CDX Learning Systems™

FUNDAMENTALS OF

Automotive Technology

Principles and Practice

SECOND EDITION

Kirk T. VanGelder

ASE Certified Master Automotive Technician & L1 & G1

Technology Educators of Oregon - President

Certified Automotive Service Instructor

Vancouver, Washington

JONES & BARTLETT
LEARNING

World Headquarters
Jones & Bartlett Learning
5 Wall Street
Burlington, MA 01803
978-443-5000
info@jblearning.com
www.jblearning.com

Jones & Bartlett Learning books and products are available through most bookstores and online booksellers. To contact Jones & Bartlett Learning directly, call 800-832-0034, fax 978-443-8000, or visit our website, www.jblearning.com.

Substantial discounts on bulk quantities of Jones & Bartlett Learning publications are available to corporations, professional associations, and other qualified organizations. For details and specific discount information, contact the special sales department at Jones & Bartlett Learning via the above contact information or send an email to specialsales@jblearning.com.

Production Credits
General Manager: Douglas Kaplan
Executive Publisher: Vernon Anthony
Content Services Manager: Kevin Murphy
Senior Vendor Manager: Sara Kelly
Marketing Manager: Amanda Banner
VP, Manufacturing and Inventory Control: Therese Connell
Composition and Project Management: Integra Software Services Pvt. Ltd.
Cover Design: Scott Moden
Rights & Media Specialist: Robert Boder
Media Development Editor: Shannon Sheehan
Cover Image (Title Page): © Umberto Shtanzman/Shutterstock
Printing and Binding: LSC Communications
Cover Printing: LSC Communications

Library of Congress Cataloging-in-Publication Data unavailable at time of printing.

6048

Printed in the United States of America
22 21 20 19 18 10 9 8 7 6 5 4 3

BRIEF CONTENTS

CONTENTS

SECTION 4 Manual Transmissions

SECTION 5 Steering and Suspension

SECTION 6 Brakes

SECTION 8 Heating and Air Conditioning

NOTE TO STUDENTS

This book was created to help you on your path to a career in the transportation industry. Employability basics covered early in the text will help you get and keep a job in the field. Essential technical skills are built in cover to cover and are the core building blocks of an advanced technician's skill set. This book also introduces "strategy-based diagnostics," a method used to solve technical problems correctly on the first attempt. The text covers every task the industry standard recommends for technicians, and will help you on your path to a successful career.

As you navigate this textbook, ask yourself, "What does a technician need to know and be able to do at work?"

This book is set up to answer that question. Each chapter starts by listing the technicians' tasks that are covered within the chapter. These are your objectives. Each chapter ends by reviewing those things a technician needs to know. The content of each chapter is written to explain each objective. As you study, continue to ask yourself that question. Gauge your progress by imagining yourself as the technician. Do you have the knowledge, and can you perform the tasks required at the beginning of each chapter? Combining your knowledge with hands-on experience is essential to becoming a Master Technician.

During your training, remember that the best thing you can do as a technician is learn to learn. This will serve you well because vehicles keep advancing, and good technicians never stop learning.

Stay curious. Ask questions. Practice your skills, and always remember that one of the best resources you have for learning is right there in your classroom . . . your instructor.

Best wishes and enjoy!
The CDX Automotive Team

ACKNOWLEDGMENTS

Editorial Board

Keith Santini
Addison Trail High School
Addison, Illinois

Merle Saunders
Nyssa, Oregon

Tim Dunn
Sydney, New South Wales
Australia

Contributors

Aims Community College
Greeley, Colorado

Addison Trail High School
Addison, Illinois

Larry Baker
Aims Community College
Greeley, Colorado

Casey's Independent Auto Repair
Vancouver, Washington

CJC Auto Parts
Lombard, Illinois

Clark College
Vancouver, Washington

Clark County Skills Center
Vancouver, Washington

John Frala
Rio Hondo Community College
Whittier, California

Fraser Automotive
Plainfield, Illinois

Brandon Fryman
Denver, Colorado

Tony Gumushian
Prairie State High School
Chicago, Illinois

Haggerty Automotive Group
Glen Ellyn, Illinois

Ed Heim
Battleground High School
Yacolt, Washington

Eileen Santini
Addison, Illinois

Les Schwab Tire Centers
Battle Ground, Washington
Orchards, Washington

McCord's Vancouver Toyota
Vancouver, Washington

Warren Tech
Lakewood, Colorado

Reviewers

Larry Baker
Aims Automotive & Technology Center

Randy Bobzien
Lincoln Educational Services

David Borelli

Charles Brecht
Waxahachie, TX

Chad Bryant
Bevill State Community College

Reuben Bruice

Jerry Clemons
Elizabethtown Community and Technical College

Jay Covey

Jerry DeLena
Middlesex County Vo-Tech High School

Robert Faoro
Moraine Valley Community College

Damon Friend
Oak Schools Technical Campus – SW Wixom

Christopher Funk
Pahrump Valley High School

Sy Gammage
Somerset South Campus

Jim Gaard
Northwest Iowa Community College

Secundino Garza
Alice Independent School District

Tony Gumushian
Prairie State College

Jim Hunnicutt
F. H. Peterson Academies of Technology
Jacksonville, Florida

Jack Ireland
Johnson County Community College

Greg Jones
Eastside Center for Applied Tech – Fayette County

James Martin
Tarrant County College South

Joel Massarello
Oakland Schools Technical Campus – Northwest

Tim McLaughlin
Western Area CTC

Ryan Monroe
Miramar College

Steve Owen
Central Nine Career Center

Doug Poteet
Elizabethtown Community and Technical College

David Sitchler
Burlington County Institute of Technology

Beth Stanberry

Ron Strzalkowski
Baker College of Flint

Jeffrey Wells
Rowanty Technical Center

Mike Wichtendahl
Kansas City Kansas Community College

Justin Winebrenner
Western Technical College

Dan Wooster
Gateway Technical College – Racine

DELL
DELL

SECTION 1

Safety and Foundation

CHAPTER 1

Careers in Automotive Technology

NATEF Tasks

There are no NATEF tasks for this chapter.

Knowledge Objectives

After reading this chapter, you will be able to:

- **K01001** Outline the history of the automobile.
- **K01002** Describe the modern automobile manufacturing process.
- **K01003** Describe the careers in the automotive service sector.
- **K01004** Describe each type of repair facility.
- **K01005** Describe the importance of automotive industry certification and ongoing training.

Skills Objectives

There are no Skills Objectives for this chapter.

▶ Introduction

Early vehicles were very basic machines with engines that were started by manually operating a crank handle and required almost continual tinkering and maintenance. As vehicle technology developed over time, the maintenance requirements have evolved as well (**FIGURE 1-1**). Although early vehicles and today's vehicles have some of the same basic systems—ignition, cooling, engine, drivetrain, suspension, and lubrication—the systems on modern vehicles are much more sophisticated and reliable. Modern vehicles travel much greater distances between maintenance appointments than their predecessors. For example, it is not unusual for a vehicle manufacturer to require maintenance at 7500- to 10,000-mile (12,000 to 16,000 km) intervals. And there is a push to extend maintenance intervals even further as new lubricants, materials, and machining processes are being developed. In fact, some manufacturers are predicting maintenance intervals of 25,000 miles (40,000 km) in the near future; in comparison, some earlier vehicles required maintenance every 1000 miles (1600 km).

Although maintenance requirements have decreased over time, the need for more substantial repairs has also decreased. It wasn't very long ago that engines would last about 100,000 miles on average. Now they routinely last well over 200,000 miles. Better metals, better machining processes, and better lubricants all extend the life of vehicle components. That doesn't mean that they don't wear out or need to be replaced, they do—just not as often as in years past. At the same time, vehicles have many more safety and convenience systems and features, making them more complex. So that has increased the types of repairs that technicians must be able to perform. In fact, many systems are now electrically or electronically operated. So technicians must have a strong understanding of electrical/electronic theory to be able to diagnose virtually all of the systems on today's vehicles.

▶ A Brief History of the Automobile

K01001

In the late 1800s, several engineers were working on the concept and design of the automobile. Karl Benz is generally acknowledged to have invented the modern automobile around 1885 (**FIGURE 1-2**). In those early days, the concept

FIGURE 1-1 Typical dealership shop.

FIGURE 1-2 Karl Benz is generally acknowledged to have invented the modern automobile around 1885.

You Are the Automotive Technician

A customer brings her 2016 V6 Dodge Minivan to the dealership for its 15,000-mile oil change. After pulling the vehicle into the bay, you set the vehicle safely on the hydraulic hoist. Next, you reference the computer to check the service history and any technical service bulletins (TSBs) or recalls for the vehicle. All previous service has been performed according to the schedule, and there are no TSBs or recalls for the vehicle. You then verify the manufacturer's scheduled maintenance recommendations for the mileage on the vehicle and find that the tires also need to be rotated and the air filter inspected. After you complete the tire rotation, you use the hydraulic hoist to raise the vehicle and then proceed to perform the oil and filter change. First, you drain and dispose of the old oil, remove and replace the filter, and then you add 4.5 quarts of new oil. When finished, you check all of the fluids, belts, hoses, and the air filter, and find them to be in good condition. You reset the oil life monitor on the vehicle, clean your area, return tools to the proper location, and process the customer's invoice with notes from your inspection, tire rotation, and oil change. The service advisor reviews the work and invoice with the customer. She thanks the customer for her business and provides her with a reminder card to return for the next scheduled maintenance appointment.

1. If you had noticed a worn belt or hose during the oil change, which type of technician would you have asked to look at the vehicle?
2. In the shop, who is responsible for keeping track of the work that is performed on customer vehicles?

of the automobile continued to develop, with many inventors producing various models. The early versions of the automobile were little more than hand-built horse carriages converted into automobiles with engines. Being hand-built, these early automobiles tended to lack uniformity, which made them expensive to buy and maintain, as well as not very reliable. Therefore, they were considered a novelty that only the wealthy could afford.

TECHNICIAN TIP

Service or maintenance intervals are also influenced by the severity of operating conditions; the more severe the conditions, the more frequent the required maintenance. Most service information gives both a normal-duty and a severe-duty maintenance schedule.

In the early 1900s, the advent of mass production made automobiles available to a wider community (**FIGURE 1-3**). Henry Ford applied two concepts that helped make the Model T affordable for the masses. The first was the concept of "interchangeability," which meant parts did not have to be custom built to match a particular car. Each part was made to the same specification so it would fit properly with its related parts. Thus, parts did not have to be produced where the vehicle was being assembled; they could be stockpiled, ready for later use. The second concept was the assembly line. Instead of the workers moving to the car as it was being assembled, the assembly line brought the car to the worker. This approach made assembly much more efficient, increasing the number of vehicles that could be built in a shift and thereby lowering production costs. Developments over the ensuing years produced more powerful and reliable automobiles, and with the further development of larger, more reliable gasoline engines, they became the predominant mode of personal transportation.

▶ Vehicle Manufacturing

K01002

Vehicle manufacturing developed from small independent makers in the late 1800s, which relied on large amounts of labor and limited automation, to the present methods, which use large-scale production lines and extensive automation. The globalization of the automotive industry has seen manufacturers sharing models and making vehicles that are sold across the world. Modern assembly lines require large-scale investments, so manufacturers must be sure that a new vehicle model will succeed in the market before they invest billions of dollars in production line retooling (**FIGURE 1-4**).

Today's vehicles are assembled on high-volume production lines, with robots used for many of the assembly processes, including welding seams. Assemblers continue to work up and down the assembly line, doing tasks that are still far too complicated for robotic assembly. Often, various parts are preassembled into unit assemblies by large-scale parts manufacturers who supply the unit assemblies to the specific vehicle manufacturer's specifications. Vehicle manufacturing is a high-volume business, so everything needs to work in the correct timing and sequence, from the supply of parts required to build a vehicle, right down to the speed at which the production line runs. Modern vehicle plants use sophisticated technology to mass-produce a product that is high quality and affordable.

This is accomplished using "just-in-time" manufacturing. With this system, vehicle manufacturers schedule the vehicles they will be producing in the order they need them built. This information is fed to the suppliers days or weeks before assembly. The supplier then produces the specified parts and unit assemblies needed by the manufacturer and delivers them shortly before assembly and in the proper order. Those parts then go in the assembly lines so that they show up for the particular vehicle they are specified for at the right time and place.

FIGURE 1-3 The assembly line was essential in the mass production of vehicles.

FIGURE 1-4 Modern assembly lines require large-scale investments in high-tech equipment.

This means that the manufacturer doesn't have to store large quantities of parts waiting for vehicles to be ordered.

Technology in Vehicles

Technology in vehicles continues to adapt and change as new research and development reach the production line. The automotive sector is very competitive and has been influenced over the years by competition between manufacturers, consumer expectations for increased technology, and, in more recent times, the need to reduce the impact on the environment. The modern vehicle contains complex electrical, electronic, and mechanical systems designed to improve efficiency, reduce emissions, and provide safety to vehicle passengers. The use of technology will continue to increase as the automotive industry responds to current and future environmental and consumer pressures.

▶ Careers in the Automotive Sector

K01003

The automotive sector provides for numerous career choices within the service, retail, and manufacturing industries. The automotive manufacturing sector provides a number of career choices, from factory workers and assemblers to design engineers and senior administrators (**FIGURE 1-5**). In the service and retail sectors, occupations range from maintenance/service technicians, light vehicle technicians, and heavy vehicle technicians to service advisors and service managers. As technical complexity has grown in modern vehicles, so has the need for more specialized job roles. For example, there is now a need for hybrid technicians, who service the systems on hybrid vehicles.

Lube Technician

Lube technicians carry out all aspects of manufacturer-scheduled maintenance activities on a range of vehicle components, such as the lubrication system (**FIGURE 1-6**). Lube technicians change oil and filters and carry out lubrication, fluid inspection, fluid service, and tire rotations. While performing these duties, lube technicians also perform a visual inspection of the vehicle, looking for any other service needs such as worn belts, hoses, and suspension system parts. When servicing vehicles, lube technicians are required to raise vehicles using hydraulic hoists or jacks and to use hand and air tools.

In addition, lube technicians may have to complete time cards or scheduled maintenance paperwork. They may be required to assist other types of technicians with dismantling or removing engine assemblies, transmissions, steering mechanisms, and other components; repairing or replacing worn or defective parts; and reassembling mechanical components. Finally, lube technicians are responsible for keeping their workspace, tools, and equipment clean and organized.

Light Line Technician

Light line technicians diagnose and replace the mechanical and electrical components of motor vehicles, such as gaskets, belts, hoses, timing belts, water pumps, radiators, alternators, and starters (**FIGURE 1-7**). In doing their job, light line technicians may be required to discuss problems with vehicle owners, operate special test equipment, and test-drive vehicles to identify faults. They need to be able to research service information, interpret wiring diagrams, and use this information to diagnose and make repairs.

In addition, light line technicians reassemble, test, clean, and adjust repaired or replaced parts or assemblies, using various instruments to make sure the parts are working properly. They also test and repair electrical systems such as lighting, instrumentation, vehicle sensors, and engine management systems. Finally, light line technicians inspect vehicles and may issue state safety certificates or list the work required before a certificate can be issued.

Heavy Line Technician

Heavy line technicians undertake major engine, transmission, and differential overhaul and repair. They may diagnose, overhaul, repair, or replace parts and assemblies (**FIGURE 1-8**). They must be able to research service information and use the information to help determine the cause of the problem and make the appropriate repairs. They also reassemble, test, clean, and adjust repaired or replaced parts or assemblies, using various test and measuring instruments and tools to make sure the parts are working properly. Some heavy line technicians are more generalized and work on a broad range of vehicles, whereas others specialize in particular areas by working on specific brand makes and models. Heavy line technicians may also specialize in particular vehicle systems, such as engines, transmissions, or final drives.

Chassis and Brake Technician

Chassis and brake technicians specialize and work primarily on the chassis and brakes of vehicles, which include steering and suspension (**FIGURE 1-9**). A chassis and brake technician inspects, diagnoses, and services the vehicle braking, steering, and suspension systems. In many cases, this also includes performing wheel alignments once any repairs have been made.

There are also separate brake technicians and chassis technicians. Chassis technicians diagnose, repair, and service steering system components and suspension systems on all types of vehicles. They diagnose faults in steering and suspension systems by speaking with the vehicle owner and test-driving the vehicle, noting its performance and comparing that to their knowledge of how the components and systems function. After diagnosis, they replace faulty components. This work could include replacing bushings and servicing wheel bearings, or checking and replacing shock absorbers or steering joints and knuckles.

Brake technicians diagnose and repair faults, replace or overhaul brake systems, and test the components of disc, drum, and power brake systems used on all types of vehicles (**FIGURE 1-10**). They diagnose faults in brake systems by speaking with the vehicle's owner and driving the vehicle and noting its performance, or by reading data from the vehicle's

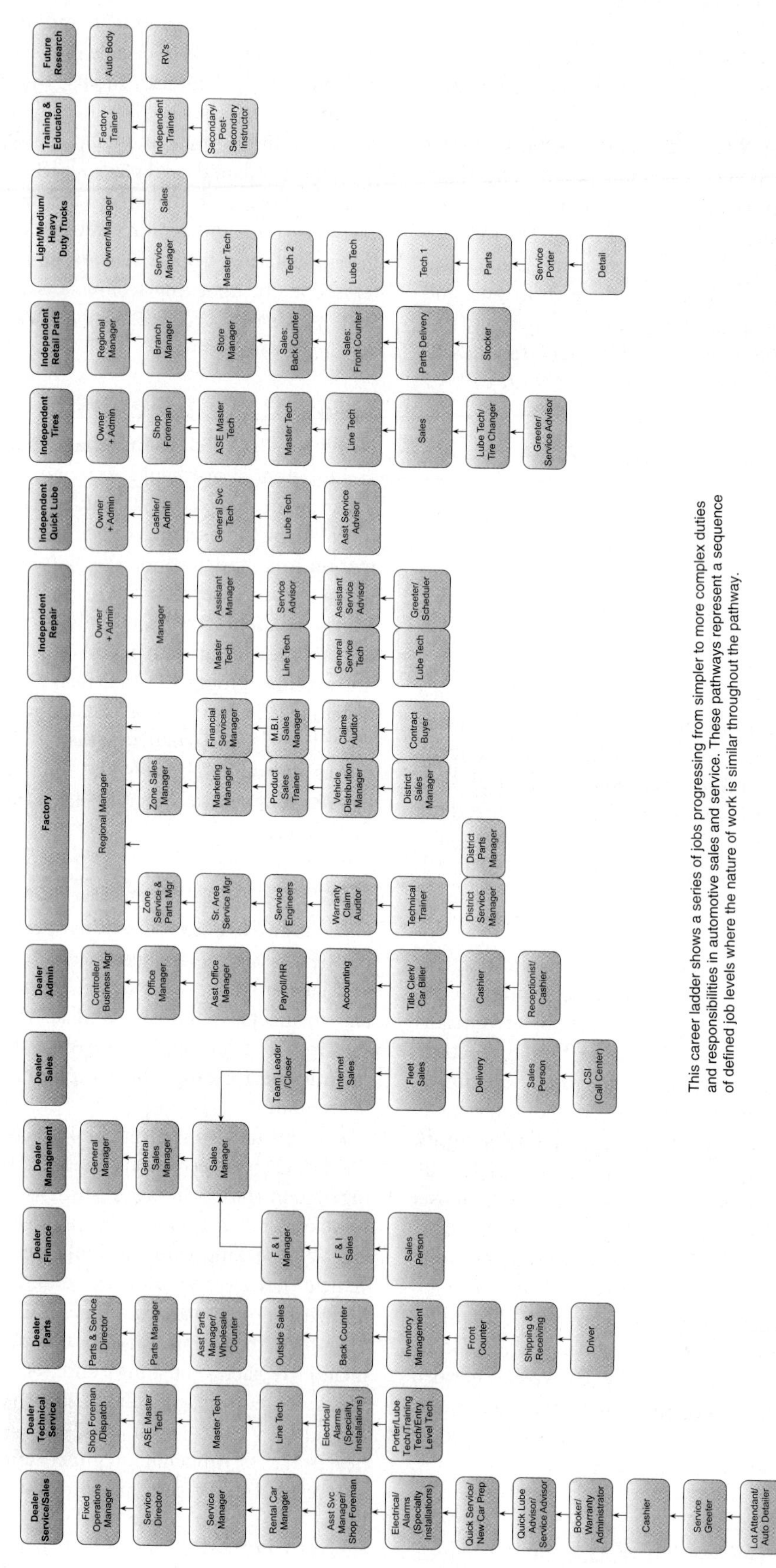

FIGURE I-5 The opportunities for challenging and rewarding careers are endless in the automotive field.

FIGURE 1-6 Lube technicians are responsible for carrying out all aspects of the standard manufacturer's scheduled maintenance.

FIGURE 1-8 Heavy line tech removing an engine.

FIGURE 1-7 Light line technicians diagnose and replace the mechanical and electrical components of vehicles.

FIGURE 1-9 Chassis and brake tech performing suspension repairs.

computer control system. This information is used along with the technician's knowledge of brake systems to properly diagnose and service the vehicle. Brake technicians also visually inspect brake units for wear, damage, or possible failure, and repair or replace the components as required. Brake technicians can measure brake drums and disc rotors to the nearest 0.0001 (0.00254 mm) to determine whether the wear or finished size meets specifications. Often, brake technicians replace leaky brake cylinders, machine rotors, and drums, when necessary, and ensure that brake systems are filled with the correct brake fluid, are bled or flushed, and are functioning properly.

Electrical/Drivability Technician

The roles of the electrical and drivability technician may be performed by a single person or by technicians who specialize in more than one area. For example, in larger shops, roles could

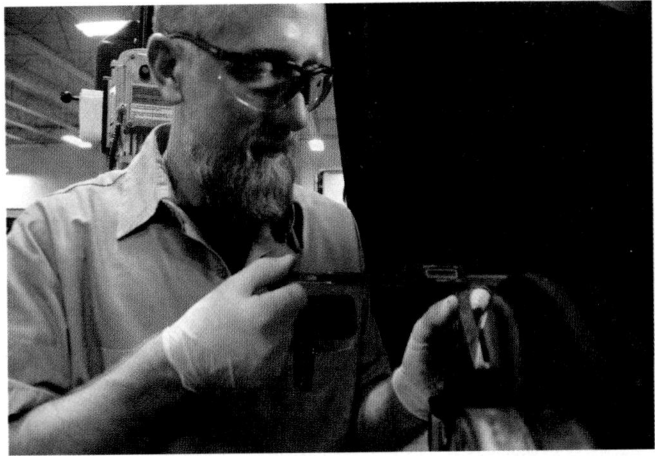

FIGURE 1-10 Brake technicians diagnose faults and repair those faults, replace or overhaul brake systems, and test the components of disc, drum, or power brake systems.

be assigned to separate electrical and drivability technicians, whereas in smaller shops, one technician could perform both roles. Often, roles cross over. For example, an electrical technician needs to understand drivability, and a drivability technician needs to understand electrical systems.

Electrical Technician

Electrical technicians diagnose, replace, maintain, identify faults with, and repair electrical wiring and computer-based equipment in vehicles (**FIGURE 1-11**). They work with computer-controlled engine management systems to identify and repair faults on electronically controlled vehicle systems such as fuel injection, ignition, anti-lock braking, cruise control, and automatic transmissions. They also install electrical components such as alternators and starter motors as well as accessories such as radios, air conditioners, driving lamps, and antitheft systems.

Often, electrical technicians use meters, oscilloscopes, test instruments, and circuit wiring diagrams to diagnose electrical faults. They test and replace faulty charging system components, starter motors, and related items such as batteries. They also repair or replace faulty ignition components, electrical wiring, fuses, and lamps and switches, often using solder equipment when repairing electrical components.

Drivability Technician

Drivability technicians diagnose and identify mechanical and electrical faults that affect the performance and emissions of vehicles. They carry out maintenance activities, replace parts, and repair both electrical wiring and computer-based equipment in vehicles. They work with computer-controlled engine management systems to service, identify, and repair faults on electronically controlled vehicle systems such as fuel injection, ignition, and electronic stability control. (**FIGURE 1-12**).

Often, drivability technicians use electronic test equipment, scan tools, pressure transducers, exhaust gas analyzers, lab scopes, meters, and circuit wiring diagrams to locate electrical, fuel, and emission systems faults. They may be required to program or reprogram (reflash) powertrain control modules (PCMs) using computerized equipment on a wide variety of vehicles; in doing so, they use updates supplied by the manufacturer to ensure that vehicles run at peak performance and within acceptable emissions limits.

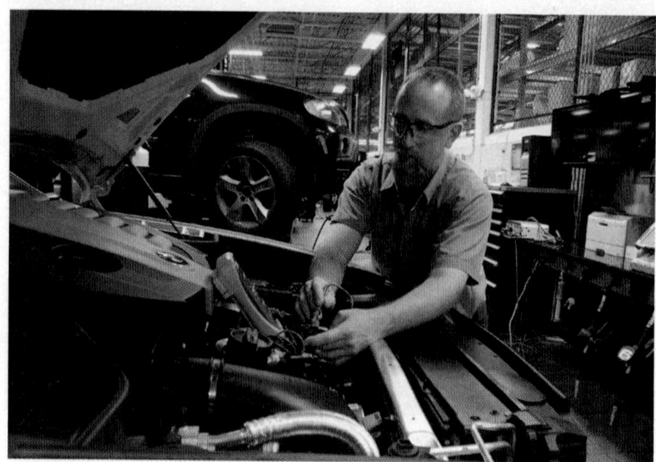

FIGURE 1-11 Electrical technicians install, maintain, identify faults with, and repair electrical wiring and computer-based equipment in vehicles.

Transmission Specialist

With the increasing complexity of modern transmissions, and the requirement for even more specialized service equipment to repair them, comes the need for transmission specialists. **Transmission specialists** diagnose, overhaul, and repair transmission units (**FIGURE 1-13**). They work on various types of manual and automatic transmissions. Transmission specialists may work on both light vehicle and heavy vehicle transmissions or may specialize in either light vehicle or heavy truck transmissions.

Transmission specialists may also work on the other components of the drivetrain, including the driveshafts and differentials. They test-drive vehicles and listen to customer concerns. They use many of the hand tools that heavy line technicians use

FIGURE 1-12 A drivability tech using a scan tool to diagnose a vehicle.

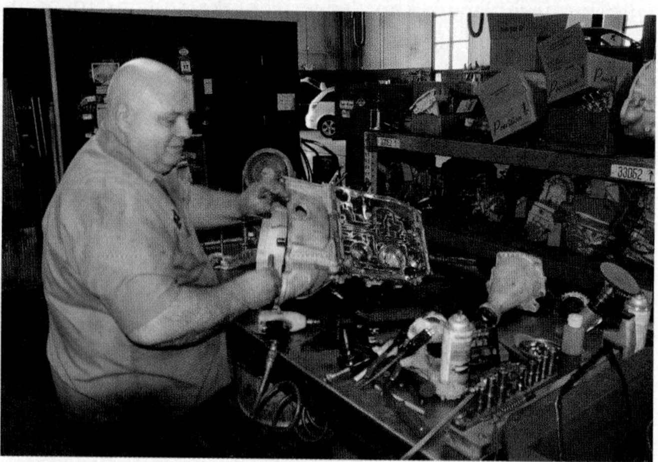

FIGURE 1-13 A transmission specialist rebuilding an automatic transmission.

and also use specialized equipment to measure tolerances, check electrical circuits, and check hydraulic pressures.

Shop Foreman

A **shop foreman** is the supervisor in a shop. Shop foremen oversee the work of all types of technicians and staff, communicate with customers and external suppliers, and handle the various administrative duties involved with operating a business. Many shop foremen are responsible for hiring and training new workers and provide regular performance reviews (**FIGURE 1-14**). They oversee technicians' work to ensure that customers receive quality repair work. Finally, the shop foreman is responsible for enforcing safety procedures at all times to avoid accidental injuries to technicians or damage to vehicles.

Service Consultant

Service consultants (advisors) work with both customers and technicians (**FIGURE 1-15**). They are the first point of contact for the customer and provide advice and assistance to customers concerning their vehicles. Service consultants book customer work into the shop, fill out repair orders, price repairs, invoice, keep track of work being performed on customer vehicles, and build customer relations in order to provide a high level of customer support. They are the interface between the technician and the customer, so good communication and organization skills are essential. A service consultant can progress to become a service manager.

FIGURE 1-14 A shop foreman training a new tech on how to perform a procedure.

Service Manager

The role of a service manager is very demanding and challenging. **Service managers** are responsible for the functioning of the entire service department. This career requires well-established skills in communicating, motivating, and creating positive work environments. The service manager often supervises many people, has to deal with customer complaints, and is accountable for the overall performance of the shop (**FIGURE 1-16**).

This job requires a high level of personal commitment and focus, exceptional people skills, excellent leadership qualities, and a high level of business knowledge. Service managers may have worked their way up through the various roles within a shop. For example, the service manager may have once been a light vehicle technician on the floor or a service consultant at the counter. Others may enter the field with backgrounds and college degrees in business management. The service manager typically reports directly to the business owner.

▶ Types of Shops

K01004

There are many facilities that cater to specific customer needs. The types of repair facilities can generally be broken down into the following categories:

- Dealership
- Independent shop
- Specialty shop
- Franchise/retailer
- Fleet shop

FIGURE 1-15 Service consultants work with both customers and technicians.

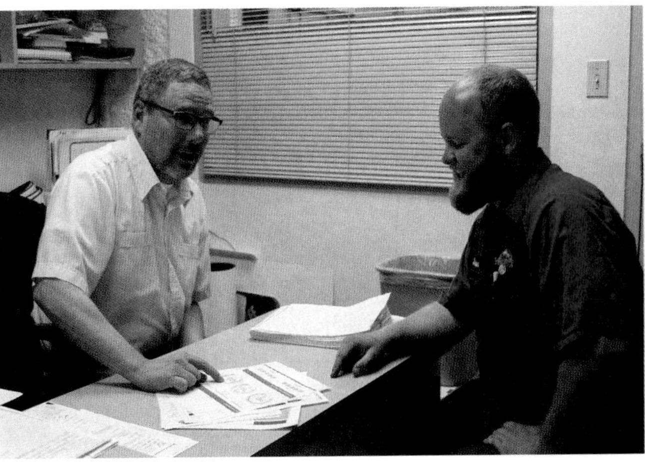

FIGURE 1-16 The service manager is accountable for the overall performance of the shop.

Each type of shop caters to a particular segment of the industry. For example, dealerships are affiliated with a specific vehicle manufacturer. The dealership sells new and used vehicles. Dealership technicians perform maintenance, service, and warranty repairs on vehicles sold by that manufacturer (**FIGURE 1-17**). In many cases, customers bring their new vehicles back to the dealership for maintenance and repair work while the vehicle is in warranty, and then may seek out other shops after that.

Because dealership technicians are working on the latest vehicles, they are right at the cutting edge of technology. Most dealerships send their technicians to manufacturer training classes regularly so that they will know how to maintain, diagnose, and repair their vehicles. Because the dealership is affiliated with the manufacturer, dealership technicians generally have instant access to the manufacturer's service information as well as the manufacturer's service representatives when they encounter a difficult diagnostic situation.

Independent shops are not affiliated with vehicle manufacturers, making it harder for independent technicians to access training on new vehicle technology (**FIGURE 1-18**). In many cases, they service a broad range of vehicles. Because vehicles are becoming increasingly specialized and complex, it is common for independent shops to limit their clientele to one of the following: European, Asian, or domestic vehicles. Some may limit their work and specialize in a particular system of the vehicle, such as brakes and alignment. In many cases, independent technicians work on vehicles that are out of warranty and are not the latest technology.

Specialty shops are usually independent shops that focus on one type of service, such as transmission service, electrical system repair, or emission system diagnosis (**FIGURE 1-19**). Because these shops offer limited services, they usually work on a variety of vehicles from a variety of manufacturers. A shop that specializes in one area, such as air-conditioning (AC), may experience a slow period during certain times of the year.

Franchises are similar to specialty shops, but they are connected to a larger parent organization that can help with marketing and provide a mechanism for warranty claims that are honored at related franchise shops across the country. Some examples include Goodyear Tire Company, AAMCO Transmissions, and Jiffy Lube (**FIGURE 1-20**).

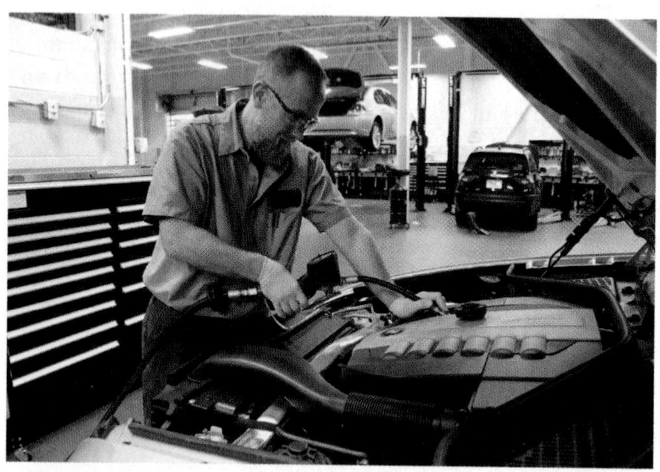

FIGURE 1-17 Dealership technicians perform maintenance, service, and warranty repairs on vehicles sold by a particular manufacturer.

FIGURE 1-19 Specialty shops focus on one type of service.

FIGURE 1-18 Independent shops are not affiliated with vehicle manufacturers.

FIGURE 1-20 Typical quick lube franchise business.

FIGURE 1-21 Fleet shop.

A fleet shop is usually connected with either a business that runs multiple vehicles or with equipment that is maintained and repaired in-house, or it could be a government agency such as a city or county would have for maintenance and repair of its vehicles and equipment.

Because fleet shops maintain and repair vehicles for the company or agency, it is more likely that the vehicles will be serviced on a regular basis (**FIGURE 1-21**).

▶ Automotive Industry Certification

K01005

The automotive service industry in the United States is generally not subject to licensure requirements. This means that a technician does not have to pass a licensure test in order to receive a license to work in the industry. At the same time, to promote professionalism and demonstrate a level of competence, many technicians become **Automotive Service Excellence** (ASE) certified (**FIGURE 1-22**). ASE is an independent, nonprofit organization dedicated to the improvement of vehicle repair through the testing and certification of automotive professionals. A few localities require technicians to be ASE certified, and many shops require it of technicians they hire. But overall, it is a voluntary certification.

To earn ASE certification, technicians are required to pass one or more ASE certification tests and have two years of qualifying work experience as a technician. ASE certification needs to be renewed every five years by taking and passing recertification tests. There are currently approximately 300,000 certified ASE technicians in the United States.

The **National Automotive Technicians Education Foundation** (NATEF) is an accrediting body for secondary and postsecondary automotive training programs (**FIGURE 1-23**). NATEF is an independent, nonprofit organization under the umbrella of the ASE. In order for programs to be accredited by NATEF, they have to demonstrate their compliance to a

FIGURE 1-22 ASE certifies technicians.

FIGURE 1-23 NATEF certifies automotive training programs.

rigorous list of standards developed by the automotive industry. Program instructors must also maintain ASE certification in the areas they teach, as well as receive a minimum of 20 hours of technical update training each year. NATEF accreditation is valid for five years, and then the program must go through a reaccreditation process.

The **Automotive Youth Educational Systems** (AYES) is an independent, nonprofit organization that is a partnership between automotive manufacturers, their dealerships, and affiliated secondary automotive programs (**FIGURE 1-24**). Because of the partnership, AYES programs receive access to new vehicle technology as well as manufacturer service information to help prepare students for working on today's vehicles

FIGURE 1-24 AYES logo.

and technology. Students receive the opportunity to intern at participating dealerships, working alongside a mentor technician. This is a great way for the students to put into practice what they are learning in the training program. The dealers and manufacturers benefit from gaining access to highly motivated students who are preparing to go to work for them. To become an AYES program, the school has to be NATEF certified and then be evaluated by AYES against another set of standards related to the commitments of carrying out the AYES training model.

Special Certification

The automotive industry has two special certification requirements for certain automotive technician specialists. To diagnose and repair emission failures on vehicles in some localities, technicians must be certified by the appropriate agency. In many cases, the technician must have ASE Advanced Engine Performance certification or have taken and passed a course specifically for emission specialists, as administered by the agency.

The second special certification required is for technicians who handle refrigerants or work on AC systems. They are required to be Environmental Protection Agency (EPA) Section 609 certified. This requires taking a training course and passing the 609 exam. Once certified, technicians are legally able to handle refrigerants and repair AC systems. It is advisable that all students obtain their 609 certification during their training program to show prospective employers when they graduate.

Ongoing Training

Even though most technicians have attended and graduated from automotive technology programs, they are never done with learning. Being an automotive technician means you will always be learning new things because vehicle technology changes at a rapid rate. In fact, it is easy to fall behind. So successful technicians attend training classes regularly to stay up to date. Dealership technicians typically have access to the manufacturer's online training classes as well as the ability to attend regional training classes (**FIGURE 1-25**). Independent technicians generally have a more difficult time accessing training, but there are both online and face-to-face classes in most areas that technicians can attend. Also, organizations like the Automobile Service Association (ASA) host annual training conferences on a regional basis. These conferences attract some of the best minds in the industry and are well attended. No matter what, successful automotive technicians are always learning. You are just beginning on that exciting and rewarding journey.

FIGURE 1-25 Dealership techs attending a regional training class.

▶ Wrap-Up

Ready for Review

- Modern vehicles are more sophisticated, with longer maintenance intervals than early vehicles.
- Karl Benz is considered the inventor of the modern automobile (in 1885).
- Application of interchangeability and the assembly line made cars more affordable for the masses.
- Today's production lines are high-volume, highly automated plants.
- Lube technicians carry out routine maintenance services.
- Light line technicians are responsible for maintenance and repair of mechanical and electrical components.
- Heavy line technicians diagnose and repair major engine or transmission problems and perform differential overhaul.
- Chassis and brake technicians repair the vehicle's chassis and brakes (respectively, or in combination), including steering and suspension.
- Electrical technicians diagnose and repair the vehicle's electrical wiring and computer-based equipment.
- Drivability technicians are responsible for diagnosing the mechanical and electrical faults that can affect the performance and emissions of vehicles.
- Transmission specialists diagnose and repair transmission units (manual and automatic).

- The shop foreman is responsible for administrative duties, supervising technicians, and ensuring customer satisfaction.
- Service consultants are the interface between the technician and the customer, so good communication and organizational skills are essential.
- Service managers run the service department and work to create a positive work environment.
- Types of repair facilities are dealerships, independent shops, specialty shops, franchises/retailers, and fleet shops.
- Automotive technicians may become ASE certified; training programs may be NATEF accredited; school programs may be AYES authorized; and technicians may choose to receive additional certification in emission failures and/or air-conditioning systems.
- Technicians require regular training to stay current with changes in vehicle technology and service procedures.

Key Terms

Automotive Service Excellence (ASE) An independent, nonprofit organization dedicated to the improvement of vehicle repair through the testing and certification of automotive professionals.

Automotive Youth Educational Systems (AYES) An independent, nonprofit organization that is a partnership between automotive manufacturers, their dealerships, and affiliated secondary automotive programs.

brake technician A technician who specializes in working on vehicle brake systems.

chassis technician A technician who specializes in working on vehicle suspension and steering systems.

drivability technician A technician who diagnoses and identifies mechanical and electrical faults that affect vehicle performance and emissions.

electrical technician A technician who diagnoses, replaces, maintains, identifies fault with, and repairs electrical wiring and computer-based equipment in vehicles.

heavy line technician A technician who undertakes major engine, transmission, and differential overhaul and repair.

light line technician A technician who diagnoses and replaces the mechanical and electrical components of motor vehicles.

lube technician A technician who carries out scheduled maintenance activities on a range of mechanical and related vehicle components.

National Automotive Technicians Education Foundation (NATEF) An accrediting body for secondary and postsecondary automotive training programs; an independent, nonprofit organization under the umbrella of the ASE.

service consultant/advisor A customer service worker who works with both customers and technicians; the first point of contact for customers seeking vehicle repairs.

service manager The shop supervisor who is responsible for the management of the service department.

shop foreman The supervisor in a shop who oversees the work of technicians and staff and communicates with customers and external suppliers.

transmission specialist A technician who diagnoses, overhauls, and repairs transmissions.

Review Questions

1. When compared to early vehicles, modern vehicles do not have:
   a. as much sophistication and reliability.
   b. as much speed.
   c. as many maintenance requirements.
   d. as many convenience systems.
2. Which of the following statements is true?
   a. "Interchangeability" meant that parts did not have to be custom built to match a particular car.
   b. Assembly line involved workers moving to the car as it was being assembled rather than bringing the car to the worker.
   c. The advent of mass production made automobiles available to only the wealthy.
   d. Efficiency in mass production is increased when parts are produced where the vehicle is being assembled.
3. Advances in vehicle manufacturing technology have decreased the need for:
   a. large-scale parts manufacturers.
   b. high-volume production lines.
   c. large-scale investments.
   d. large amounts of labor.
4. In "Just in Time" manufacturing:
   a. there is no need for scheduling.
   b. the manufacturer doesn't have to store large quantities of parts.
   c. parts are delivered a few months before assembly and in the order the supplier prefers.
   d. technology use is reduced in response to environmental pressures.
5. Which of the following would be the most likely to program or reprogram (reflash) powertrain control modules (PCMs) using computerized equipment?
   a. Transmission specialists
   b. Brake technicians
   c. Drivability technicians
   d. Light line technicians
6. All the statements below are true *except*:
   a. Service consultants work only with customers.
   b. A shop foreman is the supervisor in a shop.
   c. Service managers are responsible for the functioning of the entire service department.
   d. A shop foreman may also be responsible for hiring and training new workers.
7. A technician wants to be at the cutting edge of technology. Which of repair facility would be the best fit for him or her?
   a. Franchise
   b. Fleet shop
   c. Independent shop
   d. Dealership

8. Jiffy Lube outlets are examples of:
   a. dealerships.
   b. franchises.
   c. fleet shops.
   d. independent shops.
9. Special certification is required for technicians who handle which of the following systems?
   a. Refrigerants and AC systems
   b. Computer-controlled systems
   c. Automatic transmission systems
   d. Engine management systems
10. Which of the following organizations certify technicians in all areas of a vehicle?
   a. EPA
   b. ASE
   c. ASA
   d. AYES

ASE Technician A/Technician B Style Questions

1. Tech A says that newer vehicles require less maintenance compared to older vehicles. Tech B says that major repairs are needed more frequently on newer engines. Who is correct?
   a. Tech A
   b. Tech B
   c. Both A and B
   d. Neither A nor B
2. Tech A says that Henry Ford is credited with the invention of the gasoline engine. Tech B says that Carl Benz is credited with the invention of the automobile. Who is correct?
   a. Tech A
   b. Tech B
   c. Both A and B
   d. Neither A nor B
3. Tech A says that the production of vehicles today requires a mix of robotic and human assembly. Tech B says that most of the parts on a vehicle are preassembled into unit assemblies before they reach the assembly line. Who is correct?
   a. Tech A
   b. Tech B
   c. Both A and B
   d. Neither A nor B
4. Tech A says that only a certain few people will find jobs in the automotive industry. Tech B says the automotive industry offers numerous career choices. Who is correct?
   a. Tech A
   b. Tech B
   c. Both A and B
   d. Neither A nor B
5. Tech A says that a technician can specialize in different areas based on his or her interest and ability. Tech B says that most specialty shops, such as transmission shops, primarily work on only one manufacturer's vehicles. Who is correct?
   a. Tech A
   b. Tech B
   c. Both A and B
   d. Neither A nor B
6. Tech A says that the shop foreman is the frontline contact for customer relations. Tech B says that the service consultant typically reports directly to the business owner. Who is correct?
   a. Tech A
   b. Tech B
   c. Both A and B
   d. Neither A nor B
7. Tech A says that dealership technicians generally have access to manufacturers' training courses. Tech B says that an independent shop works on a wide variety of vehicles that requires a broad skill level in technicians. Who is correct?
   a. Tech A
   b. Tech B
   c. Both A and B
   d. Neither A nor B
8. Tech A says that AYES certifies technicians. Tech B says that having ASE certifications can make it easier to get a job. Who is correct?
   a. Tech A
   b. Tech B
   c. Both A and B
   d. Neither A nor B
9. Tech A says that the maintenance intervals of a vehicle have changed very little since the creation of the automobile. Tech B says that manufacturers are predicting 25,000-mile (40,000 km) maintenance intervals. Who is correct?
   a. Tech A
   b. Tech B
   c. Both A and B
   d. Neither A nor B
10. Tech A says that a technician can progress to different jobs within the industry. Tech B says that careers in the automotive industry include parts manager. Who is correct?
   a. Tech A
   b. Tech B
   c. Both A and B
   d. Neither A nor B

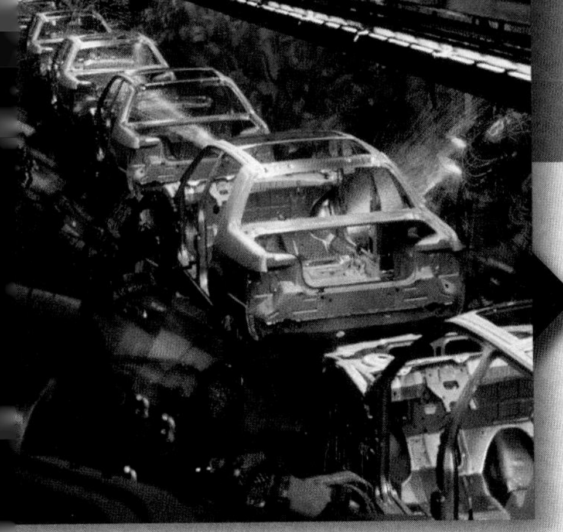

CHAPTER 2
Introduction to Automotive Technology

NATEF Tasks

There are no NATEF tasks for this chapter.

Knowledge Objectives

After reading this chapter, you will be able to:

- **K02001** Identify vehicle body types and their characteristics.
- **K02002** Describe vehicle chassis designs.
- **K02003** List the functions of common vehicle systems.
- **K02004** Describe general vehicle operation.
- **K02005** Describe drivetrain layouts.
- **K02006** Identify engine configurations.
- **K02007** Describe transmission/transaxle and axle configurations.
- **K02008** Explain how torque is applied.
- **K02009** Describe the functions of transmissions and final drives.
- **K02010** Distinguish between four-wheel and all-wheel drive.

Skills Objectives

There are no Skills Objectives for this chapter.

▶ Introduction

As you prepare to work in the automotive repair industry, you will need to learn a whole new vocabulary. Terms like "hygroscopic," "asymmetrical," "reciprocating motion," and "volumetric efficiency" are only a few examples of the new vocabulary you need to master (**FIGURE 2-1**). Healthcare professionals need to know how each of the body's systems function, the organs that make up the system, and how to diagnose an issue. Automotive technicians need to have the same level of understanding about vehicles. The good news is that automotive names aren't in Latin; however, it may seem like they are at times.

Knowing the correct automotive terminology and concepts will help you fit into your new work environment as well as to communicate accurately with customers, suppliers, and fellow employees. You will also be able to properly complete repair orders, parts requisitions, and warranty paperwork as you maintain, diagnose, and repair vehicles. This chapter will help you start the process of learning automotive terminology as it relates to vehicle types, drivetrain layouts, engine configurations, and axle arrangements.

▶ Body Designs

K02001

Vehicle bodies come in a variety of designs primarily depending on the intended function of the vehicle. But they also are designed to accommodate style, aesthetics, and,

Glossary Excerpt	
AMP	Ampere
AMP/hour	Amperes per hour. A standard measure for a rate of current flow
Amperage	An amount of current, expressed as amperes
Ampere	Usually called an amp, the unit for measuring electrical current
Ampere Turns	The unit of measurement for electrical magnetic field strength
Analog instrument	An instrument that displays measurements with a needle on a dial
Analog signal	An electrical signal that varies in amplitude within a given parameter
Anchor	A mounting point on a vehicle for a stressed, non-structural component such as a seat or seat belt
Anchor end	The end of a brake shoe that is attached to a fixed point on the backing plate
Anchor pin	The steel pin attached to the backing plate of drum brakes. Return springs are attached to the anchor pin and to the brake shoes to hold the shoes against the anchor pin in a non-applied position. In an applied position it prevents the shoes from rotating with the drums
Anion	A negative ion. Alkali, molten carbonate and solid oxide fuel cells are "anion-mobile" cells
Anode	The positively charged electrode in an electrolytic cell toward which current flows
Anodize	An electrochemical process that coats and hardens the surface of aluminum
ANSI	American National Standards Institute. A privately funded organization that promotes uniform standards in areas such as measurements
Antenna	A conductive metallic structure used for radiating or receiving electromagnetic signals, such as those for radio transmissions and television signals
Anti-foam agent	An additive that reduces foaming caused by the churning action of the crankshaft in the engine oil
Antifreeze	A liquid that mixes easily mixes with water and is used to cool the engine. It also lowers the coolant's freezing point and increases its boiling point. Coolant is typically mixed at a ratio of 50% anitfreeze and 50% water

FIGURE 2-1 A sample list of automotive vocabulary terms.

You Are the Automotive Technician

The sales manager asks you to provide training for her new vehicle sales associates in your recently expanded dealership. The first training lesson takes place in the dealership showroom where several styles of vehicles are on display. You explain the concept of the unibody design and compare it to the full-frame design. You also show examples of various drivetrain layouts and their pros and cons. The second lesson is in the shop area, designated for engine repair. You show them the various engine classifications and configurations.

1. What are the common vehicle body types and their characteristics?
2. How does the unibody design differ from the full-frame vehicle?
3. What are the benefits of the unibody design?
4. What are the various engine configurations, and how do they differ?
5. How is a transaxle different from a transmission?

most importantly, safety. Vehicle body design continues to evolve to accommodate the owners' lifestyles and personal tastes, such as the Scion XB or the M-Benz Smart Fortwo (**FIGURE 2-2**). Manufacturers also use vehicle body design as a strong marketing tool to entice buyers to purchase their particular vehicles.

Common types of body design cater to both passenger and light commercial use. Terms to describe various body designs have become part of common automotive language, although names describing the same body design type can vary from country to country. For example, a vehicle described as a sedan in the United States is called a saloon in the United Kingdom. Other types of body designs include station wagons, hatchbacks, convertibles, coupes, vans, minivans, pickups, and sport utility vehicles (SUVs).

Sedan

A **sedan** has an enclosed body, with a maximum of four doors to allow access to the passenger compartment (**FIGURE 2-3**). The sedan design also allows for storage of luggage or other items in a trunk located in the rear of the vehicle and accessible from a trunk lid. A sedan traditionally has a fixed roof; however, there are soft-top versions of sedans, which have only two doors.

Coupe

A **coupe** has only two doors. Reducing the number of doors to the passenger compartment makes the vehicle structurally more rigid. Traditionally, a coupe has two standard-size seats in front and possibly two smaller seats behind (**FIGURE 2-4**). Coupes are available in both a fixed-roof and a convertible style. They also are equipped with a trunk for storage purposes, although they can sometimes be on the small side.

Hatchback

Hatchbacks are available in three-door and five-door designs, with the odd-numbered door being a hatch that lifts up at the rear of the vehicle. This gives access to the luggage area.

FIGURE 2-2 Vehicle body design continues to evolve to accommodate the owners' lifestyles and personal tastes. **A.** Car from the 1950s. **B.** Car from the 2010s.

FIGURE 2-3 A sedan has an enclosed body, with a maximum of four doors.

FIGURE 2-4 Traditionally, the coupe has two standard-size seats in front, with two smaller seats behind.

The rear seats usually fold down to increase the luggage area (**FIGURE 2-5**). Often the rear seat is split, which provides more flexibility because only one side needs to be folded down if the other seat is required for a passenger. Hatchbacks are versatile vehicles, combining some of the benefits of both sedans and station wagons.

Convertible

A **convertible** is an automobile that can convert from having an enclosed top to having an open top by means of a roof that can be removed, retracted, or folded away (**FIGURE 2-6**). The roofing material is most often a flexible fabric such as canvas or vinyl, and most convertibles have a mechanism driven by electric motors that retract and raise the top. In some vehicles, known as hardtop convertibles, the roof can be made up of folding or fixed steel or fiberglass panels. When in place, the hard roof makes such vehicles look more like conventional fixed-roof coupe vehicles (**FIGURE 2-7**). In other vehicles, only a smaller section of the roof area is convertible, such as a T-top. Traditionally, the term "roadster" was applied to a vehicle with no permanent roof covering or side windows, but today that name is most often used to describe any convertible sports car.

> **TECHNICIAN TIP**
>
> A convertible is commonly known as a cabriolet in Europe.

Station Wagon

A **station wagon** has an extended roof that goes all the way to the rear of the vehicle. It is similar to a van, but not as tall. The extra length in the roof increases the luggage capacity. In some cases, the passenger capacity is increased with extra seats in the very rear of the vehicle. Station wagons have a large rear door for easy access, and the rear seats can usually be folded to increase storage capacity even further (**FIGURE 2-8**). Station wagons usually have fixed roofs.

FIGURE 2-5 Rear seats in hatchbacks usually fold down to increase the luggage area.

FIGURE 2-7 Hardtop convertibles have a hard roof that makes the vehicle look more like a conventional fixed-roof vehicle.

FIGURE 2-6 A convertible is an automobile that can convert from having an enclosed top to having an open top by means of a roof that can be removed, retracted, or folded away.

FIGURE 2-8 A station wagon has increased luggage capacity and a large rear door for access.

Pickup

The **pickup**, or **truck**, carries and tows cargo. Usually it has heavier-duty chassis and suspension components than a passenger car to support greater loads. Traditionally, pickups had only a single cab with two doors, which limited the number of passengers they could carry. Today's pickups have options for extended cabs or four-door versions to carry more passengers (**FIGURE 2-9**). In some cases, the four-door pickup has a reduced cargo-carrying space to accommodate the extra seating in the passenger cab.

Minivan

Minivans are generally lighter-duty vehicles with suspension systems more like passenger cars than full-size vans, which use heavy-duty pickup truck–type suspension systems. Minivans can be configured to maximize the number of seats for passengers or redesigned so that maximum cargo space is available (**FIGURE 2-10**). Also, because they are lighter duty, the fuel economy of minivans is substantially better than that of full-size vans.

FIGURE 2-9 More recent versions of pickups have options for extended cabs or four-door versions to carry more passengers.

FIGURE 2-10 Minivans can be based on common interior sedan designs or redesigned so that maximum cargo space is available.

Sport Utility Vehicle

Sport utility vehicles (**SUVs**) are popular in the United States because they can easily be used to carry out functions that would otherwise require several different vehicles (**FIGURE 2-11**). They act like both a full-size van and a pickup in that they have a heavier-duty chassis so they can carry heavier loads. This load can be in the form of passengers, luggage, or cargo. They can also tow moderately heavy loads, making them a great vehicle for family outings, as they can pull a trailer while still carrying a number of passengers and luggage.

▶ Vehicle Chassis

K02002

A **chassis** is an underlying supporting structure for vehicles—similar to the skeleton of a human—on which additional components are mounted. In a vehicle, a traditional chassis gives the vehicle structural strength as well as a platform on which to mount the engine, wheels, transmission, and all the other mechanical components. Also bolted onto this frame is the body. Originally made of wood, vehicle chassis were soon changed to an open steel ladder-frame structure, which is easier to manufacture and longer lasting (**FIGURE 2-12**).

"Body-on-frame" is the term used when a vehicle body is mounted on a rigid frame or chassis. It was the preferred way of

FIGURE 2-11 Sport utility vehicles (SUVs) are designed for flexible use while being heavier duty than minivans.

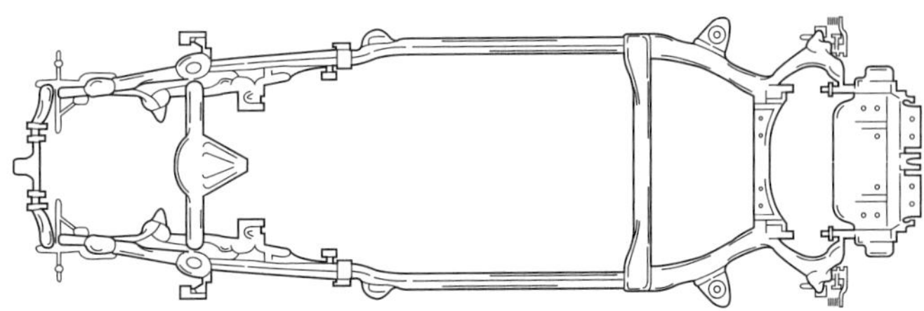

FIGURE 2-12 Steel ladder-frame chassis.

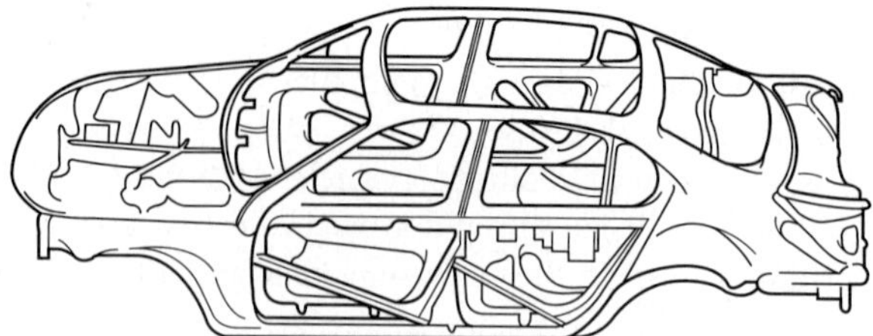

FIGURE 2-13 The unibody design.

building passenger vehicles because it allowed manufacturers to release new models of vehicles with different body styles without having to retool most of the mechanical and structural components. However, by the 1960s, most manufacturers switched to vehicle designs that either partially or wholly integrated the bodywork into a single unit with the chassis so that the vehicle body became part of the vehicle structure rather than just an external skin. This is the **unibody design**, or single shell design (**FIGURE 2-13**). The unibody design is constructed of a large number of steel sheet metal panels that are precisely formed in presses and spot-welded together into a structural unit.

The unibody design was first used in aircraft and then spread to automobiles. It became popular with vehicle manufacturers because with less of a chassis component, it was quicker to manufacture and lighter in weight. The lighter weight meant less cost in both material and labor. Another benefit of being lighter was that the vehicles became more fuel efficient.

▶ TECHNICIAN TIP

The unibody design is the predominant vehicle construction technology used today.

▶ TECHNICIAN TIP

Some high-performance racing cars today have no chassis at all. Their structural strength comes from their light, stiff, and stable body shells molded from newer lightweight materials such as carbon fiber–reinforced plastics.

Vehicle Closure Designs

A vehicle body contains many openings apart from the vehicle doors. These include the engine compartment hood, hatch and tailgate openings, the fuel door, and in some cases a battery access cover. All of these openings have to be secured and may require a remote switch or lever to be activated. In some cases, the access door is opened mechanically by the driver pulling a lever, which moves a cable and releases a latch. Other doors may use electric- or vacuum-operated solenoids to release the latch. In this case, the driver pushes a switch that sends an electric or vacuum signal to the release mechanism, which releases the door. Some rear hatch doors have a hinged window incorporated; this window offers the owner easy access to the storage space without opening the entire back door.

FIGURE 2-14 The engine compartment release may be located inside the passenger compartment under the dash, in the glove compartment, or on a doorjamb.

Engine compartment hoods on today's vehicles usually have a remote release lever to prevent unauthorized access to the engine compartment, for security reasons. The release lever is usually located inside the passenger compartment: under the dash, in the glove compartment, or on a doorjamb (**FIGURE 2-14**). Once the hood is open, it is held in the open position by one of three methods: large springs on the hinges, pressurized gas strut assemblies, or a prop rod (**FIGURE 2-15**).

▶ TECHNICIAN TIP

Some high-end sports car manufactures are no longer fitting an engine compartment release to their vehicles. The hood can only be released by the manufacturer's scan tool or by a service key fitted into a secluded opening on the vehicle body (usually behind a manufacturer's badge).

Overview of Vehicle Systems

K02003

A vehicle's components are arranged in systems that are designed to perform specific functions (**FIGURE 2-16**). Each of the systems is critical to the operation, passenger safety, and environmental impact of the vehicle. Knowing all of the systems and their purpose will aid you when we dig deeper into them. Here is a list of the major vehicle systems:

- Powertrain system—One of the most major systems on the vehicle. In fact, it contains several smaller subsystems to get its job done, which is to power the vehicle down the road as well as to provide for all on-vehicle power.
- Engine system—The powerplant of the vehicle. This system converts gasoline energy or electrical energy into mechanical energy.
- Lubrication system—Lubricates the internal components of the engine for long life and quiet operation.
- Cooling system—Regulates the temperature of the engine so it operates at the ideal temperature for efficiency and long life.

FIGURE 2-15 Mechanisms for holding hoods open. **A.** Springs. **B.** Gas Strut. **C.** Prop Rod.

- Fuel system—Stores and delivers the proper amount of fuel to the engine's combustion chambers.
- Ignition system—Creates and delivers high-voltage sparks at the right time to ignite the air-fuel mixture in the combustion chamber.
- Emission system—Controls and reduces the amount of harmful emissions that the powertrain creates.
- Transmission system—Extends the operating range by increasing the engine's torque or speed.
- Electrical system—The nerve center of the vehicle. Electricity is used in all of the vehicle's systems for both control and possibly power.
- Starting system—Used to crank the engine over so that it will start and run. This is usually performed by an electric starter motor.
- Charging system—Used to both charge the vehicle's battery as well as run the entire vehicle's electrical system when the engine is running.
- Lighting system—Used to illuminate both the outside and inside of the vehicle, including warning lights.
- Entertainment system—Provides entertainment such as audio, video, and Internet to vehicle occupants.
- Safety systems—An important system on the vehicle that is designed to protect the vehicle's occupants as well as provide a level of accident avoidance.
- SRS systems—Designed to protect occupants by both restraining them and cushioning them in an accident.

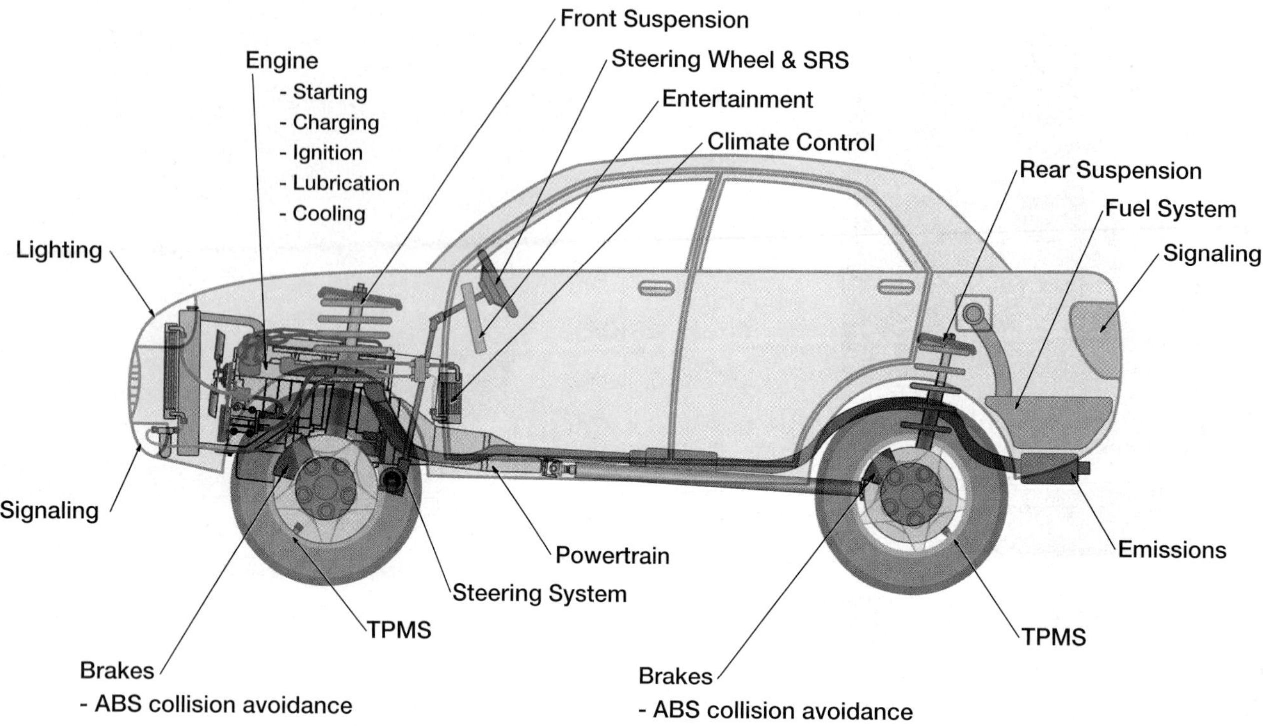

FIGURE 2-16 Vehicles are made up of a variety of systems.

- TPMS systems—Monitors tire pressure and warns the driver of tires with low pressure.
- Crash avoidance systems—Monitors the area around the vehicle and anticipates potential accidents and either alerts the driver or takes evasive actions.
- Climate control system—Clears the windshield of fog and frost and maintains a comfortable temperature and humidity inside the vehicle.
- Braking system—Used to slow or stop the vehicle. Must be capable of doing so quickly and on all types of surfaces.
- Steering system—Allows the vehicle to track straight while making it easy for the wheels to be steered by the driver, without losing control.
- Suspension system—Keeps the tires in contact with the road surface while absorbing road harshness. It also reduces vehicle body sway and dive during vehicle maneuvers.

We cover each of these systems and more in following chapters so that you will be prepared to maintain, diagnose, and repair them. But for now, in the next topic, sit back and see how some of these systems work together to drive the vehicle down the road.

▶ Vehicle Operation Overview

K02004

For a vehicle to operate, a lot of almost magical things need to happen. First, it requires a means of converting stored energy into a form of energy that can turn the wheels. In the majority of vehicles, the stored energy is in the chemical form of gasoline or diesel fuel, but newer vehicles are using natural gas, alcohol, biodiesel, hydrogen, and battery acid as their stored chemical energy source. Chemical energy can be converted into mechanical energy in two primary ways: through the operation of an internal combustion engine or through the operation of an electric motor (**FIGURE 2-17**). Both of these methods take energy in chemical form and convert it to mechanical energy by causing a shaft to rotate—the crankshaft in an internal combustion engine and the armature in an electric motor. The shaft then provides mechanical energy to move the vehicle as well as power all of the other accessories on the vehicle.

The vehicle's internal combustion engine is an engineering marvel that combines fuel with air to create a combustible mixture, which is compressed by the pistons and ignited by a spark plug in the engine cylinders (**FIGURE 2-18**). The burning, expanding gases create high pressure, which rapidly pushes the pistons down the cylinders and spins the crankshaft.

When the driver initiates the starting process by turning the ignition key to run (or using a smart key), power is sent from the battery to a variety of vehicle circuits, including energizing of the fuel pump, which pressurizes the fuel system in preparation for the engine to start. When the key is moved just a bit further to the crank position (or the start button is activated), the starter motor is energized and cranks over the engine by turning the crankshaft (**FIGURE 2-19**). The rotation of the crankshaft moves the pistons up and down, drawing air and fuel into the cylinder, which is ignited when the piston reaches the top of the cylinder. Once the cylinders start to fire off, the engine runs by itself. The engine continues to run while the ignition key is in the run position. Depressing the accelerator pedal allows more air and fuel to enter the engine, which greatly increases the engine speed and power.

FIGURE 2-17 Stored chemical energy is converted to mechanical energy to propel the vehicle down the road. **A.** Internal combustion engine. **B.** Hybrid engine with internal combustion engine and electric motor.

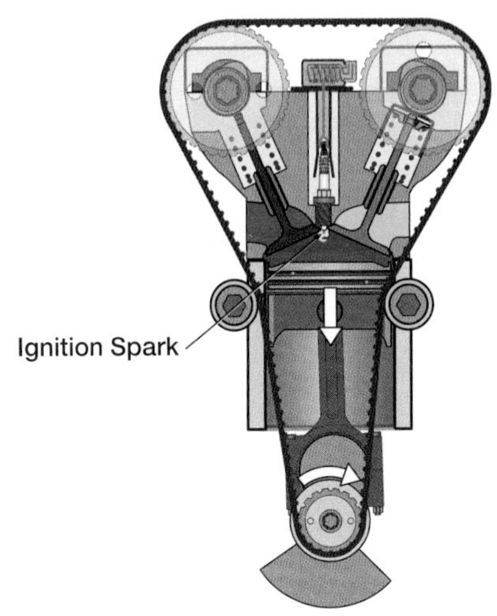

FIGURE 2-18 Internal combustion engine.

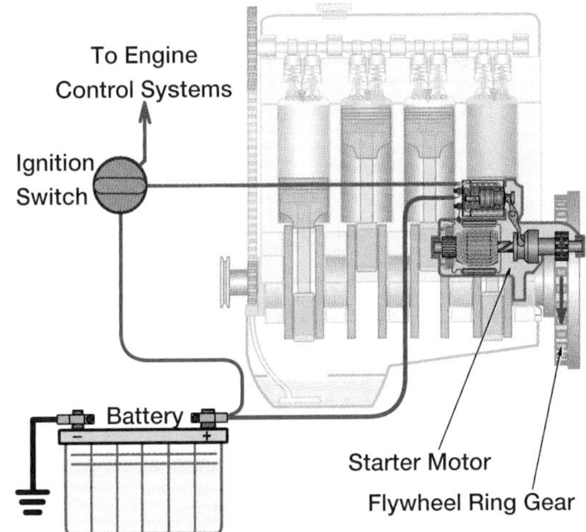

FIGURE 2-19 The starter cranks the engine over.

As the engine cranks, the engine management system monitors the vast array of engine sensors and makes critical decisions about such matters as the correct timing to fire the spark plugs and injecting and metering the proper amount of fuel (**FIGURE 2-20**). The fuel mixture is ignited by a high-voltage spark of up to 100,000 volts created by the ignition system, which causes the burning gases to expand *very* rapidly. In fact, it almost explodes. This expansion shoots each piston down the cylinder, which in turn causes the crankshaft to rotate.

As the pistons continue moving up and down, they rotate the crankshaft, turning the **flywheel** or flex plate, which is bolted to the engine crankshaft (**FIGURE 2-21**). For a manual transmission, the flywheel transmits the engine output to the transmission through a clutch; for an automatic transmission, the flex plate transmits the engine output through a device called a torque convertor.

The vehicle's transmission takes the torque from the engine and multiplies it by running it through a number of different gear ratios, which allow the vehicle to pull quickly away from a stop or cruise effortlessly down the highway (**FIGURE 2-22**). The transmission connects to a final drive assembly that divides up the power and sends it through axles to the drive wheels (**FIGURE 2-23**).

Now that the vehicle is moving at a fast speed, we need to be able to stop it. Brake assemblies are attached to the wheels to provide a means of slowing or stopping the vehicle. The driver operates them by stepping on a brake pedal, which hydraulically applies the brakes through a master cylinder. The brake master cylinder gets assistance from a power booster that amplifies the hydraulic pressure to the wheel brake units. The brake units apply brake friction pads to metal discs or drums connected to the wheels. This creates friction, which transforms the vehicle's kinetic energy into heat energy, slowing the wheel's rotation. The harder the driver presses on the brake pedal, the firmer the brakes are applied and the quicker the vehicle slows down (**FIGURE 2-24**).

The vehicle suspension system suspends the body above the wheels through flexible springs and pivoting links that allow the wheels to follow uneven roads while insulating the passengers from the bumps and dips. It also maintains the orientation of the

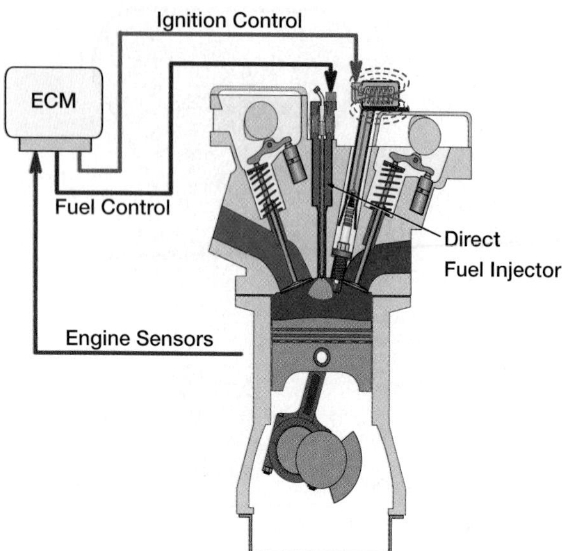

FIGURE 2-20 The engine management system controls the amount of fuel injected.

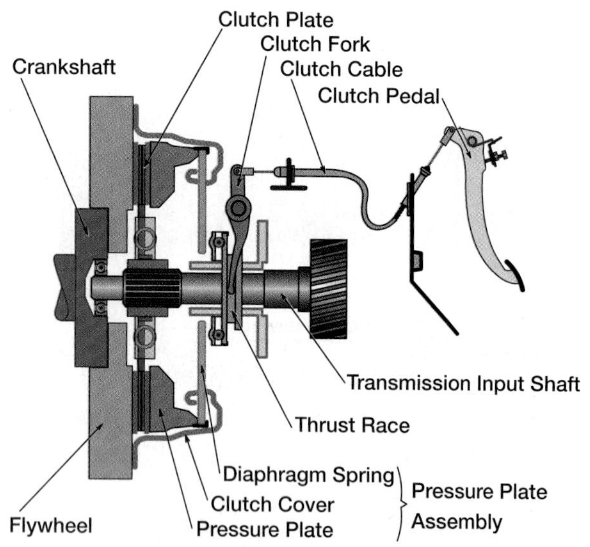

FIGURE 2-21 The crankshaft rotates the flywheel, which sends power to the clutch or torque converter.

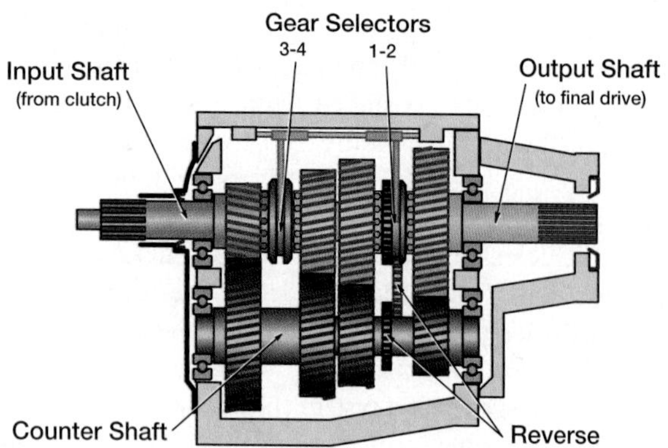

FIGURE 2-22 Transmissions transmit the engine's power through various gears.

FIGURE 2-23 Rear axle assembly sending torque from the driveshaft to the rear wheels.

wheels so the vehicle drives in a stable and predictable manner (**FIGURE 2-25**). The steering system connects the steering wheel to the road wheels so the driver can point the vehicle in the intended direction of travel (**FIGURE 2-26**). Power steering systems assist the driver in steering the vehicle's wheels and use either an engine-driven pump or electric motor to power the system.

The electrical system is the nerve system of the vehicle and is interconnected to all of the other systems. It includes the battery, which stores a supply of electricity for the purpose of starting the vehicle and operating the electrical accessories such as lights, windshield wipers, and the heater system when the engine is not operating (**FIGURE 2-27**). So that the battery is continually recharged, a charging system, driven by the engine, creates electrical energy to feed the battery and run all of the other parts of the electrical system. The electrical system is part of every other system on the vehicle, including the powertrain control system, the lighting system, accessory systems, safety systems, passenger comfort systems, and entertainment systems.

All of these systems combine to provide a safe, efficient, and enjoyable form of transportation. But because not all vehicles are designed the same way, let's look at some of the different drivetrain layouts in the next topic.

▶ Drivetrain Layouts

K02005

The **drivetrain** encompasses the major assemblies that power the vehicle down the road, including the engine, transmission/transaxle, differential, axles, and wheels (**FIGURE 2-28**). Drivetrains are designed in different layouts based on the vehicle application and manufacturer's preferences. For example, the drivetrain layout is different in a pickup used for driving up muddy mountain roads than in a high-performance sports car that runs around a smooth asphalt track. Differences in configuration between each of the drivetrain's major assemblies define the drivetrain layout.

The drivetrain layout includes three main engine mounting positions: front, mid, and rear (**FIGURE 2-29**).

FIGURE 2-24 Brakes stopping a vehicle.

FIGURE 2-25 The suspension system absorbs road shock.

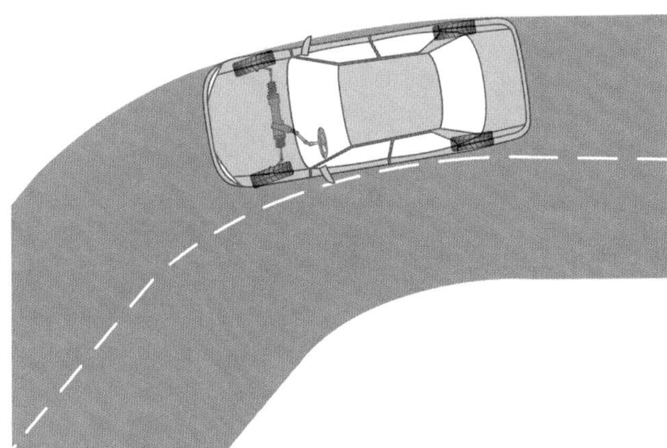

FIGURE 2-26 The steering system allows the driver to control the intended direction of travel.

FIGURE 2-27 The electrical system operates all of the systems on the vehicle.

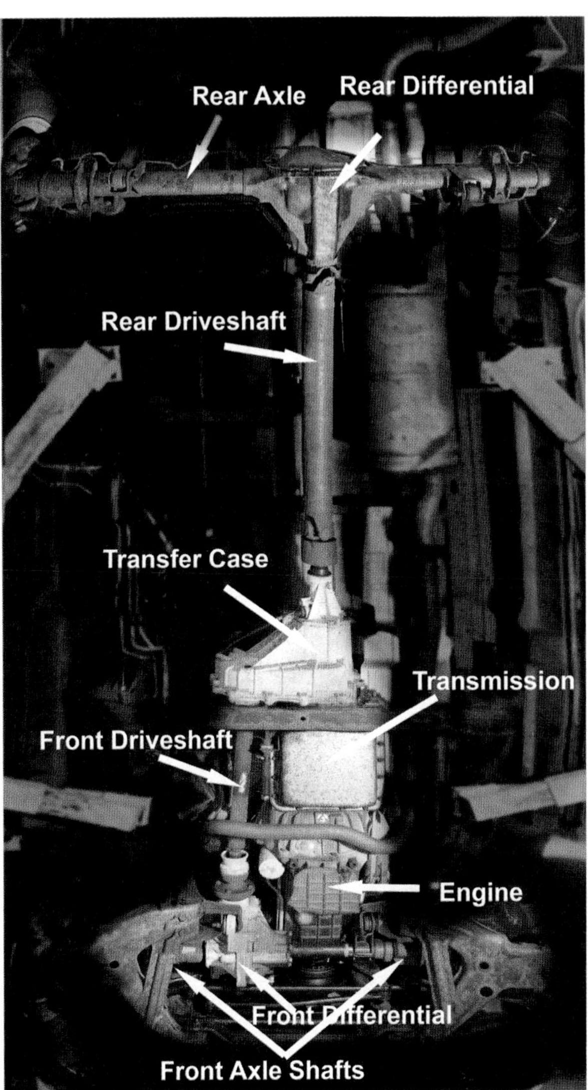

FIGURE 2-28 The drivetrain encompasses the engine, transmission, differential, axles, and wheels.

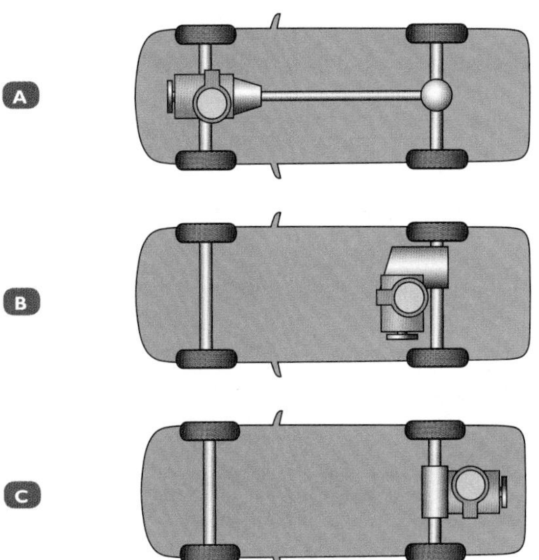

FIGURE 2-29 The three engine-mounting positions. **A.** Front. **B.** Mid. **C.** Rear.

The front engine design is the most common in everyday vehicles and has the engine mounted between the front wheels. Mid-engine vehicles have the engine mounted in front of the rear wheels and provide a more equal weight distribution between the front and rear wheels, but this takes up passenger or cargo space. Rear-engine vehicles have the engine mounted between or behind the rear wheels. Mid- and rear-engine designs are usually reserved for performance-type vehicles, although there have been some exceptions, such as the Volkswagen Beetle. Manufacturers also mount engines in one of two orientations, **longitudinal** (front to back) and **transverse** (side to side), depending on which design best fits the vehicle and the rest of the drivetrain (**FIGURE 2-30**).

The engine does not necessarily drive all four wheels. Drivetrain layouts accommodate four common drive wheel arrangements. **Front-wheel drive (FWD)** is very common in modern vehicles; in this arrangement, only the front wheels are driven by the engine. **Rear-wheel drive (RWD)** is when the engine drives only the rear wheels. Both of these arrangements are called two-wheel drive vehicles. Becoming increasingly popular is **all-wheel drive (AWD)**, with all four wheels driven by the engine all the time. The final arrangement is **four-wheel drive (4WD)**, which is slightly different from all-wheel drive. In a 4WD vehicle, the driver can select between two-wheel drive and four-wheel drive.

The combination of engine position, engine orientation, and type of drive defines the drivetrain layout. For example, using the different variations can give the following drivetrain layouts:

- Front-engine, front-wheel drive
- Front-engine, rear-wheel drive
- Front-engine, all-wheel drive/4WD
- Rear-engine, rear-wheel drive
- Rear-engine, all-wheel drive
- Mid-engine, rear-wheel drive
- Mid-engine, all-wheel drive

▶ Classifying Engines

K02006

Engines are classified by type, cylinder arrangement, number of cylinders/rotors, and total engine displacement in cubic inches or liters. Two common types of engines are piston and rotary (**FIGURE 2-31**). The piston engine uses cylindrical pistons that

FIGURE 2-30 **A.** Longitudinal engine orientation. **B.** Transverse engine orientation.

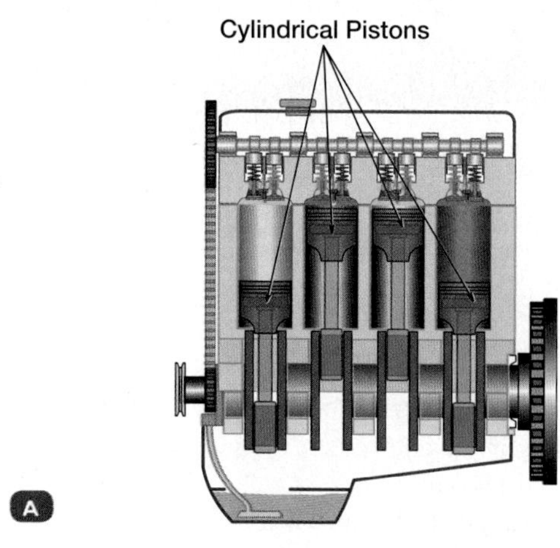

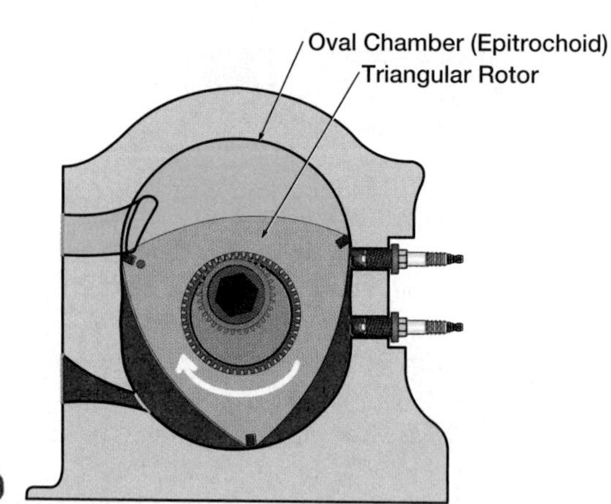

FIGURE 2-31 **A.** Piston engine. **B.** Rotary engine.

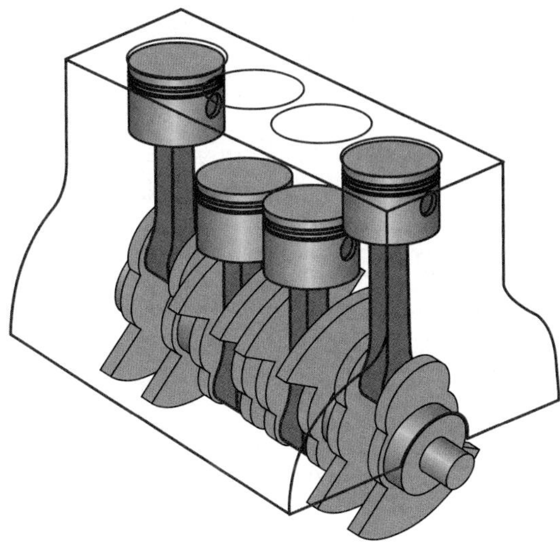

FIGURE 2-32 In-line engine.

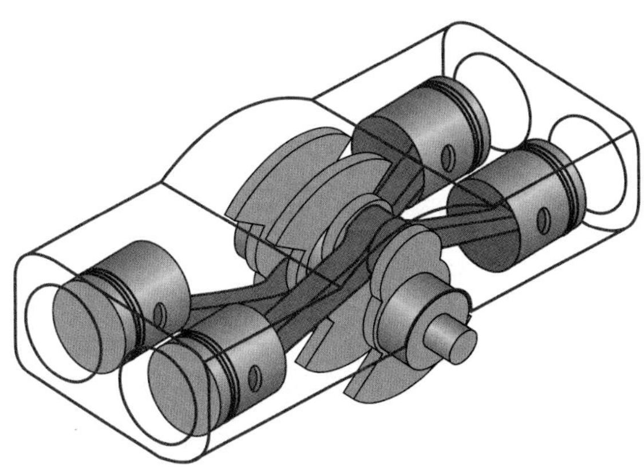

FIGURE 2-33 Horizontally opposed engine.

move up and down in cylinder bores, whereas the rotary engine uses a triangular rotor that turns inside of a roughly oval chamber. The vast majority of automotive engines are of the piston type rather than the less common rotary type, because they have a longer life and produce fewer emissions.

Piston Engines

In a **piston engine**, the way engine cylinders are arranged is called the **engine configuration**. Multi-cylinder internal combustion automotive engines are produced in four common configurations:

- In-line—The pistons are all in one bank on one side of a common crankshaft (**FIGURE 2-32**).
- Horizontally opposed—The pistons are in two banks on both sides of a common crankshaft (**FIGURE 2-33**).
- V—The pistons are in two banks on opposite sides, forming a deep V with a common crankshaft at the base of the V (**FIGURE 2-34**).
- VR and W—In a VR engine, the pistons are in one bank but form a shallow V within the bank. The W engine consists of two VR banks in a deep V configuration with each other (**FIGURE 2-35**).

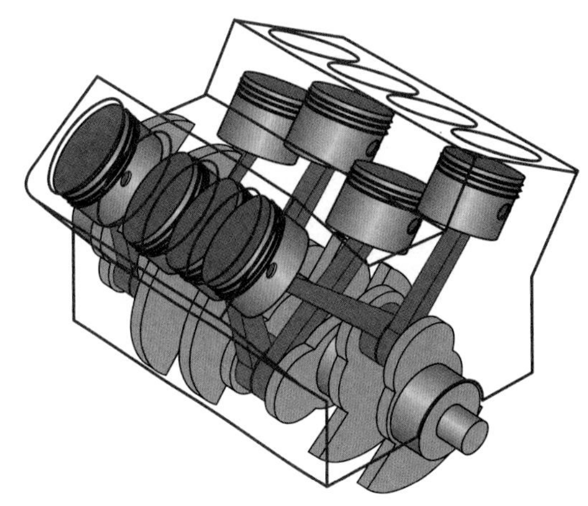

FIGURE 2-34 V engine.

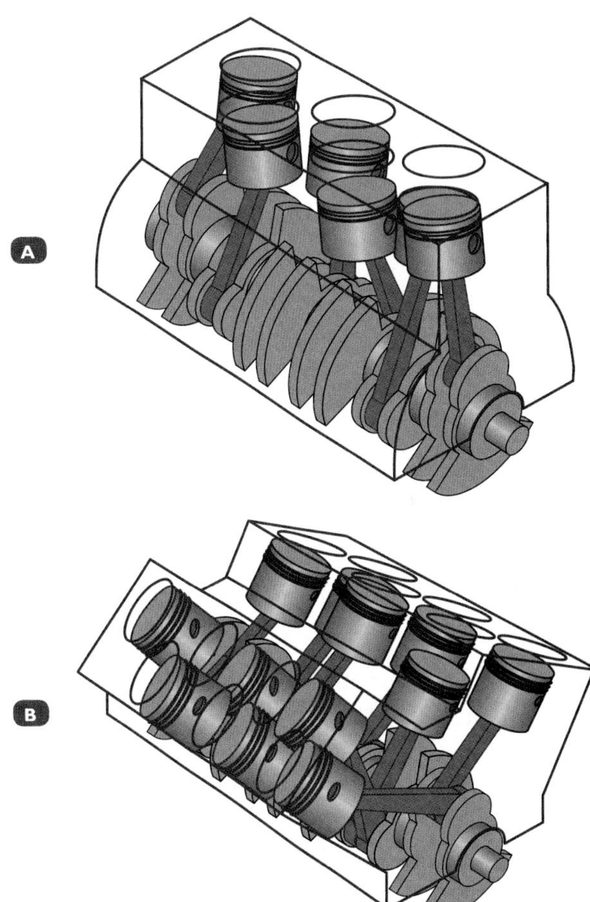

FIGURE 2-35 **A.** VR engine. **B.** W engine.

TECHNICIAN TIP

In-line, horizontally opposed, and V configurations are the most common engine configurations. Recent additions are the VR and W arrangements.

Engineers design engines with tilted cylinder banks to reduce engine height. This can reduce the height of the hood as well, which allows a more streamlined hood line (**FIGURE 2-36**).

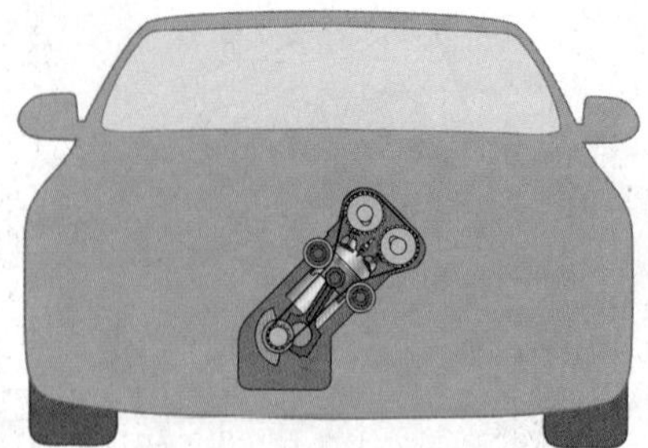

FIGURE 2-36 Tilting cylinder banks can reduce both engine height and hood height, which allows a more streamlined hood line.

Tilting can be carried to an extreme by designing the engine to lie completely on its side in the case of a horizontally opposed engine, also called a flat engine. This greatly reduces the engine height.

As the number of cylinders increases, the length of the engine block and the crankshaft can become a problem structurally and space-wise. One way to avoid this problem is by having more than one row of cylinders, as in a horizontally opposed, V, or W configuration. These designs make the engine block and the crankshaft shorter and more rigid. In vehicle applications, the number of cylinders can vary, usually up to 12. Some examples are listed here:

- In-line 4—Compact, fairly inexpensive, easier to work on, and better fuel economy than other types.
- V8—Approximately twice as powerful as the in-line 4, but with less than twice the space. Good power for its size and not overly complicated.
- Flat 6—Low-profile engine that fits well in vehicles with very low hood lines.
- W12—Most powerful compared to its overall dimensions, but more complicated and expensive than the other engines.

Common angles between the banks of cylinders are 180 degrees, 90 degrees, 60 degrees, and 15 degrees. Angles vary due to the number of cylinders and the manufacturer's design considerations.

In-line

Cylinders arranged side by side in a single row identify the **in-line engine** and can be found in three-, four-, five-, and six-cylinder configurations (**FIGURE 2-37**). There have been in-line eight-cylinder engines, but they are too long to fit into the engine bay of a conventional modern car. In-line engines can be mounted longitudinally (lengthwise) or transversely (sideways) in the engine bay. In-line engines are generally less complicated to design and manufacture because they do not have to share components with a second bank of cylinders. This lack of shared components can mean extra working room in the engine compartment when performing maintenance and repairs. As a general rule of thumb, in-line engines are easier to work on than the other cylinder arrangements.

FIGURE 2-37 Cylinders arranged side by side in a single row identify the in-line engine.

FIGURE 2-38 Horizontally opposed engines are commonly found in four- and six-cylinder configurations.

Horizontally Opposed

Horizontally opposed engines are sometimes referred to as "flat" engines and are commonly found in four- and six-cylinder configurations (**FIGURE 2-38**). They are shorter lengthwise than a comparable in-line engine, but wider than a V type. Horizontally opposed engines have two banks of cylinders, 180 degrees apart, on opposite sides of the crankshaft. It is a useful design when little vertical space is available. A horizontally opposed engine is only fitted longitudinally.

V

V engines have two banks of cylinders sitting side by side in a V arrangement sharing a common crankshaft (**FIGURE 2-39**). This compact design allows for about twice the power output from a V engine as an in-line engine of the same length. In automotive applications, V engines can typically be found in 6-, 8-, 10-, and 12-cylinder configurations. A V6 has two banks of three cylinders; a V8, two banks of four cylinders; etc. The angle of the V tends to vary according to the number of cylinders and can be found by dividing 720 degrees (two rotations of the

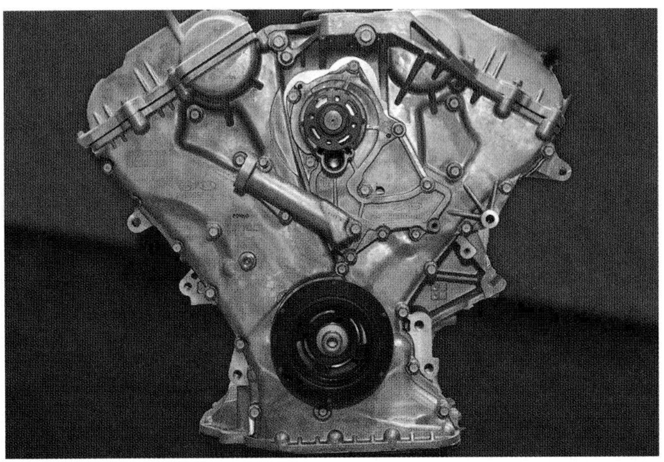

FIGURE 2-39 V engines are shorter than in-line engines of equivalent capacity.

crankshaft, which equals one complete cycle) by the number of cylinders. The natural angle for a V8 is 90 degrees. The natural angle for a V6 is 120 degrees, for a V10 is 72 degrees, and for a V12 is 60 degrees. Designing the engine around the natural angle means the engine can have a shorter length because each crankshaft throw can be shared between two cylinders. Some manufacturers vary their angles from those natural angles and use 90 degrees for a V6 and 15 degrees for a VR6 because of convenience or design requirements. Varying away from the natural angle means that the crankshaft must have one crank throw per cylinder, making the engine slightly longer. This can be accomplished by having completely separate crankshaft journals or by using splayed journals (**FIGURE 2-40**).

FIGURE 2-40 **A.** Individual crankshaft journals. **B.** Splayed crankshaft journals.

VR and W

The **VR engine** uses a single bank of cylinders, but the cylinders are staggered at a shallow 15-degree V within the bank (**FIGURE 2-41**). This design not only allows the engine to be shorter than an in-line engine but also makes it narrower than a typical V engine. The **W engine** consists of two VR cylinder banks in a deeper V arrangement to each other. This gives a very compact, yet powerful, engine design (**FIGURE 2-42**).

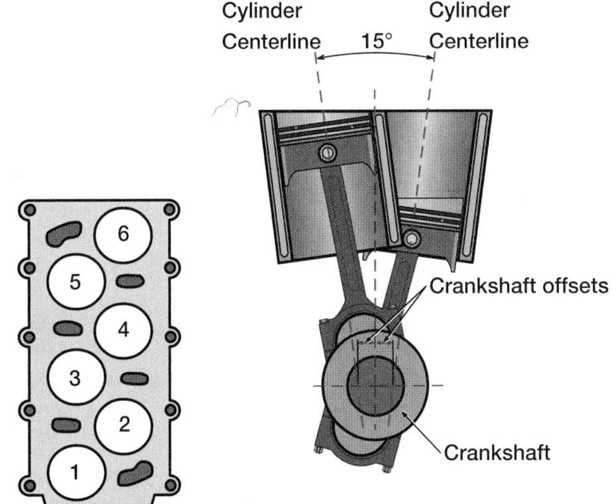

FIGURE 2-41 VR cylinder arrangement.

Rotary Engines

Rotary engines are very powerful for their size, but they do not use conventional pistons that slide back and forth inside a straight cylinder. Instead, a rotary engine uses a roughly triangular rotor that turns inside a roughly oval-shaped housing (**FIGURE 2-43**). The rotary engine has three combustion events for each rotation of the rotor. As the rotor turns, it carries the air-fuel mixture around the chambers, which are created between the tips of the rotor and the chamber wall. The rotor compresses the air-fuel mixture, and spark plugs ignite the mixture, just like a conventional engine. The rotary engine does not have intake and exhaust valves like a traditional piston engine; instead, it has exhaust and intake ports that are covered and uncovered by the rotating rotor in the chamber. This design reduces the number of parts and makes it less complicated than a piston engine. Rotary engines generally have more than one rotor, with two being most common. The design of the rotary engine provides a very compact power unit.

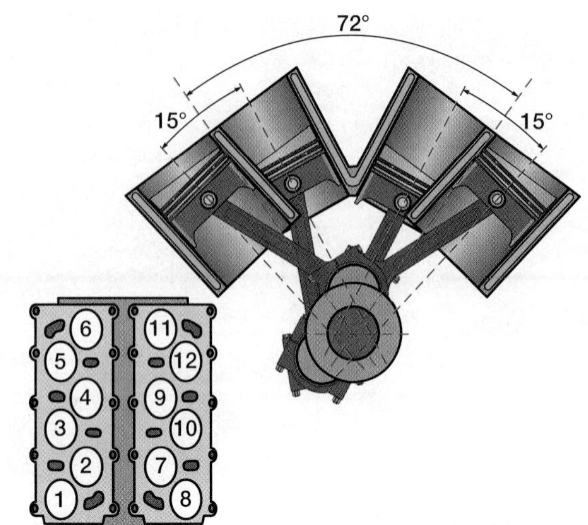

FIGURE 2-42 W engine.

FIGURE 2-43 Rotary engine.

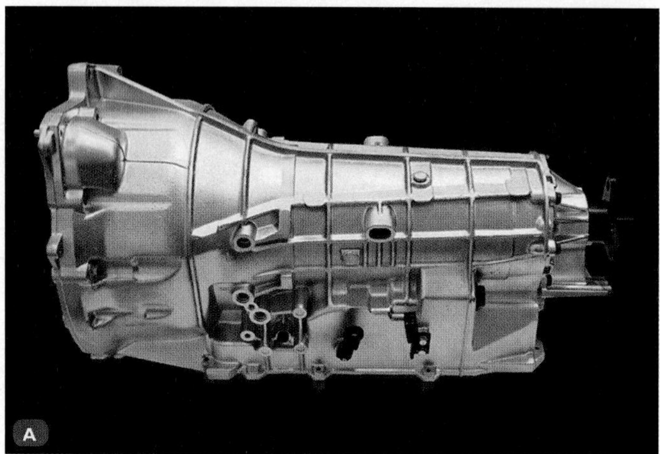

FIGURE 2-44 **A.** Transmission. **B.** Transaxle.

▶ Transmission and Axle Configurations

K02007

In most vehicles, the engine is bolted firmly to either a transmission or a transaxle (**FIGURE 2-44**). A transmission transmits engine power to a driveshaft, final drive and differential gears, and driving axles. Transmissions are usually used in front-engine, rear-wheel drive vehicles. Alternatively, a transaxle is a self-contained unit with the transmission, final drive gears, and differential located in one casing. It is usually used on front-engine, front-wheel drive vehicles or rear-engine, rear-wheel drive vehicles. It can also be used on some sports cars with front-engine, rear-wheel drive, with the transaxle connected to the engine by a driveshaft (**FIGURE 2-45**).

Live and Dead Axles

Vehicles can be described by the number of axles and driven wheels. Each axle typically has one wheel on each end of the axle. **Axles** come in two configurations: live axle and dead axle. Live axles use the engine's torque to turn the wheels (drive the vehicle) and at the same time support the weight of the vehicle. The wheels and axles on a live axle are also called drive wheels and drive axles because they propel the vehicle. Dead axles support the weight of the vehicle only while allowing the wheels to rotate freely on the axle. The wheels on dead axles are not considered drive wheels because they only support the vehicle's weight. Most light vehicles have only two axles: one live axle and one dead axle (**FIGURE 2-46**). On commercial vehicles, the load carried on a single axle is limited by law, so vehicles with extra axles are common. A heavy vehicle may have six wheels (three axles) to support the vehicle, but only two wheels (one axle) may actually drive it. The extra axle at the rear may be used only to support the weight of the vehicle and is sometimes called a lazy axle. This vehicle is called a 6 × 2 (expressed as "6-by-2") vehicle. If the lazy axle is changed to a drive axle, this becomes a 6 × 4 vehicle. Some heavy transport vehicles have an extra steering axle, which allows even more weight to be carried.

Location of Live Axles

The location of the live axle determines whether the vehicle is classified as rear-wheel drive, front-wheel drive, four-wheel drive, or all-wheel drive. If the live axle is at the front of the vehicle, it is considered a front-wheel drive vehicle. Front-wheel drive vehicles use the front wheels to pull the vehicle along. In light passenger

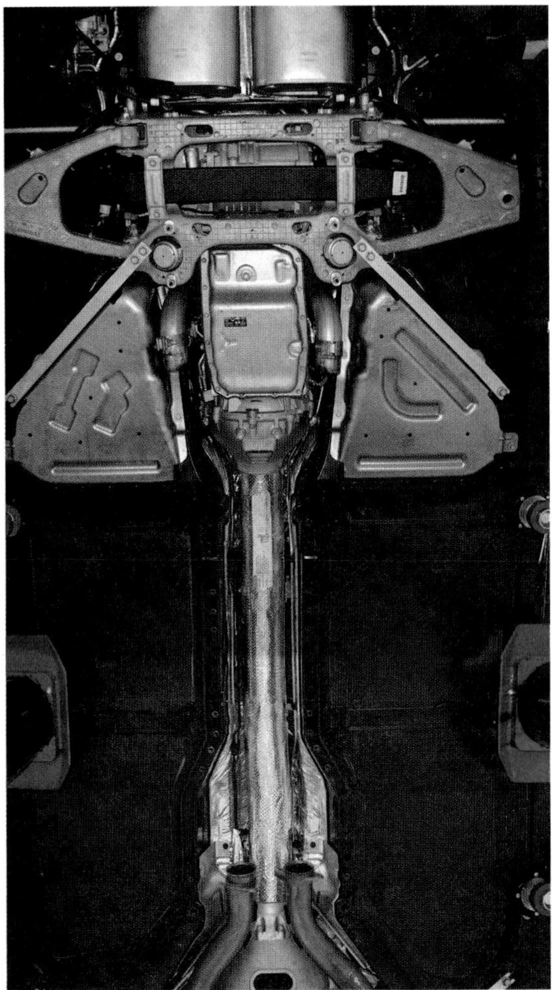

FIGURE 2-45 Front-engine, rear-transaxle arrangement used in some Corvettes.

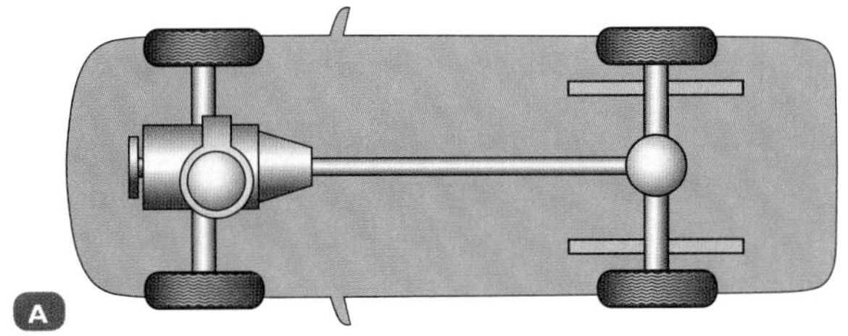

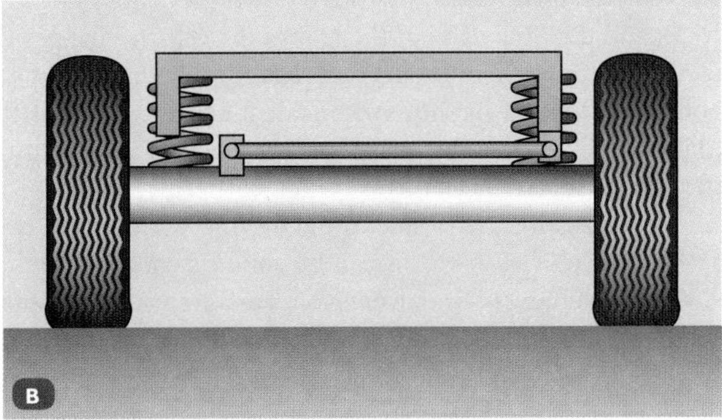

FIGURE 2-46 Most light vehicles only have two axles. **A.** Live axle. **B.** Dead axle.

vehicles, a live front axle gives it lighter body weight and increased interior room. The engine and transaxle are at the front and can be mounted transversely, with the engine parallel to the front axle, or longitudinally, where the engine is in line with the centerline of the vehicle. Front-wheel drive gives good traction on the front wheels, but less traction on the rear wheels, especially when braking or making evasive maneuvers.

If the live axle is at the rear of the vehicle, it is considered a rear-wheel drive vehicle and pushes the vehicle along. With the engine in front, this spreads the weight of the drivetrain assemblies throughout the vehicle, putting more of the weight on the rear wheels. Some rear-wheel drive vehicles have the engine at the rear, driving the wheels through a transaxle. Moving the engine to the rear allows a lower hood line, which may improve aerodynamics. The increase in weight over the rear wheels can improve their traction, but at the expense of less traction on the front wheels.

The previous examples would be called two-wheel drive vehicles, or 4 × 2. This means that there are four wheels, but only two drive wheels. On all-wheel drive vehicles, both axles are live, and all four wheels drive (4 × 4) the vehicle. Driving all four wheels requires additional components and complexity, which add weight and cost. Most 4 × 4 vehicles are more expensive and less fuel efficient than similar 4 × 2 vehicles.

Torque

K02008

Torque is the twisting force applied to a shaft. In a vehicle, torque is used to drive the vehicle down the road (**FIGURE 2-47**). Torque is developed in the engine when combustion of the air-fuel mixture causes high pressure to push the piston down the cylinder and apply a twisting force to the crankshaft. The torque in the crankshaft is then used to twist the gears in the transmission, which is then used to twist the gears in the final drive assembly, which is then used to twist the axles, which is then used to twist the wheels and tires, which drive the vehicle down the road.

Torque is also used by technicians when tightening bolts and nuts (**FIGURE 2-48**). Most of the bolts and nuts on the vehicle have a torque that they must be tightened to, which is specified by the manufacturer. This is accomplished by using a special wrench that has the ability to measure torque. Thus, torque is an important concept for technicians to understand. Torque in the Imperial system is measured by the foot-pound (ft-lb) and inch-pound (in-lb), and in the metric system it is measured by the newton meter (Nm). **TABLE 2-1** lists standard torque conversions.

Torque Measurement

The measure of torque is based on the equivalent twisting force exerted by an amount of weight (mass) applied to a perpendicular lever of a given length. The following designations are used in measuring torque:

- A foot-pound (ft-lb) is the twisting force applied to a shaft by a lever 1 foot long with a 1-pound mass on the end.
- An inch-pound (in-lb) is the twisting force applied to a shaft by a lever 1 inch (25.4 mm) long with a 1-pound (0.45 kg) mass on the end.

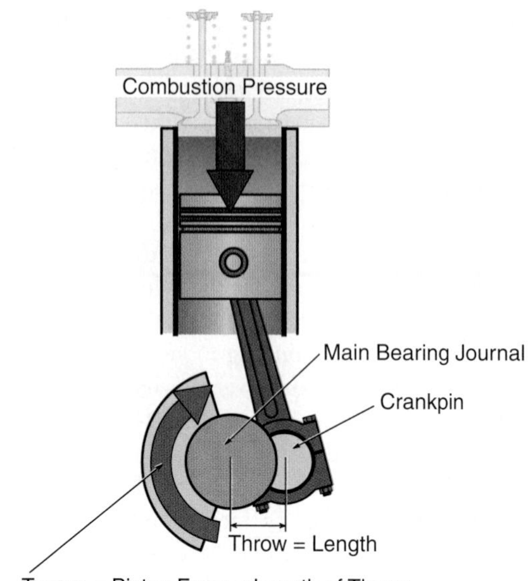

FIGURE 2-47 Torque applied to a shaft.

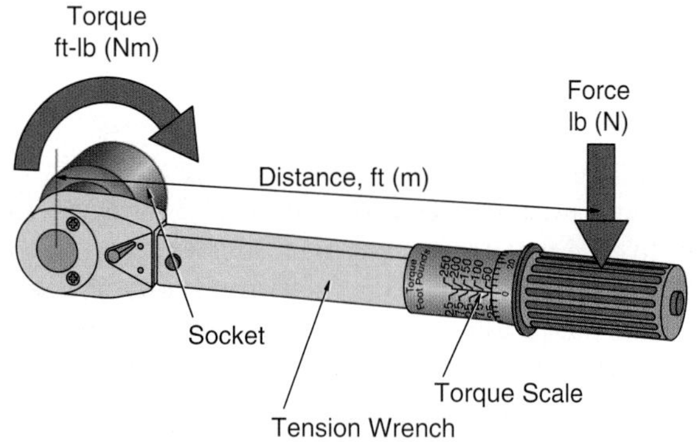

FIGURE 2-48 Torque applied to a bolt.

TABLE 2-1: Torque Conversions

12 in-lb	1 ft-lb
1 in-lb	0.08 ft-lb
1 Nm	0.74 ft-lb
1 Nm	8.8 in-lb

- A Newton meter (Nm) is the twisting force applied to a shaft by a lever 1 meter long with a force of 1 Newton applied to the end of the lever. (1 N is equivalent to the force applied by a mass of 102 g, or 0.102 kg.)

Transmissions and Final Drives

K02009

A vehicle with a manual transmission uses a clutch to engage and disengage the engine from the transmission (**FIGURE 2-49**).

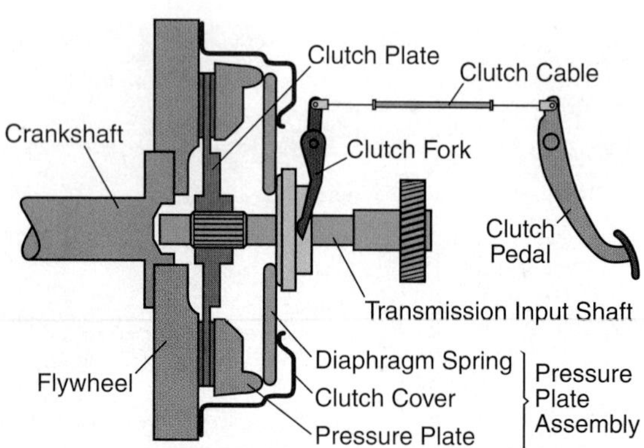

FIGURE 2-49 A vehicle with a manual transmission uses a clutch to engage and disengage the engine from the drivetrain.

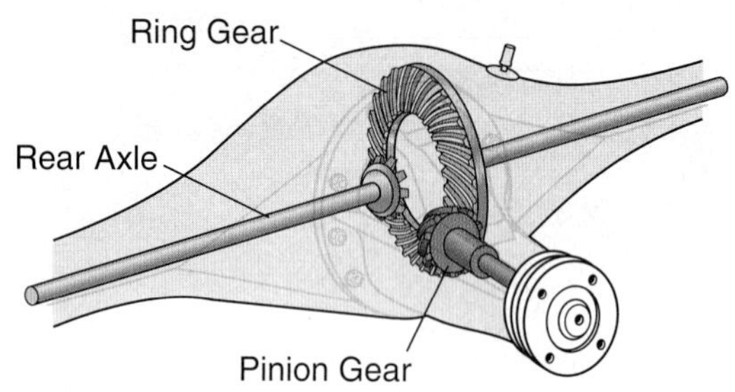

FIGURE 2-50 Final drive assembly.

Note: Differential gears and bearings removed from center of ring gear for illustration purposes.

Engine torque is transmitted through the clutch to the transmission or transaxle. The transmission or transaxle contains sets of gears that increase or decrease the torque, which allows the vehicle to have more pulling power in low gears and higher road speed in higher gears. Drive from the transmission is then transmitted to the rest of the drivetrain. The lower the gear selected, the higher the torque transmitted. A vehicle starting from rest needs a lot of torque, but once it is moving, it can maintain speed with only a relatively small amount of torque. A higher gear can then be selected and engine speed reduced.

A conventional vehicle with the engine at the front and driving wheels at the rear uses a **driveshaft** to transmit torque from the transmission to the final drive. The **final drive** provides a final gear reduction to multiply the torque before applying it to the drive axles (**FIGURE 2-50**). On front-engine, rear-wheel drive vehicles, the driveshaft is fitted down the centerline of the vehicle, and the final drive at the rear of the vehicle changes the direction of the drive by 90 degrees from the center of the vehicle out to the wheels via the axles. Inside the final drive, a **differential gear set** divides the torque to the axles and allows for the difference in speed of each wheel when cornering.

Axles transmit the torque to the driving wheels. In a rear-wheel drive vehicle, the axles can be solid or contain joints to allow for movement of the suspension. For a transaxle-equipped vehicle, each driveshaft has movable joints to allow for suspension and steering movement.

An automatic transmission or transaxle performs similar functions to a manual transmission or transaxle, except that

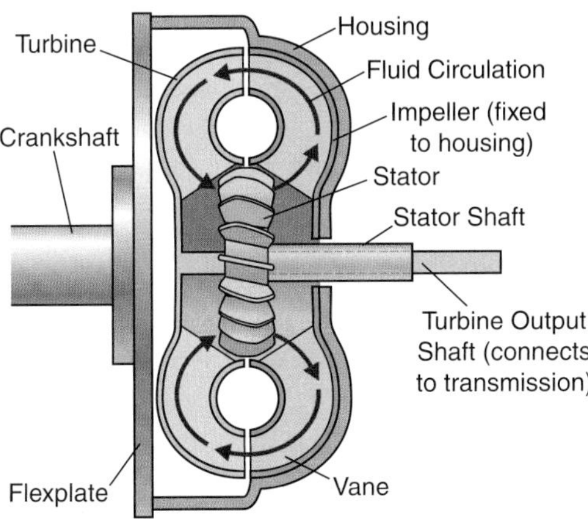

FIGURE 2-51 Torque converter.

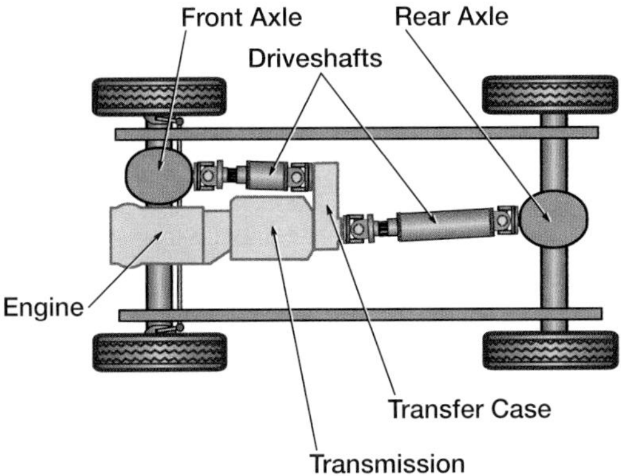

FIGURE 2-52 Four-wheel drive vehicle.

FIGURE 2-53 All-wheel drive layout using a modified transaxle.

gear selection is automatically controlled either hydraulically or electronically. The automatic transmission uses a **torque converter** (instead of a clutch), which acts as a hydraulic coupling to transfer the drive from the engine to the transmission (**FIGURE 2-51**). Automatic transmissions are covered in more depth in the Automatic Transmission section.

Four-Wheel Drive

K02010

A 4WD vehicle can drive all four wheels, so it has a driveshaft, a final drive and differential gears, and axles for both the front and rear axle assemblies (**FIGURE 2-52**). A transfer case is attached to the transmission. The transfer case controls the drive to the front and rear axles. It has a selector lever to select two- or four-wheel drive. Some transfer cases also allow for selection of high and low range, which changes the gear ratios to the wheels. This is helpful when driving in rough off-road conditions, where vehicle speed is very slow.

Part-time 4WD means the vehicle is usually driven in two-wheel drive and switched to four-wheel drive when needed by engaging the transfer case. The transfer case locks the driveshafts together and directs torque through them to both axles. When disengaged, the vehicle transfer case drives only one axle, typically the wheels on the rear axle.

All-Wheel Drive

All-wheel drive vehicles provide drive to all four wheels of the vehicle. All-wheel drive systems should not be confused with part-time four-wheel drive vehicles. The all-wheel drive cannot normally be manually disconnected or deselected. The all-wheel drive vehicle can be used on hard pavement, with drive to all wheels because the transfer case employs a center differential unit that allows the front and rear axles to rotate at slightly different speeds. There are various methods of splitting the drive between the front and rear wheels. Some transfer cases use an electronically controlled multi-plate clutch, whereas others use a **viscous coupling**. Both of these devices allow a small difference in speed between the front and rear axles when the vehicle is turning, but do not allow any great difference in speed, such as when one or more of the wheels lose traction. The driver can still temporarily lock the front and rear axles together by moving a separate lever, as in a conventional 4WD, or by moving the main gear selector. This is called a differential lock.

Some full-time 4WD sedans use a front engine and transaxle, with a driveshaft connected to drive the rear wheels (**FIGURE 2-53**). These cars are lighter and less rugged than conventional off-road types and usually operate at higher speeds. The drive to all four wheels provides better-balanced handling and traction for cornering in slippery conditions.

Regardless of the type of system fitted, the aim is the same: to provide drive to all four wheels constantly while the vehicle is in motion. The power is not necessarily split 50/50 between the front and rear wheels; torque is usually split 60% to front wheels and 40% to rear wheels. Some vehicles incorporate sophisticated traction and torque control systems to maintain effective traction of the wheels to the road under all driving conditions.

▶ Wrap-Up

Ready for Review

- Learning automotive terminology is essential to becoming an automotive technician.
- Vehicle body design types include sedan, station wagon, hatchback, convertible, coupe, van, minivan, pickup, and sport utility vehicle (SUV).
- Sedans have enclosed bodies with a maximum of four doors, whereas coupes only have two doors; either design can be fixed roof or convertible.
- The hatch of a hatchback lifts up at the rear of the vehicle to provide luggage area access.
- Convertibles can convert from enclosed top to open top via a retractable roof.
- Station wagons have increased luggage capacity because of an extended roof.
- Pickup trucks are designed to support greater loads than passenger cars.
- Minivans are lighter-duty versions of full-size vans.
- SUVs can carry and/or tow moderately heavy loads.
- A vehicle chassis is the steel structure that supports the engine, wheels, and transmission.
- The unibody vehicle design replaced the body-on-frame design, allowing for lighter-weight cars that could be manufactured more quickly.
- Vehicle openings include engine compartment hood, hatch and tailgate openings, fuel door, and battery access covers.
- Engine compartment hoods have remote release levers to prevent unauthorized access.
- Vehicles must convert stored chemical energy into mechanical energy.
- Stored energy sources include gasoline, diesel fuel, natural gas, alcohol, bio-diesel, hydrogen, and battery acid.
- Conversion of chemical energy occurs via an internal combustion engine or operation of an electric motor.
- The transmission changes gear ratios to match engine speed and output to desired road speed.
- The suspension system uses shock absorbers, linkages, and springs to connect the wheels to the chassis.
- The electrical system includes the battery, powertrain control system, lighting system, accessory systems, safety systems, passenger comfort systems, and entertainment systems.
- The drivetrain is comprised of the engine, transmission/transaxle, differential, axles, and wheels.
- Drivetrain layout can have a front-, mid-, or rear-mounted engine, which can be oriented longitudinally or transversely.
- The four common drive wheel arrangements are front-wheel drive, rear-wheel drive, all-wheel drive, and four-wheel drive.
- Engines are classified according to type, cylinder arrangement, number of cylinders/rotors, and total engine displacement.
- The two types of engine are piston (most common) and rotary.
- Piston engines are configured in one of four ways: in-line, horizontally opposed, V, or VR/W.
- Multi-cylinder engines vary in the number of cylinders; examples include in-line 4, V8, Flat 6, and W12.
- In-line engines have a single row of side-by-side cylinders.
- Horizontally opposed engines separate two banks of cylinders by 180 degrees, on either side of the crankshaft.
- V engines have two rows of side-by-side cylinders that share a crankshaft.
- VR/W engines use a much narrower angle between cylinders, which are placed in one bank on VR engines and two banks on W engines.
- Rotary engines use one or more triangular rotors.
- Axles can be live and use engine torque to turn the wheels, or they can be dead and just support the weight of the vehicle.
- Axle location determines the drive wheel arrangement.
- Torque is a twisting force that is applied to the crankshaft by movement of the pistons or to nuts and bolts by a technician's wrench.
- Manual transmissions require a clutch to transmit torque and engage/disengage the engine from the transmission.
- A conventional vehicle uses a driveshaft to transmit torque from the transmission to the final drive.
- Front-wheel drive vehicles use half shafts to transmit torque to the wheels.
- Automatic transmissions use a torque converter instead of a clutch.
- Four-wheel drive vehicles can switch from two-wheel to four-wheel drive by engaging the transfer case attached to the transmission.
- All-wheel drive vehicles cannot be switched to two-wheel drive.
- All-wheel drive vehicles provide constant drive to all four wheels, with only a small difference in speed between front and rear axles when the vehicle is turning.

Key Terms

all-wheel drive (AWD) A drivetrain arrangement in which all of the wheels drive the vehicle.

axle A shaft connected to wheels that transmits the driving torque to the wheels.

chassis The main support frame in a vehicle. It includes the running gear, such as suspension, the engine, and the drivetrain.

convertible A vehicle that converts from having an enclosed top to having an open top by a roof that can be removed, retracted, or folded away.

coupe A two-door vehicle that has seating for two people and may have a small rear seat.

differential gear set The arrangement of gears between two axles that allows each axle to spin at its own speed when the vehicle is going around a corner.

driveshaft The shaft or tube fitted with universal couplings that is connected between the transmission and other drivetrain components to transmit torque and rotation.

drivetrain A term used to identify the engine, transmission/transaxle, differential, axles, and wheels.

engine configuration The way engine cylinders are arranged—for example, V, flat, or in-line.

final drive A component that provides a final gear reduction and allows for the difference in speed of each wheel when cornering.

flywheel The heavy disc bolted to the rear of the crankshaft that smooths out the power pulses and stores energy from the power stroke for use in keeping the crankshaft rotating through the other three strokes.

four-wheel drive (4WD) A drivetrain layout in which the engine drive has either two wheels or four wheels, depending on which mode is selected by the driver.

front-wheel drive (FWD) A drivetrain layout in which the engine drives the front wheels.

hatchback A vehicle that has a shared passenger and cargo area; it typically is available in three- and five-door arrangements.

horizontally opposed engine An engine with two banks of cylinders, 180 degrees apart, on opposite sides of the crankshaft. It is also called a flat engine or a boxer engine.

in-line engine An engine in which the cylinders are arranged side by side in a single row.

longitudinal A term used to describe the front-to-back engine orientation when mounted in the engine compartment.

minivan A lighter-duty van used for carrying six to eight occupants or light cargo.

pickup A vehicle that carries cargo; it has stronger chassis components and suspension than a sedan.

piston engine An internal combustion engine that uses cylindrical pistons, moving back and forth in a cylinder, to extract mechanical energy from chemical energy.

rear-wheel drive (RWD) A drivetrain layout in which the engine drives the rear wheels.

rotary engine An engine that uses a triangular rotor turning in a housing instead of conventional pistons.

sedan A vehicle configuration that has an enclosed body, with a maximum of four doors to allow access to the passenger compartment.

sport utility vehicle (SUV) A passenger vehicle built on a light-truck chassis; it is usually equipped with four-wheel drive and capable of hauling heavier loads than typical passenger vehicles.

station wagon A vehicle configuration with four doors, with a roof line that continues into the rear cargo area and a rear door for access.

torque Twisting force.

torque converter A device that is turned by the crankshaft and transmits torque to the input shaft of an automatic transmission.

transverse A term used to describe the side-to-side engine orientation when mounted in the engine compartment.

truck A large, heavy vehicle for carrying cargo.

unibody design A vehicle design that does not use a rigid frame to support the body. The body panels are designed to provide the strength for the vehicle.

V engine A term used to describe an engine configuration that has two banks of cylinders sitting side by side in a V arrangement sharing a common crankshaft.

VR engine A term used to describe an engine configuration that uses a single bank of cylinders staggered at a shallow 15-degree V.

viscous coupling A device that acts like a limited slip clutch.

W engine A term used to describe an engine configuration consisting of two VR cylinder banks in a deep V arrangement.

Review Questions

1. A hatchback typically has:
   - **a.** 2 doors.
   - **b.** 3 doors.
   - **c.** 4 doors.
   - **d.** 6 doors.
2. A design that either partially or wholly integrates the bodywork into a single unit with the chassis is called:
   - **a.** body-on-frame design.
   - **b.** ladder-frame design.
   - **c.** retool design.
   - **d.** unibody design.
3. Which of these systems creates and delivers high voltage sparks at the right time to ignite the air-fuel mixture in the combustion chamber?
   - **a.** Emission system
   - **b.** SRS system
   - **c.** TPMS system
   - **d.** Ignition system
4. Which of these components multiplies the torque from the engine by running it through a number of different gear ratios?
   - **a.** Steering
   - **b.** Suspension
   - **c.** Transmission
   - **d.** Piston
5. Which of these does *not* define the drivetrain layout?
   - **a.** Engine position
   - **b.** Engine capacity
   - **c.** Engine orientation
   - **d.** Number of driven axles
6. Which of the following is true of a horizontally opposed configuration?
   - **a.** The pistons are in two banks on both sides of a common crankshaft.
   - **b.** The pistons are in two banks on opposite sides, forming a deep V with a common crankshaft at the base of the V.

   c. The pistons are in one bank, but form a shallow V within the bank.
   d. The pistons are in two VR banks in a deep V configuration with each other.
7. Which of the following statements is true?
   a. Both live and dead axles drive the vehicle.
   b. Both live and dead axles support the weight of the vehicle.
   c. The dead axles use the engine's torque to turn the wheels.
   d. The lazy axles use the engine's torque to turn the wheels.
8. All of the statements below are true *except*:
   a. Torque twists the gears in the transmission.
   b. Torque twists the gears in the final drive assembly.
   c. Torque turns the axles.
   d. Torque pushes the piston down.
9. Choose the correct statement.
   a. A vehicle starting from rest needs less torque.
   b. The lower the gear selected, the lower the output torque transmitted.
   c. An automatic transmission uses a torque converter instead of a clutch.
   d. The driveshaft divides the torque to the axles.
10. Which vehicles can switch from two-wheel to four-wheel drive by engaging the transfer case attached to the transmission?
   a. Front-wheel drive
   b. All-wheel drive
   c. Rear-wheel drive
   d. Four-wheel drive

ASE Technician A/Technician B Style Questions

1. Tech A says that most vehicles today are built with a ladder frame. Tech B says that most vehicles today do not have a full frame. Who is correct?
   a. Tech A
   b. Tech B
   c. Both A and B
   d. Neither A nor B
2. Tech A says that an SUV is typically a two-door convertible vehicle. Tech B says that many pickups come in two-door and four-door versions. Who is correct?
   a. Tech A
   b. Tech B
   c. Both A and B
   d. Neither A nor B
3. Tech A says that gasoline is energy in chemical form. Tech B says that chemical energy is converted to mechanical energy by the combustion process of an engine. Who is correct?
   a. Tech A
   b. Tech B
   c. Both A and B
   d. Neither A nor B
4. Tech A says that a live axle is an axle that drives the vehicle. Tech B says that a dead axle means that the axle can be steered. Who is correct?
   a. Tech A
   b. Tech B
   c. Both A and B
   d. Neither A nor B
5. Tech A says that the gear ratios in the transmission give the vehicle an extended operating range. Tech B says that the final drive assembly has various gear ratios that can be selected with a gearshift lever. Who is correct?
   a. Tech A
   b. Tech B
   c. Both A and B
   d. Neither A nor B
6. Tech A says that pushing on the brake pedal stops the vehicle by converting chemical energy into thermal energy. Tech B says that brake pedal pressure is transmitted to the brakes hydraulically. Who is correct?
   a. Tech A
   b. Tech B
   c. Both A and B
   d. Neither A nor B
7. Tech A says that one advantage of a mid-engine design is better weight distribution. Tech B says that the rear-engine design is commonly used in four-wheel drive vehicles. Who is correct?
   a. Tech A
   b. Tech B
   c. Both A and B
   d. Neither A nor B
8. Tech A says that the purpose of the battery is to charge the vehicle. Tech B says that in-line engines are generally easier to work on than V engines. Who is correct?
   a. Tech A
   b. Tech B
   c. Both A and B
   d. Neither A nor B
9. Tech A says that a piston engine has a roughly triangular shaped rotor. Tech B says that V engines have horizontally opposed cylinders. Who is correct?
   a. Tech A
   b. Tech B
   c. Both A and B
   d. Neither A nor B
10. Tech A says that the definition of torque is how far the crankshaft twists in degrees. Tech B says that torque can relate to tightness of bolts. Who is correct?
   a. Tech A
   b. Tech B
   c. Both A and B
   d. Neither A nor B

CHAPTER 3

Introduction to Automotive Safety

NATEF Tasks

- **N03001** Identify general shop safety rules and procedures.
- **N03002** Identify the location of the posted evacuation routes.
- **N03003** Identify marked safety areas.
- **N03004** Identify the location and the types of fire extinguishers and other fire safety equipment; demonstrate knowledge of the procedures for using fire extinguishers and other fire safety equipment.
- **N03005** Identify the location and use of eyewash stations.
- **N03006** Locate and demonstrate knowledge of safety data sheets (SDS).

Knowledge Objectives

After reading this chapter, you will be able to:

- **K03001** Outline safe practices in the workplace.
- **K03002** Describe how the Occupational Safety and Health Administration (OSHA) and the Environmental Protection Agency (EPA) impact the workplace.
- **K03003** Describe the difference between a shop policy and a shop procedure.
- **K03004** Identify workplace safety signs and their meanings.
- **K03005** Identify hazards in the shop environment.
- **K03006** Describe the standard safety equipment.
- **K03007** Maintain a safe air quality in the workplace.
- **K03008** List the safety precautions when working with electrical tools and equipment.
- **K0309** Explain the safety importance of the shop layout.
- **K03010** Describe how to reduce the risk of fires in the shop.
- **K03011** Describe how to use firefighting equipment.
- **K03012** Explain how to use an SDS.
- **K03013** Describe how to properly dispose of used engine oil and other petroleum products.
- **K03014** Describe shop safety inspections and hazards commonly found.

Skills Objectives

After reading this chapter, you will be able to:

- **S03001** Safely clean and dispose of hazardous dust.

▶ Introduction

Occupational safety and health is very important to ensure that everyone can work without being injured. Governments normally have legislation in place, with significant penalties for those who do not follow safe practices in the workplace. Potential hazards are in most workplaces, especially automotive shops. It is important to learn about hazards so you can identify them and take action to protect yourself and your coworkers (**FIGURE 3-1**).

Some hazards are obvious, such as vehicles falling from hoists or jacks, or tires exploding during inflation. Other hazards are less obvious, such as the long-term effects of breathing fumes from solvents. There are many things to learn about safety in the automotive shop, but it is impossible to cover every situation you will encounter. One of the most important skills to learn is the ability to recognize unsafe practices or equipment and put in place measures to prevent injuries from happening.

Occupational safety and health is everyone's responsibility. You have a responsibility to ensure that you work safely and take care not to put others at risk by acting in an unsafe manner. Your employer also has a responsibility to provide a safe working environment. To ensure the safety of yourself and others, make sure you are aware of the correct safety procedures at your workplace. This means listening very carefully to safety information provided by your employer and asking for clarification, help, or instructions if you are unsure how to perform a task safely. Always think about how you are performing shop tasks, be on the lookout for unsafe equipment and work practices, and wear the correct **personal protective equipment (PPE)**. PPE refers to items of safety equipment like safety footwear, gloves, clothing, protective eyewear, and hearing protection (**FIGURE 3-2**).

FIGURE 3-1 Identifying hazards.

▶ Safety Overview

N03001, K03001

Motor vehicle servicing is one of the most common vocations worldwide. Hundreds of thousands of shops service millions of vehicles every day. That means at any given time, many people are conducting automotive servicing, and there is great potential for things to go wrong (**FIGURE 3-3**).

It is up to you and your workplace to make sure all work activities are conducted safely. Accidents are not caused by properly maintained tools; accidents are generally caused by people.

▶ TECHNICIAN TIP

Whenever you perform a task in the shop, you must use personal protective clothing and equipment that are appropriate for the task and that conform to your local safety regulations and policies. Among other items, these may include:

- Work clothing, such as coveralls and steel-capped footwear
- Eye protection, such as safety glasses and face masks
- Ear protection, such as earmuffs and earplugs
- Hand protection, such as gloves and barrier cream
- Respiratory equipment, such as face masks and valved respirators
- If you are not certain what is appropriate or required, ask your supervisor.

You Are the Automotive Technician

You are changing the oil on a new type of vehicle for the first time. The oil pan has the drain plug on the side of the oil pan instead of the bottom of the pan. You place the drain pan directly under the drain plug like you normally do. Unfortunately, when the plug comes out, the oil shoots sideways right over the side of the drain pan. You reposition it quickly, but not before a large puddle is on the floor.

1. Why is it important to review the SDS before cleaning up a spill?
2. What is the minimum PPE that should be worn to manage this spill?
3. What are some of the health hazards of coming into frequent or prolonged contact with used engine oil?
4. If the oil were to catch fire, what fire extinguisher should be used to put the fire out?

FIGURE 3-2 Personal protective equipment (PPE) refers to items of safety equipment like safety footwear, gloves, clothing, protective eyewear, and hearing protection.

FIGURE 3-3 Accident in the shop.

Don't Underestimate the Dangers

Because vehicle servicing and repair are so commonplace, it is easy to overlook the many potential risks related to this field. Think carefully about what you are doing and how you are doing it. Think through the steps, trying to anticipate things that may go wrong and taking steps to prevent them. Also be wary of taking shortcuts. In most cases, the time saved by taking a shortcut is nothing compared to the time spent recovering from an accident (**FIGURE 3-4**).

Accidents and Injuries Can Happen at Any Time

There is the possibility of an accident occurring whenever work is undertaken. For example, fires and explosions are a constant hazard wherever there are flammable fuels. Electricity can kill quickly as well as cause painful shocks and burns. Heavy equipment and machinery can easily cause broken bones or crush fingers and toes. Hazardous solvents and other chemicals can burn or blind as well as contribute to many kinds of illness. Oil

FIGURE 3-4 Technician recovering from a broken arm due to a slip or fall.

spills and tools left lying around can cause slips, trips, and falls. Poor lifting and handling techniques can cause chronic strain injuries, particularly to your back (**FIGURE 3-5**).

Accidents and Injuries Are Avoidable

Almost all accidents are avoidable or preventable by taking a few precautions. Think of nearly every accident you have witnessed or heard about. In most cases someone made a mistake. Whether caused by horseplay, neglecting maintenance on tools or equipment, or using tools improperly, these instances lead to injury (**FIGURE 3-6**).

Most of these accidents can be prevented if people follow policies and develop a "safety first" attitude. By following regulations and safety procedures, you can make your workplace much safer. Learn and follow all of the correct safety procedures for your workplace. Always wear the right PPE, and stay alert and aware of what is happening around you. Think about what you are doing, how you are doing it, and its effect on others. You also need to know what to do in case of an emergency. Document and report all accidents and injuries whenever they happen, and take the proper steps to make sure they never happen again.

FIGURE 3-5 Poor lifting and handling techniques can cause chronic strain injuries, particularly to your back.

Evacuation Routes

N03002

Evacuation routes are a safe way of escaping danger and gathering in a prearranged safe place where everyone can be accounted for in the event of an emergency. It is important to have more than one evacuation route in case any single route is blocked during the emergency. Your shop may have an evacuation procedure that clearly identifies the evacuation routes (**FIGURE 3-7**).

Often the evacuation routes are marked with colored lines painted or taped on the floors. Exits should be highlighted with signs that may be illuminated, and should never be chained closed or obstructed (**FIGURE 3-8**).

Always make sure you are familiar with the evacuation routes for the shop. Before conducting any task, identify which route you will take if an emergency occurs.

> **TECHNICIAN TIP**
>
> Never place anything in the way of evacuation routes, including equipment, tools, parts, cleaning supplies, or vehicles.

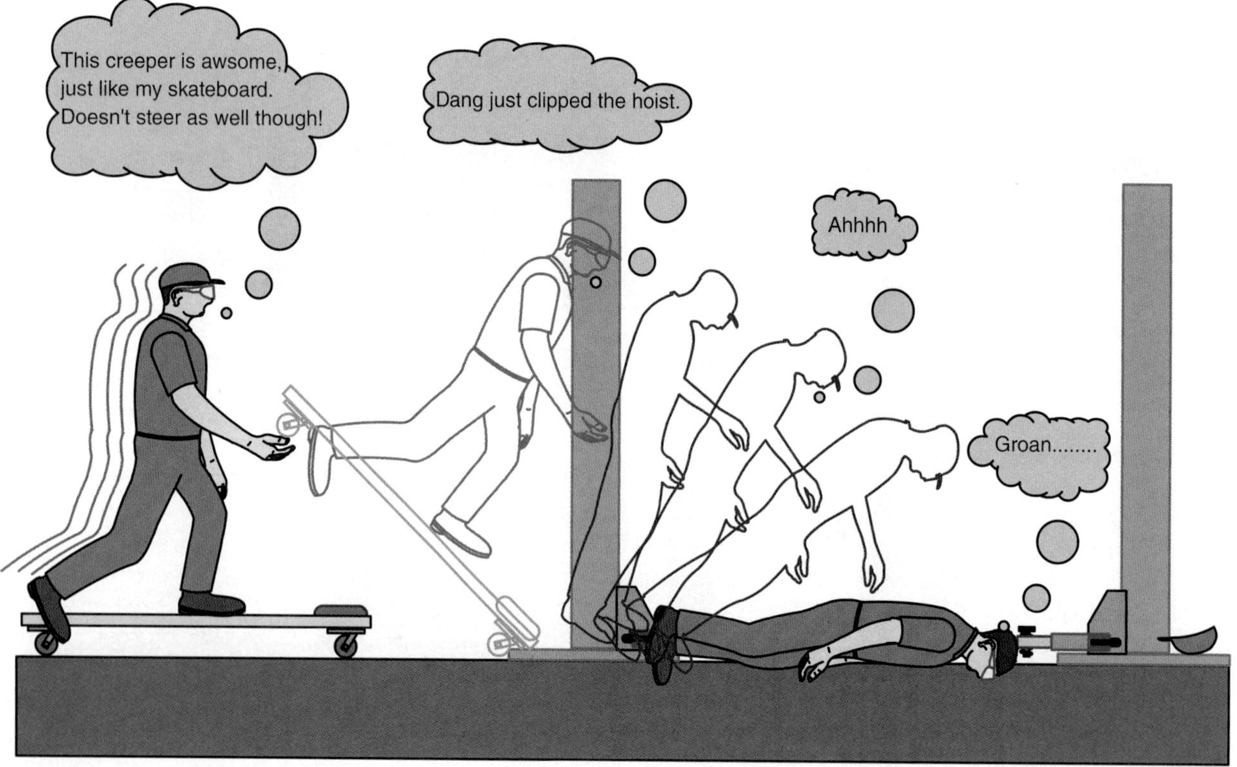

FIGURE 3-6 Accidents are usually caused by mistakes.

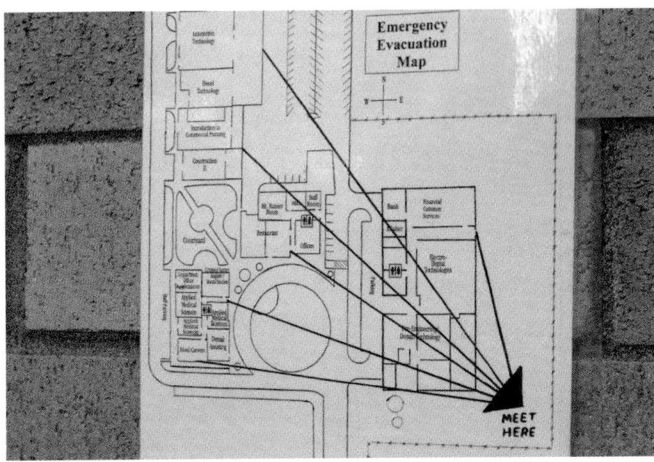

FIGURE 3-7 Your shop may have an evacuation procedure that clearly identifies the evacuation routes.

FIGURE 3-8 Exits should be marked, clear of obstructions, and not chained closed.

▶ Work Environment

The work environment can be described as anywhere you work. The condition of the work environment plays an important role in making the workplace safer. A safe work environment goes a long way toward preventing accidents, injuries, and illnesses. There are many ways to describe a safe work environment, but generally it would contain a well-organized shop layout, use of shop policies and procedures, safe equipment, safety equipment, safety training, employees who work safely, good supervision, and a workplace culture that supports safe work practices. Conversely, a shop that is cluttered with junk, poorly lit, and full of safety hazards is unsafe (**FIGURE 3-9**).

FIGURE 3-9 **A.** Relatively safe shop. **B.** Relatively unsafe shop.

OSHA and EPA

K03002

OSHA stands for the **Occupational Safety and Health Administration (OSHA)**. It is a U.S. government agency that was created to provide national leadership in occupational safety and health, and it works toward finding the most effective ways to help prevent worker fatalities and workplace injuries and illnesses. OSHA has the authority to conduct workplace inspections and, if required, fine employers and workplaces if they violate its regulations and procedures (**FIGURE 3-10**). For example, a fine may be imposed on the employer or workplace if a worker is electrocuted by a piece of faulty machinery that has not been regularly tested and maintained.

EPA stands for the **Environmental Protection Agency**. This federal government agency deals with issues related to environmental safety. The EPA conducts research and monitoring, sets standards, conducts workplace inspections, and holds employees and companies legally accountable in order to keep the environment protected. Shop activities need to comply with EPA laws and regulations by ensuring that waste products are disposed of in an environmentally responsible way, chemicals and fluids are correctly stored, and work practices do not contribute to damaging the environment (**FIGURE 3-11**).

FIGURE 3-10 OSHA inspection.

FIGURE 3-11 Correct chemical storage.

Applied Science

AS-2: Environmental Issues: The technician develops and maintains an understanding of all federal, state, and local rules and regulations regarding environmental issues related to the work of the automobile technician.

You need to keep up to date with local laws and regulations regarding environmental issues. There are large fines associated with disregarded environmental regulations. To remain informed, you can log on to local state websites to check the latest laws and regulations. The US Environmental Protection Agency (EPA) website also has up-to-date information on environmental regulations (www.epa.gov/lawsregs/). More information can be found on the specific regulations for the automotive industry at www.epa.gov/lawsregs/sectors/automotive.html. This site has a full listing of laws and regulations, compliance measures, and enforcement tactics. For additional laws that are enforced by your local state environmental agencies, go to www.epa.gov/epahome/state.htm.

AS-3: Environmental Issues: The technician uses such things as government impact statements, media information, and general knowledge of pollution and waste management to correctly use and dispose of products that result from the performance of a repair task.

When you complete a job, you must be able to identify what to do with any waste products created from the repair task. This could be as simple as knowing where and how to recycle cardboard boxes or as complex as knowing what to do with brake components that may contain asbestos. Normally, there is a table posted in the garage that details the correct measures for disposing of and storing waste material. A sample table follows:

Component	Material/Parts	Removal and Safety Information	Recommended Storage
Air-conditioning gases (refrigerant)	R12, R134a	Use approved evacuation and collection equipment required. *Note:* Requires AC license.	Use approved storage containers – reused or recycled.
Batteries	Plastic, rubber, lead, sulfuric acid	Avoid contact between sulfuric acid and your skin, clothing, or eyes.	Store off the ground in a covered area for collection by recycler.
Brake fluid	Diethylene and polyethylene glycol-monoalkyl ethers	These are corrosive and highly toxic to the environment. Drain into pan or tray.	Store in a drum in a covered area for collection by a licensed operator.
Brake shoes and pads pre-2004	Asbestos	Fibers are dangerous if inhaled.	Put in a plastic bag in a sealed container for collection by contractor.
Coolant	Phosphoric acid, hydrazine ethylene glycol, alcohols	Radiator coolant can be toxic to the environment.	Store in sealed drums in a covered area for recycling or collection by a licensed operator.
Coverings for plastic parts, and plastic bags and containers for parts shipping	Plastic-made components	Plastic components that can be recycled bear a recycling symbol: this symbol has a number inside telling the recycling company what the product is made of.	If the plastic container has a recycle code, then put it in the recycle bin. If it does not, then put in general waste. Plastic oil containers cannot be recycled.
Fuel	Unleaded, diesel	Avoid fumes. Fire hazard; keep well away from ignition sources. Siphon from tank to avoid spillage.	Store in a drum in a covered area.

Metal	Brake discs, housings made of metal (gearbox/engine case and components), metal cuttings	Some metal products can be heavy, so lift with care.	All metal components can be recycled; keep waste metal in a separate recycle bin for sale or disposal.
Oil	Engine, transmission, and differential oils	Fire hazard; keep well away from ignition sources.	Store in a drum/container in a covered area for collection by a licensed operator.
Oil filters	Steel paper fiber	—	Drain filter, then crush and store in a leakproof drum for collection.
Parts boxes and paperwork	Paper, cardboard	—	Store in a recycling bin to be taken away for recycling.
Tires	Rubber, steel, fabric	Keep away from ignition sources.	Store in a fenced area for collection by recycler.
Trim, plastic fittings, and seats	Plastic, metal, cloth	—	Store racked or binned for reuse, sale, or recycling.
Tubes and rubber components	Rubber hoses, mounts, etc.	Keep away from ignition sources.	Store in a collection bin, to be collected and recycled.
Undeployed airbags	Plastics, metals, igniters, explosives	Recommended specific training on airbags before attempting removal. Handle with care; accidental deployment can cause serious harm. If unit is to be scrapped, ensure that it is safely deployed first.	Store face up in a secure area.

Shop Policies and Procedures

K03003

Shop policies and procedures are a set of documents that outline how tasks and activities in the shop are to be conducted and managed. They also ensure that the shop operates according to OSHA and EPA laws and regulations. A **policy** is a guiding principle that sets the shop direction, and a **procedure** is a list of the steps required to get the same result each time a task or activity is performed (**FIGURE 3-12**). An example of a policy is an OSHA document for the shop that describes how the shop complies with legislation, such as a sign simply saying, "Safety glasses must be worn at all times in the shop." An example of a procedure is a document that describes the steps required to safely use the vehicle hoist.

Each shop has its own set of policies and procedures and a system in place to make sure the policies and procedures are regularly reviewed and updated. Regular reviews ensure that new policies and procedures are developed and old ones are modified in case something has changed. For example, if the shop moves to a new building, then a review of policies and procedures will ensure that they relate to the new shop, its layout, and equipment. In general, the policies and procedures are written to guide shop practice; help ensure compliance with laws, statutes, and regulations; and reduce the risk of injury. Always follow your shop policies and procedures to reduce the risk of injury to your coworkers and yourself and to prevent damage to property.

It is everyone's responsibility to know and follow the rules. Locate the general shop rules and procedures for your workplace. Look through the contents or index pages to familiarize yourself with the contents. Discuss the policy and the shop rules and procedures with your supervisor. Ask questions to ensure that you understand how the rules and procedures should be applied and your role in making sure they are followed.

▶ Standard Safety Measures

N03003

Signs

K03004

Always remember that a shop is a hazardous environment. To make people more aware of specific shop hazards, legislative bodies have developed a series of safety signs. These signs are designed to give adequate warning of an unsafe situation. Each sign has four components:

- **Signal word:** There are three signal words—danger, warning, and caution. *Danger* indicates an immediately hazardous situation, which, if not avoided, will result in death or serious injury. Danger is usually indicated by white text with a red background. *Warning* indicates a potentially hazardous situation, which, if not avoided, could result in death or serious injury. The sign is usually in black text with an orange or yellow background. *Caution* indicates a potentially hazardous situation, which, if not avoided, may result in minor or moderate injury. It may also be used to alert against unsafe practices. This sign is usually in black text with a yellow background (**FIGURE 3-13**).
- **Background color:** The choice of background color also draws attention to potential hazards and is used to provide contrast so the letters or images stand out. For example, a red background is used to identify a definite hazard; yellow indicates caution for a potential hazard. A

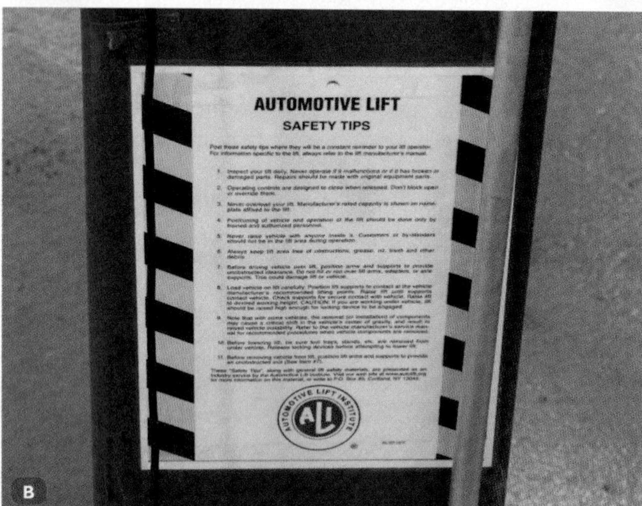

FIGURE 3-12 **A.** Policy. **B.** Procedure.

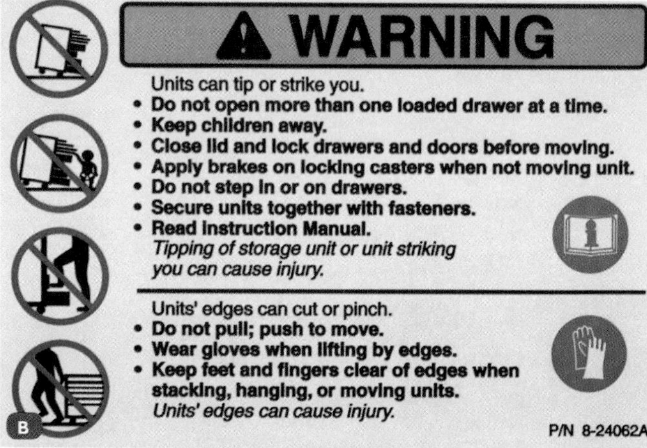

FIGURE 3-13 Signs. **A.** Danger is usually indicated by white text on a red background. **B.** Warning is usually in black text with an orange background. **C.** Caution is usually in black text with a yellow background.

green background is used for emergency-type signs, such as for first aid, fire protection, and emergency equipment. A blue background is used for general information signs.

- **Text:** The sign will sometimes include explanatory text intended to provide additional safety information. Some signs are designed to convey a personal safety message.
- **Pictorial message:** In symbol signs, a pictorial message appears alone or is combined with explanatory text. This type of sign allows the safety message to be conveyed to people who are illiterate or who do not speak the local language.

Identifying Hazardous Environments

K03005

A **hazardous environment** is a place where hazards exist. A **hazard** is anything that could hurt you or someone else, and most workplaces have them. It is almost impossible to remove all hazards, but it is important to identify hazards and work to reduce their potential for causing harm by putting specific measures in place. For example, operating a bench grinder poses a number of hazards. Although it is not possible to eliminate the hazards of using the bench grinder, by putting specific measures in place, the risk of those hazards can be reduced. A risk analysis of a bench grinder would identify the following hazards and risks: a high-velocity particle that could damage your eyesight or that of someone working nearby; the grinding wheel breaking apart, damaging eyesight or causing cuts and abrasion; electrocution if electrical parts are faulty; a risk to your hands from heat or high-velocity particles; a risk to your hearing due to excessive noise; a risk of fire if flammables are present, and a risk of entrapment of clothing or body parts through rotating machinery. To reduce the risk of these hazards, the following measures are taken: Position the bench grinder in a safe area away from where others work; make sure electrical items are regularly checked for electrical and mechanical safety; when operating the equipment, wear PPE such as protective eyewear, gloves, hearing protection, hairnets, or caps; make sure no flammables are nearby; and do not wear loose clothing that can be caught in the bench grinder.

To identify hazardous environments, follow the steps in **SKILL DRILL 3-1**.

Safety Equipment

K03006

Shop safety equipment includes items such as:

- **Handrails:** Handrails are used to separate walkways and pedestrian traffic from work areas. They provide a physical barrier that directs pedestrian traffic and also offer protection from vehicle movements (**FIGURE 3-14**).
- **Machinery guards:** Machinery guards and yellow lines prevent people from accidentally walking into the operating equipment or indicate that a safe distance should be kept from the equipment (**FIGURE 3-15**).
- **Painted lines:** Large, fixed machinery such as lathes and milling machines present a hazard to the operator and others working in the area. To prevent accidents, a machinery guard or a painted yellow line on the floor usually borders this equipment (**FIGURE 3-16**).

SKILL DRILL 3-1 Identifying Hazardous Environments

1. Familiarize yourself with the shop layout. There are special work areas that are defined by painted lines. These lines show the hazardous zone around certain machines and areas. If you are not working on the machines, you should stay outside the marked area.

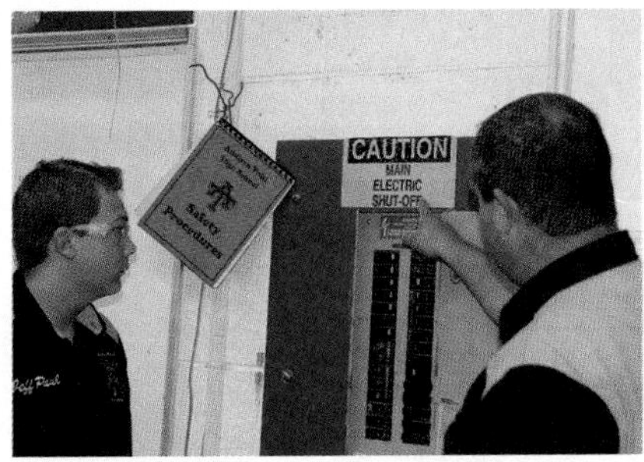

2. Study the various warning signs around your shop. Understand the meaning of the signal word, the colors, the text, and the symbols or pictures on each sign. Ask your supervisor if you do not fully understand any part of the sign.

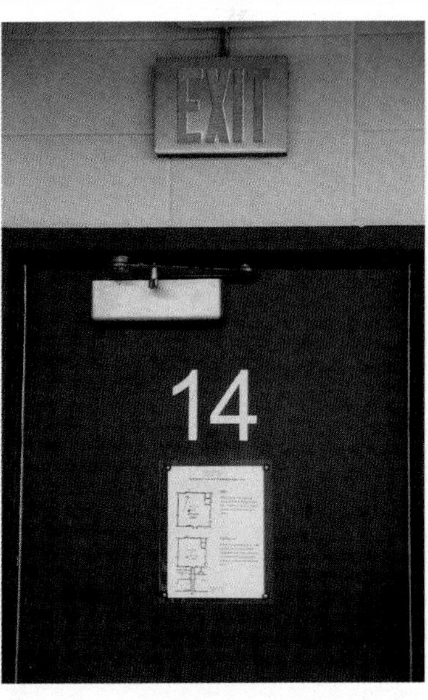

3. Identify exits. Find out where each door, window, and gate is and whether it is usually open or locked. Plan your escape route, should you need to exit in a hurry. Also know the designated gathering point and go there in an emergency.

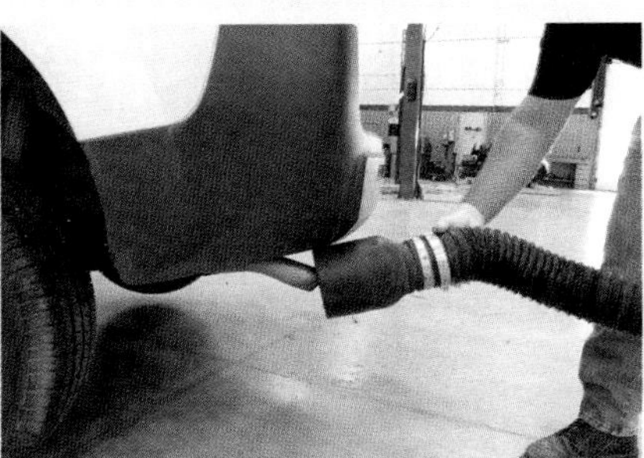

4. Check for air quality. There should be good ventilation and very little chemical fumes or smell. Locate the extractor fans or ventilation outlets, and make sure they are not obstructed in any way. Locate and observe the operation of the exhaust extraction hoses, pump, and outlets that are used on the vehicle's exhaust pipes.

5. Check the location and types of fire extinguishers in your shop. Be sure you know when and how to use each type of fire extinguisher.

SKILL DRILL 3-1 Identifying Hazardous Environments (Continued)

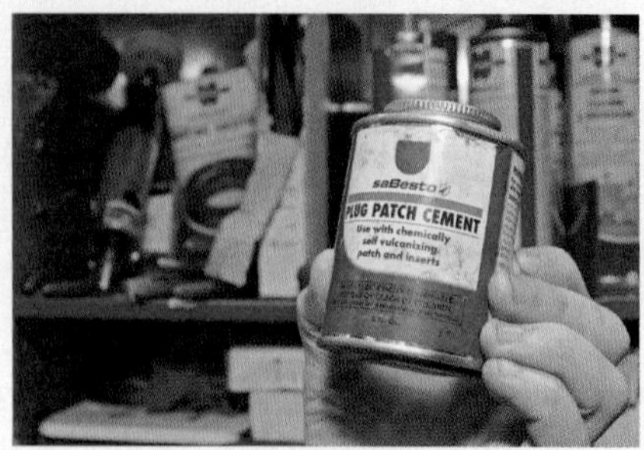

6. Identify flammable hazards. Find out where flammable materials are kept, and make sure they are stored properly.

7. Check the hoses and fittings on the air compressor and air guns for any damage or excessive wear. You must be particularly careful when troubleshooting air guns. Always disconnect an air gun from the air line before inspecting it—if you do not, severe eye damage can result.

8. Identify caustic chemicals and acids associated with activities in your shop. Ask your supervisor for information on any special hazards in your particular shop and any special avoidance procedures, which may apply to you and your working environment.

FIGURE 3-14 Handrail.

FIGURE 3-15 Machinery guard.

- **Sound-insulated rooms:** Sound-insulated rooms are usually used when operating equipment makes a lot of noise, for example a chassis dynamometer. A vehicle operating on a dynamometer produces a lot of noise from its tires, exhaust, and engine. To protect other shop users from the noise, the dynamometer is usually placed in a sound-insulated room, keeping shop noise to a minimum (**FIGURE 3-17**).
- **Adequate ventilation:** Exhaust gases and chemical vapors are serious health hazards in the shop. Whenever a vehicle's engine is running, toxic gases are emitted from its exhaust. To prevent an excess of toxic gas buildup, a well-ventilated work area is needed, as well as a method of directly venting the vehicle's exhaust to the outside. It may only take a minute or two for a poorly running vehicle to fill the shop with enough carbon monoxide to affect people's health. Chemical vapors are also a hazard and need to be vented outside (**FIGURE 3-18**).
- **Doors and gates:** Doors and gates are used for the same reason as machinery guards and painted lines. A doorway is a physical barrier that can be locked and sealed to separate a hazardous environment from the rest of the shop or a general work area from an office or specialist work area (**FIGURE 3-19**).

FIGURE 3-16 Painted safety lines.

FIGURE 3-17 Sound-insulated room.

- **Temporary barriers:** In the day-to-day operation of a shop, there is often a reason to temporarily separate one work bay from others. If a welding machine or an oxyacetylene cutting torch is in use, it may be necessary to place a temporary screen or barrier around the work area to protect other shop users from welding flash or injury (**FIGURE 3-20**).

> ▶ TECHNICIAN TIP
>
> Stay alert for hazards or anything that might be dangerous. If you see, hear, or smell anything odd, take steps to fix it, or tell your supervisor about the problem.

Air Quality

K03007

Managing air quality in shops helps protect you from potential harm and also protects the environment. There are many shop

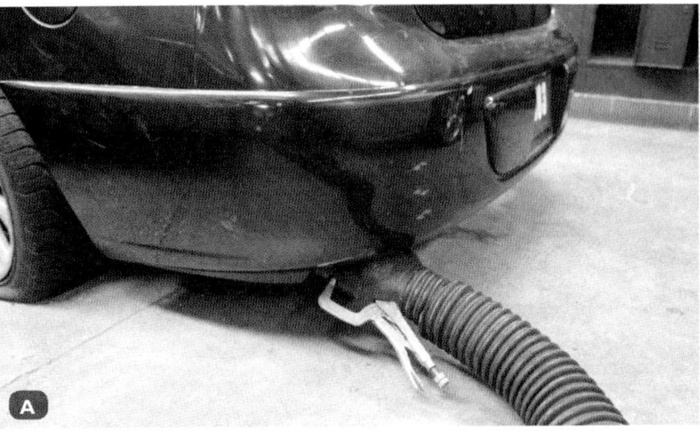

FIGURE 3-18 Exhaust extraction and shop ventilation equipment.

FIGURE 3-19 Sign on the door, used to control access.

FIGURE 3-20 Portable welding curtain.

FIGURE 3-21 Exhaust hoses should be vented to where the fumes will not be drawn back indoors.

activities, stored liquids, and other hazards that can reduce the quality of air in shops. Some of these are listed here:

- Dangerous fumes from running engines
- Welding (gas and electric)
- Painting
- Liquid storage areas
- Air conditioning servicing
- Dust particles from brake servicing

Running Engines

Running engines produce dangerous exhaust gases including carbon monoxide (CO) and carbon dioxide (CO_2). Carbon monoxide in small concentrations can kill or cause serious injuries. Carbon dioxide is a greenhouse gas, and vehicles are a major source of carbon dioxide in the atmosphere. Exhaust gases also contain hydrocarbons (HC) and oxides of nitrogen (NO_x). These gases can form smog and also cause breathing problems for some people.

Carbon monoxide in particular is extremely dangerous, as it is odorless and colorless and can build up to toxic levels very quickly in confined spaces. In fact, it doesn't take very much carbon monoxide to pose a danger. The maximum exposure limit is regulated by the following agencies:

- OSHA permissible exposure limit (PEL) is 50 parts per million (ppm) of air for an eight-hour period.
- National Institute for Occupational Safety and Health has established a recommended exposure limit of 35 ppm for an eight-hour period.

The reason the PEL is so low is because carbon monoxide attaches itself to red blood cells much more easily than oxygen does, and it never leaves the blood cell. This prevents the blood cells from carrying as much oxygen, and if enough carbon monoxide has been inhaled, it effectively asphyxiates the person. Always follow the correct safety precautions when running engines indoors or in a confined space, including over service pits, as gases can accumulate there. The best solution when running engines in an enclosed space (shop) is to directly connect the vehicle's exhaust pipe to an exhaust extraction system hose that ventilates the fumes to the outside air. The extraction system should be vented to where the fumes will not be drawn back indoors, to a place well away from other people and other premises (**FIGURE 3-21**).

Do not assume that an engine fitted with a catalytic converter can be run safely indoors; it cannot. Catalytic converters are fitted into the exhaust system to help control exhaust emissions through chemical reaction. They require high temperatures to operate efficiently and are less effective when the exhaust gases are relatively cool, such as when the engine is only idling or being run intermittently. A catalytic convertor can never substitute for adequate ventilation or exhaust extraction equipment. In fact, even if the catalytic converter were working at 100% efficiency, the exhaust would contain large amounts of carbon dioxide and very low amounts of oxygen, neither of which conditions can sustain life.

Electrical Safety

K03008

Many people are injured by electricity in shops. Poor electrical safety practices can cause shocks and burns as well as fires and explosions. Make sure you know where the electrical shutoffs, or panels for your shop are located. All circuit breakers and fuses should be clearly labeled so that you know which circuits and functions they control (**FIGURE 3-22**).

In the case of an emergency, you may need to know how to shut off the electricity supply to a work area or to your entire shop. Keep the circuit breaker and/or electrical panel covers closed to keep them in good condition, prevent unauthorized access, and prevent accidental contact with the electricity supply. It is important that you do not block or obstruct access to this electrical panel; keep equipment and tools well away so emergency access is not hindered. In some localities, 3' (0.91 m) of unobstructed space must be maintained around the panel at all times.

There should be a sufficient number of electrical receptacles in your work area for all your needs. Do not connect multiple appliances to a single receptacle with a simple double adapter. If necessary, use a multi-outlet safety strip that has a built-in overload cutout feature. Electric receptacles should be at least 3' (0.91 m) above floor level to reduce the risk of igniting spilled fuel vapors or other flammable liquids.

Portable Electrical Equipment

If you need to use an extension cord, make sure it is made of flexible wiring—not the stiffer type of house wiring—and that it is

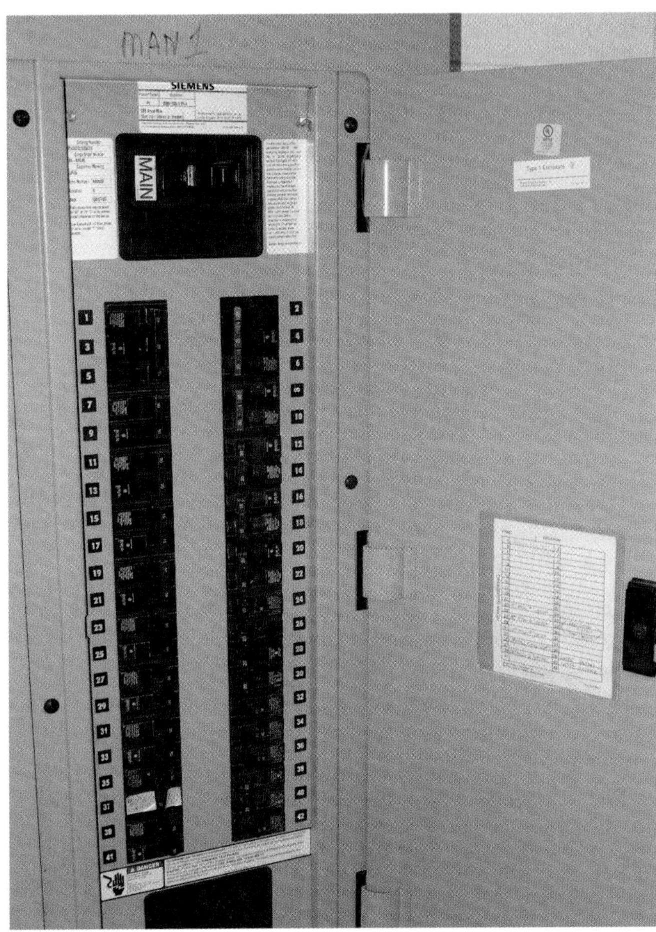

FIGURE 3-22 All electrical switches and fuses should be clearly labeled so that you know which circuits and functions they control.

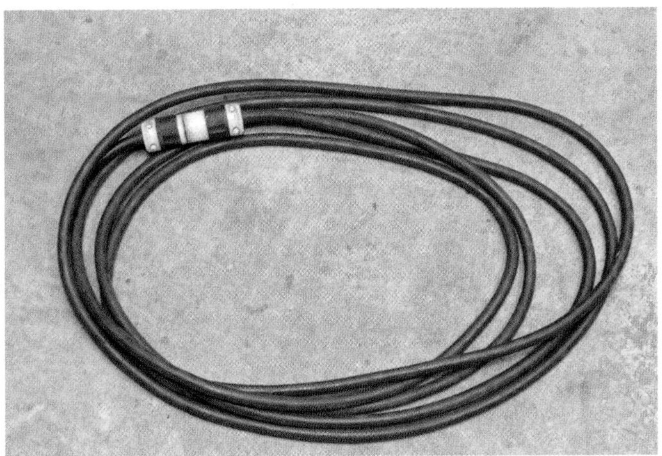

FIGURE 3-23 The extension cord should be neoprene-covered.

fitted with a ground wire. The cord should be neoprene-covered, as this material resists oil damage (**FIGURE 3-23**).

Always check it for cuts, abrasions, or other damage. Be careful how you place the extension cord, so it does not cause a tripping hazard. Also avoid rolling equipment or vehicles over it, as doing so can damage the cord. Never use an extension cord in wet conditions or around flammable liquids. Portable electric tools that operate at 240 volts are often sources of serious shock and burn accidents. Be particularly careful when using these

FIGURE 3-24 Ground prong. **A.** Okay. **B.** Missing.

items. Always inspect the cord for damage and check the security of the attached plug before connecting the item to the power supply. Use 110-volt or lower voltage tools if they are available. All electric tools must be equipped with a ground prong or be **double insulated** (**FIGURE 3-24**).

If electrical cords don't have the ground prong or aren't double insulated, do *not* use them. Never use any high-voltage tool in a wet environment. In contrast, air-operated tools cannot give you an electric shock, because they operate on air pressure instead of electricity, and therefore, they are safer to use in a wet environment.

Portable Shop Lights

Portable shop lights/droplights have been the cause of many accidents over the years. Incandescent bulbs get extremely hot and can cause burns. They are also prone to shatter, which can cut skin and damage eyes or cause fires and electric shock. In fact, in some places incandescent portable shop lights cannot be used in automotive shops and must be replaced with less hazardous lights (**FIGURE 3-25**).

One such light is the fluorescent droplight. Although this type of light stays much cooler than incandescent lights, it still can shatter, causing cuts or damage to eyes. Because of this, droplights should be fully enclosed in a clear, insulating case. They may also contain mercury, which becomes dispersed when the bulb is shattered, creating a hazardous-materials situation (**FIGURE 3-26**).

The safest portable shop light to come on the market is the LED (light-emitting diode) shop light. It uses much lower voltage, and the LED is much less prone to shattering, so it is much safer to use. It also uses a small amount of electricity to produce a large

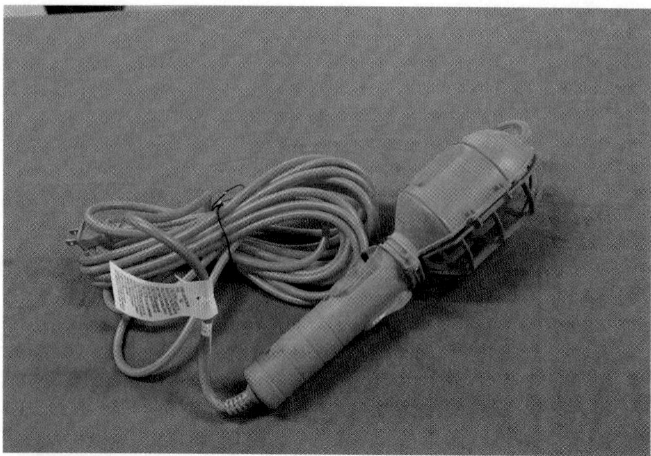

FIGURE 3-25 Incandescent droplight which should *not* be used in a shop.

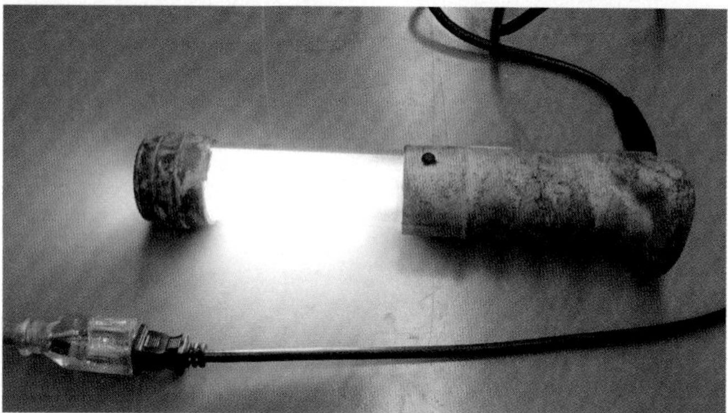

FIGURE 3-26 Fluorescent droplight.

FIGURE 3-27 Cordless LED light.

amount of light, so many of them are cordless. They may also include a magnetic base, so the light can be attached to any steel surface and then adjusted to shine where needed (**FIGURE 3-27**).

> ▶ **TECHNICIAN TIP**
>
> Try out the new LED flashlights or cordless LED shop lights. Good ones provide a lot of light, and the batteries last a long time. It is not unusual to hear about an LED flashlight that got dropped behind a toolbox while it was on, and was still illuminated days later.

Shop Layout

K0309

The shop should have a layout that is efficient and safe, with clearly defined working areas and walkways. Customers should not be allowed to wander through work areas unescorted. A good shop layout can be achieved by thinking about how the work is to be done, how equipment is used, and what traffic movements, both pedestrian and vehicular, occur within the shop.

A well-planned shop should have clearly defined areas for various activities, like parts cleaning, parts storage, tool storage, flammable liquid storage, jacking or lifting, tire service, and painting. Safety equipment such as eyewash stations, fire extinguishers, and first aid kits should be located where they are easily accessible. All flammable items should be kept in an approved fireproof storage container or cabinet, with firefighting equipment close at hand. And lastly, supplies for cleaning the shop, such as garbage and recycling containers, brooms, mops, dust pans, and even floor scrubbers, need to be strategically located so it is easy to keep the shop clean and orderly (**FIGURE 3-28**).

Preventing Fires

K03010

The danger of a gasoline fire is always present in an automotive shop. Most automobiles carry a fuel tank, often with large quantities of fuel on board, which is more than sufficient to cause a large, very destructive, and potentially explosive fire (**FIGURE 3-29**).

In fact, 1 gallon of gasoline has the same amount of energy as 20 sticks of dynamite. So take precautions to make sure you have the correct type and size of extinguishers on hand for a potential fuel fire. Make sure you clean up spills immediately and avoid ignition sources, like sparks, near flammable liquids or gases. Being aware of the following topics will help you know how to minimize the risk of fires in the shop.

> ▶ **TECHNICIAN TIP**
>
> It's not a matter of "if" but "when" a shop is going to have a fire. When shops do catch on fire, the fire can grow quickly out of control. So preventing a fire is the best action you can take. If a fire starts, react appropriately. In some

FIGURE 3-28 Shop layout can minimize accidents as well as improve working efficiency.

FIGURE 3-29 Shop fire.

cases, you just need to get out safely. In other cases, it may be safe to try to fight the fire quickly before it spreads too far. Always be aware of the potential for a fire, and plan ahead by thinking through the task you are about to undertake. Know where firefighting equipment is kept and how it works.

Fuel Vapor

Liquid fuel vaporizes rapidly, especially when spilled, and the vapor is extremely easy to ignite. Because fuel vapor is invisible and three to four times heavier than air, it can spread unseen across a wide area, filling any low spots such as service pits. So a source of ignition can be quite some distance from the original spill and still ignite it. Fuel can even vaporize from paper towels or rags used to wipe up liquid spills. These materials should be allowed to dry in the open air, not held in front of a heater element or other ignition source. Any spark or naked flame, even a lit cigarette, can start an explosive fire. Always keep flammable liquids in spill-proof containers. If you use gas cans, this means using one that automatically closes when not in use (**FIGURE 3-30**).

Always be aware of the possibility of fuel vapor buildup. If you smell gasoline or other flammable vapor, stop work and investigate the source. This may mean having coworkers stop all work and assist you. Do not ignore the presence of fuel vapors.

Spillage Risks

Spills frequently occur when technicians remove and replace fuel filters with pressurized fuel in them and the hoses. Disconnecting the hoses without properly releasing the fuel pressure causes liquid to spray a great deal, so always check the manufacturer's procedure for depressurizing the fuel system before opening it up. Spills can also occur during removal of a fuel pump or fuel tank sender unit, which are commonly located on the inside of the fuel tank. Unless the tank can be opened from the top without removing the tank from the vehicle, it is best to first empty the tank safely, to avoid spills. Spills also can occur when fuel systems are being repaired or when fuel is being drained into unsuitable containers. Avoid spills by following the manufacturer's specified procedure when removing fuel system components. Also, keep a spill response kit nearby to deal with any spills quickly. Spill kits should contain absorbent material and barrier dams to contain moderate-sized spills (**FIGURE 3-31**).

FIGURE 3-30 Spill-proof gas can.

FIGURE 3-31 Spill response kit.

And remember that fuel vapors are heavier than air and can travel a long way from the spill, so all sources of ignition need to be removed from the immediate vicinity, which can be more than 50' away.

Draining Fuel

If there is a possibility of fuel spillage while working on a vehicle, then you should first remove the fuel safely. Do this only in a well-ventilated, level space, preferably outside in the open air. Make sure all potential sources of ignition have been removed from the area, and disconnect the battery on the vehicle. Do not drain fuel from a vehicle over an inspection pit. Make sure the container you are draining into is an approved fuel storage container (fuel retriever) and that it is large enough to contain all of the fuel in the system being drained. A fuel retriever minimizes the chance of sudden large spills occurring (**FIGURE 3-32**).

You may need to use narrow-diameter hoses to bypass anti-spillage devices in the filler neck (**FIGURE 3-33**).

If removing fuel lines or fuel filters, the fuel system must be depressurized to prevent fuel spraying when the hoses are disconnected. Check the service manual for details on how best to depressurize or drain the fuel from the vehicle you are working on.

FIGURE 3-32 Fuel retriever.

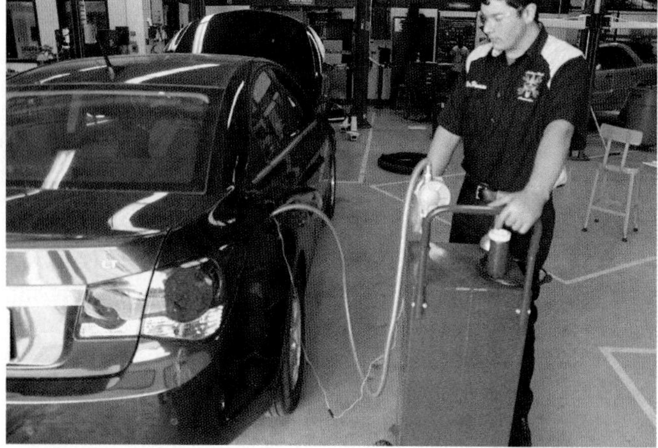

FIGURE 3-33 Fuel retriever use.

SAFETY TIP

Never weld anywhere near a gas tank or any kind of fuel line. Welding work on a tank is a job for specialists. An empty fuel tank can still contain fuel vapor and therefore would be even more dangerous than one full of liquid fuel. Do not attempt to repair a tank yourself.

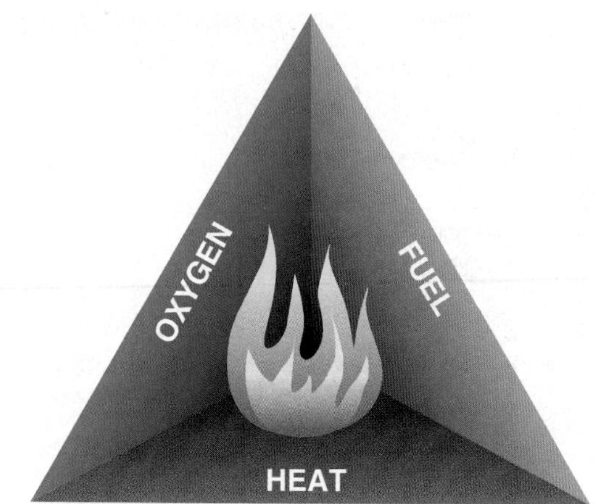

FIGURE 3-34 Fire triangle.

Extinguishing Fires

N03004, K03011

Three elements must be present at the same time for a fire to occur: fuel, oxygen, and heat (FIGURE 3-34). The secret of firefighting involves the removal of at least one of these elements. If a fire occurs in the shop, it is usually the oxygen or the heat that is removed to extinguish the fire. For example, a fire blanket when applied correctly removes the oxygen, and a water extinguisher removes heat from the fire. Fire extinguishers are used to extinguish the majority of small fires in a shop. Never hesitate to call the fire department if you cannot extinguish a fire quickly and safely.

Fire Classifications

In the United States, there are five classes of fire:

- Class A fires involve ordinary combustibles such as wood, paper, or cloth.
- Class B fires involve flammable liquids or gaseous fuels.
- Class C fires involve electrical equipment.
- Class D fires involve combustible metals such as sodium, titanium, and magnesium.
- Class K fires involve cooking oil or fat.

Fire Extinguisher Types

Fire extinguishers are marked with pictograms depicting the types of fires that the extinguisher is approved to fight (FIGURE 3-35):

- Class A: Green triangle
- Class B: Red square
- Class C: Blue circle
- Class D: Yellow pentagram
- Class K: Black hexagon

Fire Extinguisher Operation

Unless the fire is very small, always sound the alarm before attempting to fight a fire. If you cannot fight the fire safely, leave the area while you wait for backup. You need to size up the fire before you make the decision to fight it with a fire extinguisher, by identifying what sort of material is burning, the extent of the fire, and the likelihood of it spreading. Also, if the fire is in electrical wires

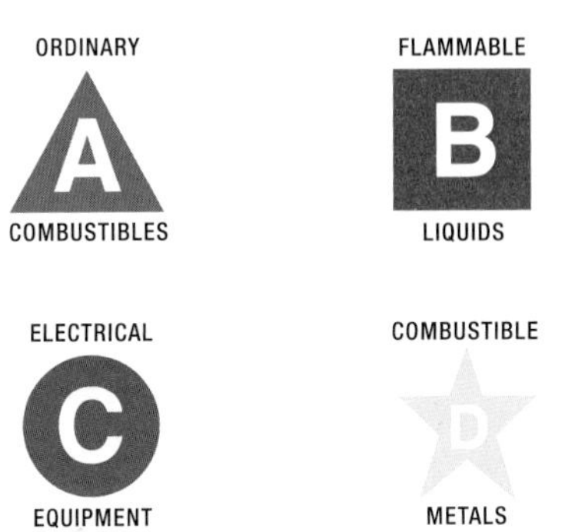

FIGURE 3-35 Traditional labels on fire extinguishers often incorporate a shape as well as a letter.

or equipment, make sure you won't be electrocuted while trying to extinguish it. To operate a fire extinguisher, follow the acronym for fire extinguisher use: PASS (pull, aim, squeeze, sweep).

- *Pull* out the pin that locks the handle at the top of the fire extinguisher to prevent accidental use. Carry the fire extinguisher in one hand, and use your other hand to:
- *Aim* the nozzle at the base of the fire. Stand about 8–12 feet (2.4–3.7 m) away from the fire, and
- *Squeeze* the handle to discharge the fire extinguisher. Remember that if you release the handle on the fire extinguisher, it will stop discharging.
- *Sweep* the nozzle from side to side at the base of the fire. Continue to watch the fire. Although it may appear to be extinguished, it may suddenly reignite. Portable fire extinguishers only operate for about 10–25 seconds before they are empty. So use them effectively (**FIGURE 3-36**).

If the fire is indoors, you should be standing between the fire and the nearest safe exit. If the fire is outside, you should stand facing the fire, with the wind on your back, so that the smoke and heat are being blown away from you. If possible, get an assistant to guide you and inform you of the fire's progress. Again, make sure you have a means of escape, should the fire get out of control. When you are certain that the fire is out, report it to your supervisor. Also report what actions you took to put out the fire. Once the circumstances of the fire have been investigated and your supervisor or the fire department has given you the all clear, clean up the debris and submit the used fire extinguisher for inspection and service.

Fire Blankets

Fire blankets are designed to smother a small fire (**FIGURE 3-37**). They are ideal for use in situations where a fire extinguisher could cause damage. For example, if there is a small fire under the hood of a vehicle, a fire blanket might be able to smother the fire without running the risk of getting powder from the fire extinguisher down the intake system. They are also very useful

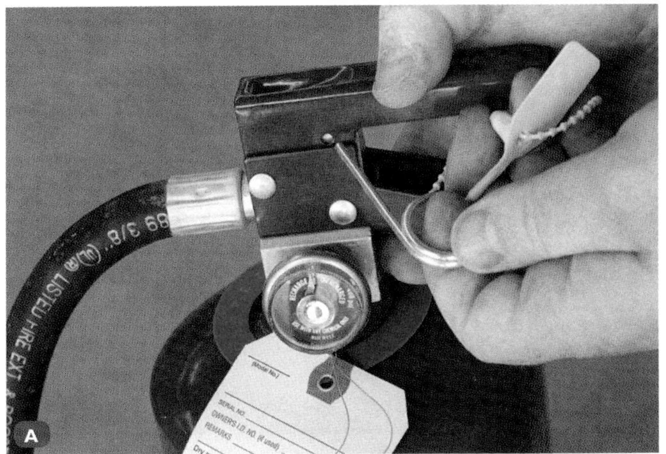

FIGURE 3-36 To operate a fire extinguisher, follow PASS. **A.** Pull. **B.** Aim. **C.** Squeeze. **D.** Sweep.

FIGURE 3-37 Fire blanket.

FIGURE 3-38 The main types of eyewashers include disposable eyewash packs and eyewash stations. Some emergency showers have an eyewash station built in.

FIGURE 3-39 Hazardous materials.

in putting out a fire on a person. The fire blanket can be thrown around the person, smothering any fire.

Obtain a fire blanket and study the use instructions on the packaging. If instructions are not provided, research how to use a fire blanket, or ask your supervisor. You may require instruction from an authorized person in using the fire blanket. If you do use a fire blanket, make sure you return the blanket for use or, if necessary, replace it with a new one.

Eyewash Stations and Emergency Showers

N03005

Hopefully you will never need to use an eyewash station or emergency shower. The best treatment is prevention, so make sure you wear all the PPE required for each specific task, to avoid as many eye injuries as possible. Eyewash stations are used to flush the eyes with clean water or sterile liquid in the event that you get foreign liquid or particles in your eye. There are different types of eyewashes; the two main types are disposable eyewash packs and eyewash stations. Some emergency or deluge showers also have an eyewash station built in (**FIGURE 3-38**).

When an individual gets chemicals in his eyes, he typically needs assistance in reaching the eyewash station by taking his arm and leading him to it. He may not want to open his eyes, even in the water, so encourage him to use his fingers to pull his eyelids open. If a chemical splashed in his eyes, encourage him to rinse his eyes for 15 minutes. While he is rinsing his eyes, call for medical assistance. One thing to note is that eyewash stations are used for flushing chemicals or debris from the surface of the eyes. Eyewash stations should not be used for punctures or cuts to the eye. In this case, cover the eye with sterile gauze, if available, and seek medical assistance.

▶ Hazardous Materials Safety

A **hazardous material** is any material that poses an unreasonable risk of damage or injury to persons, property, or the environment if it is not properly controlled during handling, storage, manufacture, processing, packaging, use and disposal, or transportation. These materials can be solids, liquids, or gases. Most technicians use hazardous materials daily, such as cleaning solvents, gasket cement, brake fluid, and coolant. In fact, there are likely to be hundreds of hazardous materials in a typical shop (**FIGURE 3-39**). Hazardous materials must be properly handled, labeled, and stored in the shop.

Safety Data Sheets

N03006, K03012

Hazardous materials are used daily and may make you very sick if they are not used properly. **Safety data sheets (SDS)** contain detailed information about hazardous materials to help you understand how they should be safely used, any health effects relating to them, how to treat a person who has been exposed to them, and how to deal with them in a fire situation. SDS can be obtained from the manufacturer of the material. The shop should have an SDS for each hazardous substance or dangerous product. In the United States, it is required that workplaces have an SDS for every chemical that is on site.

Safety data sheets (SDS) were formerly called material safety data sheets (MSDS). In 2012, OSHA changed the requirements for the hazard communication system (HCS) to conform to the United Nations Globally Harmonized System of Classification and Labeling of Chemicals (GHS). The MSDS needed to change its name and its format to fit the new standards. Whereas the original MSDS had 8 sections, safety data sheets are required to have 16 sections that provide additional details and make it easier to find specific data when needed. In addition, GHS requires all employers to train their employees in the new chemical labeling requirements and the new format for the safety data sheets.

Whenever you deal with a potentially hazardous product, you should consult the SDS to learn how to use that product safely. If you are using more than one product, make sure you consult all the SDS for those products. Be aware that certain combinations of products can be more dangerous than any of them separately.

SDS are usually kept in a clearly marked binder and should be regularly updated as chemicals come into the workplace. As of June 1, 2015, the HCS requires new SDS to be in a uniform format, and include the section numbers, the headings, and associated information under the headings below:

To identify information found on an SDS, follow the steps in **SKILL DRILL 3-2**.

Cleaning Hazardous Dust Safely

S03001

Toxic dust is any dust that may contain fine particles that could be harmful to humans or the environment. If you are unsure as to the toxicity of any particular dust, then you should always treat it as toxic and take the precautions identified in the SDS or shop procedures. Brake and clutch dust are potential toxic dusts that automotive shops must manage. The dust is made up of very fine particles that can easily spread and contaminate an area. One of the more common sources of

Section 1, Identification, includes product identifier; manufacturer or distributor name, address, and phone number; emergency phone number; recommended use; restrictions on use.

Section 2, Hazard(s) identification, includes all hazards regarding the chemical; required label elements.

Section 3, Composition/information on ingredients, includes information on chemical ingredients; trade secret claims.

Section 4, First-aid measures, includes important symptoms/ effects, acute, delayed; required treatment.

Section 5, Firefighting measures, lists suitable extinguishing techniques, equipment; chemical hazards from fire.

Section 6, Accidental release measures, lists emergency procedures; protective equipment; proper methods of containment and cleanup.

Section 7, Handling and storage, lists precautions for safe handling and storage, including incompatibilities.

Section 8, Exposure controls/personal protection, lists OSHA's permissible exposure limits (PELs); threshold limit values (TLVs); appropriate engineering controls; personal protective equipment (PPE).

Section 9, Physical and chemical properties, lists the chemical's characteristics.

Section 10, Stability and reactivity, lists chemical stability and possibility of hazardous reactions.

Section 11, Toxicological information, includes routes of exposure; related symptoms, acute and chronic effects; numerical measures of toxicity.

Section 12, Ecological information*

Section 13, Disposal considerations*

Section 14, Transport information*

Section 15, Regulatory information*

Section 16, Other information, includes the date of preparation or last revision.

**Note*: Since other Agencies regulate this information, OSHA will not be enforcing Sections 12 through 15 (29 CFR 1910.1200(g)(2)).

Source: https://www.osha.gov/Publications/HazComm_QuickCard_SafetyData.html.

SKILL DRILL 3-2 Identifying Information in an SDS

1. Once you have studied the information on the container label, find the SDS for that particular material. Always check the revision date to ensure that you are reading the most recent update.
2. Note the chemical and trade names for the material, its manufacturer, and the emergency telephone number to call.
3. Find out why this material is potentially hazardous. It may be flammable, it may explode, or it may be poisonous if inhaled or touched with your bare skin. Check the **threshold limit values (TLVs)**. The concentration of this material in the air you breathe in your shop must not exceed these figures. There could be physical symptoms associated with breathing harmful chemicals. Find out what will happen to you if you suffer overexposure to the material, either through breathing it or by coming into physical contact with it. This helps you to take safety precautions, such as wearing eye, face, or skin protection, or a mask or respirator, while using the material, or like washing your skin afterward.
4. Note the flash point for this material so that you know at what temperature it may catch fire. Also note what kind of fire extinguisher you would use to fight a fire involving this material. The wrong fire extinguisher could make the emergency even worse.
5. Study the reactivity for this material to identify the physical conditions or other materials that you should avoid when using this material. It could be heat, moisture, or some other chemical.
6. Find out what special precautions you should take when working with this material. These include personal protection for your skin, eyes, and lungs, and proper storage and use of the material.
7. Be sure to refresh your knowledge of your SDS from time to time. Be confident that you know how to handle and use the material and what action to take in an emergency, should one occur.

toxic dust is inside drum brakes and manual transmission bell housings. It is a good idea to avoid all dust if possible, whether it is classified as toxic or not. If you do have to work with dust, never use compressed air to blow it from components or parts, and always use PPE such as face masks, eye protection, and gloves.

After completing a service or repair task on a vehicle, there is often dirt and dust left behind. The chemicals present in this dirt usually contain toxic chemicals that can build up and cause health problems. To keep the levels of dirt and dust to a minimum, clean it up immediately after the task is complete. The vigorous action of sweeping causes the dirt and dust to rise; therefore, when sweeping the floor, use a soft broom that pushes, rather than flicks, the dirt and dust forward. Create smaller piles and dispose of them frequently. Another successful way of cleaning shop dirt and dust is to use a water hose. The wastewater must be caught in a settling pit and not run into a storm water drain. Many shops also have floor scrubbers that use a water/soap solution to clean the floor. These shops usually vacuum up the dirty water and store it in a tank until it can be disposed of properly.

Various tools have been developed to clean toxic dust from vehicle components. The most common one is the brake wash station. It uses an aqueous solution to wet down and wash the dust into a collection basin. The basin needs periodic maintenance to properly dispose of the accumulated sludge. This tool is probably the simplest way to effectively deal with hazardous dust because it is easy to set up, use, and store. One system that used to be approved for cleanup of hazardous dust is no longer approved for that purpose. It is called a high-efficiency particulate air (HEPA) dust collection system and used a HEPA filter to trap very small particles. But these types of hazardous dust collection systems are no longer approved.

▶ TECHNICIAN TIP

- Some vehicle components, including brake and clutch linings, contain asbestos, which, despite having very good heat properties, is toxic. Asbestos dust causes lung cancer. Complications from breathing the dust may not show until decades after exposure.
- Airborne dust in the shop can also cause breathing problems such as asthma and throat infections.
- Never cause dust from vehicle components to be blown into the air. It can stay floating for many hours, meaning that other people will breathe the dust unknowingly.
- Wear protective gloves whenever using solvents.
- If you are unfamiliar with a solvent or a cleaner, refer to the SDS for information about its correct use and applicable hazards.
- Always wash your hands thoroughly with soap and water after performing repair tasks on brake and clutch components.
- Always wash work clothes separately from other clothes so that toxic dust does not transfer from one garment to another.
- Always wear protective clothing and the appropriate safety equipment.

▶ TECHNICIAN TIP

Whenever using an atomizer with solvents and cleaners, make sure there is adequate ventilation. Wear appropriate breathing apparatus and eye protection.

To safely clean brake dust, follow the steps in **SKILL DRILL 3-3**.

Used Engine Oil and Fluids

K03013

Used engine oil and fluids are liquids that have been drained from the vehicle, usually during servicing operations. Used oil and fluids often contain dangerous chemicals and impurities

SKILL DRILL 3-3 Safely Cleaning Brake Dust

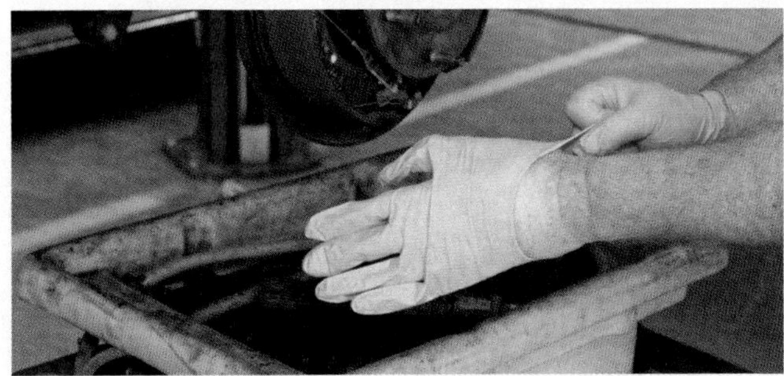

1. When performing any cleaning tasks on brake or clutch components, always wear a face mask, gloves, and eye protection.

2. Position the brake wash station under the bottom of the backing plate. When cleaning drum brakes, remove the brake drum and check for the presence of dust and brake fluid. When cleaning a clutch, position the wash station underneath the bell housing.

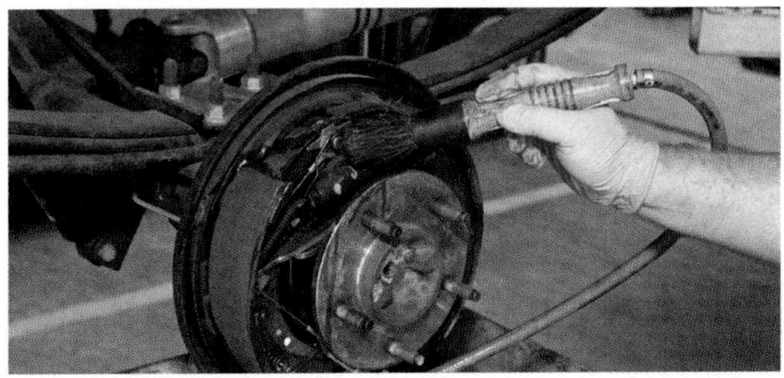

3. Turn on the wash station pump, and paint the solution over the components to wet and clean the components and remove the dust. Any toxic dust will be washed down and caught in the wash station.

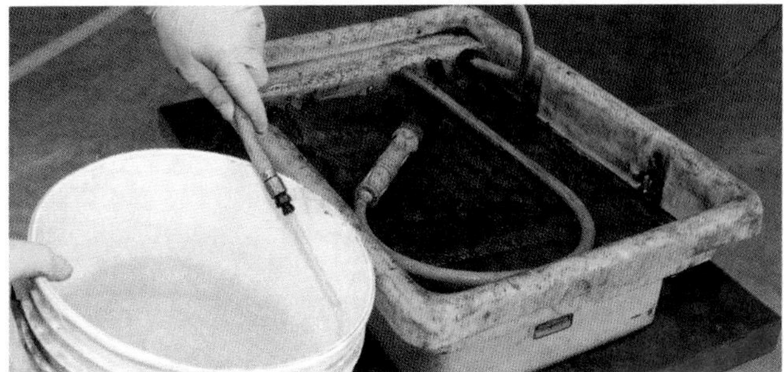

4. Periodically dispose of the residue in an approved manner.

FIGURE 3-40 Used oil and fluids often contain dangerous chemicals and need to be safely recycled or disposed of in an environmentally friendly way.

such as heavy metals that need to be safely recycled or disposed of in an environmentally friendly way (**FIGURE 3-40**).

There are laws and regulations that control the way in which they are to be handled and disposed. The shop will have policies and procedures that describe how you should handle and dispose of used engine oil and fluids. Be careful not to mix incompatible fluids such as used engine oil and used coolant. Doing so makes the hazardous materials very much more expensive to dispose of.

Generally speaking, petroleum products can be mixed together. Follow your local, state, and federal regulations when disposing of waste fluids. Used engine oil is a hazardous material containing many impurities that can damage your skin. Coming into frequent or prolonged contact with used engine oil can cause dermatitis and other skin disorders, including some forms of cancer. Avoid direct contact as much as possible by always using impervious gloves and other protective clothing, which should be cleaned or replaced regularly. Using a barrier-type hand lotion also helps protect your hands and makes cleaning them much easier. Always follow safe work practices, which minimize the possibility of accidental spills. Keeping a high standard of personal hygiene and cleanliness is important so that you get into the habit of washing off harmful materials as soon as possible after contact. If you have been in periodic contact with used engine oil, you should regularly inspect your skin for signs of damage or deterioration. If you have any concerns, consult your doctor.

▶ Shop Safety Inspections

K03014

Shop safety inspections are valuable ways of identifying unsafe equipment, materials, or activities so they can be corrected to reduce the risk of accidents or injuries. The inspection can be formalized by using inspection sheets to check specific items, or they can be general walk-arounds where you consciously look for and document problems that can be corrected (**FIGURE 3-41**).

Here are some of the common things to look for:

- Items blocking emergency exits or walkways
- Poor safety signage
- Unsafe storage of flammable goods
- Tripping or slipping hazards
- Faulty or unsafe equipment or tools
- Missing equipment guards
- Misadjusted bench grinder tool rests
- Missing or expired fire extinguishers

FIGURE 3-41 Shop safety inspection using a checklist.

- Clutter, spills, unsafe shop practices
- People not wearing the correct PPE

Formal and informal safety inspections should be held regularly. For example, an inspection sheet might be used weekly or monthly to formally evaluate the shop, and informal inspections might be held daily to catch issues that are of a more immediate nature. Never ignore or put off a safety issue.

Applied Science

AS-4: Waste Management: The technician identifies the waste products resulting from a repair task.

The most important part of identifying waste is first to determine what products are waste and what can be reused or recycled. The second is to identify what type of material the waste is and then determine what type of disposal needs to take place with the waste material.

If you replace a set of front disc brake rotors and pads, then you will have to determine where to put the waste material from this job. You will have to refer to the waste disposal table posted in the shop to determine where each item of waste should go. Most plastic bags cannot be recycled, so these items should be put in the standard waste bin. The disc brake rotors should be put in the metal scrap bin for recycling. The disc brake pads may also be put in the metal bin for recycling unless they were manufactured before 2004. If so, they may contain asbestos and need to be put in a plastic airtight bag for removal by a licensed contractor.

AS-5: Waste Management: The technician handles the disposal of materials such as automotive lubricants in accordance with applicable federal, state, and local rules and regulations.

When you carry out a major service, you may have to replace the engine oil and oil filter, coolant, and automatic transmission oil. You will have to determine where to put the waste material from this job by consulting the waste disposal guidelines posted in the shop. For example, if you refer to the sample waste disposal table shown in a previous Applied Science box, you can see that engine oil and transmission oil need to be collected and stored in a container for removal by a licensed contractor. The oil filter needs to be drained and then crushed and kept in a leakproof container to be collected by a licensed contractor. The coolant needs to be kept in a separate container and collected to be recycled by a licensed operator. Cardboard boxes that oil or air filters may come in can be put in the recycling bin for recycling.

▶ Wrap-Up

Ready for Review

- ▶ Your employer is responsible for maintaining a safe work environment; you are responsible for working safely.
- ▶ Always wear the correct personal protective equipment, such as gloves or hearing protection.
- ▶ Accidents and injuries can be avoided by safe work practices.
- ▶ Every shop should mark evacuation routes; always know the evacuation route for your shop.
- ▶ OSHA is a federal agency that oversees safe workplace environments and practices.
- ▶ The EPA monitors and enforces issues related to environmental safety.

- Shop policies and procedures are designed to ensure compliance with laws and regulations, create a safe working environment, and guide shop practice.
- Identify hazards and hazardous materials in your work environment.
- Safety signs include a signal word, background color, text, and a pictorial message.
- Shop safety equipment includes handrails, machinery guards, painted lines, soundproof rooms, adequate ventilation, gas extraction hoses, doors and gates, and temporary barriers.
- Air quality is an important safety concern.
- Carbon monoxide and carbon dioxide from running engines can create a hazardous work environment.
- Electrical safety in a shop is important to prevent shocks, burns, fires, and explosions.
- Portable electrical equipment should be the proper voltage and should always be inspected for damage.
- Use caution when plugging in or using a portable shop light.
- Shop layouts should be well planned to maximize safety.
- Fuels and fuel vapors are potential fire hazards.
- Use fuel retrievers when draining fuel, and have a spill response kit nearby.
- Fuel, oxygen, and heat must all be present for fire to occur.
- Types of fires are classified as A, B, C, D, or K, and fire extinguishers match them accordingly.
- Do not fight a fire unless you can do so safely.
- Operating a fire extinguisher involves the PASS method: pull, aim, squeeze, and sweep.
- Eyewash stations and emergency showers allow flushing of chemicals or other irritants.
- Safety data sheets contain important information on each hazardous material in the shop.
- Using a soap and water solution is the safest method of cleaning dust or dirt that may be toxic.
- Used engine oil and fluids must be handled and disposed of properly.
- Shop safety inspections ensure that safety policies and procedures are being followed.

Key Terms

double insulated Tools or appliances that are designed in such a way that no single failure can result in a dangerous voltage coming into contact with the outer casing of the device.

Environmental Protection Agency (EPA) Federal government agency that deals with issues related to environmental safety.

hazard Anything that could hurt you or someone else.

hazardous environment A place where hazards exist.

hazardous material Any material that poses an unreasonable risk of damage or injury to persons, property, or the environment if it is not properly controlled during handling, storage, manufacture, processing, packaging, use and disposal, or transportation.

Occupational Safety and Health Administration (OSHA) Government agency created to provide national leadership in occupational safety and health.

personal protective equipment (PPE) Safety equipment designed to protect the technician, such as safety boots, gloves, clothing, protective eyewear, and hearing protection.

policy A guiding principle that sets the shop direction.

procedure A list of the steps required to get the same result each time a task or activity is performed.

safety data sheet (SDS) A sheet that provides information about handling, use, and storage of a material that may be hazardous.

threshold limit value (TLV) The maximum allowable concentration of a given material in the surrounding air.

toxic dust Any dust that may contain fine particles that could be harmful to humans or the environment.

Review Questions

1. All of the statements below with respect to emergency evacuation are true *except*:
   a. There should be only one clearly identified evacuation route to avoid confusion.
   b. Exits should be highlighted with signs that may be illuminated.
   c. There should be a prearranged safe place for gathering.
   d. Exits should never be chained closed or obstructed.
2. Which of the following works toward finding the most effective ways to help prevent worker fatalities and workplace injuries and illnesses?
   a. EPA
   b. FDA
   c. OSHA
   d. ASA
3. The list of steps required to get the same result each time a task or activity is performed is known as the:
   a. shop policy.
   b. shop regulations.
   c. shop procedure.
   d. shop system.
4. The signal word that indicates a potentially hazardous situation, which, if not avoided, could result in death or serious injury is:
   a. danger.
   b. caution.
   c. warning.
   d. hazardous.
5. Shop users other than those operating the welding machine or cutting torch can be protected from welding flash or injury by:
   a. sound insulated rooms.
   b. temporary barriers.
   c. adequate ventilation.
   d. painted lines.

6. Which of these should not be used in a shop?
   a. Incandescent bulbs
   b. Extension cords
   c. Low-voltage tools
   d. Air-operated tools
7. To operate a fire extinguisher, follow the acronym for fire extinguisher use: PASS, which stands for:
   a. Press, aim, squeeze, sweep.
   b. Pull, aim, shake, sweep.
   c. Press, aim, shake, sweep.
   d. Pull, aim, squeeze, sweep.
8. All of the following statements regarding safety data sheets are true *except*:
   a. They should be consulted to learn how to use the product safely.
   b. If you are using more than one product, consult the SDS of only one of the products.
   c. They include information on chemical ingredients.
   d. All new SDS should be in a uniform format.
9. Which of these should never be used to remove dust?
   a. Floor scrubbers
   b. Water hoses
   c. Compressed air
   d. Brake wash stations
10. When disposing of used liquids:
   a. federal regulations need not be followed.
   b. incompatible liquids can be mixed together.
   c. impervious gloves are not needed.
   d. petroleum products can be mixed together.

ASE Technician A/Technician B Style Questions

1. Tech A says that exposure to solvents may have long-term effects. Tech B says that accidents are almost always avoidable. Who is correct?
   a. Tech A
   b. Tech B
   c. Both A and B
   d. Neither A nor B
2. Tech A says that after an accident you should take measures to avoid it in the future. Tech B says that it is okay to block an exit for a shop if you are actively working. Who is correct?
   a. Tech A
   b. Tech B
   c. Both A and B
   d. Neither A nor B
3. Tech A says that both OSHA and the EPA can inspect facilities for violations. Tech B says that a shop safety rule does not have to be reviewed once put in place. Who is correct?
   a. Tech A
   b. Tech B
   c. Both A and B
   d. Neither A nor B
4. Tech A says that all hazards can be removed from a shop. Tech B says that you should disconnect an air gun before inspecting it. Who is correct?
   a. Tech A
   b. Tech B
   c. Both A and B
   d. Neither A nor B
5. Tech A says that both caution and danger signs indicate a potentially hazardous situation. Tech B says that an exhaust extraction hose is not needed if the vehicle is only going to run for a few minutes. Who is correct?
   a. Tech A
   b. Tech B
   c. Both A and B
   d. Neither A nor B
6. Tech A says that if you are unsure of what personal protective equipment (PPE) to use to perform a job, you should just use what is nearby. Tech B says that air tools are less likely to shock you than electrically powered tools. Who is correct?
   a. Tech A
   b. Tech B
   c. Both A and B
   d. Neither A nor B
7. Tech A says that PASS is an acronym used to help remember how to use a fire extinguisher. Tech B says that a class A fire extinguisher can only be used to fight an electrical fire. Who is correct?
   a. Tech A
   b. Tech B
   c. Both A and B
   d. Neither A nor B
8. Tech A says that a safety data sheet (SDS) contains information on procedures to repair a vehicle. Tech B says that you only need an SDS after an accident occurs. Who is correct?
   a. Tech A
   b. Tech B
   c. Both A and B
   d. Neither A nor B
9. Tech A says that one approved way to clean dust off brakes is with compressed air. Tech B says that some auto parts may contain asbestos. Who is correct?
   a. Tech A
   b. Tech B
   c. Both A and B
   d. Neither A nor B
10. Tech A says that when cleaning brake and clutch components, the wash station should be placed directly under the component. Tech B says that you should follow state and local regulations when disposing of used oil. Who is correct?
   a. Tech A
   b. Tech B
   c. Both A and B
   d. Neither A nor B

CHAPTER 4
Personal Safety

NATEF Tasks

- **N04001** Comply with the required use of safety glasses, ear protection, gloves, and shoes during lab/shop activities.
- **N04002** Identify and wear appropriate clothing for lab/shop activities.
- **N04003** Secure hair and jewelry for lab/shop activities.
- **N04004** Utilize proper ventilation procedures for working within the lab/shop area.

Knowledge Objectives

After reading this chapter, you will be able to:

- **K04001** Identify personal protective equipment and how it should be used.
- **K04002** Describe the applications for hand protection.
- **K04003** Describe the types of ear protection.
- **K04004** Describe the types of breathing devices available.
- **K04005** Describe the applications of eye protection.
- **K04006** Apply injury protection practices.
- **K04007** Describe proper lifting techniques.
- **K04008** Describe the importance of maintaining a clean and orderly workspace.
- **K04009** Explain the procedure when approaching an emergency.
- **K04010** Explain the first aid procedures for bleeding, eye injuries, fractures, and burns.

Skills Objectives

There are no Skills Objectives for this chapter.

▶ Introduction

Personal safety is not something to take lightly. Accidents cause injury and death every day in workplaces across the world (**FIGURE 4-1**). Even if accidents don't result in death, they can be very costly in lost productivity, disability, rehabilitation, and litigation costs. Because workplace safety affects people and society so heavily, government has an interest in minimizing workplace accidents and promoting safe working environments. The primary federal agency for workplace safety is the Occupational Safety and Health Administration (OSHA). States have their own agencies that administer the federal guidelines as well as create additional regulations that apply to their state.

OSHA standard 29 CFR 1910.132 requires employers to assess the workplace to determine if hazards are present, or are likely to be present, which necessitate the use of PPE. In addition, OSHA regulations require employers to protect their employees from workplace hazards such as machines, work procedures, and hazardous substances that can cause injury. To do that, employers must institute all feasible **engineering and work practice controls** to eliminate and reduce hazards before using PPE to protect against hazards. This means that the employers have a responsibility not only to assess safety issues but to eliminate or reduce those that can be mitigated, and then provide PPE and training for those hazards that cannot be mitigated. At the same time, if you are injured, you are the one suffering the consequences, such as being out of work for a period of time, permanent disability, or death. So it is in your best interest to take responsibility for your own safety while on the job.

Engineering controls means the employer has physically changed the machine or work environment to prevent employee exposure to the potential hazard. An example might be adding

FIGURE 4-1 Accidents are costly.

You Are the Automotive Technician

It's your first day on the job, and you are asked to report to the main office, where your new supervisor gives you your PPE. Before you can begin working on the shop floor, you are given training on the proper use of PPE. Here are some of the questions you must be able to answer.

1. Which type of gloves should be worn when handling solvents and cleaners?
2. Why must safety glasses be worn at all times in the shop?
3. Why should rings, watches, and jewelry never be worn in the shop?
4. When should hearing protection be worn?
5. For what types of tasks should a face shield be worn?
6. Why must hair be tied up or restrained in the shop?
7. Which type of eye protection should be worn when using or assisting a person using an oxyacetylene welder?

a ventilation system to an area where solvent tanks are used. Work practice controls means the employer changes the way employees do their jobs in order to reduce exposure to the hazard. An example of this might be requiring employees to use a brake wash station when working on brake systems.

Personal protective equipment (PPE) is equipment used to block the entry of hazardous materials into the body or to protect the body from injury. PPE includes clothing, shoes, eye protection, face protection, head protection, hearing protection, gloves, masks, and respirators (**FIGURE 4-2**). Before you undertake any activity, consider all potential hazards and select the correct PPE based on the risk associated with the activity. For example, if you are going to change hydraulic brake fluid, put on some impervious gloves to protect your skin from chemicals.

As you go through this chapter, you will learn how to identify the correct PPE for a given activity and how to wear it safely. It is important that the PPE you use fits correctly and is appropriate for the task you are undertaking. For example, if the task requires you to wear eye protection and specifies that you should use a full face shield, do not try to cut corners and only wear safety glasses. You also need to make sure the PPE you are using is worn correctly. For example, a hairnet that does not capture all of your hair is not protecting you adequately.

▶ Personal Protective Equipment (PPE)

N04001, K04001

Protective Clothing

N04002

Protective clothing includes items like shirts, vests, pants, shoes, and gloves. These items are your first line of defense against injuries and accidents, and clothing appropriate for the task must be worn when performing any work. Always make sure protective clothing is kept clean and in good condition. You should replace any clothing that is not in good condition, as it is no longer able to fully protect you. Types of protective clothing materials and their uses are as follows:

- **Paper-like fiber:** Disposable suits made of this material provide protection against dust and splashes (**FIGURE 4-3**).
- **Treated wool and cotton:** Adapts well to changing workplace temperatures. Comfortable and fire resistant. Protects against dust, abrasion, and rough and irritating surfaces (**FIGURE 4-4**).
- **Duck:** Protects employees against cuts and bruises while they handle heavy, sharp, or rough materials (**FIGURE 4-5**).
- **Leather:** Often used against dry heat and flame (**FIGURE 4-6**).
- **Rubber, rubberized fabrics, neoprene, and plastics:** Provides protection against certain acids and other chemicals (**FIGURE 4-7**).

> ▶ **TECHNICIAN TIP**
>
> Each shop activity requires specific clothing, depending on its nature. Research and identify what specific type of clothing is required for every activity you undertake. Wear appropriate clothing for the activity you will be involved in, according to the shop's policies and procedures.

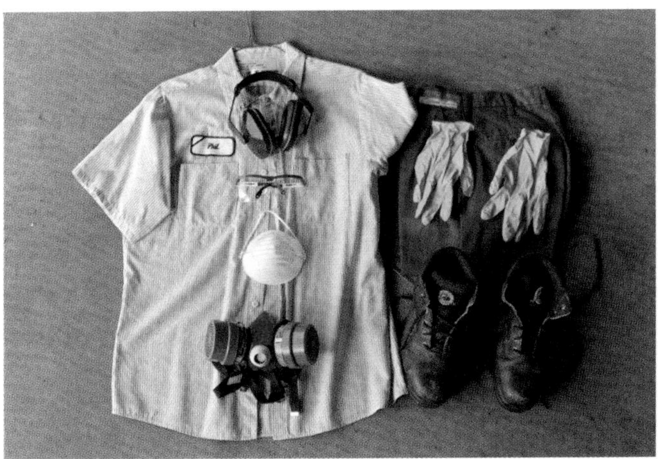

FIGURE 4-2 Personal protective equipment (PPE) includes clothing, shoes, safety glasses, hearing protection, masks, and respirators.

FIGURE 4-3 Disposable fiber suits.

FIGURE 4-4 Treated wool and cotton uniform.

FIGURE 4-5 Duck material work clothes.

Work Clothing

Always wear appropriate work clothing. Whether this is a one-piece coverall/overall or a separate shirt and pants, the clothes you work in should be comfortable enough to allow you to move, without being loose enough to catch on machinery (**FIGURE 4-8**). The material must be flame retardant and strong enough that it

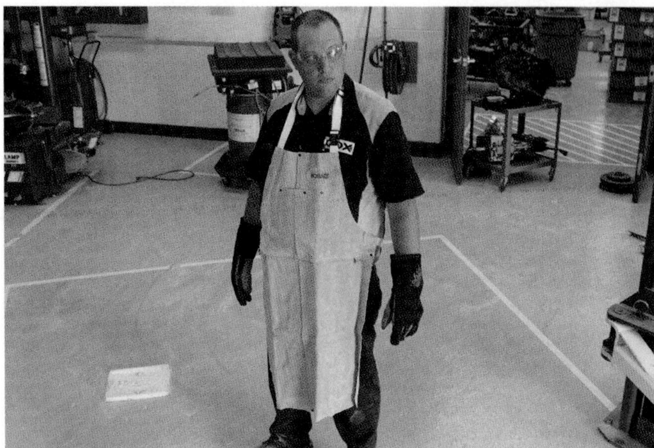

FIGURE 4-6 Leather apron.

FIGURE 4-7 Neoprene gloves.

cannot be easily torn. A flap must cover buttons or snaps. If you wear a long-sleeve shirt, the cuffs must be close fitting, without being tight. Pants should not have cuffs so that hot debris cannot become trapped in the fabric.

Care of Clothing

Always wash your work clothes separately from your other clothes to prevent contaminating your regular clothes. Start a new working day with clean work clothes, and change out of contaminated clothing as soon as possible. It is a good idea to keep a spare set of work clothes in the shop in case the ones you are wearing become overly dirty or a toxic or corrosive fluid is spilled on them (**FIGURE 4-9**).

Footwear

The proper footwear provides protection against items falling on your feet, chemicals, cuts, abrasions, punctures, and slips. They also provide good support for your feet, especially when working on hard surfaces like concrete (**FIGURE 4-10**). The soles of your shoes must be acid and slip resistant, and the uppers must be made from a puncture-proof material such as leather. Some shops and technicians prefer safety shoes with a steel toe cap to protect the toes (**FIGURE 4-11**). Always wear shoes that comply with your local shop standards.

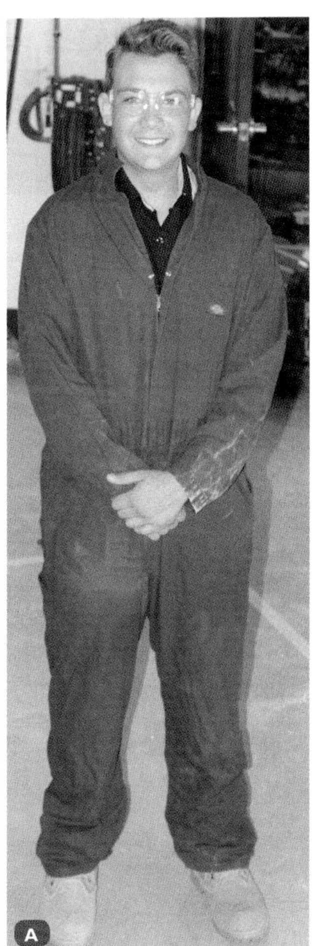

FIGURE 4-8 **A.** One-piece coverall. **B.** Shirt and pants.

FIGURE 4-9 Keep an extra set of work clothes at work in case you need them.

Headgear and Jewelry

N04003

Headgear includes items like hairnets, caps, and hard hats. These help protect you from getting your hair caught in rotating machinery and protect your head from knocks or bumps. For

FIGURE 4-10 Work shoes provide support for your feet when working on hard surfaces.

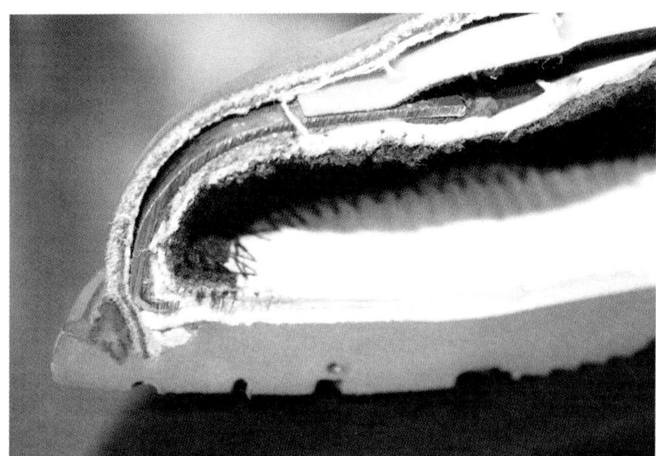

FIGURE 4-11 Steel-toed boots protect your toes from falling objects.

example, a hard hat can protect you from bumping your head on vehicle parts when working under a vehicle that is raised on a hoist (**FIGURE 4-12**). Head wounds tend to bleed a lot, so hard hats can prevent the need for visiting an emergency room for stitches.

Some technicians wear a cap to keep their hair clean when working under vehicles, or to contain hair that reaches a shirt collar. Some caps are designed specifically with additional padding on the top to provide extra protection against bumps. If hair is longer than can be contained in a cap, then technicians can either use a ponytail holder or hairnet (**FIGURE 4-13**).

When in a workshop environment, watches, rings, necklaces, and dangling earrings and other jewelry present a number of hazards. They can get caught in rotating machinery, and because they are mainly constructed from metal, they can conduct electricity. Imagine leaning over a running engine with a dangling necklace; it could get caught in the fan belt and pull you into the rotating parts if it doesn't break; not only will it get destroyed but it could seriously injure you. A ring or watch could inadvertently short out an electrical circuit, heat up quickly and severely burn you, or cause a spark that might make the battery explode. A ring can also get caught on moving parts, breaking the finger bone or even ripping the finger out of the

FIGURE 4-12 Hard hats prevent head wounds.

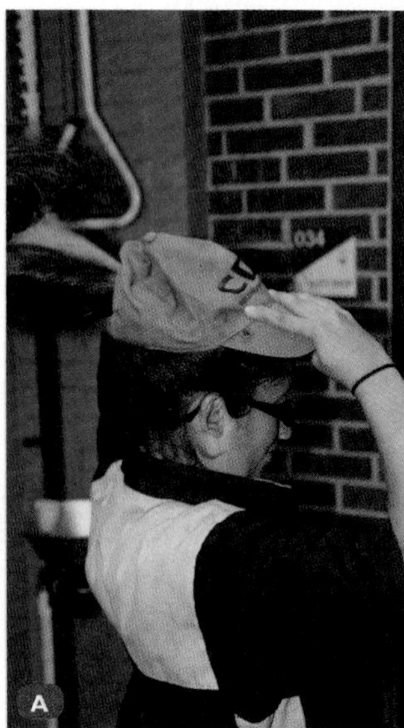

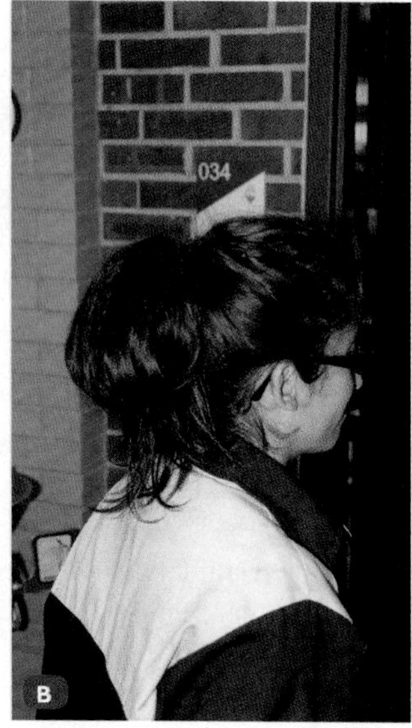

FIGURE 4-13 Containing hair. **A.** Ball cap. **B.** Pony tail.

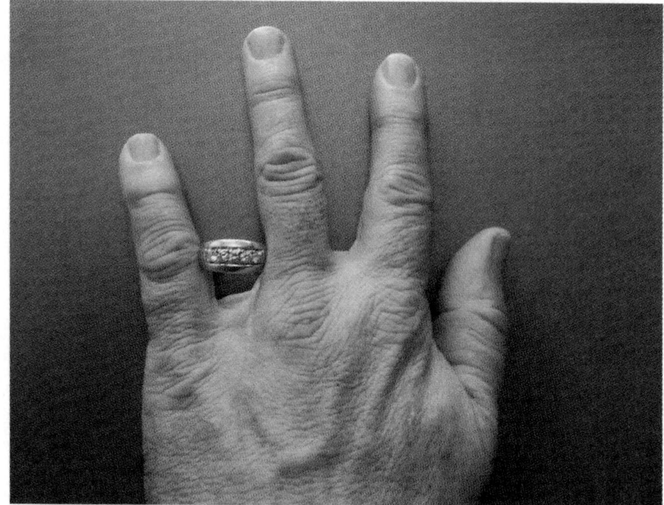

FIGURE 4-14 Finger missing because the wedding ring caught on rotating machinery.

hand (**FIGURE 4-14**). To be safe, always remove watches, rings, and jewelry before starting work. Not only is it safer to remove these items but your valuables will not get damaged or lost.

Hand Protection

K04002

Hands are a very complex and sensitive part of the body, with many nerves, tendons, and blood vessels. They are susceptible to injury and damage. Nearly every activity performed on vehicles requires the use of your hands, which provides many opportunities for injury. Whenever required, wear gloves to protect your hands. There are many types of gloves available, and their applications vary greatly as you will see below. It is important to wear the correct type of glove for the various activities you perform. In fact, when working on rotating equipment, it may be necessary for you to remove your gloves so that they don't get caught in the machinery and pull you into it (**FIGURE 4-15**).

Chemical Gloves

Heavy-duty and impenetrable chemical gloves should always be worn when using solvents and cleaners. They should also be worn when working on batteries. Chemical gloves should extend to the middle of your forearm to reduce the risk of chemicals

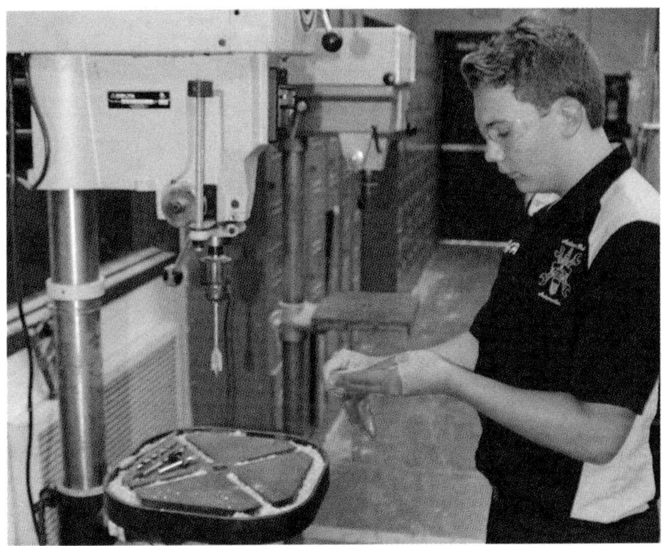

FIGURE 4-15 Removing gloves before working on a drill press.

FIGURE 4-17 Leather gloves protect your hands from burns when welding or handling hot components.

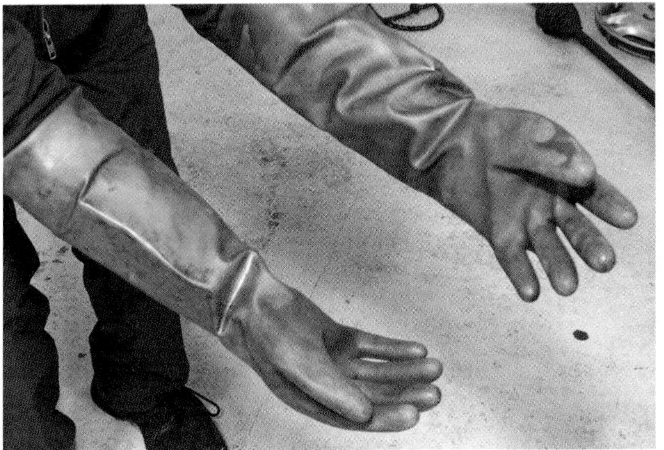

FIGURE 4-16 Chemical gloves should extend to the middle of your forearm to reduce the risk of chemical burns.

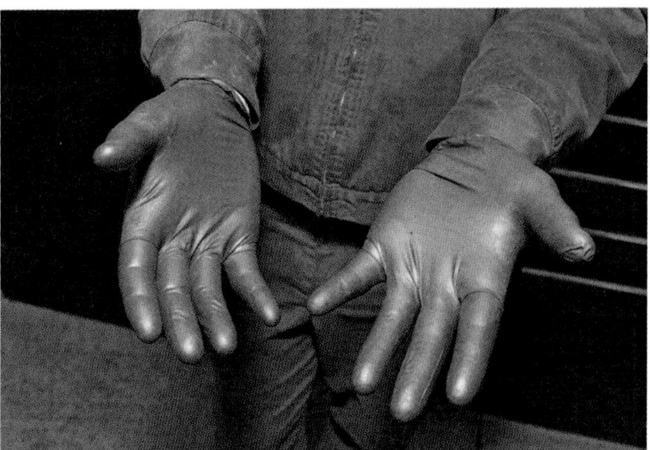

FIGURE 4-18 Light-duty gloves should be used to protect your hands from exposure to greases and oils.

splashing onto your skin (**FIGURE 4-16**). Always inspect chemical gloves for holes or cracks before using them, and replace them when they become worn. Some chemical gloves are also slightly heat resistant. This type of chemical glove is suitable for use when removing radiator caps and mixing coolant.

Leather Gloves

Leather gloves protect your hands from burns when welding or handling hot components (**FIGURE 4-17**). You should also use them when removing steel from a storage rack and when handling sharp objects. When using leather gloves for handling hot components, be aware of the potential for **heat buildup**. Heat buildup occurs when the leather glove can no longer absorb or reflect heat, and heat is transferred to the inside of the leather glove. At this point, the leather gloves' ability to protect you from the heat is reduced, and you need to stop work, remove the leather gloves, and allow them to cool down before continuing to work. Also avoid picking up very hot metal with leather gloves, because it causes the leather to harden, making it less flexible during use. If very hot metal must be moved, it would be better to use an appropriate pair of pliers.

Light-Duty Gloves

Light-duty gloves should be used to protect your hands from exposure to greases and oils (**FIGURE 4-18**). Light-duty gloves are typically disposable and can be made from a few different materials, such as nitrile, latex, and even plastic. Some people have allergies to these materials. If you have an allergic reaction when wearing these gloves, try using a glove made from a different material.

General-Purpose Cloth Gloves

Cloth gloves are designed to be worn in cold temperatures, particularly during winter, so that cold tools do not stick to your skin (**FIGURE 4-19**). Over time, cloth gloves accumulate dirt and grime, so you need to wash them regularly. Regularly inspect cloth gloves for damage and wear, and replace them when required. Cloth gloves are not an effective barrier against chemicals or oils, so never use them for that purpose.

FIGURE 4-19 Cloth gloves work well in cold temperatures, particularly during winter, so that cold tools do not stick to your skin.

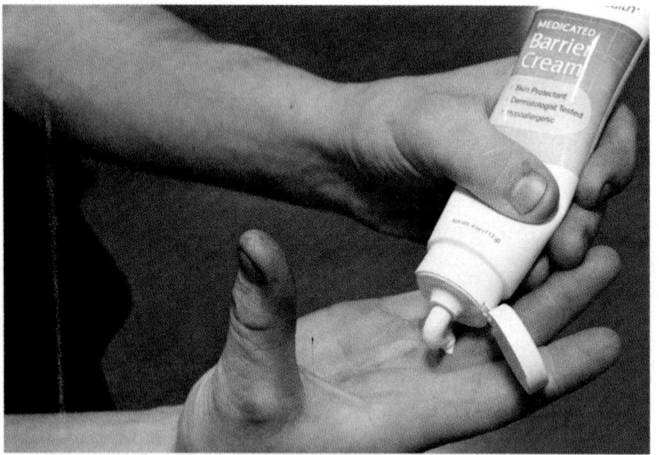

FIGURE 4-20 Barrier cream helps prevent chemicals from being absorbed into your skin and should be applied to your hands before you begin work.

Barrier Cream

Barrier cream looks and feels like a moisturizing cream, but it has a specific formula to provide extra protection from chemicals and oils. Barrier cream prevents chemicals from being absorbed into your skin and should be applied to your hands before you begin work (**FIGURE 4-20**). Even the slightest exposure to certain chemicals can lead to dermatitis, a painful skin irritation. Never use a standard moisturizer as a replacement for proper barrier cream. Barrier cream also makes it easier to clean your hands because it can prevent fine particles from adhering to your skin.

Cleaning Your Hands

When cleaning your hands, use only specialized hand cleaners, which protect your skin (**FIGURE 4-21**). Your hands are porous and easily absorb liquids on contact. Never use solvents such as gasoline or kerosene to clean your hands, because they can be absorbed into the bloodstream and remove the skin's natural protective oils.

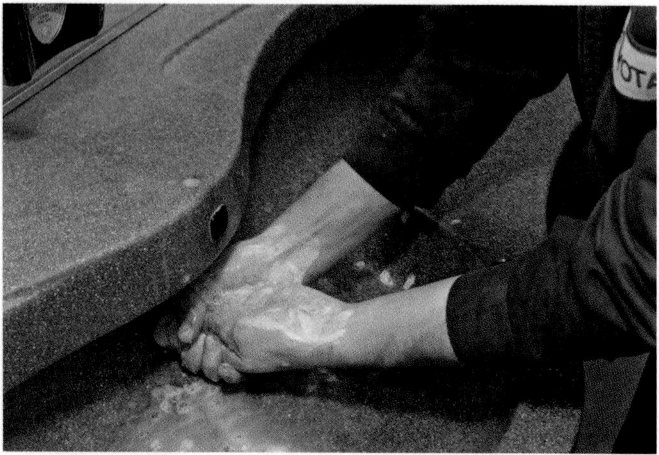

FIGURE 4-21 When cleaning your hands, use only specialized hand cleaners, which protect your skin.

Ear Protection

K04003

Ear protection should be worn when sound levels exceed 85 decibels, when you are working around operating machinery for any period of time, or when the equipment you are using produces loud noise. If you have to raise your voice to be heard by a person who is 2' away from you, then the sound level is about 85 decibels or more. Ear protection comes in two forms: One type covers the entire outer ear, and the other is fitted into the ear canal (**FIGURE 4-22**). Generally speaking, the in-the-ear style has higher noise-reduction ratings. If the noise is not excessively loud, either type of protection will work. If you are in an extremely loud environment, you will want to verify that the option you choose is rated high enough.

Breathing Devices

K04004

Dust and chemicals from your workspace can be absorbed into the body when you breathe. When working in an environment where dust is present or where the task you are performing will produce dust, you should always wear an appropriate form of breathing device. When working in an environment where chemical vapors are present, you should always wear the proper respirator. There are two types of breathing devices: disposable dust masks and respirators.

Disposable Dust Mask

A disposable dust mask is made from paper with a wire-reinforced edge that is held to your face with an elastic strip. It covers your mouth and nose and is disposed of at the completion of the task (**FIGURE 4-23**). This type of mask should only be used as a dust mask and should not be used if chemicals, such as paint solvents, are present in the atmosphere. It should also not be used when working around asbestos dust as the asbestos particles are too small for the filter to remove them, allowing then to be inhaled deeply into the lungs where their sharp tips pierce the lung's lining and become trapped. Over time, these create scar tissue in the

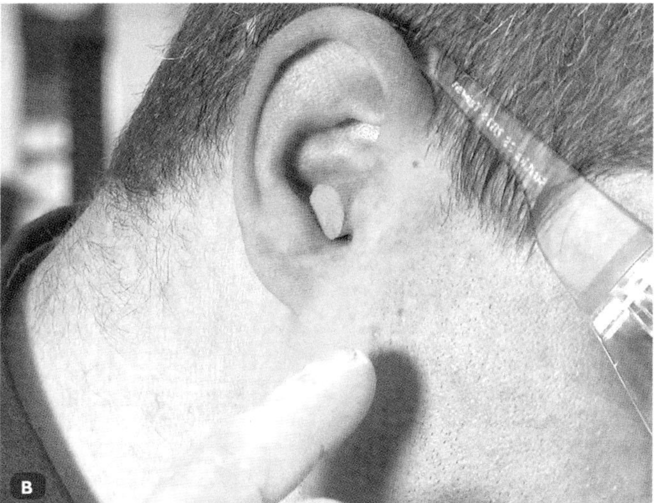

FIGURE 4-22 Ear protection comes in two forms: **A.** One type covers the entire outer ear. **B.** The other type is fitted into the ear canal.

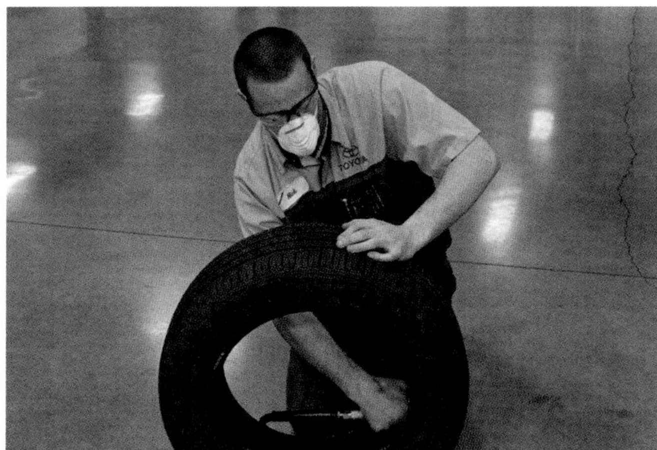

FIGURE 4-23 A disposable dust mask covers your mouth and nose and is disposed of at the completion of the task.

lungs and can potentially cause cancer or other life-threatening diseases. Dust masks and respirators should fit securely on your face to minimize leaks around the edges. This can be especially difficult to prevent if you have a beard.

Respirator

The **respirator** has removable cartridges that can be changed according to the type of contaminant being filtered. Always make sure the cartridge is the correct type for the contaminant in the atmosphere. For example, when chemicals are present, use the appropriate chemical filter in your respirator. The cartridges should be replaced according to the manufacturer's recommended replacement schedule to ensure their effectiveness. To be completely effective, the respirator mask must make a good seal onto your face (**FIGURE 4-24**).

In some situations where the environment either contains too high a concentration of hazardous chemicals or a lack of oxygen, a fresh air respirator must be used. This device pumps a supply of fresh air to the mask from an outside location (**FIGURE 4-25**). Being aware of the environment you are working in allows you to determine the proper respirator or fresh air supply system.

Eye Protection

K04005

Eyes are very sensitive organs, and they need to be protected against damage and injury. There are many things in the workshop environment that can damage or injure eyes, such as

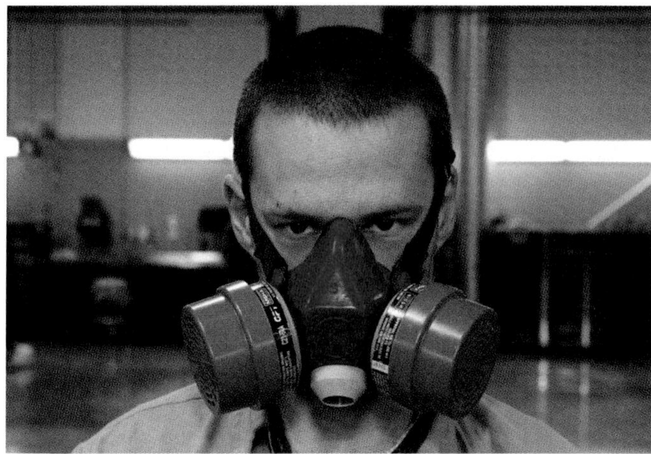

FIGURE 4-24 To be completely effective, the respirator mask must make a good seal onto your face.

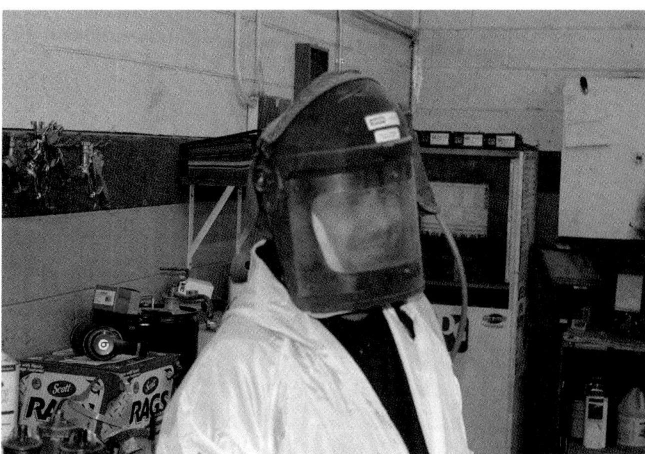

FIGURE 4-25 Fresh air respirator.

high-velocity particles coming from a grinder or high-intensity light coming from a welder. In fact, the American National Standards Institute (ANSI) reports that 2000 workers per day suffer on-the-job eye injuries. Always select the appropriate eye protection for the work you are undertaking. Sometimes this may mean that more than one type of protection is required. For example, when grinding, you should wear a pair of safety glasses underneath your face shield for added protection.

Safety Glasses

The most common type of eye protection is a pair of safety glasses, which must be clearly marked with "Z87.1" (**FIGURE 4-26**). Safety glasses have built-in side shields to help protect your eyes from the side. Approved safety glasses should be worn whenever you are in a workshop. They are designed to help protect your eyes from direct impact or debris damage (**FIGURE 4-27**).

The only time they should be removed is when you are using other eye protection equipment. Prescription and tinted safety glasses are also available. Tinted safety glasses are designed to be worn outside in bright sunlight conditions. Never wear them indoors or in low-light conditions because they reduce your ability to see clearly. For people who wear prescription glasses, there are three acceptable options that OSHA makes available:

- Prescription spectacles, with side shields and protective lenses meeting requirements of ANSI Z87.1
- Goggles that can fit comfortably over corrective eyeglasses without disturbing their alignment
- Goggles that incorporate corrective lenses mounted behind protective lenses

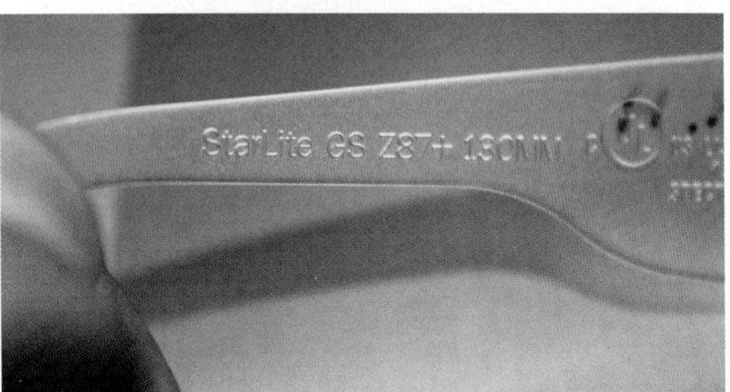

FIGURE 4-26 Safety glasses with the Z87.1 marking.

FIGURE 4-27 Safety glasses are designed to protect your eyes from direct impact or debris damage.

SAFETY TIP

You might be tempted to take your safety glasses off while you are doing a nonhazardous task in the shop like a former student of mine. He was doing paperwork in his stall while his best friend was driving pins out of the tracks of a bulldozer. Unfortunately, the head of the punch his friend was using was mushroomed. And on one hit, a fragment broke off, flew across the stall, and hit the student in the eye, blinding him permanently. So always wear safety glasses while in a work area, even if you aren't working.

Safety Goggles

Safety goggles don't provide as much eye protection as safety glasses, but they do have added protection against harmful chemicals that may splash up behind the lenses of safety glasses (**FIGURE 4-28**). Goggles also provide additional protection from foreign particles. Safety goggles must be worn when servicing air-conditioning systems or any other system that contains pressurized gas. Goggles can sometimes fog up when in use; if this occurs, use one of the special antifog cleaning fluids or cloths to clean them.

▶ TECHNICIAN TIP

Each lab/shop activity requires at least the safe use of safety glasses, clothing, and shoes, depending on its nature. Research and identify whether any additional safety devices are required for every activity you undertake.

Full Face Shield

A full face shield gives you added protection from sparks or chemicals over safety glasses alone (**FIGURE 4-29**). The clear mask of the face shield allows you to see all that you are doing

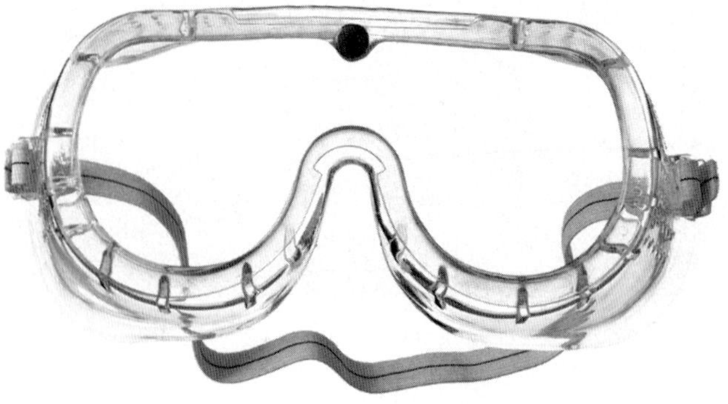

FIGURE 4-28 Safety goggles provide much the same eye protection as safety glasses, but with added protection against any harmful fluid that may find its way behind the lenses.

and helps protect your eyes and face from chemical burns should there be any splashes or battery explosions. It is also recommended that you use a full face shield combined with safety goggles when using a bench or angle grinder.

Welding Helmet

The light from a welding arc is very bright and contains high levels of ultraviolet radiation. So wear a **welding helmet** when using, or assisting a person using, an electric welder. The lens on a welding helmet has heavily shaded glass to reduce the intensity of the light from the welding arc, allowing you to see the task you are performing more clearly (**FIGURE 4-30**).

Lenses come in a variety of ratings depending on the type of welding you are doing; always make sure you are using a properly rated lens for the welder you are using. The remainder of the helmet is made from a durable material that blocks any other light, which can burn your skin similar to a sunburn, from reaching your face. It also protects you from welding sparks. Photosensitive welding helmets that darken automatically when an arc is struck are also available. Their big advantage is that you do not have to lift and lower the helmet by hand while welding.

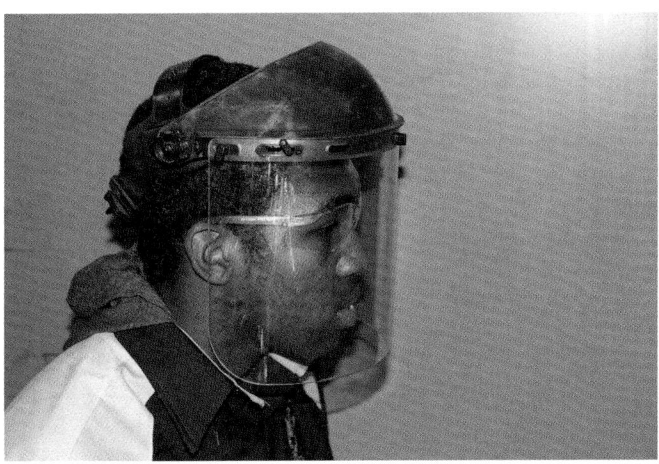

FIGURE 4-29 Full face shield.

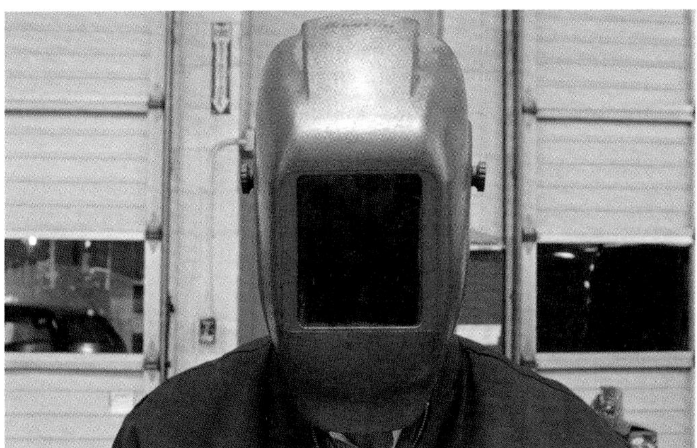

FIGURE 4-30 The lens on a welding helmet has heavily tinted glass to reduce the intensity of the light from the welding tip, allowing you to see what you are doing.

SAFETY TIP

Be aware that the ultraviolet radiation can burn your skin like a sunburn, so wear the appropriate welding apparel to protect yourself from this hazard.

Gas Welding Goggles

Gas welding goggles can be worn instead of a welding mask when using or assisting a person using an oxyacetylene welder (**FIGURE 4-31**). The eyepieces are available in heavily shaded

TABLE 4-1: Filter Lens Shade Numbers for Protection Against Radiant Energy

Welding Operation	Shade Number
Shielded metal-arc welding 1/18-, 3/32-, 1/8-, 5/32-inch-diameter electrodes	10
Gas-shielded arc welding (nonferrous) 1/16-, 3/32-, 1/8-, 5/32-inch diameter electrodes	11
Gas-shielded arc welding (ferrous) 1/16-, 3/32-, 1/8-, 5/32-inch diameter electrodes	12
Shielded metal-arc welding 3/16-, 7/32-, 1/4-inch diameter electrodes	12
5/16-, 3/8-inch diameter electrodes	12
Atomic hydrogen welding	10–14
Carbon-arc welding	14
Soldering	2
Torch brazing	3 or 4
Light cutting, up to 1 inch	3 or 4
Medium cutting, 1 inch to 6 inches	4 or 5
Heavy cutting, over 6 inches	5 or 6
Gas welding (light), up to 1/8 inch	4 or 5
Gas welding (medium), 1/8 inch to ½ inch	5 or 6
Gas welding (heavy), over ½ inch	6 or 8

Source: Reprinted from PPE Assessment, OSHA Office of Training and Education, Tab. 2, http://www.osha.gov/dte/library/ppe_assessment/ppe_assessment.html.

FIGURE 4-31 Gas welding goggles can be worn instead of a welding helmet when using or assisting a person using an oxyacetylene welder.

versions, but not as shaded as those used in an electric welding helmet. There is much less ultraviolet radiation from an oxy-acetylene flame, so a welding helmet is not required. However, the flame is bright enough to damage your eyes, so always use goggles of the correct shade rating.

▶ Injury Protection Practices

K04006

Safe Attitude

Develop a safe attitude toward your work. You should always think "safety first" and then act safely (**FIGURE 4-32**). Think ahead about what you are doing, and put in place specific measures to protect yourself and those around you. For example, you could ask yourself the following questions:

- What could go wrong?
- What measures can I take to ensure that nothing goes wrong?
- What PPE should I use?
- Have I been trained to use this piece of equipment?
- Is the equipment I'm using safe?

Answering these questions and taking appropriate action before you begin will help you work safely. If you don't know the answers to those questions, *don't* work! Stop and ask someone with more experience than yourself, or your supervisor. Also, don't count on others for your safety. You alone are primarily responsible for your own safety.

Also remember that the time to make a good decision is *before* an accident happens. Once the accident occurs, you will likely have very little to no control over the outcome. In fact, there are very few true accidents that happen all by themselves. Accidents are almost always caused by poor decisions or poor actions. Almost *all* accidents can be prevented with a little more thought, training, and practice.

Proper Ventilation

N04004

Proper ventilation is required for working in the shop area. The key to proper ventilation is to ensure that any task or procedure that may produce dangerous or toxic fumes is recognized, so measures can be put in place to provide adequate ventilation (**FIGURE 4-33**). Ventilation can be provided by natural means, such as by opening doors and windows to provide air flow for low-exposure situations. However, in high-exposure situations, such as vehicles running in the shop, a mechanical means of ventilation is required; an example is an exhaust extraction system. Parts cleaning areas or areas where solvents and chemicals are used should also have good general ventilation, and if required, additional exhaust hoods or fans should be installed to remove dangerous fumes. In some cases, such as when

FIGURE 4-32 Always think "safety first."

FIGURE 4-33 Adequate ventilation is necessary for a safe work environment.

spraying paint, it may be necessary to use a personal respirator in addition to proper ventilation.

> ▶ TECHNICIAN TIP
>
> Before starting a vehicle in the shop, make sure the correct ventilation equipment is connected properly and turned on. Also, if the system uses rubber hoses, make sure they aren't kinked.

Applied Science

AS-1: Safety: The technician follows all safety regulations and applicable procedures while performing the task.

Using a bench grinder to grind down a steel component is a simple everyday activity in shops. In terms of its potential safety implications, it carries significant risk of injury. Before beginning the task, you must ensure that the machinery is safe and ready to use. Inspect and/or adjust the guards/shields, grinding wheels, electrical cord, etc. From a personal perspective, you must make sure your clothing is suitable and safe for the task. Clothing cannot be loose, as it may get caught in the grinder; it must be made of flame-retardant material because of the risk of ignition from sparks. Long hair must be tied back or contained within a hairnet or cap because of the risk of it getting caught in rotating machinery. Various items of PPE are required to safely carry out this task: safety goggles to protect against foreign objects entering the eyes, ear protection to guard against hearing damage due to excessive noise, steel-capped boots to prevent injury from falling heavy objects, and heavy-duty gloves to protect against skin contact with sharp or hot components.

Lifting

K04007

Whenever you lift something, there is always the possibility of injury; however, by lifting correctly, you reduce the chance of injuring yourself or others. Before lifting anything, you can reduce the risk of injury by breaking down the load into smaller quantities, asking for assistance if required, or possibly using a mechanical device to assist the lift. If you have to bend down to lift something, you should bend your knees to lower your body; do not bend over with straight legs because this can damage your back (**FIGURE 4-34**). Place your feet about shoulder width apart, and lift the item by straightening your legs while keeping your back as straight as possible.

> SAFETY TIP
>
> Never lift anything that is too heavy for you to comfortably lift, and always seek assistance if lifting the object could injure you. Always err on the side of caution.

Housekeeping and Orderliness

K04008

Good housekeeping is about always making sure the shop and your work surroundings are neat and kept in good order. This habit pays off in a couple of ways. It makes the shop a safer place to work by not allowing clutter or spills to accumulate (**FIGURE 4-35**). It also makes the shop more efficient if everything is kept neat and orderly, by making it easier to find needed tools and equipment. Trash and liquid spills should be quickly cleaned up; tools need to be cleaned and put away after use; spare or removed parts need to be stored or disposed of correctly; and generally everything needs to have a safe place to be kept.

You should carry out good housekeeping practices while working, not just after a job is completed. For example, throw trash in the garbage can as it accumulates, clean up spills when they happen, and put tools away when you are finished working

FIGURE 4-34 Prevent back injuries when lifting heavy objects by crouching with your legs slightly apart, standing close to the object, positioning yourself so that the center of gravity is between your feet.

with them. It is also good practice to periodically perform a deep clean of the shop so that any neglected areas are taken care of.

Slip, Trip, and Fall Hazards

Slip, trip, and fall hazards are ever present in the shop, and they can be caused by trash, tools and equipment, or liquid spills being left lying around. Always be on the lookout for hazards that can cause slips, trips, or falls. Floors and steps can become slippery, so they should be kept clean and have antislip coatings applied to them. High-visibility strips with antislip coatings can be applied to the edge of step treads to reduce the hazard. Clean up liquid spills immediately, and mark the area with wet floor signs until the floor is dry (**FIGURE 4-36**). Make sure the workshop has good lighting, so hazards are easy to spot, and keep walkways clear from obstruction. Think about what you are doing and make sure the work area is free of slip, trip, and fall hazards as you work.

▶ First Aid Principles

The following information is designed to provide you with an awareness of basic first aid principles and the importance of first aid training courses. You will find general information about how to take care of someone who is injured. However, this information is only a guide. It is not a substitute for training or professional medical assistance. Always seek professional advice when tending to an injured person. One of the best things you can do is take a first aid class and periodic refresher courses. These courses help you stay current on first aid practices as they change. They also help remind you of the importance of preventing accidents so that first aid is less likely to be needed.

First aid is the immediate care given to an injured or suddenly ill person (**FIGURE 4-37**). Learning first aid skills is valuable in the workplace in case an accident or medical emergency arises. First aid courses are available through many organizations, such as the Emergency Care and Safety Institute (ECSI). It is strongly advised that you seek out a certified first aid course and become certified in first aid. The following information highlights some of the principles of first aid.

In the event of an accident, the possibility of injury to the rescuer or further injury to the victim must be evaluated. This means the first step in first aid is to survey the scene. While doing this, try to determine what happened, what dangers may still be present, and the best actions to take. Only remove the injured person from a dangerous area if it is safe for you to do so. When dealing with electrocution or electrical burns, make

FIGURE 4-35 Develop good housekeeping habits. **A.** Cluttered, unorganized shop. **B.** Clean, organized shop.

FIGURE 4-36 Clean up spill immediately.

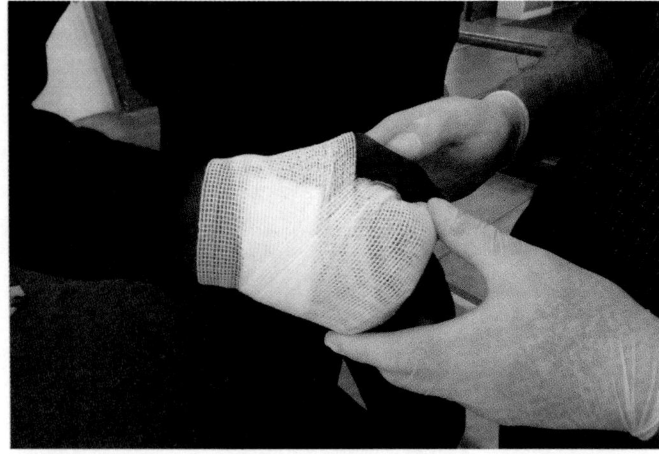

FIGURE 4-37 First aid is a valuable skill to learn.

sure the electrical supply is switched off before attempting any assistance. Always perform first aid techniques as quickly as is safely possible after an injury. When breathing or the heart has stopped, brain damage can occur within four to six minutes. The degree of brain damage increases with each passing minute, so make sure you know what to do, and do it as quickly as is safe to do so.

> **TECHNICIAN TIP**
>
> Three important rules of first aid:
>
> 1. Know what you must not do.
> 2. Know what you must do.
> 3. If you are not sure what procedures to follow, send for trained medical assistance.

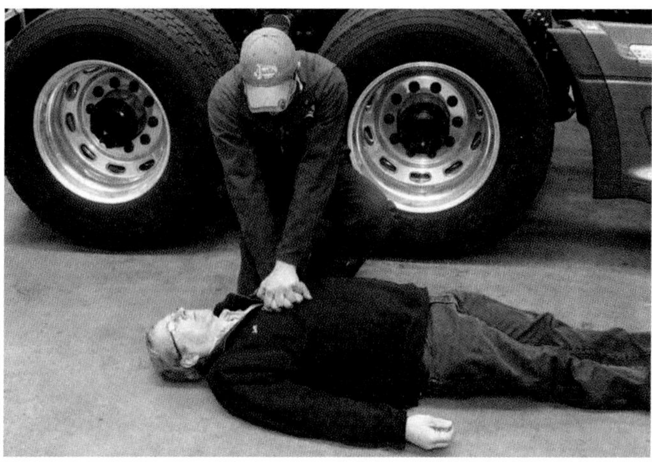

FIGURE 4-38 Chest compressions.

First Aid Concepts

K04009

Prompt care and treatment before the arrival of emergency medical assistance can sometimes mean the difference between life and death. The goals of first aid are to make the immediate environment as safe as possible, preserve the life of the patient, prevent the injury from worsening, prevent additional injuries from occurring, protect the unconscious, promote recovery, comfort the injured, prevent unnecessary delays in treatment, and provide the best possible immediate care for the injured person. When attending to an injured victim, always send for assistance. Make sure the person who stays with the injured victim is more experienced in first aid than the messenger. If you are the only person available, request medical assistance as soon as reasonably possible. When you approach the scene of an accident or emergency, do the following:

1. **Danger:** Survey the scene to make sure there are no other dangers, and assist only if it is safe to do so.
2. **Response:** Check to see if the victim is responsive and breathing. If responsive, ask the victim if he or she needs help. If the victim does not respond, he or she is unresponsive.
3. **Send for help:** Have a bystander call 9-1-1. Always have the person with the most first aid experience stay with the victim. If alone, call 9-1-1 yourself.
4. **Airway:** Open the airway by tilting the victim's head back and lifting the chin.
5. **Breathing:** Check for normal breathing by placing one of your hands on the victim's chest, and lean down so your ear is near the person's mouth. Listen and watch for breathing. If the victim is breathing, monitor the victim for further issues. If the victim is not breathing normally, start CPR.
6. **Circulation/CPR (cardiopulmonary resuscitation):** If no pulse is present, start chest compressions at the rate of approximately 100 per minute. You can either give sequences of 30 compressions to two breaths, or just compressions if you are reluctant to give mouth to mouth resuscitation. Repeat the compression and breath cycles, or compression only cycles, until an automated external defibrillator (AED) is available or emergency medical system (EMS) personnel arrive (**FIGURE 4-38**).

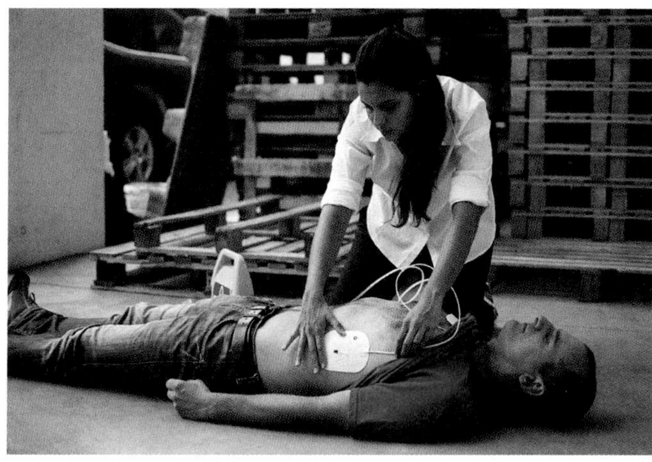

FIGURE 4-39 AED being set up.

7. **Defibrillation:** If a pulse is still not present once an AED arrives, expose the victim's chest and turn on the AED (**FIGURE 4-39**). Attach the AED pads. Ensure that no one touches the victim. Follow the audio and visual prompts from the AED. If no shock is advised, resume CPR immediately (five sets of 30 compressions and two breaths, or compressions by themselves). If a shock is advised, do not touch the victim and give one shock. Or shock as advised by AED. Follow the directions given by the AED.

First Aid Procedures

K04010

Bleeding

A wound that is severely bleeding is serious. If the bleeding is allowed to continue, the victim may collapse or die. Bleeding is divided into two categories: external and internal. **External bleeding** is the loss of blood from an external wound where blood can be seen escaping. **Internal bleeding** is the loss of blood into a body cavity from a wound with no obvious sign of blood.

Before providing first aid, make sure you are not exposed to blood. Wear latex gloves or an artificial barrier. Lay the victim down, and then apply a gauze pad and direct pressure to the wound

(**FIGURE 4-40**). Apply a pressure bandage over the gauze. If blood soaks through the bandage, apply additional dressings and pressure bandage (**FIGURE 4-41**). Call 9-1-1 if bleeding cannot be controlled. Give nothing by mouth, and seek medical aid immediately.

If an object punctures the victim's skin and becomes embedded in the victim's body, do not attempt to remove the object. Stabilize the object with a bulky dressing. Seek medical care immediately.

If the injured person has internal bleeding, it may not be immediately obvious. Symptoms of internal bleeding are bruising, a painful or tender area, coughing frothy blood, vomiting blood, stool that is black or contains bright red blood, and passing blood with urine. To assist an injured victim with internal bleeding, lay the victim down, loosen tight clothing, give nothing by mouth, and seek medical aid immediately.

Eye Injuries

Foreign objects can become embedded in the eye, or chemicals can splash into the eye. If an object penetrates and becomes embedded in the eye, do not attempt to remove it. Lay the victim down, and stabilize the object with a bulky dressing or clean cloths. Ask the victim to close the other eye, and call 9-1-1 (**FIGURE 4-42**).

If an object is loose on the surface of the eye, pull the upper lid over the lower lid. Hold the eyelid open and gently rinse with water. Examine the lower lid by pulling it down gently. If you can see the object, remove it with a moistened sterile gauze, a clean cloth, or a moistened cotton swab. Examine the underside of the upper lid by grasping the lashes of the upper lid and rolling the lid upward over a cotton swab. If you can see the object, remove it with a moistened sterile gauze or a clean cloth.

If a chemical splashes into the eyes, you may be able to flush it out using an eyewash station (**FIGURE 4-43**). Hold the eye wide open, and flush with warm water for at least 20 minutes, continuously and gently. Irrigate from the nose side of the eye toward the outside to avoid flushing material into the other eye. Loosely bandage the eyes with wet dressings. Call 9-1-1.

Fractures

A fracture is a broken or cracked bone. Always seek medical care for all fractures. There may be symptoms you are not aware of that may make the injury more complex than first thought. There are three types of bone fractures: A **simple fracture** involves no wound or internal or external bleeding; an **open fracture** involves bleeding or the protrusion of bone through

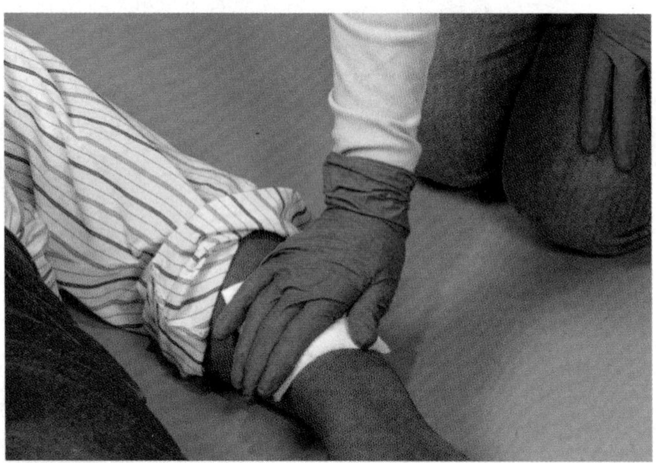

FIGURE 4-40 Apply a gauze pad and direct pressure to the wound.

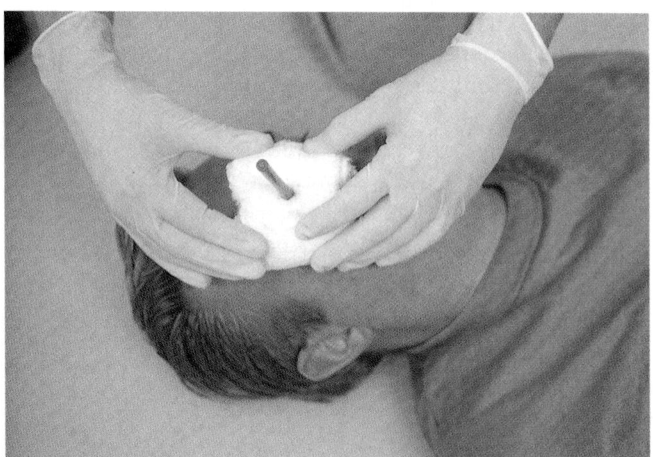

FIGURE 4-42 If an object penetrates and becomes embedded in the eye, stabilize the object with a bulky dressing or clean cloths.

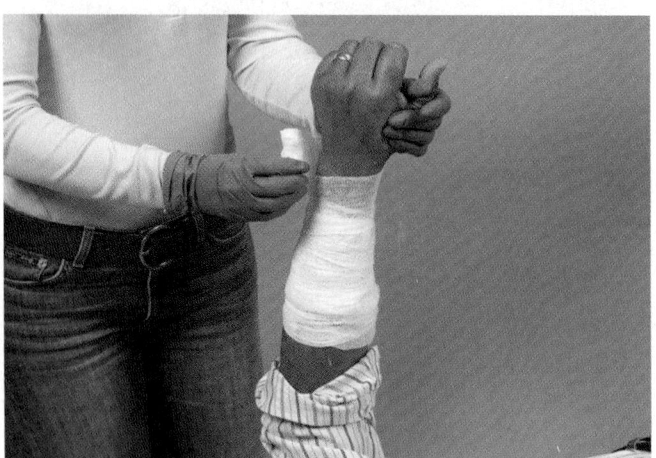

FIGURE 4-41 If blood soaks through the bandage, apply additional dressings and a pressure bandage.

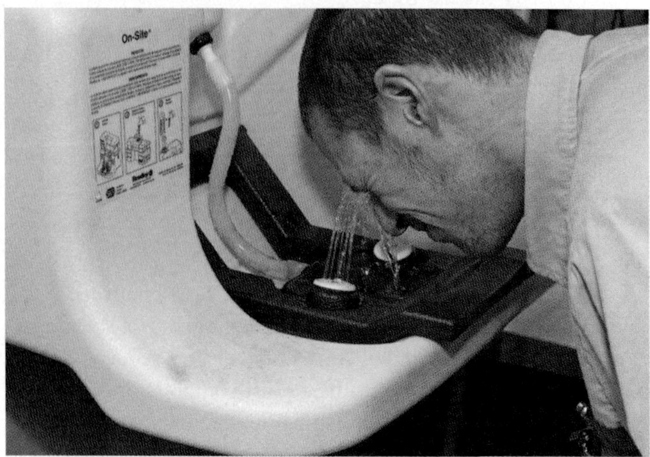

FIGURE 4-43 Flush out the eye to prevent a chemical burn.

the skin; and a **complicated fracture** involves penetration of a bone into a vital organ.

The symptoms of a fracture include hearing a snapping noise when the injury occurred, pain or tenderness at or near the injury, inability to move the limb, loss of strength in the limb, shortening of the limb or an abnormally shaped limb, swelling and/or bruising around the area, and a grinding noise if the limb is moved. Allow the victim to support the injured area in the most comfortable position. Stabilize the injured part with your hands or a splint to prevent movement. If the injury is an open fracture, do not push on any protruding bone. Cover the wound and exposed bone with a dressing. Apply ice or a cold pack, if possible, to help reduce swelling or pain. Call 9-1-1 for any open fractures or large bone fractures. Do not move the victim unless there is an immediate danger. Be aware of the onset of **shock**, which may present as the victim vomiting or fainting. Shock is when the body's tissues do not receive enough oxygenated blood.

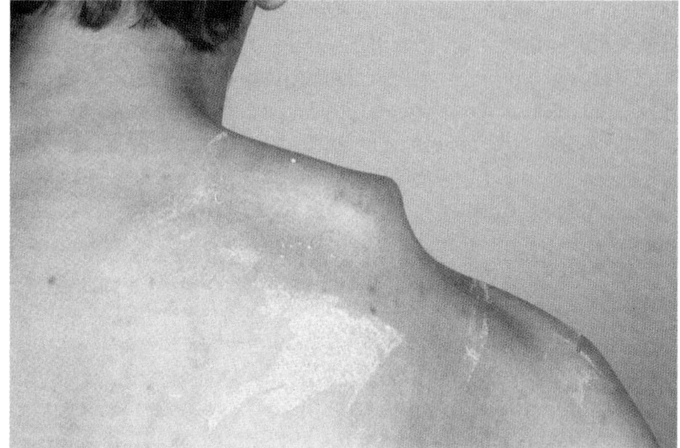

FIGURE 4-44 Dislocated shoulder.

Sprains, Strains, and Dislocations

When a joint has been forced past its natural range of movement, or a muscle or ligament has been overstressed or torn, a sprain, strain, or dislocation may occur. A **sprain** occurs when a joint is forced beyond its natural movement limit. This causes stretching or tearing in the ligaments that hold the bones together. The symptoms of a sprain include pain and loss of limb function, with swelling and bruising present. When a sprain occurs, apply covered ice packs every 20 minutes, elevate the injured limb, and apply an elastic compression bandage to the area and beyond the affected area. You should always treat a sprain as a fracture until medical opinion says otherwise.

A **strain** is an injury caused by the overstretching of muscles and tendons. Symptoms of a strain are sharp pain in the area immediately after the injury occurs, increased pain when using the limb, or tenderness over the entire muscle. The muscle may also have an indentation at the strain location. When a strain occurs, have the victim rest, elevate the injured limb, apply covered ice packs every 20 minutes, and apply an elastic compression bandage.

A **dislocation** is the displacement of a joint from its normal position; it is caused by an external force stretching the ligaments beyond their elastic limit (**FIGURE 4-44**). Symptoms of a dislocation are pain or tenderness around the area, inability to move the joint, deformity of the joint, and swelling and discoloration over the joint. If a dislocation occurs, try to immobilize the limb and seek medical attention. Do not try to put the joint back in place.

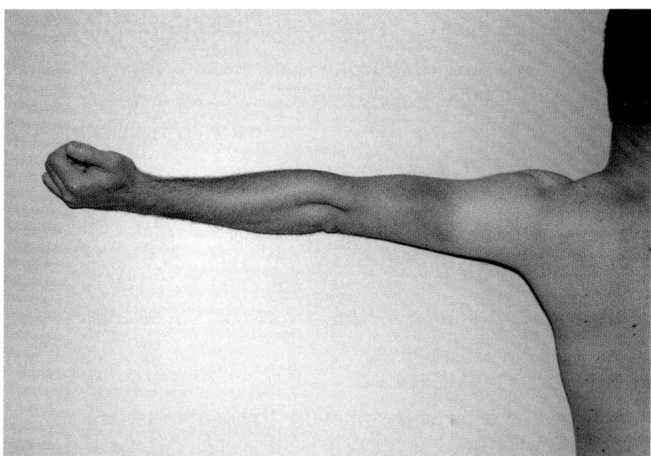

FIGURE 4-45 First-degree burn.

Burns and Scalds

Burns are injuries to body tissues, including skin, that are caused by exposure to heat, chemicals, and radiation. Burns are classified as either superficial, partial thickness, or full thickness. Superficial burns, or **first-degree burns**, show reddening of the skin and damage to the outer layer of skin only (**FIGURE 4-45**). Partial-thickness burns, or **second-degree burns**, involve blistering and damage to the outer layer of skin (**FIGURE 4-46**). Full-thickness burns, or **third-degree burns**,

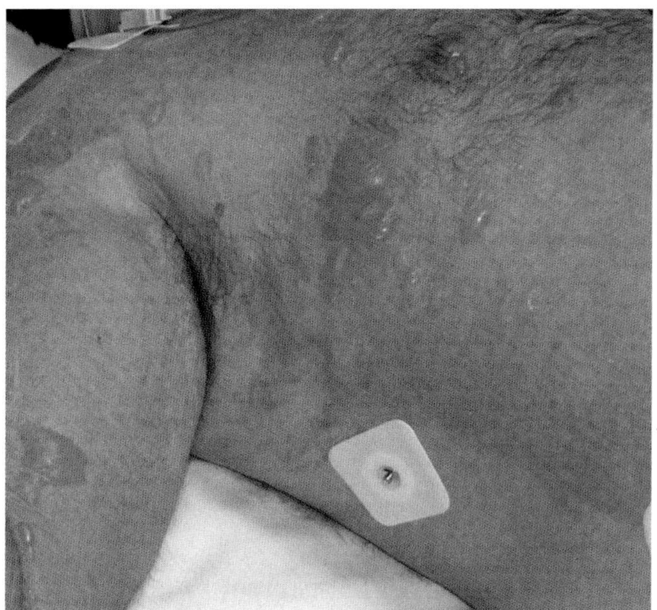

FIGURE 4-46 Second-degree burn.

involve white or blackened areas and include damage to all skin layers and underlying structures and tissues (**FIGURE 4-47**).

Burns can be caused by excessive heat, such as from fire; friction, such as from a rope burn; radiation, such as from a welding flash or a sunburn; chemicals, including acids and bases; or electricity, such as from faulty appliances. Scalds are injuries to the skin caused by exposure to hot liquids and gases. The effects of burns and scalds can include permanent skin and tissue damage, blisters caused by damage to surface blood vessels, severe pain, and shock. Remove the victim from any danger. If clothing is burning, have the victim roll on the ground using the "stop, drop, and roll" method. Smother the flames with a fire blanket or douse the victim with water. For minor burns, cool the burn with cool water until the body part is pain free. After the burn has cooled, apply antibiotic ointment. Do not apply lotions or aloe vera. Cover the burn loosely with a dry, nonstick, sterile, or clean dressing. Do not break any blisters. Give an over-the-counter pain medication such as ibuprofen. Seek medical care. Any large or third-degree burn must be treated by a qualified medical practitioner. Serious burns include skin that is blackened, whitened, or charred; a burn that is larger than 2 cm (0.79") in diameter; or a burn that is in the airway or on the face, hands, or genitals. When presented with such burns, call 9-1-1 immediately.

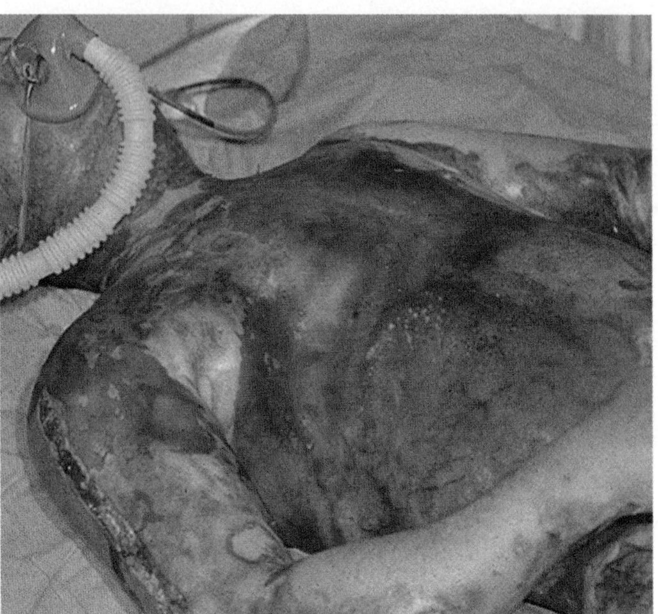

FIGURE 4-47 Third-degree burn.

Courtesy of AAOS

Wrap-Up

Ready for Review

- Personal protective equipment (PPE) protects the body from injury but must fit correctly and be task appropriate.
- Work clothing should be clean, loose enough for movement, but not baggy, and should also be flame-retardant.
- Footwear should be acid- and slip-resistant and made of puncture-proof material.
- Headgear can protect your head from bumps and should hold long hair in place.
- Hand protection includes chemical gloves, leather gloves, light-duty gloves, general-purpose cloth gloves, and barrier cream.
- Hazardous chemicals and oils can be absorbed into your skin.
- Wear ear protection if the sound level is 85 decibels or above.
- Breathing devices include disposable dust masks and respirators.
- Forms of eye protection are safety glasses, welding helmet, gas welding goggles, full face shield, and safety goggles.
- You may need two types of eye protection for some tasks.
- Before starting work, remove all jewelry and watches, and make sure your hair is contained.
- Thinking "safety first" will lead to acting safely.
- All shops require proper ventilation.
- Lifting correctly or seeking assistance helps prevent back injuries.
- Safety includes keeping a clean shop with everything put where it belongs and all spills cleaned up.
- First aid involves providing immediate care to an ill or injured person.
- Do not perform first aid if it is unsafe to do so.
- Bleeding may be internal or external; do not remove embedded objects that cause bleeding.
- Flush eyes with warm water to remove chemicals or debris.
- Fractures can be simple, open, or complicated; all fractures require medical care.
- Sprains, dislocations, and strains are possible injuries to muscles and ligaments.
- Burns can be superficial, partial thickness, or full thickness.
- Burns are classified as first, second, or third degree; large or third-degree burns require immediate medical care.

Key Terms

barrier cream A cream that looks and feels like a moisturizing cream but has a specific formula to provide extra protection from chemicals and oils.

complicated fracture A fracture in which the bone has penetrated a vital organ.

dislocation The displacement of a joint from its normal position; it is caused by an external force stretching the ligaments beyond their elastic limit.

ear protection Protective gear worn when the sound levels exceed 85 decibels, when working around operating machinery for any period of time, or when the equipment you are using produces loud noise.

engineering and work practice controls Systems and procedures required by OSHA and put in place by employers to protect their employees from hazards.

external bleeding The loss of blood from an external wound; blood can be seen escaping.

first aid The immediate care given to an injured or suddenly ill person.

first-degree burns Burns that show reddening of the skin and damage to the outer layer of skin only.

gas welding goggles Protective gear designed for gas welding; they provide protection against foreign particles entering the eye and are tinted to reduce the glare of the welding flame.

headgear Protective gear that includes items like hairnets, caps, or hard hats.

heat buildup A dangerous condition that occurs when the glove can no longer absorb or reflect heat, and heat is transferred to the inside of the glove.

internal bleeding The loss of blood into the body cavity from a wound; there is no obvious sign of blood.

open fracture A fracture in which the bone is protruding through the skin or there is severe bleeding.

respirator Protective gear used to protect the wearer from inhaling harmful dusts or gases. Respirators range from single-use disposable masks to types that have replaceable cartridges. The correct types of cartridge must be used for the type of contaminant encountered.

second-degree burns Burns that involve blistering and damage to the outer layer of skin.

shock Inadequate tissue oxygenation resulting from serious injury or illness.

simple fracture A fracture that involves no open wound or internal or external bleeding.

sprain An injury in which a joint is forced beyond its natural movement limit.

strain An injury caused by the overstretching of muscles and tendons.

third-degree burns Burns that involve white or blackened areas and damage to all skin layers and underlying structures and tissues.

welding helmet Protective gear designed for arc welding; it provides protection against foreign particles entering the eye, and the lens is tinted to reduce the glare of the welding arc.

Review Questions

1. Neoprene gloves are ideal for protection from:
   - a. dust.
   - b. heavy materials.
   - c. changing workplace temperatures.
   - d. certain acids and other chemicals.
2. All of the following may pose a problem when working on rotating equipment *except*:
   - a. gloves.
   - b. hats.
   - c. watches.
   - d. rings.
3. Heat buildup is a factor to be aware of when wearing:
   - a. leather gloves.
   - b. cloth gloves.
   - c. light-duty gloves.
   - d. chemical gloves.
4. You can use a disposable dust mask to protect yourself against:
   - a. asbestos dust.
   - b. shop dust.
   - c. chemical vapors.
   - d. paint solvents.
5. All of the following statements are true *except*:
   - a. When grinding, wear a pair of safety glasses underneath your face shield for added protection.
   - b. Never wear tinted safety glasses indoors or in low light conditions.
   - c. Wear safety glasses while in a work area only when you are working.
   - d. Safety glasses have built-in side shields to help protect your eyes from the side.
6. Carry out housekeeping activities:
   - a. only while working.
   - b. only after the job is completed.
   - c. only at the end of the day.
   - d. both while working and after the job is completed.
7. When bending down to lift something, you should:
   - a. bend over with straight legs.
   - b. put your feet together.
   - c. bend your knees.
   - d. bend your back.
8. What should you do when an object punctures the victim's skin and becomes embedded in the victim's body?
   - a. Apply direct pressure to the wound.
   - b. Remove the object immediately.
   - c. Start compression-only cycles.
   - d. Apply a bulky dressing.
9. If a chemical splashes into the eye:
   - a. apply a bulky dressing.
   - b. rinse from the ear end of the eye.
   - c. flush the eyes with warm water.
   - d. do not attempt to remove it.
10. In the case of an open fracture, you may do all of the following *except*:
    - a. push any protruding bone back in.
    - b. apply ice or a cold pack if possible.
    - c. cover the wound and exposed bone with a dressing.
    - d. stabilize the injured part with your hands or a splint to prevent movement.

ASE Technician A/Technician B Style Questions

1. Tech A says that personal protective equipment (PPE) does not include clothing. Tech B says that the PPE used should be based on the task you are performing. Who is correct?
   a. Tech A
   b. Tech B
   c. Both A and B
   d. Neither A nor B
2. Tech A says that protective clothing that is not in good condition should be replaced. Tech B says that safety glasses are adequate to protect your eyes regardless of the activity. Who is correct?
   a. Tech A
   b. Tech B
   c. Both A and B
   d. Neither A nor B
3. Tech A says that rings, watches, and other jewelry should be removed prior to working. Tech B says that you should always wear cuffed pants when working in a shop. Who is correct?
   a. Tech A
   b. Tech B
   c. Both A and B
   d. Neither A nor B
4. Tech A says that proper footwear may include both leather and steel-toed shoes. Tech B says that when working in the shop, you only need to wear safety glasses if you are doing something dangerous. Who is correct?
   a. Tech A
   b. Tech B
   c. Both A and B
   d. Neither A nor B
5. Tech A says that a hat can help keep your hair clean when working on a vehicle. Tech B says that chemical gloves should be used when working with solvent. Who is correct?
   a. Tech A
   b. Tech B
   c. Both A and B
   d. Neither A nor B
6. Tech A says that you only need to worry about external bleeding because internal bleeding doesn't leak out. Tech B says that leather gloves help protect you from hot pieces of metal. Who is correct?
   a. Tech A
   b. Tech B
   c. Both A and B
   d. Neither A nor B
7. Tech A says that one use of barrier creams is to make cleaning your hands easier. Tech B says that hearing protection only needs to be worn by people operating loud equipment. Who is correct?
   a. Tech A
   b. Tech B
   c. Both A and B
   d. Neither A nor B
8. Tech A says that dust masks should be used when painting. Tech B says that a respirator should be used when the TLV for a chemical is exceeded. Who is correct?
   a. Tech A
   b. Tech B
   c. Both A and B
   d. Neither A nor B
9. Tech A says that tinted safety glasses can be worn when working outside. Tech B says that welding can cause a sunburn. Who is correct?
   a. Tech A
   b. Tech B
   c. Both A and B
   d. Neither A nor B
10. Tech A says that you should put tools away when done using them. Tech B says that a bystander can perform first aid. Who is correct?
    a. Tech A
    b. Tech B
    c. Both A and B
    d. Neither A nor B

CHAPTER 5

Vehicle Service Information and Diagnostic Process

NATEF Tasks

- **N05001** Identify information needed and the service requested on a repair order.
- **N05002** Complete work order to include customer information, vehicle identifying information, customer concern, related service history, cause, and correction.
- **N05003** Review vehicle service history.
- **N05004** Demonstrate use of the "3 Cs" (concern, cause, and correction).

Knowledge Objectives

After reading this chapter, you will be able to:

- **K05001** Describe the purpose and use of owner's manuals.
- **K05002** Describe the purpose and use of shop manuals/service information.
- **K05003** Explain how TSBs are used.
- **K05004** Explain how service campaigns and recalls are used.
- **K05005** Describe the purpose and use of labor guides.
- **K05006** Identify the purpose and use of a parts program.
- **K05007** Describe the information and its use within a repair/work order.
- **K05008** Describe the purpose and use of service history.
- **K05009** Explain the purpose and application of VINs.
- **K05010** Identify other vehicle information labels.
- **K05011** Describe each step in strategy-based diagnosis.
- **K05012** Explain how the "3 Cs" are applied in repairing and servicing vehicles.

Skills Objectives

After reading this chapter, you will be able to:

- **S05001** Use an owner's manual to obtain vehicle information.
- **S05002** Use a shop/repair manual while conducting a service or repair.
- **S05003** Use a service information program while conducting a service or repair.
- **S05004** Use a labor guide to estimate the cost or charge of conducting a service or repair.
- **S05005** Use a parts program to identify and order the correct replacement parts for a service or repair.
- **S05006** Use service history in the repair and service of vehicles.
- **S05007** Locate the VIN and production date code.
- **S05008** Decode a North American VIN.

▶ Introduction

Over the last 100 years, motor vehicles have become increasingly comfortable and reliable through the application of technology and improved manufacturing processes. For example, the modern motor vehicle has complex computer-controlled electrical systems where older vehicles had very basic wiring with no electronic components. The increased complexity and expanded range of makes and models have created a need for timely access to relevant, complete, and accurate information to perform maintenance, diagnosis, and repair activities.

Vehicle and customer information from various sources provides the fundamental knowledge required to conduct repairs and servicing. Today, the ability to properly service vehicles is dependent on the technician's ability to research and apply technical information. Vehicle information can come from a number of sources, including vehicle identification plates, owner's manuals, shop manuals, repair orders, service and parts programs, and technical service bulletins. The various sources of information are increasingly available through online services, but can also be through printed books or manuals. It is important that you know how to research and apply this information correctly so you can be more efficient and accurate when servicing vehicles (**FIGURE 5-1**).

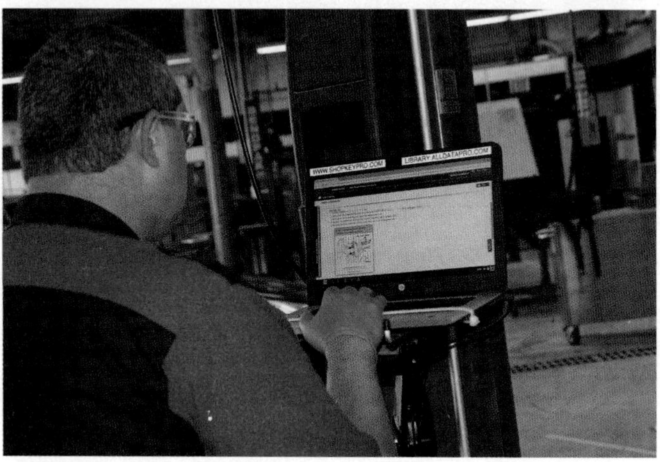

FIGURE 5-1 A technician researching service information.

▶ Owner's Manual

K05001

Manufacturers supply a vehicle owner's manual, which comes with every new vehicle purchased and is usually kept in the vehicle's glove compartment. Secondhand vehicles may or may not have the owner's manual in the glove compartment. The **owner's manual** contains information about the vehicle and is a valuable source of information for both the owner and the technician.

The information contained in the owner's manual varies for each manufacturer. A typical owner's manual includes an overview of the controls and features of the vehicle; the proper operation, care, and maintenance of the vehicle; owner service procedures; and specifications or technical data (**FIGURE 5-2**). The owner's manual also details such elements as vehicle security PIN (personal identification number) codes; warranty and service information; fuel, lubricant, and coolant capacities; tire-changing specifications; jacking and towing information; and a list of service facilities. The layout and amount of detail in an owner's manual varies according to the manufacturer and age of the vehicle.

Using an Owner's Manual

S05001

To locate the specifications for servicing a vehicle, follow the steps in **SKILL DRILL 5-1**.

▶ Shop Manual

K05002

Shop or service manuals are available for just about every make and model of every vehicle made. Service manuals come in two types—factory and after-market (**FIGURE 5-3**). Factory manuals

You Are the Automotive Technician

A well-known foreign vehicle manufacturer has recalled over 400,000 of its vehicles for a brake issue. The issue can be repaired with a software update that reprograms the vehicles' powertrain control module (PCM) to overcome the problem. But not all of the suspect vehicles need the software update. The technical service bulletin describes the steps needed to verify whether the vehicle has the fault. The manufacturer has notified their customers of a product recall through both mail and email. The dealership that you work for has been receiving many calls from customers to set up recall appointments. Today, you are working on the first vehicle for this recall.

1. Where would you locate the technical service bulletins (TSBs), service campaigns, and recalls in the dealership?
2. Why is it important to verify whether the TSB has been issued for that particular vehicle?
3. What sources would you use to look up the scheduled maintenance chart for the vehicle?
4. Write up what you would put on the repair order for each of the "3 Cs."

CONTENTS

1

FIGURE 5-2 A typical owner's manual includes an overview, a list of the controls and features, operation care and maintenance information, owner service procedures, and specifications or technical data.

are produced by vehicle manufacturers and specify the procedures to maintain, repair, and diagnose their vehicles. Usually a factory service manual is specific to one year and make of vehicle, such as a 2017 Ford Mustang.

After-market service manuals are published by independent companies for the same purpose. Usually, after-market manuals are not as detailed as factory manuals; they may not cover topics such as trim and entertainment systems. They are arranged in one of two ways: They either cover a range of years for a particular vehicle, such as 2012 to 2015 Ford Mustangs, or they cover a range of vehicles for a single year, such as 2011 General Motors vehicles.

Paper shop manuals have become less common over the years as less expensive electronic versions that can either be installed on a computer or accessed online have become available.

▶ TECHNICIAN TIP

Although factory manuals are usually more complete, it can be very costly to maintain a library of individual manuals for each of the vehicles serviced by a shop. After-market manuals help make the cost of service information more affordable. The modern vehicle is becoming very complex with all the technology now fitted to it, and this means shop manuals have also grown in complexity and size, resulting in significant expense and storage requirements.

▶ TECHNICIAN TIP

Here is an example of how much more a technician of today needs to know versus 70 years ago. In 1947, MOTOR, an organization devoted to supplying automotive data, produced a repair manual that covered the repair procedures and specifications for all of the vehicles produced by more than 20 manufacturers from 1935 to 1946. This reference was only about 1100 pages long. Today it takes MOTOR two to three 1100-page books to cover the same information for a single vehicle for one manufacturere for one year. Technicians rely on service information more now than ever before.

These online services are usually provided through a daily, monthly, or yearly subscription. Online versions are becoming very popular because they allow shops to access the information they need without having to pay for and store large numbers of shop manuals. Also, it is easier for the publisher to update information as changes or corrections are needed, so the information is generally more accurate than printed materials, which require supplemental printed updates on a periodic basis.

Typical paper and online shop manuals are broken into a number of sections that relate to systems within the vehicle—for example, engine, transmission, drivetrain, suspension,

SKILL DRILL 5-1 Using an Owner's Manual

1. Decide what information you need to know about the job and about the vehicle. For example, if your job is to change the engine oil, make sure you know the make, model, and year of manufacture of the vehicle and the type and size of the engine. In order to change the oil, you need to know the engine oil specifications, how much oil to put in, and what grade of oil to use.

2. Locate the appropriate manual. This kind of information is most readily found in the vehicle's owner's manual, which is usually kept in the glove compartment. Open the owner's manual to the first page, which is usually a table of contents to help you quickly find the information you need. Find the correct page and turn to it.

3. Locate the vehicle specifications, and identify the correct grade of motor oil for this vehicle. The table of contents lists a page number for the fuel and lubricant capacities and another page number for the refill capacities. First, turn to the page listing all of the vehicle's lubricant specifications. Find the correct specifications for the engine crankcase oil, and make a note. Next, turn to the page listing the refill capacities. You find that this eight-cylinder engine requires 5 quarts (4.7 liters) of oil.

4. Another way of finding information is to refer to the index at the back of the owner's manual. For example, look under *E* for engine, and find "Engine Oil," or look under *L* for lubricants or *O* for oil. Each item should refer you to the same page.

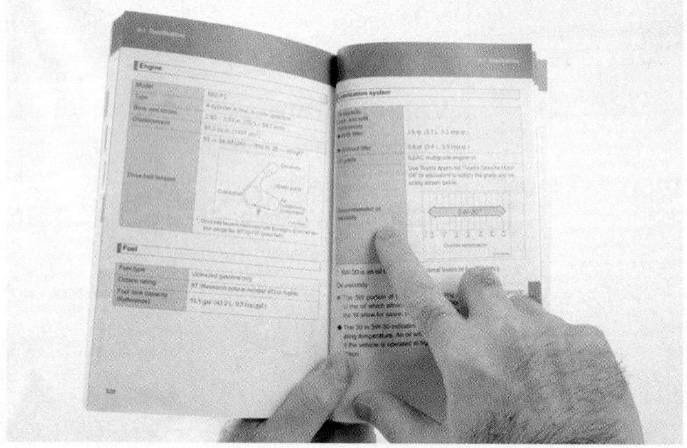

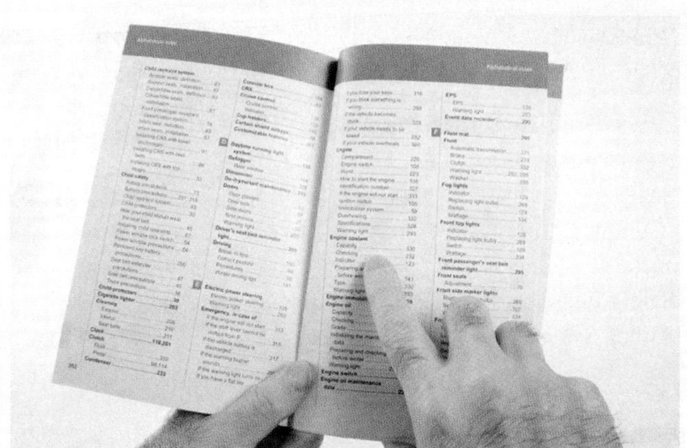

5. Once you have the specific requirements of the vehicle, you are ready to begin servicing the vehicle.

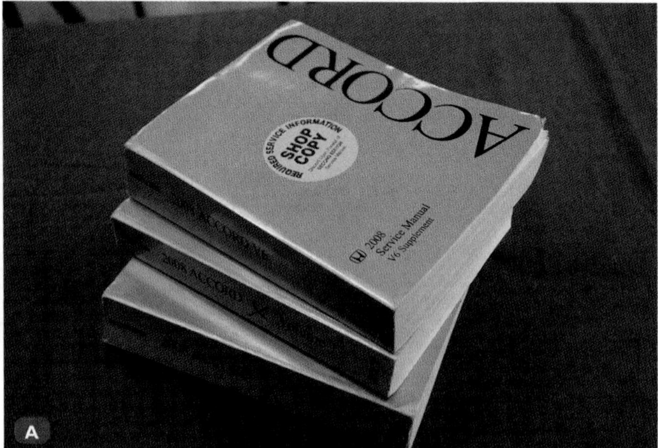

FIGURE 5-3 A. Factory manual. **B.** Aftermarket manual.

and electrical. The sections of the shop manuals are further divided into topics or subject areas; for example, in the engine section, topics could be general description, engine diagnosis, and on-vehicle service. A typical shop manual page has a task description broken into steps and diagrams or pictures to aid the technician (**FIGURE 5-4**). It is important to know that all service manuals arrange the content in their own way, so using a variety of different manuals helps you become familiar with finding the information you are looking for.

MAINTENANCE/SPECIFICATIONS

CHANGING YOUR WIPERS

The wiper arms can be manually moved when the ignition is disabled. This allows for ease of blade replacement and cleaning under the blades.

1. Disable the ignition before removing the blade.
2. Pull the arm away from the glass.
3. Left leading edge retaining block to release the blade. Swing the blade, away from with the arm, to remove it.
4. Swing the new blade toward the arm and snap it into place. Replace the retaining block at the leading edge of the wiper arm. Lower the wiper arm back to the windshield. The wiper arms will automatically return to their normal position the next time the ignition is enabled.

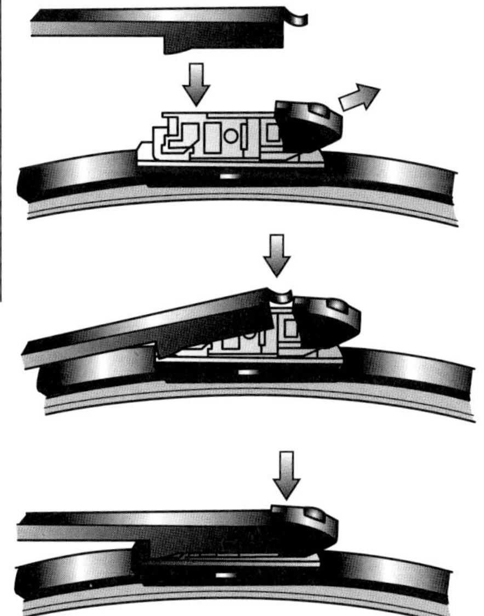

Refresh wiper blades at least twice a year for premium performance.

Poor preforming wipers quality can be improved by cleaning the blades and the windshield. See *Windows and wiper blades* in the *Cleaning* chapter.

To extend the life of wiper blades, scrape off the ice on the windshield BEFORE turning on the wipers. The ice has many sharp edges and will damage and shred the cleaning edge of your wiper blade.

FIGURE 5-4 A typical shop manual page has a task description broken into steps and diagrams or pictures to aid the technician.

Using a Shop Manual

S05002

Paper shop manuals are developed by manufacturers or after-market publishers to provide you with correct information on performing all service and repair tasks on the vehicles produced. The information found in shop manuals provides a systematic procedure and identifies special tools, safety precautions, and specifications relevant to the task. Shop manuals are organized according to vehicle systems and have indexes for quick referencing. Knowing that a water pump is part of the cooling system and that the cooling system is part of the engine system (or in some cases the HVAC system) helps you locate the specific information you need.

To identify correct service procedures using a shop manual, follow the steps in **SKILL DRILL 5-2**.

SKILL DRILL 5-2 Using a Shop Manual

1. Gather the information you need:
   - Year, make, model
   - Type and size of engine
   - Vehicle identification number (VIN)
   - Specific information you need to know
2. Find the appropriate shop manual for the make, model, and year of the vehicle you are working on.
3. Locate the correct section that contains the information you need. The first few pages of the shop manual usually contain the table of contents.
4. Locate the service procedures in the proper section. For example, if you were performing brake repairs, you would turn to the Brakes section. The text and the pictures describe how to properly perform a procedure and tell you the tools to use and how to use them.
5. Locate the vehicle specifications by consulting the specifications page in the proper section. You may need to know the type of engine of the vehicle to find the correct specifications for the vehicle you are working on.

Applied Science

AS-7: Maps/Charts/Tables/Graphs: The technician uses the information in service manual charts, tables, or graphs to determine the manufacturer's specifications for system(s) operation(s).

AS-8: Maps/Charts/Tables/Graphs: The technician uses the information in service manual charts, tables, or graphs to determine the appropriate repair/replacement procedure and/ or part.

Most service manuals have a chart that you need to consult before performing preventative maintenance. The chart lists what needs to be performed according to the vehicle's mileage. As you can see, there is a big difference between a 43,000 mile (minor) service to a 52,000-mile (major) service. On the 52,000-mile service, far more items are inspected and/or replaced. If the vehicle came into the workshop for a 43,000-mile service, you can easily read from the service chart what needs to be done.

Mileage	43,000 Miles	52,000 Miles
Maintenance Items	I: Drive belts	I: Drive belts
	R: Engine oil	R: Engine oil
	R: Engine oil filter	R: Engine oil filter
	I: Battery	I: Cooling and heater system
	I: Engine air cleaner filter	I: Engine coolant
	I: Brake pedal and parking brake	I: Exhaust pipe and mountings
	I: Brake pads and discs	I: Battery
	I: Brake fluid	R: Engine air cleaner filter
	I: Clutch fluid	I: Brake pedal and parking
	I: Brake pipes and hose	brake
	I: Power steering fluid	I: Parking brake linings and drums
	I: Steering wheel	I: Brake pads and discs
	I: Drive shaft boots	R: Brake fluid
	I: Suspension ball joint and dust covers	I: Brake pipes and hoses
	I: Tires and psi	I: Power steering fluid
	I: Rotate wheels	I: Steering wheel and linkage
	I: Seatbelt, webbing condition, buckle, and retractor mechanism operation	I: Front and rear suspension
	C: Air conditioner filter	I: Lights, horns, wipers, and washers
	I: Refrigerant amount of air conditioner	I: Seatbelt, webbing condition, buckle, and retractor mechanism operation
	I: Valve clearance	C: Air conditioner filter

Note: T = Tighten, R = Replace, I = Inspect, A = Adjust, L = Lubricate, and C = Clean.

Using a Service Information Program

S05003

Service information programs are computer applications used to provide technical information for the repair and maintenance of vehicles (**FIGURE 5-5**). In most cases, manufacturers have moved to online delivery of their service information systems through subscriptions that shops and schools purchase. In most cases, the information is laid out similarly to paper manuals. But because it is electronic, it can be searched easily. Also, publishers can link in full size illustrations, pictures, and charts.

To use a service information program, you may be required to log in to the computer and the service information program, so make sure you have the username and login available before you start. A printer is also helpful to print copies of the information so that you can use it when conducting service and repairs; alternatively, you may need to take notes.

To obtain the correct information, you need the same vehicle identification information as with a service manual. The repair order may provide you with this information, or you may have to research vehicle identification information from the vehicle. You may need to perform some initial diagnosis of the fault to further identify the specific information needed.

Information can usually be obtained by searching online for the vehicle and then selecting from the list of systems such

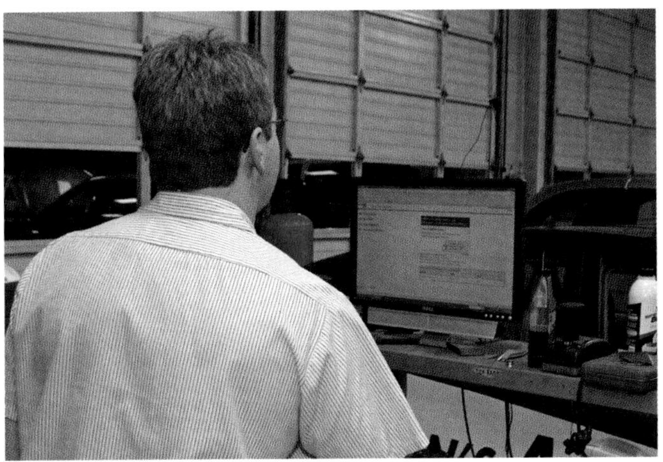

FIGURE 5-5 Computer databases provide information on procedures, parts, and service problems.

as brakes or maintenance, followed by subsystems such as disc brakes or fluid capacities. A keyword search may also be available; for example, use the keyword "service interval" to obtain a list of scheduled service intervals. Using a generic word like "engine" may return a very large list. If this occurs, the search can be narrowed further by entering more specific criteria, such as "engine oil," "water pump," or "camshaft."

The information is displayed on pages that have a mixture of text and diagrams along with explanations. Some of the diagrams may have detailed views, so you can see how parts fit together, and links may be provided to other relevant information such as a schematic diagram. Most systems contain help menus or training guides with examples to assist you in using the system, if required.

To use a service information program, follow the steps in **SKILL DRILL 5-3**.

SKILL DRILL 5-3 Using a Service Information Program

1. If necessary, start the computer and select the service information program.
2. Log in to the application, using the appropriate username and password.
3. Enter the vehicle identification information into the system in the appropriate places: year, make, model, engine, and possibly VIN.
4. Search for the information you require to perform the service or repair.
5. The search engine will provide a list of possible matches for you to select from. If the initial search does not produce what you are looking for, try changing the search criteria. Keep searching until you find the information.
6. Finally, once the general details for the item are displayed, gather the specific information on the specifications or repairs. You may need more than one piece of information.
7. Print out or write down the information needed. Put this on a clipboard, and take it with you to perform the diagnosis, service, or repair.

Applied Science

AS-11 Information Processing: The technician can use computer databases to input and retrieve customer information for billing, warranty work, and other record-keeping purposes.

Dealership service departments have access to databases run by manufacturers for the purposes of accessing warranty information, tracking vehicle servicing and warranty repair history, and logging warranty repair jobs for payment by the manufacturer. When customers present their vehicle for a warranty repair, the customer service department staff begin by consulting the database to confirm that the vehicle is within its warranty period and that the warranty has not been invalidated for any reason. Once it is confirmed that the vehicle is still under valid warranty, the repair order is passed to the technician for diagnosis and repair. Any parts required for the warranty repair must be labeled by the technician and stored for possible recall by the manufacturer.

For example, a young man comes in complaining that his vehicle is "running rough." The customer service staff confirms that the vehicle is nine months old, only has 14,500 miles, and is within the manufacturer's three-year/100,000-mile warranty period. They check the manufacturer's database to confirm that the vehicle's warranty has not been invalidated, before handing the repair order on to the technician. The technician diagnoses the fault as a defective ignition coil and fills out a warranty parts form.

Once the repair has been completed and the parts labeled, the warranty parts form and any repair order paperwork is passed back to administrative staff for processing. Processing includes billing the manufacturer for the correct, preapproved amount of time, logging the repair on the database for payment, and ensuring that all documentation is correct for auditing purposes.

Warranty Parts Form

		Vehicle Information	
Customer concern	Vehicle running rough		
Cause:	#6 ignition coil open circuit on primary winding	VIN:	1G1123456789101111
Correction:	Replaced #6 ignition coil	RO Number:	123456
Parts description:	#6 ignition coil	Date of repair:	10/04/2016

▶ Technical Service Bulletins

K05003

Technical service bulletins (TSBs) are issued by manufacturers to provide information to technicians on unexpected problems, updated parts, or changes to repair procedures that may occur with a particular vehicle system, part, or component (**FIGURE 5-6**).The typical TSB contains step-by-step procedures and diagrams on how to identify whether there is a fault and perform an effective repair.

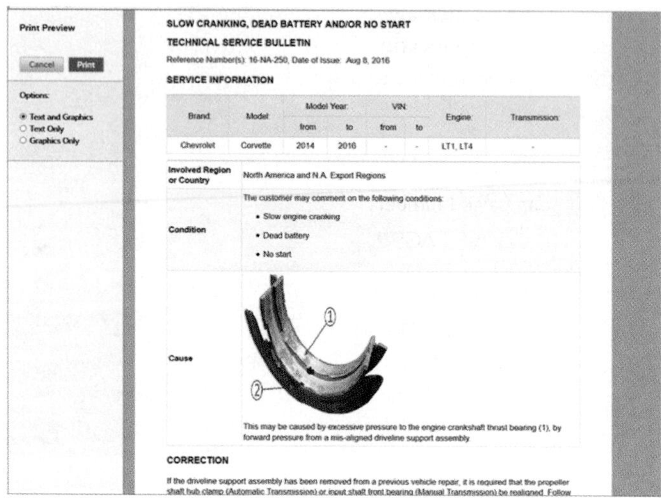

SLOW CRANKING, DEAD BATTERY AND/OR NO START

TECHNICAL SERVICE BULLETIN

Reference Number(s): 16-NA-250, Date of Issue: Aug 8, 2016

SERVICE INFORMATION

Brand	Model	Model Year: from	Model Year: to	VIN: from	VIN: to	Engine	Transmission
Chevrolet	Corvette	2014	2016	-	-	LT1, LT4	-

Involved Region or Country	North America and N.A. Export Regions
Condition	The customer may comment on the following conditions: • Slow engine cranking • Dead battery • No start
Cause	This may be caused by excessive pressure to the engine crankshaft thrust bearing (1), by forward pressure from a mis-aligned driveline support assembly.

CORRECTION

If the driveline support assembly has been removed from a previous vehicle repair, it is required that the propeller shaft hub clamp (Automatic Transmission) or input shaft front bearing (Manual Transmission) be realigned. Follow

FIGURE 5-6 Technical service bulletin.

At the time of production, manufacturers prepare service and technical information and attempt to anticipate the information the technicians need to perform service and repairs. Once the vehicle is in use, situations can arise when particular components or repair procedures may need either additional information or changes. This is where TSBs are most useful. For example, a change may be needed to update the procedure that bleeds air from the cooling system. In this situation, the manufacturer would issue a service bulletin explaining the problem and the updated procedure to bleed air from the cooling system.

Using TSBs

To use a TSB, follow these guidelines:

1. Locate where the TSBs are kept in your shop or look them up in the electronic service information system.
2. Prior to performing repairs, look through the TSBs and get to know the type of information contained in them.
3. Before working on a vehicle, it is good practice to check whether a TSB has been issued for that vehicle and type of fault or repair. This can save a lot of wasted time.
4. Compare the information contained in the TSB to that found in the shop manual. Note the differences, and if necessary, copy the TSB and take it with you to perform the repair.
5. Perform the repair following the TSB where appropriate while also referring to the shop manual.
6. If required in your shop policy, note the details of the service bulletin in the appropriate area on the repair order.

▶ Service Campaigns and Recalls

K05004

Service campaigns and recalls are usually conducted by manufacturers when a safety issue is discovered with a particular vehicle. Recalls are costly to manufacturers because they can require the repair of an entire model or production run of vehicles. Potentially, this could involve many thousands of vehicles. Depending on the nature of the problem, recalls can be mandatory and enforced by law, or manufacturers may choose to voluntarily conduct a recall to ensure the safe operation of the vehicle or minimize damage to their business and product image.

SAFETY TIP

Each country has specific laws regarding product recalls. Find out the laws in your jurisdiction.

An example of a mandatory recall is a fault within the airbag system of a vehicle that results in the airbag not deploying or deploying when it should not. In this case, the manufacturer would need to identify the problem, its cause, the vehicles affected, and the recertification requirements. A recall would then be issued and advertised in popular media. Letters would be sent from the manufacturer to known owners of the particular vehicle indicating that the vehicles should be returned for repair (**FIGURE 5-7**). Usually all costs associated with the recall are paid by the manufacturer.

Using Service Campaign Information

To utilize service campaigns or recall information, follow these guidelines. Locate where the special service messages, service campaigns/recalls, vehicle/service warranty applications, and service interval recommendations can be accessed in your shop. Look through the TSBs, service recalls, service warranty applications, and service interval recommendations, and get to know the type of information that is contained in them. Identify how they could be used in your daily tasks.

When working on vehicles, check to see if a TSB has been issued for that vehicle and type of repair. Perform service and repairs following the special service messages, service campaigns/recalls, vehicle/service warranty applications, and service interval recommendations. Fill in the required documentation as required in your shop policies. Note the details of

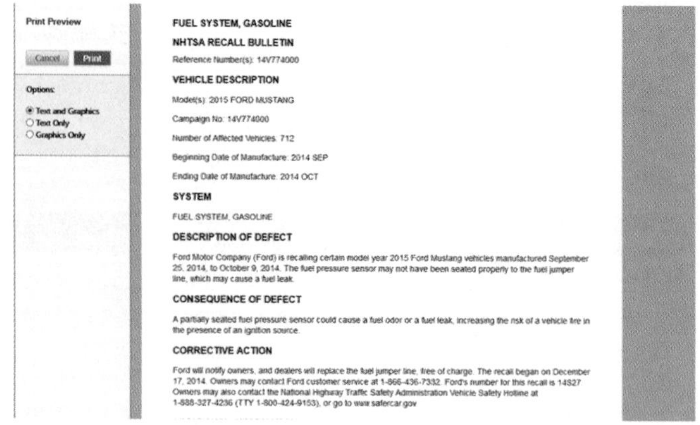

FUEL SYSTEM, GASOLINE

NHTSA RECALL BULLETIN

Reference Number(s): 14V774000

VEHICLE DESCRIPTION

Model(s): 2015 FORD MUSTANG

Campaign No: 14V774000

Number of Affected Vehicles: 712

Beginning Date of Manufacture: 2014 SEP

Ending Date of Manufacture: 2014 OCT

SYSTEM

FUEL SYSTEM, GASOLINE

DESCRIPTION OF DEFECT

Ford Motor Company (Ford) is recalling certain model year 2015 Ford Mustang vehicles manufactured September 25, 2014, to October 9, 2014. The fuel pressure sensor may not have been seated properly to the fuel jumper line, which may cause a fuel leak.

CONSEQUENCE OF DEFECT

A partially seated fuel pressure sensor could cause a fuel odor or a fuel leak, increasing the risk of a vehicle fire in the presence of an ignition source.

CORRECTIVE ACTION

Ford will notify owners, and dealers will replace the fuel jumper line, free of charge. The recall began on December 17, 2014. Owners may contact Ford customer service at 1-866-436-7332. Ford's number for this recall is 14S27. Owners may also contact the National Highway Traffic Safety Administration Vehicle Safety Hotline at 1-888-327-4236 (TTY 1-800-424-9153), or go to www.safercar.gov

FIGURE 5-7 Typical manufacturer recall announcement.

the special service messages, service campaigns/recalls, vehicle/service warranty applications, and service interval recommendations in the appropriate area on the repair order.

> **Report Receipt Date:** JAN 24, 2013
> **NHTSA Campaign Number:** 13V023000
> **Component(s):** AIR BAGS
>
> **All Products Associated with this Recall expand**
> **Manufacturer:** General Motors LLC
>
> **SUMMARY:** General Motors LLC (GM) is recalling certain model year 2012 Chevrolet Camaro, Cruze, and Sonic, and model year 2012 Buick Verano vehicles. The driver side frontal air bag has a shorting bar which may intermittently contact the air bag terminals.
>
> **CONSEQUENCE:** If the bar and terminals are contacting each other at the time of a crash necessitating deployment of the driver's frontal air bag, that air bag will not deploy, increasing the driver's risk of injury.
>
> **REMEDY:** GM will notify owners, and dealers will replace the steering wheel air bag coil, free of charge. The safety recall began on February 13, 2013. Owners may contact General Motors at 1-800-521-7300.
>
> **NOTES:** This is General Motors recall number 12261 and is an expansion of NHTSA recall 12V-522. Owners may also contact the National Highway Traffic Safety Administration Vehicle Safety Hotline at 1-888-327-4236 (TTY 1-800-424-9153), or go to www.safercar.gov

► TECHNICIAN TIP

Customer satisfaction ratings are very important to dealerships and shops. One way to impress customers is to check for recalls on every vehicle that comes in for service and inform the customer if you find anything. Some manufacturers flag a vehicle for any outstanding recalls when the VIN is entered into the dealership's computerized repair system when the vehicle is brought in for service.

► Labor Guide

K05005

Labor guides list how much time will be involved in performing a standard or warranty-related service or repair. They are regularly updated as new models are released into the market and provide a basis for making job estimates and standard charges for the customer. Flat rate servicing costs are usually derived from a labor guide. For example, if a customer wants to know how much it will cost to replace a leaking intake manifold gasket on a particular vehicle, then a technician can look up this procedure in a labor guide and find the information on the time and parts required for that particular repair on the specific vehicle.

With the advent of technology and the Internet, many providers of labor guides have started making them available online as well as in print. The online versions are paid for by subscription, which is usually a monthly or annual fee to access the information. Having access to online labor guides means the shop does not have to wait for a new version of

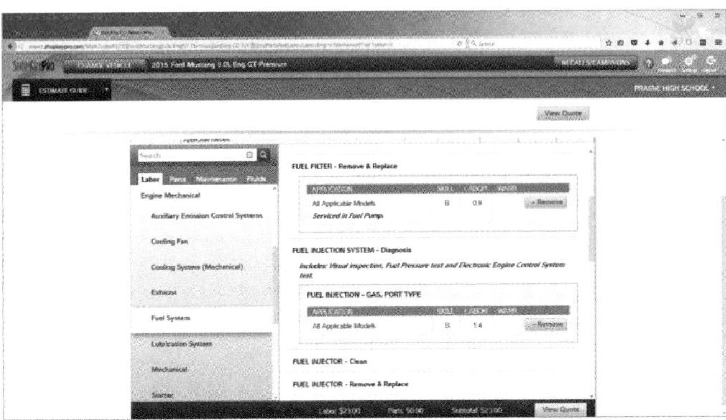

FIGURE 5-8 Online labor guide.

the print publication to become available. Online versions of labor guides also can be updated as new models of vehicles are released or updates are made available by the manufacturer (**FIGURE 5-8**).

Using a Labor Guide

S05004

The labor guide indicates how quickly an average technician can complete tasks. Experienced technicians who have performed a task many times and who are working efficiently can usually perform the job more quickly than the labor guide specifies. But because each task and vehicle has its small differences, the time given is not always completely accurate. The information contained within a labor guide is referenced in a similar manner to a repair manual or online service information system.

To use a labor guide, follow the steps in **SKILL DRILL 5-4**.

► Parts Program

K05006

Parts programs are the modern-day version of parts manuals. They are essentially an electronic version of a parts manual. Parts programs are typically accessed over the Internet, but some shops use DVDs or even paper manuals. Technicians and **parts specialists**, the individuals working at the parts counter, use these programs to identify parts and find part numbers for ordering purposes.

Parts manuals are produced for all makes and models of vehicles and are essentially a catalog of all the parts that make up a vehicle. The parts are cataloged by systems—for example, brake, engine, and transmission. Diagrams of each part are shown along with a part number, which is a unique identifying number for that particular part.

► TECHNICIAN TIP

Dealership technicians have one advantage over most independent technicians: They have an onsite parts department that stocks many of the parts needed for repairs. Many independent shops maintain a relatively small inventory of high-demand parts, such as filters, belts, and light bulbs, and use a local parts house to supply the less common parts. This can result in delays waiting for parts to arrive.

SKILL DRILL 5-4 Using a Labor Guide

1. Decide what specific labor operations you need to locate. Make sure you know the year, make, model, engine, and any other pertinent details of the vehicle.
2. Log in to the labor estimating system.
3. Enter the vehicle information into the system.
4. Find the labor operation either by working your way through the menu tree or by typing a keyword into the search bar.
5. Once you locate the labor operation, there are usually two columns that list the time. The first one is "warranty time." It is the amount of time the manufacturer would pay the shop for the operation under a customer warranty. The second time listed is the "customer pay" time. That is the amount of time a customer would be billed for and is usually 20–40% longer than warranty time. The length of time is usually listed in tenths of an hour. So 0.6 hours represent 36 minutes. Every tenth of an hour equals six minutes.
6. Check for any "combination" time that would need to be added to the base job when a related job is also being completed. Combination time recognizes that combined tasks in many cases save a lot of time over individual tasks because the customer is already being charged for part of the job in the first task. This could be time needed to flush the brakes when the main operation is to replace the brake pads.
7. Check for an "additional" time. This is extra time needed to deal with situations that occur on a relatively common basis, such as vehicle-installed options that are not common to all vehicles, like wheel locks. If you are replacing the brake pads on a vehicle with wheel locks, the customer should be charged for the extra time it takes to find the lock key and to remove and install the wheel locks (finding the lock key can itself add substantial time). Thus, "additional" time needs to be added to the base operation time if the vehicle meets the criteria for the particular "additional" situation.
8. Calculate the total time and multiply it by the shop's hourly labor rate. You now have the correct figure to estimate the charge for the particular service.

Using a Parts Program

S05005

A parts program is a computer application that is used to identify part numbers for vehicle components. Part numbers have to be identified so that correct replacement components can be ordered to replace faulty parts. The software can be installed on the computer, accessed via the Internet using a browser, or run from a CD or DVD.

To use a parts program, you need to have a basic understanding of how to start and use a computer. Usernames and passwords may be required to log in to the computer and the parts program, so make sure you have those available before you start. A printer is also helpful to print out copies of the information so that you can use it when ordering parts; alternatively, you may need to take notes.

To identify the correct part, you need to know where on the vehicle the part is installed, what system or subsystem it comes from, and vehicle identification information, such as date of manufacture, model, and engine and VIN numbers. Make sure you have this information on hand before you use the system. Searches can be conducted by keywords. If the part is for the brake system, in the search criteria box, enter "brake." Using a generic word like "brake" may return a very large list. If this occurs, the search can be narrowed further by entering more specific criteria such as "disc brake."

The parts are displayed in diagrams that are labeled and show individual parts in exploded view, making it easier to identify parts. The diagrams may number the parts and have a key on the page for reference to part numbers, or arrows may point to listed part numbers on the page. Most systems contain help menus or training guides with examples to assist you in using the software, if required (**FIGURE 5-9**).

To locate parts information on the computer, follow the steps in **SKILL DRILL 5-5**.

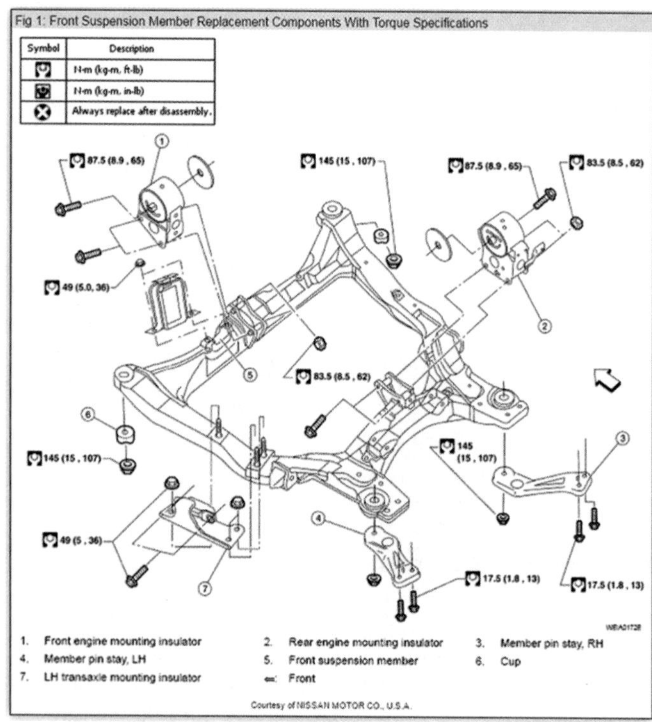

FIGURE 5-9 Typical exploded parts diagram.

▶ Repair Order Information

N05001, K05007

A **repair order**, or work order, is a form used by shops to collect information regarding a vehicle coming in for repair (**FIGURE 5-10**). They can be paper, but more and more shops use electronic repair orders. Initial information for the repair order includes customer and vehicle details, along with a brief description of the customer's concern(s). The repair order is

SKILL DRILL 5-5 Using a Parts System

1. Log in to the application, using the appropriate username and password.
2. Enter the year, make, model, and engine and VIN number information into the system in the appropriate places.
3. Search for the parts you require to conduct the service or repair.
4. The search engine will provide a list of possible matches for you to select from. If the initial search does not produce what you are looking for, try changing the search criteria. Keep searching until you find the information.
5. Gather information on the identified parts, including part numbers, location, availability, and cost.
6. Print, write down, or directly place an order for the desired parts.

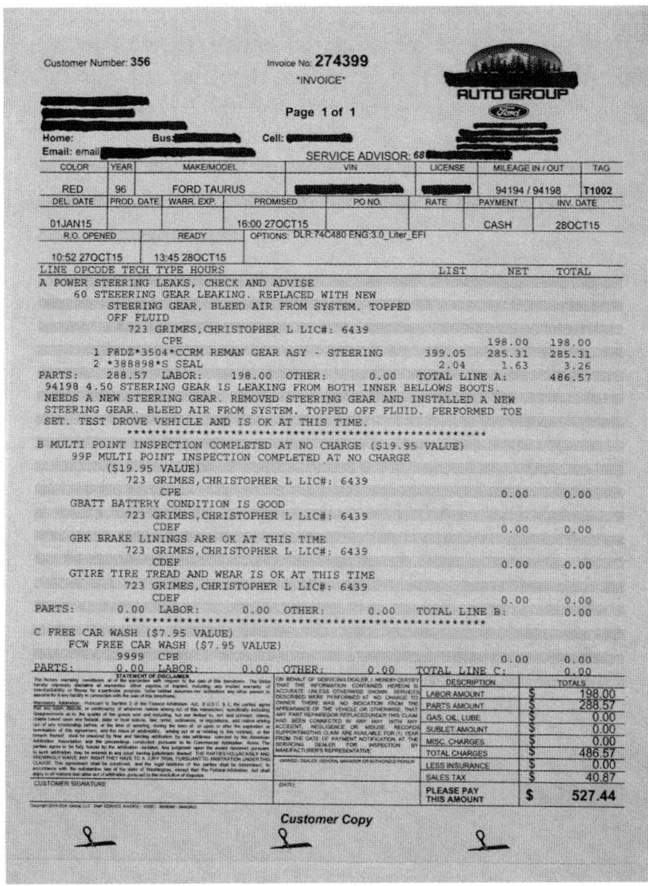

Customer Number: 356

Invoice No: 274399

INVOICE

Page 1 of 1

AUTO GROUP

Home: Bus: Cell:

Email: email

SERVICE ADVISOR: 68

COLOR	YEAR	MAKE/MODEL	VIN	LICENSE	MILEAGE IN / OUT	TAG
RED	96	FORD TAURUS			94194 / 94198	T1002

DEL. DATE	PROD. DATE	WARR. EXP.	PROMISED	PO NO.	RATE	PAYMENT	INV. DATE
01JAN15			16:00 27OCT15			CASH	28OCT15

R.O. OPENED	READY	OPTIONS: DLR:74C480 ENG:3.0_Liter_EFI
10:52 27OCT15	13:45 28OCT15	

LINE OPCODE TECH TYPE HOURS LIST NET TOTAL

A POWER STEERING LEAKS, CHECK AND ADVISE

60 STEEERING GEAR LEAKING. REPLACED WITH NEW STEERING GEAR, BLEED AIR FROM SYSTEM. TOPPED OFF FLUID

723 GRIMES,CHRISTOPHER L LIC#: 6439

CPE 198.00 198.00

1 F8DZ*3504*CCRM REMAN GEAR ASY - STEERING 399.05 285.31 285.31

2 *388898*S SEAL 2.04 1.63 3.26

PARTS: 288.57 LABOR: 198.00 OTHER: 0.00 TOTAL LINE A: 486.57

94198 4.50 STEERING GEAR IS LEAKING FROM BOTH INNER BELLOWS BOOTS. NEEDS A NEW STEERING GEAR. REMOVED STEERING GEAR AND INSTALLED A NEW STEERING GEAR. BLEED AIR FROM SYSTEM. TOPPED OFF FLUID. PERFORMED TOE SET. TEST DROVE VEHICLE AND IS OK AT THIS TIME.

B MULTI POINT INSPECTION COMPLETED AT NO CHARGE ($19.95 VALUE)

99P MULTI POINT INSPECTION COMPLETED AT NO CHARGE ($19.95 VALUE)

723 GRIMES,CHRISTOPHER L LIC#: 6439

CPE 0.00 0.00

GBATT BATTERY CONDITION IS GOOD

723 GRIMES,CHRISTOPHER L LIC#: 6439

CDEF 0.00 0.00

GBK BRAKE LININGS ARE OK AT THIS TIME

723 GRIMES,CHRISTOPHER L LIC#: 6439

CDEF 0.00 0.00

GTIRE TIRE TREAD AND WEAR IS OK AT THIS TIME

723 GRIMES,CHRISTOPHER L LIC#: 6439

CDEF 0.00 0.00

PARTS: 0.00 LABOR: 0.00 OTHER: 0.00 TOTAL LINE B: 0.00

C FREE CAR WASH ($7.95 VALUE)

FCW FREE CAR WASH ($7.95 VALUE)

9999 CPE 0.00 0.00

PARTS: 0.00 LABOR: 0.00 OTHER: 0.00 TOTAL LINE C: 0.00

STATEMENT OF DISCLAIMER

DESCRIPTION		TOTALS
LABOR AMOUNT	$	198.00
PARTS AMOUNT	$	288.57
GAS, OIL, LUBE	$	0.00
SUBLET AMOUNT	$	0.00
MISC. CHARGES	$	0.00
TOTAL CHARGES	$	486.57
LESS INSURANCE	$	0.00
SALES TAX	$	40.87
PLEASE PAY THIS AMOUNT	$	527.44

CUSTOMER SIGNATURE

(DATE)

Customer Copy

FIGURE 5-10 A print out of an electronic repair order.

the technician's main point of contact with the customer. So it is critical that the customer service staff get an accurate, complete, and yet concise record of the customer's concern(s) along with any information and details about the concern. Once the repair is complete, the technician and customer service staff complete the repair order with the exact cause of the concern and the correction required to fix the concern.

Detailed information on the repair order includes customer details such as name and address; the vehicle make, model, and year; the odometer reading; the date; customer concern information; the cause of the concern(s); the correction for the concern(s); the hours of labor; and the parts used for the repair. The repair order should always include all of the information pertaining to the customer, vehicle, and cost of repair. Repair orders are legal documents that can be used as evidence in the event of a lawsuit. Make sure the information is complete and accurate whenever filling out a repair order, and store it in an organized safe place, such as in a file cabinet or electronically on a secure computer network.

To identify the information needed and the service requested on a repair order, follow the steps in **SKILL DRILL 5-6**.

▶ TECHNICIAN TIP

Repair orders are used also to inform the customer of needed repairs or service. This usually results in the customer agreeing to the needed repair, in which case all is well. But if he or she does not agree to the repair and the vehicle is involved in an accident because of the faulty components, having the customer's initials on the repair order signifying that he or she understands the safety issues can help prevent the shop from being held liable for the accident.

SKILL DRILL 5-6 Using Repair Orders

1. Locate a repair order used in your shop.
2. Familiarize yourself with the repair order, and identify the following information on the repair order:
   a. Date
   b. Customer details: name and address, daytime phone number
   c. Vehicle details: year, make, model, color, odometer reading, VIN, and license plate number
   d. Customer concern details
3. Note any additional information that is required on your shop's repair order.
4. Following the shop procedures, determine the workflow for the tasks that are listed.
5. Use the repair order to carry out the requested service or repair. Fill in the repair order with details of the cause of the customer concern(s) and the correction(s) conducted.

Completing a Repair Order

N05002

The accounting section contains information about the methods of payment, which can be cash, credit card, or account. An account system can be set up to handle all payments related to a customer or to a company that uses your service for a number of vehicles. When a vehicle on an account system comes in, you need to record both the account number and the order number. To work out the total cost of the service, you need to know:

- The labor cost
- The cost of parts
- The tax amounts
- The cost of gas and consumables you used to service the vehicle

You also need to have the customer's authorization to carry out the service. Remember, before making changes to the estimated cost of repair beyond a certain percent or dollar amount after the initial authorization, you need to receive and document the customer's approval of the additional repair cost. Failure to do so may mean that the customer doesn't have to pay the extra cost.

▶ TECHNICIAN TIP

Always obtain the customer's authorization before servicing his or her vehicle, and also make sure you have the customer's approval before making any changes to the service invoice or the work order.

▶ Service History

N05003, K05008, S05006

Service history is a complete list of all the servicing and repairs that have been performed on a vehicle (**FIGURE 5-11**). The scheduled service history can be recorded in a service booklet or owner's manual that is kept in the glove compartment. The service history can provide valuable information to technicians conducting repairs. It also can provide potential new owners of used vehicles an indication of how well the vehicle was maintained. A vehicle with a regular service history is a good indication that all of the vehicle's systems have been well maintained, and the vehicle will often be worth more during resale. Most manufacturers store all service history performed in their dealerships (based on the VIN) on a corporate server that is accessible from any of their dealerships. They also use this vehicle service history when it comes to evaluating warranty claims. A vehicle that does not have a complete service history may not

▶ TECHNICIAN TIP

A vehicle's service history is valuable for several reasons:

- It can provide helpful information to the technician when performing repairs.
- It allows potential new owners of the vehicle to know how well the vehicle and its systems were maintained.
- Manufacturers use the history to evaluate warranty claims.

Applied Communications

AC-23: Repair Orders: The technician writes a repair order containing customer vehicle information, customer complaints, parts and materials used (including prices), services performed, labor hours, and suggested repairs/maintenance.

A repair order is a key document used to communicate with both your customers and coworkers. It is a legal contract between the service provider and the customer. It contains details of the services to be provided by you and the authorization from the customer. To make sure everyone understands clearly what is involved, a repair order should contain the following information:

- Your company or service providers—The service provider section contains the company name, address, and contact details; the name of a service advisor who is overseeing the job; and the amount of time the service technician needs to service the vehicle.
- The customer—The customer section contains the customer's name, address, and contact phone numbers.
- The customer's vehicle—The vehicle section includes details about the vehicle to be serviced. Check the vehicle's license plate before starting work. The license plate numbers are usually unique within a country. You should also record information about the vehicle's make, model, and color. This information makes it easier for you to locate the vehicle on the parking lot. You need to know the manufacture date of the vehicle to be able to order the right parts. The odometer reading and the date help keep track of how much distance the vehicle travels and the time period between each visit to the shop. The VIN is designed to be unique worldwide and contains specific information about the vehicle. Many shops do a "walk-around" with the customer to note any previous damage to the vehicle and to look for any obvious faults such as worn tires, rusted-out exhaust pipes, or torn wiper blades.
- The service operations—This section contains the details of the service operations and parts.
  - The first part is the service operation details. For example, the vehicle is in for a 150,000-mile (240,000 km) service, which can be done in three hours, resulting in approximately a $300 labor cost. The information about the chargeable labor time to complete a specific task can be found in a labor guide manual. In some workplaces, this information is built into the computer system and is automatically displayed.
  - The second part of this section is the details of parts used in the service, including the descriptions, quantities, codes, and prices. The codes for each service and part are normally abbreviations that are used for easy reference in the shop. Some shops may have their own reference code system.
  - As you do the vehicle inspection, you may discover other things that need to be replaced or repaired. These additional services can be recorded in another section. It is essential that you check with your customers and obtain their approval before carrying out any additional services.
- The parts requirements—This section lists the parts required to perform the repair.

Some repair orders also contain accounting information so they can be used as invoices.

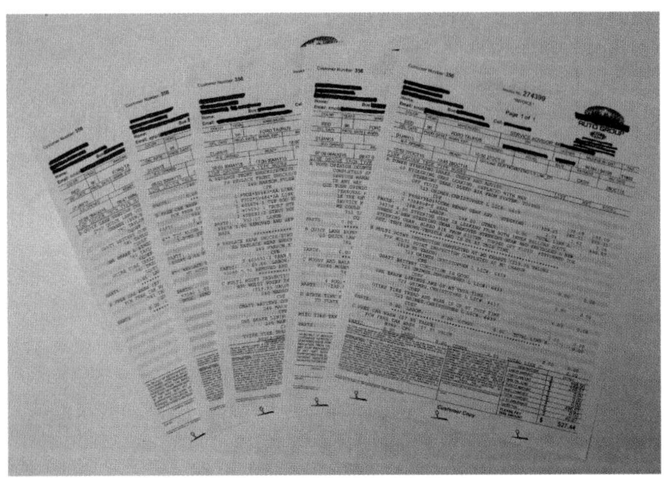

FIGURE 5-11 Print outs of completed repair orders as saved in the online repair order system.

FIGURE 5-12 A sample VIN.

be eligible for warranty claims. Independent shops generally keep records of the repairs they perform. However, if a vehicle is repaired at multiple shops, repair history is much more difficult to track and, again, may result in a denial of warranty claims.

To review the vehicle service history, follow the steps in **SKILL DRILL 5-7**.

▶ VIN and Production Date Code, and Vehicle Information Labels

K05009

Large numbers of vehicles with many variations of makes and models with different equipment levels are produced every day across the world. Vehicle information labels have become very important to technicians because they help to uniquely identify the vehicle. VIN stands for **vehicle identification number** and is a unique serial number that is assigned to each vehicle produced. This means that no two vehicles have the same VIN (**FIGURE 5-12**). Since 1981, the VIN has been made up of 17 characters. It is usually located on the front left corner of the windshield and is also inscribed on various vehicle parts. VINs can be used to check the service history of a vehicle and also are used to identify the vehicle for ordering components. Labeling the vehicle and vehicle parts with VINs also deters auto theft because it provides an easy way of uniquely identifying and tracing the vehicle and its major parts.

Production date codes and vehicle information labels also add to the identification information available on vehicles. The production date is the date of manufacture by year and month. Other information labels are fitted to the vehicle to provide ready access to information—for example, tire inflation pressures, vehicle weight, and load-carrying capacity. All of these information labels are used regularly by technicians to identify vehicles, order parts, and check service history.

Locating the VIN and Production Date Code

S05007

In order to locate a VIN and production date code, it is important to understand the principles of VINs and correctly identify the components that make up a vehicle identification number.

SKILL DRILL 5-7 Reviewing Service History

1. Locate the service history for the vehicle. This may be in shop records or in the service history booklet within the vehicle glove compartment. Some shops may keep the vehicle's service history on a computer.
2. Familiarize yourself with the service history of the vehicle.
   - **a.** On what date was the vehicle first serviced?
   - **b.** On what date was the vehicle last serviced?
   - **c.** What was the most major service performed?
   - **d.** Was the vehicle ever serviced for the same problem more than once?
3. Compare the vehicle service history to the manufacturer's scheduled maintenance requirements, and list any discrepancies.
   - **a.** Have all the services been performed?
   - **b.** Have all the items been checked?
   - **c.** Are there any outstanding items?

The VIN is a 17-character identification composed of letters and digits. The VIN is designed to identify motor vehicles of all kinds: cars, trucks, buses, motorcycles, etc. It was originally defined in the International Organization for Standardization (ISO) Standard 3779 in 1977 and was revised in 1983.

The VIN is usually located below the front left corner of the windshield and is also inscribed on the engine, the transmission, both front fenders, the hood, the doors, both bumpers, both rear quarter panels, and the trunk or hatchback. The VIN is unique worldwide, identifying the country of manufacture, manufacturer's name, division name, model, and other important information. Since 1981, all worldwide vehicle manufacturers use this numbering system. By learning to interpret the system, the identity of a vehicle or a component can be determined and verified.

Whenever a vehicle is registered or a registered vehicle is sold, a record of the VIN is kept. From this registry, information about the vehicle can be accessed, including the **title history**, which can tell you who has owned the vehicle. The registry may reveal a **salvage title**, which tells you that the vehicle has been wrecked and suffered economically irreparable damage. It also can tell you if a **lemon law buyback** has occurred. Lemon laws exist in some states to protect consumers from purchasing vehicles that have undergone several unsuccessful attempts to repair the same fault or from purchasing vehicles in which repair of the defects has caused the vehicle to be out of service for an extended time. The registry can also indicate if the vehicle has had an odometer rollback (mileage reduction), which is evidence of odometer tampering.

To locate the VIN and production date code, follow the steps in **SKILL DRILL 5-8**.

Decoding a VIN

S05008

There are two different, but essentially compatible, 17-character VIN standards: the North American VIN system and the ISO Standard 3779, which is used in most of the rest of the world. (**FIGURE 5-13**) shows how the numbers are structured.

To decode a North American VIN, follow the steps in **SKILL DRILL 5-9**.

So a vehicle with the VIN 1G1YN3DE-A5100001 can be identified as a 2010 Chevrolet Corvette ZR1 Convertible 7.0 L, V-8 LS7, SFI, and it was the first one off the production line that year.

SKILL DRILL 5-8 Locating the VIN and Production Date Code

- Locate the make of the vehicle from the body nameplate, which is usually found on the front or rear of the vehicle. Now locate the model from the body trim. The model may be a name, number, letter, or combination of these elements.
- Next locate the VIN, usually found on a plate in the upper left dashboard and often visible through the windshield. In some instances, the plate may be mounted in a different location. If the plate is not visible through the windshield, check under the hood to see if it is mounted in the engine bay area. Note each letter and number exactly as it appears on the plate.
- Decode the VIN and note the information. Each manufacturer provides a VIN decoding chart for its vehicles in its shop service information. This chart is normally found in the "general information" section of the manual or electronic service information system. Using the VIN decoding chart, write down the information or print it for later use when locating specifications or parts.

SKILL DRILL 5-9 Decoding a North American VIN

1. The VIN is 1G1YN3DE-A5100001. The first character is the country of origin. This number or letter tells you where the vehicle was manufactured. For instance, a "1" means that the vehicle was made in the United States; a "2" is for Canada; a "J" means Japan, etc. "G" stands for "General Motors."
2. The second character is usually a letter; it tells you the name of the manufacturer. The number "1" means this is a Chevrolet.
3. The third character tells you the division that made the vehicle. It could be a Pontiac, an Oldsmobile, or a GMC truck, for example.
4. The fourth and fifth characters give you the model, or series, of the vehicle. You will need a decoding chart for the details. Here we have a Corvette ZR1 Custom 3ZR Manual.
5. The sixth character describes the body type: two-door, four-door, coupe, sedan, etc. Here we have a two-door convertible.
6. The seventh character tells you the type of seat restraints fitted to the vehicle. In this case it is active manual seat belts, airbags front (driver and passenger) and front seat side.
7. The eighth character is the engine code, which provides details of the engine type, size, or displacement, and where the engine was made. Here we have a 7.0 L, LS7, gas, eight-cylinder, SFI, aluminum, GM.
8. The ninth character is the check character. It is used internally by the manufacturer.
9. The tenth character tells you the model year of manufacture. You can decode this character according to a model year identification chart, which in this example shows us that the vehicle was assembled for the 2010 model year.
10. The 11th character tells you the assembly plant or factory where the vehicle was put together.
11. The final six numbers make up the sequential number of the vehicle as it comes off the assembly line, starting at a base number, which is usually 100,000. So the first vehicle to be produced usually, but not always, has the number 100001. In our example, the vehicle was the first to come off the assembly line in that year.

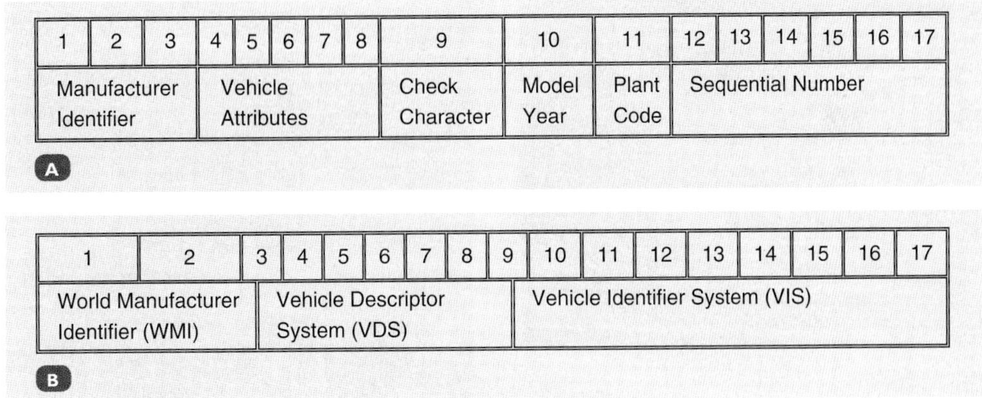

FIGURE 5-13 **A.** North American VIN system. **B.** ISO Standard 3779.

▶ Using Other Vehicle Information Labels

K05010

Vehicle Emission Control Information (VECI) Label

The **vehicle emission control information (VECI) label** is used by technicians to identify engine and emission control information for the vehicle (**FIGURE 5-14**). It is usually located in the engine compartment on either the hood, strut tower, or radiator support. It typically includes the following information:

- Engine family and displacement
- Model year the vehicle conforms to
- Spark plug part number and gap
- Evaporative emission system family
- Emission control system schematic
- Certification application

Vehicle Safety Certification (VSC) Label

The **vehicle safety certification (VSC) label** certifies that the vehicle meets the Federal Motor Vehicle Safety, Bumper, and Theft Prevention Standards in effect at the time of manufacture (**FIGURE 5-15**). It is used by technicians to identify some basic types of information about the vehicle such as month and year of manufacture, gross vehicle weight rating (GVWR), and on some vehicles, tire information. It is usually affixed to the driver's side door pillar or on the side of the door next to the pillar. It typically includes the following information:

- Month and year of manufacture
- GVWR and gross axle weight rating (GAWR)
- VIN
- Recommended tire sizes
- Recommended tire inflation pressures
- Paint and trim codes

Other Labels

Other labels include the refrigerant label, the coolant label, and the belt routing label. The **refrigerant label** lists the type and total capacity of refrigerant that is installed in the A/C system (**FIGURE 5-16**). The **coolant label** lists the type of coolant installed in the cooling system (**FIGURE 5-17**). The **belt routing label** lists a diagram of the serpentine belt routing for the engine accessories (**FIGURE 5-18**).

▶ Strategy-Based Diagnosis

K05011

Customers experience issues with their vehicles on a regular basis. Parts wear out or malfunction and cause the vehicle to not operate properly. Technicians are called on to investigate, identify,

FIGURE 5-14 VECI label.

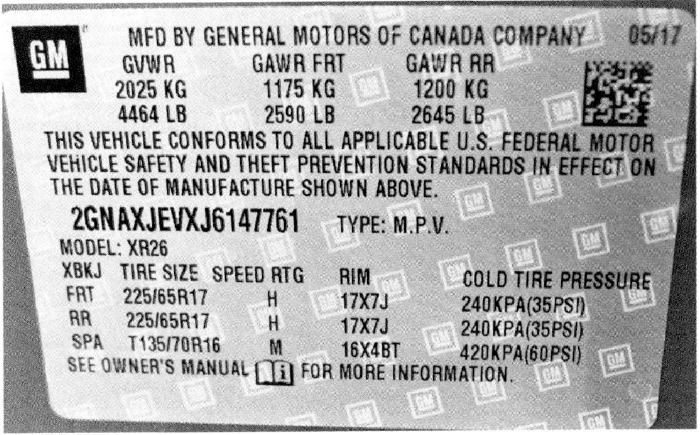

FIGURE 5-15 VSC label.

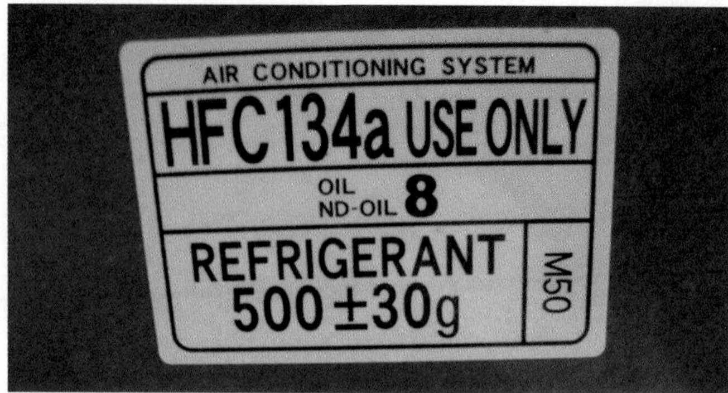

FIGURE 5-16 Refrigerant label.

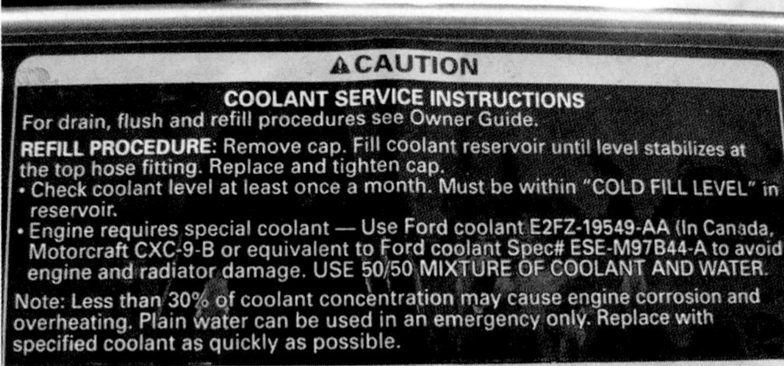

FIGURE 5-17 Coolant label.

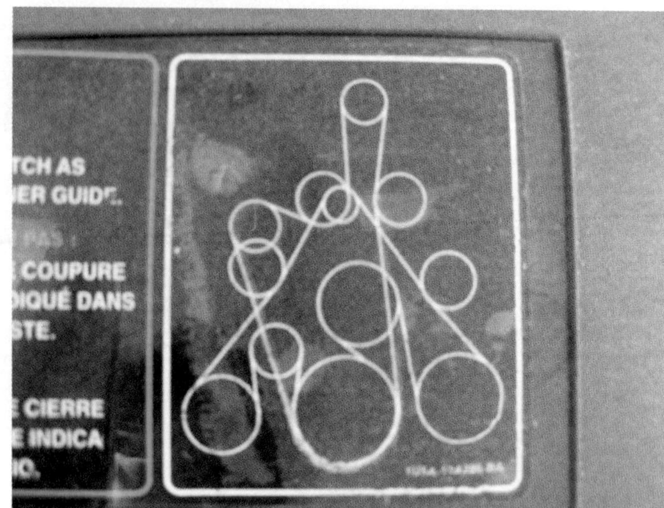

FIGURE 5-18 Belt routing label.

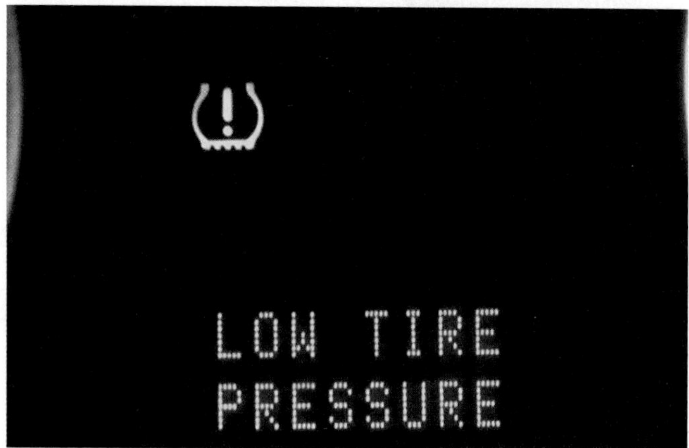

FIGURE 5-19 TPMS warning system showing the LF tire low.

and solve the cause of the malfunction along with determining what the appropriate fix is. That is called diagnosing the problem. A diagnosis is the conclusion a technician comes to after thoroughly investigating the problem. Diagnosing problems can be very challenging to perform in a timely and efficient manner. Technicians find that having a proven plan in place ahead of time vastly simplifies the process of logically and systematically (strategically) solving problems. The plan technicians use is simple to remember and consistent in its approach; and it works for the entire range of diagnostic problems that technicians will encounter. In this way, technicians have one single plan to approach any diagnostic situation they may encounter, and can be confident in their ability to resolve it. This problem-solving plan is called **strategy-based diagnosis** and involves the following steps:

Step 1: Verifying the customer's concern
Step 2: Researching possible faults and gathering information
Step 3: Focused testing
Step 4: Performing the repair
Step 5: Verifying the repair

Let's look at a simple example of the diagnostic process in action. A customer dropped off her car the day before with the following concern: "The tire pressure monitoring system (TPMS) warning lamp is on and the car pulls to the left." As a first step of "verifying the concern," when walking up to the vehicle, the technician looks at the tires and notices that the driver's side front tire appears to be quite low on air pressure. The technician then turns the ignition key to the "Run" position and observes the TPMS warning system. The display shows a warning message indicating low pressure in the left front tire. This completes Step 1 (**FIGURE 5-19**).

For Step 2, "Researching Possible Faults and Gathering Information," the technician knows from his training that the low pressure tire can cause the vehicle to pull toward the side that is low. He also knows that the TPMS warning lamp is designed to warn the driver of low pressure tires and faults in the system. Checking the tire placard reveals that the specified tire pressure is 32 psi (220.63 kPa) for this vehicle. Researching the service information, the technician finds that the air pressure needs to be adjusted after the vehicle has been parked for at least four hours. It also states that pressures under a certain point will activate the TPMS. The technician then knows that if the tire pressure is below that specified psi (kPa), the TPMS was working properly and that there is another problem causing the low tire pressure, which needs to be identified. So now it is time to test the vehicle (**FIGURE 5-20**).

For Step 3, "Focused Testing," the technician follows the specified procedure for properly testing the tire pressure, and finds that it reads 21 psi (144.79 kPa) (**FIGURE 5-21**). At this point, the technician needs to locate the cause of the low tire pressure. He does so by removing the wheel and tire from the vehicle and placing it in a tire dunk tank to locate any leaks. Bubbles are coming

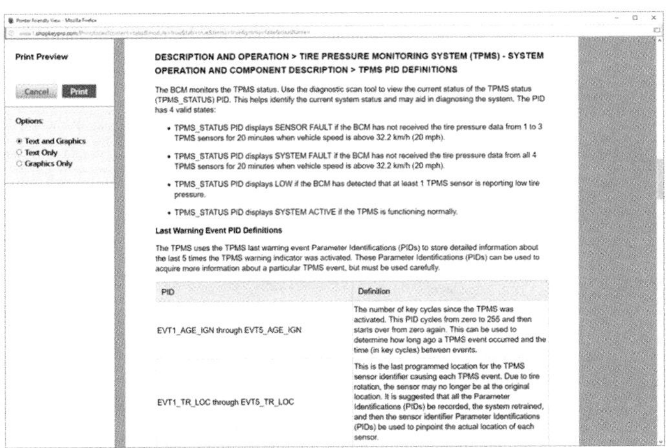

Print Preview

Cancel Print

Options:
- Text and Graphics
- Text Only
- Graphics Only

DESCRIPTION AND OPERATION > TIRE PRESSURE MONITORING SYSTEM (TPMS) - SYSTEM OPERATION AND COMPONENT DESCRIPTION > TPMS PID DEFINITIONS

The BCM monitors the TPMS status. Use the diagnostic scan tool to view the current status of the TPMS status (TPMS_STATUS) PID. This helps identify the current system status and may aid in diagnosing the system. The PID has 4 valid states:

- TPMS_STATUS PID displays SENSOR FAULT if the BCM has not received the tire pressure data from 1 to 3 TPMS sensors for 20 minutes when vehicle speed is above 32.2 km/h (20 mph).
- TPMS_STATUS PID displays SYSTEM FAULT if the BCM has not received the tire pressure data from all 4 TPMS sensors for 20 minutes when vehicle speed is above 32.2 km/h (20 mph).
- TPMS_STATUS PID displays LOW if the BCM has detected that at least 1 TPMS sensor is reporting low tire pressure.
- TPMS_STATUS PID displays SYSTEM ACTIVE if the TPMS is functioning normally.

Last Warning Event PID Definitions

The TPMS uses the TPMS last warning event Parameter Identifications (PIDs) to store detailed information about the last 5 times the TPMS warning indicator was activated. These Parameter Identifications (PIDs) can be used to acquire more information about a particular TPMS event, but must be used carefully.

PID	Definition
EVT1_AGE_IGN through EVT5_AGE_IGN	The number of key cycles since the TPMS was activated. This PID cycles from zero to 256 and then starts over from zero again. This can be used to determine how long ago a TPMS event occurred and the time (in key cycles) between events.
EVT1_TR_LOC through EVT5_TR_LOC	This is the last programmed location for the TPMS sensor identifier causing each TPMS event. Due to tire rotation, the sensor may no longer be at the original location. It is suggested that all the Parameter Identifications (PIDs) be recorded, the system retrained, and then the sensor identifier Parameter Identifications (PIDs) be used to pinpoint the actual location of each sensor.

FIGURE 5-20 Service information listing the enable criteria that activates the TPMS warning system.

FIGURE 5-21 Technician measuring tire pressure, showing 21 psi (144.79 kPa).

from a small nail embedded near the center of the tread (**FIGURE 5-22**). The technician informs the customer of the need to plug and patch the tire, and then verifies that the TPMS warning light is no longer illuminated. The customer authorizes the repair.

For Step 4, "Performing the Repair," the technician follows the proper steps to plug and patch the hole in the tire (**FIGURE 5-23**). He then inflates the tire to the recommended 32 psi (220.63 kPa), rechecks it to make sure the original leak is repaired as well as no other leaks. He then remounts the tire on the wheel, balances the wheel/tire assembly, and remounts the wheel on the vehicle, being sure to tighten the lug nuts to the proper torque.

For Step 5, "Verify the Repair," the technician turns the ignition key to the "Run" position and watches the TPMS warning light. It initially comes on for a few seconds as the system goes through its initial check. After that, the TPMS warning lamp turns off, indicating that the system is monitoring the pressure correctly (**FIGURE 5-24**). Now all the technician needs to do is test-drive the vehicle to verify that it is no longer pulling to the left, which he does, and finishes up the repair order so the customer can pay the bill and get on her way.

This strategy-based diagnostic process is used by technicians to help them logically diagnose problems. We used a very simple example of a technician using that process. As you progress further through this textbook and into other systems of the automobile, we expand on both the theory of the diagnostic strategy as well as more complicated diagnostic procedures and

FIGURE 5-22 Tire being tested for a leak and leaking from the center of the tread.

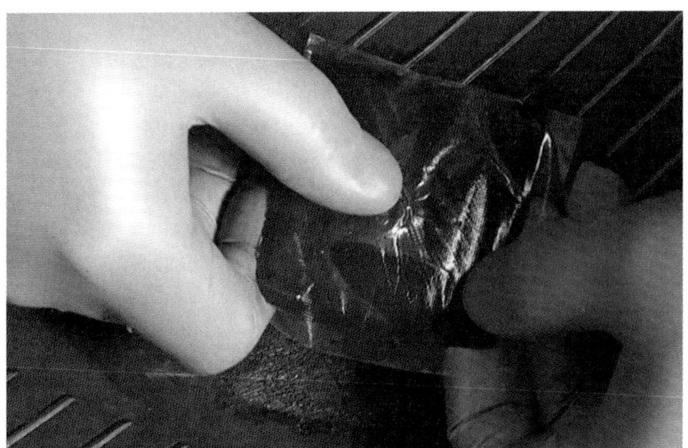

FIGURE 5-23 Tire being plug patched.

FIGURE 5-24 Technician verifying that the TPMS warning light is off.

processes. But for now, we want to introduce the process to you so that you can start to use it from the early portions of this book. Stay tuned for more details on this topic in future chapters.

▶ 3 Cs

N05004, K05012

Now that you understand how the diagnostic process works, you need a record of the repair to give to the customer so he or she can pay the bill, and to the shop so they know how much to pay you. This is done on repair orders, using the

"3 Cs," which is short for "concern," "cause," and "correction." Concern stands for the customer's understanding of the problem with the vehicle. Customers experience an issue with their vehicle and attempt to communicate that to the service advisor, who helps the customer put the concern into words that make sense and that accurately describe the actual problem. For example, a customer has trouble starting his car. He may initially communicate that the engine won't crank over. With some questioning and answers, the service advisor helps the customer identify that the engine is cranking over, but not starting. So the actual concern is "engine cranks, but won't start." That is a vastly different situation than "engine will not crank." By understanding the actual concern, the technician will avoid wasting time looking for a nonexistent problem.

Cause stands for the actual fault that is causing the concern. To determine that, you need to fully diagnose the concern, making sure you find the root cause of the concern. For example, in the previous topic, the TPMS warning lamp was on, which was the customer's concern. As we saw, the left front tire had low air pressure. This could definitely cause the TPMS lamp to be lit. If the technician had inflated the tire to the proper pressure, it would have turned the TPMS warning lamp off and solved the customer's problem, but only temporarily until the tire pressure got low again. So in that case, the technician didn't stop when he found low tire pressure; he continued his diagnosis and investigated why the tire pressure was low in the first place. This was caused by a nail that had punctured the tire. So in this situation, the root cause of the customer's concern was a nail in the tire, causing low tire pressure and the TPMS warning lamp to be on. Technicians always need to determine the root cause of a concern if the problem is to be fixed permanently. And the way you determine the root cause is by repeatedly asking yourself what caused the particular finding until you get to the root. And yes, you can go too far. In the leaking tire scenario, you could ask yourself, "Why is there a nail in the tire?", and the answer might be because the customer drove through a construction zone. Most of the time, you don't have to go that far unless the customer will encounter that situation regularly. You might consider the deeper questions if the customer comes in within a few weeks with another nail in a tire. At that point, telling him that this is the second roofing nail in his tire, he may decide that he should drive a different route home than through the construction zone.

Correction stands for the procedure and parts that will be used to fix the problem or, in fact, did fix the problem. In the preceding situation, the correction would be: "Repair tire with plug patch, inflate tire to 32 psi (220.63 kPa), and verify that the TPMS lamp is off." This step occurs once the problem and faults are fully understood and the damage assessed so that the correct parts can be replaced to complete a successful repair. Don't forget that to ensure a successful repair, the system should be thoroughly checked to verify that the fault has been corrected and is working as it should before returning the vehicle to the customer.

To apply the 3 Cs, follow the steps in **SKILL DRILL 5-10**.

SKILL DRILL 5-10 Using the 3 Cs to Document the Repair

1. Using the 3 Cs, document the repair process required for a repair order.
2. Identify and document the concern on the repair order. Obtain as much information as possible, as this will help you to understand the problem. Identify what the problem is and what vehicle systems are involved. Gather information on any recent repairs or servicing. Ask questions like, "How long has the problem been occurring? or "Does it occur at any particular time or temperature?" All of this information helps to identify the problem.
3. Identify and document the root cause of the concern. Research service information and perform tests to identify the cause of the problem. This may require a number of tests across multiple systems.
4. Review the information you collect from the tests. To review effectively, you need to understand how the systems work and interact.
5. Always work safely and use the proper tools and correct personal protective equipment (PPE).
6. Identify and document the correction required, including work activities and parts required or used. Get authorization for the repair if needed.
7. Make repairs or replace parts to complete the repair.
8. Retest the vehicle to be sure the fault has been corrected.
9. Fill in the required information and details to document the repair on the repair order.

▶ Wrap-Up

Ready for Review

- The owner's manual, usually kept in the glove compartment, provides information on how to operate the vehicle and basic maintenance to be performed.
- Manufacturers provide shop (or service) manuals for each make and model of car; these manuals provide vehicle-specific instructions on service and repair.
- Service information programs allow users to access maintenance and repair information via computer.
- After-market repair manuals are not produced by manufacturers and provide less detailed information for specific makes and models.
- Manufacturers provide technical service bulletins (TSBs) as updates to shop manuals when new problems or maintenance concerns arise for certain vehicle makes or models.
- If a safety issue is discovered on a certain make of vehicle, the manufacturer may issue a service campaign or recall.

- Labor guides provide information on service repair times and cost estimates.
- Parts programs are electronic catalogs of vehicle parts.
- Repair or work orders detail customer concern information to guide the service technician, as well as information on services they performed.
- Accounting systems track repair costs and customer methods of payment.
- A vehicle's service history consists of records of all maintenance and repairs performed on the vehicle.
- Vehicle information numbers (VINs) are unique identifiers for each vehicle produced.
- VINs are made up of 17 characters and are usually located on the front left corner of the windshield and on the engine, transmission, and other vehicle parts.
- VINs assist customers and technicians in tracking title history and reveal whether the vehicle has been wrecked or had repeated unsuccessful repairs for a particular fault.
- The two worldwide VIN systems are the North American VIN system and ISO Standard 3779.
- The VIN contains information on country of origin, manufacturer, make and model, body type, seat restraints, engine type, year of manufacture, assembly plant, and the order the vehicle came off the assembly line.
- Manufacturers also provide vehicle information labels to provide further specifications for each model of vehicle.
- The strategy-based diagnostic process helps technicians solve problem in a logical and efficient manner.
- The strategy-based diagnostic process consists of these steps:
  - Step 1: Verifying the customer's concern
  - Step 2: Researching possible faults and gathering information
  - Step 3: Focused testing
  - Step 4: Performing the repair
  - Step 5: Verifying the repair
  - The 3 Cs of vehicle repair are concern, cause, and correction.

Key Terms

belt routing label A label that lists a diagram of the routing of the belt(s) for the engine accessories.

coolant label A label that lists the type of coolant installed in the cooling system.

labor guide A guide that provides information to make estimates for repairs.

lemon law buyback A consumer protection law used in some states to identify a new vehicle that has undergone several unsuccessful attempts to repair the same fault.

owner's manual An informational guide supplied by the manufacturer; it contains basic vehicle operating information.

parts program A computer software program for identifying and ordering replacement vehicle parts.

parts specialist The person who serves customers at the parts counter.

refrigerant label A label that lists the type and total capacity of refrigerant that is installed in the A/C system.

repair order A form used by shops to collect information regarding a vehicle coming in for repair, also referred to as a work order.

salvage title Also called a branded title; a record that a vehicle has been severely damaged or deemed a total loss by an insurance company.

service campaign and recall A corrective measure conducted by manufacturers when a safety issue is discovered with a particular vehicle.

service history A complete list of all the servicing and repairs that have been performed on a vehicle.

shop or service manual Manufacturer's or after-market information on the repair and service of vehicles.

strategy-based diagnosis The process of logically and systematically (strategically) solving problems.

technical service bulletin (TSB) Information issued by manufacturers to alert technicians of unexpected problems or changes to repair procedures.

title history A detailed account of a vehicle's past.

vehicle emission control information (VECI) label A label used by technicians to identify engine and emission control information for the vehicle.

vehicle identification number (VIN) A unique serial number that is assigned to each vehicle produced.

vehicle safety certification (VSC) label A label certifying that the vehicle meets the Federal Motor Vehicle Safety, Bumper, and Theft Prevention Standards in effect at the time of manufacture.

Review Questions

1. Vehicle security PIN codes can typically be found in the:
   a. shop manual.
   b. owner's manual.
   c. technical service bulletin.
   d. after-market manual.
2. A technical service bulletin may provide information on all of the following *except*:
   a. unexpected problems.
   b. updated parts.
   c. changes to repair procedures.
   d. the scheduled maintenance chart.
3. Which of these can be used in arriving at flat rate servicing costs?
   a. Service interval recommendations
   b. Shop manuals
   c. Labor guides
   d. Service warranty applications
4. All of the following statements about a parts program are true *except*:
   a. It can only be accessed by DVD.
   b. It can be used to directly place an order.
   c. It can provide information on the cost of a part.
   d. It provides options for search by multiple criteria.
5. When a vehicle comes into a shop for repair, detailed information regarding the vehicle should be recorded in the:
   a. service booklet.
   b. repair order.

c. vehicle information label.
d. shop manual.

6. The service history of the vehicle gives information on whether:
   a. the vehicle was serviced for the same problem more than once.
   b. the vehicle is going to break down again.
   c. the vehicle meets federal standards.
   d. the vehicle has Vehicle Safety Certification.
7. All of the following statements about the VIN are true *except*:
   a. It is a unique serial number.
   b. It deters auto theft.
   c. It helps in accessing title history.
   d. It is made up of 10 characters.
8. What does the 10th character of a North American VIN denote?
   a. Type of sear restraint
   b. Model year of manufacture
   c. Body type
   d. Name of the manufacturer
9. The first step in the strategy-based diagnostic process is to:
   a. research possible faults.
   b. verify the customer's concern.
   c. perform focused testing.
   d. verify the repair.
10. Which of the following is not one of the three Cs of vehicle repair?
    a. Cause
    b. Cost
    c. Concern
    d. Correction

ASE Technician A/Technician B Style Questions

1. Tech A says that the owner's manual provides engine overhaul procedure information. Tech B says that engine oil capacity information for that specific vehicle is in the owner's manual. Who is correct?
   a. Tech A
   b. Tech B
   c. Both A and B
   d. Neither A nor B
2. Tech A says that the cause of the customer concern should be recorded on the repair order. Tech B says that as long as you replace at least one part on a vehicle with a problem, you don't need to verify the repair. Who is correct?
   a. Tech A
   b. Tech B
   c. Both A and B
   d. Neither A nor B
3. Tech A says that reviewing service history is an important part of diagnosis. Tech B says that "caution" is one of the 3 Cs. Who is correct?
   a. Tech A
   b. Tech B
   c. Both A and B
   d. Neither A nor B
4. Tech A says that service information programs are extremely helpful, as the technician can use a laptop at the vehicle for quick access to repair procedures. Tech B says that a fault from the factory within the airbag system would likely trigger a recall. Who is correct?
   a. Tech A
   b. Tech B
   c. Both A and B
   d. Neither A nor B
5. Tech A says that TSBs are typically updates to the owner's manual. Tech B says that TSBs are generally updated information on model changes that do not affect the technician. Who is correct?
   a. Tech A
   b. Tech B
   c. Both A and B
   d. Neither A nor B
6. Tech A says that labor guides are used by the service writer to quote prices for a customer. Tech B says that labor guides allow more time for repairs made under warranty. Who is correct?
   a. Tech A
   b. Tech B
   c. Both A and B
   d. Neither A nor B
7. Tech A says that strategy-based diagnosis is only used on faults you have never encountered before. Tech B says that a parts person needs to have basic computer operating skills. Who is correct?
   a. Tech A
   b. Tech B
   c. Both A and B
   d. Neither A nor B
8. Tech A says that the repair order is just a piece of paper telling the technician what to do. Tech B says that the repair order is a legal and binding contract between the customer and the repair facility. Who is correct?
   a. Tech A
   b. Tech B
   c. Both A and B
   d. Neither A nor B
9. Tech A says that customer authorization is again required if the repairs are more than a certain dollar amount or percentage above the original estimate. Tech B says that additional charges can be dealt with after the repair is complete, as the customer will be satisfied because more repairs have been completed. Who is correct?
   a. Tech A
   b. Tech B
   c. Both A and B
   d. Neither A nor B
10. Tech A says that the VIN number on a vehicle can help identify which engine is installed in the chassis. Tech B says that the digits in the VIN number can be an identifier for information pertinent to that vehicle. Who is correct?
    a. Tech A
    b. Tech B
    c. Both A and B
    d. Neither A nor B

CHAPTER 6

Basic Tools and Precision Measuring

NATEF Tasks

- **N06001** Demonstrate safe handling and use of appropriate tools.
- **N06002** Utilize safe procedures for handling of tools and equipment.
- **N06003** Identify standard and metric designation.
- **N06004** Demonstrate proper use of precision measuring tools (i.e., micrometer, dial-indicator, dial-caliper).
- **N06005** Demonstrate proper cleaning, storage, and maintenance of tools and equipment.

Knowledge Objectives

After reading this chapter, you will be able to:

- **K06001** Describe the safety procedures to take when handling and using tools.
- **K06002** Describe how to properly lockout and tag-out faulty equipment and tools.
- **K06003** Describe typical tool storage methods.
- **K06004** Identify tools and their usage in automotive applications.
- **K06005** Describe the type and use of wrenches.
- **K06006** Describe the type and use of sockets.
- **K06007** Describe the type and use of torque wrenches.
- **K06008** Describe the type and use of pliers.
- **K06009** Describe the type and use of cutting tools.
- **K06010** Describe the type and use of Allen wrenches.
- **K06011** Describe the type and use of screwdrivers.
- **K06012** Describe type and use of magnetic pickup tools and mechanical fingers.
- **K06013** Describe the type and use of hammers.
- **K06014** Describe the type and use of chisels.
- **K06015** Describe the type and use of punches.
- **K06016** Describe the type and use of pry bars.
- **K06017** Describe the type and use of gasket scrapers.
- **K06018** Describe the type and use of files.
- **K06019** Describe the type and use of clamps.
- **K06020** Describe the type and use of taps and dies.
- **K06021** Describe the type and use of screw extractors.
- **K06022** Describe the type and use of pullers.
- **K06023** Describe the type and use of flaring tools.
- **K06024** Describe the type and use of riveting tools.
- **K06025** Describe the type and use of measuring tapes.
- **K06026** Describe the type and use of steel rulers.
- **K06027** Describe the type and use of outside, inside, and depth micrometers.
- **K06028** Describe the type and use of telescoping gauges.
- **K06029** Describe the type and use of split ball gauges.
- **K06030** Describe the type and use of dial bore gauges.
- **K06031** Describe the type and use of vernier calipers.
- **K06032** Describe the type and use of dial indicators.
- **K06033** Describe the type and use of straight edges.
- **K06034** Describe the type and use of feeler gauges.
- **K06035** Describe the importance of proper cleaning and storage of tools.

Skills Objectives

There are no Skills Objectives for this chapter.

▶ Introduction

In this chapter, we explore a variety of tool and basic shop equipment topics that are fundamental to your success as an automotive technician. They provide the means for work to be undertaken on vehicles, from lifting to diagnosing, removing, installing, cleaning, and inspecting. Nearly all shop tasks involve the use of some sort of tool or piece of equipment (**FIGURE 6-1**). This makes their purchase, use, and maintenance very important to the overall performance of the shop. In fact, most tool purchases are considered an investment because they generate income when they are used. For example, if you can buy a tool for $100 that saves you two hours of working time every time you use it, you only need to use it a couple of times in order to pay for it. Then, every time after that, it pays you a bonus. This means that you need to treat tools like your own personal moneymakers. One way to do that is to always use tools and equipment in the way they are designed to be used. Don't abuse them. Think about the task at hand, identify the most effective tools to do the task, inspect the tool before using it, use it correctly, clean and inspect it after you use it, and store it in the correct location. Doing all of these things ensures that your tools will be available the next time you use them and that they will last a long time.

FIGURE 6-1 Technicians rely on tools to perform work.

▶ TECHNICIAN TIP

A sticker on the dash of a work truck I once used read: "Treat me well—your paycheck depends on it. If I can't work, neither can you!" That goes for all tools and equipment. Your ability to do work as a technician will either be enhanced or reduced by the condition of the tools and equipment you use.

▶ General Safety Guidelines

N06001, N06002, K06001

Although it is important to be trained on the safe use of tools and equipment, it is even more critical to have a safe attitude. A safe attitude will help you avoid being involved in an accident. Students who think they will never be involved in an accident will not be as aware of unsafe situations as they should be. And that can lead to accidents. So while we are covering the various tools and equipment you will encounter in the shop, pay close attention to the safety and operation procedures. Tools are a technician's best friend, but if used improperly, they can injure or kill (**FIGURE 6-2**).

Work Safe and Stay Safe

Whenever using tools, always think safety first. There is nothing more important than your personal safety. If tools (both hand and power) are used incorrectly, you can potentially injure yourself and others. Always follow equipment and shop instructions, including the use of recommended personal protective equipment (PPE). Accidents only take a second or two to happen but can take a lifetime to recover from. You are ultimately responsible for your safety, so remember to work safe and stay safe.

Safe Handling and Use of Tools

Tools must be safely handled and used to prevent injury and damage. Always inspect tools prior to use, and never use damaged tools. Check the manufacturer and the shop procedures, or ask your supervisor if you are uncertain about how to use tools. Inspect and clean tools when you are finished using them. Always return tools to their correct storage location.

To safely handle and use appropriate tools, follow **SKILL DRILL 6-1**.

You Are the Automotive Technician

After finishing work on the last vehicle of the day, you are required to return your workstation back to order. You clean, inspect, and return tools and equipment to their designated place. You wipe up any spills, according to the shop procedure, and clear the floor of any debris, to avoid slips and falls. During your workspace inspection, you determine that the insulation on the droplight cord is frayed, and there are some tools that need to be cleaned and put away.

1. What needs to happen with the droplight?
2. What should you do to with a micrometer before storing it?
3. How do you check a micrometer for accuracy?
4. Describe a double flare and how it is different from a single flare.

FIGURE 6-2 Tools used improperly can injure or kill.

SKILL DRILL 6-1 Safe Handling and Use of Tools

1. Select the correct tool(s) to undertake tasks.

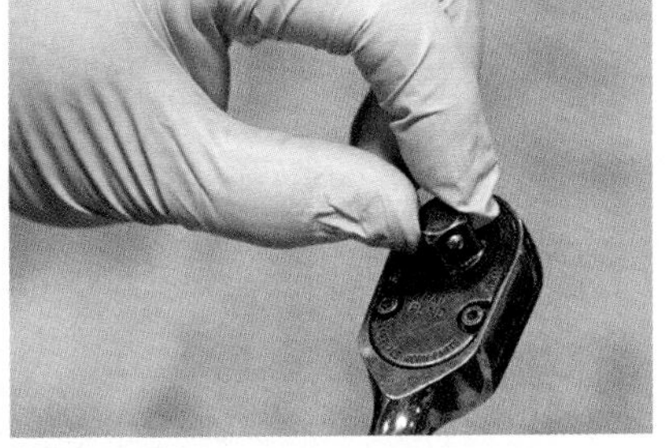

2. Inspect tools prior to use to ensure they are in good working order. If tools are faulty, remove them from service according to shop procedures.

3. Clean tools prior to use if necessary.

4. Use tools to complete the task while ensuring manufacturer and shop procedures are followed. Always use tools safely to prevent injury and damage.

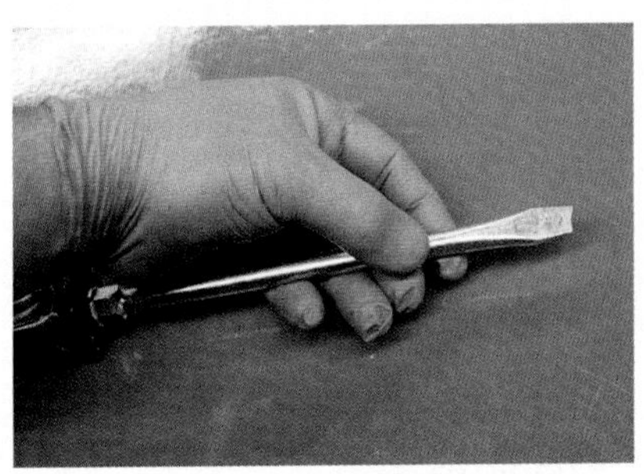

5. Ensure tools are clean and in good working order after use. Report and tag damaged tools, and remove them from service, following shop procedures.

6. Return tools to correct storage locations.

SKILL DRILL 6-2 Safe Procedures for Handling Tools and Equipment

1. Seek assistance if tools and equipment are too heavy or too awkward to be managed by a single person.

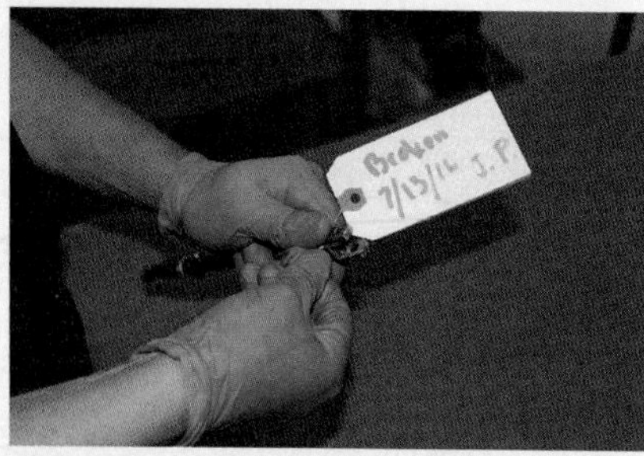

2. Inspect tools and equipment for possible defects before starting work. Report and/or tag faulty tools and equipment according to shop procedures.

3. Select and wear appropriate PPE for the tools and equipment being used.

4. Use tools and equipment safely.

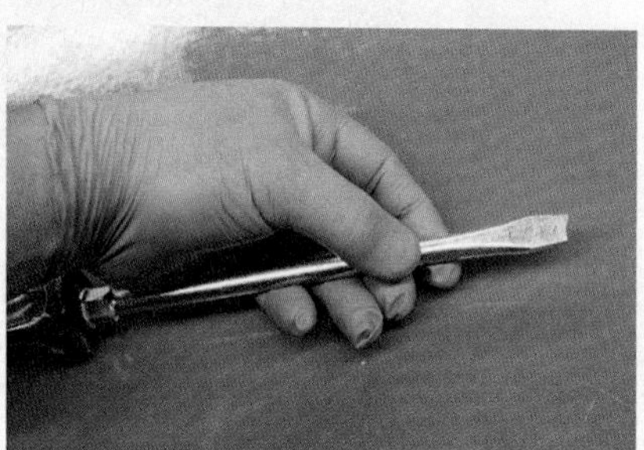

5. Check tools for faults after using them and report and/or tag faulty tools and equipment according to shop procedures.

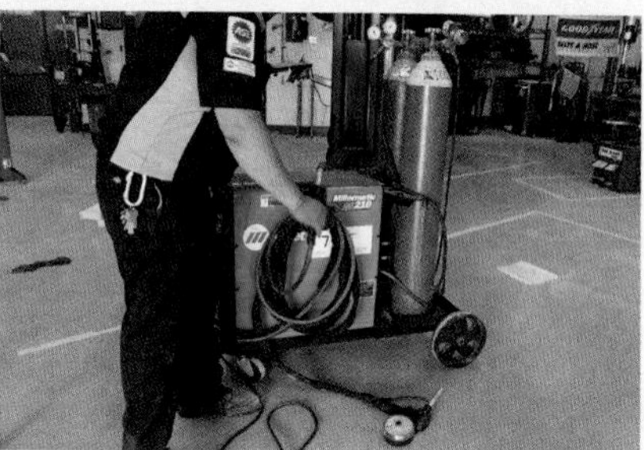

6. Clean and return tools and equipment to correct storage locations when tasks are completed.

Safe Procedures for Handling Tools and Equipment

Some tools are heavy or awkward to use, so seek assistance if required, and use correct manual handling techniques when using tools. To utilize safe procedures for handling tools and equipment, follow the steps in **SKILL DRILL 6-2**.

Tool Usage

Tools extend our abilities to perform many tasks; for example, jacks, stands, and hoists extend our ability to lift and hold heavy objects. Hand tools extend our ability to perform fundamental tasks like gripping, turning, tightening, measuring, and cutting (**FIGURE 6-3**). Electrical meters enable us to measure things we cannot see, feel, or hear; and power and air tools multiply our strength by performing tasks quickly and efficiently. As you are working, always think about what tool can make the job easier, safer, or more efficient. As you become familiar with more tools, your productivity, quality of work, and effectiveness will improve.

Every tool is designed to be used in a certain way to do the job safely. It is critical to use a tool in the way it was designed to be used and to do so safely. For example, a screwdriver is designed to tighten and loosen screws, not to be used as a chisel. Ratchets are designed to turn sockets, not to be used as a hammer. Think about the task you are undertaking, select the correct tools for the task, and use each tool as it was designed.

FIGURE 6-3 Tools extend our abilities.

TECHNICIAN TIP

Using tools can make you much more efficient and effective in performing your job. Without tools, it would be very difficult to carry out vehicle repairs and servicing. It is also the reason that many technicians invest well over $20,000 in their personal tools. If purchased wisely, tools can help you perform more work in a shorter amount of time, thereby making you more money. So think of your tools as an investment that pays for themselves over time.

Lockout/Tag-out

K06002

Lockout/tag-out is an umbrella term that describes a set of safety practices and procedures that are intended to reduce the risk of technicians inadvertently using tools, equipment, or materials that have been determined to be unsafe or potentially unsafe, or that are in the process of being serviced. An example of lockout would be physically securing a broken, unsafe, or out-of-service tool so that it cannot be used by a technician. In many cases, the item is also tagged out so it is not inadvertently placed back into service or operated. An example of tag-out would be affixing a clear and obvious label to a piece of equipment that describes the fault found, the name of the person who found the fault, and the date that the fault was found, and that warns not to use the equipment (**FIGURE 6-4**).

SAFETY TIP

Standardized lockout/tag-out procedures are a mandatory part of workplace safety regulations in most countries. Familiarize yourself with your local legislation and with the specific lockout/tag-out practices that apply in your workplace.

FIGURE 6-4 A. An example of lockout would be physically locking out a tool or piece of equipment so that it cannot be accessed and used by someone who may be unaware of the potential danger of doing so. **B.** An example of tag-out would be affixing a clear and obvious label to a piece of equipment that describes the fault found and that warns not to use the equipment.

Tool Storage

K06003

Typically, technicians have a selection of their own tools, which include various hand tools, air tools, measuring tools, and electrical meters (**FIGURE 6-5**). Often they add to their toolbox over time. These tools are kept in a toolbox that can be rolled around as needed but is typically kept in the technician's work stall. The shop usually has a selection of specialty tools and equipment that is available for technicians to use on a shared basis. These tools are located in specific areas around the shop, typically in a centralized location so they are relatively easy to access (**FIGURE 6-6**). They include specialized manufacturer tools, such as pullers or installation tools; high-cost tools such as factory scan tools; and tools that are not portable, such as hoists and tire machines. But because they are shared by everyone in the shop, it is critical that they are inspected for damage and put in their proper storage space after each use so they will be available to the other technicians.

To identify tools and their usage in automotive applications, follow these steps:

1. Create a list of tools in your toolbox and identify their application for automotive repair and service.

FIGURE 6-5 Typical technician's toolbox.

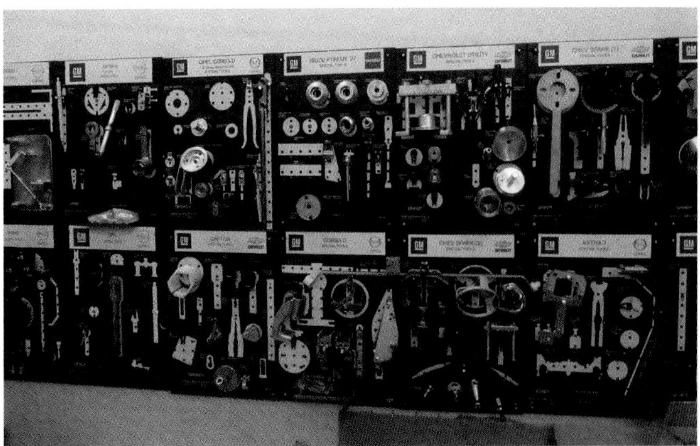

FIGURE 6-6 Manufacturer's special tools.

2. Look through the shop's tool storage areas, and create a list of the tools found in each storage area; identify their application for automotive repair and service.

Standard and Metric Designations

N06003

Many tools, measuring instruments, and fasteners come in United States customary system (USCS) sizes, more commonly referred to as "standard," or in metric sizes. Tools and measuring instruments can be identified as standard or metric by markings identifying their size on tools or the increments on a measuring tool. **Fasteners** bought new have their designation identified on the packaging. Other fasteners may have to be measured by a ruler or **vernier caliper** to identify their designation. Manufacturer's charts showing thread and fastener sizing assist in identifying standard or metric sizing.

To identify standard and metric designation, follow these steps:

1. Examine the component, tool, or fastener to see if any marking identifies it as standard or metric. Manufacturer specifications and shop manuals can be referred to and may identify components as standard or metric (**FIGURE 6-7**).
2. If no markings are available, use measuring devices to gauge the size of the item and compare thread and fastener charts to identify the sizing. Inch-to-metric conversion charts assist in identifying component designation (**FIGURE 6-8**).

▶ Basic Hand Tools

K06004

Like all tools, hand tools extend our ability to do work. Hand tools come in a variety of shapes, sizes, and functions (**FIGURE 6-9**). A large percentage of your personal tools will be hand tools. Over the years, manufacturers have introduced new fasteners, wire harness terminals, quick-connect fittings for fuel and other lines, and additional technologies that require their own different types of hand tools. This means that technicians need to continually add updated tools to their toolbox.

Speaking of toolboxes, technicians need to invest in a quality toolbox to keep their tools secure, so toolboxes should have enough capacity to hold all current and future tools. They should be built to last because a toolbox is in continuous use throughout the workday. A toolbox should have drawers that can handle the weight of the tools in it. Plus it should open and close easily; drawer slides that use bearings make them much easier to open and close (**FIGURE 6-10**). The toolbox should also be equipped with an easy-to-use locking system to secure the tools when they aren't being used.

▶ TECHNICIAN TIP

Invest in quality tools. Because tools extend your abilities, poor-quality tools affect the quality and quantity of your work. Price is not always the best indicator of quality, but it plays a role. As you learn the purpose and function of the tools in this chapter, you should be able to identify high-quality tools versus poor-quality tools by looking at them, handling them, and putting them to work.

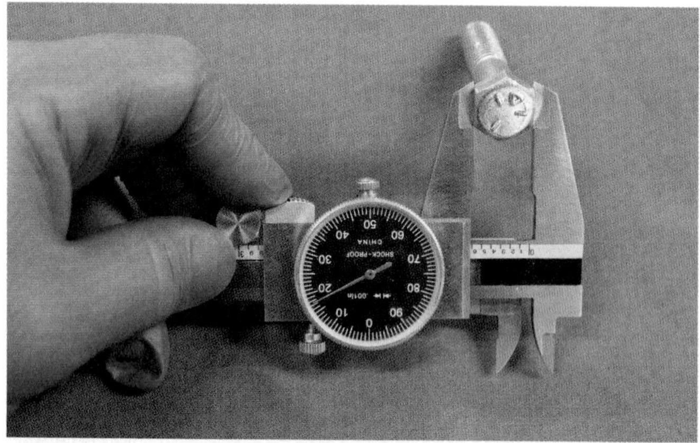

FIGURE 6-8 If no markings are available, use measuring devices to gauge the size of the item and compare thread and fastener charts to identify the sizing.

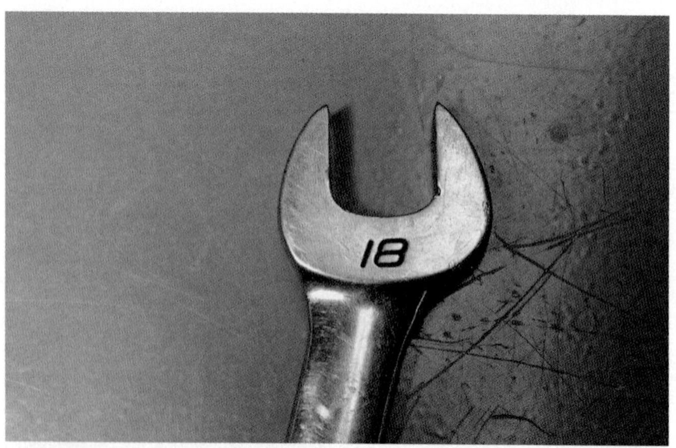

FIGURE 6-7 Manufacturer specifications and shop manuals can be referred to and may identify components as standard or metric.

FIGURE 6-9 Hand tools.

FIGURE 6-10 Toolbox drawers need to open and close easily.

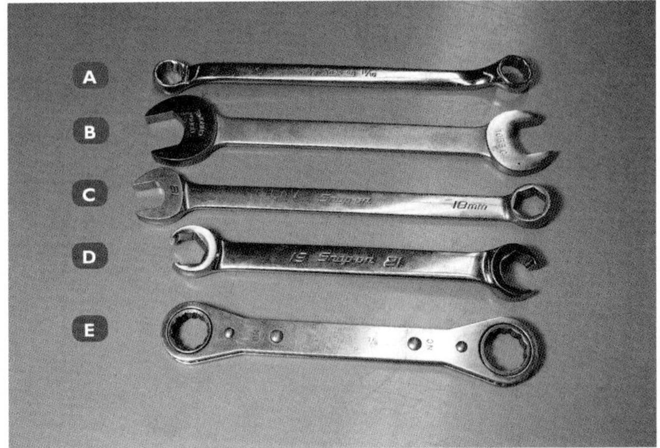

FIGURE 6-11 **A.** Box-end wrench. **B.** Open-end wrench. **C.** Combination wrench. **D.** Flare nut wrench. **E.** Ratcheting box-end wrench.

Wrenches

K06005

Wrenches are used to tighten and loosen nuts and **bolts**, which are two types of fasteners. There are three commonly used wrenches: the **box-end wrench**, the **open-end wrench**, and the **combination wrench** (FIGURE 6-11). The box-end wrench fits fully around the head of the bolt or nut and grips each of the six points at the corners, just like a socket. This is exactly the sort of grip needed if a nut or bolt is very tight, and makes the box-end wrench less likely to round off the points on the head of the bolt than the open-end wrench. The ends of box-end wrenches are bent or offset, so they are easier to grip and have different-sized heads at each end (FIGURE 6-12). One disadvantage of the box-end wrench is that it can be awkward to use once the nut or bolt has been loosened a bit, because you have to lift it off the head of the fastener and move it to each new position.

The open-end wrench is open on the end, and the two parallel flats only grip two points of the fastener. Open-end wrenches usually either have different-sized heads on each end of the wrench, or they have the same size, but with different angles (FIGURE 6-13). The head is at an angle to the handle and is not bent or offset, so it can be flipped over and used on both sides. This is a good wrench to use in very tight spaces as you can flip it over and get a new angle so the head can catch new points on the fastener. Although an open-end wrench often gives the best access to a fastener, if the fastener is extremely tight, the open-end wrench should not be used, as this type of wrench only grips two points. If the jaws flex slightly, the wrench can suddenly slip when force is applied. This slippage can round off the points of the fastener. The best way to approach a tight fastener is to use a box-end wrench to break the bolt or nut free, then use the open-end wrench to finish the job. The open-end wrench should only be used on fasteners that are no more than firmly tightened.

The combination wrench has an open-end head on one end and a box-end head on the other end (FIGURE 6-14). Both ends are usually of the same size. That way the box-end wrench can be used to break the bolt loose, and the open end can be used for turning the bolt. Because of its versatility, this is probably the most popular wrench for technicians.

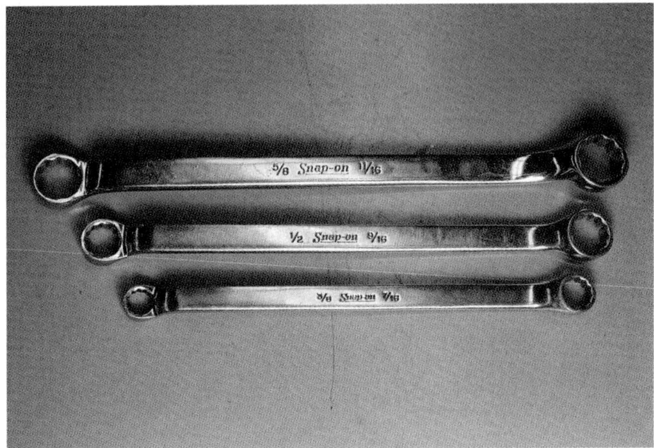

FIGURE 6-12 Box-end wrenches.

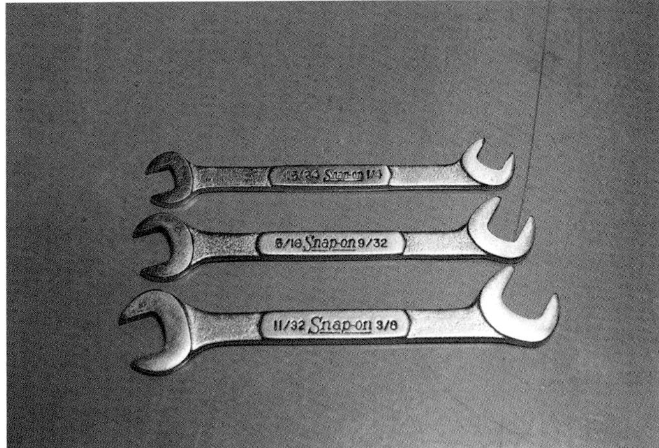

FIGURE 6-13 Types of open-end wrenches.

A variation on the open-end wrench is the **flare nut wrench**, also called a flare tubing wrench, or line wrench. It gives a better grip than the open-end wrench because it grabs all six points of

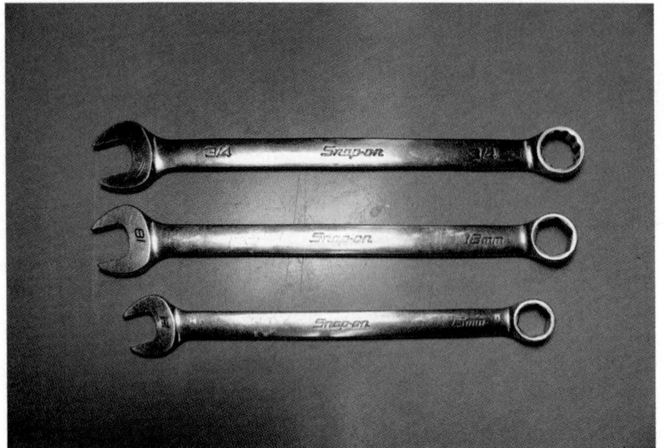

FIGURE 6-14 Combination wrenches.

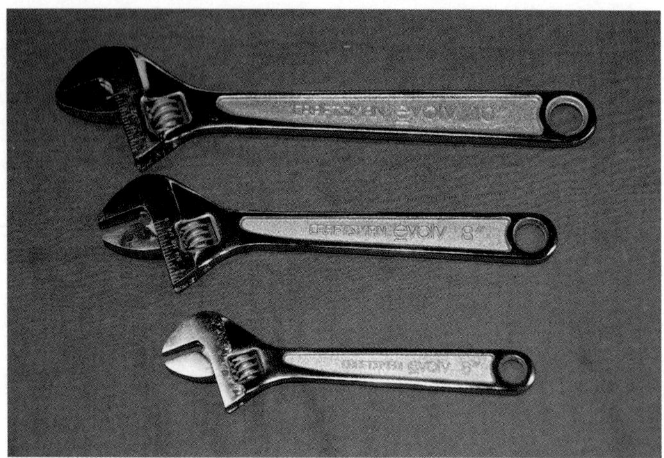

FIGURE 6-16 Open-end adjustable wrench.

FIGURE 6-15 Flare nut wrenches.

FIGURE 6-17 Ratcheting box-end wrench.

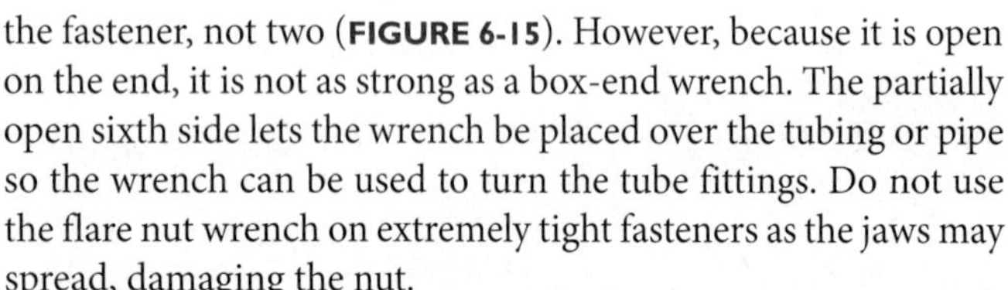

the fastener, not two (**FIGURE 6-15**). However, because it is open on the end, it is not as strong as a box-end wrench. The partially open sixth side lets the wrench be placed over the tubing or pipe so the wrench can be used to turn the tube fittings. Do not use the flare nut wrench on extremely tight fasteners as the jaws may spread, damaging the nut.

One other open-end wrench is the open-end adjustable wrench or crescent wrench. This wrench has a movable jaw that can be adjusted by turning an adjusting screw to fit any fastener within its range (**FIGURE 6-16**). It should only be used if other wrenches are not available because it is not as strong as a fixed wrench, so it can slip off and damage the head of tight bolts or nuts. Still, it is a handy tool to have because it can be adjusted to fit most any fastener size.

A **ratcheting box-end wrench** is a useful tool in some applications because it does not require removal of the tool to reposition it. It has an inner piece that fits over and grabs the points on the fastener and is able to rotate within the outer housing. A ratcheting mechanism lets it rotate in one direction and lock in the other direction. In some cases, the wrench just needs to be flipped over to be used in the opposite direction. In other cases, it has a lever that changes the direction from clockwise to counterclockwise (**FIGURE 6-17**). Just be careful to not overstress this tool by using it to tighten or loosen very tight fasteners, as the outer housing is not very strong. There is also a ratcheting open-end wrench, but it uses no moving parts. One of the sides is partially removed so that only the bottom one-third remains to catch a point on the bolt (**FIGURE 6-18**). When it is used, the normal side works just like a standard open-end wrench. The shorter side of the open-end wrench catches the point on the fastener so it can be turned. When moving the wrench to get a new bite, the wrench is pulled slightly outward, disengaging the short side while leaving the long side to slide along the faces of the bolt. The wrench is then rotated to the new position and pushed back in so the short side engages the next point. This wrench, like other open-end wrenches, is not designed to tighten or loosen tight fasteners, but it does work well in blind places where a socket or ratcheting box-end wrench cannot be used.

Specialized wrenches such as the **pipe wrench** grip pipes and can exert a lot of force to turn them (**FIGURE 6-19**). Because the handle pivots slightly, the more pressure put on the handle to turn the wrench, the more the grip tightens. The jaws are hardened and serrated, and increasing the pressure also

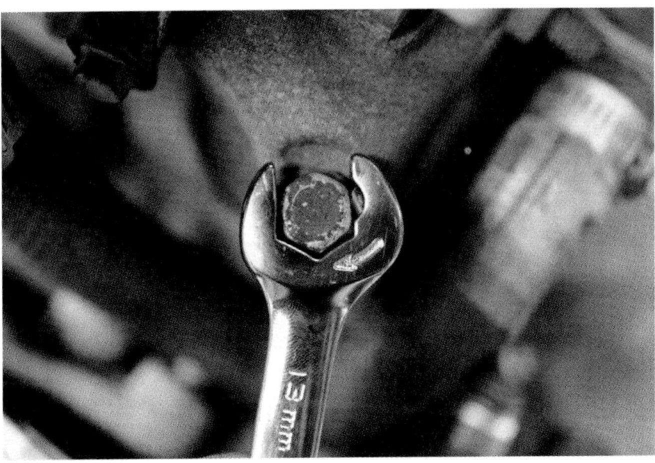

FIGURE 6-18 Ratcheting open-end wrench.

FIGURE 6-19 Pipe wrench.

FIGURE 6-20 Oil filter wrenches.

increases the risk of marking or even gouging metal from the pipe. The jaw is adjustable, so it can be threaded in or out to fit different pipe sizes. Also, they come in different lengths, allowing you to increase the leverage applied to the pipe.

A specialized wrench called an **oil filter wrench** grabs the filter and gives you extra leverage to remove an oil filter when it is tight. These wrenches are available in various designs and sizes (**FIGURE 6-20**), and some are adjustable to fit many filter sizes. Also note that an oil filter wrench should be used *only* to remove an oil filter, never to install it. Almost all oil filters should be installed and tightened by hand.

> **TECHNICIAN TIP**
>
> Wrenches (which are also known as spanners in some countries) will only do a job properly if they are the right size for the given nut or bolt head. The size used to describe a wrench is the distance across the flats of the nut or bolt. There are two systems in common use— standard (in inches) and metric (in millimeters). Each system provides a range of sizes, which are identified by either a fraction, which indicates fractions of an inch for the standard system, or a number, which indicates millimeters for the metric system. In most cases, standard and metric tools cannot be used interchangeably on a particular fastener. Even though a metric tool fits a standard sized fastener, it may not grip as tight, which could cause it to round the head of the fastener off.

Using Wrenches Correctly

Choosing the correct wrench for a job usually depends on two things: how tight the fastener is and how much room there is to get the wrench onto the fastener, and then to turn it. When being used, it is always possible that a wrench could slip. Before putting a lot of tension on the wrench, try to anticipate what will happen if it does slip. If possible, it is usually better to pull a wrench toward you than to push it away (**FIGURE 6-21**). If you have to push, use an open palm to push, so your knuckles won't get crushed if the wrench slips. If pulling toward yourself, make sure your face is not close to your hand or in line with your pulling motion. Many technicians have punched themselves in the face when the wrench slipped.

Sockets

K06006

Sockets are very popular because of their adaptability and ease of use (**FIGURE 6-22**). Sockets are a good choice where the top of the fastener is reasonably accessible. The **socket** fits onto the fastener snugly and grips it on all six corners, providing the type of grip needed on any nut or bolt that is extremely tight. Sockets come in a variety of configurations, and technicians usually have a lot of sockets so they can get in a multitude of tight places. Individual sockets fit a particular size nut or bolt, so they are usually purchased in sets.

Sockets are classified by the following characteristics:

- Standard or metric
- Size of drive used to turn them: 1/2" (12.7 mm), 3/8" (9.525 mm), and 1/4" (6.35 mm) are most common; 1" (25.4 mm) and 3/4" (19.05 mm) are less common.

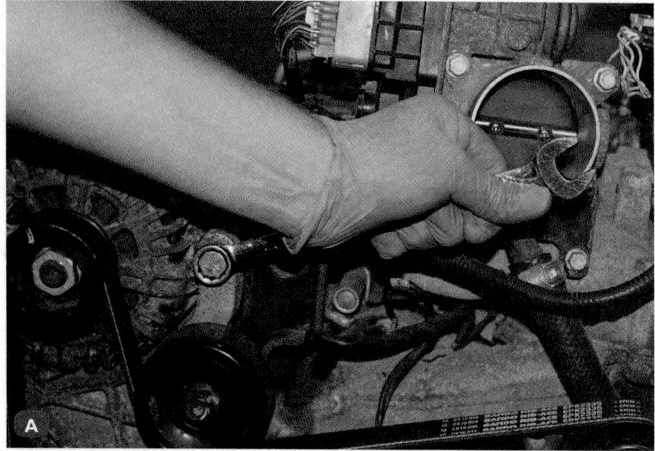

FIGURE 6-21 A. Pulling on a wrench is generally better than pushing on a wrench. **B.** Pushing regularly results in bruised or broken knuckles.

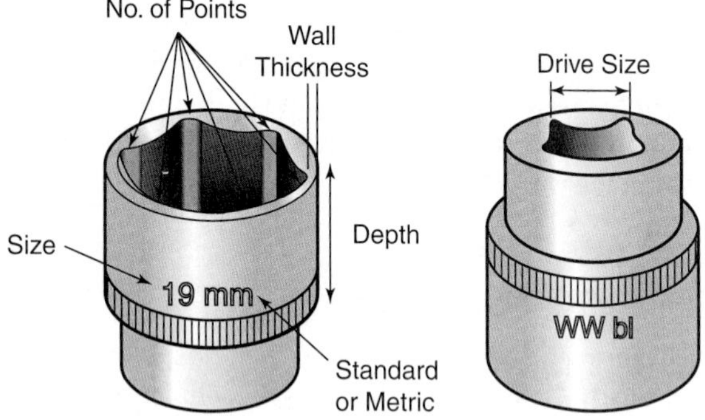

FIGURE 6-22 The anatomy of a socket.

- Number of points: 6 and 12 are most common; 4 and 8 are less common.
- Depth of socket: Standard and deep are most common; shallow is less common.
- Thickness of wall: Standard and impact are most common; thin wall is less common.

Sockets are built with a recessed square drive that fits over the square drive of the ratchet or other driver (**FIGURE 6-23**).

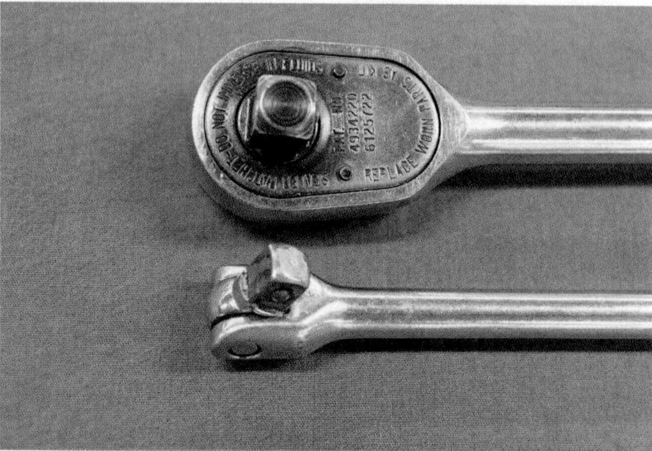

FIGURE 6-23 Sockets are designed to fit a matching drive on a ratchet.

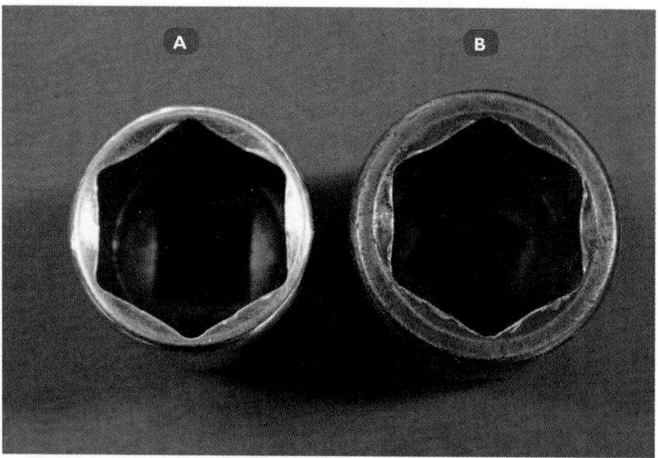

FIGURE 6-24 A. Standard wall socket. **B.** Impact socket.

The size of the drive determines how much twisting force can be applied to the socket. The larger the drive, the larger the twisting force. Small fasteners usually only need a small torque, so having too large of a drive may make it so the socket cannot gain access to the bolt. For fasteners that are really tight, an impact wrench exerts a lot more torque on a socket than turning it by hand does. Impact sockets are usually thicker walled than standard wall sockets and have six points so they can withstand the forces generated by the impact wrench as well as grip the fastener securely (**FIGURE 6-24**).

Six- and 12-point sockets fit the heads of hexagonal-shaped fasteners. Four- and 8-point sockets fit the heads of square-shaped fasteners (**FIGURE 6-25**). Because 6-point and 4-point sockets fit the exact shape of the fastener, they have the strongest grip on the fastener, but they only can fit on the fastener in half as many positions as a 12- or 8-point socket, making them harder to fit onto the fastener in places where the ratchet handle is restricted.

Another factor in accessing a fastener is the depth of the socket. If a nut is threaded quite a way down a stud or bolt, then a standard length socket will not fit far enough over the stud to reach the nut (**FIGURE 6-26**). In this case, a deep socket will usually reach the nut.

FIGURE 6-25 **A.** Six- and 12-point sockets. **B.** Four- and 8-point sockets.

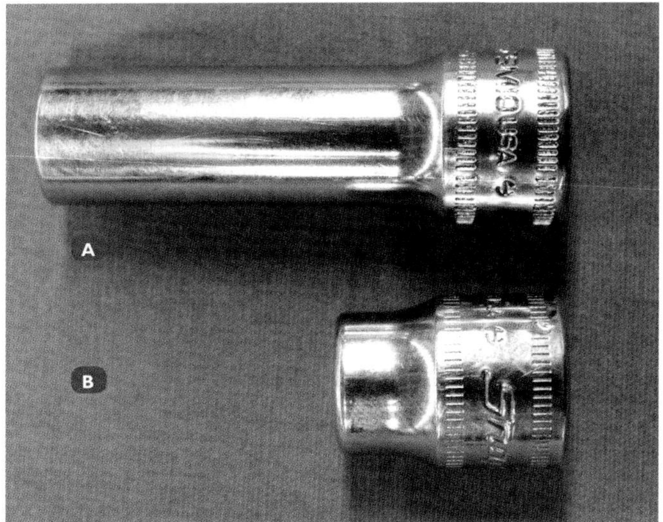

FIGURE 6-26 **A.** Deep socket. **B.** Standard length socket.

▶ **TECHNICIAN TIP**

Because sockets are usually purchased in sets, with each set providing a slightly different capability, you can see why technicians could easily have several hundred sockets in their toolbox. Having a variety of sockets allows the technician to do jobs easier and quicker, which makes them an investment. Turning a socket requires a handle (**FIGURE 6-27**).

The most common socket handle, the **ratchet**, makes easy work of tightening or loosening a nut where not a lot of pressure is involved. It can be set to turn in either direction and does not need much room to swing. It is built to be convenient, not super strong, so too much pressure could strip the ratchet mechanism. For heavier tightening or loosening, a breaker bar gives the most leverage. When that is not available, a **sliding T-handle** may be more useful. With this tool, both hands can be used, and the position of the T-piece is adjustable to clear any obstructions

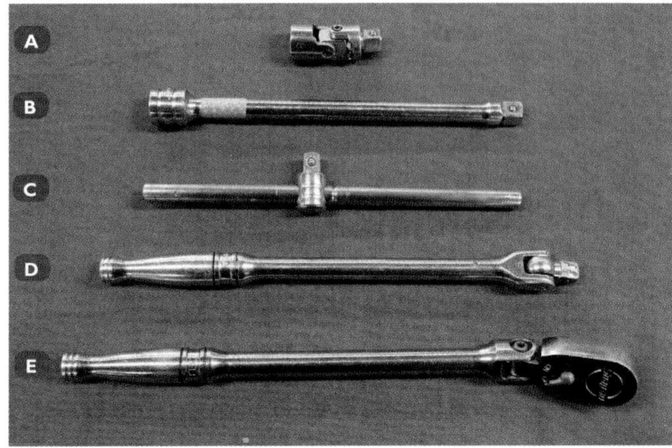

FIGURE 6-27 Tools to turn sockets. **A.** Universal joint. **B.** Extension. **C.** T-handle. **D.** Breaker bar. **E.** Ratchet.

when turning it. The connection between the socket and the accessory is made by a square drive. The larger the drive, the heavier and bulkier the socket will be. The 1/4" (6.35 mm) drive is for small work in difficult areas. The 3/8" (9.525 mm) drive accessories handle a lot of general work where torque requirements are not too high. The 1/2" (12.7 mm) drive is required for all-around service. The 3/4" (19.05 mm) and 1" (25.4 mm) drives are required for large work with high torque settings.

Many fasteners are located in positions where access can be difficult. There are many different lengths of extensions available to allow the socket to be on the fastener while extending the drive point out to where a handle can be attached. Because extensions come in various lengths, they can be connected together to get just the right length needed for a particular situation.

If an object is in the way of getting a socket on a fastener, a flexible joint is used to apply the turning force to the socket through an angle. This can allow you to still turn the socket even though you are no longer directly in line with the fastener. There are four common types of flexible joints; U-joint style, wobble extension, cable extension, and flex socket (**FIGURE 6-28**). The flex socket has the universal built into it, so it's overall length is shorter, making it able to get into tighter spaces. In some situations, you may need to use more than one flexible joint to get around objects that are in the way. This is especially true when removing some bell housing bolts on some transmissions.

A **speed brace** or speeder handle is the fastest way to spin a fastener on or off a thread by hand, but it cannot apply much torque to the fastener; therefore, it is mainly used to remove a fastener that has already been loosened or to run the fastener onto the thread until it begins to tighten (**FIGURE 6-29**).

A **lug wrench** has special-sized lug nut sockets permanently attached to it. One common model has four different-sized sockets, one on each arm (**FIGURE 6-30**). Never hit or jump on a lug wrench when loosening lug nuts. If the lug wrench will not remove them, you should use an impact wrench. The impact wrench provides a hammering effect in conjunction with rotation to help loosen tight fasteners. *Never* use an impact wrench to tighten lug fasteners. Torque all lug fasteners to the proper

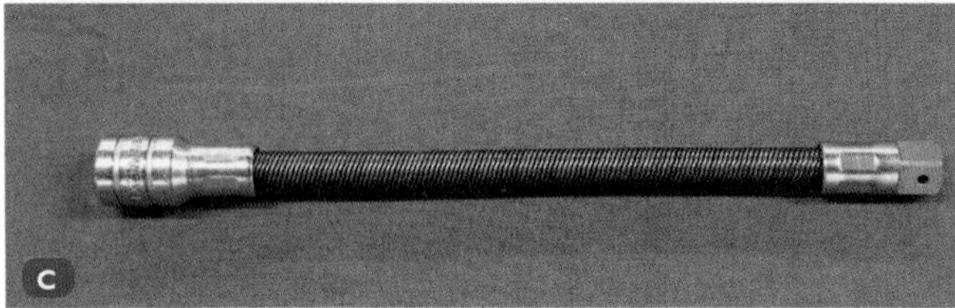

FIGURE 6-28 Flexible extensions. **A.** U-joint style. **B.** Wobble extension style. **C.** Cable extension style. **D.** Flex socket style.

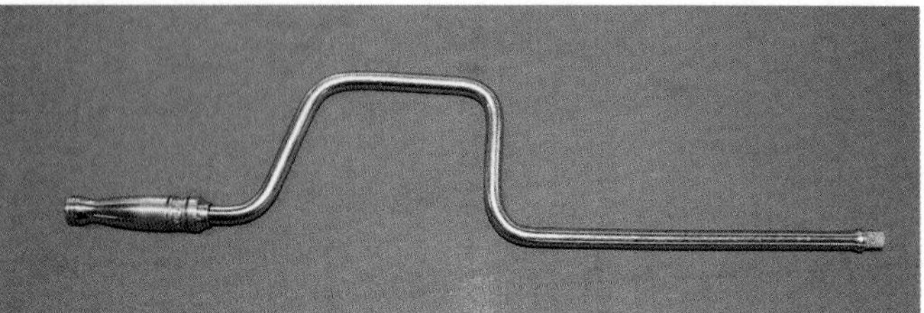

FIGURE 6-29 Speed brace.

FIGURE 6-30 Lug wrench.

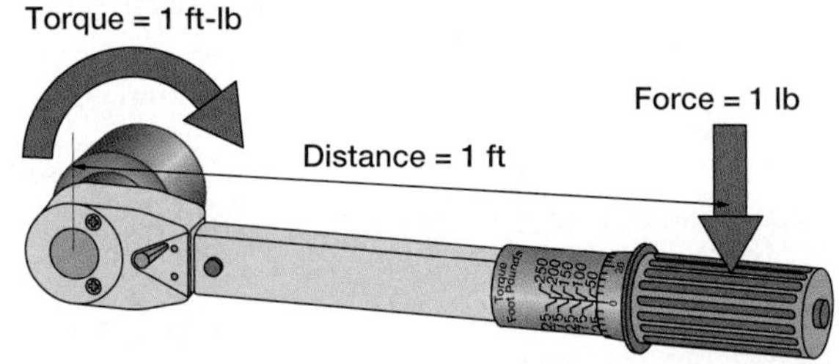

FIGURE 6-31 Torque is the measurement of twisting force. 1 ft-lb of torque.

torque with a properly calibrated torque wrench. And do it in the proper sequence.

Torque and Torque Wrenches

K06007

Bolt tension is what keeps a bolt from loosening and what causes it to hold parts together with the proper clamping force. **Torque** is the twisting force used to create bolt tension so that surfaces are clamped together with the proper force. The torque value is the amount of twisting force applied to a fastener by the torque wrench. A foot-pound is described as the amount of twisting force applied to a shaft by a perpendicular lever 1 ft long with a weight of 1 lb placed on the outer end (**FIGURE 6-31**). A torque value of 100 ft-lb is the same as a 100 lb weight placed at the end of a 1' long lever. A torque value of an inch-pound is 1 lb placed at the end of a 1" long lever. This means that 12 in-lb equals 1 ft-lb and vice versa. A newton meter (Nm) is described as the amount of twisting force applied to a shaft by a perpendicular lever 1 meter long with a force of 1 newton applied to the outer end. A torque value of 100 Nm is the same as applying a 100-newton force to the end of a 1 m long lever. One ft-lb is equal to 1.35 Nm. So torque is the measurement of twisting force.

Torque Charts

Torque specifications for bolts and nuts in vehicles are usually contained within service information. Bolt, nut, and stud manufacturers also produce torque charts, which contain the information you need to determine the maximum torque of bolts or nuts. For example, most charts include the bolt diameter, threads per inch, grade, and maximum torque setting for both dry and lubricated bolts and nuts (**FIGURE 6-32**).

A lubricated bolt and nut reach their maximum clamping force at a lower torque setting than if they are dry. In practice, most torque specifications call for the nuts and bolts to have dry threads prior to tightening. There are some exceptions, so close examination of the torque specification is critical. Also remember that the bolt manufacturer's torque chart is a maximum

Bolt Size	TPI	[illegible]sile Stress Area	Fastener Coating	Bolt Torque & Clamp Load	10,000 psi	25,[illegible]0 psi	SAE J429-Grade 2	SAE J429-Grade 5	SAE J429-Grade 8
3/8 JNC	16	0.0775		Clamp Load (Lb)	775	1,937	3,196	4,940	6,974
			Lubricated	Torque (Ft-Lb)	4	9	15	23	33
			Zinc Plated		4	11	18	28	39
			Plain - Dry		5	12	20	31	44
3/8 JNF	24	0.0878		Clamp Load (Lb)	878	2,196	3,623	5,599	7,905
			Lubricated	Torque (Ft-Lb)	4	10	17	26	37
			Zinc Plated		5	12	20	31	44
			Plain - Dry		5	14	23	35	49
[illegible]/16 JNC	14	0.1063		Clamp Load (Lb)	1,063	2,658	4,385	6,777	9,568
			Lubricated	Torque (Ft-Lb)	6	15	24	37	52
			Zinc Plated		7	17	29	44	63
			Plain - Dry		8	19	32	49	70
[illegible]/16 JNF	20	0.1187		Clamp Load (Lb)	1,187	2,968	4,897	7,568	10,684
			Lubricated	Torque (Ft-Lb)	6	16	27	41	58
			Zinc Plated		8	19	32	50	70
			Plain - Dry		9	22	36	55	78
1/2	13	0.1419		Clamp Load (Lb)	1,419	3,547	5,853	9,046	12,77
			Lubricated		9	22	37	57	80

FIGURE 6-32 Bolt torque chart.

recommended torque, not necessarily the torque required by the vehicle manufacturer for the specific application that the bolt is used for.

Torque Wrenches

A **torque wrench** is also known as a tension wrench (**FIGURE 6-33**) and is used to tighten fasteners to a predetermined torque. The drive on the end fits any socket and accessory of the same drive size found in ordinary socket sets. Although manufacturers do not specify torque settings for every nut and bolt, when they do, it is important to follow the specifications. For example, manufacturers specify a torque for head bolts.

The torque specified ensures that the bolt provides the proper clamping pressure and will not come loose, but will not be so tight as to risk breaking the bolt or stripping the threads. The torque value is specified in foot-pounds (ft-lb), inch-pounds (in-lb), or newton meters (Nm).

Torque wrenches come in various types: beam style, clicker, dial, and electronic (**FIGURE 6-34**). The simplest and least expensive is the beam-style torque wrench, which uses a spring steel beam that flexes under tension. A smaller fixed rod then indicates the amount of torque on a scale mounted to the bar. The amount of deflection of the bar coincides with the amount of torque on the scale. One drawback of this design is that you have to be positioned directly above the scale so you can read it accurately. That can be a problem when working under the hood of a vehicle.

The clicker-style torque wrench uses an adjustable clutch inside that slips (clicks) when the preset torque is reached. You can set it for a particular torque on the handle (**FIGURE 6-35**). As the bolt is tightened, once the preset torque is reached, the torque wrench clicks. The higher the torque, the louder the click; the lower the torque, the quieter the click. Be careful when using this style of torque wrench, especially at lower torque settings. It is easy to miss the click and over-tighten, break, or strip the bolt. Once the torque wrench clicks, stop turning it, as it will continue to tighten the fastener if you turn it past the click point.

The dial torque wrench turns a dial that indicates the torque based on the torque being applied. Like the beam-style

FIGURE 6-33 A torque wrench.

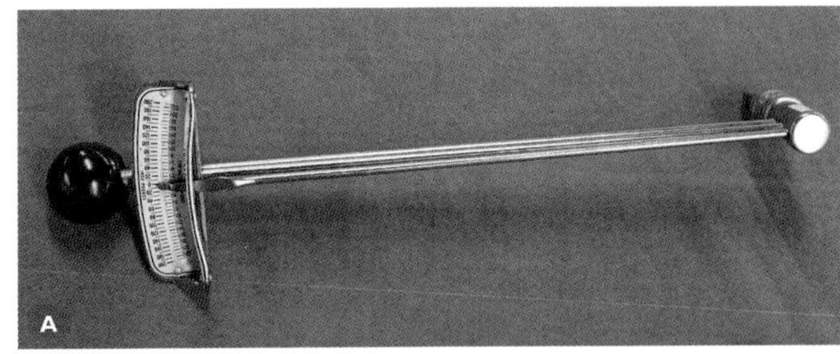

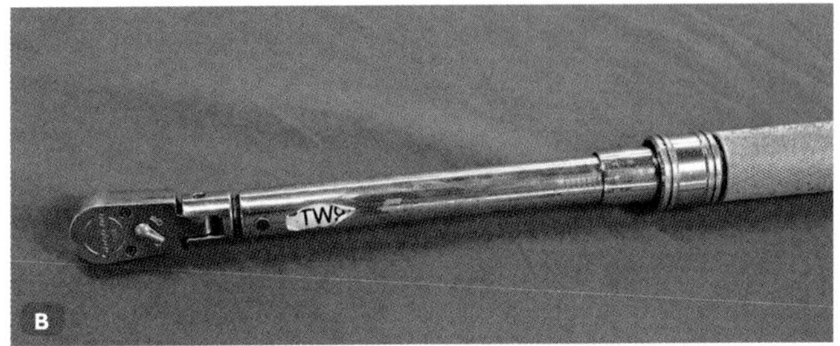

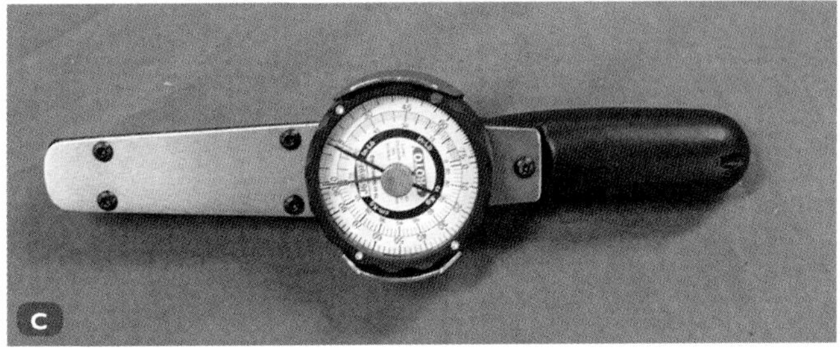

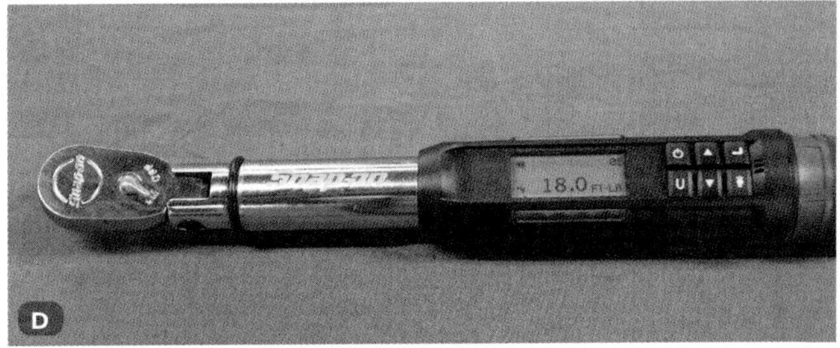

FIGURE 6-34 Torque wrenches. **A.** Beam style. **B.** Clicker style. **C.** Dial. **D.** Electronic.

torque wrench, you have to be able to see the dial to know how much torque is being applied (**FIGURE 6-36**). Many dial torque wrenches have a movable indicator that is moved by the dial and stays at the highest reading. That way you can double-check the torque achieved once the torque wrench is released. Once the

FIGURE 6-35 Torque setting scale on the handle.

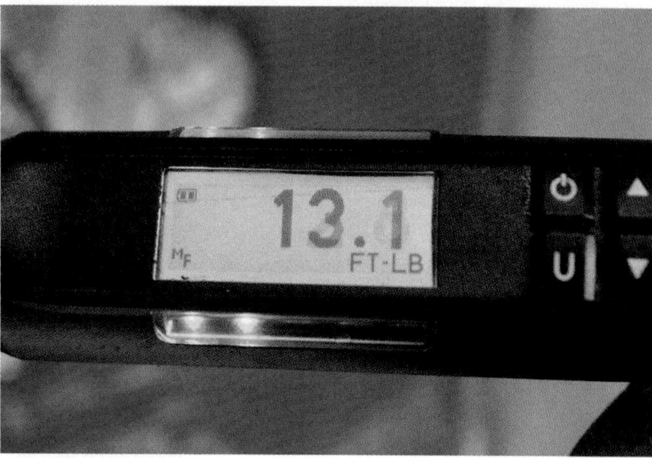

FIGURE 6-37 Digital torque wrench displaying torque.

FIGURE 6-36 Dial torque wrench reading torque.

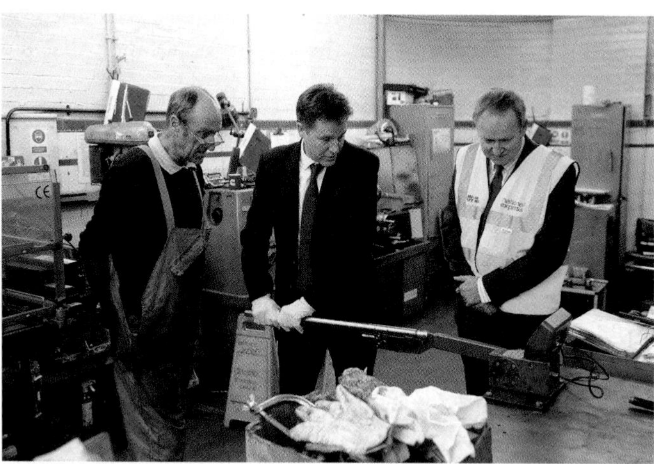

FIGURE 6-38 Checking torque wrench calibration.

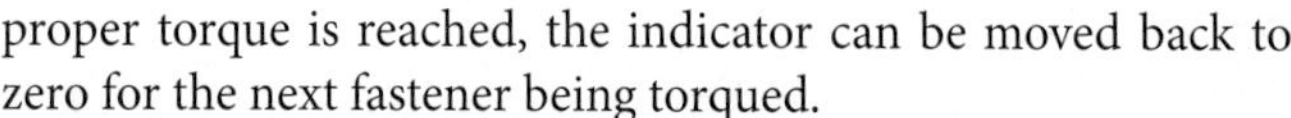

proper torque is reached, the indicator can be moved back to zero for the next fastener being torqued.

The digital torque wrench usually uses a spring steel bar with an electronic strain gauge to measure the amount of torque being applied. The torque wrench can be preset to the desired torque. It will then display the torque as the fastener is being tightened (**FIGURE 6-37**). When it reaches the preset torque, it usually gives an audible signal, such as a beep. This makes it useful in situations where a scale or dial cannot be read.

Torque wrenches fall out of calibration over time or if they are not used properly, so they should be checked and calibrated on a periodic basis (**FIGURE 6-38**). This can be performed in the shop if the proper calibration equipment is available, or the torque wrench can be sent to a qualified service center. Most quality torque wrench manufacturers provide a recalibration service for their customers.

Using Torque Wrenches

The torque wrench is used to apply a specified amount of torque to a fastener. There are various methods used by torque wrenches to indicate that the correct torque has been reached. Some give an audible signal, such as a click or a beep, whereas others give a visual signal such as a light or a pin moving or clicking out. Some provide a scale and needle that must be observed while you are torquing the fastener. To help ensure that the proper amount of torque gets from the torque wrench to the bolt, support the head of the torque wrench with one hand (**FIGURE 6-39**). When using a torque wrench, it is best not to use extensions. Extensions make it harder to support the head, which can end up absorbing some of the torque. If possible, use a deep socket instead.

Pliers

K06008

Pliers are a hand tool designed to hold, cut, or compress materials (**FIGURE 6-40**). They are usually made out of two pieces of strong steel joined at a fulcrum point, with jaws and cutting surfaces at one end and handles designed to provide leverage at the other. There are many types of pliers, including slip-joint, combination, arc joint, needle-nose, and flat-nose (**FIGURE 6-41**).

FIGURE 6-39 It may be necessary to support the torque wrench when using extensions.

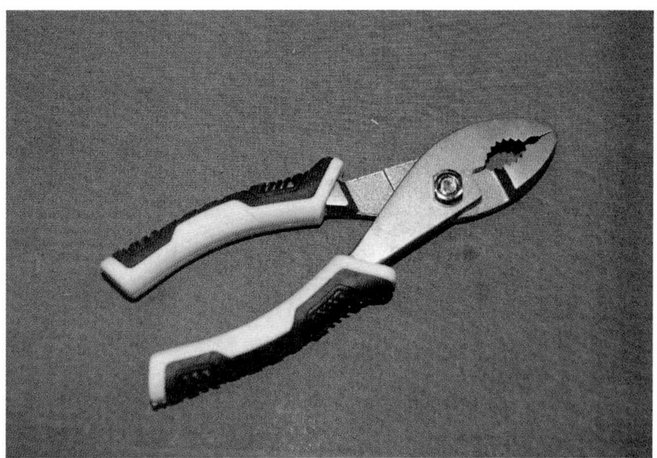

FIGURE 6-40 Pliers are used for grasping and cutting. These are slip-joint pliers.

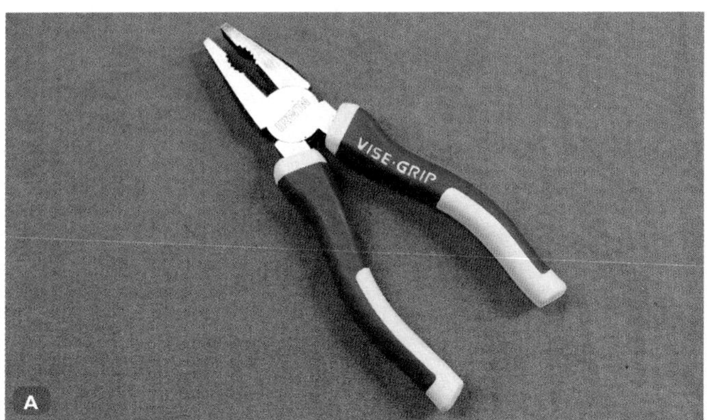

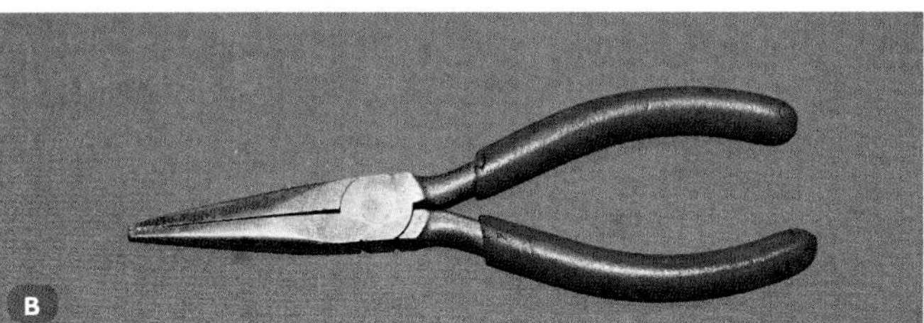

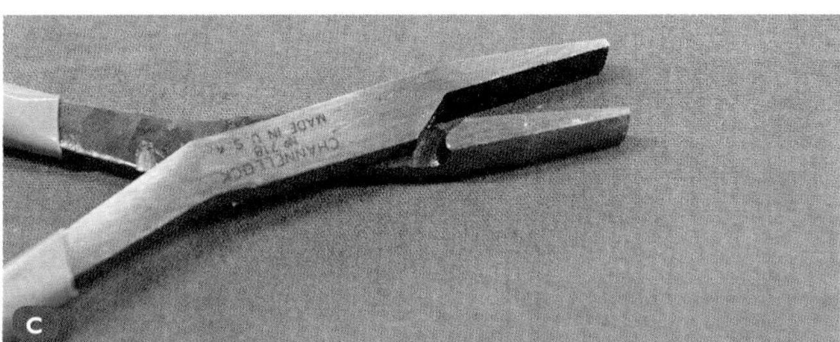

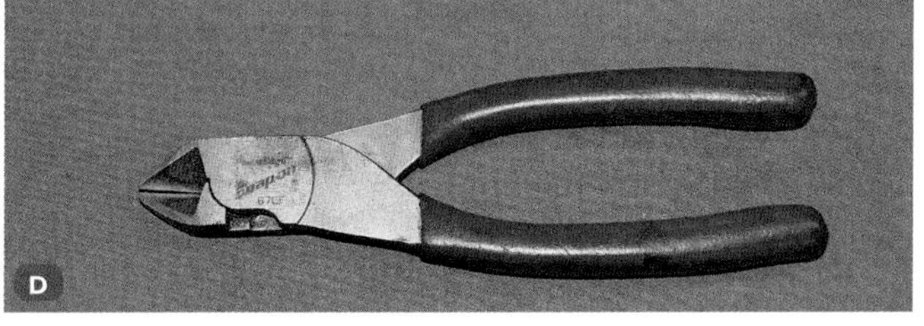

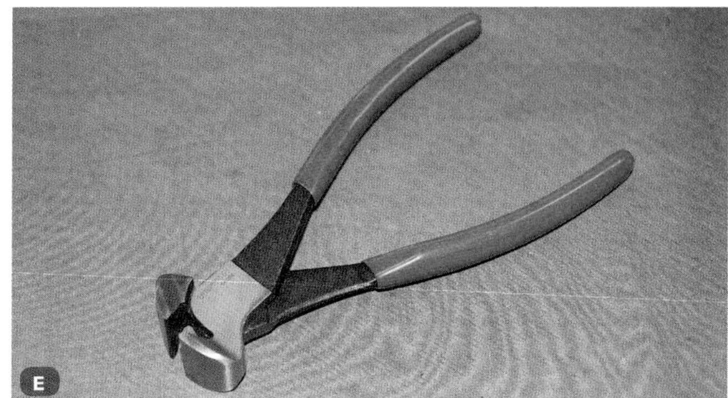

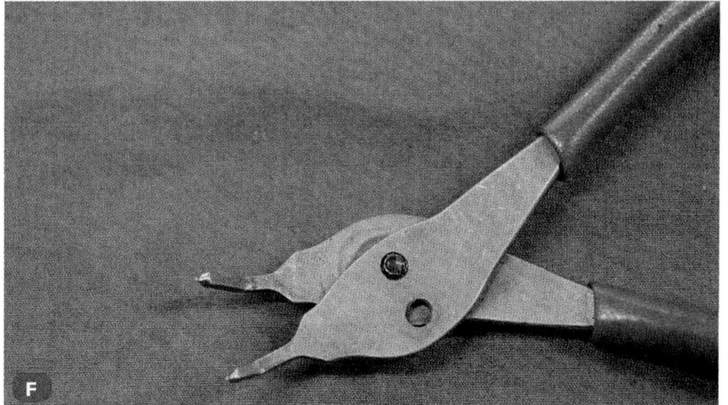

FIGURE 6-41 **A.** Combination pliers. **B.** Needle-nose pliers. **C.** Flat-nose pliers. **D.** Diagonal cutting pliers. **E.** Nippers. **F.** Internal snap ring pliers. **G.** External snap ring pliers.

Quality **combination pliers** are one of the most commonly used pliers in a shop. They pivot together so that any force applied to the handles is multiplied in the strong jaws. Most combination pliers are designed so they have surfaces to both grip and cut. Combination pliers offer two gripping surfaces, one for gripping flat objects and one for gripping rounded objects, and one or two pairs of cutters. The cutters in the jaws should be used for softer materials that will not damage the blades. The cutters next to the pivot can shear through hard, thin materials, like steel wire or pins.

Most pliers are limited by their size in what they can grip. Beyond a certain point, the handles are spread too wide, or the jaws cannot open wide enough, but **arc joint pliers** overcome that limitation with a moveable pivot. Often, these are called Channellocks™ after the company that first made them. These pliers have parallel jaws that allow you to increase or decrease the size of the jaws by selecting a different set of channels (**FIGURE 6-42**). They are useful for a wider grip and a tighter squeeze on parts too big for conventional pliers.

Another type of pliers is **needle-nose pliers**, which have long, pointed jaws and can reach into tight spots or hold small items that other pliers cannot (**FIGURE 6-43**). For example, they can pick up a small bolt that has fallen into a tight spot. **Flat-nose pliers** have an end or nose that is flat and square; in contrast, combination pliers have a rounded end. A flat nose makes it possible to bend wire or even a thin piece of sheet steel accurately along a straight edge. **Diagonal cutting pliers** are used for cutting wire or cotter pins (**FIGURE 6-44**). Diagonal cutters are the most common cutters in the toolbox, but they should not be used on hard or heavy-gauge materials because the cutting surfaces will be damaged. End cutting pliers, also called **nippers**, have a cutting edge at right angles to their length (**FIGURE 6-45**). They are designed to cut through soft metal objects sticking out from a surface.

Snap ring pliers have metal pins that fit in the holes of a snap ring (**FIGURE 6-46**). Snap rings can be of the internal or external type. If internal, then internal snap ring pliers compress the snap ring so it can be removed from and installed in

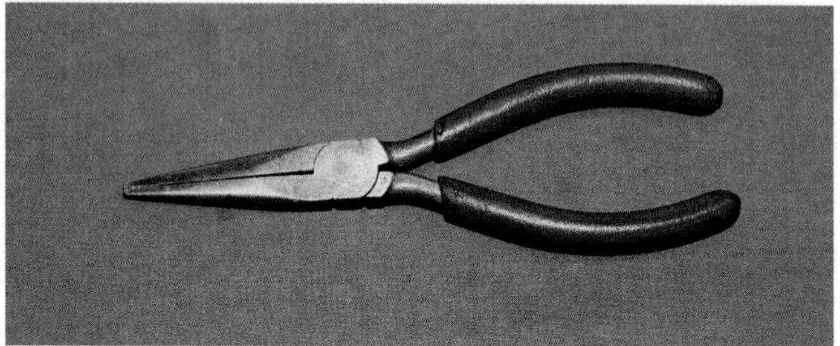

FIGURE 6-43 Needle-nose pliers.

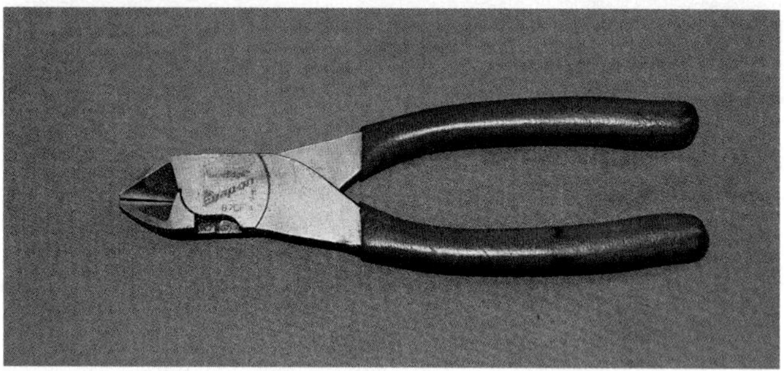

FIGURE 6-44 Diagonal side cutters.

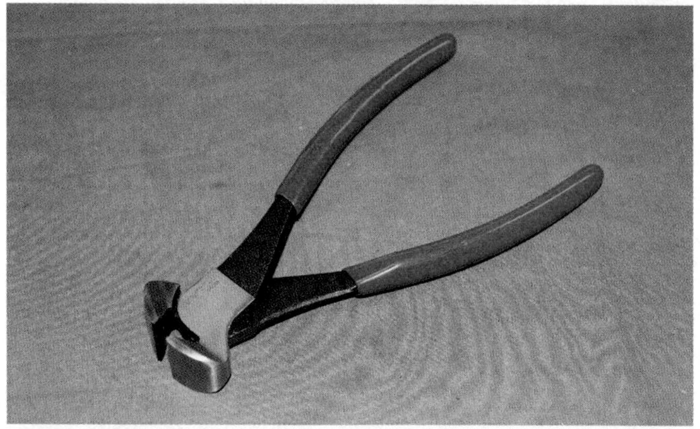

FIGURE 6-45 Nippers, or end cutting, pliers.

FIGURE 6-42 Arc joint pliers.

FIGURE 6-46 Snap ring pliers and snap ring.

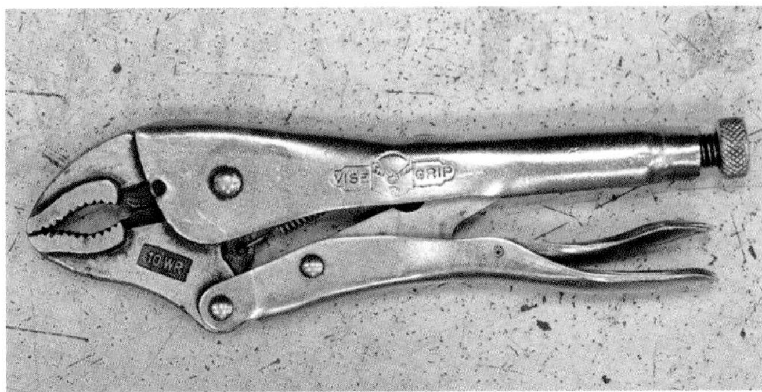

FIGURE 6-47 Locking pliers.

its internal groove. If external, then external snap ring pliers are used to expand the snap ring so it can remove and install the snap ring in its external groove. Always wear safety glasses when working with snap rings, as the snap rings can easily slip off the snap ring pliers and fly off at tremendous speeds, possibly causing severe eye injuries.

Locking pliers, also called vice grips, are general-purpose pliers used to clamp and hold one or more objects (**FIGURE 6-47**). Locking pliers are helpful by freeing up one or more of your hands when you are working, because they can clamp something and lock themselves in place to hold it. They are also adjustable, so they can be used for a variety of tasks. To clamp an object with locking pliers, put the object between the jaws, turn the screw until the handles are almost closed, then squeeze them together to lock them shut. You can increase or decrease the gripping force with the adjustment screw. To release them, squeeze the release lever, and they should open right up.

SAFETY TIP

When applying pressure to pliers, make sure your hands are not greasy, or they might slip. Select the right type and size of pliers for the job. As with most tools, if you have to exert almost all your strength to get something done, then you are using either the wrong tool or the wrong technique. If the pliers slip, you will get hurt. At the very least, you will damage the tool and what you are working on. Pliers get a lot of hard use in the shop, so they do get worn and damaged. If they are worn or damaged, they will be inefficient and can be dangerous. Always check the condition of all shop tools on a regular basis, and replace any that are worn or damaged.

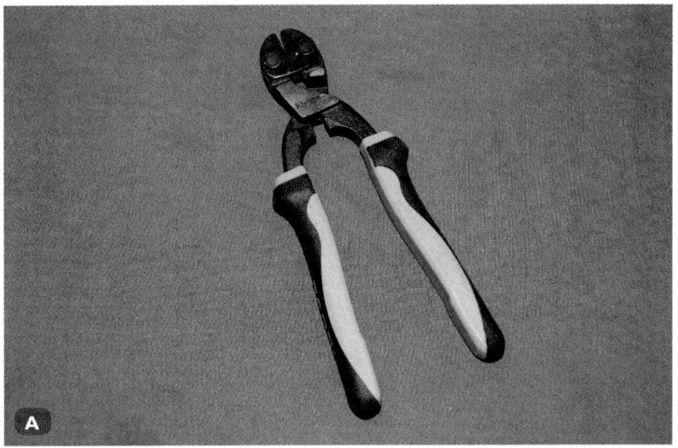

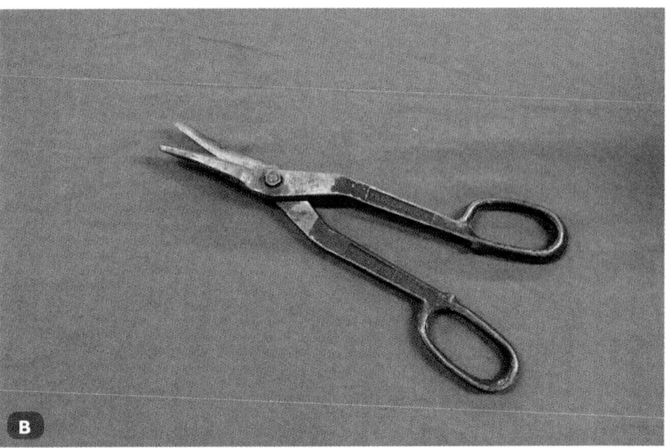

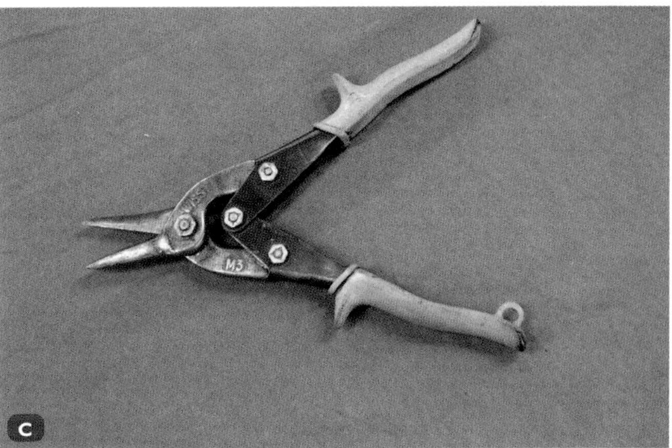

FIGURE 6-48 **A.** Bolt cutters. **B.** Tin snips. **C.** Aviation snips.

Cutting Tools

K06009

Bolt cutters cut heavy wire, non-hardened rods, and bolts (**FIGURE 6-48**). Their compound joints and long handles give the leverage and cutting pressure that is needed for heavy gauge materials. **Tin snips** are the nearest thing in the toolbox to a pair of scissors. They can cut thin sheet metal, and lighter versions make it easy to follow the outline of gaskets. Most snips come with straight blades, but if there is an unusual shape to cut, there is a pair with left- or right-handed curved blades. **Aviation snips** are designed to cut soft metals. They are easy to use because the handles are spring-loaded in the open position and double pivoted for extra leverage.

Allen Wrenches

K06010

Allen wrenches, sometimes called Allen or hex keys, are tools designed to tighten and loosen fasteners with Allen heads (**FIGURE 6-49**). The Allen head fastener has an internal hexagonal recess that the Allen wrench fits in snugly. Allen wrenches come in sets, and there is a correct wrench size for every Allen head. They give the best grip on a screw or bolt of all the

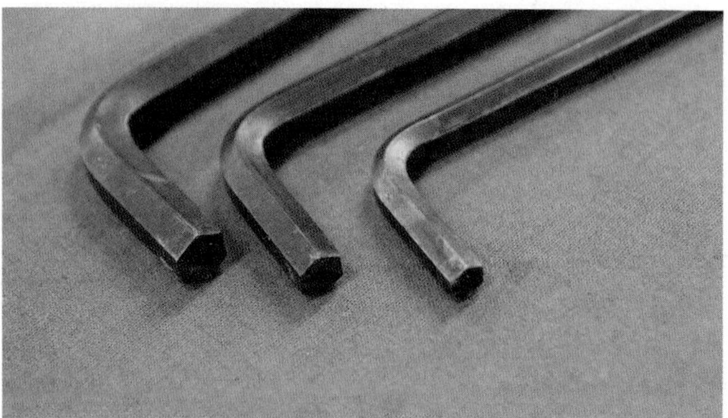

FIGURE 6-49 Typical Allen wrench head.

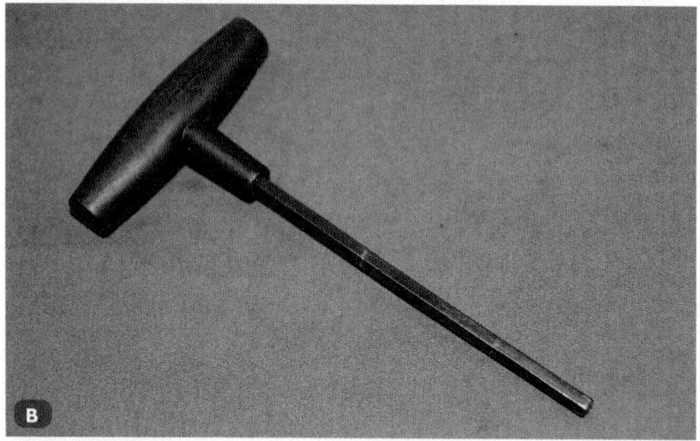

FIGURE 6-50 A. Allen socket. **B.** T-handle Allen wrench.

drivers, and their shape makes them good at getting into tight spots. Care must be taken to make sure the correct size of Allen wrench is used, or else the wrench and/or socket head will be rounded off. The traditional Allen wrench is a hexagonal bar with a right-angle bend at one end. They are made in various metric and standard sizes. As their popularity has increased, so too has the number of tool variations. Now Allen sockets are available, as are T-handle Allen keys (**FIGURE 6-50**).

Screwdrivers

K06011

The correct screwdriver to use depends on the type of slot or recess in the head of the screw or bolt, and how accessible it is (**FIGURE 6-51**). Most screwdrivers cannot grip as securely as wrenches, so it is very important to match the tip of the screwdriver exactly with the slot or recess in the head of a fastener. Otherwise, the tool might slip, damaging the fastener or the tool and possibly injuring you. When using a screwdriver, always check where the screwdriver blade can end up if it slips off the head of the screw. Many technicians who have not taken this precaution have stabbed a screwdriver into or through their hand, which can be painful as well as become infected or damage nerves.

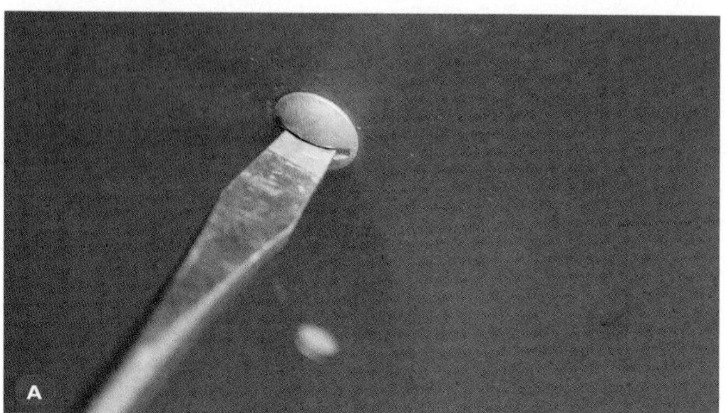

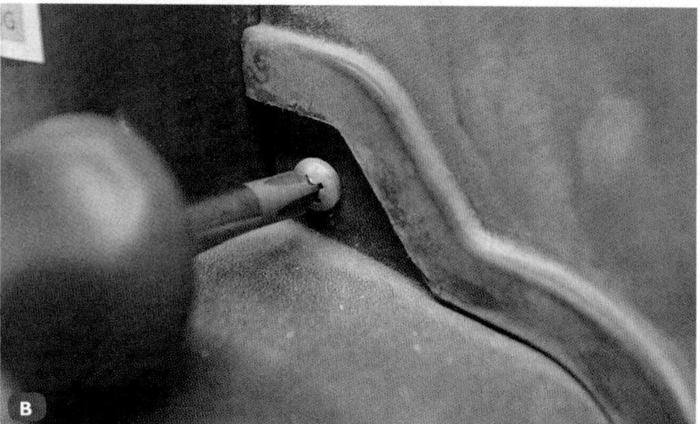

FIGURE 6-51 A. Slotted screw and screw driver. **B.** Phillips screw and screw driver.

The most common screwdriver has a flat tip, or blade, which gives it the name **flat blade screwdriver**. The blade should be almost as wide and thick as the slot in the fastener so that twisting force applied to the screwdriver is transferred right out to the edges of the head, where it has most effect. The blade should be a snug fit in the slot of the screw head. Then the twisting force is applied evenly along the sides of the slot. This guards against the screwdriver suddenly chewing a piece out of the slot and slipping just when the most force is being exerted. Flat blade screwdrivers come in a variety of sizes and lengths, so find the right one for the job. If viewed from the side, the blade should taper slightly to the very end where the flat tip fits into the slot. If the tip of the blade is not clean and square, it should be reshaped or replaced.

When you use a flat blade screwdriver, support the shaft with your free hand as you turn it (but keep it behind the tip).

FIGURE 6-52 **A.** Flat blade screwdrivers. **B.** Phillips screwdriver. **C.** Pozidriv screwdriver.

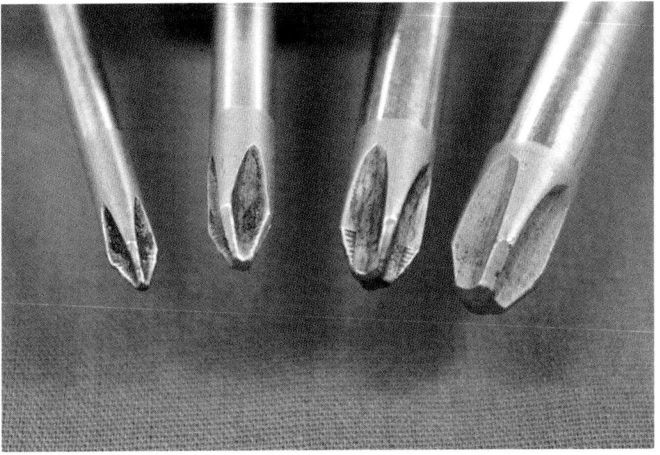

FIGURE 6-53 Four sizes of Phillips screwdrivers.

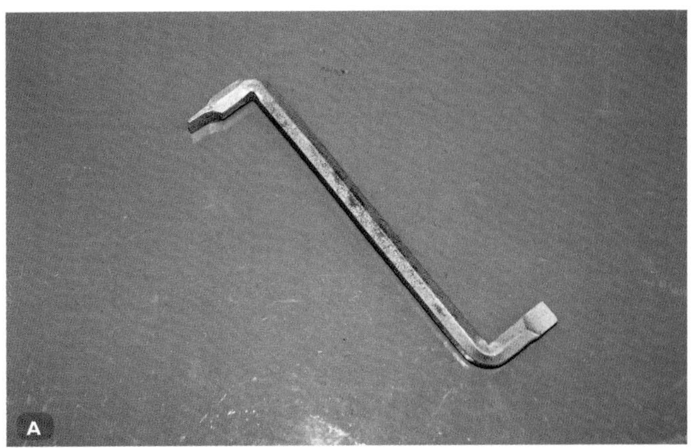

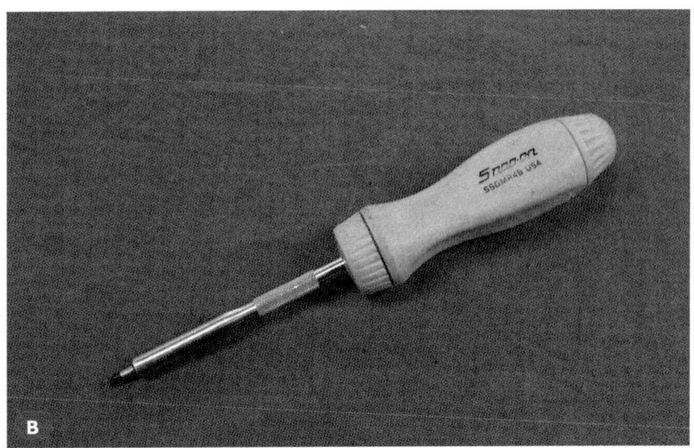

FIGURE 6-54 **A.** Offset screwdriver. **B.** Ratcheting screwdriver. **C.** Impact driver.

This helps keep the blade square in the slot and centered. Screwdrivers that slip are a common source of damage and injury in shops. A screw or bolt with a cross-shaped recess requires a **Phillips head screwdriver** or a Pozidriv screwdriver (**FIGURE 6-52**). The cross-shaped slot holds the tip of the screwdriver securely on the head. The Phillips tip fits a tapered recess, whereas the Pozidriv fits into slots with parallel sides in the head of the screw. Both a Phillips and a Pozidriv screwdriver are less likely to slip sideways, because the point is centered in the screw, but again the screwdriver must be the right size. The fitting process is simplified with these two types of screwdrivers because four sizes are enough to fit almost all fasteners with this sort of screw head (**FIGURE 6-53**).

The **offset screwdriver** fits into spaces where a straight screwdriver cannot and is useful where there is not much room to turn it (**FIGURE 6-54**). The two tips look identical, but one is set at 90 degrees to the other. This is because sometimes there is only room to make a quarter turn of the driver. Thus the driver has two blades on opposite ends so that offset ends of the screwdriver can be used alternately.

The **ratcheting screwdriver** is a popular screwdriver handle that usually comes with a selection of removable flat and Phillips tips. It has a ratchet inside the handle that turns the blade in only one direction, depending on how the slider is set. When set for loosening, a screw can be undone without removing the tip of the blade from the head of the screw. When set for tightening, a screw can be inserted just as easily.

An **impact driver** is used when a screw or a bolt is rusted/corroded in place or over-tightened, and needs a tool that can apply more force than the other members of this family. Screw slots can easily be stripped with the use of a standard screwdriver. The force of the hammer forcing the bit into the screw while turning it makes it more likely the screw will break loose. The impact driver accepts a variety of special impact tips. Choose the right one for the screw head, fit the tip in place, and then tension it in the direction it has to turn. A sharp blow with the hammer breaks the screw free, and then it can usually be unscrewed normally.

Magnetic Pickup Tools and Mechanical Fingers

K06012

Magnetic pickup tools and **mechanical fingers** are very useful for grabbing items in tight spaces (**FIGURE 6-55**). A magnetic pickup tool typically is a telescoping stick that has a magnet attached to the end on a swivel joint. The magnet is strong enough to pick up screws, bolts, sockets, and other ferrous (containing iron, making it magnetic) metals. For example, if a screw is dropped into a tight crevice where your fingers cannot reach, a magnetic pickup tool can be used to extract it.

Mechanical fingers are also designed to extract or insert objects in tight spaces. Because they actually grab the object, they can pick up nonmagnetic objects, which makes them handy for picking up rubber or plastic parts. They use a flexible body and come in different lengths, but typically are about 12–18" (305–457 mm) long. They have expanding grappling fingers on one end to grab items, and the other end has a push mechanism to expand the fingers and a retracting spring to contract the fingers.

TECHNICIAN TIP

It may be challenging getting the magnet down inside some areas because the magnet wants to keep sticking to other objects. One trick in this situation is to roll up a piece of paper so that a tube is created. Stick that down into the area of the dropped part, then slide the magnet down the tube, which helps it get past magnetic objects. Once the magnet is down, you may want to remove the roll of paper. Just remember two things: First, patience is important when using this tool, and second, don't drop anything in the first place!

Hammers

K06013

Hammers are a vital part of the shop tool collection, and a variety are commonly used (**FIGURE 6-56**). The most common hammer in an automotive shop is the **ball-peen (engineer's) hammer**. Like most hammers, its head is hardened steel. A punch or a chisel can be driven with the flat face. Its name comes from the ball peen or rounded face. This end is usually used for flattening or **peening** a rivet. The hammer should always match the size of the job, and it is usually better to use one that is too big than too small.

Hitting chisels with a **steel hammer** is fine, but sometimes you only need to tap a component to position it. A steel hammer might mark or damage the part, especially if it is made of a softer metal, such as aluminum. In such cases, a soft-faced hammer should normally be used for the job. Soft-faced hammers range from very soft with rubber or plastic heads to slightly harder with brass or copper.

When a large chisel needs a really strong blow, it is time to use a **sledgehammer**. The sledgehammer is like a small mallet, with two square faces made of high carbon steel. It is the heaviest type of hammer that can be used one-handed. The sledgehammer is used in conjunction with a chisel to cut off a bolt where corrosion has made it impossible to remove the nut.

A **dead blow hammer** is designed not to bounce back when it hits something. A rebounding hammer can be dangerous or destructive. A dead blow hammer can be made with a lead head or, more commonly, a hollow polyurethane head filled with lead shot or sand. The head absorbs the blow when the hammer makes contact, reducing any bounce-back or rebounding. This hammer can be used when working on the vehicle chassis or when dislodging stuck parts.

A **hard rubber mallet** is a special-purpose tool and has a head made of hard rubber. It is often used for moving things into place where it is important not to damage the item being moved. For example, it can be used to install a hubcap or to break a gasket seal on an aluminum housing.

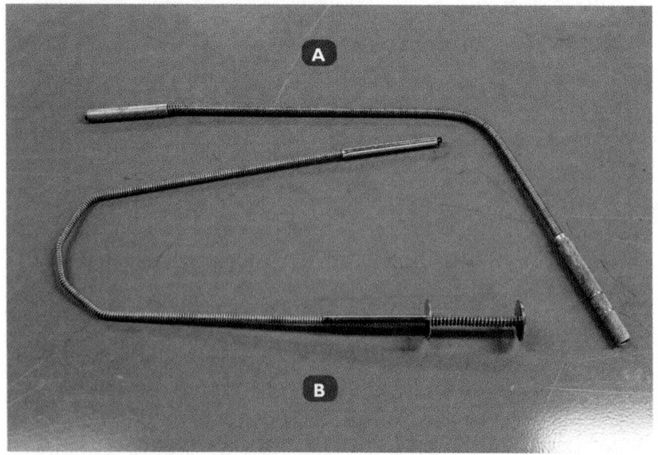

FIGURE 6-55 A. Magnetic pickup tools. **B.** Mechanical fingers.

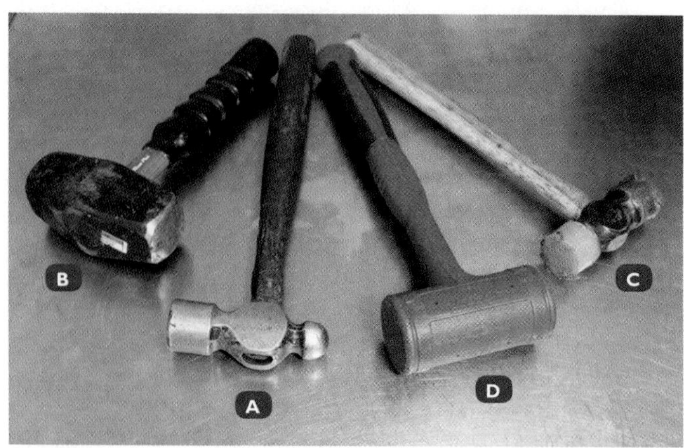

FIGURE 6-56 A. Ball peen hammer. **B.** Sledge hammer. **C.** Soft-faced hammer. **D.** Dead blow hammer.

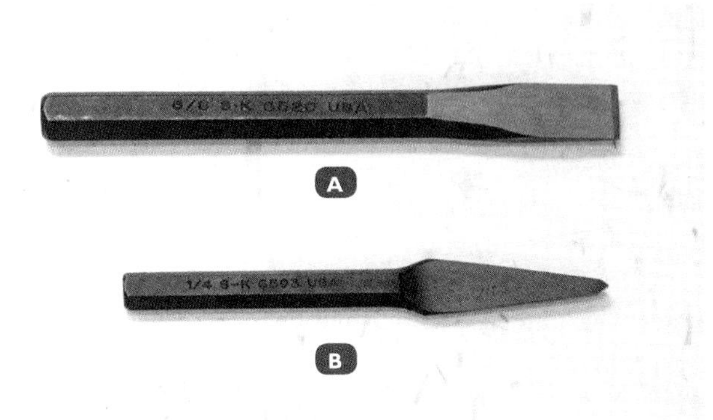

FIGURE 6-57 **A.** Cold chisel. **B.** Cross-cut chisel.

SAFETY TIP

The hammer you use depends on the part you are striking. Hammers with a metal face should almost always be harder than the part you are hammering. Never strike two hardened tools together, as this can cause the hardened parts to shatter.

SAFETY TIP

When using hammers and chisels, always wear safety glasses. Also be aware that safety glasses by themselves need to be form fitting and must not be allowed to slide down your nose, for them to be effective. It may be prudent to wear either a face shield or safety goggles for extra protection.

Chisels

K06014

The most common kind of chisel is a **cold chisel** (**FIGURE 6-57**). It gets its name from the fact it is used to cut cold metals rather than heated metals. It has a flat blade made of high-quality steel and a cutting angle of approximately 70 degrees. The cutting end is tempered and hardened because it has to be harder than the metals it is cutting. The head of the chisel needs to be softer so it will not chip when it is hit with a hammer. Technicians sometimes use a cold chisel to remove bolts whose heads have rounded off.

A variation of the cold chisel is a spring-loaded cold chisel (**FIGURE 6-58**). This chisel works really well in tight spaces where a hammer can't be swung. The chisel is made up of three parts: the chisel, the weighted hammerhead, and a spring in tension, holding the other two components together. It is operated by holding the chisel end against the part you are working on, pulling back on the hammerhead, and allowing the spring to rapidly slam the hammerhead into the end of the chisel. The force of the hammerhead hitting the end of the chisel transfers a lot of energy to the chisel. This type of arrangement is also used on some spring-loaded center punches.

A **cross-cut chisel** is so named because the sharpened edge is across the blade width. This chisel narrows down along the stock, so it is good for getting in grooves. It is used for cleaning out or even making key ways. The flying chips of metal should always be directed away from the user.

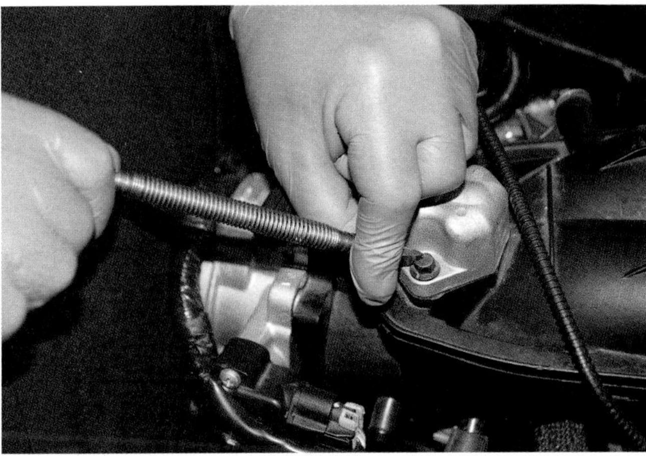

FIGURE 6-58 Spring-loaded chisels.

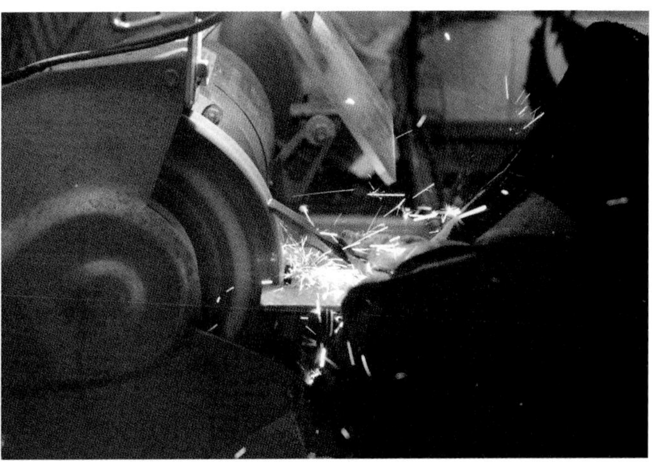

FIGURE 6-59 Dressing a chisel.

SAFETY TIP

Chisels and punches are designed with a softer striking end than hammers. Over time, this softer metal "mushrooms," and small fragments are prone to breaking off when hammered. These fragments can cause eye or other penetrative injuries to people in the area. Always inspect chisels and punches for mushrooming, and dress them on a grinder when necessary (**FIGURE 6-59**).

Punches

K06015

Punches are used when the head of the hammer is too large to strike the object being hit without causing damage to adjacent parts. A punch transmits the hammer's striking power from the soft upper end down to the tip that is made of hardened high-carbon steel. A punch transmits an accurate blow from the hammer at exactly one point, something that cannot be guaranteed using a hammer on its own.

Four of the most common punches are the prick punch, center punch, drift punch, and pin punch (**FIGURE 6-60**). When marks need to be drawn on an object like a steel plate, to help

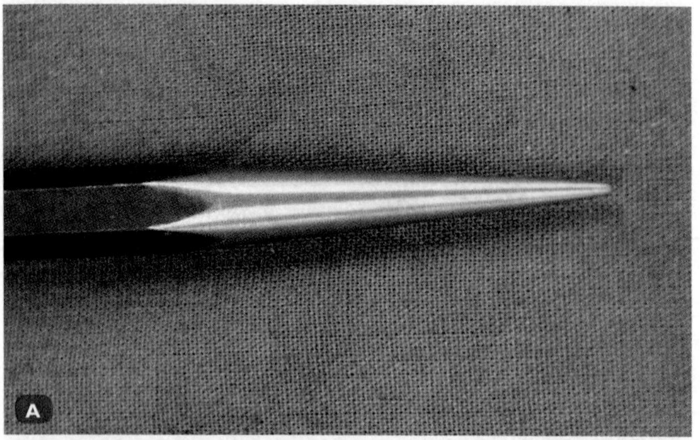
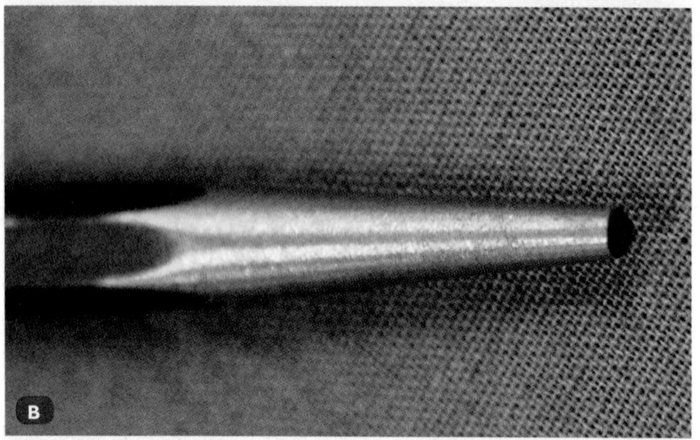
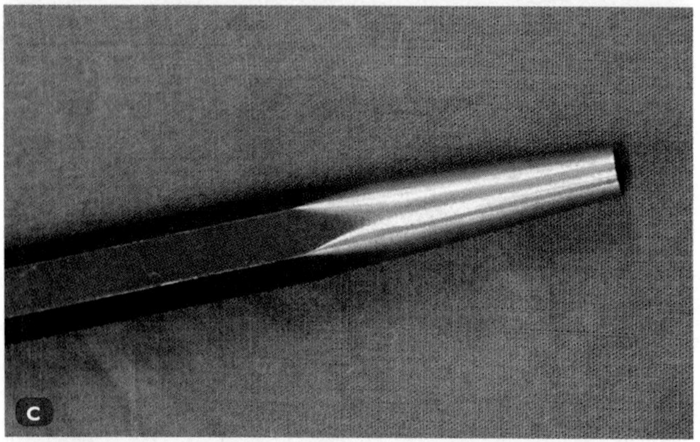
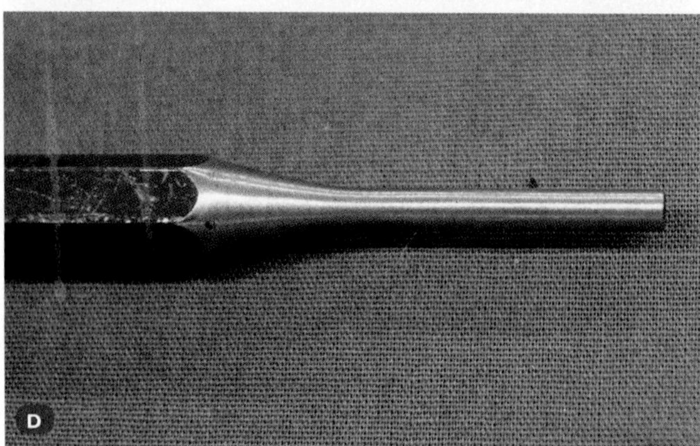

FIGURE 6-60 A. Prick punch. **B.** Center punch. **C.** Drift punch. **D.** Pin punch.

locate a hole to be drilled, a **prick punch** can be used to mark the points so they will not rub off. They can also be used to scribe intersecting lines between given points. The prick punch's point is very sharp, so a gentle tap leaves a clear indentation. The **center punch** is not as sharp as a prick punch and is usually bigger. It makes a bigger indentation that centers a drill bit at the point where a hole is required to be drilled.

Although most center punches are used with a hammer, some center punches operate automatically when the punch is pressed tightly up against the part you are punching. This type of punch has a spring and weighted hammer inside of the back end of the center punch. It is machined in such a way that pushing the center punch against a surface compresses a spring behind a movable weight (**FIGURE 6-61**). When pushed far enough, the weight is released, and the spring forces it against the center rod in the punch, causing it to indent the work surface.

A **drift punch** is also named a starter punch because you should always use it first to get a pin moving. It has a tapered shank, and the tip is slightly hollow so it does not spread the end of a pin and make it an even tighter fit. Once the starter drift has gotten the pin moving, a suitable pin punch will drive the pin out or in. A drift punch also works well for aligning holes on two mating objects, such as a valve cover and cylinder head. Forcing the drift punch in the hole aligns both components for easier installation of the remaining bolts. **Pin punches** are available in various diameters. A pin punch has a long slender shaft that has straight sides. It is used to drive out pins or rivets (**FIGURE 6-62**). A lot of components are either held together or accurately located by pins. Pins can be pretty tight, and a group of pin punches is specially designed to deal with them.

Special punches with hollow ends are called **wad punches** or **hollow punches** (**FIGURE 6-63**). They are the most efficient tool to make a hole in soft sheet material like shim steel, plastic, and leather, or, most commonly, in a gasket. When being used, there should always be a soft surface under the work, ideally the end grain of a wooden block. If a hollow punch loses its sharpness or has nicks around its edge, it will make a mess instead of a hole.

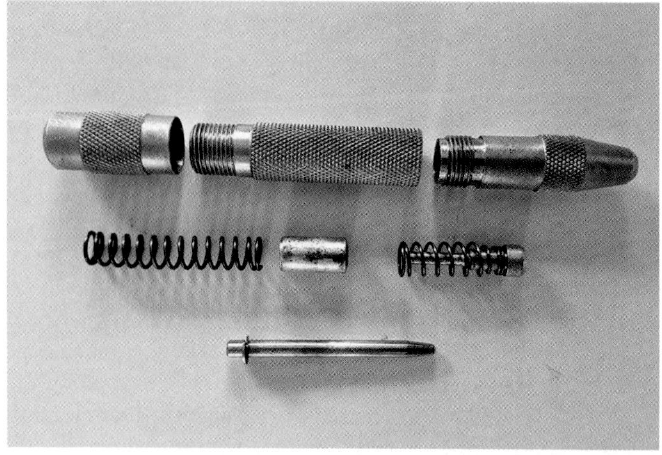

FIGURE 6-61 Internal workings of an automatic center punch.

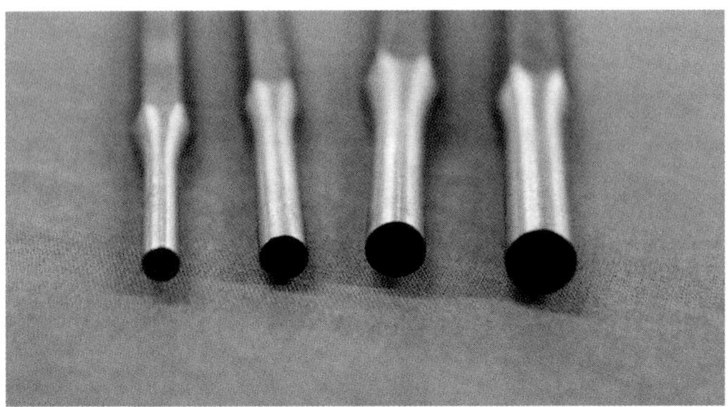

FIGURE 6-62 Various pin punches.

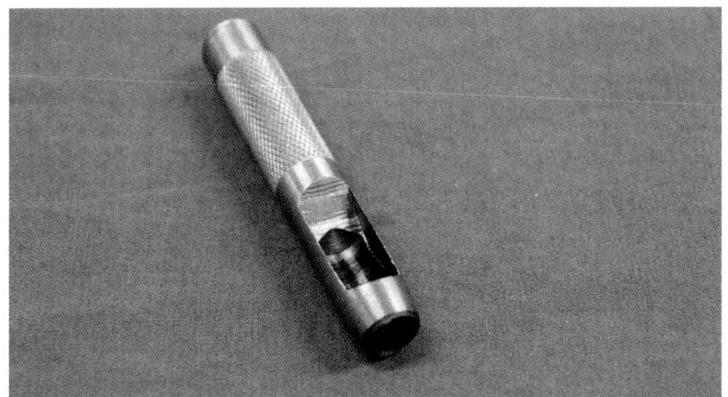

FIGURE 6-63 Wad punch.

FIGURE 6-64 Number and letter punches.

Numbers and letters, like the engine numbers on some cylinder blocks, are usually made with number and letter punches that come in boxed sets (**FIGURE 6-64**). The rules for using a number or letter punch set are the same as for all punches. The punch must be square with the surface being worked on, not on an angle, and the hammer must hit the top squarely.

Pry Bars

K06016

Pry bars are tools constructed of strong metal that are used as a lever to move, adjust, or pry (**FIGURE 6-65**). Pry bars come in a variety of shapes and sizes. Many have a tapered end that is slightly bent, with a plastic handle on the other end. This design works well for applying force to tension belts or for moving parts into alignment. Another type of pry bar is the **roll bar**. One end is sharply curved and tapered, which is used for prying. The other end is tapered to a dull point and is used to align larger holes such as transmission bell housings or engine motor mounts. Because pry bars are made of hardened steel, care should be taken when using them on softer materials, to avoid any damage.

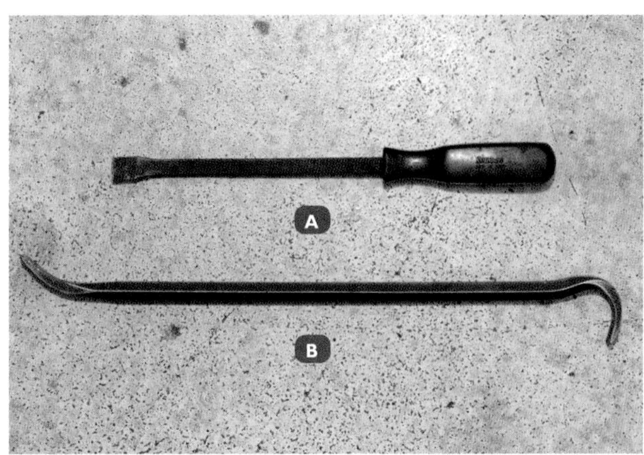

FIGURE 6-65 **A.** Pry bar. **B.** Roll bar.

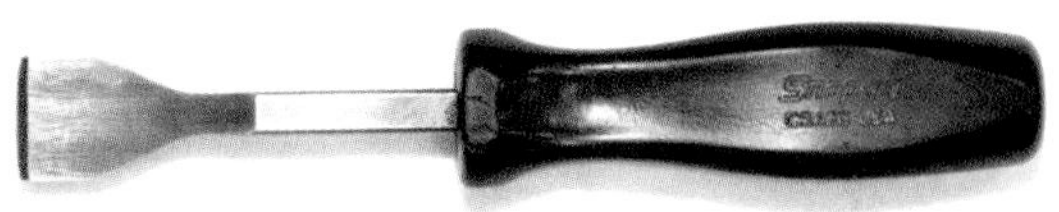

FIGURE 6-66 A gasket scraper.

Gasket Scrapers

K06017

A **gasket scraper** has a hardened, sharpened blade. It is designed to remove a gasket without damaging the sealing face of the component when used properly (**FIGURE 6-66**). On one end, it has a comfortable handle like a screwdriver handle; on the other end, a blade is fitted with a sharp edge to assist in the removal of gaskets. The gasket scraper should be kept sharp and straight to make it easy to remove all traces of the old gasket and sealing compound. The blades come in different sizes, with a typical size being 1" (25 mm) wide. Whenever you use a gasket scraper, be very careful not to nick or damage the surface being cleaned.

▶ TECHNICIAN TIP

Many engine components are made of aluminum. Because aluminum is quite soft, it is critical that you use the gasket scraper very carefully so as not to damage the surface. This can be accomplished by keeping the gasket scraper at a fairly flat angle to the surface. Also, the gasket scraper should only be used by hand, not with a hammer. Some manufacturers specify using plastic gasket scrapers only on certain aluminum components such as cylinder heads and blocks.

Files

K06018

Files are hand tools designed to remove small amounts of material from the surface of a workpiece. Files come in a variety of shapes, sizes, and coarseness depending on the material being worked and the size of the job. Files have a pointed tang on one end that is fitted to a handle. Files are often sold without handles, but they should not be used until a handle of the right size has been fitted. A correctly sized handle fits snugly without working loose when the file is being used. Always check the handle before using the file. If the handle is loose, give it a sharp rap to tighten it up, or if it is the threaded type, screw it on tighter. If it fails to fit snugly, you must use a different-size handle.

What makes one file different from another is not just the shape but how much material it is designed to remove with each stroke. The teeth on the file determine how much material will be removed (**FIGURE 6-67**). Because the teeth face one direction only, the file cuts in one direction only. Dragging the file backward over the surface of the metal only dulls the teeth and wears them out quickly. Teeth on a coarse-grade file are larger, with a greater space between them. A coarse-grade file working on a piece of mild steel removes a lot of material with each stroke, but it leaves a rough finish. A smooth-grade file has smaller teeth cut more closely together. It removes much less material on each stroke, but the finish is much smoother. On many jobs, the coarse file is used first to remove material quickly, and then a smoother file gently removes the remaining material and leaves a smoother finish to the work. The full list of grades in flat files, from rough to smooth, follows:

- **Rough files** have the coarsest teeth, with approximately 20 teeth per inch. They are used when a lot of material must be removed quickly. They leave a very rough finish and have to be followed by the use of finer files to produce a smooth final finish.
- **Coarse bastard files** are still a coarse file, with approximately 30 teeth per inch, but they are not as course as the rough file. They are also used to rough out or remove material quickly from the job.
- **Second-cut files** have approximately 40 teeth per inch and provide a smoother finish than the rough or coarse bastard file. They are good all-round intermediary files and leave a reasonably smooth finish.
- **Smooth files** have approximately 60 teeth per inch and are a finishing file used to provide a smooth final finish.
- **Dead smooth files** have 100 teeth per inch or more and are used where a very fine finish is required.

Some flat files are available with one smooth edge (no teeth), called safe edge files. They allow filing up to an edge without damaging it. Flat files work well on straightforward jobs, but some jobs require special files. A **warding file** is thinner than other files and comes to a point; it is used for working in narrow slots (**FIGURE 6-68**). A **square file** has teeth on all four sides, so you can use it in a square or rectangular hole. A square file can make the right shape for a squared metal key to fit in a slot.

SAFETY TIP

Hands should always be kept away from the surface of the file and the metal that is being worked on. It is easy for skin to get caught between the edge of the file and the piece you are working. This creates a scissor-like condition that can result in severe cuts. Also, filing can produce small slivers of metal that penetrate your skin and can be difficult to remove from a finger or hand.

A **triangular file** has three sides. It is triangular, so it can get into internal corners easily. It is able to cut right into a corner without removing material from the sides. **Curved files** are typically either half-round or round. A half-round file has a shallow convex surface that can file in a concave hollow or in an acute internal corner (**FIGURE 6-69**). The fully round file, sometimes called a rat-tail file, can make holes bigger. It can also file inside a concave surface with a tight radius.

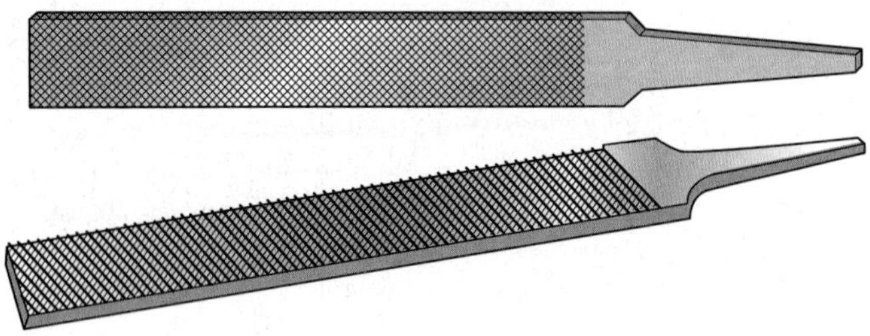

Description:
The teeth on a file.

FIGURE 6-67 The teeth on a file determine how much material will be removed from the object being filed.

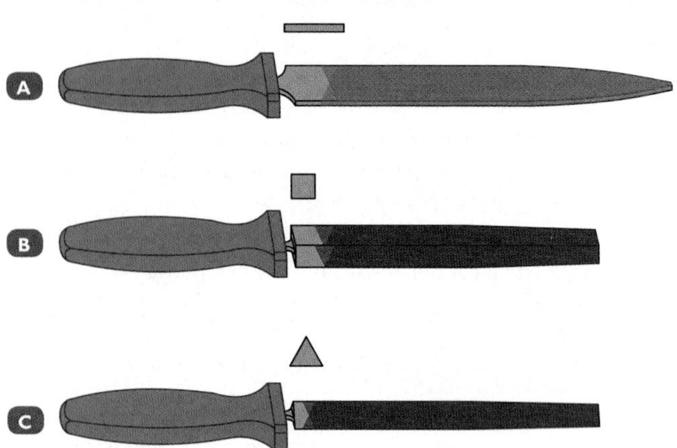

FIGURE 6-68 A. Warding file. **B.** Square file. **C.** Triangular file.

The **thread file** cleans clogged or distorted threads on bolts and studs. Thread files come in either standard or metric configurations, so make sure you use the correct file. Each file has eight different surfaces that match different thread dimensions, so the correct face must be used (**FIGURE 6-70**).

Files should be cleaned after each use. If they are clogged, they can be cleaned by using a file card, or file brush (**FIGURE 6-71**). This tool has short steel bristles that clean out the small particles that clog the teeth of the file. Rubbing a piece of chalk over the surface of the file prior to filing makes it easier to clean.

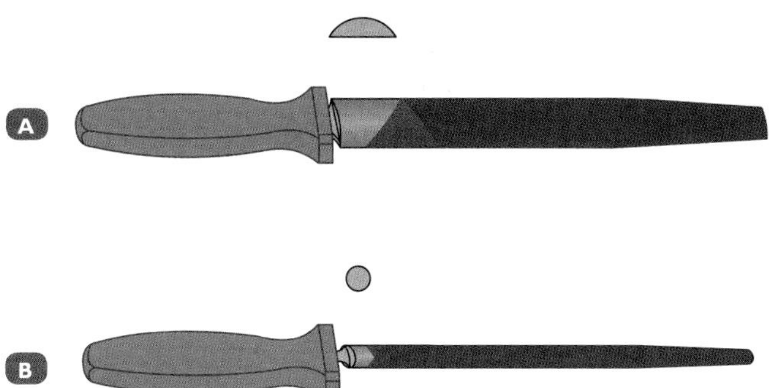

FIGURE 6-69 **A.** Half-round file. **B.** Round, or rat-tail, file.

Clamps

K06019

There are many types of vices or clamps available (**FIGURE 6-72**). The **bench vice** is a useful tool for holding anything that can fit into its jaws. Some common uses include sawing, filing, or chiseling. The jaws are serrated to give extra grip. They are also very hard, which means that when the vice is tightened, the jaws can mar whatever they are gripping. To prevent this, a pair of soft jaws can be fitted whenever the danger of damage arises. They are usually made of aluminum or some other soft metal or can have a rubber-type surface applied to them.

When materials are too awkward to grip vertically in a plain vice, it may be easier to use an **offset vice**. The offset vice has its jaws set to one side to allow long components to be held vertically (**FIGURE 6-73**). For example, a long threaded bar can be held vertically in an offset vice to cut a thread with a die.

A **drill vice** is designed to hold material on a drill worktable (**FIGURE 6-74**). The drill worktable has slots cut into it to allow the vice to be bolted down on the table, to hold material

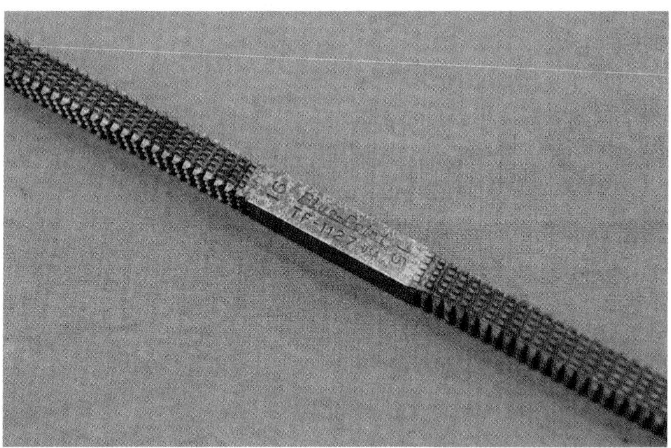

FIGURE 6-70 Thread file.

FIGURE 6-72 Bench vice.

FIGURE 6-71 File card.

FIGURE 6-73 Offset vice being used to hold a pipe.

FIGURE 6-74 Drill vice on a drill worktable.

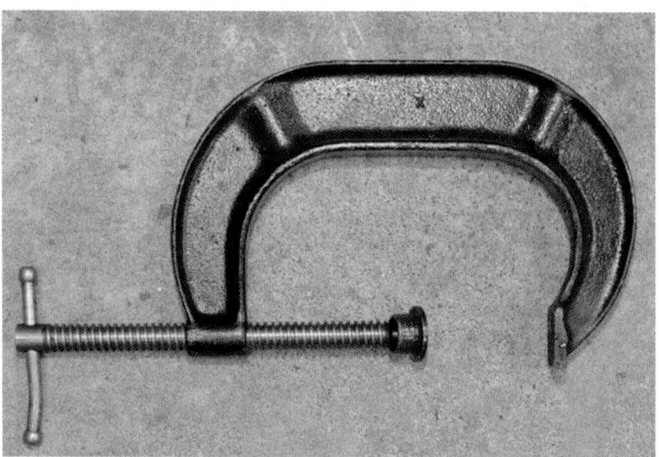

FIGURE 6-75 C-Clamp.

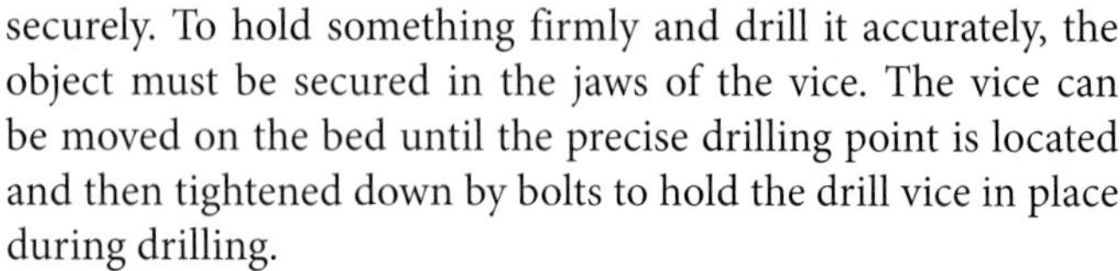

securely. To hold something firmly and drill it accurately, the object must be secured in the jaws of the vice. The vice can be moved on the bed until the precise drilling point is located and then tightened down by bolts to hold the drill vice in place during drilling.

The name for the **C-clamp** comes from its shape (**FIGURE 6-75**). It can hold parts together while they are being assembled, drilled, or welded. It can reach around awkwardly shaped pieces that do not fit in a vice. It is also commonly used to retract disc brake caliper pistons. This clamp is portable, so it can be taken to the work.

Taps and Dies

K06020

Taps and dies are used to form threads in metal so that they can be fastened together (**FIGURE 6-76**). The tap cuts female threads in a component so a fastener can be screwed into it. The die is used to cut male threads on a bolt so that it can be screwed into the female threads created by the tap. The tap and die are companion tools that create matching threads so that they can both be used to fasten things together.

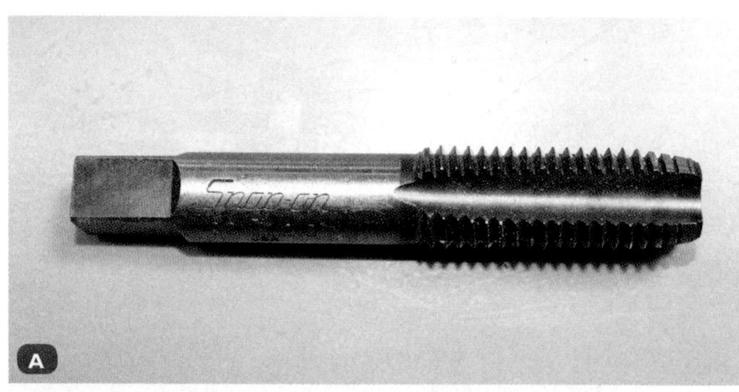

FIGURE 6-76 A. Tap. **B.** Die.

Taps

Various types of taps are designed to be used based on what you want to do and the material you are working with. The most common taps are the taper tap, intermediate tap (also called a plug tap), bottoming tap (also called a flat-bottomed tap), and the thread chaser (**FIGURE 6-77**). These are explored in more depth below.

Taper Tap A **taper tap** narrows at the tip, which makes it easier to start straight when cutting threads in a new hole. It also makes it less likely to break the tip off the tap because it removes metal in a less aggressive manner. Taps are very hard, which gives them good wear resistance but also makes them very brittle; they can break off easily. If a tap breaks in the hole, it can be very hard to remove. This makes taper taps good taps to use when starting a hole. Plus, they are easier to get started straight in the hole.

Intermediate Tap The second type of tap is an **intermediate tap**, also known as a plug tap. It is more aggressive than a taper tap, but not as aggressive as a bottoming tap. This is the most common tap used by technicians. Although it is a bit more aggressive than the taper tap, it can be used as a starter tap for a new hole.

Bottoming Tap The **bottoming tap** is used when you need to cut threads to the very bottom of a blind hole and in holes that already have threads started that just need to be extended slightly. It has a flat bottom. and the threads are the same all the way to the end. This type of tap is virtually impossible to use in a new, unthreaded hole.

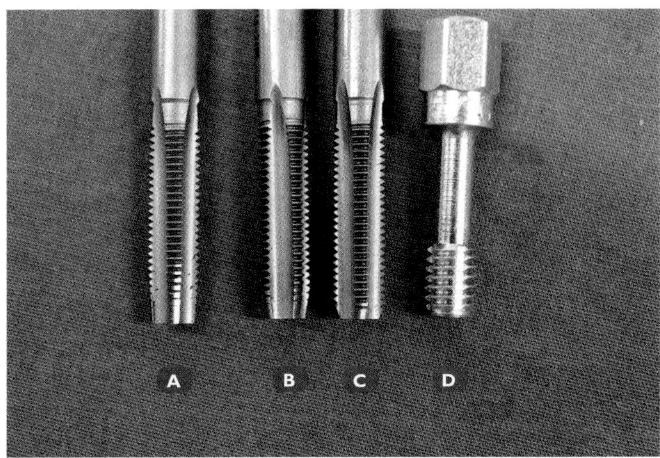

FIGURE 6-77 **A.** Taper tap. **B.** Intermediate tap. **C.** Bottoming tap. **D.** Thread chaser.

FIGURE 6-78 **A.** Die nut. **B.** Split die.

Thread Chaser **Thread chasers** are used to clean up the threads of a hole to make sure they are free from debris and dirt. They do not cut threads, but just clean up the existing threads so that the bolt doesn't encounter any excessive resistance.

Dies

A **die** is used to cut external threads on a metal shank or bolt. The threads in a die create the male companion to the female threads that have been cut into the material by a tap. Usually the dies come with a setscrew that allows the user to slightly adjust the size of the die so the threads can be cut to the right fit that matches the threaded hole. Loose-fitting threads strip more easily than they should and are not as secure. Tight-fitting threads increase the torque required to turn the bolt, thereby reducing the clamping force relative to bolt torque. So adjusting thread fit is typically accomplished by adjusting the die. Dies are hardened, which makes them very wear resistant, but very brittle. Die nuts do not have the split like the threading die does because they are used to clean up threads like the thread chasers (**FIGURE 6-78**).

FIGURE 6-79 Tap handle.

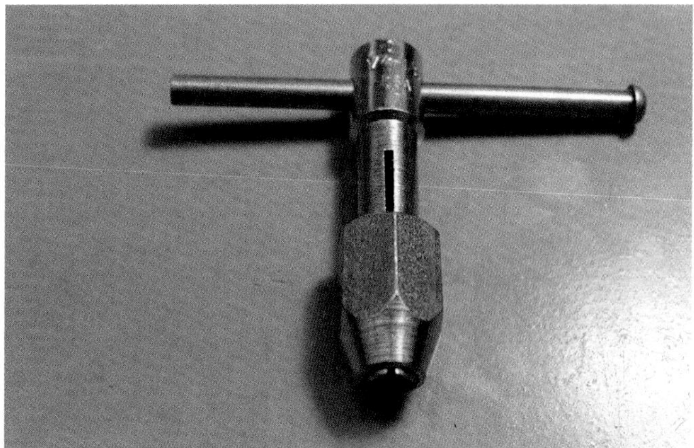

FIGURE 6-80 T-shaped tap handle.

Proper Use of Taps and Dies

The proper use of taps and dies is a major issue often overlooked in repairing a thread on a bolt or threads in a hole. Improper use of these two items causes the bolt not to thread into the hole properly, or it could even break the tool, which is something to avoid. Both of these tools are made of hardened steel, which makes them very wear resistant but also makes them very brittle. If they are used improperly they have a tendency to facture and break.

Because taps and dies both need to be rotated, special tools are used to turn them. A **tap handle** is used to turn taps. It has a right-angled jaw that matches the squared end of all the taps (**FIGURE 6-79**). The jaws are designed to hold the tap securely, and the handles provide the leverage for the operator to comfortably rotate the tap to cut the thread. To cut a thread in an awkward space, a T-shaped tap handle is very convenient (**FIGURE 6-80**). Its handle is not as long, so it fits into tighter spaces; however, it is harder to turn and to guide accurately.

FIGURE 6-81 Die stock.

Tap/Drill Recommendations

C CLEVELAND TWIST DRILL

Note: Drill size recommendations are for approximately 70-75% thread height. Drills produce a hole slightly larger than their nominal size. Size obtained will depend on drill style, machine, drilling conditions, fixturing and coating selected.

GTD GREENFIELD

MACHINE SCREW, FRACTIONAL AND METRIC SIZES

Inch Tap Size & Pitch	Metric Tap Size & Pitch	Cutting Taps Drill Size	Cutting Taps Dec. Equiv.	Forming Taps Drill Size	Forming Taps Dec. Equiv.
0-80			.0469	54	.0550
	M1.6 x 0.35	1.25	.0492	1.45	.0571
	M1.8 x 0.35	1.45	.0571	1.65	.0650
1-64		53	.0595	51	.0670
1-72		53	.0595	51	.0670
	M2 x 0.40	1.60	.0630	1.80	.0709
2-56		50	.0700	5/64	.0781
2-64		50	.0700	47	.0785
	M2.2 x 0.45	1.75	.0689	2.00	.0787
	M2.5 x 0.45	2.05	.0807	2.30	.0906
3-48		47	.0785	43	.0890
3-56		46	.0810	2.30	.0905
4-40		43	.0890	38	.1015
4-48		42	.0935	2.60	.1024
	M3 x 0.50	2.50	.0984	7/64	.1094
5-40		38	.1015	33	.1130
5-44		37	.1040	2.90	.1142
	M3.5 x 0.60	2.90	.1142	3.20	.1260
6-32		36	.1065	1/8	.1250
6-40		33	.1130	3.25	.1280
	M4 x 0.70	3.30	.1299	3.70	.1476

Inch Tap Size & Pitch	Metric Tap Size & Pitch	Cutting Taps Drill Size	Cutting Taps Dec. Equiv.	Forming Taps Drill Size	Forming Taps Dec. Equ
5/8-11			.5312	14.75	.5807
5/8-18			.5781	15.25	.6004
	M16 x 2.00	14.00	.5512	19/32	.593
	M16 x 1.50	14.50	.5209	15.25	.600
	M18 x 2.50	15.50	.6102	39/64	.609
	M18 x 1.50	16.50	.6496	17.25	.679
3/4-10			.6562	45/64	.703
3/4-16			.6875	23/32	.718
	M20 x 2.50	17.50	.6890		
	M20 x 1.50	18.50	.7283		
	M22 x 2.50	19.50	.7677		
	M22 x 1.50	20.50	.8071		
7/8-9			.7656		
7/8-14			.8125		
	M24 x 3.00	21.00	.8268		
	M24 x 2.00	22.00	.8661		
1-8			.8750		
1-12			.9219		
	M27 x 3.00	24.00	.9449		
	M27 x 2.00	25.00	.9843		
1-1/8-7			.9844		

FIGURE 6-82 Tap drill chart.

To cut a brand new thread on a blank rod or shaft, a die held in a **die stock** is used (**FIGURE 6-81**). The die fits into the octagonal recess in the die stock and is usually held in place by a thumb screw. The die may be split so that it can be adjusted more tightly onto the work with each pass of the die as the thread is cut deeper and deeper, until the external thread fits the threaded hole properly.

When tapping a hole, the diameter of the hole is determined by a tap drill chart, which can be obtained from engineering suppliers. This chart shows what hole size has to be drilled and what tap size is needed to cut the right thread for any given bolt size (**FIGURE 6-82**).

Just remember that if you are drilling a 1/4" (6 mm) or larger hole, use a smaller pilot drill first. Once the properly sized hole has been drilled, the taper tap or intermediate tap can be started in the hole. Make sure to use the proper lubricant for the metal you are tapping. Also be sure to start the tap straight. The best way to do that is to start the tap about one turn, stop, and then use a square to check the position of the tap in two places 90 degrees apart. If it isn't perfectly straight, you can usually straighten it while turning it another half turn. Again stop and verify that it is straight in two positions, 90 degrees apart. If the tap is straight, turn the tap about one full turn, and then back it off about a quarter turn. Continue cutting and backing off the tap until either the tap turns easily or you are at the bottom of the hole. Remove the tap. If you are cutting threads in a blind hole, you need to use a bottoming tap to finish the threads off. If not, clean the threaded hole and test-fit a bolt in the hole to check the threads. The bolt should turn smoothly by hand.

FIGURE 6-83 Screw extractors.

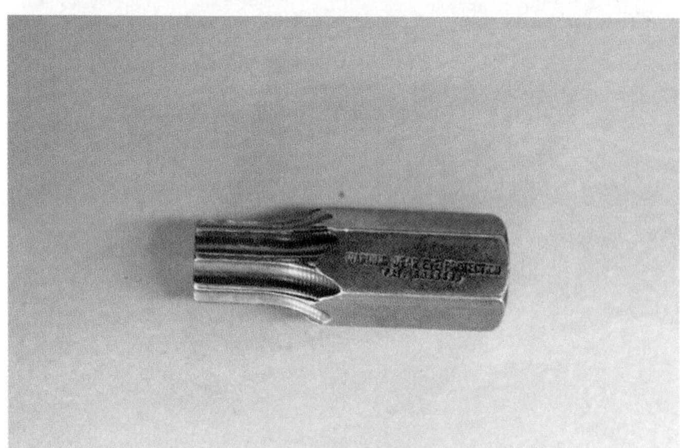

FIGURE 6-84 Straight-sided screw extractor.

Screw Extractors

K06021

Screw extractors are devices designed to remove screws, studs, or bolts that have broken off in threaded holes. A common type of extractor uses a coarse left-hand tapered thread formed on its hardened body (**FIGURE 6-83**). Normally, a hole is drilled in the center of the broken screw, and then the extractor is screwed into the hole. The left-hand thread grips the broken part of the bolt and unscrews it. The extractor is marked with two sizes: one showing the size range of screws it is designed to remove, and the other, the size of the hole that needs to be drilled. It is important to carefully drill the hole in the center of the bolt or stud in case you end up having to drill the bolt out. If you drill the hole off center, you will not be able to drill it out all the way to the inside diameter of the threads, as the hole is offset from center. This makes removal of the bolt much harder.

Some screw extractors use a hardened, tapered square shank that is hammered into a hole drilled into the center of the broken off bolt. The square edges of the screw extractor cut into the bolt and are used to grip it so it can be removed. Another type of screw extractor is the straight-sided, vertically splined, round shaft. The sides of the shaft have straight splines running the length of the extractor (**FIGURE 6-84**). It doesn't taper, so it is strong up

FIGURE 6-85 **A.** Puller. **B.** Gear puller.

and down its length. A straight hole of the correct diameter is drilled into the broken-off bolt. Then the extractor is driven into the newly drilled hole. The vertical splines, being larger than the hole, grab onto the bolt, allowing it to be unthreaded.

Pullers

K06022

Pullers are a very common universal tool that can be used for removing bearings, bushings, pulleys, and gears (**FIGURE 6-85**). Specialized pullers are also available for specific tasks where a standard puller is not as effective. The most common pullers have two or three legs that grip the part to be removed. A center bolt, called a forcing screw or jacking bolt, is then screwed in, producing a jacking or pulling action, which extracts the part. **Gear pullers** come in a range of sizes and shapes, all designed for particular applications. They consist of three main parts: jaws, a cross-arm, and a forcing screw. There are normally two or three jaws on a puller. They are designed to connect to the component either externally or internally.

The **forcing screw** is a long, fine-threaded bolt that is applied to the center of the cross-arm. When the forcing screw is turned, it applies a very large force to the component you are removing. The forcing screw typically has interchangeable feet (**FIGURE 6-86**). A tapered cone style of foot does a good job of centering the puller, but it also creates a very large wedging effect, which can distort the end of the shaft. A flat-style foot is very good for pushing against the end of a shaft, but doesn't center itself. In either situation, the wrong foot size can push on the internal threads of the shaft, damaging them. The cross-arm attaches

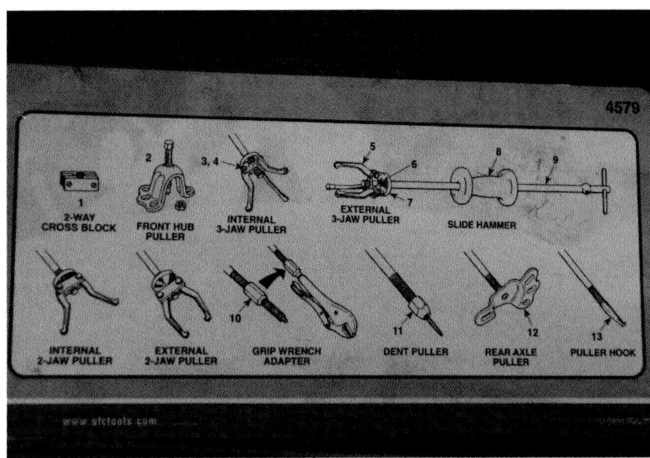

FIGURE 6-86 Interchangeable feet for the gear puller.

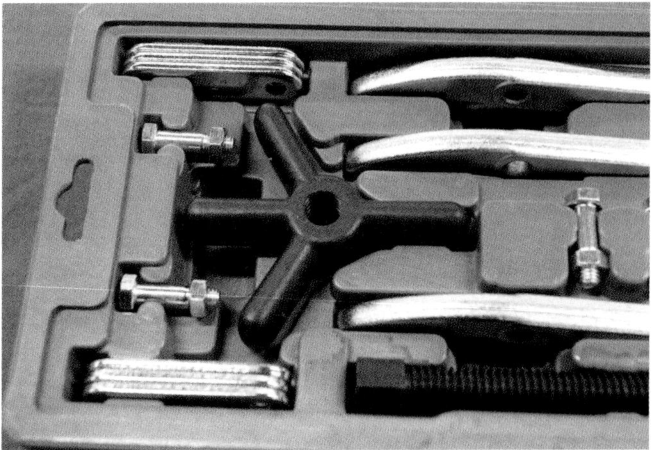

FIGURE 6-87 Four-arm puller.

the jaws to the forcing screw. If the **cross-arm** has four arms, three of the arms are spaced 120 degrees apart. The fourth arm is positioned 180 degrees apart from one arm (**FIGURE 6-87**). This allows the cross-arm to be used as either a two- or a three-arm puller.

Using Gear Pullers

Gear and bearing pullers are designed for hundreds of applications. Their main purpose is to remove a component such as a gear, pulley, or bearing from a shaft; or to remove a shaft from inside a hole. Normally these components have been pressed onto that shaft or into the hole, so removing them requires considerable force. To select, install, and use a gear puller to remove a pulley, follow the steps in **SKILL DRILL 6-3**.

SAFETY TIP

- Always wear eye protection when using a gear puller.
- Make sure the puller is located correctly on the workpiece. If the jaws cannot be fitted correctly on the part, then select a more appropriate puller. Do not use a puller that does not fit the job.

SKILL DRILL 6-3 Using Gear Pullers

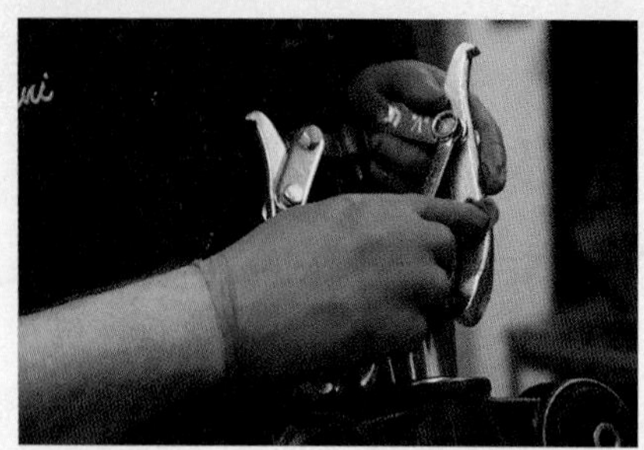

1. Examine the gear puller you have selected for the job. Identify the jaws; there may be two or three of them, and they must fit the part you want to remove. The cross-arm enables you to adjust the diameter of the jaws. The forcing screw should fit snugly onto the part you are removing. Finally, select the right wrench or socket size to fit the nut on the end of the forcing screw.

2. Adjust and fit the puller. Adjust the jaws and cross-arms of the puller so that it fits tightly around the part to be removed. The arms of the jaws should be pulling against the component at close to right angles.

3. Position the forcing screw. Use the appropriate wrench to run the forcing screw down to touch the shaft. Check that the point of the forcing screw is centered on the shaft. If not, adjust the jaws and cross-arms until the point is in the center of the shaft. Also be careful to use the correct foot on the end of the puller.

4. Tighten the forcing screw slowly and carefully onto the shaft. Check that the puller is not going to slip off center or off the pulley. Readjust the puller if necessary.

5. If the forcing screw and puller jaws remain in the correct position, tighten the forcing screw, and pull the part off the shaft.

6. You may sometimes have to use a hammer to hit directly on the end of the forcing screw to help break the part loose.

Flaring Tools

K06023

A **tube flaring tool** is used to flare the end of a tube so it can be connected to another tube or component. One example of this is where the brake line screws into a wheel cylinder. The flared end is compressed between two threaded parts so that it seals the joint and withstands high pressures. The three most common shapes of flares are the **single flare**, for tubing carrying low pressures like a fuel line; the **double flare**, for higher pressures such as in a brake system; and the ISO flare (sometimes called a bubble flare), which is the metric version used in brake systems (**FIGURE 6-88**).

Flaring tools have two main parts: a set of bars with holes that match the diameter of the tube end that is being shaped, and a yoke that drives a cone into the mouth of the tube (**FIGURE 6-89**). To make a single flare, the end of the tube is placed level with the surface of the top of the flaring bars. With the clamp screw firmly tightened, the feed screw flares the end of the tube. Making a double flare is similar, but an extra step is added before flaring the tube, and more of the tube is exposed to allow for folding the flare over into a double flare. A double flaring button is placed into the end of the tube, and when it is removed after tightening, the pipe looks like a bubble. Placing

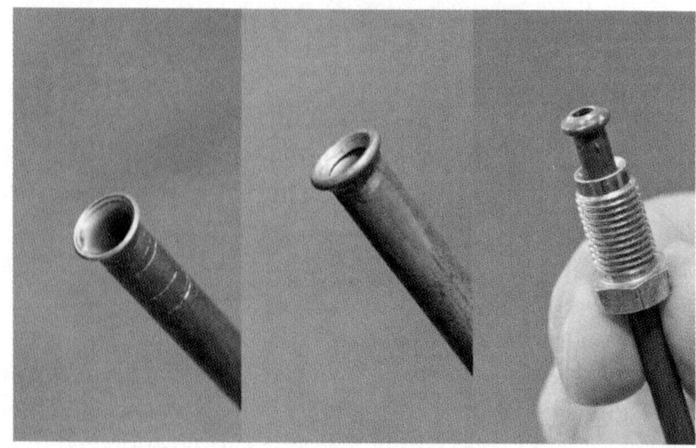

FIGURE 6-88 Single flare, double flare, and ISO flare.

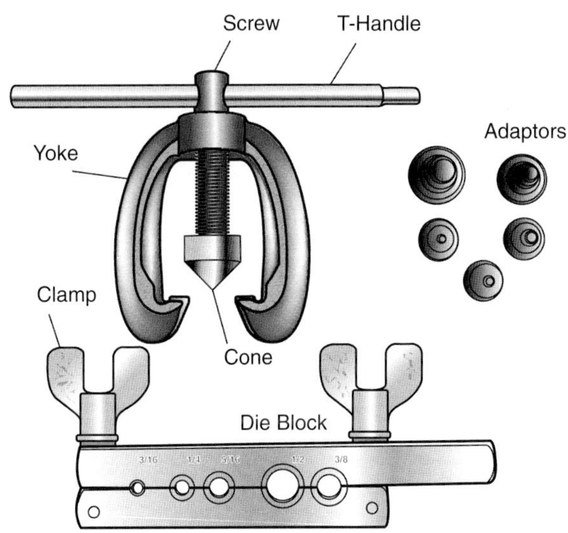

FIGURE 6-89 Components of a flare tool.

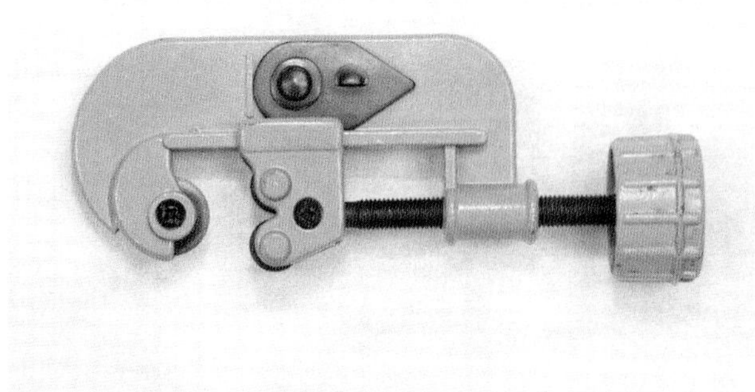

FIGURE 6-90 Tubing cutter.

the cone and yoke over the bubble allows you to force the bubble to fold in on itself, forming the double flare.

An ISO flare uses a flaring tool made specifically for that type of flare. The process is similar to that of the double-flare but differs in the use of the button, because an ISO flare does not get doubled back on itself. It should resemble a bubble shape when you are finished.

A **tubing cutter** is more convenient and neater than a saw when cutting pipes and metal tubing (**FIGURE 6-90**). The sharpened wheel does the cutting. As the tool turns around the pipe, the screw increases the pressure, driving the wheel deeper and deeper through the pipe until it finally cuts through. There is a larger version that is used for cutting exhaust pipes.

SAFETY TIP

A flaring tool is used to produce a pressure seal for sealing brake lines and fuel system tubing. Make sure you test the flared joint for leaks before completing the repair; otherwise, the brakes could fail or leaking fuel could catch on fire.

Using Flaring Tools

To make a successful flare, it is important to have the correct amount of tube protruding through the tool before clamping. Otherwise, too much of the end will fold over and leave too small of a hole for fluid to pass through. Too little and there won't be enough tube to fold over properly, and the joint won't have full surface contact.

If you are making a double flare or ISO flare, make sure you use the correctly sized button for the tubing size. The button is also used to measure the amount of tube required to protrude from the tool prior to forming it. To prevent the tool from slipping on the tube and ruining the flare, make sure the tool is sufficiently tight around the tube before starting to create the flare.

To use a flaring tool to make a flare in a piece of tubing, follow the steps in **SKILL DRILL 6-4**.

SAFETY TIP

Always wear eye protection when using a flaring tool. Make sure the tool is clean, in good condition, and suitable for the type of material you are going to flare.

Riveting Tools

K06024

There are many applications for blind rivets, and various rivet types and tools may be used to do the riveting. Rivets are used in many places in automotive applications where there is a need for a fastener that doesn't have to be easily removed. Some older window regulators, body panels, trim pieces, and even some suspension members use rivets to keep them attached without vibrating loose. **Pop rivet guns** are convenient for occasional riveting of light materials (**FIGURE 6-91**).

A typical pop, or **blind rivet**, has a body that forms the **finished rivet** and a mandrel, which is discarded when the riveting is completed (**FIGURE 6-92**). It is called a blind rivet because there is no need to see or reach the other side of the hole in which the rivet goes to do the work. In some types, the rivet is plugged shut so that it is waterproof or pressure proof.

The rivet is inserted into the riveting tool, which, when squeezed, pulls the end of the **mandrel** back through the body of the rivet. Because the **mandrel head** is bigger than the hole through the body, the body swells tightly against the hole. Finally, the mandrel head will snap off under the pressure and fall out, leaving the rivet body gripping the two pieces of material together.

Using Riveting Tools

Rivet tools are used to join two pieces of metal or other material together— for example, sheet metal that needs to be attached to a stiffening frame. To perform a riveting operation, you need a rivet gun, rivets, a drill, a properly sized drill bit, and the materials to be riveted.

SKILL DRILL 6-4 Using Flaring Tools

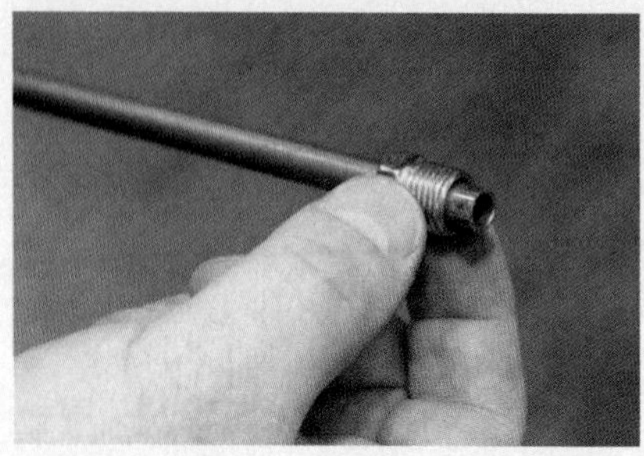

1. Choose the tube you will use to make the flare, and put the flare nut on the tube before creating the flare.

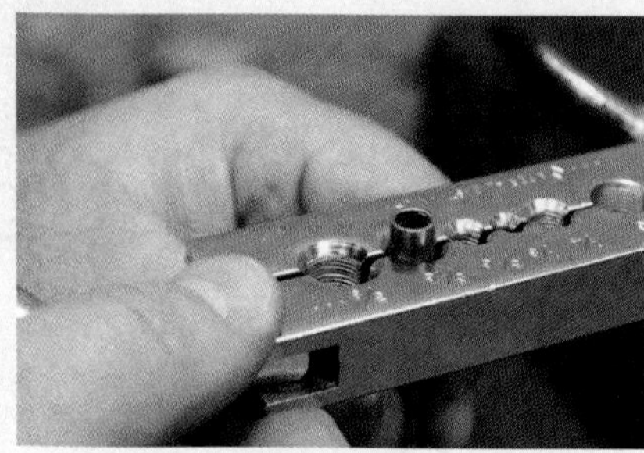

2. Match the size of the tube to the correct hole in the tubing clamp.

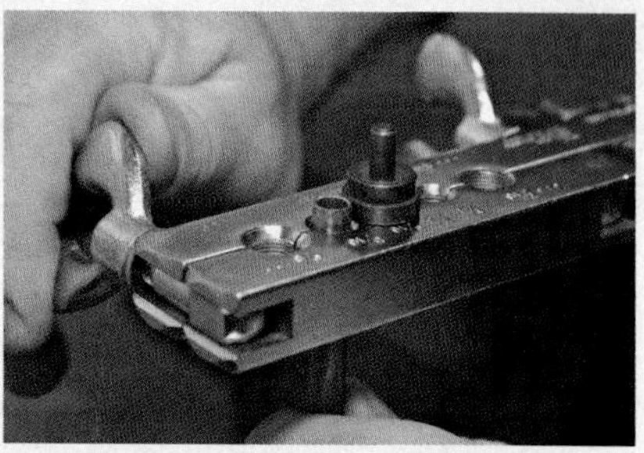

3. Holding the flaring tool, put the tube into the clamp. Position the tube so the correct amount is showing through the tool. If you are conducting a double flare, use the correctly sized button to ensure the proper amount of the tube is sticking up above the top of the clamp. Tighten the two halves of the clamp together using the wing nuts. Make sure the tool is tight enough to clamp the tube so it will not slip.

4. Put the cone and forming tool over the clamp, and turn the handle to make the flare. If you are doing a double flare or ISO flare, place the button in the end of the tube, install the cone and forming tool, and turn the handle to make the bubble. Remove the button from the tube.

5. If this is an ISO flare, inspect it to see if it is properly formed. If it is a double flare, put the cone and forming tool back on the clamp, and tighten the forming tool handle to create the double flare.

6. Remove the forming tool.

7. Remove the tube from the clamp, and check the flare to ensure it is free of burrs and is correctly formed.

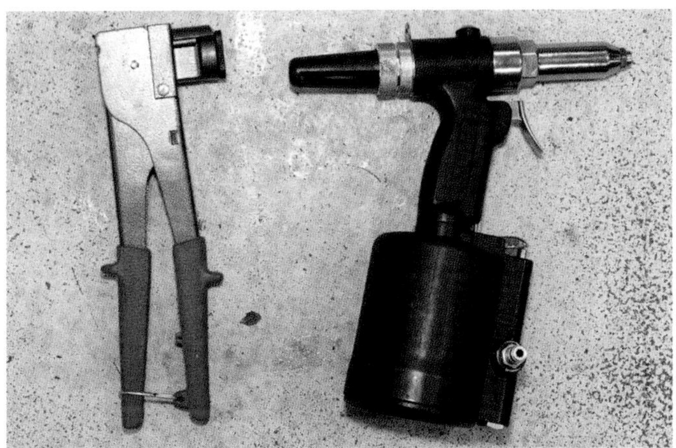

FIGURE 6-91 Pop rivet guns.

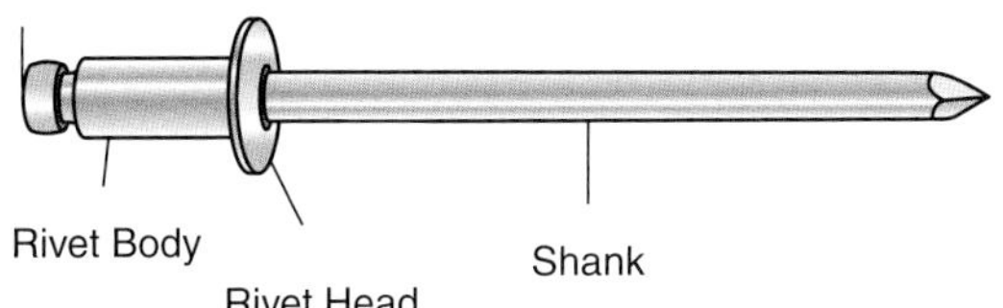

FIGURE 6-92 Anatomy of a rivet.

Rivets come in various diameters and lengths for different sizes of jobs and are made of various types of metals to suit the job at hand. When selecting rivets to suit the job, consider the diameter, length, and rivet material. Larger diameter rivets should be used for jobs that require more strength. The rivet length should be sufficient to protrude past the thickness of the materials being riveted by about 1.6 times the diameter of the rivet stem. Typically, you should select rivets that are made from the same material as that being riveted. For example, stainless steel rivets should be used for riveting stainless steel, and aluminum rivets should be used to rivet aluminum.

Pilot holes must be drilled through the metal to be riveted. Ensure that the hole is just large enough for the rivet to comfortably pass through it, but do not make it too large. If the hole is too large, the rivet will be loose and will not hold the materials securely together. When drilling holes for rivets, stay back from the material's edge to ensure that the rivets do not break through the edge of the materials being riveted. A good rule of thumb is to allow at least twice the diameter of the rivet stem as clearance from any edge.

Most rivet tools are capable of riveting various sizes of rivets and have a number of nosepiece sizes to work with different sizes of rivets. Make sure you select the properly sized nosepiece for the rivet you are using.

TECHNICIAN TIP

A rivet is a single-use fastener. Unlike a nut and bolt, which can normally be disassembled and reused, a rivet cannot. The metal shell that makes up a pop rivet is crushed into place so that it holds the parts firmly together. If it ever needs to be removed, it must be drilled out or cut off.

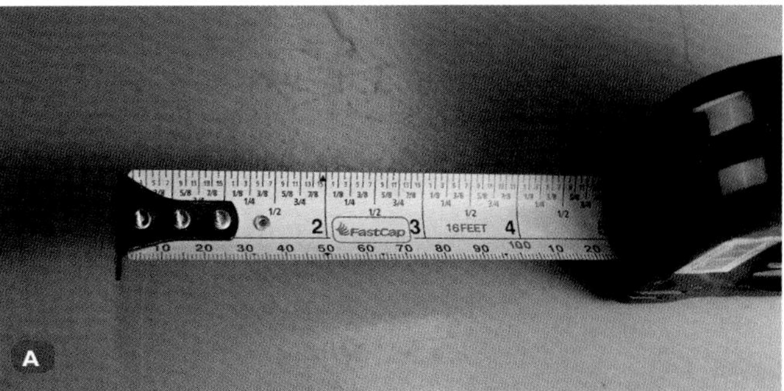

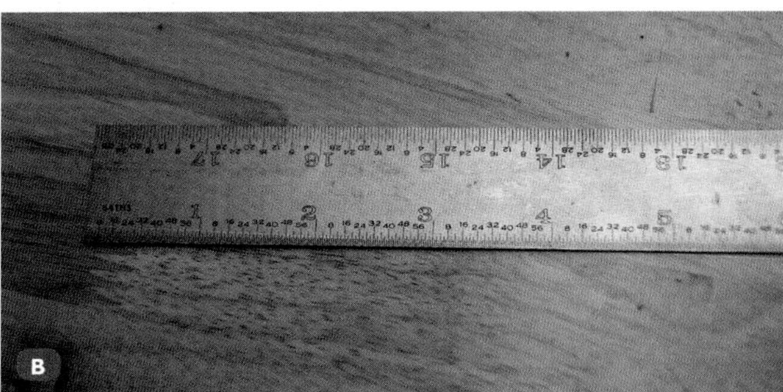

FIGURE 6-93 **A.** Measuring tape. **B.** Steel rule.

SAFETY TIP

When compressing the rivet handles, be careful not to place your fingers between the handles as they could end up pinching your fingers when the mandrel on the rivet breaks.

To use a riveting tool to rivet two pieces of material together, follow the steps in **SKILL DRILL 6-5**.

Precision Measuring Tools

K06025

Technicians are required to perform a variety of measurements while carrying out their job. This requires knowledge of what tools are available and how to use them. Measuring tools can generally be classified according to what type of measurements they can make. A measuring tape is useful for measuring longer distances and is accurate to a millimeter or fraction of an inch (**FIGURE 6-93**). A steel rule is capable of accurate measurements on shorter lengths, down to a millimeter or a fraction of an inch. Precision measuring tools are accurate to much smaller dimensions, such as a micrometer, which in some cases can accurately measure down to 1/10,000 of an inch (0.0001") or 1/1000 of a millimeter (0.001 mm).

Measuring Tapes

N06025

Measuring tapes are a flexible type of ruler and are a common measuring tool. The most common type found in shops is a thin metal strip about 0.5" to 1" (13 to 25 mm) wide that is rolled up

SKILL DRILL 6-5 Using Riveting Tools

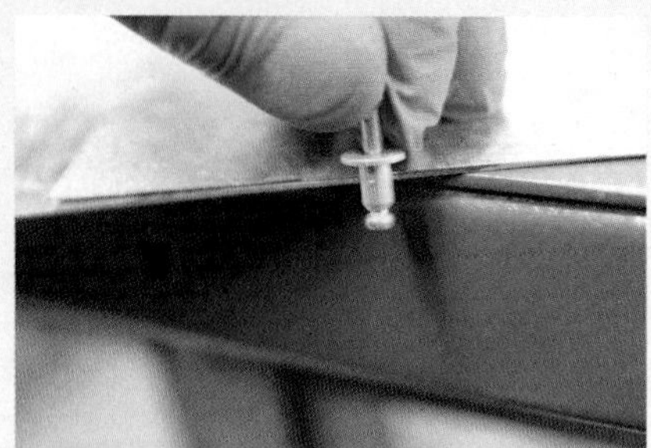

1. Select the correct rivet for the material you are riveting. Make sure the rivet is the correct length.

2. Drill pilot holes in the material to be riveted. Remove any burrs. Ensure that the pilot hole is the correct size—not too large or too small.

3. Make sure the correctly sized nosepiece for the rivet size is fitted to the rivet tool

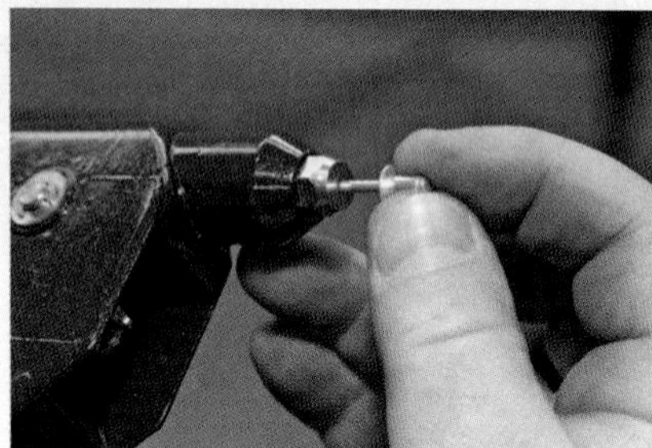

4. Insert the rivet into the gun, and push the rivet through the materials to be riveted. Hold firm pressure while pushing the rivet into the work.

5. Squeeze and release the rivet tool handle to compress the rivet. Continue this process until the rivet stem or shank breaks away from the rivet head.

6. Check the rivet joint to ensure the pieces are firmly held together.

inside a housing with a spring return mechanism. Measuring tapes can be of various lengths, with 16' or 25' (5 or 8 m) being very common. The measuring tape is pulled from the housing to measure items, and a spring return winds it back into the housing. The housing usually has a built-in locking mechanism to hold the extended measuring tape against the spring return mechanism. The hooked end can be placed over the edge of the object you are measuring and pulled against spring tension to take the measurement (**FIGURE 6-94**).

▶ TECHNICIAN TIP

The United States customary system (USCS), also called the standard system, and the metric system are two sets of standards for quantifying weights and measurements. Each system has defined units. For example, the standard system uses inches, feet, and yards, whereas the metric system uses millimeters, centimeters, and meters. Conversions can be undertaken from one system to the other. For example, 1 inch is equal to 25.4 millimeters, and 1 foot is equal to 304.8 millimeters. Tools that make use of a measuring system, such as wrenches, sockets, drill bits, micrometers, rulers, and many others, come in both standard and metric measurements. To work on modern vehicles, an understanding of both systems and their conversion is required. Conversion tables can be used to convert from one system to the other. The more you work with both systems, the easier it will be to understand how they relate to each other.

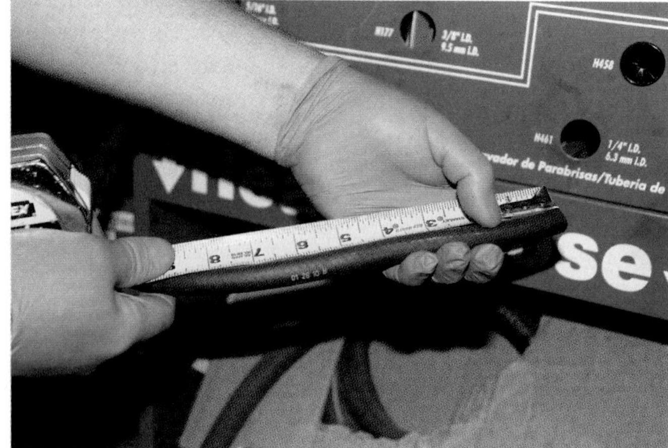

FIGURE 6-94 The hook makes it easy to take a measurement with one hand.

Steel Rulers

K06026

As the name suggests, a **steel rule** is a ruler that is made from steel. Steel rules commonly come in 12", 24", and 36" lengths. They are used like any ruler to measure and mark out items. A

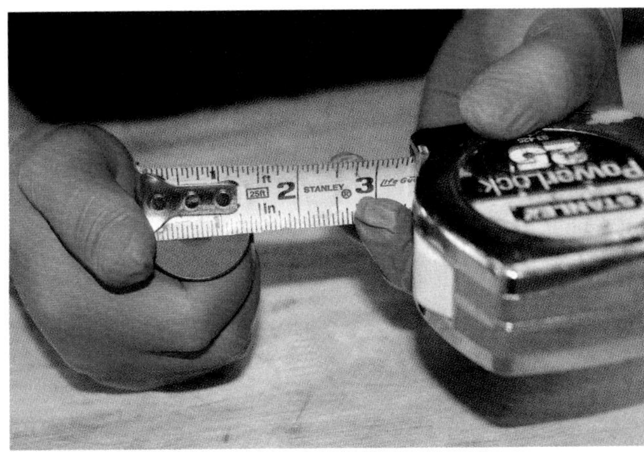

FIGURE 6-95 Tipping a steel rule on its side to get a more accurate reading.

steel rule is a very strong ruler, has precise markings, and resists damage. When using a steel rule, you can rest it on its edge so the markings are closer to the material being measured, which helps to mark the work more precisely (FIGURE 6-95). Always protect the steel rule from damage by storing it carefully; a damaged ruler will not give an accurate measurement. Never take measurements from the very end of a damaged steel rule, as damaged ends may affect the accuracy of your measurements.

> TECHNICIAN TIP
>
> If the end of a rule is damaged, you may be able to measure from the 1" mark and subtract an inch from the measurement.

Outside, Inside, and Depth Micrometers

K06027

Micrometers are precise measuring tools designed to measure small distances and are available in both inch and millimeter (mm) calibrations. Typically they can measure down to a resolution of 1/1000 of an inch (0.001") for a standard micrometer or 1/100 of a millimeter (0.01 mm) for a metric micrometer. Vernier micrometers equipped with the addition of a vernier scale can measure down to 1/10,000 of an inch (0.0001") or 1/1000 of a millimeter (0.001 mm).

The most common types of micrometers are the outside, inside, and depth micrometers (FIGURE 6-96). As the name suggests, an **outside micrometer** measures the outside dimensions of an item. For example, it could measure the diameter of a valve stem. The **inside micrometer** measures inside dimensions. For example, the inside micrometer could measure an engine cylinder bore. **Depth micrometers** measure the depth of an item such as how much clearance a piston has below the surface of the block.

The most common micrometer is an outside micrometer and is made up of several parts (FIGURE 6-97). The horseshoe-shaped part is the frame. It is built to make sure the

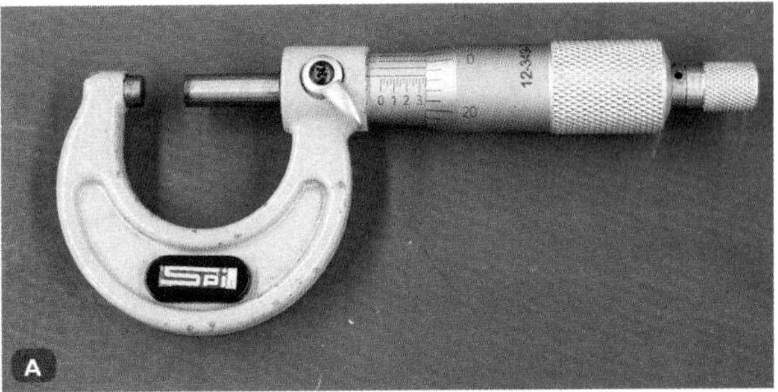

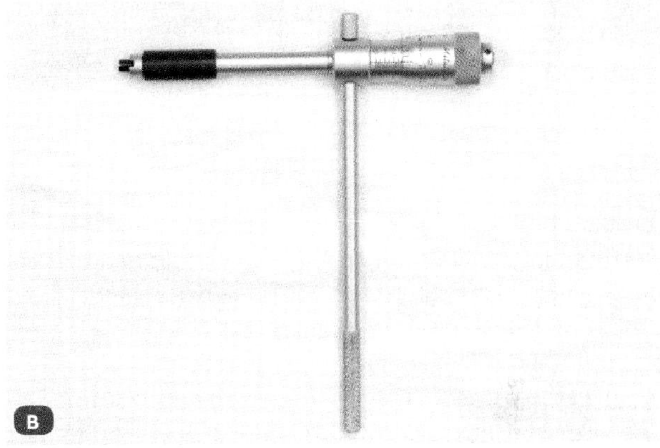

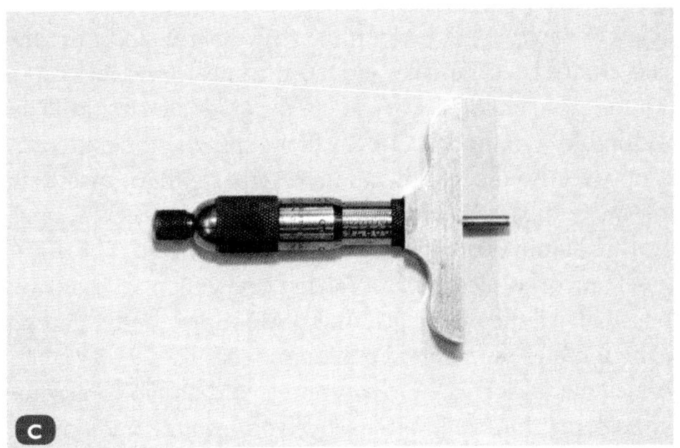

FIGURE 6-96 **A.** Outside micrometer. **B.** Inside micrometer. **C.** Depth micrometer.

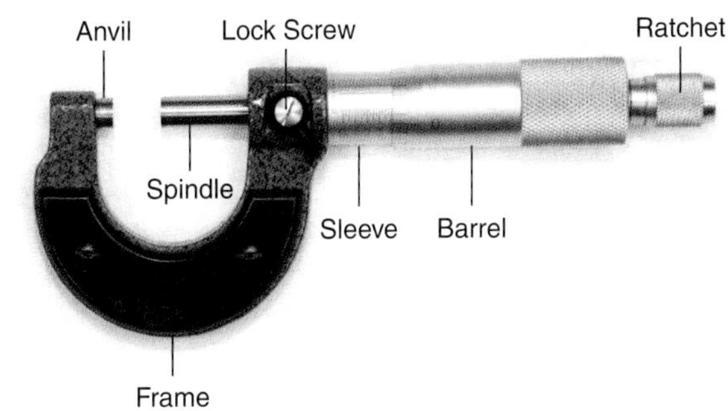

FIGURE 6-97 Parts of an outside micrometer.

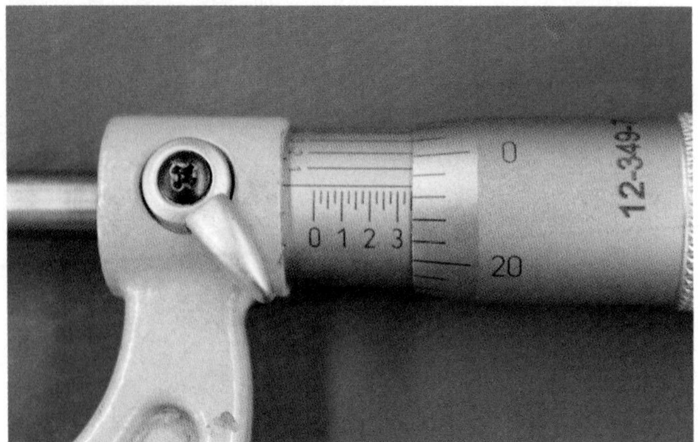

FIGURE 6-98 0.100" and 0.025" markings on the sleeve.

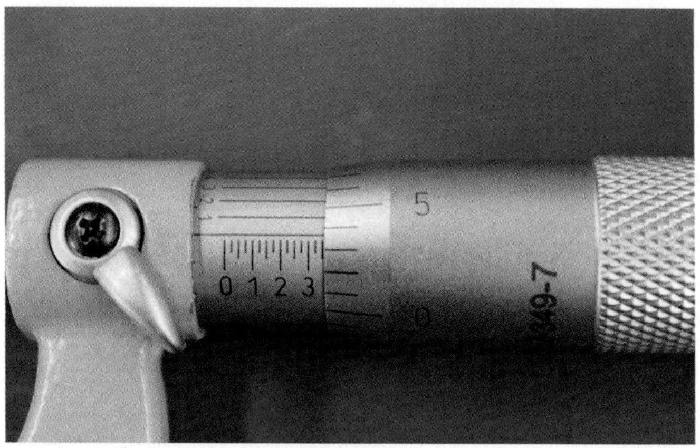

FIGURE 6-99 Markings on the thimble.

micrometer holds its shape. Some frames have plastic finger pads so that body heat is not transferred to the metal frame as easily, as heat can cause the metal to expand slightly and affect the reading. On one end of the frame is the anvil, which contacts one side of the part being measured. The other contact point is the spindle. The micrometer measures the distance between the anvil and spindle, so that is where the part being measured fits.

The measurement is read on the sleeve/barrel and thimble. The sleeve/barrel is stationary and has the linear markings on it. The thimble fits over the sleeve and has the graduated marking on it. The thimble is connected directly to the spindle, and both turn as a unit. Because the spindle and sleeve/barrel have matching threads, the thimble rotates the spindle inside of the sleeve/barrel, and the thread moves the spindle inward and outward. The thimble usually incorporates either a ratchet or a clutch mechanism, which is turned lightly by finger, thus preventing over-tightening of the micrometer thimble when taking a reading.

A lock nut, lock ring, or lock screw is used on most micrometers to lock the thimble in place while you read the micrometer. Standard micrometers use a specific thread of 40 TPI (threads per inch) on the spindle and sleeve. This means that the thimble rotates exactly 40 turns in 1' of travel. Every complete rotation moves the spindle one 40th of an inch, or 0.025" (1 ÷ 40 = 0.025). In four rotations, the spindle moves 0.100" (0.025 × 4 = 0.100). The linear markings on the sleeve show each of the 0.100" marks between 0 and 1 inch as well as each of the 0.025" marks (**FIGURE 6-98**). Because the thimble has graduated marks from 0 to 24 (each mark representing 0.001"), every complete turn of the thimble uncovers another one of the 0.025" marks on the sleeve. If the thimble stops short of any complete turn, it will indicate the number of 0.001" marks past the zero line on the sleeve (**FIGURE 6-99**). So reading a micrometer is as simple as adding up the numbers as shown below.

> **TECHNICIAN TIP**
>
> Micrometers are precision measuring instruments and must be handled and stored with care. They should always be stored with a gap between the spindle and anvil so metal contraction and expansion do not interfere with their calibration.

> **TECHNICIAN TIP**
>
> All micrometers need to be checked for calibration (also called "zeroing") before each use. A 0–1" or 0–25 mm outside micrometer can be lightly closed all of the way. If the anvil and spindle are clean, the micrometer should read 0.000, indicating the micrometer is calibrated correctly. If the micrometer is bigger than 1", or 25 mm, then a "standard" is used to verify the calibration. A standard is a hardened, machined rod of a precise length, such as 2", or 50 mm. When inserted in the same-sized micrometer, the reading should be exactly the same as listed on the standard. If a micrometer is not properly calibrated, it should not be used until it is recalibrated. See the tool's instruction manual for the calibration procedure.

To read a standard micrometer, perform the following steps (**FIGURE 6-100**):

1. Verify that the micrometer is properly calibrated.
2. Verify what size of micrometer you are using. If it is a 0–1" micrometer, start with 0.000. If it is a 1–2" micrometer, start with 1.000". A 2–3" micrometer would start with 2.000", etc. (To give an example, let's say it is 2.000".)
3. Read how many 0.100" marks the thimble has uncovered (example: 0.300").
4. Read how many 0.025" marks the thimble has uncovered past the 0.100" mark in step 3 (example: 2 × 0.025 = 0.050").
5. Read the number on the thimble that lines up with the zero line on the sleeve (example: 13 × 0.001 = 0.013").
6. Lastly, total all of the individual readings (example: 2.000 + 0.300 + 0.050 + 0.013 = 2.363").

A metric micrometer uses the same components as the standard micrometer. However, it uses a different **thread pitch** on the spindle and sleeve. It uses a 0.5 mm thread pitch (2.0 threads per millimeter) and opens up approximately 25 mm. Each rotation of the thimble moves the spindle 0.5 mm, and it therefore takes 50 rotations of the thimble to move the full 25 mm distance. The sleeve/barrel is labeled with individual millimeter marks and half-millimeter marks from the starting millimeter to the ending millimeter, 25 mm away (**FIGURE 6-101**). The thimble has graduated marks from 0 to 49 (**FIGURE 6-102**).

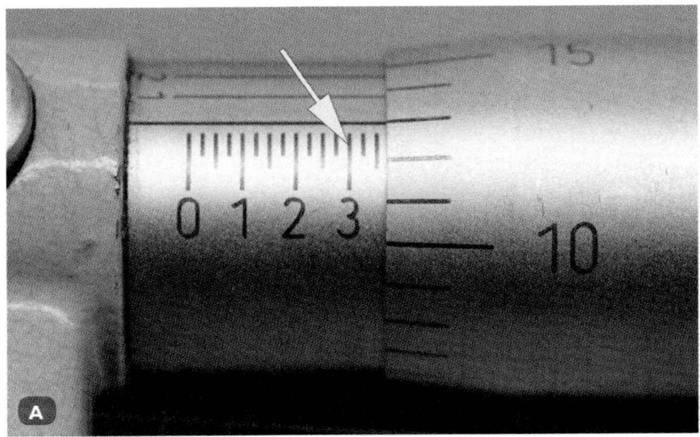

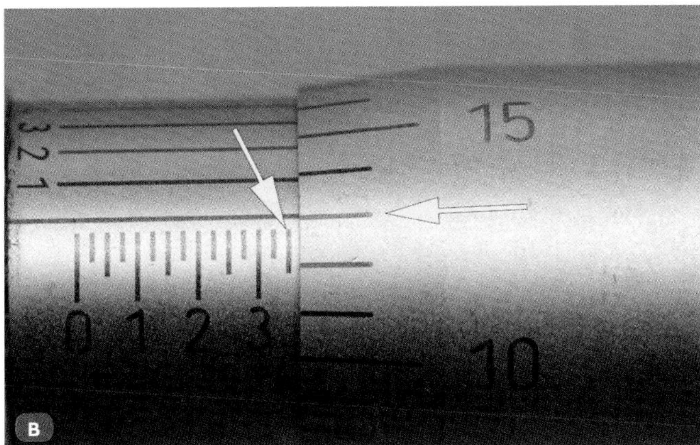

FIGURE 6-100 **A.** Read how many 0.100" marks the thimble has uncovered. **B.** Read how many 0.025: marks are uncovered, then read the number on the thimble that lines up with the zero line on the sleeve.

Reading a metric micrometer involves the following steps:

1. Read the number of full millimeters the thimble has passed (To give an example, let's say it is 23.00 mm).
2. Check to see if it passed the 0.5 mm mark (example: 0.50 mm).
3. Check to see which mark on the thimble lines up with or is just passed (example: 37 × 0.01 mm = 0.37 mm).
4. Total all of the numbers (example: 23.00 mm + 0.50 mm + 0.37 mm = 23.87 mm).

If the micrometer is equipped with a vernier gauge, meaning it can read down to 1/10,000 of an inch (0.0001") or 1/1000 of a millimeter (0.001 mm), you need to complete one more step. Identify which of the vernier lines is closest to one of the lines on the thimble (**FIGURE 6-103**). Sometimes it is hard to determine which is the closest, so decide which three are the closest, and then use the center line. At the frame side of the sleeve will be a number that corresponds to the vernier line, numbered 1–0. Add the vernier to the end of your reading. For example: 2.363 + 0.0007 = 2.3637", and 23.77 + 0.007 = 23.777 mm.

For inside measurements, the inside micrometer works on the same principles as the outside micrometer and so does the depth micrometer. The only difference is that the scale on the sleeve of the depth micrometer is backward, so be careful when reading it.

FIGURE 6-101 Metric markings on the sleeve.

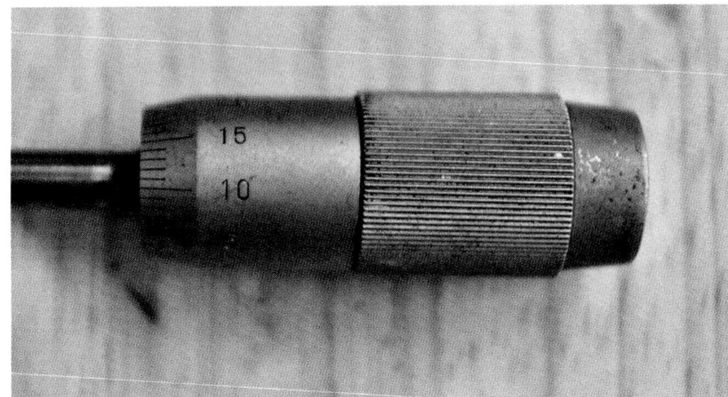

FIGURE 6-102 Markings on the thimble.

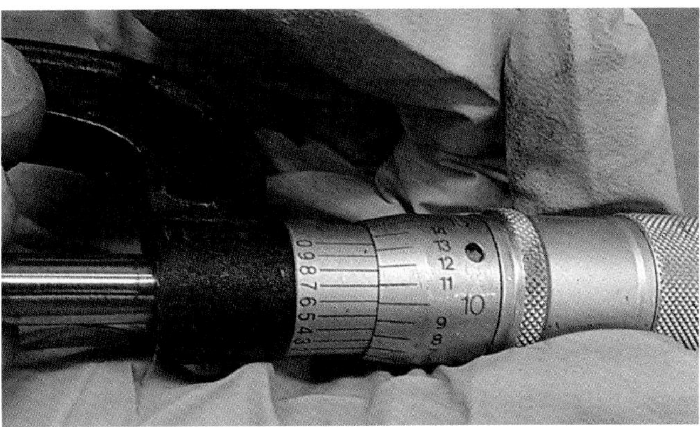

FIGURE 6-103 Vernier scale on a micrometer showing 7 on the sleeve lined up the best with a line on the thimble.

Using Micrometers

To maintain accuracy of measurements, it is important that both the micrometer and the items to be measured are clean and free of any dirt or debris. Also make sure the micrometer is zeroed before taking any measurements. Never over-tighten a micrometer or store it with its measuring surfaces touching, as this may damage the tool and affect its accuracy. When measuring, make sure the item can pass through the micrometer surfaces snugly and squarely. This is best accomplished by using the ratchet to tighten the micrometer. Always take the measurement a number of times and compare results to ensure you have measured accurately.

SKILL DRILL 6-6 Using Micrometers

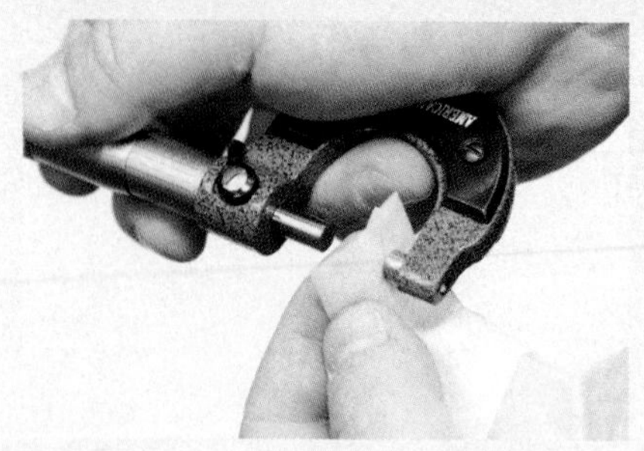

1. Select the correct size of micrometer. Verify that the anvil and spindle are clean and that it is calibrated properly. Clean the surface of the part you are measuring.

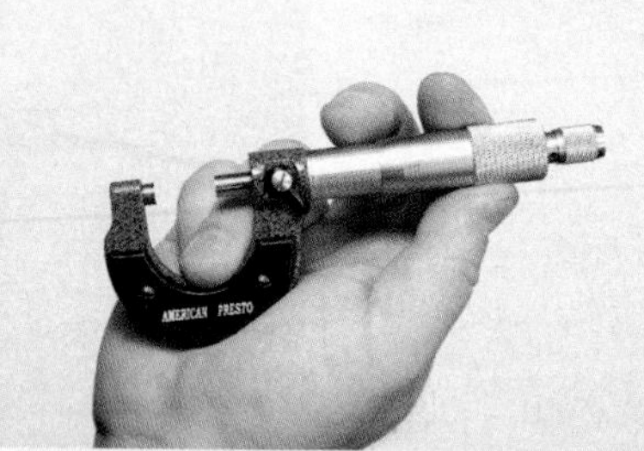

2. In your right hand, hold the frame of the micrometer between your pinky, ring finger, and the palm of your hand, with the thimble between your thumb and forefinger.

3. With your left hand, hold the part you are measuring, and place the micrometer over it.

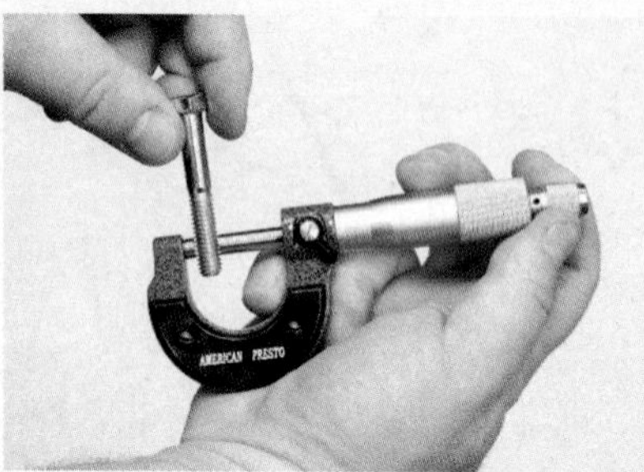

4. Using your thumb and forefinger, lightly tighten the ratchet. It is important that the correct amount of force is applied to the spindle when taking a measurement. The spindle and anvil should just touch the component with a slight amount of drag when the micrometer is removed from the measured piece. Be careful that the part is square in the micrometer so the reading is correct. Try rocking the micrometer in all directions to make sure it is square.

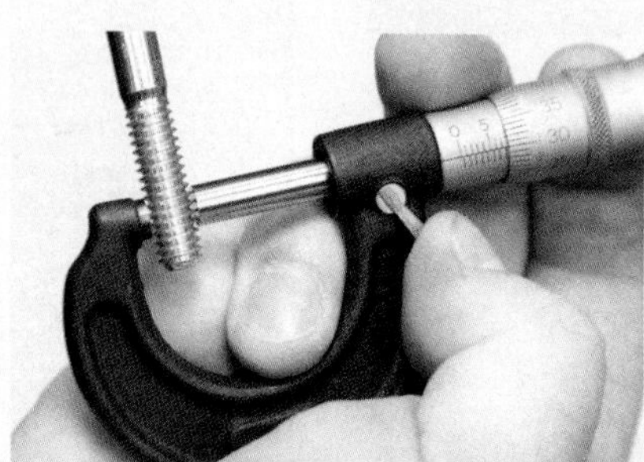

5. Once the micrometer is properly snug, tighten the lock mechanism so the spindle will not turn. Read the micrometer and record your reading.

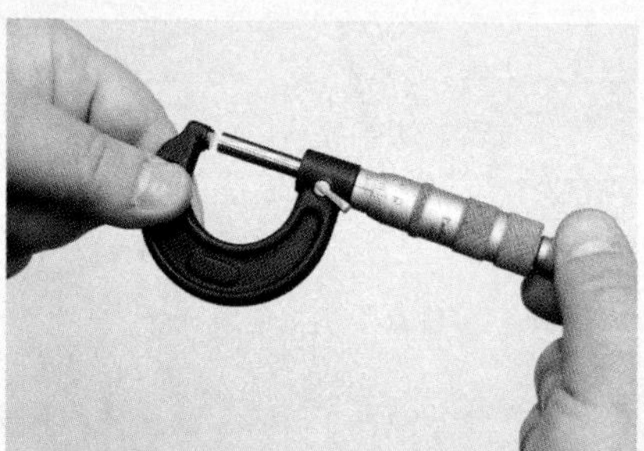

6. When all readings are finished, clean the micrometer, position the spindle so it is backed off from the anvil, and return it to its protective case.

To correctly measure using an outside micrometer, follow the steps in **SKILL DRILL 6-6**.

Telescoping Gauges

K06028

For measuring distances in awkward spots like the bottom of a deep cylinder, the **telescoping gauge** has spring-loaded plungers that can be unlocked with a screw on the handle so they slide out and touch the walls of the cylinder (**FIGURE 6-104**). The screw then locks them in that position, the gauge can be withdrawn, and the distance across the plungers can be measured with an outside micrometer or calipers to convey the diameter of the cylinder at that point. Telescoping gauges come in a variety of sizes to fit various sizes of holes and bores.

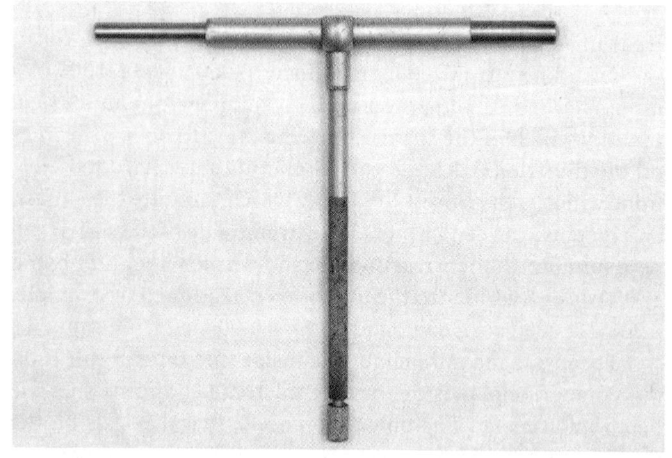

FIGURE 6-104 Telescoping gauge.

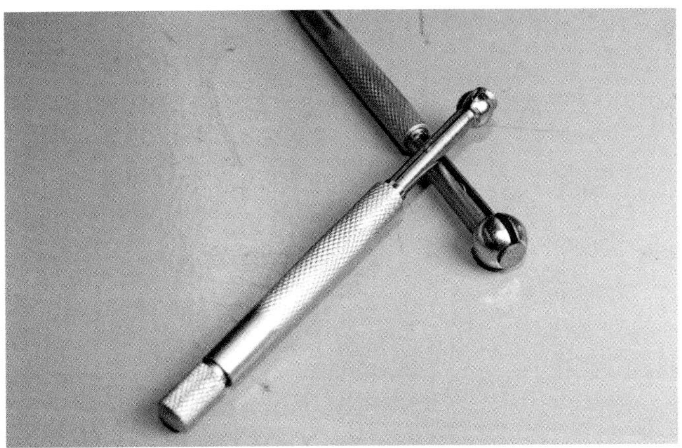

FIGURE 6-105 Split ball gauges.

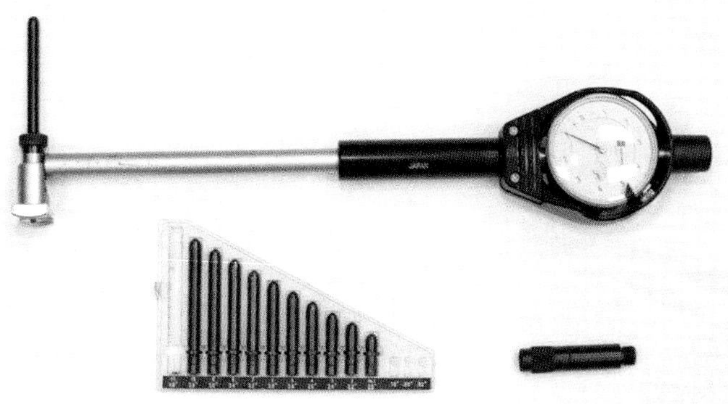

FIGURE 6-106 A dial bore gauge set.

Split Ball Gauges

K06029

A **split ball gauge** or small hole gauge is good for measuring small holes where telescoping gauges cannot fit (**FIGURE 6-105**). They use a similar principle to the telescoping gauge, but the measuring head uses a split ball mechanism that allows it to fit into very small holes. Split ball gauges are ideal for measuring valve guides on a cylinder head for wear. A split ball gauge can be fitted in the bore and expanded until there is a slight drag. Then it can be retracted and measured with an outside micrometer.

Dial Bore Gauges

K06030

A **dial bore gauge** is used to measure the inside diameter of bores with a high degree of accuracy and speed (**FIGURE 6-106**). The dial bore gauge can measure a bore directly by using telescoping pistons on a T-handle with a dial mounted on the handle. The dial bore gauge combines a telescoping gauge and dial indicator in one instrument. A dial bore gauge determines if the diameter is worn, tapered, or out-of-round according to the manufacturer's specifications. The resolution of a dial bore gauge is typically accurate to 5/10,000 of an inch (0.0005") or 1/100 of a millimeter (0.01 mm).

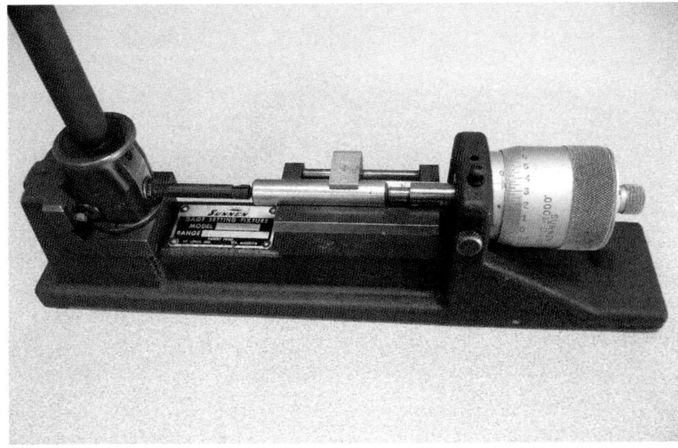

FIGURE 6-107 Dial bore gauge being calibrated to a predetermined size.

Using Dial Bore Gauges

To use a dial bore gauge, select an appropriate-sized adapter to fit the internal diameter of the bore, and install it to the measuring head. Many dial bore gauges also have a fixture to calibrate the tool to the size you desire (**FIGURE 6-107**). The fixture is set to the size desired, and the dial bore gauge is placed in it. The dial bore gauge is then adjusted to the proper reading. Once it is calibrated, the dial bore gauge can be inserted inside the bore to be measured. Hold the gauge in line with the bore, and slightly rock it to ensure it is centered. Read the dial when it is fully centered and square to the bore, to determine the correct measurement. It takes a bit of practice to get accurate readings.

Store a bore gauge carefully in its storage box and ensure the locking mechanism is released while in storage. Bore gauges are available in different ranges of size. It is important to select a gauge with the correct range for the bore you are measuring. When measuring, make sure the gauge is at a 90-degree angle to the bore and read the dial. Always take the measurement a number of times and compare results to ensure you have measured accurately.

To correctly measure using a dial bore gauge, follow the steps in **SKILL DRILL 6-7**.

Vernier Calipers

K06031

Vernier calipers are a precision instrument used for measuring outside dimensions, inside dimensions, and depth measurements, all in one tool (**FIGURE 6-108**). They have a graduated bar with markings like a ruler. On the bar, a sliding sleeve with jaws is mounted for taking inside or outside measurements. Measurements on older versions of vernier calipers are taken by reading the graduated bar scales, and fractional measurements are read by comparing the scales between the sliding sleeve and the graduated bar. Technicians often use vernier calipers to measure length and diameters of bolts and pins or the depth of blind holes in housings.

SKILL DRILL 6-7 Using Dial Bore Gauges

1. Select the correct size of the dial bore gauge you will use, and fit any adapters to it. Check the calibration and adjust it as necessary. Insert the dial bore gauge into the bore. The accurate measurement will be at exactly 90 degrees to the bore. To find the accurate measurement, rock the dial bore gauge handle slightly back and forth until you find the centered position.

2. Check the calibration and adjust it as necessary.

3. Insert the dial bore gauge into the bore. The accurate measurement will be at exactly 90 degrees to the bore. To find the accurate measurement, rock the dial bore gauge handle slightly back and forth until you find the centered position.

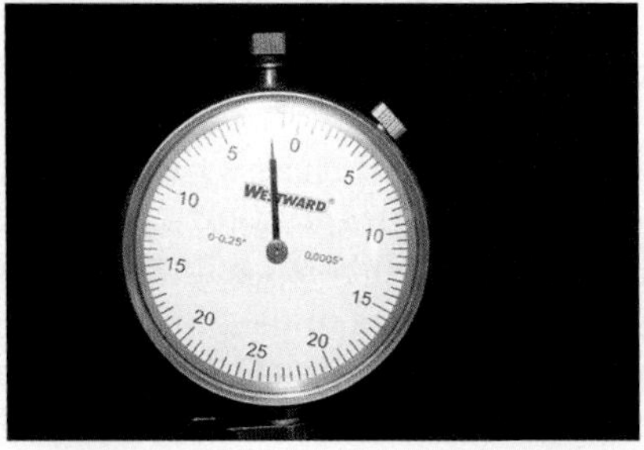

4. Read the dial to determine the bore measurement.

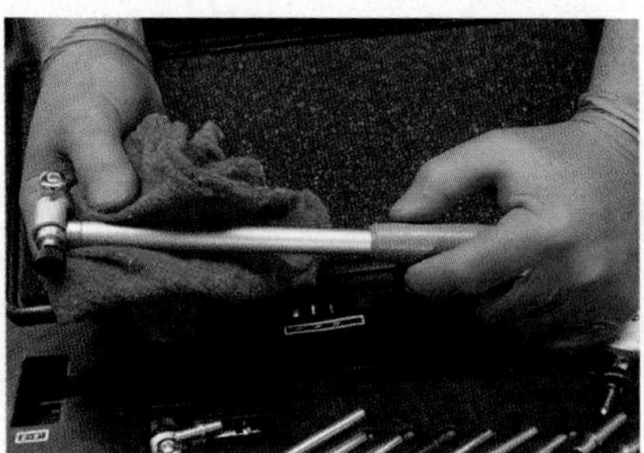

5. Always clean the dial bore gauge, and return it to its protective case when you have finished using it.

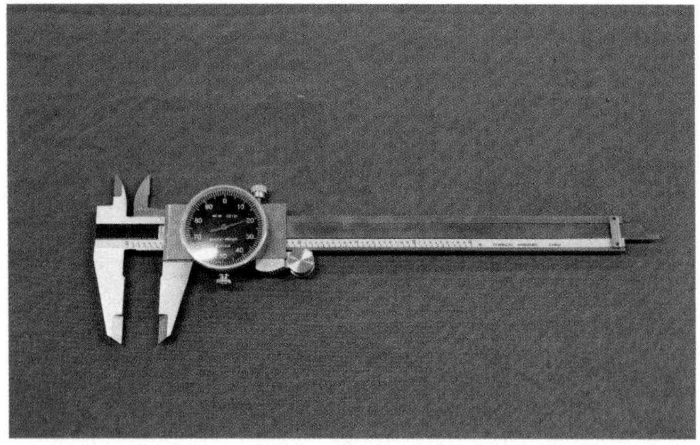

FIGURE 6-108 Vernier calipers can take three types of readings.

FIGURE 6-109 Dial vernier caliper.

Newer versions of vernier calipers have dial and digital scales (**FIGURE 6-109**). The dial caliper has the main scale on the graduated bar, and fractional measurements are taken from a dial with a rotating needle. These tend to be easier to read than the older versions. More recently, digital scales on vernier calipers have become commonplace (**FIGURE 6-110**). The principle of their use is the same as any vernier caliper; however, they have a digital scale that reads the measurement directly.

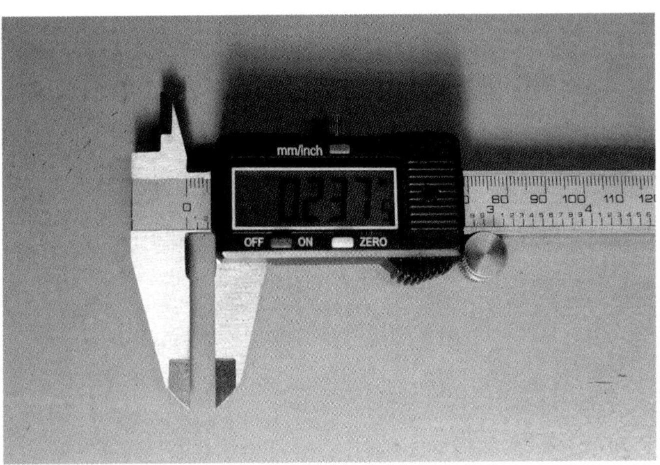

FIGURE 6-110 Digital caliper.

FIGURE 6-111 Dial indicator.

Using Vernier Calipers

Always store vernier calipers in a storage box to protect them and ensure the measuring surfaces are kept clean for accurate measurement. If making an internal or external measurement, make sure the caliper is at right angles to the surfaces to be measured. You should always repeat the measurement a number of times and compare results to ensure you have measured accurately. To correctly measure using digital vernier calipers, follow the steps in **SKILL DRILL 6-8**.

Dial Indicators

K06032

Dial indicators can also be known as dial gauges, and as the name suggests, they have a dial and needle where measurements are read. They have a measuring plunger with a pointed or rounded contact end that is spring loaded and connected via the housing to the dial needle (**FIGURE 6-111**). The dial accurately displays movement of the plunger in and out as it rests against an object. For example, they can be used to measure the trueness of a rotating disc brake rotor. A dial indicator can also measure how round something is, such as a **crankshaft**, which can be rotated in a set of **V blocks** (**FIGURE 6-112**). If the crankshaft is bent, it will show as movement on the dial indicator as the crankshaft is rotated. The dial indicator senses slight movement at its tip and magnifies it into a measurable swing on the dial. Dial indicators normally have either one or two indicator needles. The large needle indicates the fine reading of thousandths of an inch. If it has a second needle, it will be smaller and indicates the coarse reading of tenths of an inch. The large needle is able to move numerous times around the outer scale. One full turn may represent 0.100" or 1 mm. The small inner scale indicates how many

SKILL DRILL 6-8 Using Vernier Calipers

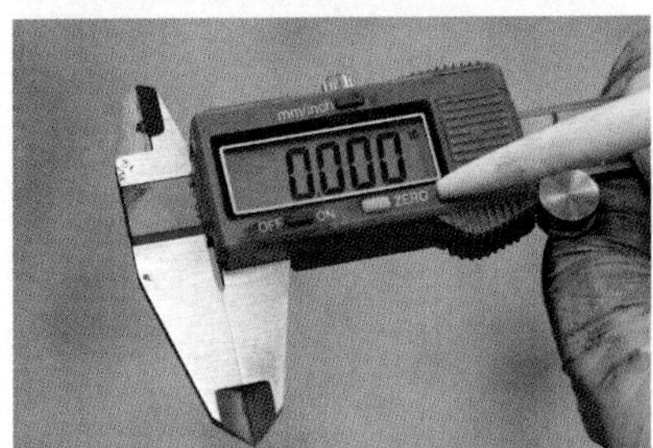

1. Verify that the vernier caliper is calibrated (zeroed) before using it.

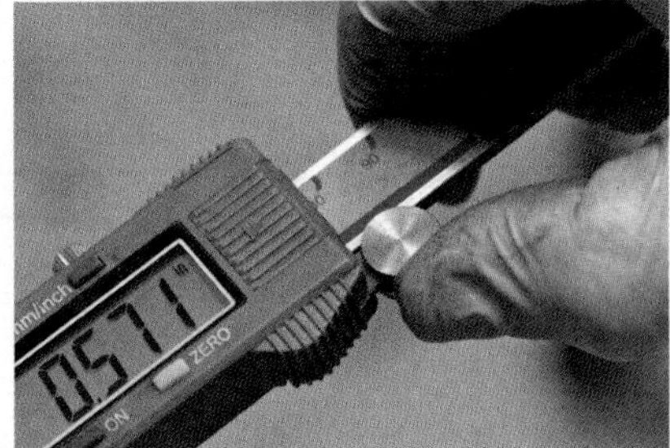

2. Position the caliper correctly for the measurement you are making. Internal and external readings are normally made with the vernier caliper positioned at 90 degrees to the face of the component to be measured. Length and depth measurements are usually made parallel to or in line with the object being measured. Use your thumb to press or withdraw the sliding jaw to measure the outside or inside of the part.

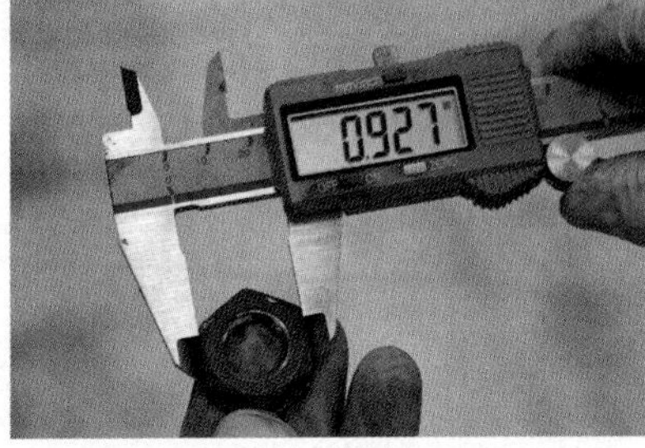

3. Read the scale of the vernier caliper, being careful not to change the position of the moveable jaw. If using a non-digital caliper, always read the dial or face straight on. A view from the side can give a considerable **parallax error**. Parallax error is a visual error caused by viewing measurement markers at an incorrect angle.

FIGURE 6-112 A dial indicator being used to measure crankshaft runout.

times the outer needle has moved around its scale. In this way, the dial indicator is able to read movement of up to 1" or 2 cm. Dial indicators can typically measure with an accuracy of 0.001" or 0.01 mm.

The type of dial indicator you use is determined by the amount of movement you expect from the component you are measuring. The indicator must be set up so that there is no gap between the dial indicator and the component to be measured. It also must be set perpendicular and centered to the part being measured. Most dial indicator sets contain various attachments and support arms, so they can be configured specifically for the measuring task.

Using Dial Indicators

Dial indicators are used in many types of service jobs. They are particularly useful in determining runout on rotating shafts and surfaces. Runout is the side-to-side variation of movement when a component is turned. When attaching a dial indicator:

- Keep support arms as short as possible.
- Make sure all attachments are tightened to prevent unnecessary movement between the indicator and the component.
- Make sure the dial indicator plunger is positioned at 90 degrees to the face of the component to be measured.
- Always read the dial face straight on, as a view from the side can give a considerable parallax error.

The outer face of the dial indicator is designed so it can be rotated so that the zero mark can be positioned directly under the pointer. This is how a dial indicator is zeroed. To correctly measure using a dial indicator, follow the steps in **SKILL DRILL 6-9**.

SKILL DRILL 6-9 Using Dial Indicators

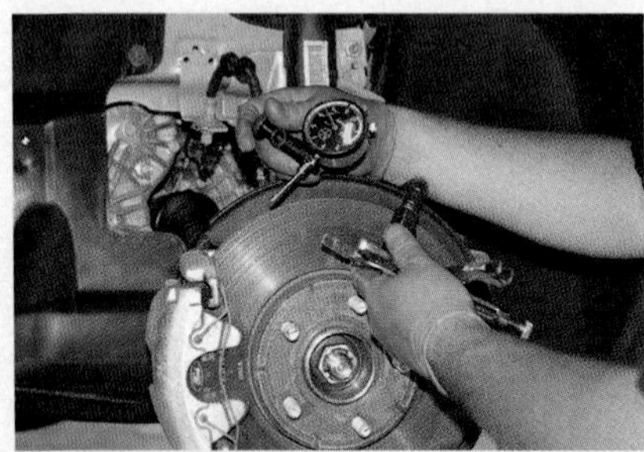

1. Select the gauge type, size, attachment, and bracket that fit the part you are measuring. Mount the dial indicator firmly to keep it stationary.

2. Adjust the indicator so that the plunger is at 90 degrees to the part you are measuring, and lock it in place.

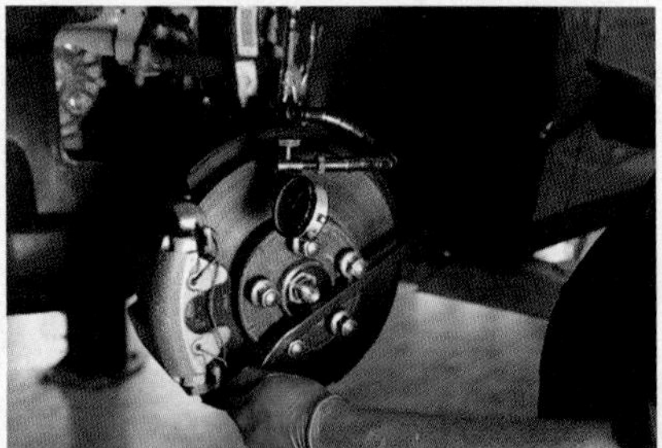

3. Rotate the part one complete turn, and locate the low spot. Zero the indicator.

4. Find the point of maximum height and note the reading. This indicates the runout value.

5. Continue the rotation, making sure the needle does not go below zero. If it does, re-zero the indicator and remeasure the point of maximum variation.

6. Check your readings against the manufacturer's specifications. If the deviation is greater than the specifications allow, consult your supervisor.

FIGURE 6-113 Straight edge being used to measure the flatness of a surface.

Straight Edges

K06033

Straight edges are usually made from hardened steel and are machined so that the edge is perfectly straight. A straight edge is used to check the flatness of a surface. It is placed on its edge against the surface to be checked (**FIGURE 6-113**). The gap between the straight edge and the surface can be measured by using feeler gauges. Sometimes the gap can be seen easily if light is shone from behind the surface being checked. Straight edges are often used to measure the amount of warpage the surface of a cylinder head has.

Feeler Gauges

K06034

Feeler gauges (also called feeler blades) are used to measure the width of gaps, such as the clearance between valves and rocker arms. Feeler gauges are flat metal strips of varying thicknesses (**FIGURE 6-114**). The thickness of each feeler gauge is clearly marked on each one. They are sized from fractions of an inch or fractions of a millimeter. They usually come in sets with different sizes and are available in standard and metric measurements. Some sets contain feeler gauges made of brass. These are used to take measurements between components that are magnetic. If steel gauges are used, the drag caused by the magnetism would mimic the drag of a proper clearance. Brass gauges are not subject to magnetism, so they work well in that situation. Some feeler gauges come in a bent arrangement to be more easily inserted in cramped spaces. Others come in a stepped version. For example, the end of the gauge might be 0.010" (0.254 mm) thick, while the rest of the gauge is 0.012" (0.305 mm) thick. This works well for adjusting valve clearance. If the specification is 0.010" (0.254 mm), then the 0.010 (0.254 mm) section can be placed in the gap. If the 0.012 (0.305 mm) section slides into the gap, then the valve needs to be readjusted. If it stops at the lip of the 0.012" (0.305 mm) section, then the gap is correct (**FIGURE 6-115**).

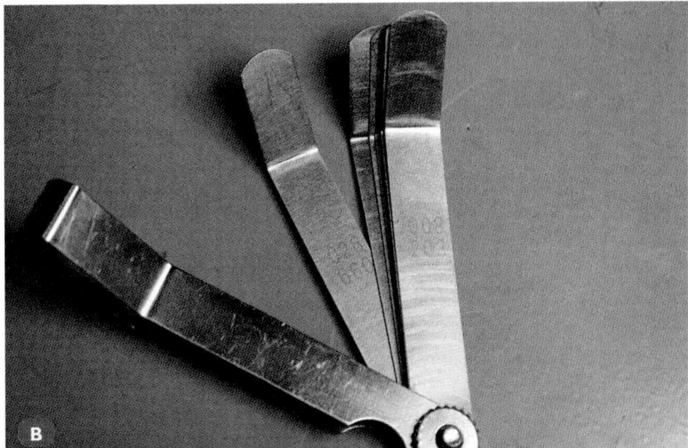

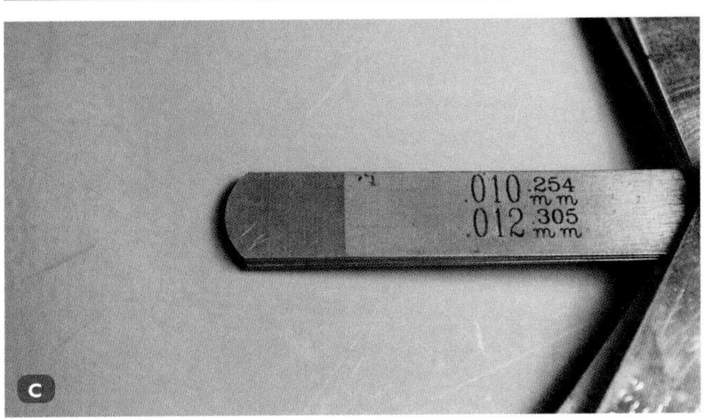

FIGURE 6-114 Feeler blade set. **A.** Straight. **B.** Bent. **C.** Stepped. **D.** Wire gauge.

FIGURE 6-115 Stepped feeler blade being used during a valve adjustment.

Two or more non-stepped feeler gauges can be stacked together to make up a desired thickness. For example, to measure a thickness of 0.029 of an inch, a 0.017 and a 0.012 feeler gauge could be used together to make up the size. Alternatively, if you want to measure an unknown gap, you can interchange feeler gauges until you find the one or more that fits snugly into the gap and total their thickness to determine the measurement of the gap. In conjunction with a straight edge, they can be used to measure surface irregularities on a cylinder head.

SAFETY TIP

Never use feeler gauges on operating machinery.

SAFETY TIP

Feeler gauges are strips of hardened metal that have been ground or rolled to a precise thickness. They can be very thin and will cut through skin if not handled correctly.

SKILL DRILL 6-10 Using Feeler Gauges

1. Select the appropriate type and size feeler gauge set for the job you are working on.

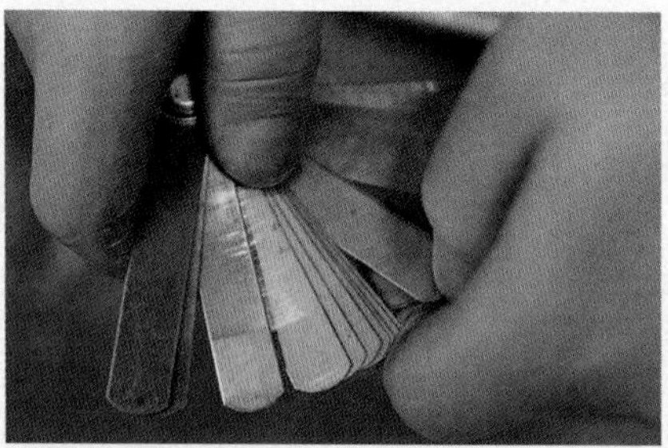

2. Inspect the feeler gauges to make sure they are clean, rust-free, and undamaged, but slightly oiled for ease of movement.

3. Choose one of the smaller wires or blades, and try to insert it in the gap on the part. If it slips in and out easily, choose the next size up. When you find one that touches both sides of the gap and slides with only gentle pressure, then you have found the exact width of that gap.

4. Read the markings on the wire or blade, and check these against the manufacturer's specifications for this component. If gap width is outside the tolerances specified, inform your supervisor.

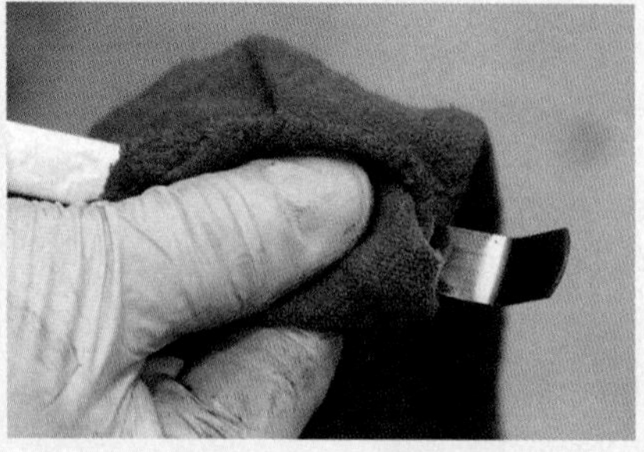

5. Clean the feeler gauge set with an oily cloth before storage to prevent rust.

FIGURE 6-116 Wire feeler gauge used to check a spark plug gap.

Using Feeler Gauges

If the feeler gauge feels too loose when measuring a gap, select the next size larger, and measure the gap again. Repeat this procedure until the feeler gauge has a slight drag between both parts. If the feeler gauge is too tight, select a smaller size until the feeler gauge fits properly. When measuring a spark plug gap, feeler gauges should not be used because the surfaces of the spark plug electrodes are not perfectly parallel, so it is preferable to use wire feeler gauges (**FIGURE 6-116**).

Wire feeler gauges use accurately machined pieces of wire instead of flat metal strips. To select and use feeler gauge sets, follow the steps in **SKILL DRILL 6-10**.

▶ Cleaning Tools and Equipment

N06005, K06035

Clean tools and equipment work more safely and efficiently. At the end of each working day, clean the tools and equipment you used and check them for any damage. If you note any damage,

SKILL DRILL 6-11 Tool and Equipment Cleaning

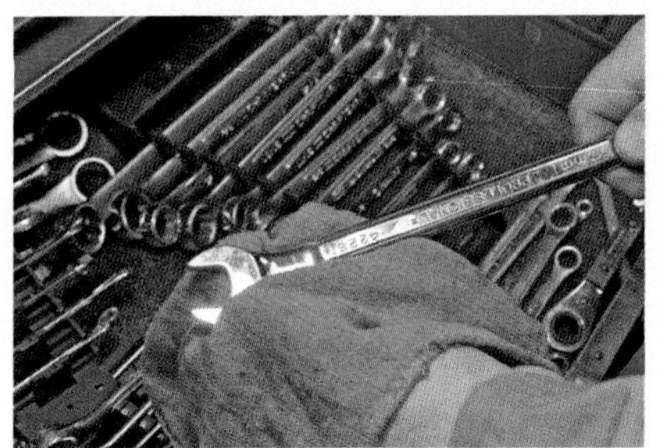

1. Clean hand tools. Keep your hand tools in good, clean condition with two types of rags. One rag should be lint-free to clean or handle precision instruments or components. The other should be oily to prevent rust and corrosion.

2. Clean floor jacks. Wipe off any oil or grease on the floor jack, and check for fluid leaks. If you find any, remove the jack from use, and have it repaired or replaced. Occasionally, apply a few drops of lubricating oil to the wheels and a few drops to the posts of threaded jack stands.

3. Clean electrical power tools. Keep power tools clean by brushing off any dust and wiping off excess oil or grease with a clean rag. Inspect any electrical cables for dirt, oil, or grease, and for any chafing or exposed wires. With drills, inspect the chuck and lubricate it occasionally with machine oil.

4. Clean air-powered tools. Apply a few drops of oil into the inlet of your air tools every day. Although these tools have no electrical motor, they do need regular lubrication of the internal parts to prevent wear.

5. Clean hoists and heavy machinery. Locate the checklist or maintenance record for each hoist or other major piece of equipment before carrying out cleaning activities.

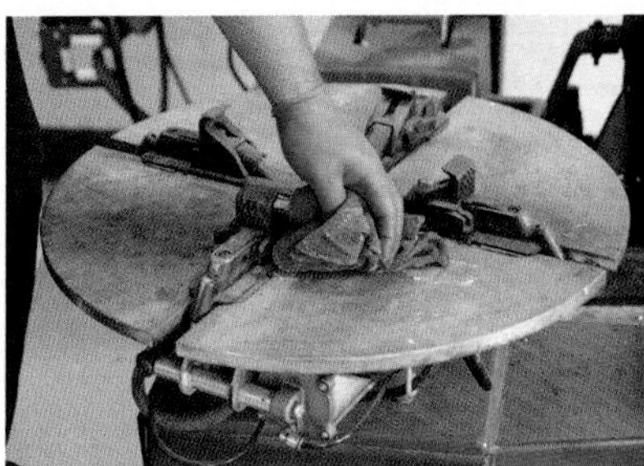

6. You should clean equipment operating mechanisms and attachments of excess oil or grease.

tag the tool as faulty and organize a repair or replacement. Electrical current can travel over oily or greasy surfaces. Be sure to keep electrical power tools clean. All shop equipment should have a maintenance schedule. Always complete the tasks described on the schedule at the required time. This helps to keep the equipment in safe working order.

Store commonly used tools in an easy-to-reach location. If a tool or piece of equipment is too difficult to return, then it will likely be left on a workbench or on the floor, where it will become a safety hazard. Keep your work area tidy. This will help you work more efficiently and safely. Keep a trash can close to your work area, and place any waste in it as soon as possible. Dispose of liquid and solid waste, such as oils, coolant, and worn components, in the correct manner. Local authorities provide guidelines for waste disposal with fines for noncompliance. When cleaning products lose their effectiveness, they need to be replaced. Refer to the supplier's recommendations for collection or disposal. Do not pour solvents or other chemicals into the sewage system. This is both environmentally damaging and illegal.

Always use chemical gloves when using any cleaning material, because excessive exposure to cleaning materials can damage skin. Also, absorbing some chemicals through the skin over time can cause permanent harm to your body. Some solvents are flammable; never use cleaning materials near an open flame or cigarette. The fumes from cleaning chemicals can be toxic, so wear appropriate respirator and eye protection wherever you are using these products.

To keep work areas and equipment clean and operational, follow the steps in **SKILL DRILL 6-11**.

SAFETY TIP

Do not use flammable cleaners or water on electrical equipment.

▶ Wrap-Up

Ready for Review

- Tools and equipment should be used only for the task they were designed to do.
- Always have a safe attitude when using tools and equipment.
- Do not use damaged tools; inspect before using, then clean and inspect again before putting them away.
- Lockouts and tag-outs are meant to prevent technicians from using tools and equipment that are potentially unsafe.
- Many tools and measuring instruments have USCS or metric system markings to identify their size.
- Torque defines how much a fastener should be tightened.
- Torque specification indicates the level of tightness each bolt or nut should be tightened to; torque charts list torque specifications for nuts and bolts.
- Torque (or tension) wrenches tighten fasteners to the correct torque specification.
- Torque value—the amount of twisting force applied to a fastener by the torque wrench—is specified in foot-pounds (ft-lb), inch-pounds (in-lb), or newton meters (Nm).
- Torque wrench styles are beam (simplest and least expensive), clicker, dial, and electronic. Each gives an indication of when proper torque is achieved.
- Bolts that are tightened beyond their yield point do not return to their original length when loosened.
- Common wrenches include box end, open end, combination (most popular), flare nut (or flare tubing), open-end adjustable, and ratcheting box end.
- Box-end wrenches can loosen very tight fasteners, but open-end wrenches usually work better once the fastener has been broken loose.
- Use the correct wrench for the situation, so as not to damage the bolt or nut.
- Sockets grip fasteners tightly on all six corners and are purchased in sets.
- Sockets are classified as follows: standard or metric, size of drive used to turn them, number of points, depth of socket, and thickness of wall.
- The most common socket handle is a ratchet; a breaker bar gives more leverage, or a sliding T-handle may be used.
- Fasteners can be spun off or on (but not tightened) by a speed brace or speeder handle.
- Pliers hold, cut, or compress materials; types include slip-joint, combination, arc joint, needle-nose, flat, diagonal cutting, snap ring, and locking.
- Always use the correct type of pliers for the job.
- Cutting tools include bolt cutters, tin snips, and aviation snips.
- Allen wrenches are designed to fit into fasteners with recessed hexagonal heads.
- Screwdriver types include flat blade (most common), Phillips, Pozidriv, offset, ratcheting, and impact.
- The tip of the screwdriver must be matched exactly to the slot or recess on the head of a fastener.
- Magnetic pickup tools and mechanical fingers allow for the extraction and insertion of objects in tight places.
- Types of hammers include ball peen (most common), sledge, mallet, and dead blow.
- Chisels are used to cut metals when hit with a hammer.
- Punches are used to mark metals when hit with a hammer and come in different diameters and different points for different tasks; types of punches include prick, center, drift, pin, ward, and hollow.
- Pry bars can be used to move, adjust, or pry parts.
- Gasket scrapers are designed to remove gaskets without damaging surrounding materials.

- Files are used to remove material from the surface of an automotive part.
- Flat files come in different grades to indicate how rough they are; grades are rough, coarse bastard, second cut, smooth, and dead smooth.
- Types of files include flat, warding, square, triangular, curved, and thread.
- Bench vices, offset vices, drill vices, and C-clamps all hold materials in place while they are worked on.
- Taps are designed to cut threads in holes or nuts; types include taper, intermediate, and bottoming.
- A die is used to cut a new thread on a blank rod or shaft.
- Gear and bearing pullers are designed to remove components from a shaft when considerable force is needed.
- Flaring tools create flares at the end of tubes to connect them to other components; types include single, double, and ISO.
- Rivet tools join together two pieces of metal; each rivet can be used only once.
- Measuring tapes and steel rules are commonly used measuring tools; more precise measuring tools include micrometers, gauges, calipers, dial indicators, and straight edges.
- Micrometers can be outside, inside, or depth.
- Learn to read micrometer measurements on the sleeve/barrel and thimble; always verify the micrometer is properly calibrated before use.
- Gauges are used to measure distances and diameters; types include telescoping, split ball, and dial bore.
- Vernier calipers measure outside, inside, and depth dimensions; newer versions have dial and digital scales.
- Dial indicators are used to measure movement.
- A straight edge is designed to assess the flatness of a surface.
- Feeler blades are flat metal strips that are used to measure the width of gaps.
- Keep work area, tools, and equipment clean and organized.

Key Terms

Allen wrench A type of hexagonal drive mechanism for fasteners.

arc joint pliers Pliers with parallel slip jaws that can increase in size. Also called Channellocks.

aviation snips A scissor-like tool for cutting sheet metal.

ball-peen (engineer's) hammer A hammer that has a head that is rounded on one end and flat on the other; designed to work with metal items.

bench vice A device that securely holds material in jaws while it is being worked on.

blind rivet A rivet that can be installed from its insertion side.

Bolt A type of threaded fastener with a thread on one end and a hexagonal head on the other.

bolt cutters Strong cutters available in different sizes, designed to cut through non-hardened bolts and other small-stock material.

bottoming tap A thread-cutting tap designed to cut threads to the bottom of a blind hole.

box-end wrench A wrench or spanner with a closed or ring end to grip bolts and nuts.

C-clamp A clamp shaped like the letter C; it comes in various sizes and can clamp various items.

center punch Less sharp than a prick punch, the center punch makes a bigger indentation that centers a drill bit at the point where a hole is required to be drilled.

cold chisel The most common type of chisel, used to cut cold metals. The cutting end is tempered and hardened so that it is harder than the metals that need to be cut.

combination pliers A type of pliers for cutting, gripping, and bending.

combination wrench A type of wrench that has a box-end wrench on one side and an open end on the other.

crankshaft A vehicle engine component that transfers the reciprocating movement of pistons into rotary motion.

cross-arm A description for an arm that is set at right angles or 90 degrees to another component.

cross-cut chisel A type of chisel for metal work that cleans out or cuts key ways.

curved file A type of file that has a curved surface for filing holes.

dead blow hammer A type of hammer that has a cushioned head to reduce the amount of head bounce.

depth micrometer A measuring device that accurately measures the depth of a hole.

diagonal cutting pliers Cutting pliers for small wire or cable.

dial bore gauge An accurate measuring device for inside bores, usually made with a dial indicator attached to it.

dial indicator An accurate measuring device where measurements are read from a dial and needle.

die Used to cut external threads on a metal shank or bolt.

die stock A handle for securely holding dies to cut threads.

double flare A seal that is made at the end of metal tubing or pipe.

drift punch A type of punch used to start pushing roll pins to prevent them from spreading.

drill vice A tool with jaws that can be attached to a drill press table for holding material that is to be drilled.

fasteners Devices that securely hold items together, such as screws, cotter pins, rivets, and bolts.

feeler gauge A thin blade device for measuring space between two objects.

finished rivet A rivet after the completion of the riveting process.

flare nut wrench A type of box-end wrench that has a slot in the box section to allow the wrench to slip through a tube or pipe. Also called a flare tubing wrench.

flat blade screwdriver A type of screwdriver that fits a straight slot in screws.

flat-nose pliers Pliers that are flat and square at the end of the nose.

forcing screw The center screw on a gear, bearing, or pulley puller. Also called a jacking screw.

gasket scraper A broad sharp flat blade to assist in removing gaskets and glue.

gear pullers A tool with two or more legs and a cross bar with a center forcing screw to remove gears.

hard rubber mallet A special-purpose tool with a head made of hard rubber; often used for moving things into place where it is important not to damage the item being moved.

hollow punch A punch with a center hollow for cutting circles in thin materials such as gaskets.

impact driver A tool that is struck with a hammer to provide an impact turning force to remove tight fasteners.

inside micrometer A micrometer designed to measure internal diameters.

intermediate tap One of a series of taps designed to cut an internal thread. Also called a plug tap.

locking pliers A type of pliers where the jaws can be set and locked into position.

lockout/tag-out A safety tag system to ensure that faulty equipment or equipment in the middle of repair is not used.

lug wrench A tool designed to remove wheel lugs nuts and commonly shaped like a cross.

magnetic pickup tools An extending shaft, often flexible, with a magnet fitted to the end for picking up metal objects.

magnetic pickup tools Useful for grabbing items in tight spaces, it typically is a telescoping stick that has a magnet attached to the end on a swivel joint.

mandrel The shaft of a pop rivet.

mandrel head The head of the pop rivet that connects to the shaft and causes the rivet body to flare.

measuring tape A thin measuring blade that rolls up and is contained in a spring-loaded dispenser.

mechanical fingers Spring-loaded fingers at the end of a flexible shaft that pick up items in tight spaces.

micrometer An accurate measuring device for internal and external dimensions. Commonly abbreviated as "mic."

needle-nose pliers Pliers with long tapered jaws for gripping small items and getting into tight spaces.

nippers (pincer pliers) Pliers designed to cut protruding items level with the surface.

nut A fastener with a hexagonal head and internal threads for screwing on bolts.

offset screwdriver A screwdriver with a 90-degree bend in the shaft for working in tight spaces.

offset vice A vice that allows long objects to be gripped vertically.

oil filter wrench A specialized wrench that allows extra leverage to remove an oil filter when it is tight.

open-end wrench A wrench with open jaws to allow side entry to a nut or bolt.

outside micrometer A micrometer designed to measure the external dimensions of items.

parallax error A visual error caused by viewing measurement markers at an incorrect angle.

peening A term used to describe the action of flattening a rivet through a hammering action.

Phillips head screwdriver A type of screwdriver that fits a head shaped like a cross in screws.

pin punch A type of punch in various sizes with a straight or parallel shaft.

pipe wrench A wrench that grips pipes and can exert a lot of force to turn them. Because the handle pivots slightly, the more pressure put on the handle to turn the wrench, the more the grip tightens.

pliers A hand tool with gripping jaws.

pop rivet gun A hand tool for installing pop rivets.

prick punch A pinch with a sharp point for accurately marking a point on metal.

pry bar A high-strength carbon steel rod with offsets for levering and prying.

pullers A generic term to describe hand tools that mechanically assist the removal of bearings, gears, pulleys, and other parts.

punches A generic term to describe a high-strength carbon steel shaft with a blunt point for driving. Center and prick punches are exceptions and have a sharp point for marking or making an indentation.

ratchet A generic term to describe a handle for sockets that allows the user to select direction of rotation. It can turn sockets in restricted areas without the user having to remove the socket from the fastener.

ratcheting box-end wrench A wrench with an inner piece that is able to rotate within the outer housing, allowing it to be repositioned without being removed.

ratcheting screwdriver A screwdriver with a selectable ratchet mechanism built into the handle that allows the screwdriver tip to ratchet as it is being used.

roll bar Another type of pry bar, with one end used for prying and the other end for aligning larger holes, such as engine motor mounts.

screw extractor A tool for removing broken screws or bolts.

single flare A sealing system made on the end of metal tubing.

sledgehammer A heavy hammer, usually with two flat faces, that provides a strong blow.

sliding T-handle A handle fitted at 90 degrees to the main body that can be slid from side to side.

snap ring pliers A pair of pliers for installing and removing snap rings or circlips.

socket An enclosed metal tube commonly with 6 or 12 points to remove and install bolts and nuts.

speed brace A U-shaped socket wrench that allows high-speed operation. Also called a speeder handle.

split ball gauge A measuring device used to accurately measure small holes.

square file A type of file with a square cross section.

steel hammer A hammer with a head made of hardened steel.

steel rule An accurate measuring ruler made of steel.

straight edge A measuring device generally made of steel to check how flat a surface is.

tap A term used to generically describe an internal thread-cutting tool.

taper tap A tap with a tapper; it is usually the first of three taps used when cutting internal threads.

tap handle A tool designed to securely hold taps for cutting internal threads.

telescoping gauge A gauge that expands and locks to the internal diameter of bores; a caliper or outside micrometer is used to measure its size.

thread file A type of file that cleans clogged or distorted threads on bolts and studs.

thread pitch The coarseness or fineness of a thread as measured by either the threads per inch or the distance from the peak of one thread to the next. Metric fasteners are measured in millimeters.

tin snips Cutting device for sheet metal, works in a similar fashion to scissors.

torque Twisting force applied to a shaft that may or may not result in motion.

torque specifications Supplied by manufacturers and describes the amount of twisting force allowable for a fastener or a specification showing the twisting force from an engine crankshaft.

torque wrench A tool used to measure the rotational or twisting force applied to fasteners.

triangular file A type of file with three sides so it can get into internal corners.

tube flaring tool A tool that makes a sealing flare on the end of metal tubing.

tubing cutter A hand tool for cutting pipe or tubing squarely.

V blocks Metal blocks with a V-shaped cutout for holding shafts while working on them. Also referred to as vee blocks.

vernier calipers An accurate measuring device for internal, external, and depth measurements that incorporates fixed and adjustable jaws.

wad punch A type of punch that is hollow for cutting circular shapes in soft materials such as gaskets.

warding file A type of thin, flat file with a tapered end.

wrenches A generic term to describe tools that tighten and loosen fasteners with hexagonal heads.

Review Questions

1. Specialty tools:
   a. should not be shared among technicians.
   b. should have tag-out practices for regular storage.
   c. should be put in the proper storage space after each use.
   d. can be used for purposes they were not designed for.
2. A set of safety practices and procedures that are intended to reduce the risk of technicians inadvertently using tools that have been determined to be unsafe is known as:
   a. tag out.
   b. shop policy.
   c. equipment storage procedure.
   d. PPE maintenance.
3. Which of these wrenches can be awkward to use once the nut or bolt has been loosened a bit?
   a. Open-end
   b. Flare nut
   c. Combination
   d. Box-end
4. Oil filters should be installed using:
   a. an oil filter wrench.
   b. a flare nut wrench.
   c. your hand.
   d. arc joint pliers.
5. When tightening lug fasteners, an impact wrench should:
   a. always be used.
   b. never be used.
   c. be used in difficult-to-access positions.
   d. never be used in high-torque settings.
6. Which of these would you use for a wider grip and a tighter squeeze on parts too big for conventional pliers?
   a. Diagonal cutting pliers
   b. External snap ring pliers
   c. Arc joint pliers
   d. Combination pliers
7. When a large chisel needs a really strong blow, use a:
   a. sledgehammer.
   b. hard rubber mallet.
   c. dead blow hammer.
   d. ball-peen hammer.
8. Which style of torque wrench is the simplest and least expensive?
   a. Clicker
   b. Dial
   c. Beam
   d. Electronic
9. The depth of blind holes in housings can best be measured using a(n):
   a. measuring tape.
   b. steel rule.
   c. dial bore gauge.
   d. vernier calipers.

10. All of the following are good practices *except*:
   a. replacing cleaning products when they lose their effectiveness.
   b. discarding solvents into the sewage system immediately after use.
   c. using chemical gloves when using any cleaning material.
   d. wearing a respirator when using toxic cleaning chemicals.

ASE Technician A/Technician B Style Questions

1. Tech A says that knowing how to use tools correctly creates a safe working environment. Tech B says that a flare nut wrench is used to loosen very tight bolts and nuts. Who is correct?
   a. Tech A
   b. Tech B
   c. Both A and B
   d. Neither A nor B

2. Tech A says that lockout is designed to secure a vehicle after all work on it has been completed. Tech B says that tag-out is used when a tool is no longer fit for use and identifies what is wrong with it. Who is correct?
   a. Tech A
   b. Tech B
   c. Both A and B
   d. Neither A nor B

3. Tech A says that torque wrenches need to be calibrated periodically to ensure proper torque values. Tech B says that when using a clicker-style torque wrench, keep turning the torque wrench ⅛–¼ turn to make sure the bolt is properly tightened. Who is correct?
   a. Tech A
   b. Tech B
   c. Both A and B
   d. Neither A nor B

4. Tech A says that when using a micrometer, a "standard" is used to hold the part you are measuring. Tech B says that the micrometer spindle should be firmly closed against the anvil prior to storage. Who is correct?
   a. Tech A
   b. Tech B
   c. Both A and B
   d. Neither A nor B

5. Tech A says that a box-end wrench is more likely to round the head of a bolt than an open-end wrench. Tech B says that 6-point sockets and wrenches will hold more firmly when removing and tightening bolts. Who is correct?
   a. Tech A
   b. Tech B
   c. Both A and B
   d. Neither A nor B

6. Tech A says that it is usually better to pull a wrench to tighten or loosen a bolt. Tech B says that pushing a wrench will protect your knuckles if the wrench slips. Who is correct?
   a. Tech A
   b. Tech B
   c. Both A and B
   d. Neither A nor B

7. Tech A says that a feeler gauge is used to measure the diameter of small holes. Tech B says that a feeler gauge and straight edge are used to check surfaces for warpage. Who is correct?
   a. Tech A
   b. Tech B
   c. Both A and B
   d. Neither A nor B

8. Tech A says that a dead blow hammer reduces rebound of the hammer. Tech B says that a dead blow hammer should be used with a chisel to cut the head of a bolt off. Who is correct?
   a. Tech A
   b. Tech B
   c. Both A and B
   d. Neither A nor B

9. Tech A says that gaskets can be removed quickly and safely with a hammer and sharp chisel. Tech B says that extreme care must be used when removing a gasket on an aluminum surface. Who is correct?
   a. Tech A
   b. Tech B
   c. Both A and B
   d. Neither A nor B

10. Tech A says that when using a file, apply pressure to file in the direction of the cut and no pressure when pulling the file back. Tech B says that file cards are used to file uneven surfaces. Who is correct?
   a. Tech A
   b. Tech B
   c. Both A and B
   d. Neither A nor B

CHAPTER 7

Power Tools and Equipment

NATEF Tasks

There are no NATEF tasks for this chapter.

Knowledge Objectives

After reading this chapter, you will be able to:

- **K07001** Explain when and how batteries are charged and jump-started.
- **K07002** Identify and describe the main categories of power tools.
- **K07003** Identify and describe the use of air tools.
- **K07004** Identify and describe the use of electric power tools.
- **K07005** Describe the application and use of soldering tools.
- **K07006** Identify and describe the use of cleaning tools and equipment.
- **K07007** Identify and describe the use of a hydraulic press.
- **K07008** Identify and describe the use of welding and cutting equipment.
- **K07009** Identify and describe the use of chassis dynamometers.

Skills Objectives

After reading this chapter, you will be able to:

- **S07001** Charge a battery.
- **S07002** Jump-start a vehicle.
- **S07003** Use an air nozzle.
- **S07004** Use an air impact wrench.
- **S07005** Use an air drill.
- **S07006** Use an air hammer.
- **S07007** Use a bench grinder.
- **S07008** Use an angle grinder.
- **S07009** Use soldering tools.
- **S07010** Use a pressure washer.
- **S07011** Use a spray wash cabinet.
- **S07012** Use a solvent tank.
- **S07013** Use a brake washer.
- **S07014** Use a sand or bead blaster.
- **S07015** Purge an oxyacetylene torch.
- **S07016** Set up an oxyacetylene torch.
- **S07017** Use an oxyacetylene torch for heating.
- **S07018** Use a welding tip for welding.
- **S07019** Use a welding tip for brazing.
- **S07020** Use a cutting torch.
- **S07021** Use a plasma cutter.
- **S07022** Use a wire feed welder.

▶ Introduction

Power tools and shop equipment use electric or air power to increase the amount of work that technicians can perform in a given amount of time (**FIGURE 7-1**). This results in more pay for both the shop and the technician, but it also comes with risk of damage to the vehicle or injury to the technician if the power tool or equipment isn't used properly. So it is important that you learn how to use these tools safely. The following topics introduce the safe working procedures of some of the common power tools and equipment used in the shop. But we can't cover all of the differences between makes and models, so always refer to the manufacturer's operator's manual, and get your instructor's permission before using any power tools or equipment.

▶ Battery Charging and Jump-Starting

K07001

Batteries are common in shops and are used in vehicles and rechargeable tools. Extreme caution should be taken when working around or with batteries, whether or not they are being charged. Batteries produce dangerous and explosive hydrogen gases, so sparks and short circuits across the battery terminals should be avoided. Always wear appropriate PPE, such as goggles, gloves, and protective clothing, when working around or with batteries. Also, some regulatory agencies require that an eyewash station be located near a battery charging station in case of a battery explosion. Check with your local authorities to determine any distance requirements.

There are many different types of **battery chargers**, and each is designed for a particular purpose and application. Battery chargers can be **fast chargers**, with high current output to charge a battery quickly. **Slow chargers** take longer to charge a battery and have lower current outputs; they put less stress on the battery, which is ideal if time is not a consideration.

Smart chargers incorporate microprocessors to monitor and control the charge rate, so the battery receives the correct amount of charge depending on its state of charge (**FIGURE 7-2**). These types of chargers are becoming more popular and ensure that the battery receives the optimal charge, promoting longer battery life. Even though a motor vehicle battery typically is 12 volts, it stores a lot of energy. The high current supply from a battery can be very dangerous. Remember, batteries have to deliver enough power to crank over a cold engine.

Batteries also produce enough power to melt a metal rod resting across the terminals. High-voltage battery packs, like those fitted to hybrid vehicles, are even more dangerous because of the potential for high voltage and current, so special precautions for dealing with high-voltage systems must be taken (**FIGURE 7-3**). Always treat batteries with care and respect.

▶ TECHNICIAN TIP

Vehicle batteries are usually lead acid types, often in a 12-volt configuration. If a vehicle requires 24 volts, then either one 24-volt battery or two 12-volt batteries can be connected in series (by connecting the positive of one battery to the negative of the other battery) to provide 24 volts.

FIGURE 7-1 Power tools increase a technician's productivity.

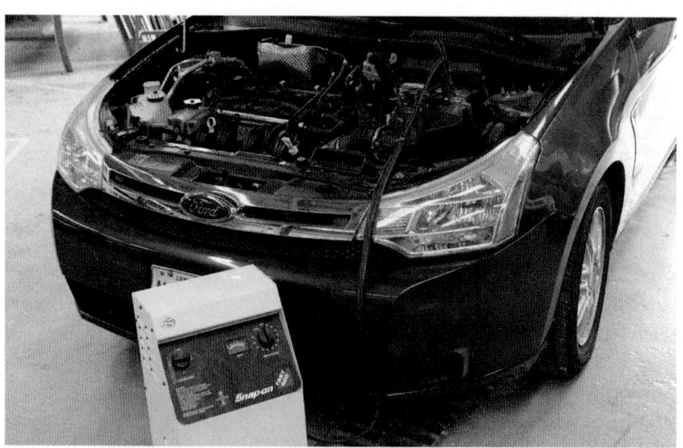

FIGURE 7-2 Smart battery charger.

You Are the Automotive Technician

The shop just agreed to accept an automotive student intern from the local college for the semester. Because you remember what it is like to be a new person in the shop, you volunteered to supervise the student and help get her up to speed. Because she will need to be familiar with certain equipment in the shop, you want to make sure you train her properly. To prepare for the training, you ask yourself the following questions, and write down the answers to make sure you cover all of the important details with the intern.

1. What are the precautions when charging batteries?
2. What are the precautions when working around compressed air?
3. What safety requirements are needed when using a bench grinder?
4. What are the steps for using a spray wash cabinet?

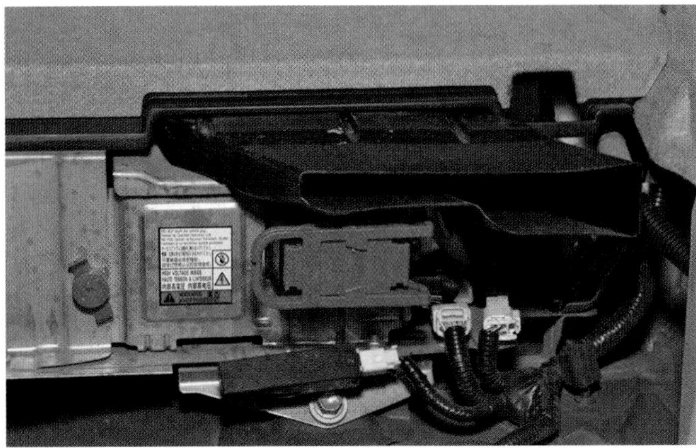

FIGURE 7-3 High-voltage battery packs, like those fitted to hybrid vehicles, are extremely dangerous because of the potential for high voltage and current. Take special precautions for dealing with high-voltage systems.

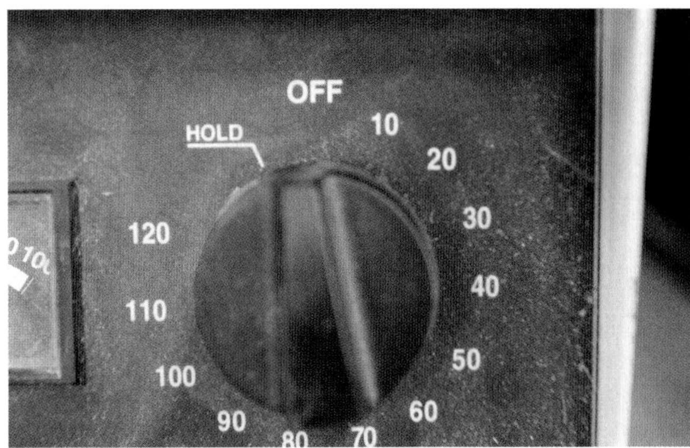

FIGURE 7-4 Switch off the charger before connecting it to or disconnecting it from the battery.

Technology changes in vehicles have increased the need for more electrical power, and in the future this will drive the need for higher-voltage battery systems. Hybrid vehicles are an example of this. Their operating voltages are typically 200–600 volts. Generally, the higher the system voltage, the more efficient the system is because the electrical current can be lower for a given amount of power. And because current determines wire size, the wires can be smaller. However, higher voltages also create a greater shock hazard, so bear this in mind when working around and with batteries.

SAFETY TIP

High voltages used in a hybrid vehicle are extremely dangerous. The voltage and current flow is several times greater than that needed to kill a person. Most hybrid manufacturers require technicians to undergo special factory training before they will allow them to service a hybrid vehicle. Also, they usually allow only very experienced technicians to undergo the training, not novices. In fact, one of the tools that Toyota requires of their shops for working on a hybrid vehicle is a nonconductive shepherd's hook. This can be used to drag a technician away from high voltage if the technician is electrocuted while working on the vehicle.

SAFETY TIP

Always remove your hand, wrist, and neck jewelry before working with batteries and electrical systems. If any of these come into contact with the battery terminals or power wire, it can cause a short circuit. You will receive painful skin burns from the very rapid heating of the metal you are wearing, or even flash burns from an arcing current. A wristwatch or ring is much harder to take off when it is red- or white-hot and burned onto your skin!

SAFETY TIP

Batteries give off hydrogen gas while they are being charged and for some time afterward. Hydrogen is a light and highly explosive gas that is easily ignited by a simple spark. Batteries are filled with **sulfuric acid**, so if the hydrogen does explode, the battery case can then rupture and spray everything and everyone nearby with this dangerous and corrosive liquid. Be very careful not to create a spark when you are connecting or disconnecting battery cables or hooking up a charger to the battery terminals. Switch off the charger before connecting and disconnecting them from the battery (**FIGURE 7-4**).

Do not try to charge a battery faster than the battery manufacturer recommends, and never use a battery load tester immediately after charging a battery. This is because both charging and rapidly discharging a battery generate heat and hydrogen. If you load-test a battery after charging it without waiting for it to cool down, you will increase the risk of distorting the plates inside the battery, thus increasing the risk of explosion if the plates end up touching each other.

How to Charge Batteries

S07001

Batteries go dead for a variety of reasons. A common cause is the driver forgetting to turn off the headlights when exiting the vehicle. Or maybe the owner went on vacation for a month, and the battery discharged slowly over that time. Because the battery only stores electricity, anything that stays on when the vehicle is not running drains the battery. The further discharged a battery is, the more it needs to be recharged.

When charging a battery, slow charging a battery is less stressful on the battery than fast charging, so, if possible, slow charge a battery instead of fast charging it. Removing the negative battery terminal while changing a battery reduces the risk of burning up any electronic devices on the vehicle, making sure that any excessive voltage from the battery charger doesn't get applied to the vehicle's electrical system. This is especially true with today's electronically intensive cars. However, disconnecting the vehicle's battery risks losing information such as radio presets and other learned data. Use a **memory saver (memory minder)**, which provides backup power to retain electronic memory settings in the vehicle's computer systems (**FIGURE 7-5**).

In some vehicles, such as in diesel pickup trucks and SUVs, manufacturers install multiple batteries to provide additional battery power when needed. Knowing how the batteries are connected together helps you determine how you connect a battery charger properly. Batteries can be connected in series (connected in line with each other, with the positive of one connected to the negative of the other) or parallel (connected side by side, with positive connected to positive and negative to negative). Two

12-volt batteries connected in series have the positive terminal of one battery connected to the negative terminal of the second battery (**FIGURE 7-6**). For example, two 12-volt batteries connected in series have a nominal output voltage of 24 volts across the most negative and most positive battery terminals. If you have a 24-volt battery charger, you can charge both batteries at once by connecting the battery charger to these same terminals. If you only have a 12-volt charger, you will have to either charge one battery at a time or reconnect them so they are connected in parallel. Just make sure you charge both batteries fully, which could take more than 12 hours each if slow charging.

Batteries connected in parallel have the negative terminals of both batteries connected to each other and the positive terminals of each battery connected to each other (**FIGURE 7-7**). The output voltage will be equal to the voltage of one battery and is taken from the positive and negative posts of either battery. For example, two 12-volt batteries connected in parallel have an output voltage of about 12 volts. In this situation, a 12-volt charger can be used to charge both batteries at the same time while the batteries are connected together. But it is likely to take about twice as long as it would if only charging one battery. After charging and reinstalling a battery, it is good practice to clean the battery terminals and posts with a battery terminal cleaner.

FIGURE 7-5 Memory savor being hooked up to a vehicle.

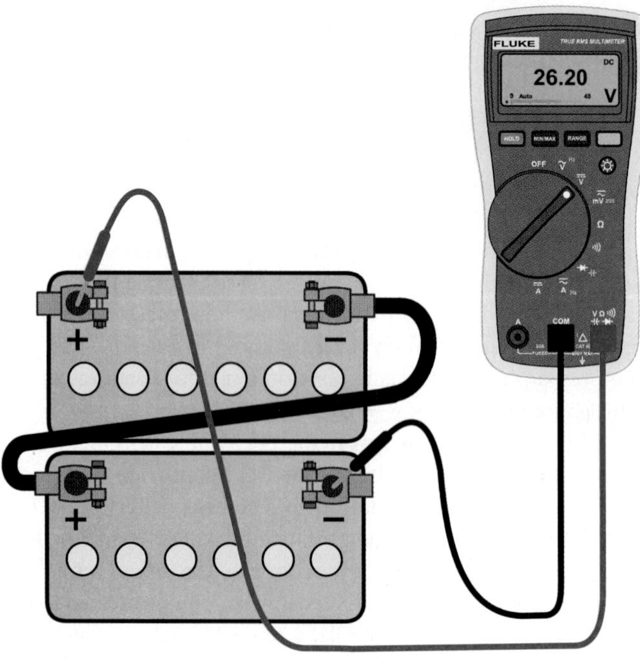

FIGURE 7-6 Batteries connected in series.

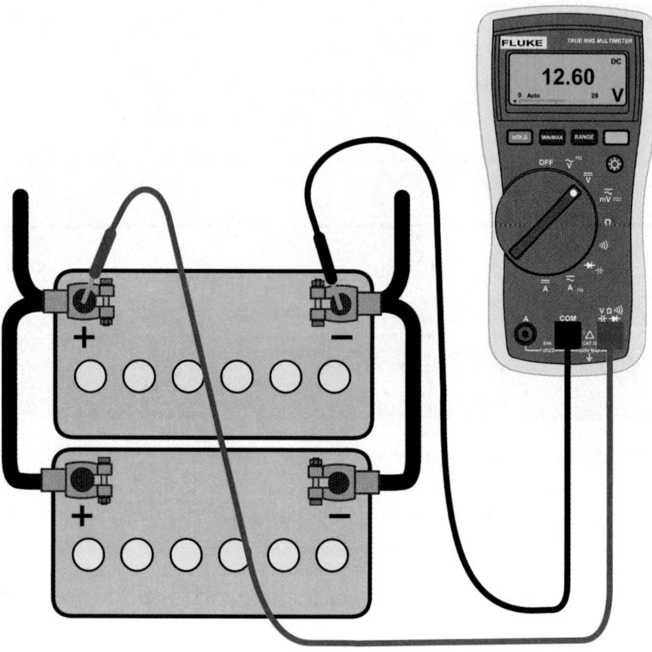

FIGURE 7-7 Batteries connected in parallel.

To correctly charge a battery using battery charging equipment, follow the steps in **SKILL DRILL 7-1**.

Jump-Starting Vehicles

S07002

Jump-starting a vehicle is the process of using one vehicle with a charged battery to provide electrical energy to start another vehicle that has a discharged or dead battery. Because starting a vehicle requires a high amount of electrical energy, jump-starting a vehicle can put stresses on both vehicles. When the discharged vehicle is being cranked, the battery voltage tends to fall very low because the battery is already discharged. This causes the alternator on the running vehicle to put out its maximum current, putting it under heavy load. But as soon as the jumped vehicle stops cranking, the voltage shoots up quickly, potentially high enough to damage electronic components in either vehicle. The same voltage spike can happen when the jumper cables are being disconnected.

Some vehicle manufacturers are now recommending that their vehicles should not be jump-started and that instead the battery should be charged or replaced. In the same way, some towing companies have policies stating that they will not jump-start certain vehicles and that they will only replace the battery or tow the vehicle to a shop to be recharged. If you do decide to jump-start a vehicle, *always* read the owner's manual for both vehicles, and follow their jump-starting guidelines. And never attempt to jump-start a frozen battery.

It is usually best to let the running vehicle charge the battery on the other vehicle for 5 to 10 minutes before trying to start

SKILL DRILL 7-1 Charging Batteries

1. Determine the voltage of the system that needs charging. If you are charging a 12-volt battery, use the 12-volt setting on the charger. If you are charging a 24-volt battery, or two 12-volt batteries connected in series, use the 24-volt setting on the charger, if it has one.

2. Identify the positive and negative terminals. Never simply use the color of the cables to determine the positive or negative terminals; use the + and – or the "Pos" and "Neg" marks.

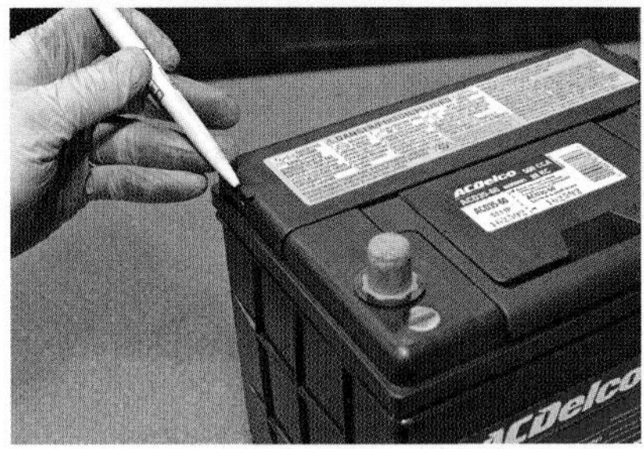

3. Inspect the battery by carrying out a visual inspection of the battery to ensure there are no cracks, holes, or damage to the casing.

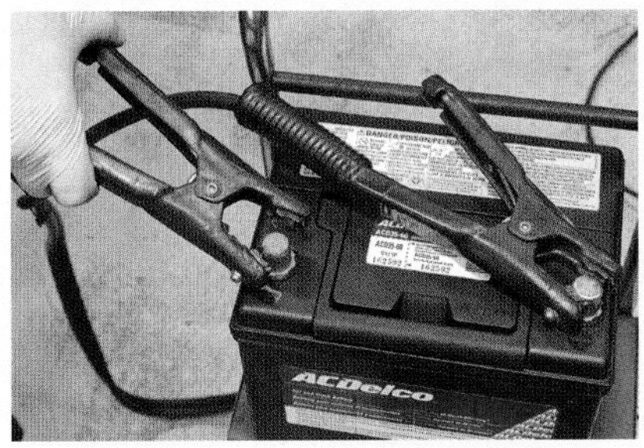

4. Verify that the charger is unplugged from the wall and turned off. Connect the red lead from the charger to the positive battery terminal. Connect the black lead from the charger to the negative battery terminal.

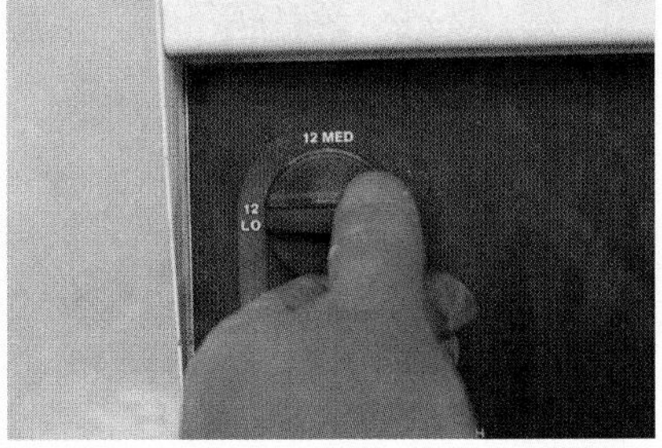

5. Check the settings on the charger, and verify that they are correct for what you are charging. Turn the charger on, and select the automatic setting, if equipped. Select the rate of charge. A fast charge should be carried out only under constant supervision.

6. Verify that the voltage and amperage the charger is putting out are proper.

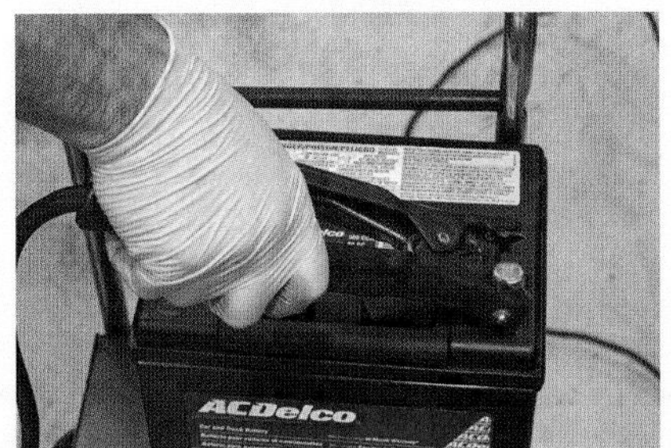

7. Once the battery is charged, turn the charger off. Disconnect the black lead from the negative battery terminal and the red lead from the positive battery terminal.

8. Allow the battery to stand for at least five minutes before testing the battery. Using a load tester or hydrometer, test the charged state of the battery.

the vehicle. Once the dead vehicle's engine has started, let both vehicles come to an idle for a moment. Do not turn off the vehicle with the discharged battery; it needs an extended amount of time to recharge. Before you disconnect the service battery from the discharged battery, it is good practice to place a load on the charged battery by turning on an accessory such as the headlights. This helps absorb any sudden rise in voltage that may occur as the load on the alternator is suddenly decreased. Another method of reducing the risk of damage to sensitive electronic devices is by using jumper leads that have a built-in or auxiliary **surge protector**.

To start a vehicle with a discharged battery, using jumper leads and a second vehicle or auxiliary battery, follow the steps in **SKILL DRILL 7-2**.

SKILL DRILL 7-2 Jump-Starting Vehicles

1. Position the charged battery close enough to the discharged battery that it is within comfortable range of your jumper cables. If the charged battery is in another vehicle, make sure the two vehicles are not touching.

2. First, connect the red jumper lead to the positive terminal of the discharged battery in the vehicle you are trying to start.

3. Next, connect the other end of this lead to the positive terminal of the charged battery or the remote terminal.

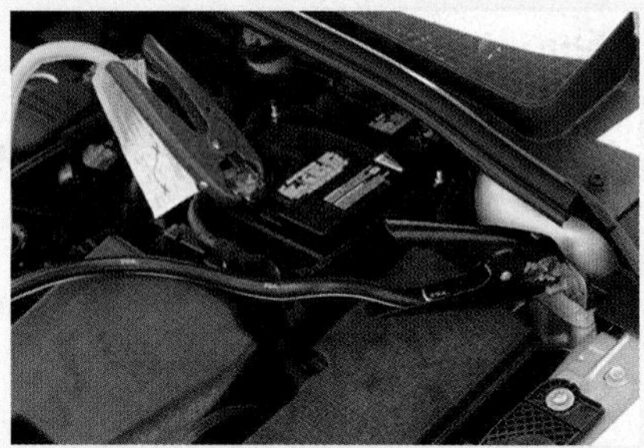

4. Then connect the black jumper lead to the negative terminal of the charged battery or the battery remote terminal.

5. Connect the other end of the negative lead to a good ground on the engine block of the vehicle with the discharged battery, and as far away as possible from the battery.

6. Do not connect the lead to the negative terminal of the discharged battery itself; doing so may cause a dangerous spark. Also, do not connect the negative lead to the body or chassis as the ground wire from the body back to the negative battery terminal is usually too small to carry current needed for jumpstarting the vehicle.

7. Try to start the vehicle with the discharged battery. If the booster battery does not have enough charge or the jumper cables are too small in diameter to do this, start the engine in the booster vehicle, and allow it to partially charge the discharged battery for several minutes. Turn on the headlights on the booster vehicle to reduce the possibility of a voltage spike damaging electronic equipment, and try starting the first vehicle again.

8. Disconnect the leads in the reverse order of connecting them.

9. If the charging system is working correctly and the battery is in good condition, the battery will be recharged while the engine is running, although it could end up overheating and damaging the alternator.

▶ TECHNICIAN TIP

Alternators are not generally designed to charge a dead battery while running several accessories. If a battery must be jump-started, it is always best to recharge the battery using a battery charger. That way, the alternator will not have to work so hard. Some technicians say it takes only 15 minutes to burn up an alternator when charging a dead battery. And if you must drive the vehicle after being jump-started, turn off as many accessories as possible, which will lessen the load on the alternator.

SAFETY TIP

When connecting jumper cables, a spark will almost always occur on the last connection you make. That is why it is critical that you make the last connection on the engine block away from the battery and any other flammables. A spark also occurs when you disconnect the first jumper cable connection, so that also needs to be the connection at the engine block.

SAFETY TIP

Make sure the hood is secured with a hood prop before going under it; otherwise, it could fall and injure someone or short out the jumper cables, causing a spark, which could result in the battery blowing up.

SAFETY TIP

- Keep your face and body as far back as you can while connecting jumper leads.
- Do not connect the negative cable to the discharged battery, because the spark may blow up the battery.
- Use only specially designed heavy-duty jumper cables to start a vehicle with a dead battery. Do not try to connect the batteries with any other type of cable.
- Always make sure you wear the appropriate PPE before starting the job. Remember, batteries contain sulfuric acid, and it is very easy to injure yourself.
- Always follow any manufacturer's personal safety instructions to prevent damage to the vehicle you are servicing.

▶ Power Tools

K07002

Power tools are typically powered by an electricity or compressed air. They can also be powered by burning of propellant, such as in a nail gun, or by a gasoline engine, such as in a portable compressor. A power tool can be stationary, such as a bench grinder, or portable, such as a portable electric drill. There are many different power tools designed to perform specific tasks (**FIGURE 7-8**). Some are corded and have to be plugged in; others are cordless and have batteries. Power tools make many tasks quicker and easier to perform, and they can save many hours of work when used and maintained correctly.

FIGURE 7-8 Power tools.

▶ Air Tools

K07003

Air Compressors and Equipment

Compressed air powers a wide range of tools and equipment in a compressed air system. Tools that use compressed air include impact wrenches, drills, grinders, pumps, grease guns, air nozzles, and jacks (**FIGURE 7-9**). The compressed air system is made up of a compressor, a pressure regulator, air hose or fixed piping, and the actual tool or item that is powered by the compressed air. The air compressor has a storage tank and is driven by an electric motor or, for more portability, a gasoline engine. In many shops, the air compressor is housed in a separate room to help isolate the noise.

Standard compressors have been, and many still are, based on piston-type compressors, which operate similarly to a piston engine, where each piston compresses the air on the compression stroke. But in the case of a compressor, the piston forces the air out of the cylinder and into a storage tank under pressure. Many newer air compressors are of the scroll compressor type, which uses a pair of rotating scrolls to compress the air and is much quieter and more efficient. With either type of compressor, a pressure regulator controls the air pressure being supplied to the distribution system while the air hose or lines transport the compressed air from the compressor to the tool. In many states, the maximum air pressure allowed in a shop's compressed air system is regulated to a certain maximum pressure to minimize accidents.

SAFETY TIP

Always respect compressed air: Wear appropriate PPE, and never use compressed air inappropriately, such as using it to blow dust off yourself.

Compressed Air Safety

Serious, sometimes fatal, injuries can be caused by compressed air being injected into the body through the skin or into a body opening, such as your mouth or ear. Do not play with air equipment,

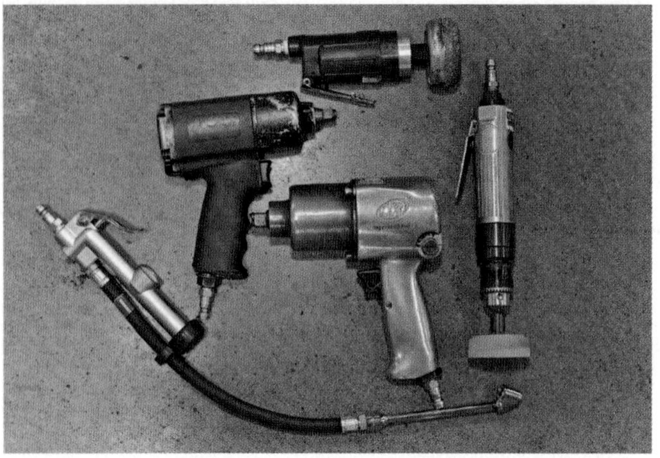

FIGURE 7-9 A wide variety of tools use compressed air, including drills, grinders, pumps, grease guns, jacks, and impact wrenches.

FIGURE 7-10 Water trap air drier.

such as blowing air at another person or yourself. Internal human blood vessels and organs will rupture at much lower pressures than those found in compressed air lines. So always handle air equipment carefully and with respect. Be extra careful when working with air equipment in a confined or awkward space, such as under a vehicle, and when clearing or cleaning the equipment, because the air pressure can blow dirt, debris, or liquids back at you with high force. Another issue to be careful of is air hoses that are starting to balloon. That means that the internal webbing in the hose has broken and is in danger of rupturing. If that happens when you are holding it, air can be injected into your skin.

FIGURE 7-11 Chiller-type air drier.

SAFETY TIP

If air, grease, or any other substance is injected into someone's body, it is not always easy to tell how much damage has been done. Always provide initial first aid, and then seek medical advice if any kind of penetration injury occurs.

Air Driers and Automatic Oilers

Air driers are fitted to compressed air systems to remove the moisture or water from the compressed air that is a result of compressing air from the atmosphere, which contains water in the form of humidity. If water gets into the air lines, it may damage the inside workings of air tools. An air drier can be a stand-alone device fitted to the compressed air system or can be incorporated into a filter/regulator system. The combination filter/regulator system removes water from the air, filters any debris that may come from the compressor tank, and regulates the line pressure from the tank.

A simple air drier can be nothing more than an inline water trap that catches water droplets that have condensed out of the air as it cooled down (**FIGURE 7-10**). This type of air drier usually needs to be drained manually on a periodic basis. However, some devices are equipped with an automatic drain system. Another type of air drier is the chiller type, which uses the principles of air conditioning to chill the hot compressed air and force condensation to occur at a higher rate (**FIGURE 7-11**). This system is much more expensive but does a much more effective job of removing all traces of moisture from the compressed air. You are likely to find this system in a body shop, where moisture in the compressed air must be removed so that it cannot mix with the paint as it is being sprayed onto the surface of the vehicle.

Compressed air tools and equipment require a regular application of a lubricating oil to reduce wear and tear. This is typically done by adding a few drops of air tool oil to the air fitting prior to use each day. This will lubricate the internal workings of the air tool (**FIGURE 7-12**). **Automatic oilers** are designed to regularly oil an air tool or air equipment so it does not have to be done manually before or during its use. Automatic oilers are usually fitted in the air hose near the air tool or equipment and regularly supply small amounts of oil into the stream of compressed air, which is then transported along with air to the tool or equipment (**FIGURE 7-13**). Automatic oilers need periodic inspection to make sure they deliver the correct amount of oil. The built-in oil reservoir also needs to be refilled with air tool oil on a regular schedule.

Air Tools

Air tools use compressed air at high pressure to operate (**FIGURE 7-14**). Air compressors in automotive shops

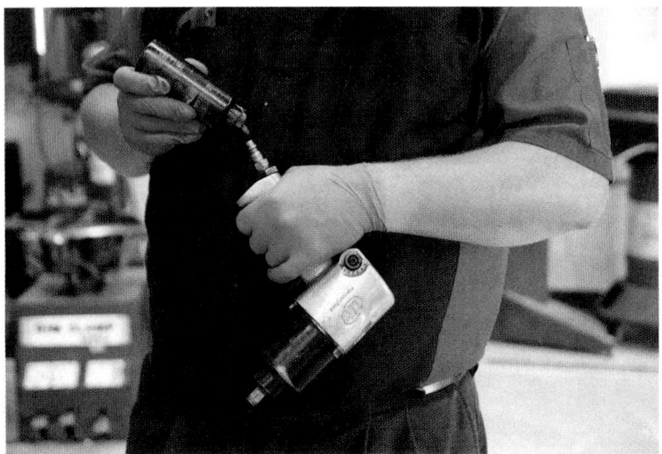

FIGURE 7-12 Manually lubricating air tools.

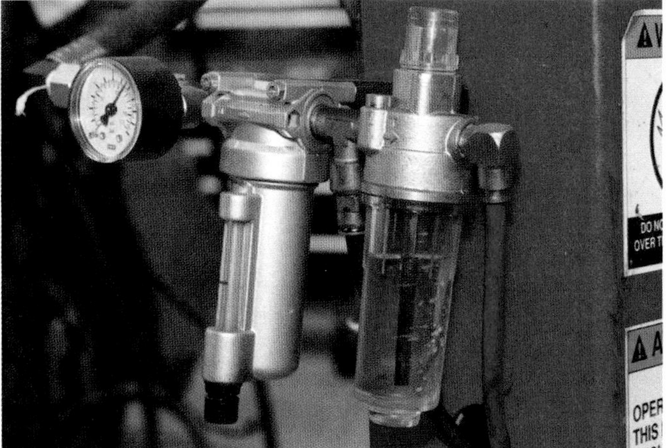

FIGURE 7-13 Compressed air automatic oiler.

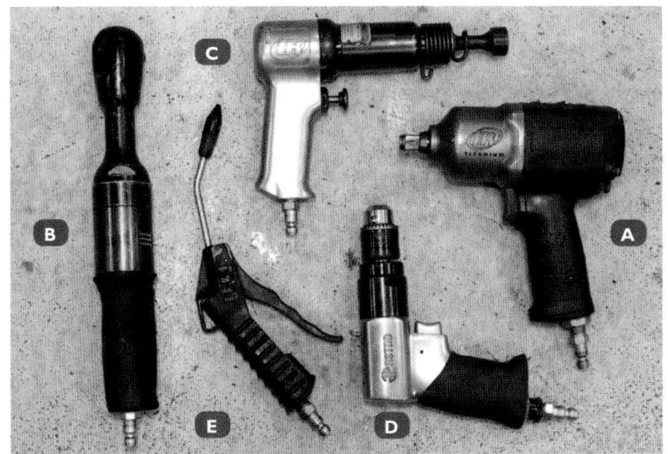

FIGURE 7-14 **A.** Air impact wrench. **B.** Air ratchet. **C.** Air hammer. **D.** Air drill. **E.** Blowgun or air nozzle.

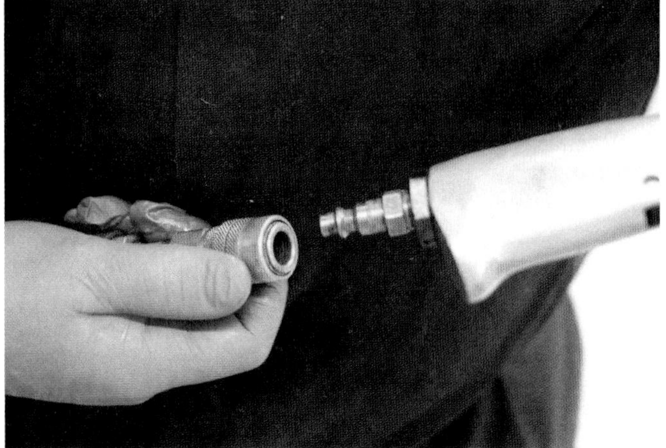

FIGURE 7-15 Quick disconnect fitting.

FIGURE 7-16 Air ratchet being used.

typically run at greater than 90 psi (621 kPa), so caution needs to be exercised when working with them. Compressed air is transported through pipes and hoses. Air tools have quick-connect fittings so that various air tools can be used on the same air hose (**FIGURE 7-15**). There are several styles of quick-connect fittings, and a shop will usually use one style throughout the entire shop.

The most common air tool in an automotive shop is the **air impact wrench**. It is sometimes called an impact gun or **rattle gun**, and it is easy to understand why when you hear one. Taking the wheels off a car to replace the tires is a typical application for this air tool. Removing lug nuts often requires a lot of torque to free the lug nuts, and air impact wrenches work well for that. The air impact wrench can be set to spin in either direction, and a valve roughly controls how much torque it applies. It should never be used for final tightening of lug nuts. There is a danger in over-tightening the lug nuts, as it can cause the lug studs to fail and the wheel to separate from the vehicle while it is moving. Another rule with the air impact wrench is that you have to use special hardened impact sockets, extensions, and joints. The sockets are special heavy-duty, six-point types, and the flats can withstand the hammering force that the impact wrench subjects them to.

An **air ratchet** uses the force of compressed air to turn a ratchet drive (**FIGURE 7-16**). It is used on smaller nuts and bolts. Once the nut is loosened, the air ratchet spins it off in a fraction of the time it would take by hand. It also works well where there isn't much room to swing a ratchet handle.

FIGURE 7-17 Air hammer being used.

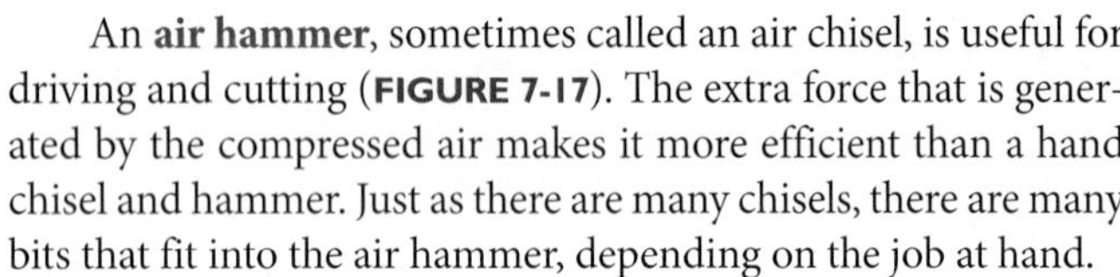

An **air hammer**, sometimes called an air chisel, is useful for driving and cutting (**FIGURE 7-17**). The extra force that is generated by the compressed air makes it more efficient than a hand chisel and hammer. Just as there are many chisels, there are many bits that fit into the air hammer, depending on the job at hand.

An **air drill** has some important advantages over the more common electric power drill. With the right attachment, it can drill holes, grind, polish, and clean parts. Unlike the electric drill, it does not run the risk of producing sparks, which is important around flammable liquids or gasoline tanks. An air drill does not trail a live electric cable behind it that could be cut, possibly causing shock and burns. It also does not get hot with heavy use.

A blowgun, or **air nozzle**, is probably the simplest air tool. It controls the flow of compressed air. It is controlled by a lever or valve that is used to blast debris and dirt out of confined spaces. Blasting debris and dirt can be dangerous, so eye protection must be worn whenever this tool is used. Noise levels are usually high, so ear protection should also be worn. It is dangerous to use an air nozzle to clean yourself off. Its blast should always be directed away from the user and anyone else working nearby. Also note that OSHA-approved nozzles lower the tip pressure by venting some of the air for safety reasons (**FIGURE 7-18**). Only use OSHA-approved nozzles for general blowing purposes; otherwise, the shop could be liable for a substantial fine.

Using Air Nozzles

S07003

An air nozzle can be a handy tool for blowing dirt and debris out of holes, for example, around spark plugs prior to removing them. It is helpful to activate the air nozzle away from the area you intend to blow off before doing it; this way you can get a feel for how the valve operates. To operate an air nozzle, pull the trigger gently, and modulate the flow of air through the nozzle. If too much air is allowed through, you may blow dirt particles back at you, so always wear your safety glasses. Also keep your mouth closed, as it helps prevent eating a bunch of dirt! To correctly operate an air nozzle, follow the steps in **SKILL DRILL 7-3**.

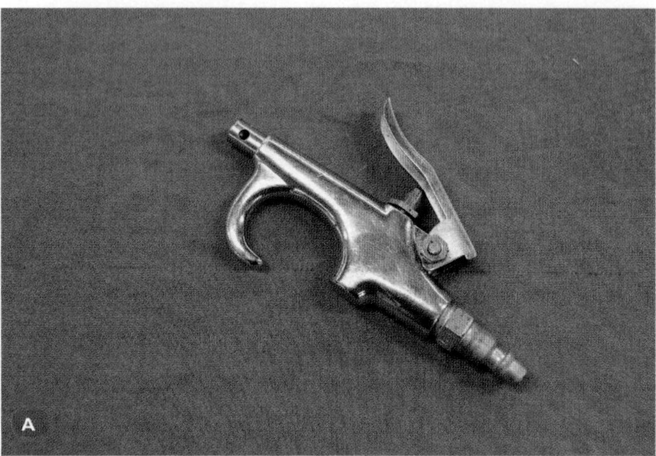

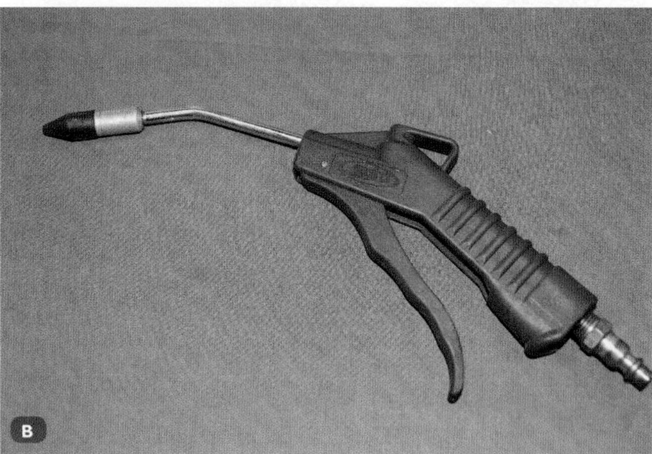

FIGURE 7-18 A. OSHA-approved air nozzle. **B.** Non-OSHA-approved air nozzle.

SAFETY TIP

- Do not use the air nozzle to clean brake dust from brake components. It will disperse the hazardous dust.
- Do not use a high-pressure air nozzle to disperse liquid solvents or fuels. A low-pressure blowing action can help these volatile materials to evaporate more quickly, but a high-pressure air jet could atomize the liquid, allowing it to form a flammable mixture.
- Do not point the air nozzle at other people.
- Never use the air nozzle to blow air over yourself or other people.
- Always wear eye protection when using air tools.
- Do not drive over the air hoses.
- Make sure the hoses are in good condition before using.

Using Air Impact Wrenches

S07004

The amount of torque an air impact wrench can produce is determined by the tool and the pressure in the air system feeding it. Because this pressure varies, there is no way of

SKILL DRILL 7-3 Using Air Nozzles

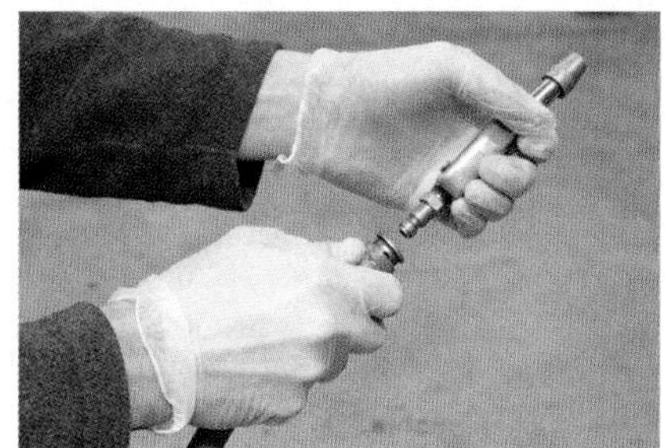

1. Fit an OSHA-approved air nozzle to the end of the air hose. Make sure there are no air leaks.

2. The air nozzle is used to blast dirt and debris out of confined spaces. To avoid injury, be sure to wear eye and ear protection whenever you use the air nozzle.

3. Do *not* use the air nozzle to dust yourself off because you risk injury. Be sure to direct the air jet away from yourself and away from anyone else who may be working nearby.

determining how much torque an impact wrench is applying to a fastener, so it is easy to over- or under-tighten fasteners. An air impact wrench can generally be used to take up the looseness in a nut or stud, but the final tightening must be performed by using a torque wrench set to the manufacturer's specifications. Every impact wrench has a control mechanism that allows it to be driven in either direction, and the ability to generally regulate the amount of torque the wrench develops. Always use six-point impact sockets when using an air impact wrench. These sockets are manufactured from a different blend of materials and have thicker walls than a standard wall socket, which could shatter during use. To correctly operate an air impact wrench, follow the steps in **SKILL DRILL 7-4**.

SKILL DRILL 7-4 Using Air Impact Wrenches

1. Select the properly sized impact gun and socket, and inspect them for damage.

2. Lubricate the gun if an automatic oiler is not installed in the system.

3. Adjust the direction of spin—forward or backward—with the selector.

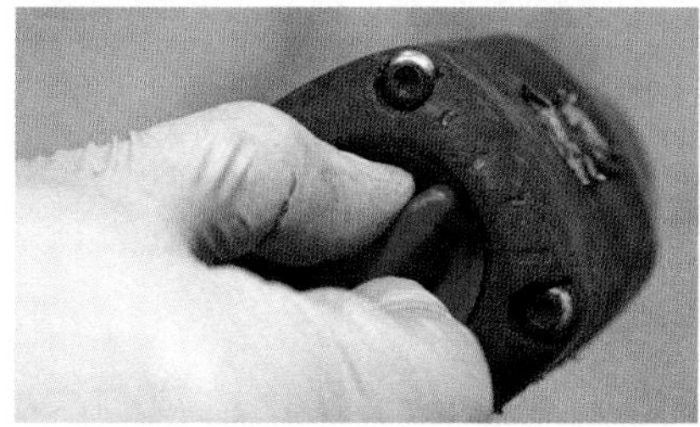

4. Turn the valve to increase or reduce the torque to match the needs of the fastener. If removing lug nuts, adjust it toward the upper middle torque setting. If running a nut back on a stud, adjust it toward the lowest torque setting. You can always readjust it if you end up needing more torque.

5. Place the impact wrench fully over the bolt or nut and give the trigger a quick squeeze to verify that the impact wrench is turning the correct direction.

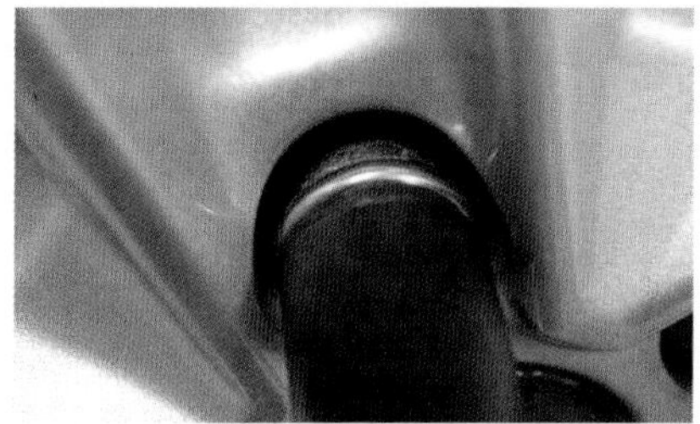

6. Continue to remove or install the fastener by squeezing the trigger only long enough to get the job done. Release the trigger before getting to the end as it takes time for the impact wrench to slow down; otherwise the nut could fly off the end of the stud, or if tightening, you could end up over tightening.

SAFETY TIP

- Air tools generally produce more noise than electric tools, so wear ear protection when using air tools.
- Always use impact sockets with impact guns.

Using Air Drills

S07005

An air drill is another tool operated by compressed air. Since they are powered by air and not electricity, they are safer to use in an environment where flammable materials are present. The amount of torque an air drill can produce is less than most electric drills, but will be determined by the pressure in the air system feeding it. Air drills are smaller and typically turn at slower speeds than electric drills. At the same time, air drills operate similarly to their electric counterparts and are equipped with the same type of drill chuck, which clamps onto the drill bit using a special key. Most air drills are of the 90-degree angle style and fit in places a typical electric drill might not. To correctly operate an air drill, follow the steps in **SKILL DRILL 7-5**.

Using Air Hammers

S07006

Air hammers act in a manner similar to a jackhammer; however, their size makes their cycling rate faster. There are a variety of attachments, so use the correct attachment for the task you are performing. This could be a chisel for cutting a bolt or a punch for driving out a broken lug stud. The attachment is held in place by a coil spring that tends to break over time, so inspect it, and never use the air hammer if it is broken. Place the tool bit against the workpiece and hold it firmly before you pull the trigger, as it is likely to jump around. To use an air hammer, follow the steps in **SKILL DRILL 7-6**.

SAFETY TIP

An air hammer's operation produces a noise level that exceeds the maximum safe exposure level for human ears. Always wear ear protection when using an air hammer.

▶ Electric Power Tools

K07004

Drills and Drill Bits

Many components make up a drill set (**FIGURE 7-19**). A portable drill can be corded or cordless. A corded drill has a cord that you have to plug into an electrical supply. The operating voltage of a drill depends on the country's electric supply. Corded drills are a good choice when moderate power is needed or if extended drilling is required. Cordless drills use their own internal batteries. When you cannot bring the work to the drill, you can take the drill to the work. But don't expect a cordless drill to be able

SKILL DRILL 7-5 Using Air Drills

1. Select the properly sized air drill and drill bit, inspect them for damage, and lubricate the gun if an automatic oiler is not installed in the system.

2. Adjust the direction of spin—forward or backward—with the selector.

3. Turn the valve to increase or reduce the torque to match the needs of the job.

4. Place the air drill into position and hold it firmly. Give the trigger a quick squeeze to verify that the air drill is turning the correct direction and does not slip.

5. Continue to operate the air drill by squeezing the trigger only long enough to get the job done.

SKILL DRILL 7-6 Using Air Hammers

1. Select the properly sized air hammer and attachment, inspect them for damage, and lubricate the gun if an automatic oiler is not installed in the system.

2. Fit the appropriate bit into the nose of the air hammer, and ensure the spring is installed correctly.

3. Place the air hammer into position and hold it firmly. Give the trigger a quick squeeze to see how the air hammer will react.

4. Continue to operate the air hammer by squeezing the trigger and watching the progress carefully. Allow the air hammer to do the work, and work slowly around the item.

to drill large holes through hard metal. Although they are very versatile, they are limited to the amount of work they can do by their power rating. The biggest drill bit that will fit into the chuck of these drills is usually marked on the body of the drill or chuck, along with the speeds at which it turns. Some portable drills have two operating speeds, but most portable drills have a variable speed rating that is determined by how much pressure is placed on the trigger and can be set to any speed within the drill's range.

A **drill press** allows for accurate drilling with more control than is offered by a portable drill, which, although convenient, can be difficult to guide accurately. A mounted drill can feed the drill bit at a controlled rate, and the worktable on the drill typically has a vise to secure the job at a constant angle to the drill bit. Also, this drill can be set to run at different drilling speeds. Most drill presses have a drill chuck that takes bits up to 0.5" (13 mm) or more in diameter.

Morse taper is a system for securing drill bits to drills. The Morse taper size changes according to drill size. The shank of the drill bit is tapered and looks like the tang of a file (**FIGURE 7-21**). It fits snugly into the drill spindle, which has a similar taper on its inside. The tang on the drill bit is located in the spindle, and it drives the drill. It is a quick way to change drills without constantly adjusting the chuck.

When a hole already drilled in sheet metal needs enlarging, a multi-fluted tapered hole drill will do the job in practically the same time it takes to say the name of this tool (**FIGURE 7-22**). A drilling speed chart is usually supplied with the drill press

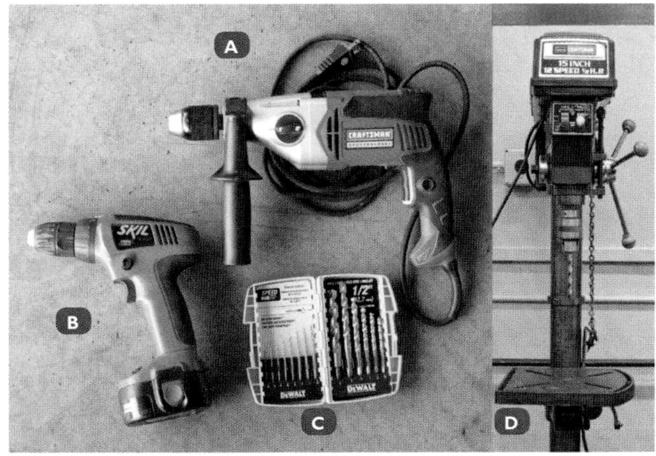

FIGURE 7-19 **A.** Portable drill. **B.** Cordless drill. **C.** Drill bits. **D.** Drill press.

and should be kept nearby for handy reference (**FIGURE 7-23**). It compares drill sizes and metals to show the proper speed. For example, to drill a 0.375" (10 mm) hole through a piece of aluminum, the drill speed should be 1800 rpm. Drilling metals is also best performed with the aid of a lubricant. The lubricant helps cool the cutting edges of the drill bit, as well as lubricate it. Each metal requires its own type of lubricant, so check a drilling guide for the metal you are working on.

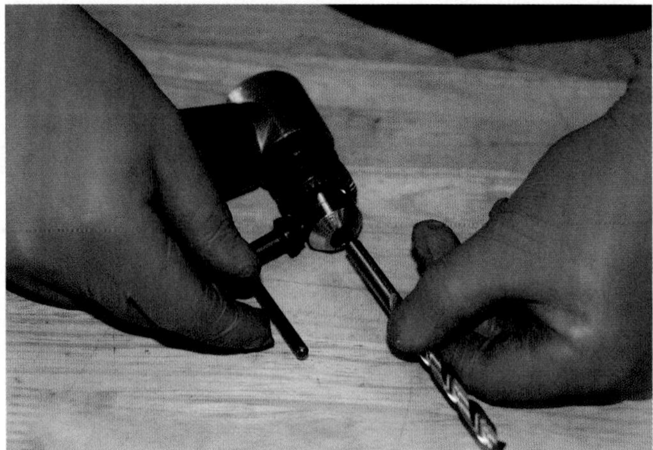

FIGURE 7-20 Twist drill bit and drill chuck.

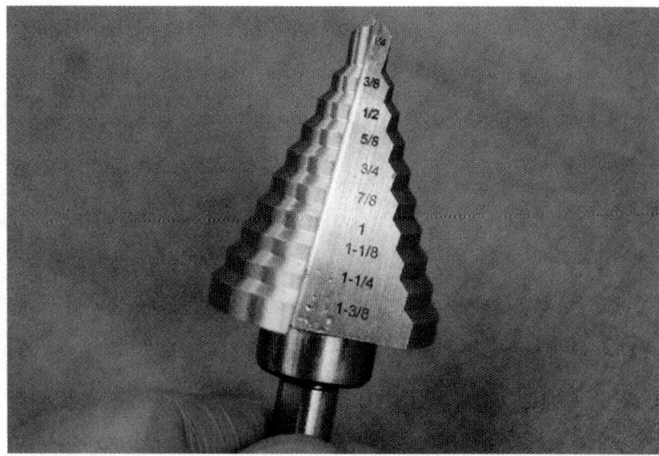

FIGURE 7-22 Multi-fluted tapered hole drill.

FIGURE 7-21 Morse taper drill bit.

Feeds and Speeds for High Speed Steel Drills, Reamers and Taps

MATERIAL	BRINELL	DRILLS			REAMERS		TAPS—S.F.M.			
							THREADS PER INCH			
		S.F.M.	POINT	FEED*	S.F.M.	FEED	3-7½	8-15	16-24	25-UP
Aluminum	99-101	200-250	118°	M	150-160	M	50	100	150	200
Aluminum Bronze	170-187	60	118°	M	40-45	M	12	25	45	60
Bakelite	—	80	60°-90°	M	50-60	M	50	100	150	200
Brass	192-202	200-250	118°	H	150-160	H	50	100	150	200
Bronze, common	166-183	200-250	118°	H	150-160	H	40	80	100	150
Bronze, phosphor, 1/2 hard	187-202	175-180	118°	M	130-140	M	25	40	50	80
Bronze, phosphor, soft	149-163	200-250	118°	H	150-160	H	40	80	100	150
Cast iron, soft	126	140-150	90°	H	100-110	H	30	60	90	140
Cast iron, medium soft	196	80-110	118°	M	50-65	M	25	40	50	80
Cast iron, hard	293-302	45-50	118°	L	67-75	L	10	20	30	40
Cast iron, chilled*	402	15	150°	L	8-10	L	5	5	10	10
Cast steel	286-302	40-50	118°	L	70-75	L	20	30	40	50
Celluloid	—	100	90°	M	75-80	M	50	100	150	200
Copper	80-85	70	100°	L	45-55	L	40	80	100	150
Drop forgings (steel)	170-196	60	118°	M	40-45	M	12	25	45	60
Duralumin	90-104	200	118°	M	150-160	M	50	100	150	200
Everdur	179-207	60	118°	L	40-45	L	20	30	40	50
Machinery steel	170-196	110	118°	H	67-75	H	35	50	60	85
Magnet steel, soft	241-302	35-40	118°	M	20-25	M	20	40	50	75
Magnet steel, hard*	321-512	15	150°	L	10	L	5	10	15	25
Manganese steel, 7-13%*	187-217	15	150°	L	10	L	15	20	25	30
Manganese copper, 30% Mn.*	134	15	150°	L	10-12	L	—	—	—	—
Malleable iron	112-126	85-90	118°	H	—	H	20	30	40	50
Mild steel, 20-30C	170-202	110-120	118°	H	75-85	H	40	55	70	90
Molybdenum steel	196-235	5	125°	M	35-45	M	20	30	35	45

FIGURE 7-23 Drill speed reference chart.

▶ TECHNICIAN TIP

Drills are also used to drive other accessories such as rotary files, screwdriver bits, and sockets.

Bench and Angle Grinders

Power grinders come in different sizes and speeds (**FIGURE 7-24**). The size of a power grinder is normally determined by the diameter of the largest grinding wheel or disc that can be fitted to it. Some grinders are fixed to a bench or pedestal, and the work is brought to the grinder; others are portable devices that can be taken to the work. Bench or pedestal grinders tend to be powered by electricity; portable ones can be electric or air powered.

A maximum safe operating speed is usually printed on **grinding wheels and discs**. This maximum speed must never be exceeded, or the wheel or disc could disintegrate. Every well-equipped shop has a solidly mounted grinder, either on a pedestal bolted to the shop floor or securely attached to the workbench. Appropriate eye protection must be worn when grinders are being used, and the wheel guards and shields must be correctly and firmly in place.

A **bench grinder (pedestal grinder)** normally has a rating giving the size of the grinding wheel it can take. Do not try to install a grinding wheel larger or smaller than is the one for which it is rated. Grinding wheels come in grades from coarse to very fine, depending on the size of the abrasive grains that are bonded together to make the wheel. They also range in hardness, depending on the abrasive used and the material that bonds the particles together. Check to find the most suitable grinding wheel for a particular grinding application.

▶ TECHNICIAN TIP

When using a bench grinder, the face of the abrasive wheel must be kept square. This is done with a dressing tool, which removes some of the abrasive compound. If the abrasive wheel is not square, use a dressing tool. Then readjust the tool rest to the proper clearance.

An **angle grinder** is usually needed when the bench grinder is not appropriate. The angle grinder uses discs rather than wheels. During grinding, the face of the disc is used instead of the edge. An angle grinder can throw sparks many feet, so direct the sparks in a safe direction, or set up a guard to catch them. Also use hearing protection whenever grinding, as it is very noisy

FIGURE 7-24 **A.** Bench grinder. **B.** Angle Grinder.

and can damage your ears. Although not as common in an automotive shop, the **straight grinder** takes conventional grinding wheels, just like the stationery grinders. However, the grinding wheel diameter is limited to about 4.75" (126 mm). In many cases, the grinder has a long shaft that moves the grinding wheel away from the motor. This makes it handy to get into recessed areas.

Hand-held cutoff wheels can be powered by electricity or air (**FIGURE 7-25**). A special thin grinding disc enables them to cut. The edge of the wheel is used for cutting, and they are useful for jobs that cannot be reached with a hacksaw.

FIGURE 7-25 Hand-held cutoff wheel.

SAFETY TIP

- Make all adjustments with the grinder stopped and unplugged.
- Stand to the side of the grinder when starting the electric motor.
- Never use a cracked or gouged grinding wheel.
- Do not operate a grinder unless it is securely mounted to the bench or floor.
- Do not grind on the side of the wheel because it may cause the wheel to shatter.
- Maintain a 2-foot (0.6 m) perimeter around the grinder free of flammables, clutter, and people.

SAFETY TIP

Ask your supervisor to demonstrate the differences between grinding wheels for soft and hard materials, and wire brush wheels. As the abrasive wheel wears down, the gap between the wheel and the tool rest increases. This creates a dangerous situation because once the gap enlarges to the thickness of the metal you are grinding, the metal can be pulled into the gap, which can cause the metal to be thrown from the grinder with much force. So always make sure the tool rest gap is set properly. This is usually about 1/8" (3.2 mm) or no more than half the thickness of the metal you are grinding, whichever is smaller.

Using Bench Grinders

S07007

When you are grinding metal, it must not be allowed to overheat, because this will adversely affect its hardness. If the metal becomes too hot and is allowed to cool slowly, it may become soft. If it is cooled quickly (quenched), it may become brittle. As you grind the metal, stop and dip it regularly into the water pot attached to the base of the grinder; this will prevent the metal from getting too hot. Some bench grinders are not supplied with a water pot. If this is the case, you need to have a water can located near the grinder so that you can cool the piece you are grinding. To set up, adjust, and use a bench grinder, follow the steps in **SKILL DRILL 7-7**.

Also, most bench grinders are equipped with a spark deflector, sometimes called a "shatter guard" which reduces sparks being thrown forward and can help contain pieces of the grinding wheel if it shatters. The shatter guard is located at the top of each grinding wheel guard opening. It needs to be periodically adjusted as the grinder wheel wears down. Refer to the manufacturer's specifications, which typically call for a gap of between 1/16 and 1/4 inch (1.59 and 6.35 mm).

Using Angle Grinders

S07008

The angle grinder uses an electric motor to drive an abrasive disc at a high speed (**FIGURE 7-26**). The grinder disc is turned at speeds that range from 5000 to 12000 rpm. The turning disc is used to grind or cut metal. The grinder size relates to the diameter of the cutting disc, which can range from 4" to 9" (102 mm

SKILL DRILL 7-7 Using Bench Grinders

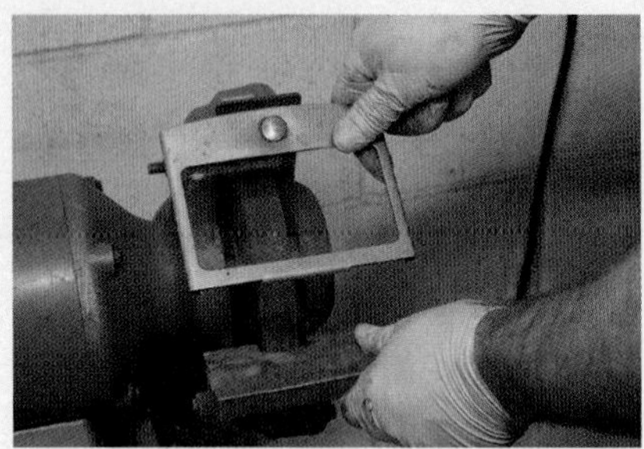

1. Before you start using the bench grinder, inspect the wheels, and check to see if the tool rest, safety shield, and shatter guard are adjusted properly. There should also be water in the pot.

2. If needed, adjust the tool rest so it has a maximum gap of 0.125" [3.2 mm] (or no more than half the thickness of the metal you are grinding), whichever is smaller, between the tool rest and wheel.

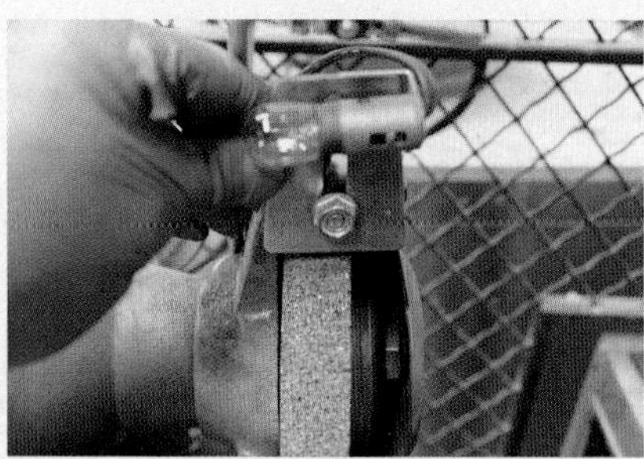

3. If needed, adjust the shatter guard to a maximum gap of 0.0625" [1.6 mm] between the guard and wheel.

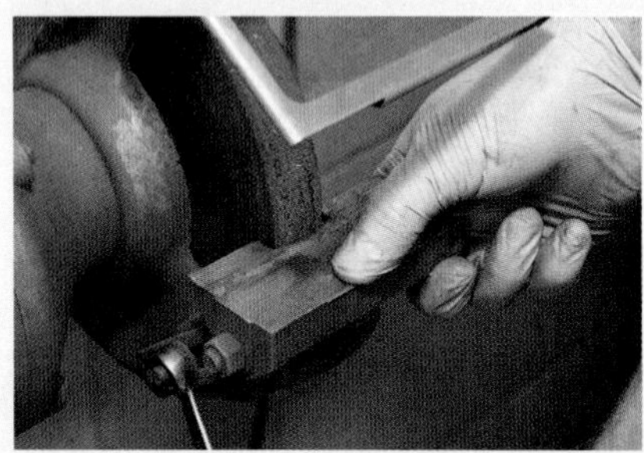

4. The tool rest should be slightly below the center of the wheel. To adjust the tool rest, locate the adjusting bolt, and loosen it with a box-end wrench. Set the tool rest at the right height and distance from the wheel, and then tighten the adjusting bolt. If you are unsure of how to do this, ask your supervisor.

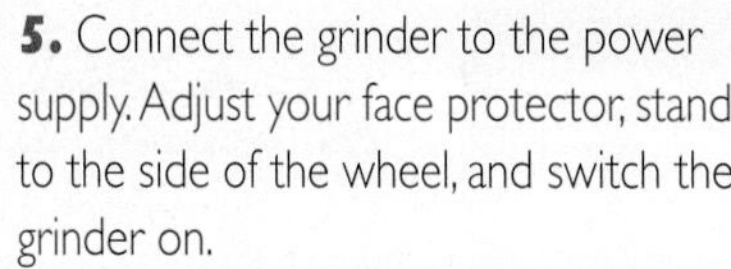

5. Connect the grinder to the power supply. Adjust your face protector, stand to the side of the wheel, and switch the grinder on.

6. After the grinder is fully up to speed, move to the front of the wheel; hold the part you are going to grind firmly onto the tool rest, and move it slowly and gently forward until it comes into contact with the wheel. The grinding wheel removes the metal it contacts. Use the full face of the grinding wheel to prevent wearing one area of the wheel.

7. Occasionally, dip the part into the water to keep it cool.

8. When you have finished, turn off the power and unplug the grinder.

to 229 mm). The size of grinder you use depends on the type of job you are doing. The smaller the grinder, the higher the speed it turns. Sanding discs and wire wheels can be fitted on the grinder, making it a versatile electric tool. An extra handle is provided that can be attached to the grinder head. This handle can be fitted to either the left, right, or top of the head to make it easy to use for left-handed as well as right-handed people.

The **abrasive disc**, or cutting wheel, is attached to the grinder by a flange and nut. The nut is specially designed to fit in a recess in the center of the pad or wheel. It is tightened by a tool that is provided with the grinder when purchased. Do not lose this wrench because it is the only tool that can tighten the nut properly. When using the grinder with cutting discs, you should always use the edge of the disc rather than the face.

FIGURE 7-26 A. The angle grinder uses an electric motor to drive an abrasive disc at a high speed. **B.** An extra handle is provided that can be attached to the grinder head.

▶ TECHNICIAN TIP

Do not confuse a grinder with a **sander/polisher**. The sander/polisher turns at lower speeds, typically 600 to 3000 rpm. It is commonly used to sand and polish paint. The pads these tools use cannot be turned at a high speed. If the polish pad were attached to an angle grinder, the higher rotational speed would cause the polishing pad to burn the paint and cause the polish pad to fly apart.

SAFETY TIP

- Always wear impact-resistant protective glasses, ear protection, and a full-face shield when using an angle grinder.
- Wear safety shoes, leather gloves, and an apron to protect your body from flying metal chips.
- Make sure the blade guard is firmly secured.
- Use the correct type of disc.
- Make sure the guard handles are secure.
- Use the correct flange or spindle nut for the type of disc being used. If you do not, the disc can shatter at high speeds and injure you.
- Angle grinders, like all portable grinding tools, need to be equipped with safety guards to protect you from flying fragments in case the disc breaks apart.
- Always follow the manufacturer's recommendations, to make sure the spindle wheel does not exceed the abrasive wheel specifications.
- Make sure there are no obvious defects or damage to the disc before you install it.

To correctly use an angle grinder, follow the steps in **SKILL DRILL 7-8.**

Soldering Tools

K07005

Solder is a mixture of metals with low melting points and is used to join metals together. Tin-lead solder has been used for soldering wires and other metals together for decades. Tin-lead solder for automotive applications consists of approximately 60% tin and 40% lead, and melts at approximately 370°F (188°C). With the environmental hazards of lead becoming an issue over the past 30 years, lead-free

SKILL DRILL 7-8 Using Angle Grinders

1. Inspect the grinding disc for any cracks or damage.

2. Check the area for any flammables and to determine where the sparks will fly. Take any precautions necessary.

SKILL DRILL 7-8 Using Angle Grinders (Continued)

3. Hold the grinder firmly with the face of the disc, not the edge, against the work. Be careful that the motor's torque does not cause the grinder to slip out of your hand. Do not press too hard. Let the grinder do the work.

4. If using the grinder with a cutting disc, use the edge of the disc, not the face.

FIGURE 7-27 **A.** Tin-lead solder. **B.** Lead-free solder.

solder was introduced in the past decade (**FIGURE 7-27**). Lead-free solder is made up of tin and copper, or of tin, copper, and silver, and usually has a melting point about 20–30°F higher than tin-lead solder. Because solder is a relatively soft compound, it is not used to make joints in situations where high stresses are involved.

In automotive applications, solder generally comes in the form of a wire. It can be solid, requiring an external **flux** cleaning agent of **rosin** if soldering electrical connections or acid if soldering non-electrical connections (**FIGURE 7-28**). The solder can also be hollow, with the rosin or acid in the core. In this case, it would be referred to as rosin-core solder or acid-core solder (**FIGURE 7-29**). Make sure you use rosin flux with electrical connections, and acid with all other connections.

The process of soldering involves heating the metals (wires) hot enough so that the solder melts and fills the spaces between the metals. When the solder cools, the solder holds the parts together and transmits electricity, if used in electrical circuits. The temperature of the soldering operation is critical. If it is not hot enough, the solder does not flow very well and does not make good contact with the metal surfaces, and tends to glob up (**FIGURE 7-30**). This makes for a weak joint and poor electrical conductivity. If it is too hot, the solder tends to run off the joint and overheat the components being soldered (**FIGURE 7-31**). In the case of electronic components, overheating can make them inoperative. Only heat the components enough to melt the solder and cause it to flow (**FIGURE 7-32**). You can also protect electronic components by using a heat dam. A heat dam absorbs some of the heat from the wires being soldered and prevents it from traveling on to the electronic component (**FIGURE 7-33**).

The heat is provided by a **soldering iron**, or gun (**FIGURE 7-34**). The heat in soldering irons is usually generated from electricity or gas. A typical soldering iron has a handle

FIGURE 7-28 **A.** Rosin flux. **B.** Acid flux.

FIGURE 7-29 **A.** Rosin core solder. **B.** Acid core solder.

FIGURE 7-30 Solder joint that was too cold.

FIGURE 7-31 Solder joint that was too hot.

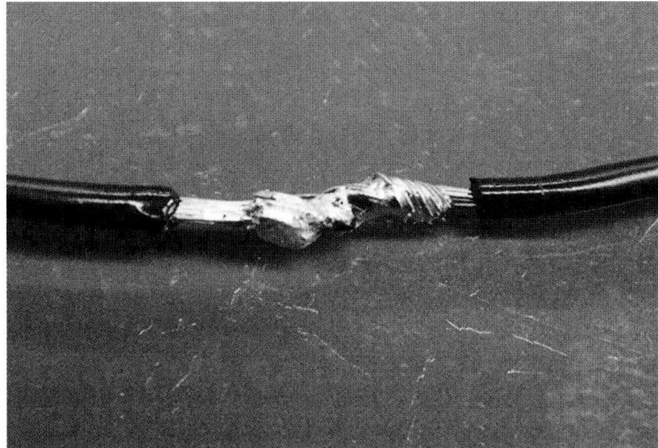

FIGURE 7-32 Good solder joint.

that is thermally insulated. The soldering tip is heated, and the heat is transferred by metal-to-metal contact from the tip into the metal to be soldered. Basic soldering irons are heated manually by a gas flame, and more sophisticated soldering irons are electrically operated and have thermal tips that are controlled

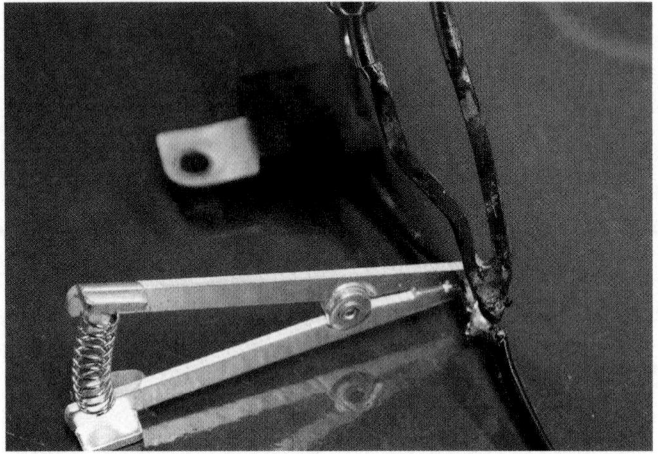
FIGURE 7-33 Soldering while using a heat dam.

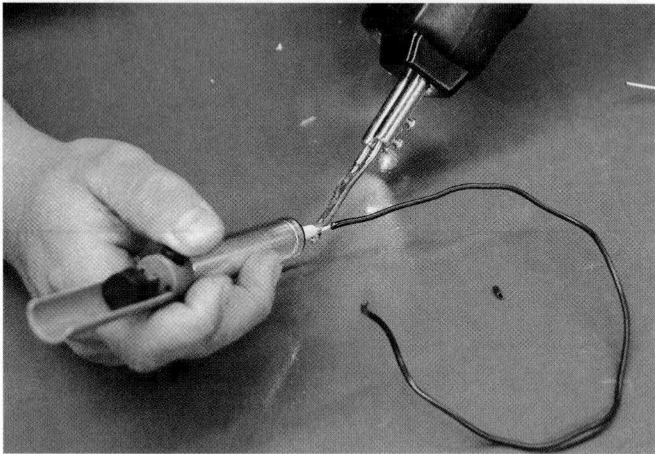
FIGURE 7-35 Desoldering tool being used.

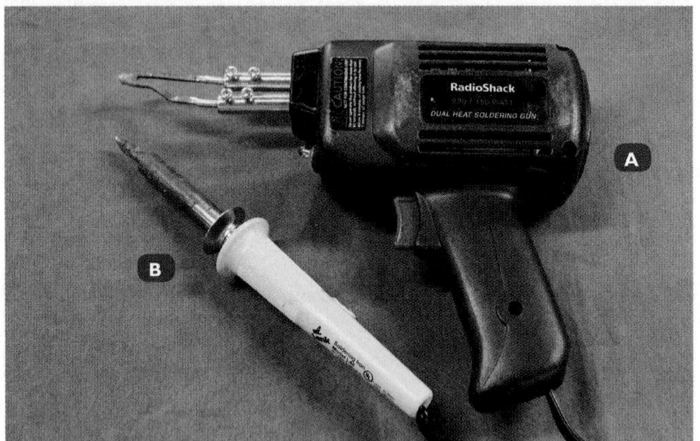

FIGURE 7-34 A. Soldering gun. **B.** Soldering iron.

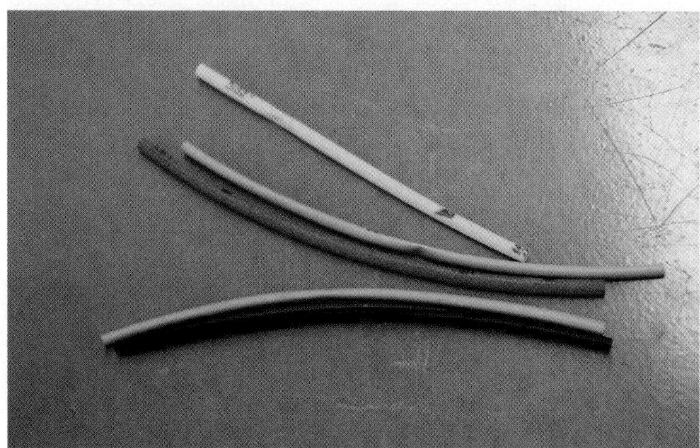
FIGURE 7-36 Heat shrink tubing.

by thermostats to maintain more accurate tip temperatures. Soldering irons can either have a fixed tip size or tips that can be interchanged with different sizes for different kinds of jobs.

Sometimes when soldering, you have to remove solder from a joint before resoldering a new part into the circuit. In this case a desoldering tool can be used to suck up the solder once it is melted, making it much easier to disassemble the joint (**FIGURE 7-35**).

Once a good solder joint has been created, you need to protect it. One of the best ways to do so is with heat shrink tubing (**FIGURE 7-36**). Heat shrink tubing is a hollow tube of insulating plastic that shrinks when it is heated. When the properly sized heat shrink tube is placed over the solder joint and heated with a heat gun, it will shrink tightly around the joint (**FIGURE 7-37**). Some heat shrink tubing is filled with a small amount of sealer that melts when heated, flows around the joint, and seals the tubing to the wire insulation.

Another way to protect solder joints is with electrical tape. When installed correctly, the tape insulates and protects the joint. The tape should be wrapped around the wire in a spiral manner while being lightly pulled tight in overlapping wraps. In most cases, the tape is fwirst wrapped down the wire (and joint), overlapping each layer by about 50%, and then back up the wire in the same fashion. Additional layers can be added for extra protection as needed (**FIGURE 7-38**).

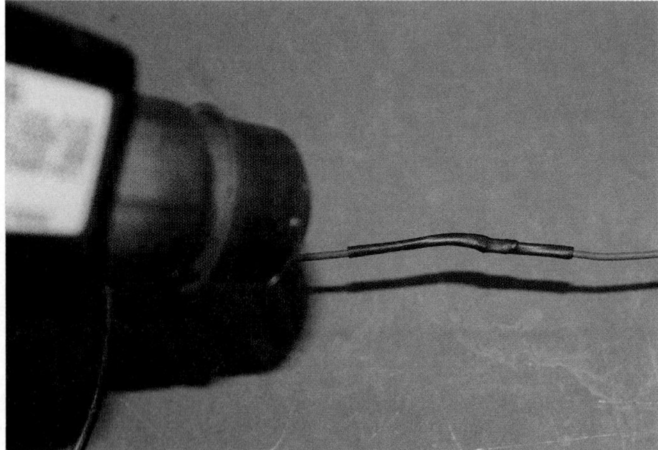
FIGURE 7-37 Heat shrink tubing being heated.

Using Soldering Tools

S07009

Apply flux to the joint if **cored solder** is not being used. Always remove any excess flux when finished. Ensure that the joint is held steady during and after the solder is applied. A dull solder

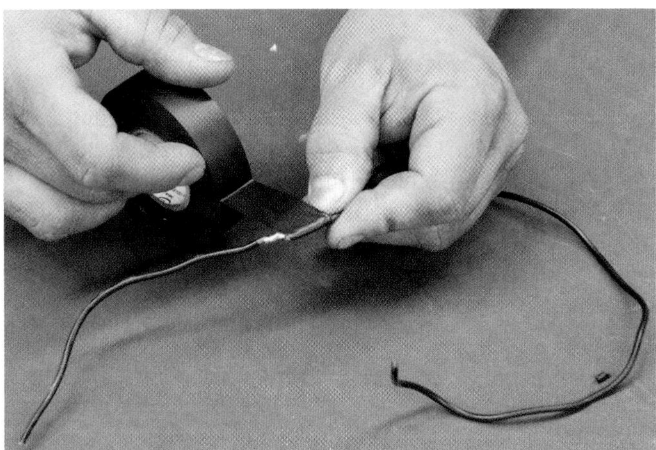

FIGURE 7-38 Wire joint being wrapped with electrical tape.

surface indicates a cold, high-resistance joint that needs to be resoldered. Select the correct tip size to heat the joint and melt the solder within a few seconds. Overheating a joint or using too small of a tip will produce a poor solder joint. To use a soldering iron to solder two pieces of wire or metal together, follow the steps in **SKILL DRILL 7-9**.

SAFETY TIP

- Always wear eye and hand protection when soldering.
- Always wipe excess solder from the iron; never flick the iron to remove excess solder.
- Make sure the tool is clean, in good condition, and suitable for the type of material you are going to solder.
- Check electrical leads and plugs for damage prior to use.

▶ Cleaning Tools

K07006

Pressure Washers and Cleaners

Pressure washers and cleaners are valuable tools for cleaning vehicles, engine compartments, and components (**FIGURE 7-39**). They can be powered by an electric motor or a gasoline engine fitted to a high-pressure pump. The pressure washer takes water at normal pressure and boosts it through the high-pressure pump to exit through a cleaning gun, which has a control trigger. The **cleaning gun** has a high-pressure nozzle that focuses high-pressure water (possibly over 2000 psi [13,790 kPa]) to quickly clean accumulated dirt and grease from components. Some pressure washers have a provision for detergent to be injected into the

SKILL DRILL 7-9 Using Soldering Tools

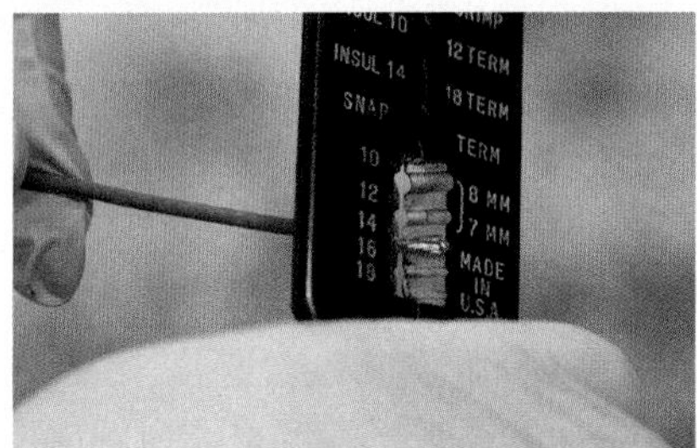

1. Prepare the materials to be soldered. Strip wires or clean metal parts before soldering.

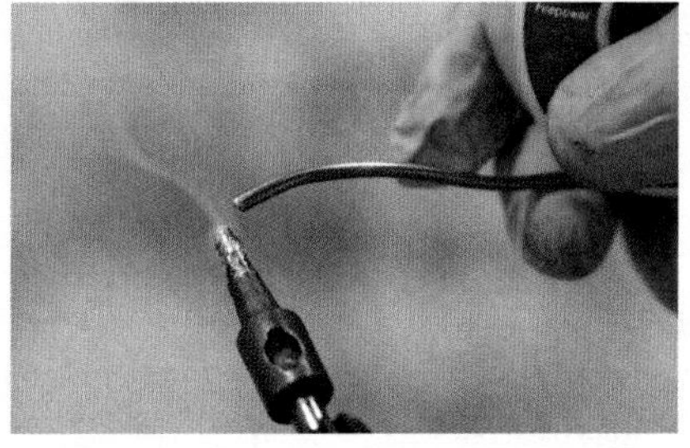

2. Prepare the soldering iron by ensuring the correctly sized tip is fitted and is clean (you may have to use a file to reshape and clean it). Tin the soldering iron tip by melting some solder to it and wiping any excess from the tip with a rag.

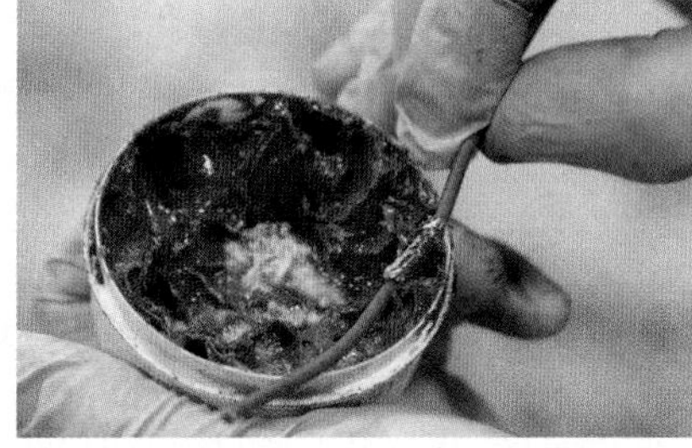

3. Apply flux to the wires or metal to be soldered. This may not be necessary if you are using cored solder.

4. Apply the hot solder iron tip to heat the joint, and then apply solder to the joint (not the soldering iron). If the solder does not melt within a few seconds, remove it and allow the joint to heat further before reapplying.

5. Once the solder has been applied, ensure the joint does not move until the solder has cooled sufficiently to set. Once cooled, inspect the joint; it should be shiny and firm.

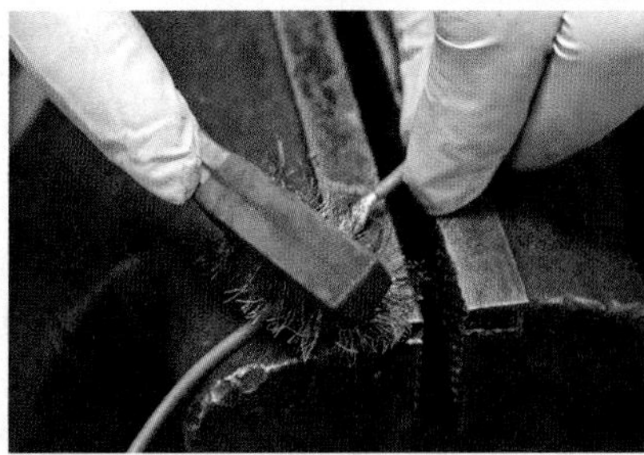

6. Clean any excess flux from the joint.

FIGURE 7-39 Pressure washer being used to clean an engine compartment.

high-pressure output to more effectively clean. Others have the ability to heat the water, in some cases hot enough to turn it to steam. Hot water and steam help loosen oil and grease buildup. Pressure washers are dangerous because of their high pressure and possibly high temperature. Always wear appropriate PPE when working with pressure cleaners—for example, goggles or face shield, protective gloves, close-fitting clothes with long sleeves and full-length pants, and leather-type boots or shoes.

Using Pressure Washers

S07010

Pressure washers and cleaners used in automotive applications are available in a range of makes and types depending on the application. There are fixed and mobile pressure washers. Familiarize yourself with the equipment prior to use; incorrect handling can result in damage to the washer or to the vehicle or components you are cleaning, in addition to health risks to yourself and your coworkers. The biggest advantage of using fluid to clean vehicles, components, and spaces is that it wets the dirt or contaminants, so no dust is created. However, the waste products must be caught and disposed of properly in either a catch basin or a wastewater settling system. The waste materials must not be released into a storm water drain.

It is imperative that you read the instructions beforehand and be familiar with the operation of the pressure washer. When cleaning exterior paintwork, extreme care must be taken to ensure that the pressure does not damage or remove paint. If in doubt, clean the area manually using a clean sponge and clean water.

It is very important to note the type of solvent or detergent being used in a pressure washer, as some vehicle components can be damaged by some solutions and should only be cleaned in wash tanks containing the correct cleaning fluid. When using high-pressure washers, it is always important not to spray in areas where water and water-based solutions can have a detrimental effect on electrical equipment such as fuse boxes, relays, and control units. If you are required to use a pressure washer in those areas of the vehicle, take precautionary measures to protect the units from high-pressure water damage by covering them with sturdy plastic bags. The damage may not become apparent for some time after the cleaning process, but it can have a catastrophic effect on the vehicle, causing system failures, which are difficult to diagnose. If using the washer results in the wheel brake units getting wet, ensure that the vehicle is driven for a short distance with the brakes slightly applied. This will remove any residual water from the brake shoes or pads through heat transfer and subsequent evaporation of the water. Depending on the outside temperature, it usually only takes several seconds to return the brakes to their proper operation.

On completion of the job, the correct disposal of contaminated materials is a top priority. Operators of cleaning equipment may be subject to prosecution for the incorrect disposal of waste materials. So always know the regulations and follow all policies and guidelines for your jurisdiction.

SAFETY TIP

- Always wear a face shield and gloves when using cleaning and washing equipment.
- Always wear safety shoes when using any washing equipment, to prevent slips on slippery surfaces.
- Always be aware of the location of safety switches located on equipment and of eyewash and first aid stations should an accident occur.
- Do not place your hand or any other part of your body in the stream of water from the high-pressure wand. Many pressure washers generate enough pressure to instantly cut skin and muscle to the bone.
- Do not aim the high-pressure wand at another person.
- Always test the temperature of the wand and the hose before you pick it up. The handle of the pressure wand is insulated to protect the user from heat, but the wand extension and the hose are not.
- If the pressure cleaner uses a heating element, turn the heater off, and allow water to flow through the wand until it has cooled, before you turn the unit off.
- If you are unfamiliar with a solvent or a cleaning agent, refer to the safety data sheet (SDS) for information about its correct use and applicable hazards.

To safely use pressure washers, follow the steps in **SKILL DRILL 7-10**.

Spray Wash Cabinets

Spray wash cabinets spray high-temperature, high-pressure cleaning solutions onto parts inside a sealed cabinet. They are automated and act like a dishwasher for parts (**FIGURE 7-40**). This significantly reduces the labor required to clean parts, because once the door is closed and the unit is turned on, the technician is free to move onto other tasks. They are available in a variety of sizes to cater to different-sized parts and provide a high level of cleaning performance. The cleaning solution is designed to effectively clean without leaving dirty residue on the parts, and most spray cabinets are fitted with a filtering system to reduce the frequency of cleaning solution changes.

SKILL DRILL 7-10 Using Pressure Washers

1. Before using the pressure washer, locate the position of safety switches, and don a face shield, work apron, and gloves. Note the location of the eyewash and first aid stations.

2. If the pressure washer is not permanently connected to the water supply, follow the manufacturer's instructions to do so, and connect the electrical plug to a power outlet protected by a ground fault circuit interrupter (GFCI). Make sure electrical connections and adjacent areas are protected from water spray by using appropriate protective shields.

3. It may be necessary to apply a degreasing agent with a spray bottle and hand brush to penetrate and soften excess dirt before you operate the pressure washer.

4. Turn on the water supply, but not the power switch. Make sure water is flowing through the washer unit by testing the flow through the pressure wand before turning the power on. It should flow freely, but not at high pressure.

5. Turn the power on, and you will hear the motor engage. Point the wand toward the ground, and test that the water now flows at high pressure.

6. Pull the trigger, and using either a circular or a sweeping motion, direct the high-pressure water onto the area to be cleaned, to remove the contaminants. Avoid getting the high-pressure spray on the exterior paintwork of the vehicle by placing the wand close to the area to be cleaned.

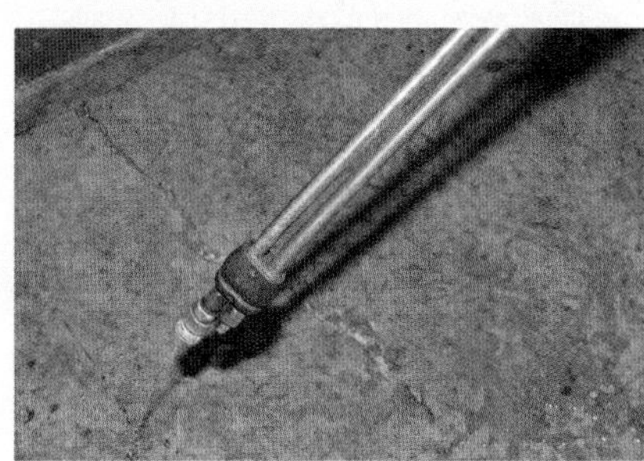

7. When the contaminants have been removed, release the trigger, and remove the wand from the cleaning area. Turn the electrical power off, and then turn the water supply off. If it is a hot water or steam cleaner, let the water run until it cools off so that you do not overheat the water in the heating coils.

10. Clean up any residual material, and place it in a bin or an environmental waste container.

8. Use an air nozzle to disperse any residual water from electrical components that have come into contact with the cleaning fluids.

9. Start the vehicle and let it run for a few moments to dry, and then remove any residual water in the engine compartment area.

FIGURE 7-40 Spray wash cabinet.

Using Spray Wash Cabinets

S07011

Spray wash cabinets used in automotive applications are available in a range of makes and types depending on the application. It is imperative that you read the instructions beforehand and be familiar with the operation of the spray wash cabinet. Incorrect handling can result in damage to the components that you are cleaning in addition to health risks to yourself and your coworkers. Spray wash cabinets must either incorporate a built-in waste recovery system or be used only where the contaminated washer fluids can be captured to enable disposal in an environmentally friendly manner. Always follow recommended safety procedures. Some spray wash cabinets use potentially dangerous chemicals at very high pressure to clean away the contaminants. So always know what chemical is being used as a cleaning agent so you can look up any precautions in the SDS.

It is also very important to note the type of cleaning agent being used, as some agents can damage some vehicle components. On completion of the job, the correct disposal of any contaminated materials is a top priority. Operators of this equipment may be subject to prosecution for the incorrect disposal of waste materials.

To use a spray wash cabinet to clean components, follow the steps in **SKILL DRILL 7-11**.

SKILL DRILL 7-11 Using Spray Wash Cabinets

1. Always refer to the manufacturer's manual for specific operating instructions. Before using the equipment, locate the position of safety switches, and put on a face shield and gloves.

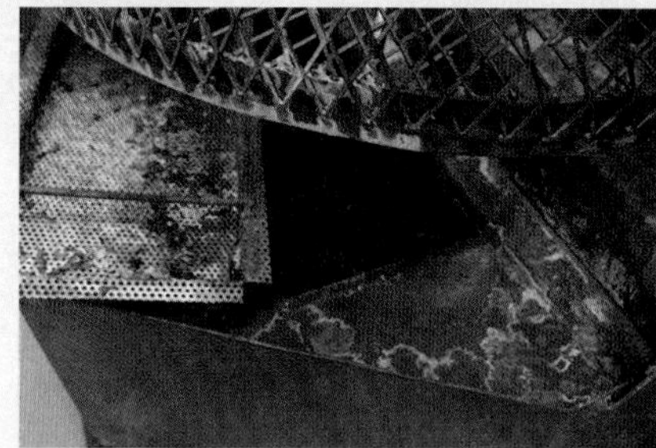

2. Check the spray wash cabinet to ensure it is operating correctly with enough cleaning fluid.

3. Open the spray wash cabinet. Place the components to be cleaned into the wash tray. Seek assistance if parts are too heavy to be handled by one person. Distribute them so each part will be cleaned effectively.

4. Close the spray wash cabinet and start the cleaning cycle.

5. Make sure the cleaning cycle has finished before you open the spray wash cabinet door. Ensure you are wearing appropriate eye and hand protection.

6. When opening the door, be careful because parts will be hot to the touch. Examine the parts to ensure they have been completely cleaned. In some models, you need to rinse the components after removing them from the washer. Follow the directions for the model you are using.

SAFETY TIP

- Do not operate the spray wash cabinet without the door securely closed.
- The spray wash cabinet uses high-pressure, high-temperature cleaning fluid, and parts will be hot after being in the spray cabinet.

Solvent Tanks

A **solvent tank** is a cleaning tank that is filled with a suitable solvent to clean parts by removing oil, grease, dirt, and grime. Solvent tanks are available in different sizes. Many solvent tanks have a pump that pushes solvent out a nozzle into a sink where it can be directed to the parts being cleaned. A brush, either on the nozzle or separate from it, can be used to loosen the grease and grime. The solvent falls back into the bottom of the solvent tank, where the heavier residue can settle to the bottom.

Other solvent tanks are designed so that parts can be immersed into the tank on racks or suspended on pieces of wire, then slowly lowered and soaked in the tank for a period of time. Some solvent tanks may have an agitation system or use a heated cleaning fluid to speed up the process. They may also have a circulation system and filters to remove debris and residue in the solvent to extend its life between changes.

SAFETY TIP

Whenever using a tank-type cleaner with solvents, make sure there is adequate ventilation, and wear appropriate breathing apparatus and eye protection.

Using Solvent Tanks

S07012

Familiarize yourself with the equipment prior to use; incorrect handling can result in damage to components that you are cleaning, in addition to health risks to yourself and your coworkers. Solvent tanks may use potentially poisonous and/or flammable chemicals to clean parts. It is very important to note the type of solvent being used and to take all necessary precautions as listed on the SDS. Also, some vehicle components can be damaged by some solvents, so ask your supervisor about what can be safely washed and what cannot. On completion of the job, the correct disposal of contaminated materials is a top priority. Operators of this equipment may be subject to prosecution for the incorrect disposal of waste materials.

To use a solvent tank to clean components, follow the steps in **SKILL DRILL 7-12**.

SKILL DRILL 7-12 Using Solvent Tanks

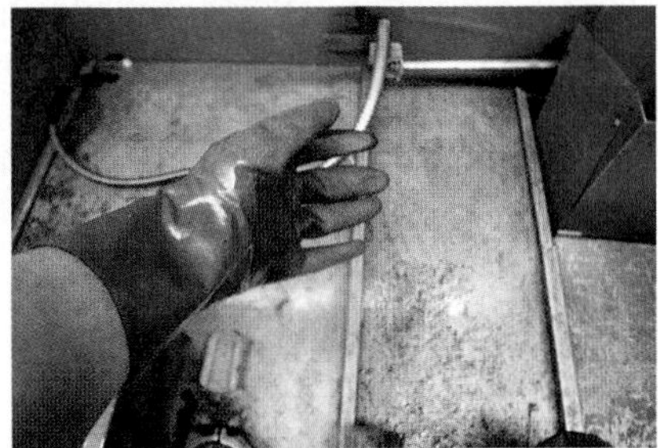

1. Always refer to the manufacturer's manual for specific operating instructions. Before using the equipment, locate the position of safety switches and put on a face shield and gloves. Note the location of the eyewash and first aid stations.

2. Lift the components to be cleaned into the wash tray or lower them into the solvent tank. Seek assistance if parts are too heavy to be handled by one person. Distribute them so each part will be cleaned effectively.

3. Start the circulating or agitating pump if equipped.

4. Soak or run the solvent onto the components in the solvent tank for the appropriate amount of time. You may need to use a brush to clean the parts. Be careful not to cause excessive splashing of cleaning fluid.

5. Examine the parts to ensure they have been completely cleaned.

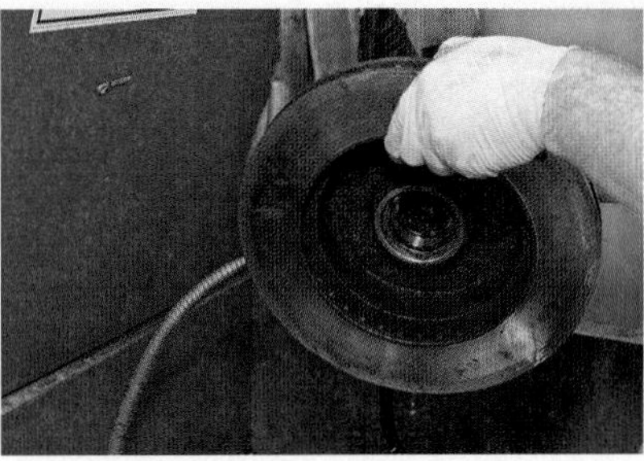

6. Remove parts carefully from the solvent tank, allowing excess cleaning solvent to drip back into the tank. Ensure contaminated cleaning fluids can be captured to enable disposal in an environmentally friendly manner.

SAFETY TIP

- If equipped, keep lids closed as much as possible, as many solvents are flammable and evaporate.
- The cleaning solution may be hot.
- Make sure there is adequate ventilation.
- Wear appropriate eye and hand protection and, if necessary, appropriate breathing apparatus.

Brake Washers

Brake washers are used to wash brake dust from wheel brake units and their components. As it is possible that the brake dust may contain asbestos, which is a cancer-causing agent, and dust in general is a lung irritant, brake washers are designed to capture the brake dust before it enters the shop environment. It does so by wetting down the dust on the brake parts and then washing it into the cleaning tray. Brake washers incorporate a built-in waste recovery system where the contaminated washer fluid can be captured to enable disposal in an environmentally friendly manner.

Brake washers are normally designed to operate at low pressure and use a range of cleaning agents. The most popular agent is an aqueous solution made up of water and a water-soluble detergent. A low-pressure air nozzle may be provided to blow excess fluid from the component into the tray area and then back to the tank by gravity. Avoid using solvent when cleaning brake components as it contaminates friction materials and may cause seals to swell. Never use kerosene as a general cleaning agent to clean brake components, as it does not clean away brake fluid, can be absorbed into lining materials, and can cause seals to swell. Soapy water is a good cleaning agent for most brake components.

SAFETY TIP

Familiarize yourself with the equipment prior to use; incorrect handling can result in damage to components that you are cleaning in addition to health risks to yourself and your coworkers. The waste products must be caught and disposed of properly.

Using Brake Washers

S07013

Brake washers are a handy piece of shop equipment to deal with hazardous dust in a quick and relatively easy manner, especially if the washer uses an aqueous (water and detergent) cleaning solution. The one thing that this solution is not suited for is cleaning greasy residue. If you need to remove grease, you might be able to first use a paper towel or grease rag to wipe it off, and then clean the brake dust with the brake washer.

To use a brake washer to clean components, follow the steps in **SKILL DRILL 7-13**.

Sand or Bead Blasters

Sand or bead blasters use high pressure to blast small abrasive particles to clean the surface of parts. The most common method of propelling the sand or glass beads is with compressed air. Sand or bead blasting can occur in a specially designed cabinet, or there are portable models that are available for use in open-air situations. The cabinets contain the blasting operation in a controlled safe environment and are best for smaller parts that can fit into the cabinet. Portable systems that do not operate within a cabinet can blast larger parts

SKILL DRILL 7-13 Using Brake Washers

1. Make sure all the washing solution is contained within the cleaning tray and returns to the reservoir. Washing solution must not enter the environment. Before using the component cleaner, make sure the solution is compatible with the component to be cleaned. An aqueous solution is an environmentally friendly solution.

2. Put on gloves and safety glasses or a face shield, and move the brake washer under the wheel brake unit to be cleaned. Make sure the waste drain is not blocked and the low-pressure air nozzle, if equipped, is operational.

SKILL DRILL 7-13 Using Brake Washers (Continued)

3. Using a semi-stiff brush, paint the solution over the components both to wet and clean the components and to remove contaminants. Continue to do so until the components are clean.

4. Use the low-pressure air nozzle to dry any excess solution from the component, and make sure the waste materials are caught in the cleaning tray and strained back into the tank below.

but do require more protection for the operator and surrounding environment.

The sand or bead blaster cabinet is fitted with a hand-operated blasting nozzle, a viewing port, and an on/off switch (often this may be a foot-operated switch), and it has openings with tough rubber gloves sealed into them to allow a technician's hands to be inside the cabinet while being protected from the abrasive sand or beads. Wet sand or bead blasters are also available and have the added advantage of reducing the amount of dust and providing additional cleaning to the part being cleaned.

Using Sand or Bead Blasters

S07014

Technicians use sand or bead blasters to clean paint, corrosion, or dirt from metal parts. Because the sand or beads are abrasive, they can remove metal from the surface of the components, so be careful what you use it on. Sand or bead blasters used in automotive applications are available in a range of makes and models, depending on application.

To use a sand or bead blaster to clean components, follow the steps in **SKILL DRILL 7-14**.

SKILL DRILL 7-14 Using Sand or Bead Blasters

1. Always refer to the manufacturer's manual for specific operating instructions. Before using the equipment, locate the position of safety switches. If not using a cabinet, then a face shield, dust mask, and gloves are required. Note the location of the eyewash and first aid stations.

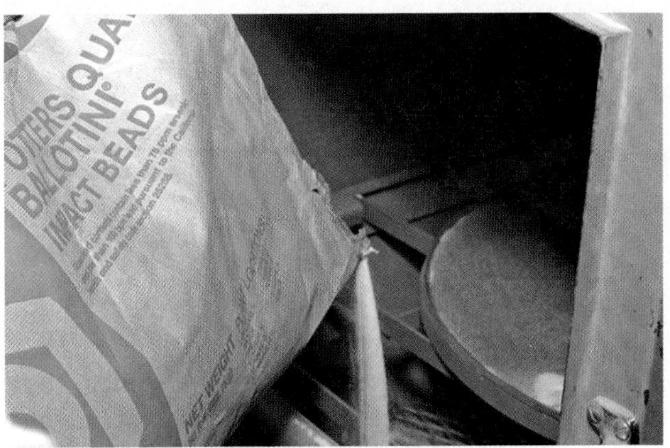

2. Check the sand or bead blaster to ensure it is operating correctly with enough cleaning sand or beads.

SKILL DRILL 7-14 Using Sand or Bead Blasters (Continued)

3. Open the sand or bead blaster. Lift the components to be blasted into the cabinet. Seek assistance if parts are too heavy to be handled by one person.

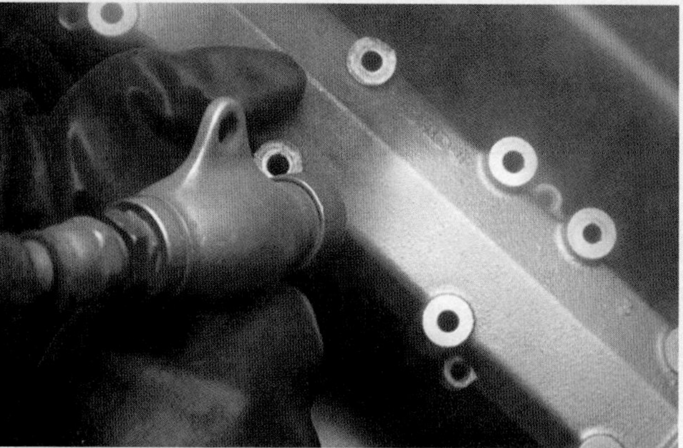

4. Close the sand or bead blaster, and start sand or bead blasting. Direct the blaster nozzle toward the parts to be cleaned. Avoid accidentally blasting the viewing port, as doing so will reduce the transparency of the port.

5. Make sure the sand or bead blasting has finished before you open the cabinet door.

6. Open the door and examine the parts to ensure they have been completely cleaned. If parts are not sufficiently cleaned, continue blasting.

7. Remove parts and clean, preferably with a liquid cleaning solution or water, to ensure all abrasive materials are removed.

SAFETY TIP

Sand or bead blasters use very fine particles at high pressure to clean away the contaminants. Also, silica from the sand particles should not be inhaled; it can cause lung damage, called silicosis, over an extended period of time.

SAFETY TIP

- Do not operate the sand or bead blaster without the door securely closed.
- Wear appropriate breathing apparatus or a dust mask.
- Always wear eye protection to prevent injury from flying particles or escaping compressed air.
- On completion of the job, the correct disposal of contaminated materials is a top priority.

Hydraulic Press

K07007

In an automobile, some parts are press fit together. This is accomplished when the engineers design an interference fit between the parts to hold them together. For example, some rear axle bearings are designed with a smaller inner diameter that is about 0.001" smaller than the outside diameter of the axle it fits over. Once installed, the bearing is clamped tightly to the axle, staying firmly in place. Over time, if the bearing fails, then it will need to be pressed off the axle shaft, using several tons of force (**FIGURE 7-41**). Depending on the size, hydraulic presses

FIGURE 7-41 Hydraulic press being used to press off an axle bearing.

can typically develop many tons of force, so they are used to perform this kind of task.

Because hydraulic presses develop so much force, they can be very dangerous to operate. The extreme forces generated can cause components to shatter with great force, resulting in shrapnel injuries or death. Parts that don't shatter can slip and be thrown a great distance at very high speed. Also, it is possible to get a body part caught under the ram. If so, it is easy for bones to be broken and crushed. So never operate a hydraulic press without specific training by and the permission of your instructor. Here are some general guidelines when using a hydraulic press:

- Never remove any guards or safety devices.
- Always wear safety glasses and a face shield.
- Never wear loose clothing. But it is good practice to wear thick clothing or a snug-fitting leather apron.
- Keep the area clear of obstacles and tripping hazards.
- Make sure all pins or locks are securely in place.
- Only use approved press plates and adapters.
- Only press in line with the hydraulic ram; don't press at an angle.
- Inspect the press for damage regularly.
- If the part doesn't move with reasonable pressure, verify that all keepers, bolts, or other fasteners have been removed from the part.

▶ Welding and Cutting Equipment

K07008

Oxyacetylene

Technicians occasionally use **oxyacetylene torches** to heat, braze, weld, and cut metal. Acetylene is a highly combustible gas, and when combined with oxygen, it produces a very hot temperature of 6300°F to 6800°F (3480°C to 3760°C). Heating is used to loosen rusted fasteners to help remove them. Brazing usesw brass filler rod, which is melted by the torch to join or patch metals. Welding is the process of joining metals by melting the metals together, usually along with a filler rod, which provides extra material to fill any gaps. Cutting involves heating metals until they are so hot that when given an extra shot of oxygen, the metal burns.

The torch consists of an acetylene cylinder, an oxygen cylinder, a pressure regulator for each cylinder, hoses, a flashback arrestor for each hose, the torch handle, and the tip. The cylinders hold the gases. Each pressure regulator has two pressure gauges. One gauge shows how much pressure is in the cylinder, and the other gauge shows how much pressure is in the line. The line (hose) pressure is adjusted on the pressure regulator by the operator. The hoses run from each regulator to the flashback arrestors on the torch handle. The acetylene hose is red, and the oxygen hose is green.

The **flashback arrestors** are spring-loaded check valves that allow flow through the hoses in one direction only—from the cylinders to the torch handle. They prevent flame from traveling back up the hose in the case of a flashback, which is when the oxygen and acetylene ignite inside the torch handle. Flashback happens if:

- the torch valves are set lower than they should be for a particular tip, which produces low gas flow out of the tip;
- if a welding spark jumps up into the tip; or
- if the torch is set with too much oxygen flowing.

The flashback arrestors may be screwed into the torch handle or built directly into it. The gas flow valves are near the base of the handle. The top of the handle is threaded so that different tips can be installed.

Different tip sizes are available depending on the size and thickness of the metal to be heated, welded, or cut (**FIGURE 7-42**).

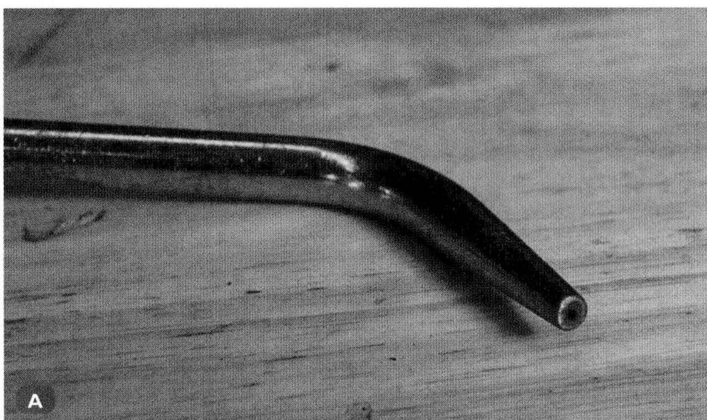

FIGURE 7-42 **A.** Welding tip. **B.** Cutting torch tip. **C.** Rose bud tip.

Larger tips allow more gas (oxygen and acetylene) mixture to flow out of the hole in the tip, which produces a larger flame. Welding tips usually have a single hole in the tip—a small hole for thin metals or a larger hole for thicker metals. Larger heating tips, usually called "rose buds," have multiple holes, typically five or more in a circle toward the outside tip face. The multiple holes allow the rose bud to create the same amount of heat as a much larger single-hole tip, but with a lower chance of flashback. The cutting torch tip uses multiple holes in a circle similar to a rose bud, but it also has a central hole for pure oxygen inside the circle of holes.

When oxyacetylene is used to cut steel, a cutting torch is fitted in place of the welding tip. The outer circle of holes provides the oxygen and acetylene for heating the metal. The center hole injects a stream of pure oxygen into the red-hot metal and causes it to burn, thereby cutting the metal. The oxygen flow is controlled by a spring-loaded lever and activated by the operator.

Neutral, Oxidizing, and Carburizing Flame Adjustment

When welding, heating, or cutting, you need to match the type of flame to the job. The torch can be adjusted to three types of flames: neutral (most common), oxidizing, and carburizing (**FIGURE 7-43**). A neutral flame is obtained once the torch is lit, by adjusting the acetylene or oxygen valves on the torch handle. With the torch lit and operating only on acetylene, the flame will be very yellow from the tip outward. As oxygen is added, the flame near the tip will turn white, and the yellow flame will shrink and turn light blue. Just when the blue flame reaches the point of no longer being visible, it is a neutral flame. If you turn the oxygen down slightly, a feathery blue cone will result, which is called a carburizing flame and is used for brazing. If you turn the oxygen up a bit more, just to where the two cones turn into one cone, you will have a neutral flame again. If you turn the oxygen up a bit more, the single cone will become sharply pointed, indicating an "oxidizing" flame. This flame tends to burn metal, and is used with a cutting torch to cut metal.

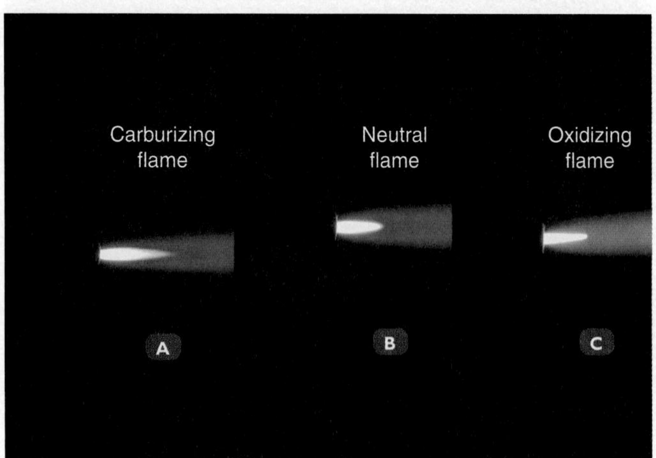

FIGURE 7-43 A. Carburizing flame. **B.** Neutral flame. **C.** Oxidizing flame.

Oxyacetylene Torch Safety

Safety needs to be first and foremost when working with an oxyacetylene torch. Oxyacetylene cylinders carry very high pressures. The acetylene pressure in a full cylinder is approximately 250 psi (1724 kPa), and the oxygen cylinder is approximately 2200 psi (15,168 kPa). If an oxygen cylinder falls over and breaks the main valve off, the cylinder will become a missile and can even go through concrete block walls, so always secure the cylinders properly to the wall or an approved welding cart. Wear a leather apron, or similar protective clothing, and welding gloves when using an oxyacetylene torch. T-shirts, nylon, and polyester blend clothing do not provide enough protection because ultraviolet light and sparks of hot metal will pass through them.

Always use proper welding goggles. Do not use sunglasses because they do not filter the extreme ultraviolet light as effectively, and the plastic used in the lenses of sunglasses will not protect your eyes from sparks. Never point the lighted flame toward another person or any flammable material. Always light the oxyacetylene torch with the striker. A cigarette lighter could explode, and a match would put your hand too close to the igniting tip. Wherever possible, use a heat shield behind the component you are heating. This prevents nearby objects from becoming hot. After heating a piece of metal, label it as "HOT" with a piece of chalk so that others will not attempt to pick it up.

Purging Oxyacetylene Torches

S07015

Purging the oxyacetylene system is the process of venting all the gas in the system, from the cylinder valve to the torch handle tip, and setting the valves so nothing can enter the system. Purging the system is required when you are finished with the torch for the day; when you are not sure if the torch was shut down properly the last time it was used; or when you are making repairs to the system.

To properly purge an oxyacetylene system, follow the steps in **SKILL DRILL 7-15**.

SAFETY TIP

- Oxygen and acetylene cylinders must be securely stored in an upright position.
- Listen and look for leaks.
- An oxyacetylene torch can produce a large amount of heat. Be aware that any objects you direct the flame toward will become hot.
- Always have a suitable fire extinguisher near your work area.
- Do not use an oxyacetylene torch near any flammable materials.

Setting Up Oxyacetylene Torches

S07016

Setting up an oxyacetylene torch means getting it ready to use. Precise steps must be followed for the sake of safety. Each of these steps is used no matter what kind of torch work you are

SKILL DRILL 7-15 Purging Oxyacetylene Torches

1. Verify that the valves are firmly closed on both cylinders. The valves normally turn off in a clockwise direction.

2. Verify that both valves on the torch handle are closed only finger tight, and no more. The seals in the torch handle valves are relatively soft and seal very easily. In contrast, the seals in the cylinder valves are solid tapered metal and need to be firmly tightened when being shut off.

3. Turn the T-handle on each pressure regulator clockwise until drag is felt, and then another half turn. This may cause the pressure to rise on the line pressure gauge.

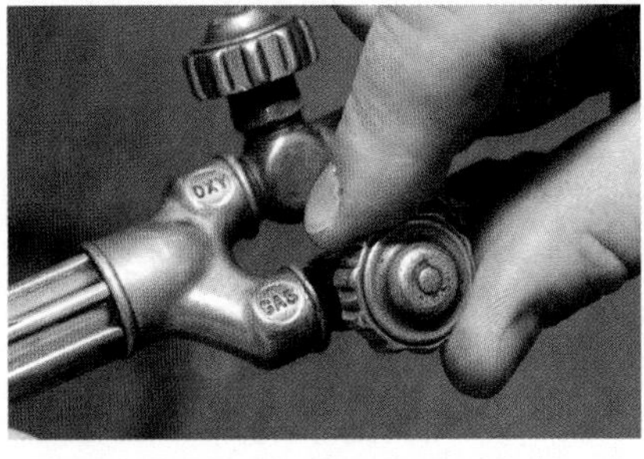

4. Open the acetylene valve on the torch handle slightly. Any acetylene in the line should purge.

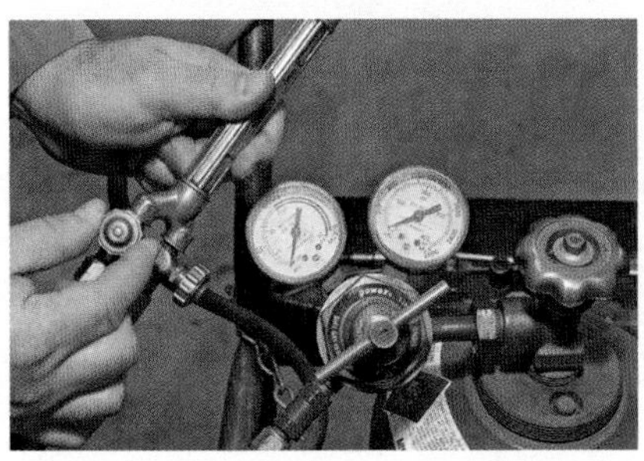

5. When both acetylene gauges read zero, close the valve on the torch handle finger tight only.

6. Back off the T-handle counterclockwise on the acetylene pressure regulator until it turns freely. If it falls out, thread it back in a few turns. Perform steps 3 through 5 on the oxygen side of the system.

performing. If in doubt, always refer the torch manufacturer's instructions.

To set up an oxyacetylene torch for heating, follow the steps in **SKILL DRILL 7-16**.

Using Oxyacetylene Torches for Heating

S07017

Technicians use an oxyacetylene torch to heat a variety of objects such as rusted bolts or brake drums that will not break loose. A torch might also be used to heat a container of water to test a thermostat in. Being able to safely use a torch is a valuable skill in a shop.

To use an oxyacetylene torch for heating to remove a ring gear from a flywheel, follow the steps in **SKILL DRILL 7-17**.

Using Welding Tips for Welding

S07018

Welding is a handy skill for technicians to have. Sometimes tools have to be modified so they will work in particular situations, or a bracket breaks and has to be welded back together. Being able to weld gives you another option in your quest to fix and repair vehicles.

To use an oxyacetylene torch for welding, follow the steps in **SKILL DRILL 7-18**.

Using Welding Tips for Brazing

S07019

Brazing allows a technician to join metals in a similar manner to soldering, but at higher temperatures of about 850°F (454°C),

SKILL DRILL 7-16 Setting Up Oxyacetylene Torches

1. First, verify that the two cylinders are secured in an upright position.

2. Verify that the valves on each of the cylinders are completely and firmly turned off. Both gauges on each cylinder should read zero. If both gauges do not read zero, you must purge the system of any gas, following the steps in Skill Drill 7-15, Purging Oxyacetylene Torches.

3. Verify that the T-handles on the pressure regulators are backed off and turn freely.

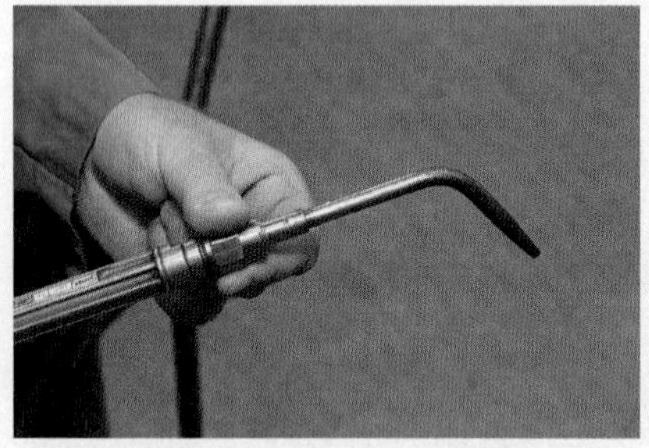

4. Select the tip you will be using, and thread it onto the torch handle by hand. Position the tip so that it faces away from you. Tighten it only moderately hand tight, as it normally uses O-rings to seal the joint.

5. Standing off to the side of the pressure gauges (in case they explode), very slowly open the valve on the top of the oxygen cylinder. When pressure stabilizes, continue to open the valve all the way until it stops.

6. While still standing off to the side of the pressure gauges, slowly open the valve on the top of the acetylene cylinder 1/4 to 1/2 turn only. This will allow for quick closing of the valve in the case of a fire or other emergency requiring quick shutoff.

7. Open the acetylene valve on the torch handle about 1/8 turn and then turn the T-handle on the acetylene pressure regulator (clockwise) until the low-pressure gauge reads 5 psi. (Never exceed 15 psi [103.42 kPa] line pressure on the acetylene side.) Lightly turn off the valve on the torch handle.

8. Open the oxygen valve on the torch handle about 1/8 turn, and then turn the T-handle on the oxygen pressure regulator (clockwise) until the low-pressure gauge reads 10 psi (68.95 kPa) (if welding or heating) or 20–25 psi (137.90–172.37 kPa) (if cutting). Lightly turn off the valve on the torch handle.

9. Before you light the torch, check the area you are working in to make sure there are no flammable materials or fluids nearby. It is absolutely vital that you wear the right safety gear: gloves and tinted gas welding goggles or gas welding helmet. Also make sure that there is plenty of slack in the welding hoses and that they are out of the way of the torch area.

SKILL DRILL 7-16 Setting Up Oxyacetylene Torches (Continued)

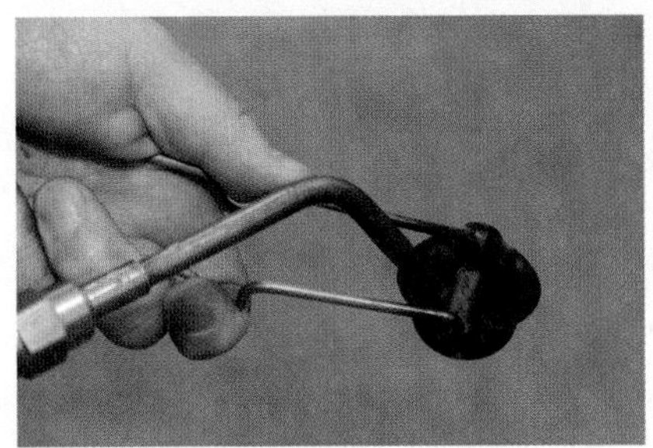

10. The tip of the torch must be pointing downward away from your body and away from the gas cylinders. Turn the acetylene valve on the torch handle slightly counterclockwise. You should hear the gas hissing. Hold the striker against the tip of the torch, with the lighter cup pointing away from you. Flick the striker to create the spark that will ignite the gas at the tip of the torch. Quickly open the acetylene valve until the sooty smoke produced by the torch disappears. Then slowly open the oxygen valve on the torch handle.

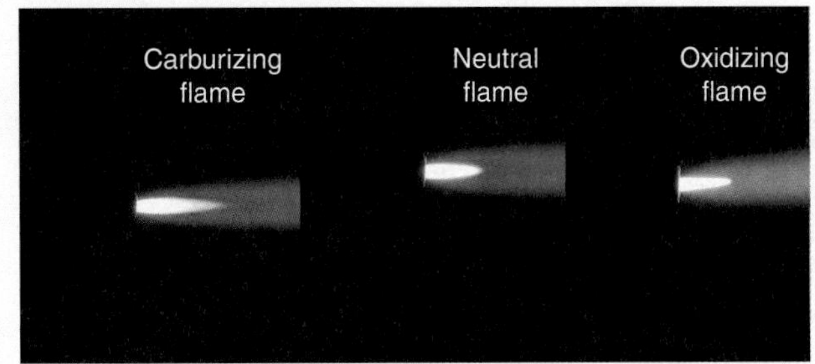

11. As you open the oxygen valve, you will see the color of the flame change. The pure acetylene flame is yellow, and it will change to blue as you add the oxygen. Continue to open the oxygen valve slowly until you can observe a single small, sharp white cone in the center of the torch flame. Do not open the oxygen valve beyond where the two flames become a single blue cone. This is the neutral flame you need for general heating and welding.

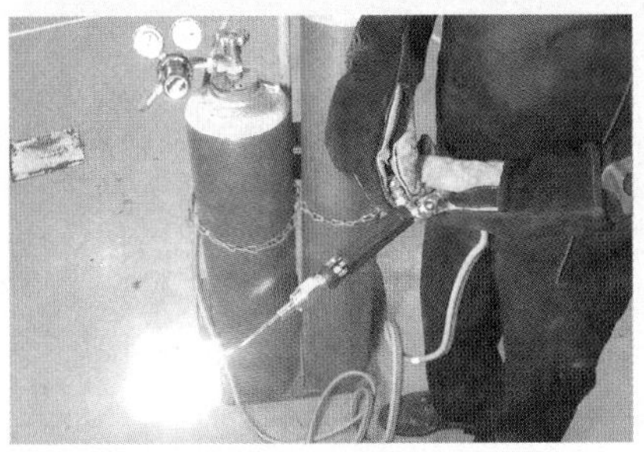

12. When you are ready to shut down the torch, quickly close the oxygen valve finger tight on the torch handle, followed by closing the acetylene valve finger tight on the handle.

13. When you are finished with the torch for the day, go through the process of purging the system (see Skill Drill 7-15, Purging Oxyacetylene Torches).

SKILL DRILL 7-17 Using Oxyacetylene Torches for Heating

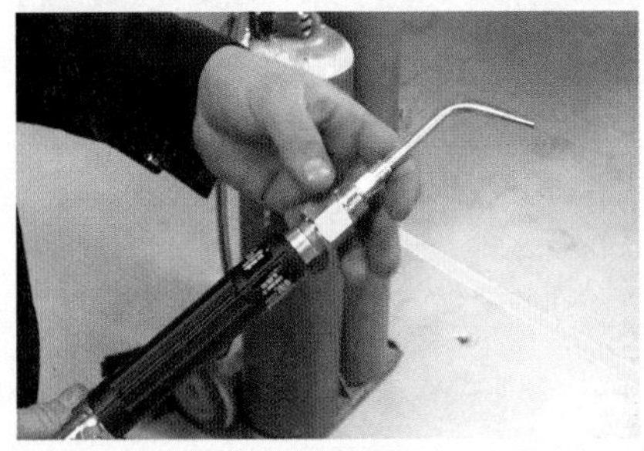

1. Select the proper tip for the amount of heating you need to perform.

2. Place the flywheel and ring gear assembly on a set of insulating spacers to elevate it from the working surface.

3. Light the torch and adjust the gas flow so that you have a neutral flame.

SKILL DRILL 7-17 Using Oxyacetylene Torches for Heating (Continued)

4. Direct the flame onto the ring gear, and apply the heat until smoke starts to appear. Stop applying the heat. At this stage, the ring gear is hot enough to remove by gently tapping with a hammer and drift punch. *Do not touch the metal with your hands.* Use welding gloves and tools that are designed for use in a hot environment.

5. When you have finished the job, you will need to shut down the equipment. Put the tip away, and return the oxyacetylene cart to its proper storage area.

SKILL DRILL 7-18 Using Welding Tips for Welding

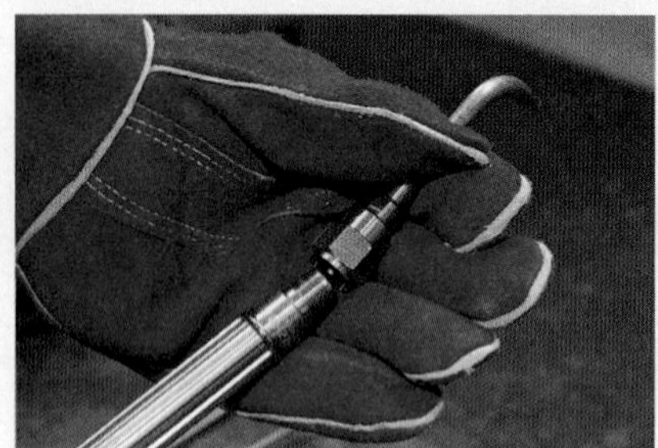

1. Select and fit the correct size tip for the weld you are about to undertake.

2. Before starting, make sure you have assembled all the materials you need to undertake the weld, including safety equipment, a fire extinguisher close at hand, welding filler rod, and the items to be welded.

3. Position the items to be welded, and secure them with clamps if necessary.

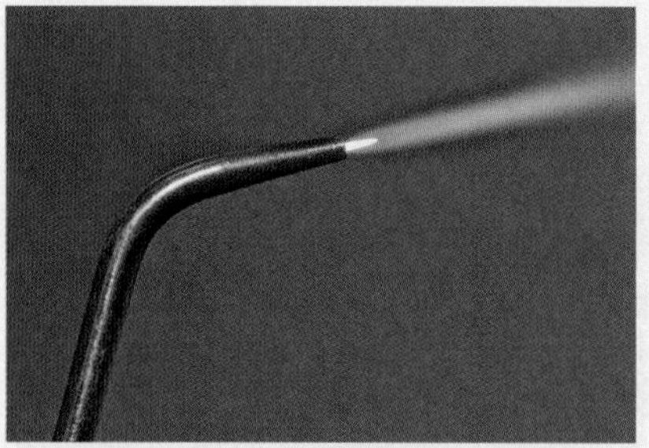

4. Light the torch and adjust the gas flow so that you have a neutral flame.

5. Direct the flame into the area to be welded, and apply the heat until the metal just starts to melt a small pool of metal. Be careful not to overheat the metals. Slowly add the welding rod into the molten pool of metal. This will cause some of the welding rod to melt and fill the gap and slightly build up the weld. If the welding rod sticks to the metal, twist it with your thumb and finger to break it loose.

6. Move the torch along the weld joint, continuously adding filler rod until the weld is complete. Avoid overheating the joint, and try to confine the heated area just to the spot you are welding.

SKILL DRILL 7-18 Using Welding Tips for Welding (Continued)

7. When the weld is complete, stop applying the heat. *Do not touch the metal with your hands.* Use welding gloves and tools that are designed for use in a hot environment.

8. When you have finished the job, you will need to properly shut down the equipment.

although not as hot as welding temperatures. Brazing typically uses brass filler rod, which melts at higher temperatures and is stronger than solder. Brazing works well when joining a heavier metal part to a thinner metal part, such as a brass fitting to a thin sheet metal oil pan. It also bonds tighter to the metals, so it is more permanent than soldering.

To use an oxyacetylene torch for brazing, follow the steps in **SKILL DRILL 7-19**.

SKILL DRILL 7-19 Using Welding Tips for Brazing

1. Select and fit the correct size tip for the braze you are about to undertake.

2. Before starting, make sure you have assembled all the materials you need to undertake the braze, including safety equipment, a fire extinguisher close at hand, brazing filler rod, appropriate flux, and the items to be brazed.

3. Position the items to be brazed, and secure them with clamps if necessary.

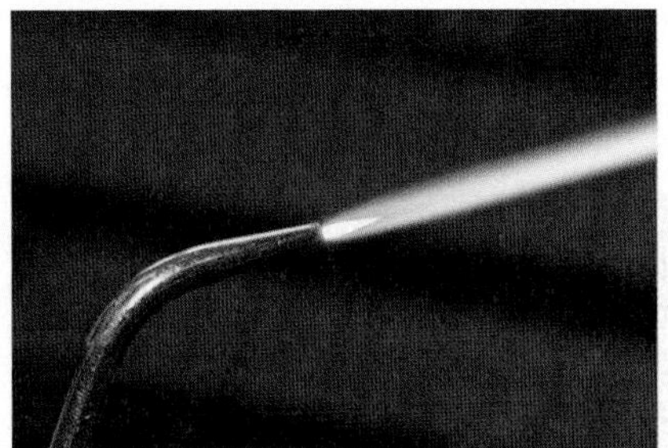

4. Light the torch and adjust the gas flow so that you have a slightly carburizing flame (not quite to neutral flame status).

5. Direct the flame into the area to be brazed, and apply the heat until the metal is cherry red. Be careful to not overheat the metals. Slowly add the flux-coated brazing rod into the space to be brazed. Some filler rods come with a coating of flux attached; others require the addition of flux. This is achieved by slightly heating the brazing rod with the welding torch and dipping the heated brazing rod into the flux.

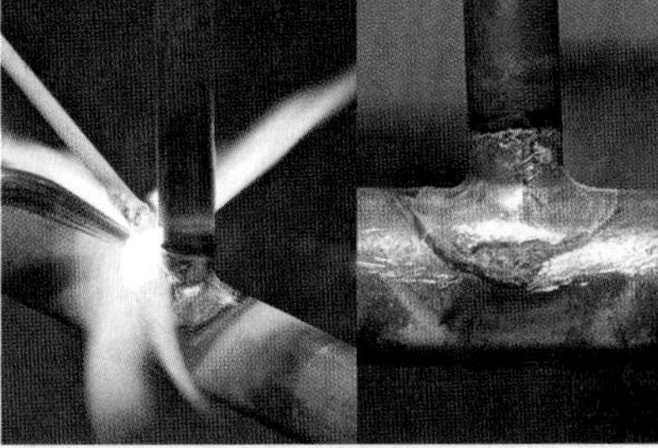

6. Move the torch along the weld joint, continuously adding filler rod until the braze is complete. Avoid overheating the joint, and try to confine the heated area to only the spot you are brazing. This will reduce the spread of the filler into unwanted areas.

7. When the braze is complete, stop applying the heat. *Do not touch the metal with your hands.* Use welding gloves and tools that are designed for use in a hot environment.

8. When you have finished the job, you will need to properly shut down the equipment.

Using Cutting Torches

S07020

Technicians use a cutting torch for various reasons. They may need to cut fasteners that cannot be disassembled due to rust and corrosion, such as exhaust pipe clamps that are frozen in place. Or they may need to cut a wrench to shorten it so it will fit into a tight space. Technicians who are very skilled with a torch can even cut out a broken-off bolt from its threaded hole. However, don't try this unless you are *very* skilled with a cutting torch.

To use an oxyacetylene torch for cutting, follow the steps in **SKILL DRILL 7-20**.

Plasma Cutters

A **plasma cutter** is a tool for cutting various thicknesses of metal. Plasma cutters work by applying a pressurized gas such as air, oxygen, nitrogen, or argon through a nozzle located in the center of a hand piece with an electrode. When power is applied to the electrode and a return electrical path is established through the metal that needs to be cut, a powerful spark is created. The spark heats the gas, changing it to plasma, a fourth state of matter. The plasma is very hot—approximately 30,000°F (16,600°C)—and moving very fast. The plasma heats the metal very quickly and blows the molten metal out to produce a clean cut through the metal. Plasma cutters have been around for quite a long time, but in the past they have been expensive, large machines reserved for large-scale manufacturing. More recently, shop-sized machines have been developed and are relatively inexpensive. This has made them very accessible and an affordable alternative to using oxyacetylene as a cutting tool for metal.

Using Plasma Cutters

S07021

If using the plasma cutter on a vehicle, make sure the battery is disconnected. Wherever possible, use a heat shield behind the

SKILL DRILL 7-20 Using Cutting Tips

1. Select and fit the correct size cutting tip for the cut you are about to undertake. Attach any cutting accessories, such as a cutting guide, to the torch.

2. Before starting, make sure you have assembled all the materials you need to undertake the cut, including safety equipment, a fire extinguisher close at hand, and the item(s) to be cut. Also clear the area of all flammable objects.

3. Position the items to be cut, and secure them with clamps if necessary.

4. Light the torch and adjust the gas flow so that you have a neutral flame. Press the oxygen cutting lever, and make sure additional oxygen is injected through the center nozzle. Further refine the gas and oxygen settings if required.

5. Direct the flame into the area to be cut, and apply the heat until the metal is cherry red. Press the oxygen cutting lever; if the metal is hot enough, doing so should instantly cause the metal under the flame to blow away.

6. Slowly draw the torch along the cut line, checking to make sure the cut is being made.

7. When the cut is complete, stop applying the heat. *Do not touch the metal with your hands.*

8. When you have finished the job, you will need to properly shut down the equipment.

component you are heating. This will help prevent nearby objects from becoming too hot. After cutting a piece of metal, label it as "HOT" with a piece of chalk so that others will not attempt to pick it up.

To use a plasma cutter for cutting, follow the steps in **SKILL DRILL 7-21**.

SAFETY TIP

- Plasma cutters require high amperage. Make sure you plug the cutter into an appropriately sized outlet.
- A plasma cutter can produce a large amount of heat. Be aware that any objects you cut will become hot.
- Always have a suitable fire extinguisher near your work area.
- Do not use a plasma cutter near any flammable materials.

Wire Feed Welders

Wire feed welders are a type of welder that has the filler rod automatically feeding into the welding joint via the hand piece. They are becoming commonplace in shops. The wire feed is controlled by a trigger on the hand piece, and its rate of feed can be adjusted on the welder to accommodate the size of weld being conducted.

SKILL DRILL 7-21 Using Plasma Cutters

1. Always refer to the manufacturer's manual for specific operating instructions. Before using the equipment, locate the position of safety switches, and put on a welding helmet and gloves. Note the location of the fire extinguishers.

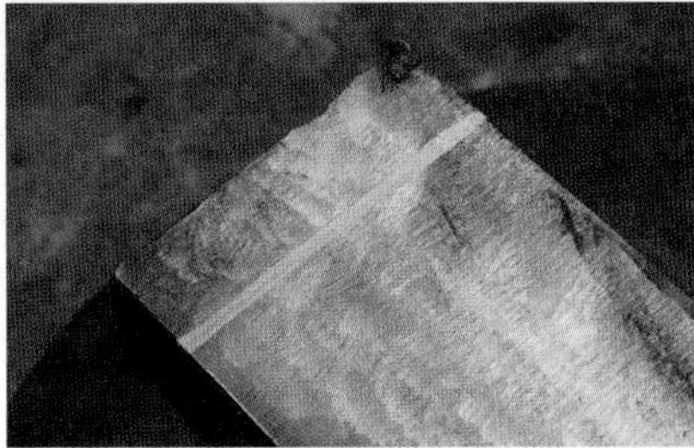

2. Ensure the cut line is free of rust and dirt. Make sure there are no flammable materials in close vicinity.

3. Connect the air hose or gas, plug the unit in, and adjust the power setting to suit the thickness of the material being cut.

4. Before starting, make sure you have assembled all the materials you need to undertake the cut, including safety equipment, a fire extinguisher close at hand, and the item(s) to be cut.

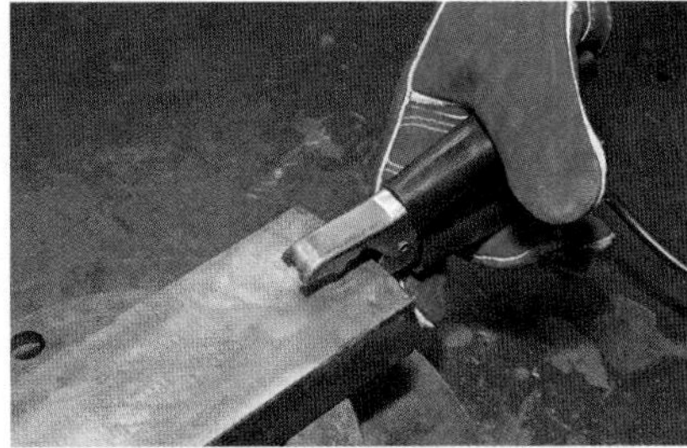

5. Connect the ground lead to a clean part of the metal to be cut, away from the cut line.

6. Turn on the plasma cutter. Put the welding helmet down; position the hand piece to cut, and then press the trigger. Check that the metal is cut all the way through. Stop and adjust the control settings if required.

7. Slowly draw the hand piece along the cut line, checking to make sure the cut is being made.

8. When the cut is complete, release the trigger. *Do not touch the metal with your hands.* Use welding gloves and tools that are designed for use in a hot environment.

9. When you have finished the job, shut down the equipment. Turn off the power and disconnect the ground lead. Put the hand piece away, and return the cutter to its proper storage area.

The welder has controls for amperage, like on a traditional arc welder, and this is adjusted depending on the thickness of metal to be welded.

Like any electric welder, the weld requires a flux or inert gas shield to prevent the oxidization of the weld. In a standard arc welder, this is provided by a coating on the filler stick. For wire feed welders, this can be provided either by an inert gas that is supplied through a nozzle to the welding tip or by a flux core in the center of the filler wire that evaporates as it is heated by the arc. This method shields the weld from oxidization while it is molten and as it is cooling. Wire feed welders are usually classified as gas or gasless. The gas wire feed welder has a provision for an inert gas bottle to be fitted to the welder to supply the shield gas to the welding tip. Gasless welders rely on the flux being in the core of the filler wire. Many wire feed welders can accommodate both gas and gasless operation.

Using Wire Feed Welders

S07022

Wear a leather apron or similar protective clothing and welding gloves when using a wire feed wielder. T-shirts, nylon, and polyester blend clothing do not provide enough protection. Ultraviolet light and sparks of hot metal will pass through them. Always use appropriate eye protection and a welding helmet. Do not use sunglasses because they do not filter the extreme ultraviolet light effectively. Also, the plastic used in the lenses of sunglasses will not protect your eyes from sparks. If using the wire feed welder on a vehicle, make sure the battery is disconnected. Wherever possible, use a heat shield behind the component you are welding. This will help prevent nearby objects from becoming too hot. After welding a piece of metal, label it as "HOT" with a piece of chalk so that others will not attempt to pick it up.

To use a wire feed welder, follow the steps in **SKILL DRILL 7-22**.

SAFETY TIP

- Wire feed welders require high amperage. Make sure you plug the welder into an appropriately sized outlet.
- A wire feed welder can produce a large amount of heat. Be aware that any objects you weld will become hot.
- Always have a suitable fire extinguisher near your work area.
- Do not use a wire feed welder near any flammable materials.

SKILL DRILL 7-22 Using Wire Feed Welders

1. Always refer to the manufacturer's manual for specific operating instructions. Before using the equipment, locate the position of safety switches, and put on a welding helmet and gloves. Note the location of the eyewash and first aid stations.

2. Prepare the area to be welded. Ensure the material is free of rust and dirt. Make sure there are no flammable materials in close vicinity.

3. Prepare the wire feed welder. Ensure enough wire is in the reel, set up the gas if using gas operation, plug the unit in, and adjust the power setting and wire feed speed to suit the thickness of the material being welded.

4. Before starting, make sure you have assembled all the materials you need to undertake the weld, including safety equipment, a fire extinguisher close at hand, and the item(s) to be welded.

5. Connect the ground lead to a clean part of the metal to be welded, as close as possible to the weld but away from the welding line.

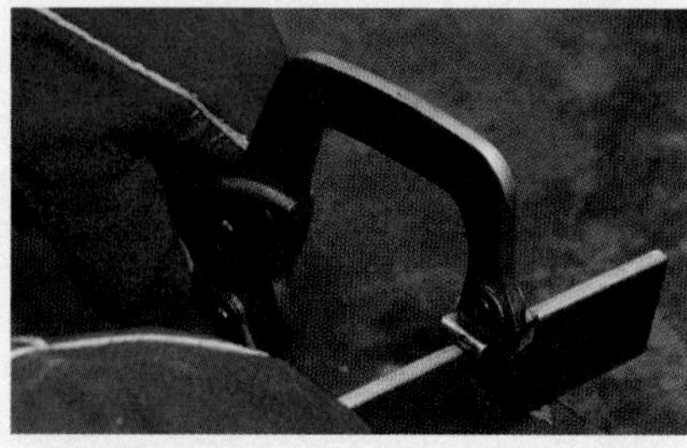

6. Position the items to be welded, and secure them with clamps if necessary.

SKILL DRILL 7-22 Using Wire Feed Welders (Continued)

7. Turn on the welder. Put on the welding helmet, position the hand piece to weld, and press the trigger. Direct the weld into the joint at about a 15-degree angle. Don't let the welding wire protrude more than about ½" (12.7 mm) from the gun. Check that the weld is satisfactory. Stop and adjust the control settings if required.

8. Slowly draw the hand piece along the line to be welded, checking to make sure the weld is being made.

9. When the weld is complete, release the trigger. *Do not touch the metal with your hands.* Use welding gloves and tools that are designed for use in a hot environment.

10. When you have finished the job, you will need to shut down the equipment. Turn off the power and disconnect the ground lead. Put the hand piece away, and return the welder to its proper storage area.

▶ Chassis Dynamometers

K07009

A **chassis dynamometer** is a device that allows a vehicle to be run at road speed while safely contained within a shop bay. Chassis dynamometers are very useful for tuning engines, testing vehicle performance, and determining output in torque and horsepower. They are also used with sensors and test instruments to create simulated load conditions and diagnose any faults while in the shop. Chassis dynamometers are usually bolted down to the floor or built into the floor and have rollers that the wheels sit on (**FIGURE 7-44**). The vehicle is chained down so it cannot move forward or backward when the accelerator is pressed down. The chassis dynamometer has a control unit that displays the controls for the dynamometer and selected engine parameters such as engine revolutions per minute (rpm), drive wheel torque, power, fuel consumption, and exhaust temperature.

FIGURE 7-44 Chassis dynamometer.

SAFETY TIP

Always make sure the vehicle is securely fastened down and the area is clear of tools, equipment, and people before using a chassis dynamometer. In many cases, the vehicle is being operated at full throttle and producing maximum horsepower. If the vehicle were ever to break free, the resulting accident would be catastrophic.

Wrap-Up

Ready for Review

- Battery chargers can be fast or slow, depending on current output; smart chargers calculate and provide the correct amount of charge needed for the battery.
- Vehicle batteries can be dangerous because of their high voltage; hybrid vehicle batteries have extremely high voltage and current flows.
- Batteries should be charged slowly, if possible.
- Jump-starting a vehicle places a stress on both vehicles; the current can damage electrical components in both vehicles.
- Compressed air systems are comprised of a compressor, a pressure regulator, an air hose or fixed piping, and the tool to be powered.
- Standard compressors use a piston to force air into a storage tank, while scroll compressors use rotating scrolls to compress air.
- Always use caution: Compressed air injuries can be fatal.
- Many compressed air systems use air driers to remove all traces of moisture from the compressed air.
- Automatic oilers provide a regular application of lubricating oil to the stream of compressed air, which then lubricates air tools and equipment.
- Chassis dynamometers allow technicians to run vehicles at road speed without leaving the shop.
- Power tools can be stationary or portable, corded or cordless, and are powered by electricity, batteries, compressed air, a propellant, or a gasoline engine.
- Drills are designed to drive a drill bit into metal (or other material) to create a hole; check drilling speed charts for proper drilling speed.
- Portable grinders are designed to grind down metals, but can also be fitted with a cutting disc to cut sheets of metal.
- Air tools use compressed, pressurized air for power; types include the air impact wrench, air ratchet, air hammer, air drill, and blowgun/air nozzle.
- Always wear eye and ear protection when using air tools; never use air nozzles on yourself or other people.
- Pressure washers/cleaners use focused, pressurized water to clean accumulated dirt and grease from vehicle components; water must be directed properly so as not to damage other parts.
- Familiarize yourself with pressure washer operating instructions and wastewater disposal regulations.
- Spray wash cabinets are designed to clean automotive parts in a sealed cabinet, much like a dishwasher.
- Solvent tanks are designed for immersion of vehicle parts to remove oil, dirt, grease, and grime; always note the type of solvent being used and take necessary precautions.
- Brake washers are designed to remove brake dust from wheel brake units and their components.
- Sand or bead blasters are designed to clean paint, corrosion, or dirt from metal parts by blasting small abrasive particles onto the surface.
- Oxyacetylene torches are designed to heat, braze, weld, and cut metal by combining acetylene with oxygen at a high temperature.
- Flashback arrestors prevent flames from traveling back up the hose in the event the oxygen and acetylene ignite inside the torch handle (flashback).
- Wear protective clothing and gear when using an oxyacetylene torch, and follow all related safety precautions.
- Plasma cutters are designed to cut various thicknesses of metal and are an alternative to oxyacetylene torches.
- A wire feed welder has a filler rod automatically feeding into the welding joint at an adjustable rate.
- Keep work area, tools, and equipment clean and organized.

Key Terms

abrasive discs (cutting wheels) Abrasive wheels or flat discs fitted to bench, pedestal, and portable grinders.

air drier A device fitted to compressed air lines to remove moisture.

air drill A compressed air–powered drill.

air hammer A tool powered by compressed air with various hammer, cutting, punching, or chisel attachments; also called an air chisel.

air impact wrench An impact tool powered by compressed air designed to undo tight fasteners.

air nozzle A compressed air device that emits a fine stream of compressed air for drying or cleaning parts.

air ratchet A ratchet tool for use with sockets, powered by compressed air.

angle grinder A portable grinder for grinding or cutting metal.

automatic oiler A device fitted to compressed air systems to oil air tools.

battery A device that converts and stores electrical energy through chemical reactions.

battery charger A device that charges a battery, reversing the discharge process.

bench grinder (pedestal grinder) A grinder that is fixed to a bench or pedestal.

chassis dynamometer A machine with rollers that allows a vehicle to attain road speed and load while sitting still in the shop.

cleaning gun A device with a nozzle controlled by a trigger fitted to the outlet of pressure cleaners.

cored solder Solder that is in the form of a hollow wire. The center is filled with flux, which is used as a cleaning agent while the solder is being applied to the metal surfaces.

drill chuck A device for securely gripping drill bits in a drill.

drill press A device that incorporates a fixed drill with multiple speeds and an adjustable worktable. It can be free standing or fixed to a bench.

fast chargers A type of battery charger that charges batteries quickly.

flashback arrestor A spring-loaded valve installed on oxyacetylene torches as a safety device to prevent flame from entering the torch hoses.

flux A liquid or paste that protects a soldering or welding joint from oxidization.

grinding wheels and discs Abrasive wheels or flat discs fitted to bench, pedestal, and portable grinders.

memory saver (memory minder) Battery backup device for vehicle computer systems.

Morse taper A system for quickly changing and securing drill bits to drills.

oxyacetylene torch A gas welding system that combines oxygen and acetylene.

plasma cutter A tool that uses electricity and compressed gas to produce a stream of high-temperature gas to cut metal.

power tools Tools powered by electricity or compressed air.

pressure washer/cleaner A cleaning machine that boosts low-pressure tap water to a high-pressure output.

rattle gun A term used to describe an air impact wrench, based on the noise it makes.

Rosin A type of liquid or paste (flux) that is in solid form contained within solder and is used to prevent oxidization.

sand or bead blasters A cleaning system that uses high pressure and fine particles of glass bead or sand.

sander/polisher A power tool with a rotating disc or head to which polishing or sanding discs can be attached.

slow charger A battery charger that charges at low current.

smart charger A battery charger with microprocessor-controlled charging rates and times.

Solder A mixture of lead and tin with a low melting point for connecting wires.

soldering irons A heating tool to heat solder and wires to produce a low-resistance joint.

solvent tank A tank containing solvents to clean vehicle parts.

spray wash cabinet A cleaning cabinet that sprays cleaning solution under pressure to clean vehicle parts.

straight grinder A powered grinder with the wheel set at 90 degrees to the shaft.

sulfuric acid A type of acid that when mixed with pure water forms the basis of battery acid or electrolyte.

surge protector An electrical protection device for preventing electrical surges.

twist drill A hardened steel drill bit for making holes in metals, plastics, and wood.

wire feed welder A welding machine that automatically feeds the filler wire by operating a trigger mechanism on a welding gun.

Review Questions

1. All of the following statements with respect to jump-starting are true *except*:
   a. Charge the dead vehicle 5 to 10 minutes before trying to start the vehicle.
   b. Jumper leads with surge protectors reduce the risk of damage to sensitive electronic devices.
   c. Read the owner's manual for both vehicles before jump-starting.
   d. Always use a fully charged battery to jump-start a frozen battery.
2. Which of these is true of an air impact wrench?
   a. It always spins in the clockwise direction.
   b. It can apply only minimal torque.
   c. It should never be used for final tightening of lug nuts.
   d. It does not need special hardened impact sockets.
3. Which of these is used for driving and cutting?
   a. Air hammer
   b. Air ratchet
   c. Air nozzle
   d. Air drill
4. When grinding metal using a bench grinder:
   a. stop and dip it regularly into the water pot.
   b. do not leave a gap between the tool rest and wheel.
   c. stand off to the side of the wheel.
   d. grind on the side of the wheel.
5. When soldering non-electrical connections, what should be used as the flux-cleaning agent?
   a. Rosin
   b. Water
   c. Acid
   d. Gasoline
6. Which of these is true of a spray wash cabinet?
   a. It removes paint under high pressure.
   b. It increases the frequency of cleaning solution changes.
   c. It does not leavy dirty residue on parts.
   d. It does not produce harmful waste.
7. All of the following are good practices when operating a hydraulic press *except:*
   a. remove guards around the press for smooth operation.
   b. wear thick clothing or a snug-fitting leather apron.
   c. only use approved press plates and adapters.
   d. never wear loose clothing.
8. Which of these is a good choice as a cleaning agent for brake components?
   a. Kerosene
   b. Soapy water
   c. Plain water
   d. Solvent

9. The advantage of using rose bud tips is that they:
   a. are best used on small heating jobs.
   b. are efficient under low pressure.
   c. can weld thicker materials.
   d. lower the chance of flashback.
10. Chassis dynamometers are very useful for:
   a. grinding down metals.
   b. cleaning accumulated dirt.
   c. testing vehicle performance.
   d. driving a drill bit.

ASE Technician A/Technician B Style Questions

1. Tech A says that it is safe to use an air nozzle to blow dust off you or your coworker. Tech B says that OSHA-approved nozzles lower the tip pressure by venting some of the air for safety reasons. Who is correct?
   a. Tech A
   b. Tech B
   c. Both A and B
   d. Neither A nor B
2. Tech A says that you should always wear ear protection when using an air hammer. Tech B says that standard wall sockets can be safely used with impact wrenches. Who is correct?
   a. Tech A
   b. Tech B
   c. Both A and B
   d. Neither A nor B
3. Tech A says that when soldering electrical wires, rosin flux must be used. Tech B says that a heat dam is used to prevent heat from traveling to the electronic component when you are soldering. Who is correct?
   a. Tech A
   b. Tech B
   c. Both A and B
   d. Neither A nor B
4. Tech A says that air tools and equipment require a regular application of a lubricating oil to reduce wear and tear. Tech B says that some compressed air systems use an inline water trap that needs to be drained periodically. Who is correct?
   a. Tech A
   b. Tech B
   c. Both A and B
   d. Neither A nor B
5. Tech A says that one of the first things to do when purging an acetylene torch is to turn off the valves on both bottles. Tech B says that when setting the line pressures on an acetylene torch, the acetylene should always be adjusted so it is above 15 psi (103.42 kPa). Who is correct?
   a. Tech A
   b. Tech B
   c. Both A and B
   d. Neither A nor B
6. Tech A says that when using a bench grinder, the tool rest should be no more than 1/8" (3.2-mm) or half the thickness of the metal you are grinding, away from the grinding wheel. Tech B says that when using a bench grinder, the piece you are grinding needs to be cooled periodically. Who is correct?
   a. Tech A
   b. Tech B
   c. Both A and B
   d. Neither A nor B
7. Tech A says that when you jump-start a vehicle, a spark typically occurs when making the last jumper cable connection. Tech B says that all four of the jumper cable connections should be made at the battery terminals. Who is correct?
   a. Tech A
   b. Tech B
   c. Both A and B
   d. Neither A nor B
8. Tech A says that an air nozzle can be used to safely blow brake dust from brake parts. Tech B says that brake wash stations use strong acid as a cleaning solution. Who is correct?
   a. Tech A
   b. Tech B
   c. Both A and B
   d. Neither A nor B
9. Tech A says that gaskets can be removed quickly and safely with a portable grinder and high-speed grinding stone. Tech B says that extreme care must be used when removing a gasket on an aluminum surface. Who is correct?
   a. Tech A
   b. Tech B
   c. Both A and B
   d. Neither A nor B
10. Tech A says that a wire feed welder is used to braze components together. Tech B says that sunglasses should be used when watching someone weld with a wire feed welder. Who is correct?
   a. Tech A
   b. Tech B
   c. Both A and B
   d. Neither A nor B

CHAPTER 8

Fasteners and Thread Repair

NATEF Tasks

- **N08001** Perform common fastener and thread repair, to include: remove broken bolt, restore internal and external threads, and repair internal threads with thread insert.

Knowledge Objectives

After reading this chapter, you will be able to:

- **K08001** Identify threaded fasteners and describe their use.
- **K08002** Identify standard and metric fasteners.
- **K08003** Describe how bolts are sized.
- **K08004** Describe thread pitch and how it is measured.
- **K08005** Describe bolt grade.
- **K08006** Describe the types of bolt strength required.
- **K08007** Describe nuts and their application.
- **K08008** Describe washers and their applications.
- **K08009** Describe the purpose and application of thread-locking compounds.
- **K08010** Describe screws and their applications.
- **K08011** Describe the torque-to-yield and torque angle.
- **K08012** Describe how to avoid broken fasteners.
- **K08013** Identify situations where thread repair is necessary.

Skills Objectives

- **S08001** Perform common fastener and thread repair.

▶ Introduction

Fasteners come in two common types: threaded fasteners and non-threaded fasteners (**FIGURE 8-1**). Both types of **fasteners** are designed to secure parts. **Threaded fasteners** are primarily designed to clamp objects together. Non-threaded fasteners are designed to hold parts together, such as preventing a component from falling off a shaft by using C-clips, cotter pins, roll pins, or other retainers. They can also be used to clamp components together, for example, with rivets. Each type of fastener has a different purpose or application, so getting to know the various kinds will help you know how to disassemble and reassemble component assemblies. We cover the major types of threaded and non-threaded fasteners below.

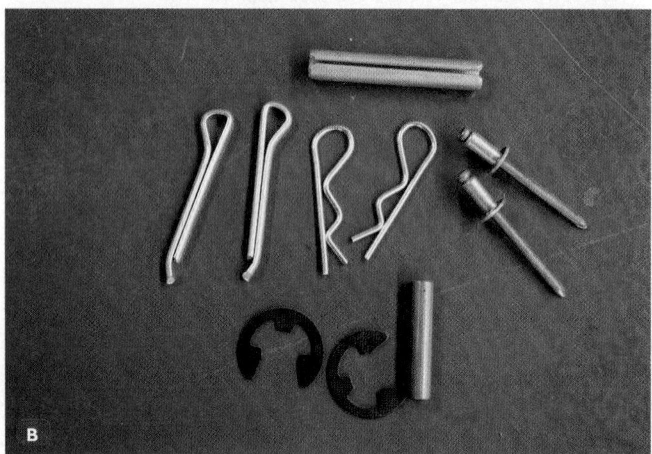

FIGURE 8-1 A. Threaded fasteners. **B.** Non-threaded fasteners.

▶ Threaded Fasteners and Torque

K08001

Threaded fasteners are designed to secure parts that are under various tension and sheer stresses. The nature of the stresses placed on parts and threaded fasteners depends on their use and location. For example, head bolts withstand tension stresses by clamping the head gasket between the cylinder head and block so that combustion pressures can be contained in the cylinder (**FIGURE 8-2**). The bolts must withstand the very high combustion pressures trying to push the head off the top of the engine block in order to prevent leaks past the head gasket. In this situation, the pressure is trying to stretch the head bolts, which means they are under tension.

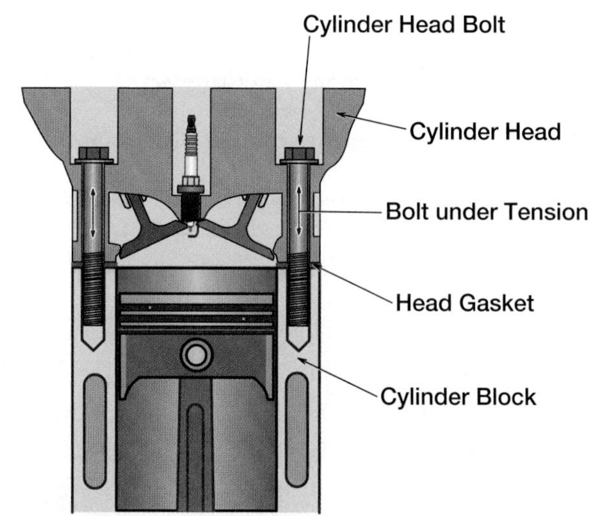

FIGURE 8-2 Bolts clamp parts together.

You Are the Automotive Technician

The new intern was helping you disassemble a leaking water pump. All was going well until one of the bolts broke off while removing it. It is evident that the shank of the bolt had been narrowed considerably due to rust and corrosion over time. You know that removing this bolt could turn the job into a real nightmare if not performed properly and carefully. It is also an opportunity to teach the intern how to perform this important task. As you are thinking about this task, several questions come to mind.

1. If part of the bolt extends past the surface it is threaded into, what can be done to try to remove it without using a bolt extractor?
2. What processes can be tried on bolts that are stuck, before they are broken off?
3. Is it okay to use a replacement bolt of a different grade from the original bolt? Why or why not?
4. What is the purpose of a thread-locking compound? Give an example of when should it be used?
5. What is the purpose of antiseize? Give examples of when it should and shouldn't be used?

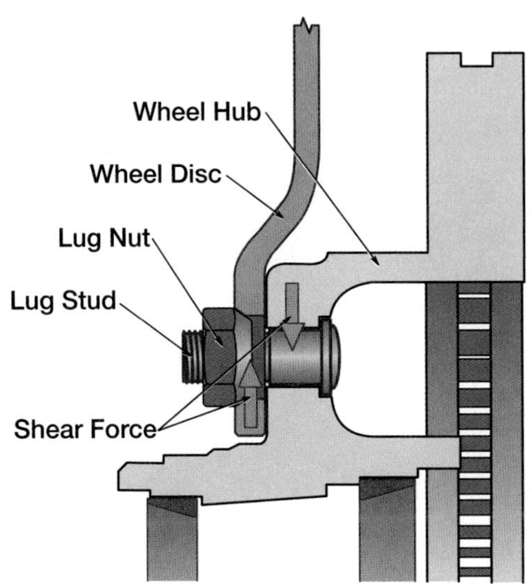

FIGURE 8-3 Bolt under sheer stresses.

Some fasteners must withstand sheer stresses, which are sideways forces trying to shear the bolt in two (**FIGURE 8-3**). An example of fasteners withstanding sheer stresses is wheel lug studs and lug nuts. They clamp the wheel assembly to the suspension system, and the weight of the vehicle tries to sheer the lug studs. If this were to happen, the wheel would fall off the vehicle, likely leading to an accident.

To accomplish their job, fasteners come in a variety of diameters for different sized loads, and hardnesses, which are defined in grades. Threaded fasteners are designed to be tightened to a specified amount depending on several factors such as the job at hand, the grade or hardness of the material they are made from, their size, and the thread type. If a fastener is over-tightened, it could become damaged or could break. If it is under-tightened, it could work loose over time. Torque is a measure of twisting force around an axis, so it is a way of defining how much a fastener should be, or is, tightened to. Manufacturers print **torque specifications** in torque charts for virtually every fastener on a vehicle.

The idea of threaded fasteners has been around since approximately 240 B.C., when Archimedes invented the screw conveyor. The screw design was adapted to a straight pin to attach different materials together. As time progressed, the materials that the screw was made from became stronger, and the sizing became more precise. Whitworth created a thread profile that had a rounded crest and rounded root in England in 1841 (**FIGURE 8-4**). Then in 1864, in the United States, Sellers created a thread profile that had a flat crest and flat root (**FIGURE 8-5**).

In order to create the threaded fasteners we use today, the flat crest and rounded root is used along with a 60-degree pitch angle (**FIGURE 8-6**). Metric and standard size bolts use the same shape of threads, excluding specialty-use bolts, which are made for a particular purpose, such as the lead screw for a vice or gear puller. The square edges of this specialty thread (**square thread**) tend to exert force more in line with the screw, where a V-thread tends to push at a 90-degree angle to the thread, not the screw (**FIGURE 8-7**).

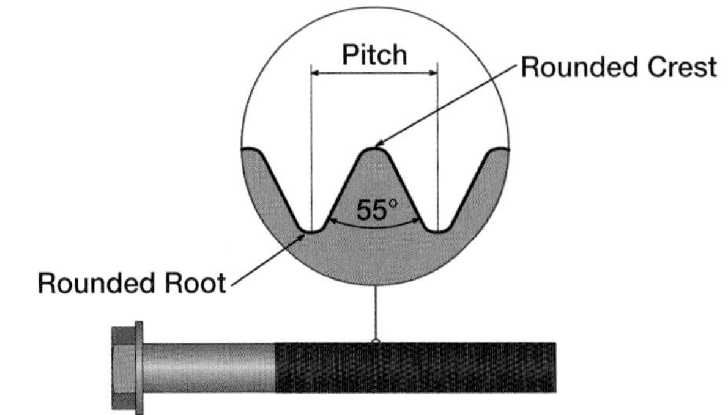

FIGURE 8-4 Whitworth thread.

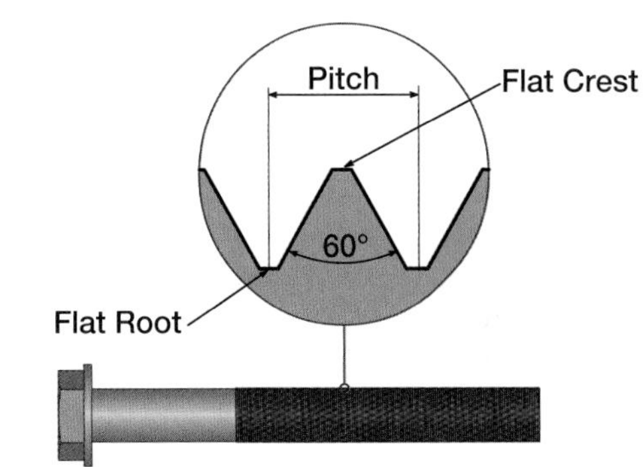

FIGURE 8-5 Sellers thread.

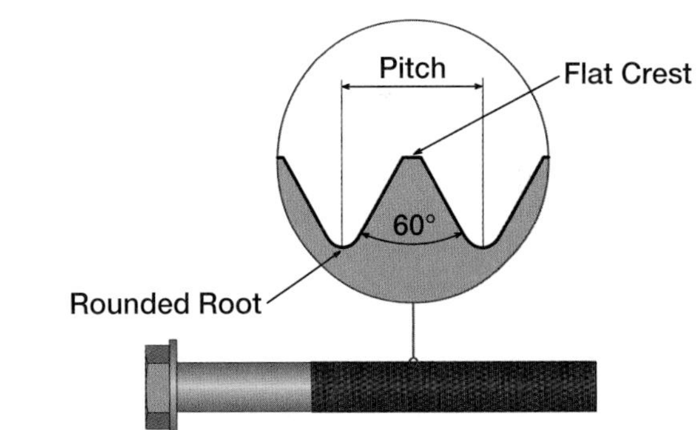

FIGURE 8-6 Current thread.

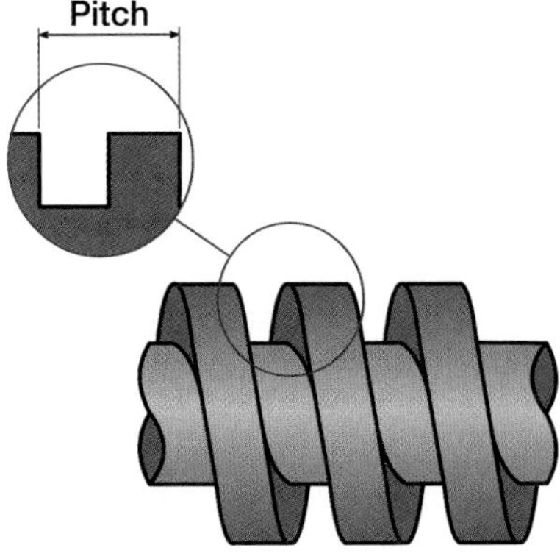

FIGURE 8-7 Square thread as used in a vice.

▶ Fastener Standardization

K08002

There are many different groups that monitor and set the standards that make the automotive industry conform to one universally recognized specification set. There are three main groups to know about. The American Society for Testing and Materials (ASTM) is a nongovernmental controlled group that tests and sets standards in all areas of industry. The International Organization for Standardization (ISO) is an independent developer of standards that are created to ensure a reliable, safe, and quality product. The Society of Automotive Engineers (SAE) was initially started to standardize automotive production and has since grown to encompass a lot of engineering disciplines creating standards and best practices literature.

FIGURE 8-8 Metric bolts can be identified by the grade number on the top of the bolt head.

Metric Bolts

The metric bolt came about from the ISO standard 898, which defines mechanical and physical properties of metric fasteners. The ISO is an independent developer of voluntary international standards. In other words, any one government or group does not control them. The ISO develops standards that apply to many different industries in many different countries, using an unbiased logical approach. The metric decimal system has been around since the late 1600s and is sequential, so people can quickly see which number is larger. Many countries have adopted this standard for their weights and measure systems, but have also retained their own customary standards. Certain industries have required more precision in measurement, so they have adopted the metric system exclusively.

The automotive industry is one that deals in precise measurement and exact specifications in every aspect of automobile design. For this reason and because most automotive companies are worldwide, the automotive industry has embraced the metric measuring system. Uniformity is very important in the production environment, as it leads to efficiencies and conformity. Because of this, most vehicle manufacturers use metric fasteners almost exclusively, but not entirely. So you need to be able to distinguish between metric and standard fasteners. Most metric fasteners can be identified by the grade number cast into the head of the bolt (**FIGURE 8-8**).

FIGURE 8-9 Standard bolts can be identified by the hash marks on top of the bolt head.

Standard Bolts

The standard measurement of bolts is a combination of the Imperial and U.S. customary measurement systems. SAE standard J429 covers the Mechanical and Material Requirements that govern this standard for externally threaded fasteners. It combines the British measuring units and the U.S. measuring units created after the Revolutionary War. Based on the British units of measure, units have been developed over the past millennia to arrive at today's standards. Instead of using a decimal number–based system like the Metric units, they are fractional based. Because the measurement system is based on a fractional scale, people unfamiliar with fractions can find using it difficult. This system is falling out of favor with many industries. As technology becomes more demanding, more precision and standardization is required for both effectiveness and efficiency. Standard bolts can generally be identified by hash marks cast into the head of the bolt, indicating the bolt's grade (**FIGURE 8-9**).

▶ Bolts, Studs, and Nuts

Bolts, studs, and nuts are threaded fasteners designed for jobs requiring a fastener heavier than a screw and tend to be made of metal alloys, making them stronger (**FIGURE 8-10**). Typical **bolts** are cylindrical pieces of metal with a hexagonal head on one end and a thread cut into the shaft at the other end. The thread acts as an inclined plane; as the bolt is turned, it is drawn into or out of the mating thread. Hexagonal **nuts** thread onto the bolt thread. The hexagonal heads

FIGURE 8-10 Bolts, studs, and nuts.

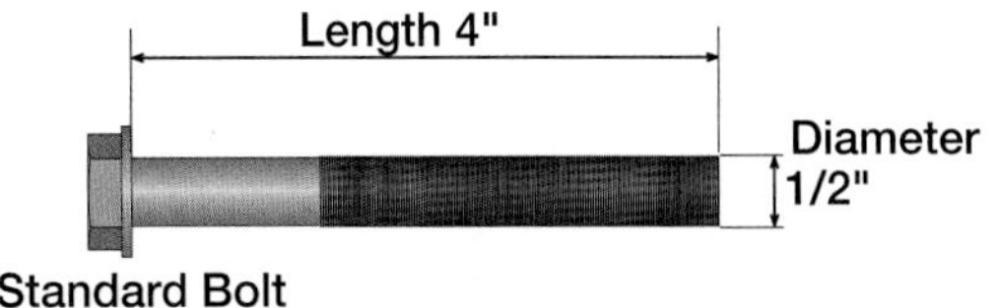

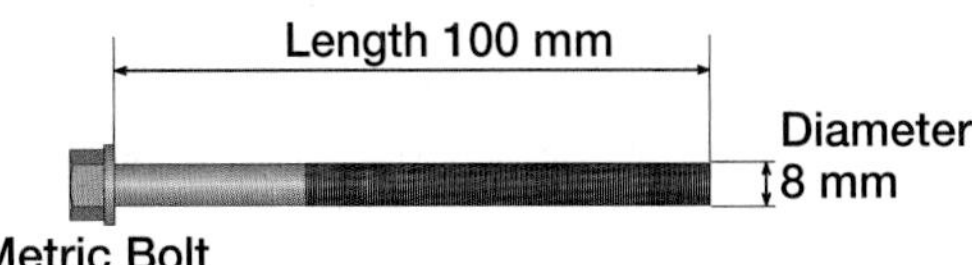

FIGURE 8-11 Bolt diameter and length.

for the bolt and nut are designed to be turned by tools such as wrenches and sockets. Note that other bolt head designs are used as well. These require special shaped tools as covered in the Hand Tools chapter.

A **stud** does not have a fixed hexagonal head; rather, it has a thread cut on each end. It is threaded into one part where it stays. The mating part is then slipped over it, and a nut is threaded onto the other end of the stud to secure the part. Studs are commonly used to attach one component to another, such as a throttle body to the intake manifold. Studs can have different threads on each end, to work best with the material they are threaded into. Coarse threads work well in aluminum; in such cases, one end of the stud may use a coarse thread to grip the threads in an aluminum intake manifold and to position the throttle body. On the other end, there may be a fine thread for pulling everything together tightly with a steel nut. Bolts, nuts, and studs have either standard or metric threads. They are designated by their thread diameter, thread pitch, length, and grade. The diameter is measured across the outside of the threads, in fractions of an inch for standard-type fasteners, and millimeters for metric-type fasteners. So a 3/8" (9.5 mm) bolt has a thread diameter of 3/8" (9.5 mm). It is important to note that a 3/8" (9.5 mm) bolt does not have a 3/8" (9.5 mm) bolt head.

Sizing Bolts

K08003

Bolts are sized either by metric or standard measuring systems. A ½" bolt does not mean that the bolt has a head that fits into a ½" socket. It means that the distance across the outside diameter of the bolt's threads measures ½" in diameter. The same goes for the metric equivalent; an 8 mm bolt is the diameter of the outside diameter of the bolt's threads. The length of a bolt is fairly straightforward. It is measured from the end of the bolt to the bottom of the head and is listed in inches or millimeters (FIGURE 8-11).

FIGURE 8-12 UNC and UNF standard bolt thread pitch.

Thread Pitch

K08004

The coarseness of any thread is called its **thread pitch** (FIGURE 8-12). In the standard system, bolts, studs, and nuts are measured in threads per inch (TPI). To determine the TPI, simply count the number of threads there are in 1" (25.4 mm). Each bolt diameter in the standard system can typically have one of two thread pitches, **Unified National Coarse Thread (UNC)** or **Unified National Fine Thread (UNF)**. For example, a 3/8–16 is coarse, and a 3/8–24 is fine. In the metric system, the thread pitch is measured in millimeters by the distance between the peaks of the threads. So course threads have larger distances, and fine threads have smaller distances. Each bolt diameter in the metric system can have up to four thread pitches (FIGURE 8-13). Consult a metric thread pitch chart, because there is no clear pattern of thread pitches for metric fasteners. These charts can be found in tap and die sets or on the Internet.

Thread Pitch Gauge

Thread pitch gauges make identifying the thread pitch on any bolt quick and easy. Without a thread pitch gauge, technicians

FIGURE 8-13 Metric thread pitches.

are stuck measuring the number of threads per inch, or the distance from peak to peak in millimeters, with a ruler. Each thread pitch gauge matches a particular thread pitch that is stamped or engraved on the gauge. To use a thread pitch gauge, find the gauge that fits perfectly into the threads on the bolt you are measuring. Just make sure you keep the gauge parallel to the shaft of the bolt; otherwise, you will get a wrong match (**FIGURE 8-14**).

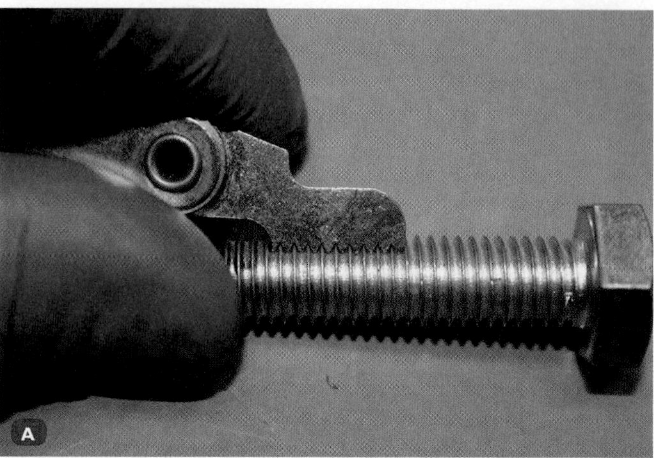

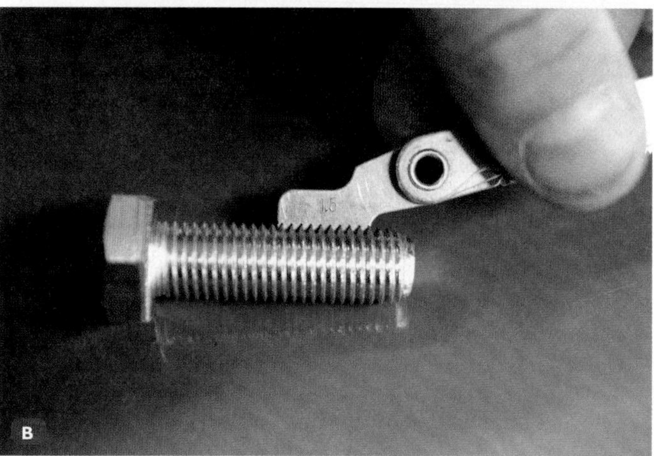

FIGURE 8-14 Thread pitch gauge being used. **A.** Standard. **B.** Metric.

Grading of Bolts

K08005

What is bolt grading? Bolt grade or class means that the fastener meets the strength requirements of that grade or classification. There are a variety of grades, so it is important to use the specified grade of bolt for each component. When bolts are tested for grade, the bolt is put in a testing tool and abused to the point of breakage so that the maximum readings can be determined to properly grade the fastener. We need to know a bolt's grade so that we are able to use the correct fastener for the application and avoid bolt failures down the road. Standard bolts are typically grade 1 to grade 8, with grade 8 being the strongest. The grade can be identified by counting the hash marks on the head of the bolt and adding two to that number. For example, there are three hash marks on the head of a grade 5 bolt and six on the head of a grade 8 bolt (**FIGURE 8-15**).

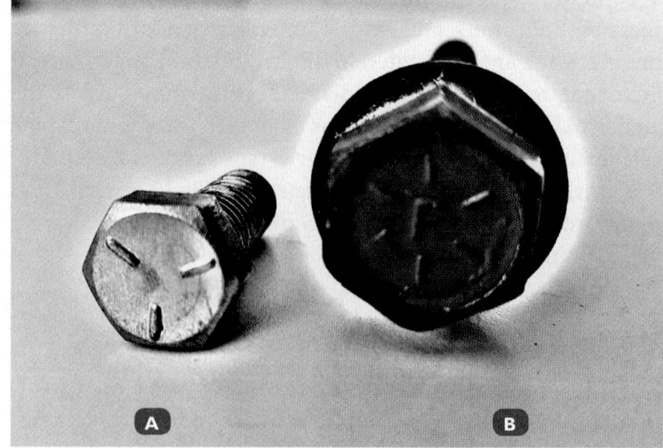

FIGURE 8-15 A. Grade 5 standard bolt. **B.** Grade 8 standard bolt.

Metric bolt grades are typically grade 4.6 to grade 12.9, with grade 12.9 being the strongest. In metric bolts, the grade number is cast into the top of the bolt head. These numbers on metric bolts have specific meanings. The number before the decimal indicates its tensile strength in megapascals (MPa), which is found by multiplying the number by 100. For example a metric 12.9-grade bolt would have a tensile strength of 1200 MPa. The number to the right of the decimal, when multiplied by 10 indicates the yield point as a percentage of the tensile strength (**FIGURE 8-16**).

Just because the higher grades are stronger doesn't mean you should use those in all applications. There are times when lower-graded bolts are more appropriate for particular applications, such as with some head bolts that need to have a little bit of give in them to allow for expansion and contraction of the cylinder head when the head gasket is clamped in place. Otherwise, the head gasket seal could fail. In other

FIGURE 8-16 **A.** Grade 8.8 metric bolt **B.** Grade 10.9 metric bolt.

situations, using the wrong grade of bolt could cause either the bolt to fail or the threads in the threaded hole to fail. So you should always use the specified grade bolt when replacing bolts with new ones.

Strength of Bolts

K08006

Tensile Strength

Tensile strength is the maximum tension (applied load) the fastener can withstand without being torn apart or the maximum stress used under tension (lengthwise force) without causing failure. Tensile strength is determined by the strength of the material and the size of the stress area. When a typical threaded fastener fails in pure tension, it usually fractures through the threaded portion, as that is the weakest area because it is also the smallest area of the bolt (**FIGURE 8-17**). Generally speaking, the higher the grade of bolt, the higher the tensile strength.

Shear Strength

Shear strength is defined as the maximum load that can be supported prior to fracture, when applied at a right angle to the fastener's axis. A load occurring in one transverse plane is known as single shear (**FIGURE 8-18**). Double shear is a load applied in two planes, where the fastener could be cut into three pieces. For most standard threaded fasteners, shear strength is not specified, even though the fastener may be commonly used in shear applications. Generally speaking, the higher the grade of bolt, the higher the shear strength.

FIGURE 8-17 Bolt failure (tension).

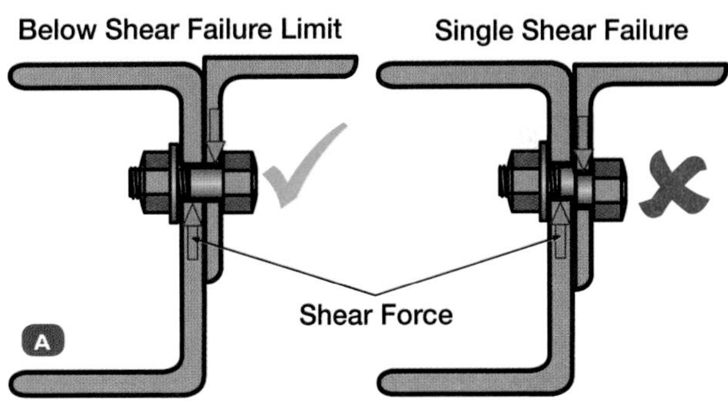

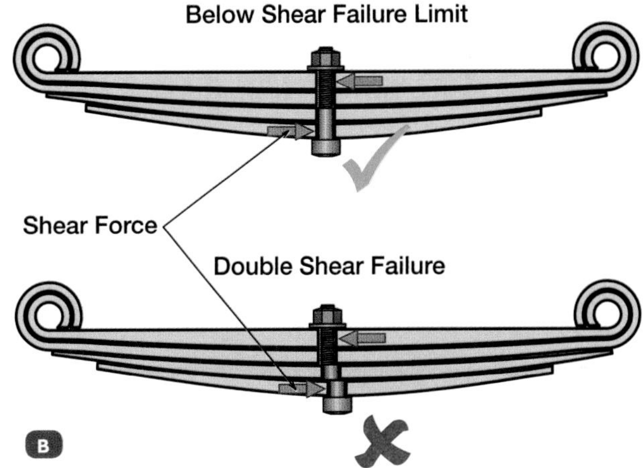

FIGURE 8-18 **A.** Single shear. **B.** Double shear.

Proof Load

The proof load represents the usable strength range for certain standard fasteners. By definition, the proof load is an applied tensile load that the fastener must support without exceeding the elastic phase, which is the point up to which the bolt returns to its original point when tension is removed.

Fatigue Strength

A fastener subjected to repeated cyclic loads can suddenly and unexpectedly break, even if the loads are well below the strength of the material. The repeated cyclic loading weakens the fastener over time, and it fails from fatigue. The fatigue strength is the maximum stress a fastener can withstand for a specified number of repeated cycles before failing. Connecting rod bolts are an example of a situation where fatigue strength would need to withstand millions of cyclic loads.

Torsional Strength

Torsional strength is a measure of a material's ability to withstand a twisting load, usually expressed in terms of torque, in which the fastener fails by being twisted off about its axis. A fastener that fails due to low torsional strength typically fails during installation or removal. Generally speaking, the higher the grade, the higher the torsional strength of the fastener.

Ductility

Ductility is the ability of a material to deform before it fractures. A material that experiences very little or no plastic deformation before fracturing is considered brittle. Think of a piece of flat glass: If you try to bend it, it breaks easily. This is an example of a material with low ductility material. Most automotive fasteners need to have some measure of ductility to avoid catastrophic failure. A reasonable indication of a fastener's ductility is the ratio of its specified minimum yield strength to the minimum tensile strength. The lower this ratio, the more ductile the fastener, meaning that the more ductile the bolt, the more it can flex or stretch without breaking. Generally speaking, the higher the grade, the lower the ductility of the fastener. So a balance between tensile strength and ductility is needed that depends on the requirements of the application.

Toughness

Toughness is defined as a material's ability to absorb impact or shock loading. Impact strength toughness is rarely a specified requirement for automotive applications. In addition to its specification for various aerospace industry fasteners, ASTM A320 specification for alloy steel bolting materials for low-temperature service is one of the few specifications that requires impact testing on certain grades.

Nuts

K08007

Nuts are screwed onto the threads of a bolt to hold something together. They have internal threads that match the threads on the same size bolt. The many different kinds of nuts are mostly application specific. We cover some of the more popular ones below. When you replace a nut with a new nut on a vehicle, you need to match the nut to the grade of bolt or stud that it is going to be screwed onto. For example, if the bolt is a grade 8, you need a grade 8 nut. By matching the grade of fasteners, you are keeping the integrity of the manufacturer's intended fastening system. If you mix different grades of bolts and nuts, one could prematurely fail, causing catastrophic failure of the component.

Locking Nuts

Locking nuts are used when there is a chance of the nut vibrating loose. Two common types of locking nuts are used in automotive applications: the nylon insert lock nut and the deformed lock nut (**FIGURE 8-19**). The nylon lock nut has a nylon insert with a smaller diameter hole than the threads in the nut. As the nut is threaded onto the bolt, the nylon insert is deformed by the bolt threads, which tends to lock the nut onto the thread of the bolt or stud so that it can't vibrate loose. In a deformed lock nut, the top thread of the nut is deformed, thus pinching the bolt threads so the nut can't vibrate loose. Both of these types of lock nuts are considered single use, meaning they should be replaced with new ones once they have been removed.

FIGURE 8-19 A. Nylon lock nut. **B.** Deformed lock nut.

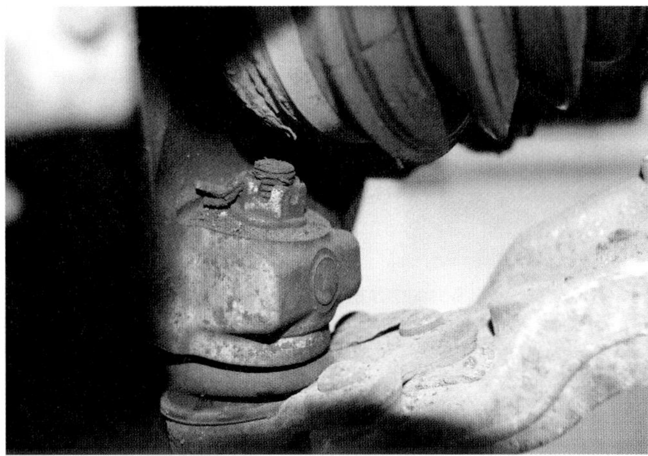

FIGURE 8-20 Castle nut used on a ball joint.

FIGURE 8-21 New cotter pins installed in the castle nuts.

Castle Nuts

Castle nuts, sometimes called castellated nuts, are used with a cotter pin to keep them from moving once they are installed. Protrusions and notches are machined into the top of the nut. The notches then can be lined up with a hole in the threaded bolt or stud (**FIGURE 8-20**). A cotter pin passes through the notch on one side of the nut, then through the hole in the stud and out the notch in the other side of the nut. The cotter pin physically prevents the castle nut from backing off the stud. These nuts are usually used on suspension components or other critical components so that they can't loosen up. It is good practice to always use new cotter pins and visually check to verify if they are installed in the castle nuts before finishing up a job (**FIGURE 8-21**).

Specialty Nuts

In the automotive industry. there are lot of specialty nuts with special characteristics—for example, an extended shoulder, a special head size, or a special material—required for a particular application. You must use the correct nut for the correct

FIGURE 8-22 A variety of specialty nuts.

application (**FIGURE 8-22**). Sheet metal nuts are a cheap alternative to a regular nut that manufactures are using on different components. Sometimes called J nuts, these nuts are folded piece of sheet metal that fits over a hole in a piece of sheet metal; they are threaded to accept a fastener, and they usually do not need to be held as they are clipped to the component. Failure to use the proper fasteners for the proper uses usually results in failure of the component.

Washers

K08008

Why are washers used underneath bolt heads and nuts? There are three main purposes of a washer:

- To distribute the force exerted on the component that it is pressing against evenly
- To prevent the surface around the hole from being worn down due to tightening of the head of the bolt or nut against a surface
- To provide a measure of locking force to keep the bolt or nut from loosening due to vibration

Not all applications require washers that provide all three functions, so make sure you know what the application requires. When selecting a washer for a bolt or nut, make sure you use the correct type, size, and grade.

Grades of Washers

Just as with nuts, you want to match the grade of washer to the nut and bolt so that you can maximize the clamping force on the component (**FIGURE 8-23**). Using dissimilar metals that have different hardness characteristics can cause the surface of the softer metal to wear away. This reduces the clamping force over time, which can cause the bolt or component to fail. Not all washers have grade ratings on them, so be careful to select the proper one when replacing a damaged or missing washer with a new one.

FIGURE 8-23 Washer with its grade rating marked on it.

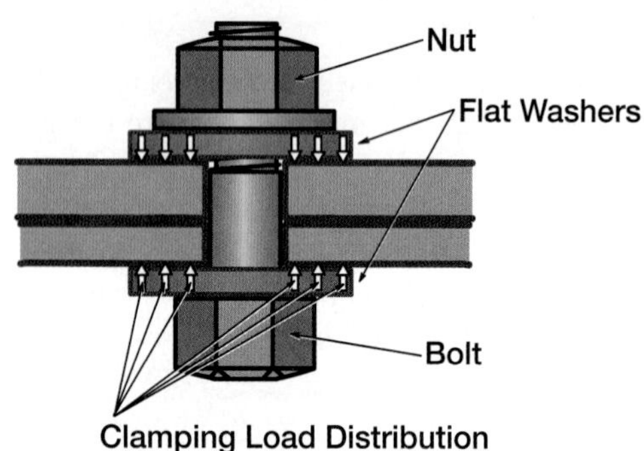

FIGURE 8-24 Flat washer spreading out the clamping force.

Flat Washers

A flat washer is a piece of steel that has been cut in a circular shape with a hole in the middle for the bolt or stud to fit through. Washers are sized and paired with a metric or standard bolt so that the pressure created by torquing the bolt down is spread out on the component that it is being used on (**FIGURE 8-24**). They prevent marring of the surface of the component around the bolt hole. They also act as a bearing surface so that the bolt achieves the expected clamping force at the specified torque. Flat washers are graded just like bolts, which means you need to match the grade on the washer with the fasteners that you are using them on.

Lock Washers

Lock washers are made from spring steel and have a slit cut in them to allow for the spring action to hold tension against the nut so that the nut will not loosen. Also, the slit is cut at an angle, with sharp edges on the top and bottom to cut into the bottom of the bolt head and the surface of the component being clamped (**FIGURE 8-25**). The sharp edges bite into the bolt and component if the bolt tries to loosen. These are used with conventional nuts and are usually used in applications that create a lot of vibration. Lock washers are usually considered to be single use and should be replaced rather than reused.

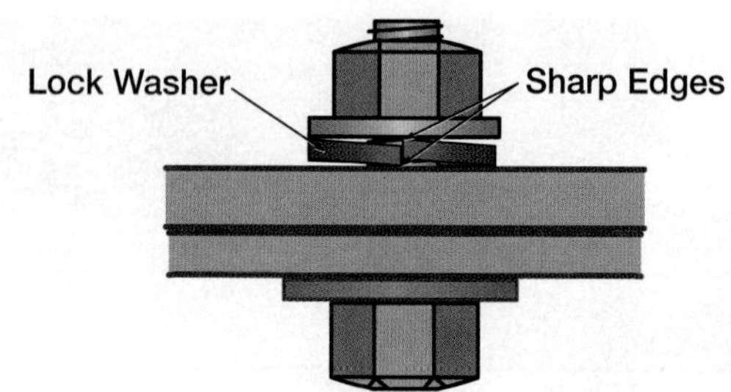

FIGURE 8-25 Lock washers prevent bolts and nuts from loosening. (Note: this nut and bolt is not fully tightened to show the action of the lock washer).

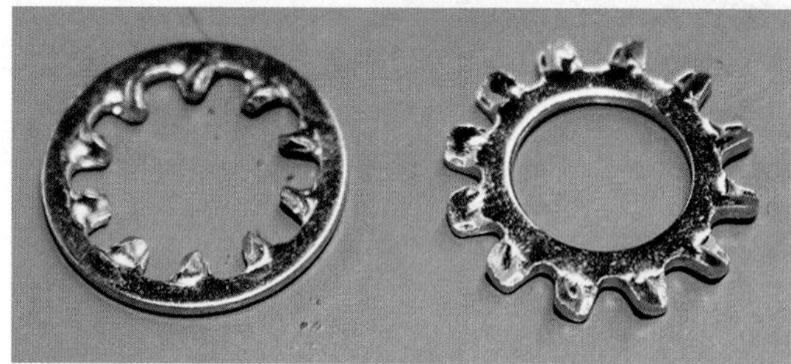

FIGURE 8-26 Star washers prevent nuts and bolts from loosening.

Star Washers

Star washers, also called toothed lock washers, are used sometimes to stop the rotation of the nut above. They work in a similar way to lock washers in that the teeth are at angles so that they bite into the bottom of the bolt head and the surface being clamped (**FIGURE 8-26**). If these washers are required for your application, they need to be reinstalled before the nut is installed. They also should be replaced with new ones rather than reused.

▶ Threadlocker and Antiseize

K08009

Thread-locking compound is a liquid that is put on the threads of a bolt or stud to hold the nut that is threaded over it. The thread-locking compound acts like very strong glue. Once the compound between the nut and bolt cures, it bonds the two together so that they do not move. A popular thread-locking compound (threadlocker) that most automotive technicians prefer is Loctite. The two main strengths of locking compound that are used in the automotive industry are blue and red (**FIGURE 8-27**). The blue allows for relatively easy removal of the nut or bolt with a socket and ratchet, whereas the red is much stronger. The red version usually requires a fair amount of heat to soften the compound so the nut or bolt can be removed. Thread-locking compound is a good safety item that technicians use on critical parts they do not want to come loose. In fact, the manufacturer may specify its use on certain fasteners, so be aware of situations that require thread-locking compound, and use the strength required.

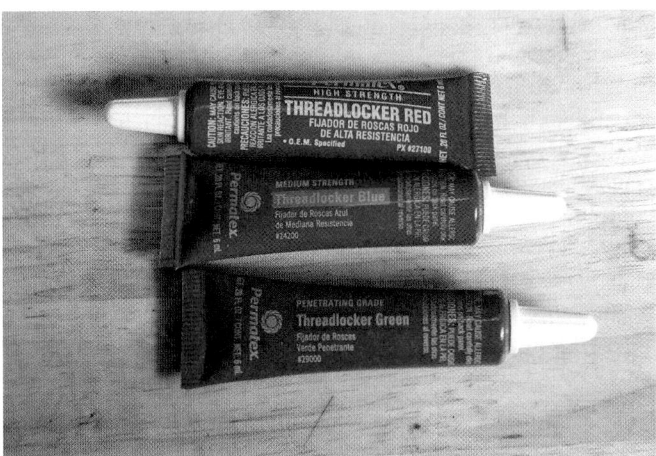

FIGURE 8-27 Thread-locking compound.

Antiseize compound is the opposite of thread-locking compound as it keeps threaded fasteners from becoming corroded together or seized due to galling. One common use of antiseize compound is on black steel spark plug threads when the spark plug is installed into an aluminum cylinder head. Antiseize compound is a coating that prevents rust and provides lubrication so that the fastener can be removed in the future. Because of that, it should *not* be used on most fasteners in a vehicle; otherwise, they could vibrate loose, causing damage or an accident. When you do use antiseize compound on specified fasteners, use a very light coating, as a little bit goes a very long way. In fact, a small can of antiseize lasts the average technician several years.

▶ Screws

K08010

Many different types of screws are found in automotive applications (**FIGURE 8-28**). A screw is very similar to a bolt except for a couple of differences: They are not hardened and they tend to be used in light-duty applications where they need to hold together components that have a low shear or tensile strength.

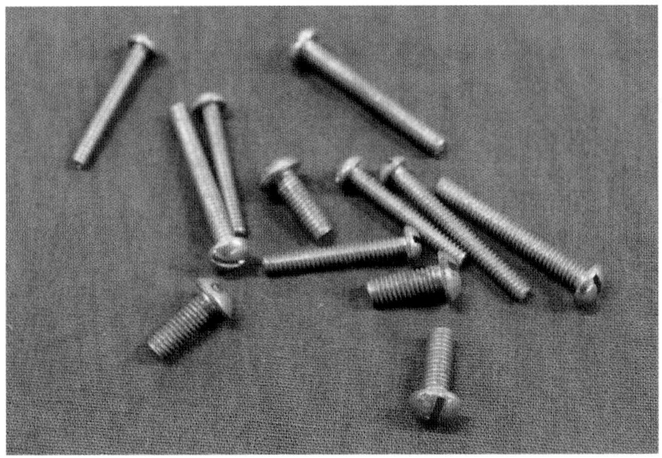

FIGURE 8-28 Variety of screws.

Machine Screws

Similar to a bolt, a machine screw is used to fasten components together, but these are usually driven by a screwdriver with a Philips or slotted head. Machine screws are usually driven into a nut or threaded hole that has been tapped to the thread pitch of the screw. As vehicles become more complex, manufacturers have started to introduce new types of driving tools for these screws. Torx, square, and Allen head screws are becoming prevalent in this fastener group (**FIGURE 8-29**).

Self-Tapping Screws

Self-tapping screws are designed to create their own holes in the material they are being driven into. They have a fluted tip so you can drill a hole into the base material, without needing a pilot hole (**FIGURE 8-30**). The threads on a self-tapping screw are very similar to those on a sheet metal screw as they grip onto the metal as you drill through it. Usually they have a cap screw–type head so they can be driven with a screw gun, but they can have Phillips, square, Allen, or Torx heads as well.

Trim Screws

Trim screws are used in applications where there is a need to hold plastic and metal ornamentation to the vehicle. They are

FIGURE 8-29 Machine screw heads.

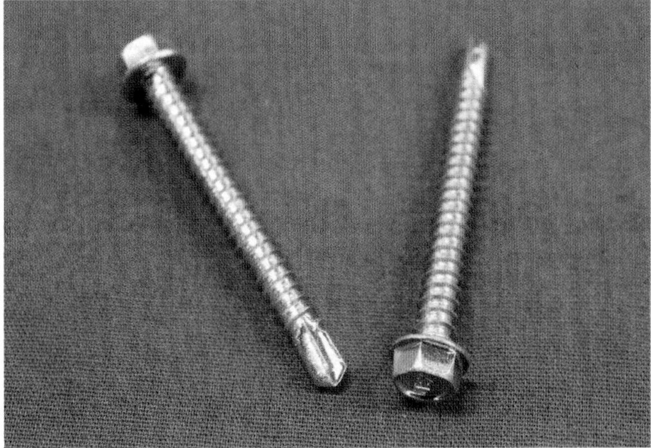

FIGURE 8-30 Self-tapping screws.

also used to hold door panels, trim pieces, and other small components to the vehicle. They are basic screws that are usually painted either the trim color or black so that they blend in with the material that they are used with (**FIGURE 8-31**).

Sheet Metal Screws

Sheet metal screws are used to attach things to sheet metal. They have a chip on the tip of the screw to push away material as it comes off (**FIGURE 8-32**). These screws are usually threaded all the way up to the head so that they can be run down all the way, clamping the pieces together.

▶ Torque-to-Yield and Torque Angle

K08009

Torque is not always the best method of ensuring that a bolt is tightened enough as to give the proper amount of clamping force. If the threads are rusty, rough, or damaged in any way, the amount of twisting force required to tighten the fastener increases. Tightening the rusty fastener to a particular torque does not provide as much clamping force as a smooth fastener torqued the same amount. This also brings up the question of whether threads should be lubricated. In most automotive cases, the torque values specified are for dry, non-lubricated threads. But always check the manufacturer's specifications.

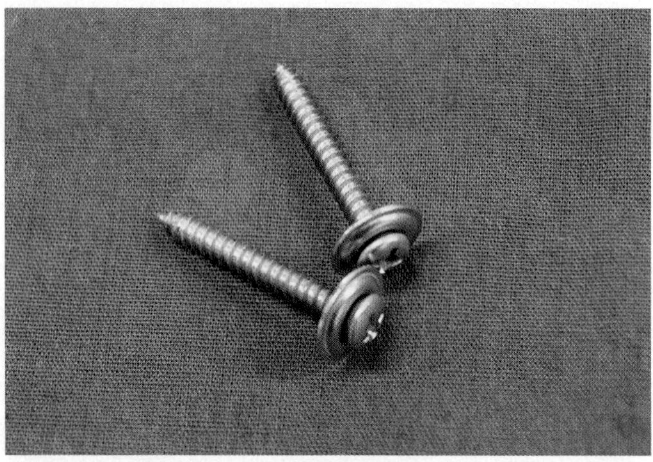

FIGURE 8-31 Trim screws.

FIGURE 8-32 Sheet metal screws.

When bolts are tightened, they are also stretched. As long as they are not tightened too much, they will return to their original length when loosened. This is called **elasticity**. If they continue to be tightened and stretch beyond their point of elasticity, they will not return to their original length when loosened. This is called the **yield point**. As threaded fasteners are torqued, they go through the following phases:

- **Rundown Phase:** Free running fastener (may or may not have prevailing torque).
- **Alignment Phase:** Fastener and joint mating surfaces are drawn into alignment.
- **Elastic Phase:** *This is the third and final stage for normal bolts!* The slope of the torque/angle curve is constant. The fastener is elongated but will return to original length upon loosening.
- **Plastic or Yield Phase:** *Over-torqued condition for normal bolts. TTY bolts are tightened just into the beginning points of this phase.* Permanent deformation and elongation of the fastener and/or joint occur. Necking of the fastener occurs.

Torque-to-yield means that a fastener is torqued to, or just beyond, its yield point. With the changes in engine metallurgy that manufacturers are using in today's vehicles, bolt technology had to change also. To help prevent bolts from loosening over time and to maintain an adequate clamping force when the engine is both cold and hot, manufacturers have adopted **torque-to-yield (TTY) bolts**. TTY bolts are designed to provide a consistent clamping force when torqued to their yield point or just beyond. The challenge is that the torque does not increase very much, or at all, once yield is reached. So using a torque wrench by itself will not indicate the point at which the manufacturer wants the bolt tightened. TTY bolts generally require a new torquing procedure called torque angle. Also, it is important to note that in virtually all cases, TTY bolts cannot be reused because they have been stretched into their yield zone and would very likely fail if retorqued (**FIGURE 8-33**).

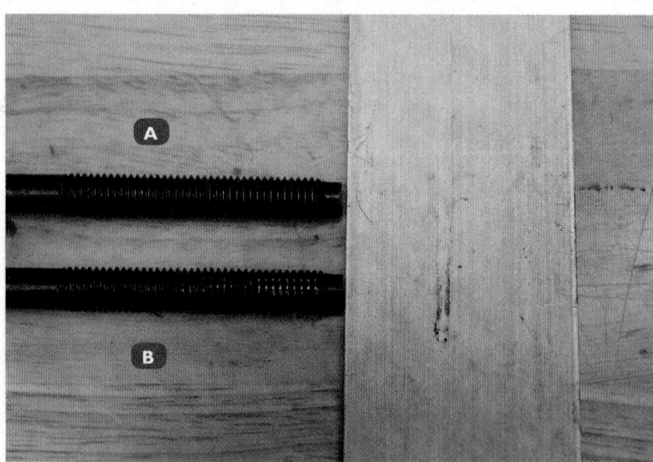

FIGURE 8-33 **A.** New TTY bolt. **B.** Used TTY bolt.

Torque angle is considered a more precise method to tighten TTY bolts and is essentially a multistep process. Bolts are first torqued in the required pattern, using a standard torque wrench to a required moderate torque setting. They are then further tightened one or more additional specified angles (torque angle) using an angle gauge, thus providing further tightening, which tightens the bolt to, or just beyond, their yield point. In some cases, after the initial torquing, the manufacturer first wants all of the bolts to be turned to an initial angle and then turned an additional angle. And in other cases, the manufacturer wants all of the bolts torqued in a particular sequence, then detorqued in a particular sequence, then retorqued once again in a particular sequence, and finally tightened an additional specified angle. So always check the manufacturer's specifications and procedure before torquing TTY bolts.

To use a torque angle gauge in conjunction with a torque wrench, follow the steps in **SKILL DRILL 8-1**.

SKILL DRILL 8-1 Using a Torque Angle Gauge with a Torque Wrench

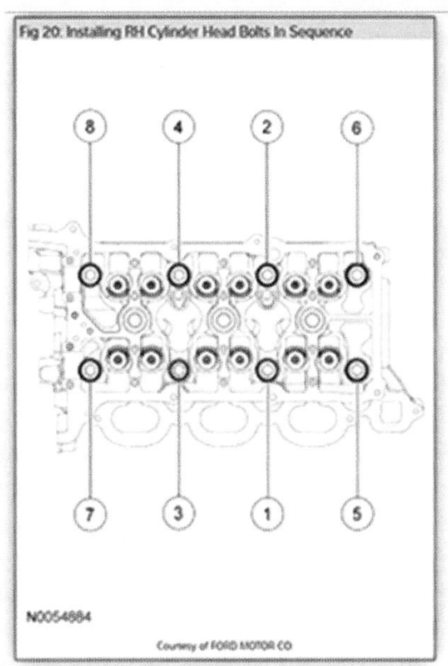

1. Check the specifications. Determine the correct torque value (ft-lb or N·m) and sequence for the bolts or fasteners you are using. Also, check the torque angle specifications for the bolt or fastener, and whether it involves one step or more than one step.

2. Tighten the bolt to the specified torque. If the component requires multiple bolts or fasteners, make sure to tighten them all to the same torque value in the sequence and steps that are specified by the manufacturer.

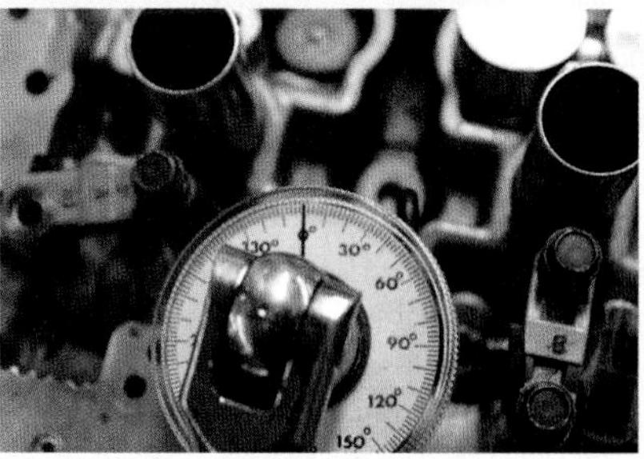

3. Install the torque angle gauge over the head of the bolt, and then put the torque wrench on top of the gauge and zero it, if necessary.

4. Turn the torque wrench the specified number of degrees indicated on the angle gauge.

5. If the component requires multiple bolts or fasteners, make sure to tighten them all to the same torque angle in the sequence that is specified by the manufacturer. Some torquing procedures could call for four or more steps to complete the torquing process properly.

▶ How to Avoid Broken Fasteners

K08012

Broken bolts can turn a routine job into a nightmare job by adding hours dealing with the broken bolt. In most cases, it is worth taking a bit of extra time doing whatever is needed to avoid breaking a bolt. One of the first things to do is to make sure that you are not dealing with a left-handed bolt (**FIGURE 8-34**). Left-hand bolts loosen by turning them clockwise. So trying to loosen them by turning them counterclockwise will actually tighten them, making it likely you will break them off.

Another cause of broken bolts is fasteners that are rusted or corroded in place. Although it is impossible to prevent all broken bolts in this case, there are several things you can do to minimize the percentage of bolts you break. One of the first things to do is use a good penetrating oil. Penetrating oil actually penetrates the rust between the nut and bolt and breaks it down, making it easier to break the nut or bolt loose. Once the penetrating oil has been given time to soak into the threads of the fastener, you may attempt, with the proper tool, to try to loosen the fastener. Note that you may want to try both loosening and tightening the bolt to help break it loose. It may be good to try using an impact wrench as well, because the hammering effect might assist in breaking it loose. Just try it a little at a time in each direction, and not with too much force.

If the fastener is still not loosening, you may need to heat up the component to help break up the rust or corrosion. The heat not only expands the metal components, it also expands the space between the nut and bolt. This also helps break the nut or bolt loose. The bolt or nut can be heated up in two primary ways. The first is with an oxyacetylene torch, but be careful because the flame from the torch also burns everything in its path. When heating up a bolt on a vehicle, there are almost always other things in the way that could be affected by the heat. The second is with an inductive heater, which uses electrical induction to heat up any ferrous metal that is placed in the inductive coil (**FIGURE 8-35**). The induction heater is preferred by many technicians for many applications, as it doesn't have a flame and it only heats up what is in the induction coil. Most induction heaters come with several sizes of induction coils that make them handy for most applications.

When heating a fastener to remove it, you usually heat it up to a dull orange color and then let it cool a bit before attempting to remove it. Once a heated fastener is removed, it must be replaced with a new bolt because the heat alters the strength of the fastener.

▶ Thread Repair

N08001, K08013, S08001

Thread repair is used in situations where it is not feasible to replace a damaged thread. This may be because the damaged thread is located in a large expensive component, such as the engine block or cylinder head of a vehicle, or because replacement parts are expensive or not available. The aim of thread repair is to restore the thread to a condition that restores the fastening integrity (**FIGURE 8-36**). It can be performed on internal

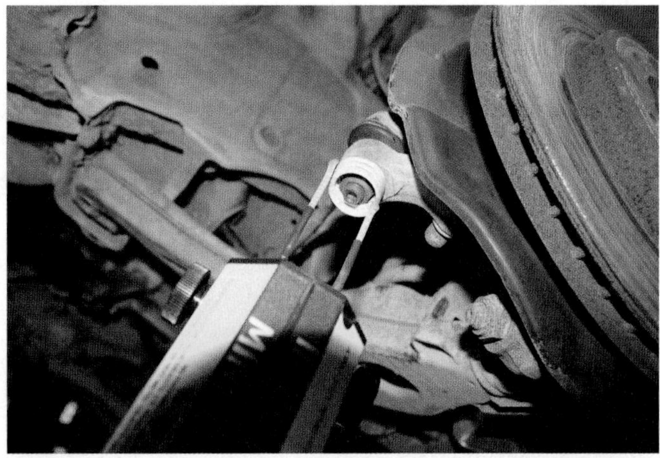

FIGURE 8-35 Heating a stuck nut with an induction heater.

FIGURE 8-34 Left-hand lug nuts are installed on some vehicles.

FIGURE 8-36 Repaired thread.

threads, such as in a housing, engine block, or cylinder head, or on external threads, such as on a bolt. However, it is usually easier to replace a bolt if the threads become damaged than repairing it.

Types of Thread Repair

Many different tools and methods can be used to repair a thread. The least invasive method is to reshape the threads. If the threads are not too badly damaged, such as when the end thread is slightly damaged from the bolt being started crooked (cross-threaded), then a thread file can be used to clean them up, or a restoring tool can be used to reshape them (**FIGURE 8-37**). Each thread file has eight different sets of file teeth that match various thread pitches. Select the set that matches the bolt you are working on, and file the bolt in line with the threads. The file removes any distorted metal from the threads. Only file until the bad spot is reshaped.

The thread-restoring tool looks like an ordinary tap and die set, but instead of cutting the threads, it reshapes the damaged portion of the thread (**FIGURE 8-38**). Threads that have substantial damage require other methods of repair.

A common method for repairing damaged internal threads is a thread insert. A number of manufacturers make thread inserts, and they all work in a similar fashion. There are two main types of thread inserts: helical thread inserts and sleeve-type thread inserts (**FIGURE 8-39**). A helical thread insert of the same internal thread pitch is used when the threads need to be replaced with the same size thread. The damaged threaded hole is drilled out bigger, tapped with a special tap to a larger size that matches the outside size of the helical insert, and then the threaded coil insert is screwed into place. The insert provides a brand new internal thread that matches the original size. A common name for this type is a Heli-Coil.

The solid sleeve type of insert also replaces the damaged thread with the same size thread as the original. The difference between this type of insert compared to the helical insert is that this one is a slightly thicker insert that resembles a sleeve or bushing, but with threads on both the inside and outside. It may also have a small flange at the top of the insert that prevents it from being installed too deeply in the hole. Like the previous insert, the damaged hole needs to be drilled out to the proper oversize. Then a special tap is used to cut new threads in the hole that match the external thread on the insert. The insert is then installed in the hole and locked into place with either thread-locking compound or keys that get

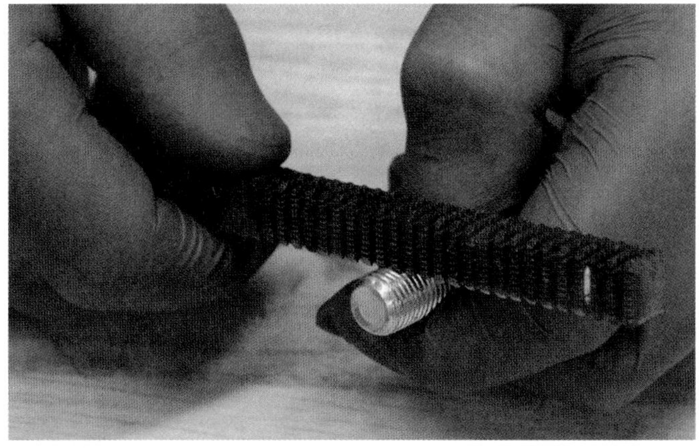

FIGURE 8-37 Threads being cleaned up with a thread file.

FIGURE 8-38 Thread-restoring tool being used.

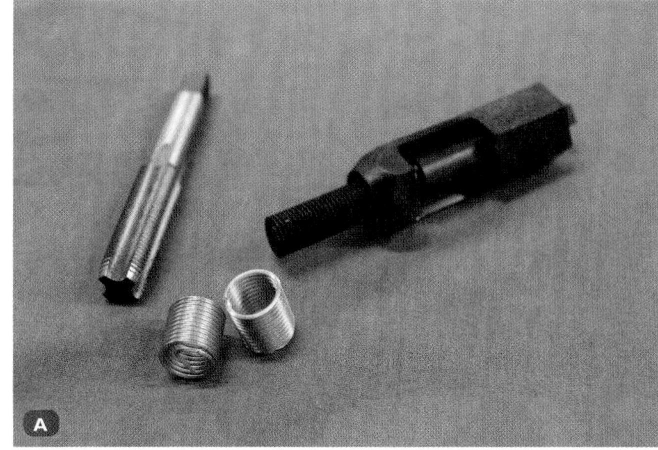

FIGURE 8-39 Thread inserts. **A.** Heli-Coil. **B.** Time-Sert.

driven down into the base metal. A common name for this type is a Time-Sert.

Self-tapping inserts are a type of insert that is driven in a similar way to a self-tapping screw, making its own threads and at the same time locking the insert into the base material. Bushing inserts are a type of threaded insert that is pressed into a blind hole with an arbor press. These types of inserts can be inserted in any material, so you must be aware of material that is soft as it may break if too much force is put to it. It has internal threads, so bolts and screws can be threaded into them.

The process of repairing a thread should first start with attempting to remove the broken bolt without damaging the threads. If this can be accomplished, then you will likely save time as well as avoid the possibility of making the problem worse, such as breaking off an easy out in the broken bolt. If the bolt can't be removed without damaging the threads, then the use of a thread insert to repair an internal thread is probably needed.

To remove a broken bolt, inspect the site. If enough of the bolt is sticking out of the surface, then a pair of pliers or locking pliers may be enough to turn and remove the bolt. The author has had quite a bit of luck using a pair of special curved jaw Channellock pliers. The curved jaw tends to get a better grip on the broken-off bolt than regular straight jaw Channellock pliers (**FIGURE 8-40**). Using penetrant or heat may help coax the bolt out. If the bolt is broken off flush with the surface, and the bolt is large enough in diameter, then you may be able to use a small center punch to turn the bolt by tapping on the outside diameter of the bolt, but in the reverse direction (**FIGURE 8-41**). If that doesn't work, or the bolt is too small in diameter, then a screw or bolt extraction tool can be tried.

Use a center punch to mark the center of the bolt to assist in centering the hole to drill. Select the correctly sized **screw extractor**, and drill the designated hole size in the center of the broken bolt to accommodate the extractor. Once the hole is drilled, insert the extractor and turn it counterclockwise (**FIGURE 8-42**). The flutes on the extractor should grab the inside of the bolt and hopefully enable you to back it out. Be careful not to exert too much force on the extractor if the bolt is extremely stuck in place. If the extractor breaks, it is almost impossible to remove it, as it is made of hardened steel that cannot be cut by most drill bits. Once the broken bolt is removed, run a lubricated tap or thread-restoring tool of the correct size and thread pitch through the hole, to clean up any rust or damage. If the screw extractor can't remove the broken bolt, then it will need to be drilled out. If the internal treads are damaged during the removal process, then the thread will have to be repaired with a thread insert, as described earlier.

To conduct thread repair, follow the steps in **SKILL DRILL 8-2**.

FIGURE 8-41 Removing a broken bolt with a center punch or cold chisel.

FIGURE 8-40 Using curved jaw pliers to remove a broken bolt.

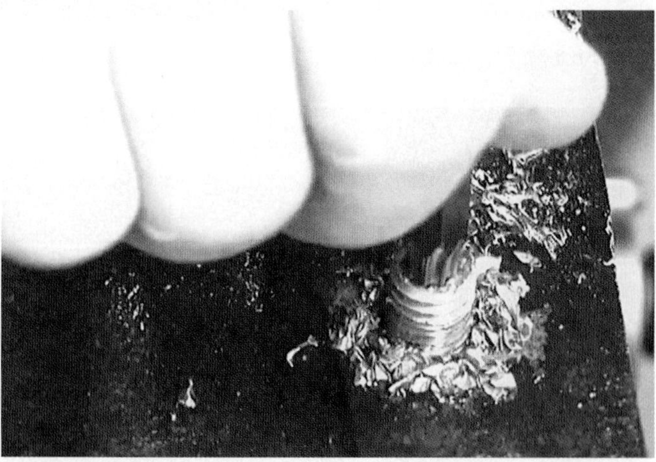

FIGURE 8-42 Removing a bolt with a screw extractor.

SKILL DRILL 8-2 Conducting Thread Repair

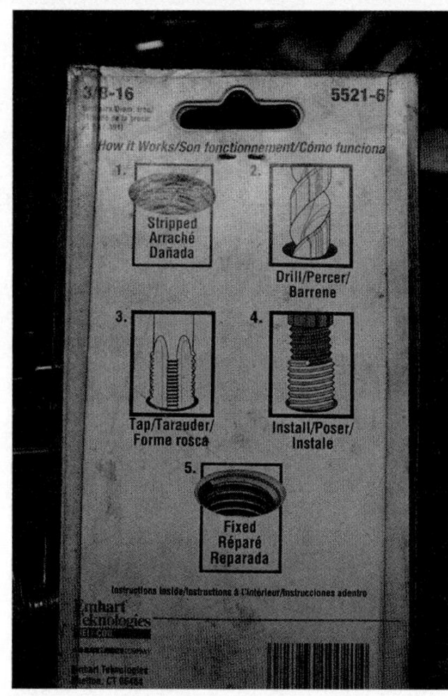

1. Always refer to the manufacturer's manual for specific operating instructions.

2. Inspect the condition of the threads, and determine the repair method.

3. Determine the type and size of the thread to be repaired. Thread pitch gauges and vernier calipers may be used to measure the thread.

4. Prepare materials for conducting the repair: dies and taps or a drill bit and drill; cutting oil, if required; and inserts.

5. Select the correctly sized tap or die if conducting a minor repair. Run the die or tap through or over the thread; be sure to use cutting lubricant.

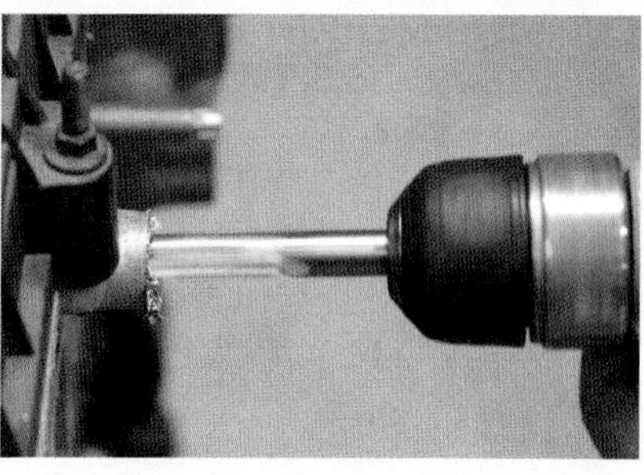

6. If using inserts, select the correctly sized insert, based on the original bolt size. Drill the damaged hole, ensuring the drill is in perfect alignment with the hole.

7. Cut the new thread, using the proper size tap. Make sure you use cutting lubricant if required.

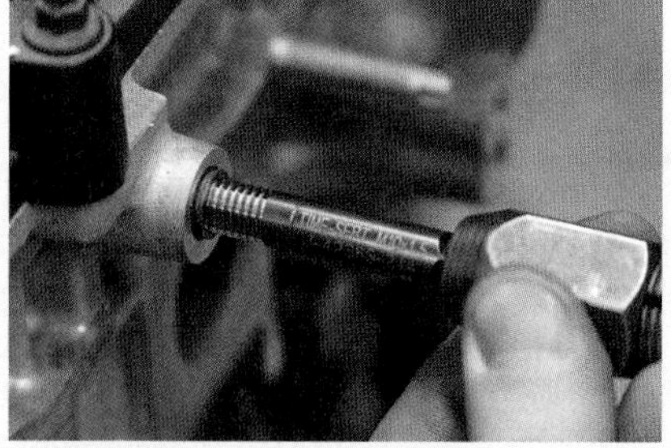

8. Using the insert-installing tool, install the insert by screwing it into the newly cut threads. Make sure the insert is secure or locked into the hole using the method specified by the manufacturer. Some inserts use a locking tab, whereas others use a liquid thread locker that hardens and holds the insert in place.

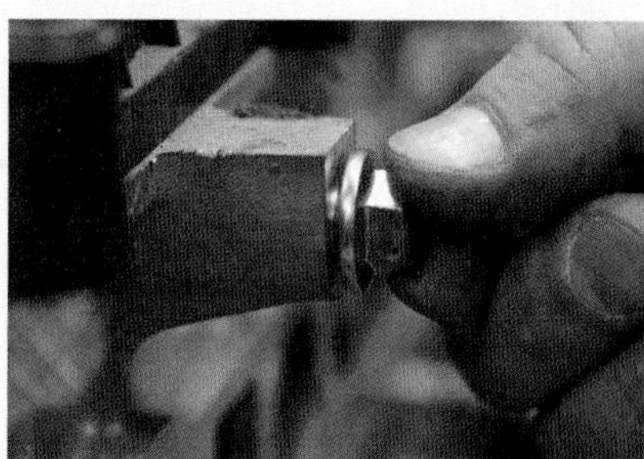

9. Test the insert to ensure it is secure and that the bolt will screw all the way in.

Wrap-Up

Ready for Review

- Threaded fasteners include bolts, studs, and nuts, and are designed to secure vehicle parts under stress.
- Torque defines how much a fastener should be tightened.
- Bolts, nuts, and studs use threads to secure each part; these threads can be in standard or metric measures.
- Thread pitch refers to the coarseness of the thread; bolts, nuts, and studs are measured in threads per inch (TPI), classified as Unified National Coarse Thread (UNC) or Unified National Fine Thread (UNF).
- Fasteners are graded by tensile strength (how much tension can be withstood before breakage).
- The SAE rates fasteners from grade 1 to grade 8; always replace a nut or bolt with one of the same grade.
- Torque specification indicates the level of tightness each bolt or nut should be tightened to; torque charts list torque specifications for nuts and bolts.
- Torque (or tension) wrenches tighten fasteners to the correct torque specification.
- Torque value—the amount of twisting force applied to a fastener by the torque wrench—is specified in foot-pounds (ft-lb), inch-pounds (in-lb), or newton meters (N·m).
- Bolts that are tightened beyond their yield point do not return to their original length when loosened.
- Torque-to-yield (TTY) bolts can be torqued just beyond their yield point, but should not be reused.
- Torque angle can be used to tighten TTY bolts and requires both a torque wrench and an angle gauge.
- Thread repair is performed to restore fastening integrity to a damaged fastener.
- Threads can be reshaped with a file, or a thread insert may be used.

Key Terms

bolt A type of threaded fastener with a thread on one end and a hexagonal head on the other.

coarse (UNC) Used to describe thread pitch; stands for Unified National Coarse.

elasticity The amount of stretch or give a material has.

fasteners Devices that securely hold items together, such as screws, cotter pins, rivets, and bolts.

fine (UNF) Used to describe thread pitch; it stands for Unified National Fine.

nut A fastener with a hexagonal head and internal threads for screwing on bolts.

screw extractor A tool for removing broken screws or bolts.

square thread A thread type with square shoulders used to translate rotational to lateral movement.

stud A type of threaded fastener with a thread cut on each end rather than having a bolt head on one end.

tensile strength In reference to fasteners, the amount of force it takes before a fastener breaks.

threaded fasteners Bolts, studs, and nuts designed to secure parts that are under various tension and sheer stresses. These include bolts, studs, and nuts, and are designed to secure vehicle parts under stress.

thread pitch The coarseness or fineness of a thread as measured by either the threads per inch or the distance from the peak of one thread to the next. Metric fasteners are measured in millimeters.

thread repair A generic term to describe a number of processes that can be used to repair threads.

torque Twisting force applied to a shaft, which may or may not result in motion.

torque angle A method of tightening bolts or nuts based on angles of rotation.

torque specifications Supplied by manufacturers and describes the amount of twisting force allowable for a fastener or a specification showing the twisting force from an engine crankshaft.

torque-to-yield A method of tightening bolts close to their yield point or the point at which they will not return to their original length.

torque-to-yield (TTY) bolts Bolts that are tightened using the torque-to-yield method.

torque wrench A tool used to measure the rotational or twisting force applied to fasteners.

yield point The point at which a bolt is stretched so hard that it will not return to its original length when loosened; it is measured in pounds per square inch of bolt cross section.

Review Questions

1. Which of these are threaded fasteners?
   a. C-clips
   b. Trim screws
   c. Cotter pins
   d. Solid rivets
2. All of the following statements are true *except*:
   a. Metric bolts can be identified by the grade number on the top of the bolt heads.
   b. Measuring units of standard bolts use a fractional number–based system.
   c. Standard bolts can be identified by the hash marks on the top of the bolt heads.
   d. Measuring units of metric bolts use a roman number–based system.
3. In an 8 mm bolt, 8 mm refers to the:
   a. length of the bolt.
   b. size of the bolt head.
   c. threads' outer diameter.
   d. number of threads per inch.

4. The easiest way to identify the TPI is to:
   a. use a vernier caliper.
   b. count the number of threads.
   c. measure distance from peak to peak.
   d. use a thread pitch gauge.
5. A bolt has 3 hash marks on its head. What is its grade?
   a. 2
   b. 5
   c. 6
   d. 8
6. Which nut is made with a deformed top thread?
   a. Lock nut
   b. J nut
   c. Castle nut
   d. Specialty nut
7. When you want a nut or bolt to hold tight such that it needs heat to be removed with a socket and ratchet, you should use a(n):
   a. lubricant.
   b. blue thread-locking compound.
   c. antiseize compound.
   d. red thread-locking compound.
8. Which of these has a fluted tip to drill a hole into the base material so that there is no need for a pilot hole?
   a. Machine screw
   b. Self-tapping screw
   c. Trim screw
   d. Sheet metal screw
9. Torque-to-yield bolts:
   a. should be tightened using torque angle.
   b. can be reused multiple times.
   c. cannot be torqued beyond the elastic phase.
   d. do not get stretched.
10. All of the following are ways to avoid breaking a bolt when removing it *except*:
   a. using penetrating oil with bolts that are rusted or corroded in place.
   b. applying maximum force in both directions.
   c. first checking whether it is a left-hand bolt.
   d. as a last resort, you may heat up the bolt that is rusted or corroded in place.

ASE Technician A/Technician B Style Questions

1. Tech A says that most metric fasteners can be identified by the grade number cast into the head of the bolt. Tech B says that standard bolts can generally be identified by hash marks cast into the head of the bolt indicating the bolt's grade. Who is correct?
   a. Tech A
   b. Tech B
   c. Both A and B
   d. Neither A nor B
2. Two technicians are discussing bolt sizes. Tech A says that the diameter of a bolt is measured across the flats on the head. Tech B says that the length of a bolt is measured from the top of the head to the bottom of the bolt. Who is correct?
   a. Tech A
   b. Tech B
   c. Both A and B
   d. Neither A nor B
3. Tech A says that a standard grade 5 bolt is stronger than a standard grade 8 bolt. Tech B says that a metric grade 12.9 bolt is stronger than a metric grade 4.6 bolt. Who is correct?
   a. Tech A
   b. Tech B
   c. Both A and B
   d. Neither A nor B
4. Tech A says that torque-to-yield head bolts are tightened to, or just past, their yield point. Tech B says that the yield point is the torque at which the bolt breaks. Who is correct?
   a. Tech A
   b. Tech B
   c. Both A and B
   d. Neither A nor B
5. Tech A says that some locking nuts have a nylon insert that has a smaller diameter hole than the threads in the nut. Tech B says that castle nuts are used with cotter pins to prevent the nut from loosening. Who is correct?
   a. Tech A
   b. Tech B
   c. Both A and B
   d. Neither A nor B
6. Tech A says that antiseize compound is a type of thread-locking compound. Tech B says that thread-locking compounds acts like very strong glue. Who is correct?
   a. Tech A
   b. Tech B
   c. Both A and B
   d. Neither A nor B
7. Tech A says that flat washers act as a bearing surface, making it so that the bolt achieves the expected clamping force at the specified torque. Tech B says that lock washers are usually considered single use and should be replaced rather than reused. Who is correct?
   a. Tech A
   b. Tech B
   c. Both A and B
   d. Neither A nor B
8. Tech A says that a common method for repairing damaged internal threads is a thread insert. Tech B says that a thread file is used to repair the threads on the inside of a bolt hole. Who is correct?
   a. Tech A
   b. Tech B
   c. Both A and B
   d. Neither A nor B

9. Tech A says that if a bolt breaks off above the surface, you might be able to remove it using locking pliers or curved jaw Channellock pliers. Tech B says that a hammer and punch can sometimes be used to remove a broken bolt. Who is correct?
   a. Tech A
   b. Tech B
   c. Both A and B
   d. Neither A nor B

10. Tech A says that when using a Heli-Coil to repair a damaged thread, the Heli-Coil is welded into the bolt hole. Tech B says that when using a solid sleeve type of insert, the damaged threads need to be drilled out with the correct size drill bit. Who is correct?
    a. Tech A
    b. Tech B
    c. Both A and B
    d. Neither A nor B

CHAPTER 9

Vehicle Protection and Jack and Lift Safety

NATEF Tasks

- **N09001** Identify purpose and demonstrate proper use of fender covers, mats.
- **N09002** Identify and use proper placement of floor jacks and jack stands.
- **N09003** Identify and use proper procedures for safe lift operation.
- **N09004** Ensure vehicle is prepared to return to customer per school/company policy (floor mats, steering wheel cover, etc.).

Knowledge Objectives

After reading this chapter, you will be able to:

- **K09001** Explain the precautions and procedures to prevent vehicle damage while conducting repairs.
- **K09002** Describe ways that vehicles can be damaged during service and their prevention.
- **K09003** Describe the application, purpose, and safe use of lifting equipment.
- **K09004** Describe the purpose and use of jacks and stands.
- **K09005** Describe the purpose and use of vehicle hoists.
- **K09006** Describe the purpose and use of vehicle inspection pits.
- **K09007** Explain how to prepare a vehicle for customer pickup.

Skills Objectives

After reading this chapter, you will be able to:

- **S09001** Use fender, seat, floor, and steering wheel covers.
- **S09002** Lift and secure a vehicle with a floor jack and jack stands.
- **S09003** Lift a vehicle using a two-post hoist.
- **S09004** Lift a vehicle using a four-post hoist.
- **S09005** Use an engine hoist to lift an engine.
- **S09006** Use a vehicle inspection pit.
- **S09007** Prepare a vehicle for customer pickup.

▶ Introduction

Whenever a shop conducts repairs, its professionalism is on display. Customers expect their vehicle to be treated with respect, repaired, and returned clean and free of any additional damage (**FIGURE 9-1**). A professional shop always takes precautions to guarantee that the customer's vehicle is treated with respect by ensuring that all work is conducted safely and efficiently. Before you undertake a task, pause for a moment to identify good work practices that will help prevent accidental damage to a customer's vehicle. For example, before starting work, install protective covers to ensure that interiors, steering wheels, and fenders are protected from accidental spills and scratches.

For your personal safety, it is important that lifting equipment such as vehicle lifts, jacks, jack stands, engine hoists, slings, and chains are inspected before each use and are well maintained. Some states require annual certification inspections of vehicle lifts and hydraulic jacks to help ensure their safety.

▶ Preventing Vehicle Damage

Vehicle Walk-Around

K09001

To avoid customer dissatisfaction, it is good practice for the service advisor to perform a vehicle walk-around with the customer. During this time, any existing damage or missing components on the vehicle should be noted on the check-in sheet or repair order and discussed with the customer (**FIGURE 9-2**). This helps prevent the customer from coming back to the shop and complaining that the vehicle was damaged in the shop. Also, this is a good time for the service advisor to visually look the vehicle over for any needed maintenance or repairs, such as worn tires and wiper blades. The service advisor should also check with the customer to make sure any valuables, medications, or firearms are not left in the vehicle during anything but the most quick maintenance tasks. You, as the technician, should also be on the lookout for any damage, needed repairs, and valuables. If found, inform your supervisor or the service advisor, and always follow company policy. Typically, the service advisor installs at least a floor cover during the walk-around. So make sure they are installed before getting into the vehicle.

FIGURE 9-1 Return the vehicle clean and free of any additional damage.

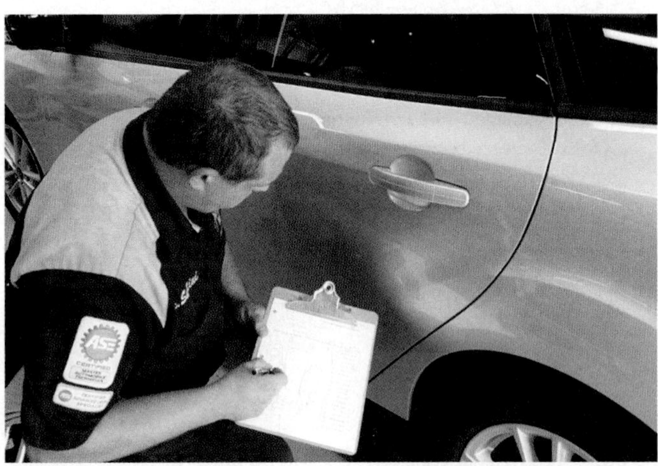

FIGURE 9-2 Vehicle walk-around with the customer.

You Are the Automotive Technician

A customer brings her 2014 BMW 325i into the auto dealership for routine service. During the vehicle walk-around, you notice a missing BMW emblem from the hood that the customer says was there when she washed the car this past weekend. You also notice that the rear tires appear to be excessively worn. Using a tread-depth indicator, you demonstrate to the customer that the tires are worn below the legal limit. You escort her to the sales department to choose new tires and wait for her vehicle. Before moving the vehicle for service, you install seat covers, a steering wheel cover, and floor mats.

1. What type of hoist is best for rotating and changing vehicle tires?
2. Why is it important to check the safe working load (SWL) rating of the lifting equipment before using it?
3. A vehicle is about to be removed from the vehicle inspection pit; what are some safety precautions you must perform before it can be removed?
4. What are some steps you need to take before returning the vehicle to the customer?

▶ TECHNICIAN TIP

The preservice vehicle walk-around with the customer can be very valuable. Occasionally, a customer is unaware of damage to his or her vehicle that was sustained before dropping the vehicle off at the shop for service. For example, maybe the customer stopped by the store to pick up a couple of things on the way to the shop. While in the store, another car sideswiped the passenger side of the vehicle. Because it was on the passenger side, the driver may not have noticed it. However, when picking up the car from the shop and looking it over to make sure nothing is damaged, the customer sees that the passenger side has been sideswiped and immediately blames the shop, saying, "My car wasn't that way when I dropped it off." A good preservice walk-around can prevent this kind of misunderstanding.

Fender, Seat, Carpet, and Steering Wheel Covers

N09001, K09002

Working on vehicles requires the use of tools, equipment, and chemicals as well as the physical process of testing and replacing parts. All of these activities have the potential to damage the vehicle if the proper precautions are not taken. For example, a dropped screwdriver or wrench may scratch the paint if fender covers are not used.

The vehicle and its components tend to be more prone to accidental damage in areas that have higher levels of service activity such as the engine bay and passenger compartment. The engine bay is the center of activity where many tools are used, batteries serviced, and oil changed. The chance of an accident in the engine bay is high if caution and care are not applied.

Protective equipment for vehicles is used to prevent damage to sensitive areas while conducting repairs and servicing (**FIGURE 9-3**). Fender covers are a protective layer used to cover the fenders when work is conducted around the engine bay. They are usually made from either a durable fabric blanket or a flexible energy-absorbing foam compound about a 0.25" (0.0006 mm) thick. They are designed to fit across the top and down the side of the fender, providing protection to those areas vulnerable to accidental damage.

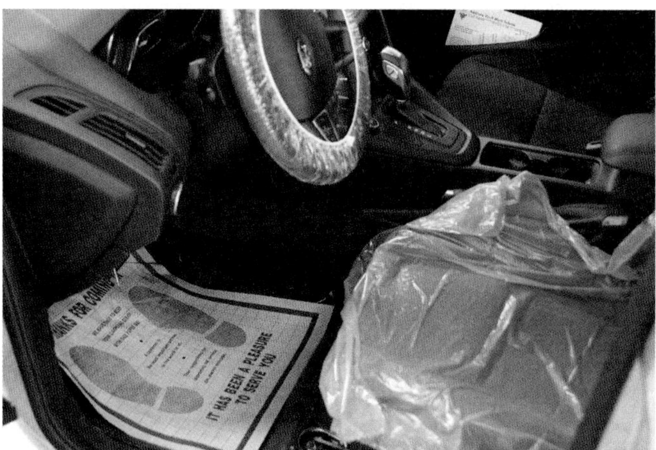

FIGURE 9-3 Always use seat and steering wheel covers, floor mats, and fender covers when working on a vehicle.

▶ TECHNICIAN TIP

Many of today's vehicles use very thin sheet metal in their fenders, making them easily dented if you lean against them or put your body weight against them. Fender covers *cannot* protect the vehicle from this kind of damage.

Seats, carpets, and steering wheels are made from materials that are sensitive to marks and damage from grease, oil, and dirt. Protectors for these areas are made from materials that provide a barrier to oil, grease, and dirt, and are designed to prevent accidental damage to the delicate materials. For example, waxed or plastic laminated carpet protectors, plastic or fabric seat protectors, and plastic steering wheel covers can be used to prevent damage and stains.

▶ TECHNICIAN TIP

Always use the appropriate protection to prevent accidental damage to vehicles. Fender, carpet, seat, and steering wheels covers should be the first things on and the last things removed when working on vehicles.

▶ TECHNICIAN TIP

Never place tools in your back pocket. If you put a screwdriver in your back pocket and then sit in the driver's seat, the screwdriver would likely poke a hole in the seat upholstery, even if you used a seat cover.

Corrosives and Greases

Corrosives and greases have the potential to cause damage to customers' vehicles. Corrosives, such as battery acids, can corrode metals, etch painted surfaces, and eat holes in fabrics. Greases, such as wheel bearing grease, can cause staining of fabric and leather materials. Corrosives and greases can also be transferred from contaminated surfaces of the vehicle onto the customer's clothing, ruining it as well. Be careful to prevent corrosives and greases from coming into contact with areas of the vehicle that aren't designed for them. This means keeping your work area clean, including your hands and uniform. Because your hands become dirty while working, it is critical that you learn not to put your hands on clean surfaces (**FIGURE 9-4**). Novice technicians have a habit of putting their hands on painted surfaces while working. This transfers grease and oil to the painted surface. Removing dirty fingerprints can be time consuming, especially on oxidized paint surfaces.

▶ TECHNICIAN TIP

Brake fluid is another liquid that can damage a vehicle. Most brake fluids soften paint quickly if allowed to get on the surface. Brake fluid on your hands can easily get onto painted surfaces if you are not careful. If brake fluid gets on a painted surface, immediately remove it with a water-soaked, clean rag, but don't scrub it, or the paint may wipe off.

FIGURE 9-4 Putting your hands on painted surfaces is a habit that needs to be stopped.

FIGURE 9-5 Special absorbent materials in granular form can be used to absorb some liquid spills, such as engine oil.

Also, be diligent to prevent spills, and if they do occur, be sure to clean them up thoroughly using appropriate methods, which can usually be found in the safety data sheets (SDS) for each material. Damage caused by corrosive agents may not be immediately visible and may lead to other problems, such as rust or corrosion, if the corrosive agents are not cleaned up and neutralized immediately. If you are handling corrosive materials and believe they have spilled, take precautions, clean the area immediately, and use a neutralizer. For example, battery electrolyte contains sulfuric acid. This material can be cleaned up by neutralizing it with an alkaline such as common baking soda and flushing the area with fresh water. Special absorbent materials in granular form can also be used to absorb some liquid spills such as engine oil (**FIGURE 9-5**). Once the engine oil is absorbed, the granules can be swept up for disposal in an environmentally safe way.

Greases and liquids can cause stains at any time, so be sure to work as cleanly as possible, and use protective covers to prevent damage. If a liquid spill does occur, clean up any excess with an absorbent material. Then use a cleaner suitable for the type of spill and the material being cleaned. For example, use upholstery cleaner for spills on vehicle seats, or carpet cleaner for carpets.

FIGURE 9-6 If you are the person guiding the driver, use large hand signals, and make sure you stand where the driver can clearly see you.

Mechanical Damage

The possibility for accidental mechanical damage is always present when a vehicle is in the shop. It can occur easily; for example, an incorrectly jacked vehicle may damage the suspension or body. Another example is hitting another vehicle while bringing a vehicle into the tight bays of the shop. The best way to minimize accidental mechanical damage is to think carefully about the task or work you are performing, plan what you intend to do, follow shop and manufacturer procedures, use tools and equipment correctly, and seek advice if you are not sure. For example, if you have to rotate a vehicle's tires, you should use the proper jacking or hoisting equipment to ensure easy and safe access to the wheels and tires.

Moving and Road Testing Vehicles

Vehicles often need to be road tested or simply moved from one place to another in the shop. Accidents can occur at this time, especially if the driver involved is inexperienced or unqualified. Only authorized, fully trained, and licensed drivers should be given the responsibility to move vehicles. Only the most skilled and experienced drivers available should be allowed to test-drive higher performance vehicles.

Have someone outside the vehicle supervise and guide any vehicle movement inside the shop, especially in restricted spaces, when reversing, and when nearing blind corners. Often someone directing you to maneuver the vehicle will use hand signals. Make sure you understand what the hand signals mean by discussing them before moving the vehicle. If you are the person guiding the driver, use large hand signals, and make sure you stand where the driver can clearly see you, but not directly in front of or behind the vehicle (**FIGURE 9-6**).

TECHNICIAN TIP

Make sure customers and visiting drivers are aware of your rules for moving cars. Keep the keys for all vehicles secure and away from the vehicles when not in use.

SAFETY TIP

Vehicles do not always run properly when they are brought in for service, which can lead to an accident. For example, a car with a hesitation problem will not move until the throttle is pressed down farther than normal. When the car does move, it lurches forward. In the tight spaces of a shop, this could cause the vehicle to run into something or someone. You have two options in this situation. The safest option is to push the vehicle instead of driving it. The second is a bit controversial—two-footing the pedals. Two-footing means that you put your left foot on the brake pedal, ready to apply the brakes, and your right foot on the throttle. Thus, as you are driving, you can use the brakes to keep the speed of the vehicle very slow. This method takes practice and goes against most driver training. But when the engine is not running properly, the standard method of using the right foot for both the throttle and brake puts the vehicle at risk of an accident.

▶ Using Protective Covers

Proper Use of Fender Covers, Seat Covers, Floor Mats, and Steering Wheel Covers

S09001

As the purpose of protective covers is to prevent damage of the vehicle, you should always inspect them prior to use. Ensure that fender covers and floor mats are clean on both sides and do not have any metal or hard objects stuck to them. Ensure that they fit securely and provide adequate protection.

To properly apply fender covers, seat covers, floor mats, and steering wheel covers, follow the steps in **SKILL DRILL 9-1**.

TECHNICIAN TIP

Some fender covers are made with a magnetic strip in them, which is designed to help hold the fender cover in place. Unfortunately, the magnet attracts metal particles, which can be held between the cover and the fender and scratch the paint. Always check these types of covers very thoroughly for metal particles.

SKILL DRILL 9-1 Applying Fender Covers, Seat Covers, Floor Mats, and Steering Wheel Covers

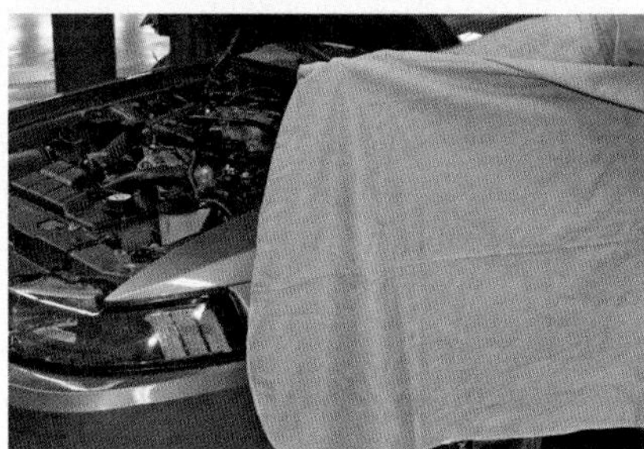

1. Select appropriate protective covers for the vehicle and type of repair. Inspect the covers for rocks, metal, or fluids that would damage the vehicle.

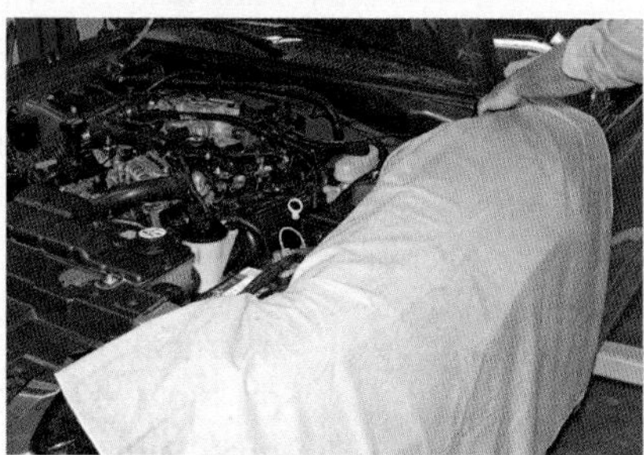

2. Position the fender covers so they provide adequate protection. Ensure that fender covers stay in position, providing protection while the vehicle is in the shop.

3. Position the seat cover so it covers both the back and bottom of the seat. Make sure it hangs over all of the edges and will stay in place.

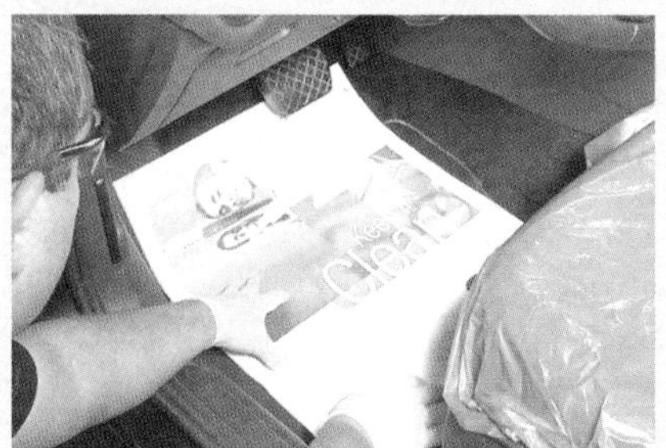

4. Install the floor mat on the floor so that it covers as much of the floor as possible. Also make sure it doesn't interfere with the throttle, brake, or clutch pedals.

5. Install the steering wheel cover so that it covers as much of the steering wheel as possible without interfering with any controls.

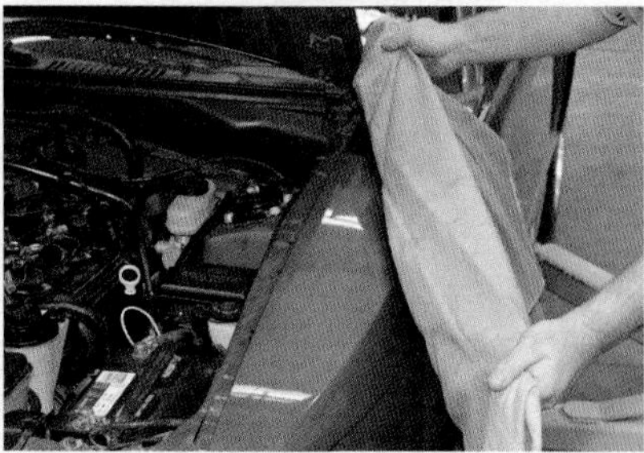

6. Remove all protective covers prior to customer pickup.

▶ Lifting Equipment

K09003

Many different types of lifting equipment may be used in a shop. Some examples include vehicle hoists, floor jacks, jack stands, engine and component hoists, chains, slings, and shackles (**FIGURE 9-7**). Lifting equipment is designed to lift and securely hold loads. Each piece of lifting equipment is designed for a specific purpose and has an operating capacity. The operating capacity is usually expressed as the **safe working load (SWL)**. For example, if the SWL is 1 ton, the equipment can safely lift up to 1 ton, or 2000 lb (907 kg). When using lifting equipment, never exceed its capacity, and always maintain some reserve capacity as an extra safety margin.

Each piece of lifting equipment should only be used for the purpose for which it is designed. For example, a vehicle hoist should only be used to lift vehicles within its capacity. Using lifting equipment incorrectly may lead to equipment failure that can cause serious injury and damage.

▶ TECHNICIAN TIP

When multiple pieces of lifting equipment are used, the SWL is limited to the lowest rated piece of equipment. For example, if a chain with an SWL of 2 tons is used with a 5-ton SWL shackle and a 3-ton SWL engine hoist, then the maximum amount of weight that can be safely lifted is 2 tons.

Testing of Lifting Equipment

Lifting equipment should be periodically checked and tested to make sure it is safe. The testing should be recorded for each piece of lifting equipment in accordance with local legislation or shop procedures, and the equipment tagged with its inspection date and SWL. Inspections should identify any damage, such as cracks, dents, marks, cuts, and abrasions, that may prevent the lifting equipment from performing as it is designed. Refer to the manufacturer's manual to find out how often maintenance inspections are recommended. The time frame is usually every 12 months in the case of hoists and lifts, but may be longer for items of lifting equipment such as chains and slings. Always check local regulations to determine the requirements for periodic testing of lifting equipment.

Check the Test Certificate

In some countries, lifting equipment is subject to statutory testing and certification. If this is the case where you work, the **test certificate** should be attached to or displayed near the lifting equipment (**FIGURE 9-8**). Before using a piece of lifting equipment, make sure the most recent inspection recorded on the test certificate is within the prescribed time limit. If it is not, the test certificate has expired, and you should notify your supervisor.

Jacks and Jack Stands

K09004

Jacks

A **vehicle jack** is a lifting tool used to raise part of a vehicle from the ground before removing or replacing components, or to raise heavy components into position. The vehicle's emergency jack can be used to raise and support the vehicle while changing a wheel on the side of the road. Vehicle jacks must not be used to support the weight of the vehicle during any task that requires you to get underneath any part of the vehicle. For those shop tasks, a vehicle jack should be used only to raise the vehicle so that it can then be lowered onto suitably rated and carefully positioned stable jack stands (**FIGURE 9-9**).

There are three main types of mechanisms that provide the lifting action for vehicle jacks: **hydraulic jacks**, **pneumatic jacks**, and **mechanical jacks**. Hydraulic and pneumatic jacks are the most common types of vehicle jacks. They can be mounted on slides or on a wheeled trolley. In hydraulic jacks, pressurized oil acts on a piston to provide the lifting action; in pneumatic jacks, compressed air lifts the vehicle. In mechanical jacks, a screw or gears provide the mechanical leverage required for lifting.

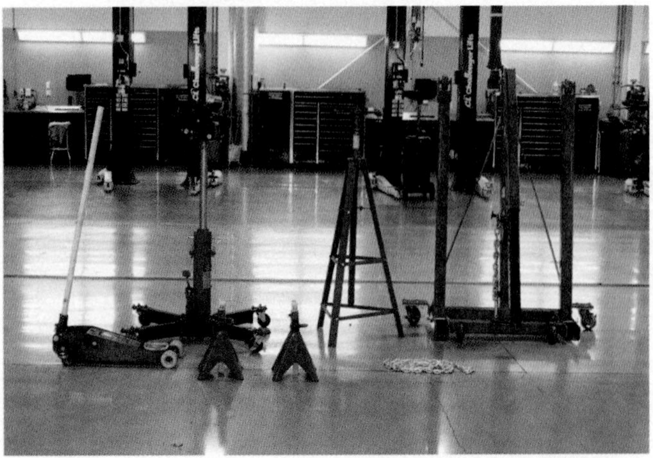

FIGURE 9-7 Some examples of lifting equipment are vehicle hoists, floor jacks, jack stands, engine and component hoists, chains, slings, and shackles.

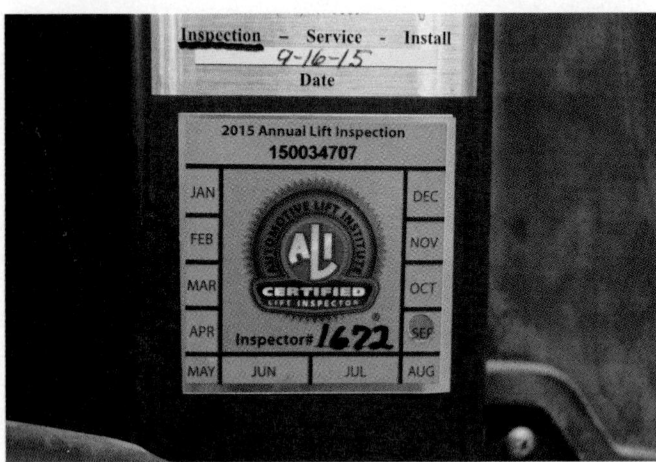

FIGURE 9-8 A vehicle hoist lift certificate.

FIGURE 9-9 A vehicle jack should be used only to raise the vehicle so that it can then be lowered onto suitably rated and carefully positioned stable jack stands.

Different jacks are available for different purposes (**FIGURE 9-10**):

- Floor jacks are a common type of hydraulic jack that is mounted on four wheels, two of which swivel to provide a steering mechanism. The floor jack has a long handle that is used both to operate the jacking mechanism and to move and position the jack. Floor jacks have a low profile, making them suitable to position under vehicles.
- High-lift (or farm) jacks are a versatile type of jack designed to lift, winch, clamp, pull, and push. They have a mechanical mechanism and are designed to provide high lift capability—for example, 36" (0.91 m) or more. Because of their high lift capability, they are often used on farms or on four-wheel drive vehicles.
- Bottle jacks are a portable jack that usually has either a mechanical screw or a hydraulic ram mechanism that rises vertically from the center of the jack as the handle is operated. They are relatively inexpensive and may be provided with vehicles for the purpose of changing flat tires.
- Air jacks use compressed air to either operate a large ram or inflate an expandable air bag to lift the vehicle. Often the air jack is fitted to a moveable platform with a long handle. Air jacks are used to lift vehicles as an alternative to floor jacks. Because a compressed air supply is required, air jacks are usually used in the shop and not for mobile operations.
- Scissor jacks are one of the most common types of jacks provided as part of the vehicle tool kit. They are used to jack a vehicle one wheel at a time. They employ a scissor action that is controlled by turning a long horizontal screw that acts on levers to raise or lower the scissor jack.
- Sliding bridge jacks are usually fitted in pairs to four-post hoists as an accessory to allow the vehicle to be lifted off the drive-on hoist runways. Operated by a hydraulic mechanism or compressed air, they use a platform mounted to a scissor-action jack to lift the vehicle along the length of the runway, thus making it more convenient to work on wheels and brakes.
- Transmission jacks are specialized jacks for lifting and lowering transmissions during removal and installation.

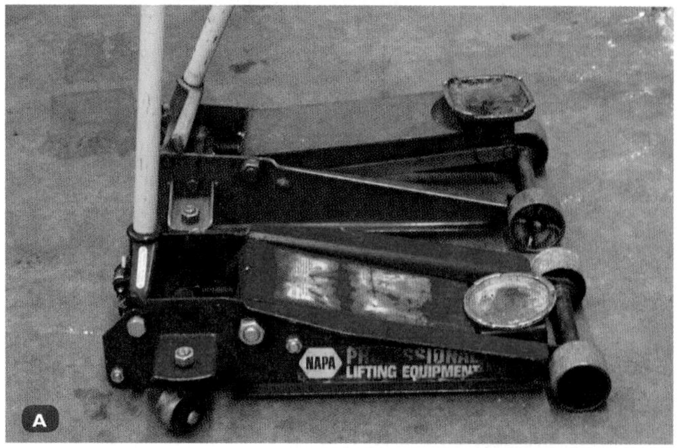

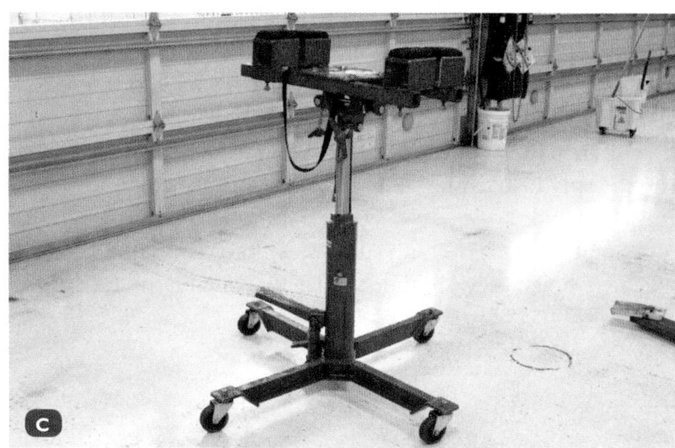

FIGURE 9-10 Jacks. **A.** Floor jacks. **B.** High-lift (or farm) jacks. **C.** Transmission jack.

Transmission jacks are usually mounted on a trolley with wheels and have a large flat plate area on which the transmission can be strapped or chained securely. They are usually operated by a hydraulic mechanism but can also be powered by compressed air.

SAFETY TIP

All vehicle jacks must always be used in accordance with the manufacturer's instructions and should be inspected on a regular basis to ensure that they are in safe working order.

▶ TECHNICIAN TIP

Make sure you always use the correct type of jack with the correct load capacity for your task.

Jack Stands

Jack stands (axle stands) are adjustable supports that are used with vehicle jacks and are designed to support the weight of the vehicle after the vehicle has been raised by a vehicle jack. Jack stands are mechanical devices, meaning they mechanically lock in place at the height selected. This makes them very dependable if they are rated strong enough for the load they are holding and if they are used properly.

Always grip jack stands by the sides to move them. Never grip them by the top or the bottom to move them, as they can slip and pinch or injure you. Check that the base of the stand is flat on the ground before lowering the vehicle onto it; otherwise, the stand might tip over, causing the vehicle to slip off. When positioned correctly, the vehicle can be lowered onto the jack stands, and the vehicle jack can be moved out of the way.

Jack stands must be used on solid, hard surfaces only. Using them on soft surfaces like dirt or even asphalt can cause the jack stand to sink in and tip over. When used properly, jack stands provide a stable support for a raised vehicle that is safer than the jack because the vehicle cannot be accidentally lowered while the jack stands are in place. To lower a vehicle that is on jack stands, it first has to be raised again so that the jack stands can be removed.

Lifting devices are also lowering devices, so it is unsafe to work underneath a vehicle that is supported only by a vehicle jack, because it could give way or be accidentally lowered. Jack stands normally come in matched pairs and should always be used as a pair (**FIGURE 9-11**). Jack stands are load rated and should only be used for loads less than the rating indicated on the jack stand.

Some shops have tall jack stands that are used along with a vehicle hoist; they are much taller than standard jack stands. Tall jack stands are used to stabilize a vehicle up on a hoist that is having a heavy component, such as a transaxle, removed or installed. Do not try to lower the vehicle with the tall jack stands still in place; doing so can cause the vehicle to slip off the hoist.

FIGURE 9-11 Jack stands normally come in matched pairs and should always be used as a pair.

Vehicle Hoists

K09005

Vehicle hoists raise whole vehicles off the ground so that a technician can easily work on the underside of the vehicle. The vehicle hoist is also useful for raising the vehicle to a height that removes the need for the technician to bend down (**FIGURE 9-12**). For example, when changing tires the vehicle can be raised to waist height to avoid excessive bending.

Vehicle hoists have a number of different designs of vehicle hoists and come in a range of sizes and configurations to meet the particular needs of the shop. For instance, some vehicle hoists are mobile, and others are designed for use where the ceiling height is limited. Some vehicle hoists can be linked together electronically so they can be used on longer vehicles such as trucks and buses.

FIGURE 9-12 Vehicle hoists allow technicians to more easily perform work under a vehicle.

Vehicle hoists are typically classified as in ground or above ground. In-ground hoists have the working mechanism in and below the concrete floor. Because most of the working parts of the hoist are below floor level, they don't have large sections above ground that can damage vehicles or impede foot traffic. One potential problem with in-ground hoists is they can leak into the ground unless they have a sealed containment chamber around the cylinders and lines to catch any fluid. Above-ground hoists have their working parts above ground. These are easier and less costly to install, but they do impede traffic, and negotiating vehicles around them can lead to accidents. Technicians need to be familiar with the operation of both types of hoists because they are both in common use.

The most common types of vehicle hoists in general use are single-post, two-post, and four-post hoists. Other types of hoists include scissor lifts, parallelogram lifts, and mobile or specialty lifts.

Safety Locks

Every vehicle hoist in the shop must have a built-in mechanical locking device so that the vehicle hoist can be secured at the chosen height after the vehicle is raised (**FIGURE 9-13**). This locking device prevents the vehicle from being accidentally lowered and holds the vehicle in place, even if the lifting mechanism fails. You should never physically go under a raised vehicle for any reason unless the safety locking mechanism has been activated.

Ratings and Inspections

All vehicle hoists are rated for a particular weight and/or type of vehicle and should never be used for any task other than that recommended by the manufacturer. In particular, a vehicle hoist should never be used to lift a vehicle that is heavier than its rated limit. In most countries, regulations require hoists to be periodically inspected, typically annually, and certified as fit for use. Before you use a vehicle hoist, check the identification plate for its rating, and make sure it has a current registration or certification label (**FIGURE 9-14**).

FIGURE 9-13 Typical mechanical lock on a two-post hoist.

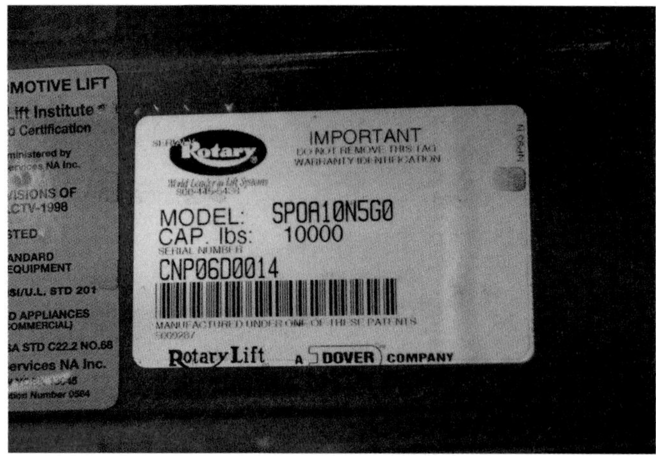

FIGURE 9-14 Typical current rating and serial number label for a hoist.

FIGURE 9-15 In-ground hoist.

Single-Post Hoist

A **single-post hoist** raises the vehicle on a platform supported by a single solid shaft located centrally under the vehicle. This type of hoist is very compact in the workshop and leaves the perimeter of the vehicle easily accessible, but the central post obscures part of the underside of the vehicle, making jobs such as transmission removal difficult or impossible. The single-post hoist is an in-ground hoist, meaning that the central post and hydraulic cylinder are in the ground as part of the concrete floor (**FIGURE 9-15**). This also means that very little of the hoist is above ground to damage vehicles or impede foot traffic in the shop. Even so, technicians need to drive over them carefully when pulling vehicles onto the lift.

Two-Post Hoist

Two-post hoists can either be in-ground or above ground and come in two configurations—symmetrical and asymmetrical. Symmetrical two-post hoists have arms that are of approximately equal length so that the vehicle is roughly centered lengthwise between the posts. This positioning creates a challenge because the posts of the hoist are usually right in the way

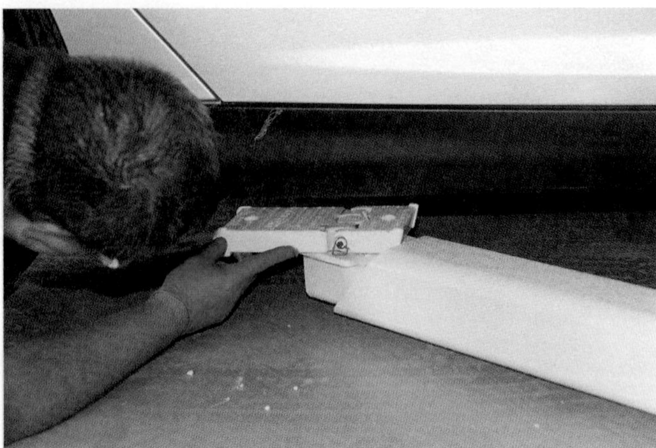

FIGURE 9-16 A two-post hoist requires careful positioning of the lift arms so that they are under appropriate lifting points, two on each side of the vehicle.

FIGURE 9-17 A four-post hoist.

FIGURE 9-18 Lifting capacity at various extensions.

of the vehicle's front doors, making it harder to get into and out of the vehicle. Asymmetrical hoists have shorter arms in front of the posts than in the rear of the posts. This allows the vehicle to be positioned farther back on the hoist, allowing better access to the doors on the vehicle. Both types of two-post hoists leave the underside of the vehicle easily accessible and also allow a technician to remove the wheels while the vehicle is raised.

Two-post hoists and single-post hoists require careful positioning of the hoist arms so that they are under the appropriate lifting points, two on each side of the vehicle (**FIGURE 9-16**). The service information for the vehicle details where those lifting points are so that the vehicle can be raised without causing structural damage.

▶ TECHNICIAN TIP

Some vehicles are equipped with an automatic leveling system, which must be disabled before lifting the vehicle. Always check the service information to see if this applies to the vehicle you are lifting.

Four-Post Hoist

A **four-post hoist** is very easy to use with most vehicles. The vehicle is driven between the four posts so that the wheels rest on two long, narrow platforms, one on each side of the vehicle (**FIGURE 9-17**). The platforms are then raised, taking the vehicle with them. The underside of the vehicle is accessible to the technician. Because the vehicle rests on its wheels on the four-post hoist, the wheels cannot be removed unless the hoist is fitted with sliding bridge jacks.

Engine Hoists and Stands

Engine hoists, or mobile floor cranes, are capable of lifting very heavy objects, such as engines, while they are being removed from or installed in a vehicle. The lifting arm of the engine hoist is moved by a hydraulic cylinder and is adjustable for length. However, extending the lifting arm reduces its lifting capacity because it moves the load farther away from the supporting frame. The supporting legs can be extended for stability, but the more the lifting arm and the legs are extended, the lower the lifting capacity of the engine hoist. The safe lifting capacity at various extensions is normally marked on the lifting arm (**FIGURE 9-18**).

The engine or component to be lifted is attached to the lifting arm by a sling or a lifting chain. The sling and lifting chain must be rated for weights in excess of the engine or component being lifted and must be firmly attached before the engine hoist is raised. When the engine or other component has been lifted and slowly and carefully moved away from the vehicle, it should be lowered onto an engine stand or onto the floor (**FIGURE 9-19**). The farther off the ground an engine is lifted, the less stable the engine hoist becomes.

SAFETY TIP

Never use a hoist to lift any weight greater than the lifting capacity of the hoist, sling, chains, or bolts.

Engine stands are a convenient device to support an engine when it is out of the vehicle and being worked on (**FIGURE 9-20**).

FIGURE 9-19 When the engine or other component has been lifted and slowly and carefully moved away from the vehicle, it should be lowered onto an engine stand or onto the floor.

FIGURE 9-20 Engine stand with engine attached.

Engine stands come in a variety of ratings, some of which are very light duty and should not be used with any but the smallest engines. When mounting an engine on the stand, use bolts of the proper strength, and ensure that they thread into the hole at least six complete turns. You should also mount the arms of the stand so that the weight of the engine is centered on the mounting head pivot. Otherwise, the engine could be top-heavy and rotate quickly, injuring you.

Vehicle Inspection Pits

K09006

Vehicle inspection pits are usually built into the floor area of a vehicle repair bay (**FIGURE 9-21**). The vehicle inspection pit allows the technician to access the underside of the vehicle without the need for a hoist or jack to raise the vehicle. Vehicle inspection pits have a center pit that is wide enough to allow a technician to move along but narrow enough to fit between the wheels of a vehicle. Stairs or steps are usually positioned at one end of the vehicle inspection pit to allow access for the technician. Lighting is installed in the vehicle inspection pit area for greater visibility. The depth of the vehicle inspection pit is fixed and is set so as to afford a general working height to the underside of the vehicle.

FIGURE 9-21 Vehicle inspection pit.

Many people have been injured falling into vehicle inspection pits, even if they have been working in or near the vehicle inspection pit for a long time. When there is no vehicle over the vehicle inspection pit, you should fence it or place boards over the top to stop people from accidentally stumbling into it. You should also restrict general access to that work area.

▶ TECHNICIAN TIP

Even low-voltage shop lights can set off an explosion. Whenever possible, use an explosion-proof lamp that will not break, even if it is dropped 6' (1.8 m). Lighting units in a vehicle inspection pit should be sealed behind toughened glass or polycarbonate and set into the vehicle inspection pit walls so that falling objects will not damage them.

SAFETY TIP

Fuel, paints, and solvents can all give off flammable vapors that are heavier than air. Like water, these vapors tend to flow down and accumulate in places like vehicle inspection pits, increasing the risk of explosion and polluting the breathable air. Unless bulkier and more expensive explosion-protected power tools are available in your shop, the only safe power tools to use in a vehicle inspection pit are air-powered, which are less likely to cause sparks that could ignite flammable gases.

▶ Using Lifting Equipment

N09002, S09002

Using Vehicle Jacks and Stands

The size of vehicle jack you use is determined by the weight of the vehicle you want to lift. Most workshops have a vehicle jack that has a lifting capacity of about 2.5 tons. If the end of the vehicle is heavier than that, or if the vehicle is loaded, you need to use a vehicle jack with a larger lifting capacity.

Make sure the jack stands are rated to support the weight you are lifting and are in good condition before you use them to support the vehicle. If they ore cracked or bent, they will not support the vehicle safely. Always use matched pairs of jack stands.

Never support a vehicle on anything other than jack stands. Do not use wood or steel blocks to support the vehicle; they may slide or split under the weight of the vehicle. Do not use bricks or concrete blocks to support the vehicle; they may crumble under its weight.

To lift and secure a vehicle with a floor jack and jack stands, follow the steps in **SKILL DRILL 9-2**.

SKILL DRILL 9-2 Lifting and Securing a Vehicle with a Vehicle Jack and Jack Stands

1. Position the vehicle on a flat, solid surface. Put the vehicle into neutral or park and set the parking brake. Place wheel chocks in front of and behind the wheels that are not going to be raised off the ground.

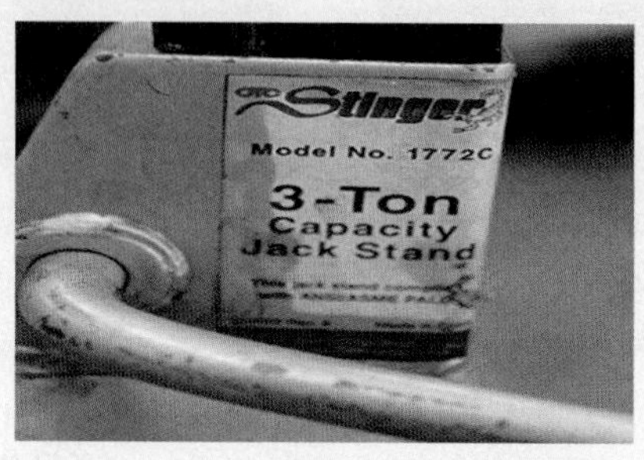

2. Check the manufacturer's labels on the jack and stands. Make sure they are rated higher than the weight you are lifting. If in doubt, ask your instructor.

3. Place one jack stand on each side of the vehicle at the same point, and adjust them so that they are both the same height.

4. Roll the floor jack under the vehicle, and position the lifting pad correctly under the frame or cross member. Turn the jack handle clockwise, and begin pumping the handle up and down until the lifting pad touches and begins to lift the vehicle.

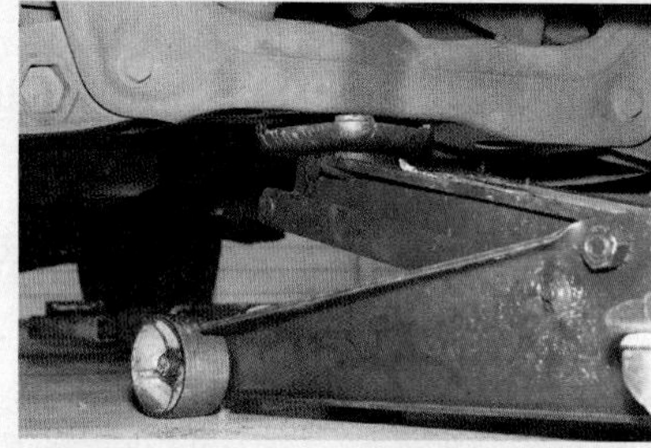

5. Stop and check the placement of the lifting pad under the vehicle to make sure there is no danger of slipping. Double-check the position of the wheel chocks to make sure they have not moved. If the vehicle is stable, continue lifting it until it is at the height at which you can safely work under it.

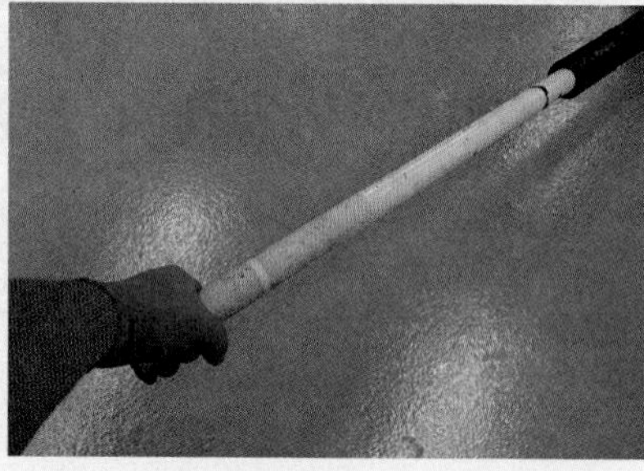

6. Slide the two jack stands underneath the vehicle and position them to support the vehicle's weight. Slowly turn the jack handle counterclockwise to open the release valve, and gently lower the vehicle onto the jack stands.

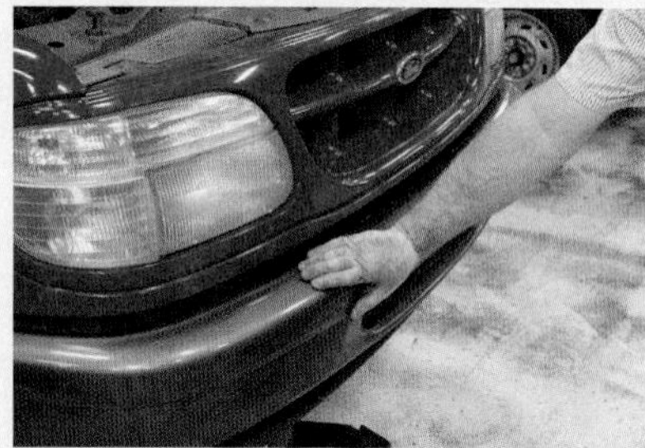

7. When the vehicle has settled onto the jack stands, lower the vehicle jack completely, and remove it from under the vehicle. Gently push the vehicle sideways to make sure it is secure. Repeat this process to lift the other end of the vehicle.

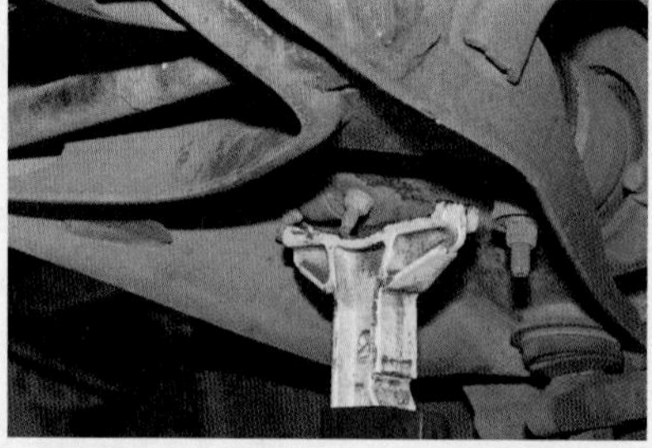

8. When the repairs are complete, use the jack to raise the vehicle off the jack stands. Slide the jack stands from under the vehicle. Make sure no one goes under the vehicle or puts any body parts under the vehicle, as the jack could fail or slip.

9. Slowly turn the jack handle counterclockwise to gently lower the vehicle to the ground. Return the jack, jack stands, and wheel chocks to their storage area before you continue working on the vehicle.

SAFETY TIP

Some independent suspension components are not strong enough to support the weight of the vehicle. Make sure you always use the specified lift points to lift the vehicle.

Using Two-Post Hoists

N09003, S09003

Before lifting any vehicle, make sure the frame is structurally sound. If you see rust or signs of major repair, lifting the vehicle with a vehicle hoist may cause damage to the vehicle or may be dangerous to you. Make sure you know exactly how to operate the vehicle hoist. Take particular care that you know exactly where the stop control is so that you can use it quickly in an emergency. Refer to the operator's manual for the correct procedure.

Check the amount of clearance under the vehicle. If any of the lifting mechanism is designed so that the vehicle is driven over it, verify that the vehicle has enough clearance. Driving a low-slung vehicle over the lifting mechanism may result in damage to the underside of the vehicle. These vehicles may require shallow ramps that raise the vehicle's wheels so it can go onto the hoist.

The lifting points on a vehicle are typically located at the same place as the jacking points. Check the vehicle's service manual if you are not sure where the lift points are. The lifting arms must be positioned under the center of the lift points so that the weight of the vehicle is distributed evenly. Check whether the vehicle requires rubber pads to protect the undercoating. If the vehicle is equipped with running boards, verify that the hoist arms will not contact the running boards before contacting the lift points. You may be able to set the hoist pads to a higher setting so that the running boards clear the hoist arms.

Make sure there will be adequate headroom above the vehicle once it is raised. Taller vehicles, especially those fitted with roof racks, may need more headroom than you think. Ask your supervisor if there is the least bit of doubt!

The vehicle hoist should be raised so you can comfortably work under it. Lock the lift in place before moving underneath or working on the vehicle. To lift a vehicle using a two-post hoist, follow the steps in **SKILL DRILL 9-3**.

SKILL DRILL 9-3 Lifting a Vehicle Using a Two-Post Hoist

1. Prepare to use the two-post hoist. Check the hoist and check the vehicle clearance. Carefully drive the vehicle so that it is centered between the two posts, left and right.

2. Also ensure that it is positioned properly, front to back, for the type of hoist and vehicle you are using. Leave the vehicle in neutral, and apply the emergency brake.

3. Position the lifting pads under the vehicle lifting points. Make sure the lifting pads are adjusted to the same height for both sides of the vehicle. Move to the operating controls, and raise the two-post hoist just far enough to come into contact with the vehicle.

4. Make sure no one is near the vehicle, and then raise the vehicle just until the wheels are a couple of inches off the floor. Check the position of the lifting pads, and shake the vehicle gently to confirm that it is stable.

5. Lift the vehicle to slightly above working height, and then lower it onto the locks or safety device.

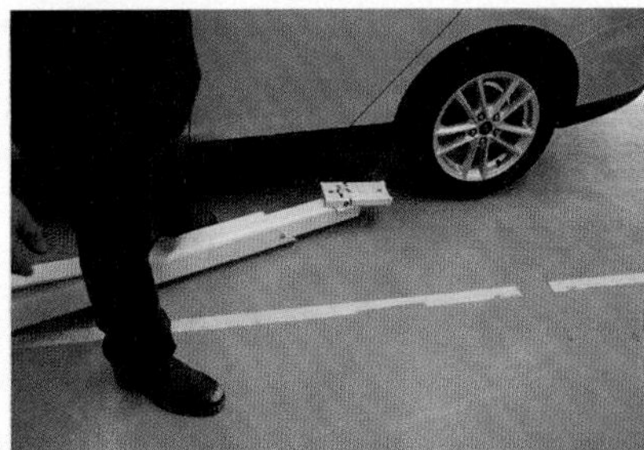

6. Before the two-post hoist is lowered, remove all tools and equipment from the area, and wipe up any spilled fluids. Raise the hoist to unlock the lift before lowering it. Make sure no one is near the vehicle before lowering it. Once the vehicle is on the ground, remove the lifting arms and drive it away.

Using Four-Post Hoists

S09004

Four-post hoists are often used to lift a vehicle for wheel alignment services and oil changes. Make sure you know how to operate the four-post hoist, taking particular care to know where the stop control is so that you can use it quickly in an emergency. Always refer to the operator's manual for the correct procedure for stopping the four-post hoist.

To lift a vehicle using a four-post hoist, follow the steps in **SKILL DRILL 9-4**.

Using Engine Hoists and Stands

S09005

Engine hoists are capable of lifting very heavy objects, which make them suitable for lifting engines. Make sure the lifting attachment at the end of the lifting arm is strong enough to lift the engine and is not damaged or cracked. When attaching the lifting chain or sling to an engine, make sure it is firmly attached and that the engine hoist is configured to lift that weight. Make sure the fasteners attaching the lifting chain, or sling, have a tensile strength that is in excess of the weight of the engine. To keep from overstressing the sling, leave enough length in the sling so that when the engine is hanging, the angle at the top of the sling is close to 45 degrees and does not exceed 90 degrees.

If removing an engine from an engine bay, lower the engine so that it is close to the ground after removal. If the engine is lifted high in the air, the engine hoist will be unstable. When moving a suspended engine, move the engine hoist slowly. Do not change direction quickly, because the engine will swing and may cause the whole apparatus to tumble.

To use an engine hoist and choose the correct attachments to lift an engine, follow the steps in **SKILL DRILL 9-5**.

SKILL DRILL 9-4 Using Four-Post Hoists to Lift a Vehicle

1. Prepare to safely use the vehicle hoist. With the aid of an assistant guiding the driver, or a large mirror in front of the hoist, drive the vehicle slowly and carefully onto the four-post hoist, and position it centrally. If the hoist has front-wheel restraints, drive the vehicle forward until the wheels lock into the brackets.

2. Get out of the vehicle, and check that it is correctly positioned on the platform. If it is, apply the emergency brake, and select first gear for a manual transmission or park for an automatic.

3. Make sure the four-post hoist area is clear. Move to the controls, and lift the vehicle until it reaches the appropriate work height. If the four-post hoist has a manual safety mechanism, lock it in place to engage whatever safety device is used.

4. Before the four-post hoist is lowered, remove all tools and equipment from the area, and wipe up any spilled fluids.

5. Remove the safety device or unlock the lift before lowering it. Make sure no one is near the area.

6. Once the four-post hoist is fully lowered, with the help of a guide you can carefully back the vehicle off the hoist.

SKILL DRILL 9-5 Using Engine Hoists and Stands

1. Prepare to use the engine hoist. Lower the lifting arm, and position the lifting end and chain over the center of the engine.

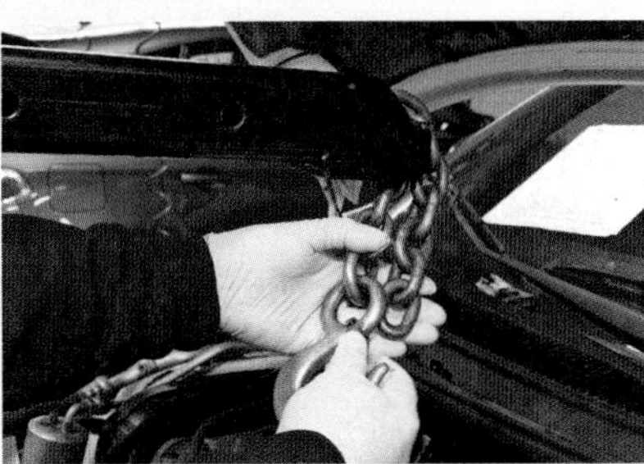

2. Inspect the chain, steel cable or sling, and bolts to make sure they are in good condition. Look carefully around the engine to determine if it has lifting eyes or other anchor points.

3. If the engine has lifting eyes, attach the sling with D-shackles or chain hooks. If you need to screw in bolts and spacer washers to lift the engine, make sure you use the correct bolt and spacer size for the chain or cable. Screw in the bolts until the sling is held tight against the engine.

4. Attach the hook of the hoist under the center of the sling, and raise the engine hoist just enough to lift the engine an inch or two (0.025–0.05 mm). Double-check the sling and attachment points for safety. The center of gravity of the engine should be directly under the hook of the engine hoist, and there should be no twists or kinks in the chain or sling.

5. Raise the engine hoist until the engine is clear of the ground and any obstacles. Slowly and gently move the engine hoist and engine to the new location.

6. Make sure the engine is positioned correctly. You may need to place blocks under the engine to stabilize it. Once you are sure the engine is stable, lower the engine hoist, and remove the sling and any securing fasteners. Finally, return the equipment to its storage area.

▶ TECHNICIAN TIP

- The load rating of the engine hoist must be greater than the weight of the object to be lifted.
- Never leave an unsupported engine hanging on an engine hoist. Secure the engine on an engine stand, or on the ground, before starting to work on it.
- If using an engine stand, make sure it is designed to support the weight of the engine and that you have the correct number of bolts to hold the engine to the stand.
- Always extend the legs of the engine hoist in relation to the lifting arm to ensure adequate stability.

Using Vehicle Inspection Pits

S09006

When using a vehicle inspection pit, you should get someone to assist in guiding you as you drive over the vehicle inspection pit. Make sure you understand each other's hand signals. Before using the vehicle inspection pit, check its lighting. Ensure that there are no obstacles in the vehicle inspection pit or its surrounding area. Do not drive or move the vehicle over a vehicle inspection pit if someone is in the vehicle inspection pit.

To use a vehicle inspection pit, follow the steps in **SKILL DRILL 9-6**.

▶ Preparation for Customer Pickup

N09004, K09007, S09007

Make sure all vehicle protection is removed prior to releasing the vehicle to the customer. After removing the vehicle protection,

SKILL DRILL 9-6 Using Vehicle Inspection Pits

1. Ensure that the vehicle inspection pit and surrounding area are clear of obstructions and you have a clear pathway to drive the vehicle over the vehicle inspection pit. Turn the lights on in the inspection pit.

2. Get someone to help guide you as you drive the vehicle over the vehicle inspection pit. Drive the vehicle slowly and carefully, and position it centrally.

3. Get out of the vehicle and check that it is correctly positioned on the platform. If it is, apply the emergency brake, and select first gear or park, depending on the transmission.

4. Perform the needed work on the vehicle, being sure to work in a safe manner, using tools approved for confined spaces.

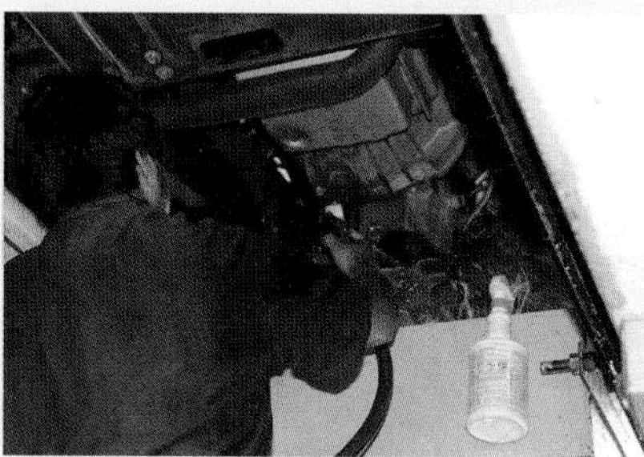

5. Before driving the vehicle off the vehicle inspection pit, remove all tools and equipment from the vehicle inspection pit area, and wipe up any spilled fluids.

6. Make sure no one is near the vehicle or in the vehicle inspection pit. Have someone guide you off the pit area.

such as fender covers, floor mats, and steering and seat covers, ensure that no damage to the vehicle has been sustained during repair. Check the vehicle for cleanliness by ensuring that all trash has been removed and no oil or grease marks are on the vehicle prior to returning it to the customer. Check that all of the windows are crystal clear; that the dashboard, knobs, steering wheel, and center console are spotless and clean; that floor mats have no dirt; and that there are no fingerprints on the door latches, fenders, or the backs of the mirrors (**FIGURE 9-22**). Now go back through the vehicle and look for tools. Wear appropriate personal protective gear such as safety glasses and gloves when working with cleaning materials.

To ensure that the vehicle is prepared to return to the customer per school/company policy, follow the steps in **SKILL DRILL 9-7**.

FIGURE 9-22 Preparing a vehicle for a customer.

SKILL DRILL 9-7 Preparing to Return Vehicle to Customer

1. Remove fender, seat, and steering wheel covers.

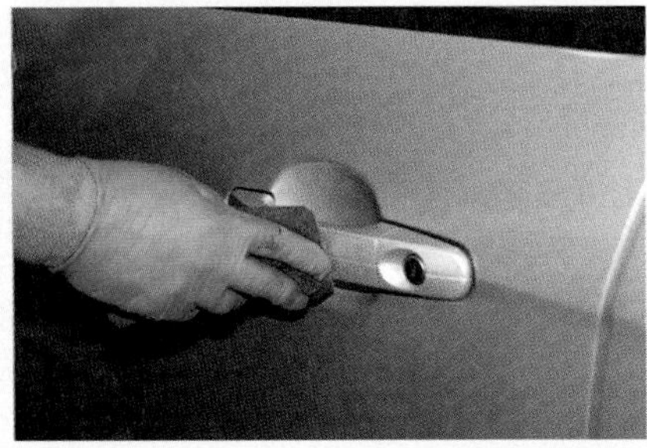

2. Clean any fingerprints from the vehicle.

3. Check for any forgotten tools.

▶ Wrap-Up

Ready for Review

- Vehicles under repair must be protected from further damage occurring while in the shop.
- A preservice vehicle walk-around ensures that the shop and customer are in agreement about existing damage to the car.
- Use protective equipment, such as seat covers, floor mats, fender covers, and steering wheel covers, to prevent damage and to protect vehicles against grease, oil, and dirt.
- Prevent corrosives and grease from getting on the surfaces of a vehicle; clean up spills thoroughly.
- Carefully use jacking or hoisting equipment to prevent mechanical damage to a customer's car.
- Only experienced, licensed drivers should test-drive vehicles.
- Moving cars in and out of the shop is another opportunity for damage to occur.
- Inspect protective covers (fender covers and floor mats) for existing damage or contamination prior to use.
- The safe working load indicates the operating capacity for lifting equipment.
- Lifting equipment includes vehicle hoists, floor jacks, jack stands, engine and component hoists, chains, slings, and shackles.
- Periodically check and test lifting equipment; consult the test certificate if available.
- Vehicle jacks can be classified by the type of lifting mechanism they use: hydraulic, pneumatic, or mechanical.
- Jack types include floor jacks, high-lift (farm) jacks, bottle jacks, air jacks, scissor jacks, sliding bridge jacks, and transmission jacks.
- Jack stands support a vehicle's weight when it has been raised; always use jack stands in pairs.
- Vehicle hoists raise the vehicle to allow technicians underside access.
- Vehicle hoists are most commonly single-post, two-post, or four-post.
- Never use a vehicle hoist without activating the safety lock or for lifting a vehicle heavier than the rated limit.
- Engine hoists can lift heavy objects out of a vehicle and onto an engine stand.
- Vehicle inspection pits allow access to the vehicle's underside without using a hoist or jack.
- Cover or fence inspection pits when not in use, to prevent others from falling in.
- Choose vehicle jacks according to size and lifting capacity.
- Do not use a vehicle hoist if the vehicle's frame is not structurally sound.
- Make sure a vehicle has enough clearance over the lifting mechanism.
- Check for damage before using an engine hoist, and make sure all components have the lifting capacity needed for the task.
- Clean the vehicle, and remove tools and vehicle protection before returning it to the customer.

Key Terms

engine hoist A small crane used to lift engines.

four-post hoist A type of hoist that the vehicle is driven onto that uses two long, narrow platforms to lift the vehicle.

hydraulic jack A type of vehicle jack that uses oil under pressure to lift vehicles.

jack stands Metal stands with adjustable height to hold a vehicle once it has been jacked up.

mechanical jack A type of vehicle jack that uses mechanical leverage to lift a vehicle.

pneumatic jack A type of vehicle jack that uses compressed gas or air to lift a vehicle.

safe working load (SWL) The maximum safe lifting load for lifting equipment.

single-post hoist A type of vehicle hoist that uses a single central platform to lift a vehicle.

test certificate A certificate issued when lifting equipment has been checked and deemed safe.

two-post hoist A type of vehicle hoist that uses two parts (one on each side of vehicle) and four arms to lift the vehicle.

vehicle hoist A type of vehicle lifting tool designed to lift the entire vehicle.

vehicle inspection pit A trench permanently fitted into the floor of the shop to allow easy work access to the vehicle's underside.

vehicle jack A tool for lifting a vehicle.

Review Questions

1. A vehicle walk-around with the customer before repair:
   a. avoids customer dissatisfaction.
   b. ensures personal safety.
   c. is mandated by federal regulations.
   d. decreases repair cost to the customer.
2. All of the following statements are true *except*:
   a. Seat covers prevent staining.
   b. The engine bay is prone to accidental damage.
   c. Avoid putting your hands on painted surfaces.
   d. Fender covers prevent denting.
3. Acid spills can be cleaned up using:
   a. absorbent granules.
   b. alkalines.
   c. water-soaked rags.
   d. solvents.
4. When multiple pieces of lifting equipment are used, the SWL is:
   a. the sum of the SWLs of the individual pieces.
   b. that of the piece with the highest rating.
   c. that of the piece with the lowest rating.
   d. the average of the SWLs of the pieces with the highest and lowest ratings.
5. Which piece of equipment is typically used to jack a vehicle one wheel at a time?
   a. Floor jack
   b. Bottle jack
   c. Scissor jack
   d. Sliding bridge jack
6. All of the following statements are correct *except*:
   a. Jack stands should be gripped by the top and bottom when moving them.
   b. Jack stands should only be used for loads less than the rating indicated on them.
   c. Jack stands should always be used as a pair.
   d. Jack stands must be used on solid, hard surfaces only.
7. Which of the following statements is true?
   a. If the end of the vehicle is heavier than the lifting capacity, use wood or steel blocks to support the vehicle.
   b. If the vehicle is loaded, use bricks or concrete blocks to support the vehicle.
   c. The size of vehicle jack to be used will be determined by the weight of the vehicle to be lifted.
   d. The lifting capacity of the vehicle jacks in most shops is about five tons.
8. Four-post hoists are most often used for:
   a. lifting engines.
   b. lifting vehicles for wheel alignment services.
   c. lifting vehicles to remove the wheels.
   d. removing an engine from the engine bay.
9. All of the following are good practices when using the vehicle inspection pit *except*:
   a. You should get someone inside the vehicle inspection pit to help guide you as you drive over it.
   b. Ensure that there are no obstacles in the vehicle inspection pit or its surrounding area.
   c. Before using the vehicle inspection pit, check its lighting.
   d. Get out of the vehicle and check that it is correctly positioned on the platform.
10. All of the following should be done post-repair *except*:
    a. go back through the vehicle and look for tools.
    b. ensure that no damage to the vehicle has been sustained during repair.
    c. look for worn tires and wiper blades.
    d. check that all fingerprints have been removed.

ASE Technician A/Technician B Style Questions

1. Tech A says that you should always deal with the customer's valuables according to company policy. Tech B says that you should protect a customer's vehicle by waxing it when you are finished working on it. Who is correct?
   a. Tech A
   b. Tech B
   c. Both A and B
   d. Neither A nor B
2. Tech A says that it is a good practice to perform a walk-around inspection of the vehicle with the customer. Tech B says that fender covers prevent the fenders from becoming dented when the engine is being worked on. Who is correct?
   a. Tech A
   b. Tech B
   c. Both A and B
   d. Neither A nor B
3. Tech A says that a vehicle jack can be used to support the vehicle while working under it. Tech B says that a jack stand automatically adjusts to the vehicle's height. Who is correct?
   a. Tech A
   b. Tech B
   c. Both A and B
   d. Neither A nor B
4. Tech A says that hoists should be inspected and certified periodically. Tech B says that safety locks do not need to be applied before working under the vehicle unless you will be working for more than 10 minutes. Who is correct?

a. Tech A
b. Tech B
c. Both A and B
d. Neither A nor B

5. Tech A says that an engine hoist can lift more weight when the legs and arm are extended. Tech B says that the bolts used to mount an engine to a stand should complete at least six turns. Who is correct?
   a. Tech A
   b. Tech B
   c. Both A and B
   d. Neither A nor B
6. Tech A says that gasoline vapors are lighter than air, so inspection pits do not have a fire hazard like above ground hoists. Tech B says that you should never exceed the lifting capacity of the hoist. Who is correct?
   a. Tech A
   b. Tech B
   c. Both A and B
   d. Neither A nor B
7. Tech A says that you should always ensure that the vehicle has enough ground clearance before driving on a lift. Tech B says that you should center the vehicle on the lift before raising it. Who is correct?
   a. Tech A
   b. Tech B
   c. Both A and B
   d. Neither A nor B
8. Tech A says that you should always inspect a hydraulic lifting device for leaks and operation before using it. Tech B says that all mechanical safety locks on a hoist should be in place before getting under the vehicle. Who is correct?
   a. Tech A
   b. Tech B
   c. Both A and B
   d. Neither A nor B
9. Tech A says that an engine sling should have an angle between 120 and 150 degrees. Tech B says that all slings and lifting chains should be inspected for damage prior to use. Who is correct?
   a. Tech A
   b. Tech B
   c. Both A and B
   d. Neither A nor B
10. Tech A says that you should have a coworker help guide you onto an inspection pit. Tech B says that the lights should be on in the pit before driving over it. Who is correct?
    a. Tech A
    b. Tech B
    c. Both A and B
    d. Neither A nor B

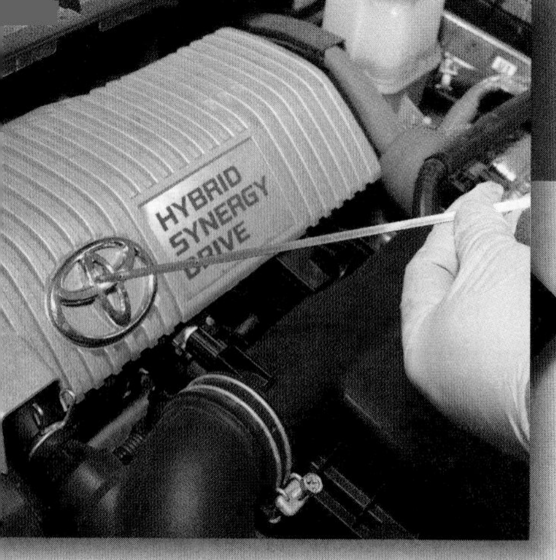

CHAPTER 10

Vehicle Maintenance Inspection

NATEF Tasks

- **N10001** Check and refill diesel exhaust fluid (DEF). (MLR/AST/MAST)
- **N10002** Inspect, service, or replace air filters, filter housings, and intake duct work. (MLR/AST/MAST)
- **N10003** Verify operation of the instrument panel engine warning indicators. (MLR/AST/MAST)

Knowledge Objectives

After reading this chapter, you will be able to:

- **K10001** Explain the systems to be inspected when performing an underhood inspection.
- **K10002** Describe what fluids are used in the typical modern automobile.

Skills Objectives

After reading this chapter, you will be able to:

- **S10001** Perform an underhood inspection.
- **S10002** Check and adjust engine oil level; perform oil and filter change.
- **S10003** Measure the freeze protection level of coolant.
- **S10004** Check the level and condition of brake and clutch fluid.
- **S10005** Check the level and condition of power steering fluid.
- **S10006** Check the level and condition of automatic transmission fluid.
- **S10007** Check the level and condition of windshield washer fluid.
- **S10008** Inspect the engine drive belts and hoses.
- **S10009** Perform an exterior vehicle inspection.
- **S10010** Inspect and replace the wiper blades.
- **S10011** Inspect the windshield.
- **S10012** Inspect the exterior lights.
- **S10013** Inspect the tires.
- **S10014** Perform an in-vehicle inspection.
- **S10015** Check the operation of the brake pedal.
- **S10016** Check the operation of the parking brake.
- **S10017** Check the operation of the horn.
- **S10018** Inspect the interior lights.
- **S10019** Perform an under-vehicle inspection.
- **S10020** Inspect under the vehicle for fluid leaks.

▶ Introduction

A regular inspection and maintenance of the many systems and components of a vehicle helps to ensure the safe and reliable operation of the vehicle (**FIGURE 10-1**). These inspections are typically performed during oil changes and include an underhood inspection, an exterior inspection, certain in-vehicle inspections, and an under-vehicle inspection. Vehicle fluid levels, lighting systems, tires, suspension, belts, and hoses are all checked. Any faults or concerns should be noted for service or repair. This chapter explains what to look for and how to perform these important services.

▶ Underhood Inspection

K10001, S10001

A check under the hood is important to the life and operation of the vehicle (**FIGURE 10-2**). The inspection should be performed at the manufacturer's recommended intervals and also prior to any long driving trip. Component damage or failure is often caused by a lack of maintenance or low fluid level in the related system. For example, low oil level in the engine can cause major damage to the engine bearings and crankshaft. Future problems also can be prevented by a thorough inspection of the underhood systems, where the discovery of a torn belt or a low fluid level may help to avoid a breakdown on the highway.

Fluids

K10002

Some of the fluids used in a vehicle, such as engine oil, are needed to keep the mechanical systems lubricated and functioning correctly. Other fluids may be safety related, such as windshield washer fluid and brake fluid. Always use the manufacturer's recommended type and amount of fluid when checking these items.

For the engine, the engine oil and coolant levels must be checked. The brake, hydraulic clutch, and anti-lock brake systems all depend on the proper level of brake fluid. Power steering fluid and transmission fluid need to be at the recommended level for these systems to operate properly. The windshield washer is an important safety feature and the fluid level should be checked as needed.

Just as the level of the fluid is important, so too is the quality of the fluid (**FIGURE 10-3**). Nearly all of the fluids in a vehicle, except some automatic transmission/transaxle fluids, get old and wear out, requiring replacement; therefore,

FIGURE 10-1 Regular inspection of a vehicle's systems helps ensure safe and reliable operation.

FIGURE 10-2 A check under the hood is vital to the life and operation of the vehicle.

You Are the Automotive Technician

You work for a car rental company that maintains their vehicles in-house. Your main task today is to perform preventive maintenance on a vehicle in the fleet that is scheduled for its next service. To perform a routine maintenance check, you follow a procedural checklist that includes changing the oil and filter; rotating the tires and inspecting brake, steering, and suspension systems; and checking/inspecting other important parts. During the inspection, you notice that the brake fluid level is low, the wiper blades smear badly, and the left rear brake light isn't working, which requires a bulb replacement.

1. What two conditions can low brake fluid level indicate?
2. How do you determine what engine oil to use?
3. What must be done after an oil change to remind the customer of the next change?
4. What is used to check the oil level after it has been changed?

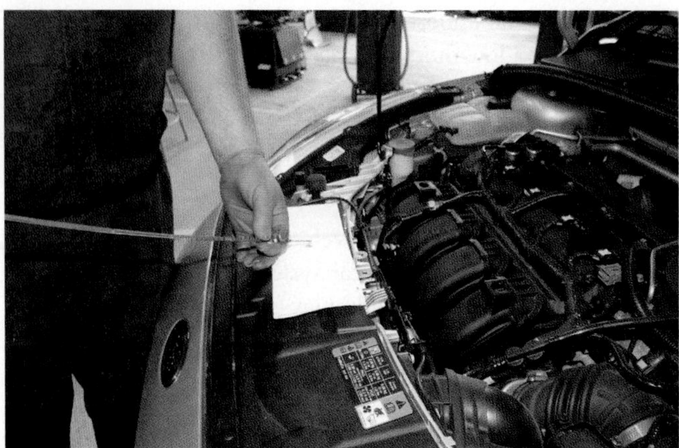

FIGURE 10-3 Checking fluid appearance.

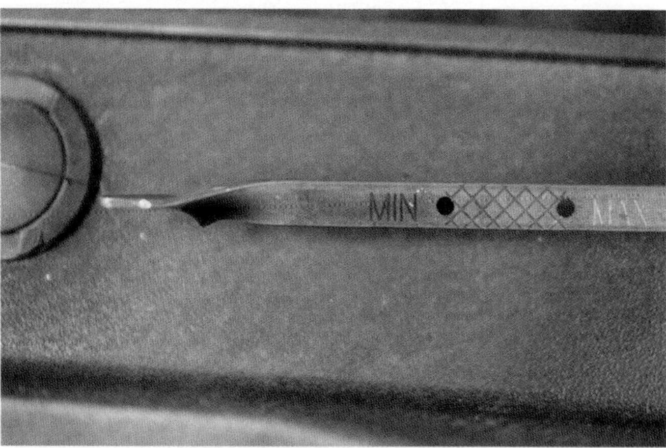

FIGURE 10-4 Checking the level of the engine oil is a vital step in the underhood inspection.

whenever you check the level of any fluid, check the quality of the fluid too. You might notice a change in color, a change in consistency, a mix of fluids, or a change in smell, such as burnt transmission/transaxle fluid. Each fluid shows its age differently, so become familiar with how to identify both good and bad fluids.

Engine Oil

S10002

The level of the oil in the engine's lubrication system is critical to the engine's operation (**FIGURE 10-4**). The engine oil is picked up by the oil pump, filtered through the oil filter, and then sent under pressure to the crankshaft and camshaft bearings. If the level is too low, the oil pump will starve for oil. If the level is too high, the oil will be struck by the crankshaft, churning it into foam. The bearings require a steady flow of oil, not air, for lubrication. If the oil is too low or too high, the engine bearings can be damaged.

The engine oil level should be checked periodically, usually at every other fuel stop as part of a preventive maintenance plan. Always check the oil level when the vehicle is on a level surface, not on a hill or slope. The oil can be checked with the dipstick, which is usually marked "oil" or brightly colored. Always check engine oil with the engine off. Wipe off the dipstick, identify the marks on the dipstick, and reinsert it fully in the tube. Then pull it out and hold it horizontally to read it. The marks on the oil dipstick usually have lines that indicate "full" and "add," or "min" and "max." The difference in quantity between the add and full marks on an engine oil dipstick is typically 1 quart (0.9 liters), but can be as much as 2 quarts (1.9 liters). Refer to the manufacturer's recommendations as noted in the vehicle owner's manual or the manufacturer's published service information.

When checking the oil level, also consider checking whether it is time for an oil change. The manufacturer specifies that the vehicle's **oil-life monitoring (OLM) system** be referenced to determine the percentage of oil life remaining, or specifies when the oil should be changed, depending on the miles traveled or period of time since the last oil change

FIGURE 10-5 Oil-life monitor.

(**FIGURE 10-5**). If the vehicle is due for an oil change, the level should still be checked first. A low reading on the dipstick could indicate that a seal or gasket is leaking or that the engine is using oil. Either of these situations requires further investigation. Refer to the chapter on Engine Lubrication for information on selecting the correct oil for the vehicle being serviced and for skill drills on how to change the oil in the engine lubrication system.

Engine Coolant

S10003

The engine **coolant** level and condition should be checked whenever the oil level is checked. The engine cooling system depends on the coolant to transfer excessive heat from the engine to the radiator. The engine temperature must be controlled to prevent overheating and to maintain proper exhaust emission levels. Some vehicles have a transparent reservoir or surge tank marked with "hot" and "cold," which allows checking the coolant without removing the cap (**FIGURE 10-6**). Some also may have an overflow tank with a tube leading from the radiator cap filler neck to a transparent overflow tank. The level can usually be seen through the side of the tank and

FIGURE 10-6 The coolant reservoir on this vehicle also includes the pressure cap.

compared to the marks. Older vehicles, which do not have an overflow tank, usually have a radiator cap located on top of the radiator that must be removed when checking the coolant level. If there is no overflow tank, then the coolant level in the radiator should be about 1.5" (38 mm) below the filler neck.

> **SAFETY TIP**
>
> Do not remove the pressure cap (radiator cap) when the engine is warm or hot. The system is under pressure, and removing the cap could allow the coolant to immediately boil and spray out, causing severe burns. Always allow the system to cool before removing the cap.

Low coolant levels can cause engine overheating, which leads to damaged cylinder head gaskets or pistons and cylinder walls. The technician should note at each maintenance inspection if any coolant needs to be added and, if so, how much. Coolant levels that are consistently low indicate a coolant leak that must be diagnosed and repaired.

Because the coolant also provides freeze protection, the coolant freeze point also should be checked during the underhood inspection. An insufficient amount of antifreeze could allow the coolant to freeze during temperatures below freezing. Water expands when frozen, so freezing of the coolant could crack the engine block, cylinder head, radiator, or other engine components. The antifreeze protection level can be checked with an antifreeze **hydrometer** **SKILL DRILL 10-1**.

A refractometer also can be used to measure the freeze protection level of coolant. A drop or two of coolant is placed on a sample plate **SKILL DRILL 10-2**.

SKILL DRILL 10-1 Using a Hydrometer to Measure the Freeze Protection Level

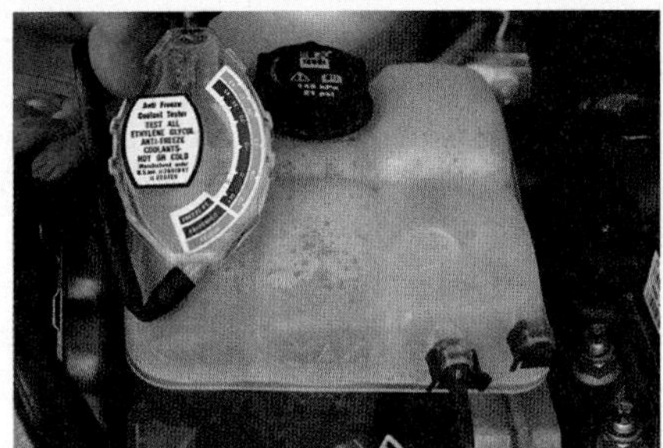

1. Select a hydrometer that is designed to be used with the type of antifreeze being tested.

2. Draw enough coolant into the hydrometer to bring it up to the "fill" line.

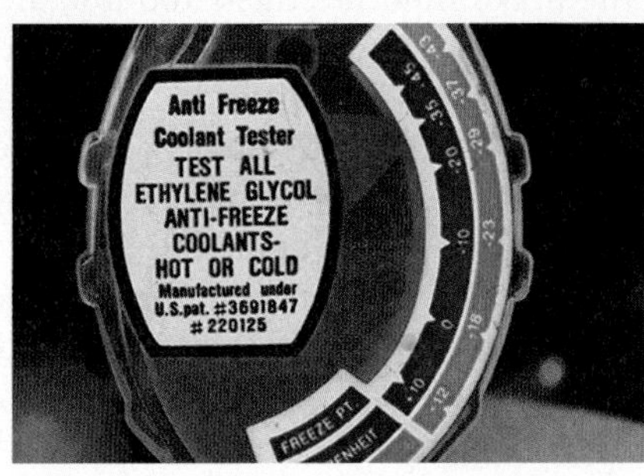

3. Hold the hydrometer vertically and read the freeze protection level.

SKILL DRILL 10-2 Using a Refractometer to Measure the Freeze Protection Level of Coolant

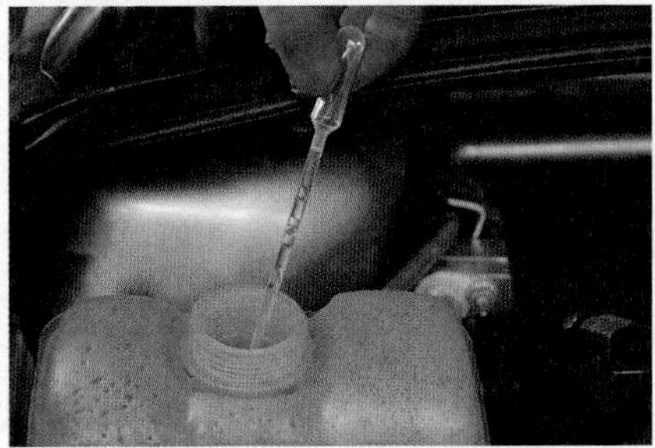

1. Remove the radiator cap and draw out a few drops of coolant.

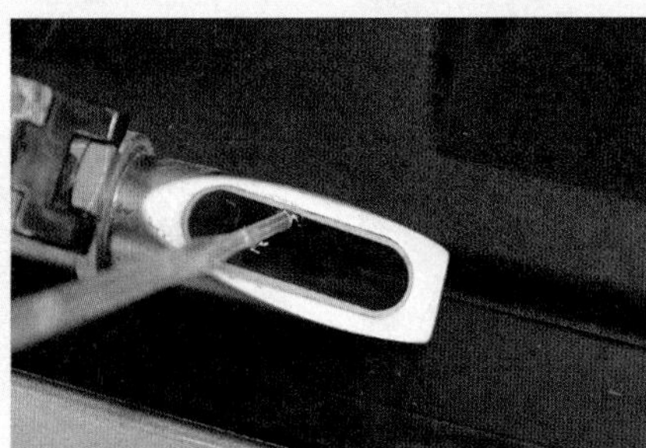

2. Place a drop or two on coolant on the sample plate.

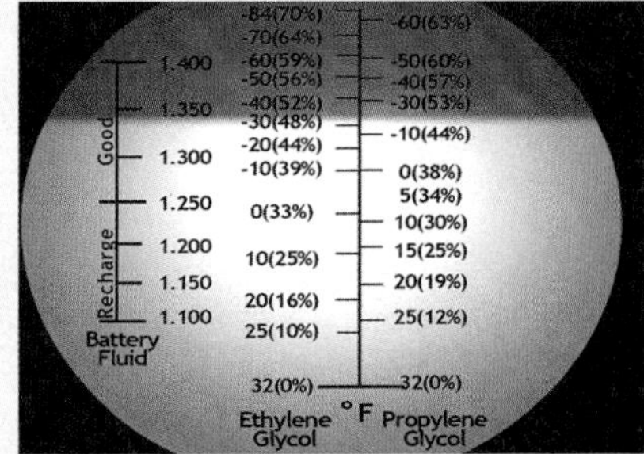

3. Close the diffuser plate, hold the refractometer level, look through the viewfinder, and read the proper scale.

Applied Math

AM-36: Ratios/Percentages: The technician can convert test readings in decimal or fractional form to a ratio or percentage form for comparison with the manufacturer's specifications, and vice versa.

A technician is servicing the cooling system of a vehicle with a 12-quart capacity per service information. Because of the cold climate of the location, the shop supervisor informed the technician to fill the cooling system with a 60/40 mix of antifreeze and distilled water. Considering that the total capacity of the system is 12 quarts, the technician needs to convert the percentages to a fractional form.

This can be done by multiplying the total quantity by each percentage. So 12 quarts × 60% is 12 × 0.60, which equals 7.2 quarts of antifreeze. And 12 quarts × 40% is 12 × 0.40, which equals 4.8 quarts of distilled water. Most climates require a 50/50 mix of antifreeze and water. This can be converted to a ratio or percentage form. Six quarts of distilled water and 6 quarts of antifreeze would be a 1:1 ratio. Concerning the percentage form, we have 50% distilled water and 50% antifreeze. This is often referred to as a 50/50 mix. Refer to the chapter on Engine Cooling for a detailed discussion on the proper type and amount of coolant for the vehicle and how to check and fill the engine cooling system.

In many cases, there are several scales that you can check, such as battery acid, ethylene glycol, and propylene glycol. Make sure you are reading the proper scale.

Brake and Clutch Fluid

S10004

A hydraulic braking system depends on a special fluid called brake fluid. Brake fluid is stored in a reservoir attached to or near to the brake system master cylinder (**FIGURE 10-7**). If the brake fluid level gets too low, air can be pulled into the hydraulic system, which causes the brake pedal to be soft, or too low. This causes the brakes to work poorly or not at all. The brake fluid should be checked whenever the oil level is checked, or at least monthly. Also check the color of the brake fluid. Most brake fluid is very clear. If it is getting dark, it is most likely becoming oxidized or contaminated and needs to be changed. Most brake fluids are **hygroscopic**, meaning they absorb water from the atmosphere. Because of this, most manufacturers recommend changing brake fluid every two to four years. However, the best process is to check the condition of the brake fluid with either a brake fluid tester or brake fluid test strips. Check service information or the vehicle owner's manual for specific recommendations.

▶ TECHNICIAN TIP

Because brake fluid test strips use the concentration of leached copper content in the brake fluid to indicate moisture contamination, hydraulic clutch systems that use plastic master cylinders and clutch lines instead of steel cannot be tested with test strips, so a brake fluid tester should be used in this situation.

Vehicles with a manual transmission may have a hydraulic clutch system that uses both a clutch master cylinder and brake fluid. Depending on the vehicle, the brake master cylinder reservoir also may supply the clutch master cylinder, or the clutch may have a separate fluid reservoir (**FIGURE 10-8**). Be sure to check both of these reservoirs for the proper fluid level. Refer to the chapter on Hydraulics and Power Brakes for an explanation of the different types of brake fluid.

FIGURE 10-7 The brake fluid is stored in a reservoir attached to or near to the brake system master cylinder.

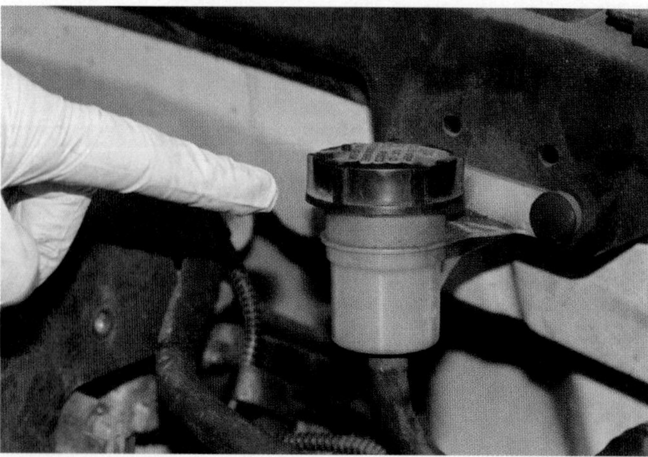

FIGURE 10-8 Check both the brake fluid reservoir and the clutch fluid reservoir in vehicles with manual transmissions.

> **TECHNICIAN TIP**
>
> Low master cylinder brake fluid levels usually indicate one of two possible issues with the system. Either there is a brake fluid leak in the system or, on vehicles equipped with disc brakes, the brake pads are worn. If the brake fluid level is low, inform your supervisor; it may be necessary to inspect the brake lining.

Power Steering Fluid

S10005

Most vehicles are equipped with power-assisted steering systems. The power for the system usually comes from an engine-driven hydraulic pump. Some vehicles, typically hybrid-electric vehicles, use either an electrically driven hydraulic pump or the electric motor directly operating the steering linkage. The pump delivers fluid under pressure to the power unit at the steering box, or rack-and-pinion, through connecting hoses and pipes. The fluid reservoir can be mounted as part of the engine-driven hydraulic pump, or it can be a separate container (**FIGURE 10-9**). The power steering fluid level must be at the proper level to avoid drawing air into the hydraulic system and to prevent fluid overflow when the engine is hot.

FIGURE 10-9 The power steering fluid reservoir. **A.** Mounted on the engine-driven hydraulic pump. **B.** Mounted separately.

FIGURE 10-10 Checking power steering fluid.

In most cases, the power steering fluid level can be checked with a dipstick connected to the filler cap. The engine should be idling and the fluid, hot. In many cases, the dipstick lists both a "cold" and a "hot" level, or a "safe" level (**FIGURE 10-10**). When checking the level, you also should check the appearance of the fluid. Dark or black fluid usually means the fluid is old and needs to be changed. The power steering fluid level should be checked as a normal part of the underhood inspection. Refer to the chapter on Servicing Steering Systems for more information on power steering fluid types and how to check them.

Automatic Transmission/Transaxle Fluid

S10006

The correct automatic transmission/transaxle fluid level (referred to as automatic transmission fluid from here on) is critical to the effective and efficient operation of the transmission. If the level is too low, slipping and shift timing faults can result. Because reverse gear usually requires a larger volume of fluid, a delayed shift into reverse could be an early indication that the fluid is low. If the fluid level is too high, the transmission fluid will churn and aerate, which can lead to low pressures, resulting in slipping clutches. Always make sure the transmission fluid level is correct.

Automatic transmission fluid level is usually checked with a dipstick located under the hood, usually near the front of the transmission. Some automatic transmissions do not have a dipstick; they are checked using a fill plug, or level plug, on the side of the transmission (**FIGURE 10-11**). If the transmission does use a dipstick, the fluid level is usually checked with the engine running, the transmission warmed up fully, and the gear selector in park or neutral, depending on the vehicle. When checking transmission fluid:

1. Wipe off the dipstick (**FIGURE 10-12**).
2. Observe the markings (**FIGURE 10-13**).
3. Reinsert the dipstick fully, hold it horizontal, and read it (**FIGURE 10-14**).

FIGURE 10-11 A few automatic transmissions are checked using a fill plug, or level plug, on the side of the transmission.

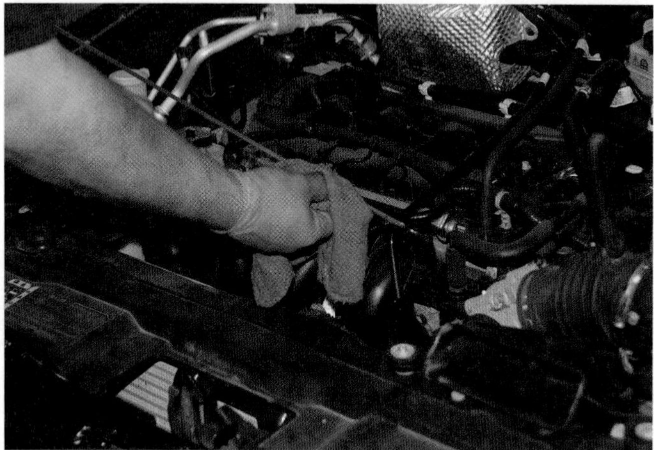

FIGURE 10-12 Wipe off the dipstick.

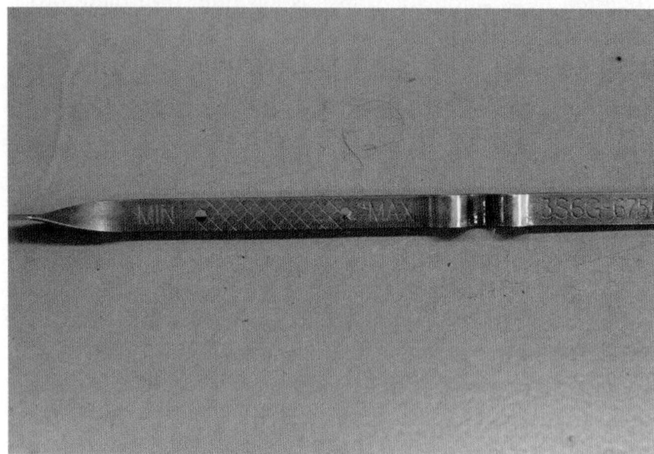

FIGURE 10-13 Observe the markings.

If adding transmission fluid, make sure you select the correct fluid, as there are a number of different transmission fluids specified for various vehicles. Fluid is added through the dipstick tube, so use a funnel. The lines between "full" and "add" are usually only about 1 pint (0.5 liters), so add only a small amount of fluid at a time, checking the fluid level in between.

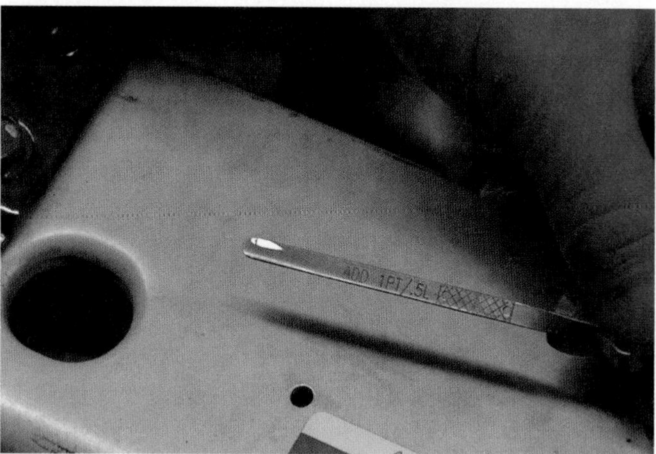

FIGURE 10-14 Reinsert the dipstick fully, hold it horizontal, and read it.

▶ TECHNICIAN TIP

Many newer vehicles—typically European models and some Asian imports—do not have a specified method of checking the transmission fluid level. The transmissions are considered sealed and lubricated for the life of the vehicle. The manufacturers have determined that the transmission fluid will not be low as long as there are no leaks; any leak requires that the transmission be repaired.

In a manual transmission/transaxle, fluid splashing up from the lower gear sets lubricates gears and bearings. Early bearing failure is the result of driving with the transmission fluid level too low. In most cases, transmission fluid level on manual transmissions/transaxles can only be checked from under the vehicle. The vehicle must be level. The level is usually accessed through a fill plug on the side of the transmission/transaxle. If fluid comes out of the fill hole when the plug is removed, allow any extra to drain out, then reinsert the fill plug. If no fluid comes out of the fill hole, carefully stick a finger in the hole and bend your finger down to feel the level of the fluid (**FIGURE 10-15**).

SAFETY TIP

Do not rotate the wheels or engine with your finger in the hole, as transmission parts could pinch or sever your finger.

If the level is within a 0.25" (0.0006 mm) of the bottom of the fill hole, the level is okay, and the fill plug can be reinserted. If the fluid level is lower than that, the proper fluid must be added until the level is even with the bottom of the fill hole. Always check the manufacturer's specifications to identify the proper fluid for the vehicle you are working on. Refer to the chapters on Servicing the Automatic (or Manual) Transmissions for more details about fluid types and checking fluid levels.

Diesel Exhaust Fluid

N10001

Some late-model diesel-powered vehicles use a fluid called **diesel exhaust fluid (DEF)**, one example being AdBlue™.

FIGURE 10-15 Checking the transmission fluid level in a manual transmission.

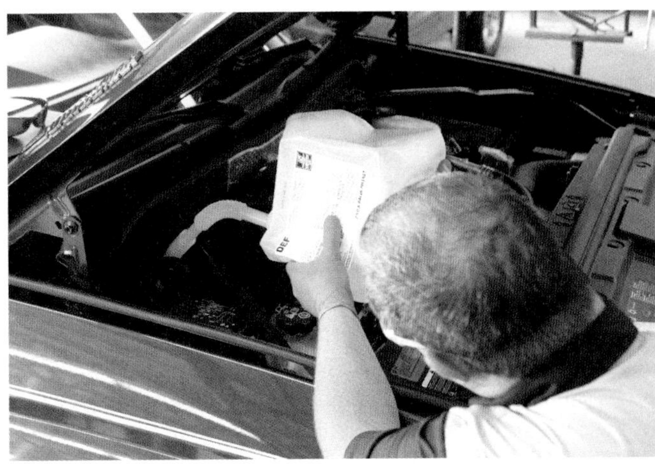

FIGURE 10-16 Adding DEF to the reservoir.

DEF is injected into the exhaust stream to reduce oxides of Nitrogen during certain driving conditions. Because DEF is consumed over time, it has to be replenished periodically, ideally during oil changes. The DEF filler cap is often located under the hood and may be colored blue (**FIGURE 10-16**). *Do not* make the expensive mistake of putting washer fluid (or any other fluid) in the DEF tank.

Windshield Washer Fluid

S10007

The windshield washer system on any vehicle is an important safety feature. Driving in muddy conditions, light mist, fog, or any other condition that causes the windshield to be obstructed requires that the washer system be ready to work when needed. The washer fluid reservoir is normally located under the hood of the vehicle (**FIGURE 10-17**). Some vehicles also may have a separate washer reservoir for the rear wiper, typically located somewhere in the rear hatch or trunk area. Refer to the owner's manual or service information to locate the filler cap.

The windshield washer fluid should be checked whenever the oil is changed or at each vehicle service. Driving in dusty or wet conditions may require that the windshield washer fluid be checked more often. It is important to use properly formulated and mixed washer fluid, especially in freezing weather. The fluid is normally purchased premixed in the proper ratio that protects against freezing.

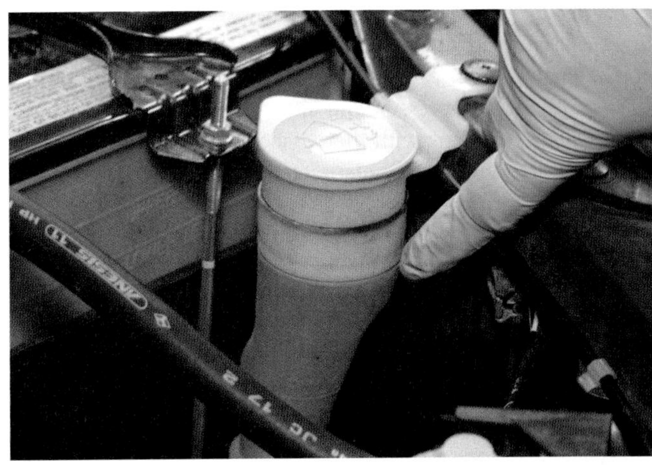

FIGURE 10-17 The washer reservoir is usually located under the hood.

To check and add to the windshield washer fluid, follow the steps in **SKILL DRILL 10-3**.

SKILL DRILL 10-3 Checking and Refilling Windshield Washer Fluid

1. Locate the front windshield wiper fluid container.

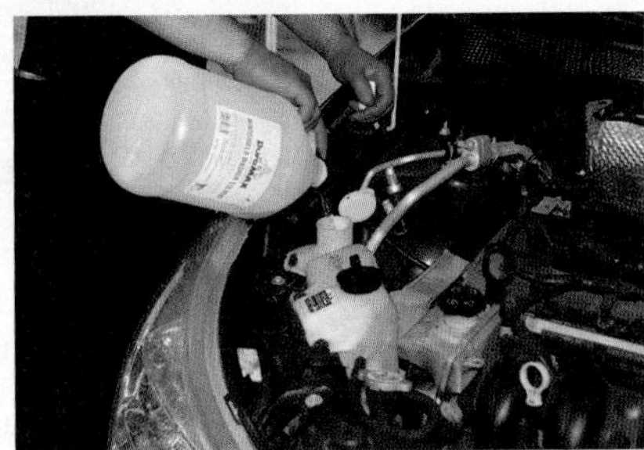

2. Check the windshield washer fluid level. If the level is low, refill the reservoir with the appropriate washer fluid.

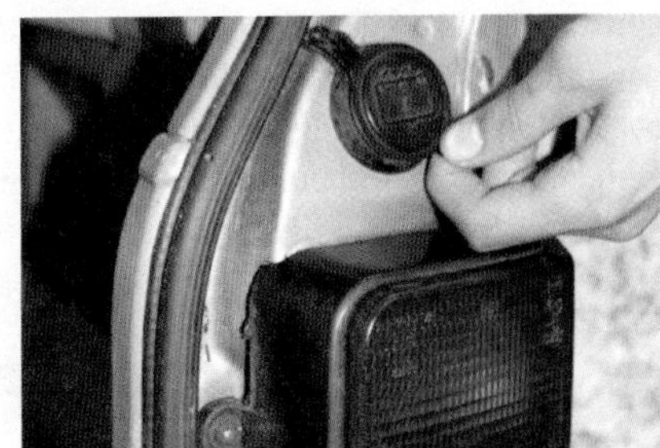

3. If equipped, check and fill the rear window washer reservoir.

> **TECHNICIAN TIP**
>
> Never use laundry or dishwashing detergent to top off the reservoir, as the chemicals in the detergent can damage the vehicle's paint.

Belts and Hoses

S10008

The engine drive belts and coolant hoses are maintenance items that should be inspected and replaced according to the manufacturer's service schedule. They do not last the life of the vehicle. Whenever the vehicle is being serviced, the belts and hoses should be checked. On most vehicles, the failure of a belt or hose when the vehicle is being driven will result in a breakdown, which could include overheating of the engine, causing damage to the engine.

Engine Drive Belts

Engine drive belts are used to operate the various accessories on the engine, such as the water pump, power steering pump, air conditioner compressor, and alternator. As the vehicle ages, these belts wear, along with their associated idler and tensioner pulleys. Some manufacturers recommend that the belts be replaced at about five years of age, as part of a preventive maintenance program. A newer belt technology is becoming more common, called stretchy belts or as one manufacturer named theirs, Stretch Fit belts. They do not use a tensioner for tensioning the belt. Their stretchiness applies an appropriate amount of tension to the belt over its useful life, which some vehicle manufacturers claim can be up to 150,000 miles. So always check the manufacturer's scheduled service guide for belt replacement intervals.

There are two main types of accessory drive belts: the V-type and the serpentine type. A V-type belt sits inside a deep V-shaped groove in the pulley. The sides of the V-belt contact and wedge in the sides of the V in the pulley. Serpentine-type belts have a flat profile with a number of grooves running lengthwise along the belt. These grooves are the exact reverse of the grooves in the outer diameter of the pulleys; they increase the contact surface area and prevent the belt from slipping off the drive pulley as it rotates (**FIGURE 10-18**).

FIGURE 10-18 Drive belts. **A.** A V-type belt fits into the V of the drive pulley. **B.** Serpentine-type belts have a flat profile with a number of grooves running lengthwise along the belt.

Check the drive belts whenever the hood is opened for service. The water pump is the most important component driven by the belt, and the engine will quickly overheat if the belt breaks or comes off. If the belts are more than five to six years old, check the manufacturer's replacement schedule in the service information. Most vehicles using a serpentine belt also have a spring-operated tensioner and pulley. This tensioner may have a built-in damper that reduces noise and vibration. Some manufacturers recommend that the tensioner be replaced along with the belt.

> **TECHNICIAN TIP**
>
> Some belts, called stretchy belts or Stretch Fit belts, do not use a method for tensioning the belt. Their stretchiness applies an appropriate amount of tension to the belt over its useful life, which some vehicle manufacturers claim can be up to 150,000 miles. You also need to know that Stretch Fit belts don't use mechanical tensioners.

FIGURE 10-19 A serpentine belt with many cracks should be replaced.

The belts should be checked for the following:

- Cracks—Cracks that exceed a certain number per inch in a belt indicate that the belt may soon fail and should be replaced (**FIGURE 10-19**).
- Oil soaking—A belt that has been soaked in oil will not grip properly on the pulleys and will slip. If the oil contamination is severe enough for this to happen, replace the belt.

- Glazing—Glazing is shininess on the surface of the belt, which comes in contact with the pulley. If the belt is very worn, the glazing could be due to the belt bottoming out, and it should be replaced. If it is not old and worn, glazing could indicate that the belt is not tight enough. Tightening the belt may be all that is necessary, depending on how bad the glazing is.
- Tears—Torn or split belts are unserviceable and should be replaced.
- Bottoming out—When a V-type belt or serpentine belt becomes very worn, the bottom of the V may contact the bottom of the groove in the pulley, preventing the sides of the belt from making good contact with the sides of the pulley grooves. This reduced friction causes slippage. A belt worn enough to bottom out should be replaced (**FIGURE 10-20**).

Hoses

The vehicle usually has at least two large radiator hoses and some smaller heater hoses. At the radiator, there is a large upper radiator hose (near the top of the radiator) and a large lower radiator hose (near the bottom of the radiator). The smaller heater hoses (usually two) run from the engine block, manifold, or water pump to connections at the heater assembly (near the firewall). The engine should be cool when inspecting the hoses. A hot engine has pressure in the cooling system that may make a soft hose feel stiff, when it really may need to be replaced. If the engine is hot, look for bulging in the hoses (**FIGURE 10-21**).

TECHNICIAN TIP

In most cases, hoses need to be felt to determine their condition. They should be neither too hard nor too soft. Also, the relative stiffness should be consistent over the length of the hose. If you feel differences in stiffness, the hose should be replaced.

Changing the Air Filter

N10002

The engine needs a free flow of clean air in order to operate correctly and with low emissions. Dust and grit in the air can be very abrasive and will severely shorten the life of the engine if not filtered out. If the filter element is not fitted correctly and does not seal properly, dirty air can bypass the filter and enter the engine without being filtered.

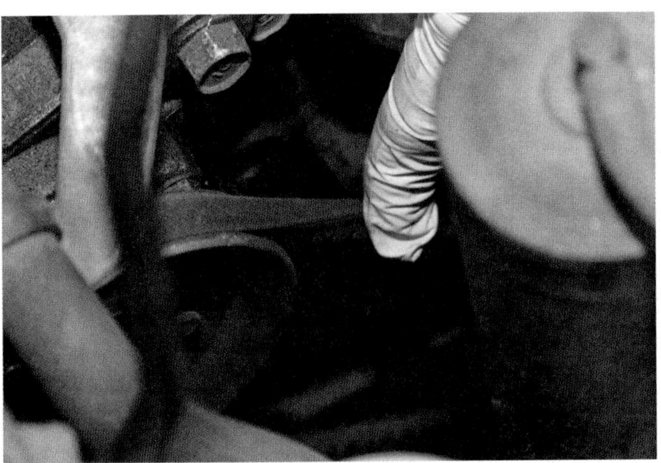

FIGURE 10-20 A V-type belt worn enough to bottom out should be replaced.

FIGURE 10-21 Inspecting coolant hoses.

Applied Math

AM-50: Deductive Reasoning: The technician can identify the specific cause of the problem by generating conclusions based on known symptoms related to the problem.

Drive belt squeal is typically a high-pitched squealing noise caused by the V or serpentine belt slipping against the pulleys. This frequently occurs only when a load is applied to the belt, so may only be noticed when turning with power steering or when air conditioning is turned on. Common causes include worn or age-hardened belts, incorrect belt adjustment, worn tensioner, incorrect length or pitch belt installed, oil contamination of the belt, or the belt pulleys being worn or incorrectly aligned. A practical example may be a car that emits a loud squealing noise occurring only when the steering wheel is turned. A technician can easily reproduce the noise in the shop and can hear that the noise is coming from the drive belt area. A visual inspection determines the condition and tension of the belt. If the engine has more than one belt, a technician can squirt some water on the edges of the belt. If the noise goes away, that is the belt that is squeaking. If the vehicle had a belt noise, make sure that you inspect the tensioner and pulleys for proper alignment and condition.

If you find one defective hose, chances are that the other hose(s) may be deteriorating in the same way and will soon need to be replaced. For this reason, most technicians generally recommend replacing all of the coolant hoses at once as a sensible precaution. This includes the upper and lower radiator hoses and both heater hoses. Also check for any small hoses that may connect components, such as between the water pump and the engine block.

The location of the air filter varies depending on the type of fuel system on the vehicle, so check the vehicle service information or owner's manual for the exact procedure. Some air filters are mounted to the top of the engine, usually found on older vehicles using a carburetor or throttle body fuel injection. The air cleaner on a multiport fuel-injected vehicle is typically located in a rectangular box within the air induction system. While inspecting the air filter, take a look at the air cleaner housing and ductwork for cracks or holes, which would allow unfiltered air to enter the engine.

TECHNICIAN TIP

The paper filter element actually becomes more efficient at filtering dirt particles as it is used. This is because the passageways become smaller as dirt is caught in them, so smaller and smaller dirt particles are caught over time. However, if the filter becomes too clogged, it will restrict air, which reduces engine power output. If it is left too long, the filter can become deformed from the pressure drop across it caused by the restricted air filter.

To change the air filter, follow the steps in **SKILL DRILL 10-4**.

▶ Exterior Vehicle Inspections

S10009

A periodic inspection of the vehicle's exterior can prevent troubles that may cause safety or operational concerns. It is much better to discover a worn tire or broken taillight lens during an inspection than when the car is broken down on the side of the road or the driver is pulled over by the police. A small percentage of owners check their own vehicle for problems, but most depend on the service technician to do it for them. Any maintenance procedure should include this inspection.

TECHNICIAN TIP

The customer or owner of the vehicle can perform an inspection while washing the vehicle. Bad wiper blades, loose trim, and other items may be noticed that are otherwise not seen.

Performing a Visual Inspection of the Vehicle's Exterior

Once a month or prior to any long trip, a vehicle should be checked for overall roadworthiness. Although the vehicle

SKILL DRILL 10-4 Changing the Air Filter

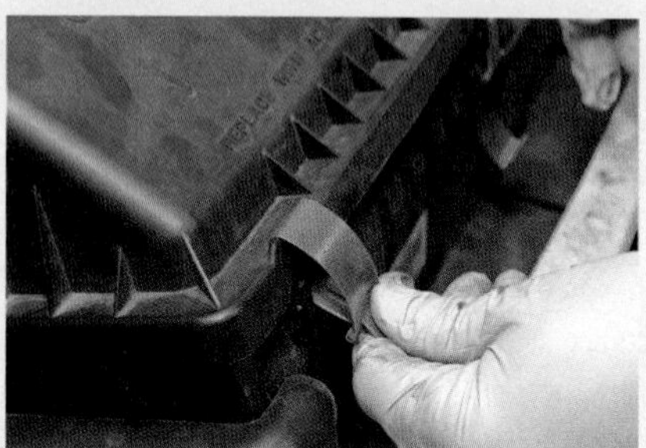

1. On fuel-injected engines, unlatch or unscrew the filter housing fasteners to remove the air filter. It may be necessary to loosen the clamps and hoses on the induction tubing to remove the filter housing cover.

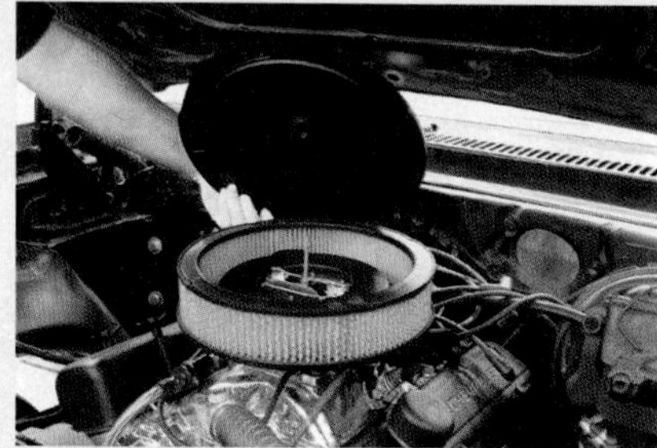

2. On carbureted or throttle body–injected engines, remove the top of the air filter by unscrewing the wing nut, and remove the air filter.

3. Inspect the air cleaner element by holding the filter element up to light and looking through it. If it is bright with no tears or cracks, it can be reused. If it is dark or damaged in any way, it needs to be replaced.

4. Clean the inside of the air filter housing, and inspect it and any ducts for cracks. If the air filter is being replaced, obtain a new air filter and compare it with the old one to ensure that they are exactly the same.

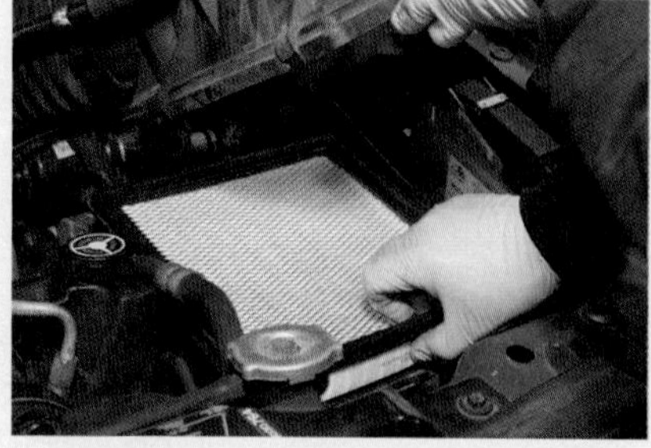

5. Place the new air filter inside the filter housing, making sure it is aligned properly on both sides.

6. Replace the cover of the air filter housing and tighten the latches, screws, or wing nut until completely closed. Reinstall any induction tubing or clamps.

owner may perform this inspection, the service technician more often does it during periodic maintenance of the vehicle. The inspection also should be made any time the vehicle is serviced at the dealer or repair shop. While doing this inspection, the technician should be sure to work in a systematic manner. Using an inspection sheet and inspecting each vehicle the same way ensures that a faulty bulb or other component is not missed.

To perform a visual inspection of the vehicle's exterior, follow the steps in **SKILL DRILL 10-5**.

Checking and Replacing the Wiper Blades

S10010

The windshield wiper blades and arms are an important safety system on every vehicle. Many states with a vehicle inspection program fail a vehicle if the wiper blades are missing, torn, or worn out. The blades, along with the washer system, help the driver to see clearly under all driving conditions.

The wiper blades should be checked as part of the exterior inspection. Usually any wiper blade that is more than a year old is ready for replacement, especially if the vehicle is parked outside. Both the blade and the wiper arm should be checked. The wiper blade should be flexible and not torn. The wiper arm should flex at the hinge and be held firmly against the windshield by the wiper arm spring. The rear wiper blade and arm are checked in the same way.

Never operate the wipers when they are dry because this may damage the blades or scratch the surface of the windshield. Never bend the arms to make better contact with the windshield. The arms are pre-tensioned by the manufacturer, and damage could result. If the arms seem to have lost their spring tension, obtain a suitable replacement.

To check and replace windshield wiper blades, follow the steps in **SKILL DRILL 10-6**.

> **TECHNICIAN TIP**
>
> Many a windshield has been broken while inspecting or replacing windshield wiper blades. The spring holds the wiper arm firmly against the windshield. If you drop the wiper arm while holding it away from the windshield, the spring will snap it against the windshield with enough force to potentially break the windshield, especially if the wiper blade is removed from the arm. You can prevent a broken windshield by making sure the arm is never allowed to slip or by placing a folded up fender cover on the windshield where the wiper blade would hit.

SKILL DRILL 10-5 Performing a Visual Inspection on the Vehicle's Exterior

1. Prepare the vehicle. Park the vehicle in a well-lit area. Turn the engine off, and unlock the doors and trunk or rear hatch.

2. Walk around the vehicle, observing any obvious items that need attention.

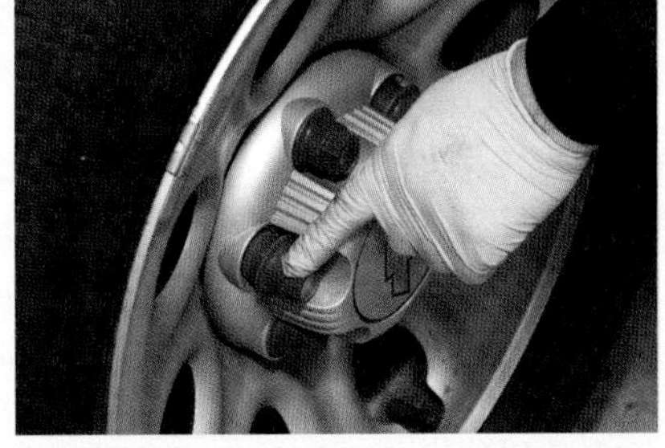

3. Check exterior component and system operation. Check the body condition to make sure all the body components are secure. Look for loose plastic trim.

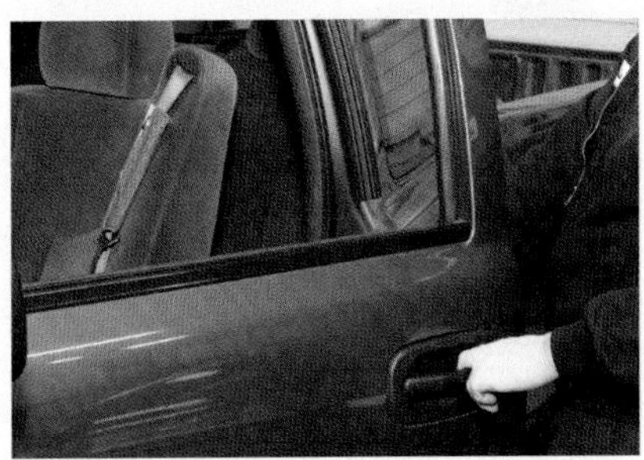

4. Open and close doors to check that they are operating correctly.

5. Push and pull on the bumpers or fenders to ensure that they are secure.

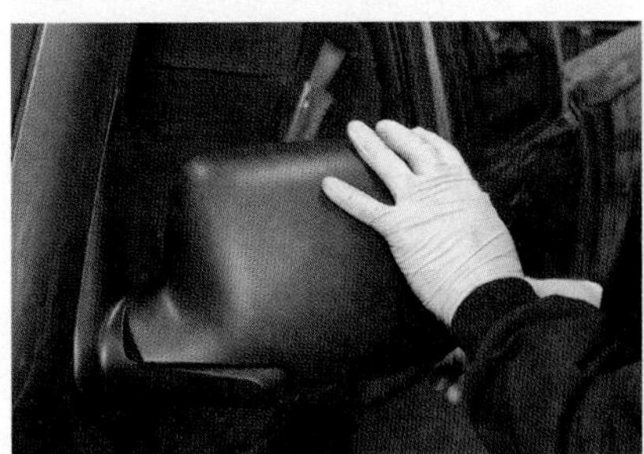

6. Inspect the external mirrors to ensure that they are secure and not broken.

SKILL DRILL 10-6 Checking and Replacing the Windshield Wiper Blades

1. Check the windshield wiper blades. Lift the wiper arm away from the windshield and inspect the condition of the blades. Look for damage or loss of resilience in the material.

2. Wet the windshield with a hose or with the washers and switch the wipers on. If the windshield is being wiped cleanly, do not replace the wiper blades. If the wiper blades are not wiping the glass evenly or are smearing, replace the blades.

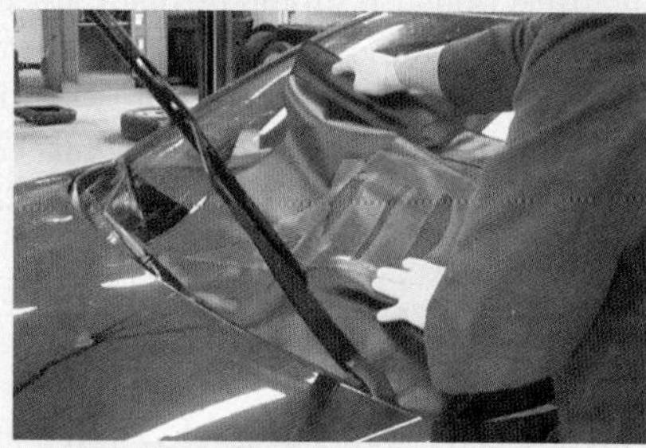

3. Place a folded up fender cover under the wiper blade you are working on to protect the windshield.

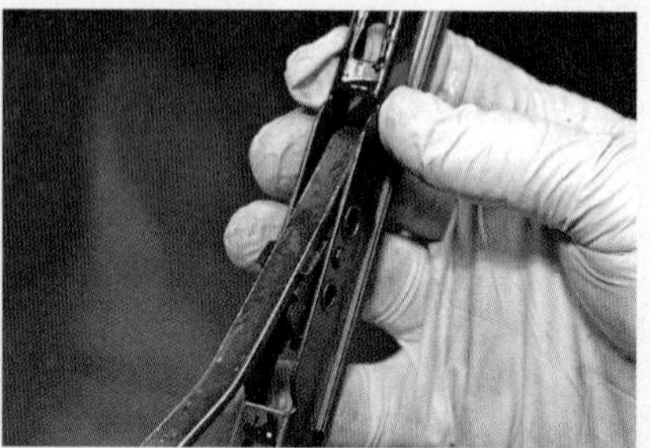

4. Remove the blade assembly.

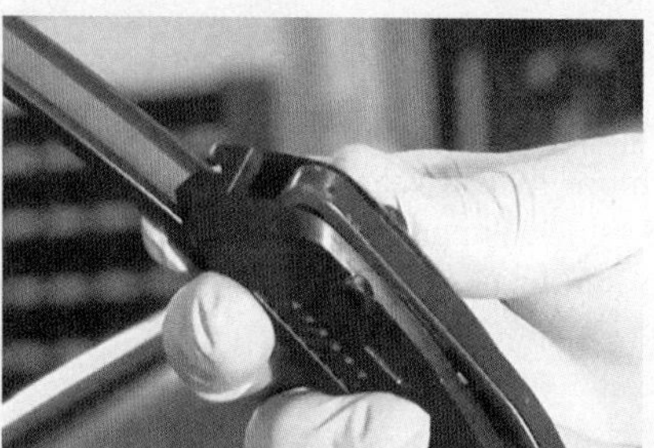

5. Obtain and install the appropriate replacement blades.

6. Once the wiper blade is installed, test them for proper operation.

Inspecting the Windshield

S10011

The windshield should be inspected during the wiper blade inspection process. Scratched, scored, or pitted glass will not wipe clean, even with new blades. Windshields may become etched or pitted, causing the wipers to function poorly. This is a safety hazard and should be repaired. The glass in some cases may be polished to repair the condition, or it may have to be replaced. Some small chips or cracks can be repaired with special resins and tools. These services can be performed by an automotive glass repair service or at some collision repair shops. Large chips or cracks longer than 3" (76 mm) may be reasons to replace the windshield.

To inspect the windshield, follow the steps in **SKILL DRILL 10-7**.

SKILL DRILL 10-7 Inspecting the Windshield

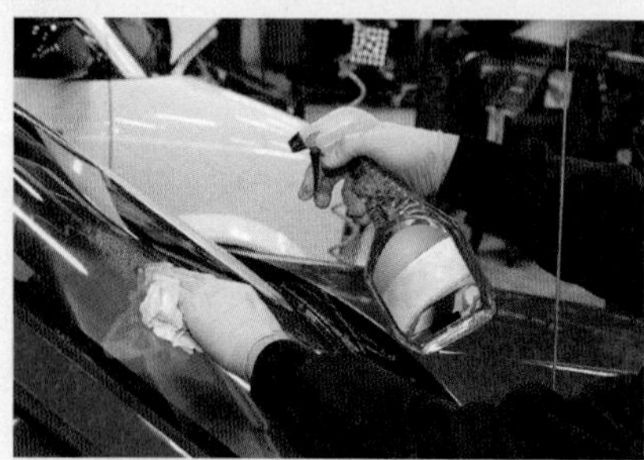

1. Prepare the windshield. First, use glass cleaner to clean the windshield thoroughly.

2. Inspect the glass. Look closely at the surface of the glass. It may help to use a flashlight or trouble light at an angle while inspecting.

3. Look for any delamination conditions present.

Inspecting the Exterior Lights

S10012

The lighting system allows the driver to see the road and sides of the road when driving at night or in poor-visibility conditions, and to signal to other drivers. These lights need to be checked periodically as they do burn out on occasion. The exterior lighting system includes the headlights, taillights, turn signals, side markers, brake lights, license plate lights, and backing lights. Some vehicles may have cornering lights, driving lights, or fog lights. Note that the rear lights may have three or more bulbs per side; be sure to check that they all are working.

To inspect the exterior lights, follow the steps in **SKILL DRILL 10-8**.

Inspecting the Tires

S10013

Tires should be inspected as part of the vehicle inspection process. They are checked for pressure, wear patterns, cuts, and tread depth (**FIGURE 10-22**). The tires and their condition are one of the most important safety considerations on the vehicle. A tire worn to a minimum tread depth may work fine on dry pavement but be dangerous in wet or snowy weather. Keep this in mind as you inspect the tires.

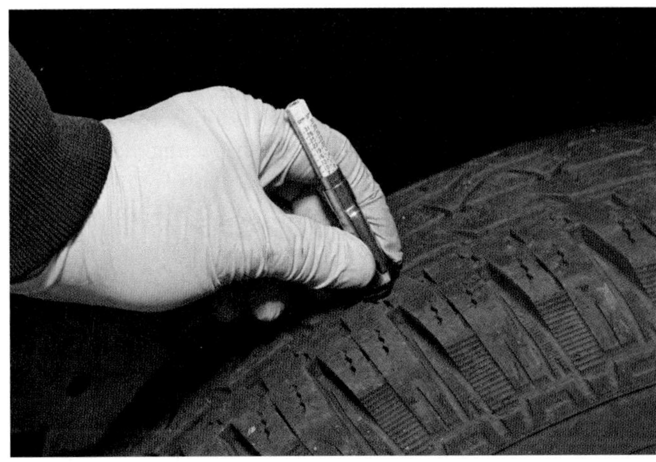

FIGURE 10-22 Tire being checked for wear patterns, cuts, and tread depth.

Tires are inflated using pressurized air or nitrogen to support the weight of the vehicle. Normal tire pressures vary from vehicle to vehicle, according to the vehicle's use. Recommended tire pressures for the vehicle are located on the vehicle manufacturer's tire placard, typically placed on the driver's side door pillar (**FIGURE 10-23**). The maximum tire pressure, located on the tire

SKILL DRILL 10-8 Inspecting the Exterior Lights

1. Park the vehicle inside or in a shaded area. Have someone stand behind the vehicle to report any problems while you turn the ignition on. Switch the lighting switch to the park light position. Check that the taillights and any side markers come on and are equal in brightness.

2. Check the rear license plate lights to be sure they are operating.

3. Put the turn signal switch in the left-turn and then in the right-turn position, and check that the signals flash equally on each side.

4. Depress the brake pedal to make sure the brake lights work. Check that the third (center) brake light works.

5. Make sure the high and low headlight beams work properly.

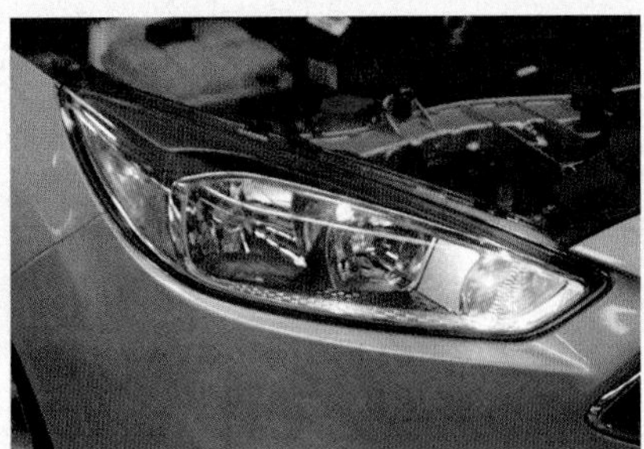

6. Make sure the park lights, side markers, turn indicators, and daytime running lamps (if equipped) are all working properly.

sidewall, is the maximum pressure for that tire, not the pressure for the vehicle (**FIGURE 10-24**). Never inflate the tire above the recommended maximum pressure, as the tire may explode. Refer to the chapter on Servicing Wheels for more information on tires, checking the tire pressure, and inspecting the tire tread.

▶ In-Vehicle Inspections

S10014

Certain in-vehicle inspections and checks should be made as the vehicle is driven into the service bay. Pay attention to the instrument cluster and warning lights. Note how the pedals feel and how the vehicle sounds when first started. Report anything unusual to the shop foreman or supervisor. These checks can be done quickly when the vehicle is first started and then completed as the vehicle is positioned in the stall.

▶ TECHNICIAN TIP

There may be concerns with the brake pedal feel or instrument panel warning lights that the customer is unaware of. The customer is used to the way the vehicle drives and feels; the technician can detect a fault when the customer may think the fault is a normal condition.

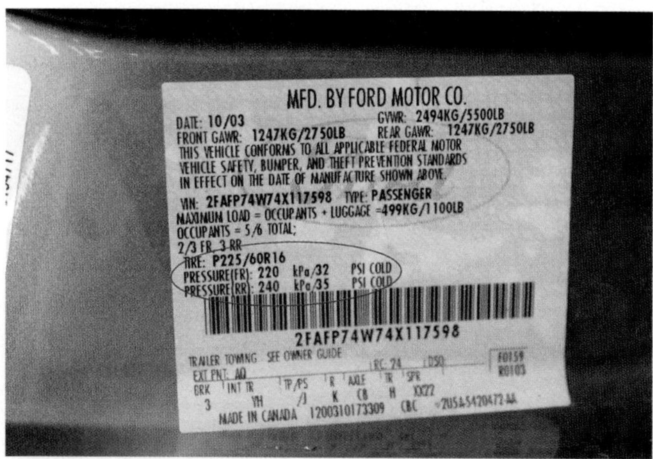

FIGURE 10-23 Typical tire placard.

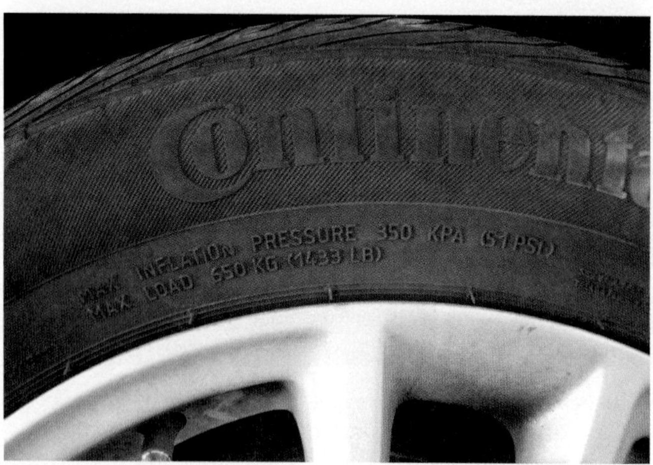

FIGURE 10-24 Maximum tire inflation pressure. (Do not exceed!)

Checking the Brake Pedal

S10015

The brake pedal acts as a lever to increase the force applied to the brake assemblies by the driver. Changes to how far the pedal travels or to its resistance—if it feels harder or softer than normal—can be an indicator of problems such as low fluid levels or even a leak in the hydraulic system. You should always check the brake pedal feel and travel before driving a vehicle into or out of the shop. Also listen for unusual brake noises when driving the vehicle into the shop. High-pitched scraping noises or heavy grinding noises could indicate a worn brake lining. The owner of the vehicle may be used to the feel of a low pedal or the sound of noisy brakes, whereas you will recognize it as an indication of a problem. See the chapter on Hydraulics and Power Brakes for more information on checking the brake pedal.

Checking the Parking Brake

S10016

All vehicles must be manufactured with a foot brake (service brake) system and a parking brake system. Most light vehicles use a foot brake that operates through a hydraulic system on all wheels, and a hand-operated parking brake that acts mechanically on the rear wheels only. The parking brake is used to hold the vehicle in position when parked. The parking brake should be checked as part of a routine safety or vehicle inspection. On vehicles with automatic transmissions, the parking brake may not work at all, and the owner would not know until it is needed. In climates with below-freezing temperatures, the parking brake cable can freeze in the applied position, making it so the parking brake will not release. In this situation, it is probably best not to test the parking brake operation. See the chapter on Disc Brake Systems for more information on checking the parking brake.

▶ TECHNICIAN TIP

If the vehicle does not have a hand- or foot-operated parking brake, check for a "P" button on the console or dash. The vehicle may be equipped with an electronically controlled parking brake.

Checking the Instrument Panel Warning Lamps

N10003

The many **instrument panel warning lamps** can indicate faults with various systems on the vehicle. If a fault is detected, the appropriate warning lamp will be illuminated. Warning lamps can also indicate proper operation of the system. Each warning system performs a self-check each time the ignition is switched on or the engine cranked. You should observe the action of the lights when starting the vehicle to determine whether any additional service may be required. For example, the amber anti-lock brake system (ABS) warning lamp will come on, then stay on for a few seconds, and then go off, indicating that the ABS control module has successfully completed a preliminary self-check.

To check instrument panel warning lamps, follow the steps in **SKILL DRILL 10-9**.

Checking the Horn

S10017

The vehicle horn is usually operated by a relay or by the vehicle **body control module** (BCM). There may be a single horn or a pair of horns, depending on the vehicle. With two horns, one sounds at a lower pitch than the other. The horns are located at the front of the vehicle, behind the grill or bumper. The horn can easily be checked before driving the vehicle into the shop.

To perform a horn check, follow the steps in **SKILL DRILL 10-10**.

Inspecting the Interior Lights

S10018

Interior lights include the courtesy lights, dome lights, and map lights. To inspect the interior lights, follow the steps in **SKILL DRILL 10-11**.

SKILL DRILL 10-9 Checking Instrument Panel Warning Lamps

1. Perform an instrument panel self-test. When the key is switched on (before starting the engine), most of the dash warning lamps will light up as a bulb check. Note any that do not light up as expected.

2. Perform an engine running check. Start the engine and observe the warning lamps. All should go off after a few seconds as the related control module runs a self-check and then commands the lamp to go off.

SKILL DRILL 10-10 Performing a Horn Check

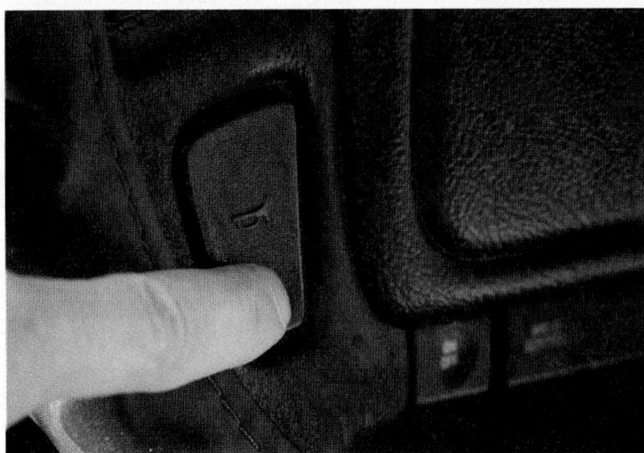

1. Check the vehicle horn. Turn on the ignition, and press the horn button. The horn should sound.

2. If the horn is not working, locate it under the hood with the help of the manufacturer's service information. Check the wiring to make sure it is connected securely.

SKILL DRILL 10-11 Inspecting the Interior Lights

1. Park the vehicle inside or in a shaded area. Using the remote key fob or door key, unlock the doors. On most vehicles, unlocking the doors causes the interior lights to come on. Check that each light works as intended. On some vehicles, the lights do not come on until the door is actually opened. Check each door on the vehicle.

2. Enter the vehicle and close the door. Wait to see that the lights go off after a time. Repeat this check with each door on the vehicle.

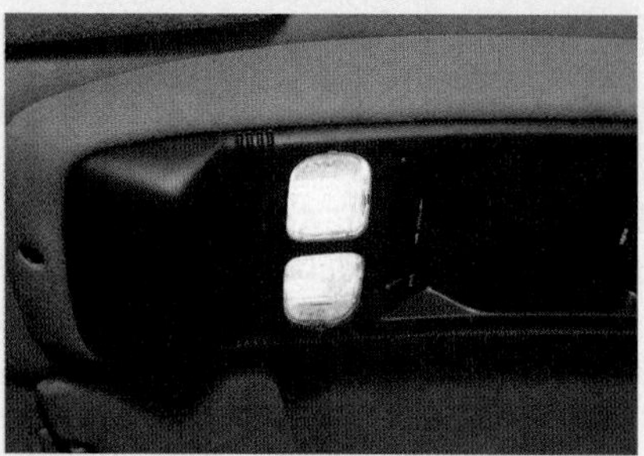

3. Operate the courtesy lights from any other switches. Check the map lights, if they exist.

▶ Under-Vehicle Inspection

S10019

The under-vehicle inspection is a systematic visual inspection of all major vehicle systems. Be prepared to write down any faults, to discuss later with your supervisor. With the vehicle safely lifted on a hoist, an under-body inspection is a good way to get a feel for the overall condition of the vehicle. Additional areas of concern are often found during the inspection, saving the customer an extra trip to the shop for repairs.

The under-vehicle inspection includes the following checks:

- The steering area: The steering area inspection includes the tie-rods and tie-rod ends, idler and pitman arm, steering rack, wheel bearings, and tires and wheels. Suspension bushings, shock absorbers, and brake hoses and lines are also inspected.
- The front-wheel drive axles: Constant-velocity joints (**CV joints**) and dust boots. Look for cracked, torn, or leaking boots.
- The engine area: Check for coolant, oil, and fuel leaks. Look for torn or cracked motor mounts, coolant hoses, and belts.
- The transmission area: This inspection includes the transmission case, clutch housing, external linkages, and wiring connectors. In the transmission area, check for fluid leaks, loose mounting bolts, and faults or looseness in the clutch mechanism or shift linkage.
- The exhaust system: Clamps and bolts may need tightening on the exhaust system or on the exhaust system-to-manifold bolts or gaskets. Check for signs of exhaust leaks, corrosion, or deterioration, including the exhaust hanger hardware. Check the condition of any heat shields.
- The parking brake cables: The cables are encased in a housing that attaches the parking brake lever or pedal to the rear brakes. Check for rusted, frozen, broken, or crushed cables.
- The driveshaft: The driveshaft transmits power from the transmission to the rear axle on rear-wheel drive vehicles. Check for any excess movement in driveshaft universal joints. Look for any dents or bending of the shaft.
- The differential, rear axle, and rear suspension area: The rear axle includes the differential and axle shafts. Look for leaks around the differential, and check the rear shock absorbers, leaf springs, brake hoses, and lines.
- The fuel tank: The fuel tank is metal or plastic, depending on the vehicle. Inspection should include the filler tube and hose, the vent and fuel delivery lines, and the fuel tank straps and protective shields. The fuel tank must be secure and fuel lines inspected for damage or abrasion.

To perform an under-vehicle components inspection, follow the steps in **SKILL DRILL 10-12**.

▶ TECHNICIAN TIP

When checking the fuel tank, any odor of gasoline indicates a leak. Keep checking until you find it. It is not normal for today's vehicles to have any odor of gasoline.

Checking for Fluid Leaks

S10020

The service bay should be clean and dry before driving the vehicle into the shop. This will help you to locate any fluid leaks that may be present on the vehicle. Active leaks leave telltale

SKILL DRILL 10-12 Performing an Under-Vehicle Components Inspection

1. Safely raise the vehicle to a comfortable working height. Work systematically in one direction. Pay particular attention to any fluid leaks.

2. Check the steering area. Locate the tie-rods and twist, push, and pull on them. Grasp the front and rear of the tire and wheel assembly and pivot it to detect lateral movement in the tie-rod end. Grasp the tire at the top and bottom and pivot it to detect movement in the wheel bearings or ball joint. Look for missing or torn rubber boots around the tie-rod ends and steering rack. Check the security of the steering box or rack mountings. Inspect rubber suspension bushings, and check shock absorbers for signs of damage or leaks. Inspect any wiring harness that is accessible for damage. Check the brake hoses and lines for signs of cracking or abrasions.

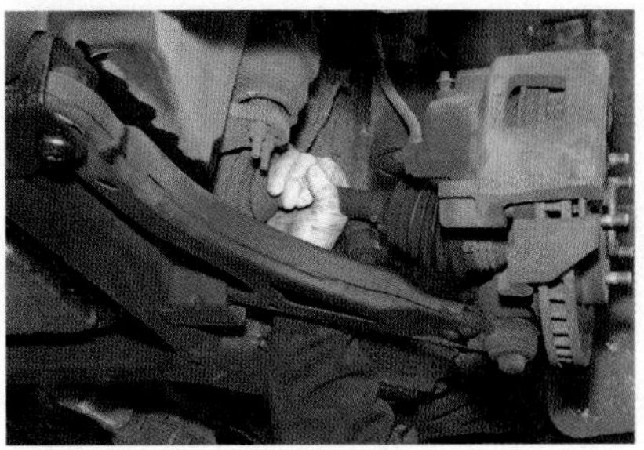

3. Check the front-wheel drive axles. On vehicles with front-wheel drive, check the drive axles for up-and-down looseness. Examine the inner and outer boots for cracks or tears.

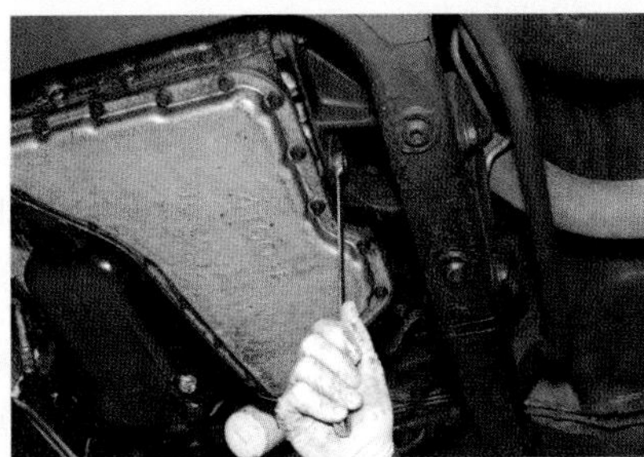

4. Check the transmission area. Look for loose mounting or cover bolts. Trace and record the source of fluid leaks. With a manual transmission, check the clutch operating mechanism for damage. For an automatic transmission, check the shift linkage for any damage. If the transmission is electronically controlled, check the wiring for damage.

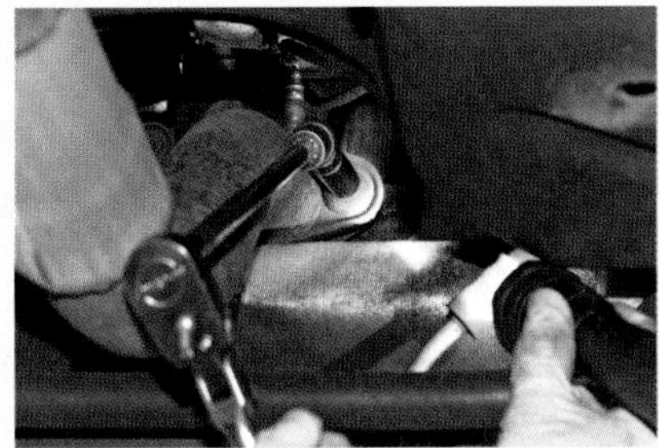

5. Check the exhaust system. Check the tightness of the flange bolts on the engine manifold pipe. Examine the catalytic converter, muffler, and resonator for signs of corrosion or deterioration. Inspect the heat shields. Check the tailpipe for corrosion, and check for looseness in the mounting brackets or hangers.

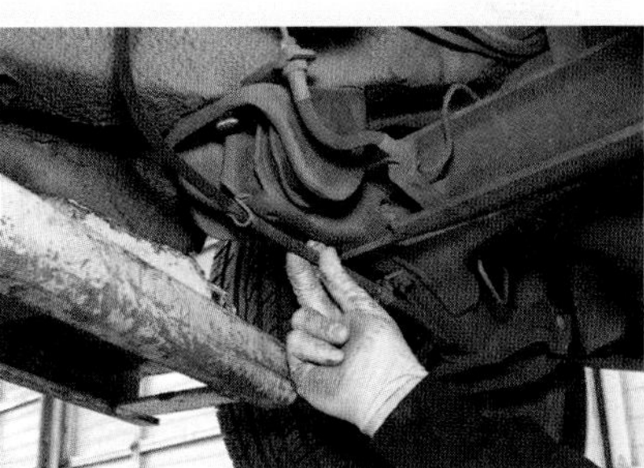

6. Check the parking brake cables. Inspect the hand brake cable to make sure it is not frayed, damaged, or binding. Look for rusted or swollen cable housings. Pull on the cables, and check that the parking brake applies.

SKILL DRILL 10-12 Performing an Under-Vehicle Components Inspection (Continued)

7. Check the drive shaft. On rear-wheel drive vehicles, inspect the drive shaft universal joints for signs of excess movement or rust. To check for wear, rotate the shaft and flange in opposite directions (there should be no movement). On four-wheel drive vehicles, repeat this procedure on the front drive shaft universals.

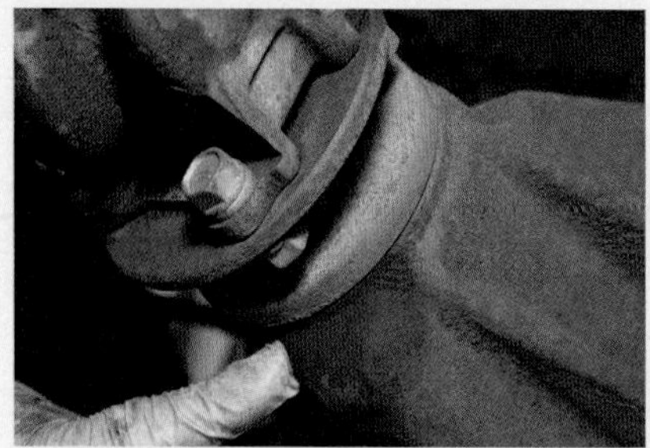

8. Check the differential and rear suspension area. Inspect the pinion shaft oil seal for any signs of leakage. Check the rear shock absorbers for signs of damage or leaks. Tighten the suspension mounting bolts. Inspect the suspension mounting bushings for signs of deterioration or damage. If the vehicle is fitted with leaf springs, inspect the leafs for any cracks or misalignment. On a vehicle with independent suspension, inspect the rear strut assemblies for physical damage or signs of fluid leaks. Inspect the brake hoses for signs of cracking or abrasion.

9. Check the fuel tank area. Tighten the fuel tank mounting bolts or retaining strap bolts. Check all the fuel lines and brake lines for signs of damage, abrasions, leaks, or rust.

drips or puddles on the clean floor, making it easier to identify what may be leaking. Fluids that may be leaking include brake fluid, transmission fluid, power steering fluid, coolant, fuel, and engine oil. Some fluids come in a variety of colors, such as red, which can be used for automatic transmission fluid or antifreeze. Become familiar with each fluid's distinctive colors, feel, or smell:

- Brake fluid—May appear clear or light amber for DOT 3 and DOT 4, whereas DOT 5 is usually purple, slightly slippery, and has an unpleasant, slightly acid-type smell.
- Automatic transmission fluid and some manual transmission fluid—Normally reddish in color, although some manufacturers may use a clear or amber color; it is very slippery and oily and has an oily smell.
- Power steering fluid—Typically red if it uses automatic transmission fluid, or clear in color if power steering fluid.
- Coolant—Normally green, orange, or yellow in color; some manufacturers (e.g., General Motors) use a coolant that is red or light red, slippery, and has a sweet smell, like syrup.
- Engine oil—Brown or black in color, very slippery and a bit thick, with an oily smell.
- Manual transmission fluid—Light brown in color, very slippery and thick (like syrup) with an oily smell.
- Gasoline—Clear in color; evaporates easily, and has a distinctive gas odor.
- Diesel—Dirty clear in color, thin, has an oily smell.

Checking for leaks can be done as part of the under-vehicle inspection, or with a light on a clean floor. The leaks are normally more visible on a warmed-up vehicle, although some coolant leaks only appear when the engine is cold. Discuss with the customer whether there are any unusual smells in the morning when the vehicle is first started.

To check for leaks, drive the vehicle into a clean, well-lit work stall. Ideally, the stall will also have a lift so that the vehicle can be further inspected if a leak is suspected. Use a flashlight or trouble light to inspect the underside of the vehicle for any drips or wet areas. With the engine running, wait a few minutes to see if any leaks appear. Turn the engine off and wait for a time to see if anything drips on the floor. Try to identify the type of fluid that is leaking and the area from which it is leaking. Remember that gravity tends to pull any leaking fluids down, so always look toward the top of the wet area to help determine the source of the leak. If fluid leaks onto a moving part, it can be thrown a good distance, so check for a common source. Lastly, a leak under pressure, such as coolant, can be sprayed a good distance from a small hole, so always use a good light to help identify the location of the leak. Finding the leak's source will tell you which component on the vehicle to inspect more closely.

▶ Wrap-Up

Ready for Review

- A check under the hood is important to the life and operation of the vehicle.
- The correct type and amount of fluid is important for reliable and safe operation of the vehicle.
- The engine drive belts and coolant hoses are maintenance items that should be inspected and replaced according to the manufacturer's service schedule.
- The engine needs a free flow of clean air from the air filter in order to operate correctly.
- A periodic inspection of the vehicle's exterior can prevent troubles that may cause safety or operational concerns.
- Once a month or prior to any long trip, a vehicle should be visually checked for overall roadworthiness.
- The windshield wiper blades and arms are an important safety system on every vehicle.
- The windshield should be inspected during the wiper blade inspection process.
- The lighting system allows the driver to see when driving at night or in poor-visibility conditions, and to signal to other drivers.
- The tires and their condition are one of the most important safety considerations on the vehicle.
- Certain in-vehicle inspections and checks should be made as the vehicle is driven into the service bay.
- The technician should always check the brake pedal feel and travel before driving a vehicle into or out of the shop.
- The parking brake should be checked as part of a routine safety or vehicle inspection.
- The many instrument panel warning lamps can indicate faults with various systems on the vehicle.
- The horn can easily be checked before driving the vehicle into the shop.
- With the vehicle safely lifted on a hoist, an under-body inspection is a good way to get a feel for the overall condition of the vehicle.

Key Terms

body control module (BCM) An onboard computer that controls many vehicle functions, including the vehicle interior and exterior lighting, horn, door locks, power seats, and windows.

Coolant An antifreeze concentrate mixed with water, called engine coolant. Most manufacturers recommend a 50/50 mixture.

CV joint An abbreviation for constant velocity, a type of universal joint used on the drive axles or half-shafts of a vehicle Usually refers to front-wheel drive vehicles.

diesel exhaust fluid (DEF) A mixture of urea and water that is injected into the exhaust system of a late-model diesel-powered vehicle to reduce exhaust oxides of Nitrogen emissions.

Hydrometer A tool that measures the specific gravity of a liquid.

hygroscopic A property of a substance or liquid that causes it to absorb moisture (water), as a sponge absorbs water. Brake fluid absorbs water out of the air; thus it is hygroscopic.

instrument panel warning lamps Lamps that illuminate to warn a driver of a fault in a system.

oil-life monitoring (OLM) system A system that displays the useful life remaining in the engine oil.

Review Questions

1. All of the following statements are true *except*:
   a. Engine oil level should be checked preferably at every other fuel stop.
   b. If the engine oil level is too high, it will foam.
   c. When checking engine oil level, hold the dipstick vertically to read it.
   d. If the oil level is too low or too high, the engine bearings can be damaged.
2. Removing the radiator cap when the engine is warm or hot may cause the coolant to:
   a. evaporate and decrease in quantity.
   b. cool down.
   c. be contaminated.
   d. boil and spray out.
3. The antifreeze protection level can be checked with an antifreeze:
   a. hydrometer.
   b. tachometer.
   c. OLM system.
   d. hydraulic pump.
4. Most brake fluids:
   a. are black in color.
   b. do not wear out.
   c. absorb water from the atmosphere.
   d. do not decrease in quantity.
5. In which of these states should the automatic transmission fluid level be checked?
   a. Engine running and the fluid cold
   b. Engine idling and the fluid hot
   c. Engine running and in drive gear
   d. Engine idling and in reverse gear
6. A delayed shift into reverse could indicate that the transmission/transaxle fluid:
   a. level is too high.
   b. is contaminated.
   c. is cold.
   d. level is low.
7. The belt causing drive belt squeal can be spotted by:
   a. replacing the tensioner.
   b. squirting some water on the belt.
   c. turning on the air conditioner.
   d. loosening the belt.

8. The following statements are true *except*:
   a. While checking, operate the wipers when they are dry.
   b. The wiper blades should be checked as part of the exterior inspection.
   c. Never bend the arms to make better contact with the windshield.
   d. The wiper blade should be flexible and not torn.
9. All of the following systems should be checked as part of an under-vehicle inspection *except*:
   a. the exhaust system.
   b. the electrical system.
   c. the driveshaft.
   d. the steering area.
10. Which of these is normally reddish in color?
   a. Engine oil
   b. Brake fluid
   c. Automatic transmission fluid
   d. Manual transmission fluid

ASE Technician A/Technician B Style Questions

1. Tech A says that engine oil should be checked with the engine idling. Tech B says that the vehicle should be on a level surface when checking the oil level. Who is correct?
   a. Tech A
   b. Tech B
   c. Both A and B
   d. Neither A nor B
2. Tech A says that coolant freeze protection can be measured with a hydrometer. Tech B says that coolant freeze protection can be measured with a refractometer. Who is correct?
   a. Tech A
   b. Tech B
   c. Both A and B
   d. Neither A nor B
3. Tech A says that most brake fluids absorb water from the atmosphere. Tech B says that brake fluid should be changed every two to four years. Who is correct?
   a. Tech A
   b. Tech B
   c. Both A and B
   d. Neither A nor B
4. Tech A says that improper handling of a windshield wiper can lead to a broken windshield. Tech B says that dishwashing detergent works well as washer fluid. Who is correct?
   a. Tech A
   b. Tech B
   c. Both A and B
   d. Neither A nor B
5. While servicing an automatic transaxle–equipped vehicle, Tech A says that some of these vehicles do not have a transaxle dipstick. Tech B says that the lines between "full" and "add" on the dipstick are usually about 1 qt (0.95 liters). Who is correct?
   a. Tech A
   b. Tech B
   c. Both A and B
   d. Neither A nor B
6. Tech A says to test new wiper blades against a dry windshield to ensure they seat properly. Tech B says that if the wiper blade don't seat properly, bend the arms to make better contact with the windshield. Who is correct?
   a. Tech A
   b. Tech B
   c. Both Tech A and Tech B
   d. Neither Tech A nor Tech B
7. Tech A says that Stretch Fit belts apply an appropriate amount of tension to the belt over its useful life. Tech B says that any cracks in a belt mean it needs to be replaced. Who is correct?
   a. Tech A
   b. Tech B
   c. Both A and B
   d. Neither A nor B
8. Tech A says that the minimum pressure the tires should be inflated to is listed on the tire sidewall. Tech B says that the specified tire pressure is listed on the tire placard, usually located on the driver's door pillar. Who is correct?
   a. Tech A
   b. Tech B
   c. Both A and B
   d. Neither A nor B
9. Tech A says that changes to how far the brake pedal travels can be an indicator of problems in the hydraulic system. Tech B says that high-pitched scraping noises or heavy grinding noises during braking could indicate a worn brake lining. Who is correct?
   a. Tech A
   b. Tech B
   c. Both A and B
   d. Neither A nor B
10. Tech A says that when the engine is started, the amber anti-lock brake system (ABS) warning lamp should come on, stay on for a few seconds, and then go off, indicating a successfully completed preliminary self-check. Tech B says that if a fault is detected in the system, the warning lamp will stay illuminated. Who is correct?
   a. Tech A
   b. Tech B
   c. Both A and B
   d. Neither A nor B

CHAPTER 11

Communication and Employability Skills

NATEF Certification Standards

Personal Standards (see Standard 7.9)

1. Reports to work daily on time; able to take directions and motivated to accomplish the task at hand.
2. Dresses appropriately and uses language and manners suitable for the workplace.
3. Maintains appropriate personal hygiene.
4. Meets and maintains employment eligibility criteria, such as drug/alcohol-free status, and clean driving record, etc.
5. Demonstrates honesty, integrity, and reliability.

Work Habits/Ethic (see Standard 7.10)

1. Complies with workplace policies/laws.
2. Contributes to the success of the team, assists others, and requests help when needed.
3. Works well with all customers and coworkers.
4. Negotiates solutions to interpersonal and workplace conflicts.
5. Contributes ideas and initiative.
6. Follows directions.
7. Communicates (written and verbal) effectively with customers and coworkers.
8. Reads and interprets workplace documents; writes clearly and concisely.
9. Analyzes and resolves problems that arise in completing assigned tasks.
10. Organizes and implements a productive plan of work.
11. Uses scientific, technical, engineering, and mathematics principles and reasoning to accomplish assigned tasks.
12. Identifies and addresses the needs of all customers, providing helpful, courteous, and knowledgeable service and advice as needed.

Knowledge Objectives

After reading this chapter, you will be able to:

- **K11001** Describe active listening.
- **K11002** Describe the components of the listening process.
- **K11003** Describe how empathy should be used when listening.
- **K11004** Describe how nonverbal feedback should be used during listening.
- **K11005** Describe how verbal feedback should be used during listening.
- **K11006** Discuss effective speaking strategies.
- **K11007** Describe each type of question and its application.
- **K11008** Describe proper telephone skills.
- **K11009** Describe how to give and receive instructions.
- **K11010** Describe how to communicate in a team.
- **K11011** Describe employability skills.
- **K11012** Describe the requirements for employment.
- **K11013** Describe the employment requirements for appearance and environment.
- **K11014** Describe the requirements for management of time.
- **K11015** Describe the customer service skills needed for employment.
- **K11016** Describe effective reading skills.
- **K11017** Describe reading comprehension.
- **K11018** Describe the process of researching and using information.
- **K11019** Use effective writing skills.
- **K11020** Describe the characteristics of good written business correspondence.

Skills Objectives

After reading this chapter, you will be able to:

- **S11001** Complete a repair order.
- **S11002** Complete a shop safety inspection form.
- **S11003** Complete a defective equipment report.
- **S11004** Complete an accident report.
- **S11005** Complete a vehicle inspection form.
- **S11006** Complete a lockout/tagout form.

▶ Introduction

Although we've been communicating all of our lives, most of us aren't aware of the listening, reading, writing, and speaking skills needed to be a good communicator. It doesn't require any extra effort to communicate well, once you know the principles behind each communication skill. Learning and applying good communication skills will save you time and help you avoid or get through tricky situations. These skills build over time, and you will find that you learn something new every day when you encounter new situations or meet new people. It's a lifelong learning process to perfect your communication skills (**FIGURE 11-1**).

Because communication is an essential workplace skill needed to function successfully in the automotive service facility, this entire chapter is dedicated to communication and employability skills. This chapter describes the steps to becoming an effective communicator, offers tips on how to be a good listener—the first step in good communication—and explains how to speak to both customers and your coworkers. Along the way, we discuss the requirement for writing and preparing documentation used in the workplace.

Note: The majority of NATEF's Applied Academic Skills for Communication are covered in the context of this chapter.

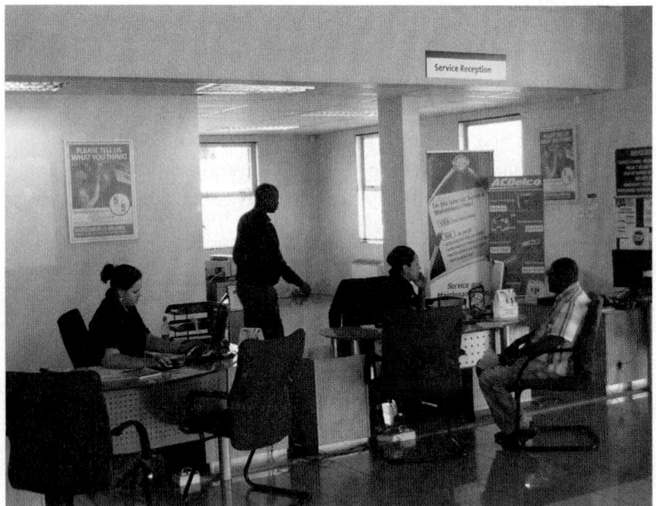

FIGURE 11-1 A service advisor listening to a customer describe their problem.

Although the applied science and math boxes found in other chapters provide examples of how to apply concepts to everyday activities in the shop, this chapter provides examples of how to apply the concepts of good communication in everyday life.

▶ Active Listening

K11001

Active listening is an essential skill. We may hear what someone is saying, but are we truly understanding what the person is trying to communicate? The listening process can be difficult to perfect but is one of the most important skills to possess when gathering information from a customer or any other person. The active listener focuses all of his or her attention on the speaker, including verbal and nonverbal messages. When appropriate, the active listener encourages the speaker to further communicate and clarify details that may have otherwise been left out.

Applied Communication

AC-15: Listening/Reading/Speaking/Writing: The technician identifies and uses effective strategies for listening, reading, speaking, and writing when dealing with customers, coworkers, and supervisors.

AC-34: Listening: The technician adapts a listening strategy that will obtain the information required for solving the problem.

The Listening Process

K11002

To be a good listener, we need to be aware of barriers that can disrupt the listening process. These barriers can be mental and physical. Mental barriers are thoughts and feelings that interfere with our listening, such as our own assumptions, emotions, and prejudices. To fully absorb what someone is telling us, we need to learn to set these feelings aside. It takes effort and isn't always easy, but it is important to keep an open mind throughout the listening process. For example, when listening to a customer who is describing a concern about his or her vehicle, it is good

You Are the Automotive Technician

During your second week on the job, your supervisor has pulled you off the shop floor and into the service department office. You will be shadowing the service manager, Bob, to gain firsthand experience communicating effectively with customers in person. The first customer has an appointment with the service department for a safety recall issue. Bob asks questions to gather all the necessary information from the customer, while making eye contact with him. As a new service technician, you will practice writing up your own work orders for Bob to critique after the customer has been taken care of. After the customer has answered all the questions, Bob reviews and reiterates to the customer his understanding of the concern, to be sure it is accurate. Bob writes up a work order, reviews the concern and agreed-on course of action with the customer, and then guides the customer to the service department lounge to wait for his vehicle.

1. What are the three types of "right questions" to ask when gathering information?
2. What are the three Cs all work orders should contain?
3. Explain why empathy is an important skill to have when dealing with upset customers.

practice to allow the customer to fully complete the thought, even if you believe you have all the information you need. Keeping an open mind is also an important first step in practicing empathy, discussed later.

As a listener and active participant, we can encourage the flow of communication by welcoming the speaker, letting him or her know that we care enough to want to understand the message, and acknowledging the speaker's feelings and concerns. To accomplish that, give the speaker your undivided attention by practicing these steps:

- Stop what you are doing.
- Remove as many distractions as possible.
- Focus on him or her.
- Look the speaker in the eye.

As the person is speaking, you also need to provide listening feedback, which indicates to the speaker that you are engaged in what he or she is saying. Feedback can be nonverbal (e.g., facial expressions, body posture) and verbal. Nonverbal language, such as nodding while listening and gesturing while speaking, is used to reinforce or add emphasis. If, while speaking, you send a conflicting message, such as looking annoyed while stating that you value the customer's opinion, the person may tend to believe the nonverbal message instead of the verbal one, even though it is only half of the total message. For example, a customer has his car, which will not start, towed in to your shop. He tells you that the cause is a bad headlight. You may be tempted to roll your eyes at this assertion, but regardless of your personal opinion, you must, as a professional, maintain a sincere and attentive attitude toward the customer. Because nonverbal communication is perceived to be more spontaneous and less conscious, most people tend to believe the nonverbal message more than the actual words expressed. In this case, rolling your eyes would likely override any "correct" words you might say.

> ▶ TECHNICIAN TIP
>
> In many situations, nonverbal communication can be more important than the verbal message itself. So make sure that your non verbal message matches the verbal message you are communicating.

Empathy

K11003

Empathy can help us to avoid selective hearing. To empathize with someone is to attempt to see the situation from that person's point of view. Empathy requires good listening skills, which include an effective use of verbal and nonverbal listening feedback, in order to take in and consider the message without applying our own biases. True empathy means breaking down or putting aside existing mental barriers. For example, when customers bring their vehicles in with a fault requiring an expensive repair that is not in their budget, we can come across as uncaring if we present them with a take-it-or-leave-it attitude. Expressing your understanding of the customer's situation and seeking to explore valid options to resolve the issue go a long way in building trust with the customer. Remember, we can empathize with people even if we do not agree with them. Also, we don't have to take responsibility for their problem. But we can at least attempt to help them find the best possible solution that will work for them. You will know that you have demonstrated empathy when no viable options were found, but the customer still thanks you sincerely for your assistance.

Empathy not only helps you understand the message better but also helps you to be less reactive in a negative way as you realize you could feel the same way in that person's situation. It may also help motivate you to find a better resolution of the situation, knowing that it could be you in the very same circumstances.

Nonverbal Feedback

K11004

> **Applied Communication**
>
> ***AC-35: Nonverbal and Verbal Cues:*** The technician uses verbal and nonverbal cues in discussion to help identify, verify, and solve problems.

Nonverbal feedback can be a very useful tool when listening. Your body position, eye contact, and facial expression can all set the direction of a conversation. We use body language to help emphasize our message, and when listening, we can use it to provide listening feedback. When listening to a speaker, try to sit or stand upright, while making eye contact (**FIGURE 11-2**). Try to avoid folding your arms, as this can be perceived as either an aggressive or defensive stance. At the same time, do not act too casual, for example, by standing with your hands in your pockets. It may seem harmless, but a customer or supervisor would likely see such a posture as a sign of disrespect or disinterest. Imagine if you were talking to a service manager during a job interview, and the manager leaned back and put his feet up on the desk. That one gesture would send you a message about the manager's level of professionalism and his respect for you and overall interest in the conversation, and in turn would have an effect on the way you viewed the meeting.

Maintaining some eye contact during the conversation, and not looking down at something or someone else, also lets the speaker know you are paying attention. If you are taking notes, be sure to look up periodically and make eye contact with the person.

Just as the speaker is noting our facial expressions, remember to pay attention to the speaker's facial expressions to get a clearer sense of the message being delivered. If, for example, a customer scrunches up her face while talking about the squealing noise the brakes are making, you can probably assume the customer would be happy if you could make that noise go away.

When we say that space is a part of nonverbal language, we are referring to the physical space between the people with whom we are communicating and ourselves, commonly known as "personal space" (Metric Equivalent for 3' is 0.76 metres.).

FIGURE 11-2 Nonverbal contact. **A.** Good. **B.** Bad.

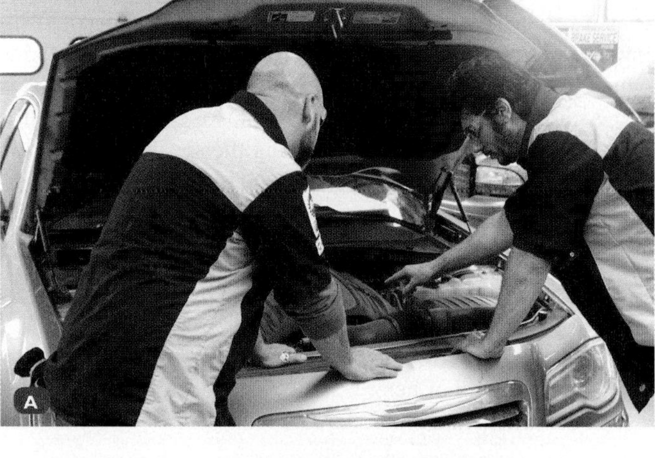

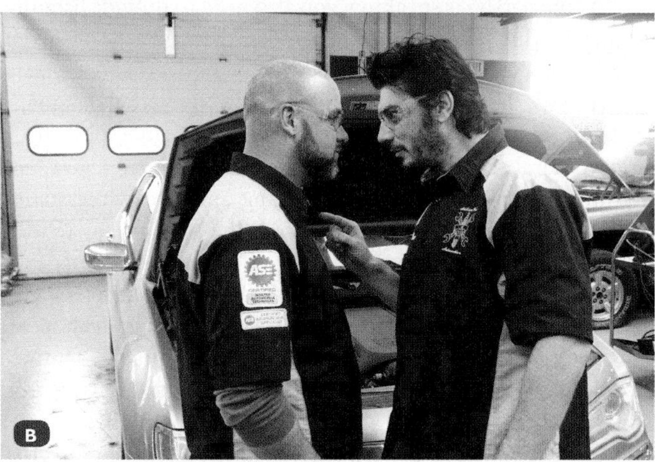

FIGURE 11-3 Personal space. **A.** About right. **B.** Too close.

How we use personal space depends on how we feel about others; it is also a consideration for another person's comfort level. Familiarity, gender, status, and culture determine the use of our personal space. Be aware of space as a nonverbal message, and adjust accordingly (**FIGURE 11-3**).

Verbal Feedback

KI1005

Verbal feedback includes very simple signals that can enhance the conversation and let the person know you comprehend. Some examples are the use of validating statements and supporting statements.

A **validating statement** shows common interest in the topic being discussed. A phrase such as "I see" or "Tell me more" indicates that you are paying attention. A validating statement also helps show empathy for the person speaking and can be a simple "I understand" or "That must be frustrating for you."

A **supporting statement** can urge the speaker to elaborate on a particular topic. Statements like "Go on" and "Give me an example" let the speaker know you would like more detail because you are genuinely interested in finding a solution. This is a necessary part of communication when you are gathering information from the customer—by asking for more details, you are likely to obtain more thorough information, giving you a better starting point for solving the problem. For example, imagine you are talking with a customer about a noise his car made while going over a bump. If you were to say, "Tell me more," the customer would know that you understood but wanted a little more information.

As an example, a customer brings in a car that will not crank. Technician A speaks with this customer and gets enough information to fill out the work order, but does not ask any clarifying questions, so doesn't gather any further details. Technician B gathers the same information as Technician A, but also uses supporting statements to gather further details that assist in making a diagnosis. What is the end result for the two technicians? Technician B is able to find out that the customer had recently gotten a new key cut for the vehicle. This led the technician to check whether the key had been programmed to the security system; it had not, causing the no start, which is exactly what was wrong. Technician A and Technician B were both able to diagnose the fault, but Technician B was able to come to a conclusion much more quickly than Technician A because of the extra information acquired by asking questions to get supporting statements.

▶ The Art of Speaking

K11006

Applied Communication

AC-28: Speaking: The technician employs communication strategies for customers, supervisors, and coworkers that will yield high-quality information for use in problem solving.
AC-29: Information–Oral: When analyzing a problem, the technician evaluates the usefulness of oral information provided by customers and coworkers.
AC-30: Information–Oral: The technician makes logical inferences and recommends solutions to problems, based on discussions with customers, coworkers, and supervisors.
AC-31: Information–Oral: The technician comprehends information gathered during discussions with customers, supervisors, and coworkers regarding problem symptoms and possible solutions.
AC-32: Information–Oral/Written: The technician analyzes information based on discussions, notes, observations, personal experiences, and data searches that will assist in solving the problem.
AC-33: Information Supplying: The technician clarifies information to customers, associates, parts suppliers, and supervisors.

Speaking is often referred to as an art because there are so many facets involved in effectively communicating and/or information gathering. Whether you are asking a customer about a vehicle or your boss for a raise, knowing how to "artfully speak" will benefit you.

Speaking is a three-step process:

1. *Think* about the message.
2. Accurately *present* the message.
3. *Check* whether the message is correctly understood. If it's not, you have to respond by rethinking and re-presenting the message, and rechecking with your listener. This process continues until you are satisfied that the listener correctly understands your message.

Think before you speak is easier said than done, but it's not impossible. In some situations, you have time to think and plan beforehand. Others require you to think on your feet. Thinking on your feet simply means you may be called upon to answer or handle a difficult situation quickly. Having to get things right in a fast-paced setting can result in rash decisions, but there are a few tactics you can use to make these situations easier:

1. *Relax.* It is not easy to do in an urgent situation, but attempting to remain calm is extremely beneficial, keeping your mind clear and helping you to embody the confidence needed in such a situation.
2. *Listen.* If the situation requires you to answer a tough question, make sure you fully understand the question before answering. You can always ask the questioner to repeat or, better yet, rephrase the question. This gives you more time to think about your answer and also another chance to read into the intent of the question. Remember that if someone is asking you a question, then this person is showing interest. Interest is a good thing!
3. *Pause.* Silence is golden—when used properly. Most people are uncomfortable with silence, but it is perfectly acceptable to take slight pauses before speaking, to organize your thoughts. This also gives you the advantage of controlling the pace of the conversation. If a situation that you have been put in charge of is slipping out of control, regain control with slight pauses and thoughtful answers. It may be beneficial to say, "Let me think about that for a second." Or you may take a few seconds to restate your understanding of the problem and the person's question. Doing so can help the other person understand that you are devoting effort to the question, while also giving you a bit of time to decide on an answer.

Before speaking, take a moment to consider that your tone of voice reveals a lot about your feelings and adds significant meaning to your message. The tone of voice includes how high or low the pitch of your voice is, how fast or slow you speak, how soft or loud, and most importantly, what your voice characteristics or emotional indications are.

Even if you disagree with what is being asserted, try to first use empathizing statements, such as "I understand" or "I can imagine how frustrating that would be," before offering a solution. For example, a regular customer brought his vehicle in last week for a malfunctioning indicator lamp (MIL). The vehicle was fixed and the codes were erased. Now the same customer is back with the MIL on again, and furious that he had to return. Regardless of whether the MIL is on for the same cause or not, you still have an upset customer to deal with.

Most times, people who are angry respond well to a statement like "I can imagine how frustrating it is for you to have to return." When you say this, remember to keep a calm and steady tone of voice. Ultimately, the goal is to de-escalate the customer's anger. Try not to argue with the customer, but rather support him or her, and offer a solution that will work for both of you. In a situation like the one given above, you could say, "I can imagine how frustrating that is for you. Let's bring the car in and see if it's the same problem as before or something different. If it's the same problem, our one-year warranty will cover it. Can I get you a cup of coffee in our waiting room while we look into it?" So, when encountering an upset customer, the goal is to de-escalate the anger, guide the situation back to rational communication, and come to an agreement about how the problem will be resolved. Always empathize, de-escalate, do not argue, and focus the discussion on finding a mutually agreeable solution, if possible.

After thinking about what we want to say, and how to say it, we can then use the second step of the process by presenting a message using verbal and nonverbal language. Remember that when we say "verbal language," we mean the actual words that are being spoken. Nonverbal language includes how we speak those words—our tone of voice, body language, and appearance—and takes into consideration the environment. Imagine you are with a friend in a quiet café, drinking coffee, and she is telling you how much she values your opinion. Now imagine that she is clenching her fists, scowling, and screaming those same words to you. It would change the message a little, wouldn't it?

The last step of the speaking process is to make sure the listener correctly understands your message. Look to see that the listener is making eye contact with you. If you perceive a lack of understanding or confusion, be prepared to repeat yourself while trying to explain it in another way or by using an example. Make sure you suppress any outward irritation you may feel at needing to do so. In stressful situations, the parties involved are often not as open-minded as in calmer moments and are therefore not as receptive to your words. This can also end with misunderstandings and expectations that go unmet. So take the extra step and ask clarifying questions or summarize the message.

▶ TECHNICIAN TIP

A calm and happy customer or coworker is normally easy to communicate with. Most people struggle with how to handle an angry one. In almost all cases, keeping a calm and steady tone of voice and a pleasant overall demeanor is the best bet. Never lose your temper; doing so will only escalate the situation and lead to an unwanted outcome.

Asking Questions

KI1007

Questioning is an important speaking skill that will help keep you out of a lot of trouble. People speak to deliver a message, but many times need more information or need to confirm the details of an agreement. Asking questions to gather more information can provide you with enough information to make good decisions. Or once you come to an agreement with another person, you can use questions to confirm those details. In either of these situations, you avoid trouble. In the first case, you gather enough information to avoid a bad decision; and in the second, you confirm the expectations that the other person had regarding the agreement, helping to avoid disappointment and loss of trust.

To use questions effectively, you need to know how to ask the right questions. You can ask three types of questions:

- Open questions
- Closed questions
- Yes-or-no questions

Each type of question is beneficial when used appropriately. Good communicators know when and how to use the appropriate type. If you have mastered each of these types of questions, you can keep a conversation going while speaking very little. These questions can be used in all kinds of situations, from casual conversations with new acquaintances to detailed conversations with customers. In dealing with customers, you should usually start with open questions to gather general information about the issue. Then, use closed questions to find out specific details. Use yes-or-no questions to further check or confirm the listener's responses or to gain the customer's agreement to authorize a repair or diagnostic procedure.

Open Questions

An open question encourages people to speak freely so you can gather facts, insights, and opinions from them. It's a good way to start a conversation with a new acquaintance or even an established customer. Open questions usually begin with these words:

- What
- How
- Why
- Could you tell me

For example, if you were questioning a customer about her visit to a repair facility, you could ask, "What type of service did you receive from XYZ Automotive?" This question opens the topic up for discussion and allows the customer to give details about her visit.

Closed Questions

If you want to know more information, you can use closed questions to establish facts and details. Closed questions usually begin with the words:

- When
- Where
- Which
- Who
- How many
- How much

This type of question requires a specific answer, and there is usually only one answer. For example, you could ask a customer, "How many times have you visited XYZ Automotive?" or, "When did you start going to them?" These closed questions allow for only one answer, without much room for discussion. These questions help you narrow the topic and guide the discussion in the direction in which you would like it to go, which is helpful when talking with a customer.

Yes-or-No Questions

Yes-or-no questions allow individuals to answer with a simple yes or no. This type of question is useful for checking or confirming a person's responses. For example, "Would you recommend them?" gets to the point and helps clarify information. Generally, you should not start out with yes-or-no questions, because they discourage further explanation or discussion. However, ending with yes-or-no questions is a good way to get confirmation. For example, after explaining the need for replacing the customer's water pump and answering the customer's questions about the job, it would be very appropriate to ask, "Can we go ahead and replace that leaky water pump for you?"

Telephone Skills

KI1008

We've learned about the important aspects of the speaking process. A phone conversation presents some different challenges.

Because we can't see each other, we can rely only upon verbal messages and some nonverbal cues, such as the tone of voice. Other nonverbal cues that we miss are body language, appearance, and the environment.

On the phone, we are limited in how we present our messages, so we need to think about our words and tone more carefully perhaps than in person. It is a good idea to plan and even write down each of the points you want to say before even picking up the phone.

Phone communication consists of three parts: greeting, exchanging messages, and finishing the call. We should always answer a phone call by first saying hello, identifying ourselves, and identifying our place of business, succinctly and clearly. An example could be, "Hello, this is John with XYZ Automotive. How may I help you?" Try to keep your greeting friendly and short.

The second part of the phone conversation is exchanging messages. This requires concise, clear communication, followed by clarifying questions and summarizing any main points. At all times, you should be polite and considerate, and remember that the most important nonverbal clue sent out over the phone is your tone of voice, as it reveals a lot about your feelings. Here are a few tips to create a good impression:

- Do not sound bored. Try to keep some inflection in your voice; do not speak in a monotone.
- Sound calm and in control, even if you were caught at a busy moment.
- No matter what, never lose your temper or patience.
- If there is a need to keep someone on hold for an extended time, offer to call the person back, and then do it.

Finish a call by confirming actions both you and the caller will take to ensure that the messages on both sides were accurately received, and end with a pleasant and friendly goodbye (**FIGURE 11-4**). Remember to thank all customers for their business, and invite them to come back. When taking a phone message for someone else, make sure you have the caller's name and organization, contact details, the date and time of the call, and a summary of the caller's message.

FIGURE 11-4 Always end calls with a friendly goodbye.

When making a phone call, use these same skills, and always have necessary information available to give to the person you are calling. For example, if you need to order parts, relevant vehicle information should be shared with your parts supplier. You usually need to have vehicle make, model, year, engine size, and transmission type, and many times you need the vehicle identification number (VIN) as well.

Giving and Receiving Instructions

K11009

Applied Communication

AC-14: Directions/Task: The technician follows all written and oral directions that relate to the applicable task or system.

A critical aspect of an efficient and well-run workshop is the ability to give clear, logical instructions and to receive instructions. Usually, these should contain information about who, what, when, where, and why, and direction on how a task should be completed. What is the job that needs to be done? Who should do it? When should the job be done? Apart from knowing what information we should include in our instructions, we also need to consider the following:

- The instructions should be clear, complete, and concise.
- The instructions shouldn't be more than the person can remember; otherwise, they need to be written down.
- Don't ramble.
- Check for understanding.

When receiving instructions, we should make sure we can understand and follow the instructions successfully. For example, the instruction "Use the tire machine to mount a set of four tires" indicates what to do. Ask follow-up questions, such as, "Should the white lettering or sidewalls be on the outside?" "Should I install new valve stems?" "Do you want me to install them back on the vehicle once they are mounted?" These help ensure we understood completely. Such questions may not be needed every time an instruction is given, but by clarifying the instructions, the quality of work will be closer to the customer's expectations and will save time and hassles in the long run.

Communication in a Team

K11010

Being part of a team can make working an enjoyable experience or a horrible experience, depending on the team. In large part, the ability of each team member to communicate effectively determines the success of the team. A high-performing team can accomplish much. As the saying goes, the sum is greater than its parts. We can achieve more when we effectively work together. Being part of a team allows us to achieve these goals:

- Learn new things from other team members.
- Share ideas, knowledge, and resources.
- Complement each other's strengths and weaknesses.
- Feel a sense of belonging.

A poorly functioning team spends a lot of time and energy bickering and blaming, and not enough time and energy being productive. Transitioning from a poorly functioning team to a high-performing team requires commitment to the team, as well as self-discipline. When all team members are committed to a set of common goals, they can contribute in positive ways to the success of the team.

Developing such a team requires good leadership skills and good skills as a follower. Good leadership skills involve setting a clear vision of the goals, empowering team members to contribute their best efforts, and recognizing each team member's strengths and weaknesses. Good leadership also provides training or mentoring to address any weaknesses in the team.

Followership is just as important as leadership. Not much would get done if everyone were a leader. Good followership skills involve being fully engaged in the team and its goals, stepping in and performing the work that needs to be done, participating fully in all decision making, and giving honest feedback.

We need to be committed to the team and team goals to make teamwork successful. Each team member should have a defined role and responsibilities that go with that role. Each member is then able to rely on the others to do their part. Think of a team as a chain; one broken link can break the whole chain. A good team player is someone who commits to being a part of that team and contributes to its success by fulfilling his or her role.

Some shops are set up to work on the team system. They typically have three to six members on a team, who all work together to accomplish the work assigned to the team. Generally the team consists of a service advisor, shop team leader, and one or two top-level techs, one or two mid-level techs, and one or two entry-level techs. The team gets paid for the number of hours of work the team produces and splits that money, based on the number of hours each person worked. So if the team works efficiently, the number of worked hours goes down, and the pay for each team member goes up. If the team does not work efficiently, the amount of work completed goes down, as does each team member's pay. Effective communication plays a big part in efficient work productivity in a team.

▶ TECHNICIAN TIP

To be able to fulfill the team commitment, people need to know what their roles and responsibilities are and what is expected of them. Leaders help facilitate that. Good communication is then required for the team to come together and work efficiently.

▶ Employability Skills

K11011

Employability skills are sometimes referred to as "job-keeping skills" and are likely the most important skills you need to acquire as a future technician. The biggest reason that technicians get fired is due to poor employability skills (**FIGURE 11-5**). Job-keeping skills can be as simple as showing up to work on time, every day, and keeping your driver's license free of tickets. Learning and practicing good employability skills will pay off over and over (**FIGURE 11-6**). Here we explore some of the main employability skills you need to master, in addition to the good communications skills we discussed previously.

Employment Requirements

K11012

Working as a technician typically requires that you be above a certain age, usually 18, because being an automotive technician is considered a hazardous occupation, especially when operating vehicle hoists and vehicles themselves. However, in some states, 16-year-olds can be employed if they are part of a formal training program. One of the biggest hurdles to being hired, or even being allowed to continue working, is the requirement of a driver's license and a clean driving record (**FIGURE 11-7**). Shops cannot employ a technician they cannot insure. And that typically means

FIGURE 11-5 Technicians getting fired for regularly showing up late.

FIGURE 11-6 Good employability skills pay off.

DRIVING RECORD HISTORY				
TYPE	VIOL/SUS	CONV/REI	HISTORY ENTRY	PTS
00	00		NONE CURRENT	0. 0
			** END OF RECORD **	

TYPE	VIOL/SUS	CONV/REI	HISTORY ENTRY	PTS
VIOL	04/11/1998	05/18/1998	RECKLESS OPERATION	6.0

City/Location.......: OH
Event Type..........: VIOLATION
State Code..........: E01
Court/Agency........: UNKNOWN
Order/Viol #........: C0126765301
Miscellaneous.......: REK OP
CODE: 3210RO NO
IMMEDIATE IMPOUND

TYPE	VIOL/SUS	CONV/REI	HISTORY ENTRY	PTS
REIN	09/25/1995	09/25/1995	FAIL TO PAY - FINE AND COSTS	

State Code..........: D53
County/Location.....: OH
Event Type..........: REINSTATEMENT
Eligible Date.......: 09/25/95
Miscellaneous.......: WITHDRAWAL EXTENT:ALL
PRIVILEGES WITHDRAWAL
LOC REFN:PENDING REIN
FEE WITHDRAWAL STATE

FIGURE 11-7 A. Good driving report. **B.** Bad driving report.

that the person cannot have more than one or two minor moving violations and no major moving violations within the last three years, such as reckless driving or driving under the influence. Because most technicians are car people who like to go fast, speeding tickets are not uncommon. Unfortunately, they can derail your career as a technician very fast. So work hard to keep your driving record clean.

Another requirement is to be drug free. Most employers require you to pass a drug test before being hired (**FIGURE 11-8**). They also may perform random drug tests periodically, and mandatory drug tests if an accident happens. So staying away from illicit drugs is a requirement for keeping your job as a technician. This can be a problem for technicians in states that have legalized marijuana. Although it may be legal in those states, it is currently illegal in the eyes of the federal government, so that can cause confusion for employers and insurance companies. If you live in a state where it is legal, know your employer's policy on marijuana so you don't get caught by surprise. Just because marijuana may be legal in your state doesn't mean you won't lose your job if you use it. Alcohol is another trap that can catch young and old technicians alike, and is caught by drug tests.

Honesty, integrity, reliability, and quality are also requirements of being a valued technician. They help you build a strong positive reputation that people can trust. And trust means people don't have to worry or wonder whether leaving their vehicle in your care was the right thing to do.

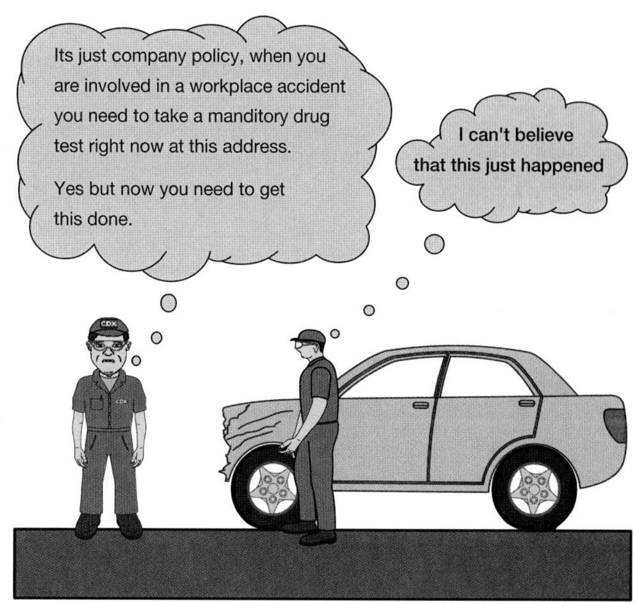

FIGURE 11-8 Passing a drug test is typically required as part of the hiring process.

- Honesty—Merriam-Webster defines "honesty" as a refusal to lie, steal, or deceive in any way. This can be as simple as telling the truth about situations. It can also mean charging people for the job you actually did, not the job you told them you did. So practice honesty; it will pay off in the long run.
- Integrity—Merriam-Webster defines "integrity" as an adherence to moral and ethical principles. This takes honesty to a higher level because it indicates an overarching commitment to being ethical in character. I could be honest by saying that the part broke while I was pressing it apart, when in fact, it broke because I didn't follow the service information, which said that a retaining ring had to be removed before pressing the component apart. A good simple summary of integrity is "doing the right thing." So if you always do the "right" thing, you won't go wrong.
- Reliability—Merriam-Webster defines "reliability" as the quality of, state of, or being fit to be trusted or relied on. It is vital to earn trust with both supervisors and customers. For most people, that means that you have to prove yourself to them by upholding your word by doing the things you agree to do in the time you say you will do it. This includes showing up to work on time and working diligently during work time. So if you agree to do something, then do it, and do it without having to be reminded. Be someone that others can rely on.
- Quality—Merriam-Webster defines "quality" as "superiority in kind." This trait should be a mark of everything you do. The old saying is true: "If it's worth doing, it's worth doing well." Customers have high expectations that shops will provide them with quality work, so shop owners expect that from their employees. Therefore, learning to do quality work will stand out to both employers and customers, and this will allow you to pay your bills.

In many training programs, learning employability skills is often overlooked because the emphasis is placed on learning technical skills. Yet employers regularly say that they can teach people who have good employability skills the technical skills, but not the other way around. So mastering employability skills along with technical skills will make you very valuable to employers.

Appearance and Environment

K11013

Appearance is the image you present of yourself to the public. All aspects of your physical appearance, including personal hygiene, clothes, jewelry, hairstyle, posture, and outward demeanor, combine to create the first impression made in any encounter. That first impression often informs the judgment others make about you, which in turn affects the level of respect and trust you achieve. It is much harder to convey your message effectively if your audience is distracted by some aspect of your appearance or does not take you seriously.

When working in a professional environment, always do your best to look professional. This starts with personal hygiene. No one enjoys being around stinky, dirty people, so personal hygiene is an important habit for all people to develop. Because technicians work in somewhat dirty conditions, and perform a fair amount of physical work, it is common to become dirty, sweaty, and stinky if you do not clean up regularly. Technicians should typically shower daily and use deodorant or antiperspirant so as not to offend fellow employees and customers. Personal hygiene also includes grooming; for men, this includes neatly shaving or trimming facial hair as well as maintaining a professional hairstyle.

Remember that while you are at work, you embody the image of your company. Shorts, improper footwear, or untucked or filthy clothes can all send negative signals to a customer. Customers are much more willing to have their vehicle serviced by someone who looks well put together than someone who does not (**FIGURE 11-9**). And that means wearing clothing that meets or exceeds your company policy. If you are a technician, that typically means wearing the shop's uniform. If that is the case, then wear it properly. That means tucked in, buttoned up, shirts and clean pants that are not drooping. If the company doesn't have an official clothing policy, then wear clothing appropriate for your job. If you are a technician, then clean, close-fitting, but not tight, work pants and shirt are in order. Also consider wearing a belt, if appropriate. Avoid wearing metal belt buckles that could damage a vehicle if you lean up against it.

If you are an employee who deals with customers, such as a service advisor, you may need to step up the level of your clothing to a shirt and tie with dress pants or slacks. If you work the parts counter, you could be required to wear either a work uniform or dress shirt and dress pants.

You should also keep an extra change of clothes at the shop in case you need to change them if they get excessively dirty. Most shops provide lockers for employees to keep their

FIGURE 11-9 When working in a professional environment, always do your best to look professional.

uniforms and street clothes in during work hours. Professional appearance means that you also take care of your clothes. Most shops provide a weekly laundry service to make sure you have clean uniforms at all times. But you have to manage getting them turned in on time so they can be picked up and cleaned. Once you get the clean uniforms back, make sure you store them properly. Don't throw them in your trunk, where they can get all wrinkled and dirty. Hang them up neatly so they will be ready to wear.

The surrounding environment is also worth consideration, as it can affect the image you are portraying. A disorganized, cluttered, and dirty area leaves a negative impression with most customers, leading them to believe that you don't care about appearances or quality. Look around you and evaluate the housekeeping. Is it clean, organized, and inviting? Or is it neglected, dirty, and gross? A little housekeeping goes a long way toward making customers feel confident in your ability to solve their problem. You also want to avoid distractions such as excessive background noise from a blaring radio, or inappropriate coworker conversations. Interruptions by coworkers, phone calls, or text messages can hinder communication. Whenever it is in your power to do so, work to keep these distractions at a minimum.

Be aware of dangers around you as well. Although most insurance policies prohibit customers in the work area, there may be occasional times when they need to see a particular issue. If you have to take customers into the shop, always escort them, and be sure to keep them safe. For example, do they need to wear safety glasses? Also, escort them back out of the shop as soon as possible, and continue the conversation in the customer write-up area.

Time Management

K11014

We know that in the work environment, time is money. This is especially true in most repair shop environments, where every minute is costing somebody something—the customer, the shop owner, and/or the technician. Punctuality is an important nonverbal message in a business environment. When you are punctual, you demonstrate a good work attitude and professionalism. Punctual means showing up on time (typically 5–10 minutes early to get ready to start work at the appointed time) (**FIGURE 11-10**). If you have an unexpected delay, such as finding a flat tire when you get into your car to go to work, make sure you call your supervisor and let him or her know what happened and when you will be in. The same applies if you need to call in sick; do so as soon as you can, so other arrangements can be made. Also, being late or sick should be very rare. Remember that customers are depending on the shop to complete their work on time, and shop managers are depending on you to get it done. Any delays or absences cause trouble for the shop management and customers. It doesn't take very many absences or late arrivals before management decides that it would be better to hire a dependable person.

Punctuality also means you complete the job when you say you will. If you tell a customer his vehicle will be finished at 3:00 P.M., you should plan on finishing it before then, just in case something goes wrong. Occasionally there are unforeseen events that keep you from completing a job on time, but those should be rare exceptions. If there is a delay, then you need to communicate that to your customers as quickly as possible so that they can make other arrangements.

Lastly, you should be giving a full measure of work for the time you are being paid. Routinely texting your friends or taking personal calls during work hours is stealing from your employer. Use your work time efficiently, just as you would want your own employees to do.

FIGURE 11-10 Showing up on time means showing up early, getting ready to work, and starting work at the appointed time.

Customer Service

K11015

Good customer service is vital in today's competitive business environment. The quality of our customer service influences people to choose us over our competitors; good service makes people feel good about continuing to buy our products or services, which is how the business gets the money to pay your wages. So customer service has a direct impact on the ability of employers to hire employees, provide wages, and offer promotions. In fact, vehicle manufacturers place great importance on the customer satisfaction index (CSI) rating at their dealerships (**FIGURE 11-11**). The CSI rating is gathered from virtually all the customers who have service work completed. The CSI rating is reported each month and used to evaluate individual technicians and the entire service facility. If the CSI rating is high, bonuses can be paid to everyone who contributed to that success. If the CSI rating slips, bonuses can be withheld, and new processes may be implemented to help restore the CSI rating.

To be able to provide good customer service, we must first understand who our customers are and then identify their needs. Internal customers, such as parts counter people, are as important as external customers. By helping our coworkers with their jobs, it ultimately helps our external customers and our organization. It is helpful to remember that it is almost always external customers who bring resources into your organization when they trade their money for your service or

FIGURE 11-11 Customer Satisfaction Index report.

product. At the same time, internal customers help us meet the needs of external customers, so treat them as respected members of the team.

Being focused on customer service means you are fully engaged in providing the highest level of service that you can. This includes clear, friendly communication to help prevent misunderstandings while establishing achievable expectations (**FIGURE 11-12**). It also means simple things like not getting grease on the vehicle's steering wheel, upholstery, or paint. And it means taking extra steps such as washing the vehicle when the work is completed or vacuuming it out so it is cleaner than when you started working on it. Keep in mind, one of the most important parts of customer service is repairing the vehicle correctly the first time. This goes a long way toward earning your customer's satisfaction.

Identifying Our Customers' Needs

Applied Communication

AC-25: Notes: The technician makes notes regarding symptoms, possible causes, and other data that aid in diagnosing and solving the problem.

AC-36: Information Requests: The technician requests specific symptom information from the customer and discusses solutions with supervisors and associates.

Different customers have different needs. The same customer may have different needs at different times. In most cases, we can divide our customers into three categories: those who are more concerned about getting things right; those who want to get things done; and those who just want to get along with people. Each of these customers is motivated by different factors. The person who wants the job done right is usually not interested in lesser-quality parts and wants you to take the time to make sure the job isn't rushed. If the diagnosis was rushed and was incorrect, this person will likely not be happy.

People who are most interested getting things done will probably be very keen to have the repairs finished on schedule. If the vehicle is not finished on schedule, this person will likely not be happy. For the person who just wants to get along with people, it is important that a level of trust is maintained. If trust is broken, it will be hard to regain. We have to identify what our customers' needs are, so we can serve them properly. In fact, it is best to address all three factors as best you can each time service is performed.

If you are the one in charge of questioning the customer about symptoms, be sure to get specific information from the customer about what happens, when it happens, and how often it happens. Take notes that will help to diagnose the problem. Then be sure that when relaying this information to someone else, the other person understands the symptoms with the same detail in which you do. For example, a customer is concerned about a particular noise in his vehicle. If the noise doesn't happen all the time, then ask when the noise happens or under what conditions the noise happens. You may need to prompt the customer with follow-up questions like "Does it happen when it is cold or hot, or when going around a corner or over bumps, or when braking or accelerating?" You may need to ask where he thinks the noise is coming from. Of course, if the noise can be reproduced, you may need to drive or ride with the customer so that he can identify the noise for you.

> **TECHNICIAN TIP**
>
> Many shops have Customer Concern sheets that prompt the customer to answer questions about the vehicle concern. These sheets can be system or symptom specific (**FIGURE 11-13**). For example, there might be one for concerns with the braking system. Another might be for unusual noises. The customer then answers the questions to the best of his or her ability. This gives the technician valuable information when planning out the steps of diagnosis or repair.

▶ Effective Reading

K11016

Every day we are faced with interpreting service information, emails, voicemails, and work orders with customer concerns. Understanding these messages without the verbal and nonverbal

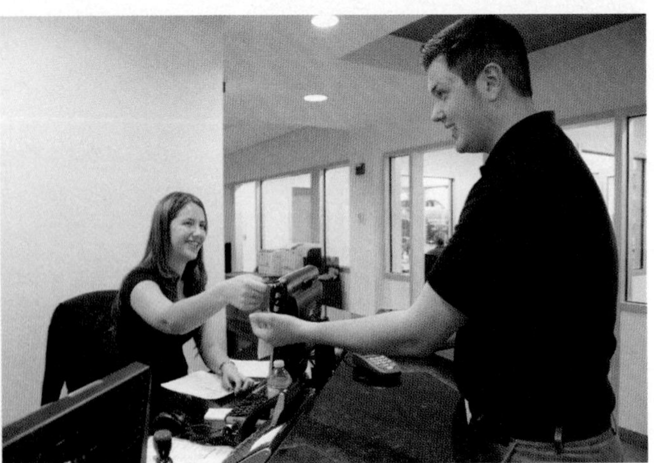

FIGURE 11-12 Focus on customer service and communicate clearly.

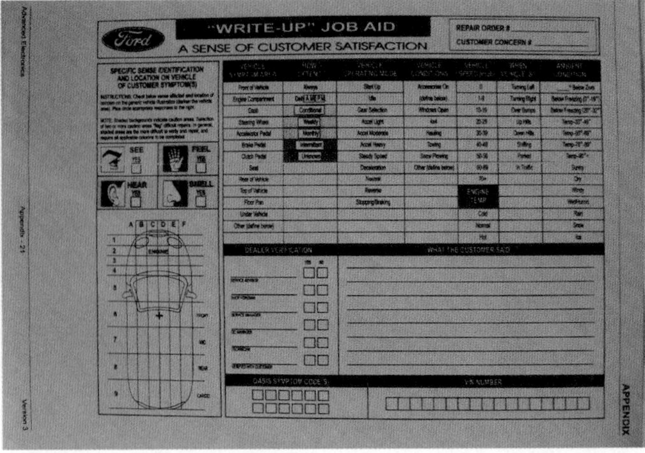

FIGURE 11-13 Customer Concern Sheet.

clues that come with face-to-face communication can be more difficult. Learning effective reading strategies and skills help you find, understand, and use information much more efficiently as you diagnose and repair vehicles.

Applied Communication

AC-1: Reading: The technician adapts a reading strategy for all written materials (e.g., customer's notes, service information, and computer/data readouts) to help identify the solution to the problem.
AC-3: Abbreviations/Acronyms: The technician identifies and uses written abbreviations and acronyms in diagnosing and solving problems.
AC-6: Charts/Tables/Graphs: The technician consults charts, tables, and graphs to determine the manufacturer's specifications to identify out-of-tolerance system components.
AC-7: Sequence: The technician consults service information to determine the appropriate sequence of procedures required for solving a specific problem.
AC-13: Skimming/Scanning: The technician reviews service information to identify problems and applies that information to appropriate repair procedures.

Reading Comprehension

K11017

Technicians are required to read a lot of information, including repair orders, service information, technical service bulletins (TSBs), and training materials. Many of these reading materials can be written at high grade levels. For example, many service manuals and TSBs are written at grade level 14 and higher. Technicians need to be proficient readers to be able to comprehend this information. Before you start reading, you need to know the purpose of the particular material and what you intend to do with the information once you understand it. In reading, the objective may be any or all of the following:

- Access a specific piece of information quickly, such as a particular specification.
- Understand the information, such as how to perform a particular series of tests on a system you are diagnosing.
- Remember the information, such as learning about new technology that a manufacturer is introducing to its vehicles, how it works, and what kind of tests are used during diagnosis.

Once we know our reading purpose, we can choose the suitable reading method, which can be selective, comprehending, or absorbing. *Selective* reading is reading only the parts we need to know. This method is useful when looking for a particular piece of information. The quickest way to use selective reading is to read through the table of contents, introduction, conclusion, headings, and index until we find what we are looking for.

When you use *comprehending* reading, you need to interpret and understand the information. Understanding what is being communicated requires careful attention to the structure of the sentences and paragraphs. You need to pay attention to the punctuation and sentence structure so you can determine the meaning of the information. In fact, you may need to read each sentence more than once, so you comprehend what is being communicated in the text.

When you need to remember information, you use the absorbing reading method. This method allows you to absorb the information into your memory. This is good for remembering information like the theory of operation of a component or system such as a fuel injector or ignition system. This type of reading takes longer than the previous two methods because the reader needs to read the information slowly, stopping to ponder it, and even reading it several times until it makes sense. You know it makes sense when you can visualize the concept in the form of a picture.

To use an *absorbing* reading method:

- Read the information slowly and methodically.
- Stop and visualize what is being said in written form.
- Reread the sentence or paragraph if needed.
- Absorb the pictured information into our memory.
- Review the information regularly so you will retain it over time.

When we talk about reading, we are not concerned only about how fast we can read but also how well we understand and retain what we have read. Unfortunately, the faster we try to read, the less we are likely to concentrate on the meaning, so these two objectives (speed and comprehension) work against each other. Reading is about practice, and the more we practice, the faster, more efficiently, and more accurately we can read.

Each of these reading strategies can be used when researching a particular problem with a vehicle. For example, if you have a vehicle that is hard to start only when it is cold outside, you might first check the service information to see if there are any related TSBs. A TSB is released when there are many of the same vehicles having the same symptoms and causes, so you can quickly skim through hundreds of TSBs, looking for specific symptoms. When a potentially relevant TSB is found, read it and interpret the information. Once you determine the appropriate TSB for the symptoms, read it carefully and work to understand the information presented, including the symptoms, cause, test procedures, and corrective actions necessary to repairing the vehicle.

▶ TECHNICIAN TIP

Charts, tables, and graphs help you to find information in a timely manner. An example is torque specifications. This information is usually presented as a chart showing the component and the required torque, usually listed in newton meters (N·m) as well as foot-pounds (ft-lb).

▶ TECHNICIAN TIP

The Society of Automotive Engineers (SAE) has a well-known list of J1930 terms, which are abbreviations used by the automotive manufacturers. In addition, each manufacturer may have additional abbreviations that differ from the SAE terms. For example, the term "GEN" (generator) is used by the SAE, and the term "ALT" (alternator) is used by several automotive manufacturers.

Researching and Using Information Sources

KI1018

Technicians spend a lot of time researching information (**FIGURE 11-14**). This could be looking for a specification needed to perform a specific step in a task, or it could be researching how a manufacturer designed the system to operate, so you can determine whether it is operating correctly or not. In some cases, research is needed to see what work has been performed on the vehicle in previous shop visits so that you will know what has and hasn't been performed previously.

Applied Communication

AC-8: Dictionary: The technician refers to a dictionary to check spelling and define unfamiliar terms.

AC-9: Text Resources: The technician uses glossaries, indexes, database menus, and tables of contents to gather the information needed for diagnosis and repair.

AC-10: Database: The technician uses databases to obtain service information.

AC-11: Operator's Manual: The technician comprehends and applies information from accompanying manuals in order to use and maintain automotive tools and equipment.

AC-12: Service (Shop) Manual: The technician uses service information in both database and print formats to identify potential malfunctions.

AC-16: Study Habits/Methods: The technician uses proven research methods when consulting the manufacturer's service information (e.g., shop manuals, service bulletins, and computer databases).

AC-17: Prior Knowledge: The technician uses prior knowledge of similar problems to determine the specific cause(s) of problems.

AC-18: Cause/Effect Relationships: The technician comprehends and uses cause-and-effect relationships presented in service information problem-solving trees.

AC-19: Definitions: The technician applies industry definitions to solve problems in automotive components and systems.

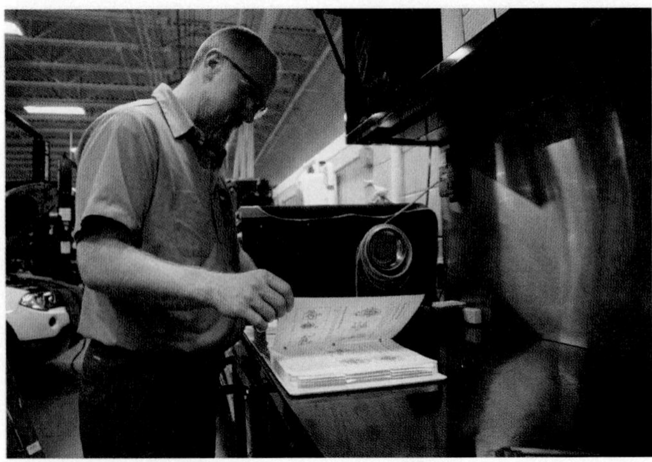

FIGURE 11-14 Researching is like conducting an investigation.

Researching is like conducting an investigation. It can be done in three steps. When we first come across a problem, we have to define what the real issue is; that requires us to know how the system or part is meant to work. Then, we look for information or clues that can help us solve the problem. Finally, we put the pieces together to determine the best solution.

If using a book for reference, start with the table of contents to find the section you are looking for. If you come across a word you do not understand, look up the definition. Most books have a dictionary or glossary of terms toward the rear of the text. When using an online resource or database, try to familiarize yourself with the site's or database's navigation first. Take note of all drop-down menus, shortcut options, and search features.

While using these types of resources, you can use the skimming technique discussed earlier to find information quickly. Try skimming for key words. For example, if you are looking for information on oil specifications for a particular vehicle, you could just skim through the resource's text, looking for the words "oil," "lubrication," or "specifications." Charts, tables, and graphs, if included in the reference material, are usually easy to skim.

Most automotive resources include diagnostic trouble charts. These provide a great way to narrow down faults without missing any steps in the procedure. Trouble trees also use the cause-and-effect approach to diagnosis. For example, if a fuse is blown, you can use the trouble tree to help you, step by step, work through the diagnosis in order to locate what caused the fuse to blow. As you gather symptoms and perform the diagnostic tests, you should take notes on your results so that you can refer back to them. Notes are also a good resource as you complete the repair order after finishing the job.

Always save the manuals for shop equipment and tools, and organize them in one place for future reference. Refer to these documents to maintain the equipment and tools. For example, most air tools require a drop or two of hydraulic oil every day, along with the use of a water separator in the air line.

When encountering problems with a new or unique system, it is often helpful to look up a definition or description of that system. This information can typically be found in either the front or the back of a book, or you may be able to search for it on computer-based information systems.

As you gain experience in the automotive industry, you will find that many vehicles have similar faults. Don't hesitate to use your prior knowledge to assist you in a new diagnostic situation. Just don't assume that all similar problems are caused by the same fault. You will want to test and verify your assumptions before suggesting a course of action to the customer.

Defining the Problem

Before you spend much time researching information, it is important that you define the problem. As you are defining the problem, make sure you narrow it down as accurately as possible. For example, if the car won't start, is this because it does not crank over at all, or does it crank over but not start? Narrowing the problem down in this way helps you filter out irrelevant information. If the problem is very broad, it is better to break it into smaller chunks and conduct a separate research activity on each one. Once the research is complete, you will have a better idea of how to proceed with determining the cause of the problem.

The next step is to look for the information that may help solve the problem. There is plenty of information readily available, but to determine how useful and reliable the information is, we need to know the sources of that information. We can obtain information from two sources: primary and secondary. The primary sources of information are people who have direct experience with the same or a similar problem. We can obtain information by interviewing those primary sources directly. However, information from primary sources can be subjective. Therefore, it is not always reliable. Information from secondary sources, or secondhand information, is compiled from a variety of sources. It is usually more reliable because it is more generic and objective.

This secondhand information is available in various formats, including print, audiovisual, and computerized (**FIGURE 11-15**). We can divide this information into different content categories, including these:

- Vehicle service information
- Automotive educational sources
- Troubleshooting

Vehicle Information

The first place to look for information about a vehicle is in the shop. Computer databases, shop manuals, aftermarket manuals, and owner manuals are good resources for basic service information, including vehicle systems and how to operate them, the locations of major components and lifting points, and vehicle care and maintenance information.

Manufacturers usually provide training videos covering specific information about a vehicle of a particular make and model. Similarly, aftermarket and components suppliers provide specific information about their products. This information is also available from their DVDs and websites.

FIGURE 11-15 Compile primary and secondary information to define the problem.

The manufacturer's service information will guide the technician in the correct sequence of procedures to correct a given problem. The basic idea is to check the simple things first. In most cases, this involves a visual inspection for obvious issues such as a vacuum line or electrical connector that is loose or disconnected. If no problem is found, then the technician should go to the next step in the service procedure, which will likely involve testing the system or components affected. The service information will continue to guide the technician through additional steps as necessary.

Educational Sources

Publishers provide extensive materials covering the basic theory of operation of automotive systems and components. These materials come in a variety of forms. Textbooks and workbooks come as paper-based materials. Then there are computer-based materials, which provide more options for interactivity and visual engagement of the student. Although there is free information available, most of these sources require a subscription to enroll in their courses.

Technical Assistance Services

For troubleshooting, there is help available on the phone and on the Internet. To use either resource, you must subscribe as a member, which normally involves paying a fee. Technical assistance hotlines put you in contact with professionals who can assist you in diagnosing a particularly difficult problem over the phone. The technical assistant has access to a variety of technical service information. Some of this information is gathered while helping other technicians with their issues. Thus, a large database of information can be accumulated, which can save a technician a lot of time when dealing with difficult diagnostic situations.

For free information, there are chat rooms and forums where questions can be posted to other technicians online, although there is no guarantee of the accuracy and availability of the resources. The websites of automotive hobbyists, enthusiasts, and car clubs are often good resources as well.

One very good online reference that is available only to professional automotive technicians and students is the International Automotive Technicians Network (www.iatn.net). This organization consists of over 85,000 active members, with more than 2 million years of combined experience, who share their knowledge with each other on over a dozen forums. They recently have begun granting free student accounts to students of member instructors. These free student accounts allow students to monitor the forums and learn from them while not allowing the students to interact with the technicians.

▶ TECHNICIAN TIP

If you are ever working on a vehicle and seem to have hit a dead end in diagnosis, don't be afraid to discuss the symptoms with other technicians or supervisors around you. Sometimes others have seen the same issue or have a bit more experience and will be able to help. Or maybe they will look at the problem from a slightly different perspective, which can give you something new to try.

▶ Effective Writing

KI1019

Technicians need to document their findings, conclusions, and any repairs on the repair order. This requires an accurate, short, but complete summary of the work you completed. In fact, the shop and ultimately the technicians get paid based on the quality of the write-up of the repair order. If there is ever a problem in the future, such as a customer filing a lawsuit against the shop over the repair, the repair order becomes a legal document that will be used by the court to determine if the shop has any liability in the situation. Thus, if the repair order is poorly written, incomplete, or inaccurate, the shop will be much more likely to lose the court case, putting it at risk of having to pay thousands of dollars—or more in the case of an injury accident. So writing is one of the most important tasks a technician does on a daily basis (**FIGURE 11-16**).

Writing Business Correspondence

KI1020

Writing for technicians generally involves completing a write-up of diagnosis and repair conclusions on repair orders and filling out parts requests. However, technicians may sometimes need to undertake a more formal business correspondence using complete paragraphs. As with any type of writing, first think about what you want to say and draft your message, and then refine and finalize your message. Just as it is important to think before speaking, you also need to think about what you want to say and what you want to achieve before composing a written message. Using your notes, sketch out a short outline of major points; this is a great place to start when writing. Then use the outline to assist you in the writing. Use complete sentences. Remember to carefully proofread your writing so that you catch and fix any spelling or grammatical errors. You should use spell-check, but don't rely on it to catch everything; spell-checkers can't tell you when you have used "two" and "to" incorrectly, or "there" and "their," or when you meant to use "they" and only typed "the"—to mention a few examples. Once you have completed the draft, it may be necessary to create a final version of the write-up.

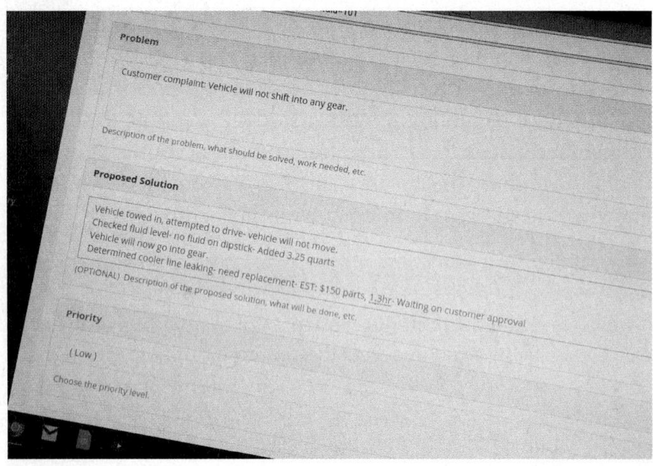

FIGURE 11-16 Technician completing an online repair order.

Applied Communication

AC-2: Information–Written: The technician can comprehend and apply the available written information needed to diagnose, analyze, and solve a problem.

AC-4: Information–Written: The technician evaluates the usefulness of available written information clearly and thoroughly when analyzing a problem.

AC-5: Information–Written: The technician makes logical inferences and recommendations based on information provided on the repair order.

AC-20: Summaries: The technician uses appropriate grammar and sentence structure when summarizing problems in reports.

AC-21: Sentences: The technician uses conventional sentence structure, spelling, capitalization, and punctuation when composing sentences for warranty reports.

AC-22: Writing: The technician adapts a writing strategy that is most appropriate for the intended audience (e.g., customers, supervisor, or fellow employees) when documenting repairs.

AC-24: Purpose: The technician adapts speaking and/or writing styles that are consistent with the purpose of the communication.

AC-26: Paragraphs: The technician composes complete paragraphs, with appropriate details, presenting accurate information regarding symptoms, diagnosis, and results when preparing warranty claims and work orders.

AC-27: Diction/Structure: The technician adapts diction and structure to the context of all verbal and written communication based on the audience, purpose, and specific situation.

When writing a business letter, after creating an outline of the main points, you must then organize and put them together. Here are a few tips:

- Structure your letter logically so readers can follow your flow of thoughts; check the content by reading it out loud.
- Use plain English when addressing a customer, avoiding complicated words or technical jargon. On the other hand, when addressing a manufacturer or warranty clerk, be very specific.
- Check your spelling and punctuation. This is easily done using spell-check or by looking up the term(s). Be aware that spell-check can misinterpret your words, so you need to double-check spelling corrections.
- Be courteous and empathize with the readers, especially when writing a letter of complaint or delivering bad news.
- Be precise, direct, and to the point. Repetition or unnecessary words waste time and can confuse readers.

There is a simple but necessary format to follow when creating a business letter (**FIGURE 11-17**). The letter should have your company name and address at the top. Most business letterheads have these details preprinted. Next include the date, followed by the recipient's name and address. Usually, you should start a letter with a greeting like "Dear Sir or Madam," or when we know the receiver's name, "Dear Mr. Mortenson." "Re:" or "Reference:" is followed by the subject or purpose of

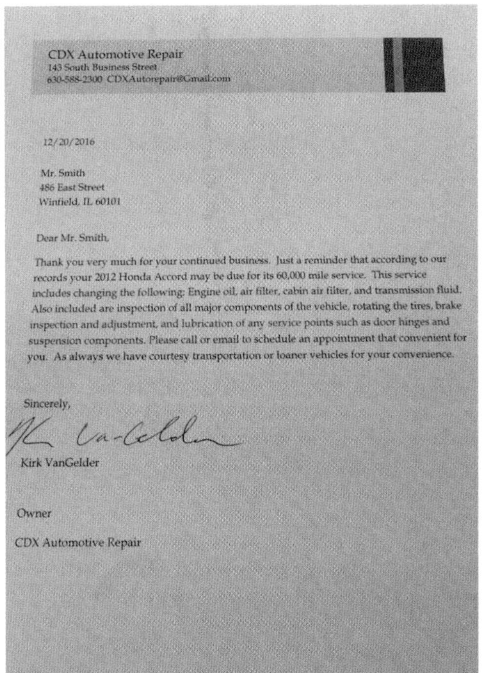

CDX Automotive Repair
143 South Business Street
630-588-2300 CDXAutorepair@Gmail.com

12/20/2016

Mr. Smith
486 East Street
Winfield, IL 60101

Dear Mr. Smith,

Thank you very much for your continued business. Just a reminder that according to our records your 2012 Honda Accord may be due for its 60,000 mile service. This service includes changing the following: Engine oil, air filter, cabin air filter, and transmission fluid. Also included are inspection of all major components of the vehicle, rotating the tires, brake inspection and adjustment, and lubrication of any service points such as door hinges and suspension components. Please call or email to schedule an appointment that convenient for you. As always we have courtesy transportation or loaner vehicles for your convenience.

Sincerely,

Kirk VanGelder

Owner

CDX Automotive Repair

FIGURE 11-17 Typical business letter.

the letter. The next part is the content that you composed in your outline. A business letter usually concludes with a complimentary closing, such as "Sincerely." Finally, a letter needs to be signed and followed by the sender's typewritten name and position.

Completing a Repair Order

S11001

If you are writing up your findings, conclusions, and repairs performed on a repair order, one thing to remember is that the information should be concise, yet complete. This is a difficult skill to develop but will improve with practice.

A work order should contain this information:

- The customer's concern: Describe the concern, how you verified the concern, and any related issues found.
- The cause of the concern: Describe what specifically is the root cause of the concern.
- The action taken, or needed to be taken, to correct the concern: Describe specifically how the concern was, or could be, resolved.

These elements constitute what is called the three "Cs": concern, cause, and correction. Remember that each of these sections needs to be written up as concisely yet completely as possible. For example:

> **Concern:** The vehicle overheats within about 10 minutes of vehicle operation from a cold start, no matter whether it is hot or cold outside. Coolant is full and protected to −30°F. Started vehicle, monitored engine temperature with temp gun. Temperature exceeded specified 195°F thermostat opening temperature, and thermostat was still not open at 205°F.

FIGURE 11-18 Shop safety inspection form being used.

> **Cause:** Thermostat faulty, not opening at specified temperature. Removed, no external damage noted.
> **Correction:** Replaced thermostat with new OE thermostat. Retested, thermostat starts to open at 193°F, which is within specifications.

While in school, you will document tasks performed in the shop on forms referred to as job sheets or tasksheets. You can use these forms to develop your ability to accurately complete the three Cs. Doing so will prepare you for the repair orders you will complete when working out in the industry.

A repair order should also contain the customer's contact information, vehicle information (e.g., make/model/VIN), parts, prices, labor time, taxes, and any recommended service that is found while repairing the vehicle. Remember that a repair order is a legal document, so treat it as if your job depends on it; it does.

Completing a Shop Safety Inspection Form

S11002

Most shops should have a safety inspection form that has to be completed on a regular basis, typically weekly or monthly, although some tasks may have to be performed daily (**FIGURE 11-18**). This form will guide you to visually inspect critical items in the shop such as automotive lifts, overhead doors, hydraulic equipment, pneumatic equipment, hoses and cords, fire extinguishers, and emergency exits. The importance of these inspections should not be taken lightly, as neglect of these items can cause safety issues and cause premature product failure. These pieces of equipment are used often and require frequent inspection.

Safety inspection forms are also a way of keeping up on routine maintenance tasks, such as changing the oil or replacing the belt of an air compressor, refilling the tire machine automatic oiler, and draining any water traps in the compressed air system. If you are the person completing the inspection, make sure you are trained to evaluate the safety and condition of the items you are inspecting. Inspect the items carefully and thoroughly so you don't miss any issues. Be sure to accurately document any issues found, and bring them to the attention of the appropriate person in the shop immediately.

Completing a Defective Equipment Report

SI1003

One way to help ensure safety in the workshop is to inspect equipment regularly and arrange for repair or replacement whenever it does not meet safety standards. Keeping equipment well maintained will help avoid equipment downtime to provide a safer work environment.

Whenever you come across any defective equipment, you should do the following:

- Tag the defective items and either place them in a secured area or secure them so no one else can use them by mistake (**FIGURE 11-19**).
- Complete a defective equipment report (**FIGURE 11-20**).
- Notify your supervisor.

Your immediate actions can help protect coworkers from accidents and injuries. If there is an accident resulting from defective equipment, you need to complete both an accident report and a defective equipment report.

On the report, you record the date when the defect was detected, the location of the defective equipment, the name of the equipment, the serial number (if possible), and a description of the defect and the action you have taken. Then sign your name as the reporter. Finally, notify your supervisor.

Completing an Accident Report

SI1004

Safety is the most important issue in the workplace. We must make a conscious decision to work safely and act responsibly to protect others and ourselves. Unfortunately, accidents do happen. When one occurs, an accident report should be completed by those involved, both the victim and witnesses, if possible (**FIGURE 11-21**). The information in the report is used to protect both employees and employers. It protects the employer and employee against false claims. And it protects future employees by calling attention to a situation that caused an accident, so hopefully measures can be taken to prevent it in the future. To ensure that the information is accurate, the accident report should be completed as soon as practically possible, while the facts of the accident are still fresh in everyone's memory.

A typical accident report includes the date and time when the accident happened, the location where the accident happened, the name of the person who was injured, the name of any witnesses, the details of the accident, any first aid treatment provided, and any medical assistance rendered. It is also important to note whether the accident will be subject to, or covered by, any insurance claims. Finally, the form needs to be signed by the person reporting the accident, and turned in to the supervisor.

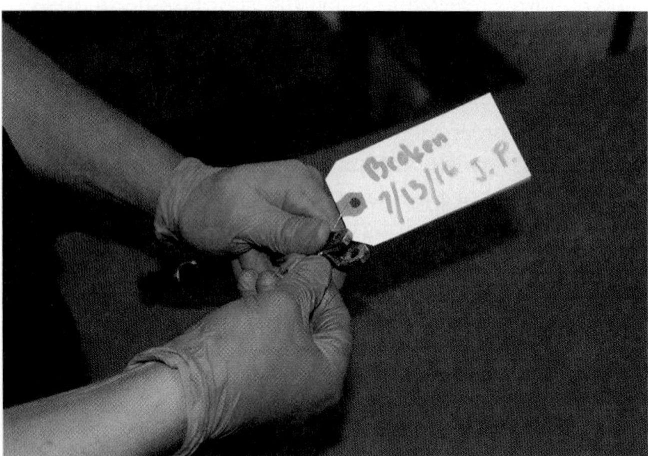

FIGURE 11-19 Tagging out and securing defective tools.

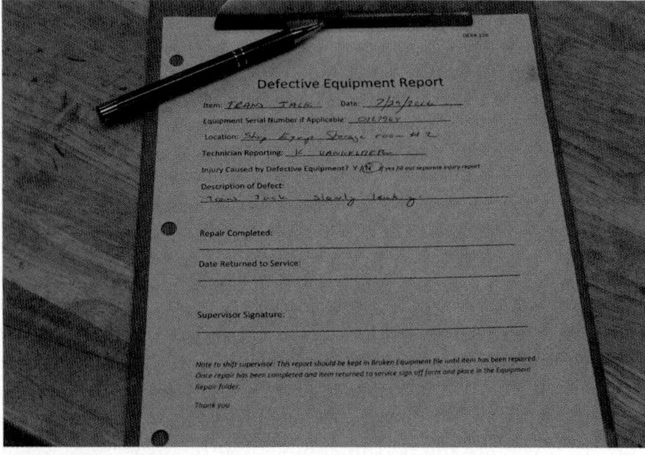

FIGURE 11-20 Completing a defective equipment report.

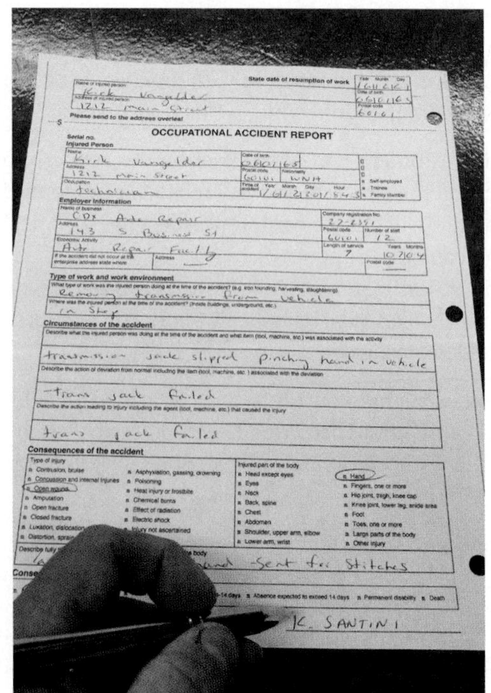

FIGURE 11-21 Tech filling out an accident report.

Completing a Vehicle Inspection Form

S11005

Technicians provide customers with a very beneficial service when a thorough inspection is performed on their vehicles. When performing an inspection, we need to check that all major components and systems are operational, secured, and safe in accordance with the vehicle manufacturer's recommendations. This means testing the operation of the electrical and mechanical systems, such as lights and brakes, and visual inspections of the components such as tires and glass.

An inspection form is a useful guide when conducting a vehicle inspection. By following the checklist, a technician can test all the components in a systematic way and ensure that they are operational or serviceable (**FIGURE 11-22**). It also becomes a record for the shop to bring a customer's attention to needed service, or in the event that a customer declines repairs, shows that the shop made the customer aware of them. Many shops use their own inspection form that lists all items tested, which are known as points. Depending on the number of points covered, the inspection may be called a 30-point inspection, a 72-point inspection, etc.

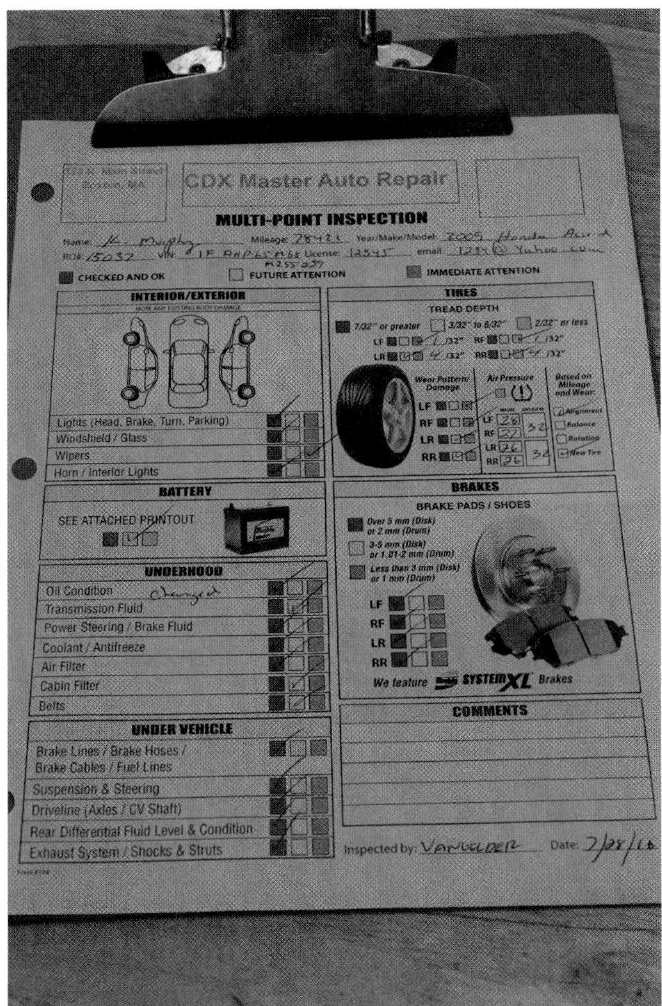

FIGURE 11-22 Technician using a vehicle inspection form.

To complete the inspection form, you should inspect all the components and systems on the checklist, including these:

- Fluids, belts, and hoses
- Steering and suspension system
- Brakes
- Drive line
- Fuel system
- Exhaust system
- Tires
- Lighting system
- Electrical system
- Visibility
- Seat belts
- General components

Once the inspection is completed, the results should be presented to the customer so he or she can decide if any repairs are to be made. This is where verbal communication comes back into play, as it is important to accurately communicate with the customer so he or she has a good understanding of what the vehicle needs, why it is important to have it repaired, and the consequences of not repairing it. This conversation should be based on a trusting relationship you have built with the customer during his or her experience with you and the shop. And as you can see, successful communication includes good verbal, nonverbal, listening, clarifying, writing, and presentation skills.

Lockout/Tagout

S11006

There are many dangers in the automotive repair facility that may need to be properly documented and tagged. One example could be a defective automotive lift. If a problem is noticed with a piece of equipment, the lockout/tagout procedure should be followed. Lockout/tagout procedures have been developed to prevent avoidable and unnecessary workshop accidents. These procedures have many functions:

- The tag notifies other users that the tool or component is dangerous to use. Any equipment that is found to be faulty must be identified so that other users are not put at risk. Write the fault, the date, and your name on the tag. Attach the tag to the tool. Smaller equipment should also be tagged and placed in a location where it is not forgotten. Notify your supervisor so that repairs or a replacement can be arranged.
- If a machine is faulty, the lockout procedure is used. Most large workshop equipment is permanently wired to the electrical supply and usually will have an isolation switch that disconnects the electrical power. A lockout tag should be placed on the isolation switch as well as the equipment. Turn the machine off at the power and master switches, attaching the lockout tag in a manner that prevents the switches from being turned on. Once again, notify your supervisor so repairs can be arranged.
- The lockout/tagout procedure is also used to notify other technicians that a vehicle is not drivable.

Your workshop will have a procedure for vehicle lockout/tagout. It may involve the technician filling out a "defective vehicle" label listing the nature of the defect, name of the technician, and date and time of the defect.

If you remove the vehicle keys, do not keep them in your pocket or on your workbench. Attach a label or tag to the keys that identify the vehicle they belong to, and store them in a secure key organizer.

If a vehicle is going through a relearn process (i.e., the vehicle's computer communicating with another computer), it may be necessary to leave the ignition on for many hours. In this case, tag the vehicle with instructions to leave the ignition on; otherwise, a passing technician may turn it off in an effort to be helpful. If a vital component has been removed for service, it may not be obvious to a casual observer, so it is necessary to tag the vehicle. The best place to tag the vehicle is in on the steering wheel or driver's window, where others trying to operate the vehicle will see it. Also, remove the ignition key and store it in a safe place.

Ask your supervisor to demonstrate the lockout/tagout process used in your workshop and to show you the location of the key organizer.

To properly identify faulty equipment, follow the steps in **SKILL DRILL 11-1**.

In conclusion, communication is a critical component of a successful business and an efficient workplace. Often overlooked in our current digital world, where face-to-face contact is getting less and less common, it is increasingly important to familiarize yourself with a wide variety of communication skills. From listening to simply filling out an inspection report, using and practicing these skills can be a big determining factor in how far you can go and how much money you make in the automotive industry. And the good news is that it doesn't even require expensive tools!

SKILL DRILL 11-1 Identifying Faulty Equipment

1. Basic workshop tools that are broken or worn should be replaced. Make sure you tag the tool as faulty or broken, and do not use it until you buy a replacement. Then discard the tool.

2. Power tools that have been identified as faulty, due to failure of parts, should also be tagged and set aside. The tool can only be used again after an authorized agent has made the repair.

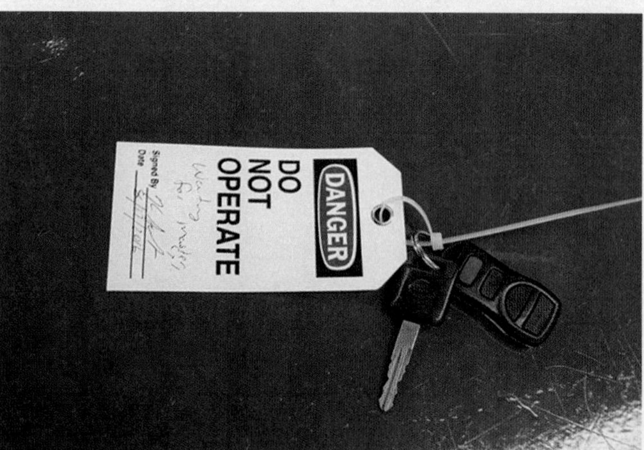

3. Isolation tags are also used on disabled vehicles or vehicles undergoing a repair. In this case, you must locate and complete the "Disabled Vehicle" warning notice. Write the license number of the vehicle and the nature of the defect. Write your name and then the date and time you completed the notice. Attach the notice to the steering wheel or driver's window.

4. Remove the keys and lock the vehicle, if appropriate. Attach a tag to the keys that identifies the vehicle they belong to. Store the keys in the key organizer, and notify your supervisor.

▶ Wrap-Up

Ready for Review

- Communication includes listening, reading, writing, and speaking skills—soft skills that every good automotive technician needs to master over time.
- To empathize with someone is to attempt to see the situation from his or her point of view.
- Nonverbal feedback helps to emphasize your message. When listening, you can use it to provide listening feedback.
- Verbal feedback includes very simple signals that can enhance the conversation and let the person know you comprehend.

- A supporting statement can urge the speaker to elaborate on a particular topic.
- Speaking is often referred to as an art because there are so many facets involved in effectively communicating and/or information gathering.
- Speaking is a three-step process: think about the message, present it, and check that it was understood.
- When encountering an upset customer, empathize, de-escalate, do not argue, and stay calm.
- There are three types of questions: open, closed, and yes-or-no.
- A critical aspect of an efficient and well-run workshop is the ability to give clear, logical instruction and to receive instruction.
- Your appearance makes the first impression and informs others' judgment of you.
- In the work environment, time is money. Always strive to be on time and ready to do your part by being present, both physically and mentally.
- Good customer service is vital in today's competitive business environment.
- Every day we are faced with interpreting service information, emails, voicemails, and work orders with customer concerns.
- Researching is an important part of troubleshooting a vehicle. First define the problem, then gather clues, and then put all the pieces together to get the conclusion.
- When a technician is writing a repair order, the three Cs must be included: concern, cause, and correction. The customer and vehicle information should also be included.

Key Terms

supporting statement A statement that urges the speaker to elaborate on a particular topic.

validating statement A statement that shows common interest in the topic being discussed.

Review Questions

1. When speaking with customers, they tend to believe your:
   a. verbal messages only.
   b. verbal messages over nonverbal messages.
   c. nonverbal messages only.
   d. nonverbal messages over verbal messages.
2. A good practice when listening to a customer is to:
   a. fold your arms.
   b. stand upright.
   c. have your hands in your pockets.
   d. look at the damaged part of the car.
3. Which of these is a good way to start a conversation with a customer?
   a. Yes-or-no questions
   b. Closed questions
   c. Open questions
   d. Validating questions
4. If there is a need to keep someone on hold for an extended time:
   a. offer to call them back.
   b. pass on the receiver to someone else to keep the caller engaged.
   c. ask them to call you back.
   d. check if they are holding the line every few minutes.
5. Which of these are basic job-keeping skills?
   a. Technical skills
   b. Employability skills
   c. Communication skills
   d. Problem-solving skills
6. A person's first impression about you is typically based on your:
   a. technical skills.
   b. educational qualifications.
   c. appearance.
   d. problem-solving skills.
7. After you have spotted a potentially relevant TSB for specific symptoms, the best reading method is:
   a. comprehending reading.
   b. absorbing reading.
   c. selective reading.
   d. skimming.
8. All of the following statements are true *except*:
   a. Customer service has a direct impact on the ability of employers to hire employees.
   b. When interacting with the customer, take notes that will help to diagnose the problem.
   c. Helping coworkers is unrelated to customer satisfaction.
   d. The same customer may have different needs at different times.
9. Which of the following sources is comparatively more subjective and therefore less reliable?
   a. Automotive educational sources
   b. People with direct experience of the same problem
   c. Technical assistance services
   d. Computer databases
10. Choose the correct statement.
    a. The spell-check feature will identify all spelling errors.
    b. Express your anger when dealing with an upset customer.
    c. Use technical jargon when addressing customers.
    d. Structure your letter logically.

ASE Technician A/Technician B Style Questions

1. Tech A says that repair orders are legal documents, so they have to be filled out accurately and carefully. Tech B says that ensuring the repair order is well written, clear, and concise promotes a professional reputation. Who is correct?
   a. Tech A
   b. Tech B
   c. Both A and B
   d. Neither A nor B

2. Tech A says that you shouldn't inspect a vehicle that is in for a repair unless the customer requests it. Tech B says that performing an inspection in addition to a repair can lead to discovery of additional concerns. Who is correct?
   a. Tech A
   b. Tech B
   c. Both A and B
   d. Neither A nor B
3. Tech A says that the proper way to listen to a customer is to maintain eye contact with the customer in between taking notes. Tech B says that talking with customers about the concern is usually a waste of time. Who is correct?
   a. Tech A
   b. Tech B
   c. Both A and B
   d. Neither A nor B
4. Tech A says that a good summary of integrity is "doing the right thing." Tech B says that if you are ever working on a vehicle and seem to have hit a dead end in diagnosis, don't be afraid to discuss the symptoms with other technicians around you. Who is correct?
   a. Tech A
   b. Tech B
   c. Both A and B
   d. Neither A nor B
5. Tech A says that maintaining an appearance of neatness is important as it conveys to the customer the idea of careful, professional technicians. Tech B says that a dirty and cluttered shop indicates to customers that the shop gets a lot of quality work done. Who is correct?
   a. Tech A
   b. Tech B
   c. Both A and B
   d. Neither A nor B
6. Tech A says you can use the skimming technique when researching information to locate it quickly. Tech B says that diagnostic trouble charts provide a great way to narrow down faults without missing any steps in the procedure. Who is correct?
   a. Tech A
   b. Tech B
   c. Both A and B
   d. Neither A nor B

7. Tech A says that an example of an open question is "What kind of problem are you having?" Tech B says that an example of a closed question is "Would you like us to go ahead with the repair?" Who is correct?
   a. Tech A
   b. Tech B
   c. Both A and B
   d. Neither A nor B
8. Tech A says that one customer who has a bad experience stemming from a misdiagnosed repair due to a technician's failure to listen has more of an effect on the repair shop than several good and happy customers. Tech B says that it is only one customer who is unhappy, and because this person paid the bill, all is well. Who is correct?
   a. Tech A
   b. Tech B
   c. Both A and B
   d. Neither A nor B
9. Tech A says that routinely texting your friends or taking personal calls during work hours is generally okay with most employers. Tech B says that being punctual means arriving early so you are ready to begin work at the appointed time. Who is correct?
   a. Tech A
   b. Tech B
   c. Both A and B
   d. Neither A nor B
10. Tech A says that the three Cs are "customer, complaint, and condition." Tech B says that the three Cs are "concern, cause, and correction." Who is correct?
   a. Tech A
   b. Tech B
   c. Both A and B
   d. Neither A nor B

0.001″
0—1″
0
10
20
30
40
50
60
70
80
90

SECTION 2
Engine Repair

CHAPTER 12
Motive Power Types—Spark-Ignition (SI) Engines

NATEF Tasks

There are no NATEF tasks for this chapter.

Knowledge Objectives

After reading this chapter, you will be able to:

- **K12001** Describe the principles of physics that allow internal combustion engines to operate.
- **K12002** Explain the difference between external combustion engines and internal combustion engines.
- **K12003** Explain the relationships among pressure, temperature, and volume.
- **K12004** Describe force, work, and power, and how they are measured.
- **K12005** Describe the operation of the four-stroke engine.
- **K12006** Explain the five events common to all internal combustion engines.
- **K12007** Explain bore, stroke, displacement, and compression ratio.
- **K12008** Describe how the Atkinson and Miller cycle engines differ from conventional four-stroke engines.
- **K12009** Identify and describe the purpose of, four-stroke engine components.
- **K12010** Describe the major components in a short block assembly.
- **K12011** Describe the functions and components of the cylinder head.
- **K12012** Describe the difference between a cam-in-block engine and an overhead cam (OHC) engine.
- **K12013** Describe the valve train, its components, and their function.
- **K12014** Describe intake and exhaust manifold.
- **K12015** Describe the operation of two-stroke engines.
- **K12016** Explain the differences between two- and four-stroke engines.
- **K12017** Describe the operation of the rotary engine.
- **K12018** Explain the basic principles of operation and components of the rotary spark-ignition engine.

Skills Objectives

There are no Skills Objectives for this chapter.

▶ Introduction

K12001

The internal combustion engine is an irreplaceable part of modern society. We rely on it to haul food and water, deliver passengers to their destinations, and even save lives. Over time, the internal combustion engine has seen many changes; however, the basics have remained similar (**FIGURE 12-1**). They have mostly just been refined. In this chapter, we cover the types of spark-ignition engines that are available, identify the major components that make up the engine, and describe how these components operate together. We also compare them briefly to diesel, two-stroke, and rotary engines.

FIGURE 12-1 A. Antique flat-head engine. **B.** Modern dual overhead cam engine.

▶ Principles of Thermodynamic Internal Combustion Engines

K12002

Thermodynamics is generally defined as the branch of physical science that deals with heat and its relation to other forms of energy such as mechanical energy. In this chapter, we discuss how heat energy is used in the internal combustion engine to produce power and make work happen.

In automotive applications, the useful effect is to move a vehicle down the road as well as provide motive (moving) power for all of the onboard systems. Engines used for motive power may be classified as external combustion or internal combustion engines. External combustion means that the fuel is being burned outside of the engine, and internal combustion means the fuel is burned inside of the engine. Two examples of the **external combustion engine** are the steam engine and the Stirling engine (**FIGURE 12-2**). Whether an engine operates on internal combustion or external combustion, they both are called "heat engines" because they run off heat energy, which is why a basic understanding of thermodynamics is critical to understand how they operate.

At one time, external combustion engines, such as steam engines, were used to power almost all equipment. Steam engines are the best example of external engines powering equipment such as farm tractors, railroad trains, automobiles, boats and ships, and more. Even a steam-powered airplane was produced, although it never became popular.

In a steam engine, steam is created in an external boiler (external combustion) and used to push a piston back and forth in a cylinder. Most steam engines applied steam alternately to each side of the piston, so the piston was powered in both directions. One problem with steam engines is that

You Are the Automotive Technician

You are working in the back shop when a salesman from the new car sales department asks if you can answer some questions for a customer. The customer's previous car was totaled in a parking lot accident, and he is very interested in a couple of cars on the lot. He has some technical questions that need to be answered, and the service manager selected you to answer the questions. You greet the customer, and he asks you the following questions:

1. "I see a nice diesel pickup truck that I could use on the farm. How does a compression-ignition engine operate differently from a spark-ignition engine?"
2. "My son would like me to buy that RX8. How does a rotary engine operate differently from a piston engine?"
3. "As I am looking at specifications on the vehicles, what is the difference between horsepower and torque?"
4. "What is meant by overhead cam?"
5. "What is an interference engine?"

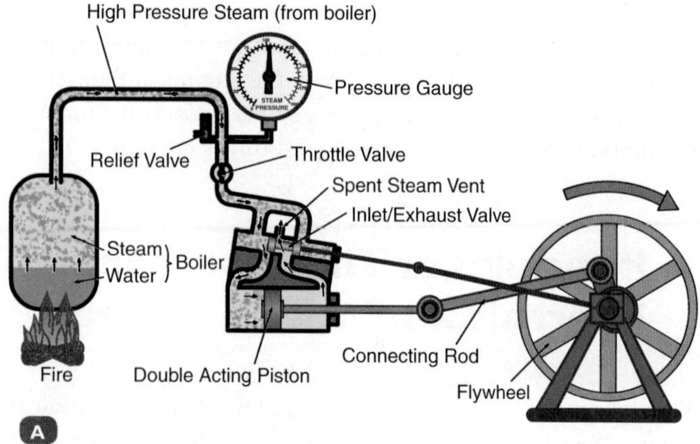

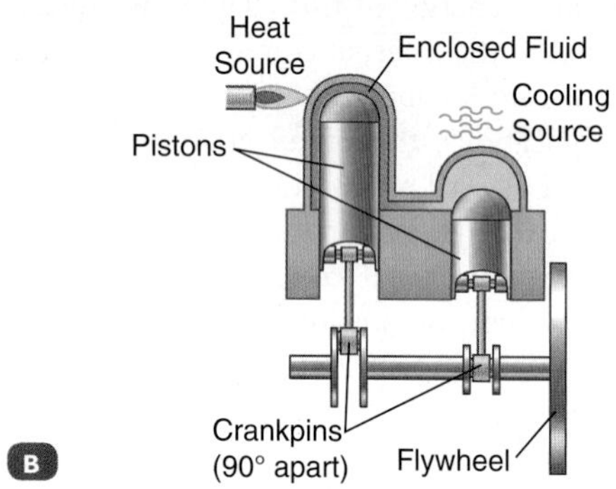

FIGURE 12-2 External combustion engines. **A.** Steam engine. **B.** Stirling engine.

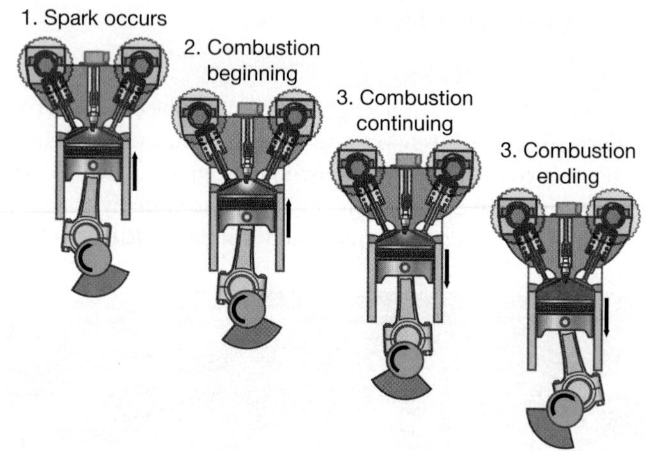

FIGURE 12-3 Combustion in an SI engine.

they take a relatively long time to generate steam pressure, so you could not just hop in a steam-powered car and take off. The boilers also presented an explosion hazard if they generated too much pressure or if the boiler weakened due to rust.

The Stirling engine is also an external combustion engine. This engine uses two cylinders with a passageway that connects both cylinders. One cylinder houses the power piston and the other, the displacer piston. The power cylinder is heated, and the air expands, pushing the piston down. The hot air is then pushed into the displacer cylinder, where it is cooled. Air moves back and forth between the cylinders to expand and cool, which drives the engine. It holds promise as an alternative source of power but has not become popular for transportation because its output cannot be easily varied. Solar-powered Stirling engines are gaining popularity as home power sources because they operate on free solar energy, which also means there are no byproducts of combustion to worry about. Stirling engines can run almost silently and therefore can be used near people without disturbing them.

The **internal combustion engine** has almost completely replaced the external combustion engine and has been around for well over a century. It is still the favored mode of power for the transportation industry, whether for on-road applications (cars, trucks, buses, etc.) or for non-road applications (tractors, trains, ships, etc.). This chapter focuses on spark-ignition internal combustion engines, as they are the most widely used engines in modern automobiles.

Refer to the Alternative Fuel Systems chapter for more information on hybrid vehicles, flex-fuel vehicles, "pure" battery-electric vehicles (BEVs), fuel cell vehicles (FCVs), and other alternative-fueled vehicles.

Gasoline (spark-ignition) and diesel (compression-ignition) engines are prime examples of internal combustion engines. Fuel is burned *inside* the internal combustion engine. According to Charles's law, when a gas is heated, it expands. Because fuel contains energy in chemical form, when it is burned in a sealed combustion chamber, it creates high pressure that pushes on a moveable piston. The moving piston produces power to do work.

The internal combustion engine (ICE) can be classified in two ways: as a reciprocating piston engine or as a rotary engine. Piston engines use a crankshaft to convert the reciprocating movement of the pistons in their cylinder bores into rotary motion at the crankshaft. The **rotary engine** uses a rotating motion rather than reciprocating motion. Both the piston and the rotary engine are described in greater detail later in this chapter.

Piston engines are either **spark-ignition (SI) engines** or **compression-ignition (CI) engines**. In SI engines, liquid fuels such as gasoline, ethanol, methanol, or butanol are compressed and ignited by an electrical spark, which jumps across the air gap of a spark plug in the combustion chamber. Timing of combustion is totally reliant on when the spark jumps across the electrodes of the spark plug (**FIGURE 12-3**). In diesel engines, air in the sealed combustion chamber is compressed so tightly that it becomes hot enough to ignite the fuel as soon as it is injected into the combustion chamber. Timing of the combustion process is reliant on when the fuel is injected; therefore, CI engines do not use spark plugs (**FIGURE 12-4**).

Refer to the Compression-Ignition Engine chapter for a more complete description of diesel engines.

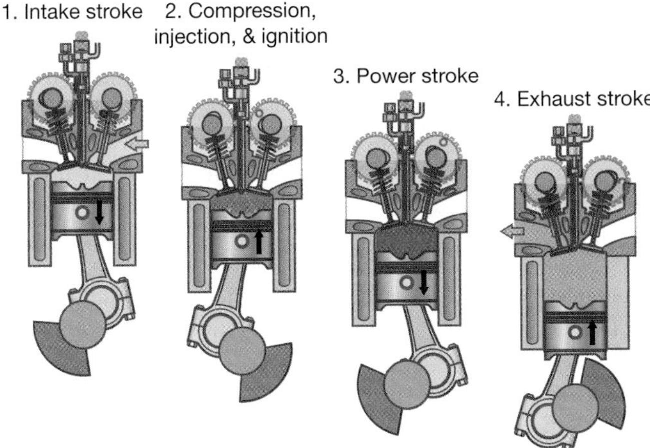

FIGURE 12-4 Combustion in a CI engine.

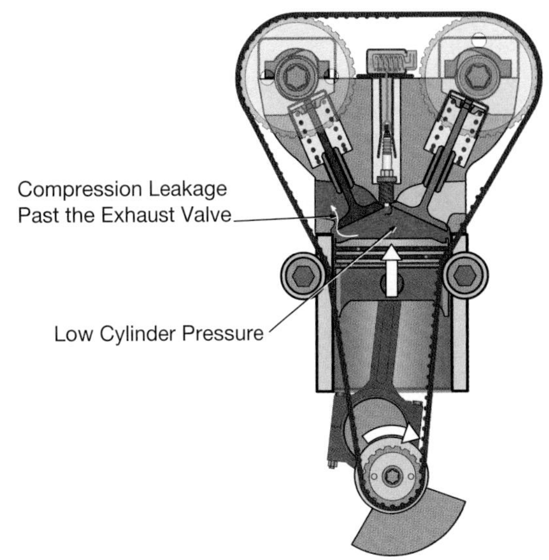

FIGURE 12-5 If the cylinder has a leak, the pressure will not rise, making combustion inefficient.

▶ Principles of Engine Operation (Physics)

K12003

Engines operate according to the unchanging laws of physics and thermodynamics. Understanding the physics and science involved with an engine will help you diagnose engine problems. For example, knowing that pressure rises when the volume of a sealed container is reduced will help you understand the need for sealing the container for maximum power.

Realize too that when molecules are tightly packed together, they produce far more expansion pressure during combustion than when they are not. Valves and ports in the engine's cylinder head(s) provide a means of sealing the combustion chamber. If the valves leak, the pressure in the cylinder will not rise as it should during the compression stroke of the piston (**FIGURE 12-5**). If there is too little pressure squeezing the air-fuel mixture together in the cylinder, the mixture will not get packed tightly enough, and less engine power will be developed when it is ignited.

A good example of the importance of compression is the burning of black powder in open air; it produces fire and smoke, but no explosion. If the same powder is wrapped tightly, it becomes a firecracker (or even a stick of dynamite)—exploding the wrapping and producing a big bang. In the same way, more power is produced when air and fuel are compressed into a tightly packed space.

Pressure and Temperature

Recall that according to Charles's law, in a sealed chamber the pressure and temperature of a gas are directly related to each other. As pressure rises, so does temperature; as pressure decreases, so does temperature. For example, think of a cylinder with a moveable plunger at one end (**FIGURE 12-6**). This plunger seals tightly in the cylinder so that no air can escape past the plunger. Installed on the other end of the cylinder are a pressure gauge and a thermometer. As we push the plunger in, the air pressure rises in the sealed cylinder as the air molecules are squeezed together. As the air molecules are squeezed together more tightly, the pressure and the temperature rise as a result of friction between the air molecules as they bounce off each other with greater force. The air in the cylinder heats up as this happens, so we see not only the pressure rise on the pressure gauge but also the temperature rise on the thermometer. The same amount of heat is in the cylinder as when uncompressed, but when it is compressed, the heat is concentrated; thus, the temperature rises.

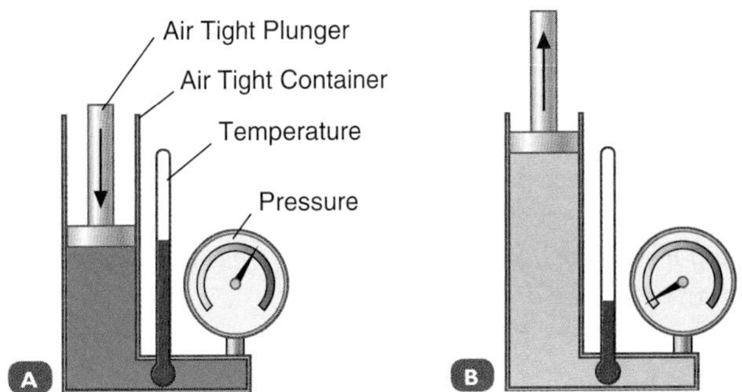

FIGURE 12-6 Pressure changes temperature. **A.** As pressure goes up, temperature goes up. **B.** As pressure goes down, temperature goes down.

▶ TEST YOUR UNDERSTANDING

When the plunger is pushed into the cylinder and the air pressure and temperature are increased, what will happen to the temperature if the plunger is held steady at that point? If you said it would stay the same because the pressure stayed the same, you would be wrong. Because heat travels from hot areas to cold areas, the heat would slowly transfer to the surrounding air until the temperatures are equalized. And yes, that would lower the pressure inside the cylinder slightly, as the temperature went down. Once the cylinder is cool, what happens to the temperature of the air inside it if the plunger is pulled way back so there is a vacuum inside? If you said it would be colder than outside air, you would be correct. Reducing pressure reduces temperature, even if that means below the outside temperature. And that is the principle we use in an air-conditioning system. To find out more, turn to the HVAC Principles chapter.

A diesel engine uses this principle to ignite the fuel injected into an engine cylinder. The air is compressed so tightly that it becomes hot enough to ignite the fuel when it is sprayed into the cylinder; this is why diesel engines are called compression-ignition engines. Again, using the example of a plunger in a cylinder, what happens when the plunger is pulled outward? Pulling out the plunger reduces gas pressure and gives the molecules more room to move; they affect each other less and temperature decreases. Thus, a drop in pressure produces a lower temperature because even though the same amount of heat is contained in the cylinder, the heat is more spread out.

Looking at a related scenario, what happens to pressure when the gas *temperature* changes? When a gas is heated, its molecules start to move more quickly and require more space. Heating a gas in a sealed container increases the pressure in the container (thermal expansion) (**FIGURE 12-7**). Cooling a gas has the opposite effect. As the molecules slow down, they demand less space. As a result of cooling a gas in a sealed container, the pressure drops.

Temperature and Energy

The temperature of a gas is one measure of how much energy it has. The more energy a gas has, the more work it can do. The heating of gas particles makes them move faster, which produces more pressure. This pressure exerts more force on the container in which the gas is located, which is how an ICE functions. Pressure is first raised quite a bit through compression and then much higher through combustion of the air-fuel mixture. Burning the air-fuel mixture increases the heat temperature inside the container tremendously, which creates the necessary pressure to produce work. The more energy the air-fuel mixture has, the more force it exerts on the piston and the more work the piston can do (**FIGURE 12-8**). This principle takes place during the power stroke and pushes the piston down the sealed container.

Latent (stored) heat energy exists in various kinds of fuels; it is released to do work when the fuel is ignited and burned. Types of fuels that contain latent heat energy include liquid fuels such as gasoline, diesel, and ethanol; gaseous fuels such as natural gas, propane, and hydrogen; and solid fuels such as gunpowder, wood, and coal. Latent heat energy is often measured and expressed as British thermal units, abbreviated Btu.

One Btu equals the heat required to raise the temperature of 1 pound (lb) of water by 1°F (17.22°C). Gasoline has a comparatively high Btu per gallon rating of around 14,000 Btu. Diesel fuel is even more energy dense, however, at around 25,000 Btu per gallon. Coal has a much lower Btu rating, which is one reason why both the home heating and the transportation (rail) industries have moved from coal to petroleum (i.e., home heating oil and diesel fuel) for better energy efficiency (transport and storage of a liquid fuel is also much easier).

Pressure and Volume

Pressure and volume are inversely related; as one rises, the other falls. A cylinder with a pressure gauge and movable piston is a good example. It contains air, and as the piston is pushed in, the inside air is forced into a smaller volume. At the same time, the

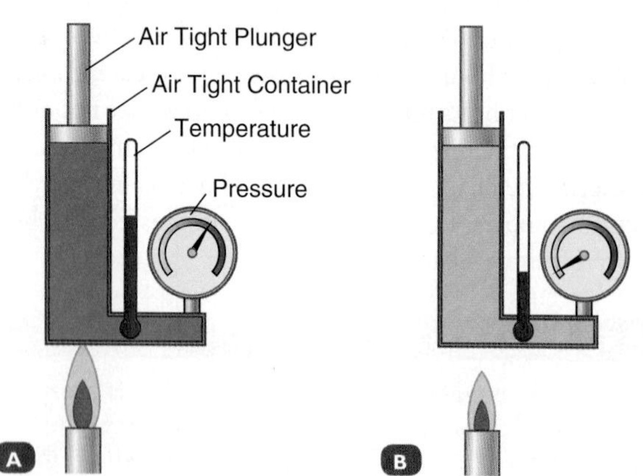
Air Tight Plunger
Air Tight Container
Temperature
Pressure
A
B

FIGURE 12-7 Temperature changes pressure. **A.** As temperature goes up, pressure goes up. **B.** As temperature goes down, pressure goes down.

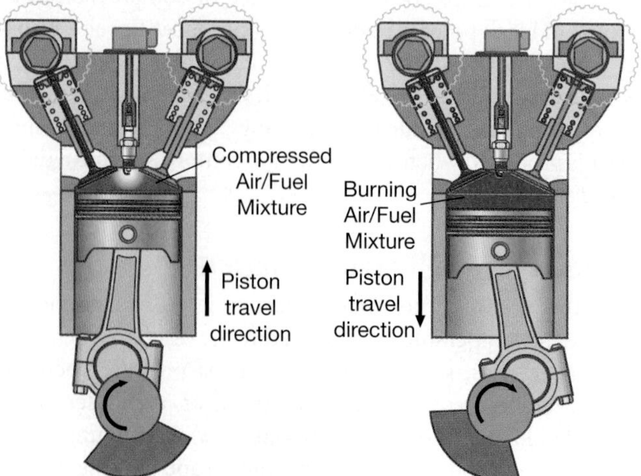
1. Compressing the air/fuel mixture raises its heat energy level.
2. Igniting the fuel increases the heat and pressure even higher.
3. The high combustion temperature causes high cylinder pressures which force the piston down the cylinder.
Compressed Air/Fuel Mixture
Piston travel direction
Burning Air/Fuel Mixture
Piston travel direction

FIGURE 12-8 Burning a compressed gas increases temperature, producing more pressure and increased force.

pressure gauge shows an increase in pressure. It is this increase in pressure that allows the pump to do its work. When the piston is pulled out, the volume occupied by the gas grows larger, and the pressure drops. A larger volume has less gas pressure, and when the volume is reduced, the gas pressure rises. Keep in mind that larger pressures are desirable to increase the amount of work done in an engine (**FIGURE 12-9**).

Force, Work, and Power

K12004

Effort to produce a push or pull action is referred to as **force**. A compressed spring applies force to cause, or resist, movement. A tensioned lifting cable applies force to cause lifting movement. Force is measured in pounds, kilograms, or newtons. When force causes movement, **work** is done (**FIGURE 12-10**).

For example, when the compressed spring or the tensioned lifting cable causes movement, work is performed. Without movement, work cannot be performed even if force is applied. *Work is equal to distance moved times force applied.* If the lifting cable of a hoist lifts a 250 lb engine 4' in the air, the amount of work done is equal to 4' times 250 lb, or 1000 foot-pounds of work. Work is measured in foot-pounds (ft-lb), watts, or joules. Work can only be accomplished when something is moved.

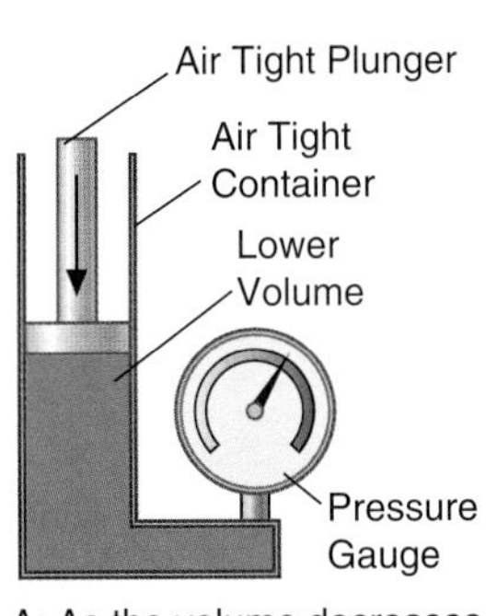
Air Tight Plunger
Air Tight Container
Lower Volume
Pressure Gauge
A: As the volume decreases, the pressure increases

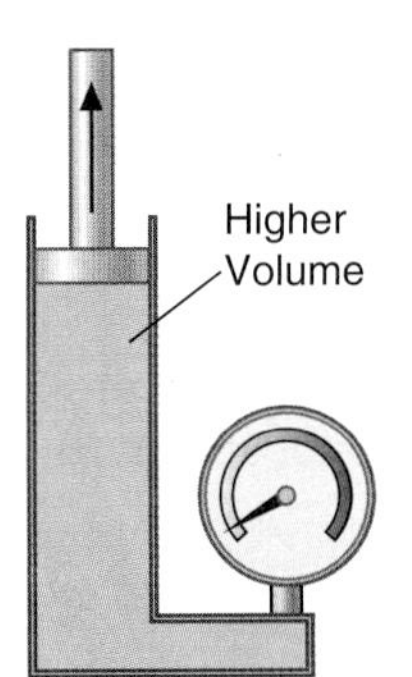
Higher Volume
B: As the volume increases, the pressure decreases

FIGURE 12-9 Volume affects pressure.

Work = distance moved × force applied
= 4 ft × 150 lb
= 600 ft-lb
4 feet
150 pounds

FIGURE 12-10 Work.

▶ TECHNICIAN TIP

Holding a heavy starter motor in place while trying to get the bolts started is not technically work; it is force. Although it seems like you are working hard, your arm muscles twitching as you stand under the vehicle and hold the part in place, you are not performing work in the true physics sense. It takes movement, along with force, to qualify as work. So lifting the starter from the ground to its position on the engine is work, but holding it there is not work. Understanding the difference between these two terms will give you a good foundation for understanding power.

The rate or speed at which work is performed is called **power** (**FIGURE 12-11**). The more power that can be produced, the more work can be performed in a given amount of time. Power is measured in ft-lb per second or ft-lb per minute. If an electric motor can lift a 600 lb weight 20' in 10 seconds, power used would be equal to:

20 (feet) × 600 (lb) ÷ 10 (seconds) = 1200 ft-lb per second.

One horsepower equals 550 ft-lb per second, or 33,000 ft-lb per minute. So 1200 ft-lb per second equals:

1200 (ft-lb) ÷ 550 (ft-lb per second) = 2.18 horsepower

The watt or kilowatt (1000 watts) is the metric unit of measurement for power, where 746 watts equals 1 horsepower. So the electric motor in the example would be developing:

2.18 (horsepower) times 746 (watts) = 1627 watts, or 1.627 kilowatts, of power

Power and Torque

Torque is described as a twisting force. Because torque is a force, movement does not have to occur to have torque. Torque is applied before or during movement. When a twist cap on a water bottle is removed, maximum torque is applied just before the cap starts to turn. When the same cap is tightened, maximum torque is applied once the cap starts to get tight. The concept of "twisting force" should always come to mind when the term "torque" is used. When a piston is pushed down a cylinder

Power = (distance moved × force applied) / time
= (4 ft × 150 lb) / 2 seconds
= 300 ft-lb / second
= 0.54 HP
4 feet in 2 seconds
150 pounds
1 Horsepower (HP) = 550 ft-lb per second

FIGURE 12-11 Formula for power.

during the power stroke, it applies force to a connecting rod linking the piston and crankshaft, causing the crankshaft to rotate. The rotational force applied to the crankshaft is called torque. The crankshaft then applies that twisting force to other components such as the transmission and front pulley.

The unit of measurement for torque in the imperial system is ft-lb; in the metric system it is newton meters. If a force of 100 lb is applied to the end of a 1' long wrench (lever) attached to a bolt, the resulting torque applied to the wrench will be 100 ft-lb. If no movement occurred, torque was still applied as a force, but no work was performed.

Work only occurs when movement happens. So applying torque (twisting force) to something while it is turning means that work is being performed.

Power means that work is being done at a certain rate, which involves time. So power is the amount of work performed in a certain amount of time. The measurement of engine power is calculated from the amount of torque at the crankshaft and the speed at which it is turning in rpm. The formula for engine horsepower is:

$$\text{Horsepower} = \text{rpm} \times \text{ft-lb} \div 5252$$

For example, if an engine creates 500 ft-lb of torque at 4000 rpm then the amount of horsepower is:

$$500 \times 4000 \div 5252, \text{ or } 380 \text{ horsepower at } 4000 \text{ rpm}$$

Because horsepower would change with rpm, it is necessary to express not only the power value but also the engine speed, in rpm, at which it occurs.

▶ TECHNICIAN TIP

Power can also be expressed in kilowatts. A kilowatt (1000 watts) is equivalent to 1000 newtons per meter per second.

The formula for computing an engine's horsepower might seem confusing, as we said earlier that 1 horsepower equals 33,000 ft-lb per minute; yet we divide the product of torque × rpm by 5252. Why do we use 5252 instead of 33,000? The reason is a bit complicated, having to do with the definition of ft-lb. When relating to *work*, ft-lb means force times distance moved. That definition works well in a lifting situation, but not so well in a twisting (torque) situation. In a *torque* situation, ft-lb means a twisting force applied to a shaft—that is, applied force times lever distance of the applied force. Thus, it is possible for torque to result in no movement, only an applied force. And we label that force in ft-lb even though for work or power to happen, movement must take place.

To calculate the twisting power of a shaft, we need a way to add *distance moved* and *time* to the equation so that power can be represented; we can't just say that 33,000 ft-lb = 1 horsepower when talking about torque. That is why rpm is included in the formula. Because rpm refers to a certain number of revolutions per minute, it includes both time and distance. But how much distance is 1 rpm? The answer relates to radians. A radian describes how many radius distances there are in the circumference of a

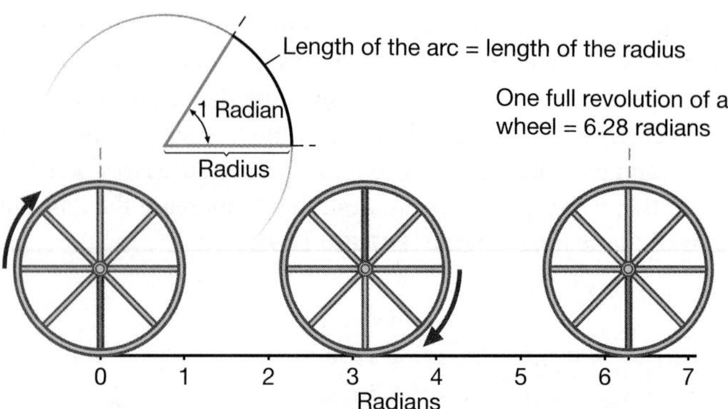

FIGURE 12-12 Radians in the circumference of a circle.

circle (**FIGURE 12-12**). Because there are 3.14 diameters in the circumference of a circle, there are twice as many, or 6.28, radius distances (radians) in the circumference of any circle. The larger the circle, the longer the radians, but still only 6.28 of them fit within the total circumference. And that is where we get our distance. Because torque relates to an equivalent amount of force a certain distance from the rotational center (radius), and there are 6.28 radians in the circumference of a circle, every revolution equals a distance of 6.28 radians.

To convert torque ft-lb to work ft-lb, there has to be movement. We can either multiply the rpm by 6.28 and use the 33,000 (ft-lb per horsepower) factor, or divide the 33,000 ft-lb by 6.28 (distance around a circle in radians) and come up with a new factor, which is 5252, that we can use with the original "torque × rpm" numbers for calculating engine horsepower. To keep the numbers more manageable, most people go with the 5252 factor. Thus, if we multiply the ft-lb of torque × rpm and divide that number by 5252, we will have calculated the engine's horsepower at that particular rpm.

Torque versus Horsepower

As torque is a twisting or turning force, **horsepower** is the rate (in time and distance) at which that force (torque) is produced. Torque alone does not mean work has been accomplished. It takes movement and time to accomplish a given amount of work in a given amount of time (horsepower). At the same time, for an engine to produce torque, it has to be running. So the rotation of the crankshaft (torque × distance) in a running engine means that work and power are occurring because torque × rpm gives us power (**FIGURE 12-13**). Also, an engine puts out varying amounts of torque and power as it is operating. Thus, if it is producing torque, it is producing power. Factors that affect an engine's torque output include engine **volumetric efficiency** (the rate of air intake and exhaust) at various crankshaft speeds and internal component friction (**parasitic losses**). In a naturally aspirated engine (nonpressurized intake system), torque peaks at the rpm where the engine's cylinders fill the most with air. Torque starts to drop as engine speed increases past peak torque rpm.

Actually, in a naturally aspirated (naturally breathing, nonpressurized) engine, air almost never completely fills the

FIGURE 12-13 An automotive chassis dynamometer measures power output of the engine at the wheels of the vehicle.

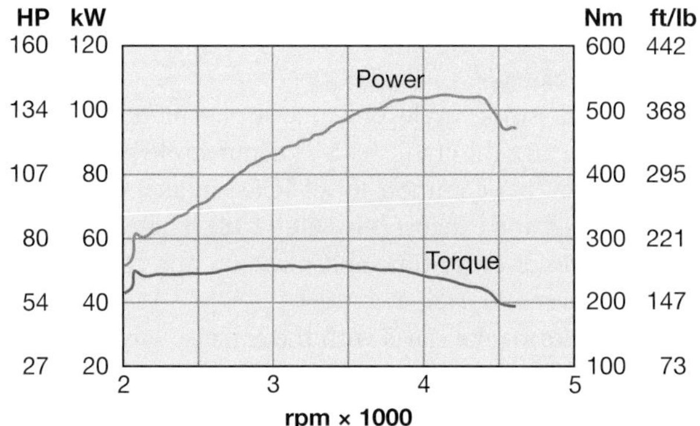

FIGURE 12-14 The relationship between torque and horsepower.

combustion chamber while the engine is running, because of the resistance of air flowing through the intake system. Peak engine torque rpm occurs at the peak volumetric efficiency (somewhere around 85% in an unmodified engine). Peak torque usually occurs at some low- to mid-rpm engine speed, depending on bore and stroke of the engine as well as intake and exhaust port size and valve timing.

Because engine rpm tends to rise faster than torque falls off (above peak volumetric efficiency), an engine's maximum horsepower occurs at a higher rpm than the peak torque rpm. At some point in the rpm range, the torque on the crankshaft drops so low that the crankshaft can no longer do additional work, and the horsepower actually starts to decrease. Remember, it is horsepower that does the work, but torque makes it happen. Engine torque (and therefore horsepower) increases can be achieved through any engine modifications that improve volumetric efficiency (**FIGURE 12-14**). In fact, a turbocharger or supercharger increases an engine's volumetric efficiency well above 100%. For example, a 1.6-liter Volkswagen diesel engine when turbocharged to its rated 11 lb of boost would be theoretically equivalent, in terms of horsepower, to a naturally aspirated 2.8-liter engine.

Applied Science

AS-97: Torque: The technician can demonstrate an understanding of how torque relates to force and angular acceleration.

Newton's second law of motion states that the change in speed of an object over a given time is proportional to the force exerted on it. We know that a train is much slower to accelerate than a car, and once it is moving, the train is far more difficult to stop. This is the basis for our understanding of the concept of torque relating to force and angular acceleration.

Physics can tell us a lot about how engines react to forces. Consider a flywheel attached to the crankshaft of an engine, which is an example of rotational motion. The torque necessary for the assembly to rotate is the product of force multiplied times distance.

In physics, torque can be thought of as a rotational force that causes a change in rotary motion. The term "twisting force" is used by most automotive technicians. To determine angular acceleration, which is measured in radians per second squared, we need to know several other items of information. We would need to know the initial and final angular velocities and time. Angular velocity describes the speed of rotation of an object that is following a specific axis. Angular acceleration is the rate of change of angular velocity with time and is measured in radians per second squared. The angular acceleration is caused by the torque, which produces the force to make it happen. This process is proportional, as stated in Newton's second law of motion.

Applied Science

AS-50: Work: The technician can explain the relationship between torque and horsepower.

Two technicians are discussing the relationship between torque and horsepower during their break time at work. Al believes that torque is the turning force at the engine's crankshaft. Bob says that horsepower is the rate at which force is produced. Both technicians are correct as they discuss the unique relationship of torque and horsepower. One cannot exist without the other.

When we compare a race car to a bulldozer, we can see two different applications of torque and horsepower. A bulldozer has lots of torque, but the engine may be operating at 2000 rpm. A race car may be operating at 8000 rpm with a lower amount of torque. Low-speed torque is needed in some situations and high rpm horsepower is needed in others. (Additional information on the relationship between torque and horsepower is described in this chapter.)

Engine Load Factor

Load factor or power range is typically used in regard to diesel engines and describes how long an engine can produce its maximum power output. An engine that is quite powerful may not be designed to produce its maximum power over long periods. A gasoline engine in a car may give good acceleration in short bursts, but it can overheat and/or fail if operated at maximum speed and power for too long. In contrast, due to its robust construction, a heavy-duty diesel engine in a large truck is designed to operate near or at its maximum speed and load for long periods without damage.

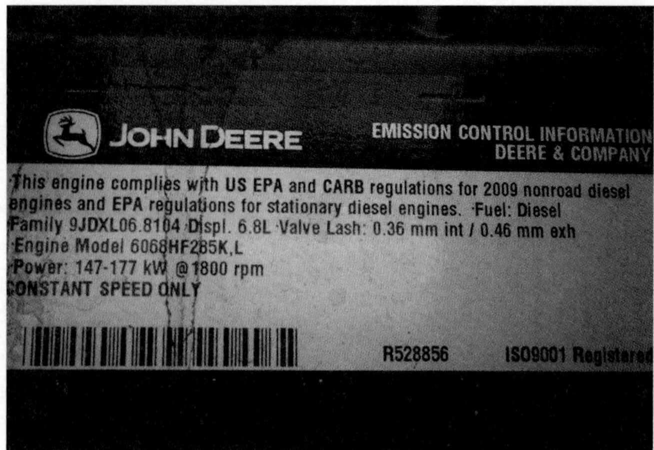

FIGURE 12-15 Example of a load factor for an engine.

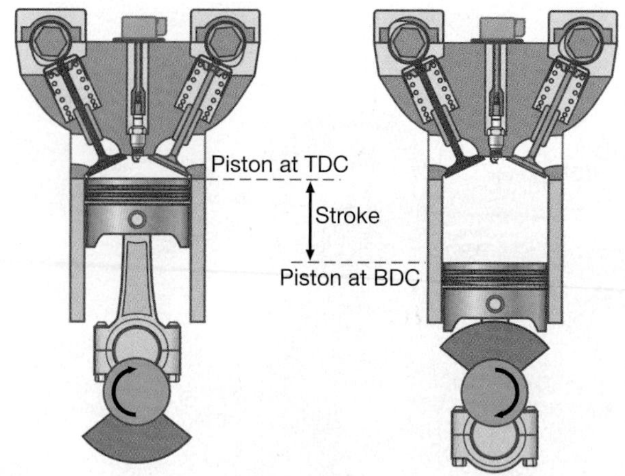

FIGURE 12-16 Piston movement from TDC to BDC or from BDC to TDC is one stroke.

One way to describe the load factor of an engine is to give its power as an average over a certain period. This is stated as a percentage and is called load factor (**FIGURE 12-15**). An engine required to operate at maximum power over 10 hours is said to have a load factor of 100%. Using that as a standard, most car engines usually only operate at about 20% to 30% during a typical daily drive. The engine is never run at 100% load for an extended time; in fact, it may never reach 100% even for a short time.

▶ Four-Stroke Spark-Ignition Engines

K12005

The SI engine used in today's vehicles operates on the four-stroke cycle principle: It takes four strokes of the piston to complete one cycle. This complete cycle is called the Otto cycle and is named after German engineer Nikolaus Otto. When the piston in a cylinder is at the position farthest away from the crankshaft, it is said to be at **top dead center (TDC)**. When the piston in the cylinder is at a position closest to the crankshaft, it is said to be at **bottom dead center (BDC)**. When the piston moves from TDC to BDC or from BDC to TDC, one **stroke** has occurred (**FIGURE 12-16**). Two or more strokes are called **reciprocating motion**, meaning an up-and-down or back-and-forth motion within the cylinder. Piston engines are therefore referred to as reciprocating engines.

Piston engines can be simple, single-piston engines such as those on lawn mowers, or they can be much more complicated, multi-piston engines, such as those in automobiles, trucks, and heavy equipment. Multi-cylinder engines come in various cylinder arrangements (**FIGURE 12-17**):

- Inline engines with pistons one behind another
- Flat engines with horizontally opposed cylinders, called boxer engines
- V-type engines with the pistons forming a V configuration
- VR or W-type engines, a modified version of the V-type engine, with cylinders in a narrower angle to one another and housed in the same cylinder bank

Basic Four-Stroke Operation

K12006

In a single four-stroke cycle, only one stroke out of four delivers new energy to turn the crankshaft. The four strokes must include the five key events common to all ICEs: intake, compression, ignition, power, and exhaust (**FIGURE 12-18**). These events occur in the same order every time and make up one complete cycle that repeats over and over.

The **intake stroke** starts with the exhaust valve closed, the intake valve(s) opening, and the piston moving from TDC to BDC. As the piston moves down, the volume above the top of the piston increases. This makes pressure inside the cylinder lower than the pressure outside the cylinder. Higher outside air pressure forces air (usually with fuel) into the cylinder. As the piston reaches BDC, the intake valve(s) closes, and the intake stroke ends.

The **compression stroke** starts near BDC when the intake valve(s) closes. The piston moves from BDC to TDC. As the piston moves up, the air-fuel mixture is compressed into a smaller and smaller volume. Compression of air caused by the moving piston causes the temperature of the air-fuel charge to rise, making ignition easier and the combustion (burning of fuel) more complete and efficient.

As the piston reaches TDC of the compression stroke, **ignition** occurs. The air-fuel mixture is ignited and burns rapidly at up to about 4500°F (2482°C). The heat of combustion causes the burning gases to expand greatly (thermal expansion), which creates very high pressure in the combustion chamber. This pressure is also applied to the entire cylinder, including the top of the piston, which is free to move down the cylinder.

The **power stroke** occurs as this extreme force moves the piston from TDC to BDC, with both valves remaining closed. The exhaust valve(s) starts to open near BDC.

At the end of the power stroke, the exhaust valve(s) opens and the **exhaust stroke** occurs as the piston moves from BDC to TDC, which pushes the burned gases out of the cylinder

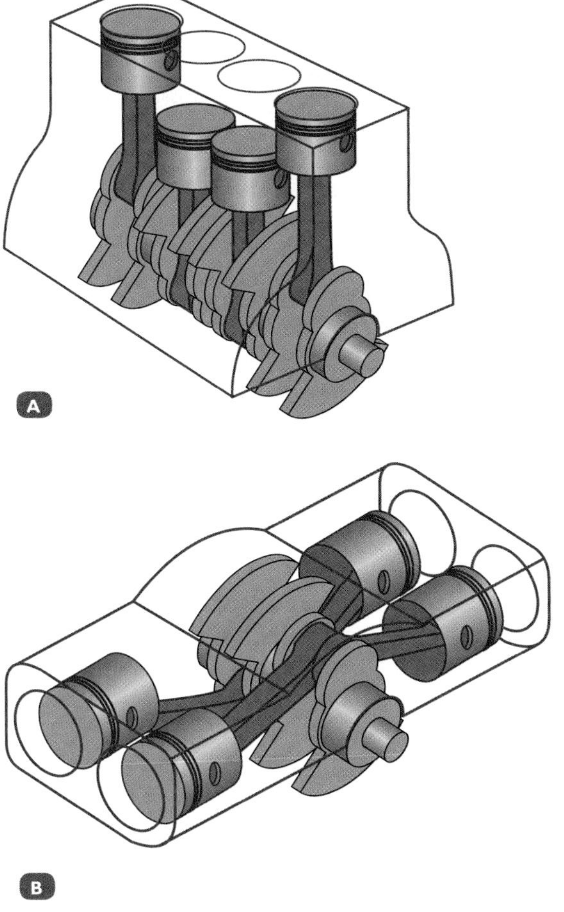

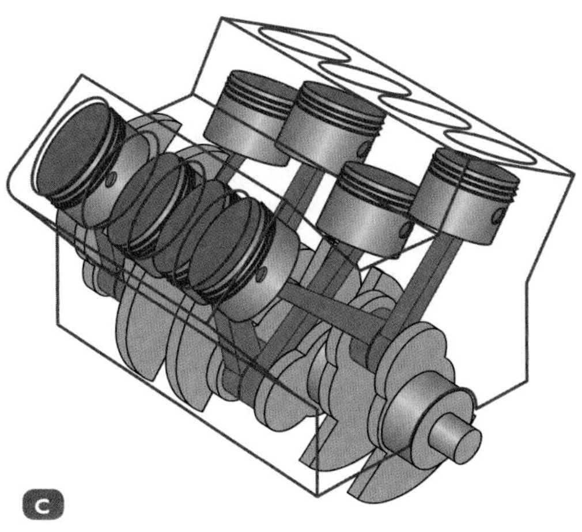

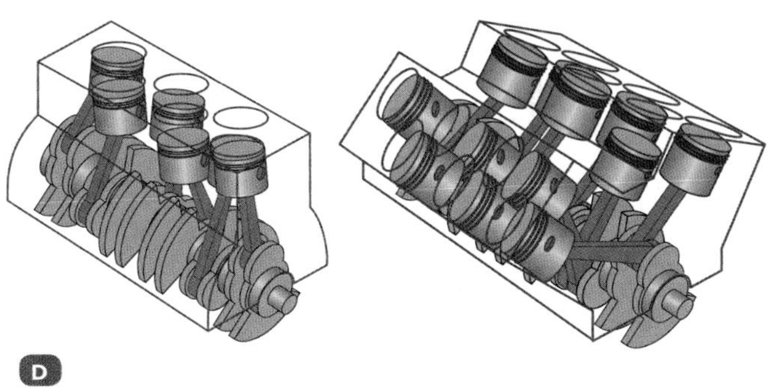

FIGURE 12-17 **A.** Inline. **B.** Flat. **C.** V-type. **D.** VR and W types.

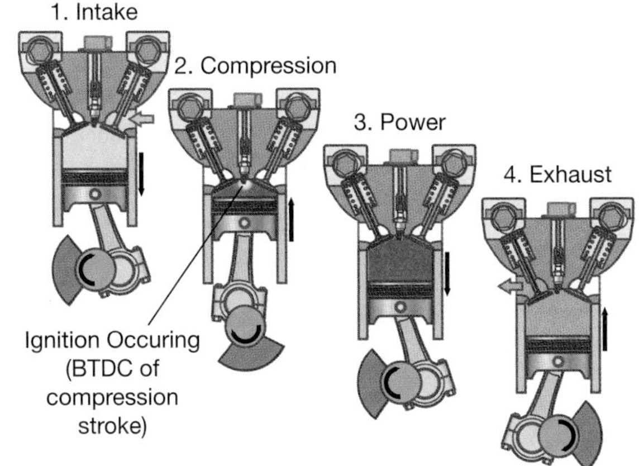

FIGURE 12-18 The basic four-stroke cycle.

through the open exhaust valve(s). When the piston nears TDC, the exhaust valve(s) starts to close and the intake valve(s) starts to open, and the four-stroke cycle starts over from the beginning. Note that the crankshaft has completed two full rotations during the four-stroke cycle. Thus, four complete strokes make one complete cycle.

Applied Science

AS-99: Rotational: The technician can explain how rotational motion can be converted to linear motion and why balance is important in rotating systems.

When working in an automotive repair facility, it is important that the technician understand basic engine operation. The text has explained much regarding each component of the engine block assembly and cylinder head. The technician should know that reciprocating motion is an up-and-down motion such as the pistons in a cylinder. The crankshaft converts this linear motion into rotational motion.

It is also important to know that the camshaft lobe converts rotary motion into linear motion. As described in the text, the lobe is the raised portion on a camshaft used to lift the lifter and open the valve. Opening the valve requires linear motion on overhead cam engines as well as cam in block engines. The text describes a **cam** as an egg-shaped piece (lobe) mounted on the camshaft. The egg shape of the cam lobe is designed to lift the valve open, hold it open, and let it close. This is the operation that converts rotating motion into linear motion.

Balance is extremely important in a rotating system for a number of reasons. In order to have a smooth running engine, with long engine life and dependability, balance is necessary. On the crankshaft, we have a vibration damper or harmonic balancer, as described in the text. This important component is carefully engineered to meet the needs of a specific engine regarding proper balance.

Engine Measurement—Size

K12007

ICEs are designated by the amount of space (volume) their pistons displace as they move from TDC to BDC, which is called **engine displacement**. So, a 5.4-liter V8 engine has eight cylinders that displace a total volume of 5.4 liters. Displacement can be listed in cubic centimeters, liters, or cubic inches. To find an engine's displacement, you need to know the bore, stroke, and number of cylinders for a particular engine.

The **cylinder bore** is the diameter of the engine cylinder. The bore is measured across the cylinder, parallel with the **block deck**, which is the machined surface of the block farthest from the crankshaft. Automotive cylinder bores can vary in size from less than 3" to more than 4".

The distance the piston travels from TDC to BDC, or from BDC to TDC, is called the **piston stroke**. Piston stroke is determined by the offset portion of the crankshaft called the **throw**. The crankshaft is described in greater detail later in this chapter. Piston stroke also varies from less than 3" to more than 4". Generally speaking, the longer the stroke, the greater the engine torque produced. A shorter stroke enables the engine to run at higher rpm to create greater horsepower. Engine specifications typically list the bore size first and the stroke length second (bore vs. stroke).

The volume that a piston displaces from BDC to TDC is called **piston displacement**. Increasing the diameter of the bore or increasing the length of the stroke produces a larger piston displacement. The formula for calculating piston displacement (**FIGURE 12-19**) is:

Cylinder bore squared × 0.785 × the piston stroke

This formula works for calculating both the standard displacement in cubic inches or the metric displacement in cubic centimeters (ccs) or liters.

For example, a 5.4-liter (329-cubic inch) V8 truck engine has a 3.55" bore, a 4.16" stroke, and eight cylinders. Using the formula for displacement:

$$3.55 \times 3.55\ (\text{bore}^2) = 12.6025 \times 0.785\ (\text{constant}) = 9.893$$
$$= 9.893 \times 4.16\ (\text{stroke}) = 41.155\text{-cubic-inch piston displacement}$$

Once you know the piston displacement, the next step to finding engine displacement is to multiply piston displacement times the number of cylinders in the engine (**FIGURE 12-20**). Continuing from the previous example:

41.155-cubic-inch piston displacement ×
8 (number of cylinders) = 329.24-cubic-inch
engine displacement

The displacement of an engine (also called engine size) can be altered by changing cylinder bore (diameter), piston stroke (length), or the number of cylinders.

TECHNICIAN TIP

An engine with the same size bore and stroke is referred to as a square engine. An engine with a larger bore than stroke is called an oversquare engine (short-stroke engine). An engine with a bore smaller than the stroke is called an undersquare engine (long-stroke engine). Oversquare engines tend to make their power at higher rpm, whereas undersquare engines tend to make their power at lower rpm.

Compression ratio compares cylinder volumes with the piston at BDC and at TDC (**FIGURE 12-21**). Maximum cylinder volume is at BDC, and minimum cylinder volume is at TDC. The ratio is given as two numbers. A compression ratio listed as 8:1 (8 to 1) means that the maximum cylinder volume is eight times larger than the minimum cylinder volume. Compression ratio is affected by changing the size and shape of the top of the piston, changing the size of the combustion chamber in the cylinder head or piston, or altering valve timing. The higher the compression ratio, the higher the compression pressures within the combustion chamber and therefore the higher the thermal expansion during combustion, making the engine more fuel efficient. But a compression ratio that is too high can cause the air-fuel mixture to be ignited by the high compression temperature before the correct time, which is when the spark occurs. This early ignition can cause damage to the engine bearings and pistons; therefore, manufacturers design their engines with an optimum compression ratio.

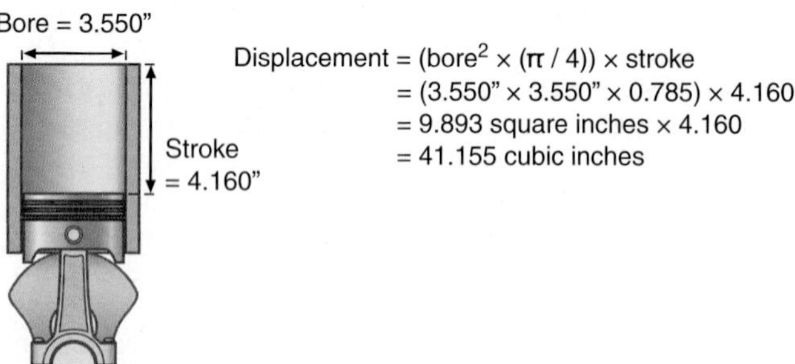
Bore = 3.550"
Displacement = (bore² × (π / 4)) × stroke
= (3.550" × 3.550" × 0.785) × 4.160
= 9.893 square inches × 4.160
= 41.155 cubic inches
Stroke
= 4.160"

FIGURE 12-19 Piston displacement.

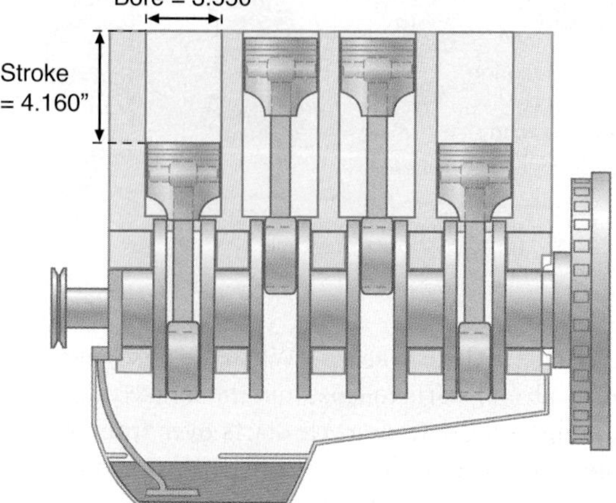
Engine displacement = (bore² × (π / 4)) × stroke × number of cylinders
= (3.550" × 3.550" × 0.785) × 4.160 × 4
= 9.893 square inches × 4.160 × 4
= 164.62 cubic inches
Bore = 3.550"
Stroke
= 4.160"

FIGURE 12-20 Engine displacement.

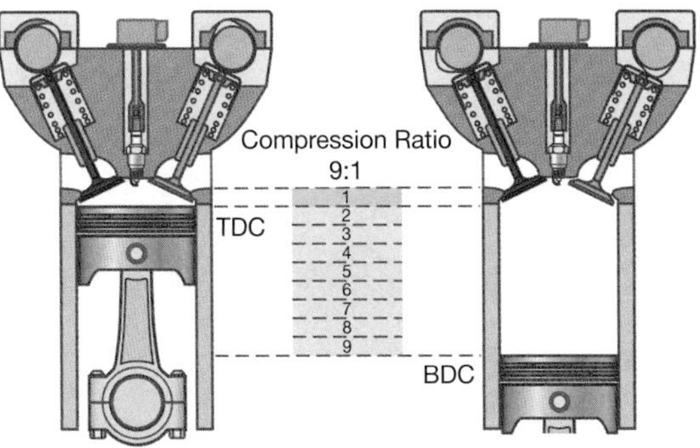

FIGURE 12-21 The compression ratio of an engine is found by taking the volume of the cylinder at BDC and comparing it to the volume at TDC. In this example, a 9:1 compression ratio is found.

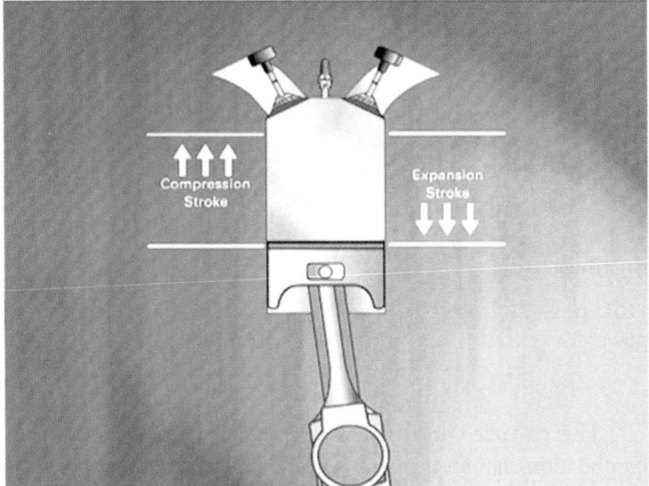

FIGURE 12-22 Conventional engine—compression and expansion strokes are the same length.

Atkinson and Miller Cycle Engines

K12008

The **Atkinson cycle** engine and the **Miller cycle** engine are both variations on the traditional four-stroke SI engine. These engines operate more efficiently but produce lower power outputs for the same displacement. In a conventional four-stroke cycle, the compression and the power (expansion) strokes are the same length (**FIGURE 12-22**). Increasing engine efficiency by increasing the stroke and raising the expansion ratio also raises the compression ratio. There is a limit to how high the compression ratio can be because raising it too much results in high enough temperatures to ignite the air-fuel mixture prematurely, causing engine damaging detonation. The Miller and the Atkinson cycles overcome this by using valve timing variations to make the effective compression stroke shorter than the expansion stroke (**FIGURE 12-23**). The effective compression stroke is shortened by delaying the closing of the intake valve at the beginning of the compression stroke. This shortens the distance that the piston has to compress the air-fuel mixture. The combustion chamber is slightly smaller so that the engine will still have a normal compression ratio. Thus, the compression pressure at ignition is still typically the same as that of a conventional four-stroke engine. The effective expansion stroke is lengthened by delaying the opening of the exhaust valve until closer to BDC, so the pressure created by the expansion of the burning gases can act on the piston longer, applying pressure to the crankshaft for a longer time and increasing efficiency.

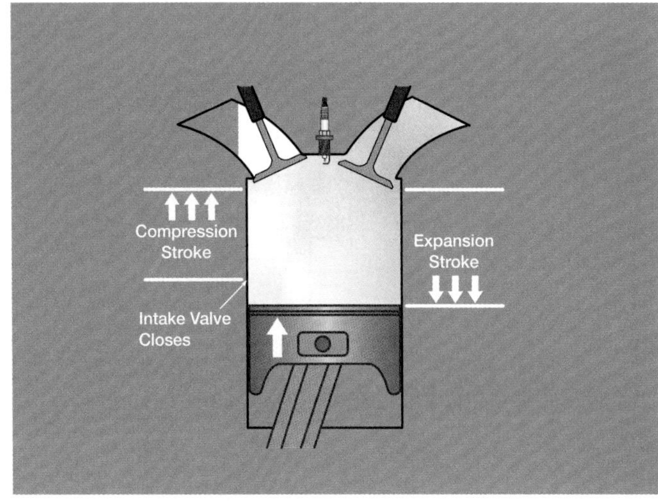

FIGURE 12-23 Miller Cycle and Atkinson Cycle engines—compression and expansion strokes are different effective lengths due to delayed intake valve closing.

Because some of the intake gases are pushed back from the cylinder into the intake manifold, Miller and Atkinson engines use a larger throttle opening for a given amount of power. This design results in lower manifold vacuum, which reduces pumping losses (parasitic drag), and increases fuel efficiency.

The Miller cycle engine adds an engine-driven supercharger to increase volumetric efficiency and boost power output when required. When the engine is operating at low load and speed, the supercharger is not needed. A clutch disengages the drive so there is no unnecessary drag on the engine. When extra power is required, the clutch is engaged, and the supercharger boosts the amount of air drawn into the engine, supercharging the cylinders.

The Atkinson cycle engine is efficient within a specific operating range (the so-called engine "sweet spot" of peak torque rpm), typically between 2000 and 4500 rpm, but its overall power output and torque are lower than a conventional ICE of the same displacement. This type of engine is less useful as the primary power source, but it is ideal in applications such as a series-parallel hybrid vehicle, where it can work in tandem with a battery-driven electric motor. This engine, combined with an electric motor, provides more torque than it can produce by itself. It is also well suited to be used to charge the high-voltage battery as it can do that within the most efficient engine rpm range.

Also, the lower maximum operating rpm allows engine components to be of lighter construction and weight as compared to a conventional ICE. Lighter and smaller components reduce friction and increase engine efficiency. In addition, the crankshaft is

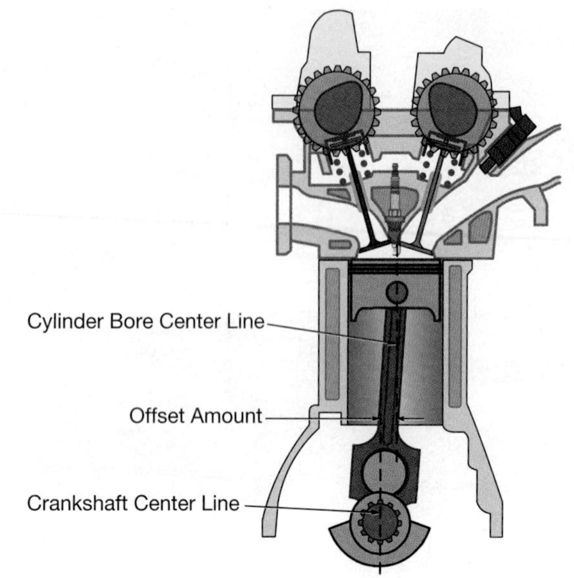

FIGURE 12-24 Crankshaft offset from cylinder bore reduces the thrust load on the piston during the power stroke.

mounted slightly off-center from the cylinder bores (**FIGURE 12-24**). This position reduces the thrust load on the piston and cylinder wall, thereby reducing power loss due to friction.

▶ Components of the Spark-Ignition Engine

K12009

The SI engine is the most widely used engine to power passenger vehicles in the United States, whereas CI engines are more common in much of the rest of the world. But both types of engines share similar components. SI engines have evolved over their 125-year life, but the fundamental principles are still the same: An air-fuel mixture is compressed in the cylinder to increase its energy; it is ignited by a high-voltage spark, and the mixture burns rapidly, causing the thermal expansion needed to push the piston down; and the exhaust gases are pushed out of the cylinder. Modern materials, machining processes, and lubricants have made these engines longer lasting, more powerful, and more environmentally friendly than ever.

Manufacturers have made incredible gains in the manufacturing of engines and engine components due to new technologies that have found their way into the automotive field. Engine blocks and cylinder heads are commonly manufactured from lightweight aluminum; valve covers and intake manifolds are being made of durable plastic materials; pistons are made of newer aluminum alloys; and in some cases connecting rods are manufactured from powdered metals, making them lighter and stronger.

The engine can be divided into a couple of main assemblies: the bottom end and the top end (**FIGURE 12-25**). The engine's so-called bottom end is where the crankshaft, bearings, connecting rods piston assemblies, and oil pump reside. The crankshaft, connecting rods, flywheel, and harmonic balancer make up the rotating assembly.

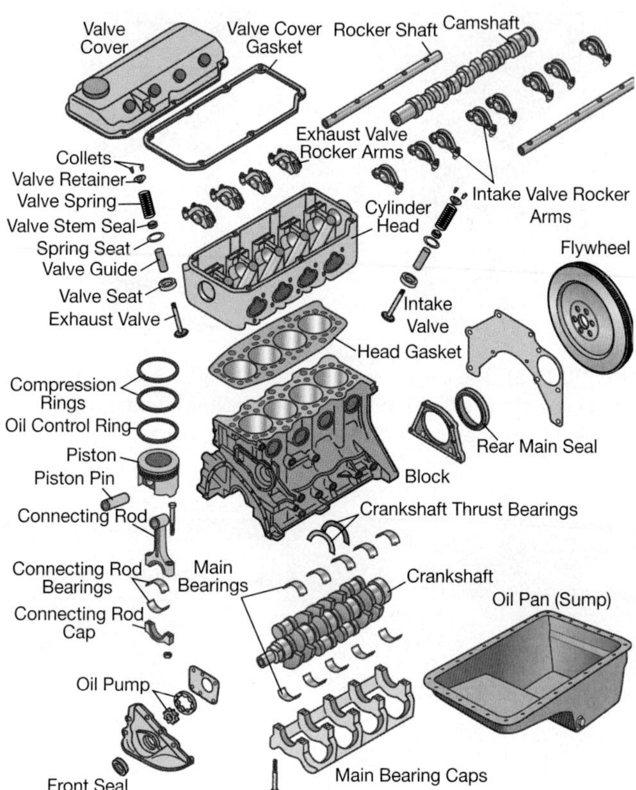

FIGURE 12-25 The engine contains many parts that work together to power the vehicle.

The so-called top end is where the cylinder heads and combustion chambers reside. The cylinder head(s) contains the overhead valves and valve train; if of the cam-in-head configuration (overhead cam), it also contains the camshaft. Each of these assemblies and components is explored further below.

Short Block and Long Block

K12010

If a rebuilt engine is needed, an engine subassembly may be purchased. A short block replacement includes all of the parts in the engine block from below the head gasket to above the oil pan (**FIGURE 12-26**). A cam-in-block engine also includes the camshaft and timing gears. An overhead-cam short block does not include the camshaft or timing gears.

A long block replacement engine includes the short block, plus the cylinder head(s), new or reconditioned valve train, camshaft and timing chain, and/or gears (or timing belt) (**FIGURE 12-27**). A long block engine still requires swapping parts from the original engine to the long block, including the intake and exhaust manifolds, fuel injection system, the starter, alternator, power steering pump, and air-conditioning compressor.

Cylinder Block, Crankshaft, and Flywheel

The cylinder block is the single largest part of the engine. The block can be made of cast iron or aluminum, which is much lighter. The block casting includes the cylinder bore openings, also known as

FIGURE 12-26 Short block assembly.

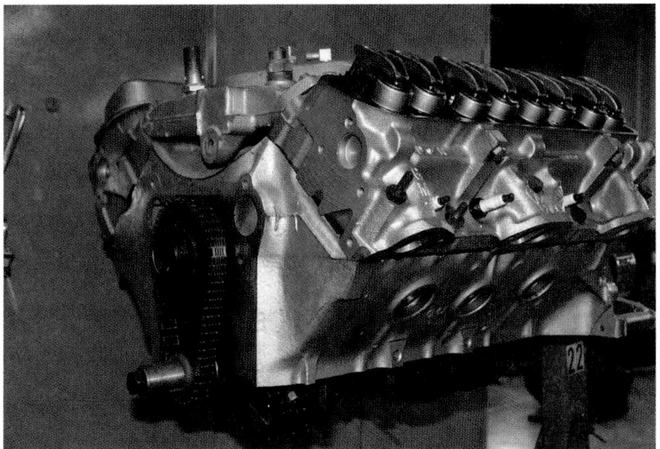

FIGURE 12-27 Long block assembly.

cylinders, which are machined into the block to allow for the fitting of pistons (**FIGURE 12-28**). The block deck is the top of the block and is machined flat. The cylinder head bolts to the block deck. Passages for the flow of coolant and lubrication are machined or cast into the block. Holes machined into the bottom of the block, called main bearing bores, have removable main caps and are used to hold the crankshaft in place. Each cap is held in place with two or more bolts. Reinforcements for strength and attachment points for related parts are also machined into the block. The lowest portion of the block is called the crankcase because it houses the crankshaft.

The oil pan completes the crankcase. On most modern engines, the main bearing caps are now a part of the engine girdle, also called a bed plate. The use of a girdle provides an even stronger design as all main bearing caps are connected and reinforce each other.

The crankshaft can be made of cast iron or forged steel, or can be machined out of a solid piece of steel. The crankshaft converts the reciprocating motion of the pistons into rotary motion at the crankshaft. The rotary motion is transferred to the engine flywheel and transmission to ultimately (in a conventional vehicle) drive the wheels. The crankshaft has main bearing journals and connecting rod journals machined and polished into it. Main bearing journals provide support for the crankshaft in the block, and the connecting rod journals provide

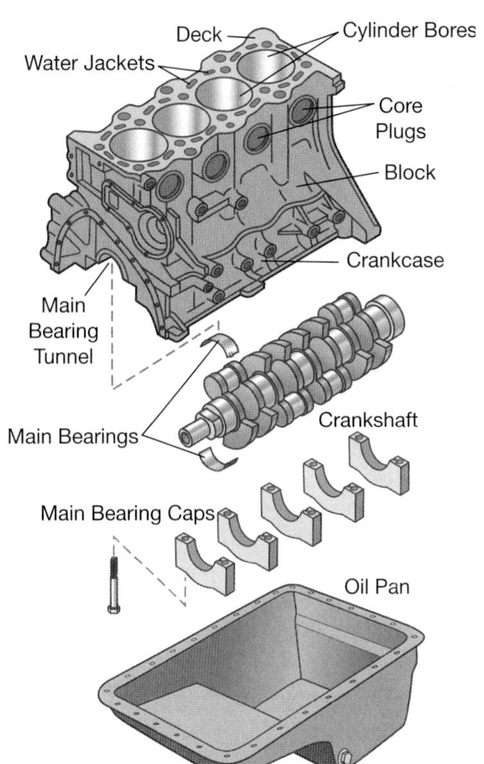

FIGURE 12-28 The block is the single largest part of the engine, with other components attached to it.

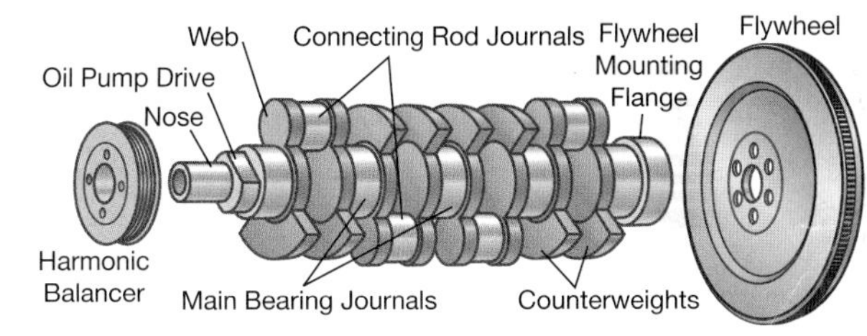

FIGURE 12-29 The basic parts of the crankshaft.

a connection point for the connecting rods (**FIGURE 12-29**). The crankshaft is supported by Babbitt-lined main bearing inserts, which fit in the main bearing saddles of the block. End movement of the crankshaft is limited by a thrust bearing at one of the main bearings.

Offset from the crankshaft centerline are the rod journals, also called throws. They are essentially the levers of the crankshaft. The longer the throw of the crankshaft, the longer the stroke of the piston and the more torque that the engine can produce. The rod journals are also machined and polished. As the crankshaft turns, the rod journals' "big ends" circle around the centerline of the crankshaft. To prevent vibration, counterweights are formed on the crankshaft. The counterweights balance the weight of the piston assembly, connecting rods, and rod journals. At the front of the crankshaft is a "snout" that provides a mount for a vibration damper (harmonic balancer) and drive gears, sprockets, or pulleys. The back of the crankshaft has a flange where the flywheel is connected by bolts or studs. The **flywheel** is a weighted assembly that stores kinetic energy from each power stroke and helps keep the crankshaft turning through non-power strokes. Vehicles with manual transmissions have a clutch assembly attached to the flywheel. Vehicles with automatic transmissions use a flexplate and torque converter assembly. The effect on the crankshaft is the same with either assembly—that of storing energy and keeping the crankshaft rotating smoothly.

Connecting Rod and Piston

A connecting rod is made of cast iron or steel in most engines, although some race cars and exotic sports cars use aluminum or titanium connecting rods to make the engine rev more quickly and higher, producing more power. The connecting rod connects the piston to the crankshaft and transfers piston movement and combustion pressure to the crankshaft rod journals. The piston end (small end) of the connecting rod follows the reciprocal movement of the piston, pivoting on a piston pin or "wrist pin" attachment. The wrist pin attaches the piston to the connecting rod and fits into a one-piece bushing in the small end of the rod. The other end (large end) of the connecting rod attaches to the crankshaft throw or rod journal through the use of a removable rod cap that bolts to the end of the connecting rod body. The connecting rod and cap are machined to allow the fit of the connecting rod bearing.

The piston is typically made of lightweight aluminum and possibly synthetic material and transfers combustion pressures to the crankshaft through the connecting rod. Pistons change direction multiple times a second (**FIGURE 12-30**). Consider that with an engine idling at 750 rpm, the piston changes direction 25 times each second. We can imagine that piston movement would be hard to see even at idle, but at a redline (maximum allowed) speed of 6000 rpm, can you imagine the stress placed on engine parts as each piston comes to a stop at TDC and at BDC 200 times per second? What's more, a 12000-rpm redline sports car or motorcycle engine would experience *400* piston reversals per second!

The top of the piston, called the piston head, is exposed to extremes of heat and pressure during combustion. Below the piston head are grooves machined into the piston that hold circular piston sealing rings. The piston rings provide a seal between the outside of the piston and the inside of the cylinder wall as the piston moves in its stroke (**FIGURE 12-31**). Usually, a total of three rings are used. The upper two rings are compression rings, which prevent combustion pressure, called **blowby gas**, from leaking past the pistons into the crankcase. The lower piston ring is an oil control ring that keeps lubricating oil on the cylinder wall and out of the combustion chamber.

Pistons can come in many styles of crowns, such as dished, domed, or flat top, depending on the compression ratio desired by the engine designer. Manufacturers today are using ceramic coatings on the pistons to provide better lubrication qualities and reduced heat absorption.

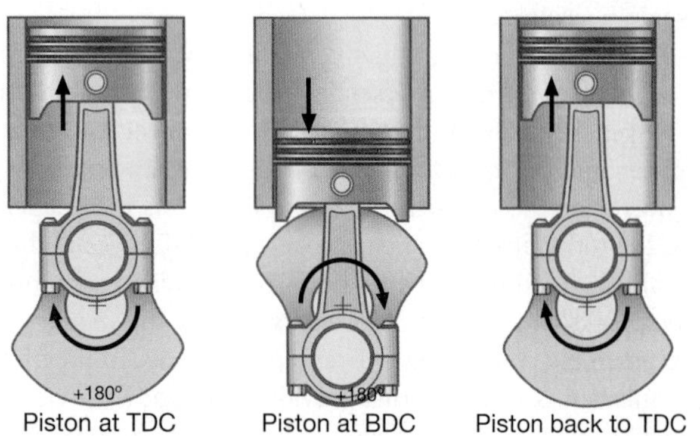

FIGURE 12-30 The piston moves two strokes during one revolution of the crankshaft.

The Cylinder Head

K12011

The cylinder head is constructed of cast iron or aluminum. Most engines are now constructed using an aluminum cylinder head, which reduces the weight of the engine. The cylinder head contains the valves and valve train (valve actuating components) of the engine. The head also includes intake and exhaust ports to which intake and exhaust manifolds are attached. The head forms the top of the cylinder and is sealed in place with the use of a head gasket. The cylinder head has a combustion chamber either cast or machined into it. Combustion chambers in the cylinder head come in several different designs, such as the wedge or the hemispherical combustion chamber, a variation that is used in most engines now (**FIGURE 12-32**).

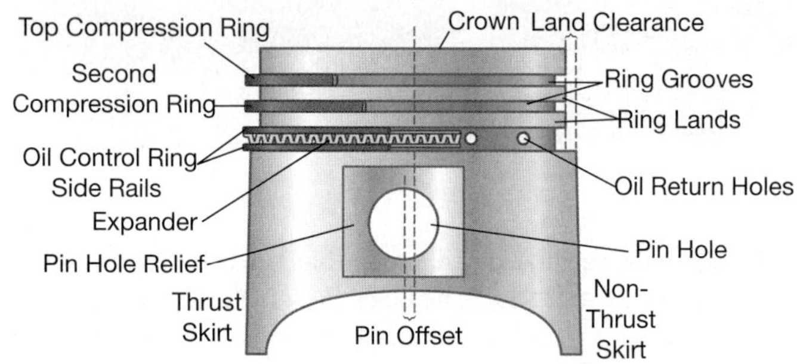

FIGURE 12-31 Piston and piston rings.

FIGURE 12-32 Combustion chambers can be designed in several configurations (wedge combustion chamber shown).

Engine Cam and Camshaft

K12012

The ICE uses so-called poppet valves. These are somewhat mushroom-shaped parts that slide up and down in the valve guides (**FIGURE 12-33**). When not actuated (closed), the valves, under pressure from the valve springs, rest on seats of hardened material such as Stellite®. Valves need a system to make them open and close. Control of the valves is accomplished through the use of cams on a common shaft. A **cam** is an egg-shaped piece (lobe) mounted on the **camshaft** (**FIGURE 12-34**). The egg shape of the cam lobe is designed to lift the valve open, hold it open, and let it close. The camshaft is timed to the rotation of the crankshaft to ensure that the valves open at the correct position of the piston. Timing the valve opening to the piston position is critical to ensure proper power output and low-emissions operation of the engine. The camshaft is turned by gears, a toothed belt, or a chain that is driven by sprockets.

Up until the 1950s, many engines had their valves installed in the engine block. Such engines are called **flat-head engines**. Some manufacturers still place the camshaft in the center of the block, but the valves are installed in the cylinder head(s). So-called **cam-in-block engines** use **pushrods** to transfer the camshaft's lifting motion to the valves by way of **rocker arms** on top of the cylinder head. **Tappets**, or "lifters," ride on the camshaft **lobes** to actuate the pushrods, rocker arms, and valves.

In most automotive engines today, however, the camshaft is mounted on top of the cylinder head. These engines are called **overhead cam (OHC)** engines (**FIGURE 12-35**). Intake and exhaust valves may all be actuated by a single camshaft, or there may be two camshafts per head, called **dual overhead cam (DOHC) engines**. One camshaft may be used to actuate all of the intake valves and another to actuate all of the exhaust valves. When separate intake and exhaust camshafts are used, there is no need for rocker arms. Most manufacturers use a lifter called a bucket lifter, placed right on top of the valve and valve spring to actuate the valve directly from the camshaft.

Camshaft lobes are designed, as described previously, to open the valve, hold it open, and allow it to close. The opening, holding, and closing of the valves are critical to ensure that the engine operates correctly. In designing the cam lobe, engineers seek a proper compromise for the application of the engine. If

FIGURE 12-33 Poppet valves in a cylinder head.

FIGURE 12-34 Cam lobes on a camshaft.

FIGURE 12-35 A. Cam-in-block engine. **B.** Overhead cam (OHC) engine.

the engine is designed to operate at a single engine rpm, then the camshaft can be designed to provide optimal power, economy, and emissions. Automotive engines do not operate at a single rpm, however, so a camshaft must be designed to provide the best balance of all speeds. High-performance engines built for racing use camshafts designed for high rpm power but would not work well for use on the street, where engines rarely stay above 3000 rpm. Newer engine designs have overcome some of these limitations by using variable valve timing, which is discussed further in the Cylinder Head chapter.

▶ Valves

K12013

A valve is used to open and close a port in a cylinder head. The intake controls the flow of air and/or fuel into the combustion chamber. The exhaust valve controls the flow of exhaust gases out of the combustion chamber and cylinder. The exposed intake port area usually must be larger than the exhaust port area to make it easier for the piston to pull air into the engine on the intake stroke. Engine vacuum created by the piston on the intake stroke is not as effective at moving air into the engine as the pressure created by the piston on the exhaust stroke is at pushing exhaust gases from the engine. The exposed port area can be increased by making the intake valve larger than the exhaust valve, or the manufacturer can use multiple intake valves. In the case of a three-valve engine, there would be two intake valves and one exhaust valve, but the intake valves would be smaller than the exhaust valve in this case.

The valve head is disc shaped, and the top of the valve head faces the combustion chamber. A machined surface on the back of the valve head is the **valve face** (**FIGURE 12-36**). The valve face seals on a hardened valve seat in the cylinder head. Located between the valve head and the valve face is a flat surface on the outer edge of the valve head, called the **valve margin**. The margin helps to prevent the valve head from melting under the heat and pressure of combustion. A shaft attached to the valve head is the **valve stem**. The stem operates in a **valve guide** in the cylinder head. The valve stem and guide work together to maintain valve alignment as the valve slides open and closed. Grooves are machined into the opposite end of the stem in order to receive locking pieces, sometimes referred to as valve **keepers**, that retain a valve spring retainer and spring on the valve. Keepers hold the retainer, preventing it from coming loose while under tension from the valve spring, and preventing the parts from coming loose under normal use. Engines may have two to five valves per cylinder.

Mechanical and Hydraulic Valve Train

The valve train is the combination of parts that work together to open and close the valves of the engine. The valve train operates off the camshaft, and the part that rides against the cam lobe is the **valve lifter**. The lifter transfers motion from the cam lobe to a pushrod, or may directly act on the valve and spring, depending on whether the cam is in the engine block or on the cylinder head. The valve lifter works as a mechanical spacer, providing a hardened bearing surface that slides across the cam lobe. If the lifter is a mechanically solid piece, it is said to be a mechanical lifter, and therefore the engine is said to have a mechanical valve train. If the lifter has a hydraulic plunger in its center, it is a hydraulic lifter, and the engine is said to have a hydraulic valve train. The hydraulic plunger allows for the expansion and contraction of components during engine warm-up and cool down.

With a hydraulic valve train, valve adjustment is made by the hydraulic lifter, which takes up any clearance automatically. If a mechanical valve train is used, valve adjustments will be necessary at periodic intervals to ensure proper clearance is maintained as parts in the valve train wear.

Roller Rockers and Lifters

Roller-equipped rocker arms are used on many new engines because they reduce friction and increase engine efficiency. They may also be used to replace the cast or stamped steel rocker arms that have been used for many years as part of a standard valve train (**FIGURE 12-37**).

FIGURE 12-36 The parts of a valve.

FIGURE 12-37 This roller rocker with a roller lifter compared to the standard rocker and flat lifter.

A typical rocker arm has a **fulcrum**, a half-round bearing, or a shaft that the rocker moves on as a bearing surface. One end of the rocker touches the top of the valve stem, and the metal slides across the valve as it pushes the valve open and then lets it close. The other end of the rocker arm connects with the pushrod. The rocker's sliding motion across the valve stem and center pivot results in friction, and friction creates loss of power and wear. To remedy this, needle bearings and a roller may be added to the end of the rocker where it contacts the valve stem, and needle bearings are added to the pivot point where the fulcrum is. This system greatly reduces friction and results in a more powerful and reliable valve train.

Rocker arms are designed to increase the amount of lift designed into the cam lobe, because they work off-center as a lever. A typical rocker arm has the center pivot closer to the pushrod end. This design causes the valve end of the rocker arm to move more than the pushrod end. Many rocker arms have a 1.5:1 lift ratio, which means that the valve moves 1.5 times farther than the pushrod. High-performance rockers have even higher ratios, such as a 1.6:1 or 1.7:1. These rockers increase the lift of the valve even more than a standard rocker arm.

The other friction loss point in the valve train is at the lifter as it rides upon the cam lobe. To reduce this friction loss, rollers with needle bearings are added to the base of the lifter on some engines. The lifter now rolls on the cam lobe profile rather than sliding on it. This modification further reduces friction in the valve train, resulting in more power delivered to the engine flywheel, and less wasted as heat. The use of rollers on the lifters also allows the use of more aggressive cam lift profiles. Such profiles would rapidly wear an ordinary flat lifter and cam lobe. This difference between cam profiles used with flat tappets versus roller lifters means camshafts and tappets should never be interchanged. Also, the roller lifter must be held in position so that it does not turn in the lifter bore and scrape on and wear the cam lobe. To prevent this, a retainer keeps the lifter from turning sideways.

Valve Clearance

Valve clearance is the amount of slack between the rocker arm or cam follower and the valve stem, or the cam and the lifter if it is a bucket-style OHC engine (**FIGURE 12-38**). If valve clearance is too large, the valves will tick and make enough noise to irritate the operator and increase wear of the valve train. If valve clearance is too small, the valve can be held open longer than it should be. As the cylinder head and valve train parts heat, they expand, so adequate clearance is needed to allow for this expansion. Insufficient valve clearance could result in burned valves. Some valves are adjustable through the use of adjusting screws, nuts, or metal shims. Other valves are nonadjustable, and the rocker arm simply bolts to the head, or the valve lifter is a preset dimension before it is installed under the camshaft if it is a bucket setup. Bucket lifters contain solid metal discs (shims) of different thicknesses. These shims are used to preset the proper valve clearance during cylinder head assembly or during a major tune-up.

FIGURE 12-38 Valve adjustments. **A.** Rocker arm screw and locknut. **B.** Rocker arm center bolt.

Valve Train Drives

The valve train is driven by the camshaft, which in turn is driven by a chain or belt (depending on the engine design) driven by the crankshaft (**FIGURE 12-39**). In older engine designs, the camshaft was driven by a gear-to-gear arrangement like that found in small lawn mower engines. The trouble with a gear-to-gear design is that it tends to be a bit noisy compared to a chain or belt, and the camshaft needs to be relatively close to the crankshaft. In any four-cycle engine design, the camshaft must rotate at half of crankshaft speed. This is accomplished by using a camshaft gear or pulley that has twice the number of teeth as the crankshaft gear. The ratio of the crank to cam gear is 2:1; thus it takes two turns of the crankshaft to turn the camshaft one turn.

The timing chain drive is louder than a belt, but the belt will not last as long as the timing chain, and if exposed to fluids or dirt will wear out even more quickly. The timing chain must have a constant supply of engine oil to lubricate the chain to keep it from wearing out quickly. Timing gears rarely jump time, but if the timing chain or belt breaks, serious engine damage could result, depending on whether it is a freewheeling engine or an interference engine.

FIGURE 12-39 Cam drives. **A.** Belt-driven OHC. **B.** Chain-driven OHC.

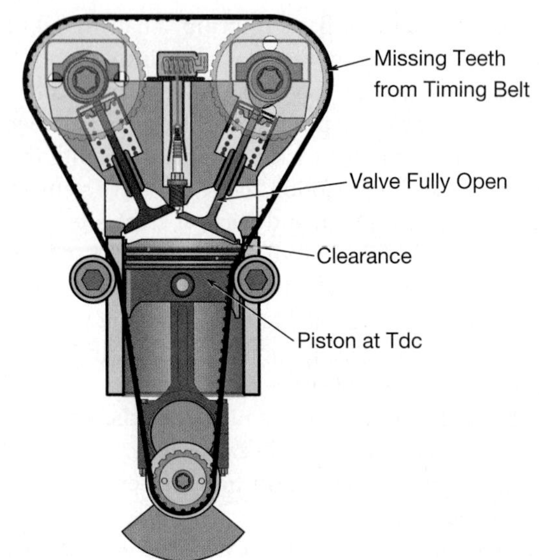

FIGURE 12-40 Freewheeling engine.

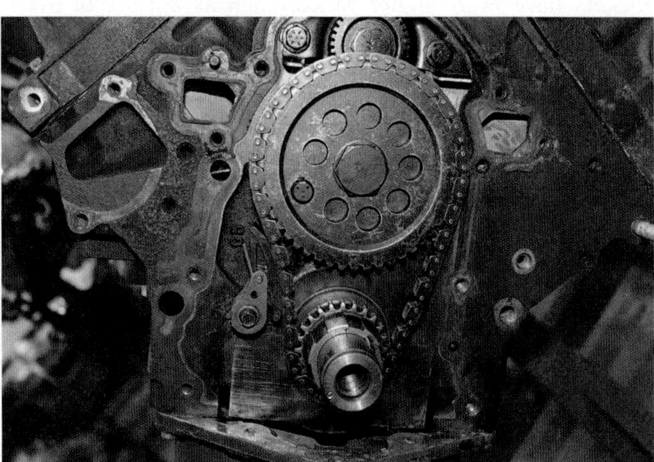

FIGURE 12-41 Cam in-block timing chain and gears.

The **freewheeling engine** has enough clearance between the pistons and the valves so that in the event the timing belt breaks, any valve that is hanging all the way open will not contact the piston, thus preventing engine damage (**FIGURE 12-40**). A broken timing belt will be an inconvenience to the customer. The engine dies and will not restart; the good news is that no mechanical engine damage usually occurs. By contrast, the **interference engine** has no clearance between the piston and valves if the valve were to open fully when the piston is at TDC. This means there is minimal clearance between the valves and pistons during normal operation. When the timing belt breaks, the valves will be tap-dancing on the piston crowns, bending the valves, with the valves possibly punching holes in the pistons. A broken timing belt on an interference engine means a huge repair expense, and in some cases the entire engine must be replaced. The manufacturer's service information will normally tell you if the engine is an interference engine.

The timing chain camshaft drive used in a cam-in-block engine is very different from that used in the OHC engine. The pushrod engine typically uses a chain behind a timing chain cover located on the front of the engine (**FIGURE 12-41**). This type of timing chain is fairly short because the camshaft is close to the crankshaft. Some pushrod engines use a timing chain tensioner to ensure that the correct timing chain tension is maintained; however, most designs do not use a tensioner. The timing chain tensioner applies pressure against the chain; as the chain wears and gets longer, the tensioner takes up play in the chain. As the timing chain stretches, cam timing may become retarded. Slack of the chain affects the positioning of the cam gear in relation to the crank gear. Retarded cam timing in this case can create undesirable engine performance problems.

The OHC engine requires a longer timing chain or belt, and in this design one or more tensioners are required (**FIGURE 12-42**). The timing chain in an OHC engine typically has hard plastic-type guides for the chain to slide on and assist the tensioner(s) with correct tracking and tension of the timing chain. The OHC timing chain must run in oil to ensure that the chain is lubricated. Without oil, the chain would wear out rapidly.

A belt system uses a toothed or cogged belt to turn toothed or cogged pulleys on the camshaft. The belt is a scheduled maintenance replacement item and needs to be replaced at

FIGURE 12-42 Timing chain, gears, and tensioner system.

the mileage or time recommended by the manufacturer (e.g., 60,000 miles [100,000 km] or five years, whichever occurs first).

Intake Manifold

K12014

The intake manifold is part of the air intake (induction) system of the engine package. It sits between the throttle body and the cylinder head(s). On a V-type engine, it usually is located between the cylinder heads. For an inline engine, it bolts to the side of the head. Intake manifolds deliver air (air with fuel on carbureted or throttle body–injected engines) to the cylinder head.

Intake manifold designs have seen many changes since the first engines. Once rectangular and boxy in shape, they are now sleek. The materials from which they are made have also changed. Intake manifolds were first manufactured in cast iron, and later in lightweight aluminum. Now, most manufacturers are using even lighter thermoplastic materials for intake manifolds (**FIGURE 12-43**). Lighter engines usually equate to better fuel economy, so manufacturers are always looking for new lighter materials to make their vehicles more fuel efficient. How smooth the inside of the intake manifold is, its inside diameter, and its length all affect how an intake manifold will contribute to engine performance. Some sportier vehicles use variable-length "tuned" intake runners, which switch between a shorter or longer path for the air to flow, depending on engine rpm. Intake manifold design can greatly affect engine performance; this topic is discussed further in the Induction and Exhaust chapter.

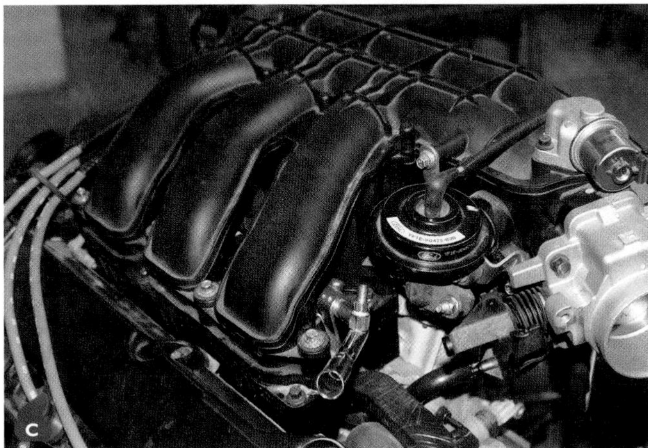

FIGURE 12-43 Intake manifolds. **A.** Cast iron. **B.** Aluminum. **C.** Thermoplastic.

Exhaust Manifold

The exhaust manifold is the output side of the engine's breathing apparatus. With a crossflow head, the exhaust manifold bolts to the cylinder head across from the intake manifold and is designed to carry exhaust gases out of the engine and deliver them to the exhaust system. As with intake manifolds, exhaust manifold design can have a great impact on the performance of the engine due to **scavenging**. So-called tuned headers, for example, direct exhaust gases from each individual cylinder using equal-length tubes (runners) to scavenge (extract) the exhaust from neighboring cylinders (**FIGURE 12-44**). This topic is discussed further in the Induction and Exhaust chapter.

Refer to the Induction and Exhaust chapter for more information about the exhaust system.

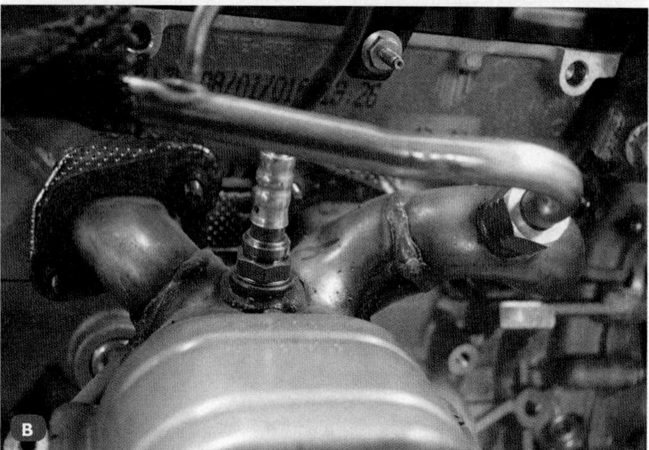

FIGURE 12-44 Exhaust manifolds. **A.** Cast iron. **B.** Tuned headers.

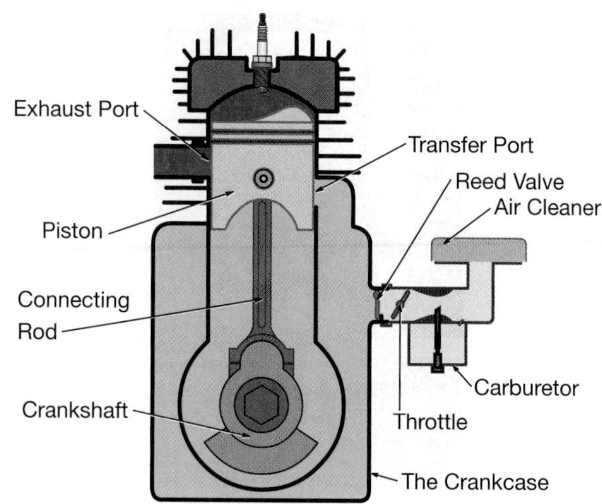

FIGURE 12-45 Two-stroke engine.

▶ Two-Stroke Spark-Ignition Engines

KI2015

Two-stroke engines are notable in their ability to produce a large power-to-weight ratio. The power capabilities of this engine come from the fact that every other stroke (TDC to BDC), instead of every four stokes, is a power stroke. However, there are drawbacks to this engine design—namely that it is a high-emissions engine. Almost all car manufacturers have moved away from two-stroke production.

Basic Two-Stroke Cycle Engine Principles

The two-stroke SI engine is different from the four-stroke SI engine. In a **two-stroke engine**, the inlet and exhaust ports are opened and closed by the movement of the piston; there are no poppet valves like those used in the four-stroke cycle engine. The two-stroke engine is still an ICE and shares the five events common to all SI engines. What is different is the method of air induction and exhaust used:

- *Intake* occurs in two parts. First, the air-fuel mixture is drawn into the crankcase as the piston moves up. It is then transferred from the crankcase to above the piston when the piston moves down.
- During *compression*, the mixture is forced into a small volume as the piston moves up.
- During *ignition*, the spark from the spark plug ignites the mixture and it burns.
- During the *power stroke*, energy released by combustion generates the force that pushes the piston down and turns the crankshaft.
- During *exhaust*, leftover gases are expelled from the cylinder when the piston is near BDC.

As in all ICEs, expanding gases drive the piston down and turn the crankshaft and flywheel, which pushes the piston back up to TDC in the cylinder (**FIGURE 12-45**).

With the two-stroke cycle engine, the crankshaft makes one revolution (two strokes) for every complete cycle. In one revolution of the crankshaft, two piston strokes occur: one down and one up. Each downward stroke of the piston is a power stroke and turns the crankshaft.

The two-stroke cycle engine differs from the four-stroke cycle engine because the upward piston movement creates suction in the crankcase to pull the air-fuel mixture into it as the piston moves from BDC to TDC (**FIGURE 12-46**). The air and fuel sit in the crankcase until the piston begins to move down, which then creates a small pressure in the crankcase, referred to as crankcase compression (**FIGURE 12-47**). As the piston moves downward with the power stroke, it uncovers a transfer port, and the fuel and air rush into the cylinder from the crankcase.

The piston begins moving up toward TDC, creating another draw of air and fuel into the crankcase. The air and fuel above the piston get squeezed in the combustion chamber, and the spark plug ignites the mixture. As the piston moves down, it also uncovers the exhaust port, and exhaust gases exit the exhaust system. As the piston moves farther down, it uncovers the transfer port again, and crankcase compression pushes the air and fuel into the cylinder once again.

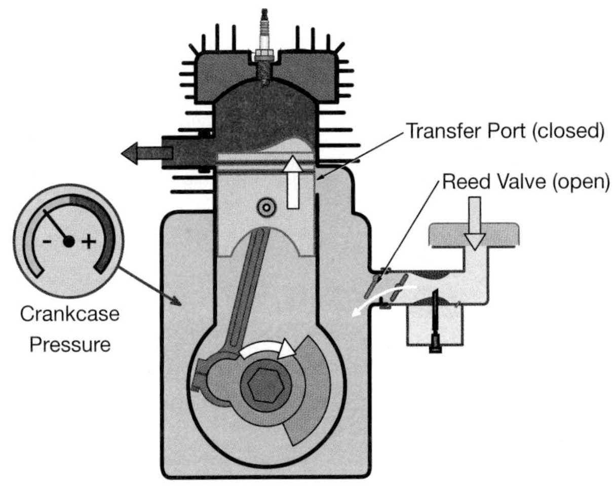

FIGURE 12-46 Two stroke—crankcase intake.

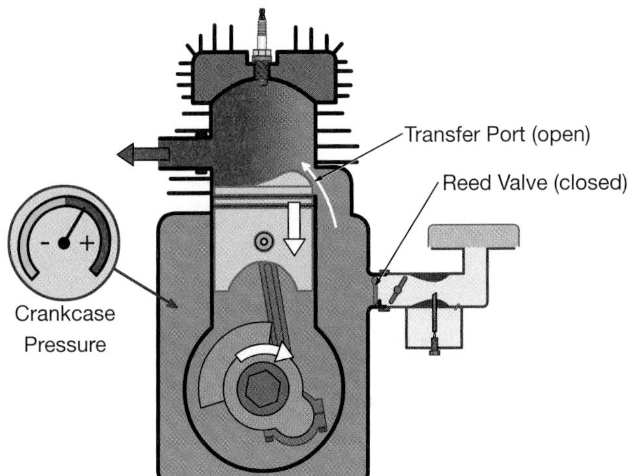

FIGURE 12-47 Two stroke—crankcase compression.

To summarize, with upward movement of the piston, compression of the air-fuel mixture is happening in the cylinder above the piston, and intake of new air and fuel is happening in the crankcase below the piston. When the piston moves down, power is being applied to the crankshaft, exhaust is happening, and crankcase compression is building to push air and fuel through the transfer port into the cylinder. The placement of the ports makes all these processes possible and eliminates the use of valves to let fuel and air in and exhaust out.

Two-Stroke Intake System

The intake system of a two-stroke engine is called a piston port system because the piston acts as a valve to cover and uncover the ports, allowing flow into and out of the cylinder. But because the crankcase alternates between vacuum and pressure as the piston goes up and down, a valve must be between the crankcase and the carburetor. Manufacturers use either a reed valve or a rotary valve for this purpose.

A **reed valve** is a small flexible metal plate that covers the inlet port (**FIGURE 12-48**). The reed valve is made of spring

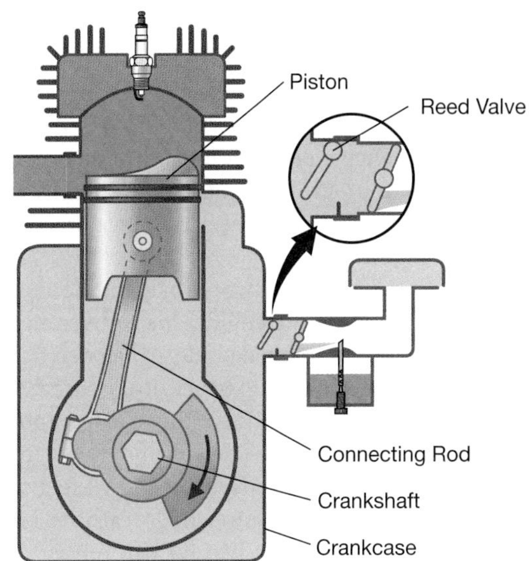

FIGURE 12-48 Two-stroke reed valve operation.

metal and can be attached to the crankcase or to the inlet port. The reed valve acts like a one-way check valve and opens and closes automatically, according to changes in pressure in the crankcase. For performance applications, such as in off-road motorcycle racing, a different reed valve may be used to change the performance characteristics of the engine.

Some two-stroke intake systems use a rotary valve. The crankcase has a port open to the carburetor, and a rotary valve that covers and uncovers that port. The rotary valve is mounted on the crankshaft. As the crankshaft turns, the rotary valve rotates with it, opening and closing the inlet port as it does so. This ensures that pressure does not push backward into the carburetor as the piston pressurizes the crankcase.

Four- and Two-Stroke Engine Differences

K12016

In a two-stroke gasoline engine, lubrication is usually performed by mixing oil with the gasoline. The oil lubricates the moving parts within the engine but also burns in the combustion chamber. This oil causes smoke to be emitted from the exhaust, which is one reason two-stroke engines have been dropped for automotive use. The exhaust emissions are beyond today's emissions regulation cutoff points.

Two-stroke engines do have some attractive benefits. The first is their lightweight, compact, and powerful package. These features make them ideal for handheld equipment such as weed cutters and chain saws. They can be run in any position because there is no worry of oil leaking out of the sump or fuel out of the carburetor. They are also simple, using fewer parts, so they are easier and less expensive to build. Pressurized oil and oil passages are not needed for the crankshaft and connecting rod bearings in two-stroke engines. Most two-stroke engines use roller bearings on the crankshaft, which are splash lubricated by the oil mixed in with the gas as it churns in the crankcase. Many two-stroke engines are air cooled, thus eliminating the water-based cooling system of the four-stroke engine. There is no oil pan with an oil pump or oil in it, nor a radiator with coolant in it. The two-stroke engine does not use any of the valve train components used by four-stroke engines, so even more weight is shed from the engine package. These are some of the benefits of the two-stroke engine.

Again, the downfall to the two-stroke engine design is the pollution released from the engine due to burning oil in the fuel and the amount of unburned fuel exhausted from the engine due to scavenging of the exhaust. It is estimated that 25% of the incoming fuel and oil leave the exhaust unburned in a typical two-stroke engine. The oil and fuel represent hydrocarbons released to the atmosphere, and wasted energy. The four-stroke engine releases fewer hydrocarbons because it controls inlet and outlet flow better. The release of less pollution makes the four-stroke engine the choice for most manufacturers trying to meet pollution requirements of the Environmental Protection Agency. Although companies continue to produce two-stroke engines and have found ways to make them pollute less, much research is still needed to improve this engine design.

▶ Rotary Combustion Spark-Ignition Engine

K12017

As we have seen from the two-stroke engine, the fewer parts that are used, the better the power production and the smaller the engine can be. The rotary engine fits into the same category of using fewer parts to produce power. The rotary combustion (RC) engine has found its way into automobiles, planes, helicopters, boats, motorcycles, lawn mowers, and other applications. Displacement has varied from tiny air-cooled models to much larger liquid-cooled, multirotor units.

The rotary engine is also called the "Wankel" engine because Felix Wankel improved it for automotive use in the 1940s. The RC engine was commercially released in 1964 in the NSU Wankel Spider and in 1967 with a two-rotor engine in the NSU RO80. Under license from NSU, Mazda successfully used the rotary engine in several vehicles, from the late 1960s all the way through the RX series (**FIGURE 12-49**). The engines were a redefining period for engine development but were never really a success for a number of other companies.

Basic Principles of the Rotary Engine

K12018

The rotary engine (aka Wankel engine) is not as common as the four-stroke or two-stroke cycle engines, but its basic principle is well accepted. The rotary engine layout is vastly different from that of a reciprocating engine. The piston engine is called a reciprocating engine because the pistons move back and forth over the same path. This reciprocating motion is converted to rotary motion at the crankshaft.

FIGURE 12-49 Wankel rotary engine.

FIGURE 12-50 Cutaway of a rotary engine.

By contrast, a rotary engine does not use a piston that reciprocates; rather, it has a rotor that—you guessed it—rotates. The rotary engine does not need to convert inefficient reciprocating motion to rotary motion because the rotor functions as the piston in the engine. In the reciprocating engine, the piston assembly must stop at BDC and move back up to TDC, then back down to BDC, etc., many times a second. The stopping and starting of the piston assembly puts tremendous pressure on the connecting rod and rod bolts. Due to inertia, the piston tries to move out of the top of the cylinder bore and through the bottom of the oil pan. The rotary engine does not have to stop/start its "piston" as it rotates. The rotor is roughly triangular in shape and turns inside of a housing. The housing works on a geometric principle called an epitrochoid curve. An **epitrochoid curve** is the circular movement around the perimeter of another circle. The rotor moves in a unique pattern to ensure that the rotor ends follow the oblong shape of the housing (**FIGURE 12-50**).

Because the rotor spins, rather than moving up and down, engine operation is relatively smooth and vibration free. Each rotor and housing is akin to that of a three-cylinder two-stroke engine because of the rotor's three-sided shape. The rotor has three working chambers; thus, for each rotation of the rotor, we get three power pulses. Low-end torque is improved to the point that (though not recommended) a four-speed transmission rotary vehicle can be driven and accelerated from a standstill in fourth gear without excessive lugging. Rotary engines can be made with one, two, or even three or more rotor housings stacked side by side.

Let's look at the basic principles of a rotary engine. Although it appears different, the rotary engine is still an ICE. Recall the five events common to all ICEs: intake, compression, ignition, power, and exhaust.

The rotary engine's intake cycle occurs when one face of the rotor passes the intake port and draws the air-fuel mixture into the working chamber through the inlet port (**FIGURE 12-51**). The turning rotor then carries it around to the spark plugs. Along the way, the volume of the working chamber decreases and compresses the mixture. The mixture is ignited and combustion occurs. Expanding gases produce a power pulse, driving the rotor farther around. When the exhaust port is uncovered, exhaust occurs as the rotor sweeps burned gases out of the housing. Each face of the rotor is a separate working chamber, so three combustion events occur for each single revolution of the rotor.

Basic Components of the Rotary Engine

The rotor is mounted in an oval-shaped housing. The housing is made of aluminum alloy, but the curved interior surface has hard chromium plating. This surface has to put up with the wear and tear of the rotor seals sliding against it as it turns in the housing (**FIGURE 12-52**). Cooling passages machined into the housing allow coolant to flow around the outside surface of the housing, to cool the engine.

There are usually two spark plugs fixed to the housing that enable combustion to occur; these are referred to as leading and trailing spark plugs. The combustion chamber design is a long trough, which can contribute to incomplete combustion due to the large surface area of the chamber and the distance the flame travels in the combustion chamber. Designers found that by installing a leading and trailing plug, a more efficient combustion process could be produced. The rotor housing also has an intake port to let air and fuel into the combustion chamber and an exhaust port to expel burned gases.

The rotor has three apexes, or points, which have seals to seal between the rotor and the rotor housing (**FIGURE 12-53**). These seals work like piston rings in a reciprocating piston engine. Side and corner seals create a seal between the rotor and the side housing to prevent combustion gases from leaking around the rotor at the apex seals. The rotor also has oil seals on the side of the rotor that keep oil from inside the rotor from finding its way into the combustion chamber.

The combustion chamber of the rotary engine is formed by hollows in the flanks of the rotor. These hollows are sometimes called bathtubs. Front and rear housings, or side housings, are bolted to each side of the rotor housing. If it is a two-rotor engine, there is an intermediate housing between the two rotors.

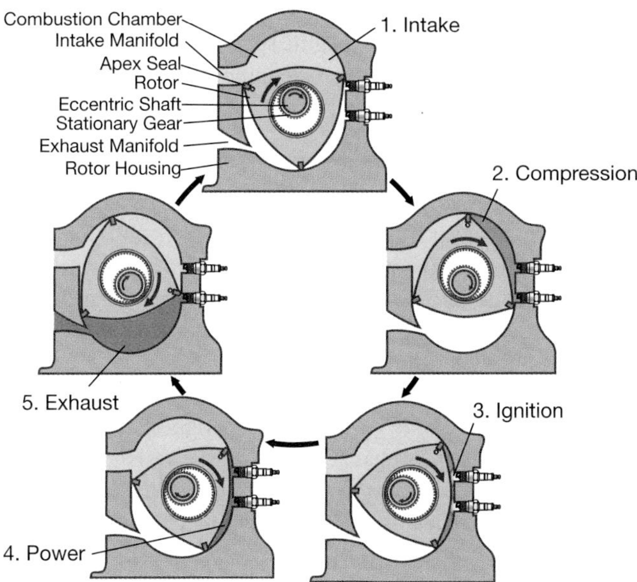

FIGURE 12-51 Operation of a rotary engine.

FIGURE 12-52 Rotary engine housing.

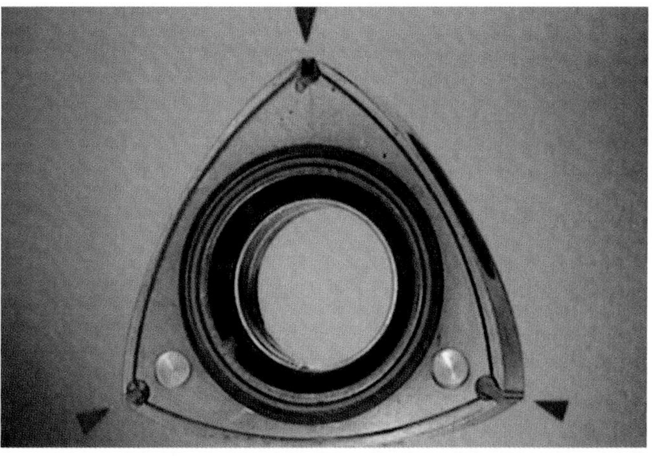

FIGURE 12-53 Rotary engine rotor and seals.

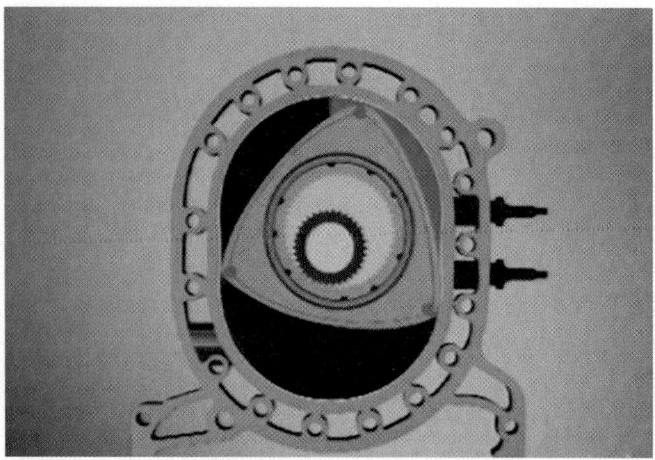

FIGURE 12-54 Stationary gear and internal rotor gear.

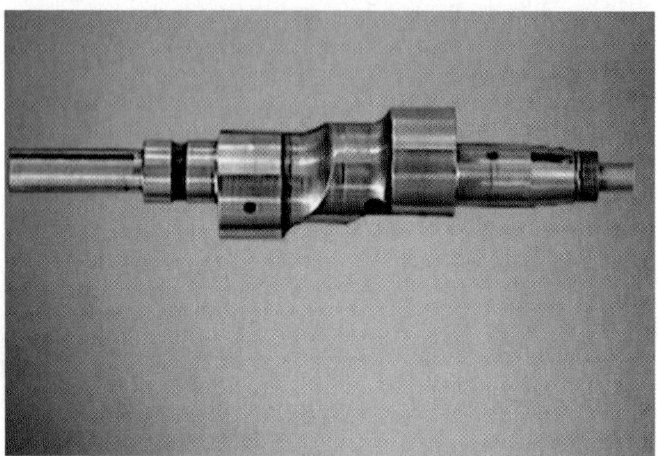

FIGURE 12-55 Rotary engine crankshaft.

An internal gear in each rotor meshes with a corresponding stationary gear in the front and rear housings (**FIGURE 12-54**).

When combustion occurs, the meshing of the teeth forces the rotor to walk around the stationary gear. The teeth being in mesh combines with an eccentric shaft to make the rotor follow the curved surface of the housing, and it gives the rotor planetary motion. The rotor is attached to the eccentric shaft at points called rotor journals. This eccentric shaft is like a crankshaft in a piston engine, but with the journals off-center (**FIGURE 12-55**). Because the rotor is off-center, the force applied to the shaft is off-center too. The whole shaft is supported by main journals, so the final output is smooth rotary motion.

Engine Power Pulses

A single-rotor rotary engine produces three power pulses per rotor rotation or one power pulse per eccentric shaft rotation. In a four-stroke engine, there is one power pulse per cylinder for every two crankshaft revolutions, and in a two-stroke engine there is one power pulse per cylinder for each crankshaft revolution. Because of the ratio of the gears in the housings and rotor, the eccentric shaft makes one revolution for each power phase. That is the same as three eccentric shaft revolutions for each rotation of the rotor. So the eccentric shaft turns at three times the speed of the rotor. A standard rotary engine typically has two rotors, offset from each other, in separate chambers, so it ends up with two power pulses per revolution of the eccentric shaft.

Renesis Rotary Engine

Because of emission and fuel economy requirements, Mazda updated the standard rotary engine to the Renesis engine. The operating cycle of the Renesis rotary engine is the same as a conventional, or Wankel, rotary engine, but with some design changes that improve fuel economy under load and compliance with current emission regulations. There is a low-output version of this engine for use with automatic transmissions and a high-output version for use with manual transmissions. The low-output Renesis engine has two intake ports: primary and secondary. The ports were enlarged and moved to the side of the housing. In this position, they can open sooner, improving power and torque, extending engine efficiency over a wider range of engine speeds, and reducing port overlap, which results in unburned fuel exiting the exhaust port.

The intake manifold has primary, secondary, and auxiliary ducts. The primary duct has no control valve; the secondary and auxiliary ducts are controlled by butterfly valves. At low engine speeds, air flows into the engine through the primary intake ducts only, keeping the air velocity in the manifold high, which provides better air-fuel mixing. At medium engine speeds and when engine load is high enough, the secondary intake ducts are opened by butterfly valves, reducing restriction and increasing airflow and torque. Intake manifolds on reciprocating piston engines now use a similar system called intake manifold runner control. The rotary engine opens an extra air duct on the air cleaner at high engine speeds, allowing more air to be drawn into the engine.

The high-output engine has three intake ports: primary, secondary, and auxiliary. At engine speeds above 6000 rpm, the auxiliary duct opens, allowing the engine to draw air in through all six ports (three intake ports per rotor), further increasing engine breathing. A butterfly valve located between each housing's main intake duct is used at speeds above 7000 rpm to shorten the effective length of the intake tubes so that pressure pulses force more air into the engine.

Low-output engines have two fuel injectors per rotor: primary and secondary. The primary fuel injectors operate at all times; the secondary injectors operate at engine speeds over 3700 rpm and when the engine load demands more fuel. The high-output engine has additional primary injectors, named "primary 2," which only operate at very high-speed and high-load conditions. The exhaust ports in a Renesis engine have also been enlarged and moved to the side of the combustion chamber housing. In this location, the exhaust ports open later than in a conventional rotary engine, and the rotor has been machined to delay the closing point. These changes deliver a longer expansion stroke and increase thermal efficiency.

Wrap-Up

Ready for Review

- Most modern vehicles use internal combustion engines.
- Internal combustion engines are typically either piston (spark ignition, using reciprocating motion of pistons) or rotary (spark ignition, using planetary motion).
- Piston engines can be spark ignition (passenger vehicles; uses a spark plug) or compression ignition (diesel vehicles; no spark plug).
- Pressure and temperature have a direct relationship in that when pressure rises, so does temperature, and vice versa.
- Internal combustion engines work by heating a gas, which increases pressure (thermal expansion), creating force to push the piston down the cylinder.
- Pressure and volume have an inverse relationship: when one increases, the other decreases.
- Force (effort) tends to cause movement, which creates work; the speed at which this happens is known as power.
- Work = distance moved × force applied.
- Power = distance × force/time in minutes.
- Engine power is measured by the amount of torque (turning effort) applied to the crankshaft multiplied by the rpm at which it is turning divided by 5252.
- Torque and power produced by an engine are called engine output.
- Horsepower refers to the speed at which torque is produced.
- Load factor refers to the period of time a vehicle can operate at maximum speed and power.
- Piston stroke refers to the distance traveled from top dead center (TDC) to bottom dead center (BDC) (or BDC to TDC).
- Internal combustion engines have either a two-stroke or four-stroke cycle.
- In a four-stroke cycle, five events must occur: intake, compression, ignition, power, and exhaust.
- The compression ratio of an engine is based on cylinder volume at BDC compared to cylinder volume at TDC, and can be affected by changes in piston stroke, piston head shape, head gasket thickness, and combustion chamber size.
- Piston displacement (the volume of movement from BDC to TCD) is calculated as bore squared × 3.14 × stroke/4.
- Engine displacement is calculated as piston displacement × number of engine cylinders.
- Two variations on the typical four-stroke spark ignition engine are the Miller cycle engine and the Atkinson cycle engine, both of which use valve timing variations to create unequal compression and expansion strokes.
- The Miller cycle engine has an engine-driven compressor that functions at high load and speed to boost power output.
- The Atkinson cycle engine is ideal for hybrid vehicles, as it has a lower power output and torque than conventional engines.
- Major components of an internal combustion engine include cylinder block, crankshaft, flywheel, connecting rod and piston, intake manifold, oil pan, oil pump, exhaust manifold, cylinder head, valve train, and engine camshaft.
- The cylinder block, the largest engine component, includes cylinder bores, coolant and lubrication passages, and the crankcase.
- The function of the crankshaft is to convert the piston's reciprocating motion into rotary (turning) motion.
- The flywheel stores energy from each piston's power stroke to smooth out the power strokes.
- The connecting rod connects the piston to the crankshaft and causes piston movement (via the crankshaft) during non-power strokes.
- Components that make up and support the piston are piston head, piston rings (compression and oil control), ring grooves, ring lands, piston skirt, pin hole, and pin boss.
- Compression and combustion gases can leak past piston rings and enter the crankcase; this is known as blowby.
- The purpose of an intake manifold is to deliver air or air and fuel to the cylinder head.
- The camshaft opens the valves and allows them to close at the right time, which ensures correct engine operation.
- Parts of the camshaft lobe include base circle, nose, cam lobe centerline, and cam lobe ramps.
- Engineers designing camshaft lobes must consider the issues of lift and duration, as well as the cam centerline and separation.
- Parts of an intake valve include head, face, margin, and stem.
- The intake valve is typically larger and tends to run cooler than the exhaust valve.
- Exhaust valves are typically smaller, allow exhaust to exit the cylinder, and tend to be run hotter than intake valves.
- Modern automotive engines have valves arranged in the cylinder head above the piston, known as an I-head arrangement.
- The valve lifter can be hydraulic or mechanical and transfers motion from the cam lobe to the valve train or directly to the valve.
- Roller rocker arms act as levers to increase the camshaft's lift of the valve.
- Valve clearance must be accurate so as not to create noise (meaning clearance is too high) or a loss of compression past the valve (clearance is too low).
- Timing of when valves open and close is vital to proper engine function; timing belts need periodic replacement and must be aligned with engine timing marks.

- Valve timing can be modified during engine design by advancing or retarding cam timing, which gives the effect of more torque and power at a lower rpm or higher rpm range, respectively.
- Variable valve timing allows cam timing to be adjusted during engine operation.
- Cam actuators can be a twisted gear arrangement or a vane-type phaser.
- The engine control module (ECM) controls variable cam timing using the following inputs: mass airflow sensor or manifold absolute pressure sensor, throttle position sensor, intake air temperature sensor, engine coolant temperature sensor, and crankshaft position sensor.
- Using a pulse width modulation signal, the cam solenoid regulates amount and direction of oil flow to the cam phaser.
- Engine design—freewheeling or interference—determines the amount of damage that will result if the timing belt breaks.
- Two-stroke engines use the piston to open and close intake and exhaust ports, allowing air-fuel to enter the cylinder.
- All events of an internal combustion engine are accomplished by a two-stroke engine within one up and one down stroke of the piston.
- Two-stroke engines use either a reed valve or rotary valve to allow air and fuel into the crankcase.
- Two-stroke engines eliminate the camshaft and valve train, making them lighter than four-stroke engines.
- Four-stroke engines release fewer hydrocarbons into the atmosphere and are more likely to meet EPA regulations.
- Rotary, or Wankel, engines use a rotor in place of a piston and move it in an epitrochoid curve to create a nearly vibration-free engine operation.
- Rotary engines have two spark plugs—leading and trailing—per rotor to enable complete combustion.
- The rotor moves along the curved surface of its housing and pushes on an eccentric shaft to produce power.
- A rotary engine cycle has four phases: intake, compression, power, and exhaust.
- Each of the three faces of the rotor act as a combustion chamber and have a power pulse every revolution of the rotor.
- The Renesis rotary engine is designed with improved fuel economy via modified intake and exhaust ports.
- A Renesis engine can be low output (for automatic transmissions) or high output (for manual transmissions).

Key Terms

Atkinson cycle An engine cycle that uses a longer effective exhaust stroke than intake stroke to reduce exhaust emissions. This type of engine is widely used in hybrid-electric vehicles.

block deck The "top" of the engine block and cylinder bore where the cylinder head is bolted on.

blowby gas The result of combustion gases leaking past the compression rings and getting into the crankcase.

bottom dead center (BDC) The position of the piston at the end of its stroke, when it is closest to the crankshaft.

cam The egg-shaped lobe machined to a shaft, used to cause opening and closing of the valves of a four-stroke cycle engine.

cam-in-block engine An engine in which the camshaft is located in the engine block rather than on the cylinder head.

camshaft The part of the engine that activates the valve train by using lobes riding against lifters.

compression-ignition (CI) engine An internal combustion engine that uses the heat of compression to ignite the compressed air-fuel mixture.

compression ratio (CR) The volume of the cylinder with the piston at bottom dead center as compared to the volume of the cylinder at top dead center, given in a ratio such as 9:1 CR.

compression stroke The stroke of the piston during which air and fuel is being compressed into a small area prior to ignition.

cylinder bore The hole in the engine block that the piston fits into.

dual overhead cam (DOHC) A design that, in a V-engine, includes four cams; also called a twin cam engine.

engine displacement The size of the engine given in cubic inches, cubic centimeters, and liters. It is found by multiplying the piston displacement by the number of cylinders the engine has. Sometimes called "swept volume."

epitrochoid curve The circular movement around the perimeter of another circle. This is the movement that the rotary engine uses to ensure that the rotor stays in contact with the housing.

exhaust stroke The stroke of piston during which the exhaust valve is open and the piston is moving from bottom dead center to top dead center, to push exhaust gas out of the cylinder.

external combustion engine An engine that runs on heat applied externally to the cylinder, for example, the steam engine.

flat-head engine An L-head engine with valves in the block.

flywheel The heavy, circular fat plate that keeps the engine rotating when power is not produced, such as on the exhaust, intake, and compression strokes.

force The effort to produce a push or pull action.

freewheeling engine An engine that has enough clearance between the piston and the valves so that in the event the timing belt or chain breaks, the valves that are hanging all the way open will not contact the piston, thus preventing engine damage.

fulcrum A half-round bearing that the rocker moves on as a bearing surface.

horsepower An amount of work performed in a given time.

ignition The lighting of the fuel and air mixture in the combustion chamber.

intake stroke The stroke of the piston from top dead center to bottom dead center, during which the intake valve is open and air is pulled into the cylinder.

interference engine An engine that has minimum clearance between the valves and the pistons during normal operation; in the event that the timing belt or chain breaks, the open valves will be contacted by the piston and bend the valves, possibly breaking the piston.

internal combustion engine An engine that burns a fuel internally and creates movement due to thermal expansion of gases.

keepers Locking devices that keep the valve retained by the valve spring seat.

lobe The raised portion on a camshaft; used to lift the lifter and open the valve.

Miller cycle An engine cycle that uses a longer exhaust stroke than intake stroke through delayed closing of the intake valve. This engine uses a supercharger to pressurize air into the cylinder when needed.

overhead cam (OHC) engine An I-head engine with the camshaft located on top of the cylinder head rather than in the block.

parasitic loss A loss of engine efficiency caused by internal friction, inefficient breathing, etc.

piston displacement The volume of air that is moved by the piston from bottom dead center to top dead center.

piston stroke The up or down motion the piston makes from one limiting position to the other.

power The rate or speed at which work is done.

power stroke The stroke during which combustion is pushing the piston from top dead center to bottom dead center in the cylinder. This stroke is where power is produced.

pushrod A tubular rod that stands between the tappet and the rocker arm in an overhead valve engine; the pushrod transfers cam motion to the rocker arm.

reciprocating motion An up-and-down motion within the cylinder.

reed valve A small flexible metal plate that covers the inlet port of a two-stroke engine and opens and closes to let air and fuel into the crankcase.

rocker arm The fulcrum that transfers pushrod endwise motion to the valve stem.

rotary engine A nonreciprocating engine with a rotor and housing instead of pistons.

scavenging The process of removing burned gases from the cylinder through the use of moving airflow pulling or extracting the gases out.

spark-ignition (SI) engine An engine that relies on an electrical spark to ignite the air and fuel mixture.

stroke The movement of an object in a straight line. The piston sees four strokes during one combustion cycle in a four-stroke cycle engine, meaning it moves up and down twice during each cycle.

tappet Another name for a valve lifter. Tappets may be fat or have rollers to ride on the cam lobes.

throw The offset area of the crankshaft where the connecting rod bolts on.

top dead center (TDC) The position of the piston when it is farthest from the crankshaft.

two-stroke engine An engine that uses only two strokes to complete its running cycle.

valve face A machined surface on the back of the valve head; this area seals onto the valve seat in the cylinder head.

valve guide An insert in the cylinder head through which the valve stem passes and moves.

valve lifter A device that transfers motion from the cam lobe to a pushrod or directly acts on the valve and spring, depending on whether the cam is in the engine block or the cylinder head; sometimes called a tappet.

valve margin The flat surface on the outer edge of the valve head between the valve head and the valve face.

valve stem The shaft that is attached to the valve head and provides the sliding surface for the valve in its guide as it opens and closes.

volumetric efficiency A ratio, given as a percentage, of the amount of air actually inducted at a given engine speed at full throttle compared to the internal engine displacement. For a normally aspirated engine (without supercharging or turbocharging), an engine's volumetric efficiency may peak at around 85%. Peak engine torque is developed at peak volumetric efficiency.

work The result of force creating movement.

Review Questions

1. When compared to internal combustion engines:
   a. the output of Stirling engines can be easily varied.
   b. steam engines take relatively less time to generate pressure.
   c. steam engines do not present an explosion hazard under too much pressure.
   d. Stirling engines run almost silently.
2. Choose the correct statement.
   a. For pressure to increase, either temperature must decrease or volume must be increased.
   b. For pressure to increase, either temperature or volume must be increased.
   c. For pressure to increase, either temperature must increase, or volume must be decreased.
   d. For pressure to increase, either temperature or volume must be decreased.
3. The effort to produce a push or pull action is referred to as:
   a. force.
   b. power.
   c. work.
   d. torque.
4. Which event occurs when the piston reaches TDC of the compression stroke?
   a. Ignition
   b. Compression stroke
   c. Exhaust
   d. Intake
5. The displacement of an engine can be altered by changing all of the below *except*:
   a. cylinder bore.
   b. type of fuel.
   c. piston stroke.
   d. number of cylinders.
6. In the Miller and Atkinson cycle engines:
   a. compression stroke is longer and expansion stroke is shorter.
   b. the combustion chamber is slightly bigger.
   c. the compression pressure at ignition is greater than in a conventional engine.
   d. compression stroke is shorter and expansion stroke is longer.

7. Which of these converts the reciprocating movement of the pistons into rotary motion?
   a. Camshaft
   b. Flywheel
   c. Crankshaft
   d. Piston ring
8. Proper power output and low-emissions operation of the engine is ensured by:
   a. using pistons with an iron coating.
   b. timing the valve opening to the piston position.
   c. keeping the combustion pressure in check by letting it into the crankcase.
   d. decreasing the number of valves per cylinder.
9. The two-stroke engine differs from the four-stroke engine in all the below aspects *except*:
   a. the events involved in the operation of the engine.
   b. the emissions produced by the engine.
   c. the method of air induction and exhaust.
   d. their production costs and size.
10. Which of the following statements is true with respect to a rotary engine?
   a. The rotor is mounted in a cylindrical housing.
   b. There is no reciprocating motion.
   c. The engine produces one power pulse per rotor rotation.
   d. The engine produces more vibration when compared to other ICEs.

ASE Technician A/Technician B Style Questions

1. Tech A says that as volume decreases, pressure increases. Tech B says that when temperature increases, pressure decreases. Who is correct?
   a. Tech A
   b. Tech B
   c. Both A and B
   d. Neither A nor B
2. Tech B says that horsepower is a measurement simply of the amount of work being performed. Tech B says that horsepower can be calculated by multiplying torque by rpm and dividing by 5252. Who is correct?
   a. Tech A
   b. Tech B
   c. Both A and B
   d. Neither A nor B
3. Tech B says that in a four-stroke engine, the piston travels to TDC four times to complete the cycle. Tech B says that in a four-stroke engine, the air-fuel mixture is ignited once every two strokes. Who is correct?
   a. Tech A
   b. Tech B
   c. Both A and B
   d. Neither A nor B
4. Tech A says that spark ignition typically occurs before TDC. Tech B says that spark ignition typically occurs during the intake stroke. Who is correct?
   a. Tech A
   b. Tech B
   c. Both A and B
   d. Neither A nor B
5. Tech A says that valve overlap occurs between the exhaust stroke and the intake stroke. Tech B says that valve overlap occurs to assist in scavenging the cylinder. Who is correct?
   a. Tech A
   b. Tech B
   c. Both A and B
   d. Neither A nor B
6. Tech A says that compression ratio is the comparison of the volume above the piston at BDC to the volume above the piston at TDC. Tech B says that scavenging of the exhaust gases occurs once the exhaust valve closes. Who is correct?
   a. Tech A
   b. Tech B
   c. Both A and B
   d. Neither A nor B
7. Tech A says that the weight of the flywheel smoothes out the engine's power pulses. Tech B says that the flexplate and torque converter perform this same function. Who is correct?
   a. Tech A
   b. Tech B
   c. Both A and B
   d. Neither A nor B
8. Tech A says that an interference engine is designed so that the pistons can hit the valves if the timing belt breaks. Tech B says that the shape of the cam lobe determines how long and far the valves are held open. Who is correct?
   a. Tech A
   b. Tech B
   c. Both A and B
   d. Neither A nor B
9. Tech A says that blowby gases occur when compression and combustion gases leak past the piston rings. Tech B says that the upper two piston rings are oil control rings. Who is correct?
   a. Tech A
   b. Tech B
   c. Both A and B
   d. Neither A nor B
10. Tech A says that the principle of thermal expansion is what pushes the piston down the cylinder on the power stroke. Tech B says that the crankshaft pulls the piston down the cylinder on the compression stroke. Who is correct?
   a. Tech A
   b. Tech B
   c. Both A and B
   d. Neither A nor B

CHAPTER 13

Engine Mechanical Testing

NATEF Tasks

- **N13001** Perform engine absolute manifold pressure tests (vacuum/boost); determine needed action.
- **N13002** Perform cylinder power balance test; determine needed action.
- **N13003** Perform cylinder cranking and running compression tests; determine needed action.
- **N13004** Perform cylinder leakage test; determine needed action.
- **N13005** Diagnose abnormal engine noises or vibration concerns; determine needed action.
- **N13006** Inspect engine assembly for fuel, oil, coolant, and other leaks; determine needed action.

Knowledge Objectives

After reading this chapter, you will be able to:

- **K13001** Explain how engine mechanical testing is part of the SBD process.
- **K13002** Identify engine testing tools and their purpose.
- **K13003** Approximate engine condition based on cranking sound.
- **K13004** Approximate engine condition using engine vacuum.
- **K13005** Describe the purpose of a cylinder power balance test.
- **K13006** Evaluate cylinder cranking and running compression.
- **K13007** Evaluate possible causes of cylinder leakage.
- **K13008** Describe the variety of engine noises and vibrations and how to locate them.
- **K13009** Describe the process for identifying various types and causes of engine fluid leaks.
- **K13010** Describe how consumption of oil or coolant not due to external leaks may be determined by the color of the exhaust.

Skills Objectives

After reading this chapter, you will be able to:

- **S13001** Perform a cranking sound diagnosis test.
- **S13002** Test engine vacuum using a vacuum gauge.
- **S13003** Test engine vacuum using a pressure transducer.
- **S13004** Perform a cylinder power balance test.
- **S13005** Perform a cranking compression test.
- **S13006** Perform a running compression test.
- **S13007** Perform a cylinder leakage test.
- **S13008** Perform a fluid leak inspection.

▶ Introduction

For an engine to operate efficiently and effectively, the mechanical condition of the engine must be in good working order. The pistons, piston rings, cylinder walls, head gasket, and valves must seal properly. If they do not, then the engine will not operate correctly, and all of the tuning in the world will not be able to fix it. So assessing the condition of the engine is a critical step in the diagnostic process, before tune-up parts are replaced. There are few things more dreaded than having to tell a customer that the $650 tune-up you just performed did not resolve the vehicle's misfiring problem caused by a burned valve and that the vehicle really needs a $1500 valve job or a $4000 engine replacement. First, the customer will not be happy that the vehicle is not fixed. Second, the repair will now cost substantially more money than the customer expected. Third, the customer now has very good reason to doubt your competence and wonder if you are correct now, when you were wrong earlier. This is a no-win situation, but it can be avoided by always diagnosing the problem instead of throwing parts at it.

Engine mechanical testing is also performed to diagnose more accurately what major engine work is needed, such as replacing a head gasket, rebuilding of the cylinder heads, or performing a full engine replacement. Understanding exactly what is wrong will allow you to better advise the customer on the appropriate repairs. It will also help you build credibility with your customers by fully understanding the situation and clearly communicating it to them in a professional manner. This chapter will help you learn the skills and procedures for performing engine mechanical tests.

▶ Engine Mechanical Testing

K13001

Engine mechanical testing uses a series of tests to assess the mechanical condition of the engine. The tests start off broad and then narrow down as each test is performed. This process will help you to first identify the location of the fault and then the cause of the fault. Having a good understanding of engine theory will help you evaluate the results of each test and provides you the information needed to know what path to follow after each step.

Mechanical testing starts with following the strategy-based diagnostic process covered earlier in this book. In this chapter, we mainly focus on only the first three steps and leave the last two steps for later chapters, when repair of the engine mechanical components is covered:

Step 1: Verify the customer's concern
Step 2: Researching possible faults and gathering information
Step 3: Focused testing
Step 4: Performing the repair
Step 5: Verify the repair

So always start by verifying the customer's concern, which involves getting as much information from the customer as possible and then observing the fault. A visual inspection of the gauges will give you general indications such as low oil pressure, overheating condition, and illuminated **malfunction indicator lamp (MIL)**. Starting the engine and listening to its operation can indicate a host of problems, from loose belts to worn main bearings. It can also reveal whether the engine is misfiring and how steady the misfire is. A visual inspection of the engine assembly for leaks will affirm the ability of the seals and gaskets to contain each of the engine's fluids and indicate other issues with the engine.

After verifying the concern and giving the vehicle a good preliminary inspection, it is time to research the vehicle's service history to see if the new concern is related to any previous work. If so, that may give you some clues as to what is causing the concern. If the vehicle service history doesn't appear to be related to the concern, gather information and put together a logical plan to attack the problem. This usually starts by researching the service information by looking up any related TSBs and the manufacturer's diagnostic flowcharts. In doing so, take your knowledge of the affected system and symptoms it is exhibiting and determine a logical starting point to begin testing. Depending on the results of your initial tests, referencing the service information and using

You Are the Automotive Technician

A customer comes into the dealership complaining that his vehicle is not running as smoothly as before and that the check engine light is flashing, which indicates a potentially catalyst-damaging fault. As an experienced, certified technician, you explain to the customer that he did the right thing by bringing in the vehicle for diagnosis. First, you locate the customer's vehicle history and write up a repair order including the customer's concerns. Second, you use a scan tool to retrieve the DTCs and find a P0304-cylinder 4 misfire detected code. Third, you hold the throttle to the floor, with the ignition switch in the off position, and then crank the engine over. The engine exhibits an uneven cranking sound, indicating a compression-related fault. Next, you research the technical service bulletins (TSBs) and find one that relates to this code and a low compression condition. The TSB indicates possible soft camshaft lobes, which can wear down and not open the valve(s) fully. To verify whether this is the case, the TSB directs you to perform a cranking compression test along with a running compression test on any misfiring cylinders. The TSB then lists the acceptable minimum pressures for each test.

1. Why did the engine have an uneven cranking sound?
2. What does a power balance test indicate?
3. What does the running compression test indicate?
4. How would you determine which cylinder is misfiring if the engine computer doesn't have that capability?

the diagnostic procedure will allow you to determine the follow-up tests needed to locate the problem. But first, let's look at the tools and equipment that are used to examine the mechanical condition of the engine.

Testing Tools

K13002

Although some seasoned mechanics have been known to use bubble gum and its wrapper to diagnose a burned valve (see the Technician Tip below), a variety of tools and equipment are available today that allow technicians to more fully assess the mechanical condition of modern engines. Each of the tools introduced here has a specific function and will be further described in the Skill Drills later in this chapter:

- Compression tester: A **compression tester** is used when a technician suspects a cylinder may have low compression. The compression tester measures the amount of compression pressure a cylinder can generate (**FIGURE 13-1**).
- Vacuum gauge: The **vacuum gauge** measures the amount of vacuum an engine can generate during various operating conditions (**FIGURE 13-2**).
- Pressure transducer and lab scope: The **pressure transducer** measures engine vacuum or pressure and displays it graphically on a lab scope. Because it is very accurate, it creates a detailed trace on the lab scope, which can be compared to a known good trace and used to determine mechanical issues in the engine (**FIGURE 13-3**).
- Cylinder leakage tester: A **cylinder leakage tester** pumps air into the cylinder and measures the percentage of air that is leaking from the cylinder. The technician can determine where the pressurized air is leaking from the cylinder by looking, listening, or feeling (**FIGURE 13-4**).
- Scan tool: The scan tool communicates to the vehicle's computers through the **data link connector (DLC)**. It displays the readings from the various sensors; retrieves trouble codes, freeze-frame data, and system monitor data; and on some vehicles performs output tests, such as a cylinder power balance test, or commands other output devices to operate (**FIGURE 13-5**).
- Stethoscope—standard and electronic: A stethoscope is used by technicians to listen to unusual noises in the vehicle. Stethoscopes come in both standard and electronic (**FIGURE 13-6**).

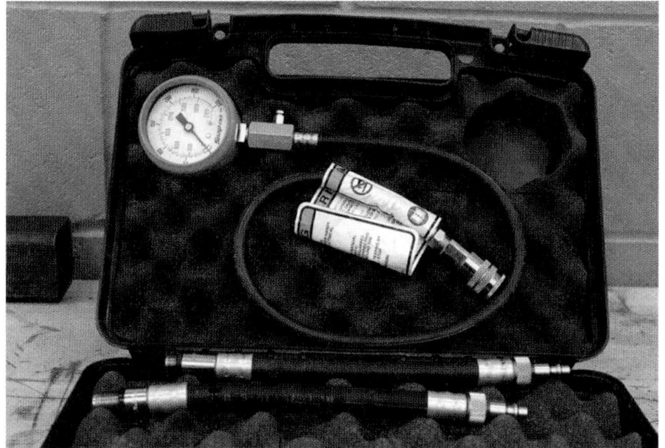

FIGURE 13-1 A compression tester.

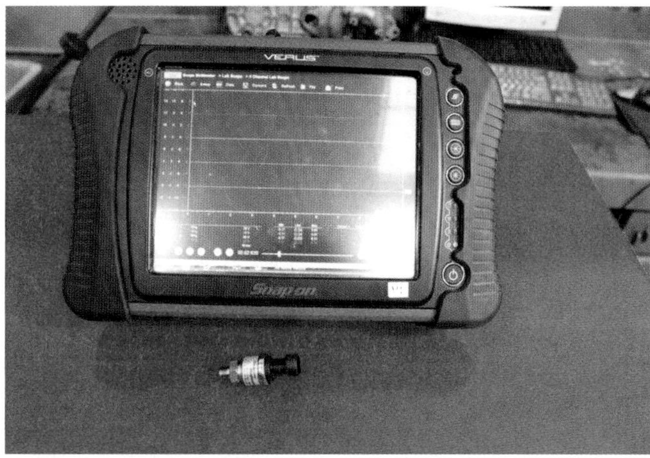

FIGURE 13-3 A pressure transducer and lab scope.

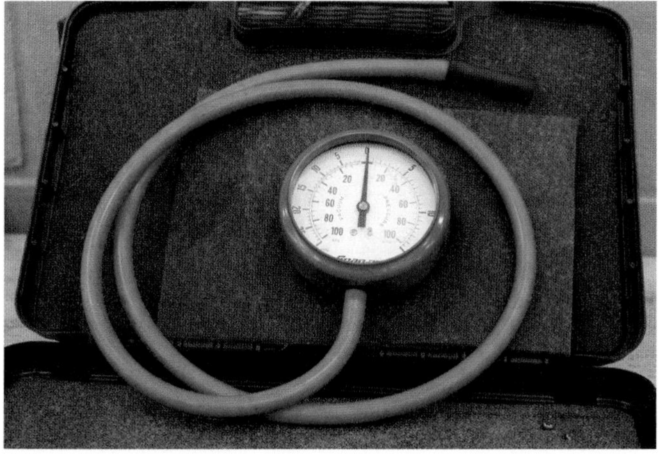

FIGURE 13-2 A vacuum gauge.

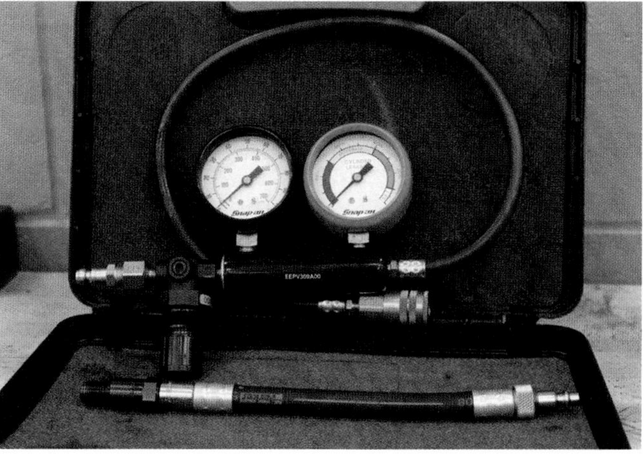

FIGURE 13-4 A cylinder leakage tester.

FIGURE 13-5 A scan tool.

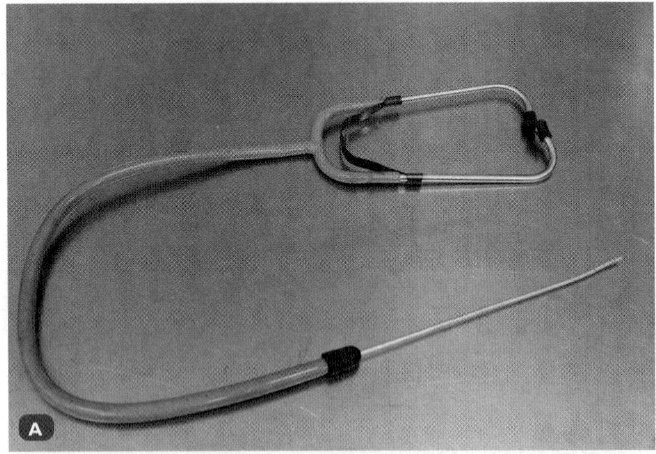

FIGURE 13-6 A. A standard stethoscope. **B.** An electronic stethoscope.

> **TECHNICIAN TIP**
>
> So how is it that a mechanic can diagnose a burned valve with just a piece of bubble gum and its wrapper? Knowing that each power pulse pushes exhaust gases out of the exhaust pipe, there should be steady pressure pulses if all of the cylinders are operating correctly. Besides listening to the engine and hearing a misfire, the technician used the gum to attach the wrapper to the end of the exhaust pipe. Each pulse pushed the wrapper away from the opening of the pipe, and each misfire tended to pull it back. If the engine had a burnt exhaust valve, the negative pressure pulse would be stronger than just a misfire, because some of the exhaust gases were pulled back into the cylinder through the burned exhaust valve, creating more of a vacuum in the exhaust pipe. The wrapper would be pulled into the exhaust pipe during the misfire in this case and blown back out during the next power pulse, resulting in a wrapper that was slapping the exhaust pipe.
>
> Suspecting that the exhaust valve was burned, the mechanic would shut off the engine, disable the ignition system, crank the engine over, and hear the telltale uneven cranking sound of a cylinder with low compression. Taking the results of both tests, the mechanic would conclude that the engine had a burnt exhaust valve. Pulling the spark plug wires off, one at a time, with the engine running resulted in one cylinder not contributing to the engine idle speed, so the mechanic knew the burnt valve was located in this cylinder.
>
> The process is similar today but involves tools and equipment that measure the results and display them numerically or graphically to more accurately assess the mechanical condition of the engine. In the scenario above, the technician would perform a cylinder leakage test on that cylinder to verify the fault.

When gathering information during an engine mechanical diagnosis, it is good to start with looking at the warning indicators. The manufacturer placed them there to warn the driver of any major issues such as low oil pressure, overheated engine, and low fuel level (**FIGURE 13-7**). The malfunction indicator light (MIL) illuminates if there are active codes in memory that can give you clues to the problem. Pulling the codes with a scan tool can reveal faults such as a cylinder that is misfiring or injector winding that is open, both of which can cause rough engine operation. There may also be an information system built into the dash that can access other information such as percent of engine oil or transmission fluid life remaining, engine oil pressure, and fuel economy. All of this information provides you with clues that can lead you to determine the proper tests and then ultimately to diagnose the root cause of the customer concern.

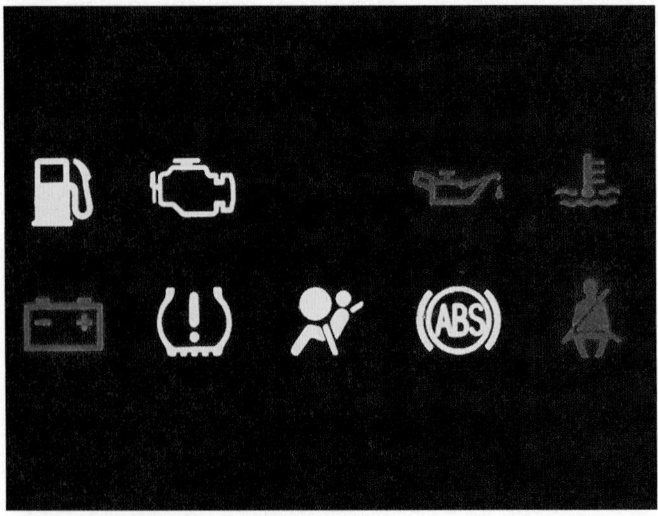

FIGURE 13-7 An example of warning indicators on the dash.

A good time to observe the gauges and warning indicators is when you are either test-driving the vehicle or as you pull it into the shop. Use that time to investigate the information provided through those indicators. One thing to remember is that the indicators can be wrong. So you may need to verify that the indicator is working properly and providing you with accurate information, instead of assuming it is correct and being led down the wrong path. You may be able to do that by other types of observation. For example, if the oil pressure is truly low, then there will likely be tapping or knocking noises coming from the engine that support this conclusion. That is what diagnosis is all about: gathering information, sorting it all out, building a focused plan of attack, testing and interpreting the results, and deciding what the next step is. And using the information provided by the warning indicators is a great starting point once you fully understand, and have verified, the concern.

Applied Science

AS-6: Operational: The technician can relate scientific terms to automotive system diagnosis, service, and repair.

The use of pressure gauges is a common procedure in most automotive repair facilities. When assigned to perform engine mechanical testing, the technician works with pounds per square inch (psi) or kilopascals (kPa) as related to engine compression gauge readings and cylinder leakage tests. A vehicle with low compression produces a number of drivability problems, including a rough running condition and low power. For a smooth-running engine, cylinder compression should be as uniform as possible between cylinders. Psi refers to the primary units of measure for pressure in the United States. The metric unit is the kilopascal, or kPa; 10 psi is equal to approximately 68.95 kPa. Most manufacturers will supply both units as technical information. If necessary, the technician can consult conversion charts to obtain the U.S. standard or metric unit that is needed.

▶ Cranking Sound Diagnosis Overview

K13003

Engine noises can give a technician valuable insight into the condition of the engine. As you have learned, compression is one of the five critical requirements for each cylinder to operate properly. If the compression in one or more cylinders is too low, then the affected cylinders will not create as much power as cylinders with the proper amount of compression. This causes the engine to run rough, in many cases misleading the customer to request a "tune-up." Yet, that will not fix the low compression issue, and the engine will still run poorly after the tune-up. This results in an unhappy customer as well as technician. To help avoid that situation, a cranking sound diagnosis can identify whether the compression is similar across all of the cylinders. If it is not similar across all cylinders, a tune-up will not fix the problem. If it is similar across all cylinders, then compression issues are not likely causing the engine to run rough, and further diagnosis of the fuel and ignition system is needed.

When you perform a cranking sound diagnosis, you should disable the engine so that it will not start. Then crank the engine over, using the key, and listen to the cranking sounds. It might take a few times to orient yourself to the sounds. Any noises you hear could be from a misaligned starter, the knock of a spun crankshaft bearing noise, an uneven cranking sound that a low-compression cylinder gives, or a fast cranking sound from a no-compression condition that is due to bent valves caused by a broken timing belt. Each of these noises can give you a clue as to where the problem may lie.

Diagnosing Cranking Sound

S13001

During engine cranking, the engine will make a rhythmic cranking sound if the compression is similar across the cylinders. As each piston comes up on compression, it loads down the starter motor and slows the cranking speed. If the compression of the engine is not the same at each cylinder, the engine will make an uneven sound. To train your ear to pick up the different compression sounds, take a vehicle with known good compression, and disable the engine so that it will not start. You can disable the engine in three possible ways:

- Use the "clear flood" mode if the vehicle is equipped with it. To do this: With the ignition key off, hold the throttle to the floor, then crank the engine. If it starts, let up on the throttle and turn the key off. Then use one of the two other methods.
- Disable the fuel pump by pulling the fuel pump fuse or relay, starting the engine, and waiting for it to die.
- Disable the ignition system (see service manual for proper procedure) so that it will crank, but not start (realize that the injectors will still spray fuel, which can wash down the cylinder walls, so it is not the best way to disable the cylinders).

You can then hold the throttle to the floor and crank the engine. Listen to the sounds it makes. Next perform the same task on a vehicle with known bad compression, and listen to the difference in cranking sounds.

To perform a cranking sound diagnosis, follow the steps in **SKILL DRILL 13-1.**

SKILL DRILL 13-1 Performing a Cranking Sound Diagnosis

1. Raise the hood and make sure it is secure.
2. Disable the engine by one of the methods described above.
3. Hold the throttle to the floor, and crank the engine over. Listen to the sound the engine makes while cranking. Determine any necessary actions.

> **TECHNICIAN TIP**
>
> Many newer vehicles are programmed with a "clear flood" capability. This effectively shuts off the fuel injectors as long as the throttle is held to the floor before the ignition key is turned to the run or crank position. When the key is turned to crank (with the throttle still held down), the vehicle's powertrain control module (PCM) will shut off the fuel injectors. This allows the engine to crank without starting. If the engine starts, it means either that the vehicle is not equipped with the clear flood mode or that fuel has leaked into the intake manifold and the problem must be diagnosed. Using the clear flood mode is a quick way of performing a cranking sound diagnosis on many vehicles.

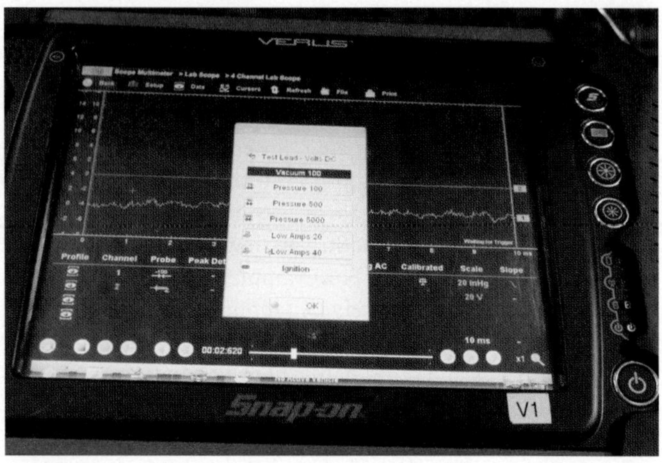

FIGURE 13-8 Vacuum trace.

▶ Vacuum Testing

N13001, K13004

A vacuum gauge is used to determine the engine's general condition. The vacuum gauge reading shows the difference between outside atmospheric pressure and the amount of vacuum (manifold pressure) in the engine. The vacuum gauge is installed in a vacuum port on the intake manifold. Always select the largest and most centrally located vacuum port. This may require using a vacuum tee so that the vacuum gauge can be connected while vacuum can still be supplied to the existing component. The typical vacuum reading for a properly running engine at idle is a steady 17" to 21" of vacuum (57.6 to 71.1 kPa) (**TABLE 13-1**). A low reading could relate to a problem with ignition, valve timing, or excessive piston, piston ring, or cylinder wall wear. A sharp oscillation back and forth in the needle or a dip in the gauge reading could relate to a problem such as a burnt or sticking valve. A low reading once the engine reaches a higher speed, such as 2500 rpm, could indicate a restricted exhaust system or faulty variable valve timing system.

With the advent of lab scope diagnostics, many technicians are using a pressure transducer to measure the engine's manifold pressure and display it graphically on a lab scope. The pressure transducer is a very accurate measuring tool; when attached to a vacuum port and hooked up to a lab scope, it gives a pressure trace that shows the low pressure created by each piston's intake stroke (**FIGURE 13-8**). If the ignition pattern for cylinder 1 is monitored on a separate trace, the vacuum pulses can be tied to specific cylinders. Technicians can view these traces and see how well each cylinder is functioning at producing and maintaining a vacuum.

Testing Engine Vacuum Using a Vacuum Gauge

S13002

During an engine vacuum test, all vacuum gauges are calibrated at sea level, and all instructions and readings are referenced to sea level. When testing above sea level, you will need to compensate. For instance, if the gauge reads 18" of vacuum at sea level, it would drop to 17" at 1000' above sea level.

The vacuum gauge reading shows the difference between outside atmospheric pressure and the amount of vacuum in the engine in a non-turbo or non-supercharged engine. The average inches of vacuum for a good running engine at idle is a steady 17" to 21" of vacuum. The average inches of vacuum at idle for a performance camshaft with a high lift and large overlap duration is around 12"–15". The amount of vacuum an engine creates relies on the piston rings, valves, ignition timing, valve timing and fuel.

To test engine vacuum using a vacuum gauge, follow the steps in **SKILL DRILL 13-2**.

TABLE 13-1: Chart of Vacuum Readings—Vacuum Gauge

Reading	Indication
17" to 21", steady needle at idle.	Good reading
Needle oscillates back and forth about 4" to 8".	Burned or constantly leaking valve
A low and steady reading.	Possible late valve timing or late ignition timing
Snap acceleration needle drops to zero and then only reads 20" to 23" of vacuum; should be much higher, around 27".	Possible worn rings
Good reading at idle. As engine speeds up to a steady 2500 rpm, the needle slowly goes down and may continue to drop.	Possible restricted exhaust system

SKILL DRILL 13-2 Testing Engine Vacuum

1. Connect the vacuum gauge to the intake manifold. You might have to use a vacuum tee. Make sure the engine is at operating temperature. Start the engine, take a reading at idle, and record the reading.

2. Snap accelerate the engine by quickly opening and closing the throttle; record the highest vacuum reading attained.

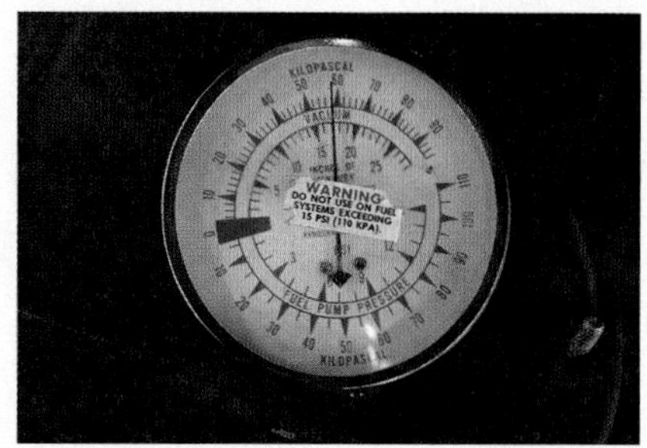

3. Hold the throttle steady at 2500 rpm and record your reading.

> **TECHNICIAN TIP**
>
> A pressure transducer can be adapted to fit in the spark plug hole to create a pressure trace of running and cranking compression. It can also be used in the exhaust system to measure the exhaust pulses. Pressure transducers are used more and more commonly when diagnosing engine problems.

Testing Engine Vacuum Using a Pressure Transducer

S13003

Using a pressure transducer and lab scope is a similar process to using a vacuum gauge, except it is much more accurate and allows you to look at the vacuum graphically. When paired with a second trace consisting of the ignition pattern for cylinder 1, any issues with the vacuum trace can be tied to a specific cylinder. By using a pressure transducer on a variety of vehicles with various issues, you will become familiar with the patterns of common faults.

To use a pressure transducer and lab scope for testing engine vacuum, follow the steps in **SKILL DRILL 13-3**.

▶ Cylinder Power Balance Test Overview

N13002, K13005

A cylinder power balance test (also called a power balance test, for short) is used for two purposes. First, it identifies which cylinder(s) is not operating properly when the vacuum test or cranking sound diagnosis test indicates a mechanical issue, or when the engine is not running smoothly. Second, it is used as a general indication of each cylinder's overall health. Every cylinder in the engine should contribute equally to the engine's power output. When a mechanical, electrical, or fuel problem occurs within an engine, the affected cylinders will not produce as much, if any, power when compared to the other cylinders. We can test to see how much each cylinder is contributing to the engine's output by disabling one cylinder at a time and measuring the rpm drop. Disabling a cylinder that is not operating correctly will not produce much, if any, rpm drop. Disabling a cylinder that is working properly, however, will produce a much larger rpm drop. The greater the difference in rpm drop, the greater the difference between each cylinder's ability to produce its share of the engine's power.

Many newer vehicles control the idle speed electronically through the PCM. In these vehicles, if something loads the engine down, such as the air-conditioning compressor turning on, the PCM will allow more air and fuel into the engine so that the idle speed stays relatively the same. In the case of a manual power balance test, disabling one of the cylinders will not produce a drop in rpm because the PCM will compensate for it. In these cases, use the PCM's built-in power balance feature, if equipped. Alternatively, in some vehicles you can disconnect the electrical connector on the idle speed control system to prevent the PCM from changing the idle speed. Unfortunately, in some vehicles if you disconnect the connector on the idle speed control system, the engine will die. On these vehicles, you can usually perform the power balance test slightly above idle, as the PCM only controls engine speed at idle. Raising the rpm can be accomplished by wedging an appropriate tool between the throttle body and the throttle linkage or by having an assistant hold the throttle steady while you perform the test.

Once the problem is isolated, you should take the next step, which is to determine whether it is a mechanical issue related to compression or an ignition or fuel-related issue.

Performing a Cylinder Power Balance Test

S13004

There are several ways that a power balance test can be performed. Knowing each of the options will allow you to choose the easiest one for the vehicle you are diagnosing. For many OBDII vehicles, you can identify which cylinders are misfiring

SKILL DRILL 13-3 Testing Engine Vacuum Using a Pressure Transducer

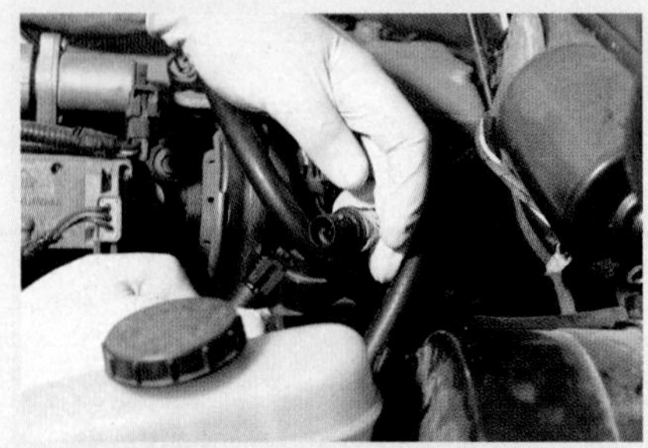

1. Connect the pressure transducer to the intake manifold and the lab scope.

2. Connect the second channel of the lab scope to the ignition system so it can identify cylinder 1.

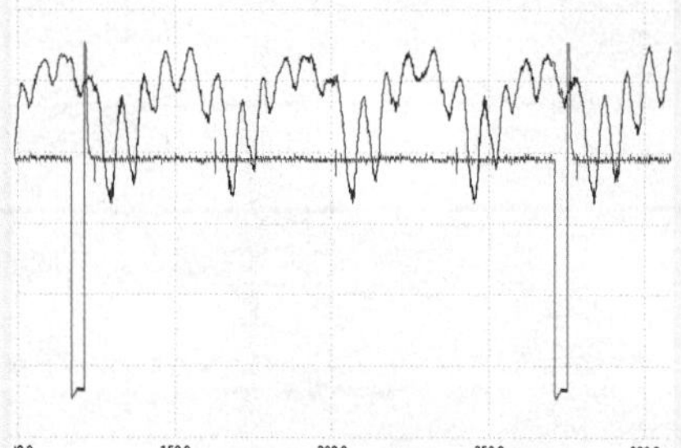

3. Start the engine and let it idle. Adjust the lab scope so that it captures the vacuum pulses for all cylinders. Observe the vacuum trace.

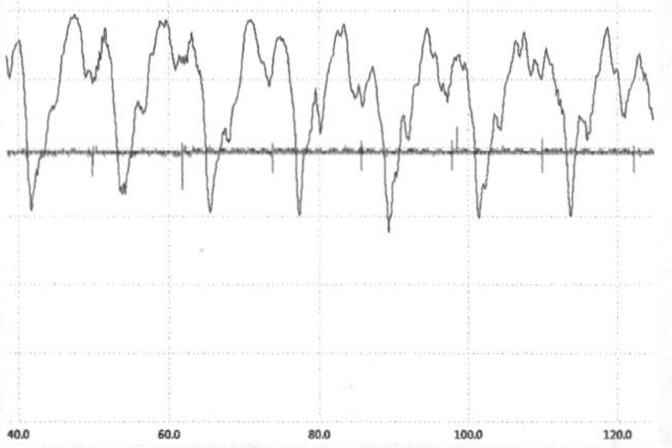

4. Snap accelerate the engine by opening and closing the throttle, and compare the trace to known good readings.

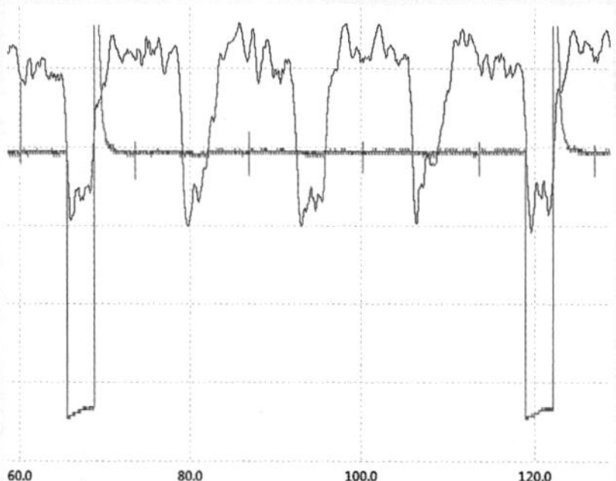

5. Hold the throttle steady at (1200–1500 rpm), and compare the trace to known good readings.

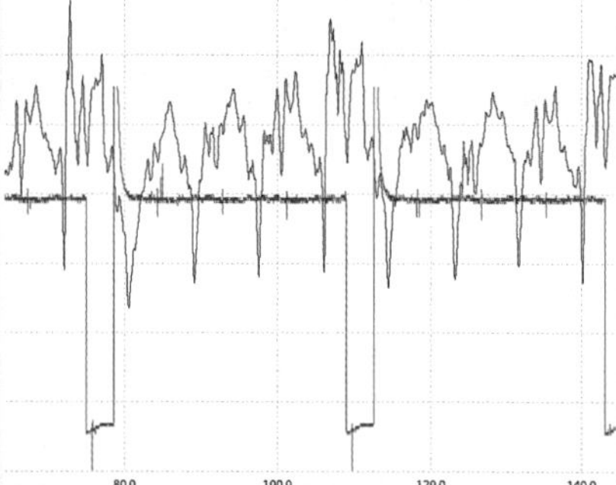

6. Then hold the throttle steady at 2500 rpm, and compare the trace to known good readings.

by using a scan tool to access stored diagnostic trouble codes (DTCs) in the PCM. Or you can access mode 6 data, which will give you information on how many misfires happen on each cylinder. In some cases, you can also use the scan tool to command the PCM to perform an automatic power balance test and report the results right on the scan tool. On most newer vehicles, using the scan tool is by far the best method of performing a power balance test.

If the system is not set up to perform the test automatically, you will have to do it manually. You will need to determine whether to disable the ignition or the fuel to each individual cylinder. If the engine has port fuel injectors that are accessible, disconnecting the electrical connector from each injector, one at a time, is the preferred method because it stops fuel from being injecting into the cylinder. Otherwise, if the ignition system is disabled on a cylinder, the fuel may still be injected but will not be burned in the cylinder. However, it will burn in the catalytic converter, which can cause it to overheat and possibly be damaged. Therefore, shutting down the fuel injector is preferable if it is an option on the engine you are working on.

If the engine has individual ignition coils on each spark plug, you can disconnect the primary electrical connector, which will shut off the spark to that spark plug. If the vehicle has coils that share cylinders (waste spark system), then you can place a 1" (25 mm) section of vacuum hose between each coil tower and spark plug wire. Then connect the alligator clip from a non-powered test light to a good engine ground, and touch the tip of the test light to each length of vacuum hose to short out each spark plug one at a time. If the vehicle is equipped with a distributor, disconnect one spark plug wire at a time from the distributor cap (it is good to use a test lead to ground the spark at the distributor cap terminal to prevent the spark from damaging the ignition module inside the distributor), which shuts down the spark for the cylinder being tested.

To perform a cylinder power balance test, follow the steps in **SKILL DRILL 13-4**.

▶ Cranking and Running Compression Tests

N13003, K13006

In a cranking or a running compression test, a high-pressure hose is hand-threaded in the spark plug hole of the cylinder to be tested and then connected to the compression gauge. The engine needs to be cranked over or started, depending on the test that is being performed. The compression gauge reads the amount of pressure that the piston is producing by

SKILL DRILL 13-4 Performing a Cylinder Power Balance Test

1. Visually inspect the engine to determine the best method to disable the cylinders. If necessary, disable the idle control system. Start the engine and allow it to idle. Record the idle rpm.

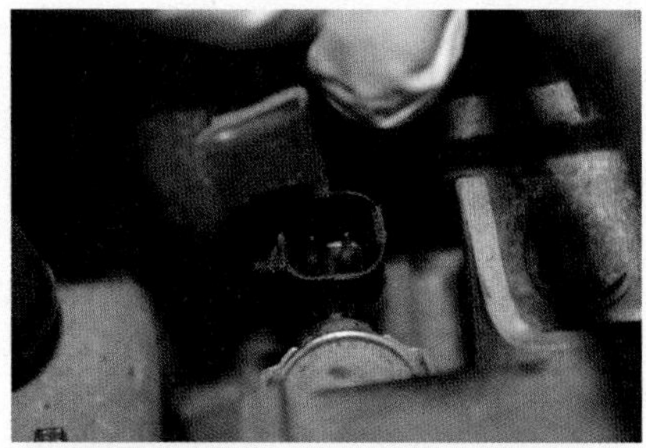

2. Using the method chosen to disable cylinders, disable the first cylinder, and record the rpm. (Do not leave the cylinder disabled for more than a few seconds.)

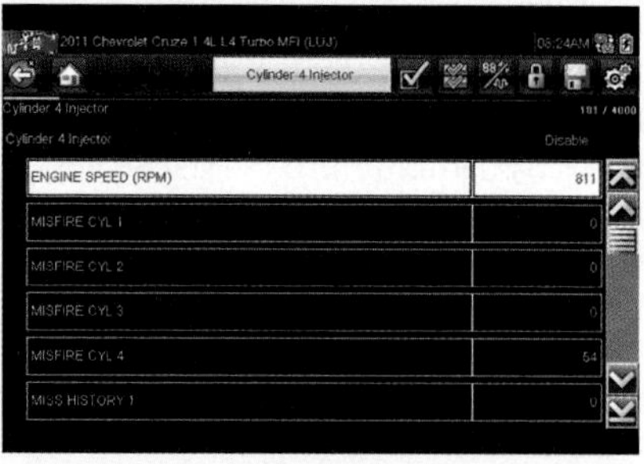

3. Reactivate the cylinder, and allow the engine to run for 10 seconds to stabilize. Repeat the steps on each of the cylinders, and record your readings. Determine any necessary action.

compressing the air in the cylinder. A cranking compression test indicates how well each combustion chamber is sealed; a running compression test indicates the engine's ability to breathe, also referred to as volumetric efficiency. You should always check factory specifications before performing these tests so that you know what to expect for results.

Performing a Cranking Compression Test

S13005

The cranking compression test is performed to determine if a misfiring or dead cylinder is caused by a compression problem. A compression test measures the air pressure as it is compressed in the cylinder. If the combustion chamber is properly sealed, the compression on all cylinders should typically measure within 10% to 15% of each other, but check the specs.

During a cranking compression test, the engine is cranked over, but not started. As the piston moves up, it compresses the air in the cylinder. The engine should crank until at least five compression pulses are observed on the compression gauge, to get an accurate reading. If the final reading is low, there could be a problem with the valves, rings, pistons, or head gasket, all of which are major engine problems

For the compression test to be completely accurate, it is recommended that the following conditions be met:

- The engine is at operating temperature.
- All of the spark plugs are removed.
- The battery is fully charged.
- The throttle is held wide open.
- At least five compression pulses are made on each cylinder (the same number of pulses for each cylinder).

All of these conditions will help the engine create maximum compression and accurate readings that can be compared to specifications. Manufacturers list the specifications differently; some examples follow:

- As the minimum pressure allowed in psi or kPa, such as 125 psi (861.8 kPa)
- As the maximum difference between cylinders in psi or kPa, such as 35 psi (241.3 kPa)
- As the maximum difference between cylinders in percent, such as 15%

At the same time, if a technician suspects that only one or two cylinders have excessively low compression (due to failing the power balance test), then only the suspect spark plugs should be removed and a compression test performed on those. If the compression is reasonably close to the specifications, then the compression is not causing the issue as it is likely caused by a fuel or ignition fault. If low compression is found, perform a wet test by placing a couple of squirts of clean engine oil in the low cylinder, and retest the compression. If the compression pressure increases substantially, the piston rings are worn. If the compression doesn't change much, a valve, the head gasket, or the top of the piston has a leak.

▶ TECHNICIAN TIP

Another way to measure compression is with a relative compression test. This test uses an inductive ammeter connected to a lab scope to measure the current flow for the starter as the engine is being cranked. As each cylinder comes up on the compression stroke, the engine is harder to turn, so the starter works harder and the current flow is higher. If every cylinder has the same relative compression, then the current spikes will be similar. If a cylinder has low compression, then that current spike will be lower than the rest. This is a quick way of checking to see if there is a compression-related problem.

To perform a cranking compression test, follow the steps in **SKILL DRILL 13-5.**

▶ TECHNICIAN TIP

When performing a compression test, the final reading is not the only thing to observe. If the piston rings are sealing well, then the first compression pulse on the gauge should be at least half as much pressure as the final reading.

When a low-compression cylinder is found, you should put a couple squirts of oil (about a tablespoon) into the spark plug hole, crank the engine a few turns, and recheck the compression. If the compression rises significantly, the problem is typically worn piston rings in that cylinder, as indicated by the higher compression due to the short-term sealing ability of the added oil. Worn piston rings are a substantial problem that typically requires rebuilding or replacement of the engine. If the oil did not make the compression rise, then the problem is likely to be a leaky valve or head gasket, or a hole in the piston. A cylinder leakage test can determine which.

A typical diagnostic test would be checking the compression of an engine after suspecting low compression. In this scenario, a customer is concerned about a four-cylinder engine that is running rough at idle. Fuel and ignition problems have been ruled out. A compression gauge will be used to measure the compression on each cylinder. The technician will record the information as the readings are taken from each cylinder, and the results could be recorded on the following chart. By consulting the manufacturer's specifications, the technician will be able to determine whether the compression meets specifications.

SKILL DRILL 13-5 Performing a Cranking Compression Test

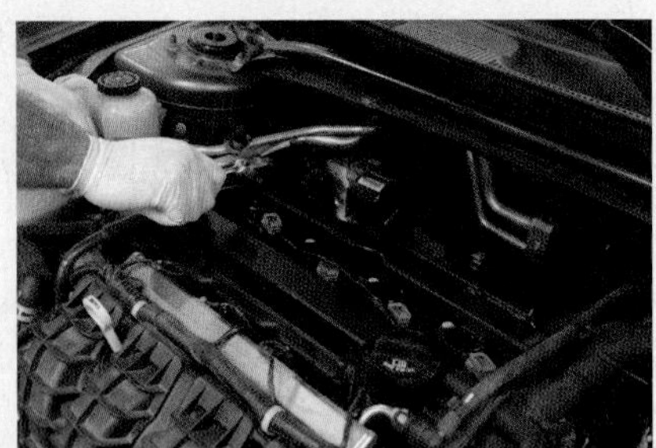

1. Remove any spark plug wires or ignition coils connected directly to the spark plugs. Disable the ignition primary system, or ground the coil wire(s).

2. Disable the injectors, or remove the fuel pump fuse or relay.

3. Remove all spark plugs.

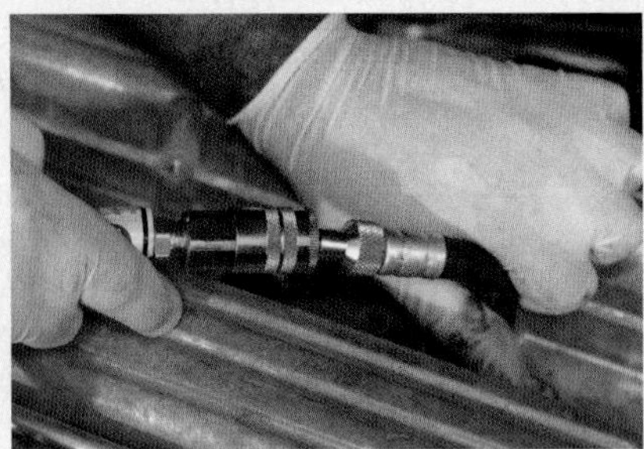

4. Connect the compression tester to the spark plug hole to be tested.

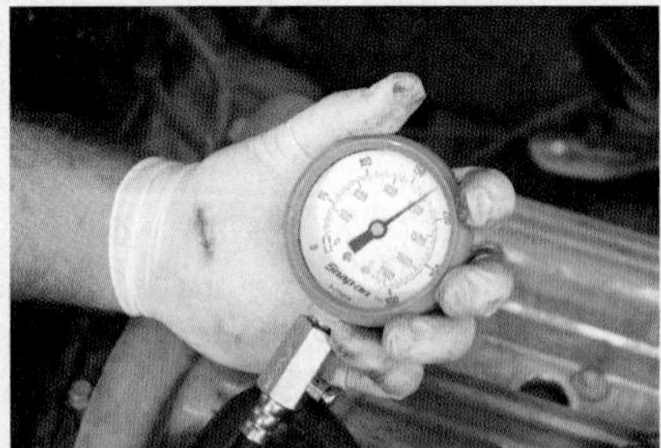

5. Hold the throttle down, and crank the engine over at least five pulses. Record the first and last needle readings. Repeat this procedure on the other cylinders.

6. Perform a wet test on any cylinders with low compression. Determine any necessary action.

Applied Math

AM-16: Charts/Tables/Graphs: The technician can construct a chart, table, or graph that depicts and compares a range of performance characteristics of various system operational conditions.

Cylinder	Cranking Compression in PSI	Running Compression in PSI
Cylinder 1		
Cylinder 2		
Cylinder 3		
Cylinder 4		

Performing a Running Compression Test

S13006

In a running compression test, the engine is running during the compression test. Unlike the cranking compression test, which checks the sealing capability of the cylinder, the running compression test checks the engine's ability to move air into and out of the cylinder. This is referred to as the engine's ability to breathe. For example, if a camshaft lobe is badly worn, less air will be entering or exiting the cylinder, depending on which valve is affected. The running compression test helps a technician to evaluate this process.

The test is performed in two parts: idle and snap throttle. During idle, because the engine is running and the throttle is relatively closed, the compression pressure will be approximately half of the cranking compression pressure. The second part of the test is a snap throttle test. With the engine idling, the throttle is snapped open and then closed very quickly (about 1 sec), which allows a big rush of air into the intake manifold. The idea is to not make the rpm change very much during the test. If the intake and exhaust system are operating correctly, then the compression tester needle will jump to about 80% of the cranking compression pressure. If the intake side of the system is restricted, then the reading will be lower than the 80% threshold. If there is a restriction on the exhaust side of the system, then the pressure will be substantially higher than the 80% threshold.

The running compression test is performed by leaving all of the spark plugs in the engine except for the one in the cylinder that you are testing. Most technicians leave the Schrader valve in the compression tester to hold pressure in the tester while performing the running compression test, although it can be hard on the Schrader valves. Always have a couple of spares handy. Also know that compression tester Schrader valves use lighter-weight springs than tire Schrader valves, so do not interchange them. The compression tester needs to be installed in the cylinder to be tested and then the engine can be started. This test can detect flat cam lobes, broken valve springs or rocker arms, carboned-up valves, or restricted intake and exhaust passageways in general. To confirm your diagnosis, perform a visual inspection of the suspect components.

To perform a running compression test, follow the steps in **SKILL DRILL 13-6**.

▶ Cylinder Leakage Test Overview

N13004, K13007

The cylinder leakage test is performed on a cylinder with low compression to determine the severity of the compression leak and where the leak is located. Compressed air is applied to the cylinder through a tester that is calibrated to show the amount of cylinder leakage as a percent of air entering the cylinder. An ideal reading is close to 0%. But because piston rings have

SKILL DRILL 13-6 Performing a Running Compression Test

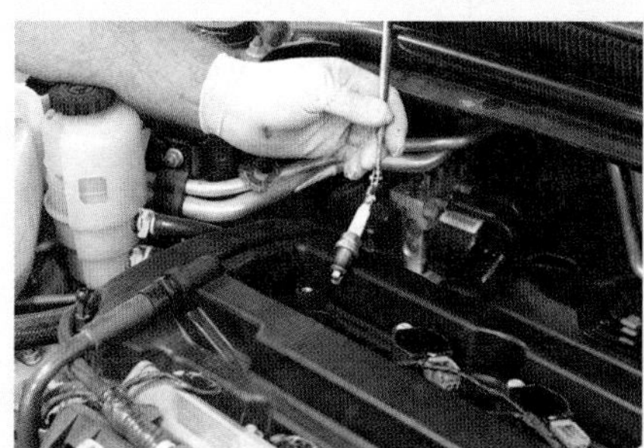

1. Remove the spark plug on the cylinder that you are testing, and ground the spark plug wire.

2. Install the proper hose and compression tester into the spark plug hole. Start the engine, allow it to idle, press and release the bleed valve, and record the reading.

3. Have your partner quickly snap the throttle open for about 1 second and then quickly close it. (Make sure the key can be turned off quickly if the throttle sticks.) Record the reading. Repeat the process on the other cylinders. Determine any necessary action.

a small gap between their ends to allow for expansion as the engine heats up, a cylinder will not be sealed 100%. There is almost always at least a small amount of leakage past the piston ring gaps. Typically, manufacturers consider up to 20% cylinder leakage past the piston rings acceptable, but the smaller the leakage, the better. Although it is okay to have a small amount of leakage past the piston rings, it is not okay to have any leakage past one of the valves or the head gasket. Leaks at these places mean the engine likely has a major mechanical engine issue.

The point of this test is to measure how much air is leaking as well as to determine the location of the leak. The gauge tells you the percentage of air leaking from the cylinder, so that is straightforward. Determining where the air is leaking from is a bit more challenging. Because there is always some air leaking past the piston rings, you will be able to hear some air leaking out of the oil fill hole when the oil fill cap is removed. If that is the only leak that you end up diagnosing, then the gauge will indicate if the leakage past the piston rings is excessive.

If an exhaust valve is burnt or warped, then you will hear leakage out of the exhaust pipe. If the intake valve is burnt or warped, then you will hear leakage out the air intake system. There should be no leakage past either of the valves, so any leak in the exhaust or intake is a bad leak. If the head gasket is blown, then you will either hear air coming out of an adjacent spark plug hole or see bubbles in the coolant when the radiator cap is removed.

▶ TECHNICIAN TIP

There are a couple of ways that a technician can get in trouble when interpreting the results of a cylinder leakage test. First, if a valve is being held open due to a piece of carbon, or not enough valve lash, then the problem could be misdiagnosed. To verify a leaky valve, remove the valve cover, and verify that the valve has the proper valve lash. Also try tapping the valve open with a soft hammer while watching the cylinder leakage gauge. If tapping on the valve stops the leakage, there was likely some carbon holding the valve slightly open. If the valve is being held open by the valve train, try adjusting the valve, and retest the valve for leakage. You should know that a valve that had too little valve lash is likely to be burnt in a very short amount of time. The longer it was operated with too little clearance, the more likely it is burnt.

Performing a Cylinder Leakage Test

S13007

There are a couple of critical steps needed to make sure the cylinder leakage test is accurate. First, the engine should be near operating temperature, which will ensure that oil has been circulated to the piston rings to help them seal. Next, it is helpful to loosen each of the spark plugs about one turn for the cylinders you will be testing and then run the engine at 1500 rpm for 10 to 15 seconds. This process helps blow out any chunks of carbon straddling the gap from the spark plug to cylinder head that break off when the spark plugs are removed. If you do not do this, it is possible for one of these chunks of carbon to get stuck between a valve and valve seat, holding the valve open slightly and producing a false reading.

Because the cylinder leakage test is usually only performed on a cylinder with low compression, remove only the spark plug for the cylinder you are testing and the spark plug for each of the cylinders next to that one. If the suspect cylinder is in the middle of the bank, then you will need to remove the spark plug on either side. If the suspect cylinder is at the end of the bank, then you will need to remove only the nearest spark plug. It also helps to remove the air cleaner assembly and radiator cap for listening purposes during the test.

Cylinder leakage is measured when the piston is on top dead center on the compression stroke. This means that you will have to turn the crankshaft to place each piston in this position before pressurizing the system. The challenge is that the piston, connecting rod, and crankshaft throw must be in near perfect alignment; otherwise, the pressure on the piston from the cylinder leakage tester will push the piston down, which turns the crankshaft. If this happens, then the intake or exhaust valve will open, depending on which way the piston ends up turning the crankshaft.

The hardest part is getting the piston exactly on top dead center. There are two primary ways to do so. One is to screw the cylinder leakage tester into the spark plug hole and then slowly turn the engine over by hand while lightly floating your thumb over the end of the hose to feel pressure and vacuum. When you feel the transition from pressure to vacuum, turn the engine in the opposite direction slightly, and stop right as the pressure stops and before the vacuum begins. It takes experience to get the feel for this. The second way is to use a plastic straw that fits down the spark plug hole and that can be pushed up by the piston without damaging the cylinder or piston. Rotate the engine by hand until the piston is as high as it will go, as indicated by the plastic straw. While the piston is on top dead center, you will not know if it is on the top of the compression stroke or the exhaust stroke until you pressurize the cylinder (unless the engine is equipped with a distributor and you can see where the rotor is pointing). If it is wrong, turn the engine one complete revolution and try it again.

To perform a cylinder leakage test, follow the steps in **SKILL DRILL 13-7.**

▶ Diagnosing Engine Noise and Vibrations

N13005, K13008

Running engines are fairly quiet considering all of the mechanical activity that happens within them. But if something starts to go wrong, noises can be one of the first indicators. Noise issues can indicate something as simple to fix as a worn accessory belt or as complicated as a spun connecting rod bearing, which would generally require rebuilding the entire engine. Understanding the engine's theory of operation and how the

SKILL DRILL 13-7 Performing a Cylinder Leakage Test

1. Remove the spark plug of the low-compression cylinder and any adjacent spark plugs. Install the cylinder leakage tester adapter hose into the spark plug hole.

2. Position the piston for the cylinder being tested at top dead center on the compression stroke.

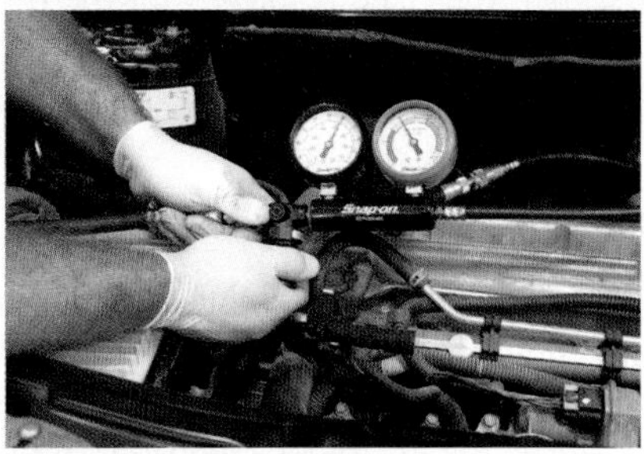

3. Connect the compressed air hose to the tester, and adjust the tester so it reads zero.

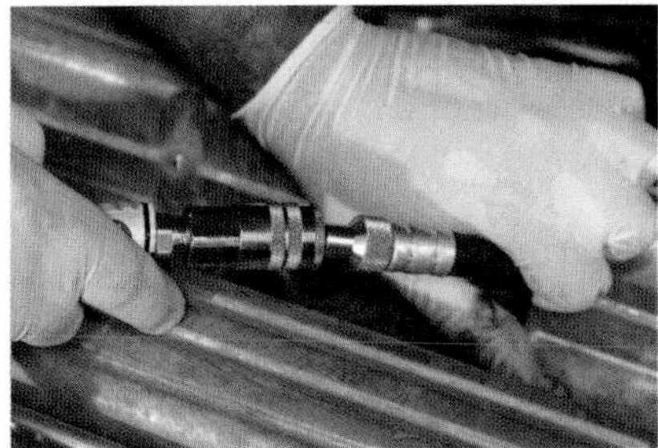

4. Connect the cylinder leakage adapter hose to the tester. Make sure the engine does not turn over. If the engine turns over, you will need to reset the piston back to the compression stroke. Record the reading.

5. Listen for leakage from the oil fill hole, the throttle body, the exhaust pipe, adjacent, cylinders, and look for bubbles in the radiator. Determine any necessary action.

individual components of an engine work will give you a good starting foundation.

Many sounds can be pinpointed through an experienced technician's previous knowledge, so investigate unusual noises and build your experience. A loud knocking noise could be from a bad main bearing or connecting rod bearing due to a worn or spun bearing. A main bearing noise is generally deeper sounding than a rod bearing. Also, a main bearing makes an evenly spaced single knock, whereas a rod bearing generally makes a double knock. A light ticking noise could be a valve lifter problem, which can be heard near the camshaft area. Or you might hear a light knocking noise that comes under slight rocking of the throttle that is caused by a collapsed piston skirt. A whirring noise can be caused by worn bearings in alternators, water pumps, and belt tensioners. One way to help locate any type of engine noise is to use a mechanic's stethoscope. Mechanical stethoscopes are the most common, but many electronic stethoscopes have settings that enhance selected sound frequencies while filtering out others, making them very handy. Place the stethoscope against stationary engine components in a variety of positions around the engine, and listen to the noises. Generally, the louder the noise, the closer you are to its source.

Depending on the noise, a stethoscope may not be appropriate. In the case of a squeaky belt, spraying water on one belt at a time will make the noise go away temporarily when you spray the one that is squeaking. Squeaks and creaks that come from linkage and joints can sometimes be sprayed one at a time with a lubricant and operated until the offending joint is found.

Vibrations can be difficult to pinpoint. Vibrations can come from the engine or the drivetrain. The best clue is to determine if the vibration occurs only when the vehicle is being driven or if it occurs with the engine running irrespective of vehicle movement. If the vibration occurs only when the vehicle is moving, suspect a component within the drivetrain or driveline, the U-joints, or even the tire balance. A good way to isolate the drivetrain is to drive the vehicle up to the speed at which the vibration is noticeable. Then place the transmission in neutral and allow the engine to idle. If the vibration is still there, then the issue is probably associated with the wheels, tires, or axles. If the vibration goes away, then raise the engine rpm while still in neutral. If the vibration reoccurs, then the issue is most likely with the engine. If there is no vibration, reengage the appropriate drive gear and accelerate moderately. If the vibration reappears, then the issue is likely with the driveshaft, U-joints, or CV joints.

▶ Fluid Leaks

N13006, K13009

Inspecting the engine assembly for fluid leaks is a common task that technicians perform during any kind of routine maintenance or service. It is also common to inspect the vehicle for leaks when the customer complains about fluid spots left on his or her driveway or if the fluid levels need to be topped off more frequently than is considered normal. It is a good practice to inspect the engine assembly for leaks before performing any engine repairs so that leaks can be taken care of at the same time.

The color of the leaking fluid can give you a clue as to the source. Black fluid could be engine oil or gear lube, reddish-orange or green fluid could be antifreeze, and red could be transmission or power steering fluid. The smell of the fluid can also provide a clue; fuel and brake fluid have very different scents. Remember that gravity will cause fluid to be pulled downward, so be sure to inspect all the way up to a place where there is no more fluid present. Also, be aware that air from the fan or wind created by the speed of the vehicle can push the fluid in the direction of the airflow.

There are several processes you may need to undertake to discover the origin of the fluid leak. The vehicle might need to be pressure washed or cleaned and reinspected for leaks. Sometimes adding a special fluorescent dye to the fluid will help locate the leak; the fluid becomes fluorescent when using an ultraviolet light and can then be more easily located. If a coolant leak is suspected, it will be easier to locate the leak if you pressure test the cooling system using a cooling system pressure tester.

Inspecting the Engine Assembly for Fuel, Oil, Coolant, and Other Leaks

S13008

Fluid leaks are a common problem that you will repair. Fluids are sealed in the various components on the vehicle, and with time, seals and gaskets begin to leak. When the vehicle begins to leak any of these fluids, customers will usually notice a puddle in their driveway or on their garage floor. The other thing customers may notice is the smell of oil or other fluids burning on the hot engine, or they may see smoke if the fluid is leaking on the exhaust pipes. When looking for the source of leaks, the color, smell, and feel of the fluid can help identify which type of fluid is leaking. Knowing the color of each of the fluids in the vehicle will help you know where to focus your inspection. Remember that gravity pulls leaking fluids down, so always look toward the highest spot that is wet, unless it is a pressure leak that is spraying or a leak onto a rotating part, such as the harmonic balancer, which would throw the fluid outward.

To inspect an engine assembly for fuel, oil, coolant, and other leaks, follow the steps in **SKILL DRILL 13-8**.

SAFETY TIP

Flammable fluids leaking onto a hot exhaust component can result in a fire. Point out any leaks found to your supervisor, and be sure to note them on the repair order. The leak poses a fire hazard, so the owner needs to be notified. Fuel leaks always pose an extreme danger; they should be fixed immediately.

To perform a fluid leak inspection by looking under the hood, follow the steps in **SKILL DRILL 13-9**.

To perform a fluid leak inspection by looking under the vehicle, follow the steps in **SKILL DRILL 13-10**.

Applied Science

AS-34: Ultraviolet: The technician can demonstrate an understanding of why dyes are added to lubricants fluoresce in ultraviolet light.

An ultraviolet dye leak detection kit can be one of the technician's most valuable tools for locating the source of a fluid leak. In the case of an engine oil leak, a special dye is added to the engine oil, and then the engine is started in order to circulate the dye. An ultraviolet light is used to search for the source of the leak as evidenced by a fluorescent yellow dye. In some cases, the detection kit may contain a special pair of glasses to assist the operator in finding the leak.

The basic scientific principle of how this works is based upon the electromagnetic spectrum. Light travels in waves called electromagnetic waves. The electromagnetic spectrum covers a wide range of components such as gamma rays, X-rays, visible light, ultraviolet light, infrared rays, and radio waves. Visible light is generally considered to be in the middle of the electromagnetic spectrum. There are many different types of electromagnetic waves that cannot be seen with the human eye. By the use of an ultraviolet light, we are able to see another part of the electromagnetic spectrum. The use of special dyes enhances our ability to find fluid leaks under the ultraviolet light source.

The ultraviolet dye leak detection kit is an excellent tool for troubleshooting difficult-to-find leaks.

AS-35: Ultraviolet: The technician can demonstrate a process for determining the source of leakage using ultraviolet light.

A vehicle is in the shop due to an engine oil leak. This is the third repair attempt to solve the customer's concern. On two previous occasions, a technician replaced the oil pan gasket. The technician was convinced that the engine oil pan gasket was the source of the leak; however, after a short period of time, the owner brought the vehicle back to the shop.

On this third repair attempt, the service manager suggested using an ultraviolet dye leak detection kit. A special dye was added to the engine lubricating oil. After running the engine for 10 minutes, the technician used an ultraviolet light to search for the leak.

After checking the engine over carefully with the ultraviolet light, the technician discovered that a slight crack in the oil pressure sending unit was the actual source of the leak. The oil was running down the block and collecting on the lip of the oil pan. The exact source of the leak was easily found as the dye showed up with a fluorescent yellow color under the ultraviolet light.

SKILL DRILL 13-8 Inspecting the Engine Assembly for Fuel, Oil, Coolant, and Other Leaks

1. Safely raise the vehicle on a lift. Verify the customer complaint of a leak by identifying which fluid is leaking. Pinpoint the location of the leak, using fluorescent dye if necessary. You may have to start the engine and look for leaks while it is running, but be very careful around moving and hot parts.

2. If a coolant leak is present, you may need to place a pressure tester on the cooling system to pressurize the system. Use a mirror and flashlight, if necessary, to look behind the engine or between components.

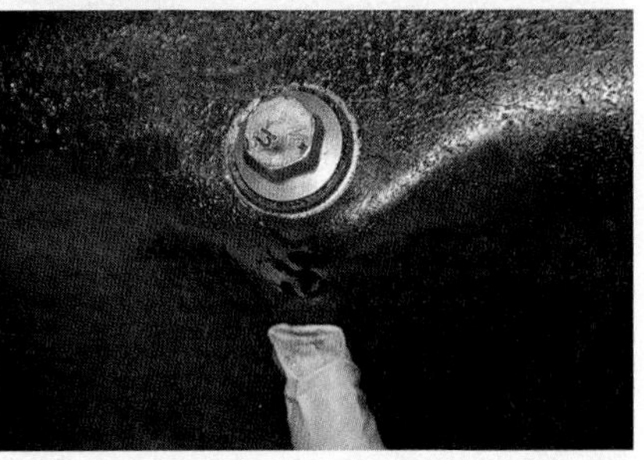

3. Repair the leak, following the manufacturer's procedures.

SKILL DRILL 13-9 Performing a Fluid Leak Inspection by Looking Under the Hood

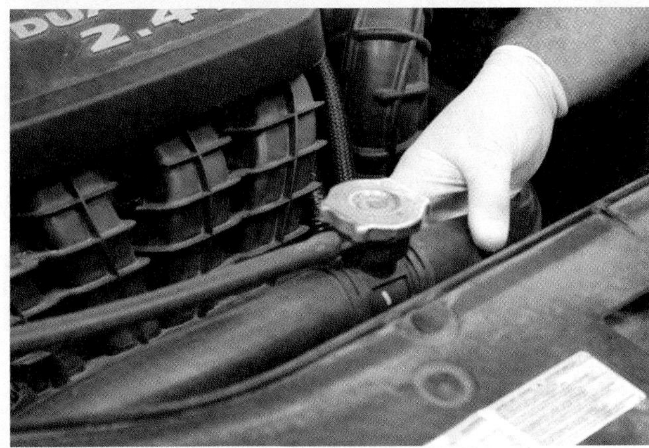

1. Raise the hood and make sure it is secure. Check for any coolant leaks. Check the radiator, radiator hoses, heater hoses, water pump, heater control valve, and any coolant lines.

2. Check for engine oil leaks, or seepage at the valve covers, intake manifold, cam seal, etc.

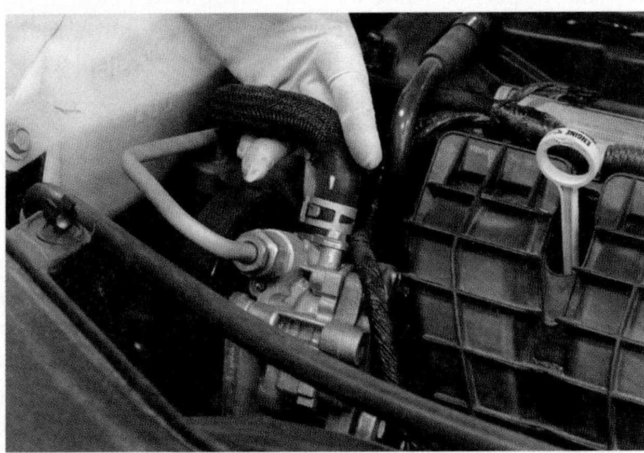

3. Check for power steering leaks at the power steering pump, lines, and steering box or rack and pinion.

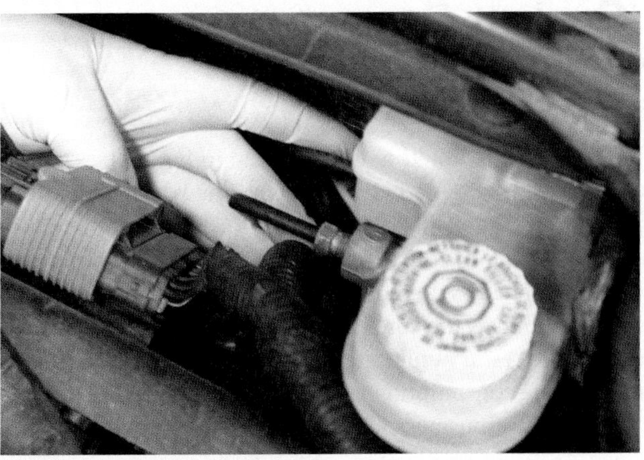

4. Check the master cylinder brake lines. Check the rear seal of the brake master cylinder, looking for signs of seepage between the master cylinder and the vacuum booster.

SKILL DRILL 13-10 Performing a Fluid Leak Inspection by Looking Under the Vehicle

1. Safely raise the vehicle, using a hoist. Check for any coolant seepage or leakage on each side of the block around the soft plugs.

2. Inspect for engine oil leaks around the front main seal, oil pan, oil filter, oil pressure switch, and rear main seal.

3. Inspect for transmission and transaxle leaks.

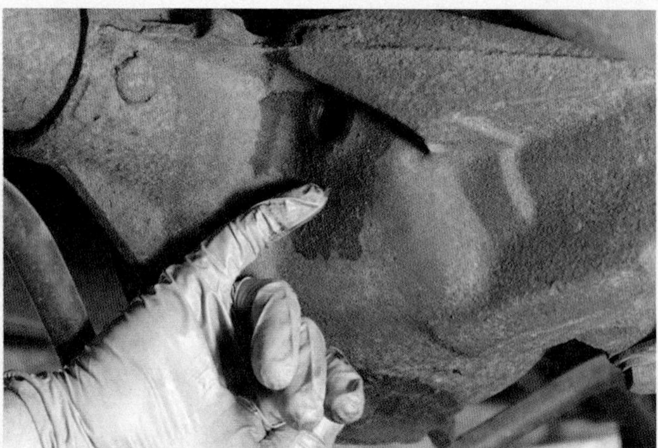

4. Inspect the front and rear differentials if equipped.

▶ Diagnosing Oil Consumption, Coolant Consumption, and Exhaust Color

K13010

Although coolant should have no consumption if the cooling system is working properly, engine oil can have a small amount of consumption and still be considered normal. Most manufacturers specify the maximum allowable consumption. In many cases, this is stated as no more than 1 quart in a certain number of miles. As long as the consumption is less than that specified, no action is necessary unless it is related to a particular customer concern. If the consumption is greater than specified, a visual is the first order of business. Consumption of oil or coolant that is not leaking externally, and therefore cannot be located with a visual inspection, may in some cases be diagnosed by the color of the exhaust. If the color has a bluish tint, it indicates that engine oil is burning. If the blue color is constant while the engine is running, it is likely that the rings are not sealing due to worn rings or cylinder walls. If the blue color is present only when you start the engine, it is likely to be a valve stem seal or valve guide issue. Black exhaust indicates an excessively rich fuel mixture. White exhaust after the engine is warmed up is an indication that coolant is leaking into the exhaust, which could mean a blown head gasket, blown intake gasket, cracked cylinder head, or leaky exhaust gas recirculation cooler on some vehicles. When coolant leaks into the exhaust, it also gives off a sweet antifreeze odor. This is the same smell as when the heater core is leaking.

Wrap-Up

Ready for Review

- Having a good understanding of engine theory will help you diagnose faults more effectively.
- Use the strategy-based diagnosis process when diagnosing engine mechanical faults.
- Engine mechanical testing tools include a vacuum gauge/pressure transducer, compression gauge, cylinder leakage tester, scan tool, and stethoscope.
- During engine cranking, the engine will make an even cranking sound if the compression is even between the cylinders.
- A vacuum gauge is used to determine the general condition of an engine.
- A cylinder power balance test is used to determine whether all of the engine cylinders are working properly, and if not, identify which ones are not.
- A cranking compression test measures how well the cylinders are sealed.
- A running compression test checks the engine's breathing ability.
- A cylinder leakage test identifies how much and where compression is leaking out of a cylinder.
- A stethoscope can be used to isolate an engine noise.
- The color of a leaking fluid can assist in identifying the source of a leak.
- Exhaust smoke can indicate the source of an engine problem.
- An odor can help diagnose an engine problem.

Key Terms

compression tester A device used to measure the amount of compression pressure a cylinder can generate.

cylinder leakage tester A device that pumps air into the cylinder and measures the percentage of air that is leaking out of the cylinder.

data link connector (DLC) The connector through which the scan tool communicates to the vehicle's computers; it displays the readings from the various sensors and can retrieve trouble codes, freeze-frame data, and system monitor data.

pressure transducer A device used to measure engine vacuum and display it graphically on a lab scope.

vacuum gauge A device used to measure the amount of vacuum an engine can generate during various operating conditions.

Review Questions

1. All of the following statements are true with respect to engine mechanical testing *except*:
   a. It helps to determine what major engine work is needed.
   b. It is an integral part of preventive maintenance.
   c. The tests start off broad and narrow down as each test is performed.
   d. Testing needs to be done before tune-up parts are replaced.
2. Which of these measures engine vacuum or pressure and displays it graphically on a lab scope?
   a. Pressure transducer
   b. Cylinder leakage tester
   c. Vacuum gauge
   d. Stethoscope
3. When performing a cranking sound diagnosis, a cylinder with no compression produces a(n):
   a. even cranking sound.
   b. slow cranking sound.
   c. uneven cranking sound.
   d. fast cranking sound.
4. When performing vacuum testing using a vacuum gauge, a steady reading of 17" to 21" of vacuum indicates:
   a. a burned valve.
   b. possible worn rings.
   c. possible late valve timing.
   d. a good reading.
5. All of the following are advantages of using a pressure transducer rather than a vacuum gauge *except*:
   a. greater accuracy.
   b. easier identification of the specific cylinder.
   c. measuring higher pressures.
   d. ability to see the levels graphically.
6. During a cylinder power balance test, when a cylinder that is not operating correctly is disabled, it:
   a. produces a large rpm drop.
   b. does not produce much, if any, rpm drop.
   c. produces the same rpm drop as when disabling a correctly operating cylinder.
   d. will not be compensated for by the PCM.
7. For the cranking compression test to be completely accurate, it is recommended that the engine:
   a. be at operating temperature.
   b. spark plugs be connected.
   c. battery be fully discharged.
   d. throttle be closed.
8. When doing the cylinder leakage test, it is acceptable to have a small amount of leakage past the:
   a. head gasket.
   b. intake valve.
   c. exhaust valve.
   d. piston rings.
9. A light ticking noise could be caused by a problem in the:
   a. main bearing.
   b. connecting rod bearing.
   c. valve lifter.
   d. piston skirt.
10. All of the following are helpful in identifying the source of a liquid leak *except*:
    a. the smell of the fluid.
    b. the part on which the fluid is dripping.
    c. the addition of a fluorescent dye.
    d. the color of the fluid.

ASE Technician A/Technician B Style Questions

1. Tech A says that a cranking sound diagnosis can be used to diagnose components in the ignition system. Tech B says that a cranking sound diagnosis can indicate differences in compression. Who is correct?
   a. Tech A
   b. Tech B
   c. Both A and B
   d. Neither A nor B
2. Tech A says that a power balance test is a good way to narrow a misfire down to a particular cylinder(s). Tech B says that a cylinder power balance test measures the volumetric efficiency of the cylinder being tested. Who is correct?
   a. Tech A
   b. Tech B
   c. Both A and B
   d. Neither A nor B
3. Tech A says that a cranking compression wet test can indicate whether the cylinder has worn piston rings. Tech B says that the throttle should be held wide open during a cranking compression test. Who is correct?
   a. Tech A
   b. Tech B
   c. Both A and B
   d. Neither A nor B
4. Tech A says that low compression on a single cylinder will cause an engine not to start. Tech B says that low compression on a single cylinder means that the engine can be fixed with a tune-up. Who is correct?
   a. Tech A
   b. Tech B
   c. Both A and B
   d. Neither A nor B
5. Tech A says that a cylinder leakage test is performed on a cylinder with low compression to determine the severity of the leak and where it is located. Tech B says that most manufacturers consider up to 50% cylinder leakage acceptable. Who is correct?
   a. Tech A
   b. Tech B
   c. Both A and B
   d. Neither A nor B
6. Tech A says that a scan tool connected to the data link connector (DLC) will report cylinder pressures during a cylinder power balance test. Tech B says that a scan tool will perform a cylinder power balance test and report whether the rings or valves have failed. Who is correct?
   a. Tech A
   b. Tech B
   c. Both A and B
   d. Neither A nor B
7. Tech A says that when performing a cylinder leakage test, the piston should be positioned at bottom dead center. Tech B says that a blown head gasket would generally leak past the exhaust valve. Who is correct?
   a. Tech A
   b. Tech B
   c. Both A and B
   d. Neither A nor B
8. Tech A says that a vacuum gauge needle that dips 4–8" rhythmically can indicate a burned valve. Tech B says that a stethoscope can be used to determine the source of unusual engine noises. Who is correct?
   a. Tech A
   b. Tech B
   c. Both A and B
   d. Neither A nor B
9. Tech A says that if the exhaust has a bluish tint, it indicates that engine is burning coolant. Tech B says that black exhaust indicates a rich fuel mixture. Who is correct?
   a. Tech A
   b. Tech B
   c. Both A and B
   d. Neither A nor B
10. Tech A says that a bad cam lobe or broken valve spring will show up during a running compression test. Tech B says that when performing a cylinder leakage test, the engine must be running during the test. Who is correct?
    a. Tech A
    b. Tech B
    c. Both A and B
    d. Neither A nor B

CHAPTER 14
Engine Lubrication Theory

NATEF Tasks

There are no NATEF tasks in this chapter.

Knowledge Objectives

After reading this chapter, you will be able to:

- **K14001** Describe the composition, functions, additives, types, and certifying bodies of lubricating oils.
- **K14002** Describe the functions of lubricating oil.
- **K14003** Describe the common additives in lubricating oil.
- **K14004** Describe the types of oil.
- **K14005** Describe the certifying bodies and their standards for engine oil.
- **K14006** Identify and describe the purpose of lubrication system components.
- **K14007** Describe wet sump and dry sump systems.
- **K14008** Identify and describe the purpose of the pickup tube assembly.
- **K14009** Describe the types and functions of oil pumps.
- **K14010** Describe the operation of the oil pressure relief valve.
- **K14011** Describe the purpose and function of oil filters.
- **K14012** Describe the purpose and function of spurt holes and oil galleries.
- **K14013** Describe the purpose and function of oil coolers.
- **K14014** Describe the purpose and function of oil indicators.
- **K14015** Describe the purpose and function of oil monitoring systems.
- **K14016** Describe the types of lubrication systems.
- **K14017** Describe the function of pressure-fed lubrication systems.
- **K14018** Describe the function of splash lubrication systems.
- **K14019** Describe the function of two-stroke premix lubrication systems.
- **K14020** Describe the function of two-stroke oil injection lubrication systems.

Skills Objectives

There are no Skills Objectives in this chapter.

▶ Introduction

Machinery like our automobiles relies on lubrication to keep the moving parts from wearing out quickly. Lubricating oil is processed from crude oil in a refinery, along with gasoline and diesel as well as with many other beneficial and useful products. Oil is much more than simply crude oil dumped into our engine's crankcase; it is heavily processed to remove impurities, and many additives are put into the processed oil to enhance its lubricating qualities.

Each moving part in the engine needs lubricating oil (**FIGURE 14-1**). The system that moves the oil through the engine is called the lubrication system. This chapter covers the theory and components of lubrication systems. It provides a solid foundation for the next chapter, which is where we discuss servicing the lubrication system and its components.

▶ Oil

K14001

Oil originates from the ground as **crude oil** (**FIGURE 14-2**). Crude oil varies in color from a dirty yellow to dark brown, to black. It can be thin like gasoline or a thick oil- or tarlike substance. Crude oil is pumped from the ground and processed into many products such as fuel for use in diesel and gasoline vehicles. Crude oil is also broken down into other products, which are used in plastics manufacturing as well as in kerosene, aviation fuel, asphalt, cosmetics, pharmaceuticals, and many other products. Many of the products refined from crude oil are used in the transportation industry. For example, **lubricating oil** is distilled from the crude oil and used as a base stock. Additives are added to the base stock to make the lubricating oil useful in engines. Other additives, such as thickening agents, are added to the base stocks and used as lubricating grease in bearings. The additives that are added to the base stock perform a variety of tasks such as keeping acids from forming, cutting down on oxidation, and maintaining the correct viscosity over a broader temperature range. We cover those qualities in more depth later.

> ▶ **TECHNICIAN TIP**
>
> Lubricating oil has been used since the invention of machinery. When metal moves on another piece of metal, the parts wear quickly without lubrication. Lubricating oil also helps to quiet the moving parts and remove heat from metal surfaces.

Functions of Lubricating Oil

K14002

Lubricating oil performs five main functions: lubricates, cushions, cools, cleans, and seals. Lubrication involves

FIGURE 14-1 Oil lubricates components inside the engine.

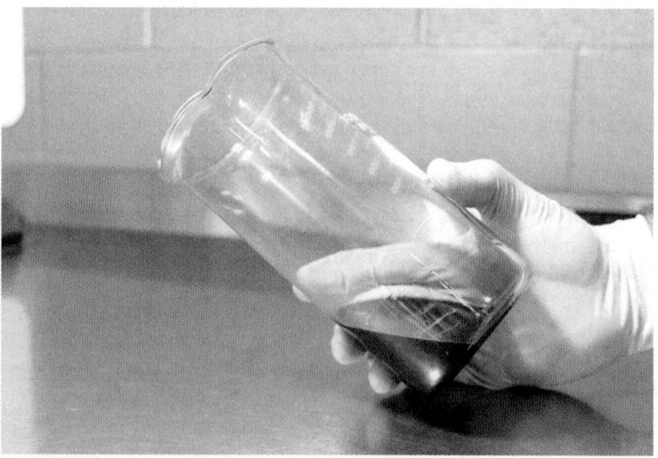

FIGURE 14-2 Crude oil straight from the ground.

You Are the Technician

A customer comes in needing a tire patched on his vehicle. When the vehicle is brought into the shop, an inspection is performed, and the oil doesn't read on the dipstick. The customer tells you that her friend added two quarts to it when he borrowed it last week, and that it regularly needs that much or more oil. She says that she hasn't been concerned about the level of the oil because the oil light is not on. She wants to know why the oil light didn't come on and what oil does that is so important. How would you answer her following questions?

1. What functions does the oil perform inside an engine?
2. Why didn't the oil pressure light indicate that the oil was below the "add" line?
3. What is viscosity?
4. What does "W" stand for in 5W30 oil?

reducing friction, protecting against corrosion, and preventing metal-to-metal contact between the moving surfaces. Friction occurs between all surfaces that come into contact with each other. When moving surfaces come together, friction tends to slow them down. Friction can be useful, as in a brake system. In the moving parts of engines, friction is a bad thing and will lead to serious damage. Friction can make metal parts so hot they melt and fuse together. When this happens, an engine is said to have seized and will need to be rebuilt or replaced.

Lubrication reduces unwanted friction as well as wear on moving parts. Clearances, such as those between the crankshaft journal and crankshaft bearing, fill with lubricating oil so that engine parts move or float on layers of oil instead of directly on each other (**FIGURE 14-3**). By reducing friction, less power is needed to move these components, and more of the engine's power can be used to turn the crankshaft instead of wasted as heat; the result is increased power to move the vehicle and better fuel economy.

How long an engine lasts depends mostly on how well it is lubricated, especially at the points of **extreme loading**, or high-wear areas, such as between the cam lobe and cam follower. At the same time, the connecting rod and crankshaft bearings take large amounts of stress as the piston transfers thousands of pounds of force to the crankshaft each time the cylinder fires. The lubricating oil between the surfaces helps to cushion these shock loads, similar to the way a shock absorber absorbs a bump in the road.

Lubricating oil also helps cool an engine. The lubricating oil collects heat from the engine's components and then returns to the **oil pan**, where it cools. The heat from the lubricating oil is picked up by the air moving over the oil pan. Many heavy-duty and high-performance vehicles have cooling fins on their oil pan or even a separate oil cooler to extract more heat from the oil, which helps the oil do an even better job of cooling critical engine components (**FIGURE 14-4**). When inspecting an engine for leaks, don't forget to check the oil cooler, if equipped.

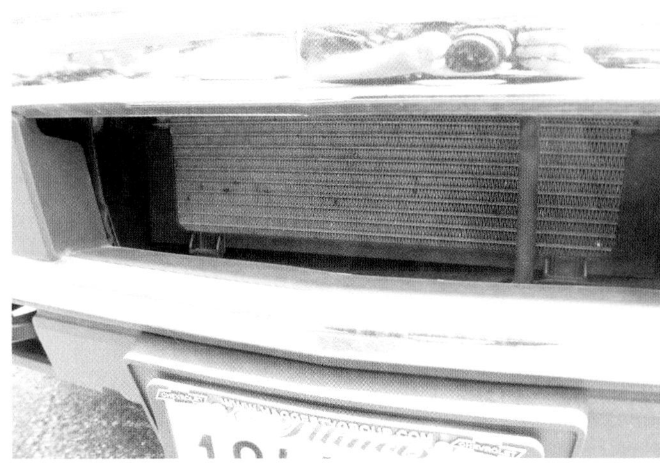

FIGURE 14-4 An oil cooler assists oil to do its job of cooling internal engine components.

Lubricating oil also works as a cleaning agent. There are additives in the lubricating oil that allow it to collect particles of metal and carbon and carry them back to the oil pan. Larger pieces fall to the bottom of the oil pan while smaller pieces are suspended in the oil and are removed when the oil moves through the oil filter. When oil is changed, most of the particles are removed with the oil filter and old oil.

The last function of oil is that it seals. It plays a key role in sealing the piston rings to the cylinder walls. Without a small film of oil between the rings and cylinder walls, blowby gases would be much higher, resulting in diluted oil, lower compression, lower power, and lower fuel economy.

Knowing the five functions of oil will allow you to better diagnose and maintain customer's vehicles, so remember them as we explore how the system works, as well as each component.

Viscosity

For oil to be able to cushion parts, it needs to have the proper viscosity. **Viscosity** is a measure of how easily a liquid flows (**FIGURE 14-5**). Low-viscosity liquid is thin and flows easily.

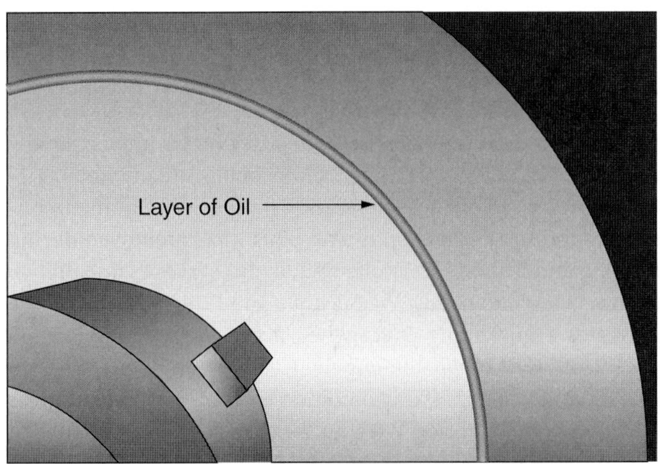

FIGURE 14-3 Clearances fill with lubricating oil so that engine parts move or float on layers of oil instead of directly on each other.

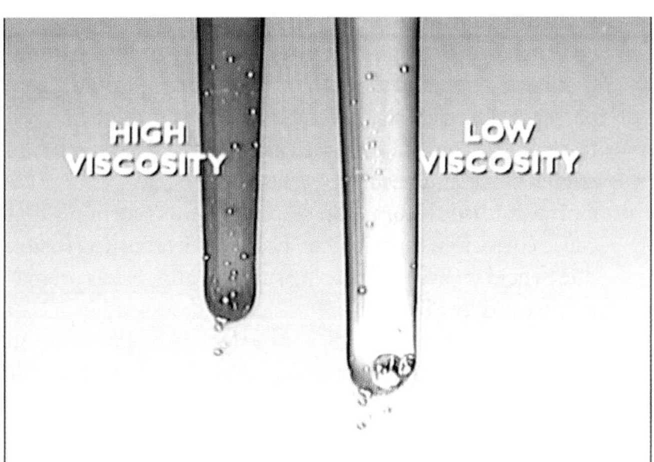

FIGURE 14-5 A. Oil with high viscosity is relatively thick. **B.** Oil with low viscosity is relatively thin.

High-viscosity liquid is thick and flows slowly. Lubricating oil must be thin enough to circulate easily between moving parts, but not so thin that it will be squeezed out easily. If it is too viscous, it moves too slowly to get to the moving parts, especially in a cold engine. As engine machining and metal technology have become more advanced, the clearances between lubricated parts have decreased. As a result, engine manufacturers have specified thinner oils for their engines so that oil can flow into the smaller clearances. The thinner oil also flows more easily, which reduces drag and increases fuel economy.

Applied Science

AS-25 Inhibitors: The technician can explain the need for additives in automobile lubricants.

Automotive lubricants are composed of base stock plus an additive package. In addition to engine oil, additives are used in lubricants for manual and automatic transmissions as well as differentials.

Oil additives are a very necessary part of modern lubricants. The improvements in modern automotive lubricants are one of the factors that enable vehicles to last longer than ever before. Oil additives consist of chemical compounds that have many beneficial functions.

Detergents are additives that help keep the oil clean. Corrosion (or rust) inhibiting additives work to prevent oxidation of engine parts. According to Wikipedia, a corrosion inhibitor is a chemical compound that, when added to a liquid or gas, decreases the corrosion rates of a material, typically a metal or an alloy.

Oil Additives

K14003

Special chemicals called additives are added to the base oil by the oil companies. Different combinations of these additives allow the oil to perform the different functions in an engine. A description of common additives follows:

- **Extreme-pressure additives** coat parts with a protective layer so that the oil resists being forced out under heavy load. This improves the cushioning effect.
- **Oxidation inhibitors** stop very hot oil from combining with oxygen in the air to produce a sticky tarlike material that coats parts and clogs the oil galleries and drain-back passages. Oil **galleries** are the passageways that carry oil through the engine. They are either cast or drilled into the engine block and head(s).
- **Corrosion inhibitors** help stop acids from forming that cause corrosion, especially on bearing surfaces. Corrosion due to acid etches into bearing surfaces and causes premature wear of the bearings.
- **Antifoaming agents** reduce the effect of oil churning in the crankcase and minimize foaming. Foaming allows air bubbles to form in the engine oil, reducing the lubricating quality of oil and contributing to breakdown of the oil due to oxidation. Because air is compressible, oil with foam reduces the ability of the oil to keep the moving parts separated, causing more wear and friction.
- **Detergents** reduce carbon deposits on parts such as piston rings and valves.
- **Dispersants** collect particles that can block the system, separate them from each other, and keep them moving. They will be removed when the oil and filter are changed
- **Pour point depressants** keep oil from forming wax particles under cold temperature operation. When wax crystals form, they result in the **gelling** of the oil and keep oil from flowing during cold start-up conditions. Gelling is the thickening of oil to a point that it will not flow through the engine; it becomes close to a solid in extreme cold temperatures.
- Base stock derived from crude oil will not retain its viscosity if the temperature gets cold enough, so viscosity improvers are added to the stock. A **viscosity index improver** is an additive that helps to reduce the change in viscosity as the temperature of the oil changes. Viscosity index improvers also keep the engine oil from becoming too thin during hot operation.

As stated above, one of the functions of lubrication is corrosion protection. Acids build up in the engine due to the accumulation of combustion byproducts and moisture. Blowby gases contain chemicals that are trapped in the oil. The chemicals react and form acids. When the engine is turned off, it begins to cool. The cooling process creates moisture that then condenses into droplets that fall into the oil and form acids. The acids attack the internal components, causing unnecessary damage. The oil contains anticorrosion additives that coat the engine surfaces, helping to protect them from the effects of the acids and water.

▶ TECHNICIAN TIP

Before multiviscosity oils, it was a normal practice for engines to need one grade of lubricating oil for summer and another for winter.

Types of Oil

K14004

Conventional Oil

Conventional oil is processed from crude oil pumped from the ground (**FIGURE 14-6**). The crude oil contains many impurities that are removed during the refining process. One of the impurities found in all crude oil is wax. This wax is removed during refining and is used for candle wax; it also serves as an additive in some food and candy. Wax is not a good thing in oil because it creates a thickening effect when it gets cold, becoming too thick to flow through the engine. Crude is broken down into mineral oil, which is then combined with additives to enhance the lubricating qualities. Without the additives, conventional oil would not work well. It would foam easily, break down quickly, and corrode the engine parts after being in the engine for a short time.

FIGURE 14-6 Conventional oil is processed from crude oil pumped from the ground.

Synthetic Oil

There are two main categories of synthetic lubricating oils: type 3, which is not a true synthetic, and type 4 (PAO), which is a true **synthetic oil**. Both types of synthetics are more costly to manufacture, as the base stocks are more highly refined or are developed in a lab and are therefore more costly to the customer. Synthetic lubricants have a number of advantages over conventional oils. They offer better protection against engine wear and can operate at the higher temperatures needed by performance engines. Synthetic oils have better low temperature viscosity, which allows the oil to be circulated through the engine more quickly during low temperature engine start-ups. Synthetics have fewer wax impurities that coagulate at low temperatures, they are chemically more stable, and they are generally thinner so they allow for closer tolerances in engine components without loss of lubrication. Modern high-performance engines run much tighter tolerances, so the need for a thinner oil that is able to hold up under higher temperatures is desirable. Some synthetics also last considerably longer, extending oil change intervals to 20,000 miles (30,000 km) or more, which benefits the environment by reducing the used oil stream and reducing the need for finding new sources of oil.

True synthetic oils are based on artificially made hydrocarbons, commonly **polyalphaolefin (PAO)** oil, which is a artificially made oil base stock—meaning it is not refined from crude oil. Synthetic oils were developed in Germany during World War II due to the lack of crude oil. Synthetic oil was used primarily in jet engines because of the high heat demands of these engines. Normal conventional oil would create heavy carbon deposits on bearings due to the extreme heat, which led to failures. Amsoil was the first synthetic to be approved by the API in 1972. Many companies now offer synthetic oils. Very few synthetic oils on the market are full PAO oils. Many of the oils allowed to be labeled as synthetic are in fact blends of processed **mineral oil** (highly refined base stock refined from crude oil) and PAO, or even just highly refined base stock, that possess lubrication qualities similar to PAOs.

Synthetic Blends

Synthetic blends give some of the benefit of the full synthetic with the cost effectiveness of conventional oil. These oils are a mix of conventional high-quality oil and full synthetic oil (**FIGURE 14-7**). They need to be changed sooner than a full synthetic but less frequently than conventional oil. The purer the base stock is after the refinement process, the longer the oil will

FIGURE 14-7 A typical synthetic blend motor oil.

last in the engine. Some manufacturers are now recommending synthetic blend oil over conventional oil due to the better protection and performance of these types of oils. Because it is half synthetic, half conventional, the full benefit of the thinner, higher performance pure synthetic is diluted, but in turn the conventional half is improved by adding oil that has no impurities. If the vehicle will be used hard, such as for hauling or towing, synthetic blends will perform better than conventional oil because of the ability of the synthetic oil to stand up to the higher heat and heavier load placed on the engine.

▶ TECHNICIAN TIP

Be sure to use at least the minimum recommended oil by the manufacturer. Manufacturers of engines spend a lot of money and time designing engines that are efficient and long lasting. Always follow the manufacturer's recommendations.

Oil-Certifying Bodies and Their Rating Standards

K14005

There are several certifying bodies for engine oil, each with its own standards. The three most common are the American Petroleum Institute (API), the American Society of Automotive Engineers (SAE), and the International Lubricant Standardization and Approval Committee (ILSAC). However, there are three others that technicians must be aware of: the Japanese Automotive Standards Organization (JASO); the Association des Constructeurs Européens d'Automobiles (ACEA), also called the European Automobile Manufacturers Association; and the original equipment manufacturers' (OEM) own standards. Let's look at them one at a time.

American Petroleum Institute (API)

The API sets minimum performance standards for lubricants, including engine oils. The API has a two-part classification: service class and service standard. The API service class has two general classifications: S for spark ignition engines and C for compression ignition engines, also referred to as "commercial." Engine oil that meets the API standards may display the API service symbol, which is also known as the API "donut." This protocol is important to understand because S-rated oil cannot be used in compression ignition engines unless they also carry the appropriate C rating, and vice versa. Be careful that the wrong oil is not used in a particular engine.

The API service standard (SA) was used in engines up until 1930, which means pure mineral oil without any additives. As engine manufacturers improved engine technology—or as government regulations changed, such as requiring reduced amounts of phosphorus—engine oil with new qualities was required, and the API would introduce a new rating level. The API SN level was added in October 2010 for 2011 gasoline vehicles. API CJ-4 was added in 2010 to meet four-stroke diesel engine requirements. New CK-4 (backward compatible) and FA-4 (not backward compatible) ratings came into effect in Dec 2016.

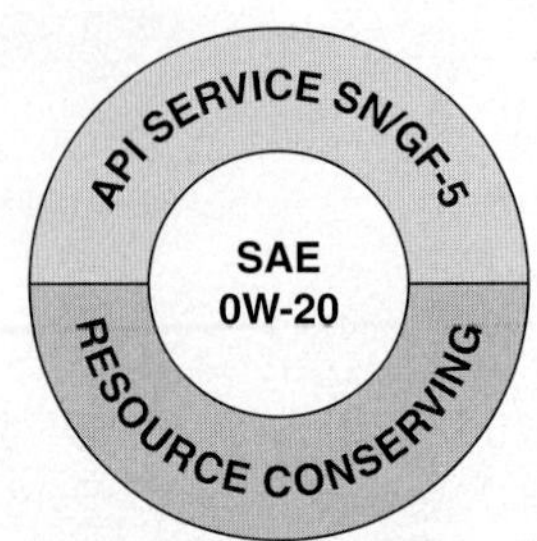

FIGURE 14-8 The API donut shows the API service class and service standard, the viscosity, the ILSAC performance rating, and the energy-conserving designation.

The API symbol is the donut symbol located on the back of the oil bottle (**FIGURE 14-8**). In the top half of the symbol is the service class—S or C—and the service standard that the oil meets. The center part carries the SAE viscosity rating for the oil. The API symbol may also carry a "Resource Conserving" or "Energy Conserving" designation if it is a fuel-saving oil. Be sure to use oil that has a correct API rating and also an energy-conserving designation in all North American vehicles.

The API also created a "starburst" symbol that can be placed on oil containers that meet the current engine protection standard and fuel economy requirements of the ILSAC, which is a joint effort of U.S. and Japanese automobile manufacturers.

▶ TECHNICIAN TIP

In most cases, higher-rated engine oils are backward compatible. This means you can use SM oil in a vehicle that requires SL. But there is one exception that some technicians have found: SN-rated oil has very low levels of phosphorus and zinc, which aids in flat tappet camshaft lubrication. So if you are working on an older engine that uses flat tappets, you probably do not want to use SN-rated oil, but SM instead.

The API classifies oils into five groups:

1. Group 1 oils are produced by simple distillation of crude oil, which separates the components of the oil by their boiling point and by the use of solvents to extract sulfur, nitrogen, and oxygen compounds. This method was the only commercial refinement process until the early 1970s, and the bulk of commercial oil products on the market are still produced by this process, such as conventional engine oils.
2. Group 2 and group 3 oils are refined with hydrogen at much higher temperatures and pressures, in a process known as **hydrocracking**. This process results in a base mineral oil with many of the higher performance characteristics of synthetic oils.
3. The more heavily hydrocracked group 3 oils have a very high viscosity index (above 120) and many, but not all, of the higher performance characteristics of a full polyalphaolefin (PAO) synthetic oil. Although not fully synthetic, these oils can be sold as synthetic oil in North America. And because

of their lower cost, many of the oils billed as "synthetic" are of group 3 base stock.

4. Group 4 oils are all of the full synthetic PAO group (most common true synthetic). This synthetic oil has more robust qualities and benefits over group 3 oils.
5. Group 5 includes all other types of synthetic oil. Typically, group 5 oils are more expensive than group 4 oils, and although they are very thermally stable, they are more commonly used in aviation and industrial applications.

▶ TECHNICIAN TIP

Over time, lubricating oil breaks down by reacting with dissolved atmospheric oxygen. In most refineries, impurities are removed by using solvent. **Hydrogenating** is a newer process that is more effective at removing impurities. Hydrogenation is the use of hydrogen during refining to assist with removing impurities. By hydrogenating the oil, and with the use of oxidation-inhibiting additives, this deterioration rate can be slowed by more than a hundredfold. Hydrogenating also reduces the presence of aromatic hydrocarbons, thereby giving more effective oxidation inhibitor action, minimizing sludge and varnish deposits, and generally avoiding other related machinery problems.

American Society of Automotive Engineers (SAE)

Engine oil producers must also meet the SAE viscosity rating for each particular oil. Engine oil with an SAE number of 30 has a higher viscosity, or is thicker, than an SAE 5 oil. Oils with low viscosity ratings, such as SAE 0W, 5W, and 10W (the "W" stands for winter viscosity), are tested at a low temperature—around 0°F (–17.8°C). These ratings indicate how the oil will flow when started cold in cold climate conditions. Oils with high viscosity ratings, such as SAE 20, 30, 40, and 50, are tested at a high temperature—around 210°F (98.9°C). These ratings indicate how the oil will flow when the engine is being used under loaded conditions in hotter situations.

Modern oils are blends of oils that combine these properties. The oils are blended with viscosity index improvers to form multigrade, or multiviscosity, oils. They provide better lubrication over a wider range of climatic conditions than monograde oils. These oils are classified by a two-part designation, such as SAE 0W-20 (**FIGURE 14-9**). In this example, when the oil was tested at 0°F (–17.8°C), it met the specifications for a viscosity of 0W weight oil, and when the same oil was tested at 210°F (98.9°C), it met the viscosity specifications for 20 weight oil. Multiviscosity oils flow easily during cold engine start-up but do not thin out as much as the engine comes up to operating temperature. These properties allow the oil to get to the components more quickly during start-up while maintaining its ability to cushion components when it is hot. Although multiviscosity oils extend the operating temperature range of the engine, always refer to the vehicle's service information to determine the correct oil viscosity to use for the climate the engine will be operated in.

For engines to meet the significantly higher fuel economy standard of 54.4 mpg by 2015, which began in 2015, engines

FIGURE 14-9 Example of multiviscosity oil.

required thinner oils that operate under more extreme conditions. Because of this, SAE, in January of 2015, introduced a new viscosity grade called SAE 16, and the following year two more grades, SAE 12 and SAE 8. These are high-temperature viscosity grades, *not* the cold temperature winter viscosity grades. For example, that would allow for engine oils such as 5W16 and 0W12. Please note that these engine oils are extremely thin and should not be used in engines that don't specify them as they are unlikely to provide adequate wear protection for engines not designed to use them.

Applied Science

AS-103: Viscosity: The technician can demonstrate an understanding of fluid viscosity as a measurement and explain how it impacts engine performance.

Viscosity is the measurement of a liquid's resistance to flow. This concept is often best understood by example. Imagine you have a small funnel that you fill with honey. You will find that the funnel drains quite slowly. If you filled the funnel with water, it would drain almost instantly. The difference is because honey has a higher viscosity than water.

International Lubricant Standardization and Approval Committee (ILSAC)

ILSAC works in conjunction with the API in creating new specifications for gasoline engine oil. However, ILSAC requires that the oil provide increased fuel economy over a base lubricant. These oils should reduce vehicle owners' fuel costs a small amount compared to an oil that does not meet the ILSAC standard. Like the API standard, ILSAC issues sequentially higher rating levels each time the standards are changed. ILSAC GF-5 replaced GF-4 and became the standard in October 2010. Engine oils that meet the highest ILSAC standard (currently GF-5) can display the API starburst symbol, which the API created to verify that the oil meets the highest ILSAC standard (**FIGURE 14-10**). The next ILSAC standard is scheduled to be called GF-6, and oils meeting the standard will be released by April 2018. This higher standard will allow engine manufacturers to keep enhancing their engines, with the goal of meeting the 2025 fuel economy standard of 54.4 mpg.

Association des Constructeurs Européens d'Automobiles (ACEA)

The ACEA classifications formulated for engine oils used in European vehicles are much more stringent than the API and ILSAC standards. Some of the characteristics the ACEA-rated oil must score high on are soot thickening, water, sludge, piston deposits, oxidative thickening, fuel economy, and after-treatment compatibility. Although some of these may be tested by the API and ILSAC, the standards are set high to achieve ACEA certification ratings. This means that the engine oil provides additional protection or characteristics that API- or ILSAC-rated oils may not match. If you are servicing a European vehicle, it is advised that you do not go by any API recommendations; instead, make sure the oil meets the recommended ACEA rating specified by the manufacturer or the manufacturer's own specification rating (**FIGURE 14-11**).

FIGURE 14-10 ILSAC starburst symbol.

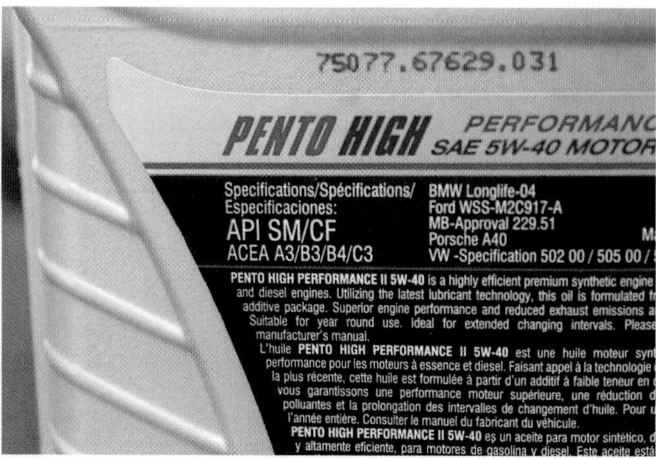

FIGURE 14-11 Oil bottle showing ACEA rating.

Japanese Automotive Standards Organization (JASO)

The JASO standards set the classification for motorcycle engines, both two-stroke and four-stroke, as well as Japanese automotive diesel engines. For four-cycle motorcycle engines, the JASO T 903:2011 came into effect in October 2011 and designates different ratings for wet clutch (MA) and dry clutch (MB). For two-stroke motorcycles, JASO M 35:2003 came into effect in October 2003. And for automotive diesel engines, JASO M355:2015 came into effect in October 2015 (**FIGURE 14-12**).

OEM-Specific Standards

As engine manufacturers continued to design new features or longer drain intervals into their engines, faster than some of the oil rating organizations could (or would) change their standards, engine manufacturers came up with their own standards. These standards are specific to individual manufacturers or even individual engines of a particular manufacturer. A few examples follow: Oil meeting Volkswagen's VW 506.00 standard is suitable for use on diesel engines (not with single injector pump), with an extended service interval of up to 31,000 miles or two years. Oil meeting General Motor's dexos1™ was specified for use starting with all 2011 GM gasoline-powered vehicles and was backward compatible in all older GM vehicles (**FIGURE 14-13**). General Motor's dexos2™ is specified for all GM vehicles equipped with Duramax diesel engines. Its viscosity is SAE 5W30 and meets the ACEA A3/B3 standard. It has a service interval of up to 18,600 miles. Oil meeting BMW's Longlife-04 standard is approved for fully synthetic long-life oil and is usually required for BMWs equipped with a diesel particulate filter.

As you can see, it is important to understand the oil requirements for the vehicle you are working on and only use the specified oil. Using the wrong oil can result in severe damage to the engine. Furthermore, using the wrong oil can void the customer's warranty, leaving the customer, or your shop, responsible for repairs. Long gone are the days of grabbing five bottles of any 10W-30 oil off the shelf and putting it into any car that rolls through the door. So always research the specified oil requirements for the vehicle you are servicing before selecting an oil to use.

FIGURE 14-12 Oil bottle showing JASO M355:2015 diesel engine oil rating.

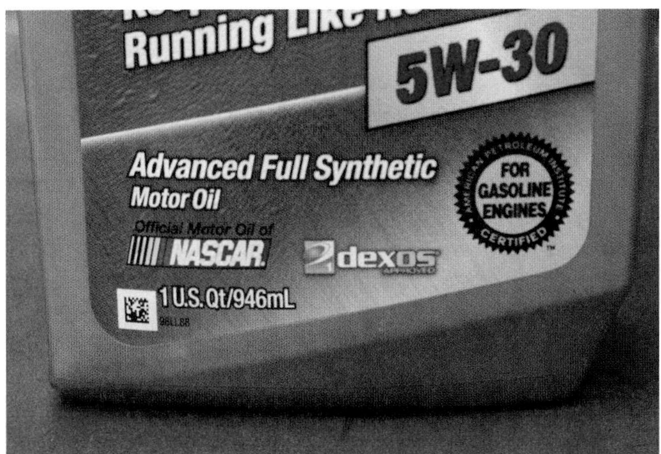

FIGURE 14-13 Manufacturer-specific oil rating.

▶ TECHNICIAN TIP

Be sure to check the owner's manual or service manual of the vehicle to ensure that you are using the correct oil rating and viscosity for the engine. Do not use just any oil that is sitting on the parts shelf.

▶ Lubrication Systems

K14006

The **lubrication system** is a series of engine components that work together to keep the moving parts inside an engine lubricated (**FIGURE 14-14**). Proper lubrication ensures that the engine runs cooler, produces maximum power, and gets maximum fuel efficiency. Lubrication also ensures that the engine will last for a long time. The lubrication system has many components that work together to deliver the oil to the correct locations in the engine. A typical lubrication system consists of an **oil sump**, an **oil pump strainer** (also called a pickup tube), an **oil pump**, a pressure regulator, **oil galleries**, an oil filter, and a low pressure warning system.

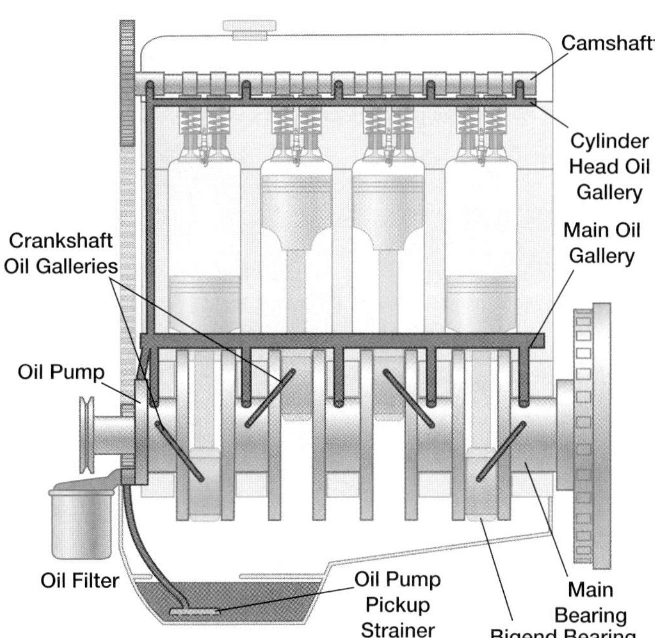

FIGURE 14-14 The lubrication system.

The oil is stored in the oil sump. Oil is drawn through the oil pump strainer from the oil sump by an oil pump. The oil travels from the oil pump to the oil filter, which removes particles of dirt from the oil. Oil moves from the filter to the oil galleries, which are small passages in the cylinder block and head(s) that direct oil to the moving parts. Oil that has been pumped to the crankshaft main bearings travels through oil-ways to the connecting rods. Oil may also be splashed from the connecting rods onto the cylinder walls, and the circulation of the oil assists with the cooling of the internal parts.

Wet Sump and Dry Sump Systems

K14007

There are two types of oil storage systems—the wet sump system and the dry sump system (**FIGURE 14-15**). In a wet sump lubrication system, the oil pan is a reservoir or storage container for the engine lubricating oil, and a collector for oil returning from the lubrication system. The oil pan is bolted to the bottom of the block and forms the lower portion of the crankcase. The oil pan is sealed to the engine with silicone sealer or an oil pan gasket. The oil pan is equipped with a drain plug that allows the oil to be drained from the engine during oil changes (**FIGURE 14-16**).

The oil pan is formed from either stamped sheet metal or cast aluminum and is shaped to ensure that oil will return to its deepest section. The oil pickup tube and strainer are located in this deep section to ensure they stay submerged in oil and to prevent air from being drawn into the oil pump. The oil pan's large external surface area helps heat transfer from the oil to the outside air. In some designs, the oil pan is an aluminum alloy casting with fins and ribs to assist in this heat transfer. A wet sump system is used on most production vehicles because of its low cost and simplicity.

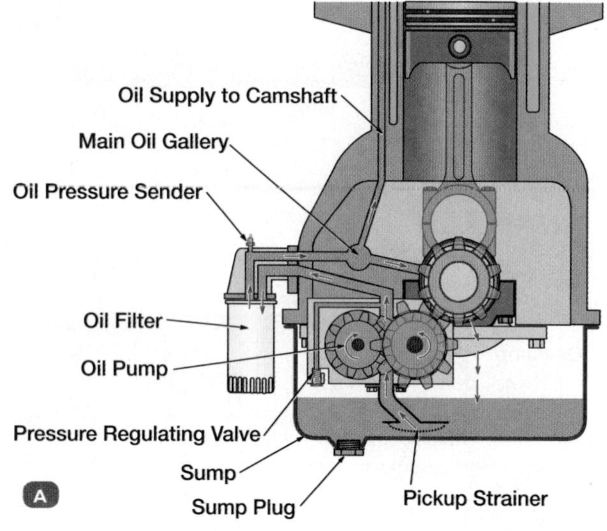

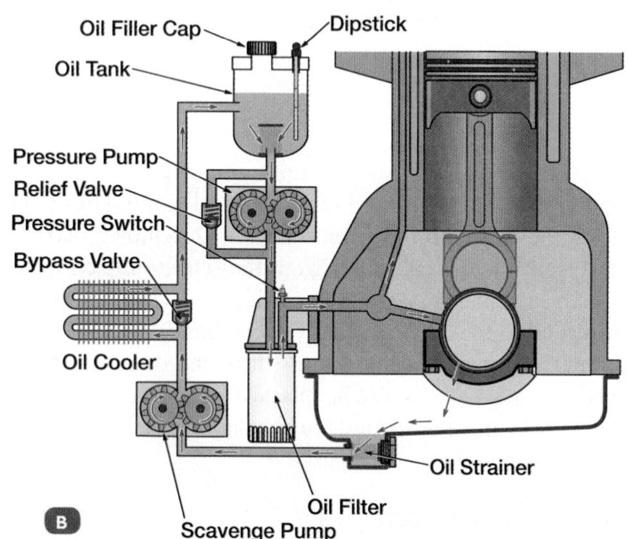

FIGURE 14-15 **A.** Wet sump. **B.** Dry sump.

Typically, a dry sump system is used in high-performance applications where ground clearance or extra capacity is needed. A dry sump system uses a shallow oil collection, which is used only to collect oil, not store it. This shallower pan allows the engine to sit lower in the engine compartment, which lowers the vehicle's center of gravity for improved handling. The dry sump system also does a better job of providing oil to the oil pump during aggressive driving conditions.

In a dry sump system, the oil is not stored under the engine in an oil pan. It collects the oil after it has circulated through the engine and directs it to the pickup strainer on the scavenge pump. The **scavenge pump** pulls oil from the sump and moves it to the oil reservoir located outside the engine. A pressure pump pulls engine oil from the oil tank, or reservoir, and delivers it to the engine at the proper pressure. Because no oil is stored in the engine in this system, a dipstick that is normally used to check oil levels in the engine is not needed. At the same time, the oil tank requires either a dipstick or a sight glass so that the oil level can be checked there.

Some high-performance vehicles have a **windage tray**, which is a shallow tray located close to the crankshaft, inside the oil pan, fitted to prevent churning of the oil by the rotation of the crankshaft (**FIGURE 14-17**). The windage tray can be made from stamped sheet metal with slots cut in it or a one-way mesh screen that allows oil to flow down through the mesh, but not vice versa. This design helps keep oil away from the rotating crank as much as possible, which prevents aeration of the oil, along with reducing drag on the spinning crankshaft. **Baffles** are also used in some oil pans on high-performance vehicles (**FIGURE 14-18**). Baffles are flat pieces of sheet metal placed around the oil pump pickup in the oil pan to prevent oil from surging away from the pickup during cornering, braking, and accelerating. Baffles can act as doors for oil flow if they are allowed to pivot. When the baffle moves one way, oil flows toward the pickup. When the baffle is moved the other way, the baffle closes and holds oil near the pickup. Baffles are commonly used on vehicles that experience strong g-forces, such as rally cars, stock cars, and drag-racing cars.

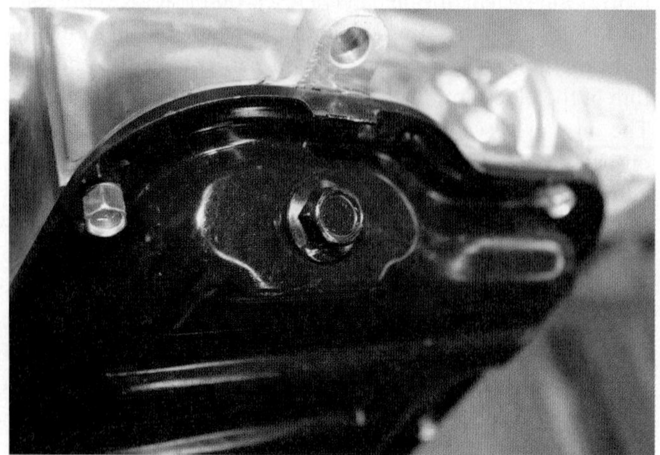

FIGURE 14-16 A drain plug allows oil to be drained during oil changes.

FIGURE 14-17 A windage tray.

FIGURE 14-18 Oil pan with baffles to keep oil near the pickup screen.

FIGURE 14-19 Pickup tube.

Pickup Tube

K14008

Between the oil pan and oil pump is a **pickup tube** with a flat cup and a wire mesh strainer immersed in the oil (**FIGURE 14-19**). The pickup tube pulls oil from the oil sump by suction of the oil pump and atmospheric pressure. A strainer on the pickup tube stops large particles of debris from entering the oil pump and damaging it. The pickup tube leads to the inlet of the oil pump on the low-pressure side. The pickup tube fits tightly into the oil pump and is usually bolted in place by a bracket to ensure that it does not fall out due to vibration. If the pickup tube were to fall out, the engine would not receive oil, as the pump would not reach down into the sump from which oil is drawn.

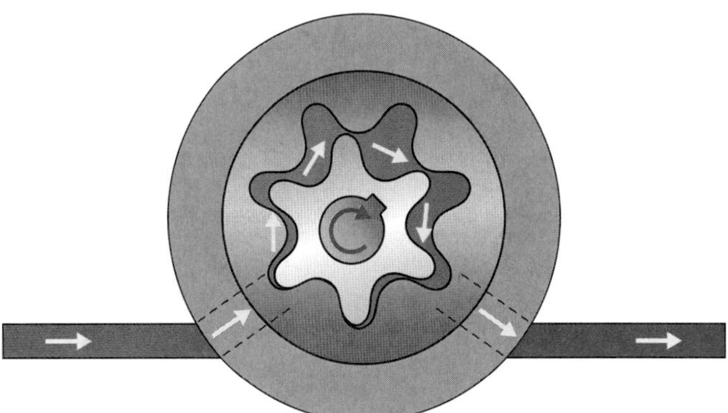

FIGURE 14-20 Rotor-type oil pump.

Oil Pump

K14009

Oil pumps force oil under pressure through the lubrication system to deliver oil to the parts that require lubrication. Oil pumps may be driven from the camshaft or the crankshaft. There are three main types of oil pumps: rotor style, gear style, and crescent style. In a **rotor-type oil pump**, an inner rotor drives an outer one; as they turn, the volume between them increases (**FIGURE 14-20**). The larger volume created between the rotors lowers the pressure at the pump inlet, creating a vacuum. Outside atmospheric pressure, which is higher, forces oil into the pump, and the oil fills the spaces between the **rotor lobes**. As the lobes of the inner rotor move into the spaces in the outer rotor, oil is squeezed out through the outlet port. In other words, oil is drawn into the spaces between the lobes on the inlet side and travels around with the lobes. The oil cannot get back to the inlet side because the lobes come together, and it is therefore forced out of the pump outlet.

In a **geared oil pump**, the driving gear meshes with a second gear (**FIGURE 14-21**). As both gears turn, their teeth separate, creating a low-pressure area. Higher atmospheric pressure outside forces the oil up into the inlet, which fills the spaces between the gear teeth. As the gears rotate, they carry oil around the chamber. As the teeth mesh again, oil is forced from the outlet into the oil gallery and toward the oil filter, where it is filtered of any particles.

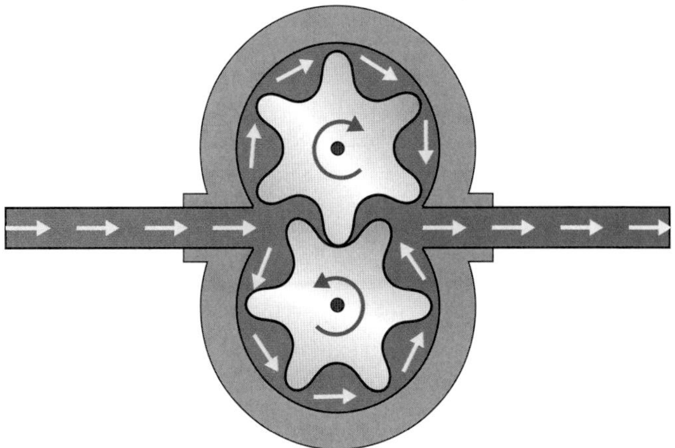

FIGURE 14-21 Geared oil pump.

The **crescent pump** uses a similar principle (**FIGURE 14-22**). It is usually mounted on the front of the cylinder block and

FIGURE 14-22 Crescent pump.

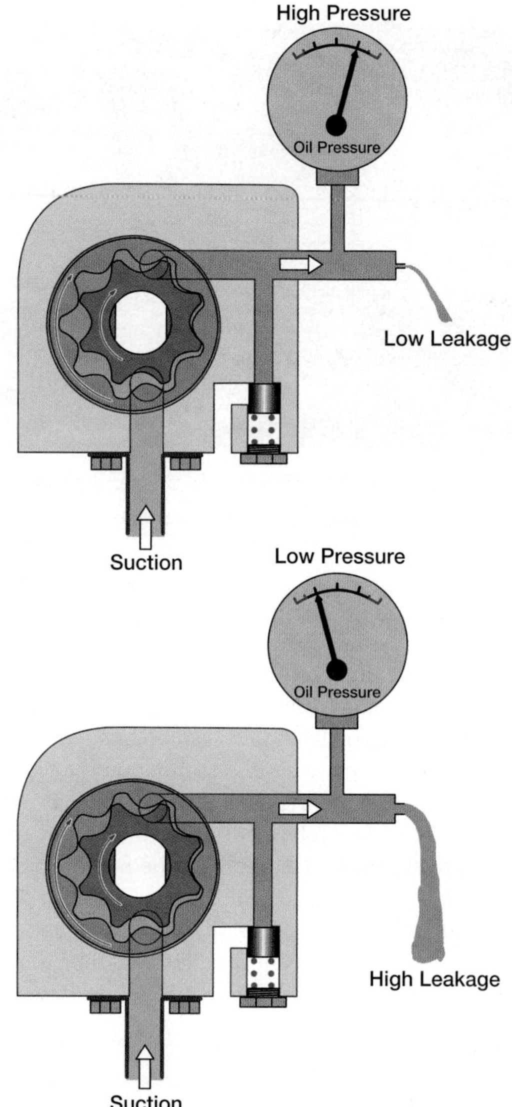

FIGURE 14-23 Oil pressure is affected by the size of the leaks in a system.

Applied Math

AM-29: Volume: The technician can use various measurement techniques to determine the volume as applicable.

Two technicians are examining an engine oil pan as it is being cleaned. It is a standard automotive pan from a V6 engine and has a rectangular shape. Randy says that the volume of the pan could be calculated by multiplying the area of the base times the height. Tom agrees with this and adds that in the U.S. system of measurement, the result would be in cubic inches, which can be converted into quarts.

At break time, they decide to measure the pan and calculate its volume in quarts. The pan measures 6' wide, 9' long, and 5.35' deep. The area of the base is 6 × 9, which is 54 square inches. Next, they multiply times the height of 5.35' for 288.9 square inches. This is approximately 5 quarts (considering one quart is equal to 57.750 cubic inches).

To calculate the example in metric units, the process would be very similar except that we would be using metric units. Centimeters would be used for the linear units to determine the base × height. There would be cubic centimeters to convert to liters.

straddles the front of the crankshaft. The inner gear is then driven by the crankshaft directly. An external toothed gear meshes with the inner one. Some gear teeth are meshed, but others are separated by the crescent-shaped part of the pump housing. The increasing volume between gear teeth causes pressure to fall, creating a vacuum, and atmospheric pressure pushes oil into the pump. Oil is then carried around between the gears and crescent before being discharged to the outlet port.

As you see, oil pumps move oil from one side of the pump to the other, and in doing so, force oil into the oil gallery. Most oil pumps are of the positive displacement design. This means that they move a given amount of oil from the inlet to the outlet each revolution. The faster the pump turns, the more the oil is pumped. But too high a speed can cause the pressure to be too high, which can erode bearings or cause oil leaks. So oil pressure must be controlled within specifications. Oil pressure is determined by two factors:

1. The size of the leaks in the system, which in the case of an engine means the amount of clearance between the bearings and the journals, and the diameter of any holes in they system (**FIGURE 14-23**)
2. The amount of oil flowing in the system, which is directly affected by the speed and size of the pump (**FIGURE 14-24**)

As you can imagine, the bearings provide a fairly consistent set of leaks. When the engine is new, the leaks are fairly small. When the engine has acquired many miles, the leaks are larger. This is why, over the life of the engine, engine oil pressure falls. In fact, low oil pressure can mean one of three things (other than a bad oil pressure gauge): (1) The oil leaks inside the engine have gotten excessive (e.g., worn bearings); (2) the oil pump is worn out and not creating as much flow as it needs to; or (3) the oil is thinner than it

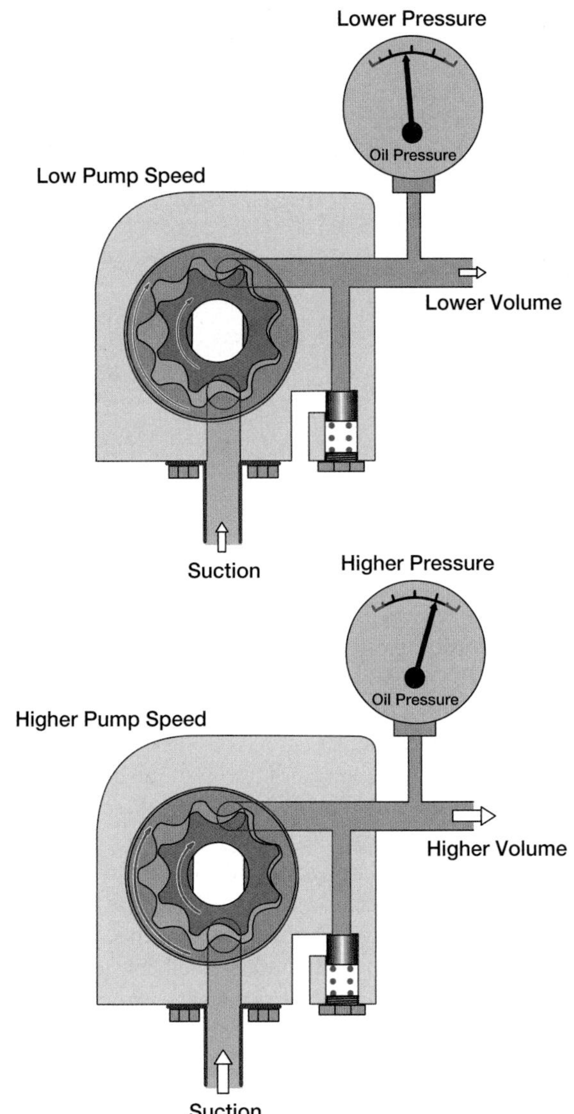

FIGURE 14-24 Oil pressure is affected by changes in the volume of oil flowing.

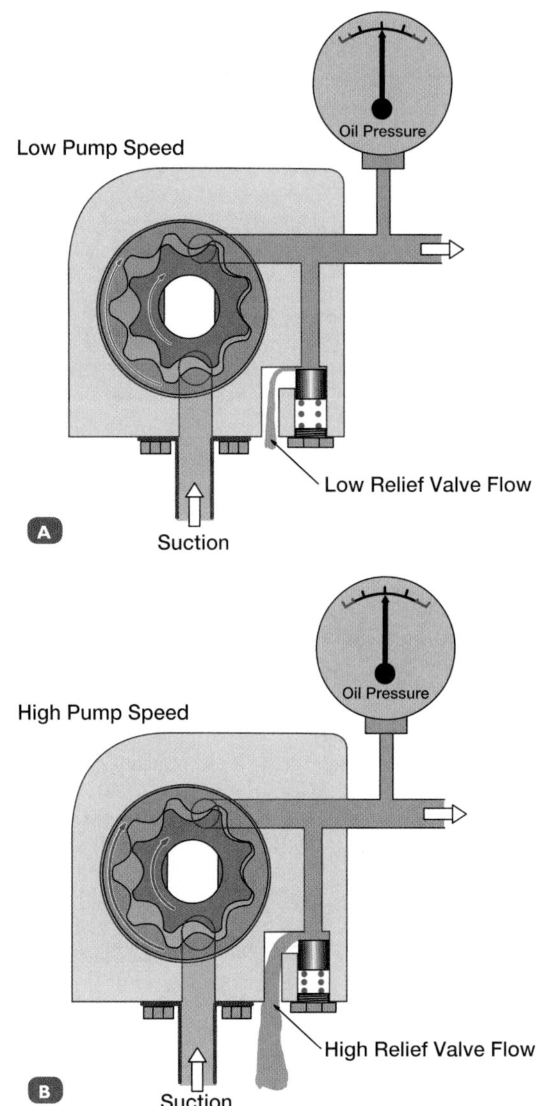

FIGURE 14-25 Oil pressure relief valve at: **A.** Low oil flow. **B.** High oil flow.

should be (e.g., saturated with gasoline from a leaky fuel injector), which causes it to pass through the leaks faster than it should. So anytime you run across low oil pressure, you should consider it to be a serious condition that needs to be diagnosed. Testing procedures are covered in the next chapter. But if the leaks inside the engine only allow a certain amount of oil to leak past them, what happens when the engine speeds up and the oil pump turns faster? If you said that the oil pressure increases, you would be correct. The next question is: What prevents the oil pressure from going too high? Keeping the engine speed low would be one answer, but this solution is not very practical. Stay tuned: We cover that issue next.

Oil Pressure Relief Valve

K14010

A normal oil pump is capable of delivering more oil than an engine needs. Extra volume provides a safety measure to ensure the engine is never starved for oil. As the oil pump rotates, and engine speed increases, the volume of oil delivered also increases. The fixed clearances between the moving parts of the engine limit the amount of oil escaping back to the oil pan, and pressure builds up in the lubrication system. An **oil pressure relief valve** stops excess pressure from developing. It is like a controlled leak, releasing just enough oil back to the oil pan to regulate the pressure in the whole system. The oil pressure relief valve contains a spring that is calibrated to a specific pressure (**FIGURE 14-25**). When the pressure is reached, the oil pressure relief valve slides open just enough to bleed sufficient oil back to the pan to maintain the preset maximum relief pressure. If the engine speed and oil flow increase, the pressure relief valve will open a bit farther and allow more oil to escape back to the sump. If the engine speed and oil flow decrease, the pressure relief valve will close an appropriate amount.

> **▶ TECHNICIAN TIP**
>
> So what is the difference between a high-pressure oil pump and a high-volume oil pump? A high-pressure pump has a stiffer pressure relief spring, which allows the higher oil pressure to build in the system. A high-volume pump has greater volume between the rotor or gear teeth, usually accomplished by making both the rotor/gear and the oil pump housing deeper. This design causes more oil to be drawn into the pump during each revolution, and therefore more oil is forced out of the pump each revolution. Generally speaking, a high-volume pump is more beneficial; because it can pump more oil, the pressure will not fall as quickly as the engine experiences wear and tear.

Oil Filters

K14011

There are two basic oil-filtering systems: full-flow and bypass (**FIGURE 14-26**). The most common **full-flow filters** are designed to filter all of the oil before delivering it to the gallery. The filter is located right after the oil pump, ensuring that all of the oil is filtered before it is sent on. The bypass filtering system is more common on diesel engines and is used in conjunction with a full-flow filtering system. The **bypass filter** is discussed later in this section.

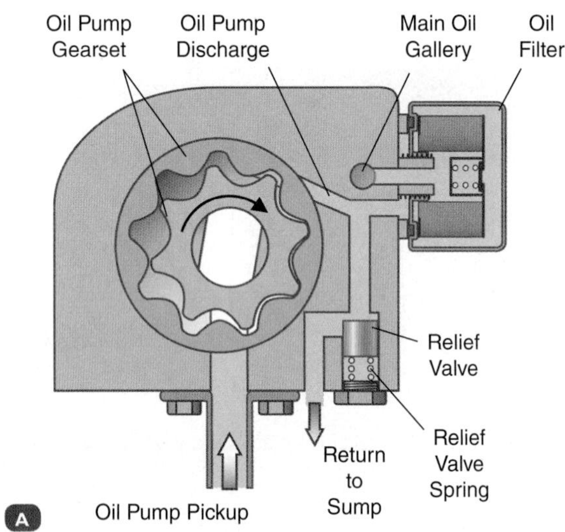

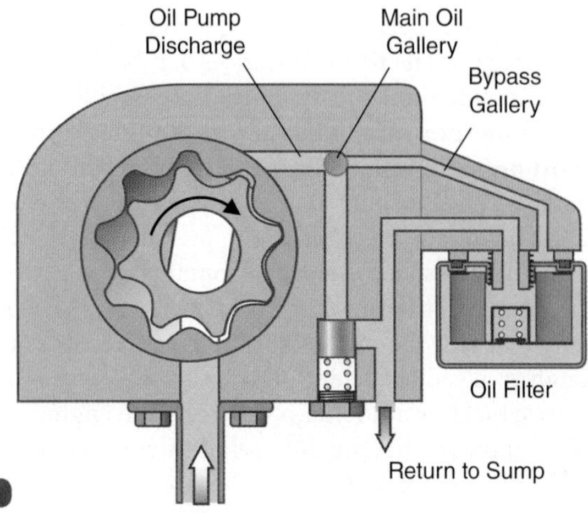

FIGURE 14-26 A. Full-flow filtering system. **B.** Bypass filtering system.

FIGURE 14-27 Pleated oil filter paper.

Oil filters use a pleated filter paper for the filtering medium (**FIGURE 14-27**). Oil flows through the paper, and as it does so, it filters out particles in the oil. Most full-flow oil filters catch particles down to 30 microns. A micron is 0.000039" (0.001 mm)—a very small particle. A human hair's thickness can be as small as 50 microns, for example. As the oil filter catches these fine particles, the paper filter element begins to clog, making it harder for the oil to flow through. As the engine is initially started cold for a few seconds, or if the filter becomes clogged, the bypass valve opens to let unfiltered oil flow to the lubricated components. The manufacturers believe it is better to have unfiltered oil flow to components than no oil at all. To prevent excessive engine wear, it is critical to change the oil filter at the manufacturer's recommended interval.

There are two common types of oil filters: spin-on and cartridge (**FIGURE 14-28**). The spin-on type is the most common. It uses a one-piece filter assembly with a crimped housing and threaded base. The pleated paper filter element is formed into the inside of the crimped housing. This kind of filter spins off with the use of an oil filter wrench and tightens by hand force only. A square-cut rubber O-ring fits into a groove in the base of the filter and seals the base of the filter to the engine block. A new O-ring comes with the filter, so it gets replaced with the filter. Be aware, though, that the old O-ring may stick to the filter adapter on the engine block. If you do not notice this and leave the old O-ring on along with the new O-ring, the old one will not be able to stay in place because it is not in any groove. As a result, it will get pushed out of place when the engine is started, and most, if not all, of the engine oil will be pumped out onto the ground. Always check for the old O-ring when removing the oil filter.

The cartridge style of oil filter uses a separate reusable metal or plastic housing and a replaceable filter cartridge. It is typically held together in one of two ways: threaded center bolt or screw-on housing. If it uses a center bolt, it will have a sealing washer between the bolt and the housing to prevent oil from leaking out. There is also a seal that fits either between the cylindrical housing and the filter adapter on the block or between the cylindrical housing and the end cap, depending on

FIGURE 14-28 A. Spin-on filter and O-ring. **B.** Cartridge paper filter, housing, and bolt.

the design. Cartridge filters must be disassembled, the housing cleaned, and the paper filter element and any O-rings or seals replaced with new ones.

Most oil filters on diesel engines are larger than those on similar gasoline engines, and some diesel engines have two oil filters. Diesel engines produce more carbon particles than gasoline engines, so the oil filter can have a full-flow element to trap larger impurities and a bypass element to collect sludge and carbon soot. In a bypass system, the bypass element filters only some of the oil from the oil pump by tapping an oil line into the oil gallery. It collects finer particles than a full-flow filter. After this oil is filtered, it is returned to the oil sump. If the bypass filter were to clog and stop oil flow, the flow of oil lubricating the engine components would not be affected.

▶ TECHNICIAN TIP

Magnets are also used as a type of filter. They attract ferrous metal particles and hold them in place until they can be cleaned off. Some manufacturers use magnetic drain plugs, which then must be inspected and cleaned off as part of an oil change. Others place a magnet to the inside or outside of the oil pan. Although this style cannot readily be cleaned, it does hold the magnetic particles in place so they cannot travel freely.

Applied Math

AM-30: Volume: The technician can determine if the existing volume is within the manufacturer's recommended tolerance.

A technician was instructed to change the engine oil and filter on a late model automobile. In addition to this, he was also to change the automatic transmission fluid and filter. The technician has access to manufacturer's service information, which he consulted before starting the tasks. The service information stated that 4.5 quarts of oil would be needed for an engine oil change with filter replacement. Concerning the automatic transmission fluid and filter change, the manual stated 9.5 quarts are needed for this service.

The technician begins working by draining the engine oil and filter. As he puts the new oil into the engine, the technician counts the number of quart containers. He discovers that after the engine was started, to fill the oil filter and shut off, it takes a total of 4.5 quarts to bring the oil level to the exact full mark on the dipstick. By this method, the technician is able to determine that the existing volume is within the manufacturer's tolerance.

At this point, the technician goes on to the transmission fluid and filter change. He drains the fluid and changes the filter. As before, the technician counts the number of quart containers necessary for filling the system properly. He pours in one quart at a time in order to take an accurate count. The technician observes that it takes 9.5 quarts to fill the system to the full line on the dipstick, with the engine running in park. By this method, the technician is able to verify that the existing volume is within the manufacturer's tolerance.

Spurt Holes and Oil Galleries

K14012

Pistons, rings, and pins are lubricated by oil thrown onto the cylinder walls from the connecting rod bearings. Some connecting rods have **oil spurt holes** that are positioned to receive oil from similar holes in the crankshaft (**FIGURE 14-29**). Oil can then spurt out at the point in the engine cycle when the largest area of cylinder wall is exposed. This oil sprays from the connecting

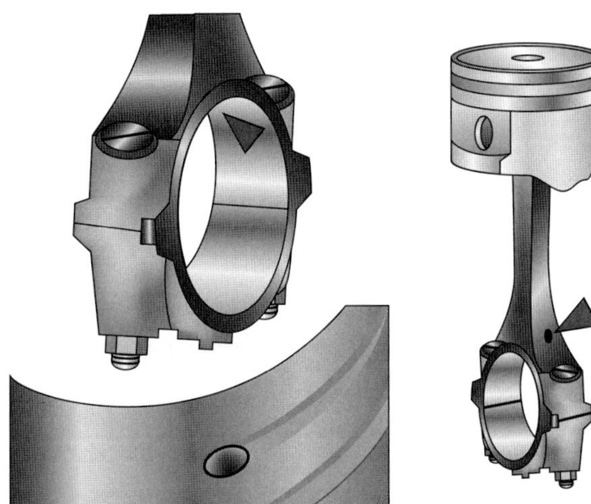

FIGURE 14-29 Some connecting rods have oil spurt holes, which are positioned to receive oil from similar holes in the crankshaft.

rod holes and lubricates the cylinder walls and piston wrist pin, and may help cool the underside of the piston. Engines with timing chains typically use one or more spurt holes that spray oil directly on the chain. Some heavy-duty engines have oil nozzles that spray oil up onto the bottom side of the piston. This is used to cool the piston and prevent it from overheating and melting down. Oil nozzles have fairly small holes that can become plugged by old, dirty oil fairly easily. So regular oil changes are critical.

Oil is fed to the cylinder head through oil galleries and on to the camshaft bearings and valve train. When oil reaches the top of the cylinder head and lubricates the valve train, it has completed its pressurized journey. The oil drains back to the oil sump through oil drain-back holes located in the cylinder head and engine block.

TECHNICIAN TIP

In many cases, the ends of the oil galleries are plugged with a threaded pipe plug or a small soft plug. During an engine rebuild, these plugs are normally removed and the galleries cleaned with stiff wire brushes. It is very important to make sure the plugs have been reinstalled with a sealer and tightened properly, so they do not leak. Many a technician has started up a newly rebuilt engine and had oil pour out of the bell housing, all because one or more of the oil gallery plugs were missing.

Oil Coolers

K14013

Engines that operate under severe conditions may use an **oil cooler** to cool the oil in the engine. There are two types of oil coolers: oil-to-water coolers and oil-to-air coolers. In some engines, an oil-to-water oil cooler and the oil filter are on the same mounting on the cylinder block (**FIGURE 14-30**). The oil cooler acts as a heat exchanger and works by transferring heat from the oil to the coolant in the cooling system. Coolant circulates through tubes in the cooler, and oil fed from the lubrication system surrounds the tubes. As the coolant circulates, heat is removed from the oil. In the oil-to-air design, the oil cooler is mounted in the airstream at the front of the vehicle. This type of oil cooler uses the flow of air passing across its fins to cool the oil circulating through it and works the same as a radiator. Most vehicles equipped with a tow package will have one of these two types of oil coolers. High-performance vehicles that will be operated at high speeds and/or temperatures can also have an oil cooler installed from the factory or as an aftermarket accessory.

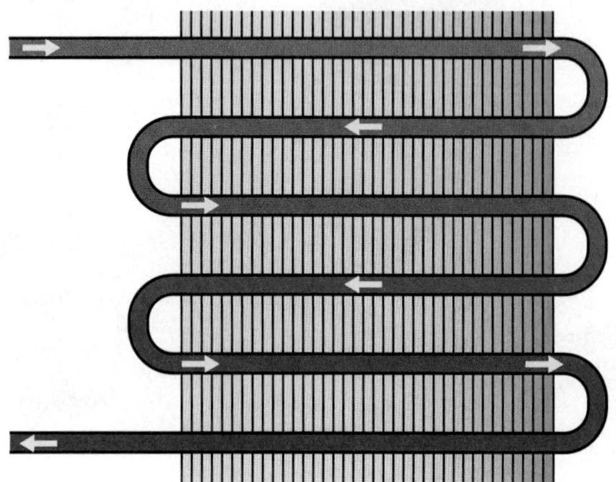

FIGURE 14-30 An oil cooler system.

Oil Indicators

K14014

A lubrication system failure can be catastrophic to the engine. Because of the damage that would happen if the lubrication system failed, a warning system is installed to let the driver know the system has failed. If oil pressure falls too low, a pressure sensor threaded into the oil gallery can activate a low oil pressure warning light, register pressure on a gauge, or turn on a low oil pressure warning message (**FIGURE 14-31**). The pressure sensor is also commonly called a sending unit because it sends a signal to the light, gauge, or message center in the dash. If the sending unit is designed as part of a warning lamp system, it is made up of a spring-loaded diaphragm and a set of switch contacts and is commonly called a pressure switch (**FIGURE 14-32**). Oil pressure is on the engine side of the diaphragm, and a spring is on the other. With the engine off and the ignition switch in the run position, the oil pressure is zero, so the spring holds the diaphragm toward the engine, and the switch contacts are closed (i.e., making contact with each other). This causes current to flow from the warning lamp in the dash through the closed switch contacts in the sending unit to ground, which turns on the warning light. When the engine is started, oil pressure increases above spring pressure, and the diaphragm is pushed away from the engine, opening the switch contacts and turning off the warning light. In some vehicles, the sending unit sends the electrical signal to the BCM, which is programmed to turn the light on below a certain pressure. This is how the system should work when everything is working normally. If the oil pressure drops below spring pressure while the engine is running, the light will come on, warning the driver of the low oil pressure condition.

If the sending unit is part of an oil pressure gauge system, it usually uses a variable resistor within the sending unit. The variable resistor is moved by the oil pressure moving the diaphragm against the spring pressure (**FIGURE 14-33**). As the pressure increases, the diaphragm is forced against spring pressure and changes the resistance of the variable resistor. This in turn changes the amount of current flowing through the oil pressure gauge and causes it to read higher. If the engine oil pressure drops, then the spring pushes the diaphragm toward the engine, again changing the resistance of the variable resistor and the current flowing through the oil pressure gauge, decreasing the pressure reading on the gauge. Some factory-installed oil pressure gauges include a warning light to warn the driver of low oil pressure. In many instances, a driver may not notice that the oil pressure gauge reading has dropped and will keep driving the vehicle, leading to engine damage. But the warning light is designed to catch the driver's attention so that he or she can stop the vehicle and investigate the cause of the low oil pressure.

FIGURE 14-31 If oil pressure falls too low, a pressure sensor in a gallery can **A.** activate a warning light, **B.** register on a gauge, or **C.** turn on a warning message.

If the sending unit is part of a driver information system, then the sending unit could be of the switch type or the variable resistance type. It usually also would include a sensor for low oil level monitoring and maybe even an oil temperature monitor. You should investigate various manufacturers' driver information systems to familiarize yourself with the different systems and strategies each manufacturer uses.

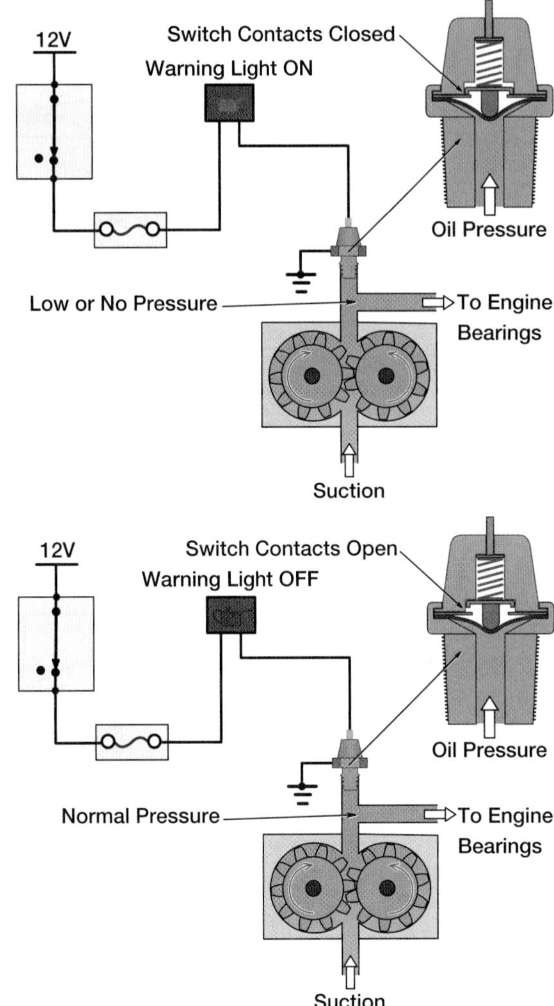

FIGURE 14-32 Oil pressure switch.

▶ TECHNICIAN TIP

Some people erroneously refer to the low oil pressure warning system as a low oil level warning system. They think this because, if the oil level gets really low, then the oil pump will draw air into the lubrication system and the oil pressure will fall, turning on the low oil pressure warning light. Unfortunately, if the oil is allowed to get that low, it is doing damage to the engine. A true low oil level warning system is designed to alert the driver when the oil level approaches the "add" mark, which is well before engine damage is being done.

▶ TECHNICIAN TIP

Many vehicle manufacturers have moved to using a switch-type sending unit with a gauge. The switch is spring loaded so that it will come on below a predetermined pressure. But it also allows current to flow through a resistor that is in series with the switch. When the oil pressure is above the spring pressure, the resistor causes the oil pressure gauge to stay mid-scale. When oil pressure falls below this setting, the contacts open, and the gauge reads low. This can confuse drivers as the oil pressure remains very steady for many years, and all of a sudden, it drops toward zero. As the engine clearances become larger, or if oil becomes thin, oil pressure drops. Once it falls low enough that spring pressure overcomes oil pressure, the gauge will read low or zero.

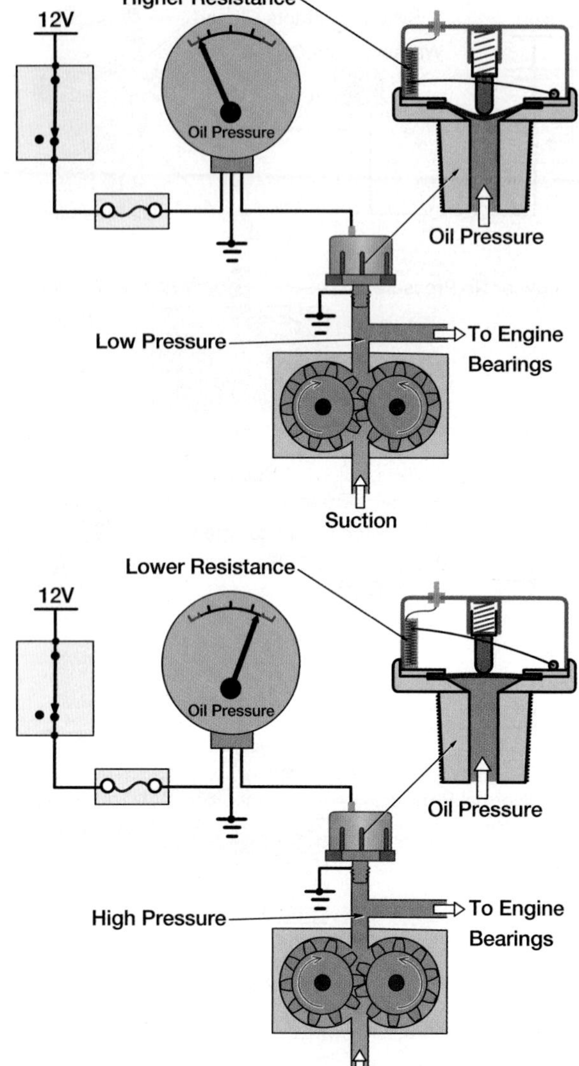

FIGURE 14-33 Oil pressure sending unit.

FIGURE 14-34 Oil change needed warning light.

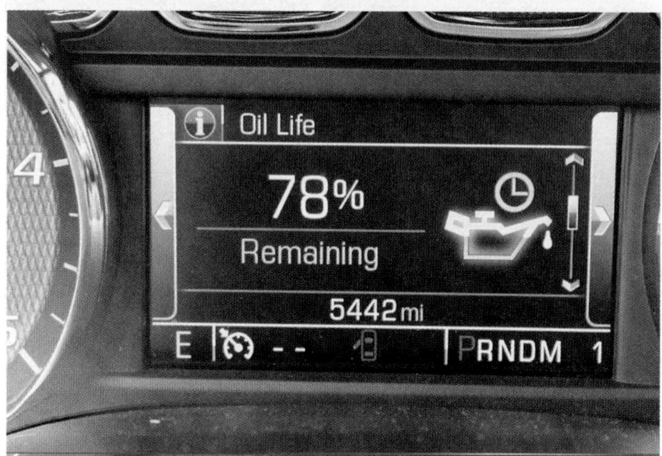

FIGURE 14-35 Oil-life monitor showing the percentage of life remaining.

Oil Monitoring Systems

K14015

Oil monitoring systems are used to inform the driver when the oil needs to be changed. There are several types of **oil monitoring systems**. Some oil systems are simply timers that keep track of mileage and will activate a warning light to notify the driver when it is time to change the engine oil (FIGURE 14-34). Other systems are very sophisticated, analyzing the conductivity of the oil through a sensor in the oil pan and monitoring changes that indicate it is time to change the oil. Depending upon the feedback from the sensor, the monitoring system computer will activate the change oil light or message to warn the driver that it is time to change the oil.

Another monitoring system, called an oil-life monitor, calculates the expected life of the oil and displays it to the driver (FIGURE 14-35). The computer receives inputs from several sensors that take into account the number of start-ups, mileage, driving habits/conditions, temperature, length of run time, and other data to calculate the remaining life of the oil, which is displayed as a percentage. When the oil is freshly changed and the oil-life monitor is reset, it will say the oil life is 100%. As the oil life wears out, the monitor will read closer to 0% oil life, informing the driver of the need to change the oil. Because it monitors the conditions the oil is operating under, the life of the oil can change drastically depending on the conditions. For example, if the vehicle were only driven in moderate temperatures for long distances, the oil would be good thousands of miles longer than a vehicle driven in stop-and-go traffic and with long periods of idling. So it is the most accurate at predicting when the oil truly needs to be changed. When changing oil, you need to reset the oil monitoring system, if equipped. Each vehicle has a specific reset procedure to turn the light or message off after an oil change.

▶ TECHNICIAN TIP

As a technician, you have to be able to find the procedure to turn the light off and reset the oil monitoring system. Refer to the service manual or owner's manual for resetting procedures. The reset procedures range from something as simple as pushing a button, to having to go through a set of steps, to having to use a dedicated tool.

▶ Types of Lubrication Systems

K14016

Pressure-Fed System

K14017

Modern vehicle engines use a **pressure, or force-feed, lubrication system** where the oil is forced throughout the engine under pressure (**FIGURE 14-36**). In gasoline engines, oil will not flow up into the engine by itself, so the oil pump collects it through a pickup tube and strainer and forces it through an oil filter and then into passageways, called galleries, in the engine block. The galleries allow oil to be fed to the crankshaft bearings first, then through holes drilled in the crankshaft to the connecting rods. The oil also moves from the galleries onto the camshaft bearings and the valve mechanism. After circulating through the engine, the oil falls back to the oil sump to cool. This design is called a wet sump lubrication system.

Diesel engines are lubricated in much the same way as gasoline engines, but there are a few differences. Diesel engines typically operate at the top end of their power range, so their internal operating temperatures are usually higher than those in similar gasoline engines. Thus, the parts in diesel engines are usually more stressed. Because diesel fuel is ignited by the heat of compression, the compression pressures (and compression ratio) are much higher than in gasoline engines. Diesel fuel has more British thermal units (Btus) of heat energy than gasoline, so it produces more heat when it is ignited, placing more stress on the engine's moving parts. Because of stress from the higher compression and combustion pressures/temperatures, parts have to be much heavier; and with heavier parts, oil must be able to handle higher shear forces. As a result, diesel oils need a different range of properties and are classified differently, usually labeled with a "C" as the first letter in the two-letter API rating system (**FIGURE 14-37**). It is also common for many diesel engines to use an oil-to-water cooler to cool the oil in the engine. The cooler and oil filter are usually on the same mounting on the cylinder block.

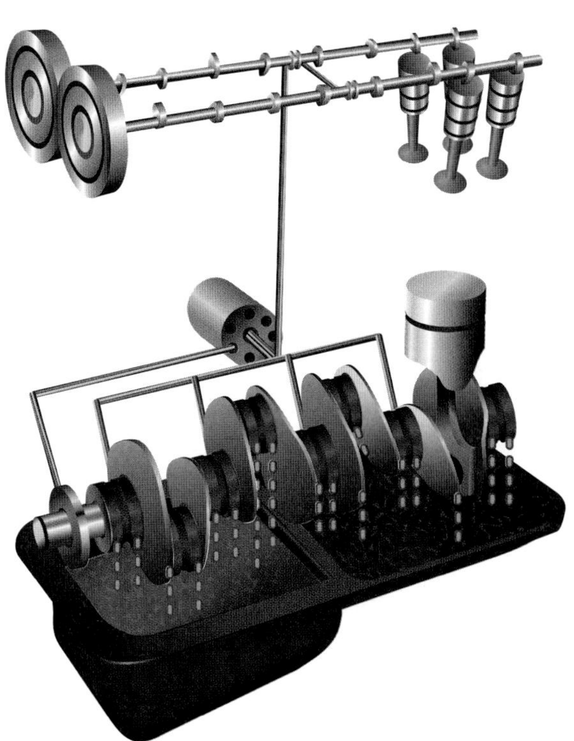

FIGURE 14-36 A pressure, or force-feed, lubrication system.

FIGURE 14-37 Diesel rated oil as designated by the "C."

Applied Science

AS-92: Friction: The technician can explain the need for lubrication to minimize friction.

Oil is a good lubricant in an engine because it has a low coefficient of friction. It creates a protective layer between two metal components, which both have a high coefficient of friction. The high coefficient of friction produces heat, causing the metal to expand, potentially creating engine wear and damage. Oil keeps the two metal components from rubbing against each other, thus preventing damage.

Splash Lubrication

K14018

Not all lubricated engine components are lubricated by the pressure-feed system. Some are lubricated by the **splash lubrication** method (**FIGURE 14-38**). In this method, the oil is thrown around and gets into spaces that need lubrication. Automotive and diesel engines use splash lubrication for lubricating the cylinder walls, pistons, wrist pin, valve guides, and sometimes the timing chain. The oil that is splashed around usually comes from moving parts that are pressure-fed; as the oil leaks out of those parts as designed, it is thrown around and provides splash lubrication to the needed components.

Most small four-stroke gasoline engines use only splash lubrication to lubricate all of the parts on the engine, including the crankshaft bearing, camshaft, and lifters. On

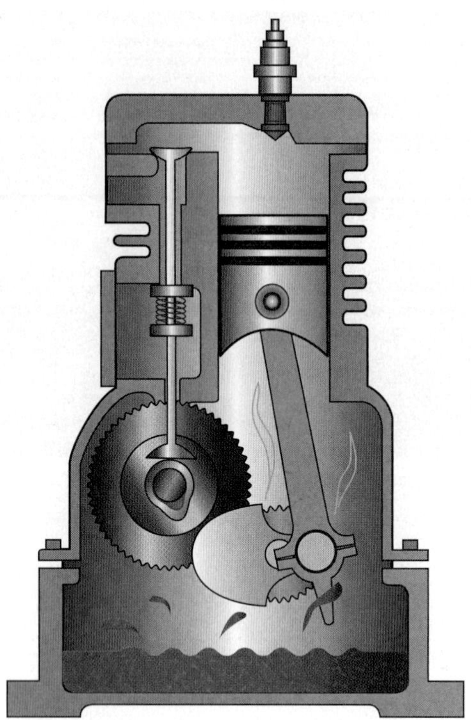

FIGURE 14-38 A splash lubrication method.

horizontal-crankshaft engines, a **dipper** on the bottom of the connecting rod scoops up oil from the crankcase for the bearings. The dipper is also able to splash oil up to the valve mechanism. Alternatively, an **oil slinger** can be driven by the crankshaft or camshaft. A slinger is a device that runs half-submerged in the engine oil. The oil is slung from the slinger upward by centrifugal force to lubricate moving parts. A similar system is used in most small vertical-crankshaft engines. Oil is also splashed up to the valve mechanism from the centrifugal force of the slinger spinning at engine speed.

Two-Stroke Engine Premix Fuel Systems

K14019

Most two-stroke gasoline engines use a specified gasoline-oil mixture for lubrication. Two-stroke oil is different than engine oil, and the two cannot be substituted for each other. They must be used according to the manufacturer's specifications. For many small two-stroke engines, the oil and fuel are premixed according to the engine manufacturer's specification. For example, an engine may require a 50:1 mixture, which is 50 parts gasoline to 1 part oil (**FIGURE 14-39**). The mixture of air, oil, and fuel passes through a sealed crankcase on its way to the combustion chamber. The crankcase thus becomes part of the fuel intake system and cannot be used as an oil sump. As the air, fuel, and oil enter the crankcase, the fuel evaporates, leaving behind enough oil to keep parts coated and lubricated. The crankshaft and connecting rod bearings in two-stroke engines typically are ball or roller types. Because the components of this kind of bearing only roll over each other, they require a lesser amount of lubrication.

FIGURE 14-39 Two-stroke engine requiring a gasoline-oil pre-mix at 40:1.

Two-Stroke Engine Oil Injection Systems

K14020

Some two-stroke gasoline engines use an oil injection system. This system does not require that the oil and gasoline be mixed manually. A small engine-driven oil pump takes oil from a tank and pumps a measured amount directly into the engine, where it mixes with the fuel and lubricates the internal engine parts (**FIGURE 14-40**). The oil pump is designed to deliver the correct amount of oil for the engine speed and the throttle setting. However, because only a small amount of oil is needed, the orifice that the oil is sprayed out of is very tiny. It is easily plugged by any debris that is allowed to fall into the oil tank when it is filled.

▶ TECHNICIAN TIP

When fueling a two-stroke engine, it is easy to forget that the gasoline must be mixed with oil. If you forget and put only gasoline in it, you will ruin it very quickly. To make matters worse, a new engine cannot tolerate this fueling error as well as an older engine. There are therefore quite a few worn-out two-cycle engines sitting around, retired long before their time.

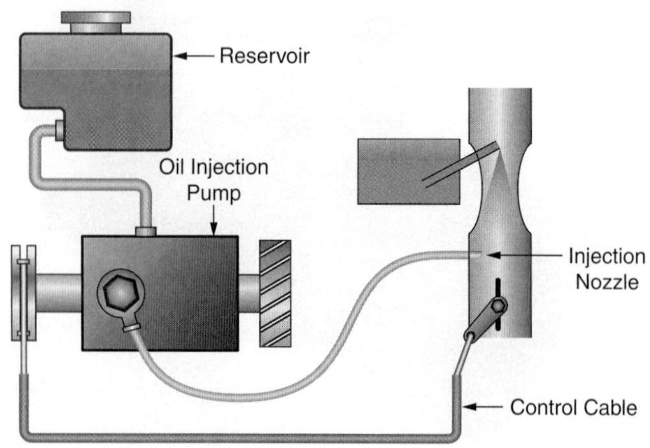

FIGURE 14-40 An oil injection system in a two-stroke engine.

▶ Wrap-Up

Ready for Review

- Lubrication oil is distilled from crude oil and has additives to prevent acid formation, reduce oxidation, maintain correct viscosity, and resist foaming.
- Functions of oil include lubricates, cushions, cleans, cools, and seals.
- Viscosity refers to how easily a liquid flows.
- Oil additives include extreme pressure additives, oxidation inhibitors, corrosion inhibitors, antifoaming agents, detergents, dispersants, pour point depressants, and viscosity index improvers.
- Engine oil also works to suppress engine noise and protect against corrosion.
- The three types of oils are conventional, synthetic, and synthetic blend.
- The three most common certifying bodies are the American Petroleum Institute (API), the American Society of Automotive Engineers (SAE), and the International Lubricant Standardization and Approval Committee (ILSAC).
- The American Petroleum Institute classifies oil into groups 1–5.
- Synthetic oil is developed in a lab, is longer lasting, operates at higher temperatures, protects better against engine wear, and is more costly to manufacture.
- Synthetic blends combine conventional and synthetic oils.
- Three other certifying bodies are the Japanese Automotive Standards Organization (JASO); the Association des Constructeurs Européens d'Automobiles (ACEA), also called the European Automobile Manufacturers Association; and the vehicle manufacturers' (OEM) own standards.
- Some vehicle manufacturers came up with their own oil rating standards.
- Components of a lubrication system include oil pan, pickup, oil pump, oil pressure relief valve, oil filter, spurt holes, and gallery.
- Oil sump systems can be wet sump or dry sump.
- Types of oil pumps are rotor type, crescent pump, and geared oil pump.
- The two most basic oil filtering systems are fullflow filters (most common) and bypass filters.
- There are two common types of oil filters: spin-on and cartridge.
- Oil is fed through passageways called oil galleries.
- There are two types of oil coolers: oil-to-water coolers and oil-to-air coolers.
- Vehicles are equipped with oil indicators and, sometimes, oil monitoring systems.
- Types of lubrication systems are splash, pressure (or force feed), two-stroke engine premix, or two-stroke engine oil injection.

Key Terms

antifoaming agents Oil additives that keep oil from foaming as it moves through the engine.

baffles Metal plates that are welded into the oil sump. These plates help to keep the oil at the oil pickup screen when the vehicle is cornering, braking, or accelerating hard. Oil will move during these conditions, and the baffles help to keep it from moving away from the pickup.

bypass filter An oil filter system that only filters some of the oil.

conventional oil Oil that is processed from crude oil; about 20% of oil is additives.

corrosion inhibitors Oil additives that keep acid from forming in the oil.

crescent pump An oil pump that uses a crescent-shaped part to separate the oil pump gears from each other, allowing oil to be moved from one side of the pump to the other.

crude oil Material pulled from the earth, originating from organic compounds broken down over time and formed into petroleum. This material is processed in a refinery to break down into various hydrocarbon substances such as diesel, gasoline, and mineral oil, among others.

detergents Oil additives that help to keep carbon from sticking to engine components.

dipper A type of splash lubricating system used in small engines. It works like a spoon scooping up oil and throwing it upward onto the crankshaft and other wear surfaces.

dispersants Oil additives that keep contaminants held in suspension in the oil, to be removed by the filter or when the oil is changed.

extreme loading Large pressure placed on two bearing surfaces. Extreme loading will try to press oil from between bearing surfaces.

extreme-pressure additive An oil additive that ensures that a protective coating is given to moving engine parts and that keeps oil from being forced out under extreme pressure. Helps oil to cushion components.

full-flow filter An oil filter installed on production cars. This oil filter cleans all oil coming from the oil pump on its way to the lubricated components.

galleries Passageways drilled or cast into the engine block or head(s), which carry pressurized lubricating oil to various moving parts in the engine, such as the camshaft bearings.

geared oil pump An oil pump that has two gears running side by side together to move oil from one side of the pump gears to the other.

gelling A thickening effect of oil in cold weather. This is not a desirable trait for lubricating oil, as it will not flow when it is gelling. Wax content in base stock mineral oil makes gelling worse.

Hydrocracking A process in which group 2 and group 3 oils are refined with hydrogen at much higher temperatures and pressures. This process results in a base mineral oil with many of the higher performance characteristics of synthetic oils.

hydrogenating A process used during refining of crude oil. Hydrogen is added to crude oil to create a chemical reaction to take out impurities such as sulfur.

lubricating oil Processed crude oil with additives to help it perform well in the engine.

lubrication system A system of parts that work together to deliver lubricating oil to the various moving parts of the engine.

mineral oil Base stock processed from crude oil in a refinery, used as the base material of all conventional oil.

oil cooler A device that takes heat away from engine oil by passing it near either engine coolant or outside air. Cooling the engine oil helps to keep it from overheating and breaking down.

oil galleries Oil passages that are drilled into the engine block and cylinder head(s). These passageways carry oil from the oil pump to critical moving parts.

oil monitoring system A system that alerts the driver when it is time to change engine oil. These systems will need to be reset for the customer after an oil change is performed.

oil pan The metal pan that covers the bottom of the engine, contains oil sump where engine oil is held.

oil pressure relief valve A valve, usually located in the oil pump, that limits the oil pressure. When oil pressure is reached, excessive pressure is bled back to the sump.

oil pump A device that pumps lubricating oil through the engine.

oil pump strainer A screen located on the oil pump pickup that keeps debris from being picked up by the oil pump.

oil slinger A device used on small engines, located on the crankshaft or driven by the camshaft. It works to fling oil up onto moving engine parts.

oil spurt holes Holes drilled into the connecting rod that spray oil up onto the cylinder walls and the piston wrist pins.

oil sump The lower part of the oil pan that holds lubricating oil for the engine. The oil pickup screen sits in this low point.

oxidation inhibitor An oil additive that helps keep hot oil from combining with oxygen to produce sludge or tar.

pickup tube A tube connected to the oil pump that acts like a straw for the oil pump to pull oil from the sump of the oil pan.

polyalphaolefin (PAO) An artificially made base stock (synthetic) used in place of mineral oil. Oil molecules are more consistent in size, and no impurities are found in this oil as it is made in a lab.

pour point depressants Oil additives that keep wax crystals from forming and causing the oil to gel during cold operation.

pressure, or force-feed, lubrication system A lubrication system that has a pump to pressurize the lubricating oil and push it through the engine to moving parts.

rotor lobes Lobes or rounded edges on rotors that squeeze oil and create pressure.

rotor-type oil pump An oil pump that uses rounded gears to squeeze oil through.

scavenge pump A pump used with a dry sump oiling system to pull oil from the dry sump pan and move it to an oil tank outside the engine.

splash lubrication A lubrication system that relies on oil being splashed onto moving parts by rotating engine parts striking the oil. These systems are typically used in small engines.

synthetic blend A blend of conventional engine oil and pure synthetic oil.

synthetic oil Synthetic oil that, in its pure form, uses artificially made base stocks and is not derived from crude oil. This oil lasts longer and performs better than normal oil. The base stock additives are similar to those in conventional oils.

viscosity The ability of a liquid to flow.

viscosity index improver An oil additive that resists a change in viscosity over a range of temperatures.

windage tray A plate that bolts onto the bottom of the main bearing saddles inside the oil pan, helping to keep the crankshaft from contacting the engine oil.

Review Questions

1. Lubricating oil performs all of the following functions *except*:
   a. cleaning.
   b. powering.
   c. sealing.
   d. cooling.
2. Which of these additives reduces carbon deposits on parts such as piston rings and valves?
   a. Oxidation inhibitors
   b. Corrosion inhibitors
   c. Detergents
   d. Dispersants
3. All of the following are advantages of synthetic oil over conventional oil *except*:
   a. lower cost.
   b. better low temperature viscosity.
   c. fewer wax impurities.
   d. operation at higher temperatures.
4. Which of these provides stringent classifications formulated for engine oils used in European vehicles?
   a. SAE
   b. API
   c. ILSAC
   d. ACEA
5. Choose the correct statement.
   a. A dry sump system stores oil in a tank outside of the engine.
   b. A wet sump system uses a scavenge pump and pressure pump.
   c. The wet sump system is the best fit during aggressive driving conditions.
   d. The dry sump system is the best fit in low performance applications.

6. Which of these is most likely to cause low engine oil pressure over time?
   a. Oil is thicker than it should be.
   b. Oil pump is creating excessive flow.
   c. Oil leaks inside the engine have become excessive.
   d. Oil pump is of the wrong type.
7. Oil nozzles have fairly small holes that can become plugged by old, dirty oil fairly easily. Therefore:
   a. change the nozzles periodically.
   b. add oil periodically.
   c. change the oil regularly.
   d. add detergents to the oil regularly.
8. Different types of oil pressure sensors/switches can do all of the following *except*:
   a. add oil to the oil pan.
   b. activate a warning light.
   c. register on a gauge.
   d. turn on a warning message.
9. Which of the following statements is true of a splash lubrication system?
   a. Oil that is splashed around usually comes from an oil pump.
   b. It can lubricate all of the parts on bigger engines.
   c. Diesel engines use only the splash lubrication system.
   d. Oil is thrown around and gets into spaces that need lubrication.
10. In a two-stroke engine oil injection system:
   a. oil and fuel are mixed manually.
   b. the oil pump is designed to deliver the correct amount of oil.
   c. oil and fuel are premixed to manufacturers' specifications.
   d. it only injects oil if you forgot to mix oil with the gas.

ASE Technician A/Technician B Style Questions

1. Tech A says that one function of oil is to clean. Tech B says that one function of oil is to cushion. Who is correct?
   a. Tech A
   b. Tech B
   c. Both A and B
   d. Neither A nor B
2. Two techs are discussing 5W20 oil. Tech A says the "W" stands for "weight." Tech B says the "W" stands for "Winter." Who is correct?
   a. Tech A
   b. Tech B
   c. Both A and B
   d. Neither A nor B
3. Tech A says that the higher the viscosity number, the thicker the oil. Tech B says that most modern vehicles use single weight oil. Who is correct?
   a. Tech A
   b. Tech B
   c. Both A and B
   d. Neither A nor B
4. Tech A says that spin-on oil filters use a square-cut rubber O-ring to seal the base of the filter to the engine block. Tech B says that some oil filters use a replaceable paper filter cartridge. Who is correct?
   a. Tech A
   b. Tech B
   c. Both A and B
   d. Neither A nor B
5. Tech A says that API's "S" rating means it should only be used in supercharged engines. Tech B says that C rated oil is certified for use for cars, and not trucks. Who is correct?
   a. Tech A
   b. Tech B
   c. Both A and B
   d. Neither A nor B
6. Tech A says that most oil pumps are of the positive displacement style. Tech B says that oil pumps are designed to deliver more oil than is needed for an engine. Who is correct?
   a. Tech A
   b. Tech B
   c. Both A and B
   d. Neither A nor B
7. Tech A says that the oil-life monitor calculates the expected life of the oil and displays it to the driver. Tech B says that an oil pressure regulator boosts the oil pressure at high engine rpm. Who is correct?
   a. Tech A
   b. Tech B
   c. Both A and B
   d. Neither A nor B
8. Tech A says that oil pressure is increased when bearing clearances increase. Tech B says that the faster the oil pump turns, the more oil that is pumped. Who is correct?
   a. Tech A
   b. Tech B
   c. Both A and B
   d. Neither A nor B
9. Tech A says that a full-flow oil filter filters all of the oil going to the bearings. Tech B says that a bypass filter bypasses the pump so that any particles in the oil won't damage the pump. Who is correct?
   a. Tech A
   b. Tech B
   c. Both A and B
   d. Neither A nor B
10. Tech A says that an oil gallery is what keeps the oil in storage at the bottom of the engine. Tech B says that spurt holes spray oil onto some parts such as timing chains. Who is correct?
   a. Tech A
   b. Tech B
   c. Both A and B
   d. Neither A nor B

CHAPTER 15
Lubrication System Service

NATEF Tasks

- **N15001** Perform engine oil and filter change; use proper fluid type per manufacturer specification. (MLR/AST/MAST).
- **N15002** Perform oil pressure tests; determine needed action. (AST/MAST)
- **N15003** Inspect, test, and replace oil temperature and pressure switches and sensors. (AST/MAST)
- **N15004** Inspect auxiliary coolers; determine needed action. (AST/MAST)

Knowledge Objectives

After reading this chapter, you will be able to:

- **K15001** Identify and describe the purpose of lubrication system tools.

Skills Objectives

After reading this chapter, you will be able to:

- **S15001** Perform lubrication system inspection, maintenance, diagnosis, and repair.
- **S15002** Check the engine oil.
- **S15003** Drain the engine oil.
- **S15004** Replace oil filters.
- **S15005** Refill the engine oil.
- **S15006** Diagnose lubrication system faults.
- **S15007** Inspect auxiliary coolers; determine necessary action.

▶ Introduction

This chapter covers the methods and specialized tools involved to ensure that you can properly maintain, diagnose, and correct common problems associated with the lubrication system.

▶ Maintenance and Repair

S15001

The engine lubrication system requires periodic inspection, maintenance, diagnosis, and repair to provide the intended protection for the internal workings of the engine and its related essentials (**FIGURE 15-1**). Inspection involves checking the level, condition, and service life of the oil. This should be performed by the operator every couple of fuel tank fill-ups or monthly. It also involves technicians inspecting the engine for oil leaks during oil changes. Maintenance involves changing the oil, filter, and resetting the oil-life monitor, if applicable. This is performed according to the manufacturer's maintenance schedule. Diagnosis involves testing the oil pressure, determining the cause of low or high oil pressure, identifying the cause of oil leaks, and diagnosing oil pressure and level warning system faults. We will cover each of these tasks in this chapter.

FIGURE 15-1 Entry-level technician performing an oil and filter change.

When working on the lubrication system, always wear appropriate PPE to help protect you. Also remember that waste oil is a hazardous material that needs to be stored and disposed of properly. Review your state and local requirements for disposing of waste oil and oil filters. Engine oil can be extremely hot, leading to severe burns. And if it spills on the floor, it only takes a dime-sized spot to cause slips and falls, which are some of the most common injuries in a shop. So always clean up spills immediately.

Tools

K15001

The tools for lubrication system maintenance and repair include (**FIGURE 15-2**):

- A variety of oil filter wrenches to remove the oil filter.
- A set of wrenches and a socket set to remove the oil drain plug and any engine covers.
- A mirror and a quality light for leak testing.
- A special socket to remove many of the oil pressure switches or sensors.
- A digital multimeter (DMM) to test wiring and sensors of the oil warning light or gauge.
- A pressure gauge for testing oil pressure
- A sensor substitution box

Checking the Engine Oil

S15002

Checking the engine oil level regularly is necessary and should be performed during every fuel fill-up or every other fuel fill-up, depending on the age and condition of the vehicle. There is a danger of damaging the engine if the engine oil level drops too low. The oil level can be low either because there is an oil leak or because the engine is consuming oil by burning it. Burning engine oil could be due to worn piston rings, worn valve guides, or a malfunction with the positive crankcase ventilation (PCV) system. Oil leaks can occur at the various gaskets, seals, and oil-pressurized components such as oil pressure sending units. Checking oil level should also be part of any pre-trip check or

You Are the Automotive Technician

A customer brings his five-year-old vehicle into your shop for an oil and filter change. The vehicle is right at the recommended 7500-mile interval. He said he has noticed some oil spots on his garage floor, and the oil level was a bit below the "add" mark this morning when he checked it. He is concerned that he has an oil leak. As you start the vehicle to pull it onto the hoist, the oil light comes on when the ignition key is turned on. It goes off once the engine starts, proving that the oil pressure warning light circuit is working. With the engine running, you find that a drop of oil forms every so often on the end of the oil pressure switch, indicating it is leaking and needs to be replaced. None of the engine seals and gaskets show any signs of leakage.

1. What must you look for when removing a spin-on oil filter?
2. When performing oil and filter changes, what must you do with most oil drain plug gaskets?
3. What is the process for tightening spin-on oil filters?
4. Describe the two types of engine oil coolers.
5. What things must be done after changing the oil and verifying there are no oil leaks?

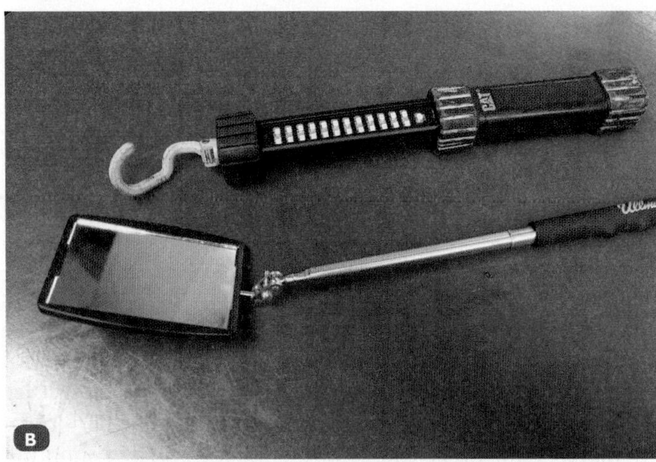

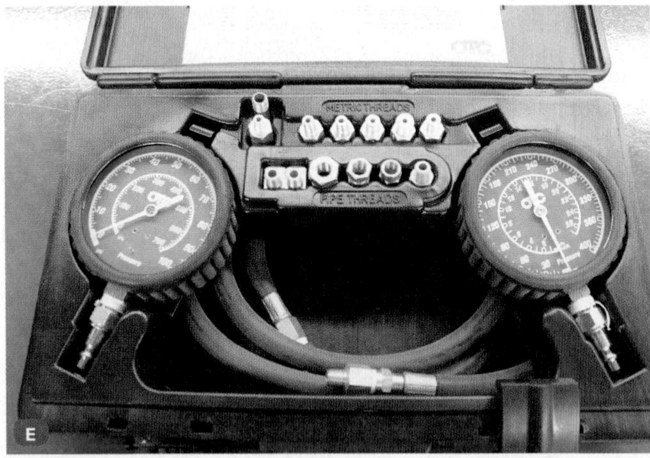

FIGURE 15-2 Tools for lubrication system maintenance. **A.** Oil filter wrenches. **B.** Mirror and quality light. **C.** Special socket. **D.** Digital multimeter (DMM). **E.** Pressure gauge. **F.** Sensor substitution box.

part of a pre-delivery inspection on a new car at the dealership before handing the vehicle over to the customer.

Oil level is typically checked with a dipstick located under the hood. The dipstick sits firmly in a tube so that the lower end of it is positioned in the oil in the oil pan. It's marked so that when the engine is filled to the proper level, oil will coat the dipstick up to the "full" line (**FIGURE 15-3**). A "safe" or "add" line is usually also included, which is typically at the point where the engine needs 1 quart of oil added to make it full. The oil level should always be maintained between the two lines.

When checking oil, always make sure the vehicle is on a level surface and the engine is off before taking a reading. If you do not, you will get inaccurate readings. Also wipe the dipstick off and reinsert it fully before removing it and reading it. When reading it, hold the dipstick horizontal so the oil will not run down the stick. Typically, the amount of oil needed to raise the

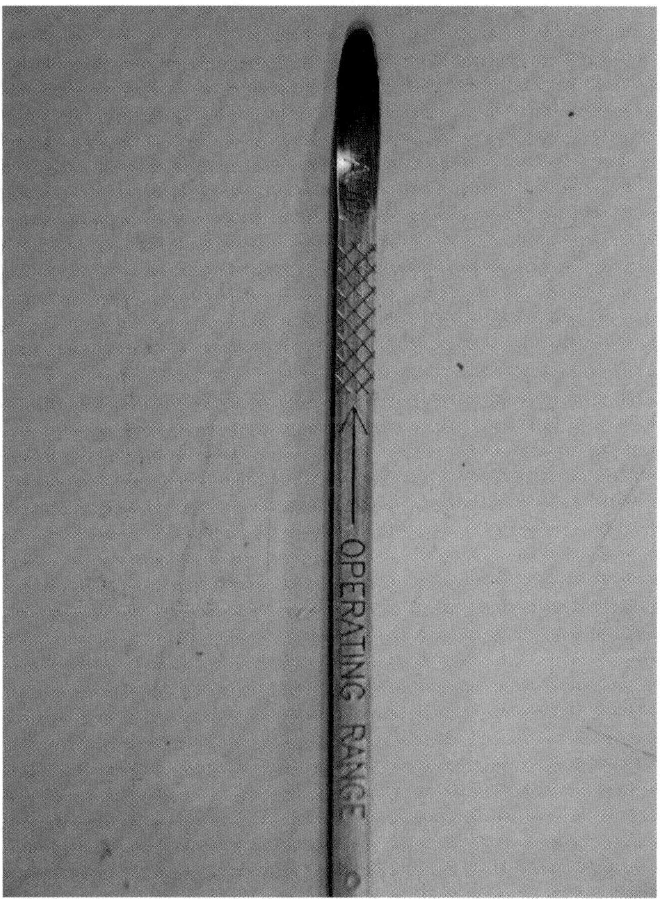

FIGURE 15-3 A dipstick showing "full" and "add" marks.

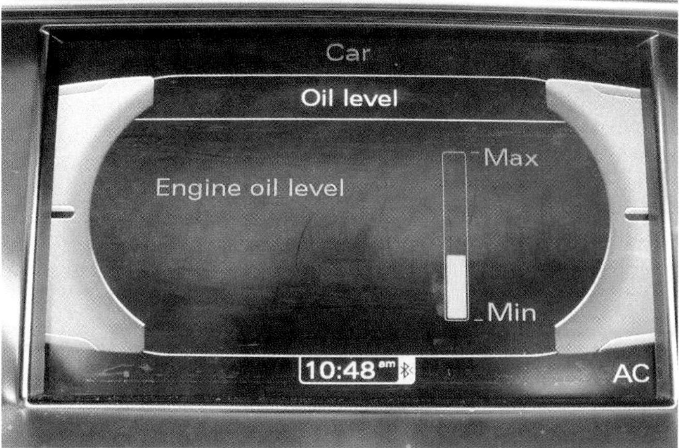

FIGURE 15-4 Instead of a dipstick, some vehicles use an oil level sensor and display the level on the instrument panel.

oil level from the bottom of the "safe" mark on the dipstick to the "full" mark is about a quart. This amount may vary, so always check the service manual to determine the correct quantity.

Some newer vehicles are not equipped with an engine oil dipstick. Instead, they may use an oil level sensor that reports the oil level to an indicator on the dash (**FIGURE 15-4**). In this way, the driver can monitor the oil level from the driver's seat, without even opening the hood.

Always install the recommended amount and type of oil specified in the service manual. Although fresh oil is translucent and oil that needs to be replaced often looks black and dirty, it is usually difficult to assess the condition of engine oil simply by its color. Oil loses its clean, fresh look very quickly but may still have a lot of life left in it.

The best guide to knowing when to change the oil is the vehicle's oil-life monitor, if the vehicle is so equipped. It will tell you the percentage of oil life remaining before a change is needed. If the vehicle is not equipped with an oil-life monitor, check the oil change sticker on the windshield, or ask the owner for oil change records to determine if the oil needs to be changed. This can ensure that the oil is not changed too often, which would be an unnecessary expense for the customer.

▶ TECHNICIAN TIP

Our job as technicians is to provide high-quality work only when it is truly needed by the customer. Part of being a professional is letting customers know when they do and when they do not need a service performed. Informing the customer of what needs to be done and what does not, along with an explanation, helps to build trust with the customer and often results in the customer returning to your shop for future repairs.

When checking the level of the oil, use the dipstick to wipe some oil onto a clean white paper towel. Hold a light over the towel, and move it back and forth to see if light reflects from metal. If metal is in the oil, there is substantial wear happening inside the engine that will require major engine diagnosis and repair.

If the oil on the dipstick looks milky gray, it is possible that there is water (or coolant) being mixed into the oil (**FIGURE 15-5**). This could indicate a serious problem somewhere inside the engine, such as a leaking head gasket or a cracked head, and you should report this to your supervisor immediately. Engine operating conditions can also influence the oil's condition. For instance, continuously stopping and starting the engine with very short operating cycles can cause condensation to build up

FIGURE 15-5 Oil that looks milky gray could indicate that coolant or water is being mixed into the oil. This could be caused by a bad head gasket, cracked head, or leaky oil cooler.

inside the engine. An extreme case of this will cause very rapid oil deterioration and will require frequent oil changes. The oil of a vehicle that is running too rich or that has a leaking fuel injector will smell like fuel and will be very thin. These problems can ruin an engine quickly, as oil will not adequately lubricate the moving surfaces in the engine. Lastly, if you had to add oil to the engine, do not forget to reinstall the filler cap after topping off the oil.

To check the engine oil, follow the steps in **SKILL DRILL 15-1**.

TECHNICIAN TIP

Make sure the hood is secure with a hood prop rod, if necessary. Always make sure you wear the appropriate personal protection equipment before starting the job. It is very easy to think that nothing can happen on a basic job like checking the oil level.

SAFETY TIP

If the engine has been running, be careful not to burn your hand or arm on the exhaust manifold or any other hot part of the engine when reaching for the dipstick. Remember, the dipstick and the oil on it will also be hot. Dripping oil from the dipstick will smoke or burn if it falls on any hot engine surfaces.

Oil Analysis

Oil will suspend particles as the engine wears. Analysis of the engine oil is a useful way to see what parts are wearing in the engine. The military as well as many companies use oil analysis to ensure that the engine oil is changed at the appropriate interval. A small tube is slipped down the oil dipstick tube all the way to the oil pan sump. A vacuum device is hooked to this tube to

SKILL DRILL 15-1 Checking the Engine Oil

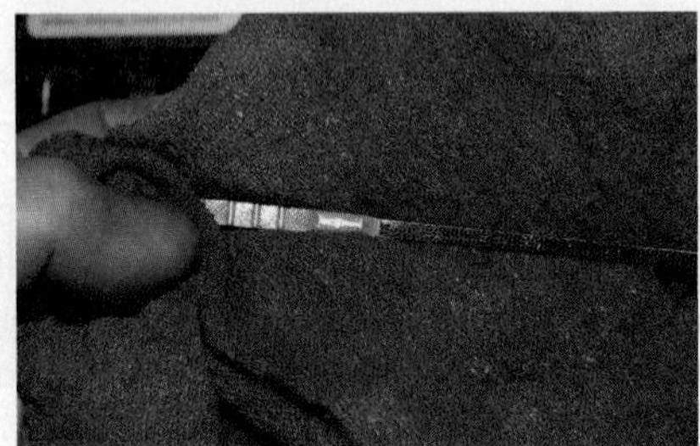

1. Locate the dipstick. With the engine off, remove the dipstick, catching any drops of oil on a rag, and wipe it clean. Observe the markings on the lower end of the stick, which indicate the "full" and "add" marks or specify the "safe" zone.

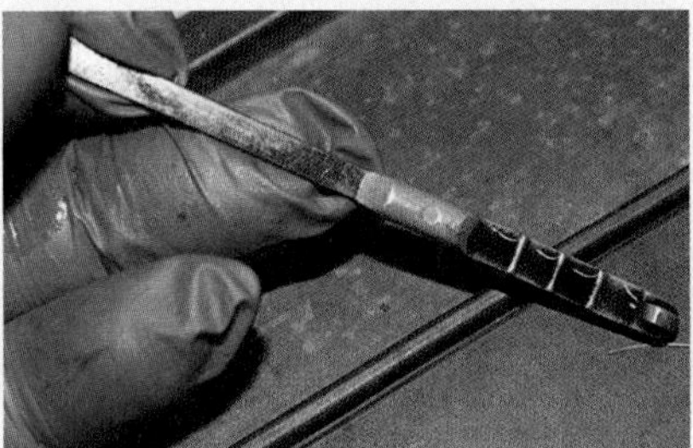

2. Replace the dipstick and push it back down into the sump as far as it will go. Remove it again, and hold it level while checking the level indicated on the bottom of the stick. If the level is near or below the "add" mark, then you will need to determine if the engine just needs topped up to the full level with fresh oil or replaced with new oil and oil filter.

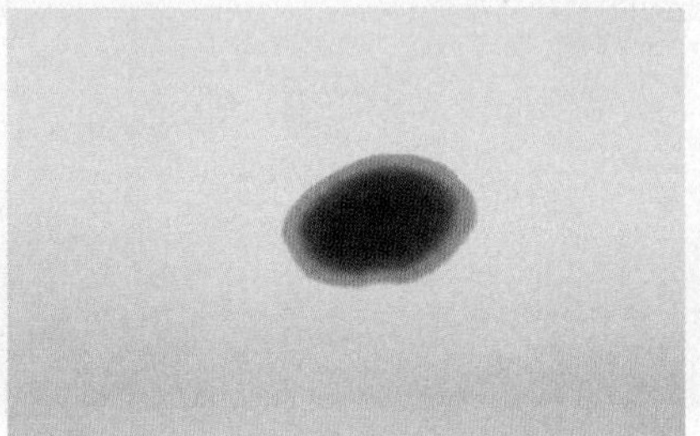

3. Check the oil for any conditions such as unusual color or texture. Report these to your supervisor.

4. Check the oil monitoring system, oil sticker, or service record to determine if the oil needs changed. (Some oil monitoring systems show the percentage of life left in the oil.)

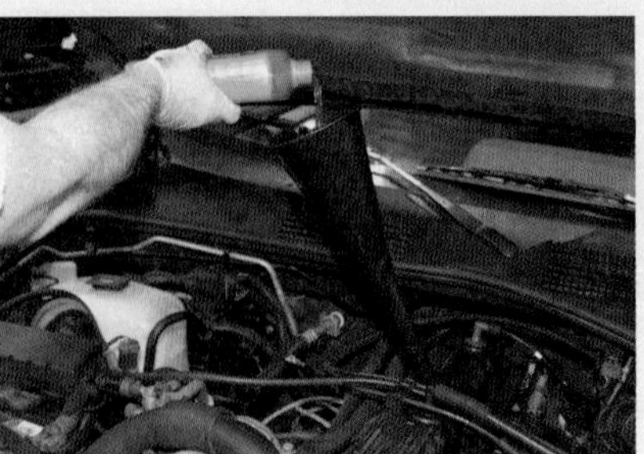

5. If additional oil is needed, estimate the amount by checking the service manual guide to the dipstick markings. Unscrew the filler cap at the top of the engine, and using a funnel to avoid spillage, turn the oil bottle so the spout is on the high side of the bottle, and gently pour the oil into the engine. Recheck the oil level.

6. Replace the oil filler cap, and check the dipstick again to make sure the oil level is now correct.

pull a sample of oil from the pan to a collection container. This sample is labeled with the vehicle information and sent to a lab. The lab thoroughly analyzes the oil and reports the findings back to the shop. The report lists physical properties such as viscosity, condensed water, fuel dilution, antifreeze, acids, metal content, and oil additives. Each of these attributes can be used to determine the condition (remaining life) of the oil as well as the engine. When oil analysis is performed for a particular engine on a regular basis, issues can usually be addressed before they become catastrophic. Oil analysis is typically used in heavy vehicle applications because they may use 3 or more gallons (11.4 or more liters) of engine oil. Some race teams also analyze engine oil to ensure that the engine is not being damaged.

Oil and Filter Change: Draining the Engine Oil

N15001, S15003

Draining the engine oil is performed any time the oil and filter are to be changed or whenever the oil pan needs to be removed for service work. Over time, the oil becomes dirty and the additives wear out. Also, as the engine wears, very small pieces of metal get suspended in the oil. Removing the old contaminated oil helps to make the engine last longer. Always follow the manufacturer's oil change interval, and remember that normal use and severe use have different oil change intervals.

When draining the oil, several precautions are necessary. First, engine oil is normally changed after the engine is fully warmed up. This helps to stir up any contaminants, making them easier to flush out with the draining of the oil. However, that means that the oil can be 200–300°F (93–149°C), so use disposable gloves so that you don't burn yourself.

Second, make sure you locate the correct drain plug—or in some cases, the correct two drain plugs. Some vehicles have a drain plug on the transmission/transaxle that can be mistaken for the oil drain plug. If in doubt, look it up or ask your supervisor to point it out.

Third, many drain plugs are either angled off the bottom radius of the oil pan or almost sideways at the bottom side of the pan. This means when the drain plug is removed, hot oil will shoot sideways. Always make sure you take into consideration what path the oil will take. Oil can shoot out pretty far. The lower the drain pan is compared to the drain plug, the harder it is to judge the distance the oil will spray.

Fourth, the drain plug gasket needs to be inspected and possibly replaced. The gasket can be a single-use gasket made of plastic, aluminum, or fiber. This type of gasket should be replaced during every oil change because it is crushed to conform to any irregularities of the pan and drain plug. Because it is crushed, it will not conform as easily the next time it is tightened. This means that someone may think it needs to be tightened excessively to prevent seepage. Over-tightening can strip the threads on the oil pan, especially if it is made of aluminum, or it can also strip the threads on the drain plug itself. Replace a non-silicone drain plug gasket every time it is removed. If it is of the integrated silicone, long-life style of drain plug, the silicone seal retains its ability to seal for quite a few oil changes as long as it hasn't been abused. So seals rarely need to be replaced.

Fifth, drain plugs need to be tightened to the proper torque. If a drain plug were to loosen and fall out, all of the oil would leak out pretty quickly, ruining the engine if the vehicle is not stopped immediately. At the same time, if drain plugs are tightened too tight, the threads in the oil pan can be stripped, requiring replacement of the oil pan, especially if it is an aluminum oil pan. In some cases, the threads can be replaced with a thread insert or oversize drain plug. But neither of those is ideal as they are costly for the shop or customer.

To drain the engine oil, follow the steps in **SKILL DRILL 15-2**.

TECHNICIAN TIP

When removing the drain plug, be sure to use the proper-sized box wrench or socket. Always inspect the drain plug gasket for damage before reusing; some manufacturers require the use of a new drain plug gasket each time the drain plug is removed. Always look up and torque the drain plug to proper specifications, as over-tightening could damage the threads of the plug and oil pan, and under-tightening could cause an oil leak or the plug to vibrate loose and fall out, resulting in the loss of all the engine oil.

TECHNICIAN TIP

Refer the customer to the owner's manual to determine when the oil should be changed. That could be based on a mileage interval, time interval, or oil-life monitor. You may also need to verify whether the vehicle is operated under normal or severe conditions. Severe conditions may include driving in low temperatures, stop-and-go driving, dusty conditions, short trips, towing, etc. Intervals for severe driving conditions are sooner than normal driving intervals.

Oil and Filter Change: Replacing Oil Filters

S15004

Oil filters are designed to filter out particles that find their way into the oil. The filter catches particles that result from a variety of sources such as carbon from combustion or small metal flakes that result from normal engine wear. The engine oil suspends some of the particles while the heavier particles fall to the bottom of the oil pan. The oil filter is designed to catch the particles that are flowing with the oil. It is critical to change the oil filter at the manufacturer's recommended mileage to help ensure that it does not become clogged, which would restrict the oil flow.

When changing a spin-on oil filter, be careful of a few things. First, an oil filter wrench is only used to remove an oil filter. It is never used to install an oil filter. Installation must be performed by hand.

Second, the O-ring that seals the spin-on oil filter to the block tends to stick on the engine block when the filter is being removed. This can lead to double-gasketing, which occurs when the new O-ring in the new filter is installed over the old O-ring. Because the groove in the filter is only deep enough to hold the new O-ring, the old one is not held in place. Once the engine is started, the oil pressure pushes it out of place and oil is pumped very quickly out of the engine and onto the floor. This not only makes a huge mess and wastes good oil, but if you don't realize it has happened, the engine could be damaged in only a minute or two of running.

SKILL DRILL 15-2 Draining the Engine Oil

1. Before you begin, obtain the oil drain container (and make sure it has enough room for the oil to be drained), verify that you have the correct oil and oil filter, and ensure that the engine oil is up to operating temperature.

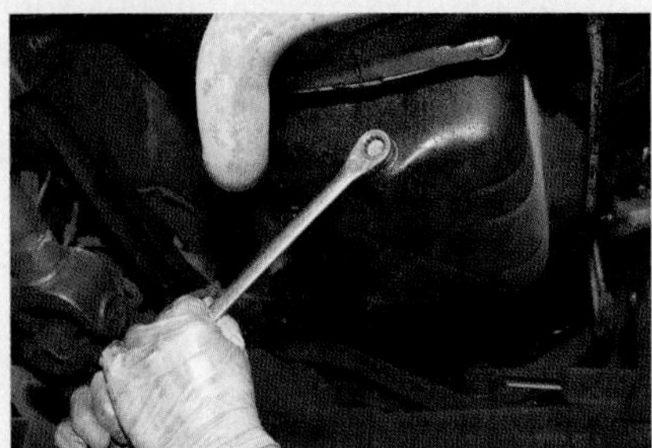

2. Identify the location of the oil drain plug. Some vehicles have two drain plugs, draining separate sump areas. Use a box wrench or socket to remove and replace the drain bolt. Be careful not to remove the transmission drain plug by mistake.

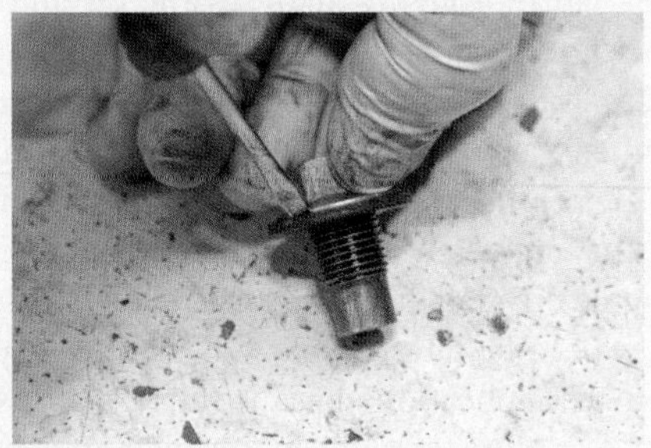

3. Position the drain pan so it will catch the oil. Remove and inspect the drain plug and gasket; replace as necessary.

4. Allow the oil to drain while you are dealing with the drain plug, gasket, and oil filter (see Skill Drill 15-3).

5. Screw in the drain plug all the way by hand, and then tighten it to the torque specified by the manufacturer. Wipe any drips from the underside of the engine.

6. Safely dispose of the drained oil according to all local regulations.

Always check that the old O-ring was removed with the filter. If it was not, reach up and peel it off the filter mounting adapter.

Third, when installing the spin-on oil filter, smear a bit of oil on the surface of the O-ring. Doing so lubricates it so that it will spin with the oil filter as it is being tightened. Failure to lube the O-ring can cause it to bind and roll out of the oil filter groove when the filter is being tightened, again causing a substantial oil leak.

Last, when installing the spin-on oil filter, it can be hard to see how it is going on the threaded filter adapter. If you get it cross-threaded, it will leak just like a double-gasketed O-ring. Plus you are likely to damage the threads on the adapter, making it harder to install an oil filter in the future. To prevent this problem, always start the filter by turning it with your fingers. Once you suspect it is started, stop and try to lift the filter off the adapter. If it comes off, it has not started. Try to start it by finger again; then try to lift it off. If it does not lift off, then it has started onto the threads. Now count the turns that the filter spins on. If it is not cross-threaded, it should go on at least five full turns before the gasket contacts the filter mounting surface. If it doesn't spin on at least five full turns by hand, unscrew it and try it again. Once you get it started properly and run down by finger, you can then tighten it by hand the appropriate amount as specified by the filter manufacturer, typically about three-quarters to one turn after the gasket contacts the adapter.

Increasingly, manufacturers are returning to using cartridge filters because it is easier to properly dispose of the oil. It also reduces the amount of waste generated from used spin-on filters. When changing a cartridge filter, be aware of the following situations. First, the only parts that get replaced are the paper filter cartridge and the O-rings or seals. All of the other parts are reused, so do not damage them or throw them away during disassembly.

Second, some cartridge filters are near the bottom of the engine, whereas others are on top. Use the service information to help you locate the filter. If the filter is on the top of the engine, there is a good chance that you will need to prefill the cartridge before installing the filter end cap. Again, check the service information for the vehicle you are working on.

Third, be careful with tightening a cartridge filter, as it is easy to crack or damage the housing—especially if it is plastic. Always follow the manufacturer's torque procedure.

Last, because the filter housing is being reused, it is important that it is clean before being reinstalled. You may need to wash it out in a clean solvent tank. Just be sure that you also remove any solvent residue before reinstalling it.

SKILL DRILL 15-3 Replacing a Spin-On Filter

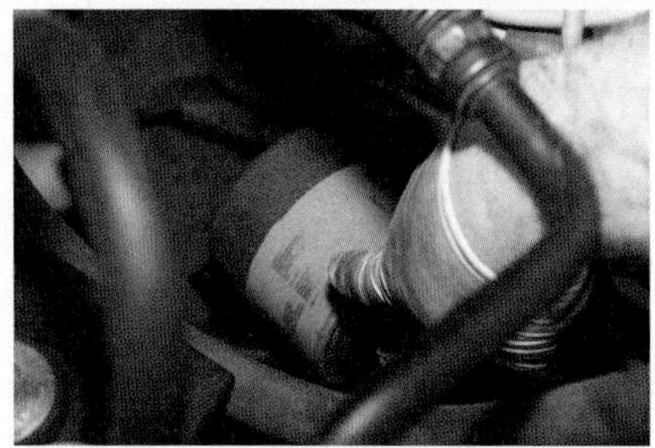

1. Check for new filter availability. Locate the filter being changed. It will usually be located on the side of the engine block or at an angle underneath the engine. Select the proper oil filter wrench.

2. Position a drain pan to catch any oil that will leak from the filter.

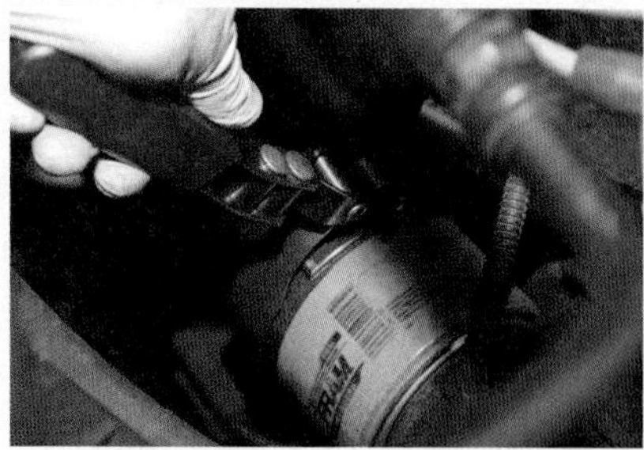

3. Remove the filter.

4. Clean the seating area on the oil filter adapter so that its surface can seal properly. Make sure the O-ring from the removed filter is not still stuck to the mounting surface like in this picture.

5. Confirm you have the correct replacement filter. Smear a little oil on the surface of the new O-ring.

6. Screw in the filter until it just starts, and ensure that it cannot be pulled off. Then turn the filter by hand until the filter lightly contacts the base. Be careful not to cross-thread the oil filter.

Replacing Spin-On and Cartridge Filters

To replace a spin-on filter, follow the steps in **SKILL DRILL 15-3**.

To replace a cartridge filter, follow the steps in **SKILL DRILL 15-4**.

Oil and Filter Change: Refilling Engine Oil

S15005

Refilling an engine's oil supply is necessary when performing an oil and filter change. It may also be necessary to refill the engine oil after a lubrication system part has been replaced, if the oil was drained. Always add the recommended amount of oil and grade of oil listed in the service information. After adding the required amount of oil, be sure to start the engine to build oil pressure and fill the oil filter, and then shut the engine off to check for leaks and the level on the dipstick. Fill the oil to the max line and no farther.

To refill the engine oil, follow the steps in **SKILL DRILL 15-5**.

▶ Diagnosis

Common Issues

S15006

Some of the common issues in lubrication systems found in the repair shop tend to stem from the same common variable—the customer. Changing the engine oil at the recommended time is the most important maintenance item the customer can have performed. Failure to change oil results in sludge formation, which clogs the oil passageways and keeps oil from lubricating critical parts. When bearings fail to get enough lubricant, they begin to wear and then will seize, resulting in the need for a new or rebuilt engine. An engine replacement is a vastly more expensive alternative to the relatively inexpensive oil change the customer did not want to pay for or did not have time for. As technicians, we need to educate customers on the importance of maintaining their vehicles.

Another common issue is oil leaks. You will need to be able to diagnose where that puddle of oil on the customer's garage floor is coming from. Oil will run down the engine from valve

SKILL DRILL 15-4 Replacing a Cartridge Filter (Replaceable Element)

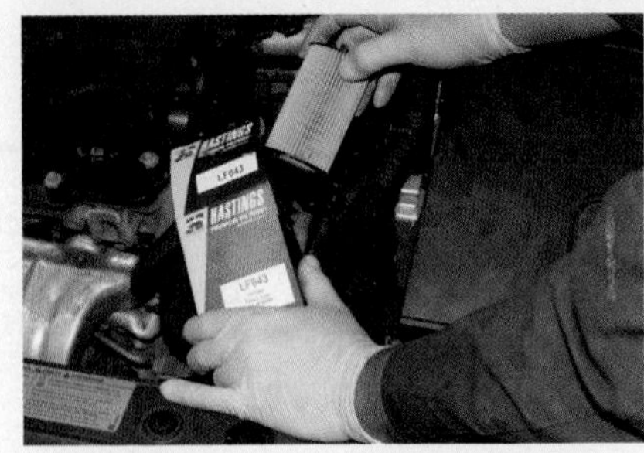

1. Before removing a cartridge-style oil filter, make sure a suitable replacement filter is available.

2. The filter may be located on the side of the engine block, at an angle underneath the engine, or on the top of the engine. If the filter is located in a plastic end cap housing, be sure to use the special tool that the service information calls for.

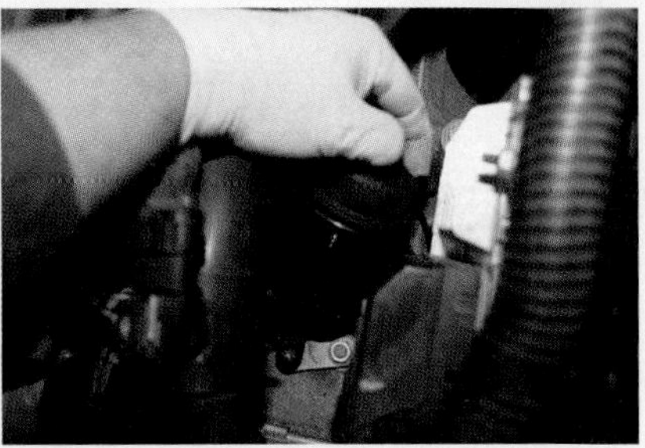

3. Position a drain pan to catch any oil that leaks from the filter. Unscrew the filter cartridge retaining bolt or oil filter cap, and remove the cartridge filter and housing or end cap.

4. Remove the filter and clean the housing or end cap as necessary, and replace any O-rings or gaskets on the assembly.

5. Smear a little oil on the surface of any new O-rings. Install the new filter cartridge back into its housing.

6. If the cartridge is on the top of the engine, the service information may direct you to pour a specified amount of oil into the filter cavity before the end cap is installed. Screw the cartridge bolt or end cap back in place, and tighten to the specified torque. Be careful not to cross-thread the oil filter bolt or end cap.

covers or the camshaft seals and may appear to be coming from a lower point. Remember to look over the engine closely and check all mating surfaces. Oil leaking from seals of rotating parts will typically sling oil in a circle; for example, a crankshaft seal will sling oil around the front of the engine in a circular pattern.

Another common area to look for leaks is on the oil pressure sending unit. The sending unit uses a diaphragm, which can develop a hole over time. When this happens, oil will typically leak out of the body either at the crimp or at the terminal end while the engine is running. If you watch it while it is running, you will usually see it drip. Just be aware that if the sending unit is at a downward angle, then oil from another location can possibly run down to it and then drip off it. So always look above the sending unit. If there is little to no oil above it, the sending unit is bad.

Noises are another lubrication issue found in the repair shop. Low oil pressure can result in valve lifters not pumping up, creating valve train noise. The use of an oil pressure gauge will show if the noise is related to low oil pressure or if it is a result of worn parts with improper running clearances. Low oil pressure can be caused by a low oil level or too thin of oil in the crankcase. It could also be caused by worn bearings or oil pump. Some engines even use an automatic cut-out that turns off the engine if oil pressure falls too low.

Oil can be low for a couple reasons: It may have leaked away or it may have burnt off. An engine can leak its lubricating oil from seals or gaskets that are intended to seal the oil in. Burning the engine oil can be caused by worn piston rings or by worn valve guides that let oil into the combustion chamber or by an improperly operating PCV system.

Too little oil in the engine is a problem, but so is too much oil. The crankshaft can whip it into foam and cause leaks by flooding the seals. Another problem with excessive oil is the **drag** produced when the crankshaft hits the oil in the oil pan and tries to stop the crankshaft from turning. This condition will cause a loss of power and result in more fuel consumption. Additionally, the foaming caused by the crankshaft hitting the oil will result

SKILL DRILL 15-5 Refilling Engine Oil

1. Research the correct grade and the quantity of oil you will need. Position the spout on the high side of the bottle. Pour the oil into the funnel slowly enough to avoid the risk of blowback or overflow. Fill the engine only to the level indicated on the engine dipstick. Replace the filler cap.

2. Start the engine and check the oil pressure indicator on the dash. If the oil pressure is inadequate, stop. Do not continue to run the engine.

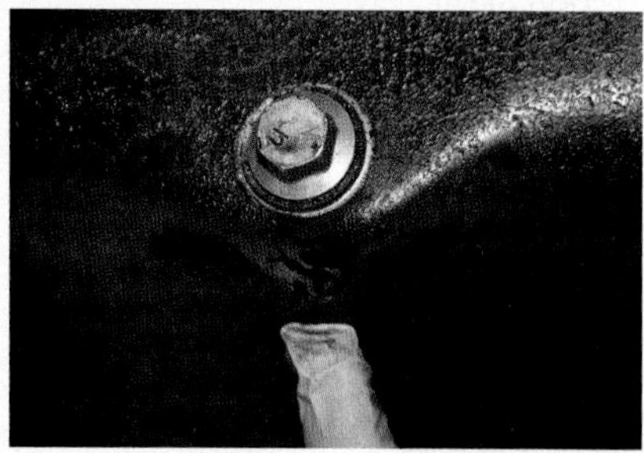

3. If the oil pressure is good, turn the engine off and check underneath the vehicle to make sure no oil is leaking from the oil filter or drain plug.

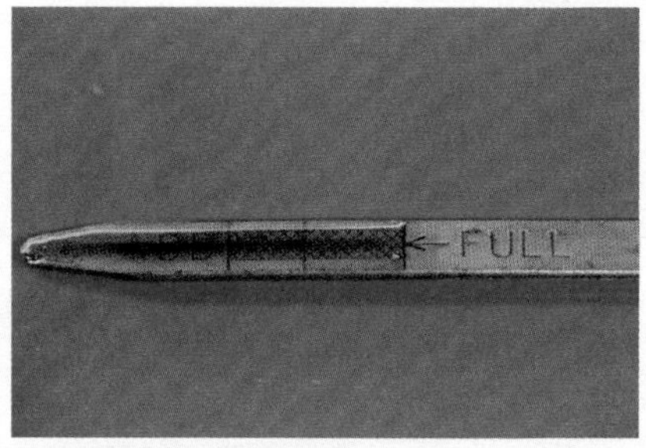

4. With a level vehicle, check the oil level again with the dipstick. It may be necessary to top off the engine by adding a small quantity of oil to compensate for the amount absorbed by the new filter. Do not overfill.

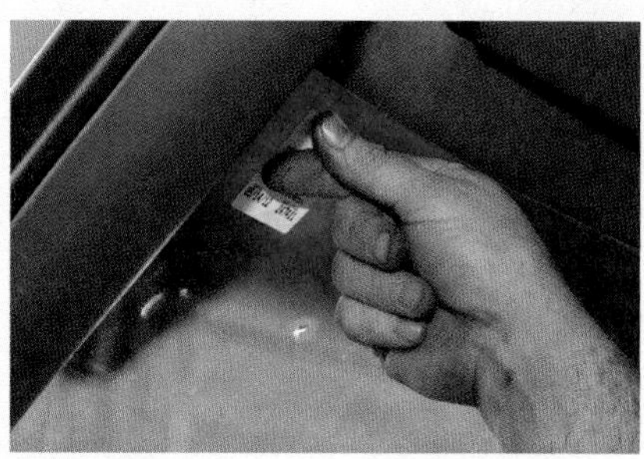

5. Refer to the owner's manual or the service information, and install a static sticker.

6. Reset the maintenance reminder system to remind the owner when the next oil change is due.

in starving the engine of precious lubricating oil since the pump will send these air bubbles to the moving parts and damage them.

> **TECHNICIAN TIP**
>
> Remember that oil will sometimes blow upward when the vehicle is driving down the highway or blowback under the vehicle, so always try to imagine how oil is being thrown around. The use of fluorescent dye may be necessary if a leak is covering more than one area of the engine. It may also be necessary to pressure wash the undercarriage and retest for a large leak.

> **TECHNICIAN TIP**
>
> In the case of finding a leak, always remember that you need to find the root cause of the problem. Gaskets and seals are only designed to withstand a certain amount of pressure before they leak. If the engine has excessive blowby or the PCV system is not working properly, pressure can build up in the crankcase, causing oil to leak past gaskets and seals. A quick way to check for crankcase pressure is to start the engine and then remove the crankcase fresh air intake tube or hose from the engine and cover the engine side of it. If a vacuum develops, the crankcase is not pressurizing. If pressure develops, further diagnosis of the PCV system and piston rings is necessary.

Tools for Diagnosing the Lubrication System

The tools used for diagnosis of the lubrication system include the following (**FIGURE 15-6**):

- An oil pressure gauge, with various size adapters that screw into the pressurized oil passage, is used to measure the engine's oil pressure.
- A mechanic's stethoscope is used to pinpoint noises from the engine. The stethoscope allows you to verify where the noise is coming from, such as the valve train or the main bearings. Stethoscopes can also be electronic versions with headphones, which allow technicians to isolate separate sound frequencies.
- A DMM is used to test the electrical circuits of oil pressure sensors and gauges.
- Fluorescent dye and an ultraviolet light are useful tools to find oil leaks. Fluorescent dye specially made for engine

oil is added to the crankcase. The engine is run to circulate the oil, and then a black light is used to pinpoint the location of the leak.
- Oil pressure sending unit sockets are useful tools for removing and installing oil pressure sending units.
- A resistance substitution box is used to diagnose gauge problems by substituting precise resistances in the gauge circuit and comparing the gauge reading to the specifications.

Testing the Oil Pressure

N15002

Oil pressure testing must be performed whenever low oil pressure is suspected and the oil level is within the "safe" range. If the oil pressure gauge in the dash of the vehicle shows low, you need to verify that it is correct with a manual test gauge. If the test gauge indicates a different pressure from the vehicle's gauge, you need to test the gauge according to the manufacturer's procedure.

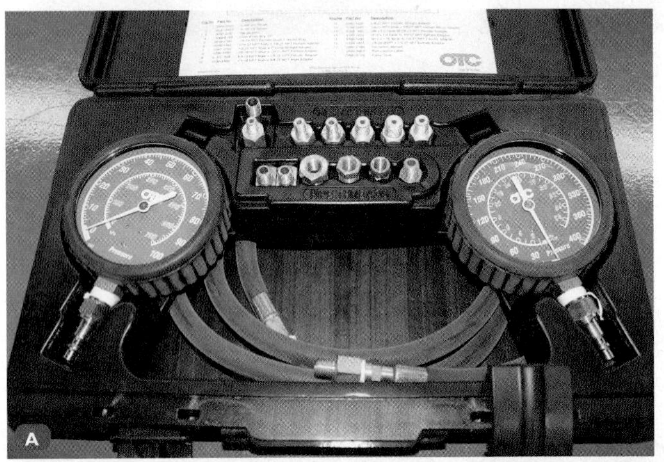

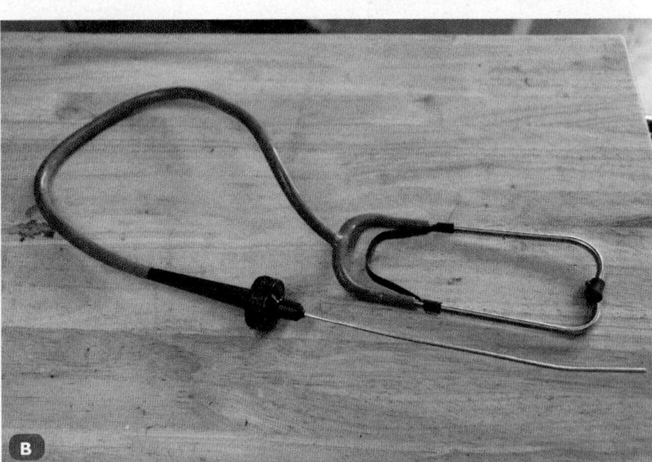

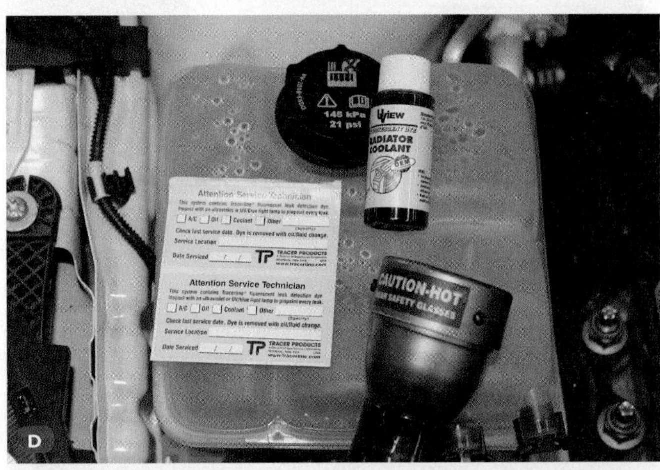

FIGURE 15-6 Tools used for diagnosis of the lubrication system. **A.** Oil pressure gauge. **B.** Mechanic's stethoscope. **C.** DMM. **D.** Fluorescent dye and ultraviolet light. **E.** Oil pressure sending unit sockets. **F.** Resistance substitution box.

(If an engine is noisy, it may be necessary to check oil pressure to verify that the engine is repairable. An engine that has been operated for long periods of time with no or low oil pressure will have extensive damage and in almost all cases will require complete engine replacement or overhaul. So if the noise is accompanied by low oil pressure, the low oil pressure must be diagnosed first. If the noise is accompanied by normal oil pressure, then further diagnosis of the noise is warranted.)

You need a quality oil pressure gauge of the correct pressure reading and correct threaded adapters. The oil pressure gauge is usually installed where the oil pressure sensor (sometimes called an oil pressure sending unit or oil pressure switch) is located. Remove the oil pressure sensor, and install the oil pressure gauge in its place. If oil pressure is lower than specifications, refer to service information for additional information. There may be excessive clearance in the engine bearings (which causes oil to bleed off and pressures to decline), and/or the oil pump may be worn. If the oil pressure is higher than specifications, suspect a sticking oil pressure regulator. Be sure to follow service information to troubleshoot this condition.

To perform oil pressure testing, follow the steps in **SKILL DRILL 15-6**.

SKILL DRILL 15-6 Testing the Oil Pressure

1. Locate the oil pressure sensor by using the component locator in the service information. It is typically near the oil filter.

2. Place an oil drain pan under the engine to catch any oil when you remove the oil pressure sensor.

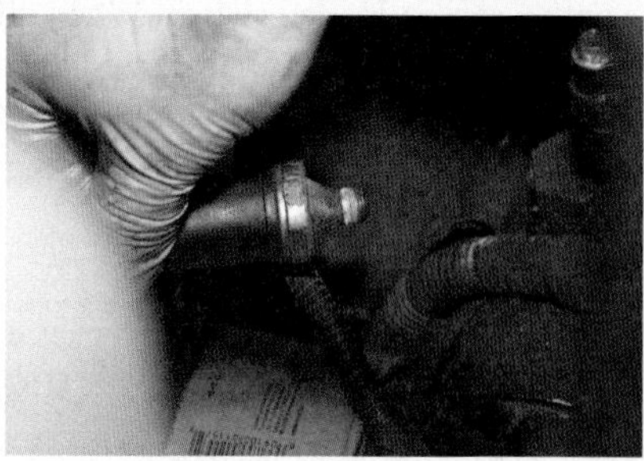

3. Disconnect the wire harness connector, and remove the oil pressure sensor from the engine, using the recommended tool in the service information.

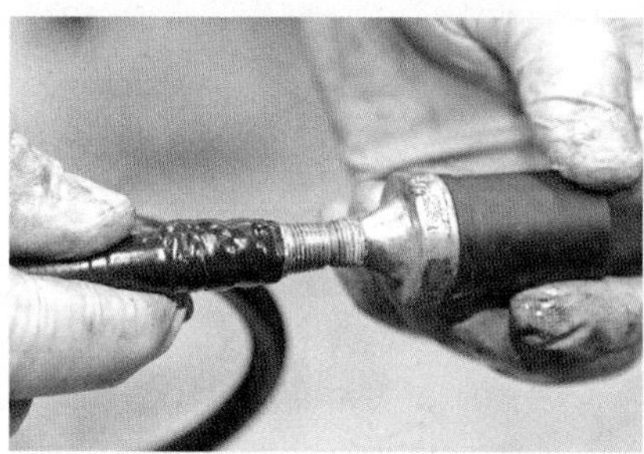

4. Match the adapter thread to the thread on the pressure sensor to ensure that the correct adapter will be screwed into the engine.

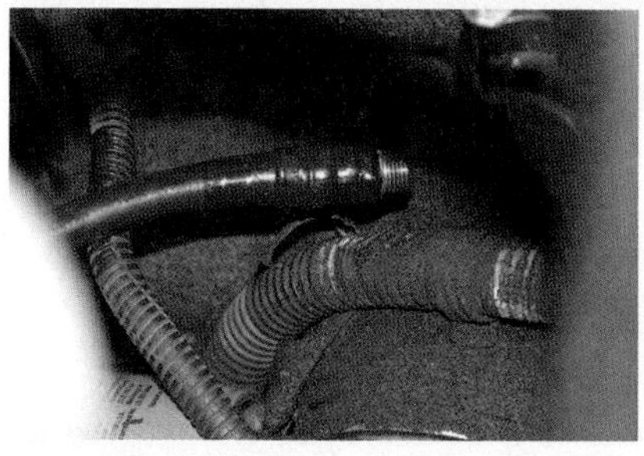

5. Carefully thread the adapter into the engine, ensuring that you do not cross-thread the adapter.

6. Connect the oil pressure gauge, then start and completely warm up the engine. Compare the readings to the specifications in the service information.

7. Remove the manual pressure gauge and adaptor. Using a thread sealant, reinstall the oil sending unit, and torque to specifications.

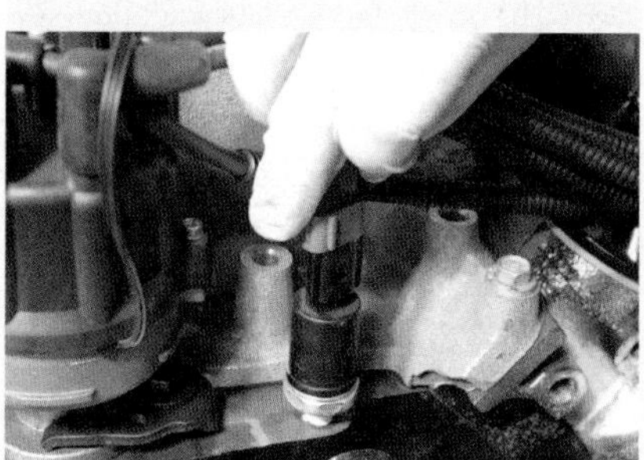

8. Install the wire harness connector, and start the engine. Inspect for oil leaks and proper pressure gauge operation.

SAFETY TIP

The lubrication system contains pressures up to approximately 80 psi (552 kPa), which will shoot oil a good distance. Be sure to wear safety glasses, and be careful if the engine oil is hot, as it could cause burns.

Inspecting and Testing Oil Sensors

N15003

Oil sensors feed oil pressure information back to the gauge or warning light to issue the operator a warning if oil pressure goes too low. Many oil pressure sensors are typically variable resistors that change resistance based on pressure changes. The variable resistance causes the gauge to read differing amounts, depending on the variable resistance. Others have a fixed resistance in line with a spring-loaded switch. When oil pressure is above the spring pressure, the switch closes and the gauge indicates a midpoint reading. When oil pressure is below the spring pressure, the switch opens and the gauge reading falls toward the low pressure side. Testing oil pressure sensors usually involves substituting the specified resistance with a resistor substitution box in place of the sensor and observing the reaction of the oil pressure gauge. Be sure to follow service information to diagnose the sensor you are working with. Use care with hot engines, as serious burns can result.

To inspect the oil sensors, follow the steps in **SKILL DRILL 15-7**.

Inspect Auxiliary Coolers

N15004, S15007

Remember from the previous chapter that one of the functions of oil is to cool engine components. Normally, oil is cooled by radiating heat from the oil pan to the atmosphere. It is easy to see that additional oil cooling might be necessary on vehicles that experience heavy loads or those that are turbocharged. Auxiliary coolers are used to provide extra cooling of the engine oil on these types of vehicles. They can be of the oil-to-water type, or

SKILL DRILL 15-7 Inspecting Oil Sensors

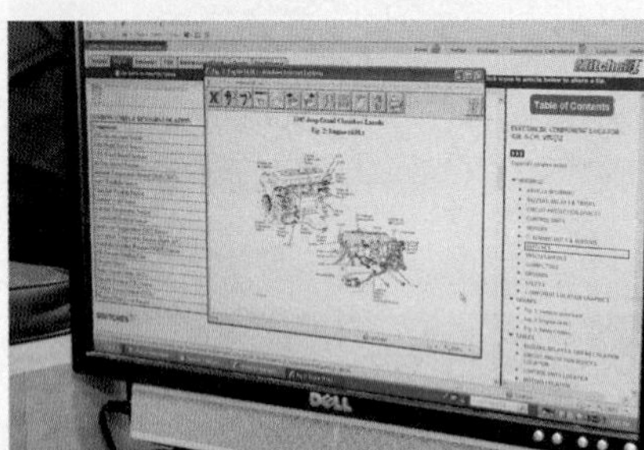

1. Locate the oil sensor by using the service information.

2. Locate the oil pressure switch or sensor. Disconnect the connector to the switch or sensor. If it is a sensor, determine which wire is the feedback to the gauge or warning light. You may need to use a connector end view to ensure that the correct wire is found.

3. Set the variable resistance to the specified setting.

4. Find the voltage feed wire color by using the wiring diagram for the vehicle, and use a variable resistor to connect between the voltage feed wire and the signal return wire.

5. Check the oil pressure gauge in the dash. The gauge should be at the specified reading.

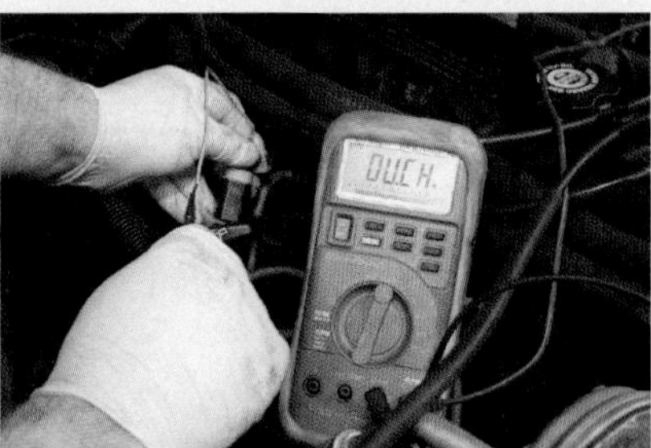

6. Replace the oil sensor if the gauge responds correctly. Double-check the oil level after replacement, as the oil level may have dropped. If the gauge does not respond correctly, then test the wiring and gauge.

the oil-to-air type (**FIGURE 15-7**), both of which remove excess heat from the engine oil and help prevent it from overheating.

When inspecting oil-to-water type coolers, look for mixing of the fluids: oil in the coolant, and coolant in the oil, both of which indicate a leaking cooler heat exchanger or cooler seal. If a mixing of fluids is observed, the cooler heat exchanger needs to be removed and pressure tested to determine if it is leaking. In most cases, a leaking cooler needs to be replaced with a new one.

When inspecting oil-to-air type coolers, the cooler should be inspected for external leaks. This is similar to locating any other type of external leak, where you identify any seepage or other leak and trace it to its origin. If the origin can't be determined easily, you may need to add the appropriate fluorescent dye to the engine oil, operate the engine for the specified amount of time, and use an ultraviolet light and special glasses to identify the source of the leak in the cooler or lines.

FIGURE 15-7 Typical engine oil coolers.

▶ Wrap-Up

Ready for Review

- The engine lubrication system requires periodic inspection, maintenance, diagnosis, and repair to provide the intended protection for the internal workings of the engine.
- Maintenance and repair procedures include draining the engine oil; replacing the oil filter; refilling engine oil; and inspecting, testing, and replacing oil pressure and temperature switches.
- Lubrication repair tools include oil filter wrench, wrenches and socket set, mirror, good-quality light, digital volt-ohmmeter, and a special socket for removing oil pressure switches or sensors.
- When checking oil level, hold the dipstick horizontal so the oil does not run down the stick.
- Sending oil out for analysis is a good way to evaluate the condition of the oil.
- Oil should typically be changed when it is hot so that more of the contaminants can be flushed out.
- Always check the service information when changing engine oil, to determine the correct quantity and type of oil.
- The drain plug gasket needs to be inspected and possibly replaced.
- Drain plugs need to be tightened to the proper torque.
- Always check that the old O-ring was removed with the filter.
- Install a static sticker, and reset the maintenance reminder system to remind the customer of the next oil change.
- Tools used to diagnose the lubrication system include oil pressure gauge, oil pressure sensor socket, digital volt-ohmmeter, sensor substitution box, and fluorescent dye.
- Diagnosis of the lubrication system may include oil analysis; checking engine oil; oil pressure testing; inspecting oil sensors; and inspecting the engine assembly for fuel, oil, coolant, and other leaks.
- Common lubrication system issues are infrequent oil changes, oil leaks, and low oil pressure.
- If the vehicle's oil pressure gauge reads low, verify this with a test gauge.
- If the oil pressure is good, but the gauge reads wrong, use an ohmmeter to test the sending unit and a resistor substitution box to test the gauge.
- When inspecting oil-to-water type coolers, look for mixing of the fluids: oil in the coolant, and coolant in the oil.
- When inspecting oil-to-air type coolers, the cooler should be inspected for external leaks.

Key Term

drag The slowing down of the crankshaft when it strikes the oil in the crankcase.

Review Questions

1. The tools for lubrication system maintenance include all of the following *except*:
   a. a compression tester.
   b. an oil filter wrench
   c. a sensor substitution box.
   d. a pressure gauge.
2. Which of these is the best guide to knowing when to change the oil?
   a. Dipstick
   b. Oil-life monitor
   c. Oil level sensor
   d. Pressure gauge

3. If the oil on the dipstick looks milky gray, it is possible that there is water (or coolant) being mixed into the oil. This could have been caused by any of the following *except*:
   a. a bad head gasket.
   b. a cracked head.
   c. a leaky oil cooler.
   d. a leaky fuel injector.
4. Choose the correct statement with respect to oil change.
   a. Normal use and severe use have the same oil change intervals.
   b. When draining oil, the oil should be cold.
   c. The drain plug should be tightened with maximum force to prevent oil leak.
   d. The engine should be fully warmed up when draining oil.
5. When installing a spin-on oil filter, failure to lube the O-ring will:
   a. lead to double gasketing.
   b. damage the threads on the adapter.
   c. cause it to bind and roll out of the oil filter groove.
   d. cause the oil pressure to be too high.
6. During replacement, when compared with a spin-on filter, in a cartridge filter:
   a. all parts get replaced.
   b. it is easier to properly dispose of the oil.
   c. a greater amount of waste is generated.
   d. reuse is minimal.
7. Excessive oil results in all of the following *except*:
   a. low oil pressure.
   b. high oil pressure.
   c. drag.
   d. foaming.
8. Which of these is used to test the electrical circuits of oil pressure sensors and gauges?
   a. A resistance substitution box
   b. An oil pressure gauge
   c. A florescent dye check kit
   d. A stethoscope
9. If oil pressure is higher than specifications:
   a. there may be excessive clearance in the engine bearings.
   b. the oil pump may be worn.
   c. the engine is noisy.
   d. the oil pressure regulator might be sticking.
10. Low oil pressure reading on the instrument panel can be caused by all of the following except:
   a. worn crankshaft bearings.
   b. worn oil pump gears and housing.
   c. oil pressure regulator spring too strong.
   d. Faulty oil pressure sensor.

ASE Technician A/Technician B Style Questions

1. Tech A says that when changing the spin-on oil filter, you need to verify that the old O-ring was removed with the filter. Tech B says that the O-ring on the new filter should be dry when the filter is screwed on. Who is correct?
   a. Tech A
   b. Tech B
   c. Both A and B
   d. Neither A nor B
2. Tech A says that the engine oil drain plug typically needs to be replaced with a new one at each oil change. Tech B says that engine oil drain plugs must be tightened with a torque wrench when they are installed. Who is correct?
   a. Tech A
   b. Tech B
   c. Both A and B
   d. Neither A nor B
3. Tech A says that the oil pressure will typically be low if the oil level is at the "add" line on the dipstick. Tech B says that worn engine bearings could cause low oil pressure. Who is correct?
   a. Tech A
   b. Tech B
   c. Both A and B
   d. Neither A nor B
4. Tech A says that it takes about 1 quart of oil to raise the oil level from "add" to "full." Tech B says that it takes about 1 pint to raise it that much. Who is correct?
   a. Tech A
   b. Tech B
   c. Both A and B
   d. Neither A nor B
5. Tech A says that if the oil pressure reads low, the oil pressure sending unit may need to be replaced. Tech B says that if the oil pressure is low, the oil pressure should be measured using a separate pressure tester. Who is correct?
   a. Tech A
   b. Tech B
   c. Both A and B
   d. Neither A nor B
6. Tech A says that when draining the engine oil, the engine should be fully cold for best results. Tech B says that the engine should be running when checking the engine oil level. Who is correct?
   a. Tech A
   b. Tech B
   c. Both A and B
   d. Neither A nor B
7. Tech A says that after changing the engine oil and filter, you should apply a static sticker to remind the owner when the next oil change is due. Tech B says that you should reset the maintenance reminder system to remind the owner when the next oil change is due. Who is correct?
   a. Tech A
   b. Tech B
   c. Both A and B
   d. Neither A nor B
8. Two techs are discussing engine oil leaks. Tech A says that oil pressure sending units are common leaks. Tech B says that oil leaking from seals of rotating parts typically slings oil in a circle. Who is correct?
   a. Tech A
   b. Tech B
   c. Both A and B
   d. Neither A nor B

9. Tech A says that many oil pressure sensors are typically variable resistors that change resistance based on pressure changes. Tech B says that a sensor substitution box can be used in place of the sensor while observing the reaction of the oil pressure gauge on the dash. Who is correct?
   a. Tech A
   b. Tech B
   c. Both A and B
   d. Neither A nor B

10. Two techs are discussing oil coolers. Tech A says that when inspecting oil-to-water type coolers, look for mixing of the fluids: oil in the coolant, and coolant in the oil. Tech B says that when inspecting oil-to-air type coolers, look for oil in the coolant only. Who is correct?
    a. Tech A
    b. Tech B
    c. Both A and B
    d. Neither A nor B

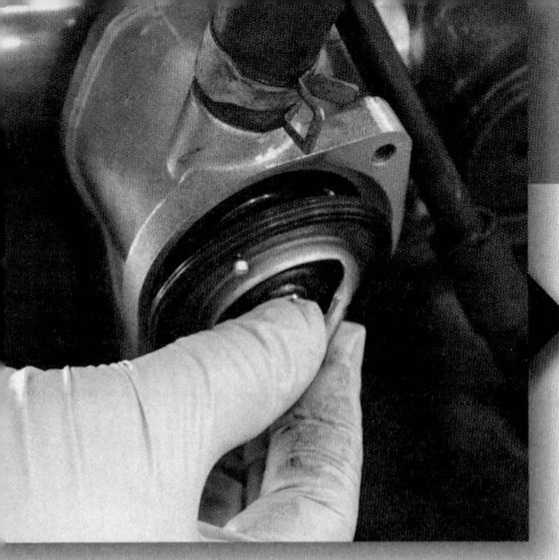

CHAPTER 16

Cooling System Theory

NATEF Tasks

There are no NATEF tasks for this chapter.

Knowledge Objectives

After reading this chapter, you will be able to:

- **K16001** Describe the methods of heat transfer.
- **K16002** Describe cooling system configurations and their operation.
- **K16003** Describe engine coolant and its required properties.
- **K16004** Describe how pressure affects the boiling point of coolant.
- **K16005** Describe the cause and concerns of electrolysis in a cooling system.
- **K16006** Describe how centrifugal force is used in a cooling system.
- **K16007** Describe the operation of a rotary engine cooling system.
- **K16008** Identify and describe the purpose of each of the cooling system components.
- **K16009** Describe the types of coolant and their properties.
- **K16010** Describe the purpose, function, and components of a radiator.
- **K16011** Describe the purpose and function of the surge tank and recovery system.
- **K16012** Describe the purpose and function of the thermostat.
- **K16013** Describe the purpose and function of the water pump.
- **K16014** Describe the types, purpose, and function of cooling fans.
- **K16015** Describe the purpose and function of coolant hoses.
- **K16016** Describe the types, purpose, and function of drive belts and tensioners.
- **K16017** Describe the type, purpose, and function of temperature indicators.
- **K16018** Describe the purpose and function of water jackets and core plugs.
- **K16019** Describe the purpose and function of the heater core, control valve, and air doors/actuators.

Skills Objectives

There are no Skills Objectives for this chapter.

▶ Introduction

Cooling systems play a critical role in the life span of the engine (**FIGURE 16-1**). Typically, a great deal of focus is placed on engine lubrication and the maintenance of the engine's lubrication system, but little focus is placed on the engine's cooling system. Yet, because the Department of Transportation has stated that cooling system failure is the leading cause of mechanical breakdowns on the highway, it deserves our attention. The cooling system can have a huge effect on the lubrication system and can also affect engine emissions and fuel economy, thus making its role a significant one. In fact, a myth regarding the cooling system is that it keeps the engine cool. That is not true. The purpose of the cooling system is to allow the engine to reach its ideal operating temperature as quickly as possible and keep it there during all operating conditions. When the cooling system fails, it can fail in two ways.

The most common failure most people are familiar with is allowing the engine to overheat. This is usually evident by steam pouring out of the engine compartment, and can lead quickly to major engine damage. The second way the cooling system fails is by not allowing the engine to reach its ideal operating temperature, either at all or too slowly. This may be evident by the heater not getting very warm or the defrost not working very well. Although this doesn't cause damage immediately, the engine does wear out much more quickly than one that warms up properly. Low engine temperatures also cause poor fuel economy and high vehicle emissions.

FIGURE 16-1 Cooling systems are the leading cause of mechanical breakdowns on the highway.

This chapter explains how the principles of heat transfer are used to keep the engine within its optimum temperature range. We also explore how each of the cooling system components contributes to that goal.

▶ Heat Transfer

K16001

Heat is thermal energy. It cannot be destroyed; it can only be transferred. It always moves from areas of higher temperature to areas of lower temperature. This principle is applied to the transfer of heat energy from engine parts to **ambient** air, using the coolant as a medium to carry it. To control this heat transfer, it is necessary to understand how heat behaves. Heat travels in only three ways:

1. From one solid to another, by a process called **conduction**.
2. Through liquids and gases, by a process called **convection**, whereby heat follows paths called convection currents.
3. Through space, by **radiation** (**FIGURE 16-2**).

Heat Transfer in Internal Combustion Engines

The internal combustion engine relies on the heat of combustion to produce torque to move the vehicle and power accessories. Unfortunately, much of the heat produced during combustion is not used productively and must be removed to avoid overheating of the engine. No matter how efficiently fuel burning occurs, and no matter the size of the engine, the heat energy generated never completely transforms into

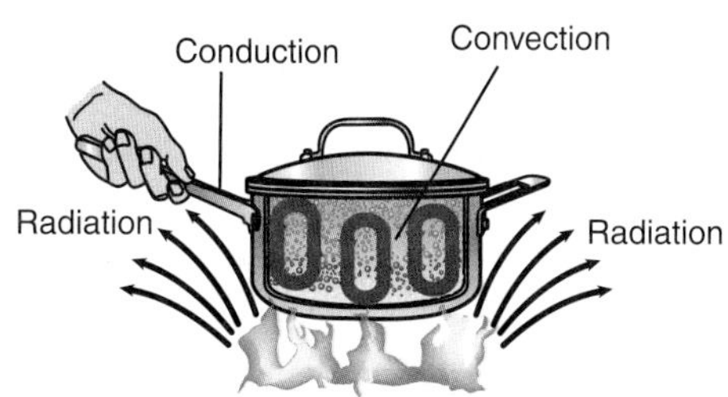

FIGURE 16-2 Conduction is the movement of heat energy through solids; convection is the movement of heat through liquids or gases; and radiation is the movement of heat through space.

You Are the Automotive Technician

You are discussing cooling system theory with a fellow technician. She said that her teacher taught the cooling system theory in a way that made sense to her, and she feels confident that she knows how they operate. You told her that you feel confident as well and that she should try to stump you with some questions. Answer her following questions:

1. What are the three ways that heat transfers from one place to another?
2. How does a thermostat regulate engine temperature?
3. How does pressure affect the boiling point of coolant?
4. What are the different types of fans, and how do they operate?

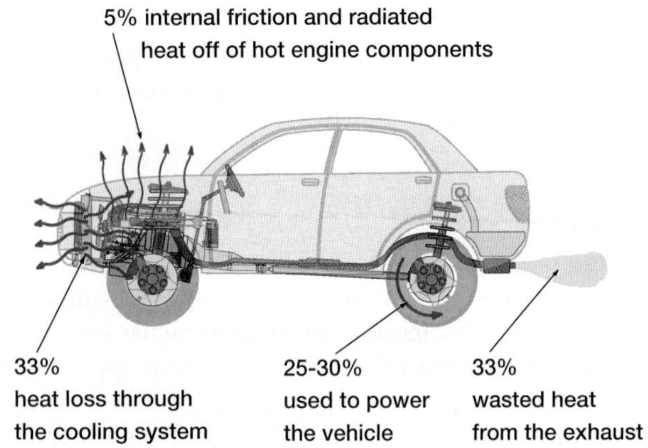

FIGURE 16-3 Heat loss in a gasoline engine.

kinetic energy (**FIGURE 16-3**). Some heat energy remains unused for powering the vehicle and is typically wasted in three ways:

- About 33% is wasted by being dumped straight out of the exhaust to the atmosphere (some of this wasted energy can be recovered by a turbocharger).
- About 33% is wasted by the cooling system, which prevents overheating of the engine components.
- About 5% is wasted by internal friction and from radiating off hot engine components straight to the atmosphere.

This leaves only about 25% to 30% of the original energy for powering the vehicle and its accessories. However, turbocharged and diesel vehicles can attain as much as about 35% efficiency. Turbochargers recover some of the energy lost from the exhaust, whereas diesel engines operate at higher compression ratios. These factors increase engine efficiency greatly.

Applied Science

AS-27: Heat: The technician can demonstrate an understanding of the effect of heat on automotive systems.

The extreme heat created by burning fuel in the internal combustion engine causes a rapid rise of pressure, which drives the pistons (see the Motive Power Types—Spark-Ignition (SI) Engines chapter). Some of the heat from the combustion process must be dissipated so that it does not overheat or melt engine parts. This is the job of the engine cooling system, which helps the engine to quickly come up to its specified operating temperature and to maintain that temperature in spite of ambient weather conditions or engine load. If the engine is not allowed to warm up properly, other systems will not operate properly. For example, the engine coolant temperature greatly affects the engine air-fuel mixture (fuel trim) and thus fuel economy and emissions. Operation of the evaporative emission system and other systems is also affected by engine temperature, so it is important that the cooling system be maintained so it operates correctly.

Applied Science

AS-30: Fusion/Vaporization: The technician can demonstrate an understanding of how heat causes a change in the state of matter.

Vaporization and fusion are two ways in which the application of heat can cause matter to change states. Vaporization refers to the transitional phase from liquid to vapor, whereas fusion refers to the transitional phase from solid to liquid. Both concepts have applications in the shop.

Pressurized cooling systems in motor vehicles are designed to prevent vaporization, in this case from boiling. Water boils at 212°F (100°C) at atmospheric pressure at sea level. When coolant in the cooling system is pressurized, its boiling point is raised, allowing the system to continue to cool the engine at temperatures over 212°F (100°C). Vaporization must be avoided because, unlike liquids, vapors do not absorb as much heat from engine components. Most cooling systems have pressure limits of 14 to 15 pounds per square inch (psi) (97 to 103 kilopascals [kPa]), effectively raising the boiling point of the coolant by approximately 45°F (25°C).

In contrast, water turns from a liquid to a solid (freezes) at 32°F (0°C). When water freezes, it expands approximately 10%. Pressure from ice is strong enough to break cast iron, which would cause leaks in the cooling system, causing extensive damage to the engine. Antifreeze is used to lower the freezing point of water to prevent freezing of the water and antifreeze mixture, called coolant.

▶ Principles of Engine Cooling

K16002

From the second law of thermodynamics, we know that heat always moves from hot to cold. These principles are put to use in the automotive cooling system to keep everything working properly. The automotive cooling system provides a means of transferring heat from the hot parts of the engine to ambient air. This is accomplished through a group of parts working together to circulate coolant throughout a sealed system, to carry heat away from those engine parts. Heat can then be released to outside air away from the vehicle or to air entering the passenger compartment, which can be used for comfort and safety purposes.

Manufacturers use various cooling system configurations, but the basic concept is to transfer excessive heat from hotter to cooler environments or materials. Regardless of the parts used by the manufacturer, the job is the same; maintain the ideal operating temperature. The ideal operating temperature for most engines is somewhere around 200°F (93°C), give or take 20°F (11°C), depending on the vintage of the vehicle. The purpose of the cooling system is to allow the engine to warm up to its optimum temperature as quickly as possible, and then transfer any excess heat energy away from the engine, thereby maintaining that ideal temperature and preventing the engine from overheating.

There are two types of cooling systems: liquid cooled and air cooled. Most automotive engines are liquid cooled. A liquid-cooled system uses coolant, a liquid that contains special antifreezing and anticorrosion chemicals mixed with water to transfer the heat throughout the system. In general terms, a water pump causes coolant to circulate through passages called **water jackets** in the engine block and cylinder head. The coolant picks up heat along the way and then gives up much of that heat when it passes through a radiator (**FIGURE 16-4**).

As air moves over and through the radiator, the heat energy is transferred from the coolant to the ambient air. The lower-temperature coolant is then returned to the engine to absorb more heat energy and continue the cycle.

In modern vehicles, radiators are low and wide to allow the hood to sit lower to the ground, for aerodynamics. Because of this design, modern vehicles use a water pump to circulate coolant through the system.

The circulation of the coolant is controlled by the thermostat, which opens and closes coolant flow from the engine to the radiator (**FIGURE 16-5**). When the engine is cold, coolant circulates through the engine (and heater core) only. Once the engine warms up, the thermostat opens and allows coolant from the engine to flow through the radiator. In this manner, a thermostat is used to regulate the coolant temperature. The thermostat initially blocks coolant flow to the radiator, keeping coolant circulating in the engine, where it heats up quickly and does not lose heat from the warming engine. Once operating temperature is reached, the thermostat starts to open and allows coolant to flow to the radiator. The thermostat does not normally open completely, but operates somewhere between its fully closed or fully open position to maintain an optimum engine operating temperature regardless of engine load.

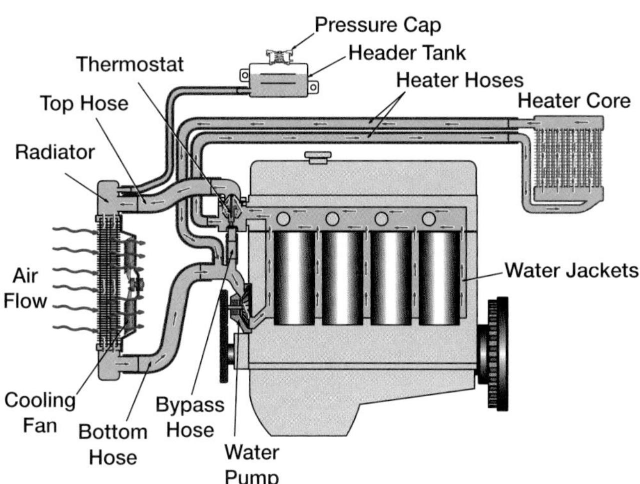

FIGURE 16-4 Simple cooling system.

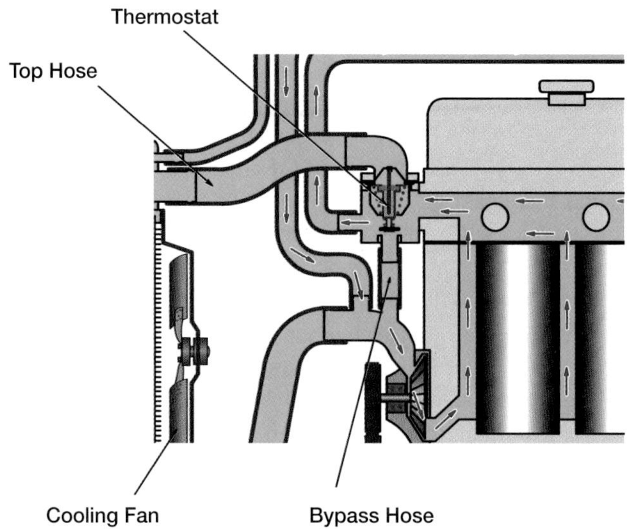

FIGURE 16-5 The thermostat controls the flow of coolant.

Some newer vehicles use a coolant heat storage system (CHSS). This system uses a vacuum-insulated container, similar to a thermos bottle. The storage container holds an amount of hot coolant and maintains the temperature for up to three days when the engine is shut down (**FIGURE 16-6**). When the vehicle is started the next time, a small electric water pump preheats the engine by circulating this hot coolant through the engine, which greatly reduces hydrocarbon exhaust emissions during start-up and shortens warm-up.

Air cooling is common on smaller internal combustion engines. Most engines use cooling fins around the cylinders and heads. Their design makes the exposed surface area as large as possible. Many air-cooled engines also use a fan to direct air through and over the cooling fins, which increases the cooling capacity. Air-cooled engines have fallen out of favor in passenger vehicles because they are harder to maintain at a stable temperature, thereby increasing emissions output. When air-cooled engines are used in automobiles, they are usually not exposed to the air; rather, the engine is housed in an enclosed engine bay. Still, for a vehicle moving at highway speed, airflow over the engine may be high enough to prevent overheating. At lower speeds or during idling, heat may build up and overheat the engine. One way to remove excess heat from the engine is to use a fan, along with shrouds and ducts, to direct air over or even between the cylinders. An engine that uses these components is called a "forced draft" air-cooled engine. Some air-cooled engines are "open draft" air-cooled, which requires the engine to be moving through the air to have sufficient airflow (**FIGURE 16-7**).

Vehicle Coolant

K16003

Coolant is a mixture of water and antifreeze, which is used to carry heat from the engine. If an engine did not have the heat removed from it, it would fail very quickly. Coolant absorbs the heat from the engine by convection. Because coolant contacts the hot metal directly, it is a very effective heat sink. The coolant is important for three reasons:

1. It prevents an engine from overheating while in use.
2. It keeps the engine from freezing while not in use in cold climates.
3. It prevents corrosion of the parts in the cooling system.

Water alone is by far the best and most plentiful coolant there is, as it can absorb a larger amount of heat than most other

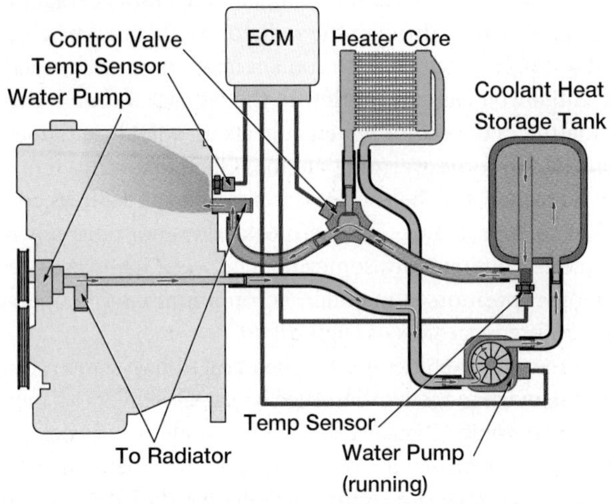

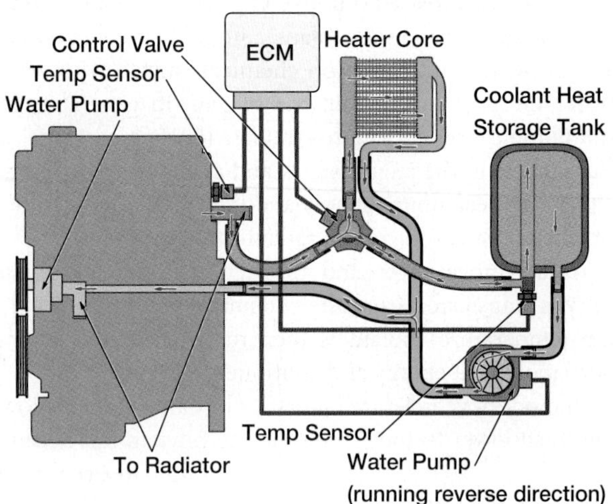

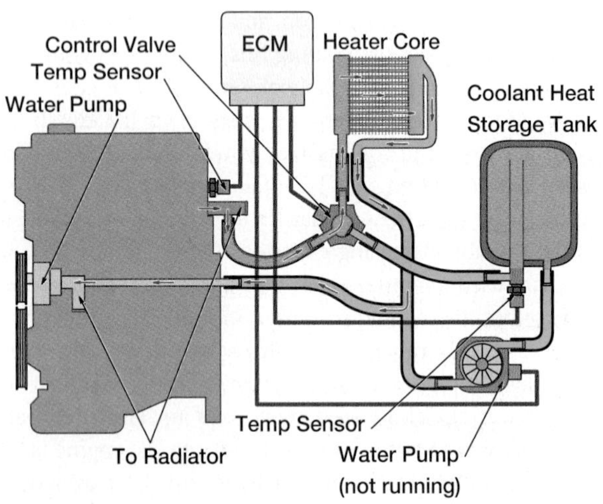

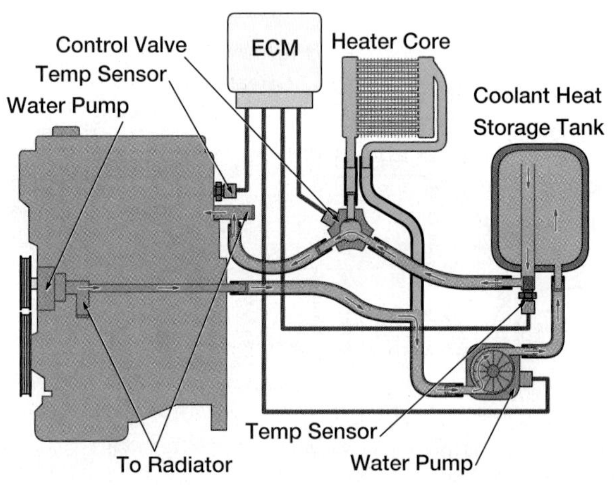

FIGURE 16-6 Toyota coolant heat storage system (CHSS).

FIGURE 16-7 A cylinder from an air-cooled Volkswagen engine. Note the cooling fins around the cylinder.

liquids. But water has some drawbacks. It freezes if its temperature drops below 32°F (0°C) (the temperature at which water becomes a solid). As water freezes, it expands approximately 10% as it becomes a solid. If it expands in the coolant passages inside the engine, these passages—typically made of cast aluminum or cast iron—cannot flex to allow expansion and will break. This causes blocks and heads to crack, which allow coolant to leak internally inside the engine, rendering the engine unrepairable in most cases.

Another thing to realize about using water alone as a coolant is that water is corrosive and causes metal to rust. Think about a piece of unpainted metal that is lying outside in the rain. Rust and corrosion build up quickly on that metal simply because of the reaction (oxidation) of the metal, water, and oxygen. Antifreeze prevents corrosion and rusting through anticorrosion additives (called corrosion inhibitors) mixed into the solution. Another important note on water is that water contains minerals and will potentially lead to excessive deposits even when added to antifreeze. Because of this, most manufacturers recommend using distilled water for cooling systems.

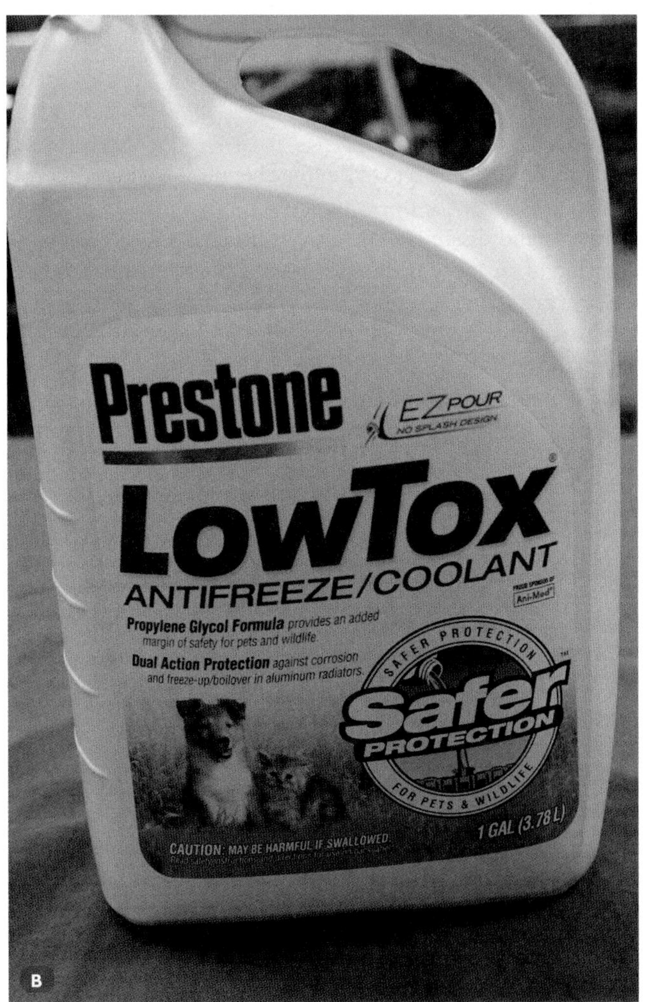

FIGURE 16-8 A. Ethylene glycol. **B.** Propylene glycol.

Antifreeze is made from one of two base chemicals—ethylene glycol or propylene glycol—plus a mixture of additives to protect against corrosion and foaming (**FIGURE 16-8**). Ethylene and propylene glycol may achieve a maximum very low freezing point of −70°F (−57°C) when mixed with the appropriate amount of water. **Ethylene glycol** is very toxic to humans and animals. **Propylene glycol** is not toxic and is used in nontoxic antifreezes. Both of these antifreezes actually freeze around 0°F (−18°C) if not mixed with water, so water is a necessary part of coolant. The freezing point of coolant will vary depending upon how much water is added to the antifreeze (**FIGURE 16-9**). Because antifreeze does not absorb heat as effectively as water, it should not be mixed at a ratio higher than 65% antifreeze and 35% water. Using a higher proportion of antifreeze will actually reduce the cooling quality of the mixture and raise the operating temperature of the engine.

The most common coolant mixture is a 50/50 balance of water and antifreeze, making it an ideal coolant for both hot and cold climates and providing adequate corrosion protection. Also, when antifreeze is added at a 50% mixture, the boiling point increases from 212°F (100°C) to around 228°F (109°C). As you can see, these are extremely beneficial characteristics of antifreeze as manufacturers continue to build engines that are more powerful, create more heat, and operate at higher temperatures.

Antifreeze can be purchased as full strength (100%) or as a 50/50 premix with water. Full-strength antifreeze that you buy from the dealer or parts store consists of three parts: glycol (around 96%), corrosion inhibitors and additives (around 2–3%), and water (around 2%). Glycol, as discussed previously, keeps the freezing point low and the boiling point high. Corrosion inhibitors and additives prevent corrosion and erosion, resist foaming, ensure coolant is compatible with cooling system component materials and hard water, resist sedimentation, and balance the acid to alkaline content of the antifreeze. Water is added to blend the inhibitors with the glycol.

Antifreeze is an amazing chemical that performs a monumental task in the operation of our vehicles. It works so well that it is often overlooked for maintenance by the customer. However, because the additives wear out and become less effective over time, coolant does need to be changed at recommended intervals. Doing so reduces the possibility of engine damage and

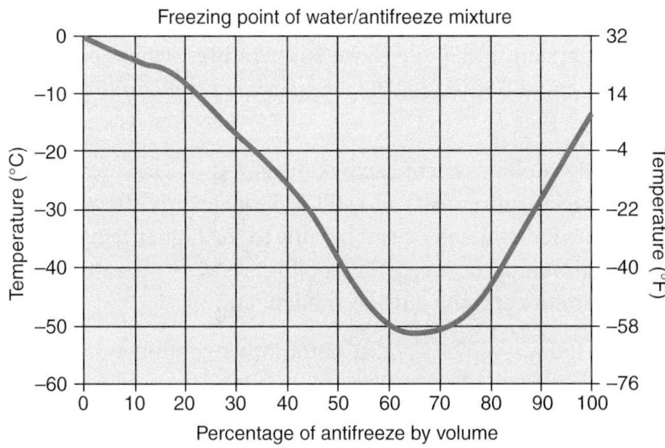

FIGURE 16-9 Freezing point of antifreeze and water solution.

failure over time. Likewise, lubrication enhancers, which keep the water pump seals functioning properly, wear out and need to be replaced.

Boiling Point and Pressure

K16004

The **boiling point** of a liquid is the temperature at which it begins to change from a liquid to a gas. Water at sea level atmospheric pressure of 14.7 psi (101.4 kPa or 1 atmosphere [atm]) boils at 212°F (100°C). Atmospheric pressure decreases as elevation is increased. Because atmospheric pressure is lower at higher elevations (such as in the mountains), the boiling temperature of a liquid in an unsealed system is lower. Think of it as *lower pressure = lower boiling point*. Conversely, raising the pressure has the opposite effect; it raises the boiling point of a liquid. Stated another way:

- Liquid in a vacuum = lower boiling point.
- Liquid under pressure = higher boiling point.

Thus, the boiling point of a liquid varies depending upon the surrounding environmental pressure.

This principle is used to enable water in the engine's cooling system to remain a liquid at temperatures well above the normal boiling point of 212°F (100°C). The pressurized coolant in the cooling system boils at a higher temperature than when the radiator pressure cap is removed (or if the system has a leak). This is because for every psi the pressure is raised on coolant, the boiling point is raised about 3°F. (1.67°C).

SAFETY TIP

Because the pressurized coolant in the cooling system boils at a higher temperature than when the radiator pressure cap is removed, you should ***never*** remove a radiator pressure cap on a hot engine. If the coolant is above 212°F (110°C), all of the coolant will turn to steam as soon as the pressure is released, pushing superheated steam out of the radiator, potentially scalding you.

Over the years, manufacturers intentionally have raised the operating temperature of their engines to increase efficiency and reduce emission. This required raising the pressure in the cooling system to handle those temperatures, which can be as high as about 250°F (121°C). Pressurizing the cooling system can mean that:

- The cooling system can be downsized (for less weight and space requirement) and still cool the engine effectively.
- The radiator can be smaller due to the higher temperature differential between the cooling system's operating temperature and the outside ambient air.

In today's vehicles, most automotive cooling systems are pressurized at 13–21 psi (89.6–144.8 kPa). However, some factory radiator caps are rated as high as 24 psi (165.5 kPa), and high-performance systems can go as high as 34 psi (234.4 kPa) (**TABLE 16-1**).

TABLE 16-1: Boiling Point of Water at Various Pressures

(Boiling point increases 3°F per each psi of increased pressure.)
13 psi : 212 + 39 = 251°F (121.6°C)
15 psi : 212 + 45 = 257°F (125°C)
19 psi : 212 + 57 = 269°F (131.6°C)
21 psi : 212 + 63 = 275°F (13°5C)
24 psi : 212 + 72 = 284°F (140°C)

TECHNICIAN TIP

The term "psig" is another way of saying the pressure that is read on a normal American pressure gauge. The "g" stands for gauge. So gauge pressure is the amount of pressure measured above 1 atm. A compound pressure gauge typically reads both above and below atmospheric pressure. It measures psig above zero, and inches of vacuum (or mm of mercury) below zero.

In contrast, the term "psia" means ***absolute pressure***, which does not include atmospheric pressure. In other words, a psia gauge would read 14.7 psig at sea level. If you want to know the absolute pressure (in psia), just add gauge pressure (psig) to atmospheric pressure (usually considered 14.7 psi at sea level).

Electrolysis

K16005

Electrolysis is the process of pulling chemicals (materials) apart by using electricity or by creating electricity through the use of chemicals and dissimilar metals. Electrolysis is used in manufacturing to create some metals, gases, and chemicals. You may have performed electrolysis in a science class by passing electrical current through water to break the hydrogen out of the water. Electrolysis is what takes place in the automotive battery. Two dissimilar metals are submerged in an acid solution, resulting in the reaction that is called electricity. Another common experiment in science class is the potato battery, made from two different types of metal stuck into the potato and used to power a digital clock or a lightbulb.

Electrolysis can occur in places where it is not desired and can have undesirable effects. For instance, in automotive cooling systems, electrolysis is possible when the coolant breaks down and becomes more acidic. Many types of metal are used in the engine, such as cast iron, aluminum, copper, and brass. Introducing an acid solution into a mix of metals produces electricity. When this electricity is produced, the movement of the electrons begins to erode away the metal in the system (**FIGURE 16-10**). Eventually, pinholes are created in the thinner, softer cooling system components (typically, the heater core or aluminum cylinder head). To combat electrolysis, customers should have regular scheduled maintenance performed on their vehicle, including flushing out the old coolant and replacing it with new coolant. Electricity can also appear in the cooling system due to faulty grounds on accessories or even the starter motor circuit. Electricity is known to follow the path of least resistance and will be

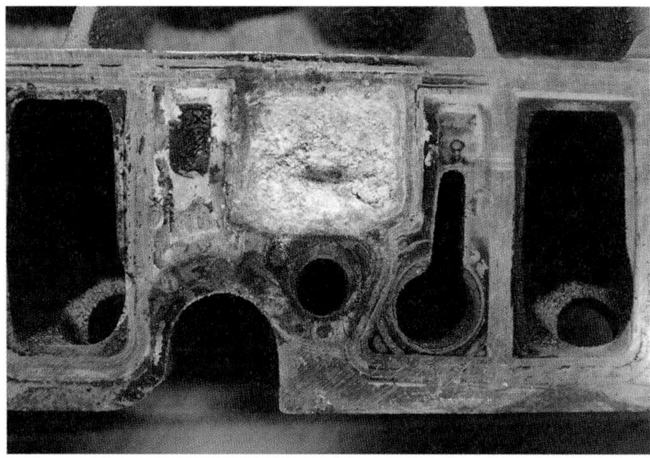

FIGURE 16-10 Erosion caused by electrolysis.

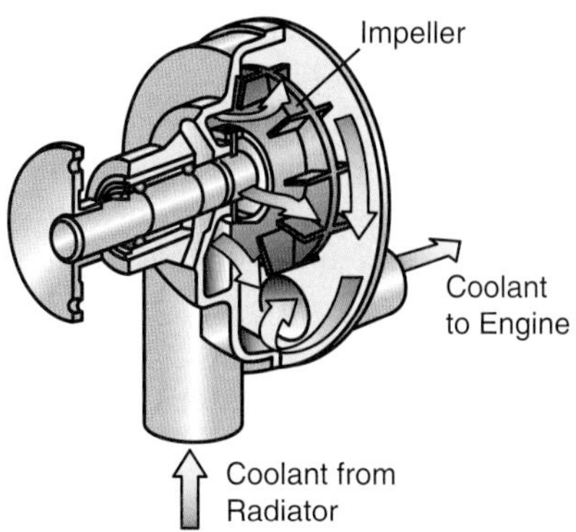

FIGURE 16-11 When coolant enters the center of this pump and the internal rotor spins, centrifugal force moves the liquid outward.

partially carried through the coolant, where it can erode metals, if that path is easier than its intended path. We discuss how to test for electrolysis later in this chapter and also revisit the topic of bad grounds.

Centrifugal Force Is Used to Circulate Coolant

K16006

Centrifugal force is a force pulling outward on a rotating body. For example, if you take a tennis ball, tie it to a string, and swing it around you, centrifugal force pulls the tennis ball outward, making the string taut. Another example of centrifugal force occurs when a vehicle turns a curve. Centrifugal force resists the turning of the vehicle and tries to keep the vehicle moving in a straight line, creating a sliding condition if centrifugal force is great enough. Centrifugal force can be useful in some cases, such as in the water pump. When coolant enters the center of the water pump and the internal rotor spins, centrifugal force moves the liquid outward toward the outlet (**FIGURE 16-11**). Centrifugal force pushes the coolant into the engine block and head passageways that surround the cylinders. The coolant then travels through the radiator to be cooled.

If a vehicle's specifications call for a total of 16 quarts of coolant, then a 50/50 mixture would consist of 8 quarts of antifreeze mixed with 8 quarts of distilled water. A 10-quart system would require 5 quarts of each, and so forth. The challenge in ensuring a 50/50 mix is the unknown mixture that remains in the cooling system, because not all of the coolant will drain out. If the system has been flushed with clean water and drained, then to end up with a 50/50 mix, the technician would first add the entire 50% of antifreeze and then top this off with clean water, leaving a 50/50 mix. If the coolant was flushed out with a 50/50 mix, then the technician can premix coolant and top the cooling system off with that. After topping off the cooling system, be sure to run the engine to circulate the coolant, and then check the mixture's freeze point with a hydrometer or refractometer.

Coolant can now be purchased premixed for convenience, but you will be paying for 50% water, plus the cost of shipping it, so it is usually less expensive to mix your own coolant.

Applied Science

AS-101: Proportion Mixtures: The technician can correctly mix fluids using proportions.

As discussed earlier, engine coolant is composed of water and antifreeze. The normally recommended mixture of these two liquids is 50/50—50% water and 50% antifreeze—which provides freeze protection to about –34°F (–37°C). Automotive cooling system capacity typically is given in quarts or liters, with smaller vehicles requiring perhaps 5 quarts (or liters) and larger SUVs with air conditioning requiring up to 20 or more quarts (or liters).

▶ TECHNICIAN TIP

Preventing overheating is one function of the cooling system. It also helps the engine reach its best operating temperature as soon as possible. Every engine has a temperature at which it operates best. If operated below or above this temperature, ignition and combustion problems may occur. Likewise, the thickness (viscosity) of the lubricating oil in the engine, and in other operating systems, may be affected. An MIL (malfunction indicator lamp) may illuminate if the thermostat has failed and does not allow the engine to reach its normal operating temperature within a specified time. Because of these situations, proper thermostat operation is a very important part of engine maintenance.

Coolant Flow—Normal and Reverse Flow

In a water-cooled cooling system with normal flow, the flow of coolant starts at the water pump. Cold coolant is moved through the engine by the water pump and starts to warm up as the engine begins to run. The pump relies on centrifugal force of the impeller (rotor) to force coolant through the cooling passageways of the engine. The coolant flows around the cylinders where combustion is taking place and picks up excess heat. It then moves upward through the cylinder head, where it passes over the top of the combustion chamber in the head (the hottest part of the engine) and flows around the valve guides, all the while picking up more heat as it passes those hotter surfaces.

From the head, it will move to the thermostat, which works like a trapdoor: If it is closed, the coolant will continue to circulate within the engine, flowing through a bypass hose or passage to move back down to the water pump, which was the starting point of its journey again. The thermostat is meanwhile sensing the coolant's temperature, and at a specified temperature it begins to open (**FIGURE 16-12**). As the thermostat opens, the coolant begins to flow through the radiator hose and to the radiator. Coolant enters the radiator's inlet tank and then the radiator core, where it flows through small tubes that have heat-dissipating fins on the outside. The tubes and fins transfer the coolant's heat to the outside air. As it cools, coolant moves to the cool side of the radiator to begin its journey back to the engine through the other radiator hose. The coolant's return flow to the engine is aided by the suction created by the water pump impeller.

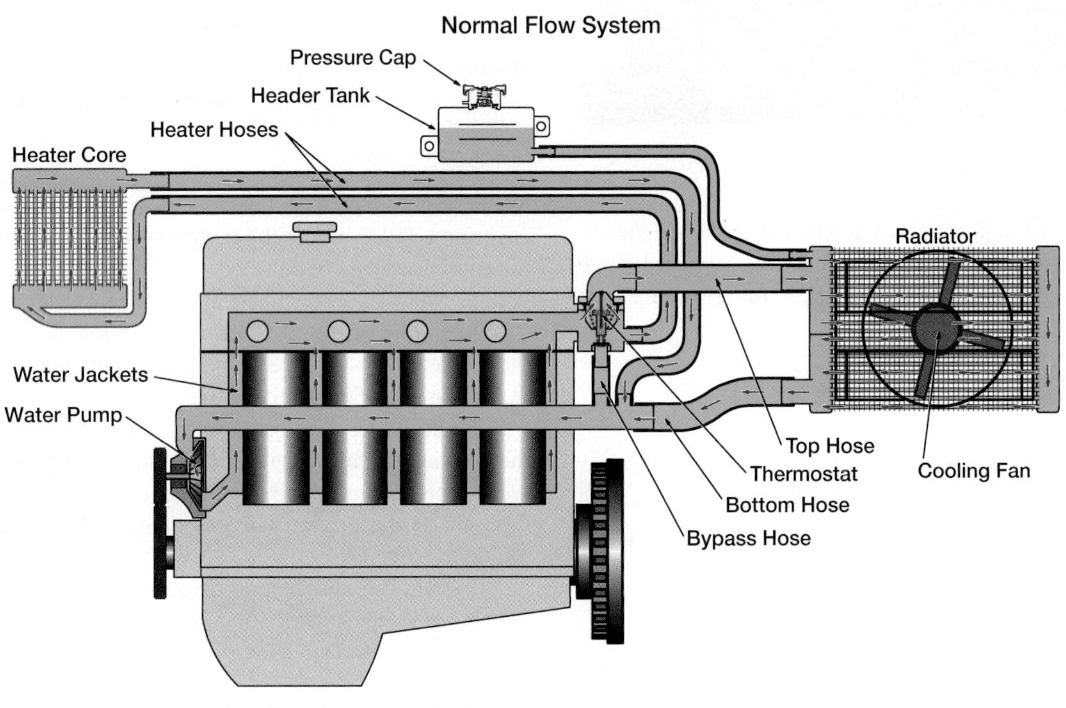

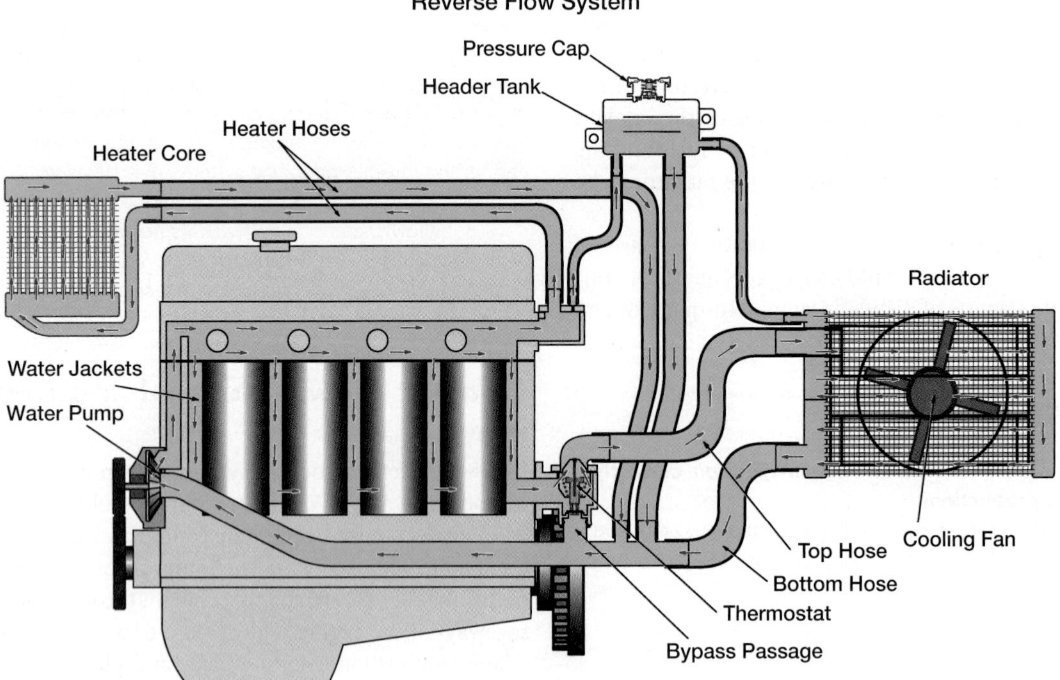

FIGURE 16-12 Coolant flow in a normal flow system and in a reverse flow system.

One problem with this normal-flow system was discovered on race cars, which develop more heat than a standard automobile. Because race car engines run hotter, and cylinder heads in general run hotter than engine blocks, the heads would get too hot and fail. The hottest part of any engine is the cylinder head because this is where the combustion chamber is located. High cylinder head temperatures tend to increase chances of detonation and failure of head gaskets. As a cylinder head heats more than the cylinder block, it will expand further and slide across the head gasket more. When the head does this repeatedly, the gasket is more likely to fail.

In the normal-flow cooling system, coolant goes through the engine block first and then moves to the hottest part—the cylinder head. Engineers found that if they changed the flow design, they could keep the cylinder head and block closer to the same temperature by pushing coolant through the head first, thus making the head, valves, and head gaskets last longer. Therefore, in some reverse-flow cooling systems, coolant flows to the cylinder heads first and then through the engine block.

In the typical reverse-flow design, coolant starts from the radiator and flows through the radiator outlet hose to the thermostat and then to the water pump. The thermostat is located between the inlet and outlet side of the engine to help regulate cold coolant more closely and to help reduce temperature shock to engine components. Coolant moves to the cylinder head first, where it can do the most cooling, then moves through the block and back to the radiator through the other radiator hose. Because the flow through the radiator is the same, it is simply the flow through the engine that has changed. This design differs only in the way coolant flows, head to block; all the components are generally the same. In other reverse-flow designs, coolant may flow from the thermostat in the lower radiator hose to the bottom of the radiator, then up to the top of the radiator and back to the cylinder heads, through the block, through the thermostat, and on to the lower radiator.

One problem with the reverse-flow design was easily fixed: As coolant moved through the hottest part of the engine, steam tended to form and get stuck at the head cooling passages, as this is the highest part of the engine. Because the engine will overheat if gas pockets get stuck in the water passageways, the solution was to drill holes in the head for steam to escape and to run a tube from the head back to a **surge tank**, where the steam turns back into coolant and is recycled through the system. The surge tank is discussed in greater detail later in the chapter.

Rotary Engine Cooling System

K16007

The cooling system of the rotary engine is not much different from the piston engine's cooling system. The engine itself is extremely different because it does not use pistons but instead uses a triangular part called a rotor that moves with combustion and turns the crankshaft. The cooling system uses a standard radiator, thermostat, and radiator hoses. Water flows from the radiator to the water pump, then into the **rotor housing**, which is essentially the engine block of the rotary engine; this is where the rotor moves. The rotor housing has passageways (water jackets) cast and machined into it, which go around the rotor housing and allow heat to be pulled from around the rotor, where combustion is taking place (**FIGURE 16-13**). After coolant moves through the water jackets, it finds its way to the thermostat, where it returns to the radiator.

FIGURE 16-13 Rotary engine coolant passages.

▶ Cooling System Components

K16008

The primary components of a vehicle cooling system are:

- *Coolant*: Coolant is the liquid used to prevent freezing, overheating, and corrosion of the engine.
- *Radiator*: The radiator is usually made of copper, brass, or aluminum tubes with copper, brass, aluminum, or plastic tanks on the sides or top for coolant to collect in. Air is drawn through the radiator to transfer heat energy to ambient air. The fins on the tubes of the radiator give more surface area for **heat dissipation**—the spreading of heat over a large area to ease heat transfer.
- *Thermostat*: The thermostat regulates coolant flow to the radiator. It opens at a predetermined temperature to allow

coolant flow to the radiator for cooling. It also enables the engine to reach operating temperature more quickly for reduced emissions and wear.

- *Recovery system*: The recovery system uses an **overflow tank** to catch any coolant that is released from the pressure cap when the coolant heats up. It works like a catch can.
- *Surge tank*: This pressurized tank is piped into the cooling system. Coolant constantly moves through it. It is used when the radiator is not the highest part of the cooling system. Remember, air collects at the highest point in the cooling system.
- *Water pump*: This pump is used to force coolant throughout the cooling system in order to transfer heat energy. The water pump is typically driven off the engine timing belt or accessory belt. On some engines, it is driven by the camshaft timing chain.
- *Cooling fan*: This fan forces air through the radiator for faster heat transfer. Cooling fans can be driven by a belt or by an electric or hydraulic motor. The fan can be controlled by viscous fluid or thermostatic sensors, switches, and relays.
- *Radiator hoses*: These hoses are used to connect the radiator to the water pump and engine. They are usually made of formed, nylon-reinforced rubber. Some radiator hoses use coiled wire inside them to prevent hose collapse as the cooling system temperature fluctuates.
- *Heater hoses*: These hoses connect the water pump and engine to the heater core. They carry heated coolant to the heater core to be used to heat the passenger compartment.
- *Drive belts*: These belts provide power to drive the water pump and other accessories on the front of the engine. Three types are used: V-belts, serpentine (also called multi-groove) belts, and toothed belts.
- *Temperature indicators*: Temperature indicators provide information to the operator about engine temperature. The temperature gauge indicates engine temperature continuously. A temperature warning indicator comes on only when the engine is overheating, to warn the operator that engine damage will occur if the vehicle is driven much farther.
- *Water* jackets: Water jackets are passages surrounding the cylinders and head on the engine where coolant can flow to pick up excess heat. They are sealed by replaceable core plugs.
- *Heater core*: The heater core is a small radiator used to provide heat to the passenger compartment from the hot coolant passing through it. The amount of heat can be controlled by a heater control valve.
- *Auxiliary coolers*: Auxiliary coolers are used to cool automatic transmission fluid, power steering fluid, exhaust gas recirculation (EGR) gases, and compressed intake air. Each of these coolers transmits heat either to the cooling system or directly to the atmosphere. Because there is such a wide variety of auxiliary coolers, refer to the manufacturer's service information for how to properly inspect the auxiliary cooler for leaks and proper operation.

Coolant Types

K16009

As mentioned earlier, water-cooled engines must be protected from freezing, boiling, and corrosion. Water absorbs a larger amount of heat than most other liquids. But it freezes at a relatively high temperature, and it is corrosive. Mixing antifreeze with water provides an adequate coolant solution by lowering the freezing point of water, raising the boiling point of water a bit, and providing anticorrosive properties.

There are several types of coolants available for use in the liquid-cooled automobile engine. The recommended coolant depends on the original equipment manufacturer's (OEM) recommendation. This is influenced by the metallurgy of the engine parts and the length of time or mileage that the manufacturer has determined between scheduled services.

It is important to note that brands and types of coolant (antifreeze) differ from one manufacturer to another. Some believe coolant can be identified according to its color, which may be anything from green or purple to yellow/gold, orange, blue, or pink. OEM cooling system designs and coolant recommendations have changed in recent years, so the color of coolant is no longer a reliable way to identify a particular type of antifreeze. Mixing types of antifreeze can cause a reaction that turns the chemicals in antifreeze to sludge that plugs up the passages in the system, including the radiator and heater core. Always read the container label and follow OEM coolant recommendations.

Most coolant types start with a base of ethylene glycol and add specific corrosion inhibitors, lubricants, and other additives, which all determine the type of coolant it is. Each coolant has antifoaming and antiscale additives. Maintaining the proper coolant acid/alkaline pH balance is also critical (which also determines when to perform coolant replacement).

Ethylene glycol is a toxic chemical that works very well as an antifreeze. Ethylene glycol mixes well with water and has a low viscosity, allowing it to circulate easily through the cooling system. Propylene glycol performs essentially the same as ethylene glycol except it is not as toxic. In fact, propylene glycol antifreeze is sold as a nontoxic coolant. The types of ethylene glycol or propylene glycol antifreeze/coolants available, based on the categories of corrosion inhibitors used in them, are as follows:

- Inorganic acid technology (IAT)
- Organic acid technology (OAT)
- Hybrid organic acid technology (HOAT)
- Poly organic acid technology (POAT)

The first category of coolant, IAT, is an early designed chemical formula that became available in the 1930s and was green in color. This coolant is still in use today. It contains phosphate and silicate as corrosion additives. Phosphate protects iron and steel parts, and silicate keeps aluminum from corroding. IAT coolant needs to be changed every two years or 24,000 miles (39,000 km), as the additives break down over that time.

The second category, OAT, is a longer-lasting coolant. Called extended-life coolant, it is designed to be changed at five years or 150,000 miles (241,000 km), a giant increase in the change interval from IAT coolant. OAT coolant was introduced in North America around 1994 and was intended for certain vehicles of that year and newer vehicles that followed. One example of OAT is Dex-Cool, the orange coolant used by GM. The anticorrosion additives in OAT coolant do not break down as quickly, which explains the longer service time. The primary additives in OAT coolant are organic acids, such as sebacate. These coolants do not use the additives used in IAT coolants.

The third category, HOAT, is a coolant that contains a mixture of inorganic and organic additives. This type of coolant can use silicate and organic acid; it is the best of both worlds of coolants. Some manufacturers found that without silicate, problems arose if the system was not properly serviced, such as oxidation of the coolant, leading to breakdown of the corrosion inhibitors and, ultimately, failure of engine parts and gaskets. Other tests indicate that silicates cause premature water pump failures. Less silicate is present in HOAT coolant than in IAT coolants, so this coolant is still considered extended life and has to be changed at five years or 150,000 miles (241,000 km). An example of this type of coolant is the yellow coolant used by Ford.

The fourth category, POAT, is a relatively new coolant that contains a proprietary blend of corrosion inhibitors. It is a very long-life coolant, providing up to seven years or 250,000 miles (402,000 km) of protection. It is claimed that it is compatible with most other types of coolant, but check the manufacturer's current service information.

Within the OAT or HOAT classifications, manufacturers may specify different corrosion additives and colored dyes. It is important that you service the vehicle with the coolant that is called for by the manufacturer. In most cases, you should never mix two or more types of coolants and/or refill with a type other than that originally used in a vehicle. Doing so will likely compromise the cooling system's service life and cause premature failure of the system.

Radiator

K16010

The **radiator** is located in the front part of the vehicle, under the hood, where maximum airflow can pass through it. Its actual location under the hood depends on the engine configuration, the available space, and the shape or line of the hood itself. The radiator consists of top and bottom tanks or side tanks and a core. The radiator core allows the coolant to pass through it in either a vertical downflow or a horizontal cross-flow direction. In addition, the radiator core serves as a good conductor of heat away from the engine.

The materials used in the radiator must be good heat conductors, such as brass, copper, or aluminum. Brass or copper is often used for tanks when combined with a brass or copper core. Modern vehicles often use plastic tanks combined with an aluminum core. This design saves weight and cost while still providing good heat transfer.

The core consists of a number of cooling tubes that carry coolant between the two tanks. The tubes can be in a horizontal (cross-flow) design or a vertical (down-flow) design (**FIGURE 16-14**). In a **down-flow radiator**, the cooling tubes run top to bottom, with the tanks on the top and bottom. In a **cross-flow radiator**, the cooling tubes are arranged horizontally, with one tank on each side. Because of this arrangement, the same amount of cooling area can be achieved without the need for a very tall radiator; instead, it is wide and short. This design feature allows the hood profile to be lower, allowing for better aerodynamics, which improves fuel economy and safety by increasing the driver's vision in front of the vehicle. The function of both types of radiator configurations is the same, which is to cool the coolant before it reenters the engine.

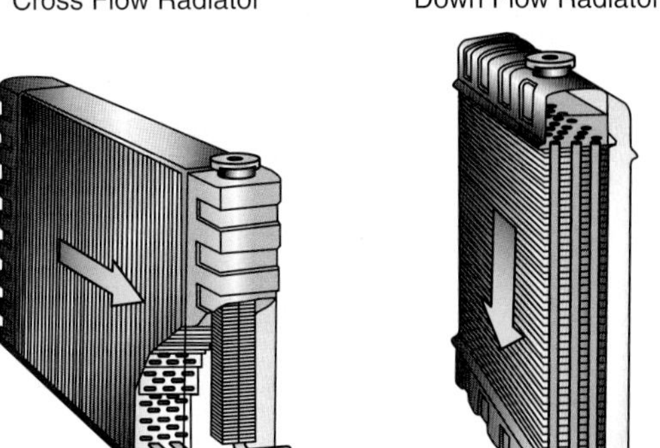

FIGURE 16-14 Cross-flow radiators have cooling tubes mounted horizontally, whereas down-flow radiators have tubes mounted vertically, requiring a taller hood profile.

In both the cross-flow and the down-flow design, the core is built of the same components. Although having only cooling tubes exposed to airflow would cool the coolant somewhat, radiators are designed with heat-dissipating fins to allow more surface area to dissipate heat, which is then released to the air (**FIGURE 16-15**). Coolant touches tube walls, and fins touch the tubes, so heat is removed from the coolant by conduction into the cooling fins, then by convection at the surface of the fins. Air rushing by the fins carries the heat away. Liquid coolant emerges cooler at the outlet of the radiator, as the coolant was able to give up much of its heat to the atmosphere.

Most radiators come with a method of draining coolant from them. This drain plug can take several forms. Older vehicles used a petcock (**FIGURE 16-16**). This usually has a T-shaped metal knob that needs to be turned "IN" to open it up. Many a new technician has tried to turn it "OUT" only to break it off, creating a problem that is not easy to fix. Some drain plugs are plastic quarter-turn valves. If turned too far, these drain plugs will break off. These drain plugs cannot always be found as a replacement part, and replacement of the entire radiator may be necessary. The last type of common drain valves are simple plastic threaded plugs. They typically are only tightened by finger and use a rubber sealing washer. Remember that radiator petcocks and drain plugs can be sources of coolant leaks, so don't overlook them.

FIGURE 16-15 Radiator fins help dissipate heat.

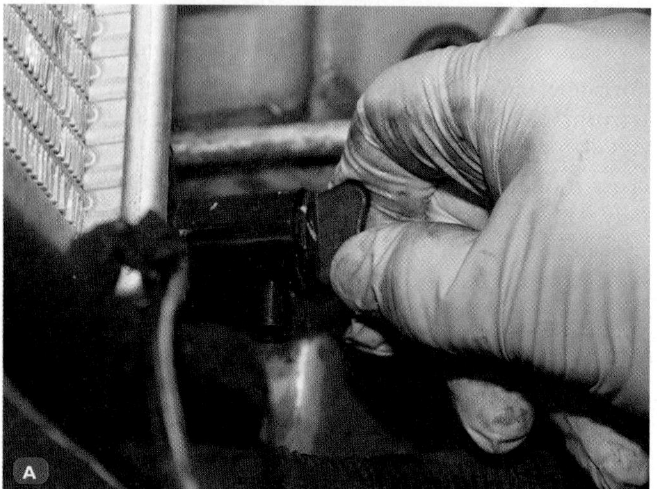

FIGURE 16-16 Radiator drain plugs. **A.** Petcock. **B.** Plastic drain plug.

Radiator Shrouding

In the interest of fuel economy (less drag coefficient), vehicle hood lines have become more streamlined, and engine compartments smaller and more crowded. These changes in turn make it more difficult for ram air to flow through the radiator and around the engine. Airflow is therefore greatly reliant on shrouding above, behind, and below the radiator to direct ram air to the radiator and from it (**FIGURE 16-17**). Shrouds are also used around the radiator fan to help draw air through the entire radiator core, not just in front of the fan blades. Also, the fan shroud prevents air from simply circulating around the tips of the fan blades instead of being drawn through the radiator. If the shrouding is removed and not replaced, such as after service has been performed, the engine will not be cooled as efficiently as it should be and will likely overheat, especially on hot days.

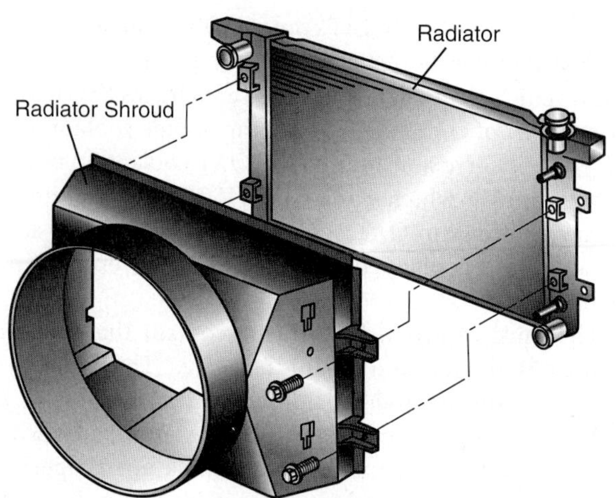

FIGURE 16-17 Radiator shrouding.

Applied Science

AS-31: Insulation: The technician can explain the role of insulation in maintaining temperatures.

Certain parts of the vehicle's heating and cooling system, as well as the passenger compartment, must be insulated in order to perform properly. Air-conditioning refrigerant lines, heating ducts, and other areas are insulated to prevent the loss of, or absorption of, heat. Even the hood itself may be insulated, not only to reduce the amount of heat radiated from the engine but also to quiet engine noise from being heard outside the vehicle or inside the passenger compartment.

Applied Science

AS-32: Radiation: The technician can demonstrate an understanding of heat transfer that involves infrared rays.

At one end of the light spectrum, just beyond visibility, lies the infrared portion of light. Infrared is a form of heat energy, and it can be detected by special instruments. Thermography involves the use of heat-sensing instruments to detect heat sources. The amount of radiation emitted by an object increases with temperature; therefore, thermography allows one to see variations in temperature. In the automotive trade, an infrared "temp gun" is pointed at objects to determine their temperature. This method is useful for detecting cylinders that are contributing less power to the engine, finding leaks in air-conditioning systems, etc. On the dashboard of many vehicles lies an infrared-sensing "sun-load sensor," which helps automatic HVAC systems to regulate cabin temperature and enables automatic headlight dimming.

Radiator Pressure Cap

If coolant boils, it can be as damaging to an engine as having it freeze. Boiling coolant changes the coolant from a liquid into a gas in the engine water jackets. No liquid is left in contact with the cylinder walls or head, causing heat transfer to stop or slow to the point of overheating. Without cooling, the pistons will seize as temperatures soar with each combustion cycle, and engine meltdown takes place.

One way to prevent coolant from boiling is to use a radiator pressure cap. As coolant temperature rises, the coolant expands, and pressure in the cooling system rises. The increased pressure raises the boiling point of the coolant. Engine temperature keeps rising, and the coolant expands further. Pressure builds against a spring-loaded valve in the radiator cap until, at a preset pressure, the valve opens and releases a small amount of coolant. When this valve opens, the coolant leaves the radiator through the overflow tube to the overflow container (**FIGURE 16-18**). In past designs, the overflow tube would drop the coolant onto the ground, which was an environmental hazard. By releasing coolant into a coolant overflow container, the coolant can be pulled back into the system once it cools.

Coolant loses temperature in the radiator when the engine is off. When coolant cools, it contracts, and with this contraction, it creates a vacuum (low pressure) in the radiator. A vacuum valve is located in the radiator cap that will open and allow coolant to be pulled back into the radiator from the overflow container. Because of this design, no coolant is lost as with the older systems that dropped coolant onto the ground. The vacuum valve in the cap also stops low pressure from developing in the radiator and causing collapsed radiator hoses as a result of atmospheric pressure on the outside of the hose crushing the hose against the vacuum in the system.

Surge Tank

K16011

Some vehicles are equipped with a surge tank. The surge tank has coolant constantly running through it and is located higher than the top of the radiator. If the radiator sits lower than any other part in the cooling system, then gas will collect in whatever component is highest. On some vehicles, this may be the heater core. To solve this problem, engineers have installed a surge tank and placed it so it is the highest component in the system. Thus, any gas in the system will make its way to this tank, ensuring that only liquid coolant circulates through the system. The surge tank has at least one line in and one line out. It is usually made of hardened plastic, which allows for a visual checking of the fluid level through the plastic. This tank is usually where the cooling system is filled or topped off with coolant. In many vehicles with a surge tank, the pressure cap is mounted on the surge tank instead of the radiator (**FIGURE 16-19**).

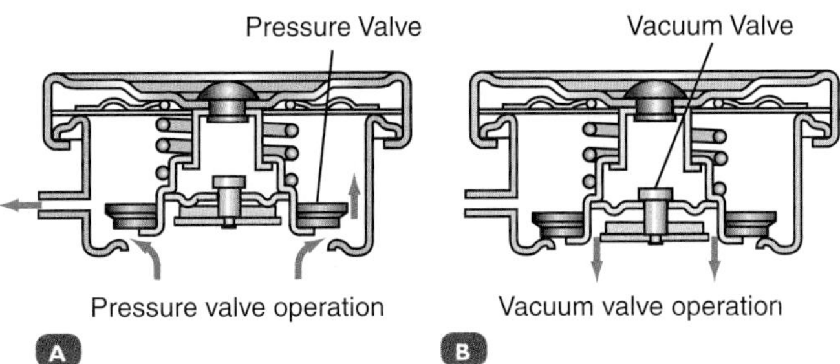

FIGURE 16-18 A. Pressure valve being forced open. **B.** Vacuum valve being pulled open.

Recovery System

A coolant recovery system maintains coolant in the system at all times. The recovery system consists of an overflow bottle, a sealed radiator pressure cap, and a small hose connecting the bottle to the radiator neck (**FIGURE 16-20**). As engine temperatures rise, the coolant expands. Pressure builds against a valve in the radiator cap until, at a preset pressure, the valve opens. Hot coolant flows out of the radiator, through the connecting hose, and into an overflow bottle. As the engine cools, the coolant contracts, and pressure in the cooling system drops below atmospheric pressure. Atmospheric pressure in the overflow bottle opens the vacuum valve in the radiator cap, and overflow coolant flows back into the radiator. Like water, air contains oxygen, which reacts with metals to form corrosion. With use of a recovery system, no coolant is lost, and excess air is kept out of the system.

Thermostat and Housing

K16012

The **thermostat** is located under the thermostat housing. The thermostat regulates the flow of coolant, allowing coolant to flow from the engine to the radiator when the engine is running at its operating temperature. The thermostat prevents coolant from flowing to the radiator when the engine is cold, to allow the engine to warm up more rapidly, thus reducing engine wear and emissions.

The thermostat is a spring-loaded valve that is controlled by a wax pellet located inside the valve (**FIGURE 16-21**). As the temperature of the coolant rises, the wax pellet melts and expands, forcing the spring-loaded valve open at a preset temperature. As the valve opens, coolant is allowed to flow through it. The thermostat works like a door to control movement of coolant. When the engine is cold, the door is closed; when hot, the door opens. Some dashboard temperature gauges show a slight swing of engine temperatures as the thermostat cycles slightly toward open or closed.

Some engines are designed such that the coolant bypass passage is directly under the thermostat. In those situations, the thermostat may have a flat disc attached to the bottom of it, which moves along with the thermostat. When the thermostat

FIGURE 16-19 A surge tank removes air and gases from the coolant. This is where you normally fill the cooling system.

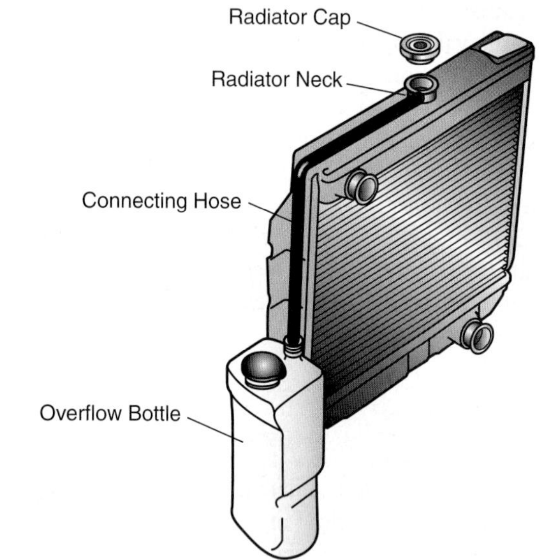

FIGURE 16-20 A coolant recovery system.

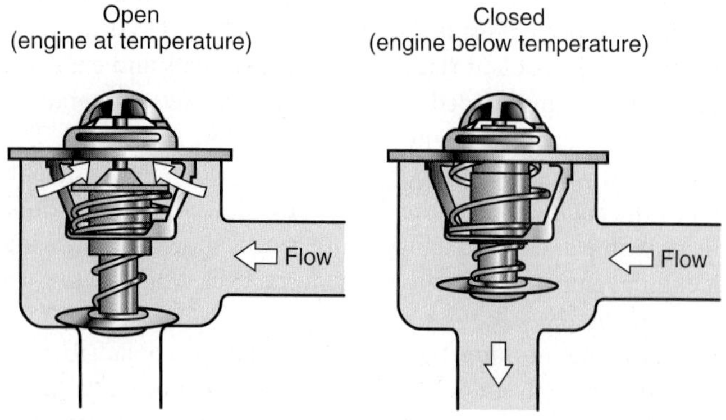

FIGURE 16-21 The thermostat has a moving valve that is controlled by a wax pellet. When the wax is cool, the valve stays closed; as temperature increases, the wax melts and forces the valve open.

fully opens, the flat disc blocks off the bypass passage so that all coolant must flow through the radiator. Yet, when the thermostat partially closes, the bypass passage will be partially open. This helps give more effective cooling when the thermostat is fully open.

Most thermostats have a small hole on one side of the thermostat valve that allows any air in the system to move past the closed thermostat when the valve is closed (**FIGURE 16-22**). This is especially helpful when the cooling system has been drained and is being filled. The hole usually contains a little pin, called a jiggle valve or jiggle pin, to help break the surface tension of the coolant and allow any air to flow slowly through the hole. Air trapped in the cooling system is thus able to slowly find its way to the uppermost part of the cooling system, where it can be bled. When installing the thermostat, the jiggle valve should be in the uppermost position so that as much air as possible can be bled from the system.

The thermostat and housing are normally located on the outlet side of the coolant flow from the engine. However, on some engines they are located on the inlet side of the engine. The thermostat is identified by being located in the housing connected to the inlet radiator hose. The reason for installing the thermostat on the inlet side is to better control the amount of cold water that rushes into the engine, which can create a temperature shock to the engine.

Some vehicles include a manual bleed valve on the thermostat housing or within a high part of the cooling system (**FIGURE 16-23**). After the vehicle cooling system has been serviced and refilled, the technician should carefully open the bleed valve to vent any trapped air to the atmosphere. It is good practice to bleed the system again once the engine has warmed up.

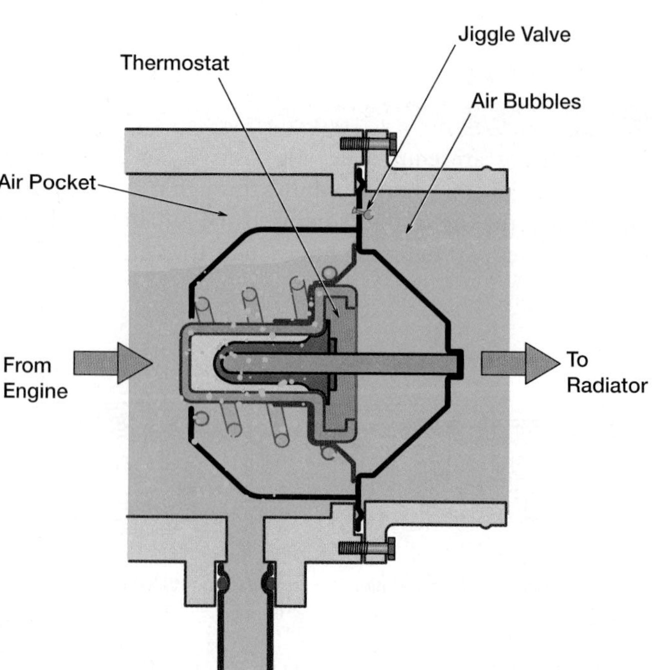

FIGURE 16-22 Thermostat with a jiggle valve for bleeding air from the system when refilled.

FIGURE 16-23 Coolant bleed valve.

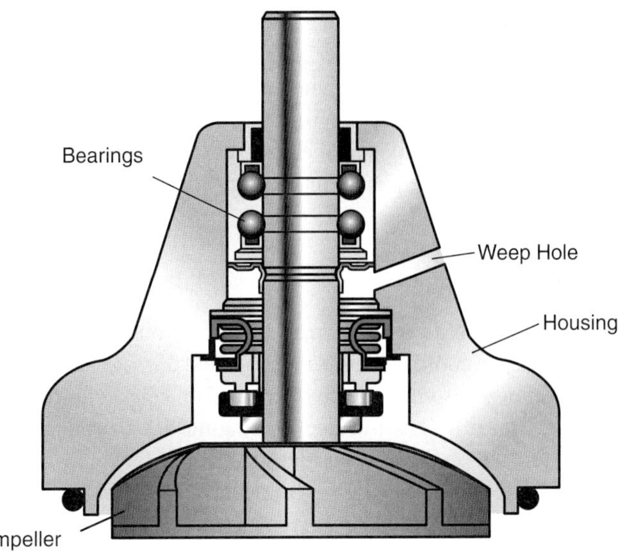

FIGURE 16-24 Cutaway view of water pump and weep hole.

Water Pump

K16013

The water pump is usually belt driven from a pulley on the front of the crankshaft. The engine drives the water pump using either an accessory V-belt or the timing belt. Some newer vehicles use an electrically driven water pump. Internally, the water pump has fanlike blades on an impeller or rotor, which is turned by a shaft connected to the pulley. The shaft rides on a heavy-duty double-row ball bearing, for long life and to withstand belt tension. The shaft is sealed where it enters the pump chamber. Most water pumps have a small weep hole that sits between the seal and the bearing and vents any coolant that leaks past the seal, to be drained to the outside of the engine (**FIGURE 16-24**). Checking this hole for signs of coolant leakage is part of the process of inspecting the engine for leaks.

The water pump is usually located at the front of the engine block; a hose typically connects it to the output of the radiator, where the relatively cool coolant emerges. Coolant enters the center of the pump. As the impeller rotates, it catches the coolant and flings it outward with centrifugal force. This type of water pump is called a centrifugal pump, meaning that it creates movement of the coolant due to centrifugal force, but does not cause much buildup of pressure. Coolant is driven through the outlet into the water jackets of the engine block and to the cylinder heads. Coolant can be directed to critical hot spots, such as around the exhaust ports in the cylinder head, to stop localized overheating. The cylinder head gasket has holes of various sizes to help control how much coolant flows to these and other locations of the head. If the head gasket is not properly installed, cylinder head damage may result from localized overheating.

Replacing a water pump is a fairly common task that used to be performed mostly as a result of noise or leak issues on accessory-belt-driven water pumps (**FIGURE 16-25**). But now, with many water pumps being driven by timing belts, it is common to replace the water pump along with the timing belt at the belt's recommended replacement schedule (**FIGURE 16-26**). When changing this type of pump, the timing belt has to be removed, so following the manufacturer's

FIGURE 16-25 Replacing a water pump.

FIGURE 16-26 Water pumps are typically driven by timing belts.

procedure is necessary to prevent damage to the valves and pistons. Once the drive belt is removed from either style of water pump, the removal of the pump involves the unbolting of the pump, carefully prying it off the mating surface, cleaning the mounting surface, and rebolting it back in place with the appropriate gasket or O-ring.

Cooling Fan

K16014

Cooling fans are used to provide airflow through the radiator core for engine cooling. This is most needed during slow driving or stop-and-go driving. There are two main categories of cooling fans: engine driven and electric. Engine-driven cooling fans may be located on the water pump shaft or in a few cases may be attached directly to the engine crankshaft. Most vehicles today use one or more electrically driven engine cooling fans, which are mounted directly to the radiator. On a fan that is driven by the engine, engine horsepower is needed to directly drive the fan, requiring extra fuel even when the fan is not needed. Such fans are also noisy and dangerous to work around.

It takes a fair amount of energy to turn a fan, which ultimately comes from the crankshaft, consuming energy that could be used to drive the vehicle down the road. Yet, the fan is not needed during cold engine operation. It is also not normally needed when the vehicle is traveling above about 35 mph (miles per hour; 56 kph [km per hour]) as the airflow is strong enough to cool the radiator without using the fan. For years, vehicle manufacturers have sought ways to reduce the energy the fan uses. One type of fan design (called a flex fan) uses flexible steel or plastic blades that straighten out and lessen their **pitch** as engine speed increases (**FIGURE 16-27**). This design increases fuel economy and reduces noise.

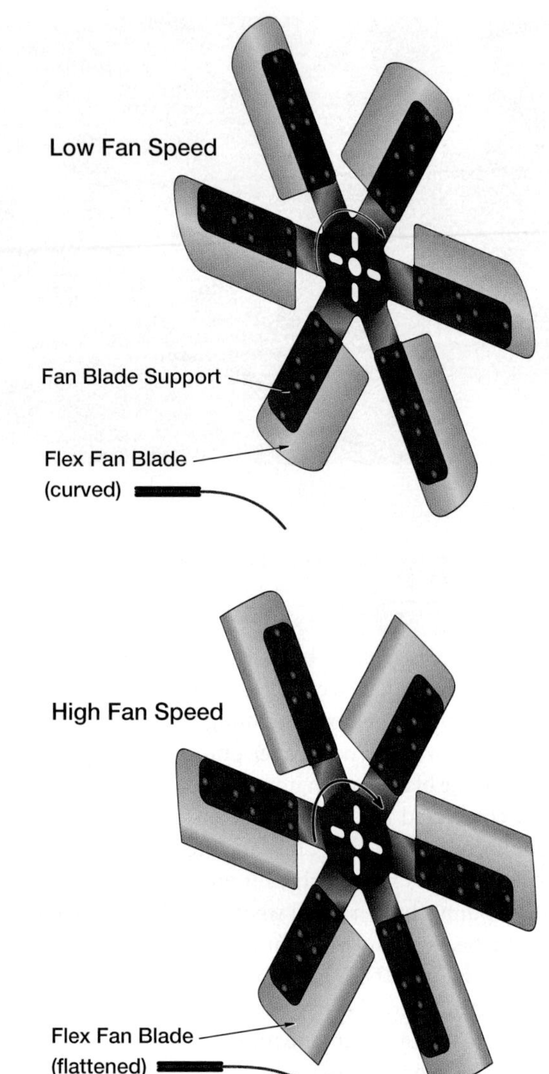

FIGURE 16-27 Flex fan. Blades flatten out at higher rpm.

Another type of engine-driven fan uses a **viscous coupler** to connect the water pump pulley to the cooling fan. The viscous coupler is called a fan clutch, as it can engage and disengage the fan from the pulley (**FIGURE 16-28**). The fan clutch typically uses two discs that have closely fitted interwoven rings and grooves. When viscous silicone oil is allowed to fill the small spaces between the rings and the grooves, it transmits torque from one disc to the other and is used to transmit torque across the two halves of the hub. A bimetallic spring on the front moves as air temperature from the radiator changes. This spring is attached to a valve that turns and allows more silicone oil to flow into the coupler, thereby transmitting more of the pulley's speed to the fan to move more air. As air temperature decreases from air flowing through the radiator, the bimetallic spring turns the valve, and the silicone oil moves out of the coupler back to the reservoir located in the clutch body. This causes the fan to slow and move less air. This type of fan is driven at all times by the accessory belt. The benefit of using a viscous clutch is to increase fuel economy by being able to cycle on and off.

A variation of the clutch fan is the solenoid-controlled fan clutch, which operates in the same manner as the thermostatic fan clutch. The clutch uses oil to control the speed of the fan in relation to engine speed (allows slip to slow fan down) and temperature of the air moving across the radiator. The only difference in the operation is that the bimetallic spring is removed, and an electric solenoid is installed on the front of the clutch. This solenoid is controlled by the powertrain control module (PCM), based on engine coolant temperature. As temperature increases, more oil is allowed to flow to the viscous coupler, and fan speed increases. As temperature decreases, oil is directed back to the reservoir in the body of the clutch, and fan speed decreases. On this type of fan, a feedback sensor provides actual fan speed, or rpm information, back to the PCM.

As fan designs evolved to increase efficiency, and as more vehicles moved to front-wheel drive, which turns the front of the engine away from the radiator, the electric fan was introduced (**FIGURE 16-29**). Now it is by far the most common and simple type of cooling fan. It can be turned on and off easily whenever it is needed, and it can operate at full speed even though the engine is idling. This versatility makes it very efficient. It only runs when the engine is above the ideal operating temperature. In addition, because it can run at full speed independent of the engine, it can move plenty of air to cool the engine effectively.

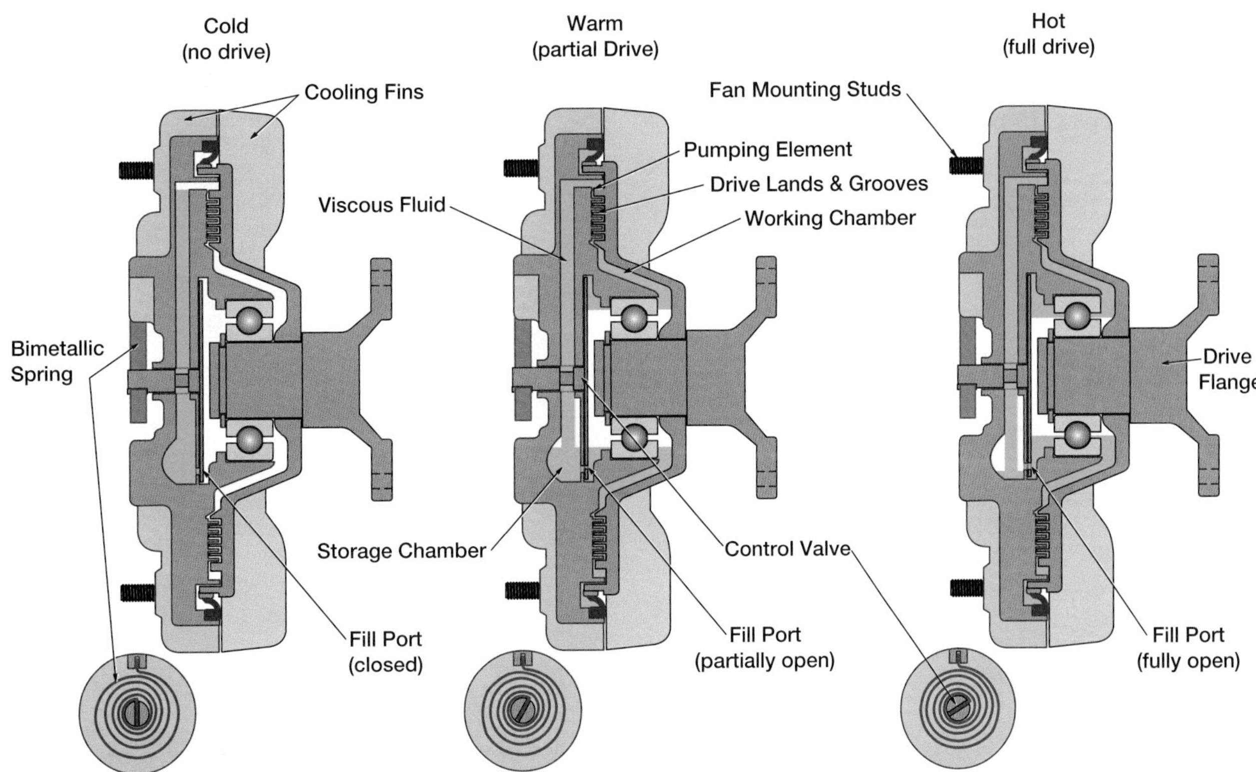

FIGURE 16-28 Viscous fan clutch.

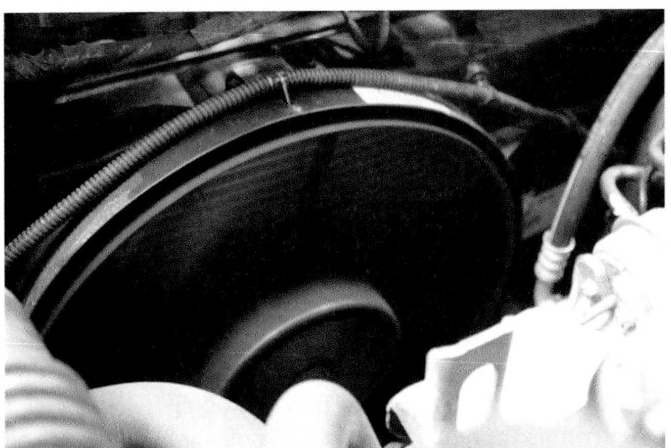

FIGURE 16-29 Electric cooling fan.

Electric fans use an electric motor to turn an attached fan. In most cases, the fan blades are made of plastic, making them safer than metal blades. The blades can be designed to either push or pull air through the radiator, so mounting options are increased. The electric fan is controlled by one of two methods:

- A control module, such as the PCM, is used to energize a fan relay to turn the fan on and off.
- A thermo control switch to either directly turn the fan on and of, or through a fan relay.

The PCM knows the temperature of the engine by means of the coolant temperature sensor. The PCM can then supply either power or ground to the fan relay, energizing the relay and fan.

The **thermo-control switch** is a temperature-sensitive switch that is mounted into a coolant passage on the engine, or into the radiator. When engine temperature gets hot enough (say, 215°F [102°C]), the thermo-control switch closes and either sends power directly to the fan or causes a relay to be activated, which turns on the fan. Once the engine coolant temperature cools back down, the switch opens and the fan stops.

Thermo-control switches often operate on the bimetallic strip principle. These consist of two different metals or alloys laminated back to back. As different metals and alloys heat and cool, they expand and contract at different rates. That means that if two different metals are joined, and then heated, the greater expansion of one forces the whole strip to flex into a curved shape. As the strip changes shape, it can be designed to complete an electric circuit by closing a switch, which turns the fan on (**FIGURE 16-30**).

With an electric fan, the electricity to run it comes from the alternator, which comes from the engine, which comes from the gasoline fuel. Because an electric fan typically needs to operate only part of the time, fuel is saved whenever it is off. Some manufacturers use multiple fans and control them separately; others use multispeed fans to provide only enough fan operation to keep the coolant at the proper temperature.

Another type of cooling fan is the hydraulically operated fan. In many cases, it uses power steering fluid from the power steering pump to power the fan (**FIGURE 16-31**). Because the power steering pump can create substantial power, hydraulically driven fans can be used to draw a large amount of air through the radiator. They are sometimes used on vehicles with heavy trailer towing capacities, as well as on ordinary vehicles. The system

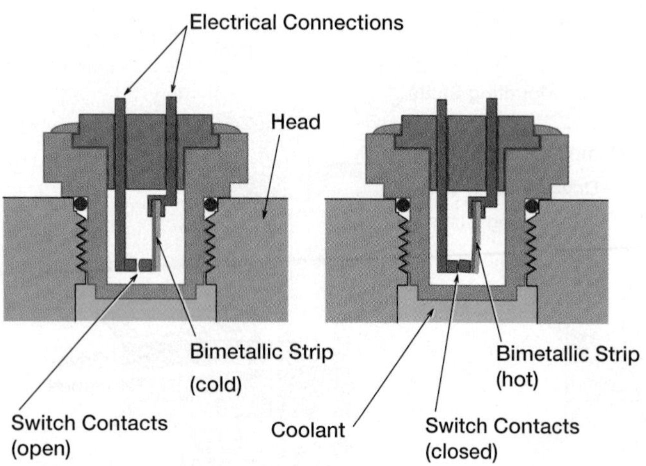

FIGURE 16-30 Bimetallic strip.

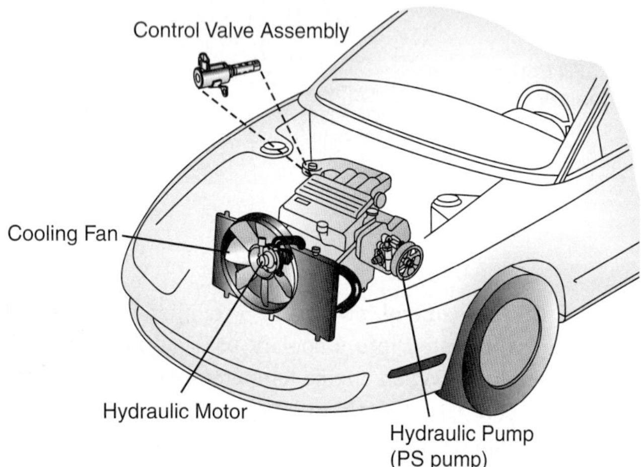

FIGURE 16-31 Hydraulically operated cooling fan.

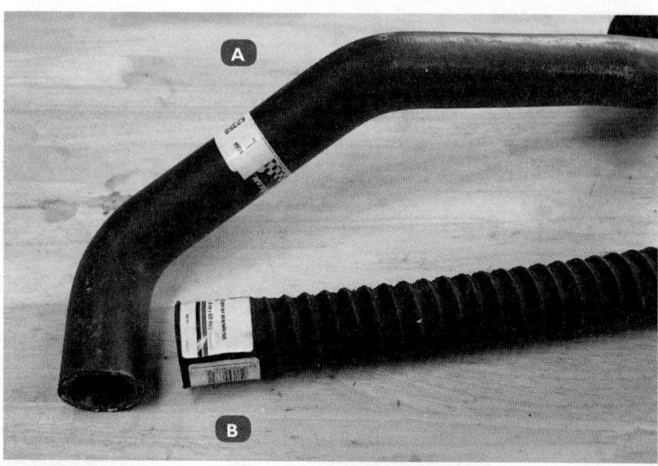

FIGURE 16-32 Radiator hoses. **A.** Molded. **B.** Flexible.

FIGURE 16-33 Hose clamps.

consists of the power steering pump, a fluid control device, the hydraulic fan motor, and high-pressure connecting hoses. The fan is typically controlled by a pulse-width-modulated solenoid valve. The solenoid controls how much hydraulic fluid is directed to the fan motor. That way the PCM can vary the signal to the solenoid valve, which controls the amount of hydraulic fluid to the fan motor, which determines how fast it spins. One benefit of the hydraulically controlled fan is that it can be operated at near full speed and force, even at idle, similar to an electric fan. Yet it can tap into more engine power than the electric fan, so it is more heavy duty.

Radiator Hoses

K16015

On most vehicles there are two radiator hoses: the upper hose and the lower hose, also called the inlet hose and the outlet hose. **Radiator hoses** are rubber hoses that are subject to high pressure and temperature; they are therefore reinforced with a layer of fabric, typically nylon, to give them strength and prevent them from ballooning, and yet still be flexible. They are often molded into a special shape to suit the particular make and model of vehicle (**FIGURE 16-32**). Some radiator hoses, especially lower hoses, have a spiral wire inside to keep the hose from collapsing during heavy acceleration when the water pump is drawing a lot of water from the radiator. The reason radiator hoses need to be flexible is that engines are mounted on flexible engine mounts to reduce noise and vibration. Because the radiator is mounted to the vehicle body, which does not move with the engine, the hoses must be flexible.

The top radiator hose is typically attached to the thermostat housing, which allows the heated coolant to enter the inlet side of the radiator. The bottom or lower radiator hose is connected between the outlet of the radiator and the inlet of the water pump. The radiator hoses are held in position by clamps. These can be spring clamps, wire wound clamps, or worm drive clamps (**FIGURE 16-33**). Radiator hoses deteriorate over time and use. They can also be damaged by oil or fuel leaking on them. Thus, they need to be inspected and changed periodically. Many vehicle manufacturers recommend radiator and heater hose replacement approximately every four to five years or 48,000 to 60,000 miles (77,000 to 97,000 km). When servicing hoses, be sure to reinstall the clamps correctly or the seal between the hose and the component will leak. Every component has a raised ridge built into it. The hose clamp has to clamp on the inside of this ridge, but not on top of it. If the hose clamp is on top of the ridge, the hose is likely to pop off once the cooling system becomes pressurized (**FIGURE 16-34**).

Heater Hoses

The heater hoses carry a smaller volume of coolant than do the radiator hoses, so they are smaller in diameter than radiator hoses. Some heater coolant hoses have special shapes and must be ordered specially for the vehicle application being serviced. Other heater hoses are straight and can be replaced by heater hose that is supplied on a roll. The construction of the heater hose is the same as the radiator hose, with a reinforcing material embedded into it.

There are two hoses for the heater core: one inlet hose and one outlet hose. The heater hose directs hot coolant to the heater core to provide heat to inside the passenger compartment. The heater hoses are sealed and retained by the use of a hose clamp. As with radiator hoses, be sure to install hose clamps correctly.

Some coolant hoses are made of silicone, which is designed to withstand heat better and last longer than the standard rubber hose. Be sure to inspect hoses whenever you are servicing the customer's vehicle.

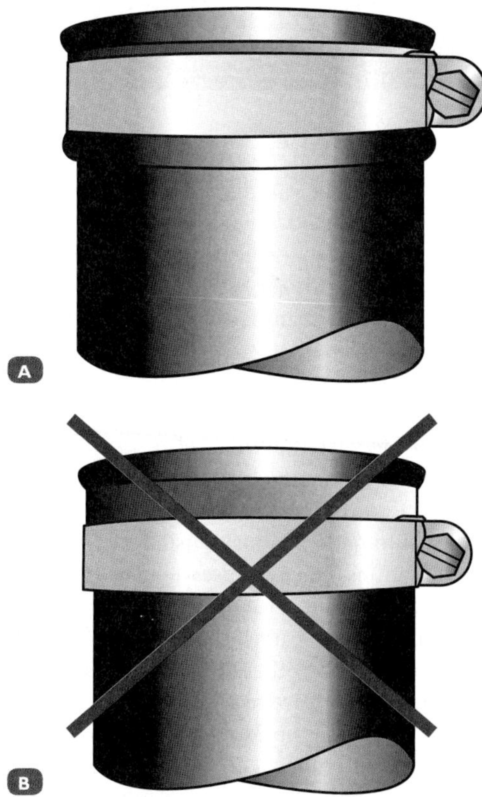

FIGURE 16-34 A. Proper installation of radiator hose clamp. **B.** Improper installation.

Other Coolant Hoses

There are additional coolant hoses that carry hot coolant through the cooling system, which are often ignored when checking hoses (**FIGURE 16-35**). As part of a maintenance inspection, the technician inspects the upper and lower radiator hoses along with the heater hoses, but may forget about these other hoses, which can fail if neglected. It is critical that these hoses be checked. Any break in the system will dump the coolant and cause the engine to quickly overheat.

One of the additional hoses is the bypass hose. The bypass hose is typically located on the water pump and connects to the intake manifold on many V-configured engines, such as a V6 or V8. This hose allows the water pump to circulate the water in the engine when the thermostat is closed. This hose is made of the same materials as the radiator hoses and heater hoses. Another coolant hose that may be used is a throttle body coolant line. This hose runs from the intake up to the throttle body to keep the throttle body from freezing during cold outside temperatures when there is high moisture content in the air. The cold wet air being pulled through the throttle plate may create ice, restrict airflow, and cause the throttle to stick. Running coolant through the throttle body eliminates this problem. There are many possible configurations of additional cooling system hoses, such as to remote oil coolers, turbochargers, or even some alternators. In a compressed natural gas vehicle, coolant hoses are routed to the CNG pressure regulator under the vehicle to keep the regulator warm. A similar setup may be used for the engine's idle air control motor and more. The point is to be sure to inspect all of the coolant hoses used on the vehicle.

Drive Belts

K16016

Typically, the water pump is turned by a belt that is driven by the crankshaft. This belt may be part of the accessory drive belt system found on the front of the engine. If the belt is located on

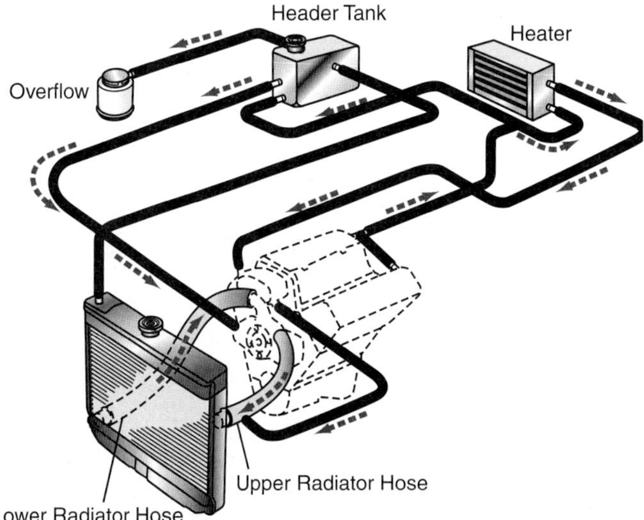

FIGURE 16-35 The cooling system may have multiple flexible coolant hoses, as seen in this cooling system schematic.

the front of the engine, it may be tensioned by a separate tensioner or by simply moving a component such as the alternator on slotted bolt holes. Some engines use the camshaft timing belt or chain as the drive for the water pump.

There are four types of drive belts:

- *V-type:* A V-type belt has a wedge-shaped interior and sits inside a corresponding groove in the pulley (**FIGURE 16-36**). The sides of the V-belt wedge in the sides of the pulley.
- *Serpentine:* A serpentine-type belt, also called a multi-groove V-belt, has a flat profile with a number of small V-shaped grooves running lengthwise along the inside of the belt (**FIGURE 16-37**). These grooves are the exact reverse of the grooves in the outer edge of the pulleys; they increase the contact surface area, as well as prevent the belt from slipping off the pulley as it rotates. The serpentine belt is used to drive multiple accessories and therefore saves underhood space forward of the engine. It winds its way around the crankshaft pulley and some or all of the engine-driven accessories. Most serpentine belts use a spring-loaded tensioner to maintain proper tightness of the belt and prevent it from slipping.
- *Stretch belt:* A stretch belt looks like an ordinary serpentine belt but is found on vehicles without a tensioner. It is made of a special material that allows it to stretch just enough to be installed over the pulleys but then shrink back to its original size, which is shorter than the distance around the pulleys. This stretchiness keeps the belt properly tensioned. Stretch belts require special tools to install and are usually cut off when being removed.
- *Toothed belt:* The toothed belt has teeth on the inside that are perpendicular to the belt and fit inside the teeth of a gear (**FIGURE 16-38**). Timing belts are always toothed belts to keep the camshaft running exactly half the speed of the crankshaft. To save on labor cost, these belts are generally replaced whenever a water pump replacement is required; likewise, the water pump is generally replaced whenever the timing belt is changed.

FIGURE 16-36 V-type belt and pulley.

FIGURE 16-37 Serpentine belt and pulley.

FIGURE 16-38 Toothed belt and pulley.

The technician must be careful when replacing drive belts to avoid tensioning them too much, or too little. If excessive tension is placed on the belt, the bearings on the accessories or tensioner can be overloaded, get hot, and fail due to excessive working load. Follow the manufacturer's service information for correct tensioning.

Tensioners

Tensioners are used to keep the drive belt at the proper tension around the pulleys to ensure the least amount of slippage without causing damage to component bearings. Tensioners can be either manual or automatic. Manual tensioners come in a wide variety of configurations. One type uses a pulley that is adjusted by turning a tensioning bolt (**FIGURE 16-39**). When the bolt is tightened, the pulley moves against the belt with increased tension; if the bolt is rotated the other direction, tension decreases. The tensioner is locked in place by tightening the nut either on the front of the pulley or on the adjustment slot.

A spring-loaded automatic tensioner is typically used with serpentine accessory belts (**FIGURE 16-40**). This type of tensioner is very simple to operate and adjusts itself, so there is no chance of getting it too tight. However, be sure to note the routing of the serpentine belt when servicing it; there are many

FIGURE 16-39 Manual tensioner.

FIGURE 16-41 Timing belt tensioner—spring style.

FIGURE 16-40 A spring-loaded automatic tensioner is typically used with serpentine accessory belts.

FIGURE 16-42 Timing belt tensioner—oil-actuated style.

pulleys to route around, and it can become confusing if you do not have a routing picture. Automatic tensioners can wear out and lose spring tension, which causes belt slippage or unusual noises. Their pivot point also wears out, which causes misalignment of the tensioner pulley on the belt. This can cause the belt to wear out more quickly, slip, or make noises. Also, a seized front engine drive component such as a tensioner can cause a no-crank or slow crank engine condition that mimics a seized engine. Simply loosening or removing the belt may help to diagnose the problem.

If the water pump is driven by the timing belt, one of two types of tensioners may be used. One type is the spring-loaded tensioner, in which a spring sets the tension and a bolt locks the tensioner into position (**FIGURE 16-41**). The other type is the oil-actuated tensioner (**FIGURE 16-42**). The oil-actuated tensioner uses oil pressure from the engine to provide additional tension on the belt. Regardless of the style of tensioner, belt tension can be checked with a belt tension tool and compared to specifications, which are typically published in the service information.

Temperature Indicators

K16017

Temperature indicators can come in two forms: a temperature gauge or a temperature light located in the instrument cluster (**FIGURE 16-43**). The two forms are sometimes used in conjunction. Overheating can heavily damage an engine, so a warning indicator is necessary. A temperature warning light is a good indicator of an overheating condition, but it cannot indicate a condition where the engine stays below operating temperature, which causes excess engine wear, increased emission output, and decreased fuel economy. A temperature gauge indicates to the driver whether the temperature is normal, below normal, or above normal. But drivers can forget to monitor it, which means that the engine could overheat without the driver noticing. Thus, a warning light in addition to a temperature gauge affords the best assurance.

Temperature gauges and warning lights both operate from a signal sent from a coolant temperature sensor located on the engine in a coolant passage. When engine coolant gets hotter

FIGURE 16-43 **A.** Temperature gauge. **B.** Temperature warning lamp.

than it should, the sensor causes the warning light or message to turn on to alert the driver that the engine is overheating. The temperature gauge also uses a sensor, which is designed to continuously indicate the temperature of the engine. The sensor for either type of warning device sits in engine coolant so that an accurate reading is always given—that is, as long as the coolant is not low.

There may also be a low coolant indicator that shows when engine coolant level is low. This system works by having a low level sensor in the surge tank or overflow bottle that turns on a warning light in the instrument cluster if the coolant level falls below an acceptable level. If the low coolant indicator illuminates, there is a good chance that the cooling system has a leak that must be located.

Water Jackets

K16018

Coolant passages such as water jackets are cast into the block and heads during the manufacturing process. They are designed to allow coolant to circulate around the tops and sides of the cylinders and are critical for the transfer of excess heat energy (**FIGURE 16-44**). It is crucial to the efficiency of heat transfer to keep the coolant passages free of scaling and buildup that can restrict heat transfer and coolant flow by ensuring proper cooling system maintenance. As part of proper maintenance, the technician must drain, flush, and refill the cooling system according to the recommended maintenance intervals. (Also see earlier reference to head gasket design.)

FIGURE 16-44 Cutaway view of the water jacket in the cylinder head.

Core Plugs

Core plugs are also known as soft plugs or expansion plugs. These aluminum, brass, or steel plugs are designed to seal the openings to the water jackets that were left from the casting process when the casting sand was removed (**FIGURE 16-45**). Under some conditions, the core plugs *might* pop out if the engine coolant is allowed to freeze—that is, if the proper mixture of antifreeze was not used. Because water expands when it freezes, the block or heads can crack internally or externally near the coolant passages. Sometimes the core plug will be pushed out and the coolant will leak out before the block cracks; however, that is not what they are designed for, so do not rely on soft plugs to protect the engine from freezing. Also, core plugs can rust out and start leaking coolant, so do not forget to inspect them when trying to locate a coolant leak.

Heater Core

K16019

The heater core is simply a small radiator that is mounted inside the heater box in the passenger compartment. As air is blown past the fins of the core, heat energy is radiated to the air and used to heat the passenger compartment for comfort (**FIGURE 16-46**). The heater core connects to the engine's cooling system and is supplied with hot water by circulation of the water pump. Typically, hot water enters the bottom of the heater core and exits the top; thus, the hot water flows from the bottom to the top of the heater core, so more heat can be pulled from the coolant. Heater cores are typically constructed of aluminum, brass, or copper.

Heater Control Valve

If used, the heater control valve is mounted in one of the heater hoses that supply coolant to the heater core. This valve controls the flow of coolant to the heater core to control the temperature

FIGURE 16-45 Core plugs are designed to seal the openings left from the casting process of the block and heads.

FIGURE 16-46 Heater core.

of the air desired by the operator (FIGURE 16-47). The climate control panel, which is adjusted by the operator, controls this valve. This feature is discussed further in the Electronic Climate Control chapter.

Air Doors and Actuators

The heater box consists of many air doors, which are plastic or metal flaps that seal off parts of the air box to control airflow (FIGURE 16-48). The air doors are moved by one of three methods: cable, vacuum actuator, or electric actuator, called a stepper motor (FIGURE 16-49). The **actuator** is a device that is electrically or vacuum controlled and is used to physically move doors within the heater box to control airflow. This system is discussed in detail in the Electronic Climate Control chapter.

The layout and function of the doors depend on the design of the system. Most systems flow air through the evaporator at all times, whereas air can be diverted around the heater core when heat is not wanted. Also, for best defrost operation, the air is directed first over the evaporator to remove any moisture from the air; then it is directed over the heater core to warm it up. This process results in dry, warm air to more quickly defog or defrost the windshield.

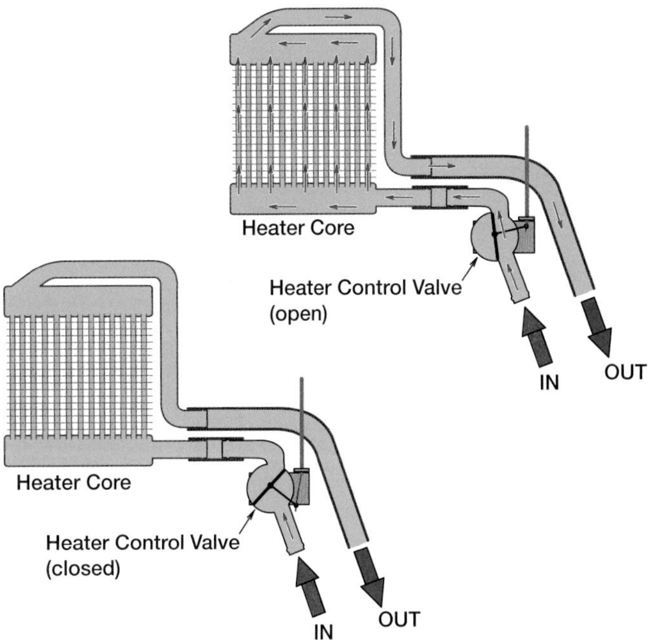

FIGURE 16-47 Heater control valve.

FIGURE 16-48 Typical heater box and air doors.

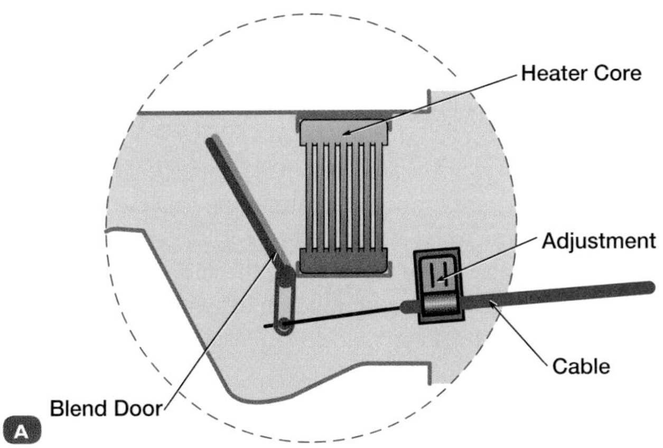

FIGURE 16-49 Air door operating mechanisms. **A.** Cable. **B.** Vacuum actuator. **C.** Electric actuator.

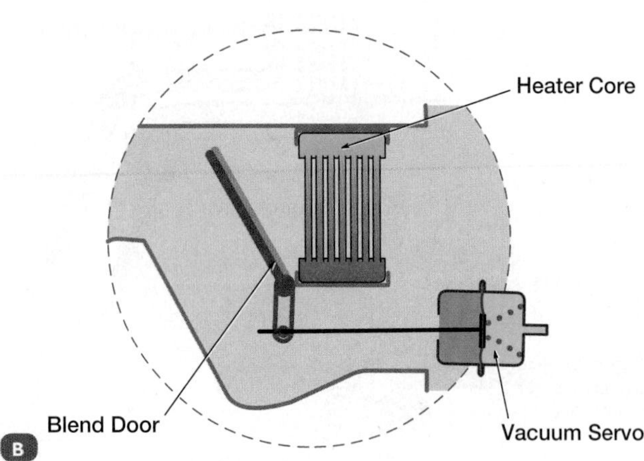

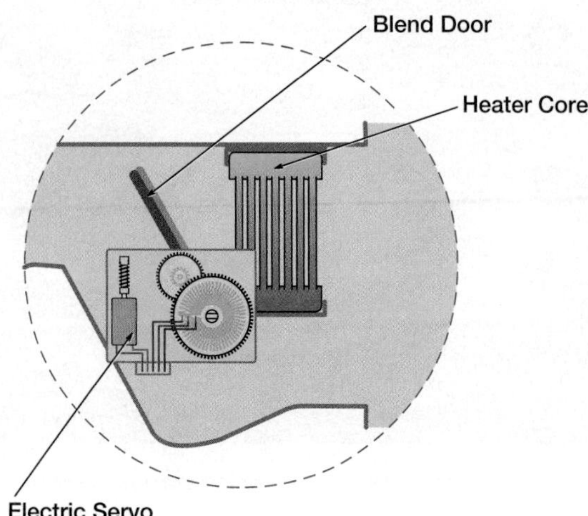

FIGURE 16-49 *(Continues)*

▶ Wrap-Up

Ready for Review

- Most cooling systems rely on coolant, a special mix of chemicals (antifreeze) and water.
- Coolant absorbs heat from the engine, is cooled in the radiator, and flows back to the engine to absorb more heat.
- Heat travels from hot to cold in one of three ways: conduction, convection, or radiation.
- Coolant works to keep an engine from overheating and from freezing.
- Coolant must contain antifreeze to prevent the water content from freezing and to reduce corrosion.
- Antifreeze contains either ethylene glycol (toxic) or propylene glycol (nontoxic).
- The combination of water and antifreeze lowers the freezing point and raises the boiling point of both components.
- Manufacturers can create more efficient combustion by raising the engine's operating temperature.
- Increasing the pressure on coolant raises its boiling point 3°F (~5.4°C) per psi (6.9 kPa).
- Radiator caps maintain a specified pressure throughout the cooling system, generally 13–21 psi (89.6–144.8 kPa).
- Changing coolant regularly prevents acid buildup and electrolysis.
- The stationary parts of the cooling system (heater core and radiator) are connected to the engine via radiator and cooling hoses.
- Modern vehicles have replaced the thermo-siphon process for moving coolant through the engine with a water pump that uses centrifugal force to circulate coolant through the system.
- An engine thermostat regulates coolant circulation, keeping it in the engine until the engine reaches operating temperature.
- Engineers have developed a reverse-flow cooling system in which coolant is first pushed through the cylinder head, thereby better equalizing temperature between the block and head, which extends the life of the head gasket.
- Engines with the reverse-flow cooling system typically have a surge tank to capture steam and reconvert it to coolant.
- Rotary engines use similar cooling systems to piston engines, with a radiator, thermostat, radiator hoses, and water jackets.
- The radiator's function is to allow coolant to pass through it and to transfer heat away from the engine to the atmosphere.
- Cooling tubes in the radiator core run in a vertical (down-flow) or horizontal (cross-flow) design.
- Radiator pressure caps contain a spring-loaded valve to allow excess coolant to pass into the overflow container, and a vacuum valve to allow coolant to be pulled from the overflow container back into the radiator when the engine cools down.
- A surge tank is situated as the highest component so that it collects any air present in the system and allows for easy air removal.
- In a recovery system, coolant flows into an overflow container and then back into the radiator so that no coolant is lost.
- The thermostat's valve is controlled by a wax pellet that melts and expands, forcing the valve open against spring pressure.
- Thermostats installed on the inlet side of the engine better control the amount of cold water flowing into the engine.
- The water pump uses centrifugal force to drive coolant into the water jackets.

- A fan clutch is driven by an accessory belt and uses a viscous fluid to control speed changes of the fan, determined by air temperature from the radiator.
- A solenoid-controlled fan clutch replaces the bimetallic spring with an electric solenoid controlled by the powertrain control module.
- Electric fans only operate when needed, typically at low vehicle speeds or when additional airflow is needed to prevent overheating.
- Radiator hoses must be correctly clamped to the radiator assembly, or a leak will develop.
- Some heater hoses contain a coolant control valve to regulate coolant flow when the driver adjusts the temperature of the heater.
- The three types of accessory drive belts are V-belts, serpentine belts, and toothed belts.
- To ensure minimal slippage, drive belts are tightened around the pulley by tensioners.
- Either a temperature gauge or a temperature light can function as a coolant temperature indicator.
- Coolant passages are critical to the transfer of heat energy, so coolant must be serviced regularly to prevent the passages from becoming clogged.
- The heater box contains air doors and actuators that work to control passenger compartment airflow.
- IAT coolant is the standard green coolant used in many vehicles and must be changed every two years or 24,000 miles (39,000 km).
- OAT coolant is an extended-life coolant that should be changed every five years or 150,000 miles (241,000 km).
- HOAT coolant (yellow) combines inorganic and organic additives and is also an extended-life coolant.
- POAT coolant, the newest type, is a very long-life coolant that should be changed every seven years or 250,000 miles (402,000 km).

Key Terms

actuator A device that is electrically or vacuum controlled and is used to physically move doors within the heater box to control airflow.

ambient the air surrounding the vehicle; atmospheric air.

boiling point The temperature at which a substance begins to change from a liquid to a gas.

centrifugal force A force pulling outward on a rotating body.

conduction Movement of heat energy through solids.

convection Movement of heat energy through gases or liquids.

coolant control valve A valve that blocks off coolant flow to keep hot water from entering the heater core when less heat is requested by the operator.

cross-flow radiator A radiator that uses cooling tubes that run horizontal with tanks on each end. This design allows lower hood profile for better vehicle aerodynamics.

down-flow radiator A radiator that uses cooling tubes that run vertical. This design requires a higher hood profile.

electrolysis The process of pulling metals apart by using electricity or by creating electricity through the use of chemicals and dissimilar metals.

ethylene glycol A chemical used as antifreeze that provides the lower freezing point of coolant and raises the boiling point. It is a toxic antifreeze.

heat dissipation The spreading of heat over a large area to increase heat transfer.

overflow tank A tank used to catch any coolant that is released from the radiator cap (works like a catch can).

pitch The angle of a fan blade. A steeper pitch draws more air, and a shallower pitch draws less air.

propylene glycol A chemical used as antifreeze. It is labeled as a nontoxic antifreeze.

radiation The movement of energy through space, such as the movement of energy from the sun to the earth.

radiator A device that takes hot coolant and cools it by passing heat energy to the surrounding air.

radiator hoses Rubber hoses that connect the radiator to the engine. Because they are subject to pressure, they are reinforced with a layer of fabric, typically nylon.

rotor housing The engine block of the rotary engine. The rotor moves within it.

surge tank A sealed tank that captures coolant coming from the head that has turned to steam and then changes the steam back to coolant to be reused by the cooling system.

thermo-control switch A temperature-sensitive switch that is mounted into a coolant passage on the engine or into the radiator to control electric fan operation.

thermostat Located under the thermostat housing, the thermostat regulates the flow of coolant, allowing coolant to flow from the engine to the radiator when the engine is running at its operating temperature.

viscous coupler Called a fan clutch, a hub that connects the water pump drive to the cooling fan using a temperature-sensitive viscous fluid to cause the fan to turn faster as the temperature of the air pulled through the radiator increases.

water jackets The passages in the engine block and cylinder head that surround the cylinders, valves, and ports.

Review Questions

1. The movement of heat through space is known as:
   a. conduction.
   b. radiation.
   c. convection.
   d. evaporation.
2. The coolant heat storage system:
   a. uses a condenser.
   b. prevents the engine from overheating on hot days.
   c. increases hydrocarbon exhaust emissions.
   d. preheats the engine when the vehicle is started the next time.

3. A coolant serves all of the following purposes *except*:
   a. preventing an engine from overheating while in use.
   b. keeping the engine from freezing while not in use in cold climates.
   c. reducing friction in the moving parts of the engine.
   d. preventing corrosion of the parts in the cooling system.
4. When pressurizing coolant:
   a. its boiling point increases.
   b. the cooling system becomes heavier.
   c. the radiator becomes bigger.
   d. engine efficiency decreases.
5. Which of these is used to force coolant throughout the cooling system in order to transfer heat energy?
   a. Water pump
   b. Thermostat
   c. Radiator
   d. Cooling fan
6. Coolants based on which of these technologies contain a proprietary blend of corrosion inhibitors and are claimed to be compatible with most other types of coolant?
   a. IAT
   b. HOAT
   c. POAT
   d. OAT
7. The surface area that dissipates heat is increased by the use of:
   a. soft plugs.
   b. fins.
   c. hood insulation.
   d. drain plugs.
8. The surge tank:
   a. regulates coolant temperature.
   b. prevents coolant leak.
   c. reduces engine noise.
   d. removes gas from the system.
9. All of the following statements are true of a thermostat *except*:
   a. It regulates the flow of coolant.
   b. It is a spring-loaded valve controlled by a wax pellet.
   c. When the engine is cold, the valve is open; when hot, the valve closes.
   d. When installing the thermostat, the jiggle valve should be in the uppermost position.
10. Which hose allows the water pump to circulate the water in the engine when the thermostat is closed, but not when it is open.
   a. Bypass hose
   b. Throttle body coolant line
   c. Overflow hose
   d. Radiator hose

ASE Technician A/Technician B Style Questions

1. Tech A says that the cooling system is designed to keep the engine as cool as possible. Tech B says that heat travels from cold objects to hot objects. Who is correct?
   a. Tech A
   b. Tech B
   c. Both A and B
   d. Neither A nor B
2. Tech A says that the thermostat is open until the engine warms up, and then it closes. Tech B says that a faulty radiator cap can be the cause of boiling coolant. Who is correct?
   a. Tech A
   b. Tech B
   c. Both A and B
   d. Neither A nor B
3. Tech A says that when 100% antifreeze is used in the cooling system, the protection level is approximately –70°F. Tech B says that 100% antifreeze will cool the engine better than 100% water. Who is correct?
   a. Tech A
   b. Tech B
   c. Both A and B
   d. Neither A nor B
4. Tech A says that some drive belts are of a stretch fit design and are not adjustable. Tech B says that automatic belt tensioners never wear out. Who is correct?
   a. Tech A
   b. Tech B
   c. Both A and B
   d. Neither A nor B
5. Tech A says that the design of a surge tank system helps to purge air from the cooling system. Tech B says that the overflow tank adds coolant to the radiator when the engine is hotter than normal. Who is correct?
   a. Tech A
   b. Tech B
   c. Both A and B
   d. Neither A nor B
6. Tech A says that when you find coolant hoses collapsed after the engine cools down, the radiator pressure cap has likely failed. Tech B says that the thermostat may have a bleed valve that should be accurately positioned when the thermostat is replaced. Who is correct?
   a. Tech A
   b. Tech B
   c. Both A and B
   d. Neither A nor B
7. Tech A says that electric cooling fans are used to cause a large airflow over the radiator when the engine is cold. Tech B says that electric cooling fans are used to cause a large airflow over the radiator at low vehicle speeds. Who is correct?
   a. Tech A
   b. Tech B
   c. Both A and B
   d. Neither A nor B
8. Tech A says that there are a number of coolant types that each has its own life span. Tech B says that mixing of coolants is generally okay, as all coolants use the same chemical base. Who is correct?
   a. Tech A
   b. Tech B

c. Both A and B
d. Neither A nor B

9. Tech A says that coolant leaking out of the water pump weep hole means the pump is operating normally. Tech B says that the water pump uses centrifugal force to circulate coolant. Who is correct?
   a. Tech A
   b. Tech B
   c. Both A and B
   d. Neither A nor B

10. Tech A says that raising the pressure on the cooling system raises the boiling point of the coolant. Tech B says that air-cooled engines operate within a more consistent and ideal temperature range. Who is correct?
   a. Tech A
   b. Tech B
   c. Both A and B
   d. Neither A nor B

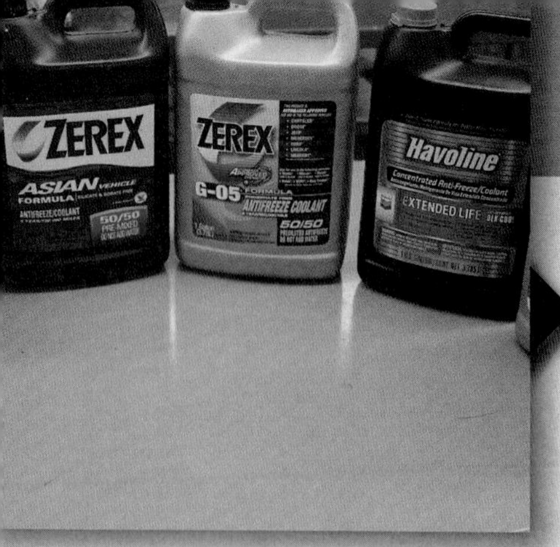

CHAPTER 17

Cooling System Service

NATEF Tasks

- **N17001** Inspect and/or test coolant; drain and recover coolant; flush and refill cooling system; use proper fluid type per manufacturer specification; bleed air as required. (MLR/AST/MAST)
- **N17002** Inspect, replace, and/or adjust drive belts, tensioners, and pulleys; check pulley and belt alignment. (MLR/AST/MAST)
- **N17003** Inspect engine cooling and heater system hoses and pipes; determine needed action. (MLR/AST/MAST)
- **N17004** Remove, inspect, and replace thermostat and gasket/seal. (MLR/AST/MAST)
- **N17005** Identify causes of engine overheating. (AST/MAST)
- **N17006** Perform cooling system pressure and dye tests to identify leaks; check coolant condition and level; inspect and test radiator, pressure cap, coolant recovery tank, heater core, and galley plugs; determine needed action. (MLR/AST/MAST)
- **N17007** Verify engine operating temperature; determine needed action. (MLR/AST/MAST)
- **N17008** Inspect and test fan(s), fan clutch (electrical or mechanical), fan shroud, and air dams determine needed action. (AST/MAST)
- **N17009** Inspect and test heater control valve(s); perform needed action. (AST/MAST)
- **N17010** Remove and replace radiator. (AST/MAST)
- **N17011** Determine procedure to remove, inspect, reinstall, and/or replace heater core. (AST/MAST)

Knowledge Objectives

There are no knowledge objectives in this chapter.

Skills Objectives

After reading this chapter, you will be able to:

- **S17001** Diagnose cooling system faults.

▶ Introduction

Now that you understand the function and operation of the cooling system, this chapter covers the maintenance required to keep it functioning properly over the life of the engine. And if the cooling system does fail for some reason, we show you how to efficiently diagnose the root cause of the fault. This chapter prepares you to properly perform cooling system service on a variety of vehicles.

▶ Preventive Maintenance

Preventive maintenance of the cooling system is critical for long life and reliability of the engine. Failure to perform required maintenance will result in cooling system failure, which can lead to breakdowns and major engine damage. Manufacturers publish the required maintenance for the vehicle in the owner's manual and in the service information (**FIGURE 17-1**). The maintenance schedule lists which services are due at a particular mileage or date. With standard IAT coolant, a maintenance schedule will say "every two years or 30,000 miles, whichever comes first." Belts and hoses have similar inspection and maintenance requirements. These service intervals are critical to follow, so always check the maintenance schedule for the vehicle you are working on.

EPA Guidelines

Ethylene glycol antifreeze, the most common antifreeze, is highly toxic to humans and animals. Because of this, the Environmental Protection Agency (EPA) has strict regulations for the handling and disposal of vehicle coolant. Coolant should never be dumped into a storm drain or down a shop floor drain. Coolant is a poison and should only be poured into an approved container and either be recycled in-house or removed by a licensed recycler (**FIGURE 17-2**).

Care must be taken when a spill occurs when servicing a vehicle; it should be cleaned up promptly according to EPA regulations.

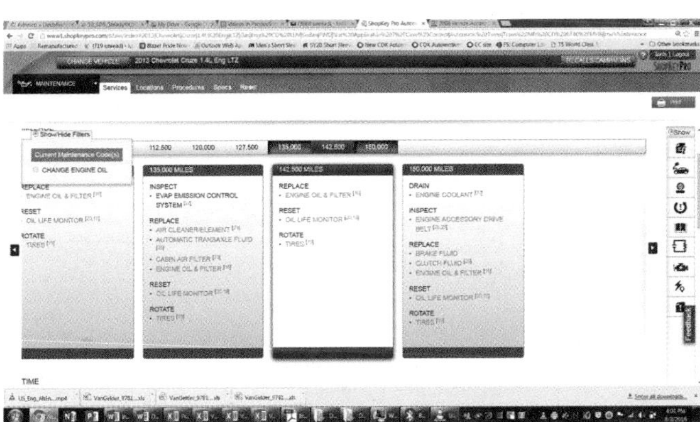

FIGURE 17-1 Maintenance schedule for the cooling system on a sample vehicle.

FIGURE 17-2 Used coolant being stored for recycling.

You Are the Automotive Technician

A customer brings his 2012 Honda Accord into your dealership complaining about the coolant boiling over on hot days. The engine gauge gets higher than normal, but never to the red zone, when steam comes out from under the hood. You ask some clarifying questions and she gives these answers; it happens most during stop and go traffic on her way home from work, which is at the hot time of the day. She isn't sure if the electric cooling fan is coming on or not since she is focused on traffic. She says that the quick lube shop takes care of her fluids, so she doesn't know if it is using coolant or not. She agrees to let you diagnose it, so you take it back to the shop and verify that the coolant level is near the full mark. You then take a minute to remember the cooling system theory your auto instructor taught you and plan your steps for diagnosis.

1. What are the possible causes for coolant that boils over below the red zone?
2. How can engine operating temperature be verified?
3. What two factors determine the boiling point of engine coolant?
4. What are some of the most likely causes of coolant that boils over when in the red zone?

Taking a little extra time to place a catch pan under the component being removed can prevent a spill and save time in the long run. Also, coolant should never be mixed with oil or other liquids; separate catch pans should always be used and marked accordingly.

Measuring Freeze Protection

N17001

The use of a hydrometer or refractometer is necessary when testing the freeze protection of the coolant in the cooling system. The customer may request this service as part of a winterization package performed by the shop. Any time coolant is replaced in the cooling system, the freeze point should be verified.

The hydrometer is a tool that measures the specific gravity of a liquid. When coolant is drawn into the hydrometer, a float will rise to a certain level depending upon the density of the coolant. Antifreeze has a higher specific gravity than water, so the higher the float rises in the liquid, the greater the percentage of antifreeze in the mix. One drawback to hydrometers is that they are typically antifreeze specific. That means you need one for ethylene glycol and one for propylene glycol, as the specific gravities of the two chemicals are different. Another drawback is that as the temperature of the coolant goes up, the specific gravity goes down. Some hydrometers have a built-in thermometer and a chart, allowing you to compensate for the temperature of the coolant.

A refractometer can also tell the proportions of antifreeze and water in the coolant mix (or the level of freeze protection) by measuring a liquid's specific gravity. It works by allowing light to shine through the fluid. The light bends in accordance with the particular liquid's specific gravity. The bending of the light displays on a scale inside the tool, indicating the specific gravity of the fluid. One nice thing about a refractometer is that it has a scale for both types of antifreeze and reads the freeze point accurately.

To use a hydrometer to test the freeze point of the coolant, follow the steps in **SKILL DRILL 17-1**.

To use a refractometer to test the freeze point of the coolant, follow the steps in **SKILL DRILL 17-2**.

SKILL DRILL 17-1 Using a Hydrometer to Test the Freeze Point of the Coolant

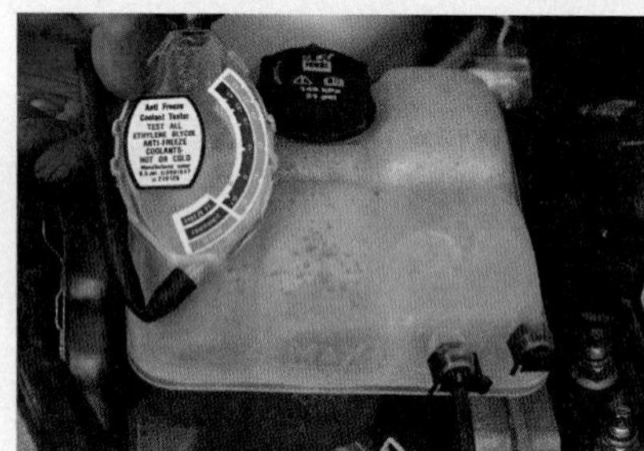

1. Remove the pressure cap. Place the hydrometer tube in the coolant, and squeeze the ball on top.

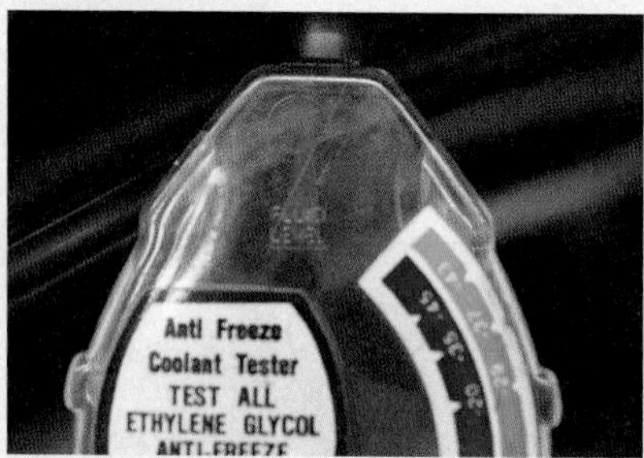

2. Release the ball to pull a coolant sample into the hydrometer. Verify it is above the minimum level in the tester.

3. Hold the tool vertical. Read the scale to verify the freeze protection of the coolant. Return coolant sample to the radiator or surge tank.

SKILL DRILL 17-2 Using a Refractometer to Test the Freeze Point of the Coolant

1. Remove the pressure cap. Be sure the cooling system is cool first.

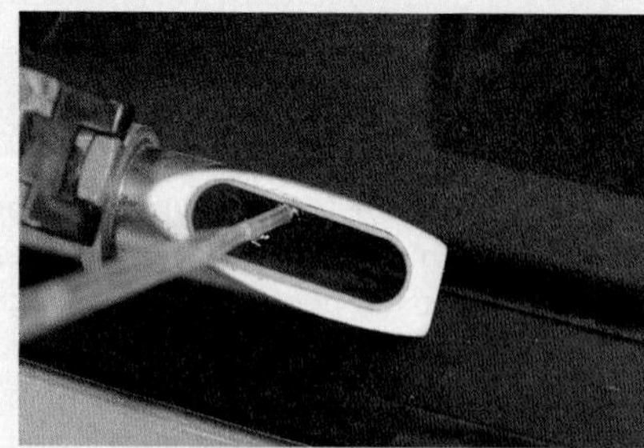

2. Determine the type of antifreeze, and verify that the refractometer is designed to be used with it. Place a few drops of coolant on the sample plate on the top of the tool.

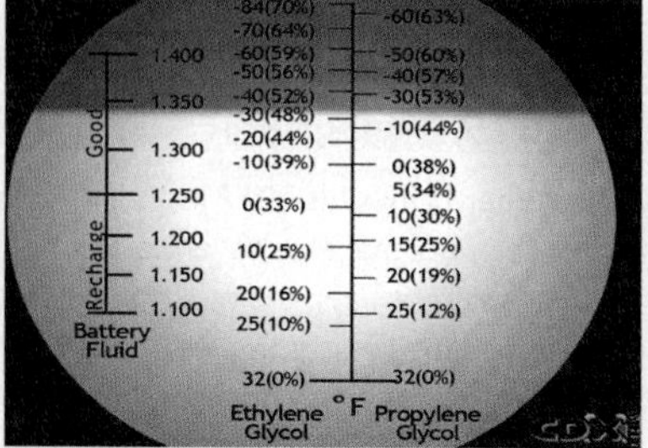

3. Hold the refractometer roughly level under a light, look through the viewfinder, and read the scale to verify the freeze protection of the coolant..

Testing the Coolant pH

The pH testing of coolant is subject to a fair amount of speculation. There isn't any hard and fast specification for an acceptable pH level. Vehicle manufacturers, coolant suppliers, and coolant test strip companies seem to have a wide range of opinion of what pH is acceptable and what is not. The best we can say is that most entities suggest that the pH should be somewhere between 9.5 and 10.5 when installed new in the vehicle, and that they recommend a flush below approximately 8.5. As corrosion inhibitors break down over time, the solution of water and antifreeze becomes more acidic. As the acid level builds, so does corrosion and electrolysis. At the same time, if the pH level exceeds 11.0, aluminum components in the cooling system start to corrode. So there is a happy middle range pH level you should shoot for.

Measuring the pH can be accomplished with test strips that turn color based on the level of acidity in the coolant or with electronic testers that measure the pH of the coolant directly. pH testing of coolant can help determine if the corrosion inhibitors are still working in the antifreeze. But the pH level of the water you use and the antifreeze chosen will affect the pH level at which you should recommend coolant replacement.

To test the coolant pH, follow the steps in **SKILL DRILL 17-3**.

Testing for Electrolysis

As the search for greater fuel efficiency continues and lighter-weight nonconductive materials find their way into our engines and vehicles, problems have arisen that result in the need for new training of technicians. **Electrolysis** is the reaction of different metals to an acid solution to produce electricity. It can result in negative effects on the cooling system. In this case, electrolysis can erode metals from inside the cooling system, leading to damage.

Electrolysis can also be due to faulty grounds in the electrical system. If a circuit has a faulty ground, the current will try to find its way back to the battery in whatever way it can. And that can include sending some of the current flow through the coolant. For example, if the engine ground is dirty and has an excessive voltage drop, then when the engine is cranked over, some of the current could flow through the coolant to another ground. Current flowing through the coolant can erode metal surfaces in the cooling system. In this case, the coolant is not at fault, and flushing will have little to no effect on the condition. Repairing the failed ground is the solution in this case. To verify whether electricity is finding its way into the cooling system, voltage can be measured when various electrical loads are operated to see if there is any stray voltage in the coolant. If voltages are over 0.3 volts when the load or loads are activated, then you will have to perform electrical diagnosis of the circuit being tested.

To perform electrolysis testing for a suspected coolant problem, follow the steps in **SKILL DRILL 17-4**.

Checking and Adjusting Coolant

Checking and adjusting coolant implies two separate tasks. First, is the level at the full mark? Second, is the level of freeze protection appropriate for the climate the vehicle is operated in? Checking coolant level should be part of every oil change so that any leaks can be identified before they become more serious. You may also need to check or adjust coolant level if the customer's low coolant indicator comes on. There are usually two correct level marks on the reservoir because the coolant in the system expands and contracts according to changes in engine temperature. The coolant level should not be below the lower mark when the vehicle is cold. It should be near the upper mark when the coolant is hot. If the coolant is indeed low, testing for a coolant leak will be necessary.

Testing of the coolant's pH can be performed with a test strip or electronic tester. Its freeze protection level can be tested with a coolant hydrometer or refractometer. If either of these tests shows that the coolant does not meet specifications, it will have to be flushed and replaced.

To check and adjust coolant, follow the steps in **SKILL DRILL 17-5**.

SKILL DRILL 17-3 Testing the Coolant pH

1. Ensure that the coolant is relatively cool before removing the radiator cap.

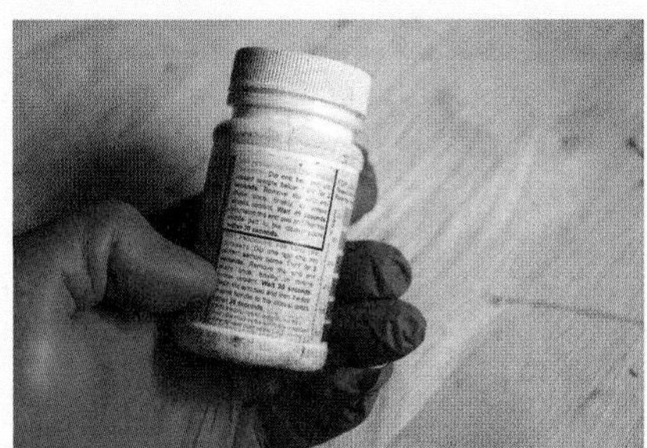

2. If using a pH test strip kit, read the instructions. Know how long to dip the strip and how long to wait to compare it to the scale. Some strips measure the freeze point, so know the time for that as well.

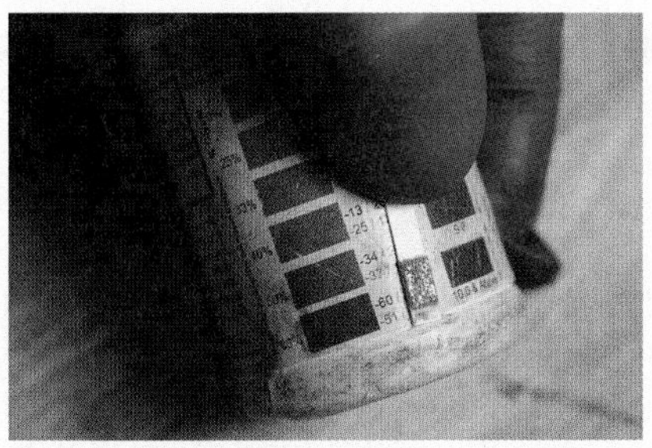

3. Dip litmus paper into the coolant, and wait the amount of time directed by the instructions. Compare the color of the litmus paper to the color scale on the kit.

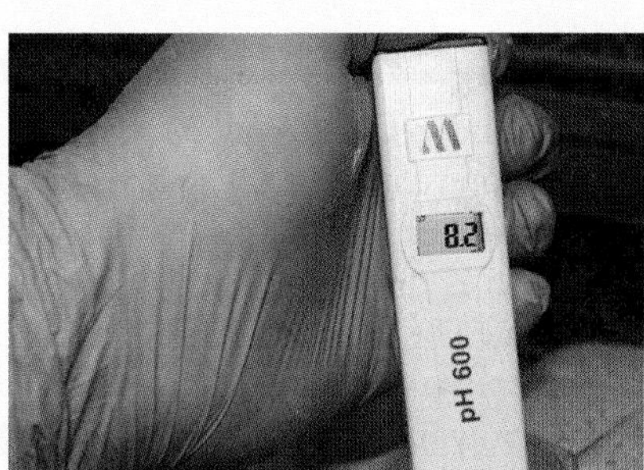

4. If using an electronic pH tester, turn on the tester, and immerse it in the coolant. Read the meter. Some cooling system experts suggest a coolant flush if the pH level is below 8.5.

SKILL DRILL 17-4 Performing Electrolysis Testing

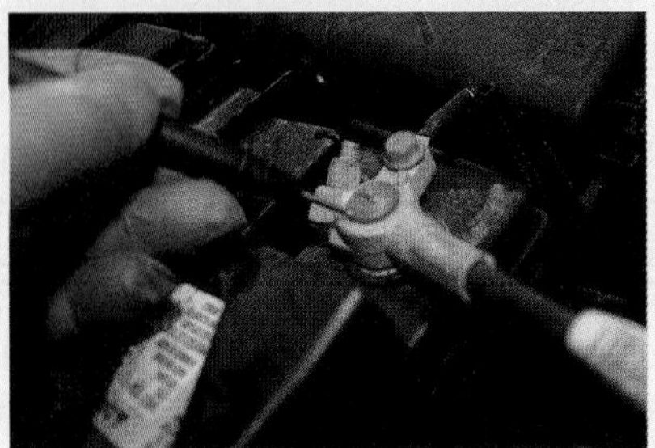

1. Connect the black lead of the voltmeter to a good engine ground.

2. Hold the red lead of the voltmeter in the coolant in the radiator.

3. Observe the voltage reading. If it is greater than 0.3 volts (300 millivolts), flush the cooling system, refill, and retest. If less than 0.3 volts, crank the engine and read the meter again. If greater than 0.3 volts, check the vehicle ground connections.

SKILL DRILL 17-5 Checking and Adjusting Coolant

1. Check the level of coolant in this reservoir; if the engine is hot, the level should be visible near the upper mark. If the engine is cold, it should be at or above the lower mark.

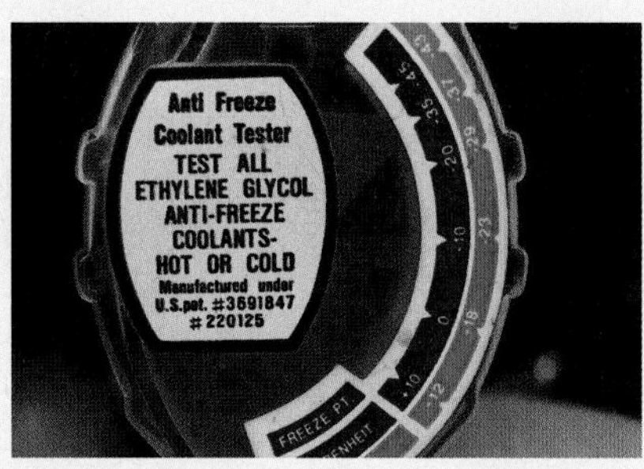

2. Before adding new coolant, check the specific gravity of the coolant in the system with a coolant hydrometer or refractometer.

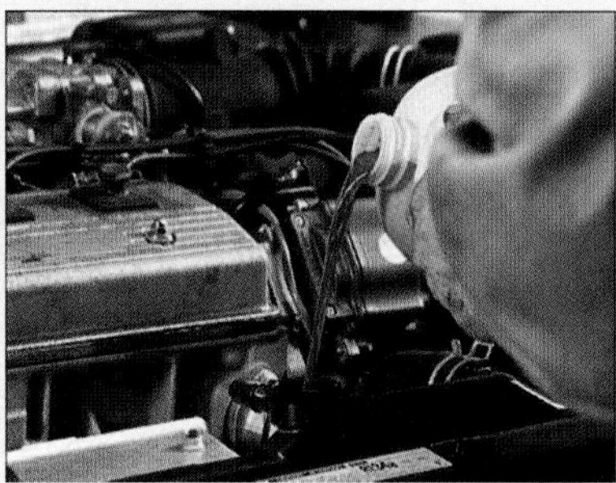

3. Depending on the freeze protection of the coolant, add either 50/50 premixed coolant, straight antifreeze, or straight water as appropriate to bring it to the proper level. If the level was low, find the cause of the loss.

Draining and Refilling Coolant

Draining and refilling coolant is necessary if the customer requests a preventive maintenance service to flush the cooling system. Another reason you may need to perform this task is if corrosion is discovered in the system or if the coolant was contaminated by an incorrect fluid being added to the radiator, such as by mistakenly adding power steering fluid to the coolant reservoir. Any time you have to replace a part of the cooling system, you will have to drain and refill the coolant. The discovery of extremely acidic coolant also requires flushing the cooling system.

When draining the coolant, it is helpful to remove the radiator cap when the petcock is open. To drain the cooling system further than the petcock can do itself, some engines are equipped with block drains. This allows more coolant to drain from the block. When refilling the system, it is critical that any trapped air be bled from behind the thermostat. This can be accomplished by loosening a bleed screw if so equipped or by loosening the highest hose in the cooling system. Fill the radiator or surge tank until coolant comes out the hose. Lastly, always dispose of used coolant in an environmentally approved manner.

Your shop may have a coolant flushing machine, which will change the steps of the following skill drill. If you are using a flushing machine, follow the directions for the machine you are using.

To drain and refill coolant, follow the steps in **SKILL DRILL 17-6**.

SKILL DRILL 17-6 Draining and Refilling Coolant

1. Locate the radiator drain plug, if equipped, and place a catch pan marked for coolant underneath the drain valve. Drain the radiator into the catch pan.

2. Remove the block drain plugs, and allow the coolant to drain into the pan.

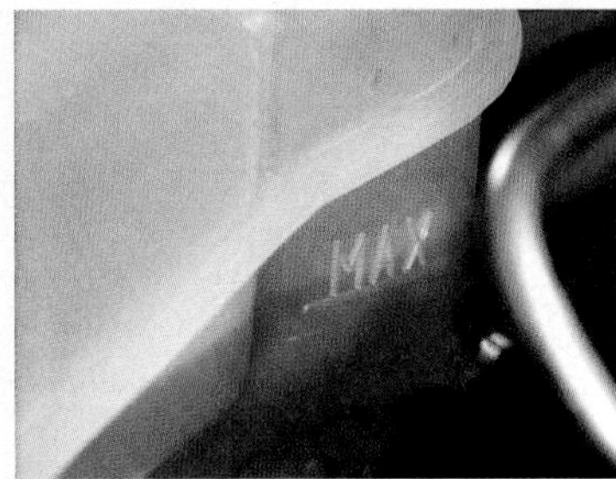

3. Refill the cooling system with the proper coolant mix after closing the drain plugs. Start the engine and verify the proper level. Dispose of coolant in an approved way.

Flushing the Coolant

Coolant flushing is necessary either as a preventive maintenance service performed by most shops or because the coolant is worn out or contaminated. It is necessary if the corrosion inhibitors in the coolant are found to be worn out or if some type of contamination is found. Flushing of the coolant is performed in one of two ways: manually or with a flushing machine. Most shops have a coolant flushing machine; if you are using this machine, follow the directions on the machine as they are all different. In this example, we perform the manual flush.

To flush the coolant, follow the steps in **SKILL DRILL 17-7**.

Inspecting and Adjusting an Accessory Drive Belt

N17002

Inspecting the engine drive belt should be part of any maintenance inspection. Drive belts stretch over time with use and may require adjustment. Adjusting the drive belt may be necessary if the vehicle is not equipped with an automatic tensioner. Never try to inspect belts with the engine running. When adjusting belts, if they are too loose, they will squeal or chirp and slip. If the belt is too tight, it will put extra force against the bearings on the accessories being turned, which can cause them to wear out prematurely. Conditions to look for on a drive belt include:

- *Cracks:* Cracks in a belt used to indicate that immediate replacement was needed. With today's belts, many manufacturers tolerate a certain number of cracks per inch. Check the manufacturer's specifications before recommending a belt be replaced (**FIGURE 17-3**).
- *Oil contamination:* A belt that has been soaked in oil will not grip properly on the pulleys and will slip. If the oil contamination is severe enough for this to happen, diagnose and repair the cause of the oil contamination, and replace the belt.
- *Glazing:* Glazing is shininess on the surface of the belt, which comes in contact with the pulley. If the belt is worn, the glazing could be caused by the belt "bottoming out" in the pulley, and it should be replaced. If it is not old and worn, glazing could simply indicate that the belt is not tight enough. Tightening the belt may be all that is necessary, depending on how bad the glazing is.

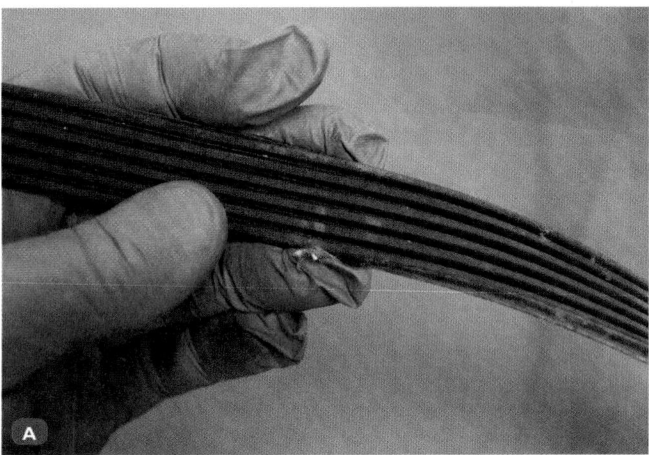

FIGURE 17-3 A. Acceptable belt. **B.** Unacceptable belt.

SKILL DRILL 17-7 Flushing the Coolant

1. Locate the radiator drain plug, if installed. Drain the coolant into the catch pan.

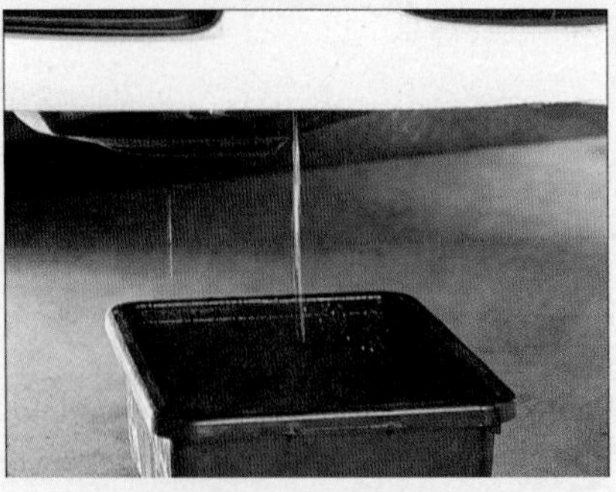
2. Remove any engine drain plugs in the block and allow the block to drain into the catch pan.

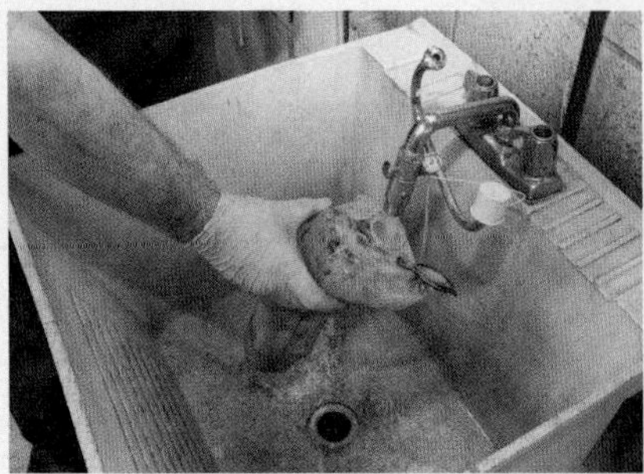
3. Remove the surge tank or overflow tank, and clean thoroughly with hot water.

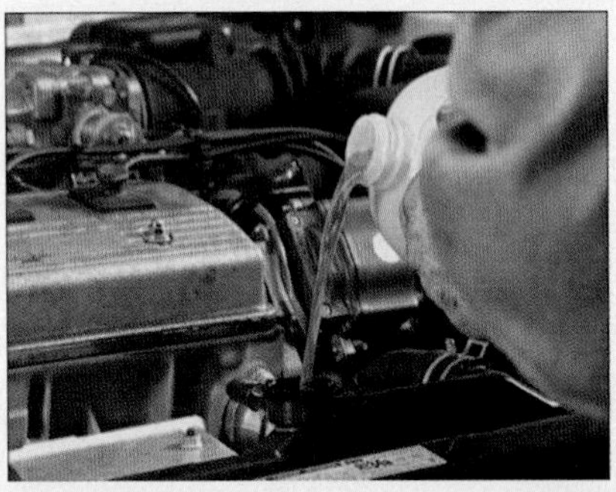
4. Reinstall the tank and all drain plugs, and refill the radiator with clean water, and if desired, a coolant flushing additive. Start the engine and allow it to warm fully.

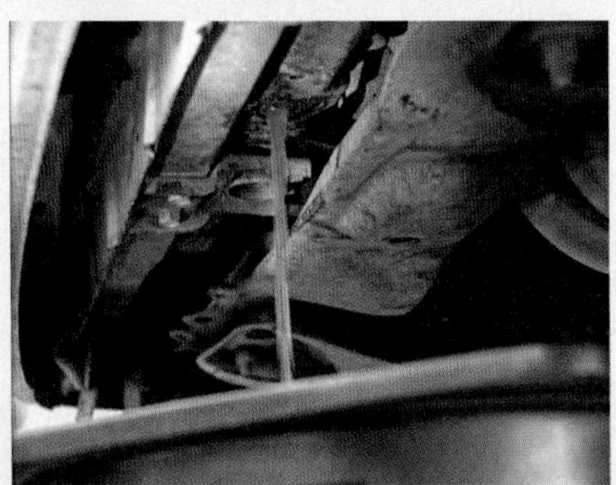
5. Drain water from the radiator and engine block. Replace all drain plugs.

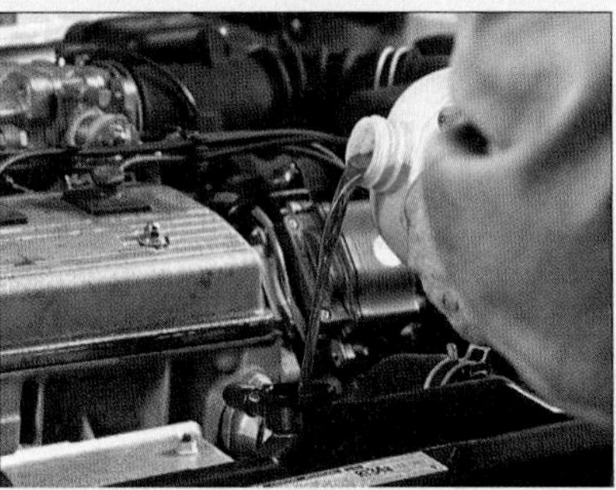
6. Refill the radiator with a 50/50 mix of antifreeze and water. Bleed any air from the system, using a vacuum bleeding system or by running the engine with the cap off and keeping the system full.

- *Tears:* Torn or split belts are unserviceable and should be replaced immediately.
- *Bottoming out:* When V-type or serpentine-type belts become very worn, the bottom of the V-shape may contact the bottom of the groove in the pulley, preventing the sides of the belt from making good contact and wedging with the sides of the pulley groove. This reduced friction causes slippage. A belt worn enough to bottom out should be replaced. V-belts can be inspected for the depth of the belt in the pulley visually. A serpentine belt can be checked with a plastic wear gauge. If the plastic tool is even with the top of the belt ribs, or lower, the belt needs to be replaced.
- *Pulley wear:* Always inspect the pulley when inspecting the belt. A worn pulley will slip and squeal.
- *Too wide*: When a V-belt that is too wide is used, the belt sits above the edges of the pulley. As the sides of the belt wear, a step develops in the side of the belt. After a while, this step runs on the top of the pulley, reducing the grip on the sides of the pulley. This can commonly cause the belt to squeal even though it is tight. Replacing the belt with one of the proper width will solve the issue.

To inspect and adjust an accessory drive belt, follow the steps in **SKILL DRILL 17-8**.

▶ TECHNICIAN TIP

Manual belt tension versus automatic belt tension: Many vehicles require the technician to manually adjust the tension on the belt. Other vehicles have an automatic spring tensioning system. There are several different types of tension gauges, so follow the operating instructions for the tool. If you do not have a tension gauge, you can estimate the tension by pushing the belt inward with your thumb. If it is correctly tensioned, you should be able to deflect the belt only about half an inch for each foot of belt span between pulleys.

Replacing an Accessory Drive Belt

Replacement of an engine drive belt may be necessary when the belt is cracked, glazed, separating, and getting ready to

SKILL DRILL 17-8 Inspecting and Adjusting an Accessory Drive Belt

1. If a V-belt, twist the belt so that you can see the underside of the V shape.

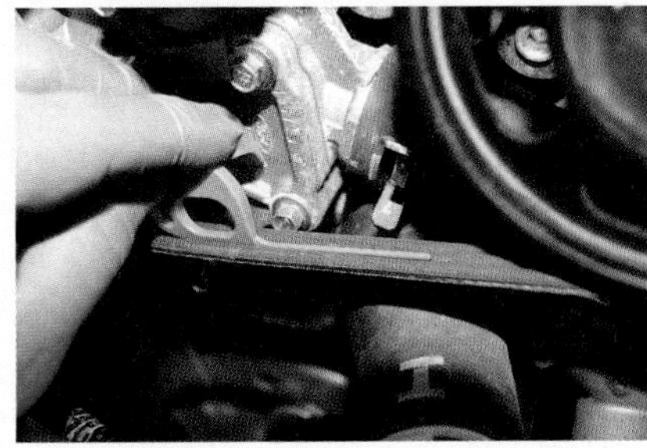

2. If a serpentine belt, check the ribs with a serpentine belt gauge.

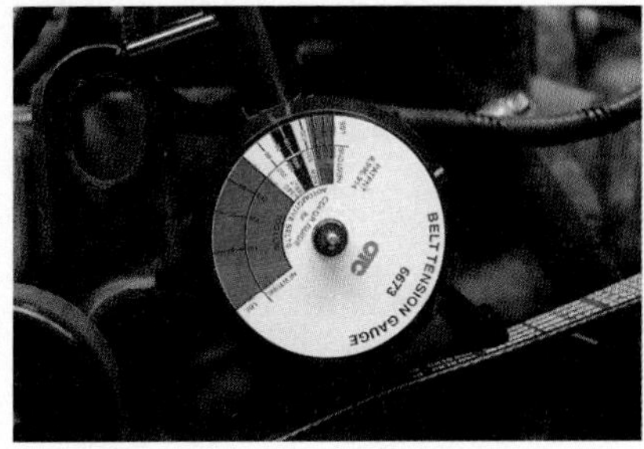

3. Check the belt tension by attaching the belt tension gauge to the belt, and measure the tension. On manually adjusted belts, loosen the locking fastener.

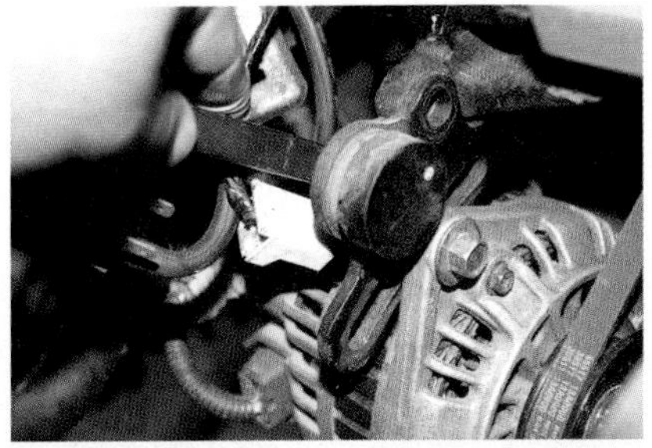

4. For the slotted style, use a pry bar to carefully pry the adjustable component until the belt tension is adjusted. Tighten the locking fastener.

5. On the style that uses a tensioning screw, tighten the tensioning screw until the belt is properly tensioned. Tighten the locking fastener.

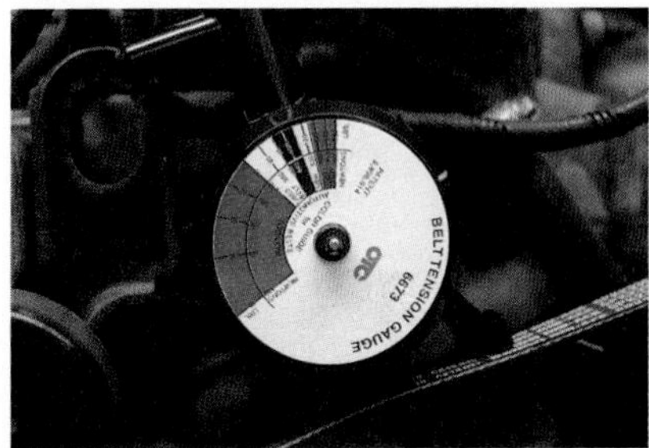

6. Check belt tension with the belt tension gauge. Remove the gauge. Start the engine and verify proper operation.

fail. Verifying that the pulley system is not damaged or misaligned will be the first step if noise is a problem. Perform a pulley alignment check as part of inspecting the drive belt system if noise is a concern or if belts are being thrown off. Checking pulley alignment is done with a straightedge across the face of the pulleys or with a special laser that fits in the grooves of the pulley. If it is a serpentine belt system, the pulley edges should be within 1/16" (1.6 mm) of alignment with each other.

Another reason to replace belts is if the customer requests belt replacement as part of the vehicle's preventive maintenance.

Before removing the belt, you need to verify two things. First, verify that the replacement belt is the correct belt. Compare both the width and the length of the belt to specifications. And second, verify the routing of the belt. Some serpentine belts wrap around about 10 different pulleys. So it is easy to get the routing wrong, which could end up turning an accessory backward. Most vehicles have a label under the hood that has a belt routing diagram. Find that and compare it to the routing of the existing belt. If the belt routing diagram isn't available or is unreadable, create your own diagram on a piece of paper. That can save you a big headache. When replacing serpentine belts it is easy for the belt not to seat in the proper grooves in one or more pulleys. So always check each pulley to make sure the belt is centered in the pulley.

To replace a standard accessory drive belt, follow the steps in **SKILL DRILL 17-9**.

Replacing a Stretch Fit Belt

Although Stretch Fit belts can last up to 10 years or 150,000 miles, you should still be familiar with the procedure to remove and replace them. First, because they are designed to be operated without a tensioner, there is no way to loosen the belt to remove it. So Stretch Fit belts are typically cut with diagonal side cutters to remove them. This means that you need to replace any Stretch Fit belt with a new one whenever the belt needs to be removed for any reason.

To install a new Stretch Fit belt, you need a special installation tool. This tool grabs onto the crankshaft pulley and levers the belt over the side of the pulley as the crankshaft is turned. This presents several safety concerns. First, you need to turn the engine over by hand because the tool is on the crankshaft pulley and would be thrown if the engine happened to start while cranking it over. So don't use the starter to crank the engine to install the belt. But doing it by hand means you *must* have the ignition switch OFF when turning the engine. You don't want it to start, or even kick when you

SKILL DRILL 17-9 Replacing an Accessory Drive Belt

1. On a manually adjusted belt, locate and loosen the adjustment locking fastener. Loosen the belt and remove it.

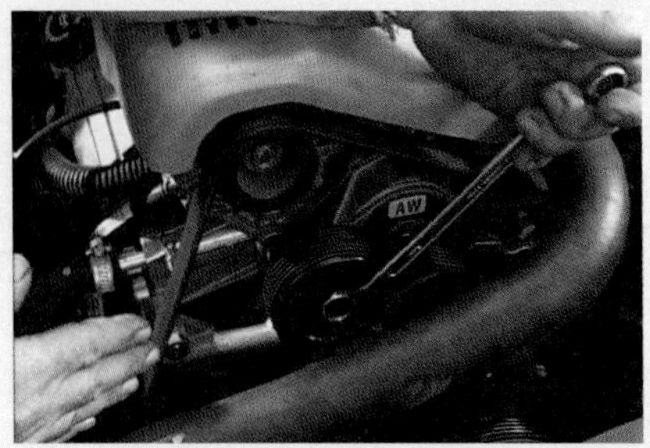

2. On an automatically adjusted belt, locate the spring-loaded idler pulley, install an adjuster tool, and pry the tensioner back. Remove the belt.

3. Inspect the belts and pulleys for wear and damage.

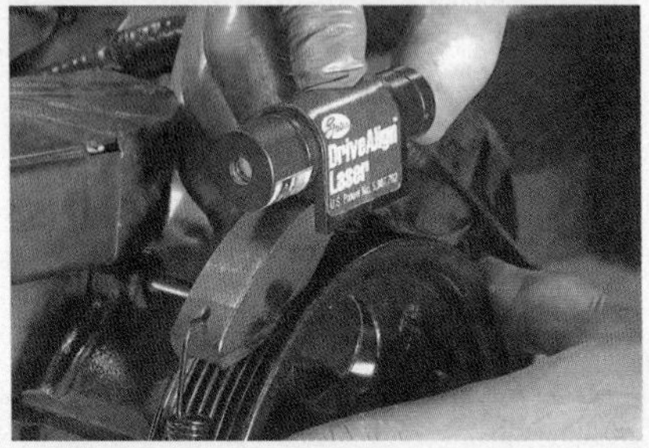

4. Check all drive pulleys for alignment using a straightedge or laser.

5. Select the correct replacement belt. Install the V-belt or serpentine belt, if equipped. Make sure the belt is properly routed around the pulleys and fully seated in each pulley.

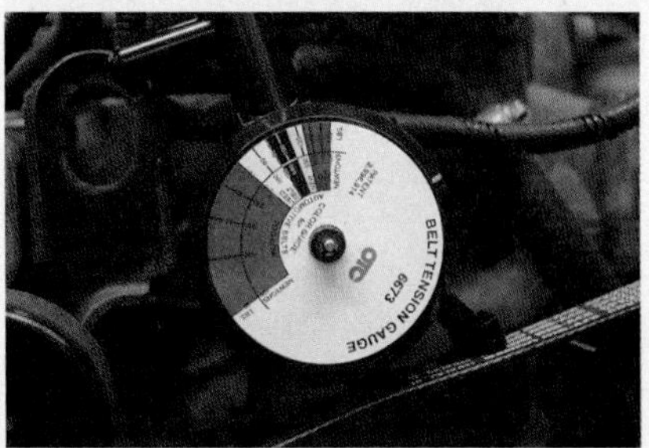

6. Correctly tension the new belt. Start the engine and verify proper operation. Stop the engine and recheck the tension.

are turning it by hand. Use a socket and ratchet on the front crankshaft bolt to slowly turn the crankshaft while installing the belt.

To install a Stretch Fit drive belt, follow the steps in **SKILL DRILL 17-10**.

Inspecting and Replacing a Coolant Hose

N17003

Checking and replacing coolant hoses is a very important job of the technician. If the hose is deteriorating, it will eventually burst, and coolant will be pumped out of the engine, resulting in overheating. Coolant hoses should be checked anytime the vehicle is in the shop for a maintenance inspection. If you find one deteriorated radiator hose, chances are that the other hose(s) may be deteriorating in the same way and will soon need to be replaced. For this reason, most technicians generally replace both radiator hoses at once as a sensible precaution. Do not forget that there may be many hoses in the cooling system, and you need to inspect all of them to ensure proper function of the cooling system. A flashlight may be helpful when inspecting the coolant hoses so that you can clearly see if the hose is starting to bulge, crack, or become damaged. Radiator hose problems include:

- *Swollen hose:* This hose has lost its reinforcement and is swelling under pressure. It may soon rupture; typically you will see a bubble protruding from the side of this hose. Replace this hose immediately (**FIGURE 17-4**).
- *Hardened hose:* This hose has become brittle and will break and leak. Verify hardening by squeezing the hose and comparing it to a known good hose (**FIGURE 17-5**).
- *Cracked hose:* This hose has cracked and will soon start to leak. Verify cracking by a visual inspection (**FIGURE 17-6**).
- *Soft hose:* This hose has become very weak and is in danger of ballooning or bursting. Verify softening by squeezing the hose and comparing it to a known good hose (**FIGURE 17-7**).

SKILL DRILL 17-10 Installing a Stretch Fit Drive Belt

1. Gather the correct replacement belt and installation tool.

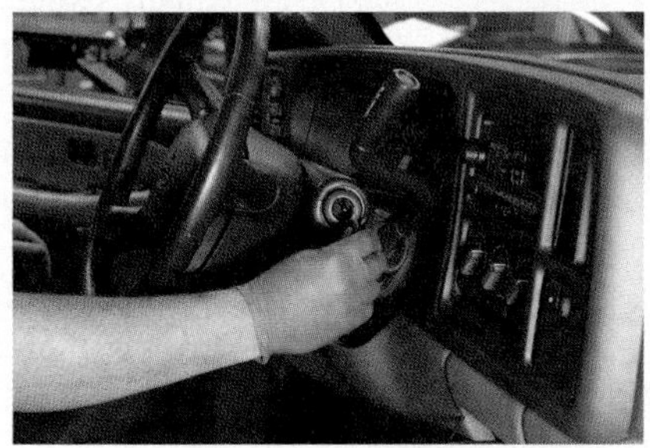

2. Make sure the ignition key is OFF.

3. Cut the old belt with a pair of diagonal side cutters.

4. Position the belt fully on all of the pulleys except the crankshaft pulley.

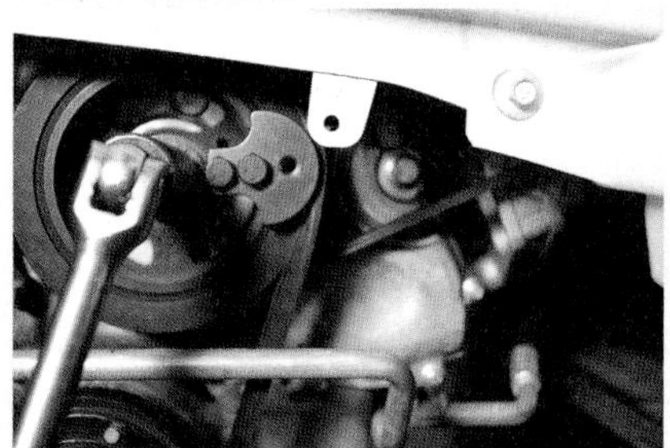

5. Position the installation tool and belt so that the belt is snug against the pulley and tool. Using a socket, turn the crankshaft slowly to stretch the belt over the side of the pulley.

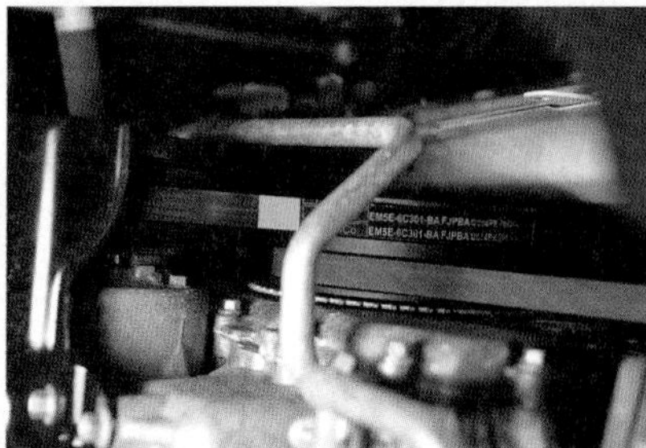

6. Verify that the belt is centered in all of the pulleys properly.

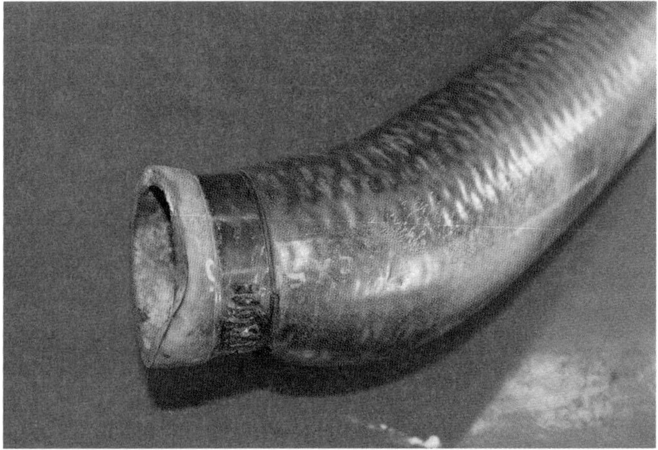

FIGURE 17-4 Swollen hose due to broken cords.

FIGURE 17-5 Hardened/brittle hose due to old age and heat.

▶ TECHNICIAN TIP

"You can't diagnose bad if you don't know good." This saying is very appropriate when checking radiator hoses. It is a good idea to check the feel of radiator and heater hoses on a variety of new vehicles to become familiar with what new hoses feel like. Then it will be easy to identify the bad ones.

Hose clamps come in several forms and require different tools to properly remove or replace. These clamps secure the hose to the component it is connected to. Be sure to follow proper installation instructions to prevent future leaks from this seal. Types of clamps include:

- *Gear or worm clamp:* Adjust with a screwdriver or nut driver.
- *Banded or screw clamp:* Adjust with a screwdriver.
- *Wire or spring clamp:* It is not adjustable. It is fitted and removed with special hose clamp pliers, which have grooved jaws (**FIGURE 17-8**).

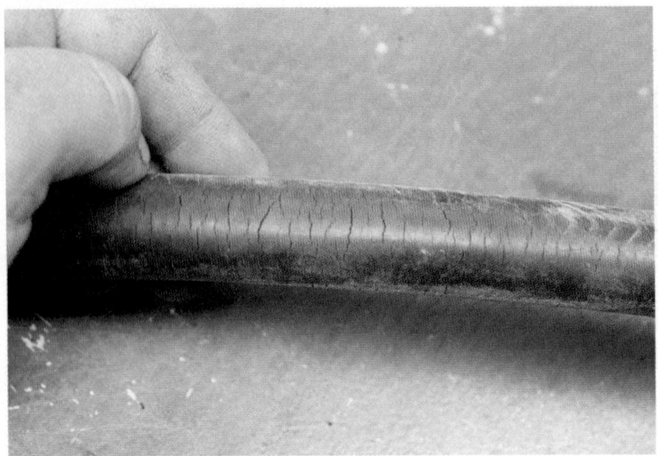

FIGURE 17-6 Cracked hose due to old age.

FIGURE 17-7 Soft hose due to old age or contamination from oil or gas.

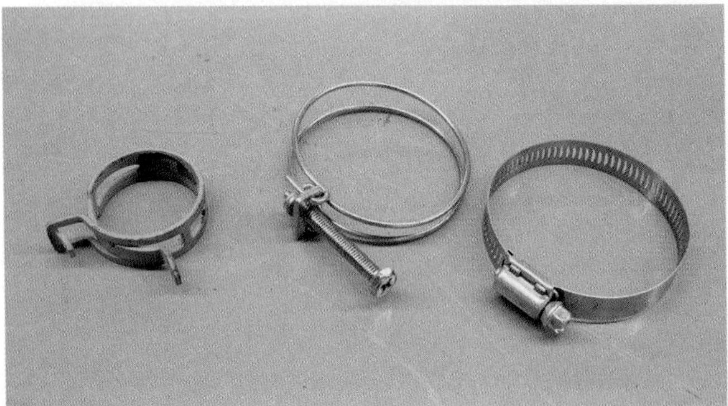

FIGURE 17-8 **A.** Spring clamp. **B.** Banded or screw clamp. **C.** Gear or worm clamp.

Clamps are not expensive, so it is good practice to install new ones at the same time as new hoses. Even if not corroded, the old clamps may have become distorted when being removed from an unserviceable hose. The time and money associated with replacing clamps is very small compared to the hassle the customer would experience if a used hose clamp fails.

When removing hoses from the radiator and heater core, be very careful that you don't damage either one, as the tubes are usually either thin metal or plastic. If metal, then the tubes are soldered to the radiator or heater core. If you try to twist the hose off, you can twist the tube out of the solder joint. You also shouldn't try to pry the hose off with a screw driver as that can dent the tube. which can cause leaks. So the best way to remove hoses is to slit them and carefully peel them off if they are stuck and don't come off easily.

To check and replace a coolant hose, follow the steps in **SKILL DRILL 17-11**.

SAFETY TIP

Never try to assess the serviceability of a coolant hose while the engine is hot. Let it cool so that you can handle the hoses comfortably and safely. Always ensure the engine is turned off before attempting to check the coolant hoses, and always wear the appropriate personal protection equipment before starting the job. It is easy to hurt yourself, even when the most exhaustive protection measures are taken.

Removing and Replacing a Thermostat

N17004

It will be necessary to remove and replace a thermostat in the event that the thermostat is found to be faulty, creating either an overheating concern or an underheating concern. You may want to suggest to the customer a thermostat replacement during a cooling system flush or water pump replacement as a precautionary maintenance step. Before starting a repair or service task on the cooling system, allow sufficient time for the system to cool adequately before opening the pressurized system.

Drain at least 50% of the coolant in the system to avoid spills. Once the thermostat has been removed, clean any old gasket material and corrosion that has built up on both sealing surfaces where the thermostat seats. Properly position the thermostat air bleed valve (if equipped). Always install a new gasket and/or O-ring seal when installing the thermostat. Make sure the thermostat is facing the correct direction (sensing bulb toward the engine) and that it is fully seated in the groove and stays there, before installing the housing. In most cases, if the thermostat falls out of its recessed groove, one of the thermostat housing ears will break off when the bolts are tightened, requiring replacement. Also, the thermostat itself is likely to be damaged. Tighten the housing bolts to the correct torque. Use the manufacturer's procedure to properly bleed all air from the cooling system.

To remove and replace a thermostat, follow the steps in **SKILL DRILL 17-12**.

▶ Diagnosis

S17001, N17005

The cooling system is an often overlooked part of the automobile. It does its job very effectively and is rarely thought of until problems arise. The lack of periodic service of the cooling

SKILL DRILL 17-11 Checking and Replacing a Coolant Hose

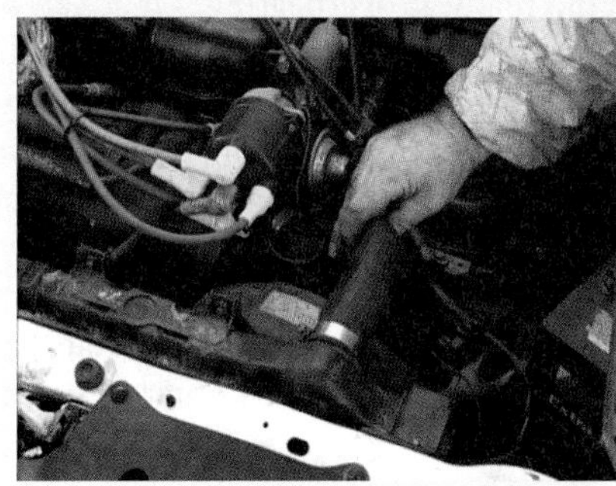

1. Inspect the hoses by squeezing them, and visually inspect the clamps. Remove the hose clamp. Carefully pull or cut the hose off the fittings.

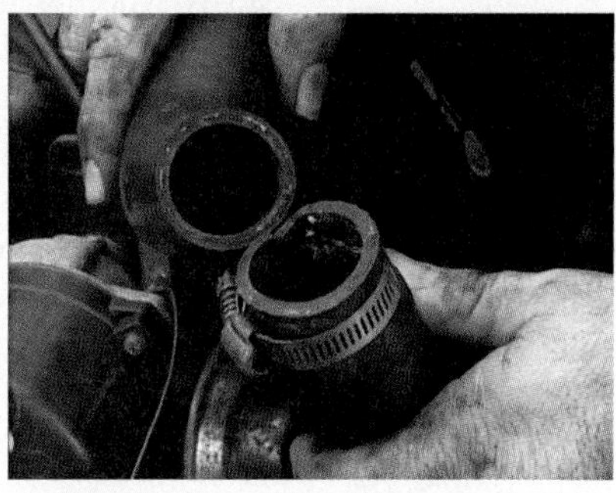

2. Verify that the replacement hose is correct. Cut to length if necessary. Reinstall the new hose all the way into position. Install new clamps.

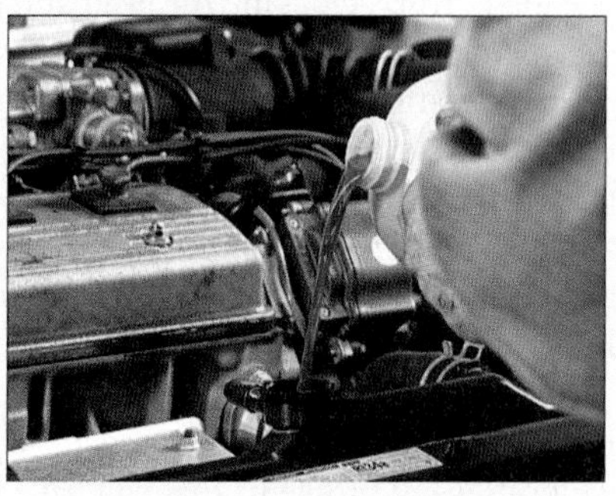

3. Refill coolant, verify correct level, and pressure check for leaks.

system according to OEM maintenance schedules is the leading reason that cooling systems fail. Properly performed cooling system maintenance will ensure that the cooling system continues to perform as designed.

Common failures of the cooling system can require repairs ranging from simple to complex. An example of a failure that would require a simple repair is a radiator hose clamp that has not been installed properly and is creating a leak. An example

SKILL DRILL 17-12 Removing and Replacing a Thermostat

1. Drain the coolant from the cooling system, using an approved catch container. Remove the thermostat.

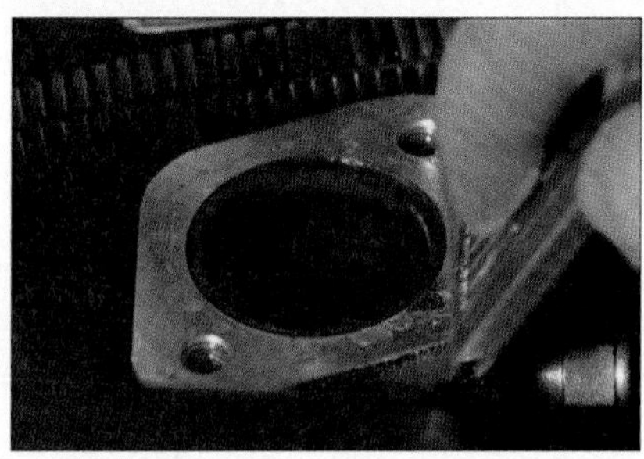

2. Inspect the thermostat housing, and clean any gasket material from the mating surfaces. Clean and inspect the thermostat groove.

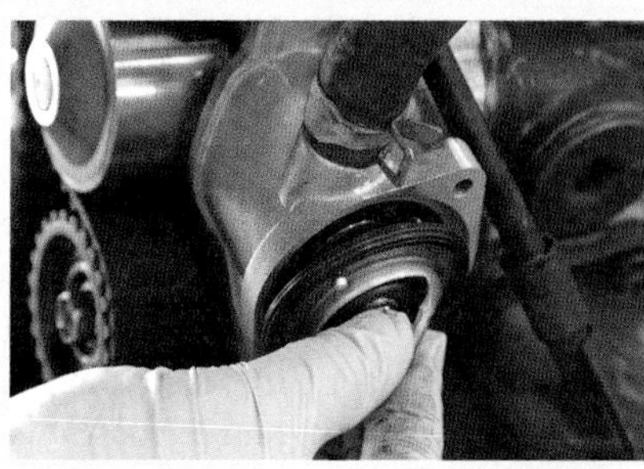

3. Install the new thermostat facing the correct direction and with the bleed valve in the correct position.

4. Install the gasket or O-ring and thermostat housing.

5. Torque to housing specifications.

6. Refill the cooling system to the proper level.

of a more complex repair is a vehicle that needs a new head and block because the antifreeze protection level was too low and the coolant froze, cracking the head and block.

Corrosion can build up to the point that normal cooling system operation cannot be restored without replacing multiple parts, including the engine itself, due to corrosion pitting of the engine block or cylinder head, leading to an internal coolant leak. Corrosion can also coat and/or block coolant passages and lead to an engine that runs hotter than normal or overheats.

Another common customer complaint is that the heater does not produce heat when turned on. This issue can be the result of several problems that will have to be diagnosed by the technician. Often the culprit is that the thermostat is stuck open and will not let the engine come up to operating temperatures. If the thermostat fails, it can fail in either the open or the closed position. If it sticks closed, the result will be an overheating condition because coolant cannot get to the radiator to be cooled. If it sticks open, the engine will not warm up fully, leading to excessive engine wear, higher emissions, and reduced fuel economy.

Overheating conditions can have several causes, such as low coolant level due to a leak, a stuck-closed thermostat, a faulty radiator cap, clogged radiator tubes or fins, an inoperative cooling fan, a water pump impeller that is eroded or slipping on the shaft, or a blown head gasket, to name a few. Understanding how the cooling system works, as well as the manufacturer's test procedures, is necessary to successfully diagnose and repair these faults.

When diagnosing a cooling system problem, you need to use your senses to help determine what is wrong. Your sense of touch can allow you to feel for hot and cold parts in the system. For example, if the engine is overheating, but the upper radiator hose is cold, that could indicate a thermostat that is stuck closed. Your sense of smell could pick up the scent of antifreeze leaking into the passenger compartment from a leaky heater core. Your sense of hearing can tell you that the accessory belt is loose and the water pump is slipping. And your sense of sight could see the telltale stream of coolant leaking out of the water pump weep hole or a rusted-through soft plug. However, it is advisable not to use your sense of taste when diagnosing cooling system issues.

If the problem cannot be determined by your senses alone, you may have to resort to tools or equipment to help you locate the issue. A cooling system pressure tester allows you to pressurize the cooling system and radiator cap to check for any visible leaks. Starting the engine and using an exhaust analyzer to sniff the vapors coming out of the neck of the radiator can indicate if there is a small leak from one of the combustion chambers, which is leaking combustion gases into the cooling system. Using a cylinder leakage tester and pressurizing each cylinder can help locate which cylinder has the combustion leak, due to either a blown head gasket or a cracked head. If the leak is a small external leak and the source cannot be located easily, you may need to use a fluorescent dye and ultraviolet light to make it stand out better.

An infrared temperature gun is a useful tool that can be used to measure the operating temperature of the engine to verify whether the engine really is overheating. Or it can be used to check the radiator for cold spots, which would indicate blockage within the core of the radiator. A scan tool provides a quick way to monitor the operating temperature, verify any cooling system diagnostic trouble codes (DTCs), or command the electric fan to come on. As you can see, understanding how cooling systems operate and observing what is happening, as well as having a few tools available, can take you a long way toward diagnosing customers' cooling system concerns.

Tools

A properly trained and experienced technician uses special tools for diagnosing and servicing the engine cooling system, including the following (**FIGURE 17-9**):

- *Coolant system pressure tester:* Used to apply pressure to the cooling system to diagnose leakage complaints. Under pressure, coolant may leak internally to the combustion chamber, intake or exhaust system, or the engine

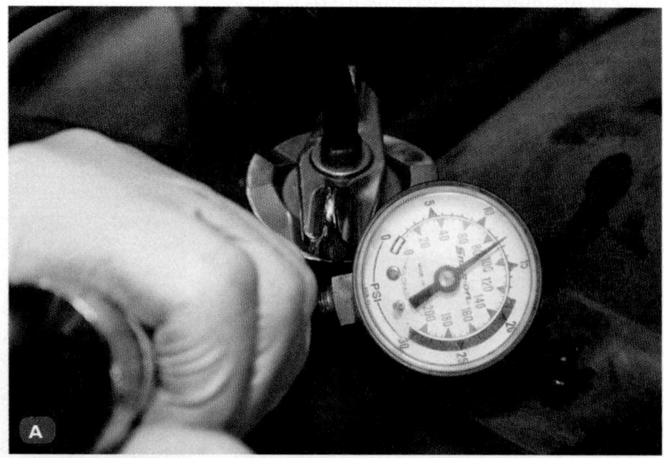

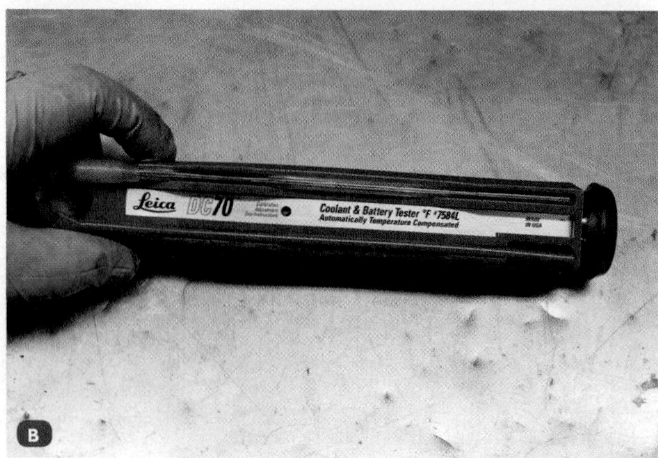

FIGURE 17-9 Common tools needed to service cooling systems. **A.** Pressure tester. **B.** Refractometer.

lubrication system. It can also leak externally to the outside of the engine.

- *Hydrometer:* Used to test coolant mixture and freeze protection by testing the specific gravity of the coolant. You must use a hydrometer specifically designed for the antifreeze you are testing.
- *Refractometer:* Used to test coolant mixture and freeze protection by testing the fluid's ability to bend light. This tester can be used with any type of antifreeze.
- *Coolant pH test strips:* Used to test the acid-to-alkalinity balance of the coolant.
- *Coolant dye kit:* Used to aid leak detection by adding dye to coolant and using an ultraviolet light source (black light) to trace the source of the leak; the dye glows fluorescent when an ultraviolet light is shined on it.
- *Infrared temperature sensor:* A noncontact thermometer used to check actual temperatures and variations of temperature throughout the cooling system, to help pinpoint faulty parts and system blockages.
- *Thermometer:* Used to check the temperature of air exiting the heating ducts.
- *Voltmeter:* Used to check for electrical problems such as cooling fan and temperature gauge issues.
- *Belt tension gauge:* Used to check belt tension.
- *Serpentine belt wear gauge:* Used to check whether the serpentine belt grooves are worn past their specifications.
- *Hose clamp pliers:* Used to safely remove spring-type radiator clamps.
- *Borescope:* Used for examining internal passages for evidence of a coolant leak.
- *Scan tool:* Used to activate the cooling fan through bidirectional controls for testing; to monitor cooling sensor operation and to command air door actuators when testing low heat complaints; and to read DTCs related to cooling system operation.
- *Cooling system flush machine:* Used to flush coolant backward through the system with cleaners that remove corrosion buildup and old coolant. Most of these machines have their own pump, so the vehicle does not have to run to perform the flush.
- *Exhaust gas analyzer:* Used to detect exhaust gases that are finding their way into the cooling system due to a leaking head gasket or damaged head or block. Be careful not to allow liquid coolant to be picked up by the analyzer probe.

Visual Inspection

Many times a visual inspection of the cooling system will give you a good indication of any issues. Check the level of the coolant in both the overflow bottle and radiator. At the same time, check the condition of the coolant to see if it is cloudy or contaminated. Also check the belt condition and for the proper tension, and check hoses for any leaks or wear. If the sweet aroma of coolant is detected when the engine is first started, check for leaking coolant at the heater box drain, which would indicate a possible cracked or rotted leaking heater core. Check the engine exhaust for excessive white smoke. A head gasket failure or a cracked head or valve seat may cause coolant to leak into the combustion chamber and be seen as white smoke, especially when combustion pressures are high when accelerating. Also start the engine with the radiator cap off, and look for bubbles from combustion in the radiator. Bubbles would indicate a leak in the combustion chamber, likely at the head gasket. Disassembly and inspection would need to occur to verify that it is the head gasket and not a cracked head or block.

SAFETY TIP

When working around the cooling system, care must be taken, particularly if the engine is at operating temperature, as the coolant may be hot enough to scald. Always allow the system to cool before removing the radiator or pressure cap, and use extreme caution when removing the radiator cap. If you must remove the radiator cap from a hot system, wear protective gloves and eyewear and place a cloth fender cover or other large rag on top of the radiator cap before releasing it slowly, to the first (safety) point, to prevent the pressure inside from erupting.

Testing the Cooling System for Leaks

N17006

Pressure testing the cooling system for leaks is usually an effective way to locate leaks because it causes coolant to leak out much more quickly, making leaks easier to locate. But before you pressurize the system, make sure it is topped off with coolant or water. Otherwise, the leak could be above the coolant level and only leak air, which is much harder to observe. Topping off leads to quicker pressurization because it removes air, which is compressible. Also, if the system is full of liquid, the pressure reading on the gauge will fall faster if there is a leak in the system.

The pressure tester puts pressure on the cooling system, and with pressure applied, leaks generally show up easily, as identified by coolant coming from the source. The use of a droplight and a mirror may be necessary to see behind the engine or in tight areas. Don't forget to check the heater core in the passenger compartment. If the pressure gauge indicates losing pressure, ensure that the tester is installed correctly; if it still loses pressure, ensure that the tool is working properly. Once the tool is verified and you cannot find an external leak, you can test to see if the cooling system is leaking internally into the engine. Check engine oil for evidence of coolant. It will have a milky appearance on the dipstick. If coolant is not leaking into the oil, it could be leaking into the cylinder. Remove the spark plugs and look for coolant being burned in the cylinders, as evidenced by a color-stained spark plug insulator. A combustion chamber experiencing a coolant leak will appear to be steam cleaned. A borescope can be used to inspect the cylinders through the spark plug holes.

Normally the engine should be off when carrying out any visual inspection of the system or when you connect test equipment such as the pressure tester. However, it is possible that the leak only occurs when the engine is running, such as around the water pump seal, making it necessary to run the engine while testing. If you do have to run the engine after the tester has been installed and pressurized, make sure to watch the pressure gauge, and release excess pressure as the engine heats up. When the engine is running, make sure you keep well away from any rotating or hot parts.

Also remember to pressure test the radiator cap, as a leak at the cap prevents the cooling system from building pressure. This lowers the boiling point, which can cause the coolant to boil at normal operating temperatures. Many a technician has been fooled into thinking that the vehicle had a serious overheating problem after changing the thermostat, water pump, etc., when all it needed was a new radiator cap.

Most pressure testers are hand-operated and come with a number of adapters to fit a variety of cooling systems. Adapters are used to connect the tester to the radiator, surge tank, or radiator cap.

To test the cooling system pressure, follow the steps in **SKILL DRILL 17-13**.

> **TECHNICIAN TIP**
>
> If you need to replace a pressure cap, use only a cap with the correct recommended pressure. If a cap with a lower pressure rating is installed, it could cause the coolant to boil over. Alternatively, a higher-rated cap will increase the pressure in the system and could result in a hose, radiator, or heater core bursting.

Verifying Engine Operating Temperature

N17007

Technicians need to verify engine operating temperature whenever the customer complains about an overheating issue, an underheating issue such as inadequate heat from the heater, or poor fuel economy. Engine operating temperature can be verified in a couple of ways. First, if the vehicle is equipped with a dash-mounted temperature gauge, it provides a good indication of the engine's temperature, as long as it is working correctly. If the gauge reads too cold or too hot, or even if it is reading normally but you suspect the engine temperature is not as indicated, you can verify it by using either an infrared noncontact temperature gun or a scan tool. The temperature gun, as described earlier, measures the amount of heat energy (temperature) of an object. Just realize that some objects do not conduct heat as well as others. If you can point the temperature gun at a metal component, it will produce a more accurate reading. If pointed toward the engine's thermostat housing (or next to it), with the engine fully warmed up, a close approximation of the engine's operating temperature can be measured and compared to the specifications. Another way to verify the engine temperature is to use a scan tool to access the data for the engine coolant temperature sensor, and compare that to both the dash gauge and the temp gun readings. Comparing all three readings is good practice as it will not only verify the engine operating temperature but also the temperature reading on the temperature gauge on the dash as well as the coolant temperature signal that the vehicle's PCM receives.

Note that a vehicle with an electric fan usually has two listed temperatures: one temperature at which the fan should turn on and another temperature at which it should turn off. The normal operating temperature is anywhere between those temperatures in this situation.

To verify the engine operating temperature, follow the steps in **SKILL DRILL 17-14**.

Inspecting and Testing the Cooling Fan

N17008

The cooling fan is a critical part in the cooling system. When coolant moves through the engine, it picks up heat. Even if the coolant is picking up the heat from the engine and moving it to the radiator,

SKILL DRILL 17-13 Testing the Cooling System Pressure

1. Inspect for leaks. Verify specified cooling system pressure, install the radiator cap on the pressure tester, and pressurize the cap to the correct pressure. It should hold pressure at approximately the rated pressure and vent at slightly above the rated pressure.

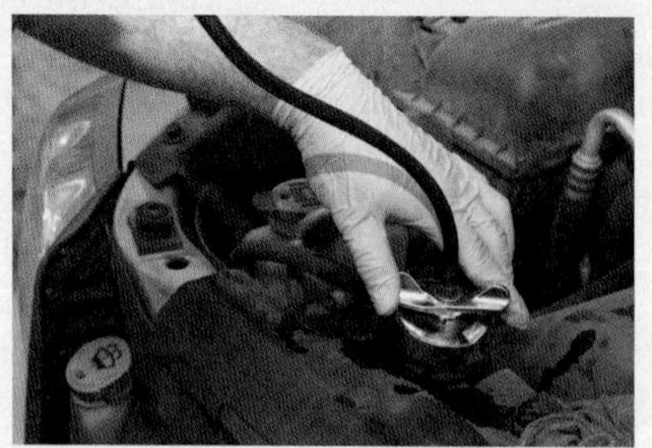

2. Top off the radiator with coolant or water, and install the tester. Pressurize the system to the specified cap pressure.

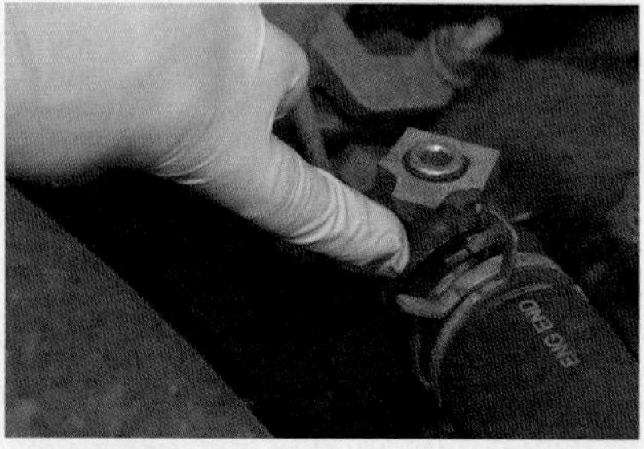

3. Watch the pressure reading for a drop while performing a visual check for any leaks. Check heater hoses, soft plugs, and any heater cores; determine necessary action.

SKILL DRILL 17-14 Verifying Engine Operating Temperature

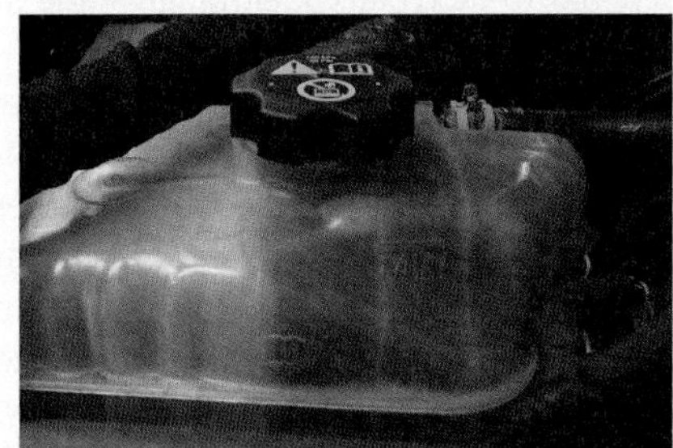

1. Verify that the coolant level is correct before starting the engine. If low, check for the presence of a leak before measuring the operating temperature.

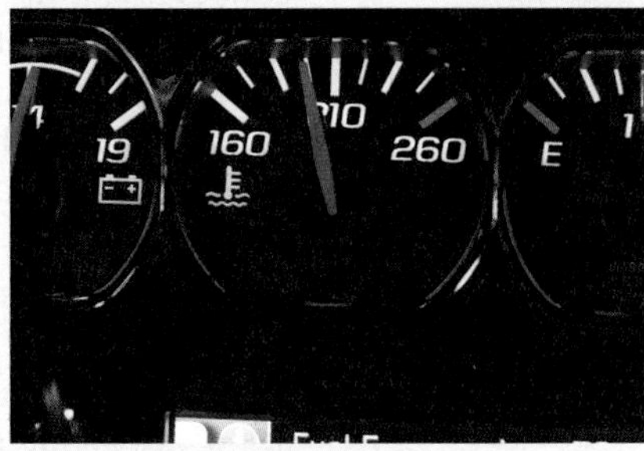

2. Start the engine and allow it to reach full operational temperature, monitoring the temperature along the way in case it starts to overheat.

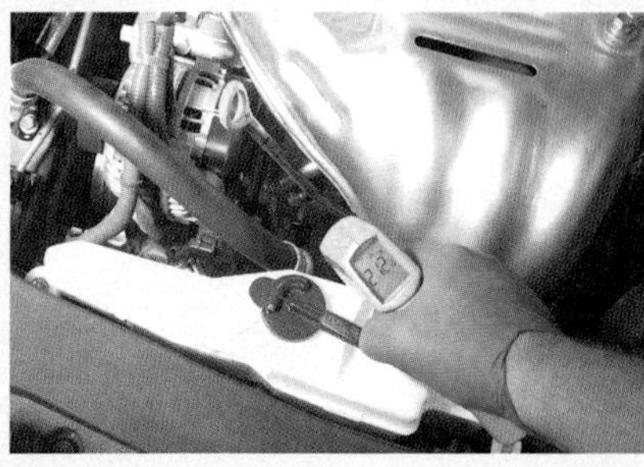

3. Using an infrared temperature gun, test the temperature of the engine near the location of the thermostat or coolant temperature sensor. Compare to specifications.

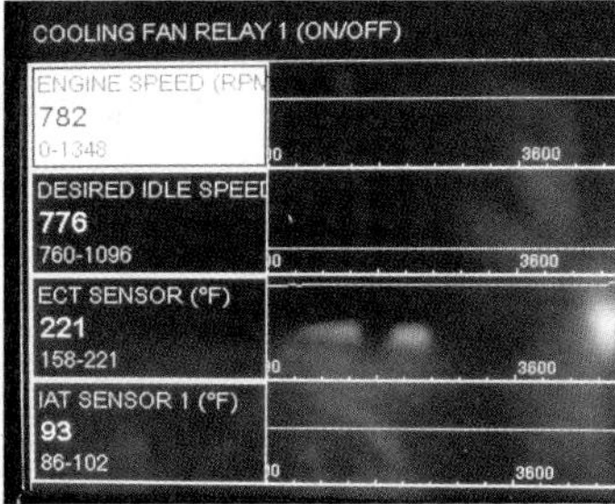

4. Using a scan tool, retrieve the engine coolant temperature sensor temperature reading.

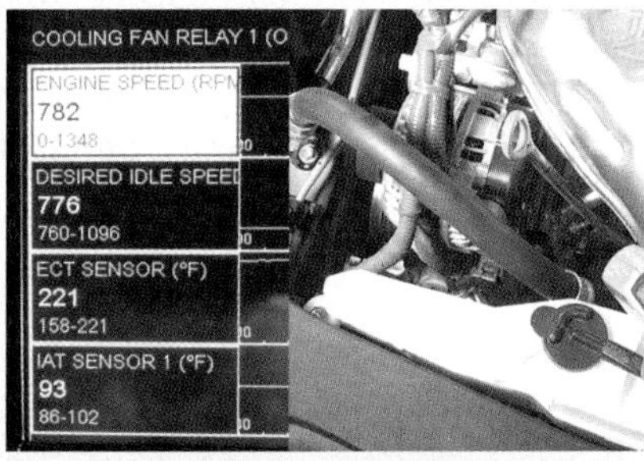

5. Be aware that there will be some temperature difference because the temperature sensor is sitting in the coolant, and the temperature gun is measuring the surface temperature near the sensor.

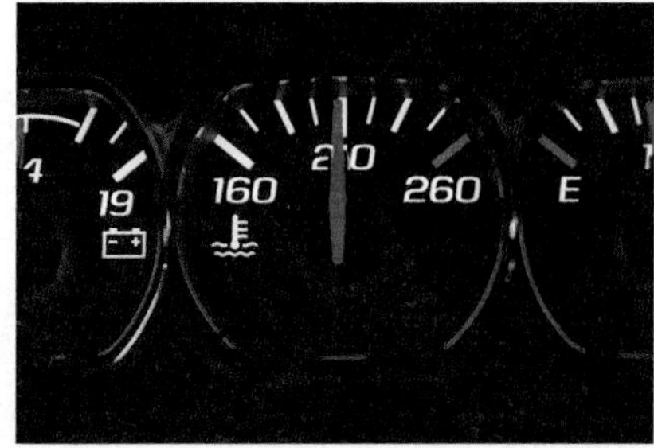

6. Compare the results from the temperature gun, the coolant temperature sensor, and the vehicle's temperature gauge; this gives you a good idea of the engine's operating temperature.

air still has to move across the radiator to remove the heat from the coolant. The cooling fan must operate to pull air across the radiator at times when airflow is low, such as when sitting in traffic. If the cooling fan stops working and the vehicle sits in traffic, engine heat will continue to build, possibly until the coolant's boiling point is reached. If reached, the pressure relief valve in the radiator cap will vent, releasing steam from the radiator and producing a hissing sound. The customer may note that a temperature warning light has come on and/or the temperature gauge has moved toward its full hot position. If the engine is operated in this condition, it will be damaged. It is possible to even create damage to the automatic transmission if the engine gets too hot, as the transmission fluid is typically cooled through the cooler located in the radiator.

When testing the fan, start by determining what type of fan the vehicle is equipped with—mechanical fan, clutch fan, electric fan, or hydraulic fan. Mechanical fans are the easiest to inspect, typically requiring only a visual inspection. Check the condition and tension of the drive belt, the condition of the pulley, and the condition of the fan blades themselves. Especially look for cracks in any metal fan blades, as they can cause the blade to be ripped off and thrown with deadly force. It is also possible that someone previously installed the fan blades backward, which lowers the effectiveness of the fan (**FIGURE 17-10**).

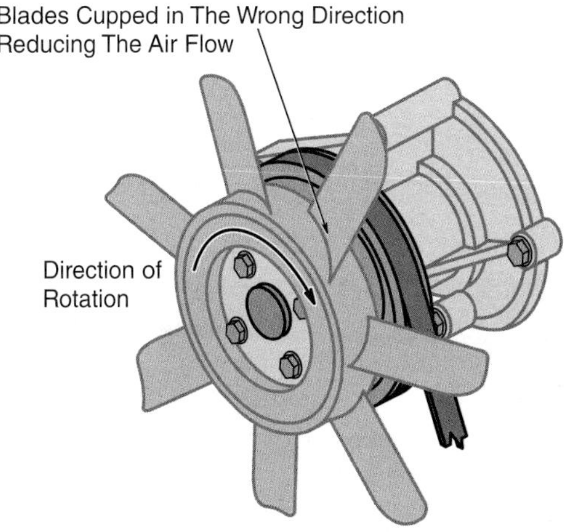

FIGURE 17-10 The mechanical fan has been installed backward on this vehicle, making it much less efficient at moving air.

A clutch fan is inspected in the same way as the mechanical fan, but the clutch also must be tested. First, with the engine off, the bearings in the fan clutch can be tested by trying to move the blades frontward and backward, and then rotating them to see if there is any play in the bearings in any direction (**FIGURE 17-11**).

FIGURE 17-11 Testing a fan clutch for excessive play.

To test the operation of the fan clutch, some manufacturers specify placing cardboard in front of the radiator to block most of it. The engine is then started with the vehicle in the shop, while monitoring the engine temperature and watching and listening to see if the fan clutch engages at the proper engine temperature. If it does not, it most likely has to be replaced, although some fan clutches can be refilled with silicone oil.

Electric fans are quite reliable. If they do cause trouble, it is usually because they don't come on when they are supposed to. Because electric fans are electrically operated, diagnosing them is quite different from diagnosing mechanical fans. Be aware that some electric fans can come on automatically, even when the key is off, so be careful when working around them. In fact, you should unplug the fan's electrical connector before physically inspecting the fan assembly. Just be sure to reconnect it when you are finished. Once the fan is disconnected, check that the fan is free to turn and does not catch or bind on any of the shrouding. If it is free to rotate, then the electrical system and the fan motor itself will have to be tested. Remember, the electric fan only operates when the engine is at the upper end of the engine's operating temperature, so it shouldn't run all the time.

To inspect and test the electric cooling fan, follow the steps in **SKILL DRILL 17-15**.

SKILL DRILL 17-15 Inspecting and Testing the Cooling Fan

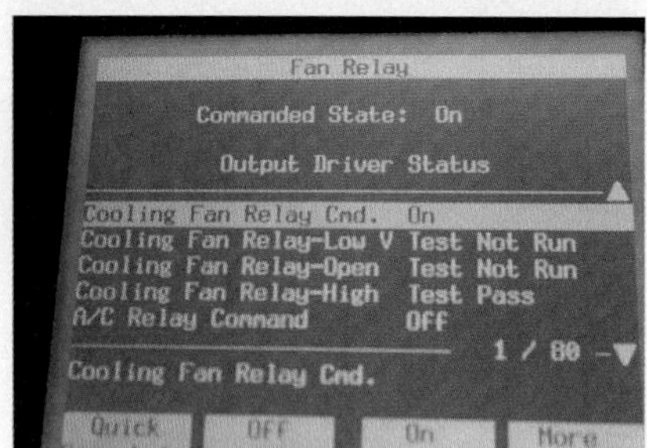

1. Install the scan tool. Find the bidirectional controls, and command the fan on. Verify fan operation.

2. Use a temperature gun to measure the operating temperature, and compare its reading to the coolant temperature displayed on the scan tool. If substantially different, diagnose the issue and determine necessary actions.

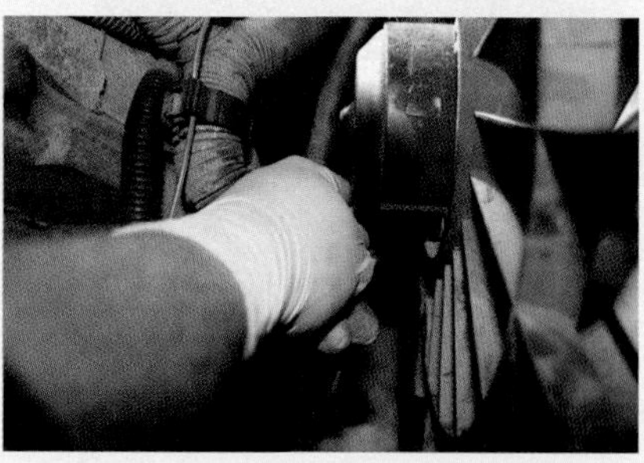

3. If the fan does not operate, wiggle test the fan wires.

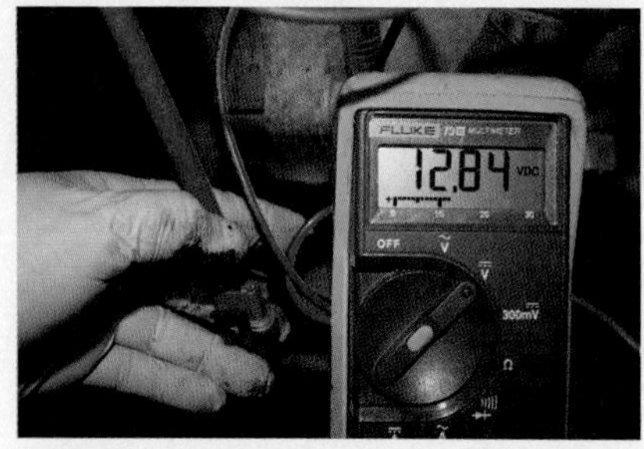

4. If it still does not work, measure the voltage to the input terminal of the fan. If substantially less than battery voltage is present, use a wiring diagram of the fan circuit to diagnose the cause.

5. If voltage is present to the input terminal, measure the voltage drop on the ground side of the fan.

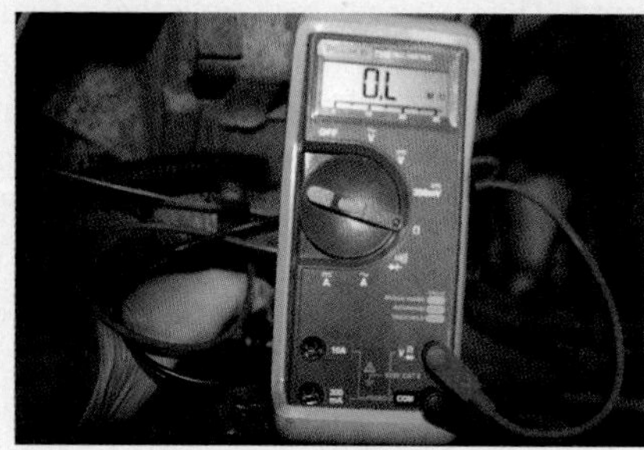

6. If the voltage drop is acceptable on the ground side, measure the continuity of the fan motor with an ohmmeter. If out of specs, replace the fan motor.

Inspecting and Testing Fans, Fan Clutch, Fan Shroud, and Air Dams

The need to inspect and test fans and fan shrouds will arise when the customer complains of overheating. Air dams are located under the vehicle and are designed to direct airflow into the radiator as the vehicle is moving down the road. If the air dam is damaged or missing, the engine temperature may be higher than originally designed while driving on the highway. Air dams are easily damaged by customers who hit curbs with the front of the car when parking. So always be on the lookout for damaged air dams.

If the vehicle only overheats while stopped, it could be an issue with the fan shroud, which directs airflow through the radiator. The shroud is what surrounds the cooling fan and allows a maximum amount of air to be moved by the cooling fan. If the shroud is missing, air will be pulled from around the fan blades rather than pulling air through the radiator. Sometimes a customer will complain about a drumming noise when accelerating or going up hills. This could be caused by the engine fan hitting the fan shroud because of broken motor mounts. If this is the case, motor mount replacement is required. A visual inspection of these components will verify whether they are in proper position and operating as designed.

A mechanical or electric fan can fail, which will also result in air not being pulled through the radiator. A quick test to verify that air is moving through the radiator is to hold a piece of paper in front of the radiator and see if suction pulls the paper in. If it does not, then air is not being pulled through the radiator. Follow the service manual information on diagnosing the fan equipped on your vehicle.

To inspect the shroud/air dam and test the electric cooling fan, follow the steps in **SKILL DRILL 17-16**.

SKILL DRILL 17-16 Inspecting the Fan Shroud/Air Dam and Testing the Electric Cooling Fan

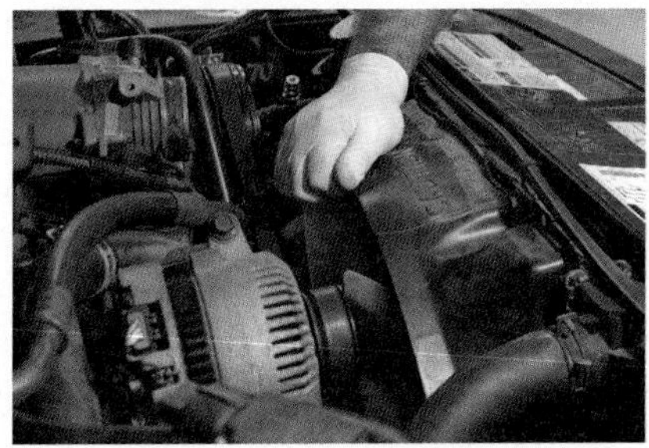

1. Visually inspect all shrouds, ducts, and air dams around the radiator and fan for damaged or missing pieces. If equipped with a fan clutch, rotate the fan and check it for looseness, play, or binding. If bad, replace the fan clutch.

2. If the fan clutch feels free, place cardboard or a fender cover in front of most of the radiator. Start the engine, monitor the engine temperature with a temperature gun, and verify that the fan cycles on at the proper temperature. If not, replace the fan clutch.

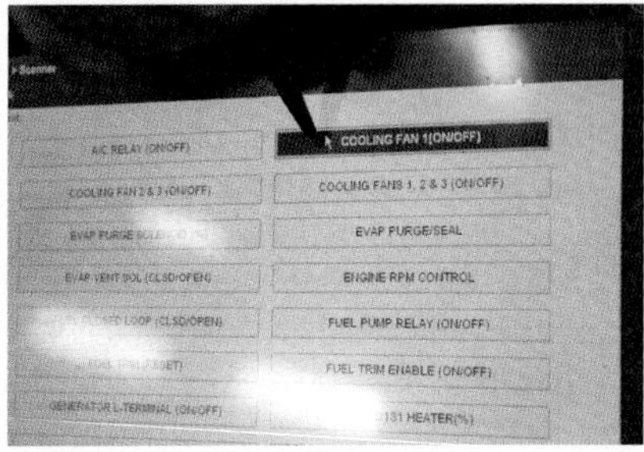

3. If equipped with an electric fan, connect a scan tool to the data link connector under the dash. Locate the bidirectional command menu, and command the cooling fan on. This activates the relay and should send power to the fan.

4. Verify that the fan is running properly. Verify that air is flowing across the radiator. If the fan is not operating, locate the cooling fan relay, remove the relay, and check for the presence of power and ground at the appropriate terminals in the relay block. Two terminals should have battery positive voltage, and two terminals should have battery negative voltage. If not, diagnose the faulty circuit.

5. If all terminals read properly, test the relay winding for continuity. If out of specifications, replace the relay.

6. If okay, install a relay socket tester, reinstall the relay into the relay box, and then activate the electric fan. Measure the voltage drop on each leg of the circuit and the relay contacts themselves. Trace the path that indicated a problem, and locate the voltage drop or defective component. Determine the necessary actions based on your findings.

Inspecting and Testing Heater Control Valves

N17009

Heater control valves control the flow of coolant to the heater core so that the operator can control heater output. If this valve becomes stuck open, the operator will not be able to turn down the heat, or the air conditioning will not be as cold as before. If the valve sticks shut, the customer will complain of no heat in the vehicle. The problem can be with the valve or the control system. The valve can get plugged up with contaminants in the coolant, or the lever can slip on the shaft that turns the valve. If it is cable operated, the cable can slip, or even corrode in place. If it is vacuum operated, the diaphragm can get a hole in it, the vacuum hose can leak or fall off, or the vacuum controller can quit working. If it is electrically operated, the motor can seize or the circuit can go bad. Heater control valves can also leak coolant, so they should be inspected for leaks. In fact, one of the first indicators of low coolant in a vehicle is that the heater stops blowing hot air, or blows hot air intermittently. Always verify that the coolant is full before spending too much time tracking down any kind of heater issues.

To test the heater control valve, follow the steps of the **SKILL DRILL 17-17**.

Removing and Replacing a Radiator

N17010

A radiator may need to be replaced due to several reasons. One reason is an overheating condition caused by partially plugged radiator tubes, or plugged radiator fins. If the radiator fins are blocked by debris, you will probably need to remove the radiator to carefully clean the fins with a dry brush or garden hose. Just make sure you don't bend the fins over or damage them in any way. If the radiator tubes are plugged, the radiator will have to be removed and sent out to a radiator shop for cleaning. A good flush of the entire cooling system will be necessary as further clogging will be likely, especially in the heater core.

Another reason to remove the radiator is for leaks traced to the radiator through the use of the coolant pressure tester or coolant dye and an ultraviolet light. Some radiators can be repaired by a radiator shop, but a cost analysis must be performed to see if doing so is more cost effective for the customer than buying a new radiator.

When replacing the radiator, make sure the coolant catch tray is large enough to catch any spills and has the capacity to hold all the coolant from the system. If you are replacing the coolant, dispose of the old coolant properly in accordance with environmental and legislative requirements. If reusing the old fluid, keep it stored in a covered and uncontaminated container. Most situations will require replacement of the coolant. However, if, for example, the vehicle just had new coolant installed, you may want to catch and reuse the coolant.

Inspect the cooling system hoses and clamps. Replace them if worn or damaged. When removing the hose from the radiator fitting, do *not* twist it; doing so can damage the fitting or even rip

SKILL DRILL 17-17 Inspecting and Testing Heater Control Valves

1. Verify that the cooling system is topped off, and bring the engine to full operating temperature. Locate the heater control valve.

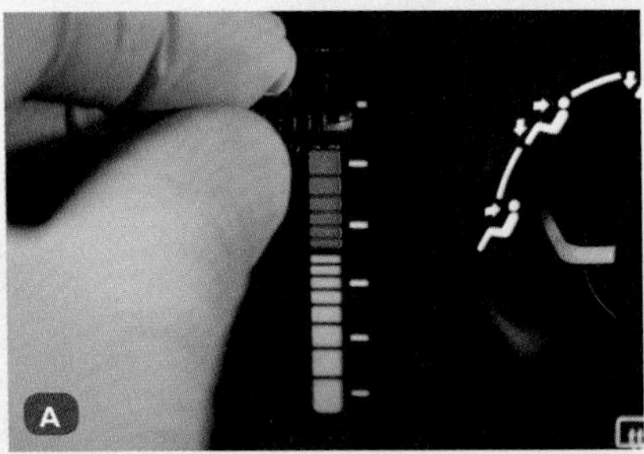

2. Move the temperature control to hot, and measure the vent temperature. It should be hot.

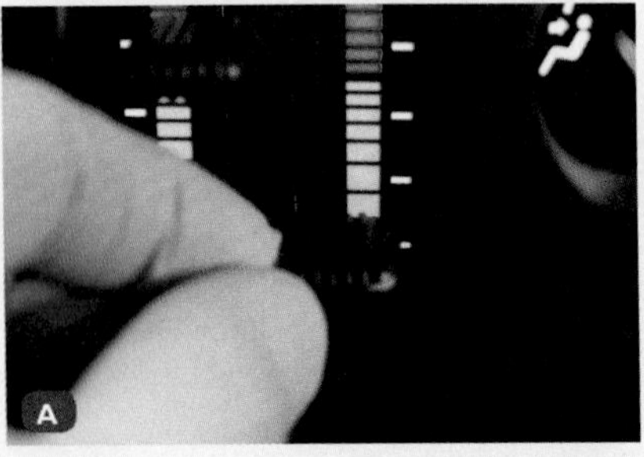

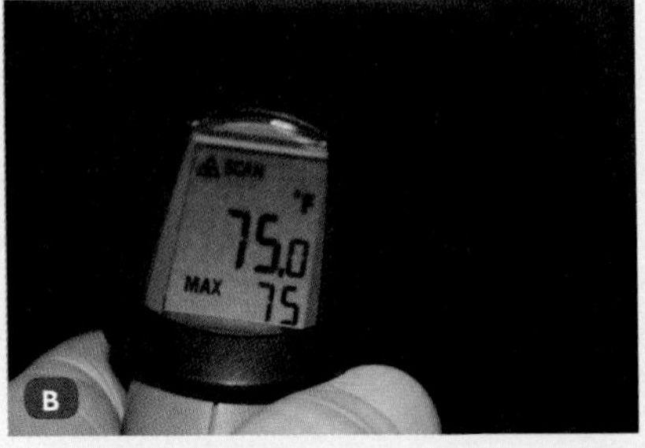

3. Move the temperature control to cold and measure the vent temperature. It should be cool.

it out of the radiator. If it does not easily release, you might be able to carefully work a tool between the hose and the fitting, breaking it loose all the way around. If the hoses are to be replaced, you can carefully slit them with a knife or hose-cutting tool and peel them off the fitting, which is less likely to damage the fitting.

When replacing hoses, reinstall them all the way onto the fittings. Make sure the clamps are installed just inside the flared or barbed segment, not on top of it. Placing a clamp on top of the flare or barb will likely result in the hose being blown off the fitting once the engine warms up and the system is pressurized. Many vehicles have automatic transmission cooler lines attached to the radiator. Remember to disconnect these lines when removing the radiator, and always reinstall them before refilling the system with coolant. Remember when you disconnect these lines that transmission fluid will leak out. Do not mix coolant and oil! Using a separate catch container will ensure that coolant and oil are not mixed.

Refill the system with the correct coolant at the proper antifreeze/water ratio. It is advisable to pressure test the system to check for leaks upon completion of the job. Start the engine, warm it up until the thermostat has opened, and check for proper operation of the cabin heater. Check for proper coolant level after it cools sufficiently.

To remove and replace a radiator, follow the steps in **SKILL DRILL 17-18**.

Removing, Inspecting, and Reinstalling the Heater Core

N17011

Removal of a heater core may become necessary if the customer notices a puddle of coolant on the carpet in the vehicle, usually on the passenger side. Typically, the customer may have the smell of coolant inside the vehicle or may complain of a film on the inside of the windshield, which upon inspection is coolant that is being discharged with the air when the customer uses the defroster. The removal of a heater core is typically a difficult task, as sometimes it may be necessary to pull the entire dash from the vehicle. It is highly recommended to take pictures of the process as you go, to make reassembly easier. Follow the manufacturer's service information to perform this task. The procedure in the following skill drill is generic and may be substantially different from the one for the vehicle you are working on.

To remove, inspect, and reinstall the heater core, follow the steps in **SKILL DRILL 17-19**.

Applied Science

AS-48: Centrifugal/Centripetal: The technician can explain the relationship of centrifugal/centripetal force to the functioning or a failure of a rotating system.

An automotive instructor is presenting a class on the vehicle's cooling system. He has an assortment of new and used water pumps to serve as training devices. The instructor begins by stating that the water pump operates on the principle of centrifugal force. As our text indicates, **centrifugal force** is a force pulling outward on a rotating body. *Centripetal* force is just the opposite. Wikipedia states that *centrifugal* is from a Latin word meaning "to flee the center." *Centripetal* is from a word meaning "to seek the center." These are forces which are opposite in direction.

The impeller is the rotating portion of the water pump which is equipped with fins to direct the coolant outward, based on the principle of centrifugal force. The correct flow rate is based on the design of the water pump's impeller and the restrictions in the cooling system, namely the thermostat.

During class, new and used water pumps are inspected to find out the condition of the impeller. Erosion damage is visible on several of the used pumps in the area of the impeller fins. Erosion is a physical wearing away of the metal caused by contaminants in the coolant. These contaminants include dirt, grit or sediment circulating in the coolant. The instructor stresses proper cooling system maintenance to prevent premature failure of the water pump and other system components.

SKILL DRILL 17-18 Removing and Replacing a Radiator

1. Place a catch pan below the radiator, and remove the drain plug. Drain the coolant from the system. Replace the drain plug, and dispose of the drained coolant.

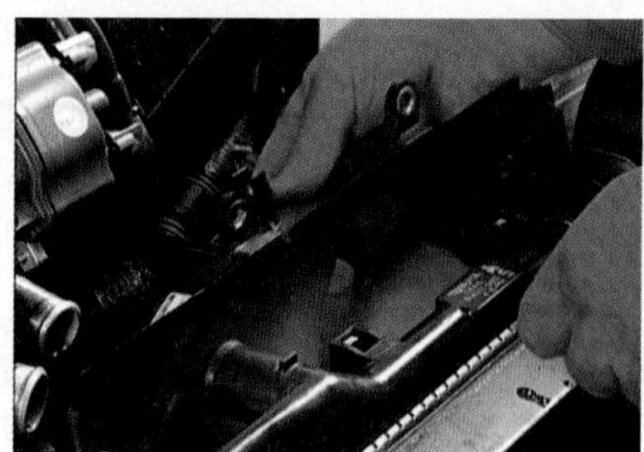

2. Unscrew any shrouds or covers from the radiator. Remove the bolts or screws that hold the radiator in position in the engine bay, and lift the radiator from its location.

3. Inspect the radiator, then place it into position and replace the securing bolts or screws. Refit the shrouds or covers. Rotate the fan and belts by hand to check that the covers do not restrict movement. Attach the coolant hoses to the radiator. Refill the system, and verify the correct level.

SKILL DRILL 17-19 Removing, Inspecting, and Reinstalling the Heater Core

1. Find service information on the removal and replacement of the heater core for the vehicle you are working on. Determine whether the air-conditioning refrigerant needs to be recovered.
2. Drain the coolant from the radiator.
3. If the air-conditioning evaporator needs to be removed, recover the air-conditioning refrigerant.
4. Remove necessary dash panels and covers, making sure to collect screws in a magnetic tray or plastic bag. Mark screws to remind you where they were removed from. This becomes more critical the longer the vehicle is apart; it can be hard to remember after a couple of weeks have gone by.
5. Remove heater core hoses (and remove air-conditioning lines to the evaporator, if necessary).
6. Locate all bolts securing the air box to the bulk-head of the vehicle, and remove, if necessary.
7. Pull the air box from the vehicle, if necessary.
8. Remove the bolts holding the air box halves together, and separate the halves.
9. Remove the heater core.
10. Now is a good time to clean the air box of any debris or gunk in the bottom around the drain; doing so could prevent a vehicle from coming back at a later date for an air-conditioning odor complaint. Inspect all air box doors to ensure there are no cracks or dry-rotting rubber or foam. If found, the doors will need to be replaced.
11. Install the new heater core, and reverse the steps to reinstall the air box. Properly refill the cooling system, and recharge the air-conditioning system.

▶ Wrap-Up

Ready for Review

- Cooling systems need regular maintenance to ensure engine longevity.
- Common cooling system diagnostic concerns include coolant leaks, thermostat that is stuck closed or open, faulty water pump, inoperative cooling fan, or failed head gasket.
- Test pH concentration of the coolant to ensure the antifreeze acid inhibitors are working correctly.
- Measure the voltage of the coolant to determine whether electrolysis is occurring due to acidic coolant or faulty grounds.
- Always follow EPA regulations for handling and disposing of coolant.
- Coolant condition and level should be checked regularly; coolant should be at the correct level mark (upper or lower) according to engine temperature.
- When removing and replacing a radiator, be sure to properly catch, handle, and dispose of or replace the coolant.
- Do not attempt to remove or replace the thermostat until the engine has cooled for at least 30 minutes.
- Potential engine drive belt problems can be belts that are cracked, oil soaked, glazed, torn, or bottomed out.
- Coolant freeze point can be verified by a hydrometer or refractometer, both of which indicate the specific gravity of the fluid, revealing the ratio of antifreeze to water.
- Removing and reinstalling the heater core can be challenging; taking pictures as you go is highly recommended for ease of reassembly.

Key Terms

centrifugal force A force pulling outward on a rotating body.

electrolysis The process of pulling metals apart by using electricity or by creating electricity through the use of chemicals and dissimilar metals.

ethylene glycol A chemical used as antifreeze that provides the lower freezing point of coolant and raises the boiling point. It is a toxic antifreeze.

Review Questions

1. A refractometer is used to test:
   a. coolant pH.
   b. the freeze protection of the coolant.
   c. electrolysis.
   d. coolant level.
2. Draining and refilling the coolant is necessary in all of the following cases *except* when:
   a. a customer requests it.
   b. the pH is out of specification.
   c. the coolant is contaminated.
   d. the coolant level is low.
3. Choose the correct statement.
   a. Belts should be inspected when the engine is running.
   b. Tears or split belts are serviceable.
   c. A bottomed out belt will slip and squeal.
   d. Always replace belts with cracks..
4. The best way to remove hoses from heater cores is:
   a. to slit them and peel them off.
   b. pry them off with a screwdriver.
   c. twist the hose off.
   d. to heat them mildly.
5. A borescope is used:
   a. for examining internal passages for evidence of a coolant leak.
   b. to check belt tension.
   c. to check the temperature of air exiting the heating ducts.
   d. to activate the cooling fan through bidirectional controls.

6. When testing the cooling system for leaks using a pressure tester:
   a. ensure there is only 50% coolant or water.
   b. the radiator cap need not be tested.
   c. do not adjust the pressure when the engine heats up.
   d. look for both internal and external leaks.
7. The engine's operating temperature can be verified using all of the following *except* a:
   a. dash gauge.
   b. refractometer.
   c. scan tool.
   d. temperature gun.
8. If the vehicle only overheats while stopped, it is likely an issue with the:
   a. heater core.
   b. thermostat.
   c. cooling fan.
   d. coolant.
9. When removing a radiator:
   a. do not disconnect automatic transmission cooler lines.
   b. inspect the cooling system hoses and clamps.
   c. do not drain the coolant.
   d. use the same container to hold transmission fluid and coolant.
10. All of the following should be checked for coolant leaks when pressure testing a cooling system *except:*
   a. Soft plugs.
   b. Heater core.
   c. Thermostat housing.
   d. clutch fan.

ASE Technician A/Technician B Style Questions

1. Tech A says that most coolants are designed to last the life of the vehicle. Tech B says that a hydrometer can be used to test the freeze point of coolant. Who is correct?
   a. Tech A
   b. Tech B
   c. Both A and B
   d. Neither A nor B
2. Tech A says that the thermostat may have an air bleed that must be positioned properly when the thermostat is replaced. Tech B says that the thermostat housing must be flush against the mating surface or the ear may break off when tightening the bolts. Who is correct?
   a. Tech A
   b. Tech B
   c. Both A and B
   d. Neither A nor B
3. Tech A says that coolant leaks can be located by performing a pressure test on the system. Tech B says that when pressure testing the cooling system, you should pressurize it to at least 10 psi (69 kPa) above the radiator cap pressure rating. Who is correct?
   a. Tech A
   b. Tech B
   c. Both A and B
   d. Neither A nor B
4. Tech A says that the radiator cap should be pressure tested as part of a cooling system diagnosis. Tech B says that the water pump weep hole could be the source of a coolant leak. Who is correct?
   a. Tech A
   b. Tech B
   c. Both A and B
   d. Neither A nor B
5. Tech A says that the engine operating temperature can be tested with an infrared temp gun. Tech B says that it can be tested with a scan tool. Who is correct?
   a. Tech A
   b. Tech B
   c. Both A and B
   d. Neither A nor B
6. Tech A says that an electric fan that runs all the time can cause the engine to overheat. Tech B says a refractometer is used to turn a clutch fan on and off. Who is correct?
   a. Tech A
   b. Tech B
   c. Both A and B
   d. Neither A nor B
7. Tech A says that one coolant test is a coolant electrolysis test. Tech B says that a thermostat that is stuck open will cause the engine to run cooler than it should. Who is correct?
   a. Tech A
   b. Tech B
   c. Both A and B
   d. Neither A nor B
8. Tech A says that there are a number of coolant types and each has its own life span. Tech B says that mixing of coolants is generally okay because all coolants use the same chemical base. Who is correct?
   a. Tech A
   b. Tech B
   c. Both A and B
   d. Neither A nor B
9. Tech A says that when replacing cooling system hoses, it may be best to slit the hose rather than trying to twist it off. Tech B says that cooling system hoses only need to be changed when they start leaking. Who is correct?
   a. Tech A
   b. Tech B
   c. Both A and B
   d. Neither A nor B
10. Two technicians are discussing cooling system hose replacement. Tech A says to make sure that the hose clamps get installed just inside the flared or barbed segment of the fitting, not on top of it. Tech B says that it is good practice to install new hose clamps at the same time as new hoses. Who is correct?
   a. Tech A
   b. Tech B
   c. Both A and B
   d. Neither A nor B

CHAPTER 18
Engine Removal and Replacement

NATEF Tasks

- **N18001** Remove and reinstall engine on a newer vehicle equipped with OBD; reconnect all attaching components and restore the vehicle to running condition. (MAST)

Knowledge Objectives

After reading this chapter, you will be able to:

- **K18001** Describe the purpose of individual engine removal tool.
- **K18002** Explain the safety concerns when removing and installing engines.

Skills Objectives

After reading this chapter, you will be able to:

- **S18001** Disconnect the topside components.
- **S18002** Relieve fuel pressure and remove fuel lines.
- **S18003** Disconnect the battery.
- **S18004** Disconnect electrical components and wire harnesses.
- **S18005** Disconnect the front accessories.
- **S18006** Disconnect and/or remove radiator.
- **S18007** Remove accessories and brackets.
- **S18008** Disconnect the underside components.
- **S18009** Drain remaining fluids.
- **S18010** Disconnect exhaust system components.
- **S18011** Remove starter.
- **S18012** Disconnect torque converter.
- **S18013** Disconnect transmission/transaxle.
- **S18014** Remove the engine.
- **S18015** Remove engine using an engine hoist.
- **S18016** Remove engine using a subframe jack.
- **S18017** Install the engine.
- **S18018** Reconnect motor mounts.
- **S18019** Perform start-up and break-in of the engine.
- **S18020** Perform prestart engine checks.
- **S18021** Perform initial engine break-in.
- **S18022** Perform DTC inspection and necessary module updates.

▶ Introduction

N18001

Once a major engine failure has been diagnosed, the engine often must be removed for repair or replacement. Typically, if well maintained, an engine will last many years and 200,000 miles (322,000 km) or more; however, abuse or defective components can severely shorten the life of an engine. If an engine has failed, the customer needs to take the vehicle to a clean, professional, and well-equipped shop for repair or replacement (**FIGURE 18-1**).

Generally, an engine lets you know if there is a serious mechanical problem. It produces excessive noise or may not even crank over. Usually, before the engine fails, there have been warning signs such as dash lights indicating low oil pressure, high operating temperature, or malfunctions in the powertrain. The **malfunction indicator light** (**MIL**) is a dash light that indicates the presence of diagnostic trouble codes (DTCs) in the powertrain control module (PCM). DTCs indicate faults such as misfiring engine cylinders, lean cylinder banks, and engine not warming up as designed.

A major mistake made by technicians when removing and replacing an engine is rushing to get the job done. To perform this task well, technicians must be organized, careful, and methodical in their work. One of the most important steps in the performance of this task is to research the appropriate service information, preferably the manufacturer's service information, and TSBs. It is important to follow the steps listed in the service information in order and step by step. Keep in mind the cost of replacing damaged components caused by sloppy or hasty workmanship. Some of the electronically controlled components in today's vehicles can cost thousands of dollars.

FIGURE 18-1 An engine being replaced in a clean, organized shop.

▶ Engine Removal Tools

K18001

The tools you will need to remove and replace an engine include special service tools, a vehicle lift, an engine hoist, an engine stand, jacks, and general shop hand tools. Additionally, a scan tool is important on all but vintage vehicles, to check and clear trouble codes, relieve pressure from fuel systems, and monitor engine and transmission systems after starting the replacement engine.

Many common hand tools are used during engine removal and replacement that were covered in the tools chapter, so we won't cover those here. In this chapter, we cover the more specialized tools and equipment that relate to engine removal and replacement.

- **Engine Hoist**—An engine hoist is necessary if the engine will be pulled from the top of the engine compartment (**FIGURE 18-2**). Make sure the engine hoist is rated for the capacity you will be lifting.

FIGURE 18-2 Using an engine hoist to remove an engine.

You Are the Automotive Technician

Today you will be removing and replacing the engine of a 2014 Ford Fusion with 187,000 miles (300,947 km) on it. Before you begin the job, you review the customer work order. The customer has complained of dash warning lights illuminating and noises coming from under the hood. You confirm the customer's complaints and diagnose the engine as having low oil pressure. The engine also exhibits a "rod knock" noise, indicating an excessively worn or spun connecting rod bearing. These conditions verify the need to replace the engine. You gather the tools needed to remove and reinstall an engine, including any special service tools, a vehicle lift, an engine hoist, jacks, and general shop hand tools.

1. What are the considerations prior to disconnecting the battery?
2. Why is it critical that all hoses, wires, and cables be disconnected from the engine before starting to lift it out?
3. What precautions are needed when installing a torque converter into the transaxle?
4. What are the important checks to make before starting the engine after installation?

- **Engine sling or chain**—An engine sling is commonly used when removing and installing engines from the top (**FIGURE 18-3**). It works well, especially if it can be adjusted by turning a threaded shaft. A strong chain can also be used. Make sure the chain exceeds the weight capacity that it will be supporting. Some engines have pulling brackets that allow the sling or chains to be attached by hooks when pulling the engine. The hooks too must exceed the weight capacity of the engine being pulled.
- **FWD engine hoist**—An FWD engine hoist is useful when removing the engine and transaxle from below the vehicle when dropping the entire cradle (**FIGURE 18-4**).
- **Engine support bar**—This bar is used to support the engine during engine mount removal in a front-wheel drive (FWD) vehicle or when removing the transaxle without removing the engine (**FIGURE 18-5**).
- **Transmission jack**—A transmission jack can come in handy if removing the engine only on a rear-wheel drive (RWD) vehicle (**FIGURE 18-6**).

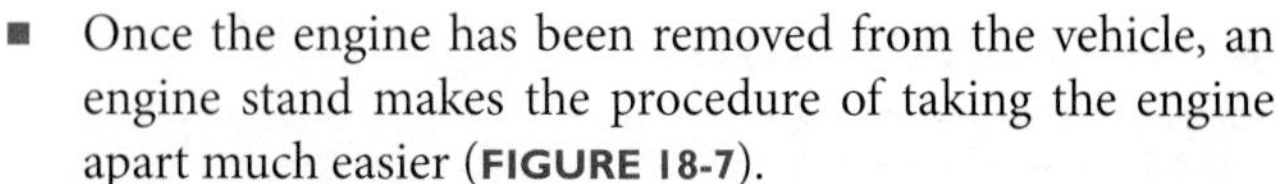

- Once the engine has been removed from the vehicle, an engine stand makes the procedure of taking the engine apart much easier (**FIGURE 18-7**).

FIGURE 18-5 An engine support bar for supporting the engine on a front-wheel (FWD) drive vehicle.

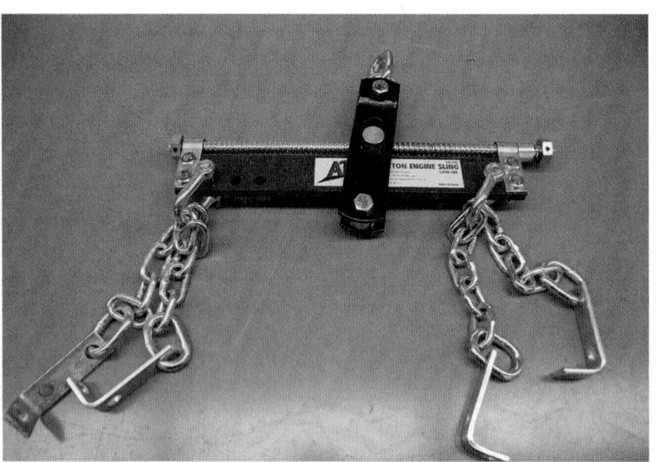

FIGURE 18-3 An engine sling used with an engine hoist.

FIGURE 18-6 A transmission jack used to support the transmission when only the engine is being removed on an RWD vehicle.

FIGURE 18-4 An FWD engine hoist is used to remove the engine and transaxle, including the cradle.

FIGURE 18-7 An engine stand.

▶ Preparation and Safety

K18002

There are many engine installation configurations, including FWD, RWD, and all-wheel drive designs. Although it is impossible to go into detail for every vehicle ever made, they all have similar engine removal procedures. In this chapter, a typical RWD and a typical FWD vehicle are used. Always refer to the service information for the specific vehicle for step-by-step disassembly procedures.

Most engines in RWD vehicles are best removed from the top of the engine compartment, whereas many engines in FWD vehicles can be removed from the top or the bottom of the engine compartment (**FIGURE 18-8**). If you are removing from the top of the engine compartment, the engine can be separated from the transmission and removed by itself, or in some cases, the engine and transmission are removed as a unit. If the engine is removed from the bottom of the engine compartment, the engine and transaxle are usually removed as a unit attached to the engine cradle, which bolts the engine and transmission to the vehicle (**FIGURE 18-9**).

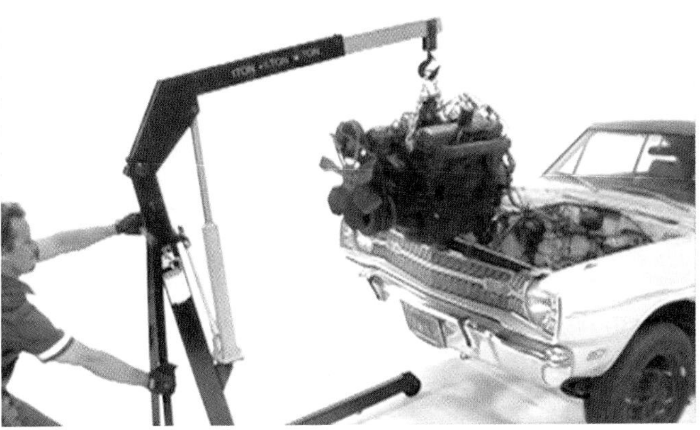

FIGURE 18-8 An RWD engine being removed from the top.

FIGURE 18-9 An FWD engine being removed from under the vehicle.

Note: For the purposes of this text, the term "transmission" may include "transaxle," and vice versa, but not always. We leave it up to the reader to determine its use, based on the vehicle being worked on.

SAFETY TIP

Engine removal and installation can be dangerous; therefore, it is important to put safety first in each step of the process. Vehicles and engines are heavy. During the removal and installation procedure, one of those heavy items is lifted high enough to clear the other. If the engine slips, it can crush or pinch fingers, hands, and feet. Many parts have sharp edges that can cut through skin and tendons, which gloves can help protect. Fluids and chemicals can spray into your eyes and mouth, or they can be absorbed through your skin, so always use personal protective equipment (PPE). Fluids on the floor can cause slips and falls, so clean up messes quickly. It is important to follow the vehicle manufacturer's guidelines to the letter. Always respect and implement shop safety standards and regulations.

▶ Topside Disconnection

S18001

If removing the engine from the top, the first step is to remove the hood. With the hood removed, you have greater access to the engine compartment and greater visibility because more light can get in. If the vehicle is FWD, and the engine is coming out from underneath the vehicle, or if the hood opens at least 90°, you may not need to remove the hood. Before removing the hood, mark the hood bracket where the bolts attach the hood to the hinge (**FIGURE 18-10**). By doing this, you will have a reference mark showing the exact place to reinstall the hood.

After referencing the appropriate service information, begin the engine removal process from the topside of the engine. Disconnect everything necessary for engine removal that is within reach from the top of the engine. One of the first things to do is to relieve any fuel pressure in the fuel system. Once that is complete, disable the electrical system by disconnecting the battery. Doing so reduces the risk of fire or explosion due to an inadvertent spark near exposed gases during the engine removal process. It also protects the electrical/electronic circuits from short-circuiting. Before the fluids are properly

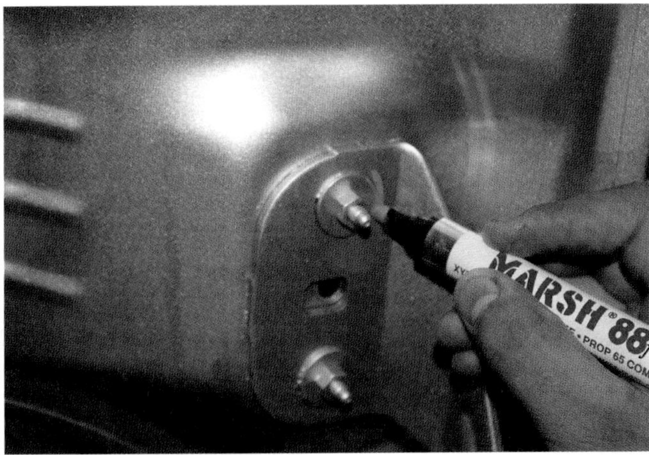

FIGURE 18-10 Marking the hood bracket.

drained, you need to avoid removing anything that would cause a fluid leak.

Fuel and Fuel Lines

S18002

Most vehicles are fuel injected. Typically, a fuel injection supply line stays pressurized even when the engine is not running; therefore, you need to relieve the pressure in the fuel system before disconnecting the lines. This should be done in two steps. First, loosen the gas cap so any pressure in the tank is vented. Second, relieve the pressure inside the fuel lines. Check the service information for proper procedures to release the fuel system pressure.

If the vehicle is in running condition, one example of relieving the pressure is to remove the fuel pump fuse while the engine is running, and wait until the engine runs out of fuel. If the fuel-injected vehicle is not running, install a fuel pressure gauge, press the relief valve while the relief hose is in a fuel container, and relieve the pressure (**FIGURE 18-11**). Once the fuel pressure has been relieved, you can disconnect the fuel line. For most fuel-injected vehicles, you need a fuel disconnect tool (**FIGURE 18-12**). Be sure to block off and seal the fuel supply line from the fuel tank. This is accomplished by using a vacuum hose plug. Just make sure it is compatible with gasoline.

Battery

S18003

Most modern vehicles have 10 or more computers monitoring and controlling nearly every operational system of the vehicle. In fact, an automobile could have more than 70 separate microprocessors onboard. Considering the sophistication of the onboard electronics, it is easy to understand why the vehicle's electrical/electronic system is vulnerable to damage from short-circuiting or grounding during the removal and installation process. Disconnecting the battery before disconnecting any other component from the engine helps avoid damage. Some electronic components, such as radios and navigation devices, have an antitheft system built into them. If power is removed from the component, it will not work until the proper unlock code is keyed into it. Therefore, before disconnecting the battery, make sure you have the unlock codes available. These codes usually come with the owner's manual.

FIGURE 18-11 A fuel gauge used to relieve fuel pressure.

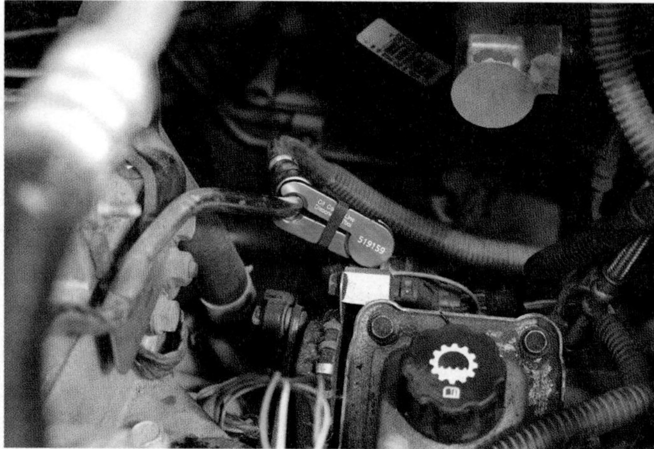

FIGURE 18-12 A fuel injection disconnect tool being used.

FIGURE 18-13 A technician disconnecting the negative terminal of the battery.

When disconnecting the battery, remove the negative terminal of the battery first (**FIGURE 18-13**). The negative side is always disconnected first because it is already grounded and is considered safe. After disconnecting the negative side of the battery, disconnect the positive terminal at the battery. As a note of caution, do not lay tools or anything else on the battery because an accidental connection could be made between the positive and the negative posts of the battery, which could cause the hydrogen gases around and inside the battery to explode.

Electrical Components

S18004

Before removing the engine from the vehicle, you must disconnect many electrical wires and connectors from the engine, bearing in mind that they must be reconnected after installation. This can be time-consuming and tedious work. Take your time; be methodical and careful during this process. If your work is rushed and sloppy, this job could become your worst nightmare.

To begin, look closely at the electrical wire harnesses to evaluate the best area to disconnect them. Often, it is easier to disconnect main harnesses at their central connector instead of at the end of each branch, so you should investigate this option (**FIGURE 18-14**). Before disconnecting the electrical wire harness plug ends, it is a good idea to take pictures (**FIGURE 18-15**). If you take pictures for later identification, examine the pictures before disconnecting anything. You want to be sure you can identify connector and wire locations based on the pictures. For vehicles built in the 1980s and earlier, you might want to label the plug-in connectors (**FIGURE 18-16**). For late-model vehicles, the electrical plug ends are not interchangeable. However, pictures will help you remember the general locations and the way they are routed. When in doubt, take pictures and label them to save yourself time and aggravation later.

FIGURE 18-16 Wires and connectors labeled and ready to be disconnected.

FIGURE 18-14 Many times it is easier and faster to disconnect the wiring harness at a central connector rather than at the end of each branch.

FIGURE 18-15 Taking pictures is a good way to remember how wires are routed and what they connect to.

▶ Front Accessories Disconnection

S18005

Continuing through the steps of engine removal, you will likely need to remove the radiator and at least some of the accessories. The air-conditioning compressor, power steering pump, and related mounting brackets may need to be removed from the engine before the engine is removed from the vehicle (**FIGURE 18-17**). After checking the appropriate service information, you can determine if the air-conditioning or power steering systems must be opened for the engine to be removed. If the air-conditioning system has to be opened, the refrigerant must be removed and stored by a properly certified technician (EPA 609 certified) using the appropriate recycling machine. If the power steering circuit needs to be opened, the fluid must be drained first to avoid spills.

Radiator

S18006

Make sure the engine is cool, and safely remove the radiator cap so that coolant will be able to flow freely out of the drain plug. At the bottom of the radiator, there should be a drain valve or plug (**FIGURE 18-18**), but if not, you will need to disconnect the lower radiator hose. Place the drain pan underneath the valve, and open the valve. If there is no valve, place the pan under the lower radiator hose. Next, disconnect the lower radiator hose until all the coolant has been drained, and then disconnect the upper radiator hose. For RWD vehicles, disconnect the fan shroud, unbolt the radiator, and remove it. For most FWD vehicles, determine whether there is enough room to remove the

FIGURE 18-17 At least some accessories typically have to be removed before the engine can be removed.

FIGURE 18-18 Draining the radiator.

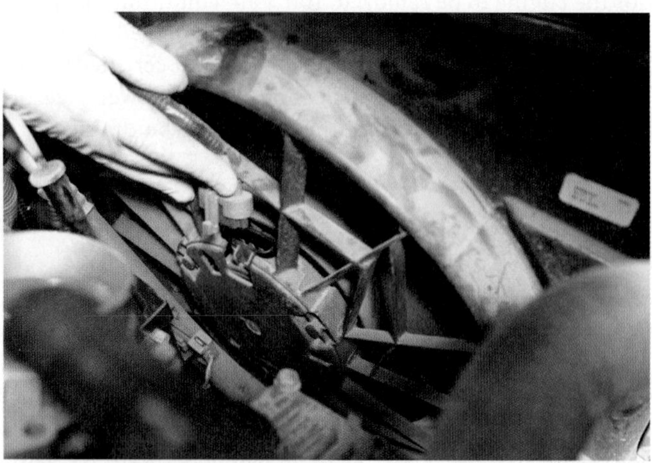

FIGURE 18-19 Removing the electric fans.

radiator without removing the electric fans, and if not, remove the electric fans (**FIGURE 18-19**), unbolt the radiator, and remove the radiator (**FIGURE 18-20**).

FIGURE 18-20 Removing the radiator.

Accessories and Brackets

S18007

Air-conditioning compressors and power steering pumps are examples of accessory components that are mounted to the engine with brackets. Often, these brackets have several pieces because they are not only used to mount but also to adjust the mounted components. Therefore, if the brackets must be removed, keep as many of the components and attachment bolts together as possible, to make reassembly easier.

Take off only the brackets that are necessary to remove the engine. Again, pictures are helpful in reminding you of the locations and positioning of the brackets. When removing the air-conditioning system, it is best to keep the system sealed, using mechanics' wire to secure the compressor and hoses up and out of the way (**FIGURE 18-21**). If you are removing the air-conditioning pump and have to disconnect the air conditioning, use the proper air-conditioning recovering equipment. Remove the power steering pump, take it off the bracket, and wire it out of the way (**FIGURE 18-22**). It is usually not necessary to disconnect the hydraulic lines. The alternator is another accessory that typically has to be removed. Carefully unbolt any nuts that hold wires on, and replace the nuts onto their studs so you won't lose them. Disconnect any wire connectors that plug into the alternator, and then unbolt the alternator from the brackets.

> ▶ TECHNICIAN TIP
>
> Only technicians who are EPA 609 certified are allowed to handle refrigerants, including breaking open any air-conditioning components. In order to be certified, you need to study a booklet, take a test, and pay the test fee. If you pass, you will be certified for life. This is a very cost-effective certification to obtain.

▶ Underside Disconnection

S18008

When preparing to raise the vehicle and make the necessary disconnections for removing the engine, start by reviewing the manufacturer's information. By following the steps for the

FIGURE 18-21 Wiring the air-conditioning compressor and hoses out of the way.

FIGURE 18-23 Missed hoses, lines, wires, and cables prevent the engine from being removed. Make sure to disconnect or remove everything the first time.

FIGURE 18-22 Remove the power steering pump as an assembly, if possible, and wire it out of the way.

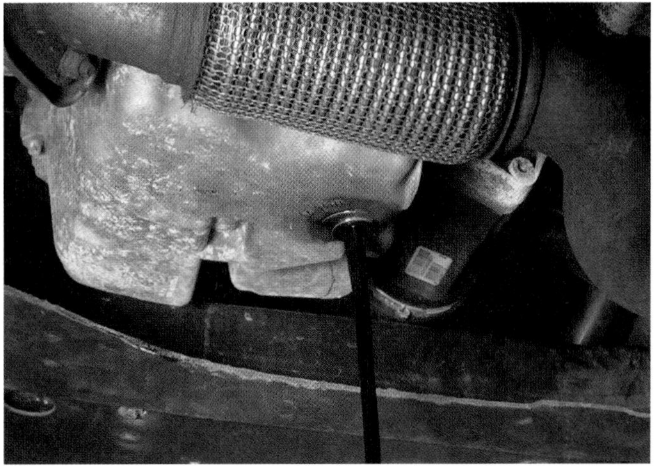

FIGURE 18-24 Draining the engine oil.

specific vehicle you are working on, you can avoid missing components, bolts, hoses, and wires that must be disconnected before the engine can be removed. When the engine is loose, hanging from the engine hoist and unsupported by the vehicle, it is not the time to discover that you missed removing something under the vehicle (**FIGURE 18-23**). Remember to take your time, and work carefully and methodically as you make the underside disconnections.

Draining Remaining Fluids

S18009

Because most fluid drain points are located under the vehicle, now is the time to drain fluids not previously drained. Drain the engine oil (**FIGURE 18-24**). If the transmission or transaxle will be coming out with the engine, you need to drain the transmission/transaxle fluid as well. If not already drained, drain the cooling system. Look for related units that may need to be drained, such as the power steering fluid if the pump is being removed from the lines.

Exhaust System

S18010

Exhaust systems are not quite as simple as they once were. Currently produced vehicles could have four or six oxygen sensors, a catalytic converter, and a turbocharger driven by the exhaust gases (**FIGURE 18-25**). Light-duty diesel-powered vehicles are now equipped with particulate filters. It is important to understand the exhaust system and related components on the vehicle before removing the exhaust system.

To begin the exhaust removal process, identify the oxygen sensors that need to be removed or disconnected. Disconnect the sensors (**FIGURE 18-26**). Check the exhaust pipe and component routing to determine the best places to disconnect the system, and be certain that the exhaust pipes will not restrict engine removal. Disconnect the exhaust pipe from the exhaust header, or manifold (**FIGURE 18-27**). On most vehicles, there usually is enough room to remove the engine with the stock exhaust header in place. Nonetheless, always follow the exhaust removal guidelines in the manufacturer's information or other reliable service information.

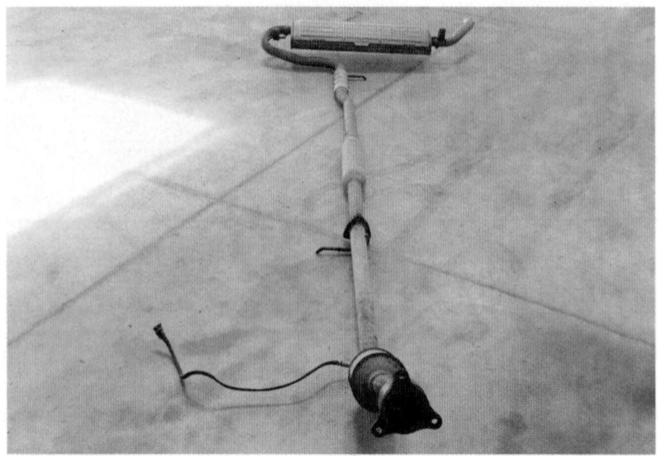

FIGURE 18-25 Exhaust assembly with oxygen sensors.

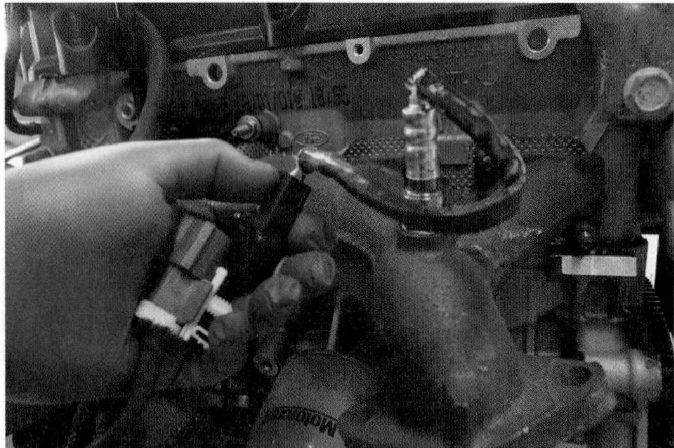

FIGURE 18-26 Disconnect any oxygen sensors.

FIGURE 18-27 Removing the exhaust pipe from the vehicle.

Starter Removal

S18011

The starter motor is normally located where the engine and the transmission/transaxle are mounted together. That mounting point is called the bell housing, and the starter is usually bolted to it (**FIGURE 18-28**). Unless the engine and transmission are being removed together, the starter has to be removed before the engine is removed. It is important to understand that the positive battery cable connects to a terminal on the starter; therefore, it is imperative that the battery be disconnected before removing the cable and wires from the starter. To remove the starter motor, disconnect the starter electrical wires, and mark where they go (**FIGURE 18-29**). Then, while holding the starter firmly (it is heavy), remove the starter bolts, and lower the starter from the engine or bell housing (**FIGURE 18-30**).

FIGURE 18-28 The starter motor is bolted to the bell housing.

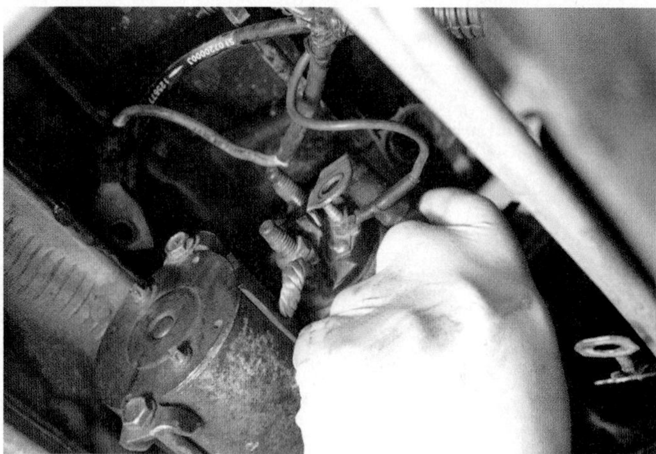

FIGURE 18-29 Removing the starter wires.

Automatic Transmissions: Torque Converter Disconnection

S18012

Remove the inspection cover on the bell housing to access the torque converter bolts on RWD and FWD vehicles if the engine is being removed from the top. For FWD vehicles in which the engine is being removed from the bottom, the engine and **subframe**, also called a cradle, will likely be removed together. Disconnecting the torque converter is not needed at this time, because the engine, transaxle, axles, and subframe are removed at the same time.

FIGURE 18-30 Removing the starter bolts.

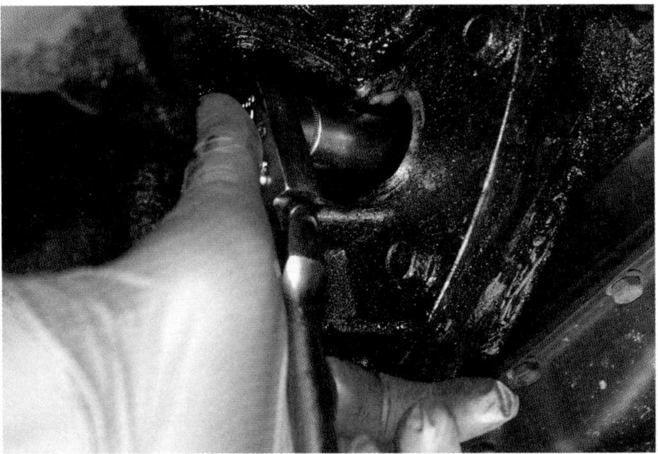

FIGURE 18-31 Unbolt the torque converter bolts or nuts.

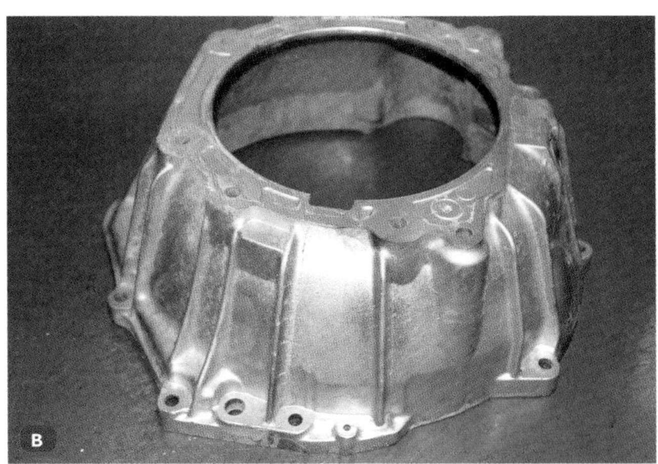

FIGURE 18-32 Manual transmissions. **A.** Integral bell housing. **B.** Separate bell housing.

After the inspection cover is removed, turn the flex plate so that one of the torque converter bolts is accessible. If the bell housing does not have an inspection cover, then you may have to access the flex plate by going through the starter mounting hole. Either turn the flex plate by hand, or turn the bolt on the front of the crankshaft, to align the flex plate bolts or nuts for easy access, and unbolt the torque converter bolts or nuts (**FIGURE 18-31**). Once the bolts or nuts have been removed, push the torque converter back into the bell housing so it will not fall out during engine removal.

Transmissions/Transaxles, Automatic and Manual

S18013

If the engine is being removed from the bottom of the vehicle using an FWD engine hoist, do not disconnect the transaxle at this time. If the engine is being removed from the top of the vehicle without the transmission/transaxle, you ultimately will need to unbolt the transmission/transaxle, but do not try to separate the transmission/transaxle from the engine until the transmission is supported and the engine is supported by the hoist.

If you are working on a manual transmission on an RWD vehicle, determine if it can be separated from the bell housing or if the bell housing is integral with the transmission case (**FIGURE 18-32**). Determine if the bell housing should be unbolted from the engine or transmission. Support the transmission, and then unbolt it or the bell housing at this time, but do not yet separate the two.

▶ Engine Removal

S18014

When removing an engine, it is always helpful to have two other people assisting you. Most importantly, it is much safer to have someone readily available to assist in an emergency or to get the appropriate assistance, if needed. Also, it is good to have one person operating the hoist while two others guide the engine as it is being lifted or lowered from the vehicle (**FIGURE 18-33**). Before removing the engine, it is good work practice to do a final check for anything that has to be disconnected that was missed earlier.

FIGURE 18-33 One person operating the engine hoist and one or two people guiding the engine.

FIGURE 18-34 Attaching the sling or chain to the engine. Make sure to use the proper bolts and see that they thread in far enough.

Engine Hoist/Boom Lift—Rear-Wheel Drive

S18015

Using an engine sling or chain, connect it to the proper lifting points, making sure the sling or chain and attaching bolts are rated for the engine weight. Use at least grade 5 bolts, and make sure they screw in at least five full turns (**FIGURE 18-34**). Connect the chain to the engine hoist, and center it on the chain as well as you can. Start by taking up the chain's slack; do not try lifting the engine, because the engine mounts are still connected. Unbolt the engine motor mounts, and begin lifting the engine out of the vehicle (**FIGURE 18-35**). A floor jack may be required to support the transmission. Lift the engine high enough to clear the radiator support, but no higher, to help prevent the hoist from tipping over (**FIGURE 18-36**). Once you have the engine out of the vehicle, either install it on an engine stand, or set it on the floor, leaving tension on the chain.

FIGURE 18-35 Remove or unbolt the motor mounts.

Engine Subframe Jack—Front-Wheel Drive

S18016

On an FWD engine, position the FWD engine hoist under the engine and subframe, according to the service information. After the hoist is positioned in place, strap the engine and transaxle to the hoist (**FIGURE 18-37**). Unbolt and remove the upper motor mounts (**FIGURE 18-38**). Remove the front brake calipers, and secure them to the vehicle, or remove the front brake lines to the calipers. Remove the cradle bolts (**FIGURE 18-39**), and lower the subframe scissor jack very carefully (**FIGURE 18-40**). After the engine and transaxle have been lowered, you can begin disassembly. Remove the engine from the transaxle by disconnecting the torque converter bolts or nuts first. Make sure the transaxle and subframe are secured to the hoist. Secure the chain of the engine hoist/boom lift to the engine, following the vehicle's service information. Take out the chain's slack; do not try to lift the engine yet, because the transaxle and engine mounts are still attached to the engine. Remove the transaxle bolts, and remove the engine motor mounts. Raise the engine about an inch, and start to separate the engine from the transaxle while starting to raise the engine.

FIGURE 18-36 The engine being raised.

FIGURE 18-37 The engine of an FWD vehicle strapped to a scissor jack.

FIGURE 18-38 Unbolt the motor mounts.

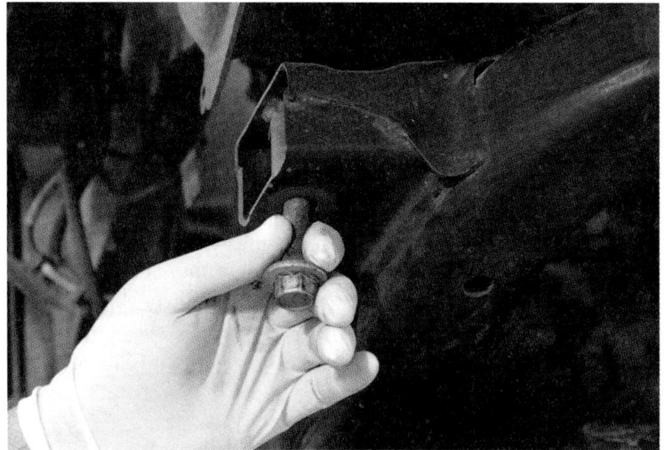

FIGURE 18-39 Remove the cradle bolts.

After the engine has been separated from the transaxle, lower the engine to the floor. Once the RWD or FWD engine is on the floor, remove the flex plate and the exhaust, if they are in the way. The engine is ready for the engine stand; make sure the engine stand is rated for the proper weight of the engine

FIGURE 18-40 Lower the scissor jack.

FIGURE 18-41 Bolt the engine to an engine stand.

and accessories, and use a minimum of grade 5 bolts threaded at least five complete turns into their holes. Mount the engine to the engine stand, making sure the engine's center of gravity is balanced on the engine stand (**FIGURE 18-41**). Finally, drain any remaining oil and fluids.

▶ Engine Installation

S18017

Just as in the engine removal process, have assistants help you install the engine. First, organize the work site. Begin by making sure the area is clean and clear of obstructions. To be prepared, arrange the bolts and engine mounts needed to secure the engine so they will be convenient. Additionally, lay out tools that will be needed for the initial installation of the engine. Time spent organizing the work site ensures a safe and smooth engine installation.

To reinstall the engine, basically reverse the process used to remove it. However, there are some additional procedures. For vehicles equipped with automatic transmissions, it is a good idea to consider changing the front transmission seal while the torque converter is removed (**FIGURE 18-42**). It only takes a few

FIGURE 18-42 Consider changing the front transmission seal even if it hasn't been leaking. It may prevent a costly comeback.

FIGURE 18-44 The torque converter should produce three distinct clunks as each component lines up inside it.

FIGURE 18-43 Fill the torque converter about halfway with the proper transmission fluid.

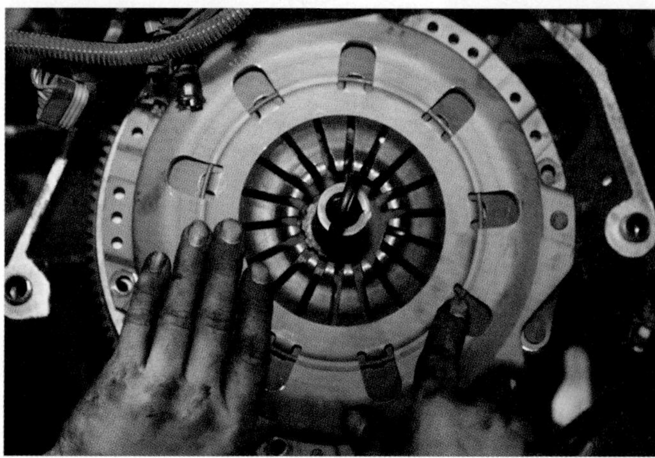

FIGURE 18-45 Use a clutch alignment tool to align the clutch disc with the flywheel.

minutes and avoids having to remove the transmission again if it were to leak. Also be sure to fill the torque converter with the proper transmission fluid. Lay the torque converter down facing up, and pour enough transmission fluid into the torque converter to fill it about halfway (**FIGURE 18-43**). When reinstalling the torque converter, push in and turn the torque converter, making sure it engages into the front pump (**FIGURE 18-44**). You should feel three distinct clunks. If the torque converter is not being replaced, make sure it is engaged into the front pump. When mating the engine up to the transmission, make sure the torque converter does not bind against the flex plate. If it does, remove the engine, and reengage the torque converter into the transmission. Some converters have studs that protrude out and are connected with nuts. The converter will have to be rotated to match the holes in the flex plate for the two to mate up.

For vehicles equipped with manual transmissions, the clutch disc has to be aligned with the clutch pilot hole, using a clutch alignment tool. First verify that the correct pilot bearing or bushing is installed in the rear of the crankshaft. Once you know the correct pilot bearing is in place, you can use an alignment tool that has the same splines as the clutch disc and fits snugly in the pilot bearing (**FIGURE 18-45**). Align the clutch disc, and install the pressure plate.

When installing a manual transmission to the engine, you sometimes need to turn the splined input shaft to align it with the clutch disc splines. A good way to do this is to leave the transmission in fourth gear and then turn the output shaft. Also, the front of the input shaft has to align with the flywheel pilot bearing or bushing. Make sure the bell housing holes are aligned with the engine's dowel pins, and tighten all of the bell housing bolts evenly.

▶ TECHNICIAN TIP

You can make alignment pins by taking two bolts that are longer than the bell housing bolts, cutting off their heads, and chamfering the cut ends. Screw these by finger into two of the bell housing holes in the block. The pointed ends will guide the bell housing into alignment against the block. Once a couple of bell housing bolts are threaded into the block, remove the guide pins, and replace them with the proper bolts.

FIGURE 18-46 Installation of engine mounts.

FIGURE 18-47 Going through engine prestart checks can help ensure proper break-in of the engine.

Engine Motor Mounts

S18018

After the transmission is bolted to the engine, the engine has to be mounted to the vehicle frame or cradle. To do so, attach the motor mounts that connect the engine to the frame or mounting platform (**FIGURE 18-46**). To mount the engine, lower the engine carefully, aligning the motor mounts. On some vehicles, the exhaust header must be aligned to the exhaust pipe at this time. Tighten the motor mounts when the engine and mounts are in place. Reinstall all parts and components in reverse order of disassembly.

▶ Engine Start-Up

S18019

Engine start-up after installation is a critical time to ensure proper engine operation and longevity. For the start-up process to be successful, a number of tasks must be performed before turning the key. These items are addressed in the Prestart Engine Checks topic. During the initial start-up, you need to keep your eyes and ears open, listening for unusual noises and looking for fluid leaks. You also need to monitor the temperature and oil pressure gauges, and shut the engine down immediately if the gauges indicate overheating or low oil pressure. If the engine does not start up right away, do not crank the engine excessively, because the starter can be damaged from excessive heat. Also, the lack of proper lubrication from the slow-turning oil pump can cause premature engine damage. Crank the engine for no more than 10 seconds at a time. If it doesn't start in that time, diagnose it to find out why. Once the engine fires up, the camshaft may need a specific process to properly break it in. Some new camshaft break-in procedures require the engine to start immediately and to run at 1200 to 2000 revolutions per minute (rpm) for 20 or 30 minutes to provide proper lubrication and break-in. Let's look at each of these topics in more depth.

Prestart Engine Checks

S18020

The goal of the prestart engine checks is to make sure that the engine is ready to be run for 10–30 minutes during the break-in period without fear of damaging it. Forgetting to fill the engine with the proper amount of engine oil will ruin all your good engine rebuilding work in just a few seconds. The steps listed here are designed to help you make sure that the engine break-in is successful (**FIGURE 18-47**). Before starting the engine after installation, you need to make several checks, including the following:

- Check all fluid levels, especially the engine oil and coolant levels. If a vacuum fill method of filling the cooling system was used, the cooling system will be full. If this method was not used, the system will look full, but likely there will be air pockets in the system. If there is an air pocket at the thermostat, it may prevent the thermostat from getting up to temperature and opening, potentially causing the brand new engine to overheat and seize.
- Check other fluids, including automatic transmission fluid, power steering fluid, brake fluid, and diesel exhaust fluid, if appropriate.
- Check that the battery terminals are clean, connected firmly, and that the battery is charged.
- Make sure that there is enough good fuel in the tank to run the engine for 30 minutes. If the fuel has been sitting for more than about three months, the fuel should be drained and new fuel added.
- Make sure the fuel is supplied to the injection system by turning the key to the run position, and check that there are no fuel leaks at the connections and fittings.
- With the key on (engine off), check that the gauges and warning lights operate properly.
- If the ignition timing is adjustable, make sure to set the static timing close enough to start. Have a timing light hooked up and the distributor slightly loose. Once the engine starts, check the timing, and adjust it if necessary (**FIGURE 18-48**).
- Check all of the auxiliary components, such as the air conditioner and the power steering pump, for proper installation and hose routing.
- Check the serpentine belt and/or other belts and pulleys for proper installation and tension.

FIGURE 18-48 If the ignition timing is adjustable, make sure it is set close enough to start, and then hook up a timing light to make an adjustment when the engine first starts.

FIGURE 18-49 Monitor the gauges to make sure there are no problems with oil pressure and engine temperature.

- Determine that electrical connectors and vacuum hoses are properly connected and routed. Verify that hoses and wires are not sitting on the exhaust manifolds.
- Double-check that the throttle linkage is free to operate, is routed correctly, and returns freely to the closed position by itself.
- Connect the exhaust removal system to the exhaust pipe.
- Verify that the brake pedal is not spongy.
- If it is an automatic transmission, verify that the gear shift indicator points to the proper gear as the gear shift is moved.
- Verify that the vehicle will only crank in Park or Neutral if it has an auto transmission.
- Verify that the clutch switch only allows the engine to be cranked when the clutch pedal is depressed.
- Apply the parking brake firmly.
- It is also good to have a fire extinguisher nearby and ready to be used.
- Have someone else to help you while you fire it up for the first time.

Initial Break-In

S18021

To ensure that the piston rings are seated properly and that the camshaft and other internal components are not damaged, it is important to follow the manufacturer's break-in procedures. If required during the initial engine break-in period, run the engine at high idle (1200–2000 rpm). Observe the temperature and oil pressure gauges (**FIGURE 18-49**). If the engine begins to run hot or the oil pressure drops, shut the engine off immediately. Also, monitor the engine for external leaks. If any are found, shut the engine down, and repair the leak. Note that it is normal for a new engine to smoke out the exhaust for the first minute or two as excess assembly lube and oil burns away. But it should clear up within a short time. Also, it is not uncommon for the engine to smoke on initial fire-up due to small amounts of oil or paint that got on the exhaust manifolds. But again, that should clear up within a few minutes. If it is getting worse over time, shut down the engine, and determine the problem. You don't want to catch the engine on fire.

Run the engine at fast idle until it reaches and holds normal operating temperature or for the prescribed break-in time, and then shut down the engine. Check the engine oil and oil level to ensure that there is no evidence of coolant in the oil. If the oil looks milky or if the oil level has risen above the full line, there may be an internal coolant leak.

Engine Timing Adjustment and Diagnostic Trouble Code Inspection

S18022

After starting the engine, you need to determine whether the engine ignition timing is set correctly if it is adjustable. There are almost as many methods of setting the timing as there are vehicle models. Again, you need to consult the appropriate service literature to determine the correct method.

You will notice that there are significant differences in timing methods from one decade to another. On vehicles from the 1970s and earlier, the primary method for setting the timing is to adjust the distributor while using a timing light. For 1980s models, you might need to disconnect a computer spark control wire or ground a particular wire or disconnect a sensor, and then set the timing by adjusting the distributor while using a timing light. On most vehicles made since the mid- to late 1980s, the computer controls the timing, and you do not need to manually set it.

Another important task to complete after the initial break-in is connecting a diagnostic scan tool to the **data link connector (DLC)**, checking for **diagnostic trouble codes (DTCs)**, and monitoring the sensor data (**FIGURE 18-50**). If you find that a DTC was set during the engine break-in period, diagnose the

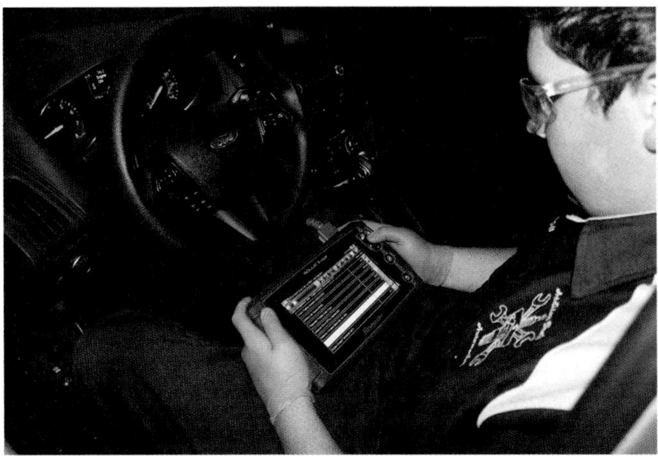

FIGURE 18-50 Use a scan tool to monitor engine data and DTCs.

condition. Once corrected, run the engine until it is fully warm again, and check for any DTCs. Note that some DTCs require two or more trips to set, so check for codes again after the test-drive is completed.

Once the engine is timed properly and the engine runs without turning on the MIL, and the engine has reached operating temperature and there are no sign of leaks or other problems, take the vehicle for a test-drive. Many rebuilders suggest that you drive it for 30 minutes without exceeding 50 mph (80 kph) and 2000 rpm. This allows the piston rings to seat and ensure good seating of other internal components such as cam lobes and valves. Once back at the shop, recheck all of the fluid levels, and look for any leaks. Also check for any loose or missing components. If everything is in working order, clean the vehicle and return it to the customer.

▶ Wrap-Up

Ready for Review

- Before the engine fails, there are usually warning signs such as dash warning lights indicating low oil pressure, high operating temperature, or a malfunctioning emission control system.
- It is important to follow the steps listed in the service information, in order and step by step.
- The tools needed to remove and replace an engine include special service tools, a vehicle lift, an engine hoist, engine sling or chain, jacks, and engine support bar.
- Make sure the engine hoist and sling are rated for the weight you will be lifting.
- Once the engine has been removed from the vehicle, use an engine stand to support the engine while taking it apart.
- There are many engine installation configurations. They include FWD, RWD, and all-wheel drive designs.
- Engine removal and installation can be dangerous; therefore, it is important to put safety first in each step of the process.
- Determine whether the engine is being removed from the top or the bottom of the vehicle.
- Typically, a fuel injection supply line stays under pressure even when the engine is not running. Therefore, you need to relieve the pressure on the fuel system before disconnecting the fuel line.
- Modern vehicles typically have 10 or more computers monitoring and controlling virtually every operational system of the vehicle.
- Disconnect the battery before disconnecting any other component from the engine.
- When disconnecting the battery, remove the negative terminal from the battery first.
- If an air-conditioning system must be opened, the refrigerant will have to be removed and stored, using the appropriate recycling machine.
- The boiling point of engine coolant is increased by the coolant properties, and the cooling system pressure is maintained by the pressure cap.
- Make sure the engine is cool before draining the radiator.
- Air-conditioning compressors and power steering pumps are examples of accessory components that are mounted to the engine.
- Currently produced vehicles could have four or six oxygen sensors, a catalytic converter, and a turbocharger driven by the exhaust.
- The starter motor is normally located where the engine and the transmission are mounted together.
- On RWD vehicles equipped with an automatic transmission, the torque converter has to be unbolted from the flexplate.
- The transmission and engine must be supported before trying to separate them.
- When removing an engine, it is always helpful to have two other people assisting you.
- Time spent organizing the work site will ensure a safe and smooth engine installation.
- Make sure the engine stand is rated for the proper weight of the engine and accessories, and use a minimum of grade 5 bolts threaded at least five complete turns into their holes.
- When reinstalling the torque converter, push in and turn the torque converter, making sure three distinct clunks are felt.
- Engine start-up after installation is a critical time for engine operation and longevity.

- Make sure that the engine is ready to be run for 10–30 minutes during the break-in period, without fear of damaging it.
- To ensure that the piston rings are seated properly and that the camshaft and other internal components are not damaged, it is important to follow appropriate break-in procedures.
- An important task after you have started the engine is to connect a diagnostic scanner to the data link connector (DLC) and check for diagnostic trouble codes (DTCs).

Key Terms

data link connector (DLC) An under-dash connector for connecting a diagnostic scanner.

diagnostic trouble code (DTC) A code set by the computer indicating system or component malfunction.

engine support bar Used to support the engine during engine mount removal in an FWD vehicle or when removing the transaxle without removing the engine.

FWD engine hoist Useful when removing the engine and transaxle from below the vehicle when dropping the entire cradle.

malfunction indicator light (MIL) A dash light usually indicating the presence of a diagnostic trouble code or malfunction.

subframe A mount attached to the vehicle that is used to support the engine and transaxle assembly.

Review Questions

1. Which of these tools is used when removing the engine from a rear wheel drive vehicle?
   a. Scan tool
   b. Transmission jack
   c. Engine support bar
   d. FWD engine hoist
2. Removing engines are likely to pose all of the below safety hazards *except*:
   a. crushing fingers due to engine slips.
   b. fluids and chemicals spraying into your eyes.
   c. sharp edges of parts causing cuts.
   d. flash burns from arc welders.
3. During topside disconnection:
   a. always remove the hood irrespective of the degree to which it opens.
   b. maintain the fuel pressure in the fuel system.
   c. disable the electrical system by disconnecting the battery.
   d. retain the fluids in the engine.
4. When disconnecting a battery:
   a. have the unlock codes of the antitheft system available.
   b. remove the positive terminal of the battery first.
   c. lay the tools on the battery for easy use.
   d. ensure that all other components are disconnected.
5. When removing accessory components that are mounted to the engine:
   a. remove all the brackets.
   b. always disconnect the hydraulic lines.
   c. place all attachment bolts separate from the components.
   d. take pictures to remind you of bracket locations and positions.
6. During the exhaust system removal process, the exhaust pipe and component routing should be checked to:
   a. prevent leaks.
   b. improve their efficiency.
   c. determine the best places to disconnect.
   d. avoid emission of toxic gases.
7. For front-wheel drive vehicles in which the engine is being removed from the bottom, all of the below may be removed along with the engine *except*:
   a. transaxle.
   b. radiator.
   c. cradle.
   d. axles.
8. Until the transmission is supported and the engine is supported by the hoist, do not try to:
   a. drain engine fluids.
   b. remove the radiator.
   c. separate the transmission from the engine.
   d. disconnect the battery.
9. When reinstalling the torque converter, push in and turn the torque converter. We can make sure it fully engages into the front pump when we feel:
   a. four distinct clunks.
   b. three distinct clunks.
   c. two distinct clunks.
   d. single clunk.
10. As part of prestart engine checks, it is essential to do all of the below *except*:
    a. checking all fluid levels, especially the engine oil and coolant levels.
    b. checking that the battery terminals are clean and connected firmly.
    c. checking all auxiliary components for proper installation.
    d. checking the ignition timing with a timing light.

ASE Technician A/Technician B Style Questions

1. Tech A says that one of the most important steps in performing an engine removal and installation procedure is to research the appropriate service information. Tech B says that researching service information is a waste of time. Who is correct?
   a. Tech A
   b. Tech B
   c. Both A and B
   d. Neither A nor B
2. Tech A says that engines in most RWD vehicle are removed from the top. Tech B says that before an engine fails, there are usually warning signs such as dash warning lights indicating low oil pressure, high operating temperature, or abnormal engine noises. Who is correct?
   a. Tech A
   b. Tech B

c. Both A and B
d. Neither A nor B

3. Tech A says that when disconnecting the battery, remove the positive terminal from the battery first. Tech B says that it is often easier to disconnect main wire harnesses at their central connector instead of at the end of each branch. Who is correct?
   a. Tech A
   b. Tech B
   c. Both A and B
   d. Neither A nor B
4. Tech A says that a fuel injection system automatically relieves pressure when the engine is turned off. Tech B says that you need to relieve fuel pressure from most fuel-injected vehicles before disconnecting the lines. Who is correct?
   a. Tech A
   b. Tech B
   c. Both A and B
   d. Neither A nor B
5. Tech A says that if refrigerant removal is necessary, air-conditioning refrigerant must be removed with an approved recycling machine. Tech B says that if refrigerant removal is needed, it is much better to release it into the atmosphere because modern refrigerants are safe. Who is correct?
   a. Tech A
   b. Tech B
   c. Both A and B
   d. Neither A nor B
6. Tech A says to remove the radiator cap so that coolant will be able to flow more freely out of the drain plug. Tech B says that if a technician were to remove a cooling system pressure cap from a hot engine, the sudden reduction in the coolant boiling point could boil the water violently and scald the technician. Who is correct?
   a. Tech A
   b. Tech B
   c. Both A and B
   d. Neither A nor B
7. Tech A says that after starting the engine, it should be revved higher than 4000 rpm for a few minutes to break in the rings. Tech B says that after initial break-in, connect a diagnostic scan tool to the DLC and check for DTCs. Who is correct?
   a. Tech A
   b. Tech B
   c. Both A and B
   d. Neither A nor B
8. Tech A says that a technician needs to make sure an engine hoist is rated for the lifting capacity needed for the engine being removed. Tech B says that when installing a clutch in an engine with a manual transmission, a clutch alignment tool is needed. Who is correct?
   a. Tech A
   b. Tech B
   c. Both A and B
   d. Neither A nor B
9. Tech A says to check all fluid levels before starting a newly installed engine, especially the engine oil and coolant levels. Tech B says that once started, the engine should be checked for external leaks. Who is correct?
   a. Tech A
   b. Tech B
   c. Both A and B
   d. Neither A nor B
10. Tech A says that a torque converter should engage at three different levels as it is being installed in a transmission. Tech B says that the torque converter requires an alignment tool when installing it in the transmission. Who is correct?
    a. Tech A
    b. Tech B
    c. Both A and B
    d. Neither A nor B

CHAPTER 19
Cylinder Head Components

NATEF Tasks

- **N19001** Remove cylinder head; inspect gasket condition; install cylinder head and gasket; tighten according to manufacturer's specification and procedures. (AST/MAST)
- **N19002** Clean and visually inspect a cylinder head for cracks; check gasket surface areas for warpage and surface finish; check passage condition. (AST/MAST)
- **N19003** Inspect valves and valve seats; determine needed action. (MAST)
- **N19004** Inspect valve guides for wear; check valve stem-to-guide clearance; determine needed action. (MAST)
- **N19005** Inspect valve springs for squareness and free height comparison; determine needed action. (MAST)
- **N19006** Inspect pushrods, rocker arms, and rocker arm pivots and shafts for wear, bending, cracks, looseness, and blocked oil passages (orifices); determine needed action. (AST/MAST)
- **N19007** Inspect and/or measure camshaft for runout, journal wear, and lobe wear. (MAST)
- **N19008** Inspect camshaft bearing surface for wear, damage, out-of-round, and alignment; determine needed action. (MAST)

Knowledge Objectives

After reading this chapter, you will be able to:

- **K19001** Identify cylinder head configuration and components.
- **K19002** Describe unique features of an OHV cylinder head.
- **K19003** Describe unique features of a SOHC cylinder head.
- **K19004** Describe unique features of a DOHC cylinder head.
- **K19005** Describe individual cylinder head components.
- **K19006** Describe the various cylinder head designs.
- **K19007** Explain the function of the valve assemblies.
- **K19008** Describe valves and their related components.
- **K19009** Describe common valve arrangements.
- **K19010** Describe the operation of intake and exhaust valves.
- **K19011** Describe valve spring assemblies and their purpose.
- **K19012** Describe valve guides and seals.
- **K19013** Describe valve seats and their purpose.
- **K19014** Explain the role of the camshaft in engine operation.
- **K19015** Define the terms used in camshaft specifications.
- **K19016** Describe how valve timing affects engine operation.
- **K19017** Explain how and why valve timing is altered in variable valve timing engines.
- **K19018** Explain the operation of valve train components.
- **K19019** Describe the operation of the different types of lifters.
- **K19020** Describe the purpose and types of pushrods.
- **K19021** Describe the purpose and types of rocker arms.
- **K19022** Explain the function and application of gaskets, seals, and sealants.
- **K19023** Describe the purpose and types of head gaskets.
- **K19024** Describe the purpose and types of seals.
- **K19025** Describe the purpose and types of sealants.
- **K19026** Describe the process of removing, inspecting, and diagnosing cylinder heads and related components.

Skills Objectives

After reading this chapter, you will be able to:

- **S19001** Reassemble a cylinder head.

▶ Introduction

The modern internal combustion engine consists of many parts that were present in the original engines of a century ago. Over time, the form and efficiency of these parts have changed significantly, though their basic function has not (**FIGURE 19-1**). The cylinder head, in particular, is an example of what many years of improvement on a solid design can yield. Today's cylinder heads are more durable and efficient than any of their past counterparts. In this chapter, the modern cylinder head will be explored, including the theory behind its many designs and its role in the combustion process.

The cylinder head assembly is the "end cap" for the combustion chamber. It provides a means for admitting air and fuel to the combustion chamber and exhausting gases after combustion. The cylinder head is also an integral part of the engine cooling system; it allows coolant to flow around the top of the combustion chamber, removing excess heat and preventing overheating of the engine.

The camshaft works with the cylinder head and **valves** to regulate the flow of the air-fuel mixture into, and the exhaust gases out of, the combustion chamber by opening and closing the valves at the proper times. The camshaft may be located either in the cylinder head or adjacent to it in the cylinder block (**FIGURE 19-2**). Each location has its own pros and cons. Engines with a camshaft located in the block have been around for decades and therefore are inexpensive to continue manufacturing. But they do not allow for varying of the valve timing, which is used to reduce emissions and increase engine

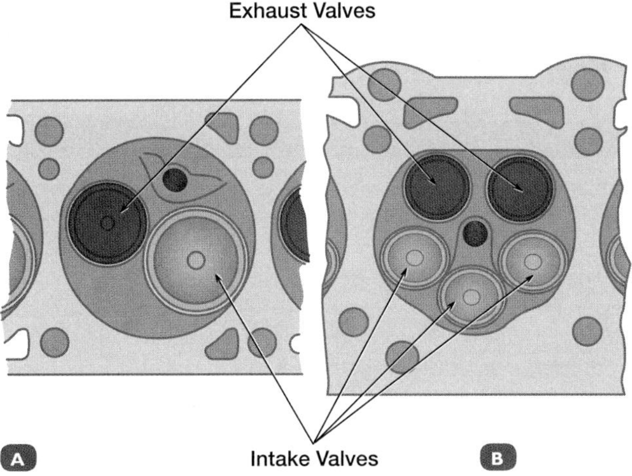

FIGURE 19-1 Cylinder head design has improved significantly over the past 50 years. **A.** 1970s OHV 2-valve cylinder head. **B.** 2010s OHC 5-valve cylinder head.

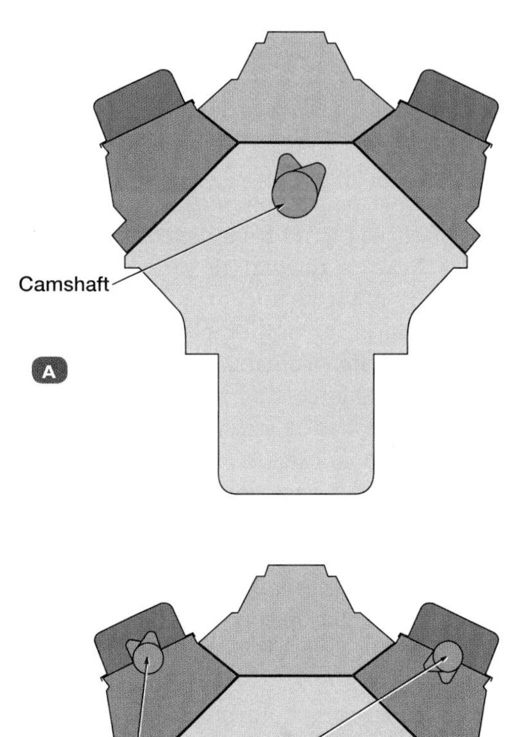

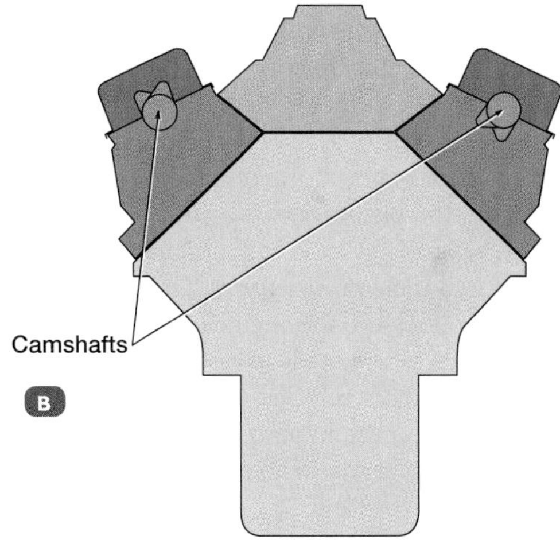

FIGURE 19-2 A. In-block camshaft position. **B.** Overhead camshaft position.

You Are the Automotive Technician

You have diagnosed a blown head gasket on a customer's 2012 Shelby GT500, which is run regularly on the local road racetrack. Because of the stresses from racing and because the vehicle is nearing 75,000 miles, he would like you to disassemble, clean, and inspect the cylinder heads while you have them off, to see if they need to be rebuilt. He agrees with your suggestion to have them checked for flatness and cracks. He is considering a high-performance camshaft, which he understands would create more power. But he wants to know what "duration," "lift," and "lobe separation" means, and asks if you could explain them to him so he can decide which camshaft to use.

1. What do the terms "duration," "lift," and "lobe separation" mean?
2. How are heads checked for cracks?
3. What two locations are checked for flatness on heads equipped with overhead cams?
4. How are hydraulic lifters and solid lifters different from each other?

efficiency. Engines with camshafts in the cylinder head are usually bulkier and require a more complicated drive system, so they are usually more expensive. Nonetheless, overhead camshafts make the engine more efficient and powerful, so they are used in more and more engines.

▶ Cylinder Heads (Gasoline Engines)

K19001

There are five events in the operation of any internal combustion engine: intake of the air-fuel mixture, compression of the air-fuel mixture, ignition of the air-fuel mixture, a power stroke triggered by combustion of the air-fuel mixture, and the exhaust of burned gases from the combustion chamber. The **cylinder head** is integral to each of these steps. It controls the entry of the air-fuel mixture through the intake valve, allows compression of the air-fuel mixture by capping the top of the cylinder, contains a spark plug that ignites the air-fuel mixture, forces the pistons down the cylinder in a power stroke by not allowing the expanding gases to escape, and contains an exhaust valve to release the burned gases from the cylinder once the piston reaches the bottom of the cylinder. Because of the vital roles that the cylinder head plays, cylinder head design has a huge influence on the engine's efficiency and power output.

To contain the combustion pressures, a head gasket is firmly clamped between the cylinder head and engine block with head bolts or studs (**FIGURE 19-3**). The head gasket is a thin metal sheet, often made of layers of different materials, with holes cut to accommodate the piston openings and passageways for oil and coolant. The head gasket keeps combustion pressure, coolant, and oil inside their proper areas. The cylinder head also distributes air and fuel into the engine's cylinders.

Cylinder heads can be made of cast iron or aluminum alloy (**FIGURE 19-4**). In today's vehicles, you will find most cylinder heads are made of aluminum alloy because of its light weight. The lighter cylinder heads help reduce the vehicle's overall weight, which helps increase fuel economy. Aluminum also conducts heat away from the combustion chamber more rapidly than cast iron. Thus, with an aluminum-alloy head, the heat of combustion can be conducted to the coolant faster, reducing the chance of localized hot spots. At the same time, engine efficiency is reduced when heat is removed too quickly from the combustion chamber, so aluminum cylinder heads may be designed with smaller water jackets to prevent them from overcooling the engine.

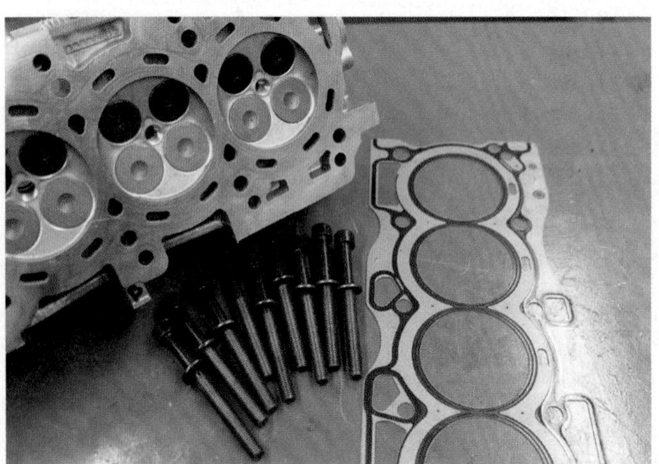

FIGURE 19-3 A cylinder head with head gasket and head bolts.

FIGURE 19-4 Cylinder heads: **A.** Aluminum. **B.** Cast iron.

> **▶ TECHNICIAN TIP**
>
> Soft plugs (or core plugs) have been used to plug the holes left over from the casting process. For years the term "freeze plugs" has also been used; however, this term is inaccurate. The cylinder head was never designed to withstand freezing with water in the cooling passages, and the soft plugs were never designed to prevent damage by popping out during a freeze. If the engine freezes and the soft plugs do pop out, more than likely the cylinder heads will have already cracked. The crack is typically found in the thin portions of the water jacket.

Overhead Valve Engine

K19002

In an **overhead valve (OHV) engine**, the valves are positioned in the cylinder head assembly, directly over the top of the piston (**FIGURE 19-5**). The valves are operated by a camshaft that is located in the cylinder block. A valve lifter (tappet) rides on the camshaft lobe and transmits the movement of the lobe to the pushrod. Valve lifters can be solid, requiring periodic adjustment, or hydraulic, which are self-adjusting. One valve lifter is located on each lobe of the camshaft. As the camshaft rotates, the lifter rises and transfers the motion to the pushrod. The pushrod is a metal, usually hollow, rod that transfers the motion from a lifter to the rocker arm. The pushrod then moves a rocker arm, which pivots near the center and pushes the valve open. This is a relatively simple design that is inexpensive to build.

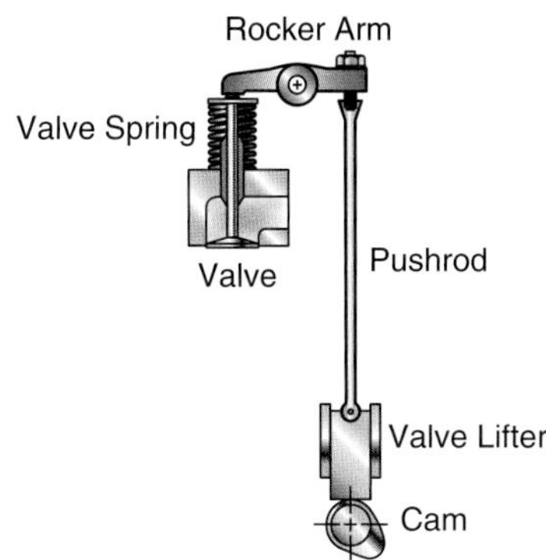

FIGURE 19-5 Overhead valve (OHV) engine.

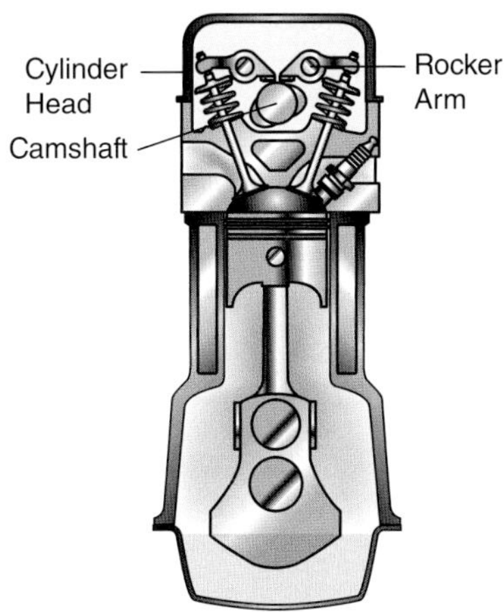

FIGURE 19-6 Overhead cam (OHC) engine.

Single Overhead Cam Engine (SOHC)

K19003

In a single OHC engine, the valve train works by having one camshaft mounted either between the rows of intake and exhaust valves on cross-flow heads, or directly over top of the valves on non-cross-flow heads. On cross-flow heads, rocker arms that pivot near the center, or cam followers that pivot on one end, ride on each cam lobe and directly operate each valve (**FIGURE 19-6**). Some systems incorporate hydraulic lash adjusters right in the rocker arms or in the cam follower pivot so that the valve train will be self-adjusting. On non-cross-flow heads, the cam may be positioned right over the top of the valves and operate them through a bucket lifter, which may also have a built-in lash adjuster. This design makes for a very compact valve train, although it calls for a more extensive cam drive as well as the need for enhanced oiling at the top of the engine for the camshaft, lifters, and followers.

Dual Overhead Cam Engine

K19004

The **dual overhead cam (DOHC) engine** contains two camshafts per cylinder head (**FIGURE 19-7**). Typically, one camshaft operates the **intake valves** and the other operates the **exhaust valves**. In most cases, the camshaft sits right on top of the valves and operates the valves directly through a bucket lifter, with no rocker arms or cam followers. Bucket lifters may have a built-in lash adjuster or be mechanically adjusted. This design makes adjusting and servicing the valves more difficult because the camshaft usually has to be removed to gain access. Offsetting the camshaft and using cam followers allows easier access. Having dual camshafts also allows intake and exhaust valve timing to be controlled separately from each other, which can reduce emissions, increase power, and increase engine efficiency or mileage.

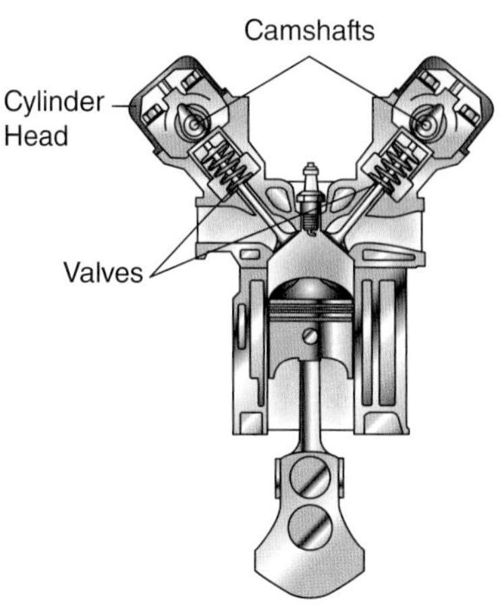

FIGURE 19-7 Dual overhead cam (DOHC) engine.

Cylinder Head Components

K19005

The cylinder head is composed of many distinct parts. At its core is the bare cylinder head casting, the name given to the base onto which all other parts are attached. Other parts of the cylinder head include (**FIGURE 19-8**):

- *Intake valves:* The air-fuel mixture enters the combustion chamber through these valves.
- *Exhaust valves:* The exhaust gases exit the combustion chamber through these valves.
- *Valve springs:* These coil springs return valves to their fully closed positions and hold them closed, after being opened.
- *Valve retainer:* This thick, washer-shaped piece of metal with a tapered center hole is fitted over the top of the valve stem and holds the top of the valve spring to keep pressure on the valves.
- *Valve keepers:* These tapered devices keep a valve spring's retainer attached to the valve stem.
- *Valve stem seals:* These seals allow only enough oil past them to lubricate the valve guide.
- *Rocker arm:* This lever actuates a valve by pivoting near the center and pushing on the tip of the valve stem to open it.
- *Plugs (e.g., galley, coolant):* Much like seals, plugs are used to block off passages where oil and coolant flow and are

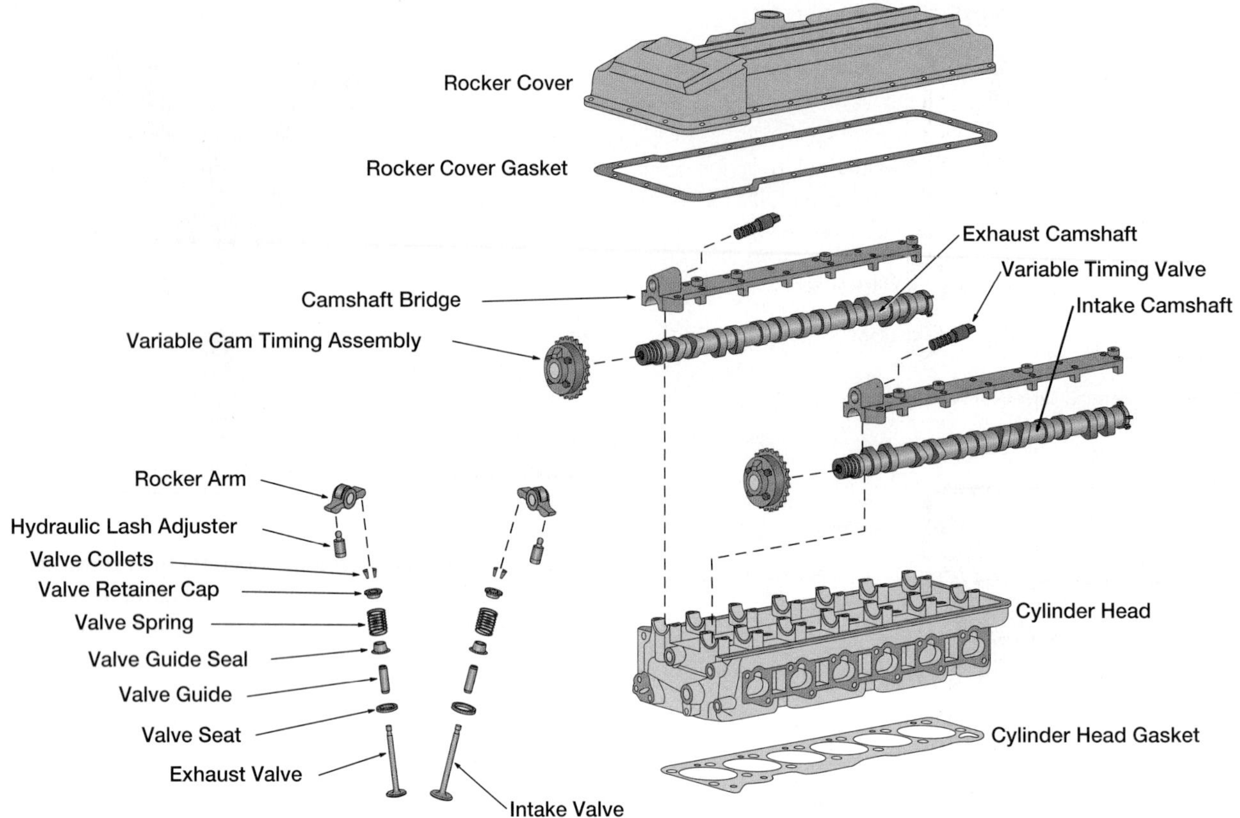

FIGURE 19-8 Exploded view of a DOHC cylinder head and components.

primarily used to block holes used in the original casting and machining of the head.

- *Camshaft (only fitted in the cylinder heads of OHC and DOHC engines):* This device provides the force to open the valves.
- *Cam seals:* These seals keep oil from leaking past a rotating camshaft.

Cylinder Head Design

K19006

The combustion chamber is located inside the engine where fuel and air are compressed by the piston and then ignited by the spark plug. The combustion chamber is generally formed into the cylinder head during the casting process. The movable floor of the combustion chamber is formed by the top of the piston. The overall efficiency of the combustion chamber is determined by its design, including the shape of the combustion chamber, the positioning of the valves, the location of the spark plug, the placement of the water jackets near the combustion chambers, and the shape of the top of the piston. Some engines use two spark plugs per combustion chamber, but this is not very common (**FIGURE 19-9**).

FIGURE 19-9 A cylinder head with two spark plugs per cylinder.

Today's cylinder heads are designed to help improve the swirl and turbulence of the air-fuel mixture. This helps the fuel mix with the air and prevents fuel droplets from settling on the surfaces of the combustion chamber. Swirl is initially created by the shape and angle of the **intake port**. Turbulence is increased in the **quench**, or **squish, area**, which is the area between the piston head at top dead center and the cylinder head. Its name comes from the squishing of the air-fuel mixture into a small "charge." The piston squeezes the air-fuel mixture and forces it with great speed out of the squish area. This creates turbulence, which helps to thoroughly mix the heavier fuel and lighter air.

Cylinder Head Shapes

Hemispherical Combustion Chamber

The **hemispherical cylinder head** has been around since approximately 1901. Several early designs were by a Belgian carmaker by the name of Pipe in 1905. In the early 1960s, Chrysler Corporation trademarked the name Hemi™, which is perhaps the most widely known hemispherical combustion chamber engine created.

A hemispherical combustion chamber is shaped like a semicircle and has the intake valve on one side of the chamber and the exhaust valve on the other (**FIGURE 19-10**). The valves are tilted away from each other. This design provides good cross-flow of the gases and is known as a cross-flow cylinder head. The air-fuel mixture enters on one side, and exhaust gases exit out the other side. Positioning the valves with the spark plug just off center of the combustion chamber leaves more room for relatively large valves and ports, which helps the engine breathe.

With the spark plug located near the center of the chamber, the air-fuel mixture starts to burn at the plug and then continues burning outward in all directions. This moving **flame front** is called **flame propagation**. With the hemispherical design, the flame front, or front edge of the burning air-fuel mixture, has less distance to travel than in some other designs, which gives rapid and effective combustion.

One of the drawbacks of the original hemispherical combustion chamber engine was that it required a complex valve train design that included two rocker arm shafts per head, with uneven rocker arm lengths. In later engines, manufacturers overcame this complex valve train design by using the OHC design. Thus, the benefits of the hemispherical combustion chamber were combined with the benefits of the OHC cylinder heads. But hemispherical combustion chambers are limited to two valves per cylinder.

In today's DOHC engines, it is very common to see multi-valve, pent-roof cylinder heads. Pent-roof cylinder heads are similar to hemispherical heads in the fact that they are of the cross-flow design, but typically with four or five valves (**FIGURE 19-11**). The combustion chamber is shaped more like a fairly flat rectangular pyramid than a hemisphere. Also, typically the spark plug is more centrally located in the top of the combustion chamber. The addition of multiple intake and/or exhaust valves on the cylinder increases the speed at which intake and exhaust gases can be moved into and out of the combustion chamber. This concept is similar to having a room full of people and trying to evacuate them with just one door versus multiple doors: The more doors, the faster the flow. This increased flow increases engine power output.

FIGURE 19-10 A hemispherical cylinder head and combustion chamber.

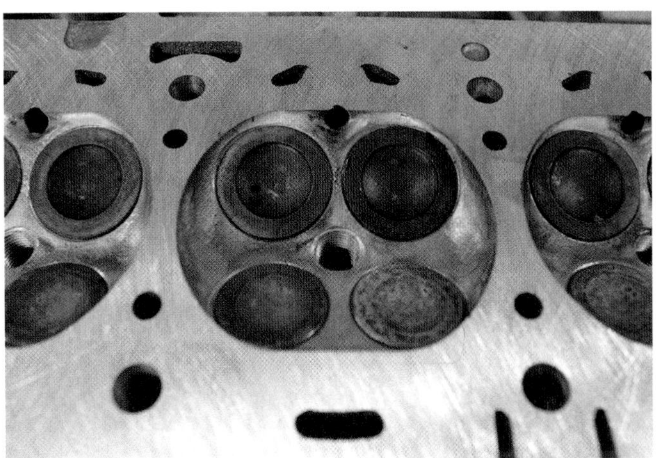

FIGURE 19-11 Pent-roof cylinder heads have four or five valves per cylinder.

Oval Combustion Chamber (Bathtub)

The bathtub combustion chamber is oval shaped, like an inverted antique bathtub (**FIGURE 19-12**). The valves are mounted vertically and are side by side, making the valve train simple in its design. On some of the newer designs, the surface of the head overlaps the piston and forms two squish areas: a large area near the spark plug and a smaller one on the opposite side.

Wedge Combustion Chamber

The **wedge combustion chamber** tapers away from the spark plug, which is at the thick end of the wedge. The valves are typically arranged in a straight line positioned along the tapered angle, which puts them at an angle to the top of the piston (**FIGURE 19-13**). The wedge design has a smaller combustion chamber surface area than the other types of cylinder heads, and that means less of a boundary area where fuel cannot burn as easily, which lowers hydrocarbon (unburned fuel) emissions.

Because the flame is directed toward the small end of the wedge, damage caused by detonation is reduced. This wedge design speeds up the velocity of the flame front. If the velocity of the flame front is not fast enough, the pressure building up in the combustion chamber causes the remaining gases to spontaneously ignite, which is called detonation. Detonation creates a second flame front that collides with the first flame front. A very

FIGURE 19-12 Oval (bathtub) combustion chambers.

FIGURE 19-14 Gas direct injections.

FIGURE 19-13 Modern wedge combustion chamber.

FIGURE 19-15 Port size affects the power range of the engine. **A.** Small port = good low rpm torque. **B.** Large port = good high rpm torque.

intense pressure wave and audible detonation or pinging sound can be heard. Extreme detonation can damage the engine by concentrating high temperatures and pressures, which can melt a hole in the top of the piston.

Gas Direct Injection Cylinder Head

The latest cylinder head design is the **gas direct injection (GDI) cylinder head**. GDI refers to gasoline injected directly into the combustion chamber just above the piston. GDI heads are typically of the multi-valve pent-roof design with the spark plug centrally located and the fuel injector offset to one side (**FIGURE 19-14**). Instead of mixing the fuel/air in the intake manifold a GDI engine mixes in the combustion chamber. Some GDI systems spray fuel into the combustion chamber during the intake stroke and use the fuel to actually cool the incoming air, which increases its density and resistance to detonation. Other GDI systems have the ability to spray fuel not only on the intake stroke but also near the top of the compression stroke, thus preventing detonation and allowing a lean air-fuel mixture to be burned by concentrating the sprayed fuel very close to the spark plug. The computer has control of the amount and timing of the injection sequence. It varies based on a variety of engine conditions, such as engine load, speed, and temperature. With the GDI cylinder head design, some manufacturers have been able to remove the throttle plate and control engine speed and power by controlling the fuel by itself. This reduces the engine's pumping losses of trying to pull air past the throttle plate. Overall, GDI can improve fuel economy up to about 35% and reduce emissions, especially carbon dioxide.

Intake and Exhaust Ports

The size and shape of intake and exhaust ports in the cylinder head can affect engine output by affecting the speed of the air flowing through them (**FIGURE 19-15**). It takes a certain amount of speed to maintain good turbulence, air-fuel mixing, and column airflow. If airflow is too slow, fuel will fall out of the air and not much ram air effect will be available. If airflow is too fast, the airflow will be restricted, which reduces the total amount of air-fuel mixture that can enter the combustion chamber.

The use of smaller intake and exhaust ports allows the engine to develop more torque at low engine speeds. By comparison, larger intake and exhaust ports allow greater airflow during the engine's high-speed operation, giving the engine more torque at higher engine speeds. Engines with smaller ported heads, meaning a cylinder head with smaller ports, work well in applications such as four-wheel drive vehicles, which are typically operated at lower engine speeds on trails. Larger port heads work well in engines that operate at high speeds and loads such as in circle-track racing. To get the advantage of both small and large ports, newer engines use two or three intake valves and intake ports on each cylinder. At low rpm, airflow is supplied through only one intake port. At higher rpms, airflow is supplied through all of the intake ports. This gives high velocity airflow at both low and high rpms resulting in more efficiency and power across a broader rpm range.

TECHNICIAN TIP

In most cylinder heads, intake valves are larger than exhaust valves. In a naturally aspirated engine, the engine relies totally on atmospheric pressure to push air into the cylinder. A drop in pressure occurs as the piston moves down the cylinder during the intake stroke. Atmospheric pressure pushes air and gas through the intake manifold, filling the void where the piston was. The average atmospheric pressure at sea level is 14.7 pounds per square inch (psi), or 101 kilopascals (kPa). The exhaust valves are smaller because removing the exhaust gas is a much easier task due to the piston creating pressure as it moves up the cylinder. The exhaust gases are actually pushed out by the piston, which creates a greater force on the exhaust gases than atmospheric pressure pushing air into the cylinder.

Valves and Valve Trains

K19007

The valve train is responsible for moving air and fuel through the engine. This means that all of the valve train components must be opened and closed at precise times. In fact, the optimum points in the cycle that the valves open and close change with engine speed (rpm) and load. Newer engines have valve trains that are computer controlled and can adjust valve timing while the engine is operated.

The term "valve train" incorporates all of the parts between the drive of the camshaft and the valves (**FIGURE 19-16**). Valve train configuration is determined by the manufacturer, so valve trains come in a variety of configurations. The following sections will familiarize you with each of the components and layout.

Valve Sizes and Components

K19008

Valves come in different sizes depending on the engine. Valves experience enormous stress even in normal conditions. In a vehicle that is driven at 2500 rpm, each valve opens and closes approximately 21 times per second, and at 6000 rpm, each valve opens and closes 50 times per second. The opening and closing is controlled by the **camshaft lobe** and valve spring. The **valve train** includes all of the components that are driven from the camshaft drive, including the lifters, pushrods, rocker arms or cam followers, valves, valve springs, valve keepers, and valve retainers (**FIGURE 19-17**).

The valve spring is compressed when the cam lobe pushes the valve open. The valve spring pushes the valve to its fully closed position as the cam lobe returns to its base circle. Valve keepers, also called valve retaining devices, are usually half-circle, wedge-shaped pieces that lock onto a groove, or grooves, near the top of the valve stem to keep the valve spring retainer and spring retained on the valve while in the head. The valve retainer is similar to a thick washer, with a tapered center hole that mates with the wedge-shaped valve keepers. The retainer is placed near the top of the valve stem and keeps the valve spring compressed between it and the surface of the cylinder head. The retainer works in conjunction with the valve keeper to retain the valve in the cylinder head.

The components of the valve include the following (**FIGURE 19-18**):

- **Valve tip**: The portion of the valve stem that comes in direct contact with the rocker arm, cam follower, or bucket lifter, which opens the valve.
- **Keeper groove**: The specific notch in the valve stem into which the keeper fits securely.

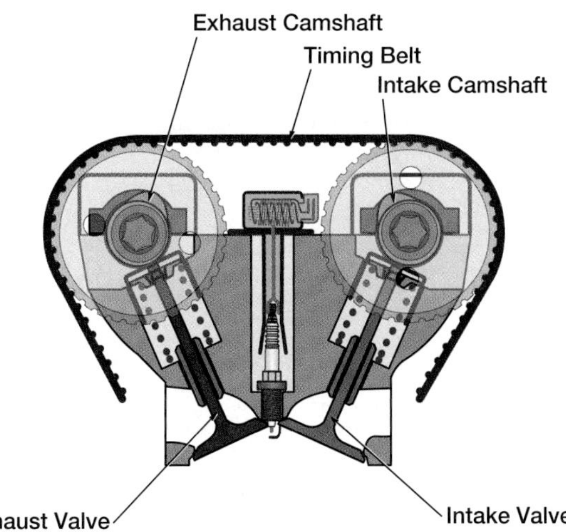

FIGURE 19-16 Typical valve train (dual overhead cam design).

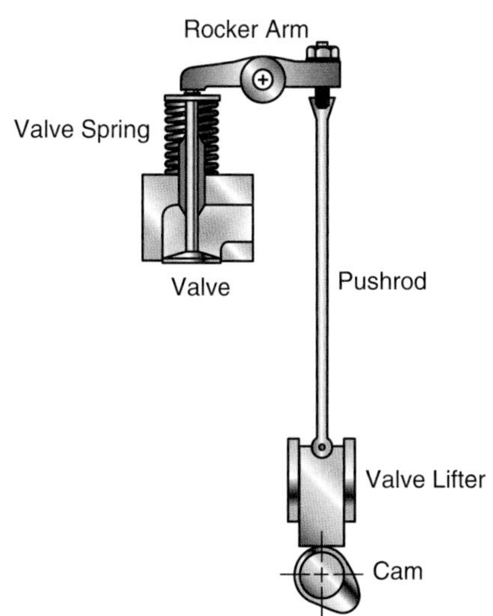

FIGURE 19-17 Valve train (cam in block design).

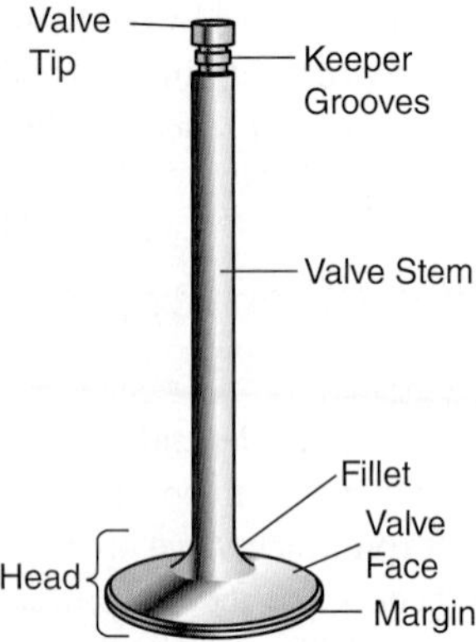

FIGURE 19-18 Parts of the valve.

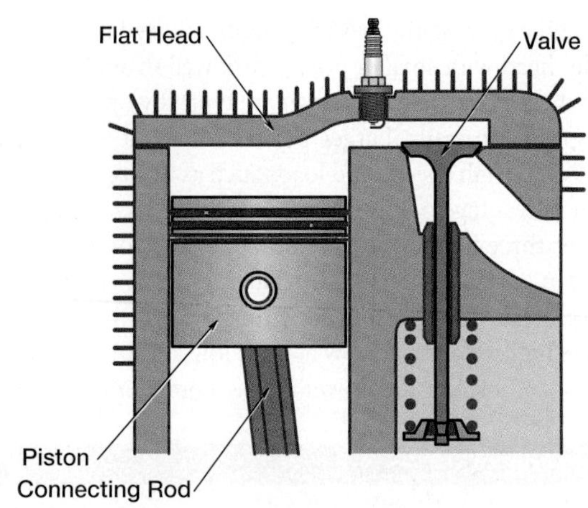

FIGURE 19-19 Flat-head design.

- **Valve stem**: A long shaft-like portion of the valve assembly that allows the valve to move freely inside a valve guide and contains the valve groove(s) into which the keeper is inserted.
- **Valve head**: The mushroomed portion of the valve, which contains the face and margin.
- **Valve face**: The portion of the valve that does the actual sealing off of the port. It is commonly ground to about a 45-degree angle to provide a positive seal during the combustion process.
- **Valve margin**: The portion of the valve between the valve face and the valve head.

FIGURE 19-20 Four-valve arrangement.

> **TECHNICIAN TIP**
>
> The valve margin gets thinner as the valve face is machined during a valve job. If it is too thin, the valve will burn in a short time. Always measure the margin and compare it to specifications before reusing it.

Valve Arrangement

K19009

In early engine designs, the valves were positioned with the valve head facing up; they were located in the engine block off to one side of the cylinder. This design was called the **L-head** (flat head) (**FIGURE 19-19**). In today's engines, the valves are located in the cylinder head above the pistons (OHV).

In OHV engines, there are several different valve arrangements, ranging from one intake valve and one exhaust valve to a five-valve arrangement. A three-valve arrangement usually has two intake valves and one exhaust valve. A four-valve arrangement usually has two intake and two exhaust valves (**FIGURE 19-20**). A five-valve arrangement usually has three intake and two exhaust valves (**FIGURE 19-21**). By adding more than one intake or exhaust valve, volumetric efficiency can be increased while keeping the speed of the air-fuel mixture high, leading to greater power output.

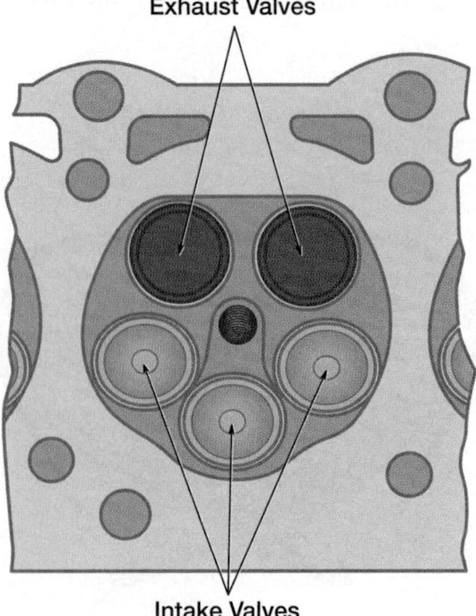

FIGURE 19-21 Five-valve cylinder head.

TECHNICIAN TIP

It has always been said that the intake valve is larger than the exhaust valve. When we had only one intake valve and one exhaust valve per cylinder, that was an accurate statement. Now, with multi-valve heads, that is not always the case. In many three-valve and five-valve engines, the extra valve is an intake valve. In these situations, because there are more intake valves than exhaust valves, the intake valves might be smaller. Therefore, be careful when identifying the valves; these newer multi-valve designs might trip you up.

Types of Valves

K19010

Valves control the flow of air and fuel into and out of the cylinder. There are many different valve designs with variations to suit the unique features of the engine into which they are installed. Some engines, like those in heavy construction vehicles, need valves with strength and longevity to handle long, grueling work. Other engines, such as those in a race car, require valves that are lightweight and capable of withstanding high stress situations, usually for shorter longevity.

The vast majority of automotive engines use a **poppet**, or **mushroom, valve**. It is called this because it "pops" up and down to open and seal its port. It also resembles a mushroom, with a narrow stem and larger head. It fits into a port in the cylinder head. Its valve face makes a gas-tight seal against the valve seat (**FIGURE 19-22**). The intake valve opens and closes the intake port to allow air and fuel to enter the combustion chamber and to seal it during the compression, power, and exhaust strokes. The exhaust valve covers the exhaust port during the intake, compression, and power strokes.

Because the valves are exposed to both the high-combustion temperatures and the hot exhaust gases, they can become red-hot under high-power conditions, so they have to be able to withstand high temperatures. At the same time, they must be able to stand up to both the **spring pressure** trying to pull the valve apart and the **combustion pressure** trying to force it into the valve seat. Valve tensile failure causes the valve to stretch or even be pulled apart by the action of the valve spring. When tensile failure of a valve first occurs, it generally causes the valve stem to stretch, which results in a cylinder misfire due to loss of compression from the valve not fully closing. In an extreme case, it could cause the head to separate from the stem and drop into the combustion chamber, causing catastrophic engine damage.

Exhaust valves are made of nickel-based alloys to help withstand the extreme heat. The problem with getting the exhaust valve too hot is that it can either melt, or it will deform the sealing surface of the face as it slams closed, resulting in a burned valve and loss of compression. The majority of the valve cooling happens when the hot valve face sits in the cooler valve seat in the head, although some of the cooling takes place through the stem and across into the valve guide. Some exhaust valves are manufactured to have hollow stems that are half filled with sodium during the manufacturing process to help transfer heat away from the head of the valve. Generally, sodium-filled valves cannot be repaired and must be replaced when problems occur.

Various surface treatments are used to help the valve resist wear, burning, and corrosion. Intake valves are made of steel mixed with chromium or silicon to make them more resistant to corrosion, and manganese and nickel to improve their strength (**FIGURE 19-23**). The valve face may also be coated in a material called Stellite™, which is a mixture of chromium and cobalt that can hold up to higher heat. Intake valves must also withstand spring pressure and combustion pressure, just as the exhaust valves do. However, the intake valve does not get as hot as the exhaust valve does, for two reasons. First, it is closed when exhaust gases are expelled from the exhaust, so hot gases do not flow around it. Second, the cool intake charge flows around it on the intake stroke, which helps to cool it. In high-performance applications, the intake and exhaust valves can be made of stainless steel or titanium, both of which are suited to extreme conditions.

FIGURE 19-22 A poppet, or mushroom, valve, which makes a gas-tight seal against the valve seat.

FIGURE 19-23 Intake valves are made of steel mixed with chromium or silicon to make them more resistant to corrosion, and manganese and nickel to improve their strength.

Valve Springs

K19011

The **valve spring** closes the valve and holds it firmly against the valve seat. Valve springs in automotive applications are almost always of the coil spring variety. Valve springs have several specifications they must meet. One is the *valve spring closed pressure*, which is a measure of the amount of pressure the spring exerts on the valve to close it tightly against the valve seat. Another specification is *valve spring open pressure*, which is a measure of the amount of pressure the valve spring is exerting to overcome the inertia of the valve train as the valve reaches its fully opened position. Both of these specifications are dependent on the type of valve train used in the engine and are measured in pounds or kilograms. The *spring-installed height* is another important specification that must be met. It indicates how tall the spring is when everything is assembled in the cylinder head. This measurement is important because spring length affects the amount of pressure the spring exerts on the valve train.

The valve could have one spring or multiple springs per valve (**FIGURE 19-24**). Usually in engines that operate at high rpm, the manufacturer uses multiple springs to close the valve. This is done to increase spring pressure and prevent **valve float**, a condition that occurs when the valves cannot be closed (sealed) by the valve springs as fast as is needed before the next stroke. Two springs can close a valve faster than one spring can.

It is important not to have overly strong valve springs (too much pressure), as excessive pressure could cause premature wear on the other valve train components. This is especially true for the cam lobes and lifters as well as the pushrods and rocker arms. Insufficient spring pressure is also problematic, as it causes valve float, which can allow the piston to hit an open valve on some engines.

FIGURE 19-24 Single- and dual-valve springs.

Valve Spring Retainer

The **valve spring retainer** is used along with valve keepers to hold the valve spring in place, centered on the valve stem (**FIGURE 19-25**). The valve spring retainer has a tapered center hole that provides a mating surface for the valve keepers. Valve retainers are usually made from steel; however, aluminum and titanium are used in some high-performance engines.

Some exhaust valve spring retainers have a built-in device called a valve rotator. These retainers provide a rotation function to slightly rotate the exhaust valve in one direction or the other each time it is opened, thus preventing hot spots and carbon buildup, and evening out the wear on the valve face and valve seat. They are found only on exhaust valves, as the intake valve is cooled by incoming fuel and air.

FIGURE 19-25 Valve spring retainers.

Valve Keepers

Valve keepers (also called valve locks) lock the valve stem to the retainer (**FIGURE 19-26**). The valve keepers are usually a split taper design that matches the valve retainer taper, with a protruding ridge or rings that fit into a groove on the valve stem. This creates a wedging effect that holds the retainer firmly to the valve stem. The valve keepers are hardened to withstand the pounding action of the opening and closing of the valves.

FIGURE 19-26 Valve keepers lock the valve stem to the retainer.

Valve Guides

K19012

Valve guides are the hollow metal passageways in the cylinder head in which the stem of the valve rides (**FIGURE 19-27**). The valve guide allows the valve to move only up and down while

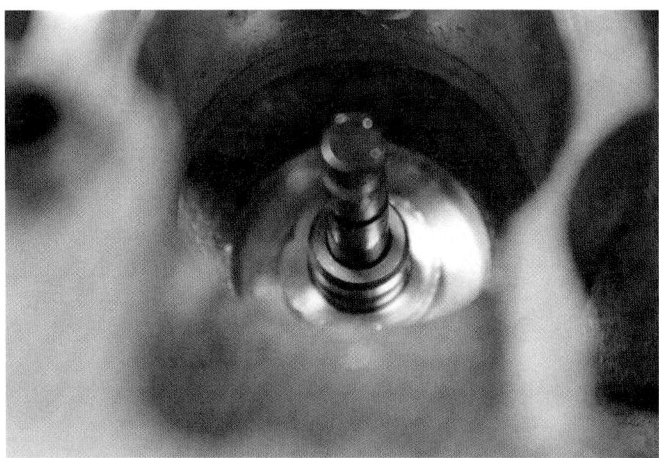

FIGURE 19-27 Valve guides position the valves properly in the head.

FIGURE 19-28 Insufficient valve stem clearance leads to improper lubrication, which can gall or stick the valve.

holding it in the same position side to side. To give long life, the valve guide must be able to retain lubrication for the valve stem. The valve moves back and forth inside the guide approximately 21 times a second at 2500 rpm, so lubrication is critical to reduce wear on the guide and stem. Clearance between the valve stem and the valve guide must be within the manufacturer's specifications. When valve stem clearance is too small, the valve stem is not able to be lubricated properly and can gall or stick (**FIGURE 19-28**). When valve stem clearance is too large, the engine suffers from premature wear on the valve stem, valve face, and valve seat. If the stem wear is bad enough, it can cause an intermittent loss of compression and misfires. Also, oil will be drawn down the intake valve guides, causing oil to enter the intake airstream. Oil will also seep down the exhaust port when the engine is off, or be drawn past the exhaust valve stem during engine deceleration. Any oil burned in the combustion chamber or exhaust system will result in excessive oil consumption, which appears as a bluish-white exhaust smoke.

Integral Valve Guide

The **integral valve guide** (non-replaceable valve guide) is cast into the cylinder head when the cylinder head is formed and

FIGURE 19-29 Integral valve guide cast and machined into the cylinder head.

is machined into the casting (**FIGURE 19-29**). Because it is constructed as part of the cylinder head, it cannot be removed and replaced. Instead, an integral guide can be repaired by reaming it oversize and installing a thin wall bronze liner. Once installed, the inside of the liner is burnished by forcing a specially sized ball bearing through it. A machinist can also repair the integral guide by using a head shop to machine the valve guide (boring out) oversize and then installing a thick wall guide, which can be made of iron or bronze. The guide is then machined to the proper size. Integral guides must be repaired when the valve guide has worn past its service limits. For example, if a valve guide's size is stated in the service information as 0.250" (6.35 mm) plus or minus 0.0005" (0.0127 mm), and the guide is worn to 0.252" (6.4008 mm), then the guide is 0.0015" (0.0381 mm) too worn to be put into service.

Non-Integral Valve Guide

The **non-integral valve guide** (replaceable valve guide) is manufactured separately from the cylinder head and press fit into a specially machined hole in the cylinder head that was formed as the head was cast. Non-integral guides can be removed and replaced, as needed, when they are excessively worn. They are typically made of cast iron or bronze—robust materials that can withstand the process of being pressed or driven into the cylinder head as well as provide a durable surface for the valve stem (**FIGURE 19-30**). The non-integral valve guide has an interference fit to the cylinder head and is usually driven into place by using a valve guide driver and air hammer or small sledge hammer.

Valve Stem Seals

Because valve stems require a small amount of running clearance as well as a small amount of lubrication, it is possible for an excessive amount of oil to leak past the valve stems into the intake and exhaust ports. The intake valve is more likely to pass oil through its guide than the exhaust valve because the intake port has low manifold pressure (vacuum) due to the intake

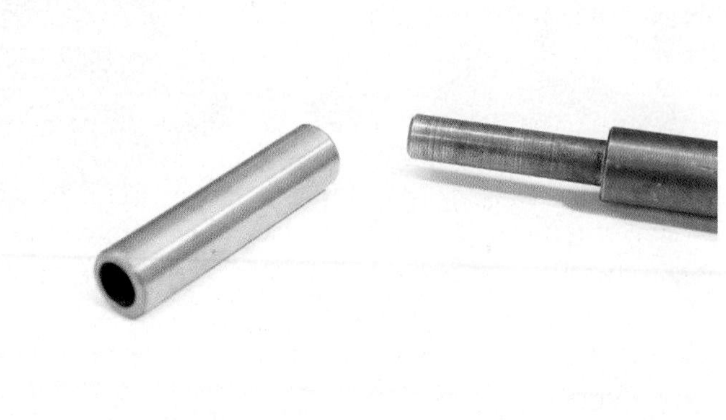

FIGURE 19-30 Non-integral valve guides are typically made of cast iron or bronze, providing a durable surface for the valve stem.

FIGURE 19-31 An umbrella-style valve stem seal.

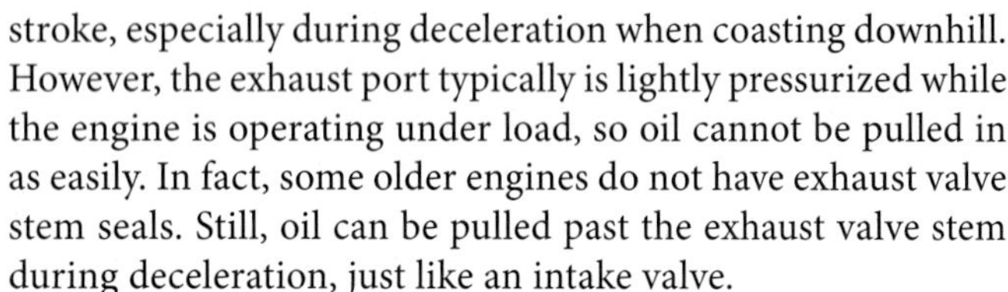

stroke, especially during deceleration when coasting downhill. However, the exhaust port typically is lightly pressurized while the engine is operating under load, so oil cannot be pulled in as easily. In fact, some older engines do not have exhaust valve stem seals. Still, oil can be pulled past the exhaust valve stem during deceleration, just like an intake valve.

Worn valve seals and/or guides can also allow oil to seep down the stems and puddle in the port, causing a puff of blue smoke during start-up. In an engine with the proper valve guide clearance, this oil leakage is reduced, but not eliminated. To control the amount of oil that lubricates the valve stems, valve stem seals are used on both the intake and the exhaust valves. These seals fit around the valve stem and usually seal to the top of the valve guide. Valve stem seals are designed to allow just enough oil to get past them to lubricate the valve stems and guides. If valve stem clearance is excessive or if the valve seals wear out or become damaged, excessive oil will pass through the guide, resulting in high oil consumption, excessive smoke emitted from the exhaust pipe, and possibly fouling of the spark plugs.

Cylinder heads can have one of three basic types of valve seals:

- **Umbrella-style valve stem seals** fit tightly around the top of the valve stems and ride up and down with the valves (**FIGURE 19-31**). The umbrella is a larger diameter than the valve guide, so oil drips off the umbrella outside of the guide. Umbrella-type seals are generally found on older (1980s and earlier) American-built V8 engines. They are not typically found in modern engines.
- **O-ring valve stem seals** fit on top of the valve in a groove just below the valve spring keepers (**FIGURE 19-32**). The O-ring seals the oil from leaking between the valve stem and the valve retainer. These seals were used primarily on General Motors V8 engines from the 1980s and earlier. O-ring seals are not typically found in modern engines.
- **Positive valve stem seals** are used in today's engines (**FIGURE 19-33**). The positive seal fits tightly on the valve guide as the valve stem moves up and down inside of the seal. This type of seal is the most effective because it forms a positive seal between the guide and the valve stem. Positive valve stem seals have a small groove cut into the top of the seal that rides on the stem. This design keeps a small amount of oil against the valve stem and only allows a very small film of oil past it to lubricate the valve guide.

FIGURE 19-32 An O-ring valve stem seal.

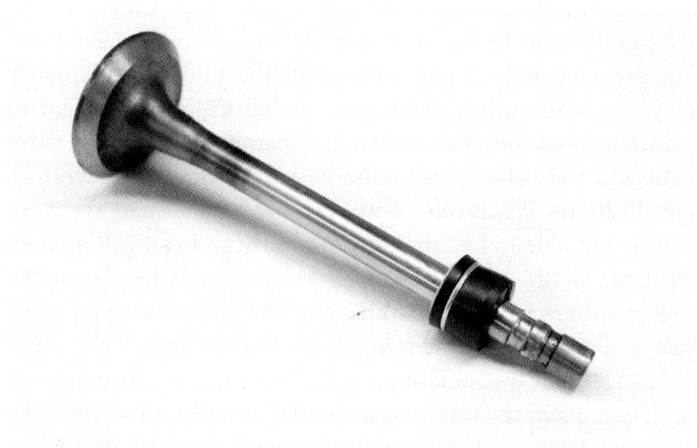

FIGURE 19-33 A positive valve stem seal.

Valve Seats

K19013

The **valve seat** is the part of the cylinder head that mates with the face of the valve when it is fully closed. It is typically made of hardened steel in order to withstand the extreme temperatures and pressures encountered during engine operation. Valve seats must be compatible with today's gasoline. Use of today's unleaded gasoline in an engine with non-hardened valve seats will cause the seats to slowly wear away and sink deeper into the head, a condition known as valve seat recession.

Valve seat inserts are made of metal rings that match the shape of the valve face in order to form a strong and complete seal (**FIGURE 19-34**). They are usually made of an iron alloy because the alloy is stronger than iron itself and provides a more durable surface. Most late-model, or "newer," vehicles require hardened seats in both the intake and the exhaust ports due to the extreme temperatures and pressures that newer engines must handle. With government air quality standards constantly rising, achieving a cleaner burn for an internal combustion engine means a hotter burn and higher pressure and thus a more intense combustion process. Valves, valve seats, pistons, and cylinders all have to be able to meet these increased demands.

Valve seats are either ground or cut (machined) into the proper shape and **concentricity** so that the valve face and seat seal tightly. Valve seat grinding is when the seat is ground to the proper angle with a special grinding stone. Valve seat machining uses a special hardened cutter to cut the proper angles into the seat.

Some engines use valves that have a 30-degree valve face angle. Others have a 45-degree valve face angle. In some cases, the intake valve angles are different than the exhaust valve angles. The differences in angles simply represent the manufacturer's engine design; the exact angle has minimal affect on the cylinder head function as long as the valve and seat fit tightly enough to seal the force of combustion in the combustion chamber. In many cases, the valve and seat are machined about ½–1 degree different from each other for sealing purposes. For instance, the valve is cut to a 45-degree angle, and the seat is cut to a 46-degree angle. This is done so that the angles will meet at a "point" toward the outside of the valve face and seat. This difference is called an **interference angle** (**FIGURE 19-35**). An interference angle allows for a quick **bedding-in** of the valve face to the valve seat on new engines. Bedding-in is the process of the valve face wearing into the valve seat during the initial breaking-in of the engine. This provides a long-lasting, leak-proof seal.

TECHNICIAN TIP

Up until the early 1970s, gasoline contained tetraethyl lead, which was used to inexpensively boost the octane. Lead also lubricated the valve seats by providing a cushioning effect, which allowed the valve seats to be machined into the base cast iron material and still give good life. However, when the installation of catalytic converters was mandated for emission control purposes in the mid-70s, lead could not be used in fuel anymore, as lead would contaminate and kill the converter. This situation required the use of hardened valve seats, which resist wearing out quickly.

Camshaft

K19014

The camshaft provides the motion to open and close each of the valves. The camshaft is a shaft with offset eccentrics, called cam lobes (**FIGURE 19-36**). The cam lobe determines when each valve opens, how far it opens, and how long it stays open. The

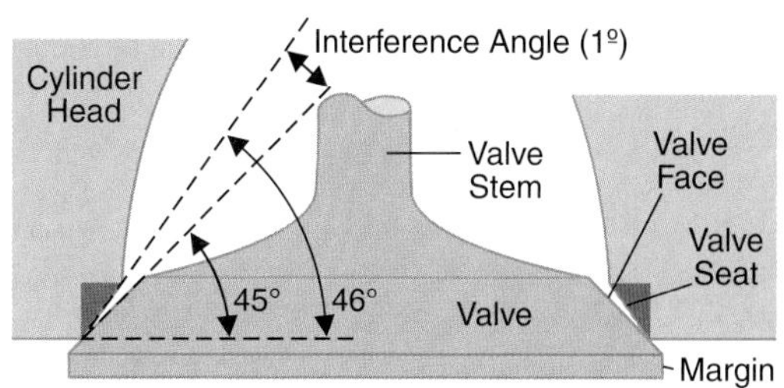

FIGURE 19-35 Valve-to-valve seat interference angle.

FIGURE 19-34 Valve seat inserts are common on aluminum cylinder heads.

FIGURE 19-36 Typical camshaft.

profile of the cam lobe determines these functions. The base circle takes up about half of the entire profile. When the lifter is on the base circle, the valve is closed. The nose is the highest point of the cam lobe and determines how far the valve opens. The difference between the base circle and nose is called lobe lift. The opening ramp is what opens the valve. The opening ramp starts at the base circle and ends at the nose. The closing ramp is the opposite of the opening ramp and allows the valve to close properly. It starts at the nose and ends at the base circle.

Camshafts are ground for specific applications and may or may not be symmetrical, meaning that the opening and closing ramps may or may not be identical. Also, different camshaft profiles create different engine operating characteristics. Two terms that you need to understand are "lift" and "duration." Lift refers to how far the cam lobe will lift the lifter, cam follower, or bucket lifter. The greater the lift, the further the valve opens, and the easier it is for air to enter the cylinder. However, the greater the lift, the faster the valve has to move from closed to open, and the harder the valve springs have to work at closing the valve. Duration refers to how long the valve is held open. The longer it is held open, the more time there is for air to enter the cylinder. Higher-duration camshafts are usually used in engines that operate at higher rpm. We look at these concepts in more depth next.

Camshaft Specifications

K19015

As seen previously, the **base circle** of the cam lobe is the rounded bottom part of the egg shape; this is where the lifters rest when the valves are closed (**FIGURE 19-37**). The shape of the camshaft lobes themselves affects the power characteristics of the engine. One specification of the cam lobe is the lift of the cam lobe. **Lift** is the amount the lifter or cam follower moves with the cam. The more the valve is lifted off its seat, the more air that can enter the engine. However, more lift creates more pressure on the valve train components due to pushing the valve further and compressing the valve spring further. Too much lift can create coil bind in the valve springs. **Coil bind** occurs when the coils of the spring touch each other. This causes the spring or the camshaft and lifter to wear excessively or break. Another problem with too much valve lift is too little valve-to-piston clearance. Too much valve lift can cause the valve to hit the piston and ruin the engine.

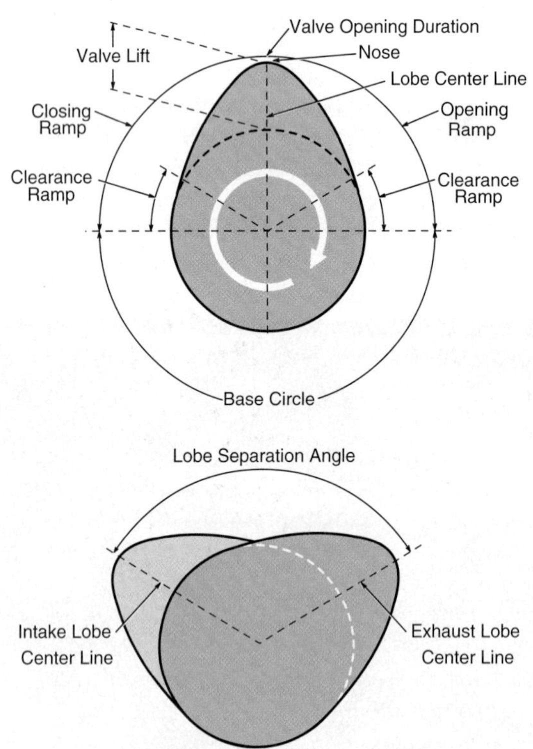

FIGURE 19-37 The camshaft lobe controls the opening and closing of the valves and affects performance greatly.

Valve duration is another specification engineers use when designing a cam lobe. **Duration** is the amount of time the valve stays open, given in degrees of *crankshaft* rotation (not camshaft rotation). The longer the valve is kept open, the more air will be able to move into and out of the engine. **Cam lobe centerline** is where the cam lobe is located in relation to TDC of the engine in degrees. There is a cam lobe centerline specification for both the intake and exhaust lobes. Cam centerline and duration can be used to determine when the valves will open in degrees of crankshaft rotation. **Cam lobe separation** refers to how far the centerline of the intake lobe is offset from the centerline of the exhaust lobe. Generally speaking, the wider the separation, the better the camshaft is for building power at lower rpm. Narrowing the separation builds power at higher rpm. The smaller the number of degrees of lobe separation, the more valve overlap there is. Valve overlap simply means that the intake and exhaust valves are open at the same time.

The **cam lobe ramp** is where the rise of the cam lobe starts from the base circle to the top of the lobe. The ramp is where the valve starts to lift and is on the opposite side from where it starts to close. The degrees of rotation the valve remains open and the cam lobe separation are specifications measured in crankshaft degrees, not in camshaft degrees. Valve lift, duration, lobe separation, and overlap are features determined by the grind (profile) of the camshaft.

High-performance engines have more degrees of valve overlap than street engines because of longer cam duration. Increased valve overlap and lift increase top-end rpm power due to **column inertia** (ram effect), but they reduce low-speed power and idle quality, again due to a loss of column inertia. Valve overlap specifications will not be given if you have a DOHC engine because the exhaust and intake lobes run on separate shafts. So-called street cams, three-quarter race, and full race cams have progressively more lift and duration, all tailored to meet the operating conditions of the engine. For street engines, too much lift or duration may not be beneficial because the resulting lower intake manifold vacuum causes the engine to idle roughly.

When comparing cam specifications, it is important to know that there are two listings for duration specifications: advertised duration and duration at 0.050" lift (**FIGURE 19-38**). Advertised duration measures the duration from where the lifter first starts to move. Duration at 0.050" lift measures the

Specifications for a 2000- to 5000-rpm street camshaft:

	Intake		Exhaust
Lift	0.440		0.440
Duration at 0.050″	222°		222°
Cam lobe centerline	109° atdc		116° btdc
Lobe separation angle (camshaft degrees)		113° between lobe centerlines	
Degree valve opens at 0.050″	2° btdc		48° bbdc
Degree valve closes at 0.050″	40° abdc		–6° atdc

Specifications for the same camshaft to be used up to 6500 rpm:

	Intake		Exhaust
Lift	0.500		0.515
Duration at 0.050″	256°		270°
Cam lobe centerline	108° atdc		116° btdc
Lobe separation angle (camshaft degrees)		112° between lobe centerlines	
Degree valve opens at 0.050″	20° btdc		71° bbdc
Degree valve closes at 0.050″	56° abdc		19° atdc

FIGURE 19-38 An aftermarket performance camshaft includes a specifications card that is used during engine assembly to ensure correct timing. The top chart shows the specifications of a street camshaft, and the bottom chart a higher performance camshaft. Notice the differences in valve lift and duration.

FIGURE 19-39 Valve timing marks on the crank and cam gear.

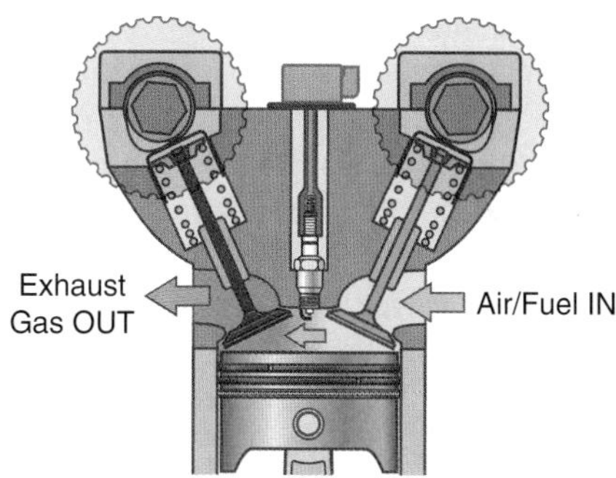

FIGURE 19-40 The intake valve starts to open while the piston is still moving up on the exhaust stroke. The exhaust valve is closing, but not closed until the piston starts moving back down.

duration starting when the lifter hits 0.050" and ending when it returns to 0.050". This specification is a better indicator of the camshaft's profile and how it will affect the operation of the engine, because by 0.050" lift, the cam is well into its lift profile.

Valve Timing

K19016

Valve timing is critical to the proper operation of the ICE. Many performance gains can be made by simply changing valve timing, but it all depends on what rpm range the power is desired to be. On fixed valve timing engines, valve timing is a mechanical setting, and once set, it cannot be changed without physically changing the timing components in some way. On variable valve timing engines, the valve timing can be modified to provide the best operating conditions for the engine speed, power, efficiency, and emissions. Variable valve timing is discussed further in a later topic.

On fixed valve timing engines, the technician sets the timing components to marks the factory has made on the crank and cam gears (**FIGURE 19-39**). It is critical for the technician to ensure that the timing marks are lined up correctly when installing the timing components. The intake and exhaust valves must open at the correct point in the rotation of the crankshaft so that the pistons will be in the correct position and not strike the valves while they are open.

Four-Stroke Cycle Engine Valve Timing

Theoretical valve timing is used when explaining the operation of a four-stroke engine and assumes that the valves open and close instantly at TDC and BDC. In reality, the valves open and close sooner or later than this. Because of inertia and the amount of time it takes the valves to fully open and close, we have to begin the opening times earlier and delay the closing time. Thus, we begin the intake stroke by opening the intake valve as the piston is still moving up and the exhaust valve is still slightly open, but closing (**FIGURE 19-40**).

As air flows into the cylinder, it gains velocity in the intake manifold, which creates a column of air that has inertia and keeps the airflow moving, or racing, into the cylinder. If we keep the intake valve open after BDC, air continues to rush in because of the inertia of the airflow. This is called column inertia; as a column of air flows into the combustion chamber, it creates inertia, which keeps air flowing until its inertial energy is spent. This principle of airflow allows the intake valve to be left open even when the piston is starting back up in the cylinder. The inertia causes air molecules to pack together more tightly, especially at higher engine rpm, thus creating a ram-air effect.

The same situation happens on the exhaust stroke. The exhaust valve begins to open shortly before the piston reaches BDC. This timing allows the remaining pressure in the cylinder to escape before the piston starts moving back up. More importantly, it allows time for the exhaust valve to open far enough so that at higher rpm there is enough time to effectively evacuate the cylinder. When the piston travels back toward TDC, the exhaust gases are pushed rapidly out of the cylinder, causing a column effect in the exhaust system. The intake valve begins to open slightly before TDC. The exhaust valve is still open past TDC (overlap), with exhaust gases rushing out; this helps provide the scavenging effect discussed earlier. All airflow that happens in the engine, whether in or out, is helped by the column inertia effect.

Consider again the definition of valve overlap described earlier. In the real world, even standard camshaft grinds (valve timing) include some degree of valve overlap at TDC. Yet in traditional engine designs, camshaft design and valve timing are a compromise for street engine use. Racing engines are designed with greater valve lift, longer valve duration, and increased valve overlap in order to maximize the effect of column inertia encountered at higher rpm.

At cranking speeds, there is not much column inertia, so having the valve open for a long time would reduce engine intake vacuum and make the engine difficult to start because the intake air would move back out past the intake valve as the piston travels up on the compression stroke. With no column inertia to speak of, leaving the intake valve hanging open during the initial portion of the compression stroke lowers the effective compression.

The timing of the valves is controlled by the physical shape and position of the cam lobes on the camshaft. The entire cam can be moved forward or backward in relation to the crankshaft position to make the events happen sooner or later in the piston stroke. Aftermarket performance companies manufacture crank gears with offset keys or keyways that allow the cam to be advanced or retarded manually to gain additional power at the desired rpm (**FIGURE 19-41**). Advancing cam timing closes the intake valve sooner, giving more cranking compression, again because of low column inertia. This results in more torque and power in the low rpm range. Retarding cam timing closes the intake valve later giving more torque and power at a higher rpm range. Because the column inertia is greater at high rpm, more air can be pushed into the engine if the valve is held open later in the four-stroke cycle. This allows the timing to be tailored for a distinct engine purpose (peak torque rpm range). On turbocharged or supercharged engines, minimal valve overlap is typically used, as air forced into the engine by the turbo or supercharger would be forced out of the exhaust during overlap reducing efficiency. Note that cam duration is built into the cam lobe profile and cannot normally be changed while the engine is running.

FIGURE 19-41 After-market timing gear with offset keyways to adjust the cam timing on the engine.

Variable Valve Timing

K19017

We can see from the previous section that the ability to control the advance and retard of the cam timing is a huge benefit to engine performance. Retarded cam timing provides greater engine torque when large amounts of air are flowing into the engine at higher rpm. Advanced timing is useful when the engine needs high torque during lower engine rpm operations. In past applications, an engine designer would have to decide which valve timing best served the operating range of the engine and then design the camshaft for that purpose.

For optimum engine performance at both high and low rpm, what if an electronic module could manage optimum valve timing for such widely differing engine demands? Well, today's automobiles do that in two primary ways; the first uses **phasers** to adjust the position of one or both camshafts hydraulically while the engine is operating. The other type uses camshafts with two lobes per valve: one mild lobe for low rpm, and one aggressive lobe for high rpm. We look at the phaser style first.

Camshafts are advanced or retarded through the use of cam phasers, or actuators. The phaser typically takes the place of the standard cam gear or pulley and uses engine oil pressure to move (some advance, some retard) the camshafts when commanded by the **powertrain control module (PCM)**. One type of actuator, or phaser, uses a twisted gear arrangement and a return spring (**FIGURE 19-42**). When oil pressure pushes against the gear mechanism, the cam rotates and advances. When oil pressure is released, the return spring brings the cam back to the fully retarded position. At rest, these types of actuators may be either fully retarded or fully advanced.

The other type of actuator used is the **vane-type phaser**. This phaser uses vanes similar to those used in a vane-type oil pump to move the camshaft (**FIGURE 19-43**). Oil pressure moves into the phaser and pushes on one side of the vane to advance the cam timing. When oil pressure is switched to the other side of the vane, the camshaft is retarded. Cam timing is adjusted according to what the PCM desires based on its input sensors. If oil pressure is held constant, the cam phaser will likewise hold cam timing constant. The cam phaser uses a mechanical stop to limit maximum valve timing advance or retard.

The second style of variable valve timing uses separate cam lobes for each valve, as mentioned earlier. At low rpm, the mild lobe actuates the valve, allowing good low rpm power. At some point in the rpm range, the PCM causes the aggressive lobe to actuate the valve, improving higher rpm power. Switching between cam lobes can happen in several ways. The first is by a sliding pin that locks the two cam followers together at higher rpms, causing the aggressive lobe to operate the valve

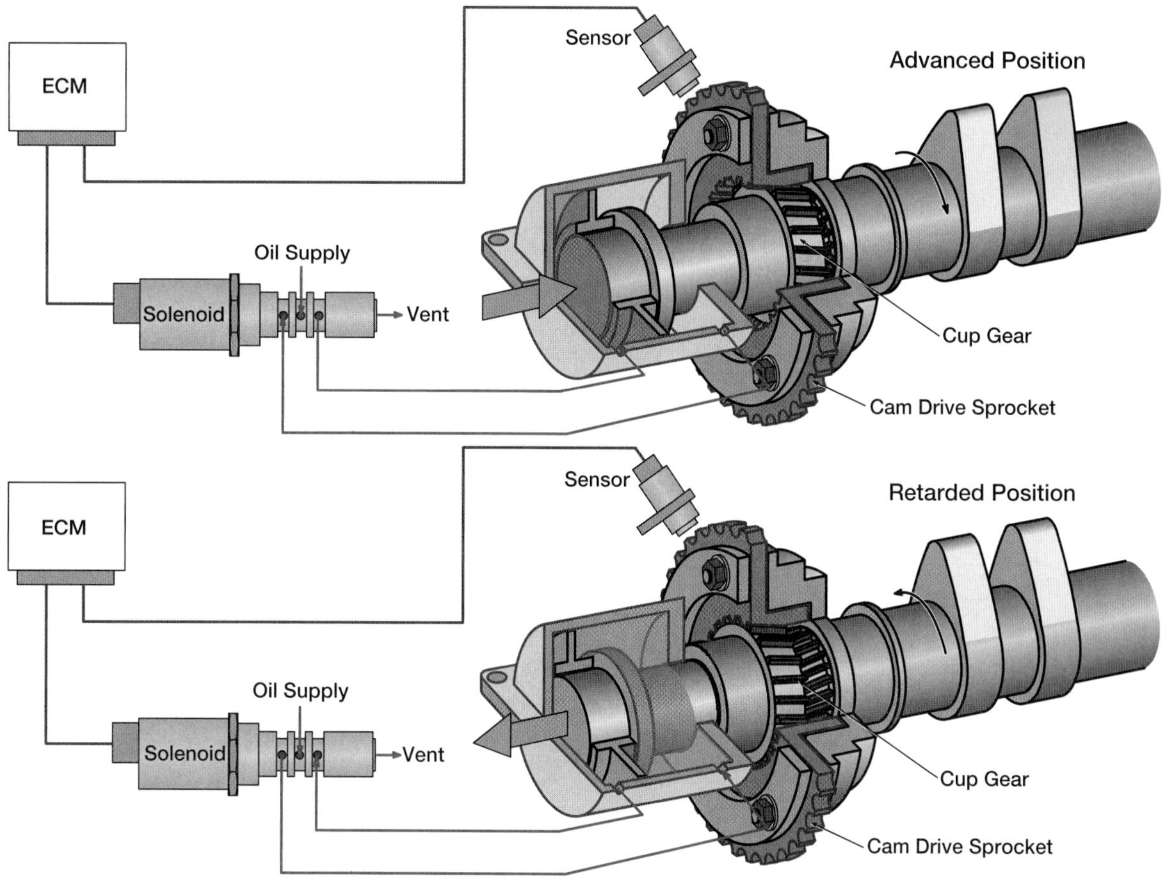

FIGURE 19-42 Twisted gear cam phaser.

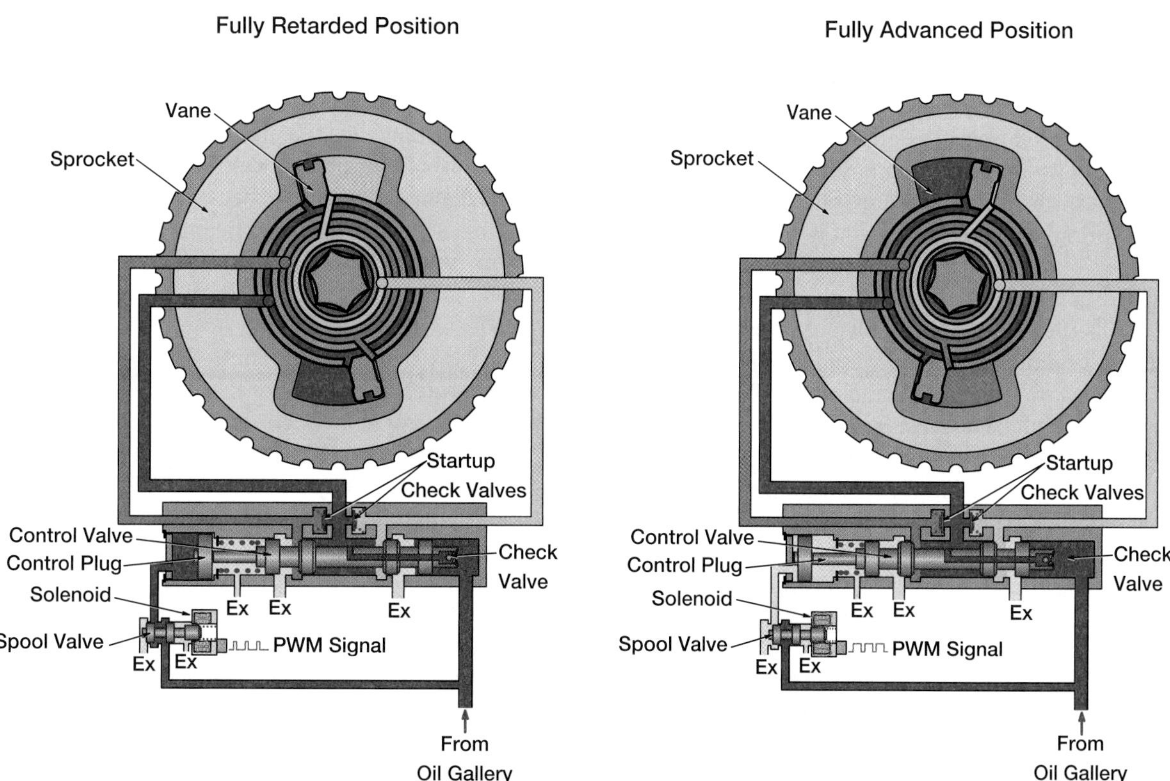

FIGURE 19-43 Vane-type cam phaser.

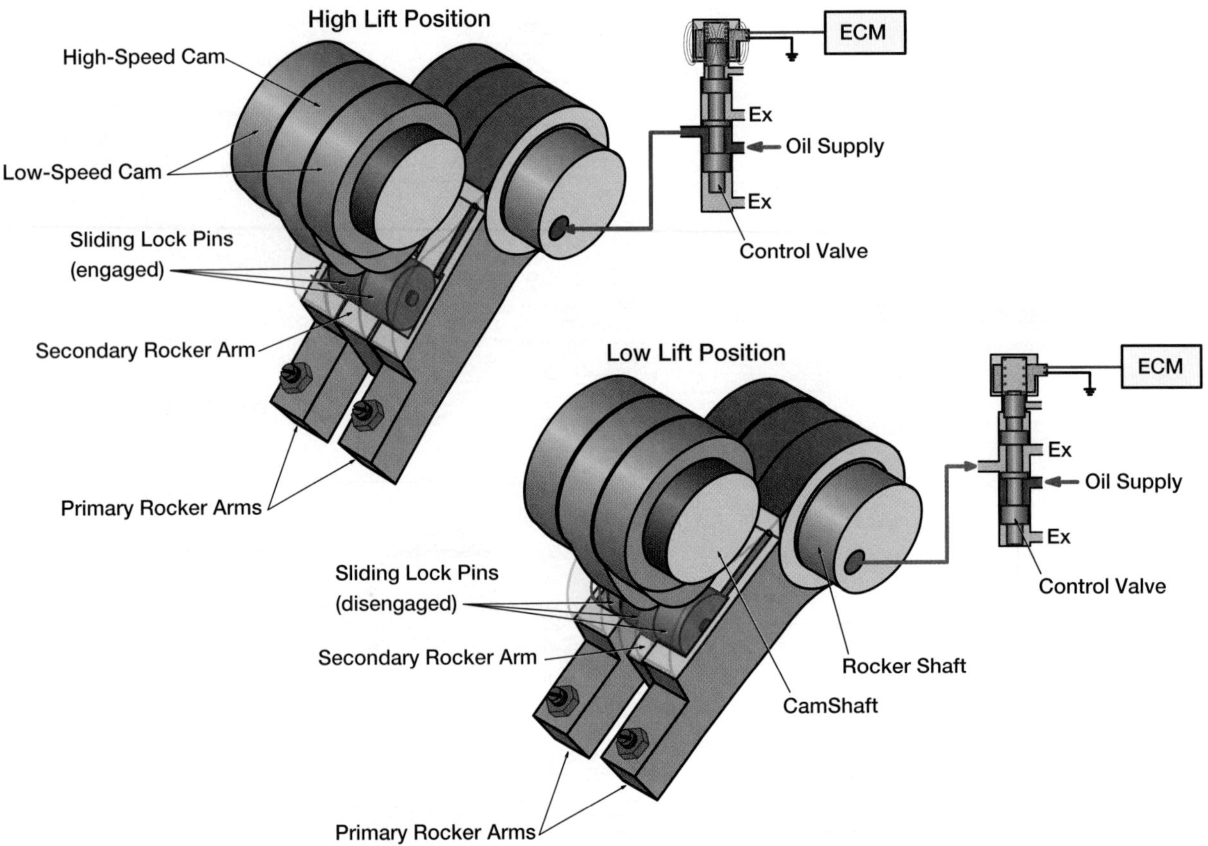

FIGURE 19-44 Variable valve timing through the use of two cam lobes per valve. In this situation, a sliding pin locks both cam followers together, which allows the aggressive cam lobe to operate the valve.

(**FIGURE 19-44**). When the pin is removed, the mild lobe operates the cam follower to open and close the valve. The second type also uses two cam lobes, but only one cam follower for each valve (**FIGURE 19-45**). The cam lobes are free to slide sideways to line up with the cam follower. A pair of solenoids operate plungers, which drop down into slotted ramps to push the cam lobes in one direction or the other. The cam lobes are held at either extreme by a spring-loaded detent assembly. Each plunger only remains in the slot for about half a revolution of the camshaft, just enough to push the lobes to one side or the other. Each plunger is used to push the lobes in only one direction.

Variable valve timing is primarily used on OHC engines. If the cam is a single cam with exhaust and intake lobes, then both intake and exhaust will be affected by rotating the camshaft to the advanced or retarded position. In a DOHC engine, some engine designers use a phaser on the exhaust camshaft only. The phaser used with this type of engine is typically not intended for increased engine power, but rather for better exhaust gas recirculation. If the exhaust valve is held open later, then it will hang open into the intake stroke, which can result in exhaust gases being pulled back into the engine. This dilutes the incoming air-fuel mixture with inert exhaust gas to help reduce oxides of nitrogen emissions. Because **inert gas** does not react chemically (does not burn), it causes the combustion gases to burn more slowly, which reduces the peak combustion temperature. Keeping the peak temperature below 2500°F (1371.1°C) reduces the amount of oxides of nitrogen, or NOx, created. Refer to the Emission Control chapter for more information on NOx.

Controlling NOx in this way allows for the elimination of the exhaust gas recirculation (EGR) valve on some vehicles. In other DOHC engines, the phaser can be installed only on the intake camshaft for greater engine power, or it can be installed on both the intake and the exhaust camshafts. The newest cam phaser in use relies on an electric motor–driven phaser to vary the cam timing (**FIGURE 19-46**); this type of phaser is more accurate, as it does not rely on the oil being a consistent viscosity or squeaky clean.

Variable cam timing is controlled by the PCM. There are several inputs used to ensure that the ECM is able to control the timing accurately. These inputs are the same inputs used by the PCM to control fuel delivery and ignition system timing. The PCM also must know oil temperature to ensure that the oil viscosity is not too thick or thin. Based on input from an oil temperature sensor, the PCM either allows or disables variable valve timing; the viscosity of the oil must be correct for accurate valve timing. The PCM must know the amount of engine load to determine the need for an advance or retard of cam timing. To calculate engine load, the PCM primarily uses the following inputs:

- Mass airflow sensor or manifold absolute pressure sensor
- Throttle position sensor
- Intake air temperature sensor

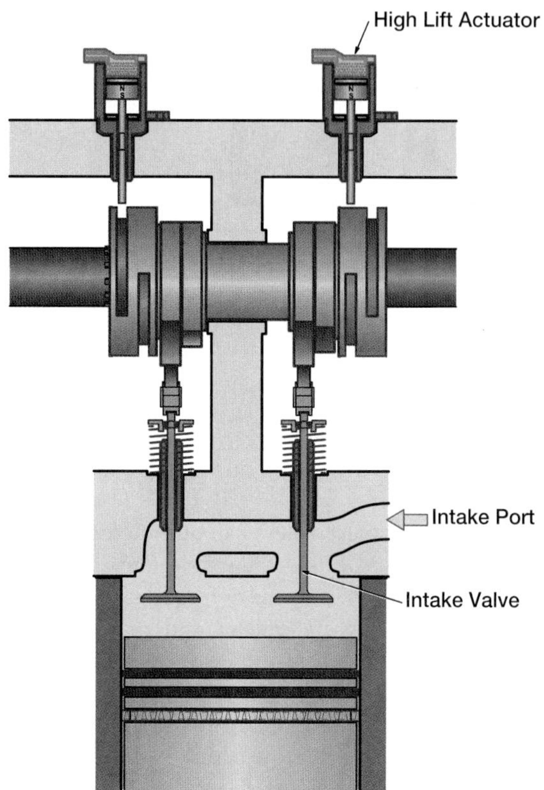

FIGURE 19-45 Sliding cam lobes used as variable valve timing. Each plunger is used to push the cam lobes in only one direction.

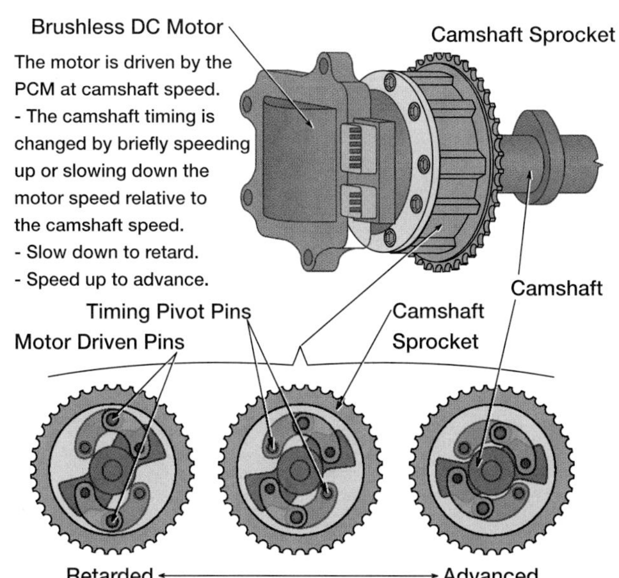

FIGURE 19-46 Electric motor-driven cam phaser.

- Engine coolant temperature sensor
- Crankshaft position sensor

The PCM relies on feedback to ensure that the cams are rotating as commanded; the cam position sensor (CMP) is used for this function. The output of the PCM is delivered to the cam timing solenoid. The cam solenoid allows oil pressure to move into the cam phaser. The solenoid is turned on and off with a **pulse-width modulation (PWM)** signal. PWM is the variable rapid (in milliseconds) time-based on/off switching of a DC signal. The longer the solenoid device is turned off, the less it opens; the longer it is turned on, the more it opens. In this way, the solenoid can regulate the amount, and direction, of oil flow to the cam phaser to alter camshaft/valve timing.

▶ TECHNICIAN TIP

PWM is used to control outputs such as fuel injectors, shift solenoids, torque converter clutches, and even interior vehicle lighting.

Types of Lifters

K19018, K19019

There are two types of lifters: solid and hydraulic. **Hydraulic valve lifters** are constructed as a centrally located plunger inside a hollow cylindrical body (**FIGURE 19-47**). Engine oil fills the inside of the lifter body, pushing the plunger up until all the play in the valve train is removed. Because they

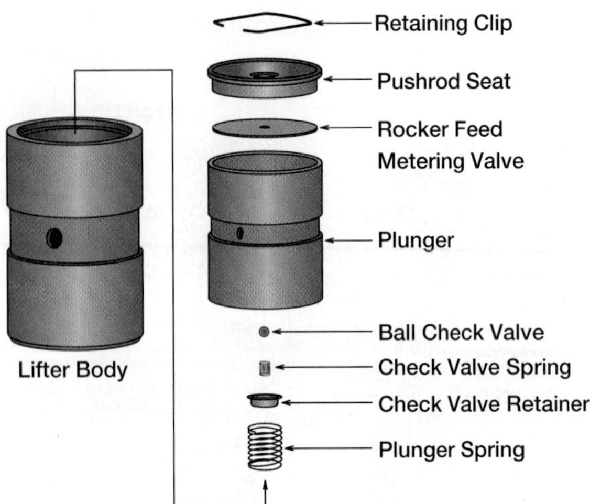

FIGURE 19-47 Hydraulic lifter—exploded view.

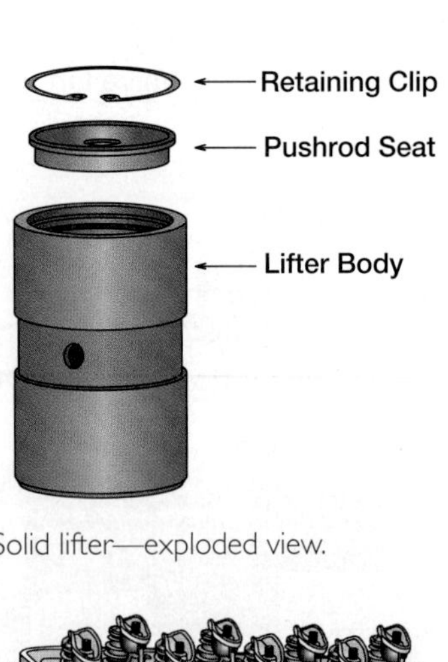

FIGURE 19-48 Solid lifter—exploded view.

are self-adjusting, these lifters can automatically accommodate changes caused by part wear or temperature fluctuations. Hydraulic lifters are more frequently used in today's engines because they are quieter and require no periodic maintenance.

Solid valve lifters are constructed as a hollow cast iron cylinder, capped on both ends (**FIGURE 19-48**). They function by simply transferring the action of the cam lobe to the pushrod. Because there is no means for changing the length of the solid lifter, as seen in the oil-filled plunger action of the hydraulic lifter, they are not able to self-adjust to changes caused by wear or thermal conditions. As a result, all adjustments must be performed manually. These types of lifters require frequent valve adjustments and are typically noisier than their hydraulic counterparts. An advantage of the solid lifter is that it can be made lighter than a hydraulic lifter, so it can be used at higher engine rpm without contributing to valve float.

The valve lifters can be located in four different places: in the engine block, in the cylinder head, in the rocker arms, or between the cam lobe and the valve tip (bucket style) (**FIGURE 19-49**). Both solid and hydraulic lifters can be of either the **flat tappet** or the roller tip design. The flat tappet style has a nearly flat surface that rides directly on the camshaft lobe. The lifter is slightly offset to one side of the cam lobe. This causes the lifter to rotate as it rides up the lobe, instead of sliding up the lobe. This rotating action reduces friction and spreads out the wear over the face of the lifter. The lifter face-to-camshaft lobe contact is one of the highest load areas in the engine. Using a roller tip lifter reduces this friction. Instead of sliding on the camshaft lobe as a flat tappet does, the roller rolls along the lobe with less friction. Because the roller must remain in rolling contact with the cam lobe, the lifter must incorporate a method of preventing it from rotating in the lifter bore. This can be accomplished by connecting two adjacent lifters together with a pivoting link, or it could use a pin on the side of the lifter that rides in a groove in the lifter bore of the block or head.

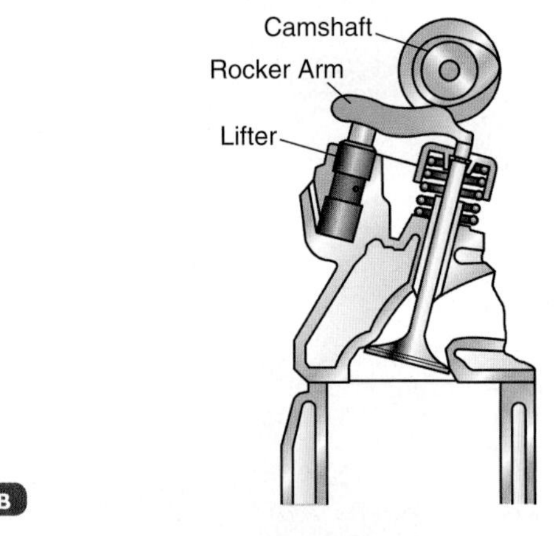

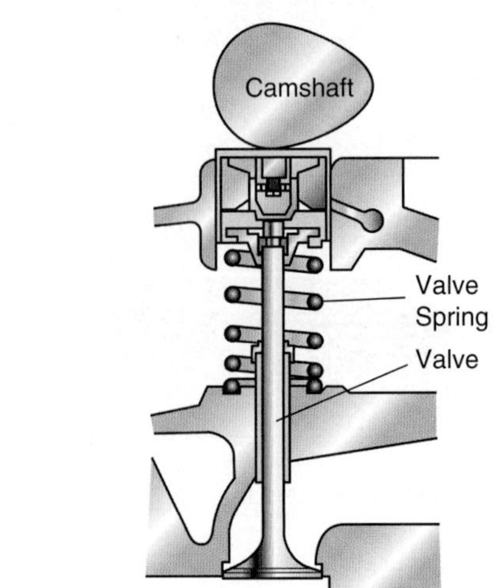

FIGURE 19-49 **A.** Engine lifter in the engine block. **B.** Engine lifter in the cylinder head. **C.** Engine lifter in between the cam and the valve tip (bucket style).

In OHC engines, you will find no rocker arms. Instead, these engines use **camshaft followers** (**FIGURE 19-50**). A camshaft follower performs the same basic function as a rocker arm: It opens a valve. The camshaft follower has a pivot on one end and a slider or roller near the center that directly contacts the camshaft lobe. As the camshaft turns, the roller rolls over the surface of the camshaft. The lobes push the followers away, transmitting the movement to the valve and pushing it down (open).

If the camshaft is located on top of and centered over the valves, the engine will use **bucket-style lifters** (**FIGURE 19-51**). These lifters have a lifter assembly that is mounted on top of the valve and are hollowed out, like a bucket. The hollow portion is placed upside down over the top of the valve stem and provides a wear surface between the camshaft and the valve stem. They perform the same job as all lifters: transmitting movement from the cam lobe to the tip of the valve. Bucket-style lifters may be hydraulic or mechanical. To adjust the valve clearance in the mechanical lifter, you must change the wafer installed on the top of the bucket. Some technicians refer to wafers as spacers or pucks. These wafers are replaced to obtain the proper valve clearance between the camshaft and the valve stem. They come in different thicknesses to achieve the correct valve clearance. Hydraulic bucket lifters are self-adjusting in the same manner as a standard hydraulic lifter and therefore do not require periodic maintenance.

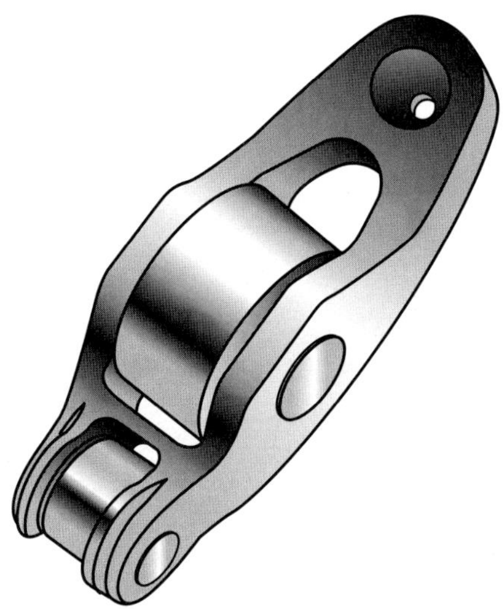

FIGURE 19-50 Camshaft follower.

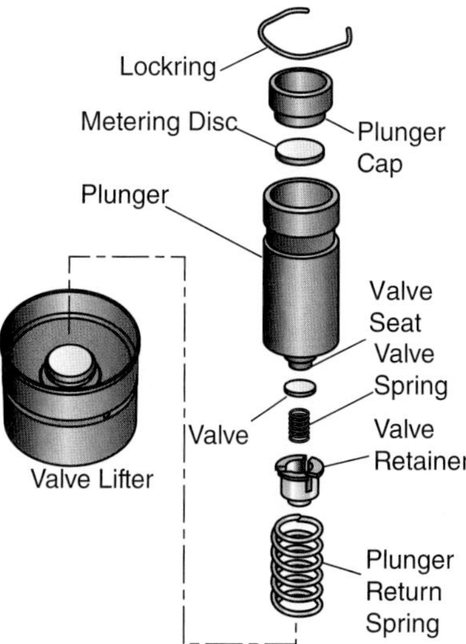

FIGURE 19-51 Bucket-style lifter.

Applied Science

AS-14: Distance/Length: The technician can use precision measuring devices to determine if wear and adjustments are within the manufacturer's tolerances.

The use of precision measuring devices is common within shops and takes place in some form in the maintenance of almost all mechanical systems on a vehicle. Looking only at engines, two common examples are the use of an inside micrometer to measure the inside diameter of camshaft bores in alloy cylinder heads to determine wear from the manufacturer's specification, and the use of feeler gauges to measure valve clearances when carrying out adjustments during servicing.

Pushrods

K19020

OHV engines incorporate **pushrods**. The pushrod transfers the motion from the valve lifter to the rocker arm. Pushrods vary in design, depending on the engine. Some pushrods are hollow inside and transfer oil to the rocker arms. Some engines have different lengths of pushrods; the exhaust is a different length than the intake. Pushrod ends come in several different styles: ball to ball, ball to cup, and ball to adjustable tip (**FIGURE 19-52**). A ball-to-ball pushrod is smooth and rounded at each end. One end fits into a "cup" or dished-out piece in the valve lifter, and the other fits into a cup in the rocker arm assembly. A ball-to-cup pushrod has a smooth, rounded ball at one end that sits in a cup on the valve lifter and a cup at the other end into which a smooth, rounded ball end of the rocker arm assembly fits. The ball-to-adjustable screw pushrod is the

FIGURE 19-52 Three types of pushrod ends.

same as the ball-to-ball pushrod except that there is a screw thread and lock nut near the end closest to the rocker arm that, when turned, lengthens or shortens the pushrod.

Rocker Arms

K19021

The **rocker arm** pivots near the middle, which results in a change of the direction of movement. The camshaft lobe pushes the rocker arm up on one end, and the other end pushes the valve down to open. Rocker arms can be used on OHV or OHC engines. In OHV engines, each valve has one lifter, one pushrod, one rocker arm, and one lobe on the camshaft dedicated to its opening and closing. In OHC engines, each valve has one rocker arm and one lobe on the camshaft. The rocker arm comes in different rocker arm ratios depending on the engine design. The ratio refers to the amount of movement on the valve side of the rocker arm in comparison with that on the opposite side. For example, a 1.5:1 rocker arm will move a valve 1.5 times the lift and speed of the camshaft lobe. The most common ratios are 1.5:1, 1.6:1, and 1.7:1.

Rocker arm ratios allow manufacturers to use smaller lift camshafts. That is, a camshaft with smaller lobes than needed to generate a specified amount of lift can be used because the rocker arm ratio magnifies the camshaft's lift. There are different styles of rocker arms, including stamped, cast, or forged (**FIGURE 19-53**). Stamped rocker arms are "stamped" from a piece of metal. They are not a high-strength or high-horsepower rocker arm and are intended for regular passenger vehicles. Cast rocker arms are cast in a die with molten metal. Metal alloys can be used, making cast, forged, or billet rocker arms that are more durable than their non-alloy counterparts. These rocker arms are generally used in sports cars and other high-performance vehicles. Forged rocker arms use localized pressing forces to form a rocker arm from steel or a steel alloy. The forging process enables the metal to withstand extreme pressure, heat, and wear. These rocker arms are used in high-torque, high-horsepower applications such as race cars and hot rods.

In many pushrod engines, the valves are positioned in a line on one side of the cylinder head and the pushrods on the other side of the head, with the rocker arms positioned between them. To aid in mounting and aligning all the rocker arms, the rocker arms may be mounted on a common rocker shaft (**FIGURE 19-54**). Or in the case of **canted valves** used in hemispherical-shaped and some wedge-shaped combustion chambers, the rocker arms may be individually mounted to the cylinder head (**FIGURE 19-55**). In OHC engines, typically the intake valves are on one side of the cylinder head and the exhaust valves on the other side of the head, with the camshaft sitting between them. Rocker arms or cam followers would then be positioned between the camshaft and the valves on each side of the camshaft.

With the push for increased fuel efficiency, many manufacturers use roller rocker arms on their engines. Roller rocker arms use rolling pivots at the valve end and the center of the rocker arm (**FIGURE 19-56**). The rolling action at the pivot

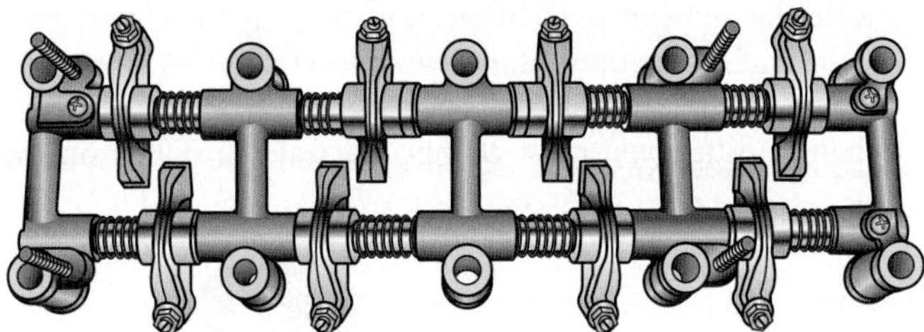

FIGURE 19-54 Rocker arm shaft assembly.

FIGURE 19-55 Rocker arms mounted directly to the cylinder head.

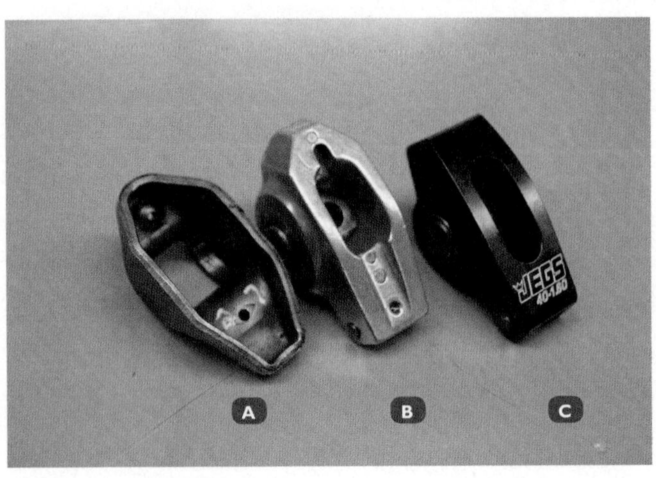

FIGURE 19-53 Rocker arm types. **A.** Stamped. **B.** Cast. **C.** Forged.

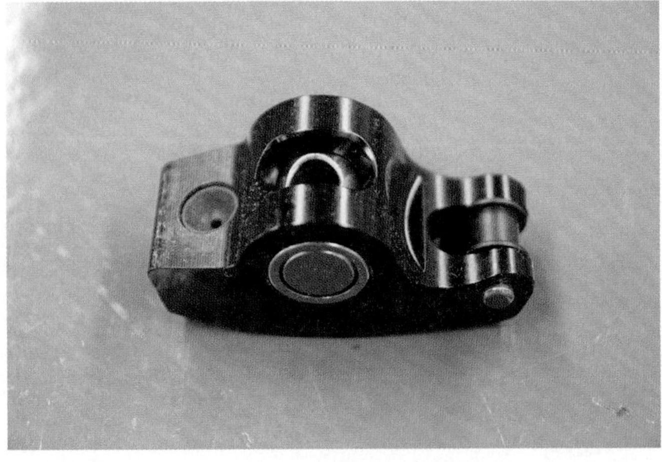

FIGURE 19-56 Roller rocker arms.

points reduces friction as well as wear on the rocker arm. The reduced friction gives a small increase in fuel efficiency and performance.

▶ Gaskets

K19022

Gaskets form a seal by being compressed between stationary parts where liquid or gases could pass. Gaskets are generally designed to be single use and therefore cannot be reused. They can be made of soft materials such as cork, paper, or various types of rubber (**FIGURE 19-57**). They can also be made of soft alloys and metals such as brass, copper, aluminum, and stainless steel. Such materials may be used individually or in some cases as blends to produce the required functional material.

With the increasingly hostile environment that gaskets are required to seal, and with the reduction in the use of asbestos, enhanced gasket materials have been developed. One of the ways to make engines run cleaner is to raise the temperature at which they operate. Today's cleaner engines typically run 30–40°F (-1.11–4.44°C) hotter than their counterparts of 30 years ago. Thus, the gaskets used in modern engines need to be made of more robust materials, such as vulcanized rubber or EPDM (ethylene propylene diene monomer [M-class] rubber).

Modern gasket manufacturers are producing improved material combinations such as nitrile-cork blends to deal with the extreme demands placed on modern engines, such as higher temperatures, higher compression, longer engine service life, and longer manufacturer warranties. Such combinations can better deal with these issues while maintaining compressibility and minimizing wicking.

Some gasket materials are designed to swell once they are put into service, thus increasing sealing ability. For instance, when oil inside a valve cover penetrates the edge of the gasket material, it may be designed to swell by approximately 30%. This swelling effect increases the sealing pressure between the head and the valve cover sealing surfaces and helps to seal potential leaks.

Choosing which material and design to use depends on the substance to be sealed (i.e., oil, fuel, exhaust gases), the pressures and temperatures involved (i.e., combustion gases, coolant), and the materials the mating surfaces to be sealed are made of (i.e., what materials the gasket will seal against—aluminum, plastic, cast iron, etc.). The best way to determine which gasket material is appropriate for an engine is to refer to the manufacturer's service information.

Head Gaskets

K19023

The head gaskets seal and contain the pressures of combustion within the engine, between the cylinder head and the engine block. They also seal oil and coolant passages between the engine block and the cylinder head.

Some high-temperature head gaskets are called **anisotropic** in nature. Anisotropic means that the gasket is designed to conduct heat laterally to transfer heat from the combustion chamber to the coolant faster. These gaskets are normally constructed with a steel core. Special facing materials are added to both sides of the gasket core to provide a comprehensive seal under varying expansion conditions.

On some engines, the head gasket provides or adjusts the proper clearances between the piston and the cylinder head by the thickness of the gasket. The service information (or repair information) for a vehicle will specify how to select the proper thickness of gasket and why that thickness is needed. Sometimes it is as simple as looking for a mark denoting the thickness of the old gasket and using that to order the new gasket.

Other head gaskets incorporate stainless steel **fire rings** to help contain heat and pressure within the cylinder. Fire rings are steel rings built into the cylinder head gasket (**FIGURE 19-58**). The rings provide extra sealing on the top of the cylinder to help seal in the high-combustion pressures (hence the name "fire ring"). For high-performance use, such as in race cars, some

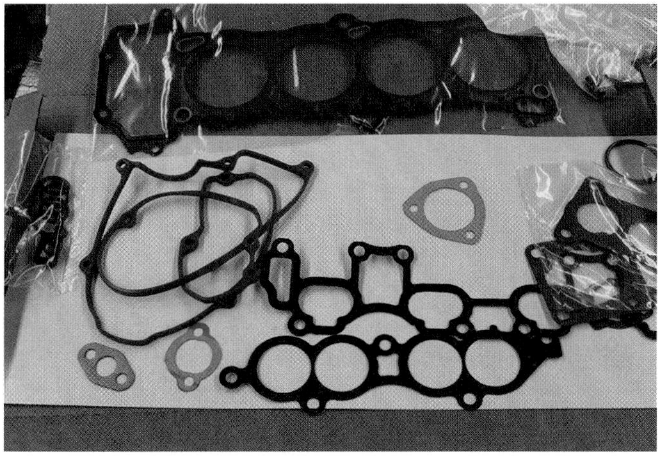

FIGURE 19-57 Various gaskets.

FIGURE 19-58 Fire ring built into a head gasket.

engine builders use soft metal O-rings, which fit in shallow grooves cut in the head around the cylinders and passageways to seal the compression and fluids (**FIGURE 19-59**). The O-rings are crushed in place when the head bolts are torqued into place. Because the O-rings fit in grooves, they are a very effective but expensive way to seal the heads.

For many late-model vehicles, the preferred head gasket is a **multilayer steel (MLS) head gasket** (**FIGURE 19-60**). These gaskets offer a wide range of benefits, such as strategically placed sealing beads that help eliminate leak paths, extra strong layers that provide superior combustion sealing, and a stainless steel material that maintains its shape despite thermal expansion and scrubbing between the engine block and the cylinder head. Many MLS and other head gaskets have an added silicone-based outer coating on both sides of the material layers to provide additional cold sealing ability during start-up and warm-up.

Seals

K19024

Gaskets cannot be used around a rotating part, such as where the camshaft protrudes through the front of the cylinder head, because they would quickly wear out and leak. To seal the rotating parts of an engine, **oil seals** are needed. Oil seals are round seals made of rubber or a rubber-type compound of silicone, EPDM, or another durable, flexible material, placed in a metal housing (**FIGURE 19-61**). These seals are typically driven into a machined bore around a rotating part. A metal spring, called a **garter spring**, is wrapped circularly around the inside of the seal and applies a small, constant pressure to keep the sealing lip in contact with the rotating part it is sealing. The most widely used seal for rotating parts is the **lip-type dynamic oil seal** (**FIGURE 19-62**). This seal is a precisely shaped, dynamic rubber lip. Like other oil seals, the lip-type dynamic oil seal also uses a garter spring to help keep the lip of the seal in contact with the shaft.

> **TECHNICIAN TIP**
>
> As a general rule, oil seals must be replaced with new ones when they are removed or when a component is overhauled or replaced.

A similar sealing principle is used to seal the valve stem to prevent oil from entering the engine combustion chamber. Like the oil seals, the valve stem seal is pressed onto the valve guide. It allows the valve stem to be wiped almost clean of oil

FIGURE 19-59 O-rings used instead of a head gasket.

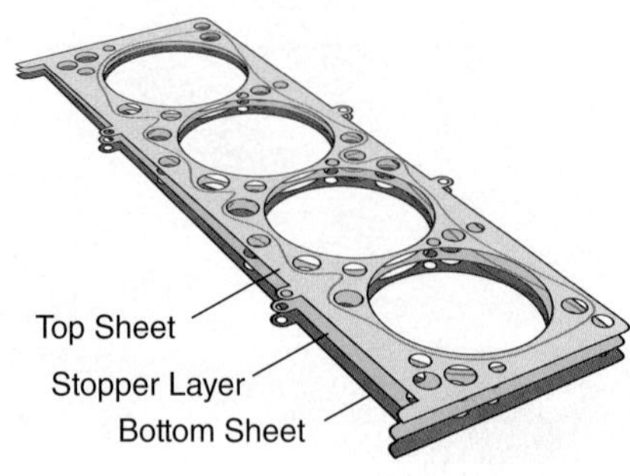

FIGURE 19-60 Multilayer steel (MLS) head gasket.

FIGURE 19-61 Oil seals.

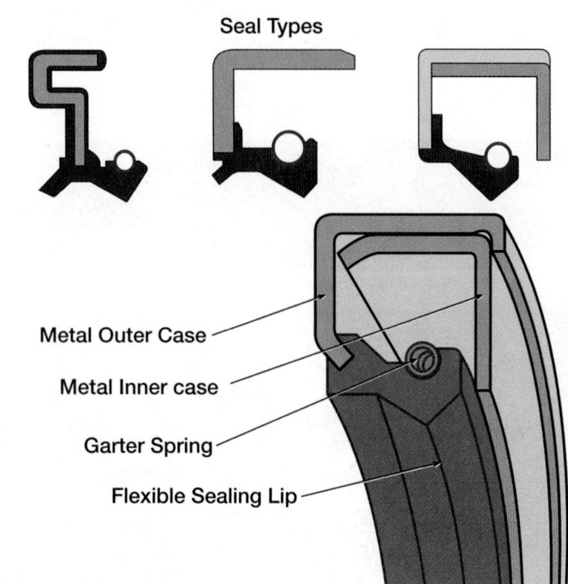

FIGURE 19-62 Lip-type dynamic oil seal.

on its opening trip, keeping a minimum amount of oil pulled down between the valve stem and the guide, for lubrication purposes.

O-rings, a simple sealing device consisting of a rounded ring of rubber or plastic, are also used to seal components. In some cases, they are used instead of gaskets, such as on thermostat housings (**FIGURE 19-63**). They can also be used to seal stationary and slowly rotating or sliding shafts. The O-ring seals the two surfaces. In many cases, the housing supplies the force to keep the ring in direct contact with the shaft. O-rings are generally effective at sealing high pressures where the differential speed between the opposing surfaces is minimal. In contrast, a lip seal is effective at sealing low pressures, but the differential speed between the opposing surfaces can be substantial. As the lip-type seal wears, the garter spring holds tension on the seal, keeping it against the part it is sealing. The O-ring seal has no mechanism for self-adjustment. Once worn, its sealing ability is compromised.

Applied Science

AS-13: Pulleys: The technician can explain how pulleys can be used to increase an applied force over distance.

To increase applied force over distance, a multiple or moveable pulley system must be used. This theory relies on the concept of mechanical advantage, which is the multiplication of applied force through the use of a mechanical device.

For a workshop example, imagine that you are trying to lift an engine block from the workshop floor, using a rope and a single pulley attached to the ceiling. To lift the engine block off the floor, you must pull down on the rope with a force greater than the weight of the engine block. To keep the block suspended in the air, your exerted force must equal the weight of the block. The mechanical advantage of this system is 1. If you were to add a second pulley attached to the engine block, the weight of the block is now supported by both you and the section of rope between the upper and the lower pulleys, providing an increased mechanical advantage. The force you apply to the rope is now multiplied by the factor of mechanical advantage to exert more force to lift the block.

FIGURE 19-63 O-ring used to seal a thermostat housing.

Sealants

K19025

Sealants are used to seal two surfaces that are in stationary contact with each other, such as between a gasket and the sealing surface of the part being sealed. Or it could be between two mating surfaces without a gasket between them. There are different sealants for various applications. And there are applications in which sealants are not to be used. For example, sealants are almost never applied to head gaskets.

To aid in gasket sealing, there are many different brands and types of sealants. One of the most popular types is **room temperature vulcanizing (RTV) silicone**. This adhesive is able to set, or "vulcanize," at room temperature (**FIGURE 19-64**). When RTV is applied, it has the consistency of a gel, and as it sets, it becomes rubbery. RTV should only be applied thick enough to fill the gaps with a small amount of squeeze-out. If too much is used, it will squeeze out from between the surfaces and form large globs of material on the inside of the engine that can break off and plug up the oil pump screen, starving the engine of oil. Another caution when using RTV silicone is to make sure you use oxygen sensor–safe RTV on engines equipped with oxygen sensors. Regular RTV gases will be drawn through the PCV system and burned in the combustion chamber. The resulting burned silicone gases will then coat the surface of the oxygen sensor, rendering it inoperative.

Some applications require a non-hardening gasket sealant so that the sealant will "give" when the metal parts expand and contract under normal heating and cooling cycles. Additionally, this type of sealant is very forgiving with gasket surfaces that are less than perfect (e.g., scratched or pitted). Gasoline-proof hardening sealants are also available, but they have little to no give once dry and are only used in extreme situations such as to help hold a cracked component together when a new component is not available. Always follow the manufacturer's recommendations for application.

A variety of different chemicals are available to hold gaskets in place during assembly and to aid in sealing. The most popular adhesive to hold a gasket in place is a form of

FIGURE 19-64 Room temperature vulcanizing (RTV) silicone material being applied.

FIGURE 19-65 Gasket adhesive being used to hold a gasket in place.

contact cement that comes in either an aerosol can or squeeze tube (**FIGURE 19-65**). Gasket adhesive is capable of adhering to many different surfaces, including felt, cork, metal, paper, rubber, and asbestos gaskets. In an automotive application, contact cement has been used on **valve cover gaskets**, fuel pump gaskets, and intake and exhaust manifold gaskets. However, adhesive sealants are not typically used on most newer vehicle models. Before using any sealant, check the vehicle manufacturer's service information to be sure the product can be used with your application.

▶ Cylinder Head Diagnosis

K19026

The cylinder head is subject to a variety of problems that can affect engine function. Cylinder head gaskets wear over time, cracking and creating holes through which coolant and compression can leak. Cracks in the cylinder block or head also may leak compression gases into the cooling system. The continual exposure to heat and cold (thermal cycles) over time may warp the cylinder head, causing secondary problems such as compression, coolant, and oil leaks. Other relatively common cylinder head problems include burnt or warped valves and stripped spark plug hole threads.

Diagnosis of cylinder head problems begins with confirming the customer's concern, such as checking to see if the engine is actually misfiring, overheating, etc. If it is misfiring, then a scan tool, scope, or power balance test can be used to identify which cylinders are misfiring. If the cause of the misfire is determined to be compression related, then compression and cylinder leakage tests should be performed.

If the engine is overheating due to a cylinder head or head gasket problem, then a cylinder leakage test or combustion gas detection test should be performed on the suspect cylinders. If the cylinder leakage test or combustion gas detection test indicates a combustion gas leak into the cooling system, the head will need to be removed so the head gasket and head can be inspected for leaks. Refer to the service information to identify the applicable procedures for these tests. Most cylinder head problems can be repaired by a cylinder head machinist. We cover most of those processes in the Engine Machining chapter.

▶ Removing Cylinder Heads

N19001

Removing a cylinder head is considered a major task and is usually time intensive; therefore, cylinder head removal is done only after extensive testing and with a certainty that the problem lies in the cylinder head itself, its associated parts, or a part requiring removal of the cylinder head to access it (such as a piston). Once the diagnosis has been made, you need to follow the steps as listed in the service information for the vehicle you are working on, as every engine is different. In fact, some manufacturers require that their head bolts be loosened in a specified pattern in specified steps and at certain temperatures. Failure to follow the manufacturer's instructions can lead to damage to the head.

Applied Science

AS-93: Metallurgy: The technician can explain the critical need for metals of different hardnesses in automotive parts.
A typical automotive application of metals with different hardnesses is the relationship between crankshaft journals and crankshaft bearings in engines. Crankshafts are typically manufactured from cast iron or steel and are frequently surface-hardened to provide a hard surface for bearing contact, providing resistance to wear. Crankshaft bearings are surfaced with soft materials, typically babbitt, copper-lead, or aluminum alloys.

In normal running, a pressurized film of oil is maintained between the crankshaft and the bearing. Two essential properties are achieved through the use of soft surface materials—embeddability and conformability. Embeddability refers to the ability of the soft surface to absorb foreign materials forced between the crankshaft and the bearing, helping to prevent crankshaft wear. Conformability allows the bearing surface to move or shift to conform to a worn or damaged crankshaft journal.

To remove a cylinder head, follow the steps in **SKILL DRILL 19-1**.

▶ SAFETY TIP

When lifting heavy cylinder heads, you will probably need to lean over the fender or front of the vehicle. This puts a large amount of strain on your lower back. Always get help when lifting cylinder heads and other heavy components.

Cleaning, Disassembling, and Inspecting a Cylinder Head

N19002

The cylinder head must be cleaned after removal to make the inspection and repair process more efficient. You cannot have a good repair on a cylinder head if there are chunks of sludge

SKILL DRILL 19-1 Removing Cylinder Heads

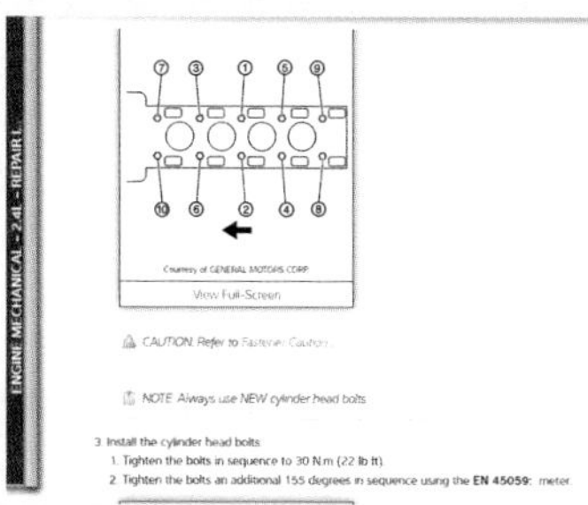

1. Determine whether the head bolts are torque-to-yield (TTY) bolts. If so, you need to discard them and replace them with new TTY bolts upon reassembly. Determine whether the head bolts need to be loosened in a specified sequence.

2. Before the heads come off, put an identifying mark on at least one of the heads. Follow the specified procedure to remove all of the head bolts, and set them aside.

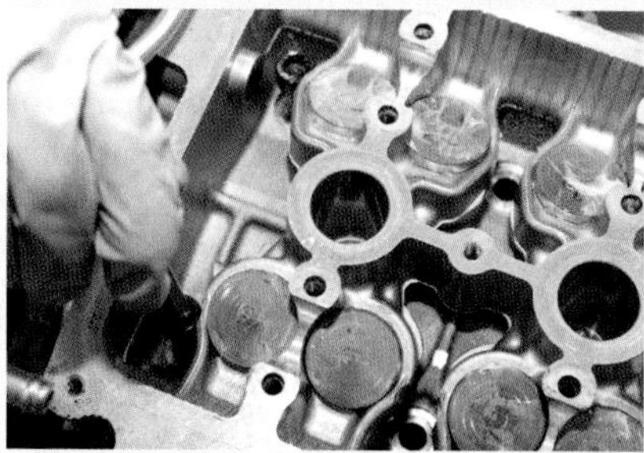

3. If dealing with a head gasket that is stuck to the head and block surface, reinstall two corner head bolts three or four turns in their respective holes.

4. If the engine is mounted on an engine stand, check that the safety pin is in the engine stand. Insert a pry bar or a long breaker bar handle into one of the intake port openings of the head and give it a firm push.

5. Remove the safety bolts from the head, and carefully lift it.

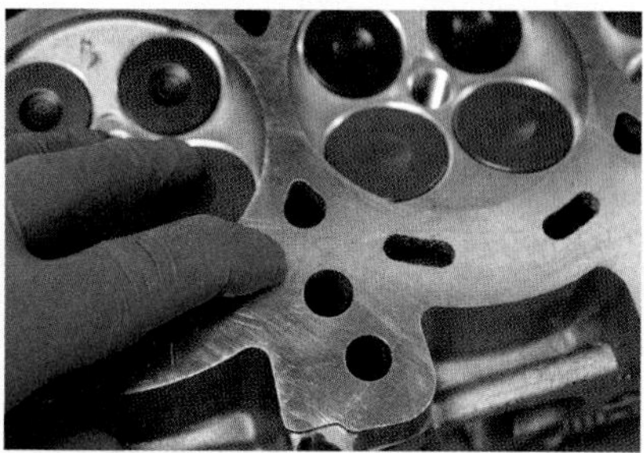

6. Inspect the cylinder head visually for any unusual conditions.

and grime concealing damage or constantly falling into the area where you are working. Initial cleaning of the cylinder head usually involves removing any gasket material by mechanically cleaning it with a scraping device such as a gasket scraper. If so, most aluminum heads must be scraped with a nonmetallic tool such as a plastic or nylon scraper made for the job.

To disassemble the head, place the head in a cylinder head holding fixture. These fixtures are usually designed to hold the cylinder head off of the work bench, allowing you to gain access for valve removal. Tap the valve retainers lightly with a hammer and socket to break the valve keepers loose from the retainers. Then use a valve spring compressor to compress the valve spring on each valve (one at a time), and remove the valve keepers. After the keepers are removed, the valve can be removed and all parts kept in order of the way they came off. That way, if they are going to be reused, their wear surfaces will match.

All parts you take off of the cylinder head need to go into a cylinder head disassembly container or small labeled bins. These will help you keep track of what valve and spring goes where for later installation. When in doubt, write it down and label it!

In most cases, once the head has been disassembled, it will also need to be cleaned in a spray tank, hot tank, cold tank, ultrasonic cleaner, or specially designed oven. Follow the equipment manufacturer's recommendations when using a particular type of cleaner.

When checking the cylinder head for flatness, lay the calibrated straightedge parallel with the combustion chambers across the already-cleaned cylinder head deck surface. Starting at the top of the cylinder head, try to fit a 0.001" (0.025-mm) feeler gauge under any part of the mating surface between the calibrated straightedge and the cylinder head. This process is repeated 5 to 10 times at various places—5 or so times with the straightedge parallel to the combustion chambers and another 5 or so times with the straightedge running at a 90-degree angle to the combustion chambers. Also be sure to cross the straightedge from one side to the other, making an X across the head with the two measurements. This approach will show any twisting of the head from end to end.

If you can slide a 0.001" feeler gauge underneath it, then you need to find out how big the deformation (warpage) is. Keep going up in value—0.002" (0.051 mm), 0.003" (0.0762 mm)—until the feeler gauge will no longer fit underneath it. The last feeler blade that will go under the straightedge is the reading. Compare the reading to the service information warpage limit. Typical allowed warpage is 0.003" or less.

Clean the cylinder head in an approved manner. Be aware that aluminum heads cannot generally be cleaned in a hot tank or high-temperature oven. Follow the manufacturer's recommended procedure for thoroughly cleaning the head.

If the head is made of cast iron, you can locate small cracks with a Magnaflux® tool. The Magnaflux® process involves placing a strong electromagnet across each part of the head surface at 90-degree offsets and lightly dusting the head with a magnetic powder. If there is a crack present, the powder will adhere to the crack because the crack creates a north and south magnetic field at that point, and the powder sticks to it.

If the head is made of aluminum, the Magnaflux® process cannot be used because the metal is nonferrous. Spray or wipe a penetrating die over the surface to be checked, followed by a light coat of developer. Once the developer dries, any cracks should show up as colored lines.

Pressure checking can be used on both cast iron and aluminum heads, but is usually available only at well-equipped machine shops. The cooling system passages are blocked off and filled with compressed air. They are then either immersed in water and inspected for leaks, or sprayed with a soapy solution and inspected for bubbles.

▶ TECHNICIAN TIP

It is always a good idea to measure and record the installed valve height for at least one exhaust valve and one intake valve. This specification can be hard to locate for some vehicles. Since hydraulic lifters can only compensate approximately 0.060" to 0.080", knowing the installed height will be invaluable.

To clean and inspect a cylinder head, follow the steps in **SKILL DRILL 19-2**.

SKILL DRILL 19-2 Cleaning and Inspecting a Cylinder Head

1. Place your cylinder head in a cylinder head holding fixture. Tap the valve retainers lightly with a hammer and socket to break the valve keepers loose from the retainers.

2. Use a valve spring compressor to compress the valve spring on each valve (one at a time), and remove the valve keepers, then remove the valve. Place all parts you take off into a cylinder head disassembly board or small labeled bins.

3. Clean the gasket mating surfaces. On a cast iron head, cleaning is usually done with a flexible abrasive pad or by hand with a gasket scraper. If it is an aluminum head, use only plastic or nylon gasket scrapers.

4. Using a straightedge, measure the cylinder head warpage by setting the straightedge against the surface you are checking and seeing what size feeler blade will fit between the straightedge and the head surface.

5. Clean the cylinder head. Be aware that aluminum heads cannot generally be cleaned in a hot tank or high-temperature oven.

6. If the head is made of cast iron, you can locate small cracks with a Magnaflux® tool. If there is a crack present, the powder will adhere to the crack.

SAFETY TIP

When cleaning and scraping a cylinder head, be very careful: Many of the surfaces are sharp and can severely cut you.

Inspecting Valves and Valve Seats

N19003

Valves and valve seats must operate in very hostile conditions, so they need to be inspected closely whenever the cylinder head is removed from the engine. Valves can burn or warp, which leads to misfire issues. The valve tip can also become worn from lack of lubrication, leading to a ticking sound, and the valve stems can wear in the area that moves inside the valve guide leading to burning of engine oil. When valves are removed from the cylinder head, the valves and seats should be inspected visually for cracks, burn marks, signs of leaks, and damage to the valve face, stem, and tip. If any of these issues are found, the valve will need to be replaced and the seat will need to be either replaced or repaired, depending on the damage. If there are no apparent issues, then the valves and seats should be measured and compared to specifications.

The valve stem diameter should be measured in the area that rides in the valve guide with a micrometer and compared to specifications. Some valves come with a tapered valve stem where the stem closest to the head end of the valve is slightly narrower to allow for expansion of the stem on today's hotter-running engines. The margin of the valve will have to be measured with a machinist's rule to make sure it is within specifications. The valve seat width also must be measured and compared to specifications.

To inspect valves and valve seats and determine any necessary actions, follow the steps in **SKILL DRILL 19-3**.

Inspecting Valve Guides and Valve Stem-to-Guide Clearance

N19004

Valves and valve guides are subject to high temperatures and pressures that tend to wear their sides. If they wear much at all, oil can be pulled down the valve stem and burned in the combustion chamber or exhaust. If the wear continues, the valve can move around in the guide so far that the head of the valve does not close properly and catches on the top of the valve seat. This causes the valve to leak and leads to intermittent misfires in that cylinder.

Valves and guides can be measured in two ways: with a micrometer or with a dial indicator. If using a micrometer, a ball gauge is used to measure the top, center, and bottom of the valve guide. The ball gauge is placed in the valve guide in each of those positions and expanded until both sides of the gauge touch the inside of the guide. It is then removed from the guide, measured with a micrometer, and recorded. The center reading is then subtracted from the ends to determine how much the

SKILL DRILL 19-3 Inspecting Valves and Valve Seats

1. Inspect each valve for signs of burning; warpage; or excessive face, stem, or tip wear. If found, replace the valves with new ones.

2. Inspect each valve seat for signs of burning, leakage, or excessive wear. If found, the seat must be machined or replaced.

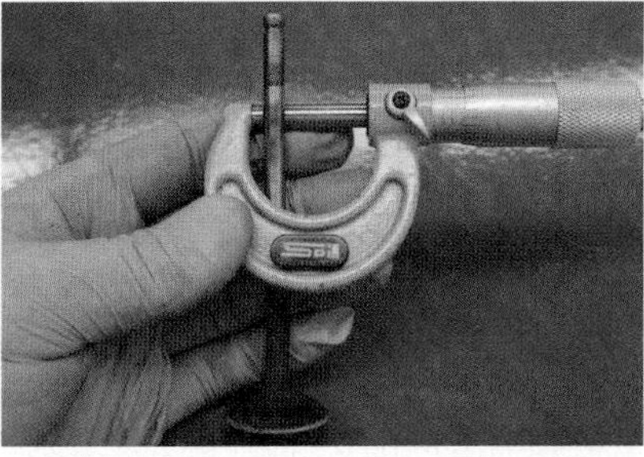

3. Using a micrometer, measure each valve stem diameter in three places where the valve rides in the valve guide, and average your answers.

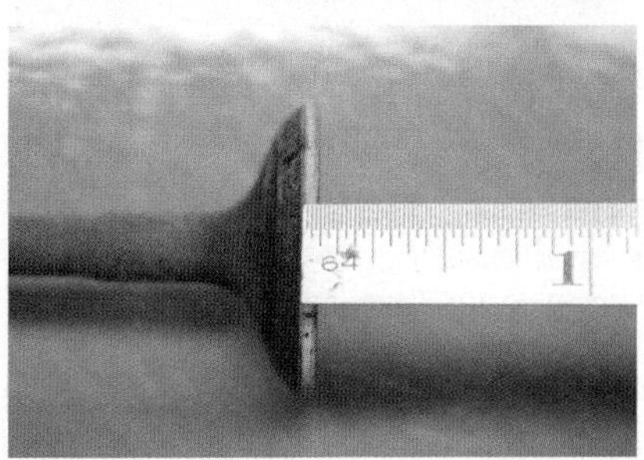

4. Measure the valve margin with a machinist's rule and record your readings.

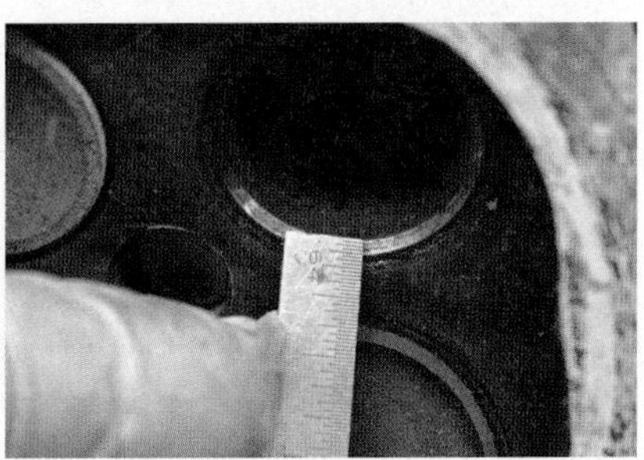

5. Measure the width of the valve seats and record your readings.

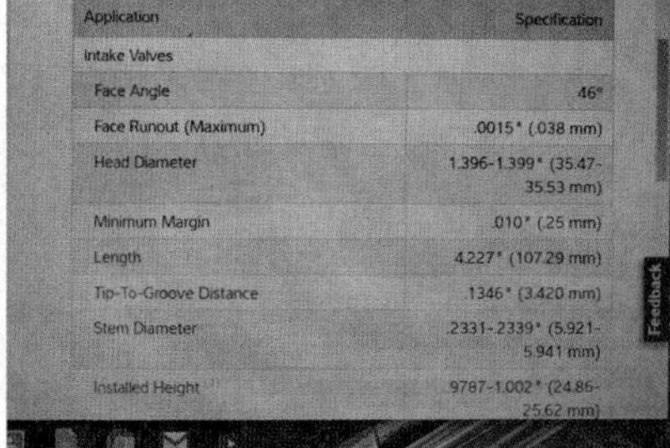

Application	Specification
Intake Valves	
Face Angle	46°
Face Runout (Maximum)	.0015" (.038 mm)
Head Diameter	1.396-1.399" (35.47-35.53 mm)
Minimum Margin	.010" (.25 mm)
Length	4.227" (107.29 mm)
Tip-To-Groove Distance	.1346" (3.420 mm)
Stem Diameter	.2331-.2339" (5.921-5.941 mm)
Installed Height	.9787-1.002" (24.86-25.62 mm)

6. Compare readings to the specifications and determine needed actions.

guide is tapered. The smallest reading (usually the center reading) is used to determine the overall size of the guide. The valve stem is also measured and recorded. If the valve stem is within specifications, then the difference between the valve guide and the valve stem is the valve-to-guide clearance, and all of the readings can be compared to the manufacturer's specifications.

A dial indicator is used with the valve installed in the valve guide. The dial indicator is mounted on the head and placed against the side of the valve stem near the top of the valve. With the valve at its normally open height, rock the head of the valve stem back and forth, observing the movement on the dial indicator, and compare to specifications.

To inspect valve guides for wear, check valve stem-to-guide clearance, and determine any necessary actions using a micrometer, follow the steps in **SKILL DRILL 19-4**.

To inspect valve guides for wear using a dial indicator, follow the steps in **SKILL DRILL 19-5**.

Inspecting Valve Springs

N19005

Using the correct size and strength of valve spring is critical to engine operation. Valve springs must be able to quickly and firmly force the valves closed. It is important to always measure the valve springs and confirm that they meet the specifications before installing them. The three key steps in testing a valve spring are (1) checking for squareness, (2) measuring spring height, and (3) measuring the installed pressure. Checking for squareness is a test to ensure that the spring is not damaged and can compress and load the valve evenly. Measuring spring height is simply checking that the spring is indeed the correct height and has not fatigued. A valve with uneven pressure or the incorrect installed pressure will cause the valve to wear the valve stem and guide due to the side-to-side pulling force exerted by the "bent" spring.

Measuring the valve spring's installed pressure involves determining how much pressure the spring can produce when it is at its installed height and making sure it meets specifications. This pressure is measured by compressing the spring to its installed height with a valve spring tester. The tester will show the pressure exerted by the spring. This tester looks like a drill press without the drill and is typically a freestanding or bench-mounted unit.

To inspect valve springs, follow the steps in **SKILL DRILL 19-6**.

Reassembling a Cylinder Head

S19001

Once all work on the valves, springs, and head is completed, the cylinder head will have to be cleaned again and then reassembled.

SKILL DRILL 19-4 Inspecting Valve Guides for Wear, Checking Valve Stem-to-Guide Clearance, and Determining Necessary Actions

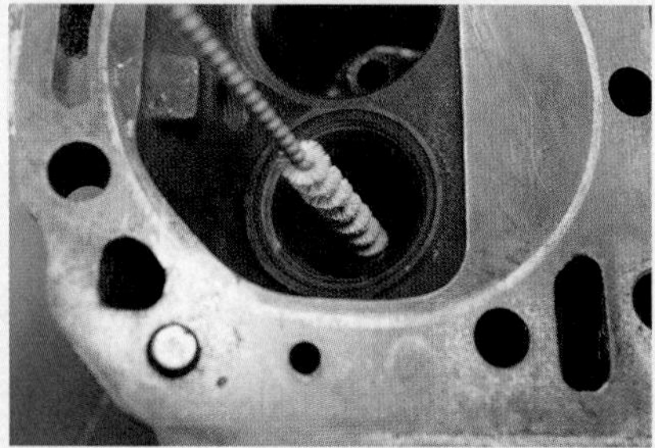

1. Clean the guide with solvent and a valve guide brush if not already cleaned.

2. Insert the ball gauge into the top of the guide and expand it until it touches both sides with a slight amount of drag when rotated to the largest diameter.

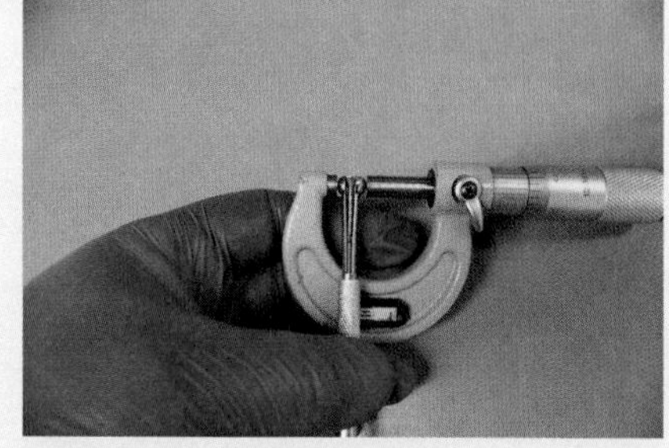

3. Carefully pull the ball gauge out of the guide without disturbing the setting, and measure it with a micrometer. Record your reading.

4. Repeat this procedure on the center and bottom portions of the valve guide, and record your readings.

5. Measure the minimum valve stem diameter at the top, center, and bottom of the wear area, and record your readings.

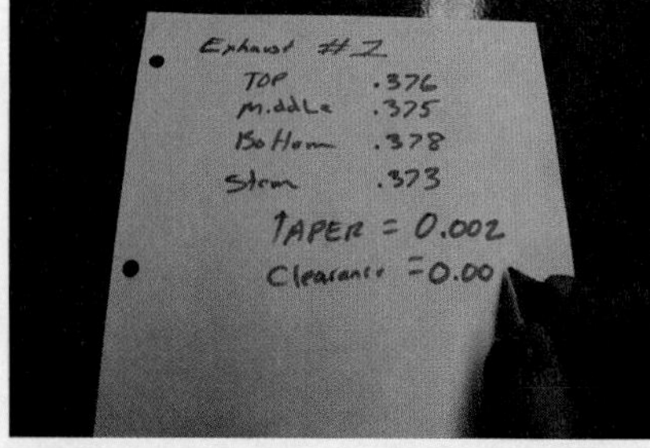

6. Calculate the taper and valve clearance; compare all of the readings to specifications, and determine any necessary actions.

SKILL DRILL 19-5 Inspecting Valve Guides for Wear Using a Dial Indicator

1. Verify that the valve guide is clean. Reinstall the valve in the guide at the fully open length.

2. Set the dial indicator so it can measure the side-to-side movement of the head of the valve.

3. Move the valve side to side, and measure the movement with the dial indicator. Record your reading. Compare to specifications and determine any necessary actions.

SKILL DRILL 19-6 Inspecting the Valve Springs

1. Set the spring on a perfectly flat surface. Place a combination square against the spring, and rotate the spring as you measure any gaps around it using a feeler gauge. Gaps should not exceed 0.0625" (1.6 mm). Replace all suspect springs. Repeat this process for all springs.

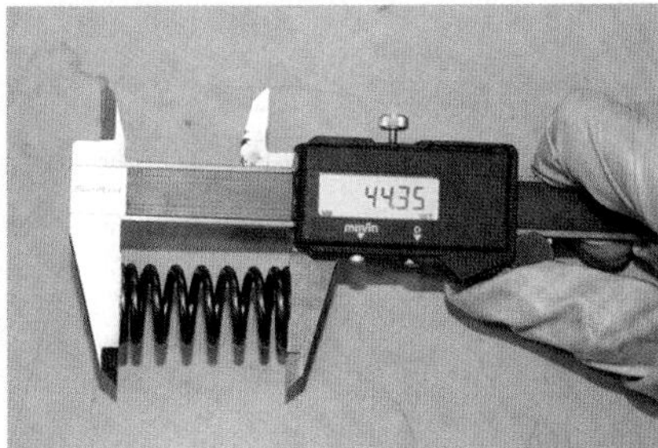

2. Using a vernier caliper, measure the spring free height, and compare your measurements with the specifications provided in the service information. Replace the spring if it is out of specification. Repeat this process for all springs.

3. To measure the spring's installed pressure, install the valve spring on the valve spring tester. Pull the side handle and compress the spring to the installed height specified in the service information while reading the tester. Record the force that the spring produced and compare that to the specifications. Do not use springs out of specification.

The following is a generic version of cylinder head reassembly and is not meant to supersede any manufacturer's cylinder head reassembly instructions. You should always check for the most current information on the vehicle you are working on through the manufacturer's recommended service information.

To reassemble a cylinder head, follow the steps in **SKILL DRILL 19-7**.

Inspecting Pushrods and Rocker Arms

N19006

Because pushrods and rocker arms operate under large sliding forces, they tend to experience wear over time and need to be inspected at each point of contact. On pushrods, that means each end. The balls or sockets typically lose their roundedness and become slightly pointy. Also, there is typically a small flat spot on the top of the pushrod. If the pushrod is worn, the flat spot will be smaller than it should be or worn off altogether. Pushrods also need to be perfectly straight; otherwise, they are subject to bending, which would render them inoperative. Rolling them on a piece of glass or perfectly flat surface plate will indicate if they are bent.

Rocker arms have three potential wear surfaces to inspect: the pushrod end, the valve end, and the pivot. The pushrod end of the rocker arm mates with the end of the pushrod, so they usually wear similarly. The valve end of the rocker arm can also wear, creating a groove in the face. The pivot can mate to a ball-shaped pivot, a T-shaped pivot, or a pivot shaft. Look at both the pivot and the mating surface on the rocker arm to see if there is excessive wear. If any of the three surfaces of the rocker arm are excessively worn, the rocker arm and pivot will have to be replaced.

To inspect pushrods, rocker arms, and rocker arm pivots and shafts for wear, bending, cracks, looseness, and blocked oil

SKILL DRILL 19-7 Reassembling a Cylinder Head

1. Ensure that all parts and passageways have been cleaned, checked for defects, and replaced if necessary. If any defects are found, correct them now.

2. Dip the valve stem that is about to be installed in clean engine oil or use the appropriate assembly lubrication.

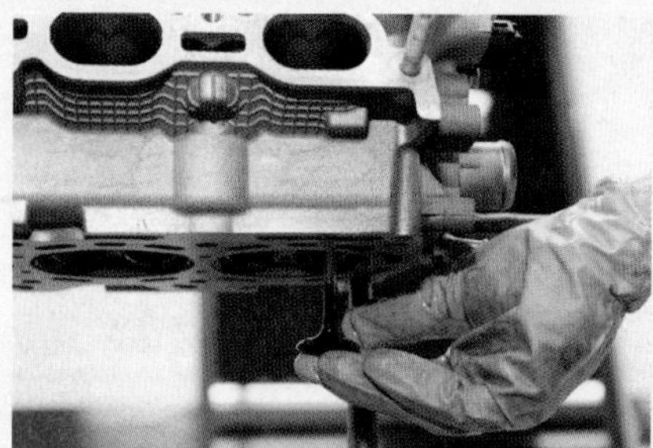

3. Insert the corresponding valve into the valve guide, and place the protective sleeve over the grooves near the end of the valve stem.

4. Dip the valve seal in clean engine oil and install it down over the exposed top of the valve guide assembly, using an installer if necessary.

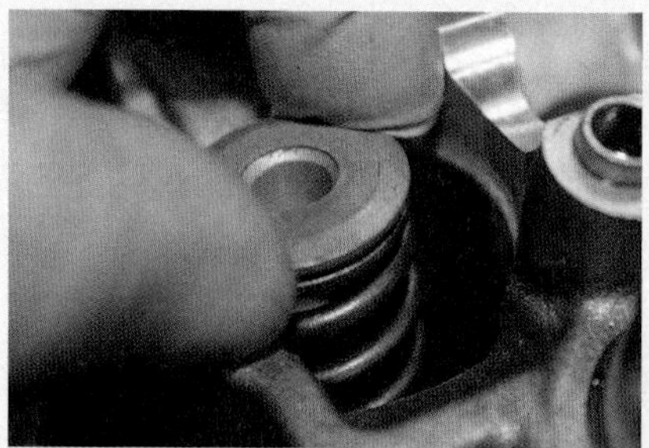

5. Remove the protective sleeve and place the valve spring and retainer over the installed valve.

6. Compress the valve spring and retainer with the valve spring compressor, and install the valve keepers. Be sure the keepers are locked into their groove before releasing pressure on the valve spring compressor. Repeat this process for the remaining valves.

passages (orifices) and determine any necessary actions, follow the steps in **SKILL DRILL 19-8**.

Inspecting and Measuring Camshaft

N19007, N19008

The camshaft experiences heavy loads on the cam lobes and journals, which need to be inspected and measured if the camshaft is being reused. A good inspection reveals whether the cam needs to be measured. Look for obvious indicators of excessive wear such as heavy scoring on cam lobes or a bluish color indicating overheating. If there is not obvious wear, take a closer look at the cam lobes. Can you see worn layers, similar to tree rings, in the surface? If so, the hardened surface of the cam lobe is worn through and needs to be replaced.

Camshafts can also warp, which can put extra wear on the camshaft journals as well as on the camshaft bearings. The cam journals first should be measured for diameter and out-of-round with a micrometer and compared to specifications. If the journals do not meet the specifications, the camshaft will have to be replaced, and the other measurements can be skipped.

The camshaft warpage and camshaft lobes are best measured with a dial indicator and set of V-blocks. The camshaft is placed on the V-blocks so that two of the camshaft journals are each sitting in a V. Some manufacturers specify the two center journals, whereas others specify the outer two journals. The dial indicator is used first to check for warpage; it is placed on either the center journal or the outer journal, whichever is not sitting in a V. The dial indicator must be at 90 degrees to the journal. Once set properly, the camshaft is carefully rotated and the dial indicator watched, noting the maximum variation. This process is repeated on each of the other journals and compared to specifications. If the cam is not warped, then use the dial indicator to measure the maximum lift on each cam lobe and compare to specifications.

To inspect and/or measure the camshaft for runout, journal wear, and lobe wear, follow the steps in **SKILL DRILL 19-9**.

Inspecting Camshaft Bearings

Camshaft bearings support the camshaft; therefore, their alignment and size are critical to the operation of the camshaft and

SKILL DRILL 19-8 Inspecting Pushrods, Rocker Arms, and Rocker Arm Pivots and Shafts

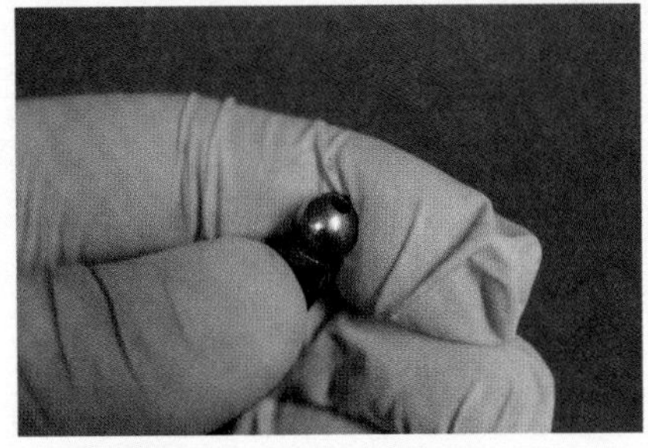

1. Inspect the ends of the pushrods for excessive wear. If the pushrod is hollow, blow compressed air through it to verify it is not plugged.

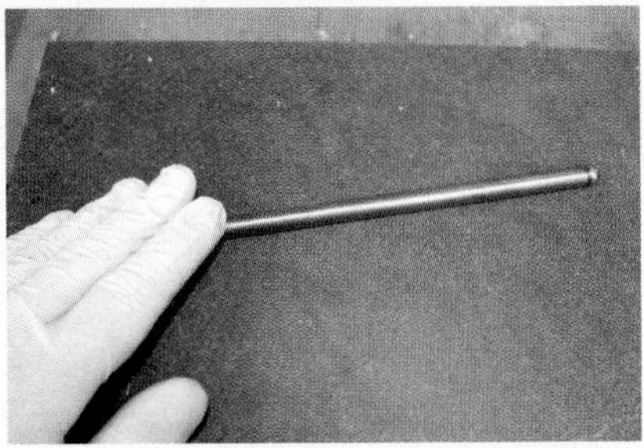

2. Roll each pushrod on a piece of glass or surface plate to check if it is bent. If the pushrods are hollow, blow compressed air through them.

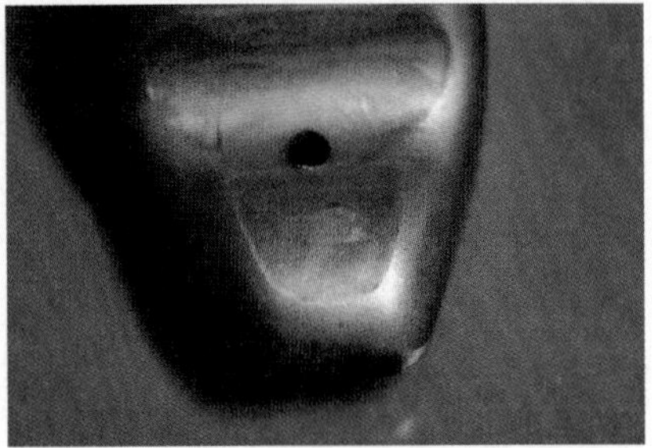

3. Inspect the valve end of the rocker arms for excessive wear or damage.

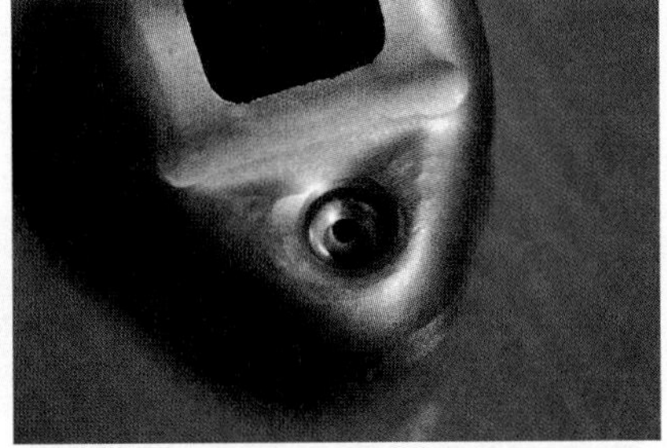

4. Inspect the pushrod end of the rocker arms for excessive wear or damage. If they are equipped with oil passages, blow compressed air through them.

5. Inspect the center pivot surface of the rocker arms for wear or damage.

6. Inspect the pivots or rocker arm shafts for wear or damage. If the rocker arm shafts are equipped with oil passages, blow compressed air through them.

engine oil pressure. In cast iron heads, the cast iron is too hard of a surface for the camshaft journals to ride on, so bearing inserts are used in the camshaft towers. Most aluminum heads allow the camshaft journal to ride directly on the aluminum because it is a softer material and provides an able bearing surface. The bearing diameter must be measured with a dial bore gauge or inside micrometer and compared to specifications. If the diameter is oversized and the head uses replaceable bearings, then the bearing will have to be replaced with new ones. If the head uses a bolt-on aluminum cap as a bearing, then a machinist can remachine the bore to its original size, but the camshaft will sit just a bit lower in the head, which can affect valve timing. If the head uses solid cam towers as camshaft bearings and the bores are oversized or damaged, in many cases the head will have to be replaced; however, in some cases, oversized camshafts or bushings are available.

The last thing to check is for camshaft bore alignment. This can be checked by laying a straightedge along the bottom of the camshaft bearings and using a feeler blade to measure any space between the straightedge and the camshaft bearing surface. Or, if the camshaft is straight and not undersized, it can be installed and checked for binding. If there is no binding, the cam bearing alignment is okay.

To inspect the camshaft bearing surface for wear, damage, out-of-round, and alignment, and then determine necessary actions, see **SKILL DRILL 19-10**.

Installing the OHC and DOHC Camshaft

Now that the valves are installed, it may or may not be time to install the camshaft(s) and valve actuators. In some engine designs, the camshaft(s) prevent access to tighten the head bolts; therefore, the camshaft(s) should be installed after the head has been properly torqued into place on the block. In other engines, the camshaft and valve actuators can be installed while the head is off the engine without interfering with torquing the head bolts. Always check the manufacturer's procedure to know when to install the camshaft(s).

Bucket-Style Lifter Heads

Assembling the camshaft and valve actuators is fairly straightforward on a head that uses bucket-style lifters, but mistakes can be made. Some engines use the same bucket lifters for intake and exhaust; other engines use separate ones. Verify that the proper bucket lifter is installed in the correct hole. Some bucket

SKILL DRILL 19-9 Inspecting and/or Measuring the Camshaft

1. Inspect the camshaft journals and lobes for obvious damage or wear. If found, replace the camshaft. Install the camshaft on a set of V-blocks so that two specified cam journals are resting in the Vs.

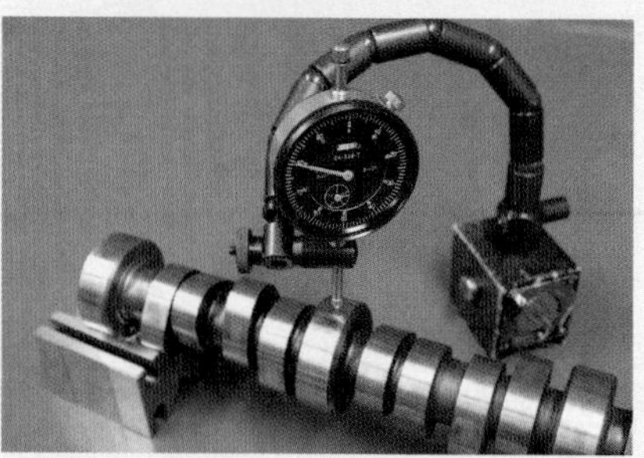

2. Set the dial indicator so it is engaged on one of the other cam journals perpendicularly.

3. Rotate the camshaft until the dial indicator reads the lowest point and zero the dial.

4. Rotate the camshaft until the dial indicator reads the highest point and record the reading. Continue rotating the camshaft to verify that the needle does not go below zero. If it does, zero it again, and remeasure the runout. Perform this process on all of the remaining journals, and compare to the runout specifications.

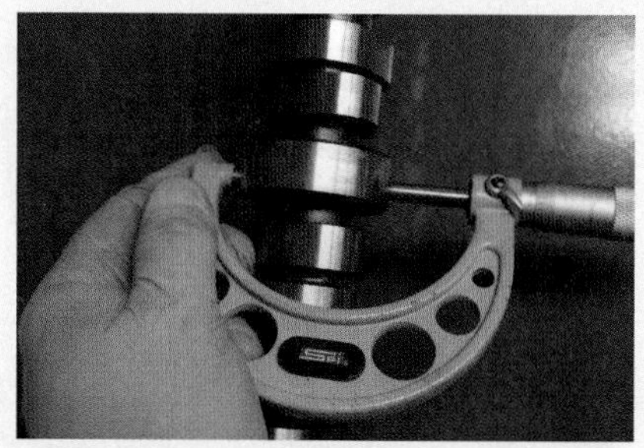

5. Using a micrometer, measure each journal in two places 90 degrees apart to determine any out-of-round condition. Record your readings, and compare to the specifications.

6. Reset the dial indicator so it is engaged on one of the cam lobes perpendicularly.

7. Rotate the camshaft until the dial indicator reads the lowest point, and zero the dial.

8. Rotate the camshaft until the dial indicator reads the highest point and record the reading. Continue to rotate the camshaft to verify that the needle does not go below zero. If it does, zero it again, and remeasure the lobe.

9. Measure each of the remaining cam lobes, and compare your readings to specifications.

SKILL DRILL 19-10 Inspecting Camshaft Bearings

1. Inspect the camshaft bearing surfaces for obvious wear or damage. If observed, the bearing surfaces will have to be machined or replaced.

2. If the cam bearings bolt onto the head, bolt them on, and torque them to specifications.

3. Place the dial bore gauge vertically in each camshaft bearing, and measure the diameter. Record your reading.

4. Place the dial bore gauge horizontally in each camshaft bearing, and measure the diameter. Record your reading, and calculate any out-of-round.

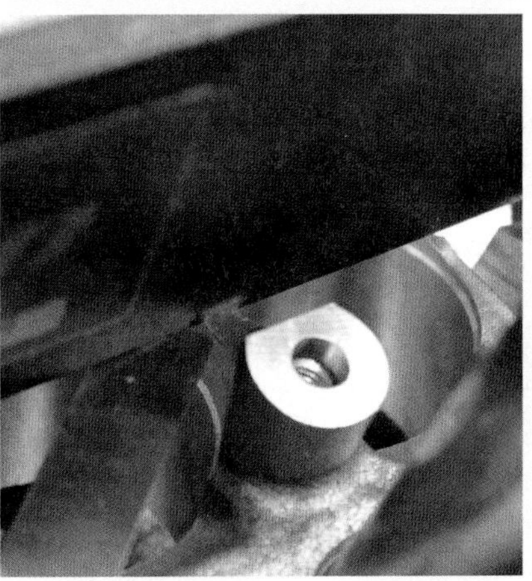

5. Lay a straightedge within the camshaft bearings, and measure any clearance between the straightedge and the bearings with a feeler blade.

6. If you have already verified that the camshaft is straight and the journals are the specified size, slide the camshaft into the bearing bores, and check for any binding. If there is no binding, the bearing alignment is okay.

lifters use an integrated hydraulic lifter, and no adjustment is necessary if the valve installed height is correct. Other engines use solid bucket lifters, which typically use shims to adjust the valve clearance. The clearance is usually adjusted after the camshaft is installed and torqued into place.

When installing the camshafts on a bucket lifter head, you must be careful to install the intake camshaft on the intake valves and the exhaust camshaft on the exhaust valves; otherwise, the engine is likely to not run or to run improperly as the valve timing will be way off. Another challenge is to get the timing marks of the timing gears aligned to the manufacturer's specifications. In many cases with dual cams, once the cams and valve actuators are in place and any valve adjustments are made, a special tool is used to hold the cam gears in the proper position while waiting for installation of the head on the block.

To install the camshaft in heads with bucket-style lifters, follow the steps in **SKILL DRILL 19-11**.

Cam Follower/Rocker Arm–Style Heads

Assembling a head that uses cam followers or rocker arms may begin with the installation of the camshaft, or the rocker arms may have to be installed first, depending on the design. If the manufacturer requires assembly to begin with the camshaft, lubricate the camshaft journals, and slide the camshaft into place in the bearings, making sure the cam can be turned freely. If it binds, you either have a warped camshaft or one or more bearings are too tight and will need to be scraped with a bearing scraper. Once the camshaft is free to turn, install any lifters into their bores, preferably not pumped up. You should then be able to slide the camshaft into place on the tips of the valves and lifters.

If the head uses a rocker arm shaft instead of cam followers, position the rocker arm shaft onto the engine with the rocker arms sitting on top of each valve and up against each cam lobe. Once everything is in place, you should be able to bolt the rocker arm shaft down slowly. Only turn each bolt about one turn at a time, to help draw it down evenly. Once all of the bolts are seated, torque each one, following the specified order to the specified torque. This could take multiple passes, torquing each bolt in order.

To assemble cam follower/rocker arm–style heads, follow the steps in **SKILL DRILL 19-12**.

SKILL DRILL 19-11 Installing the Camshaft in Heads with Bucket-Style Lifters

1. Lubricate the top of the valve stem and the bucket lifters with clean engine oil or assembly lube.

2. Install the bucket lifters on top of the valve in the same hole they were removed from originally, if they are being reused.

3. Lubricate the cam bearings, journals, and lobes with clean oil or assembly lube.

4. Install the camshaft in the bearings, and place the bearing caps in their original position. Tighten cam bearing bolts finger tight.

5. Tighten each cam bearing bolt about one turn at a time, following the specified sequence. Once all of the bolts have the caps against the cylinder head, torque the bolts to the specified torque.

6. If the camshaft is driven by a timing belt, lubricate the camshaft seal(s) and install them around each camshaft.

SKILL DRILL 19-12 Assembling Cam Follower/Rocker Arm–Style Heads

1. Lubricate the rocker arm pivot points and the valve tips, using engine assembly lube or clean engine oil.

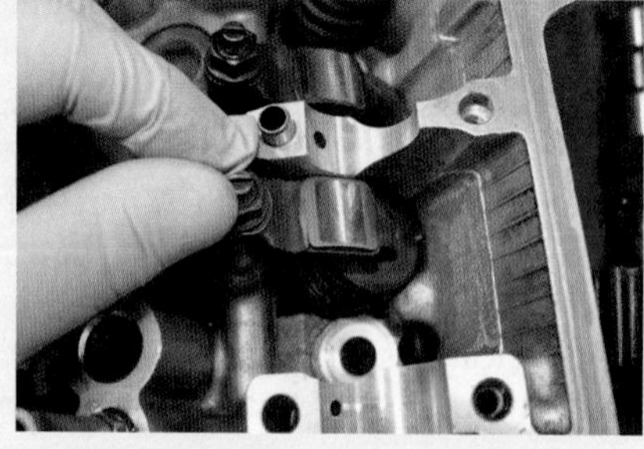

2. Install the rocker arms into the pivot points and over the tip of the valve.

3. Lubricate the rocker arm where the camshaft lobe will ride using engine assembly lube or clean engine oil. Lubricate the camshaft bearings, journals, and cam lobes with engine assembly lube or clean engine oil. Install the camshaft into the camshaft bearings.

SKILL DRILL 19-12 Assembling Cam Follower/Rocker Arm–Style Heads (Continued)

4. Install the camshaft support caps. Tighten the support cap bolts finger tight, and torque the bolts to the manufacturer's specification.

5. Use a feeler gauge to measure the valve clearance, following the manufacturer's recommended procedure.

6. Adjust the valve clearance to the manufacturer's specifications. Be sure to retighten the adjustment jam nut to prevent the clearance from changing.

▶ Wrap-Up

Ready for Review

- Functions of the cylinder head include admitting air and fuel to the combustion chamber, releasing exhaust after combustion, and allowing coolant to flow around the top of the combustion chamber.
- The cylinder head forms the top of the combustion chamber and is securely fastened with a head gasket between the head and block.
- Cylinder heads may be aluminum alloy (lighter weight, boosts fuel economy) or cast iron (heavier, but lower engine operating temperatures).
- Engines may be overhead valve, overhead cam, or dual overhead cam.
- Cylinder head parts include base cylinder head casting, intake valves, exhaust valves, valve keepers, valve retainers, valve springs, oil seals, coolant seals, rocker arms, plugs, and camshaft.
- Cylinder heads are designed to improve the swirl and turbulence of the air-fuel mixture moving into the combustion chamber.
- A hemispherical (HEMI) combustion chamber has intake and exhaust valves located on opposite sides to create cross flow of gases.
- The amount of air an engine can take in and remove is referred to as "breathing."
- The greater the air-fuel mixture, the more heat is generated, and thus the more engine power.
- Dual overhead cam engines often have multi-valve cylinder heads to create quicker, more efficient air-fuel movement.
- Cylinder head shapes can be oval (bathtub), wedge, hemispherical, or gas direct injection (GDI).
- Smaller intake and exhaust ports are generally better for engines that operate at low speeds (four-wheel drive), whereas larger intake and exhaust ports are ideal for engines operating at high speeds.
- Components of the valve train are driven from the camshaft and include lifters, pushrods, rocker arms or cam followers, valves, valve springs, keepers, and retainers.
- Components of the valve assembly are valve tip, keeper grooves, valve stem, valve face, and valve margin.
- Valves are located in the cylinder head and are configured in two-, three-, four-, and five-valve arrangements.
- Most engines are fitted with poppet (mushroom) valves.
- Exhaust valves are designed to withstand higher temperatures than intake valves.
- The valve spring is designed to close the valve and exert pressure to hold it firmly against the valve seat.
- Valve-spring retainers and valve locks (keepers) hold the valve spring in place.
- Valve guides are integral (non-replaceable) or non-integral (replaceable) and are designed to limit a valve's movement to up and down and to retain lubrication for the valve stem.
- Valve stem seals are designed to prevent oil leakage; the three basic types are umbrella style, O-ring, and positive.
- Valve seats form a strong seal with the valve face and must withstand high temperatures and extreme pressure.
- Valve train refers to all the parts that contribute to the opening or closing of the intake and exhaust valves.
- Types of valve trains correspond to engine types: overhead valve, overhead cam, or dual overhead cam.
- The camshaft provides the motion to open and close each of the valves.
- Valve timing can be fixed or variable.
- Lifters can be solid (manually adjusted) or hydraulic (self-adjusting) and are located in the engine block, cylinder head, rocker arms, and between the cam and valve tip.
- Camshaft followers are used in overhead camshaft engines in place of rocker arms.
- Bucket-style lifters transmit movement from the cam lobe to the tip of the valve.

- Pushrods, found in overhead valve engines, transfer the motion from the valve lifter to the rocker arm; pushrod types are ball-to-ball, ball-to-cup, and ball-to-adjustable tip.
- Rocker arms are categorized by the ratio of movement on the valve side to movement on the pushrod side.
- Gaskets are designed to create a seal when compressed between two stationary parts.
- Gaskets are single use only and must withstand engine heat.
- Functions of a head gasket are to seal and contain combustion pressure, seal oil passages, and control coolant flow.
- Anisotropic head gaskets conduct heat laterally.
- Oil seals can seal the rotating parts of an engine and are made of rubber or a rubber-type compound comprised of silicone, EPDM, or other flexible material.
- The lip-type dynamic oil seal is the most commonly used type.
- O-rings are used to seal stationary or slowly rotating shafts.
- Sealants are designed to seal together two surfaces in stationary contact with one another.
- Always check manufacturer recommendations for the proper type of sealant.
- Common cylinder head problems include wear, cracks, leaking coolant or compression gases, warped cylinder head, burnt or warped valves, and stripped spark plug hole threads.
- Thermal stress is the main cause of cylinder head cracks.
- Many cylinder head repairs require the skill of a certified machinist.
- Only remove the cylinder head after extensive testing and certainty that removal is required.
- Always clean the cylinder head and check for flatness after removal.
- Check valves and valve seats for cracks, burn marks, and signs of leakage.
- Valve guides should be checked for excessive wear, using a micrometer or dial indicator.
- Test valve springs for squareness, and measure both spring height and installed pressure.
- Inspect all parts when reassembling the head to make sure there are no problems.

Key Terms

anisotropic An object that has unequal physical properties, along its various axes. Used in head gaskets to pull heat laterally from the edge surrounding the combustion chamber to the water jacket.

base circle The rounded bottom part of the camshaft (off the lobe) where the valves remain closed or at rest.

bedding-in The process of a valve wearing into the valve seat and creating a positive seal around the whole diameter.

bucket-style lifter A bucket-shaped lifter assembly that sits on top of the valves and is operated directly by the camshaft. It can be hydraulic or mechanical.

cam lobe centerline The location of the cam lobe in relation to top dead center of the engine in degrees.

cam lobe ramp The rise of the lobe from the base circle to the top of the lobe, which is where the valve starts to lift, on the side opposite of where it starts to close.

cam lobe separation The number of degrees between the centerline of the intake lobe and the centerline of the exhaust lobe; this with cam duration determines the amount of valve overlap.

camshaft follower A slider or roller placed in direct contact with the lobes of the OHC camshaft that pushes on the tip of the valve to open it.

camshaft lobe The eccentric "egg-shaped" portion of the camshaft that pushes on the valve lifter or camshaft follower.

canted valves A valve arrangement in the cylinder head where the valves are at an angle to the cylinder bore. Canting the valves can make for a straighter path for air to flow into and out of the intake and exhaust ports.

coil bind A result of excessive valve lift. When the coils of the spring touch each other, excessive wear on the cam lobe and bending the pushrod.

column inertia The principle that as a column of air flows, it creates inertia, which keeps air flowing until its inertia energy is spent; sometimes referred to as a "ram effect" when using tuned intake or exhaust systems.

combustion chamber The area of an engine in which the air-fuel mixture is burnt. It consists of the bottom of the cylinder head and the top of the piston.

combustion pressure The force exerted by the expanding gases during the burning process. It is what causes the internal combustion engine to operate.

concentricity A term used to describe a valve seat where the valve seat and valve stem share a common center. In this design, the valve can seal against the seat no matter how it is rotated.

cylinder head The part of the engine that is bolted to the engine block and caps off the top of the combustion chamber.

dual overhead cam (DOHC) engine An engine design that is like the overhead cam engine but with two camshafts used per cylinder head; one operates the intake valves and the other operates the exhaust valves.

duration The amount of time the valve stays open, given in degrees of rotation of the crankshaft.

exhaust valve The valve through which exhaust gases are forced out of the combustion chamber.

fire rings Steel rings integrated into the cylinder head gasket nearest the combustion chambers that provide extra sealing to seal in the high combustion pressures.

flame front The front edge of the burning air-fuel mixture in the combustion chamber.

flame propagation The movement of the flame through the combustion chamber during the combustion process.

flat tappet A camshaft specifically designed to push on flat bottom lifters (nonroller types). It typically has a higher rolling resistance than the roller-style camshafts.

garter spring A metal spring wrapped circularly around the inside of a lip seal to keep it in constant contact with the moving shaft.

gas direct injection (GDI) cylinder head A cylinder head designed to accept a gasoline injector directly in the combustion chamber instead of in the intake manifold.

gasket Any material used on an engine to seal a fluid or gas into a particular area.

head gasket A thin piece of material, often a multilayered, bimetallic sheet used to seal the cylinder head assembly to the engine block.

hemispherical cylinder head A combustion chamber that is hemispherical in shape with the valves in a cross-flow arrangement and the spark plug near the center of the cylinder head directly over the top of the piston.

hydraulic valve lifter A small mechanical cylinder with a hydraulically operated internal piston used to automatically take up the slack (valve clearance) in the valve train.

inert gas A gas that will not react chemically.

intake port The port through which the air or air-fuel mixture travels from the throttle body area to the combustion chamber.

intake valve The valve through which air and fuel enter the combustion chamber.

integral valve guide A valve guide machined into the cylinder head during cylinder head construction.

interference angle The built-in differences in the angles between the valve seat and the valve face for the purpose of providing quick wearing-in of the surfaces; there usually must be between 0.5 and 1 degree of difference.

keeper groove A groove machined into the top of the valve stem near the tip of the valve that is used to "lock in" the valve keepers to help retain the valve spring onto the valve assembly.

L-head A type of four-stroke internal combustion engine having both intake and exhaust valves located in one side of the engine block, which are operated by lifters actuated by a single camshaft. It is sometimes called a flat head because there are no valves in the head.

lift The amount the valve will open. The more the valve lifts off its seat, the more air can get into and out of the engine.

lip-type dynamic oil seal A seal with a precisely shaped dynamic rubber lip that is held in contact with a moving shaft by a garter spring. An example would be a valve seal or camshaft seal.

multilayer steel (MLS) head gasket A gasket composed of multiple layers of steel and coated with a rubberlike substance that adheres to metal surfaces. They are typically used between the cylinder head and the cylinder block.

non-integral valve guide Also called a replaceable valve guide, a valve guide that is pressed into the cylinder head after the head has been constructed. These guides can be removed and replaced.

oil seal Any seal used to seal oil in and dirt, moisture, and debris out.

O-ring A small, donut-shaped, rubber ring used to seal connections between components.

O-ring valve stem seal A seal used where the valve spring retainer attaches to the valve stem to seal the juncture and prevent excessive oil from leaking into the valve guide.

overhead cam (OHC) engine Any engine in which the camshaft is housed inside the cylinder head and directly actuates the valves through rocker arms or cam followers or by pushing directly above the valve stem on a bucket.

overhead valve (OHV) engine An engine in which the valves are positioned in the cylinder head assembly, directly over the top of the piston.

poppet (mushroom) valve A cam-operated, spring-loaded, mushroom-type valve used to control intake into, and exhaust out of, the combustion chamber.

positive valve stem seal A valve seal located at the top of the valve guide that is a more positive seal than an umbrella or O-ring seal.

pulse-width modulation (PWM) A digital on/off electrical signal used as a variable control for devices such as solenoids.

pushrod A long, often hollow, metal tube that transfers the force from the valve lifter to the rocker arm assembly.

quench or squish area The narrow area between the top of the piston at top dead center and the cylinder head. It derives its name from the squishing of the air-fuel mixture into a small "charge."

rocker arm A lever that actuates a valve by pivoting near the center and pushing on the tip of the valve to open it.

room temperature vulcanizing (RTV) silicone A silicone adhesive that sets up, or "vulcanizes," at room temperature.

solid valve lifter A non-hydraulic valve lifter.

spring pressure Pressure exerted by a metal coil, usually measured in pounds or kilograms.

umbrella-style valve stem seal A seal that surrounds the top of the valve stem to prevent excessive oil from leaking into the valve guide.

valve A device used to control the flow of air and fuel into the combustion chamber and exhaust gases out of the combustion chamber.

valve cover gasket A gasket used to seal the valve cover (also called a rocker arm cover or cylinder head cover) to the cylinder head assembly.

valve face The portion of the valve that does the actual sealing to the valve seat.

valve float A condition that occurs when the valves are moving so fast that the spring tension is not great enough to fully seat the valves as designed.

valve guide A hole machined into the cylinder head or a hollow metal tube pressed into the cylinder head in which the valve stem rides, holding the stem in the proper position within the cylinder head.

valve head The portion of the valve that is exposed to the combustion chamber and contains the valve face and margin.

valve keeper A device used to keep the valve spring retainers attached to the valve while in the cylinder head.

valve margin The portion of the valve between the valve face and the valve head.

valve seat An integral part of the head, or circular metal rings that are pressed into the cylinder head, that makes up the mating surface where the valve face sits when it is closed.

valve spring A metal coil spring that returns valves to their fully closed positions after being opened.

valve spring retainer A washer-shaped piece of metal positioned near the top of the valve stem that holds the top of the valve spring to keep pressure on the valve while in the cylinder head.

valve stem The long shaftlike portion of the valve.

valve tip The end of the valve stem against which the rocker arm, cam follower, or hydraulic bucket-style lifter directly presses on to open the valve.

valve train A system encompassing all of the parts used in the opening and closing of the valves.

wedge combustion chamber A combustion chamber design where the valves are often directly lined up beside each other in a row and positioned at an angle over the pistons, forming a wedge-shaped combustion chamber.

Review Questions

1. The smaller water jackets in aluminum cylinder heads:
   a. reduce the heat generated by combustion.
   b. increase the chance of localized hot spots.
   c. prevent them from overcooling the engine.
   d. prevent them from overheating the engine.
2. As the camshaft rotates, the lifter rises and transfers the motion to the:
   a. camshaft lobe.
   b. pushrod.
   c. rocker arm.
   d. tappet.
3. Which of these is primarily used to block holes used in the original casting and machining of the head?
   a. Valve springs
   b. Cam seals
   c. Valve stem seals
   d. Soft plugs
4. The technician observes that the valves are arranged in a straight line positioned along the tapered angle. This typical arrangement is found in which of the following cylinder head designs?
   a. Wedge combustion chamber
   b. Oval combustion chamber
   c. Hemispherical combustion chamber
   d. Gas direct injection cylinder head
5. Which component in the valve train assembly determines the opening and closing of valves?
   a. Camshaft lobe
   b. Valve keeper
   c. Valve spring
   d. Valve retainers
6. Valve springs are checked for all of the following EXCEPT:
   a. Valve spring squareness
   b. Valve spring installed pressure
   c. Valve spring length
   d. Valve spring thickness
7. When the valve stem clearance is excessive, there is:
   a. a valve that is burned.
   b. more oil passing through the guide.
   c. less smoke emission from exhaust pipe.
   d. less oil consumption.
8. All of the statements below are true EXCEPT:
   a. The timing of the valves is controlled by the physical shape and position of the cam lobes on the cam shaft.
   b. Racing engines are designed to minimize the effect of the column inertia encountered at higher rpm.
   c. Valve timing is critical to the proper operation of the ICE.
   d. On variable valve timing engines, the valve timing can be modified to provide the best operating conditions.
9. Which of the following is a simple sealing device used to seal stationary and slowly rotating or sliding shafts?
   a. Garter spring
   b. Lip-type dynamic oil seal
   c. O-rings
   d. Multilayer steel (MLS) head gasket
10. All of the statements below are true EXCEPT:
    a. Cylinder head gaskets can be repaired and reused.
    b. Most cylinder head problems can be repaired by a cylinder head machinist.
    c. Cast iron cyilnder heads can be magnafluxed to locate cracks.
    d. The cylinder head is subject to a variety of problems that can affect engine function.

ASE Technician A/Technician B Style Questions

1. Tech A says that variable valve timing allows the engine's piston displacement to be increased or decreased. Tech B says that in-block cams usually include variable valve timing. Who is correct?
   a. Tech A
   b. Tech B
   c. Both A and B
   d. Neither A nor B
2. Tech A says that DOHC engines use two camshafts in each head. Tech B says that cylinder head warpage can be measured with a straightedge and feeler blade. Who is correct?
   a. Tech A
   b. Tech B
   c. Both A and B
   d. Neither A nor B
3. Tech A says that the head gasket is designed to provide a seal between the head and the block. Tech B says that most head gaskets require that you coat them with special sealers before installation. Who is correct?
   a. Tech A
   b. Tech B
   c. Both A and B
   d. Neither A nor B

4. Tech A says that pent-roof heads are of the cross-flow design, but typically with four or five valves. Tech B says that GDI heads typically have the spark plug centrally located and the fuel injector offset to one side. Who is correct?
   a. Tech A
   b. Tech B
   c. Both A and B
   d. Neither A nor B
5. Tech A says that an interference angle is the difference in angle between the valve face and valve seat. Tech B says an interference angle is the point where the cam lobe contacts the lifter. Who is correct?
   a. Tech A
   b. Tech B
   c. Both A and B
   d. Neither A nor B
6. Tech A says that soft plugs allow the head or block to expand and contract without cracking. Tech B says that soft plugs are designed to prevent damage to the head or block whenever coolant freezes. Who is correct?
   a. Tech A
   b. Tech B
   c. Both A and B
   d. Neither A nor B
7. Tech A says that valve seals are designed to prevent combustion gases from entering the valve train area. Tech B says that valve retainers hold the valve in the spring. Who is correct?
   a. Tech A
   b. Tech B
   c. Both A and B
   d. Neither A nor B
8. Tech A says that valve springs close the valves. Tech B says that turbulence in the combustion area creates a better burn of the air-fuel mixture. Who is correct?
   a. Tech A
   b. Tech B
   c. Both A and B
   d. Neither A nor B
9. Tech A says that when the lifter is on the base circle of the cam lobe, the valve is closed. Tech B says that the cam-in-block lifter rides directly against the valve. Who is correct?
   a. Tech A
   b. Tech B
   c. Both A and B
   d. Neither A nor B
10. Tech A says that when cleaning the gasket mating surface on a cast iron head a metal gasket scraper can be used. Tech B says that when cleaning the gasket mating head on an aluminum head, use only plastic or nylon scrapers. Who is correct?
    a. Tech A
    b. Tech B
    c. Both A and B
    d. Neither A nor B

CHAPTER 20

Engine Block Components

NATEF Tasks

- **N20001** Disassemble engine block; clean and prepare components for inspection and reassembly. (MAST)
- **N20002** Deglaze and clean cylinder walls. (MAST)
- **N20003** Inspect crankshaft for straightness, journal damage, keyway damage, thrust flange and sealing surface condition, and visual surface cracks; check oil passage condition; measure end play and journal wear; check crankshaft position sensor reluctor ring (where applicable); determine needed action. (MAST)
- **N20004** Inspect main and connecting rod bearings for damage and wear; determine needed action. (MAST)
- **N20005** Identify piston and bearing wear patterns that indicate connecting rod alignment and main bearing bore problems; determine needed action. (MAST)
- **N20006** Inspect and measure cylinder walls/sleeves for damage, wear, and ridges; determine needed action. (MAST)
- **N20007** Inspect and measure piston skirts and ring lands; determine needed action. (MAST)
- **N20008** Inspect auxiliary shaft(s) (balance, intermediate, idler, counterbalance and/or silencer); inspect shaft(s) and support bearings for damage and wear; determine needed action; reinstall and time. (MAST)

Knowledge Objectives

After reading this chapter, you will be able to:

- **K20001** Describe the block and its related components.
- **K20002** Describe how engine blocks are manufactured.
- **K20003** Describe the engine block construction.
- **K20004** Describe the function of the oil pan.
- **K20005** Describe the features and function of the cylinders.
- **K20006** Describe the features and function of the pistons.
- **K20007** Describe the features and function of the piston rings.
- **K20008** Describe the features and function of the connecting rods.
- **K20009** Explain the function of the rotating assembly.
- **K20010** Describe the features and functions of the crankshaft.
- **K20011** Describe the features and function of friction bearings.
- **K20012** Describe the features and function of flywheels, flexplates, and harmonic balancers.
- **K20013** Describe the features and function of crankshaft seals.
- **K20014** Describe the purpose and function of balance shafts.

Skills Objectives

After reading this chapter, you will be able to:

- **S20001** Remove the oil pan.
- **S20002** Drain the block.
- **S20003** Disassemble the block underside.
- **S20004** Remove the oil pump and tray.
- **S20005** Label and remove the piston assemblies.
- **S20006** Remove the crankshaft and bearings.
- **S20007** Remove passage plugs.
- **S20008** Clean the components.
- **S20009** Prepare cylinder block for reuse.
- **S20010** Inspect the cylinder block.

▶ Introduction

The engine block (also called the cylinder block or simply the block) is the main structural member of the internal combustion engine. Although many different engine designs are available, in all configurations the engine block acts as the framework for the entire engine and provides for the mounting of all the components. In the early days of the automobile, the distinct parts of the engine, such as the cylinders, intake and exhaust passages, and crankcase, were constructed as separate pieces that were then bolted or welded together. Today, engine blocks are cast in one piece, with machined surfaces and holes added for the placement and attachment of the engine components. The process of constructing a single block into which all parts fit is more efficient from a manufacturing standpoint. It also gives the block more strength and rigidity.

This chapter discusses the engine block and its related components as well as the procedures for inspecting them, and explains how all the components work together to generate the power that drives the vehicle.

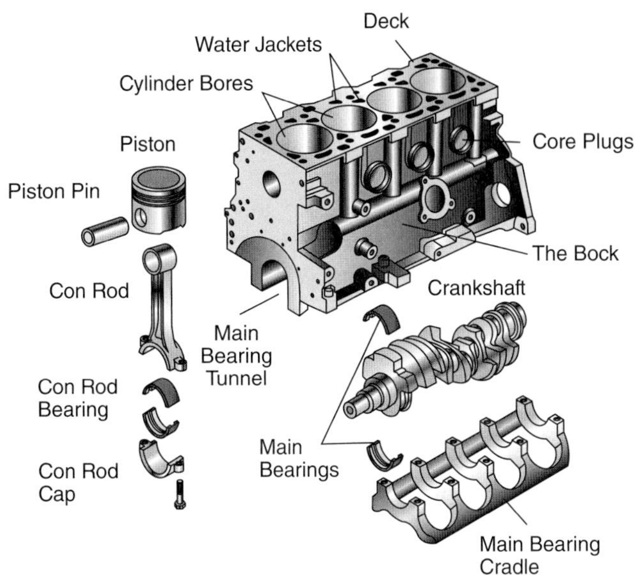

FIGURE 20-1 The engine block of an overhead cam engine.

▶ Engine Block

K20001

The engine block is the largest part of the engine; it is the main supporting and aligning structure for all of the engine parts (**FIGURE 20-1**). Within the engine block are the cylinders and pistons; the top of the piston and the matching cylinder head form the combustion chamber. The rapidly expanding gases produced by combustion pressurize the combustion chamber and push down on the piston during the power stroke, forcing it down the cylinder until it reaches the bottom of its travel. This movement causes the crankshaft to spin, which pushes the piston back up the cylinder in preparation for the next combustion event. The crankshaft also powers the rest of the drivetrain through the flywheel or flexplate and the accessories from the crankshaft pulley.

For optimum performance and engine life, the engine block must be designed, cast, and machined within precise tolerances, meaning that the components must be designed and connected together within narrow specifications. These precise specifications are critical to prevent leakage of air and fuel, withstand the thermal stresses of combustion, and accommodate shocks and vibration applied to the engine during the power pulses.

The upper section of the engine block contains the cylinders and pistons. The lower section of the engine block forms the crankcase, which supports the main bearings, crankshaft, camshaft (some engines), and balance shafts (some engines). A few engine designs, like the various Subaru models, have a **horizontally opposed engine**, sometimes called a boxer engine. In these engines, the cylinders are positioned side to side, rather than up and down, and the crankcase is located between the cylinders (**FIGURE 20-2**).

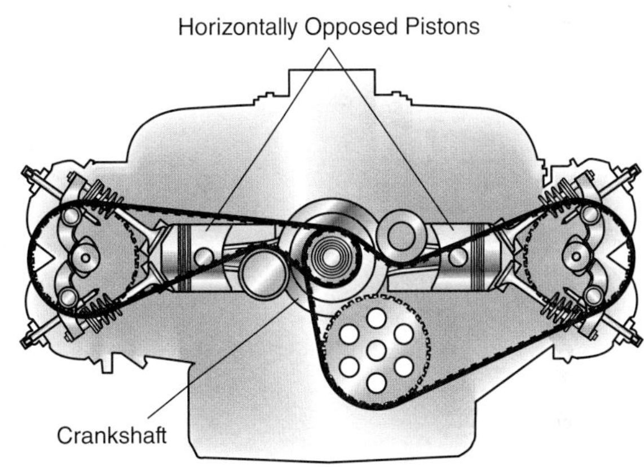

FIGURE 20-2 A horizontally opposed engine.

You Are the Automotive Technician

Today you are disassembling an engine that has more than 250,000 miles on it. The vehicle still runs well but doesn't have as much power as it used to. The fuel economy is also not as good as it once was, and the vehicle is burning more oil than specified. The owner would prefer that this engine be rebuilt rather than replaced, because it has lasted so long, but only if that makes economic sense. She has agreed to pay you to disassemble the engine, measure the critical components of the engine, compare them to specifications, and estimate the cost to rebuild it.

1. What measurements will you need to perform on the block and its internal components?
2. How will you determine whether the block has any cracks?
3. What conditions would require the cylinders to be bored?
4. What are the two methods of measuring crankshaft bearing clearance?
5. What precautions should be taken when disassembling an engine in which the major components may be reused?

Engine Block Manufacturing

K20002

In the metal casting foundry process, different types of materials and techniques are used to create steel or cast iron and aluminum engine blocks. The most common methods are the green sand, shell, and **lost foam** processes. In all three processes, hot molten metal is poured into a mold. The engine block can be cast in one piece from gray iron, or it can be alloyed with other metals such as nickel or chromium. The specific type of metals or metal used depends on many factors, including the intended use of the engine block, the type of environment it will be run in, and the cost.

The casting process begins by making foam models of the block (**FIGURE 20-3**). Different engine block foam models are used depending on the specific engine type desired, such as V8, V6, or four-cylinder in-line. The foam models have voids for the **water jackets** and cylinders. The foam models are placed into molds and packed tightly with sand, which fills the voids and supports the foam (**FIGURE 20-4**). The foam has two risers that become the entry and exit points for the molten metal, which is poured into the casting mold from a tub or ladle. This is done without the use of pressure, a process known as **gravity pouring**. The molten metal instantly burns up the foam model, allowing the space to be filled with the metal. Once the metal cools, the core sand is removed through holes in the sides and ends of the engine block casting, leaving cavities for the cooling and lubricating passages.

These holes are sealed with **core plugs**, which may also be called **Welch plugs** or **soft plugs**. The plugs are rounded discs or cups that are used to seal the openings created in the casting process for removal of core sand (**FIGURE 20-5**). The discs can be flat, concaved, or threaded to fit the particular opening. For years, the term "freeze plugs" has also been used; however, the term is inaccurate. The engine block was never designed to withstand freezing of the water in the cooling passages. Water expands approximately 10% as it freezes into ice. If the engine temperature drops below freezing, the pressure of the expanding ice may cause the block to crack, even if the core plugs pop out.

Once the engine block casting is inspected and cleaned, it goes through a series of machining processes to bore, flatten, and thread the various surfaces and holes (**FIGURE 20-6**). Parts of the engine block that typically require machining include the

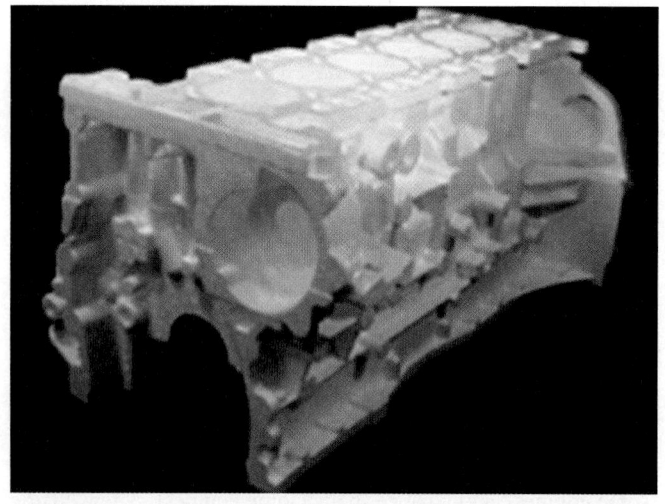

FIGURE 20-3 Foam core of an engine block.

FIGURE 20-5 Soft plug being installed in a block.

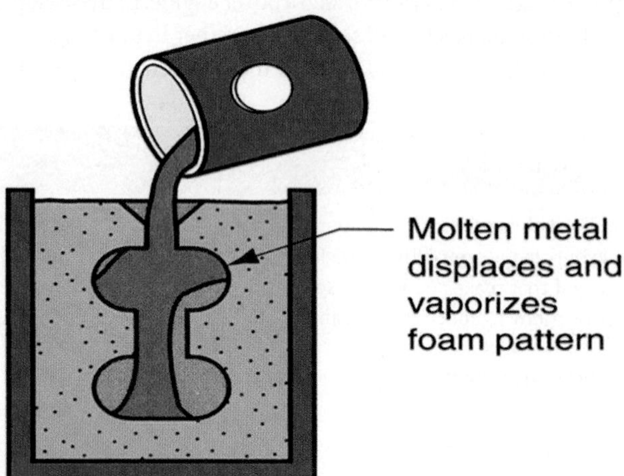
Molten metal
displaces and
vaporizes
foam pattern

FIGURE 20-4 Foam core being placed into a mold.

FIGURE 20-6 Engine block being machined.

FIGURE 20-7 Aluminum block.

top of the engine block (also called the **deck**); cylinders; front of the block; rear of the block; bottom of the block, including the crankshaft main bearing and thrust areas; and any other area to which another engine part is attached, such as the cam bearing or balancer shaft journals. Other finishing processes include honing the cylinders, drilling holes, and cutting threads.

As more manufacturers try to make vehicles lighter and more fuel efficient, more engine blocks are being cast from aluminum (**FIGURE 20-7**). An engine block made of aluminum alloy is lighter than one made of cast iron. The lighter aluminum engine provides a higher power-to-weight ratio and also provides the vehicle with greater fuel efficiency.

Engine Block Construction and Crankshaft Support

K20003

The engine block assembly, or lower end as it is sometimes called, consists of the following: block, oil pump, oil galleries and plugs (oil passage plugs), water jacket and plugs, pistons and rings, piston pins, connecting rods and caps, rod bearing inserts, main bearing caps, main bearing inserts, the thrust bearing, the crankshaft, rear seals (lip or rope seal), the camshaft (for non-overhead cam engines), a vibration damper, and a flywheel or flexplate. The engine block also has smooth surfaces in certain areas to which the mating parts, such as the cylinder heads, oil pan, timing cover, water pump, and transmission, are attached with a precise and tight fit.

The engine lubrication system also begins in the engine block assembly, with a system of oil passages that feed oil to the main and rod bearings, cam bearings, and cylinder heads. Oil pressure is developed by an attached oil pump, and the resistance to flow of the oil is provided by the tight spaces the oil is pumped into. The oil pump may be attached on either the front or the bottom of the block, generally near the crankshaft (**FIGURE 20-8**). The oil pump may be driven directly by the crankshaft or by a gear off the camshaft. A mounting surface for attachment of the oil filter is usually near the oil pump.

FIGURE 20-8 Locations of oil pumps. **A.** Crankshaft. **B.** Crankcase.

FIGURE 20-9 Ribs, webs, and fillets.

In casting the engine block, **ribs**, **webs**, and **fillets** are used to provide rigidity to the casting while reducing overall engine weight (**FIGURE 20-9**). These structures are cast into the engine block when it is formed, but are constructed into the shape of ribs, webs, or fillets. They are placed near areas that encounter high levels of stress and vibration, reinforcing and adding strength to the engine block without adding excessive bulk.

The crankshaft is supported in the engine block's main bearing bores by **main bearing caps**. The main bearing bores and caps provide support for the crankshaft. The caps can be fastened using a two- or four-bolt design. The shape is the same for each design; only the number of bolts—two or four—varies. The main bearing caps can be inset within the block. The bolts can be fastened through the top of the main cap with two or four bolts or studs (**FIGURE 20-10**). They can also be fastened with two bolts or studs on top and two bolts through the side of the block and into the side of the main bearing cap (**FIGURE 20-11**).

Some engines use an integrated **main cap girdle** (**FIGURE 20-12**) that has all the main caps cast in a single supporting structure. Others use a main cap girdle that is separate from the main caps. It acts as a cover to the main bearing caps and links the caps together by connecting each individual cap to all of the others in order to reinforce their strength. The **girdle** also helps position the main bearings in the engine block assembly, supplies engine block rigidity and structure, and prevents lateral or vertical movement.

The Oil Pan

K20004

The oil pan seals off the bottom of the crankcase and holds oil for the engine lubrication system (**FIGURE 20-13**). It radiates oil heat to the outside ambient air and may include cooling fins (heat sinks) to effectively transfer heat to the airstream flowing below the vehicle. The oil pan bolts onto the bottom of the engine block and has undergone many changes since the first engine design. The oil pan was originally designed as part of the engine block in cast iron but was later changed to lighter stamped steel. Most are now designed of lightweight aluminum or thermoplastic. The oil pan has become more of a structural part of the engine and helps to reinforce the engine block.

On many engines, the oil pan houses the oil pump, which is the heart of the engine in that it supplies critical lubrication for the internal moving parts of the engine. The screened oil pump pickup lies near the bottom of the oil pan and is used to screen out any larger particles that may contaminate the oil.

FIGURE 20-10 Typical main bearing cap and bolts.

FIGURE 20-12 Integrated main cap girdle.

FIGURE 20-11 Typical cross-bolt main bearing cap and bolts.

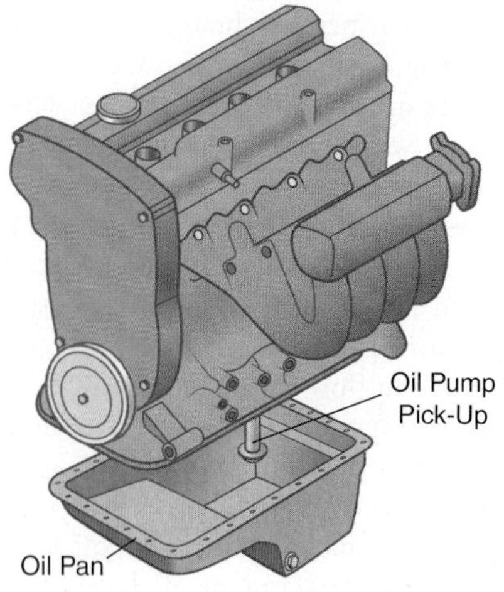

FIGURE 20-13 The oil pan caps off the bottom of the crankcase.

Cylinders

K20005

Cylinders are the large holes in the block in which the pistons move up and down. Cylinders can be machined directly into the cylinder block as in the case with most cast iron blocks, or they can be made of a separate steel or cast iron sleeve in the case of some aluminum blocks (**FIGURE 20-14**). In the case of a cast iron block, the cast iron provides a relatively hard wear surface for the pistons and piston rings. In blocks made from softer materials such as aluminum, cylinder sleeves are used to provide a durable surface for pistons and piston rings to ride against.

Hardened materials can better withstand the continuous scuffing forces of the piston and piston rings, especially under extreme temperatures. Most aluminum engine blocks use cast iron cylinder sleeves that are cast into the engine block and cannot be removed or replaced. The sleeves have grooves on the outside that form ridges; once the cylinder heads are bolted on, these ridges lock the cylinder to the engine block and prevent the cylinder from shifting during engine operation. They also increase the surface area of the sleeve, which assists in the transfer of heat away from the sleeve to the cylinder block and ultimately the water jacket. The movement of combustion heat away from the engine block is critical in preventing damage from overheating such as when a piston seizes to the cylinder wall.

There are three main types of cylinder sleeves: dry, dry flanged, and wet (**FIGURE 20-15**). The **dry sleeve** gets its name because it is not in direct contact with the water jacket. Dry sleeves are held in place by an **interference fit** with the cylinder walls. An interference fit is a fastening between two parts that is activated by friction after the two parts are pushed together. In this type of fit, one part slightly interferes with the space that the other occupies, such as a slightly oversized piece being pushed into a slightly undersized hole. The wall of the dry sleeve is usually between 3/32" and 1/8" thick (2.4–3.2 mm)—just thick enough to create the necessary friction for a strong interference fit within the engine cylinder block. When pressed into the engine block, the cylinder sleeve is smooth at the top and has full surface contact through the engine cylinder. The top of the cylinder is flush with the top of the engine block and is difficult to see. Once in place, dry sleeves become a permanent part of the cylinder block.

A **dry flanged sleeve** is similar to a normal dry sleeve in that it too is not in direct contact with the water jacket; however, it differs from the traditional cylinder sleeve in its installation. The dry flange sleeve is not held in place by an interference fit, but has a flange at the top of the cylinder that is used to lock the sleeve into the engine block. The flange fits in a matching recessed area at the top of the engine cylinder block. The cylinder head then clamps the flange in place so it cannot move. However, this type of sleeve must be thicker than a normal dry sleeve as it is not an interference fit.

In a **wet sleeve**, the outer surface is in direct contact with coolant in the water jacket that surrounds the cylinder. It is called a wet sleeve because coolant from the cooling system circulates against the wet sleeve's outer surface. This design helps speed up heat transfer between the wet sleeve and the coolant because there is direct contact between the wet sleeve and the water jacket. The wet sleeve is sealed at the top and bottom to prevent coolant leaks. The recessed fit of the sleeve forms the top seal. At the bottom, one or two neoprene seals fit tightly between the outside of the sleeve and the block. Proper sealing of the wet sleeve is important

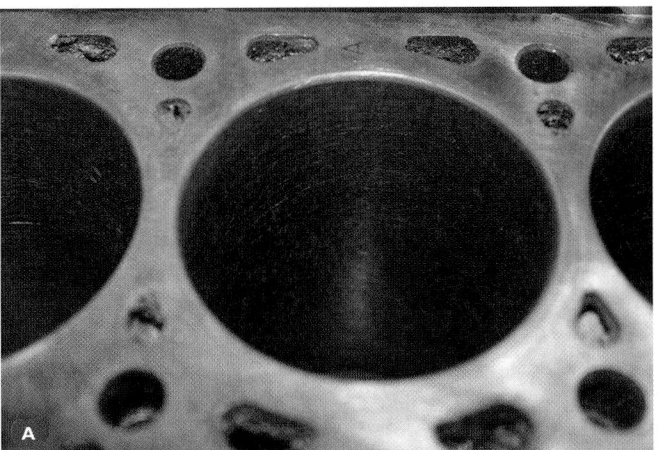

FIGURE 20-14 A. Integral cast iron block and cylinder. **B.** Cast aluminum block and cast iron sleeve.

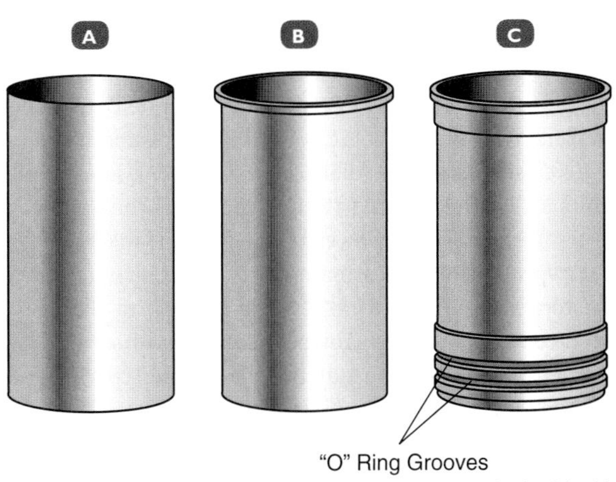

FIGURE 20-15 Cylinder sleeves. **A.** Dry. **B.** Dry Flanged. **C.** Wet.

to prevent coolant from leaking into the combustion chamber at the top and the crankcase at the bottom. Because the sleeve is in direct contact with the water jacket, corrosion can be a problem. Corrosion can insulate the wet sleeve from the coolant, creating a layer that slows down heat transfer to the water jacket.

Applied Math

AM-31: Length/Volume/Weight: The technician can determine the degree of conformance to the manufacturer's specifications for length, volume, weight, and other appropriate measurements in the standard and metric systems.

An automotive engine is being overhauled in a repair facility. Various measurements are being taken to determine the degree of conformance to the manufacturer's specifications. The technician uses an inside micrometer to measure the diameter of the cylinders. The measurement is 3.800 inches (96.50 mm). This information verifies to the technician that the engine matches the specifications in the service information.

In order to measure the cylinders with a more exact procedure, a dial bore gauge is used. This gauge is a special tool used to measure cylinder bore for taper and out-of-round conditions. The dial bore gauge is able to measure cylinders to within 1/10,000 of an inch. Service information for this engine allows a maximum out-of-round of 0.0004" (0.010 mm). The maximum allowable taper is 0.0005" (0.013 mm). If a cylinder is worn beyond these limits, the piston rings will not seal properly, and the likely solution would be to bore the cylinders oversize. In some cases, this could be 0.030" or 0.060" oversize, and matching pistons would be installed. By comparing the measurements to manufacturer's specifications, the technician is able to determine the degree of conformance.

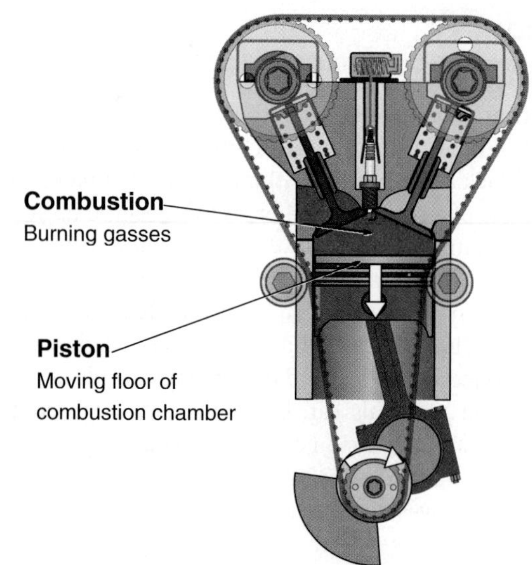

FIGURE 20-16 The piston is the moveable floor of the combustion chamber that extracts energy from the burning gases.

FIGURE 20-17 Piston assembly.

Pistons

K20006

The **piston** fits in the cylinder like a close-fitting, movable plug. The clearance between the piston and the cylinder wall is typically within a few thousandths of an inch, and on some current model engines, it can be less than a thousandth of an inch. A good way to think of the piston is as the "movable floor of the combustion chamber" (**FIGURE 20-16**). When the air-fuel mixture burns in the combustion chamber, it creates pressure that pushes the piston down on the power stroke. The job of the piston is to extract the thermal energy from the burning fuel and convert it into mechanical energy, which can be used to power the vehicle. Accordingly, it is critical that the piston fit the cylinder with close tolerance so that the pressurized gases cannot easily leak past it. The piston, with its connecting rod and bearing, transfers the force of the **power stroke** to the crankshaft.

The piston comprises many parts, together referred to as the **piston assembly** (**FIGURE 20-17**). The top of the piston is referred to as the head or crown. It comes in different designs, depending on the specific function or engine type into which it will be placed. The piston skirt is the area around the side of the piston that slides against the engine cylinder walls and helps keep the piston straight in the cylinder. The skirt has two thrust faces 90 degrees to the piston pin. These skirt thrust faces are usually longer than the piston pin sides of the skirt, so there is more surface area for the piston and cylinder wall to spread the load over. Also, the curved cutouts of the skirt prevent the crankshaft counterweight from hitting the piston skirt when the piston moves down the cylinder. Some skirt designs also include a small groove called the expansion slot that allows piston expansion.

The **piston pin (wrist pin)** is a tubular structure that connects the piston to the top of the connecting rod. The pin passes through the piston skirt into an area called the piston boss, a heavily reinforced area designed to help the pin withstand the pressures exerted on it, and is joined to the top of the connecting rod (**FIGURE 20-18**). Because of this extra material in the piston boss, the piston is not perfectly round. It is slightly oval in shape to allow for the extra expansion of the metal in the piston boss area as the piston heats up from combustion heat as the engine is operating. When the engine is at operating temperature, the piston becomes round. When it is cold, the piston is narrower across the wrist pin diameter than across the thrust surfaces of the piston.

The piston pin (wrist pin) is made of steel and is hollow in the center to reduce weight, thereby reducing the engine's

FIGURE 20-18 Wrist pin connects the piston to the connecting rod.

FIGURE 20-20 A rod heater being used to expand the small end of the rod.

FIGURE 20-19 Floating wrist pin held in place by snap rings.

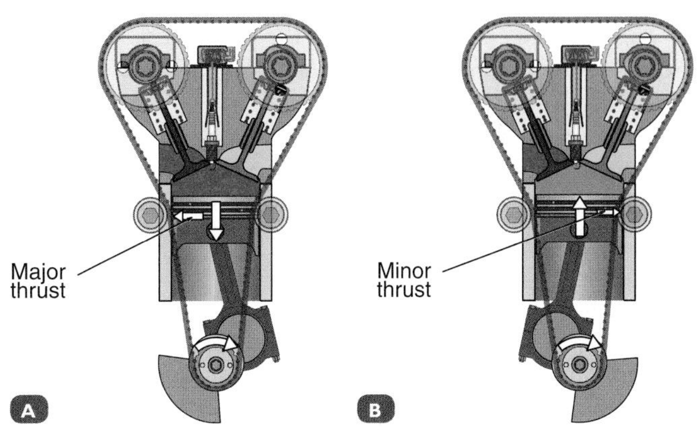

FIGURE 20-21 **A.** Major thrust. **B.** Minor thrust.

reciprocating mass (**FIGURE 20-19**). The wrist pin design can be a **floating pin** or a **press-fit pin**. In both cases, the pin is slightly smaller than the pin bore in the piston, and the piston is free to move on the pin. With the floating pin design, the piston pin floats in both the piston and the rod, with the piston pin being held in place by snap rings on each side of the piston (figure 20-19).

With the press-fit pin design, the piston is held in place by a wrist pin that is pressed into the connecting rod. The small end of the connecting rod uses an interference fit with the wrist pin to lock the pin firmly in place. The wrist pin is installed into the connecting rod by pressing it in or by heating the end of the rod with a rod heater (**FIGURE 20-20**). The rod heater heats up the small end of the connecting rod, expanding the size of the hole just enough for the pin to slide into the connecting rod. When it cools, the pin is firmly locked into the rod.

Each piston has a major thrust face and a minor thrust face (**FIGURE 20-21**). The side of the piston that is forced into the cylinder wall on the power stroke is the major thrust face. The side of the piston that is forced into the cylinder wall on the compression stroke is the minor thrust face. Most piston pin bores are not centered in the piston; they are set slightly off center, typically with the pin offset toward the major thrust side of the piston (**FIGURE 20-22**). This slightly cocks the piston in the bore, which keeps the piston from rocking as it transitions between the minor thrust face to the major thrust face.

FIGURE 20-22 Piston pin offset.

If the clearances between the piston and the cylinder wall become too great, the transfer from one piston skirt to the other can cause a condition known as **piston slap**. Piston slap causes an audible noise that is most noticeable when the pistons are cold and at their smallest diameter. Piston slap occurs when the piston is forced to the thrust side of the cylinder at the top of the power stroke. Offsetting the wrist pin in the piston keeps the piston slightly cocked in the cylinder, which helps reduce piston slap.

Pistons are made of aluminum alloy for weight savings. Aluminum by itself is too soft and expands too rapidly to stand up to the heat and pressure of combustion. If aluminum is mixed with other materials, such as silicon, aluminum alloys are produced that are strong and lightweight and that can withstand the heat and pressures of combustion and resist expansion. If expansion can be kept low, then the piston clearance can be tighter, helping to reduce blowby.

Control of Heat in Pistons

The piston must stand up to great heat and pressure. All pistons expand as they heat up. Because there is more metal near the wrist pin, this area tends to expand more. To allow for this, many pistons are machined into a slightly oval shape. This process is called cam grinding, and the piston is called a cam-ground piston. As an oval piston heats up to operating temperature, it expands and becomes round.

Other methods that allow, and help to control, piston expansion involve using steel struts or ribs, expansion slots in the skirt of the piston, or slots called heat dams that restrict movement of the heat. In the first method, steel struts or ribs are built into the piston assembly. Their sole purpose is to control expansion by slowing heat flow and stiffening the piston. However, they cause more piston weight. In the second method, expansion slots are cut vertically into the piston skirt. As the piston expands, the slots are slowly closed. The third method uses a horizontal slot cut directly beneath the piston head (**FIGURE 20-23**). The slot acts like a dam or retardant, preventing heat from traveling down from the top of the piston.

▶ TECHNICIAN TIP

A piston becomes round once it heats up to operating temperature. But what if it overheats? Then the piston boss areas will continue to expand and grow faster than the rest of the piston. This makes the piston wider at the piston bosses, which can cause it to expand larger than the cylinder bore. If this happens, then the piston will be forced against the cylinder walls and quickly seize in the cylinder, requiring rebuilding or replacement. Low-mileage engines have smaller clearances and are therefore more prone to this problem. The best prevention is ensuring proper piston clearance and maintaining the cooling system.

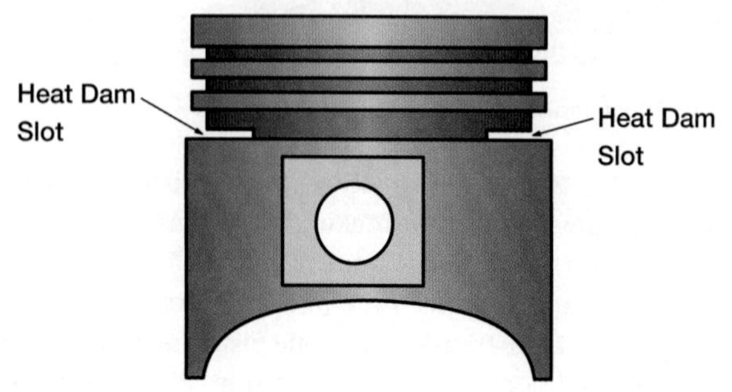

FIGURE 20-23 Horizontal slots in the piston act as heat dams.

Piston Head Shape

The shape of the top of the piston depends on the shape of its combustion chamber and its compression ratio. Combustion chambers vary in shape depending on the manufacturer and type of engine for which they are intended. The top of the piston forms the bottom of the engine's combustion chamber. The compression ratio can be changed by using a piston with a different head design or a cylinder head with a different chamber size. This is because the sizes of the piston head and chamber affect the ratio between the cylinder's volume at BDC and the volume at TDC within the combustion chamber. A taller piston head takes up some of the available volume in the combustion chamber as compared to a shallower piston head (**FIGURE 20-24**). The automotive engineer or manufacturer will select the correct piston head shape for the performance level desired.

In designing an engine, the automotive engineer creates the design with a particular piston head in mind. The piston head must be able to work in accordance with the design of the rest of the combustion chamber and the valve train assembly to obtain the engineer's goals for engine output and performance. The top of the piston may be flat, concave, dome shaped, or recessed (**FIGURE 20-25**), and it can have one to five valve relief areas. Valve relief areas are used to prevent pistons from hitting the valves, or vice versa. Depending on the needs and shape of the engine, valve relief areas may be used on just one side of the piston head or on both.

There are two processes in manufacturing pistons: forging and casting. The forged piston is the stronger and more costly of the two. In forging a piston, metal molecules are compressed into a denser area. This makes the piston stronger and more desirable for diesel engines and racing applications. With the stronger forged piston, the higher compression ratios that create hotter combustion gases and higher pressures can be used with less piston damage.

There are two versions of the cast piston: (1) cast aluminum with steel struts and (2) cast aluminum with 16% to 19% silicon, which is called a **hypereutectic piston**. Steel struts are used to help

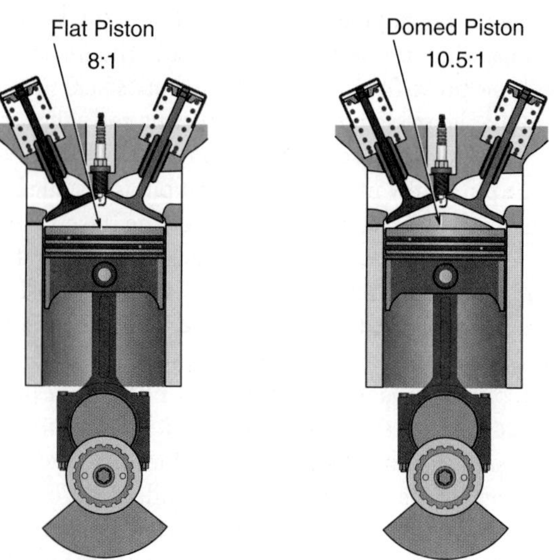

FIGURE 20-24 Piston head design affects compression ratio.

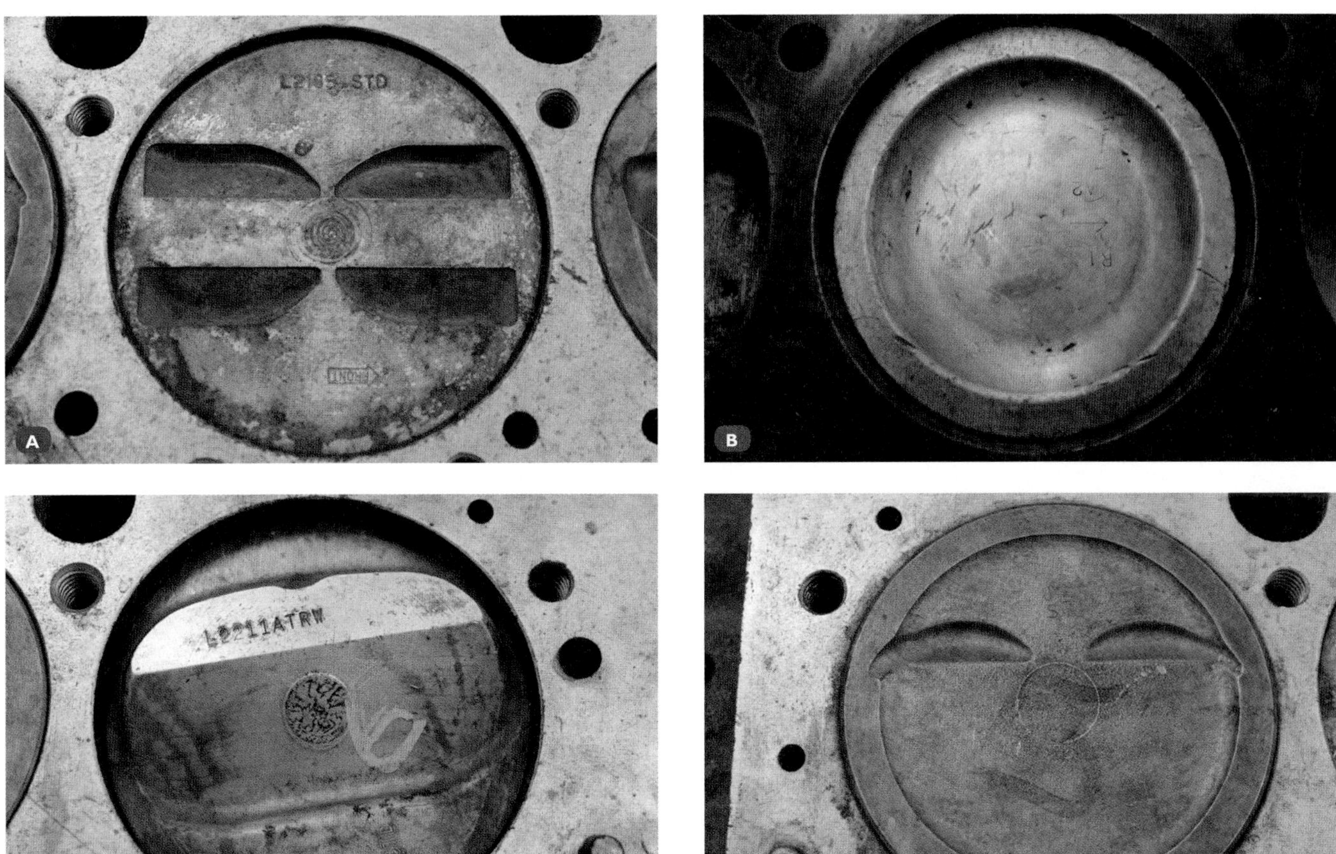

FIGURE 20-25 Piston shapes. **A.** Flat. **B.** Concave. **C.** Dome. **D.** Recessed.

control the flow of heat from the combustion process and can be embedded into the piston assembly to help control piston expansion. Hypereutectic pistons are stronger than pure aluminum pistons and are used in many applications. Although the cost of hypereutectic pistons is considerably less than forged pistons, the trade-off is that forged pistons are stronger than hypereutectic pistons. Another advantage of hypereutectic pistons is that the clearance between the piston and the cylinder is extremely small, which allows for a tighter fit of the piston assembly inside the cylinder as compared to all other piston types. The smaller clearance allows (requires) the use of lower viscosity oils that can provide lubrication between the closer running components.

Piston Coatings

It is not uncommon today to see pistons with a coating on them (**FIGURE 20-26**). The most popular is a black **moly**, which is a molybdenum disulfide coating on the piston skirt that helps reduce friction between the piston skirt area and the cylinder wall. Ceramic coating is also available. Some manufacturers and aftermarket companies coat the top of the piston with a ceramic coating, which creates a thermal barrier. The ceramic coating prevents the piston metal from absorbing heat as quickly, which allows the piston to operate at a lower temperature and withstand greater combustion temperatures without deforming. It also increases the efficiency of the engine by retaining more heat in the combustion chamber rather than transferring it through the piston to the cooling system. Designing the exhaust system to take advantage of the faster exhausting of gases due to the increase in exhaust gas temperature can aid in the scavenging effect and help the engine breathe more efficiently.

Compression Ratio

Compression ratio relates to how tightly the air-fuel mixture is compressed in the combustion chamber. The tighter the air-fuel mixture, the more power that can be extracted from a given amount of air and fuel. Unfortunately, increased compression ratios can cause engine damage. Increased compression ratios increase compression pressures, thereby raising the temperature of the mixture within the combustion chamber as it is being compressed. If the mixture gets too hot, it will ignite by itself before it should, which is known as preignition or detonation. This is very hazardous to the engine, so compression ratios must not be raised too high. Typical compression ratios for stock gasoline engines run between 8:1 and 10.5:1. High-performance vehicles can be around 12:1 or higher. Diesel engines run around 20:1.

To calculate the compression ratio, divide the volume of the combustion chamber above the piston at BDC by the volume of the combustion chamber above the piston at TDC. For example,

FIGURE 20-26 Piston coatings. **A.** Moly coating. **B.** Ceramic coating.

a cylinder that has a total combustion chamber volume of 50 cubic inches (819.35 cc) at BDC and 5 cubic inches (81.93 cc) at TDC has a compression ratio of 50 to 5, or 10 to 1, which is written 10:1. In other words, the piston is compressing 50 units of volume into 5 units of volume (**FIGURE 20-27**).

Compression ratio can be raised in several ways. First, a taller piston or one with a dome will reduce the volume of the combustion chamber at both TDC and BDC, thus raising the ratio. Another option is to use cylinder heads with smaller chambers so that the mixture is squeezed into a smaller space. Likewise, machining the cylinder head surface or using a thinner head gasket will increase the compression ratios slightly. Also, using a crankshaft with longer throws increases the engine's stroke, which increases how high and low the piston travels in the cylinder, raising the compression ratio but also changing engine size. Finally, boring out the cylinders and using larger pistons makes the cylinder volume larger so the compression ratio increases, but this method also changes engine size.

What we have just discussed is theoretical compression ratio, which does not take into consideration valve timing. We have assumed that the intake valve closed at BDC, which would mean that compression could start at BDC. In reality, intake valves close after BDC, so compression cannot start until the intake valve is fully closed. Complete closure could be substantially after BDC in the case of a high-performance camshaft, and the piston has already moved partially up the cylinder, lowering the effective compression ratio. So the effective compression ratio is the volume above the piston when the intake valve just closes, compared to the volume above the piston at TDC.

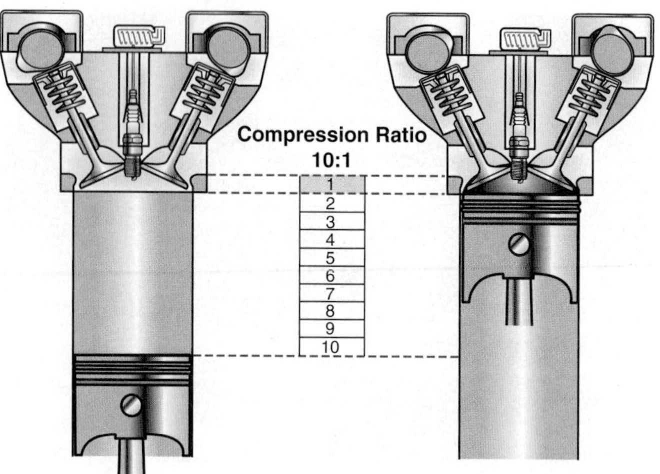

FIGURE 20-27 Compression ratio.

FIGURE 20-28 The compression rings are the top two rings.

Piston Rings

K20007

Piston rings are thin, metallic, C-shaped rings set into grooves located around the piston skirt; they fill the space between the piston and the cylinder wall. Generally, three to four rings are used per piston. The top two rings are called **compression rings** and are designed to seal and create a close fit between the piston and the cylinder wall (**FIGURE 20-28**). The other rings are called oil control rings and are specially designed to hold a small amount of lubrication oil to lubricate the piston and cylinder walls. They also scrape excessive oil from the walls and release

FIGURE 20-29 Piston ring grooves and ring lands.

FIGURE 20-30 Piston rings have an end gap so they can be installed on the piston.

it back to the crankcase as needed. The piston rings are typically splash lubricated by the oil released onto the cylinder wall from the oil leaking out of the connecting rod bearing. Rings can be square, tapered, flat, or slanted on the ring surface.

Piston rings ride in ring grooves in the piston. The most common arrangement has two thin **compression ring grooves** and one larger **oil ring groove**. The sides of the grooves, called **ring lands**, support the piston rings (**FIGURE 20-29**). The **piston ring grooves** are located between the top of the piston and the piston pin. The width and depth of the piston ring groove dictate the width and depth of the piston rings that should be used.

Compression Rings

Compression rings must seal against compression loss during the compression and power strokes. Compression loss results in low compression pressures and excessive amounts of blowby gases into the crankcase, leading to reduced combustion efficiency, less engine power, more emissions, and poor fuel economy. The compression ring is typically made of cast iron. It is subject to wear due to its movement against the cylinder wall. To enhance the longevity of the compression rings, the rings may have a coating made of chromium, nitride, plasma, molybdenum, or a ceramic material.

Piston rings are not a complete circle; they are split and must have a specified piston ring end gap. This **ring gap** allows the piston rings to be expanded so they can fit over the ring lands when the rings are being installed on the piston (**FIGURE 20-30**). The gap also allows the piston rings to expand when they heat up as the engine is operating. The small gap should almost close when the engine is up to full operating temperature, but should not close entirely. Insufficient end gap results in the compression ring binding, breaking, twisting, or scoring the cylinder walls when the compression ring expands as the engine warms up. Conversely, if the gap is too large, then compression pressure will be lost through the gap, which leads to excessive blowby and lost efficiency. Thus, technicians measure the piston ring end gap to make sure the gap is within the specified tolerances (**FIGURE 20-31**).

The piston ring end gaps should be staggered around the piston when the piston rings are installed. Staggering prevents a straight path for the compression or gases to escape. If the path is too straight and smooth, the speed of gas escape is far too rapid, increasing the blowby gases.

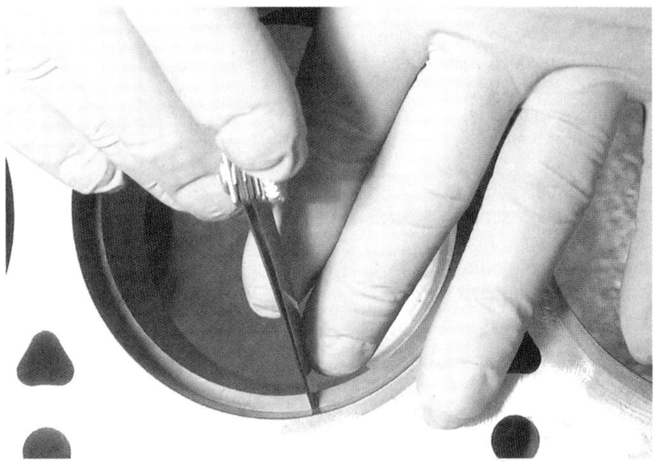

FIGURE 20-31 Technician measuring piston ring end gap.

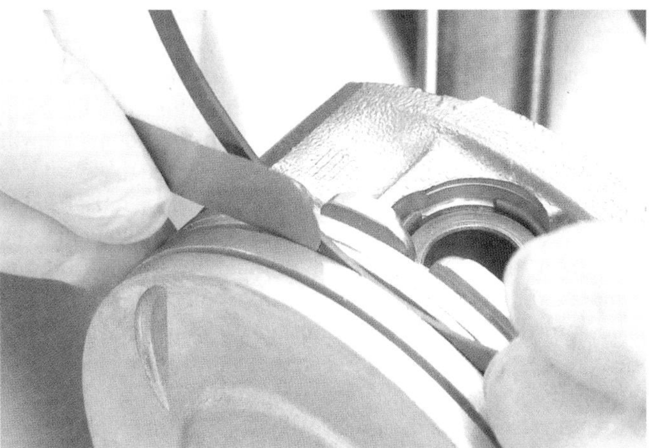

FIGURE 20-32 Piston ring side clearance.

There must also be side clearance between the compression rings and the ring lands (**FIGURE 20-32**). The clearance prevents

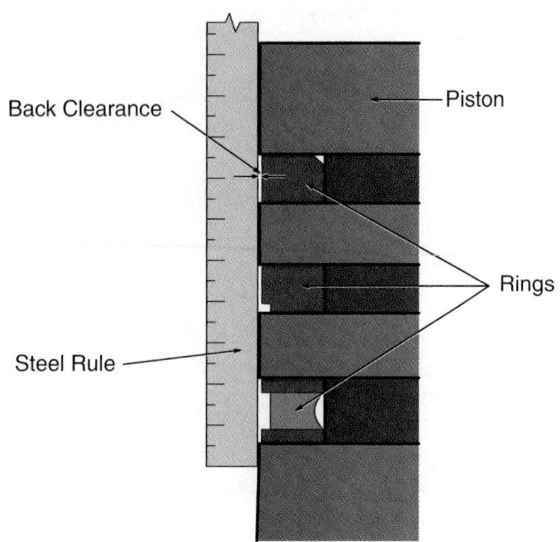

FIGURE 20-33 Piston ring back clearance.

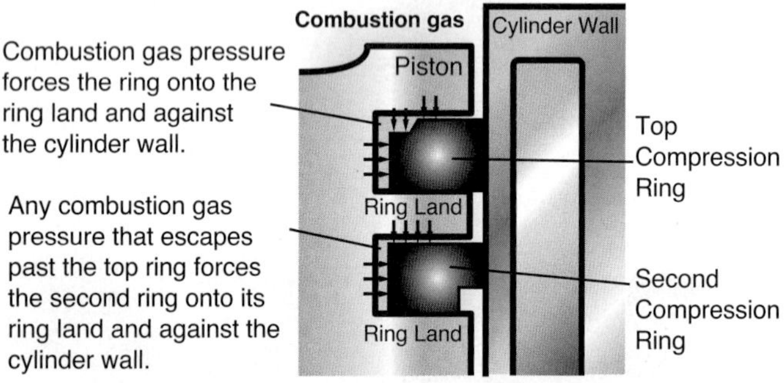

FIGURE 20-34 Combustion pressure forcing the top compression ring against the cylinder.

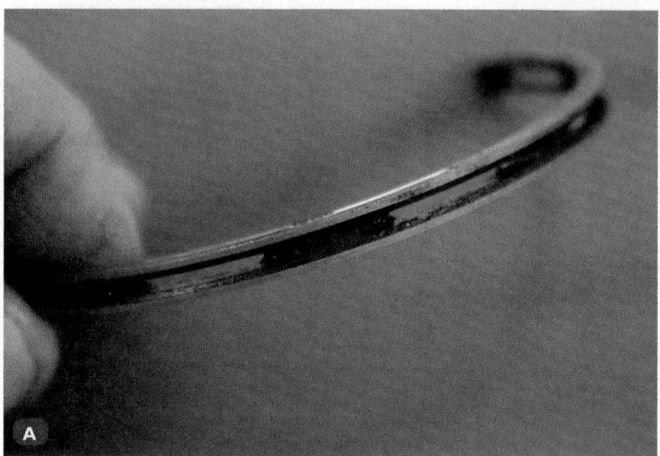

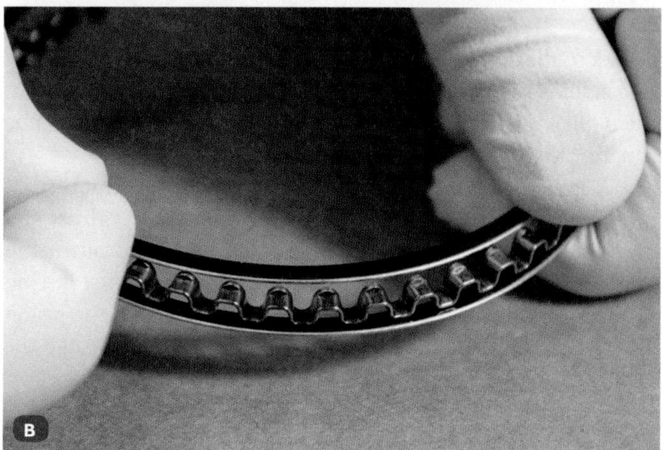

FIGURE 20-35 Oil rings. **A.** One-piece design. **B.** Three-piece design.

the piston ring from becoming stuck in the groove when it expands due to combustion heat. The ring must also have **back clearance** inside the piston groove to accommodate heat expansion of the piston (**FIGURE 20-33**).

The compression rings are built to have static pressure, with their diameter larger than the piston, thus creating **ring tension** so that the piston ring is forced against the cylinder wall to help provide a good seal. It must not have too much tension or it will wear out prematurely, but it needs to have enough tension to maintain an adequate seal. The compression rings must be able to seal under all conditions found in the combustion chamber—vacuum during the intake stroke and high pressure during the compression and power strokes. Some compression rings and pistons are designed so that combustion pressure can get behind the compression ring and force it out against the cylinder wall during the power stroke, increasing its sealing effectiveness (**FIGURE 20-34**). This design also allows the compression rings to have a smaller amount of static pressure, which reduces piston ring drag and improves fuel efficiency. It also means that most compression rings have a top and bottom and will not function correctly if installed improperly.

Oil Rings

Oil control rings prevent excessive oil from working up into the combustion chambers. Oil rings have two rails separated by a groove around the outer surface of the ring, giving them a U-shaped cross section that holds a small amount of oil and directs excess oil back to the crankcase. The oil control ring is designed to leave a film of lubricating oil less than a thousandth of an inch thick on the cylinder wall. It can be a one-piece ring that depends on its own tension to hold it against the cylinder walls, or it can be a three-piece ring with two side rails and an **expander**, which provide the tension for the side rails (**FIGURE 20-35**). Both styles have slots or holes in the oil ring and the piston ring groove. These holes allow oil return to the crankcase, and in some pistons they also direct oil to the wrist pin. This movement of oil helps with the cooling process. On modern engines, most oil rings are made of steel. The expander is made of thin steel with a series of **crimps**, or folds, in the metal that give it an outward spring force. The expander has a top and a bottom groove that hold the solid **side rails** and together make up the oil control ring. The side rails depend on the expander to hold them against the cylinder walls.

Connecting Rods

K20008

The **connecting rod** connects the piston to the crankshaft. It has a small end that houses the piston pin and connects it to the piston assembly. The middle of the rod, called the beam area, transmits the power flow to the bottom of the rod. At the bottom, or big end, the connecting rod houses the bearing inserts and attaches firmly around the crankshaft (**FIGURE 20-36**).

In order for the connecting rod to fit over the crankshaft rod journal, the big end is split in half; the bottom half called the **rod cap** is removable. The connecting rod's big end is connected

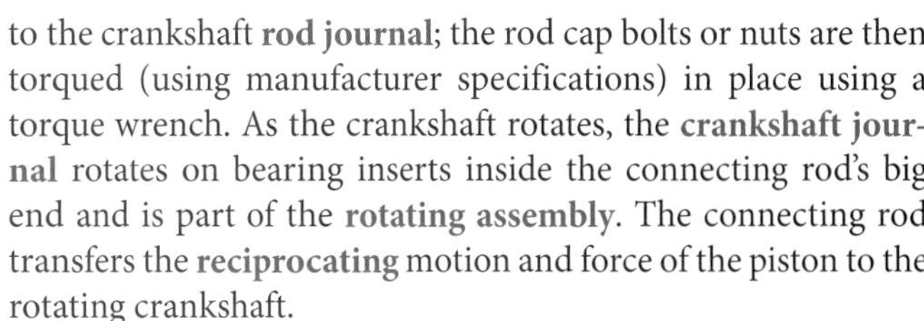

to the crankshaft **rod journal**; the rod cap bolts or nuts are then torqued (using manufacturer specifications) in place using a torque wrench. As the crankshaft rotates, the **crankshaft journal** rotates on bearing inserts inside the connecting rod's big end and is part of the **rotating assembly**. The connecting rod transfers the **reciprocating** motion and force of the piston to the rotating crankshaft.

Most stock connecting rods are shaped like an I-beam down the length of the connecting rod, but some aftermarket performance connecting rods have an H shape. The H-shaped rod is the stronger of the two in high rpm applications. Most connecting rods have extra metal on each end (balance pads) for balancing.

There are four types of connecting rods: cast, forged, aluminum, and powdered metal (**FIGURE 20-37**). Regardless of how the connecting rod is manufactured, when the rod cap is taken off, it must go back on the same connecting rod; it is not interchangeable. It must also be reinstalled in the exact orientation. The best process to make sure the caps and connecting rods do not get mixed up is to mark them close to the parting line.

The big end of the rod is where the **rod bearings** are inserted. On most engines, there is a notch in the connecting rod and a notch in the rod cap, although some newer rods don't use notches. If specified, rod bearings have matching **tangs** on the end; these tangs are inserted in the notch of the rod and the

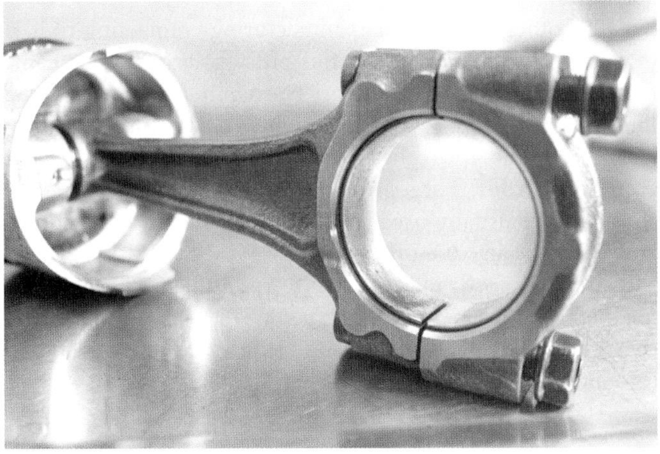

FIGURE 20-36 Connecting rod.

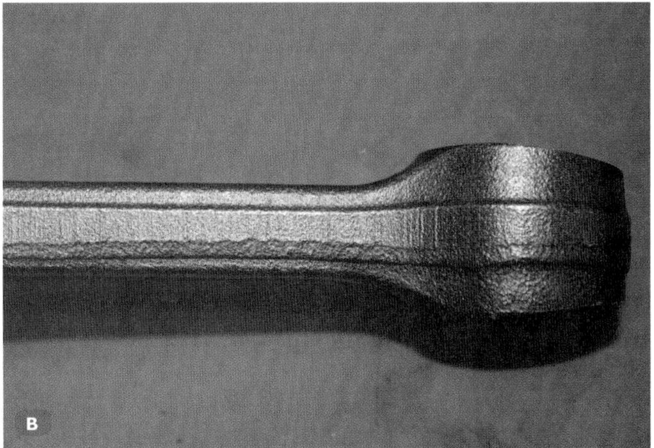

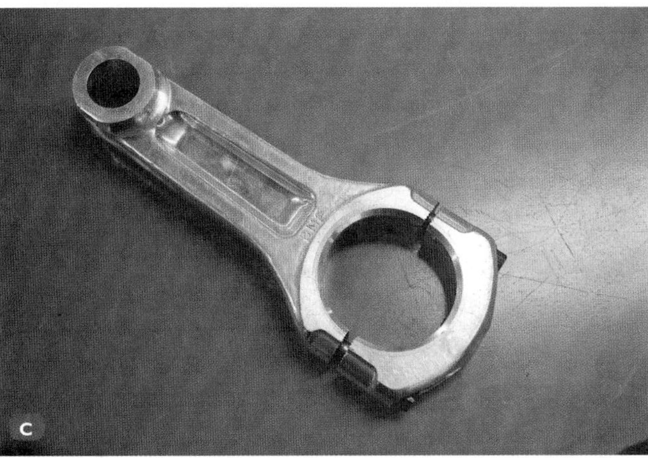

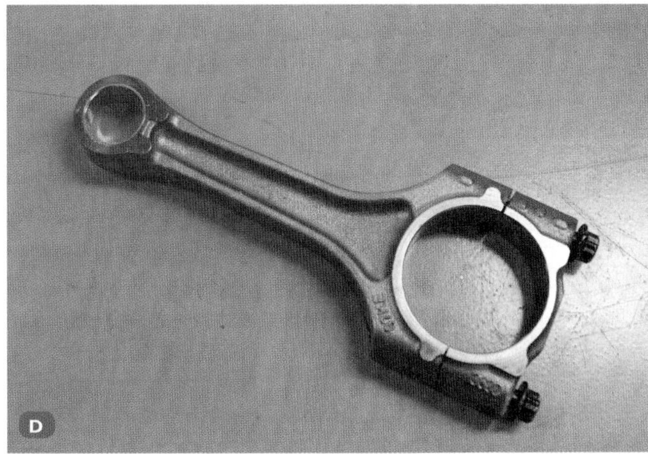

FIGURE 20-37 Connecting rods. **A.** Cast. **B.** Forged. **C.** Aluminum. **D.** Powdered metal.

rod cap (**FIGURE 20-38**). The tang in the bearing shell of the rod cap butts up against the parting surface of the rod. The tang in the rod butts up against the rod cap. When the rod cap is torqued into place, the tangs hold the rod bearing halves in place and prevent them from spinning inside the big end of the rod.

Some connecting rods and rod bearings have a hole that sprays oil out to lubricate the cylinder wall, wrist pin, and camshaft when the hole lines up with the lubricating hole in the rod journal. When installing the rod bearings, the lubricating hole in the connecting rod must line up to the hole in the bearing shell (**FIGURE 20-39**).

> ▶ TECHNICIAN TIP
>
> Although most bearings are held in place by tangs, some bearings are held in place by a hole in the bearing, which fits over a short pin in the rod or cap. Failure to line up the hole and pin results in a locked up crankshaft that won't turn.

In the manufacturing of connecting rods, most are made using a metal casting process wherein molten metal is poured into a mold. Down the center of the connecting rod mold, there is an imperfection in the shape of a line separating one half of the rod from the other. This mold line is referred to as a **casting line** and can be used to determine whether a rod was created by the metal casting process (**FIGURE 20-40**). The **cast rod** is the most common connecting rod style. Rods manufactured by this process are inexpensive, yet strong enough to meet the requirements for most small truck, passenger car, and small engine applications.

The other method for manufacturing connecting rods is through a forging process. In this process, large presses hammer the metal into the rod shape. The hammering squeezes the molecular structure of the metal tightly, creating an extremely strong, durable rod. These rods are referred to as **forged connecting rods**. Forged rods are much stronger than cast rods and are usually used in high-performance applications. However, because the process is more labor intensive, forged rods are more costly to manufacture, resulting in a higher price for the buyer. To identify this type of rod, look on the side of the **rod beam** for a wide flat line where the rod was forged (**FIGURE 20-41**).

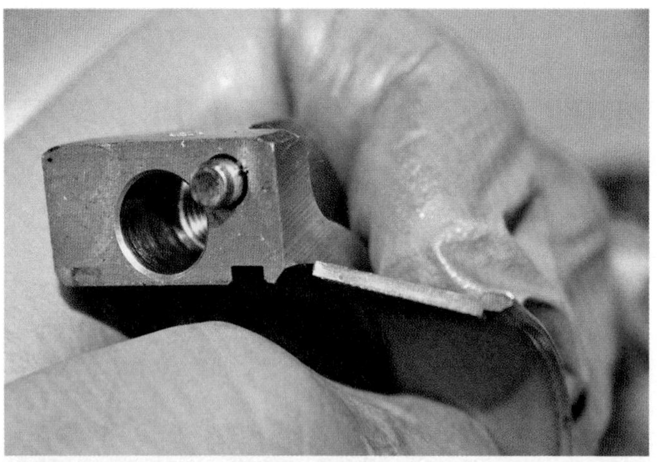

FIGURE 20-38 Many rods and bearings use notches and tangs to align the bearing and hold it in place.

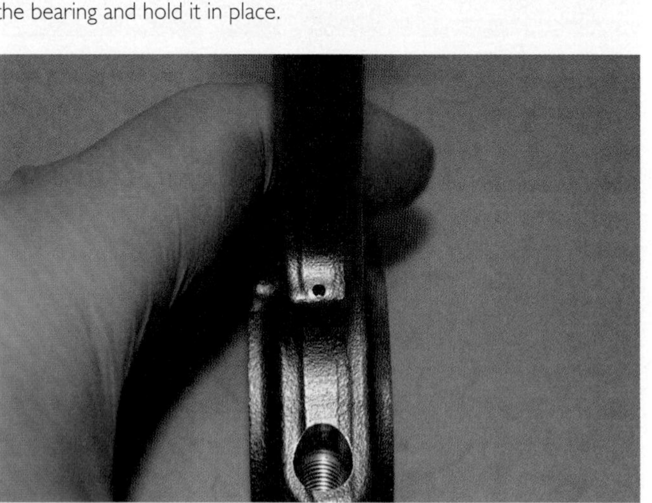

FIGURE 20-39 Small hole in the rod must line up with the hole in the rod bearing; used for lubricating the cylinder wall and wrist pin.

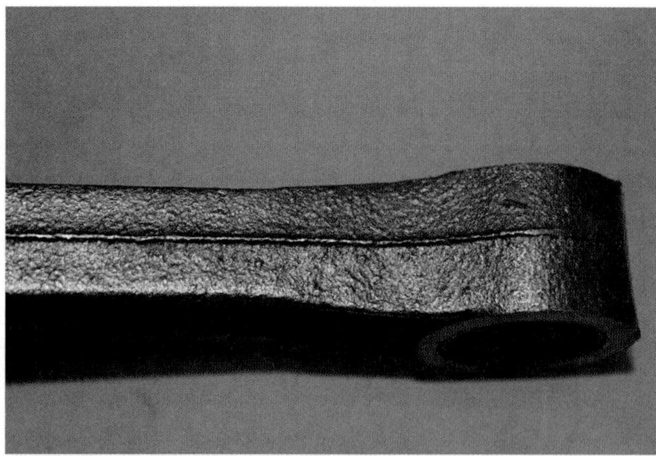

FIGURE 20-40 Parting line on a cast rod.

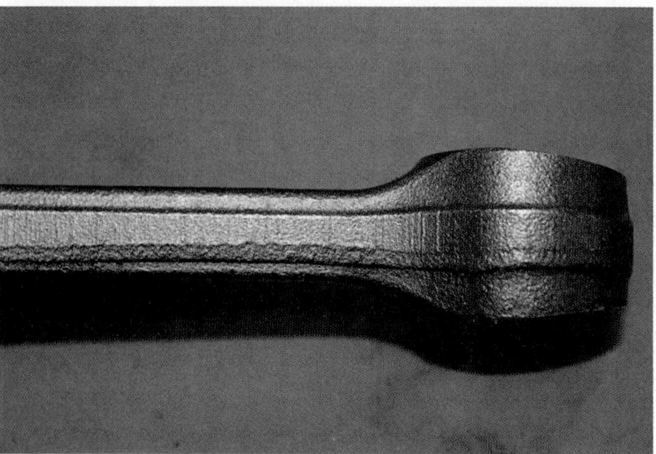

FIGURE 20-41 Forged rod showing the forging line.

The **aluminum rod** is used in some drag race cars. It is lighter and wider than cast or forged rods. The extra width strengthens the rod, enabling it to withstand the pressures and forces applied to the connecting rod and crankshaft assembly during the short, intense duration of a drag race. Frequent replacement of aluminum connecting rods is necessary because of their lack of durability. They are not appropriate for endurance race cars, aircraft, or general-purpose vehicles.

The **powdered metal rod** is the most recent technology. The rod starts with a precise amount of powdered metal, which includes iron, copper, carbon, and other agents. The powder is placed in a die and forged (pressed) under great pressure and then put through a **sintering process**, where the rod is heated up until it almost melts. Sintering is a process used in metal fabrication. It uses high temperatures, ranging from about 2000–2500°F (1093–1371°C), to melt several powdered metals into one high-quality, **metallurgically bonded** metal alloy. It is an old process that provides excellent surface finish and allows for the creation of hard-to-make parts. Additionally, the end alloy is stronger than the original individual metals. Some of the benefits are that the rod is lighter than the typical cast and forged rod and that approximately 40% less metal is used, compared to a standard forged rod. The powdered metal rod meets or exceeds the tensile and yield strength of forged rods, and the cost is considerably less.

The powdered metal rod is then fracture split at the big end, giving more adhesion at the parting line of the two halves of the connecting rod assembly (**FIGURE 20-42**) than the machined cap and rod. This is also a time saver, eliminating the machining involved on the parting line between the rod and the rod cap. With the fracture split design, the rod cap is a precision fit on the rod, and the rod bolt is not relied upon to align the cap to the rod. The powdered metal rod is used on most newer engines. Also, because the fractured surfaces would be ruined by machining, fracture rods cannot be rebuilt if they are out of specification.

▶ TECHNICIAN TIP

Do not use a punch or stamping die to mark powdered metal rods and caps as this will cause them to fail. Use paint or permanent marker instead.

FIGURE 20-42 Mating surfaces of a powdered metal fracture split rod.

Applied Math

AM-19: Proper Operation: The technician can determine the sequence of arithmetic operations needed to arrive at a solution when comparing system measurements with the manufacturer's specifications.

An engine is being overhauled at an automotive repair facility. After the engine has been disassembled, the technician uses precision measuring tools to evaluate the components. The camshaft is being checked with an outside micrometer regarding the journal diameter. The technician finds that all journals are 1.7875" (45.40 mm) in diameter.

Using service information, it is discovered that the specifications for the journal diameter should be 1.7865–1.7885" (45.37–45.42 mm). The proper operation for this task is to compare the technician's measurement of 1.7875" (45.40 mm) to the manufacturer's specification. The result shows a very good camshaft regarding the journal diameter: It is exactly between the minimum and maximum specifications. The manufacturer has allowed a measurement of 0.002" (0.050 mm) between the minimum and maximum diameters, and this part is 0.001" (0.050 mm) between each specification.

▶ Crankshaft Assembly and Related Parts

K20009

The crankshaft assembly and related parts are responsible for transferring the power created inside the cylinder and combustion chamber to the external drives of the engine (**FIGURE 20-43**). The piston and connecting rod assembly inside the cylinder are connected to the crankshaft. The movement of the piston and connecting rod assembly driven by the force of combustion turns the crankshaft. Rotation of the crankshaft drives the flywheel/flexplate, and then the drivetrain, which rotates the wheels. The crankshaft also transmits power to the front of the crankshaft assembly, which drives the valve train assembly through a series of belts, chains, and gears. The power to the front of the crankshaft assembly also drives the pulley configuration that connects to the power steering pump, air-conditioning compressor, water pump, and alternator.

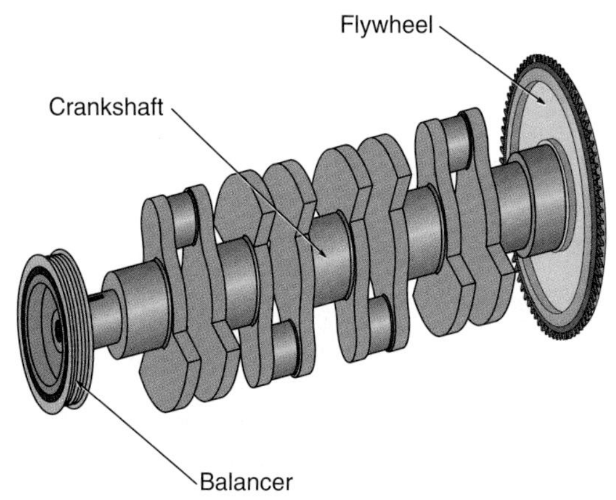

FIGURE 20-43 Power is provided out the front and back of the crankshaft.

Crankshaft

K20010

The main role of the **crankshaft** is to convert the up-and-down strokes of the pistons into rotary motion. It is situated in the **crankcase**, which is the lower part or bottom of the engine block. Connecting rods are attached to offset journals called **throws** (**FIGURE 20-44**). This is where the reciprocating motion of the piston is changed into a rotary motion, which is then used to power the vehicle.

The crankshaft now also plays a role in creating the spark for the combustion process. In newer engine models, the crankshaft houses the reluctor for the crankshaft position sensor, which is a device that lets the engine computer know the position of the crankshaft in order to help time the spark and fuel injection events (**FIGURE 20-45**). Teeth and notches may also be embedded in the crankshaft, or shields and windows used in a Hall-effect type of sensor may serve the same purpose. Many current engines also drive the oil pump off the front of the crankshaft.

Crankshafts are a one-piece casting or forging, or they can be machined out of a **billet**, which is a solid piece of metal. The raw crankshaft is then machined into the desired shape for the particular engine into which it will be installed. If the crankshaft journals are worn or damaged, they can generally be refinished to a smaller size and reused with undersized bearings—that is, bearings that are thicker.

FIGURE 20-44 Crankshaft rod journals (throws).

FIGURE 20-45 Reluctor for the crankshaft position sensor.

Applied Math

AM-20: Place Value: The technician can interpret standard or metric units when conducting precision measurements.

A technician and his helper are rebuilding an automotive engine. It has been disassembled, and the task is to measure the components with micrometers. The helper is familiar with use of a common micrometer, which can measure parts to within one-thousandth of an inch. The micrometer set that is being used is capable of measuring parts to within one-ten-thousandth of an inch.

To assist the helper, the technician mentions the term "place value." Place value is the value of the position of a digit in a number. For example, in the number 2.4137", the "3" represents thousandths of an inch. Most automotive engine specifications are listed with four numbers to the right of the decimal. This indicates the ten-thousandths position. In our example, the "7" is four numbers to the right of the decimal, which is the ten-thousandths position.

Concerning the metric system, the unit of linear measurement is the millimeter or centimeter. Understanding place value of this system is essential. A typical specification in the metric system would be 45.428 mm, which is equal to 1.7885". In metric units, engine specifications extend three numbers to the right of the decimal for linear measurement in millimeters.

Main Journals and Connecting Rod Journals

Main bearing journals are the part of the crankshaft assembly that mate to the main bearings found at the bottom of the engine block. Main bearing journals are very smooth and perfectly round so that they can rotate inside of the softer main bearings. The main bearing journals are positioned in a straight line down the center of the crankshaft and are used to position the crankshaft assembly into the lower part of the engine block. The **main journals** are ground to manufacturer specifications from the **crank core** (the rough, unfinished crankshaft assembly that has just left the foundry or forging area) or billet. The center of the main journals is referred to as the centerline of the crankshaft.

The crankshaft is held in the block by main bearing caps which mate to **journal saddles**, located at the bottom of the engine block (**FIGURE 20-46**). The main bearing cap and the

FIGURE 20-46 Main bearing cap and saddle.

FIGURE 20-47 Main journals rotate in main bearings.

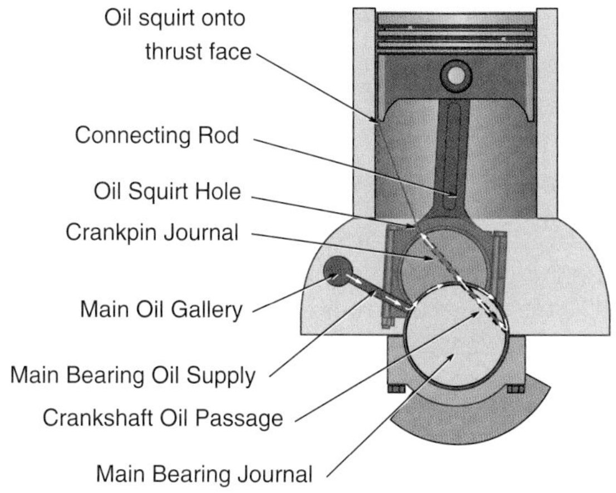

FIGURE 20-48 Oil passage used to lubricate the rod bearings.

FIGURE 20-49 Splayed crankshaft from a 90-degree V6 engine.

FIGURE 20-50 Crankshaft counterweights.

journal saddle create a perfectly round bore that clamps bearing inserts in place to support the crankshaft journals. The main bearing inserts are secured by main bearing caps and bolts.

As discussed earlier, the number of journal saddles in an engine is generally dependent on the length of the engine block. A common V8 has five main journals on the crankshaft; many straight in-line six-cylinder engines have seven main journals; and a V6 typically has only four main journals.

The crankshaft main journals rotate in bearing inserts called **main bearings** (**FIGURE 20-47**). Cross-drilled passages in the crankshaft carry lubricant under pressure from the main bearings to the adjacent crank throws (rod throws) (**FIGURE 20-48**). The crank throws (also called rod journals) are offset from the centerline of the crankshaft; the amount of offset determines the **piston stroke**. The crankshaft design in some 90-degree V6 engines differs significantly from that of 60-degree V6 engines. In order for a 90-degree V6 to have evenly spaced power pulses, the crankshaft must have connecting rod journals that are splayed (**FIGURE 20-49**). A splayed journal is designed with two connecting rod journals, each one slightly offset from the other. The crank pin, or journal, appears split, but both journals are actually located between the crankshaft throws or counterweights. This arrangement creates a 90-degree V6 engine with evenly spaced power pulses and is called an even fire V6. It is smoother running than an odd fire V6, which has uneven power pulses because it does not use splayed connecting rod journals.

Counterweights are formed on the crankshaft to balance the components of the rotating assembly that are mounted inside and attached to the engine block, including the rod journals, rod bearings, connecting rods, pistons, piston pin, and rings (**FIGURE 20-50**). The balancing of the rotating assembly is done by drilling out weight or by adding weight to the counterweights. A heavy metal called **Mallory metal** is added to the counterweights during manufacturing to balance the counterweights. Mallory is approximately 117% heavier (more than twice as heavy) than steel.

Crankshafts are classified as either internally balanced or externally balanced (**FIGURE 20-51**). Internally balanced crankshafts are balanced solely from the counterweights of the crankshaft. No external counterweights are needed. Externally balanced crankshafts are not heavy enough, so counterweights need to be positioned on the flywheel or flexplate at the rear of the crankshaft and the harmonic balancer at the

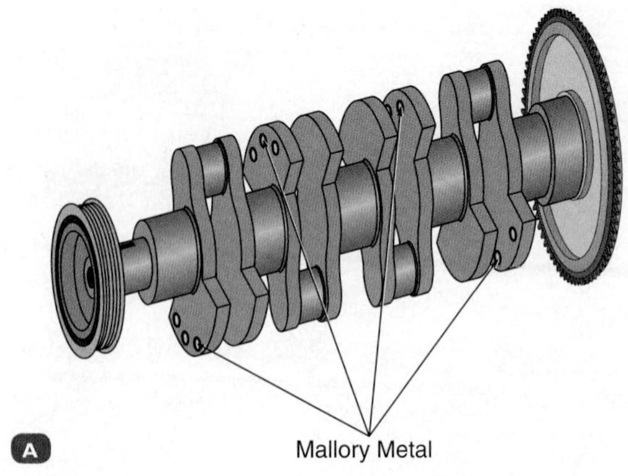

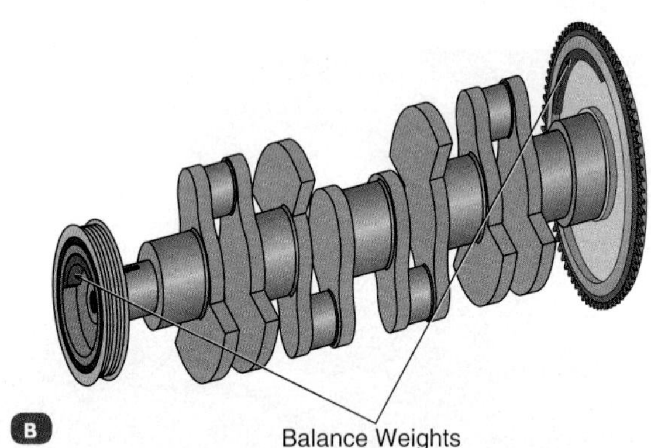

FIGURE 20-51 **A.** Internally balanced crankshaft. **B.** Externally balanced crankshaft.

front. On an externally balanced engine, it is critical to use the proper externally balanced flywheel or flexplate and harmonic balancer for the engine, and they need to be installed in the proper orientation. Check the service information carefully.

Most crankshafts use **induction-hardened** journal surfaces to give good wear qualities. Induction hardening is a process for hardening the outer surface of a metal structure by rapid heating and cooling. Using the induction hardening process allows the crankshaft to be reground on the main and rod-bearing surfaces without grinding through the hardened layer. Regrinding is used to resize, smooth, and polish the main and rod journals so they can be reused with undersize bearings when the engine is being rebuilt. It is used as an alternative to replacing the entire crankshaft assembly.

Some heavy-duty and performance crankshafts use a **nitridization process**, which takes longer and is more costly but results in an extremely hard surface. **Nitriding** is a surface-hardening heat treatment that introduces nitrogen into the surface of steel at a temperature range of approximately 930–1020°F (499–549°C). Nitriding creates a surface that is harder than that produced by the induction hardening process, but the hardened layer is not as thick. The crankshaft will have to be re-nitrided if the main journals or connecting rod journals need to be reground due to wear.

Applied Math

AM-37: Mentally: When comparing the observed measurement with the manufacturer's specifications, the technician can mentally compute whether the observed measurement meets specifications.

A technician is assigned to overhaul a V8 engine. He is currently measuring the crankshaft, which has been removed from the block and cleaned. The manufacturer's service information lists the minimum diameter for the main bearing journals as 2.6576", maximum taper of 0.0002", and maximum out-of-round of 0.0003". While measuring each journal in four places with an outside micrometer, the technician compares each reading to the minimum diameter specification and mentally computes whether it meets it. Also, the technician compares the two readings used to measure the amount of taper and mentally computes the difference between them to come up with the amount of taper, which again is compared to specifications. The same process is performed with the out-of-round measurement. The two measurements are taken, the difference is calculated mentally, and that difference is compared to specifications. These measurements are then performed on each of the other crankshaft journals.

Crankshaft Bearings

K20011

Bearings are used to help support the crankshaft in the block as well as the connecting rods on the connecting rod journals. Crankshaft bearings are of the friction bearing type as compared to the antifriction type. This means that the surfaces are in sliding contact with each other rather than rolling contact, such as in a wheel bearing. Connecting rod bearings and the crankshaft main bearings are of the split-sleeve type, which means they are in two halves, called **bearing inserts**, or half-shell bearings (**FIGURE 20-52**).

Bearing inserts are made from Babbitt metal bonded to a steel shell (**FIGURE 20-53**). The shell gives rigidity to the bearing, and the Babbitt metal is relatively soft and provides the bearing wear surface. Bearings, along with lubricant, are used in engines to support and protect rotating parts and allow them to turn freely. A very thin layer of oil separates the bearing insert from the surface of the crankshaft journals. The oil acts as a shock absorber and is designed to keep the journal from making contact with the surface of the bearing.

In some bearing shells, a built-in tang fits into a matching machined notch in the bearing saddle and cap (**FIGURE 20-54**). The tang aligns the bearing in the saddle and cap by butting up against the machined faces of the saddle and cap. Some newer engines don't use tangs and notches, as they are held in place only by bearing crush, explained later. In a main bearing, the saddle has a drilled passage for supplying oil to the bearing and journal.

It is important to pick the proper bearing half with the matching oil hole, as it must align to the oil passage in the main saddle. The lower half is carried in the main bearing cap. The main cap is held in place on the main saddle by main bearing cap bolts. Refer to the manufacturer's specifications for torque procedures.

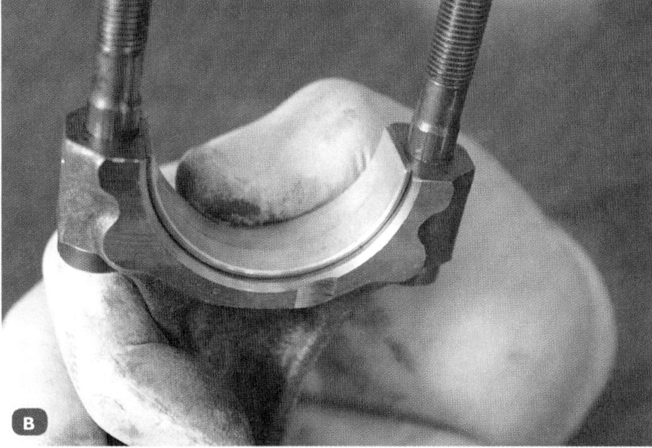

FIGURE 20-52 Bearing inserts. **A.** Main bearings. **B.** Rod bearings.

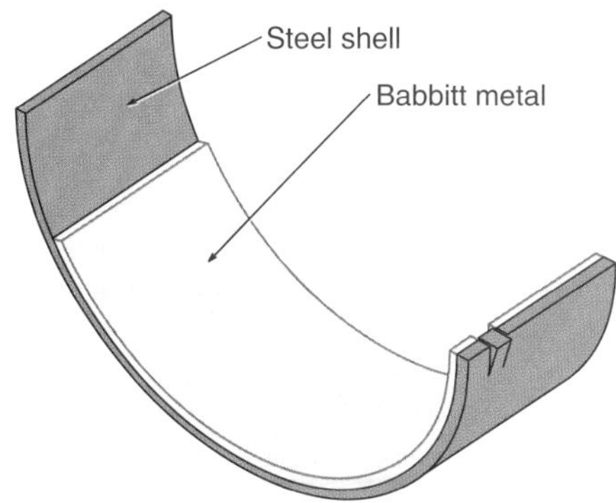

FIGURE 20-53 Bearing construction.

In a connecting rod bearing, the bearing's upper half is carried in the big end of the connecting rod, which may have a notch for the bearing tang. The lower half is carried in the connecting rod cap, which may also contain a notch for the bearing tang. The cap is bolted to the connecting rod and torqued to specifications. Refer to the manufacturer's specifications for torque procedures.

The crankshaft requires a thrust bearing, which limits the end play movement of the crankshaft. These bearings can be in the form of bearing flanges that are part of one of the main bearings, giving each main bearing insert a U shape. This U-shaped bearing insert fits over a precisely machined main bearing saddle and cap (**FIGURE 20-55**). The thrust bearing will only fit on the specified bearing support. Alternatively, a separate two-piece **thrust bearing** can be fitted into a machined recess in each side of the bearing support and cap, and acts as the thrust bearing for the crankshaft (**FIGURE 20-56**). In the same way that the

FIGURE 20-54 Bearing shell tang and mating notch in the bearing cap.

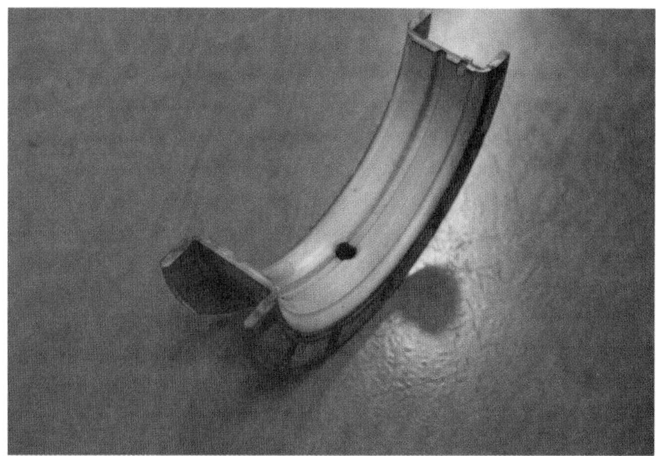

FIGURE 20-55 Thrust bearing.

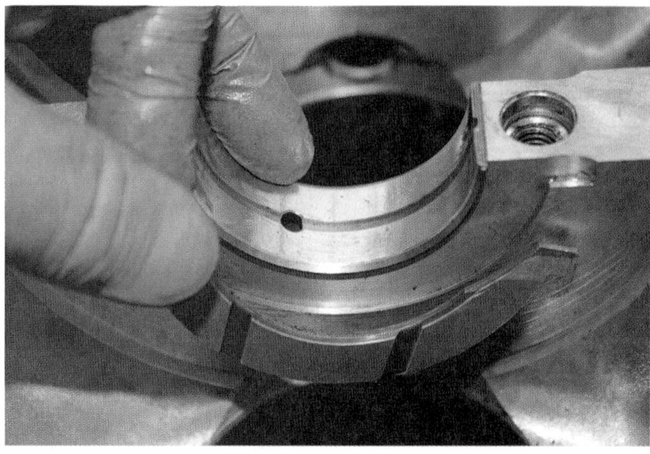

FIGURE 20-56 Two-piece thrust bearing and positioning.

bearing-to-journal clearance is important, so is the clearance between the bearing flange (or thrust bearing) and the mating surface on the crankshaft. The thrust bearing clearance prevents excessive end play (forward or backward movement) of the crankshaft, especially in a vehicle equipped with a manual transmission/transaxle and clutch, because the crankshaft thrust surfaces bear all of the force of compressing the pressure plate springs when the operator pushes down on the clutch pedal.

Oil flows through an oil gallery in the cylinder block. Each main bearing receives its own oil under pressure from the oil hole in the main bearing saddle, which connects directly to the oil gallery. Some top bearing inserts have a groove around the middle. The groove provides a space for oil to flow around the journal, which helps provide more lubrication for the bearing. The grooved bearing also helps extend the time that the rod bearing receives pressurized oil, as the oil hole in the crankshaft lines up to the bearing groove for half of the rotation of the crankshaft. Passageways drilled in the crankshaft carry oil from the main bearing journals to the rod journals (**FIGURE 20-57**).

Oil flow maintains an oil cushion between the crankshaft and the bearing as the engine is operating. But when the engine is first started up, especially when it is cold, it takes a few seconds to pump oil to the bearings, which can result in the crankshaft touching the bearing surface for a short time. The oil flow also carries away heat and particles that could cause wear. If the cushion of oil did not exist, the bearing would overheat and fail. Engine manufacturers specify the clearance required between the bearing material and the crankshaft. This clearance specification helps maintain the balance between adequate oil pressure and adequate oil flow.

All main bearings and rod bearings are held in place by what is called **bearing crush** (**FIGURE 20-58**). This term refers to the fact that the bearing inserts are slightly larger than the bore they fit into and are therefore slightly crushed into position when tightened into place. The proper bearing crush is designed into the bearing when they are made by allowing the bearing insert to slightly protrude from the parting line of the bearing support and cap. When the cap is tightened properly, the bearing inserts are forced very tightly into the bore, which helps hold them in place.

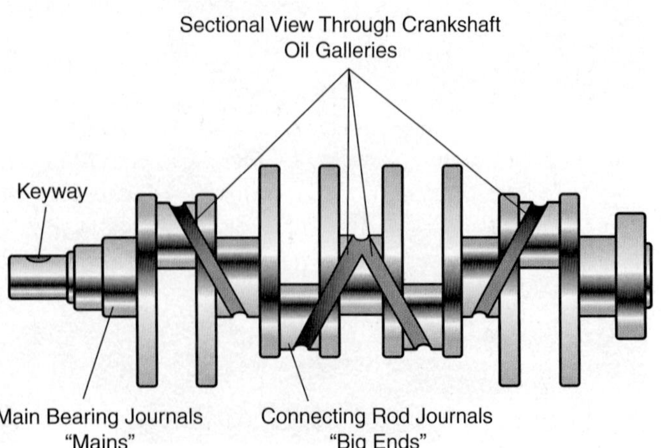

FIGURE 20-57 Oil passageway drilled in the crankshaft.

> **TECHNICIAN TIP**
>
> In some high-performance and race applications, rather than using an oil groove in both halves of the bearings, the crankshaft is cross-drilled in the main journal of the crankshaft. This allows oil to be sent through the crankshaft to the rod journal during the full revolution of the crankshaft while only using a groove in one of the two bearing inserts. This technique allows more bearing surface on the load side of the main bearing. By allowing more bearing surface on the load side of the bearing insert, the bearing can support a greater load.

Main bearings and rod bearings come in various sizes. The most common bearing size is a standard size (std), which means that it fits the same size journal as originally manufactured (**FIGURE 20-59**). Undersized bearings can be purchased for many engines in the following sizes: 0.001" (0.025 mm), 0.010" (0.25 mm), 0.020" (0.51 mm), and 0.030" (0.76 mm). If you are replacing the crankshaft, it is important to measure the crankshaft journals and check to make sure they are within the specifications set out by the crankshaft supplier. The clearance between the crankshaft journal and the bearings is crucial and must be within the manufacturer's specifications. If there is too little clearance, the oil will not be able to flow through the

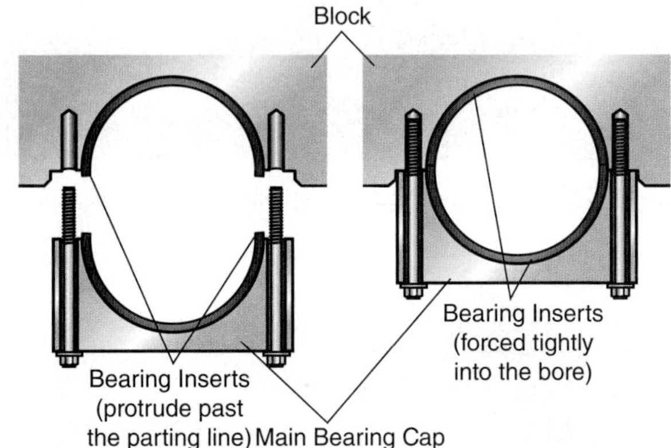

FIGURE 20-58 Bearing crush (exagerated for clarity).

FIGURE 20-59 Main and rod bearing size is stamped on the back of the bearing.

clearance in the bearing, which can cause the bearing material to overheat and possibly to seize the journal. If the clearance is too large, the oil pressure will be lowered, and the oil will be squished out of the clearance in the bearing, which allows the journal to come into direct contact with the bearing, creating a knocking noise and accelerated wear.

> ▶ TECHNICIAN TIP
>
> You might be wondering why bearings would be called undersized when in reality they are thicker than a standard-sized bearing. The reason is that the crankshaft journal is typically machined to give it a new wear surface; as a result, the journal becomes undersized. The matching bearings are then referred to as undersized bearings.

Bearing Materials

Bearings require a difficult mix of properties; they must be hard enough to resist wear, but soft enough not to damage the crankshaft. One bearing material, called Babbitt metal, is an alloy that can include metals such as tin, lead, aluminum, and copper. This mix of metals creates a unique combination of hardness and softness.

The two most popular **engine bearings** in use are the **trimetal** and the **bimetal** types. Both bearings have their distinct advantages. Most trimetal bearings are made of a lead, tin, and copper overlay that rides next to the crankshaft, followed by a nickel barrier, and next a high-fatigue strength copper and lead lining, which is bonded to a steel backing (**FIGURE 20-60**). The biggest advantage of the trimetal bearing is that it has a higher load capacity; thus, it is ideal for high-performance and racing engines. Most bimetal bearings are made of an aluminum, tin, and silicon alloy that is bonded on a steel backing. There are also some aluminum bearings that are made of an aluminum alloy that contain tin and copper, which is bonded to a steel backing. Some of the advantages of the aluminum bearings are that they are less expensive and contain no lead (which is an environmental concern).

Some vehicle manufacturers and some high-performance engine builders are now using coated bearings to reduce friction, handle dry starts, and withstand higher peak loads, while still increasing bearing life. Coatings such as polyamide or PTFE (Teflon) are applied to the wear surface of the bearing. The coatings are applied in a very thin layer, generally around 0.00025" to 0.00030" (6 to 7.5 mic). But even at this minimal thickness, it is a benefit. This technology is also useful in vehicles with start–stop capability, as the coatings provide better protection over the much higher number of start cycles.

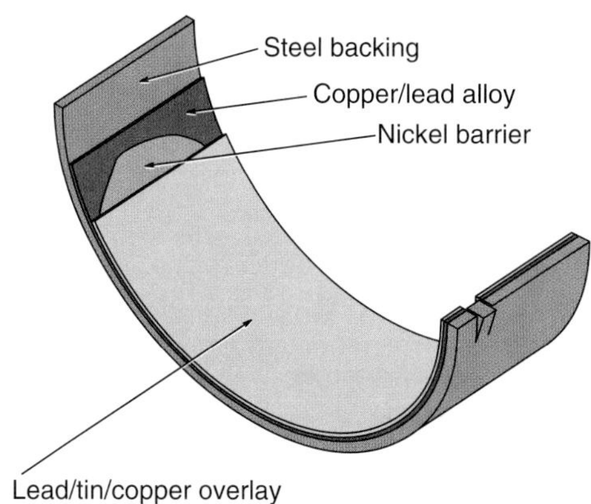

FIGURE 20-60 Trimetal crankshaft bearing.

Bearing Wear

Bearing wear can be caused by many sources, but the four most common causes seen by the automotive technician are dirt, insufficient oil, improper assembly, and operational faults. Dirt can enter the engine through the oil system or air intake system or from engine rebuilding. Particles of dirt can become sandwiched between the bearing and the bore, appearing as a distortion on the bearing. Conditions of insufficient oil arise from low engine oil levels, clogged oil passageways, or a bad oil pump. Without proper lubrication, the parts will grind against each other, creating extreme wear in the bearing and journal. Improper assembly can arise from reversed caps on the journals, insufficient clearance allowance, misalignments, or bearing inserts placed in the wrong position. The wear patterns on the bearing differ depending on the exact assembly error. Operational faults that lead to bearing wear include engine lugging at low rpm, over-revving the engine, etc. As with improper assembly, the exact pattern of wear varies by issue (**FIGURE 20-61**).

Flywheel, Flexplate, and Harmonic Balancer

K20012

The **flywheel** is a heavy metal disc that bolts to the rear of the crankshaft of all engines equipped with a manual transmission/transaxle. The flywheel is very heavy, and its momentum helps smooth out engine operation. It maintains crankshaft momentum during the non-power strokes of the piston. Essentially, it stores up rotational energy, which ensures less fluctuation in crankshaft speed between the engine's power strokes. The flywheel is bolted to the rear of the crankshaft (**FIGURE 20-62**). It links the crankshaft to the manual transmission/transaxle, through the clutch assembly. The size of the flywheel varies based on the manufacturer's priorities. Heavier flywheels do a better job of smoothing out the power pulses but do not allow the engine to accelerate as quickly. Lighter flywheels allow much quicker engine acceleration, but they do not smooth out the power pulses as well.

The flywheel rim carries the **ring gear**. The teeth on the ring gear mesh with the **started teeth** and are used to crank over the engine. The ring gear is typically attached to the flywheel by an interference fit, so no bolts or welds are used to hold it on. The ring gear is installed by heating it up until it has expanded enough to fit over the edge of the flywheel. Once it cools, it shrinks enough to firmly lock it in place. The face of the flywheel is machined smooth and flat and acts as one of the friction surfaces for the clutch.

Flexplate

In automatic transmission engines, a **flexplate** is used instead of a flywheel (**FIGURE 20-63**). Unlike the flywheel, a torque converter is bolted to the flexplate. Like the flywheel, the flexplate

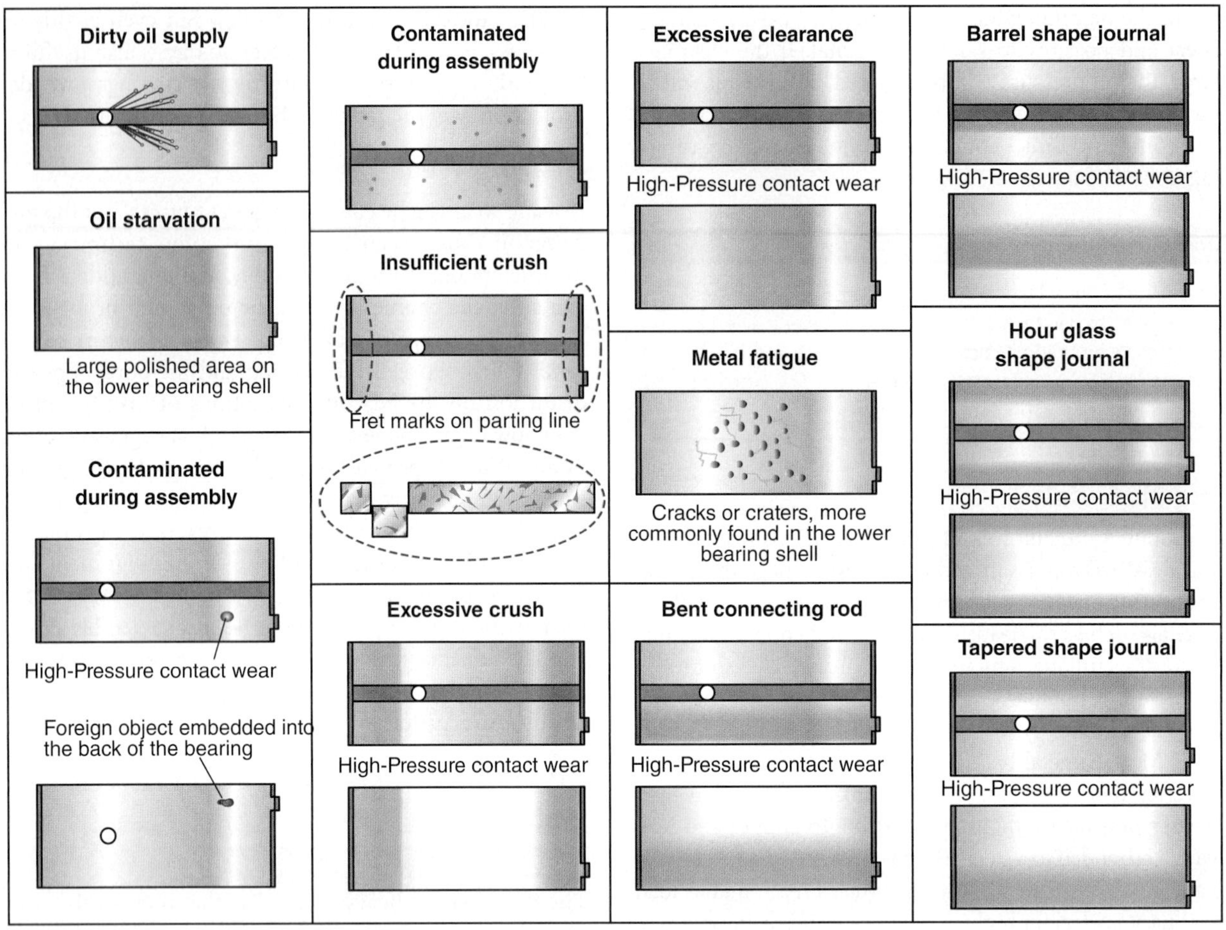

FIGURE 20-61 Bearing wear patterns.

FIGURE 20-62 Flywheel, with ring gear, installed on the crankshaft.

FIGURE 20-63 Flexplate with ring gear.

is constructed as a round disc and mounted to the end of the crankshaft. The flexplate is lighter than a conventional flywheel because the weight of the torque converter performs the function of the flywheel. Because the torque converter is quite heavy, the flexplate can be constructed to be much lighter and thinner than a flywheel. The torque converter is bolted to the flexplate. The flexplate often has a ring gear welded to its edge because of the flexplate's light weight. If an interference fit were used, the flexplate would warp under the pressure created by the fit. Some manufacturers weld the ring gear to the outside of the torque converter and use a smaller diameter flexplate.

Harmonic Balancer

The **harmonic balancer** is also known as the crankshaft damper. It is located on the front of the crankshaft assembly, and its function is to reduce vibration from the crankshaft and piston assembly. When the crankshaft deflects or twists slightly (vibrates) from the power pulses, the harmonic balancer absorbs the vibrations and reduces their severity. Minimizing vibration is critical in preventing damage to the crankshaft. Excessive torsional vibration can cause a crankshaft to break in two.

A harmonic balancer is composed of two elements: a mass and an energy-dissipating system, mounted to a central hub. The mass is a heavy circular steel disc that absorbs and releases the energy from the twisting of the crankshaft. A rubber or **fluid dissipater** is fitted between the circular mass and the hub (**FIGURE 20-64**). The dissipater allows the mass to rotate at a more constant speed than the crankshaft. This helps to even out the torsional vibrations of the crankshaft. Over time, the rubber material can get brittle and hard, letting the mass of the harmonic balancer slip. When the mass slips on the rubber, or if the rubber is brittle and cracked, the harmonic balancer cannot provide the needed dampening effect and needs to be replaced with a new or remanufactured one.

The harmonic balancer is attached to the crankshaft using a **Woodruff key** or a square key and **center bolt**. The Woodruff key is a metal half-moon–shaped piece that protrudes from a slot near the end of the crankshaft (**FIGURE 20-65**). It fits into a slot in the harmonic balancer center hub and prevents the harmonic balancer from spinning on the end of the crankshaft. Some manufacturers use a square-cut key instead of the Woodruff key. The center bolt is threaded so that it will clamp the front side of the harmonic balancer to the front of the crankshaft.

The **accessory drive pulleys** are bolted to the front hub of the harmonic balancer. In modern vehicles, there typically is one pulley and one serpentine drive belt attached to the balancer to operate the various accessories. The pulley and belt transfer the power output needed for various systems in the vehicle. Older vehicles generally contained multiple pulleys and belts to operate systems that are now powered electronically or through other means.

FIGURE 20-64 Harmonic balancer.

> **TECHNICIAN TIP**
>
> The harmonic balancer may have counterweights for an externally balanced engine. Be sure to replace it with the correct harmonic balancer, or the engine will be out of balance and have a bad vibration.

Crankshaft Seals

K20013

Crankshaft seals are located at each end of the crankshaft assembly where it projects through the crankcase. The seals help contain the lubricating oils inside the lower part of the crankcase and prevent them from leaking out. A rear main seal is located just after the rear main bearing (**FIGURE 20-66**). In engines that use a **timing chain** or **timing gear**, the front main seal is located in the front timing cover. In engines equipped with a timing belt, the front main seal is located in a housing bolted to the front of the engine block (**FIGURE 20-67**) and sits between the block and the timing belt. Most crankshaft seals today are a circular one-piece style with a sealing lip. Some seals use a small spring, called a **garter spring**, around the inside of the sealing lip to tension the lip against the sealing surface of the crankshaft or harmonic balancer.

FIGURE 20-65 A Woodruff key protrudes from a half-moon–shaped slot in the crankshaft and mates with a full length slot in the harmonic balancer.

FIGURE 20-66 Rear main seal.

FIGURE 20-67 Front main seal.

FIGURE 20-68 Balance Shaft.

Balance Shaft

K20014

Some in-line four-cylinder and 90-degree V6 engines use one or two **balance shafts** to reduce vibrations from engine operation. Their asymmetrical design, or lack of symmetry in cylinder configuration, causes these engines to have an inherent **second-order vibration**. The second-order vibration is at twice the engine rpm. It cannot be eliminated no matter how well the internal parts are balanced. This vibration is produced because of the even firing of the four-cylinder and the uneven firing of some 90-degree V6 engines. The descending and ascending pistons are not always completely opposed in their acceleration.

To address this problem, engine manufacturers have developed a balance shaft with counterweights spaced so as to cancel out the inherent vibrations. The balance shafts are mounted in the engine block, either below the crankshaft, next to the crankshaft, or above the crankshaft, depending on design. The shafts are typically driven at twice the engine rpm, usually by a chain, gear, or belt from the crankshaft. The weights on the shaft are timed to cancel the second-order vibrations in the engine (**FIGURE 20-68**).

FIGURE 20-69 Disassembling an engine block.

▶ Disassembly of Engine Block Components

N20001

Disassembly of the engine block components is usually performed only when major repairs or rebuilding is needed. In many cases, a new or rebuilt engine is used, rather than rebuilding the existing engine, so that the vehicle can be returned to the customer more quickly. Usually, if the decision is made to repair or rebuild the engine, the engine is removed from the vehicle and disassembled once it is out, although some tasks, such as replacing head gaskets, are usually performed with the engine still installed. Disassembling the engine is a very large task, so make sure it is the proper course of action before proceeding (**FIGURE 20-69**).

If the engine block assembly is being rebuilt, then the cylinder head(s) and, if it is an overhead cam engine, the harmonic balancer, and timing belt or chain(s) have already been removed following the steps given in the Cylinder Head Components chapter. The exact process for disassembling an engine block varies from engine to engine because each manufacturer's engines are assembled differently. Therefore, only an overview of the process is presented here. Always refer to the manufacturer's service information, and follow the steps exactly.

Removing the Oil Pan

S20001

To gain access to the rotating assembly of the engine, the oil pan has to be removed. If the engine has a lot of miles on it, was poorly maintained, or had extensive damage, there will probably be a layer of oil sludge and possibly metal particles or other foreign objects in the bottom of the oil pan. In order to keep this evidence intact, do not rotate the engine block upside down yet.

To remove the oil pan, you need to remove the bolts holding the oil pan onto the block and store them in a bin or bag. If you are removing the oil pan with the engine side up, leave two bolts in opposite corners, threaded in about halfway. Insert a putty knife or gasket scraper between the engine block and the oil pan, and start to pry the pan loose (**FIGURE 20-70**). Use care to avoid bending the pan while prying. When you separate the oil pan enough to rest on the catch bolts, support the oil pan from the bottom, and remove those remaining two bolts. Pull the oil pan away carefully, working around any obstructions such as the oil pump screen or any baffles.

Draining the Engine Block

S20002

Now that the oil pan has been removed, it is time to remove any remaining fluids from the engine block. It is not uncommon to find some leftover oil and coolant inside the engine block once the heads and oil pan have been removed. The coolant often becomes trapped inside the water jacket passages, especially if you were unable to remove one of the drains plugs earlier in the disassembly process. Leftover oil is also frequently found in trapped oil passageways. It is a good idea to always place a catch tray underneath the engine when it is rolled over to catch these fluids as they drain out of the engine (**FIGURE 20-71**).

FIGURE 20-70 Break the oil pan loose.

Disassembling the Block Underside

S20003

The block underside refers to the area located in the oil pan and timing chain part of the engine cylinder block assembly. These areas are fully covered by the timing chain cover and the oil pan cover, which are stamped sheet metal assemblies or cast aluminum assemblies.

Removing the Oil Pump and Tray

S20004

Some oil pumps are located inside the crankcase and have to be removed before the pistons; others are located on the front of the block and must be removed along with any timing components. On a crankcase-mounted oil pump, you may be able to unscrew one bolt and simply pull off the oil pump. Other oil pumps have two bolts and may also have a few other bolts that hold the pickup tube and screen in place (**FIGURE 20-72**). As the oil pump is removed, you may see the plastic collar that holds the oil pump drive rod onto the oil pump. If the collar is broken, you will need to replace it.

If the engine has a windage tray (to prevent oil from splashing onto the crankshaft), remove it now, along with any dipstick tube or extension that might be inside the oil pan area. With the oil pan removed, it is easier to check for evidence of engine problems. If there were knocking noises heard coming from deep inside the engine, then it may be visible or apparent that one of the rod bearings has disintegrated and/or spun.

Labeling the Connecting Rods and Caps

S20005

The connecting rod caps must be replaced in the same orientation as they are removed. If they are not already labeled, label them in two places: above and below the parting line of the cap and rod body. It is a good idea to also label the main caps of the block in the same way. One way to label these components is with a number punch set, but you could use a regular center punch. *Note: Do not* stamp or center punch powdered

FIGURE 20-71 Catch any remaining fluids in a catch tray.

FIGURE 20-72 Unbolt and remove the oil pump and pickup tube.

FIGURE 20-73 Number both halves of the connecting rods.

FIGURE 20-74 Remove the ridge with a ridge reamer if the ridge is significant.

metal rods; doing so could destroy them. Clean the side of a powdered metal rod with a solvent, and use paint to mark it instead (**FIGURE 20-73**).

You need to rotate the crankshaft over during disassembly, so you may want to use a crankshaft turning tool that allows you to turn the crankshaft assembly. Or you can also reinstall the harmonic balancer and turn the engine over by grabbing the harmonic balancer and turning the engine over by hand. Do not use pliers or vise grips because they will damage the crankshaft and/or Woodruff key.

Removing the Piston Assemblies

Removing the pistons is a multistep process that requires making some decisions before starting the process. The two most basic decisions are deciding if the pistons are going to be reused or replaced and if the ridge at the top of the cylinder needs to be reamed off. The answer to those questions will determine the order in which you perform the steps and the amount of care you use to remove the pistons.

It may be obvious that the pistons will have to be replaced, such as when the cylinders are worn enough that they need to be bored oversize or when the engine exhibits enough wear requiring piston replacement. When a motor is running, the piston rings slowly wear away the cylinder wall where they ride and create what is called a ridge at the top of the cylinder. Higher mileage and harder worked engines develop more severe ridges than other engines. Carbon buildup also adds thickness to the ridge of a cylinder, but this can be removed with a razor blade. A thick ridge will make it more difficult to push the piston out.

The ridge at the top of the cylinder is too big if it catches a fingernail after the carbon has been scraped off. If the ridge is significant and you are planning on reusing the pistons, you will need to remove the ridge with a tool called a **ridge reamer** (**FIGURE 20-74**). The ridge reamer is a cutting tool that will cut into the metal ridge and carbon buildup at the top of the cylinder and remove the lip. If the ridge is not removed, the piston rings could break as they move across the ridge. The piston itself could be damaged if it is forced over the ridge when removing the

FIGURE 20-75 Position a piston at BDC and remove the connecting rod nuts.

piston. Also, care must be taken not to damage the connecting rod journal on the crankshaft when removing the piston. Using rod bolt covers will help protect the connecting rod journal.

> ▶ TECHNICIAN TIP
>
> It is easy for the ridge reamer to remove more than the ridge, which could damage the cylinder wall. If you know the pistons will be replaced with new ones, it may be less hazardous to risk breaking the old, useless pistons than to risk damaging the cylinder with the ridge reamer.

Removing the Rod Caps

To remove the rod caps, start by positioning the number 1 piston at BDC. Then remove the nuts or bolts for the connecting rod cap on the number 1 piston (**FIGURE 20-75**). Next, remove the rod cap. You can do this by tapping the end of the connecting rod on the parting line or by tapping the bolts that are pressed into the rod body with a brass hammer (**FIGURE 20-76**). When the cap separates, pull it off by maneuvering it back and forth (**FIGURE 20-77**). Inside the cap, there should be a bearing insert. If the bearing insert is not in the cap, it will be stuck to the rod journal of the crankshaft, so remove it from there.

Preparing to Remove the Piston Assemblies

Before you remove the pistons, it is important to put some protecting sleeves onto the rod bolts so they cannot scratch the rod journal or cylinder walls as the pistons are pushed out (**FIGURE 20-78**). To prevent the piston from falling out of the engine block and onto the ground when you push it out of the cylinders, thread two long bolts partially into the block on either side of the cylinder, and then stretch some rubber bands or twist a piece of wire between them to catch the piston (**FIGURE 20-79**).

Removing the Piston Assemblies

Make sure the rod journal of the crank is at the centerline of the cylinder bore. Place a wood or plastic dowel of about 1/2" (13 mm) in diameter and 2' (0.6 m) in length inside the bottom of the piston. Several hits to the dowel in the right spot should cause the piston to slide out (**FIGURE 20-80**). In extreme cases where the piston might be frozen in its bore by corrosion, a wooden dowel might not be strong enough. In that case, it may be necessary to switch to a metal rod of some sort and then apply more force with the hammer. Be careful not to scratch the cylinder walls with the edges of the connecting rod as you slide

FIGURE 20-76 Loosen the rod cap by tapping on the side of the rod at the parting line, or on the bolts.

FIGURE 20-77 Remove the rod cap.

FIGURE 20-78 Place protective covers on the rod bolts to protect the rod journal.

FIGURE 20-79 Use bolts and rubber bands to catch a piston.

FIGURE 20-80 Use a long dowel or hammer handle to tap a piston out of the cylinder.

the piston out. Also be careful to avoid damaging the rod journal surface on the crankshaft.

Remove the other pistons in the same way. Line up the rod journal with the bore centerline again, and tap the dowel until the piston pops out. You may have to hit the piston with enough force that it breaks the rings to get it past the ridge. In that case, have someone there to catch it so you do not cause more damage than necessary.

▶ TECHNICIAN TIP

It is not uncommon for a piston ring or even a piston skirt to break when worn pistons are being forced out. Piston assemblies have to be replaced if the technician determines that the cylinder should be machined or that the piston itself has been damaged while inside the engine cylinder block. The cylinders should be inspected for a smooth surface and measured for out-of-round and taper. If these measurements are not within specifications, then the cylinder will have to be remachined, and new pistons will have to fit the resized cylinders.

Removing the Crankshaft and Bearings

S20006

Before removing the main bearing caps, make sure they are marked for reinstallation (**FIGURE 20-81**). To do so, use a number punch or marking paint. To remove the crankshaft bearing caps, you may need some additional leverage to loosen the bolts that hold the crankshaft caps. Some engine blocks have two bolts per cap. Others have four. Remove the bolts (**FIGURE 20-82**). The cap may not come away easily, especially if it has long alignment dowels. Make sure the bolts are not threaded in at this point. Hold them up halfway out of their bolt holes, and tap lightly side to side on the bolt heads. You will need to alternate sides until the cap is clear to be removed. The lower half of the main bearing inserts may come away with the cap as you remove it, or they may stick to the crankshaft journal (**FIGURE 20-83**).

To remove the crankshaft on an application with a one-piece rear main seal, you first need to remove the rear main seal and housing (**FIGURE 20-84**). This may require removing the engine from the engine stand to gain access to the rear of the block. Depending on the maker of engine stand, and the engine block, sometimes the engine stand adapter will prevent the ease of removing the crankshaft assembly due to the end

FIGURE 20-81 Make sure the main bearing caps are marked. If not use a paint stick or number stamp to mark them.

FIGURE 20-82 Remove the main bearing cap bolts.

FIGURE 20-83 Remove the main bearing caps.

FIGURE 20-84 Remove the rear main seal and housing.

part of the crankshaft sticking out past the engine block and the adapter covering it. Crankshafts are both heavy and awkwardly shaped, so be very careful lifting it out of the engine block, or ask for assistance (**FIGURE 20-85**).

When the crankshaft is removed from the engine cylinder block, sometimes the oil has become sticky, and the crankshaft bearings may get stuck to the crankshaft assembly. In this case, you must examine the crankshaft thoroughly and remove the sticky bearings. Also examine the engine cylinder block saddles, where the crankshaft lies inside the block, to see if there are any bearings left inside the engine block. Next, remove any bearing inserts left in the engine cylinder block. Use your fingers to push each half-shell to the side from the center, opposite the bearing tang (**FIGURE 20-86**). If necessary, use a wooden hammer handle to help remove the bearing. The thrust bearing is a special bearing for the crankshaft assembly. In addition to being the largest bearing, the thrust bearing can be identified by the sides it has on it. These sides are to cover the engine cylinder block saddles and prevent bearing movement.

Bearings with visible wear patterns, especially if the copper core is showing through in some areas, are not reusable and should be replaced. If you are not sure of their reusability, take them to a machine shop specialist for advice. If the engine has a rope seal or a two-piece rear main seal, pull out both halves with needle-nose pliers, a scratch awl, or a flat-blade screwdriver.

Removing Passage Plugs

S20007

Remove the water jacket core plugs. The water jacket core plugs cannot be easily pulled out, but they can usually be tilted so that Channellock pliers can grab them and pry them out. Use a punch to carefully drive one side of the core plug toward the water jacket. This should tilt them (**FIGURE 20-87**). If not, keep tapping one side in, but be careful to hang on to it because some core plugs can fall into the water jackets and drop down out of sight. Grip one side of the plug with Channellock pliers, and pry it back out (**FIGURE 20-88**).

If you have a stuck water jacket drain plug, now is the time to remove it. Drill right through the plug, then heat it up and let it cool again to try to break some of the threads loose. Insert an

FIGURE 20-85 Lift out the crankshaft.

FIGURE 20-87 Carefully hammer one side of the core plug until it tilts or pushes into the water jacket.

FIGURE 20-86 Remove the bearing inserts from the bearing saddles in the block.

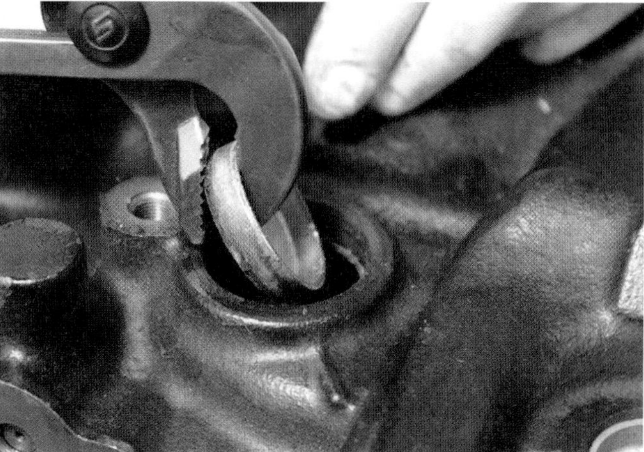

FIGURE 20-88 Grip the core plug with channel lock pliers and pry it out.

FIGURE 20-89 Remove the block from the engine stand and remove the adapter.

FIGURE 20-90 Remove the oil gallery plugs.

extractor tool into the open hole, and use a wrench to break it free and unthread it. Next, take the engine block off the stand. Lower the engine block carefully onto a bench, and remove the engine stand adapter (**FIGURE 20-89**). Remove the oil gallery plugs (**FIGURE 20-90**). If they are square-type plugs, only use the proper hardened steel tool to remove them. Do not use the square part of a socket wrench. If they are stuck, do not force them to the point of stripping them. Use an oxyacetylene torch to heat up frozen plugs. Get them very hot, and then let them cool, after which they may be loose enough to undo. If not, or if they strip, a machining specialist may have to remove them for you.

With the rear plugs out, you can pass a metal rod through the oil gallery and hammer out the front plugs from the rear, if they are the hammer-in type (**FIGURE 20-91**). Hammer-in types do not have any threads showing, but rather a smooth surface on the sides and top of the plug. A threaded plug has visible threads on the side and a square recessed area on top, where a drive from a ratchet would be placed during installation or removal.

Each engine block has its own particular engine block plugs. Plug type is decided by the manufacturer of the engine cylinder block, based on the engine cylinder block application. Check the service information to locate all of the plugs.

FIGURE 20-91 Remove the front oil plugs.

Go around the engine block, taking notes as you remove any plug, adapter, fitting, or sensor that might hinder the cleaning process.

Cleaning the Components

S20008

Before the parts are thoroughly inspected and measured, they must be cleaned. The cleaning process of the engine cylinder block can vary depending on what is available to the technician. Most shops use a spray wash cabinet, which operates much like a dishwasher and usually uses environmentally friendly degreasers and hot water (**FIGURE 20-92**). Some machine shops dip the major engine parts for several hours in a hot or cold caustic solution. The strong corrosive chemicals remove mineral deposits in the water jackets, carbon deposits, oil, greases, and paints.

Depending on the condition of the components, everything from the engine block and disassembled heads to the crankshaft and all the covers may have to be put into the spray wash for degreasing. After that, the heads and engine block will have to be hot tanked, blasted, or baked until they are perfectly clean, and any of the external covers must be bead blasted to be ready for paint later on. It is much easier to inspect and properly assess the condition of the parts when everything is clean. *Note:*

FIGURE 20-92 Place the components to be cleaned in a spray wash cabinet.

FIGURE 20-93 With the parts clean, visually inspect them for any obvious damage or wear that would require replacement.

It is of utmost importance to remember that, after beading, all of the beads will have to be thoroughly cleaned out and off the parts. Otherwise, beads will get stuck in parts with baffles and passageways, and will quickly destroy a newly rebuilt engine when they become dislodged and circulate inside the engine.

▶ Engine Block and Component Inspections

S20009

Now that the engine has been disassembled and cleaned, all of the components must be inspected thoroughly to assess their condition. A quick visual inspection should have been made as each part was removed from the engine, and any obvious damage noted. With the parts clean, inspect each part more thoroughly for any obvious damage that you missed before (**FIGURE 20-93**). If found, determine if the part needs to be replaced or can be repaired. If there are no obvious issues, perform any measurements or structural tests to verify that the parts are in useable or repairable condition. To do so, you need to know how to use the various tools involved, including measuring tools, crack detection tools, and various testers.

Inspecting the Engine Block

S20010

The engine block should be inspected for cracks after the cleaning process is finished and the block is thoroughly dried. You want to determine whether the block is cracked before performing any machining work so that the crack can be repaired or the block replaced; otherwise, time and materials will likely be wasted. Cracks can occur anywhere in the block, inside or outside, so you need to thoroughly test all of the surfaces of the block. Cracks in cast iron parts can be detected by a process called magnetic particle inspection, more commonly referred to as **magnafluxing**. In this process, an electromagnet is attached to the part, which is then sprayed with a special magnetic powder or liquid that sticks to the magnetic field created at any cracks, thereby making them visible.

Another option is to use a leak detection dye. The product used could be a fluorescent dye such as **Zyglo™**, which is sprayed on the surface being checked, soaks into cracks, and is wiped off. A special black light is then shone on the surface, and the dye appears in any cracks. Another type of dye detection system uses a two-part process. First the dye is applied and wiped off. Then a white developer is sprayed onto the surface. If the dye has soaked into any cracks, it will bleed out of the crack and stain the white developer. Leak detection dyes can be used on nonferrous metals such as aluminum.

Another task when inspecting the block is to check the deck surface, where the head gasket sits, for warpage. A straightedge and feeler gauge is used to measure any warpage. Check the deck surface in several different angles across the length of the deck. Most machine shops will resurface an engine block with more than 0.0025" (0.064 mm) warpage. Check the manufacturer's specifications for maximum warpage.

To inspect the engine block for visible cracks, follow the steps in **SKILL DRILL 20-1**.

Deglazing and Cleaning Cylinder Walls

N20002

If the cylinder measures within specifications, it will need to be prepared for the new piston rings. The cylinder wall will probably have a polished "glaze" on its surface that must be removed. Removing the glazed surface with an abrasive stone of the proper grit will give the cylinder walls a nice crosshatch pattern, which allows a small amount of oil to be retained in the fine scratches. This oil helps to lubricate the piston rings during engine operation, and the fine crosshatched scratches allow the new piston rings to seat against the surface of the freshly honed cylinder wall. The recommended crosshatch for a cylinder wall is approximately 45 degrees.

A cylinder hone can be rigid, flexible, or ball style. A rigid hone uses fixed rectangular abrasive stones and is used on a cylinder after it has been bored to give a perfectly cylindrical shape with the proper finish. A flexible hone uses spring-loaded rectangular stones that can more easily follow

SKILL DRILL 20-1 Inspecting the Engine Block

1. Perform a visual check for cracks. If more detection is needed, magnaflux the block if it is cast iron.

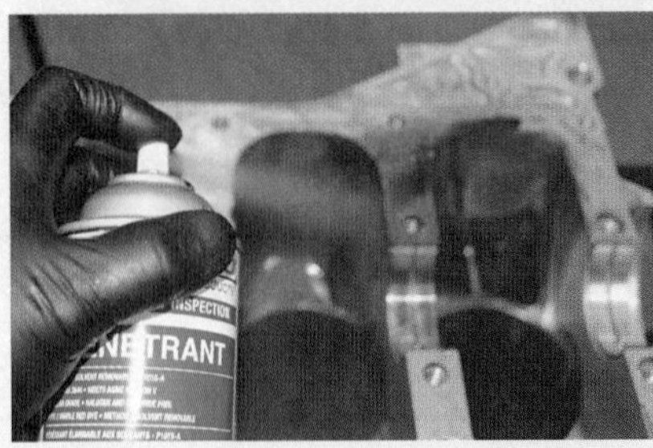

2. If the block is aluminum use a dye check such as Zyglo™.

3. Check all plug areas for obstruction, blockage, and cleanliness.

4. Clean the deck surface using an appropriate gasket scraper, wire brush, or bristle discs for cast iron.

5. Use nonmetallic and nonabrasive scrapers on aluminum sealing surfaces.

6. Check the deck surface for warpage, using a straightedge and a feeler gauge.

the contours of a minimally used cylinder; thus, it is used to deglaze a cylinder that will be reused. A ball hone uses round abrasive stones on the ends of flexible metal spokes. A ball hone also works well on a minimally worn cylinder that will be put back into service. When removing either type of hone, the hone should still be spinning slowly. Don't pull the hone out when it is stopped, as it will wear grooves in the side of the cylinder walls. When honing is complete, the block should be washed in a spray cabinet; then the cylinders should be cleaned with soap and hot water, using a cylinder brush. Use a white rag to verify that the cylinders are clean. Once clean, dry the block and spray the machined surfaces with an anti-rust protectant.

To deglaze and clean a cylinder wall, follow the steps in **SKILL DRILL 20-2**.

Inspecting the Crankshaft

N20003

A crankshaft can be faulty in several ways. First, it could be warped, which would cause the journals to be forced against the bearings, wearing them out prematurely. Second, the main and rod journals could be excessively worn, leading to too much bearing clearance, which would cause low engine oil pressure. Third, the bolt holes, keyways, and sealing surfaces could be damaged, causing their own set of issues. These situations mean that a thorough inspection along with precise measurements must be made to ensure the crankshaft can be put back in service.

There are two methods you can use to check the crankshaft for straightness.

- **Crankshaft in block:** With the pistons and rods removed, make sure all main caps are numbered. Pull the number 1 main cap, and clean the main journal on the crankshaft. Using a dial indicator with a 0.001" (0.025 mm) scale and a magnetic base, attach the dial indicator on the pan rail, and center the dial indicator over the number 1 main journal. Preload the dial indicator on the main journal, and move the face on the dial indicator to zero. Rotate the crankshaft one full rotation, watching the dial indicator, and record your findings. Repeat the process to check the center main. Determine any necessary actions based on your findings.
- **Crankshaft out of block:** With the crankshaft removed, make sure the main journals are clean. Position the crankshaft on two V-blocks, using the first main journal and the last main journal. Using a dial indicator with a 0.001" (0.025 mm) scale, attach the dial indicator on the bench with a magnetic base or clamp. Position the dial indicator over the number 1 journal. Preload the dial indicator on the main journal, and set the face on the dial indicator to

SKILL DRILL 20-2 Deglazing and Cleaning a Cylinder Wall

1. Using a drill, attach the proper-sized flex hone.

2. Carefully insert the hone, and apply honing oil to the cylinder walls.

3. Using the drill with the attached flex hone, stroke up and down in the cylinder that you are honing for approximately 10 to 15 seconds at a speed at which you get a good crosshatch.

4. Use a rag to wipe the cylinder out so you can inspect the cylinder walls.

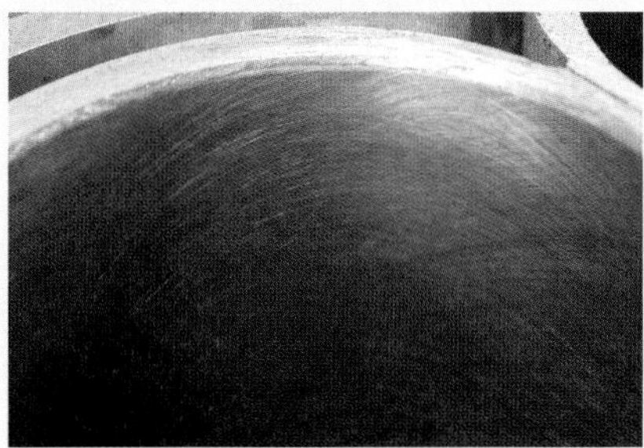

5. Inspect the crosshatch on the cylinders. Repeat until a good crosshatch pattern is achieved.

6. Clean the block in a spray cabinet, followed by using soap and water with a cylinder brush. Finish by drying and applying protectant.

zero. Rotate the crankshaft one full rotation, watching the dial indicator. Record your findings. Repeat the process on the center main.

Visually inspect the journals for grooves. The journals should be smooth to the touch. Use an outside micrometer to measure the rod and main journals. Both rod and main journals should have measurements taken 90 degrees apart. Also, rod journals should be measured across the face of the journal to check for taper. Compare your findings to factory specifications.

Visually inspect the square-cut or Woodruff key and keyway. If the square key or Woodruff key is out, make sure the keyway is straight with sharp edges. The square-cut or Woodruff key must fit snugly. Visually inspect the thrust flange, bolt holes, and rear seal surface to make sure there are no nicks or scratches. Visually inspect the crankshaft for cracks. If you need a more detailed inspection, the crankshaft should be checked by magnafluxing.

Probe the oil passages with a small plastic tube (a spray can nozzle extension works well), making sure the oil passages from the mains into the rod journals are clear. Also check the crankshaft position sensor reluctor ring for any damage, when applicable.

To inspect the crankshaft, follow the steps in **SKILL DRILL 20-3**.

Inspecting Main and Connecting Rod Bearings

N20004

The main and connecting rod bearings must be inspected to determine whether they are in good condition or need replacing. Both normal wear and traumatic events can damage the bearings, resulting in a knocking noise and/or low oil pressure. Typically, a visual inspection is all that is needed to determine that they need to be replaced, as excessive wear will be apparent. If the bearings look to be in good shape, then the bearing clearance should be measured to determine if it meets the manufacturer's specifications. This will be covered in a later skill drill.

If you are considering reusing the main and rod bearing inserts, be sure to keep them together, and mark them once they are removed. The main caps and connecting rods should have already been marked with numbers to avoid mixing them up. Remove each main bearing insert from the main cap and main housing support, one at a time. If the bearing insert will not come out using your fingers, use a wooden hammer handle to push the bearing on the opposite side of the tang at the parting line; the bearing should slip out.

Clean each main bearing, and mark the back of each half-shell with a permanent marker with the correlating bearing support

SKILL DRILL 20-3 Inspecting the Crankshaft

1. Check the crankshaft for straightness with a dial indicator.

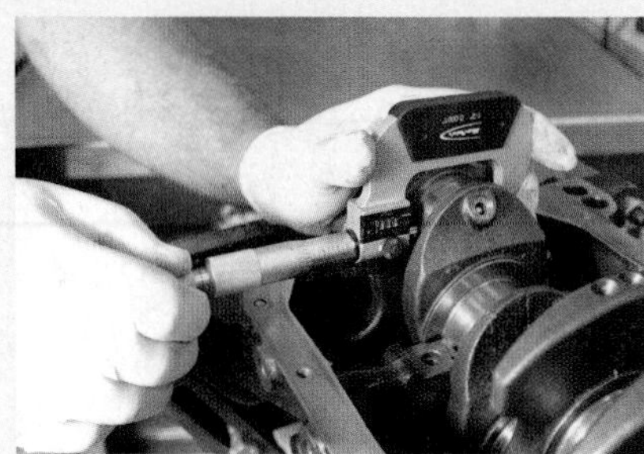

2. Measure the rod and main journals using a micrometer, and record your findings.

3. Visually inspect the square-cut or Woodruff key and keyway for damage.

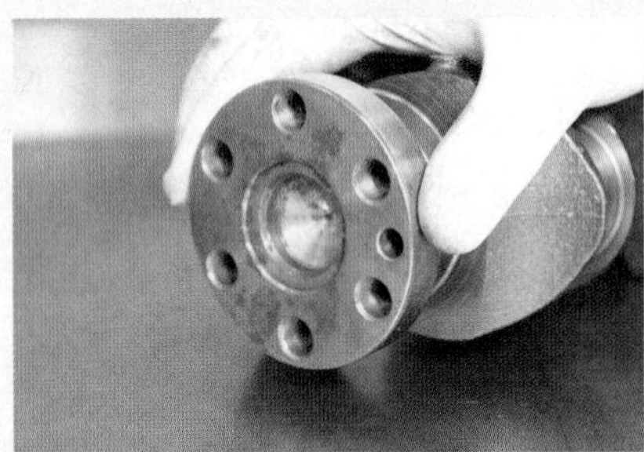

4. Visually inspect the thrust flange, bolt holes, and rear seal surface for nicks or scratches.

5. Visually inspect the crankshaft for cracks; magnaflux, if necessary.

6. Make sure the oil passages are clear. Check the crankshaft position sensor reluctor ring to make sure there is no damage.

number. It is important to know what journal the bearings were located on, to identify any unusual wear patterns. Visually check that there are no grooves or embedded foreign matter. Check with the bearing wear chart to see if the bearings are wearing normally or if the wear pattern indicates that the bearings need replacement. Repeat the process for the connecting rod bearings.

Before you inspect the thrust bearing, go back to your crankshaft end play findings. If it was excessive, you should be able to determine why by checking for damage or wear on either the front or the rear thrust bearing surface. If the rear surface of the thrust bearing is worn and the vehicle is equipped with an automatic transmission, the wear is usually caused by a torque converter ballooning, which causes pressure on the thrust bearing. If it is a manual transmission, the wear could be caused by someone holding the clutch pedal down excessively or by a pressure plate with too much spring pressure.

To inspect main and connecting rod bearings, follow the steps in **SKILL DRILL 20-4**.

SKILL DRILL 20-4 Inspecting Main and Connecting Rod Bearings

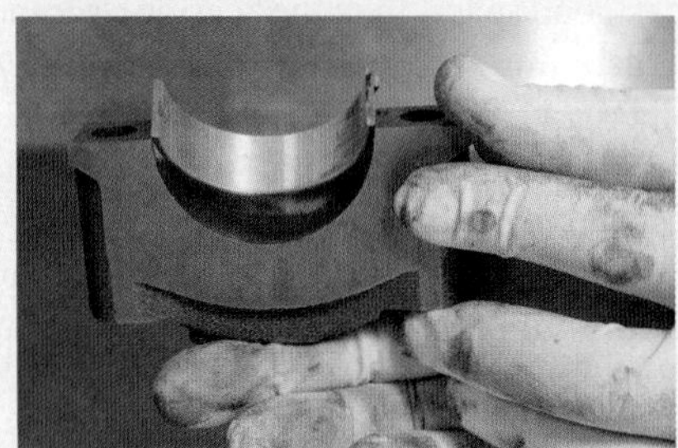

1. Take each main bearing insert out of the main cap and main bearing support, one at a time.

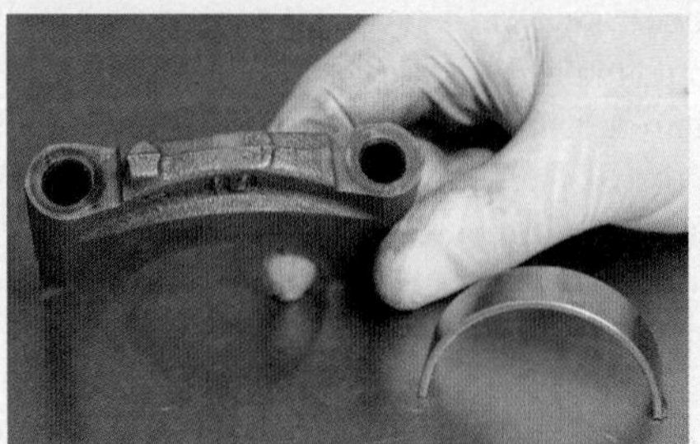

2. Clean each main bearing and mark the back of each half-shell with a permanent marker with the correlating bearing support number.

3. Check that there are no grooves or embedded foreign matter. Check with the bearing wear chart to see whether the wear pattern indicates that the bearings need replacement.

Identifying Piston and Bearing Wear Patterns

N20005

Examining the piston and bearing wear patterns is useful because the patterns serve as a diagnostic tool for determining whether the parts are operating normally or if there are possible alignment issues or engine operating problems with the engine block and its internal components. The technician must compare the wear patterns with those provided in the manufacturer's wear chart (**FIGURE 20-94**). The wear chart correlates the wear pattern with the contributing problem, such as a bent connecting rod, damaged crankshaft assembly, lubrication system, or cooling system.

There should be an even wear pattern across the full face of the rod and main bearings. Greater wear on one side of a rod bearing than the other is an indication of a bent connecting rod. A bent rod also typically shows an increased wear pattern on diagonally opposite sides of the piston. If uneven wear is observed, have a machine shop check the rod for straightness.

Greater wear on one side of a main bearing than the other is an indication that the main line that supports the crankshaft is no longer straight. If uneven wear is observed, inspect the crankshaft for straightness and determine if it needs to be straightened or replaced.

To identify piston and bearing wear patterns that indicate connecting rod alignment and main bearing bore problems, follow the steps in **SKILL DRILL 20-5**.

Removing and Replacing the Piston Pin

Removing and replacing the piston pin is typically performed when new pistons are being used or if the rods are being rebuilt or replaced. Removing and inspecting the pins is important in order to see if the piston or the piston pins have developed any abnormal wear during the operation of the engine. Full floating pins can be removed by hand once the snap rings have been removed. Semi-floating pins must be pressed apart with a hydraulic press; this is covered in the Machining Chapter.

To remove and replace the floating piston pin, follow the steps in **SKILL DRILL 20-6**.

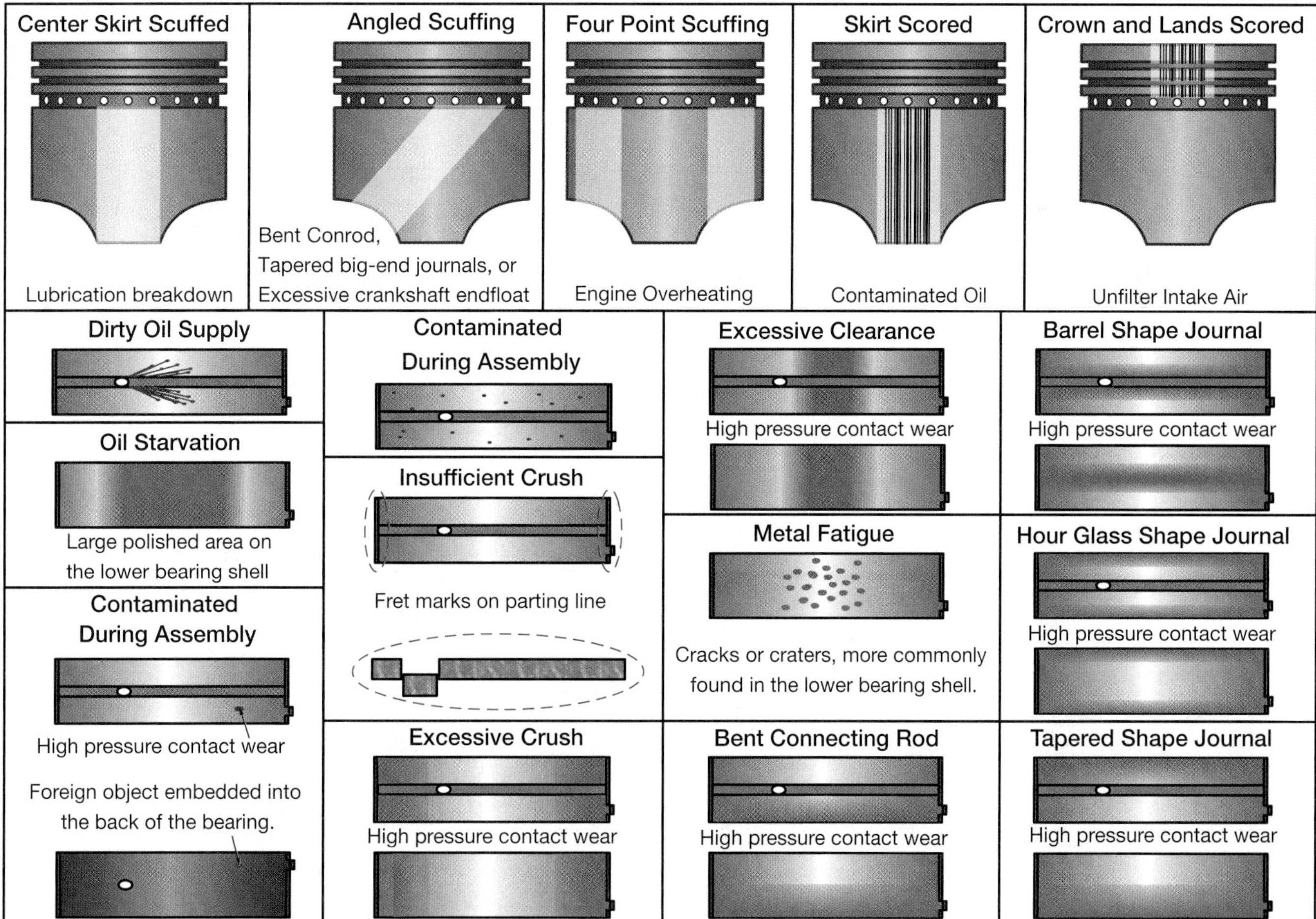

FIGURE 20-94 Piston and bearing wear chart.

SKILL DRILL 20-5 Checking Piston and Bearing Wear Patterns

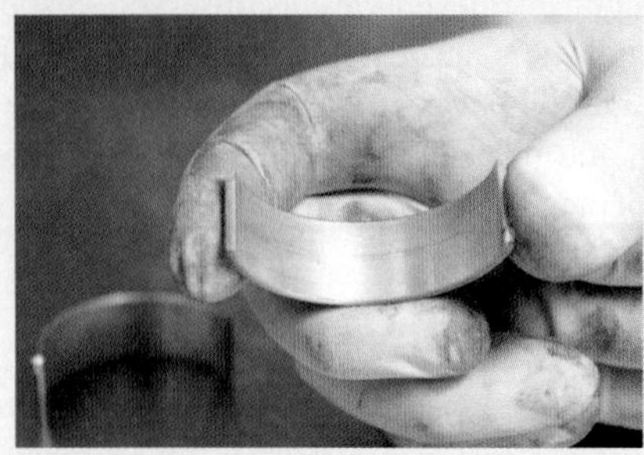

1. Inspect the rod bearings. There should be an even wear pattern across the full face of the bearing.

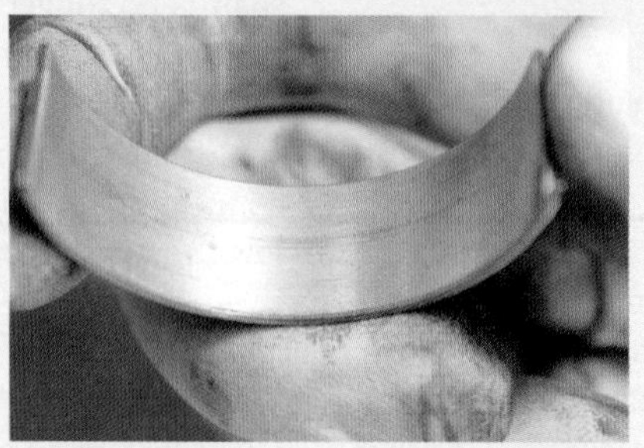

2. Inspect the main bearings. There should be an even wear pattern across the full face of the bearing.

3. Inspect the piston assembly for wear patterns that indicate bent or damaged components.

Inspecting and Measuring the Cylinder Wall/Sleeves

N20006

The surfaces of the cylinder bores wear over time as the pistons travel up and down the cylinders millions of times over their life. The cylinder bore in both sleeved and non-sleeved engine blocks must be perfectly round and of a uniform diameter from top to bottom. Careful measurement is required to determine if the cylinders can be reused or if they will need to be replaced (sleeves) or bored oversize. The tool that most quickly measures the cylinder bores is a dial bore gauge. It uses a dial indicator gauge connected to a handle. Once it is calibrated to a certain size, it can be inserted into the cylinder and a quick

SKILL DRILL 20-6 Removing and Replacing the Floating Piston Pin

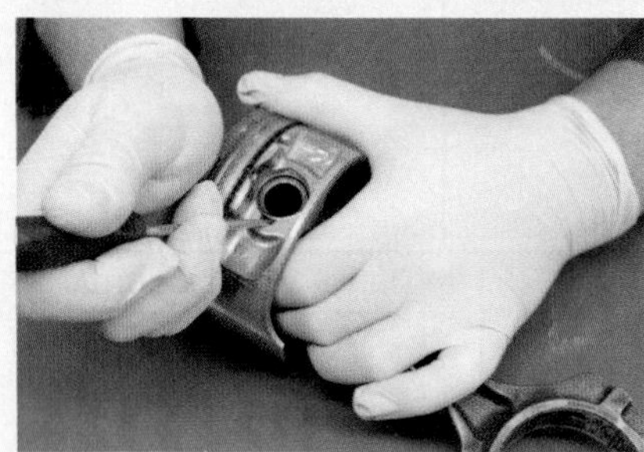

1. Using the proper snap ring pliers that fit the holes in the snap ring, remove both snap rings.

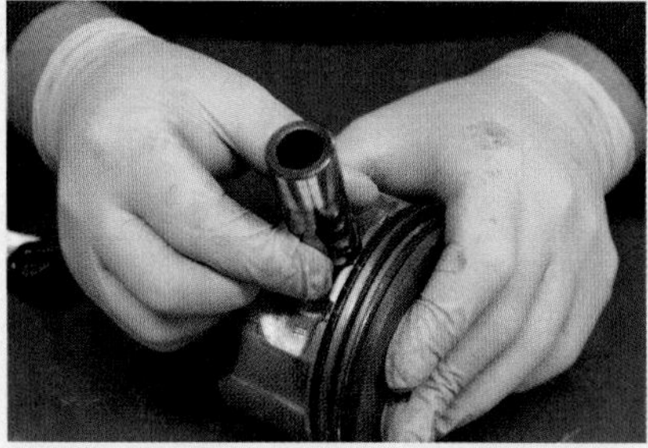

2. Push on the wrist pin to separate the wrist pin from the piston and connecting rod.

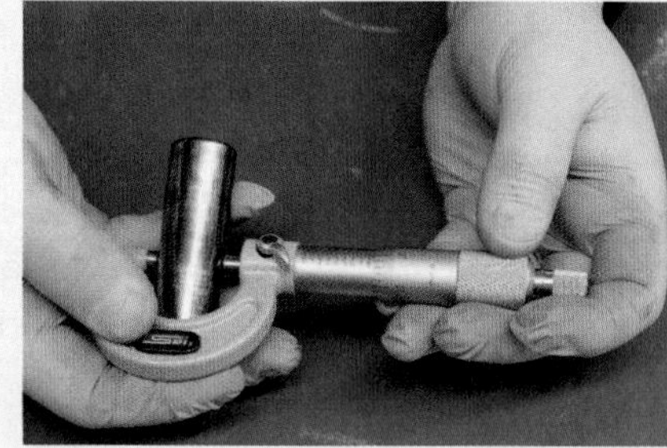

3. Using an outside micrometer, measure the piston pin outside diameter, and record your findings.

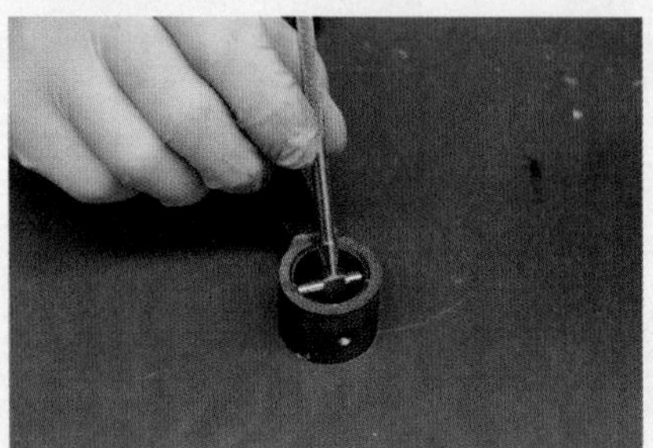

4. Using a bore gauge or inside micrometer, measure the inside of the small end of the connecting rod. Record the clearance.

5. Install the floating pin design. Lubricate the wrist pin hole on the piston and the small end of the connecting rod along with the wrist pin. Using your snap ring pliers, install one snap ring on the piston.

6. Position the piston so it is in the proper relationship to the connecting rod (see service information).

SKILL DRILL 20-6 Removing and Replacing the Floating Piston Pin (Continued)

7. Place the small end of the connecting rod between the piston pin bosses. Using your fingers or a wrist pin driver, push the wrist pin through the piston and small end of the rod until it stops against the snap ring.

8. Install the second snap ring, and move the piston and connecting rod, rocking the rod back and forth to ensure that it moves freely and smoothly.

measurement made. Another tool that is just as accurate, but not as easy and quick, is the inside micrometer.

To inspect and measure cylinder walls/sleeves for damage and wear, follow the steps in **SKILL DRILL 20-7**.

Inspecting and Measuring Piston Skirts

N20007

Pistons, like cylinder walls, wear over time. If they are to be reused, they must be carefully inspected, measured, and compared to specifications first. It does no good to put new piston rings on a piston that has too much wear at the ring lands, as the rings will leak oil and compression. If the piston skirt is worn, the piston may make a knocking noise due to too much clearance between the skirt and the cylinder wall.

To inspect and measure piston skirts, follow the steps in **SKILL DRILL 20-8**.

Inspecting Balance Shafts and Support Bearings

N20008

On engines equipped with balance shafts, it is important to inspect them to verify whether they are in a reusable condition. Unless something catastrophic happened in the engine, they usually don't cause problems. When inspecting them, you need to measure the bearing journals for undersize, out-of round, or poor surface finish, and compare your readings to specifications. Also inspect the support bearings for wear and compare to specifications.

SKILL DRILL 20-7 Inspecting and Measuring the Cylinder Walls/Sleeves

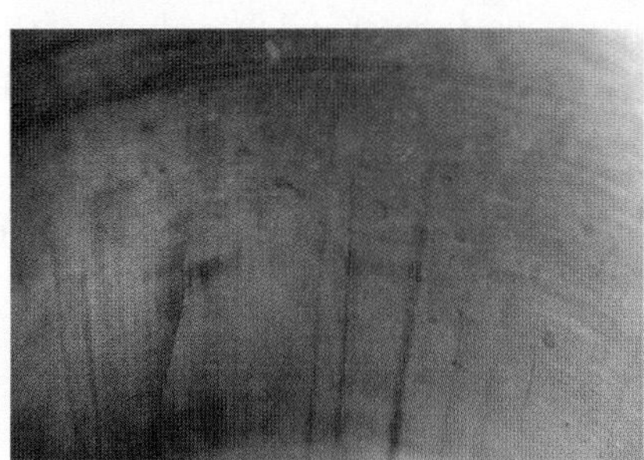

1. If there is obvious scoring or damage, then the block will have to be machined.

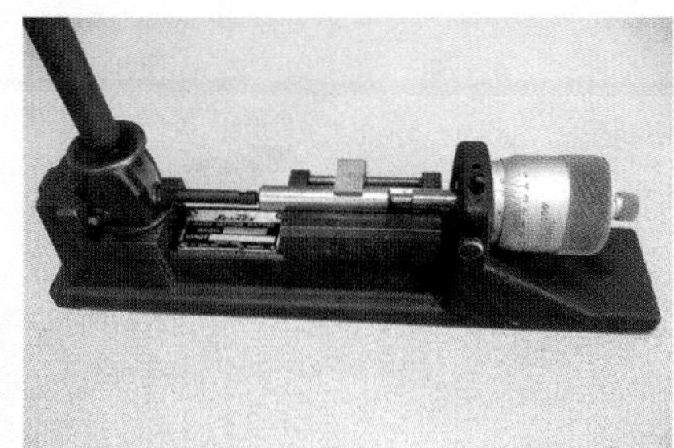

2. If there is no obvious damage, set a dial bore gauge to the specified standard cylinder bore size.

3. Measure each cylinder at the top, center, and bottom, 90 degrees across the thrust side and pin side at each level. Compare to specifications.

SKILL DRILL 20-8 Inspecting and Measuring Piston Skirts

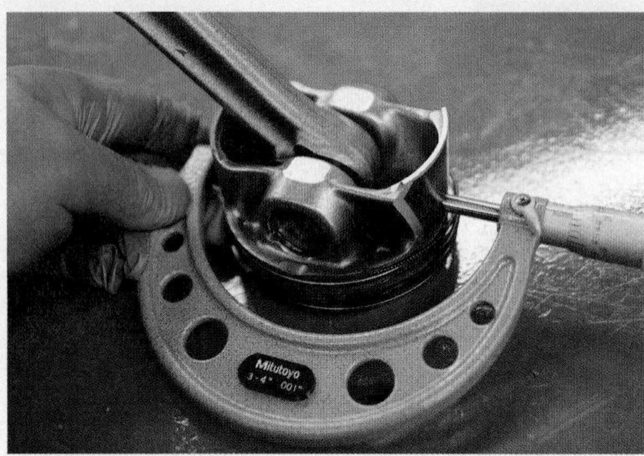

1. With the piston inverted, measure the diameter of the piston skirt even with the center of the wrist pin, and compare to specifications. Replace the pistons if necessary.

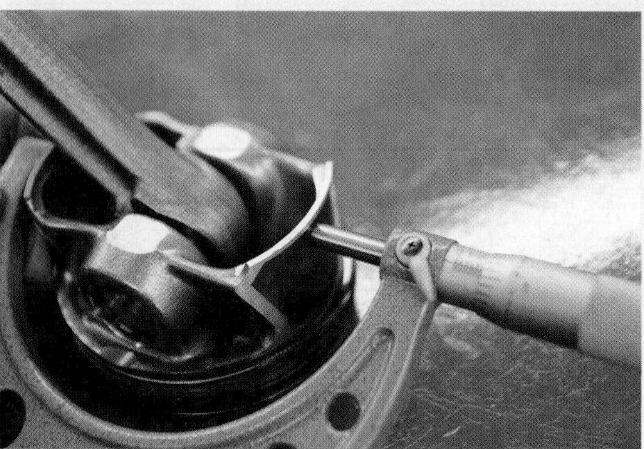

2. Measure the diameter of the bottom of the piston skirt. It should never be smaller than the previous measurement. If the measurement is smaller, then the piston has a collapsed skirt, and the piston must be replaced.

▶ Wrap-Up

Ready for Review

- The engine block is the main structure of the internal combustion engine; it provides the engine's framework and a mounting surface for all components.
- The engine block is machined in one piece, with precise specifications that help to prevent air and fuel leakage, withstand the thermal stresses of combustion, and accommodate any resulting shocks and vibration from power pulses.
- Engine block casting must accommodate cylinders and water jackets (passageways around the cylinders); this can be done with core sand or foam shapes.
- Manufacturers are increasingly making aluminum alloy engine blocks, as they are lighter weight and increase fuel efficiency.
- The engine block assembly (lower end) includes oil galleries and plugs, water jacket and plugs, pistons and rings, piston pins, connecting rods and caps, rod bearing inserts, main bearing caps, main bearing inserts, the thrust bearing, crankshaft, rear seals, camshaft (in block cam), a vibration balancer, and smooth surfaces to accommodate mating parts.
- Ribs, webs, and fillets provide reinforcement and strength to high-stress areas of the engine block.
- The main bearing caps, which provide crankshaft support, are sometimes supported by a main bearing cap girdle (some of which are integrated).
- Cylinders are either machined into the cylinder block or are made of a cast iron sleeve.
- Cylinder sleeves are dry (not in contact with a water jacket; held by an interference fit), dry flange (also not in contact with a water jacket; locked into engine block with a flange), or wet (direct contact with water jacket coolant).
- The piston fits into the cylinder and is responsible for transferring power stroke force to the crankshaft.
- The piston assembly includes a head, skirt, pin, connecting rod, piston boss, compression rings, oil rings, piston ring grooves, ring lands, and wrist pin.
- Piston slap occurs when there is too great a clearance between the piston and the cylinder wall.
- Cam-ground pistons are slightly oval shaped to accommodate heat expansion.
- Compression ratio refers to the difference between the volume in the combustion chamber above the piston when the piston is at bottom dead center and the volume of the combustion chamber above the piston when the piston is at top dead center.
- The compression ratio can be changed by altering the shape of the piston head (e.g., flat, concave, dome shaped, recessed, machined reliefs).
- Stock compression ratios typically run between 8:1 and 10.5:1 and can be intentionally increased if necessary.
- Pistons can be manufactured via forging (expensive, but stronger) or casting (cast aluminum or hypereutectic).
- Piston skirts may be coated with ceramic or molybdenum disulfide to reduce friction with the cylinder wall.
- Piston compression rings must have proper end gap, back clearance, and ring tension.
- Connecting rods are H- or I-beam shaped, and types include cast, forged, aluminum, and powdered metal

(most recent technology); rod caps are not interchangeable among the different types.

- Connecting rods (cast, forged, aluminum, and powdered metal) connect the piston to the crankshaft.
- The crankshaft is designed to transform piston stroke power into rotary motion.
- The crankshaft assembly contains main bearing journals that help to center the crankshaft at the bottom of the engine block, where they mate with main bearings.
- Crankshafts may be internally or externally balanced, using counterweights.
- Crankshaft bearings are split in two parts; some have a tang that fits into the bearing saddle and cap. Bearing crush is used to hold them in place.
- A thrust bearing is used to limit end play movement of the crankshaft.
- An oil hole in the main bearing saddle allows oil from the oil gallery to lubricate the main bearings.
- Bearings can be undersized or standard and are generally either trimetal or bimetal. Newer bearings may be coated with polyamide or PTFE (Teflon™).
- An engine will have either a flywheel (manual transmission) or flexplate and torque converter (automatic transmission) to store rotational energy.
- The harmonic balancer (crankshaft damper) reduces vibration from the piston and crankshaft assembly.
- Crankshaft seals are designed to contain lubricating oil.
- Balance shafts reduce engine vibrations and may use counterweights to reduce second-order vibrations.
- When disassembling the engine, refer to manufacturer's service information.

Key Terms

accessory drive pulley A round circular metal disc attached to the front of the engine crankshaft with a rubber/fiber belt that helps operate automotive accessories.

aluminum rod A lightweight, strong, either cast or forged metal shaft that connects the piston assembly to the crankshaft assembly.

back clearance The area or space behind the piston rings when the rings are in the piston ring grooves.

balance shaft A metal shaft attached and located inside the engine cylinder block assembly to counteract crankshaft vibrations.

bearing crush The force created to seat the bearing by the extra bearing material when the ends of the bearing inserts touch each other and are forced against each other.

bearing inserts Components made out of soft metal materials that are replaceable and come in pairs; also called half-shell bearings.

billet The roughly shaped steel piece that has been cast or forged but has not undergone its final machining.

bimetal Aluminum, tin, and silicon alloy metals placed together with steel to form one piece of material and used for bearing materials.

casting line A by-product of the metal casting process that forms a line found on parts that are cast in metal.

cast rod A connecting rod that is created by the metal casting process.

center bolt A fastener or bolt found in the middle of the harmonic balancer or cam sprocket that attaches it to the crankshaft or the camshaft.

compression ring A metal ring found inside the grooves on the side of the engine piston.

compression ring groove A square groove found on the outside of the piston skirt in which the piston ring fits.

connecting rod A cast or forged metal rod that connects the engine pistons to the engine crankshaft.

core plug A metal cap for the holes that are used to remove core sand used during the casting process.

crankcase The bottom area of the engine cylinder block where the crankshaft is located.

crank core The rough, unfinished crankshaft assembly that has just left the foundry or forging area.

crankshaft A part mounted in the lower side of the engine block that has offset journals that rotate to change the up-and-down motion of the pistons into rotary motion at the crankshaft.

crankshaft journal A metal part of the crankshaft that is positioned offset to the main bearing journal on the crankshaft.

crimp The bent shape of the oil ring expander that allows it to provide outward force on the oil rings.

deck The surface area at the top of the engine block against which the cylinder head seals.

dry flanged sleeve A metal cylinder with a flange embedded into the block pressed into an engine cylinder to give it a new wear surface. The flanged sleeve is held in by the cylinder head.

dry sleeve A metal cylinder that is pressed into an engine cylinder block that does not come in direct contact with coolant. It is held in place with an interference fit.

engine bearing A specially designed metal piece that supports circular moving parts.

expander The part of an oil control ring that holds the ring against the cylinder wall.

fillet The radius portion of an inside corner that reduces stress at the corner.

flexplate A metal part attached to the rear of the crankshaft that connects to the torque converter in a vehicle with an automatic transmission.

floating pin A round metal pin that has a very small clearance but is free to float in the piston and connecting rod.

fluid dissipater A silicone fluid component that allows the mass of a fluid type harmonic balancer to rotate at a more constant speed than the crankshaft, to help even out the torsional vibrations of the crankshaft.

flywheel A heavy round metal disc attached to the end of the crankshaft to smooth out vibrations from the crankshaft assembly and provide one of the friction surfaces for a clutch disc used on manual transmission/transaxle applications.

forged connecting rod A strong metal connecting rod hammered into shape by large presses used to connect the piston to the crankshaft.

garter spring A round circular spring used in seals to provide tension on oil seal lips.

girdle A metal device connected to the bottom of the engine cylinder block to create strength for the main bearing caps.

gravity pouring The casting process used for creating metal parts.

harmonic balancer A round metal disc connected to the front of the crankshaft that smoothes out torsional vibrations created by the crankshaft.

horizontally opposed engine An engine with a centrally located crankshaft and pistons on each side.

hypereutectic piston An aluminum piston that is cast with a 16% to 19% silicon content and that has minimal expansion rates.

induction hardening A metal hardening process that uses an electromagnetic field to quickly heat up the surface of a metal part so it can be quenched to provide an appropriate amount of hardness.

interference fit A fastening between two parts that is activated by friction after the two parts are pushed together.

journal saddle A semicircular cut located at the bottom of the engine block that is used to support the engine crankshaft.

lost foam A metal casting process that uses a foaming process to create the pattern of the part being cast in metal.

magnafluxing An electromagnetic process used to locate cracks in ferrous engine blocks and cylinder heads and other ferrous metal parts.

main bearing Metal bearing insert located in the main saddles that support the main journals.

main bearing caps Sturdy caps placed over the main journals of the crankshaft assembly and fastened with either two or four bolts.

main cap girdle A metal attachment that strengthens and connects all the main bearing caps together. It can be integral from the main bearing caps or separate.

main journal The smooth machined area of the crankshaft assembly that allows the crankshaft to rotate within the main bearings.

Mallory metal A tungsten alloy of copper and nickel that is used as a metal substitute added to the crankshaft counterweight for balancing purposes.

metallurgical bonding A process of sintering metals until they are fused together as one metal.

moly The abbreviation for the lubricant called molybdenum disulfide.

nitriding A metal surfacing process that uses nitrogen ammonia gas to create a very thin, highly hardened surface.

nitridization process The nitriding metal surfacing process in which nitrogen ammonia gas is used inside a furnace and heated up and then cooled.

oil control ring The widest of the metal rings on the piston head assembly that controls oil flow to and from the cylinder walls.

oil ring groove A metal groove cut into the piston head assembly designed to hold the oil control ring, which lubricates the cylinder wall and pistons.

piston The round metal plug found inside the engine cylinder that moves up and down inside the cylinder.

piston assembly All of the parts of the piston including the piston, piston rings, and piston pin.

piston pin (wrist pin) A round circular metal manufactured part that attaches the piston assembly to the connecting rod assembly.

piston ring A metal ring that is placed in a square groove around a piston for sealing purposes.

piston ring groove A square-cut groove located on the piston, designed to hold a metal piston ring.

piston slap An engine noise caused by excessive clearance between the piston skirt area and the cylinder wall.

piston stroke The distance the piston travels through the engine cylinder from top dead center to bottom dead center.

powdered metal rod A connecting rod manufactured through the process of heat, compression, and forging of powdered metals into a connecting rod.

power stroke The travel of the piston following ignition of the air-fuel mixture as the expanding gases of combustion push the piston down the cylinder.

press-fit pin A description of how a piston pin is placed inside a connecting rod by pressing the pin into the connecting rod.

reciprocation The back-and-forth movement of the piston assembly inside the cylinder.

rib A design feature created in the metal casting/forging process that places extra metal area on a part for strength and durability, yet keeps the weight of the engine block low.

ridge reamer A tool used to remove the metal lip on top of the cylinder walls caused by engine wear.

ring gap The distance between the ends of the piston rings when the rings are seated against the inside of the engine cylinder.

ring gear A large circular gear that is typically mounted on the flywheel or flexplate and mates to the starter gear, for the purpose of cranking the engine over.

ring land The area between the piston ring grooves found on the piston head assembly.

ring tension The built-in force created inside the piston rings; it is caused by the piston rings being made bigger than the cylinder walls in order to generate a scraping action against the cylinder walls.

rod beam The area of a connecting rod between the big and small ends.

rod bearing A bearing insert located in the big end of the connecting rod.

rod cap The very bottom part of the connecting rod that retains the rod bearing inserts.

rod journal Also called the crankshaft rod throw, an area machined to a very smooth finish that contains an oil hole to provide oil to lubricate the surfaces of the crankshaft and rod bearing inserts.

rotating assembly The assembly of the crankshaft, connecting rod, and piston that is found inside the engine cylinder block.

second-order vibration Vibration that occurs at twice the engine rpm.

side rail The thin portion of the oil control ring that is used to scrape oil off the cylinder walls.

sintering process A metal hardening process in which the metal is fused together without melting.

soft plug A thin-walled, metal, cup-shaped disc designed to be pressed into a machined passageway in the block for the purpose of plugging it.

starter teeth Machined teeth located on the ring gear for meshing with the starter drive teeth located in the starter motor.

tang A part of the bearing insert that helps to lock the bearing insert into the bearing saddles and caps.

throw The part of the crankshaft that is offset for the connecting rods to mate with; also called the connecting rod journal.

thrust bearing A main bearing insert that either has integrated flanges or separate flanges that provide bearing surfaces that prevent forward or lateral movement of the crankshaft assembly.

timing chain A steel chain connecting the crankshaft assembly to the camshaft assembly.

timing gear A sprocket attached to the crankshaft assembly and the camshaft assembly.

trimetal The use of three different types of metals to build up a bearing insert to give it long-lasting wear characteristics.

water jacket A passageway for coolant to flow inside the engine block that is formed when the block is cast.

web A reinforced area of a metal part formed during the casting process to create strength and durability.

Welch plug A soft, round, metal disc pressed into the engine block to plug a water jacket or oil passageway.

wet sleeve A replaceable steel cylinder installed into the block that provides a wear surface for the piston and rings. It is in direct contact with coolant on its outside surface.

Woodruff key A half-round, rectangular key that is placed into a half-round, rectangular opening in the nose of the crankshaft or other round shafts found on the engine.

Zyglo™ A crack detection method used on engine blocks and cylinder heads that uses a liquid fluorescent penetrant and a fluorescent light to find cracks in ferrous metal.

Review Questions

1. The bottom of the crankcase is sealed off by the:
   - **a.** oil pump.
   - **b.** crankshaft bearings.
   - **c.** oil pan.
   - **d.** flexplate.
2. When removing the pistons from the block:
   - **a.** remove them out the bottom of the block.
   - **b.** remove them after removing the crankshaft from the block.
   - **c.** use large pliers to pull them out of the cylinders.
   - **d.** check for an excessive amount of cylinder ridge.
3. Which of these functions is related the most to a piston?
   - **a.** It converts thermal energy into mechanical energy.
   - **b.** It transfers movement of the cam lobe to the valve.
   - **c.** It prevents the engine from overheating.
   - **d.** It reduces vibrations during engine operation.
4. Compression rings are designed to:
   - **a.** seal the piston and the cylinder wall.
   - **b.** scrape excessive oil from the walls and release it back to the crankcase as needed.
   - **c.** to hold a small amount of lubrication oil to lubricate the piston and cylinder walls.
   - **d.** decrease compression to prevent engine damage.
5. Choose the correct answer.
   - **a.** The top end of the connecting rod is where the large rod bearings are inserted.
   - **b.** Powdered metal rods have a vert smooth surface finish.
   - **c.** Forged connecting rods are made of metallurgically bonded metal alloy.
   - **d.** Connecting rods transfer motion and force from the piston to the crankshaft.
6. Which of these converts the reciprocating motion of the piston into rotary motion?
   - **a.** The upper part of the engine block
   - **b.** The piston rings
   - **c.** The balance shafts
   - **d.** The crankshaft
7. All of the below are likely causes of bearing wear *except*:
   - **a.** insufficient oil.
   - **b.** improper assembly.
   - **c.** overcooling.
   - **d.** poor cleaning during rebuild.
8. The flexplate and fly wheel are similar in that both:
   - **a.** have a torque converter bolted to them.
   - **b.** are mounted to the end of the crankshaft.
   - **c.** are of similar weight.
   - **d.** are used in automatic transmission engines.
9. The balance shaft serves to:
   - **a.** increase the compression ratio.
   - **b.** decrease cylinder wear.
   - **c.** increase fuel efficiency.
   - **d.** reduce vibrations from engine operation.
10. Increased wear pattern on diagonally opposite sides of the piston indicates:
    - **a.** a bent connecting rod.
    - **b.** problems in the main bearing bores.
    - **c.** problems in the rod bearing inserts.
    - **d.** wear in the rear seals.

ASE Technician A/Technician B Style Questions

1. Tech A says that the upper portion of the block contains the cylinders and pistons. Tech B says that the main bearings support the crankshaft. Who is correct?
   a. Tech A
   b. Tech B
   c. Both A and B
   d. Neither A nor B
2. Tech A says that most pistons use one compression ring and one oil control ring. Tech B says that most pistons are made of steel to withstand the heat of combustion. Who is correct?
   a. Tech A
   b. Tech B
   c. Both A and B
   d. Neither A nor B
3. Tech A says that each main bearing in the engine is fitted with a thrust bearing. Tech B says that thrust bearings limit the end play of the crankshaft. Who is correct?
   a. Tech A
   b. Tech B
   c. Both A and B
   d. Neither A nor B
4. Tech A says that using a thicker head gasket would lower the compression ratio. Tech B says that the effective compression ratio is affected by valve timing. Who is correct?
   a. Tech A
   b. Tech B
   c. Both A and B
   d. Neither A nor B
5. Tech A says that a piston is manufactured slightly oval in shape. Tech B says that the piston pin is slightly oval in shape to allow for heat expansion and lubrication. Who is correct?
   a. Tech A
   b. Tech B
   c. Both A and B
   d. Neither A nor B
6. Tech A says that cylinder walls are manufactured with a slight taper so that the piston rings will make a tighter seal at TDC. Tech B says that ceramic piston coatings assist in the control of piston temperature, which allows for tighter bore clearances and greater engine efficiency. Who is correct?
   a. Tech A
   b. Tech B
   c. Both A and B
   d. Neither A nor B
7. Tech A says that fractured rods have a unique mating surface and that only the cap that was fractured from that rod is the one that will fit that rod. Tech B says that a powdered metal rod is less expensive, lighter, and as strong as a forged rod. Who is correct?
   a. Tech A
   b. Tech B
   c. Both A and B
   d. Neither A nor B
8. Tech A says that when grinding main journals, an undersized main bearing will be required. Tech B says that the main bearing journals should be measured for minimum diameter, out-of-round, and taper. Who is correct?
   a. Tech A
   b. Tech B
   c. Both A and B
   d. Neither A nor B
9. Tech A says that a failed harmonic balancer can cause a crankshaft to break. Tech B says that the crankshaft rod journals are typically measured using a dial indicator. Who is correct?
   a. Tech A
   b. Tech B
   c. Both A and B
   d. Neither A nor B
10. Tech A says that main and rod bearings are glued in place with special bearing cement. Tech B says that main and rod bearings are held in place by a precise amount of "bearing crush." Who is correct?
   a. Tech A
   b. Tech B
   c. Both A and B
   d. Neither A nor B

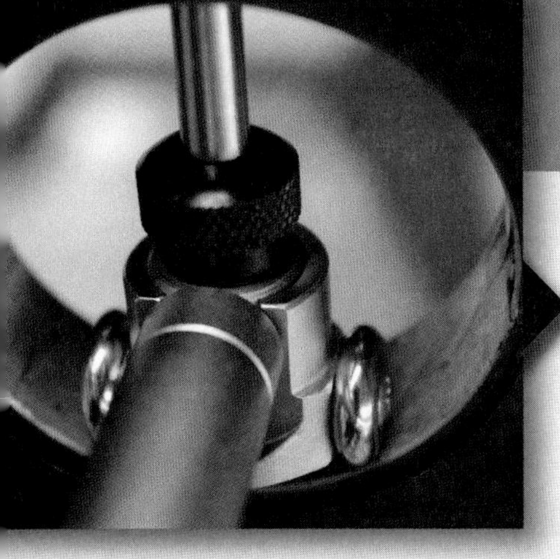

CHAPTER 21 Engine Machining

NATEF Tasks

There are no NATEF tasks for this chapter.

Knowledge Objectives

After reading this chapter, you will be able to:

- **K21001** Explain the purpose of engine machining.
- **K21002** Explain the cylinder block machining processes.
- **K21003** Explain how surface finish is important to the head gasket sealing surfaces.
- **K21004** Explain the procedure for boring and honing the cylinder bores.
- **K21005** Explain the process for resizing the main bearing bores.
- **K21006** Describe the process of machining crankshafts and connecting rods.
- **K21007** Describe the crankshaft refinishing process.
- **K21008** Explain the process of resizing the connecting rods.
- **K21009** Explain the process for machining the cylinder heads.
- **K21010** Explain the process of detecting cracks.
- **K21011** Explain the process of pressure testing a cylinder head.
- **K21012** Explain the process of servicing overhead cam bearings.
- **K21013** Explain the valve grinding process.
- **K21014** Explain valve guide service.
- **K21015** Explain valve seats service.
- **K21016** Describe the process of balancing an engine.

Skills Objectives

There are no Skills Objectives for this chapter.

▶ Introduction

In the past 20 years, the field of automotive technology has changed dramatically. More often than not, when the vehicle is under warranty, the vehicle manufacturer takes failed parts back for analysis and sends the technician brand new parts. For example, if a connecting rod failed and caused catastrophic engine damage, the manufacturer would ask for the engine back for a complete analysis of why it failed. The technician's responsibility would be to replace the broken engine with a newly supplied unit.

If the vehicle is not under warranty, it is still usually less expensive to replace most engine assemblies than to tear down and repair them. Typically, once the engine is replaced, the broken engine is sent to the supplier from which the new engine was purchased for a "core return." The parts store sends the broken engine back to the manufacturer to have it professionally rebuilt and all worn parts replaced or remachined (**FIGURE 21-1**). The automotive technician of today does not necessarily need to know how to proficiently perform each of the procedures listed in this chapter, but he or she does need to know that these processes exist and understand what is possible to repair and what is not. The ability to recognize failures of engines that are beyond repair will help the technician to know whether it is more cost effective for the customer to replace rather than repair.

▶ Engine Machining

K21001

Engine machining is done any time the tolerances of an engine component must be adjusted by adding or removing material. Typically, this process is performed by taking precise measurements and comparing those measurements to a set of factory specifications that are found in service information for that particular vehicle (**FIGURE 21-2**). Engine machining is performed when an engine exhibits a problem, and the technician determines that problem is coming from a component of the engine that is not operating within specifications. For example, a customer complains of a noise when the engine is started up cold. Due to the tone and frequency of the noise and the fact that the volume of the noise goes away as the engine warms up, the technician diagnoses the noise as excessive piston-to-cylinder wall clearance, which is called piston slap. The technician

FIGURE 21-1 A cylinder being machined as part of the rebuilding process.

FIGURE 21-2 Taking precise measurements is part of a quality rebuild.

You Are the Automotive Technician

You are an engine machinist at the local machine shop. A customer wants to know about the machining processes your shop can handle. The 1970 Boss 429 is up for its first rebuild. It spun a main bearing and wore a few thousandths of an inch off of the number 3 crankshaft main bearing journal as well as the main bearing saddle in the block. She wants to know what the options are for repairing those issues. She also said the cylinder bores are worn 0.015–0.016' oversize and would like to know how much larger you would have to bore it so she can order new pistons.

1. How can the crankshaft be saved, even though it is worn badly?
2. What process would you use to repair a main bearing bore that is a few thousandths out of line?
3. To what size would you most likely bore the cylinders out to?
4. What is the purpose of using a torque plate?
5. Once the cylinder has been initially bored, what process is used to finish the surface of the cylinders?

disassembles the engine, checks the piston-to-cylinder wall clearance, and finds that the pistons have excessive clearance. The pistons are replaced, and the block is bored oversize to accommodate the larger pistons. The engine is reassembled, and the technician verifies the repair by testing the vehicle on a cold start-up.

Most technicians and shop owners in today's competitive market have chosen to sublet, or send out, all of their machining work to a machine shop. The main task of machine shops is to restore engine parts to factory specifications. The amount of money it takes to purchase the equipment necessary to perform these machining functions makes it financially unwise for most automotive shop owners to purchase these tools and train their employees on them, when they will only use them once in a while. It is best to leave machining to Master Certified Machinist professionals certified by Automotive Service Excellence (ASE)—although ASE no longer offers Machinist certifications.

TECHNICIAN TIP

Replacing the pistons can change the engine's balance. Take engine balance into consideration when selecting new pistons, or consider having the engine rebalanced during the machining process. See the Engine Balancing section later in this chapter.

Engine Cylinder Block Machining

K21002

Engine Cylinder Block and Cylinder Head Mating Surface Finish

K21003

Surface roughness refers to the irregularities found in the surface finish. These irregularities are the result of the machining process employed to refinish the surface. **Surface roughness average (Ra)** is a measure of how rough a surface is at the microscopic level and is rated as the arithmetic average of the surface valleys and peaks. For precise and consistent results, surface finish should be within a specified range or maximum level of Ra. The Ra is the measurement used to classify how rough a certain mating surface is (**FIGURE 21-3**). A 30 Ra finish is smoother than a 60 Ra finish.

Today's cylinder heads are much lighter and less rigid and use a multilayer steel (MLS) gasket; as a result, the surface needs to be smoother than in the past. A cast iron head must be between 60 and 100 Ra, and an aluminum head must be approximately 30 Ra. The surface must be smooth enough for initial cold seal when starting the engine, and the gasket requires a certain amount of support so the gases from the combustion chamber cannot distort the head gasket. If the surface is too rough (too much bite) and the cylinder head is aluminum and the block cast iron, then sideways shearing from the

FIGURE 21-3 Surface roughness. **A.** 60 Ra. **B.** 30 Ra.

aluminum cylinder head during expansion and contraction will occur, causing the gasket to fail. Always check the manufacturer's specifications.

The engine block can be resurfaced by a surface grinder using a stone wheel or a broach-style surfacer, applying a **cubic boron nitride (CBN) cutter** for cast iron and a **polycrystalline diamond (PCD) cutter** for aluminum. This procedure is performed because the cylinder head and the engine cylinder block need to be perfectly flat at the mating surface where they are bolted together. If they are not perfectly flat, leaks or ruptures will quickly occur.

The desired procedure in surfacing the **engine block's deck** is to surface the deck parallel to the **mainline** where the crankshaft is installed. Surfacing the engine block this way makes the deck height on each cylinder the same. It is important that the deck height be the same at each cylinder to ensure that all pistons have the exact same amount of space between the cylinder head and the cylinder block, to maintain equal compression values.

To check the engine block, measure the distance from where the bearing insert is installed on the mainline to the top of the engine block on each end of the deck surface. Some surface machines use a precision bar that runs through the mainline, and the measurements and settings are taken from the precision bar (**FIGURE 21-4**).

The computer numerical control (CNC) machine is a tool that uses computer programs to automatically execute a series of machining operations. These programs are usually completely computer automated, requiring little input from the operator. With some of the newer CNC-style machines, the measurements are taken from the top of the **engine block dowel pin hole**. The engine block dowel pinhole is where the engine block dowel pins are fixed (**FIGURE 21-5**). These pins are guides that allow exact alignment of the cylinder head and cylinder head gasket when reassembling the engine assembly.

During the machining process, the engine block dowel pins are removed and the dowel pinholes (which are precision-drilled

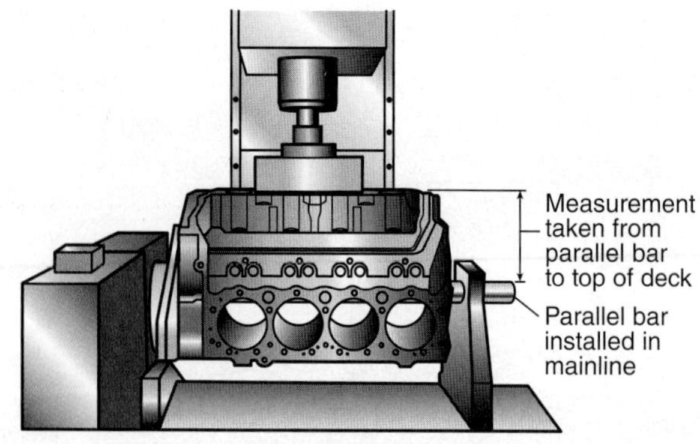

FIGURE 21-4 Measuring the deck height from the setting fixture.

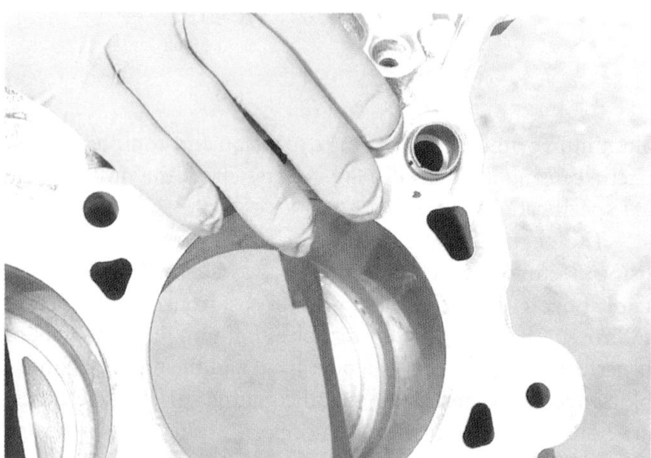

FIGURE 21-5 Engine block dowel pinholes and dowels.

Applied Math

AM-38: Estimate/Exact Value: The technician can distinguish the need to use a specified value versus an estimated value, depending upon the system malfunction.

In the case of certain mechanical failures, it may be possible to determine whether or not a component requires replacement based on estimation, rather than measurement and comparison to specified values. One example is lobe wear on camshafts. Worn cam lobes can cause different symptoms depending on the extent of the wear.

Mildly worn lobes result in a loss of valve lift but still open and close the valve at the correct time with correct lash adjustment. Engine operation at low speed is normal but higher RPM performance is lost due to insufficient valve opening restricting air flow. Excessive valve train noise may be evident, particularly at low speed.

More severe wear can result in the lobe wearing to nearly circular and results in late valve opening, early closure, and dramatically reduced lift. The engine may exhibit only a slight loss of idle quality but will misfire on the affected cylinder at speeds much above idle. Severe wear is noticeable to the eye when comparing against other lobes. Cams are generally case hardened, resulting in only a thin hard layer on their surface. Once the hardened surface is worn through, the lifter will run on the softer base metal, causing rapidly accelerated wear.

holes) are used as depth markers. There are usually two or more of these holes in an engine block deck surface. The machinist will measure the depth of each hole to determine exactly how much metal to take off to get the holes to the exact same depth. In this process, the engine block surface will be resurfaced to be completely flat and true.

Boring the Cylinder Bores

K21004

If the cylinders are damaged, have too much **taper** (meaning they are smaller in diameter at one end), or are out of round, then they should be rebored and honed to fit oversized pistons. Oversized pistons need to be used following boring because the boring process enlarges the cylinder walls, leaving the regular piston too small to be effectively reused. In some cases, the cylinder **bore** might be damaged beyond boring, worn beyond the limits for an oversized piston, or cracked. To repair the cylinder, it must be **bored out** (machined) and a cylinder sleeve installed to fill the extra space created in the boring process. Dry cylinder sleeves require the cylinder to be bored out and the sleeve to have an **interference fit** of approximately 0.002" to 0.0025" (0.05 to 0.06 mm). Dry cylinder sleeves are pressed into the existing bore and do not come into contact with an engine coolant passage (**FIGURE 21-6**). Wet cylinder sleeves are installed and do come into contact with engine coolant. Using sleeves allows the original size of pistons to be used, if desired, restoring the factory bore dimensions.

▶ TECHNICIAN TIP

Boring an engine increases its displacement slightly. To calculate an engine's displacement, you must find the swept volume of the cylinder, based on the bore and **stroke** dimensions. The formula is bore squared, times stroke, times 0.785, times number of cylinders.

Boring of the engine cylinders to a larger size requires the use of a boring machine. The boring machine typically used by engine machine shops is semiautomated (**FIGURE 21-7**). The semiautomatic machine cuts the engine block cylinders to their new oversized dimensions. To start, the engine block must be correctly mounted underneath the cutting bit. The machinist tells the boring machine what engine is being bored and performs preliminary adjustments to the machine, such as inserting the correct cutting bit for the application to be serviced. After all preliminary setup steps are completed, the machine centers itself in the first bore and begins the cutting process. After the first bore is finished, the machine automatically moves to the next cylinder bore, centers itself, and begins the boring process all over again.

When boring the cylinder for a dry sleeve with no top lip, the cylinder wall must be bored to leave a lip on the bottom. This is done by not running the boring bar all the way through the cylinder, so the cylinder sleeve cannot slip in the cylinder bore.

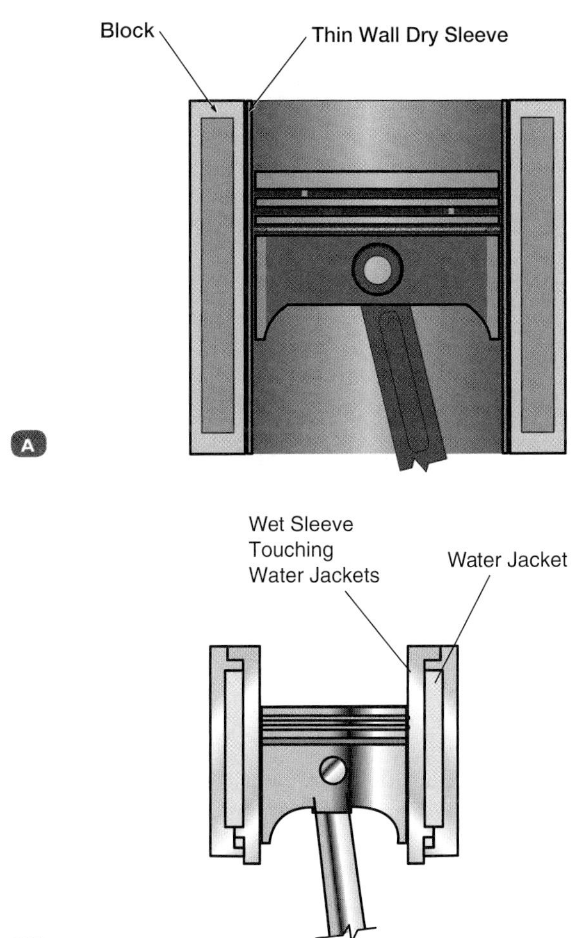

FIGURE 21-6 Cylinder sleeves. **A.** Dry sleeve. **B.** Wet sleeve.

FIGURE 21-7 Semiautomated boring bar.

If the cylinder can be bored oversize without needing a sleeve, usually the cylinder is bored to either 0.030" (0.76 mm) or 0.060" (1.5 mm) oversize, although there are two other relatively common oversizes of 0.020" (0.51 mm) and 0.040" (1.0 mm). For example, if the standard bore size is 4.000" and the cylinder measures 4.016" at the top and 4.009" at the bottom, the machinist will likely bore out the cylinder to 4.030". Before boring begins, the exact diameter of the new pistons must be determined to know how much material to leave in the cylinders to hone out for final clearance. Be sure to refer to the manufacturer's specifications to see how large a particular engine can be bored oversize.

A **boring bar** is another name for a boring machine used for boring cylinders. It usually includes fingers for centering it in the cylinder. There are three styles of boring bars: the portable boring bar, which can be mounted to the top of the engine cylinder block or to a boring machine stand, the semiautomated boring bar machine, and the CNC-style boring machine.

All boring bars fracture the surface as they cut the material away from the cylinder walls, so there must be enough material left so the fractured metal can be honed to final size. The boring machine simply cuts away metal with a sharp bit, leaving microscopic hills and valleys that must be ground down to prevent premature cylinder wear. This grinding is done with a cylinder hone. Typically, the portable boring bar needs to leave more material for honing, approximately 0.005" (0.13 mm), whereas the more rigid machines with the optional cutters and spindle speeds need only 0.0015" to 0.002" (0.04 to 0.05 mm). The portable boring bar should not be used on modern thinner-walled engine blocks, as cylinder distortion can take place from the weight of the boring bar bolted to the engine block.

The boring process leaves a corkscrew pattern on the walls of the cylinder. Hone stones are used to smooth out the surface of the cylinder walls, leaving a smooth, unidirectional finish also known as a **crosshatch** pattern. This pattern appears as a series of very fine, large Xs ground into the walls. Most hones are self-tensioning, which alleviates the need for the technician to determine the amount of tension needed. The crosshatch pattern is made at either a standard 45- or 60-degree angle. Refer to the manufacturer's specifications for the engine you are working on.

A good crosshatch helps to trap a small amount of oil in the surface of the cylinder walls, where it is used to lubricate the piston rings. Achieving the desired crosshatch is a matter of adjusting the speed and stroke of the cylinder hone. As the cylinder hone is moved up and down inside the cylinder, it is also moving in a circle. This up-and-down circular motion allows the hone stones to make the crosshatch on the cylinder walls. For a 45-degree pattern, move the stones at a normal pace. If a more aggressive or steeper-angled X pattern is needed, the process must be sped up. It is not an exact science except when an actual automatic cylinder hone machine is doing the work.

The faster or more rapid the hone is stroked in the cylinder, the higher the angle of the crosshatch becomes. If the angle gets too high, then the cylinder wall will allow oil to run down the crosshatch and not provide enough lubricant to the rings and pistons, causing scoring and leading to low compression. If the angle gets too flat, then too much oil will be held on the crosshatch, and the oil will be burned.

Applied Math

AM-11: Decimals: The technician can multiply decimal numbers to determine conformance with the manufacturer's specifications.

During the overhaul of a V-6 engine, it is discovered that the cylinders are worn beyond manufacturer's specifications. The technician determines that the cylinders must be bored 0.060" (1.52 mm) oversize. The current displacement of the engine is 231 cubic inches (3.79 liters). Our task is to use a formula to determine the new cubic inch displacement of the engine.

The bore of the engine is 3.860" (98.04 mm) after it was enlarged and the stroke is 3.4" (86.36 mm). For the first step, we will calculate the cubic inch displacement of one cylinder, and then multiply times six cylinders. The volume of a cylinder is determined by multiplication of pi (3.14) * radius squared * length of stroke. In our scenario, this is 3.14 * 3.725" (2403 mm) = 11.69 square inches (7542 square mm) * 3.4" (86.36 mm) = 39.768 cubic inches (0.651 liters).

This is the volume or cubic inch displacement of one cylinder. When multiplied by six cylinders, we have a "new" displacement of 239 cubic inches (3.9 liters). This new displacement is the result of boring the cylinders and installing new pistons that are 0.060" (1.52 mm) oversize. Now each cylinder will meet manufacturer's specifications because excessive cylinder taper and out-of-round conditions have been corrected. This engine should give many miles of trouble-free service.

Applied Math

AM-39: Equal/Not Equal: The technician can distinguish when a measurement is not equal to the manufacturer's specifications or tolerances.

When assessing an engine's machining requirements for reconditioning, one of the most important requirements is to measure the cylinder bores to determine if boring, or in more extreme cases, sleeving, is required. Cylinder bores are first inspected for obvious damage, then they are measured for wear, out-of-round, and taper.

Using an accurate cylinder bore gauge, measurements are taken at the upper, middle, and lower sections of the bore in two directions, along and across the block. The amount of bore wear is calculated by taking the maximum and minimum of the six numbers and calculating the difference between them.

If the factory bore measurement is 78 mm (3.079"), with a specified wear limit of 0.15 mm (0.006"), and the difference between a bore's taper measurements is 0.19 mm (0.007"), the block must be rebored and fitted with oversized pistons. If measured wear is below the specified limit, 0.10 mm (0.004") in this example, the engine could be reassembled with the original pistons, assuming no other damage.

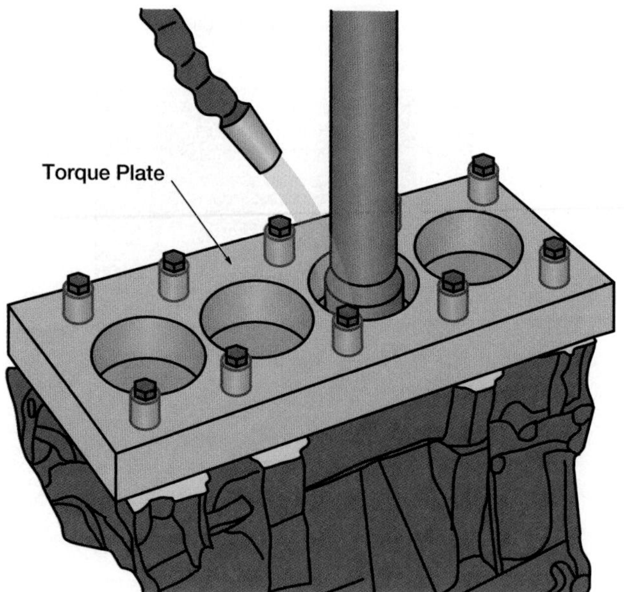

FIGURE 21-8 A torque plate bolted on a block to simulate a head torqued in place while being torqued.

Torque Plate

When the cylinder head is torqued, it exerts force around the cylinder wall and can distort the machined dimensions, especially on lighter-weight engine blocks. To eliminate this problem, a **torque plate** is installed that simulates the cylinder head torqued in place when the cylinder is bored and honed (**FIGURE 21-8**). The torque plate is used in performance machine shops and is also being used more often on the new style of engines that have lightweight cores made of thinner aluminum, which are more prone to cylinder shifting. Some cylinders can distort as much as 0.0015" (0.04 mm) when the cylinder head is bolted down. One high-performance industry, NASCAR, has even gone to the trouble of not only using a torque plate but also circulating hot water through the block to ensure that the cylinder bores are the size they will be at full temperature. This process is known as hot honing. The use of hot honing is not cost effective for the typical engine, as the machine is very expensive to purchase and operate.

Honing Cylinders

There are several options to consider before **honing**. If there are no imperfections or grooves, and the cylinder walls need only to be deglazed, a **ball hone** is used. If the cylinder was bored oversize, then the cylinders have to be honed to final size with a rigid hone. A ball hone consists of stone balls that are attached to the ends of flexible metal rods. When this assembly is inserted into the cylinder and spun with a hand drill and moved back and forth, the abrasive stones wear the "glazing" off the cylinder walls and give the cylinder walls a nice new crosshatched pattern (**FIGURE 21-9**). Glazing refers to the finish produced when the crosshatch scratches have been worn away, leaving the cylinder smooth and shiny and unable to retain oil. This condition reduces compression, as oil helps the rings to seal, and can also result in oil moving past the rings and burning in the combustion chamber. The honing process with a

FIGURE 21-9 The ball hone is used by hand with a drill motor to restore the crosshatch pattern in the engine cylinder.

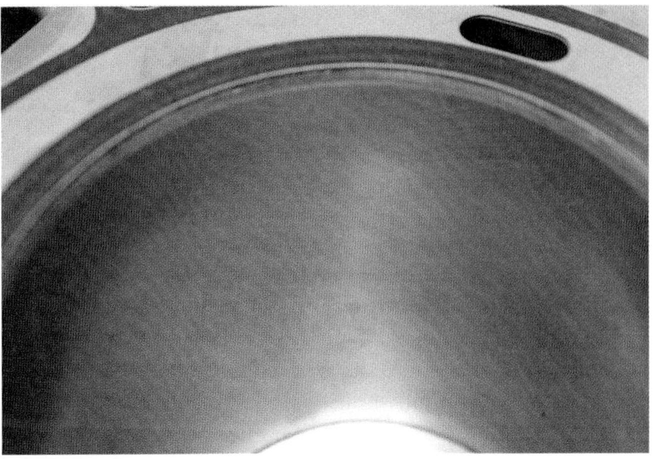

FIGURE 21-10 The proper crosshatching pattern ensures that the piston rings will seat properly and not result in oil burning.

ball hone is done only if the cylinder does not need to be bored and is in good shape in regard to taper and wear. The ball hone does not restore cylinder size; it simply reapplies the crosshatching. Producing the correct pattern takes practice to get just the right speed of the stone moving up and down in the cylinder in relation to the spinning speed.

After the boring process, a rigid hone is used for the smoothing-out process of the cylinder walls and is performed as the final step in refinishing the cylinder. A crosshatched pattern allows the proper amount of oil to stay on the cylinder walls, which allows the piston rings to seat into the cylinder walls and ensures a proper seal so as to not lose compression or burn oil (**FIGURE 21-10**).

Honing is typically performed by a honing machine (**FIGURE 21-11**). This machine consists of three or more elongated smooth stones attached to an arbor. This assembly is inserted inside a finished/bored cylinder and is spun (either by a computer-controlled honing machine or by a machine operated by a machinist) until the proper surface finish and final bore size of the cylinder are reached. Honing is typically performed by the engine rebuilder or machinist.

Once the cylinders are at the proper dimension and have the correct crosshatch pattern, it is time to clean and lubricate them. First, wash the block in a spray cabinet; then scrub the cylinders with warm, soapy water and a stiff brush to remove any metal and stone particles. Next, take a clean, lint-free rag and pour some clean oil on it. Rub the oil into the cylinder walls. If any discoloration appears, rewash the cylinders, and recheck them with a clean, oiled rag. When no dirt or discoloration occurs, the oiled cylinders are ready when it is time to install the pistons.

FIGURE 21-11 Cylinder being honed on a honing machine.

Main Bearing Bores

K21005

Premature bearing failure and crankshaft failure can occur with an out-of-line main bearing bore. Proper bore alignment must be ensured between all the main bearing bores. The bore alignment can be checked for straightness by using a straightedge with a feeler gauge (**FIGURE 21-12**). The engine main bearing bores can become misaligned due to the stresses released over time from the initial casting process. Over time, engine blocks can distort with temperature changes, and the bearing bores can become misaligned. If this happens, engine bearings will be excessively worn, particularly the center bearing because of the bore shift.

FIGURE 21-12 Checking main bearing bore straightness with a straightedge and feeler blade.

The main crankshaft bores must be perfectly circular to ensure that they are not out of round or egg shaped. Just like engine block cylinders, these bores can wear into an oblong or egg-shaped hole, which can damage the crankshaft. Out-of-round is checked with a bore gauge tool or inside micrometer.

To correct a bore alignment issue, the engine block has to be **line bored**, or **line honed**. In this process, the main journals must first be made smaller. This is done by removing the main caps and cutting the mating surface down about 0.003" to 0.005" (0.076 to 0.13 mm) (**FIGURE 21-13**). After all of the main caps are cut, the engine block mating surface is cleaned to remove dust or debris. The main caps are torqued back into place, the line hone or line bore machine is set up, and they are resized to the original main bearing bore specification. The manufacturer's specifications for the main bearing bore diameter must be followed to ensure that the proper bearing crush is maintained. Line boring ensures that the bores line up properly and are perfectly round (**FIGURE 21-14**).

FIGURE 21-13 Cutting main caps.

▶ TECHNICIAN TIP

Whenever the crankshaft centerline is refinished, the centerline of the crankshaft moves slightly upward. If the centerline moves too far, then cam timing issues can occur. It can also cause the compression ratio to be increased slightly. So always make sure that only a minimal amount of material is removed during line boring.

▶ Crankshaft Grinding and Polishing

K21006, K21007

The **crankshaft journals** have to be inspected for scratches, pits, cracks, and grooves when rebuilding the engine. The bearings need a true, clean surface to run on, or they will be torn up on the very first running of the engine. When imperfections cannot be polished out to 0.0005" (0.013 mm) or less, the crankshaft needs to be reground. Regrinding involves placing the crankshaft in a crankshaft grinding machine (**FIGURE 21-15**). The lathe spins the crankshaft while a special bit shaves or grinds the metal away.

Most crankshafts are **induction hardened**. This process hardens the metal by raising the temperature of the part with heat or magnetic fields and then suddenly cooling it in water, oil, or another chemical. Using the induction hardening method allows the crankshaft to be reground on the mains and rod bearing journals without having to redo the hardening process, which is important because the hardening will be as deep as 0.08" (2.03 mm). Some heavy-duty and performance cranks use **nitriding**. Nitriding is the process of slowly heating the crankshaft in a sealed container with the oxygen removed. Nitrogen is introduced and penetrates the surface. Nitriding produces a harder, more durable surface than induction hardening but only penetrates approximately 0.02" (0.51 mm). The crankshaft will have to be rehardened if the mains or rod journals must be reground. Heating and hardening cause the crankshaft to grow, which the machinist must take into account when performing the initial metal removal before hardening.

Crankshafts are usually ground to 0.010" (0.25 mm), 0.020" (0.51 mm), or 0.030" (0.76 mm) undersize, meaning they are

FIGURE 21-14 Line honing a mainline.

FIGURE 21-15 A crankshaft being ground.

ground slightly smaller than their original size. When choosing a crankshaft main or rod journal bearing, it must match the amount of material that was removed during grinding. For example, a crankshaft ground to 0.010" undersize must match up to a 0.010" undersize bearing, usually marked ".010 US" on the back of the bearing shell (**FIGURE 21-16**). The undersize bearing is actually thicker to make up for the smaller crankshaft journal. The crankshaft main and rod journals can be ground to different sizes, but all rod journals must be cut to the same size. Likewise, all the main journals must be the same size as each other. For example, the crankshaft could be a 0.010/0.020, which equals 0.010" rods and 0.020" mains.

The crankshaft includes a **fillet area** (the area where the rod bearing journal or main bearing journal meets the crankshaft) that is radiused, from the crankshaft journal up along the crankshaft counterweight (**FIGURE 21-17**). The **radius** is important because it helps strengthen the crankshaft. The whole "radius" process gets rid of sharp edges, which create stress points inside the crankshaft. If a crankshaft is going to break, it will usually break on a sharp edge or corner. These fillet areas help to reduce those stress areas.

When a crankshaft cannot be restored because the grooves are too deep, or a bearing has spun on the journal, then it is a common practice to use spray welding or submerged arc welding. Spray welding adds materials to the journal by melting and spraying molten metal onto the crankshaft journal surface so that there is material that can be ground. Submerged arc welding uses a steady stream of flux to protect the molten metal from oxygen as the filler material melts and binds to the crank surface. Crankshafts are usually ground the opposite direction of the crank's rotation and then polished the same direction of the crank's rotation to counteract the corkscrew effect of grinding.

The oil passage holes in the crankshaft need to be chamfered before polishing the journals. **Chamfering** is cutting into the top of the oil passage hole (which is usually right in the center of the bearing's operating surface on the crankshaft journals) with a chamfering bit (**FIGURE 21-18**). This process takes any sharp edge at the top of the oil hole and makes it more like a 45-degree angle, which allows the edge to not dig into the bearing the way a 90-degree edge would. It also alleviates the 90-degree stress point on the crankshaft. Crankshaft polishing is performed on a crankshaft polisher (**FIGURE 21-19**). The crankshaft polishing machine rotates the crankshaft slowly, and a polishing belt slides over the journals to clean up and polish the surface. The polishing belt is usually 320 grit or finer.

FIGURE 21-16 Undersize bearing.

FIGURE 21-18 Crankshaft oil passage chamfer.

FIGURE 21-17 The radiused fillet area of a crankshaft reduces stress points in the crankshaft.

FIGURE 21-19 Crankshaft being polished.

Connecting Rod Resizing

K21008

Before any machining of connecting rods can be performed, the pistons must be removed. If the rods are of the full-floating style, then the snap rings can be removed from the piston, and the pins can be worked out of the piston and rod. If the rods are of the semi-floating style, then the piston pins have to be pressed out of the piston and rod with a small hydraulic press. When pressing piston pins, be sure to use the proper piston support tool to cradle the piston in order to avoid damaging it during removal and installation (**FIGURE 21-20**).

Rod—Big End

The big end of the connecting rod is made up of two halves. Each mating half can have either machined surfaces or, in the case of powdered metal connecting rods, **fracture split** surfaces, which are identified by the ragged finish of the parting surfaces on the big end (**FIGURE 21-21**). The ragged fracture line ensures better rod cap alignment on assembly because the fractured surfaces fit together in perfect alignment. The connecting rod's big end has to be checked for roundness. This measurement is just as important as measuring the roundness of the crankshaft main bearing journals and main bearing bores. Any time the old rod bolts are removed and new bolts are installed, the size of the "big end bore" may become distorted due to the press fit of the bolts. Accordingly, the size of the big end of the connecting rod should be rechecked and machined if necessary.

To resize the connecting rod's big end, the bolts are removed (**FIGURE 21-22**) and the cap and rod mating surfaces are ground approximately 0.0015" to 0.0025" (0.038 to 0.14 mm). Then the rod cap is bolted back onto the rod and the bore machined back to the proper size. The challenge with powdered metal rods is that because the fractured surface is critical to their alignment, they cannot tolerate having those surfaces ground smooth. Thus, they cannot be resized, unless bearing inserts of a larger diameter are available for that particular engine.

Once the rod mating surfaces have been machined lightly and bolted back together, either a rod honing machine or a rod boring machine can be used for resizing (**FIGURE 21-23**).

FIGURE 21-20 A semi-floating piston pin being pressed out of the rod and piston.

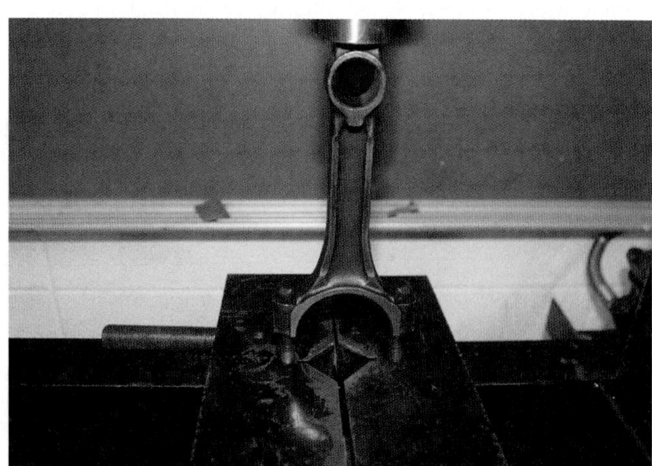

FIGURE 21-22 Removing the connecting rod bolts.

FIGURE 21-21 Big end connecting rods. Powdered metal with fracture split faces.

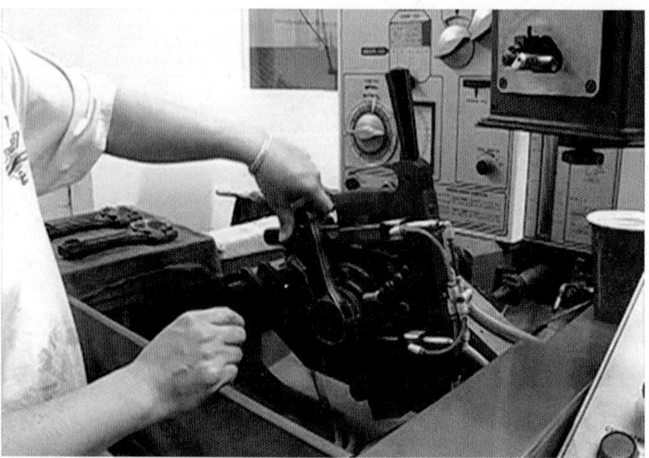

FIGURE 21-23 Connecting rod big end being resized.

A machinist will frequently measure the size of the connecting rod to ensure that it does not become oversized during this process. The big end of the connecting rod must be resized to factory specifications to achieve the proper amount of bearing crush when torqued. The torquing process actually "squeezes" the bearings, causing them to maintain just the right amount of grip on the connecting rod while maintaining the correct oil clearance for the engine crankshaft to spin smoothly within them.

Rod—Small End

Connecting rods may have pressed-in wrist pins or, in some engines, a floating wrist pin. With the floating type of wrist pin, the pin can spin in both the piston pin bosses and the small end of the connecting rod. The wrist pin is typically held in place with a snap ring on each side of the piston pin boss. Some floating wrist pin designs use a bushing or bearing on the small end of the connecting rod. A small hydraulic press is used to press the bushing out of the rod and install a new one. Some manufacturers require the bushing to be burnished (rubbed until smooth or glossy) when installed. A rod honing machine or rod boring machine can be used for this procedure.

The pressed-in wrist pin is either press fit into the piston pin boss or more often press fit into the small end of the connecting rod. Wrist pins can be installed in two ways: pressed in or with a rod heater. When pressing piston pins in, use the correct tool to support the piston; otherwise, the piston will be damaged (**FIGURE 21-24**). Lube the wrist pin and inside of the rod and piston bores with assembly lube. Carefully press the wrist pin into place and to the proper position. When using a rod heater, only heat the rod to the minimum amount needed (**FIGURE 21-25**). Before heating, set the pin stop to the proper depth so that the pin can be inserted quickly. Use the pin tool to push the pin into the piston and rod, all the way to the stop, in one smooth push (**FIGURE 21-26**).

FIGURE 21-24 Pressing a wrist pin into a rod and piston.

FIGURE 21-25 When using a rod heater, only heat the rod to the minimum amount needed to install the pin.

FIGURE 21-26 Use a pin tool to push the pin into the piston and rod.

▶ Cylinder Head Repair

K21009

Cracks in cylinder heads are a serious problem that is relatively common because of their lightweight construction. The main cause of cylinder head cracks is thermal stress. When metal is heated, it expands. Although cylinder heads and head bolts are designed to withstand a certain amount of expansion and stretch, the rise in engine operating temperatures due to cooling system faults can push a cylinder head, gaskets, and fasteners beyond their design limits. Over time, expansion and contraction can cause the metal to deform, causing cracks to form. Another cause of cracks is caused by coolant freezing. Water expands approximately 10% when it turns to ice, so it creates extreme forces that can crack cast iron and aluminum. Depending on the location of the crack, coolant, oil, or combustion gases may leak out of the cylinder head into the surrounding engine parts. Pressure testing the cylinder block and heads is a reliable method for identifying hard-to-see cracks.

FIGURE 21-27 A cylinder head being resurfaced.

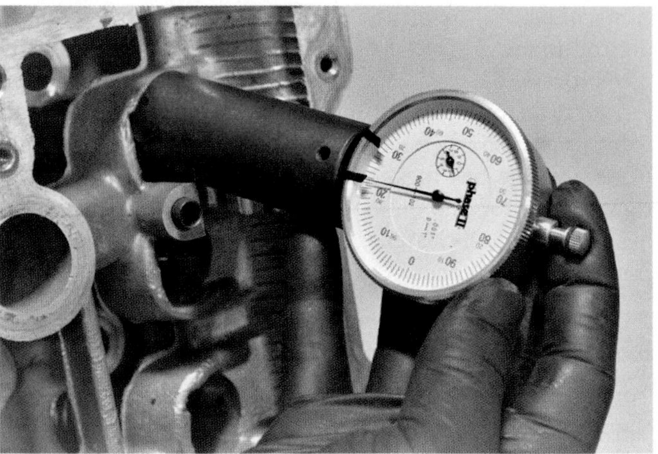

FIGURE 21-28 Before machining the valves or seats, take an installed height measurement for reference when installing the valves.

The presence of a crack does not automatically condemn the cylinder head to the scrap bin. In some cases, repair methods have proven successful and less costly than replacement. Small cracks can be fixed by pinning (also known as cold repair), a process of drilling holes in a crack and installing overlapping pins to fill it that are then peened over to seal and blend the surface. Pinning sounds simple enough, but it should only be performed by a highly trained machinist. Furnace welding can also be used to repair cylinder head cracks. This process involves heating the cylinder head to extreme temperatures (about 1300°F [704°C]) and welding iron into the crack with an acetylene torch. This process requires even more skill than pinning and should not be attempted by anyone who is not an expert.

Heat stress and pressure stress can also cause the head to warp so that the sealing surface is no longer perfectly flat. Resurfacing the cylinder head restores a perfectly flat and specified surface finish to the head for placement of a new gasket (**FIGURE 21-27**). Resurfacing is best performed by milling the surface but can also be performed by grinding with the proper head-surfacing machine. The resurfacing process must be precise and to the proper surface finish, with no more material removed than is necessary to clean up the surface, and is best left to a skilled machinist.

If you are dealing with an OHC head, then there are two surfaces that must be in alignment—the head surface and the camshaft journal bores. If the head surface is warped, it is likely that the camshaft bore is also warped. Because the camshaft is straight, if the camshaft bore is warped, the camshaft will bind in the journal bearings. This means that a warped OHC head should first be straightened so that the journal bore is aligned before the head surface is machined. Straightening a head is usually performed by bolting a head down to a straightening plate with shims placed between the head and the plate so that the head will be under tension. Then the head is heated up not more than 500°F (260°C) and allowed to cool slowly. The head is again checked for flatness, and if it is flat, the head surface and camshaft journals can then be machined.

Damaged valves and valve seats can sometimes be reconditioned and reused rather than replaced. However, the price of new valves has dropped significantly over the years, making replacement the more common option for newer vehicles. For older vehicles for which replacement parts are not readily available, reconditioning is necessary. Reconditioning involves cleaning, fault detection, and machining. The process requires special equipment and a skilled and knowledgeable operator.

Cylinder head machining is fairly time intensive, so make sure you are fully prepared before beginning. When disassembling a nonadjustable valve-type cylinder head, it is recommended that before machining the valves or seats, the valve installed height measurement be taken and recorded for future reference (**FIGURE 21-28**). This information is needed during the machining and assembly process. Also, before any machining can begin after the disassembly and cleaning of the cylinder head, the cylinder head should be **magnafluxed** or dye checked to locate any external cracks. To check for internal cracks, pressure testing is performed.

Crack Detection

K21010

If a cylinder head or block is made of cast iron, then it can be magnaflux tested using a dry or wet method. The dry method consists of using an iron dusting powder that is available in different colors. A permanent magnet or electromagnet creates a magnetic field that causes a high concentration of magnetic flux that flows through the cast iron cylinder head. At any cracks in the surface, a north and south magnet is created on each side of the crack, which attracts the iron powder when applied (**FIGURE 21-29**). The machinist will notice a nice solid line of powder sticking to the crack. The wet method uses a liquid iron and fluorescent dye mixture. It is used along with an electromagnet to pull the dye into any cracks, and then a black light is used to illuminate it.

A dye check can also be used. It works by soaking the metal surfaces with a special dye that wicks down into any surface cracks. The surface is then wiped dry, and either an ultraviolet light is used to illuminate any cracks or a white chemical developer is sprayed onto the surface, and the dye bleeds into the developer, making the crack visible (**FIGURE 21-30**). Dye checking works on both ferrous and nonferrous metals, including aluminum.

FIGURE 21-29 Magnafluxing a cast iron cylinder head with cracks.

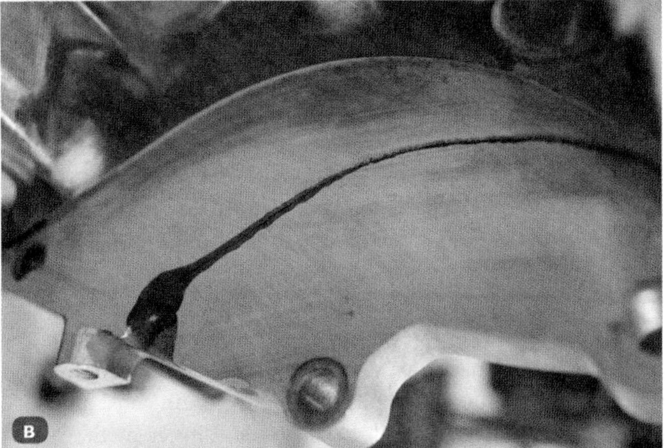
FIGURE 21-30 A. Fluorescent dye and ultraviolet light showing a crack. **B.** Fluorescent dye and white developer showing the crack.

Pressure Testing

K21011

Pressure testing is normally used on aluminum cylinder heads because the magnaflux process cannot be used. There are two different styles of pressure testers. In the first, a plate is cut out for the combustion chamber areas and used to seal off the water jackets, except for one where air pressure is injected (**FIGURE 21-31**). The cylinder head is checked for leaks by submerging it in water or by using a soapy water formula that is sprayed all over the cylinder head with a squirt bottle. The technician looks for bubbles, which result from an air leak. In the other type of pressure tester, the cylinder head is placed on a pressure-testing bench. The surface of the bench is usually made of soft rubber, which acts as the seal around the bottom of the cylinder head. Pressure is then introduced into the combustion chamber via the spark plug hole, and the machinist notes if any pressure is lost or if a "hissing" or leaking noise is heard coming from around the valves or any part of the casting.

FIGURE 21-31 A head being pressure tested.

Overhead Cam Bearings

K21012

Overhead cam engines have camshafts that typically ride directly in bores in the cylinder head. There are no removable bearings. The cam rides right on the machined aluminum and is lubricated by a thin coating of oil that is fed to the surface through an oil galley hole. Oil pressure must make its way up from the bottom of the engine to the top of the cylinder head to lubricate the camshaft. If oil pressure falls, then the camshaft lacks lubrication and can score the camshaft bores in the cylinder head. This is a problem because if the surface is damaged, there are no bearings to replace.

Another potential problem occurs if the engine has overheated and the cylinder head is warped; in this case, the cam bores can become misaligned, just as the main bearing bores for the crankshaft can become misaligned. If the cam bores are misaligned enough, it can cause the camshaft to bind up and possibly break.

When machining a head surface for a warpage concern, do not forget to check the cam bore alignment. To do so, use a straightedge designed for this purpose, or use a straight camshaft and make sure it turns freely in the bores. If the cam bores are misaligned, then the head most likely will have to be straightened. A machine shop can do this using heat and pressure to pull the head back into line. Once the head is straight, the cam bores can be rechecked.

If the head has removable cam bearing caps (**FIGURE 21-32**), then the bore can be machined in a process similar to line

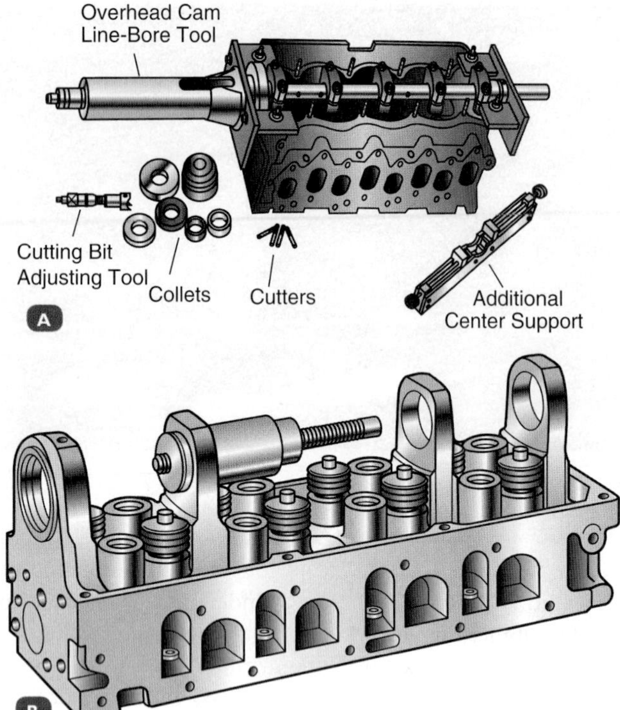

FIGURE 21-32 **A.** Cam bores using removable bearing caps. **B.** Cam bearing with one-piece cam towers.

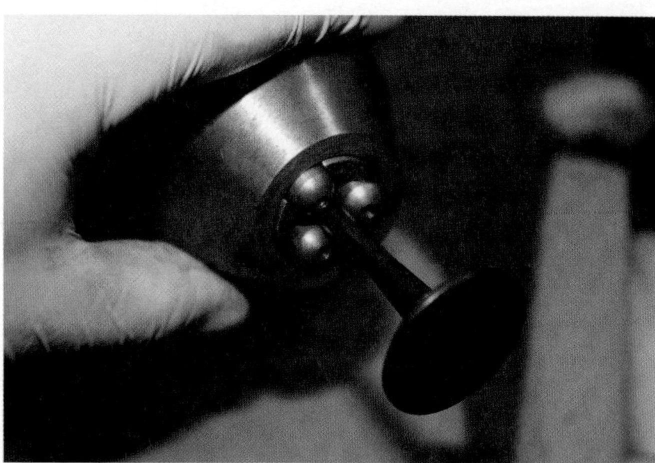

FIGURE 21-33 Installing the valve in a collet.

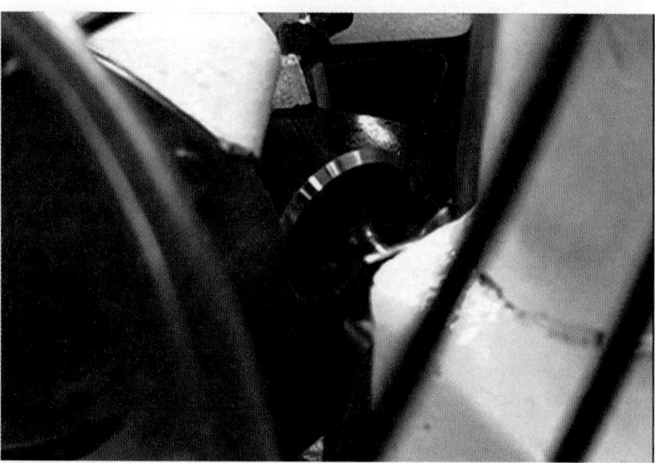

FIGURE 21-34 Grinding the valve.

boring of the main bearing journals. Simply cut a small amount of metal off of the bottoms of the cam bore caps and remachine the bore back to its original specification. If the bore has one-piece cam towers, then the machinist may have to see if a camshaft with oversized cam journals can be purchased for the engine. Oversized cam journals can sometimes be found for an engine in 0.015" or 0.030" (0.38 or 0.76 mm) sizes. This means that the original bore size will have to be machined 0.015" to 0.030" larger. Another possible option is to find replacement bearing shells or inserts made by some aftermarket companies, to repair the cam bore. These bearing inserts work similarly to a standard camshaft bearing. The last repair option would be to either spray weld the cam journals to a larger size and machine them to the desired size, or weld the cam bores and remachine them to the proper size. This process is very costly and is done only when a replacement head cannot be easily found.

Valve Grinding

K21013

If the valves are in reusable condition, both the valve face and the valve tip will have to be ground. If the engine uses adjustable valves, then both surfaces can be ground at this time. If the engine uses non-adjustable valves, then the tips must be ground after the valve face and valve seat have been ground so that the installed valve height can be adjusted by grinding the valve tip to the proper length.

The valve stem must be within factory specifications, which is why the valve stems and margins should be checked first. There are multiple styles of valve grinding benches or valve refacing benches; follow the manufacturer's specific procedures. The valve stem end is first chamfered so that when inserted in the **collet** that holds the valve, it is held centered (**FIGURE 21-33**). The valve grinding machine can be set up to grind the valves at any angle, typically specified by the manufacturer. Always check the specifications, and set the machine to the proper valve face angle. When the machine is turned on, the valve rotates in one direction, and the grinding stone rotates in the other. The valve is slowly fed into the grinding stone; do not take too much off at a time, and make sure the face of the valve moves across the face of the grinding stone from side to side until the valve face is refinished and true (**FIGURE 21-34**).

When grinding the valves, make sure not to remove any more metal than necessary, to avoid making the margin thinner. This margin, if allowed to become too thin, can cause the edge of the valve to burn or melt because the thinner metal cannot dissipate the heat as well. The smaller the valve margin, the hotter the edge of the valve becomes and the harder it is to dissipate the heat to other parts of the cylinder head. That is why the thickness of the valve margin is important.

Valve Guides

K21014

Before grinding the valve seats, the valve guides must be checked and repaired to ensure they are within factory specifications. The valves and valve stems must also be checked to verify that they are reusable. If the valves are worn beyond specifications, they must be replaced. If the valve guides are worn, they have to be repaired. There are different procedures to repair valve guides in an integral cylinder head (non-replaceable guide) or a non-integral cylinder head (replaceable guides). Each procedure is described next.

For the integral cylinder head (non-replaceable guide), there are four methods of repairing the valve guides:

1. **Knurling** the cylinder head using a diamond knurl, a tool offered by Sunnen Products. Knurling is when a bit with a spiral groove (unique to the cylinder head that is being worked on, because of its specific diameter) is threaded through or run through the valve guide. This tool leaves behind raised ridges on either side of the groove and effectively shrinks the size of the valve guide because of these raised ridges. This process takes the current material and squeezes it out, closer to the area where the valve stem rides. The knurled guide is then reamed so that it is back within specifications. Reaming removes a small amount of the material displaced by the knurling tool. The knurling process is not used by most reputable machine shops anymore because it is typically considered a "quick fix" and would usually only last for 20,000 to 40,000 miles. Knurled guides do not stand up to the same abuse and wear and tear that an insert or liner would typically stand up to.
2. Reaming the guide oversize and using a valve with an oversized valve stem. Valves with oversized stems are usually very expensive and may not be available for all engines.
3. Installing a thin-walled bronze liner (**FIGURE 21-35**). The valve guide is reamed out with a special ream, and an oversized thin-walled bronze liner is installed and is burnished to size.
4. Installing a thick-walled guide that is made of bronze or heat-treated cast iron. Thick-walled guides can usually be driven out with an air hammer, but in some cases, they may need to be machined out with a head shop. Thick-walled valve guides are generally installed with an air hammer (**FIGURE 21-36**).

Hardened Seat Inserts

K21015

Hardened seats are valve seats that have been heat treated to make them more robust and capable of withstanding the severe demands of today's engines and fuels. Hardened valve seats are used to prevent valve seat regression. Regression refers to the valve actually sitting farther up inside the cylinder head due to the missing material that has been worn away from the valve seat over the life of the vehicle. Hardened seats are also necessary in aluminum heads, as the aluminum is relatively soft.

Installing hardened seats is also a common practice on early cast iron heads that have seats that were not induction hardened from the factory. Lead was added to gasoline, starting in the 1920s, as an antiknock additive. It also proved to be an antiwear additive for valve seats. The lead acts as a chemical buffer, keeping the hot valves from "microwelding" themselves to the seat. After lead was taken out of gas, if the exhaust valve seat was not induction hardened or a hardened seat insert was

> **TECHNICIAN TIP**
>
> Valve guide knurling is a procedure used to resize the inner diameter (ID) of a valve guide. The process creates spiral grooves down the length of the valve guide inner bore. The material displaced by the creation of the grooves reduces the valve guide ID. This process can be used to extend the life of a valve guide. Generally, it is preferred to replace the guide instead of knurling, because the replacement will last for longer than the knurled guide. In situations where replacement is not possible, valve guide knurling can put the cylinder head back to factory specification and return the component to service.

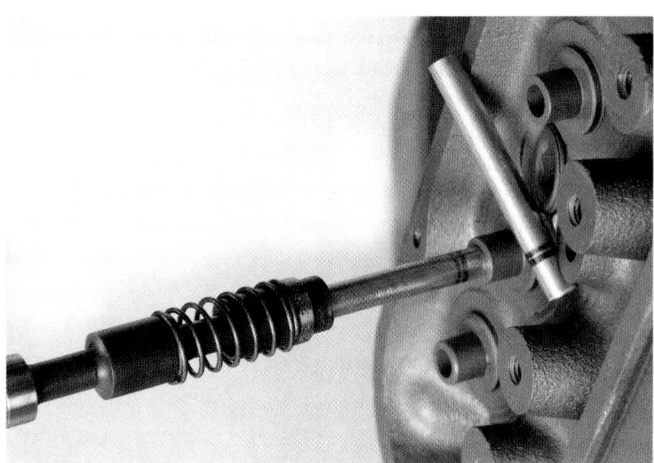

FIGURE 21-35 Thin-walled guide being installed.

FIGURE 21-36 Thick-walled valve guide being installed.

not installed, the valve seat would deteriorate and fail. This would result in the valve sinking into the head, quickly causing valve seat failure.

Valve seats can be of the integral type or the replaceable type. Integral valve seats are used in some cast iron heads by induction hardening the valve seat area. Replaceable valve seats can come in many different materials—nickel alloy, chrome alloy, cobalt alloy, stainless steel, or even cast iron. Valve seats intended for compressed natural gas (CNG) or propane engines may be made of tungsten steel, with additives blended for lubricating qualities. Check with the manufacturer to determine which hardened seat is recommended. This will vary depending on which fuels are being used.

To install a hardened seat in an integral type head, a head shop is used to cut a recessed cavity in which the seat will fit (**FIGURE 21-37**). If the head uses replaceable valve seats, the seats will have to be removed by either prying, heating the head in a cylinder head oven, or even welding the seat, which causes the seat to contract when the weld cools. New hardened seats are pressed or hammered into place with an interference fit (**FIGURE 21-38**).

Grinding/Cutting Valve Seats

Valve seats must be newly refinished to create a near-perfect sealing area for the valves when they are installed into the cylinder head. Refinishing removes minor pits and wear that appear over time. There are two basic ways to refurbish the **valve seats**. The first way is to grind the valve seat with stones (**FIGURE 21-39**). When using grinding stones, a **three-angle grind** is typically used to properly form the valve seat. The three-angle grind provides three different angles—one to create the surface of the valve seat, one to narrow the valve seat from the top, and one to narrow the valve seat from the bottom. Using the top- and bottom-narrowing stones allows the width and height of the valve seat to be properly created and positioned (**FIGURE 21-40**). The second way to refurbish valve seats is with a cutter. Valve seat cutters are available in different forms. One style uses individual cutters with a particular size and angle, similar to how the grinding stones work. The other type of cutter is a single cutter made to cut all three angles at once (**FIGURE 21-41**).

Valve seat contact is critical to the operation of the valve because much of the valve's cooling occurs where the valve face contacts the valve seat. Usually, this is slightly outward from the center of the valve face. This contact area is what seals the valve

FIGURE 21-37 Cutting the recessed cavity for a hardened valve seat.

FIGURE 21-39 Grinding a valve seat.

FIGURE 21-38 Installing a hardened valve seat.

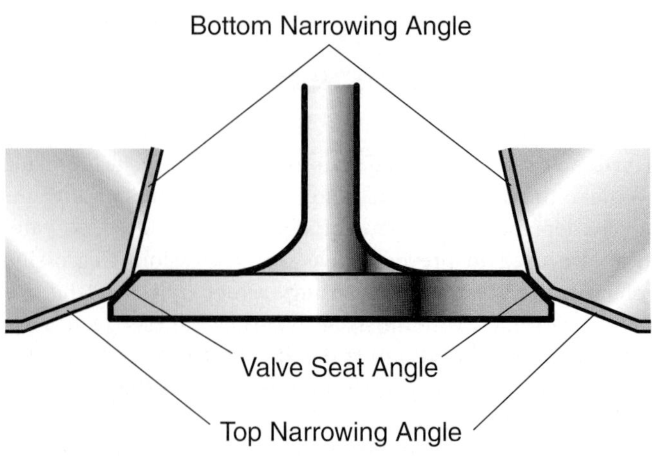

FIGURE 21-40 Valve seat with proper angles.

FIGURE 21-41 Cutting a valve seat with a valve seat cutter.

FIGURE 21-42 Checking valve seat contact. It should show an unbroken narrow line toward the outer portion of the valve face.

to the cylinder head and makes a completely sealed combustion chamber. Valve seat contact must be checked to ensure that the valves are sealing at the proper location on the valve and also to ensure that the valves are actually sealing. A good way to check the valve seat contact on the valve is to use Prussian blue or a dry-erase marker. Put Prussian blue or a dry-erase marking on the face of the valve. Install the valve in the cylinder head and open and close the valve rapidly against the valve seat. Remove the valve and inspect the valve face. There should be a small area of no contact on the top of the face and a small area of no contact on the bottom of the face (**FIGURE 21-42**).

Valve seat runout is how much the valve seat is out of round, called valve seat concentricity. Valve seat runout should be checked with a special runout gauge. The seat runout gauge uses a pilot shaft to slide tightly into the valve guide. The tool is able to spin on the pilot shaft, and the top has a dial indicator. As you spin the dial indicator against the seat contact surface, runout can be determined (**FIGURE 21-43**). The ideal runout is zero; however, most engines will function well with less than 0.002" (0.05 mm). For an idea of how much that is, the thickness of a sheet of notebook paper is typically about 0.004" (0.1 mm).

Valve guide-to-valve stem clearance should typically be less than 0.0025" (0.064 mm), depending on the engine application. Any more than this will affect valve sealing. Excessive valve guide clearance will also cause valve stem seals to leak, as the valve stem will move and wear the seal quickly. If the clearance is too small, the valve could seize or hang up in the cylinder head assembly, causing a cylinder misfire and possibly catastrophic damage if the piston hits the open valve at top dead center. Also, if the valve hangs open on pushrod engines, the pushrod can be dislodged from the rocker and lifter, and on some applications, the lifter may come out of its bore, resulting in a loss of oil pressure.

When the valve and seat are first machined, their surfaces need to wear together to create an effective seal. To help the valves and seats initially wear together, the valves are ground or cut at a half or one degree less than the valve face angle. This allows the valve face and seat to meet at a very narrow ring near the top of the valve face and creates what is called an **interference angle**. For instance, if the valve face is cut to a 44-degree slope, then the seat will be cut to a 45-degree slope, thus allowing the two metal pieces to meet at a point (**FIGURE 21-44**). The valve face and seat would then wear in fairly quickly to create a tight seal across the complete seat width.

FIGURE 21-43 A valve seat being checked for runout.

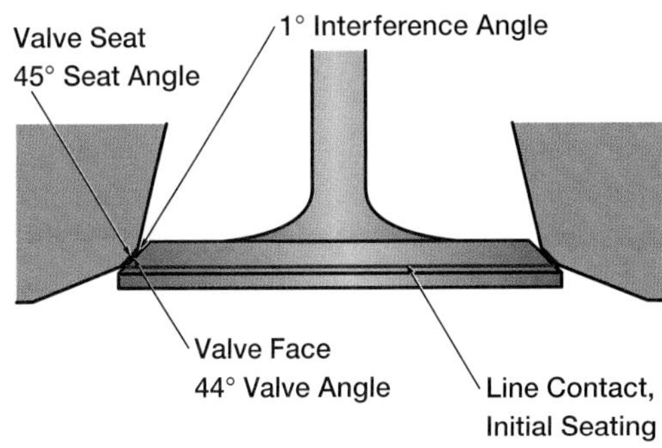

FIGURE 21-44 An interference angle allows the valve to seal well initially while wearing in to a full seat over a short period of time.

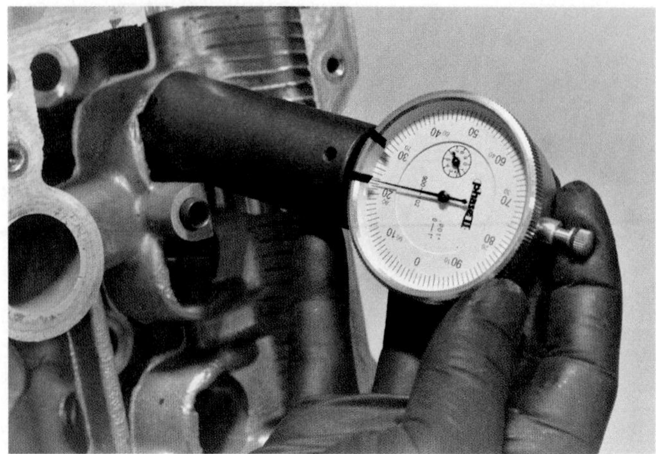
FIGURE 21-45 Valve stem height being measured.

FIGURE 21-46 Valve spring height being measured with a valve spring height tool.

Another consideration for the engine machinist to remember is that as material is cut from the valve seat and the valve face, the valve will move higher up in the head. This change in position can result in a valve that is held open when the valve train is installed. A critical measurement is valve stem height after the machining of the valve and valve seat (**FIGURE 21-45**). Refer to the manufacturer's specifications, or ensure an initial reading is taken before valve work is done, if no specifications are published.

Valve stem height is measured with the valve in the head and seated fully. The machinist then measures from the tip of the valve stem to the bottom of the valve spring pocket. If the valve stem height is not correct, then the machinist will use the valve grinding machine to remove the required amount of material from the tip of the valve stem. Care must be taken not to remove too much material, as doing so could cause excessive valve clearance leading to valve train noise. The machinist also must be sure not to machine the valve stem of sodium-filled exhaust valves; if these types of valves are used, then replacement of the seat and valve may be required to restore original engine specifications. Sodium reacts violently with water, so be aware of what you are working with.

Also remember that when the valve stem moves up in the head, so does the retainer groove on the valve stem. This movement affects valve spring installed height, which affects spring pressure. If the spring installed height is too high, then pressure will not be as high, and the valve could hang open at high rpm and create lots of problems, such as being contacted by a piston. Valve spring installed height is restored by using shims under the valve spring to bring the bottom of the spring up. Measurement of the valve spring height is from the bottom coil of the spring to the top coil of the spring (**FIGURE 21-46**). Be sure to shim correctly because if the spring installed height is too short, coil spring binding could occur when the valve reaches full lift, and this will result in valve spring breakage or valve train damage. There should be a minimum of 0.060" (1.52 mm) of clearance between coils of a valve spring when fully compressed, but like all rule-of-thumb specifications, be sure to follow the manufacturer's specifications if available.

FIGURE 21-47 Surfacing a cylinder head.

Cylinder Head Surface

The cylinder head surface is critical because of the combustion pressures, water passages, and, sometimes, oil passages. The cylinder head must be flat and surfaced to the proper Ra. The cylinder head warps much more easily than the engine block because of its design; it does not have the same supporting structure. A precision ground straightedge and feeler gauge are used to determine flatness.

Before surfacing the cylinder head, check the manufacturer's cylinder head thickness specifications to make sure the cylinder head has enough material to be surfaced. If the cylinder head is warped more than 0.003" (0.076 mm), it is recommended the cylinder head be surfaced; always check the manufacturer's specifications. The cylinder head can be surfaced by a surface grinder using a stone wheel, or a **broach-style surfacer** using a CBN cutter for cast iron or a PCD for aluminum (**FIGURE 21-47**). A cylinder head with overhead cams usually cannot be machined as much as an adjustable pushrod engine, as the timing chain or timing belt tension will be affected. Again, refer to specifications before machining.

> ▶ TECHNICIAN TIP
>
> If an overhead cam head is warped, the cam bearing bores should be checked for straightness before surfacing the head. If the cam bores are not straight, then surfacing the head will leave the cam journals out of line when the head is bolted onto the block. This will cause binding of the cam in the bearings. If the cam journals are not in line, then the head will have to be straightened before resurfacing.

▶ Engine Balancing

K21016

Almost all production engines on the road today, as well as 99.9% of all vehicles that a technician encounters on a daily basis, have been "prebalanced" at the factory. Because of manufacturing tolerances, that balancing is relatively close, but not perfect. In a standard engine rebuild, no further balancing should normally be necessary. Still, if you are rebuilding a performance engine, or you just want the extra comfort of knowing that the engine is within much tighter tolerances, consider having the engine balanced. This process is usually only done in a machine shop. Engine balancing is performed to ensure that all weight imbalances in the engine are neutralized by taking away or adding the proper amount of weight in the proper places.

Some engines are internally balanced, and others are externally balanced. An internally balanced engine is balanced completely from the inside. This means the piston assemblies, rods, and crankshaft are balanced so that their rotating forces cancel out. In some in-line four-cylinder engines and 90-degree V6 engines, manufacturers may use an extra counterbalance shaft to smooth out engine vibrations (**FIGURE 21-48**). Counterbalance shafts are only added at the factory or by the original equipment manufacturer, and they must be timed with the crankshaft.

FIGURE 21-48 Internally balanced engine with a counterbalance shaft.

An externally balanced engine has external weights added on the harmonic balancer and on the flex plate or flywheel (**FIGURE 21-49**) These weights work in conjunction with the internal counterweights to balance the rotating mass of the engine. The challenge of externally balanced engines is that if the harmonic balancer or flywheel ever needs to be replaced, the new component will not match the rotating assembly, and the engine will not be as balanced.

FIGURE 21-49 Engine balancing. **A.** Externally balanced harmonic balancer. **B.** Externally balanced flywheel.

If an engine has a balance issue during initial creation and testing, the factory will calculate the necessary amount of counterforce needed as well as exactly where it should be applied. This process has worked very well for manufacturers over the years and helps explain why our vehicles are as smooth and quiet as they are, even though the engine parts are moving very fast in various directions within the engine.

For engine balancing purposes, the parts of the crankshaft assembly are divided into two categories: **rotating weight** and **reciprocating weight**. Rotating weight is the amount of weight that is moving in a circular motion. It includes everything from the center of the connecting rod down to the connecting rod cap and all components of the crankshaft. Reciprocating weight is the amount of weight that is moving up and down. It is everything from the middle of the connecting rod upward, including the piston, wrist pin, and rings. Even the estimated weight of the oil on the piston and rings is added to the reciprocating weight. The crankshaft spins, so it is part of the rotating mass, and so are the components that spin with it, such as the rod bearings, the crank gear, the flex plate or flywheel, and the harmonic balancer. The pistons, rings, and wrist pins move up and down, so they are part of the reciprocating weight. With the connecting rods, the small end reciprocates while the big end rotates.

The first step in balancing is ensuring that all pistons and connecting rods weigh the same amount (**FIGURE 21-50**). All pistons are weighed and compared, the lightest piston is identified, and some of the aluminum from the heavier pistons is removed so that all pistons are made equal in weight. The weight is then recorded. The same procedure is performed on the wrist pins. If there are wrist pin retainers, one is weighed and recorded.

The rotating rod (the big end) is weighed on the connecting rod balance fixture (**FIGURE 21-51**). All of the rods must weigh the same, so weight is removed as needed on the balance pad on the big end to match the weight of the lightest rod. The weight is recorded. The reciprocating end of the rod (the small end) is weighed using the connecting rod balance fixture. All small ends must weigh the same. Some of the material on the balance pad is ground away as needed on the small end so all the rods match the lightest rod, and then the weight is recorded.

Once the pistons and connecting rods are balanced and weighed, the bob weights are bolted to the rod journals on the crank to mimic this weight (**FIGURE 21-52**). The reciprocating

FIGURE 21-50 A piston being weighed.

Applied Science

AS-16: Balance: The technician can measure and balance rotating systems.

The bottom-end components of an engine function as a unit, balancing them as an assembly ensures reliable and vibration-free operation. Bottom-end balancing occurs in two stages: the static balancing of reciprocating components and the dynamic balancing of the crankshaft.

The first stages of the balancing operation address the weight of the connecting rods, pistons, and associated components. Connecting rods are mounted in a balancing fixture, and their ends are weighed. First, all big ends are weighed, the lightest is determined, and metal is machined from the heavier rods until they match the lightest. This procedure is repeated for the small ends. Next, the pistons and their associated wrist pins, clips, and rings are weighed, and the lightest identified. Again, machining is carried out to remove metal from the heavier pistons until they match the lightest.

The weights of the balanced connecting rods and piston assemblies are used to calculate the rotating and reciprocating weights. **Bob weights** are used to simulate the weight of the reciprocating components and are fitted to the crank journals. The crankshaft is mounted in a specialized balancing machine and spun at optimum speed. A computer indicates the amount of imbalance present and at which points on the crankshaft weight needs to be added or removed. Weight may be removed by machining metal from existing counterweights or adding heavy metal. In some situations, flywheels, clutch pressure plates, and harmonic balancers may be fitted to the crankshaft during dynamic balancing.

FIGURE 21-51 The big end of the rod being weighed.

FIGURE 21-52 Bob weights being bolted to the crank to simulate the weight of the pistons and rods when balancing the crankshaft.

weight is balanced first, using a bob weight card (a generic card for recording specific weights as they are added and removed) to record the weights.

The bob weight card will say "×2" on the rod's big end since there are two rods sitting on one journal in a V8 application. We must multiply the weight of the big end by two to come up with the weight of the bob weight we will bolt to the journal. An estimate of the oil in the crank is added to the bob weight card. The rotating weight (the big end of connecting rod, the main bearings, and a bit of oil) is added to 50% of the reciprocating weight (wrist pin, small end of the connecting rod, and piston). This calculation will give the total weight needed for the bob weight.

For example, say that the total calculated bob weight is 1260 grams, which includes the lead shot (adapters). The bob weights will be adjusted to weigh 1260 grams and bolted onto the rod journals. The Woodruff key (a piece of metal used to position the harmonic balancer on the crankshaft), the timing chain crankshaft gear, the harmonic balancer, and the flywheel or flex plate are installed. The crankshaft is spun, and using a strobe light, the **engine balancer** shows the operator the heavy spot on the crankshaft and how much weight to remove (**FIGURE 21-53**). The balancer works very similarly to a wheel and tire spin balancer.

The crankshaft is balanced by either adding weight in a certain area or removing weight from a certain area. Weight is removed by selectively drilling the already present counterweights on the crankshaft assembly. Weight is added by drilling holes and inserting Mallory metal plugs (**FIGURE 21-54**). Mallory is about twice as heavy or dense as regular steel; for example, for every 10 ounces of steel you drill out, you can fit 20 ounces of Mallory metal into the resulting hole. All weights should be recorded in grams because it is the unit of weight used by manufacturers and in most shops.

FIGURE 21-53 A crankshaft being spun while it is being checked for balance.

FIGURE 21-54 Mallory metal plugs being installed in a crankshaft.

▶ Wrap-Up

Ready for Review

- Most automotive shops and technicians send their engine machining work to a specialty machining shop where components are restored to factory specifications.
- The measure of surface roughness is based on surface valleys and peaks and is specified as surface roughness area (Ra).
- The deck height of an engine block must be identical at each cylinder in order to maintain equal compression values.
- The engine cylinder block dowel pinholes are used as alignment guides for the cylinder head and cylinder head gasket.
- Damaged cylinders must be rebored to fit oversized pistons.
- Boring bars, which position and align a single-point tool, can be portable, semiautomated, or CNC style.
- After boring, the cylinder walls must be honed to the proper surface texture and crosshatch pattern.
- A ball hone may be used to reapply the cylinder's crosshatch pattern.
- Most hones are self-tensioning but must be adjusted to the correct speed to achieve the desired angle of crosshatch.
- A torque plate may be used to keep the cylinder true to the machined dimensions when honed.
- Main bearing bores must be properly aligned to prevent premature bearing and/or crankshaft failure.
- The engine block must be line bored or line honed to correct a bore alignment issue.
- Oil clearance can be checked quickly with Plastigauge or by subtracting the size of the main bearing journal from the size of the main bearing bore with bearings installed.
- Crankshaft journals can be machined undersize if they aren't damaged too badly.

- Small imperfections in crankshaft journals can sometimes be polished out.
- Crankshaft rod journals must all be cut to the same size, as must all main journals.
- A crankshaft can be spray welded to add enough surface material for grinding.
- Crankshaft oil passage holes must be chamfered before journal polishing begins.
- The big end of a connecting rod must be resized to ensure proper rod cap alignment; this can be done with a rod honing machine or a rod boring machine.
- Before disassembling a non-adjustable valve-type cylinder head, measure and record the valve installed height.
- Cylinder heads should be magnafluxed, using the wet or dry method, to check for cracks.
- Aluminum cylinder heads cannot be magnafluxed; therefore, they must either be dye checked or pressure tested via water or a pressure-testing bench.
- Overhead cam engines risk camshaft damage if they lose lubrication.
- If the engine overheats, the head can warp, which can cause excessive wear on the camshaft bearings or even break the camshaft.
- Cam bores must always be aligned with each other.
- The four methods of repairing valve guides are knurling, reaming the guide oversize and using a valve with an oversized valve stem, installing a thin-walled bronze liner, or installing a thick-walled guide (bronze or cast iron).
- When grinding valve faces, be careful not to grind away too much valve margin.
- Hardened valve seats are installed to prevent the valve from sinking into the head.
- Valve seats may be made of nickel alloy, chrome alloy, cobalt alloy, stainless steel, cast iron, or tungsten steel (for CNG or propane engines).
- Valve seats can be refurbished with stones or cutters.
- Check valve seat contact, valve seat runout, and valve guide–to–valve stem clearance.
- It is critical to measure valve stem height after machining to ensure that the valve will close once the valve train is installed.
- Measure valve spring height and adjust to correct height with shims.
- Cylinder head resurfacing may be performed if the head deck surface is found to be warped or if there is excessive pitting.
- The majority of production engines are "prebalanced" before leaving the factory.
- Engines can be internally balanced (crankshaft) or externally balanced (flywheel and harmonic balancer).
- When balancing a crankshaft assembly, the parts are divided into rotating (moving in a circular motion) and reciprocating (moving up and down) weight categories.
- Equipment needed to simulate balance of a V8 engine includes an engine balancer, a connecting rod balance fixture, a gram scale, gram calibration weights, bob weights, lead shot, a disc sander, and an equipment lathe or milling machine.

Key Terms

ball hone An assembly of metal rods, with abrasive stones attached, which is inserted into the cylinder and spun to break the glazing off the cylinder walls and make a new crosshatched pattern.

bob weights Weights used during the balancing process that are used to mimic the exact weight of the piston and connecting rod weights for that particular rod bearing journal.

bore A tool used to make or enlarge a hole in or through an object.

bored out An object that has been bored.

boring bar A long bar used to position and align a single-point tool, such as an engine cylinder block boring machine, for boring operations.

broach-style surface A tool used in resurfacing the engine block.

chamfering The process of cutting into the top of the oil passage hole with a specialized chamfering tool or regular drill bit to make the top of the oil hole more like a 45-degree angle, to keep the edge from digging into the bearing the way a 90-degree edge would.

collet A cone-shaped gripping adapter that encloses and grips a rod or shaft when inserted into the sleeve of a lathe or other machine.

crankshaft journals Smooth, curved, machined areas where other parts fit to the crankshaft.

crosshatch A pattern of lines placed at angles to each other, appearing as a series of Xs across a surface.

cubic boron nitride (CBN) cutter An engine block resurfacing tool used on cast iron parts.

engine balancer An expansive bench unit, complete with computer controls, a lathe and drill press for machining metal for the balancing procedure, and all necessary adapters for most applications of crankshaft assemblies.

engine block deck The portion of the engine cylinder block on which the head gasket lies and to which the cylinder head is bolted.

engine block dowel pin hole The hole where the engine cylinder block dowel pins are fixed.

fillet area The area of the crankshaft that meets up with the rod bearing journal or main bearing journal.

fracture split A method used on powdered metal connecting rods where the big end of the connecting rod is broken in half and can be identified by a ragged finish of its parting surfaces.

hardened seat A valve seat that has undergone a heat-treating process to become more robust and capable of withstanding the severe demands of today's vehicles and drivers.

honing The smoothing-out process of the cylinder walls performed after the boring process. This is the final step in refinishing the cylinder; it creates the proper crosshatched pattern necessary for the piston rings to seal and seat into to give the combustion chamber the proper seal.

induction hardening A process for hardening metal by using a magnetic field to quickly heat up the surface and then suddenly cooling it in water, oil, or another chemical.

interference angle The angle formed between the valve face and the valve seat; usually it is ½ to 1 degree.

interference fit A process used for holding parts in place where the inside diameter of the outer piece is slightly smaller than the outside diameter of the inner piece.

knurled A process in which a special bit with a spiral groove is threaded through the valve guide to create raised ridges on either side of the groove, effectively shrinking the size of the valve guide.

line bored/line honed A process for correcting a mainline that is out of line.

magnafluxing A process for checking the cylinder head for cracks, using a magnetic powder that sticks to any cracks.

mainline The imaginary centerline down the engine cylinder block.

nitriding A metal surface-hardening process using nitrogen and high temperatures.

polycrystalline diamond (PCD) cutter An engine block resurfacing tool used on aluminum parts.

pressure testing A process for checking the cylinder heads for cracks, using air pressure.

radius A straight line extending from the center of a circle to its edge or from the center of a sphere to its surface.

reciprocating weight The amount of weight that is moving up and down. It is everything from the middle of the connecting rod upward, including the piston, wrist pin, and rings.

rotating weight The amount of weight that is moving in a circular motion, including everything from the center of the connecting rod down to the connecting rod cap and all components of the crankshaft.

stroke The movement of the piston up or down.

surface roughness average (Ra) A measurement used to classify how rough a particular mating surface is at the microscopic level.

taper An object that is smaller in diameter at one end.

three-angle grind A process of grinding the valve seats on the cylinder head so that air can pass through them more smoothly and quickly. Also used to narrow and position the seat.

torque plate A 2" (51 mm) thick plate that is bolted where the cylinder head is fastened on the engine block during the boring process, simulating head bolt torque.

valve seat The surface against which a valve seals when the valve is closed.

Review Questions

1. When resurfacing an engine block surface, the machinist determines exactly how much metal to take off by measuring the:
   a. depth of the dowel pinholes.
   b. surface roughness of the cylinder head.
   c. height of the dowel pins.
   d. thickness of the head gasket.
2. Choose the correct statement regarding crosshatch.
   a. It does not allow oil to be trapped on the surface of the cylinder walls.
   b. The desired pattern can be achieved by adjusting the speed and stroke of the cylinder hone.
   c. If a steeper-angled X pattern is needed, move the stones up and down at a slower pace.
   d. If the angle gets too flat, the compression will be too low.
3. The ball hone:
   a. restores cylinder size.
   b. simply reapplies the crosshatching.
   c. is done when there is considerable wear and tear.
   d. does not remove glazing.
4. What must be done when correcting a main bore alignment issue?
   a. the crank journals must first be made smaller.
   b. the cylinders must be bored oversize.
   c. the pistons must be replaced with oversize pistons.
   d. main caps should be cut.
5. During crankshaft grinding, the:
   a. main and rod journals can be ground to different sizes.
   b. rod and main journals must be cut to the same undersize.
   c. main journals that are worn are the only ones that should be ground.
   d. crankshaft is ground slightly bigger than its original size.
6. A small crack in a cylinder head can be fixed by:
   a. boring it oversize.
   b. milling the surface.
   c. welding.
   d. sealing.
7. Which of these can be used to check for internal cracks?
   a. Magnaflux testing using dry method
   b. Dye checking
   c. Pressure testing
   d. Magnaflux testing using wet method
8. If the cam bores are misaligned, you should first:
   a. machine the bore in a process similar to line boring of the main bearing journals.
   b. straighten the head if it is warped.
   c. weld the cam bores and remachine them to the proper size.
   d. simply increasing the bore size.
9. During valve grinding, the:
   a. valve and grinding stone rotate in the opposite directions.
   b. grinding stone moves side to side.
   c. valve and grinding stone rotate in the same direction.
   d. valve is held in place and does not move.
10. Knurling is:
    a. reaming the guide oversize and using a valve with an oversized valve stem.
    b. when a bit with a spiral groove is threaded through or run through the valve guide.
    c. installing a thin-walled aluminum liner.
    d. installing a thick-walled guide that is made of aluminum.

ASE Technician A/Technician B Style Questions

1. Tech A says that failure analysis will assist the technician in determining whether it is more cost effective to replace rather than repair an engine. Tech B says that when an engine fails under warranty, the manufacturer may want it back for complete analysis of why it failed. Who is correct?
   a. Tech A
   b. Tech B
   c. Both A and B
   d. Neither A nor B
2. Tech A says that Ra is critical in the machining process to ensure head gasket sealing. Tech B says that a Ra rating of 0–1 is the best sealing surface. Who is correct?
   a. Tech A
   b. Tech B
   c. Both A and B
   d. Neither A nor B
3. Tech A says that full floating pistons use keepers to retain the piston pin. Tech B says that some piston pins are press fitted in the rod. Who is correct?
   a. Tech A
   b. Tech B
   c. Both A and B
   d. Neither A nor B
4. Tech A says that when measurement of the cylinder bore is out of specifications, just install an oversized piston. Tech B says that 0.030" over and 0.060" over are two common piston sizes. Who is correct?
   a. Tech A
   b. Tech B
   c. Both A and B
   d. Neither A nor B
5. Tech A says that after boring the cylinders oversize, the cylinder bore will have to be honed. Tech B says that you should always bore out the block as big as possible. Who is correct?
   a. Tech A
   b. Tech B
   c. Both A and B
   d. Neither A nor B
6. Tech A says that the main bore alignment can be checked with a straightedge and feeler blade. Tech B says that the main bore can be straightened by resurfacing the bottom of the main bearing caps and then resizing the main bore. Who is correct?
   a. Tech A
   b. Tech B
   c. Both A and B
   d. Neither A nor B
7. Tech A says that valves must be replaced if the margin is below the minimum specified. Tech B says that the valves must be replaced if the stems are not within specifications. Who is correct?
   a. Tech A
   b. Tech B
   c. Both A and B
   d. Neither A nor B
8. Tech A says that some valve seats can be replaced. Tech B says that newer style valve seats are made of soft materials, so they don't create as much wear on the valves. Who is correct?
   a. Tech A
   b. Tech B
   c. Both A and B
   d. Neither A nor B
9. Tech A says that main bearing clearance can be calculated by knowing the inner diameter of the main bearing and the outer diameter of the main journal. Tech B says that rod bearing and main bearing clearance can be measured with Plastigauge. Who is correct?
   a. Tech A
   b. Tech B
   c. Both A and B
   d. Neither A nor B
10. Tech A says that when balancing an engine, bob weights are used to simulate the weight of the engine block. Tech B says that balancing a crankshaft can include removing or adding weight to the crankshaft. Who is correct?
    a. Tech A
    b. Tech B
    c. Both A and B
    d. Neither A nor B

CHAPTER 22
Engine Assembly

NATEF Tasks

- **N22001** Assemble engine block. (MAST)

Knowledge Objectives

After reading this chapter, you will be able to:

- **K22001** Describe the application and use of various types of adhesives/sealers.

Skills Objectives

After reading this chapter, you will be able to:

- **S22001** Perform all preassembly actions and measurements.
- **S22002** Prepare the crankshaft and block for preassembly.
- **S22003** Perform piston, ring, and block clearance checks.
- **S22004** Install the pistons on the connecting rods.
- **S22005** Perform a temporary test buildup and related measurements of the engine.
- **S22006** Install main bearings, and measure the clearances.
- **S22007** Install the crankshaft, and measure the clearance.
- **S22008** Assemble the piston and rod assemblies.
- **S22009** Measure rod bearing clearance.
- **S22010** Check piston-to-deck and piston-to-valve clearance.
- **S22011** Perform final assembly of the short block.
- **S22012** Clean and paint the engine.
- **S22013** Install soft plugs and oil gallery plugs.
- **S22014** Install the crankshaft.
- **S22015** Install the piston rings.
- **S22016** Install the pistons in the block, and measure side clearance.
- **S22017** Install the cylinder head and valve train.
- **S22018** Install the head gaskets.
- **S22019** Install the cylinder heads.
- **S22020** Assemble the valve train, and adjust the valves.
- **S22021** Install the oil pickup, oil pump, and oil pan, and prime the engine.
- **S22022** Install the oil pan and timing cover.
- **S22023** Install the external components.
- **S22024** Install the intake manifold.
- **S22025** Install the other external components.
- **S22026** Install the accessories.
- **S22027** Inspect and replace camshaft drive belts and chains.
- **S22028** Establish camshaft position sensor indexing.
- **S22029** Inspect and replace valve stem seals on an assembled engine.

▶ Introduction

N22001

Engine assembly is a process that must be carried out in a very clean, organized, and uncluttered environment and with the utmost attention to detail (**FIGURE 22-1**). Failure to heed this advice or carelessness during this process will result in an engine that may not live for more than a few minutes after being started up, so organization is critical. In this chapter, we first perform a test buildup of the engine to measure all of the critical clearances and to make sure all of the parts work together properly. Next, if everything is within specifications, the engine will be carefully disassembled, cleaned once more, and painted so it is ready for final assembly. Following the procedures laid out in this chapter, along with the service manual's procedures for the engine you are rebuilding, will result in a long-lasting engine that runs well.

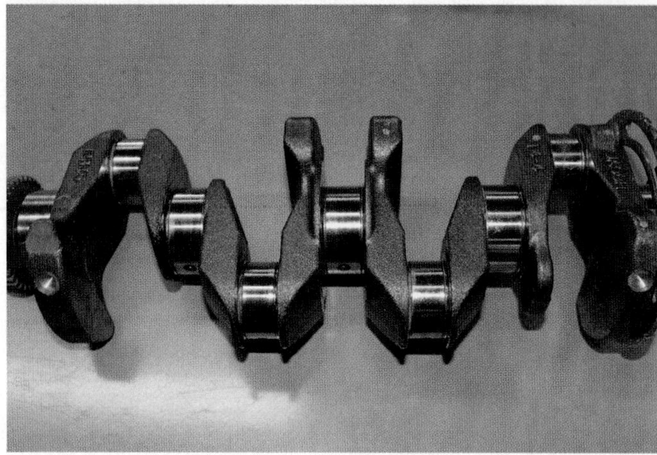

FIGURE 22-2 A crankshaft that needs to be evaluated before reuse or machining.

▶ Engine Preassembly

S22001

Before final assembly of the engine, some preparatory work is necessary. It is good practice to examine all clearances and surface finishes prior to assembly, especially if the engine has been machined (**FIGURE 22-2**). In most cases, measurements will be taken, but keep your eyes open for any defects that would affect the engine operation. We also show you how to measure piston-to-valve clearance in case you are using a nonstock camshaft. All of this helps to avoid issues once the engine is reassembled and running.

FIGURE 22-1 An engine being assembled in a clean, organized workspace.

Crankshaft and Block Preparation

S22002

The engine block should have been cleaned and sprayed with rust inhibiter at the machine shop. It is good practice to reclean the block and crankshaft passages. Using a small brush or other similar tool, ensure that all oil galleries and passages are clean. You may need to run a matching thread chaser down all of the holes in the block to make sure they are clean (**FIGURE 22-3**). Spray the block with a rust inhibiter, and wrap the engine with a plastic bag until you are ready to assemble the engine.

Clean the threads in the rear flange of the crankshaft with a thread chaser. Clean the crankshaft, using an industrial parts cleaner. After the crankshaft has been washed, take a small wire brush and make sure all oil passages are clean (**FIGURE 22-4**). Next, flush the passages with hot water. After the entire crankshaft has been cleaned and rinsed with hot water, blow it dry, and spray an antirust inhibitor over the crankshaft.

You Are the Automotive Technician

You have successfully disassembled, cleaned, and machined the engine parts and checked all necessary clearances. Prior to the final reassembly of the engine, you performed a temporary test buildup to verify that no additional machining was needed. After assembling the block and heads, you need to install the timing belt and set the timing for all four camshafts.

1. Why is it important to perform a temporary test buildup of the rotating assembly prior to final assembly?
2. How do you measure piston-to-valve clearance?
3. What damage is possible if you install the timing belt/chain with the wrong camshaft timing?
4. How are rod and main bearing clearances measured?
5. How is the main thrust bearing centered?

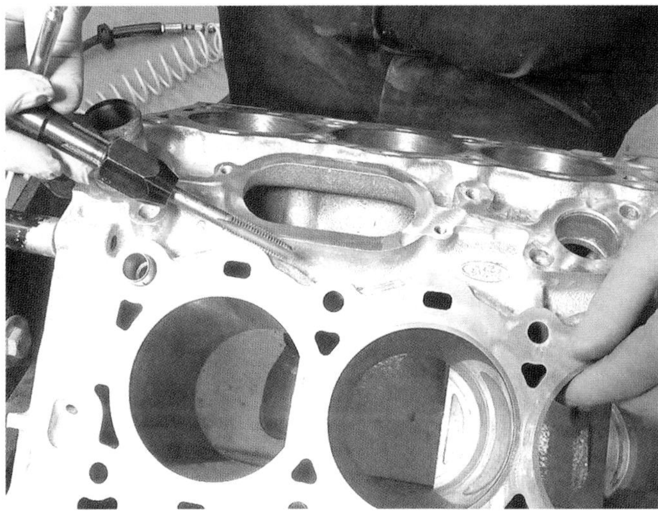

FIGURE 22-3 Threaded holes being chased.

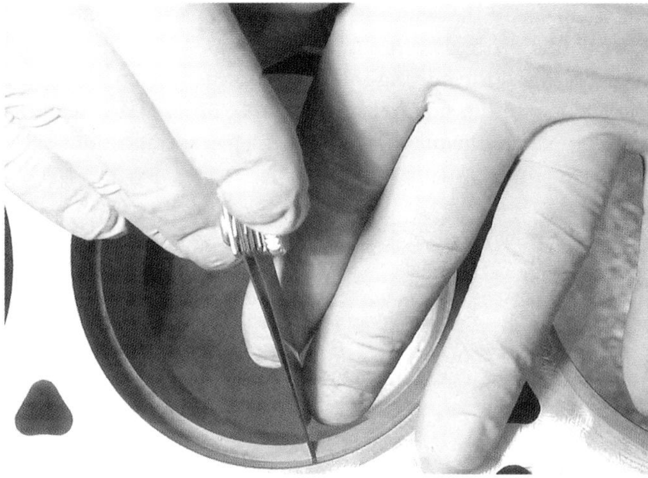

FIGURE 22-5 The pistons and rings must have the proper clearance for the engine to have a long life.

FIGURE 22-4 Crankshaft oil passages being cleaned.

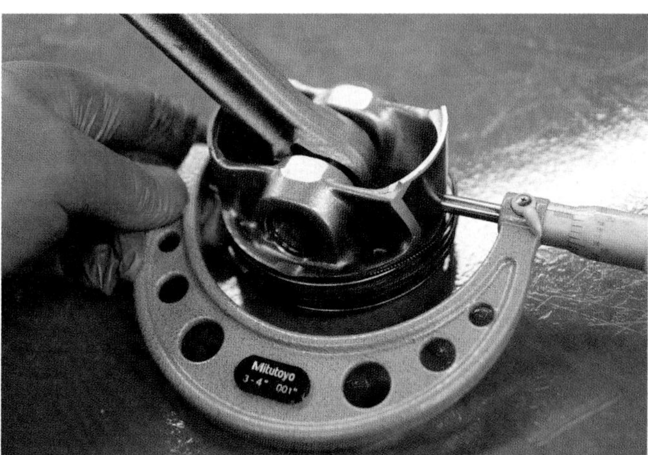

FIGURE 22-6 Measuring a piston.

Piston, Ring, and Block Clearance Checks

S22003

For the engine to have a long, trouble-free life, the pistons and rings must have the proper clearance with the cylinder bore. Measuring the piston-to-cylinder wall clearance and the piston ring end gap ensures that the machining was performed properly and that the parts you have are fitted properly to the cylinders (**FIGURE 22-5**). Be sure to check the piston-to-cylinder wall clearance and the piston ring end gap before installing the soft plugs and oil gallery plugs. This step could prevent you from having to remove and reinstall the gallery plugs and soft plugs if additional honing is needed. If the block were to need further machining, the plugs would have to be removed again.

Piston-to-Cylinder Wall Clearance

Using an outside micrometer, measure the piston skirt at the wrist pin centerline, and record the piston size. Repeat this process for all the pistons, and record your findings (**FIGURE 22-6**). Set the cylinder dial bore gauge to the piston size, using the bore gauge setting fixture. Recheck the bore gauge with the outside micrometer, and make sure the dial is set to zero. With the bore gauge set to the exact piston size, the dial will indicate the cylinder size with a plus or minus reading, as compared to the piston. This reading indicates the amount of oil clearance for that piston and cylinder. For example, if the cylinder bore is larger than the piston by 0.002" (0.051 mm), the dial will indicate a reading of + 0.002 (**FIGURE 22-7**). Check each cylinder wall with the bore gauge, adjusting the bore gauge if the pistons differ in size, and record your findings. Make sure your findings are within the manufacturer's specifications. If the clearance is not correct, contact the machine shop to explore the options for correcting the issue.

Measuring Piston Ring End Gap

Before installing piston rings on the piston, you need to check the ring end gap. Carefully slide the top compression ring into the cylinder bore about an inch down into the cylinder bore. Turn a piston upside down, and use the head of the piston to square up the ring in the cylinder bore (**FIGURE 22-8**). Using a feeler gauge, measure the width of the ring end gap (**FIGURE 22-9**). Repeat this process with the other compression ring and oil control ring. If the oil ring is the three-piece version, then only the end gap of the side rails is measured. Record each reading, and compare to the manufacturer's ring gap specifications.

FIGURE 22-7 Measuring the piston-to-cylinder clearance using a bore gauge.

Installing the Piston on the Connecting Rod

S22004

Refer to the manufacturer's specifications to determine which direction the piston is installed on the connecting rod (**FIGURE 22-10**). This is important, as many engines use a slightly offset pin on the piston. In addition, the rod may have an oil "spit" hole that helps to lubricate the bottom of the cylinder walls; if the oil hole is on the wrong side, the oil will not be sprayed at the right time or place. The two most popular types of piston pins are the full-floating and the semi-floating pins, also called wrist pins.

Floating Pin Design

The full-floating pin is a slide-fit in both the piston and the connecting rod bore. The pin is retained by a pair of snap rings. Note that some snap rings are directional, which means they must be installed with the correct side facing the wrist pin. Otherwise, the snap ring can work its way out of its groove and score the cylinder wall, destroying the engine. Install one snap ring on one side of the piston (**FIGURE 22-11**), and then lubricate the pin area of the piston,

FIGURE 22-8 Squaring up piston rings.

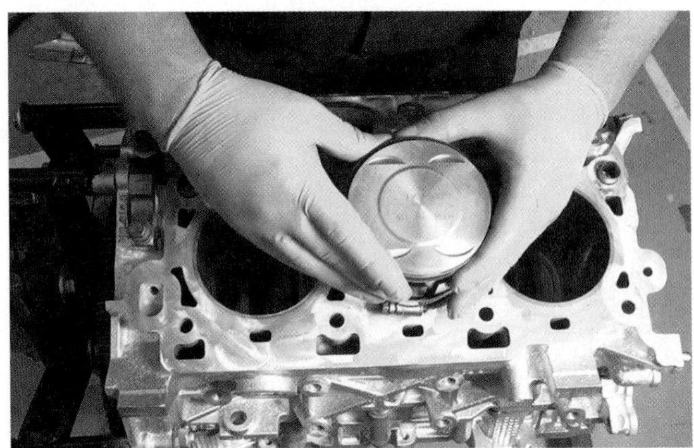

FIGURE 22-10 On most pistons, the notch goes toward the front of the engine.

FIGURE 22-9 Checking compression ring end gap.

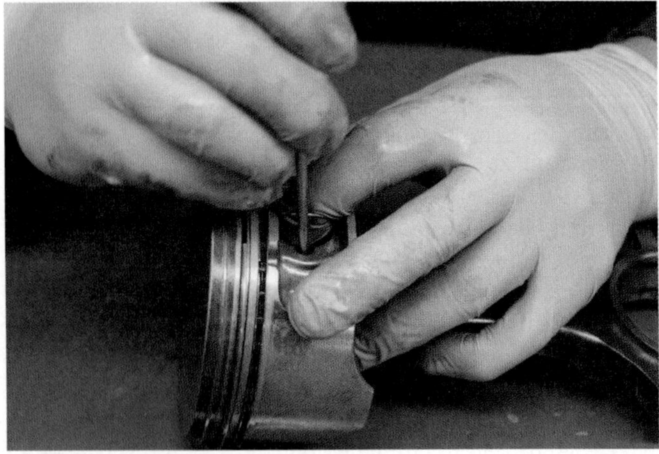

FIGURE 22-11 Installing the snap ring.

the connecting rod, and the pin. Next, slide the piston pin in place, and secure the pin with the other snap ring (**FIGURE 22-12**).

Semi-Floating Pin Design (Interference Fit)

Most engines have semi-floating or wrist pins, in which case the machine shop will have already attached the pistons to the rods. The semi-floating pin moves freely in the piston bore but is press fit in the connecting rod bore. The preferred method for installing the wrist pin is to use a rod heater. Adjust the setting fixture where the piston pin needs to stop (**FIGURE 22-13**). When installed correctly, the wrist pin will be centered in the connecting rod (**FIGURE 22-14**). Make sure the rod heater is hot; lubricate the piston pin bore with a thin film of automatic transmission fluid (ATF). The small end of the connecting rod will have a slight glow from heat; slide the piston pin in using a mandrel or pin press and let cool.

Temporary Test Buildup

S22005

If you are using any non-stock components in your rebuild such as high compression pistons or larger camshaft, it is very important to perform a temporary test buildup of the rotating assembly and heads. This buildup does not need to include the piston rings or a rear main seal for the crankshaft, but you should install the cylinder heads and the valve train. Do not skip this section of the rebuild; it will determine whether additional machining or exchange of any parts will be necessary.

FIGURE 22-12 Sliding the piston pin into the piston and rod.

FIGURE 22-13 Adjusting the pin stop on the rod heater.

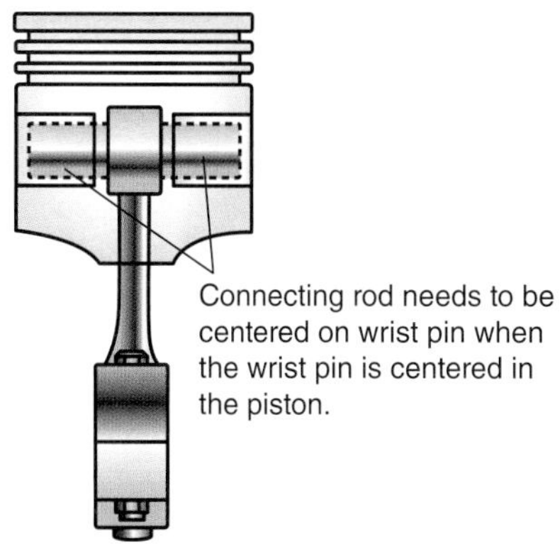

Connecting rod needs to be centered on wrist pin when the wrist pin is centered in the piston.

FIGURE 22-14 Wrist pin centered in the connecting rod.

All of your components should now be refurbished, new, or thoroughly cleaned and inspected. The decks, cylinders, and main bores of the block should all be refurbished to specifications. The pistons should be clean or new, and the rods refurbished on both the big and the small ends. All of the crankshaft journals should have been inspected and reground if necessary. The camshaft and lifters, as well as the timing set, should have passed inspection or been replaced with new parts. The cylinder heads and all related valve train components should also be clean, refurbished, or new. For this preassembly, have the head and intake gaskets as well as the oil pan and valve cover gaskets nearby. Also have the main and rod bearings ready for when you install their related components.

Installing Main Bearings

S22006

Before installing the main bearings, you need to make sure the bearings are the proper size. The crankshaft could have been turned undersize by 0.010" (0.25 mm), 0.020" (0.51 mm), or 0.030" (0.76 mm). Make sure the bearings match the crank size. You can do this ahead of time by checking the stamp on the back of the bearings or by looking at the mark on the bearing box. The back of the bearing insert and the bearing cap and saddle should be clean and dry.

> **▶ TECHNICIAN TIP**
>
> The back of the bearing insert and the bearing cap and saddle should all be clean and dry when installing the bearings. Do not apply grease or oil to the back side of the bearing shell. Even the smallest bit of dirt on the bearing back can cause a raised spot on the bearing, which leads to scoring of the crankshaft.

To insert the bearing if the bearing has a tang, start the **tang** of the bearing into the matching slot on the main bearing saddle. Make sure the bearing with the oil hole is inserted in the block and not the main bearing cap. Use your fingers to push the bearing insert into place (**FIGURE 22-15**). Check to make sure the oil hole is aligned with the main oil hole when inserted (**FIGURE 22-16**).

The thrust bearing can be installed only in its proper saddle. This location can be in the center of the block or at the rear main saddle of the block. You will need to look for the cutout area for the thrust-bearing flange to fit. If the thrust bearing is not a flange thrust bearing, grease will have to be applied to the back side of the bearing to hold the bearing in place until the crankshaft is installed or the main cap is installed.

Checking Main Bearing Clearance

It is critical that the main bearing clearances be within the manufacturer's tolerances. The specifications usually have a range of approximately plus or minus 0.0005" (0.0127 mm), so accurate measurement is required. There are several methods used to check clearance. If you want to check the clearance prior to installing the crankshaft, you can use an inside and outside micrometer or a bore gauge (**FIGURE 22-17**). If you want to check the bearing clearance with the crankshaft installed, then you can use a plastic gauging material (**Plastigauge**) Each of those is discussed next.

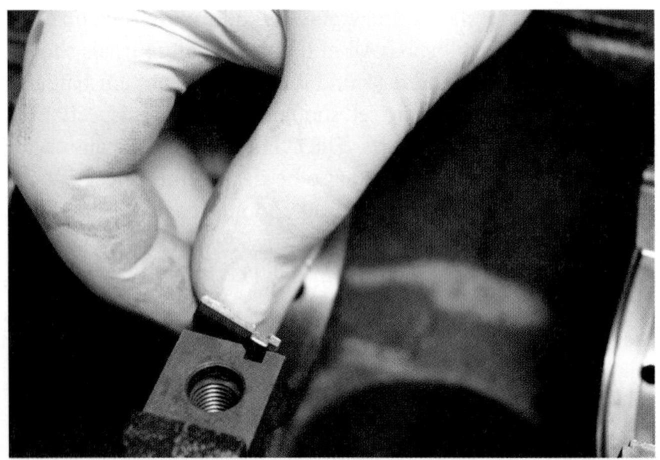

FIGURE 22-15 Installing main bearing insert.

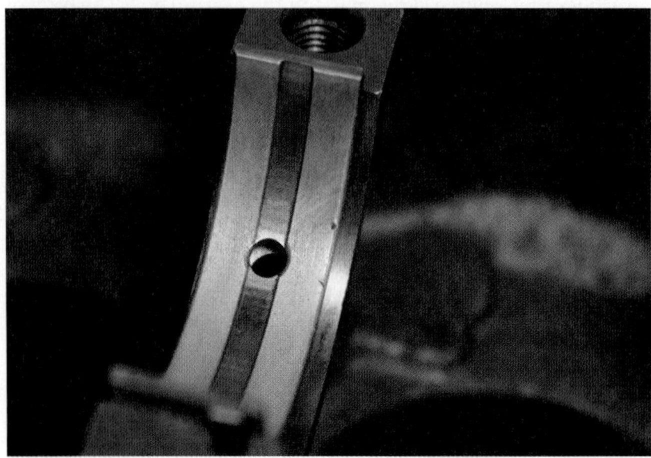

FIGURE 22-16 Oil hole lined up.

FIGURE 22-17 Using a bore gauge to measure a cylinder.

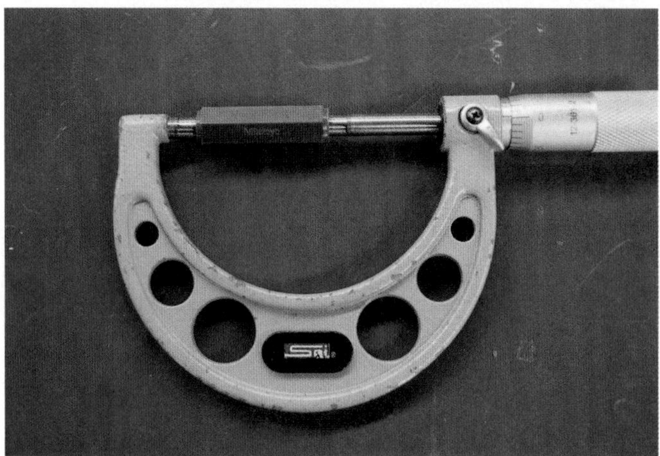

FIGURE 22-18 A micrometer being checked for accuracy with a standard.

Using a Micrometer

Before using a micrometer to measure the crankshaft, it is important to make sure the crankshaft is clean and free of oil and grease. Also make sure that the micrometer is zeroed using a standard (**FIGURE 22-18**). Use the **outside micrometer** to measure each main bearing journal, and record your findings (**FIGURE 22-19**). With all the main caps and bearings installed and torqued (crankshaft is not installed at this time), use an inside micrometer or snap gauge to measure the inside diameter of the main bearing bore (**FIGURE 22-20**). Take care not to damage the bearing surface. This procedure does take some experience, but when performed correctly, it is very accurate.

Once the inside micrometer or snap gauge is set to the correct size, read the inside micrometer, or use the outside micrometer to measure the outside of the snap gauge, and record your findings. Subtract the crankshaft main journal readings from the main bearing bore readings, and record your findings. The difference will be in thousandths and ten-thousandths of an inch, giving you the bearing clearance. Compare this with the manufacturer's specifications.

FIGURE 22-19 Measuring crankshaft journals with an outside micrometer.

FIGURE 22-20 Measuring the inside of a main bearing.

Using a Bore Gauge

When using a bore gauge, it is important to make sure the crankshaft is clean and free from oil or grease. You need to measure the main journals with an outside micrometer and record your findings. Next, use a bore gauge setting fixture to set the bore gauge according to the crankshaft main journals measurements, and zero the dial. With all the main bearings in place and the main caps torqued to the proper torque setting, use the bore gauge to measure the clearance (**FIGURE 22-21**). The reading on the bore gauge is the bearing clearance. Record the clearance, and compare the reading to the manufacturer's specifications.

Using Plastigauge

Plastigauge is an extruded plastic thread that comes in different sizes, depending on the expected clearances to be checked (**FIGURE 22-22**). Because it is soft, it can be squished between the crankshaft journal and bearing insert without damaging either surface. And because it is a precise diameter, the width it squishes out to indicates the clearance between those parts. The wider it is squished, the smaller the clearance, and vice versa. A measuring gauge is printed on the wrapper.

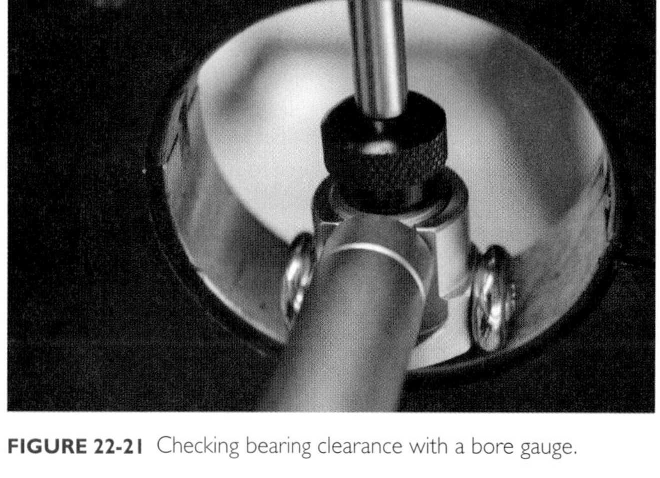

FIGURE 22-21 Checking bearing clearance with a bore gauge.

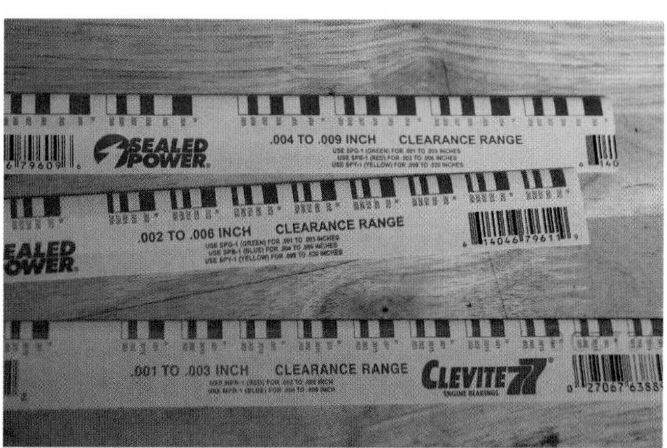

FIGURE 22-22 Plastigauge comes in several sizes to match various applications.

FIGURE 22-23 Plastigauge being placed on a main journal in preparation to measuring bearing clearance.

For most main bearings, the 0.0005" to 0.003" (0.013 to 0.076 mm) sized Plastigauge, which is green in color, is used. A piece of Plastigauge just short of the width of the bearing is laid across the width of the crankshaft main journal (**FIGURE 22-23**). The proper main cap is then torqued to manufacturer's specifications, making sure the crankshaft does not turn. If the crankshaft turns during this measurement, the gauge strip will smear, giving an inaccurate reading.

After the main cap has been torqued into place, the cap is then removed. The plastic gauge strip will be stuck to the crankshaft journal or the bearing insert. The paper wrapper that the Plastigauge came in has a gauge printed on it for determining the actual bearing clearance, which is then compared to the manufacturer's specifications. This process is repeated for all the main journals.

Test-Installing the Crankshaft

S22007

Testing crankshaft bearing clearances is done to ensure that all clearances are correct. Any problems can be noted and corrected before the rest of the engine is assembled. After the main bearing inserts are installed, the crankshaft is installed, and then the clearance between the crankshaft journals and the main bearings is measured. This skill drill uses the Plastigauge method to check clearances. If the main journals of the crankshaft were cut undersize, make sure the new main bearing inserts match the undersized journals. Before you open the sealed bearing insert package, double-check the labels to make sure they are the correct type and size.

To test-install a crankshaft, follow the steps in **SKILL DRILL 22-1**.

Thrust Bearing and Crankshaft End Play

Crankshaft end play is controlled with the thrust bearings. Checking the thrust bearing end play can be done with a feeler

SKILL DRILL 22-1 Test-Installing the Crankshaft

1. Wipe out the saddles of the block and the cap-to-block mating surfaces. Press the bearings into place by hand. The bearing insert with the oil hole fits in the saddle.

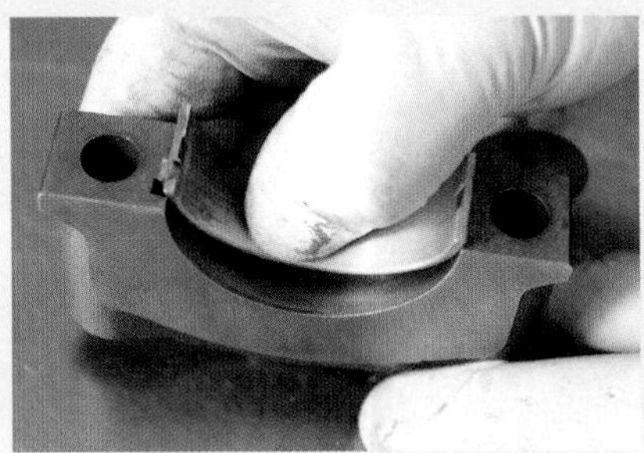

2. Clean the main bearing caps, and install the bearing inserts in the same manner. Wipe a small amount of oil on each bearing insert and on the thrust surface of the thrust bearing.

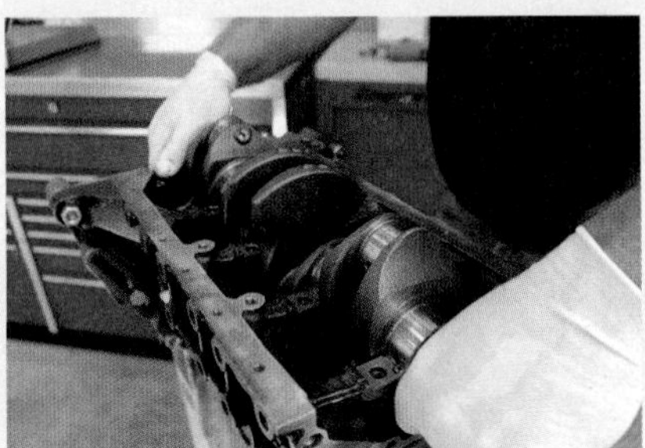

3. Lift the crankshaft and set it into place. Ask for lifting help if it is needed, and then gently set the crank into the block.

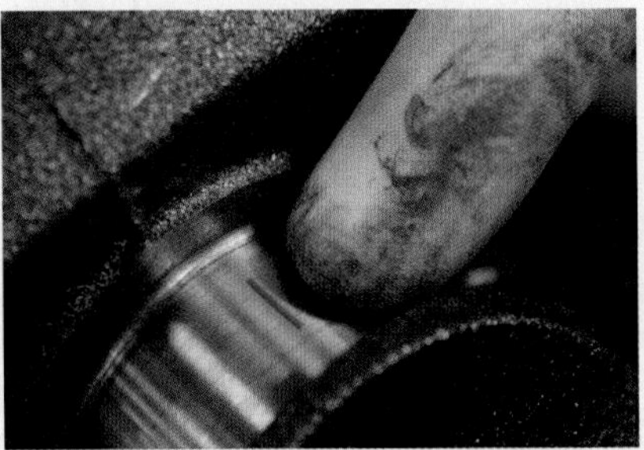

4. Lay one strip of Plastigauge on top of each of the main journals of the crank.

5. When all the bolts are in and hand-tightened, torque the main bearing cap bolts to the manufacturer's specification.

6. Remove the main bearing caps. Be careful not to move or rotate the crankshaft.

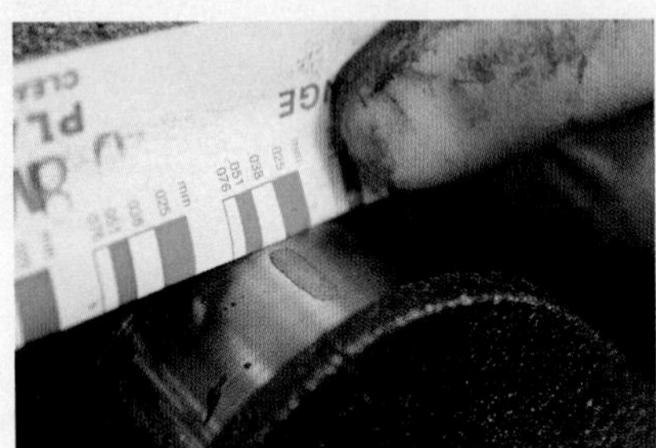

7. Determine the main bearing clearance by placing the measuring strip that came with the Plastigauge next to the squished plastic.

8. Remove the Plastigauge. Install all of the main caps finger tight.

9. Align the thrust bearing by using a dead blow hammer to tap the crankshaft forward and backward, and torque the caps. Verify that the crank is very easy to turn by hand.

FIGURE 22-24 Checking thrust bearing clearance using a feeler gauge.

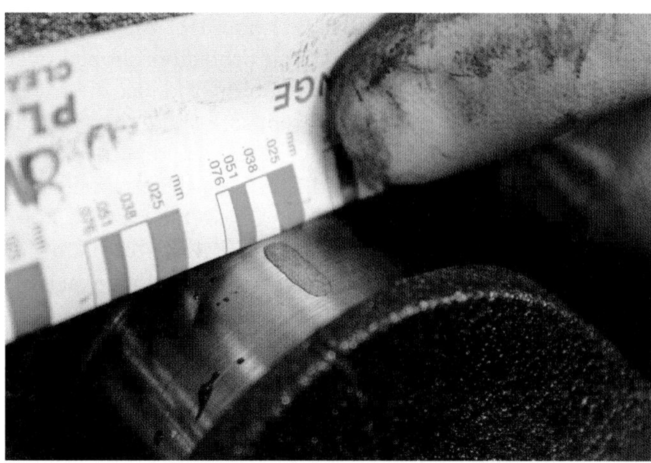

FIGURE 22-26 Rod bearing clearance using Plastigauge.

FIGURE 22-25 Checking the crankshaft end play using a dial indicator.

gauge after the crankshaft is installed. Push the crankshaft forward, and slide a feeler gauge on the back side of the thrust bearing (**FIGURE 22-24**). Check the manufacturer's specifications; an average for the thrust bearing end play is 0.003" to 0.004" (0.076 to 0.1 mm). If checking using a dial indicator, set up the indicator and zero the dial indicator with the crankshaft pushed all the way back against the rear thrust. Next, carefully push the crankshaft forward, and read the amount of end play on the dial indicator (**FIGURE 22-25**).

Test-Installing the Piston and Rod Assemblies

S22008

The same method used to check main bearing clearances can be used to check the rod bearing clearance. If a micrometer or bore gauge is used, these checks can be done before the pistons are assembled on the connecting rods. A rod vise is needed to hold the connecting rod while it is being torqued. Be sure to record all of the connecting rod clearances, and check your findings against the manufacturer's specifications.

Checking Rod Bearing Clearance

S22009

It is easier to check the bearing clearance when using the Plastigauge method if the connecting rod is installed on the piston. After the piston has been installed on the connecting rod, you can check rod clearance by test-installing the pistons without the piston rings installed. Be careful, however, because the piston could fall out of the cylinder bore and become damaged. Once the piston and rod are in place, you need to use the same process that was followed previously for checking main bearing clearance. Cut the Plastigauge the same length as the bearing insert width, and set the Plastigauge on the rod journal. Torque the connecting rod nuts or bolts to the manufacturer's specifications. Take the rod cap off, and compare to the paper graph for your clearance measurement (**FIGURE 22-26**).

The piston and rod assemblies are test-fitted next. This allows you to check rod bearing clearances and internal clearances to ensure there are no problems. The piston and rods should already have been assembled. A preassembly check of the piston bearing clearance can be done more quickly if you leave the piston rings off during the test-install. Leaving them off will also make it easier to rotate the crank for the rest of the checks. Make sure everything is clean. Wipe any dust off the rod journals, and then wipe out the cylinder bores with a clean, dry paper towel. Do not use a shop rag; they tend to leave cloth fibers on the cylinder walls. When the bores are clean, wipe an even coat of fresh engine oil over the insides of the cylinder bores. Turn the crankshaft so that the rod journals farthest forward are facing up (engine upside down), and lock the crankshaft.

Place the Plastigauge strip on the bearing in the connecting rod, not the bearing in the rod cap. It is placed on the bearing in the connecting rod so that it will not get squished further when the rod cap is being removed from the rod. Thus, placing the Plastigauge strip on the bearing in the connecting rod gives you the most accurate reading. Some technicians use a very thin layer of oil to keep it from falling off, but that can take up a small amount of the clearance, so take that into consideration. Bolt protectors slipped over the rod studs make it easier to line up the rod bearing to the journal of the crank and help prevent scratching the cylinder wall or nicking the crank journal. Do not let go of the piston; because there are no rings to hold it in place, it will fall right out. Keep your hand on the rod, and have the cap and bolts nearby so that your other hand can install the cap. Do not forget to line up the rod and the cap properly. When you have at least one nut threaded on, you can let go and make the nuts hand tight.

Set the wrench to one-third of the final torque setting for the rod nuts. Next, torque the nuts to the two-thirds measurement, and then finish tightening them all at full torque. Immediately afterward, loosen the rod nuts. The most important thing is to make sure not to rotate the crank at all while you are tightening and removing the caps.

Remove the pistons, and check bearing clearance. Sometimes the caps can be worked off by hand. Usually, you have to hit the bolts to break the caps loose. That is why you put the gauge in the rod and not the cap. If the gauge strip were in the cap, it would be distorted when the cap is tapped for removal. Keep a hand on the piston or the rod; because there are rings to hold the piston in place, it will fall right out of the cylinder when it breaks loose. Install the protector boots, and remove the piston. If the clearance is correct, clean the plastic off the bearing or the crank. Now lubricate the bearing, slip the piston back into its cylinder, and snug the rod nuts. Next, rotate the crank at least twice, slowly. If everything is clear, rotate the crankshaft so that the next set of rod journals are facing up. Continue installing and measuring the rest of the pistons. When all of the pistons are installed, check to make sure they rotate smoothly.

To test-install pistons, follow the steps in **SKILL DRILL 22-2**.

Checking Piston-to-Deck Heights

S22010

Most engines have a specification for piston height as compared to the cylinder block deck. It is important to check this dimension, especially if the deck of the block has been machined or if new, nonstock pistons are being installed.

SKILL DRILL 22-2 Test-Installing Pistons

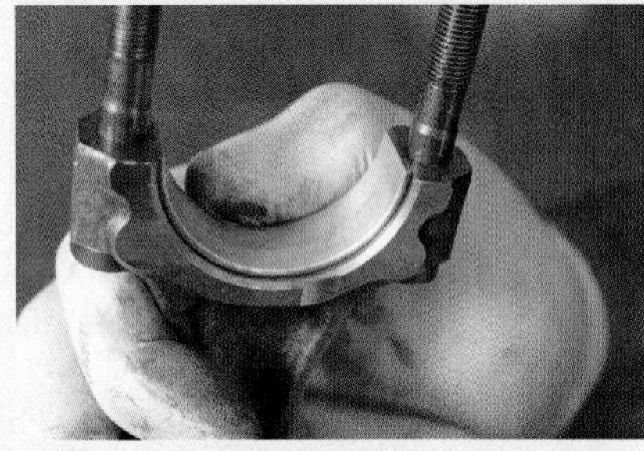

1. With the caps separated from the rods, install the new rod bearings after you verify the parts numbers to ensure they match the sizes on the crankshaft.

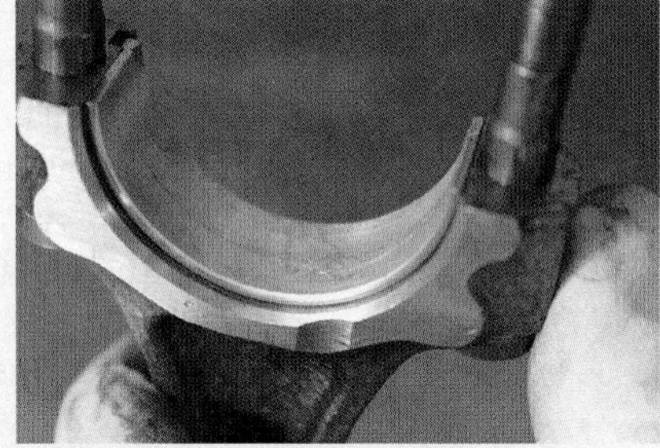

2. Stick the Plastigauge in place with a film of oil. Put the gauge in the rod and not the rod cap.

3. Install protector boots over the rod studs. Torque the rod nuts to manufacturer's specifications, in three increments. Immediately after, loosen the rod nuts.

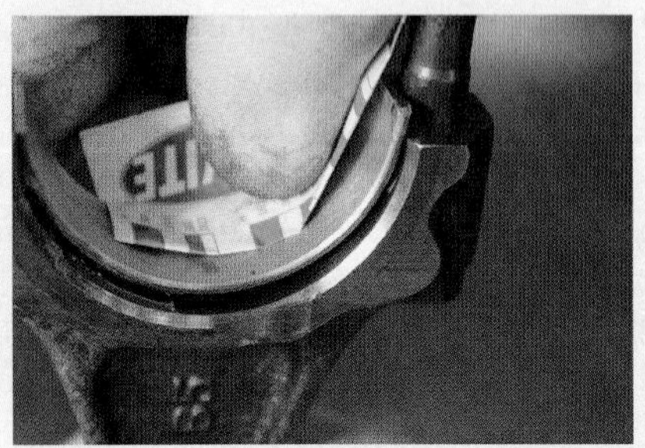

4. Remove the pistons, and check bearing clearance. If the clearance is correct, reinstall the piston and rod, rotate the crankshaft, and repeat the procedure for the next set of pistons.

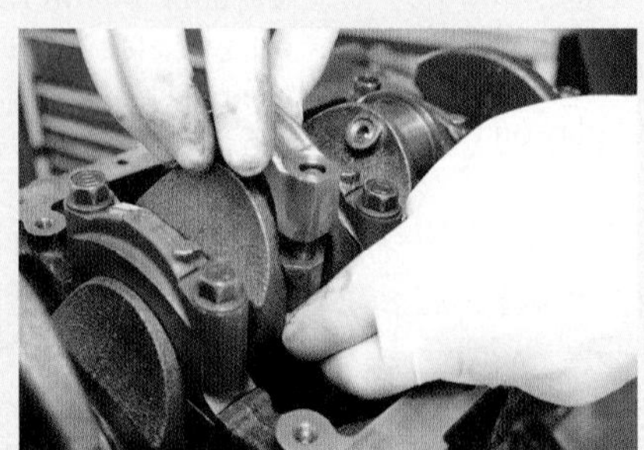

5. When the caps are back on, torque them again in three increments.

6. Continue installing and measuring the rest of the pistons.

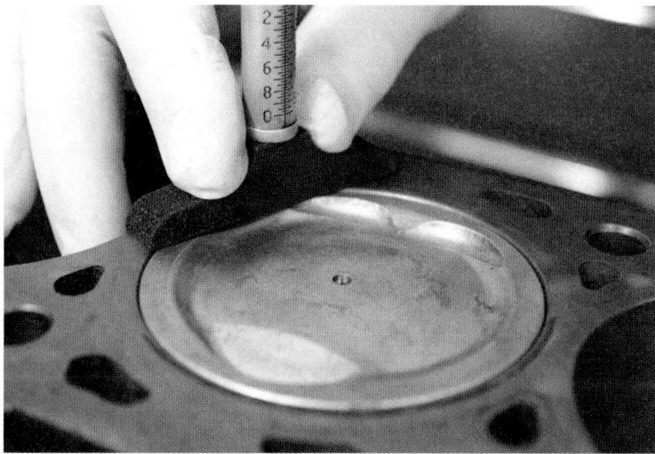

FIGURE 22-27 Checking piston-to-deck height.

FIGURE 22-28 Valve impression in clay on top of piston.

Improper height will affect the engine's compression ratio and could cause interference issues if the piston rises above the deck. Using a depth gauge micrometer, you can measure the piston height at top dead center (TDC), referenced to the block's deck (**FIGURE 22-27**). Record your findings, and compare to the manufacturer's specifications.

Checking Piston-to-Valve Clearance of a Performance Camshaft

If a performance camshaft is used and there is a higher lift and more duration in the cam profile than in a stock camshaft, it is recommended to check piston-to-valve clearance. At the end of the exhaust stroke, as the piston travels to the top of the cylinder (TDC), the exhaust valve will be just closing and the intake valve just beginning to open. There needs to be enough clearance so that the pistons do not hit the valves, resulting in bent valves and damaged pistons.

To check this clearance, the crankshaft, piston, and rod must be installed in the block. After installing the piston and rod, temporarily install the cylinder head, along with the head gasket and valve train. Just remember that if hydraulic lifters are used, they will need to be pumped up first, and lighter test springs should be used in place of the stock valve springs so that the valves will open fully. The clearance can be checked by using one of two methods: the clay method or the dial indicator method. Both methods show you the minimum amount of clearance between the piston and the valve. The clay method gives an imprint of where the valve and piston have an interference issue so that you will be better able to determine how to fly cut the pistons if necessary. The dial indicator method allows you to check the clearance at various positions of the piston.

Clay Method

Before beginning, determine the use of solid or hydraulic lifters. If solid lifters, use new ones with plenty of assembly lube on the bottom of the lifter and cam lobes. If using hydraulic lifters, you will need to remove the stock springs and install the lightweight springs in their place. Have the right pair of hydraulic lifters for that cylinder nearby.

Make two balls, using soft clay, one for each valve. Stick them to the piston in the upper valve reliefs (if present) that are cut into the top of the piston. Cover them with a little oil to keep the valves from sticking to the clay when they make an imprint in it. Make sure the pistons are positioned somewhere in mid-stroke. Lay the head in place with the head gasket, install a few bolts in the cylinder head, and tighten them by hand. If using a wrench, do not tighten the bolts very much—no more than 15 ft-lb (20.3 N·m). Most head gaskets must be replaced once they have been torqued even once.

Install the valve train, and adjust the valves to the specified lash, if applicable. If the engine uses hydraulic lifters, set the clearance to zero lash. Carefully rotate the crank until you see the exhaust and intake valves open and close two times each. If you feel solid resistance, *stop*; the piston may be contacting the valve.

Now remove the valve train and head. When you remove the head, the two balls of clay now have valve imprints pressed into them (**FIGURE 22-28**). Cut each piece of clay, and remove half. Measure the distance from the piston to the top of the valve imprint with a ruler or the depth end of a dial caliper. Compare this measurement with the specification in the service information. If it is greater than the minimum allowed, there is very little danger of your valves contacting the piston once the engine is heated up and running, unless the valves float at high engine revolutions per minute (rpm). If the clearance is smaller than specified, the pistons will need to be fly cut by a machine shop, or a smaller camshaft must be installed. Clean any leftover clay off the piston.

To perform a piston-to-valve clearance check using the clay method, follow the steps in **SKILL DRILL 22-3**.

Dial Indicator Method

Using a dial indicator involves test-assembling one cylinder of the engine and using a dial indicator to measure the amount

SKILL DRILL 22-3 Checking Piston-to-Valve Clearance (Clay Method)

1. Assemble the valve train for the number 2 cylinder. If using hydraulic lifters, install lightweight valve springs.

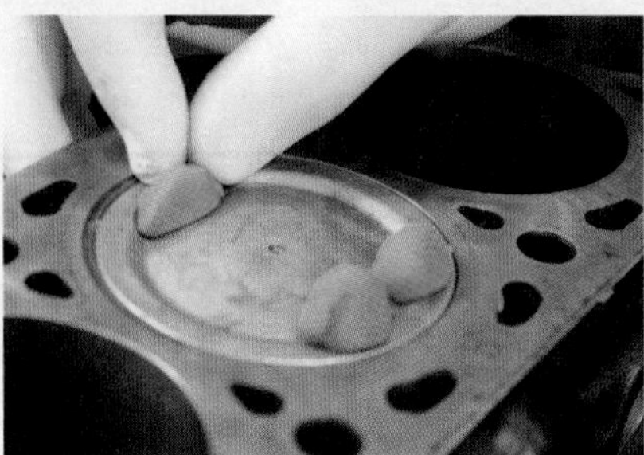

2. Make balls using soft clay, one for each valve. Stick them to the piston in the valve reliefs that are cut into the top of the piston, and coat the clay with clean engine oil.

3. Make sure the crank and cam are positioned properly, most likely with the timing marks lined up on each. Place the head gasket and cylinder head on the block. Tighten all the head bolts to about 15 ft-lbs (20.3 N·m).

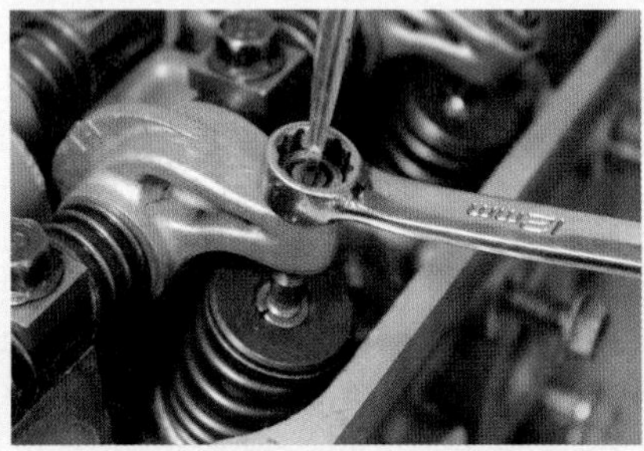

4. Install the valve train, and adjust the valves to the specified lash, or if using hydraulic lifters, zero lash.

5. Carefully rotate the crank until you see the exhaust and intake valves open and close two times each.

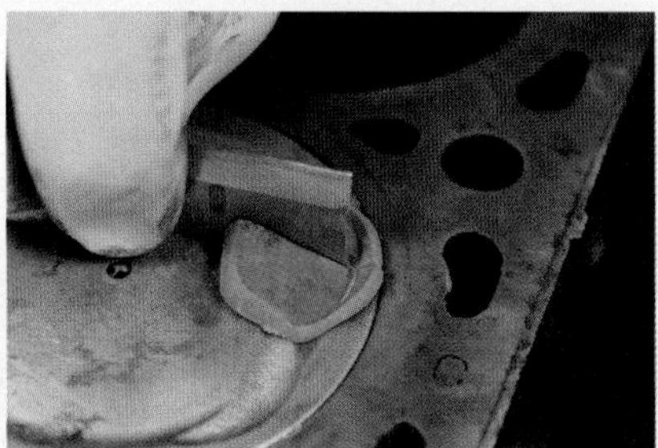

6. Remove the valve train and head. Cut each piece of clay, and remove half. Measure the distance from the piston to the top of the valve imprint with a ruler or the depth end of a dial caliper.

of clearance between the piston and the valves when the piston is near TDC. Because the dial indicator will be mounted so it can measure valve movement, and the valve will have to be compressed by hand, the stock valve springs cannot be used. Lighter-duty test springs will have to be used in their place. It is also helpful to know where the piston is in relation to TDC, so the use of a **degree wheel** or the harmonic balancer and timing cover with timing marks will help identify the crankshaft position.

Once the test piston and valve train are assembled for the test, and the dial indicator is set perpendicularly on the spring retainer of the intake valve, you are ready for the test (**FIGURE 22-29**). The crankshaft is turned until the piston is about 20 degrees before TDC of the exhaust stroke (between the end of the exhaust stroke and the beginning of the intake stroke). At that point, the dial indicator is zeroed (**FIGURE 22-30**), and the valve is carefully pushed down at least 0.250" (6.4 mm), if possible (**FIGURE 22-31**). If the valve contacts the piston, the distance measured by the dial indicator is the valve-to-piston clearance. Write down the measurement, and recheck in 5-degree increments until about 20 degrees after TDC. Once you are finished with the first valve, repeat this process for the exhaust valve. Note that many rebuilders recommend a minimum piston-to-valve clearance of at least 0.080" (2.0 mm).

FIGURE 22-29 Dial indicator positioned on the intake valve spring retainer.

FIGURE 22-30 At about 20 degrees before TDC, zero the dial indicator.

FIGURE 22-31 Lightly push the valve down, and read the dial indicator if the valve hits the piston.

▶ Final Assembly of the Engine Rotating Components (Short Block)

S22011

Now that the test-fitting of the components is done, and assuming that all measurements are within the manufacturer's specifications, the block is ready for final assembly. If everything fits, completely disassemble the engine again, making sure the valve train parts all go back to their places in the organizer tray, and the gaskets go back into their packages to stay clean. The pistons should go back in their tray, and any bolts you remove should go back to the correct storage bins or bags.

When the crankshaft is out of the block again, wipe off the main bearings and the rod bearings, and put them in bins away from any dust that could collect on them. Use extreme caution when you remove the camshaft, as it is very easy to gouge the new bearings.

▶ TECHNICIAN TIP

The engine block, including the assembled crankshaft, rods, pistons, camshaft, and timing gear set (in overhead valve [OHV] engines), is referred to as a **short block**. A short block has all of the rotating parts of the engine installed—everything except the heads, covers, manifolds, accessories, and oil pan. A rebuilt engine that includes the heads, covers, timing components (in overhead cam [OHC] engines), oil pan, and sometimes the flywheel is referred to as a **long block**.

Painting the Engine

S22012

The engine has to be painted to protect it from rust and corrosion. It also gives the whole rebuilding job a professional, finished look. Painting is especially necessary if the cast iron castings have been cleaned such that all of the original paint is removed. The various stamped or cast metal covers should also be painted. The engine can be painted before or after installing the oil gallery and soft plugs (see Skill Drill 22-5).

To paint the engine and covers, follow the steps in **SKILL DRILL 22-4**.

Installing the Soft Plugs and Oil Gallery Plugs

S22013

To help ensure that the engine has no leaks when started up, all the oil gallery plugs and soft plugs have to be installed. Make sure all oil galleries are clean and clear of debris. You can check this visually with a flashlight, or blow air through the passageways. Next, any threaded plugs such as the rear oil gallery plugs should be coated with a sealer, preferably a liquid Teflon or a similar non-hardening sealer. Do not use Teflon tape; pieces can break off inside and cause an oil system clog. These plugs are usually a tapered pipe thread, so do not over tighten them.

The soft plugs need an appropriate non-hardening sealer applied to the soft plug and the block's sealing edge. Soft plugs are hammered into place in the block and in some heads. Use an appropriate driver, and keep them straight when driving them in place. On the small oil gallery soft plugs, tap them in about $^{1}/_{16}$" below the surface, and use a chisel to slightly indent the surface of the block around each opening in at least two places. This will keep the plugs from backing out when the oil pressure builds.

Always check for miscellaneous plugs. Almost every engine has some hidden plug, like an oil gallery plug, in an odd place. Carefully review your disassembly notes and the service information to make sure you do not miss any plugs.

To install the engine plugs, follow the steps in **SKILL DRILL 22-5**.

Installing the Crankshaft

S22014

With the main caps off the block, make sure the main bearings are clean; denatured alcohol works well for cleaning the

SKILL DRILL 22-4 Painting the Engine and Covers

1. Prepare the engine parts for painting by building an empty shell of the long block, and then paint everything. Install all the bolts you want to have painted, but make sure you tighten at least four bolts for the heads and the intake. You do not need any gaskets or internal parts. The covers will mask off all the big openings, but for all the other holes, use some masking tape and cut around the edges with a razor blade.

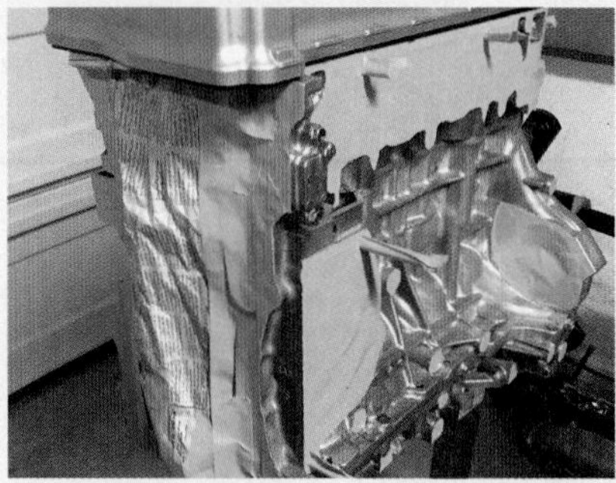

2. When the masking is done, if necessary, wipe all the surfaces to be painted with a degreaser or lacquer thinner to prepare the surface, and remove any oil residue. Put some old spark plugs in the heads.

3. *In a well-ventilated area* and wearing a painter's mask, spray the engine with your chosen color of engine paint. You can also spray a coat or two on any accessories that you want to protect and brighten up with some color.

SKILL DRILL 22-5 Installing Engine Plugs

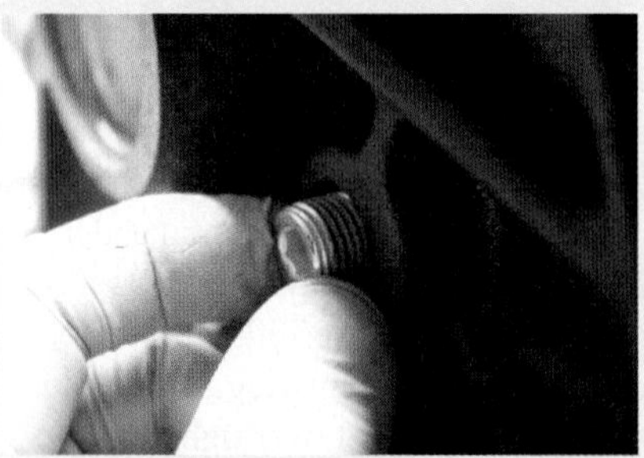

1. Start with the bare block sitting on the floor with the rear area facing up. Apply a small amount of sealer to the threaded oil gallery plugs, and screw them in.

2. Clean the edges of the rear water jacket holes before you install the soft plugs.

3. Coat the soft plug with a non-hardening sealer. Wipe it thin and even, being careful not to leave any bare spots.

4. Carefully tap in each soft plug, using an impact socket and hammer.

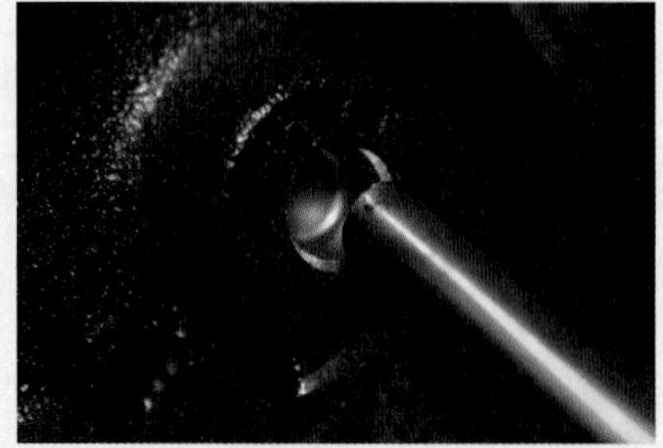

5. Turn the block over so that the front is facing up, and select a punch that is smaller than the inner diameter of the front oil gallery plugs. Wipe sealer evenly around each plug and tap it in until it is $^{1}/_{16}$" (1.6 mm) below the rim.

6. Turn the block on its side. Clean, apply sealer, and install the soft plugs in the side of the block, making sure not to miss any.

bearings. Most bearing manufacturers sell assembly lubricant. Using this, lubricate both the main journals of the crankshaft and the main bearings with the bearing assembly oil.

If the engine has a two-piece main seal, install the upper half before lowering the crankshaft into place. Offset the ends of the seal so that one end sits slightly higher than the block surface and the other end slightly lower than the block surface. That way the parting seam on the seal is not exactly lined up with the block surface, which will help prevent leaks. On a neoprene rubber seal, the pointy lip must point toward the inside of the engine. When you press it into place, use the plastic protector shim between the seal and the block to keep the rubber back of the seal from being damaged by the sharp edge of the block.

For engines with an older type of rope seal, work the rope into the groove of the block, and then use a large socket to tap it deeper into the groove. Cut off the excess flush with the block, using a razor knife. Put a little oil or assembly lube on the lip of the rope so the seal will not burn out on start-up. For all types of engines, it is very important to put a dab of silicone sealer on each side of the seal where the rear cap meets the metal of the block. If you do not apply this sealer, oil will leak from the rear of the block. Spread a little silicone sealer on the ends of neoprene or rope seals as well.

Install the crankshaft in the main saddles, and make sure it spins freely. Next, install the main caps. If the bearings have tangs, make sure they are installed squarely in their grooves. Place the main caps carefully into their recesses in the block and seat them with a light tap, making sure the caps are installed in their correct positions. When installing a girdle-style main cap design, all the caps are anchored together and installed at once. If called for by the manufacturer, apply a light coat of oil on the main cap bolt threads and the head of the bolt where it makes contact with the main cap. Snug the bolts down by finger. Hit the back of the crank and the front of the crank with a dead blow hammer to line up the thrust bearing. Torque the main caps following the manufacturer's procedure. This is likely a multistep process.

For engines with a one-piece rear main seal, you need access to the rear of the block. Many engines use an adapter housing that bolts to the back of the block. The rear main seal is fitted in after the adapter is sealed and bolted into place. To install the seal, you will probably have to take the block out of the stand so that you can access the rear area of the block. You then need to lift the engine back onto the stand after installing the seal. Alternatively, you can leave the main seal out and install it later, just before installing the flywheel or flex plate—just don't forget it.

To install a crankshaft, follow the steps in **SKILL DRILL 22-6**.

SKILL DRILL 22-6 Installing the Crankshaft

1. Install the main bearing inserts, and double-check that the oil holes line up with the ones in the block.

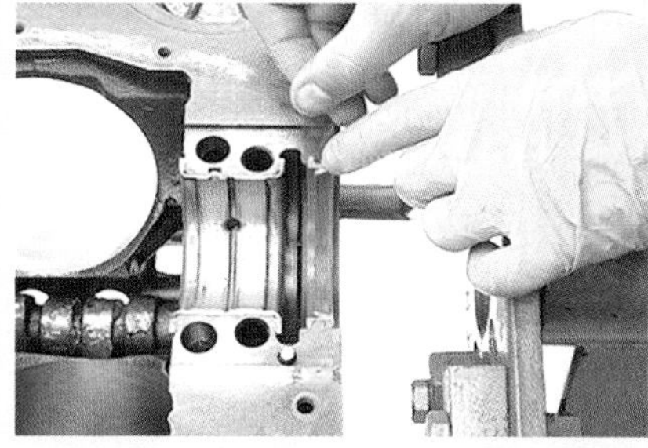

2. If the engine has a two-piece seal, install the upper half before lowering the crankshaft into place. For engines with an older type of rope seal, work the rope into the groove of the block, and then use a large socket to tap it deeper into the groove.

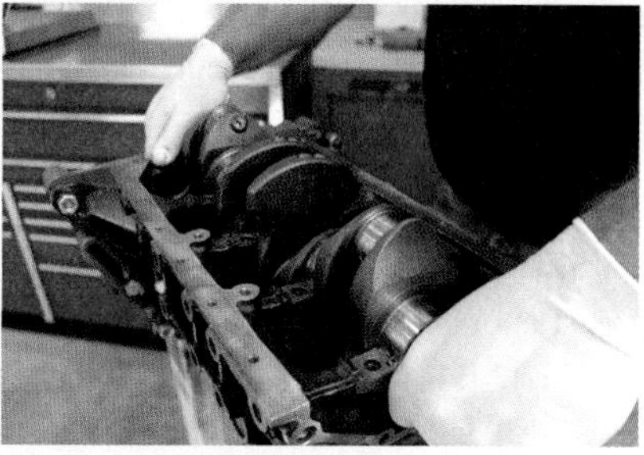

3. Carefully lay the crankshaft in place. Get someone to help you if the crankshaft is heavy. Next, lubricate each of the caps and put them in place.

4. Place the main cap bolts into their respective holes.

5. Carry out the final torquing of the main caps, following the manufacturer's procedure.

6. Mount the adapter plate, and install the one-piece main seal.

Installing Piston Rings

S22015

The ring kit should have a label that says what each ring in the package is and where it gets installed. The first ring to go on is the oil expander ring, which slips into the bottom groove. Most engines use an oil control ring set made up of an expander ring and a pair of thin scraper rails. Position the parting line of the expander to the center of the piston pin. Make sure the ends of the expander ring do not overlap. Now install the first of the oil scraper rails, which go above and below the oil expander ring. Start it about 45 degrees to one side of the expander gap, and slide it in between the roof of the groove and the top of the expander ring. When the edge of the ring is in, slowly work it clockwise around the piston. Slow down when you get to the other side of the piston. The edge of the ring could scratch the side of the piston, so grasp it firmly and carefully pull it slightly away from the piston; then drop the end of it in place above the expander ring. Start the other scraper rail about 45 degrees to the other side of the expander gap, and slide it into the bottom slot. Work this ring around in the opposite direction, guiding it past the top scraper and into the bottom groove as you go. Pull it away slightly at the end, and drop into its groove. Double-check that the expander ring gap is not overlapped. It is important to rotate the rings now and make sure nothing is stuck or crooked.

The packaging or instruction sheet should tell you which compression rings are which. Find the ring for the lower groove first. Look closely at the ring. There is usually a label on one side of the ring, such as a dot, a **pip mark**, a "T," or some other marking. The marked side of the ring typically, but not always, faces up. Hold the ring expander with one hand, and steady the ring with the other. Expand it only enough to get the ring to slip down past the first ring groove and into its place in the second groove. It should rotate freely. Using the same technique as earlier, expand the top compression ring just enough, with the top mark facing up and lower the ring into its place. Once all of the rings have been installed, position them around the piston as specified by the manufacturer.

To fit the piston rings, follow the steps in **SKILL DRILL 22-7**.

Installing Pistons in the Block

S22016

Installing the pistons requires a few special steps. First, the piston rings must be aligned in their proper positions to minimize blowby gases. All of the pistons, rings, and cylinder walls must be lubricated with assembly lube or clean engine oil. Once the piston rings are aligned and lubricated, they need to be compressed with a ring compressor so they will fit into the cylinder. Make sure the rings collapse into the ring lands. If any ring overlaps onto the surface of the piston, it will be broken as the piston is inserted. It is also important to use bolt protectors on the rod bolts so that the rod bolts do not nick the rod journals on the crankshaft. Also, a liberal amount of bearing assembly lube should be placed on both halves of the connecting rod bearing inserts and the rod journal on the crankshaft. When the piston is being installed, the crankshaft should be positioned so its journal is at bottom dead center (BDC), and the front of the piston should be facing the front of the engine.

The piston will most likely have to be tapped into the cylinder with the handle end of a hammer or other soft tool. It is important to make sure the connecting rod is aligned to the crank journal so it slides over without gouging the surface. If you feel even the slightest resistance, stop and check the compressor. If you see any of the oil scraper rings or compression rings popping out, stop and immediately pull the compressor off. If you are tapping very slowly and a ring keeps popping out, try hitting a little harder in the beginning so that the piston will move quickly past the chamfer at the block deck. Once the piston is in, remove the compressor, and tap lightly to move the rod down while guiding it over the rod journal.

When you install the rod cap, the numbers you marked on the connecting rod and cap during disassembly should match. Also, if the bearing inserts have tangs, they should be on the same side as each other. Tighten the rod bolts about

SKILL DRILL 22-7 Fitting the Piston Rings

1. Install the oil control rings. The first ring to go on is the oil expander ring, which slips into the bottom groove. Make sure the ends of the expander ring do not overlap. Install the scraper rails positioned 45 degrees to each side of the expander gap.

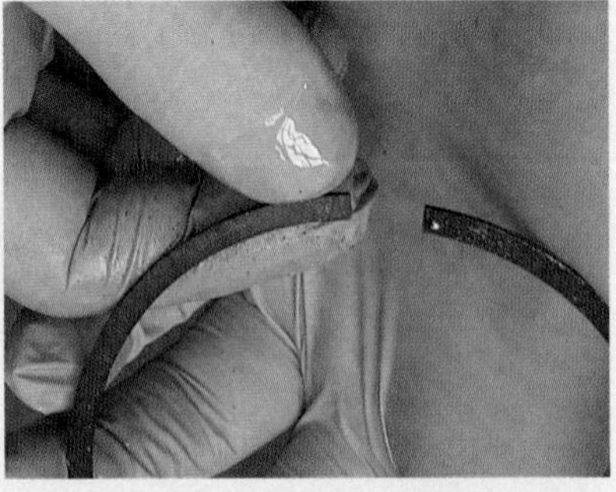

2. Find the ring for the lower compression ring first. There is usually a label on one side of the ring, such as a dot, a pip mark, a "T," or some other marking. Position as specified.

3. Use a ring expander to install the compression rings, and position the gaps as specified.

one-third of the specified torque, and make sure that the crank turns appropriately and is not binding up. Do this after each rod is installed. If the crank binds up, loosen the bolts on the last rod, and check the bearings for any damage.

To install the pistons, follow the steps in **SKILL DRILL 22-8**.

Measure Connecting Rod Side Clearance

After all of the pistons and rods are installed, use a feeler gauge to check the rod side clearance (**FIGURE 22-32**). Refer to the manufacturer's specifications. If there is not enough side clearance, make sure the large radius of the connecting rod is toward the rod fillet area on the crankshaft. If the radius of the rod is not facing the fillet area, either the connecting rod was installed on the piston wrong or the piston is in the cylinder backward. If you check and everything is installed correctly, then the crankshaft was machined wrong. If there is too much clearance, either the crankshaft was machined wrong or the connecting rod's big end is too narrow and should be replaced.

SKILL DRILL 22-8 Installing the Pistons

1. Install and lubricate the rod bearing inserts in the rod and cap. Also install the rod bolt protectors.

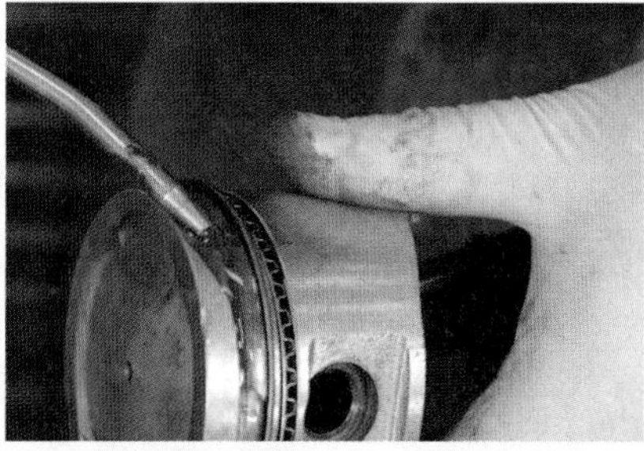

2. Set the positions of the ring gaps in the piston, and dip the head of the piston in clean engine oil.

3. Fit the piston into the ring compressor, and tighten it until it is firmly against the piston.

4. Check to make sure the crankshaft rod journal is at BDC, and the forward mark on the piston is facing the front of the engine. Tighten the compressor. Tap the compressor down until it is square against the block.

5. Use the handle of a hammer to tap the piston in place. Hit it just hard enough to make it slide past the grip of the ring compressor into the cylinder. If you feel even the slightest resistance, stop and check the compressor.

6. Guide the connecting rod onto the rod journal. Lubricate the cap that goes with the rod, and line up the marks you stamped on the sides of the rod and cap. Torque the rod nuts to one-third the final torque.

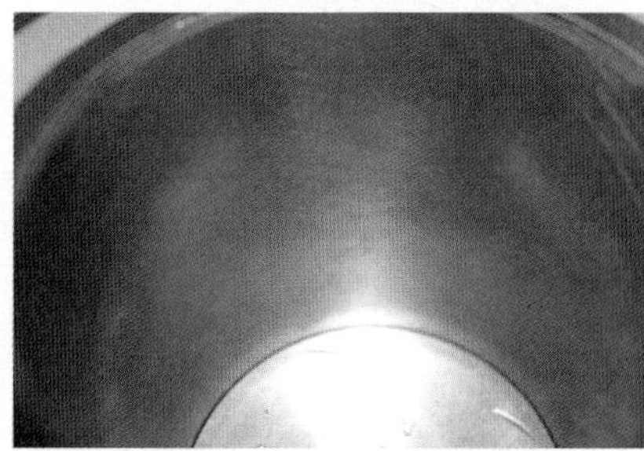

7. If the rod bearings are binding, the crank will be hard to turn. If a ring broke while you were putting in the pistons, you should notice some scratching on the walls. If you see a scratch, pull the piston out and check the rings.

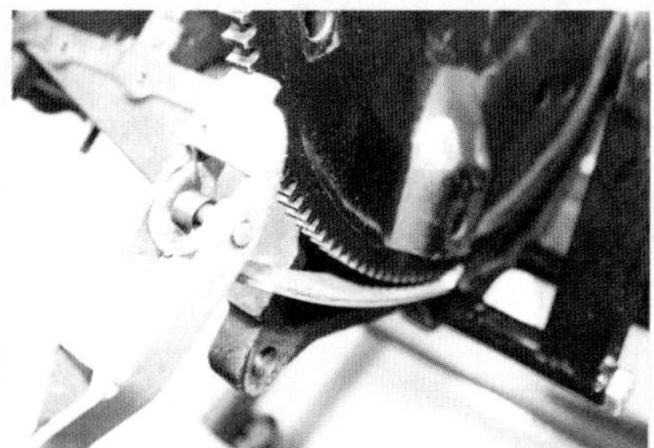

8. Rotate the next rod journal to its bottom position. When the ring compressor is tight, tap on it again to make sure it is square with the block. Then tap the piston into the cylinder.

9. Using a crank turning tool, rotate the first rod journal up. Torque the rod bolts the last two steps. After all the bolts are torqued, rotate the crankshaft to verify that the bearings are not dragging.

FIGURE 22-32 Checking rod side clearance.

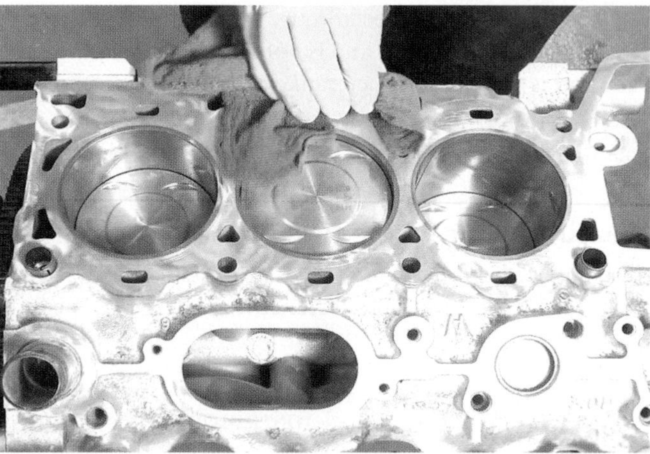

FIGURE 22-33 Make sure the deck and head surfaces are perfectly clean.

▶ Cylinder Head Installation

S22017

With the pistons and rods installed, the next step is to install the cylinder head or heads (V-type engines). Clean the block deck surface and the decks of the heads with the appropriate cleaner (spray brake cleaner), making sure to leave no fuzz or lint on the surface (**FIGURE 22-33**). The head and block need to be clean and dry, with no particles that may cause the gasket to leak.

If the head bolts are of the torque-to-yield design, then they will need to be replaced with new head bolts. Never reuse torque-to-yield bolts as they will likely fail in use. If the head bolts are reusable, retrieve them from their storage bin, and make sure the threads are clean. If any of the head bolts go into holes that extend into the water jacket, it is essential to put some non-hardening sealer on the threads of the bolts to keep the coolant from leaking past the threads. Some oil or assembly lubricant will also be needed on the underside of the bolt heads.

FIGURE 22-34 Checking to see that the head gasket holes line up properly.

Applied Science

AS-102: Adhesives/Sealants: The technician can demonstrate an understanding of how surface processes and cohesive/adhesive forces determine the effectiveness of glues, tapes, and sealants.

When working in an automotive repair facility, the technician uses adhesives and sealants on a regular basis. Every part that requires a gasket or sealant must be installed correctly to avoid a costly mistake resulting in an additional teardown to correct.

Sealant chemistry often uses the terms "cohesive" and "adhesive." Cohesive comes from a Latin word that means "to cling together." An adhesive is a material designed to hold two surfaces together. The adhesion/cohesion theory is part of the science of adhesion. Adhesion is the force that holds substrates together in opposition to stresses exerted to pull the substrates apart. Cohesion is the attraction of particles within the adhesive that holds the mass together. The combination of adhesion and cohesive strength determines sealing effectiveness. In most cases, surface preparation determines how well and for how long a bond will hold. Grease, oil, and dirt reduce the bond. All surfaces should be properly prepared per instructions of the sealant being used.

Installing Head Gaskets

S22018

Before installing a head gasket, look for "This side up" or "Front" labels on the head gaskets. If there are no labels, it should not matter which side goes up. Double-check that for every hole in the block or heads there is a corresponding hole in the gasket (**FIGURE 22-34**).

Most engines use a composition-type gasket that does not require sealant on either side. Some older engines use a metal shim gasket that requires an even coat of a high-tack gasket sealer to be sprayed on both sides or high-heat aluminum paint to be sprayed lightly on both sides. This type of gasket coating should dry before the gasket is installed.

Installing the Head

S22019

If you have not done so already, install the safety pin in the stand to keep the engine from spinning upside down. Gently lay the head on the head gasket that is on the block, using the alignment dowels to guide the head into place (**FIGURE 22-35**). Make

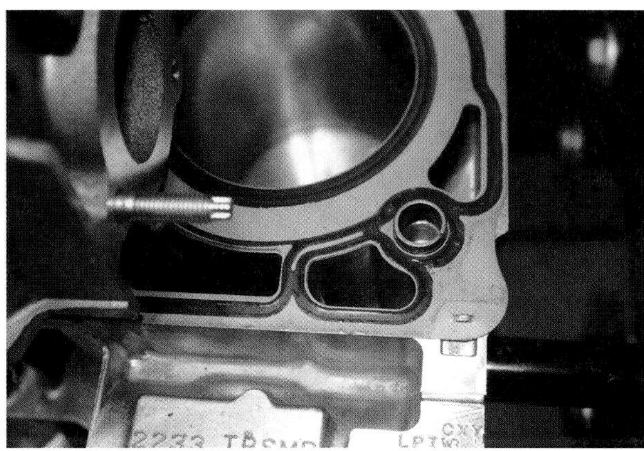

FIGURE 22-35 Align cylinder head on dowel pins.

sure to lay it down flat and not to slide it on the gasket. Thread in the head bolts by hand. If you have a few different sizes of head bolts, make sure the right bolts go in the right places. Also, remember to use a non-hardening sealer on any bolts that thread into the water jackets.

Torquing Head Bolts

Go through each bolt on both heads, running the bolts down with a ratchet and socket until they are just snug. If the bolts are standard bolts, torque in increments recommended by the manufacturer. The manufacturer will also specify the sequence that the bolts must be torqued. Many engines use either a **torque angle** or a **torque-to-yield** method when tightening the head bolts. The torque angle method is a more precise way of torquing a standard fastener to a predetermined tension. When using this method, the bolts are torqued to a minimal specification, say 20 ft-lb, and then tightened a further specified number of degrees, say 90 degrees. This process tends to cause each fastener to create an equal clamping pressure, regardless of any imperfections on the threads that would affect torque readings.

The torque-to-yield method uses the same process except that the fastener is stretched to, or near, its yield point, which means it is at the point where it will not return to its original length when loosened. Thus, the fastener will provide the best overall clamping pressure over a wide range of conditions, such as thermal expansion of the cylinder head and head gasket relaxation. Torque-to-yield bolts cannot be reused; they must be replaced with new fasteners each time they are disassembled.

Sometimes it is hard to keep track of which bolts have been torqued and which have not. Make a note of how many total bolts you have, and count as you go. If you get to what you think is the last bolt in the sequence and your count is one short, then you will know that you missed one. You should double-check regardless, especially when you get to the final torque specification.

To install the heads, follow the steps in **SKILL DRILL 22-9**.

SKILL DRILL 22-9 Installing the Heads

1. Clean all the head bolt threads and bores. Use sealer on any bolts that extend into the water jacket. Lubricate the threads and under the heads on other bolts, if specified.

2. Clean the deck surface of the block and head, and install the head gaskets. Look for any "This side up" or "Front" labels on the head gaskets.

3. On OHC cam engines, make sure the crankshaft is set to the specified position before installing the head.

4. On OHC engines, make sure the camshaft is set to the specified position before installing the head. Failure to do so could lead to damaged or bent valves.

5. Gently lay the head over the alignment dowels of the block, and start threading in the bolts. If torque-to-yield head bolts are used, tighten each bolt in the specified sequence to the specified torque. Torquing may need to be performed in several stages.

6. Install the torque angle gauge, and tighten each bolt in the specified sequence to the specified angle. If torquing standard head bolts, torque them in the specified sequence in three increments, resetting the torque wrench and increasing by one-third each time.

Assembling Valve Train and Adjusting Valves

S22020

If working on an OHC engine, the valve train components in the head should have been set up from the machine shop. On OHV engines, the pushrods and rocker assemblies have to be installed. If the valves are of the adjustable type, they have to be adjusted.

When assembling the valve train components, make sure everything is clean. Wipe them down, or wash them if necessary. If the pushrods are hollow, blow them out with compressed air (**FIGURE 22-36**). Lube the pivot points of the pushrods, rocker arms, and valve tip with assembly lube (**FIGURE 22-37**). Make sure that everything is seated properly in its mating part; otherwise, you can bend or break components (**FIGURE 22-38**). Install the valve train components following the specified procedure and order.

Once the valve train is assembled, it is time to adjust the valves. If the engine uses mechanical lifters, feeler blades are used to measure and set the valves to the specified clearance. Position the first cylinder on TDC compression. Select the proper thickness feeler blade for the valve you are checking. Insert the feeler blade between the rocker arm and the valve stem (**FIGURE 22-39**). If the feeler blade is too loose or too tight, loosen the locking nut (**FIGURE 22-40**). Turn the adjustment screw just enough so that there is slight drag on the feeler blade when you move it between the rocker arm and valve stem (**FIGURE 22-41**). Once you have the proper clearance, hold the adjusting screw still, and tighten the lock nut (**FIGURE 22-42**). Retest the valve clearance. If it changed while tightening the lock nut, readjust the clearance again. Do this for the companion valve on that cylinder. Then reset the next cylinder to TDC on compression, and adjust the valves on it.

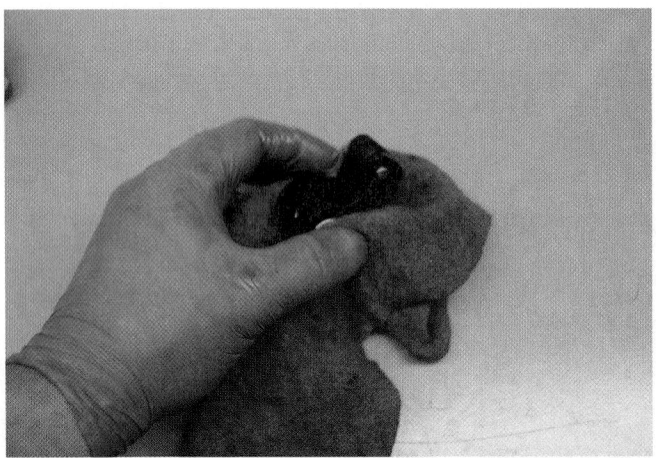

FIGURE 22-36 Before assembling the valve train parts, make sure they are clean.

FIGURE 22-37 Lube each of the pivot points on the components with assembly lube.

FIGURE 22-38 Assemble the components, making sure they are seated properly.

FIGURE 22-39 Place the specified feeler blade between the rocker arm and valve stem.

Installing Timing Chain and Gears or Timing Belt—OHC Engines

On OHC engines, the timing chains or belts are installed after the cylinder head is installed. They also use a tensioner to keep the belt or chain properly tensioned. This design can make it harder to get the timing correct, as the tensioner pulls all of the slack to one side of the belt or chain, which causes the cam to rotate slightly. Thus, the cam timing can be out of specifications. There are many different timing set configurations for these engines

FIGURE 22-40 Loosen the lock nut.

FIGURE 22-41 Turn the adjustment screw to obtain the proper clearance. There should be slight drag on the feeler blade.

FIGURE 22-42 While holding the adjusting screw, tighten the lock nut.

(**FIGURE 22-43**). It is essential that the correct service information be used when setting the cam-to-crankshaft timing on these types of engines. Also, if the engine uses a crankshaft-mounted oil pump, it will be necessary to install it before the timing chain or belt. Follow the service information in this case.

FIGURE 22-43 A dual overhead cam arrangement can have a complicated cam drive arrangement. Be sure to follow the service information when installing it.

FIGURE 22-44 Typical OHC timing chain—roller style.

Chain Drive

OHC chain drives typically are of the roller chain style, which looks a lot like a bicycle chain (**FIGURE 22-44**). Roller chain has a long life, usually lasting the life of the engine as long as the engine oil is maintained properly. Chains need lubrication to operate, so engine oil is usually sprayed onto the chain from the front of the block. The timing chain area of the engine is sealed from the outside. OHC engines using timing chains have guides and chain tensioners to keep the chain in place, in addition to two or more sprockets. Refer to the service information for timing mark locations and the specific instructions for installing these.

Chain tensioners are installed with the tensioner in the collapsed position (**FIGURE 22-45**). When installing the chain, you might be directed to count the chain links between the mark on the crank sprocket and the mark on the cam sprocket to get the right timing. Usually the timing links are a different color to make alignment easier. Align the links to the chain gear pip marks on both the crank and the cam sprockets (**FIGURE 22-46**). Install any chain guides, and then release the chain tensioner to take up the

FIGURE 22-45 Chain tensioner in collapsed position.

FIGURE 22-46 Chain links in position on the timing sprocket marks.

FIGURE 22-47 Timing marks aligned.

FIGURE 22-48 Timing belt tensioner installed.

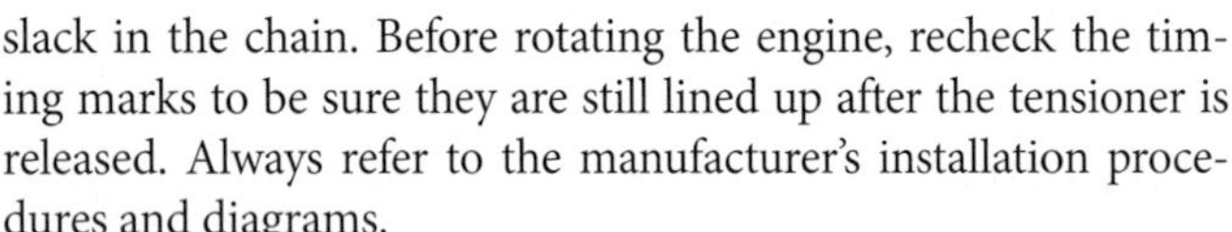

slack in the chain. Before rotating the engine, recheck the timing marks to be sure they are still lined up after the tensioner is released. Always refer to the manufacturer's installation procedures and diagrams.

Belt-Driven Timing Sets

Belt-driven timing sets use a flexible belt to transmit drive from the crankshaft to the camshaft. The belt runs more quietly than a chain and does not require lubrication. But timing belts do wear out and must be changed periodically, typically between 50,000 and 100,000 miles (80,000 to 160,000 km), depending on the vehicle manufacturer. Because they do not need to be lubricated, they are generally behind a timing belt cover that is not sealed to the front of the engine. The cover acts more as a belt guard and to protect it from dirt and debris.

As with the timing chain type of engine, always refer to current service information for timing mark locations. Install the cogged gears onto the camshaft if not already installed, being sure not to forget the keys, if equipped. They should have already been aligned when the cylinder head was on the bench, but it is best to double-check for the proper alignment of the timing marks. Be careful not to move the camshaft very much, as the valves could have an interference fit to the pistons. If the camshaft does need to move substantially, turn the crankshaft so it is approximately 20 to 30 degrees before or after TDC. This will pull all of the pistons down from the top of the cylinder so that the valves will not hit the pistons when the cam is turned. With the timing marks aligned (**FIGURE 22-47**), install the timing belt with as much of the slack as possible on the tensioner side of the belt, and then install the belt tensioner (**FIGURE 22-48**). Verify that all timing marks are aligned. Always refer to the manufacturer's installation procedures on timing belts and belt tension.

▶ TECHNICIAN TIP

On dual overhead cam (DOHC) engines, it is important to install the correct cam gear on each camshaft. Failure to do so will cause the cam timing to be off, which could lead to valve and piston damage as well as an engine that will not run or run properly. The same thing can happen if the camshafts are installed in the wrong place.

FIGURE 22-49 Oil pump installed.

FIGURE 22-50 Bolt-on style oil pickup tube.

Installing the Oil Pickup Tube and Oil Pump

S22021

The oil pump may be located in the oil pan, on the side of the block, or in the timing cover (**FIGURE 22-49**). It is typically driven by the crankshaft or camshaft. It is recommended that you pack the oil pump with some assembly lube. Doing so helps to provide the suction that is needed when the engine is started in order to get the oil moving through the pickup tube and into the oil pump. Be sure to install it to the manufacturer's specifications and torque.

In many cases, the oil pickup tube is not part of the oil pump. The tube is installed as a separate component. Oil pickup tubes come in three different versions—a screw-in style that screws into the oil pump or engine block, a bolt-on type (**FIGURE 22-50**), and a press-in style (**FIGURE 22-51**). The screw-in style has a pipe thread and requires a non-hardening sealer to ensure that the oil pump does not pull air around the threads. The bolt-on type is bolted onto the pump or block and usually uses a gasket for sealing purposes. The press-in style pickup tube is held in place with an interference fit, but the oil pump needs to be installed first. Be sure to install the oil pump drive gear or shaft before installing the pump, since it cannot be installed later. Once the oil pump is mounted, the pickup tube can be installed.

FIGURE 22-51 Press-in style oil pickup tube.

If the strainer is of the press-fit style, it is usually good practice to adjust it so it is at the proper height inside the bottom of the oil pan. This can be done by installing the pickup tube in the pump just enough so that it is fairly solid but can still twist. Rotate the strainer so that it is obviously going to touch the bottom of the oil pan. Then set the oil pan in place, which will rotate the strainer until the pan contacts both block rails fully. Lift off the oil pan, and measure the height of the strainer from the block. Adjust the strainer until it has between $\frac{1}{4}$" and $\frac{3}{8}$" (6.4–9.5 mm) of clearance from the bottom of the pan. Then tap the pickup tube into the oil pump. Some engine builders tack-weld the pickup assembly to the oil pump to prevent the pickup tube from falling out or loosening up. If you do weld it in place, make sure the pump is removed from the engine when you weld it and that the pump cover is removed from the pump so that there is no chance the welding process will hurt the gears.

Adhesives and Sealers

K22001

When installing engine covers, manifolds, and oil pans, sealers are often used to ensure there are no leaks. There are many different engine applications and engine designs, which require various sealers (**FIGURE 22-52**). Refer to the service information for your specific engine. There are also many types of adhesives and brands of room temperature vulcanizing (RTV) oxygen-safe silicone produced. Always follow the manufacturer's instructions when deciding which adhesive or sealer to use.

Adhesives

To help hold the gaskets in place, you can use a product by 3M called Trim Adhesive™, which is contact cement in a liquid or spray form. It does a good job of holding a gasket to a smooth metal surface while the parts are being assembled (**FIGURE 22-53**). In many cases, the gasket adhesive is placed on both the gasket and metal sealing surface. The adhesive is allowed to tack up, and then the gasket is pressed into place and held there by the adhesive. The spray type has a spray nozzle that can be controlled and directed

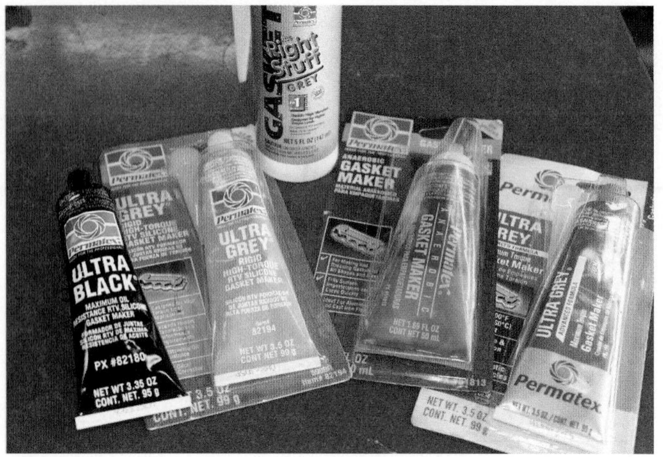

FIGURE 22-52 Various sealers and adhesives used by technicians.

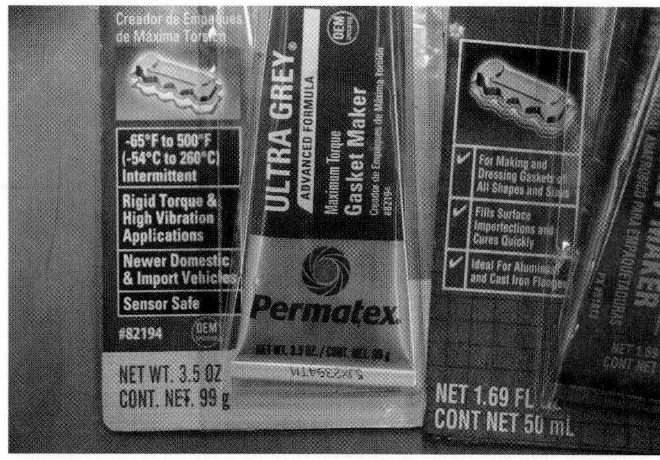

FIGURE 22-54 A variety of silicone sealers are available, depending on the temperature and fluid the sealer must stand up to.

FIGURE 22-53 Gasket adhesive being applied to a gasket.

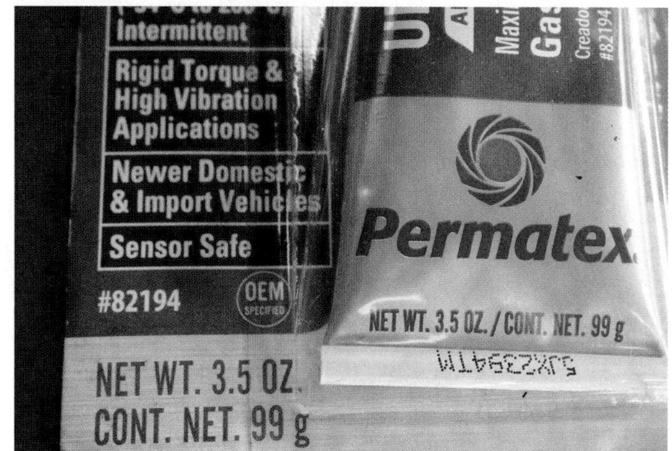

FIGURE 22-55 Oxygen sensor safe RTV.

FIGURE 22-56 Once the adhesive is smeared around, remove the gasket and let it tack up.

in a specific fan direction. However, use caution when you apply it around areas that you do not want any spray adhesive to get inside or on. Use a piece of light cardboard or paper shop rags to catch any overspray.

Silicone Sealers

There are many different silicone sealers, which are based on the temperature range the silicone can withstand or the type of fluid it is designed to seal (**FIGURE 22-54**). RTV is used to help seal fiber gaskets and gasket joints and sometimes on surfaces designed to be assembled with no gasket. When working on the engine, make sure the RTV is labeled oxygen sensor safe (**FIGURE 22-55**). Otherwise, it could harm the oxygen sensors on the engine by coating them with a silicone film, which, after a short time of engine operation, would negatively affect the operation of the oxygen sensor. Be sure to use the appropriate product and follow its instructions.

Applying Gaskets with Adhesive

For cork, felt, or neoprene gaskets, a light coat of adhesive is used on each surface of the block and the gasket. Avoid getting overspray into the engine, which can be done in two different ways: One way is to spray the gasket only with a heavy coat, and then put the sprayed side of the gasket on the block surface, and slide it around to smear the adhesive on the block surface. Then pull the gasket off, making sure there is adhesive on both surfaces (**FIGURE 22-56**). Let both the gasket and the block area dry. Once dry, position the gasket and make sure it is aligned, because you will not be able to move the gasket after you set it in place. The other way is to spray the gasket and then spray the block carefully, not allowing any overspray to get into the engine. You might have to use thin cardboard or paper towels to block any overspray. Let the adhesive dry, and position the gasket, making sure all holes are aligned. Once in place, the gasket will not move.

Applying RTV Silicone

Using silicone RTV on some gaskets is debatable. Most technicians agree that RTV can be used on paper gaskets. Some technicians say that cork and cork-rubber gaskets should be installed dry. Other technicians say that neoprene gaskets should use an adhesive rather than RTV, as RTV can make the neoprene slippery, causing the gasket to slip out of place during assembly. Do not use RTV on neoprene gaskets with multi-sealing edges. These gaskets are designed so that when one edge fails, another edge will still seal. If you use RTV on this style of gasket, the RTV fills up the multi-sealing edges and leaves only one sealing edge. Also, RTV should be used in appropriate amounts. Using too much RTV will cause excess RTV to be squeezed out from between the surfaces and form ribbons inside the engine (**FIGURE 22-57**). This excess RTV can then tear away and be carried by oil or coolant throughout the system and clog up the oil pickup screen or passages in the lubrication or cooling system.

Applying RTV Where the Manufacturer Specifies No Gaskets

On some engines, the manufacturer has designed some sealing surfaces that do not require a gasket, only RTV. When using only RTV and no gasket, a moderately larger bead of RTV is required to seal between the two surfaces (**FIGURE 22-58**). An example of this application is an oil pan; the manufacturer may have designed the engine to use no gasket. This is another reason that checking the manufacturer's procedures is so important.

FIGURE 22-57 Using too much RTV can lead to ribbons of RTV inside the engine that can break away and plug up passageways.

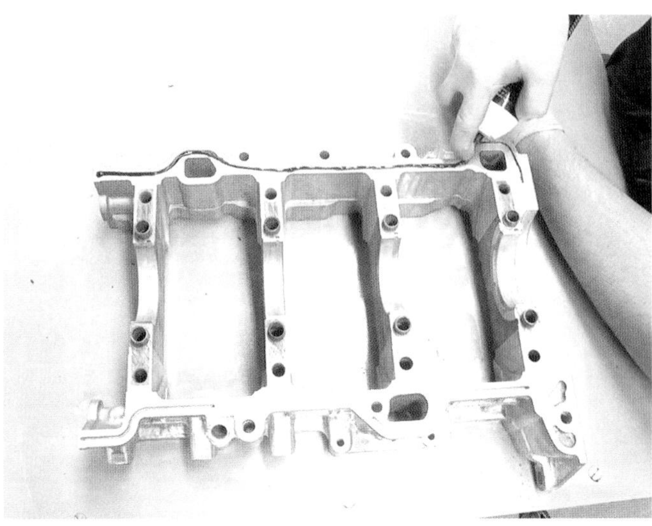

FIGURE 22-58 RTV is used instead of gaskets on some applications.

Installing the Oil Pan and Timing Cover

S22022

A certain sequence must be followed during installation of oil pans and timing covers as the parts can overlap one another. For example, in some engines, the timing cover also forms part of the mating surface for the oil pan, so the timing cover must be installed first (**FIGURE 22-59**). In other arrangements, the bottom of the timing cover has a half-moon shape with a deep groove that holds a neoprene gasket in place, against which the oil pan fits (**FIGURE 22-60**). In this case, the parts are more easily assembled if the timing cover is installed first. If the wrong sequence is used, the parts may not line up properly, which could lead to a fluid leak. Refer to the manufacturer's installation specifications for the particular engine you are working on.

To install the timing cover and oil pan, follow the steps in **SKILL DRILL 22-10**.

FIGURE 22-59 On this engine, the timing cover forms a flat surface for the oil pan gasket.

FIGURE 22-60 A timing cover that uses a half-moon neoprene gasket as part of the oil pan seal.

Oil Priming (Pre-oiling) the Engine

Priming the engine is very important; you want to make sure all moving parts receive oil as soon as the engine is started. There are a couple ways to prime the engine. One way is to spin the oil pump with a drill attached to the pump drive rod. To do this, it is handy to have a mechanical oil gauge installed in an oil gallery, the correct amount of oil installed in the engine, and a heavy-duty reversible drill. If the oil pump is driven by the distributor, put the oil pump primer rod into the distributor hole. The primer rod will engage onto the oil pump drive rod so you can spin the oil pump from the outside of the engine. The drill should be spinning in the same direction that the distributor rotates when the engine is running. Also rotate the crankshaft 90 degrees every 30 seconds or so, to fill all of the passageways in the crank and valve train. Leave the valve cover or covers off so that you can make sure all the oil is getting to the upper valve train. While priming the engine, watch the oil pressure gauge to make sure it reads

SKILL DRILL 22-10 Installing the Timing Cover and Oil Pan

1. On engines equipped with timing chains, install the timing cover to the front of the engine. Refer to service information to determine the proper sequence of assembly.

2. Make sure the rod bolts are tight; then check your notes and the manual to make sure you have not forgotten to reinstall other parts such as a windage tray. Clean the block rail and the edge of the oil pan with some solvent. Dab some gasket sealant at a few points along the block rail to help the gasket stay lined up over the bolt holes. Dab some silicone sealer at the four points where the rear main seal and the front timing cover seal arch up, away from the block rail. Lay the block rail gaskets in place. If they do not fit perfectly in the corners, you may need to cut a few small pieces with some scissors. You should always follow the manufacturer's instructions, but in most cases you will not need any sealant for rubber or cork gaskets. Gasket sets can come with a few choices for the oil pan front and rear seals. Make sure to pick the right-size gasket from the kit that will match your timing cover seal and oil pan. The same thing applies to the rear cap seal. Once you have the right seal, press it firmly into place.

If you have a cork rear cap gasket, it is a good idea to coil up the gasket and hold its shape together with a rubber band. Doing so will give a pre-curl to the gasket, making it easier to install in the curved slot on the rear cap. With all types of engines, after the front and rear oil pan seals are in place, you should put another four dabs of silicone sealer in each of the corners where the oil pan will curve from the block rail to the arch of the front or rear seal.

SKILL DRILL 22-10 Installing the Timing Cover and Oil Pan (Continued)

3. Lower the oil pan carefully into place. Start with the four corner bolts, which will make it easier to line up all of the others. Thread in the rest of the bolts a few turns by hand. You may need to tap on the pan a little to get some of the bolts to start in their holes, but make sure you do not dislodge the gasket or push it inside the block.

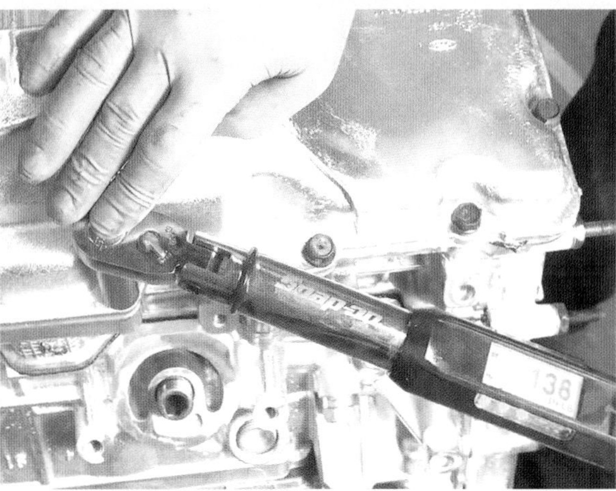

4. When the bolts are all seated hand tight, switch to a torque wrench. These bolts have a low torque rating, so make sure your wrench is accurate at low settings. These fasteners are usually torqued in inch-pounds (in-lb). Work around the pan from the center bolts in a spiral outward to the front and rear bolts. With almost all oil pan gaskets, it is a good idea to let the pan sit for a minute after you torque the final setting. Then go back through all the bolts again. You will probably find they need a third torquing to achieve the torque specification. If your engine has a thin gap between the oil pan and the timing cover, it is a good idea to lay a thin bead of silicone sealer along this lip. Spread the silicone evenly with a finger.

within the normal measurement. Even at normal drill speeds, the oil gauge should read that it has plenty of pressure.

When oil pressure builds inside the engine, oil will make it through the galleries and into the crankshaft and the lifters. After oil goes through the lifters, it will move up the pushrods and will start to drip out of the holes in the rocker arms. Make sure this happens for each of the rockers. Because there are no valve covers installed, be aware that oil might spill out of the head when the rockers get full.

Another way is to use a pressure bleeder tapped into the engine oil galleries. You will need to release all pressure in the bleeder and empty any oil in the engine, and then fill the bleeder with the correct amount and proper rating of oil you will be using. Close the valve on the hose and pressurize the bleeder. Install the bleeder hose in an oil gallery; then install a mechanical oil gauge, and make sure the valve cover is off. Open the valve and make sure the oil reaches all of the valve train. Watch the oil pressure gauge on the engine; it should have a reading close to the pressure bleeder's setting.

Once the engine has been pre-oiled, install the valve covers. Rubber-style gaskets should not require any sealant, but may require a gasket adhesive. If you have cork gaskets and want to use some kind of sealer, spray one side of the gasket with a high-tack aerosol gasket sealant. Stick the sealed side to the valve cover so that the gasket stays with the cover. When you place the gaskets, make sure the gaskets contact the edges of the heads on all sides so that you do not have any oil leaks once the engine starts. The valve cover bolts do not have a very high torque rating, so be careful not to torque them beyond specifications.

To pre-oil the engine, follow the steps in **SKILL DRILL 22-11**.

▶ External Engine Components and Accessories

S22023

The engine is now ready for installation of the external components, such as the manifolds, sensors, water pump, thermostat, alternator, power steering pump, air-conditioning compressor, or any other component removed during disassembly (**FIGURE 22-61**). Make sure all components are installed in the proper order, as many parts fit behind or over other parts. Also be careful to install and properly tighten all fasteners. You should not have any parts left over that were not replaced with new parts. If you do have leftover parts, determine where they came from.

Installing the Intake Manifold

S22024

With the heads installed, the intake manifold is the next component to be installed. New gaskets will be required, which may or may not require additional sealants. Also, the intake manifold bolts must be torqued to the proper specifications, in the proper sequence. It is usually a good idea to tighten the bolts in three steps,

SKILL DRILL 22-11 Pre-Oiling the Engine

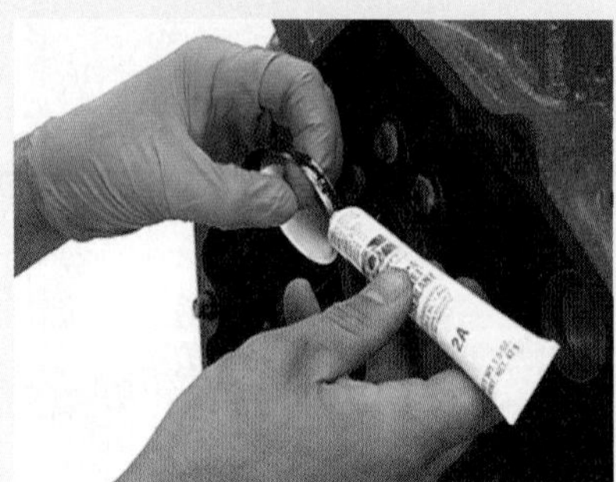

1. Install the rear camshaft plug and rear main seal, if not installed yet.

2. Install the oil filter. Lubricate the rubber seal of the oil filter with a little engine oil, and thread it on hand tight. Also install a manual pressure gauge.

3. Fill the crankcase with the type and quantity of engine oil recommended in the repair information. If the oil pump is driven by the distributor, use a drill to spin the oil pump.

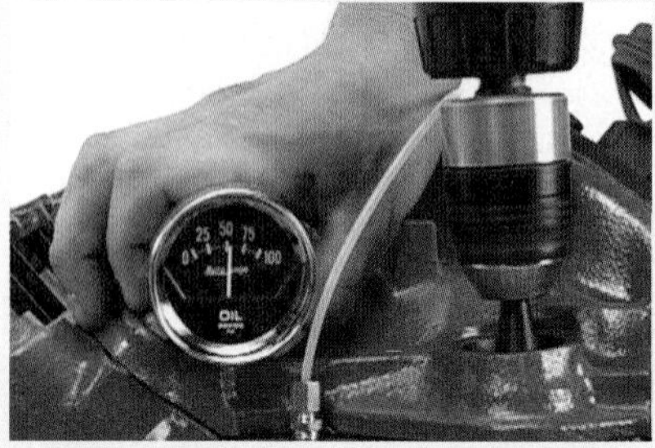

4. Check that oil is reaching the valve train. You may need to rotate the engine two turns to get oil to flow to each valve.

5. If the oil pump is driven by the crankshaft, use a pressurized pre-oiler to pressurize the lubrication system.

6. Once the engine has been pre-oiled, install the valve covers.

one-third of the torque at a time. Refer to service information for the proper torque values and sequence for the manifold bolts. An error here could result in a coolant, oil, or vacuum leak that may be difficult to diagnose and repair after the engine is in the vehicle.

On V-type engines, put some RTV sealant in the four corners where the heads meet the block. Put a thin layer of gasket sealer around the water jacket ports and their mating surfaces on the intake gaskets, if equipped. When the sealer on the gaskets is tacky, line up the bolt holes, and stick the gaskets in place. Then lightly spread some more gasket sealer on the tops of the gaskets, around the water jacket ports. On any engine, if the decks of the heads or the block, or both, were machined, the heads will sit a little bit lower, and the gap between the intake manifold and the block rail will be smaller. Some gasket sets have rubber or cork gaskets that will be too thick to allow the intake to seat properly. Such gaskets are not recommended. In this case, to seal the block rail to the intake, run a solid bead of silicone sealer along the front and rear upper rails of the block. Run the bead evenly from gasket to gasket. This is a good idea for many engines, whether the decks have been machined or not, because it usually provides a much better seal.

FIGURE 22-61 External components being installed.

On most V-type engines, the intake manifold bolts thread directly into the oil-soaked area under the valve covers. Treat the threads of these bolts with non-hardening sealer to avoid oil leaks. It is sometimes hard to use a socket to get to the bolts near the plenum of the intake. In such cases, use the box end of a wrench. The intake bolts usually have a tightening sequence and specified torque similar to the heads, but the sequence usually starts with the middle bolts near the plenum. If you cannot get to all of the bolts with a torque wrench, tighten one of the bolts to the first torque increment. Then put a box-end wrench on the same bolt, and feel with your hand how tight it is. Now you can tighten the first few inaccessible bolts for each of the three increments, and you should end up with all of the bolts torqued to a similar rating.

SKILL DRILL 22-12 Installing the Intake Manifold

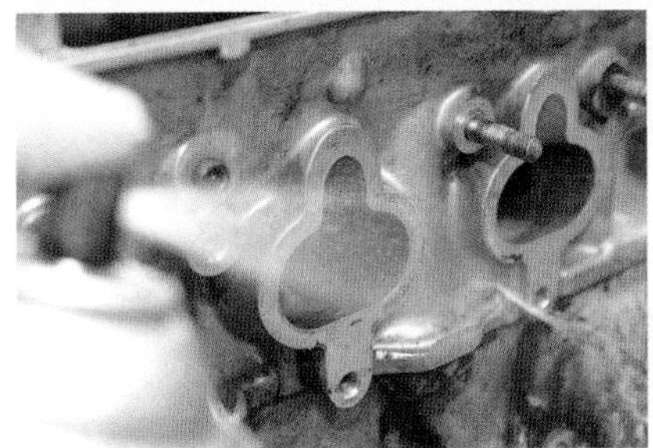

1. Clean the manifold and head sealing surfaces with a solvent cleaner.

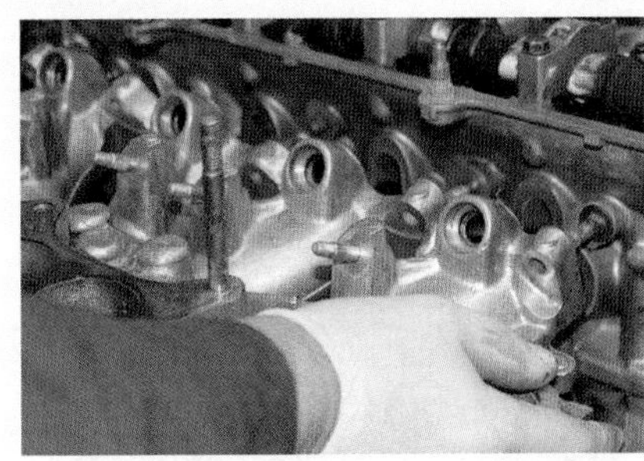

2. Make sure the manifold is ready to install, and then set the intake manifold in place.

3. Torque the manifold bolts to specifications.

To install the intake manifold, follow the steps in **SKILL DRILL 22-12.**

Installing the Remaining External Components

S22025

The remaining external components may include the balancer, exhaust manifolds, sensors, water pump (on some engines), thermostat, alternator, or any number of other components that will complete the engine assembly. The fasteners of these components are often different from one another, so it is easy to get them confused, which can lead to stripped bolt holes or damaged fasteners. This is where organization during disassembly pays off.

When installing the balancer, fit it onto the nose of the crank and line it up with its key and keyway. For a press-on balancer, do not hammer it onto the crank. Use a quality harmonic balancer installing tool, and thread it into the same threads as the center bolt of the crank. Draw up the installer nut and start turning. If it seems to bind as it is going in, check it or pull it off and start again. If it starts to go crooked and you do not adjust it, you could end up snapping off the nose of the tool in the end of the crankshaft. When it gets close or is all the way in, the crank will start to spin. If it is not fully seated, install the flex plate, and disable the rotation of the crank to draw it on the last little bit.

Exhaust gaskets usually do not require any sealant. If the gasket has a metallic finish on one side, that side will typically face toward the manifolds. Some engines do not use exhaust gaskets. Torque all bolts to the proper specifications.

To install the engine externals, follow the steps in **SKILL DRILL 22-13.**

Assembling the Accessories

S22026

As a last step, there are various accessories that have to be installed on the engine. Depending on the vehicle, these may be installed before or after the engine is installed in the vehicle. Once again, check your notes from the disassembly. The engine must be brought back to the same state it was in prior to coming out of the chassis. Replace any sensors, plugs, or brackets. You might even go as far as reinstalling the fuel injection system or the carburetor.

All engines will need the flex plate or flywheel torqued back into place. But don't forget to first install the spacer plate that goes between the engine and transmission. On most engines, it must be installed before the flywheel or flex plate.

Many engines require that new bolts be used each time the flywheel is installed; check service information for the engine you are working on. Use a flywheel wrench to keep the crank from moving while you tighten the bolts. Automatic transmission engines will only need the flex plate installed for now.

For engines with a flywheel and clutch assembly, the clutch pressure plate and disc should be installed. After the flywheel is torqued into place, set the friction disc into a centered position. A clutch alignment tool will center this disc where it needs to be while you fit the pressure plate on and tighten its screws. When the pressure plate is torqued onto the flywheel, the disc will be pinched into its proper position and will not move. Now pull out the alignment tool and put it away.

Decide what else you need to bolt to the engine. Keep in mind that you might still need to roll the engine through a doorway or past some other obstacle to get it to the hoisting location, but the aim is to be close to the completed assembly, ready for reinstallation in the vehicle.

To assemble the accessories, follow the steps in **SKILL DRILL 22-14.**

▶ Inspecting and Replacing Camshaft Drive Belt/Chain

S22027

The camshaft and drive belt/chain assembly make up an important system for the internal combustion engine. They link the valve train system to the crankshaft, and determine the right time in the engine cycle for the valves to open and close. Due to operational wear, these parts need to be serviced to make sure the engine continues to operate correctly. Timing belts need to be replaced according to the manufacturer's scheduled service. Timing chains usually last much longer than timing belts, but may need to be replaced especially if the engine oil is neglected.

If it is unknown whether a timing belt has been replaced, you may be able to visually inspect it to see whether it looks new. But unless it looks very new, it could be old enough that it is in danger

SKILL DRILL 22-13 Installing the Engine Externals

1. Thread a harmonic balancer installing tool into the same threads as the center bolt of the crank. Use the bolt to pull the balancer onto the crankshaft.

2. Install the thermostat in the correct direction. Put gasket sealer on both sides of the thermostat gasket and install the housing. O-ring seals typically do not require sealer.

3. Prepare the threads of the exhaust manifold bolts by cleaning them with a wire brush or thread chaser. Coat the threads with plenty of antiseize lubricant. Torque bolts to the proper specifications.

SKILL DRILL 22-14 Assembling the Accessories

1. Install the miscellaneous accessories, checking your notes from the disassembly.

2. Install the flex plate, and torque it into place. Use a flywheel wrench to hold the flex plate. For engines with a clutch assembly, use a clutch alignment tool to align the clutch disc.

3. Torque the front crankshaft center bolt. Make sure you have the crank held still with a flywheel wrench while you tighten this bolt.

of breaking. So if in doubt, it would be better to encourage the customer to authorize replacement. On a timing chain, unless the timing chain is making noise, or a symptom on the vehicle indicates a timing chain issue, it isn't necessary to change it unless you are doing work that deep already.

To inspect and replace the camshaft and drive belt/chain, follow the steps in **SKILL DRILL 22-15**.

Establishing Camshaft Position Sensor Indexing

S22028

Camshaft position sensors must be positioned correctly so that they accurately signal the position of the camshaft to the powertrain control module and the ignition module. Any misadjustment of the sensor will cause the fuel system timing to be off. Most newer engines do not have an adjustable camshaft position sensor, so indexing is usually needed only on sensors that can be manually adjusted.

The adjustment can typically be made in one of two ways: A special tool can be used to properly align the sensor, or a lab scope can be used to compare the camshaft position sensor adjustment to the crankshaft position sensor. If one channel of the lab scope is connected to the camshaft position sensor signal wire and another channel is connected to the crankshaft position sensor signal wire, then when the engine is running, the camshaft position sensor can be adjusted so they align. This is also a great way to measure timing chain or belt stretch on a vehicle with nonadjustable camshaft sensors. The greater the two signals are out of line, the more stretch in the belt or chain.

To establish camshaft position sensor indexing, follow the steps in **SKILL DRILL 22-16**.

▶ Inspecting and Replacing Valve Stem Seals on an Assembled Engine

S22029

It is sometimes necessary to replace the valve stem seals on an assembled engine. This is performed when the engine is in relatively good condition, but the valve seals are causing the engine to burn oil that is getting past the valve seal and valve guides.

SKILL DRILL 22-15 Inspecting and Replacing the Camshaft and Drive Belt/Chain

1. Determine all specifications for timing chain or belt tension assembly according to the manufacturer. Remove required fluids. Remove all components that cover the timing chain/belt assembly, such as the harmonic balancer, water pump, alternator, and power steering pump.

2. Remove the camshaft timing chain/belt cover. Inspect the cover for wear marks, and replace if worn. Look for causes for the worn cover.

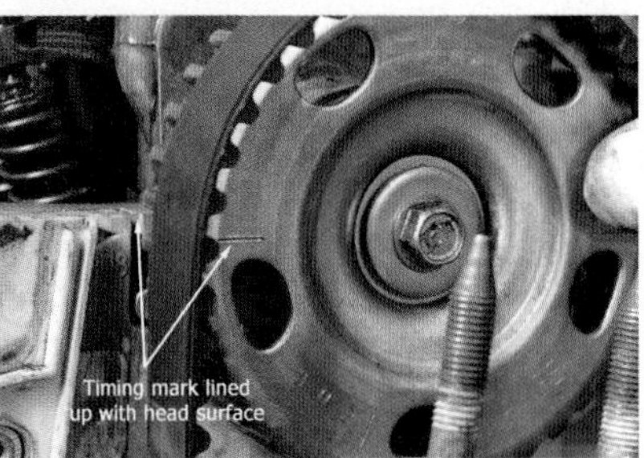

3. With the timing chain/belt cover off and the timing gears and belts/ chains in full view, turn the engine over manually to line up the timing marks of the crankshaft and camshaft sprockets with the appropriate marks on the block and head.

4. Inspect the belt/chain and measure for wear, replace if not within specifications. Measure the clearance between tensioners, if applicable, and replace if not within specifications.

5. Remove the chain or belt following the specified procedure. Use the proper cam holding tool, if specified, to prevent damage to valves.

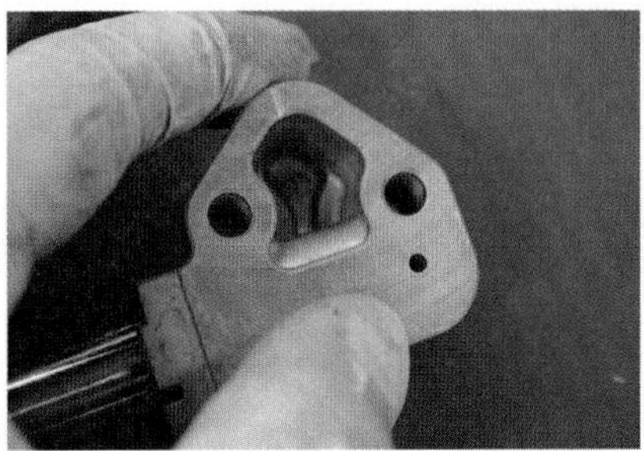

6. If the tensioner is oil operated, check the oil passages to the tensioner for clogs or buildup of dirt sludge, and clean or replace according to recommendations.

7. Inspect any guide pulleys for smooth rotation on their bearings, and replace if damaged. With the chain/belt removed, inspect the cam sprockets visually for wear, cracking, and damage. Inspect sprockets for backlash and end play, if applicable.

8. On engines with any type of variable valve timing, check components for worn and damaged parts on the gears, inspect any oil control devices for leaks, and perform other manufacturer recommended tests.

9. Reassemble the timing chain/belt assembly. Turn the crankshaft two complete revolutions by hand, and recheck the timing marks and belt/chain tension. Reassemble components following the specified procedure.

The seals can be changed once the engine is disassembled far enough to remove the valve springs. Removing the valve springs presents a problem. With the valve springs off, there is nothing to hold the valves from falling down into the cylinders. Thus, compressed air, typically from a cylinder leakage tester, is used to hold the valves closed while the valve spring is off. Careful work is required around the valve so the pressure seal is not broken and the valve dropped into the cylinder. Once the valve springs are off, the seal can be replaced with a new one. When you do so, use a special sleeve that fits over the valve keeper grooves and protects the seal as it is installed.

To replace valve stem seals on an assembled engine and to inspect valve spring retainers, locks/keepers, and valve lock/keeper grooves, follow the steps in **SKILL DRILL 22-17**.

SKILL DRILL 22-16 Establishing Camshaft Position Sensor Indexing

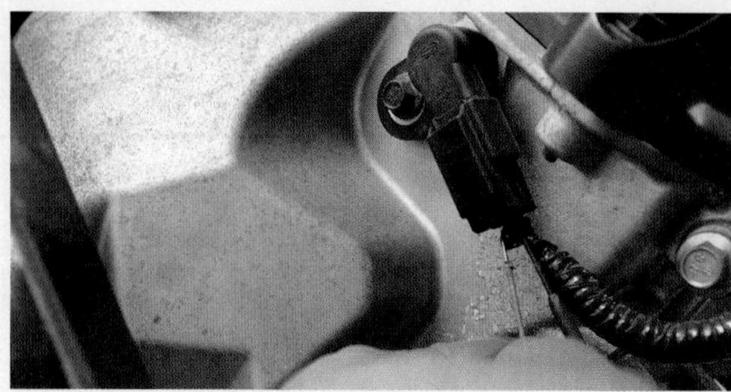

1. Attach a lab scope to the crankshaft position sensor signal wire and the camshaft position sensor signal wire.

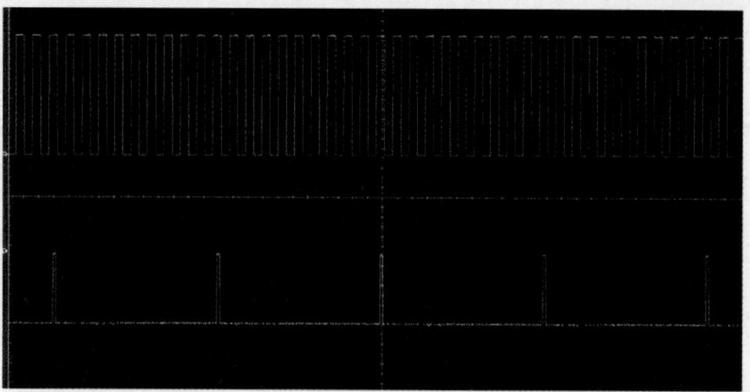

2. Start the engine, and observe the lab scope. The readings should line up. Adjust the camshaft position sensor, if adjustable, to bring it in line with the crankshaft position sensor.

SKILL DRILL 22-17 Replacing Valve Stem Seals and Inspecting Valve Spring Retainers, Locks/Keepers, and Valve Lock/Keeper Grooves

1. Remove any engine components that are in the way of removing the valve cover(s). Remove the valve cover(s).

2. Remove the rocker arms or cam followers.

3. Rotate the engine so the cylinder you are working on is at TDC on the compression stroke, and pressurize the cylinder. If the engine rotates, reset it in the proper position.

4. Use a socket and hammer to tap on the valve retainers.

5. Use a spring compressor to compress the valve spring, and remove the keepers and valve spring assembly. Remove the old valve seal.

6. Install the protective sleeve over the valve stem. Lubricate the valve stem, and install the valve seal into position. The valve stem may need to be pushed onto the valve stem boss.

SKILL DRILL 22-17 Replacing Valve Stem Seals and Inspecting Valve Spring Retainers, Locks/Keepers, and Valve Lock/Keeper Grooves (Continued)

7. Reinstall the valve spring and keepers. Repeat the process on the other valve(s) on that cylinder; then repeat the process for the valve seals on the other cylinders. Reassemble the engine and check.

▶ Wrap-Up

Ready for Review

- ▶ Preparation work necessary prior to engine assembly includes recleaning the engine block and crankshaft passages; ensuring piston, ring, and block clearances are within specifications; and determining the correct direction for piston installation on the connecting rod.
- ▶ Engine assembly includes the following:
  - Check piston-to-cylinder wall clearance by using an outside micrometer.
  - Check piston ring end gaps prior to installing piston rings.
  - Piston pins are either full-floating or semi-floating pins (wrist pins).
  - If non-stock parts are used, it is a good idea to perform a temporary test buildup of the rotating assembly (including cylinder heads and valve train).
  - Ensure that main bearings are properly sized to the crank and that the bearing cap and saddle are clean and dry.
  - Main bearing clearance must meet the manufacturer's specifications and can be measured using an inside micrometer, outside micrometer, bore gauge, or Plastigauge.
  - Check thrust bearing end play, as this controls crankshaft end play.
  - Check rod bearing, internal, and piston-to-deck clearances.
  - Check piston-to-valve clearance using either the clay or dial indicator method.
    - The clay method clearance check uses a valve impression in clay to indicate where the valve and piston may have interference.
    - The dial indicator method checks clearance at various piston positions.
  - Paint the engine to protect it from rust and corrosion.
  - Install new soft plugs and oil gallery plugs.
  - Install the crankshaft, first ensuring that main bearings are clean and properly lubricated.
  - Carefully install piston rings with any marks facing the correct way.
  - Install pistons after all piston prep work is finished, including properly positioning piston rings; lubricating pistons, rings, and cylinder walls; compressing piston rings; and lubricating connecting rod bearing inserts and rod journal.
  - Check rod side clearance with a feeler gauge, following piston and rod installation.
  - Install the cylinder head(s) after ensuring the head and block are clean and dry.
  - Use new torque-to-yield head bolts, or ensure reusable bolts have clean threads.
  - Ensure the head gasket is properly positioned, and apply sealant if necessary.
  - Torque head bolts to the manufacturer's specifications.
  - Install the timing belt (or chains) following cylinder head installation; ensure it is properly tensioned.
  - Ensure timing marks on chains and belts are aligned according to the manufacturer's specifications.
  - Install the oil pump and oil pickup tube.
  - Oil pickup tubes may be one of three types: screw in, bolt on, or press in.
  - Only use adhesives and sealers that meet service requirements for the specific engine you are working on.

- Room temperature vulcanizing (RTV) silicone should not be used on neoprene gaskets with multi-sealing edges, but is safe for paper gaskets.
- Priming the engine ensures all moving parts receive oil once the engine is started.
- The engine can be primed by spinning the oil pump with a drill attached to the pump drive or by using a pressure bleeder tapped into the engine oil galleries.
- The following external engine components should be installed in the proper order: manifolds, sensors, water pump, thermostat, alternator, balancer, power-steering pump, and air-conditioning compressor.
- The first external component to install is the intake manifold.
- Install engine accessories after installation of the external components.
- The camshaft drive belt/chain assembly may need to be replaced periodically. Always check the service information for replacement schedules.
- Camshaft position sensors must be positioned correctly so that they accurately signal the position of the camshaft to the powertrain control module and the ignition module.

Key Terms

degree wheel A disc with 360 one-degree markings near its outer edge; it bolts to the front of the crankshaft and is used to check valve and cam timing.

long block An engine assembly that includes the short block and adds the valve train cylinder heads, timing cover, oil pan, and, in some cases, manifolds.

outside micrometer A precision measuring instrument meant to measure the outside of components. It is usually accurate to 0.0001" (0.0025 mm).

pip mark A small indent or dimple on the piston ring that indicates which side of the ring is installed upward. It is also used on some timing sprockets.

Plastigauge® The trademarked name for a plastic gauging material used to check the clearances between two surfaces.

short block An engine assembly that includes the engine block, camshaft, timing set, pistons, rods, and crankshaft installed.

tang A raised or slightly bent corner on some engine bearing shells that fits into a matching recess in the bearing mounting surface to keep the bearing from rotating in the bore.

torque angle A tightening procedure in which a bolt is torqued to a set torque value and then tightened using a measured angle instead of a torque value. Example: A bolt is torqued to 40 ft-lb and then tightened another 90 degrees.

torque-to-yield A tightening procedure in which a bolt is designed to be slightly elastic when tightened; the elastic bolt retains an even pressure on the head gasket.

Review Questions

1. Most engines use a composition-type head gasket that:
   a. requires special head gasket sealer on both sides of the gasket.
   b. needs antiseize applied to both sides.
   c. can be reused more than once.
   d. does not require sealant on either side.
2. The preferred method for installing a semi-floating wrist pin is to use a:
   a. sealant.
   b. hammer.
   c. rod heater.
   d. welding machine.
3. All of the below statements are true with respect to a temporary test buildup EXCEPT:
   a. It should always include the piston rings or a rear main seal for the crankshaft.
   b. It will determine if additional machining will be necessary.
   c. It should include the cylinder heads and the valve train.
   d. It will determine if exchange of any parts will be necessary.
4. Which of these should be used to check the bearing clearance with the crankshaft installed?
   a. Micrometer
   b. Bore gauge
   c. Plastigauge
   d. Feeler gauge
5. When installing the crankshaft, to prevent oil leak from the rear of the block:
   a. paint the surface in a double coat.
   b. use a dab of silicone sealer.
   c. use plugs to close any openings.
   d. temporarily solder together the parts in the rear of the block.
6. When the piston is installed:
   a. the piston rings must be aligned in their proper positions to minimize blowby gases.
   b. ensure lubricants or oil is not present.
   c. the crankshaft should be positioned so its journal is at top dead center.
   d. hammer the piston hard into the cylinder until you do not feel any resistance.
7. RTV should not be used on:
   a. paper gaskets.
   b. fiber gaskets.
   c. thermostat housing gaskets.
   d. neoprene gaskets with multi-sealing edges.
8. When establishing camshaft position sensor indexing:
   a. use a labscope.
   b. use a micrometer.
   c. use Plastigauge.
   d. use a feeler blade and straight edge.
9. When using this torque angle method of tightening head bolts, the bolts are torqued to a minimal specification, say 20 ft-lb, and then:
   a. loosened and reinstalled in a different hole.
   b. bent over so they don't loosen.
   c. locked in place with either a cotter pin or safety wire.
   d. tightened a further specified number of degrees.

10. When replacing valve seals, which of these is used to hold the valves closed while the valve spring is off?
    a. A special sleeve
    b. Compressed air
    c. An adhesive
    d. A wrench

ASE Technician A/Technician B Style Questions

1. Tech A says that measuring piston ring end gap in the cylinder bore will verify that the rings are properly sized for the bore. Tech B says that measuring piston-to-cylinder bore clearance determines the compression ratio of the cylinder. Who is correct?
    a. Tech A
    b. Tech B
    c. Both A and B
    d. Neither A nor B
2. Tech A says that semi-floating wrist pins can be pressed into the rods. Tech B says that a rod heater can be used to install semi-floating wrist pins. Who is correct?
    a. Tech A
    b. Tech B
    c. Both A and B
    d. Neither A nor B
3. Tech A says that the bearing tang's purpose is to be used as a balance pad that can be ground, if needed, to lighten the bearing. Tech B says that bearing tangs are designed to prevent the bearing from spinning in its bore. Who is correct?
    a. Tech A
    b. Tech B
    c. Both A and B
    d. Neither A nor B
4. Tech A says that the wider the Plastigauge is squished out, the smaller the bearing clearance. Tech B says that you should turn the crankshaft one full turn after the bearing cap is torqued in place when using Plastigauge. Who is correct?
    a. Tech A
    b. Tech B
    c. Both A and B
    d. Neither A nor B
5. Tech A says that after installing piston rings on the piston, the ring gaps should be lined up with each other. Tech B says that the oil control ring typically is positioned closest to the head of the piston. Who is correct?
    a. Tech A
    b. Tech B
    c. Both A and B
    d. Neither A nor B
6. Tech A says that rod bearings are held in place by the torque of the rod bolts crushing the bearing into the big end of the rod bore. Tech B says that one of the connecting rod bearings act as a thrust bearing for the engine. Who is correct?
    a. Tech A
    b. Tech B
    c. Both A and B
    d. Neither A nor B
7. Tech A says that when installing a performance camshaft, the cam-to-lifter clearance should be measured and recorded. Tech B says that when installing a performance camshaft, piston-to-valve clearance should be checked. Who is correct?
    a. Tech A
    b. Tech B
    c. Both A and B
    d. Neither A nor B
8. Tech A says that when installing the rod caps on a rod, the bearing tangs should be on the same side of the rod. Tech B says that when installing the piston and rod into the block, rod bolt protectors should be used. Who is correct?
    a. Tech A
    b. Tech B
    c. Both A and B
    d. Neither A nor B
9. Tech A says that torque-to-yield head bolts are special bolts that are reusable. Tech B says that head bolts should be torqued in only one step to provide the proper bolt stretch. Who is correct?
    a. Tech A
    b. Tech B
    c. Both A and B
    d. Neither A nor B
10. Tech A says that the use of some RTV sealants to seal components on an engine can damage the oxygen sensor. Tech B says that it is common to use a heavy-duty reversible drill to spin the oil pump to pre-lube the engine. Who is correct?
    a. Tech A
    b. Tech B
    c. Both A and B
    d. Neither A nor B

SECTION 3

Automatic Transmissions

CHAPTER 23 Automatic Transmission Fundamentals

NATEF Tasks

There are no NATEF tasks for this chapter.

Knowledge Objectives

After reading this chapter, you will be able to:

- **K23001** Describe the types and operation of automatic transmissions.
- **K23002** Describe the operation of each type of automatic transmission.
- **K23003** Describe the operation of a torque converter.
- **K23004** Describe the purpose and function of a torque converter.
- **K23005** Identify the components of a torque converter, and describe their function.
- **K23006** Describe the various modes of operation of a torque converter.
- **K23007** Identify the components used to lock up a converter, and describe their function.
- **K23008** Describe the purpose and function of a transmission heat exchanger.
- **K23009** Describe automatic transmission geartrain components and their operation.
- **K23010** Describe the operation of the automatic transmission geartrain.
- **K23011** Identify and describe automatic transmission gear set styles.
- **K23012** Describe planetary gear sets and their operation.
- **K23013** Describe the methods used to hold and drive automatic transmission gears.
- **K23014** Describe the components of an automatic transmission and their function.
- **K23015** Describe the qualities and additives of automatic transmission fluid.
- **K23016** Describe purpose, types, and function of front pumps.
- **K23017** Describe the purpose and function of the flexplate and ring gear.
- **K23018** Describe the purpose and function of the case, extension housing, and pan.
- **K23019** Describe each type of gasket and seal.
- **K23020** Describe the purpose and function of the parking pawl assembly.
- **K23021** Describe the purpose and function of standard and compound planetary gears.
- **K23022** Describe the different types of thrust washers and thrust bearings used in an automatic transmission.
- **K23023** Describe the bushings used in an automatic transmission.
- **K23024** Describe the purpose and function of automatic transmission holding devices.
- **K23025** Describe the purpose and operation of brake bands and servos.
- **K23026** Describe the purpose and function of multidisc clutches.

Skills Objectives

There are no Skills Objectives for this chapter.

▶ Introduction

Automatic transmissions were once considered an expensive option on automobiles. Today, the vast majority of vehicles sold in the United States include automatic transmissions as standard equipment. In fact, it now mostly comes down to how many speeds a customer prefers, not whether they want an automatic or manual transmission. Some passenger vehicle manufacturers produce automatic transmissions with up to nine forward speeds, and they are likely to increase that number further over time (**FIGURE 23-1**).

Although many technicians will never rebuild transmissions, a working knowledge of automatic transmission fundamentals is necessary for vehicle diagnosis purposes due to the interrelated nature of the transmission with the rest of the vehicle. Technicians may encounter malfunctions that appear to be transmission problems, only to discover after transmission replacement that the problem is still present and caused by an entirely different system. For example, a plugged catalytic converter can cause a transmission to shift erratically, or a bad vehicle speed sensor (VSS) can prevent a transmission from shifting at all.

In some cases, a transmission can be easily repaired without a complete teardown and rebuild if the technician has a good understanding of how a transmission operates (**FIGURE 23-2**).

Unfortunately, though, technicians will often immediately condemn a transmission and replace it with a rebuilt unit. This may seem like an easy option to the technician; just replace the transmission, and the problem is likely to go away. But that can be much more expensive than a smaller targeted repair for the customer. Also, a customer who is able to drive away with a $300 repair is likely to become a much more loyal customer than one who is sold a complete transmission for $3000 and discovers later that he or she was taken advantage of. For these reasons, the more knowledge and experience you have with transmissions, the better you will be able to diagnose and repair vehicles for customers, which will lead to more job security and satisfaction. In this ASE area, we cover the various types of automatic transmissions, theory of operation, service and repair procedures, and diagnosis of hydraulically and electronically controlled transmissions. In this chapter we will cover the theory of operation.

Function of an Automatic Transmission

K23001

An automatic transmission has two major functions that separate it from a manual transmission. First, the transmission can select and shift gears without input from the driver. This

FIGURE 23-1 Nine-speed ZF transmission.

FIGURE 23-2 Sometimes transmission concerns can be fixed with simple repairs.

You Are the Automotive Technician

In your second week as a new technician, you overhear two older technicians discussing automatic transmissions. Both of the technicians have been in the industry for a long time, but they have had very little training, having learned most of what they know on the job. They seem to agree on much of their understanding about how automatic transmissions operate, but they have several points of disagreement. Because you have been through an automotive technology training program, they ask you to explain some things to them.

1. How does a torque converter transmit more torque to the transmission than the engine creates?
2. How can the planetary gears in an automatic transmission always be in mesh, and yet the transmission can change from one gear ratio to other gear ratios?
3. How do multidisc clutches, bands, and one-way clutches operate?

function is accomplished within the geartrain of the transmission with the aid of the hydraulic and electronic control systems. Second, the transmission can automatically couple and uncouple from the engine when needed, much like a clutch on a manual transmission vehicle, but without any input from the driver. This is the function of the torque converter. Without the ability to automatically connect and disconnect the transmission from the engine, a vehicle would stall when it came to a stop, much like a manual transmission vehicle will do if the driver forgets to depress the clutch pedal. These two functions separate automatic transmissions from manual transmissions and are discussed in depth over the next several chapters.

Types of Automatic Transmissions

K23002

There are several types or classifications of automatic transmissions: conventional transmissions, transaxles, dual clutch, continuously variable, hybrid transmissions, and a Honda/Saturn type of transmission. Each of these types of automatic transmissions has elements common to the conventional transmission, but we explain only the major differences here.

A conventional automatic transmission uses one or more planetary gear sets to create several gear ratios needed to drive the vehicle. Conventional automatic transmissions are typically connected to the engine through a torque converter. The torque converter is covered in detail later in the chapter. The gears inside a conventional transmission are constantly meshed to each other. Holding devices, such as clutches or bands, will either stop or drive the rotation of gears in a planetary gear set in order to create the needed gear ratios.

On most rear-wheel drive (RWD) vehicles, the differential and final drive gear are located separate from the transmission in the rear axle assembly. The differential allows for the vehicle to make turns without scrubbing or dragging the tires, and the final drive increases the torque to the wheels. In the 1970s, most manufacturers developed front-wheel drive (FWD) vehicles for several reasons: improved traction over RWD vehicles, less overall vehicle weight, and increased passenger compartment space. Having the engine, transmission, differential, and final drive all driving the front wheels required a different arrangement. That arrangement is the **transaxle**, which is a combination of a transmission, differential, and final drive gear all in one compact unit (**FIGURE 23-3**).

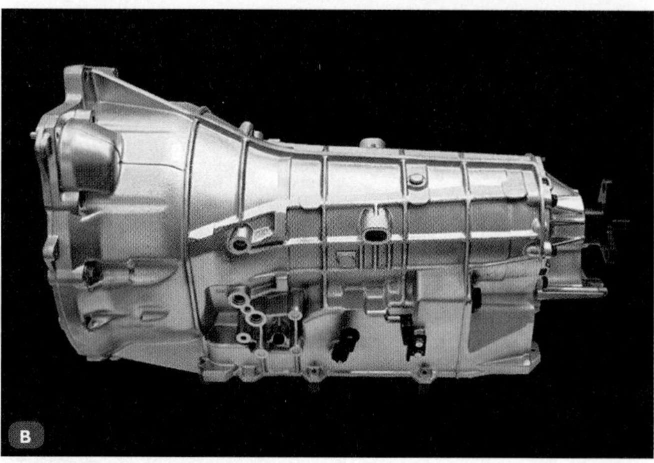

FIGURE 23-3 **A.** Transaxle. **B.** Transmission.

In most cases, the transaxle bolts directly to the engine, although some RWD vehicles, such as late-model Chevy Corvettes, use a transaxle assembly located in the rear of the vehicle and the engine is located in the front. The Corvette uses this transaxle arrangement to better distribute weight between the front and rear wheels to improve handling. A fairly long shaft connects the engine and transaxle together.

Dual-clutch transmissions are a newer type of automatic transmission that uses two clutches, either wet or dry, in place of a **standard torque converter** to connect the engine to the transmission. A wet clutch is one that is immersed in hydraulic oil or transmission fluid rather than being dry like a conventional manual transmission clutch. Each clutch is connected to different driving gears. Clutch 1 is connected to gears 1, 3, and 5, and clutch 2 powers gears 2, 4, and 6 (**FIGURE 23-4**). The transmission operates by locking the individual driving gears to the output shaft in order to change the speed of the output

FIGURE 23-4 A Ford Focus dual-clutch transmission. Gears 1, 3, and 5 are on one shaft and 2, 4, and 6 are on the other shaft. Note that this is more like a manual transmission than an automatic transmission.

shaft. Their operation is similar in some ways to that of a manual transmission, but the changing of gears is controlled by the vehicle computer, not the driver.

Continuously variable transmissions (CVTs) are transmissions that do not use typical gears as in other transmissions. Rather, CVTs commonly use two pulleys that change diameter in response to vehicle load and speed. There is a large, heavy metal belt connecting the two pulleys (**FIGURE 23-5**). When the vehicle is starting from a stop, the input pulley, which is coupled to the engine, has a small diameter; the output pulley, which is coupled to the drive wheels, has a large diameter. This size difference creates a large amount of torque multiplication as the smaller input pulley turns approximately three times for every one turn of the output pulley. When the vehicle reaches cruising speed, the input pulley has compressed, causing the diameter of the pulley to increase, and the output pulley is reduced in diameter. At this point, the output pulley spins up to three times for every revolution of the input pulley. Using this system, the transmission does not have fixed gear ratios like a conventional transmission; instead, it is able to vary the gear ratio infinitely within the changing diameters of the pulleys. This variability allows the engine to operate in its most efficient rpm range to save fuel.

Hybrid vehicles use drivetrains that can be categorized as series or parallel hybrid drivetrains. In a series drivetrain, the power going into the transmission from the engine is supplemented with power from an electric motor. In a **series hybrid drivetrain**, the electric motor is not typically able to propel the vehicle on its own. This electric motor is often placed between the engine and the transmission.

A **parallel hybrid drivetrain** has two or more devices to power the transmission. The engine can mechanically send power through the transmission to the wheels, or one or more electric motors can send power through the transmission to the wheels. Either the electric motors or the gasoline engine usually propels the vehicle along, individually or in combination.

A **series-parallel hybrid drivetrain** is designed so that it can function as both a series hybrid and a parallel hybrid. It uses what is called a **power-splitting transmission (PST)** (**FIGURE 23-6**). This transmission is sometimes referred to as a type of CVT transmission, but it uses a planetary gear set to obtain an infinitely adjustable transmission. It allows power to be applied to the planetary gear set (epicyclic gearing) through three sources: the internal combustion engine (ICE) and two motor/generators. This design is an ingenious application of the planetary gear set, as in its normal use one member is held and another is driven. But in the case of a hybrid vehicle, where you have more than one power source, two components can be driven. Moreover, they can be driven at different speeds, and in the case of the electric motors, they can be driven in reverse. This ability to drive more than one member at various speeds allows the planetary gear to act like a CVT. In this configuration, the planetary gear set acts as a power divider and uses all three power sources or two generators to balance the overall needs of the system, including vehicle speed, traction motor battery charge, and ICE efficiency. This system is operated by sophisticated computer controls. Hybrids and CVT transmissions are covered in greater detail in the chapter on that topic.

FIGURE 23-5 The pulleys and metal belt of a CVT transmission.

FIGURE 23-6 A power-splitting transmission (PST) uses a planetary gear set to obtain an infinitely adjustable transmission.

Many Honda and Saturn vehicles use an automatic transmission that more closely resembles a manual transmission, as they do not use planetary gear sets. These transmissions are sometimes called **dual-shaft transmissions** because they have a main shaft and a countershaft, like in a manual transmission (**FIGURE 23-7**). In these transmissions, the synchronizer assemblies that would be in a manual transmission are replaced with multidisc clutch packs to engage the individual gears onto the main shaft.

▶ Torque Converters

K23003

Torque converters transmit engine torque to the transmission. They allow the engine to idle at a stop even though the transmission is still in gear. As engine speed increases, they transmit greater amounts of torque, driving the vehicle. We explain their operation in depth a bit later. Torque converters are generally robust components that don't give a lot of trouble, but if they do, it is important to have a basic working knowledge of them in order to facilitate proper vehicle diagnosis. And when you

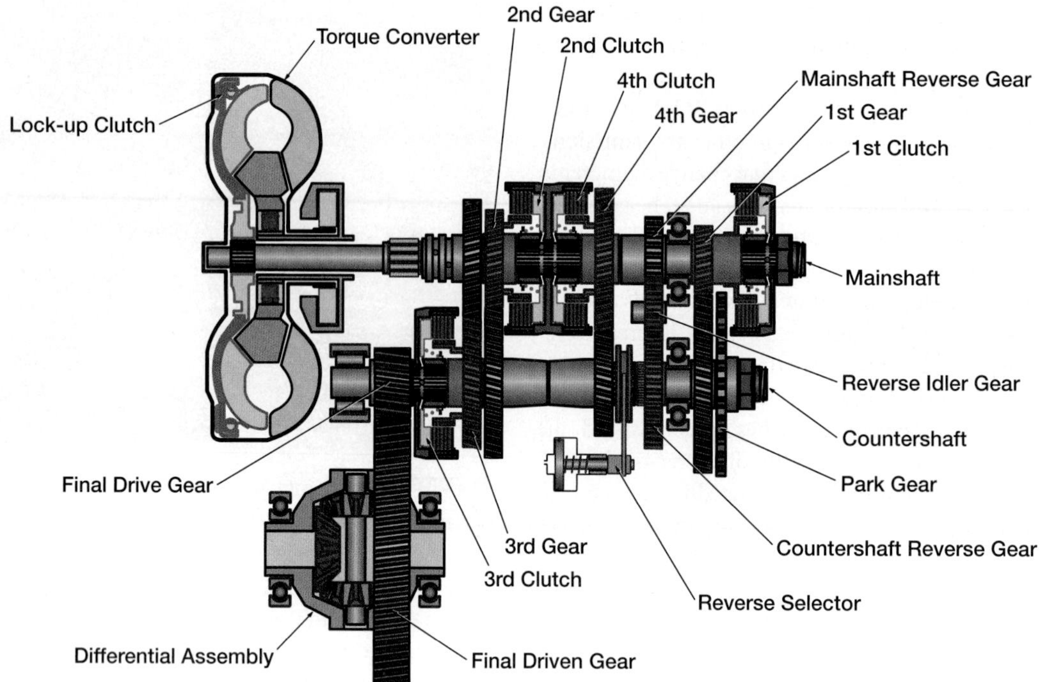

FIGURE 23-7 A Honda automatic transmission with a main shaft and countershaft.

do find one that is faulty, you will typically only replace it, not repair it, as they are usually welded together from the factory. So most of the time torque converters only get replaced when the transmission is being replaced or rebuilt, although occasionally the torque converter itself fails.

Understanding how the torque converter operates will help you successfully diagnose problems that may occur. When lock-up torque convertors came out in the early 1980s, one manufacturer had a very large percentage of these torque convertors coming in for warranty replacement. The problem, however, was that the customer complaint was not rectified after the torque convertor was replaced. After careful diagnosis, the problem turned out to be a faulty spark plug wire that would only appear once the lock-up torque convertor engaged. The manufacturer released several technical service bulletins to address this concern. If the technicians had been better able to diagnose the system, they might have accurately diagnosed the problem sooner, and for much less money.

As stated, most light vehicle torque convertors cannot be disassembled by a traditional shop, or even a transmission repair facility, as they are a welded component. Most shops simply order a remanufactured torque convertor when rebuilding the transmission. On some heavy-duty vehicles, the torque convertors can be disassembled and rebuilt in a regular shop because the two halves of the housing are bolted together instead of welded (**FIGURE 23-8**).

FIGURE 23-8 A. A light vehicle torque convertor that is welded together. **B.** A heavy-duty truck torque converter that is bolted together.

Torque Converter Principles

K23004

The purpose of a vehicle transmission is to transmit engine torque to the driving wheels. In a manual transmission, engine torque is controlled by the driver by opening or closing the throttle and modified by manually selecting the appropriate gear with the shifter mechanism. In an automatic transmission, the driver still controls the engine torque with the throttle, but in this case gear selection is automatically controlled by the transmission. The transmission also works in concert with the torque converter to modify the engine's torque, as we explain later (**FIGURE 23-9**).

The torque converter is mounted between the engine and the transmission, in the same place as a manual transmission clutch (**FIGURE 23-10**). The torque converter effectively connects the engine flywheel to the input shaft of the transmission. The torque converter does the same job as a manual clutch, transmitting engine torque to the input shaft of the transmission. But it can do one thing a manual transmission clutch cannot do—multiply torque from the engine to the input shaft of the transmission under certain driving conditions. This is one reason that vehicles equipped with automatic transmissions make good tow vehicles. The torque converter can multiply the engine torque when starting from a standstill, which is just what is needed when pulling a boat trailer out of the water on a steep boat ramp. How can it do that, you ask? Well, keep reading as we explain how each of the torque converter components work together to multiply torque under certain conditions.

FIGURE 23-9 A typical torque converter that has been removed from a vehicle.

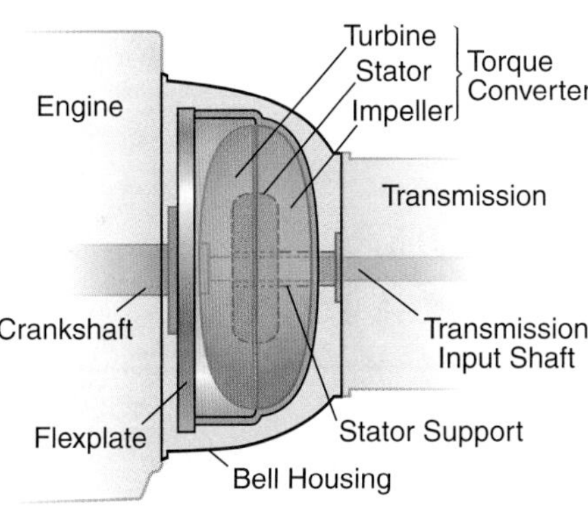

FIGURE 23-10 A torque converter mounted between the engine and the transmission.

Torque Converter Components

K23005

Early automatic transmissions used a **fluid coupler**, which was the precursor to a torque converter. A fluid coupler is basically two fans facing each other (**FIGURE 23-11**). One fan is the driving fan and is driven by the engine. The second fan is attached to the input shaft of the transmission. When the fluid is thrown off the driving fan, it hits the driven fan, causing the second fan to begin to spin. When the second fan spins, the input shaft of the transmission also begins to spin.

The device is called a fluid coupler because none of the converter components are physically connected to the others; rather, the force of the fluid flow causes the transfer of power. This fluid coupler acts as an automatic clutch. At engine idle speeds, it allows the engine to operate while the vehicle is stationary and the transmission is in a drive range, without the driver having to shift the transmission to neutral or step on a clutch pedal. Unfortunately fluid couplers were very inefficient, so they haven't been used since the 1950s.

Modern transmissions no longer use fluid couplers, but the principle of the torque converter operation is still similar to the fluid coupler, but with an extra component. In its simplest form, a single-stage torque converter has three elements: the impeller, the turbine, and the stator (**FIGURE 23-12**). All three have angled or curved vanes and are contained in a single housing. They

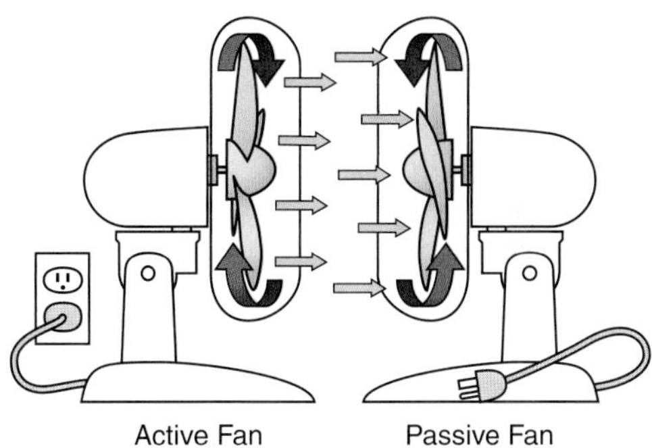

FIGURE 23-11 One fan driving a second fan functions much like a fluid coupler, with air as the medium rather than transmission fluid.

FIGURE 23-12 A torque converter that has been cut open to reveal the impeller, stator, and turbine.

are separated from each other by thrust bearings but maintain a close relationship for efficient torque transmission.

The impeller has a large number of vanes attached to the converter housing to form the driving member (like the fan that is plugged in for the fluid coupler) (**FIGURE 23-13**). The vanes rotate with the housing, and because the housing is bolted to the engine's flexplate, the impeller rotates at engine speed. Each impeller vane has a slight curvature and is set radially in the case. The impeller causes fluid to flow inside the converter.

The turbine is similar in construction to the impeller, but with more vanes and a greater curvature. The direction of curvature of the turbine vanes is opposite to that of the impeller vanes. The turbine is free to rotate in the housing, and the center hub has splines that mate with splines on the input shaft of the transmission (**FIGURE 23-14**). The turbine is turned by fluid leaving the impeller. This then rotates the transmission input shaft.

Both the impeller and the turbine are fitted with a guide ring, which helps to secure the vanes in position. It also reduces turbulence as fluid is flowing in a circle from the impeller to the turbine and back to the impeller again, and it improves torque converter efficiency. So far we have what could be called a fluid coupler. The next component (stator) is what turns this assembly into a torque converter.

The stator has a small set of curved blades attached to a central hub and is positioned between the impeller and the turbine. The center hub is mounted on a one-way clutch splined to a stator support shaft (**FIGURE 23-15**). The stator support shaft is attached firmly to the transmission case and does not rotate. The one-way clutch allows the stator to rotate only in the same direction as the impeller. Trying to rotate the stator in the opposite direction locks the stator on the support shaft and holds it stationary.

FIGURE 23-13 An impeller brazed to the housing of the torque converter.

Converter Operation

K23006

When the engine starts, the transmission pump rapidly fills and pressurizes the converter with transmission fluid. The impeller is driven by the engine and turns at crankshaft speed. The turbine is splined to the transmission input shaft. When a gear is selected by the driver, the input shaft becomes locked to the output shaft of the transmission through the various gears, bands, and clutches in the transmission. The spinning impeller uses centrifugal force to throw fluid outward; the fluid follows the case and strikes the turbine vanes and then goes back around the guide ring in a forward direction. This is because of the shape of the housing and the curvature of the vanes. With the engine idling, little torque is transferred from the impeller to the turbine, as the fluid flow is too slow. When the engine accelerates, higher impeller speed discharges the fluid across and against the turbine vanes with greater force, causing the turbine to begin to spin and transferring torque to the input shaft of the transmission.

The fluid, still at high velocity, now flows between the turbine vanes. The fluid leaves the turbine in a direction opposite to impeller rotation due to the curvature of the turbine vanes. In a fluid coupler, this fluid flow would then strike the impeller in the opposite direction of impeller rotation, reducing efficiency and torque. In a torque converter, the stator redirects the fluid so that it reenters the impeller in the same direction as impeller rotation (**FIGURE 23-16**). The energy that was left in the fluid

FIGURE 23-14 A turbine mounted inside the converter case. Note that the turbine is free to spin inside the housing on bearings.

FIGURE 23-15 A stator removed from a torque converter.

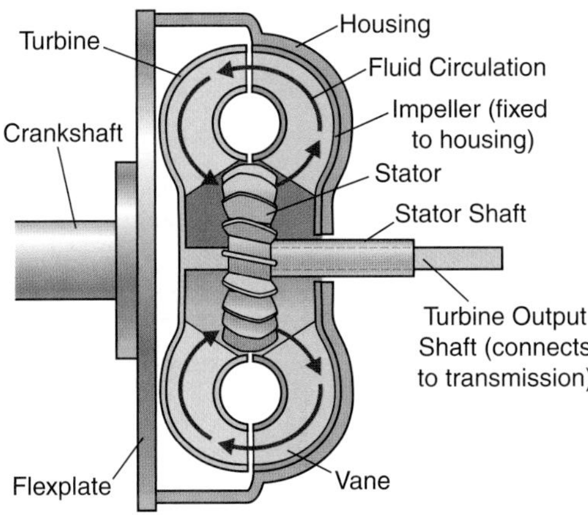

FIGURE 23-16 Fluid flow through a torque converter.

after turning the turbine is recycled and helps to spin the impeller, which allows the engine to turn faster. This process creates torque multiplication. During torque multiplication, the stator remains stationary while the turbine and impeller spin.

Torque Multiplication

Torque multiplication exists only when there is a difference in speed between the impeller and the turbine. The amount of torque multiplication depends on load. When the turbine is stalled, it has a maximum torque multiplication of about 2.5:1 for most passenger vehicles. **Stall** is an operating condition where the turbine is stationary and the engine throttle is wide open, making the rotational speed of the impeller as high as possible.

Stall can be approximated when a vehicle moves from rest, up an incline, towing a heavy trailer. Instantaneously, as the vehicle begins to move, maximum torque multiplication occurs because there is a large difference between the speed of the impeller and the turbine. At a stall speed of 2100 rpm, if the torque at that speed is 200 ft-lbs (271.16 N·m), the torque input to the transmission could be as high as 500 ft-lbs (677.91 N·m) which is 2.5 times the input torque. This torque enters the transmission, where it is multiplied further by the planetary gears, and multiplied again at the final drive assembly before being applied to the vehicle's wheels. This torque multiplication is what allows a vehicle its pulling power moving away from a stop.

SAFETY TIP

Stall speed can be tested by performing a stall test (brake torque) in the shop. To perform this test, place the transmission in low gear, apply the brake pedal fully, and then carefully push the throttle open all the way. The maximum rpm the engine achieves is the stall speed. This can be a dangerous test and should only be performed under your supervisor's direct guidance.

This multiplication tapers off as the turbine speed begins to match the speed of the impeller. When turbine speed reaches around 90% of impeller speed, torque multiplication falls to zero. Torque multiplication is then about 1:1. This is known as the coupling point. At the coupling point, fluid flow from the turbine vanes is relatively low, but it is at high speed, following the direction of the torque converter rotation. The rapidly turning turbine discharges its fluid against the back of the stator blades. This force unlocks the stator's one-way clutch, allowing the stator to rotate with the fluid. The stator, turbine, and impeller all rotate as one unit at this time. Unlocking the stator prevents the stator from being in the way of the fluid flow.

Fluid Flow

The rotating impeller carries the fluid with it inside the converter housing. The fluid is rotating around with each of the three converter components. This is known as **rotary flow**. At the same time, centrifugal force moves the fluid outward, away from the converter axis. During torque multiplication, the shape of the converter case makes the fluid flow in a circular motion, through the impeller, turbine, and stator. This is known as **vortex flow**. Combining these two fluid flows produces a progressive circular, or spiraling, motion. This is known as the spiral flow (**FIGURE 23-17**).

In a stalled converter, fluid flows at high velocity from the revolving impeller through the stationary turbine and stator. This condition results in a fast-moving vortex flow and high torque multiplication. When the turbine starts to rotate and increases in speed, the centrifugal force on the fluid in the turbine opposes the high-velocity flow from the impeller, reducing vortex flow and torque multiplication. At coupling point, the vortex flow of fluid is slight, and there is no torque multiplication. The converter now acts as a fluid coupler.

During hill climbing, the turbine slows, and at the same time the driver increases engine power, causing the impeller to speed up. The combination of these two factors causes an increase in vortex flow. This again causes torque multiplication until either the turbine speeds up or the impeller slows down. The converter automatically adjusts its output, within design limits, to meet driving requirements.

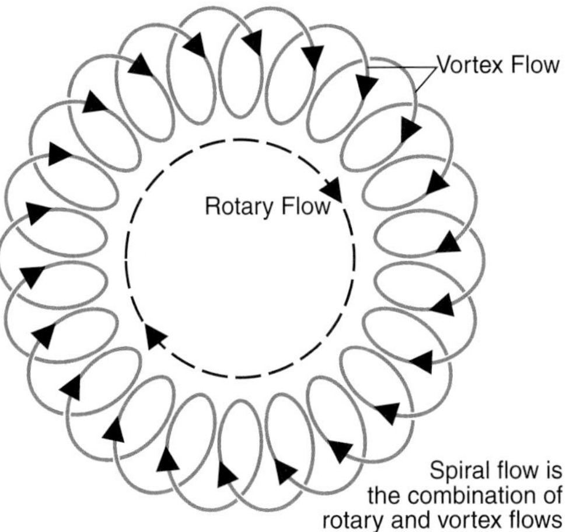

FIGURE 23-17 Rotary fluid flow, vortex flow, and spiral flow.

Lock-Up Converters

K23007

Under a drive condition, the impeller and turbine generally never achieve a 1:1 speed ratio. The turbine almost always spins slightly slower than the impeller. This means that the converter is typically slipping approximately 5% to 10% when comparing engine speed to transmission input speed. During the late 1970s and early 1980s, manufacturers added a lock-up function to the torque converter in their vehicles, to deal with this slippage condition and increase a vehicle's fuel economy.

In a lock-up converter, the impeller and turbine are locked together when conditions are suitable, to provide a 1:1 drive from the engine to the transmission input shaft. Lock-up normally occurs at higher road speeds and when the vehicle is under light load. Inside the front of the torque converter, there is a large piston that has friction material bonded to its surface near its outside diameter (**FIGURE 23-18**). When the piston is hydraulically applied, the piston and friction material are pressed against the front housing of the torque converter, locking the turbine to the housing (**FIGURE 23-19**). This provides a 1:1 connection from the engine to transmission, with no slippage between components. The lock-up torque converter also helps to reduce transmission fluid temperatures; in some cases this requires an ATF warmer system.

Torsional damper springs are built into the piston assembly (**FIGURE 23-20**). When the clutch is engaged, these springs dampen engine and drive line torsional vibrations. Without these damper springs, every power pulse of the pistons would be felt by the driver as excessive engine vibration.

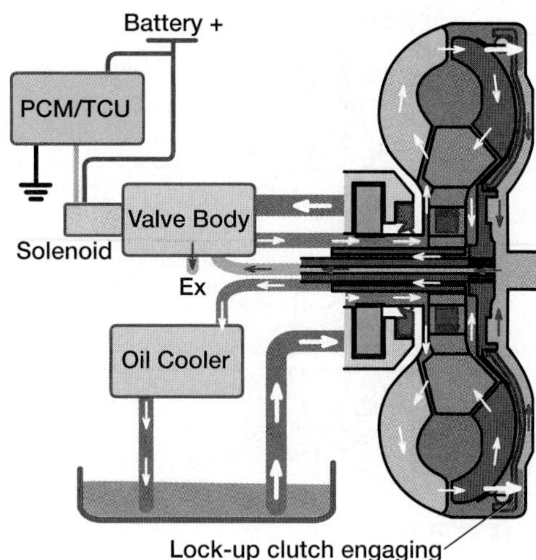

FIGURE 23-19 The torque converter clutch is hydraulically applied and forced against the front housing of the torque converter, locking the two together.

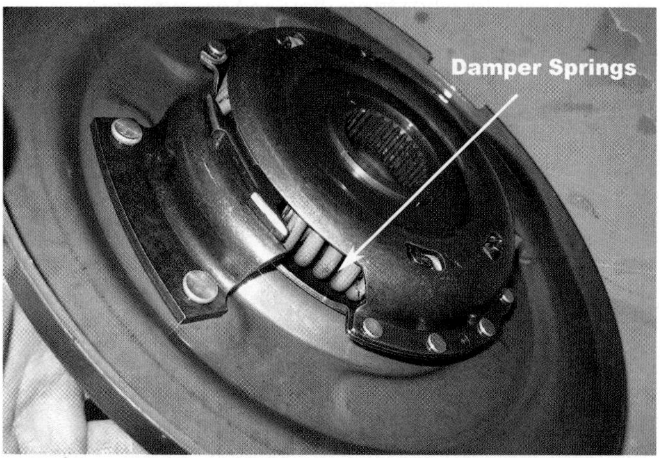

FIGURE 23-20 Lock-up pistons use torsional springs to absorb power pulses from the engine.

Heat Exchanger

K23008

Torque converter slip and loss of power through the transmission produce heat, which must be dissipated. At stall, a lot of engine output is converted into heat, bringing the transmission fluid operating temperature closer to its boiling point. Excessive temperature can produce cavitation bubbles in the fluid, which reduce converter efficiency. It can also shorten the life of the transmission fluid greatly. Too hot and it oxidizes; too cold and it does not lubricate as well. Most automatic transmission vehicles use a heat exchanger, sometimes called a transmission cooler, usually in outlet tank of the radiator. In this way, the transmission fluid can be cooled, but the engine coolant at the outlet side of the radiator will prevent it from overcooling the transmission fluid (**FIGURE 23-21**).

FIGURE 23-18 A lock-up torque converter piston with friction material bonded to the back of the piston.

Fluid flows from the pump to the valve body, where part of the fluid flows to the converter and the rest to operate the bands and clutches. Because the converter creates a lot of heat, the fluid then flows through the heat exchanger before returning to the transmission, where it is used to lubricate the planetary gears and bushings in the transmission. Heavy-duty vehicles or those with towing packages installed have an extra external transmission cooler installed in front of the radiator to help dissipate the added heat from towing (**FIGURE 23-22**).

FIGURE 23-21 Cutaway showing the transmission cooler inside the radiator tank.

FIGURE 23-23 A deep sump finned transmission pan.

FIGURE 23-22 An external transmission cooler (the smaller one behind the tubes) is typically used on vehicles equipped with towing packages.

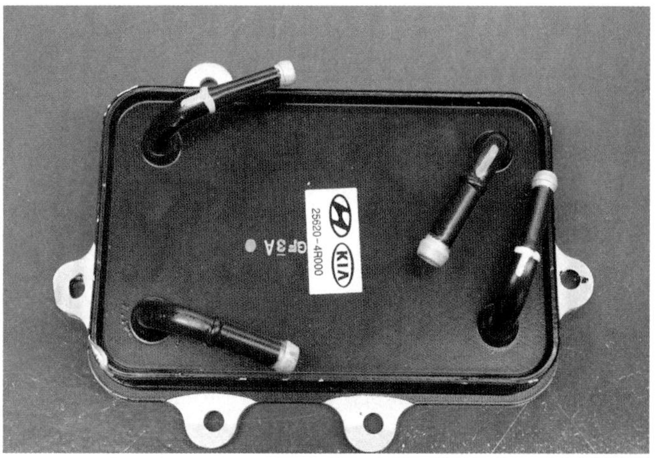

FIGURE 23-24 A transmission fluid warmer from a Hyndai Sonata.

When using an external transmission fluid cooler, it is generally agreed that it is best to run the transmission fluid through the external cooler first and then through the cooler in the radiator. Doing so prevents overcooling of the transmission fluid.

Another method of improving cooling on heavy-duty vehicles or performance vehicles is to install a finned aluminum transmission pan. This finned pan allows air that is passing under the vehicle to draw heat away from the fluid in the transmission pan. Often a pan of this type is a deeper pan, allowing the transmission to hold an extra quart or two of fluid (**FIGURE 23-23**).

Some manufacturers incorporate a transmission fluid warmer that uses engine coolant to warm up the automatic transmission fluid (ATF) quickly and keep it from cooling too much. These warmers are often found on vehicles with CVTs. They help reduce the viscosity of the ATF, which reduces the friction losses of the thicker ATF. At least one manufacturer routes coolant to the warmer assembly, which provides heat to warm up the ATF. The warmer assembly uses a thermostatic valve to direct the flow of ATF through the heated warmer assembly when the ATF is cold or to bypass the heated warmer assembly if the ATF is hot (**FIGURE 23-24**).

▶ Gear Train—Principles of Operation

K23009, K23010

The geartrain of a modern automatic transmission generally consists of the planetary gears, shafts, transfer chains or gears, bearings, bushings, and the final drive unit on a transaxle. These components are often referred to as the "hard parts" of a transmission, and they would generally not be replaced during a traditional rebuild unless they show signs of wear, with the exception of some bearings and bushings.

Most automatic transmissions use hydraulic pressure to apply individual clutches and bands in order to hold or drive components of one or more planetary gear sets. A planetary gear set is a device with several intermeshed gears assembled in a compact design. The planetary gear set has a **sun gear** in the center, with smaller **planetary gears** revolving around the sun gear (**FIGURE 23-25**). These planetary gears are held together in a **planet carrier**. The planetary gears revolve inside a larger ring gear that wraps around the outside of

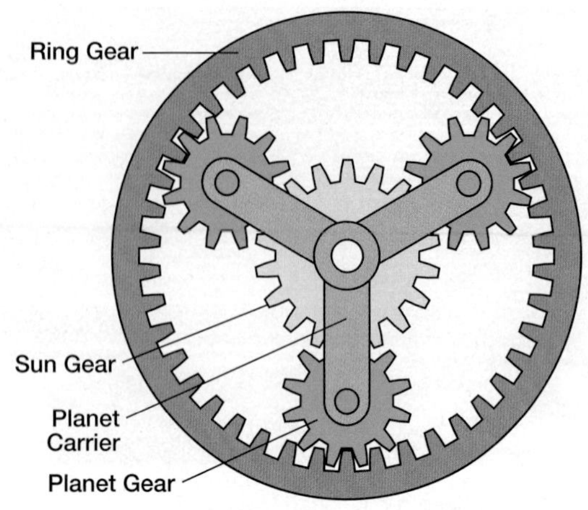

FIGURE 23-25 Simple planetary gear set.

the whole planetary gear set. Various gear ratios can be created in a planetary gear by holding and driving the different members. We explain that in more depth later in the chapter. Combining two or more planetary gear sets (compound planetary gear sets) creates the needed gear ratios, or speeds, for modern transmissions.

Gear Ratio/Torque Multiplication

Gear ratio refers to the difference in diameter between the driving (input) gear and the driven (output) gear. When you loosen a bolt, what tool do you use? Do you use a socket alone? No, that would not give you much leverage. Instead, you place a long ratchet or breaker bar onto the end of the socket to increase the amount of leverage. This increase in leverage is the same as torque multiplication in a gear ratio.

FIGURE 23-26 shows a gear with 8 teeth driving a gear with 24 teeth. The smaller drive gear will be able to turn three times before the large driven gear turns once. In this scenario, we have created a gear ratio of three to one (3:1). This gear ratio will create three times more torque, but the driven gear will turn at three times less speed and distance, otherwise known as gear reduction. If we were to put 100 ft-lb (136 N·m) of torque into this gear system, we would get 300 ft-lb (407 N·m) of torque out, but at one-third the speed and distance.

If we have a gear with 15 teeth driving a gear with only 10 teeth, we will have a gear ratio of 0.67:1. This is an **overdrive** ratio (**FIGURE 23-27**), and it results in less torque being transmitted to the wheels but will increase wheel speed. Any ratio of less than 1:1 is considered an overdrive ratio. Overdrive ratios are typically used to increase fuel economy while traveling at highway speeds, by slowing the engine speed.

A typical automatic transmission has four or more forward gear ratios and one reverse. Transmission gear ratios vary by transmission and manufacturer but are often approximately as listed here:

- First gear—3:1, resulting in a large increase in torque, but a major decrease in speed
- Second gear—1.75:1, still resulting in a torque increase with a smaller speed decrease
- Third gear—typically 1:1, or direct, ratio, meaning that power is sent directly through the transmission with no torque increase or speed increase
- Fourth gear—0.80:1, resulting in an overdrive ratio in order to increase fuel economy
- Reverse gear—2:1, resulting in a torque increase with a speed reduction

Early transmissions were simple two-speed automatics with a low gear and a direct gear. As engine size in vehicles decreased, the two gear ratios were no longer adequate to propel the vehicles being produced. Three-speed automatic transmissions became the norm through the early 1980s, when four-speed automatics were installed in vehicles to improve fuel economy. Now with an even greater need to increase fuel economy, manufacturers have turned to adding more gear ratios to their transmissions. Many vehicles are leaving the factory today with six- and seven-speed transmissions, with one manufacturer even using a nine-speed automatic. Even with all of these gears, the basic principle of gear reduction and overdrive remains the same.

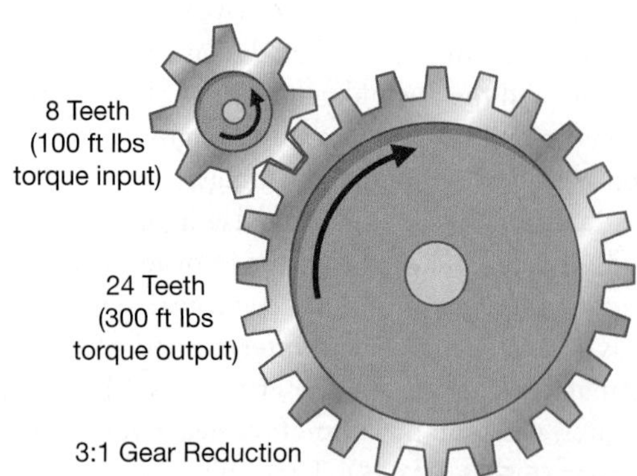

FIGURE 23-26 Having a smaller input gear with 8 teeth driving a larger output gear with 24 teeth results in a 3:1 gear reduction.

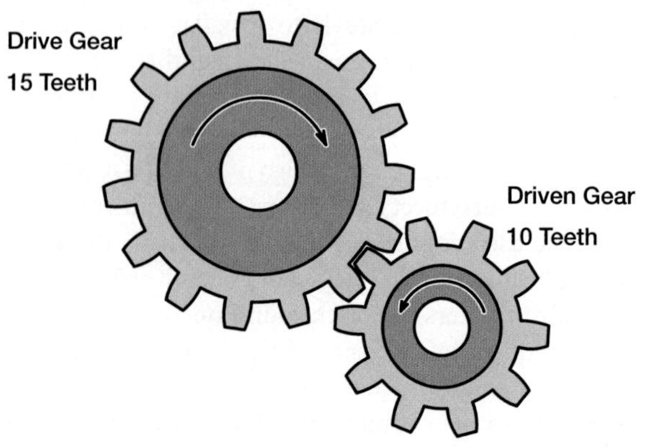

FIGURE 23-27 Having a larger input gear with 15 teeth and a smaller output gear with 10 teeth results in a 0.67:1 overdrive.

Gear Set Styles

K23011

Individual gears in automatic transmissions are typically one of three types: helical cut, spur (also known as straight-cut gears), or hypoid gears (**FIGURE 23-28**). **Spur gears** were used in early transmissions due to ease of manufacturing and lower cost. Spur gears also tend to be much louder and do not offer as much strength as the other types of gears; for this reason, manufacturers usually use helical gears for passenger vehicles. At the same time, spur gears don't create thrust forces, so that makes them useful in high-torque applications such as heavy off-road diesel equipment.

Helical-cut gears (or simply "helical gears") are cut on a spiral around the axis of the shaft. Helical gears offer the benefit of always having more than one gear tooth in contact with the driving gear, thus increasing the strength. Helical gears are quieter than spur gears but have more thrust motion due to the spiral cut of the gear teeth. Because of this increased thrust, helical gears in an automatic transmission must have a method to control the amount of thrust movement, such as **thrust washers** or thrust bearings.

Hypoid gears are often used in the final drive of FWD transaxles and also in conventional final drives on RWD vehicles. Hypoid gears are often used to change the direction of power flow by 90 degrees. They are a type of helical gear in which the axes of the two gears are not aligned (**FIGURE 23-29**). These gears therefore have the thrust action of helical gears along with a scraping action between the teeth. Because of this scraping action and very high pressure on the gear teeth, hypoid gears require a special lubricant to prevent damage.

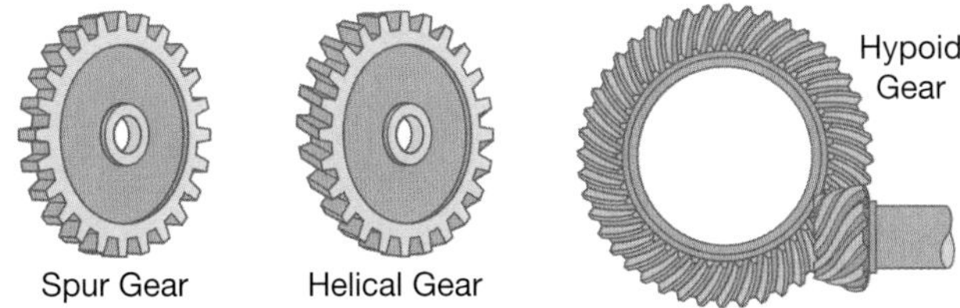

FIGURE 23-28 Spur, helical, and hypoid gears.

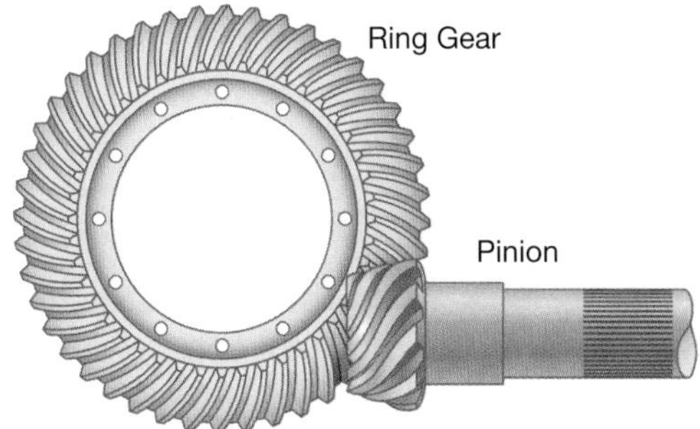

FIGURE 23-29 A hypoid gear arrangement.

Planetary Gear Sets

K23012

A simple planetary gear set is the backbone of most modern automatic transmissions. One of their biggest benefits is that they are in constant mesh. And because they can provide multiple gear ratios, a variety of speeds can be obtained without having to take them out of mesh or put them back in mesh. Planetary gear sets use either helical gears or straight-cut gears. A simple planetary gear set contains a sun gear in the center, with multiple revolving planetary gears around it. The planetary gears are held in place by the planet carrier. On the outside edge of the planetary gears is the ring gear. By holding one of the components (planet carrier, sun, or ring) and driving one of the other components, we can create up to six separate gear ratios—four in the forward direction (two reduction and two overdrive) and two in the reverse direction (one reduction and one overdrive). If any two components are locked together, we end up with another speed—1:1, called direct drive.

Just because a planetary gear can obtain four forward gears and two reverse gears does not mean the manufacturer can use each of those ratios. Manufacturers have to design the planetary gear sets to get the required number of gear ratios as well as gear ratio splits to work with the particular application. The benefit of designing transmissions with planetary gears is that if you need more gear ratios than you have, you can combine multiple planetary gear sets to create the needed ratios. We cover these compound planetary gear sets later in the chapter.

Holding/Driving Gears

K23013

As we stated earlier, to create a gear ratio, we need to hold one member of a planetary gear set and drive another. There are four basic holding devices used in transmissions today: bands,

Applied Math

AM-43: Direct/Inverse Variation: The technician can solve problems requiring the use of fractions, decimals, ratios, and percentages.

When working out gear combinations, it may be necessary to calculate gear ratios based on counting the number of gear teeth. This is simple in the case of differentials and manual transmissions, but the planetary gear sets used in automatic transmissions require more involved calculations.

In the case of a four-speed automatic transmission with an overdrive top gear obtained by driving the planet carrier, the sun gear (30 teeth) held, and the ring gear (72 teeth) driven, the formula to calculate the gear ratio would be:

$$1/(1 + \text{sun/ring})$$

The effective ratio is therefore calculated as:

$$1/(1 + 30/72), \text{ or } 0.71:1.$$

Different combinations of driving, driven, and held gears require different formulae to calculate ratios.

multidisc clutches, one-way clutches, and dog clutches. These holding devices will either lock two members of the planetary gear set together or lock a member to the transmission housing to prevent it from spinning.

The **band** is a friction-lined steel belt that wraps around the outside of a drum inside the automatic transmission (**FIGURE 23-30**). When a servo applies pressure to one end of the band, it closes the diameter of the band, squeezing the drum and preventing it from turning. The drum is then locked to one member of a planetary gear set.

Multidisc clutches are unique in that they not only can be used to hold a member of the planetary gear set, but they are also the only type of component that currently is used to drive a member of the planetary gear set (**FIGURE 23-31**). In a multidisc clutch, individual friction discs, often called frictions, are stacked between smooth steel plates, often called steels. Both the frictions and the steels have a hollow center, allowing them to wrap around the outside of a hub or drum connected to a member of the planetary gear set. The steels typically have splines, or teeth, on the outside edge to hold them in place inside a drum, and the frictions have splines or teeth on the inside edge to hold them to a member of the planetary gear set. When the steels and frictions are forced together by a hydraulic piston, the separately revolving steels and frictions match speed and lock the drum or hub to the member of the planetary gear set.

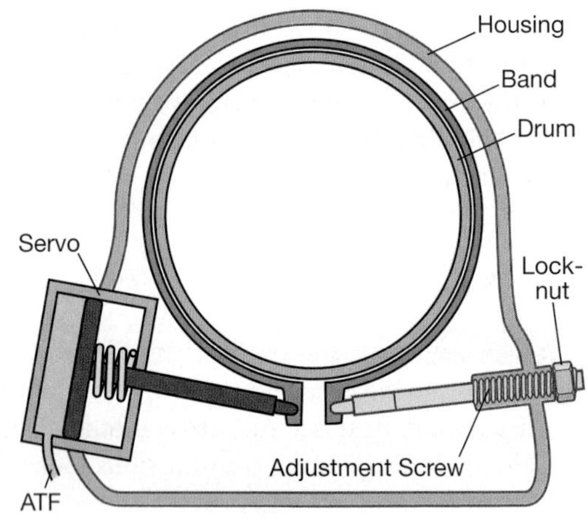

FIGURE 23-30 A typical transmission band.

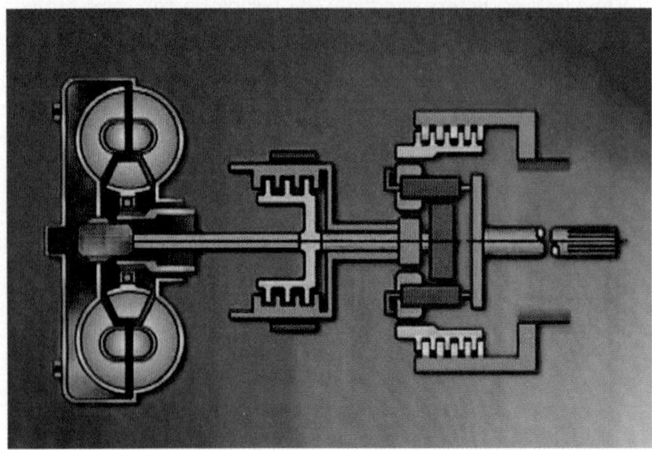

FIGURE 23-31 Two multi-disc clutches shown here.

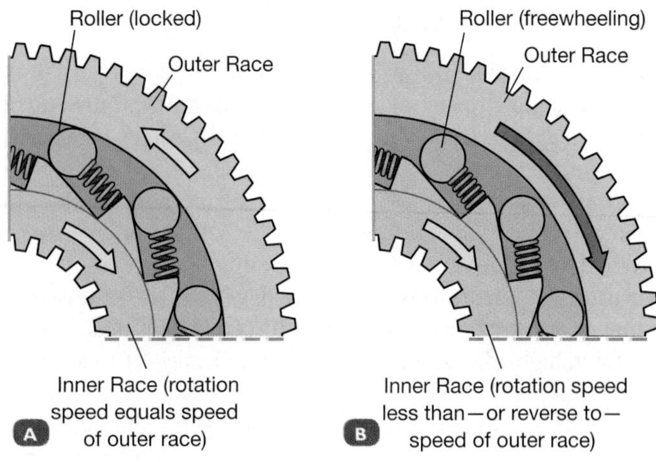

FIGURE 23-32 One-way clutches. **A.** A one-way roller in the locked position. **B.** A one-way roller in its freewheeling position.

One-way clutches can be either one-way rollers or sprags. Both the one-way roller and the sprag operate on the same principle, but sprags have a greater torque-holding capacity. On both the one-way roller and the sprag, an inside hub is allowed to spin freely in one direction, but spinning the hub in the opposite direction causes the rollers or sprags to become wedged in place (**FIGURE 23-32**). This wedging action prevents rotation of the hub. In the one-way roller clutch, each roller has a waved compression spring that pushes the roller toward the narrow end of the wedge. Rotating the hub in the opposite direction locks the outer race onto the rollers wedged between the two races. Thus, a one-way clutch allows rotation in one direction but prevents rotation in the opposite direction.

Dog clutches are two-piece mechanisms that can be engaged or disengaged from each other. Their interlocking teeth can hold or drive components (**FIGURE 23-33**). They do not rely on friction to lock up, but rather provide a direct mechanical connection between the two components. When disconnecting them from each other, one is moved away from the other.

▶ Automatic Transmission Components

K23014

Automatic Transmission Fluids

K23015

ATF is a specialized fluid that has been designed for a specific job. The ATF must be able to transfer heat from the internal components of the transmission to the transmission cooler in order to prevent damaging the internal seals of the

FIGURE 23-33 Dog clutches have teeth on each side that interlock with each other, providing a direct mechanical connection between the two.

transmission. The fluid must also lubricate the internal gears, bearings, and bushings of the transmission, yet have enough of a **coefficient of friction** to allow the clutches to hold and not slip. Coefficient of friction is the force required to move two sliding surfaces over each other. Some ATF fluids are typically dyed red for easy identification, although there are other colors used as well. But all ATF contains additives such as those listed here:

- *Rust and corrosion inhibitors:* These additives prevent the internal parts of the transmission from developing rust and corrosion on metal components. Rust and corrosion can affect the shift quality and longevity of the transmission. As rust particles break off, they become an abrasive in the fluid, causing increased wear.
- *Friction modifiers:* Manufacturers add friction modifiers to the fluid to ensure that the fluid has the proper coefficient of friction in order to produce the desired shift quality. A fluid with a lower coefficient of friction produces softer, longer shifts, causing an increase in clutch slippage. A fluid with a higher coefficient of friction causes shorter, harsher shifts, reducing clutch slippage but increasing drive line shock.
- *Seal conditioners:* These additives are designed to help protect the seals inside a transmission and cause the seals to swell slightly to help prevent leaks and clutch slippage.
- *Detergents:* ATF has a large amount of detergent to prevent dirt and other foreign particles from becoming trapped inside the transmission. The detergent causes the dirt and other particles to be attracted to the fluid so they transfer with the fluid. When the fluid passes through the filter, the large particles become trapped in the filter; the small particles are removed during the next transmission fluid change.
- *Antifoam:* These additives help prevent foaming of the transmission fluid. When moving parts spin through a fluid, they tend to produce air bubbles. These air bubbles can quickly multiply and become foam. Foam is compressible and can cause a transmission to slip because insufficient pressure is applied to the clutches.
- *Viscosity modifiers:* These additives are similar to the engine oil additives that allow us to have multiviscosity engine oils such as 5W-30. The additives allow the fluid to remain thin when the temperature is cold, and they prevent the fluid from becoming too thin as the transmission fluid warms up.

▶ TECHNICIAN TIP

When building a performance transmission, it is beneficial on some transmissions to increase the number of individual rollers or sprags in order to increase holding power. Common performance transmissions have multiple versions of one-way rollers and sprags.

ATFs can be mineral oil based or a synthetic lubricant (**FIGURE 23-34**). Many late-model vehicles recommend a specific synthetic ATF. Originally, most manufacturers used and recommended General Motors and Ford Motor Company fluids. In the past few decades, most auto manufacturers have developed their own fluids for use in their transmissions. This practice has required repair facilities and quick lube shops to carry many different types of fluids. Some shops carry a few major types of fluids and use those in every vehicle, but that is not recommended as it may cause undesirable transmission operation and may void the transmission warranty. Several vehicle manufacturers have published technical service bulletins (TSBs) related to incorrect fluid use and the negative effects on the transmission. Research all applicable information, including service history and precautions.

On most vehicles, ATF needs to be changed periodically to remove dirt and contaminants from the transmission. Technicians used to tell if the transmission fluid needed to be changed by

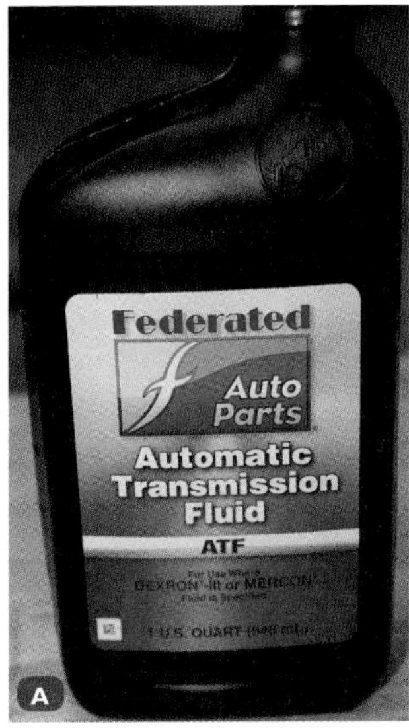

FIGURE 23-34 A. Mineral-based ATF. **B.** Synthetic-based ATF.

looking at and smelling it. If the fluid was dark or smelled burnt, it needed to be changed. You cannot determine the condition of modern ATF by looks and smell. Some newer synthetic fluids have a burnt smell when they are brand new and are often darker than mineral-based ATF. Most manufacturers specify the recommended change interval in miles or months, although some manufacturers have specified no fluid changes for the life of the transmission and may not even provide a method of checking the fluid.

Front Pump

K23016

The hydraulic system of an automatic transmission requires a continuous supply of fluid under pressure. This is accomplished by the **front pump**. The pump is driven by the torque converter when the engine is running. The pump driving member is engaged by tangs, machined flats, or slots on the converter hub or snout (**FIGURE 23-35**).

Pump output always exceeds normal operating demands. Pressures are controlled by a pressure relief valve on older vehicles or a computer-controlled pressure control solenoid on late-model vehicles. Excess fluid returns to the pump inlet or is dumped into the oil pan, where it can be reused (**FIGURE 23-36**).

There are four types of pumps used in automatic transmissions: the gear, or "crescent," pump; the rotor pump; the vane pump; and a variable displacement pump. All of these begin pumping as soon as the engine is cranked, no matter which position the gear selector is in.

Gear pumps have been used in automatic transmissions for a long time and are classified as a positive displacement pump. This means that the pump produces the same amount of flow each time the pump turns. Gear pumps are extremely reliable and rarely fail during operation. A gear pump has an outer driven gear and an inner driving gear (**FIGURE 23-37**). The inner gear is driven by the torque converter. The outer gear is meshed with the inner gear and is turned by the movement of the inner gear. A crescent is placed on one side of the pump between the two gears. As the gears turn toward the crescent, they are pushed away from each other. This creates a low-pressure area at the front of the crescent and draws fluid in the spaces between the gear teeth. As the gears continue to turn past the crescent, fluid is moved from the low-pressure side of the pump to the high-pressure side. When the two gears remesh, the space between the gear teeth becomes smaller, forcing the fluid out. Because it is a positive displacement pump, restricting the fluid flow after the pump can create very high pressure, so a pressure regulator is used to maintain proper pressure. This pressure is used to apply bands and clutches. Most gear pump housings and gears are made of cast iron, which resists wear. Often during a transmission overhaul, the gear pump does not need any service other than replacing the pump bushing and seal.

A **rotor pump** is another type of positive displacement pump. Rotor pumps have two rotors, an inner rotor and an outer rotor (**FIGURE 23-38**). The inner rotor is driven by the torque converter. The outer rotor is driven by the inner rotor. Each rotor has multiple lobes on it. There is no crescent in a rotor pump, but the principle of operation is similar to that of a gear pump. When the two rotors move away from each other, they create a low-pressure area that draws fluid in. When the rotors move closer together, the fluid is forced out under pressure.

FIGURE 23-36 A pressure relief valve on a TH200 front pump. The relief valve sends pressure back to the oil pan at (approximately 175 psi [1207 kPa]).

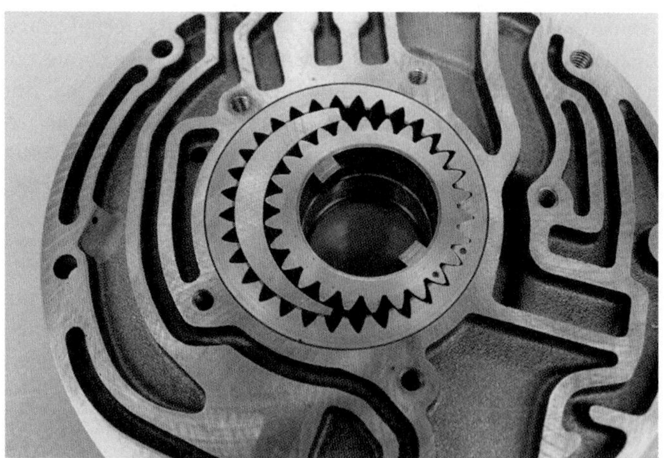

FIGURE 23-37 A typical gear pump.

FIGURE 23-35 A front pump showing the converter drive engaged in the pump.

FIGURE 23-38 A rotor pump.

▶ TECHNICIAN TIP

Automatic transmissions since the 1960s have been equipped with only a front pump. This means that the vehicle cannot be push-started because there is no way to turn the pump without the engine running. If the pump is not turning, then the bands and clutches cannot be applied to drive the torque converter, meaning the engine will not turn. Some older transmissions were equipped with an additional pump located in the rear of the transmission and turned by the output shaft. This design allowed the transmission to create pressurized fluid to apply the bands and clutches if the vehicle was rolling, making it possible to push-start the vehicle.

A **vane pump** has a center rotor that is driven by the torque converter. The rotor has vanes located around its outside circumference (**FIGURE 23-39**). The vanes are free to move in and out of the rotor. Centrifugal force causes the vanes to be pressed against the pump housing, creating a seal. The pump housing is slightly larger than the rotor, with the rotor mounted off-center in the pump housing. On the pump inlet side, the space between the pump vanes and the pump housing increases in volume as the pump turns. This creates a low-pressure area and draws the fluid into the pump. As the rotor spins past its midpoint, the volume between the vanes and the housing starts to decrease. This causes an increase in pressure, forcing the fluid out of the pump outlet.

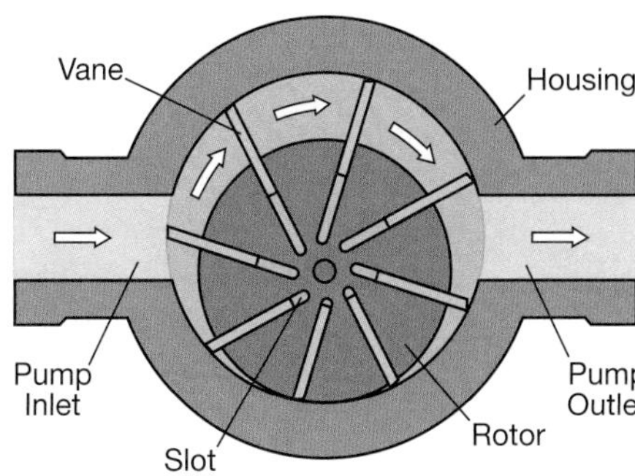

FIGURE 23-39 A vane pump.

▶ TECHNICIAN TIP

On many transmissions, vane pumps are available with different numbers of vanes. When rebuilding a transmission and replacing the pump, compare the number of vanes on the original pump to the new one; it is possible to replace the stock pump with one that has more vanes when building a performance transmission.

In recent years, the vane pump has become popular due to the invention of a variable displacement version of the vane pump (**FIGURE 23-40**). A **variable displacement vane pump** reduces the pumping load on the engine when the transmission requires less fluid to be pumped. In the variable displacement pump, the pump housing can pivot in relation to the rotor. As the housing moves in one direction, the volume change between the inlet side of the pump and the outlet side decreases. This decreased volume change results in less fluid being pumped and at a lower pressure. When the pump housing moves in the opposite direction, the amount of volume change increases, resulting in more fluid flow and higher pressure. Usually, transmissions do not require maximum pressure and volume of fluid. By reducing the pump load, manufacturers are able to improve fuel efficiency.

Front pumps should be inspected when catastrophic engine damage occurs, such as a thrown connecting rod. The sudden stopping of the engine, and the connected front pump, can damage the pump. It would be wise to remove and inspect the pump while the engine is out of the vehicle. Technicians have been known to replace an engine only to find that the front pump makes knocking noises after the new engine is installed and started.

Flexplate and Ring Gear

K23017

On an automatic transmission there is no flywheel; rather, there is a thin lightweight steel flexplate. The flexplate bolts to the rear crankshaft flange in the same manner as

FIGURE 23-40 A variable displacement vane pump from a General Motors 4L60E transmission. Note the moveable housing and the large spring.

a flywheel would. The torque converter is bolted to the flexplate with three or more bolts. Because there is no thrust force of a pressure plate as in a manual transmission, the flexplate can be significantly lighter than a traditional flywheel (**FIGURE 23-41**). Also, the large mass of the torque converter dampens the engine pulses in a similar manner to that of a flywheel.

Wrapped around the outside edge of the flexplate is a ring gear that is used when starting the vehicle. The starter pinion gear moves out to contact the ring gear, and the starter motor cranks the engine over when the driver turns the ignition switch to the crank position. This ring gear is not typically a serviceable, separate part of the flexplate. If the ring gear wears out, the entire flexplate will typically have to be replaced. Ring gears on flywheels are often available as separate parts from the flywheel and can be replaced. In some applications, manufacturers weld the ring gear to the torque converter instead of the flexplate (**FIGURE 23-42**).

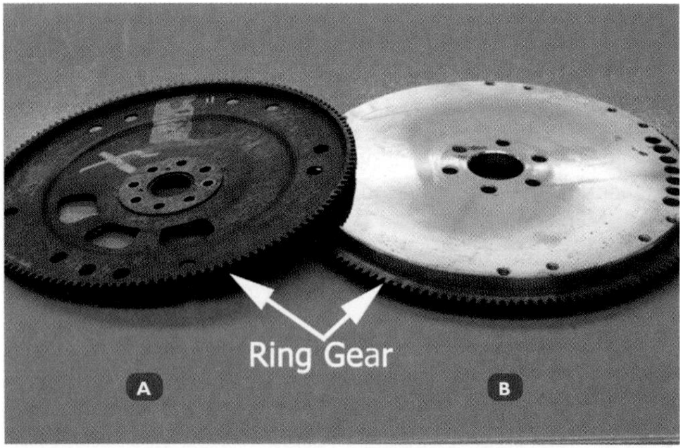

FIGURE 23-41 A. An automatic transmission flexplate and ring gear. Note the ring gear around the outside edge of the flexplate that is used when starting the vehicle. **B.** A manual transmission flywheel and ring gear.

> **TECHNICIAN TIP**
>
> When a faulty engine or transmission is suspected because of knocking noises, double-check that the flexplate bolts are tight. Loose bolts on the flexplate can sound like a knocking engine. Also, a cracked flexplate can cause a knocking noise that can sound like a defective rod bearing. In some cases, the entire center of the flexplate can break free of the rest of the flexplate, preventing the engine from cranking over when the starter is engaged. It is easy for a technician to think that the engine has a broken crankshaft because the front pulley does not turn when the engine is cranked, when in reality the outer portion of the flexplate is spinning on the broken center piece.

Front Pump Seal, Bushing, and Pump Drive

On the front pump, there is a seal between the pump and the torque converter. This seal is similar in appearance to an axle seal you would find on a wheel bearing and is designed to keep dirt out of the inside of the transmission and keep low-pressure transmission fluid from leaking out the front of the transmission.

Behind the seal is the front pump bushing. The bushing is a friction-type bearing similar to a camshaft bearing. It supports the torque converter hub and should always be replaced when overhauling a transmission, replacing a failed torque converter, or replacing a leaking front pump seal. A special bushing installation tool is needed to correctly install the front pump bushing without damaging it (**FIGURE 23-43**).

Inside the front pump opening where the torque converter enters the pump, there are two tangs, or flats, that drive the pump. The end of the torque converter hub or snout has matching splines, flats, or slots that engage into the front pump (**FIGURE 23-44**). These components drive the pump when the engine is spinning. It is critical that the torque converter drive tangs or slots engage the pump tangs or flats properly when installing the torque converter. If the tangs or flats are not engaged properly and the transmission is reinstalled into the vehicle, the front pump will be destroyed when the engine is first started. Keep in mind that more than one student has failed to install the torque converter properly, which has earned the student the additional experience of removing the transmission again, replacing a front pump, and cleaning the filings out of the cooler and transmission pan.

FIGURE 23-42 Torque converter with a welded-on ring gear.

FIGURE 23-43 A front pump bushing properly installed. Note the front pump seal in front of the bushing.

FIGURE 23-44 **A.** Splines. **B.** Flats. **C.** Slots.

Case, Extension Housing, and Pan

K23018

Modern automatic transmission cases are made of lightweight aluminum and alloys. Early transmission cases were sometimes made of cast iron. When a transmission is new, the manufacturer typically leaves the transmission case uncoated. When the transmission is rebuilt, the rebuilders often paint the transmission a particular color. This serves two purposes: (1) It helps identify a transmission that was previously rebuilt, and (2) it helps seal the transmission after it has been cleaned. Aluminum can become porous after excessive cleaning of the case with detergents and solvents. The paint helps to fill in any case porosity.

On the rear or output side of the transmission, there is an aluminum housing called the **extension housing**. The extension housing is usually bolted to the transmission case with a **gasket** between the housing and the case. The extension housing may contain the vehicle speed sensor drive gear or speedometer drive gear and/or a governor assembly to measure vehicle speed (**FIGURE 23-45**). Inside the end of the extension housing is the extension housing bushing that supports the driveshaft. The extension housing on most four-wheel drive pickups and SUVs has a surface that bolts the transmission up to the transfer case.

On the bottom of most transmissions is a sheet metal or aluminum pan. Inside this pan, the transmission filter and possibly the transmission valve body are located. Most FWD transaxles have a second pan called the side pan, which houses the valve body (**FIGURE 23-46**). The bottom transmission pan is often

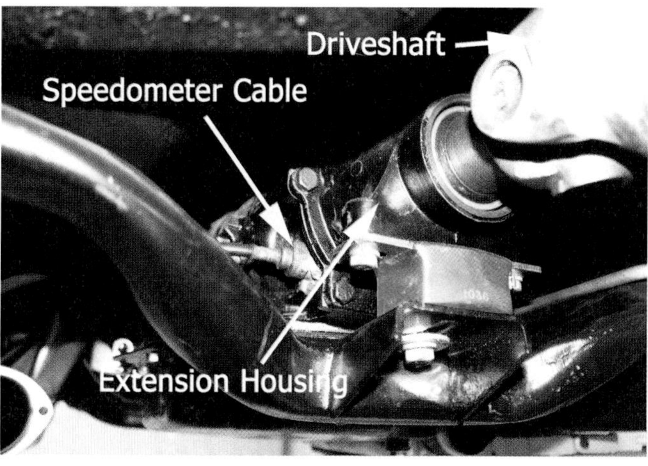

FIGURE 23-45 A typical extension housing with the speedometer cable connected.

FIGURE 23-46 A common FWD transaxle showing the bottom oil pan and side pan.

removed during routine transmission fluid service to change the filter and clean a magnet, if the vehicle is so equipped, which catches small particles of steel and iron that are floating in the fluid. The transmission pan holds about half the transmission fluid when the vehicle is off. When the vehicle is running, the pump draws fluid from the transmission pan to supply lubrication and pressure to the different parts of the transmission.

Gaskets and Seals

K23019

Gaskets are used throughout the automatic transmission. Some gaskets are made from paper, fiber, or cork material that must be replaced every time a component is removed. Other gaskets are a reusable type with neoprene inserted into a plastic housing. Reusable gaskets are fairly common for transmission pan gaskets where the part is frequently removed for routine service. Some reusable gaskets have built-in torque limiters. These torque limiters are metal sleeves built in the gasket around the bolt holes to help prevent over-tightening of the gasket (**FIGURE 23-47**).

There are many types of internal seals for an automatic transmission (**FIGURE 23-48**). These internal seals are often the root cause of a transmission slipping and eventually failing. All seals need to be properly lubricated when assembling the transmission, to prevent premature failure and to help with installation. Seals are lubricated with ATF, petroleum jelly, or automatic transmission assembly lubricant. Do not use grease to lubricate the seals, as grease can clog up the internal passages of the transmission when the vehicle is put back into service. A description of seals used in a transmission system follows:

- *Square-cut seals:* Square-cut seals can be made from neoprene rubber or Teflon. Teflon seals require special handling and are discussed separately. Square-cut seals are similar to the seal found inside a brake caliper. They are sometimes located on the piston of a multidisc clutch.
- *Lip seals:* Lip seals are made from neoprene. A lip seal has greater sealing ability than a square-cut seal because when pressure is applied to the back of the seal, it forces the seal out against the inside of the clutch drum. These seals are similar to the cup seals located inside a drum brake wheel cylinder. Lip seals can be easily damaged by careless installation and infrequent fluid and filter changes. They are typically used inside multidisc clutch assemblies on the piston.
- *Locking seal:* The locking seal is made from cast iron and is similar to a piston compression ring except that the ends overlap and lock together. Locking seals are typically used between the front pump and the rotating clutch drums.
- *O-rings:* O-rings are made from neoprene rubber and are often found sealing external parts of the transmission, such as a speedometer cable assembly or governor cover. O-rings can also be used on accumulators and servo covers.
- *Teflon seals:* Teflon seals come in several styles—continuous, butt cut, scarf cut, and step joint. The continuous seal provides the best sealing action and should be used whenever possible, but it does require special tools for proper installation. Teflon seals have replaced the locking seals on most modern transmissions and are found sealing the front pump to input shafts and clutch drums.

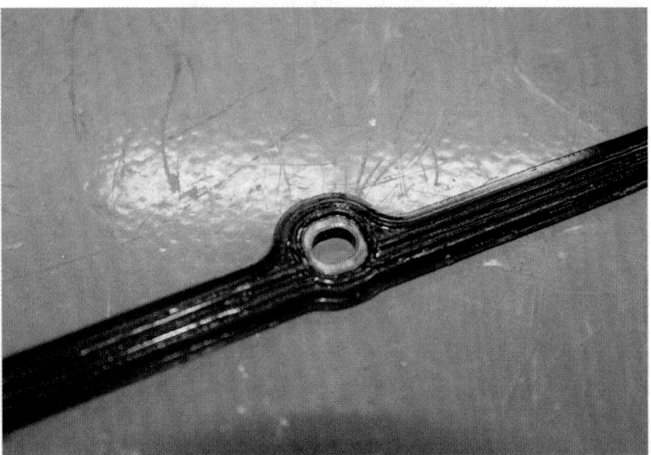

FIGURE 23-47 Some reusable gaskets have built-in torque limiters that prevent over-tightening of the gasket.

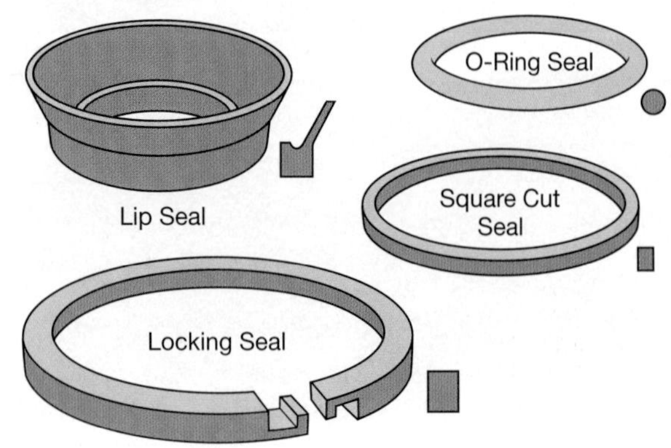

FIGURE 23-48 Typical seals used in an automatic transmission.

Parking Pawl Assembly

K23020

Most automatic transmissions for passenger vehicles include an internal parking mechanism. When the driver shifts the transmission into park, a lever, called a parking pawl, is forced into notches cut into a hardened steel drum on the output shaft of the transmission (**FIGURE 23-49**). This parking pawl prevents rotation of the output shaft when it is engaged. The parking pawl is often operated through a spring in case the driver accidentally places the vehicle into park while it is still moving. In this scenario, the vehicle will make a loud ratcheting noise as the parking pawl bounces against the rotating notches of the output shaft drum. The spring is designed to help minimize damage to the parking pawl and the output shaft. The spring also allows the driver to place the vehicle shift selector into park even when the parking paw does not quite align with the notches in the drum. If this occurs, the gearshift is placed in park, but the vehicle may need to roll an inch or two for the parking pawl to engage.

FIGURE 23-49 A parking pawl assembly on a late-model CVT transmission.

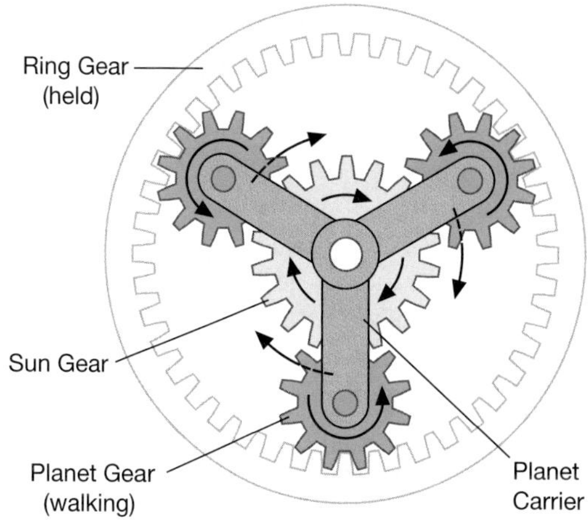

FIGURE 23-50 A planetary set with the ring gear held so the planet gears are "walking" around the sun gear.

TECHNICIAN TIP

Often, Teflon seals are placed in boiling water prior to assembly to help expand the seal, and the component on which the seal is being installed may be placed in the freezer to contract the part.

Planetary Gears

K23021

As mentioned earlier, most automatic transmissions use planetary gears that are constantly in mesh with one another. A basic planetary gear set has a sun gear, which meshes with planet gears, also called planet **pinions**. The planet pinions, in sets of three or more, rotate on bearings in a planet carrier. The planet carrier equally spaces the planets around the sun gear. It also locates them so they can mesh with an internally toothed ring gear. This design means the planet pinions are always in mesh with the sun gear and the ring gear.

The planetary gears' motion is described as either **"walking"** or **idling**. Walking means that if either the sun gear or the ring gear is held stationary, the opposite driving member rotates the planetary gears in the carrier. This rotating motion turns the planet carrier in the same direction as the driving member. Planetary gears always turn in the same direction on their pins as the planet carrier while they walk around a stationary sun gear. They always turn in the opposite direction on their pins while walking inside a stationary ring gear (**FIGURE 23-50**).

Idling refers to the rotation of the planetary gears on their pins whenever the planet carrier is stationary (**FIGURE 23-51**). Torque is transmitted from the sun gear to the ring gear or from the ring gear to the sun gear via the planetary gears and the stationary carrier.

In both cases, the driven member is turned in the opposite direction from the driving member. To provide the ratios available from the gear set, one or more of the components must be held or released. This is normally done by a band, a multidisc clutch, one-way clutches, or dog clutches. In practice, a combination of these devices is normally used to hold individual members of the planetary gear set.

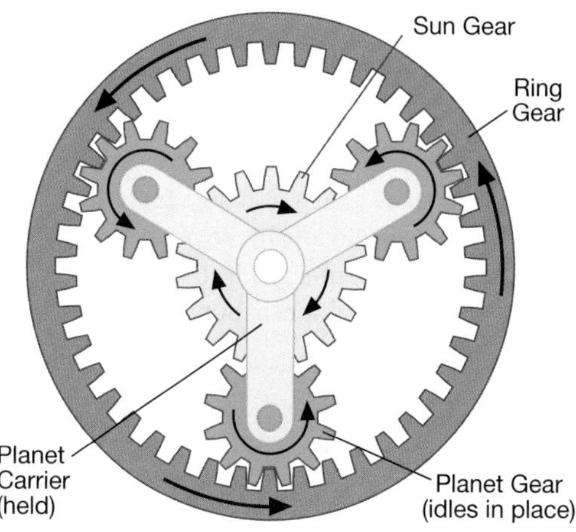

FIGURE 23-51 A planetary set with the planet carrier held. The planet gears idle in place.

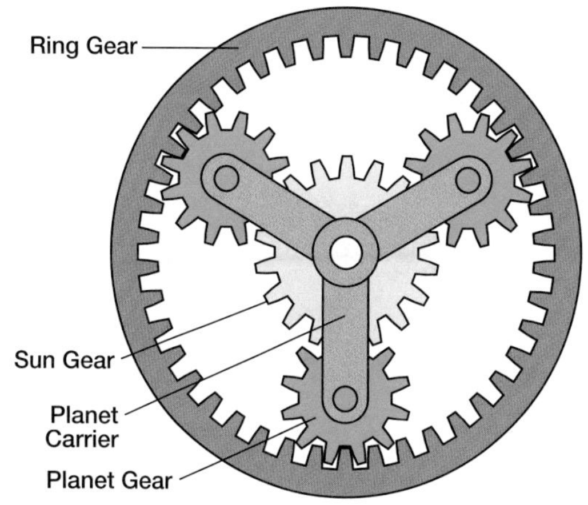

FIGURE 23-52 Neutral - Sun gear driven, planet carrier - output shaft, ring gear free.

Neutral is obtained when none of the members are held nor any two members locked together. Just driving one of the members by itself will result in no output torque. In **FIGURE 23-52**, the sun gear is attached to the input shaft from the turbine. The planet carrier is attached to the output shaft. The ring gear is not held. When the input shaft rotates the sun gear, the planetary gears idle in the carrier and turn the ring gear in the opposite direction to engine rotation. This free rotation of the gears provides a neutral state. No torque is transmitted to the output shaft.

For low gear, we use a holding device to hold the ring gear stationary. Now when the sun gear rotates, the planetary gears can no longer idle. They must walk around the inside of the stationary ring gear. The planet carrier must move with them in the same direction as engine rotation, but at a slower speed and with an increase in torque.

A direct drive, or 1:1 ratio, is obtained by locking together any two members of the gear set. A multidisc clutch is used for this purpose. The outer drum and steels are attached to the ring gear. The frictions and inner drum are attached to the planet carrier. When fluid under pressure is directed onto the clutch piston, the frictions and steels are locked together, thus locking the planet carrier to the ring gear. Now when the sun gear is rotated, the planetary gears can neither idle nor walk. The whole gear set turns as one unit to give direct drive.

For reverse, the ring gear is attached to the output shaft, and the planet carrier is held stationary. Rotating the sun gear causes the planetary gears to idle in the planet carrier. This turns the ring gear and the output shaft in the opposite direction to engine rotation, at a reduced speed, but increased torque, resulting in reverse (**FIGURE 23-53**).

Compound Planetary Gear Sets

The simple gear set illustrates basic principles, but it cannot obtain all the ratios required in a modern vehicle. In practice, **compound planetary gear sets** are used. There are a variety of planetary gear arrangements, but there are two that are commonly used. One common type of compound planetary gear set is the **Simpson gear set**. A Simpson gear set has two planetary gear sets that share a common sun gear (**FIGURE 23-54**). The sun gear is free to rotate on the output shaft. It is constantly meshed with the planets of both sets.

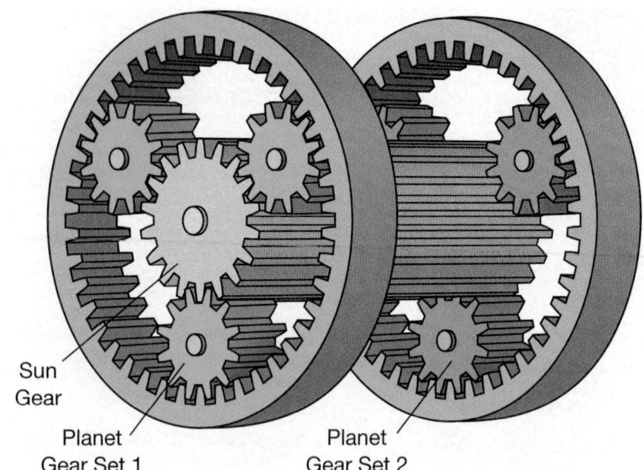

FIGURE 23-54 A Simpson gear set. Note the single sun gear driving two sets of planets.

They in turn mesh with their individual ring gears. Both sets are also connected by each having a part splined to the output shaft—the ring gear of the rear set and the planet carrier of the front set. The input shaft is isolated from the gear set by one or more clutches, depending on the gear and range engaged. The Simpson gear set is capable of producing three forward gears along with a reverse.

A second common type of compound planetary gear set is the **Ravigneaux gear set**, which uses two different-sized sun gears connected to long and short pinion gears (planets) and one ring gear. All gears are always in mesh, just as in a simple gear set. The primary sun gear meshes with short primary pinions. The secondary sun gear meshes with long secondary pinions (**FIGURE 23-55**). Both sets of planet pinions are mounted in a common planet carrier, and they mesh in pairs. The outer secondary pinions mesh with the ring gear, which is attached to the output shaft. The Ravigneaux gear set is capable of producing four forward gears—two with gear reduction, a direct, and an overdrive—along with reverse.

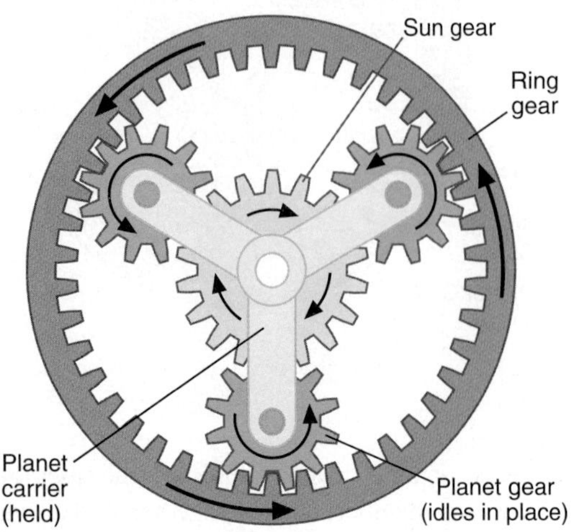

FIGURE 23-53 A simple planetary gear set with the planet carrier held to create reverse.

FIGURE 23-55 A Ravigneaux gear set showing the two sun gears, with the long and short pinions.

Thrust Washers and Thrust Bearings

K23022

Forward and backward motion of components inside a transmission must be controlled. Typically, when there are two parts spinning next to each other, a thrust washer or thrust bearing is used to control that motion.

Thrust washers are a friction type of bearing similar to engine main bearings. They can be made of plastic or steel coated with a bronze bearing material. Thrust bearings must have a small amount of clearance for proper oil lubrication. A typical manufacturer specification for a thrust bearing clearance is 0.005" (0.13 mm). Thrust washers need to be checked during a rebuild and are sometimes replaced if they are worn (**FIGURE 23-56**).

Thrust bearings are a type of flat needle roller bearing with the rollers placed radially around the center of the bearing. These bearings are typically called Torrington bearings. Thrust bearings must have less clearance than thrust washers. If a thrust bearing has excessive clearance, it will suffer impact damage.

Bushings

K23023

Most automatic transmissions use bushings around spinning components, rather than roller bearings. These bushings are a friction-type bearing, similar to an overhead valve camshaft bearing. One of the most critical bushings to be replaced during an overhaul of an automatic transmission is the front pump bushing. The front pump bushing is used to support the torque converter hub and sees the largest amount of wear in the transmission because it is in use anytime the engine is running (**FIGURE 23-57**).

Another bushing that is prone to wear is the output shaft bushing (**FIGURE 23-58**). It supports the driveshaft on a RWD vehicle and is located in the extension housing. The output shaft bushing is subject to higher amounts of wear both from the spinning of the driveshaft and from the in-and-out action of the driveshaft as the suspension moves up and down. This movement can also allow small particles of dust and dirt to move past the extension housing seal and cause more wear to this bushing. The remaining bushings inside the transmission should be checked, but not necessarily replaced, during an overhaul of the transmission.

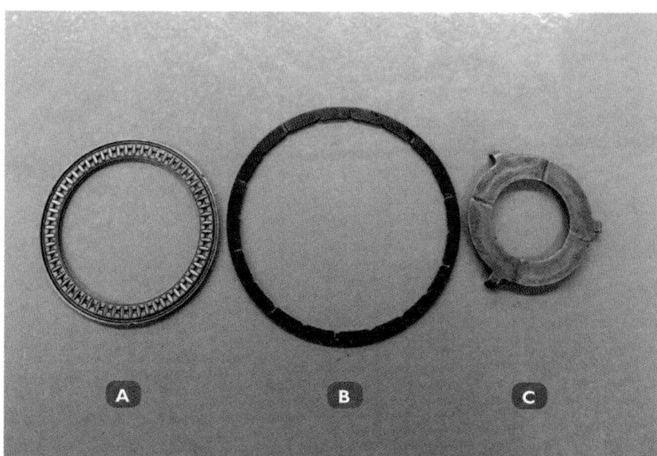

FIGURE 23-56 **A.** A typical thrust bearing (bearing has been opened to show individual rollers). **B.** A plastic thrust washer. **C.** A typical thrust washer. Note the copper color, as the outer coating of the washer has been worn away.

FIGURE 23-57 A front pump bushing removed from the transmission.

FIGURE 23-58 Output shaft bushing.

> **TECHNICIAN TIP**
>
> A worn extension housing bushing will often present as a vibration at high speeds. Some extension housing bushings are only serviceable with the entire extension housing.

▶ Holding/Driving Devices

K23024

Components of the planetary gear set need to be either held, driven, or allowed to freewheel to get the required gear ratio. These jobs are performed by bands and clutches (**FIGURE 23-59**). A band is specifically used to hold one member of the planetary gear set stationary. In other words, it locks

FIGURE 23-59 **A.** Band. **B.** Multidisc clutch. **C.** One-way clutch. **D.** Dog clutch.

the member to the transmission case. There are three types of clutches: multidisc, one-way, and dog clutches. Multidisc clutches are used for three purposes. The first is like the band; the clutch holds one member of the planetary gear set stationary. The second is to drive a member of the planetary gear set. The last purpose of a clutch is to lock one rotating member to another rotating member. For example, the clutch can lock two members of the planetary gear set together as they are spinning, to create direct drive. One-way clutches hold a component when it is trying to turn in one direction and allow it to freewheel when turning the other direction. Dog clutches are two-piece mechanisms that can be engaged or disengaged from each other. They have interlocking teeth that can hold or drive components (**FIGURE 23-60**). They are most often used to hold a planetary gear component to the transmission case.

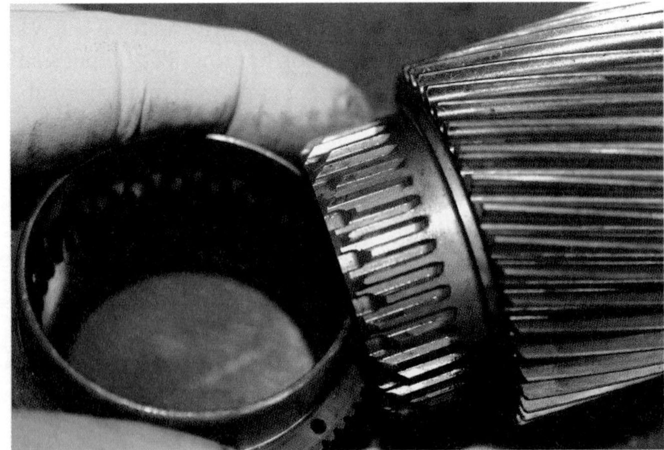

FIGURE 23-60 Dog clutches have teeth on each side that interlock with each other, providing a direct mechanical connection between the two.

Automatic Transmission Brake Bands

K23025

All bands inside an automatic transmission are externally contracting types. They have a thin layer of friction material bonded onto spring steel or cast steel backing. One end of the band rests against a stop, or strut, in the transmission case. The other end is operated by linkage from a hydraulically operated servo. When the servo pushes on the linkage, the band is squeezed tightly around the drum, stopping the drum from spinning and holding whichever planetary gear component the drum is attached to. Bands come in a variety of configurations (**FIGURE 23-61**). The most common is the one-piece band, which is one solid band with friction

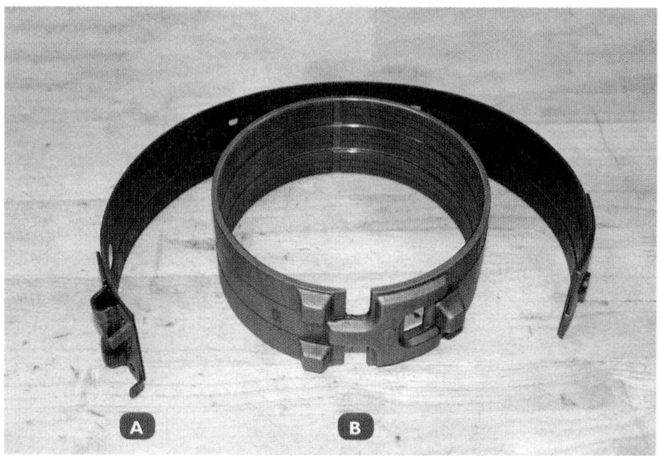

FIGURE 23-61 Brake bands. **A.** One-piece band. **B.** Multi-wrap band.

lining on it. Other bands come in a multi-wrap configuration, which uses a band made up of three parallel straps that give increased gripping power as well as allow fluid to be squeezed out of the slots for quicker band application.

Servos

Servos are used in automatic transmissions to convert the hydraulic pressure acting on a piston to a mechanical force that is then applied to a brake band. The simplest servo consists of a cylinder containing a piston, a piston return spring, and a pushrod. Servos are designed for the particular application, so they come in a variety of configurations.

Fluid under pressure is directed onto one side of the piston in the closed cylinder, moving it to apply the band. When this fluid is "dumped," or exhausted from the supply line, the spring returns the piston and flexible band to their released position. Some servos are the "pressure release" type. This means that the servo is applied in the normal way by fluid pressure acting on one side of the servo piston. Release is obtained by applying fluid at the same pressure to the opposite side of the servo piston (**FIGURE 23-62**). Once pressure equalizes, the return spring pushes the piston to its release position. This method of releasing a servo allows the speed of release to be carefully controlled.

Sometimes servos can be accessed from the outside of the transmission, which means they can be serviced without removing the transmission from the vehicle (**FIGURE 23-63**). Some servos have multiple pistons to stage the application of the band for different gears. A common example of this is the General Motors 4L60 and 4L60E transmissions. These transmissions are used in RWD vehicles and have a second and fourth gear band. The servo is a selective apply servo based on the gear required.

Servo Springs, Seals, and Linkage

To aid in the release of a servo, a large spring is typically located behind the servo piston. The servo piston typically has either a neoprene-type seal or a Teflon seal to prevent fluid pressure leaks.

The linkage connected to the servo needs to be checked for proper length. If the linkage is too short when the servo applies, the band will slip. If the linkage is too long, the band will drag when the servo is released. On older vehicles, the band and servo linkage was adjustable, and adjustment was required when the transmission fluid was serviced. Often the technician would adjust the band adjustment bolt to a specified torque, loosen the bolt a specified number of turns, then lock it in place with a jam nut (**FIGURE 23-64**).

Multidisc Clutches

K23026

Multidisc clutches can be used to hold or drive members of the planetary gear set. Multidisc clutches have a set of driving plates and a set of driven plates, collectively called a clutch pack. Both types of plates are made from steel, with one having friction material bonded to both sides. The plates without friction materials are called steels, and the ones with friction material are called frictions. Because they both operate in the transmission fluid, they are called wet clutches.

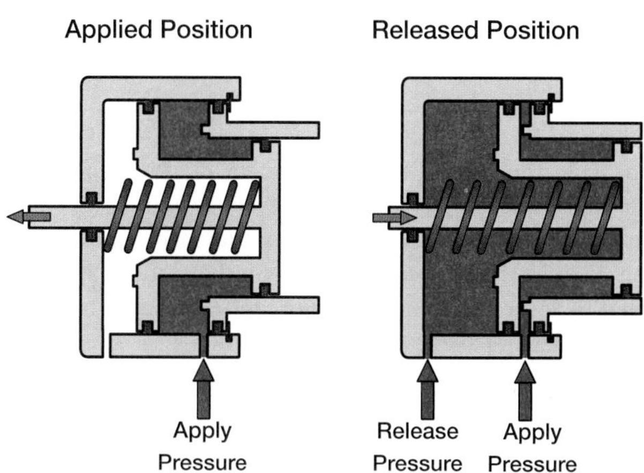

FIGURE 23-62 Pressure release servo.

FIGURE 23-63 A servo being removed from the transmission.

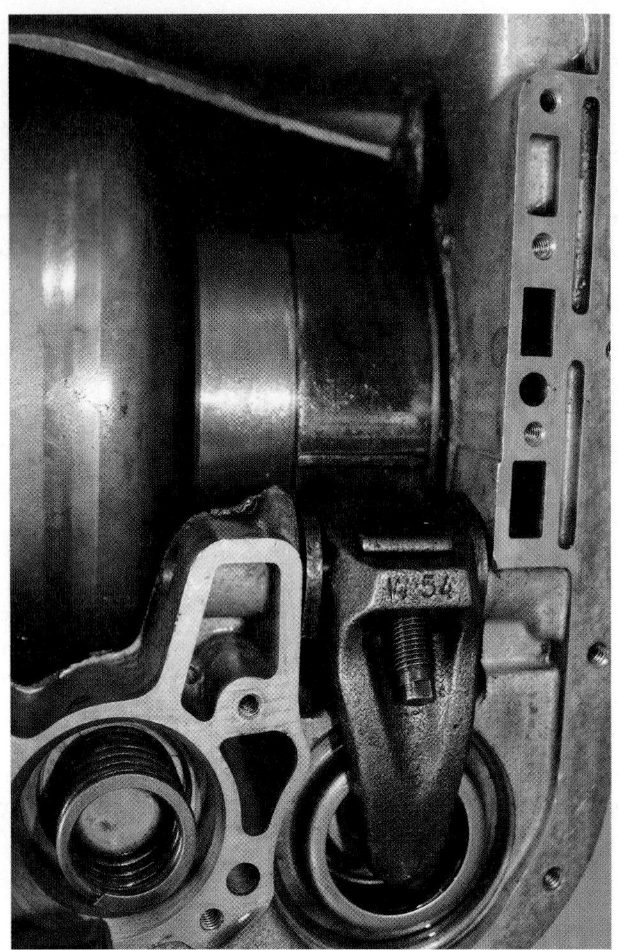

FIGURE 23-64 A band with an adjustment bolt inside the transmission pan.

FIGURE 23-65 Clutch plates. **A.** A burned clutch. **B.** A good in-service clutch. **C.** A new clutch. Note that the clutch on the left has lost all of its friction material.

FIGURE 23-66 A piston removed from a drum on a multidisc clutch pack.

The friction material may be plain or grooved. Grooving allows the fluid to escape from between the plates when the clutch engages and creates a firmer, quicker shift. Using a plain or smooth friction material results in a softer, delayed shift. Over time, as the clutch engages and disengages, the friction material slowly wears away (**FIGURE 23-65**). If the clutch slips due to low transmission fluid level or a faulty clutch seal, the friction material will be destroyed very quickly.

Three to five frictions and steels are commonly used in a multidisc clutch. The frictions and steels are pressed together for engagement by a hydraulically operated piston. It moves in a short cylinder inside a clutch drum, or in a cylinder machined in the transmission housing. The steels are flat and have external splines or teeth that sit in matching grooves in the clutch drum.

The inner driven plates are the frictions, and their splines, or teeth, mate with splines on the outside of a hub located in the center of the clutch pack. The hub is splined and attaches to a shaft or component. The frictions and steels are placed alternately in the clutch pack and are loaded into the clutch drum against the relatively thick pressure plate (**FIGURE 23-66**).

The number of plates installed determines the torque capacity of the clutch. Installing more plates increases the torque capacity, but in all cases the clutch must maintain its specified clearance so that the plates can separate from one another while in the released position. Thinner frictions and steels can be installed to increase the total number of clutch plates and increase the torque capacity.

Releasing of the clutch can be helped by one large coil spring, a pack containing many small springs, or a diaphragm spring. When a diaphragm spring is used, it also gives extra leverage to increase the applied force on the clutch pack and to provide the release force. The outer circumference of the diaphragm spring rests against a step in the clutch drum. The piston pushes against the inner circumference of the spring to apply the clutch. The force exerted on the clutch pack is multiplied by leverage from the diaphragm as it pivots against the pressure plate (**FIGURE 23-67**).

Trapped fluid can accumulate at the outer portion of the clutch cylinder. Due to centrifugal force, at high speed the trapped fluid can produce enough pressure to cause partial application of the clutch. A centrifugal relief valve in the clutch piston or clutch drum releases fluid trapped in the cylinder

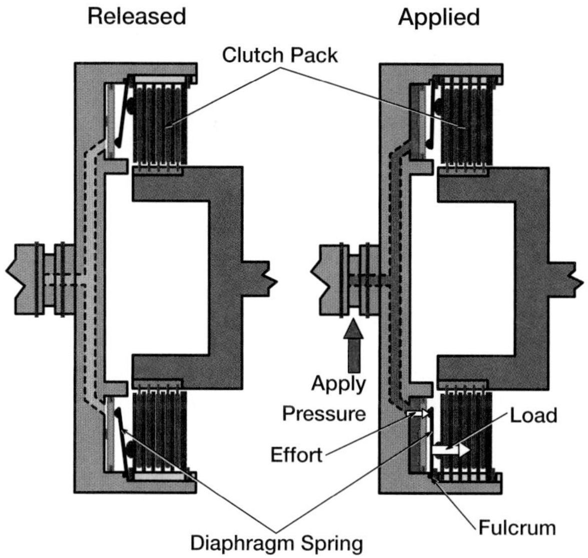

FIGURE 23-67 A diaphragm spring also uses leverage to increase pressure on the clutch pack.

when the clutch is released (**FIGURE 23-68**). This prevents partial application of the clutch during high-speed operation, when the particular clutch should be disengaged, thus preventing unwanted wear of the clutch discs.

Plate Clearance

In the released position, the plates must retain a set clearance. This clearance allows them to separate from one another to allow oil to flow between the clutches and the steels in order to avoid excessive or premature wear. Clutch pack clearance must be checked when rebuilding an automatic transmission (**FIGURE 23-69**). Most transmissions give a range for clutch pack clearance. As mentioned previously, greater clearance results in harsher shifts, and less clearance produces slow, smooth shifts. When the clutch pack has more clearance, the piston will develop more speed as it is applied, creating a harsh engagement with very little clutch pack slippage. If a clutch pack has less clearance, the piston will have a slower apply speed, resulting in a smoother shift as the clutches slowly apply.

▶ TECHNICIAN TIP

Always check TSBs, Automatic Transmission Rebuilders Association (ATRA) publications, or Automatic Transmission Service Group (ATSG) manuals for revised clutch pack clearances. Some transmission specifications are revised after the transmission is put into service. The revised specifications are important for the longevity of the transmission rebuild.

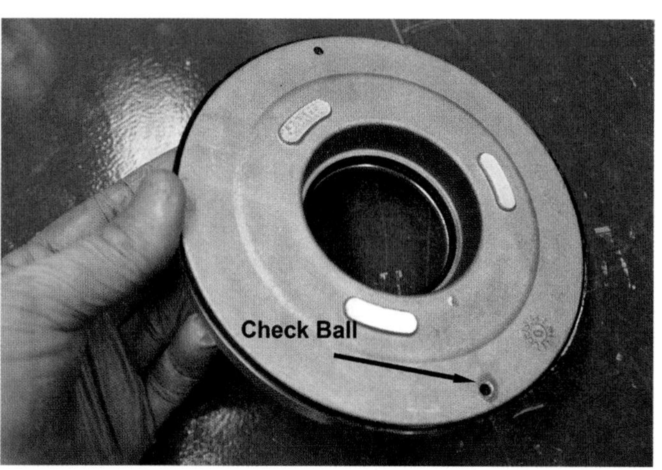

FIGURE 23-68 Centrifugal relief valve in a clutch piston.

FIGURE 23-69 A technician checking clutch pack clearance with a set of feeler gauges.

▶ Wrap-Up

Ready for Review

- Automatic transmissions are now standard equipment in vehicles.
- Most automatic transmissions contain planetary gear sets.
- Automatic transmissions have four or more forward gear ratios and one reverse ratio.
- Typical forward gear ratios are as follows: first gear, 3:1; second gear, 1.75:1; third gear, 1:1; and fourth gear, 0.80:1.
- Adding more gear ratios to the transmission boosts fuel economy.
- Automatic transmissions select and shift gears without driver input and can couple or uncouple from the engine as necessary.
- Types of automatic transmissions include conventional, transaxle, dual clutch, continuously variable, dual shaft, and hybrid (series or parallel).

- Conventional automatic transmissions have planetary gear sets, connect to the engine via torque converter, and use holding devices to stop one or more parts of the planetary gear.
- Transaxle transmissions add a differential and final drive gear to a standard transmission.
- Dual-clutch transmissions use two wet clutches instead of a standard torque converter.
- Continuously variable transmissions use two moveable pulleys rather than typical gears.
- Honda and Saturn vehicles use dual-shaft transmissions that have a main shaft and a countershaft as well as multidisc clutch packs.
- Hybrid vehicles often use series transmissions, in which an electric motor supplements power to the transmission.
- A hybrid vehicle may use a parallel hybrid transmission, in which either the engine or an electric motor can power the transmission.
- Automatic transmission gears are typically helical (most commonly used), spur (straight cut), or hypoid.
- Planetary gear sets use either helical- or straight-cut gears and are comprised of three components: planet carrier, sun gear, and ring gear.
- Bands, multidisc clutches, and one-way clutches are the three basic holding devices used in transmissions.
- Automatic transmission fluid (ATF) transfers heat from the internal transmission components to the transmission cooler.
- ATFs are oil-based or synthetic and may contain these additives: rust and corrosion inhibitors, friction modifiers, seal conditioners, detergents, antifoam, and viscosity modifiers.
- Components of an automatic transmission include the front pump; flexplate and ring gear; front pump seal, bushings, and pump drive; case, extension housing, and pan; gaskets and seals; and parking pawl assembly.
- The front hydraulic pump ensures a continuous pressurized supply of fluid to the transmission.
- The four types of automatic transmission front pumps are gear/crescent, rotor, vane, and variable displacement.
- On automatic transmissions, the flywheel is replaced with a flexplate, with a ring gear wrapped around the outside edge.
- Gaskets may be made of paper, fiber, cork, or neoprene (which may be reusable).
- Types of internal seals include square cut, lip, locking, O-ring, and Teflon.
- Compound planetary gear sets are used to create all the ratios needed for modern vehicle travel.
- Two types of compound planetary gear sets are Simpson and Ravigneaux.
- Thrust washers and thrust bearings are used to control forward and backward motion of transmission components.
- Bushings should be checked during transmission overhaul; the front pump bushing or the output shaft bushing will likely need replacing.
- Automatic transmission brake bands use friction material and servo pressure to stop the drum from spinning.
- Multidisc clutches are used to couple planetary gear members to input shaft, output shaft, or the case.
- Clutch release may be assisted via a large coil spring, a pack of small springs, or a diaphragm spring.
- Check clutch pack clearance when rebuilding an automatic transmission.
- A torque convertor works with the transmission to select the proper gear to meet the speed and power demands of the vehicle.
- A single-stage torque converter contains an impeller, a turbine, and a stator.
- Torque multiplication is a state that occurs when the impeller is turning substantially faster than the turbine. The greater the difference in speed the greater the torque multiplication.
- A torque converter clutch is used to lock the turbine to the impeller when the vehicle is operating under light loads, to improve fuel economy.

Key Terms

band A metal band with friction material bonded to one side. The band is contracted around a drum to stop the drum from spinning.

coefficient of friction A value assigned to materials to describe the amount of friction when two objects slide against each other.

compound planetary gear set Two or more simple planetary gear sets connected together.

continuously variable transmission (CVT) A transmission that does not have conventional set gear ratios, but is instead infinitely variable between the transmission's lowest gear ratio and highest gear ratio.

dual-clutch transmission A type of automatically shifting manual transmission in which two separate input shafts are connected to their own clutch. Shifting of the gears alternates between the two input shafts.

dual-shaft transmission An automatic transmission that more closely resembles a manual transmission, as they do not use planetary gear sets.

extension housing A component of the automatic transmission housing that covers the output shaft of the transmission. The extension housing also supports the end of the driveshaft and may hold components such as the vehicle speed sensor, speedometer drive assembly, and governor assembly.

fluid coupler A type of hydraulic coupling used on vintage vehicles to connect and transfer power from the engine to the transmission.

front pump A hydraulic pump used to supply lubrication oil and hydraulic pressure to the different components inside the automatic transmission.

gasket A rubber, cork, or paper spacer that goes between two parts to seal the gap between the parts.

gear pump A type of front pump used on some vehicles that uses two rotating gears to force fluid out of the pump.

gear ratio The ratio of the size or teeth of one gear compared to the size or teeth of a mating gear.

helical-cut gear A type of gear in which the teeth are cut in a spiral down the axis of the gear.

hypoid gear A type of helical gear used to change the direction of motion 90 degrees. The axis of the input gear does not line up on the centerline of the output gear.

idling A condition in which a gear is spinning, but not moving.

multidisc clutch A type of holding device used by an automatic transmission to stop the movement of one component of a planetary gear set. It uses several thin friction discs and thin steel plates that are squeezed together when hydraulic pressure is applied to a piston in the clutch.

one-way clutch A type of holding device used by an automatic transmission to stop the movement of one component of a planetary gear set. It allows free spinning in one direction but will lock up when the part attempts to spin in the opposite direction.

overdrive Any gear ratio that results in a torque reduction with a speed increase. Overdrive is used on vehicles to reduce the engine speed when traveling at highway speed in order to save fuel.

parallel hybrid drivetrain A type of hybrid transmission in which power can flow from either a gasoline engine or an electric motor or any combination of the two.

pinions The small gears in a planetary gear set that revolve around the sun gear, also known as planetary gears.

planet carrier The device that holds the planet gears in place, keeping them equally spaced.

planetary gears The small gears in a planetary gear set that revolve around the sun gear; also known as pinion gears.

power-splitting transmission (PST) A type of hybrid transmission that splits the power flow going to the wheels from one or more electric motors and an internal combustion engine.

Ravigneaux gear set A type of compound planetary gear set that uses two different sun gears and two different-diameter planets while only using one ring gear.

rotary flow A type of fluid flow in a torque converter in which fluid flows around the centerline of the torque converter in a circle.

rotor pump A type of front pump found in automatic transmissions. The rotor pump has two different-sized rotors that mesh together to force fluid out of the pump.

series hybrid drivetrain A type of hybrid transmission in which power flows from the engine through an electric motor. The electric motor supplements the power from the engine to the wheels.

series-parallel hybrid drivetrain A type of hybrid drivetrain that can function as both a series hybrid and parallel hybrid. That means the gasoline engine can turn a generator that can be used to power an electric motor. The gasoline engine can also drive the vehicle directly through the transmission. And the electric motor can work in parallel with the gasoline engine to drive the vehicle.

servo A hydraulic device consisting of a piston inside a cylinder that is used to apply a band. The servo typically has a spring that releases the servo when it is no longer needed.

Simpson gear set A type of gear set with two planetary gear sets that share a common sun gear.

spur gear A type of gear in which the teeth of the gear are cut in a straight line down the axis of the gear.

stall Often referred to as stall speed, the maximum rpm difference between the turbine and the impeller in the torque converter.

standard torque converter A hydraulic coupling device consisting of an impeller, turbine, stator, and housing; located between the engine and the transmission.

sun gear The center gear of a planetary gear set around which the other gears rotate.

thrust washer A small steel washer coated with bearing material. The thrust washer is used to control forward and backward movement of a part in an automatic transmission.

transaxle A type of transmission typically used in FWD vehicles in which the transmission also includes the differential and final drive gear assembly.

vane pump A type of front pump in an automatic transmission that uses small vanes placed radially around a center hub. The vanes may be forced out against the housing of the pump by centrifugal force. The volume between the vanes varies as the pump turns, resulting in fluid being forced out of the pump.

variable displacement vane pump A type of front pump that uses a vane-style pump that can pivot so that its displacement can be varied.

vortex flow State in which the fluid in the torque converter is traveling between the turbine, stator, and impeller.

Review Questions

1. Which of these transmissions uses two, either wet or dry, clutches in place of a standard torque converter to connect the engine to the transmission?
   a. Continuously variable transmissions
   b. Power-splitting transmissions
   c. Dual-clutch transmissions
   d. Parallel hybrid transmission
2. A torque converter does all of the following *except*:
   a. enables the transmission to automatically couple and uncouple from the engine when needed.
   b. allows the engine to idle at a stop, even though the transmission is still in gear.
   c. multiplies torque from the engine to the input shaft of the transmission under certain driving conditions.
   d. transmits lower amounts of torque when driving away from a stop.
3. When there is a large difference between the speed of the impeller and the turbine:
   a. maximum torque multiplication occurs.
   b. the vehicle is coasting.
   c. torque multiplication falls to zero.
   d. fluid flow from the turbine vanes is relatively low.
4. What is achieved when the transmission fluid first runs through the external cooler and then through the cooler in the radiator?
   a. It increases the viscosity of the transmission fluid.
   b. It prevents overcooling of the transmission fluid.
   c. It draws heat away from the fluid in the transmission pan.
   d. It prevents overheating of the transmission fluid.

5. Any gear ratio of less than 1:1 is considered an overdrive ratio. Overdrive ratios are typically used to:
   a. increase engine speed.
   b. increase fuel economy.
   c. increase engine noise.
   d. increase engine wear.
6. Planetary gears are held in place by a:
   a. sun gear.
   b. ring gear.
   c. planet carrier.
   d. turbine.
7. Which of the below additives allows the automatic transmission fluid to remain thin when the temperature is cold and prevents the fluid from becoming too thin as the transmission fluid warms up?
   a. Seal conditioners
   b. Detergents
   c. Antifoam
   d. Viscosity modifiers
8. Which of these reduces the pumping load on the engine when the transmission requires less fluid to be pumped?
   a. Gear pump
   b. Variable displacement vane pump
   c. Crescent pump
   d. Rotor pump
9. All of the below can be used to lubricate seals *except*:
   a. grease.
   b. petroleum jelly.
   c. automatic transmission assembly lubricant.
   d. automatic transmission fluid.
10. Which of these converts the hydraulic pressure acting on a piston to a mechanical force that is then applied to a brake band?
   a. Seals
   b. Springs
   c. Servos
   d. Pinions

ASE Technician A/Technician B Style Questions

1. Tech A says that the higher the numerical gear ratio (4:1), the more torque will be applied to the wheels. Tech B says that the lower the numerical gear ratio (2:1), the more speed can be obtained at the wheels. Who is correct?
   a. Tech A
   b. Tech B
   c. Both A and B
   d. Neither A nor B
2. Tech A says that spur gears create more thrust motion than helical gears. Tech B says that helical cut gears are noisier than straight-cut gears. Who is correct?
   a. Tech A
   b. Tech B
   c. Both A and B
   d. Neither A nor B
3. Tech A says that the impeller is splined to the input shaft. Tech B says that the stator is fixed solidly to the torque converter housing. Who is correct?
   a. Tech A
   b. Tech B
   c. Both A and B
   d. Neither A nor B
4. Tech A says that a parallel hybrid drivetrain has two or more different devices to propel the vehicle. Tech B says that power-splitting transmissions use two adjustable pulleys and a heavy metal belt. Who is correct?
   a. Tech A
   b. Tech B
   c. Both A and B
   d. A nor B
5. Tech A says that most automatic transmissions use planetary gear sets. Tech B says that planetary gear sets are often combined together to create the needed gear ratios for a transmission. Who is correct?
   a. Tech A
   b. Tech B
   c. Both A and B
   d. Neither A nor B
6. Tech A says that the torque converter hub drives the front pump in an automatic transmission. Tech B says that the turbine shaft drives the front pump. Who is correct?
   a. Tech A
   b. Tech B
   c. Both A and B
   d. Neither A nor B
7. Tech A says that multidisc clutches are used to drive or hold planetary gear components. Tech B says that bands are used to drive planetary gear components. Who is correct?
   a. Tech A
   b. Tech B
   c. Both A and B
   d. Neither A nor B
8. Tech A says that bands allow components to freewheel in one direction, and lock up in the other direction. Tech B says that some servos are the "pressure release" type. Who is correct?
   a. Tech A
   b. Tech B
   c. Both A and B
   d. Neither A nor B
9. Tech A says that the front pump only operates when the turbine is rotating. Tech B says that a torque converter clutch locks the turbine to the converter housing when applied. Who is correct?
   a. Tech A
   b. Tech B
   c. Both A and B
   d. Neither A nor B
10. Tech A says that the torque converter can multiply the amount of torque transmitted from the engine to the transmission. Tech B says that the torque converter allows the engine to idle when the vehicle stops in traffic. Who is correct?
   a. Tech A
   b. Tech B
   c. Both A and B
   d. Neither A nor B

CHAPTER 24
Hydraulic Fundamentals

NATEF Tasks

- **N24001** Diagnose transmission/transaxle gear reduction/multiplication concerns using driving, driven, and held member (power flow) principles. (AST/MAST)

Knowledge Objectives

After reading this chapter, you will be able to:

- **K24001** Explain the principle of Pascal's law and how it relates to the operation of an automatic transmission.
- **K24002** Identify the hydraulic system components and their function.
- **K24003** Describe the types and function of automatic transmission filters.
- **K24004** Describe the types and function of front pumps.
- **K24005** Describe the types and function of pressure regulators.
- **K24006** Describe the types and function of application devices.
- **K24007** Describe the types and function of hydraulic valves.
- **K24008** Describe the types and function of spool valves.
- **K24009** Describe the types and function of solenoid valves.
- **K24010** Describe the types and function of check valves.
- **K24011** Describe the types and function of orifices and accumulators.
- **K24012** Describe the use of clutch and band application charts.

Skills Objectives

There are no Skills Objectives for this chapter.

▶ Introduction

Automatic transmissions function using the principles of hydraulic operation to apply and release different clutches and bands. This system achieves the different gear ratios needed for vehicle operation. Using hydraulics to apply clutches and bands is a good choice because fluids are noncompressible and therefore transmit force effectively. Another benefit of hydraulics is that fluids, having no shape of their own, take the shape of the container. This allows force to easily be applied in a variety of directions (**FIGURE 24-1**). A good understanding of how the hydraulic system operates is necessary to properly diagnose and repair modern automatic transmissions. The basic hydraulic principles and components covered in this chapter are common to most automatic transmissions, whether they are modern computer-controlled transmissions or older hydraulically controlled transmissions.

▶ Pascal's Law

K24001

Pascal's law is one of the fundamental physics principles behind the operation of an automatic transmission. Simply put, Pascal's law states that if we place fluid in a confined space, such as a transmission hydraulic system, pressure will be transferred equally in all directions inside the space. For example, in the transmission, if the front pump produces 100 pounds per square inch (psi) of hydraulic pressure, that pressure is transmitted to

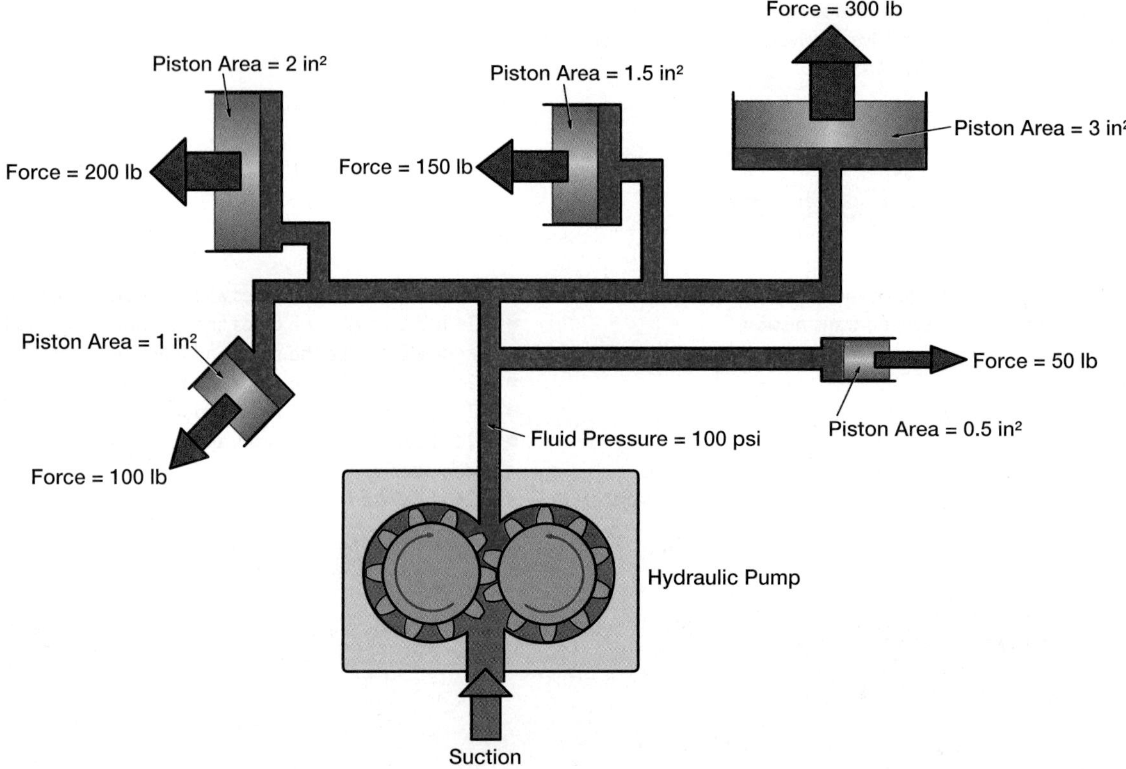

FIGURE 24-1 Hydraulic force can be applied in a variety of directions.

You Are the Automotive Technician

A customer recently purchased a 2012 Ford Taurus with an automatic transmission and stops by your shop to ask about her car. She informs you that the car hasn't been running smoothly lately, and she has seen dark red fluid from the car on the garage floor. Also, she isn't sure if the transmission fluid has ever been checked or serviced. First, you see there is no vehicle service history in the shop's computer. Next, you check the transmission fluid level and find it looks dirty, but not burnt, and is a bit below the safe mark. You notify the customer that you will need to complete a thorough inspection of the vehicle to diagnose the car's problem. The customer agrees to leave the vehicle with you for further inspection.

1. If the vehicle slipped only in one or two specific gears, how would you diagnose which component(s) may be faulty?
2. How does the pressure regulator operate?
3. How would a transmission act if a lip seal tore on a multidisc clutch piston?
4. What does a band do? And how is it actuated?

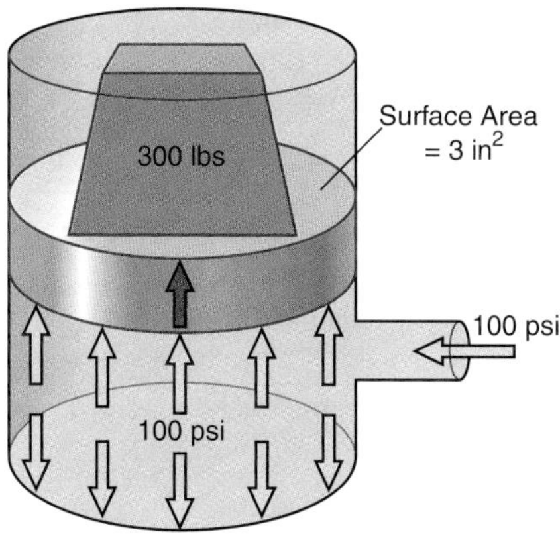

FIGURE 24-2 Hydraulic pressure of 100 psi applied to a 3-square-inch piston results in 300 lb of output force.

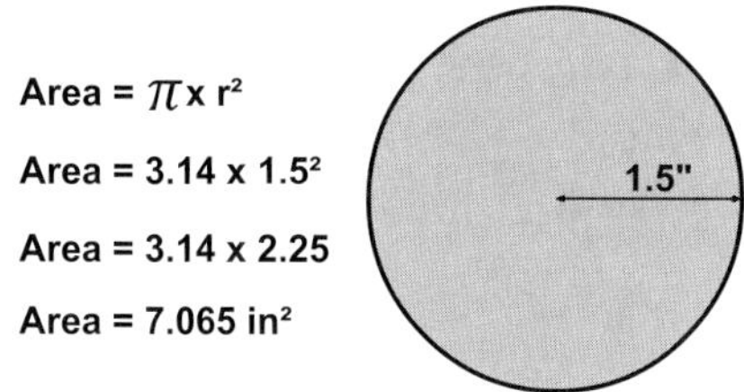

FIGURE 24-3 Area = π × r².

the valve body and then on to a clutch or servo. There will be 100 psi (689 kPa) in the passages and 100 psi (689 kPa) in the clutch or servo.

We are also able to multiply force using Pascal's law. For example, if we have 100 psi (689 kPa) and apply that pressure to a piston, whose surface area is 3 square inches (in^2), the resulting force will be equivalent to 300 lb (100 psi × 3 in^2 = 300 lb) (**FIGURE 24-2**).

Another example: Suppose we apply 200 psi (1379 kPa) to a piston that has an area of 4 in^2. The resulting force will be 800 lb (200 psi × 4 in^2 = 800 lb). Before we can calculate the amount of force created by a piston, we must calculate the area of the piston. A piston is classified as a circle. To calculate the area of a circle, we need to use the formula $\pi \times r^2$. Recall that $\pi = 3.14$, and r = the radius of the circle (in other words, the center of our piston to the outside edge). For example, a piston that has a radius of 1.5" (3" diameter) would have an area of 7.065 in^2 ($\pi \times 1.5^2 = 7.065$ in^2) (**FIGURE 24-3**).

The principles of Pascal's law allow us to apply an enormous amount of force to holding and driving devices inside a transmission. In one typical transmission, the hydraulic system is regulated to approximately 175 psi (1207 kPa) of pressure. That 175 psi (1207 kPa) is applied to a clutch piston with a radius of 3" (6" diameter). To calculate the total clamping force of the piston, we need to calculate the area of the piston: $\pi \times 3^2 = 28.26$ in^2. So applying the 175 psi to a 28 in^2 piston gives a total clamping force compressing the clutch pack together of 4900 lb (2223 kg) (**FIGURE 24-4**)!

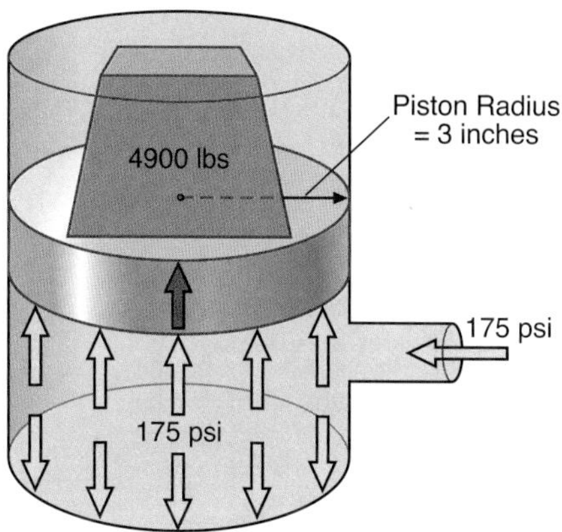

FIGURE 24-4 A piston with a radius of 3" under a pressure of 175 psi will create 4900 lb of output force.

▶ Hydraulic System Components

K24002

The hydraulic system of an automatic transmission includes a sump, filter, a hydraulic pump (front pump), a pressure regulation device, hydraulic application devices for clutches and bands, and several types of control valves and accumulators (**FIGURE 24-5**). Each of the components works with the others to direct fluid under the proper pressure to the appropriate places in the transmission. Learning the function and purposes of each of these hydraulic components will help you to diagnose transmission-related problems, as well as inspect and test the components when replacing them.

Filter

K24003

The pump (and the rest of the transmission) must be provided with a steady stream of clean transmission fluid. Any dirt or particles in the fluid can cause excessive wear to the inside components of the transmission. This dirt can also cause any of the precision valves inside the transmission to become stuck. Vehicle transmissions can use either internal or external filters (**FIGURE 24-6**). Some vehicles have both an internal and an external filter, such as the Allison series of transmissions found in General Motors heavy-duty pickup trucks.

An external filter looks very similar (or almost identical) to an engine oil filter (**FIGURE 24-7**). The filter can be changed without removing the transmission oil pan. A vehicle that used an external filter was the (now discontinued) Saturn S series, as well as many Subaru models.

▶ TECHNICIAN TIP

Many beginning technicians or students have accidentally removed an external transmission filter instead of the oil filter when performing an oil change. Always double-check that the filter you are removing is the correct one.

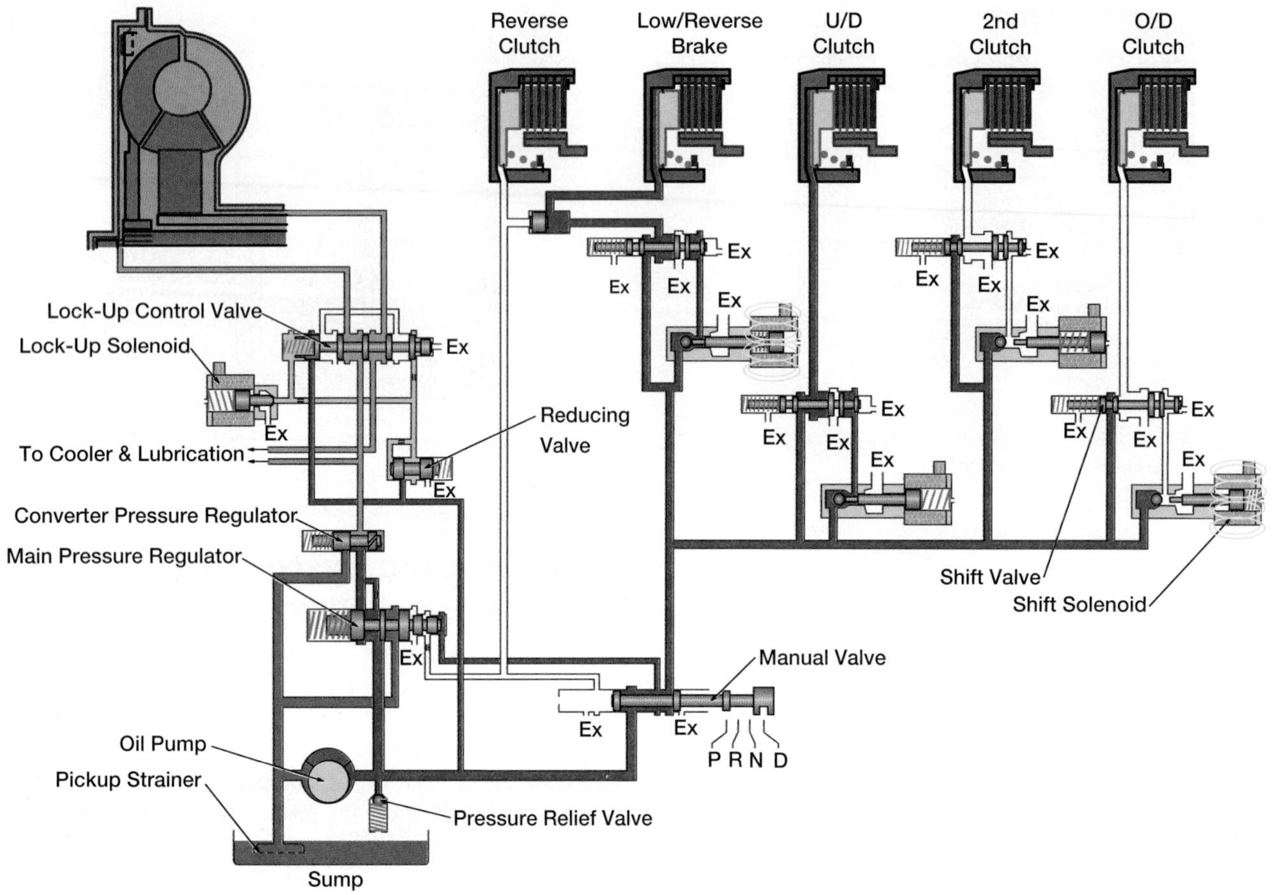

FIGURE 24-5 Typical automatic transmission hydraulic system.

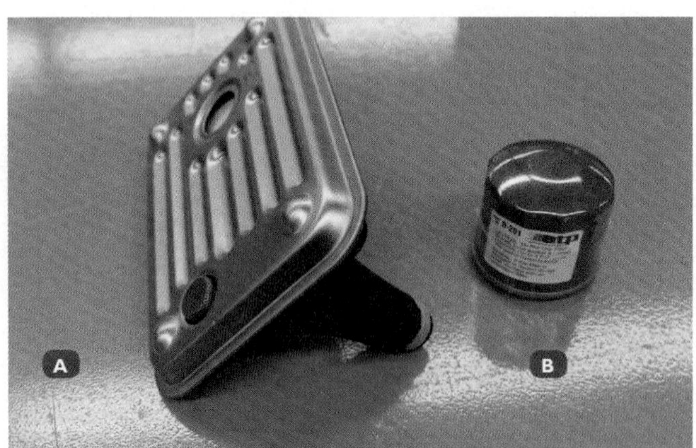

FIGURE 24-6 Transmission filters. **A.** A typical internal transmission filter. **B.** A typical external filter.

FIGURE 24-7 External transmission filter.

Internal filters are much more common than external filters. Replacement of an internal filter requires removing the transmission pan. Internal filters can be one of two types: a **surface filter** or a **depth filter**. A surface filter is often a simple fine-mesh screen made out of metal or a plastic such as polyester. The screen catches particles of dirt as the fluid is drawn through the filter. Surface filters do not have the ability to trap a large amount of material. Most surface filters can stop particles between the size of 50 and 100 microns. One micron is 0.000039" (0.0001 mm), so the smallest particle a surface filter would be able to stop is 0.00195" (0.05 mm)—approximately the thickness of an average human hair.

Depth filters have a thick filter medium that the fluid must pass through. As the fluid passes through the filter, particles become trapped throughout the medium. They use felt or other filter media to trap dirt. Depth filters are able to hold a much larger volume of dirt and particles. This allows them to be effective for a longer period of time. They can also filter out much smaller particles than a surface filter. High-quality depth filters can filter out particles as small as 10 microns, or 0.00039" (0.001 mm).

Inside the transmission, there is often a magnet stuck to the transmission pan. This magnet is designed to help catch ferrous (iron-containing) particles that could cause severe damage to the internal surfaces of the pump as well as the seals, bearings, and bushings within the transmission. This magnet needs to be cleaned off when the transmission fluid and filter are serviced.

Transmission filters should be changed according to the manufacturer's recommended service interval, using high-quality parts and the correct fluid. Changing the filter and fluid on a regular schedule can greatly increase the life of a transmission. This task is covered later in the chapter.

Front Pump

K24004

Modern automatic transmissions have a hydraulic pump referred to as a front pump. They are called front pumps because they are located at the front of the transmission (closest to the engine). In pre-1970s vehicles, some transmissions had a front pump and a rear pump—hence how the specific term front pump came to be. The front pump is turned by the torque converter hub, which is directly turned by the flexplate and crankshaft (**FIGURE 24-8**). Thus, the pump is turning whenever the engine is turning, and hydraulic pressure is available inside the transmission when the engine is operating.

Front pumps can be one of four types of pumps: gear, rotor, vane, and a variable displacement style of the vane pump. Each of these was discussed in the Automatic Transmission Fundamentals chapter. Each type of pump has a different design, but all serve the same purpose: providing hydraulic pressure to all of the hydraulically operated components, supplying lubrication oil to internal transmission components, and circulating fluid to help cool the transmission.

FIGURE 24-8 The front pump is turned by the torque converter hub, which is driven by the crankshaft.

Pump housings can be made from cast iron or aluminum. Cast iron housings tend to resist wear very well and require replacement less often than aluminum pump housings; however, cast iron is much heavier than aluminum, and most manufacturers have switched to aluminum to reduce vehicle weight and improve fuel economy.

Pressure Regulation

K24005

Fluid pressure in an automatic transmission needs to be regulated for several reasons. The first is to apply enough clamping pressure to allow clutches and bands to hold without slipping. The second reason is to help control the quality of the shift based on engine load and rpm. And the third reason is for efficiency purposes by reducing the pressures during cruise when less clutch and band holding power is needed. Without regulation, fluid pressure coming out of the front pump will vary greatly as engine speed increases and decreases while the customer drives the vehicle. When the engine speed is low, the pump output is low, resulting in low pressure. When engine speed increases, the volume of fluid and the pressure coming out of the front pump increase. Front pumps are designed with the capacity to provide more than enough fluid volume and pressure at low speeds to ensure proper transmission operation; therefore, we need the capability of reducing the pressure to the correct amount. The fluid pressure or line pressure can be controlled either by using a mechanical pressure control regulator valve or by using an electronic pressure control solenoid.

A **mechanical pressure regulator valve** is a simple mechanism that uses a spring and a **spool valve** to dump excess fluid back to the transmission pan (**FIGURE 24-9**). A spool valve is a type of valve commonly used in automatic transmissions. The spool valve got its name because of its resemblance to a spool of thread or fishing line. The spool valve operates back and forth inside a bore in the valve body or the front pump. The larger ends of the spool, called lands, fit inside the bore with a very small amount of clearance, typically less than 0.001" (0.0254 mm) (**FIGURE 24-10**). Such a limited clearance means that even the smallest dirt or debris in a valve body can cause a valve to stick in its bore. All internal transmission work must be performed in a clean environment.

This snug fit prevents most fluid from slipping past the land. The clearance between the land and the bore must be just great enough to prevent the valve from sticking as it moves back and forth.

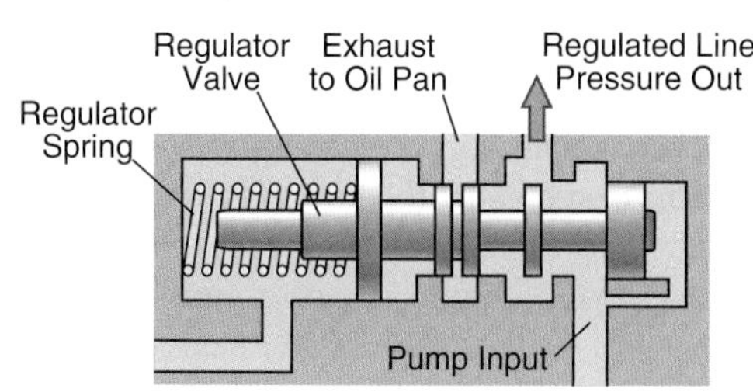

FIGURE 24-9 A typical mechanical pressure control regulator valve.

FIGURE 24-10 A typical spool valve installed in its bore.

FIGURE 24-11 Pressure regulation with a variable displacement pump. Note the spring has been removed to show the pump in a low volume position.

TECHNICIAN TIP

The author's transmission instructor had a device that demonstrated the clearances of valves in a valve body. It consisted of a precision machined spindle that fit into a housing with a precision machined bore. The clearance between the two was less than 0.0001" (0.00254 mm). The bore went clear through the housing. With the housing vertical, the instructor put the spindle about halfway into the bore and a neoprene stopper in the bottom of the bore. In this position, the spindle was supported by the air trapped in the bottom of the bore, as the air could not easily escape between the spindle and the bore. The instructor would proceed to spin it, and it would spin on the cushion of air for more than two minutes. He would then repeat the experiment by touching the surface of the spindle that went inside the bore with his fingers and reassembling it. This time, the spindle would not spin for more than a few seconds because of the oils from his fingerprints. This made clear the importance of treating valves and valve bodies very carefully and keeping them clean.

Fluid coming from the front pump is forced against one side of the spool valve. The other side of the spool valve has a spring that pushes back against the fluid. The spring pressure is set at the factory to a predetermined value, which determines the regulated fluid pressure. When the pump pressure acting on the spool valve becomes greater than the spring pressure, the spool valve moves slightly, opening an exhaust port, and excess fluid pressure is exhausted back to the transmission pan. As the fluid pressure drops, the spring forces the spool valve back into position to block the exhaust port in the bore, thereby raising the fluid pressure. This process happens continuously as the vehicle is driven.

One way to think of a pressure regulator valve is to picture a garden hose turned on all the way. If the end of the hose is open, the water flows out of the hose easily, and the pressure in the hose is very small. But when you place your thumb over the end of the hose and block the flow of water, pressure in the hose increases. You can vary the pressure in the hose by increasing or decreasing your thumb pressure, thus controlling the flow of water out of the hose. This is the same way a standard pressure regulator valve works, except the valve is controlled by spring pressure. In many cases, spring pressure can be assisted by applying hydraulic pressure on the spring side of the valve or by using linkage to increase the spring pressure.

TECHNICIAN TIP

When thinking about pressure in a hydraulic (closed) circuit, it is good to remember that we cannot have fluid pressure without flow and resistance to flow. Without flow, there is no pressure, and without resistance to flow, there is no pressure.

On a vehicle with a variable displacement pump, some of the output fluid is sent to the moveable housing of the pump. This pressure causes the housing to move to one side, as shown in **FIGURE 24-11**. When the housing moves, the pump's volume is reduced. The balance between pump pressure and spring pressure causes the pump's displacement to be controlled so that the specified pump pressure is maintained.

TECHNICIAN TIP

High-performance rebuild kits often contain a stiffer pressure-regulating spring to increase pump pressure and create more clutch- and band-holding power as well as firmer shifts. On older mechanical pressure-regulated transmissions, it was a common practice to install small washers behind the spring, or to use a stiffer spring, to increase the regulated pump pressure on high-performance transmissions.

On computerized transmissions, an **electronic pressure control (EPC) solenoid** is used to regulate line pressure. Most EPC solenoids are a pulse-width modulated valve that provides a certain duty cycle (**FIGURE 24-12**). Pulse-width modulated means that they are pulsed on and off continuously a certain number of times per second. A 50% duty cycle means that the valve is open 50% of the time and closed 50% of the time.

EPC solenoids are usually designed so that they reduce pressure when they are operating. The solenoid is normally closed, resulting in no fluid being returned to the transmission pan (**FIGURE 24-13**). If the solenoid is pulsed on 25% of the time,

FIGURE 24-12 An electronic pressure control (EPC) solenoid mounted on a valve body from a General Motors 4L60E transmission.

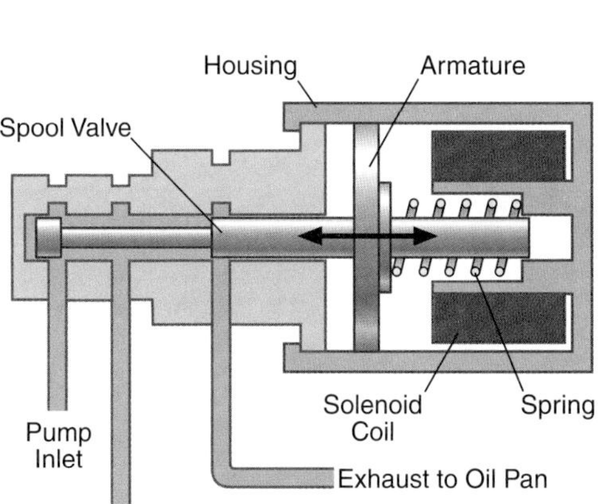

FIGURE 24-13 A typical EPC solenoid and the fluid passages going to and from the valve.

line pressure is reduced by approximately 25%. The higher the pulse width, the lower the line pressure. If the solenoid fails, the transmission will have maximum line pressure to prevent transmission damage and allow the vehicle be driven to a shop for service.

Application Devices

K24006

In an automatic transmission, there are four basic types of holding devices used to hold a member of the planetary gear set, as discussed in the Automatic Transmission Fundamentals chapter. These holding devices are a multidisc clutch, a band, a one-way roller or sprag, and a dog clutch. Both multidisc clutches and bands are hydraulically operated, so we go into more detail about them here.

As discussed earlier, the multidisc clutch is often located inside a drum (**FIGURE 24-14**). It may also be mounted inside the case, using the case as the drum (housing). At the bottom of the drum, after you have removed the clutches and steels, you will find a snap ring and one or more piston return springs (**FIGURE 24-15**). By properly compressing the return spring(s) to remove the snap ring, you can see a large piston that fits snugly inside the drum. This piston typically has a lip seal around the outside edge to seal the piston to the drum and prevent fluid and pressure leakage (**FIGURE 24-16**). There is another lip seal

FIGURE 24-14 A piston removed from a multidisc clutch.

FIGURE 24-15 Snap ring holding the return spring(s) in place.

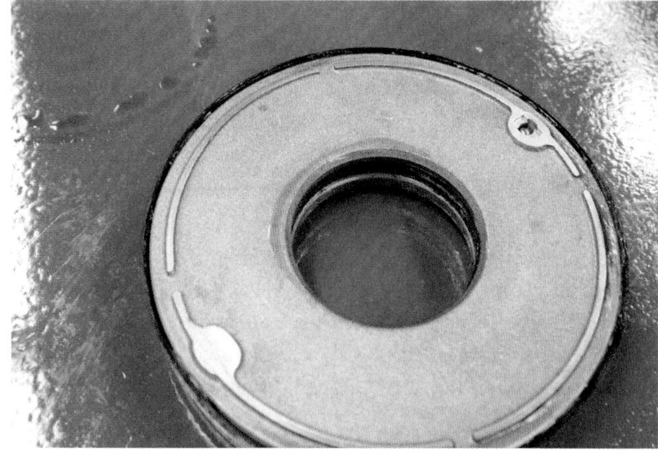

FIGURE 24-16 Inner and outer lip seals installed in the piston.

FIGURE 24-17 A piston from a multidisc clutch that has cracked, leading to a failed transmission.

around the center of the piston. Often, transmission failure can be traced back to failed lip seals that have been leaking and preventing the piston from applying the full pressure needed to lock up the multidisc clutch.

When the transmission sends fluid pressure to the multidisc clutch piston, the piston is forced outward by the fluid. The average piston has an approximate surface area of 12 in^2 (197 mm^2). If the pump applies 150 psi (1034 kPa) of pressure to the piston, the piston will apply a force of 1800 lb (816 kg) to squeeze the friction discs and steels together. If there is a leak in the system, the amount of force exerted will be greatly reduced and can result in clutch slippage.

When rebuilding a clutch pack, the piston needs to be carefully inspected for cracks (**FIGURE 24-17**). Cracks in the piston will cause a pressure loss, resulting in a clutch pack that will fail. New lip seals should always be installed before the piston is reinstalled in the drum. Seals should be properly lubricated with transmission assembly lube or petroleum jelly to prevent damaging the seals. Reinstalling the piston into the drum takes some patience and practice to prevent ripping the lips of the seals. After rebuilding a multidisc clutch, always check the operation of the clutch, using the shop's compressed air to actuate the piston.

When the transmission stops sending fluid pressure to the clutch pack piston, the piston return spring(s) pushes the piston back into its seat in the drum. Once the pressure is relieved, typically a check ball in the piston or drum opens and allows the remaining fluid to be thrown out of the drum as it is rotating. The fluid then is returned to the transmission pan.

> **TECHNICIAN TIP**
>
> Operating the clutch with compressed air to check it for leaks is called "air checking." If a seal was damaged during installation, there will be a large air leak from inside the clutch, requiring disassembly and replacement of the damaged seal. A word of warning: only air check a clutch with all of the clutch plates and snap rings installed; this approach will prevent the piston from traveling too far and coming out of its bore, leaking, and then cutting the lips of the seals between the piston and the edge of the bore. It can be very difficult disassembling the clutch if this happens.

Bands are applied by applying fluid pressure to a servo. The servo is a piston inside a bore. The edge of the piston is sealed to the bore using a square-cut or lip seal. As the fluid pressure pushes on the piston, the piston pushes a linkage to contract the band. An average servo piston is approximately 2" (51 mm) in diameter, with a surface area of approximately 3 in^2 (49 mm^2). This means if we apply the 150 psi (1034 kPa) pump pressure to the servo piston, the piston will exert a force of 450 lb (204 kg) on the band. A large spring around the linkage returns the piston to its seat in the bore when the pressure is released. Some transmissions use a staged, or dual-piston, servo to apply a band. This dual piston allows for greater force to be applied to the band by increasing the effective surface area of the pistons. Some servos operate a lever with an offset fulcrum that multiplies the force from the piston to the band.

> **TECHNICIAN TIP**
>
> When building performance transmissions, some standard servos can be swapped out for high-performance servos, which increase the surface area of the piston. An example of this is the General Motors 4L60E transmission, where the servo can be replaced with a servo from the Corvette version of the transmission.

▶ Valve Types and Functions

K24007

The majority of automatic transmission control valves are located inside the **valve body**. The valve body is a housing made of aluminum or cast iron that holds many different spool valves, check valves, springs, and orifices to make the transmission operate properly. The valve body may be located in the transmission oil pan or in a side pan, as on many front-wheel drive transaxles (**FIGURE 24-18**).

Spool Valves

K24008

Spool valves are used quite often in automatic transmissions. As we discussed earlier, spool valves resemble the spool that thread or fishing line comes on. The spool valve has larger lands located on the valve that fit tightly inside a bore in the valve body. A spool valve may have several lands placed in different spots on the valve. Smaller-diameter sections, called valleys, are located between the lands. The face of each land is machined square and sharp to the land diameter, and some valves have a stem or extension on one end to locate a regulating or return spring or to limit valve travel (**FIGURE 24-19**). The sharp corners on the lands give the valve a self-cleaning action as it moves back and forth in the bore.

FIGURE 24-18 **A.** Transmission valve body location. **B.** Transaxle valve body location.

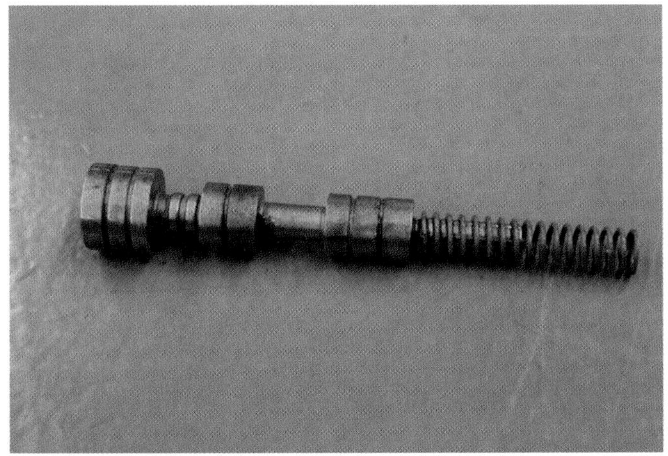

FIGURE 24-19 A spool valve and return spring that have been removed from the transmission valve body.

Fluid is able to flow through the valleys easily. As the land of the spool valve is pushed, either by a spring or by hydraulic pressure, the land moves to either block fluid flow or allow fluid flow in a passage in the valve body. When you are servicing a transmission, spool valves must be handled with care because they have very tight clearances, and any damage will make them unusable. Because the clearances are so small, always use lint-free towels when working around automatic transmissions. A small piece of lint can cause a valve to stick in position and not operate, affecting the operation of the transmission.

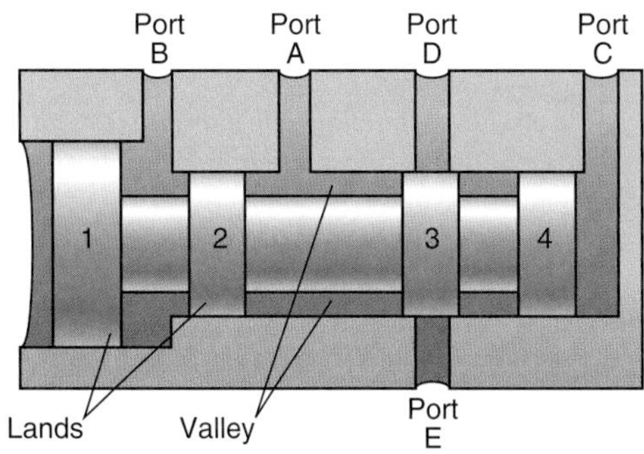

FIGURE 24-20 A typical spool valve with different-sized lands.

Spool valves can be positioned manually, by springs, or by pressure, and many applications have lands of different diameters. In FIGURE 24-20, the different diameters result in a larger face area on land 1 than on land 2. This is known as a differential area. When fluid under pressure enters at port B, this differential area creates a differential force. This force moves the valve to the left because the force acting on land 1 is greater than the force on land 2. Alternatively, the valve could be actuated by admitting fluid at port C. In both cases, the valve movement allows previously blocked fluid at port D to flow between lands 3 and 4 and either exhaust at port E or be directed to some other part of the system through port E. If fluid pressure is removed from port C, the valve will remain stationary. Even if fluid is admitted at port A, no movement can occur because the surface areas of lands 2 and 3 are equal, and both generate an equal force. The valve can be moved back to its original position only by introducing spring pressure or fluid pressure on the left side of land 1.

Shift valves are spool valves that direct the flow of hydraulic oil to a clutch or band on upshift when they are moved by hydraulic pressure called governor pressure, or, on a modern transmission, when the computer commands the shift valve to move directly. Shift valves are labeled in a manner such as 2–3 upshift or 3–2 downshift to distinguish their function.

Pressure-regulating valves are similar to the line pressure regulation valve, except these valves reduce line pressure even further for specific components. Some typical places where manufacturers use additional pressure regulation valves to lower line pressure are the transmission cooler and the torque converter. These pressure regulators are often located in the transmission valve body, but they can also be located inside the front pump.

The **manual valve** is a type of spool valve directly connected to the gearshift mechanism in the vehicle. When the driver selects a drive gear, the manual valve is positioned in the valve body to allow line pressure to push fluid to the proper valves and components inside the transmission. The manual

Applied Science

AS-49: Pressure: The technician can demonstrate an understanding of the concept of pressure in relation to the concept of force.

"Pressure" and "force" are two terms that share some similarities and whose meanings are often confused. Force can be simply defined as a push or a pull (or both) that tends to make an object change its state of motion. Pressure is a physical quality, defined as the amount of force per unit of area applied to the surface of an object.

Let's say that we have a hydraulic circuit in an automatic transmission that activates a clutch piston. The piston has an area of 10 sq/in (64.5cm^2), and when the circuit is activated, a pressure of 75 psi (517 kPa) is applied to the piston. The force generated by the piston to apply the clutch is determined by multiplying the piston area by the pressure applied; in this case, 75 psi (517 kPa) of pressure results in 750 lb (3.3 kg) of force holding the clutches together.

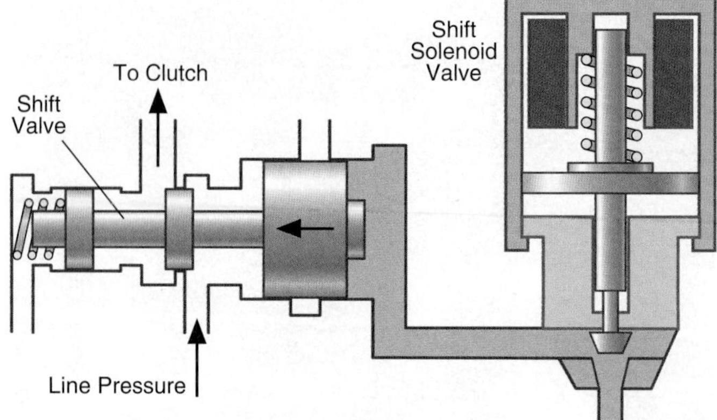

FIGURE 24-22 A shift solenoid that sends line pressure to a secondary shift valve.

FIGURE 24-23 A transmission check ball installed in a transmission valve body.

valve is often connected to the gearshift through a metal linkage inside the transmission that is connected to a cable on the outside of the transmission (**FIGURE 24-21**). Some older models used complicated linkages from the gearshift lever to the transmission that were prone to wear and often needed adjustment. Refer to the Hydraulically Controlled Transmission chapter for more detail on governors.

Solenoid Valves

K24009

A solenoid valve is an electromagnetically operated valve. When electrical current is passed through the solenoid, a plunger is pulled up, opening the valve and allowing fluid to flow. On modern computer-controlled transmissions, shifting is accomplished by one of two methods. In the first method, the computer commands a solenoid valve to open, allowing line pressure to push fluid past the solenoid valve and move a spool-type shift valve which sends fluid to the clutch or band (**FIGURE 24-22**). In the second method, the computer commands a solenoid valve to allow line pressure to be directly sent to the clutch or band to complete the shift.

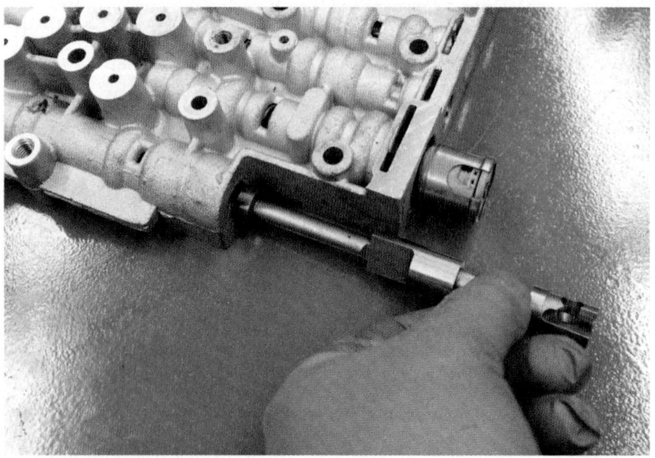

FIGURE 24-21 A manual valve installed in an automatic transmission.

Check Valves

K24010

Check valves can be one of two types in an automatic transmission: one-way check valves or two-way check valves (also known as shuttle valves) (**FIGURE 24-23**). In a one-way check valve, fluid flows in one direction, causing the valve to unseat and allowing hydraulic oil to flow freely. Fluid flow in the opposite direction causes the valve to close, blocking the passage **FIGURE 24-24**. Check valves can have springs inside to help with the closing of the valve, or they can be a simple check ball design, which uses gravity or fluid flow to close the valve.

Two-way check valves are often used in the release circuit for application devices. For example, some transmissions have one small passage that supplies oil to an application device, but have two passages that allow oil to release from the application device.(**FIGURE 24-25**) If we did not have a check ball or check

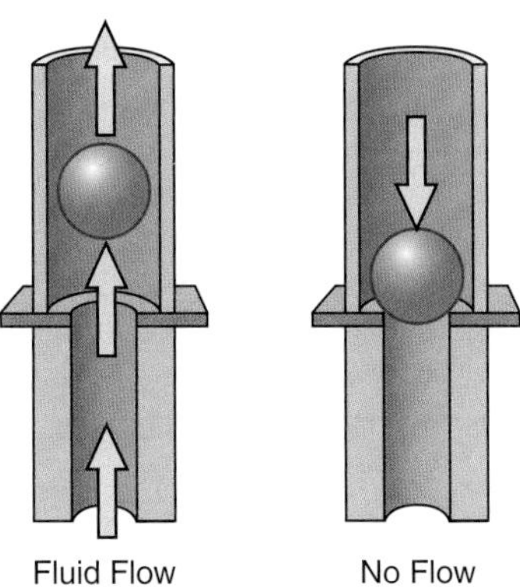

FIGURE 24-24 One-way check valve operation.

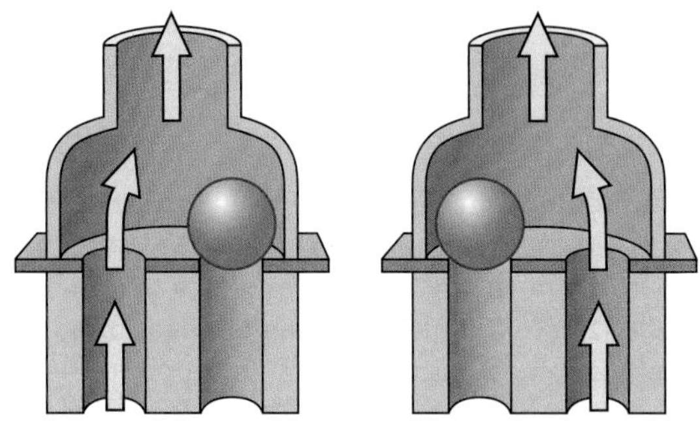

FIGURE 24-25 A two-way check valve.

Applied Science

AS-86: Action/Reaction: The technician can demonstrate an understanding of the action/reaction of fluids involving valves or pistons.

According to Newton's third law of motion, for every acting force there is an equal (in strength) and opposite (in direction) reacting force. This principle is applied in several areas within automatic transmission hydraulic circuits.

For gear changes to occur at optimum times, the transmission must balance engine load and road speed. In hydraulically controlled transmissions, engine load may be sensed by means of a throttle valve in the transmission, connected directly to the accelerator pedal. When changes occur in engine load, the throttle valve raises or lowers its output pressure correspondingly. Road speed is monitored via a governor valve, which is sensitive to centrifugal force and is mounted on the transmission's output shaft. The governor valve increases or decreases its output pressure in proportion to output shaft speed.

Shift valves are exposed to both throttle pressure and governor pressure on opposite sides of a piston. The position of the shift valve is dependent on their relationship. If governor pressure overcomes throttle pressure (+spring pressure), an upshift will result. If throttle pressure (+spring pressure) overcomes governor pressure, a downshift occurs.

Two-way check valves also commonly use a check ball (**FIGURE 24-25**). In the two-way check valve, there are two fluid inlet ports and one fluid outlet port. Thus, two different control devices can share the outlet passage.

valve in the second passage, the fluid pressure would simply dump out the exhaust port rather than applying the clutch or band. By having a second larger passage for the release of oil from the clutch or band, we can exhaust the oil pressure more quickly to prevent clutch dragging.

Orifices and Accumulators

K24011

Orifices are often used in conjunction with check balls and spool valves to restrict fluid flow for timing purposes or to prevent valve fluctuations. An orifice is a precision hole. The orifice is smaller in diameter than the passage or line feeding the orifice. It is designed to prevent sudden or undue movement of a valve. It is much like a section of highway where a few lanes must merge together. This reduction in the size of the passage (or the highway) causes the fluid to be delayed and reduces pressure after the orifice (just as after the construction zone on the highway, there is much less traffic on the road around you). Vehicle manufacturers often build orifices into automatic transmission valve bodies or into the **separator plate** installed between the valve body halves or between the valve body and the transmission case (**FIGURE 24-26**). The orifices control the flow of hydraulic fluid between the valve body and the transmission.

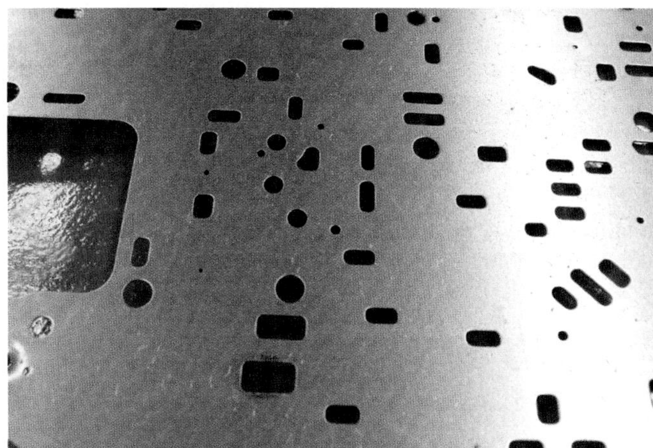

FIGURE 24-26 Orifice used to control the flow of fluid in a separator plate.

Orifices are often used in the passage feeding oil pressure to a clutch or band. The narrow section or orifice slows the application of the gear to prevent the transmission from having a harsh application.

▶ TECHNICIAN TIP

If you purchase a high-performance rebuild kit for a transmission, often the kit will come with a different valve body separator plate with larger holes. These larger holes are actually larger orifices that will result in less of a delay in clutch application. Some high-performance kits specify that you use a specific drill bit size to enlarge specific holes in the spacer plate. These modifications affect the operation of the transmission, so only experienced transmission specialists should perform this work.

Accumulators are another hydraulic device found in automatic transmissions (**FIGURE 24-27**). Accumulators are used to dampen the application of clutches and bands (**FIGURE 24-28**). Without an accumulator in an application device circuit, the transmission might apply the clutch or band harshly enough to cause the customer to complain about the shift quality. Or the transmission might shift harshly enough to break the tires loose from the pavement during each transmission shift. Harsh shifting increases stress on the rest of the drivetrain inside and outside of the transmission. Accumulators contain a piston installed in a bore. The piston has rubber or Teflon seals around its outside edge, to seal the piston to the bore. There is a large spring behind the accumulator piston. When the shift valve sends line pressure to the clutch or band, fluid is also sent to the side of the accumulator piston opposite the spring. The clutch or band begins to apply progressively as the accumulator spring is completely compressed by the piston.

FIGURE 24-27 An automatic transmission accumulator being removed from a transmission case.

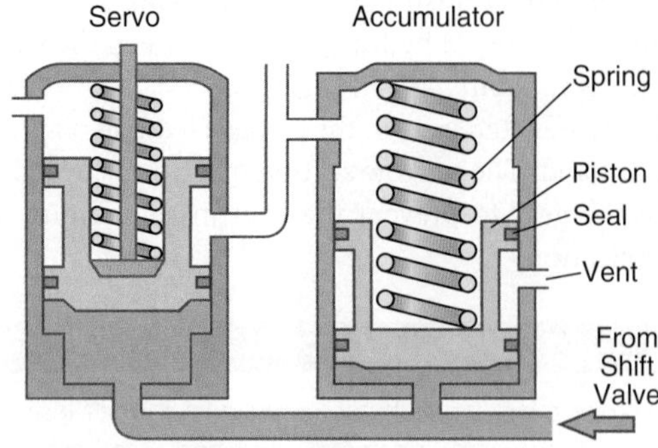

FIGURE 24-28 Accumulator operation in a servo circuit.

▶ **TECHNICIAN TIP**

Often when building high-performance transmissions, specific accumulator springs are removed to improve the shift quality of a transmission that is installed behind a high-performance, high-horsepower engine. Without the spring installed in the accumulator, the piston will remain in the bottom of the bore so that the piston cannot move. Without the piston movement, the clutch or band will apply much more quickly, but will result in a harsh shift.

Clutch and Band Application Charts

N24001, K24012

When diagnosing an automatic transmission problem, it is important to first verify the customer complaint by taking the vehicle on a road test, if possible. If the vehicle transmission will not function at all, this is not possible. After you have verified the customer complaint, pay attention and document when the problem occurs. For example, if you place the gearshift into 2 rather than D, does the problem still occur? Does the problem only occur during hard acceleration, or during light acceleration as well? Use this information when you look at the manufacturer's clutch and band application chart, sometimes referred to as a truth chart. The chart lists specifically which clutches, bands, and other components are used for the different functions of the transmission.

TABLE 24-1 is a sample of a clutch and band application chart for a transmission. Using this chart, we can start our diagnosis of a failed transmission *before* we pull the transmission. By using this information before beginning the repair, we can diagnose the cause of the fault and give the customer a better estimate of the repair costs. We may also determine that it is not even necessary to remove the transmission from the vehicle in order to repair the problem.

Using the table, what are the possible problems if the transmission operates only in third gear, no matter which forward range the gear selector is in? If you said that it must be a problem with the shift solenoids, you are correct. By looking at the table, we can determine that neither shift solenoid is used in third gear.

What if the vehicle has no reverse but moves forward? When you look at the table, you will see that there are several components of the transmission in use. If the transmission seems to drive forward fine, could the problem be a bad 1–2 shift solenoid, or a 2–3 shift solenoid? If you answered no, you are correct. If it were one of the shift solenoids, other gears would also show a problem. Now you have reduced the number of possible components to two: the low reverse clutch and the reverse input clutch. If you were to put the vehicle into first, you would be able to diagnose further. When in first, the transmission is supposed to hold onto the engine and not freewheel or coast when you decelerate. If the transmission holds on deceleration in first, you can eliminate all of the possible choices except the reverse input clutch. You now have a very good idea of what has failed inside the transmission before you have even turned a wrench.

TABLE 24-1 Sample Application Chart for a Transmission

Gear Range (Selector Lever Position)	1–2 Shift Solenoid	2–3 Shift Solenoid	Forward Clutch	Forward Sprag	Low Roller Clutch	2–4 Band	3–4 Clutch	Overrun Clutch	Low Reverse Clutch	Reverse Input Clutch
Park	×	×								
Neutral	×	×								
Reverse	×	×							×	×
D (overdrive), First gear	×	×	×	×	×					
D (overdrive), Second gear		×	×	×		×				
D (overdrive), Third gear			×	×			×			
D (overdrive) Overdrive	×		×			×	×			
D (drive), First gear	×	×	×	×	×					
D (Drive), Second gear		×	×	×		×				
D (drive), Third gear			×	×		×	×			
2, First gear	×	×	×	×	×			×		
2, Second gear		×	×	×		×		×		
1, Low gear	×	×	×	×	×			×	×	

▶ Wrap-Up

Ready for Review

- Hydraulic operation is used to apply and release different clutches and bands in automatic transmissions.
- Pascal's law states that if fluid is placed in a confined space, then pressure will be equally transferred in all directions within that space.
- Force can be multiplied using Pascal's law.
- Vehicle transmissions use filters (internal, external, or both) to ensure clean transmission fluid.
- Internal filters (most common) can be surface or depth filters.
- Surface filters have a fine-mesh screen but cannot trap a large amount of material.
- Depth filters use thicker filter media, trap a larger volume of material, and are effective longer than surface filters.
- A magnet stuck to the transmission pan is designed to catch iron-containing particles.
- Components of a hydraulic system are the hydraulic pump (front pump), the pressure regulation device, hydraulic application devices, control valves, and accumulators.
- Types of front pumps are gear, rotor, vane, and the variable displacement style of the vane pump.
- Purposes of the hydraulic/front pump are to provide hydraulic pressure to all hydraulically operated components, supply lubrication oil to internal transmission components, and circulate fluid to cool the transmission.
- Pump housings can be cast iron (wear resistant, require less frequent replacement) or aluminum (lighter weight).
- A feeler gauge is used to measure pump gear-to-pump housing clearance, driving gear teeth-to-crescent clearance, and pump gear-to-straightedge on pump halves clearance.
- Gear pumps can be checked for wear by hand, while the pump housing should be checked with a straightedge and feeler gauge.

- Rotor pumps are checked in a similar fashion to gear pumps.
- Vane pumps are checked for wear by comparing rotor thickness to specifications.
- Check vane pump housing and vanes for wear; reinstall vanes in the same direction they were removed.
- Automatic transmission fluid pressure is controlled via either a mechanical pressure control regulator valve or an electronic pressure control (EPC) solenoid.
- A mechanical pressure regulator valve uses a spring and a spool valve to ensure that excess fluid spills into the transmission pan.
- Variable displacement pumps rely on movement of the housing to reduce volume and regulate pressure.
- Computerized transmissions use EPC solenoids to regulate line pressure.
- EPC solenoids are generally pulse-width–modulated valves that are continuously pulsed on and off and reduce pressure during their "on" cycle.
- The multidisc clutches and bands used to stop a member of the planetary gear set are hydraulically operated.
- The transmission sends fluid pressure to the multidisc clutch piston, forcing the piston to transfer that force into squeezing friction discs and steels together; any leaks will reduce the amount of force.
- When fluid pressure is applied to a servo piston, it in turn exerts force on a band.
- Types of automatic transmission valves are spool, solenoid, and check.
- Spool valves have lands and valleys that are used to block or allow flow of transmission fluid.
- Differential diameters of the spool valve's lands create differential force when receiving pressurized fluid.
- Types of spool valves include manual and shift valves.
- On computer-controlled transmissions, the computer can direct a shift solenoid to send line pressure to a secondary shift valve.
- Check valves can be one-way or two-way valves (shuttle valves).
- Check valves can allow fluid to flow past them, but fluid flowing in the opposite direction will close the valve, blocking the fluid.
- Check valves may contain a spring or a check ball (common in two-way check valves).
- Orifices are precision holes that work to restrict fluid flow for timing purposes or to prevent valve fluctuations.
- Accumulators are designed to dampen the application of clutches and bands.
- Automatic transmission problem diagnosis should start with a road test. Verify the customer's complaint, and document the circumstances under which the problem occurs.
- Use a manufacturer-provided clutch and band application chart to assist in diagnosis and estimate repair costs.

Key Terms

accumulator A device used to reduce the speed of clutch or band application to help prevent harsh shifting.

check valve Also known as a shuttle valve, a valve that allows the flow of hydraulic oil in one direction only. It is typically used to allow oil to escape quickly from an application device when shifting gears.

depth filter A hydraulic filter that has thick filter media to trap dirt and other particles of various sizes as they pass through the filter.

electronic pressure control (EPC) solenoid A solenoid used to regulate line pressure on a computerized transmission.

manual valve A type of spool valve directly connected to the gearshift mechanism in the vehicle; when the driver selects a drive gear, the manual valve is positioned in the valve body to allow line pressure to push fluid to the proper valves and components inside the transmission.

mechanical pressure regulator valve A spool valve and spring assembly that is used to control the amount of line pressure in a transmission.

orifice A precisely sized hole used to reduce the speed and pressure of hydraulic oil flowing to a clutch or band.

Pascal's law One of the fundamental physics principles behind the operation of an automatic transmission, this law states that if fluid is placed in a confined space, pressure will be transferred equally in all directions inside the space.

separator plate Sometimes called a spacer plate, a thin sheet metal plate installed between the valve body and the transmission case. Orifices can be installed in the separator plate, and check valves can work with holes in the plate.

shift valve A spool valve used to control oil flow to a clutch or band for a particular gear.

spool valve A type of valve commonly used in automatic transmissions that resembles the spool that thread or fishing line comes on.

surface filter A filter that is a simple screen mechanism to catch dirt and other particles in the hydraulic oil as it passes through.

valve body An aluminum or cast iron housing inside the transmission that houses the majority of the valves that control transmission operation.

Review Questions

1. Pascal's law states that if we place fluid in a confined space, pressure will:
   a. be transferred equally in all directions inside the space.
   b. be higher at the end of the space.
   c. be lower at the end of the space.
   d. vary at various points inside the space.
2. The hydraulic system of an automatic transmission includes all of the following *except*:
   a. a sump or pan.
   b. a filter.
   c. a front pump.
   d. a bushing.

3. Depth filters may be made of:
   a. metal.
   b. felt.
   c. plastic.
   d. polyester.
4. A pressure regulator valve ensures that:
   a. minimal clamping pressure is applied on clutches and bands at all times.
   b. engine load and rpm do not affect transmillsion shift points.
   c. the pressures are reduced during cruise, when less clutch and band holding power is needed.
   d. fluid pressure and line pressure stay uniformly high during all driving conditions.
5. Which of these is used to regulate line pressure on computerized transmissions?
   a. EPC solenoid
   b. Multidisc clutches
   c. Check valves
   d. Accumulators
6. When rebuilding a clutch pack, the piston needs to be carefully inspected for:
   a. clearance.
   b. expansion.
   c. contraction.
   d. cracks.
7. Which of these valves is directly connected to the gearshift mechanism in the vehicle?
   a. Shift valve
   b. Manual valve
   c. Solenoid valve
   d. Check valve
8. Which of these is often used in conjunction with check balls and spool valves to restrict fluid flow for timing purposes?
   a. Lip seal
   b. Accumulator
   c. Orifice
   d. Pressure regulator valve
9. Accumulators are used to:
   a. dampen the application of clutches and bands.
   b. direct fluid under the proper pressure to the appropriate places in the transmission.
   c. regulate line pressure.
   d. prevent valve fluctuations.
10. A clutch and band application chart helps to:
   a. obtain a clearer problem description from the customer.
   b. better record the problem on a repair order.
   c. identify a problem *before* you pull the transmission.
   d. verify which transmission you are working on.

ASE Technician A/Technician B Style Questions

1. Tech A says that band and clutch application charts can help to diagnose faults in a transmission. Tech B says that 100 psi of pressure on a 2-square-inch piston results in 50 lb of force. Who is correct?
   a. Tech A
   b. Tech B
   c. Both A and B
   d. Neither A nor B
2. Tech A says that one benefit of using hydraulic pressure is that it is not affected by leaks in the system. Tech B says that hydraulic pressure is the same in all directions within a closed container. Who is correct?
   a. Tech A
   b. Tech B
   c. Both A and B
   d. Neither A nor B
3. Tech A says that a mechanical pressure regulator exhausts excess fluid back to the transmission pan. Tech B says that if the transmission pan is removed, the magnet must be replaced. Who is correct?
   a. Tech A
   b. Tech B
   c. Both A and B
   d. Neither A nor B
4. Tech A says that one type of front pump is a variable displacement pump. Tech B says that the front pump is turned by the torque converter hub. Who is correct?
   a. Tech A
   b. Tech B
   c. Both A and B
   d. Neither A nor B
5. Tech A says that spring tension against a spool valve regulates system pressure. Tech B says that a 50% pulse width modulation means the valve is closed 50% of the time. Who is correct?
   a. Tech A
   b. Tech B
   c. Both A and B
   d. Neither A nor B
6. Tech A says that transmission failures can be caused by failed lip seals preventing full pressure from being applied to the clutches. Tech B says that orifices are designed to speed up the movement of a valve. Who is correct?
   a. Tech A
   b. Tech B
   c. Both A and B
   d. Neither A nor B
7. Tech A says that park is obtained by mechanically locking one band and one clutch together. Tech B says that spool valves tend to be self-cleaning. Who is correct?
   a. Tech A
   b. Tech B
   c. Both A and B
   d. Neither A nor B
8. Tech A says that multi-plate clutches must have a specified amount of clearance when they are installed. Tech B says that there should be no clearance when they are installed because the clutch would slip. Who is correct?
   a. Tech A
   b. Tech B

c. Both A and B
d. Neither A nor B

9. Tech A says that bands are a type of hydraulic valve. Tech B says that servos are devices used to allow free movement in one direction and lock up in the other direction. Who is correct?
   a. Tech A
   b. Tech B
   c. Both A and B
   d. Neither A nor B

10. Tech A says that accumulators are used to filter contaminates. Tech B says that accumulators are used to soften clutch engagement. Who is correct?
    a. Tech A
    b. Tech B
    c. Both A and B
    d. Neither A nor B

CHAPTER 25

Hydraulically Controlled Transmission

NATEF Tasks

There are no NATEF tasks for this chapter.

Knowledge Objectives

After reading this chapter, you will be able to:

- **K25001** Describe purpose and function of hydraulically controlled transmission components.
- **K25002** Describe line pressure and how it is controlled.
- **K25003** Describe the purpose and function of the governor valve.
- **K25004** Describe the purpose and function of the vacuum modulator.
- **K25005** Describe the purpose and function of the throttle valve.
- **K25006** Describe the purpose and function of the kickdown valve.
- **K25007** Identify and describe the function of shift valves.
- **K25008** Describe the purpose and function of the secondary regulator valve.
- **K25009** Describe the operation of the hydraulically controlled transmission.
- **K250010** Describe the operation of the transmission in park.
- **K250011** Describe the operation of the transmission in first gear.
- **K250012** Describe the operation of the transmission in second gear.
- **K250013** Describe the operation of the transmission in third gear.
- **K250014** Describe the operation of the transmission in reverse.

Skills Objectives

There are no Skills Objectives for this chapter.

▶ Introduction

In fully hydraulically controlled transmissions, the timing and speed of shifts are completely controlled by hydraulic and/or mechanical devices (**FIGURE 25-1**). Although none of the automobile manufacturers in the United States are still producing fully hydraulically controlled transmissions, it is nonetheless important for students and technicians to understand how these transmissions operate. This is because on a modern computer-controlled transmission, the electronic sensors and actuators that replaced the hydraulic and mechanical control devices perform the exact same function: controlling the flow of hydraulic fluid to and from the transmission components. In fact, both hydraulic and electronic transmissions with multi-plate clutches and bands use hydraulic pressure to apply the bands and clutches. The only difference between the two types is how the fluid is controlled. We go into detail about each of the parts used to control the hydraulically controlled transmission in this chapter. Electronically controlled transmissions are covered in the Electronically Controlled Transmission chapter.

FIGURE 25-1 A hydraulically controlled valve body.

▶ Hydraulically Controlled Transmission Concepts and Components

K25001

The hydraulic controls in a hydraulically controlled transmission work together to apply gears along with causing upshifts and downshifts at the proper times (**FIGURE 25-2**). For the controls to work properly, a supply of pressurized automatic transmission fluid is required. This pressurized fluid is supplied by the pump and maintained at the proper pressure by the pressure regulator valve. The pressure controlled by the pressure regulator valve is called line pressure and is what applies the bands and clutches as well as supplies fluid to the governor and throttle valves. Next, We explore each of the valves, the jobs they perform, and their operation.

Line Pressure

K25002

In a hydraulically controlled transmission, clutches and bands are applied using **line pressure**, which is created by regulating hydraulic pressure that is generated in the front pump. A pressure regulator valve regulates that pressure before it can be used by the clutches and bands. However, because the line pressure must vary according to engine load, other valves are used to act on the pressure regulator valve. We look at these now in greater depth.

Controlling Line Pressure

The Hydraulic Fundamentals chapter describes the operation of the front pump and pressure regulator. For the purposes of that chapter, the operation of the pressure regulator is simplified to help you build your understanding of the way an automatic transmission operates. Now we need to go into greater detail as to how the pressure regulator controls line pressure. In a transmission, line pressure should not remain constant through all driving ranges and vehicle demands. The line pressure must

You Are the Automotive Technician

A customer has come into your shop, requesting a custom transmission rebuild for his modified, high-performance Corvette. You will be working alongside your coworker, who has extensive experience in transmission rebuilds. Before you begin the rebuild, she asks you to prepare the workspace and make sure you have the two most essential tools ready: a transmission jack and a vehicle lift. Your coworker explains to you that this process will take longer than a traditional transmission rebuild because of the nature of a performance rebuild. This is in part because the adjustments to the valve body and governor will have to be done slowly and road tested after each adjustment. The valve body is one part you will need to carefully disassemble so that you can clean and inspect each of the valves and valve bores. The repair order states that the transmission had some performance modifications when it was rebuilt six years ago and that the vehicle has had several high-performance aftermarket components installed.

1. What are some of the modifications to the governor that can change performance?
2. Why would high-performance engine components change how you perform a rebuild on a transmission?
3. What modifications can be made to accumulators, and how would shift feel be changed?
4. If the transmission has an adjustable vacuum modulator, what would happen to modulator pressure if you adjust the spring a bit tighter?

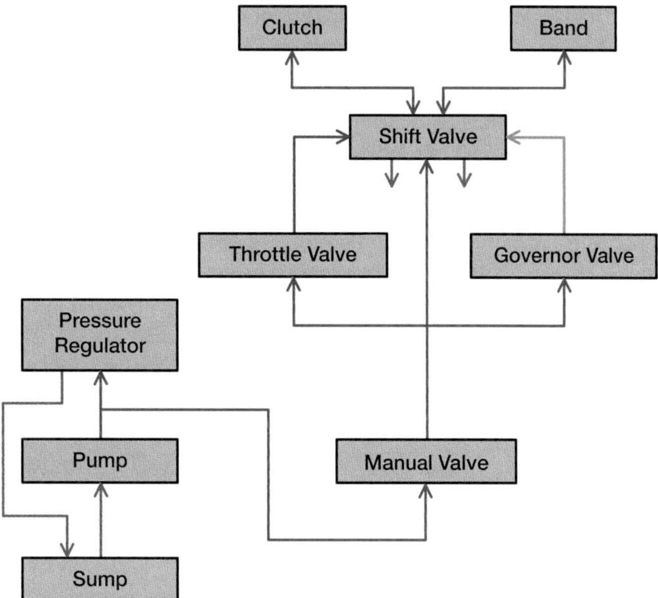

FIGURE 25-2 Automatic shifting happens based on the interaction of opposing pressures acting on valves.

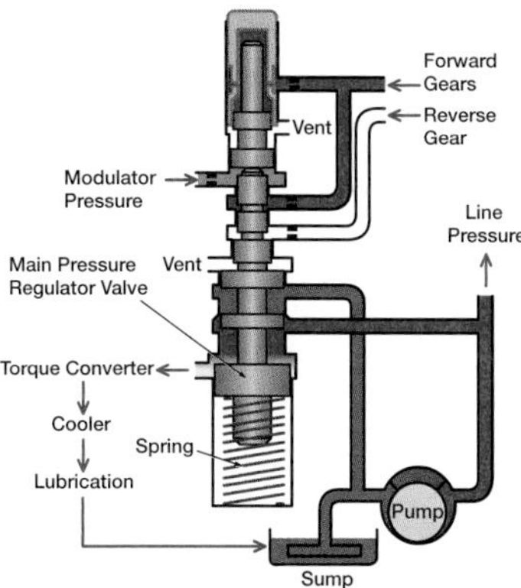

FIGURE 25-3 A diagram of a common pressure regulator valve at its crack point in a hydraulically controlled transmission.

increase when needed to help create greater holding power in the clutches and bands, and decrease when the additional pressure is not needed, so fuel economy can be improved. Also, if the line pressure is too high during normal cruising speeds, the customer may complain about a transmission that shifts harshly.

FIGURE 25-3 shows the standard pressure regulator discussed in the Hydraulic Fundamentals chapter but with a few additional hydraulic passages. As explained in that chapter, the pump supplies pressure to port A. Line pressure flows out of port B. When the pump pressure exceeds the spring pressure, the valve moves to the left, exposing the exhaust port (port C) and allowing excess fluid pressure to be returned to the transmission pan. As the throttle is opened further, line pressure must be increased to help hold the clutches and bands with the additional engine force being produced. To increase the line pressure, throttle valve pressure is allowed to flow into port F, thus pushing the throttle valve. By boosting valve pressure to the right and adding additional pressure to the spring force keeping the pressure regulator valve closed, the line pressure is increased.

Transmission pressures vary greatly among manufacturers and with different transmissions, but line pressure with the throttle closed is as low as 75 psi (517 kPa) and could be almost 200 psi (1379 kPa) when the throttle is wide open. To test the operation of the throttle valve boost valve, you must install a pressure gauge to the transmission line pressure test port. Disconnect the throttle valve cable or linkage, and start the vehicle. With the vehicle idling in gear, measure the line pressure with the throttle valve closed and again with the throttle valve all the way open. Compare your results with the manufacturer's specifications. The same test can be performed on a transmission equipped with a vacuum modulator. A vacuum pump is applied to the vacuum modulator. As the vacuum changes, so should line pressure. As vacuum increases, line pressure should decrease. Check the manufacturer's specifications.

Often line pressure has to be increased in the reverse and L1 gear positions, or "1" range. The torque multiplication from the torque converter is greatest when starting from a stop, especially when in reverse. The combination of starting from a stop along with the reverse gear's lower ratio means the clutches and bands need to be held tighter to prevent slippage. In Figure 25-18, you can see that there is another valve called the **reverse boost valve**. When the transmission is placed in reverse, the manual

Applied Science

AS-94: Dynamics: The technician can explain the forces and motions involved in a hydraulic system.

Hydraulic systems operate on a very simple principle: In a simple hydraulic system consisting of two pistons connected by a hydraulic line, force that is applied to one point in the system is transferred to all other points by means of a noncompressible fluid. The amount of force applied and the amount of piston movement are governed by the relative sizes of the pistons.

valve allows hydraulic pressure to flow to port E. This pressure is added to the spring pressure acting against the pressure regulator valve, which increases the line pressure. Often line pressure at WOT in the reverse gear is near the pump's maximum pressure. It is not uncommon to see line pressures approach 300 psi (2068 kPa) in reverse at WOT. If the transmission is placed in the L1 gear position again, the manual valve will allow line pressure to flow to port E. This additional pressure is added to the pressure-regulating spring, as was done in reverse, thus increasing line pressure to approximately the same pressure that we had in reverse.

Governor

K25003

The **governor** is a hydromechanical valve that produces a variable pressure based on vehicle speed, otherwise known as ground speed. This variable pressure is called **governor pressure**. Governors can be installed in the transmission in one of two places: mounted in the transmission case or mounted on the output shaft of the transmission (**FIGURE 25-4**). Both types are driven by the output shaft so that the governor pressure they create is proportional to output shaft speed, which is directly related to vehicle speed.

The governor is supplied with line pressure and sends governor pressure out. It uses centrifugal force to vary the governor pressure in proportion to vehicle speed. Visualize what happens when you place some water in a bucket and quickly swing the bucket in a circle, from your side to up over your head and then back down, repeatedly. The water stays inside the bucket even when the bucket is upside down, as long as the bucket continues moving fast enough in a circle. The force acting on the water is centrifugal force. If you swing the bucket more slowly, the water will fall out of the bucket. If you increase the speed at which you swing the bucket, the amount of outward force on the bucket will increase (**FIGURE 25-5**). The governor uses the same principle to convert line pressure to governor pressure, as we will discover.

The transmission case–mounted governor is free to spin inside a precision bore that is lubricated with transmission fluid. At the inboard end of the governor is a gear that engages with another gear that is pressed onto the output shaft. These gears are often made of plastic and can have issues with wear on high-mileage vehicles. They are replaceable on most transmissions. On the opposite end of the governor, the outboard end, there is a pair of weights that are free to pivot (**FIGURE 25-6**). At low speed, these weights are tucked in next to the governor shaft by either spring pressure or line pressure. As the vehicle speed increases, centrifugal force causes these weights to move outward on their pivots, against the spring or hydraulic pressure, in proportion to the vehicle speed. As these weights move outward, they move the governor valve within its housing, causing governor pressure to increase.

The governor valve inside the governor shaft moves back and forth in its bore in response to the movement of the weights on the outboard side of the governor. Line pressure is supplied to the valley of the valve from port C in (**FIGURE 25-7**). When the vehicle is stopped, line pressure seats the governor valve in the bore all the way to the left. With the valve seated, governor pressure is exhausted out port A, where the fluid will be returned to the transmission pan. As vehicle speed increases, centrifugal force causes the weights to fly out on their pivots, moving the governor valve to the right, against line pressure. This reduces

FIGURE 25-4 A case-mounted governor on a General Motors 4L60 transmission. The cover has been removed to show the governor.

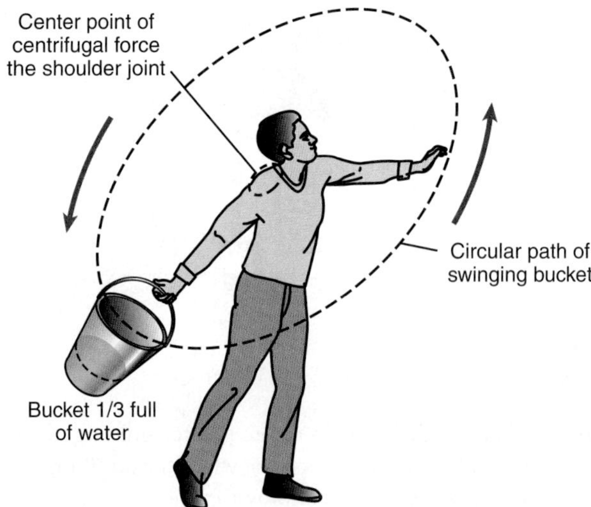

FIGURE 25-5 Centrifugal force keeps the water in the bucket as a person swings the bucket upside down.

FIGURE 25-6 A governor that has been removed from a transmission.

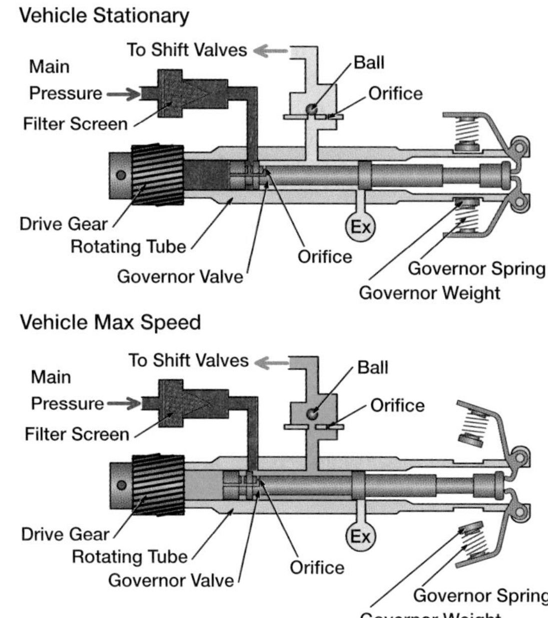

FIGURE 25-7 A hydraulic diagram of a governor valve.

the amount of fluid being exhausted, which increases governor pressure. The farther the valve moves to the right, the greater the governor pressure, and the less pressure that is exhausted back to the pan.

Springs behind the weights help to calibrate the governor to the particular engine that is installed in the vehicle. Increasing the strength of the springs causes an increase in governor pressure, which causes the vehicle to shift earlier.

When rebuilding a transmission, the governor assembly needs to be cleaned thoroughly. If the valve has dirt and old clutch material stuck in it, it may become jammed, preventing the transmission from shifting properly. Some governors have a screen or filter installed in line to help prevent debris from becoming trapped in the valve. This screen needs to be cleaned or replaced during transmission overhaul.

▶ TECHNICAN TIP

When rebuilding a transmission for a high-performance vehicle, an experienced transmission rebuilder often modifies or replaces the governor with a different governor. If manual shifting is desired, the governor may be removed entirely. Placing weaker springs in the governor causes a delay in the transmission shift. Also, drilling small holes in the governor weights reduces their weight, and in turn the centrifugal force, thereby delaying the shift. Any modifications should be made carefully and slowly; it is very easy to remove too much material. Removing too much material can cause the transmission's shift point to be beyond the engine's red line.

The second type of governor is mounted directly on the output shaft. This type of governor can have one or two valves that are moved by centrifugal force. Line pressure is applied to the governor valve through port B (**FIGURE 25-8**). Line pressure moves the valve fully to the left in its bore. When the valve is pushed to the left, line pressure is exhausted out port A and returned to the transmission pan. As vehicle speed increases, centrifugal force causes the weights on the right side of the governor to move to the right. This moves the governor valve to the right, restricting fluid out of the exhaust port. Governor pressure then builds in port C. The farther the valve moves to the right, the higher the governor pressure.

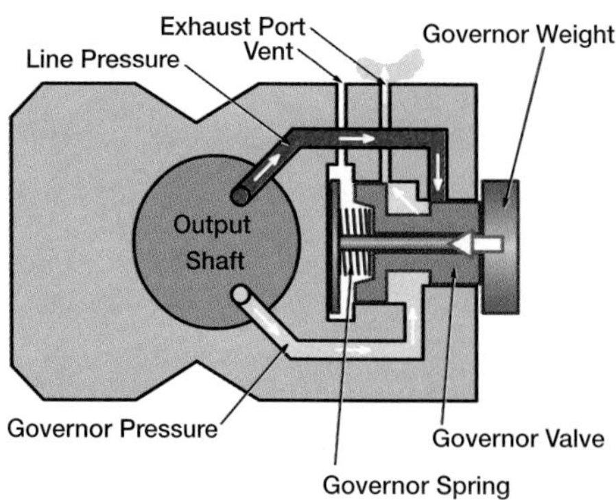

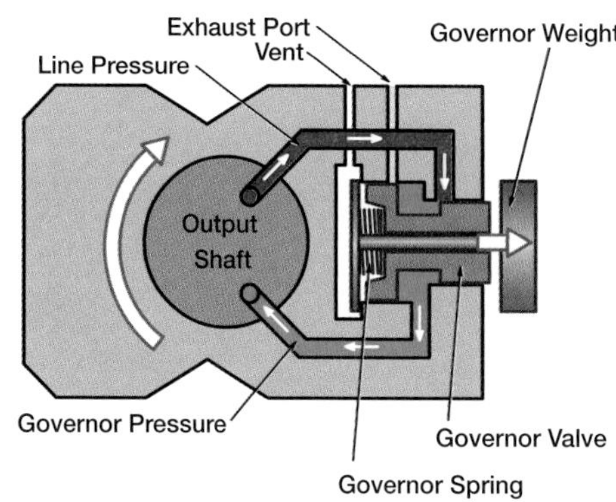

FIGURE 25-8 A hydraulic diagram of an output shaft–mounted governor.

Typically, governors have trouble accurately measuring vehicle speed when the vehicle is moving slowly. To reduce this problem, manufacturers can install larger springs in the governor, but this affects the high-speed operation of the governor. Some manufacturers fix this problem by using a **staged governor**. A staged governor is one in which the assembly uses two valves: a primary valve and a secondary valve (**FIGURE 25-9**). The primary valve begins moving as soon as the vehicle begins moving, creating a small amount of governor pressure. The secondary valve begins moving at a higher speed, creating additional governor pressure. This two-stage governor enables more precise control of governor pressure and therefore the transmission shift as well.

As we consider the operation of each valve, it is interesting to note that if a transmission has only a governor valve and no throttle valve or vacuum modulator valve, the transmission would shift at the same speeds regardless of the engine load or throttle position. In other words, the transmission will shift at the same speed if the throttle is held wide open or just barely open. Once governor pressure exceeds spring pressure, the shift occurs (**FIGURE 25-10**).

FIGURE 25-9 A cutaway of a staged governor that contains a primary valve piston and a secondary valve piston. Note the size difference of the two valves.

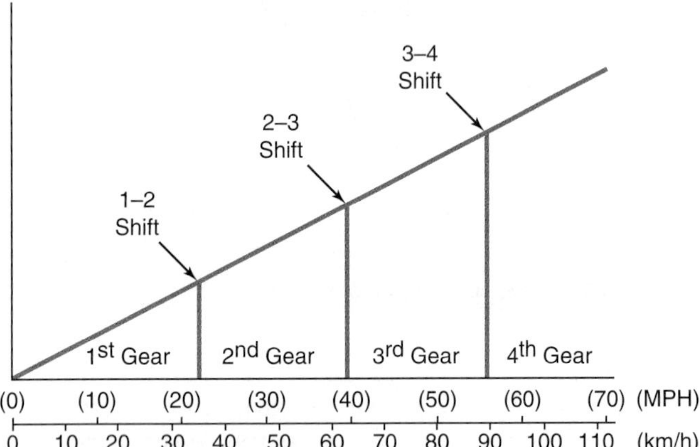

FIGURE 25-10 A shift chart with a governor only.

FIGURE 25-11 A vacuum modulator installed in a General Motors TH400 transmission.

Vacuum Modulator

K25004

An automatic transmission needs to vary its shift points and line pressure based on engine load. When a vehicle is under heavy load, such as when towing a trailer or going up a hill, the transmission delays the shift to allow the engine to operate at higher rpm, creating more power. At the same time, the clutches and bands have to be applied with more force to prevent slipping. A manifold vacuum can be used to determine the amount of engine load. When an engine is operating under a light load, the engine manifold vacuum is high (manifold pressure is low). When the engine is operating under a heavy load, the engine manifold vacuum is low or even zero (manifold pressure is high, near 14.7 psi, or 101 kPa). On many hydraulically controlled transmissions, this vacuum is sent to a **vacuum modulator** (**FIGURE 25-11**). The vacuum modulator converts this engine vacuum signal into a hydraulic pressure signal called **modulator pressure**. When engine vacuum is high, the modulator produces a low modulator pressure, and when engine vacuum is low, the modulator produces a high modulator pressure.

FIGURE 25-12 shows a vacuum modulator assembly. You can see that the modulator has a rubber diaphragm that is pushed to the bottom of the housing by a large spring when there is no engine vacuum. With no line pressure, this also moves the modulator valve to the right. As the engine is started, engine vacuum is applied to the vacuum port on the modulator, and line pressure pushes the valve to the left, opening the exhaust port (port C) and lowering modulator pressure. During wide-open throttle, when there is no vacuum applied to the modulator, the spring pushes harder against the modulator valve and line pressure. This closes the exhaust port (port C) and raises governor pressure. All pressure is sent out port B as modulator pressure. The modulator pressure is proportional to engine load, or inversely proportional to engine vacuum.

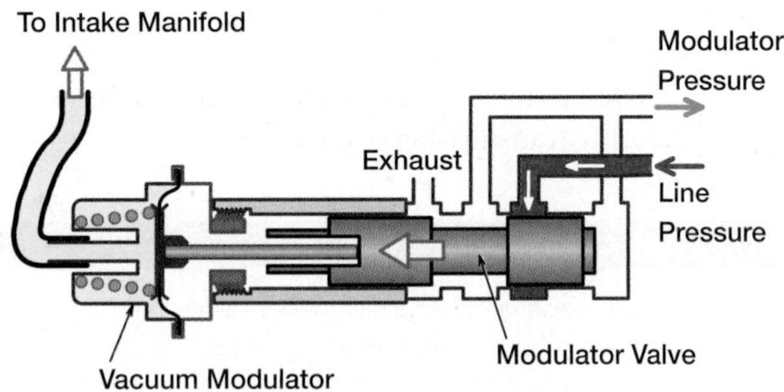

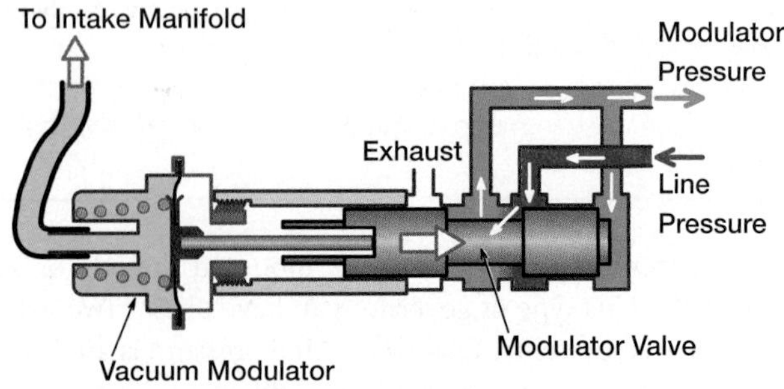

FIGURE 25-12 A cross-sectional view of a vacuum modulator.

Vacuum modulators should be replaced when rebuilding a transmission. They can often be replaced without pulling the transmission from the vehicle and can be a simple, low-cost transmission repair. The rubber diaphragm can become ripped or torn in service. If the diaphragm becomes ripped, the customer may report that the vehicle shift is extremely delayed and harsh, often shifting only near the engine red line regardless of the driving condition or throttle position. The customer may also report white smoke coming from the vehicle's exhaust pipe. This white smoke is from automatic transmission fluid being sucked up from the modulator and burned inside the engine.

The vacuum line leading from the engine to the vacuum modulator on the transmission is also susceptible to damage. The steel section of the line can become rusted, causing a vacuum leak and a delayed shift, and the rubber connectors on the ends of the steel line can deteriorate over time.

Replacement vacuum modulators are often adjustable to fine-tune the shift timing (**FIGURE 25-13**). If you remove the vacuum hose from the modulator, you will see a small screw inside, if the modulator is adjustable. Turning the screw puts more or less spring pressure on the diaphragm and, depending on the direction the screw is turned, advances or delays the shift. When adjusting the modulator, adjustments should be made in small increments to prevent shifting problems.

Throttle Valve

K25005

Some hydraulically controlled transmissions use a **throttle valve** in place of a vacuum modulator or in addition to a vacuum modulator. Throttle valves, along with many of the other valves, are located inside the transmission valve body. The throttle valve is connected to the engine throttle linkage by a cable or, in older vehicles, a system of mechanical linkages (**FIGURE 25-14**). As the throttle opening increases, **throttle valve pressure** increases. Increased throttle pressure delays the upshift, whereas decreased throttle pressure allows upshifts at lower vehicle speeds (**FIGURE 25-15**).

When the throttle is closed, throttle valve pressure is reduced in the following way. As shown in **FIGURE 25-16**, line

FIGURE 25-13 A replacement adjustable vacuum modulator.

FIGURE 25-14 An engine throttle body with cables for the accelerator pedal, cruise control, and transmission throttle valve.

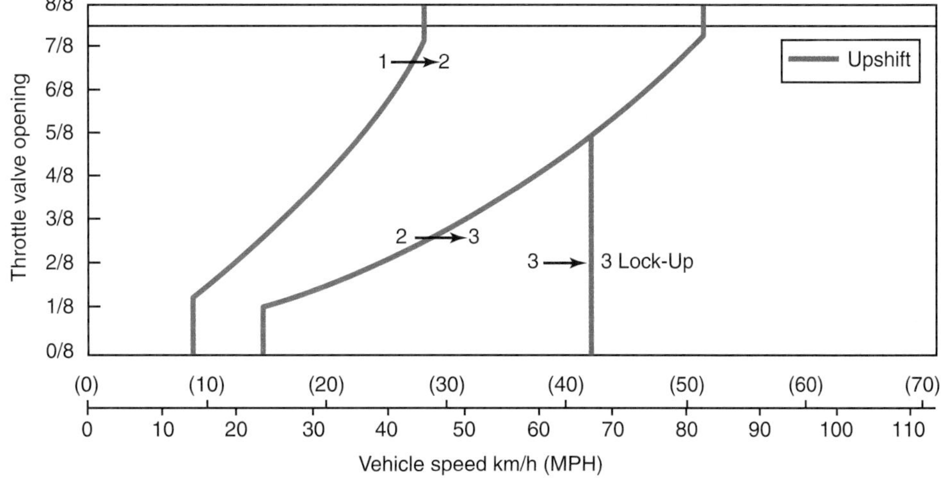

FIGURE 25-15 A shift chart with a governor and throttle valve or modulator.

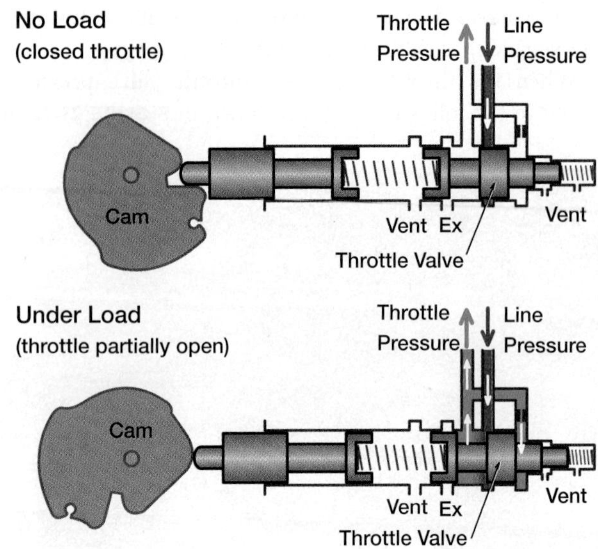

FIGURE 25-16 A hydraulic diagram of a throttle valve.

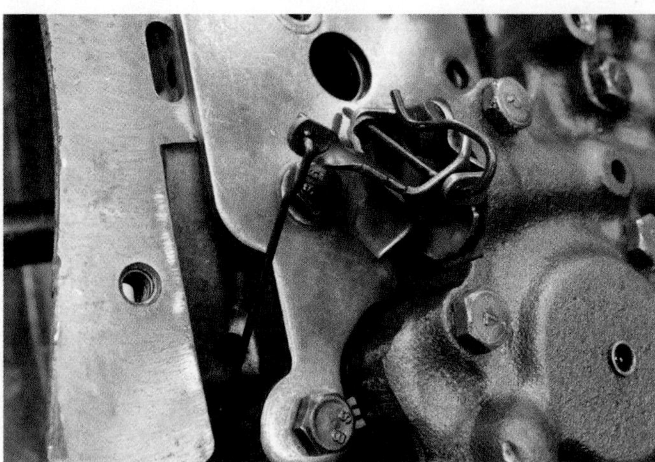

FIGURE 25-17 A kickdown valve connection inside the transmission.

pressure is supplied to the inlet port. When the throttle is closed, the spring causes the throttle valve to remain pushed all the way to the right in its bore. This allows all of the fluid pressure to be exhausted out port B and returned to the transmission pan. As the throttle is opened, the throttle valve moves to the left, blocking part of the exhaust port and opening part of the throttle valve pressure port, port C.

The pressure exiting the throttle valve pressure port is proportional to the amount of valve movement and throttle position. Throttle valve cables and linkages are a critical adjustment on a hydraulically controlled transmission. A common generic adjustment specification is that the cable should have no slack when the throttle is fully opened. If the cable is adjusted with too much slack, the throttle valve pressure is too low, resulting in early shifts and poor downshifts. If the cable is adjusted so that there is no slack before the throttle reaches wide-open throttle (WOT), the throttle valve in the transmission could become damaged. The vehicle will also not have the capability to operate at WOT because the throttle will not be able to be opened fully. Follow the manufacturer's procedure for adjusting the throttle valve cable or linkage.

Kickdown Valve

K25006

With most transmissions, if you apply the accelerator pedal all the way to the floor, the transmission will downshift to accelerate faster. On older, hydraulically controlled transmissions, this downshift was often accomplished through a **kickdown valve**. In some transmissions, the kickdown valve is called a **detent valve**. The kickdown valve can be operated by a cable (**FIGURE 25-17**), linkage, or electric solenoid that moves a valve inside the transmission when the accelerator is pressed all the way down (i.e., WOT). This kickdown valve opens a passage, allowing full line pressure to pass the kickdown valve and increase throttle pressure. The transmission will then downshift if the vehicle speed isn't too high for the lower gear.

On vehicles with a kickdown solenoid, a switch is either installed under the accelerator pedal or attached to the throttle linkage on the engine. When the throttle is opened all the way, the electric switch closes, allowing current to flow to the kickdown electric solenoid inside the transmission. The solenoid then opens a valve to allow line pressure to increase throttle pressure, forcing a downshift if appropriate.

Shift Valves

K25007

A shift valve typically refers to a spool valve with multiple lands on it. Remember that the lands are the larger part of a spool valve that fits with minimal clearance into the bore to prevent fluid from flowing around the valve. The spool valve is located inside the transmission valve body. The shift valve will either block or allow the passage of line pressure to a particular clutch or band in order to apply the gear. A typical shift valve is shown in **FIGURE 25-18**. In the figure, notice the calibrated spring on the left side of the valve. The spring keeps the shift valve against the right side of the bore until the valve needs to move. Line pressure enters the spool valve at port A and is free to flow out of port B to the clutch or band. Governor pressure is applied to port C of the spool valve. The governor pressure pushes against the calibrated spring at the left side of the valve. Modulator pressure or throttle valve pressure is applied to port D of the valve. When the valve moves far enough to the left, port B is covered, causing the clutch or band to be released. Port E becomes exposed and line pressure is allowed to flow to the clutch or band for the next gear range.

Secondary Regulator Valve

K25008

Many hydraulically controlled transmissions regulate the pressure to the torque converter, cooler, and lubrication circuit so that it is lower than line pressure; therefore, they use a secondary

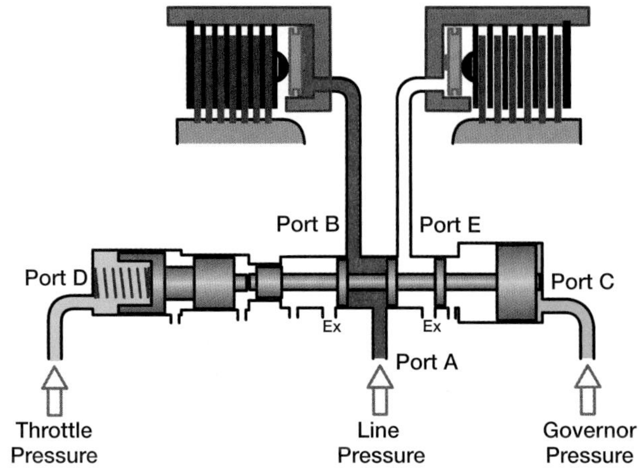

FIGURE 25-18 A hydraulic diagram of a simple shift valve.

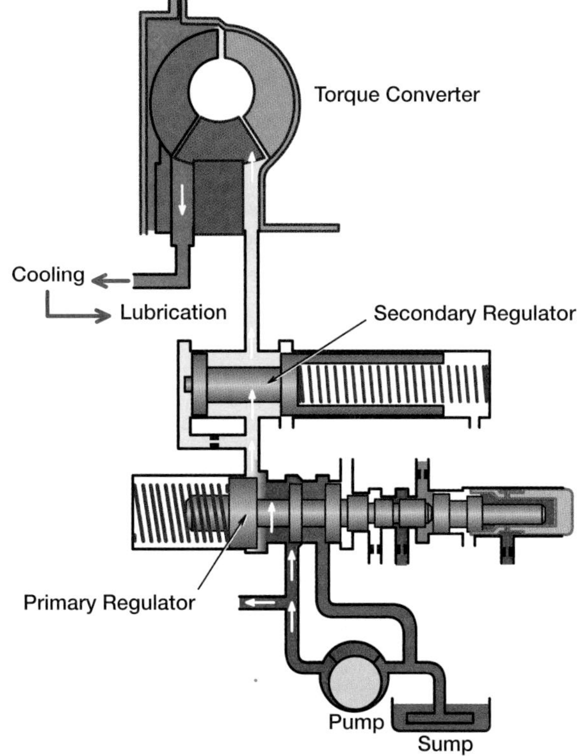

FIGURE 25-19 A secondary pressure regulator valve showing a hydraulic circuit.

regulator valve. Typically, the regulator valve uses a simple spring-loaded valve that dumps fluid above the intended pressure back to the oil pan (**FIGURE 25-19**). This pressure is used to fill the torque converter, thereby circulating fluid through and out of the converter. Transmission fluid leaving the torque converter can be extremely hot, so it goes next to the transmission cooler. After leaving the cooler, it is ready to enter the transmission and lubricate the planetary gears and bushings. In some designs, the throttle pressure or modulator pressure acts on the secondary regulator valve to increase the converter pressure so that the converter clutch can be applied more firmly under moderate load conditions.

▶ Operation of the Hydraulically Controlled Transmission

K25009

Park

K25010

When the driver starts the vehicle, the front pump draws fluid up from the transmission pan, through the filter, and into the pump. The pump then sends fluid out under pressure to the pressure regulator valve, the throttle valve or vacuum modulator valve, the governor, and the manual valve. This release of fluid creates line pressure. The pressure regulator valve keeps the pressure at a regulated level by exhausting excess pressure back to the transmission pan. With the manual valve in the park position, line pressure is not able to flow past the valve to any other components of the transmission (**FIGURE 25-20**). Line pressure that is sent to the governor is allowed to be exhausted back to the transmission pan when the vehicle is not moving and the governor is not creating any pressure. Line pressure sent to the throttle valve, or vacuum modulator, is mostly exhausted to the transmission pan when the throttle is closed and the vehicle is under no load.

Drive: First Gear

K25011

When the driver shifts the vehicle into drive, the manual valve moves (**FIGURE 25-21**), allowing line pressure to flow to port A of the 1–2 shift valve. The 1–2 shift valve is held to the right (downshift) side of its bore by the spring. This allows the line pressure from port A to flow freely from port B. Port B is connected to the clutch or band that applies first gear. As the vehicle accelerates, the governor sends fluid pressure to port D. This fluid pressure compresses the shift spring and moves the shift valve to the left. The vacuum modulator or throttle valve applies fluid pressure to port C. This pressure adds force to the spring in order to resist the pressure from the governor. Remember that the vacuum modulator and the throttle valve create fluid pressure that varies with engine load or throttle opening. The higher the engine load, or the more the throttle is opened, the greater the pressure that develops. Governor pressure increases linearly as vehicle speed increases. When the governor pressure exceeds the spring pressure plus the vacuum modulator or throttle valve pressure, the shift valve moves to the left (upshift position), resulting in fluid flow being cut off from port B. When the fluid pressure stops flowing from port B, the clutch or band for first gear is disengaged.

Drive: Second Gear

K25012

As the driver proceeds, fluid pressure is still flowing into port A. The governor pressure has exceeded the spring pressure and vacuum modulator, or throttle valve, pressure, and the shift valve moves to the left (upshift position), compressing the shift

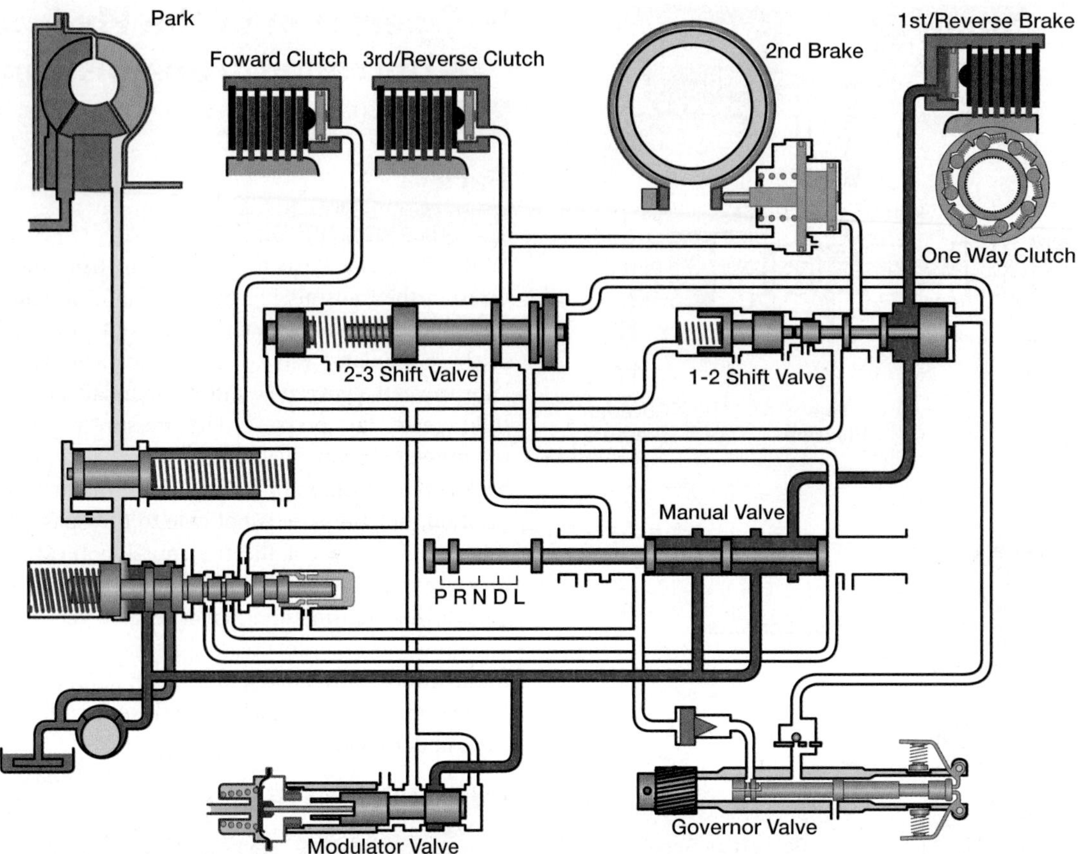

FIGURE 25-20 A simplified hydraulic diagram of a transmission in park.,

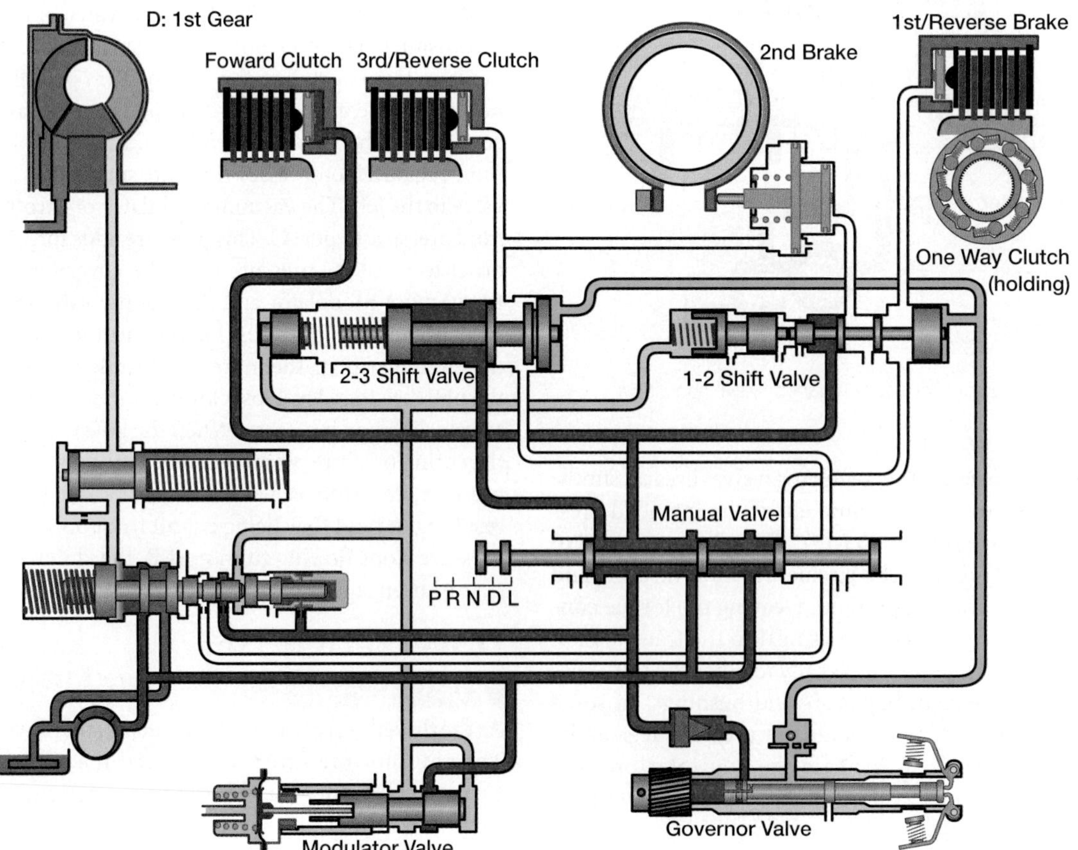

FIGURE 25-21 A simplified hydraulic diagram of a transmission in first gear.

spring. Port B becomes blocked, cutting off fluid flow to the clutch or band that is used for first gear. As the valve moves, port E becomes exposed, and now line pressure is allowed to flow from port A to port E. Port E sends line pressure to the 2–3 shift valve, as shown in (**FIGURE 25-22**). The 2–3 shift valve allows line pressure to flow from port A of the 2–3 shift valve to port B of the 2–3 shift valve. If the vehicle speed decreases, governor pressure is reduced, resulting in the shift valve moving back to the right, blocking port E of the 1–2 shift valve and re-exposing port B of the 1–2 shift valve. This causes the vehicle to downshift back into first gear. The same downshift would happen if the vehicle were cruising down the road and the driver applied the accelerator pedal far enough to overcome the governor pressure, resulting in a downshift. The shifting of the hydraulically controlled transmission is caused by the constant push-pull between the pressure of the governor and pressure of the shift spring and vacuum modulator or throttle valve pressure.

Drive: Third Gear

K25013

As the driver continues to accelerate, governor pressure continues to increase. This increased fluid pressure moves our 2–3 shift valve farther to the left, blocking port B of the 2–3 shift valve. This prevents line pressure from going to the clutch or band for second gear as shown in **FIGURE 25-23**. The spring at the left of the 2–3 shift valve is stiffer than the spring found in the 1–2 shift valve. Line pressure has been supplied to port A of the 2–3 shift valve since the transmission shifted into second gear. Governor pressure is applied to port D of the 2–3 shift valve. (Note that some transmissions use a combined 1–2, 2–3 shift valve.) The vacuum modulator or throttle valve applies pressure to port C. When the governor pressure becomes great enough to overcome the spring pressure and the modulator/throttle valve pressure in the 2–3 shift valve, line pressure is allowed to flow to port E. Port E sends line pressure to the clutch for third gear. If the vehicle slows down, governor pressure is reduced, causing the valve to move back to the right and a downshift to second gear. The same will happen if the driver applies pressure to the accelerator pedal, as modulator or throttle valve pressure adds additional force to the shift spring. This will push the shift valve back to the right and return the vehicle to second gear. This process, as mentioned previously, is called kickdown. Transmissions are designed to prevent a downshift if it exceeds the maximum rpm that an engine is designed for. If the vehicle speed is not too high in third gear, and the driver steps on the accelerator hard enough, the transmission can downshift from third to second to first in one shift.

Reverse

K25014

In reverse, line pressure is allowed to flow from the manual valve directly to the clutches and/or bands that are used for reverse (**FIGURE 25-24**). Governor pressure and modulator

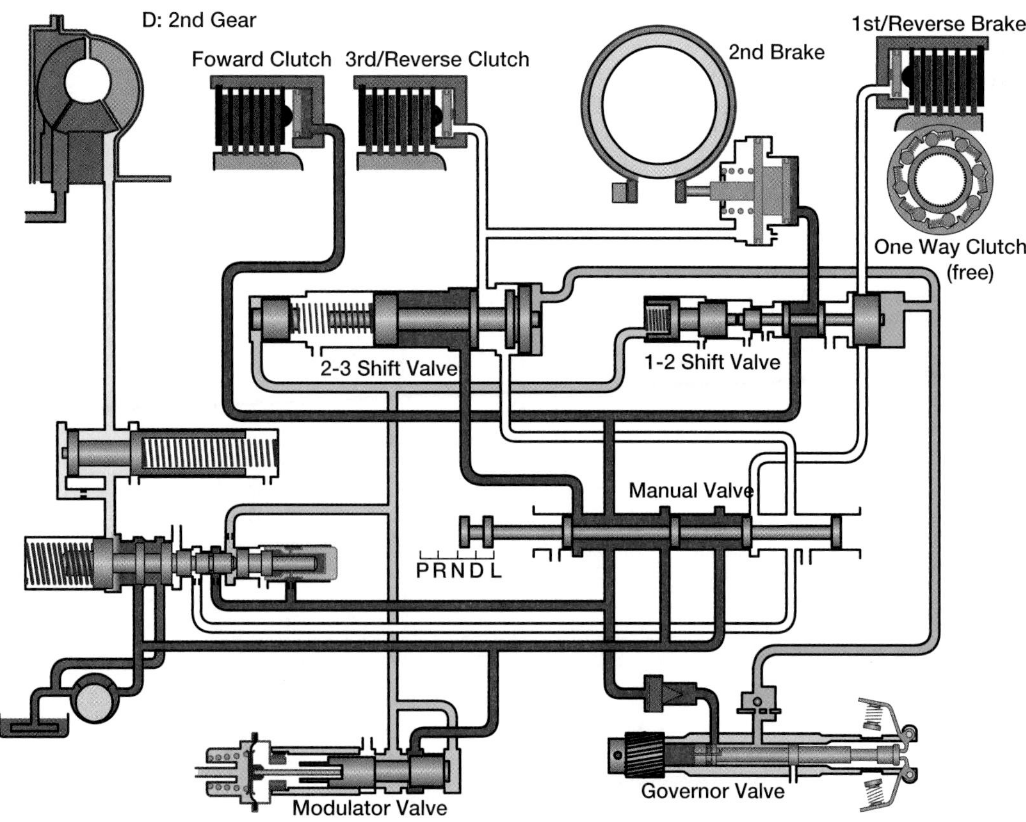

FIGURE 25-22 A simplified hydraulic diagram of a transmission in second gear.

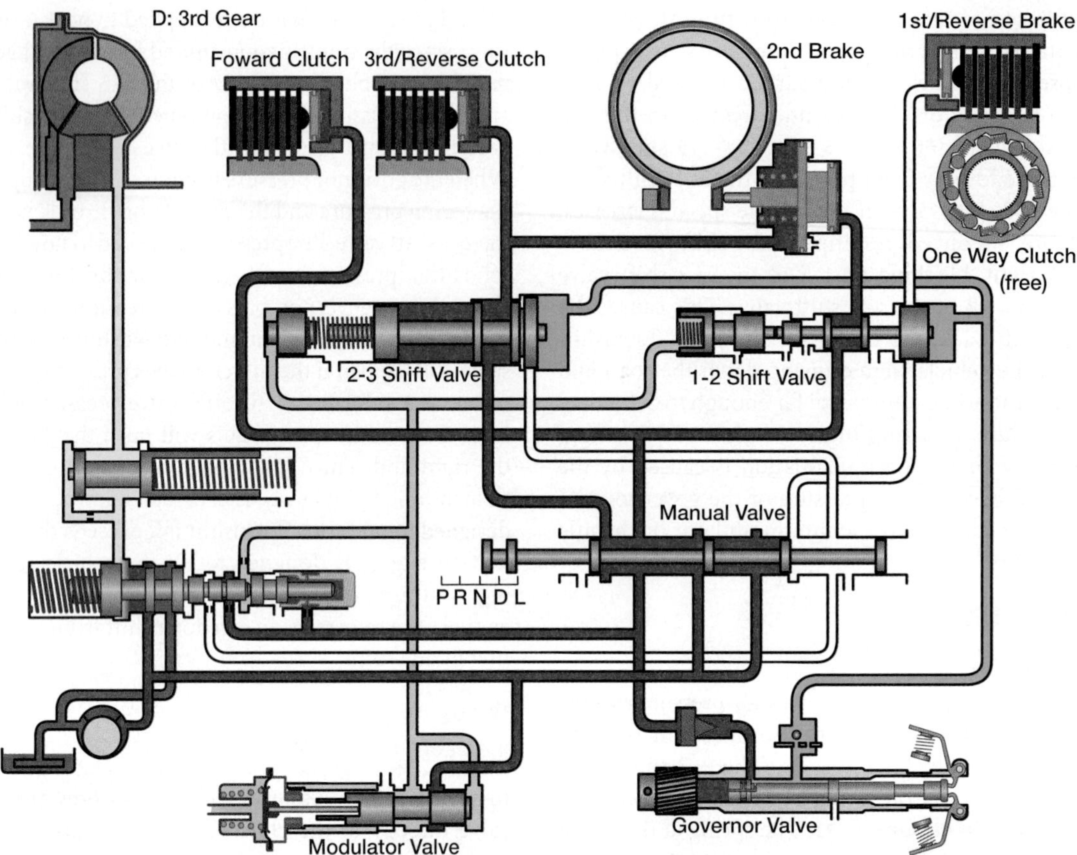

FIGURE 25-23 A simplified hydraulic diagram for a transmission in third gear.

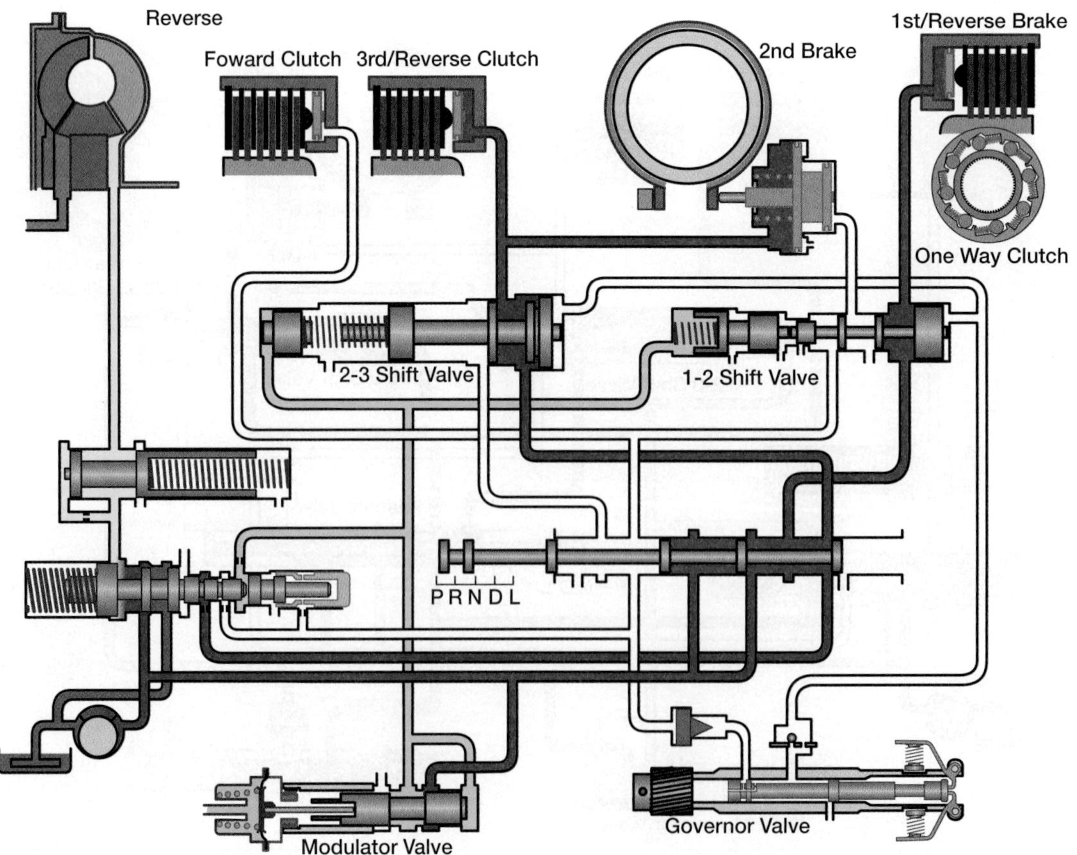

FIGURE 25-24 A simplified hydraulic diagram for a transmission in reverse.

pressure/throttle valve pressure are not used in reverse because there is only one speed available. These pressures are instead exhausted directly to the transmission pan while in reverse. Reverse does often require greater line pressure, so the manual valve will direct additional fluid under pressure to port D of the pressure regulator valve. This results in additional force added to the spring pressure in the regulating valve, which causes a much higher line pressure for reverse.

▶ Wrap-Up

Ready for Review

- ▶ Hydraulically controlled transmissions are no longer produced in the United States.
- ▶ Governors are hydromechanical valves that produce governor pressure proportional to vehicle speed.
- ▶ A governor can be located on the output shaft or in the transmission case.
- ▶ A staged governor uses a primary valve at lower speeds and a secondary valve at higher speeds.
- ▶ Vacuum modulators convert manifold vacuum into modulator pressure.
- ▶ Modulator pressure and engine vacuum have an inverse relationship.
- ▶ A throttle valve may be used in place of, or in addition to, a vacuum modulator.
- ▶ Kickdown valves (detent valves) respond when the accelerator is completely depressed to downshift the transmission and increase acceleration.
- ▶ Line pressure is the hydraulic pressure used to apply clutches and bands.
- ▶ Line pressure is controlled by the pressure regulator.
- ▶ More line pressure is needed as engine load is increased, and in reverse.
- ▶ A shift valve is a spool valve with multiple lands that is designed to block or allow line pressure to pass to a band or clutch in order to apply the gear.
- ▶ When the manual valve is in the park position, line pressure cannot flow to any of the bands or clutches.
- ▶ Shifting the vehicle into first gear moves the manual valve and allows line pressure to flow.
- ▶ As vehicle speed increases, governor pressure increases linearly.
- ▶ Governor pressure is reduced as vehicle speed decreases.
- ▶ Shifting of a hydraulically controlled transmission is due to the push–push between the governor pressure on one side and the shift spring and throttle pressure on the other.
- ▶ Transmissions can prevent a downshift that would exceed the engine's maximum rpm.
- ▶ Reverse only uses one speed; therefore, governor pressure and modulator/throttle valve pressure are not required.

Key Terms

detent valve See *kickdown valve.*

governor A mechanical device that creates a pressure using centrifugal force. The pressure is proportional to vehicle speed.

governor pressure The pressure created by the governor, which is used to make the shift valves upshift and is proportional to vehicle speed.

kickdown valve A type of spool valve that is connected to the throttle on the vehicle. The valve is used to force a downshift when the throttle is opened all the way, assuming the governor pressure is below a specified point. Kickdown valves are also called detent valves.

line pressure A hydraulic pressure that is used to apply bands and clutches, and is regulated by the pressure regulator valve.

modulator pressure A pressure created by the vacuum modulator that is proportional to engine load. Modulator pressure is used to delay transmission upshifting based upon engine load. It may also be used to raise line pressure to more firmly apply bands and clutches under higher engine loads.

reverse boost valve A component of the pressure regulator valve that increases line pressure when the vehicle is in reverse.

staged governor A governor in which the assembly uses two valves: a primary valve and a secondary valve.

throttle valve A type of spool valve that is connected to the throttle on a vehicle. The throttle valve creates a pressure proportional to throttle opening and is used to delay upshifting based on throttle opening.

throttle valve pressure The pressure created by the throttle valve that is proportional to throttle opening.

vacuum modulator A device on a hydraulically controlled transmission that converts engine manifold vacuum into an engine load signal called modulator pressure.

Review Questions

1. Harsh-shifting transmission could be caused:
   - **a.** when the line pressure is too high during normal cruising speeds.
   - **b.** when the line pressure is too low during low cruising speeds.
   - **c.** when the line pressure is too high during high cruising speeds.
   - **d.** if the line pressure is too low during normal cruising speeds.
2. The device that varies incoming line pressure using centrifugal force is known as:
   - **a.** vacuum modulator.
   - **b.** reverse boost valve.
   - **c.** governor.
   - **d.** throttle valve.

3. The vacuum modulator converts the engine vacuum signal into:
   a. engine pressure.
   b. line pressure.
   c. governor pressure.
   d. modulator pressure.
4. All of the following statements are true *except*:
   a. Throttle valves are located inside the transmission valve body.
   b. As the throttle opening increases, throttle valve pressure decreases.
   c. Increased throttle pressure delays the upshift.
   d. Decreased throttle pressure allows upshifts.
5. Which of the following has multiple lands designed to block or allow line pressure to pass to a band or clutch in order to apply the gear?
   a. Kickdown valve
   b. Throttle valve
   c. Modulator valve
   d. Shift valve
6. When the manual valve is in park position, the line pressure:
   a. is able to flow past the valve to all other components of the transmission.
   b. is sent to apply the parking pawl.
   c. is not able to flow past the valve to any other components of the transmission.
   d. increases to full pressure so it will be ready to apply a band or clutch.
7. If an automatic transmission only had a governor, and no throttle valve or vacuum modulator, what is the shift most lilely to be like?
   a. No difference since the throttle valve and vacuum modulator valve don't affect shift timing.
   b. The transmission would shift later with greater engine load.
   c. The transmission would shift earlier with greater engine load.
   d. It would shift at the same point no matter what the engine load is.
8. In second gear with low governor pressure, the line pressure is allowed to flow from the 2-3 shift valve to:
   a. the 3rd gear clutch or band.
   b. the 2nd gear clutch or band.
   c. the 1st gear clutch or band.
   d. the reverse gear clutch or band.
9. When the governor pressure becomes great enough to overcome the spring pressure and the modulator/throttle valve pressure in the 2–3 shift valve, line pressure is allowed to flow to:
   a. the 3rd gear clutch or band.
   b. the 2nd gear clutch or band.
   c. the 1st gear clutch or band.
   d. the reverse gear clutch or band.
10. Which gear needs higher line pressure, and the modulator/throtle valve pressure are not used?
   a. first gear.
   b. second gear.
   c. third gear.
   d. reverse.

ASE Technician A/Technician B Style Questions

1. Tech A says that a vacuum modulator converts manifold vacuum into an engine load signal. Tech B says that the manual valve is moved by governor pressure to force a shift. Who is correct?
   a. Tech A
   b. Tech B
   c. Both A and B
   d. Neither A nor B
2. Tech A says that governor pressure is controlled by torque converter rpm. Tech B says that governor pressure is controlled by output shaft speed. Who is correct?
   a. Tech A
   b. Tech B
   c. Both A and B
   d. Neither A nor B
3. Tech A says that reciprocating force controls governor pressure. Tech B says that centrifugal force controls governor pressure. Who is correct?
   a. Tech A
   b. Tech B
   c. Both A and B
   d. Neither A nor B
4. Tech A says that modifying governor weights will alter transmission shifting. Tech B says that excessive modification of the governor weights may cause the transmission to not shift. Who is correct?
   a. Tech A
   b. Tech B
   c. Both A and B
   d. Neither A nor B
5. Tech A says that the converter and cooler operate at line pressure. Tech B says that line pressure causes the manual valve to move during upshifts and downshifts. Who is correct?
   a. Tech A
   b. Tech B
   c. Both A and B
   d. Neither A nor B
6. Tech A says that the throttle valve is controlled by a cable or linkage hooked to the engine throttle linkage. Tech B says that a misadjusted throttle valve cable can prevent wide-open throttle (WOT) on an engine. Who is correct?
   a. Tech A
   b. Tech B
   c. Both A and B
   d. Neither A nor B

7. Tech A says that line pressure is reduced during light throttle conditions to improve fuel economy. Tech B says that as the engine speeds up in rpm, the transmission line pressure gradually reduces to maintain clutch pack pressure. Who is correct?
   a. Tech A
   b. Tech B
   c. Both A and B
   d. Neither A nor B
8. Tech A says that when governor pressure on one side of a shift valve overcomes spring pressure and throttle pressure on the other side, an upshift occurs. Tech B says when reverse boost pressure on one side of the reverse valve overcomes governor pressure on the other side, shift into reverse occurs. Who is correct?
   a. Tech A
   b. Tech B
   c. Both A and B
   d. Neither A nor B
9. Tech A says that line pressure in an automatic transmission is controlled by engine rpm. Tech B says that modulator pressure changes with engine vacuum. Who is correct?
   a. Tech A
   b. Tech B
   c. Both A and B
   d. Neither A nor B
10. Tech A says that in reverse, governor pressure regulates line pressure. Tech B says that in reverse, modulator pressure regulates line pressure. Who is correct?
   a. Tech A
   b. Tech B
   c. Both A and B
   d. Neither A nor B

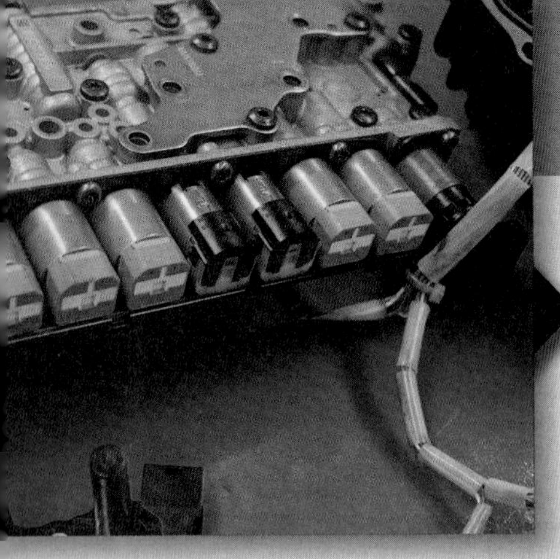

CHAPTER 26 Electronically Controlled Transmission

NATEF Tasks

There are no NATEF tasks for this chapter.

Knowledge Objectives

After reading this chapter, you will be able to:

- **K26001** Describe the purpose and function of electronically controlled transmission components.
- **K26002** Describe the purpose and function of transmission sensors.
- **K26003** Describe the purpose and function of the vehicle speed sensor.
- **K26004** Describe the purpose and function of the input shaft speed sensor.
- **K26005** Describe the purpose and function of the transmission oil temperature sensor.
- **K26006** Describe the purpose and function of transmission pressure switches.
- **K26007** Describe the purpose and function of the line pressure sensor.
- **K26008** Describe the purpose and function of the manual lever position switch.
- **K26009** Describe the purpose and function of the overdrive switch.
- **K26010** Describe the purpose and function of transmission-related engine sensors.
- **K26011** Describe the purpose and function of the throttle position sensor.
- **K26012** Describe the purpose and function of the accelerator pedal position sensor.
- **K26013** Describe the purpose and function of the engine coolant temperature sensor.
- **K26014** Describe the purpose and function of the manifold absolute pressure sensor.
- **K26015** Describe the purpose and function of the mass airflow sensor.
- **K26016** Describe the purpose and function of the crankshaft position sensor.
- **K26017** Describe the purpose and function of the brake on/off switch.
- **K26018** Describe the purpose and function of the manual upshift/downshift control system.
- **K26019** Describe the purpose and function of the electronic actuators.
- **K26020** Describe the purpose and function of the electronic pressure control solenoid.
- **K26021** Describe the purpose and function of the shift solenoids.
- **K26022** Describe the purpose and function of the torque converter clutch solenoid.
- **K26023** Describe the operation of the electronically controlled transmission system.
- **K26024** Describe the purpose and operation of the powertrain control module/transmission control module.
- **K26025** Describe the purpose and function of electronic shift programs.

Skills Objectives

There are no Skills Objectives for this chapter.

▶ Introduction

Transmissions began changing from hydraulically controlled to electronically controlled during the late 1980s and early 1990s. Currently, there are no auto manufacturers in the United States still producing hydraulically controlled transmissions for use in new vehicles. This change is partially because of the increased need for greater fuel economy and the additional gears required to achieve it. More accurate control of vehicles' emissions is another reason why manufacturers have gone to electronically controlled transmissions. The computer can better predict engine load changes when the vehicle computer is in charge of the shift timing. This keeps emissions under better control by adjusting the engine settings in preparation for the shift. On a conventional hydraulically controlled transmission, the computer can react to the increased load only *after* the transmission shifts.

Fully computer-controlled transmissions have also allowed the manufacturers to install lighter transmissions into vehicles because the computer controls can be used to reduce the amount of torque flowing through a transmission as it shifts (**FIGURE 26-1**). Note that computer-controlled means that the computer is controlling hydraulic valves in most cases, thereby directing hydraulic pressure to activate the various clutches and bands within the transmission. So while electronically controlled transmissions are controlled electronically, they still use hydraulic pressure to apply and release the bands and clutches.

▶ Electronically Controlled Transmission Components

K26001

Transmission Sensors

K26002

As discussed in the Hydraulically Controlled Transmission chapter, in a fully hydraulically controlled transmission, the shift points are controlled by hydraulic and mechanical devices such as springs and valves. In the fully electronically controlled transmission, the shift points are controlled by the vehicle's powertrain control module (PCM) or a **transmission control module (TCM)**. Most vehicle manufacturers have integrated the TCM into the construction of the PCM, so we use the abbreviation PCM for this chapter. The PCM receives various inputs from the engine and transmission to determine the shift timing and the firmness of the shift. We come back to the PCM after we learn about some of the inputs and outputs that the PCM uses (**FIGURE 26-2**).

FIGURE 26-1 A typical electronically controlled transmission removed from a vehicle.

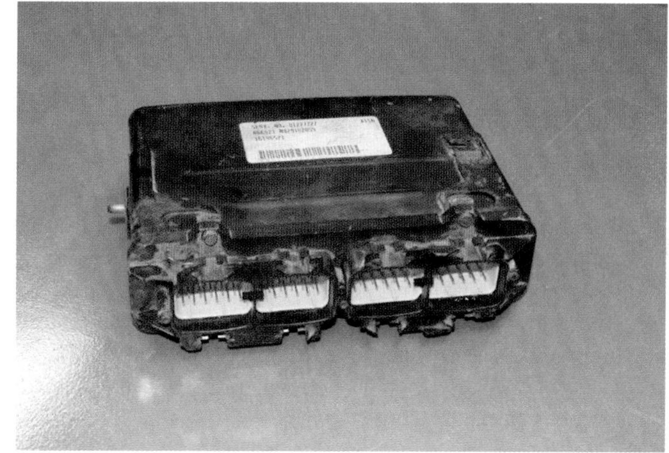

FIGURE 26-2 A typical vehicle PCM that has been removed from a vehicle.

You Are the Automotive Technician

A customer brings her vehicle into the transmission shop complaining that it has not been shifting properly for the last few days. The customer is concerned that she may need to have the transmission replaced, which can be a very expensive repair. After gathering all of the customer's information, you review the service history and technical service bulletins for this vehicle. You then tell the customer that you need to verify the fluid level and condition and then perform a test-drive to verify the problem and gather information. The fluid level is within the safe zone but is a bit dirty. During the test-drive, you notice that the speedometer is erratic, during which the transmission shifts erratically. You also verify that there is no apparent slippage in any of the gears. After the test-drive, you explain to the customer that the erratic speedometer could indicate a faulty vehicle speed sensor (VSS), which can cause the transmission to shift erratically. She agrees to allow you to continue to diagnose the issue, and says that if the VSS is causing the shifting problem, she would like it fixed, along with changing the fluid and filter based on your recommendation.

1. How could a malfunctioning speedometer be related to the transmission shifting issue?
2. Although the technician didn't feel any slippage, how is slippage monitored in some electronically controlled transmissions?
3. What is limp-in mode? And how does it operate?
4. How can shift solenoid charts help you diagnose transmission faults?

FIGURE 26-3 A VSS installed in a transmission. The case on the transmission has been disassembled to show the sensor through the case and the reluctor wheel.

Vehicle Speed Sensor (VSS)

K26003

In a hydraulically controlled transmission, vehicle speed, or ground speed, is measured by the governor. The governor produces a hydraulic pressure that moves shift valves inside the transmission. In a computerized transmission, the governor is replaced by a **vehicle speed sensor (VSS)**. This VSS is sometimes called an output shaft speed sensor (**FIGURE 26-3**).

The VSS is often also responsible for sending the vehicle speed information to the vehicle speedometer. Sometimes this information is sent to the PCM and the PCM relays the information, or they may both receive the same signal. On early electronically controlled transmissions, sometimes the manufacturer would use an electronic VSS *and* a mechanical speedometer cable because the vehicle was still using a mechanical speedometer in the instrument panel.

> **TECHNICIAN TIP**
>
> When a customer states that the automatic transmission will not shift, perform some basic diagnostic tests before replacing a transmission. A simple diagnostic test is to observe whether the speedometer is working. If the speedometer is not working, there is probably a problem with the VSS or its wiring.

Most VSSs are a type of sensor called a magnetic pickup, or magnetic reluctance, sensor. The sensor assembly consists of an iron core wrapped with fine copper wiring. There is also a reluctor wheel, which has raised "teeth" located around the circumference of the wheel. **FIGURE 26-4** shows the parts of the magnetic pickup sensor. The reluctor wheel is attached to the output shaft of the transmission and spins with the output shaft. As the teeth of the reluctor wheel get closer to the iron core, a positive voltage is produced. As the tooth moves away, the voltage becomes negative. This process repeats over and

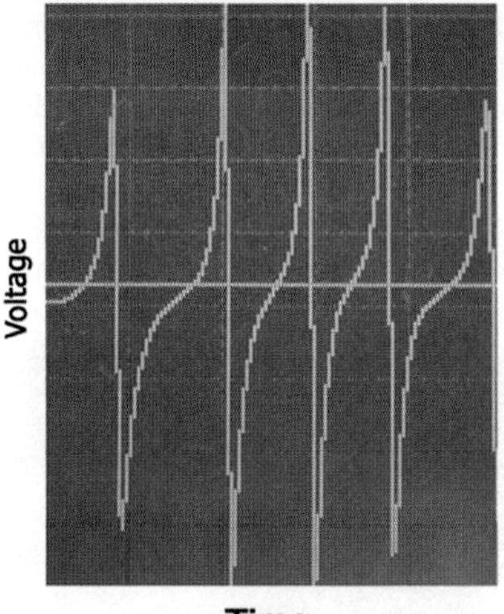

FIGURE 26-4 A. A typical VSS magnetic pickup sensor. **B.** Typical VSS output signal.

over again as the reluctor wheel spins, creating an AC voltage. This signal is sent to the PCM, where the vehicle speed can be determined from the signal frequency. The magnetic pickup sensor produces its own voltage while the reluctor wheel is spun. These sensors do not need a separate wire to supply them with a reference voltage.

> **TECHNICIAN TIP**
>
> Magnetic pickup sensors do not need to be installed in the vehicle to be tested. If you place a voltmeter on the leads of a magnetic pickup sensor and spin the output shaft of a transmission, you should get an AC voltage reading. To a certain point, the faster the shaft is spun, the higher the voltage.

Another type of VSS is called a **reed switch**. There is a thin movable blade switch inside the sensor and a magnet installed in a rotating part, called the rotor, on the output shaft (**FIGURE 26-5**). As the output shaft spins, the rotor spins. When the magnet comes near the reed switch, the switch closes, allowing current to flow through the switch. When the

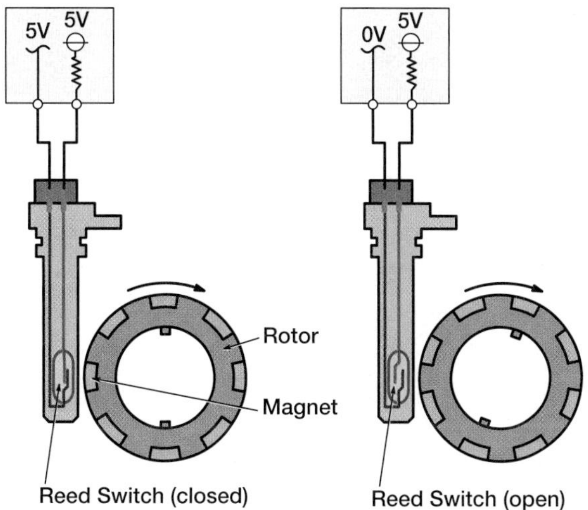

FIGURE 26-5 A reed switch VSS. As the magnet in the rotor passes the reed switch, the switch closes, allowing current to flow.

FIGURE 26-6 A typical input shaft speed sensor.

magnet moves away from the switch, the switch opens, breaking the current flow. This opening and closing of the switch creates a DC square wave pattern that the PCM can use to determine vehicle speed. Reed switches need to be supplied with electricity to operate. The amount of voltage that the sensor is supplied with varies among vehicle manufacturers. Some of these switches apply a ground to a wire coming from the PCM; others simply allow a voltage to pass from a voltage source, through the reed switch, to the PCM.

Input Shaft Speed Sensor

K26004

Many transmissions also incorporate an **input shaft speed sensor**, also called a turbine speed sensor because the input shaft is splined to the turbine of the torque converter (**FIGURE 26-6**). The input shaft speed sensor is very similar to a VSS. These sensors can also be a magnetic pickup type or a reed switch type. Both types send their signal as an input to the PCM.

Some PCMs compare the input shaft speed sensor to the output shaft speed sensor to determine gear ratios. For example, if the input shaft is spinning at 3000 revolutions per minute (rpm) and the transmission is in first gear (3:1 reduction), then the output shaft should be spinning at 1000 rpm. The PCM can also use this information to determine if a clutch or band is slipping. For example, if the output shaft speed sensor is only reading 900 rpm, the PCM knows that first gear is slipping (**FIGURE 26-7**) and will trigger a diagnostic trouble code (DTC) such as a P0731 gear ratio error in first gear.

Some manufacturers compare the input shaft speed sensor to the output shaft speed sensor to measure the amount of time required to complete the shift from one gear to another. Chrysler vehicles often use this information and list it as the clutch volume index (CVI). This CVI number can be used by the PCM to determine how worn the clutch or band is and to adjust shift timing and pressures accordingly.

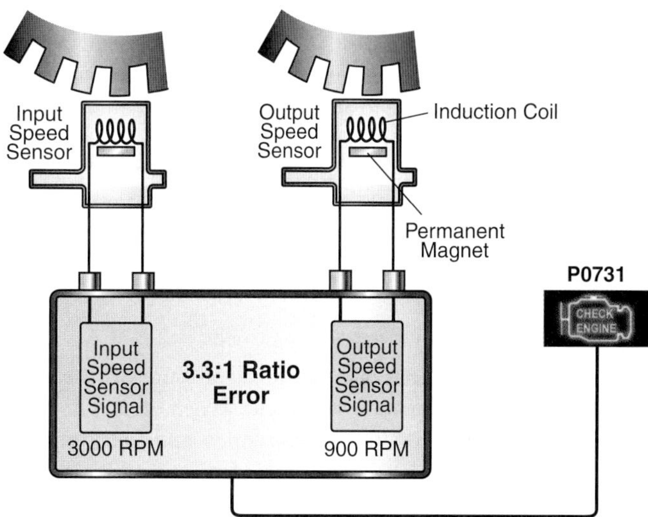

FIGURE 26-7 Input shaft speed sensor versus output shaft speed sensor.

Transmission Fluid Temperature Sensor

K26005

Another input the PCM uses when controlling the shifting of an electronically controlled transmission is the **transmission fluid temperature (TFT) sensor**. Most TFT sensors are a type of resistor called a thermistor. A thermistor changes its resistance based upon its temperature. The PCM will supply a reference voltage to the sensor and measure the amount of voltage returned to the PCM through the TFT sensor (**FIGURE 26-8**).

The PCM varies shift timing and pressures based upon the TFT sensor because cold transmission oil is thicker than hot transmission oil and does not flow as easily through small orifices and passages in the transmission. So, if the transmission oil is cooler, the PCM will allow higher engine rpm before shifting the transmission and will also allow more time for the transmission to complete the shift. If the transmission oil is hot, the PCM may increase line pressure to compensate for internal leaks, as thinner oil flows past valves and seals more easily. The

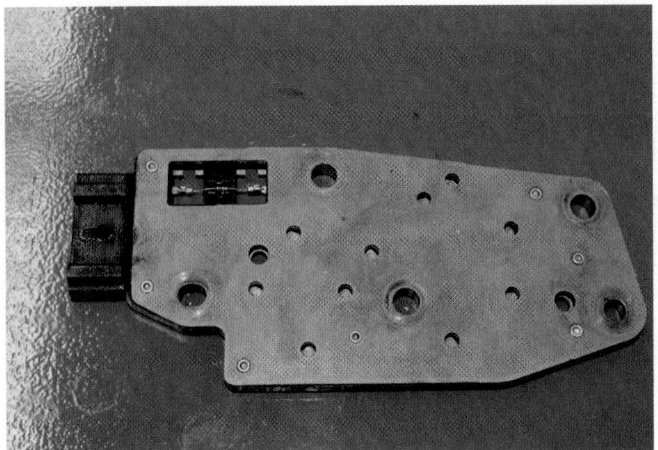

FIGURE 26-8 A General Motors TFT sensor and pressure switch assembly.

FIGURE 26-9 A typical line pressure sensor.

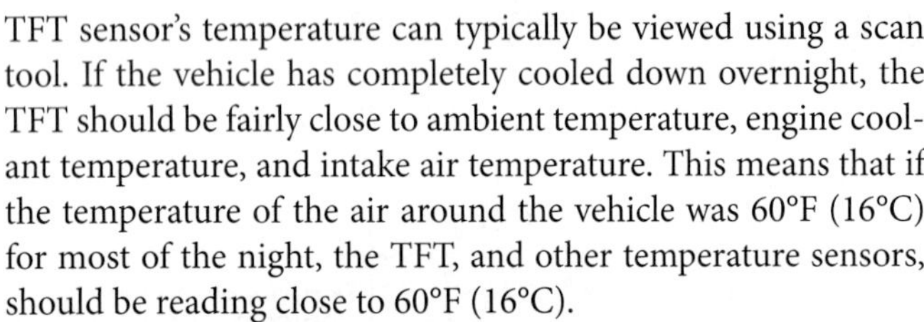

TFT sensor's temperature can typically be viewed using a scan tool. If the vehicle has completely cooled down overnight, the TFT should be fairly close to ambient temperature, engine coolant temperature, and intake air temperature. This means that if the temperature of the air around the vehicle was 60°F (16°C) for most of the night, the TFT, and other temperature sensors, should be reading close to 60°F (16°C).

Transmission Pressure Switches

K26006

Inside many transmissions, there are small pressure switches. The switches are a simple open or closed type. When there is no pressure in a hydraulic circuit, the switch is open, and when pressure is applied, it closes. The PCM uses this information to determine if a particular hydraulic circuit of the transmission has pressure, such as the circuit for the **torque converter clutch** (TCC) lock-up system. The pressure switches can also be used to determine in what gear the transmission is operating. Figure 26-12 shows a common General Motors pressure switch used to verify which gear the transmission is operating in.

Line Pressure Sensor

K26007

The PCM receives voltage signals on some transmissions from a **line pressure sensor** (**FIGURE 26-9**). The line pressure sensor is a pressure transducer, which is a type of resistor that changes its resistance based upon pressure. The sensor is supplied with reference voltage from the PCM. The sensor varies the voltage based upon the line pressure and sends the signal back to the PCM. The PCM can then interpret the signal and vary the line pressure inside the transmission in order to adapt the shift timing based upon line pressure.

Manual Lever Position

K26008

The **manual lever position** (MLP) **switch** is sometimes called a transmission range switch or a neutral safety switch. It is a simple multi-position switch that sends a signal to the PCM regarding which gear the manual lever is in. This information is used to prevent the vehicle from starting in any gear except Park or Neutral. The PCM also uses this information to turn on the reverse lights and to control the selection of gears and automatic shifting.

The MLP switch is often mounted directly on the transmission case where the shift linkage connects to the transmission (**FIGURE 26-10**). On some vehicles, this switch is adjustable and will have to be adjusted after replacing the transmission, the shift cable, or the switch.

Overdrive Switch

K26009

The overdrive switch is operated by the driver to manually select or deselect overdrive (**FIGURE 26-11**). Many manufacturers recommend not using overdrive when pulling a trailer or heavy loads because doing so can overload the transmission. Therefore, the driver can activate the overdrive switch (sometimes labeled as "Tow/Haul"), which sends a simple on/off signal to the PCM indicating that the driver wants to disable overdrive. The PCM then modifies the shifting pattern accordingly, locking out overdrive. The driver activates the overdrive switch again when overdrive is desired. The PCM then allows shifting into overdrive when appropriate. Some vehicles automatically reset the overdrive switch to allow overdrive when the vehicle is restarted.

Engine Sensors

K26010

The PCM also uses a variety of engine sensors in determining how best to control the transmission. Each of these sensors is covered in more depth in the Engine Management System chapter. But we mention them and their basic operation here as well.

Throttle Position Sensor

K26011

The **throttle position sensor** (TPS) is a critical sensor for an electronically controlled transmission (**FIGURE 26-12**). The TPS

FIGURE 26-10 A manual lever position (MLP) switch mounted on the outside of the transmission. Note the manual lever linkage passing through the middle of the switch.

FIGURE 26-11 An overdrive switch used by the driver to manually select or deselect overdrive.

FIGURE 26-12 A TPS mounted on the side of a throttle body assembly.

is a type of variable resistor called a **potentiometer**, which varies its resistance based on its position. A volume knob on a stereo is another example of a potentiometer.

The TPS is mounted on the side of an engine throttle body assembly, and the inside of the TPS moves with the position of the throttle valve. The PCM supplies the TPS with a steady reference voltage and monitors the voltage returned through the potentiometer. Most TPSs have a relatively high resistance at closed throttle, dropping the reference voltage, which results in a low-voltage signal being returned to the PCM through the signal return wire. Usually this voltage is around 0.5 volt at closed throttle. The sensor's resistance steadily drops as the throttle valve is opened. This drop causes the voltage signal returned to the PCM to increase to approximately 4.5 volts at wide-open throttle.

▶ TECHNICIAN TIP

If a customer brings a vehicle in and complains about erratic shifting and a vehicle hesitation problem, it is a good idea to check the TPS first. It is difficult for the PCM to determine whether there is a problem with the TPS because the sensor is directly tied to human reaction. Often, if the PCM sees the voltage drop out on a TPS signal, it believes the driver released the accelerator. This makes it difficult to determine sensor failure unless the sensor fails completely, resulting in either 0 volts or reference voltage being returned to the PCM.

Accelerator Pedal Position Sensor

K26012

A sensor that is similar to the TPS is the accelerator pedal position sensor (**FIGURE 26-13**). It is used on drive-by-wire vehicles to determine the driver's intent as related to acceleration/deceleration. Beyond indicating the position of the accelerator pedal, the signal can also be used to determine how quickly or slowly the pedal is being pressed or released. Quick application of the pedal can indicate the need for more engine power and the corresponding need for a downshift and a delay of the upshifts. Quick release of the pedal can indicate the start of a panic stop. So the PCM can start to anticipate taking actions that would help avoid an accident, such as downshifting and applying the brakes.

Engine Coolant Temperature Sensor

K26013

The PCM will look at the **engine coolant temperature (ECT) sensor** not only for starting the engine but also for transmission operation. An ECT sensor is another thermistor, a resistor that varies its resistance based upon temperature, like the TFT sensor. The ECT sensor is installed in an engine coolant passage, often near the thermostat housing on an engine (**FIGURE 26-14**). The PCM sends out reference voltage to the ECT sensor, which varies the voltage returned to the PCM through the thermistor, allowing the PCM to determine the temperature of the engine from the voltage.

The PCM varies shift timing and pressures based upon engine temperature. When the engine is cold, the PCM delays transmission shifts to allow the engine to warm up faster and allows for more consistent engine and transmission operation.

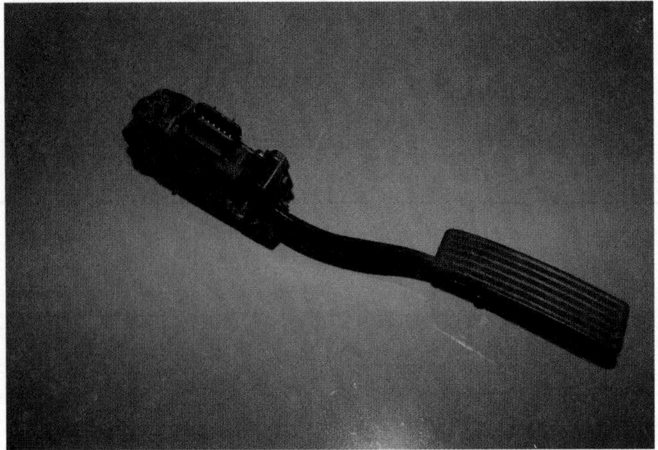

FIGURE 26-13 Accelerator pedal position sensor.

FIGURE 26-14 An ECT sensor installed near the thermostat housing.

> ▶ TECHNICIAN TIP
>
> An ECT sensor cannot accurately read engine temperature when the coolant level is low. Before condemning the ECT sensor, make sure the coolant level is up to the proper level and that the thermostat is operating properly on the vehicle.

Manifold Absolute Pressure Sensor

K26014

The PCM needs to know how much load the engine and transmission are under. One method of determining engine load is through the use of a **manifold absolute pressure (MAP) sensor**. The MAP sensor can be either directly bolted to the intake manifold or connected through a vacuum hose (**FIGURE 26-15**). Most modern MAP sensors use a type of flexible silicon chip within a sealed chamber; the chip changes resistance when manifold pressure flexes the chip. The varying resistance then changes the output signal based on the manifold pressure. When the engine has more vacuum, the chip is pulled down, changing its resistance. The PCM supplies a reference voltage to the sensor, which is reduced as it travels through the potentiometer and is then returned to the PCM. Because manifold pressure is an indication of engine load, the PCM can use this voltage signal to monitor engine load.

FIGURE 26-15 A common MAP sensor installed on an intake manifold.

Manifold pressure is a good indicator of engine load and was used by the vacuum modulator on our hydraulically controlled transmission. When an engine is operating under light load, engine manifold pressure is low. When the engine is operating under heavy load, engine manifold pressure is high. As the manifold pressure changes, so do the shift points. When the engine is under heavy load, the shifts are delayed, and vice versa. The engine load is also used to determine when the TCC should be applied and released. The TCC cannot transmit very much torque without slipping, which would burn it up quickly. Thus, when the engine is under more load than the TCC is designed for, the PCM turns the TCC off.

Although many vehicles still have a MAP sensor on the intake manifold, most of these sensors on vehicles built after 1996 are not the primary sensor used for determining engine load, as you will see in the next section. MAP sensors are now used to test if certain emissions-related components are functioning properly, as they will cause changes in manifold pressure if they are.

> ▶ TECHNICIAN TIP
>
> MAP sensors were originally connected to the intake manifold through a vacuum hose. Over time these hoses deteriorate and can develop small holes and cracks in the lines, or they get soft and collapse. Vacuum hoses should be replaced periodically as they start to show signs of deterioration. A vacuum hose that has deteriorated will typically affect fuel economy and result in a late, harder shift, as the PCM thinks the vehicle is under greater load than it really is.

Mass Airflow Sensor

K26015

To meet more stringent emission standards, most vehicles come equipped with a **mass airflow (MAF) sensor**, rather than a MAP sensor, to measure engine load. MAF sensors are a much

FIGURE 26-16 A MAF sensor installed in the air intake tube on a vehicle. It is important to make sure all of the clamps are securely fastened.

FIGURE 26-17 A crankshaft position (CKP) sensor installed on the front of a crankshaft.

more accurate and faster means of measuring engine load. The MAF sensor actually measures the mass of the air coming into the engine, so the PCM can accurately inject the proper amount of fuel. MAF sensors measure airflow in grams per second (gps). The more airflow, the harder the engine is working. As engine speed increases, so does airflow. The MAF sensor is typically installed in the air intake hose leading to the throttle body assembly. Although some MAF sensors are built into the throttle body assembly, others are mounted on the air cleaner assembly (**FIGURE 26-16**).

Many different types of MAF sensors are currently being installed on vehicles, and you need to check the specific vehicle's service information for testing methods. It is important to check that all of the air intake hoses are tightly secured to the engine so that the MAF sensor will measure all of the air entering the engine. Any air leaks *after* the sensor will result in an inaccurate reading by the MAF sensor, resulting in early shifting and poor vehicle drivability. Some MAF sensors require periodic cleaning, as dirt and contaminants can build up on the sensing surface inside the sensor. Clean these components carefully, using a cleaner recommended for MAF sensor cleaning. Note that some manufacturers specifically say not to clean their MAF sensors and to instead replace them if they are not reading correctly. Be careful to check the manufacturer's recommendation regarding whether the MAF sensor can be cleaned. And if it can, follow the manufacturer's directions precisely.

Crankshaft Position Sensor

K26016

The PCM uses the **crankshaft position (CKP) sensor** to determine engine speed (**FIGURE 26-17**). CKP sensors can be a magnetic pickup-style sensor, a Hall-effect sensor, or an optical-style sensor. All three types of sensors serve the same function: to send a pulsed signal to the vehicle PCM to determine engine speed.

The PCM compares the CKP sensor signal to the input shaft speed sensor in the transmission to determine torque converter slippage between the impeller and the turbine. It is also used as the PCM operates the TCC to monitor the engagement and slippage of the TCC. Most TCCs are pulsed on slowly to prevent a harsh engagement that drivers might complain about. By observing the difference in speed, the PCM can precisely control how long it takes for the TCC to become fully locked.

Brake On/Off Switch

K26017

Another input the PCM looks at is the brake on/off switch (**FIGURE 26-18**). The brake switch is used to control the brake lights on the vehicle when the driver applies the brakes, but the PCM also uses this information for transmission control. The brake on/off switch is a normally open switch that closes when the driver applies the brake pedal, allowing current to flow to the stop lamps and to the PCM. In some vehicles, the brake on/off switch is separate from the brake light switch, so always check the service information to make sure you are testing the correct switch.

> ▶ TECHNICIAN TIP
>
> After replacing a transmission, make sure the MLP switch is adjusted correctly *before* starting the vehicle. If the switch has not been adjusted properly, when the technician continues to add hydraulic oil to top off the transmission, the vehicle may suddenly have enough transmission fluid to engage the clutches and begin moving.

The PCM uses the brake on/off switch input to disengage the TCC. As soon as the driver applies any pressure to the brake pedal, the TCC must be released to prevent excessive drag on the engine and a possible stall. On some newer vehicles, the brake switch is also used to put the transmission into neutral while idling at a stoplight. As long as the driver has the brake pedal applied, the vehicle will remain in

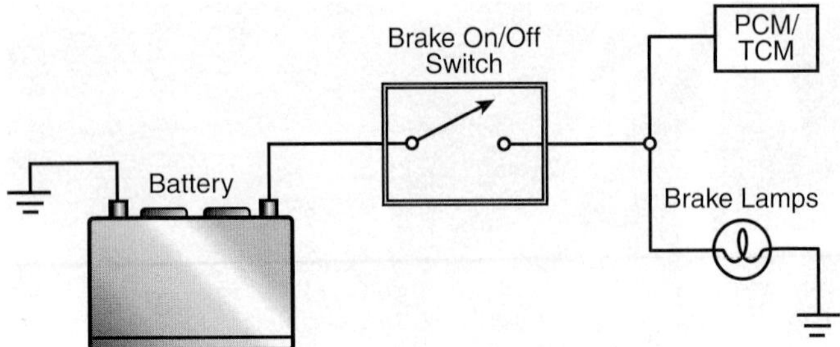

FIGURE 26-18 A common brake on/off switch wiring diagram.

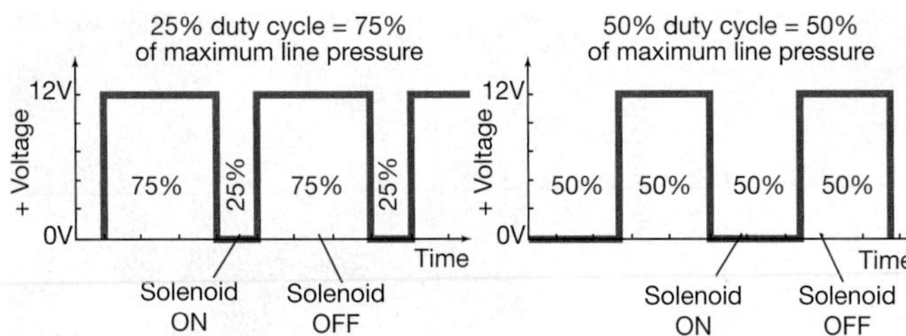

FIGURE 26-19 A pulse-width-modulated pattern of an electronic pressure control (EPC) solenoid (ground side switched).

neutral to save fuel. As soon as the driver releases the brake pedal, the transmission reengages into gear.

Manual Upshift/Downshift Control System

K26018

Many electronically controlled transmissions are equipped with a manually selectable upshift/downshift control system. When the driver selects manual control, he or she can either press an up/down button or move the gearshift lever forward/backward to shift the transmission into a higher or lower gear. This is a great feature when driving on hilly roads, especially with heavy loads since the driver can see what is coming next, and prepare for it. For example, if the vehicle is travelling at freeway speeds and a steep hill is ahead, the driver can shift down a gear or two at the bottom of the hill to increase the engine's power to maintain vehicle speed. Currently, vehicles cannot anticipate upcoming elevation gains or losses, so allowing the driver to control this if they like, is a helpful feature. It can also increase fuel economy and reduce wear and tear on both the engine and transmission in these situations.

▶ Powertrain Control Module Outputs

K26019

Electronic Pressure Control Solenoid

K26020

On electronically controlled transmissions, line pressure is controlled using an electronic pressure control (EPC) solenoid. Most EPC solenoids are controlled by a digital signal that rapidly cycles on and off to control pressure. This on /off cycling can be classified in two ways; pulse-width modulation and duty cycle. Pulse width is generally measured in milliseconds (mS) of on-time per cycle, and duty cycle is generally measured in percent of on-time per second. A pulse width of 25 mS means that the solenoid would be activated for 25 mS for one cycle. A duty cycle of 50% means the solenoid is on 50% of a second and off the other 50% of the second. If we were to look at the voltage to the EPC solenoid using a DSO, we would see a square wave pattern like that shown in (**FIGURE 26-19**).

EPC solenoids are installed in the line pressure circuit and exhaust some of the line pressure back to the transmission pan (**FIGURE 26-20**). If the line pressure is 200 pounds per square inch (psi) and we have an EPC duty cycle of 50%, the line pressure will be approximately 100 psi. Roughly 50% of the line pressure will be exhausted back to the pan. If higher line pressure is required, the PCM will reduce the duty cycle. At a 25% duty cycle, the line pressure would be approximately 150 psi, or 75% of the available line pressure in this example.

Most EPC solenoids are designed so that if there is a problem with the PCM, the wiring, or the EPC solenoid, the vehicle will default to maximum line pressure. The EPC solenoid will be forced closed by an internal spring, and no line pressure will be exhausted to the pan. When using a scan tool, you will notice that some manufacturers give you the amount of amperage being used by the EPC solenoid, rather than the duty cycle. This amperage will change as the duty cycle of the solenoid is changed. **FIGURE 26-21** shows an EPC solenoid electrical diagram.

Shift Solenoids

K26021

Shift solenoids are typically an electromagnetic type of valve that is either open or closed. These solenoids can directly control line pressure to an application device, such as a clutch or band, or they may control a shift valve, which in turn directs fluid to the clutch or band. Many four-speed automatic transmissions use two shift solenoids to control the transmission. **TABLE 26-1** lists the shift solenoid operation on a typical four-speed automatic transmission. By turning the different solenoids on and off, all of the gears needed can be hydraulically applied.

Based on the information in **TABLE 26-1**, what gear would the vehicle start in if both solenoids failed? Notice that only third gear does not use either shift solenoid. If the vehicle has a problem with the PCM, the shift solenoids, or the wiring, the customer may complain about a very poor-performing vehicle that will only start in third gear. This is often called limp-in mode, as the vehicle would still be drivable with a shift solenoid failure.

FIGURE 26-22 is a simplified diagram for a common Chrysler 41TE transmission, found in many of the Chrysler minivans. In this transmission, the solenoids directly operate

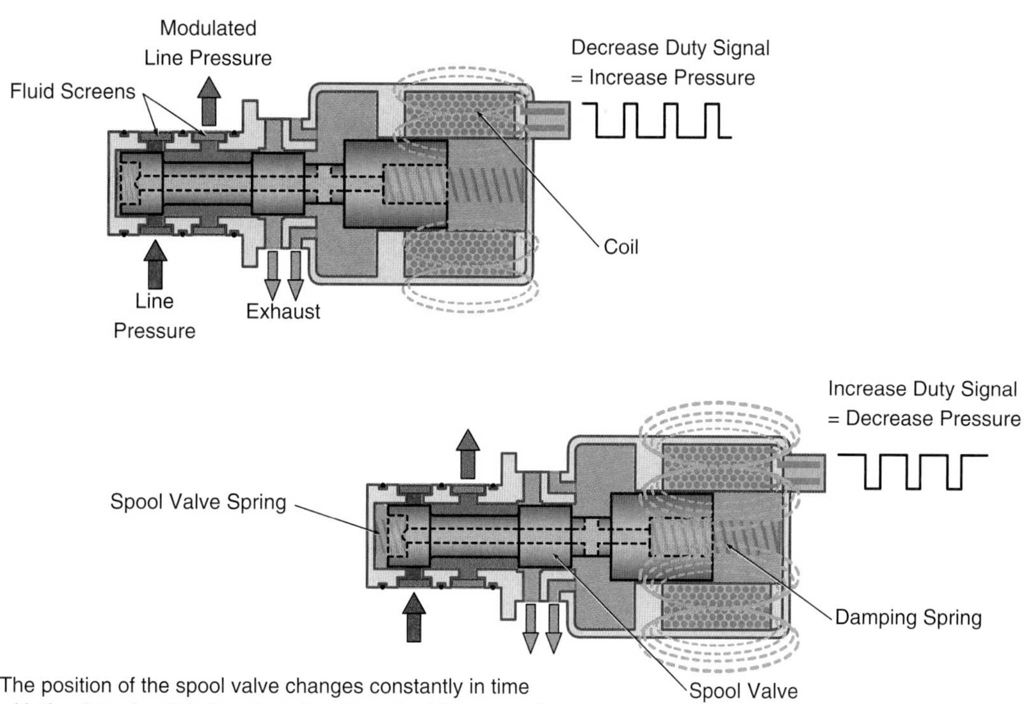

FIGURE 26-20 An EPC solenoid controls how much fluid is exhausted back to the pan to control line pressure.

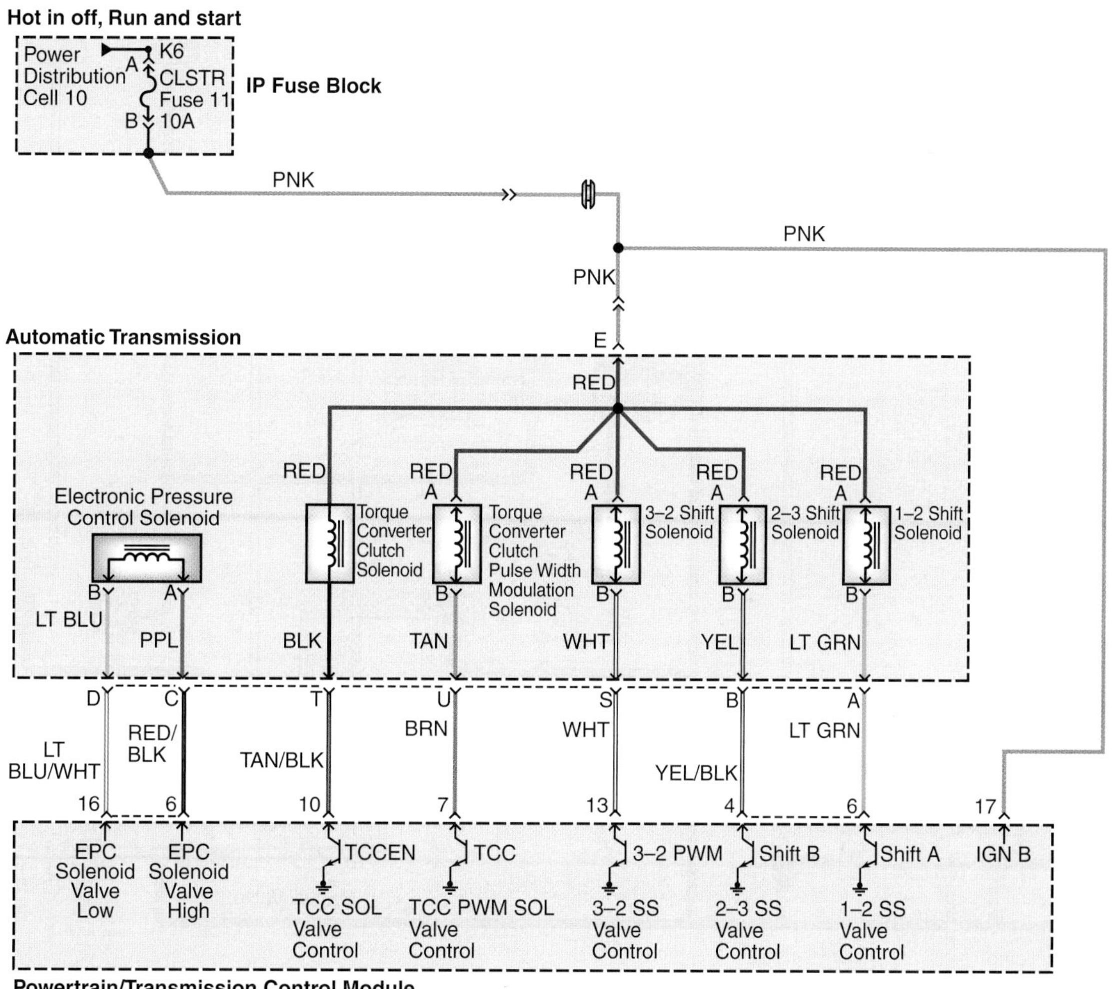

FIGURE 26-21 An EPC solenoid electrical diagram.

TABLE 26-1 Shift Solenoid Operation on a Typical Four-Speed Automatic Transmission

PCM Commanded Gear	1–2 Shift Solenoid	2–3 Shift Solenoid
Park, neutral, reverse	On	On
First	On	On
Second	Off	On
Third	Off	Off
Fourth	On	Off

the clutches to control the shifting of gears. This particular transmission has four solenoids. They are labeled 1, 2, 3, and 4 on the diagram. Solenoid 1 controls the underdrive clutch, solenoid 2 controls the overdrive clutch, solenoid 3 controls the low reverse clutch, and solenoid 4 controls the 2–4 clutch. Looking at **FIGURE 26-23**, you can see that solenoids 1 and 3 are being used. Line pressure is being routed through the solenoids to the underdrive clutch and the low reverse clutch to engage first gear.

In Figure 26-23, you can see the transmission in second gear. For this transmission to operate in second gear, the PCM commands solenoids 1 and 4 on. This allows line pressure to flow to the underdrive clutch through solenoid 1. Line pressure is allowed to flow through solenoid 4 to the 2–4 clutch. With both the underdrive clutch and the 2–4 clutch engaged, the vehicle is in second gear.

FIGURE 26-24 shows the same transmission in third gear. In third gear on this vehicle, the PCM commands solenoids 1 and 2 on. Solenoid 1 again supplies line pressure to the underdrive clutch, and solenoid 2 supplies line pressure to the overdrive clutch. With both the underdrive clutch and the overdrive clutch engaged, the vehicle is in third gear.

FIGURE 26-25 shows fourth gear operation. For fourth gear in this transmission, solenoids 2 and 4 need to be engaged by the PCM. Solenoid 2 allows line pressure to flow to the overdrive clutch, and solenoid 4 allows line pressure to flow to the 2–4 clutch. When both the 2–4 clutch and the overdrive clutch are engaged, the vehicle is in fourth gear.

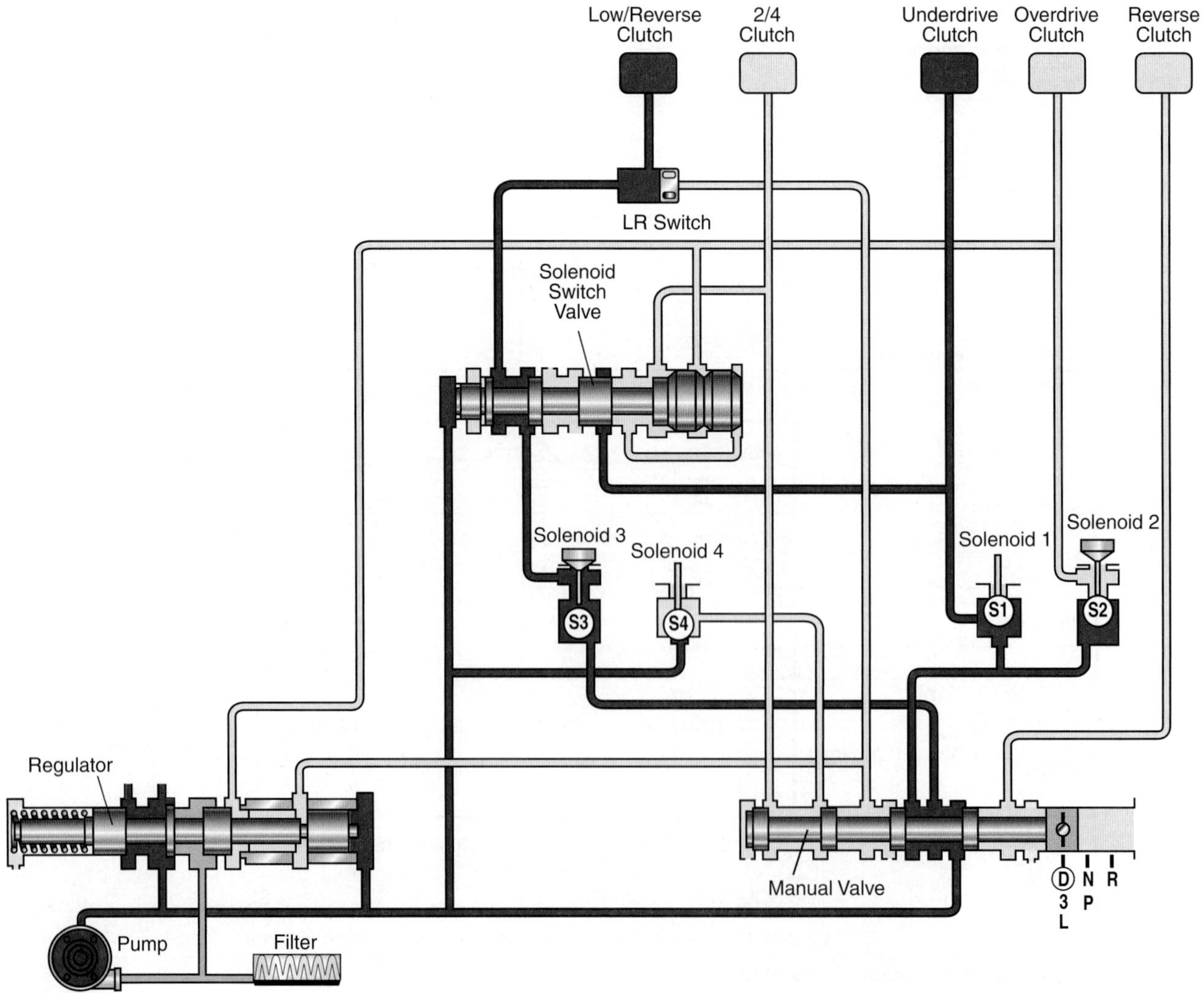

FIGURE 26-22 A simplified diagram of a Chrysler 41 TE transmission in first gear.

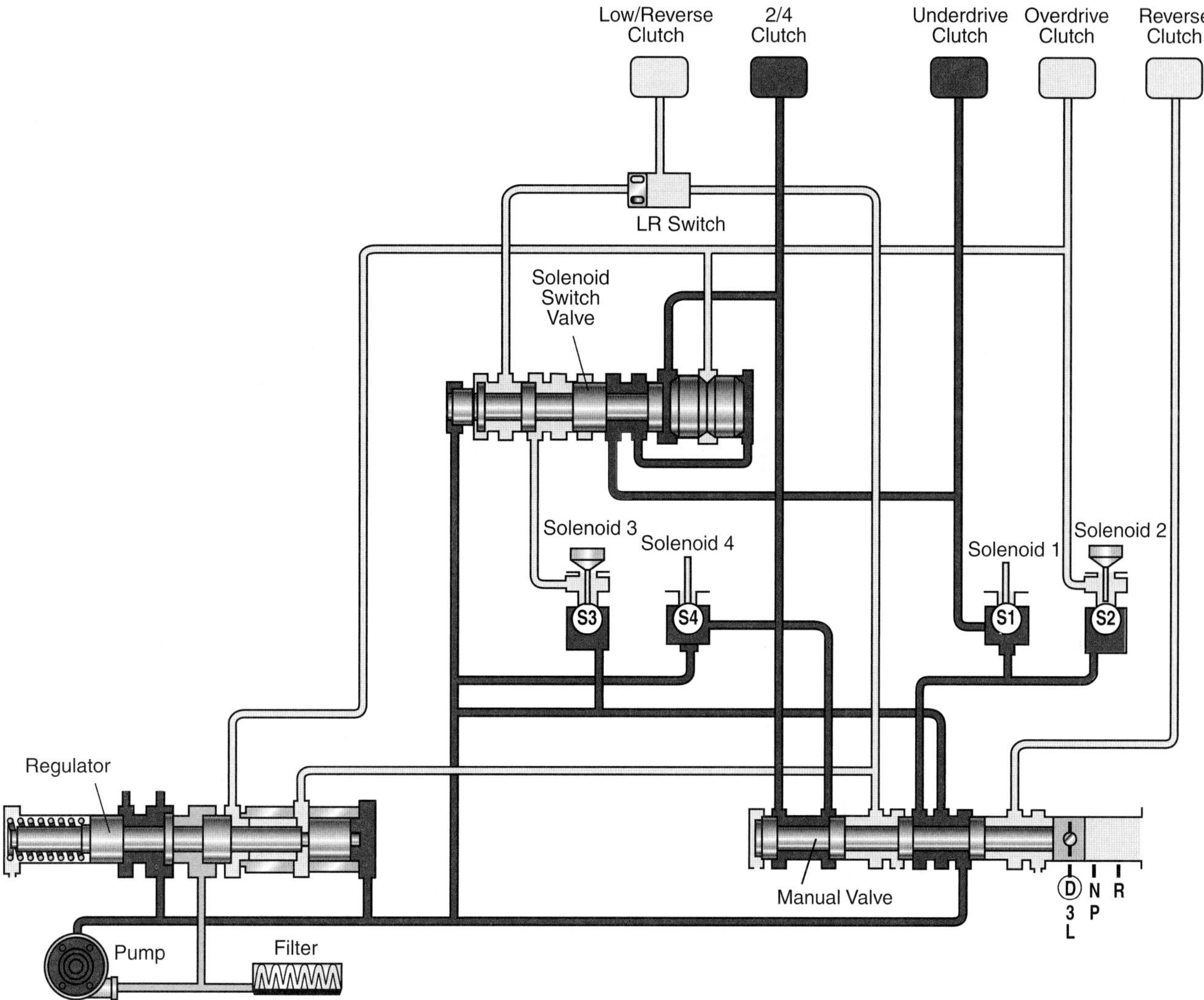

FIGURE 26-23 A simplified diagram of a Chrysler 41 TE transmission in second gear.

In many transmissions, the solenoids do not directly operate the clutches and bands. Rather, the solenoid allows hydraulic oil to flow to a spool valve. The spool valve then moves, allowing line pressure to flow past and engage the clutch or band. **FIGURE 26-26** is a simplified hydraulic diagram for a transmission. In the figure, the transmission is in first gear. The manual valve allows line pressure to flow to all four of the pressure control valves, and to the four shift solenoids. The PCM has energized solenoids 1 and 3. Solenoid 1 is now able to flow line pressure to pressure control valve 1, causing the spool valve to move to the left. When the spool valve moves to the left, line pressure is allowed to flow to the low reverse band. Solenoid 3 allows line pressure to flow to shift valve 3, causing the spool valve to move to the left. As this valve moves to the left, line pressure is allowed to flow to the underdrive clutch. When the low/reverse band and the underdrive clutch are engaged, the vehicle is in first gear.

FIGURE 26-27 shows the same transmission in second gear. For second gear, the PCM energizes solenoids 2 and 3. The manual valve allows line pressure to flow to all four shift valves and all four solenoids. Opening solenoid 2 allows line pressure to be applied to shift valve 2, causing the spool valve to move to the left. This movement allows line pressure to flow past shift valve 2 and to the second band. Opening solenoid 3 allows line pressure to be applied to the right side of shift valve 3, causing the valve to move to the left. As the spool valve moves to the left, line pressure is allowed to flow through the shift valve to the underdrive clutch. When the underdrive clutch and the second band are applied, the vehicle is in second gear.

The same generic transmission, this time in third gear, is shown in **FIGURE 26-28**. Looking at the diagram, we can see that the PCM has energized solenoids 3 and 4. Again, line pressure has been supplied to all four shift solenoids and the four shift valves. Opening solenoid 3 allows line pressure to flow to the right side of shift valve 3, causing the spool valve to move to the left. When the valve moves to the left, line pressure is allowed to flow through the spool valve to the underdrive clutch. When solenoid 4 is engaged by the PCM, line pressure flows to the right side of shift valve 4, again moving the spool valve to the

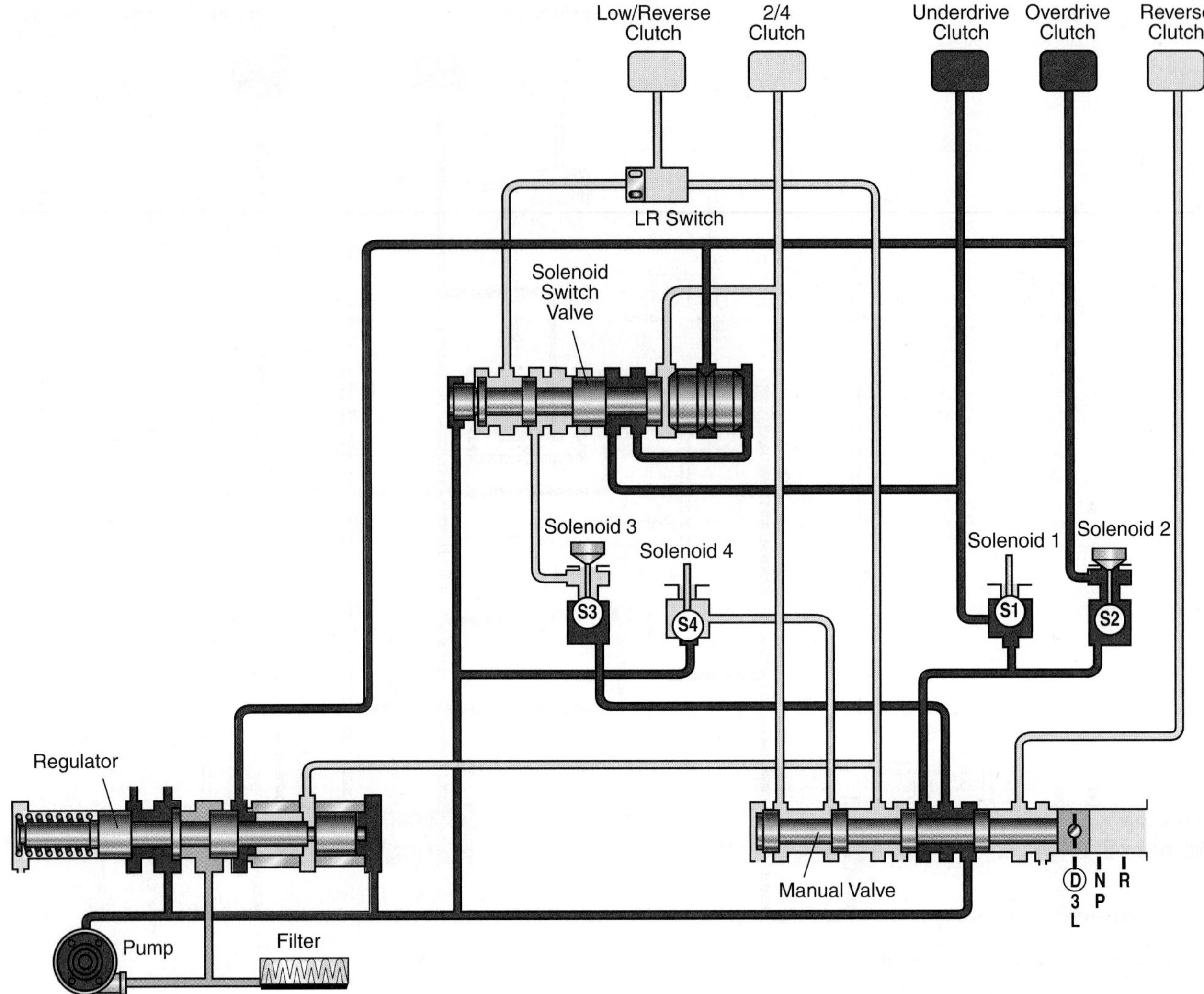

FIGURE 26-24 A simplified diagram of a Chrysler 41 TE transmission in third gear.

left. Line pressure is then allowed to flow to the overdrive clutch. When both the underdrive clutch and the overdrive clutch are engaged, this transmission is in third gear.

In **FIGURE 26-29**, we have the same transmission once more. This time the PCM has commanded operation of fourth gear. The PCM has energized solenoids 2 and 4. Each of these solenoids then allows line pressure to its respective shift valve. The line pressure causes the spool valve to move to the left, allowing line pressure to flow from shift valve 2 to the second band, and from shift valve 4 to the overdrive clutch. When both the second band and the overdrive clutch are engaged, this transmission is in fourth gear.

In the hydraulic diagrams for these examples, the direct solenoid shift type and the indirect solenoid shift type have been simplified to ease understanding. Having grasped the basic concept of how these systems control the transmission, you can follow a more complicated hydraulic diagram. On many transmissions, the manufacturers will use multiple valves to ensure proper operation of the clutches and bands, and to prevent the transmission from being in two gears at once. If a transmission were allowed to be in two gears at once, the results would be disastrous for the transmission, as it would tend to lock up and the weakest part would burn out or break.

Torque Converter Clutch Solenoid

K26022

The PCM controls the TCC (torque converter clutch) lock-up solenoid (**FIGURE 26-30**). Early torque converters always had some slippage, so in the late 1970s and early 1980s manufacturers added a lock-up clutch to the torque converter. When the PCM determines that the torque converter should be locked up (i.e., the vehicle is operating under an appropriate load, at an appropriate speed, in the appropriate gear, and at the proper engine temperature, etc.), it engages the TCC lock-up solenoid. The solenoid allows line pressure to be applied to the TCC and locks up the turbine to the impeller, providing direct drive through the torque converter.

Some TCC solenoids are a pulse-width–modulated solenoid. Remember that pulse-width modulation means

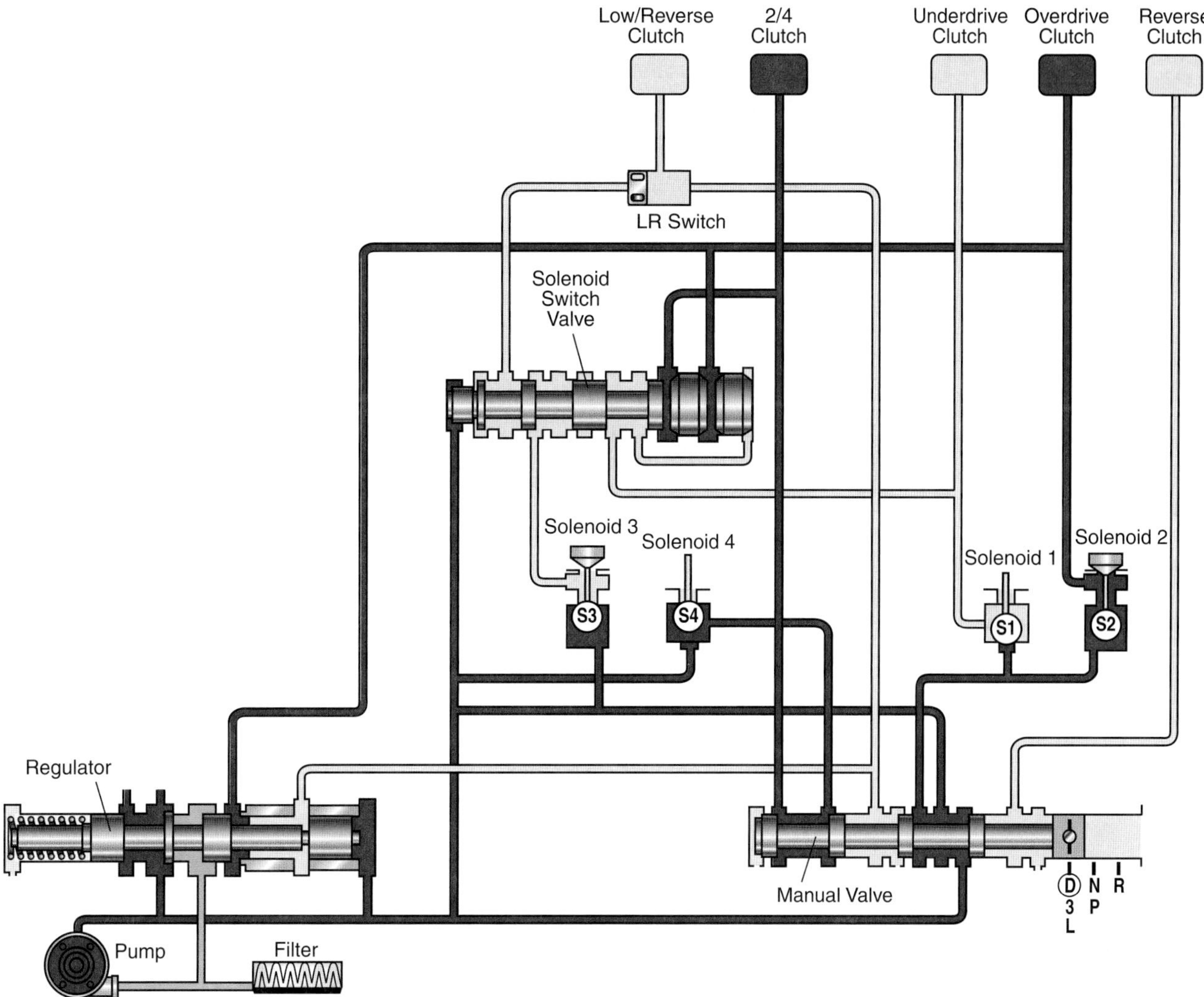

FIGURE 26-25 A simplified diagram of a Chrysler 41TE transmission in fourth gear.

that the solenoid is only on for a percentage of the time. For a solenoid with a 25% pulse-width modulation, the solenoid is on 25% of the time and off 75% of the time. Manufacturers slowly pulse the TCC on to prevent a harsh engagement that a customer might complain about. The PCM monitors the CKP sensor and the input shaft speed sensor to determine the amount of torque converter slippage and control the lock-up of the TCC.

▶ TECHNICIAN TIP

Always test-drive a vehicle to verify the reported problem. One customer brought in a turbocharged vehicle he had recently purchased, complaining that the turbocharger was not functioning correctly. After a quick test-drive, it was determined that the turbocharger was not the cause of the poor performance; rather, the transmission solenoids had failed and the vehicle only had third gear. Replacing the solenoids fixed the problem and the customer was happy. If the technician had listened to the customer and replaced the turbo unit, the vehicle would still have had very poor performance.

▶ Powertrain Control Module/Transmission Control Module

K26023, K26024

The PCM or the TCM is the brain in the system. In basic terms, it comprises a storage memory section, a data processor section, an input sensor control and processing section, and a section for output drivers. The storage memory contains the software for the computer, as well as a place to retain DTCs, freeze-frame data, and learned data. Software is loaded into the memory, which is the basis for processing data. The software contains all of the information needed for the processor to interpret the sensor information and make decisions based on the programmed information. It also contains the instructions for performing diagnostic tests on the circuits and PCM itself. In many cases, the software can be updated by uploading software updates, called a relearn process. With software that can be updated, manufacturers can design fixes for many of the issues that may arise once their vehicles go into operation. This makes it easier, quicker, and less expensive to resolve some common

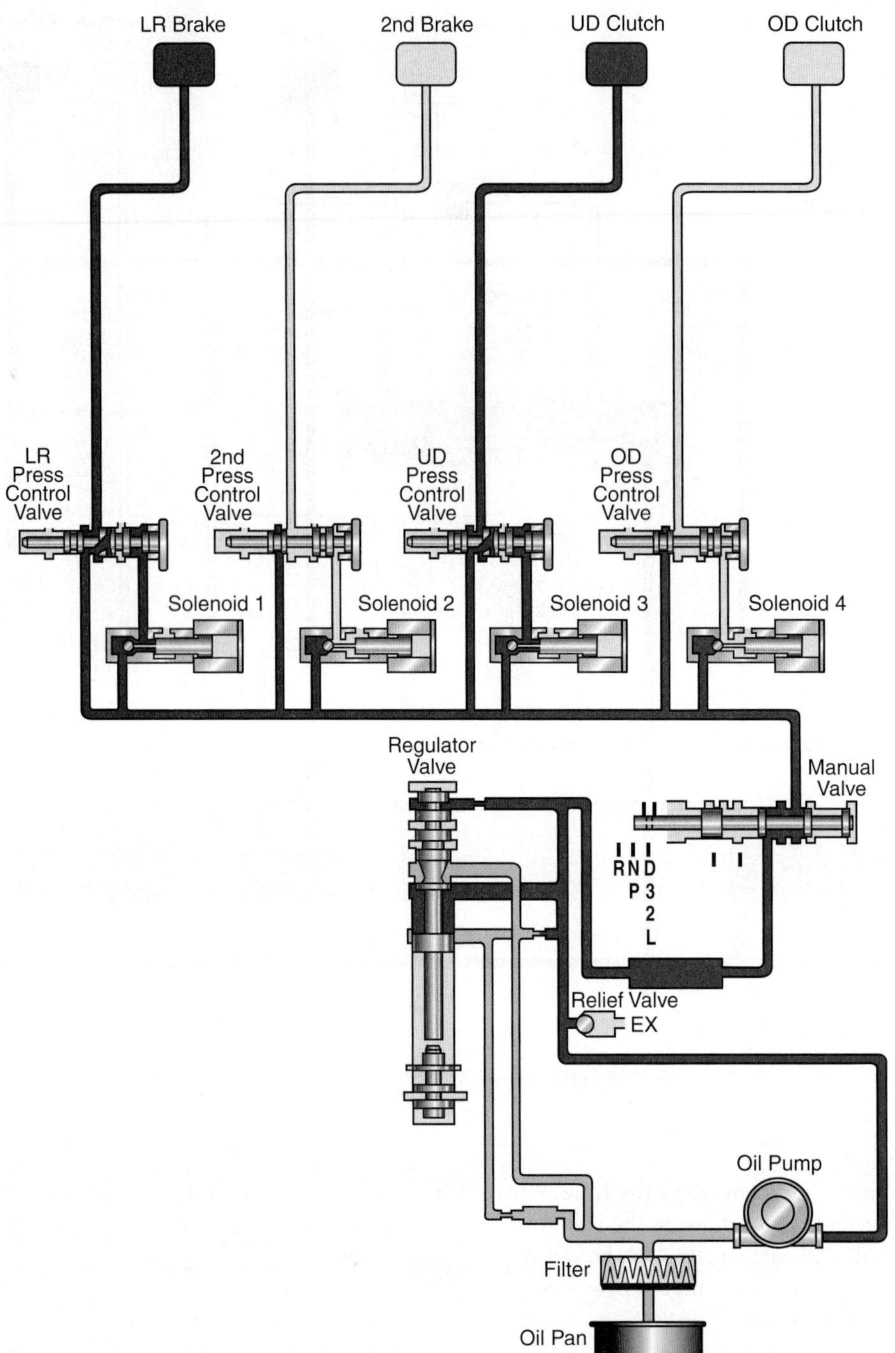

FIGURE 26-26 A simplified hydraulic diagram for a transmission in which the shift solenoids indirectly operate the clutches or bands in first gear.

customer or drivability issues rather than replacing the PCM with an updated part.

The processor and drivers are hardware related, so they make up the physical portion of the PCM. The input sensor section is responsible for sending the proper reference voltage to many of the sensors. It also receives the sensor signals and processes them into information that the main computer processor can use. The processor processes the data and compares them to data maps in the memory. Once the sensor data has been analyzed, the processor looks up the appropriate data map and determines the necessary actions. The processor then sends the appropriate commands to each of the drivers, which are electronic switching or control devices that control each of the actuators in the transmission. The processor also commands the diagnostic tests on the system and evaluates the results compared to the information stored in memory. The output drivers typically send either simple on/off signals or pulse-width–modulated signals to the actuators, depending on the actuator being controlled.

Refer to the chapters on Servicing Steering and Suspension Systems for more detailed information on the PCM.

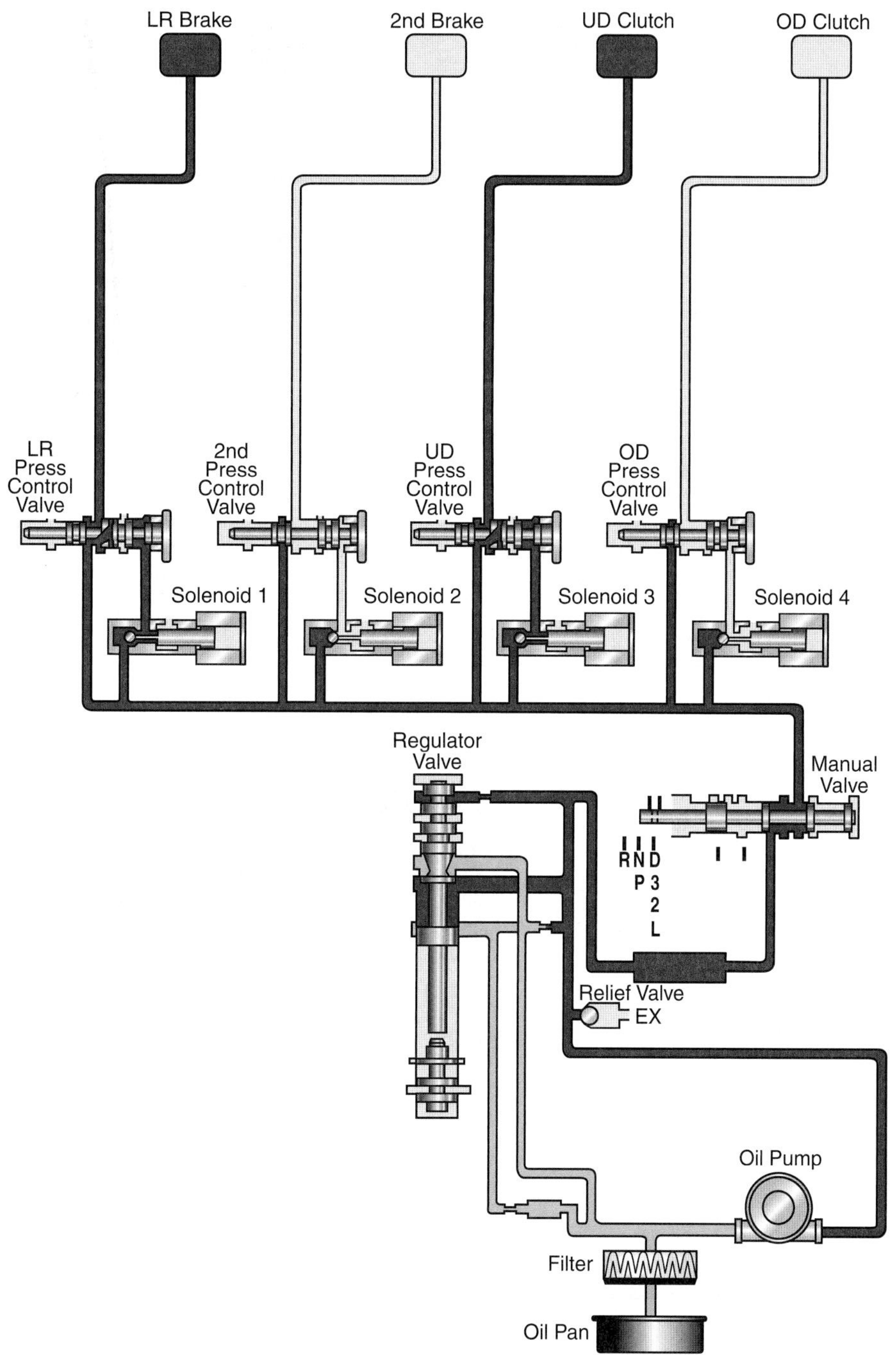

FIGURE 26-27 A simplified hydraulic diagram of a transmission in second gear that uses indirect solenoids to control the shifts.

Electronic Shift Programs

K26025

Most vehicles can identify and adapt to a driver's individual style and to environmental conditions. In time, the PCM learns how a particular driver uses the vehicle. For example, does the driver always accelerate hard or more gradually?

The PCM monitors the TPS sensor along with the VSS to determine a driver's driving pattern. For a driver who has a more aggressive driving pattern, the PCM will delay the shift points to allow the engine to operate closer to its peak rpm range. The PCM will also increase the line pressure to give firmer, quicker shifts. During deceleration, the transmission will downshift at a higher rpm, so the engine will be at an ideal rpm, for maximum performance after the downshift.

For a more conservative driver, the PCM will allow the transmission to shift earlier to keep the engine's rpm lower and

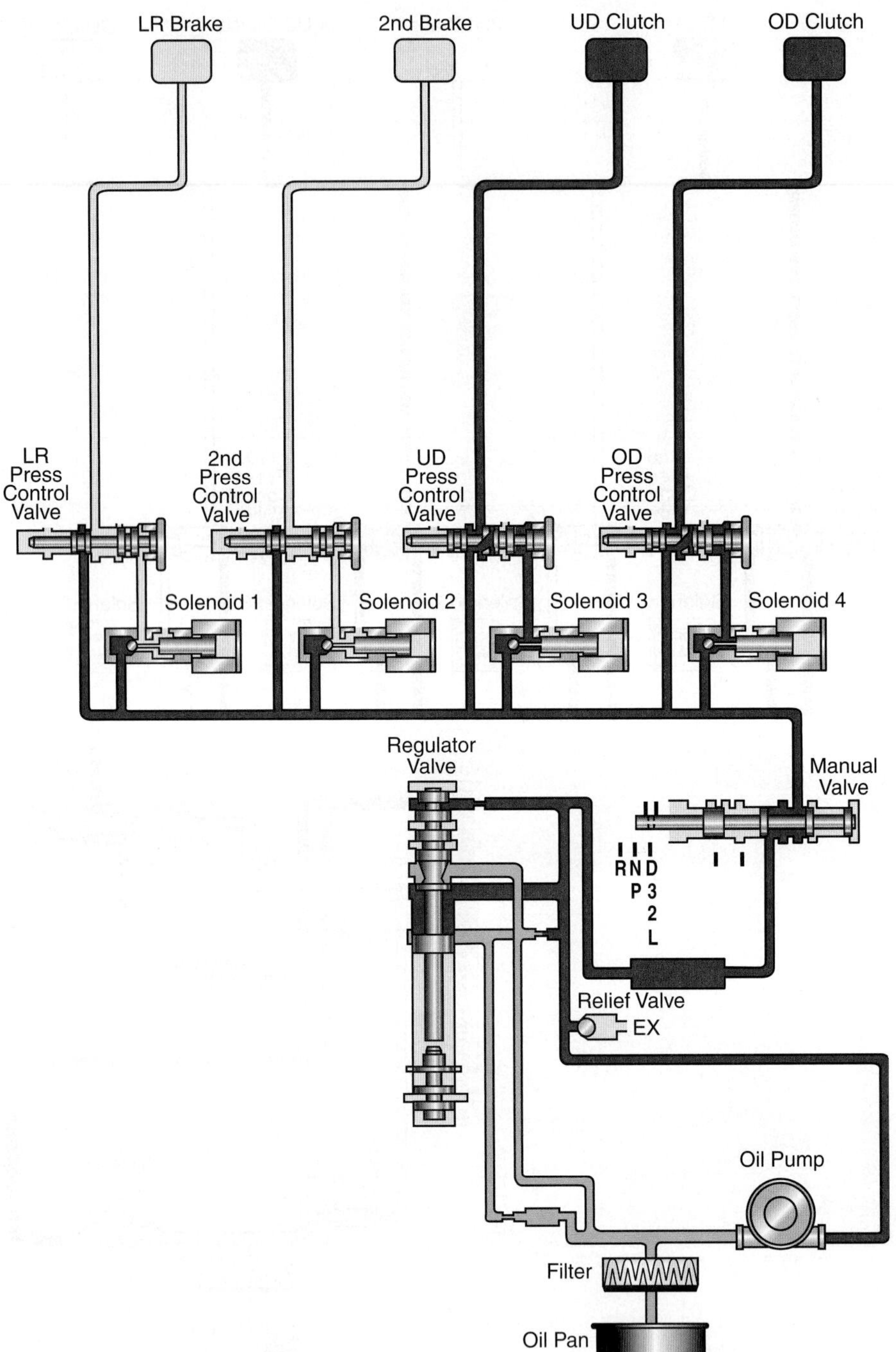

FIGURE 26-28 A simplified hydraulic diagram of a transmission in third gear that uses indirect solenoids to control the shifts.

help improve fuel economy. The PCM will also reduce the line pressure to soften the shifts and allow for a more comfortable drive.

Some vehicles have a mode selection switch on, or near, the gearshift lever to change the shift characteristics. The mode selector might have modes such as sport or normal, or sport or economy (**FIGURE 26-31**). On some hybrid vehicles, a brake mode, or regenerative braking mode, is available (**FIGURE 26-32**). This mode is used when going down steeper hills and increases the amount of regeneration that is commanded. It provides greater braking power than in the normal mode.

Because a driver's driving responses can vary considerably within a short time, depending on traffic and road conditions, response patterns are analyzed continuously. If a driver's requirements change suddenly after a stop, the driving style identification circuits within the control unit must be able to adapt.

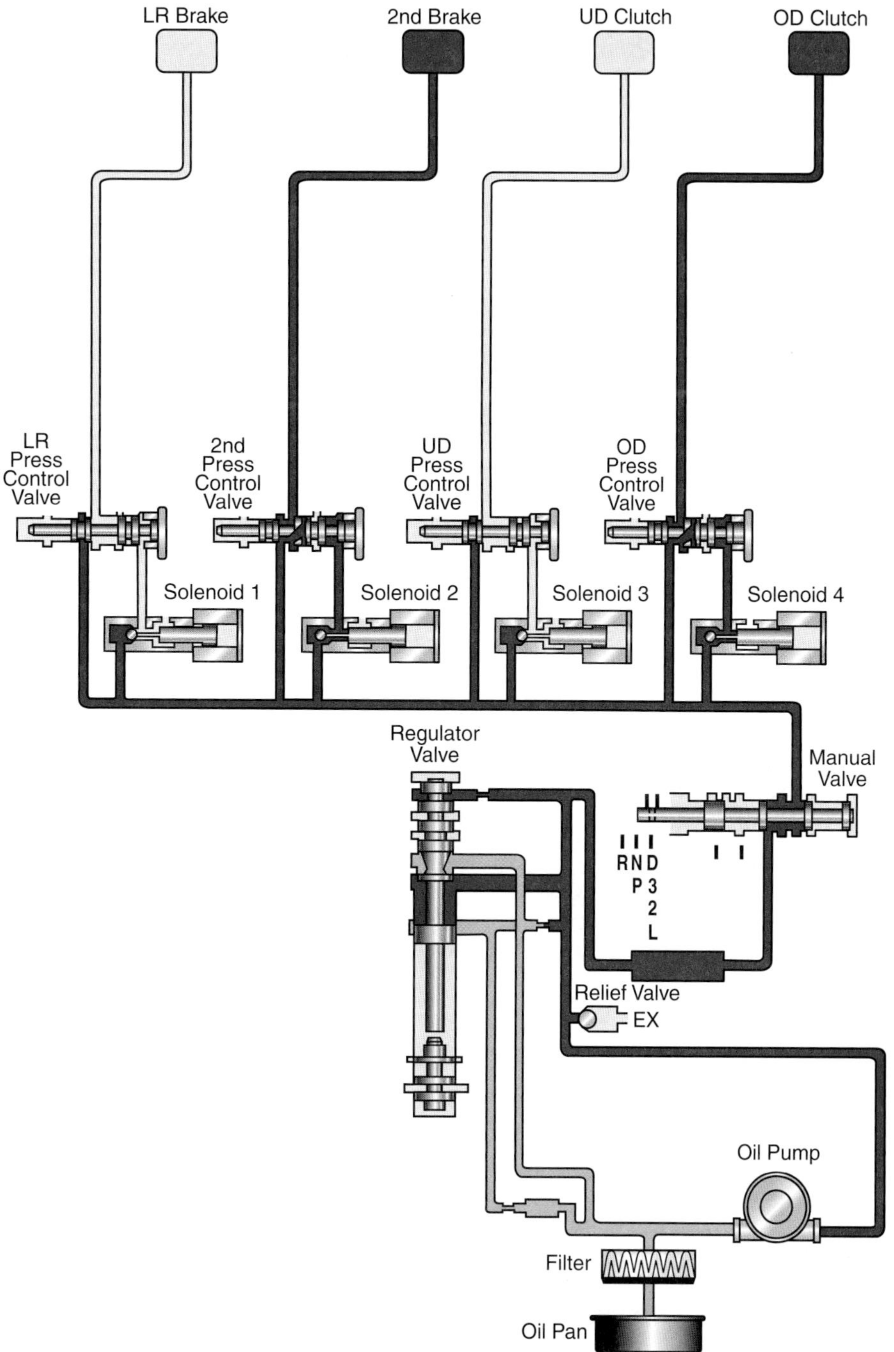

FIGURE 26-29 A simplified hydraulic diagram of a transmission in fourth gear that uses indirect solenoids to control the shifts.

> **▶ TECHNICIAN TIP**
>
> If the adaptive memory is lost due to clearing of the memory, or disconnection of the battery on some vehicles, the customer may come back and complain, "What did you do to my transmission? It doesn't work the same as before!" To prevent this, you can either use a memory minder while the battery is disconnected, or you can test-drive the vehicle to help it relearn a driving style. The other alternative is to educate drivers and tell them that they will have to drive the vehicle for a few days for it to relearn their driving style.

Environment Identification

Identification of environment means identifying unusual ambient conditions in which the vehicle is operating. Slippery conditions require a gear ratio that enhances traction and the stability of motion of the vehicle. On vehicles equipped with traction control, one method of providing improved traction is to change the gear ratio of the transmission. Wheel slip monitoring via the anti-lock braking sensors may confirm poor tire grip, and a winter shift program can be initiated by the PCM

FIGURE 26-30 A lock-up torque converter solenoid being removed from a General Motors 4L60E transmission.

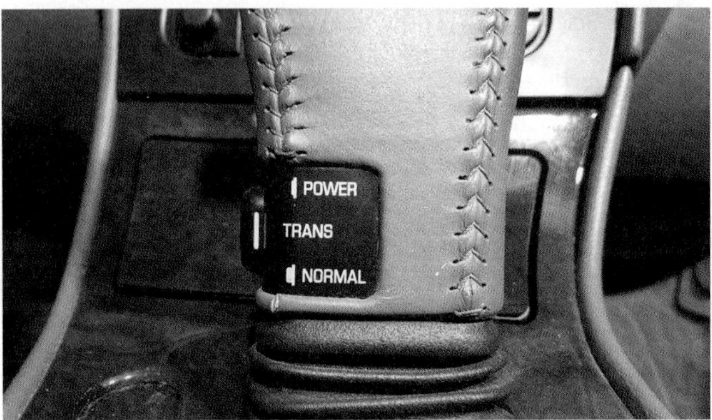

FIGURE 26-31 An economy/power button located on the shifter handle.

FIGURE 26-32 Regeneration mode (B) on this Prius Hybrid vehicle is activated by the shift selector.

on a vehicle with traction control. When traction control is activated, if a tire starts to spin, the PCM will command the brake for that wheel to apply in order to slow down the wheel so power can be given to the other wheel. The PCM typically commands a higher transmission gear, and thus reduced engine torque, when the vehicle is just starting from a stop. This helps to prevent the tires from losing traction.

Electronic Transmission Fault Detection

On electronically controlled transmissions, the PCM must be able to monitor the transmission for proper operation. One of the first methods of fault detection is a basic circuit test. On solenoids and switches, the PCM can test the continuity of those circuits and tell whether they are open, shorted to ground, or shorted to power. If the test value is out of range, the PCM will set a DTC and turn the malfunction indicator lamp (MIL) on.

A more sophisticated type of fault detection is when the PCM monitors the input shaft speed sensor and compares it to the VSS to determine gear ratios. If the ratio is not correct, the PCM will set a DTC for a ratio error and turn the MIL on.

Newer PCMs also have logic programmed into them. No longer will the PCM only look for opens, shorts, and ratio errors. It will now also apply logic to the operating conditions it sees. For example, suppose the PCM is sent a signal from the CKP sensor of 650 rpm, an MLP signal of neutral, and a TPS signal of 75% throttle. The PCM has the logic built into its programming to realize that it is not possible to have the vehicle in neutral, an engine speed of only 650 rpm, and the throttle at 75% open. The PCM would trigger a code for the TPS. Before PCMs were programmed with this type of logic, the PCM could not make these kinds of determinations.

Torque Management

On many modern electronically controlled transmissions, the PCM can manage the torque output of the engine and vary the amount of torque being sent to the transmission. By having a completely electronically controlled transmission and engine, the PCM is able to reduce the amount of torque being produced by the engine right before the transmission shifts. Much of the wear and tear in an automatic transmission occurs during the relatively short time that the transmission is shifting between gears. By reducing the amount of torque being transmitted to the transmission during this time, transmission wear is greatly reduced, and smoother shifting is experienced. Manufacturers have also found that by using torque management to reduce the amount of torque during shifting, they can install a smaller transmission in the vehicle. For example, a manufacturer might be able to install a transmission that was designed to handle less torque than otherwise would be necessary, without affecting the transmission's longevity.

To manage the engine torque, the PCM might be programmed to cut back on the ignition timing of the engine right before the transmission shifts. As soon as the transmission has completed the shift, the PCM returns the ignition timing to normal. During normal operation, the spark plug is fired anywhere between about 10 degrees and 45 degrees of crankshaft revolution *before* the piston reaches top dead center. By retarding the ignition timing, the engine torque is dramatically decreased, reducing the strain on the transmission. All of this happens in less than 1 second, so the driver rarely feels any loss of power. If anything, the driver might notice that the transmission shifts more smoothly than a vehicle without torque management.

Limp-in Mode

Most electronically controlled transmissions have a fail-safe, or **limp-in mode**. Limp-in mode means that even if the computer for the transmission or the part of the PCM that controls the transmission fails, the vehicle will still be drivable. The transmission will have only one forward gear. Whichever gear does not require any shift solenoids will be the limp-in gear. On many transmissions, third gear is designed to be the limp-in gear. However as vehicle manufacturers have started adding more gear ratios to automatic transmissions, this fail-safe mode has been disappearing. Some vehicle manufacturers will prevent the vehicle from moving at all if the PCM/TCM detects a problem with the transmission.

Also, on most transmissions, if the PCM fails, the EPC solenoid will mechanically default to maximum line pressure to prevent clutch and band slippage, which would severely damage the transmission. This results in additional holding power for the clutches and bands. Failure of the transmission in any of these ways should turn on the MIL, notifying the driver of the failure and allowing him or her to safely drive the vehicle to a repair facility.

▶ Wrap-Up

Ready for Review

- US auto manufacturers switched to electronically controlled transmissions by the late 1980s.
- Shift points in an electronically controlled transmission are controlled by the powertrain control module (PCM) or transmission control module (TCM).
- A vehicle speed sensor (VSS) relays vehicle speed information to the speedometer.
- Types of VSS are magnetic pickup and reed switch.
- An input shaft speed sensor is similar to a VSS and relays its signal to the PCM.
- Shift timing and pressure are regulated based on input from the transmission fluid temperature (TFT) sensor.
- Transmission pressure switches are used to determine when pressure is in a particular circuit.
- Transmission line pressure is adjusted based on feedback from the line pressure sensor.
- The throttle position sensor (TPS) is a variable resistor (potentiometer) that moves with the position of the throttle.
- The engine coolant temperature (ECT) sensor relays information to the PCM, enabling the PCM to vary shift timing and pressures based on engine temperature.
- The manifold absolute pressure (MAP) sensor allows the PCM to monitor how much load the engine and transmission are under.
- Modern MAP sensors (installed after 1996) are primarily used to determine functionality of emissions-related components.
- Mass airflow (MAF) sensors measure the mass of air (in grams per second) entering the engine, enabling the PCM to inject the correct amount of fuel.
- Types of crankshaft position (CKP) sensors are magnetic pickup, Hall effect, and optical.
- The PCM relies on the CKP sensor to determine engine speed.
- The PCM reads signals from the brake on/off switch to know when the driver has applied the brake pedal.
- The manual lever position (MLP) switch may be referred to as the transmission range switch or a neutral safety switch.
- The PCM reads the manual lever position to prevent the vehicle from starting unless it is in park or neutral.
- An electronic pressure control (EPC) solenoid is used to control line pressure.
- Shift solenoids are used to direct line pressure directly to an application device or via a shift valve.
- Shift solenoids may allow hydraulic oil to flow to a spool valve rather than directly operate clutches and bands.
- A torque converter clutch (TCC) lock-up solenoid is engaged once vehicle speed has stabilized.
- The PCM is designed to adapt to a driver's individual driving pattern.
- The PCM can also identify and adapt to environmental conditions.
- The first step in a PCM's fault detection system is a basic circuit test.
- The PCM monitors gear ratios as a method of fault detection.
- The PCM can reduce transmission wear by torque reduction during gear shifting.
- The PCM manages engine torque via retarding engine ignition timing prior to transmission shifts.
- Limp-in mode refers to the vehicle still functioning in the event of a failure of the transmission computer or part of the PCM that controls the transmission.

Key Terms

crankshaft position (CKP) sensor A sensor used by the PCM to monitor engine speed. It can be one of three types of sensors—Hall effect, magnetic pickup, or optical.

engine coolant temperature (ECT) sensor A sensor that changes resistance based upon coolant temperature; also known as a thermistor.

input shaft speed sensor A sensor inside the transmission that measures the rpm of the input shaft; also called a turbine shaft sensor.

limp-in mode A transmission operating mode in which limited computer controls are needed to operate for the purpose of getting the vehicle to a shop.

line pressure sensor A variable resistor sensor used to monitor line pressure. It sends a signal back to the PCM where it can be translated into a psi reading.

manifold absolute pressure (MAP) sensor A vacuum sensor that is attached to the intake manifold by a passageway or vacuum hose. The sensor measures engine intake manifold pressure to determine engine load and sends a corresponding signal to the PCM.

manual lever position (MLP) switch A switch that is used by the PCM to tell which gear range the driver has selected with the shift lever.

mass airflow (MAF) sensor A sensor located in the air intake system that is used to measure the mass of the air flowing into the engine.

potentiometer A type of variable resistor that increases or decreases its resistance as it is turned, which creates a varying voltage signal.

reed switch A type of speed sensor that uses a magnetic field to open and close a movable set of contacts. It is used with a rotating magnet to measure rpm of a shaft and send the signal to the PCM.

shift solenoid An electromechanical device used to control oil flow to bands and clutches in an automatic transmission to help shift the transmission.

throttle position sensor (TPS) A type of potentiometer used by the PCM to measure throttle angle.

torque converter clutch (TCC) A hydraulically operated clutch located inside the torque converter that applies at predetermined conditions and stops torque converter slippage.

transmission control module (TCM) A computer module that controls the transmission operation. It may be integrated into the PCM.

transmission fluid temperature (TFT) sensor A type of variable resistor used inside the transmission to monitor oil temperature.

vehicle speed sensor (VSS) A sensor used by the PCM to measure vehicle speed. It is often located in the transmission extension housing.

Review Questions

1. All of the following statements are true with respect to vehicle speed sensor (VSS) *except*:
   - **a.** It is responsible for sending the vehicle speed information to the vehicle speedometer.
   - **b.** It is sometimes called an output shaft speed sensor.
   - **c.** Most VSSs are a type of sensor called a magnetic pickup, or magnetic reluctance, sensor.
   - **d.** It replaces the vacuum modulator on a computerized transmission.
2. All the following statements are true *except*:
   - **a.** In a fully hydraulically controlled transmission, the shift points are controlled by vehicle's power train control module (PCM).
   - **b.** In the fully electronically controlled transmission, the shift points are controlled by the vehicle's power train control module (PCM) or transmission control module (TCM).
   - **c.** In the fully electronically controlled transmission, the shift points are controlled by the transmission governor.
   - **d.** On a hydraulically controlled transmission, the computer can react to the increased load only *after* the transmission shifts.
3. Transmission pressure switches are:
   - **a.** open when there is pressure.
   - **b.** used to determine what gear the transmission is operating in.
   - **c.** present in the dashboard.
   - **d.** used to determine the temperature of the transmission fluid.
4. Which of the following is a pressure transducer?
   - **a.** Electric-only propulsion
   - **b.** Manual lever position (MLP) switch
   - **c.** Overdrive switch
   - **d.** Line pressure sensor
5. Which of the following is installed in an engine coolant passage, often near the thermostat housing on an engine?
   - **a.** MAF sensor
   - **b.** ECT sensor
   - **c.** MAP sensor
   - **d.** CKP sensor
6. All of the following statements are true *except*:
   - **a.** The PCM or the TCM is the brain in the system.
   - **b.** The PCM monitors the TPS sensor along with the VSS to determine a driver's driving pattern.
   - **c.** Slippery conditions require a gear ratio that enhances the traction of the vehicle.
   - **d.** Increasing the amount of torque being transmitted to the transmission reduces transmission wear.
7. Which of the following is used by the PCM to tell which gear range the driver has selected with the shift lever?
   - **a.** Manual lever position switch
   - **b.** Reed switch
   - **c.** Line pressure sensor
   - **d.** Throttle position sensor
8. Choose the correct statement with respect to limp-in mode.
   - **a.** It is when the transmission can only operate in one forward gear.
   - **b.** It is a sensor used by the PCM to monitor engine speed.
   - **c.** It is a sensor that changes resistance based upon coolant temperature.
   - **d.** It is a type of speed sensor that uses a magnetic field to open and close a movable set of contacts.
9. An electromechanical device used to control oil flow to bands and clutches in an automatic transmission is the:
   - **a.** torque converter clutch.
   - **b.** throttle position sensor.
   - **c.** shift solenoid.
   - **d.** potentiometer.

10. All the following statements describing VSS are true *except*:
    a. The types of VSS are magnetic pickup and reed switch.
    b. An input shaft speed sensor is similar to a VSS and relays its signal to the PCM.
    c. It relays vehicle speed information to the speedometer.
    d. It is primarily used to determine functionality of emissions-related components.

ASE Technician A/Technician B Style Questions

1. Tech A says that electronically controlled automatic transmissions improve fuel economy over hydraulically controlled transmissions. Tech B says that electronically controlled transmissions also reduce emissions. Who is correct?
    a. Tech A
    b. Tech B
    c. Both A and B
    d. Neither A nor B
2. A vehicle comes in only operating in 3rd gear. Tech A says the transmission will need to be replaced. Tech B says that the transmission may be operating in limp-in mode. Who is correct?
    a. Tech A
    b. Tech B
    c. Both A and B
    d. Neither A nor B
3. Tech A says that reed switches produce an AC signal. Tech B says that a magnetic VSS can be diagnosed by observing the pattern on a lab scope and comparing it to a known good pattern. Who is correct?
    a. Tech A
    b. Tech B
    c. Both A and B
    d. Neither A nor B
4. Tech A says that transmissions with an input shaft speed sensor and a vehicle speed sensor (VSS) can use this information to determine shift solenoid failure. Tech B says the input shaft speed sensor is also called a turbine speed sensor. Who is correct?
    a. Tech A
    b. Tech B
    c. Both A and B
    d. Neither A nor B
5. Tech A says that delayed shifts could be caused by a malfunctioning TPS. Tech B says that the crankshaft position sensor is used to indicate engine speed. Who is correct?
    a. Tech A
    b. Tech B
    c. Both A and B
    d. Neither A nor B
6. Tech A says that the ECT sensor and the TFT sensor cause a delay in shifting when they indicate cold temperatures. Tech B says that the ECT and the TFT should read substantially different temperatures after sitting over night. Who is correct?
    a. Tech A
    b. Tech B
    c. Both A and B
    d. Neither A nor B
7. Tech A says that the EPC solenoid controls line pressure. Tech B says that if an EPC fails, it will default to minimum line pressure. Who is correct?
    a. Tech A
    b. Tech B
    c. Both A and B
    d. Neither A nor B
8. A vehicle stalls when coming to a stop. After testing the transmission it is found that the TCC solenoid is sticking in the on position. Tech A says that the vehicle probably needs a new torque convertor. Tech B says that the TCC solenoid needs to be replaced. Who is correct?
    a. Tech A
    b. Tech B
    c. Both A and B
    d. Neither A nor B
9. Tech A says that pressure switches are used in electronically controlled transmissions to indicate which gear the transmission is operating in. Tech B says that the brake on/off switch signals the PCM to disengage the TCC. Who is correct?
    a. Tech A
    b. Tech B
    c. Both A and B
    d. Neither A nor B
10. Tech A says that electronically controlled automatic transmissions can vary their gear ratios based upon winter weather conditions. Tech B says that on a vehicle with traction control, the transmission will change shift programming to assist when slippery conditions are encountered. Who is correct?
    a. Tech A
    b. Tech B
    c. Both A and B
    d. Neither A nor B

CHAPTER 27
Servicing the Automatic Transmission/Transaxle

NATEF Tasks

- **N27001** Check fluid level in a transmission or a transaxle equipped with a dipstick. (MLR/AST/MAST)
- **N27002** Check fluid level in a transmission or a transaxle not equipped with a dipstick. (MLR/AST/MAST)
- **N27003** Check transmission fluid condition; check for leaks. (MLR)
- **N27004** Diagnose fluid loss and condition concerns; determine needed action. (AST/MAST)
- **N27005** Drain and replace fluid and filter(s); use proper fluid type per manufacturer specification. (MLR/AST/MAST)
- **N27006** Identify and interpret transmission/transaxle concerns, differentiate between engine performance and transmission/transaxle concerns; determine needed action. (AST/MAST)
- **N27007** Diagnose noise and vibration concerns; determine needed action. (MAST)
- **N27008** Diagnose electronic transmission/transaxle control systems using appropriate test equipment and service information. (MAST)
- **N27009** Perform stall test; determine needed action. (AST/MAST)
- **N27010** Perform pressure tests (including transmissions/transaxles equipped with electronic pressure control); determine needed action. (MAST)
- **N27011** Inspect for leakage; replace external seals, gaskets, and bushings. (MLR/AST/MAST)
- **N27012** Inspect, replace, and align powertrain mounts. (MLR/AST/MAST)
- **N27013** Perform lock-up converter system tests; determine needed action. (AST/MAST)
- **N27014** Inspect, adjust, and/or replace external manual valve shift linkage, transmission range sensor/switch, and/or park/neutral position switch. (MLR/AST/MAST)

Knowledge Objectives

After reading this chapter, you will be able to:

- **K27001** Describe the strategy-based diagnostic process as it relates to automatic transmissions.
- **K27002** Describe the safety and precautions needed when diagnosing automatic transmissions.
- **K27003** Describe the methods of testing automatic transmissions during diagnosis.
- **K27004** Describe the process of test-driving a vehicle during diagnosis of the transmission.

Skills Objectives

After reading this chapter, you will be able to:

- **S27001** Perform maintenance tasks on an automatic transmission.
- **S27002** Perform transmission in-vehicle diagnosis.
- **S27003** Perform in-vehicle transmission repairs.

▶ Introduction

Most automotive technicians will not perform complete rebuilding of automatic transmissions and transaxles in their shops because of their complexity and requirement of special tools. Nevertheless, it is very important for technicians to have a good understanding of how the transmission operates and the correct testing and service procedures. A technician who does not have the ability to diagnose a transmission failure may be doomed to transmission replacement. Or worse, this technician may replace the transmission only to find that the presenting problem still exists. If a customer brings a vehicle to a shop for a transmission problem and is told that the vehicle needs a new transmission for $3500, that customer may then take the vehicle to another shop for a second opinion. What if that other shop actually does some diagnostic work and finds that the transmission is only leaking from a gasket and is low on fluid, or a shift solenoid is faulty? The total bill may be less than $300, and the customer has now found a new, reliable shop. Not only that, but the customer may tell other potential customers not to take their vehicles to the original shop—all because of that one critical experience.

Note, when we say "transmission," we use the term generically to include transaxles also. When we use the term "transaxle," we generally only mean transaxle. Readers should use their understanding of transmission and transaxle theory to understand the intended meaning. For example, if we are discussing ring and pinion backlash, that can only apply to a transaxle.

▶ General Transmission Maintenance

S27001

Most automatic transmissions require some periodic preventive maintenance. This maintenance is designed to reduce transmission failure and save the customer money in the long run. It should include a visual inspection for leaks and a check of the transmission for proper fluid level. Periodic maintenance often also includes transmission fluid and filter replacement, which allows for the inspection of debris or contaminants in the bottom of the transmission pan. On older vehicles, the bands also needed periodic adjustment.

Checking Transmission Fluid

N27001, N27002

On transmissions with a dipstick, transmission fluid should be checked at least at every oil change. Checking the fluid level in an automatic transmission is one of the first steps in diagnosing a transmission problem. Without the proper fluid level, the transmission will not shift properly and may suffer internal damage when the fluid level is too low or too high. If the fluid level is low, it is important to properly identify the source of fluid loss, which we cover in more depth when we discuss the individual components.

For most, but not all, automatic transmissions, the fluid level is checked with the vehicle idling in park or neutral and the transmission at operating temperature. Check the service information for the vehicle you are working on. The vehicle must be on level ground to accurately measure the fluid level. When reading a transmission dipstick, it is important to realize that on most transmission dipsticks it only takes a pint of fluid to raise the level from the bottom of the crosshatched area or add mark to the full mark, unlike on the engine oil dipstick, which typically requires 1 quart (**FIGURE 27-1**). In recent years, vehicle

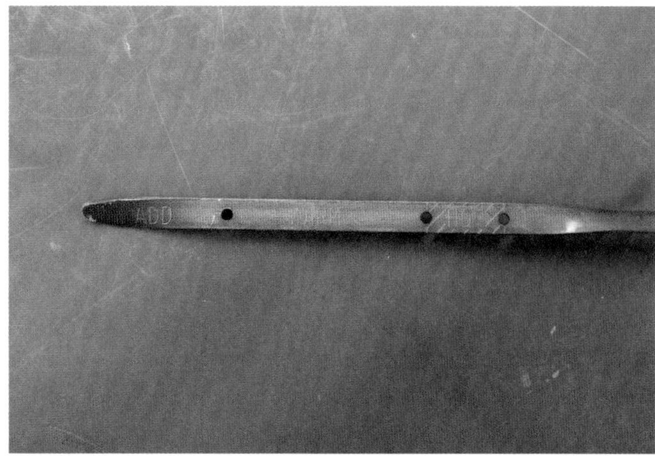

FIGURE 27-1 A typical transmission dipstick. This transmission typically requires 1 pint to bring the fluid level from the bottom of the crosshatched area to the full mark.

You Are the Automotive Technician

Today a customer visits your shop with a transmission concern in a 2011 Chevy Impala. He tells you that it takes about five seconds for it to go into gear the first thing in the morning, and sometimes the engine rpm flares on an upshift. You start by checking the fluid level and find it well below the safe mark. It leaves a moderately dark center when placed on a white paper towel, but no metal filings. You top it off with about 1.5 quarts of the proper transmission fluid to get it ready for a test-drive. During the test-drive, you notice that the malfunction indicator light (MIL) is illuminated and the torque converter clutch doesn't engage, but the customer says the flaring on upshifts is gone. Everything else appears normal.

1. What might a visual inspection reveal on this vehicle?
2. What test(s) should be performed based on the results of the test-drive?
3. Assuming the tests reveal that the transmission does not need to be rebuilt, what service(s) should be recommended to the customer?

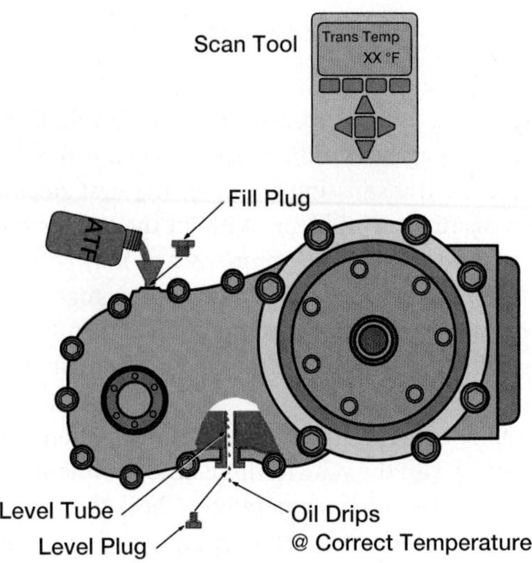

FIGURE 27-2 On transmissions with the fill tube on the bottom of the pan, once the fluid is at the correct level, additional fluid spills out.

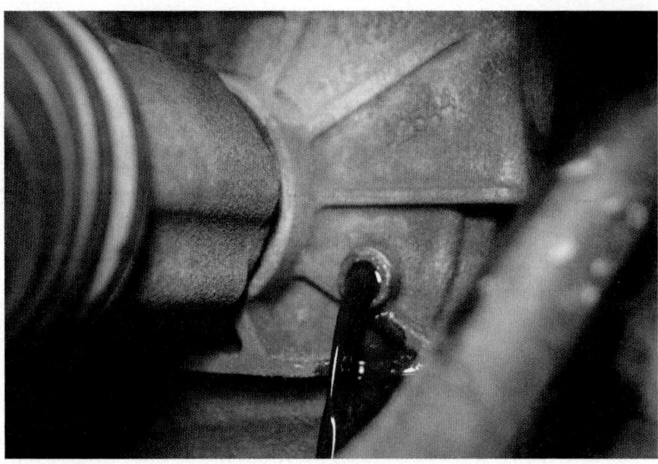

FIGURE 27-3 On transmissions with a fill plug on the side of the transmission, fill until transmission fluid just starts to come out.

manufacturers have been eliminating the transmission dipstick on some of their vehicles. This prevents the vehicle owner from installing the incorrect transmission fluid or overfilling the transmission. It also provides one less entry point for dirt and contaminants to enter the transmission.

On transmissions without a dipstick, it is critical to check the service information for the proper procedure for checking the fluid level. Many late-model Ford and Toyota vehicles have what looks like a drain plug installed in the transmission pan. With the vehicle running, this plug is removed, and fluid is forced up into the transmission through this hole. The hole has a tube attached to it inside the transmission (**FIGURE 27-2**). Once the fluid level is at full, the fluid will simply drain back out, and then the plug can be reinstalled.

General Motors vehicles often have a threaded plug, located on the transmission case, that you remove while the vehicle is running, in order to check the fluid level (**FIGURE 27-3**). Again, fluid is added until transmission fluid flows out of the hole, and then the plug is reinstalled.

Locating Leaks

N27003, N27004

When fluids leak, they travel downward due to gravity and typically toward the back of the vehicle because of air movement during driving. This fluid travel can make finding a leak more challenging, especially if the vehicle has been leaking for a while. If the transmission has a large amount of oil and dirt on it, locating the leak might first require cleaning the transmission case with an engine degreaser or pressure washer. Another option if the leak is hard to locate is to place a leak detection dye in the transmission fluid. After running the transmission for a few minutes, any leaking dye will be visible with an ultraviolet light.

Transmissions can leak from a variety of places. Some places to check are the transmission pan, area around the entrance of the filler tube to the transmission, extension housing gasket, output shaft seals, speedometer/vehicle speed sensor seal, selector shaft seal, area around the electrical connectors that go into the transmission case, front pump seals, fluid cooler lines, and fittings. Also, if the vehicle has a vacuum modulator, remove the vacuum hose from the modulator and see if there is any transmission fluid in the hose. If there is, the modulator is bad. Be sure to remove the radiator cap (with vehicle cold) and check for any transmission fluid in the radiator indicating a leaky cooler.

To diagnose fluid loss and condition concerns, follow the steps in **SKILL DRILL 27-1**.

Replacing Fluid and Filters

N27005

The most common transmission work that the average technician performs is a visual inspection along with fluid and filter replacement. At the minimum, transmission fluid should be changed according to the manufacturer's maintenance schedule. If the vehicle is used for towing or operates in dusty environments, the fluid should be changed more often. Some manufacturers recommend fluid and filter changes every 25,000 miles (40,000 km), whereas others advise against changing the fluid for the life of the vehicle because the transmission is sealed—and you will find many recommendations in between these.

As discussed in the Automatic Transmission Fundamentals chapter, it used to be possible to check the quality of transmission fluid by looking at its color and smelling it. If the fluid was a darkish red or had a burnt smell, it was time for fluid replacement. Some modern transmission fluids are a darker red when they are brand new and even have a slightly burnt smell to them. For this reason, it is important to check the fluid in a new way. Take a few drops of transmission fluid and place them on a clean paper towel. The fluid will disperse on the paper towel, but any contaminants will remain where the fluid was placed. If the fluid spot on the paper towel has a darker center, it is a sign that there are contaminants in the transmission fluid, and it should be changed (**FIGURE 27-4**).

SKILL DRILL 27-1 Checking Fluid Level and Inspecting Fluid Loss

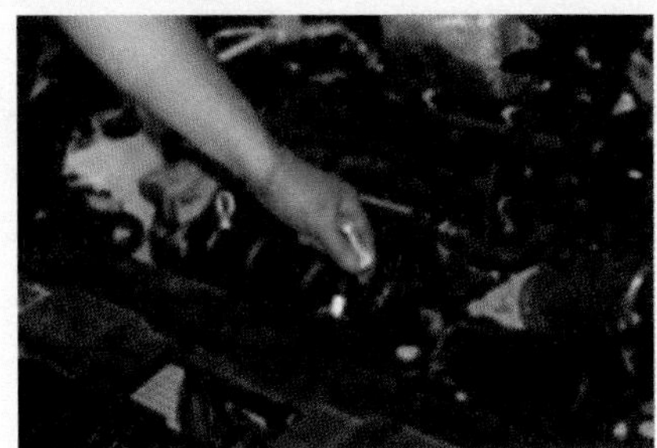

1. Look up the procedure for checking the transmission fluid level in the appropriate service information, and check the level.

2. Inspect the transmission for signs of leakage.

3. If the transmission has a large amount of transmission fluid or engine oil covering it, pour a leak detection dye in the transmission fluid, and use an ultraviolet light to look for fresh transmission fluid leaking.

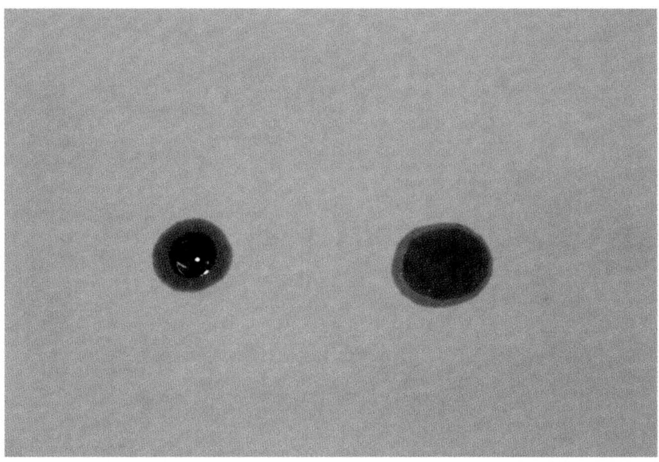

FIGURE 27-4 Place a drop of transmission fluid on a clean paper towel. If the fluid is dirty, the center of the drop will appear darker.

When changing the fluid and filter there are a few things to be careful of:

- Don't lose any parts. On some transmissions, check balls and springs can be held in place by the filter. Removing the filter frees these parts to fall on the floor or in the drain pan.
- Tighten any filter and pan bolts to the specified torque to avoid stripping the aluminum threads or breaking the bolts.
- Research the proper transmission fluid type and capacity (which is approximated for a filter change).

TECHNICIAN TIP

Some manufacturers install a replaceable in-line filter in the cooler line. Be aware of that when you research the replacement procedure for the transmission filter. Also, some transmissions have external filters that look like an atypical engine oil filter.

To replace the fluid and internal filter, follow the steps in **SKILL DRILL 27-2.**

Transmission In-Vehicle Diagnosis

S27002, K27001

To properly diagnose a transmission (and not simply replace it), you need to follow the strategy-based diagnostic process:

Step 1: Verify the Customer's Concern

Step 2: Researching Possible Faults and Gathering Information

Step 3: Focused Testing

Step 4: Performing the Repair

Step 5: Verifying the Repair

So start with verifying the customer's concern by thoroughly reading the comments on the repair order, speaking with the service advisor, or talking directly with the customer. Once you feel you have a good understanding of the customer's impressions regarding the concern, perform any visual inspection needed, such as checking transmission fluid level or listening to any unusual noises. If it is safe, test-drive the vehicle to verify the concern and gain any further clues to the fault. Make sure you drive the vehicle in a safe place and manner. Also, operate the vehicle in the same conditions that the customer described.

The next step is to research the service information to compare the symptoms to any known faults or potential faults. This may involve researching any past service or repairs, technical service bulletins, or the service information for the particular transmission. As you are performing research, start to develop a plan of attack for diagnosing the fault. This involves evaluating

SKILL DRILL 27-2 Draining and Replacing Fluid and Filter

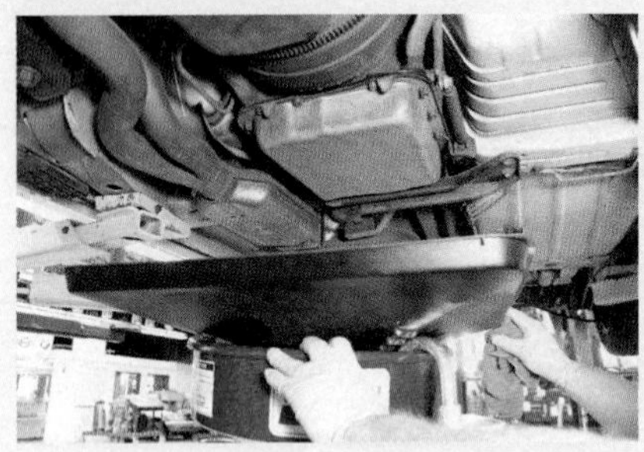

1. Safely raise and support the vehicle on the hoist, Place a drain pan with a large transmission drain funnel under the transmission pan. If the transmission has a drain plug, remove it, and place it to the side to prevent losing it. *Be careful: The transmission fluid will be hot.* If the transmission does not have a drain plug, loosen and remove all of the bolts except for two or three that are next to each other in a corner.

2. Hold the pan against the transmission with one hand while loosening the remaining bolts a few turns. Slowly allow the transmission pan to tilt downward so transmission fluid flows into the drain pan. Once most of the fluid has been drained, remove the remaining bolts the rest of the way.

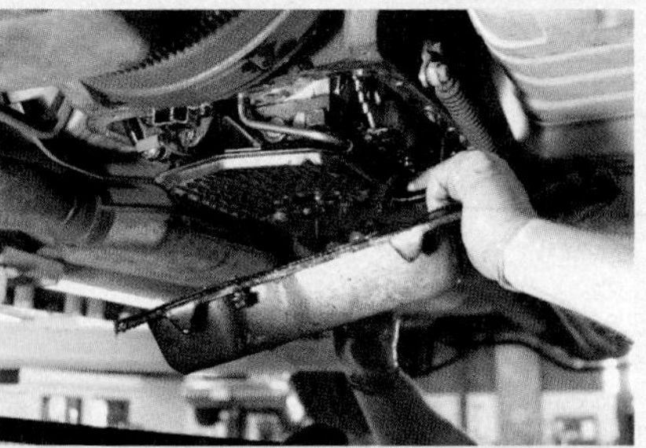

3. Lower the pan and inspect it for nonferrous metal and old clutch material. Metal is sparkly, and clutch material is a dark residue in the bottom of the pan. The presence of some clutch material is normal. Also, inspect the magnet in the pan (if equipped). A small amount of metal is normal.

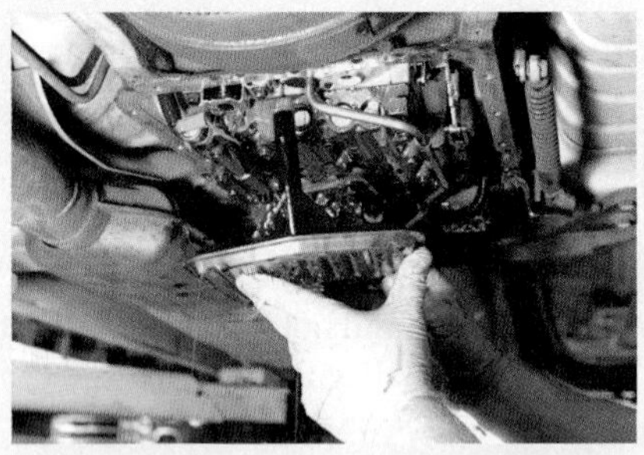

4. If equipped, remove any bolts or clips holding the transmission filter in place, and lower the filter.

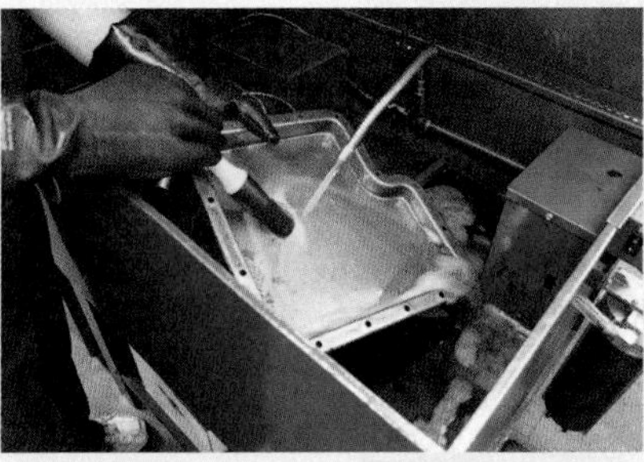

5. Clean the pan and the magnet thoroughly in a parts washer. Dry the pan with a lint-free shop towel or compressed air. Check that the sealing surface of the pan is flat. See your supervisor if it is not.

6. Compare the new filter and transmission gasket to the old one to make sure they are correct. If the filter kit came with a new filter gasket or grommet, install the new one to prevent air leaks and possible transmission damage.

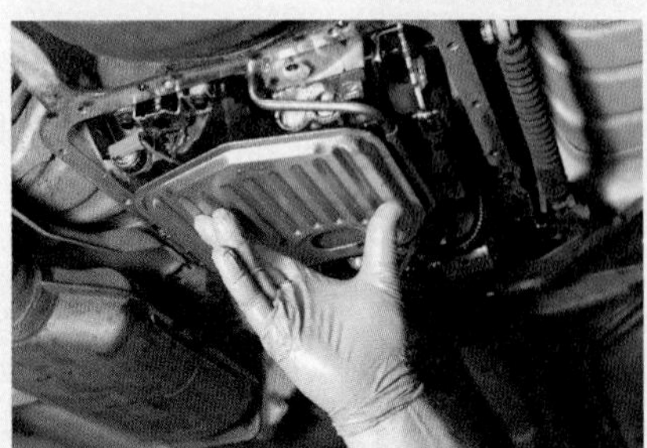

7. Install the new filter, and torque any retaining bolts to the manufacturer's specifications.

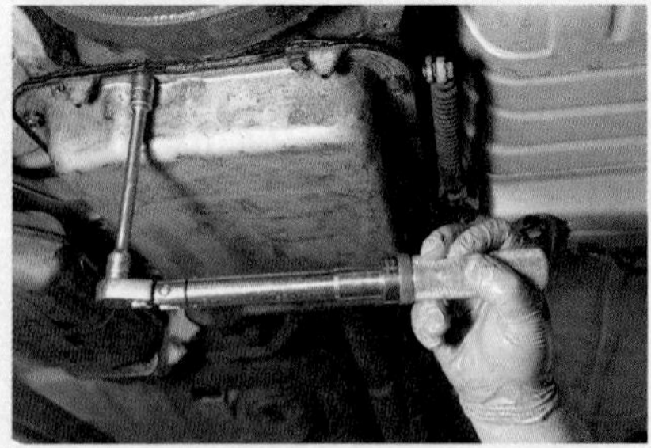

8. Put the new gasket on the transmission pan, and place the pan onto the transmission. Start all of the bolts before tightening any of them. Make sure the gasket remains in the correct place. Torque all of the bolts to specifications.

9. Lower the vehicle and install about 75% of the correct new fluid. Start the vehicle and check the fluid level. Add fluid as necessary to bring it to the bottom of the safe or add mark. While your foot is firmly depressing the brake pedal, place the gear selector in each of the gear ranges. Check the fluid level and top off as necessary. Raise the vehicle, and check for any leaks.

and prioritizing the individual tests you will perform based on these factors:

- Which is the most logical test based on symptoms and specified service procedures?
- Which test eliminates the most possibilities or narrows them down substantially?
- How easy/quick is the test due to access, tool/equipment availability, etc?

You need to consider each of the above factors to determine your initial plan of action. Also, as you implement the plan you need to modify it based on the results you get. For example, if a no-shifting concern could be caused by either a hydraulic or an electrical issue, it would probably be easier and faster to test the electrical signals to the shift solenoids by commanding them on and off, one at a time, than to hook up pressure gauges to test pressures. If the scan tool is able to command the shift solenoids on and the transmission responds appropriately, then your diagnostic plan can be modified toward testing of the electronic control unit and sensors that it uses to decide when to shift. If commanding the solenoids on results in no reaction from the transmission, then your plan will have to be modified toward seeing if the electrical signals are getting to the transmission. As you can see, diagnosis is a process that involves logic, knowledge, accurate service information, appropriate tools, and adequate time to be successful.

Now it is time to implement the focused testing plan you just created. Remember that in-vehicle diagnosis enables you to test the transmission in a variety of ways and with a variety of tests that are impossible to reproduce once the transmission is removed from the vehicle. Scan tools and pressure gauges are often required to perform those tests. The goal is twofold: (1) Diagnose whether the fault is inside or outside of the transmission, and (2) if it is inside the transmission, then diagnose the faulty components before removing the transmission and tearing it down. Performing the repair and verifying if it corrected the problem are covered in the next chapter.

Safety and Precautions

K27002

Most of the testing on an automatic transmission requires that the vehicle be running, or even running in a drive gear. This presents a number of safety issues. Use extreme caution around moving parts of the vehicle, to prevent injury. Often, transmissions are located right next to hot exhaust components, such as the catalytic converter and exhaust pipes. Also, there are many sharp edges under a vehicle, such as cotter pins and sheet metal shrouds. Wearing insulated gloves and arm sleeves can help prevent a technician from being severely burned or cut by these components.

Some tests require the vehicle to be "driven" while in the air on a hoist. Vehicle speed should never be allowed to exceed 45 mph (72 kph) on the speedometer, as one wheel's speed may be double that due to the operation of the differential assembly. That means that even though the speedometer reads 45 mph (72 kph), if one drive wheel is stopped, the other wheel is spinning at 90 mph (145 kph). At high speeds, tires can fly apart and send pieces of tire across the shop. Tires that are not in perfect balance can cause the vehicle to shake, which puts it in danger of falling off the hoist.

Often, a road test will have to be completed to properly diagnose the transmission. You may need to use a scan tool to monitor electronic components inside and outside of the transmission during a road test. Distracted driving can lead to an accident; always use the "movie" feature of a scan tool or a data logger to record the information while you test-drive a vehicle, or have another person drive the vehicle while you monitor the scan tool data. Also, you should never drive with a scan tool in front of an airbag in case of an airbag deployment.

Methods of Testing

K27003

When diagnosing an automatic transmission, whether it is a hydraulically controlled transmission or an electronically controlled transmission, there are several tests that a technician performs. Not all of these tests are done on every transmission or for every problem, nor are they always done in the same order. The order and appropriateness of the test depends on your strategy-based diagnostic procedure. But here are some of the most common tests used in transmission diagnosis:

- *Fluid check:* Although we have already discussed how to check transmission fluid, having the proper level of transmission fluid is critical to the proper operation of the transmission and should be double-checked any time there is a possible transmission problem. The fluid in the transmission has to be at the proper level and in good condition and must be the fluid specified by the manufacturer. It cannot be stressed enough that the wrong fluid or low fluid can cause many issues with an automatic transmission.
- *Road test:* A road test is typically required when the transmission is not so bad that the vehicle cannot easily be driven. In some cases, just driving the vehicle enough to get it in the shop is enough to start the diagnostic process. If the vehicle is brought in on a tow truck, then a test-drive is not an option. If the vehicle can be driven, then a test-drive will give you lots of information and help to verify the customer complaint. What a customer describes as a slipping transmission may not actually be a slipping transmission. The best way to accurately verify the customer complaint is usually by performing a road test.
- *Visual inspection:* A quick check under the hood or under the vehicle often identifies problems that can be causing undesirable transmission operation. For example, on a hydraulically controlled transmission, is the throttle valve cable attached to the throttle body? Did a vacuum hose fall off that leads to a vacuum modulator? On an electronically controlled transmission, an air intake hose after the mass airflow sensor can cause the transmission to shift erratically. After an oil change, this hose may be left disconnected. Or maybe the customer backed into a curb and kinked the exhaust pipe, causing a restriction that affects

the operation of both the engine and the transmission. A quick visual inspection of the entire vehicle can point out issues that relate to the customer concern.

- *Scan of computer for codes (on electronically controlled transmissions):* Scanning the powertrain control module (PCM) will not necessarily find the exact problem, but it may give the technician a general direction of focus. All codes, not just transmission codes, should be documented and evaluated. Often, a code for the engine may be affecting the transmission operation, or vice versa. Once codes are obtained, the diagnostic procedure can be researched in the service information to determine the best way to continue with the diagnosis.
- *Stall test:* A **stall test** checks the torque converter operation and also checks the transmission for slippage by operating the engine at full throttle with the transmission in gear and the brakes applied. It heavily stresses the torque converter and transmission, so any weaknesses that are present will be reflected in the results of the stall test. Because the stall test stresses the drivetrain so much, many manufacturers do not recommend performing a stall test anymore. It is important to check the service information for the particular vehicle.
- *Hydraulic pressure test:* The **hydraulic pressure test** checks the pump operation, the pressure regulator, and the seals and gaskets inside the transmission. Some transmissions provide a test port for line pressure only, whereas others provide ports for each individual gear, line pressure, modulator pressure, etc. The results can be compared against the manufacturer's specifications. Some electronically controlled vehicles have a PID, which can show pressure(s) on a scan tool. Just remember that this information is coming from a pressure sensor and is interpreted by the PCM before being transmitted to the scan tool. Thus, you may have to verify hydraulic pressure manually with a pressure gauge.
- *Leak check:* Leak checking the transmission is often done during many other routine services, as mentioned earlier. If a transmission problem is suspected or a low fluid level is found, the transmission should be inspected for leaks. This check can be a visual inspection, or it can be aided by a special fluorescent dye that is added to the transmission fluid.
- *Pan inspection:* **Pan inspection** was discussed as part of the general service of the transmission, under fluid and filter replacement, but removing the transmission pan to check for debris in the pan can provide a good indicator of transmission failure as well as which part of the components failed. Check the pan for excessive clutch material or metal particles. A small amount of clutch material and metal particles in the pan is acceptable due to normal wear and tear over time.
- *Air checking:* **Air checking** or testing uses compressed air to check the different circuits and components in the hydraulic system. Air checking can indicate if a clutch or band is holding pressure and moving. It can also identify a leak in the circuit or unusual application of a band or clutch.
- *Cooler flow test:* A **cooler flow test** uses a flowmeter on the cooler circuit of the hydraulic system of the transmission to help identify poor pump performance, restriction in the cooler, converter clutch problems, and worn spool valves in the valve body.
- *Noise, vibration, and harshness test:* The **noise, vibration, and harshness (NVH) test** describes a process that technicians have always done—listening for noise and feeling for vibrations and harshness. Now there are test tools, called NVH analyzers, that can help pinpoint noises in a vehicle while the vehicle is driven.

A technician uses all of the tests just described, in conjunction with the service information, to make a comprehensive diagnosis of the transmission. By comparing the transmission operation during a test-drive to the correct clutch and band application charts in the service information, the technician can pinpoint the best method for further testing, including using scan tools, checking on noises heard during the test-drive, pressure testing, air checking, etc. Another critical step in diagnosis is checking for technical service bulletins (TSBs). In many cases, the manufacturer has identified common problems that occur with a particular vehicle and has provided updated programming, parts, or test procedures to verify the problem. This assistance from the manufacturer can save a technician a lot of time and the customer a lot of money. If a technician knows how the system works and has access to the right service information, knowledge of the proper tests, and the correct tools, he or she should be able to create a successful strategy-based diagnosis plan.

Test-Driving

K27004, N27006, N27007

Test-driving, or performing a road test, is a critical part of the diagnostic procedure. Transmissions operate much differently when they are placed under a load, such as driving up a hill, than they do sitting in a shop on a hoist. A transmission might operate perfectly fine when operated on the hoist but have a major amount of clutch slippage when taken on a road test. Very few repair shops have access to a chassis dynamometer that would allow them to "test-drive" the vehicle in the shop, so a road test is a great option. Test-driving the vehicle must be done in a safe manner to prevent an accident. If you need to monitor test equipment such as a scan tool or pressure tester, then request the assistance of another technician to operate the vehicle safely.

During the test-drive, it is critical to place the transmission in the same operating conditions in which the customer stated the problem occurred. Did the customer state that the problem occurred during hard acceleration? Or maybe the problem only occurs when the transmission is cold or during shifts.

When test-driving, it is important to test the transmission in many different operating conditions and record the results for further analysis. Test the transmission during light, medium, and hard acceleration. Place the transmission in different gear ranges. For example, does the problem occur only in the overdrive range, or does it still occur in the L2 range? While test-driving, listen for noises and vibrations that the customer may interpret as a transmission problem but that in fact indicate a problem with another component, such as a U-joint, transmission mount, or engine fault.

Applied Math

AM-53: Word Problem: The technician can evaluate symptoms of problems with a customer or associate technician and identify any relevant missing data required to solve the problem.

Perhaps the most important element of automotive diagnostics is to obtain all of the important information about the problem from the customer. Most customers do not understand the technical operation of their vehicles; they can only describe the symptoms of a problem that they can feel, see, or hear. They may have additional relevant information but may not think to include it. It is up to the technician to ask questions that will guide the customer to provide all the information that could help with the diagnosis.

An example might be this complaint: "The car sometimes seems to rev too hard and then bang into gear." As technicians, we may think this "sounds" like an automatic transmission problem, but to guide our diagnostic process, we need more information from the customer. Clarifying questions in this scenario may include the following:

- Does the reported problem occur when the vehicle is cold or after driving for a long time?
- Has the transmission been serviced recently?
- How long has the reported problem been occurring?
- Is the vehicle used to carry or tow heavy loads?

Answers to these questions guide us in different directions when looking at a problem and may help us reproduce the required conditions to get a fault to occur during testing. Sometimes it can be useful to take the customer along during a road test, to provide further clarification or have him or her reproduce the fault.

Scanning the Transmission Control Module/PCM

N27008

A high-quality or factory scan tool can be an invaluable asset when diagnosing an electronically controlled transmission. A generic scan tool can read codes, but a factory scan tool can also view live data from the sensors and actuators inside the transmission. Further, many factory scan tools give the technician **bidirectional control** of the transmission. Bidirectional control means the technician is able to command different solenoids and actuators on and off to check their operation. For example, the torque converter clutch (TCC) could be commanded on while the engine is idling in drive with the brakes on. This should kill the engine if the TCC solenoid is operating and the clutch is in good shape. Activating this solenoid and observing the reaction gives you valuable information that helps determine the next steps in the diagnosis process.

Any trouble codes should be researched in the appropriate service information to find the correct diagnostic procedure. Following the diagnostic procedure step by step is critical in a successful repair of the vehicle. The diagnostic procedure often requires the use of a digital volt-ohmmeter (DVOM) to aid in the diagnosis of wiring, switches, solenoids, sensors, and actuators.

To diagnose electronic transmission control systems using appropriate test equipment and service information, follow the steps in **SKILL DRILL 27-3**.

Stall Testing

N27009

A transmission stall test is used to test the torque converter's operation and to check for slippage in the transmission. Most manufacturers do not recommend stall testing on late-model vehicles, as it can damage the transmission. Check with the service information before performing a stall test on any vehicle. Also, a stall test can be dangerous if it is not performed properly.

SKILL DRILL 27-3 Scanning the TCM/PCM

1. Install a scan tool onto the vehicle's data link connector (DLC). For the location of the DLC, consult the service information.

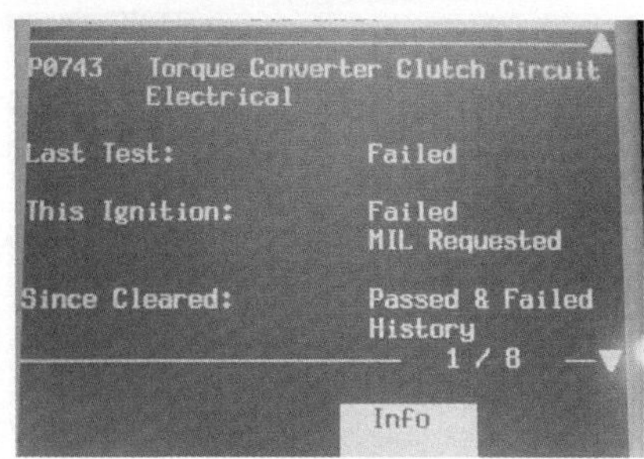

2. Retrieve any diagnostic trouble codes (DTCs) from the vehicle. Record the codes.

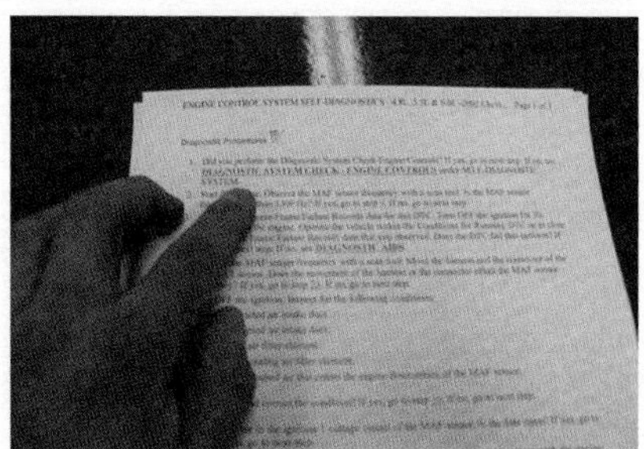

3. Using the service information, research the diagnostic procedure for any codes found. Follow the diagnostic procedure step by step until you have completed the diagnostic procedure and found the cause of the DTC.

Make sure no one is near the front or rear of the vehicle while you are performing the test. A stall test puts maximum load on the engine, transmission, and brake system. If any one of these systems fails during the test, it can have serious consequences. Do not perform a stall test if there is an engine problem, such as worn engine bearings, broken powertrain mounts, or overheating. The same goes for the brake system. Also, a stall test can make a marginal transmission fail. During a stall test, a large amount of heat will be produced. Place a large fan in front of the vehicle to blow air through the radiator. Some other rules to follow while performing a stall test are listed here:

- Check all engine and transmission fluid levels before performing a stall test.
- The vehicle should be at normal operating temperature before performing the stall test.
- *Do not perform the test for more than five seconds*, as severe transmission damage can occur.
- Allow the engine and transmission to cool down for about 30 seconds between tests of the different gear ratios. When the test is complete, allow the engine to run for a few minutes to allow the transmission fluid to cool.

When performing a stall test, place the vehicle in a gear with the emergency brake fully applied and the brake pedal firmly held down. Depress the accelerator until the engine reaches its maximum speed (the vehicle should not be moving). Record the engine rpm, and compare to the manufacturer's specifications. Engine rpm that is lower than the specified rpm indicates one of two problems: Either the engine has a problem and is not producing enough power or the torque converter has a malfunction internally. Engine rpm that is high in all gears can be a sign of a faulty pump, an internal pressure leak, a faulty pressure regulator, low fluid, or a clogged filter. If the engine rpm is high in only one gear, check the service information to determine which clutch or band may be slipping, and, if possible, air check that component.

To perform a stall test, follow the steps in **SKILL DRILL 27-4**.

Testing Hydraulic Pressure

N27010

Automatic transmissions, whether they are hydraulically controlled or electronically controlled, use hydraulic devices inside the transmission to apply, drive, or hold mechanical devices.

SKILL DRILL 27-4 Stall Testing

1. Look in the service information to determine whether a stall test is recommended for this vehicle. If it is, follow the procedure. Place wheel chocks in front of and behind the wheels to help prevent vehicle movement.

2. Check engine and transmission fluid levels. If the vehicle does not have a tachometer, place a tachometer on the engine. Place a large fan in front of the vehicle's radiator to help cool the radiator and transmission fluid.

3. Start the vehicle and allow it to reach operating temperature. Firmly apply the emergency brake, depress the brake pedal, and place the vehicle into a gear.

4. Gradually press the throttle all the way to the floor to increase the engine rpm to the stall speed for a maximum of five seconds while firmly holding the brake pedal, and note the rpm.

5. Release the throttle and place the vehicle into neutral. Record the stall speed. Cool the engine down for 30 seconds. Continue testing the remaining gear ranges (R, D4, D3, D2, D1). Compare your results with the manufacturer's specifications.

This means that to perform a proper diagnosis, a technician often has to perform a pressure test of the hydraulic system. Use the appropriate service information to determine which test ports the manufacturer has supplied and the specified pressures. It is sometimes useful to install multiple pressure gauges into the different test ports so that you can monitor the different pressures as the transmission is operated.

Manufacturer pressure specifications vary greatly depending on the transmission and its application. Most manufacturers specify a range for the pressure readings in particular gear ranges. Generally, line pressure should be highest in the reverse gear range. Always select a gauge that can read higher than the expected pressure you will be reading. That said, if you are expecting a reading of 120 psi (827 kPa) and you select a gauge that reads up to 400 psi (2758 kPa), you may not get an accurate reading. A gauge that reads up to 150 psi (1034 kPa) would be more accurate.

On electronically controlled transmissions, the procedure might involve using a factory scan tool to actuate the electronic pressure control (EPC) solenoid in order to check the PCM's ability to vary the line pressure to meet the demands of the transmission. Low line pressure can be a sign of a weak or worn front pump, a bad pressure regulation system, or internal leaks in the transmission. Follow the diagnostic procedure found in the service information. High line pressure can be a sign of a stuck pressure regulator valve or failed EPC solenoid.

To perform pressure tests (including transmissions/transaxles equipped with EPC), follow the steps in **SKILL DRILL 27-5**.

▶ In-Vehicle Transmission Repair

S27003

Many automatic transmission repairs can be completed without removing the transmission from the vehicle. Transmission repairs that are completed with the transmission still in the vehicle are often less expensive than pulling the transmission for a complete rebuild because removal and reinstallation of the transmission will take a few hours at a minimum. Examples of typical in-vehicle transmission repairs include replacement of the external gaskets and seals, vehicle speed sensor, shift solenoids, extension housing bushing, and powertrain mounts; repair of the valve body; and adjustment of some servos and bands.

SKILL DRILL 27-5 Performing Pressure Tests

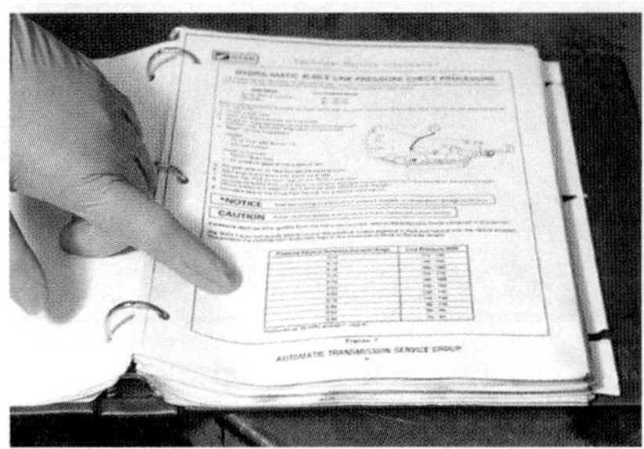

1. Research the transmission's specified hydraulic pressures. Verify the correct transmission fluid level in the transmission. Place a drain pan under the transmission and remove the correct pressure test port plug(s).

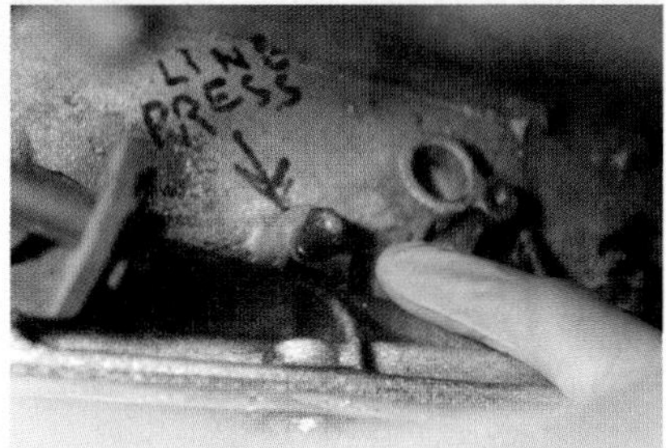

2. Install a transmission pressure tester(s) capable of measuring the maximum pressure into the test port(s) on the transmission.

3. Start the engine, and place the vehicle in the specified operating conditions to monitor the pressures (e.g., transmission hot, in drive, idling). Record the pressure(s).

4. Shut off the engine, remove the transmission pressure tester(s), seal the threads, and reinstall the test port plug(s).

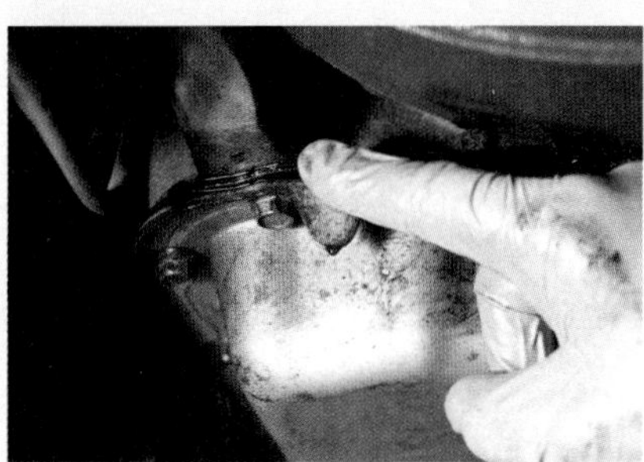

5. Clean off any transmission fluid that dripped onto the transmission, restart the vehicle to check for leaks, and top off the fluid if necessary.

Inspecting and Repairing an External Leak

N27011

Transmissions have many external seals and gaskets that can be replaced without removing the transmission from the vehicle. Examples include the extension housing seal, the vehicle speed sensor seal, the pan gasket, and the extension housing gasket. A transmission fluid leak can result in failure of the transmission due to a low fluid level. As a technician, it is important to fix any external leaks on the transmission to prevent this failure. In the case of a leaky gasket (pan or extension housing), the leak may be caused by loose bolts. If so, retighten them to the specified torque. But *never* tighten them beyond specifications.

Some seals, such as the extension housing seal, may require the replacement of the extension housing bushing to properly repair the leak. With a worn bushing, the driveshaft may move up and down excessively, flexing the seal and allowing transmission fluid to flow out of the transmission. Replacement of an extension housing bushing requires the removal of the driveshaft and the extension housing. The bushing can then be driven out using a bushing driver set.

Most seals on the outside of a transmission require specialized tools to be removed and replaced without removing the transmission from the vehicle. Some of these specialty tools are universal types; others are specific for a particular model of transmission.

To inspect for leakage and replace external seals, gaskets, and bushings, follow the steps in **SKILL DRILL 27-6**.

Inspecting and Repairing Powertrain Mounts

N27012

The **powertrain mounts** hold the engine and transmission in the proper position in the vehicle. If the transmission or engine is allowed to move out of position, all of the components that connect the engine and transmission to other parts of the vehicle go out of alignment. In the case of a transmission

SKILL DRILL 27-6 Inspecting for Leakage and Replacing Seals, Gaskets, and Bushings

1. Safely raise and support the vehicle on a hoist. Carefully inspect the front pump, output shaft, selector shaft seals, pan gasket, side pan gasket (if equipped), and extension housing gasket for leaks.

2. Inspect the extension housing bushing by moving the driveshaft up and down. If there is excessive movement in the driveshaft, the extension housing bushing must be replaced. Place a drain pan under the extension housing and remove the driveshaft from the vehicle.

3. Remove the extension housing from the vehicle. On some older vehicles, the seal and bushing can be replaced while it is still in the vehicle, using a specially designed puller. Remove the extension housing seal using the correct tool.

4. Use the correct-sized bushing driver to remove the extension housing bushing.

5. Use the correct-sized bushing driver to carefully drive the bushing into place. Take note of any lubrication holes that need to be lined up before installation.

6. Use the correct seal installation tool to install the new seal in the housing. Lubricate the edge of the seal with clean transmission fluid. Reinstall the extension housing and the driveshaft.

equipped vehicle, the shift linkage, or throttle linkage, could become bound or out of place, resulting in a safety concern. With the shift linkage out of place, the gear selector indicator might indicate one gear, but the transmission is in another gear. This could cause an accident, especially if the driver expects the vehicle to be in neutral or reverse when it is actually in drive. Worn powertrain mounts can also cause drive axles or the driveshaft to be out of alignment, causing a vibration. Worn powertrain mounts should always be replaced with the correct part for the vehicle.

▶ TECHNICIAN TIP

A technician was once called by a customer who was on a trip, towing his large boat, and was experiencing a strange phenomenon. It seemed that the vehicle had developed a mind of its own. As soon as the vehicle would start up a hill, the gas pedal would drop to the floor, and the vehicle would go to full throttle. If the driver turned the engine off at the top of the hill and restarted the engine, everything worked fine until the next hill, when the gas pedal would once again drop to the floor. The technician walked the customer through the process of testing the powertrain mounts, one of which was torn, allowing the engine to move and pulling the throttle linkage open and the pedal to the floor.

Most powertrain mounts are of the rubber style, which are molded from precisely engineered rubber and with specific shapes and voids (**FIGURE 27-5**). Rubber mounts are subject to cracks and tears that occur over time due to their constant flexing. Their life is also reduced by being saturated with oil, power steering fluid, or gasoline, which can soften the rubber.

Powertrain mounts can also be of the hydraulic style (**FIGURE 27-6**). Hydraulic mounts use silicone hydraulic fluid that is squeezed back and forth between two chambers, similar to the design of a shock absorber. They use a metered orifice between the chambers to control the flow of fluid, which dampens the engine pulsations. Some hydraulic mounts use a computer-controlled electronic control valve that can

FIGURE 27-5 Rubber style powertrain mounts.

FIGURE 27-6 Hydraulic style powertrain mount.

open an alternate passageway or vary the size of the orifice. This changes the rate of flow between the chambers, thereby allowing the computer to control the amount of dampening that is needed for various operating conditions. Hydraulic mounts should be inspected for leaking and excessive movement. The solenoid valve can also be tested for continuity or commanded to operate with a scan tool. Hydraulic-style powertrain mounts are typically much more expensive than rubber mounts and may be available only through the manufacturer.

Often, in the event of a broken powertrain mount, customers complain of a loud thump when they apply the accelerator in drive or reverse, and again when they apply the brake pedal. This loud thump may be the engine and/or transmission being pulled up from the broken mount due to the engine's torque, and then being set back down when the brake is applied.

Powertrain mounts wear out over time, so they need to be inspected periodically. Because leaking fluids can soften the rubber in the mounts over time, it is wise to encourage customers to have leaks repaired before they damage other parts.

To inspect, replace, and align powertrain mounts, follow the steps in **SKILL DRILL 27-7**.

Testing the Lock-Up Torque Converter

N27013

The lock-up torque converter has been used by manufacturers since the late 1970s to help improve fuel economy when the vehicle is cruising at highway speeds. The lock-up torque converter contains a TCC that forces the turbine and impeller to match speeds. This system has several possible components that can fail, resulting in a loss of the lock-up ability or a converter that will not unlock. The system has a mechanical friction clutch that is actuated by a hydraulic piston and control circuit (**FIGURE 27-7**). The hydraulic control system uses an electrical solenoid valve to direct fluid away from, or to, the TCC assembly.

SKILL DRILL 27-7 Inspecting, Replacing, and Aligning Powertrain Mounts

1. Look up the correct procedure for checking and replacing powertrain mounts in the service information. Safely raise and support the vehicle on a hoist to inspect the powertrain mounts under the vehicle.

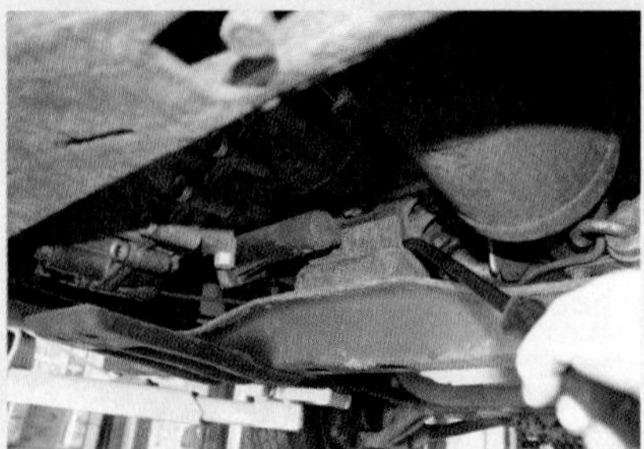

2. Use a pry bar to carefully push up on the engine and transmission while watching the powertrain mounts. If the rubber section of the powertrain mount separates from the metal bracket, or if the rubber is torn, replace the mount.

3. Lower the vehicle, apply the brake, and start the engine. Place the vehicle into gear, and apply the throttle slowly. Watch for excessive engine movement on the mounts. Repeat in reverse to check the opposite mounts.

4. Use an engine support fixture, engine hoist, or transmission jack to raise the engine or transmission just far enough that the weight is off the powertrain mount. Be *very* careful not to cause the vehicle to shift on the hoist!

5. Remove the bolts securing the mount to the transmission or engine, and then the bolts securing the mount to the frame of the vehicle.

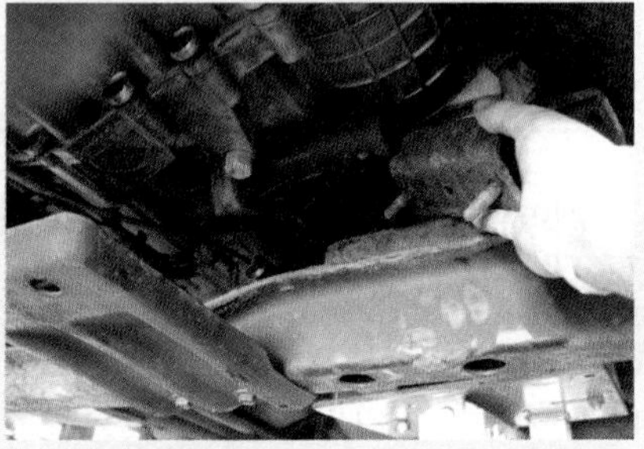

6. Remove the old mount, and compare it to the new mount.

7. Place the new mount in the correct position, and lower the jack slightly. Be careful to keep your fingers away from any pinch points. Reinstall the bolts and torque them to specifications.

8. Lower the engine or transmission back down. Reinstall all components that were removed to access the powertrain mount.

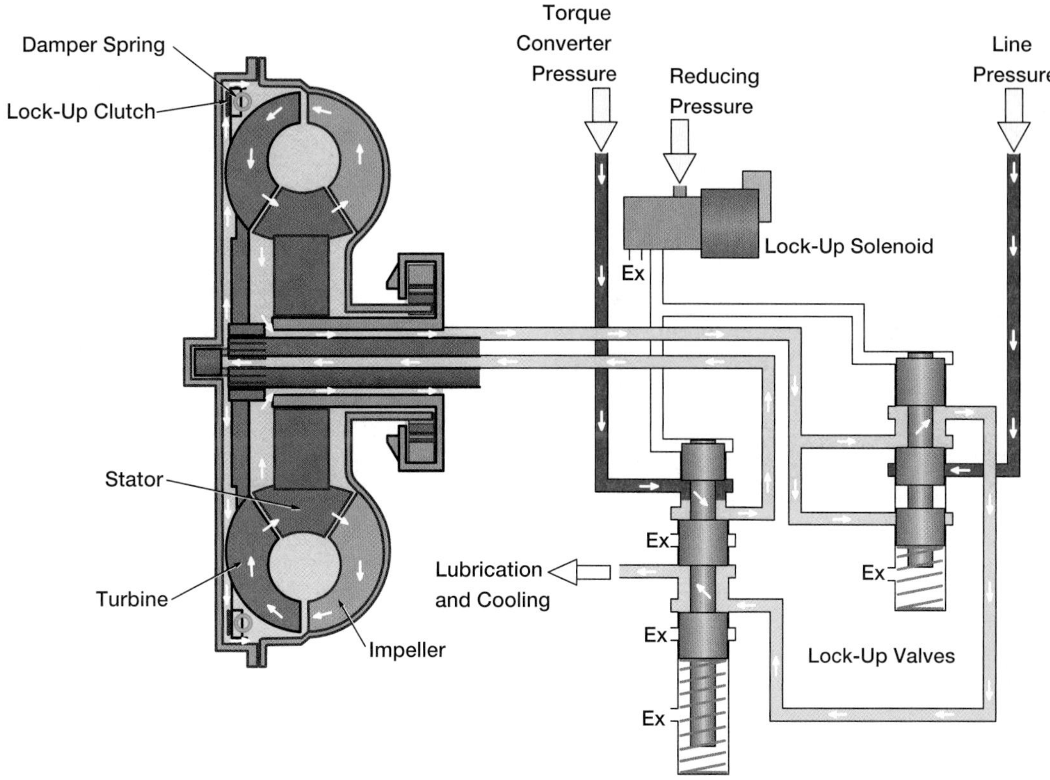

FIGURE 27-7 A typical lock-up converter cutaway.

On electronically controlled transmissions, the operation of the TCC solenoid is monitored by the PCM and can, but will not always, set DTCs if a fault is present. For testing of a TCC on an electronically controlled transmission, a factory scan tool is invaluable. If the PCM has set any codes related to the TCC circuit, look up the proper diagnostic procedure in the appropriate service information. Follow the procedure step by step.

Usually on hydraulically controlled transmissions, the PCM will not set a DTC except in the case of an open circuit or a shorted solenoid for the TCC circuit. On a hydraulically controlled transmission, test-drive the vehicle to see whether you can feel the engagement of the lock-up torque converter. The scan tool might give data as to when the solenoid has been energized to apply the TCC solenoid, but many hydraulically controlled transmissions were built before auto manufacturers gave the technician much diagnostic data through the scan tool. If the TCC is not engaging, always check with the service information for a diagnostic procedure. Before beginning any diagnostics, check to make sure the system has power. Next, inspect and test the solenoid's ground circuit. If the power and grounds check out, test the solenoid with an ohmmeter. If the solenoid's resistance is within specifications, remove the solenoid from the transmission. Apply power and ground to the solenoid while attempting to blow air through the solenoid. If the solenoid operates properly, the problem is probably in the torque converter itself. If the solenoid does not allow air to flow, place it in a clean container of automatic transmission fluid, and electrically cycle the solenoid to see if you can remove any blockage from the solenoid. If this does not work, it will be necessary to replace the solenoid.

▶ TECHNICIAN TIP

When diagnosing a TCC circuit, take the time to double-check that the brake light circuit is working properly. It is not unheard of for a TCC system to fail because the brake light switch is sticking, preventing the operation of the system. This type of failure will not set a DTC, as the PCM believes the customer is applying the brake pedal. The TCC cannot be applied at the same time as the brake pedal because the two systems would be fighting each other.

▶ TECHNICIAN TIP

In some vehicles, especially older General Motors vehicles, when the transmission was hot, the engine would want to die when coming to a stop. The TCC solenoid would stick on, causing the TCC to not release. This caused the engine to die because the torque converter could not freewheel while the vehicle was stopped. In this case, the TCC solenoid required replacement with an updated part.

To perform lock-up converter system tests and determine necessary actions on electronically controlled transmissions, follow the steps in **SKILL DRILL 27-8**.

SKILL DRILL 27-8 Performing a Lock-Up Converter Test

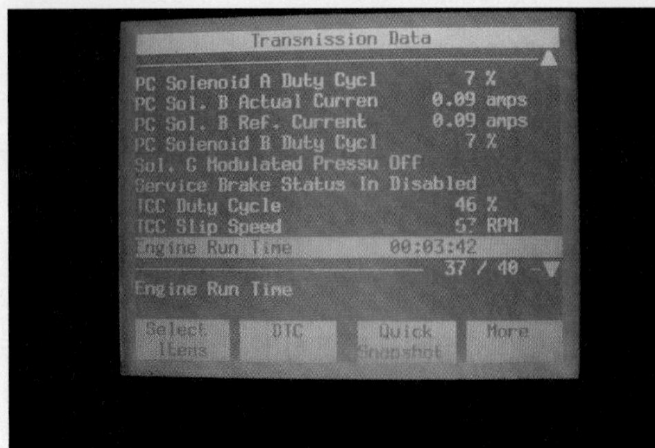

1. Scan the vehicle to identify any TCC-related DTCs the PCM has, and record. Follow the diagnostic procedure for the DTC. Test-drive the vehicle and observe the operation of the TCC circuit and the TCC slippage on the scan tool. If not within specifications, check for loose wiring or connections.

2. Check the TCC circuit for power and ground at the TCC connector. If full power and ground are not available at the TCC connector, locate the open or high resistance, using the voltmeter to measure voltage drop.

3. If power and ground are available at the TCC connector on the transmission, measure the resistance of the solenoid through the TCC connector. If out of specifications, the transmission will have to be drained and the pan removed to test the wires and solenoid. If the solenoid is not within specification, replace the solenoid.

4. If power and ground are available to the TCC connector, and the TCC resistance is within specifications, the TCC solenoid will have to be removed and checked to see if the valve is actually opening and closing. Apply power and ground to the solenoid while attempting to blow air through the solenoid. If the solenoid operates properly, the problem may be in the TCC itself.

Adjusting the Shift Linkage and Transmission Range Sensor

N27014

When servicing an automatic transmission, it may be necessary to adjust the shift linkage along with the transmission range sensor. If these are not adjusted properly, the vehicle may not start, or it may start in gear, resulting in a major safety hazard. If the linkage is not adjusted properly, the transmission position indicator (PRNDL indicator) will not be set correctly; the customer will not know which gear the vehicle is in; and the backup, or reverse, lights may not work. Also, the neutral safety switch requires proper linkage adjustment so that the vehicle starts only in the park and neutral positions. Make sure the brake pedal is firmly applied when checking that the vehicle starts only in the park and neutral positions.

To inspect, adjust, and replace the manual valve shift linkage, transmission range sensor/switch, and park/neutral position switch, follow the steps in **SKILL DRILL 27-9**.

SKILL DRILL 27-9 Inspecting, Adjusting, and Replacing Manual Linkage

1. Look up the proper service procedure in the appropriate service information. Place the gear selector in the park position.

2. Safely raise and support the vehicle on a hoist (if necessary to access the transmission range switch and manual valve linkage). Disconnect the shift linkage from the transmission.

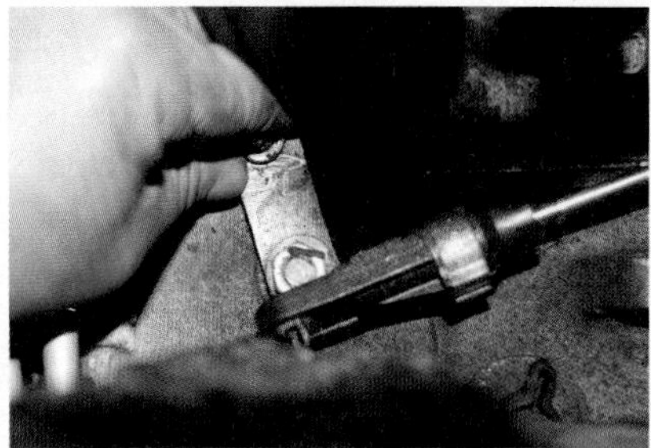

3. Check that the manual lever is in the park position. The lever should snap into position.

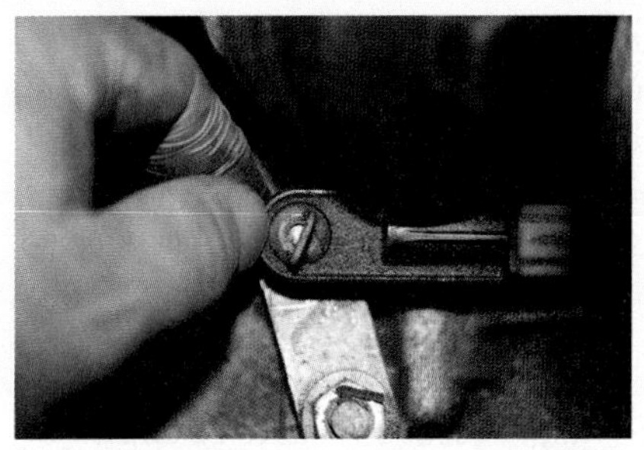

4. The shift linkage should fit right onto the manual valve with no pulling on the linkage or the manual valve.

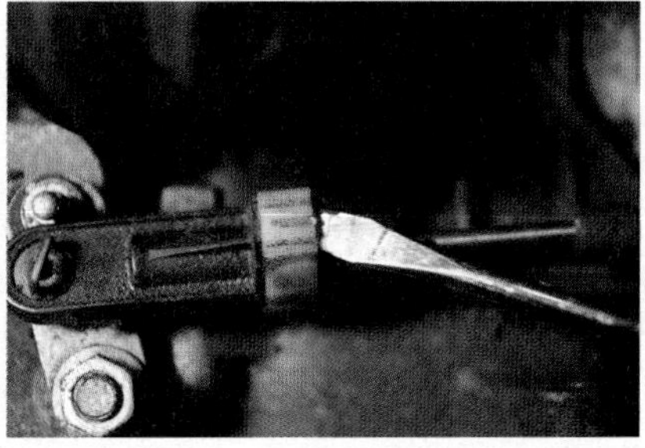

5. If the linkage does not line up, loosen the adjustment on the shift linkage and adjust the linkage so that it will install properly on the manual valve. Tighten the adjustment on the shift linkage.

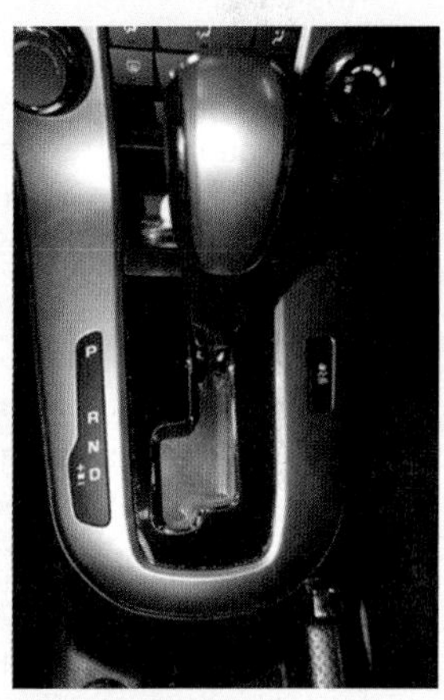

6. Double-check that the PRNDL indicator still indicates the vehicle is in park.

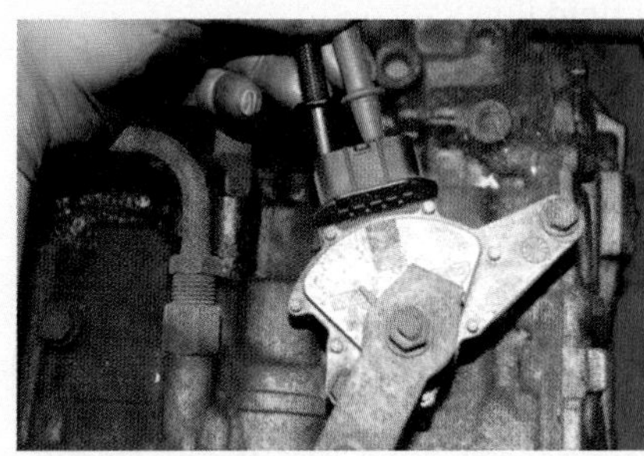

7. Use an ohmmeter to check that the neutral safety switch/transmission range switch has continuity on the correct terminals.

8. If not, loosen the switch and adjust its position, if possible. If continuity is never obtained or is obtained in every gear, replace the switch.

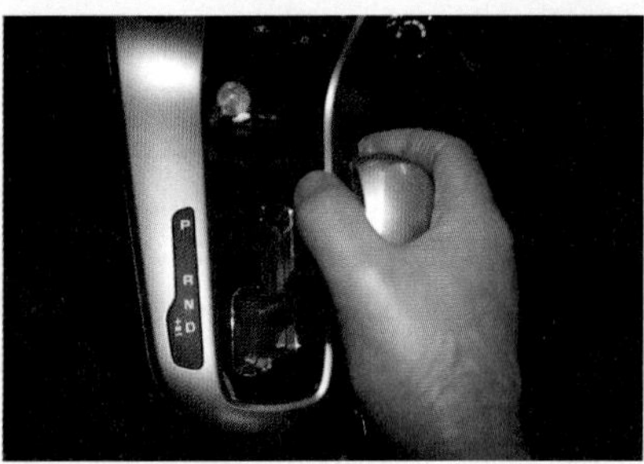

9. Run the shifter through all of the gear ranges, checking for proper operation.

▶ Wrap-Up

Ready for Review

- ▶ Automatic transmissions require periodic maintenance, which includes checking transmission fluid and replacing fluid and filters.
- ▶ If transmission fluid levels are low, identify the source of fluid loss.
- ▶ Visually inspect transmission fluid, and replace as necessary.
- ▶ Diagnosis of transmission problems requires a thorough transmission inspection.
- ▶ Use extreme caution and wear proper protective clothing and equipment when diagnosing a transmission in a running vehicle.
- ▶ Diagnostic transmission tests include the fluid check; road test; visual inspection; scan of the computer for codes; stall test; hydraulic pressure test; leak check; pan inspection; air check; cooler flow test; and noise, vibration, and harshness test.
- ▶ Always perform a diagnostic test-drive under a range of operating conditions.
- ▶ Using a factory scan tool allows technicians bidirectional transmission control.
- ▶ Stall testing involves determining whether the engine can achieve manufacturer-specified rpm in each gear.
- ▶ Hydraulic system pressure testing should be performed via manufacturer-supplied test ports.
- ▶ In-vehicle transmission repairs include replacement of the vehicle speed sensor, extension housing bushing, and powertrain mount; repair of the valve body and the external gaskets and seals; and adjustments of the servo and bands.
- ▶ Leaks from gaskets or seals may require bushing replacement to properly repair the leak.
- ▶ Repair of powertrain mounts is critical for keeping the engine and transmission in proper position.
- ▶ Hydraulically controlled transmission vehicles will likely require a test-drive to determine functionality of the torque converter clutch.
- ▶ Adjust the shift linkage and transmission range sensor as part of automatic transmission servicing.

Key Terms

air checking The use of compressed air to check clutch and servo operation on transmissions during and after assembly.

bidirectional control The ability to command different solenoids and actuators on and off to check their operation.

cooler flow test The placement of a specialty measuring device into the transmission cooler line to measure fluid flow to the cooler. Low cooler flow can be a sign of other issues with the pump and lubrication system.

hydraulic pressure test The use of a hydraulic pressure gauge to measure the amount of hydraulic pressure produced in each gear range.

noise, vibration, and harshness (NVH) test A test to measure for any audible noises, vibrations, and harsh operation. It can be completed by the technician with or without the aid of an NVH tester. The tester is used to pinpoint the exact frequencies of the noise and vibrations.

pan inspection The process of removing the transmission pan to check for clutch material, metal, and other debris or contaminants.

powertrain mount A rubber or metal bracket used to secure the engine and transmission into the vehicle. Some vehicles use hydraulic or electrohydraulic powertrain mounts.

stall test A test that involves raising the engine rpm to wide-open throttle while the brake is firmly applied and the transmission is in gear. The test is used to check torque convertor and transmission operation on some vehicles.

Review Questions

1. As part of general transmission maintenance, all of the following should be done *except*:
   a. checking transmission fluid.
   b. disassembling and checking the transmission.
   c. locating leaks.
   d. replacing filters.
2. When diagnosing problems in the transmission:
   a. assume the fault.
   b. ignore the customer's point of view.
   c. take a quick decision based on visual inspection.
   d. use accurate service information.
3. When using a scan tool to monitor electronic components inside and outside of the transmission during a road test, which of the following precautions has to be taken?
   a. Wear insulated gloves and arm sleeves.
   b. Check if tires are in balance.
   c. Do not use the movie feature of a scan tool.
   d. Do not drive with a scan tool in front of an airbag.
4. When a customer describes a slipping transmission as a complaint, what test is the best way to verify?
   a. Fluid test
   b. Visual inspection
   c. Road test
   d. Test converter fluid flow
5. Which test stresses the drivetrain so much that many manufacturers do not recommend performing it?
   a. Hydraulic pressure test
   b. Stall test
   c. Fluid test
   d. Road test
6. To check the pump operation, what test should the technician conduct?
   a. Leak check
   b. Air checking
   c. NVH test
   d. Hydraulic pressure test
7. All of the following can be replaced during in-vehicle transmission repairs *except*:
   a. vehicle speed sensor.

   b. shift solenoids.
   c. sun gear.
   d. extension housing bushing.
8. Having access to what would allow the technician to test drive the vehicle in the repair shop?
   a. Scan tool
   b. Pressure tester
   c. Chassis dynamometer
   d. Modulator
9. All of the following are very critical while doing a test drive *except*:
   a. avoiding testing the transmission in different gear ranges.
   b. recording the results for further analysis.
   c. testing the transmission during light, medium, and hard acceleration.
   d. testing the transmission in many different operating conditions.
10. Which of these is an invaluable tool for testing a TCC on an electronically controlled transmission?
   a. Pressure gauge
   b. Factory scan tool
   c. NVH analyzer
   d. Infrared temp gun

ASE Technician A/Technician B Style Questions

1. Tech A says that the fluid level in many automatic transmissions is checked with the engine idling and the transmission in park. Tech B says that some manufacturers don't provide a dipstick to check the automatic transmission fluid (ATF) level. Who is correct?
   a. Tech A
   b. Tech B
   c. Both A and B
   d. Neither A nor B
2. Tech A says that increasing the transmission fluid level from the add mark to the full mark typically requires a quart of fluid, just like with engine oil. Tech B says that when checking for automatic transmission fluid leaks, you should also remove the radiator cap and look for ATF in the radiator, Who is correct?
   a. Tech A
   b. Tech B
   c. Both A and B
   d. Neither A nor B
3. Tech A says that when checking the fluid level on the dipstick, always read the highest level on either side. Tech B says that when reinstalling the pan, start all of the bolts before tightening any of them. Who is correct?
   a. Tech A
   b. Tech B
   c. Both A and B
   d. Neither A nor B
4. Tech A says that when changing the fluid and filter, the presence of a small amount of clutch material in the pan is normal. Tech B says that if the pan gasket is leaking, you should tighten the pan bolts about a quarter turn at a time until the leak stops. Who is correct?
   a. Tech A
   b. Tech B
   c. Both A and B
   d. Neither A nor B
5. Tech A says that powertrain mounts hold the powertrain firmly in place and prevent the engine from moving at all. Tech B says that the life of rubber powertrain mounts is reduced by being saturated with oil from a leak. Who is correct?
   a. Tech A
   b. Tech B
   c. Both A and B
   d. Neither A nor B
6. Tech A says that performing a stall test can reveal slippage within the torque converter or transmission. Tech B says that a stall test can damage some transmissions and should not be performed on all transmissions. Who is correct?
   a. Tech A
   b. Tech B
   c. Both A and B
   d. Neither A nor B
7. Tech A says that using a scan tool to momentarily activate the TCC, with the engine idling in drive and the brakes applied, is a typical troubleshooting task. Tech B says that diagnosis is a waste of time on a faulty transmission because it will be rebuilt anyway. Who is correct?
   a. Tech A
   b. Tech B
   c. Both A and B
   d. Neither A nor B
8. Tech A says that when checking line pressure, the engine should be off. Tech B says that low line pressure can be caused by a worn pump. Who is correct?
   a. Tech A
   b. Tech B
   c. Both A and B
   d. Neither A nor B
9. Tech A says that low line pressure can be caused by a plugged transmission vent. Tech B says that high line pressure is caused by a restricted filter. Who is correct?
   a. Tech A
   b. Tech B
   c. Both A and B
   d. Neither A nor B
10. Tech A says that commanding shift solenoids on and off, one at a time, is an example of focused testing. Tech B says that test driving the vehicle is an example of verifying the customer's concern. Who is correct?
   a. Tech A
   b. Tech B
   c. Both A and B
   d. Neither A nor B

CHAPTER 28
Rebuilding the Automatic Transmission/Transaxle

NATEF Tasks

- **N28001** Remove and reinstall transmission/transaxle and torque converter; inspect engine core plugs, rear crankshaft seal, dowel pins, dowel pin holes, and mounting surfaces. (AST/MAST)
- **N28002** Inspect converter flex (drive) plate, converter attaching bolts, converter pilot, converter pump drive surfaces, converter end play, and crankshaft pilot bore. (AST/MAST)
- **N28003** Inspect, leak test, flush, and/or replace transmission/transaxle oil cooler, lines, and fittings. (AST/MAST)
- **N28004** Measure transmission/transaxle end play and/or preload; determine needed action. (MAST)
- **N28005** Disassemble, clean, and inspect transmission/transaxle. (MAST)
- **N28006** Inspect, measure, and reseal oil pump assembly and components. (MAST)
- **N28007** Inspect and measure planetary gear assembly components; determine needed action. (MAST)
- **N28008** Inspect bushings; determine needed action. (MAST)
- **N28009** Inspect, measure, and/or replace thrust washers and bearings. (MAST)
- **N28010** Inspect one-way clutches, races, rollers, sprags, springs, cages, retainers; determine needed action. (MAST)
- **N28011** Inspect oil delivery circuits, including seal rings, ring grooves, and sealing surface areas, feed pipes, orifices, and check valves/balls. (MAST)
- **N28012** Inspect case bores, passages, bushings, vents, and mating surfaces; determine needed action. (MAST)
- **N28013** Inspect clutch drum, piston, check-balls, springs, retainers, seals, friction plates, pressure plates, and bands; determine needed action. (MAST)
- **N28014** Measure clutch pack clearance; determine needed action. (MAST)
- **N28015** Air test operation of clutch and servo assemblies. (MAST)
- **N28016** Inspect, measure, clean, and replace valve body (includes surfaces, bores, springs, valves, switches, solenoids, sleeves, retainers, brackets, check valves/balls, screens, spacers, and gaskets). (MAST)
- **N28017** Inspect, measure, repair, adjust, or replace transaxle final drive components. (MAST)
- **N28018** Assemble transmission/transaxle. (MAST)

Knowledge Objectives

There are no knowledge objectives in this chapter.

Skills Objectives

After reading this chapter, you will be able to:

- **S28001** Install and seat the torque converter to engage drive/splines.
- **S28002** Perform final checks after installation.

▶ Introduction

Once it has been determined that an internal transmission fault is the cause of the customer concern and that the faulty component can't be accessed while in the vehicle, the transmission will have to be removed, disassembled, and repaired (**FIGURE 28-1**). In most cases, if the transmission has internal issues, the entire transmission is rebuilt at that time. This is because in most situations the transmission must be mostly disassembled to repair the faulty part. And while you are in there, it usually makes sense to perform a full rebuild. Proper removal and installation are just as critical as proper rebuilding techniques and practices. Poor workmanship during the removal and installation can result in new problems and reflect poorly on the reputation of the shop.

▶ Safety and Precautions

When removing and installing a transmission, as always, it is important to wear safety glasses while working in the shop. Transmission fluid, rust, and dirt tend to drip down and can get into your eyes. Also, the transmission is heavy and needs to be supported properly with an approved transmission jack while removing it from the vehicle. The transmission's weight is great enough that during its removal the sudden redistribution of weight can cause the vehicle to shift enough that it slips on or even falls off the hoist. It is important to anticipate these weight changes when setting the vehicle on the hoist. When moving the transmission around the shop, get another person to help put the transmission on a cart or workbench. Never try to move the transmission by yourself.

FIGURE 28-1 If a transmission has an internal failure, it is usually best to perform a full rebuild on it.

SAFETY TIP

Transmission fluid is slippery, and even after you wipe it off your hands, a film remains. This can lead to the transmission slipping out of your hands while removing it or installing it. In the same way, transmission fluid is very slippery on a smooth concrete floor. Wear oil-resistant shoes, and keep the floor clean to prevent slips and falls.

▶ Removing/Replacing the Transmission

N28001

Removing a transmission from a vehicle is not a small job and will often take several hours to complete. Removal of the transmission should be carried out only after a thorough diagnosis of the transmission in the vehicle. You should know what the problem is before removing the transmission. For example, suppose the low/reverse clutch hydraulic circuit leaks. You can then spend a little extra time inspecting the related parts to identify what caused the leak so that you can correct it permanently. Some systems become much more difficult, if not impossible, to test once the transmission has been removed from the vehicle, requiring additional inspection time and focus.

To remove and reinstall the transmission and torque converter, and to inspect the engine core plugs, rear crankshaft seal, dowel pins, dowel pin holes, and mating surfaces on a typical rear-wheel drive vehicle, follow the steps in **SKILL DRILL 28-1**.

You Are the Automotive Technician

Today you will be working with your mentor, rebuilding an automatic transmission out of a 2009 Ford F150. It has been diagnosed with low line pressure and severe slippage in all gears. In fact, it is so bad that the vehicle cannot be test-driven. The visual inspection confirmed that the fluid was in the "safe" zone, but it is very dark and smells badly burnt. Before removing the transmission, your mentor asks the following questions:

1. What components do you believe are possibly bad?
2. How is a crescent-style front pump inspected to make sure it meets specifications?
3. What things should you look for when inspecting a multidisc clutch assembly?
4. What is air testing, and why is it important?
5. When installing the torque converter, what must happen and why?

SKILL DRILL 28-1 Removing and Reinstalling the Transmission

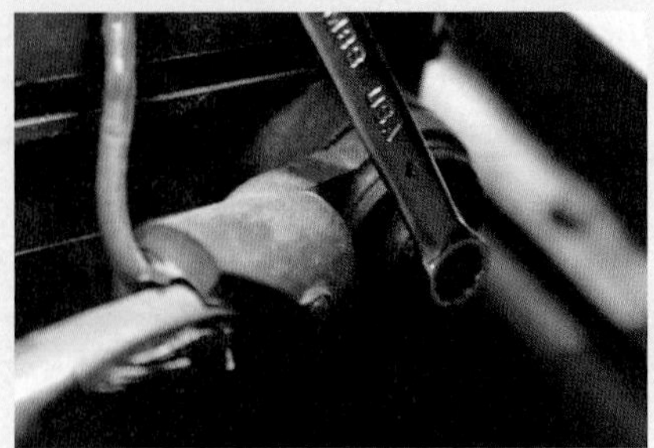

1. Look up the procedure for removal of the transmission. Disconnect the negative battery cable. Place tape around the cable end to prevent it from accidentally touching the battery terminal.

2. Drain the transmission fluid. Reinstall the transmission pan after the transmission has been drained.

3. Remove the driveshaft from the vehicle.

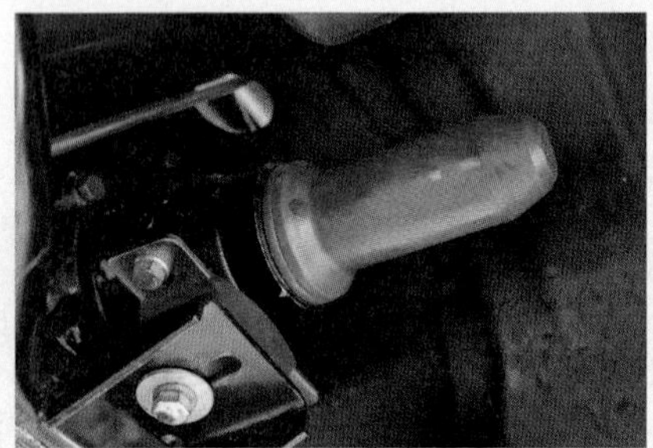

4. Install an output shaft cap over the end of the transmission output shaft to prevent fluid from leaking out while the transmission is removed.

5. Remove the speedometer cable/vehicle speed sensor harness and vacuum line to the vacuum modulator (if equipped) from the transmission.

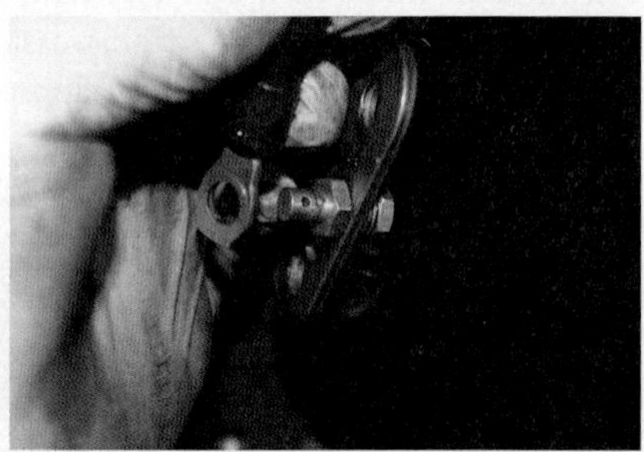

6. Remove the TV cable (if equipped), shift linkage, and any electrical connectors.

7. Remove the throttle valve cable. Remove the starter motor and the exhaust and catalytic converter if necessary to gain access.

8. Remove the dust cover from the torque converter (if equipped) and the bolts connecting the torque converter to the flywheel.

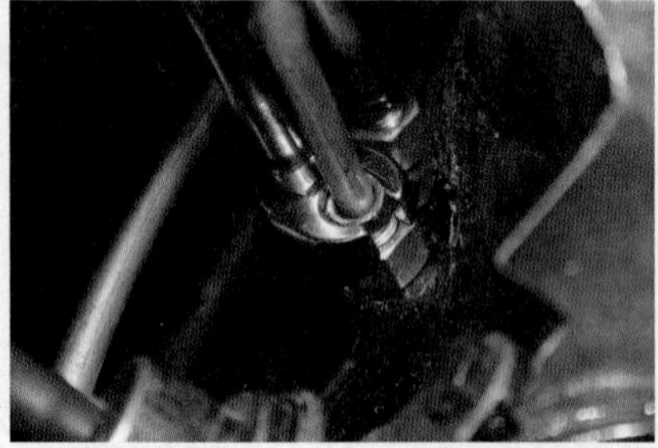

9. Disconnect and plug the transmission cooler lines and the transmission vent hose (if equipped).

SKILL DRILL 28-1 Removing and Reinstalling the Transmission (Continued)

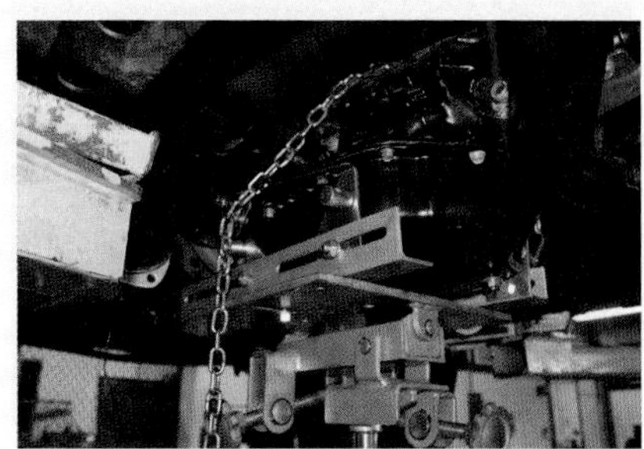

10. Remove the transmission dipstick tube. Install the transmission jack onto the transmission. Secure the safety chain around the transmission. Raise the transmission jack slightly to support the transmission.

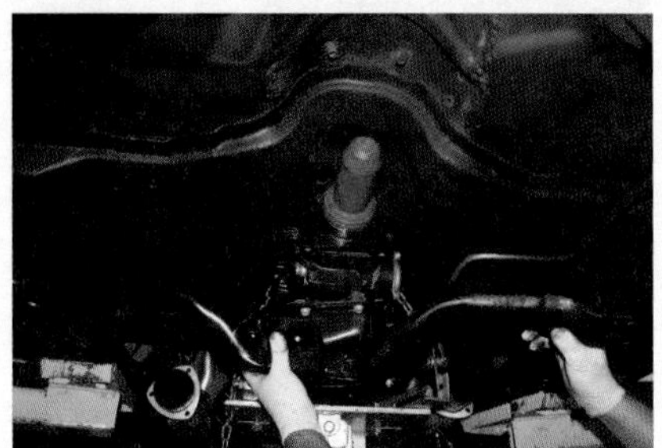

11. Remove the transmission powertrain mount. Remove the transmission frame cross member.

12. Lower the transmission slightly. Loosen and remove the bolts connecting the bell housing to the engine. Pull the transmission away from the engine until it clears the flexplate and the alignment dowels. Lower the transmission and move it to where it will be repaired. Inspect the crankshaft rear main seal, engine core plugs, dowel pins and holes, and mating surfaces. To reinstall, reverse the order of removal.

Inspecting the Torque Converter and Flexplate

N28002

The flexplate on the engine is usually a trouble-free component, but it needs to be checked when the transmission has been removed. Flexplates can crack, resulting in knocking noises and front pump damage. When the transmission is removed, inspect carefully around the bolt holes that mount the flexplate to the crankshaft and the flexplate to the torque converter. Cracks tend to start at one of the bolt holes and spread outward. Also inspect the bolt holes for elongation of the hole. If there are cracks or elongation of the holes, the flexplate and the bolts must be replaced. While inspecting the flexplate, inspect the starter ring gear for damage. If the ring gear has damage or excessive wear, replace the flexplate, and inspect the starter drive gear for damage. Some manufacturers weld the ring gear onto the converter, and if damaged teeth are noted, the converter will have to be replaced.

The bolts or nuts securing the torque converter to the flexplate have a difficult job. All of the engine's power pulses and torque are transmitted through three or four bolts to the torque converter. Some manufacturers recommend that if the torque converter is replaced, the bolts should be replaced also. If the bolts will be reused, check the threads carefully for damage and wear. If any are found to be damaged, replace all of the torque converter bolts. If the converter uses nuts, inspect the threads of the studs on the converter for thread damage as well. If the studs are damaged, the converter must be replaced. Also inspect the mounting pads on the converter for any damage.

Many shops automatically install a new or rebuilt torque converter when rebuilding a transmission. If you are not replacing the torque converter, the torque converter pump drive should be visually inspected. Check that the pump drive tangs or lugs on the converter are not damaged. If the torque converter was installed incorrectly during a past service, the pump drive lugs (and the pump) may have become damaged. Also check the snout of the converter where the seal and bushing rides for excessive wear.

Before reusing a torque converter, the internal end play of the converter must be checked. Thrust bearings inside the converter can be subjected to wear over time. A dial indicator is often used for checking the converter end play. This is accomplished by placing a special tool inside the turbine. The tool expands to grab the turbine. By pulling up and down on the tool, the movement can be measured with a dial indicator. A typical specification for torque converter end play is 0.030" to 0.050" (0.76 to 1.27 mm).

To inspect the converter flexplate (drive plate), converter attaching bolts or studs, converter pilot, converter pump drive surfaces, converter end play, and crankshaft pilot bore, follow the steps in **SKILL DRILL 28-2**.

Inspecting and Cleaning the Transmission Cooler

N28003

Before installing a new or rebuilt transmission, it is critical to clean and inspect the transmission fluid cooler and lines. If a transmission suffers catastrophic failure, metal particles and

SKILL DRILL 28-2 Inspecting the Converter and Flexplate

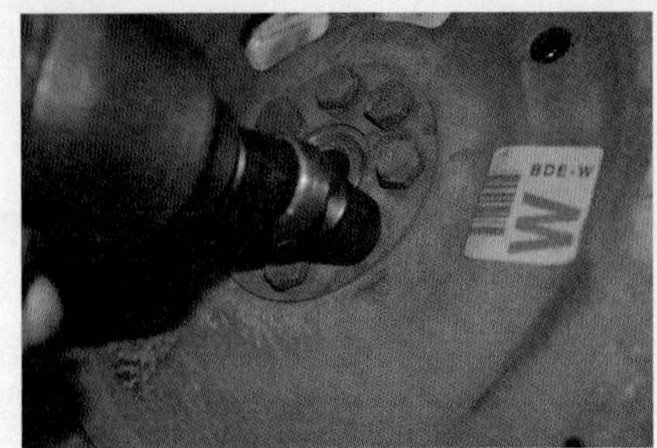

1. Remove the bolts securing the flexplate to the crankshaft.

2. Inspect the bolts and bolt holes for the flexplate to crankshaft and flexplate to torque converter.

3. Inspect the torque converter pilot, crankshaft pilot bore, and rear main seal for signs of damage.

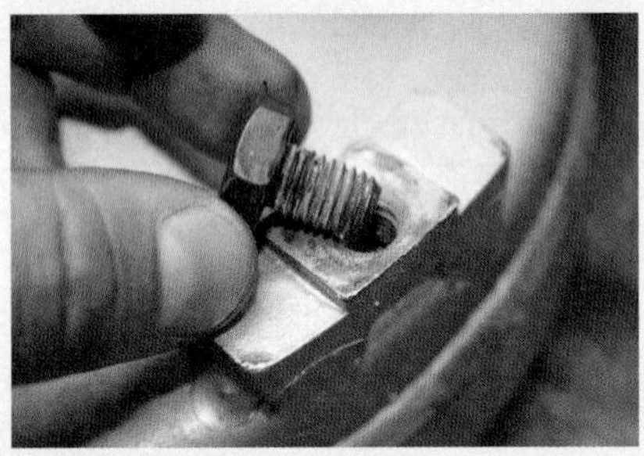

4. Inspect the torque converter mounting pads for damage.

5. Inspect the pump drive tangs for damage.

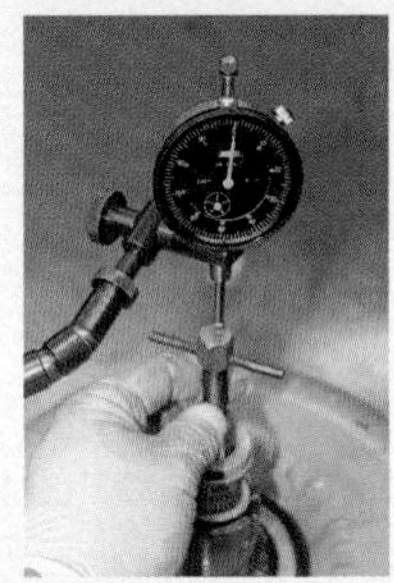

6. Install the required tool into the turbine to check for torque converter end play. Use a dial indicator to measure the amount of turbine movement.

old clutch material can become stuck inside the fluid cooler or lines. Even if this material is not stuck in the cooler, the cooler is still full of old contaminated transmission fluid. This contaminated fluid will contaminate the new transmission as soon as the vehicle is started. Some transmissions have a separate filter located in the cooler lines; these should be replaced, not cleaned. Also be aware that some manufacturers install a check valve assembly in one of the cooler lines to prevent back drainage of the fluid in the converter when the engine is off. This valve can catch debris, which can cause a restriction or even make the check valve stick closed. If this happens, no transmission fluid will get to the planetary gears for lubrication, ruining them fairly quickly. There is debate about whether these check valves can be cleaned adequately because the lines cannot be reverse flushed, so follow the manufacturer's recommendations.

Transmission coolers can be cleaned in one of two ways. The first method is to use a dedicated cleaning system with a pump that will pulse clean solvent through the transmission cooler and back to the cleaning machine. Dirt and debris are removed and returned to the cleaning system with the solvent, and removed with a filter. The second method of cleaning the transmission cooler involves using an aerosol flushing kit available at many parts stores. The aerosol can is a one-time use system and must be replaced for every transmission. When performing either method, hook up to the transmission cooler so that the solvent will flow backward through the cooler to help dislodge any stuck debris. Cooler lines with check valves cannot be reverse flushed.

A transmission fluid cooler that is leaking could cause the new transmission to fail from low fluid level or it could cause antifreeze to be drawn into the cooler from the radiator, contaminating the transmission fluid. Many times, the transmission fluid will flow into the radiator, causing the antifreeze to become contaminated with transmission fluid and the transmission to suffer from low fluid level.

To inspect, leak test, and flush or replace transmission cooler lines and fittings, follow the steps in **SKILL DRILL 28-3**.

SKILL DRILL 28-3 Inspecting and Flushing Cooler Lines

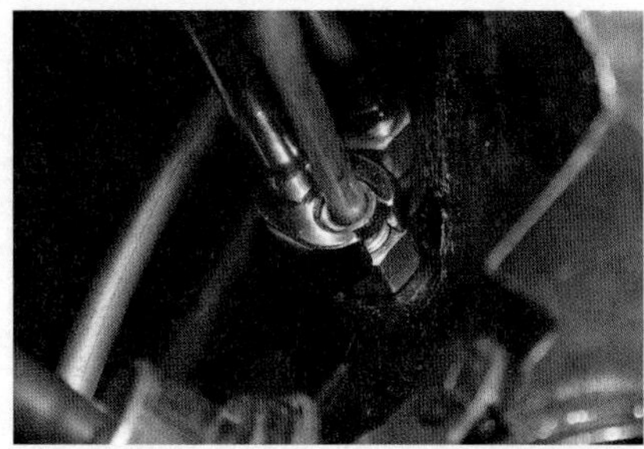

1. Look up the recommended transmission cooler service method. Remove the fluid cooler lines from the transmission if the transmission is still in the vehicle.

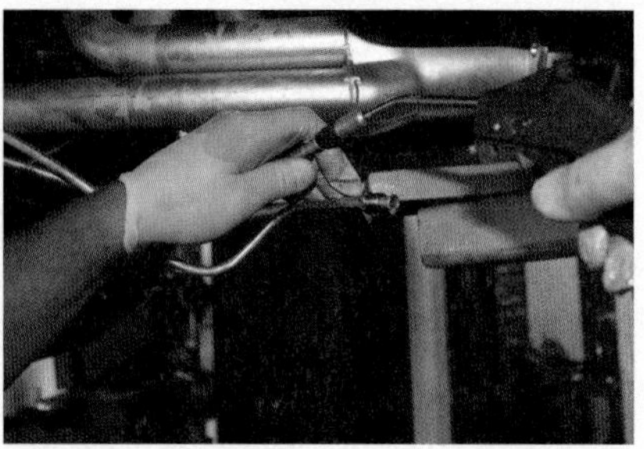

2. Using compressed air, blow into one cooler line while catching the residue in a container as it comes out the other line. Switch directions and repeat.

3. Install the cooler flush machine or aerosol can lines onto the transmission cooler lines so that the flow is in the reverse direction.

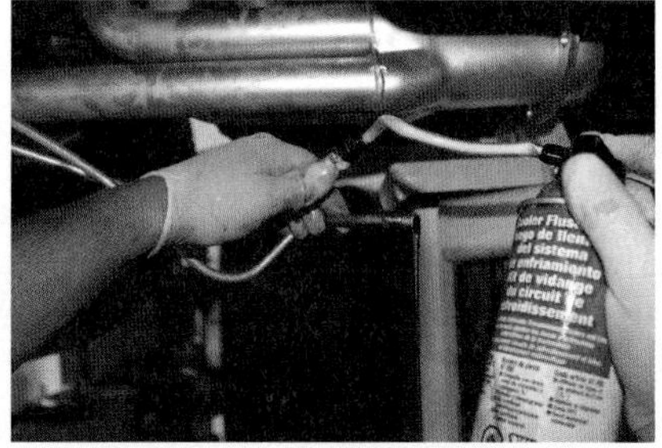

4. Start the flush machine or aerosol can, and allow it to run the recommended time. If necessary, switch directions on the lines so they can be flushed in the other direction.

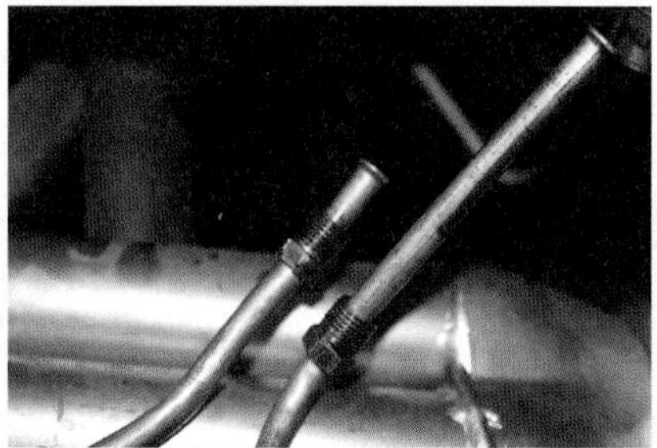

5. Remove the flush machine, and blow out the lines again so no residue remains inside the lines.

6. Reinstall the lines onto the transmission, or cap them if the transmission is removed from the vehicle. After properly filling the transmission with fluid, start the vehicle and inspect the lines and fittings for any signs of leakage. Check inside the radiator for signs of the transmission cooler leaking into the radiator.

Checking Bearings for Preload and End Play

N28004

When working with bearings, there are two terms you need to become familiar with: **preload** and **free play**. Preload refers to the amount of force placed on a bearing *before* any load is on the bearing. It is the over-tightening of a bearing to the point that it becomes increasingly difficult to spin. In most scenarios, we do not want much preload on a bearing, as it will cause bearing failure. The opposite of preload is free play. Free play refers to the amount of clearance between the bearings and the bearing races. If a bearing has too much free play, it will suffer from impact damage as the bearings hammer back and forth against the bearing races.

A critical measurement of a transmission is its end play. End play is similar to free play except that it measures the amount of free play not only in the bearing but also in the gear train inside the transmission. The end play is checked by placing the transmission so that the input shaft is vertical. A dial indicator is placed on the end of the input shaft. Depending on the transmission, either the input shaft will have to be pulled up, which allows the amount of end play to be measured on the dial indicator, or the output shaft will have to be pushed up, which moves the dial indicator. For some transmissions, there is a special fixture that can be used to pull on the input shaft to check for end play.

If the end play on a transmission is insufficient, it may cause preload on the thrust bearings inside the transmission, or dragging of the clutch packs, resulting in premature wear. If the end play is excessive, the inside of the transmission may suffer from impact damage and possible harsh shifting. Transmissions with incorrect end play may have been reassembled incorrectly. Many transmissions have a variety of select fit thrust washers of different thicknesses available that can be installed in the transmission to adjust the end play so it is within specification.

To measure transmission end play or preload, follow the steps in **SKILL DRILL 28-4**.

SKILL DRILL 28-4 Measuring Transmission End Play or Preload

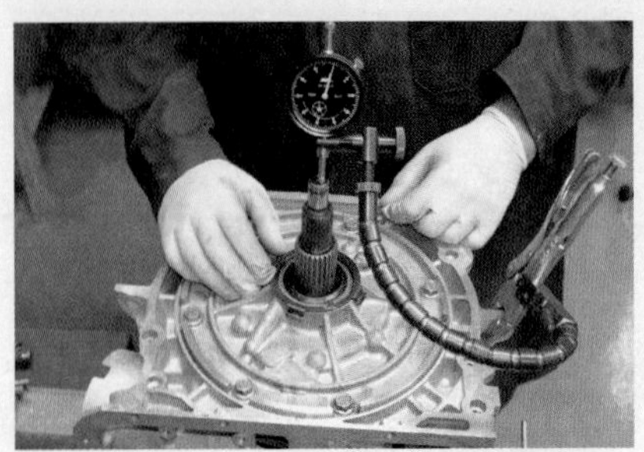

1. Research the specified procedure, and note any special tools required. Mount the transmission in a fixture with the input shaft vertical. Install a dial indicator onto the top edge of the input shaft. Set the indicator needle to zero.

2. Pull up on the input shaft or push up on the output shaft, depending on the transmission, and record the dial indicator movement. This measurement is the input shaft end play.

3. Release the input shaft. The dial indicator should return to zero. If it does not, reset and measure again.

▶ Disassembly and Inspection

N28005

A transmission that will be repaired or rebuilt has to be carefully disassembled. Many students quickly tear apart a transmission and begin cleaning the parts, which is not the best method because they miss clues to any faults present. While disassembling the transmission, discovering clues as to why a transmission failed in the first place enables you to correct any problems. Without finding and correcting these problems, the rebuilt transmission is doomed to fail again from the same cause(s).

When disassembling a transmission, it is good to disassemble it into its unit assemblies, such as front pump, clutch assemblies, planetary gear sets, valve body, and shafts. Once the transmission is disassembled to this level, you can work on one assembly at a time. This methodical approach helps to prevent missed steps and to locate any faults. And it makes it easier to remember how the parts go back together because you are only working on one assembly at a time.

▶ TECHNICIAN TIP

Some transmissions come with multiple configurations of the different clutches. Always carefully check that you have the correct, matching parts for the job. For example, one particular transmission has a clutch that comes in two configurations, with one having one less spline than the other. The difference is very difficult to see until you attempt to put the transmission back together, and the parts do not fit with one another.

Initial Disassembly and Safety

Transmissions can have many sharp parts and edges on castings that may cut you. Be careful when handling these parts. Also, many parts have large return springs that must be compressed with specialty tools. Always follow the manufacturer's recommended disassembly and service procedures to prevent injury.

Small parts can be easily lost if the technician is not careful. Place these small parts in trays or plastic bags, and label them for easy identification later. Also, many parts are round and can roll off a workbench easily, so always place them flat. If that is not possible, place them in a tray with sides that will prevent them from rolling off the edge. Parts that fall on the shop floor can be damaged, requiring replacement of the part.

To disassemble, clean, and inspect the transmission, follow the steps in **SKILL DRILL 28-5**.

Front Pump Service

N28006

The front pump is the heart of the transmission. Without a properly functioning front pump, the transmission cannot operate. The pump needs to be disassembled and checked for wear. Often, if a pump is severely worn, a technician will purchase a new or rebuilt pump for the transmission. When inspecting front pumps, there are two general categories of pumps that have different inspection procedures. The first category includes the crescent type and rotor type of pumps. The second category is the vane pump. Inspection procedures for each of these categories of pumps is explored here.

Servicing the Crescent or Rotor Pump

The crescent and rotor pumps are very simply designed pumps. Often during a transmission overhaul, the gear or rotor pump does not require any service other than replacing any seals and the front pump bushing. However, it is still important to check the pump for wear to prevent any problems when the transmission is put back into service. Most manufacturers specify a clearance measurement between the outer pump gear or rotor

SKILL DRILL 28-5 Disassembling, Cleaning, and Inspecting the Transmission

1. Follow the disassembly procedure after mounting the transmission in a transmission fixture. Remove the transmission pan and all external components, including any removable bell housing. On a transaxle, remove the side transmission pan (if equipped).

2. If equipped, disconnect electrical connectors from the solenoids and sensors inside the transmission, and inspect the connectors. If equipped, remove the wiring harness from inside the transmission.

3. If equipped, remove the shift solenoids, EPC solenoid, and TCC solenoid from the transmission.

4. Unbolt and remove the valve body assembly. Remove the manual valve from the valve body to prevent damaging it. Place the valve body on a flat surface to prevent it from becoming warped.

5. Remove the bolts holding the front pump, and use the recommended puller to remove the pump.

6. Remove the front clutch and the band, drum, and input shaft assembly.

7. Remove the planetary sets. *Note:* There may be a snap ring holding the planetary set to the output shaft.

8. Remove the extension housing assembly. Remove the governor assembly, if equipped.

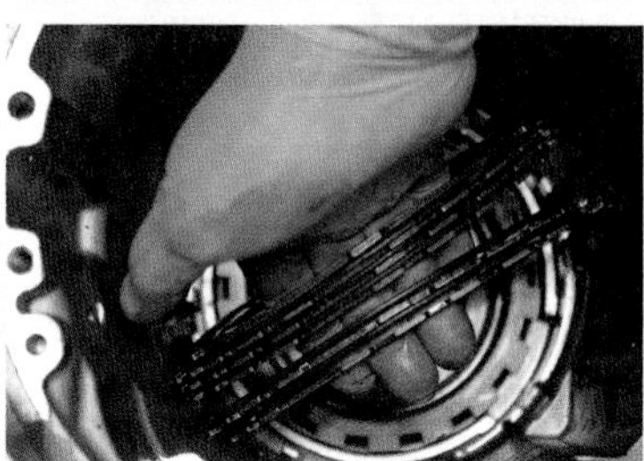

9. Remove the output shaft and any clutches, bands, and one-way rollers that are with it.

10. Remove any servos from the transmission case.

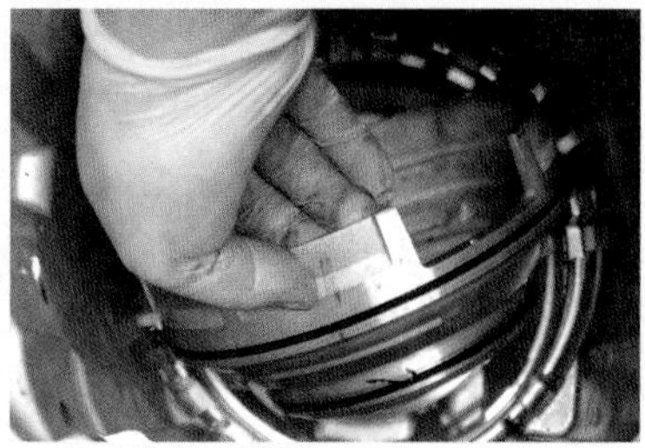

11. Remove any remaining clutch pack pistons. After all electrical components are removed from the transmission, clean the transmission case.

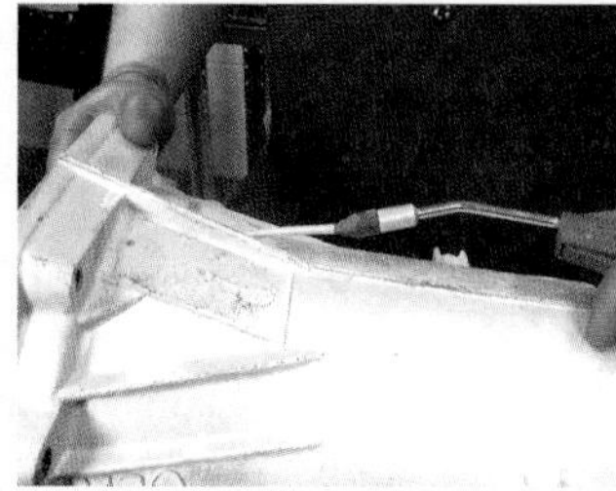

12. Inspect all threaded holes, bores, and splines for damage. Fluid passages should be blown out using compressed air to check for blockages.

FIGURE 28-2 Pump gear-to-housing clearance being measured with a feeler gauge.

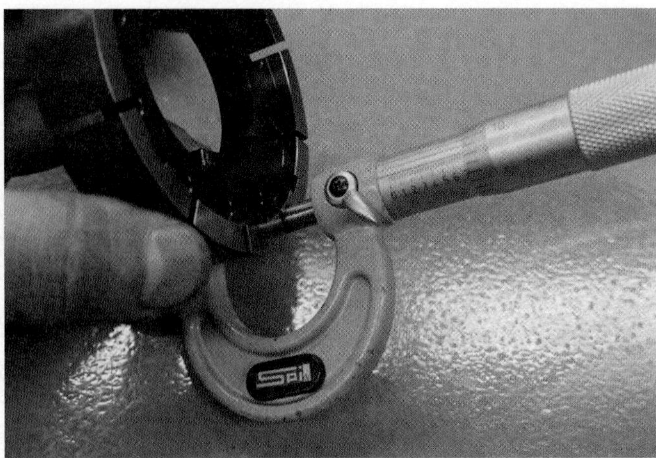

FIGURE 28-3 A vane pump's rotor thickness being measured to check for wear.

and the pump housing. This clearance can be measured with a small feeler gauge (**FIGURE 28-2**).

On a crescent pump, manufacturers also usually specify a clearance measurement between the teeth of each gear and the crescent. The last measurement is completed by placing a precision straightedge across the pump halves with the two pump gears installed. The clearance between the pump gears and the straightedge is then measured with a feeler gauge. Some manufacturers also specify an inner-gear-tooth to outer-gear-tooth clearance that must be checked with a feeler gauge.

A generic method for quickly checking a gear pump on a high-mileage vehicle is to run your hand across the face of the pump gears. If you feel a sharp edge on the gear, it is worn out and will require replacement. When the gears for a gear pump are machined at the factory, a slight bevel is machined into the edge of the gears. As the gear wears, this bevel wears away, leaving a sharp edge.

The housing of the pump also must be checked for wear. This can be accomplished with a straightedge and a feeler gauge. Often, a pump on a transmission that has suffered from infrequent fluid changes winds up with a small piece of dirt trapped between the gear and the housing, causing a groove to be ground into the pump housing. These grooves can be felt by running your fingernail across the surface of the pump housing. If you can feel your fingernail catch on a groove, the groove is at least 0.001" (0.0254 mm) deep. Light scratches and grooves can be removed on some pumps by rubbing a whetstone across the pump housing while running solvent over the pump in the solvent tank. A whetstone is a type of flat stone used to sharpen knives and other cutting tools.

The rotor pump has two rotors that mesh together, one inside the other. As with the gear pump, most manufacturers specify rotor-to-pump housing clearance as well as rotor lobe–to–rotor lobe clearance. The same methods for quickly checking a gear pump can be used on the rotor pump.

To inspect, measure, and reseal a removed pump assembly and components for a crescent pump (a rotor pump is similar), follow the steps in **SKILL DRILL 28-6**.

Servicing the Vane Pump

The vane pump and the variable displacement vane pump are quite a bit different from the gear pump and rotor pump. The center section of the vane pump is the rotor. The rotor has vanes that stick out from grooves radially spaced around the rotor. The vanes seal against the housing, which is offset to the rotor. Vehicle manufacturers typically recommend measuring the thickness of the rotor and comparing it to specifications (**FIGURE 28-3**).

The housing of the vane pump, like that of the gear pump and rotor pump, must be checked for wear, grooves, or scoring. The vanes have to be inspected for excessive wear. They must be reinstalled in the same direction that they were removed. If the vanes are flipped, rapid wear will result when the transmission is placed back into service. The vane pump typically has sealing rings on the rotor that have to be replaced during service. The sealing rings are usually made from cast iron or Teflon.

A variable displacement vane pump uses a rotor and vanes just like the standard vane pump. However, the variable displacement vane pump uses a spring-loaded movable pump housing that pivots in such a way that the volume of the pumping chambers can be increased or decreased based on fluid pressure. This design allows the pump chambers to have a large volume during low engine speeds, and smaller chambers during higher engine speeds. This results in less pumping losses and reduces fluid foaming in a pressure regulator.

To inspect, measure, and reseal a removed pump assembly and components for a vane pump, follow the steps in **SKILL DRILL 28-7**.

Inspecting and Measuring Planetary Gear Sets

N28007

Planetary gear sets are generally very durable and usually do not require service when rebuilding a transmission, but they must be checked properly to prevent a comeback related to the transmission. When checking planetary gear sets, there are several considerations.

SKILL DRILL 28-6 Inspecting, Measuring, and Resealing the Pump (Crescent Pump)

1. After looking up the proper service procedure, remove the front pump thrust washer or bearing, gaskets, sealing rings, and converter seal.

2. Unbolt the two halves of the pump and disassemble. Mark the gears so they can be reassembled in the proper position. Remove the gears and any valves from the pump.

3. Use a bushing driver to remove the front pump bushing.

4. Clean and blow-dry the pump. Inspect the sealing ring grooves for damage and wear. Install the gears into the pump housing in their original orientation.

5. Using a feeler gauge, measure the clearance between the outer gear and the housing. Measure the clearance between the inner gear teeth and the crescent. Measure the clearance between the outer gear teeth and the crescent. Compare all measurements to specifications.

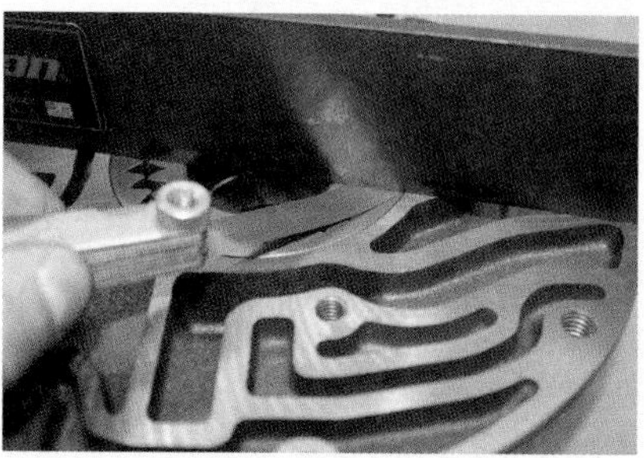

6. Use a straightedge and feeler gauges to check for clearance between the pump cover and the gears. Compare to specifications.

7. Remove the gears again. Using a bushing driver, carefully install the new front pump bushing and a new converter seal.

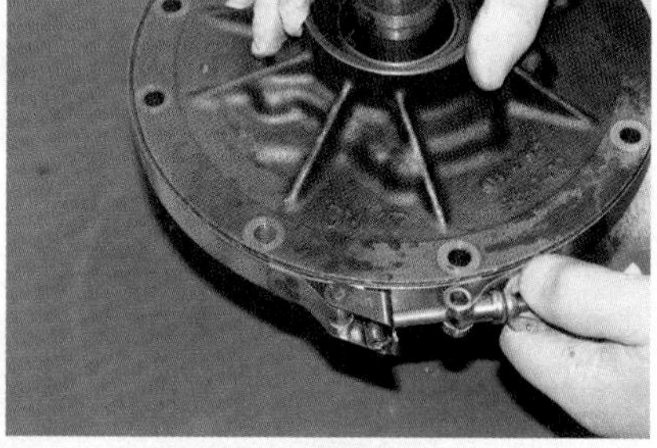

8. Lubricate the gears with transmission fluid or transmission assembly lube, and reinstall them in the proper position. Place the two halves of the pump together, and use the appropriate tool to line up pump halves. Torque the pump housing bolts to specifications.

9. Install new sealing rings and a new pump gasket onto the pump.

SKILL DRILL 28-7 Inspecting, Measuring, and Resealing the Pump (Vane Pump)

1. Research the procedure in the appropriate service information; then remove the front pump thrust washer or bearing, gaskets, sealing rings, and converter seal.

2. Mark the halves of the pump for reassembly, and then unbolt the two halves and disassemble.

3. Remove the vanes from the rotor, being careful to note their orientation, so you can reinstall them in the same direction.

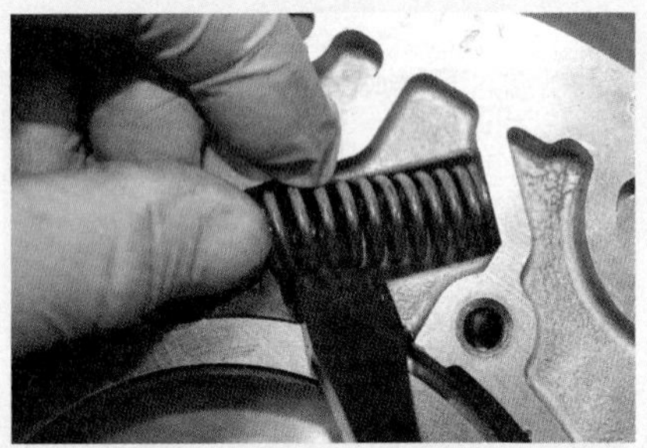

4. Remove the rotor, pressure spring, slide, pivot, and sealing rings.

5. Use a bushing driver to remove the front pump bushing. Clean the pump thoroughly in a solvent tank, and then blow-dry.

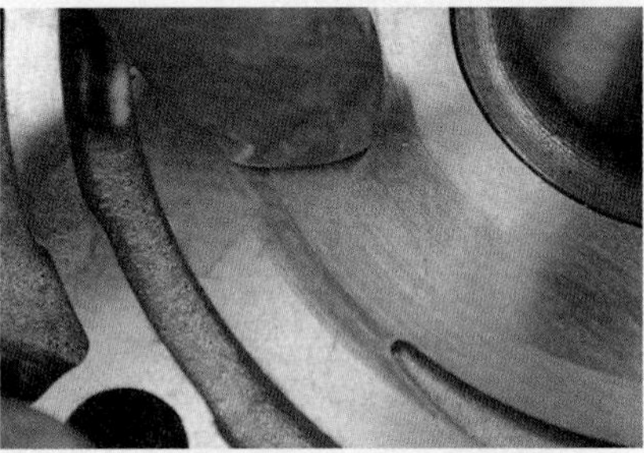

6. Inspect the sealing ring grooves for damage and wear. Inspect the housing and the slide where the rotor and vanes ride for deep scratches and wear.

7. Measure the thickness of the rotor, and compare to specifications.

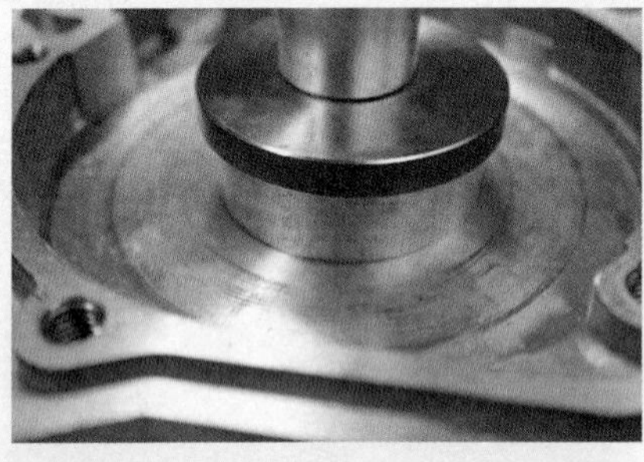

8. Using a bushing driver, carefully install the new front pump bushing.

9. Using a seal driver, install a new converter seal.

10. Lubricate the sealing rings, slide, and pivot with transmission fluid or assembly lube, and install the sealing rings, slide, pivot, and pressure spring into the pump housing.

11. Lubricate the rotor and vanes, and install the rotor and vanes in the correct original direction.

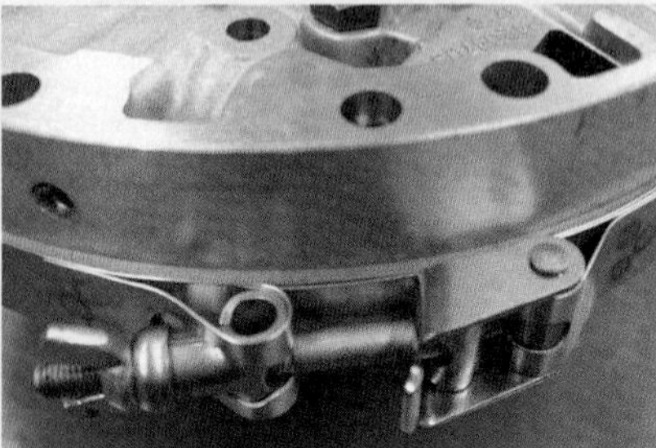

12. Place the two halves of the pump together, and start the bolts, but do not tighten. Follow the manufacturer's procedure to line up the pump halves. Torque the bolts to specification.

SKILL DRILL 28-7 Inspecting, Measuring, and Resealing the Pump (Vane Pump) (Continued)

13. Install new sealing rings on the front pump hub. If ready to install the pump, place a new gasket onto the pump.

The first check is for discoloration of the planetary gears or the planet carrier. Discoloration can be a sign of a failed lubrication circuit in the transmission and must be repaired before the discolored gears are replaced and the transmission is placed back into service.

The next inspection is of each gear in the planetary set, for cracked or worn teeth. If any gear is found to be damaged, the companion gears must also be replaced to prevent noisy gears after the rebuild.

Check each planetary pinion gear by spinning it. The gear should spin freely without binding. Check that the gear is not loose on its shaft. The end play of each of the planetary pinion gears must be checked with a feeler gauge. If the clearance is not correct, the planetary set will have to be disassembled, if it is a serviceable type, and thrust bearings of a different thickness must be installed.

To inspect and measure removed planetary gear assembly components, follow the steps in **SKILL DRILL 28-8**.

Inspecting Bushings

N28008

Bushings inside the transmission do not necessarily require replacement except for the front pump bushing and the extension housing bushing. All of the bushings inside a transmission do, however, require inspection to prevent later problems. On high-mileage transmissions, it may be necessary to purchase a complete bushing replacement kit and replace every bushing inside the transmission.

Each bushing inside the transmission should be inspected for scratches, scoring, and large pits. Place the shaft into the bushing. Wiggle the shaft back and forth. More than about 0.003" (0.076 mm) of clearance requires replacement for most bushings. You can also check the amount of clearance with a wire-style feeler gauge.

Always use a bushing driver to install new bushings into a transmission. Make sure any lubrication holes are correctly lined up, or severe transmission damage could result. Many

SKILL DRILL 28-8 Inspecting and Measuring Planetary Gear Assemblies

1. Look up the proper service procedure for the planetary gear set you are working on. Inspect the planetary gears, carrier, sun gear, and ring gear for discoloration due to lack of lubrication.

2. Inspect the teeth of each gear in the planetary gear set for wear and damage. Check the operation of the pinion gears on their shafts. Check for binding and looseness.

3. Using a feeler gauge, measure the end play of the planetary pinions, and compare to the manufacturer's specifications.

bushings are a split ring bushing that must be installed carefully to prevent splitting the new bearing apart. More than one student or beginning technician has had to purchase a second bushing to replace the first one that split on installation.

To inspect bushings on a disassembled transmission, follow the steps in **SKILL DRILL 28-9**.

Inspecting Thrust Bearings and Washers

N28009

Thrust bearings and washers have to be inspected during a transmission rebuild. If the thrust bearings or washers are worn, lateral movement of components inside the transmission can occur, affecting their operation.

Thrust bearings should be cleaned, and the races and rollers inspected carefully for pitting, scoring, and scratches. If any damage or wear is found, the entire thrust bearing assembly should be replaced. Because any wear means the thrust bearing must be replaced, it doesn't typically have to be measured for thickness unless it is used as a selective fit component to adjust end play.

Thrust washers can wear thinner without showing appreciable signs of wear. So a micrometer should be used to check the thickness of a thrust washer, and the thickness should be compared to the manufacturer's specifications. A small degree of scratching on the face of a thrust washer is acceptable. When reinstalling thrust bearings or washers in a transmission, apply a small amount of transmission assembly lube. This will help hold the bearing or washer in place while related parts are assembled.

To inspect, measure, and replace thrust washers and bearings on a disassembled transmission, follow the steps in **SKILL DRILL 28-10**.

SKILL DRILL 28-9 Inspecting Bushings

1. Research the proper procedure and specifications for checking bushings. Wipe the bushing dry with a lint-free shop towel.

2. Inspect each bushing in the transmission for pitting, scoring, or scratches.

3. Measure the inside diameter of the bushing, and compare to the manufacturer's specifications. Replace any bushings that are not within specifications and/or that show signs of wear.

SKILL DRILL 28-10 Inspecting, Measuring, and Replacing Thrust Washers and Bearings

1. Research the procedure and specifications for inspecting the thrust bearings and washers. Inspect each thrust bearing race and roller for pitting, scoring, and scratches. Replace any bearings that show signs of wear.

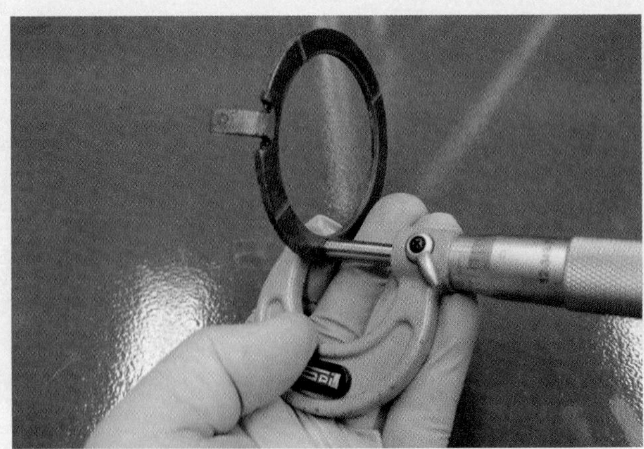

2. Measure the thickness of each thrust washer, using a micrometer, and compare to the manufacturer's specifications. Replace any washers that are not within specifications.

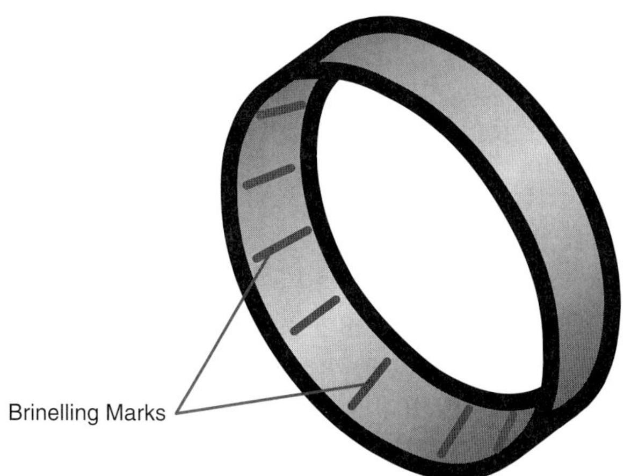

FIGURE 28-4 A bearing race showing brinelling.

Inspecting One-Way Roller and Sprag Clutches

N28010

Both one-way roller clutches and sprag clutches must be able to stop the rotation of an object in one direction but allow the object to spin freely in the opposite direction. One-way roller clutches are easy to disassemble, but sprag clutches have the sprags permanently connected to a guide, so they typically cannot be fully disassembled.

On a one-way roller clutch, the unit should be disassembled to inspect each individual roller. The roller should be checked for pitting, scoring, and brinelling, such as in a wheel bearing. Brinelling is damage found in the race from a high bearing load (**FIGURE 28-4**). The race should be checked to make sure it is smooth and has no nicks or pits in it. Rollers or races that show any signs of damage must be replaced, and the lubrication circuit for the one-way roller should be carefully inspected for blockage. The folded springs should be checked for cracks. Often the ends of the springs will crack off, leaving a shorter spring. The springs should all look the same. A one-way roller will not work if assembled in the wrong direction, so be careful to assemble it in the correct direction.

Sprag clutches can be only partially disassembled. Check the faces of the sprags and the races for any damage. Sprag clutches also must be assembled in the correct direction. Sprag clutches must be serviced as a complete unit if any damage is found.

After visually inspecting one-way roller and sprag clutches, they should be reassembled. They both should spin freely in one direction and lock up in the opposite direction. To inspect the one-way roller or sprag clutch and related components in a disassembled transmission, follow the steps in **SKILL DRILL 28-11**.

SKILL DRILL 28-11 Inspecting One-Way Roller and Sprag Clutch Assemblies

1. Research the proper service procedure for one-way roller or sprag clutch assemblies, and disassemble them.

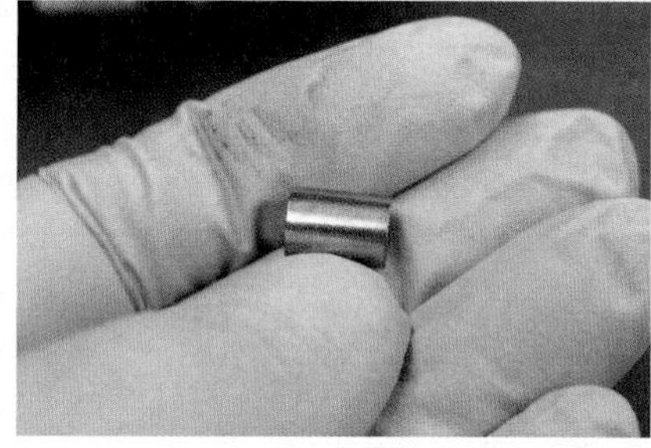

2. Inspect each roller and race for signs of pitting, scoring, or brinelling.

3. Inspect the folded springs on the one-way roller for cracks or broken springs.

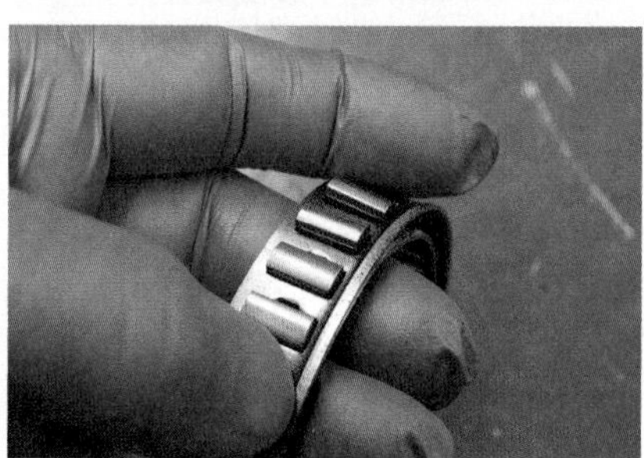

4. Inspect the faces of the sprags and the race for signs of scoring or damage. If any damage is found, replace the one-way roller or sprag.

5. Reassemble the one-way roller or sprag, lubricate with transmission fluid, and test the operation of the one-way roller or sprag.

Inspecting Fluid Delivery Circuits

N28011

Fluid flow in an automatic transmission is critical to the operation of the transmission. Any debris from a transmission failure can become trapped in a fluid passage, preventing the flow of fluid for lubrication or clutch/band application. It could also dislodge and move into the bearings and bushings, where it could cause damage. All fluid passages should be checked during a rebuild of the transmission.

The proper service information is critical for checking the operation of the fluid delivery circuits. A technician needs to use a transmission case diagram to identify the fluid passages through the case. Compressed air should be blown through each fluid passage in the transmission case, the valve body, the input and output shafts, and the front pump.

Feed pipes should be checked for cracks and damaged ends. Sealing rings should be replaced or inspected for damage, along with their corresponding sealing surface. Often on high-mileage vehicles, a groove may become worn in the sealing surface, requiring replacement of the part for proper sealing.

Check valves and/or check balls should be inspected for proper operation. Plastic check balls tend to wear and are often replaced during a rebuild. Metal check balls can cause wear to the valve body or separator plate. Check the area around the check ball for damage or enlargement of the hole in the separator plate.

To inspect fluid delivery circuits, including sealing rings, ring grooves, sealing surface areas, feed pipes, orifices, and check valves/balls in a disassembled transmission, follow the steps in **SKILL DRILL 28-12**.

Servicing the Transmission Case, Servo, and Accumulator

N28012

After a thorough cleaning of the transmission case and all components, an inspection of the transmission case is necessary. Inspect the inside bore of the transmission along with any splines. If any damage is found, the entire transmission case may have to be replaced. On high-mileage vehicles, the clutches and steels can sometimes wear grooves into the sides of the case. These grooves can create spots where the clutches can hang up during their application and release.

The sealing surfaces of the transmission case need to be checked with a precision straightedge to ensure that the surface has not become warped. A gasket is designed to fill in small imperfections in the sealing surface, but if the surface has deep gouges or scratches, the gasket will be unable to do its job. If the sealing surface of the transmission has become damaged, it may be possible to machine it back to within an acceptable limit.

As was covered earlier in the chapter, any bushings in the transmission should be checked for wear. The exceptions are the

SKILL DRILL 28-12 Inspecting Fluid Delivery Circuits and Components

1. Look up the proper service procedure for the fluid delivery circuit. Use compressed air to blow through each of the fluid delivery circuits to check for blockages.

2. Inspect each sealing ring and corresponding sealing surface for damage and wear.

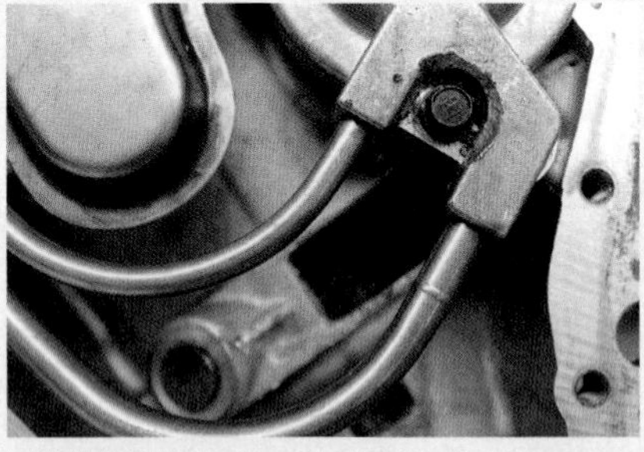

3. Inspect the fluid feed pipes for cracks and damaged ends.

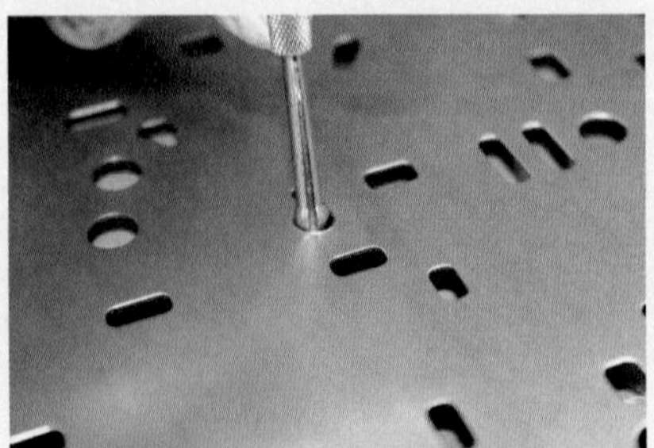

4. Measure the orifices in the valve body and separator plate for proper sizing. Compare to the manufacturer's specifications.

5. Inspect the check balls and check ball housing for signs of wear. Replace any components showing signs of wear.

front pump bushing and the extension housing bushing, which should always be replaced when rebuilding a transmission.

As the transmission and the fluid heat up during the vehicle's operation, the transmission fluid expands. This expansion would cause an increase in the pressure found inside the transmission case if we did not have a method to vent the pressure. Nearly every transmission has a vent that must be checked when rebuilding the transmission. If the vent has become clogged, the transmission will suffer from external leaks as the buildup of pressure pushes fluid out in places other than the appropriate transmission vents. Often the vents in off-road vehicles will become plugged with dried mud, creating transmission leaks.

Servo and accumulator bores have to be checked for scoring. Scores or scratches in the bore allow hydraulic fluid to flow past the servo or accumulator seal and cause a loss of pressure. Check the piston on both an accumulator and a servo for cracks. Aluminum pistons are more prone to cracking. Look closely; sometimes the top of the piston is stuck in the housing after it has completely separated from the rest of the piston. Check that the piston fits tightly on its pin. If the center hole of the piston is worn, the piston will be able to rock in the bore as it is applied, resulting in binding. Check the bore for the pin. The pin should not be allowed to wobble in the bore. Check the spring because often springs in accumulators can become cracked or broken, requiring replacement. Also check the spring seat to make sure a groove has not been worn.

To inspect case bores, passages, bushings, vents, and mating surfaces and determine any necessary actions, and to inspect servo and accumulator bores, pistons, seals, pins, springs, and retainers in a disassembled transmission, follow the steps in **SKILL DRILL 28-13**.

Inspecting Bands and Drums

N28013

Bands should be inspected for wear if they are not being replaced during an automatic transmission rebuild. They should be inspected for cracks and chipping in the friction material. Also, the band can become burned during service or glazed like a brake lining. If any of these conditions are identified, the band should be replaced.

The drum should also be checked for wear. Place a straightedge across the drum where the band rides, and check that the drum is straight. Use a feeler gauge to measure the amount of warpage on the drum, and compare your measurement to the manufacturer's specifications. Inspect the drum for any discoloration due to band slippage. A drum that is not badly scored can be refinished by running emery cloth across the drum while it is spun on a machinist lathe.

To inspect bands and drums in a disassembled transmission, follow the steps in **SKILL DRILL 28-14**.

Rebuilding and Testing the Clutch Assembly

When rebuilding an automatic transmission, all clutch assemblies should be rebuilt. Part of the rebuilding process is a thorough disassembly and inspection of all of the components in the assembly. Because most transmissions have multiple clutches, it is good practice to rebuild only one at a time, so parts do not get interchanged.

The clutch drum should be checked for cracks and broken or worn splines. The splines should be straight with no worn edges. On high-mileage transmissions, you may find that the clutch discs become stuck as they move up and down on splines. If the drum has a check ball, make sure the check ball is free to move. The sealing ability of the check ball must also be checked. Place some solvent in the bore of the drum, and check the sealing operation of the check ball. If the solvent leaks out, the drum has to be replaced.

The spring retainer and springs should all be checked for straightness, and each spring's height should be checked. Any springs that are bent should be replaced to prevent the clutch from becoming stuck when it is disengaged.

Check the friction discs for loose or flaking friction material. Friction material that is discolored was overheated and should be replaced. The friction discs must be flat. Warpage is a sign of overheating or improper installation. The steels should be checked for flatness, and their thickness should be measured. Compare your measurements against the manufacturer's specifications. The pressure plate should also be checked for warpage and replaced if necessary.

Inspect the clutch piston for cracks and warpage. Replace all of the lip seals on the clutch piston. If the piston has a check ball, verify that the ball is free to move. If the ball will not move, use solvent and a small wire to free up the check ball.

To inspect the clutch drum, piston, check balls, springs, retainers, seals, and friction and pressure plates on a disassembled transmission, follow the steps in **SKILL DRILL 28-15**.

Measuring Clutch Pack Clearance

N28014

After rebuilding and reassembling a clutch pack, it is necessary to measure the amount of clearance between the plates with a feeler blade. If the clutch pack has too little clearance, the clutch will drag when it is disengaged, causing the frictions to be burned up. If the clutch pack has too much clearance, the clutch may slip or have a very harsh engagement. If the clutch pack clearance is not correct, there are several methods manufacturers use to change the amount of clearance. One method is to change the thickness of the steels in the clutch pack. Another method is to change the pressure plate to one with a different thickness. And a final method is to change the thickness of the snap ring holding the pressure plate in the clutch pack. It is necessary to check which method the manufacturer specifies for that particular clutch pack.

To measure clutch pack clearance, follow the steps in **SKILL DRILL 28-16**.

Air Checking

N28015

After a clutch or servo has been completely rebuilt, it should be tested. Clutch packs that are rebuilt outside of the transmission are typically air tested before installing them back in the transmission. Clutch packs that are rebuilt inside the transmission are typically air tested once the clutch pack is fully assembled in the

SKILL DRILL 28-13 Inspecting the Transmission Case and Servo/Accumulator Bores and Components

1. Research the service procedure for the transmission case, servos, and accumulators. Inspect the transmission case bore for excessive wear. Check the splines inside the bore for wear and grooves. Inspect all hydraulic passages in the transmission case, using compressed air.

2. Inspect and measure the clearance on all bushings inside the transmission case.

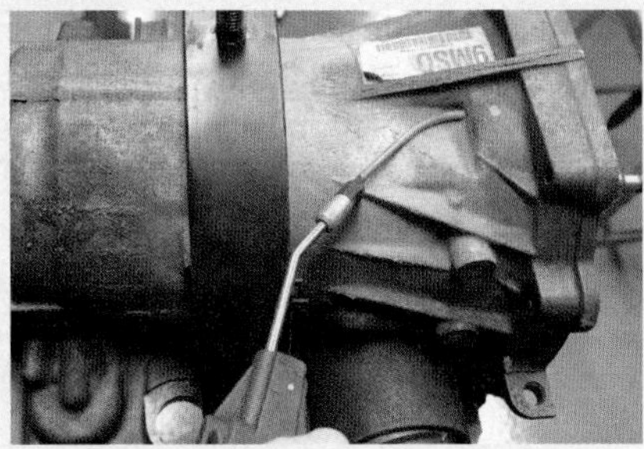

3. Check the operation of the transmission vent, using compressed air.

4. Use a precision straightedge and feeler blades to check the surfaces of the pan gasket and the valve body mounting for flatness. Inspect gasket sealing areas for nicks and scratches.

5. Inspect the servo and accumulator bores and seals for wear or damage.

6. Inspect the servo and accumulator pistons for cracks and loose-fitting pins and/or retainers.

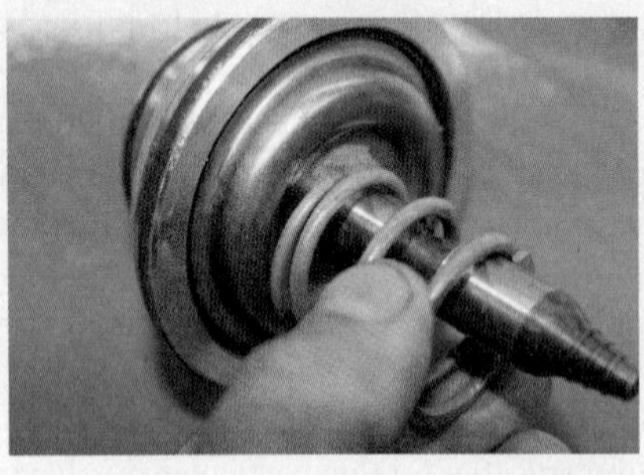

7. Inspect the servo and accumulator springs for cracks and wear grooves in the spring seats.

case. Servos are typically air tested after the associated band is installed and adjusted as the transmission is being assembled.

Using compressed air is a good method of determining whether the clutches and servos will function properly. However, never air test a clutch without the clutch pack assembled and the snap ring fully engaged. Doing so will cause the piston to travel too far, and the sealing lips will pop out of their bore. Once that happens, the seals cannot seal, and the return springs will pinch the seal between the piston and the bore, damaging the seal.

To air test the operation of clutch and servo assemblies, follow the steps in **SKILL DRILL 28-17**.

Servicing the Valve Body

N28016

To properly rebuild an automatic transmission, the valve body must be completely disassembled. There are commercially made

SKILL DRILL 28-14 Inspecting Bands and Drums

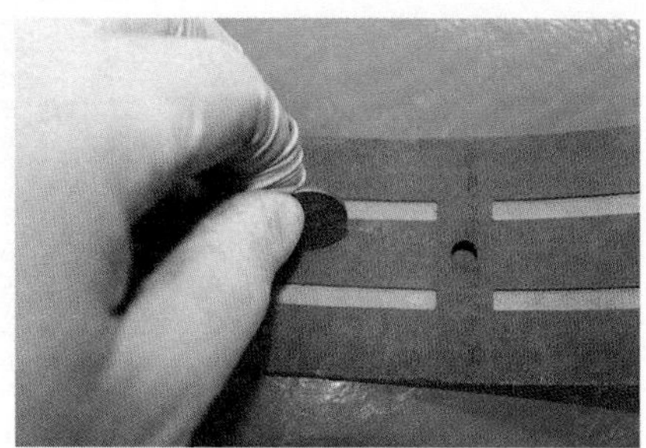
1. Research the service procedure and specifications for the band. Inspect the band friction material for cracks, chips, flaking, glazing, and burn marks.

2. Inspect the band's strut (if equipped) for wear. Inspect the drum for discoloration and scoring where the band rides.

3. Use a straightedge and feeler gauge to check the drum for straightness. Compare your measurements with the manufacturer's specifications.

SKILL DRILL 28-15 Inspecting Clutch Assemblies

1. Research the manufacturer's procedure and specifications for rebuilding the clutch pack. Disassemble and clean the clutch pack assembly.

2. Inspect the clutch drum for spline damage, and inspect the check balls for proper operation.

3. Inspect the clutch piston for warpage and cracks. Check operation of the piston check ball (if equipped).

4. Inspect the springs for length and squareness. Check the spring retainer for warpage.

5. Inspect the friction discs for damaged or deteriorating friction material.

6. Inspect the steels for warpage. Measure the thickness and compare to the manufacturer's specifications. Inspect the pressure plate for warpage and cracks.

7. Replace all seals and lubricate with transmission assembly lube.

8. Using the proper seal tools, carefully reinstall the clutch piston.

9. Install the clutch return spring assembly.

SKILL DRILL 28-15 Inspecting Clutch Assemblies (Continued)

10. Use the correct clutch spring compression tool to slowly compress the spring assembly, and install the snap ring.

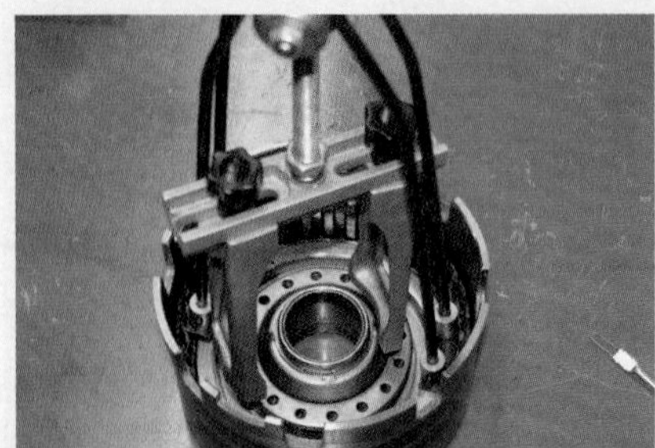

11. Slowly release the clutch spring tool and remove.

12. Install the friction discs and steels back into the clutch pack in the correct order according to the service manual.

13. Install the wave washer or Belleville spring (if equipped) into the clutch pack.

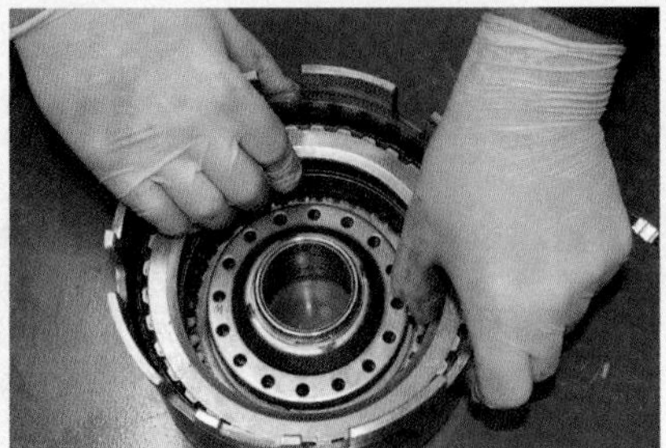

14. Install the pressure plate into the clutch pack. Make sure the pressure plate is facing the correct direction, and reinstall the large snap ring.

SKILL DRILL 28-16 Measuring Clutch Pack Clearance

1. Research the manufacturer's procedure and specifications for clutch pack clearance. Locate the feeler gauge that fits between the pressure plate and the first friction with some drag as it is removed.

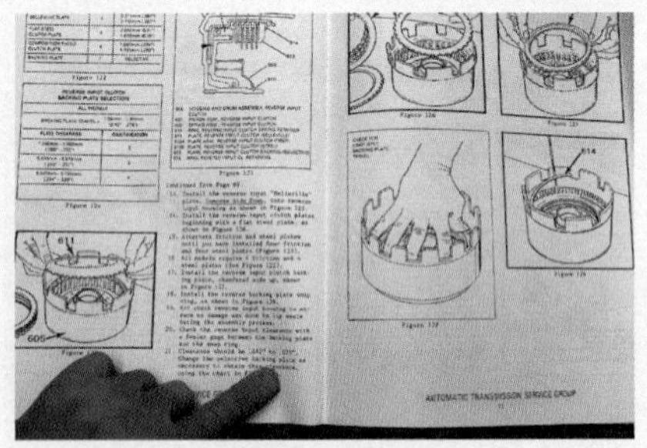

2. Compare your measurement to the manufacturer's specifications, and adjust, if necessary, by measuring the selective fit component and determining the correct replacement.

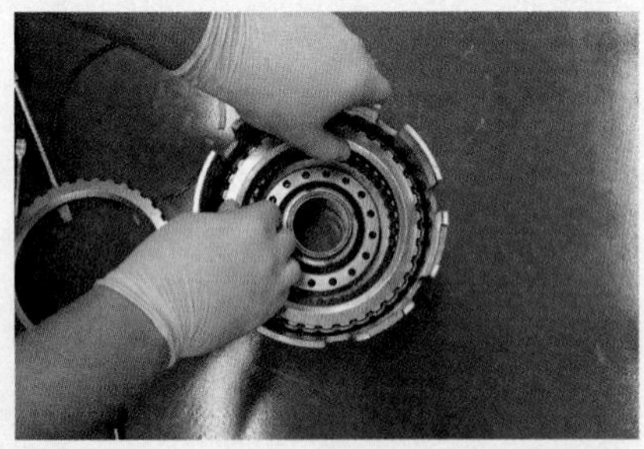

3. Install the selective fit component, and remeasure the clearance.

fixtures to keep all of the different valves and springs in order. Disassemble the valve body, and clean all of the components in the solvent tank or parts washer. Be careful not to lose any parts, and remember how, and where, each valve goes so that you can reassemble it properly.

Each valve in the valve body must be checked for wear. On modern electronically controlled transmissions, many of the valves pulse back and forth slightly during vehicle operation. This motion results in more wear on the valves than was seen on hydraulically controlled transmissions. Also, modern transmissions often use aluminum valves, which wear faster than the older steel valves. The valve bore should be measured to check for valve clearance. Use a telescoping gauge and micrometer to measure the clearance of the valve to the bore. Compare the clearance to the manufacturer's specifications. If the clearance is too great, the valve body will need to be replaced or a service kit will be needed. Many common transmissions have repair kits for the valve body assemblies that involve reboring the valve body bore and installing a sleeve or larger valve. Many transmission rebuild shops will replace

SKILL DRILL 28-17 Air Testing Clutch and Servo Assemblies

1. Research the manufacturer's recommended procedure to air check the clutch or servo. Apply compressed air to the fluid inlet hole for the clutch or servo.

2. Ensure that the clutch or servo is holding properly with the air pressure applied, with limited air leakage.

3. Release the air pressure and observe that the clutch or servo releases properly.

every valve body on transmissions that have common valve body problems.

The surface from the valve body to transmission case should be checked for warpage, using a precision straightedge and feeler blades. If the valve body is warped, it is recommended that it be replaced with a reconditioned unit.

Each valve should be reinstalled and checked to make sure it moves freely in its bore. If the valve hangs up, the bore can be lightly sanded using a small roll of crocus cloth. Steel valves can be rotated on a wet sharpening stone to remove any small burrs.

To inspect, measure, clean, and replace the removed valve body, follow the steps in **SKILL DRILL 28-18**.

SKILL DRILL 28-18 Inspecting, Measuring, Cleaning, and Replacing the Valve Body

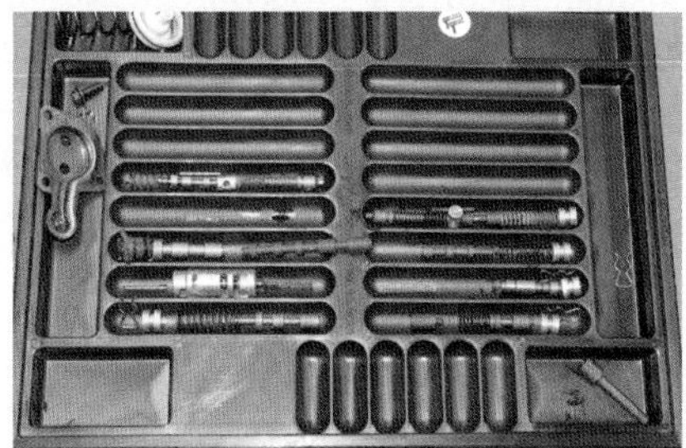

1. Research the proper service procedure for the valve body. Disassemble the valve body completely, keeping valves, springs, sleeves, and other components in order. Clean the valve body, valves, sleeves, and springs in a solvent tank.

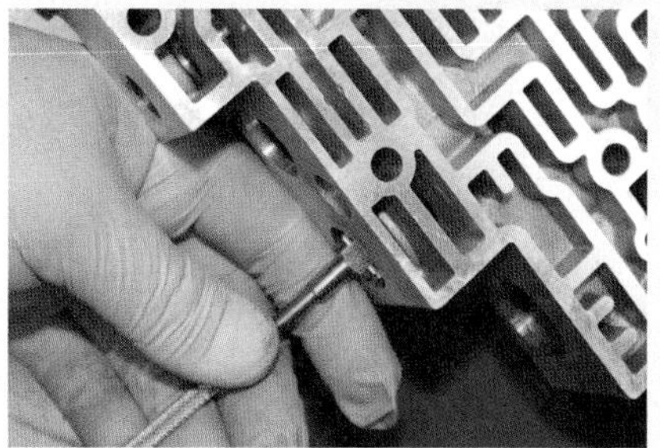

2. Inspect valves and valve bores for wear. Measure valve bores for clearance, and compare to manufacturer's specifications.

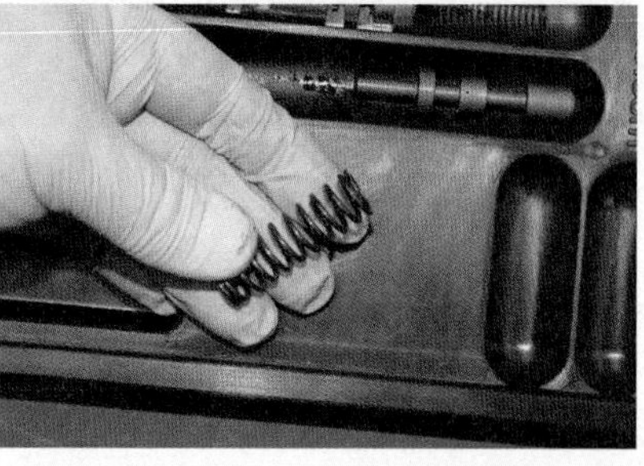

3. Inspect springs for damaged or distorted coils.

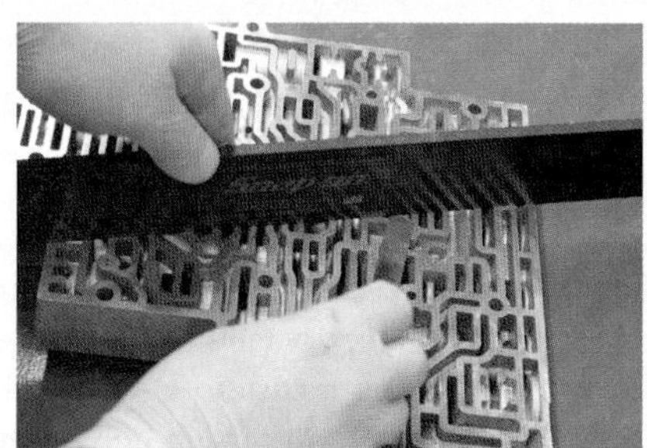

4. Check valve body to transmission case surface for warpage.

5. Check that all check balls and screens are functioning and are in the correct place. Reinstall valves, springs, and sleeves in the valve body, and check that they move freely.

6. Reinstall all plates and tighten all valve body screws and bolts to the proper torque.

FIGURE 28-5 Power transfer. **A.** Gear style. **B.** Chain style.

▶ Transaxle Types and Components

N28017

Transaxles are a transmission with an integrated final drive unit such as that found in the differential of a rear-wheel drive vehicle. These transaxles have a few more components than a traditional transmission and have a few more tests involved with their diagnosis. In transaxles, the power from the transmission portion of the transaxle needs to be transferred to the final drive portion of the transaxle. This power transfer can occur in one of two methods—a gear style or a chain style (**FIGURE 28-5**).

Gear Style

In the gear-style transaxle assembly, power is transferred from the transmission planetary gears to the differential through a set of helically cut gears. One of the helical gears is splined to the output shaft of the transmission, and the other is splined to a transfer shaft. The transfer shaft transmits the power to the ring gear of the differential case. These gears have to be checked for wear when servicing the transmission. If excessive wear is found on either gear, both gears in the set will require replacement, as they are a matched set.

Chain Style

Many transaxles use a large **Morse-style chain** and sprockets to connect the output shaft of the transmission to the differential assembly. Some transaxles use a chain to connect from the torque converter to the input shaft of the transmission, as in many General Motors transaxles. When rebuilding the transmission, the chain and sprockets have to be checked for wear. Many manufacturers specify a limit for the amount of chain deflection allowed. If the chain is replaced, both sprockets should also be replaced.

To inspect the transaxle drive, link chains, sprockets, gears, bearings, and bushings, follow the steps in **SKILL DRILL 28-19**.

Final Drive Assembly

The transaxle final drive assembly has to be inspected for wear during a rebuild. The measurements and adjustments are similar to those that would be completed during a differential overhaul. Each gear in the final drive unit should be inspected for worn, chipped, or broken teeth. The bearings should be inspected for scoring or overheating. If the unit is a planetary gear set style, the pinion gears have to be checked for proper operation in the same manner as they were when the planetary gear sets in the transmission were checked.

Output shaft end play and side gear backlash must be checked in order to prevent seizing of the gears or excessive noise. When assembling the final drive unit, measure bearing preload, following the manufacturer's procedure. Changing the preload and backlash often involves changing selective thickness shims.

To inspect, measure, repair, adjust, or replace transaxle final drive components, follow the steps in **SKILL DRILL 28-20**.

▶ Assembling the Transmission

N28018

After taking the time to rebuild each of the individual transmission components, the transmission has to be carefully assembled. Check the service information for the correct procedure before beginning assembly. All components should be lubricated with the correct automatic transmission fluid or transmission assembly lube as they are installed in the transmission. Transmission parts should not have to be forced together. They may require lots of wiggling and turning, but if you try to force them together, you will succeed only in damaging the parts. All gaskets and seals should be replaced at this time, if they have not already been replaced.

Take the time to look up torque specifications and torque sequences. The manufacturer published them for a reason. Failure to torque parts correctly can result in an early failure of the transmission and a second visit from an unhappy customer. Many bolts also have a specific location and should be installed only where they belong. A misplaced valve body bolt can cause catastrophic damage when the vehicle is started, if the bolt protrudes into any of the rotating parts of the transmission from the valve body.

To assemble the transmission, follow the steps in **SKILL DRILL 28-21**.

SKILL DRILL 28-19 Inspecting Transaxle Drive Assembly

1. First, look up the manufacturer's recommended procedure and specifications for checking the transaxle drive components. Measure the amount of chain slack, and compare to the manufacturer's specifications (if equipped).

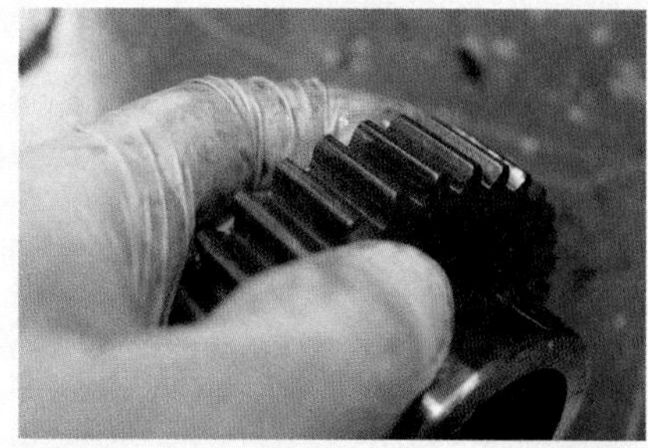

2. Inspect the sprockets for tooth wear (if equipped).

3. Inspect the transfer gears for excessive wear (if equipped).

4. Inspect the bearings for damage and wear to the needle bearings and their races.

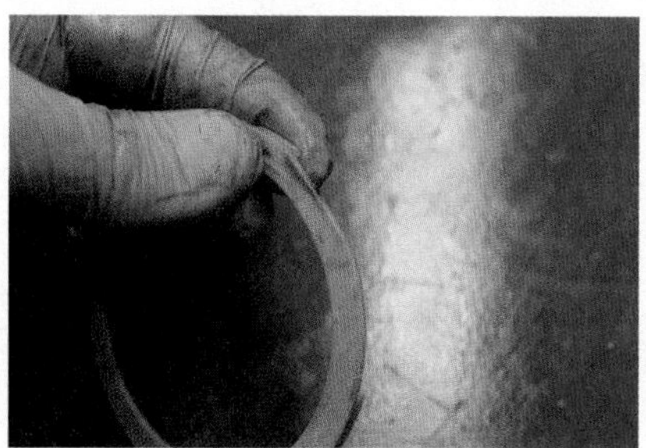

5. Inspect the bushings for scoring, pitting, or wear. Replace any components worn beyond the manufacturer's specifications.

SKILL DRILL 28-20 Inspecting, Measuring, and Repairing Transaxle Final Drive Components

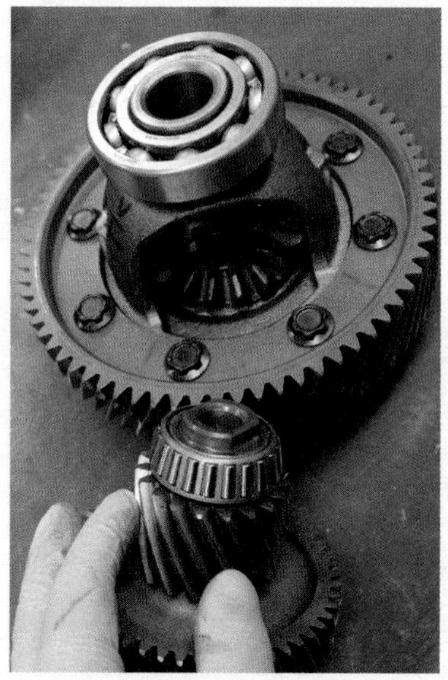

1. First look up the manufacturer's recommended procedure and specifications for the final drive unit. Inspect all gears in the final drive unit for chipped or broken teeth. Inspect the planetary gear assembly (if equipped).

2. Use a dial indicator to check the side gear backlash, and compare the backlash to the manufacturer's specifications.

3. If the backlash is too great, install a thicker shim behind the side gear. If the backlash is too little, install a thinner shim behind the side gear.

SKILL DRILL 28-20 Inspecting, Measuring, and Repairing Transaxle Final Drive Components (Continued)

4. Install the final drive assembly into the transaxle and torque the retaining bolts. Use a dial indicator to measure end play on the differential assembly, and compare to specifications. If the end play is not correct, change the selective shim behind the bearings to change the end play.

5. Reinstall the final drive and check end play. Measure the rotating torque of the final drive unit, and compare to specifications.

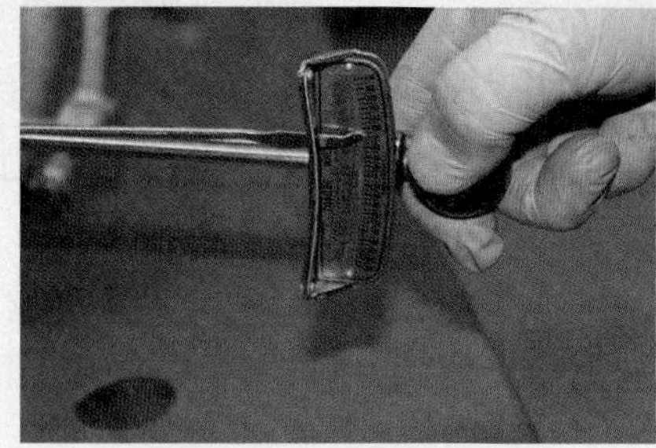

6. If the torque is above the recommended amount, replace the shim with a slightly thinner shim. If it is lower than the amount recommended by the manufacturer, replace the shim with a slightly thicker one.

SKILL DRILL 28-21 Assembling the Transmission

1. First look up the proper assembly procedure and specifications. Install the rear clutch piston and return springs.

2. Install the rear thrust bearing and output shaft into the transmission.

3. Install any rear clutch or band assembly into the transmission.

4. Install the planetary sets and any snap rings into the transmission.

5. Install the clutches, drums, and bands with the input shaft. Make sure all components are fully engaged; otherwise, there will be no end play, and components will be damaged.

6. Install the front pump. Make sure the pump lines up correctly and that there is a small amount of end play. Torque the bolts to specifications.

SKILL DRILL 28-21 Assembling the Transmission (Continued)

7. Check the transmission end play (refer to Skill Drill 28-4). Adjust any bands if adjustable with the proper torque wrench.

8. Install any check valves and the valve body spacer plate.

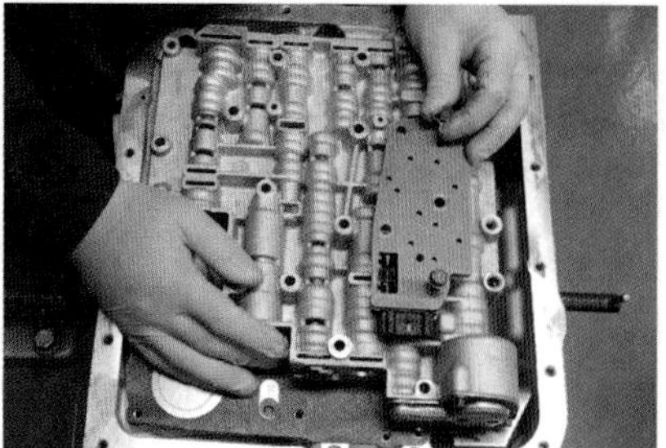

9. Install the valve body, and torque the bolts to specifications, following the manufacturer's specified sequence.

10. Install the manual valve and manual valve linkage.

11. Install the shift solenoids, EPC solenoid, and sensors into the valve body. Install the wiring harness.

12. Install the extension housing.

13. Install the transmission filter and transmission pan. Install the bell housing, if removed.

14. Install all external sensors, plugs, and covers. Remove the transmission from the transmission vise. Prepare to install it in the vehicle.

Installing a Torque Converter

S28001

Installing the torque converter properly is often where beginning students and technicians have a problem. Sometimes it is difficult to get the torque converter fully seated when installing it into the transmission. It takes practice to feel when each of the three separate engagements of the mating parts occurs. The first engagement is the pump drive. Aligning the pump gears with the slots or flats on the torque converter gets them close. The second and third engagements are the stator support to the stator and the input shaft to the turbine. A torque converter that is not properly installed will cause catastrophic damage to the front pump when the vehicle is first started. If the transmission was rebuilt by a transmission shop, this damage will not be covered under the warranty.

To install and seat the torque converter to engage drive/splines, follow the steps in **SKILL DRILL 28-22**.

Transmission Final Checks

S28002

After the transmission is installed properly into the vehicle, add about 75% of the total transmission fluid to the transmission.

SKILL DRILL 28-22 Installing and Seating the Torque Converter

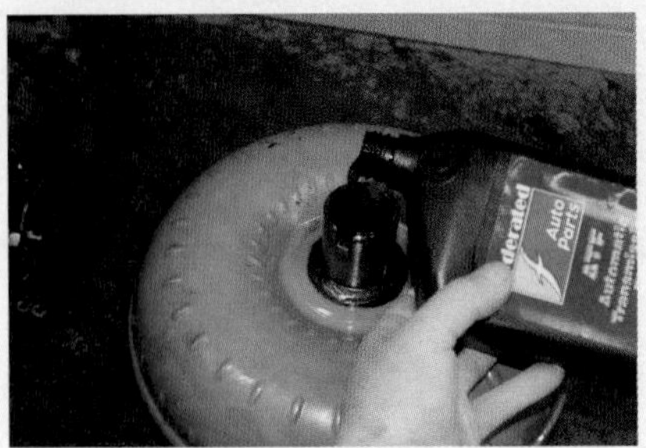

1. Look up the manufacturer's recommended procedure for installing the torque converter. Prefill the new or rebuilt torque converter about halfway with clean automatic transmission fluid.

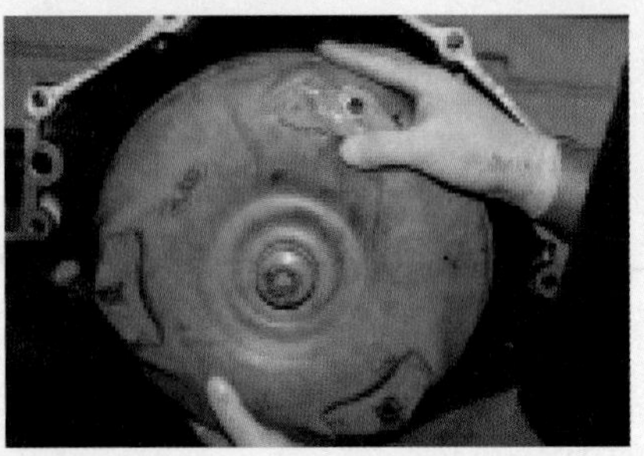

2. Install the torque converter onto the input shaft of the transmission. Spin and wiggle the torque converter to install it properly. You should feel the torque converter catch three times.

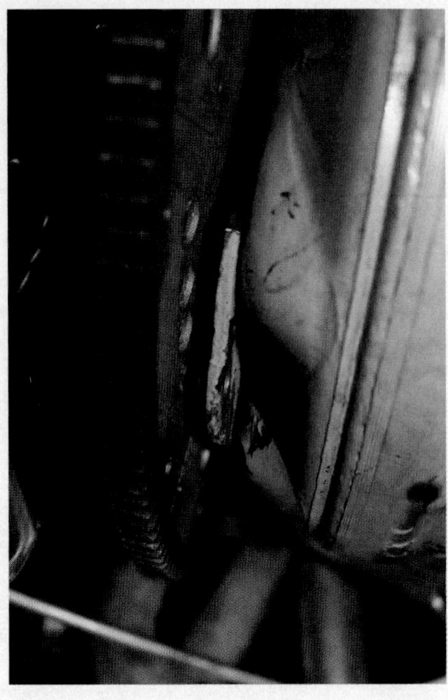

3. Install the transmission into the vehicle. If the torque converter will not spin, or is bound up against the flexplate, stop. Remove the transmission. The torque converter is not installed properly.

Firmly apply the brakes and start the engine, and check the fluid level. Add fluid as necessary to bring it to the bottom of the safe or add mark on the dipstick. Make sure you clear any transmission codes that are in the PCM, if equipped. Double-check that all of the adjustments are correct, and then test-drive the vehicle. If required, make sure to look up any procedures for resetting the adaptive memory after a rebuild, so the PCM will reprogram its shift timing and pressures for the rebuilt transmission.

After the test-drive, double-check the fluid level. Make sure there are no leaks, and all of the mounting bolts are properly torqued.

▶ Wrap-Up

Ready for Review

- Always follow safety precautions and wear personal protective clothing and equipment when removing the transmission for repair.
- Thoroughly inspect the flexplate, even though it is unlikely to need repair.
- If not replacing the torque converter, be sure to inspect the bolts, pump drive, pump drive tangs, pump drive lugs, and converter snout, and check the end play.
- Always clean and inspect the transmission oil cooler and lines prior to installing a new or repaired transmission.
- Check bearings to ensure preload or end play are at optimum levels.
- Disassemble a transmission in a stepwise fashion to avoid missing transmission failure clues.
- Take care during transmission disassembly not to lose or damage any parts.
- Check planetary gear sets for discoloration, cracked or worn teeth, and end play.
- Inspect the smaller transmission parts, including bushings, thrust bearings, and washers.
- Disassemble and inspect one-way rollers for signs of damage.
- Visually inspect sprags, as they cannot be completely disassembled.
- All components of the oil delivery circuits must be inspected for debris, damage, and proper operation.
- Inspect the transmission case for damage or warping, and ensure the vent is unclogged.
- Inspect bands for damage, wear, or burns; inspect the drum for wear and warpage.

- Always rebuild clutch assemblies during transmission rebuilding, and thoroughly inspect all components for damage and wear.
- Measure clutch pack clearance after reassembly.
- Disassemble the transmission valve body, and inspect each valve for wear.
- Both gear-style and chain-style transaxle assemblies must be inspected for wear.
- Carefully reassemble the transmission in the correct sequence, with all components properly lubricated.
- Install the torque converter and ensure that all three components (oil pump drive, stator support, and turbine input shaft) engage with their mating parts.
- When finished, test-drive the vehicle, and double-check the fluid level to ensure there are no leaks.

Key Terms

free play The amount of movement between two mating parts.

Morse-style chain A heavy-duty chain constructed of many links and held together with pins. It is used in some transaxles and transfer cases to transmit torque from one component to another.

preload Further pressure applied to bearing-supported parts after all the free play is taken up.

Review Questions

1. All of the following statements are true with respect to the bolts and nuts securing the torque converter to the flexplate except:
   a. All of the engine's power pulses and torque are transmitted through three or four bolts to the torque converter.
   b. If any bolts are found to be damaged, replace only the damaged bolts.
   c. If the converter uses nuts, inspect the threads of the studs on the converter for thread damage
   d. If the studs are damaged, the converter will need to be replaced.
2. Transmission coolers can be cleaned by:
   a. an aerosol flushing kit.
   b. compressed air.
   c. detergents.
   d. rust inhibitors.
3. When disassembling a transmission, it is good to disassemble it into its unit assemblies because it helps in all of the following except:
   a. preventing missed steps.
   b. preserves the warranty.
   c. locating faults.
   d. remembering how the parts go back together.
4. Discoloration of the planetary gears or the planet carrier is most likely a sign of:
   a. incorrect clearance.
   b. a gear that is loose on its shaft.
   c. slightly over full fluid level.
   d. a failed lubrication circuit.
5. Which of these should be used to identify fluid passages through the case when inspecting fluid delivery circuits?
   a. Scan tool
   b. Mirror and light
   c. Transmission case diagram
   d. Telescoping gauge
6. A drum that is not badly scored (lightly worn) should:
   a. still be replaced.
   b. be refinished using emery cloth.
   c. be used as is.
   d. be used with more lubrication.
7. If the clutch pack has too little clearance, the clutch:
   a. will drag when it is disengaged.
   b. may slip.
   c. will have a very harsh engagement.
   d. piston needs to be replaced.
8. How many engagements should you feel when installing a torque converter?
   a. 1
   b. 2
   c. 3
   d. 4
9. Choose the correct statement.
   a. Planetary gear sets are generally replaced with new ones during a transmission rebuild.
   b. Fluid flow in an automatic transmission does not affect the operation of the transmission.
   c. Bands should always be replaced with new bands during a transmission overhaul.
   d. When rebuilding an automatic transmission, all multi-disc clutch assemblies should be rebuilt.
10. After the transmission is installed properly into the vehicle, approximately what percentage of the total transmission fluid should be added to the transmission?
    a. 100%
    b. 50%
    c. 75%
    d. 25%

ASE-Type Technician A/Technician B Style Questions

1. Two technicians are discussing transmission end play. Tech A says that if the end play is insufficient, it may cause premature wear on the thrust bearings. Tech B says that transmissions with incorrect end play may have been reassembled incorrectly. Who is correct?
   a. Tech A
   b. Tech B
   c. Both A and B
   d. Neither A nor B
2. Tech A says that one crescent-type front pump measurement uses a feeler blade between the outer gear and pump body. Tech B says that a micrometer is used to measure the width of the crescent. Who is correct?
   a. Tech A
   b. Tech B

c. Both A and B
d. Neither A nor B

3. Tech A says that to determine whether roller type thrust bearings need to be replaced, they should be measured with a micrometer. Tech B says that to determine whether thrust washers are worn excessively, they should be measured with a micrometer and compared to specifications. Who is correct?
   a. Tech A
   b. Tech B
   c. Both A and B
   d. Neither A nor B
4. Tech A says that a one-way roller clutch will work fine if assembled in either direction. Tech B says that sprag clutches must be serviced as a complete unit if any damage is found. Who is correct?
   a. Tech A
   b. Tech B
   c. Both A and B
   d. Neither A nor B
5. Tech A says that when rebuilding a multidisc clutch, the steels should be checked for flatness, and their thickness should be measured. Tech B says that it is common to reuse the clutch piston's lip seals. Who is correct?
   a. Tech A
   b. Tech B
   c. Both A and B
   d. Neither A nor B
6. Tech A says that clutch pack clearance should be checked with a micrometer. Tech B says that any clearance in the clutch pack means that the clutch discs are worn out. Who is correct?
   a. Tech A
   b. Tech B
   c. Both A and B
   d. Neither A nor B
7. Tech A says that after rebuilding a clutch, it needs to be air checked before the clutch plates are installed. Tech B says that clutches that are rebuilt outside of the transmission are typically air tested before installing them back in the transmission. Who is correct?
   a. Tech A
   b. Tech B
   c. Both A and B
   d. Neither A nor B
8. Tech A says that when assembling a transmission, the components should be lubricated with the correct automatic transmission fluid or transmission assembly lube. Tech B says that the transmission end play needs to be measured after the front pump is installed. Who is correct?
   a. Tech A
   b. Tech B
   c. Both A and B
   d. Neither A nor B
9. Tech A says that the torque converter should be bolted to the flexplate before installing the transmission. Tech B says that when installing the torque converter, three separate engagements with their mating parts must be felt. Who is correct?
   a. Tech A
   b. Tech B
   c. Both A and B
   d. Neither A nor B
10. Tech A says that when inspecting the torque converter, you need to check the snout of the converter where the seal and bushing rides for excessive wear. Tech B says that the internal end play of the converter must be checked. Who is correct?
   a. Tech A
   b. Tech B
   c. Both A and B
   d. Neither A nor B

CHAPTER 29
Hybrid and Continuously Variable Transmissions

NATEF Tasks

- **N29001** Describe the operational characteristics of a hybrid vehicle drivetrain. (MLR/AST/MAST)
- **N29002** Describe the operational characteristics of a continuously variable transmission (CVT). (MLR/AST/MAST)

Knowledge Objectives

After reading this chapter, you will be able to:

- **K29001** Describe the purpose and function of idle stop systems.
- **K29002** Describe the purpose and function of torque smoothing systems.
- **K29003** Describe the purpose and function of regenerative braking.
- **K29004** Describe the purpose and function of torque assist.
- **K29005** Describe the purpose and function of electric-only propulsion.
- **K29006** Describe the operational characteristics of various models of hybrid vehicles.
- **K29007** Describe the purpose and function of belt alternator starter systems.
- **K29008** Describe the purpose and function of the Honda Integrated Motor Assist system.
- **K29009** Describe the purpose and function of the Honda two-motor hybrid powertrain.
- **K29010** Describe the purpose and function of the Toyota and Lexus hybrid powertrain.
- **K29011** Describe the purpose and function of the Ford hybrid powertrain.
- **K29012** Describe the purpose and function of the two-mode hybrid powertrain.
- **K29013** Describe the function of the VDP CVT.
- **K29014** Describe the function of the toroidal CVT.

Skills Objectives

There are no Skills Objectives for this chapter.

▶ Introduction

With the growing need for increased fuel economy, lower carbon dioxide emissions, and a smaller global footprint, manufacturers have been working to create more efficient powertrain systems. Automobile manufacturers have had hybrid vehicles and vehicles with continuously variable transmissions (CVT) in mass production for more than 20 years (**FIGURE 29-1**). These vehicles are showing up in dealer and non-dealer repair facilities in greater numbers every day. And because of their unique safety and operational characteristics, it is important that technicians become familiar with their system design and operation.

▶ Hybrid Drive Systems

N29001

A **hybrid drive system** is defined as a system that uses two or more power sources, such as an internal combustion engine (ICE) and an electric motor, to propel the vehicle (**FIGURE 29-2**). There are many different styles and systems of hybrid vehicles being sold today. We cover several of the more common hybrid vehicles on the market at the time of this writing—primarily gasoline–electric hybrids—and how they operate. Also expect various other hybrid arrangements, such as diesel–electric hybrids and fuel cell–electric hybrids, in the near future.

Before exploring how a hybrid operates, it is important to first discuss the purpose of a hybrid drive system. Hybrid vehicles have several different functions that separate them from conventional vehicles, such as idle stop and regenerative braking. Not all hybrids have every one of these functions, but often they have most of them. Hybrids may be considered mild hybrids or full hybrids based on the number of hybrid functions they use. Knowing what functions a hybrid vehicle is equipped with and how the vehicle operates is necessary for two reasons: first, so you will know whether it is operating correctly, and second, so you will know what precautions you need to take in order to work on it safely.

Idle Stop

K29001

Idle stop is one of the most common functions of a hybrid vehicle, and when used with an integrated starter/generator, qualifies the vehicle for mild hybrid status. When the driver stops the vehicle, such as at a stoplight, the powertrain control module (PCM) shuts off the engine. Shutting off the engine reduces fuel

FIGURE 29-1 Inside view of a continuously variable transmission.

FIGURE 29-2 Inside view of both electric motors from a power-splitting transmission out of a Toyota Prius.

You Are the Automotive Technician

Recently the Express Delivery company you work for has converted its fleet from diesel to hybrid vehicles. Hybrid vehicles have lower carbon dioxide emissions, increased fuel economy, and a smaller global footprint. These vehicles often use very high-voltage batteries that present a severe shock hazard, which can easily lead to death if handled improperly. Your supervisor has asked you to complete a scheduled maintenance service on one of the hybrid vehicles.

1. Why is it important to fully understand the hybrid system and its operation before attempting any repairs on a hybrid vehicle?
2. Describe "regeneration."
3. Describe the operation of a variable-diameter pulley CVT.
4. When checking a front or rear turn signal bulb, why would it be unsafe to turn the ignition key on and then go check to see if the bulb is working?

FIGURE 29-3 A picture of the driver information center on a Toyota Prius, showing the vehicle in an idle stop situation. The engine is not running in this mode.

consumption and carbon dioxide emissions. By using high-voltage electric motors, the PCM is able to crank the vehicle over very quickly and smoothly, creating an almost instant start as soon as the driver presses down on the accelerator. This cranking speed can be as high as 1000 rpm.

Stopping the engine at stoplights or when idling creates several problems that need to be addressed, such as the operation of the heating and cooling systems when the engine is stopped. In a conventional vehicle, the water pump and air-conditioning compressor are driven by an engine belt. This is not possible on a vehicle with the idle stop function (**FIGURE 29-3**).

Many hybrid vehicles that use idle stop have a small, electric transmission hydraulic pump. This hydraulic pump is used to prevent a delay in the engagement of the transmission when the vehicle is restarted.

Refer to the Principles of Heating and Air-Conditioning Systems chapter for more information on the heating and cooling system in hybrid vehicles.

Torque Smoothing

K29002

Torque smoothing refers to the ability of the hybrid vehicle to use an electric motor to help smooth out the power pulses of the ICE and create a flatter torque output curve. This feature is especially helpful on small three- and four-cylinder engines, which tend to produce stronger engine pulsations at lower engine speeds. When the ICE is on a compression pulse, the electric motor can be used to increase the amount of torque during compression (**FIGURE 29-4**). As the ICE is on a power stroke, the electric motor creates less torque. This smoothes out the variations in crankshaft speed during each revolution. As the ICE's rpm increases, the torque smoothing pulses are needed less and less.

Regenerative Braking

K29003

Accelerating a vehicle from a stoplight requires a large amount of energy. Yet, when a conventional vehicle is being stopped, most of the kinetic energy of the vehicle's movement is converted into heat by the friction of the brake pads against the

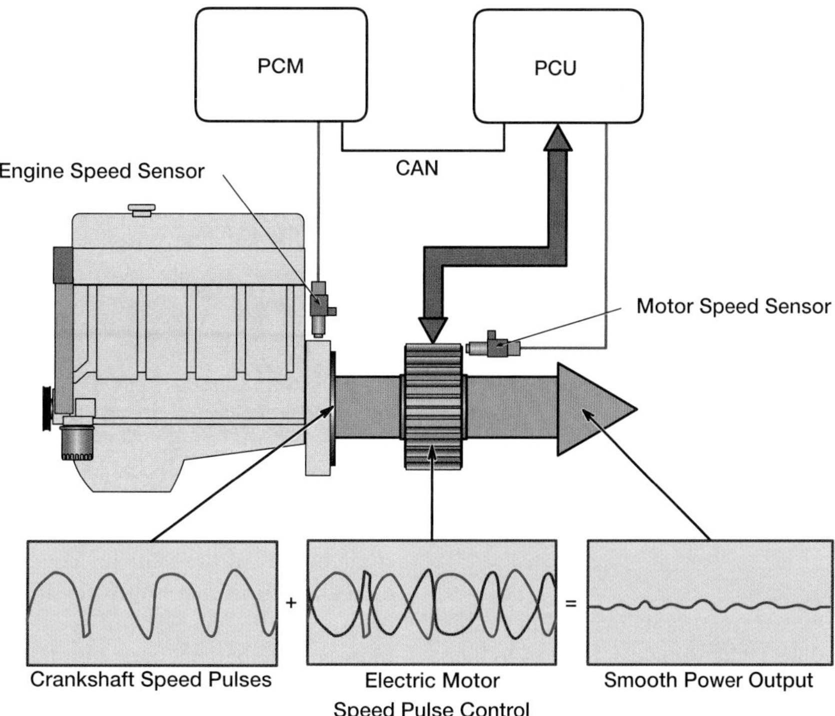

FIGURE 29-4 Torque smoothing uses the electric motor to smooth out the compression and power pulses.

FIGURE 29-5 The driver information screen from a Toyota Prius, showing power flow from the electric motor to the battery (regeneration).

FIGURE 29-6 A picture of the driver information center from a Toyota Prius showing the ICE and the electric motor being used to propel the vehicle (torque assist).

brake rotors; therefore, all of this braking energy is lost on a conventional vehicle. Most hybrid vehicles use **regenerative braking** to ensure that not all of the energy that was used to accelerate the vehicle is lost during braking.

On a hybrid vehicle, when the driver initiates a stop, the electric motor becomes a generator. The kinetic energy of the vehicle's movement is used to turn the generator and create electricity to charge the high-voltage battery. The harder the driver steps on the brake pedal, the more electricity is generated. It is important to remember that the more electricity that a generator is required to produce, the harder it is to turn the generator. This action causes the vehicle to slow down and eventually stop with the help of a conventional braking system, if needed (**FIGURE 29-5**).

Regenerative brakes can only develop a certain amount of stopping power. If the driver needs more braking power than the regenerative brakes can supply, mechanical brakes are available as backup. However, they cannot recapture any of the kinetic energy. Judicious use of the brakes on a hybrid vehicle, such that the mechanical brakes are not activated, results in the best fuel economy.

Continuous driving, such as on the interstate, does not offer the regenerative braking system an opportunity to energize the battery. It is during stop-and-go driving that hybrids benefit most from this system, often resulting in a significantly higher gas mileage (mpg) rating during city driving than on the highway. The regenerative braking system also makes hybrids ideal for delivery vehicles and city transit buses that operate in stop-and-go conditions all day long.

Torque Assist

K29004

An electric motor is capable of creating its maximum torque as soon as it begins spinning. This torque can be used to help propel the vehicle from a stop, creating **torque assist** for the ICE, which creates its maximum torque much higher in the rpm range. By using an electric motor with an ICE, the overall displacement of the ICE can be reduced. For example, a

FIGURE 29-7 A driver information center on a Toyota Prius, showing the electric motor alone propelling the vehicle.

vehicle that might have needed a 2.0-liter engine to operate adequately might only need a 1.4-liter engine to perform the same way when it is combined with the hybrid electric motor. This reduction in size of the ICE results in a fuel savings, not only during city driving but also on the highway (**FIGURE 29-6**).

Electric-Only Propulsion

K29005

On many full hybrids, the vehicle can operate at low speeds using the electric motor only. This is especially helpful in slow moving or stop and go traffic to greatly lower emissions output of the vehicle. This cuts down on the amount of air pollution that builds up around cities. When a driver presses on the accelerator pedal, the PCM commands the high-voltage battery pack to apply electricity to the electric motor until the vehicle reaches speeds up to approximately 40 mph (64 kph) (depending on the model), at which point the PCM starts the ICE to continue accelerating the vehicle (**FIGURE 29-7**).

SAFETY TIP

Extreme caution should be exercised when working around hybrid vehicles. These vehicles often use very high voltages that can cause serious harm or death. Only begin working on a hybrid vehicle after you have had the proper safety training and thoroughly understand the system and its operation, and have the proper safety equipment. Always check with the appropriate service information before attempting any repairs on a hybrid vehicle. In addition to the high voltages, many systems have very strong permanent magnets that attract any ferrous metal that gets close to them and can pinch your fingers or hand between the part and the magnet. The magnets often cannot be removed without a special tool.

▶ Hybrid Electric Vehicle Models

K29006

Currently, hybrid vehicles are available from almost every vehicle manufacturer, with new models being added every year as the demand for more fuel-efficient vehicles increases (**FIGURE 29-8**). In this section, we cover some of the more common hybrid vehicle models used by the different manufacturers.

Belt Alternator Starter

K29007

Belt alternator starter (**BAS**) vehicles have been produced primarily by the General Motors Corporation (**FIGURE 29-9**). The BAS unit is a belt-driven alternator and starter motor. Conventional starter motors and alternators operate on nominal 12 volts, whereas the BAS system uses a 42-volt battery.

When the engine is running, the belt spins the BAS motor, which in turn produces electricity, like a conventional alternator, to charge the high-voltage battery. When the engine is stopped, the BAS motor can crank the vehicle over very quickly through the belt drive. This design allows the vehicle to use the fuel savings of the idle stop function.

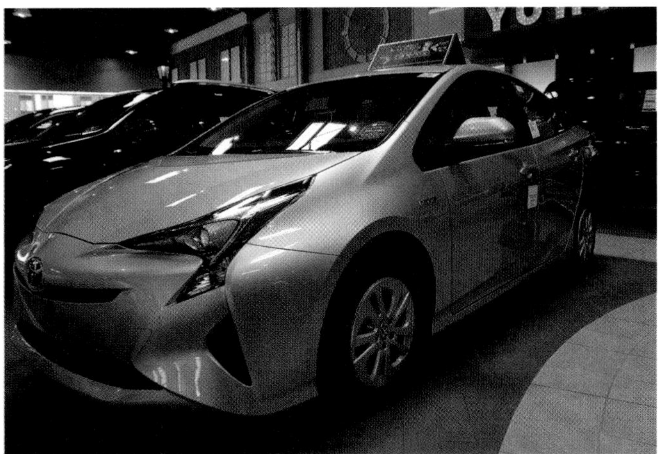

FIGURE 29-8 Toyota Prius is the most common hybrid vehicle.

FIGURE 29-9 A BAS system installed on a Chevrolet Malibu.

As the vehicle is slowing down, the BAS motor is used to produce a large amount of current to charge the battery while also slowing the vehicle down due to the extra drag of the alternator. This allows the vehicle to capture some of the fuel savings of the regenerative braking. Note that this system is not designed to power the vehicle down the road independent of the ICE.

The transaxle used in vehicles with the BAS systems is relatively the same as that used in non-hybrid counterparts. The only major difference, other than PCM programming, is the addition of an electrically operated transaxle hydraulic pump. This pump allows a smooth acceleration as the engine restarts after an idle stop. Without this electric pump, there would be a momentary lag as the transaxle reengaged after the engine was restarted.

Honda Integrated Motor Assist

K29008

Honda hybrids have used a system called the **integrated motor assist (IMA)**. The IMA system uses a thin electric motor in place of a conventional flywheel or flexplate. This electric motor is used to supplement the gasoline engine's power when accelerating. The Honda IMA system is classified as a parallel hybrid because the electric motor is operating at the same time as the gasoline engine. The Honda IMA system uses between 144 and 158 volts, depending on the year of the vehicle and the specific model of hybrid. Currently, Honda IMA hybrids cannot drive using the electric motor only because the IMA is not able to disconnect from the gasoline engine. Therefore, in order for electricity to be produced, the gasoline engine must be spinning. (It does not have to be using gasoline, though; it can be coasting.)

The electric motor is used to charge the high-voltage battery pack when the vehicle is decelerating, converting some of the kinetic energy of the vehicle motion into electrical energy rather than wasting some of the energy in heat during braking. The electric motor is also used to rapidly restart the engine after the driver releases the brake pedal.

The Honda IMA system does also use a conventional 12-volt starter to initially start the vehicle when it is cold out or when the high-voltage battery is not charged. The transaxle used in combination with the IMA system can be a conventional automatic transmission, a manual transmission, or, more typically, a CVT. CVTs are covered later in this chapter. All three types of transmissions have been modified to be smaller so they can fit with the IMA assembly between the engine and the transmission. Both conventional transmissions and CVTs use an auxiliary electric transaxle hydraulic pump to keep the clutches engaged when the vehicle is stopped at a light. Other transaxle modifications include a stronger lock-up torque converter, which allows the drive wheels to run the IMA assembly during deceleration in order to charge the high-voltage battery pack. Gear ratios are different in the transaxles used in conjunction with IMA assembly than in conventional vehicles to maximize the fuel savings and performance of the hybrid functions (**FIGURE 29-10**).

Honda Intelligent Multi-Mode Drive (i-MMD) Two-Motor Hybrid Powertrain

K29009

The Honda two-motor hybrid system uses two electric motors (propulsion motor and generator motor) and an ICE to power the vehicle. It is considered a full hybrid that can be classified as a series-parallel hybrid because either the gasoline ICE or the electric propulsion motor can propel the vehicle, or both can be used together, either in series or parallel. One especially unique feature of this powertrain is that there is no real transmission. It is a direct drive powertrain with no shifts, or a continuously variable geartrain. It utilizes a single-speed gearbox, where power sources come together from the ICE and propulsion motor to drive the vehicle (**FIGURE 29-11**). To do that, the propulsion motor is permanently coupled to the direct drive system, and the ICE is intermittently coupled to the drive system by a lock-up clutch. The generator motor is permanently coupled to the ICE and acts as a generator used to both charge the high-voltage battery and power the propulsion motor. It also acts as a starter motor for the ICE during idle/stop operation and coupling with the drive system.

FIGURE 29-10 An electric motor from a Honda IMA system.

Operation at low vehicle speeds is provided entirely by the propulsion motor, which can be powered electrically by either the high-voltage battery or power from the generator motor that is driven by the ICE, or both, if needed. At slow vehicle speeds, the ICE will operate as needed to charge the high-voltage battery and provide power to the propulsion motor. Above approximately 40 mph (64 kph), where the ICE is more efficient, the generator motor spins the ICE up to speed, and it is coupled to the drive system through a lock-up clutch. Above this speed, the ICE can propel the vehicle by itself, as well as power the generator motor to charge the battery, or power the propulsion motor, depending on what is needed at the time.

During deceleration, the propulsion motor is turned by the wheels and regenerates power back to the battery to recapture braking energy that would otherwise be wasted. Reverse is obtained simply by reversing the direction of the propulsion motor, which then drives the vehicle in reverse. This reduces the need for additional parts and complexity.

Toyota and Lexus Hybrids

K29010

Toyota and Lexus hybrids use a power-splitting device that is considered a type of CVT. These hybrids use voltages from 200 to 650 volts, depending on the application. The Toyota and Lexus hybrids are full hybrids that can be classified as a series-parallel hybrid because either the gasoline ICE or the electric motor/generator can propel the vehicle, or both can be used together.

SAFETY TIP

Because the vehicle can be driven by the electric motor only, and the ICE can be started by the PCM at any time that the key is turned on or the system is powered up by the start button, it is very important that you follow the manufacturer's procedure for testing the vehicle. This imperative could be illustrated by a simple turn signal bulb inspection. You would normally turn the ignition switch to the run position, turn on the turn signal, and then go look at the front and rear turn signal bulbs. Unfortunately, with a hybrid system, the vehicle could start on its own and travel across the shop, putting others and the vehicle in danger. Follow the manufacturer's procedures exactly when working on a hybrid vehicle.

The transaxle contains two electric motor/generators, a final drive gear set, a differential, and a planetary gear set. Each of the electric motor/generators and the ICE are connected to a separate part of the planetary gear set, as shown in **FIGURE 29-12**.

The ring gear of the planetary gear set is connected to the final drive through a drive chain as in many front-wheel drive

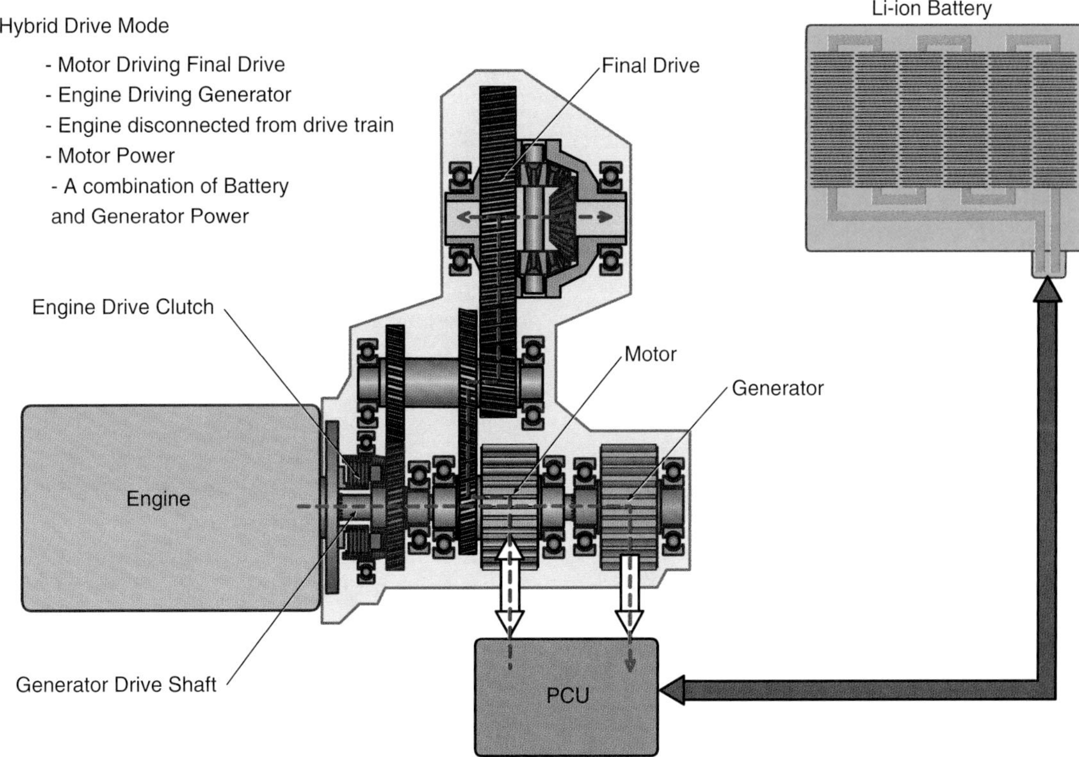

FIGURE 29-11 Honda two-motor hybrid powertrain schematic.

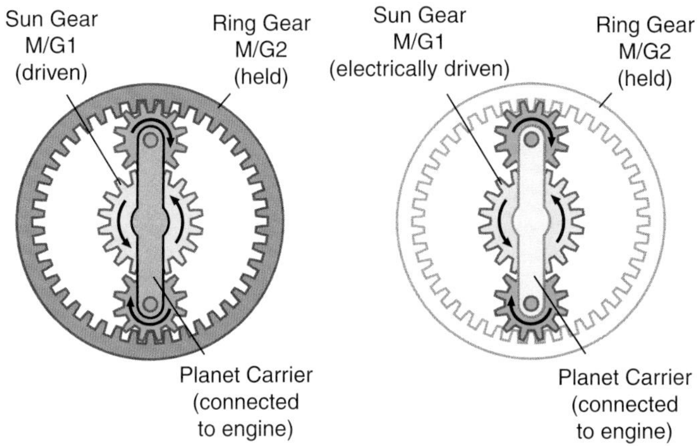

FIGURE 29-12 The planetary operation as the engine is started.

FIGURE 29-13 The ring gear is attached to MG2 and the drive chain. The sun gear is attached to MG1.

(FWD) transaxles (**FIGURE 29-13**). The ring gear is also directly connected to motor/generator 2 (M/G2). The sun gear of the planetary gear set is attached to the motor/generator 1 (M/G1). M/G1 is used as a generator to charge the high-voltage battery pack and to crank the engine over. The planet carrier of the planetary gear set is attached to the ICE.

When starting the gasoline engine, M/G1 spins the planetary carrier, which is attached to the ICE, in order to crank the engine. With the vehicle stopped, M/G2 is basically locked and prevented from turning.

When the engine starts, M/G1 begins spinning. M/G1 switches to generator mode and begins charging the high-voltage battery pack. M/G2 then begins to spin, and the power from the ICE is added to M/G2 to spin the ring gear and propel the vehicle (**FIGURE 29-14**).

Toyota and Lexus hybrids are able to propel the vehicle in electric-only mode at low speeds. During electric-only mode, the ICE is shut down, effectively holding the planet carrier from spinning. High voltage is supplied to M/G2, which spins the ring gear and drives the wheels. M/G1 is allowed to spin freely (**FIGURE 29-15**). In reverse, the current flow to M/G2 is reversed and causes the motor to spin in the opposite direction. In reverse, the ICE is not used to propel the vehicle.

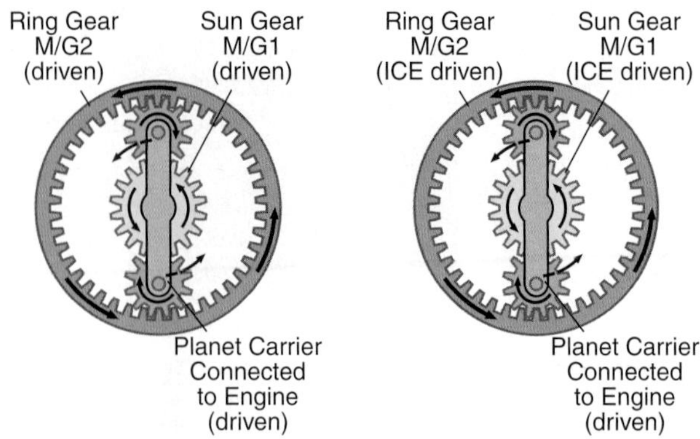

FIGURE 29-14 The planetary operation as the vehicle is moving and M/G2 and the ICE are moving the vehicle.

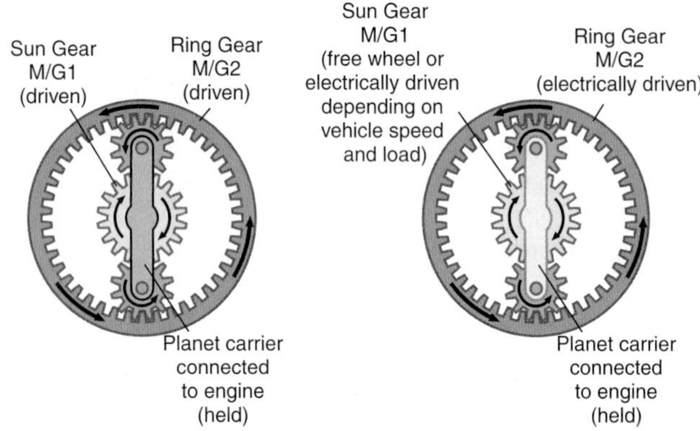

FIGURE 29-15 The planetary operation as M/G1 is propelling the vehicle without the aid of the ICE.

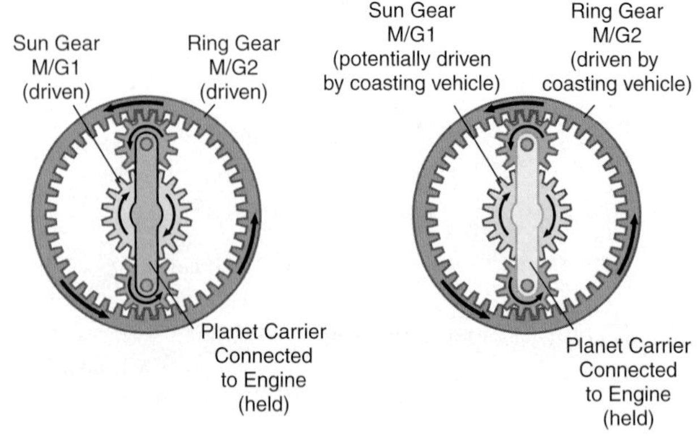

FIGURE 29-16 The planetary gear operation during deceleration.

During deceleration at low speeds, the engine is shut off to hold the carrier. M/G1 is allowed to freewheel while M/G2 is driven by the wheels. M/G2 is switched to generator mode, and the power generated is used to recharge the high-voltage battery pack (**FIGURE 29-16**).

The vehicle's PCM controls the amount of current produced from M/G2 during deceleration to prevent wheel lock-up and skidding. During electric-only operation, the PCM continuously monitors the battery voltage and restarts the ICE as needed. There is no torque converter or clutch mechanism. There is, however, a damper assembly that is used to cushion the power pulses from the engine and reduce the drive line shock.

Ford Motor Company Hybrids

K29011

The Ford Escape and Fusion hybrids (along with the Lincoln and Mercury versions) are very similar in operation to the Toyota system. The Ford transaxle also uses two electric motor/generators that, in conjunction with an ICE, split power through a planetary gear set. The only major difference between the Ford and the Toyota is that rather than having the two electric motor/generators directly connected to the ring gear and the sun gear, they are attached through a set of transfer gears (**FIGURE 29-17**).

Two-Mode Hybrid

K29012

The **two-mode hybrid** was a joint venture by the General Motors, Chrysler, and BMW companies to create a hybrid system that could be used on trucks and larger luxury vehicles. The system is mostly housed in what looks like a conventional truck transmission housing (**FIGURE 29-18**). It uses a system

FIGURE 29-17 A cutaway of a Ford hybrid transmission. Notice M/G1 and M/G2.

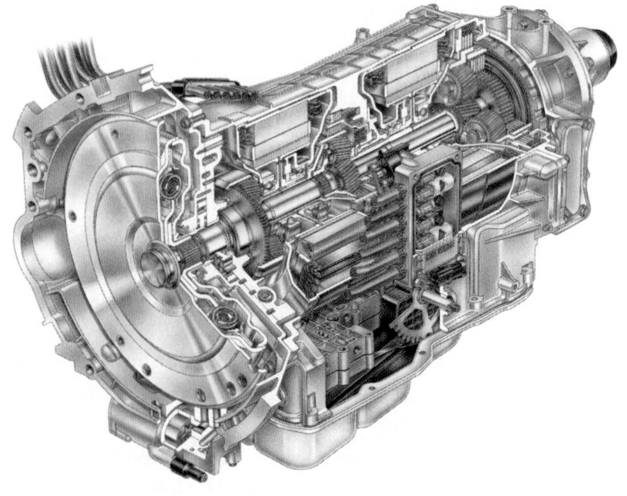

FIGURE 29-18 A typical two-mode hybrid transmission.

voltage of 300 volts to power the two electric motor/generators housed inside the transmission case, which connect to either three or four planetary gear sets, depending on the application. The electric motors/generators and ICE operate in a continuously variable fashion somewhat similarly to the Toyota power-splitting transmission, except that the Toyota version only uses one planetary gear set.

The transmission gets its name from the two operational modes (**FIGURE 29-19**). The first mode is used at low speeds and light engine load. During mode 1, the vehicle can be propelled by the ICE, the electric motor/generator, or a combination of the ICE and electric motor/generator. This functionality qualifies it as a full hybrid system. When operating in mode 1, one of the electric motor/generators is used to keep the high-voltage battery charged while the second motor/generator is used to assist in propelling the vehicle. The vehicle can be driven by the electric motor only if the PCM determines that the high-voltage battery has enough charge to operate the vehicle. If at any time the PCM determines that the battery voltage is too low, or if the driver demands more power, the ICE will be restarted automatically, using the one motor/generator to crank the engine.

In mode 1, the electric motor/generators, in combination with the planetary gear sets, allow the transmission to operate as a type of CVT, keeping the engine operating in its most efficient rpm range. During deceleration, both electric motor/generators are switched to generator mode to recharge the battery.

The second mode is used for higher vehicle speed and/or load. During the second mode, the gasoline engine is running to propel the vehicle. When maximum power is needed, both electric motor/generators are used to help propel the vehicle. When less power is needed or the PCM determines that the battery voltage is low, one of the motor/generators is switched over to generate power to recharge the battery. The PCM phases in and out each of the planetary gear sets to create four distinct gear ratios. This also keeps the electric motor/generators operating in their efficient rpm range.

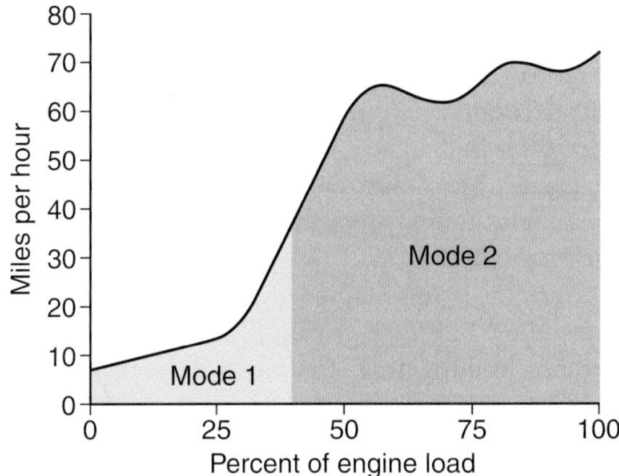

FIGURE 29-19 A chart showing the operation of the two modes of a two-mode hybrid.

During the second mode, the PCM employs different fuel-saving techniques, such as shutting down cylinders and using variable cam timing, to achieve the highest fuel economy without sacrificing power when it is needed. When the vehicle's ICE is operating and cylinders are shut down to save fuel, the electric motor/generators are used to smooth out the power demands (torque smoothing) and compensate for the reduced power from the ICE.

▶ Continuously Variable Transmission (CVT)

N29002

A CVT has no fixed gear ratios; the transmission can infinitely change the gear ratio within its operational design. Changes to the gear ratio occur in a smooth stepless progression to suit speed and load conditions. This design allows the engine to operate in its most efficient rpm operating range for fuel economy or performance, depending on the driver's demands. This means that the engine can operate within a fairly narrow rpm range, even though the vehicle speed is being varied substantially.

Types of CVTs

There are three basic types of CVTs commonly used in production vehicles (**FIGURE 29-20**). The first type is often referred to as an **electronic continuously variable transmission** (**ECVT**) and is commonly found in full hybrid vehicles. The second type is the most common CVT, called a **variable-diameter pulley** or Reeves drive CVT. It is used on both hybrid and standard vehicles. The last type of CVT found in automobiles is a toroidal or roller-based CVT. It is much less common than the other two.

We have already covered the ECVT because it is the transmission/transaxle found in Toyota/Ford hybrid vehicles. The GM two-mode transmission also uses planetary gears in a partial continuously variable manner. By using the combination of two electric motor/generators and an ICE driven through a specially designed planetary gear set(s), the manufacturer is able to create an infinite number of gear ratios.

Variable-Diameter Pulley CVT

K29013

The **variable-diameter pulley (VDP)** CVT is one of the most common types of CVTs being used in vehicles today. The VDP system operates using two variable diameter pulleys with either a steel or a rubber belt between them. Some of the early CVTs in Europe used a stiff rubber belt, as do many snowmobiles and all-terrain vehicles (ATVs). The rubber belt has a limited service life in an automobile, so most manufacturers use a steel drive belt. Each of the pulleys has two movable drive faces called sheaves. These sheaves can be moved inward or outward, relative to each other, to change the effective diameter of the pulley (**FIGURE 29-21**).

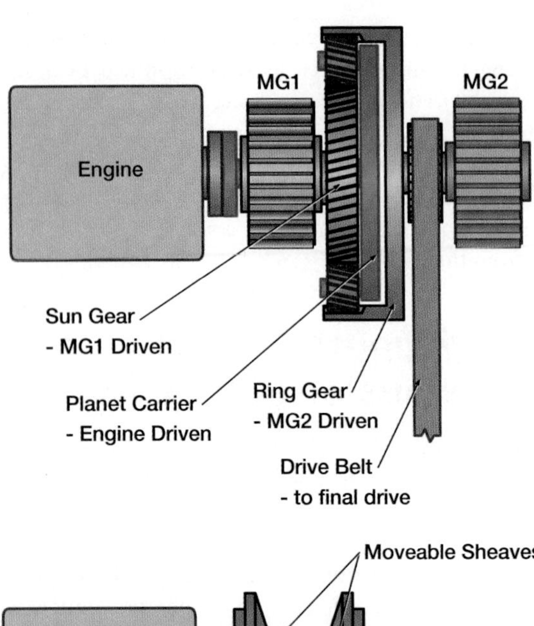

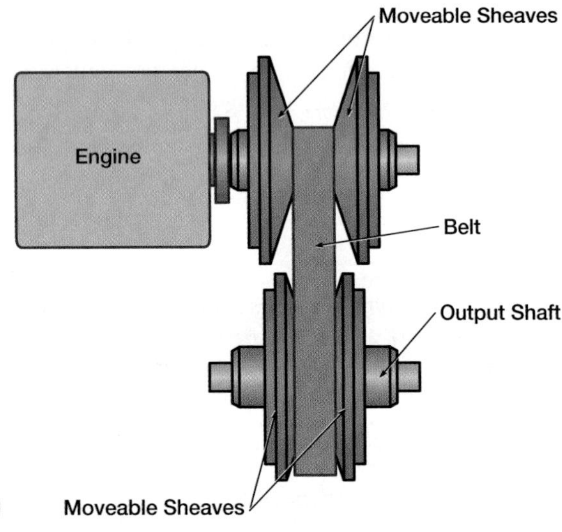

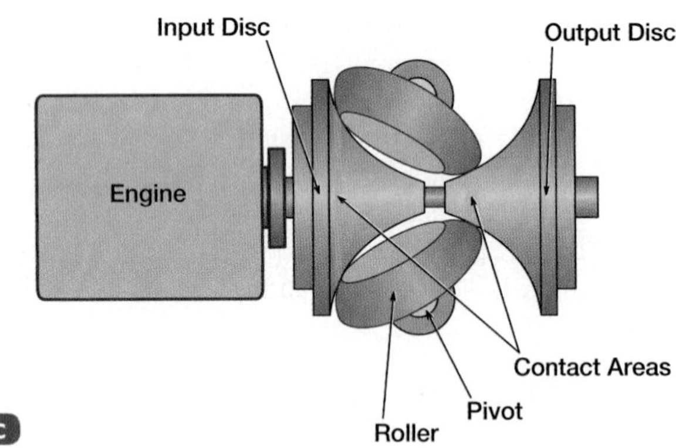

FIGURE 29-20 **A.** Electronic continuously variable transmission (ECVT). **B.** Variable-diameter pulley or Reeves drive CVT. **C.** Toroidal or roller-based CVT.

When the vehicle starts from a stop, the input pulley has a small diameter while the output pulley has a large diameter. This size difference creates a high torque multiplication (low speed) gear ratio to start the vehicle from a stop. The drive ratio from a stop is approximately 3:1 through the pulleys. As the vehicle gains speed, the input pulley diameter increases while the output pulley diameter decreases (high speed).

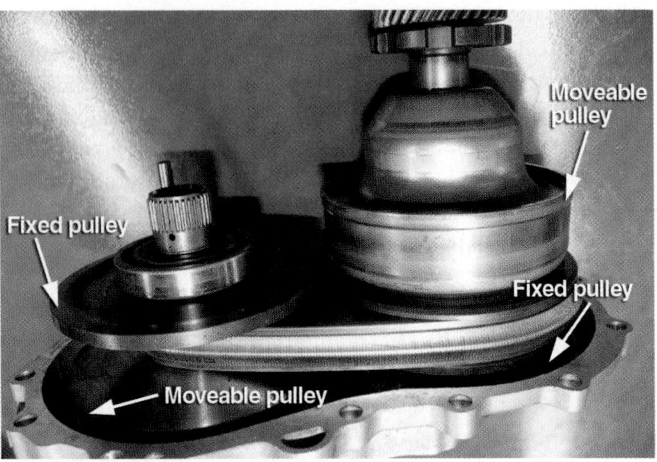

FIGURE 29-21 A VDP CVT. Note the small-diameter input pulley on the left and the large-diameter output pulley on right. The two sheaves that are moveable are labeled. The other two sheaves remain stationary.

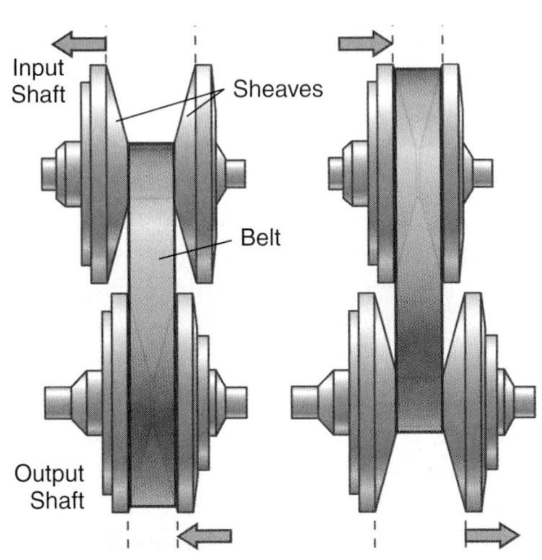

FIGURE 29-22 An illustration of the changing sizes of the input and output pulleys.

The input pulley diameter is changed by applying hydraulic oil to one of the pulley sheaves to push the two sheaves closer together. The belt is forced to ride higher on the faces of the pulley, changing the effective diameter of the pulley, as shown in Figure 29-22. When the two pulleys have the same diameter, the gear ratio of the pulleys is 1:1. As vehicle speed increases, the input pulley continues to increase in diameter, and the output pulley decreases in diameter, to create an overdrive ratio of 1:2, or even greater on some transmissions. Remember that this ratio does not include any final drive gearing, which reduces the actual gear ratio of the CVT.

Belt tension is maintained by large springs in the output shaft pulley. The springs cause the sheaves of the output pulley to be close together, creating a large-diameter pulley. As the input pulley diameter increases, the springs in the output pulley are compressed, decreasing the diameter of the output pulley. The diameter of the pulleys is controlled to keep the engine operating in its most efficient rpm range (**FIGURE 29-22**). The PCM monitors the speed of the two pulleys to maintain the correct gear ratio. Some CVTs also incorporate a sensor to measure the position of one of the input pulley sheaves. This allows for greater control and monitoring of the CVT.

Some manufacturers have programmed into the PCM regular shift points, at which the pulley sheaves move to predetermined positions to create the feel of separate conventional shifts. Some manufacturers also allow the driver to manually shift through these predetermined pulley diameters while driving the vehicle.

Currently, most manufacturers do not recommend repairing internal components of a CVT. Instead, they recommend the transmission be replaced as a complete unit with a new or rebuilt transmission. CVTs can be rebuilt, but many parts are difficult and costly to obtain. Specialty tools are required to compress the output pulley and reinstall the belt.

The steel belt is made up of hundreds of transversely mounted steel plates that are held in place with several steel bands running longitudinally around their edges. The transverse steel plates grab the sides of the pulley and in turn are pushed by the pulleys. The longitudinal steel bands are used to hold the belt together and keep the steel plates lined up. These bands hold the plates closely together in a very strong and flexible ring. The plates are guided by the bands, but not attached to them. Drive is transmitted by compressing the plate elements rather than relying on tension in the band. Each block leaving the primary pulley pushes the blocks ahead of it to the secondary pulley, where the bands keep the blocks in contact with the pulley faces. The blocks are compressed on the drive side and float loosely along the bands on the return side.

The large number of plates that are in contact with the pulleys keeps the surface pressures low on the belt, allowing high torque to be transmitted. Both the plates and pulleys are made of very hard materials to resist wearing out quickly. The belt and pulleys are lubricated by transmission fluid sprayed directly onto them at high pressure. A typical steel belt is shown in **FIGURE 29-23**.

FIGURE 29-23 A steel CVT belt that has been removed from a Nissan transmission.

CVTs require special transmission fluid, and many CVTs have a special transmission oil heater built into them. This heater is used to warm up the transmission oil so the transmission can operate properly during cold weather operation. Filter and fluid change intervals must be adhered to in order to prevent damage to the transmission.

Low and Reverse

Most CVT manufacturers use multidisc clutch packs and a set of planetary gears to allow the transmission to operate in reverse. These clutch packs and planetary gears are typically located on the input pulley shaft, to change the direction of rotation of the pulleys and belt (**FIGURE 29-24**).

Ford Motor Company also uses these planetary gear sets to create a low ratio when it is requested by the driver (**FIGURE 29-25**). Ford advertises their CVTs as two-speed CVTs. When the driver selects low, the planetary gear set is used to create a greater torque multiplication for higher pulling power and greater traction with limited vehicle speed. This is helpful when driving in rough terrain or when starting from a stop on a steep hill.

Applied Science

AS-98: Simple Machines: The technician can demonstrate an understanding of how cams, pulleys, and levers are used to multiply forces or change the direction of force in a mechanical system.

Simple machines, such as cams, pulleys, and levers, have become much improved and refined over the years. Looking back into history, we can understand that these three mechanical units have transformed the way that we perform many tasks.

Cams have the ability to change rotary motion to linear motion; for example, a camshaft in an engine. Cams have also been used in printing presses, textile machinery, and machine shop equipment of various types.

Pulleys can be used to change the direction of a force or provide a mechanical advantage with a multiple part line. As an example, raising a flag on a flagpole requires the use of a pulley.

Levers are one of the basic tools that may have been used in prehistoric times. The Greek mathematician, Archimedes, described the use of levers in 260 B.C. Examples of levers include pry bars, pliers, claw hammers, and tongs.

As described in our text, the variable-diameter pulley (VDP) continuously variable transmission is one of the most common types of CVT being used in vehicles today. The VDP system operates using two variable diameter pulleys with either a steel or rubber belt between them. The rubber belt has a limited service life in an automobile, so most manufacturers use a steel drive belt. Each of the pulleys has two movable drive faces, called sheaves. These sheaves can be moved inward or outward, relative to each other, to change the effective diameter of the pulley.

A CVT is a transmission that can change through an infinite number of gear ratios. This infinite number is limited only between the minimum and maximum capabilities of the particular unit. This contrasts with other mechanical transmissions that offer a fixed number of gear ratios. Fuel economy is enhanced by the engine being allowed to run at its most efficient rpm for a range of vehicle speeds.

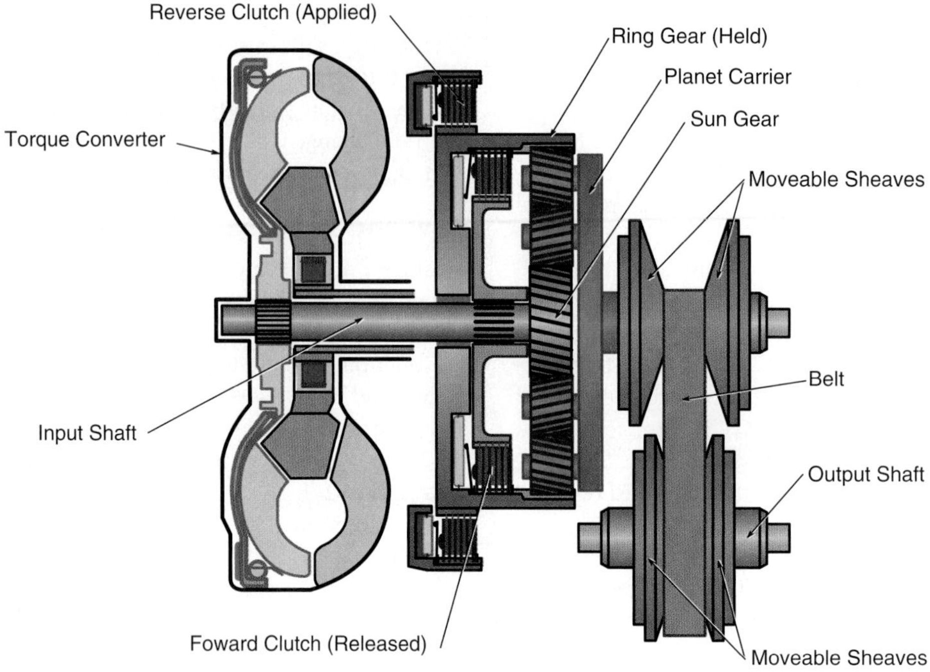

FIGURE 29-24 A planetary gear controlled by clutches allows some CVTs to have reverse and low gears.

FIGURE 29-25 Control switch for selecting low gear ratio.

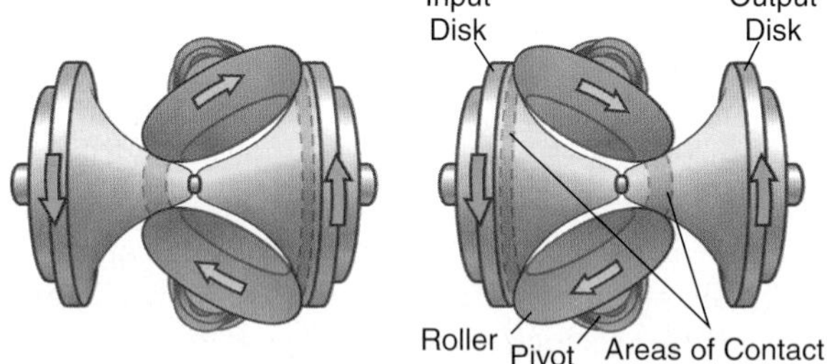

FIGURE 29-26 An illustration of a toroidal CVT. The transmission on the left is in a high torque multiplication ratio, and the transmission on the right is in an overdrive ratio.

Toroidal CVT

K29014

The **toroidal CVT** design currently is limited in production because of high manufacturing costs. The toroidal, or roller, design uses two curved discs—an input disc and an output disc. Power is transferred from the input disc to the output disc through a set of variable-angle rollers (**FIGURE 29-26**). When the rollers are tilted down toward the center axis of the input disc, the opposite side of the roller moves up toward the outer edge of the output disc. This positioning creates a high gear ratio of approximately 4:1 to begin accelerating the vehicle. As the vehicle accelerates, the roller angle changes. When the rollers are perpendicular to the input and output discs, the gear ratio is 1:1. As the vehicle continues to accelerate, the rollers angle in the opposite direction, toward the center axis of the output disc. When the roller is near the center axis of the output disc, its opposite side is near the outer edge of the input disc, creating a gear ratio of approximately 1:4. Remember that these ratios do not include any final drive gearing in their calculations.

The toroidal CVT is currently more costly to manufacture than a belt-style CVT, but it can handle larger amounts of torque. A special transmission fluid is needed for these transmissions to prevent wear to the rollers and discs while providing the correct coefficient of friction for the rollers to operate.

▶ Wrap-Up

Ready for Review

- Vehicles with continuously variable transmissions (CVTs) are becoming more prevalent.
- A system that uses two or more power sources is a hybrid drive system.
- A hybrid vehicle may have the following functions: idle stop, torque smoothing, regenerative braking, torque assist, and electric-only propulsion.
- Idle stop refers to engine shutoff when the driver stops (but does not turn off) the vehicle.
- A hybrid vehicle that uses an electric motor to smooth out internal combustion engine (ICE) power pulses is using torque smoothing.
- Most hybrid vehicles use braking power to generate electricity to charge the battery (regenerative braking).
- Torque assist refers to using torque to help propel the vehicle when additional torque is needed.
- Common hybrid vehicle systems include belt alternator starter (BAS), Honda integrated motor assist (IMA), two-mode hybrid, Toyota and Lexus hybrids, and Ford hybrids.
- A BAS system uses a 42-volt battery and relies on the belt drive to help quickly crank the vehicle over following an idle stop.
- The IMA system replaces the conventional flywheel with a thin electric motor to supplement the engine's torque during acceleration.
- The two-motor hybrid system uses two electric motor/generators and an ICE connected to a direct drive and operates as a series-parallel hybrid.
- The two-mode hybrid system uses approximately 300 volts to power the two generators inside the transmission case.
- The Toyota, Lexus, and Ford hybrid systems are series-parallel systems because the ICE and the electric generator can propel the vehicle, either individually or together.
- In Toyota, Lexus, and Ford hybrid systems, the PCM monitors battery voltage; if voltage is low, the ICE is engaged.
- Ford hybrid systems have two electric motors attached through a set of transfer gears.
- A CVT is able to change gear ratios to suit vehicle speed and load conditions for optimum fuel economy and performance.
- Types of CVTs are the electronic continuously variable transmission (ECVT), variable-diameter pulley (VDP) CVT, and toroidal (or roller-based) CVT.
- The most common CVT in vehicles today is the VDP.
- A VDP has two pulleys (input and output) with movable drive faces that can be adjusted to change the pulleys' effective diameter.
- The output pulley of a VDP utilizes large springs to maintain belt tension.
- Manufacturers recommend replacing CVTs as a whole, rather than repairing internal components.
- Change CVT fluid and filters on a regular basis to maintain transmission functionality.
- Toroidal CVTs use curved discs (input and output) with a set of variable-angle rollers instead of pulleys.
- Toroidal CVTs are costly to manufacture, and production is limited.
- In some VDP CVTs, multidisc clutch packs and planetary gear sets are designed to allow transmission operation in reverse and to create a low gear ratio.

Key Terms

belt alternator starter (BAS) A type of hybrid drive system that uses a belt-driven alternator/starter that operates on 42 volts.

electronic continuously variable transmission (ECVT) A type of hybrid transmission that often uses two electric motors in combination with an ICE. The two electric motors and the ICE transfer power through a planetary gear set, allowing an infinite amount of gear ratios.

hybrid drive system A drive system that uses two or more propulsion systems such as electric motors and an ICE.

integrated motor assist (IMA) A Honda hybrid drive system that uses a moderate-sized electric motor installed between the engine and the transmission.

regenerative braking A type of braking in which the kinetic energy of the vehicle's motion is captured rather than being lost to heat in a conventional braking system.

toroidal CVT A type of CVT that uses moveable rollers in contact with input and output drive discs. The rollers transfer power from one drive disc to the other. Their position determines the effective gear ratio.

torque assist Use of an electric motor to supplement the engine's torque whenever additional torque is needed, allowing for a smaller ICE to be used.

torque smoothing A process that uses an electric motor to smooth out engine power pulses when an ICE is operating at low rpm or when the vehicle is using fuel management techniques such as cylinder deactivation.

two-mode hybrid A type of hybrid drive system in which there are two distinct modes of operation. In one mode, the electric motor can propel the vehicle and be used for regenerative braking; in the second mode, the electric motors can be used to assist the engine while the engine uses fuel management techniques such as cylinder deactivation.

variable-diameter pulley (VDP) A type of CVT that uses two pulleys with moveable sheaves, allowing the effective diameter of the pulleys to change, resulting in variable gear ratios.

Review Questions

1. Which function uses a small, electric transmission hydraulic pump that prevents a delay in the engagement of the transmission?
   a. Torque smoothing
   b. Idle stop
   c. Regenerative braking
   d. Electric-only propulsion
2. What is helpful on small three- and four-cylinder engines with respect to compression and power impulses?
   a. Electric-only propulsion
   b. Belt alternator starter
   c. Torque assist
   d. Torque smoothing
3. All of the following statements are true *except*:
   a. regenerative braking system turns the generator and creates electricity to charge the high-voltage battery.
   b. regenerative brakes can develop great amount of stopping power.
   c. regenerative braking system is an opportunity to recharge the battery.
   d. regenerative braking system makes hybrids ideal for delivery vehicles and city transit buses.
4. Which of these functions makes it so that the overall displacement of the ICE can be reduced?
   a. Torque smoothing
   b. Torque assist
   c. Electric-only propulsion
   d. Regeneration
5. All of the following systems are considered as full hybrid that can be classified as a series-parallel hybrid *except*:
   a. Integrated motor assist (IMA)
   b. Honda two-motor hybrid systems.
   c. Toyota and Lexus hybrid systems.
   d. Ford hybrid powertrain.
6. During idle-stop:
   a. The engine idles while the vehicle is stopped.
   b. the engine stops while the vehicle is stopped.
   c. the high-voltage battery is charged when the vehicle is stopped.
   d. the engine charges the high-voltage battery while it idles.
7. All of the following statements with respect to i-MMD are true *except*:
   a. it is a direct drive powertrain with no shifts.
   b. its unique feature is that there is no real transmission.
   c. it utilizes a four-speed gear box.
   d. the propulsion motor is permanently coupled to the direct drive system.
8. Torque smoothing:
   a. adds power when high torque is required.
   b. use an electric motor to help smooth out the power pulses of the ICE.
   c. is used to smooth out road vibration at high speed.
   d. is used during regeneration to smooth out the electrical pulsations going to the high-voltage battery.
9. Which system uses the combination of two electric motor/generators and an ICE driven through a specially designed planetary gear set(s), which creates an infinite number of gear ratios?
   a. Honda Ingetrated motor assist (IMA).
   b. Two-mode hybrid
   c. Toyota and Lexus hybrids
   d. Honda Intelligent Multi-Mode Drive
10. The variable-diameter pulley CVT:
   a. As the vehicle gains speed, the input pulley diameter increases while the output pulley diameter decreases.
   b. As the vehicle gains speed, the input pulley diameter decreases while the output pulley diameter increases.
   c. prevents high torque multiplication.
   d. has no effect on the gear ratio.

ASE Technician A/Technician B Style Questions

1. Tech A says that hybrid vehicles have two power sources such as an internal combustion engine and an electric motor. Tech B says that most hybrid electric vehicles utilize regenerative braking to help improve fuel economy. Who is correct?
   a. Tech A
   b. Tech B
   c. Both A and B
   d. Neither A nor B
2. Tech A says that it is critical to have the proper safety equipment before working on a hybrid vehicle. Tech B says that hybrid vehicles use safety interlocks to prevent technician injury so it is no longer necessary to use protective gloves. Who is correct?
   a. Tech A
   b. Tech B
   c. Both A and B
   d. Neither A nor B
3. Tech A says that hybrid vehicles do not need a transmission since they have two power sources. Tech B says that some hybrid transmissions have a small electric fluid pump for when the transmission is in idle/stop mode. Who is correct?
   a. Tech A
   b. Tech B
   c. Both A and B
   d. Neither A nor B
4. Tech A says that BAS vehicles use an alternator to help slow the vehicle as well as act as a starter. Tech B says that the BAS alternator can be used to propel the vehicle independent of the ICE. Who is correct?
   a. Tech A
   b. Tech B
   c. Both A and B
   d. Neither A nor B
5. Tech A says that the transmission in a Toyota hybrid has a separate gear to provide reverse. Tech B says that Toyota hybrids use a VDP CVT transmission. Who is correct?
   a. Tech A
   b. Tech B

c. Both A and B
d. Neither A nor B

6. Tech A says that on a vehicle with a VDP CVT, when the vehicle starts from a stop, the input pulley has a small diameter while the output pulley has large diameter. Tech B says that VDP CVTs typically use a separate electric motor for reverse. Who is correct?
   a. Tech A
   b. Tech B
   c. Both A and B
   d. Neither A nor B
7. Tech A says that some CVTs require a heating system to warm the transmission fluid during cold weather operation. Tech B says that CVTs use conventional automatic transmission fluid. Who is correct?
   a. Tech A
   b. Tech B
   c. Both A and B
   d. Neither A nor B
8. Tech A says that VDP CVTs use a steel belt made up of hundreds of transversely mounted steel plates. Tech B says that some VDP CVTs are programmed to give the feel of shift points. Who is correct?
   a. Tech A
   b. Tech B
   c. Both A and B
   d. Neither A nor B
9. Tech A says that two-mode hybrid vehicles uses two motor/generators along with an ICE to propel the vehicle. Tech B says that at low speeds, the two-mode hybrid vehicle can operate off one motor only. Who is correct?
   a. Tech A
   b. Tech B
   c. Both A and B
   d. Neither A nor B
10. Tech A says that the Honda IMA system uses a thin electric motor in place of a conventional flywheel or flexplate. Tech B says that the Honda IMA system can operate on the electric motor only. Who is correct?
   a. Tech A
   b. Tech B
   c. Both A and B
   d. Neither A nor B

SECTION 4
Manual Transmissions

CHAPTER 30
Manual Transmission/ Transaxle Principles

NATEF Tasks

There are no NATEF tasks for this chapter.

Knowledge Objectives

After reading this chapter, you will be able to:

- **K30001** Give an overview of manual drivetrains and their history.
- **K30002** Explain the principles of operation of manual transmissions.
- **K30003** Explain the principles of mechanical advantage.
- **K30004** Explain the principles of gear ratios.
- **K30005** Explain the principles of power flow.
- **K30006** Describe the operation of a manual transmission drivetrain.
- **K30007** Explain the operation of a vehicle equipped with a manual transmission.
- **K30008** Describe the purpose and function of manual drivetrain components.
- **K30009** Describe the purpose and function of shafts, gears, and bearings.
- **K30010** Describe the purpose and function of the clutch system.
- **K30011** Describe the purpose and function of the transmission/ transaxle.
- **K30012** Describe the purpose and function of the transfer case.
- **K30013** Describe the purpose and function of the final drive assembly.
- **K30014** Describe the purpose and function of drive axles.

Skills Objectives

There are no Skills Objectives for this chapter.

▶ Introduction

In this chapter, we explore the history and principles of the modern manual transmission **drivetrain** system. Some of those principles involve certain aspects of physics, such as mechanical advantage and gear ratios. Other principles require an understanding of the nomenclature and theory of operation of the major drivetrain assemblies. Having a good understanding of these principles will give you a framework on which to hang the higher-level theory and concepts covered in later chapters.

A good starting place is to understand what is meant by "drivetrain." The drivetrain consists of the component assemblies that transmit power from the engine all the way to the drive wheels. The manual transmission is at the heart of the drivetrain and receives power from the engine by way of the clutch assembly. The clutch can be operated by the driver to disconnect and connect the transmission from the engine. The transmission provides a range of shiftable gears for the driver to select when operating the vehicle. In a front-engine, rear-wheel drive (RWD) vehicle, the transmission sends power to the final drive assembly, which then changes the direction of the twisting force 90 degrees so it can be sent out the axles to the wheels and tires. In a front-engine, front-wheel drive (FWD) vehicle, the transaxle contains both the transmission and the final drive assembly in a common unit and sends the power out the axles to the front wheels and tires (**FIGURE 30-1**). In all-wheel drive vehicles (AWD), the transmission or transaxle sends power to all four wheels and tires.

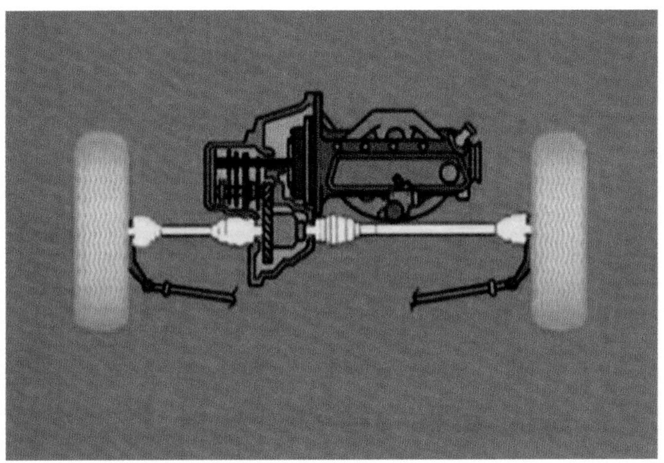

FIGURE 30-1 Front-wheel drive layout.

▶ The History of Manual Transmissions

K30001

In 1877, a patent for an FWD carriage with a one-cylinder engine was obtained by George Selden. The "claim to fame" of that vehicle was the transmission (**FIGURE 30-2**). The power from the engine drove a set of bevel gears, which in turn drove a shaft and a pulley. Leather belts were used on a pulley, and a geared wheel on the axle to make it move. One small wheel on the engine got the car going by meshing with the ring gear on one of the drive wheels. The big wheel on the axle then made the car move along at a staggering—at the time—20 miles per hour.

This early engine had one belt-driven high gear for speed and one belt-driven low gear for increased torque. If the car needed to climb a hill, the driver had to stop, get out, and change the belt to the lower gear. This setup was similar to a 10-speed bicycle, where the bike has a shifting mechanism to physically move a chain from one gear to another.

In 1894, a couple of Frenchmen, Louis René Panhard and Émile Levassor, designed an FWD multi-gear manual transmission (**FIGURE 30-3**). When they tried to put on a demonstration of their new transmission, the engine in their demo vehicle

FIGURE 30-2 The engine and transmission, which were on the front axle, drove this early vehicle.

You Are the Automotive Technician

A Boy Scout troop is working on a merit badge regarding vehicles and vehicle maintenance. They come into your shop to ask you questions that their scoutmaster cannot answer. You greet them, show them around the shop, and introduce them to your fellow employees, who tell them about their specialties. Once that is done, they gather together to ask you some specific questions about terms they heard but didn't understand. How would you answer their questions?

1. What is the difference between a transmission and transaxle?
2. What is the purpose of a final drive?
3. What is a transfer case, and what does it do?
4. What are spur gears and helical gears?

FIGURE 30-3 Panhard and Levassor's attempt at a front drive axle.

encountered problems, and they were unable to make it move under its own power. One year later, Panhard and Levassor successfully demonstrated their new multi-gear transmission. Their transmission and clutch arrangement is the prototype for most manual transmissions today in that it used a clutch-driven, three-speed, sliding gear transmission. When the driver wanted to shift gears, he or she would push the clutch pedal to disengage the engine. Then the driver would move the shifter lever, which slid the previous gears out of mesh and the newly selected gears into mesh. This process allowed the vehicle to move at higher and lower speeds. This sliding gear arrangement became the basis for nearly all manual transmissions since then, with manufacturers developing enhancements to the design.

Before 1898, vehicles were either belt or chain driven. Because the chain or belt was exposed to the elements, this design required frequent maintenance. It could also be dangerous if someone got too close to it. In 1898, Louis Renault connected an engine to a transmission and created a live rear axle by using a metal axle shaft supported by bushings. Renault then adapted a differential-type rear axle that was based on an idea that American C. E. Duryea had back in 1893. The differential assembly had a number of gears set in such a way as to allow each wheel to turn at its own speed when going around a corner. This alleviated the problems of loss of traction and rapid tire wear due to scuffing of tires when making turns with a solid axle.

By 1904, most of the carmakers had adopted the sliding gear manual transmission design in one form or another. Many improvements have been made since then, including the invention of the **gear synchronizer** (syncromesh), which applies a friction device between the gear and the shaft to match the gear speed to the shaft speed. As a result, the gear and shaft spin at the same speed, which allows gear selection to be made without "grinding" of the gears. In 1928, Cadillac introduced the first **synchromesh transmission**. Porsche improved the design in 1952 by using moly-coated steel rings, which were more efficient and reliable and were licensed by many of the manufacturers of the time. In the early 1960s, BorgWarner introduced the modern cone-style synchronizer, which is still widely used today.

▶ TECHNICIAN TIP

In the time span between the sliding gear transmission and the synchromesh transmission, there were other attempts to make it easier for the driver to change gears. In 1907, the Ford Model T was equipped with a transmission that used constantly meshed planetary gear sets. These sets consist of a central gear—called a sun gear—that is meshed with multiple planetary pinion gears, which are meshed inside a ring gear. Although this gear arrangement is not common in manual transmissions, it is widely used in automatic transmissions today.

▶ TECHNICIAN TIP

All designs of transmissions since Panhard and Levassor have always had one goal in mind: to make shifting easier for the driver. The easiest transmission to shift is the automatic transmission, which is strictly an American innovation.

▶ Principles of Operation of Manual Transmissions

K30002

Mechanical Advantage

K30003

Mechanical advantage is defined as the amplification of the input force by trading distance moved for greater output force. For example, imagine using a 10' lever to move a large rock. If you place a short section of log (the pivot point) 2' from the end of the lever nearest the rock, you would have a mechanical advantage of 8 to 2 (**FIGURE 30-4**). Ratios are normally expressed as the equivalent of the first number to 1; thus, 8 to 2 would be expressed as 4 to 1, which is written as 4:1. Assuming no frictional loss, a lever with this mechanical advantage would exert four times as much force against the rock as the amount of force being applied by the person to the other end of the lever. So we could say the lever gives a mechanical advantage of 4:1.

Given this 4:1 mechanical advantage, if a person pushes the lever down with a force of 100 lb, the force the lever generates against the rock is 400 lb. By using a lever, the person achieves more output force than he or she puts in. However, the person also is moving the long side of the lever four times as far as the rock side of the lever. So although the output force is four times greater than the input force, the input distance moved is four times greater

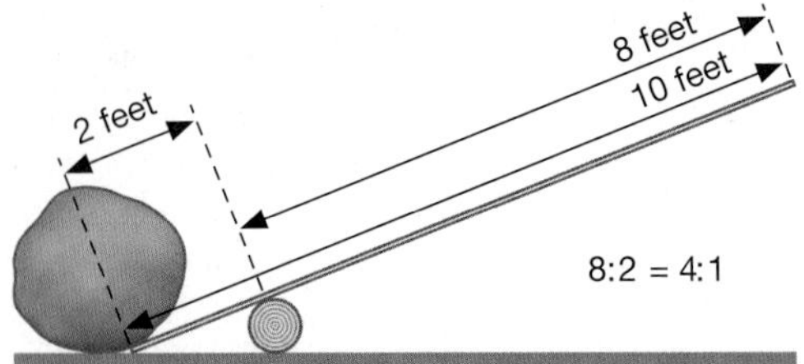

FIGURE 30-4 Using a 10' lever to move a rock, with the pivot point 2' from the end nearest the rock, gives a mechanical advantage of 8 to 2 (4:1).

than the output distance. Also, the input speed is four times faster than the output speed. Thus, the total amount of work on each end of the lever is the same (work = force × distance); it is simply rearranged to give an increase of the output force. The lesson here is that mechanical advantage can be used to generate a larger output force, but it requires an increase in the input distance and speed.

This concept applies directly to manual transmissions. Each set of gears, called a **gear set**, in the transmission provides a certain amount of mechanical advantage and has a specific speed differential associated with it. For example, in low gear, where the smallest gear in the transmission turns the largest gear in the transmission, one complete turn of the small gear turns the larger gear only a small portion of a full turn. This provides mechanical advantage and is called gear reduction. At the same time, because the small gear is turning much faster than the output gear, the output speed is slow compared to the input speed. The driver can select the transmission gear set that best matches the operating conditions under which he or she is driving. When driving on rough and hilly off-road terrain, then a lower gear works well. When driving on a smooth, straight, and level freeway, a higher gear is best.

Gear Ratios

K30004

How do gear sets create mechanical advantage? They do so by having different ratios. Gears with the same number of teeth have a ratio of 1:1, meaning that this set of gears has no mechanical advantage. At the same time, there is no difference of speed between the gears; they each turn at the same speed because they have the same number of teeth. A gear set in which the input gear has half as many teeth as the output gear has a ratio of 2:1. Gear ratios can be found by comparing the number of teeth each gear has (**FIGURE 30-5**). If the drive gear (the input gear) has fewer teeth than the driven gear (the output gear), then the gear set will have mechanical advantage because the smaller drive gear has to rotate faster than the larger driven gear, giving an increase in output force.

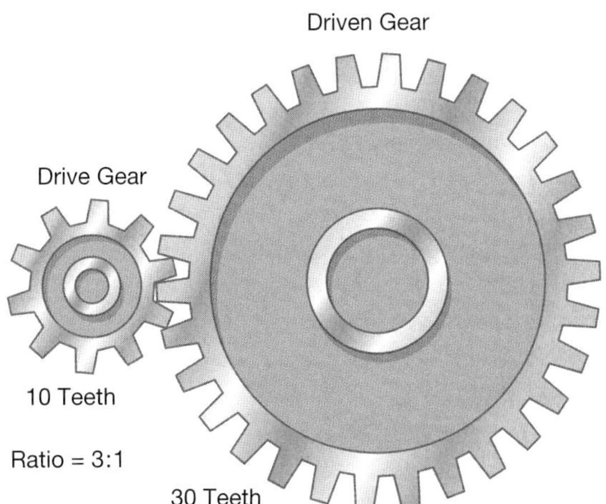

FIGURE 30-5 The mechanical advantage of two gears with a different ratio, which provides torque to get moving.

All gear ratios are calculated by the formula:

$$\frac{\text{Driven}}{\text{Drive}},$$

or the number of driven gear teeth divided by the number of drive gear teeth. For example, if the drive gear has 10 teeth and the driven gear has 30 teeth, then the mechanical advantage is 3:1. In this case, the driven gear has three times the output force (torque), but only one-third of the **rotational speed**. In other words, if the drive gear is turning at 100 revolutions per minute (rpm), the driven gear is turning at 33.3 rpm, but with three times the torque. If you were to make the drive gear larger by increasing it to 15 teeth and then you connected it to a driven gear with 30 teeth, the gear ratio would drop to 2:1, but the speed of the driven gear would increase to one-half as fast as the drive gear. In other words, if the drive gear is turning at 100 rpm, the driven gear is now turning 50 rpm, but with only two times the torque.

As the gear ratio decreases, the output speed increases. This is what happens inside a transmission each time the driver selects a different gear ratio. The output torque and speed can be varied as necessary, based on the speed and load of the vehicle. When the vehicle speed is low, the driver selects a lower gear (which has a high gear ratio) so that the engine can be used to accelerate the vehicle. As the engine speed increases, the driver selects the next gear so that the vehicle can be accelerated further. This continues to happen until the vehicle reaches the speed desired by the driver.

Speed and torque output is the result of the selected gear ratio and allows the driver to tailor the engine speed and power to the driving conditions. Controlling an automobile over different terrains and weather conditions is accomplished by using the various gear ratios that are appropriate for those conditions. The ultimate task of the manual transmission is to allow the vehicle to easily move away from a stop by multiplying the torque through use of a low gear, which has a high gear ratio, and once moving, to allow the driver to sequentially select higher gears (lower gear ratios) for higher vehicle speeds, while keeping the engine speed appropriate.

Refer to the Manual Transmissions/Transaxles Diagnosis and Maintenance chapter for more information on gear ratio.

▶ TECHNICIAN TIP

Following manufacturer-recommended shifting speed guidelines will ensure good fuel economy. In fact, many vehicles have an indicator on the dash that signals the driver when an upshift or downshift would provide better fuel economy.

Power Flow

K30005

Power flow is defined as the path in which power is transmitted through a series of components. Power flows through the entire drivetrain. The engine in the vehicle is typically the main

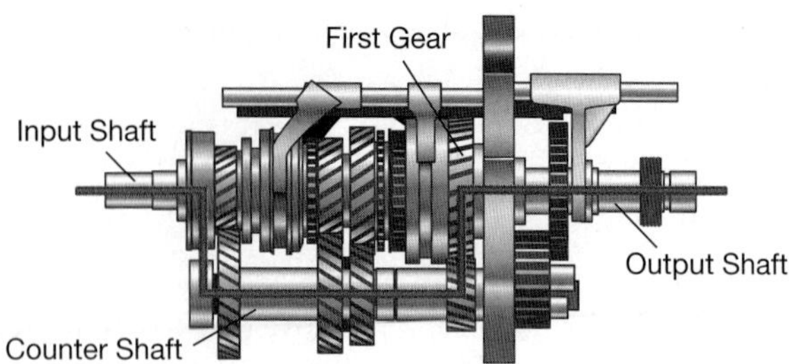

FIGURE 30-6 Power flow in a transmission.

source of all mechanical power for the vehicle. The power that the engine creates flows from the crankshaft to the flywheel and to the clutch assembly. When the clutch pedal is released (in the up position), the clutch transmits power from the flywheel to the input shaft of the transmission/transaxle. Power leaves the transmission/transaxle output shaft and flows to the final drive assembly, where it changes directions and flows through the axles and out to the wheels. This is a very simple and generic explanation of the path in which power flows from the engine to the tires.

When it comes to the manual transmission, power flows into the transmission through the input shaft, where a gear is used to transfer power to a meshed gear on the countershaft. The countershaft has a number of gears that mesh with gears on the output shaft (**FIGURE 30-6**). The gear the driver selects determines the power flow from the countershaft to the output shaft. Because of the various gear sets and driver selection, the transmission/transaxle has multiple paths along which the power can flow. This is covered in more depth in the next chapter.

▶ Manual Transmission Drivetrain Overview

K30006

The manual transmission allows the driver to control the power flow in the drivetrain. The **clutch system** is the medium by which the driver can connect and disconnect the engine from the transmission, resulting in the vehicle's forward or rearward movement. On the driver's side, there are usually three pedals on the floor, which, from the driver's right to left, are as follows: the **accelerator pedal**, used for acceleration or deceleration of the engine; the **brake pedal**, used to stop the vehicle's motion; and the **clutch pedal**, which operates the clutch system. Pushing down on the clutch pedal disconnects the clutch components and allows the selection of the forward or reverse gears within the transmission. Letting the clutch pedal back up reengages the clutch components and transmits engine power to the transmission.

The term **transmission** relates to layouts where the transmission sends power externally to the final drive assembly, and is usually used in RWD. The term transaxle relates to layouts where the transmission and final drive are integrated into a common assembly, and is usually used on FWD vehicles.

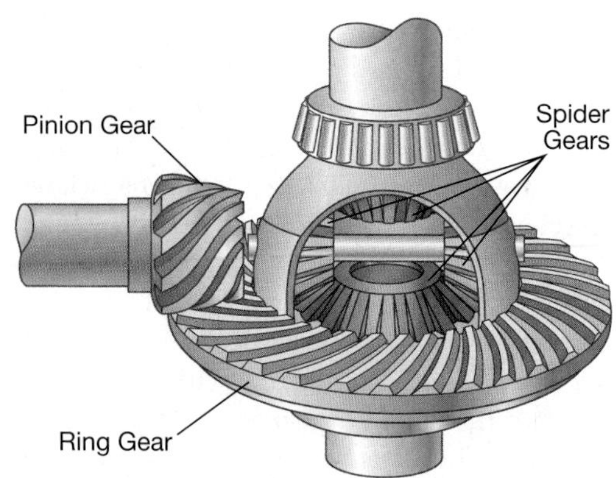

FIGURE 30-7 Typical final drive assembly.

The **final drive assembly** gives the final gear reduction to the drivetrain and powers the drive wheels through axles (**FIGURE 30-7**). The final drive assembly takes the torque from the transmission and increases it further by means of mechanical advantage. This has an effect on the torque applied by the wheels to the ground, the speed of the engine, and the fuel economy. If the final drive has a high gear ratio (higher numerical number), then the vehicle will have more pulling power, but the engine will operate at a higher rpm at any vehicle speed, thereby decreasing the fuel economy. Manufacturers match the final drive gear ratio to the vehicle based on the engine, vehicle weight, and expected use of the vehicle.

When a vehicle travels around a corner, the outside wheel travels a greater distance than the inside wheel and thus turns at a faster rate. If both wheels were connected to a solid axle, the tires would scuff on the road surface when turning a corner. In modern automobiles, the final drive assembly sits between the ends of two separate axles. The final drive assembly incorporates a set of **differential gears** arranged so that they sit between the two axles. This connects them in such a way that they allow each axle to rotate at its own speed while going around a corner. At the same time, the final drive assembly powers both axles through the differential gear assembly.

On RWD vehicles, the **driveshaft** transmits power from the transmission to the final drive assembly (**FIGURE 30-8**). The driveshaft uses **universal joints**, which allow the driveshaft to change angles due to the movement of the suspension relative to the body. Because FWD vehicles incorporate the final drive in the transaxle, no external driveshaft is needed. Both FWD and RWD vehicles use **drive axles** to power the wheels. The axles can be solid, as is the case with many RWD vehicles, or flexible **half-shafts**, which use **constant-velocity (CV) joints**, as is the case with FWD vehicles (**FIGURE 30-9**). FWD axles must be able to change length as the vehicle goes over bumps and dips, and must allow for the wheels to be steered. This is provided by different types of CV joints.

Operation

K30007

The driver steps on the clutch pedal, which releases the clutch and disconnects the engine from the transmission. Depressing

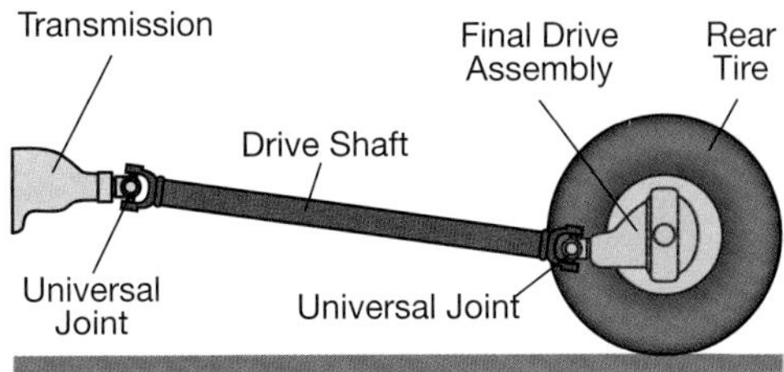

FIGURE 30-8 A typical driveshaft.

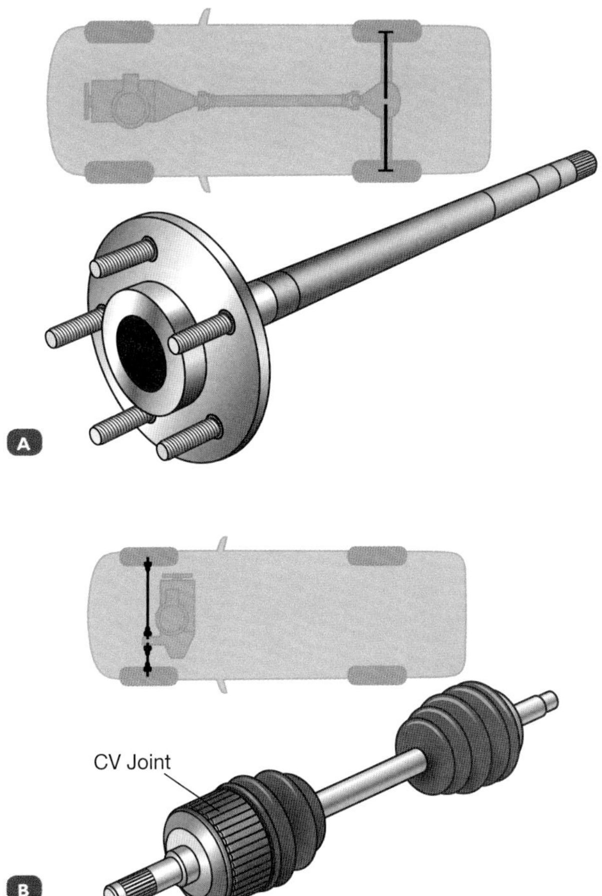

FIGURE 30-9 Drive axles. **A.** Solid drive axle. **B.** Half-shaft drive axle.

the clutch pedal also closes the contacts of the **clutch safety switch**. The closed clutch safety switch allows the starter motor to crank the engine over when the driver turns the key to the "crank" position. The engine then starts and idles. With the clutch pedal fully depressed, the driver moves the gearshift lever to select first gear or reverse gear. Once in gear, the driver can slowly release the clutch pedal and progressively connect the engine to the transmission.

With the transmission in first gear, the transmission is in the lowest gear; in this gear, the vehicle cannot travel very fast, but it provides the best pulling power. Torque is then sent to the final drive, where its speed undergoes a further gear reduction, which increases the torque again before it is sent to the axles. The axles turn the wheels, which are connected to the ground and cause the vehicle to move. When the driver wants to change gears, he or she pushes the clutch pedal and moves the gearshift lever to the next gear position. Once the gear is selected, the driver again slowly releases the clutch pedal to reconnect the engine and transmission and continues driving the vehicle.

▶ Manual Drivetrain Components

K30008

Today's manual transmissions/transaxles are precisely machined, mechanical wonders packed into a relatively small package (**FIGURE 30-10**). Each of the components plays a critical role in allowing the driver to easily operate the transmission. Understanding the main components gives you a good overview of the transmission, which will help you understand how the individual components fit into the overall function of the transmission. Also, understanding the internal workings provides you with a foundation to draw upon when diagnosing transmission problems. We now cover some of the main components of the manual transmission.

FIGURE 30-10 Manual transmissions are mechanical wonders with many precisely machined gears and shafts.

Shafts, Gears, and Bearings

K30009

Inside a manual transmission is a series of shafts, gears, and bearings that make it possible for the driver to select the preferred gear for the road conditions. The availability of several different gear ratios within the transmission allows the vehicle to accelerate from a dead stop almost effortlessly, as well as travel at a variety of speeds without over-revving the engine.

Shafts are used to support gears and are machined precisely to accommodate bearings and individual gears. The surfaces are smooth, allowing the bearings and gears to rotate on a thin film of oil (**FIGURE 30-11**) in order to reduce metal-to-metal friction and prevent catastrophic failure from overheating. Some surfaces on the shaft are **splined** (parallel grooves in a shaft that mate with a component with matching grooves) to allow other parts of the transmission such as synchronizers to be held stationary on the shaft. Other machining may create raised areas on the shafts such as shoulders, which provide a firm support for gears and bearings to butt up against. Alternately, grooves may be cut into the shafts to accommodate **snap rings** and thrust washers, which are used for holding the gears in the proper position on the shaft once the gears are installed. The ends of the shafts have machined surfaces to allow for bearings. Bearings are used to center and hold the shafts in alignment so the gears rotate smoothly when in mesh with their mating gears. Friction bearings provide sliding metal-to-metal contact between components and are usually made of brass or bronze. Nonfriction bearings provide rolling metal-to-metal contact between components and usually use ball or roller-type bearings.

Gears are round parts with teeth cut on the outside perimeter in order to mate with one or more other gears to achieve a specific gear ratio. The mated gears are called a gear set. Each set is cut on the same angle or pitch to allow for proper contact of the gears with each other. Manual transmission gear teeth are usually of two configurations: spur or helical. Spur gears have straight teeth that are parallel to the shaft the gear turns on (**FIGURE 30-12**). They can be noisy during operation. However, they don't create axial forces like helical gears do, which can place high stress on the bearings the case, so they are used in heavy-duty vehicles such as off-road equipment. Helical gears have angled teeth that are cut on an angle to the shaft (**FIGURE 30-13**). They operate more quietly and are more commonly used in passenger vehicles.

Bearings are necessary to maintain the position and alignment of the shafts and gears while under two types of loads: radial and axial. **Radial loads** are perpendicular to the shaft and occur because the gears have a tendency to push each other apart as torque is transmitted between them while driving (**FIGURE 30-14**). **Axial loads** are in line with the shaft and occur in transmissions because the cut of the helical gear teeth, which are at an angle to each other, causes a fore and aft load (**FIGURE 30-15**). Bearings also allow shafts and gears to have greater rotational speeds while also minimizing metal-to-metal friction, which contributes to increased fuel economy and longer transmission life. Bearings can be of the friction or nonfriction type. Both are used in transmissions to resist overheating and keep shafts aligned.

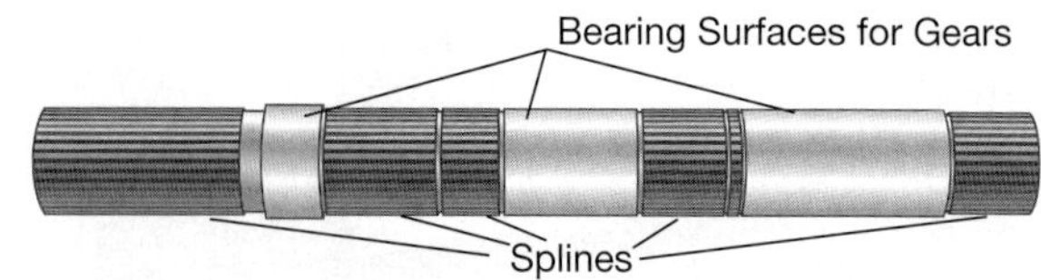

FIGURE 30-11 Shafts are used to support gears and are machined precisely to accommodate bearings and individual gears.

FIGURE 30-12 Spur gears have straight teeth. They can be noisy during operation.

FIGURE 30-13 Helical gears are cut on an angle for better coverage. They operate quietly.

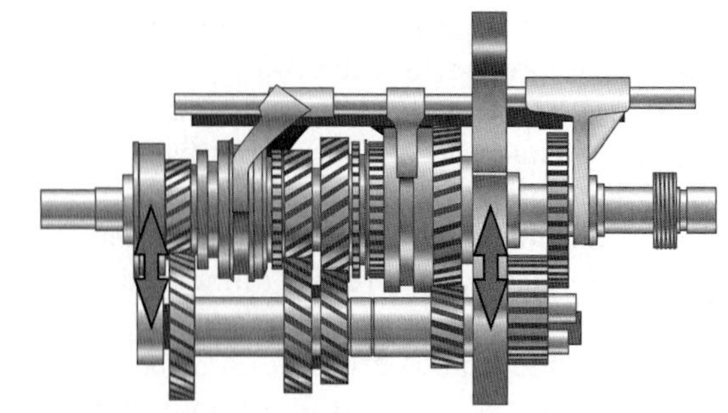

FIGURE 30-14 Radial-loaded bearing.

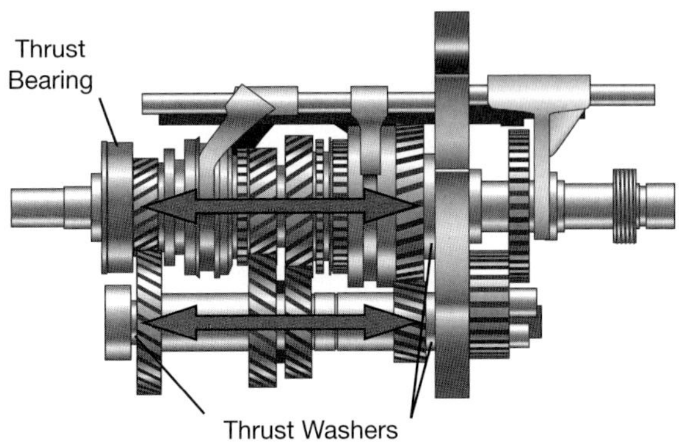

FIGURE 30-15 Axial-loaded thrust washer.

> **TECHNICIAN TIP**
>
> Manual transmission failures occur, which is why a proper preventive maintenance program is critical to the life of all parts inside the transmission. Proper oil levels, types, and viscosities contribute to longevity of these units.

Clutch System

K30010

In a manual transmission vehicle, the component that locks the engine and transmission together is the clutch system. The engine is connected to the clutch, and the clutch is connected to the transmission. The clutch is designed so it can connect and disconnect. Imagine a wall-mounted dimmer switch for a ceiling light in a house. When the dimmer switch is in the off position, no electrical power is transmitted to the light, resulting in no illumination of the room. When the clutch pedal is in the depressed or down position, no power is transmitted to the transmission, resulting in no forward movement of the vehicle. When the dimmer switch is in the full on position, electrical power is transmitted to the lightbulb, resulting in illumination. Similarly, when the clutch pedal is in the released or up position, power is transmitted to the transmission, and forward or rearward movement of the vehicle can be achieved.

Much like the dimmer switch, the clutch pedal requires movement to connect and disconnect power to the transmission many times repeatedly (**FIGURE 30-16**). This movement can cause the clutch components to wear (just as wear can be present in the dimmer switch over time). When the dimmer switch is turned on, there is a drag of amperage on the power source as the light draws power to illuminate a room. This is also what happens when the clutch is released—a drag on the engine is required to get the vehicle to move from the stationary position. This transition from no power to power creates heat, which is detrimental to both the dimmer switch and the clutch system. If the clutch is only partially released, there will be only a partial transmission of the power through the clutch, just like power being reduced by the dimmer switch when it is left in a partially on position. The difference is that the dimmer switch is designed to handle the resulting heat, but the clutch is not.

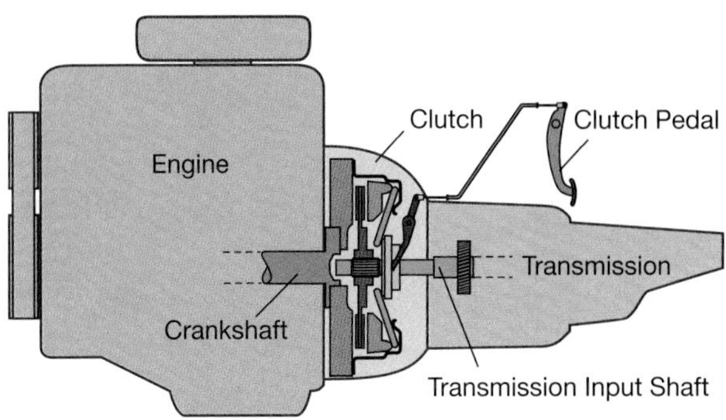

FIGURE 30-16 Clutch components are designed to connect and disconnect the power to the transmission.

> **TECHNICIAN TIP**
>
> The clutch system may have to be inspected and maintained at regular intervals depending on the driver's habits. Customers who tend to slip the clutch, rest their foot on the clutch pedal, or drive aggressively put more wear on the clutch components. It is always a good practice to follow the preventive maintenance schedule to catch a problem before it becomes serious.

Transmission/Transaxle

K30011

Since the invention of the automobile, the need for a device to control engine power and torque for various applications has been satisfied by the transmission. As vehicle design has evolved over the decades, the transaxle was developed (**FIGURE 30-17**). The only difference between the transmission and the transaxle is that the transaxle incorporates the final drive assembly in its construction, making the transmission and final drive a single unit. In a vehicle with a transmission, the final drive is a separate unit.

> **TECHNICIAN TIP**
>
> Most manual transmissions/transaxles are manually shifted, but some are electronically controlled. This type of transmission will be discussed in detail in the Manual Transmissions/Transaxles Basic Diagnosis and Maintenance chapter. In these systems, a computer plays a part in how and when the shift will occur. These new electronic systems require the use of a scan tool to diagnose faults in the transmission as well as to program all parameters for transmission shifting.

> **TECHNICIAN TIP**
>
> The manufacturer determines the gear ratios in the transmission/transaxle based on the power-to-weight ratio, the vehicle size, the application, and the intended use. This concept is be discussed in depth in the Manual Transmissions/Transaxles Basic Diagnosis and Maintenance chapter.

Transmissions/transaxles are rated by the manufacturers for how much twisting force (torque), measured in ft-lb, they can handle. This rating system ensures that the transmission/transaxle is strong enough for the engine it is used with. Automobile

FIGURE 30-17 Transmissions/transaxles. **A.** Typical RWD transmission. **B.** Typical FWD transaxle.

FIGURE 30-18 A typical transfer case used for four-wheel drive (FWD).

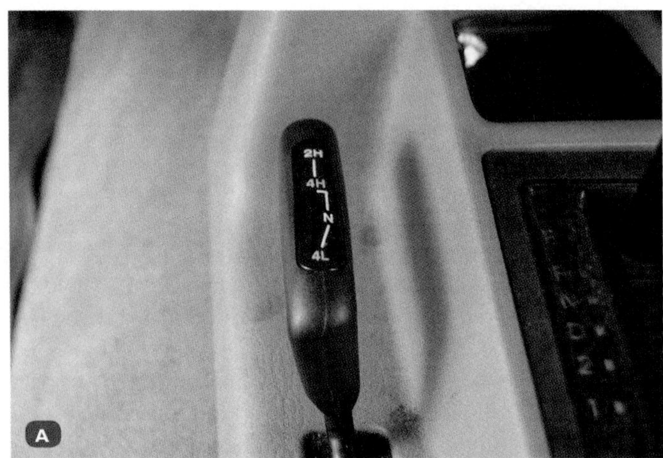

FIGURE 30-19 Transfer case control can be: **A.** manually controlled or **B.** electronically controlled.

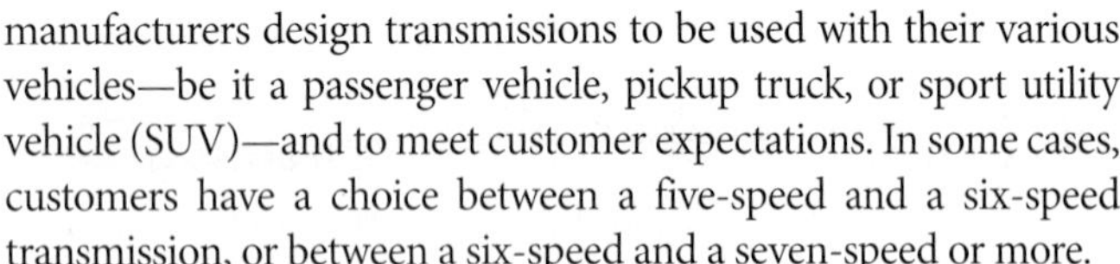

manufacturers design transmissions to be used with their various vehicles—be it a passenger vehicle, pickup truck, or sport utility vehicle (SUV)—and to meet customer expectations. In some cases, customers have a choice between a five-speed and a six-speed transmission, or between a six-speed and a seven-speed or more.

Transfer Case

K30012

The **transfer case** is typically mounted to the rear of the transmission. Its purpose is to transfer power to both the front and rear axles, providing 4WD so the vehicle can move with more stability in inclement weather and compromised road conditions. Steering ability is also increased, making the vehicle track better in slippery conditions.

The transfer case takes the power from the transmission and directs it to one or both axles, depending on the mode selected (**FIGURE 30-18**). This system requires that the front wheels be powered; thus, a front differential and two front axles were added. A second driveshaft is necessary to provide power to the front axle assembly.

The transfer case and the manual transmission are similar in operation in that they both have a series of shafts, gears, and bearings and include shift mechanisms. The transfer case can be made of cast iron to provide strength, or magnesium or aluminum to lessen the weight. It also is filled with lubricant to reduce friction, increase life span, and maintain quiet operation of the gears by providing a thin film of oil between them.

Transfer cases can be operated manually or electronically (**FIGURE 30-19**). Mechanically shifted transfer cases have a shift lever that is usually next to the transmission gearshift. The transfer case shift lever is also usually shorter, as it is not used as frequently as the transmission gearshift. Electronically shifted transfer cases use either an electronically controlled vacuum actuator or an electric motor to shift the transfer case. Sensors may be used to operate a light that informs the driver which mode is selected.

> **TECHNICIAN TIP**
>
> There are two types of FWD systems—full-time FWD and part-time FWD. These will be explained in greater depth in the Drivetrain Components chapter. Most manufacturers of SUVs incorporate some form of FWD or AWD as a selling/safety feature to the public. Early FWD vehicles were not very fuel efficient due to heavy and bulky designs. Over time, manufacturers have made fuel efficiency and safety priorities.

Differential and Final Drive

K30013

The term "differential" is used in two different ways. The first is the most technically accurate and refers to the components inside the differential housing that allow the axles to turn at different speeds when the vehicle is cornering or turning (**FIGURE 30-20**). This is called the differential assembly. Because the outside tire must travel farther when cornering than the inside tire, it must turn at a faster speed. The differential assembly allows this to happen; otherwise, the tires would bind, skip, hop, and slide when going around a corner, which causes them to wear out quickly and could easily cause a loss of vehicle control.

The second way technicians use the term "differential" is as another name for the complete final drive assembly (**FIGURE 30-21**). The final drive assembly provides the final gear reduction necessary for drivetrain operation. The gear reduction happens because the drive gear (called a pinion gear) is much smaller than the driven gear (called the ring gear). This difference in size results in an increase in mechanical advantage.

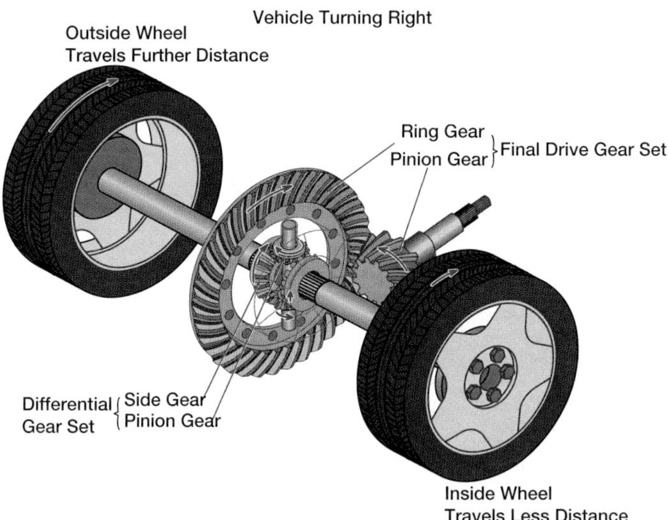

FIGURE 30-20 The differential assembly allows the wheels on an axle to rotate at different speeds when cornering.

In addition, the final drive assembly takes the power from the transmission, turns it 90 degrees, and sends it out the axles to the tires. Then the differential assembly allows the tires to travel around corners without binding while the final drive is powering them. Because the differential assembly is housed within the final drive assembly, many technicians simply call the entire final drive assembly the differential. Any time you hear the term "differential," you need to ask whether the speaker means the final drive assembly or the differential assembly.

Differential assemblies also come in two varieties. First, there is the **open differential assembly**. With the open differential assembly, power is supplied to both wheels equally only when each tire maintains traction, which is not likely on ice and snow. If one wheel is stuck in the snow or ice, the other wheel cannot supply power because of the nature of the open differential assembly, which supplies power only to the slipping wheel. The second type is the **limited slip differential assembly**, which allows both of the rear wheels to supply power to the ground even if one wheel loses traction. This allows the wheel with traction to drive the vehicle forward or backward.

There are two types of rear axle assemblies used in modern vehicles: the solid rear axle and the independent rear axle. The **solid rear axle** uses a solid axle housing, which means movement of one wheel affects the other wheel (**FIGURE 30-22**). As the wheel on one side of the vehicle hits a bump, the wheel on the other side pivots, affecting the ride of the vehicle. An **independent rear axle** uses a final drive assembly that is firmly bolted

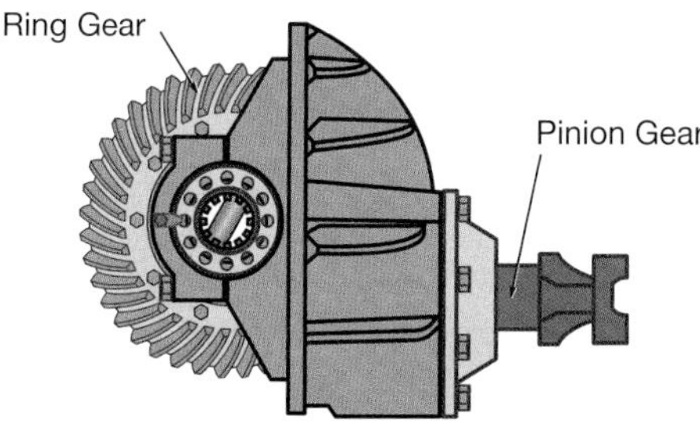

FIGURE 30-21 A final drive assembly sometimes inaccurately referred to as a differential.

FIGURE 30-22 Solid rear axle assembly.

FIGURE 30-23 Independent rear axle assembly.

FIGURE 30-24 Live axles drive wheels.

FIGURE 30-25 Dead axles hold wheels in position only; they don't drive the wheels.

to the vehicle frame or unibody while each wheel is allowed to move independently of the other wheel, which results in a more comfortable ride and better handling (FIGURE 30-23).

> TECHNICIAN TIP

If the vehicle is equipped with traction control, the traction control system can help overcome the loss of traction by applying the brakes to the wheel that is slipping. This results in more power being sent to the wheel that is not slipping.

> TECHNICIAN TIP

Preventive maintenance is important because the axles are filled with special lubricants and additives that promote long life and maintain quiet operation of the gears. Maintenance schedules help technicians know when each component requires maintenance and when visual and physical inspection are due, thus preventing premature failure.

Drive Axle

K30014

The drive axle is the component that supplies the power from the final drive to the wheels. It is part of the **drive axle assembly**. There are two types of axles—live and dead. A **live axle** powers the wheels attached to it (FIGURE 30-24). **Dead axles** allow the wheels to freely rotate on the axle assembly and do not drive the wheels (FIGURE 30-25).

There are two types of live axles. The first type is the **independent suspension drive axle** (FIGURE 30-26). It uses one half-shaft axle for each of the two wheels. These axles must flex to allow suspension movement, so they incorporate what are called constant-velocity (CV) joints. CV joints connect the transaxle final drive to the wheels. CV joints are known for their flexibility in allowing greater operating angles; they are used to allow steering the front wheels while also supplying them with power. Rubber boots are on each end to contain the lubricant inside each CV joint to maintain its integrity.

The second type of live axle in use today is the **solid axle** (FIGURE 30-27). These axles are fitted within a strong

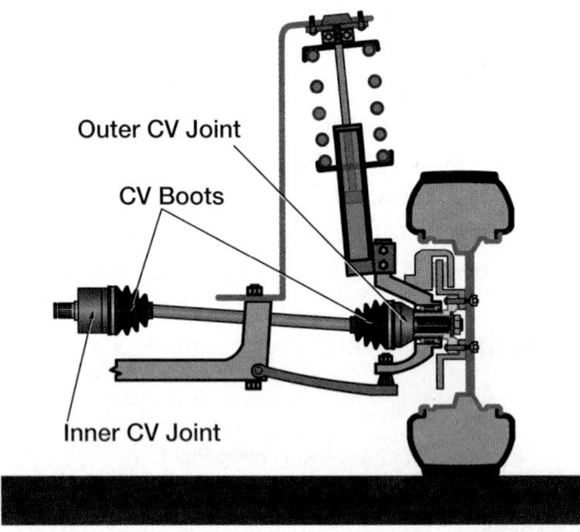

FIGURE 30-26 Independent suspension drive axle.

FIGURE 30-27 Solid axle fits inside a solid axle housing.

nonflexible housing. The outer ends of the axles have flanges with studs and nuts that are used to secure matching-sized wheels and tires. The other end of the axle has splines that slide into the differential side gears inside the differential housing. This is where the solid axle gets its power. There are three types of solid live axles used today in RWD vehicles: the semi-floating axle, the three-quarter floating axle, and the full floating axle. All three types are discussed in detail in the Drivetrain Components chapter.

▶ TECHNICIAN TIP

A good preventive maintenance program will keep on top of all leaks and fluid losses. Always consult manufacturer specifications for correct fluid types.

▶ Wrap-Up

Ready for Review

- ▶ The drivetrain comprises component assemblies that transmit power from the engine to the wheels.
- ▶ Early engines had belt-driven gears that required drivers to manually change the belt.
- ▶ Some form of the multi-gear transmission, with clutch operation, has been used since its invention in 1894.
- ▶ A gear synchronizer matches the gear speed and the shaft speed to allow gear selection without "grinding" the gears.
- ▶ Mechanical advantage refers to gaining greater output force by increasing input distance, such that the total work on either end is equal (work = force × distance).
- ▶ Each gear set creates mechanical advantage by the ratio of teeth of the input gears to that of the output gears.
- ▶ Gear ratios are calculated by the number of driven gear teeth divided by the number of drive gear teeth; as the gear ratio decreases, the output increases.
- ▶ Power flow refers to the path by which power is transmitted from one component to another.
- ▶ The manual transmission drivetrain consists of the clutch system, transmission/transaxle, and final drive assembly.
- ▶ When depressed, the clutch pedal disconnects the engine from the transmission and allows the driver to change gears.
- ▶ Releasing the clutch pedal reconnects the engine to the transmission.
- ▶ The main components of a manual transmission are shafts, gears, bearings, and the clutch assembly.
- ▶ Shafts support the gears, and bearings maintain the position and alignment of shafts.
- ▶ Helical gears have angled teeth; spur gears have straight teeth and are stronger than helical gears.
- ▶ Transaxles have final drive assemblies incorporated into their construction.
- ▶ Transmissions/transaxles are rated as to how much torque they can handle.
- ▶ The transfer case transfers power to both axles to provide four-wheel drive.
- ▶ Differential refers to the components that allow the axles to turn at different speeds when a vehicle is cornering.
- ▶ The final drive assembly houses the differential and provides the final gear reduction necessary for drivetrain operation.
- ▶ Differentials can be open or limited slip.
- ▶ Drive axles supply power from the final drive to the wheels.
- ▶ An axle can be live (powers the wheels attached to it) or dead (allows the wheels to rotate freely).
- ▶ The two types of live axles are independent suspension and solid (which can be semi-floating, three-quarter floating, or full floating).

Key Terms

accelerator pedal The foot-operated pedal used by the driver to increase and decrease the amount of power the engine develops.

axial load The load applied in line with a shaft. It can be controlled with thrust bearings.

clutch pedal The foot-operated pedal used by the driver to engage and disengage the clutch.

clutch safety switch An electrical switch that is operated by the clutch pedal and keeps the starter motor from cranking the engine over until the clutch is fully depressed.

clutch system A mechanically operated assembly that connects and disconnects the engine from the transmission.

constant-velocity (CV) joints Joints commonly used in FWD vehicles to allow flexibility of the axle while turning.

dead axle Axle that supplies no power to the wheels.

differential gears Gears situated in the final drive assembly that are meshed together and with both axles, allowing the wheels to rotate at different speeds when turning a corner.

drive axle assembly The components that make up the drive axle, including the axles, final drive assembly, bearings, and axle housing.

drive axle An axle that provides power to a wheel.

driveshaft The hollow tube with flexible joints on each end that transmits power from the transmission to the final drive unit.

drivetrain The component assemblies that transmit power from the engine all the way to the drive wheels.

final drive assembly An assembly used to power the drive wheels and allow the wheels to rotate at different speeds as the vehicle turns.

gear set Two or more gears that are in mesh with each other.

gear A relatively round, rotating part with internal or external teeth that are designed to mesh with another gear for the purpose of transmitting torque.

gear synchronizer An assembly in the transmission that is used to bring two unequally spinning shafts or gears to the same speed when upshifting or downshifting.

half-shaft An axle that has CV joints on each end and that fits between the transaxle and wheel. Typically, one is used on each side of a vehicle.

independent rear axle A type of rear suspension system that allows each wheel on the axle to move independently of the other.

independent suspension drive axle A type of suspension that allows each wheel on a drive axle to move independently of the other.

limited slip differential assembly A differential assembly that uses a clutch assembly or gear assembly to allow a limited amount of slip between the two axles. It is used to increase drive wheel traction in slippery conditions.

live axle An axle that provides power to the wheels.

mechanical advantage The process of using a device to get more output force than the amount of input force, with the trade-off being that the input distance is proportionately longer than the output distance.

open differential assembly A differential assembly that allows both axles to turn at their own speed when turning a corner, but is dependent on the traction of the tires to deliver torque to the ground. If one wheel has no traction, all of the engine's torque will be used at that wheel, causing it to simply spin.

power flow The path that power takes from the beginning of an assembly to the end. In a transmission, power flow changes as different gears are selected by the driver.

radial load The load that is perpendicular to a shaft, usually controlled by bearings or bushings.

rotational speed The speed at which an object rotates, measured in revolutions per minute (rpm).

shaft The long, narrow component that carries one or more gears or has gears machined into it.

snap ring The spring steel, C-shaped ring that is fitted in a groove and holds gears, bearings, and shafts in place.

solid axle A type of axle that is not flexible, with splines on one end to fit the final drive unit and a flange on the other end to power the wheel.

solid rear axle A type of axle that has a one-piece axle housing, so that the action of hitting a bump with one wheel affects the other wheel.

splined Typically, a shaft and gear that have parallel grooves machined in them so they mate with each other and lock together rotationally.

spur gears Gears with straight-cut gear teeth.

synchromesh transmission A modern transmission that uses gear synchronizers to match the speeds of gears and shafts during upshifts and downshifts.

transfer case An assembly used in FWD vehicles to transmit power to either two wheels only or all four wheels.

transmission An assembly that houses a variety of gear sets that allow the vehicle to be driven at a wider range of speeds and terrain conditions than would be possible without a transmission.

universal joint A cross-shaped joint with bearings on each leg, where one set of parallel legs is connected to the end of one shaft and the other set of parallel legs is connected to the end of a second shaft. This arrangement allows the shafts to operate at shallow angles to each other.

Review Questions

1. The manual transmission receives power from the engine by way of the:
   a. drive shaft.
   b. clutch assembly.
   c. gear set.
   d. final drive.
2. Mechanical advantage can be used to generate a larger output force, but it requires a(n):
   a. increase in input distance and speed.
   b. decrease in input distance and increase in speed.
   c. increase in input distance and decrease in speed.
   d. decrease in input distance and speed.
3. A 4:1 gear ratio results in ________ output torque and ________ output speed than a 2:1 gear ratio.
   a. lower; lower
   b. higher; higher
   c. lower; higher
   d. higher; lower
4. In manual transmission, which of these determines the path of the power flow from the countershaft to the output shaft?
   a. Gears
   b. Transmission/transaxle
   c. Engine
   d. Input shaft
5. The driver can increase and decrease the amount of power the engine develops using the:
   a. accelerator pedal.
   b. differential gears.
   c. clutch pedal.
   d. final drive assembly.
6. All of the following statements with respect to manual transmission are true *except*:
   a. gears can be applied only when the clutch is depressed.
   b. when in gear, releasing the clutch connects the engine to the transmission.
   c. releasing the clutch pedal closes the contacts of the clutch safety switch.
   d. the transmission provides the best pulling power in the lowest gear.
7. Which of these are used for holding the gears in the proper position on the shaft once the gears are installed?
   a. Parallel grooves in a shaft
   b. Shoulders
   c. Snap rings and thrust washers
   d. Bearings

8. Which of the following statements is correct with respect to the difference between a transmission and a transaxle?
   a. A transmission is typically used in front-wheel drive vehicles.
   b. A transaxle is typically used in rear-wheel drive vehicles.
   c. The final drive assembly is incorporated in transmission construction.
   d. The final drive assembly is incorporated in transaxle construction.
9. All of the following statements are true *except*:
   a. the transfer case is typically mounted to the rear of the transmission.
   b. the transfer case can only be operated mechanically.
   c. the transfer case takes the power from the transmission and directs it to one or both axles.
   d. the transfer case system requires that the front wheels be powered at all times.
10. Choose the correct statement.
   a. The differential assembly does not allow the wheels on an axle to rotate at different speeds.
   b. Technically, "differential assembly" is the other name for "final drive assembly."
   c. The two varieties of differential assembly are open and closed.
   d. Without the differential assembly, the tires would bind, skip, hop, and slide when going around a corner.

ASE Technician A/Technician B Style Questions

1. Tech A says that friction bearings are made up of balls and rollers. Tech B says that nonfriction bearings are in sliding contact between moving surfaces. Who is correct?
   a. Tech A
   b. Tech B
   c. Both A and B
   d. Neither A nor B
2. Tech A says that fifth gear provides more torque to the wheels than second gear. Tech B says that fourth gear provides more wheel speed than first gear. Who is correct?
   a. Tech A
   b. Tech B
   c. Both A and B
   d. Neither A nor B
3. Tech A says that the countershaft is splined to fit into the hub in the clutch disc. Tech B says that the output shaft has gears that are meshed with the countershaft gears. Who is correct?
   a. Tech A
   b. Tech B
   c. Both A and B
   d. Neither A nor B
4. Tech A says that bearing axial loads are in line with the shaft. Tech B says that bearing radial loads occur because the gears have a tendency to push each other apart as torque is transmitted between them. Who is correct?
   a. Tech A
   b. Tech B
   c. Both A and B
   d. Neither A nor B
5. Tech A says that the final drive is used to provide an increase in the rotational speed of the axles. Tech B says that the final drive is used to provide an increase of twisting force to the axles. Who is correct?
   a. Tech A
   b. Tech B
   c. Both A and B
   d. Neither A nor B
6. Tech A says that a gear set that has a drive gear with 9 teeth and a driven gear with 27 teeth has a gear ratio of 3:1. Tech B says that the drive gear is also called the output gear. Who is correct?
   a. Tech A
   b. Tech B
   c. Both A and B
   d. Neither A nor B
7. Tech A says that a transfer case is designed to transfer torque from the engine to the transmission. Tech B says that a transfer case is used on FWD vehicles. Who is correct?
   a. Tech A
   b. Tech B
   c. Both A and B
   d. Neither A nor B
8. Tech A says that dead axles allow the wheels to freely rotate on the axle assembly and do not drive the wheels. Tech B says that the ring gear and pinion gear are part of the final drive assembly. Who is correct?
   a. Tech A
   b. Tech B
   c. Both A and B
   d. Neither A nor B
9. Tech A says that an open differential allows both of the rear wheels to supply power to the ground even if one wheel loses traction. Tech B says that a live axle powers the wheels attached to it. Who is correct?
   a. Tech A
   b. Tech B
   c. Both A and B
   d. Neither A nor B
10. Tech A says that the differential assembly provides a means for the inside and outside wheels to turn at different speeds when going around a corner. Tech B says that the differential assembly provides smooth shifts and reduces gear "grinding" by matching gear speeds. Who is correct?
   a. Tech A
   b. Tech B
   c. Both A and B
   d. Neither A nor B

CHAPTER 31
The Clutch System

NATEF Tasks

- **N31001** Inspect clutch pedal linkage, cables, automatic adjuster mechanisms, brackets, bushings, pivots, and springs; perform needed action. (AST/MAST)
- **N31002** Bleed clutch hydraulic system. (AST/MAST)
- **N31003** Diagnose clutch noise, binding, slippage, pulsation, and chatter; determine needed action. (AST/MAST)
- **N31004** Inspect flywheel and ring gear for wear, cracks, and discoloration; determine needed action. (AST/MAST)
- **N31005** Measure flywheel runout and crankshaft end play; determine needed action. (AST/MAST)
- **N31006** Inspect and/or replace clutch pressure plate assembly, clutch disc, release (throw-out) bearing, linkage, and pilot bearing/bushing (as applicable). (AST/MAST)

Knowledge Objectives

After reading this chapter, you will be able to:

- **K31001** Explain the purpose and function of a clutch.
- **K31002** Describe the purpose and function of the clutch components.
- **K31003** Describe the purpose and function of the flywheel.
- **K31004** Describe the purpose and function of pressure plates.
- **K31005** Describe the purpose and function of clutch discs.
- **K31006** Describe the purpose and function of the throw-out bearing and clutch fork.
- **K31007** Describe the purpose and function of the pilot bearing.
- **K31008** Describe the purpose and function of clutch operating mechanisms.
- **K31009** Describe the purpose and function of cable operating mechanisms.
- **K31010** Describe the purpose and function of hydraulic clutch mechanisms.
- **K31011** Describe the purpose and function of linkage operated systems.
- **K31012** Describe the safety and hazard concerns related to clutch repair.
- **K31013** Identify common tools used in clutch repair.
- **K31014** Describe the required clutch preventive maintenance procedures.

Skills Objectives

After reading this chapter, you will be able to:

- **S31001** Perform clutch maintenance and repair.
- **S31002** Perform a clutch replacement.
- **S31003** Remove and replace a transmission/transaxle.
- **S31004** Inspect the engine block, crankshaft, crankshaft seal, and soft plugs.

▶ Introduction

The clutch is a mechanical device located in the bell housing, which sits between the engine and the transmission (**FIGURE 31-1**). The clutch allows the driver to engage and disengage the engine from the transmission while operating the vehicle. The driver controls the operation of the clutch with his or her foot through the clutch pedal. The clutch has a series of components that help to make driving an interactive experience, allowing the driver to control the shifting of gears in the vehicle. An automatic transmission does not use a manual clutch, giving the driver less control over transmission operation other than to put it in or out of gear.

▶ Clutch Principles

K31001

The purpose of the clutch is to allow the driver to disconnect and progressively connect the engine to the transmission. It progressively transmits torque from the engine to the transmission. It also allows the driver to disconnect the transmission from the engine for the purpose of shifting between gears while accelerating or decelerating the vehicle.

Automotive manual transmission clutches are dry clutches, as opposed to wet clutches in automatic transmissions, which run in a lubricating fluid (**FIGURE 31-2**). Dry clutches use the friction between the clutch surfaces to transmit torque from the engine to the transmission. The amount of torque a clutch can transmit depends on the amount of friction between the clutch disc and the mating surfaces of the flywheel and pressure plate. This friction is dependent on four variables: the coefficient of friction of the clutch disc facings, the diameter of the clutch, the number of clutch discs in the clutch assembly, and the total spring force clamping the parts together.

Increasing the friction of the clutch disc increases the torque-carrying ability but makes the clutch grab, which makes it more difficult to start from a stop. Increasing the diameter

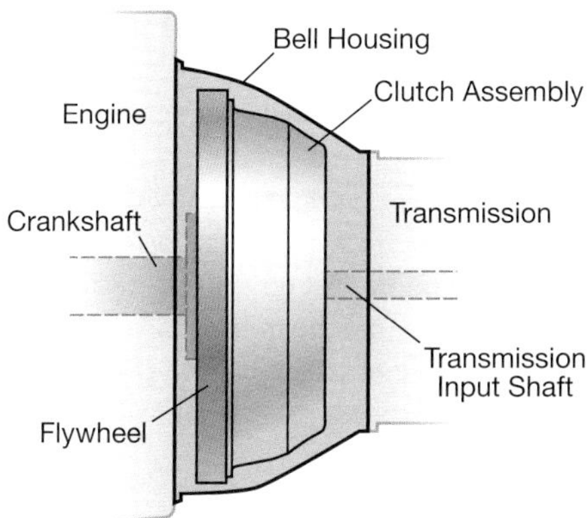

FIGURE 31-1 The clutch engages and disengages the engine from the transmission.

FIGURE 31-2 Types of clutches. **A.** Dry clutch from a manual trans. **B.** Wet clutch from an auto trans.

You Are the Automotive Technician

A customer brings her 2008 Ford Mustang into the shop for a clutch inspection. She has noticed the clutch slipping when she is in fourth gear, driving on the highway. You inform the customer that it would be beneficial to perform a road test to properly diagnose this problem. She agrees, and you verify that the clutch slips when applying throttle while going up a hill. You drive back to the shop and inform the customer that you need to perform some additional diagnosis in the shop and will let her know what you find.

1. What things should be done to confirm the clutch is worn out before removing the transmission?
2. Why is a test-drive important to help diagnose clutch system issues?
3. What are the main components to the clutch assembly?
4. What are the different clutch operating systems?

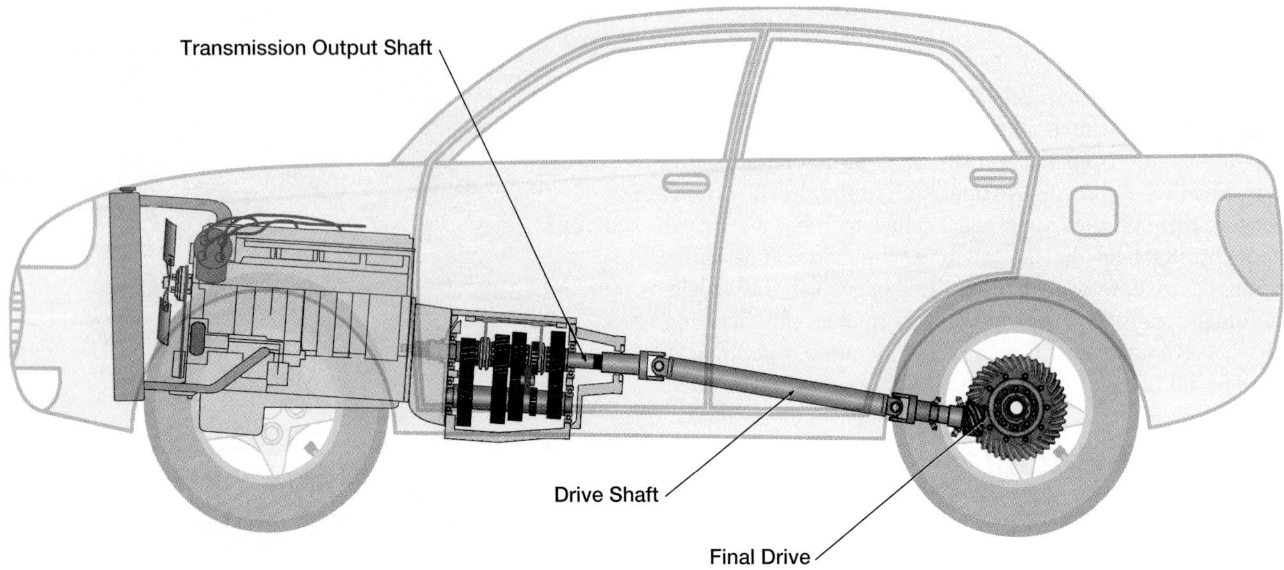

FIGURE 31-3 The transmission output shaft turns with the wheels because they are directly connected.

of the clutch gives it more leverage, which increases its torque capacity but takes up more area. Increasing the number of clutch discs increases torque capacity but is more complicated due to the extra parts. Increasing the spring force clamping the parts together increases torque capacity but takes more foot pressure to operate the clutch pedal. Manufacturers balance all of these factors when designing a clutch for a particular application. This also comes into play if the engine is modified to develop more horsepower. The clutch may need to be upgraded as well.

▶ TECHNICIAN TIP

Two or more clutch plates can be used to form a **multi-plate clutch**, increasing the number of facings and the torque capacity. This design is useful where a reduction in clutch diameter is advantageous or where increasing the spring strength is undesirable because it would increase pedal effort and driver discomfort.

Input and Output Shaft Speed

Because vehicles are dependent on shifting between various gears to be able to drive down the road at a variety of speeds, it is important to understand how the transmission input and output shaft speeds play a critical role in shifting. First, you need to understand that output shaft torque and speed can each be increased, but not at the same time. In low gear, the input shaft speed is relatively high while the output shaft speed is relatively low. This results in an increase of torque from the output shaft. However, the speed of the output shaft (and the speed of the vehicle) is relatively low. This setup makes for good acceleration from a stop or while driving at slow speeds. As higher transmission gears are selected, the output shaft speed increases with each higher gear. However, the torque of the output shaft is lessened with each higher gear, allowing the vehicle to travel at higher speeds without over-revving the engine, but with decreased torque.

The second role that the input and output shaft speeds play is during the actual shift. When transitioning from one gear to another, the relative speeds of the shafts must change to allow the gears to shift from one set to another. In order for a gear to be selected, at least one of the shafts must be able to turn freely for this speed change to happen. The input shaft is able to connect to and disconnect from the engine's flywheel through the operation of the clutch, thus allowing the input shaft's speed to change during a shifting of the gears. Because the output shaft is connected directly to the wheels through the drivetrain components, it cannot be disconnected during a shift (**FIGURE 31-3**). This means that any time the vehicle is moving, the output shaft is turning. The faster the vehicle speed, the faster the rotation of the output shaft. This has an impact on shifting because the speeds of both shafts must be such that the individual gears can be selected. The clutch plays the role of engaging and disengaging the input shaft, so the input shaft speed can be changed as needed.

▶ Clutch Components

K31002

The main components of a clutch assembly are the **flywheel**, **clutch disc**, **pressure plate**, **throw-out bearing**, **clutch fork** and **pilot bearing** (**FIGURE 31-4**). The flywheel bolts onto the rear of the crankshaft, and the pressure plate bolts onto the flywheel. Most light vehicles use a single-plate clutch disc with two **friction facings** attached to a central hub and splined to engage the **transmission input shaft**. The friction facings on the clutch disc are clamped between the flat surfaces of the engine flywheel and the spring-loaded pressure plate. With engine rotation, the flywheel and pressure plate rotate together with the clutch disc. The flywheel and pressure plate are the drive unit, and the clutch disc is the driven unit. Engine torque is transferred from the flywheel and pressure plate through the friction facings of the driven clutch disc to the splines of the input shaft and into the transmission.

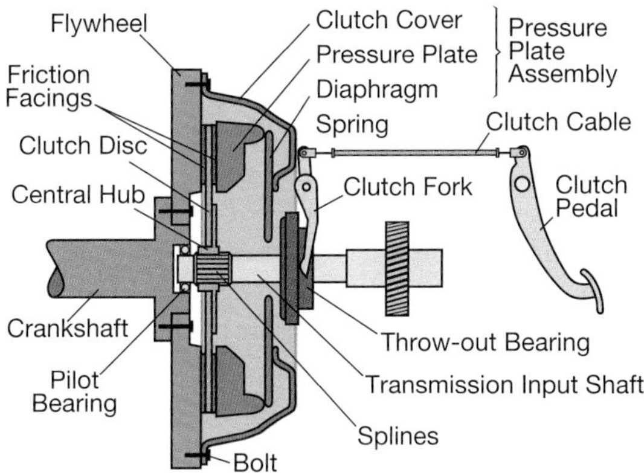

FIGURE 31-4 A standard light vehicle clutch.

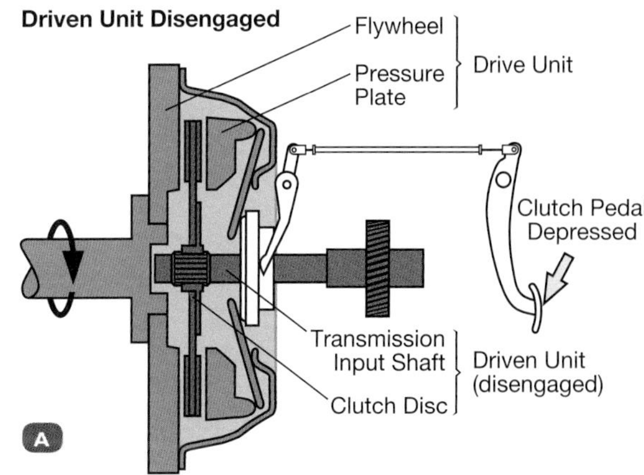

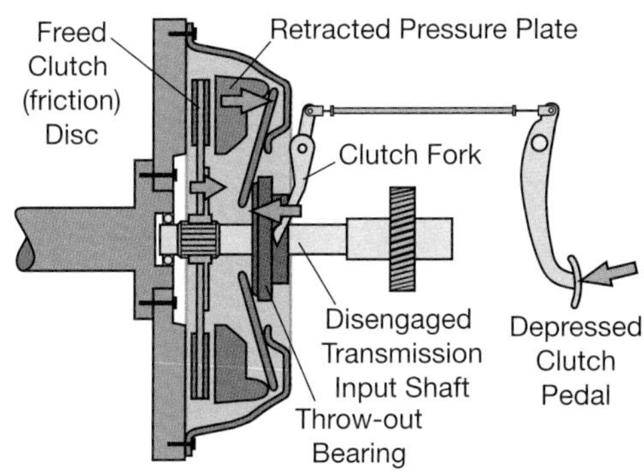

FIGURE 31-5 Depressing the clutch pedal retracts the pressure plate against the force of its springs or diaphragm and frees the friction disc from its clamping action.

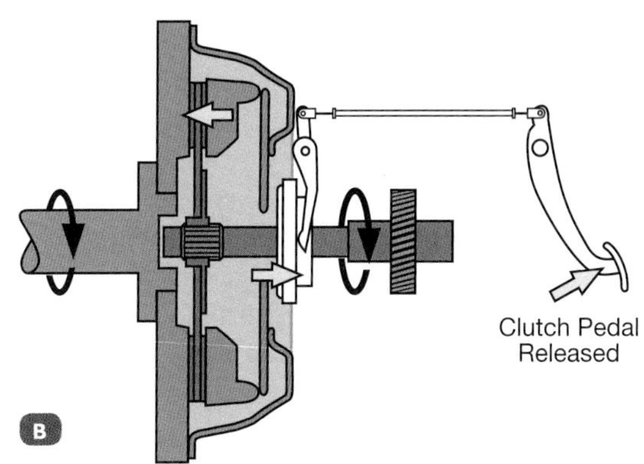

FIGURE 31-6 Releasing the clutch pedal reapplies the clamping force and reconnects the engine and transmission, firmly clamping them together to continue rotating as a unit.

Pushing the clutch pedal operates the **release mechanism**, which controls the flow of torque between the engine and the transmission. Depressing the clutch pedal retracts the pressure plate against the force of its springs and frees the friction disc from its clamping action (**FIGURE 31-5**). Releasing the clutch pedal reapplies the clamping force and reconnects the engine and transmission by firmly clamping the clutch disc between the pressure plate and the flywheel, allowing them to rotate as a unit (**FIGURE 31-6**).

Flywheel

K31003

The main purpose of the flywheel is to smooth out the power pulses from the pistons during the power strokes. It also provides a friction surface for the clutch disc and a mounting surface for the pressure plate (**FIGURE 31-7**). The flywheel is quite heavy. It is usually made of cast iron so that it can store energy from each power pulse from the engine; it uses that energy to keep the crankshaft turning through the intake, compression, and exhaust strokes. However, a heavy flywheel means that it slows the engine's acceleration. A lighter flywheel does not smooth out the power pulses as effectively, but it works well on a drag-racing car, for example, because it allows the engine to accelerate faster.

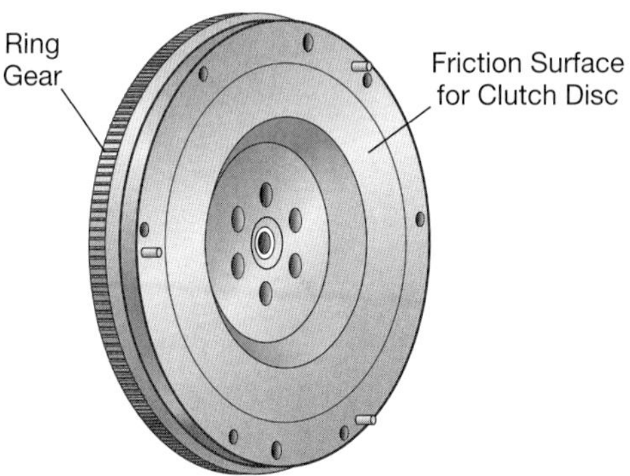

FIGURE 31-7 The flywheel.

The flywheel is bolted to the rear of the crankshaft and allows the crankshaft to mate with the transmission via the clutch system. It also incorporates the **flywheel ring gear**, which enables the starter motor drive gear to crank the engine over. In most cases, the ring gear is made of hardened steel and is press fit onto the outer edge of the flywheel.

> **TECHNICIAN TIP**
>
> If replacement of the starter was performed due to failure, the teeth on the ring gear should be inspected thoroughly as well. If the teeth on the ring gear are damaged, they will likely destroy the matching teeth on the starter motor.

Types of Flywheels

There are two main types of flywheels: the single one-piece flywheel and the dual mass flywheel. The single one-piece flywheel is by far the most common on light vehicles and is what most people think of when they think of a flywheel. Its one-piece construction makes it simple, inexpensive, and reliable. It usually has a starter ring gear pressed onto its outer edge and a machined mating surface for the clutch disc. The center of the flywheel has machined holes for bolts to mount it firmly to the flywheel. In some cases, the bolt pattern is equal, so it can be mounted in any position. In other cases, where the flywheel is used as the primary method of balancing the engine, it may have offset bolt holes, so it can be mounted in only one position on the crankshaft. It is generally designed for a particular application, so do not try to use a flywheel from one type of engine on another type.

Some flywheels are of the stepped style (**FIGURE 31-8**). In this design, the friction surface of the flywheel is recessed, and the outer diameter, against which the pressure plate is bolted, is raised. The depth of the step is critical for proper operation of the clutch assembly. If it is too deep, the clutch will slip or not engage. If it is too shallow, the clutch will be stiff and may not disengage. Both stepped and flat flywheels can be resurfaced if the wear or defects are minor.

> **TECHNICIAN TIP**
>
> Refinishing the flywheel moves the pressure plate toward the engine and away from the throw-out bearing, increasing free play. If too much material is removed from the flywheel surface, some release mechanisms will not be able to compensate for the loss, and the clutch will not fully release. Also, as the flywheel is machined thinner, the clutch center hub may contact the flywheel bolts. So even though most manufacturers do not list a minimum thickness for the flywheel, know that it can cause problems if it becomes too thin.

FIGURE 31-8 Stepped flywheels have a recessed friction surface for the clutch disc.

The dual mass flywheel improves the engine's fuel economy by smoothing out the power pulses and focusing them in the direction of engine rotation. Its inner workings help absorb engine vibrations, thus minimizing gear rattle and putting less strain on the drivetrain components. This also makes for smoother shifting. There are two basic types of dual mass flywheels. The first is composed of a primary and a secondary flywheel with a series of torsion springs and cushions. The second uses a planetary gear and torsional springs (**FIGURE 31-9**).

In the first type of dual mass flywheel, a friction ring is located between the inner and the outer flywheel that allows the inner and the outer flywheel to slip. This feature is designed to alleviate any damage to the transmission when torque loads exceed the vehicle rating of the transmission. The friction ring is

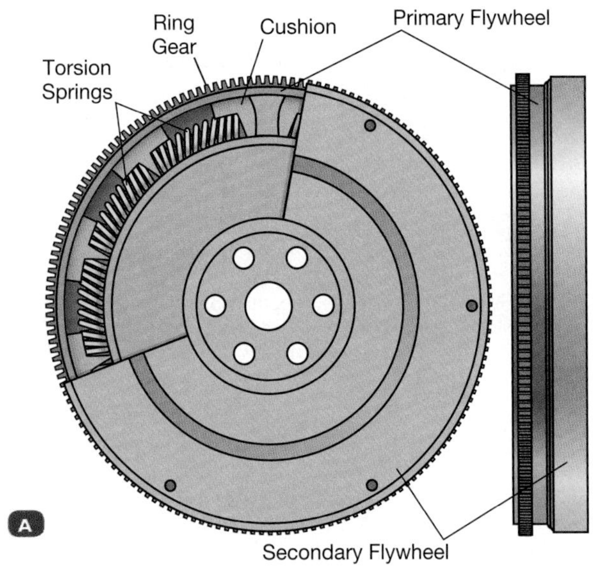

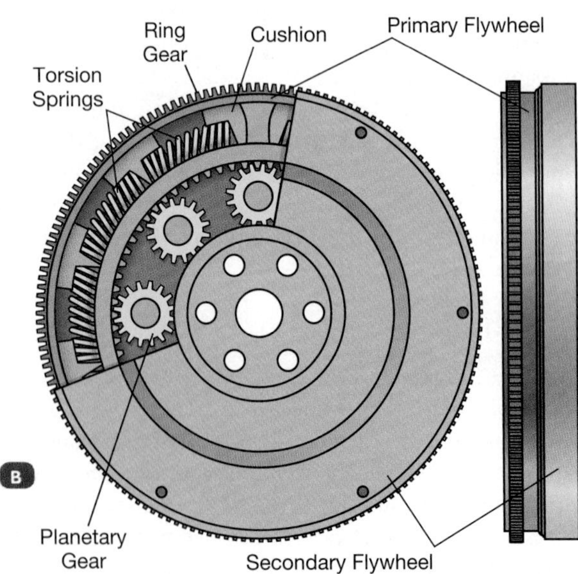

FIGURE 31-9 There are two types of dual mass flywheels. **A.** The first is composed of a primary and a secondary flywheel with a series of torsion springs and cushions. **B.** The second uses a planetary gear and torsional springs.

the weak spot in the system and can wear out if excessive engine torque loads are applied. This type of dual mass flywheel also has a center support bearing that carries the load between the inner and the outer flywheel, and is fitted with damper springs to absorb shocks.

The second type of dual mass flywheel incorporates planetary gearing along with torsion springs. It is designed for engines with stronger vibrations at lower engine speeds. Some manufacturers use this style for their high-performance vehicles to gain greater driving and shifting comfort. Because of the increased dampening effect at lower engine speeds, the engine can be idled at fewer revolutions per minute (rpm), which reduces fuel consumption slightly.

Dual mass flywheels are most often fitted to light-duty diesel trucks with standard manual transmissions, and to higher performance luxury vehicles. However, dual mass flywheels are now being used in economy-type vehicles to dampen vibrations in the drivetrain. The function of the dual mass flywheel is to absorb torsional crankshaft vibrations, which are twisting forces created in opposite directions. A twisting force happens in one direction when a piston is on the compression stroke and in the opposite direction on the power stroke. This vibration is magnified in diesel engines, which have higher compression ratios than gasoline engines. By minimizing the torsional vibration, the dual mass flywheel eliminates any potential damage to the transmission gear teeth. If the dual mass flywheel were not used, the torsional vibration could cause increased wear or even chipping of the transmission gears.

The dual mass flywheel construction relocates the light-duty torsional damper from the clutch disc in a one-piece flywheel to the engine flywheel on the dual mass flywheel style. This repositioning and heavy-duty torsional damper dampens engine torsional vibrations much more effectively than is possible with standard clutch disc dampening technology.

▶ TECHNICIAN TIP

Dual mass flywheels are designed to provide maximum isolation of the frequency below the engine's operating rpm, usually between 200 and 400 rpm. They are also most effective during engine startup and shutdown.

Pressure Plates

K31004

The pressure plate provides the clamping force to clamp the clutch disc between the pressure plate and the flywheel so that torque can be transmitted between those parts. The force is generated by one or more very strong springs. When the driver pushes on the clutch pedal, the spring(s) in the pressure plate are what the driver is pushing against. Pushing the clutch pedal compresses the spring(s) and removes the clamping force from the clutch disc. Releasing the clutch pedal allows the spring(s) to apply their clamping force to the clutch disc again. Most automotive pressure plates are of either the diaphragm spring style or the coil spring style.

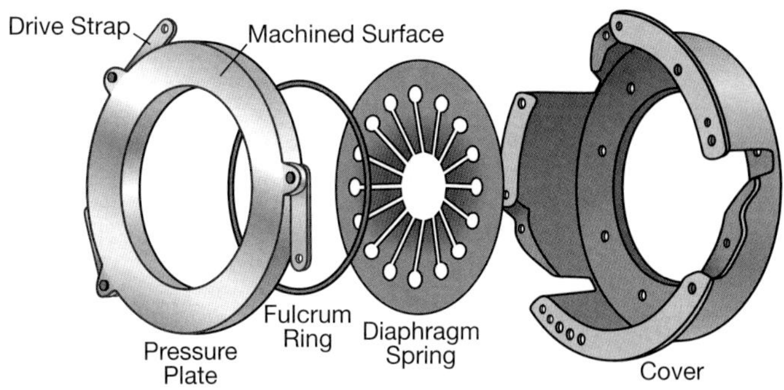

FIGURE 31-10 A diaphragm pressure plate consists of a pressed steel cover, a pressure plate with a machined flat surface, a number of spring steel drive straps, and the diaphragm spring.

Diaphragm Pressure Plate

In light vehicles, the pressure plate is normally a diaphragm type and is serviced as an assembly, which means it is not designed to be disassembled and repaired (**FIGURE 31-10**). A **diaphragm pressure plate** consists of a pressed steel cover, a pressure plate with a machined flat surface, a number of spring steel drive straps, and the diaphragm spring. This diaphragm is located inside the clutch cover on two **fulcrum rings** held in place by a number of rivets passing through the diaphragm. The pressure plate is connected to the cover by the spring steel drive straps, which are riveted to the cover at one end, and two or more projecting lugs on the plate at the other. Retraction clips hold the pressure plate in contact with the outer edge of the diaphragm. During clutch operation, the throw-out bearing pushes on the diaphragm levers. The diaphragm pivots on the fulcrum rings and pulls the outer edge of the diaphragm away from the flywheel. The retraction clips pull the pressure plate away from the clutch disc. Diaphragm clutches are known for their lighter feel when the clutch pedal is fully down, making them more comfortable to drive.

Coil Spring Pressure Plate

The **coil spring pressure plate** uses coil springs to create the clamping pressure and uses release levers to release the clutch disc (**FIGURE 31-11**). Typically, three or four release levers are used, depending on whether the application is in a car or a truck. The release levers control the movement of the friction portion of the pressure plate. They pivot on the pedestals that are part of the pressure plate housing. When the release bearing pushes the levers toward the flywheel, they pivot and pull the friction portion of the pressure plate back toward the driver, which then relinquishes its clamping pressure on the clutch disc. When the clutch pedal is released, the levers return to their rest position, the clamping force is restored to the clutch disc, torque is transmitted, and the vehicle moves forward.

An advantage of the coil spring pressure plate is that the more coil springs there are, the tighter the clamping force and torque capacity. A disadvantage is that if the clutch disc overheats, the springs can become weak, and the clamping force is compromised. This slippage causes more heat, which weakens

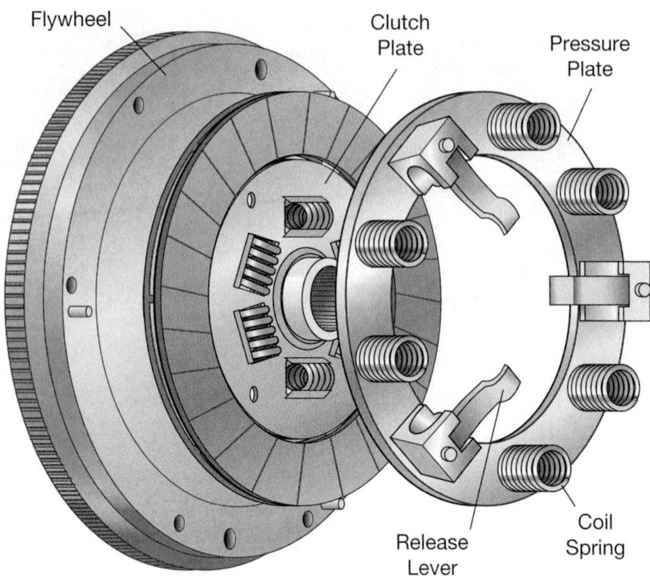

FIGURE 31-11 The coil spring pressure plate uses coil springs to create the clamping pressure and uses release levers to release the clutch disc.

the coil springs even further. Once a clutch starts slipping, it likely will have to be replaced very soon. Additional disadvantages are that the spring pressure is less evenly spaced on many coil spring pressure plates, and there are fewer release levers, which can cause uneven wear on the friction surface over time.

> TECHNICIAN TIP
>
> The coil spring–type pressure plate generally requires more pedal pressure to hold the pedal down, which makes it less comfortable to drive. This is why diaphragm pressure plate clutches, which require less pedal effort to hold the pedal down, are more desirable for light-duty passenger vehicles.

Clutch Disc

K31005

The clutch disc is also called a **driven center plate** or a friction disc (FIGURE 31-12). The clutch disc provides the friction material needed to transmit engine torque from the flywheel and pressure plate to the input shaft of the transmission. On the clutch disc, the friction facings are riveted to waved spring steel segments, which are themselves riveted to a steel disc. The central alloy-steel splined hub is separate from the steel disc. Drive is transmitted from the steel disc to the hub through heavy torsional coil springs or rubber blocks. This arrangement dampens torsional vibrations from the engine. It also absorbs shock loads imposed on the drive line by sudden or violent clutch engagement. A molded friction washer between the hub and the spring retaining plate also acts as a damper.

Waved spring steel segments located between the friction facings cause the facings to spread apart slightly when the clutch

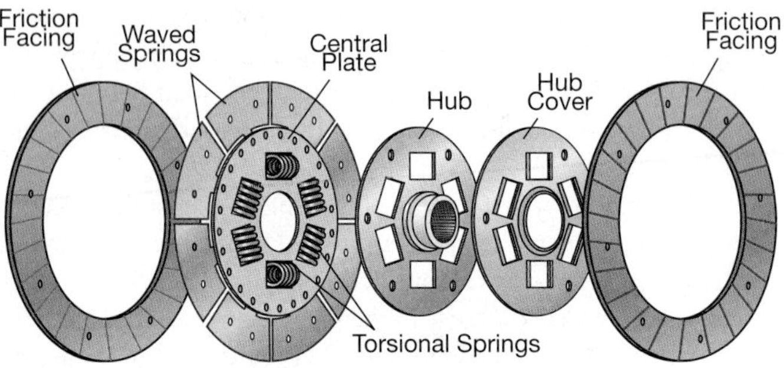

FIGURE 31-12 The clutch disc components.

is disengaged and to compress as it is engaged (FIGURE 31-13). These waved springs allow a progressive application of the pressure plate clamping force as the waved springs are being compressed when the pedal is being released. This results in a smoother engagement of the clutch when starting from a stop.

> TECHNICIAN TIP
>
> Some high-performance clutch discs do not use wavy springs between the clutch facings, which makes them more robust. This removes the progressive nature of the clutch action, making them much more "grabby."

Throw-Out Bearing and Clutch Fork

K31006

The clutch throw-out bearing and clutch fork work together (along with the clutch linkage) to compress the pressure plate springs when the clutch pedal is pressed. Because the pressure plate rotates with the engine flywheel when the engine is running, the throw-out bearing must be able to rotate with the

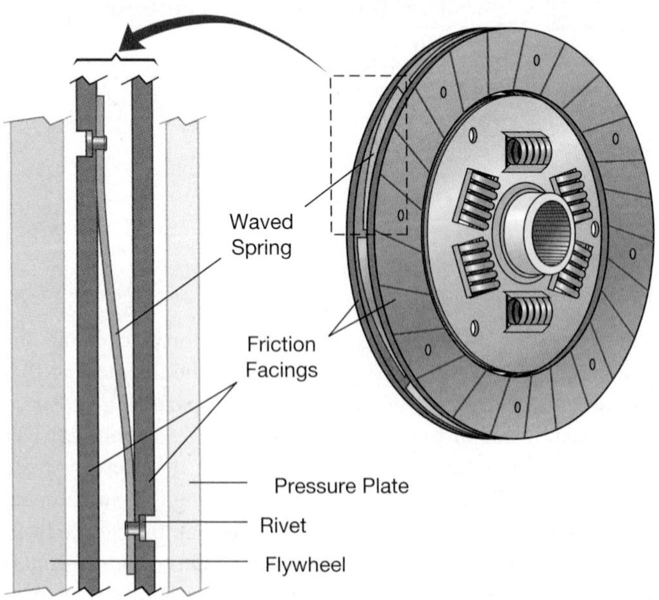

FIGURE 31-13 Waved springs allow progressive application of the clutch.

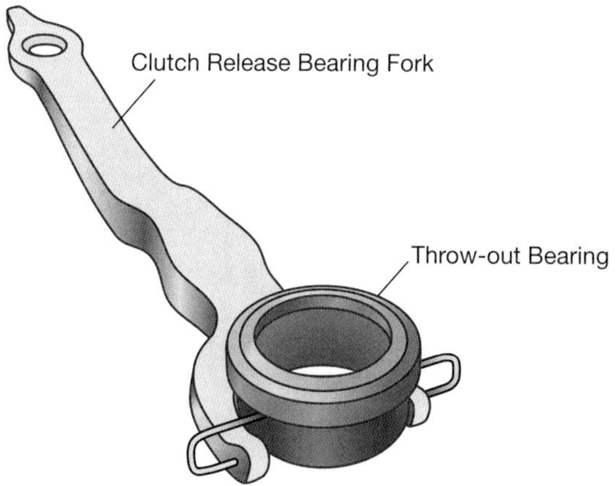

FIGURE 31-14 A clutch release bearing fork.

FIGURE 31-15 Collar-style slave cylinder operates the throw-out bearing directly, without a clutch fork.

pressure plate while the clutch fork remains stationary. Thus, the throw-out bearing must include a thrust bearing as part of its assembly. Throw-out bearings are usually **thrust-type angular-contact ball bearings** that are pressed onto a **carrier**. The carrier slides on the sleeve of the **front bearing retainer** that extends from the front of the transmission. This is considered a **push-type clutch** design, as the throw-out bearing pushes on the levers of the pressure plate. However, there are pull-type designs used on heavy truck applications. The bearing carrier is located on the clutch release bearing fork (**FIGURE 31-14**).

Moving the clutch release fork (clutch fork) brings the bearing thrust face into contact with the pressure plate levers, which causes it to rotate against the linear motion of the clutch fork and absorb the rotary motion of the levers. The thrust-type angular contact ball bearing is packed with lubricant during manufacture and requires no periodic maintenance during its service life, as long as it is not abused and the clutch free play is maintained.

The clutch fork is usually made of stamped steel or cast iron. It pivots either in the center or at the end inside the bell housing. The pivot is generally screwed into the bell housing and is usually replaceable. The pivot should be inspected for wear and lubricated whenever the clutch is replaced. The release bearing engages in tabs in the clutch fork and is usually held in place by clips.

Not all release bearings are operated by a clutch fork. Some are operated directly by collar-style slave cylinders. These are sometimes called central or concentric slave cylinders because they are donut shaped and fit around the input shaft (**FIGURE 31-15**). This style uses hydraulic pressure to directly push the release bearing against the pressure plate fingers without the use of a clutch fork.

Pilot Bearing

K31007

The pilot bearing is essentially an alignment support bearing for the snout of the input shaft to ride on (**FIGURE 31-16**). Because it takes two bearings to support a rotating shaft, the pilot bearing is the front bearing for the input shaft. The other end of the input shaft is supported by the transmission input bearing. The pilot bearing can be a brass or bronze bushing, a needle-type bearing, or a roller-type bearing (**FIGURE 31-17**). Larger vehicles, such as trucks, may use a ball bearing type. Some of these bearings are installed in the cavity on the end of the crankshaft, or they may be placed or pressed into the center of the flywheel. To access the pilot bearing, should it fail in any way, the transmission and clutch assembly must be removed. Some vehicles do not require a pilot bearing because of the construction of the input shaft. On this type of shaft, the input shaft is long enough (typically front-wheel drive) so it is supported on the front and back of the transaxle case by support bearings (**FIGURE 31-18**).

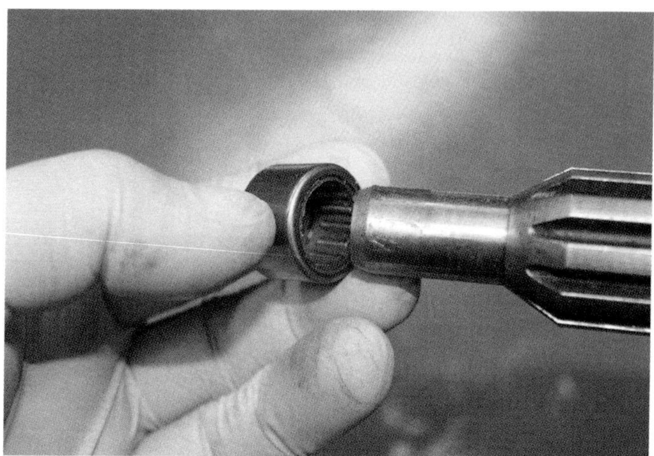

FIGURE 31-16 Pilot bearing and input shaft.

▶ Clutch Operating Mechanisms

K31008

Movement of the clutch pedal is transferred through an operating mechanism to the clutch assembly on the rear of the flywheel. This mechanism may be one of two types of mechanical

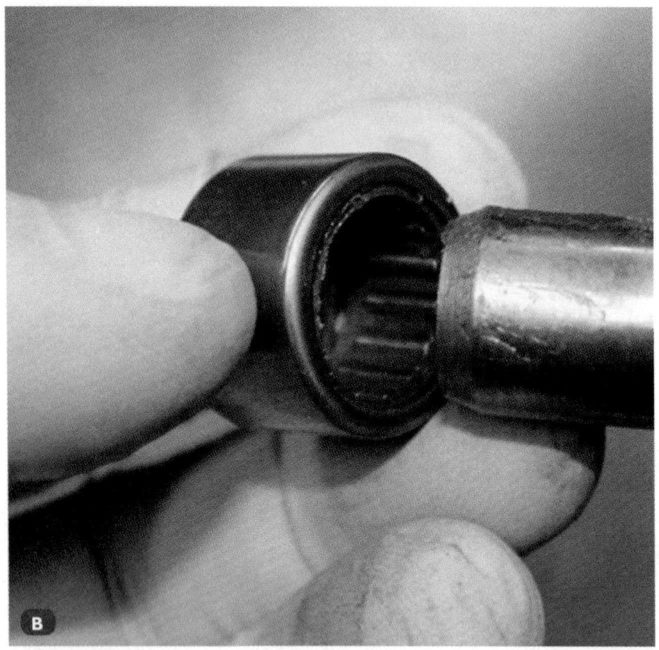

FIGURE 31-17 A. Brass or bronze bushing. **B.** Needle-type bearing. **C.** Roller-type bearing.

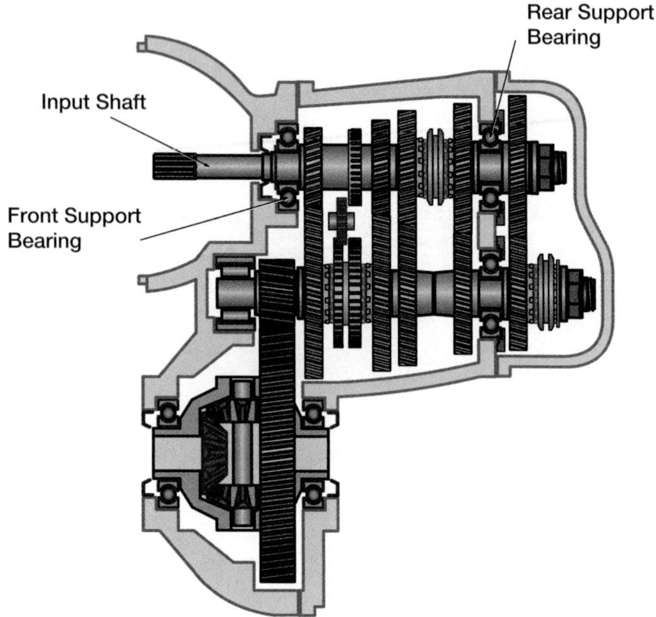

FIGURE 31-18 A vehicle that doesn't use a pilot bearing because its two bearings are located in the transmission.

systems or a hydraulically operated system. The two types of mechanical systems are the linkage style and the cable style. Linkage-style systems use a system of links, rods, and levers between the clutch pedal and the clutch fork. Cable-operated linkage uses a strong cable in a flexible housing. This style offers more flexibility and is easier to install in the factory. Hydraulic systems use a series of steel, plastic, or reinforced rubber lines, along with fluid, to accomplish movement for engagement and disengagement of the clutch system.

▶ TECHNICIAN TIP

Regardless of the operating mechanism, it is always a good idea for drivers to make sure, when shifting gears, to completely remove their foot from the clutch pedal to preserve the life of the throw-out bearing and friction facings.

Cable Mechanisms

K31009

Cable-operated clutch control systems are easily installed in the vehicle during manufacture and take up less overall engine compartment room. The outer cable housing is fixed to the pedal support inside the vehicle and to the transmission bell housing in the engine compartment. The inner cable connects between the upper end of the clutch pedal and an external lever on the end of the clutch fork (**FIGURE 31-19**). This lever is part of the clutch fork, which operates the throw-out bearing. Depressing the clutch pedal transfers the movement through the cable, and the throw-out bearing thrusts against the levers on the pressure plate, pushing the clutch pressure plate into the released position.

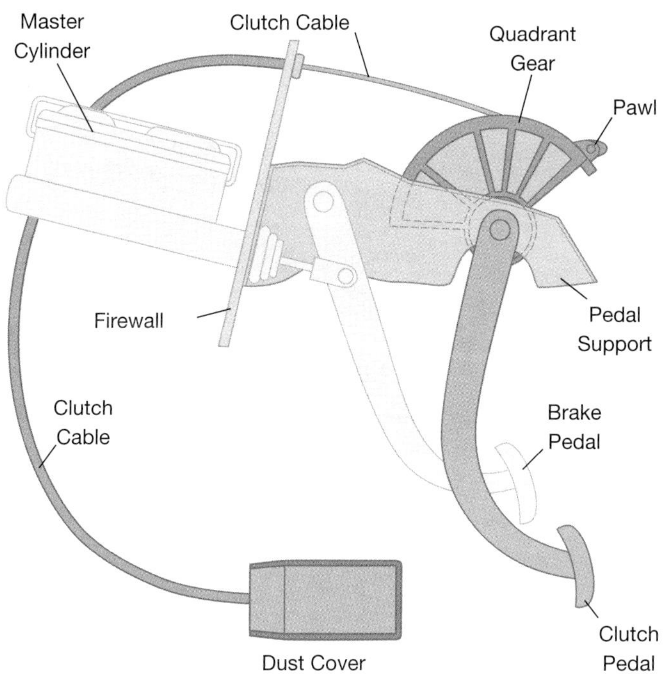

FIGURE 31-19 A cable-operated clutch.

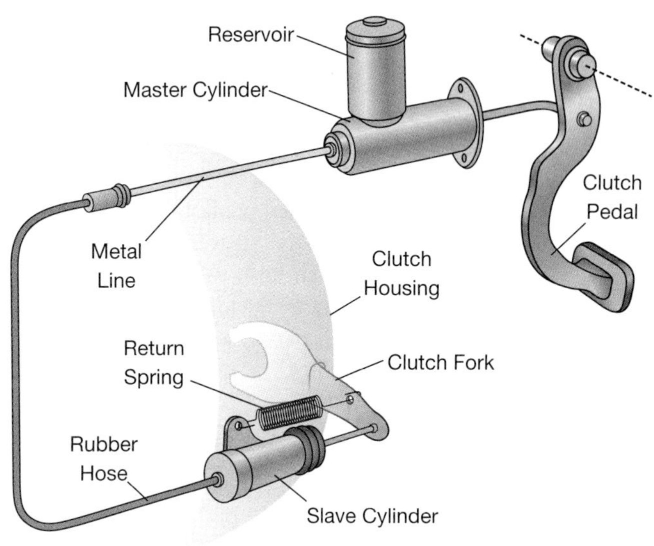

FIGURE 31-20 A hydraulic clutch control.

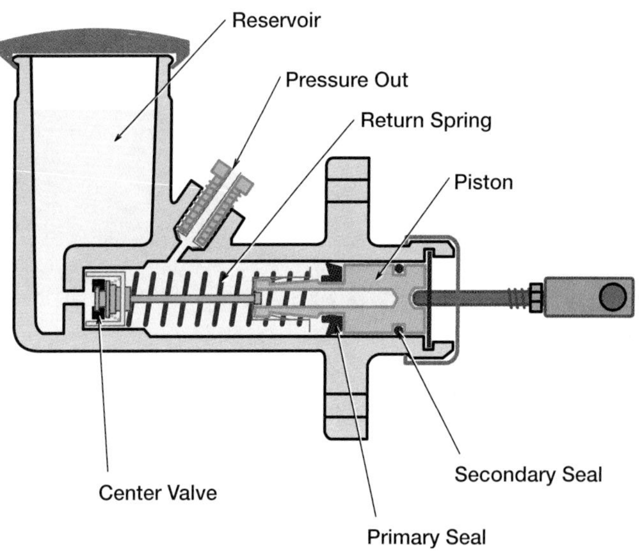

FIGURE 31-21 Cutaway view of a clutch master cylinder.

An adjustment on the clutch cable provides for the specified amount of free play to be maintained between the throw-out bearing and the pressure plate levers when the clutch pedal is in the released position. This free play prevents constant contact of the bearing with the pressure plate levers and subsequent rotation of the bearing when the engine is running, which would cause it to wear prematurely as well as produce a rotating sound. Some cable-operated clutches are adjusted manually by turning a threaded nut or collar to obtain the proper free play. Some vehicles use a **quadrant ratchet**, which automatically adjusts the clutch pedal free play as needed when the pedal is lifted by the driver's toe.

Hydraulic Clutch Mechanisms

K31010

In hydraulic clutch release mechanisms, the clutch pedal acts on a master cylinder connected by a hydraulic tube and flexible hose to a slave cylinder mounted on (external type) or in (internal type) the transmission bell housing (**FIGURE 31-20**). The **slave cylinder** operates the clutch fork (external type) or directly on the throw-out bearing (internal type) on some vehicles. With the clutch pedal in the released position, the center valve in the master cylinder is clear of the inlet port, and fluid is free to flow to or from the reservoir into the cylinder (**FIGURE 31-21**). This allows for expansion and contraction of the fluid as it heats and cools.

When the clutch pedal is initially depressed, the master cylinder piston moves forward, taking the valve assembly with it. The center valve closes off the inlet port from the reservoir, trapping fluid in the cylinder bore. Further piston movement displaces fluid through the outlet port and into the connecting lines to act on the slave cylinder piston. The movement of the slave cylinder piston is commonly transferred through a push-rod to the clutch release fork to operate the clutch. In other configurations, the slave cylinder is located directly behind the throw-out bearing and pushes directly on it (collar style). When the clutch pedal is released, displaced fluid returns to the master cylinder, and the center valve returns to being slightly clear of the inlet port, allowing excess fluid to return to the reservoir. Most hydraulic clutches are self-adjusting because the clutch fork causes the slave cylinder piston to return as far as necessary, venting any excess hydraulic fluid to the master cylinder reservoir. This system automatically compensates for any clutch disc wear and makes it so the clutch, in most cases, does not have to be adjusted.

Linkage-Operated Systems

K31011

Older technology systems used a series of links and levers with an equalizing mechanism called a bell crank (equalizer bar) that rotated and was attached to the engine and the vehicle frame (**FIGURE 31-22**). The bell crank pivoted in plastic or nylon bushings and wore out over time. Adjustments were made at the end of the link connected to the clutch fork, using a threaded rod and lock nuts. As the clutch plate wore, adjustments were necessary to maintain proper clutch pedal free play, so the clutch would operate correctly.

▶ TECHNICIAN TIP

You might think that because a clutch plate wears over time, the clutch free play would increase. In fact, it decreases. As the lining wears, the pressure plate levers move backward toward the throw-out bearing, reducing clutch pedal free play. If the clutch disc wears enough, all of the free play can be lost, meaning that even with the driver's foot off the clutch pedal, the linkage is holding some pressure on the lining. This reduces the pressure plate clamping force and, if bad enough, can cause the clutch to slip, leading to a burnt clutch requiring replacement.

▶ Maintenance and Repair

S31001

Clutch Safety and Hazards

K31012

Clutch maintenance and repair can be hazardous if the proper safety precautions are not taken. Wear appropriate clothing and eye protection while servicing all clutch components (**FIGURE 31-23**). Gaining access to work on clutches generally requires a lift or jack stands. Always make sure the vehicle is properly secured on jack stands or lifts before performing any type of service. Because transmissions are heavy and awkward to remove and install, use an approved transmission jack along with the appropriate tie-down adapters. In many cases, it is best to get the assistance of another technician when removing or installing a transmission. This gives you an additional set of hands to aid in preventing an injury. Always be mindful of spills, and clean the area as necessary to avoid any slips or falls.

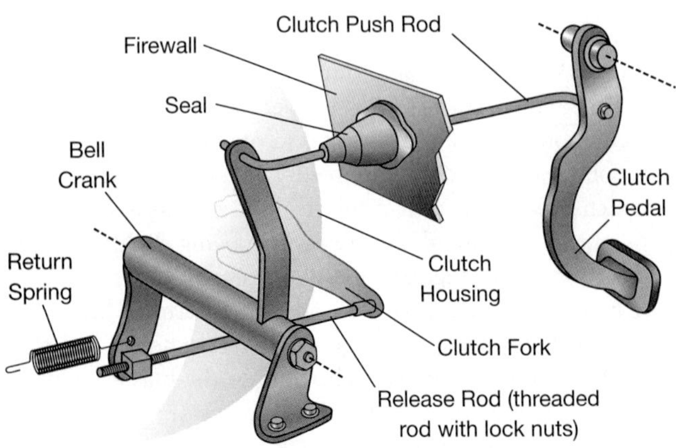

FIGURE 31-22 A linkage-operated system.

FIGURE 31-23 Protect yourself by wearing appropriate clothing and eye protection while servicing clutch components.

SAFETY TIP

Due to designs of the manufacturer, it may be necessary to maneuver the transmission in awkward positions. Make sure the transmission is secured to the transmission jack to avoid injuring yourself.

Asbestos is a carcinogen. In fine particles, it can be breathed into the lungs, making it a health hazard. It was commonly present in older clutch disc linings. Modern technology has generally replaced asbestos by using organic compound resins with imbedded copper filings and some ceramic materials. Use equipment approved by the Occupational Safety and Health Administration (OSHA) to remove clutch dust and debris, and follow all local, state, and federal regulations for particulate disposal.

Never use compressed air to blow off clutch parts such as the flywheel, pressure plate, clutch plate, transmission, or engine bell housings. Although it is possible that the dust does not contain asbestos, even non-asbestos dust is hazardous and can damage lungs. Blowing the dust off parts puts it up in the air where it can be inhaled. OSHA-approved methods entail using a soap and water solution brushed onto the component to loosen and remove the dust, where it is rinsed off and collected in a container to be properly disposed of according to federal, state, and local laws.

Above all, work in a clean and orderly manner whenever servicing a clutch. Keeping things clean will help prevent grease and dirt stains, and keeping things organized will help ensure nothing is forgotten during reassembly. The end result will show the customer what kind of technician you are. Remember that your main objective is to keep the customer satisfied!

Tools

K31013

The following are the main tools used to maintain and repair the clutch system (**FIGURE 31-24**):

- **Transmission jack**—A tool used to support a transmission during removal and installation.
- **Dial indicator**—A tool used to measure runout or end play in clutch service.
- **Clutch alignment tool**—A tool used to center the clutch plate between the flywheel and the pressure plate during the installation of the pressure plate.
- **Flywheel wrench**—A lever-type tool that grabs onto the ring gear teeth and holds the flywheel while the flywheel bolts are being torqued.
- **Clutch wash station**—Typically an air-operated soap-and-water cleaning station that is portable and used to

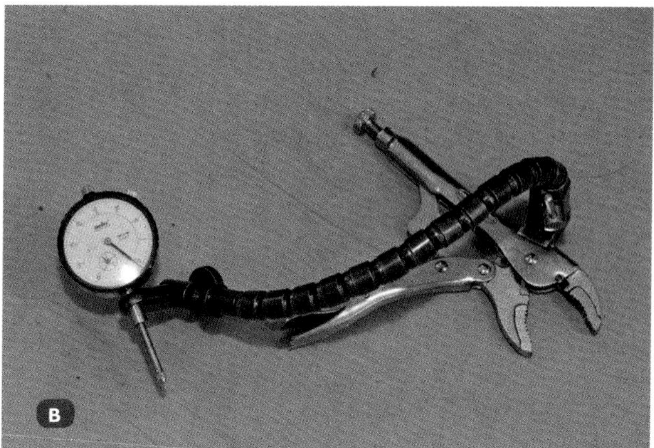

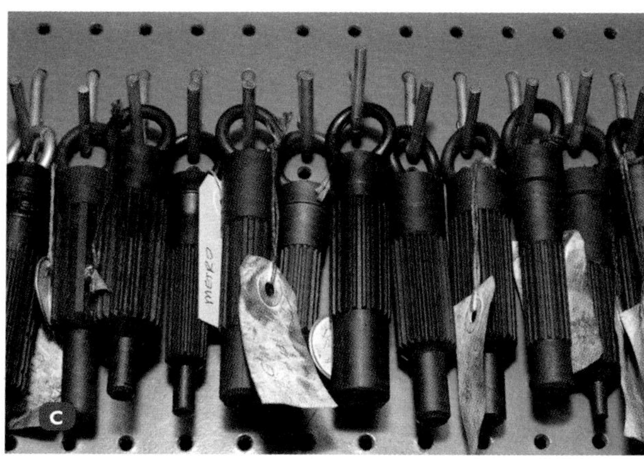

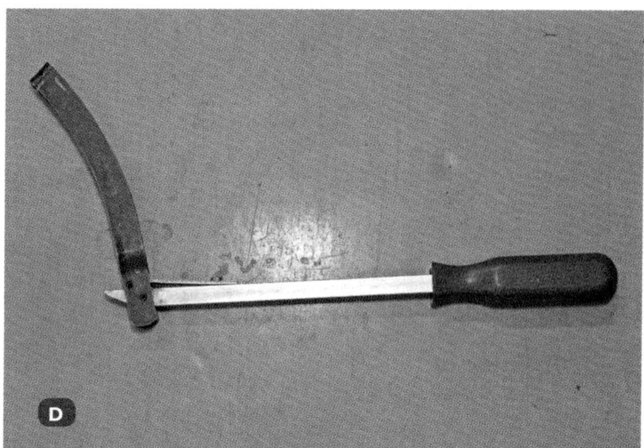

FIGURE 31-24 A. Transmission jack. **B.** Dial indicator. **C.** Clutch alignment tools. **D.** Flywheel wrench. **E.** Clutch wash station.

clean potential asbestos debris from bell housings and clutch parts.

Preventive Maintenance

K31014

Regular preventive maintenance should be performed on the clutch per manufacturer-specified intervals. Clutch pedal free play should be checked and compared to specifications. Maintaining the specified clutch pedal free play is critical to maintaining clutch life. This can be accomplished by following the specified adjustment procedures for the clutch system you are maintaining.

The operation of the clutch pedal and return springs should be checked for binding and excessive movement. If a clutch switch is used on the clutch pedal, check that it is functional by verifying that it prevents the engine from cranking over while the clutch pedal is released. Mechanical linkage should be inspected for damage and lubricated with specified lubricants at all pivot points.

Hydraulic clutches should be checked for moisture-contaminated fluid in a manner similar to checking brake fluid (**FIGURE 31-25**). Clutch fluid is often more neglected than even brake fluid, so there is a good chance that the clutch fluid needs to be flushed if the vehicle is more than a few years old. Hydraulic clutch lines should be checked for leaks at the master cylinder and the slave cylinder. If accessible, pull the boot back on the slave cylinder and inspect it for signs of leakage. Check that all hydraulic lines are not kinked or leaking at their connections. Make sure all hydraulic components are secure in their mountings.

> **TECHNICIAN TIP**
>
> Be careful of the pressure that comes out of the slave cylinder, as it may splash or spray into your eyes. Also, brake fluid will eat paint, so when handling brake fluid, cover fenders and clean up any spilled fluid with generous amounts of water.

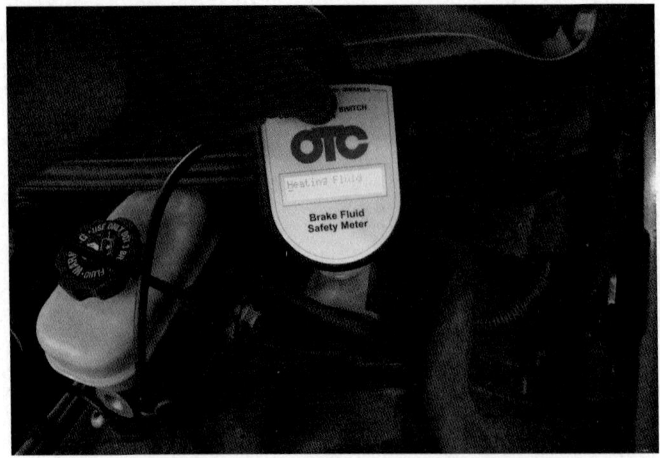

FIGURE 31-25 Checking brake fluid for excess moisture.

Checking and Adjusting a Mechanical Clutch

N31001

As the clutch wears, the friction disc becomes thinner. This results in the pressure plate release levers moving closer to the release bearing and the clutch linkage losing its operational clearance. Some clutches are self-compensating for wear, whereas others require checking and adjusting. You must refer to the manufacturer's shop information to find out exactly where any adjustment should be made.

It is important to check the clutch linkage mechanism for proper operation and correct the adjustment (free play) periodically. It is common to do so during every routine maintenance service. If the clutch pedal has too little free play, the throw-out bearing could remain in contact with the pressure plate levers and prevent the pressure plate springs from applying full pressure. This would result in an incomplete clamping of the disc, leading to premature wear or failure of the clutch assembly, and would require replacement. Also, the pedal will have to be released a long way before the clutch starts to engage. If the adjustment has too much free play, the clutch pedal might not have enough travel to fully release the pressure plate when the pedal is pushed down, causing the gears to clash (grind) when shifting and resulting in heavy synchronizer wear. With this condition, the clutch will start to engage right from the floor when releasing the pedal.

To check and adjust a linkage style clutch, follow the steps in **SKILL DRILL 31-1**.

Checking, Adjusting, and Bleeding a Hydraulic Clutch

N31002

It is important to check the clutch hydraulic system and components for proper operation and correct adjustment. It is common to do so during periodic routine maintenance service. If the hydraulic clutch system is improperly maintained, clutch operation could be compromised in a similar manner as the operation of a mechanical linkage clutch, resulting in pressure plate, friction disc, and transmission synchronizer failure. Also, improper or old fluid in the hydraulic system can cause master cylinder and slave cylinder damage and leaks.

To check and adjust a hydraulic clutch, follow the steps in **SKILL DRILL 31-2**.

> **TECHNICIAN TIP**
>
> Make sure there are no floor mats or other obstructions that will affect the operation of the clutch pedal.

Bleeding a Hydraulic Clutch System

In the case of a hydraulic clutch system failure, it may be necessary to bleed the air from the system. Bleeding is also needed whenever any hydraulic component is replaced or the hydraulic

SKILL DRILL 31-1 Checking and Adjusting a Linkage Style Clutch

1. Following the specified procedure, inspect the clutch linkage parts for damaged, worn, bent, or missing components. Look for signs of binding, looseness, and excessive wear. Operate the clutch pedal, and inspect all components under the dash.

2. Check the clutch linkage components under the hood for the same signs of wear or damage as the components under the dash.

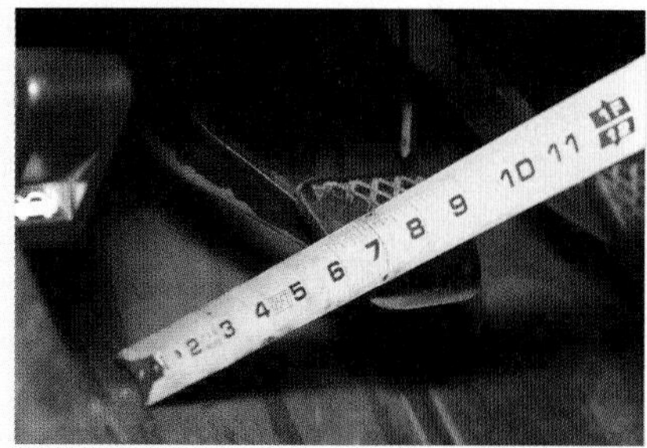

3. Measure the clutch pedal height. Compare your reading to the specifications, and determine any necessary actions to correct any fault.

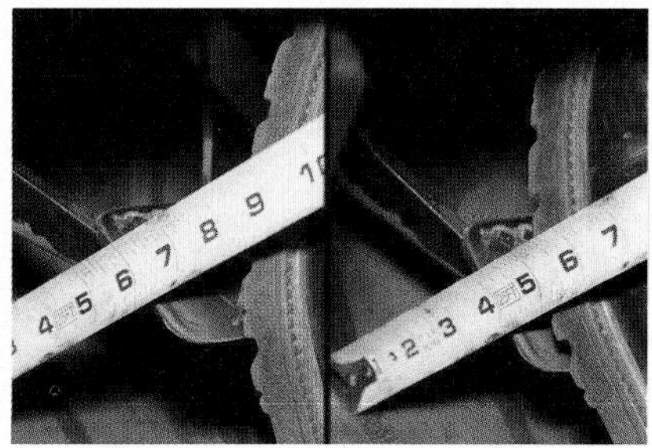

4. Measure the clutch pedal free play. Perform any adjustments as necessary, following the manufacturer's procedure.

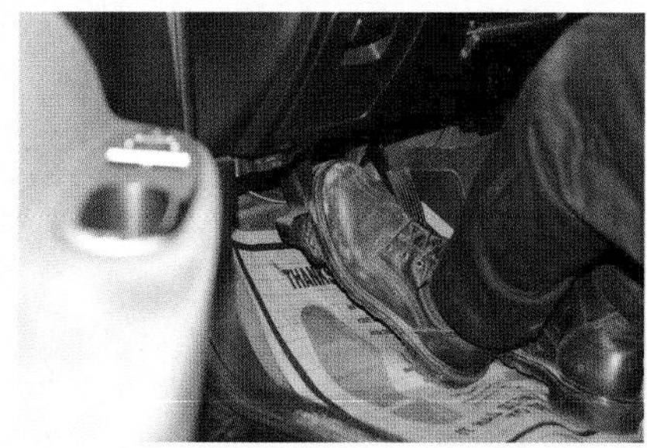

5. Start the vehicle and depress the clutch. The clutch should engage at the proper height and have the proper free play. Make a gear selection to ensure the gears do not clash going into mesh.

fluid becomes unfit for use due to age or contamination. Not all systems are fitted with a bleeder screw, because of how the system is constructed. Also, it may be necessary to bleed the system from the line entering the slave cylinder.

Research the procedure and specifications for bleeding the hydraulic clutch system. There are three types of bleeding: gravity bleeding, manual bleeding, and pressure/vacuum bleeding. Determine the proper method of bleeding to use by consulting the manufacturer's specifications.

Gravity bleeding uses gravity to push fluid and air from the master cylinder and lines out through the slave cylinder bleeder screw. In most vehicles, the clutch master cylinder is quite a bit higher than the slave cylinder. The weight of the fluid can therefore be used to supply the pressure to push fluid and air out of the system.

To bleed/flush a hydraulic clutch system using the gravity method, follow the steps in **SKILL DRILL 31-3**.

The manual bleeding method uses the clutch master cylinder to push fluid and air from the system. The procedure usually requires an assistant to hold the clutch pedal down while the other person opens the bleeder valve on the slave cylinder. This forces air and old fluid from the system. You and the assistant need to coordinate actions. The assistant will have to hold the pedal down until you have the bleeder screw closed, and should also allow the pedal to rise slowly to allow fluid to be drawn from the reservoir into the clutch master cylinder bore. Allowing the pedal to rise too fast can create a vacuum that draws air into the cylinder past the rubber seals. The assistant should avoid pumping the pedal, which could result in a single air bubble breaking up into smaller bubbles and becoming distributed throughout the system. This will require more time to bleed the system completely.

To bleed or flush a hydraulic clutch system using the manual method, follow the steps in **SKILL DRILL 31-4**.

SKILL DRILL 31-2 Checking and Adjusting a Hydraulic Clutch

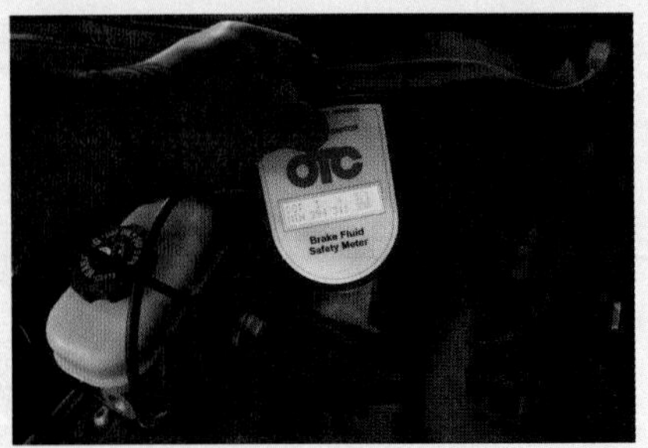

1. Inspect the clutch master cylinder for correct fluid level, and test the quality of the fluid.

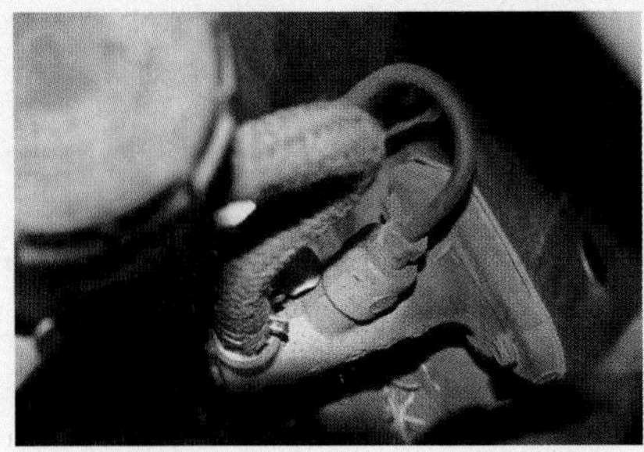

2. Inspect all line connections to the master cylinder.

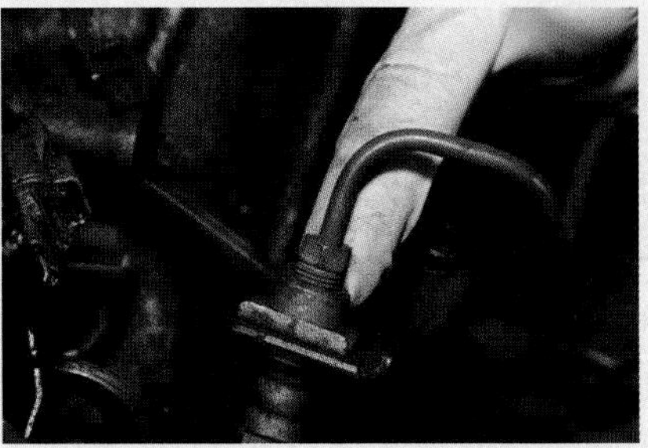

3. Check that all hydraulic lines are not kinked or leaking at their connections. This will require that the system be repaired and bled of any air. Check all rubber hoses for dry rot, bulges, or leaks. Make sure all hydraulic components are secure in their mountings.

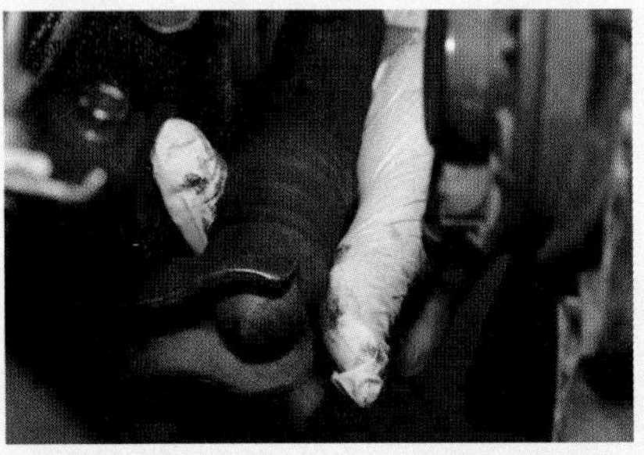

4. Check the boot on the slave cylinder for seepage, which may indicate a leaking slave cylinder piston seal.

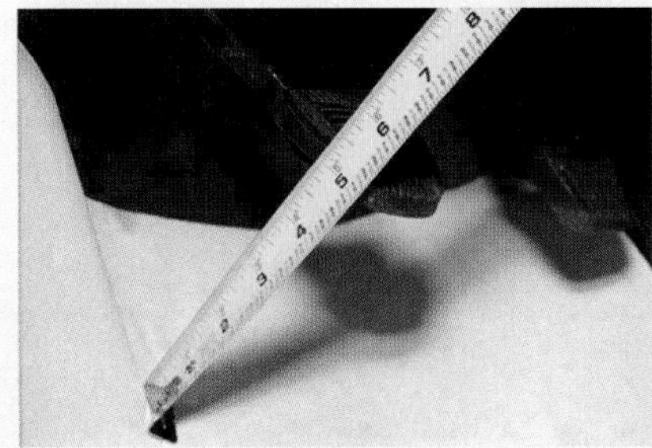

5. Check clutch pedal height. Measure clutch pedal free play, using a tape measure. Compare your readings to the specifications and determine any necessary actions to correct any fault.

SKILL DRILL 31-3 Bleeding/Flushing Hydraulic Clutch System Using the Gravity Method

1. If the fluid needs to be flushed, use a suction gun or old antifreeze tester to suck the fluid out of the clutch master cylinder reservoir.

2. Fill the reservoir with the specified fluid from a freshly opened bottle.

3. Open the bleeder screw on the slave cylinder.

SKILL DRILL 31-3 Bleeding/Flushing Hydraulic Clutch System Using the Gravity Method (Continued)

4. Allow air and fluid to drain from the system into a container.

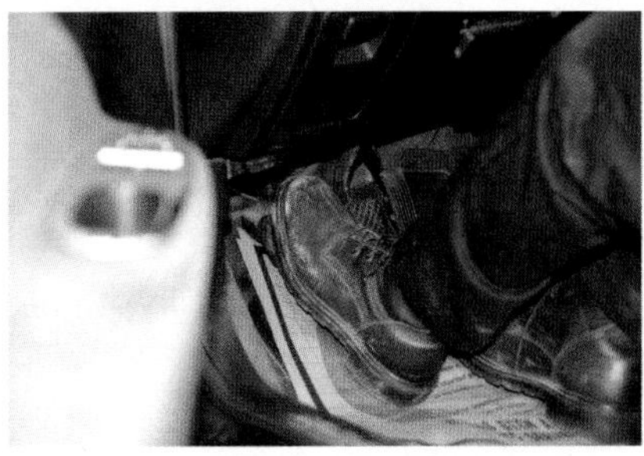

5. Keep the master cylinder filled. Once all air and old fluid are removed, close the bleeder screw, and operate the clutch pedal to check for normal operation.

6. After bleeding the clutch hydraulic system, fill the master cylinder to the correct level with the specified type of brake fluid.

SKILL DRILL 31-4 Bleeding or Flushing a Hydraulic Clutch System Using the Manual Method

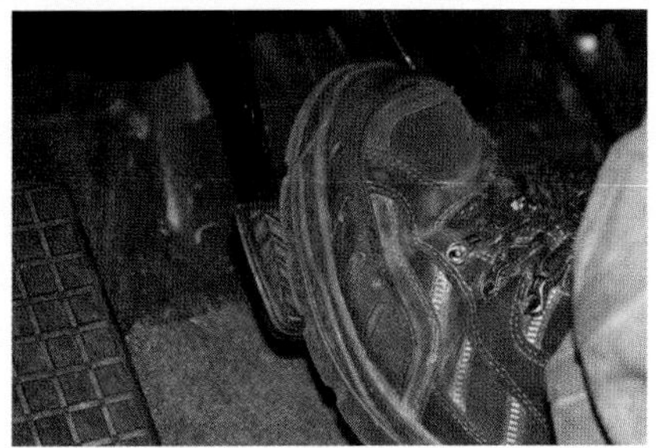

1. Remove all old fluid from the reservoir. Fill it with the specified fluid. Have an assistant depress the clutch pedal slowly.

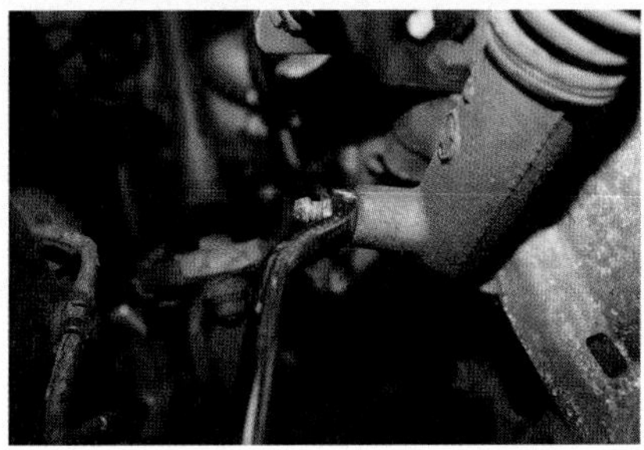

2. Open the bleeder valve on the slave cylinder, and let fluid run out into a container. When all of the fluid stops flowing, close the bleeder valve, and slowly release the pedal. Repeat this process until all air and old fluid are removed from the system.

3. After bleeding the clutch hydraulic system, check for correct pedal feel, and fill the master cylinder to the correct level with the specified type of brake fluid.

The pressure or vacuum bleeding method is the most common method of bleeding a hydraulic clutch in a shop. It uses pressure or vacuum to push or pull fluid and air from the system. This method works well for systems that tend to trap air in the hydraulic system that cannot be bled manually, as it keeps the fluid moving continuously through the system. It does require special bleeding tools or equipment to create the pressure or vacuum.

To bleed a hydraulic clutch system using the pressure method, follow the steps in **SKILL DRILL 31-5**.

▶ TECHNICIAN TIP

In some applications, such as four-wheel drive and all-wheel drive vehicles, it may be quicker to remove the engine than it is to remove the transmission to gain access to the clutch assembly. Look up the flat rate time for removing the engine, and compare that time to the transmission removal time.

▶ Diagnosis

N31003

Clutch diagnosis is usually only performed when the driver complains about abnormal operation, sounds, or chatter that could be related to the clutch. Clutch diagnosis generally starts with getting as much information from the customer as possible. That is typically followed by a test-drive to verify the concern

SKILL DRILL 31-5 Bleeding a Hydraulic Clutch System Using the Pressure or Vacuum Method

1. Remove all old fluid from the reservoir. Fill it with the specified fluid.

2. Hook up the pressure or vacuum bleeding tool to the vehicle with the correct adapters.

3. Apply pressure or vacuum to the system.

4. Open the bleeder screw, and allow the fluid and air to be purged from the system. Repeat this process as necessary.

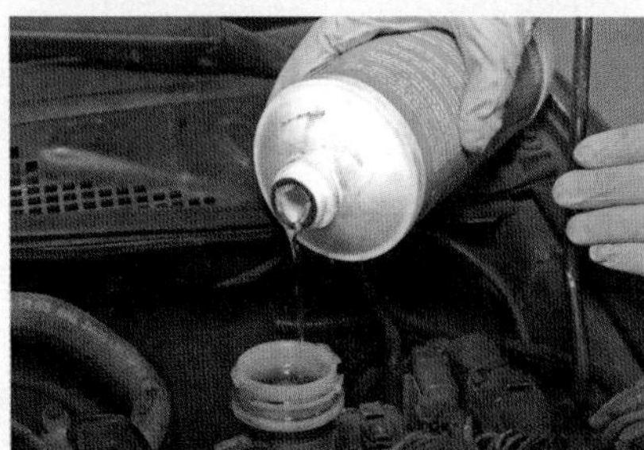

5. After bleeding, check for correct pedal feel, and fill the clutch master cylinder to the correct level with the specified type of brake fluid.

and gather information about the fault. Depending on the condition present, you may be able to determine what is causing the condition based on the test-drive alone. In some cases, further inspection needs to happen to pinpoint the cause. This could be as simple as inspecting the linkage or could be more complex, requiring the removal of the transmission for further inspection of the clutch components.

Often, when a customer has a clutch-related concern, the vehicle has been towed to the shop, and a test-drive is not possible. If the vehicle cannot be driven, you will have to verify the symptoms as best you can, and then perform a visual inspection. For example, if the customer complains that the clutch pedal is spongy and he or she can't put the vehicle in gear when the engine is running, you may need to check the level of fluid in the hydraulic clutch master cylinder. If it is low, you will likely need to repair that before test driving the vehicle and checking out the operation of the clutch to determine whether there are any other faults in the clutch assembly (**FIGURE 31-26**).

Once you have verified and understand the symptoms, research them in the appropriate service information to determine the possible causes and steps to diagnose the concern. Be sure you identify the root cause of the concern. For example, if the clutch disc is badly worn, it could be caused by weak pressure plate springs or misadjusted clutch free play. Just replacing the clutch

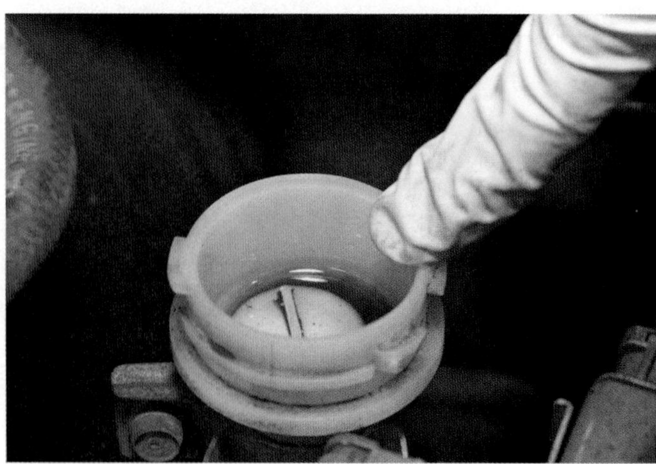

FIGURE 31-26 A technician is checking the level of fluid in the clutch master cylinder reservoir before a test-drive.

disc will result in the clutch disc wearing out quickly again. So always determine all of the necessary actions to correct the fault.

Common Issues

Clutches typically have several relatively common faults that occur when there is an issue. The more familiar you are with clutch fundamentals and the manufacturer's diagnostic procedures, the more likely you are to successfully diagnose the cause of the customer complaint. Some common clutch-related complaints include noises, pedal pulsations, slippage, clutch binding or drag, clutch vibration, and clutch chatter. These issues are explored here individually.

Noises

Clutch-related noises come from three general places: the pilot bearing, the release bearing, and the transmission input shaft bearing. A worn pilot bearing can make a howling or squeaking sound. The noise generally can be heard by fully depressing the clutch pedal with the engine idling and the transmission in first gear. This causes the pilot bearing to rotate on the input shaft, which produces noise if the bearing is faulty.

A typical throw-out bearing noise may be more pronounced when the clutch pedal is partially or fully depressed and may disappear when the clutch pedal is released. A growling noise or whirring noise is prevalent when the throw-out bearing is going bad. Thus, forcing the throw-out bearing into the pressure plate levers by slightly depressing the clutch pedal will reveal the difference between a worn pilot bearing and a worn throw-out bearing. A worn throw-out bearing generally will make noise starting with contact of the throw-out bearing with the pressure plate levers, whereas a pilot bearing will not start making noise until the clutch disengages. Noise that occurs when the engine is idling in neutral and the clutch pedal is in the released position may indicate a worn transmission input shaft bearing, misalignment of the transmission, low transmission fluid level, or abuse by the operator.

Pedal Pulsations

Pedal pulsations are continual up-and-down pulsations due to release bearing contact from diaphragm lever misalignment or coil spring pressure plates that have lever misalignment. Pedal pulsations also may be due to abuse of the clutch, such as heavy slippage while trying to pull a heavy boat and trailer up a steep boat ramp. As a result of the clutch abuse, the flywheel and/or pressure plate may overheat and acquire hot spots.

Excessive slippage from a misadjusted clutch can contribute to overheating of the clutch components and pedal pulsations. To check for this, place your foot on the clutch pedal and lightly push down. Usually the pedal pulsations can be felt as soon as the release bearing starts to contact the pressure plate levers. Further diagnosis will require removal of the clutch assembly and replacement of worn and damaged parts.

Slippage

A **slippage** condition can usually be identified while driving the vehicle in a gear that is direct drive, typically fourth gear, while accelerating. Observe both the engine speed and the vehicle speed. They should both increase proportionately. If there is slippage, the engine rpm will increase much faster than the vehicle speed. In this case, the clutch is unable to transmit full engine torque, so it slips, which is indicated by the increased engine speed.

As the clutch disc starts to wear, it becomes thinner. This results in less clamping pressure from the pressure plate and less clutch pedal free play. This in turn results in an engine rpm increase as slippage starts to occur, which will decrease the amount of torque being transferred to the transmission. To properly diagnosis this problem, a road test is required. You also may be able to feel the slippage as each gear is being shifted and the throttle is being depressed.

Another way to diagnose this condition can be performed while the vehicle is at rest. Engage the parking brake, and shift the transmission into fourth gear. Slowly let the clutch pedal release. If the clutch disc and pressure plate are still working properly, the engine should stall. If the engine continues to run, check the clutch adjustment, and then repeat the same procedure. If it fails to stall and continues to run, the assembly must be replaced, and the flywheel should be checked for warpage and hot spots. This condition may require flywheel resurfacing or replacement.

Clutch Binding

Clutch binding occurs when the clutch does not quite disengage when the clutch pedal is depressed, causing a dragging of the clutch assembly. Clutch binding causes a grinding noise when shifting into and between gears because the clutch does not fully disengage. Trying to shift into a gear while the gears are being powered by the binding clutch is very difficult and results in the grinding noise. It also can make it difficult to put the transmission into neutral.

Applied Math

AM-49: Conditionals: The technician understands that if the described problem has certain conditions (symptoms), then there are a limited number of probable solutions.

In all automotive systems, effective diagnosis requires an operational understanding of the system in which the problem has occurred. A typical workshop scenario is a customer concern regarding a slipping clutch. The presence of the fault is generally confirmed by a road test before starting work on the vehicle.

As outlined in the chapter, a typical clutch system is made up of only a limited number of components, so the slipping condition can be attributed to only a limited number of causes. Potential causes could include worn clutch plate friction material; over-adjustment, causing the release bearing to ride on the pressure plate springs preventing full engagement; weak pressure plate springs, creating insufficient clamping force; or oil-soaked clutch plate friction material, reducing friction (although this generally results in a clutch chatter condition first).

The diagnostic process should start with the less labor-intensive checks first. For example, you should check the clutch adjustment and check for oil leaks around the bell housing area before removing the transmission from the vehicle to physically inspect the clutch assembly.

If clutch binding is present, it will be evident when trying to engage a gear with the engine running. If the transmission shifts easily when the engine is off, but is difficult with the engine running, suspect a binding clutch condition. Clutch binding may result from a damaged clutch disc, a warped clutch disc, or rusted input shaft splines and clutch hub. A bad input shaft bearing or pilot bearing also may cause clutch dragging. Loose transmission bolts cause misalignment of the input shaft, and clutch dragging will occur. A twisted input shaft will affect the engagement and disengagement of the clutch disc. This condition may result from driver abuse or possibly from the driver's foot slipping off the clutch pedal due to rain, snow, or ice. Other possible causes of clutch binding include a clutch linkage that is bent or binding because of excessive wear, or a damaged or rusted firewall that flexes and does not allow the clutch to be released fully. The firewall can sometimes be reinforced by bolting a metal plate to it.

Clutch Vibration

Clutch vibration can be very annoying to drivers, so you need to be familiar with how to diagnose it. First, isolate the type of vibration that is present. **Torsional vibrations** may be the result of engine parts that are starting to go bad. A vibration damper or flywheel that is out of balance can cause torsional vibrations. If this is the case, the vibration will be felt throughout the rpm ranges of the engine. If it is a clutch disc that is out of balance (usually due to clutch lining that has broken off), the vibration will be worse when the clutch pedal is released and the transmission is in neutral. **Driveline vibrations** may come from worn u-joints or CV joints, along with motor mounts that are worn and are allowing the engine to contact the frame or cross member. A thorough inspection and diagnosis of all of these conditions should be performed before removal of the clutch assembly parts for inspection.

Clutch Chatter

Clutch chatter should not be taken lightly, as the result may be clutch failure. **Clutch chatter** is a shuddering feeling as the clutch pedal is being released. If there is uneven wear of the friction disc or friction surfaces of the pressure plate and flywheel, you may feel chatter as the pressure plate starts to contact the friction disc. Oil on the friction disc from either a leaking rear main seal or front transmission seal can also contribute to this condition.

▶ TECHNICIAN TIP

When diagnosing vehicle problems, it is sometimes helpful to let the owner drive the vehicle while you observe the way he or she operates the vehicle. All too often we take for granted that people drive the way we expect them to. Some problems can be solved only by observing the owner's driving habits, and then educating him or her. A technician tells a story of an elderly lady who had recently purchased a new vehicle. She brought it back several times, complaining of a vibration. The technician test drove and inspected the vehicle each time and found nothing wrong. Finally, he had the customer drive the vehicle while he rode along. He was quite surprised when she put it in gear, pushed down the throttle pedal, and spun the tires a good distance, making the wheels hop and creating the vibration she was describing. The technician took her back to the sales department and had her exchange the high-performance version of the vehicle for a less powerful version that fit her needs better. Don't forget—our job is to diagnose almost any problem and find appropriate solutions, and as long as we have all the data available, this can be accomplished.

Too much crankshaft end play from a severely worn crankshaft thrust bearing may also create clutch chatter. As the engine is running, the crankshaft can move back and forth, and a shuttering may be felt. Verification of this may require the removal of the engine oil pan to inspect the thrust bearing. Clutch chatter can also be caused by binding clutch linkage, worn or broken motor mounts, misalignment of the bell housing, a bent or warped friction disc, the friction disc hub binding on the input shaft, or loose friction disc facings. A thorough inspection of each component is required to determine the cause of the chatter.

▶ TECHNICIAN TIP

As with all test-drives, obey all speed laws, and treat the customer's car better than your own.

▶ Removing a Transmission/Transaxle

S31002, S31003

Removal of the transmission is usually required only when the clutch is not operating correctly; the flywheel ring gear is damaged; there is a major concern with the transmission; or access to the rear of the engine is needed for leaks involving core plugs, oil gallery plugs, or the crankshaft rear main seal. This job can be time consuming and requires the ability to remember how components and parts go back together again. Depending on the vehicle, it can be difficult to gain access to the fasteners and other components necessary to remove and reinstall the transmission. Because the transmission is heavy and awkward, use a transmission jack with a solid support system during removal and installation. Also, many of the parts related to the transmission, clutch, and flywheel have very sharp edges, so wear leather or nylon gloves, and be careful when grabbing onto parts.

▶ TECHNICIAN TIP

Core plugs seal the holes in the block as part of the casting process and are located around the outside of the cylinder block and head(s), including behind the flywheel. Core plugs can rust from either side. They may look good on the outside, but they may be almost rusted through from the inside. A variety of core plugs are available if replacement is required. Be aware of hot coolant that may be leaking from the core plugs if the engine has recently been running.

To remove and reinstall a transmission/transaxle, follow the steps in **SKILL DRILL 31-6**.

SKILL DRILL 31-6 Removing and Reinstalling a Transmission/Transaxle

1. Research the procedures and specifications. Ensure that bolts, clips, and fasteners are kept in containers. Make sure the vehicle is secure on the lift. If necessary, drain the transmission fluid to avoid spills.

2. Disconnect the driveshaft or axles and secure them from hanging. Disconnect all wires, tubes, and hoses, and inspect for damage. Disconnect the clutch and shifter linkage, and inspect for wear or damage. Secure the transmission with a transmission jack, and remove any transmission mounts.

3. Following the specified procedure, remove the transmission/transaxle from the vehicle. At this time, you may want to overhaul the transmission or perform other tasks such as replacing the clutch assembly or rear crankshaft main seal.

4. Prepare the transmission to be reinstalled. Ensure that the release (throw-out) bearing and clutch fork are properly installed, and the clutch disc is centered in the pressure plate. If specified, lubricate the pilot bushing/bearing.

5. Lightly lubricate input shaft splines to ensure that there is no binding upon entering the clutch disc hub. Position all wires, hoses, and tubes out of the way and in their specified positions. Secure them temporarily if necessary.

6. Following the specified procedure, reinstall the transmission/transaxle. Tighten all fasteners to the proper torque. If the vehicle is equipped with a cable- or linkage-style clutch release system, reinstall and adjust it for the proper free play.

7. If the vehicle is equipped with a hydraulic clutch, reinstall the slave cylinder and bleed and/or adjust if necessary.

8. Refill the transmission/transaxle to the proper level with the specified fluid.

9. Check that all electrical wires, connectors, linkages, and other removed components are properly installed and adjusted. Place exhaust hose(s) over the exhaust pipe(s), and set the parking brake.

Inspecting the Flywheel and Ring Gear

N31004

Anytime the clutch assembly has been removed, it is a good idea to conduct a thorough visual inspection of the flywheel and ring gear. The flywheel should be checked for bluing, hot spots, and cracks. These conditions can result from possible driver abuse or maladjustment of the clutch, causing the clutch disc to overheat. Also inspect the flywheel friction surface for cupping with a straightedge and feeler blades. Cupping can happen because the pressure plate exerts a pulling effort on the outside diameter of the flywheel, which then allows the outer portion to cup over time; in addition, the outer edge of the friction surface wears more quickly than the inner edge. A straightedge can help you identify this situation quickly.

Some flywheels are designed with a step, meaning that the friction surface is lower than the outer surface against which the pressure plate is bolted. The depth of the step is critical to the operation of the pressure plate. If it is too deep, the pressure plate will not develop enough clamping force to keep the clutch from slipping. If it is not deep enough, the clutch might not be able to be released. In any case, the depth of the step must be within the manufacturer's specifications. This becomes more important to check if the flywheel has been resurfaced. The machinist should have made sure the depth is correct, but you need to double-check it.

In regard to resurfacing flywheels, most vehicle manufacturers as well as clutch suppliers suggest that the flywheel be resurfaced whenever the clutch is replaced. This gives a new, flat, and consistent friction surface that helps prevent clutch chatter. It is good practice to check flywheel runout once a resurfaced flywheel is reinstalled on the crankshaft, so you can identify any machining errors.

Ring gears must be inspected for missing teeth and excessive wear on the teeth. If any of these conditions are present, the ring gear should be replaced. Many ring gears are press fit onto the flywheel; others are a permanent part of the flywheel. If a ring gear is press fit onto the flywheel, it can be removed by heating it up carefully with an acetylene torch to a few hundred degrees Fahrenheit and using a hammer and punch to quickly drive it off the flywheel. To reinstall it, a good method is to cool the flywheel in a freezer for 30 minutes and to place the ring gear in an oven at about 300°F. Then, using welding gloves, quickly take them out, set the flywheel on a solid surface and slide the ring gear over the flywheel. Use a hammer and flat punch to seat the ring gear against the flange on the flywheel.

To inspect the flywheel and ring gear (with the transmission removed), follow the steps in **SKILL DRILL 31-7**.

SKILL DRILL 31-7 Inspecting the Flywheel and Ring Gear

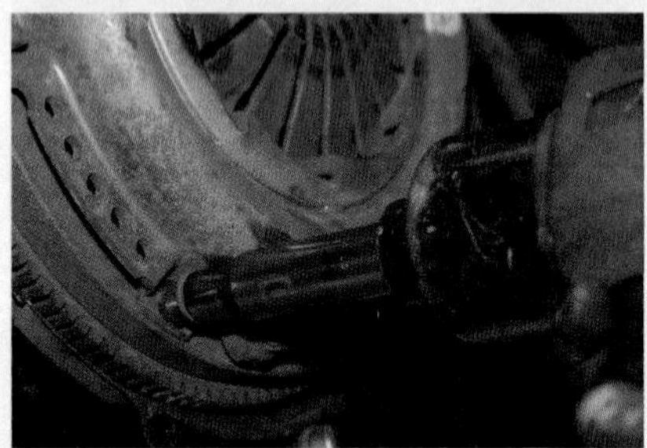

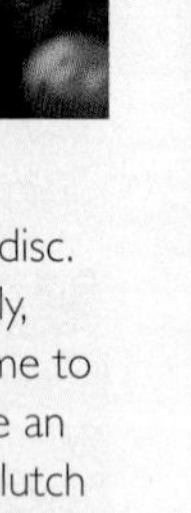

1. Follow the specified procedure to remove the pressure plate and clutch disc. Remove the pressure plate bolts evenly, backing each bolt out one turn at a time to avoid warping the pressure plate. Have an assistant hold the pressure plate and clutch disc as they are being removed.

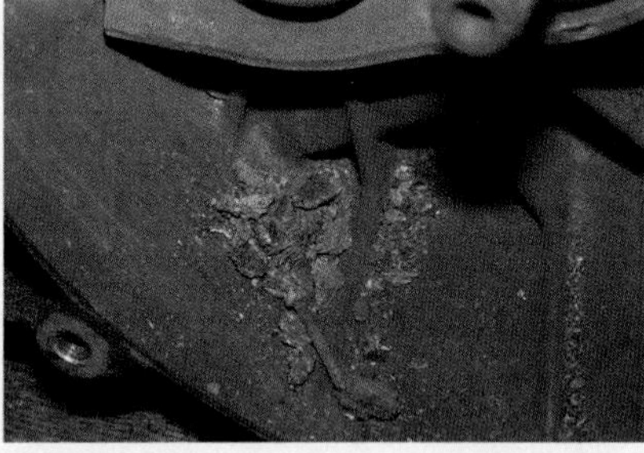

2. Clean up any clutch dust and debris, using an approved method for disposing of hazardous dust.

3. Inspect the flywheel for wear, hot spots, bluing, and cracks. Use of a straightedge can give a preliminary check for flatness.

4. Inspect the ring gear for wear, chipped teeth, and cracks. If teeth are worn in one area, it may also be necessary to replace the starter drive.

5. Check that the ring gear is secure on the flywheel.

6. Inspect the starter drive, as it may have been damaged from a faulty ring gear.

Measuring the Flywheel Runout and Crankshaft End Play

N31005

The flywheel must rotate true. This means that the surface of the flywheel must not wobble, which would indicate excessive runout. It is always a good idea to check and measure the flywheel runout in the case of catastrophic clutch failure or routine clutch replacement any time the transmission is removed. Flywheel runout can be measured with a dial indicator, which is held in position against the flywheel friction surface. When the flywheel is rotated, the runout should be within specifications, near zero. If the runout is excessive, you will need to use the dial indicator to measure the crankshaft runout to see if that is causing the flywheel runout.

Crankshaft end play is the amount of forward and rearward movement of the crankshaft and must be within the specifications for the vehicle you are working on. Crankshaft end play must be checked if any irregular clutch noises have been detected. Too much crankshaft end play could result in an engine's crankshaft thrust bearing becoming worn past its specified tolerance. Crankshaft end play can also result in clutch chatter when the clutch pedal is being released.

Use a dial indicator to measure crankshaft end play at the end of the crankshaft. If the engine oil pan has been removed, you also can measure the thrust bearing clearance with a feeler gauge. Also check runout of the pilot bearing cavity, as this could indicate that the pilot bearing may not be held in position as is required for proper operation; if so, damage to the clutch assembly may result.

To measure flywheel runout and crankshaft end play, follow the steps in **SKILL DRILL 31-8**.

SKILL DRILL 31-8 Measuring Flywheel Runout and Crankshaft End Play

1. Following the specified procedure, measure flywheel runout. Mount the dial indicator base on the engine block, and place the dial on the flywheel friction surface.

2. Rotate the crankshaft using a flywheel wrench or a socket and breaker bar. Find the high spot on the flywheel, and zero the dial indicator. Continue rotating the crankshaft until the dial indicator reads the greatest amount of runout. Record your reading, and compare with specifications.

3. If flywheel runout is excessive, measure the runout on the crankshaft flange with a dial indicator. To get an accurate reading, make sure that you keep the crankshaft pushed forward against the thrust bearing while rotating it.

4. Crankshaft end play should also be checked with a dial indicator to make sure it is not excessive; if it is, it should be addressed before replacing the clutch. Mount the dial indicator on the engine block surface, and set the dial on the rear face of the crankshaft flange.

5. Carefully pry the crankshaft back and forth, using a pry bar to pry on a bolt threaded into one of the crankshaft bolt holes. Record your reading, and compare to manufacturer specifications.

Inspecting the Engine Block

S31004

While the flywheel is removed, it is a good practice to inspect the rear of the engine block for any damage to the mating surfaces and alignment dowels. This should include inspecting the threads of all bolt holes to make sure they are not stripped. Also check for leaking coolant and/or engine oil, and check the integrity of the soft plugs. Now is the time to address any issues in this area, as it requires substantial disassembly time to access it if you have to go back in later to fix something.

▶ TECHNICIAN TIP

The alignment dowels ensure that the transmission and engine line up correctly. They can be easily damaged when removing or installing the transmission, so inspect them for damage whenever the transmission has been removed.

Many newer engines use a full-circle rear main seal around the crankshaft flange at the rear of the block. This can be replaced only by removing the transmission and flywheel. If the clutch and flywheel have been removed, now would be an ideal time to inspect and change the full-circle rear main seal.

To inspect the engine block, follow the steps in **SKILL DRILL 31-9**.

Installing a Clutch Assembly

N31006

Whether you are replacing a clutch because of a failure or removing the clutch and reinstalling it after replacing an engine, thorough inspection of all components is critical to ensure that the clutch will operate properly and last a long time. Installing the clutch assembly requires a few precautions. First, the clutch disc is directional, meaning that it must be installed with the correct face toward the flywheel. New clutch discs usually are marked with a sticker or paint indicating the "flywheel side" or "engine side." This marking cannot usually be read on used discs, so always mark the clutch disc when you remove it if you will possibly be reinstalling it. Second, alignment of the clutch disc is required prior to bolting down the pressure plate so that the input shaft can align with the clutch disc hub and pilot bearing without binding. Use of a pilot shaft or clutch

SKILL DRILL 31-9 Inspecting the Engine Block

1. Mark the flywheel-to-crankshaft position. Remove the flywheel to access the rear of the engine.

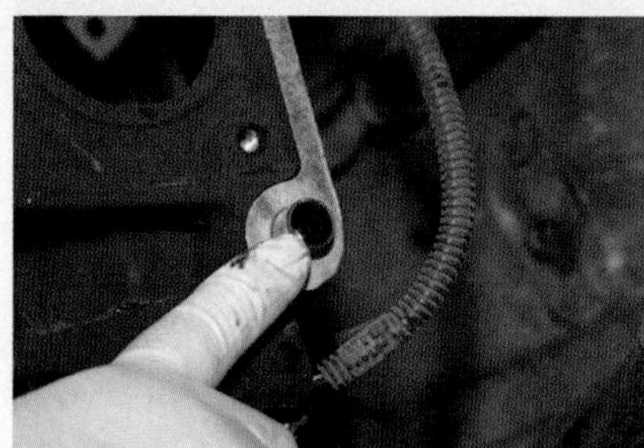

2. Inspect the engine block for obvious defects.

3. Inspect the core plugs. Test their integrity by tapping on them lightly with a punch. Replace the core plugs if they are leaking or show signs of deterioration.

4. Check the rear crankshaft oil seal for seepage, as this could leak onto the clutch assembly.

5. Inspect the clutch (bell) housing for elongated holes or broken ears due to loose bolts.

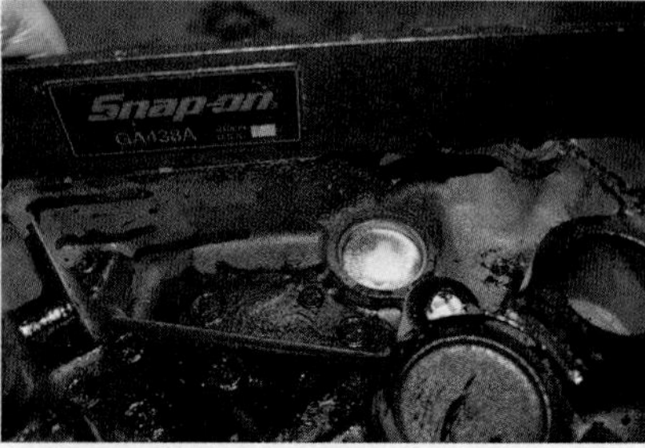

6. Inspect the transmission/transaxle case mating surfaces for flatness. Inspect the alignment dowels for proper fit, sizing, and wear. Check the engine block for stripped bolt holes.

alignment tool helps ensure proper clutch disc alignment. With the clutch disc inserted, under *no circumstances* should you remove the pilot shaft until after the clutch pressure plate has been installed and torqued to specification. If it is removed prematurely, misalignment of the clutch disc will occur, and installation of the transmission will be extremely difficult or impossible.

Upon installation of the pressure plate, insert bolts and tighten about a turn at a time in a multistep, crisscross pattern, which will allow the pressure plate cover to be seated slowly and without distortion. Double-check all torque values to ensure proper installation.

To install a clutch assembly, follow the steps in **SKILL DRILL 31-10.**

SKILL DRILL 31-10 Installing a Clutch Assembly

1. Research the procedure. Inspect the pressure plate assembly for worn springs or diaphragm and also for hot spots. Inspect the clutch disc for loose dampening springs or rubber blocks, and check the condition of the rivets.

2. Check the release (throw-out) bearing and linkage for damage or binding. Check the pilot bushing/bearing and its cavity for damage. Replace any worn parts with new or remanufactured components.

3. Following the specified procedure, install the pilot bushing/bearing, if removed. Lubricate it with the specified lubricant, if applicable.

4. Following the specified procedure, install the flywheel. Be sure to torque the flywheel bolts in the proper sequence.

5. Following the specified procedure, install the clutch disc and pressure plate. Use an appropriate clutch alignment tool to ensure proper positioning of the clutch disc.

6. Tighten the pressure plate bolts down evenly by tightening each bolt, a turn at a time, until the pressure plate is evenly seated on the flywheel. Tighten each pressure plate bolt to the specified torque and sequence. Install the release (throw-out) bearing and linkage in the bell housing.

▶ Wrap-Up

Ready for Review

- The clutch is designed to engage and disengage the engine from the vehicle's transmission.
- The driver controls the shifting of gears via the clutch while the vehicle is in operation.
- Dry (automotive) clutches rely on friction to transmit torque from the engine to the transmission.
- There are four variables affecting the clutch torque transmission: amount of friction between the clutch disc and mating surfaces of the flywheel and pressure plate;

diameter of the clutch; number of clutch discs in the clutch assembly; and total spring force clamping the parts together.

- Speed and output shaft torque have an inverse relationship: when one is increased, the other is decreased.
- The clutch disengages the input shaft from the engine flywheel in order to switch gears.
- The output shaft cannot be disengaged because it is directly connected to the wheels.
- The clutch assembly is composed of flywheel, clutch disc, pressure plate, throw-out bearing, clutch fork, and pilot bearing.
- Flywheels are designed to provide a friction surface for the clutch disc and dampen the power pulses from pistons during power strokes.
- Types of flywheels are single one-piece flywheel (light-duty vehicles) and dual mass flywheel (diesel trucks and luxury vehicles).
- A dual mass flywheel can be made of a primary and secondary flywheel with a series of torsion springs and cushions or can use a planetary gear and torsional springs.
- The pressure plate uses springs to clamp the clutch disc between the pressure plate and flywheel, allowing it to transmit torque.
- Pressure plates use diaphragm spring pressure (light-duty vehicles) or coil spring pressure.
- A clutch disc (also known as a driven center plate or friction disc) allows engine torque to transmit from the flywheel and pressure plate to the transmission input shaft by providing necessary friction material.
- The clutch throw-out bearing and clutch fork compress the pressure plate springs.
- Pilot bearings can be brass or bronze bushing, needle type, or roller type, and provide alignment support for the input shaft.
- Clutch operating mechanism styles can be linkage, cable, or hydraulic.
- Cable-operated clutch control systems must be adjusted to ensure adequate free play between the throw-out bearing and pressure plate levers.
- Hydraulic clutch control systems rely on a slave cylinder to operate the clutch fork.
- Clutches can be single plate (light-duty vehicles) or multi-plate (heavy-duty vehicles).
- Common clutch-related problems are noises, pedal pulsations, slippage, clutch binding or drag, clutch vibration, and clutch chatter.
- Clutch vibrations may be torsional or driveline.
- Proper clutch diagnosis will likely require a test-drive.
- Take proper safety precautions during clutch maintenance and repair; older clutch disc linings may contain asbestos.
- Tools needed to maintain and repair a clutch system include: transmission jack, dial indicator, clutch alignment tool, flywheel wrench, and clutch was station.
- Perform regular preventive clutch maintenance per the manufacturer's guidelines.
- Check and adjust the clutch linkage mechanism (mechanical or hydraulic) during routine maintenance service.
- Hydraulic clutch systems may need bleeding if the system fails; when a component is replaced; or if the hydraulic fluid becomes unfit for use.
- Hydraulic clutch systems may be bled via gravity, manually, or via the pressure/vacuum method.
- Removing a transmission and transaxle is difficult; remember how components go back together again.
- Inspect the flywheel, ring gear, and engine block whenever the clutch assembly has been removed.
- Measure the flywheel run out and crankshaft end play anytime the transmission is removed.
- Be sure to properly align the clutch disc when installing a clutch assembly.

Key Terms

carrier The part of the throw-out bearing assembly that holds the bearing.

clutch binding A condition in which the clutch disc is dragging, leading to grinding gears during gear shifts and possibly clutch chatter.

clutch chatter A condition in which the clutch shudders when the clutch pedal is released and the vehicle starts to move forward.

clutch disc The center component of the clutch assembly, with friction material riveted on each side. Also called a clutch plate or friction disc.

clutch fork The part of the clutch linkage that operates the throw-out bearing.

coil spring pressure plate A type of pressure plate that uses coil springs to provide the clamping force.

crankshaft end play The amount of forward and rearward movement of the crankshaft in the main bearings. Crankshaft end play is controlled by the engine main bearing thrust bearing.

diaphragm pressure plate A slightly conical, spring steel plate used to provide the clamping force for the clutch assembly.

driveline vibrations Rotational fluctuations caused by out-of-balance, misaligned, worn, or bent driveline components.

driven center plate The friction disc that is held firmly against the flywheel by a pressure plate and that transfers power from the flywheel to the transmission input shaft.

flywheel A heavy metal disc bolted to the crankshaft that is used to smooth out the engine's power pulses and keep the engine moving through the non-power strokes. Also provides the mating surface for the clutch disc and pressure plate.

flywheel ring gear Large, round, externally toothed gear that is usually press fit to the outer diameter of the flywheel and used along with the starter to crank the engine over.

friction facing The material riveted to each side of the clutch disc that mates to the flywheel and pressure plate. Used to provide friction and a wear surface for the clutch assembly.

front bearing retainer The housing that bolts the input shaft bearing in place on the front of the transmission.

fulcrum ring A steel ring that is used as a pivot point for the diaphragm spring in the pressure plate.

multi-plate clutch A clutch assembly that is comprised of two or more clutch plates and used to increase the torque-carrying capacity of the clutch.

pilot bearing The bearing or bushing that supports the front of the transmission input shaft.

pressure plate The assembly that applies and removes the clamping force on the clutch disc.

push-type clutch A typical clutch system, used in modern vehicles, where the clutch fork pushes the release bearing forward to release the friction facing from the pressure plate.

quadrant ratchet The device used in some cable-operated clutches to provide self-adjustment as the clutch disc wears. Some quadrant ratchets adjust if you lift up on the clutch pedal.

release mechanisms Components that operate the clutch. Usually included are the throw-out bearing and the clutch fork. Some manufacturers include the operating system.

slave cylinder The component in a hydraulically operated clutch that converts hydraulic pressure to mechanical movement at the clutch fork.

slippage A condition in which two surfaces in firm contact with each other slide.

throw-out bearing The part of the clutch release mechanism that imparts clutch pedal force to the rotating pressure plate levers.

thrust-type angular-contact ball bearing A type of bearing that uses a deep groove in the bearing races where the ball bearings ride; this design is for thrust conditions.

torsional vibrations The speeding up and slowing down of a shaft, which happen at a relatively high frequency. Crankshafts have torsional vibrations due to the power pulses of the pistons.

transmission input shaft The shaft that brings engine torque into the transmission.

Review Questions

1. All of the following statements are true with respect to clutch principles *except*:
   a. increasing the friction of the clutch disc decreases the torque-carrying ability.
   b. increasing the diameter of the clutch increases its torque capacity.
   c. increasing the number of clutch discs increases torque capacity.
   d. increasing the spring force clamping the parts together increases torque capacity.
2. Which of these parts compress the pressure plate springs?
   a. The clutch fork and the pilot bearing
   b. The clutch throw-out bearing and the pilot bearing
   c. The clutch throw-out bearing and clutch fork
   d. The clutch fork and the clutch disc
3. Which of these components smoothens out the power pulses from the pistons during the power strokes?
   a. Pressure plate spring
   b. Flywheel
   c. Clutch fork
   d. Throw-out bearing
4. The clamping force to clamp the clutch disc, allowing it to transmit torque, is provided by the:
   a. flywheel.
   b. clutch pedal.
   c. pressure plate.
   d. clutch fork.
5. A clutch disc allows engine torque to transmit from the flywheel and pressure plate to the:
   a. transmission input shaft.
   b. front bearing retainer.
   c. axles.
   d. pilot bearing.
6. All of the following statements are true *except*:
   a. the throw-out bearing and clutch fork work together.
   b. the throw-out bearing rotates with the pressure plate.
   c. the clutch fork rotates with the pressure plate.
   d. the throw-out bearing includes a thrust bearing in its assembly.
7. All of the following are types of clutch operating systems *except*:
   a. hydraulic style.
   b. cable style.
   c. linkage style.
   d. clutch fork.
8. In some cable-operated clutch vehicles, which device provides self-adjustment as the clutch disc wears?
   a. Single-plate clutch
   b. Clutch fork
   c. Quadrant ratchet
   d. Bell crank
9. What is clutch pedal free-play?
   a. The distance the clutch pedal moves the master cylinder piston.
   b. The distance the clutch pedal moves before the master cylinder piston moves.
   c. The distance from the floor where the clutch engages.
   d. The distance the clutch pedal moves for the clutch to be fully disengaged.
10. Which of these tools is used to center the clutch plate between the flywheel and the pressure plate during the installation of the pressure plate?
    a. Flywheel wrench
    b. Dial indicator
    c. Clutch alignment tool
    d. Transmission jack

ASE Technician A/Technician B Style Questions

1. Tech A says that the pressure plate friction surface rides on the flywheel friction surface to transmit torque. Tech B says that the flywheel can either be flat or stepped. Who is correct?
   a. Tech A
   b. Tech B
   c. Both A and B
   d. Neither A nor B

2. Tech A says that insufficient clutch pedal clearance (free play) can cause gear clashing when shifting. Tech B says that when the engine is idling and the clutch pedal is released, the friction disc should stop rotating. Who is correct?
   a. Tech A
   b. Tech B
   c. Both A and B
   d. Neither A nor B
3. Tech A says that hot spots on the flywheel are a result of excessive heat. Tech B says that a pulsation in a clutch pedal could be due to uneven clutch pressure plate levers. Who is correct?
   a. Tech A
   b. Tech B
   c. Both A and B
   d. Neither A nor B
4. Tech A says that a leaking rear main seal can cause clutch damage. Tech B says that the throw-out bearing rides directly on the clutch disc. Who is correct?
   a. Tech A
   b. Tech B
   c. Both A and B
   d. Neither A nor B
5. Tech A says that the flywheel runout can be checked with a dial indicator. Tech B says that when checking flywheel runout, it is good practice to also check crankshaft end play. Who is correct?
   a. Tech A
   b. Tech B
   c. Both A and B
   d. Neither A nor B
6. Tech A says that clutch slippage can be a result of excessively strong pressure plate spring(s). Tech B says that when replacing the friction disc, it is good practice to also replace the pressure plate. Who is correct?
   a. Tech A
   b. Tech B
   c. Both A and B
   d. Neither A nor B
7. Tech A says that the pilot bearing can be a needle-style bearing. Tech B says that the pilot bearing can be a brass bushing style. Who is correct?
   a. Tech A
   b. Tech B
   c. Both A and B
   d. Neither A nor B
8. Tech A says that a bad pilot bearing can cause a whirring noise when the clutch pedal is released. Tech B says that many ring gears are press fit onto the flywheel. Who is correct?
   a. Tech A
   b. Tech B
   c. Both A and B
   d. Neither A nor B
9. Tech A says that waved springs between the clutch facings results in a smoother engagement of the clutch when starting from a stop. Tech B says that the heavy torsional coil springs in the clutch disc clamp the clutch disc between the flywheel and pressure plate. Who is correct?
   a. Tech A
   b. Tech B
   c. Both A and B
   d. Neither A nor B
10. Two technicians are discussing clutch operating systems. Tech A says that most hydraulic clutches need to be adjusted periodically. Tech B says that cable clutches use a slave cylinder to operate the clutch fork. Who is correct?
   a. Tech A
   b. Tech B
   c. Both A and B
   d. Neither A nor B

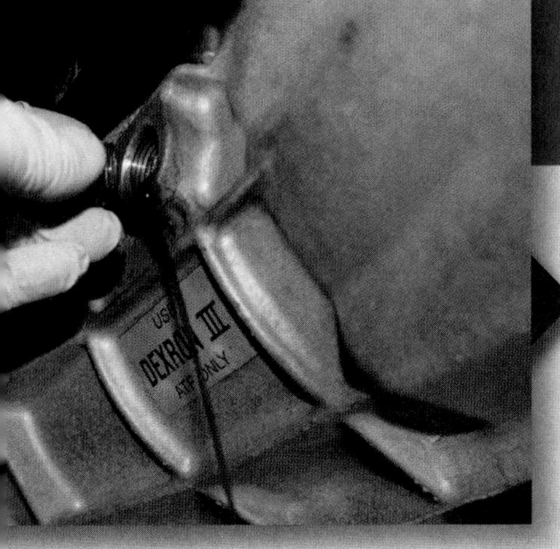

CHAPTER 32

Manual Transmissions/ Transaxles Basic Diagnosis and Maintenance

NATEF Tasks

- **N32001** Describe the operational characteristics of an electronically controlled manual transmission/transaxle. (MLR/AST/MAST)
- **N32002** Check fluid condition; check for leaks; determine needed action. (MLR/AST/MAST)
- **N32003** Drain and refill manual transmission/transaxle and final drive unit; use proper fluid type per manufacturer specification. (MLR/AST/MAST)
- **N32004** Diagnose noise concerns through the application of transmission/transaxle power flow principles. (MAST)
- **N32005** Diagnose hard shifting and jumping out of gear concerns; determine needed action. (MAST)
- **N32006** Diagnose transaxle final drive assembly noise and vibration concerns; determine needed action. (MAST)

Knowledge Objectives

After reading this chapter, you will be able to:

- **K32001** Describe the geartrain and the way it is shifted.
- **K32002** Describe gear terminology and calculate gear ratios.
- **K32003** Describe gear selection and mechanisms.
- **K32004** Describe the purpose and function of bearings, thrust washers, gaskets, and seals.
- **K32005** Describe the purpose and function of bearings and thrust washers.
- **K32006** Describe the purpose and function of seals and gaskets.
- **K32007** Describe the differences in design between a transmission and a transaxle.
- **K32008** Describe the internal operation of transmissions/transaxles.
- **K32009** Describe the internal operation of a transmission.
- **K32010** Describe the internal operation of a transaxle.

Skills Objectives

After reading this chapter, you will be able to:

- **S32001** Perform preventive maintenance on manual transmissions/transaxles.
- **S32002** Diagnose common concerns on transmissions/transaxles.
- **S32003** Diagnose drivetrain concerns.
- **S32004** Diagnose electronically controlled transmission/transaxle.

▶ Introduction

Manual transmissions have been around since the invention of the automobile. The ability of the technician to drive, service, repair, and maintain the manual transmission/transaxle is a critical skill. The manual transmission/transaxle is not the most common transmission used in automobiles but there are enough used to create a need for technicians that can service these devices. Transmissions have made improvements in gear ratios, weight reduction, and fuel saving features such as closer gear ratios and more gear selections (**FIGURE 32-1**). Proper diagnosis of the manual transmission/transaxle will be a required skill of the service technician.

Note, when we say *transmission,* we use the term generically to include transaxles also. When we use transaxle, we generally only mean transaxle. Readers should use their understanding of transmission and transaxle theory to determine the intended meaning.

▶ Manual Transmission Geartrain

K32001

Gear Ratios

K32002

In light vehicle applications, the engine operates over a wide range of speeds, or revolutions per minute (rpm), but it produces maximum torque only within a relatively narrow rpm range (approximately 1200 rpm to 3000 rpm). A relatively large twisting force must be applied to the wheels to accelerate the vehicle from a stop. Once the vehicle is up to speed, the required twisting force is generally much less, although this force must vary to overcome wind resistance along with **gradient resistance** while going up hills and **rolling resistance** while driving down the road. Gradient resistance is the resistance caused by inclines, and rolling resistance is the resistance caused by the tires' friction on the road surface. Rolling resistance increases due to front-end alignment defects, road crown, and driving aggressiveness of the driver. Because of the rpm and torque limitations of the engine, a multi-gear transmission is used to provide adequate acceleration at slow speeds while not over-revving the engine at cruising speeds. Manual transmissions typically have three to six forward gears along with a reverse gear and a neutral position, which disconnects the engine from the driving wheels, allowing the engine to idle at a stoplight.

FIGURE 32-1 The internal workings of a manual transmission.

When two gears are in mesh, one is a driven (output) gear. The other, providing the turning action, is the drive (input) gear. If the gears have the same number of teeth, they will create a one-to-one **gear ratio** (**FIGURE 32-2**). In **one-to-one** rotation, both the output and the input gears turn at the same speed, and the turning effort of the output gear equals the effort applied by the input gear.

Calculating the Gear Ratio

Gear ratios of different gear sizes can be calculated by dividing the number of teeth on the driven gear by the number of

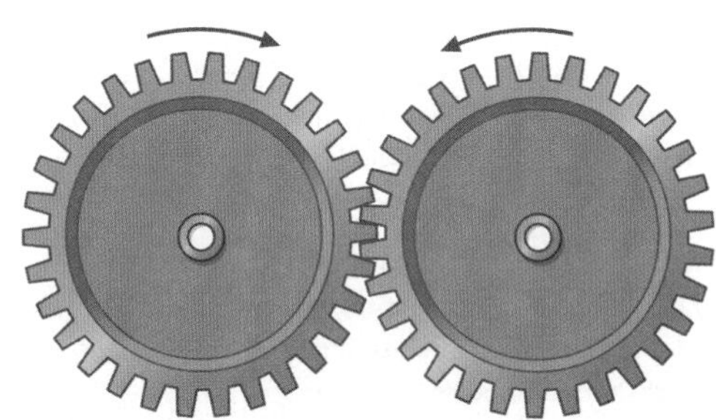

FIGURE 32-2 Gears with the same number of teeth rotate at the same rate of speed, resulting in direct drive.

You Are the Automotive Technician

A customer brings her 2013 Acura RDX 6-speed into the shop for a manual transmission inspection. She informs you that when she is shifting into 3rd gear she sometimes hears a loud grinding noise. You carefully document the issue, review her vehicle service history, check for technical service bulletins, and then take her vehicle for a test-drive. You notice that 3rd gear does grind, mostly on downshifts, but also a bit on upshifts. There are no other unusual gear or bearing noises. You bring the vehicle into the shop and inspect the lubricant level and condition. Both are OK. You also check the shift linkage for worn bushings or pivots. All are in good shape. You notify the customer that the blocker ring on the 3rd gear synchronizer is worn and needs to be replaced. You also recommend that the transmission should be rebuilt at the same time, since the entire transmission has to be disassembled to get to the blocker ring. She has several questions for you:

1. What is a blocker ring and what does it do?
2. How do you know it isn't worn gears that are making the grinding noise?
3. Why doesn't it grind when shifting into other gears?

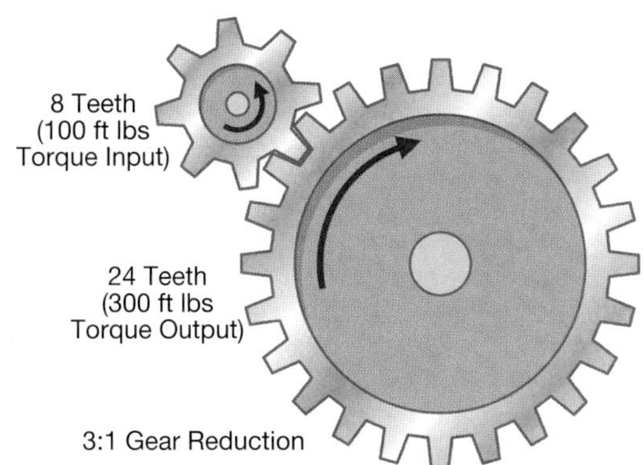

FIGURE 32-3 Gear ratios are determined by the size difference of each gear or the number of teeth on each gear.

teeth on the drive gear (**FIGURE 32-3**). For example, if the driven gear has 30 teeth and the drive gear has 10 teeth, the gear ratio is 30:10, which reduces to 3:1. In this case, the drive gear has to turn three times to turn the driven gear once. In continuous rotation, the driven gear turns three times slower than the drive gear.

Driving a large gear with a smaller gear results in a **gear reduction**, just as a bicycle with a large gear in the back and a small gear in the front gives you a very low gear ratio. If we think of the radius of the gear as a lever, then a large gear being turned by a small gear results in more leverage and therefore more torque. We turn the input gear quickly but the output gear speed is very low, resulting in higher torque. For example, with a gear ratio of 3:1, if input torque is 100 ft-lb (135.6 Nm, then output torque is three times that, or 300 ft-lb (406.7 Nm). Different gear ratios are used inside the transmission to achieve varied torque to accelerate the vehicle. The increase of torque is referred to as **torque multiplication**.

In reverse, three gears are in mesh and the input and output gears are meshed with an idler or intermediate gear commonly used in reverse. When two gears are in mesh, the output gear turns in the opposite direction of the input gear. When an idler gear is used between the input and output gears, the output gear turns in the same direction as the input gear. This means the output gear in a three-gear arrangement will rotate in the reverse direction as compared to the output gear on a two-gear arrangement. The **idler gear** simply transfers motion and does not have an effect on the gear ratio of the input to output gears, which is why it is called an idler gear (**FIGURE 32-4**).

The terms input gear and drive gear have the same meaning. The same goes for output gear and driven gear. For the sake of simplicity, we will generally refer to the gears as input and output. Note that other textbooks or service information may use the alternate terms.

▶ TECHNICIAN TIP

Gear ratio theory will help you to understand how power is distributed and how power and speed vary as the vehicle travels.

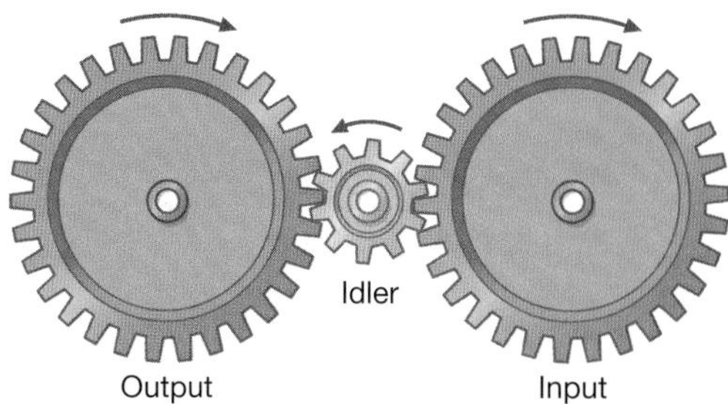

FIGURE 32-4 The idler gear is a gear used to change the direction of the rotation of shafts and is used to provide reverse.

Applied Math

AM-35: Standards: The technician can demonstrate conformance to standards defined by the industry and/or manufacturer for the system being analyzed.

A vehicle requires a remanufactured transmission. The owner of the vehicle has indicated to the automotive repair shop service advisor that he would like to inspect the transmission before it is installed. As the owner compares the original five-speed transmission with the remanufactured unit he has some concerns. The part numbers on the housing are not exactly the same and the customer has some doubts about the remanufactured unit being correct.

Using service information for the customer's vehicle, the technician asks the customer to verify the correct gear ratios for the remanufactured transmission. With the transmission on a workbench, the technician places the shifter into first gear. The service information states that the gear ratio should be 3.5:1 for first gear. This means that the engine would rotate 3.5 times as compared to the output shaft of the transmission rotates 1 time. Using a marker, the technician places reference marks on the input and output shafts of the transmission. The technician rotates the input shaft exactly three and one half turns as the customer observes the output shaft turning exactly one turn. This action demonstrates conformance to the standards of the manufacturer's service information. The same procedure was conducted to indicate a ratio of 2:1 for second gear, 1.4:1 for third gear, 1:1 for fourth gear, and 0.76:1 for fifth gear. After verifying that all gear ratios were correct, the owner was convinced that the remanufactured transmission was correct for his vehicle.

Compound Geartrains

In keeping with the gear ratio theory, compound geartrains have two or more pairs of gears in constant mesh so that they rotate together (**FIGURE 32-5**). They are found inside the transmission. Compound geartrains enable a variety of gear ratios, also accounting for torque output and speed requirements.

On a simple compound geartrain, the gear on the input shaft meshes with a larger gear on a countershaft or cluster gear, which is usually the bottom shaft in the manual transmission (**FIGURE 32-6**). The countershaft has a smaller gear on it, in mesh with the output shaft gear. The rotation of the input shaft

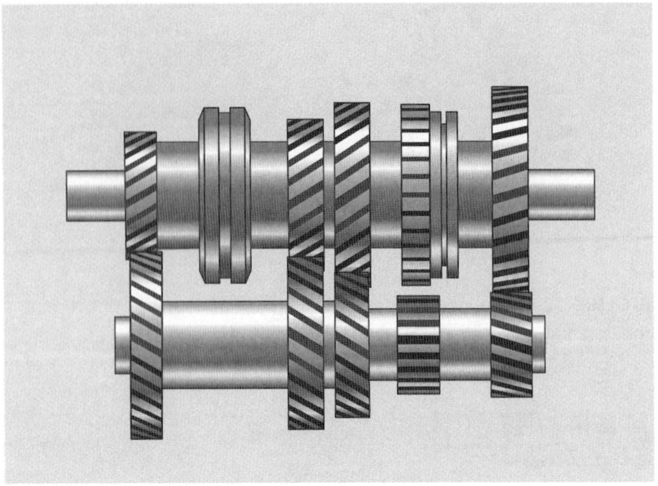

FIGURE 32-5 Compound geartrains have two or more gear sets in constant mesh at all times.

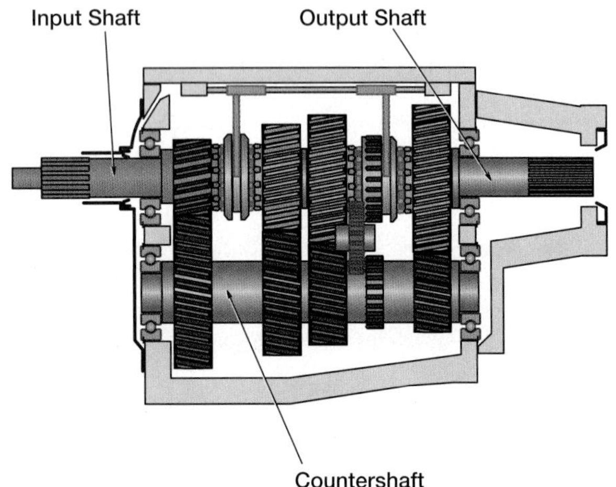

FIGURE 32-6 Simple compound geartrain. **A.** Input shaft. **B.** Countershaft. **C.** Output shaft.

is transferred through the large gear on the countershaft, then along the countershaft to the smaller gear, and then to the output shaft gear. The output gear turns in the same direction as the input gear, but at a reduced ratio depending on the relative sizes and ratios of the gears. Because two pairs of gears are involved, their ratios are compounded, or multiplied together. This is how the overall geartrain ratio is calculated.

The input shaft gear, with 12 teeth, drives its mating gear on the countershaft, which has 24 teeth. This creates a gear ratio of 2:1. This ratio (2:1) is then multiplied by the output ratio of the countershaft to the output shaft, which has a ratio of 3:1. Multiplying these two calculations together results in a gear ratio of 2:1 times 3:1 = 6:1 between the input and the output of the compound gear, resulting in a speed reduction and a corresponding increase in torque.

▶ TECHNICIAN TIP

Gear reductions in the lower gears of manual transmissions can be provided by compound geartrains. The typical ratios are first gear = 4.41:1, second gear = 2.63:1, third gear = 1.61:1, fourth gear = 1:1, and fifth gear = 0.87:1.

FIGURE 32-7 This type of gearshift lever is mounted in the top of the transmission and is sometimes called a top loader.

Fourth gear is normally a gear ratio of 1:1, or direct drive. The input and output shafts are mechanically locked together. There is no torque multiplication, but there is an increase of vehicle speed from third gear. A fifth gear is normally an overdrive ratio, typically with a value of about 0.87:1. The output shaft of the transmission turns faster than the input, but the output torque is reduced. Now output speed is again increased over fourth gear, and fuel mileage is usually increased, as the engine is turning at a slower speed. Another way to describe overdrive is to say the output speed of the transmission is faster than its input speed. Many vehicles employ a sixth gear, which can be an additional overdrive gear to reduce the engine speed even further, increasing fuel mileage. A sixth gear may have an overdrive ratio of approximately 0.75:1.

A further gear reduction is always provided by the final drive assembly. The final drive ratio has to be included when calculating overall gear reduction. It can be calculated by using the preceding compound gear ratio math to find the overall gear ratio. The overall gear ratio is the gearbox ratio multiplied by the final drive ratio. For example, a gearbox ratio of 3:1 with a final drive ratio of 4:1 results in an overall ratio of 12:1. This means 12 revolutions of the crankshaft result in one turn of the road wheels. Assuming 100% efficiency, the torque applied is 12 times the engine torque.

Gear Selection

K32003

A **gearshift lever** allows the driver to manually select gears via a gearshift mechanism. The gearshift lever can be mounted separate from the gearbox on the floor, the center console, or the steering column, and is connected to the gearbox by a rod-style linkage or cables. It can also be mounted in the top of the gearbox (called a top loader transmission), where it acts through a ball pivot in the top of the extension housing (**FIGURE 32-7**). In this design, the lower end of the gearshift lever fits into a socket in a control shaft inside the transmission extension housing and is used to move the shift rails. Typically, on the bottom of the shift lever, below the ball portion, is a flattened end that fits into U-shaped cutouts in several shafts.

FIGURE 32-8 The selector gate.

FIGURE 32-9 The shift fork is the part that fits into the square groove in the synchronizer sleeve.

Shifting Mechanisms and Components

The flattened portion at the bottom of the shifter fits into U-shaped cutouts in several shafts called selector shift rails. The **selector shift rail** is the set of shafts that are connected to pieces called shift forks. **Shift forks** are the parts that actually move the components that change the gears selected inside the transmission. There is typically a shift fork for every two gear selections; thus, for a five-speed transmission, there are three shift rails. When we move one shift rail forward or backward, we engage one of two gear possibilities. We must have a way to then move between the other two shift rails. This is where the U-shaped cutouts come in. The U-shaped cutouts that are located on the selector shift rails are referred to as the **selector gate** (**FIGURE 32-8**). To select one of the other two shift rails, the shift lever must be moved to the side to engage the U-shaped cutout in one of the other shift rails and then pulled back or pushed forward. To select the opposite shift rail, the lever is moved to the other side and pulled back or pushed forward.

The selector shift rails are supported in the casing and are moved backward or forward by the gearshift lever. One selector shift rail moves to engage first and second gear; another operates third and fourth; and the other operates fifth and reverse. Note that some five-speed transmissions use four shift rails, which means fifth gear uses a shift rail separate from the reverse shift rail. A shift fork is attached to each selector shift rail. It sits in a square section groove on the outside of the **synchronizer sleeve** (**FIGURE 32-9**). A synchronizer sleeve is part of the synchronizer assembly and is what engages the dog teeth on the selected gear and transmits torque from the gear to the output shaft. The **synchronizer** is discussed in greater detail later in this chapter.

The fifth gear/**reverse shift fork** engages in a groove in the synchronizer sleeve of the reverse gear synchronizer. The operation is no different from the other gear selections. The shift lever engages the correct shift selector rail and moves the attached shift fork, which moves the correct synchronizer sleeve forward or backward. For the example we have used in this material, you have a total of three shift forks—one for first and second gear, one for third and fourth, and one for fifth and reverse.

FIGURE 32-10 The transmission detent mechanism helps to hold the transmission in the gear.

Each selector shift rail has a **detent mechanism**, usually in the form of a spring-loaded steel ball held in the casing (**FIGURE 32-10**). The ball engages in a groove in the rail when the gear lever is in neutral and in a similar groove when a gear is selected. This helps hold the gear in the selected position along with the synchronizer **spring-loaded keys** that hold the synchronizer collar in the neutral position (**FIGURE 32-11**). The detents also give the driver a feel of when the shift is complete.

When a gear is changed or selected, an **interlock mechanism** prevents more than one gear from being engaged at the same time (**FIGURE 32-12**). Some manufacturers have combined the interlock and detent mechanisms into one unit. The interlock mechanism's function in ensuring that two gears cannot be selected at the same time is vital; if two gears could be selected, then the transmission main shaft or countershaft would break due to the binding condition created. In Figure 32-12, the interlock mechanism is made up of two steel balls that are located in the casing, between the center and the outside shift rails, and in alignment with grooves machined in the selector shift rails. A small plunger is captive in a hole drilled in the center rail. In the neutral position, the balls and plunger can move laterally. But

FIGURE 32-11 Spring-loaded keys aid in holding the sleeve in the neutral position.

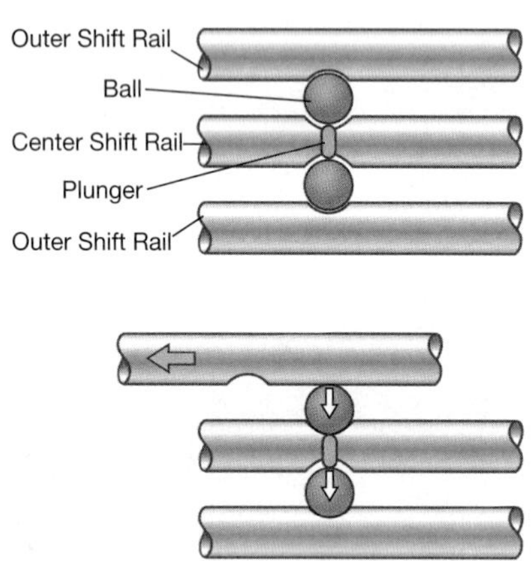

FIGURE 32-12 The interlock mechanism keeps two gears from being selected at the same time.

when an outside rail is moved, the ball is pushed sideways to contact the plunger and move the opposite ball into the groove of the other outside rail. The two balls held in the casing prevent movement of the interlocked rails and hold them in the neutral position. When the center rail is moved, the balls move into the grooves of the outside rails and hold them in a similar manner.

▶ TECHNICIAN TIP

All gearboxes are fitted with interlock and detent mechanisms, although designs vary according to manufacturer. Some manufacturers combine the two.

Many newer manual transmissions are equipped with **electronic reverse lockout** systems, which use an electronic solenoid to prevent the transmission from being shifted into reverse while the vehicle is in forward motion. When the vehicle is moving, a signal from the powertrain control module (PCM) energizes the reverse **lockout solenoid** at a speed of around 5 to 10 mph (8 to 16 kph) and above. Once the cam and lock pin is operated by the electronic solenoid, it is impossible for the manual transmission to be shifted into reverse gear. However, at very low speeds, the solenoid is not energized and the return spring keeps the lock pin away from the interlock, enabling the transmission to be shifted into reverse.

FIGURE 32-13 Steering wheel-mounted paddles are being used to shift manual transmissions electronically.

Paddle Shifters

The ability to shift gears and operate the clutch electronically or by **hydraulic actuators** such as a clutch slave cylinder is now facilitated by buttons or **paddles** on the steering wheel (**FIGURE 32-13**). Some of these paddle units can be operated in a fully automatic mode. Most paddle shifters are installed by the manufacturers to control an automatic transmission, but some manufacturers are using them to control a manual transmission. The shifter discussed here is strictly for manual transmissions. The paddles control switches that send signals to the powertrain control module/transmission control module (PCM/TCM) indicating the driver's desire to either upshift or downshift to the next gear. The PCM/TCM then sends a signal to the correct actuator to engage and disengage the appropriate gears. Paddle-shifted transmissions typically can only be shifted sequentially. In other words, you cannot skip gears like you can do with a typical manual transmission. These transmissions are further explored later in the chapter.

▶ Bearings, Thrust Washers, Gaskets, and Seals

K32004

Bearings and Thrust Washers

K32005

In a manual transmission, the gears and shafts are supported on bearings or bearing surfaces to reduce rolling friction. This design allows for smooth rotation and also maintains the alignment of the components because of rotational forces that work against each shaft due to the cut of the teeth on each gear. The type of bearing used at each locating point depends on the load that must be sustained. Radial loads try to force the gears and shafts apart. The bearing therefore carries the load along its radius. Thrust loads are applied along the length of the components, so the bearing must cater to side thrusts. In many cases, these loads are combined, and some bearings designed

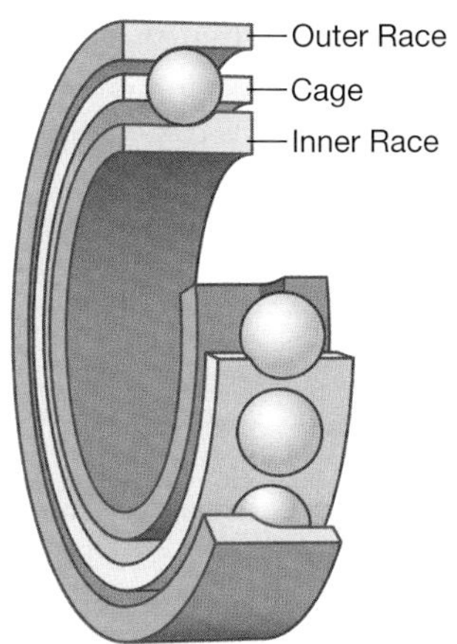

FIGURE 32-14 Single-row ball bearings are usually used to support the input shaft in the front of the transmission case.

FIGURE 32-15 The pocket bearing is located in the end of the input shaft and supports the front of the main shaft.

for radial loads also accept light thrust loads. This description applies to most ball bearings.

Single-row, deep-groove ball bearings normally support the input shaft and main shaft in conventional rear-wheel drive transmissions (**FIGURE 32-14**). They can also be used at each end of the countershaft. The helical gearing creates radial and thrust loads as the gears try to force each other apart and move endwise. A circlip or snap ring, in a groove on the outside of the bearing, can locate each bearing and shaft in the casing, which maintains the alignment of the gears and engagement hubs.

Pocket bearings are used in transmissions and are roller bearings that sit in the pocket of the input shaft (**FIGURE 32-15**). The input shaft is separated from the main shaft in order to transmit power to the countershaft. The main shaft is supported by the output bearing in the rear of the housing, and the front of the shaft is supported by the pocket bearing, which is held in place in the rear of the input shaft. The two shafts must be able to turn at differing speeds, so the pocket bearing reduces friction between the two shafts. This style of bearing is used only in rear-wheel drive transmissions.

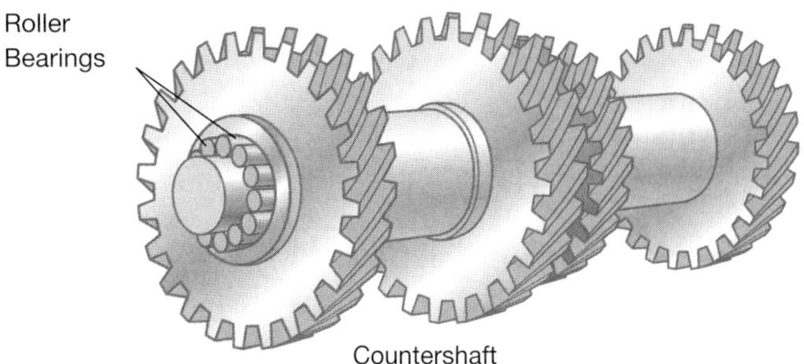

FIGURE 32-16 Roller bearings are used to support radial loads.

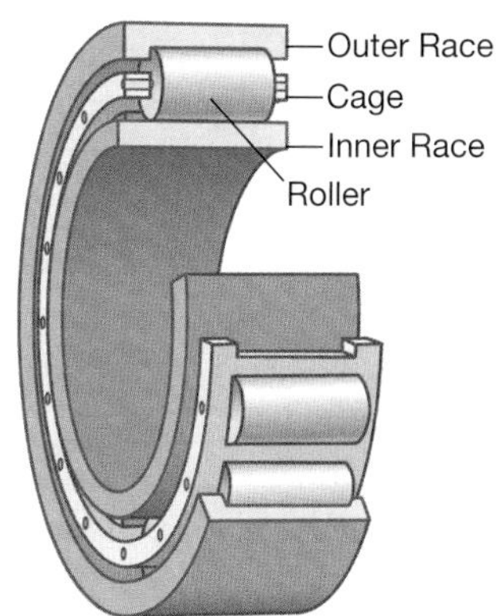

FIGURE 32-17 Caged roller bearings can be used to support gears rotating on an output shaft.

FIGURE 32-18 Thrust washers act as wear surfaces and can be used to adjust end play.

Roller bearings are common where loads are purely radial (**FIGURE 32-16**). **Needle roller bearings** can be caged or can be a loose number assembled to operate directly on a hardened gear or shaft (**FIGURE 32-17**). Typical applications are pilot bearings, pocket bearings, and flat thrust bearings. Some caged-style roller bearings are also used inside the freely rotating gears on the main shaft.

Thrust washers are designed to fit between parts that rotate against each other. They protect against thrust loads and act as a wear surface, reducing friction between the rotating parts (**FIGURE 32-18**). They are often of a selective thickness to allow for end play adjustments. Thrust washers are plain bearings, meaning they have no rollers. They are usually bronze, but can be steel, with a bronze facing on one side. They can be dimpled or grooved to allow for improved lubrication when under constant rotation.

A plain copper bushing is also used as a bearing wear surface in the extension housing to support the front of the yoke on the driveshaft. Machined and hardened surfaces on the main shaft can provide a bearing surface for the freely rotating gears. Flanges and snap rings on the main shaft locate the gears and hold them in place against any end thrust forces. A plain copper or bronze bearing is also used for the inside of the reverse idler gear to rotate on the idler shaft.

Tapered roller bearings are normally used in pairs and can sustain heavy radial and thrust loads in either direction (**FIGURE 32-19**). They are commonly used in transaxles, where they support the front and rear of the primary and secondary shafts, and the final drive assembly. **Selective thickness shims** are washers that can be used to provide for adjustment of end play or preload (**FIGURE 32-20**).

Oil Seals and Gaskets

K32006

The bearings are lubricated by being submerged in gear fluid. Gear fluid is splashed inside the casing while the transmission shafts rotate. Lubrication is extremely important for long gearbox life. Where the input and output shafts extend from the casing, oil seals prevent fluid leakage and the entrance of dirt and moisture into the transmission. If fluid levels are too high, pressure can build up and damage the seals. Low fluid levels cause overheating of all parts and may cause major damage to the wear surfaces of all of the internal components within a transmission.

Garter spring lip seals are normally used (**FIGURE 32-21**). They can be single or double lipped, depending on the application. They can also be **hydrodynamic seals**, with helical flutes molded into the sealing lip (**FIGURE 32-22**). The flutes or swirls create a pumping action as the input or output shaft rotates, causing any fluid at the sealing edge to be drawn back inside the casing. Because the flutes are molded in one direction during manufacture, care must be taken to ensure they are installed with the correct direction of shaft rotation.

O-ring seals are typically used for static locations such as the speedometer drive gear housing, while gaskets are used to seal the mating surfaces between the casing and the retaining

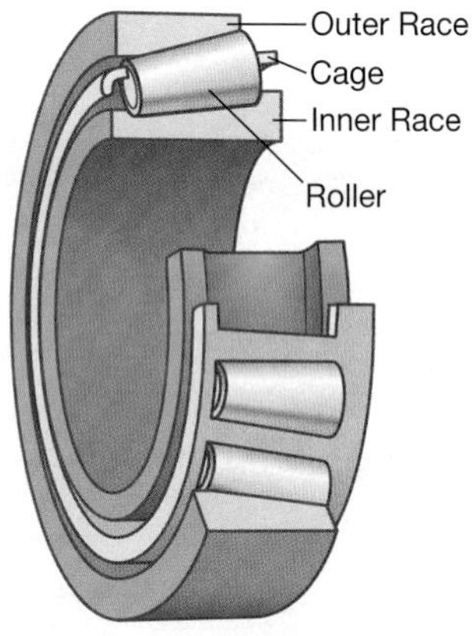

FIGURE 32-19 Tapered roller bearings are used to sustain radial and thrust loads.

FIGURE 32-20 Shims are used for correcting preload adjustments.

FIGURE 32-21 Seals prevent fluid leakage and keep dirt from entering the transmission.

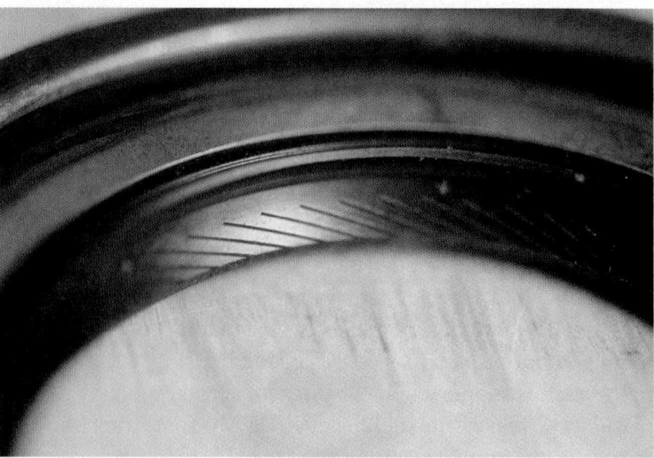

FIGURE 32-22 The helical flutes in the seal surface help push the fluid back into the transmission.

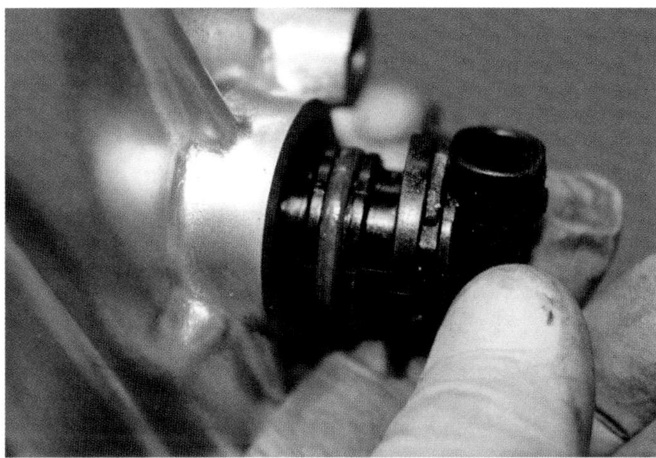

FIGURE 32-23 O-ring seals are typically used for static locations.

FIGURE 32-24 A rubberized cork transmission pan gasket.

plates or housings (**FIGURE 32-23**). In some cases, the gaskets that seal flat surfaces may be used by the manufacturer as selective thickness shims or spacers, in addition to their sealing function.

Cork gaskets are cut out of a large sheet of cork and rubber mixture. They compress and provide a sealing action for case covers on the transmission (**FIGURE 32-24**). Some transmissions use a bead of room temperature vulcanizing (RTV) sealant to form the gasket. "Room temperature vulcanizing" means the liquid gasket material will cure at room temperature. This type of gasket costs less to manufacture because gaskets do not have to be produced for each transmission.

▶ TECHNICIAN TIP

Some manufacturers use reusable silicone gaskets in some applications that can be used more than once.

▶ Transmission Design

K32007

In a manual transmission for a rear-wheel drive vehicle, the geartrain is built on three shafts—the input shaft, the countershaft, and the main shaft (sometimes referred to as the output shaft) (**FIGURE 32-25**). The front of the input shaft is supported by the pilot bearing in the rear of the crankshaft. The rear of the input shaft is supported by the input shaft bearing at the front of the transmission. The input shaft gear meshes constantly with a mating gear on the countershaft. The countershaft is supported by bearings on each end. Other gears on the counter shaft are in constant mesh with mating gears on the main shaft, or output shaft. The main shaft is supported by bearings on either end of the shaft—in the front by the pocket bearing in the rear of the input shaft and on the rear by the rear main shaft bearing.

The gears are able to freely rotate on the main shaft until they are locked to the main shaft when a particular gear is selected. Each freewheeling main shaft gear has an external toothed section (commonly called dog teeth) on one side. The teeth face an internally toothed engagement sleeve and **blocking ring**, located on a central hub (**FIGURE 32-26**). The central hub is splined to the main shaft.

The engagement sleeve or sliding collar can slide in either direction to engage the dog teeth on the appropriate gear. Engagement of the dog teeth locks the gear through the sleeve and hub to the main shaft. Before engagement of the gear

FIGURE 32-25 Input shaft.

FIGURE 32-26 The synchronizer sleeve engages the dog teeth on the mating gear.

occurs, a synchromesh device called the synchronizer assembly matches the speed of the gear with the sleeve so that the sleeve can engage the dog teeth, transmitting power from the gear to the sleeve, and ultimately to the main shaft. For more on how the synchronizer works, see the Synchronizer section later in this chapter.

The gears constantly in mesh have their teeth cut on a **helix** at an angle to the gear centerline. This design reduces gear noise and distributes the load more evenly, as several teeth are in contact at any time (**FIGURE 32-27**).

FIGURE 32-28 Spur gears use straight-cut teeth for more coverage and heavy-duty operation.

▶ TECHNICIAN TIP

In the early days of automotives, gear designs were all spur-type gears and did not have synchronizers to enable a smooth shift. These gear arrangements required special clutch operation called double-clutching when shifting from one gear to another. **Double-clutching** refers to depressing the clutch pedal twice for each gear change—once to enter neutral and again to select the next gear. The clutch pedal is then released a second time, and the vehicle can be accelerated. The reason for double-clutching was to match the engine speed with the gear speed to ensure a relatively smooth engagement of the gears, thereby avoiding gear clashing and transmission damage. Double-clutching a transmission was not an easy task, as the clutch pedal pressure was usually very high in an older vehicle to ensure the disc did not slip (spring technology was not the same as it is today). In today's transmissions with synchronizers, there is no need to double-clutch to change gears; the synchronizer automatically matches the gear speed to the shaft speed as the gear shift is being performed.

FIGURE 32-29 A. Sliding reverse idler. **B.** Constant-mesh reverse gear. Both cause the output shaft to rotate in the reverse direction.

Teeth on some reverse gears are straight-cut, or spur, gears, which are cut parallel to the gear centerline and in most cases are very noisy when rotating in mesh (**FIGURE 32-28**). The noise made by spur-cut gears is demonstrated by the whining of the transmission when a car is backed up rapidly. Many newer manual transmissions are now using helical-cut reverse gears.

When the reverse gear is selected, one of two things occurs. In one design, the reverse idler gear connects and slides into its two mating gears on the countershaft and main shaft. This results in an opposite rotation of the output shaft as compared to the forward direction, propelling the vehicle in the reverse direction. The reverse idler rotates on a shaft fixed in the casing. It transfers the drive from the countershaft to the main shaft. In the second design, reverse is engaged by a constant-mesh reverse gear that uses a synchronizer sleeve to lock the reverse gear to the main shaft (**FIGURE 32-29**).

FIGURE 32-27 Helical gears have teeth that are cut on a helix for quieter operation.

Transaxle Designs

In transaxle designs, the drive is transferred through the clutch unit to a primary or input shaft. In this design, the primary shaft carries gears of different sizes, meshed with gears on the secondary shaft. Each pair of gears typically has one gear fixed to the primary shaft and one free to rotate on bearings on the secondary or output shaft.

An engagement sleeve and hub assembly, with a synchromesh unit facing each gear, is on the secondary shaft between first and second, and similarly between third and fourth.

The engagement sleeve moves to engage with the dog teeth of the relevant gear when a gear is selected. This locks the gear to the shaft so that drive can be transmitted.

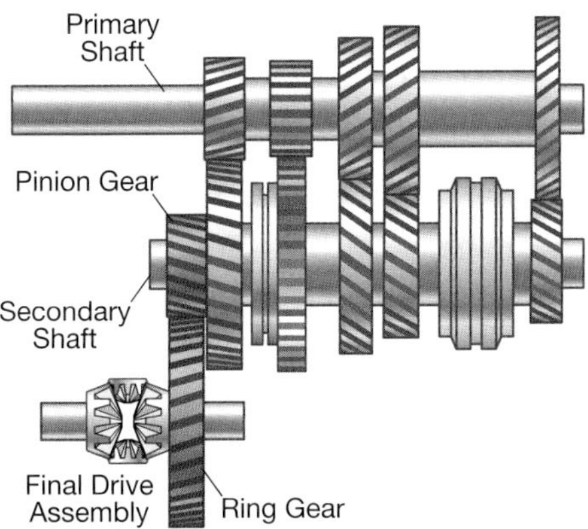

FIGURE 32-30 In a transaxle, the pinion gear is on the end off the secondary shaft and engages the ring gear in the final drive assembly.

FIGURE 32-31 The fifth gear is attached to the output shaft to provide higher vehicle speeds at lower engine rpm, which increases fuel economy.

A drive pinion gear fixed to the end of the secondary shaft is constantly in mesh with a final drive ring gear (**FIGURE 32-30**). The drive is transferred through these gears to the differential unit.

A reverse idler transmits the drive from the primary to the secondary shaft when reverse is selected. One design of a five-speed transaxle has an extra fixed gear on the primary shaft, meshed with a mating free gear on the secondary shaft (**FIGURE 32-31**). The free gear has its own engagement sleeve and synchromesh unit beside it. Selecting fifth gear locks the gear to the shaft. The gear on the primary shaft is larger than the gear on the secondary, so an overdrive ratio is provided. The secondary shaft will then rotate faster than the primary shaft.

▶ TECHNICIAN TIP

Having the gearbox and final drive gears in one casing provides a compact and relatively quiet unit and fits well into front-wheel drive vehicles.

▶ Transmission/Transaxle Operation

K32008

Transmission Operation

K32009

In a rear-wheel drive manual transmission, the splines on the front of the input shaft engage with the splines of the clutch-driven plate. With engine rotation and the clutch pedal released, the input shaft transfers its motion through the countershaft to rotate the gears on the main shaft. In neutral condition, the engagement sleeves and hubs splined to the main shaft are stationary. No drive is transmitted to the main/output shaft.

Depressing the clutch pedal disconnects the engine from the input shaft, allowing an engagement sleeve to be moved into engagement with the external dog teeth on the gear selected. Engaging the dog teeth locks the gear to the main shaft. When the clutch is released, the drive is transmitted to the input gear along the countershaft to the gear selected. Because this gear is now locked to the main shaft, the main shaft rotates and transfers the drive to the final drive unit.

The speed ratio and the torque transferred depend on which gear is selected. In a four- or five-speed transmission, the first gear, the smallest on the countershaft, meshes with the largest gear on the main shaft, producing first gear (**FIGURE 32-32**). First gear is always the lowest of the gear ratios in forward. Remember, reverse often has a lower gear ratio because the vehicle needs high torque more than fast speeds when moving in reverse. A low gear ratio also gives the greatest amount of torque multiplication.

In second gear, power flow is through the next adjacent gear (**FIGURE 32-33**). The difference in speed between input and output shafts is reduced. Fewer turns of the input shaft are required to give one turn of the output shaft, but there is less torque multiplication.

Refer to the Manual Transmission/Transaxle Principles chapter to learn more about power flow.

When the engagement sleeve for third gear moves into place, drive is transmitted through the meshing gears at an even higher gear ratio (**FIGURE 32-34**). In fourth gear, the input and output

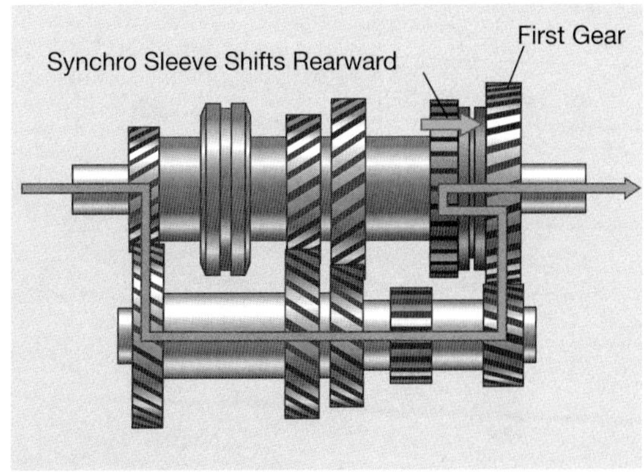

FIGURE 32-32 Power flow for first gear in a transmission.

shafts are directly locked together (**FIGURE 32-35**). This provides a **direct drive** with a 1:1 ratio, with no reduction through the countershaft gears. The input and output shafts turn at the same speed, and there is therefore no torque multiplication.

The countershaft gears and their corresponding gears on the main shaft continue to rotate, as the countershaft is constantly in mesh with the input gear. They do not transmit any drive, but the countershaft still provides lubrication by its movement, which picks up and flings transmission fluid within the housing. With the vehicle in motion, these gears turn at different speeds in relation to the main shaft and to the engagement sleeves, which are locked to the main shaft by the mating splines on the sleeve and synchronizer hub. A transaxle operates in a similar manner, transferring drive through gears on the primary and secondary shafts, which then transfer drive to the final drive assembly.

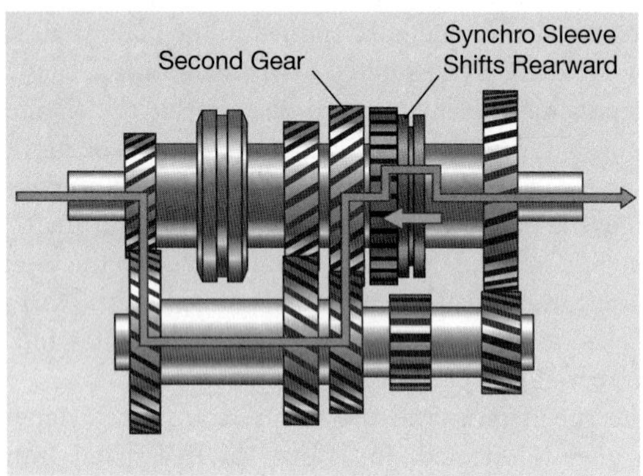

FIGURE 32-33 Power flow for second gear in a transmission.

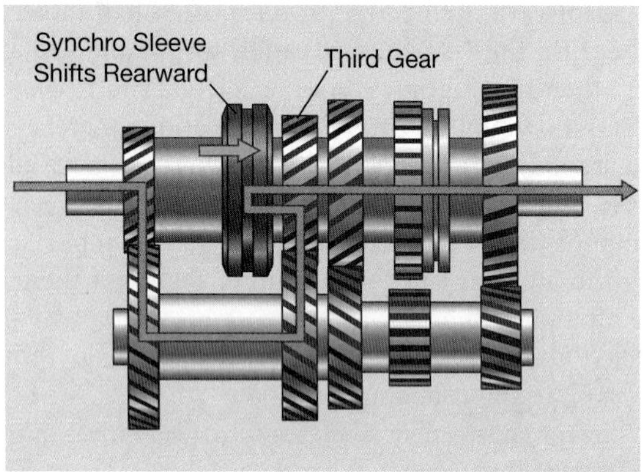

FIGURE 32-34 Power flow for third gear in a transmission.

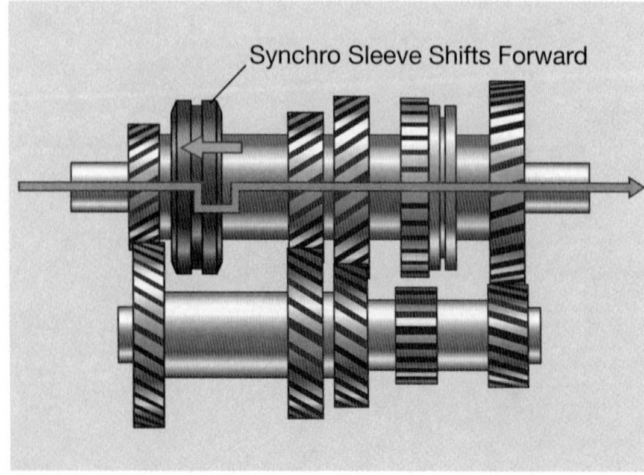

FIGURE 32-35 Power flow for fourth gear in a transmission.

Role of Electronics

For the most part, historically, manual transmissions have not required the use of electronics. But as with all vehicle systems, electronics are starting to play more of a roll. Vehicle speed sensors (VSS) are mounted in the transmission case and are used to generate a signal, based on the vehicle's speed, which is sent to the PCM and/or speedometer. A sensor may also be used to tell the PCM/TCM what gear is selected. The PCM uses that information along with data from other sensors to control various aspects of the vehicle, including the speedometer reading (on some vehicles) as well as some electronic shift routines, which are covered in the Manual Transmission/Transaxle Principles chapter. Another innovation in technology is the dual-clutch, six- or seven-speed transmission designed by several manufacturers, which makes shifting easy and effortless by incorporating electronics into the mix. See the Dual-Clutch Manual Transmissions section later in this chapter for more information.

Baulk-Ring Synchromesh Unit

A **baulk-ring synchromesh unit**, commonly called a blocker ring synchromesh unit, is used to synchronize the speeds of gear and shaft before engagement (**FIGURE 32-36**). Once the speeds of the gear and shaft are matched, the engagement sleeve slides on its splines on the hub and engages with the dog teeth on the gear, effectively locking them together. The synchronizer assembly makes this possible. Three metal inserts, with a center ridge, fit into slots in the hub. A radial spring at each end holds them out so that the ridge on each insert sits in a groove inside the sleeve and holds each one centered on the hub (**FIGURE 32-37**). When the synchronizer is assembled, the hub is splined to the main shaft, and the engagement sleeve is splined to the hub. A

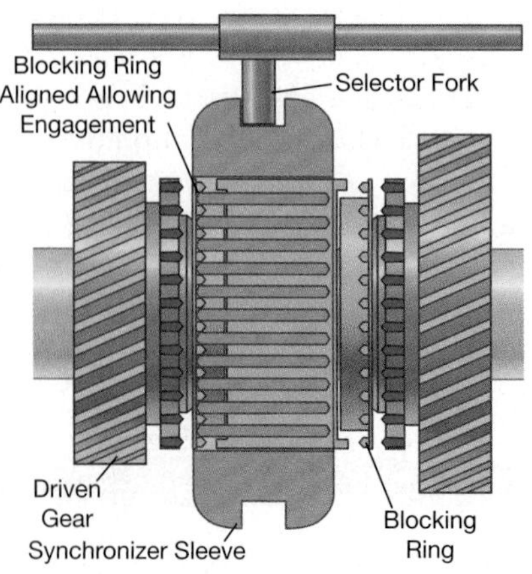

FIGURE 32-36 Blocker ring teeth line up with the dogteeth to allow smooth engagement of the sleeve.

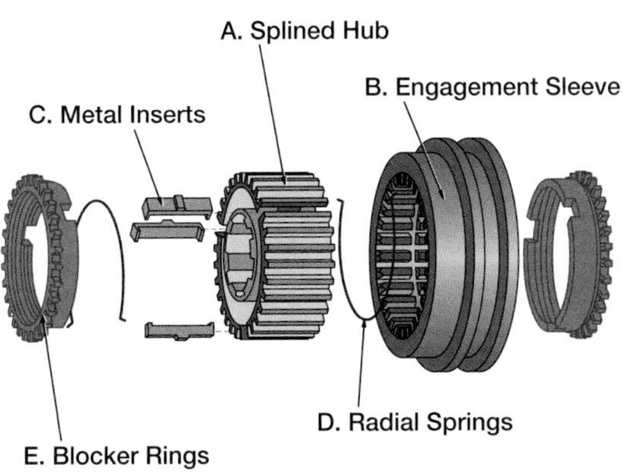

FIGURE 32-37 Synchronizer components. **A.** Splined Hub. **B.** Engagement Sleeve. **C.** Metal Inserts. **D.** Radial Spring. **E.** Blocker Ring.

FIGURE 32-38 Fine groves are machined on the conical surface of the blocker ring. They wear out over time.

bronze blocker ring with recesses to accommodate the ends of the inserts is located in each end of the hub, and a conical inner surface faces a matching steel cone on the gear.

Fine grooves are machined on the conical surface of the blocker ring and help with cutting through the oil film on the cone-shaped surface on the gear (**FIGURE 32-38**). The teeth on the outer edge of the blocker ring are the same size as the dog teeth on the gear and the splines on the sleeve. The recesses in the blocker ring are wider than the inserts to allow the blocker ring to move radially. Then the teeth on the blocker ring can be rotated so they are out of alignment with the teeth on the sleeve.

When the sleeve is moved to select a gear, the spring-loaded inserts move the blocker ring into contact with the conical face of the gear. The grooves in the face of the blocker ring start to cut through the oil film. The difference in speed between the two components causes the blocker ring to be dragged around with the gear to the limit of the recesses, where it is held by the inserts.

Because the teeth of the blocker ring are now out of alignment with the teeth of the sleeve, the sleeve is blocked from moving over the ring into engagement (**FIGURE 32-39**).

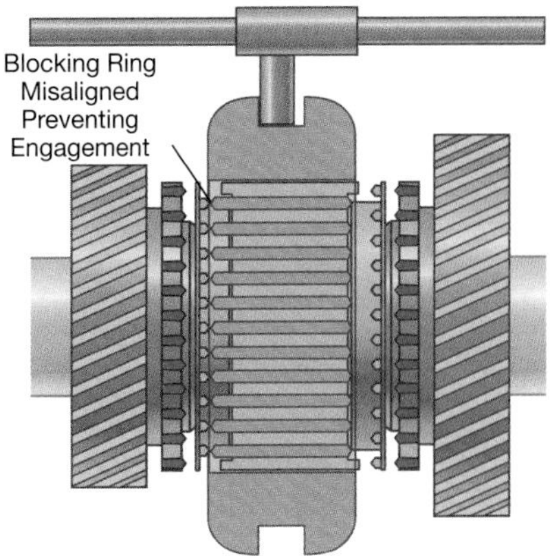

FIGURE 32-39 Blocker ring teeth out of alignment, which allows the blocker ring to bring the gear to the same speed.

The force exerted by the driver on the gear lever, shift rails, and shift fork now presses the sleeve against the teeth of the blocker ring and increases the force of the conical face of the blocker ring against the cone of the selected gear until the friction causes the two components to rotate at the same speed. The blocker ring teeth can now come into alignment, and the sleeve slides over them and into engagement without grinding. This step is assisted by a **chamfer** (a tapered cut) on the teeth of the gear, blocker ring, and sleeve, which helps to guide the sleeve into position over each of those parts (**FIGURE 32-40**). Smooth and rapid gear changes are thus ensured.

Transaxle Power Flow

K32010

In transaxles, the synchromesh unit plays a similar role as in a transmission. They are located between the end of each engagement sleeve and the gear. In this arrangement, the engagement hub and sleeve assembly for first and second gears is typically on the secondary shaft, as is the third and fourth speed hub. The fifth gear engagement hub is at the end of the secondary shaft,

FIGURE 32-40 The teeth are chamfered to allow easier engagement. This wears away over time and causes hard engagement.

farthest from the main drive pinion. When first gear is selected, the engagement sleeve locks the first gear to the secondary shaft. When the clutch is released, the drive is transferred through the fixed first gear on the primary shaft to the locked gear on the secondary shaft. The secondary shaft rotation transfers the drive through the main drive pinion to the final drive ring gear and the differential case.

In first gear, the smallest gear on the primary shaft drives the largest gear on the secondary shaft, resulting in the lowest gear ratio—in this case, approximately 3.4:1 (**FIGURE 32-41**). In second gear, a higher gear ratio is obtained, as the fixed second gear on the primary shaft is larger. Fewer turns of the primary shaft are required to obtain one turn of the secondary shaft—in this case, resulting in an approximate ratio of 1.95:1 (**FIGURE 32-42**).

In third gear, the engagement sleeve locks the freely rotating third gear to the secondary shaft (**FIGURE 32-43**). Drive is transferred from the fixed gear on the primary shaft to the locked gear on the secondary shaft, and then to the final drive. The drive transfer is similar for fourth gear, but the gears are the same size (**FIGURE 32-44**).

FIGURE 32-41 Power flow for first gear.

FIGURE 32-42 Power flow for second gear.

The fifth gear ratio is provided by transferring the drive through the fixed fifth gear on the primary shaft to the engaged fifth gear on the secondary shaft. This is often an overdrive ratio, with the driving gear larger than the driven gear, making the driven gear rotate at a higher speed (**FIGURE 32-45**).

FIGURE 32-43 Power flow for third gear.

FIGURE 32-44 Power flow for fourth gear.

FIGURE 32-45 Power flow for fifth gear.

FIGURE 32-46 Power flow for reverse gear.

For reverse, the reverse idler moves to mesh with a reverse gear on the primary shaft and the matching gear teeth on the outside of the first and second gear engagement sleeve (**FIGURE 32-46**). Drive transfers through the idler to the splined secondary shaft sleeve. The idler between the shafts makes the secondary shaft rotate in the reverse direction. This reverses the direction of the final drive ring gear and thus the road wheels.

Dual-Clutch Manual Transmissions

N32001

A dual-clutch transmission is sometimes referred to as a semi-automatic or automated manual transmission because it uses a geartrain more similar to a manual transmission than an automatic transmission. Still, dual clutches are controlled by the PCM/TCM, so there is no clutch pedal. The dual-clutch arrangement also requires two input shafts. Each clutch powers one of the input shafts (**FIGURE 32-47**). The dual-clutch transaxle uses electric solenoid valves to hydraulically engage and disengage each clutch. This transaxle has dual input shafts that are connected by gears to two main shafts containing the synchronizers. One shaft has gears two, four, and six, and the other shaft has gears one, three, and five. The transaxle control module is able to move the shift forks by electric actuators to select each gear (**FIGURE 32-48**). Only one gear on one input shaft transmits power at a time. While one gear is engaged on one input shaft, the control module commands an actuator on the other input shaft to engage the next sequential gear. Because the clutch on the second input shaft is disengaged, the gear can be selected, ready, and waiting to be applied for either an upshift or downshift. This type of transaxle provides a very quick shift, as all it takes to upshift is releasing one clutch and activating the other. This makes for extremely fast and accurate shifts.

The dual clutches are typically hydraulically operated wet clutches similar to clutches in an automatic transmission, but dry clutches are coming into more use (**FIGURE 32-49**). The PCM/TCM controls their application and release by sending hydraulic pressure to each clutch to apply it and venting hydraulic pressure

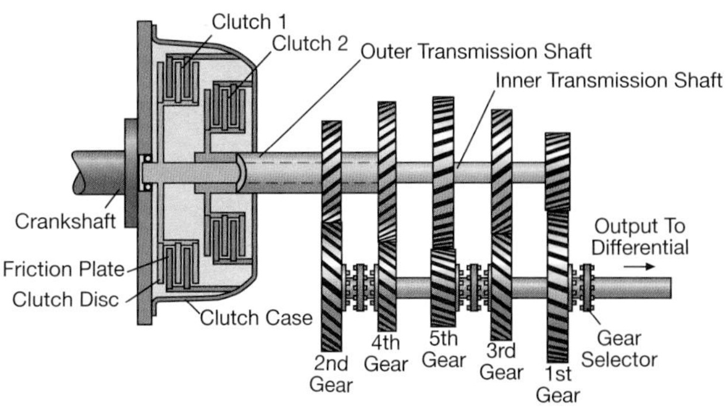

FIGURE 32-47 Dual-clutch transmission layout.

FIGURE 32-48 Shift actuator for a dual-clutch transmission.

to release it. Hydraulic pressure is controlled by solenoid valves that react to the signals sent from the PCM/TCM. The PCM/TCM monitors data such as engine rpm, rpm of each input shaft, gear selected, vehicle speed, throttle position, brake pedal position, and other sensor information. Based on this information, the PCM/TCM sends the appropriate signals to the solenoid valves.

▶ Preventive Maintenance

S32001

Preventive maintenance is designed to extend the service life of the transmission/transaxle and drivetrain components. Although some newer vehicles have fewer requirements for transmission maintenance, due to self-adjusting clutches and fill-for-life lubricants, those that do require maintenance must be serviced according to the manufacturer's schedule to ensure long life. It usually consists of checking and replacing lubricants on a regular basis, lubricating linkage pivot points, performing adjustments such as the clutch adjustment, and inspecting all drivetrain components for damage or wear.

Failure to perform required maintenance of transmissions can result in severe damage to the internal components, which are expensive to repair. For example, operating the transmission with substantially low lubricating fluid, because it had a small leak and never got checked during engine oil changes, can ruin

FIGURE 32-49 Dual clutch transmission clutches. **A.** Wet clutch. **B.** Dry clutch.

FIGURE 32-50 Checking the fluid level.

the transmission long before its time. Don't overlook the maintenance; your customer expects it to be performed properly and at the specified time (**FIGURE 32-50**).

Lubrication

When it comes to lubrication, virtually every component in the manual transmission requires lubrication according to manufacturer specifications even after the warranty runs out. Clutches, the transmission, the final drive, driveshafts, and drive axles need to be checked or lubricated to maintain their integrity on a regular basis. Proper lubricants will ensure that all manual transmission components last a long time. Manufacturers have used a variety of lubricants for their particular transmissions over the years. The most common lubricant is gear lube. However, some manufacturers specify a specific weight of engine oil be used. Some manufacturers even specify automatic transmission fluid be used in their manual transmission. Considering the amount of stress put on the manual transmission, it is important to maintain all lubricants on a regular basis.

The transmission/transaxle and differential lubricant/oil levels keep all rotating parts cool by removing heat and transferring it to the case, which can dissipate the heat to the atmosphere. The lubricant also supplies a thin film of oil between close tolerances of gears, shafts, and bearings to reduce metal-to-metal contact. This oil film prolongs the life of these parts. Improper lubricant/oil levels could cause overheating and unwanted failures to occur.

Components such as CV joints require different types of semisolid lubricants or grease. Considering the amount of torque that CV joints must endure, special lead-based grease is used in most applications. Grease possesses a property called **thixotropy**. This term refers to its ability to be a solid but, while under stress, to flow or become thin in order to lubricate properly. Many types of semisolid lubricants are in use today. Always remember to consult the manufacturer's service information for correct usage.

Tools

When performing service and repair operations on a manual transmission, you need to be familiar with several tools (**FIGURE 32-51**):

- Suction gun—used to suck up gear lube and fill the transmission.
- Dial indicator—used to check input or output shaft end play, which is set by selective thickness shims. The selective thickness shims are measured with an outside micrometer.
- Hydraulic press—may be used to tear down and reassemble the transmission for service.
- Bearing splitter—used to press bearings from the transmission shafts or to remove synchronizer sleeves.
- Various gear pullers and bearing pullers—tools that are manually operated.
- Snap ring pliers—required to remove snap rings that may be used to retain gears or synchronizers.
- Seal installer—used to ensure that seals are not damaged when being installed into transmission housings.

Checking Gearbox Fluid

N32002

As a preventive maintenance task, fluid level of the manual transmission should be checked when other tasks are performed, such as engine oil and filter change or lubrication of steering parts,

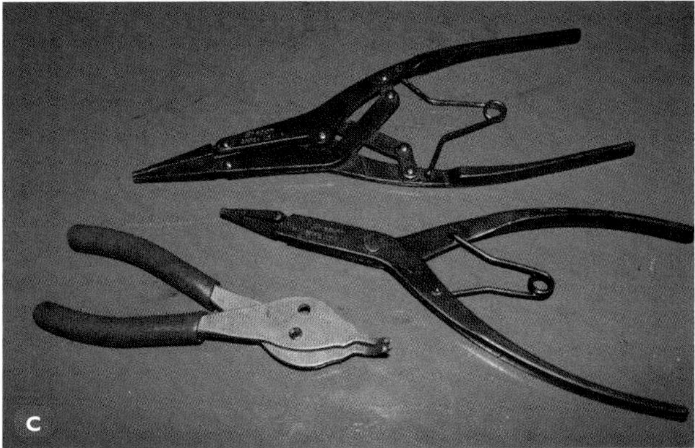

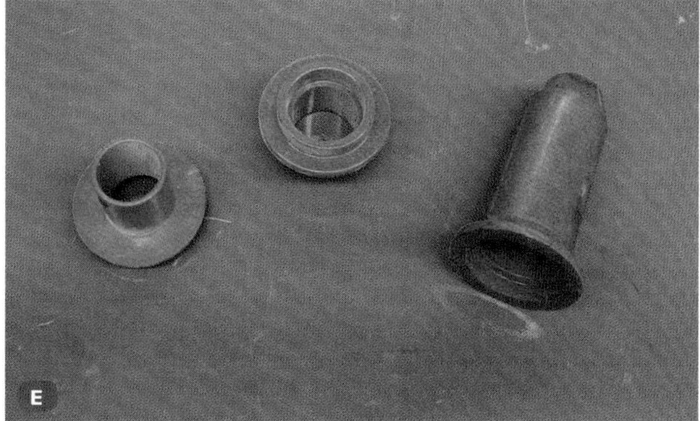

FIGURE 32-51 Properly servicing manual transmissions requires a variety of tools. **A.** Hydraulic press. **B.** Bearing splitter. **C.** Snap ring pliers. **D.** Bearing puller or gear puller. **E.** Seal installer.

according to the manufacturer's periodic intervals. Also, not all manual transmissions use gear lube; some use engine oil of a specified viscosity, and others even use a certain type of automatic transmission fluid. So make sure you check the service information to determine the correct type and quantity of fluid.

Fluid level and condition should be checked as part of a thorough visual inspection anytime a transmission fault is suspected. Condition and level of the transmission fluid can tell a technician a lot about what is going on inside. For example, if metal is found floating inside the transmission fluid, then the technician knows that the transmission will have to be pulled apart or replaced. In the event of a suspected leak, the transmission fluid level should be checked and topped off once repairs are complete, to prevent catastrophic failure. Make it a habit to visually inspect the transmission every time it is in for an oil change or other service so you can catch issues before they become severe.

TECHNICIAN TIP

It is always a good idea to keep up with preventive maintenance on a regular basis. Remind the customer that preventive maintenance is the road to long life for the vehicle and can help save thousands of dollars in repairs.

To check and adjust the fluid level, follow the steps in **SKILL DRILL 32-1**.

Identifying the Cause of Fluid Loss

Performing preventive maintenance on a regular basis helps ensure that fluid leaks and concerns are identified and addressed before they become a major problem. Diagnosing leaks and concerns requires that you put the vehicle on a hoist or jack stands. Noises and shifting concerns can be signs of fluid leaks or problems, so always inspect the transmission if these symptoms are present. If a leak is found, in some cases repairing fluid leaks requires removal of the transmission.

Fluid leaks may be a result of the following conditions:

- Too much fluid in the transmission
- Leaking seals, including the following:
  - Half-shaft axle seals
  - Input shaft retainer O-rings or lip seals
  - Speed sensor seals or O-ring
  - Backup light switch O-ring
  - Shifting lever shaft seals
  - Transmission rear seal
  - Side covers or access plates
- Improper fluid type
- Transmission case porosity or cracks
- Missing, loose, or stripped case bolts
- Damaged gaskets for case halves
- Loose drain or fill plugs

To identify the cause of fluid loss in a transmission, follow the steps in **SKILL DRILL 32-2**.

Checking and Adjusting Differential/Transfer Case Fluid

Checking and adjusting the differential/transfer case fluid level is very similar to the process for checking the transmission fluid level, but there are some differences. First, some differential/final drive assemblies, such as limited-slip and posi-traction assemblies, require specially designed additives or fluid. Transfer cases may also require special lubricants as specified by the manufacturer. Always check the service information for the vehicle you are working on. Differentials and transfer cases generally have fill plugs that can be used when checking the fluid level in the same way a transmission does. However, they often do not have a drain plug. In these cases, either a specific bolt may need to be removed or a cover may need to be unbolted and removed. Follow the service information.

SKILL DRILL 32-1 Checking the Fluid Level of a Manual Transmission

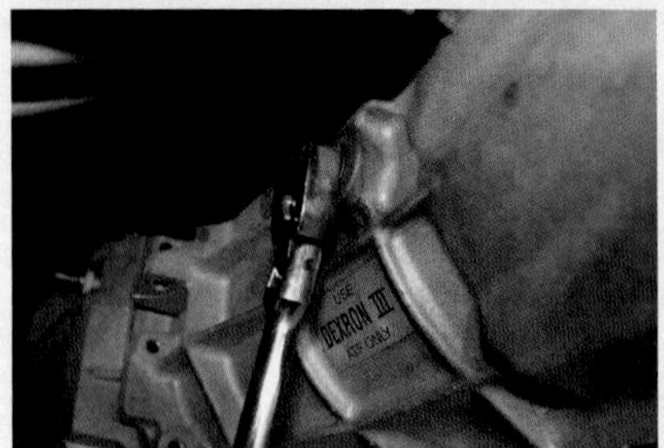

1. Safely raise and support the vehicle on the lift so that it is level. Inspect the transmission for leaks. Remove the filler plug, using the proper tool. Inspect the filler plug and fill hole for thread damage, and replace or repair if necessary.

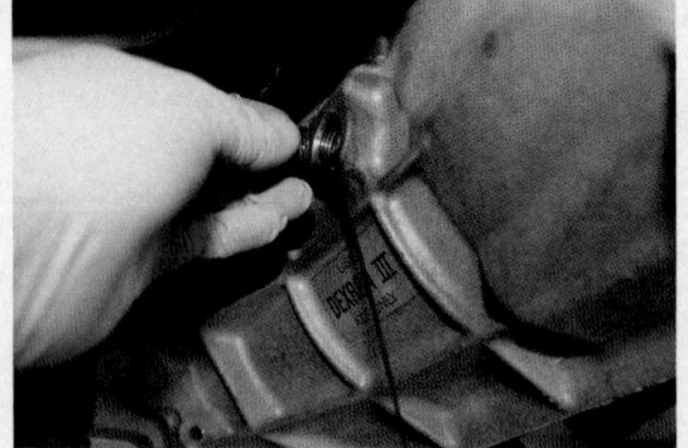

2. If the gearbox fluid begins to run out as the filler plug is removed, let the gearbox fluid seek its own level before reinstalling the filler plug. The gearbox fluid level should be at the bottom of the filler plug hole.

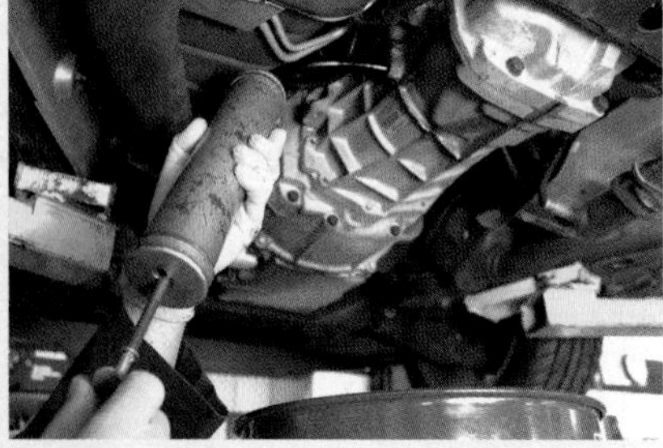

3. If the fluid level is low, refill with the specified fluid, reinstall the filler plug, and wipe the area around the filler plug hole with a clean shop towel. Tighten the filler plug to the specified torque.

SKILL DRILL 32-2 Identifying the Cause of Fluid Loss in a Transmission

1. Safely raise and support the vehicle. Look for leaks in the transmission bell housing to the engine block (front seal).

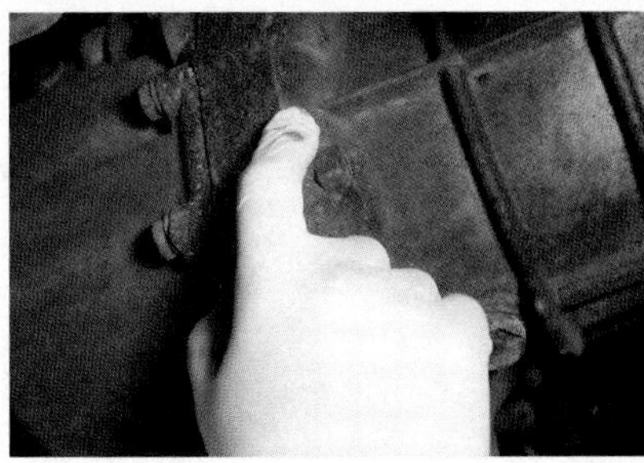

2. Look for leaks in the transmission breather outlet. Look for leaks in all case gasket areas.

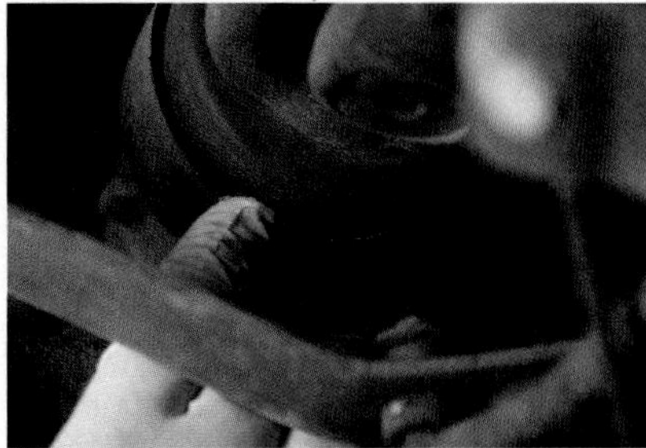

3. Look for leaks in the rear tail shaft seal.

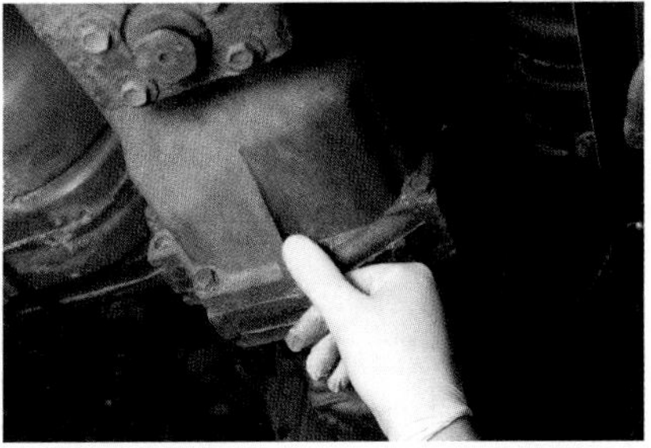

4. Inspect for transmission case defects, such as cracks and porosity.

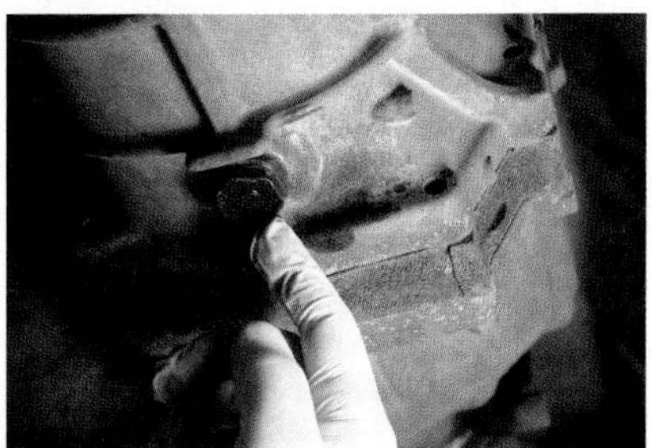

5. Look for leaks at the drain and fill plugs.

6. Check the gear fluid level. If excess level is found, let the excess drain into a container. If the fluid is more than ¼" (6 mm) below the bottom of the threaded hole, add new fluid of the correct viscosity and type.

To check and adjust the differential/transfer case fluid level, follow the steps in **SKILL DRILL 32-3**.

Changing Manual Gearbox and Final Drive Fluid

N32003

At the manufacturer's specified intervals, the gearbox fluid must be changed in those vehicles in which manufacturers require it. Gearbox fluid left unchanged will break down from heat and lose its capacity to lubricate all rotating parts, causing premature bearing and gear failure. This potential damage also makes it critical to maintain proper gearbox fluid levels. If the level becomes too low, bearings and gears will starve for lubrication; if levels are too high, aeration of the gearbox fluid will be present and cause premature damage. Regular change intervals help to ensure that none of these conditions occur.

Manual transmissions use fluids specified by the manufacturer. Do not assume that all transmissions use the same fluid. They don't. Some use gear lube of differing viscosities. Some use engine oil of a specified viscosity. And some even specify automatic transmission fluid in the transmission. Therefore, always verify that you are using the specified fluid in a particular vehicle. One other thing to keep in mind: Some transaxles use a separate reservoir for the final drive, which is separate from the transaxle reservoir. And in some cases, they use different types of fluid. Always check the manufacturer's specifications before changing out the fluids.

To prepare for a gearbox fluid change, it is a good idea to have the gearbox fluid warm. When it is warm, it will flow better to ensure proper draining. If possible, drive the vehicle until the engine is at operating temperature, and then change the gearbox fluid. Do not touch the gearbox fluid. It will be hot and could cause burns.

To change the gearbox fluid, follow the steps in **SKILL DRILL 32-4**.

Inspecting, Replacing, and Aligning Powertrain Mounts

The purpose of the powertrain mounts (motor mounts) is to support the engine and transmission as it is placed in the vehicle. They also isolate torsional vibrations created by the running engine and vibrations from the transmission, and allow for flexing as torque is achieved in various gear selections. Powertrain mounts are subject to the elements and are known to break, crack, chunk (broken pieces), and become oil soaked (**FIGURE 32-52**).

The powertrain mounts should be inspected when a customer is complaining of gear shifting issues or clunking

SKILL DRILL 32-3 Checking and Adjusting the Differential/Transfer Case Fluid Level

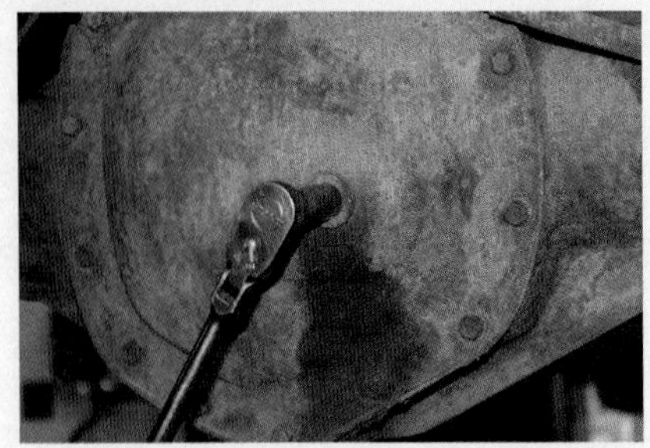

1. Safely raise and support the vehicle. Inspect the differential and transfer case for leaks. Position a clean drain pan under the filler plug. Remove the filler plug, using the proper tool. Inspect the filler plug threads for thread, and replace if necessary.

2. If the fluid begins to run out as the filler plug is removed, let the fluid seek its own level before reinstalling the filler plug. The fluid level should be at the bottom of the filler plug hole.

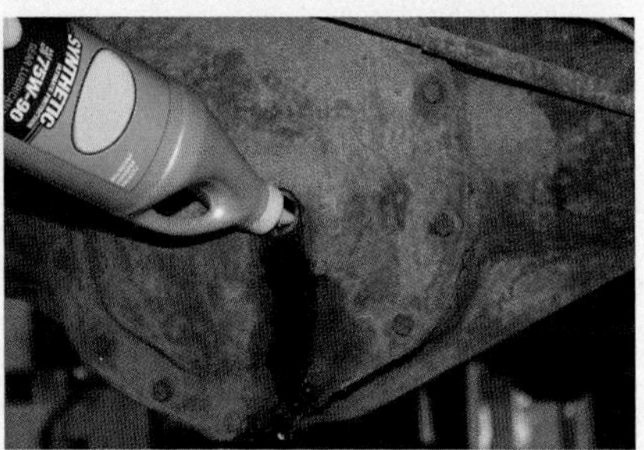

3. If the fluid level is low, refill with the specified fluid, reinstall the filler plug, and wipe the area around the filler plug hole with a clean shop towel. Tighten the filler plug to the specified torque.

SKILL DRILL 32-4 Changing the Gearbox Fluid

1. Safely raise and support the vehicle, using an approved lift. Obtain a clean drain pan to put the used fluid in.

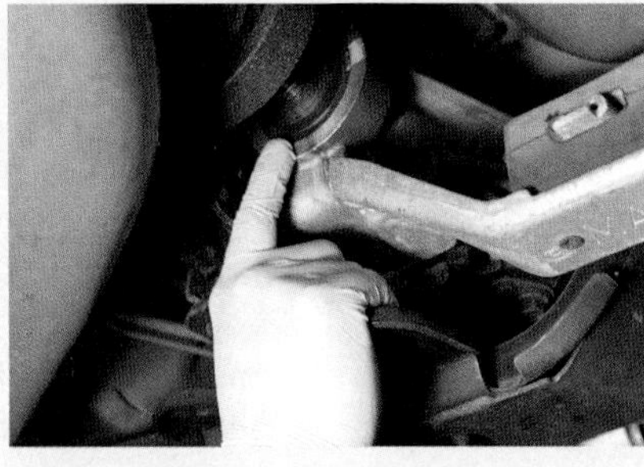

2. Inspect the transmission for leaks.

3. Remove the drain plug from the bottom of the transmission, being careful of the hot gearbox fluid. Let gearbox fluid drain until it has stopped running. If necessary, drain the final drive assembly.

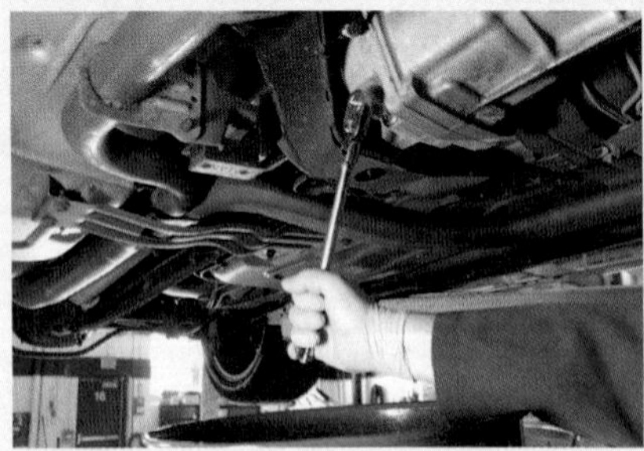

4. Replace the drain plug(s) and tighten to specification, and remove the fill plug(s).

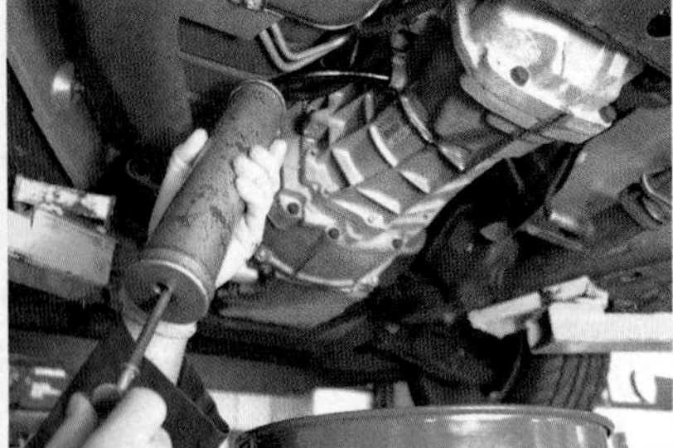

5. Refill the transmission and final drive to the proper level, using manufacturer-approved gearbox fluid. Replace the fill plug(s), and tighten to the proper torque. Wipe away any spillage. Road test the vehicle. If necessary, put the vehicle back on the lift, and check for any leaks.

FIGURE 32-52 Faulty powertrain mount that needs to be replaced.

sounds when accelerating or decelerating. The powertrain mounts can also create issues with constant-velocity (CV) axles as a result of axles running at incorrect angles. This can result in vibrations when driving or possible early failure of the CV axles. They can also cause hard shifting and other shift-related symptoms. To inspect and service powertrain mounts, refer to the Servicing the Automatic Transmission/Transaxle chapter.

Inspecting and Servicing Sensors and Switches

Although a manual transmission is being shifted manually, electronics are used to a small degree in the process of selecting gears on some transmissions. A backup light switch is used to let people know that the driver is reversing the vehicle. A speed sensor is used to tell the computer how fast the driver is traveling down the highway.

With electronic shifting mechanisms, some electronic switches and solenoids are involved. Some vehicles use a hydraulic fluid pressure sensor or a gear-selecting sensor. Some of these transmission shifting systems require the use of a scan tool when troubleshooting. Use a scan tool that can communicate with the vehicle's transmission control module.

To inspect and service sensors and switches, follow the steps in **SKILL DRILL 32-5**.

▶ Diagnosis

S32002

Diagnosing Common Noise Concerns

N32004

In many cases, manual transmissions will start making noises as they begin to fail. Listening to, and evaluating, noises is an important part of transmission diagnosis. It is usually a good practice to take the vehicle out on a road test to verify the customer's concern and to listen to any unusual noises. Some noises may be a normal characteristic of that vehicle, such as the whining noise that is produced in reverse due to the spur-cut gears. Obviously, most noises are a valid concern, so a noise is always worth looking into.

When listening to noises, listen for when they occur, along with the type of noise. For example, is the noise only evident or is it worse while the transmission is in neutral, or when in a particular gear, or only when driving down the road? The key is to be sure the customer has given you all the information about when the noise is happening. Be aware that some noises may not emanate from the manual transmission, but may be coming from other sources such as the engine, transfer case, and/or differential. In fact, even

SKILL DRILL 32-5 Inspecting and Servicing Sensors and Switches

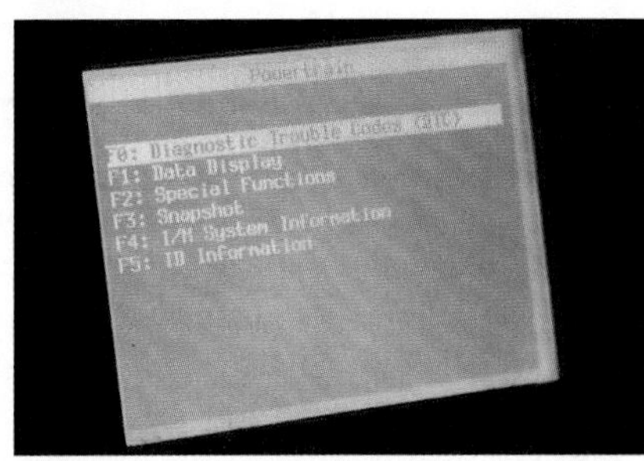

1. Hook up a scan tool to the diagnostic link connector to read information or any codes that may be present. Go to the transmission section on the scan tool screen.

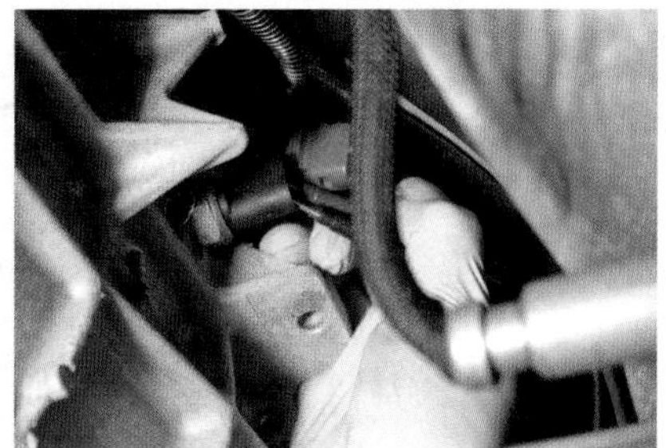

2. Perform recommended tests with the scan tool. If necessary, remove the switch or solenoid, and use an ohmmeter to check the resistance specification, if available; replace as necessary.

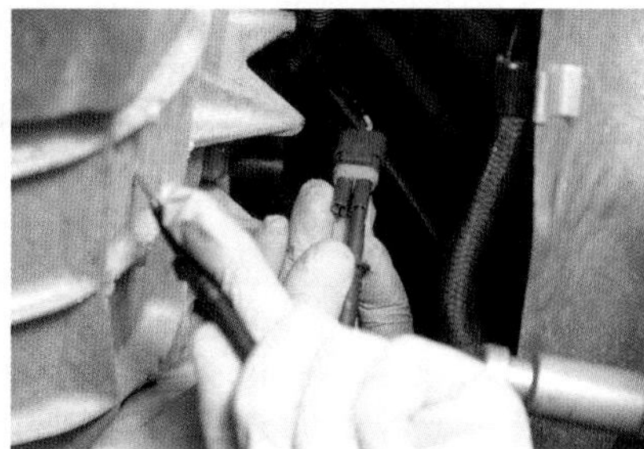

3. Check for voltage to the switch or solenoid. Check all wiring and connectors for excessive voltage drops, as high resistance can cause failures as well. Once all repairs are made, verify that the repair is successful.

bad motor mounts can cause noises that can be mistaken for noises coming from the transmission area.

Most transmission noises stop or become much quieter when the vehicle is brought to a standstill; therefore, test-driving the vehicle will help you determine the source of the noise. Transmissions can create many different types of noises. The following paragraphs will help you to diagnose the common noises associated with the manual transmission.

Grinding or clashing noises during upshifts and downshifts are fairly common on manual transmissions. In some cases, it could be as simple as the driver not fully disengaging the clutch between shifts, and may be solved by training the customer to push the clutch pedal all the way to the floor. A related cause could be the clutch having too much free play and needing an adjustment or replacement of worn parts. Or if it is a hydraulic clutch, the fluid may have air in the system, preventing the clutch from fully releasing.

Grinding noises could also be caused by a component such as a worn synchronizer blocker ring that cannot grip the gear and match the gear speed to the shaft speed. The synchronizer blocker rings have ridges on them that grab the gear cone. If the ridges are worn, the synchronizer cannot grip the gear as well and will slip. Another cause could be that a shift rail is binding, causing two gears to try to engage at the same time, which creates grinding noises.

A less common cause of grinding noises relates to the alignment of the bell housing. If the bell housing of the transmission is misaligned, the clutch disc will be out of alignment with the flywheel and pressure plate, which can cause the clutch disc to drag, making it harder for the synchronizers to match the speed for shifting into gear. This results in grinding noises.

Another noise that can come from the transmission is a rough growling noise while driving down the road. This type of noise can be caused by a couple of faults—worn bearings or worn gears. If the bearings on the input shaft, countershaft, or main shaft are worn, either front or rear, a growling noise will result when driving. If the noise is caused by a set of worn gears, it will normally be loudest in that gear while driving under a load. However, if it is the input gear and mating counter gear, then it will be loud in all gears, but not direct drive, as direct drive doesn't go through the counter shaft. It may be necessary to drive the vehicle on a lift to pinpoint the location of the noise. Teardown of the transmission may be necessary to diagnose the exact bearing or gear issue.

Clicking noises in any gear selection are typically related to a tooth of a gear that is damaged on the main shaft, input shaft, or countershaft. The clicking will increase with speed or may change with gearshifts. Pinpointing the exact gear that is chipped usually requires a teardown of the transmission.

If there is a rough growling noise in neutral, you need to know whether the clutch is engaged or disengaged. Think about what you know is spinning when the engine is running and the transmission is in neutral. The flywheel is spinning, and if the clutch pedal is held down, then the input shaft is not spinning; thus the pilot bearing is operating between the turning flywheel and the stationary input shaft. In this case, the growling noise is likely caused by a worn pilot bearing. If the clutch pedal is released, then the input shaft and countershaft are spinning, but the output shaft will be stationary, as no gears are locked to the main shaft, and they are freewheeling. By thinking through and visualizing what is moving inside the transmission, you can figure out that the noise must be related to the input shaft front bearing or one of the countershaft bearings. Although a countershaft bearing or input shaft bearing would normally be louder when driving in a reduction gear.

If the customer concern is that there is a growling noise in all gears except direct drive, then focus on the parts that are under load such as the input bearing, countershaft bearings, pocket bearing in the back of the input shaft, and the individual gears themselves. If the noise gets much quieter when the shafts are turning at the same speed (fourth gear), then it is related to the power flow through the countershaft, which could be any of the bearings on all three shafts. Noises in transaxle vehicles would be diagnosed in the same manner, first knowing how the transaxle operates and then applying that logic to when and what is causing the noise.

These concerns are not necessarily all the possible customer concerns related to manual transmissions and transaxles. Be sure to refer to the manufacturer's service information for all transmission-related problems. Do not forget to check technical service bulletins (TSBs) for any information related to the concern you are diagnosing. Many concerns could have been solved quickly and efficiently if TSBs had been checked before beginning service work on a transmission and not finding anything wrong inside.

Diagnosing Hard Shifting

N32005

During the life of the vehicle, components wear, and **hard shifting** can occur. Hard shifting is when the shifter does not move smoothly into the desired gear and requires excessive force by the driver to force it into gear. The hard shifting may also be accompanied by grinding gears. Hard shifting concerns can be the result of an incorrectly adjusted clutch or most other clutch problems. If the clutch does not disengage fully, then the input shaft will still spin, preventing the synchronizers from matching the speed between the gear and the output shaft. A grinding noise and hard shifting will result. The easiest way to diagnose this issue is to lift the drive wheels of the vehicle and place the transmission in first gear with the clutch pushed in. If the wheels spin when the clutch is pushed in, then the input shaft is still spinning. Check the clutch adjustment at this point, or check for travel issues in the hydraulic clutch due to air in the system, or incorrect slave cylinder or master cylinder installation.

Another reason for hard shifting is the clutch pedal binding. If the clutch does not move freely, it can result in a clutch that stays partially engaged. Remember, anything keeping the clutch from fully disengaging can result in hard shifting, so if the bell housing to transmission is misaligned or loose, then hard shifting can result.

Shift rails can be another reason that the transmission shifts hard. If the shift rails will not move freely, then hard shifting will result, and more force will have to be placed on the shifter by the driver. With this in mind, it is easy to see that if the synchronizers or shift forks are binding, this will also result in hard shifting. One test to see if the synchronizers are able to do their job is to try shifting the transmission into gear with the engine running, and compare that to how hard it is to shift into the same gear with the engine not running. Any difference means the synchronizers cannot adequately do their job. At the same time, the synchronizers may be overworked by a dragging clutch.

A simple issue that can cause hard shifting is created by maintenance work that is performed improperly on a manual transmission. Let's say that a customer changes transmission fluid as part of the preventive maintenance, but the customer is used to a different vehicle that uses an 80W-90 gear oil in the transmission. The customer drains the fluid and installs 80W-90 in the transmission, which requires 10W-30 motor oil. The thick gear oil will result in shift linkage that is sluggish and hard to move. The fix for this is simple: Remove the thick oil and replace with the required fluid. If the transmission will not shift into one or more gears, then check the square portion of the shift lever for wear, or check for worn shift linkages. Check for the detent system sticking. The detent ball must move up out of the way as the shift rail moves forward. If the detent ball is stuck down and will not move up, then the transmission may not be able to shift into that related gear(s). If we go inside the transmission, there may be an issue with the dog teeth on the front of the gear being worn to the point that they will not accept the synchronizer sleeve as the driver moves it into mesh with the worn teeth.

If the transmission is stuck in one gear, the technician needs to check for a stuck detent mechanism. A detent that is stuck could keep the shifter from moving to select the next gear. Another cause of this concern would be a binding synchronizer that does not allow movement of the shifter. If a gear has broken dog teeth, it is possible that the synchronizer has become jammed and will not allow changing gears. The technician's first step would be to disconnect the linkage to the transmission, if equipped, and verify that the shifter will now move freely. The issue could simply be a cable that is frozen or a linkage that is binding. It would be very bad for the technician to tear down a transmission only to find that a stuck shift cable was the cause of the customer's concern.

If the transmission is jumping out of gear, then the technician should first check for shift linkage, shift cable, or shifter boot binding. A shifter boot has been known to cause gears to pop out as the engine is accelerated or decelerated. Remember, the engine and transmission twist with torque; if the boot is pulled tight, it could push the shifter out of gear. Worn or broken powertrain mounts can also pull the transmission out of gear if it uses an external shifter as the engine moves when applying a load. Be sure to check TSBs for any issues of redesigned shifter boots. Gears popping out can also be caused by issues inside the transmission. Be sure to eliminate all causes outside the transmission before pulling the transmission down for inspection.

The synchronizer is used to ensure proper engagement of the gear without grinding. The blocking ring of the synchronizer grips the cone of the gear to slow or speed the gear to shaft speed, and it helps the synchronizer sleeve move over and lock into the gear dog teeth. The dog teeth have a back cut that ensures that the synchronizer sleeve locks into mesh with the teeth. If the dog teeth are worn, the synchronizer sleeve will not lock into place and can pull back off with torque from the engine. If the transmission has too much shaft end play, the moving the shafts can cause the shift fork to walk the synchronizer sleeve out of mesh with the gear dog teeth.

Diagnosing Transaxle Noises and Vibrations

N32006

Transaxles can exhibit similar noises as a transmission when the issue is related to the bearings or gears, so diagnose them in a similar manner, realizing that the geartrain and power flow are slightly different. Because the transaxle includes a final drive assembly that transmits power out of the transaxle to the half-shaft drive axles, it can have noises that are different from a regular transmission. The transaxle's secondary shaft drives the final drive ring gear. The ring gear then transfers torque and engine power to the differential case. The transaxle case houses the ring gear with its counterpart, the pinion gear. The differential axle side gears drive the half shaft axles that are splined to the front axle hubs. The main difference between the differential in a rear-wheel drive vehicle and the differential in a front-wheel drive transaxle is the direction of power flow. The rear-wheel drive differential power flow changes 90 degrees between the drive pinion gear and the ring gear. This 90-degree change in direction is not necessary with most front-wheel drive vehicles. Noise and vibration can be a concern in the front-wheel drive differential, as this usually can be felt in the steering wheel assembly.

Transaxle final drive assembly noise and vibrations can be caused by multiple things. Be sure to check for TSBs and service information for the vehicle you are working on. Differential noise can be related to ring gear bearings, which could have been damaged by low transaxle fluid levels or fluids of the wrong viscosity type. Bearings can be damaged by bent half-shafts and can create vibrations while driving. If the transaxle ring and pinion assembly has too little clearance, then a whining noise can result. This clearance is referred to as backlash and is typically set with selective thickness shims. Backlash is a critical inspection item upon assembly, and care should be taken to ensure it is set to the manufacturer's specifications. If backlash is excessive, then a clunking noise can result as the vehicle accelerates or decelerates.

Other issues related to transaxles include noises from damaged teeth on the ring and pinion gears. Typically, a clicking noise will result from chipped teeth. Customers who race their front-wheel drive vehicle may chip teeth due to the excessive shock placed on gear teeth. Constant-velocity (CV) joints can create vibration if the engine cradle assembly is shifted or bent. This will result in CV joints running at incorrect angles and therefore binding, which creates vibration. If either of the outer CV joints is damaged, then clicking will result when the steering wheel is turned completely to one side, especially while accelerating. If you notice the CV boot is torn, it is likely that water and dirt has already ruined the joint.

Diagnosing Drivetrain Concerns

S32003

Customer concerns related to the drivetrain will typically be evident while the vehicle is moving. The first step in diagnosis is to verify the customer concern. If the concern cannot be verified at the shop, then a test-drive with the customer may be necessary.

A customer interview is critical to the diagnosis to obtain all of the information concerning the symptoms and to ensure a proper and lasting repair. Service consultants may use a check sheet to help the customer express all of the conditions related to the concern. The check sheet asks the customer a series of questions to help determine when the noise is occurring, whether the noise is louder on acceleration or on deceleration, or whether the noise is in a particular gear. If a check sheet is not available, then the consultant should ask similar questions to ensure the technician has as many details as possible to aid in diagnosis. If the service consultant has failed to obtain the proper information, it may be necessary for the technician to interview the customer personally or go on a test-drive with the customer.

Depending on the depth of the problem, researching for TSBs or recalls may play a critical role in proper diagnosis and repair. Perform any necessary diagnostic tests according to the manufacturer's specifications, to identify the cause of the problem. The troubleshooting procedure in the service information will help the technician to diagnose the issue. For example, let's say a noise is found only in third gear on a pickup truck. The technician verifies that the noise is in third gear. The service information states that the technician should note the rpm range in which the noise occurs. The technician notes that the rpm is 3000 when the concern is occurring. The service information tells the technician to ensure that the noise is not occurring when the transmission is in neutral at that rpm. The technician finds that the noise does occur at 3000 rpm in neutral. This finding tells the technician that the concern is not in the transmission, but in the engine. Further diagnosis of the engine is needed at this point.

A visual inspection should always be performed when diagnosing the manual transmission. Manufacturers have designed prescribed tests to aid in finding a solution to the complaint. Once all data have been collected and analyzed, determine the correct action to make the repair.

Diagnosing an Electronically Controlled Manual Transmission

S32004

Manual transmissions are now being equipped with shift mechanisms using new electronic technologies. These shift mechanisms are fully electronic and are controlled by a series of modules that provide better control of gear engagement while monitoring optimum shift force at time of the shift. By controlling this shift force, which relies on engine, transmission, and vehicle operation, shift mechanisms are able to vary shift forces to complete a gear shift. These forces account for synchronizer clearances and shift rail end forces for shifting into gear smoothly. Special electronic shift routines ensure that shift rails operate at maximum efficiency and shifts are made at the right times.

Transmission gear selection is done by shift solenoids, actuators, and modules, which are controlled by the PCM/TCM according to the vehicle data, programmed software, and driver commands. The PCM/TCM then provides an output signal to operate the appropriate shift solenoid or actuator for the shift to occur. Clutch operation can be mechanical or solenoid operation. The shift lever itself is not connected to any linkage; rather, it is a shift-by-wire or electronic connection to the transmission, and information comes from the control module. When shifts are made, the computer determines the correct time to allow the transmission to shift. This type of transmission also has fail-safes built into the computers, in case the computer sees a problem in the system.

To diagnose an electronically controlled manual transmission, consult the manufacturer's service information to identify all possible problems that might result from a fault in the system. Start by locating the description and operation section to define operational characteristics of the electronic manual transmission. Understanding how the system works will help you avoid misdiagnosing the fault, as well as point you in the right direction for diagnosing it. Once you know how the transmission should be acting, verify that it is not acting properly. Use a scan tool to read any codes that may be present. You can also look at any freeze-frame and data-stream information while operating the vehicle. Doing so will give you further clues as to the cause of the fault. If the vehicle has codes, follow the troubleshooting chart for the code found. Be sure to check for codes that are related to the customer concern. Do not chase codes that are unrelated to the customer concern. For example, if the customer disconnected the battery with the ignition key on, multiple codes might be set. None of these codes is likely to be related to a noise issue.

▶ Wrap-Up

Ready for Review

- The manual transmission allows a driver to manually vary engine torque and output to the drivetrain.
- The ability of the technician to drive, service, repair, and maintain the manual transmission/transaxle is a critical skill.
- A manual transmission/transaxle allows the driver to directly vary the gear ratio, by manual selection, between the engine and the driving road wheels.
- The components of a manual transmission can vary depending on the use and size of the transmission.
- The gear ratio is the number of turns of the input gear necessary to achieve one turn of the output gear.
- Gear ratio theory helps you understand how power is distributed and how power and speed vary as the vehicle travels.
- In keeping with the gear ratio theory, compound geartrains have two or more pairs of gears in constant mesh so that they rotate together; they are found inside the transmission/transaxle.
- In direct drive the input shaft and output shaft are locked together, and the turning effort of the output gear equals the effort applied by the input gear.
- All gearboxes have shift rails and shift forks. The gearshift lever/mechanism can be located on the top, on the side, or on the column.
- The interlock mechanism is the device that ensures two gears cannot be selected at the same time.
- Many newer manual transmissions are equipped with electronic reverse lockout systems, which use an electronic solenoid to prevent the transmission/transaxle from being shifted into reverse while the vehicle is in forward motion.
- In a manual transmission/transaxle, the gears and shafts are supported on bearings or bearing surfaces, to reduce rolling friction.
- Roller bearings are used to support radial loads.
- Caged roller bearings can be used to support gear rotating on an output shaft.
- Thrust washers act as wear surfaces and can be used to adjust end play.
- Tapered roller bearings are used to sustain radial and thrust loads.
- Transmission problems include nose, hard shifting, jumping out of gear, no gear selection, and fluid leaks.
- Lubrication type and condition are critical factors in transmission performance and reliability.
- Having the gearbox and final drive gears in one casing provides a compact and relatively quiet unit and fits well into front-wheel drive vehicles.
- Historically, manual transmissions have not required the use of electronics, but as with all vehicle systems, electronics are starting to play a more common role.
- Manual transaxles problems can be diagnosed with the guidance of the manufacturer's diagnostic tables.
- A simple issue that can cause hard shifting is created by maintenance work that is performed improperly on a manual transmission/transaxle.
- Always adjust shift linkage following the manufacturer's methods.
- Always refer to the manufacturer's prescribed maintenance schedules and fluid changes to ensure the unit lasts the life of the vehicle.
- It is always a good practice to inspect powertrain mounts whenever the vehicle is on the lift or if the customer complains of unusual shifting or noises present.
- Although a manual transmission/transaxle is being shifted manually, there is a bit of electronics used in the process of selecting gears on some transmissions.

Key Terms

baulk-ring synchromesh unit A synchronizer used in all current transmissions. It uses a blocking ring. Also called a blocker ring synchromesh unit.

blocking ring A synchronizer part that increases or decreases a gear's speed to match shaft speed so that the synchronizer sleeve can lock the gear to the shaft. Also called a baulk ring.

chamfer A rounded or angled edge located on the blocking ring, with ridges machined into it to help grab the cone surface of the gear.

detent mechanism The mechanism that holds or helps hold the shift rail into position to ensure that the gear does not pop out when selected and to let the driver feel when a shift is completed.

direct drive A condition in which the input shaft and output shaft are locked together and turning at the same rate of speed.

double-clutching A technique used to shift gears when a non-synchronized transmission is used. The driver must push the clutch in multiple times to change gears.

electronic reverse lockout An electronic solenoid used to prevent accidental engagement into reverse gear while the vehicle is moving forward.

garter spring lip seal A seal that has a spring to maintain pressure on the seal's surface to prevent leakage.

gear ratio The ratio of the number of turns that a drive gear must complete to turn the driven gear one turn. Typically, this is a calculation of the driven gear to the drive gear. Ratios are listed, for example, as 2:1 or 4.3:1.

gear reduction The use of a small gear to drive a large gear. The result is in an increase in torque, but a decrease in speed.

gearshift lever The lever that the driver uses to shift the transmission.

gradient resistance Resistance encountered when a vehicle travels up a hill, requiring torque to be applied to overcome it.

hard shifting A shifting problem in which the shifter will not move smoothly into the desired gear, requiring excessive force by the driver to set it into gear.

helix The curve created by a smooth spiral and used in the angle of gear teeth and coil springs.

hydraulic actuator A hydraulically controlled cylinder that engages or disengages the clutch pedal.

hydrodynamic seal Oil seal flutes or swirls that are part of the oil seal and that create a pumping action to return oil to the transmission as the shaft rotates, enabling lubrication.

idler gear A gear used in between two gears to change the direction of the rotation of the driveshaft or drive axles in the transmission.

interlock mechanism A mechanical device that prevents engagement of two different gears at the same time.

lockout solenoid The cam and lock pin operated by an electronic solenoid that keeps the reverse gear in lockout until it is selected.

needle roller bearing Typically a small-diameter pin rolling bearing that can be held in a cage or placed into the inside diameter of a hole.

O-ring seal A seal that is a complete circle. If a cutaway is done, it will also be a complete circle. These seals are usually set into a groove that is machined into the housing or part that does not move and is intended to be sealed.

paddle A shifting mechanism or electronic control usually attached to the steering wheel.

pocket bearing A roller in the rear of the input shaft that supports the front of the main shaft.

reverse shift fork A shift fork used to engage the reverse gear.

roller bearing A long cylindrical roller held in position by a cage.

rolling resistance Resistance that is present from tires contacting the road and wind resistance against the vehicle while rolling down the highway.

selective thickness shim A shim of a prescribed thickness that is used to control shaft clearances in a transmission.

selector gate The U-shaped cutaway in shift shafts that the shifter lever fits into.

selector shift rail A rod that is attached to the shift fork. These rails move when the shifter lever is moved against them.

shift fork A mechanism that moves the synchronizer sleeve to lock the gear to the main shaft.

spring-loaded key A part of the synchronizer that helps hold the synchronizer collar in position.

synchronizer An assembly that allows for the selection of gears without grinding by matching the speed of the two assemblies.

synchronizer sleeve The sleeve that slides to lock the selected gear to the main shaft. The synchronizer sleeve is part of the synchronizer assembly.

thixotropy The ability of a semisolid grease to flow when agitated or stressed.

torque multiplication The increase of torque.

Review Questions

1. Choose the correct statement with respect to direct drive rotation.
   - **a.** The input gear turns faster than the output gear.
   - **b.** The output gear turns faster than the input gear.
   - **c.** The input shaft and output shaft are locked together.
   - **d.** The turning effort of the output gear is higher than the effort applied by the input gear.
2. For a gear ratio of 4:1, the output torque is 300 ft-lb. What would be the input torque?
   - **a.** 1200 ft-lb
   - **b.** 600 ft-lb
   - **c.** 150 ft-lb
   - **d.** 75 ft-lb
3. Which of the below engages with the dog teeth on the selected gear and transmits torque from the gear to the output shaft?
   - **a.** Synchronizer sleeve
   - **b.** Selector gate
   - **c.** Shift forks
   - **d.** Shift rails
4. Which of the following statements are true?
   - **a.** pocket bearings sit in the pocket of the input shaft.
   - **b.** pocket bearings support the front of the input shaft.
   - **c.** pocket bearings are generally not used in rear-wheel drive transmissions.
   - **d.** pocket bearings are a type of ball bearing.
5. What type of bearing is best used to sustain radial and thrust loads?
   - **a.** Cylindrical roller bearing
   - **b.** Tapered roller bearing
   - **c.** Needle roller bearing
   - **d.** Pin roller bearing
6. All of the following statements are true except:
   - **a.** gear fluid is splashed inside the casing while the transmission shafts rotate.
   - **b.** oil seals prevent fluid leakage.
   - **c.** oil seals prevent dirt and moisture from entering into the transmission.
   - **d.** the helical flutes in the seal prevent the fluid from going back into the transmission.
7. In transaxle designs, the drive is transferred through the clutch unit to a:
   - **a.** primary shaft.
   - **b.** secondary shaft.
   - **c.** counter shaft.
   - **d.** alternate shaft.
8. Which of these is used to synchronize the speeds of gear and shaft before engagement?
   - **a.** Vehicle speed sensor
   - **b.** Blocker ring synchromesh unit
   - **c.** Sliding reverse idler
   - **d.** Constant-mesh reverse gear
9. Choose the correct answer.
   - **a.** All manual transmissions use the same lubricating fluid.
   - **b.** Ensure that the gearbox fluid is not warm during gearbox fluid change.

c. If gearbox fluid levels are too low, aeration of the fluid will be present.
d. Inspect the transmission for leaks when changing gearbox fluid.

10. Which of these could be related to hard shifting?
a. Power flow through the countershaft
b. A gear that is damaged on the main shaft
c. Fluid level to high.
d. The alignment of the bell housing

ASE Technician A/Technician B Style Questions

1. Tech A says that in a conventional transmission, the gears freewheel around the main shaft until they are locked to it by the synchronizer. Tech B says that when not engaged, main shaft gears slide away from the counter gears until they are no longer in mesh. Who is correct?
a. Tech A
b. Tech B
c. Both A and B
d. Neither A nor B

2. Tech A says that the countershaft assembly can be made up of several gears machined out of a single piece of steel. Tech B says that the countershaft assembly is driven by the gear on the input shaft and drives the gears on the main shaft. Who is correct?
a. Tech A
b. Tech B
c. Both A and B
d. Neither A nor B

3. Tech A says that the speeds at which two meshed gears turn depend on the number of teeth on each gear. Tech B says that if the driving gear has fewer teeth than the driven gear, the speed of the driven gear is faster. Who is correct?
a. Tech A
b. Tech B
c. Both A and B
d. Neither A nor B

4. Tech A says that an overdrive gear ratio means the input gear turns faster than the output gear. Tech B says that overdrive ratios provide less torque output than underdrive ratios. Who is correct?
a. Tech A
b. Tech B
c. Both A and B
d. Neither A nor B

5. Tech A says that if a noise is caused by a set of worn gears, it will normally be loudest in that gear while driving under a load. Tech B says that if the bearings on the input shaft, countershaft, or main shaft are worn, either front or rear, a growling noise will result when stopped in gear. Who is correct?
a. Tech A
b. Tech B
c. Both A and B
d. Neither A nor B

6. Tech A says that some manual transmissions use two input shafts and two clutches. Tech B says that many manual transmissions now use planetary gear sets. Who is correct?
a. Tech A
b. Tech B
c. Both A and B
d. Neither A nor B

7. Tech A says that blocker rings prevent the vehicle from rolling when parked. Tech B says that engine torque is transferred from the dog teeth through the synchronizer sleeves. Who is correct?
a. Tech A
b. Tech B
c. Both A and B
d. Neither A nor B

8. Tech A says that if gearbox fluid begins to run out as the filler plug is removed, the technician must add the same amount back in. Tech B says that blocker rings have fine grooves that cut through the oil on the tapered cone of the matching gear. Who is correct?
a. Tech A
b. Tech B
c. Both A and B
d. Neither A nor B

9. Tech A says that a broken detent spring can cause the transmission to pop out of gear. Tech B says that a broken interlock pin can cause the transmission to go into two gears at the same time? Who is correct?
a. Tech A
b. Tech B
c. Both A and B
d. Neither A nor B

10. Tech A says that some manufacturers specify automatic transmission fluid to be used in their manual transmissions. Tech B says that some manufacturers specify engine oil to be used in their manual transmissions. Who is correct?
a. Tech A
b. Tech B
c. Both A and B
d. Neither A nor B

CHAPTER 33 Manual Transmission Overhaul

NATEF Task

- **N33001** Disassemble, inspect, clean, and reassemble internal transmission/transaxle components.

Knowledge Objectives

After reading this chapter, you will be able to:

- **K33001** Describe the things you need to know before rebuilding a transmission.
- **K33002** List and identify the tools used for rebuilding transmissions.
- **K33003** Describe the things necessary to prepare for disassembly of the transmission.

Skills Objectives

After reading this chapter, you will be able to:

- **S33001** Inspect transmission case and components.
- **S33002** Inspect the gaskets and seals.
- **S33003** Inspect and reinstall the synchronizer assembly.
- **S33004** Inspect and service lubrication devices.
- **S33005** Inspect and service the shift cover.
- **S33006** Inspect and service shift linkages.
- **S33007** Remove and replace the transaxle final drive.
- **S33008** Remove and service transaxle final drive pinion gears.
- **S33009** Reassemble the transmission.
- **S33010** Measure and adjust transmission end play/preload.

▶ Introduction

K33001

As the transmission wears, it will develop behaviors that are of concern to the customer, such as popping out of gear, making growling noises, or becoming difficult to shift. Although these can be due to normal wear and tear, a transmission that has been operating with insufficient gear oil will wear extremely quickly and require rebuilding or replacement. The manual transmission must be diagnosed properly to verify that the concern is indeed a fault inside the transmission and not something external that could be easily fixed, such as a bound-up shift boot that is causing the transmission to pop out of gear. You should be positive that the issue requires the teardown and inspection of the internals of the manual transmission before removing the transmission. This chapter helps guide you through the teardown and replacement of faulty components after a proper diagnosis has been performed.

Rebuilding a transmission can be a large undertaking considering all of the makes and models produced for the automotive market. Rebuilding a transmission often requires specialty tools that are specific to each transmission. The cost of these tools is usually quite high. The differences between each transmission also require you to have access to transmission repair manuals for specifications and instructions. The most important part of any rebuild/overhaul is paying attention to detail and making sure you locate the cause(s) of a transmission failure, such as worn synchronizer rings, a broken detent spring, or a worn bearing. Gear oil can be sent to a local laboratory to be analyzed for contamination or chemical breakdown. This analysis can help determine the origin of a transmission failure; for example, the wrong type of oil may have been used in the transmission or some type of incorrect additive may have been added to the oil. Locating the cause of transmission failure is critical to making an accurate and lasting repair.

Note, when we say "transmission," we use the term generically to include transaxles also. When we use "transaxle," we generally only mean transaxle. Readers should use their understanding of transmission and transaxle theory to understand the intended meaning. For example, if we are discussing ring and pinion backlash, that can apply only to a transaxle.

▶ Tools and Equipment

K33002

When performing a rebuild of a transmission, proper tool usage is vital. Manufacturer specifications call for certain tools and equipment to be used to rebuild these units. Some special tools and equipment may be required to disassemble and reassemble modern manual transmissions, such as special pullers or electronic devices. Remember, the proper tool for the proper job makes the job go smoothly and easily. Wherever possible, consult the manufacturer's publications. Some of the tools that may be needed for rebuilding the transmission are listed here (**FIGURE 33-1**):

- Dial indicator
- Outside micrometer
- Bearing press tools
- Seal driver tools
- Bearing preload tools to assist in finding correct free play
- Engine support tool
- Bushing driver kit
- Slide hammer and adapters
- Bearing splitter
- Gear pullers
- Depth micrometer
- Snap ring pliers of the proper size for the transmission
- Transmission holding fixture

Some of these tools are necessary for one transmission, but not for another. Refer to the manufacturer's unit repair manual for the necessary tools to repair the unit you are working on.

Failures inside the transmission may simply indicate replacement of wear items, such as seals, gaskets, bearings, and synchronizer blocker rings, or may indicate the failure of hard parts, such as gears, synchronizer assemblies, and input, output, or countershafts. Rebuild kits can be purchased in various levels. The most basic kits may include only gaskets and seals. Other kits may include only bearings. Higher-level kits may include gaskets, seals, and synchronizer rings. Top-level kits include gaskets, seals, synchronizer rings, bearings, and sometimes even shafts. It is possible that the kit you need will not be available and that you will have to order individual parts as needed.

You Are the Automotive Technician

Today you are working alongside Dave, a senior technician, to rebuild a manual transmission. Dave has several years of experience doing this complicated repair and you are assisting him. Dave has informed you that Mrs. Marston's Jeep makes grinding noises when shifting gears. The fluid level is fine, but gold flakes in the fluid indicate worn synchronizer assemblies, requiring a transmission rebuild. Dave carefully checks the service information prior to transmission disassembly and explains the importance of patience, organization, and cleanliness during the rebuilding process. As Dave has the transmission disassembled into its subassemblies, he disassembles the main shaft and shows you the worn blocker rings and how the fine grooves are mostly worn away. The dog teeth on the gears and collars on first and second gear are worn due to repeated grinding and will require replacement.

1. Why is it important to be organized during a transmission replacement?
2. Why are gold flakes in the fluid an indicator of worn synchronizers?
3. Why should manual transmissions subassemblies be disassembled one by one?

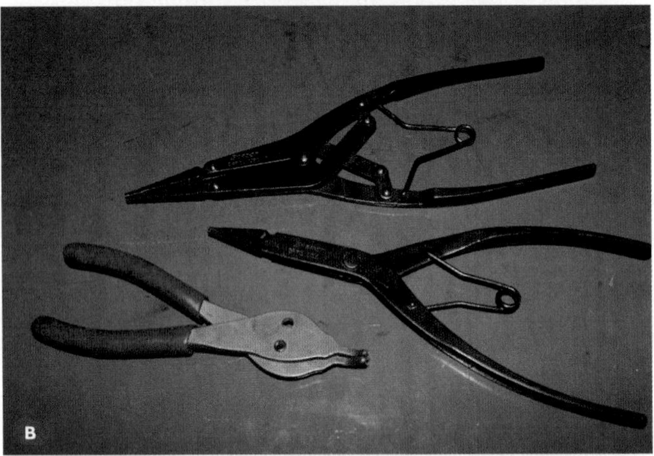

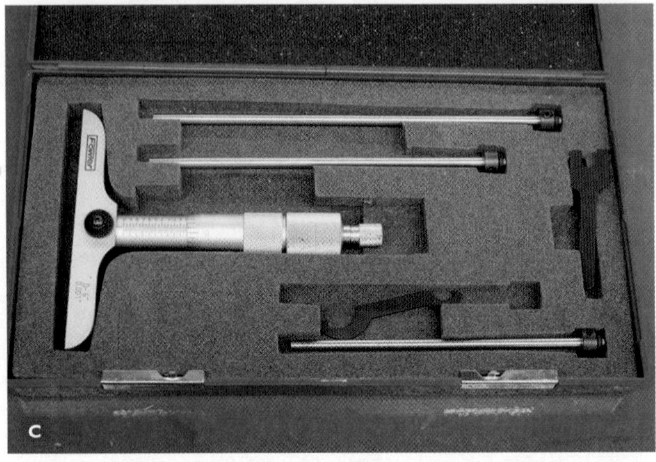

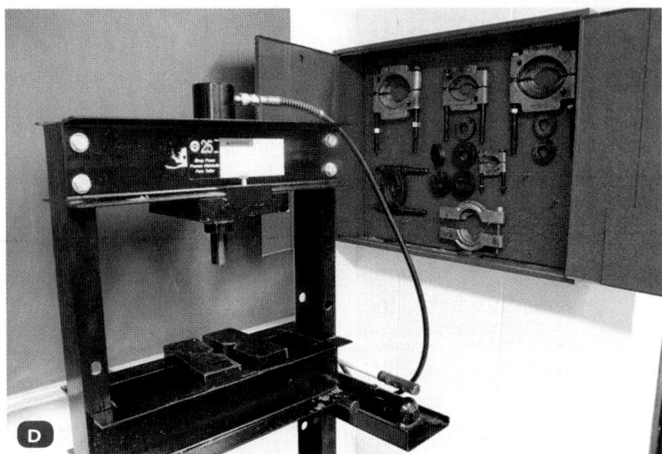

FIGURE 33-1 Depending on the transmission being serviced, specialty tools may be necessary. Some of the basic tools needed are pictured here.

▶ Preparing for Disassembly

K33003

The disassembly of a transmission is not something to be performed by a novice technician without any training. Many clearances must be checked as the transmission is torn down and again as it is reassembled. Specialty tools must be used to ensure that damage is not done to the parts that do not need to be replaced. The goal of the rebuild is to restore the transmission to like-new condition for the customer at the agreed-upon price. Damaging components because of lack of knowledge and skill is not keeping the customer's or shop owner's interests in mind. It is necessary to use a unit repair manual to perform this task properly.

Among the first tasks when doing any kind of service to a manual transmission is obtaining and reading all manufacturer publications and service information. The manufacturer's information should provide the technician with a good idea of how the transmission is disassembled before the technician actually tears the unit down. It is wise to take some time to look over the procedures in the manual, ensure that the proper tools are available, and check for any technical service bulletins related to the customer concern you are addressing.

Diagnosis

If it is necessary to disassemble a manual transmission because of an internal problem, a complete and thorough diagnosis is warranted to complete an error-free repair. This diagnosis involves a thorough understanding of the customer complaint, which may require an interview with the customer to determine the events that led to the failure and bring to light any symptoms that were present before the failure occurred.

Once the concern is fully understood, the technician needs to verify the concern, usually by test-driving the vehicle, and diagnose the possible causes. A troubleshooting chart should help you track the customer concern to the specific fault inside the transmission. For example, if you find that the customer does not fully depress the clutch to shift between gears, then the popping-out-of-gear concern may be related to the dog teeth on the gear being worn due to the grinding that would happen as the driver forces the synchronizer sleeve into mesh. Following are some additional areas of diagnosis that you should consider before tearing a transmission down. Once all the information is obtained, you may begin the disassembly process knowing that you have a good understanding of what may have happened and what you expect to find.

▶ Disassembly Procedures

N33001

For example purposes, this disassembly procedure is based on the Borg-Warner T-5 transmission. First, remove the manual transmission according to the manufacturer's guidelines. Once the manual transmission is removed, place it on a clean and organized bench. If not already performed, drain the transmission fluid by removing the drain plug as required. Remove any electronic sensors or wiring that may be attached to the transmission case. Perform a thorough inspection of the outside of the transmission, looking for any obvious signs of the failure or any other damage.

In most cases, manual transmissions come apart in main assemblies, such as the shift cover, extension housing, input shaft, output shaft, and countershaft. But the parts can be close fitting, requiring patience in determining how they will come apart. Some parts may be held together only by the positioning of the assemblies, and when you start to remove them, these loose parts will fall off. Take your time as each assembly is removed. Have a number of parts trays available to store parts during the disassembly. Labeling unknown parts is a good idea to ensure that they are reassembled correctly. Pay close attention to detail here, as organization during disassembly is a critical part of any repair.

▶ TECHNICIAN TIP

There may be a check ball under the speedometer gear. Make sure to secure it as the gear is removed.

To disassemble a manual transmission, follow the steps in **SKILL DRILL 33-1**.

SKILL DRILL 33-1 Disassembling a Manual Transmission

1. Remove the retaining bolts and remove the shift lever from its seat. On some manual transmissions, it is best to place the gearshift in neutral prior to removing the shift lever.

2. Remove the bolts that retain the tail shaft housing; then remove the housing.

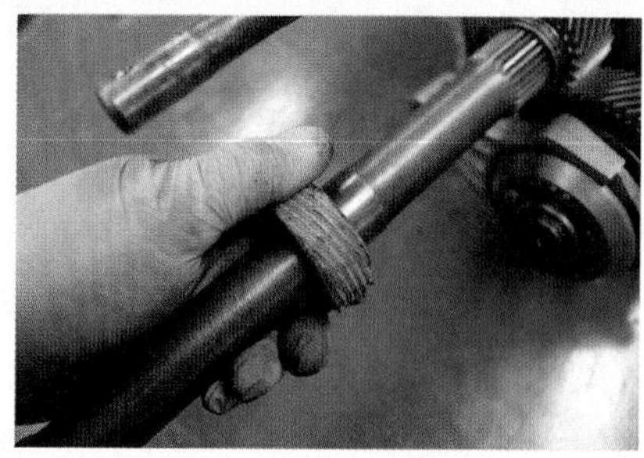

3. Remove the speedometer gear and retainer, and slide it off the output shaft. Some gears are pressed on, so you may have to use a gear puller to remove these.

4. Remove the shift mechanism housing bolts, and remove the shift mechanism housing from the top of the manual transmission.

5. Remove the input shaft bearing retainer bolts, and remove the retainer from the transmission case.

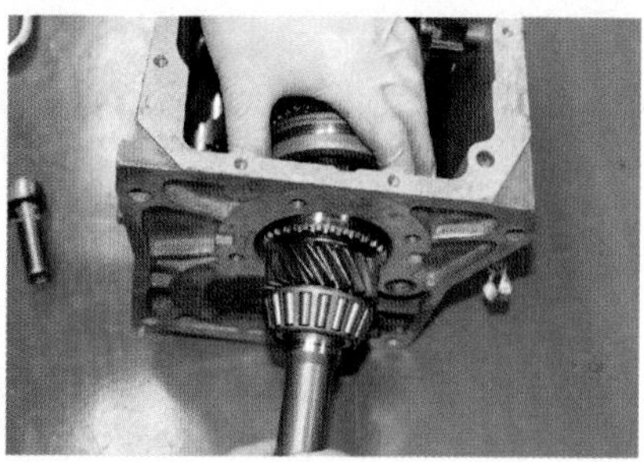

6. Rotate the input shaft so the dog teeth cutout is lined up with the gear on the cluster shaft for easy removal. Pay attention to the roller bearings in the end of the input shaft. Now pull the input shaft from the transmission case.

SKILL DRILL 33-1 Disassembling a Manual Transmission (Continued)

7. Remove the fifth gear driven gear from the countershaft by removing the synchronizer and snap ring for access.

8. Remove the rear bearing race from the transmission case with a brass drift and hammer.

9. Tilt the output shaft on an upward angle to enable removal of the shaft from the transmission case. Before removal, make sure the third gear synchronizer is engaged to make clearance for removal of the output shaft. Remove the output shaft from the case.

10. Remove the reverse idler gear shaft retaining roll pin from the idler shaft. Remove the idler shaft and gear from the transmission case.

11. To remove the countershaft from the transmission case, it may be necessary to put the transmission case in a press to remove the rear countershaft bearing. Press out the bearing, and then remove the countershaft from the transmission.

12. The front pocket bearing (which is a caged roller-type bearing) must also be removed in order to be replaced upon overhaul.

13. To disassemble the main shaft/output shaft:

- Position the shaft in a soft-jaw vise. Remove the first snap ring that holds third and fourth gears in place.
- Remove the third and fourth gear sets and roller bearing cages.
- Remove the second snap ring, and remove the second gear.
- Remove the first and second/reverse synchronizer sliding collar.

14. Disassemble the output shaft:

- Mount the output shaft in the press to remove fifth gear, which is pressed onto the shaft. This will also remove first gear and the rear output shaft bearing.

15. Remove the output shaft pocket bearings from the inside of the input shaft. Place the input shaft in the press, and press off the front bearing.

▶ TECHNICIAN TIP

Pay close attention to the magnet in the bottom of the transmission case. The amount of filings found on it indicate gear and bearing wear.

Transmission Case and Components Inspection

S33001

When **overhauling** a transmission, all transmission case components should be cleaned and inspected to make sure they can be reused. A close inspection of all vital parts is necessary to complete a manual transmission. If there is any doubt as to whether a case component is reusable, then it should be replaced. You have an important cost analysis job to perform for the customer. If the transmission rebuild will cost more than a remanufactured unit from a parts supplier, then the customer should be informed of the option of using a remanufactured unit. Remanufactured transmissions come with a warranty from the remanufacturer and will ensure that the customer is covered against any warranted defects for the term of the warranty, typically up to one year. This may present a significant savings to the customer and ensure that the repair facility performing the repair is providing the best value for the customer.

To inspect a transmission case, follow the steps in **SKILL DRILL 33-2**.

SKILL DRILL 33-2 Inspecting a Transmission Case

1A. To inspect the main transmission case:

- Check flat mating surfaces for warpage, using a straightedge and a feeler gauge to ensure straight surfaces.

1B. Check bores for concentricity or roundness and any specific defects. Check that all plug hole threads are in good shape and are able to accommodate the plug without stripping.

1C. Check the transmission case for cracks and porosity. Check that all attaching ears are in good shape and are not elongated.

1D. Check the condition of the magnet in the bottom of the transmission case, and replace as necessary. Make sure old gasket material has been removed.

2A. To inspect the output or tail shaft housing:

- Inspect the mating surface for flatness, using a straightedge and feeler gauge.

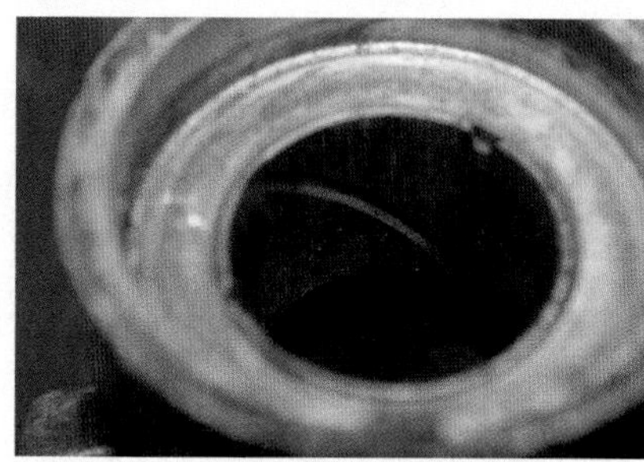

2B. Inspect the output shaft bushing, and replace as necessary.

SKILL DRILL 33-2 Inspecting a Transmission Case (Continued)

2C. Check the speedometer opening for any damage, and install a new O-ring when assembling the transmission.

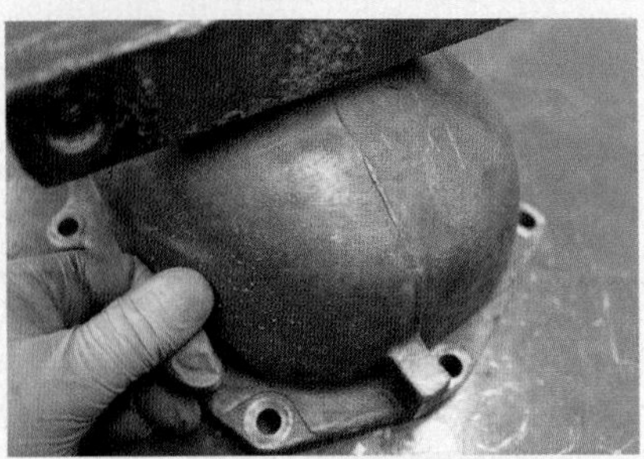

2D. Inspect the tail shaft housing for cracking and/or porosity problems.

2E. Inspect the tail shaft seal surface, and replace the seal when reinstalling it on the transmission case. Inspect the housing vent to make sure it is not clogged.

3A. To inspect the input shaft retainer:

- Inspect the retaining collar that the release bearing rides on for imperfections.

3B. Inspect the bearing cone surface for galling or cracking, or out-of-round conditions.

3C. Inspect the front seal surface, and replace the front seal as necessary.

3D. Check to make sure the bolt holes are not elongated or damaged from bolts that were loose.

Next, inspect the manual transmission by following the steps in **SKILL DRILL 33-3**.

Inspecting the Gaskets and Seals

S33002

When rebuilding a transmission, it is always a good practice to replace the gaskets and seals. A thorough inspection of all gasket surfaces is important to prevent premature leakage of transmission fluids. Imperfections of seal surfaces can cause improperly installed seals and transmission failures by allowing oil to seep past the imperfections.

Before inspecting the gaskets, clean all gasket surfaces to make sure the surface will accept the gasket and prevent leaks. The transmission mating surfaces, such as the main case and output shaft housing, should be checked with a straightedge and feeler gauge to make sure both surfaces are flat, to ensure a tight seal. This is especially important to prevent leakage while the vehicle is in operation.

SKILL DRILL 33-3 Inspecting the Manual Transmission

1. Inspect the front seal surface on the input shaft for damage.

2. Inspect the input shaft:

- Inspect the input shaft gear for excessive wear or damage on the gear teeth.
- Inspect the dog teeth for any broken, cracked, or unusual distortions. The dog teeth are pointed to ensure complete synchronous hub engagement.
- Inspect the input shaft splines for wear and twisting.
- Inspect the synchronizer smooth surface for any brinelling or unusual wear.
- Inspect the pilot bearing surface of the input shaft for any wear.

3. Inspect the output shaft gear and components:

- Inspect necessary gears for tooth wear and broken or cracked dog teeth; check the blocking ring surface and ensure that it is smooth.
- Inspect caged needle bearings that are used inside the rotating output shaft gears.
- Check washers and snap rings for tension and possible distortion.
- Inspect smooth surfaces that gears rotate on for out-of-round and brinelling.
- Inspect splines that gears slide over for any wear that would inhibit proper installation of the gears.
- Inspect the output shaft hub for imperfections that will hinder synchronizer collar operation.
- Inspect output shaft pilot surface for brinelling from the input pocket bearing.

4. To inspect the countershaft gears and shaft:

- Inspect the bearing ends of the countershaft for imperfections.
- Inspect gears for wear, cracking, or chipping.
- Inspect gears that are not part of the shaft for wear, cracks, and chipping.

5. To inspect the reverse idler gear:

- Inspect the reverse gear brass bushing for excessive wear.
- Inspect the idler shaft for brinelling or wear.

The front bearing retainer and the main transmission case should also be checked with a straightedge and feeler gauge. If this area starts to leak, it could contaminate the clutch friction disc surface and cause a catastrophic failure of the clutch system. In addition, consider checking the shift tower surface for flatness with a straightedge and feeler gauge, as a leakage of fluid can occur in this area.

All seal surface openings should be checked for **concentricity** with a dial caliper or bore gauge. If you are using a dial caliper, then expand the dial caliper to touch the surfaces of the bore and get a size reading. Now move the dial caliper to a position 90 degrees from the original, and ensure that the reading is the same or very close. Refer to the manufacturer's information for concentricity of the case bores. If no specifications are given, inspect for scoring or cracks, and replace if present. Sealants should be applied where necessary and according to the manufacturer's specifications.

To perform a gasket and seals inspection, follow the steps in **SKILL DRILL 33-4**.

SKILL DRILL 33-4 Inspecting Gaskets and Seals

1. Clean all gasket surfaces as required. Use a straightedge and feeler gauge to check the main case and output shaft housing, the front bearing retainer and main case, and the shift tower surface on the main case.

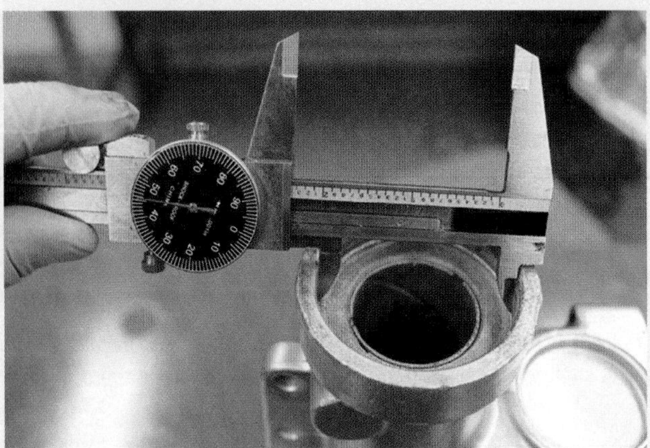

2. Measure seal surface openings with a dial caliper. When appropriate, apply all necessary sealants as required.

Inspecting and Reinstalling the Synchronizer Assembly

S33003

Synchronizer assemblies allow for the smooth shifting of the transmission. When overhauling a manual transmission, it is important to inspect for damage if you are reusing the synchronizer assembly. If the synchronizer assembly is damaged, replace all parts or the entire component, as necessary.

A good place to start when inspecting for synchronizer wear and damage is when you drain the transmission fluid from the transmission. Any gold-colored particles may indicate that the **blocking rings** are excessively worn and may need to be serviced. Sloppy or loose shifting, or jumping out of gear can indicate damaged synchronizer units as well as other shift mechanism problems. A thorough inspection of all synchronizer parts is critical to make sure the transmission gear selections will be smooth, quick, and easy.

To inspect and reinstall a synchronizer hub, follow the steps in **SKILL DRILL 33-5**.

SKILL DRILL 33-5 Inspecting and Reinstalling a Synchronizer Hub

1. Inspect the synchronizer hub for any malformations such as chipping, cracking, or broken parts. Such damage may require replacement of the output shaft in some cases.

2. Inspect the synchronizer sleeve for any visible defects such as chipping or cracking.

3. Check the inserts for any defects. If bent or cracked, replace the synchronizer inserts by sliding the synchronizer sleeve off the synchronizer hub. The keys and springs will fall out. It is necessary to disassemble the synchronizer for inspection.

SKILL DRILL 33-5 Inspecting and Reinstalling a Synchronizer Hub (Continued)

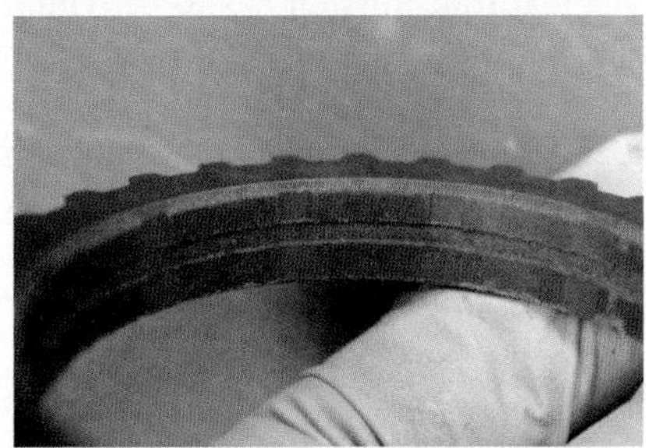

4. Inspect the blocking rings for wear, which can be seen on the friction side of the ring. There are usually serrations or friction material that are visible, but if missing or worn, replace the blocking rings as a pair. On transmissions that give blocking ring clearance specifications, measure the clearance with a feeler blade.

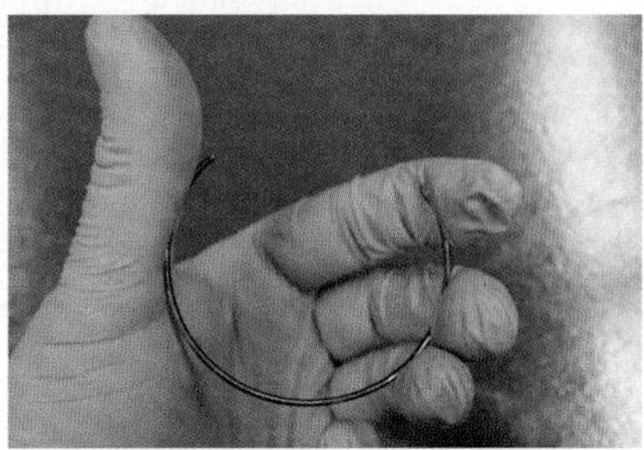

5. Inspect the insert springs for tension, and replace as necessary.

6. To assemble the synchronizer hub:Lubricate the synchronizer hub.

- Install the insert springs, paying attention to orientation.
- Lubricate the inserts and install in slots in the synchronizer hub.
- Insert the blocking rings into the hub, and align the slots in them so as to accommodate the inserts.
- Align the synchronizer sleeve, paying attention to grooves that may line up with the inserts, and slide into position.
- Reinstall the synchronizer on the output shaft, or if already a part of the output shaft, check for proper operation.
- Check synchronizer clearance, using a feeler gauge.

Inspecting and Servicing Lubrication Devices

S33004

Some manual transmissions use an **oil slinger** behind the input shaft. The oil slinger ensures that oil is directed toward the front bearing, and oil is returned to the transmission case. The oil slinger should be inspected for any defects or damage such as scratches or bending of the slinger that might have occurred during the transmission failure.

Sometimes it is necessary to use an oil pump to lubricate a transmission in order to maintain a safe temperature inside the case. Some manufacturers, such as the Ford Motor Company, use the ZF six-speed transmission. This manual transmission has an oil pump with an external oil cooler used to carry heat out of the transmission, similar to an automatic transmission during heavy-duty usage of the vehicle. During a rebuild, it is always good practice to inspect or replace the oil pump parts when necessary. Loss of oil pressure could result in transmission damage.

To inspect and service an oil pump and related lubrication devices, follow the steps in **SKILL DRILL 33-6**.

SKILL DRILL 33-6 Inspecting and Servicing an Oil Pump

1. Start by removing the oil pump from the transmission case, following the specified procedure. Inspect all gears for excessive wear and clearances, and replace as necessary.

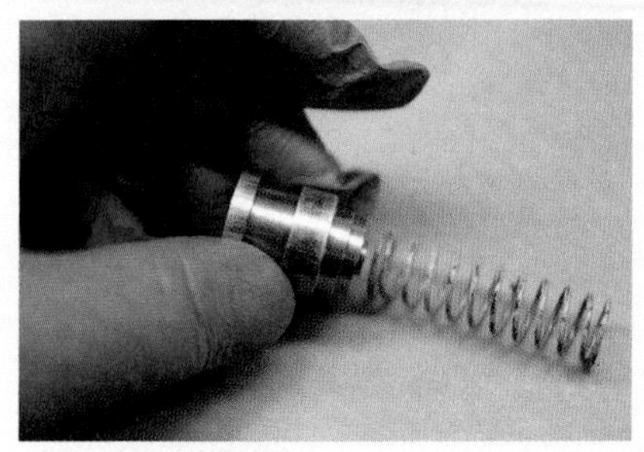

2. Inspect the regulator piston for wear and etching in the bore.

3. Replace any parts that have wear or etching. Reassemble the oil pump in the transmission case and check clearances. Any oil slingers should be replaced as necessary with the gaskets and seals.

Inspecting and Servicing the Shift Cover

S33005

The shifting mechanism of a manual transmission takes a lot of abuse from the driver. Shift lever bushings, shafts, pivot points, and seals wear over time and may require service. To prevent failure in the future, it is always a good practice to inspect and service the shift mechanism when rebuilding a transmission. Upon disassembly of the transmission, the shift cover and associated parts should be inspected along with the rest of the transmission parts. Shift mechanisms may also be part of the reason the manual transmission has to be rebuilt. Disassembly and cleaning should be the first priority before inspection.

When removing the shift cover and mechanism, make sure the manual transmission is in neutral. This will make it easier to install the shift cover when the reassembly is complete.

To inspect and service the shift cover, follow the steps in **SKILL DRILL 33-7.**

SKILL DRILL 33-7 Inspecting and Servicing the Shift Cover

1. If an O-ring is present that is used to seal the cover to the output shaft housing, it should be replaced.

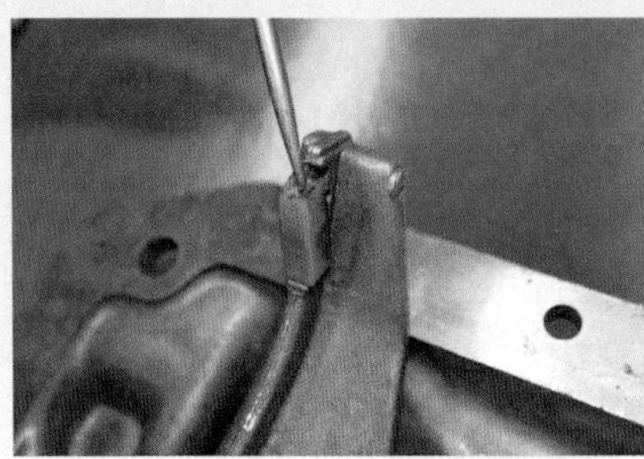

2. Check all shifter forks for wear in the gear selector slots. If the gearshift fork inserts are worn or cracked, they should be replaced before the installation of the shift cover.

3. Check the gear selector interlock sleeve for excessive wear and binding. If necessary, clean or replace.

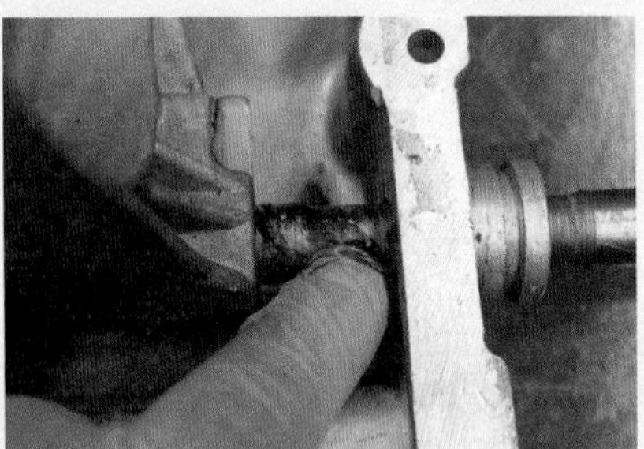

4. Lubricate all main shift control shafts with petroleum jelly before installation.

5. As the shift rails are being installed, check all detent springs and balls for damage, and replace as necessary.

6. If roll pins are used to secure shift forks or sleeves, inspect for tightness and replace as necessary by driving the pin out and replacing with a new one. If the new pin fits loosely in the case, the case may be damaged and will have to be replaced or repaired.

7. Secure a new shift cover gasket, usually by using a dab of RTV sealer, and install the shift cover. Some transmissions use RTV without an additional gasket, as does the transmission in the picture. Follow the manufacturer's directions to ensure a leak-free repair. Make sure the shift forks line up with the corresponding gears.

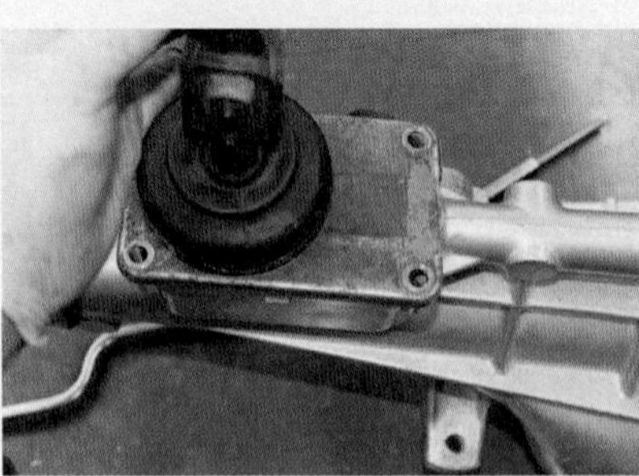

8. Install the shift lever, and check that all gears can be shifted without binding.

Inspecting and Servicing Shift Linkages

S33006

On removal of the transmission, it is necessary to remove the shift linkage and associated parts. Any adjustments that were present will have to be performed again because the shifter cables or linkages will have to be loosened. It is always a good practice to check all components of the shifting mechanism and replace any parts that are damaged or broken. Upon inspection, it may be possible to find out what caused the failure of the transmission to begin with; for example, if the transmission shift linkage is keeping one gear partially engaged, the dog teeth of the transmission can be damaged.

To inspect and service shift linkages, follow the steps in **SKILL DRILL 33-8**.

Removing and Replacing the Transaxle Final Drive

S33007

Sometimes it is necessary to remove the transaxle final drive, whether because of a failure or because of a noise problem emanating from that area, such as gear or bearing noise. The procedure for this task is specific to each vehicle. Because bearing preload and gear contact patterns are crucial to the longevity of the transaxle, be sure to carefully follow the specified procedure in the service information.

SKILL DRILL 33-9 is based on a Ford MTX transaxle and is used solely as an example of the process.

Removing and Servicing Transaxle Final Drive Pinion Gears

S33008

After transaxle disassembly, the transaxle final drive is exposed for service. The pinion gear is part of the output shaft of the transaxle. The pinion gear is what drives the differential ring gear, which is located inside the transaxle assembly. If it needs to be replaced, it will also be necessary to replace the ring gear as well. These are, as in a rear-wheel drive vehicle, a matched set. Remove the differential case, following the manufacturer's procedures.

To remove and service the transaxle final drive pinion gears, follow the steps in **SKILL DRILL 33-10**.

Reassembling the Transmission

S33009

After inspection and replacement of faulty components, the transmission is now ready for reassembly. If it was found

SKILL DRILL 33-8 Inspecting and Servicing Shift Linkages

1. Lay all the associated parts out on a bench, if possible, to get a clear picture for inspection. Clean off the dirt and grease in order to properly inspect.

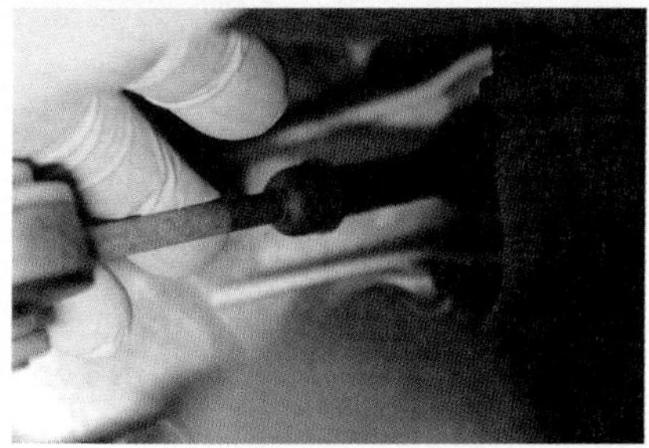

2. Inspect the cable linkage for fraying and stretching.

3. Inspect the mechanical linkage for worn or elongated holes where the bushings ride.

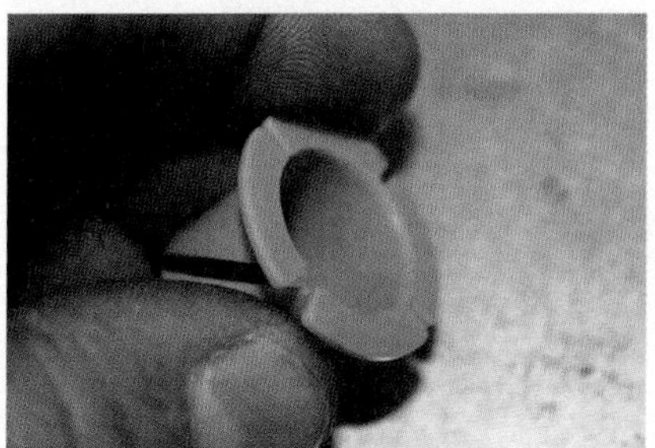

4. If the vehicle has plastic or brass bushings, inspect them for excessive wear and replace as necessary.

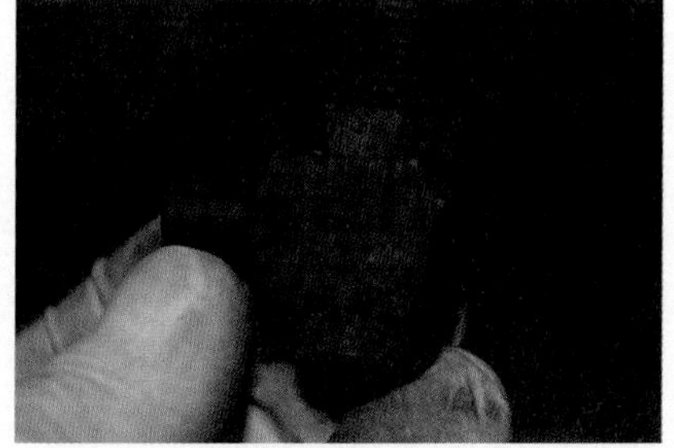

5. Inspect linkage rod ends for excessive wear from abuse. Inspect the rods for abnormal shape, which can occur as a result of driver abuse.

6. Inspect mounting brackets for excessive wear and looseness, and replace as necessary. Check all sensors and solenoids.

SKILL DRILL 33-9 Removing and Replacing the Transaxle Final Drive

1. Following the manufacturer's specifications, disassemble the transaxle case and shafts as necessary to gain access to the final drive assembly. Parts of the final drive include the output shaft, which has the pinion gear molded on to it, and the differential case.

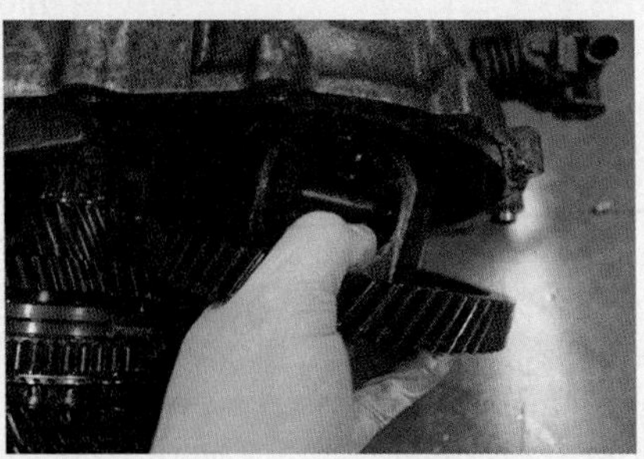

2. Remove the output shaft from the transaxle case and inspect. See Skill Drill 33-2 for more information. Remove the differential case from the transaxle housing. Be careful of the spider gear orientation, and do not let them fall out of the differential case. In some cases, there is no retaining pin to hold them in place.

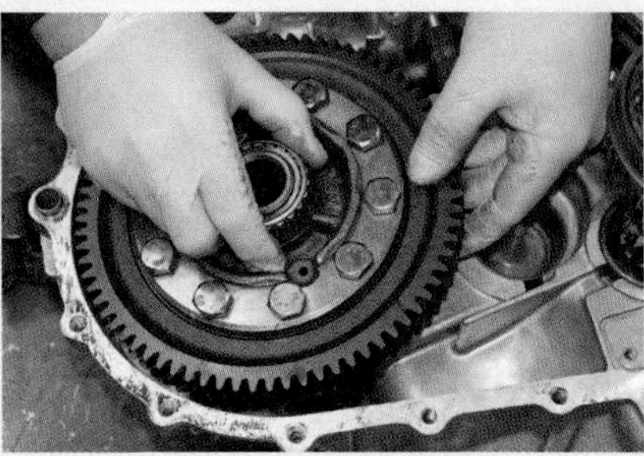

3. After performing any needed service, reinstall the differential case in the transaxle housing, seating the bearing in its race.

4. Reinstall the output shaft in the transaxle case and mate with all other shafts.

5. Measure the preload and correct by using shims under the case bearings of the differential by using a dial indicator, following the manufacturer's procedure. Reassemble the transaxle's major components.

SKILL DRILL 33-10 Removing and Servicing Transaxle Final Drive Pinion Gears

1. Remove the ring gear from the differential case. If inspection shows that it should be replaced, remove by drilling the rivets out of the ring gear itself for replacement.

2. Start disassembly of the differential case by removing the roll pin from the pinion shaft lock pin.

3. Remove the differential lock pin by sliding it out of the housing.

SKILL DRILL 33-10 Removing and Servicing Transaxle Final Drive Pinion Gears (Continued)

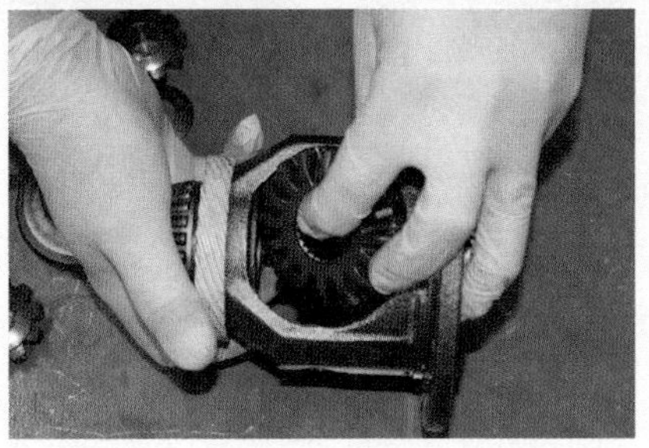

4. Remove the differential pinion side gears and differential pinion thrust washers.

5. Remove the differential spider gears and differential spider gear thrust washers.

6. Use a push-type bearing puller and a step plate to remove the differential case bearings.

7. Remove the speedometer drive gear.

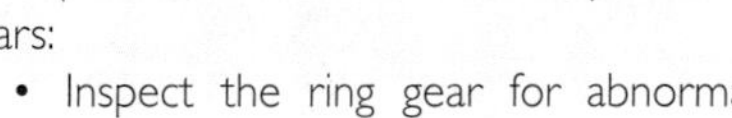

8. Inspect the transaxle final drive pinion gears:

- Inspect the ring gear for abnormal wear or cracking.
- Inspect differential spider gear sets as well as the side gears for abnormal wear and cracks or scratches or galling in the thrust washer area.
- Inspect thrust washers; if unusable, replace as necessary.
- Inspect locking pins for wear and distortion.
- Inspect the thrust surfaces in the differential case for damage in the thrust washer areas.
- Inspect the differential case for any damage, and replace as necessary.
- Check bearing surfaces on the differential case for any problems.

9. Reassemble the differential case, following the manufacturer's procedures. Install a new ring gear with either bolts or new rivets.

10. Reinstall the differential case in the transaxle case. Measure any necessary clearances.

to be cheaper to use a replacement transmission, then the transmission may need to be reassembled for a core return to the manufacturer. It will likely be sent out to a rebuilder for refurbishing and then resold as a remanufactured transmission. In most cases, rebuilders require you either to give them your old transmission or pay a core charge, which allows the rebuilder to purchase another core for rebuilding purposes.

You may need to use some of the special tools previously mentioned to reassemble the transmission. Take extreme care to ensure the transmission is assembled correctly, and the proper clearances and bearing end play/preloads have been set.

To reassemble the major components of a manual transmission, follow the steps in **SKILL DRILL 33-11**.

SKILL DRILL 33-11 Reassembling the Major Components of a Manual Transmission

1. To reassemble the input shaft assembly:
- Install a new input shaft bearing using a hydraulic press by pushing the bearing down on to the shaft.
- Pack and install the pocket bearings.
- Install the spacer on top of the pocket bearing.

2. To reassemble the output shaft assembly:
- Install the output shaft in a hydraulic press.
- Install the first gear and blocking ring and the rear bearing and fifth gear.
- Press all items into their proper position.
- Install the fifth gear snap ring.
- Install the speedometer check ball and speedometer gear according to the manufacturer's specifications.
- Remove the output shaft from the press. Install the output shaft in a vise.
- Assemble the first and second gear synchronizer assembly.
- Install the second gear with a spacer and snap ring.
- Install the third gear along with the fourth gear synchronizer.
- Install the third gear retaining snap ring.

3. To reassemble the countershaft:
- Press all gears back onto the countershaft shaft.
- Press all bearings back onto the ends of the countershaft shaft.

Measuring and Adjusting End Play/Preload

S33010

End play and preload adjustments are a necessary part of the disassembly and assembly of the transmission. End play is the movement back and forth of the input or output shaft. It is typically set by using preselected spacers or washers to ensure the correct clearance is present. The need for end play and preload adjustments can be an indicator of washer wear or bearing failure, which causes the shaft to walk back and forth in the case. This movement leads to misalignment of shift forks, making shifting difficult or popping the transmission out of gear. Axial or radial clearances (i.e., end play and preload adjustments) can be used to determine whether gears and shafts are to be replaced or reused.

Most manufacturers require that certain end play and preload adjustments be done upon assembly of transmissions. Preload ensures that enough pressure is placed on the bearing to keep the shaft from moving around, but not so much that the bearing will run hot and begin to fail. Most of these adjustments require the use of a dial indicator or feeler gauge to obtain proper clearances. Follow all manufacturer specifications and procedures upon disassembly and assembly to ensure a proper and lasting repair.

If one of the transmission shafts is found to have too much free play, then a thorough inspection must be performed to ensure that there is not wear on a component, such as a thrust bearing, causing the fault. If there is no issue with components, then check for thrust washer or shim, and install a proper-sized thrust washer, if needed.

Measuring end play is critical to ensure that the transmission stays in service for a long time. Incorrect end play and bearing preload will result in failure of the transmission.

To measure and adjust end play/preload, follow the steps in **SKILL DRILL 33-12**.

SKILL DRILL 33-12 Measuring and Adjusting End Play/Preload

1. Once the transmission assembly is together, turn the transmission case on its end, and mount a dial indicator on the end of the output shaft. Rotate the dial indicator to zero.

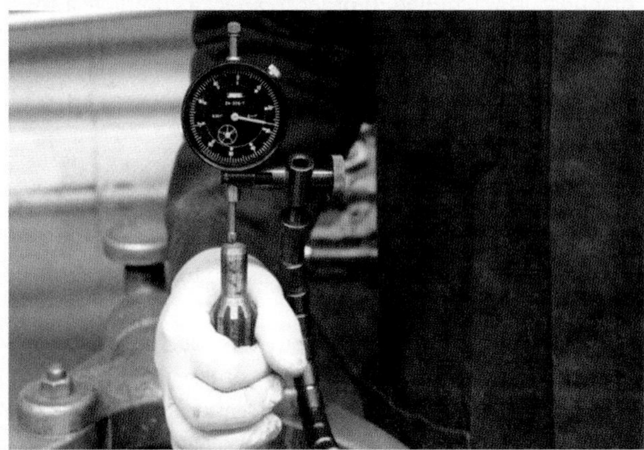

2. Grab the output shaft and raise it upward until the end play is totally removed; doing so will give you an accurate reading. Record the measurement and compare with the manufacturer's specifications. Most are usually around 0.005" to 0.025" (0.127 to 0.635 mm).

3. If an adjustment is necessary, remove the front bearing retaining cover. Remove the front bearing race, which may expose any existing shims.

4. Install the correct shims to bring the end play into specifications. Reinstall the front bearing retainer and gasket. Torque the bolts to the manufacturer's specifications.

▶ Wrap-Up

Ready for Review

- As the transmission wears, it may begin to develop concerns for the customer such as popping out of gear, making growling noises, or becoming difficult to shift.
- A transmission that has been run low on gear oil will wear extremely quickly and will result in the need for rebuilding or replacement.
- Rebuilding a transmission/transaxle can be a large undertaking, considering its complexity and all of the makes and models produced for the automotive market.
- The most important part in any rebuild or overhaul is to pay attention to detail and make sure you locate the cause/s of a transmission/transaxle failure, such as worn synchronizer rings, broken detent spring, or worn bearing.
- Some special tools and equipment may be required to disassemble and reassemble today's modern manual transmissions, such as special pullers or electronic devices.
- Failures inside the transmission can simply indicate replacement of wear items such as seals, gaskets, bearings, and synchronizer blocker rings; or the failure of hard parts, such as input, output, or counter shafts, gears, and synchronizer assemblies.
- The rebuilding of a transmission/transaxle is not something to be performed by a novice technician without any training.
- Among the first things to consider when doing any kind of service to a manual transmission/transaxle is obtaining and reading all manufacturer publications and service information.

- If it is necessary to disassemble a manual transmission because of an internal problem, a complete and thorough diagnosis is warranted in order to affect a complete and error-free repair.
- Once the concern is fully understood, the technician needs to verify the concern, usually by test-driving the vehicle, and diagnose the possible causes.
- When overhauling a transmission/transaxle, all transmission case components should be cleaned and inspected to make sure that they can be reused.
- You have an important cost analysis job to perform for the customer.
- When rebuilding a transmission/transaxle, it is always a good practice to replace the gaskets and seals.
- Before replacing the gaskets, clean all gasket surfaces to make sure that the surface will accept the gasket and prevent leaks.
- The transmission/transaxle mating surfaces, such as the main case and output shaft housing, should be checked with a straight edge and feeler gauge to make sure that both surfaces are flat, to ensure a tight seal.
- The front bearing retainer and the main transmission/transaxle case is another area that should be checked with a straight edge and feeler gauge.
- The shift tower surface is also an area to consider and check for flatness with a straight edge and feeler gauge, as this may cause a leakage of fluid.
- When overhauling a manual transmission, it is important to inspect the synchronizer assemblies for damage before reusing them.
- A good place to start when inspecting for synchronizer wear and damage is when you drain the transmission fluid from the transmission/transaxle.
- Any gold-colored particles may indicate that the blocking rings are excessively worn and may need to be serviced.
- The oil slinger should be inspected for any defects or damage such as scratches or bending of the slinger that might have occurred during the transmission failure.
- During a rebuild, it is always good practice to inspect or replace the oil pump parts, if equipped.
- To prevent failure in the future, it is always a good practice to inspect and service the shift mechanism when rebuilding a transmission.
- Shift linkage adjustments that were present will have to be performed again, as the shifter cables or linkages must be loosened.
- Sometimes it is necessary to remove the transaxle final drive, whether there was a failure or for service because of a noise problem emanating from that particular area, such as gear or bearing noise.
- If it was found to be cheaper to use a replacement transmission, then the transmission may have to be reassembled for a core return to the manufacturer.
- When reassembling all components, it is a good practice to lubricate parts as they are being installed.
- End play and preload adjustments are a necessary part of the disassembly and assembly of the transmission/transaxle.
- End play and preload adjustments can be an indicator of washer wear or bearing failure, which causes the shaft to walk back and forth in the case.
- Measuring end play is critical to ensure that the transmission stays in service for a long time.
- It is highly important that the technician always follow the manufacture's service and maintenance procedures as well as safety precautions.

Key Terms

blocking ring The ring on the synchronizer assembly that slows or speeds up the gear to match the main shaft speed, which allows smooth shifts. Also called a blocker ring.

concentricity The roundness of a hole. If the hole is not round, it can also be referred to as out of round.

end play Referring to the input or output shaft, fore and aft movement in the transmission.

oil slinger A device that rides partially in the transmission fluid; as it spins, it flings oil to lubricate the internal workings of the transmission.

overhauling The process of refurbishing the transmission to like-new condition.

Review Questions

1. Which of the following could cause the transmission to make noise only when the clutch pedal is depressed?
   - **a.** Worn input shaft bearing
   - **b.** Worn synchronizer rings
   - **c.** Throwout bearing
   - **d.** Worn counter shaft bearing
2. Which of these tools is required to pull differential case bearings?
   - **a.** Push-type bearing puller
   - **b.** An old bearing race
   - **c.** Chain-style vice grips
   - **d.** Snap ring pliers
3. Which tool is required to remove rear bearing race from the transmission case?
   - **a.** Brass drift and hammer
   - **b.** Seal driver
   - **c.** Gear pullers
   - **d.** Impact screw driver
4. The selective thickness shims are measured with a(n):
   - **a.** hydraulic press.
   - **b.** outside micrometer.
   - **c.** dial indicator.
   - **d.** depth micrometer.
5. All of the following statements with respect to disassembly of the transmission are true *except*:
   - **a.** obtain and read all manufacturer publications and service information.
   - **b.** diagnose the possible causes.
   - **c.** ensure proper tools are available for disassembly and assembly.
   - **d.** ensure that all transmission case components are replaced.

6. All of the following statements with respect to overhauling a transmission are true *except*:
   a. transmission case components should be cleaned and inspected.
   b. when in doubt about the reusability of a case component, do not replace it.
   c. perform a cost analysis for the customer.
   d. if rebuilding will cost more than a remanufactured unit, the customer should be informed.
7. Choose the correct statement.
   a. Always reuse gaskets and seals.
   b. Sealants should not be used in transmissions.
   c. Gasket surfaces don't need to be cleaned because the new gasket will seal any residual gasket material.
   d. Check the main case and output shaft housing with a straightedge and feeler gauge.
8. A good place to start when inspecting for synchronizer wear and damage is when the:
   a. transmission fluid is drained from the transmission.
   b. transmission is being assembled.
   c. vehicle is running.
   d. transmission endplay is measured and adjusted.
9. Choose the correct statement.
   a. The transaxle final drive can be removed without disassembling the transaxle.
   b. The pinion gear is part of the input shaft of the transaxle.
   c. The pinion gear drives the differential ring gear, located inside the transaxle assembly.
   d. Do not replace the ring gear if the pinion gear has to be replaced.
10. All of the following statements referring to transmission end play/preload are true *except*:
   a. end play and preload adjustments are a necessary part of the disassembly and assembly of the transmission.
   b. end play is the movement back and forth of the input or output shaft.
   c. Washer wear or bearing failure causes the shaft to walk back and forth in the case.
   d. axial or radial clearances should not determine whether gears and shafts are to be replaced or reused.

ASE Technician A/Technician B Style Questions

1. Tech A says that broken or worn powertrain mounts can cause a transaxle to have shifting problems. Tech B says that poor shift boot alignment cannot cause a transaxle to jump out of gear. Who is correct?
   a. Tech A
   b. Tech B
   c. Both A and B
   d. Neither A nor B
2. Tech A says that a leaky transmission output shaft seal is likely to cause clutch plate contamination. Tech B says that a worn transmission rear bushing may contribute to output shaft seal leakage. Who is correct?
   a. Tech A
   b. Tech B
   c. Both A and B
   d. Neither A nor B
3. A car jumps out of gear into neutral when decelerating or going down hills. Tech A says that the shift lever and internal gearshift linkage should be checked. Tech B says that a detent spring could be broken. Who is correct?
   a. Tech A
   b. Tech B
   c. Both A and B
   d. Neither A nor B
4. Tech A says that as long as the dog (clutch) teeth of the synchronizer are rounded, the synchronizer can be reused. Tech B says that the synchronizer sleeve should slide smoothly on its splines. Who is correct?
   a. Tech A
   b. Tech B
   c. Both A and B
   d. Neither A nor B
5. A transmission is being assembled, but the input shaft does not slide all the way in. Tech A says this is normal and to use the bolts to pull the front bearing retainer into place. Tech B says that you should use a hammer to tap the bearing retainer into place. Who is correct?
   a. Tech A
   b. Tech B
   c. Both A and B
   d. Neither A nor B
6. A rough growling noise occurs when the vehicle is moving and the transmission is in the forward or reverse gear. The noise is not heard when the transmission is in neutral. Tech A says that the output shaft (main shaft) bearings may be faulty. Tech B that says the pilot bearing may be faulty. Who is correct?
   a. Tech A
   b. Tech B
   c. Both A and B
   d. Neither A nor B
7. Tech A says that the input shaft normally has to be installed before the main shaft. Tech B says that the countershaft is normally installed before the main shaft. Who is correct?
   a. Tech A
   b. Tech B
   c. Both A and B
   d. Neither A nor B
8. Tech A says that the condition of blocking rings should be checked by measuring the clearance between the blocking ring and the face of the gear. Tech B says that the blocking ring should be visually inspected for worn teeth and worn (smooth) internal surfaces. Who is correct?
   a. Tech A
   b. Tech B
   c. Both A and B
   d. Neither A nor B

9. Transaxle inspection is being discussed. Tech A says that in some cases, a magnet is installed in the case to catch any ferrous metal particles, and it should be inspected. Tech B says that tapping on the gears with a ball-peen hammer and listening for a ringing sound is a good way to determine if a gear is reusable. Who is correct?
   a. Tech A
   b. Tech B
   c. Both A and B
   d. Neither A nor B

10. Transmission end play is being discussed. Tech A says that a micrometer should be used to measure the inward and outward movement of the output shaft. Tech B says that an angle gauge should be used to measure output shaft rotation. Who is correct?
   a. Tech A
   b. Tech B
   c. Both A and B
   d. Neither A nor B

CHAPTER 34
Driveshafts, Axles, and Final Drives

NATEF Tasks

- **N34001** Check fluid condition; check for leaks; determine needed action. (MLR/AST/MAST)
- **N34002** Check shaft balance and phasing; measure shaft runout; measure and adjust driveline angles. (AST/MAST)
- **N34003** Diagnose universal joint noise and vibration concerns; perform needed action. (AST/MAST)
- **N34004** Measure drive axle flange runout and shaft end play; determine needed action. (AST/MAST)
- **N34005** Inspect and replace drive axle wheel studs. (MLR/AST/MAST)
- **N34006** Remove and replace drive axle shafts. (AST/MAST)
- **N34007** Inspect and replace drive axle shaft seals, bearings, and retainers. (AST/MAST)
- **N34008** Inspect, service, and/or replace shafts, yokes, boots, and universal/CV joints. (MLR/AST/MAST)
- **N34009** Diagnose noise and vibration concerns; determine needed action. (MAST)
- **N34010** Diagnose drive axle shafts, bearings, and seals for noise, vibration, and fluid leakage concerns; determine needed action. (MAST)
- **N34011** Diagnose noise, slippage, and chatter concerns; determine needed action. (MAST)
- **N34012** Diagnose constant-velocity (CV) joint noise and vibration concerns; determine needed action. (AST/MAST)
- **N34013** Inspect and replace companion flange and/or pinion seal; measure companion flange runout. (AST/MAST)
- **N34014** Inspect ring gear and measure runout; determine needed action. (MAST)
- **N34015** Remove, inspect, and/or reinstall drive pinion and ring gear, spacers, sleeves, and bearings. (MAST)
- **N34016** Disassemble, inspect, measure, and/or replace differential pinion gears (spiders), shaft, side gears, side bearings, thrust washers, and case. (MAST)
- **N34017** Measure and adjust drive pinion depth. (MAST)
- **N34018** Measure and adjust drive pinion bearing preload. (MAST)
- **N34019** Reassemble and reinstall differential case assembly; measure runout; determine needed action. (MAST)
- **N34020** Check ring and pinion tooth contact patterns; perform needed action. (MAST)
- **N34021** Clean and inspect differential case; check for leaks; inspect housing vent (MLR/AST/MAST)
- **N34022** Measure rotating torque; determine needed action. (MAST)

Knowledge Objectives

After reading this chapter, you will be able to:

- **K34001** Describe rear-wheel drive and front-wheel drive layout.
- **K34002** Describe the layout and function of rear-wheel drive systems.
- **K34003** Describe the layout and function of rear-wheel drive systems.
- **K34004** Describe the purpose and function of driveshafts, axles, and half-shafts.
- **K34005** Describe the types, purpose and function of axles and half-shafts.
- **K34006** Describe the types, purpose and function of rear-wheel drive solid axles.
- **K34007** Describe the types, purpose and function of half-shafts.
- **K34008** Describe the purpose and function of joints and couplings in driveshafts and axles.
- **K34009** Describe the types, purpose and function of universal joints.
- **K34010** Describe the types, purpose and function of constant-velocity joints.
- **K34011** Describe the purpose and function of final drives/differentials.
- **K34012** Explain the purpose and function of rear-wheel final drives.
- **K34013** Explain the purpose and function of limited slip differentials.
- **K34014** Describe the purpose and function of front-wheel drive differentials.

Skills Objectives

After reading this chapter, you will be able to:

- **S34001** Perform driveline and axle inspection and repair.
- **S34002** Perform final drive and axle diagnosis.
- **S34003** Perform final drive inspection and repair.
- **S34004** Inspect and reinstall limited slip differential components.

▶ Introduction

K34001

In previous chapters, we have covered how the engine's torque is modified and delivered through the transmission, whether it is a manual or an automatic type. Now the torque must get delivered to the appropriate wheels. That job is performed by the remainder of the drivetrain components, such as driveshafts, transfer cases, final drives, differentials, and drive axles. All of these assemblies are made up of subcomponents that allow them to perform their task. Depending on the vehicle, the arrangement of these assemblies can vary. For example, most passenger vehicles are driven by two wheels, whereas most off-road vehicles are driven by all four wheels. This chapter discusses the various assemblies and components of the drivetrain for two-wheel drive vehicles: their layout, function, how to diagnose them, and ultimately how to maintain and repair them.

FIGURE 34-1 A typical RWD solid axle assembly with the final drive assembly enclosed in one housing.

▶ Rear-Wheel Drive Layout

K34002

In a rear-wheel drive (RWD) vehicle, the engine and transmission power the rear wheels to drive the vehicle down the road. In a conventional RWD vehicle, the engine and transmission are mounted longitudinally at the front. Drive from the engine is transmitted to a rear axle assembly by a drive (propeller) shaft. **Beam-type** or solid rear axle assemblies enclose the final drive gears, differential gears, and axle shafts into one housing (**FIGURE 34-1**). In vehicles with an independent rear suspension, the final drive unit is mounted on the chassis frame, and drive is transferred to each road wheel through external driveshafts (half-shafts), which have flexible joints on each end. Some RWD vehicles have rear- or mid-mounted engines along with a transaxle and external drive half-shafts to drive to the road wheels.

In beam axle or solid rear axle applications, suspension action makes the final drive assembly rise and fall relative to the vehicle frame. This happens as a vehicle goes over bumps and various types of terrain. This movement produces continuous change in the distance from the transmission output shaft to

You Are the Automotive Technician

After a long, icy-cold Michigan winter, Mrs. Simpson visits her local repair shop because she has started to notice an unusual noise that sometimes comes from the front of the vehicle. She drives her vehicle to and from work on Highway 94, which has multiple potholes as a result of the heavy trucking industry and harsh winter weather. You complete an interview with Mrs. Simpson and take her 2006 Toyota Camry for a test-drive, during which you notice that the vehicle makes a popping noise when turning corners. Your preliminary diagnosis is a worn constant-velocity (CV) joint, but you will need to further inspect the vehicle in the service bay. The visual inspection of the underside of the vehicle confirms your preliminary diagnosis: the right front CV joint boot is torn and dripping water. There is also dirt mixed with the little bit of grease still inside the boot.

1. What is likely causing the customer's concern? And what will you recommend to correct it?
2. What are the three different drive axle arrangements used with an RWD solid axle? Which drive axle is most commonly used on medium to heavy trucks?
3. What is the difference between a final drive unit and the differential assembly?

the final drive pinion, and in the angle between the driveshaft and its connections. In addition, the pinion nose is forced up on acceleration by torque and down when the brakes are applied on deceleration.

Despite these movements, the driveshaft must transfer the drive smoothly. Change in length is accommodated by a sliding coupling built into the driveshaft or through the use of a **slip yoke** on the front of the driveshaft. The slip yoke has splines that mate to the splines on the transmission output shaft. The slip yoke slides in and out on the output shaft, allowing the length changes needed as the suspension moves up and down. If the vehicle uses a **sliding spline driveshaft**, a two-piece driveshaft is joined in the middle with splines. The driveshaft can slide on itself to increase and decrease in length. Universal joints (or simply U-joints) are fitted at the front and rear ends of the driveshafts to allow for up-and-down suspension movement. The engine crankshaft and driveshaft rotate in line on this arrangement.

As shown in Figure 34-1, the centerline of the RWD vehicle is not in line with the axle lines; this means the power needs to be turned 90 degrees to get it to the rear wheels. A ring gear and pinion assembly is commonly used for this purpose. The ring and pinion gear set allows the transfer of power 90 degrees (**FIGURE 34-2**). It also reduces the rotational speed of the axles relative to the driveshaft, which increases the torque applied to the wheels.

▶ Front-Wheel Drive Layout

K34003

In an FWD vehicle, the front wheels are driven by a transversely or longitudinally mounted engine. **Longitudinally** means the front of the engine is facing the front of the vehicle, while **transverse** mounting means the front of the engine is facing the side of the vehicle; that is, the engine is facing one of the fenders. Most front-engine FWD vehicles use a transverse-mounted engine coupled to a transaxle. Power is transmitted in a straight line from the engine through the transmission and final drive (differential) gearing and out to the front axles (half-shafts) (**FIGURE 34-3**). This design eliminates the need to transfer power 90 degrees as with an RWD arrangement.

FWD power reaches the front wheels through the transaxle. In transverse applications, the transaxle is normally mounted at the rear of the engine and a primary shaft engages with the splines of the clutch center plate. When a gear is selected, the drive is transferred to a secondary shaft, and through a secondary shaft pinion, to a helical ring gear attached to the differential case. Drive is then transferred through the differential gears to each driveshaft and each front wheel.

On an FWD vehicle with a longitudinally mounted engine, the power must be turned 90 degrees to drive the axles (**FIGURE 34-4**). This is accomplished by the design of the transaxle, which is mounted behind and sometimes under the engine. Power is transmitted through the clutch in a rearward direction. It is then typically transferred by a large chain or gears to the rest of the geartrain and final drive assembly located under the engine. The final drive turns the power 90 degrees so half shafts can transmit power to the wheels.

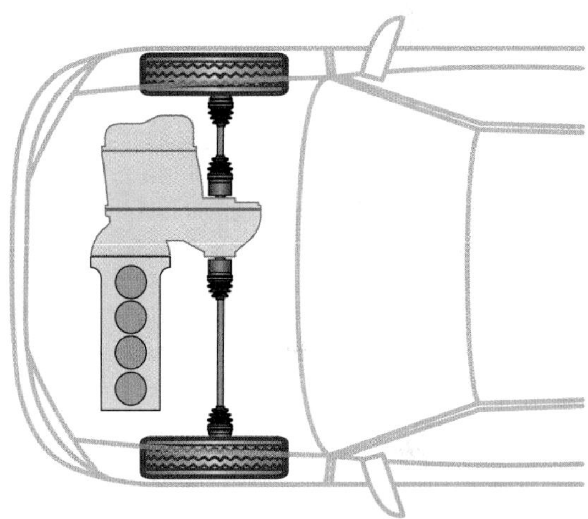

FIGURE 34-3 Powerflow in a transverse engine is in line with the engine, all the way to the wheels.

FIGURE 34-2 The ring and pinion gear set allows power to be transferred ninety degrees from the power source.

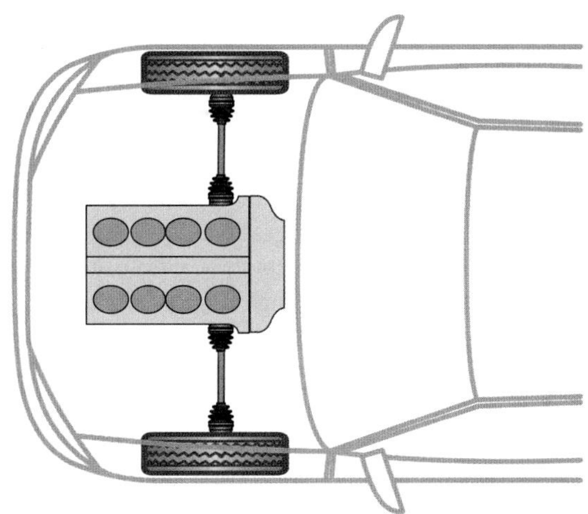

FIGURE 34-4 Powerflow in a longitudinally mounted engine must turn 90 degrees to power the wheels.

▶ Driveline Subassemblies and Components

K34004

The driveline subassemblies are made up of driveshafts, axle shafts, and half-shafts. Each shaft transfers torque from one component to the next. Terminology of the components varies depending on the type of drivetrain. For example, on most RWD vehicles, there is one driveshaft from the transmission to the final drive assembly and two axle shafts—one from each side of the final drive to one of the rear wheels. On FWD vehicles, there are two half-shafts (axles)—one on each side of the transaxle to each of the front wheels. Four-wheel drive (4WD) and AWD drive (AWD) vehicles use combinations of driveshafts and half-shafts depending on if the axle shafts are independent or non-independent. We will explore each of these types of shafts and the joints that allow them to transfer torque.

Driveshafts—Rear-Wheel Drive

The driveshaft is a device that transfers torque from one component to another, such as from the transmission output shaft to the final drive assembly (**FIGURE 34-5**). The driveshaft is typically made from metal tubing material. The driveshaft can be constructed from steel or aluminum and can come in various sizes and lengths depending upon the application. The driveshaft can be a short one-piece assembly with a U-joint at each end, or in some vehicles with a long wheel base, such as a school bus, the driveshaft can be made up of as many as four separate segments. The driveshaft typically requires the use of joints (described later in this chapter) to allow power to transfer smoothly as the driveshaft follows the movement of the suspension system.

U-joint yokes are typically welded to each end of the tube. The front U-joint connects to the driveshaft's internally splined slip yoke, which engages with splines on the transmission output shaft. The front slip yoke on the driveshaft slides over the output shaft in the transmission and is supported on the outside by a bushing located inside the rear transmission housing. This design enables the slip yoke to move longitudinally and vary the shaft length with suspension movement. The rear U-joint mates with a **companion flange** on the differential pinion shaft. The companion flange is a splined flange that transmits power to the pinion gear as discussed in the differential section.

Some applications use a two-piece driveshaft that is able to flex in the middle. The front section of the driveshaft is supported at its rear end by a center bearing, called the carrier bearing (**FIGURE 34-6**). The carrier bearing is bolted to a mounting bracket on the vehicle frame that supports the center of the driveshaft. Since the maximum length of a single section of driveshaft tubing is approximately 72" (183 cm) to prevent twisting of the tube from the torque output, two or more sections may be used. The other reason to use the two-piece driveshaft is because the angle of the U-joint may be too great if the driveshaft were just a single piece. The use of three, or more U-joints breaks up the excessive angle between the transmission and the rear axle. The U-joint is capable of working at a maximum angle of 3 degrees; beyond this, vibration and additional wear will occur. Be sure to check the driveline angles when diagnosing this concern. Provision is normally made for adjusting the alignment of the two shafts, using shims under the carrier bearing.

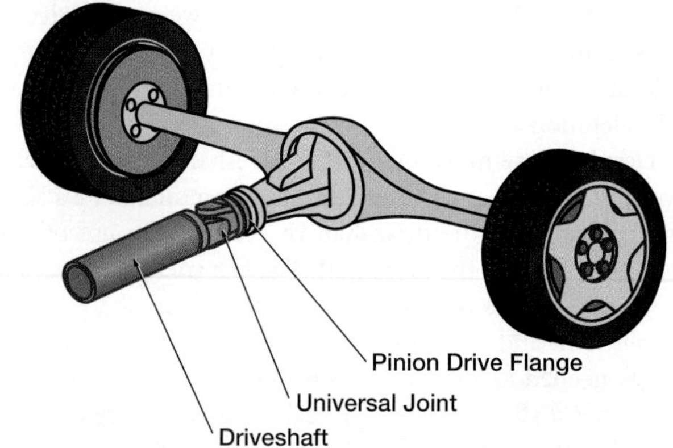

FIGURE 34-5 Rear-wheel driveshaft.

FIGURE 34-6 A carrier bearing supports the center of a two-piece driveshaft.

When working with a two-piece driveshaft or a splined driveshaft, it is critical that the yokes on both halves of the driveshaft be phased correctly, which means the pivot points of the U-joints are lined up in the same exact orientation from one shaft to the other. If the driveshaft halves are not assembled correctly, then vibration will result. With a one-piece driveshaft, the yokes are welded to the tube and cannot be moved. When working on a two-piece driveshaft, be sure to mark the location of each driveshaft half with a marker or white paint mark to ensure the yokes line up (**FIGURE 34-7**).

Axles and Half-Shafts

K34005

There are two types of axles used today—the dead axle and the live axle (drive axle). The dead axle does not provide any drive capabilities and is used on the rear of FWD vehicles, and the front of RWD vehicles. It works the same as a trailer axle, allowing the wheels to spin freely and follow the drive axle. The purpose of live axle is to transfer torque to the wheels and tires, propelling the vehicle forward or backward.

FIGURE 34-7 The driveshaft yoke alignment is critical for smooth operation.

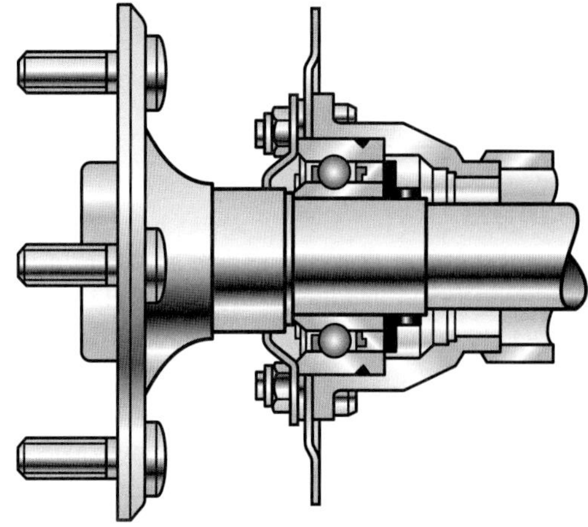

FIGURE 34-8 Semi-floating axle—bearing located between the axle and housing.

RWD Solid Axles

K34006

Drive axles used with an RWD solid axle assembly come in three varieties:

1. **Semi-floating axle**: On the semi-floating axle, the axle shafts are splined to the differential side gears, and the outer bearing is between the outer end of the axle shaft and the inside of the axle housing (**FIGURE 34-8**). Wheel studs are pressed into the axle flange at the end of this type of axle. The axles support the weight of the vehicle and are also subject to bending forces as the vehicle corners. If the vehicle with a semi-floating axle were to hit a curb and break or snap the axle shaft, the wheel would come off the vehicle.
2. **Three-quarter floating axle**: On this type of axle, there is a single roller bearing between the hub and the outside of the axle housing (**FIGURE 34-9**). The axle flange is bolted to the housing and stabilizes the wheel vertically, while the vehicle's weight is supported by the hub and the bearing. The ¾-floating axle was used on older vehicles, such as Chrysler vehicles and pickup trucks.
3. **Full floating axle**: On the full floating axle, there are two tapered roller bearings between the hub and the outside of the axle housing (**FIGURE 34-10**). This arrangement helps to isolate the weight of the vehicle on the hub and bearings, not on the axle itself. Since full floating axles can carry more weight, they are commonly used on medium- to heavy-duty pickup trucks. This type of axle floats between the axle side gears and the wheel hub. Torque is delivered to the wheel hub by the flange on the end of the axle that is bolted to the hub. Heavy-duty applications sometimes use conical wedges to keep the full floating axle centered on the hub. The flange requires a gasket or room temperature vulcanizing (RTV) sealer to seal the oil in the axle housing from leaking out while driving down the highway. The inner axle seal is part of the hub and can be serviced only by the removing the wheel and hub assembly.

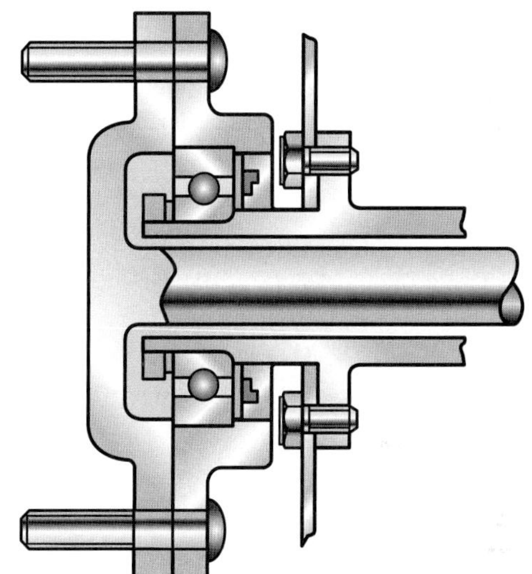

FIGURE 34-9 A ¾-floating axle—one bearing between the outside of the axle housing and the hub.

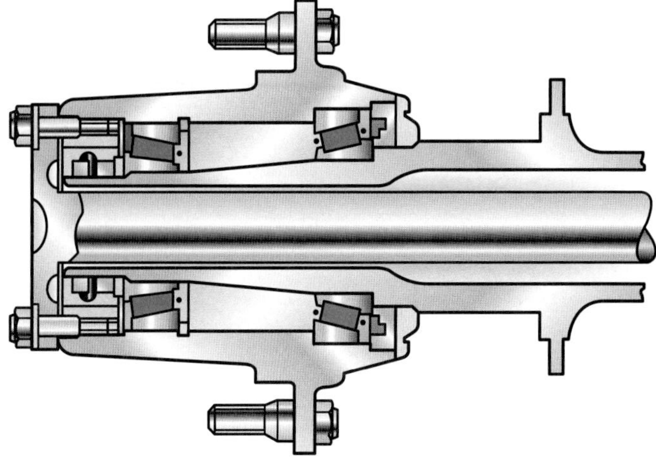

FIGURE 34-10 A full floating axle—two bearings between the axle housing and hub.

▶ TECHNICIAN TIP

Because of their design, full floating axles can be removed without removing the wheels; thus, the vehicle may not even need to be lifted off the ground.

Axle Flanges

In most cases, the flange on the end of the axle in either the semi-floating or the full floating axle is a molded part of the axle (**FIGURE 34-11**), whereas the ¾-floating axle may or may not have a built-in flange as part of it. It is always good practice to check runout on the flange in case of any vibration complaints. If flange damage is diagnosed, replacement of the entire axle shaft will often be required. Also, lug studs are pressed into the axle flange and used to secure the wheel onto the vehicle. It is good practice to inspect the studs for damaged threads, stretching, or even breaking off. Lug studs that are damaged can usually be replaced.

Refer to the Disc Brake Systems chapter for the procedure to replace lug studs.

Axle Seals

Axle seals can be of the single-lip or double-lip design depending on the application (**FIGURE 34-12**). The purpose of axle seals is to contain the lubricating fluid in the final drive assembly, to maintain quiet operation, and to seal contaminants out of the sealed cavity. Too much oil inside the axle housing can cause the oil to be forced against the axle seal, which can lead to seepage past the axle seals. Failure of any of the axle shaft bearings or differential bearings can cause leaking of the axle seals as well. Worn bearings allow the axle shafts to move, and the shaft movement will wear the seals quickly and create gaps that the fluid can leak out of.

Sometimes the seal surface on the axle becomes worn or nicked. An aftermarket repair kit called a **speedy sleeve** is available for the seal surface on some axles (**FIGURE 34-13**). The speedy sleeve is a thin metal sleeve that fits tightly over the seal surface of the axle, providing a new, undamaged surface for the seal to ride against. The seal is designed slightly oversize to maintain lubrication integrity.

FIGURE 34-11 Axle flange with lug studs.

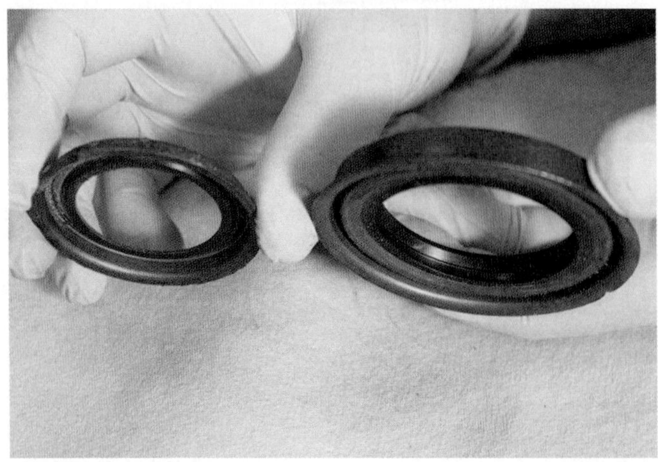

FIGURE 34-12 Single-lip and double-lip seals.

▶ TECHNICIAN TIP

A common issue you might find is the leaking of the axle grease onto the rear drum brakes. Typically, this will be found as part of a state inspection or during a routine brake system inspection. A customer may complain of the rear brakes grabbing. The sticking could be the result of axle gear oil creating a thin, sticky layer on the surface of the brake lining. This in turn makes the brake material grab the drum brake surface and can create a lock-up of the brake, especially at slow speeds.

FWD/AWD Axles/Half-Shafts

K34007

In FWD vehicles and AWD vehicles, the driveshafts transfer the drive directly from the final drive inside the transaxle to the front wheels. FWD vehicles typically use axles called half-shafts, which have an inner and an outer CV joint (**FIGURE 34-14**).

An FWD transaxle typically does not place the final drive directly in the center of the vehicle; because the final drive is bolted to the end of the transversely mounted engine, it is offset to one side. Thus, the half-shafts are usually different lengths.

FIGURE 34-13 The speedy sleeve restores a damaged surface so that the seal has a clean smooth surface to ride against.

FIGURE 34-14 Typical half shaft and CV joints.

FIGURE 34-15 Intermediate shaft.

The use of different-sized axle shafts can cause a problem known as torque steer. **Torque steer** is when the vehicle pulls to one side during hard acceleration. When half-shafts are not equal lengths, more torque is applied to the side with the short half-shaft, creating the pulling condition. In addition to the unequal axle length, the angles at the CV joints are different, which also results in the pulling concern and can create vibration. To combat this condition, many manufacturers use an intermediate shaft (**FIGURE 34-15**). The intermediate shaft is a short section of shaft that typically has a bearing pressed onto it (similar to a carrier bearing). The intermediate shaft makes it so that both half-shafts are the same length from left to right. In some cases the manufacturer uses a longer half-shaft on one side, and a rubber dynamic damper may be fitted to help absorb vibrations, although this does not reduce torque steer issues.

▶ Joints and Couplings

K34008

The driveshaft and half-shafts require a flexible joint at either end to allow for angular changes as the suspension travels up and down. There are two common types of joints used in this way—the U-joint and the CV joint. Both types of joints allow the shaft to transmit torque through a change of drive angle. The U-joint does so with an increase and decrease in velocity as the joint rotates every 90 degrees. A CV joint maintains the same velocity as it goes through its rotation. While both can transfer torque through an angle, the CV joint can do so at a much greater angle than the U-joint, which is why it is used on front half-shafts to accommodate large steering angles. U-joints are typically used on driveshafts because they generally need to accommodate only small angles, although some manufacturers use CV joints to provide a smoother ride, or on off-road vehicles that have been lifted and experience a greater amount of suspension travel. A large amount of torque is transmitted through the driveshaft and half-shafts, so joints can wear over time, making inspection and diagnosis important.

Universal Joints

K34009

A **universal joint (U-joint)** is a cross-shaped flexible joint. There are caps that fit over the ends of the cross. Needle bearings fit between the ends of the cross and the caps, allowing the caps to rotate smoothly. The U-joint is considered a non-CV joint. The most common type is a **Hooke's joint** (cross-and-roller joint) (**FIGURE 34-16**). The Hooke's joint consists of a steel cross with four hardened bearing journals, mounted on needle rollers in hardened caps, which locate the cross in the eyes of the yokes. The cross swivels in the yokes as the drive is transferred across the joint. However, the swiveling conflicts with the rotation, and in each revolution the velocity of the yoke changes every 90 degrees due to the angles that the driveshafts are operating on. This change in the angular velocity increases as the angle that the joint operates on increases.

There are two planes of stop/start as the joint flexes due to the cross shape of the joint. As the shaft rotates, it changes the velocity of the shaft up and down and then side to side, creating a speeding up and slowing down of the joint. The sharper the angle of the joint, the more pronounced the velocity change becomes, which may cause a vibration complaint from customers. The effect of the changing velocity can be minimized by having identical non-CV joints at each end of the propeller shaft in phase (in alignment). This is done by having the yokes of the

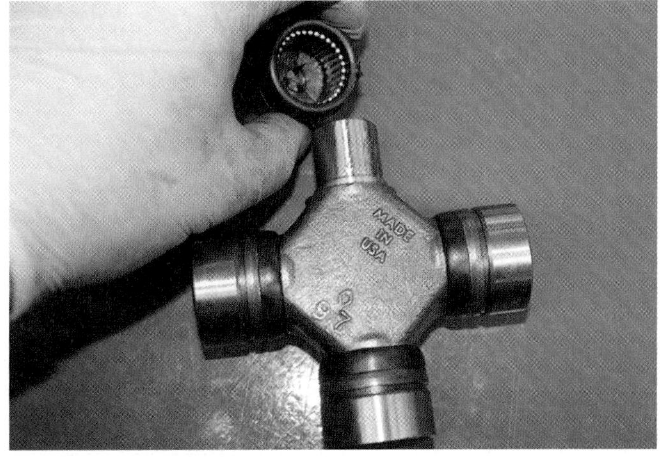

FIGURE 34-16 Cross-and-roller U-joint.

driveshaft in line with each other. An increase in the velocity of the front yoke is canceled out by a similar decrease in the velocity of the rear yoke in each revolution. In some designs, engineers fit a CV joint to the front end of the driveshaft, which allows for an increase of angles of the shafts without creating vibration.

A **double Cardan joint** greatly reduces the change in velocity of a single Cardan joint by using the second joint to cancel out the changes in velocity of the first joint. A double Cardan joint is considered by most technicians to be a CV joint under normal driveline angles. The double Cardan joint uses two Cardan joints housed in a short carrier (**FIGURE 34-17**). Each cross has either a centering ball or a socket that is connected to its back side and joins with its mating component on the other cross. The ball-and-socket assembly keeps each Cardan joint at an equal angle, thereby canceling the change in velocity.

FIGURE 34-17 A typical double cardan joint.

Constant-Velocity Joints

K34010

For independent suspension vehicles, the external drive axle shafts (half-shafts) to each road wheel most likely use CV joints at their connecting points. CV joints allow for more torque transfer than U-joints due to their larger bearing surfaces. Also, because of their construction, CV joints can operate at greater angles than U-joints.

Often, a sliding spline or a **plunge-type joint** is used as the inner half-shaft joint to accommodate for changes in shaft length when traveling on different types of terrain. Plunge-type joints allow smooth power flow while allowing the joint to slide in and out, effectively increasing and decreasing the length of the axle shaft during up and down suspension travel. One type of plunge CV joint is the tulip tripod joint. The **tulip/tripod joint** has three equally spaced fingers shaped like a star. On the ends of the star are three round bearing surfaces that sit on needle bearings on each finger. The round bearing surfaces move on the fingers by needle bearings. The outer race has three straight grooves that run from side to side. This configuration allows in-and-out movement of the shaft while allowing flexing (**FIGURE 34-18**).

The **fixed-type joint** does not slide to allow for shaft lengthening or shortening; it simply allows for angle changes as the suspension moves. The fixed joint is typically used on the outboard side of the half-shaft. One type of fixed joint that you will service as a technician is the **Rzeppa joint**. The Rzeppa joint has an inner race, six steel ball bearings, a bearing cage, and an outer race. The ball bearings, retained by a spherical cage, are carried in angular grooves in each race. These balls transfer the drive from one race to the other. The inner race is splined to the axle shaft while the outer race is splined to the

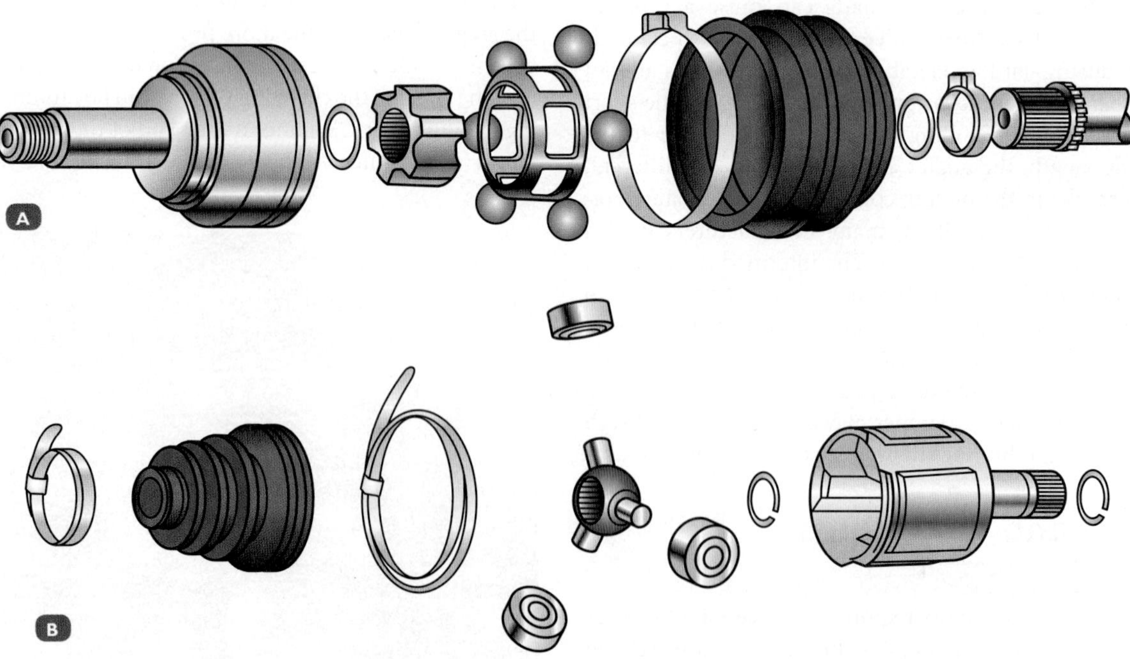

FIGURE 34-18 The CV joint comes in several forms. The Rzeppa joint **(A)** and the tulip tripod joint **(B)** are two types that are commonly found on half-shafts.

FIGURE 34-19 CV boot.

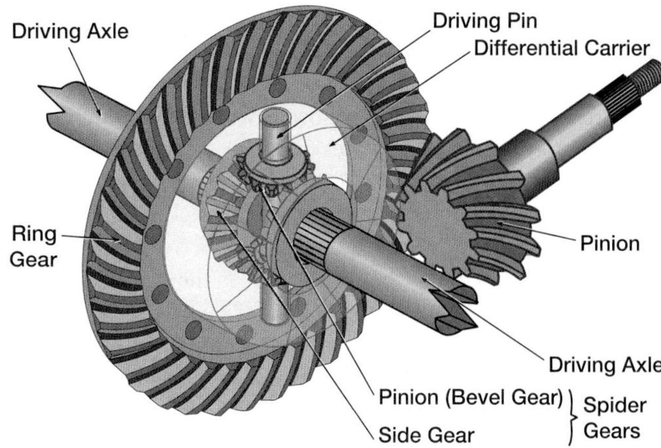

FIGURE 34-20 Differential gears are housed in the center of the ring gear.

wheel hub. The Rzeppa joint has been modified so that it will allow a limited amount of plunge capability

CV Joint Lubrication and Inspection

Each CV joint must be lubricated. Special bearing grease is used to keep the bearing surfaces moving freely. To keep the grease inside the joint, and to keep dirt and debris out, an accordion-type synthetic rubber boot, called a CV boot, is attached to the outside of the joint (**FIGURE 34-19**). The CV boot should be inspected during routine maintenance, such as during an oil change, for cracks or splits. If a boot is leaking, traces of lubricant thrown on nearby components will be noted. In some cases, split-type boots can be used as replacements. Otherwise, the half-shaft will need to be removed to replace the defective boot. If any leakage is noted, the boot should be replaced since dirt and water can enter the joint, causing the joint to ultimately fail.

Although CV joints are larger than U-joints, they are still subject to wear. The indication of excessive wear is typically a clicking or popping noise when the vehicle is driven with the wheels fully turned in one direction or the other. The noise is caused by the drive balls being forced into and out of the wear grooves. Severe cases are indicated by noise at lesser steering angles.

SAFETY TIP

Some CV joint greases use lead to help cushion the high metal-to-metal contact loads within a CV joint. Always wash your hands thoroughly after working with any CV joint grease.

▶ Final Drives/Differentials

K34011

The terms "final drive" and "differential" are confused by technicians. Specifically, the term differential gets used in two different ways. The first is the most technically accurate and refers to the components inside of the differential housing that allow the axles to turn at different speeds when the vehicle is cornering or turning (**FIGURE 34-20**). These components make up what is called the differential assembly. Since the outside tire must travel farther around a corner than the inside tire, it must turn at a faster speed. The differential assembly allows this to happen; otherwise the tires would bind, skip, hop, and slide when going around a corner, which would then cause them to wear out quickly.

FIGURE 34-21 Final drive assembly.

The second way technicians use the term "differential" is as another name for the complete final drive assembly. The final drive assembly provides the final gear reduction necessary for drivetrain operation, and it houses the "differential assembly." The final drive takes the power from the transmission and sends it to the wheels (**FIGURE 34-21**). And the differential assembly within it allows the tires to travel around corners without binding while the final drive is powering them. Since the differential assembly is housed within the final drive assembly, many technicians improperly call the entire final drive assembly the differential. Thus, any time you hear the term "differential," you need to ask whether the person means the final drive assembly or only the differential assembly.

Final drives can be found in axles either at the front or rear of the vehicle or can be found in the transaxle of an FWD vehicle. The speed reduction gears in the final drive are called the ring and pinion gears. Various "rear axle" ratios are available to suit the demands of the vehicle use. Lower ratios are used to increase low-speed pulling power, while higher ratios are used to improve high-speed fuel economy, such as on the freeway. Some "rear axles" allow for final drive ratios to be quickly changed, such as for racing on different track surfaces (dirt, paved, etc.) or track sizes. If the vehicle uses a front or rear axle, the driveshaft attaches to the final drive. (In the case of an FWD transaxle, half-shafts attach to the final drive.)

On some of the final drive assemblies manufactured today, the anti-lock brake system (ABS) wheel speed sensor is incorporated in them. Speed sensors read wheel speed and are part of the final drive gear set. If the vehicle is equipped with traction control, the traction control system can overcome the loss of traction by applying more power to the wheel that is not slipping by applying the brake on the wheel that is slipping.

Differential assemblies come in two varieties. First there is the open differential, which means power is supplied to both wheels equally only when each tire maintains traction, which on ice and snow is nearly impossible. If one wheel is stuck in the snow or ice, the other wheel can supply virtually no power due to the nature of the open differential assembly. The second type is the limited slip differential, which allows, under the same conditions, power to be delivered to both of the rear wheels to supply power to the ground to drive the vehicle.

Preventive maintenance of the final drive assembly, regardless of its design, is important, as these axles are filled with special lubricants that promote long life and maintain quiet operation of the gears. Maintenance schedules tell technicians when each component requires maintenance or inspection and thereby help prevent premature failure. More information on final drive assemblies will be provided in subsequent sections in this chapter.

Rear-Wheel Final Drive

K34012

In a conventional RWD vehicle, such as a pickup truck, a solid axle assembly incorporates and encloses the final drive gears, differential, and axle shafts in one housing. A ring gear and a pinion gear transfer power through 90 degrees and provide a final gear reduction to the driving road wheels. **Hypoid bevel gears** are normally used for this purpose. Hypoid gears are a special design of spiral bevel gears, with the centerline of the pinion below the centerline of the ring gear (**FIGURE 34-22**). This design reduces the height of the driveshaft tunnel, allowing for a flatter vehicle floor pan. The tooth shape also provides a greater area of tooth contact, and therefore greater strength.

In an RWD vehicle, the ring gear is bolted to the differential carrier, which is supported in the axle housing or case by tapered roller side bearings, retained by bearing caps and bolts. In some axle designs, threaded adjusting rings engage with threads in the housing and will push against the side bearings of the carrier. These threaded rings are used to set the carrier bearing preload and a backlash clearance between the ring and pinion. **Backlash** is the amount of movement between the pinion gear teeth versus the ring gear teeth. If you hold the ring gear and turn the pinion, you will have a slight clearance back and forth, which is the backlash. On other axle designs, shims are used to push against the side bearings to adjust bearing preload and backlash.

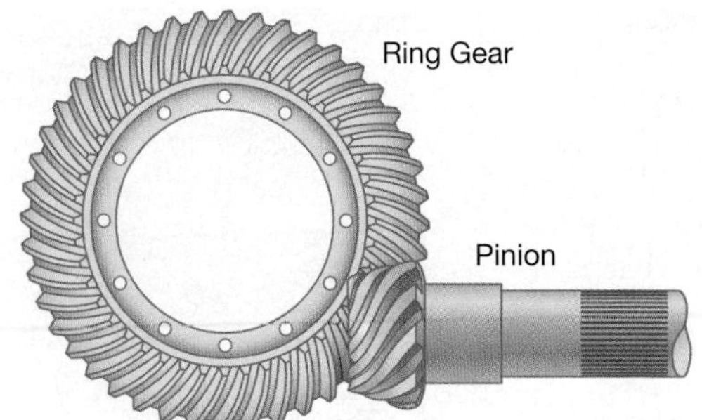

FIGURE 34-22 A hypoid gear arrangement.

The two smaller bevel gears, or pinions, are mounted on a driving pin that passes through the carrier (**FIGURE 34-23**). Two **side gears** mesh with the pinions and are in recesses in the differential carrier. The drive axles are splined to these side gears. With the ring gear bolted to the carrier, the splines on the axle shafts are engaged with the splines in the differential side gears, and drive is transferred through the ring gear to the differential carrier pins, which carry the pinion gears with the carrier. The pinion gears carry the side gears with them, turning the axles and wheels. The pinion and side gears are also called spider gears.

When the vehicle travels in a straight line, the ring gear rotates the case. The driving pin and pinion gears rotate end over end, carrying the side gears with them. The side gears, which are splined to the axles, then turn the drive axles. There is no relative motion between the pinion gears and the side gears; each side gear turns at the same speed.

As soon as the vehicle turns from a straight-ahead position, the inner wheel slows down and its side gear turns more slowly than the differential case. The turning effort applied to the driving pin allows the pinion gears to rotate slowly on their pin. They walk around the inner side gear while still being turned end over end (**FIGURE 34-24**).

This rotation of the pinion gears makes the outer side gear and its road wheel speed up while the inner side gear and its road wheel slow done in rotation by an equivalent amount. The outer side gear then turns faster than the case. This provides an equal torque to each drive axle while allowing for their rotational speed difference.

Limited Slip Differentials

K34013

Limited slip differentials allow normal differential action under normal driving conditions, but when road conditions are not normal, the limited slip differential reduces or prevents differential action so that a wheel cannot spin freely. Drive is maintained to both wheels. There are two main types of limited slip differentials—clutch style and gear style. The clutch-style

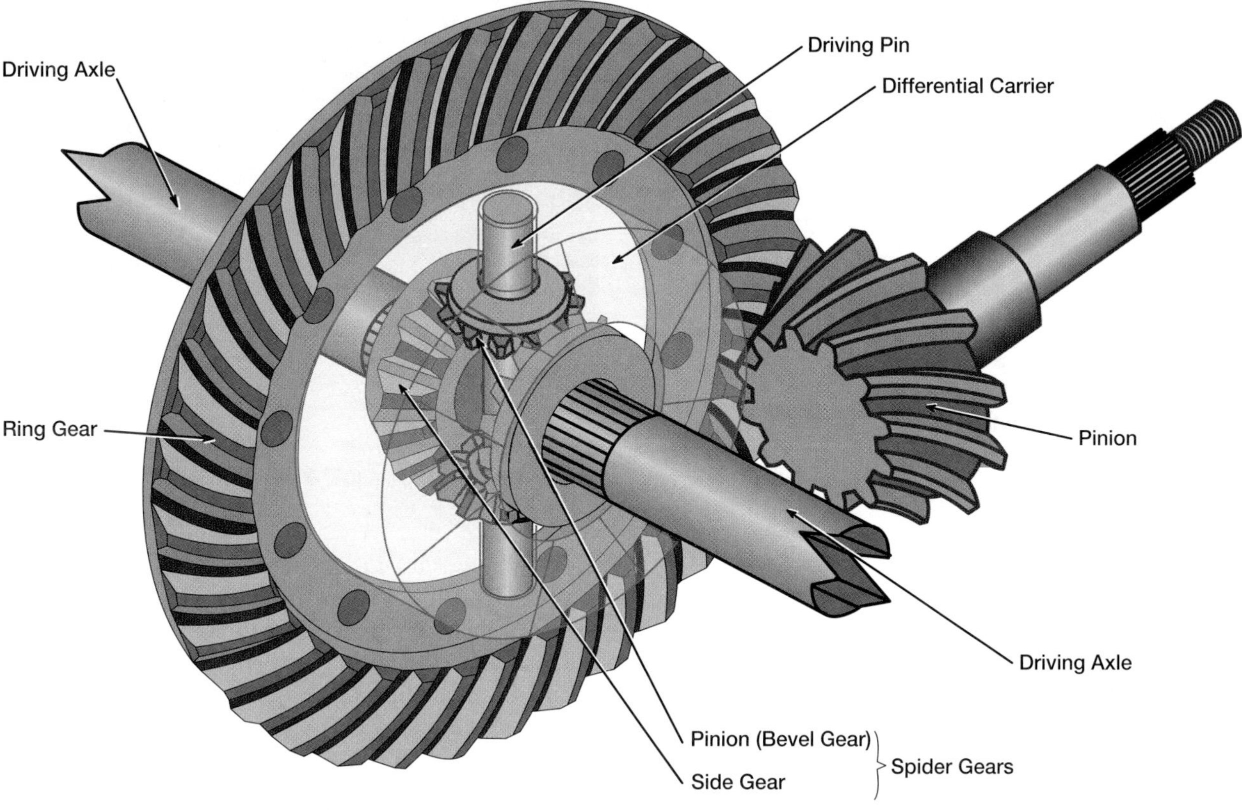

FIGURE 34-23 Differential pinion gear assembly.

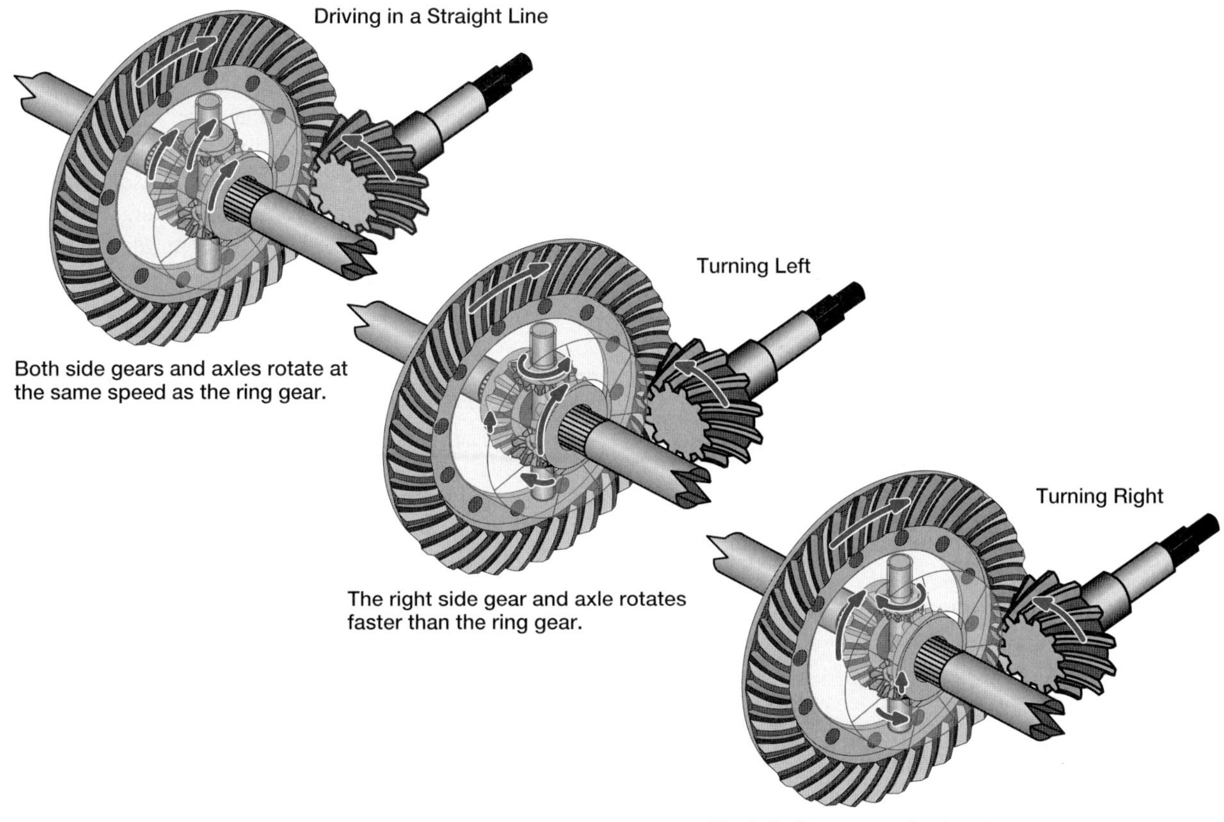

FIGURE 34-24 The differential assembly allows the axles to rotate at different speeds during cornering.

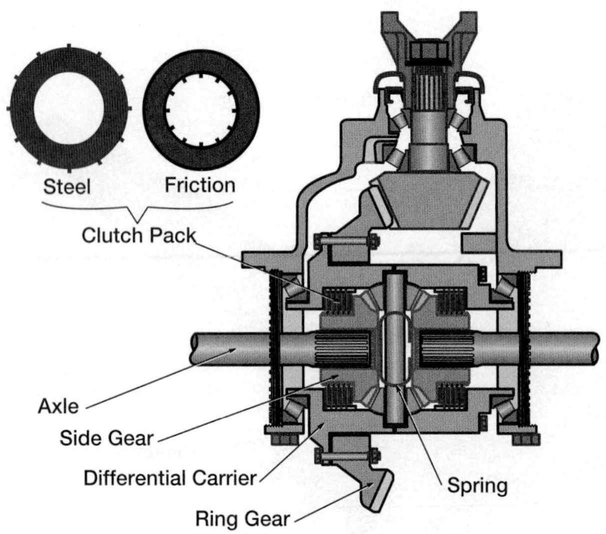

FIGURE 34-25 Clutch-style limited slip differential.

limited slip differential uses a multi-plate clutch pack between each side gear and the differential case (**FIGURE 34-25**). Each clutch pack has two different types of flat steel plates, placed alternately in the pack. One type has internal splines that mate with splines on the side gear pressure ring. The other has driving lugs that locate in slots in the differential carrier.

The outside plate is dished, or cup shaped, to provide initial tension on the clutch pack when the two halves of the carrier are assembled. Four differential pinions are mounted on two driving pins, at right angles to each other, so that they mesh with the side gears. The pinion shafts are relieved, so as not to make contact at their intersection. The ends of the shafts have two flat surfaces forming wedge shapes, which fit into similar wedges in the carrier.

In straight-ahead driving, the driving force through the ring gear to the differential carrier causes the pinion shafts to rotate end over end. This transfers the drive through the pinion gears and the side gears to the axles. There is no relative motion between the gears; however, the resistance at the road wheels forces the pinion shafts up the incline formed by the wedges in the case. As a result, the piston shafts are forced apart and the pinion gears exert a greater force on the side gears and on the clutch packs. The force locks the side gears to the carrier, preventing any sudden spinning of either wheel. Under normal operating conditions, the driving torque is transmitted equally to each axle shaft and wheel, but when patches of loose gravel or mud are encountered, the ratio of torque delivered depends on the traction available at each road wheel.

The greatest amount of torque will be transmitted to the wheel with the most traction. When turning a corner, the limited slip differential gives normal differential action and permits the outer wheel to turn faster than the inner wheel. At the same time, the differential applies the major driving force to the inner wheel, improving stability and cornering.

Torsen Style

Some manufacturers use a gear-style limited slip differential design rather than the usual clutch-type design (**FIGURE 34-26**).

FIGURE 34-26 A Torsen helical gear limited slip differential.

This design was used to improve vehicle stability and tire traction. Some use gear-based, or torque-biased or torque-sensing (Torsen), units. The heart of these limited slip differentials is the parallel-axis helical gear set. The idea behind the Torsen differential is to multiply the torque available from the wheel that is losing traction and turn it over to the slower turning wheel with better traction. Torque application to the wheel with better traction begins because of the resistance between the sets of gears in mesh.

Helical-geared limited slip differentials respond very quickly to changes in traction. They also do not bind from friction in turns and do not lose their effectiveness since there are no clutches, like in a clutch-style unit. The unit described here is the Torsen differential, which is a very popular limited slip differential. Another manufacturer, Eaton, uses a similar design. The side gears of the differential are cut in a worm gear configuration. The pinion gears are also cut with a worm gear cut, and they are splined to each other. When one wheel begins to slip under torque, the worm gears create a locking situation that transfers power to the wheel with greater traction.

▶ TECHNICIAN TIP

Oil changes are required less frequently with gear-style units because there are no clutch discs to wear as the vehicle makes turns, which contaminates the gear oil.

FWD Differentials

K34014

The final drive and differential action is the same for FWD vehicles as it is for RWD differentials (**FIGURE 34-27**). The only real difference with an FWD vehicle is that the final drive and differential are located inside the transaxle and are driven by the secondary shaft, which is connected to the pinion drive gear. Some automatic transaxles use a chain to drive the differential assembly.

FIGURE 34-27 A typical FWD differential located in the transaxle case.

▶ Driveline and Axle Inspection and Repair

S34001

Inspecting Fluid Leakage

N34001

Fluid loss and leaks are a part of the aging process of a vehicle. Fluid leakage concerns are important and should be checked for on a regular basis. Axle seals, pinion seals, and gaskets used to seal the housing are usually the likely places to investigate. Seals are subject to expansion and contraction through heat and cold changes that take place on a regular basis. Overfilling of the differential will create a pressure buildup in the differential from foaming of the liquid that the breather cannot handle, so the weakest place to vent pressure will be through the axle seal itself.

A lack of lubrication will cause the axle bearing to heat up and damage the seal. Brake heat that comes from a dragging brake can also cause an axle seal to leak. A bent axle or axle bearing failure will add to the cause of a failure of the seal. Preventive maintenance can identify some of these concerns and should be practiced on a regular basis.

To inspect fluid leakage, follow the steps in **SKILL DRILL 34-1**.

Checking Driveshaft Joints, Phasing, Angles, Balance, and Runout

N34002, N34003

Vehicles today are put through a tremendous variety of driving styles and scenarios and terrains. Delivering torque to a driveshaft is a violent action due to different engine sizes and torque. Driver abuse can also play an important role in driveshaft diagnosis. Each component of the drivetrain can develop a small amount of play as components wear. One of the most common places for wear is in the U-joint. As the U-joints wear, vibrations can occur and major damage can

SKILL DRILL 34-1 Inspecting Fluid Leakage

1. Put the vehicle on an approved lift, and make sure it is secure. Visually inspect around the housing where the axle seats for any seepage.

2. Inspect the pinion flange for any seepage.

3. If necessary, remove the rear wheels and install a dial indicator on the axle flange to check for any distortion.

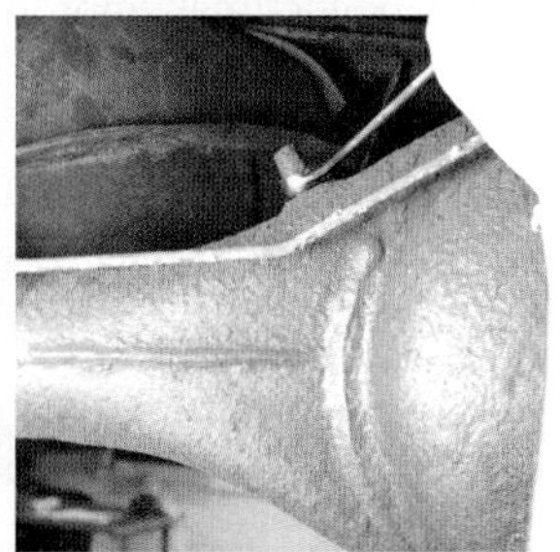

4. Check and clean the breather or vent for any obstructions that may cause a pressure buildup to occur.

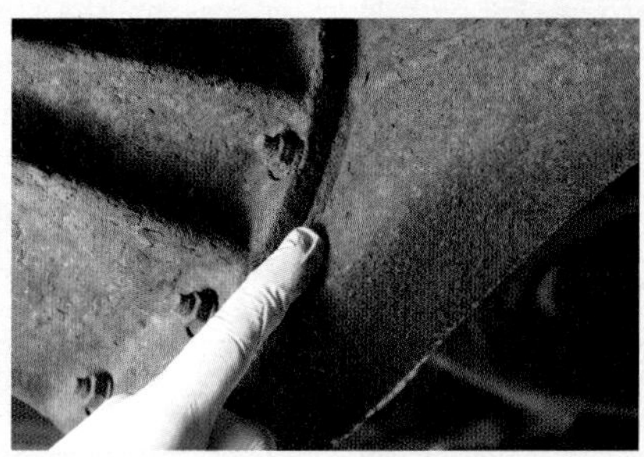

5. Check the differential rear cover for a leaking gasket, if so equipped, and tighten if loose or replace as necessary.

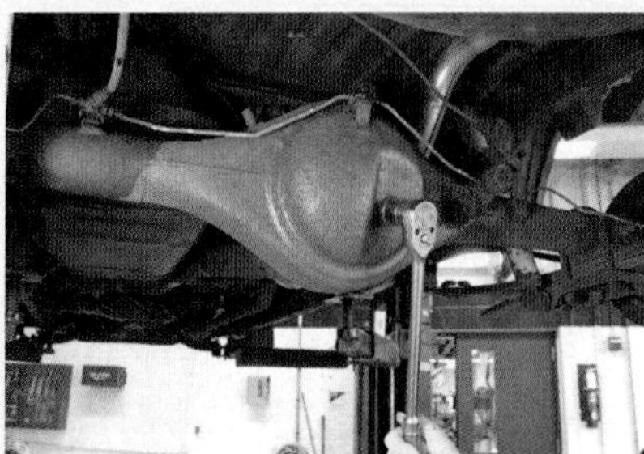

6. Check the fluid level for lack of fluid or overfilling of the differential, as either one can indicate that a leak is present.

happen if the joint were to break. Regular inspection of the driveline should be performed when any vehicle service is performed.

U-joint phasing is important when multiple shafts are involved. If you are going to disassemble a multi-piece driveshaft, then it is critical to mark the parts in relation to each other. If the parts are not marked during disassembly, then an out-of-phase condition could result, and vibration will occur even if the shaft is in balance. Along with balance and phasing, runout and concentricity should be checked. If any damage has occurred, then the damaged parts will need to be replaced.

Driveline angularity is critical to making sure U-joints are working at their proper angle. Driveline angularity is the relationship of the driveshaft to the component that the driveshaft attaches to, as measured in degrees. For example, the angle of the rear axle assembly companion flange should be perpendicular to the transmission output shaft. If any deviation is present, the U-joints will bind, which will cause both a vibration and additional wear on the joints. An angle gauge or a protractor should be used to measure these angles (**FIGURE 34-28**). If the angles are out of specifications, follow the manufacturer's procedure to bring the components into alignment.

To check driveshaft joints, phasing, angles, balance, and runout, follow the steps in **SKILL DRILL 34-2**.

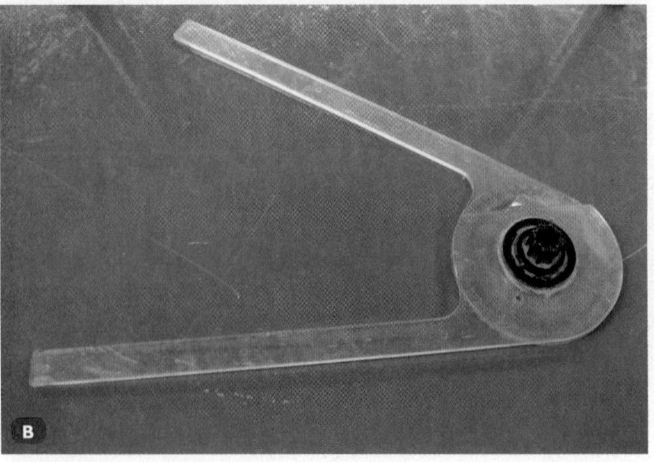

FIGURE 34-28 A. An angle gauge. **B.** A protractor.

Inspecting and Servicing Center Support Bearings

The center bearing is an important part of the driveline. The bearing is usually supported by a rubber insert similar to a powertrain mount which isolates noise and vibration from the chassis or body. The rubber can tear or deteriorate in the same way powertrain mounts do. The center bearing is subject to weather and the same abuses as a U-joint. Whenever an under-vehicle service is performed, the center bearing should be inspected for any type of damage that can happen along the way. If this bearing is equipped with a grease zerk, it should be lubricated whenever an oil change and filter and chassis lubrication is performed. Without lubrication, it could cause a major vibration and a catastrophic failure to occur.

To inspect and service center support bearings, follow the steps in **SKILL DRILL 34-3**.

Measuring Drive Axle Flange Runout and Shaft End Play

N34004

Once axles have been serviced or replaced, it is a good practice to measure axle flange runout and axle end play. Runout is important to make sure the brake drum or rotor and the tire and wheel assembly will rotate properly and not cause any kind of vibrations. If all repairs have been done properly, runout and end play should be in specifications. Some procedures may differ due to the manufacturer's protocol. Always consult the manufacturer's guidelines and service information to ensure proper procedures for runout and end play.

To measure drive axle flange runout and shaft end play, follow the steps in **SKILL DRILL 34-4**.

Inspecting and Replacing Wheel Studs and Lug Nuts

N34005

Lug nuts and studs are what hold the drive wheels in place and have been known to come loose or go bad over time. In some states, the law requires an inspection of the brakes that involves removal of the tires to gain access to the brake system. Lug nuts must be removed to do this. Inspection of the studs should be done in order to look for stretching or thread defects on these studs (**FIGURE 34-29**). If any defects or thread damage is present, the stud should be replaced. Sometimes improper torquing of the lug nuts can cause the studs to fail or lose a wheel when driving. Under- or over-tightening can result in these conditions. Uneven tightening can cause warping, which can cause problems with the brakes. *Note:* Refer to the Servicing Disc Brakes chapter for the procedure to replace lug studs.

▶ TECHNICIAN TIP

Never use an air gun to torque lug nuts without using a torque stick followed by a torque wrench that is set to the manufacturer's torque specifications; otherwise, lug and stud damage can be present and result in loss of a wheel and tire assembly while driving down the highway.

SKILL DRILL 34-2 Checking Driveshaft Joints, Phasing, Angles, Balance, and Runout

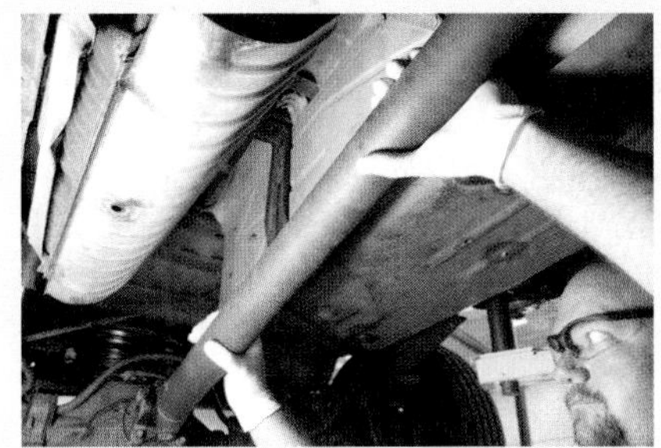

1. Raise the vehicle on a lift, making sure it is secure. Inspect the entire driveline by grasping the driveshaft and checking for abnormal play.

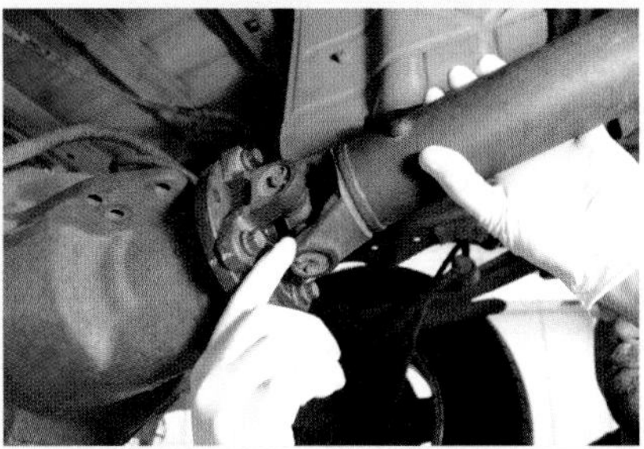

2. Move the shaft fore and aft and radially. Play indicates worn U-joints.

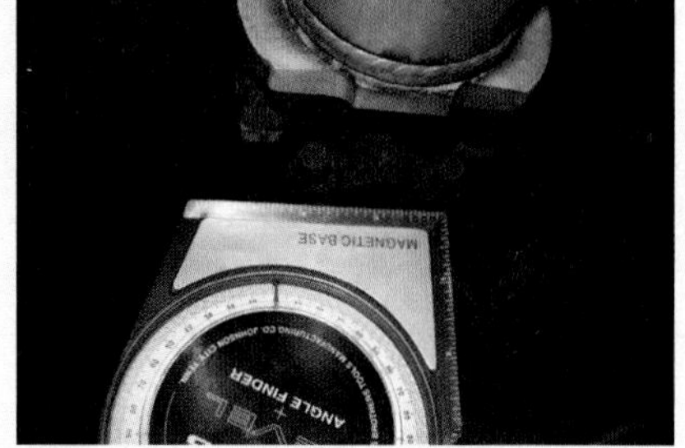

3. Start the vehicle and safely allow it to run in drive at slow speed on the lift. Watch and listen to the joints. Any noise or off-centeredness indicates damaged U-joints. Check that the joints are in phase by checking the U-joint yoke on each end of the driveshaft with a level.

4. Slowly raise the speed and watch for balance issues on the driveshaft. If any balance issues are present, slow the speed so the vehicle does not slip on the hoist. If all is well, stop the engine and measure the runout with a dial indicator, and check the driveshaft for concentricity.

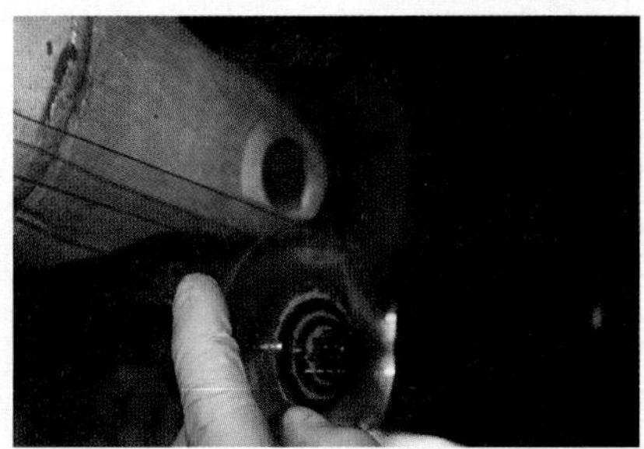

5. Check the angularity using a protractor, and make all necessary adjustments. Shims may be required.

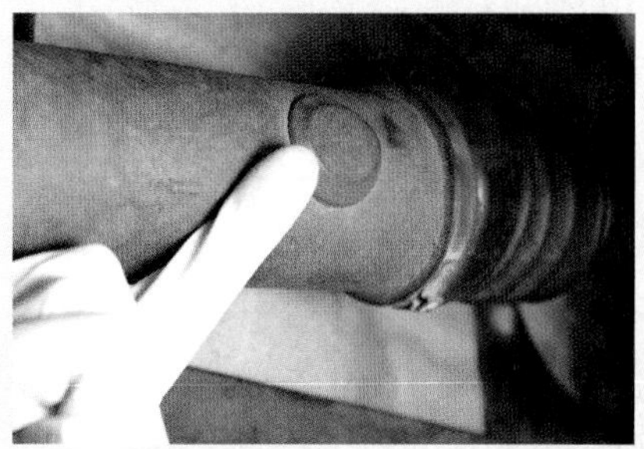

6. Check the driveshaft for any missing weights that are used to balance the shaft. Usually missing weights will be evident by the welds that are made when the driveshaft is balanced.

SKILL DRILL 34-3 Inspecting and Servicing Center Support Bearings

1. Safely raise the vehicle on an approved lift. Inspect the center bearing components for any major defects, such as looseness or noises.
2. Inspect the center bearing for proper mounting.
3. Inspect the bearing mount rubber insert for dry rotting and cracking.
4. If the bearing must be replaced, follow the manufacturer's specifications and procedures for proper installation of a new bearing. Typical bearing replacement may go as follows: mount the driveshaft in an approved vise.
5. Mark the shafts so they may be properly phased when put back together. Separate the two driveshafts.
6. Remove the U-shaped metal mounting bracket. Remove the rubber mount from around the center bearing.
7. Remove any snap rings or circlips that may be holding the bearing in place.
8. Use an appropriate puller or press to remove the bearing from the driveshaft.
9. Check the splines for the slip yoke for any defects. Check the slip yoke on the mating shaft for wear and defects.
10. Press on a new bearing. Reinstall any necessary snap rings or circlips.
11. Install a new rubber mount around the new bearing. Reinstall the mounting bracket.
12. Put the two shafts back together, paying attention to driveshaft phasing.
13. Remove the driveshaft from the vise, and reinstall it into the vehicle.

SKILL DRILL 34-4 Measuring Drive Axle Flange Runout and Shaft End Play

1. Measure axle flange runout with the tires, wheels, and brake drum or rotor removed. Start by mounting a dial indicator on the backing plate, and set the dial gauge on the flange of the axle; zero out the indicator to obtain a correct reading.
2. Slowly rotate the axle, and observe the runout as it is turning.
3. Compare with the manufacturer's specifications. If it does not match, replace the axle.
4. Once runout has been completed, mount the dial on the rim perch of the axle, and zero out the indicator.
5. Pry the axle in and out, and obtain an end play reading.
6. Compare to the manufacturer's specifications. If it does not match, the axle may have to be removed, and the axle groove should be inspected (integral axle type), or the bearings in the housing should be inspected or replaced (removable carrier type).

FIGURE 34-29 Inspecting lug studs. **A.** Stretched lug stud. **B.** Cross-threaded lug stud.

Removing and Replacing Drive Axle Shafts

N34006

Sometimes it is necessary to remove the drive axles with seal leakage or differential damage, or possibly a bent or damaged axle. Depending on the type of differential (integral or removable carrier), the removal process differs between the two. If it becomes necessary to remove the axles, make sure to follow all manufacturer guidelines and procedures explicitly.

To remove integral-type differential axles, follow the steps in **SKILL DRILL 34-5**.

To remove the removable carrier-type differential axles, follow the steps in **SKILL DRILL 34-6**.

Inspecting and Replacing Drive Axle Shaft Seals, Bearings, and Retainers

N34007

While the axle is out, depending on the type of final drive, you can service the seals, bearings, and retainers. Follow the manufacturer's procedures for the removal of the drive axles. If axles will be replaced, it is a good practice to replace the bearings and seals. Some retainers can be reused if clearances are still within specifications on integral-type final drives. Clearances on the end of the axle shafts for snap rings should be checked, as the axle has a tendency to move axially and cause the end of the axle shaft groove to become wide enough to drop the snap ring into the housing; this will cause the axle to walk out of the housing while the vehicle is traveling down the highway. Check the manufacturer's specification for axial clearances between the clip and the axle groove.

To replace the integral axle housing seal and bearing, follow the steps in **SKILL DRILL 34-7**.

To replace the removable carrier seal and bearing, follow the steps in **SKILL DRILL 34-8**.

Inspecting Half-Shaft Components

N34008

It is important to make the proper diagnosis before entering into any type of repair. If a repair of the CV joints is necessary, follow all manufacturer guidelines and procedures.

Service of CV joints depends on the type of failure that has occurred. For example, if the axle is clicking when turning sharply, then the joint or the entire half-shaft will need to be replaced. If the CV boot is torn, it can be replaced without replacing the entire joint or shaft. Some can be easy and some can be extremely hard, but a thorough inspection is a good practice to diagnose whether to replace the half-shaft in its entirety or to replace and service one CV joint or boot.

To inspect half-shaft components, follow the steps in **SKILL DRILL 34-9**.

> **TECHNICIAN TIP**
>
> Some CV joints do not require a retaining ring and just need to be driven off with a soft metal hammer, preferably a brass one.

SKILL DRILL 34-5 Removing Integral-Type Differential Axles

1. Put the vehicle on a lift. Remove the wheels and tires. Drain the differential fluid in a clean pan if reusing; otherwise, refill with fresh fluid.

2. Remove the rear differential cover bolts. Clean and remove the old gasket material to allow for the new gaskets.

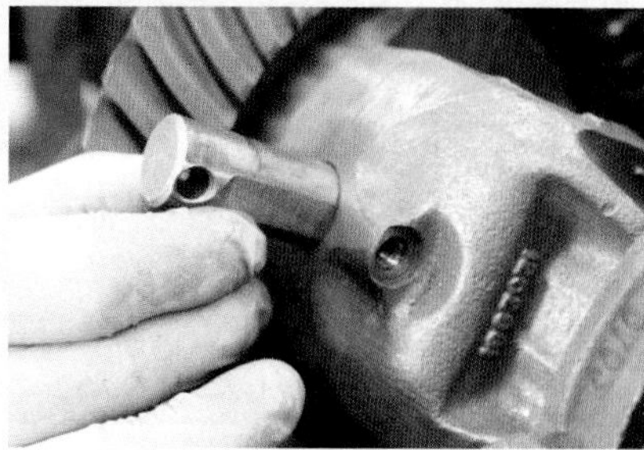

3. Remove the lock pin from the differential lock pin, and remove the pin.

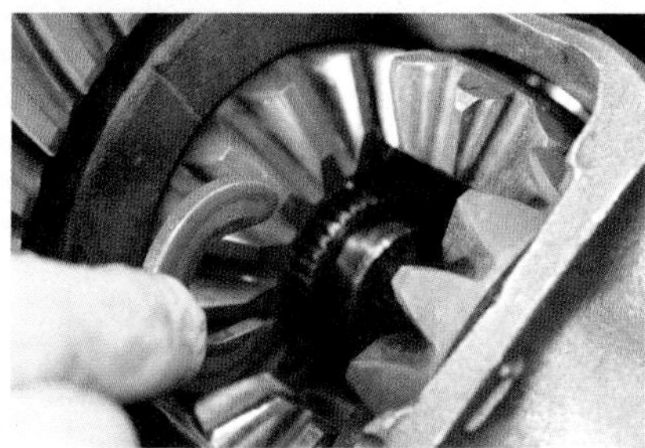

4. Slide one axle inward to remove the circlip from the end of axle. Then remove the other clip from the opposite axle. Slide the old axles out of the differential housing.

5. Replace the axle seals, along with new outer axle bearings, to avoid any leaks. Slide the new axles into the housing, paying attention to the axle side gear splines. Make sure they align properly.

6. Reinstall the circlips in the grooves on the new axles.

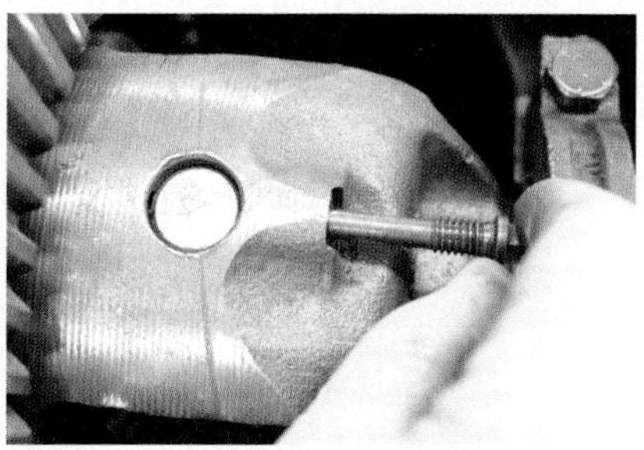

7. Slide the lock pin back into place, and reinstall the lock pin retainer. Tighten it to specifications.

8. Reinstall the differential cover with a new gasket, and tighten the bolts to specifications.

9. Refill the axle housing with appropriate type fluid recommended by the manufacturer. After a road test, put the car back on the lift, and inspect for any leaks.

▶ Final Drive and Axle Diagnosis

S34002

Noise/Vibration/Harshness and Other Issues

N34009

Drivetrain noises and vibrations can occur at any point in the life of the vehicle. As a vehicle starts to age, things begin to wear out. Differential gears and bearings can wear, putting adjustments out of specifications. U-joints or bent rear axles can contribute to noise or vibration conditions. Proper diagnosis is a priority if a reliable repair is to be performed. A thorough customer interview and road testing is essential.

Road testing the vehicle for vibrations will reveal the type of vibration that is present. First eliminate any torsional vibrations by starting the engine and running it at various revolutions per minute (rpm). If a vibration is felt, it may be engine noises or vibrations causing the problem, rather than the drivetrain.

SKILL DRILL 34-6 Removing Carrier-Type Differential Axles

1. Put the vehicle on an approved lift and secure it. Remove the tires and wheels. Remove the axle flange retaining bolts by rotating the axle and aligning the hole in the flange with the bolt to be loosened.

2. Pull the axles from the differential housing, using a puller if necessary.

3. Replace the axle seals and bearings (see Skill Drill 34-7).

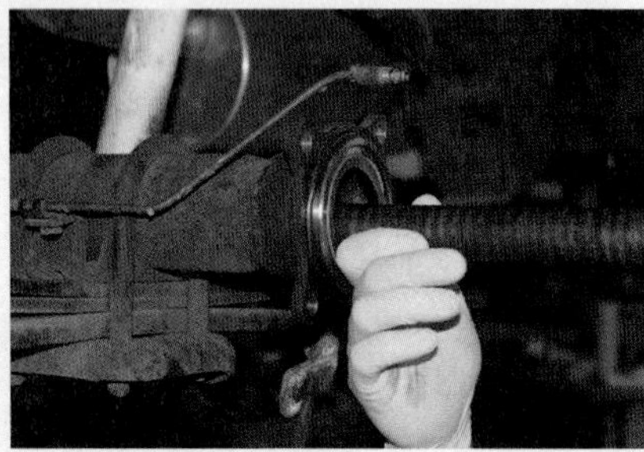

4. Install the new axles, and turn until splines are aligned and feel like they have fallen into place.

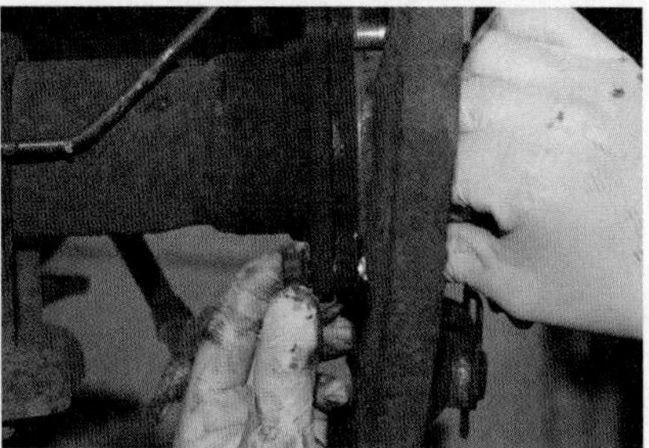

5. Replace the axle retaining bolts, and torque to the manufacturer's specifications. Reinstall the tires and wheels. Refill or top off the fluid level. After a road test, put the vehicle on the lift, and inspect for any leaks.

SKILL DRILL 34-7 Replacing the Integral Axle Housing Seal and Bearing

1. Remove the axle.

2. On some models, remove the bearings and seal by putting the axle in a press. On some models, pull the bearing out of the housing.

3. Replace all seals and inspect the seal surface of the axle.

SKILL DRILL 34-8 Replacing the Removable Carrier Differential Seal and Bearing

1. Remove the bolts of the axle retainer, and then remove the axle.

2. Press the bearing and seal off.

3. Remove the seal and bearing from the bearing housing.

4. Install new bearing and seal into the bearing housing.

5. Inspect the axle for any damage. Press the bearing assembly back onto the axle.

6. Push the axle back in the housing. Reinstall the flange bolts, and torque.

SKILL DRILL 34-9 Inspecting Half-Shaft Components

1. Clamp the entire half-shaft into a soft-jawed vise, and make sure it is secure. Remove the retaining clamps from the CV boot. Slide the boot down the shaft, paying attention to the condition of the boot.

2. Wipe out as much grease as possible to be able to access the retaining ring from the CV joint itself. If there is a retaining ring present, remove the retaining ring with the appropriate tool. When the retaining ring is removed, remove the CV joint from the half-shaft, and inspect the splines on the end of the half-shaft. This also applies to the other end of the half-shaft.

SKILL DRILL 34-9 Inspecting Half-Shaft Components (Continued)

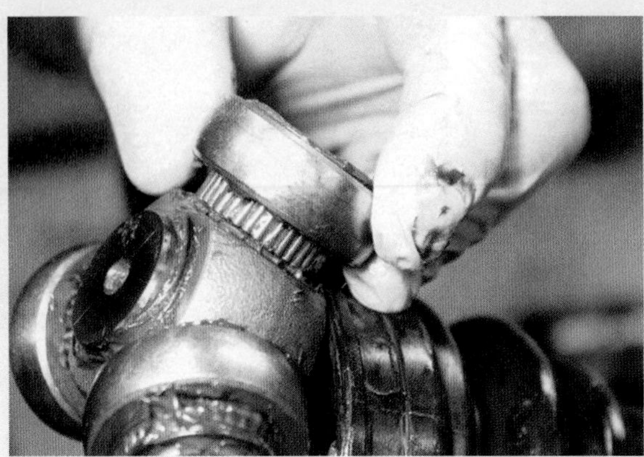

3. Inspect the old joint to gain an accurate assessment of the failure to prevent a reoccurrence of this failure. Reinstall the new CV joint onto the shaft splines as required.

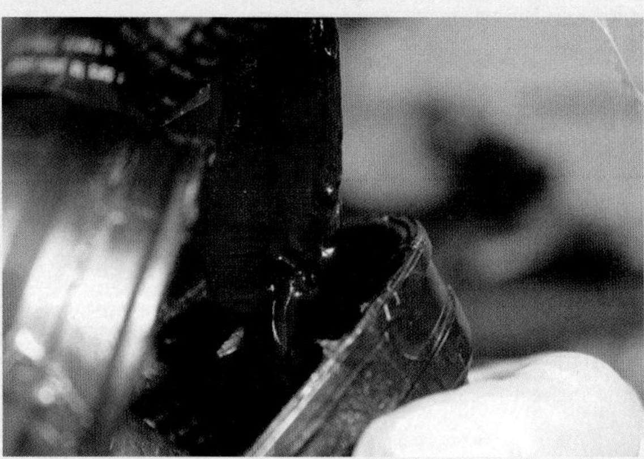

4. Apply the lubrication grease that comes with the new CV joint. Tighten the boot clamp to ensure that grease will not be lost. This applies to both sides of the half-shaft. Reinstall the half-shaft, following the manufacturer's guidelines.

Once the engine is eliminated as an issue, road test the vehicle at different speed ranges. Axle and wheel bearing noises tend to change in severity when steering side to side. Pinion bearing noises remain constant as the vehicle drives straight down the highway. Differential gears (spider gears) can make a whining or popping noise as the vehicle makes turns or travels around curves in the road. A clunking noise on acceleration and deceleration can be caused by worn ring and pinion gears and also tends to make a variable whirring noise when decelerating from 40 mph (64 kph) to a complete stop.

Knocking noises may indicate excessive clearance between gears such as the spider gears or axle side gears. Limited slip differentials may also contribute to noises in the differential. Rapid clicking or chattering noises may be coming from the clutch packs or cones that are part of the limited slip configuration. A fluid change may be required to eliminate these noises, or a friction modifier may need to be added.

Vibrations may result from a bent axle, bent driveshaft, or bent tire and rim combination. A thorough interview of the customer is important when diagnosing the complaint, as he or she can generally give you clues to what is causing the fault. The problem may be a result of damage from hitting a curb or an extremely large pothole. A thorough road test of the vehicle by operating the vehicle at different speeds and conditions will verify the driver's complaint.

Diagnosing Axle Shafts, Bearings, and Seals

N34010

Axle shafts and bearings are subject to radial and axial forces that under the right conditions can cause damage to these components. Conditions such as accidents can cause axles to bend or break, and bearings to fail. Seals can develop leaks, which can cause bearings to become dry, overheat from a lack of lubrication, and become damaged. When bearings fail, they generally will cause a whirring or growling noise that can increase and disappear as the vehicle turns from side to side. Lack of lubrication also results in failure of all the differential bearings, requiring a complete overhaul if not addressed.

Fluid leaks can be caused by a plugged breather in the housing, which causes a pressure buildup within the housing. They can also be caused by overfilling the final drive assembly, which exerts additional pressure on the seals. The weakest points on the final drive assembly are the seals. Seals are in constant contact with a rotating shaft. If pressure is allowed to build in the system, this pressure will force the seal onto the shaft harder, increasing the wear. Also, if the seal does start to leak, the pressure will force the fluid past it much more quickly than it otherwise would. Leaks can also be a result of dragging brakes, which create heat and destroy the seal. Sometimes vintage vehicles sit for long periods without being driven, and the seals dry out or shrink as a result of disuse.

Diagnosing Noise, Slippage, and Chatter in Limited Slip Differentials

N34011

Noise diagnosis requires a thorough road testing under all conditions necessary to recreate the problem. The type of noise in most cases tells the technician where the noise is coming from. Whining noises while swerving or cornering usually come from the differential pinion gears and pinion shaft. Damage to these components or scoring of the pinion shaft can cause this noise to occur. This problem can happen when the one of the wheels is turning faster than the other for an extended period of time—usually the result of mismatched tire sizes on the same axle or too much wheel spinning on one wheel. A "hmm, hmm" noise or chuckle noise can be the result of too much clearance between the ring gear and pinion gears that occurs during deceleration at around 40 mph (64 kph) or less. Harsh acceleration or driver abuse can contribute to this condition.

A knocking or clunking noise is usually the result of parts that are broken inside the final drive. Chipped or broken ring and pinion gears, broken spider or axle side gears, or damaged bearings can be the problem. Driver abuse or potholes contribute to this issue by making one wheel leave the ground for a split second and then grab quickly to take up clearances suddenly. The result is chipped or cracked gear teeth. Limited slip differential growling or chattering can result from clutch parts that are worn; in some cases, changing the limited slip lubricating oil fixes the problem. Slippage can result from worn clutches in the limited slip differential.

To perform a diagnosis, thoroughly interview the customer to gain information on when or how the condition first started. Thoroughly road test the vehicle under all conditions necessary to recreate the condition being diagnosed. Consult the service information, including technical service bulletins (TSBs), to help with diagnosis. Follow all manufacturer guidelines when performing the repair. Once the repair is complete, road test the vehicle to make sure the condition no longer exists.

Applied Math

AM-42: Everyday Occurrences: The technician compares the performance outcome of a normally operating system with the anticipated outcome of an everyday occurrence, such as comparing system tolerances to the manufacturer's specifications.

When an AWD vehicle with properly inflated tires goes around a corner, the process is very different from when that same vehicle has a substantially low tire. Low tire pressure reduces the rolling circumference of the tire. It is like having a smaller tire on one side. The difference in wheel speeds must be absorbed in the transfer case. If a viscous coupling or an electric clutch is used, it will be subjected to extreme torsional stress due to the uneven rotational speeds. The customer will notice binding or chirping of the tires, or there may be a banging or popping noise from the drivetrain. This condition will result in a damaged viscous coupling or electric clutch after a short time of driving in this situation.

The solution for this problem is to maintain all tires at the manufacturer's specified tire pressure. The tire pressure specification is usually found on the driver's side doorjamb. Another source of information for correct tire pressure is the owner's manual.

Starting in 2008, a tire pressure monitoring system, or TPMS, was required by law on all passenger cars and light trucks. The TPMS monitors the air pressure in the tires, and the driver is alerted of low tire pressure by a dashboard warning. This system is not meant to be a substitute for checking the pressures on a regular basis, as the TPMS warning may not come on until the air pressure drops by 25% or more below the correct pressure.

Diagnosing CV Joint Issues

N34012

CV joints have to withstand and transmit tremendous amounts of force when the vehicle is being driven. Grease helps to lubricate and cool the joint, which prevents or reduces wear. CV joints use rubber boots to contain the grease in the joint while still allowing the joint to pivot. Sometimes the CV joint boots crack or become torn. The grease then runs out, and contaminants can get in. This combination of problems causes the joint to overheat and wear out, resulting in a failed CV joint that must be replaced. This is one reason that periodic inspection of the CV boots are important. If caught soon enough, the boot and grease can be replaced before the joint is damaged. One indicator of CV joint wear is clicking or popping noises when cornering. The ball bearings roll into and out of the worn spot in the bearing races, making the popping noise. CV joints that make this noise have to be replaced; they cannot be repaired. A thorough road test along with a visual inspection on a lift may be warranted in order to diagnose CV joint problems.

To diagnose a potential issue with a CV joint, talk to the customer to fully understand his or her concern. Verify the concern with a test-drive or visual inspection, depending on the concern. You should then have enough information to make a diagnosis of the fault.

To diagnose CV joint issues, follow the steps in **SKILL DRILL 34-10**.

▶ Final Drive Repair

S34003

Inspecting and Replacing the Companion Flange and Pinion Seal

N34013

The pinion companion flange is what the driveshaft bolts up to and connects the transmission to the differential. Inspection from time to time may reveal any problems that may be present. Customer concerns can also warrant an inspection of the pinion companion flange. It is held on by a large nut that also sets the preload for the pinion gear on crush sleeve types. If replacement is necessary, consult the service information for complete replacement procedures.

SKILL DRILL 34-10 Diagnosing CV Joint Issues

1. Thoroughly road test the vehicle to verify the customer complaint. Safely raise the vehicle on an approved lift, and make sure it is secure.
2. Visually inspect all four CV joints.
3. Look for broken or ruptured boots. Look for cracked or dry-rotted boots, as they are subject to all types of weather.
4. Manually move the axle shaft up and down, looking for unnecessary play or bad bearings. These conditions can cause a vibration coming from the side that is bad. They will also cause a clicking sound emanating from the side that is affected.
5. Look for any type of axle damage, such as a bent axle from a recent accident. Axle damage causes a vibration from the affected shaft as well.

To inspect and replace the companion flange and pinion seal, follow the steps in **SKILL DRILL 34-11**.

> TECHNICIAN TIP

When installing any type of seal, it is a good practice to lubricate the seal and spring with grease to prevent the spring from popping off upon installation. Failure to do so may result in seal leakage and flange damage.

Inspecting the Ring Gear and Measuring Runout

N34014

When disassembling a differential case, it is good practice to measure ring gear runout before disassembly and after reassembly. This will indicate if there is a problem with the differential case or the ring gear itself, which can be an indicator of reuse or replacement. Any excessive runout could be the result of debris caught between the ring gear and the differential case upon assembly of the unit or possibly bad bearings on the differential case housing.

To inspect the ring gear and measure runout, follow the steps in **SKILL DRILL 34-12**.

Removing, Inspecting, and Reinstalling the Drive Pinion and Ring Gear

N34015

The overhaul process for a set of differential gears, including the ring gear and pinion gear, requires an overhaul kit. In the kit are new bearings and gaskets along with spacers and washers. A new crush sleeve may need to be purchased, as these are not reusable. If shims are underneath the bearing race cups or pinion gear, they will have to be added to or subtracted from in order to make any necessary adjustments. All bearings should be replaced along with the bearing races that they mate with.

To remove, inspect, and reinstall the drive pinion and ring gear, follow these step in **SKILL DRILL 34-13**.

> TECHNICIAN TIP

Lubricate the new bearings in the appropriate type of fluid used by the manufacturer.

SKILL DRILL 34-11 Inspecting and Replacing the Companion Flange and Pinion Seal

1. Raise the vehicle on an approved lift. Inspect the pinion companion flange for any visible defects. Inspect the pinion seal for leakage, and replace as necessary.

2. Remove the driveshaft to expose the retaining nut. Remove the large retaining nut. Slide the flange off the pinion gear spline, using a soft hammer or puller, and inspect the splines for any damage.

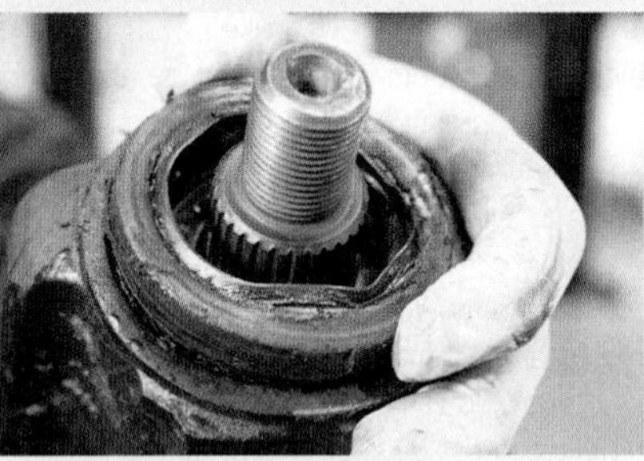

3. Remove the companion flange seal with the proper removal tool. Inspect seal surface for any damage. Replace the seal, using the appropriate driver.

4. Reinstall the companion flange, paying attention to the spline alignment if necessary. Install the retaining nut, and torque it to the proper specifications.

5. Using a dial indicator, check the runout of the companion flange, and compare it to the manufacturer's specifications.

6. Reinstall the driveshaft, paying attention to the alignment of the slip yoke. Check and refill the differential fluid. Road test the vehicle, and put it back on the lift to double-check your work.

SKILL DRILL 34-12 Inspecting the Ring Gear and Measuring Runout

1. Clean off the flat side of the ring gear to remove any excess oil. Attach a dial indicator to the differential housing near the area of the ring gear. Zero out the indicator so an accurate reading can be taken.

2. Rotate the ring gear by installing a socket and a ratchet to the pinion companion flange nut. Load the ring gear up by wedging a screwdriver or equivalent in order to get an accurate reading.

3. Rotate the ring gear slowly, paying attention to the dial, and compare the reading with specifications. Determine any necessary actions.

SKILL DRILL 34-13 Removing, Inspecting, and Reinstalling the Drive Pinion

1. Remove the pinion bearing by installing it in a press with a bearing splitter. Pump the handle and watch the bearing start to move and come off.

2. Install the new pinion bearing (and any proper shims) by pressing the new bearing back on in the same fashion.

3. Install the bearing splitter on the differential case bearings, and press these off using a press.

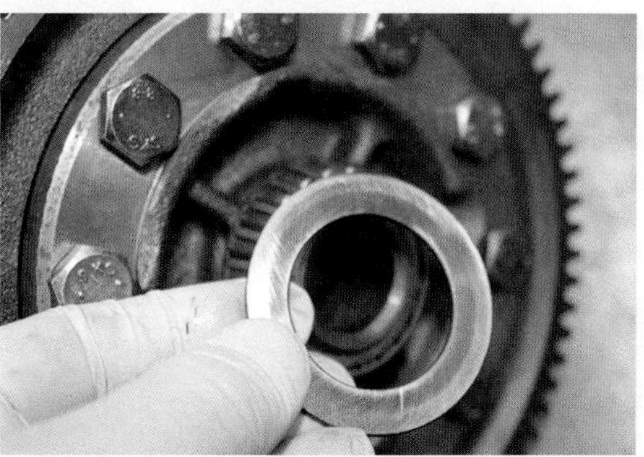

4. Install the new bearings (and any necessary shims), using the press, and make sure they are seated correctly.

Disassembling, Inspecting, Measuring, and Adjusting the Differential Pinion Gear Assembly

N34016

In the case of a catastrophic failure or just an upgrade of parts for performance purposes, it is extremely important to follow the correct procedures to replace or adjust the ring and pinion gears. The pinion-to-ring-gear backlash should be measured and recorded for future reference. It will be helpful should it be necessary to reinstall the same gear set back in the housing. Once the final drive assembly is removed from the housing, the differential assembly can be disassembled, inspected, and measured. Removal of some carriers requires that the axle housing be spread using the proper housing spreader tool recommended by the manufacturer. Remove the ring gear and differential case to access the differential side gears and pinion (or spider) gears for service.

To disassemble, inspect, measure, and adjust the differential pinion gear assembly, follow the steps in **SKILL DRILL 34-14**.

SKILL DRILL 34-14 Disassembling, Inspecting, Measuring, and Adjusting the Differential Pinion Gear Assembly

1. Disassemble the differential case by removing the ring gear bolts and separating the ring gear from the case. Then remove the spider gears, axle side gears, and thrust washers from the case.

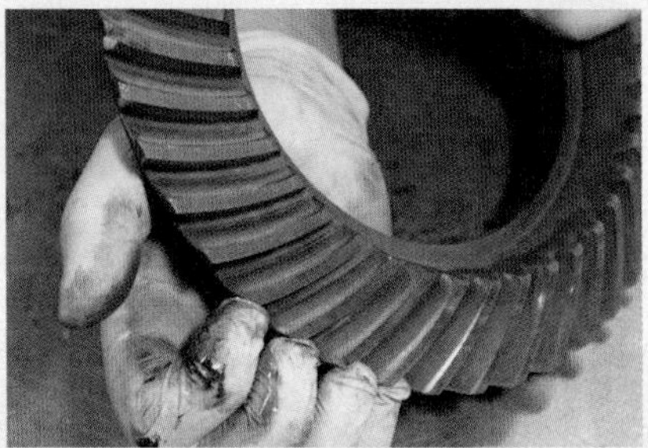

2. Inspect all gears for chipping and discoloration; this determines whether the gears should be renewed.

3. Inspect all thrust washers for chipping or galling and discoloration to determine whether they should be replaced.

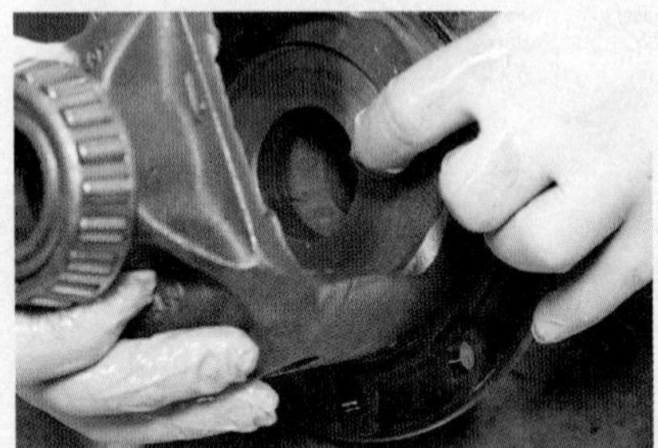

4. Inspect the differential case for any defects that would cause a problem for new or used gears and washers, especially the thrust washer surfaces.

5. Inspect the ring gear surface for trueness, using a straightedge, and inspect the ring gear mounting holes for elongation.

6. Reassemble the differential case, using new or used gears and thrust washers, and check the clearances between the gear sets.

7. Using a feeler gauge, check the clearances between the axle side gears and the spider gears. Check the clearances between the axle side gears and the differential case. Reinstall the ring gear and torque the bolts to manufacturer specifications.

Measuring and Adjusting Drive Pinion Depth

N34017

The purpose of measuring pinion depth is to ensure that the pinion gear teeth are making correct contact with the ring gear teeth. The depth must be correct in order for the ring to be driven properly, or there could be a catastrophic failure. When these spiral bevel gears rotate, they try to separate each other by a twisting motion, which makes the gears want to walk away from each other. The bearings keep them in check as they rotate, but the separation effect is always present. Correct depth and gear contact will keep separation at a minimum.

To adjust pinion depth, follow the steps in **SKILL DRILL 34-15**.

Measuring and Adjusting Drive Pinion Preload

N34018

Adjusting pinion bearing preload is necessary to make sure the conical bearings that support it are seated and positioned correctly. The pinion rotates and drives the ring gear. The forces on the pinion gear are tremendous in that the driveshaft, which is connected to it, transmits the power to the pinion gear, then to the ring gear, and then to the axles and tires. Pinion bearing preload positions the bearings such that the pinion gear is held firmly in place but still allows smooth rotation of the gear.

To measure and adjust drive pinion preload, follow the steps in **SKILL DRILL 34-16**.

SKILL DRILL 34-15 Measuring and Adjusting Drive Pinion Depth

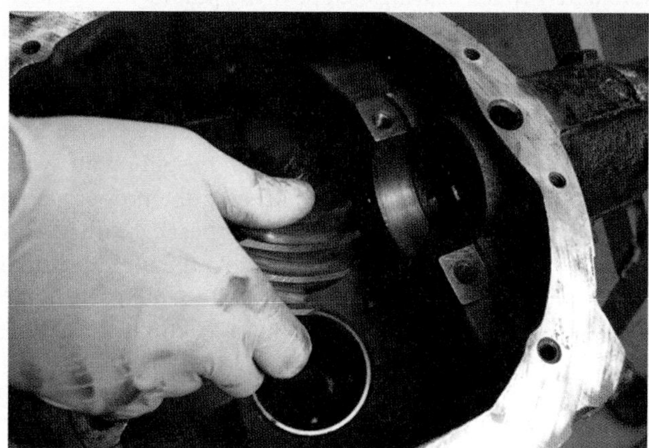

1. Install the pinion gear bearing race and any shims that were present.

2. Use the factory-designated tool to set the depth of the pinion gear in the housing. After the measurement is made, install the proper shims to meet the required height. Reassembly of the pinion can be performed for the next adjustment, called pinion bearing preload.

SKILL DRILL 34-16 Measuring and Adjusting Drive Pinion Preload

1. Reinstall the pinion into the housing with a new bearing, making sure to install the crush sleeve or shims.

2. Install the front bearing over the splined end of the gear.

3. Reinstall the pinion seal.

SKILL DRILL 34-16 Measuring and Adjusting Drive Pinion Preload (Continued)

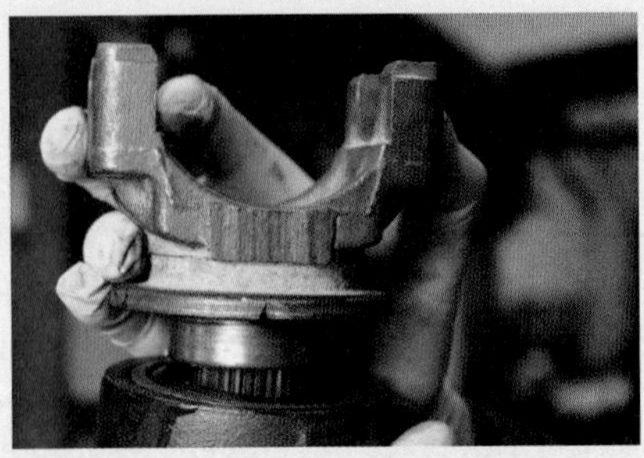

4. Reinstall the companion flange, lining up the splines.

5. Reinstall the nut and washer. Torque the nut to the manufacturer's specifications.

6. Check the preload, using an inch-pound torque wrench or pull scale, looking for the correct specification.

> **TECHNICIAN TIP**
>
> Never over-tighten (exceed the preload). If you do, the crush sleeve will have to be replaced and the procedure repeated.

Reassembling and Reinstalling the Differential Case, and Measuring Runout

N34019

Now that the differential case has been serviced, the assembly can be reinstalled. This should be done carefully, with attention given to the installation specifications. Upon installation, put the new differential case bearing races on the bearings and any spacers that were removed. These spacers determine backlash, which is addressed in Skill Drills 34-18 and 34-19.

To reassemble and reinstall the differential case assembly and measure runout, follow the steps in **SKILL DRILL 34-17**.

Differential Adjustments

The tapered roller bearings support the pinion in the carrier. The companion flange splines engage with the pinion splines to transfer the drive torque from the driveshaft. A collapsible spacer, called a **crush sleeve**, between the bearings allows a specified preload to be set (**FIGURE 34-30**). The crush sleeve provides a means of maintaining a preset torque on the pinion nut, which holds the companion flange while keeping the proper amount of bearing preload so that the pinion gear bearings are not running too tight or too loose. The depth that the pinion sits at in the axle housing is controlled by shims. The depth of the pinion determines how the pinion gear teeth mesh with the ring gear.

The ring gear is bolted to the differential carrier, which is supported in the axle housing or case by tapered roller side bearings, retained by bearing caps and bolts. In some axle designs, threaded adjusting rings engage with threads in the housing and push against the side bearings of the carrier. These threaded rings are used to set the bearing preload and

SKILL DRILL 34-17 Reassembling and Reinstalling the Differential Case Assembly, and Measuring Runout

1. Use a factory-recommended housing spreader to reinstall the differential case as a unit. Insert the differential case back in the housing, paying attention to the spacers that were present.

2. Reinstall the bearing caps on the housing, paying attention to the marks made as they were removed. Torque to specifications.

3. Attach a dial indicator to the housing, set the dial on the ring gear, and check the runout. Set the dial to zero and rotate the ring, using the companion flange nut and a socket and ratchet. Check specifications for proper runout.

FIGURE 34-30 Pinion with crush sleeve.

a backlash clearance between the ring and pinion gears. If you hold the ring gear and turn the pinion, you should feel a slight clearance back and forth. On other axle designs, shims are used to push against the side bearings to adjust bearing preload and backlash.

Measuring and Adjusting Side Bearing Preload and Ring and Pinion Gear Backlash

Setting backlash on a set of differential gears is critical in that if it is too loose or tight, a catastrophic failure will happen for sure. Carrier bearing preload should be adjusted at the same time as a backlash setting. Side bearing preload limits the amount of lateral movement that the ring gear can travel, whereas backlash sets the closeness of the ring gear into the pinion gear. If this is an integral differential setup, backlash and bearing preload are usually set using shims. Removable center section carriers usually have bearing cone adjusters that are turned to set preload and backlash.

To adjust backlash of an integral differential, follow the steps in **SKILL DRILL 34-18**.

To adjust backlash of a removable carrier differential, follow the steps in **SKILL DRILL 34-19**.

SKILL DRILL 34-18 Adjusting Backlash of an Integral Differential

1. Make sure the bearing preload is set, and attach a dial indicator to the housing. Set the dial indicator on the teeth of the ring gear, and zero out the gauge.

2. Rock the ring back and forth, and take a reading. Compare to the specifications. If adjustment is required, the ring gear will have to be removed and spacers of the required thickness will have to be installed.

3. Repeat the measurement process.

SKILL DRILL 34-19 Adjusting Backlash of a Removable Carrier Differential

1. Install the ring gear/differential assembly and bearing cones and caps.

2. Snug the bolts on the bearing caps.

3. Thread the bearing adjusting nuts into the cap and housing threads, being careful not to cross-thread the adjusters (slight bearing preload for test).

SKILL DRILL 34-19 Adjusting Backlash of a Removable Carrier Differential (Continued)

4. Attach a dial indicator to the housing. Set the dial indicator on the ring gear teeth, and zero out the dial indicator. Rock the ring gear back and forth, and take a measurement.

5. Set the backlash to the manufacturer's specifications by alternately turning the adjuster nuts until the reading falls into the manufacturer's reading. Torque the bearing cap bolts to specifications.

Checking Ring and Pinion Tooth Contact Patterns

N34020

Once a differential gear set has been replaced, checking the gear tooth contact pattern is a good practice to make sure there is proper contact between the ring and pinion. On the ring gear tooth, there are several parts that need to be identified. The first thing to identify is the coast side of the ring gear tooth and the drive side of the ring gear tooth. The drive side of the ring gear tooth is the convex side that torque is applied to when the power is being transferred through the gear teeth. Coast is when the vehicle is coasting and contact is being made on the concave side of the ring gear teeth. The pinion drive gear is now the driven gear in coast mode. The ends of the tooth are the toe and the heel, with the heel being the wider part of the tooth farther from the gear centerline. Looking at the tooth where the contact is made, the top of the tooth surface is called the face; the middle of the tooth is the flank; and the bottom is called the undercut. These parts must be understood, as the gear contact pattern needs to be in the center or the flank of the gear tooth. It is a good idea to check the tooth contact pattern before disassembly, as this will create a comparison between the old and the new tooth pattern and allows the technician to see if the pinion depth is set correctly (**FIGURE 34-31**).

This pattern represents the distance between the ring and pinion gears and ensures quiet operation while they are in mesh. Any necessary adjustments should be made in the advent that the contact pattern is not acceptable. A contact pattern is made by using some type of grease or Prussian blue marking compound to make the pattern stand out. Paint several teeth, and turn the companion flange nut while holding back on the ring gear river to prevent the ring gear backlash from affecting the tooth contact pattern.

To check contact patterns, follow the steps in **SKILL DRILL 34-20**.

FIGURE 34-31 Understanding the tooth pattern is important in ensuring that setup of the ring and pinion has been correctly performed.

Cleaning, Inspecting, and Lubricating the Differential Housing

N34021

If a catastrophic failure has occurred, then it will be necessary to clean the differential housing. Start by disassembling all major parts to expose the bare housing. Use an approved parts cleaner or pressure washer to clean the housing. The entire tubes must be flushed of all debris and particles that are present. Doing so will ensure that the new parts being used will not be contaminated and cause a premature failure. Inspect the housing for any major defects that may be present from any type of failure that may have occurred. Reassemble all major parts and components as described in previous Skill Drills.

To clean and inspect the differential housing and to refill it with lubricant, follow the steps in **SKILL DRILL 34-21**.

SKILL DRILL 34-20 Checking Ring and Pinion Tooth Contact Patterns

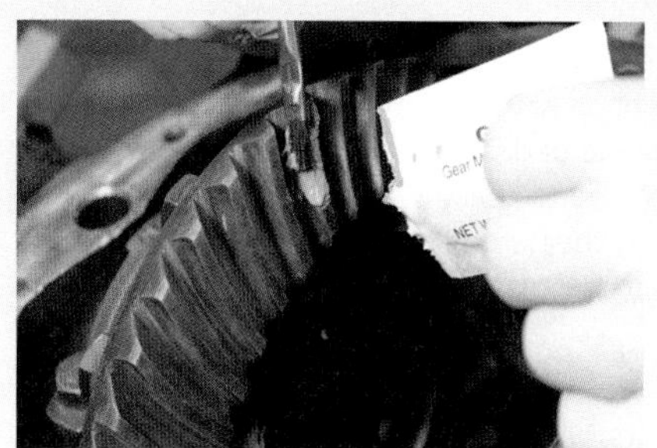

1. Paint several teeth with a suitable compound to make the pattern stand out in order to read it for any corrective actions that may be necessary.

2. Rotate the ring gear a full 360 degrees with a ratchet and a socket, using the companion flange nut.

3. Wedge the ring gear with a screwdriver in order to limit the backlash from affecting the pattern. Read the pattern, compare it to the specifications, and determine the necessary course of action if it is not satisfactory. Adjust the pattern as necessary and clean the components thoroughly before reassembling.

SKILL DRILL 34-21 Cleaning and Inspecting the Differential Housing, and Refilling with the Correct Lubricant

1. Disassemble all major parts and components.

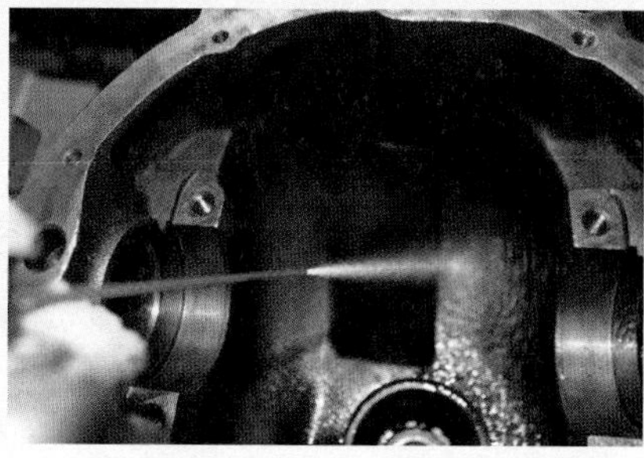

2. Flush and clean using proper procedures and equipment.

3. Inspect the housing for any defects. Reassemble all parts and components. Refill with manufacturer-specific differential fluids or additives.

Inspecting and Reinstalling Limited Slip Differential Components

S34004

If the differential clutches must be replaced because of a catastrophic failure, then all components should be thoroughly inspected to ensure a lasting repair. Inspection of the housing is described in Skill Drill 34-21. Inspect the differential casing for any damage that may require replacement of the housing. Inspect all axle side gears and differential pinion gears and shafts for damage, and replace as necessary.

Start by placing the new clutch facings in the differential lubricant, and soak them thoroughly. If they are put into the housing dry, a growling noise on turns will be present, and damage to the clutches can occur prematurely. Once they have been soaked, install them in proper order by putting in a steel plate first, then a clutch facing, and then alternating them. Install the side gears; then install the pinion gears into the housing with all pinion shafts in proper order. Once this process is completed, the assembly may be reinstalled into the axle housing. Make sure the differential is filled with the proper oils and additives to ensure quiet operation. Not all limited slip assemblies are assembled the same way; always consult the manufacturer's guidelines and procedures.

To inspect and reinstall limited slip differential components, follow the steps in **SKILL DRILL 34-22**.

Measuring Rotating Torque

N34022

Measuring rotation torque on a limited slip differential tests the condition and adjustment of the clutch pack. If the clutch discs are worn, rotating torque will be lower than specifications. If

SKILL DRILL 34-22 Inspecting and Reinstalling Limited Slip Differential Components

1. Inspect all potentially reusable components for damage that may require replacement of the parts.
2. Soak all of the clutch friction facings prior to assembly to ensure quiet operation of the differential.
3. Insert the clutches into the housing, paying attention to the orientation of the steel plates and the friction facings.
4. Install the side gears against the clutches with the flat side facing the clutch friction facings.
5. Install the pinion gears and pinion shafts in the order they were removed, paying attention to any roll pins or retaining pins that were present at the time of disassembly.
6. Install the unit into the housing, along with a new gasket, and perform any necessary measurements.
7. Fill the differential with the manufacturer-recommended oils and additives.
8. Road test the vehicle to ensure a proper repair.

SKILL DRILL 34-23 Measuring Rotating Torque

1. Place the vehicle in neutral, raise one drive wheel a few inches off the ground, and secure it with a jack stand.
2. Bolt the torque wrench adapter to the raised wheel.
3. Use the torque wrench to turn the wheel. Rotating torque is the torque required to keep the wheel turning, not the breakaway torque. Compare to specifications.

rotating torque is above specifications, the vehicle may try to go straight when steered on slightly slippery surfaces, or chirp the tires on dry surfaces. Rotating torque is measured with the vehicle in neutral, one drive wheel off the ground and the other on the ground. It is easiest to use a beam style torque wrench when measuring rotating torque.

To measure rotating torque, follow the steps in **SKILL DRILL 34-23**.

▶ Wrap-Up

Ready for Review

- Without a set of gears, flexible joints, and shafts to transfer the power from the engine and transmission, the vehicle would not be able to propel itself up hills, handle turns, or hold itself back as it goes down hills.
- In a conventional RWD vehicle, the engine and transmission are mounted longitudinally at the front.
- Vehicles with rear- or mid-mounted engines normally use a transaxle and transfer the drive to the road wheels by independent half-shafts.
- The front-engine FWD vehicle typically, but not always, uses a crosswise-mounted (transverse) engine coupled to a transaxle.
- CV joints allow for smoother transfer of power and allow for the vehicle to turn more tightly without the joint binding.
- CV joints are most often used on the half-shafts of FWD vehicles and may be mounted at the end of the drive axles in some RWD vehicles.
- The most common type of joint is the universal joint, or simply U-Joint (the correct term for it is a Hooke's joint).
- The driveshaft itself allows power transfer from one component to another, such as from the transmission output shaft to the differential and drive axles.
- The rear-wheel driveshaft transfers the power from the transmission to the final drive at the rear of the vehicle.
- Because the maximum length of a driveshaft is approximately 72" (183 cm), to prevent twisting from the torque output, two or more sections may be used.
- The front section of the two-piece driveshaft is supported at its rear end by a center bearing, called the carrier bearing.
- The U-joint is capable of working at a maximum angle of 3 degrees; beyond this, vibration and damage will result from U-joint binding.
- When working with a two-piece driveshaft or a splined driveshaft, it is critical that the yokes of the driveshaft (the portion of the shaft that has the holes in it for the U-joint end caps to fit into) line up with each other from one end of the driveshaft to the other.
- In FWD vehicles, the driveshafts transfer the drive directly from the differential inside the transaxle to the front wheels.
- An FWD transaxle typically does not place the differential directly in the center of the vehicle; instead, it is offset to one side because it is bolted to the end of the transversely mounted engine.
- Torque steer is caused by the flexing of the longer half-shaft, resulting in lower torque to one side than the

other, creating the pulling condition. To combat this condition, many manufacturers use an intermediate shaft.

- The differential provides the means of transferring power from the driveshaft (propeller) to the drive wheels while allowing the vehicle to turn smoothly.
- Final drives can be found in axles either at the front or rear of the vehicle or can be found in the transaxle of an FWD vehicle.
- A ring gear and pinion gear located in the final drive transfer power through 90 degrees and provide a final gear reduction to the driving road wheels.
- The transfer case or power take-off (PTO) allows the transfer of power to the front and rear axles.
- Limited slip differentials allow normal differential action under normal driving conditions, but when road conditions are not normal, the limited slip differential reduces or prevents differential action so that a wheel cannot spin freely.
- Helical-geared limited slip differentials respond very quickly to changes in traction. They also do not bind from friction in turns and do not lose their effectiveness as wear develops in the clutch-style units.
- There are three different axle setups, semi-floating, full floating, and ¾-floating axle assemblies. The difference is in the arrangement of the wheel bearings.
- The purpose of axle seals is to keep the required amount of oil in the differential to maintain quiet operation of the differential and related parts and to ensure lubrication for the bearings and gears.
- Proper diagnosis of vibration, noise, and other problems can keep a vehicle running and lasting longer. Always ask clarifying questions when a customer complains of noises or vibrations.

Key Terms

backlash The amount of movement between the pinion teeth and the ring gear teeth.

beam-type axle A rear-wheel drive axle assembly that has a solid tube incorporating the differential gears

companion flange A splined flange that transmits power from the driveshaft to the pinion gear.

constant-velocity (CV) joint A joint used to transmit torque through wider angles and without the change of velocity that occurs in U-joints.

crush sleeve A collapsible spacer between the bearings that provides a means of maintaining a preset torque on the pinion nut.

double Cardan joint A type of joint that uses two Cardan joints housed in a short carrier and that reduces the change in velocity of a single Cardan joint by using the second joint to cancel out the changes in velocity of the first joint.

driveline angularity The relationship of the driveshaft to the component that the driveshaft attaches, measured in degrees of angle.

fixed-type joint A joint that does not slide to allow for shaft lengthening or shortening; it simply allows for angle changes as the suspension moves.

full floating axle An axle that does not support any weight; if removed, the vehicle will still roll on its wheels.

helical-geared limited slip differential A type of differential that responds very quickly to changes in traction and that does not bind from friction in turns or lose its effectiveness since there are no clutches.

Hooke's joint A joint that consists of a steel cross with four hardened bearing journals, mounted on needle rollers in hardened caps, which locate the cross in the eyes of the yokes. The cross swivels in the yokes as the drive is transferred across the joint.

hypoid bevel gear A special design of spiral bevel gear, with the centerline of the pinion below the centerline of the ring gear.

longitudinal The orientation of the engine in which the front of the engine is facing the front of the vehicle. It is most commonly found in rear-wheel drive vehicles.

plunge-type joint The inner joint on the half shaft that allows for changes in shaft length.

Rzeppa joint A type of fixed constant-velocity joint that has an inner race, six steel ball bearings, a bearing cage, and an outer race.

semi-floating axle An axle that carries the weight of the vehicle; if removed, there is no way to connect the wheel to the vehicle.

side gear A gear that is splined to the axle shaft and meshes with the spider gears and allows the axles to rotate at their own speeds when cornering and turning.

sliding spline driveshaft A two-piece driveshaft that is joined in the middle with splines. The driveshaft can slide on itself to increase or decrease in length.

slip yoke Part of a two-piece driveshaft that is splined and allows for a change in length of the shaft as the suspension compresses and rebounds.

speedy sleeve An aftermarket repair kit that consists of a thin metal sleeve that fits tightly over the seal surface of the axle, providing a new, undamaged surface for the seal to ride against.

three-quarter floating axle An axle on which there is only one wheel bearing that bears the weight of the vehicle, but the axle prevents the wheel from tipping inward or outward.

torque steer A condition in which the vehicle pulls to one side during hard acceleration.

transverse The orientation of the engine in which the front of the engine is facing the side of the vehicle.

tulip/tripod joint A constant-velocity joint that has three equally spaced fingers shaped like a star. This configuration allows in-and-out movement of the shaft while allowing flexing.

universal joint (U-joint) A cross-shaped flexible joint on which caps fit over the ends of the cross. Needle bearings fit between the ends of the cross and the caps, allowing the caps to rotate smoothly.

Review Questions

1. All of the following statements describing rear-wheel drive systems are true *except*:
   a. the ring and pinion gear set allows the transfer of power 90 degrees.
   b. the engine and transmission are transversely mounted at the front.
   c. drive from the engine is transmitted to a rear axle assembly by a propeller shaft.
   d. splines of the slip yoke mate to the splines on the transmission output shaft.
2. Choose the correct statement.
   a. Most front-engine front-wheel drive vehicles use a longitudinal-mounted engine.
   b. Front-wheel drive power reaches the rear wheels through the transaxle.
   c. In a transverse-mounted engine, the front of the engine is facing the front of the vehicle.
   d. Power is transmitted in a straight line from the engine through the transaxle and out to the front axles.
3. What is done to prevent twisting of long driveshafts from the torque output?
   a. Thicker metal is used.
   b. It is held in place by reinforcements.
   c. The length of the tube is decreased.
   d. Two or more sections may be used.
4. The front section of the two-piece driveshaft is supported at its rear end by a:
   a. differential bearing.
   b. carrier bearing.
   c. U-joint.
   d. pocket bearing.
5. Which of the following axles uses a single roller bearing between the hub and the outside of the axle housing?
   a. Semi-floating axle
   b. Full floating axle
   c. Floating axle
   d. Three-quarter floating axle
6. Choose the correct statement.
   a. The drive shafts are connected directly to the final drive gears inside the transaxle.
   b. Half-shafts have an inner and an outer CV joint.
   c. Front-wheel drive transaxle typically places the final drive directly in the center of the vehicle.
   d. The intermediate shaft makes it so that both half-shafts are of different length from left to right.
7. All of the following statements with respect to universal joints are true *except*:
   a. needle bearings fit between the ends of the cross and the caps, allowing the caps to rotate smoothly.
   b. in Hooke's joints, change in the angular velocity increases as the angle of the joint decreases.
   c. a double Cardan joint is considered as a CV joint under normal drive line angles.
   d. the ball-and-socket assembly keeps each Cardan joint at an equal angle, thereby canceling the change in velocity.
8. Which type of joint has an inner race, six steel ball bearings, a bearing cage, and an outer race?
   a. Rzeppa joint
   b. Fixed-type joint
   c. Tulip joint
   d. Universal joint
9. Choose the correct statement.
   a. The differential assembly provides the final gear reduction necessary for drivetrain operation.
   b. The differential assembly takes the power from the transmission and sends it to the wheels.
   c. The differential assembly is housed within the final drive assembly.
   d. The differential assembly is another name for the complete final drive assembly.
10. All of the following statements referring to limited slip differentials are true *except*:
    a. the limited slip differential gives normal differential action and permits the outer wheel to turn faster than the inner wheel.
    b. the clutch-style limited slip differential uses a multi-plate clutch pack between each side gear and the differential case.
    c. the Torsen differential is to multiply the torque available from the wheel that is losing traction and turn it over to the slower turning wheel with better traction.
    d. helical-geared limited slip differentials do not respond very quickly to changes in traction.

ASE Technician A/Technician B Style Questions

1. Tech A says that axle seals are a common source of solid axle fluid leaks. Tech B says that pinion seals are a common source of solid axle fluid leaks. Who is correct?
   a. Tech A
   b. Tech B
   c. Both A and B
   d. Neither A nor B
2. Tech A says that the angle of the rear axle assembly companion flange should be perpendicular to the transmission output shaft. Tech B says that U-joints should be 45 degrees out of phase from one another to minimize vibration. Who is correct?
   a. Tech A
   b. Tech B
   c. Both A and B
   d. Neither A nor B
3. Tech A says that a driveshaft center bearing is used on a vehicle with a short driveshaft. Tech B says that a driveshaft center bearing has a rubber insert that can deteriorate. Who is correct?
   a. Tech A
   b. Tech B
   c. Both A and B
   d. Neither A nor B
4. Tech A says that if a CV boot is torn, it can typically be replaced without replacing the entire joint or shaft. Tech B

says that worn CV joints can make a clicking noise when driving while making a tight turn. Who is correct?
   a. Tech A
   b. Tech B
   c. Both A and B
   d. Neither A nor B
5. Tech A says that the gear tooth contact pattern can be taken using Prussian blue. Tech B says that the convex side of the ring gear is the coast side. Who is correct?
   a. Tech A
   b. Tech B
   c. Both A and B
   d. Neither A nor B
6. Tech A says that excessive ring gear runout could be the result of debris caught between the ring gear and the differential case during assembly. Tech B says that excessive ring gear runout could indicate bad bearings on the differential case housing. Who is correct?
   a. Tech A
   b. Tech B
   c. Both A and B
   d. Neither A nor B
7. Tech A says that the clutch-style limited slip differential uses a multi-plate clutch pack between each side gear and the differential case. Tech B says that if the clutch discs are worn, rotating torque will be higher than specifications. Who is correct?
   a. Tech A
   b. Tech B
   c. Both A and B
   d. Neither A nor B
8. Tech A says that pinion bearing preload is measured with a dial indicator. Tech B says that pinion bearing preload positions the bearings such that the pinion gear is held firmly in place but still allows smooth rotation of the gear. Who is correct?
   a. Tech A
   b. Tech B
   c. Both A and B
   d. Neither A nor B
9. Tech A says that on the full floating axle, there are two tapered roller bearings between the hub and the outside of the axle housing. Tech B says that a semi-floating axle, the axle supports the weight of the vehicle and is also subject to bending forces as the vehicle corners. Who is correct?
   a. Tech A
   b. Tech B
   c. Both A and B
   d. Neither A nor B
10. Tech A says that U-joints can make a chirping or squeaking nose when they are going bad. Tech B says that U-joints are typically used on FWD axles to provide torque through all angles of steering. Who is correct?
   a. Tech A
   b. Tech B
   c. Both A and B
   d. Neither A nor B

CHAPTER 35

Four-Wheel Drive/ All-Wheel Drive

NATEF Tasks

- **N35001** Diagnose noise, vibration, and unusual steering concerns; determine needed action. (MAST)
- **N35002** Identify concerns related to variations in tire circumference and/or final drive ratios. (AST/MAST)
- **N35003** Diagnose, test, adjust, and/or replace electrical/ electronic components of four-wheel drive systems. (MAST)
- **N35004** Inspect, adjust, and repair shifting controls (mechanical, electrical, and vacuum), bushings, mounts, levers, and brackets. (AST/MAST)
- **N35005** Disassemble, service, and reassemble transfer case and components. (MAST)
- **N35006** Inspect locking hubs; determine needed action. (AST/MAST)

Knowledge Objectives

After reading this chapter you will be able to:

- **K35001** Describe the purpose, function, and difference between four-wheel drive and all-wheel drive systems.
- **K35002** Describe four-wheel drive layout.
- **K35003** Describe all-wheel drive layout.
- **K35004** Describe the purpose and function of 4WD and AWD transfer cases.
- **K35005** Describe the purpose and function of the 4WD transfer case.
- **K35006** Describe the purpose and function of the AWD transfer case.
- **K35007** Describe the purpose and function of 4WD driveshafts, axles, and locking hubs.
- **K35008** Describe the purpose and function of 4WD driveshafts.
- **K35009** Describe the purpose and function of drive axles.
- **K35010** Describe the purpose and function of 4WD locking hubs.

Skills Objectives

- **S35001** Perform diagnosis and repair on 4WD components.
- **S35002** Remove and reinstall transfer case.

▶ Introduction

K35001

In the last chapter, we covered front-wheel drive (FWD) and rear-wheel drive (RWD). In this chapter, we will cover four-wheel drive (4WD) and all-wheel drive (AWD). The ability to drive all of the wheels enhances the vehicle's ability to connect the engine's power to the ground. This is very handy when traveling on slippery on-road or off-road surfaces, as power can be applied to the ground through more wheels. In some performance vehicles, AWD is used in conjunction with traction control and stability control to enhance acceleration and handling characteristics when driving aggressively. In this situation, power can be distributed to each of the wheels individually, based on their loading and the directional stability of the vehicle. We cover traction control and stability control in the Electronic Brake System chapter. In this chapter, we cover the mechanical systems used to get power to each of the four wheels.

▶ Four-Wheel Drive Layout

K35002

Vehicles with part-time 4WD are designed for optional off-road use. Four-wheel drive can be selected as needed for abnormal surfaces (mud, snow, ice, etc.) and can be disconnected for normal road surfaces. Selection is made by a manual lever or by electronic control via a push button. In such applications, the engine and transmission are normally mounted longitudinally at the front. A transfer case is mounted to the rear of the transmission. The transfer case is a device designed to split the power from the transmission to both the front and the rear drive axles when 4WD is selected (**FIGURE 35-1**). Driveshafts are used to move power from the transfer case to the front and rear drive axles. The transfer case gives the driver control of four possible power transfer options:

1. The selection most often used is two-wheel drive (2WD), as most driving is done on clear non-slippery roads. Two-wheel drive sends power to the rear wheels only; the front axle is not being supplied with any power.

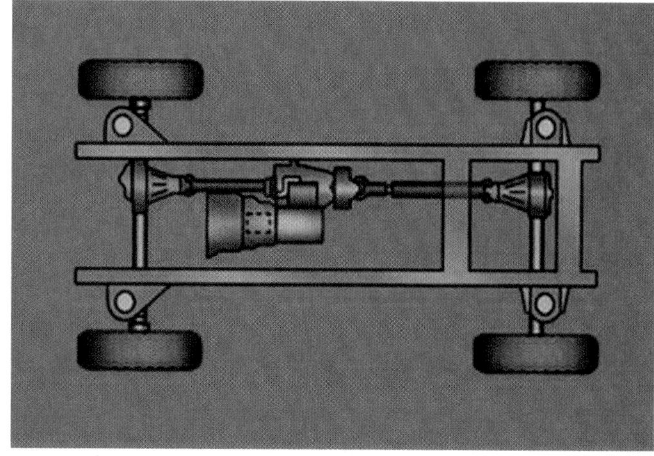

FIGURE 35-1 The transfer case is a device designed to split the power from the transmission to both the front and the rear drive axles when 4WD is selected.

2. The next selection is typically neutral, which allows the vehicle to be towed if it is stuck in mud or snow, or towed down the highway with the engine off, such as behind a motor home; no power is being transmitted to either axle in this selection.
3. Another selection is 4WD *high range*, which transmits power to the front and rear drive axles at normal drive speed.
4. The last selection is 4WD *low range*. This position provides a gear reduction inside the transfer case and results in higher torque to the front and rear drive axles. This position causes the vehicle to move slowly while the engine turns at a higher rpm. Thus, to avoid over-revving the engine, the low range should be selected only when driving at slow vehicle speeds.

▶ TECHNICIAN TIP

A neutral position on the transfer case allows an accessory, such as a cable winch or a hydraulically driven dump bed, to be driven from a **power take-off (PTO)** gear mounted to the side of the transfer case.

You Are the Automotive Technician

Mrs. Smith was out four-wheeling pretty hard in her restored 1987 Bronco, when it made a loud bang and the front wheels lost their drive. She was able to drive it to your shop and asked you to diagnose it. You turn the freewheeling hubs to the lock position, and they engage normally when doing that. You test-drive the vehicle on the dirt road behind the shop and verify that the front wheels are not being driven. You visually inspect the drivelines and axles, and there is some additional play in one of the outer CV joints, but otherwise, normal play from the wheels to the transfer case. You drain the transfer case, and several chunks of metal that look like they are from the drive chain come out with the fluid.

1. What are some precautions when removing the transfer case?
2. What is the purpose and function of the transfer case drive chain?
3. Why are CV joints used instead of U-Joints on the outer ends of front axles?
4. How do 4WD locking hubs connect and disconnect the axle and the wheel?

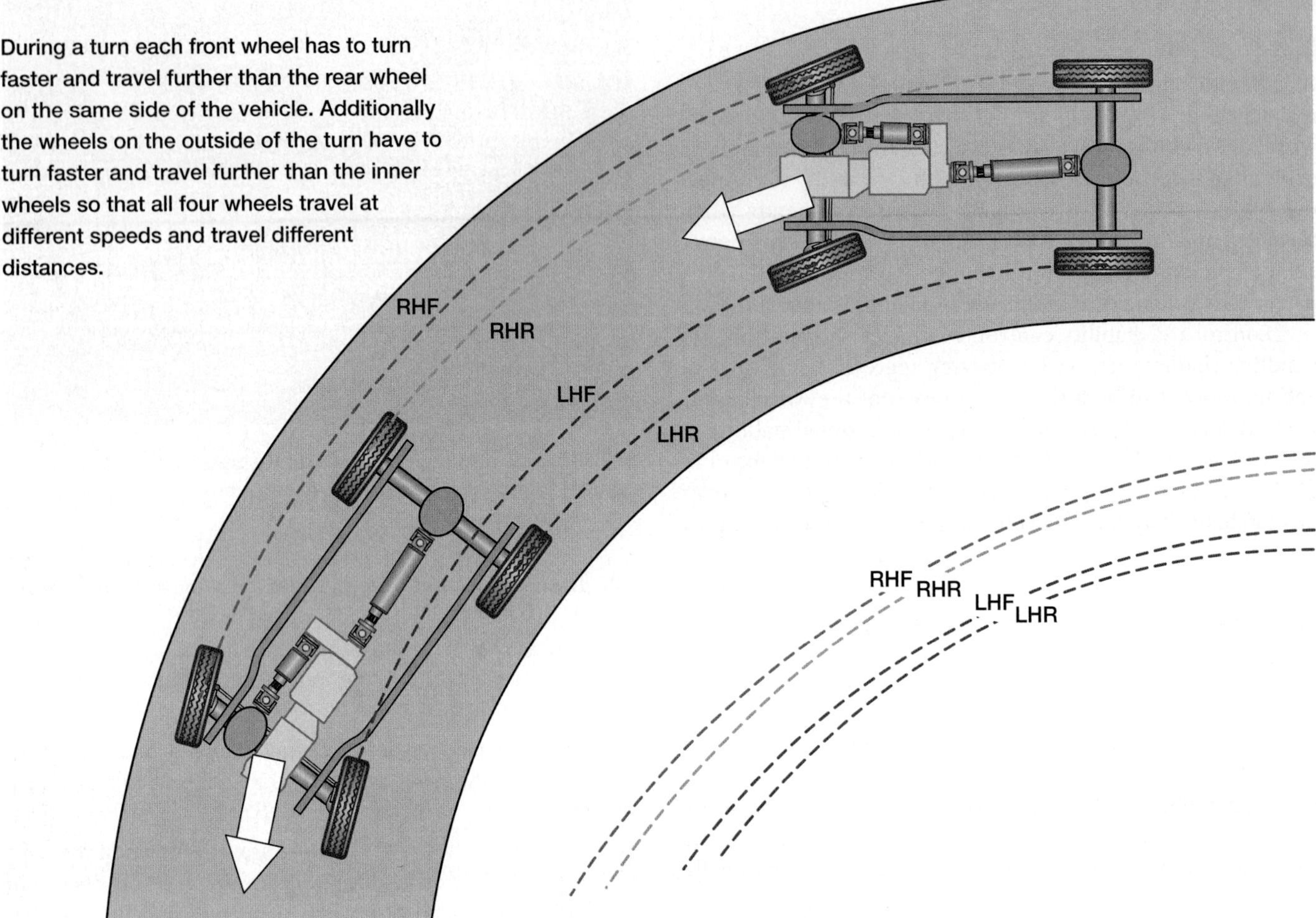

During a turn each front wheel has to turn faster and travel further than the rear wheel on the same side of the vehicle. Additionally the wheels on the outside of the turn have to turn faster and travel further than the inner wheels so that all four wheels travel at different speeds and travel different distances.
RHF
RHR
LHF
LHR
RHF
RHR
LHF
LHR

FIGURE 35-2 The front wheels travel on a larger arc than the rear wheels, making the front wheels travel faster.

The beam-type or solid axle housings enclose the ring gear and pinion and differential gears, and driveshafts supply power from the transfer case. The front drive axle must allow for steering, and to do this, front wheel hubs are mounted to upper and lower ball joints, or pins and bearings. To transfer power to the movable hubs, U-joints must be used in the axle shafts that are splined to the hubs. Newer designs in pickup trucks and SUVs use a constant-velocity (CV) joint—on one or both ends of each axle. CV joints allow for smoother transfer of power and also allow the vehicle to turn more tightly without the joint binding. CV joints also allow for differences in working angles, a condition that would have caused vibrations from standard U-joint setups.

Four-Wheel Drive Differentials

In part-time 4WD vehicles, differentials are fitted to both front and rear axle assemblies. When a 2WD range is selected, the drive is transferred through the rear final drive and the differential gears to the rear axle shafts and road wheels, but not to the front wheels. The differential gears allow the rear wheels to rotate at different speeds when the vehicle is turning, while continuing to transmit an equal turning effort to each wheel. When 4WD is engaged, the drive is transmitted through both front and rear final drive assemblies, and differential action occurs in both.

However, when making a turn, the front wheels travel a greater distance than the rear wheels. This causes a difference in the rotational speeds of the front and rear wheels due to the differing arcs that each wheel is tracing (**FIGURE 35-2**). Because front and rear driveshafts are locked together at the transfer case, the difference in speed cannot be absorbed in the transmission, and the drivelines can be subjected to torsional binding because they are forced to rotate at the same rate of speed. The customer will notice binding of the drivetrain. If the binding is severe, the tires may chirp, or a popping noise from the drivetrain may occur.

In off-road conditions, differences in speed between the front and rear driveshafts can be absorbed through slippage of the tires on the ground—be it mud, snow, or ice conditions. On firm road surfaces, only 2WD should be engaged. Using 2WD prevents excessive tire wear and possible damage to the drivelines or other drivetrain components. It also increases performance and fuel economy.

TECHNICIAN TIP

Differences in wheel speed can also arise from any condition that causes uneven tire circumferences, such as unmatched tires, unevenly inflated tires, or uneven wear. Whenever excessive binding is experienced in a 4WD vehicle, it is a good idea to check for these issues. Once you have determined that all the tires match and are inflated properly, use a tape measure to measure the circumference of each tire. Manufacturers' recommendations for the maximum difference in tire circumferences typically vary from 0.125" to 0.5" (3.2 to 12.7 mm). This difference can easily occur when one tire is damaged enough that it has to be replaced, but the other tires are partially worn. Replacing just one tire may not be a viable option unless the new tire is shaved to match the wear of the other tires.

▶ All-Wheel Drive Layout

K35003

AWD vehicles are widely used today because they offer both the convenience of not having to select 4WD when needed and the necessary traction when on-road (or occasionally off-road) conditions warrant it. Such vehicles are intended for lighter duty and are sometimes referred to as "boulevard SUVs."

In some AWD vehicles, only two of the wheels are normally powered, typically the front wheels. In this case, the RWD is applied automatically when needed to maintain traction. The driver is not normally aware of when this takes place, but has the peace of mind that it happens when needed. In other AWD vehicles, all four wheels are powered all of the time, but the torque may be split unevenly when driving under normal conditions. We cover that in more depth later in the chapter.

▶ Transfer Cases

K35004

Transfer cases are used in 4WD and AWD vehicles to split power to the front and rear drive axles (**FIGURE 35-3**). The transfer case, or in some cases a power take-off (PTO), performs this task. A transfer case is used on vehicles that are primarily RWD; a PTO (also called a power transfer unit [PTU]) is used on vehicles that are primarily FWD. The transfer case is typically bolted to the rear of the transmission, and the PTO is bolted to the side of the transaxle. In either situation, transfer cases and PTOs perform the same task, splitting power between the front and rear axles.

FIGURE 35-3 A part-time 4WD transfer case.

Four-Wheel Drive Transfer Case

K35005

In a typical 4WD vehicle, the transfer case is bolted to the rear of the transmission (**FIGURE 35-4**). The output shaft of the transmission is splined to the input shaft of the transfer case. Power comes into the transfer case through the input shaft and moves to the rear output shaft and, if 4WD is selected, also to the front output shaft. A variety of styles of transfer cases are used in 4WD vehicles, depending on the manufacturer. Some transfer cases use a large chain to allow power to move to the front output shaft. The chain rides on bearings on the input shaft and will typically freewheel unless 4WD is selected. When 4WD is selected, the mode fork,

FIGURE 35-4 A typical transfer case bolted to the rear of the transmission.

which is similar to a shift fork found in a manual transmission, moves a synchronizer sleeve to the input shaft chain sprocket. This locks the chain sprocket to the input shaft and transfers power to the front output shaft. Power then moves out of the transfer case front output shaft to the front driveshaft and to the front axle. In this style of transfer case, the function of 4WD low and neutral is possible through the use of a range fork and a planetary gear set. When the planetary gear carrier is held stationary, a lower gear ratio results than when the sun gear is held.

Another style of transfer case uses a gear drive. The transfer case described here was used by Toyota in their Land Cruiser model. Gear drive transfer cases tend to be noisier than chain drives and are not used as much as they once were. Drive is transmitted from the transmission to the transfer case through the input shaft. The input shaft has a gear that mates to an idler gear that mates to a drive gear on the output shaft. In 2WD, power comes into the transfer case by the input shaft through the idler gear to the output shaft and to the rear driveshaft. The idler shaft contains two gears: a high-speed gear and a low-speed gear. The low-speed gear is free to rotate on bearings and is only locked to the shaft when the shift fork moves to engage the low-speed gear. When in four-wheel high range, power is transferred from the input gear to the idler to the output shafts, which are locked together by the high and low shift fork. This transfer case uses a differential assembly on the output shaft to allow for slowing of the front axle when turning in tight corners. Some vehicles use an electronic center lock to lock this differential in the event that the driver is stuck and needs full lock of front and rear axles.

All-Wheel Drive Transfer Case

K35006

Many vehicles today are using transverse front-mounted engines with AWD. In an AWD vehicle, both the front drive axle and the rear drive axle propel the vehicle. Depending on the system, the majority of engine drive torque is normally delivered to just one set of wheels, but upon demand, all four wheels receive drive torque and share the load. The transfer case powers both the front and the rear driveshafts, which power both final drive assemblies. Axles may be either solid or half-shafts, depending on whether the vehicle uses an independent suspension, or not.

The AWD transfer case is bolted to the output of the transaxle (**FIGURE 35-5**). Drive is transferred from the transmission output shaft to an internally splined helical gear on the transfer case input shaft. The internally splined helical gear is constantly meshed with the larger of two idler gear pinions. They are constantly meshed with the high- and low-output pinions on the transfer case output shaft. The low-range output pinion is bolted to a differential case that contains the third set of differential gears. The high-range pinion can rotate on needle roller bearings on one side of the differential case.

The idler gear pinions are also separated, with the smaller low-speed gear free to rotate on needle rollers on the idler shaft formed with the high-speed gear. A common selector fork links a splined engagement sleeve on the idler, and a similar sleeve on the output moves the fork to engage the high range and locks the high-output pinion to the differential case.

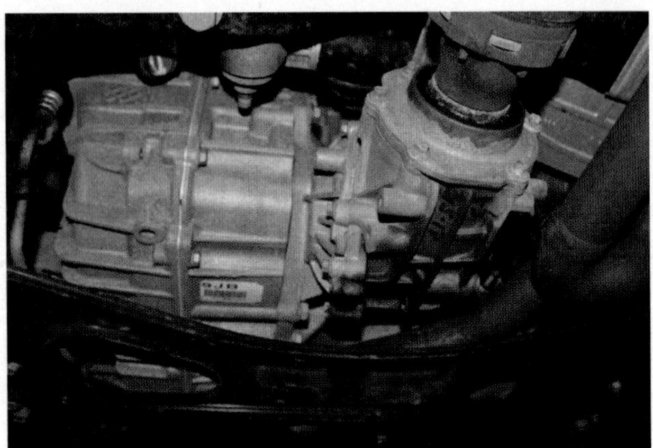

FIGURE 35-5 The AWD transfer case is bolted to the output of the transaxle.

Drive is transmitted at a one-to-one ratio, through the larger pinion of the idler gear and the locked output pinion, to the differential case. It is then transmitted through the differential gears to the front and rear output shafts, which drive the front and rear driveshafts. Engaging low range moves the selector fork in the opposite direction. This locks the low idler gear pinion to the idler shaft.

From the transmission output, drive is transmitted to the large idler pinion, and through the smaller locked pinion to the low-range output pinion, bolted to the differential case. This double reduction provides a low-output speed ratio of about 2.48 to 1 to both front and rear shafts, which can be used to gain traction in snow and ice or even mud. In addition, in some vehicles, when low range is selected, a geared electric motor automatically operates an engagement sleeve splined to the front output shaft. It engages mating splines on the differential case. This locks out all differential action between the front and rear output shafts; thus, it should only be used in off-road situations so that both the front and rear shafts rotate at the same rate of speed.

In high range, the differential lock is disengaged. Differential action is provided for by the differential gears located in the output case and splined to the output shafts. This design allows for a difference in speed between front and rear wheels when the vehicle is on firm road surfaces.

Transfer Case Differential Action

On AWD vehicles, when high range is engaged for driving on firm road surfaces, differences in speed between the front and rear wheels while turning and driving into curves must be provided. This can be accomplished by use of a third set of differential gears in the transfer case output assembly. However, the third differential operation means that if a front wheel spins on a slippery surface, virtually no drive is transmitted to the rear wheels.

There are several ways manufacturers have addressed this slipping condition. One of the easiest is to use the traction control system to apply the brake unit at the wheel that is slipping, which increases the amount of torque to the other wheels that have more traction. Other manufacturers use a **viscous coupling** unit between the front and rear driveshafts to act as a limited slip differential and drive both axles if one is slipping

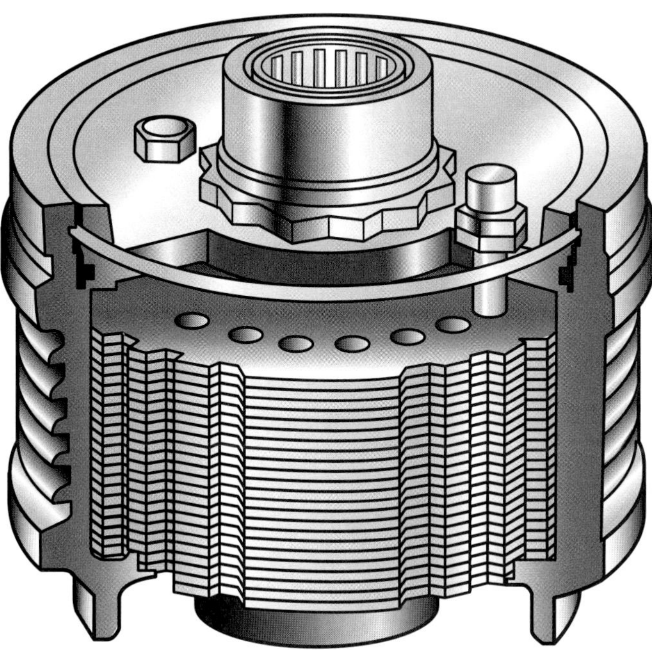

FIGURE 35-6 This viscous coupling allows slippage to keep the rotational speeds of the front and rear axles separate.

(**FIGURE 35-6**). A viscous coupler uses two sets of clutch plates that are alternated front to back. One set of plates is splined to the outer housing (one driveshaft), and the other set of plates is splined to the inner housing (other driveshaft).

In normal driving conditions, the third differential operates to provide the small degree of differential action necessary. In this condition, there is only a minor speed difference between the viscous coupling plates, so the viscous coupling allows them to move without much resistance. However, when a wheel spins, there is a higher level of third differential action and a greater speed difference between the front and rear driveshafts. This speed difference also applies to the viscous coupling plates. Fluid in the unit is sheared by the relative motion between them. This shearing action causes the silicone to heat up and become much thicker (higher viscosity), transferring torque to the wheels that are not slipping.

The degree of locking achieved depends on the severity of the wheel slippage occurring. Only under severe conditions is 100% locking achieved. When the plates are fully locked, the fluid starts to cool, and the cycle of shearing and locking starts all over again.

Some manufacturers control the differential action electronically through the use of an electromagnetic clutch to lock the driveshafts together (**FIGURE 35-7**). This device squeezes the clutches together by applying a magnetic force on the clutch discs. In some models, the clutch can be pulse-width modulated, which controls the amount of torque that is being transferred. The use of a control module is necessary to operate this clutch.

Another type of differential used in transfer cases is the Torsen style, which uses gears similar to the version used in final drives. This style is good for splitting torque to each axle as long as the tires maintain traction, but it is not as effective in low-traction conditions.

The last type of coupling device is the mechanical differential lock. This type locks the driveshafts together using a mechanical sleeve that slides over the splines on both shafts. The differential lock is generally activated by a switch on the dash that operates either an electric or a vacuum motor, which engages or disengages the locking sleeve. This type of locking mechanism does not allow for slippage in the transfer case, so it should be used only on slippery surfaces, and disengaged before driving on high-traction surfaces.

Electrical Components—Four-Wheel Drive/All-Wheel Drive

Today's vehicles almost exclusively use computer-controlled transmissions with solenoids and actuators built in. The transmission control module (TCM) manages line pressures, shift functions, etc., and even incorporates a "self-healing" adjustment capability along with the ability to memorize driver preferences/behaviors. Pulse-width modulated shift patterns are used to save transmission component wear. The TCM knows

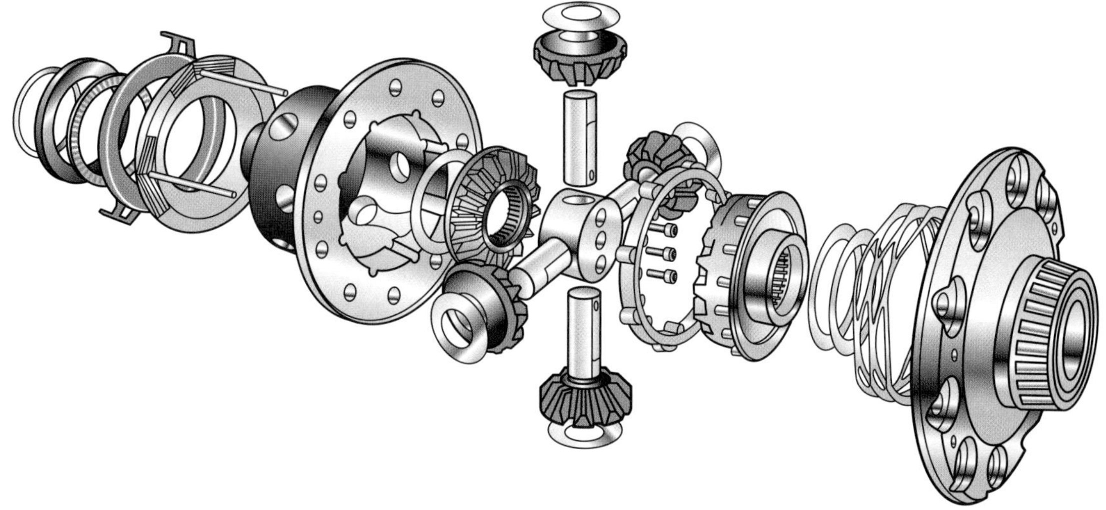

FIGURE 35-7 Some manufacturers control the differential action electronically through the use of an electromagnetic clutch to lock the driveshafts together.

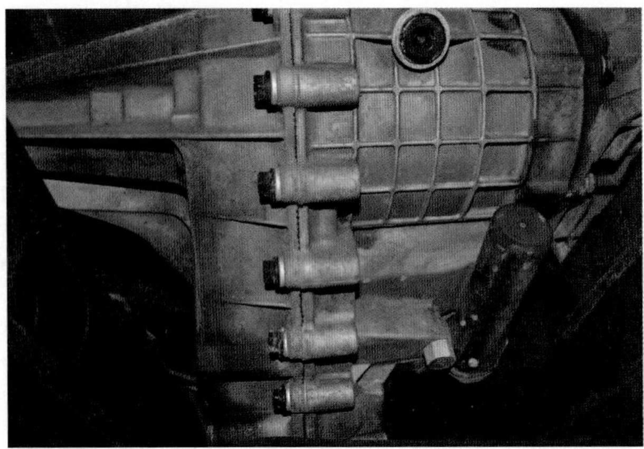

FIGURE 35-8 The electronically controlled transfer case shown here uses a control module to lock the front and rear axles automatically as needed, and monitors wheel speed to determine the torque needed.

FIGURE 35-9 Front driveshaft with slip joint.

when to switch from 2WD to 4WD, along with many other functions; all of these are managed by the TCM (**FIGURE 35-8**).

▶ Four-Wheel Drive Driveshafts and Axles

K35007

Driveshafts—Four-Wheel Drive

K35008

In 4WD applications, driveshafts transfer the drive from the transfer case (torque splitter) to the final drive units at the front and rear axle assemblies. Each driveshaft has a U-joint at each end to transmit the drive through varying angles, and a sliding joint (slip joint) incorporated into the shaft to accommodate changes in driveshaft length by allowing the driveshaft to slide on the splines of its counterpart (**FIGURE 35-9**).

Drive Axles

K35009

Whether a vehicle is 2WD, 4WD, or AWD, axles deliver torque from the differential(s) to the drive wheels. Rear axles in 4WD vehicles are just like solid rear axles in 2WD vehicles. In some 4WD vehicles, the front axle driveshafts are enclosed within an axle housing. They are splined at their inner end to the differential side gears, and to the front wheel hubs at their outer end. A CV joint in each driveshaft allows for greater steering angle movement on turns while continuing to transmit drive to the front wheels (**FIGURE 35-10**). Other front axle shafts are exposed in similar manner of a FWD vehicle (**FIGURE 35-11**). They have CV joints on each end.

FIGURE 35-10 Enclosed front axle from a 4WD vehicle. This type is enclosed in an axle housing.

Four-Wheel Drive Locking Hubs

K35010

On part-time 4WDs, the front wheels and axle driveshafts are splined together through the differential side gears. When the vehicle is operating in 2WD range, the front axle driveshafts, final drive assembly, and front driveshaft all turn with the road wheels but do not transmit any drive. The drag from turning all of these components with the wheels while driving down the highway causes reduced fuel economy and vehicle performance. Freewheeling hubs or manually **locking hubs** may be installed at the factory or as an aftermarket add-on to the front hubs. Freewheeling hubs disconnect the hubs from the axle shafts and prevent rotation of the FWD components (**FIGURE 35-12**). This reduces wear and noise and improves performance and fuel economy when driving in a 2WD range.

When "lock" is selected on a manually locking hub, it acts on a sleeve. The sleeve has external splines that engage matching splines on the hub body and internal splines to engage splines on the axle shaft (**FIGURE 35-13**). The sleeve slides in and out to lock the two splined components together. In a 4WD range, the drive is transmitted from the axle shaft, through the sleeve, to the hub and road wheel. Unfortunately, in inclement weather conditions, the driver must get out of the vehicle and turn the locking hub knobs located on the front wheels to lock in the front drive axles.

When "free" mode is selected, the compressed spring pushes the sleeve out of engagement, leaving the wheel free to rotate

FIGURE 35-11 Exposed front axle on a 4WD vehicle.

FIGURE 35-12 Freewheeling hubs disconnect the hubs from the axle shafts and prevent rotation of the FWD components.

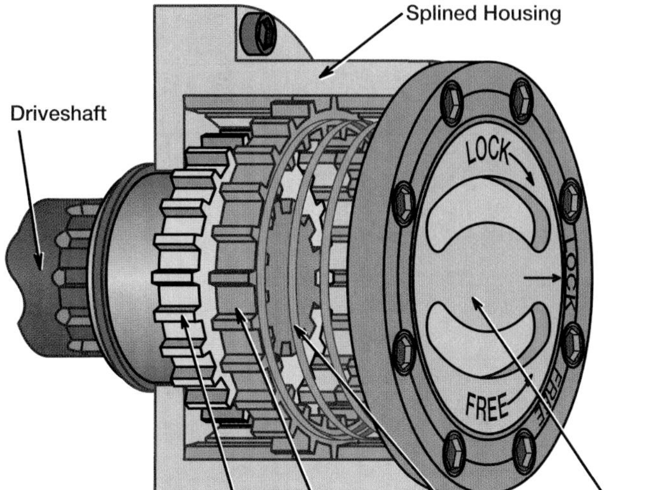

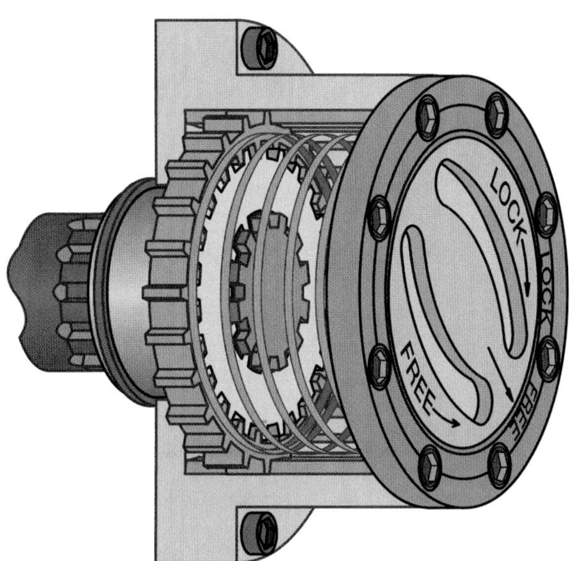

FIGURE 35-13 Freewheeling hub assembly.

without turning the axle shaft. Most vehicles require slightly reversing the vehicle in order to disengage the sleeve, as binding can happen. Reversing the motion of the axle shafts will unbind the sleeve.

Another type of hub is the automatic hub. The automatic hub is factory installed on many newer vehicles. The automatic hub has a spring-loaded sleeve that moves out to engage the hub when power is applied to the front axle. So if 4WD is selected and the vehicle is driven forward, the sleeve will lock the hub to the axle. When 4WD is deactivated, the front axle is not transmitting power, so the spring automatically pushes the sleeve out of engagement and the hub freewheels.

Diagnosing Noise and Unusual Steering Concerns

S35001, N35001

In 4WD vehicles, steering concerns can be present because of the arrangement of the axles and steering linkage. Most pickups and SUVs use a parallelogram-type steering, and any stresses caused by the 4WD system will be applied to the linkage. This can cause vibrations and shimmying problems. Stress may come in the form of driving in 4WD mode with the hubs in the locked position on a dry road. Front axles and U-joints experience stress from the same practice.

Wheel bearings and hubs are also subject to wheel rim offsets that throw off the wheel's scrub radius and cause premature failure of the bearings and hubs. Tire types and profiles as well as tire pressure can also be a cause of unwanted noise or vibrations. Periodic inspection of the front steering/driving axle is always a good practice.

Identifying Concerns Related to Tire Circumference and/or Final Drive Ratios

N35002

When it comes to vehicle computer systems, everything hinges on the computer monitoring the vehicle for proper operation. Sensors such as those for the ABS rely on accurate information while the vehicle is traveling down the highway. Any deviation from the programmed routines in the electronic control module (ECM) may cause a malfunction indicator lamp (MIL) or other system warning light to illuminate, alerting the driver of a system malfunction. Tire sizes and final drive gear ratios must be maintained within the manufacturer's specifications. Tire wear can also have an effect on the ECM routines to report a malfunction. Wheel speed is monitored and, if out of specifications, might cause an MIL to illuminate. Installing a spare tire of a different circumference on the same axle could cause the light to come on. When checking tire circumference, verify that the tire pressure is at the specified pressure; then measure the circumference around the center of the tire. Many manufacturers require the circumference of *all* tires to be within 0.25" (6.4 mm) of each other.

Installing the wrong gear ratios in the final drive will cause the drivetrain to bind when 4WD is called for. Final drive gear ratios in a 4WD unit are almost always the same in the front as in the rear. The only exception is in vehicles that use a slightly different ratio front to rear so as to apply more torque to one set of wheels than the other. In this instance, a viscous coupler is used to provide the needed slippage and prevent binding. To check the final drive for correct ratios, look for a tag under one of the mounting bolts. It may list the ratio or provide you with a code that will allow you to look up the ratio. If there is no tag, you may have to remove the inspection cover. The number of teeth of the pinion and ring gear is often written on the ring gear's outer perimeter so the ratio can be calculated.

To identify concerns related to variations in tire circumference and/or final drive ratios, follow the steps in **SKILL DRILL 35-1**.

Diagnosing, Testing, Adjusting, and Replacing Electrical Components of 4WD Systems

N35003

In older, mechanically controlled 4WD vehicles, visual inspection of all mechanical components is the best way to diagnose customer concerns. This includes the transfer case linkage and the locking hubs. If the fault cannot be identified by an external inspection, you may have to isolate the fault to a particular area or unit and then disassemble it to inspect the internal components. Today, diagnosis of the electrical aspects of 4WD and AWD vehicles is done with a digital volt-ohmmeter (DVOM) and a scan tool. With the inception of the second generation of on-board diagnostic standards (OBDII), many of the older mechanical components have been replaced by the electronic and electrical components common on today's vehicles. Typical electrical components are ECMs, sensors, and solenoids (also known as electronic actuators).

Sensors provide input signals to control modules or computers. Modules are preprogrammed with instructions on what to do under certain conditions, and if possible, they compensate for any unwanted conditions and/or store a code and alert the driver by illuminating an MIL of a fault. Computers can send output signals to actuators. Electric actuators are electric motors or solenoids that cause a desired action to be performed, such as engaging 4WD, or locking the axles together.

If the 4WD system is electronically controlled, typical diagnosis might start with connecting a scan tool and reading any current codes or real-time data to make an accurate assessment of the customer's concern. A quick visual inspection of each of

SKILL DRILL 35-1 Identifying Concerns Related to Variations in Tire Circumference and/or Final Drive Ratios

1. Check all tires for mismatches or unusual wear, and set the tire pressures to specifications.
2. If the vehicle is equipped with ABS and you can access the data stream, connect the appropriate scan tool and test-drive the vehicle. Compare each wheel speed sensor reading to the others. They should be within 0.2 mph (0.3 kph) of one another.
3. If necessary, measure the circumference of the tires and compare to each other. They should be within 0.25" (6.4 mm) of each other.
4. Check the differentials for correct ratios, which may require removal of the inspection cover. The number of teeth of the pinion and ring gear is often written on the ring gear's outer perimeter so that the ratio can be calculated.
5. Consult the manufacturer's service information for any necessary procedures to further diagnose the fault.

the exposed connectors on the circuit should be performed, as a corroded, dirty, or loose connection can cause a component not to operate correctly. From there, a DVOM is used to measure the voltage at the component. If the voltage is not correct, then the voltage drop on each side of the affected circuit must be measured to determine which side of the circuit has the voltage drop. Generally, each side of a control circuit should have less than a 0.5-volt drop. Signal circuits should have less than a 0.1-volt drop. If all voltage readings are within specifications, you will need to inspect and test the electrical component's resistance to see if it has failed.

Resistance checks look for high resistance within the component that would not allow enough current to flow to allow the component to operate. This type of problem is common in today's electronic age. Resistance testing requires the component to be disconnected from the circuit and tested for its specified resistance. If the resistance is outside of specifications, the part is faulty and will have to be replaced. Make sure you have determined the cause of the fault so that it does not occur again. Upon replacement of any electrical component, there may be some necessary adjustments, such as to servo length or linkage or to the part's orientation. Always follow all manufacturer recommended procedures for correct installation and adjustments.

Inspecting, Adjusting, and Repairing Shifting Controls

N35004

There are various types of transfer case shifting controls in use today, ranging from the older mechanical and vacuum controls to the newer electronically motorized control systems. Periodic inspection of all of these components is always a good practice. Mechanical shifting controls can wear out over time, and electrical problems may develop due to corrosion or looseness. Furthermore, vacuum controls have rubber diaphragms in them, and heat and moisture can cause them to deteriorate over time.

Mounting bushings and grommets wear out as they are used over and over again. Inspection of these parts is critical to prevent problems along the way. Mounting brackets are used to hold components in place. They are subject to weather and will rust, wear, or bend under constant usage. These problems usually show up over time and should be inspected under normal preventive maintenance practices.

To inspect, adjust, and repair shifting controls, follow the steps in **SKILL DRILL 35-2**.

Removing and Reinstalling Transfer Case

S35002

In some cases, it may be necessary to remove the transfer case, such as when there has been a failure of the transfer case or of the transmission in a 4WD vehicle. Four-wheel drive vehicles can be used for various purposes and must endure a variety of stress situations depending on the type of weather, climate, or terrain they are subjected to. For example, snowplowing can put stress on the transfer case, as can off-roading. This can lead to wear or damage that requires removal of the transfer case.

One thing to remember when removing a transfer case is that it is both heavy and awkward. In many cases, its shape is such that it wants to tip over if lifted by a jack placed under the transfer case. So make sure that you use a proper jack that will secure it properly. Failure to do so can result in the transfer case falling off the jack, leading to severe injury.

To remove and reinstall the transfer case, follow the steps in **SKILL DRILL 35-3**.

SKILL DRILL 35-2 Inspecting, Adjusting, and Repairing Shifting Controls

1. Raise the vehicle with an approved vehicle hoist, and make sure it is secure. Check the manufacturer's publications for proper procedures for inspection of any special parts.
2. Check for any codes that may be present by using a scan tool on the vehicle.
3. Perform a visual inspection of all mechanical linkages or mounting brackets for any damage or excessive wear of bushings or grommets.
4. Make any necessary adjustments to linkages or cables if they are loose or have been replaced.
5. Inspect any hoses that are dry rotted or cracked on any vacuum-operated components.
6. Refer to the service information for proper diagnosis of any electrical components that are in question.
7. After all repairs have been made, road test the vehicle to ensure that it was repaired properly.

SKILL DRILL 35-3 Removing and Reinstalling the Transfer Case

1. Remove the driveshafts and attaching bolts.
2. Remove all electronic wiring and shift linkage. *Note:* It may be necessary to remove any skid plates or protective plates from the frame and transfer case.
3. Remove the attaching bolts for the transmission, and store them in a safe place.
4. Using a transmission jack, slowly back the transfer case away from the transmission.
5. Perform all necessary operations or repairs according to the manufacturer's specifications and operations.
6. Refill the oil to its proper level if it was drained.
7. Lubricate the splines of the transfer case input shaft. Align it with the transmission output shaft splines, and slide it into position.
8. Reattach the transfer case bolts, and torque them to specifications.
9. Attach all wiring and connections as required.
10. Reinstall all linkage that was removed, and adjust as necessary.
11. Align and reinstall the front and rear driveshafts.
12. Road test the vehicle to ensure proper operation.

Disassembling, Servicing, and Reassembling the Transfer Case and Components

N35005

Most transfer cases are fairly durable, so they don't usually need more than routine maintenance. If there are any unusual bearing or gear noises coming from inside the transfer case, it will most likely have to be disassembled and inspected to determine the cause. As always, check the manufacturer's service information for all necessary procedures when servicing the transfer case.

Once the transfer case has been removed, inspect the outer casing for any cracks or oil leaks or other signs of damage, as these indicate what types of repairs you will need to perform. This task is based on a typical Ford transfer case, which we use for teardown procedures.

To disassemble a transfer case and related components, follow the steps in **SKILL DRILL 35-4**.

To reassemble the transfer case and related components, follow the steps in **SKILL DRILL 35-5**.

Inspecting Front Wheel Bearings and Locking Hubs

N35006

To connect the front wheels with the front drive axles on a 4WD vehicle, some sort of hub locking mechanism must be used. These vehicles have either automatic locking hubs or manual locking hubs. All of these systems require periodic service and inspection to maintain functionality and prevent early failure of these parts.

To inspect front wheel bearings and locking hubs, follow the steps in **SKILL DRILL 35-6**.

SKILL DRILL 35-4 Disassembling the Transfer Case and Components

1. Remove the drain plug, and drain all oil before the disassembly.

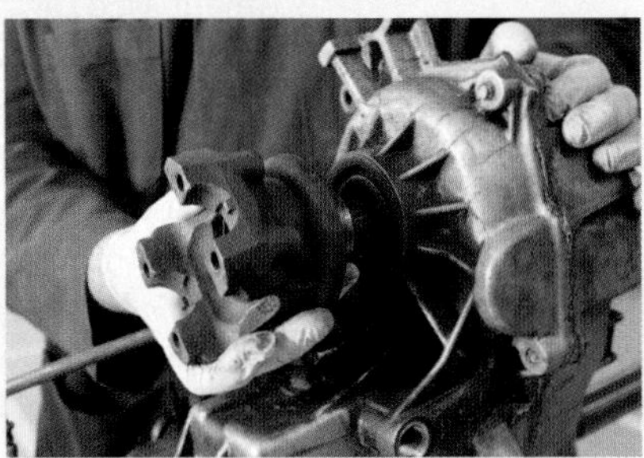

2. Remove the output flange yokes.

3. Remove the 4WD dash light switch. Unplug the shift harness connector. Remove the speed sensor.

4. Remove the front case retaining bolts. Separate the halves of the transfer case.

5. Remove snap rings. Remove the reluctor wheel.

6. Remove the oil pump assembly.

7. Remove the driven and drive sprockets with the chain.

8. Remove the output shaft assembly.

9. Remove the 4WD shift fork and collar.

SKILL DRILL 35-4 Disassembling the Transfer Case and Components (Continued)

10. Remove the high/low shift fork and collar.

11. Remove the planetary gear set.

12. Clean and inspect all components for wear or damage. Replace any defective parts.

SKILL DRILL 35-5 Reassembling the Transfer Case and Components

1. Reinstall the planetary gear set into the case. Reinstall any snap rings holding the planetary gear set.

2. Reinstall the high/low shift fork and collar into the case.

3. Reinstall the 4WD shift fork and collar into the case. Make sure the shift rod is installed in the correct position in the case.

4. Reinstall the output shaft into the case. Make sure the shift collars line up properly on the shaft.

5. Reinstall the chain and driven sprocket.

6. Reassemble the oil pump assembly and pickup tube.

7. Apply a thin bead of RTV to one case half. Reassemble the two halves of the case, and torque.

8. Reinstall all electrical sensors and switches.

9. Reinstall output flange yokes and torque to specification.

SKILL DRILL 35-6 Inspecting Front Wheel Bearings and Locking Hubs

1. Raise the front axle off the ground, and secure it with jack stands. Loosen the lug nuts, and remove the tire and wheel assembly.
2. Remove the screws from the manual locking hub, and pull the hub from the axle.
3. Inspect the O-ring on the back side of the cover, and if necessary replace it.
4. Remove the wire ring from the hub, and pull the retaining plate out of the hub.
5. Remove the snap ring from the axle.
6. Remove the locking sleeve from the hub.
7. Remove the bearing retaining nut(s) from the hub assembly to expose the bearings, and remove the hub and rotor assembly for inspection.
8. Knock out the seal and rear bearing for inspection. Replace the seal with a new one when reassembling.
9. Clean and inspect the bearings closely for brinelling and bluing due to heat concerns, and replace as necessary.
10. Pack and install new bearings and races using the proper bearing race driver.
11. Install the new seal assembly. *Note:* When replacing the seal, be sure to put some grease on the spring inside the seal to prevent it from coming off when the seal is driven into the hub.
12. Reinstall the hub and rotor on the axle, and place the outer wheel bearing in the hub.
13. Thread the bearing adjusting nut on by hand.
14. Set the wheel bearing preload following the manufacturer's specifications, using the proper tool.
15. Install the locking washer over the adjusting nut.
16. Thread the lock nut up against the locking washer, and tighten to the specified torque. If necessary, bend the lock tang up against the locking nut.
17. Install the locking sleeve.
18. Install the retaining plate and wire clip.
19. Install a new O-ring on the hub cover, and reinstall the hub cover on the axle.
20. Tighten the hub cover screws to the manufacturer's specifications.
21. Road test the vehicle to ensure that it operates properly.

▶ Wrap-Up

Ready for Review

- ▶ The transfer case or power take-off (PTO) allows the transfer of power to the front and rear axles.
- ▶ The transmission control module (TCM) manages line pressures, shift functions, etc., and even incorporates a "self-healing" adjustment capability, along with the ability to memorize driver preferences/behaviors.

Key Terms

locking hub Four-wheel drive front axle hubs that are manually locked or unlocked by turning the knob on the hub.

power take-off (PTO) A device attached to the transmission that is gear driven and can be used to run accessories such as winches and towing equipment. It can also refer to the gears that send power to the rear axle in a predominantly FWD vehicle.

viscous coupling An silicone clutch assembly used in AWD differentials to provide a slight amount of differential action for control of axle rotational speeds.

Review Questions

1. Choose the correct statement.
   - a. A transfer case is mounted to the front of the transmission.
   - b. Drive shafts are used to move power from the transfer case to the front and rear drive axles.
   - c. When making a turn, the rear wheels travel a greater distance than the front wheels.
   - d. The ring and pinion gears allow the front wheels to rotate at different speeds.
2. Which of the following is a device designed to split the power from the transmission to both the front and the rear drive axles when four-wheel drive is selected?
   - a. Clutch
   - b. Transfer case
   - c. Driveshaft
   - d. CV joint
3. Which of these perform the same task of splitting power between the front and rear axles?
   - a. Transfer cases and PTOs
   - b. Output shaft and PTOs
   - c. CV joints and PTOs
   - d. Output shaft and CV joints
4. Choose the correct statement.
   - a. Chain drive transfer cases tend to be noisier than gear drives.
   - b. When the planetary gear carrier is in motion, a lower gear ratio results.
   - c. Power comes into the transfer case directly from the engine flywheel and moves to the rear input shaft.
   - d. The idler shaft contains two gears: a high-speed gear and a low-speed gear.
5. Which of the following devices squeezes the clutches together by applying a magnetic force on the clutch discs?
   - a. Torsen-style clutch
   - b. Mechanical differential lock
   - c. Electromagnetic clutch
   - d. Viscous coupling
6. All of the following statements are true *except*:
   - a. axles may be either solid or half-shafts, depending on whether the vehicle uses an independent suspension or not.
   - b. a viscous coupler uses two sets of clutch plates that are alternated front to back.

c. the PTO determines when to switch from two-wheel drive to four-wheel drive.
d. the degree of locking achieved depends on the severity of the wheel slippage occurring.

7. In four-wheel drive applications, which of the following components transfers the drive from the transfer case (torque splitter) to the final drive units at the front and rear axle assemblies?
   a. Drive shafts
   b. Drive axle
   c. Electromagnetic clutch
   d. Vacuum motor
8. All of the following statements are true *except*:
   a. when lock is selected on a manually locking hub, it acts on a sleeve.
   b. the sleeve slides in and out to lock the two splined components together.
   c. freewheeling hubs disconnect the hubs from the axle shafts and prevent rotation of the front-wheel drive components.
   d. in a four-wheel drive range, the drive is transmitted from the transfer case to the transmission.
9. Under which of these conditions must the driver get out of the vehicle and turn the locking hub knobs located on the front wheels to lock in the front drive axles?
   a. Hot weather conditions
   b. Dry weather conditions
   c. Inclement weather conditions
   d. Heavy traffic
10. All of the following statements are true *except*:
   a. the automatic hub is factory installed on many newer vehicles.
   b. when four-wheel drive is activated, the front axle is not transmitting power.
   c. when "free" mode is selected, the compressed spring pushes the sleeve out of engagement.
   d. reversing the motion of the axle shafts will unbind the sleeve.

ASE Technician A/Technician B Style Questions

1. Tech A says that when making a turn, the front wheels travel a greater distance than the rear wheels. Tech B says that in off-road conditions, differences in speed between the front and rear driveshafts can be absorbed through slippage of the tires on the ground. Who is correct?
   a. Tech A
   b. Tech B
   c. Both A and B
   d. Neither A nor B
2. Tech A says that AWD vehicles have a differential in the middle of the drivetrain. Tech B says that many AWD vehicles incorporate a viscous clutch in the center differential. Who is correct?
   a. Tech A
   b. Tech B
   c. Both A and B
   d. Neither A nor B
3. Tech A says that all front wheel hubs lock automatically. Tech B says that all front wheel hubs lock manually. Who is correct?
   a. Tech A
   b. Tech B
   c. Both A and B
   d. Neither A nor B
4. Tech A says that some front differentials are operated by a vacuum motor. Tech B says that some front differentials are operated by electric motors. Who is correct?
   a. Tech A
   b. Tech B
   c. Both A and B
   d. Neither A nor B
5. Tech A says that part-time 4WD does not use a transfer case. Tech B says that AWD vehicles can safely drive with all four wheels on pavement. Who is correct?
   a. Tech A
   b. Tech B
   c. Both A and B
   d. Neither A nor B
6. Tech A says that constant-velocity joints are used on some 4WD vehicles. Tech B says that a slip joint is part of a driveshaft. Who is correct?
   a. Tech A
   b. Tech B
   c. Both A and B
   d. Neither A nor B
7. Tech A says that electric actuators are used for things like engaging 4WD or locking the axles together. Tech B says that sensors send signals directly to electric actuators to cause them to operate. Who is correct?
   a. Tech A
   b. Tech B
   c. Both A and B
   d. Neither A nor B
8. Tech A says that some transfer cases have a chain inside them to transfer torque to the wheels. Tech B says that some transfer cases have a belt inside them to transfer torque to the wheels. Who is correct?
   a. Tech A
   b. Tech B
   c. Both A and B
   d. Neither A nor B
9. Tech A says that the PCM/TCM compensates for different size tires as long as the tires on the front match each other, and the tires on the rear match each other. Tech B says that many manufacturers require the circumference of ALL tires to be within 0.25" (6.4 mm) of each other. Who is correct?
   a. Tech A
   b. Tech B
   c. Both A and B
   d. Neither A nor B
10. Tech A says that freewheeling hubs disconnect the hubs from the axle shafts. Tech B says that freewheeling hubs disconnect the wheels when steering around corners. Who is correct?
   a. Tech A
   b. Tech B
   c. Both A and B
   d. Neither A nor B

SECTION 5

Steering and Suspension

CHAPTER 36
Wheels and Tires Theory

NATEF Tasks

There are no NATEF tasks in this chapter.

Knowledge Objectives

After reading this chapter, you will be able to:

- **K36001** Explain the principles of tire distortion and center of gravity.
- **K36002** Describe the functions and features of wheels and tires.
- **K36003** Describe purpose, function, and types of a wheel and wheel lug nuts/studs.
- **K36004** Describe the components of a vehicle's tire.
- **K36005** Decipher the tire markings on the sidewall of each tire.
- **K36006** Describe the common safety features found on today's tires.
- **K36007** Describe the purpose, function, and types of tire pressure monitoring systems.
- **K36008** Describe the purpose and function of run-flat tires.
- **K36009** Describe the purpose and function of self-sealing tires.
- **K36010** Describe the purpose and function of space saver spare tires.

Skills Objectives

There are no Skills Objectives in this chapter.

▶ Introduction

The wheel and tire assembly is the only point of contact between the vehicle and the road surface. The entire vehicle literally rests on the wheel and tire assembly. The contact between the tire and the road surface plays a large role in determining how a vehicle handles and rides (**FIGURE 36-1**). Today's newer, lightweight vehicles are more sensitive than ever to minor tire or wheel issues. Poorly maintained wheels and tires decrease effective handling, fuel economy, and ride quality and increase the potential for accidents or breakdowns.

▶ Principles

K36001

Tires and wheels work together to allow the vehicle to roll smoothly down the road. Tires by themselves have very little rigidity; they flex and deform whenever a force is put against them. On the other hand, wheels are very rigid and have no way of absorbing the unevenness of the road surface, so driving would be annoying were they used by themselves. Also, because wheels have a low coefficient of friction, traction would be almost nonexistent. But when you pair a tire with a wheel, you get the rigidity needed to provide directional control (from the wheel), and you get flexibility and traction to grip the road surface while smoothing out the bumps in the surface of the road (from the tire).

FIGURE 36-1 Wheels and tires provide the connection between the vehicle and road surface.

Tires on most passenger vehicles are called pneumatic tires because they are filled with pressurized air, which gives them support. Some vehicles, such as forklifts, may use solid rubber tires. These tires are not susceptible to punctures, so they are used where tire hazards are present such as garbage dumps or construction sites. Early pneumatic tires were called tube-type tires because they used an inner tube inside the tire to seal the air in the tire/wheel assembly. Recent vehicles use tubeless tires; these tires are designed to seal tightly to the wheel in such a way that the pressurized air does not leak out.

The air pressure gives the tire its shape when the weight of the vehicle is sitting on it. Air pushes on the entire inside surface of the tire, pushing it outward. Tires on many vehicles are inflated to about 32 pounds per square inch (psi). This means that the air pressure is pushing against every square inch of the inside of the tire. It would not be unreasonable to say there is approximately 1200 square inches of surface area inside a typical passenger car tire. So 32 psi multiplied by 1200 square inches equals 38,400 lb of force trying to expand the tire. So why doesn't the tire expand like a balloon? This is because it has lots of strong reinforcing strands molded into it, which gives it strength while still allowing it to be flexible. So the air inside the tire pushes outward on the tire, stretching the strands tightly, which prevents the tire from ballooning (**FIGURE 36-2**). Also, the relatively high pressure in the tire stiffens the tire so that it will support the weight of the vehicle. This design also explains the appearance of a tire that is substantially low on air pressure: The bottom will appear flat, and the sidewall relatively soft. The air pressure is not fully stretching the cords in the tire.

Tire Distortion

During cornering, centrifugal force acts on a vehicle to produce a **side force**. The side force is the pressure on the wheel that pushes it toward the outside or inside of the rim as the vehicle makes a turn. In most instances, the contact of the tire with the surface of the road produces enough friction to prevent this force from actually pushing the wheel and tire sideways across the road surface. However, when the roads are slippery, or the vehicle is traveling too fast, the tire is unable to grip the road surface adequately to generate enough friction to overcome the side force, resulting in the wheels and vehicle skidding sideways during a turn. Without the resistance created by friction, side force will cause the vehicle to continue in a straight line.

You Are the Automotive Technician

A customer brings her 2014 Mustang GT into your shop and needs to buy a new set of tires. She races the car regularly at the local road race track, so she wants to buy tires that will help her out there. She has several questions for you as she looks at the various options.

1. Would lower profile tires be helpful on a road race track? Why or why not?
2. Her tire placard on the driver's door shows 32 psi (~221 kPa), but the tire shows a maximum pressure of 40 psi (~226 kPa). Which pressure should she use?
3. Would bias-ply tires give her better traction than radial tires? Why or why not?
4. What types of tires could she use if she wanted to avoid having to change tires out on the track if she got a puncture?

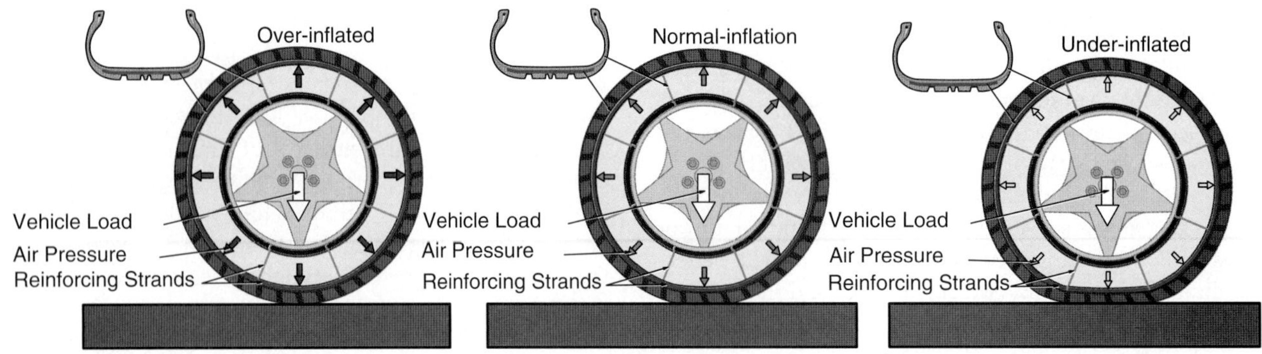

FIGURE 36-2 The strands in the tire are stretched tight by the air pressure inside the tire allowing it to support the weight of the vehicle. At the same time, the strands keep the tire from ballooning.

FIGURE 36-3 Tires distort during cornering, providing an opposing force that allows a vehicle to make a turn.

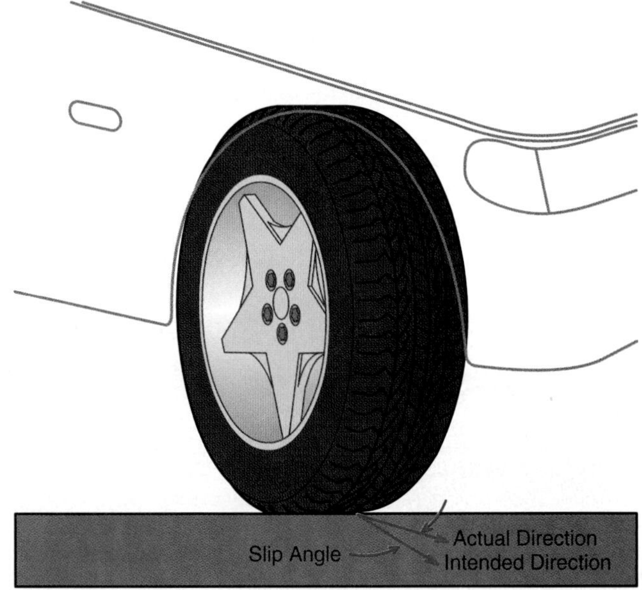

FIGURE 36-4 Slip angle is the difference in angle between the direction the tire is pointed and the direction the tire is going.

The tire provides this opposing force by being able to distort while still gripping the road (**FIGURE 36-3**). Because the tire's construction makes it elastic, it exerts a force, called **cornering force**, that acts between the tread and the road surface. It allows the rubber to distort from its normal position. The tire's sideways distortion makes the vehicle follow a path at an angle to the direction the road wheel is pointing (**FIGURE 36-4**). This is called the **slip angle**. As the cornering force increases, so does slip angle.

When a vehicle is being driven into a turn with a decreasing radius, both slip angle and cornering force increase until a point is reached where the tire slides. At this point, the only resistance comes from sliding kinetic friction across the road surface. The tire grips again only when the vehicle has slowed or the wheels are not turned so sharply—that is, when the side force is reduced to a level the tire can withstand without skidding.

Because both front and rear tires develop a slip angle in a turn, the vehicle's path is determined by the steering of the front tires and the slip angles of both the front and the rear tires. These slip angles depend on the weight distribution within the vehicle, the wheelbase, the tire track, and the overall length of the vehicle. The weight distribution is affected by whether the engine is front, mid, or rear mounted. Another factor is whether the vehicle is front, rear, or all-wheel drive.

During cornering, centrifugal force puts more weight on the outside wheels. Acceleration puts more weight on the rear wheels. Deceleration or braking puts more weight on the front wheels. In a turn, centrifugal force tries to push the vehicle away from the corner. This is resisted by the cornering force of the tires. The tires' slip angles may not be equal, due to the cornering, deceleration, or acceleration forces acting on them. When the front slip angles are larger than the rear slip angles, the vehicle is said to be in an **understeer** condition, which is referred to as the vehicle "pushing" in the corners (**FIGURE 36-5**). Understeer results in the front of the vehicle being pushed toward the outside of the corner. At

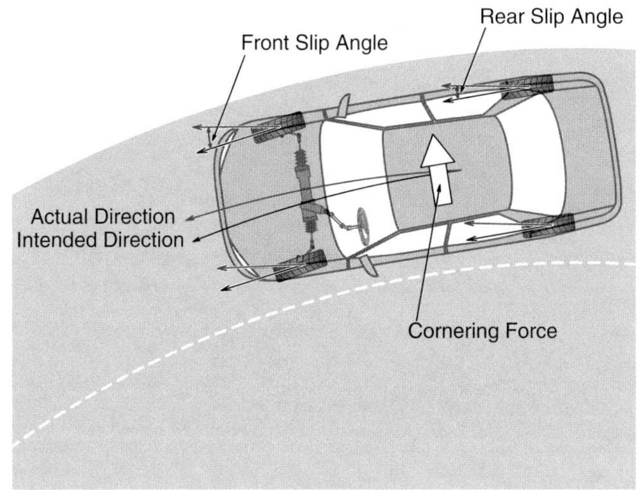

FIGURE 36-5 Understeer occurs when the front wheel slip angle is larger than the rear slip angle.

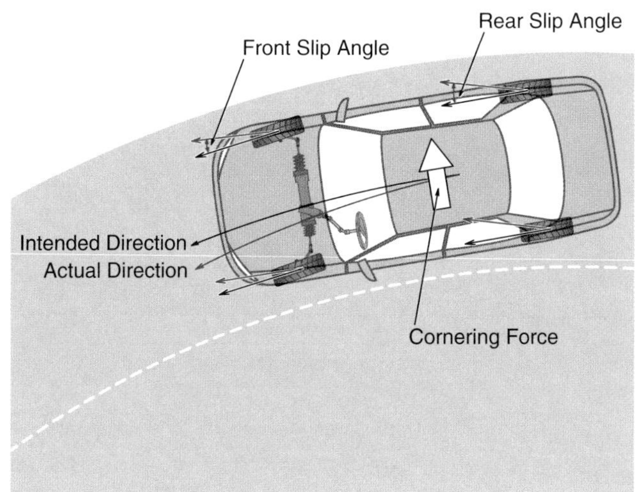

FIGURE 36-6 Oversteer occurs when the rear wheel slip angle is larger than the front slip angle.

its worst, it causes the vehicle's front wheels to slide toward the outside of the corner and into the curb or off the track. **Oversteer** is when the rear slip angle is larger than the front slip angle; it is referred to as the vehicle being "loose" in the corners (**FIGURE 36-6**). Oversteer tends to cause the rear of the vehicle to slide toward the outside of the corner. In this situation, the rear wheels lose traction, typically resulting in the vehicle spinning out. When both the front and the rear tires have equal amounts of slip angle, the vehicle is said to have **neutral steer**, in which case the vehicle tends to go in the direction that the front tires are pointing.

Center of Gravity

The center of gravity is the balance point of the entire vehicle. Its actual position depends on location of the engine and transmission (**FIGURE 36-7**). It is always located above the road surface and between the tires. When a vehicle is cornering, this is the

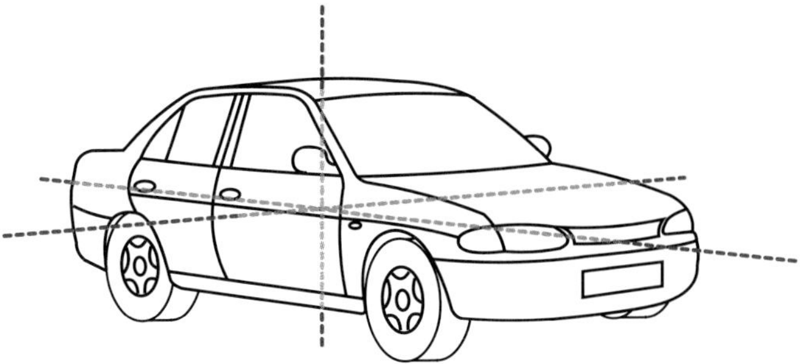

FIGURE 36-7 The vehicle's center of gravity.

point through which all centrifugal force is assumed to act. Its position is determined by the load carried by the front and rear wheels—that is, by how weight is distributed. In a typical rear-wheel drive vehicle, the weight distribution is approximately 60% fore and 40% aft; 60% of the weight is carried on the front wheels, 40% on the rear, making the center of gravity closer to the front than the rear. On a typical front-wheel drive vehicle, the weight distribution is approximately 75% fore and 25% aft. Lateral, or side-to-side, weight distribution can be expressed in the same way and is affected by things like the location of the fuel tank and battery.

Every vehicle has static weight distribution, whether it is at rest or traveling in a straight line at a steady speed. This distribution is changed laterally by centrifugal force when the vehicle is turning and in a fore-and-aft direction during acceleration or braking.

▶ TECHNICIAN TIP

The height of the center of gravity is determined by the height of the center of the mass above the road surface.

▶ Wheels and Tires

K36002

Wheels

K36003

Wheels are usually made from pressed steel or cast aluminum alloy. They are lightweight, yet strong enough to withstand normal operational forces. Alloy wheels are popular because of their appearance and because they are lighter than steel wheels. Aluminum is a better conductor of heat, so alloy wheels can dissipate heat from the brakes and tires more effectively than steel. Alloy wheels are often called mag or magnesium wheels, but wheels made of magnesium are rarely used on vehicles.

The terms "wheel" and "rim" are often used synonymously. Technically, the **rim** of a wheel is the outer circular lip of the metal on which the inside edge of the tire is mounted (**FIGURE 36-8**). The purpose of the rim is to hold and seal the tire to the wheel.

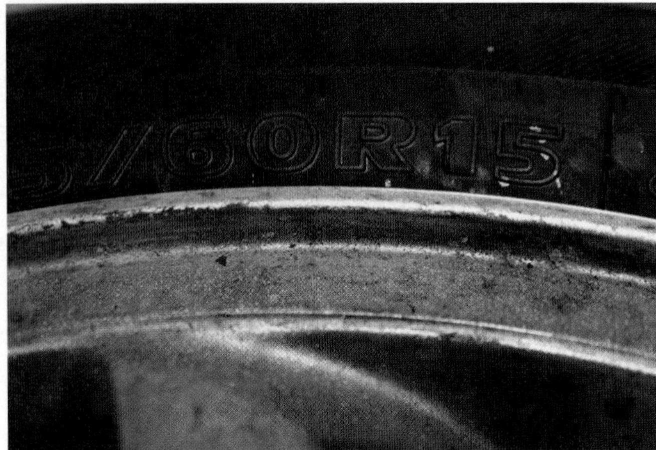

FIGURE 36-8 Rim of the wheel.

FIGURE 36-9 Drop-center rim.

Rim width is the distance across the inside of the **rim flanges** at the **bead seat**. The bead seat is the edge of the rim that creates a seal between the tire bead and the wheel. The rim flange is the exterior lip that holds the tire in place. The rim diameter is the distance across the center of the rim from bead seat to bead seat. The width of the rim and the diameter are traditionally stated in inches, although some are stated in millimeters.

> TECHNICIAN TIP

Rims are stamped with a code. For example, a rim designated "7 JJ by 14" would refer to a rim measuring:

- 7" across the rim flanges;
- 14" in diameter from bead seat to bead seat;
- and with the flange profile conforming to a JJ code.

Most passenger vehicle **wheel rims** are of the drop-center design. In drop-center rims, the inner section sits lower than the sides (**FIGURE 36-9**). This design allows for tire removal and mounting. The drop-center rim is made in one piece and is permanently fastened to the wheel disc. When inflated, the tire is locked to the rim by tapering the bead seat toward the flange, or more commonly by safety ridges or humps close to the flange.

> TECHNICIAN TIP

Most wheels have ventilation holes in the flange so that air can circulate to the brakes.

An important feature in a rim is the location of the drop center in the rim. The drop center can be closer to the front of the wheel or the rear of the wheel, depending on the desired position of the wheel relative to the axle flange. In most stock wheels, the drop center is closer to the front side of the wheel. In **deep dish wheels**, the drop center is closer to the rear of the wheel (**FIGURE 36-10**). Because the drop center is crucial to removing the tire from the wheel, the side of the wheel that is closest to the drop center is the side the tire should be removed from. In fact, it is virtually impossible to remove the tire from the side of the wheel farthest from the drop center.

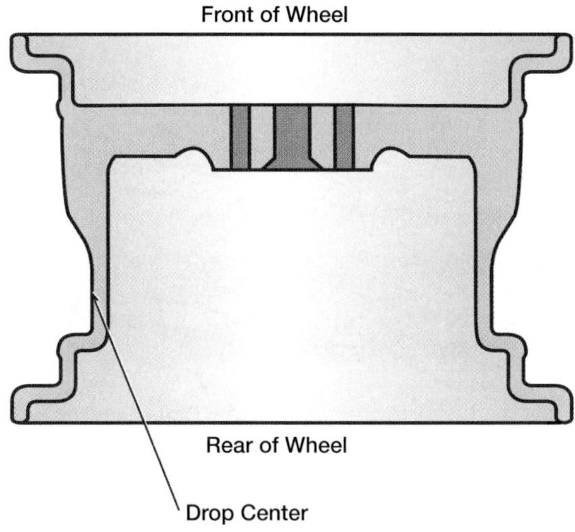

FIGURE 36-10 Deep dish wheel.

With safety in mind, manufacturers and legislators are insisting that passenger cars and light trucks be equipped with safety-type drop-center rims. **Safety-type drop-center rims** have a slight ridge, or hump, at the inside edge of the bead ledges. This ridge helps to hold the tire beads in place when the tire goes flat. In the event of sudden deflation or blowout, safety ridges help prevent the tire from moving down into the drop center. This helps maintain control of the vehicle while the driver is braking.

Rims can be made from two sections of pressed steel—a flange or disc that is drilled for the wheel fasteners and the rim. The rims for passenger vehicles are usually the **steel-disc type**. The disc may be solid, but the major manufacturers place holes in them to reduce the disc's weight and to provide ventilation for cooling the brakes. The disc is either welded or riveted to the rim and bolted to the axle or axle flange. Additional types of rims can be found in **TABLE 36-1**.

TABLE 36-1: Types of Rims

Type	Purpose
Steel rims	Rims that can be painted or chromed. Generally used on bottom-of-the-line vehicles.
One-piece alloy rims	Rims that are constructed in one piece, including the well, to facilitate the mounting and dismounting of the tire.
One-piece alloy billet	Rims that are machined out of a solid piece of alloy.
Two-piece alloy rims	Rims that are constructed as two separate pieces of aluminum, one being the center piece and the other a mounting flange, that are welded together.
Multi-piece alloy rims	Rims that are constructed by welding together several separately molded pieces.
Custom rims	Rims that are built to suit the specific, unique needs and requirements of the customer.
Spinning rims (spinners)	Rims that are constructed with a piece that rotates independent from the wheel. High-speed roller bearings placed within the rim create the piece's movement.
Split rims	Rims that are constructed of a single long piece of metal formed into a circle, but whose ends are not welded together (a "split" ring). The split allows removal of the outside flange so the tire can be demounted. This type of rim is common on tractor trailers and other big-wheeled trucks.
Semi-drop well rims	Rims constructed of a solid rim to allow removal of the outside flange, as in the split rim.
Drop well rims	Rims constructed with a trough in the center of the rim that is smaller in diameter than the edge by the lip where the bead of the tire seals. The trough allows the tire to be mounted on the rim.
Safety rims	Rims designed to hold the tire bead in place on the rim in the event of inadequate tire pressure.

▶ TECHNICIAN TIP

The tire must be an exact fit on the rim to fulfill a number of functions:

- To ensure that the narrow contact area between the beads of the tire and the rim will seal the air in a tubeless tire
- To transfer all the forces between the tire and the wheel, without slipping or chafing
- To ensure that the friction between the tire and the rim prevents the tire from turning on the rim

SAFETY TIP

The rim width and diameter can also be stated in millimeters. Metric rims are not interchangeable with imperial rims; nor can you fit a metric tire to an imperial rim. Conversely, you cannot fit an imperial tire to a metric rim. Doing so could result in personal injury or a vehicle accident.

Applied Math

AM-27: Distance: The technician can measure distance using a variety of devices to determine conformance to the manufacturer's specifications and tolerances.

A technician may be called upon to measure distances with a variety of devices to determine conformance of components to manufacturer's tolerances and specifications. Some examples may include measuring:

- Tire tread with a depth gauge
- Brake pad thickness with a ruler
- Vehicle ride height with a tape measure
- Brake disc thickness with a vernier caliper
- Engine valve clearance with feeler gauges
- Crankshaft journal thickness with a micrometer
- Wheel runout with a dial gauge

Wheel Offset

The offset of a wheel is the distance from its hub mounting surface to the centerline of the wheel (**FIGURE 36-11**). Offset is important because it is typically used to bring the tire centerline into close alignment with the larger inner wheel bearing, and it reduces load on the stub axle. This requires the inside of the wheel assembly to be shaped so that there is space for the brake assembly, especially disc brakes, which are typically larger in diameter than drum brakes.

The offset can be either zero, positive, or negative:

- **Zero offset**: The plane of the hub mounting surface is even with the centerline of the wheel. Vehicles manufactured in the mid-1970s through the 1980s were typically built with zero offset wheels.
- **Positive offset**: The plane of the hub mounting surface is shifted toward the outside or front side of the

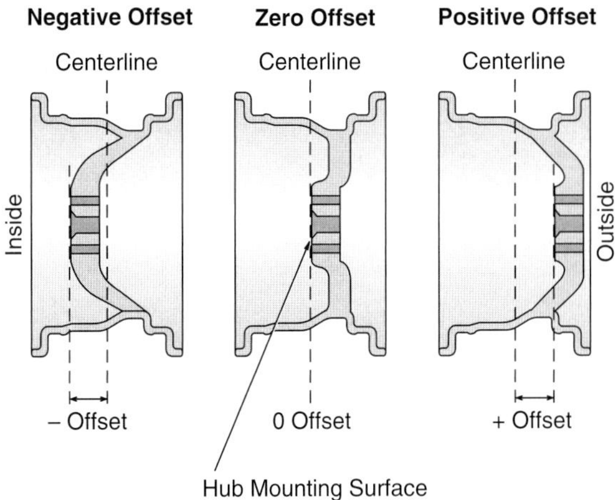

FIGURE 36-11 Zero, positive, and negative offset.

wheel. Positive offset wheels are generally found on front-wheel drive vehicles and the newer rear-wheel drive vehicles.

- **Negative offset**: The hub mounting surface is toward the brake side or back of the wheel's centerline. Older model vehicles and specialized high-performance vehicles with deep dish wheels typically have wheels with negative offset.

Wheel Studs and Lug Nuts

Wheels are fastened to the rims by **wheel studs** and **lug nuts**. Wheel studs and lug nuts are highly stressed by loads from the weight of the vehicle and the forces generated by its motion. Wheel studs and nuts are made from heat-treated, high-grade alloy steel. The threads between the studs and the nuts are close fitting and accurately sized. All wheel nuts must be tightened to the correct torque with the proper sequence; otherwise, the wheel could break free from the hub. You must use the correct type of wheel retaining stud or nut—which includes **tapered seat, flat seat with washer, flat seat without washer**, or **ball seat** (**FIGURE 36-12**)—for the wheel. Tapered seat lug nuts have a tapered end that is placed toward the rim and fit into a matching taper in the rim to help center the wheel on the lug studs. Flat seat studs are flat at the point where they contact the wheel. Those with washers have a washer affixed to the lug nut, enabling it to turn independent from the hex part of the lug nut.

When a wheel assembly is changed, it is imperative that the mating surfaces of the wheel inner hub and the axle flange mounting face are clean and free of dirt, mud, and rust. After the wheel assembly has been fitted, the **wheel retaining nuts** that hold the wheel onto the vehicle are torqued to the manufacturer's specifications. Many shops request that the customer return and have the lug nut torque rechecked within a few days or about 100 miles. If the nut or stud is under-tightened, then it is possible that they will loosen up while driving, and the wheel could fall off. If the nut or stud is over-tightened, then there is a good chance that the studs will be weakened, causing them to break because of overstressing and fatigue fractures. Over-tightening can also cause the brake rotors to warp.

Most lug nuts and studs are right-hand threaded, which means they tighten when turned clockwise. However, some manufacturers use lug nuts and studs with left-hand threads on the right side (passenger side) of the vehicle, and right-hand lug nuts and studs on the left side (driver side) of the vehicle. The manufacturers claim that left-hand lug nuts are less likely to

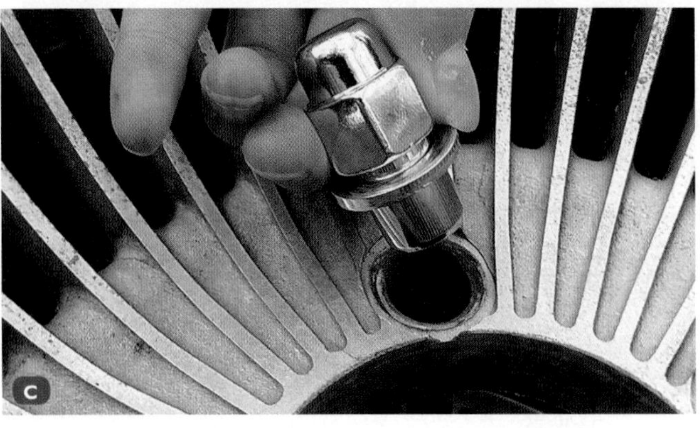

FIGURE 36-12 **A.** Tapered seat lug nut and wheel. **B.** Flat seat with integrated washer and wheel. **C.** Flat seat with separate washer and wheel. **D.** Ball seat.

back off when used on the right side of the vehicle. This may or may not be true, but properly torqued lug nuts do not generally loosen up, no matter if they are left-hand or right-hand. Left-hand lug nuts can usually be identified by an "L" stamped on the end of the lug stud (**FIGURE 36-13**). When you place the socket or lug wrench over the lug nut, make sure you check for the "L." If there is one, then you will need to turn the lug nut clockwise to loosen it.

TECHNICIAN TIP

Don't be like the apprentice technician who was trying to get a vehicle ready for a brake job for the journeyman technician. When the journeyman came into work, the apprentice was just taking off the last wheel, and the journeyman could hear the impact wrench hammering away and then the impact wrench zing to full speed. By the time he got to the apprentice, he was on the next to last lug nut. The journeyman stopped him. The apprentice said that all of the lug nuts were rusted on the studs on the right side of the car, and they all broke off when he tried to loosen them. The journeyman took the impact wrench, switched directions on it, and proceeded to easily remove the lug nuts. The apprentice was quite embarrassed, especially when the other technicians in the shop started calling him "Lefty."

If a lug nut does not start to loosen quickly when you are trying to remove it, always stop and try to turn it the other direction. It could be a left-hand lug nut, which you will have to turn clockwise to remove. And even if it is a right-hand lug nut, sometimes tightening it just a bit helps break loose any rust so that it comes off more easily.

The **wheel center** is the part of the wheel containing the holes for the lug studs. It usually has a machined hole in the center, which accurately centers the wheel rim on the axle. In some cases, the fit of the centering hole is snug enough that it can rust in place and stick to the axle. This requires penetrating oil, a hammer, or even heat from a torch to break it loose. Always seek the assistance of your supervisor if you encounter a wheel rusted to the axle flange. The wheel center also provides the required offset from the centerline of the wheel to the face of the mounting flange.

FIGURE 36-13 Left-hand lug nut identified by the "L" on the lug nut.

Bolt Pattern

The bolt pattern refers to the number and spacing of the lug nuts or wheel studs on the wheel hub and the wheel rim (**FIGURE 36-14**). As the studs are most often evenly spaced, the number of studs determines the pattern. For example, some smaller cars have three studs, some may have four studs, but most passenger cars have five. Pickup trucks and large SUVs can have as many as 6, 8, or 10 studs. The exact number and pattern of studs vary depending on the vehicle type and manufacturer design.

TECHNICIAN TIP

Along with the bolt pattern configuration is **pitch circle diameter (PCD)**. PCD is the diameter of a circle drawn through the center of the wheel's bolt holes. It is a fixed measure set during manufacture that cannot be altered. PCD is measured in both inches and millimeters, and it also indicates the number of studs or bolts on the wheel. One of the most common configurations has four studs and a PCD of 100 mm, hence the size 4 × 100 designation.

Tires

K36004

Tires are hollow, donut-shaped structures designed to provide traction to the wheel assembly and act as a cushion to absorb shock from road surfaces (**FIGURE 36-15**). The air in the tire supports the vehicle's mass, and the tire tread provides frictional contact with the road surface so the vehicle can maneuver safely.

The tire itself is generally composed of the tread, plies or cords, sidewalls, inner liner, and bead. The tread is the exterior rubber portion of the tire that comes in contact with the road. It is configured in a variety of patterns based on the application of the tire, such as for driving in mud and snow. The plies

FIGURE 36-14 The bolt pattern refers to the number and spacing of lug nuts or wheel studs on the wheel hub on the wheel rim.

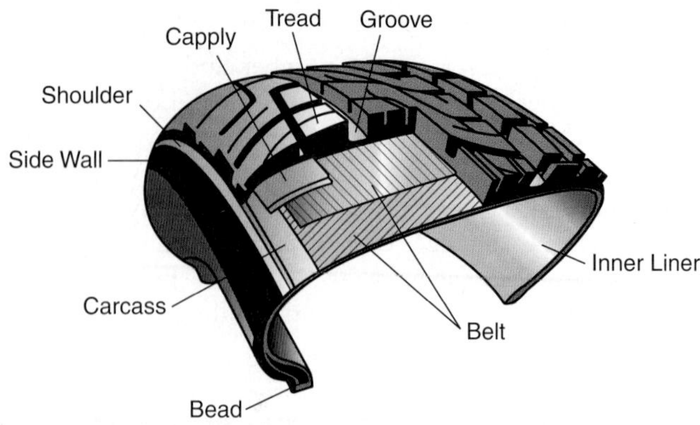

FIGURE 36-15 The tire itself is generally composed of the tread, sidewalls, inner liner, and bead.

or cords are the reinforcing material that gives the tread and sidewall their ability to hold their shape. The sidewalls are the lightly reinforced sides of the tire that provide lateral strength to the tire and prevent it from ballooning. They do not come into direct contact with the road. The inner liner is a covering of the casing material and seals the air in the tire. Beads are bands of steel wire coated in rubber that give the bead area the stiffness to hold the bead against the bead seat and to seal to the wheel rim. Tires can be inflated through a valve assembly located in the wheel.

Modern tires are made from a range of materials. The rubber used is mostly synthetic, with carbon black added to increase strength and toughness. When used in the tread, this combination gives the tire a long life. Natural rubber is weaker than the synthetic version, so it is used mainly in sidewalls.

Cords of synthetic strands or fabric have high **tensile strength**. Because of their high strength, the cords resist stretching but are flexible under load. The cords are placed in parallel and impregnated with rubber to form sheets called plies or belts. Plies have high strength in one direction and are flexible in other directions. When cotton was used as a cord, the number of plies, or layers in a tire, was a measure of the tire's strength. Newer cord materials use fewer plies. A modern steel-belted radial tire with a six-**ply rating** may have six plies in the tread, but just two plies in its sidewall. Having fewer plies makes the tire more flexible. Higher numbers of

Applied Math

AM-28: Distance: The technician can use standard and metric measurement instruments to determine correct sizes and distances.

In practice, both standard and metric measurement instruments must be used to determine correct sizes and distances because specifications may be published in either scale. Some common examples include measuring the diameter of wheel rims, which are almost universally measured in inches, and measuring the PCD of wheel studs, which is usually quoted in millimeters.

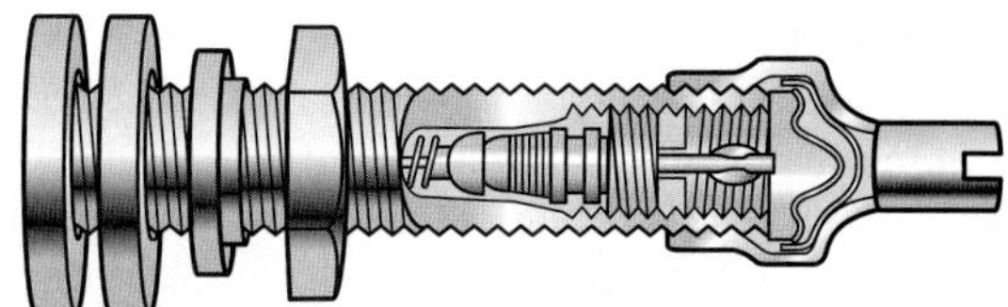

FIGURE 36-16 Components of a tire valve.

plies make a tire's response to bumps harsher, but the tire can withstand punctures much better.

The bead of the tire is made of a cord of high-tensile steel coated with rubber. The ends of the plies are wrapped around each bead, which is then wrapped in rubber to stop chafing of the plies and also to seal the bead against the rim. The length of the wire used for the bead determines the rim diameter of the tire. Belts that reinforce the tread area of the tire are typically made of braided, high-tensile steel wire, but they can also be made of rayon or polyester. The inner liner of a tubeless tire is made of soft rubber. The inner liner must be flexible and airtight.

Tire Valve Stems, Cores, and Caps

The **valve stem** is a specially designed part that allows a tire to be inflated and then automatically closes to prevent air from escaping. A **valve core** is threaded into the valve stem, and it is used on virtually all automobile tires. The valve stem is a rubber or steel piece that attaches the valve to the rim of the tire. The valve core is a **Schrader valve** (a spring-loaded, one-way valve) (**FIGURE 36-16**). A **valve stem cap** is used to keep debris out of the valve stem, which could cause the valve stem core to fail. The valve stem cap is removed to check air pressure and to inflate or deflate the tire. Some valve stem caps come with a built-in tool for tightening or loosening the valve core. In wet climates, the valve stem cap keeps mud, water, and ice from entering the valve stem. The valve stem cap can also slow a leak from a valve core, but will not stop a leak completely. In newer vehicles, a green valve stem cap indicates that the tire is filled with nitrogen and, to prevent contamination, should only be topped off with nitrogen.

Types of Tire Construction

Although there are two types of tire construction—**bias-ply** and **radial**—radial tires are by far the most common. In fact, one tire retailer estimates that 98% of the tires sold today for passenger vehicles are radial tires.

Bias-Ply Tires The bias-ply tire is the older form of tire and is still in use on some trailers and off-road vehicles, primarily because of a slightly lower cost and their more durable construction. Bias tire construction uses body ply cords extended diagonally from bead to bead at 30- to 40-degree angles, with successive plies laid at opposing angles, resulting in a crisscross pattern onto which tread is applied. This design provides a strong, stable casing, but with relatively stiff sidewalls. However, during cornering, stiff sidewalls can distort the tread and partially lift it off the road surface, which reduces the friction between the road and the tire.

> **TECHNICIAN TIP**
>
> The bias-ply tire's stiff sidewalls can make tires run at a higher temperature. This is because, as the tire rotates, the body ply cords in the plies flex over one another, causing friction and heat. A bias-ply tire that overheats can wear prematurely. At the same time, stiffer sidewalls can resist punctures from sharp rocks or other objects better than a radial tire.

Radial Tires Radial tire construction uses the same body ply cords, but they are laid across the tread extending from bead to bead, so the cords end up parallel to each other at approximately 90 degrees to the centerline of the tread. Radial tires are extremely durable and maintain good traction with the road, providing better steering control. Most passenger cars now use radial tires, as do most four-wheel drive vehicles and heavy transport vehicles.

Radial ply tires have much more flexible sidewalls because of their construction (**FIGURE 36-17**). They use two or more layers of **casing plies**, the cords that give the tire shape and strength, with the cord loops, which are loops of high-strength material (typically polyester) inlaid with rubber, running radially from bead to bead. The sidewalls are more flexible because the casing plies do not cross over each other; however, a bracing layer of two or more steel belts running horizontally must be placed under the tread to strengthen and stabilize the tire. The cords of the bracing layers may be of fabric or steel and are placed at 12 to 15 degrees to the circumference line. This design forms triangles where the belt cords cross over the radial cords. The stiff bracing layer links the cord loops together to give stability when accelerating or braking, and it prevents any movement of the cords during cornering. The cord plies flex and deform only in the area above the road contact patch. There are no heavy plies to distort, and flexing of the thin casing generates little heat, which is easily dispersed. A properly inflated radial tire runs cooler than a comparable bias-ply tire, increasing tread life. Also, a radial tire has less rolling resistance as it moves over the road surface, increasing fuel economy.

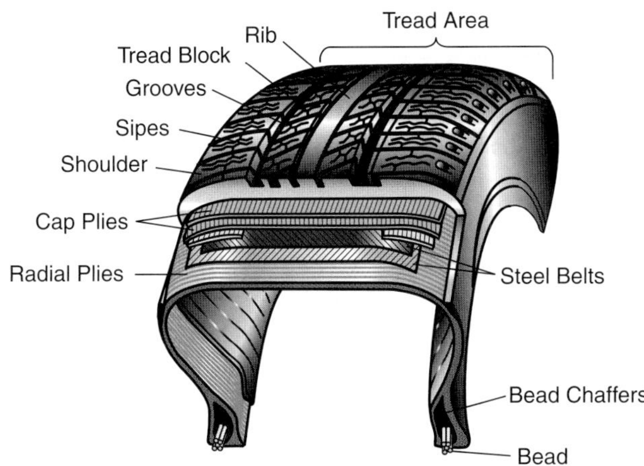

FIGURE 36-17 Radial tires.

> **TECHNICIAN TIP**
>
> The sidewalls of radial tires bulge where the tire meets the road, making it difficult to estimate the inflation pressure visually. Tire pressure needs to be checked with an accurate tire gauge. Using the correct inflation pressures extends the life of the tire and is vital for safety. Sidewalls of an under-inflated tire flex too far, which causes the center section of the tread to be pushed up and away from the road surface. This causes wear at the shoulders of the tire. In an over-inflated tire, the sidewalls are straightened, which pulls the edges of the tread away from the road and causes wear at the center of the tread.

Tread Designs

Differing tread patterns give manufacturers the ability to design tires for special applications, such as rainy, snowy, muddy, highway, or high-performance driving. Manufacturers spend a lot of money trying to design better tread patterns than their competitors. And when they discover a good design, it can be very profitable. Tire tread patterns can be classified with these general characteristics (**FIGURE 36-18**):

- **Directional tread patterns** are designed to provide a range of attributes during particular driving conditions, such as driving in wet weather where there is a possibility of hydroplaning (sliding across water). The predominate tread pattern is mounted in such a manner as to provide maximum moisture dissipation, thereby creating a drier surface for the tire to adhere to. Specifically, these tread patterns actually pump water out from under the tire, bringing it in direct contact with the surface of the road. This type of tire can only be mounted to the wheel so that it revolves in a particular direction to correspond with the tread pattern. An arrow on the tire sidewall indicates the designed direction of forward travel.
- **Nondirectional tread patterns** are designed in such a way that the tire can be mounted on the wheel for any direction of rotation. This allows the tire to be installed on either the left or the right side of the vehicle and to be mounted in multiple directions. These tires are used for general applications, when the customer may not be looking for a high-performance tire. They are usually cheaper to manufacture than the specialized tires, and they allow tire stores to stock fewer tires.
- **Symmetric tread patterns** have the same tread pattern on both sides of the tire and are usually nondirectional tires. As the name indicates, they tend to feature treads that are similar in continuous designs across both sides of the tire tread as well as around the tire. A main feature of most symmetric tires is that they are nondirectional and thus can be fitted in either direction.

FIGURE 36-18 Tread designs: **A.** Directional tread pattern. **B.** Nondirectional tread pattern. **C.** Symmetric tread pattern. **D.** Asymmetric tread pattern. **E.** Directional and asymmetric tread pattern.

- **Asymmetric tread patterns** have a tread pattern that is different from one side of the tire to the other. They are designed to provide good grip when traveling straight and in turns. Asymmetric tires are labeled "outside" on the side of the tire that faces outward from the vehicle.
- **Directional and asymmetric tread patterns** are both directional and asymmetric, which means the tire is designed to rotate in only one direction and has one side that must face outward to ensure that the tire performs as designed under operating conditions. These tires tend to be used on high-performance vehicles.

FIGURE 36-19 Typical sidewall markings.

FIGURE 36-20 Maximum pressure that the tire is designed for. Always make sure this pressure is higher than the recommended pressure on the tire placard.

▶ Tire Markings

K36005

Tires come in a variety of sizes and ratings to accommodate the wide range of vehicles and driving situations. To help select the most appropriate tire for a particular application, tires use common sizing and rating systems. All tires meeting legislative codes must have the following information clearly marked on the sidewall (**FIGURE 36-19**):

- Manufacturer or brand name
- International Organization for Standardization (ISO) tire class:
  - P—passenger
  - LT—light truck
  - C—commercial
  - T—temporary use as a spare wheel
  - No letter designation before the section width—either a European metric tire or an off-road tire
- Section width: measured in millimeters from sidewall to sidewall
- Aspect ratio: the height of the sidewall expressed as a percentage of the section width
- Type of tire construction: R for radial; blank for bias-ply
- Wheel diameter: diameter of wheel from bead seat to bead seat, usually measured in inches
- Speed rating designation: the maximum speed for which a particular tire is rated. Vehicles should not be driven in excess of their speed ratings, with the exception of Z-rated tires, which are rated for speeds above 149 mph (240 kph). *Note:* Virtually all tires are rated for higher speeds than legal speed limits. Tire speed ratings do *not* give you legal cover for speeding! Speed ratings are as follows:
  - Q—up to 100 mph (161 kph)
  - R—up to 106 mph (171 kph)
  - S—up to 112 mph (180 kph)
  - T—up to 118 mph (190 kph)
  - U—up to 124 mph (200 kph)
  - H—up to 139 mph (224 kph)
  - V—up to 149 mph (240 kph)
  - W—up to 168 mph (270 kph)
  - Y—up to 186 mph (299 kph)
  - Z—149 mph (240 kph) and over
- Maximum air pressure: the maximum pressure that the tire can be inflated to (**FIGURE 36-20**). This does not indicate the vehicle manufacturer's recommended inflation pressure. Always refer to the **tire placard** on the vehicle to determine proper inflation pressures.
- Load index: the maximum amount of weight that a tire can safely carry at the maximum rated tire pressure. When replacing tires, verify that the load index meets the vehicle manufacturer's weight rating for front and rear axles.
- Uniform Tire Quality Grading (UTQG) system:
  - Tread wear grade: an approximation of how long a tire will last when compared to another tire from the same manufacturer. For example, a tire rated at 450 will last approximately three times as long as one rated 150.
  - Traction grade: a representation of a tire's wet traction characteristics. From highest to lowest: AA, A, B, and C.
  - Temperature grade: a representation of a tire's ability to resist and dissipate heat. From highest to lowest: A, B, and C.
- Department of Transportation compliance symbols and serial numbers, including date of manufacture code:
  - First two letters following the letters "DOT" identify the tire manufacturer and manufacturing plant.
  - The third and fourth letters are codes that indicate the tire's size.
  - The final three or four letters are codes indicating manufacturer-specified characteristics.
  - Week of manufacture: The first pair of digits represent the week of the year in which the tire was manufactured, starting in January.
  - Year of manufacture: The last two digits represent the year of manufacture; for example, "17" indicates a tire manufactured in 2017.

> **TECHNICIAN TIP**
>
> Radial tires are marked with the section width in millimeters, but with the rim diameter in inches.

Tire Sizes and Designations

The size of a tire must be appropriate for the vehicle application and intended use. The vehicle manufacturer designates the recommended size and load rating of the tires to be used on each particular vehicle. The bead diameter must match the rim diameter. The section width must be suitable for use on the rim and large enough to have a suitable load-carrying capacity for the vehicle. The overall tire size must allow sufficient clearance between the tire and the chassis and body components. The section width of the tire is measured in millimeters, from sidewall to sidewall, by the tire manufacturer when installed on a standard wheel, inflated to its recommended pressure and without any load on it. Section width varies from manufacturer to manufacturer, so do not assume tires with the same section width are all the same.

The **aspect ratio** of a tire is the ratio of its height to its width. It is usually given as a percentage. Information on tire aspect ratio is now included in the sidewall marking. The lower a tire's aspect ratio, the wider the tire is in relation to its height. An aspect ratio of 75 means the height of the sidewall is 75% as much as the section width. Generally, the higher the aspect ratio, the smoother the ride will be, but there will be more flex during cornering.

Low-profile tires have very short sidewalls and can be difficult to remove and install (**FIGURE 36-21**). They have an aspect ratio as low as 25. The low-profile tire improves cornering performance but sacrifices a smooth ride, and there is a greater danger of wheel and suspension damage when hitting potholes. Often, low-profile tires have a higher speed rating, as there is less centrifugal force trying to throw the tire apart.

FIGURE 36-21 Tire profile. **A.** Standard-profile tire. **B.** Low-profile tire.

Tire Ratings for Tread Wear, Traction, and Temperature

One of the markings on the sidewall of a tire is a **Uniform Tire Quality Grading (UTQG)** rating (**FIGURE 36-22**). As discussed previously, the tire's UTQG rating provides information on three aspects of the tire's durability and operational characteristics: tread wear, traction, and temperature. The **tread wear grade** comes from testing the tire in controlled conditions. The higher the number, the longer the life expectancy of the tread as compared to another tire from the same manufacturer. Because no one vehicle will be subjected to exactly the same surfaces and at the same speeds as the controlled conditions, the number can only be an indicator of expected tread life in normal conditions. The rating is based on a percentage of the projected wear life. For instance, when looking at two tires from the same manufacturer, one tire rated at 400 has a projected life of four times that of a tire rated at 100.

FIGURE 36-22 Typical UTQG ratings on a tire sidewall.

> **TECHNICIAN TIP**
>
> There are many factors that influence wear, such as vehicle speed, road surface, climate, vehicle wheel alignment, and the driving characteristics of the driver. The rating can only be an indication of the anticipated wear characteristics of the tire in controlled conditions.

A **traction grade** is a letter-based indicator system. The rating is based on the tire's ability to stop a vehicle on wet concrete and asphalt in a straight-line situation. It does not indicate the tire's cornering ability. The tire traction indicators are rated from highest to lowest as AA, A, B, or C. It is important to note that the relevant rating does not indicate hydroplaning resistance; dry or snow traction capacity; or cornering capability in wet, dry, or snow conditions.

The **temperature grade** of a tire is a letter based on a test performed by the tire manufacturer and overseen by the US government. The goal of the test is to determine how well a tire stands up to heat and measures how well the tire dissipates heat. The US government uses the UTQG criteria for tire temperature ratings. The tires are graded from C (least tolerant of heat dissipation) to A (most tolerant of heat dissipation). Although a C-graded tire runs hotter, it is not necessarily unsafe. It is important to remember that these ratings are based on standardized test conditions, and the tests do not reflect tires that are operated in overloaded, under-inflated, and/or misaligned conditions. It should also be noted that one tire might be rated a low A and another a high B, so the actual operating performance differences might be relatively small.

It is not uncommon for there to be differences in UTQG ratings within a given tire design. Sometimes a particular vehicle manufacturer requires certain properties for the tires supplied to its vehicles, which can affect the ratings both positively and negatively. Sometimes there are differences between small sizes and large sizes of tires in a given design. All of these aspects can affect the actual rating that is put on the sidewall.

Tire Date of Manufacture Coding

The **US Department of Transportation (DOT)** inspects everything and anything pertaining to transportation—including tires. As part of DOT regulations, there must be a tire manufacture date code stamped on the sidewall of every tire. In fact, it is illegal to sell a tire intended for use on a public road within the United States without a DOT stamp. Some manufacturers mold this code on only one sidewall, so you might need to get under the vehicle and look at the inward-facing side of the tire, where you will find a three- or four-digit code. This code denotes when the tire was manufactured. For safety concerns and as a rule of thumb, you should never use tires more than six years old. The rubber in tires degrades over time, irrespective of whether the tire is being used or not. Reading the DOT code is relatively simple. A three-digit DOT code was used for tires manufactured before 2000. For example, "176" means that the tire was manufactured in the 17th week of the sixth year of the decade, or 1986.

For tires manufactured in the 1990s, there may be a little triangle after the DOT code. For example, a tire manufactured in the 17th week of 1996 might have the code "176." After 2000, the code was switched to a four-digit code. For example, "30 11" means the tire was manufactured in the 30th week of 2011 (**FIGURE 36-23**).

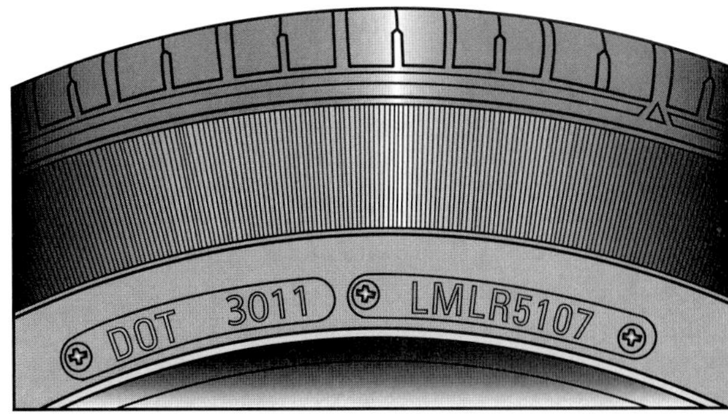

FIGURE 36-23 The DOT tire date manufacturing code is a four-digit code.

▶ Tire Safety Features

K36006

Flat tires are one of the leading causes of vehicle breakdowns. Generally speaking, vehicles cannot be driven with one or more flat tires. Attempting to do so will ruin a potentially repairable tire very quickly. Manufacturers address this situation in various ways. The most common is through the use of a **tire pressure monitoring system (TPMS)** to monitor the air pressure in each tire and warn the driver if one or more tires is low. Another tactic that manufacturers use is run-flat tires. These tires are designed so that they can still be driven for a reasonable amount of time when the tire pressure is low or empty. Similar to the run-flat tires are self-sealing tires, which resist leaks and help maintain air pressure. The last safety feature involves the use of a spare tire, which is usually of the space-saving type, also called a temporary tire. It is designed to be used for reduced speeds and distances so that a driver can get the vehicle to a service facility. Each of these safety features is discussed further in the following sections.

Tire Pressure Monitoring Systems

K36007

Maintaining proper tire pressure is essential for the safety and performance of a vehicle. It also plays a significant role in decreasing fuel consumption, reducing CO_2 emissions, and extending tire life. All tires lose inflation over time and, as many modern vehicles have extended service intervals, tires can become dangerously under-inflated without regular checking by the vehicle driver. In addition to increased fuel consumption and tire wear, long periods of driving with low tire pressures can cause additional stress on the tire sidewalls. This results in increased operating temperatures that can lead to premature tire failure. Tires operating with low pressures can also affect the vehicle's handling and performance. In a worst-case scenario, under-inflation can lead to a tire blowout or tread separation. Because of these situations, the Transportation Recall Enhancement, Accountability, and Documentation (or TREAD) Act called for all new passenger vehicles under 10,000 lb (4536 kg) to be equipped with a TPMS by October 1, 2007.

The automated tire pressure monitoring system (TPMS) provides a means of reliable and continuous monitoring of the vehicle tire pressure and is designed to increase safety, decrease fuel consumption, and improve vehicle performance. A TPMS monitors the tires for low air pressure and alerts the driver when one or more tires are lower than (or in some cases, higher than) the designated thresholds. This alert can be an illuminated warning lamp or a chime. A TPMS can be fitted to all vehicles using conventional and run-flat tires. With some TPMS systems, drivers can monitor the tire pressures and temperatures from the driver's seat to ensure that their tires are properly inflated under all operating conditions. The TPMS is designed to ignore normal pressure variations caused by changes in ambient temperature.

There are two basic configurations used to monitor the vehicle's tire pressures: direct and indirect. A **direct TPMS** directly measure the tire pressure and sometimes pressure via a sensor that is installed inside each wheel, which helps protect it from damage. There is a sensor with an antenna in each wheel that wirelessly relays the information it senses to receivers located within the vehicle (**FIGURE 36-24**). The receivers send the signal to the control unit. The sensor is able to respond to a drop in pressure of as little as 2 psi (14 kPa). The control unit sends an appropriate signal to the driver's information circuit or, in some vehicles, the onboard computer, which in turn illuminates a display in the vehicle's multifunction screen to warn the driver of low tire pressure in a certain wheel. An audible and visual warning instantly alerts the driver, allowing time for the vehicle to stop or be driven to a service station. The sensors are powered by an internal battery that is designed to last between 5 and 10 years. The use of a **centrifugal switch** in the sensor allows the sensor to go to sleep when the vehicle stops, which extends battery life. When the battery goes dead, it has to be replaced, which usually means that the entire sensor must be replaced because the battery is typically sealed inside the sensor.

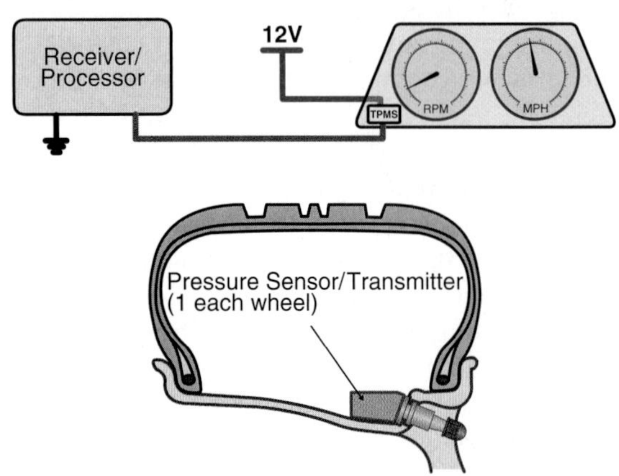

FIGURE 36-24 The tire pressure monitoring system (TPMS).

TECHNICIAN TIP

In vehicles with a direct TPMS, the dashboard may show:

- The required tire pressure
- The actual tire pressure
- The tire pressure status
- The temperature of the tire

In some vehicles, the driver can use the display control buttons to check the status of each tire.

A direct TPMS can be of two types: one-way communication or two-way communication. In one-way communication, the TPMS sensor can only transmit to the receiver; it cannot receive any information. In this type of system, the sensors usually use a centrifugal switch to turn them on and off to conserve the battery energy. In a two-way communication TPMS, the sensor can receive as well as transmit signals. This allows the control unit to send signals to wake up or cause the sensor to sleep, thus extending the sensor's battery life. The two-way communication system is more complex and expensive than the one-way system. The wheel sensors are located inside the tire and fastened in some manner to the wheel. Many wheel sensors are integrated into a one-piece design along with the valve stem, whereas others are a separate unit screwed into the valve stem (**FIGURE 36-25**). The latter style uses a band or strap that fastens around the wheel in the drop center and holds the sensor in place. In a band-style wheel sensor, the sensor is typically located opposite of the valve stem (**FIGURE 36-26**). When removing and replacing a tire on a wheel, you must be aware of the sensor's location and take steps to avoid damaging it.

TECHNICIAN TIP

Most wheels with TPMS sensors use an aluminum valve stem, which requires a nickel-plated valve core to prevent corrosion. Also, the valve stem acts as the antennae for the TPMS sensor on some applications.

An **indirect TPMS** indirectly monitors tire air pressure. The most prevalent indirect tire pressure monitoring systems in use today utilize the wheel speed anti-lock braking systems (ABS)

FIGURE 36-25 Typical pressure sensors used in TPMS.

FIGURE 36-26 Band-type TPMS sensor.

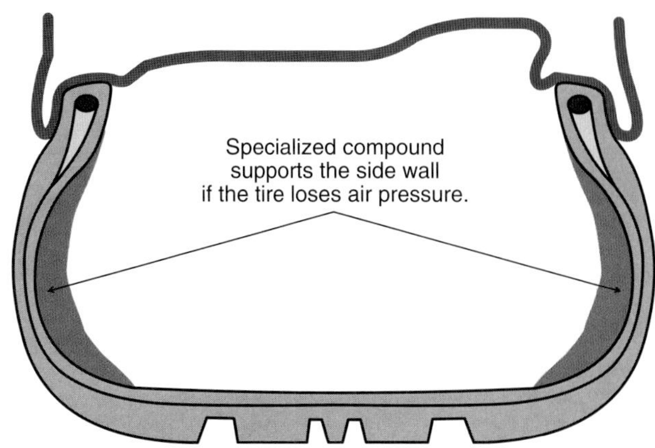
Specialized compound
supports the side wall
if the tire loses air pressure.

FIGURE 36-27 Run-flat tire construction.

to measure the difference in the rotational speed of the four wheels. A wheel that is rotating faster than the others indicates that the tire has lower pressure than the other tires. For example, if a tire loses pressure, then its rolling radius is reduced, which increases the speed of rotation. Because it is rolling faster than the other tires, the wheel speed sensor will send a slightly faster wheel speed signal for that wheel to the ABS control unit. The control unit monitors the changes in wheel speed, and when a low tire pressure failure is detected, it sends a signal to the low tire pressure light and/or chime on the dashboard.

Run-Flat Tires

K36008

The major safety benefit of **run-flat technology** is that it enables a driver to maintain vehicle control if a tire suffers a rapid pressure loss when in motion. In addition, run-flat tires enable the driver to continue the journey within specified speed and distance limits (about 50 mph [80 kph] and at least 50 miles 80 [km]). This alleviates the need to replace the wheel on the side of the road or in an unsafe area. But as with any tire, the run-flat tire cannot continue to be driven if the sidewall has been compromised or blown out. In these instances, the tire needs to be changed. A TPMS is normally mandatory for all run-flat technology applications to monitor the drivability of the vehicle's run-flat tires.

Tire manufacturers also maintain that run-flat technology usually saves weight and space by eliminating the need to carry a spare tire. However, because of their construction, they are generally between two to three times heavier than their conventional counterparts. This adds unsprung weight (weight not held up by the vehicle's springs) to the vehicle, affects the suspension, and can increase fuel consumption.

Because of the extra materials used in the construction of run-flat tires, they are also more expensive to purchase. In addition, run-flat tires are usually harsher riding and noisier in operation, which can be a disadvantage in some applications, such as in luxury cars that are expected to have a quiet, smooth ride. From a manufacturing perspective, the free space created by eliminating a standard spare wheel gives the vehicle manufacturer a range of additional design opportunities, such as providing more storage space in the trunk.

The design features of run-flat technology generally focus on two aspects of operational use: rigidity and heat resistance. The objective is for the tire to support the vehicle's weight when it is rotating with a total air loss. The sidewall is constructed with reinforced rubber and is thicker than in conventional tires, enabling it to carry the vehicle's weight at zero pressure (**FIGURE 36-27**).

The bead shape configuration of the run-flat tire is largely unchanged from conventional tires to enable compatibility with conventional rims. However, the bead wire is normally wider and reinforced to ensure a secure fit on the specialized rim, even at zero pressure. A special bead filler with low heat generation is used as part of the run-flat technology construction to help prevent excess heat buildup that can be generated with zero pressure.

The rim used with some run-flat technology tires is referred to as an **EH2 rim**. EH2 is an abbreviation for double-extended hump. These rims have a wider, or more extended, safety hump than a standard safety rim, to accommodate the wider and more square-shaped tire bead. The run-flat technology tire bead also has a larger diameter than the standard safety bead of a conventional radial ply tire.

▶ TECHNICIAN TIP

Tire manufacturers call their run-flat safety tire by various terminologies, but all have the same characteristics. Some run-flat tires are known as:

- Run-Flat Technology (RFT) tires
- **Extended Mobility Technology (EMT)** tires
- Continuous Mobility Technology (CMT) tires
- Zero-pressure (ZP) tires

EMT tire sidewalls can be six times thicker than the sidewalls of traditional tires. As a result, the manufacturers claim that run-flat tires can be driven at speeds of about 50 mph (80 kph) for at least 50 miles (80 km) in a deflated condition before being damaged. These tires are not indestructible. Major damage that slices the tire casing can still result in complete tire failure.

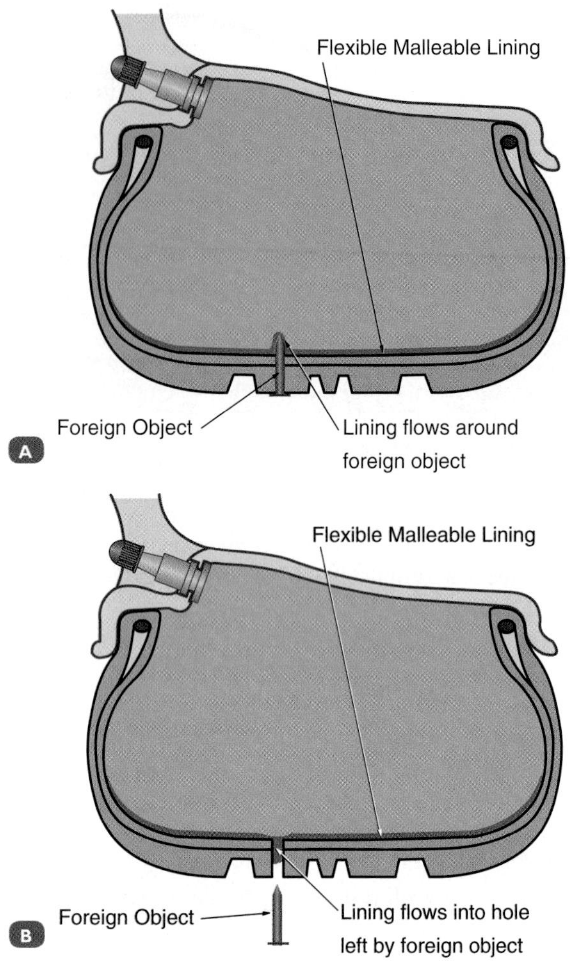

FIGURE 36-28 Self-sealing tire protects against small punctures.

Self-Sealing Tires

K36009

Self-sealing tires were introduced recently to commercial passenger cars. One self-sealing tire manufacturer states that its self-sealing tire is designed to repair most small tread-area punctures instantly and permanently. Self-sealing tires feature standard tire construction with the addition of a flexible and malleable lining inside the tire in the tread area (**FIGURE 36-28**). This lining serves as a puncture sealant that can permanently seal most punctures from nails, bolts, or screws up to 3/16" (or 5 mm) in diameter. Self-sealing tires first provide a seal around the object when the tire is punctured and then fill in the hole in the tread if the object is removed. Most drivers never even know that they just had a puncture. Because these tires can still leak air from between the bead and the rim or the valve stem and the valve core, they still require a TPMS to detect any low-pressure conditions on 2008 and newer vehicles.

Space-Saver Spare Tires

K36010

Space-saver spare tires are designed for emergency use only. They are designed to get the driver to a service center to have the regular tire fixed or purchase a new one. When provided with the vehicle as part of the original manufacturer's equipment, most manufacturers warn not to exceed 50 mph (80 kph) and not to go farther than 50 miles (80 km) on the space-saver spare tire. Some vehicles have miniature or collapsible space-saver spare tires as spares. These normally require a specially charged canister for inflation when being installed. Other vehicles have small, temporary spare tires that have been inflated normally with a compressed air supply, but, because of their small size, to a much higher pressure than normal road tires (**FIGURE 36-29**).

FIGURE 36-29 Typical space-saver tire.

▶ Wrap-Up

Ready for Review

- Poorly maintained wheels and tires decrease effective handling and may lead to steering and suspension problems.
- Tire distortion refers to the tire's cornering force countering the side force that occurs when a vehicle corners, to create a slip angle.
- Center of gravity refers to the balance point of the vehicle, which is determined by location of the engine and transmission.
- Wheel offset refers to the distance from the hub mounting surface to the centerline of the wheel.
- Wheel offset can be zero, positive, or negative.
- Tire and wheel assemblies must be balanced to prevent both static and dynamic imbalance.
- Tire and wheel components include the wheel or rim, wheel studs and nuts, wheel center, and tires.
- A Schrader valve is a one-way valve used in a valve stem.

- Wheels usually have a deep well—a widened area on one side of the wheel.
- Types of rims include steel, one-piece alloy, two-piece alloy, multi-piece alloy, custom, spinning, split, semi-drop well, drop well, and safety.
- Wheel studs and lug nuts fasten the wheels to the rims.
- Wheel retaining studs or nuts can be tapered seat, flat seat with washer, flat seat without a washer, or ball seat types.
- Tires provide the wheel with coverage and protection and absorb shock from road surfaces.
- Tires are composed of treads, sidewalls, inner liners, and beads.
- Synthetic fabric cords are used to create plies, giving the tire strength and flexibility.
- Tires are most commonly either cross-ply or radial (used by most passenger vehicles).
- Tire pressure monitoring systems (TPMS) come in two types: direct and indirect.
- Direct TMPS systems use pressure sensors in each wheel and transmit the pressure readings wirelessly.
- Indirect TPMS systems use wheel speed sensors to indicate wheel speed, which increases as tires lose pressure.
- TPMS sensors can be part of the tire's valve stem or mounted with a band around the inside of the wheel.
- TPMS sensors may have a centrifugal switch that turns them on only when the vehicle is moving.
- Most TPMS systems need to be manually calibrated to identify which pressure sensor is mounted inside each wheel.
- Run-flat tires allow a vehicle to be driven for up to 50 miles if it experiences a sudden loss of pressure.
- A TPMS is normally mandatory for all run-flat technology applications.
- Self-sealing tires have a lining that serves as a puncture sealant that can permanently seal most punctures from nails, bolts, or screws up to 3/16" (or 5 mm) in diameter.
- Space-saver spare tires are designed to get the driver to a service center to have the regular tire fixed or purchase a new one.
- Some vehicles have miniature or collapsible space-saver spare tires as spares. These normally require a specially charged canister for inflation when being installed.
- Other vehicles have small, temporary spare tires that have been inflated to a much higher pressure than normal road tires because of their smaller size.

Key Terms

aspect ratio The ratio of sidewall height to section width of a tire.

asymmetric tread pattern A tread pattern that differs on each side and therefore is usually directional.

ball seat

bead seat The part of the wheel that the tire seals against.

bias-ply A tire constructed in a latticed, crisscrossing structure, with alternate plies crossing over each other and laid with the cord angles in opposite directions.

casing plies A network of cords that give the tire shape and strength; also known as casing cords.

centrifugal switch A switch that is only activated when centrifugal forces are placed on a vehicle.

cornering force The force between the tread and the road surface as a vehicle turns.

deep dish wheel A wheel with negative offset, which gives the outside of the wheel a deep dish appearance. Deep dish refers to the side of the wheel that is farthest from the drop center.

direct TPMS A type of automated tire pressure monitoring system that measures tire pressure and possibly temperature via a sensor installed inside each wheel.

directional and asymmetric tread pattern A tread pattern that is both directional and asymmetric, which means the tire is designed to rotate in only one direction and has one side that must face outward to ensure that the tire performs as designed under operating conditions.

directional tread pattern A tread pattern designed to pump water out from under the tire; each tire must be placed in a particular spot on the vehicle.

drop center A wheel design with part of the center section of the wheel a smaller diameter than the rest. It is used for mounting and demounting the tire.

EH2 rim The specialized rim design that is used with some run-flat tires.

Extended Mobility Technology (EMT) Tires with thick sidewalls that allow the tire to be driven on even when it has no air pressure.

flat seat with washer A type of lug nut that is flat where it bolts to the wheel and has a washer affixed that allows it to turn independent of the hex part of the lug nut.

flat seat without washer A type of lug nut that is flat where it bolts to the wheel.

indirect TPMS A type of automated tire pressure monitoring system that uses the anti-lock braking system of a vehicle to measure the difference in the rotational speed of the four wheels, to determine tire pressure.

lug nuts Nuts that secure the wheel onto the wheel studs.

negative offset A condition in which the plane of the hub mounting surface is positioned toward the brake side or back of the wheel centerline.

neutral steer A condition in which both the front and the rear tires of a vehicle are experiencing the same slip angle.

nondirectional tread pattern A tread pattern that is nonspecific, allowing the tire to be placed on any wheel of a vehicle.

oversteer A condition in which a vehicle's front slip angles are larger than the rear slip angles. This vehicle is said to be "pushing" in the corners.

pitch circle diameter (PCD) The diameter of the imaginary circle drawn through the center of the wheel bolt holes.

ply rating A rating system that denotes the number of belt layers, or plies, that make up the tire carcass. In radial tires, ply rating denotes the relative strength of the plies, not the actual number of plies.

positive offset A condition in which the plane of the hub mounting surface is positioned toward the outside or front of the wheel centerline.

radial tire A tire with two or more layers of casing plies and cord loops running radially from bead to bead.

rim the outer circular lip of the metal on which the inside edge of the tire is mounted.

rim flanges The outside edge of the wheel that helps keep the tire from popping off the wheel.

rim width The distance across the rim from one rim flange to the other.

run-flat technology A tire design that allows the vehicle to keep moving under driver control, following a puncture or rapid loss of pressure.

safety-type drop-center rim A rim designed with a slight hump at the inside edge of the bead ledges to hold the tire beads in place during a flat tire.

Schrader valve A one-way valve used in a valve stem.

self-sealing tire A tire constructed with a flexible and malleable lining inside the tire around the inner tubeless membrane. The lining can seal small tread-area punctures instantly and permanently.

side force The pressure on the wheel that pushes it toward the outside or inside of the rim as the vehicle makes a turn.

slip angle A tire's sideways distortion that makes the vehicle follow a path at an angle to the direction the road wheel is pointing.

steel-disc–type rim A plain steel wheel that is typically covered by a hubcap.

symmetric tread pattern A tread pattern with the same tread pattern on both sides of the tire; typically nondirectional.

tapered seat A type of lug nut with a tapered end toward the rim that helps center the wheel on the wheel studs.

temperature grade A standardized grading system that indicates the extent to which heat is generated or dissipated by a tire.

tensile strength A measure of how strong a material is as it is being pulled apart.

tire inflation pressure The level of air in the tire that provides it with load-carrying capacity and affects overall vehicle performance.

tire placard A metal, vinyl, or paper tag permanently affixed to a vehicle that indicates the appropriate tire size and inflation pressure for the vehicle.

tire pressure monitoring system (TPMS) An automatic sensor system in most modern vehicles that alerts drivers of tire pressure problems.

tire valve The valve through which air is inserted into a tire to inflate it.

traction grade A standardized grading system that indicates how well a tire will maintain contact with the road surface when wet.

tread wear grade The number imprinted on the sidewall of a tire by the manufacturer, as required by the National Highway Traffic Safety Administration (NHTSA), that indicates the tread life of a tire's tread.

understeer A condition in which a vehicle's front wheels turn more sharply than the vehicle's direction when the vehicle is steered around a corner. This vehicle is said to be "loose" in the corners.

Uniform Tire Quality Grading (UTQG) A standardized grading system, established by the National Highway Traffic Safety Administration (NHTSA), designed to provide tire buyers with a comparative measure of a tire's tread life, traction, and temperature characteristics.

US Department of Transportation (DOT) A federal agency that regulates transportation safety in the United States, including vehicles' wheels and tires. The DOT requires a code—a series of letters and numbers—to be stamped into the sidewall of every tire made for public use in the United States. These codes contain information such as the date of manufacture and the plant where the tire was manufactured.

valve core The one-way spring-loaded valve that screws into the valve stem; it allows air to be pumped into a tire and prevents it from flowing out.

valve stem A rubber or steel piece that attaches the tire valve to the rim.

valve stem cap A cap that fits tightly onto the valve stem to prevent debris from clogging it and acts as a secondary seal.

wheel center The part of the wheel containing the holes for the lug studs.

wheel retaining nuts Lug nuts used to hold the wheel on the hub.

wheel rim The part on which the tire is mounted. Also called a "wheel" or "rim."

wheel studs The threaded fasteners that attach the wheel to the vehicle.

zero offset A condition in which the plane of the hub mounting surface is even with the centerline of the wheel.

Review Questions

1. All of the following statements are true *except*:
   - **a.** the front slip angles are larger than the rear slip angles in an understeer condition.
   - **b.** the rear slip angles are larger than the front slip angles in an understeer condition.
   - **c.** the rear slip angles are larger than the front slip angles in an oversteer condition.
   - **d.** The front and the rear tires have equal amounts of slip angles in a neutral steer condition.
2. In a typical rear-wheel drive vehicle, the weight distribution is approximately:
   - **a.** 75% fore and 25% aft.
   - **b.** 60% fore and 40% aft.
   - **c.** 40% fore and 60% aft.
   - **d.** 25% fore and 75% aft.

3. Which type of rims have a slight ridge, or hump, at the inside edge of the bead ledges that helps to hold the tire beads in place when the tire goes flat?
   a. Multi-piece alloy rims
   b. Spinning rims
   c. Semi-drop well rims
   d. Safety-type drop-center rims
4. Which type of lug nut is placed toward the rim and fits into a matching taper in the rim to help center the wheel on the lug studs?
   a. Tapered seat
   b. Flat seat with washer
   c. Flat seat without washer
   d. Ball seat
5. Which of the following tread designs actually pumps water out from under the tire, bringing it in direct contact with the surface of the road?
   a. Symmetric tread pattern
   b. Asymmetric tread pattern
   c. Directional tread pattern
   d. Nondirectional tread pattern
6. Which of the following speed rating designations is rated for speeds above 149 mph (240 kph)?
   a. V
   b. Z
   c. Y
   d. W
7. Which of the following allows a vehicle to be driven for a reasonable amount of time when the tire pressure is low or empty?
   a. TPMS
   b. Run-flat tire
   c. Self-sealing tire
   d. Space-saving tire
8. All of the following statements are true in the case of TPMS *except*:
   a. it monitors the tires for low air pressure and alerts the driver when pressure in one or more tires is lower than the designated thresholds.
   b. the TPMS adds air to tires which has slow leaks so that tire pressure is maintained.
   c. the indirect TPMS utilizes the anti-lock braking systems (ABS) to measure the difference in the rotational speed of the four wheels to detect low pressure in tires.
   d. the direct TPMS uses a sensor with an antenna in each wheel that wirelessly relays the information it senses to receivers located within the vehicle.
9. All of the following statements are true in describing run-flat tires *except*:
   a. run-flat tires are not compatible with conventional rims.
   b. run-flat tires enable the driver to continue the journey within specified speed and distance limits.
   c. run-flat tires support the vehicle's weight when it is rotating with a total air loss.
   d. run-flat tire sidewalls are normally much thinner than the sidewalls of traditional tires.
10. A tire constructed with a flexible and malleable lining inside the tire around the inner tubeless membrane is known as a:
   a. bias-ply tire.
   b. radial tire.
   c. self-sealing tire.
   d. tubeless tire.

ASE Technician A/Technician B Style Questions

1. Tech A says that most wheels have a drop center or deep well that is used in installing a tire on the wheel. Tech B says that the drop center or deep well is used to prevent the tire from coming off the wheel in the case of low tire pressure. Who is correct?
   a. Tech A
   b. Tech B
   c. Both Tech A and B
   d. Neither Tech A nor B
2. Tech A says that all wheels must be torqued to prevent wheels from loosening up and falling off. Tech B says that all wheels must be torqued to prevent over-tightening, which can weaken lug studs and warp brake rotors. Who is correct?
   a. Tech A
   b. Tech B
   c. Both Tech A and B
   d. Neither Tech A nor B
3. Tech A says that tires are marked with a date code listing the date the tires should be discarded. Tech B says that any tires with a three-digit date code should be discarded because they are too old to be safely in service. Who is correct?
   a. Tech A
   b. Tech B
   c. Both Tech A and B
   d. Neither Tech A nor B
4. Tech A says that a TPMS system can save fuel over time. Tech B says that a TPMS will help prevent blowouts. Who is correct?
   a. Tech A
   b. Tech B
   c. Both Tech A and B
   d. Neither Tech A nor B
5. Tech A says that the center of gravity on a vehicle is located at the vehicle body's lowest point. Tech B says that having a low center of gravity makes a vehicle handle better. Who is correct?
   a. Tech A
   b. Tech B
   c. Both Tech A and B
   d. Neither Tech A nor B
6. Tech A says that oversteer is when the front slip angles are larger than the rear slip angles. Tech B says that understeer is when the rear slip angles are larger than the front slip angles. Who is correct?
   a. Tech A
   b. Tech B

c. Both Tech A and B
d. Neither Tech A nor B

7. Tech A says that the "P" in "P225/70, R15" means "the tire's inflation pressure in kPa." Tech B says that the "T" in "T135R16" means "temporary." Who is correct?
   a. Tech A
   b. Tech B
   c. Both Tech A and B
   d. Neither Tech A nor B
8. Tech A says that a direct TPMS system uses a pressure sensor located in each wheel. Tech B says that an indirect TPMS system uses wires to connect the pressure sensor located in each wheel to the PCM. Who is correct?
   a. Tech A
   b. Tech B
   c. Both Tech A and B
   d. Neither Tech A nor B
9. Tech A says that radial ply tires have much more flexible sidewalls than bias-ply tires because of their construction. Tech B says that bias-ply tires have a more durable construction than radial tires. Who is correct?
   a. Tech A
   b. Tech B
   c. Both Tech A and B
   d. Neither Tech A nor B
10. Tech A says that most lug nuts and studs are right-hand threaded, which means they tighten when turned clockwise. Tech B says that some lug nuts and studs are left-handed, which means they tighten when turned counterclockwise. Who is correct?
   a. Tech A
   b. Tech B
   c. Both Tech A and B
   d. Neither Tech A nor B

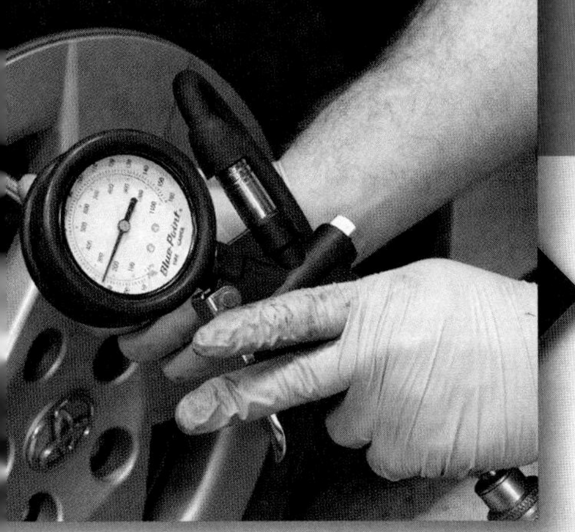

CHAPTER 37
Servicing Wheels and Tires

NATEF Tasks

- **N37001** Inspect tire condition; identify tire wear patterns; check for correct tire size and application (load and speed ratings), and adjust air pressure as listed on the tire information placard/label; determine needed action. (MLR/AST/MAST)
- **N37002** Rotate tires according to manufacturer's recommendation including vehicles equipped with tire pressure monitoring systems (TPMS). (MLR/AST/MAST)
- **N37003** Dismount, inspect, and remount tire on wheel; balance wheel and tire assembly. (MLR/AST/MAST)
- **N37004** Demonstrate knowledge of steps required to remove and replace sensors in a tire pressure monitoring system (TPMS) including relearn procedure. (MLR/AST/MAST)
- **N37005** Dismount, inspect, and remount tire on wheel equipped with tire pressure monitoring system sensor. (MLR/AST/MAST)
- **N37006** Diagnose wheel/tire vibration, shimmy, and noise; determine needed action. (AST/MAST)
- **N37007** Diagnose tire pull problems; determine needed action. (AST/MAST)
- **N37008** Inspect tire and wheel assembly for air loss; perform needed action. (MLR/AST/MAST)
- **N37009** Repair tire following vehicle manufacturer approved procedure. (MLR/AST/MAST)
- **N37010** Measure wheel, tire, axle flange, and hub runout; determine needed action. (AST/MAST)
- **N37011** Identify indirect and direct tire pressure monitoring system (TPMS); verify operation of instrument panel lamps. (MLR/AST/MAST)

Knowledge Objectives

After reading this chapter, you will be able to:

- **K37001** Describe the types of tire maintenance tasks and their purpose.
- **K37002** Describe the purpose and process of maintaining proper tire inflation.
- **K37003** Describe the purpose and process of rotating tires.
- **K37004** Describe the purpose and process of balancing wheels.
- **K37005** Describe the purpose and process of maintaining TPMS sensors.
- **K37006** Identify the tools used for servicing tires.
- **K37007** Describe some of the common issues with wheels and tires.

Skills Objectives

After reading this chapter, you will be able to:

- **S37001** Perform tire maintenance and repair procedures.
- **S37002** Perform wheel and tire diagnosis.

▶ Introduction

Paying attention to the wheels and tires can also assist in identifying problems with other vehicle systems, particularly in steering and suspension systems. The wear patterns on a tire correlate with wear or damage to particular vehicle components. Overlooking the wheel and tire assembly may result in missed faults, leading to larger issues and associated costs. It is critical that you be familiar with each part of the wheel and tire assembly and possess the skills necessary to identify and correct problems associated with them. Also, you need to understand the various tire markings and ratings so that you can properly service wheels and tires, as well as answer customer questions when they need repair or replacement.

▶ Tire Maintenance Preliminaries

K37001

A periodic inspection of the tires is necessary to ensure a long life (**FIGURE 37-1**). The tire pressure should be checked and adjusted on vehicles that do not have direct TPMS. The tread area is inspected for cuts, flat spotting, or irregular wear patterns. The tread depth is inspected to ensure that it is above the built-in wear indicators, which indicate that the tire is at the end of its legal life and should be replaced. The sidewalls are inspected for cuts or gashes and any signs of damage from impacting solid objects such as curbs. At the same time, vehicle and tire manufacturers recommend that the tires be rotated during each oil change. Tire balance is performed when the tires are new and any time there is a customer concern regarding vibrations that are diagnosed to be caused by tire balance faults or if tire wear indicates an out-of-balance condition.

FIGURE 37-1 Whenever a vehicle comes into the shop, check the tires' air pressure and look for signs of wear.

Wheel alignment is another service that is performed on occasion. The alignment is normally checked and adjusted in cases such as when there is abnormal tire wear, when steering/suspension parts are replaced, when the vehicle does not handle or drive correctly, and after an accident that could affect the alignment. So the goal of an alignment is to restore proper handling along with extending the life of the tires. Most tire shops recommend an alignment when new tires are purchased to ensure that the tires don't wear out prematurely. This is usually a fairly easy sell, especially if the customer came in to replace tires that wore out due to an alignment issue.

Proper Tire Inflation

K37002

Tire inflation pressure is the amount of air pressure in the tire that provides it with load-carrying capacity and affects the overall performance of the vehicle. Vehicle manufacturers determine the tire inflation pressure based on the vehicle's designed load limit, which is the greatest amount of weight a vehicle can safely carry, and the vehicle's tire size. The proper tire pressure for the vehicle is referred to as the recommended cold inflation pressure and is measured in psi or kPa. You will find this information on the tire placard, expressed in psi or kPa (**FIGURE 37-2**). The tire placard is typically located on the A- or B-pillar door frame, in the glove compartment, or on the fuel filler flap.

Because tires are designed to be used on more than one type of vehicle, tire manufacturers list the maximum safe inflation pressure on the tire sidewall. This number is the highest amount of air pressure that should ever be put in the tire. The maximum tire pressure should never be exceeded, even when the vehicle's tire placard indicates a higher pressure. As explained

You Are the Automotive Technician

A customer brings her 2010 Dodge Caravan to the dealership with the tire pressure light on. The customer tells you that the tires are filled with nitrogen, and on the drive to the dealership, the van pulled to the right. You bring the vehicle into the shop and notice that the right front tire looks low, and the pressure reads 22 psi. You inspect the tire and notice a small screw embedded near the center of the half-worn tread. You mark a big "X" across the tire (centered on the screw) to mark the hole. Air leaks out as you pull the screw out of the tire. You remove the wheel from the vehicle in preparation for repairing the tire.

1. How can you identify whether tires have been inflated with nitrogen?
2. What is the proper method of repairing a tire?
3. What are some conditions when a tire should not be repaired?
4. What happens if you repair a high-performance speed-rated tire?
5. What must be done once the tire is repaired and back on the vehicle?

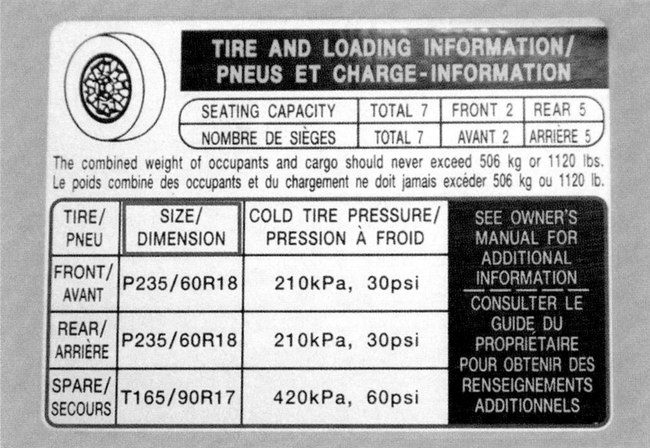

FIGURE 37-2 Typical tire placard.

earlier, 32 psi of air pressure can generate over 30,000 lb of force on a tire. If the maximum tire pressure is exceeded, this generates even greater forces and can cause the tire to explode violently. Too many young technicians have lost their life when they over-inflated a tire and it blew up in their face. This is one of the more dangerous situations for inexperienced technicians. It is also why it is good to use a tire cage when inflating a tire (**FIGURE 37-3**). If the tire blows up, the explosion will affect your hearing, and you might be hit by some debris, but the cage is likely to contain most of the larger chunks of the tire.

To get an accurate pressure reading, the tires must be checked when cold. The term "cold" does not relate to the outside temperature. Rather, a cold tire is one that has not been driven on for at least three hours or less than 1 mile (1.6 km). While driving, tires get warmer because of the heat of friction created as the tire materials flex against each other. As tires rotate and the sidewalls flex due to variations in the road surfaces, they generate heat as the layers of plies are forced to accommodate the sidewall flexing. Flexing results in heat. The heated tire materials, in turn, heat the air within the tire. As the air within the tire warms, it expands, causing the air pressure within the tire to increase. To obtain an accurate tire pressure reading, then, you must measure tire pressure when the tires are cold or compensate for the extra pressure in warm tires. Some manufacturers supply this information, but others do not.

> **TECHNICIAN TIP**
>
> According to the U.S. Department of Energy, under-inflated tires can lower gas mileage by 0.3% for every 1 psi drop in pressure in all four tires. So running them 10 psi low reduces gas mileage approximately 3%. Under-inflation wastes approximately 2 billion gallons of fuel each year in the United States alone.

Nitrogen Fill

Oxygen is harmful to rubber and other tire materials. The oxygen reacts with the rubber through oxidation and causes the rubber to lose its flexibility and sealing ability. This allows

FIGURE 37-3 After installing a new tire, it is good to inflate it in a tire cage.

oxygen to permeate the rubber and degrade it further over time. Also, as the inner liner oxidizes, more air molecules can pass through it, causing an increased rate of pressure loss. In addition, moisture in shop air causes metal to rust or corrode, which can clog the valve stems, causing them to leak. In recent years, some manufacturers have been filling their tires with pure or nearly pure nitrogen in hopes that the problems associated with oxygen-filled tires can be avoided (**FIGURE 37-4**). Nitrogen is an inert gas that does not react with the rubber compounds in the tire. Nitrogen generators also remove any moisture from the nitrogen gas, so corrosion effects are reduced on TPMS sensors and other metal components of the wheels and tires.

Advantages for Nitrogen Fills

Although both nitrogen and oxygen can permeate rubber, nitrogen does so at a much slower rate. It might take three months to lose 1 psi (14 kPa) with nitrogen, compared to just a month with normal air. In addition, nitrogen is far less reactive. It does not cause rust and corrosion on steel or aluminum, and it does

FIGURE 37-4 A nitrogen filling station.

not degrade rubber. Wheel surfaces stay smooth and clean, and rubber remains supple and resilient.

Because tires operate most efficiently when they are at the correct pressure, nitrogen filling can better maintain a tire's performance over time. It will also increase the vehicle's fuel efficiency and tire life by reducing the effects of pressure loss due to permeation. Proper tire inflation also helps prevent accidents by reducing the possibility of blowouts and maintaining maximum tire traction with the road surface.

The air that we breathe and that is typically used to inflate tires chemically consists of 78% nitrogen, 21% oxygen, and 1% other. When using pure nitrogen as an inflation gas, the composition of nitrogen increases from 78% to near 100%. To meet the standards for proper nitrogen fill means that the nitrogen level in the tire must be at 95% or higher. This typically requires two or three inflations and deflations to remove the oxygen. Nitrogen-filled tires can be identified by the green valve stem caps that are placed on the valve stem when the tire is inflated (**FIGURE 37-5**). Although tire manufacturers generally support the use of nitrogen, they do not generally mandate its use based on its ability to better retain a consistent tire pressure over a period of time.

FIGURE 37-5 Nitrogen-filled tires can be identified by the green valve stem caps.

Tire Rotation Pattern

K37003

Tire rotation service is the removal and relocation of each tire/wheel assembly on the vehicle. Each tire on the vehicle can wear differently. Regular tire rotation promotes uniform tread wear, extending the life of the set of tires. Uniform tread wear can also boost the vehicle's fuel economy and increase the vehicle's performance. Each manufacturer designates the proper rotation sequence, depending on whether the vehicle is front-wheel drive, rear-wheel drive, or all-wheel drive; the type of tires (e.g., directional); and whether a spare tire is involved in the rotation (**FIGURE 37-6**). In general, a four-tire rotation with nondirectional tires can have one of three rotation patterns: forward-cross, rearward-cross, and X pattern. Directional tires require keeping the tires on the same side of the vehicle and generally use a front-to-rear pattern. If the directional tires are differently sized from front to rear, then the tires will have to be dismounted and remounted on the wheels from the other side of the vehicle. If the vehicle uses a full-sized spare tire, then the pattern should be one of two 5-tire rotations, which are variations of the four-tire forward-cross or rearward-cross pattern. Generally, it is recommended to rotate the tires on vehicles approximately every 5000 miles (8000 km) or with every oil change.

Wheel Balance

K37004

Wheels and tires are mass-produced so many times that there are small (or sometimes big) imbalances manufactured into them. Also, they are not always perfectly round (concentric), but can still be sold as long as they are within the tolerances specified by the manufacturer. To complicate matters, tolerance stacking can happen. For example, the wheel hub may have an out-of-round tolerance of 0.005" (0.13 mm); the wheel may have a tolerance of 0.005" (0.13 mm); and the tire may have a tolerance of 0.025" (0.64 mm). If all of the tolerances stack up (one on top of another), then the wheel will experience balance and runout issues. Wheels and tires that are not balanced or that

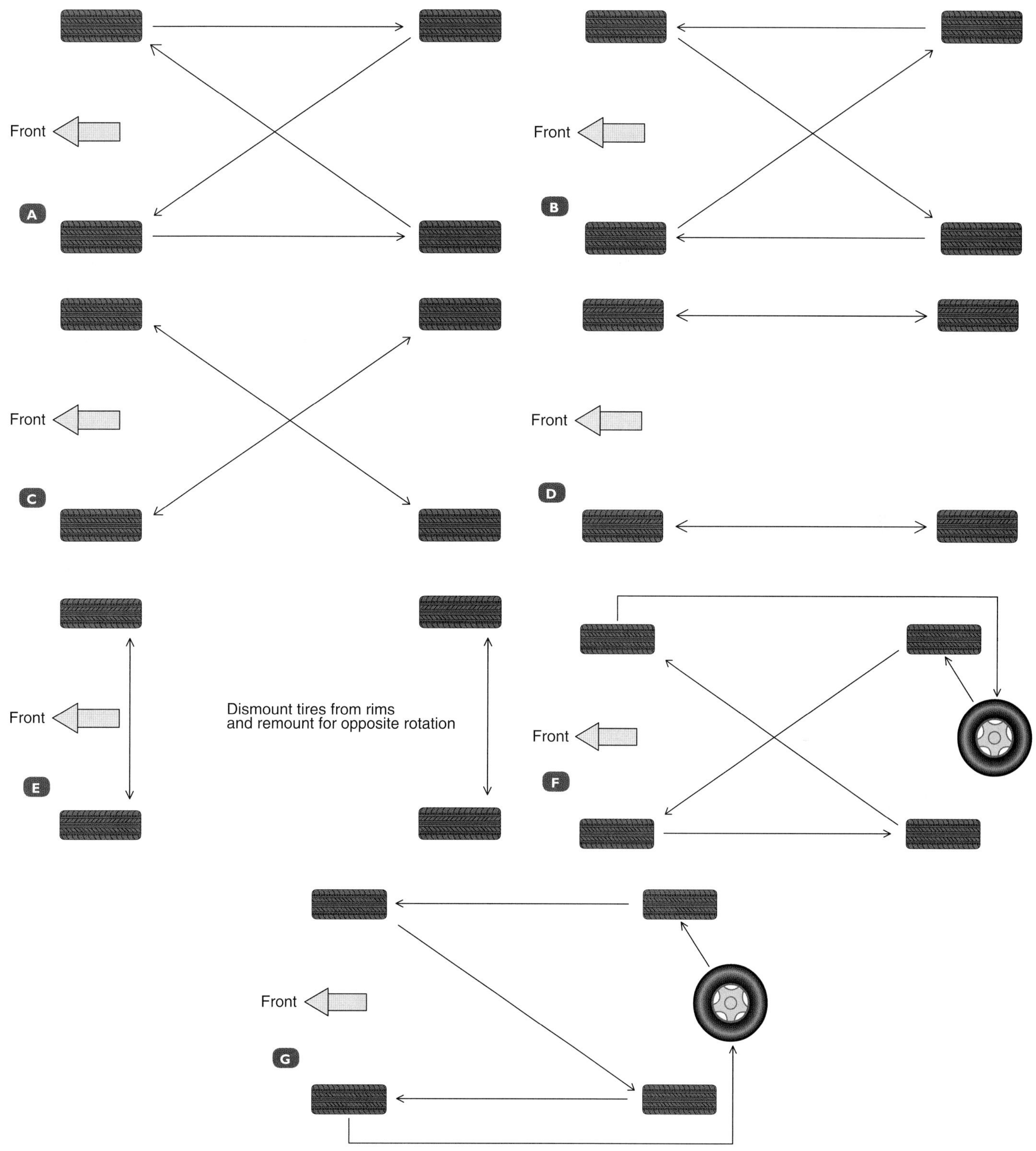

FIGURE 37-6 A. Forward-cross pattern. **B.** Rearward-cross pattern. **C.** X pattern. **D.** Front-to-rear pattern. **E.** Side-to-side pattern from the other side of the vehicle. **F.** Five-tire forward-cross pattern. **G.** Five-tire rearward-cross pattern.

are out of balance generally produce an uncomfortable vibration and result in premature wearing of suspension and steering components as well as uneven tire wear. An out-of-balance tire will cause the vehicle to vibrate at certain speeds, usually between 50 and 70 mph (80 to 113 kph). A tire is out of balance when one section of the tire is heavier or stiffer than the others. One ounce of imbalance on a front tire is enough to cause a noticeable vibration in the steering wheel at highway speeds.

There are two ways wheels can be balanced: off car and on car. Although most shops use off-car balancers, if the brake rotor, drum, or hub is out of balance, then a balanced wheel cannot compensate for that, so on-car balancers are used to balance the entire rotating assembly. The only problem with on-car balancing is that the wheel will need to be rebalanced when it is rotated to another position on the vehicle. This is one reason that most shops use off-car balancers (**FIGURE 37-7**).

Tires and wheels can be out of balance in three ways: static, dynamic, and road force. **Static imbalance** is when the imbalance is centered across the width of the tire (**FIGURE 37-8**). Static balancing does not take into consideration that the tire has width; thus, static balancing can address the imbalance only as it relates to how much weight is needed to counterbalance the imbalance X inches away from the center of the axle. Therefore, it can be effective only when dealing with an imbalance condition that is centered within the width of the tire. When performing a static balance, the tire's imbalance is measured when the tire is stationary. Static imbalance tends to cause the tire to move purely up and down. Static balancing is no longer an acceptable method of balancing tires on today's vehicles.

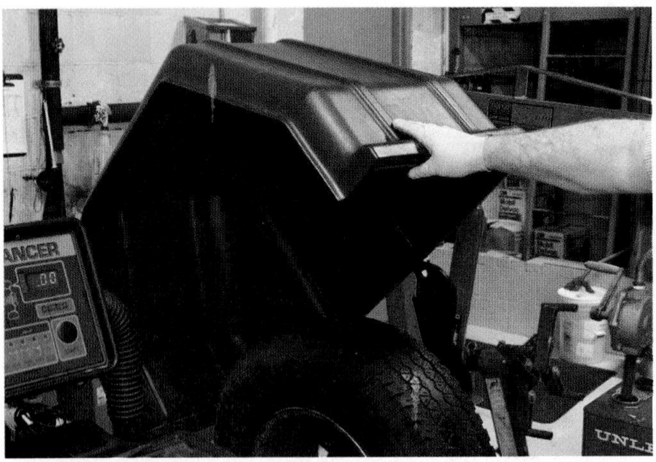

FIGURE 37-7 Off-car balancer.

Dynamic imbalance is when the imbalance is off center of the tire width (**FIGURE 37-9**). Dynamic balancing performed when the tire is rotating. It takes into consideration that the tire has width; therefore, the imbalance is not only a certain weight and distance from the centerline of the axle, but it can also be anywhere within the width of the tire. Dynamic imbalance can cause the tire to move side to side as well as up and down. Dynamic balancing is performed by placing specific amounts of weight on each side of the rim to provide the exact counterbalance needed. It can effectively account for an imbalance located anywhere within the volume of the tire/wheel assembly. Dynamic balancing is used in most shops today.

It is possible that the tire and wheel can be perfectly in balance statically and dynamically, yet the vehicle still experiences a shimmy or shake that feels just like an out-of-balance tire. This problem is caused by **road force imbalance**. Road force imbalance occurs when the wheel or tire is not concentric or when the tire's sidewall has uneven stiffness (**FIGURE 37-10**). As the tire rotates and contacts the surface of the road, any issues with concentricity or sidewall stiffness will push up and down on the vehicle and simulate a tire that is out of balance. Road force balancing a tire gives the best ride quality to owners as it takes into consideration all balance factors related to the wheel and tires.

When dynamically balancing a tire using an off-car balancer, the wheel and tire assembly is mounted on the balancer. It is then spun up to speed so that the precise location and amount of the imbalance can be identified. The balancer then shows the technician where and how much weight to add on each side of the wheel. The wheel is spun up again to ensure that the weights have been affixed in the correct location and there is no more imbalance. If a road force balancer is used, then the tire is run up against a roller, and any uneven forces are measured. If an uneven force is encountered, the machine will usually direct you to rotate the tire on the wheel and spin it again to see if there are any concentricity issues with the wheel. It can also tell you where the best location is for the tire on the rim, to reduce road

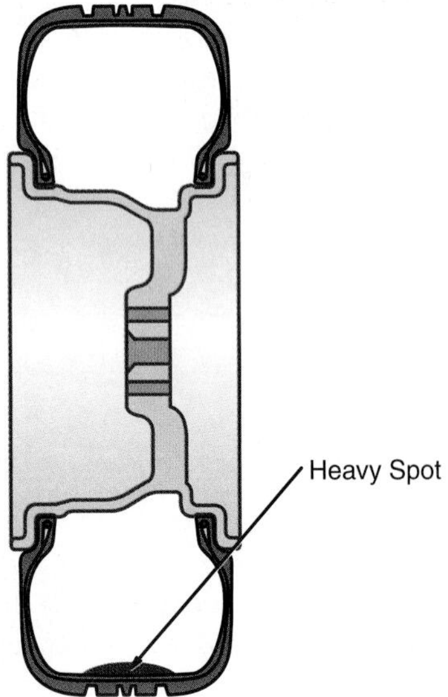

FIGURE 37-8 Static imbalance—the imbalance is centered across the width of the tire.

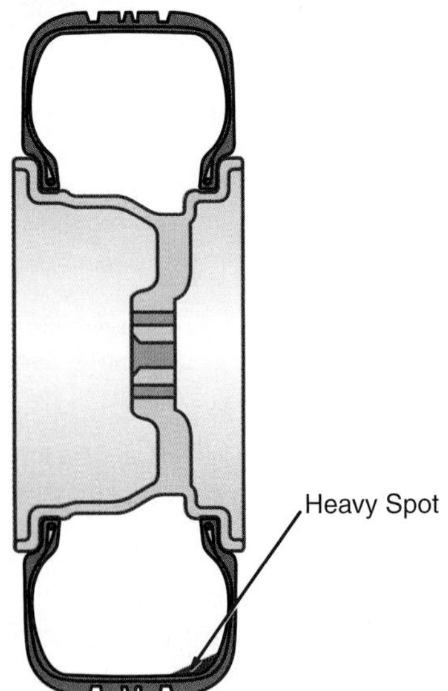

FIGURE 37-9 Dynamic imbalance—the imbalance is off-center of the tire's width.

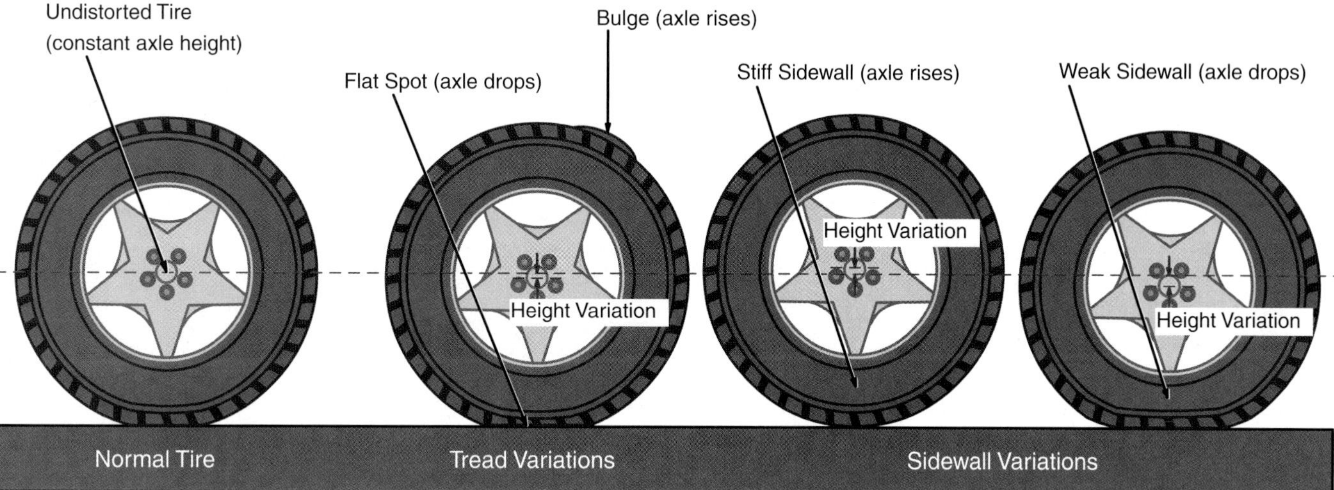

FIGURE 37-10 Road force imbalance—when the sidewall is either stiffer, or taller, in one place.

force fluctuations. After this process is completed, the wheel is reinstalled on the vehicle, and the wheel nuts are torqued to the manufacturer's specification.

> ▶ TECHNICIAN TIP
>
> Not all shimmies and shakes are caused by tire balance issues. Sometimes belts inside the tire carcass break, creating a lump in the tire. Every time the lump hits the road, the tire pushes back. This really is just an extreme case of road force imbalance.

The majority of top-quality tires hold their balance reasonably well as long as the vehicle is not driven over potholes. If the driver should notice a vibration that was not there the day before, it is possible that one of the balancing weights fell off. If the driver feels the vibration mostly in the steering wheel, the problem is most likely in a front wheel. If the vibration is mostly in the seat, the problem is probably in one of the rear tires.

Applied Science

AS-47: Force, Balance/Unbalanced: The technician can demonstrate an understanding of the role of balanced and unbalanced forces on linear or rotating vehicle assemblies.

When a tire is fitted to a wheel, two imperfectly weighted components are assembled together to form a heavy, large-diameter rotational component. The chances of this assembly having perfect weight distribution are extremely small; hence the importance of adding weight in specific areas to balance the assembly.

A car wheel rotates on a central axis, supported by bearings. When a wheel turning at high speed is out of balance, either statically or dynamically, forces are created perpendicular to the rotational axis. These forces are typically experienced as a vibration through the steering wheel if the front wheels are out of balance, although in extreme cases a vibration may also be felt through the vehicle's body if the rear wheels are out of balance. When a wheel is in perfect balance, no forces exist to interrupt the smooth rotation around the central axis.

Wheel weights come in a variety of styles to fit different rim configurations, so you will need to verify the type of weight needed for a particular wheel (**FIGURE 37-11**). Weights come in 0.25-ounce or 5-gram increments. Although wheel weights have been primarily made of lead for decades, several states have outlawed them because of the potential environmental hazards of lead. In those states, steel weights coated with zinc or another protective layer are being used instead of lead weights. It is good practice to always use new wheel weights when balancing a tire. Do not reuse the old wheel weights, as they are likely to be thrown from the wheel when driven at freeway speeds.

TPMS Maintenance

K37005

TPMS sensors need to be treated carefully when tires are being removed and reinstalled. Sometimes normal corrosion can damage the TPMS sensors, but more likely, a careless technician can damage the TPMS sensors by not removing the tire properly. If a sensor is damaged during a tire change, then the entire TPMS sensor unit will generally need to be replaced.

The TPMS sensor unit is a sealed component. Some models are powered by a small, sealed, disposable battery. These batteries are designed to last for 5 to 10 years without replacement. When the battery does expire, the sensor will have to be

FIGURE 37-11 Wheel weights come in a variety of styles, materials, and coatings. So make sure you use the proper one for the wheel you are working on.

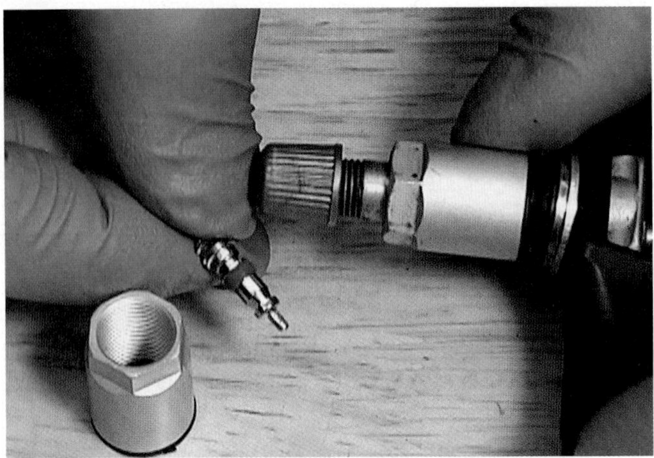

FIGURE 37-12 A nickel-plated valve core.

FIGURE 37-13 A technician checking tire pressure during a routine inspection.

removed from the rim and checked. In most instances, the sensor must be replaced as a unit. The cost of replacing a battery compared to the sensor may not be economically viable, and new TMPS sensors may have to be installed.

Many TPMS sensors use nickel-plated valve cores in an aluminum valve stem (**FIGURE 37-12**). Never use brass valve cores or unplated brass caps; doing so can cause galvanic corrosion between the two different metals. Many TPMS sensors must be torqued into place. This torque is a very small amount measured in inch-pounds. Be careful, as they are easily stripped or broken. Always use new sealing washers with TPMS sensors when removing them from a rim.

▶ Maintenance and Repair

S37001

Tires are one of the most maintenance-intensive parts of a vehicle. They require regular visual inspections, pressure inspections, and rotations. Visual inspections of the tires and wheels should be made whenever a vehicle comes into a shop for any work (**FIGURE 37-13**). It is common to check for any unusual wear patterns and for any embedded objects or other signs of damage. If the vehicle is not equipped with direct TPMS, then the operator should be encouraged to check the pressure monthly. Another task is rotating the tires. Most manufacturers recommend that the tires be rotated at each oil change, to even out the wear and extend the life of the tires. Research the vehicle and service information for more information.

Tools

K37006

To service tires in an efficient manner, you must have the appropriate tools and equipment. The basic equipment list should consist of (**FIGURE 37-14**):

- Tire pressure gauge—used to check the air pressure in tires.
- Tread depth gauge—used to measure the tire's tread depth.
- Valve stem tool—used to remove and install tubeless valve stems in rims.
- Valve core tool—used to remove and install valve cores in valve stems.
- TPMS torque wrench—used to tighten TPMS sensor to the valve stem.
- Tire-changing machine capable of handling the range of tires that the shop stocks—in a shop that handles run-flat tires, the tire-changing machine must be designed to handle the more robust bead and sidewalls.
- Tire dunk tank—used to locate leaks in tires and rims.
- Tire spreader—used to spread the sidewalls of a tire for easier access during tire patching.
- Air tire buffer—used to lightly buff the inside surface of the tire as preparation for tire patching.
- Patch stitching tool—used to apply pressure to the patch when positioning it.
- Tire inflation cage—used to contain the tire and rim during tire inflation in the event of a tire explosion.
- Wheel balancing machine capable of handling the range of tires that the shop stocks.
- Wheel weight hammer—used to remove and install wheel weights onto rims.
- TPMS reset tool—used to reset the TPMS (not required on all TPMS-equipped vehicles).

Common Issues

K37007

Common tire and wheel issues include:

- *Air loss:* The most common issue with tires is air loss. Tires normally lose a small amount of air over time and require periodic refilling. Punctures occur that cause leaks of various sizes. Valve stems and tire beads can also allow air to leak from the tire.
- *An out-of-balance tire or wheel:* The wheel and tires must be balanced, with the weight equally distributed throughout. When a tire rotates, any points of unequal weight will cause the tire to wobble, placing stress on the shocks, bearings, and wheel assembly. Additionally, the vehicle will vibrate

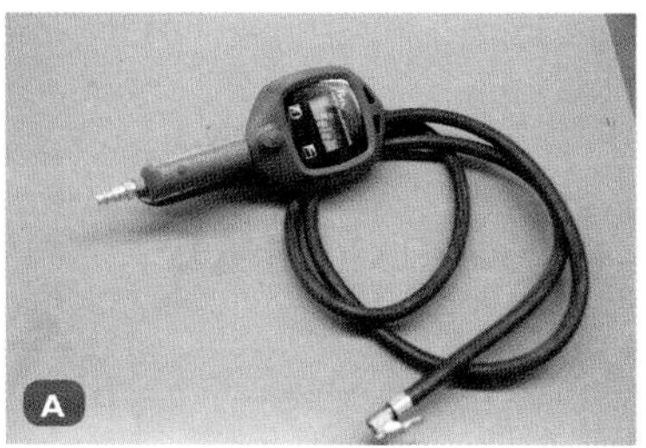

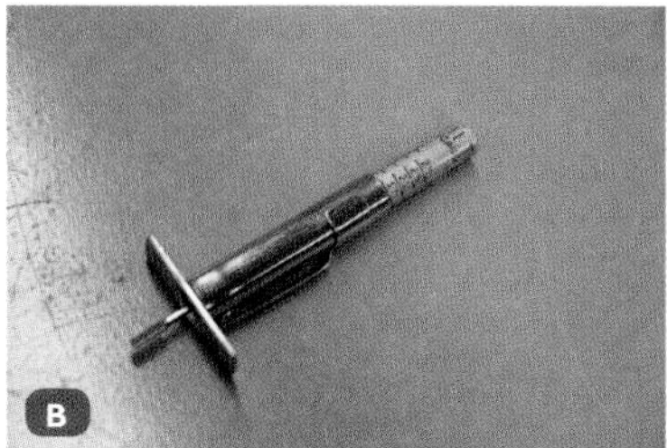

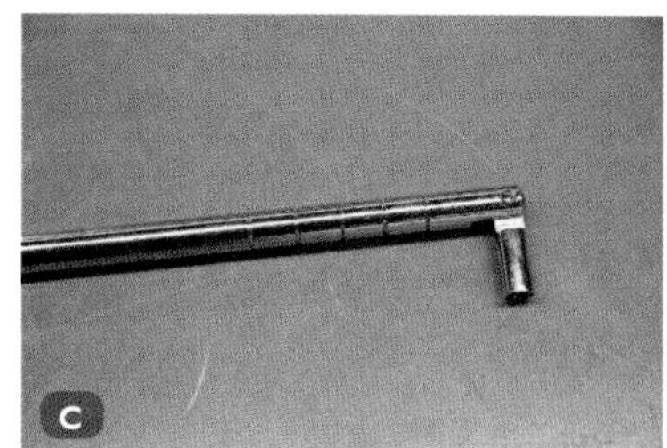

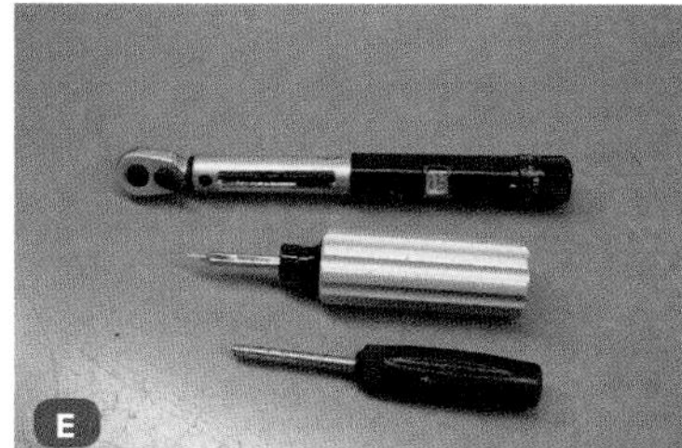

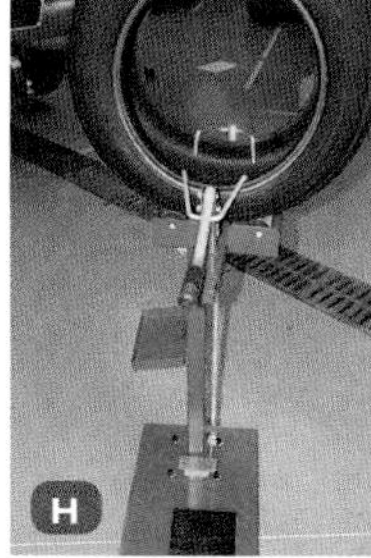

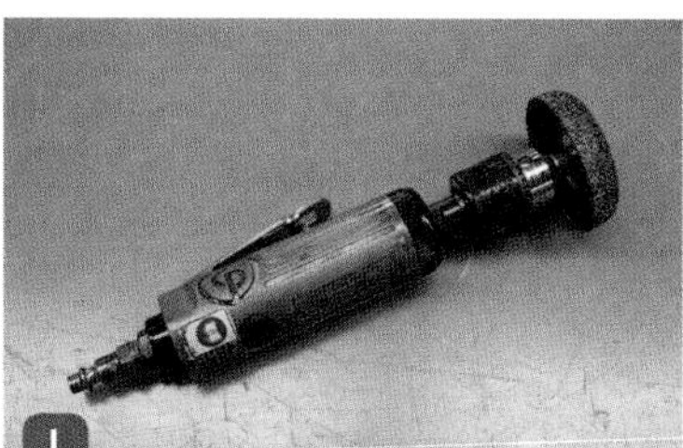

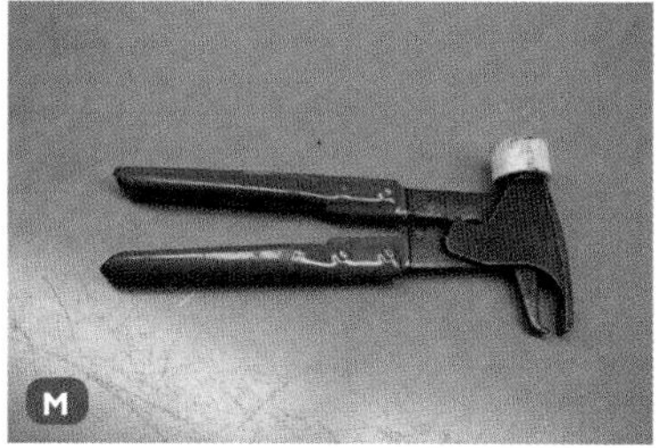

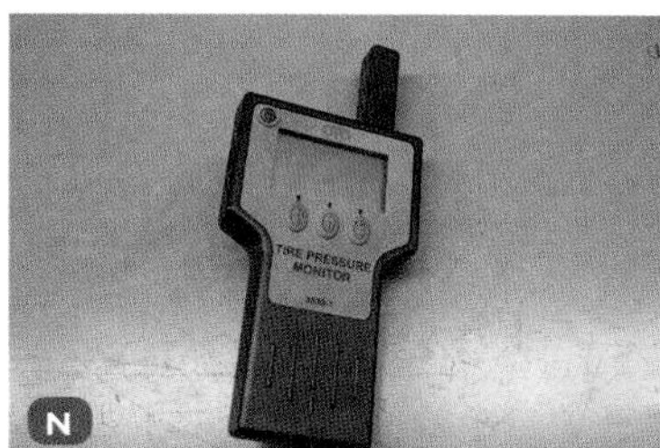

FIGURE 37-14 Typical tire repair tools and equipment.

at speeds of about 35 mph (56 kph) and above, it will have a rough ride, and the steering wheel may vibrate. A wheel balancer is used to identify and correct wheel imbalances.

- *Excessive loaded radial runout on the tire, wheel, and hub assembly:* With radial runout, the tire tread moves up and down. It is caused by incorrect manufacture or by damage to the tire, such as a broken belt. Correction involves replacing the tire.
- *Excessive lateral runout on the tire, wheel, and hub assembly:* Runout occurs when a part of the wheel assembly becomes bent or was manufactured improperly. The result is a wobble. When lateral runout occurs, the only method for fixing it is to replace the bent or improperly manufactured component.
- *Wheel trim imbalance (if fitted):* The term "wheel trim" refers to any pieces attached to a wheel that are not necessary for actual wheel function, such as a hubcap or trim ring. Imbalance or wear in trim pieces is corrected by removing or replacing the part.
- Heavy pulling of a vehicle to either the left or the right while the customer is driving is an indication of these possible problems:
  - Mismatched tire sizes or pressures
  - Tire with broken or misaligned belts
  - Out-of-alignment wheels
  - Worn suspension or steering components
  - A dragging front brake assembly

Using a Tire Pressure Gauge

N37001

There are two main types of **tire pressure gauges**: fixed workshop gauges and portable pocket-size gauges (**FIGURE 37-15**). The three most popular types of pocket tire pressure gauges are the pencil type, the dial type, and the digital type. The pencil type looks similar to a pencil and contains a graduated sliding extension that is forced out of the sleeve by air pressure when it is attached to the tire valve. The dial type has a similar chuck to the pencil type but includes a graduated gauge and needle. The digital type can look like any of the others, but it gives a digital reading of the pressure and is generally the most accurate. Some digital pressure gauges can also read the temperature.

Each tire pressure gauge measures pressures in pounds per square inch (psi), kilopascals (kPa), or **bars**.

- One bar is equivalent to 14.5 psi or 100 kPa.
- One psi is equivalent to approximately 7 kPa. Some tire pressure gauges have scales for more than one unit of measurement.

The tire pressure will vary from vehicle to vehicle, and also depends on its use and driver preference. Recommended tire pressures are located on the vehicle's tire placard, usually located on the driver door pillar. The maximum tire pressure is stamped on the tire sidewall. Never inflate the tire above the maximum pressure listed on the sidewall. The tire may explode, or the wheel rim may give way and cause a blowout, which can easily be fatal.

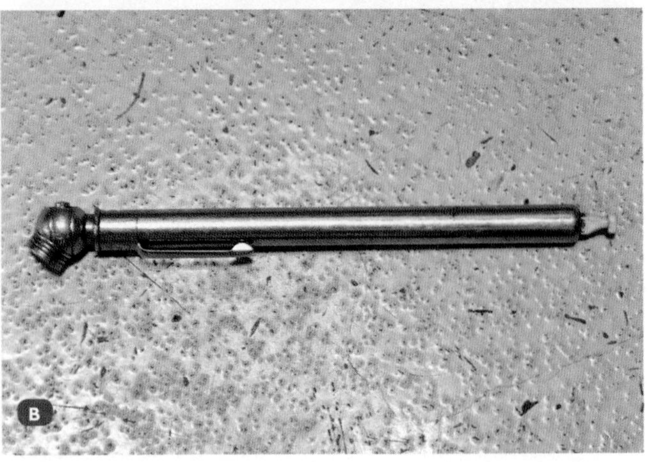

FIGURE 37-15 Tire pressure gauges. **A.** Fixed workshop gauge. **B.** Portable pocket-size gauge.

> **TECHNICIAN TIP**
>
> A common mistake when inflating tires is to use the pressure stamped on the sidewall instead of the one listed on the tire placard. The pressure branded on the sidewall is the maximum pressure that the tire is designed to withstand and should *never* be exceeded, even when the placard lists a higher pressure! So you should always first check the placard and then verify that it is not higher than the maximum pressure stamped on the sidewall. If it is, someone installed underrated tires on the vehicle.

> **TECHNICIAN TIP**
>
> Pocket-type tire pressure gauges are inexpensive and generally more accurate than the gauges provided by service stations. Service station gauges are often damaged by weather, misuse, or being run over. There may also be a significant difference in readings between the tire pressure gauge of one service station and the gauge from another service station. If the same pocket-type tire pressure gauge is always used to check tire pressures, then there will be no variation of readings.

To use a tire pressure gauge, follow the steps in **SKILL DRILL 37-1**.

> **TECHNICIAN TIP**
>
> If you check the tire pressures after the vehicle has been driven and the tires are warm or hot, *do not* release this excess pressure. If you bleed the tire pressure down to the manufacturer's recommendation, it will be under-inflated when the tire is cold or at normal operating temperature. This could cause premature wear on the tires and handling issues with the vehicle. Most tire manufacturers recommend checking tire pressures before the vehicle has been driven more than 1 mile (1.6 km).

Adjusting Tire Pressure

Even brand-new high-quality tires lose air over time, so tire pressure should be adjusted periodically. It is good practice to check tire pressure at least monthly to catch a leaking tire. Vehicles that are 2008 model year and newer are required to be equipped with a TPMS, designed to monitor the pressures in each individual tire and alert the driver of a problem if a tire pressure is outside of the specified limits. These vehicles may have a specific tire inflation and TPMS reset procedure required by the manufacturer, so make sure you investigate that before inflating the tire.

Many tires are filled with nitrogen instead of regular air to reduce the amount of air that is lost over time as well as reduce oxidation of the rubber on the inside of the tire. It is best to fill these tires only with nitrogen to avoid introducing oxygen into the tire, but regular compressed air can be used in an emergency. Nitrogen-filled tires can usually be identified by a green

SKILL DRILL 37-1 Using a Tire Pressure Gauge

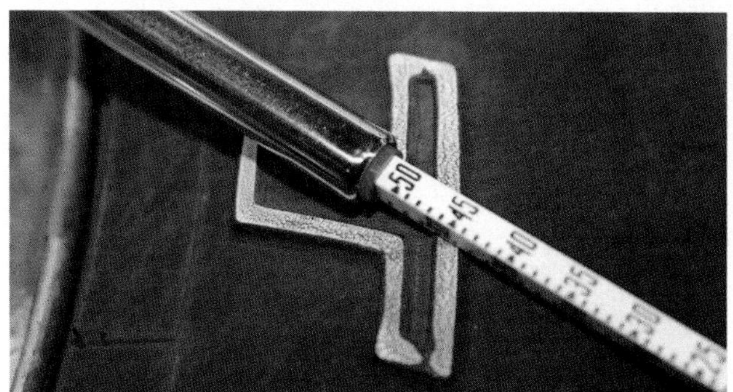

1. Remove the valve cap from the tire valve. Fit the pencil gauge to the valve. Make sure the graduated sleeve is seated into the gauge body, and then push the tire gauge chuck firmly onto the head of the valve. Read the scale and add up the numbers.

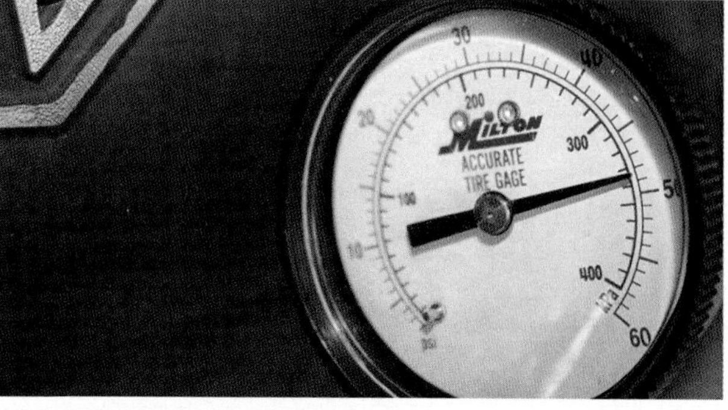

2. Attach the dial pressure gauge to the top of the valve. Adjust your hand pressure and angle so that no air escapes from the valve. When the needle has jumped, remove the dial pressure gauge from the valve, and read the dial.

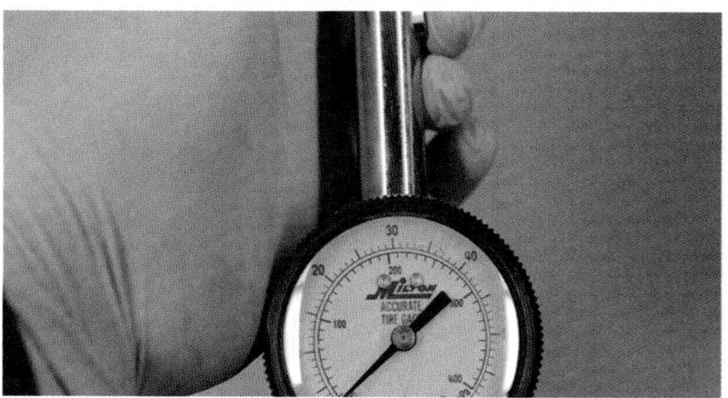

3. Reset the dial pressure gauge to zero by pressing the button on the neck of the dial.

4. Repeat the procedure for all wheels. Remember to replace the valve cap on each wheel as you go.

cap on the valve stem. Nitrogen-filled tires are best topped off at the manufacturer's dealership or a tire store.

When checking or adjusting tire pressure, check the recommended tire size and pressures on the tire placard, usually located on the driver's side door pillar or surrounding location. Compare that to the tire size and maximum load-carrying capacity as stamped on each tire, as customers sometimes install mismatched tires. If the tire markings do not meet the specifications of the vehicle, ask your supervisor for direction. Never allow a vehicle to be operated with tires that do not meet the vehicle manufacturer's specifications. Doing so can have serious safety repercussions. Also check the sidewall markings on the tire for the maximum operating pressure to make sure it can be inflated to the pressure listed on the tire placard. Always check the pressure when the tires are cold. Remove the cap from the valve stem on the first tire. Use a reliable tire gauge to check the air pressure in the tire. A pocket-type pencil gauge is ideal for this purpose. A gauge attached to the tire inflator is less likely to be accurate as it is more vulnerable to damage.

To adjust the tire pressure using a tire pressure gauge, follow the steps in **SKILL DRILL 37-2**.

Checking for Tire Wear Patterns

Tires come in a wide variety of tread patterns. Patterns differ based on the manufacturer and the tire's intended purpose. For example, tires designed to wick water away from the road surface will have deep grooves angled back toward the side of the road. Racing tires will have no grooves, as the smooth tire surface grips dry road surfaces tightly, increasing friction. Regardless of the exact pattern, all four tires should be inspected regularly to ensure they are wearing evenly. Irregular wear patterns are indicative of a problem. Common irregular wear patterns encountered include feathering, one-sided wear, cupping, center wear, and edge wear.

Feathering is observed as a rib with a slightly rounded edge on one side and a sharp edge on the other. This condition can be difficult to identify visually, so the technician should run a hand across the tire in both directions, feeling for the sharp edges. Feathering is most commonly a result of the tires set with excessive toe-in or toe-out. If the tire's sharp edge is toward the outside of the treads, then the tire is toed out. If there are sharp edges toward the inside of the treads, then the tire is toed in. One-sided wear refers to ribs on one side of the tire wearing out faster than those on the other side. This type of wear indicates that the wheels are not properly aligned. Cupping is the appearance of dips around the edge of the tread, usually on just one side of the tire. It occurs when one or more suspension parts are worn or bent. Center wear is when the ribs in the middle of the tire wear faster than those on each side. It results from driving

SKILL DRILL 37-2 Adjusting Tire Pressure Using a Tire Pressure Gauge

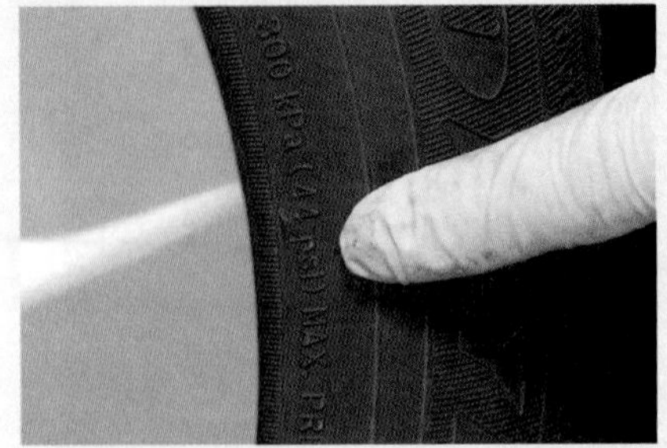

1. Park the vehicle so you can reach all four tires with the air hose. Check the tire placard and tire sidewall markings. Make sure the tires meet the pressure and load carrying specifications.

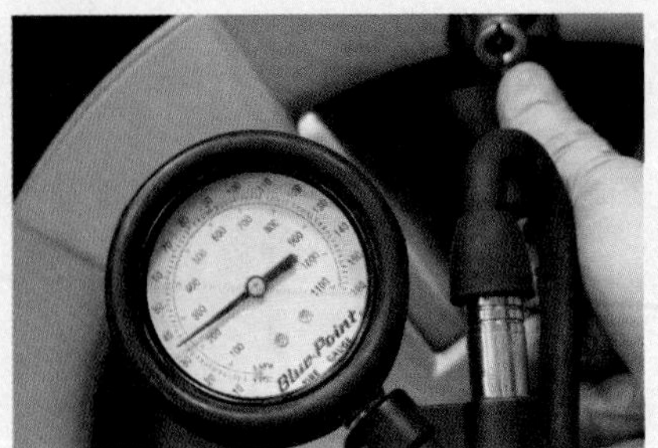

2. Remove the cap from the tire valve on the first tire. Use a reliable tire pressure gauge to check the air pressure in the tire.

3. If you need to add air, use short bursts with the air hose so you do not over-inflate the tire. Recheck the tire pressure after filling it, and replace the cap on the tire valve. Repeat the process for the other tires.

on over-inflated tires. Edge wear occurs when the ribs on the outer edges of the tire wear out faster than those in the middle of the tire (the reverse of center wear). It indicates that the vehicle has been driven with the tires consistently under-inflated, or the driver regularly corners the vehicle at excessive speeds. Most tires have wear indicator bars incorporated into the tread pattern. Inspect the wear indicator bars. Tires should have at least 1/16" (2 mm) of tread remaining. The wear indicator bars are normally set at this depth. If the tread is worn down to that level or below, the tires are unserviceable and must be replaced.

To check for tire wear patterns, follow the steps in **SKILL DRILL 37-3**.

SKILL DRILL 37-3 Checking for Tire Wear Patterns

1. Inspect the tires for embedded objects in treads and remove them. If anything penetrates the tread, mark the hole with a tire crayon.

2. Look for signs of wear on all tires, including the spare. Check the air pressure in the tires (see Skill Drill 37-2).

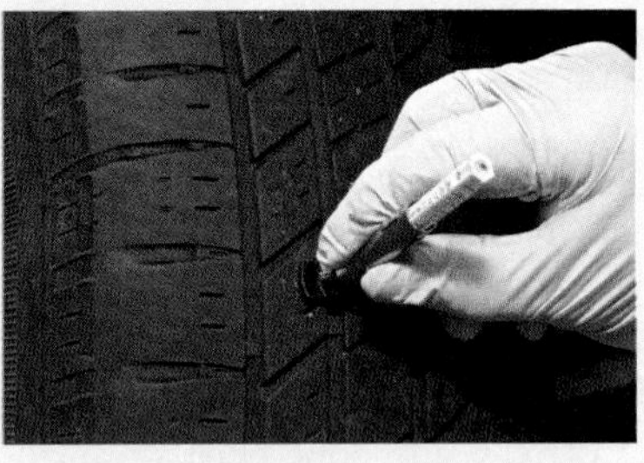

3. Check the tread wear depth. Inspect the wear indicator bars. Tires should have at least 1/16" (2 mm) of tread remaining. If the tread is worn down to that level or below, the tires are unserviceable and must be replaced.

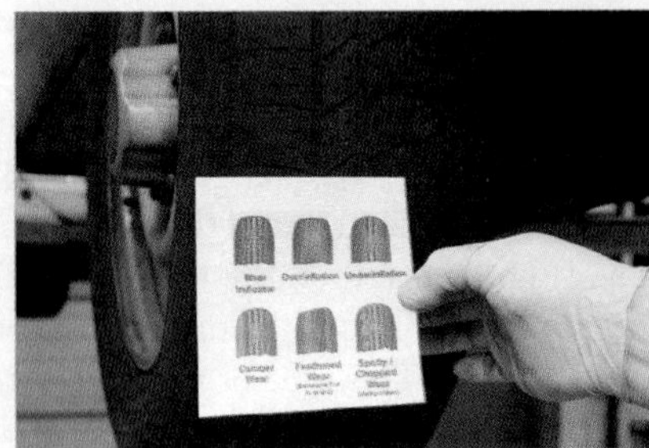

4. Check the tread wear patterns with the vehicle's service information to indicate the types of wear that have occurred.

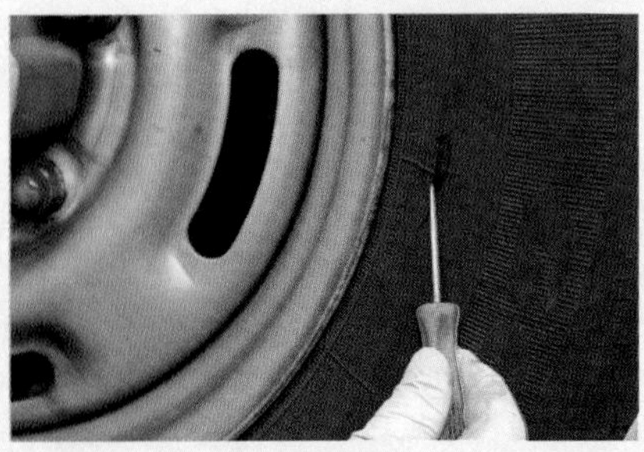

5. Inspect the sidewalls of the tires for signs of weather cracking and gouges from impacts with blunt objects. Carefully examine the tread area for separation. This is usually identified as bubble sunder the tread area.

6. Spin the wheel and see if it is running true. If it is wobbling as it rotates, report it to your supervisor.

Rotating the Tires

N37002

Rotating the tires to new positions on the vehicle helps to even out the tire wear, which extends their useful life. Manufacturers recommend certain tire rotation patterns for their vehicles, so always research this information for the vehicle you are working on. Many manufacturers recommend tire rotations at every oil change. This means that the tires generally should be rotated at intervals of approximately 5000 to 10,000 miles (8000 to 16,000 km). When the tires are removed, it is a good time to measure brake lining thickness and look for any leaks or damage to the brake assembly.

When reinstalling the lug nuts, thread them on by hand at least two full turns. Make sure the correct side of the lug nut is facing the wheel; this is usually the tapered side that matches the taper in the wheel. Do not put the nut or stud into the socket of an impact wrench and power them on directly. This practice can lead to the lug nuts or studs going on cross-threaded. Once the lug nuts are started onto the studs, an impact wrench can be used to run the lug nuts down lightly. Do not tighten them with the impact wrench.

Tighten the lug nuts to the correct torque in the proper tightening sequence, as specified in the service information. Typically, the lug nuts are tightened in a crisscross sequence on four stud wheels and in a star arrangement for five stud wheels until all of the nuts are torqued, but there are exceptions, so refer to the manufacturer's recommendations. Normally, the lug nuts should be tightened to 50% of the recommended torque for the first stage and then to 100% of the specification at the second stage, using the same tightening sequence for the particular wheel.

> **TECHNICIAN TIP**
>
> Lug nuts and lug bolts are designed with a specific grade (i.e., strength or amount of holding force) indicating a certain amount of stretch. Why do they stretch? Through proper torque, which stretches the bolt, stretching is what allows the threads of the stud or bolt to tightly mate to prevent them from working loose. Also, the torque listed for lug nuts is a dry torque, meaning that no lubricant should be used. Using a lubricant will cause the lug nuts to be over-tightened, which will likely result in failure.

As with all types of wheels, retorquing lug nuts per the manufacturer's specifications is typically recommended between 25 and 100 miles (40 to 160 km) after the initial tightening. Always refer to the owner's manual for proper factory specifications, which take precedence over any listed recommendations.

To rotate the tires, follow the steps in **SKILL DRILL 37-4**.

SKILL DRILL 37-4 Rotating the Tires

1. Prepare the vehicle by removing any hubcaps or lug nut covers. If using hand tools, break loose the lug or wheel nuts while some of the weight is still on the ground, and then raise the vehicle to a comfortable working position.

2. Remove the lug nuts and place them in a convenient place such as the arm of the hoist.

3. Remove the tire and wheel from the vehicle. Rotate the tires to the new specified position.

4. Reinstall the lug nuts by hand, at least two full turns, making sure the correct side of the lug nut is facing the wheel. Do not put the nut or stud into the socket of an impact wrench and power them on directly. Tighten the lug nuts to the correct torque in the proper tightening sequence, as specified in the service information.

▶ TECHNICIAN TIP

The tire pressures should be checked when the tires are cold (~70°F [21°C]). On average, the pressure in a tire will increase or decrease by about 1 psi for each 10°F (or 12.5 kPa for each 2°C) the tire is above or below its normal operating temperature.

Dismounting a Tire

N37003

Tires generally have to be removed from a rim for only a few reasons: replacing old tires with new tires, patching a leaky tire, and possibly switching between snow tires and regular tires. Because the tire is held in place by the air pressure forcing the bead of the tire into the wheel flange, repeatedly removing a tire from its wheel risks damaging the sealing surface of the bead. So it is best to remove the tire only when absolutely necessary.

Dismounting a tire is usually performed on a tire machine, which is very powerful, so use extreme care and closely follow the manufacturer's procedure. Tire machines are strong enough to break the bead loose from the rim as well as hold the rim while the bead is forced over the flange during removal and installation. This means the tire machine has several operations that you must become familiar with. Because different machines have different operating parameters, make sure you understand the manufacturer's specified operating procedure.

The turntable jaws on the tire changer can hold the rim by grasping it from the outside or the inside. When mounting an alloy rim, it is normal that the rim be clamped from the outside, whereas steel rims are clamped from the inside. Always check the instruction manual for the tire changer you are using for the correct method of clamping a rim.

To dismount a tire, follow the steps in **SKILL DRILL 37-5**.

SKILL DRILL 37-5 Dismounting a Tire

1. Before removing the tire, check to see if the wheel is equipped with a TPMS sensor. If it is, follow the manufacturer's procedures. Inspect the tread and sidewalls for any sharp cords sticking out that could injure you. If there is any damage, the tire should be discarded.

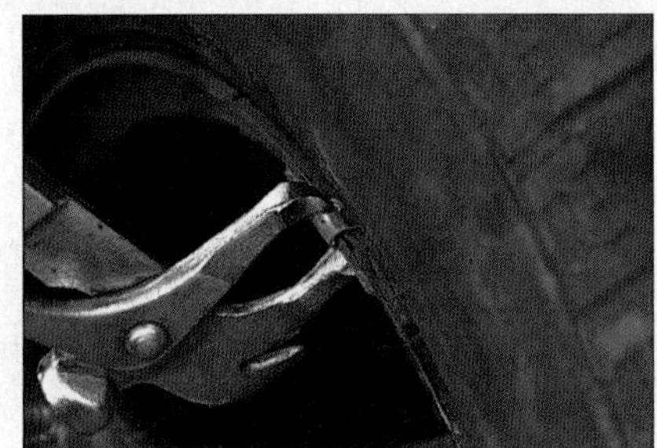

2. Check the wheel for any balance weights, and pry them off with the wheel weight tool.

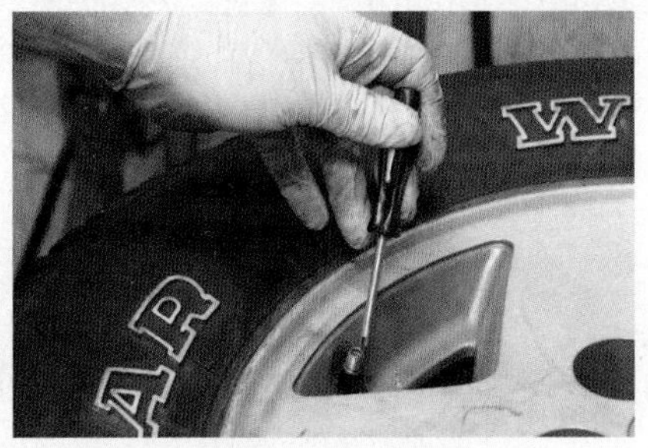

3. Locate the valve stem, unscrew the dust cap, and store it for later use. Using the valve core tool, unscrew the valve core carefully.

4. Once all the air has been removed from the tire, locate the wheel in the bead breaker, with the outside of the rim facing toward the blade. Locate the blade close to the edge of the rim while keeping your hands at a safe distance. Activate the bead breaker, which will force the tire bead away from the edge of the rim and over the safety ridge.

5. Release the blade, turn the wheel one-third to one-half of a turn, reposition the blade, and release this section of the tire as well. Release the blade. Roll the tire away from the machine, and reposition it with the inside of the rim facing toward the blade. Repeat the bead-breaking procedure on all four tires.

6. Set the wheel on the turntable with the shallow dish side of the wheel facing up. Position the wheel and tire assembly on the turntable, and activate the jaws so the wheel is centered in the jaws. Activate the turntable to verify that the wheel is centered and securely held.

SKILL DRILL 37-5 Dismounting a Tire (Continued)

7. Lubricate the top bead with tire lubricant. Make sure you get the lube on the flat bead seat, not just the side of the bead.

8. Position the bead remover against the edge of the rim; if necessary, adjust it so it has the proper clearance between the rim and the roller.

9. Use the tire lever to pry the tire bead over the bead remover knuckle while pushing down on the sidewall on the opposite side of the tire.

10. Activate the turntable so the bead is guided off the rim. Once the bead is removed, stop the turntable, lift the tire slightly, and remove the tube, if fitted. Guide the lower bead into the drop center, and using the tire lever, pry the lower bead over the knuckle and activate the turntable. The tire will come off the rim. If replacing the tire with a new one, remove the valve stem by either unscrewing it or using a valve stem tool, and discard the valve stem.

SAFETY TIP

For your safety, follow these tips:

- An inflated tire is a pressure vessel that must be treated with care and respect. Always fully deflate the tire before performing any repair tasks.
- Keep your hands clear of the bead breaker when it is operating. It applies great force to the tire that will cause you a severe injury if your hand is trapped.
- Always use correct lifting techniques when lifting a tire on and off the tire changer and vehicle.
- The tire tread comes in contact with many unknown substances that are transferred directly to your hands. For this reason, it is recommended that you wear protective gloves when handling tires. If you do not wear gloves, wash your hands after the tire change is completed and before eating.

Replacing a Valve Stem

Valve stems come in a few styles: rubber press-fit valve stems, screw-in valve stems, and TPMS sensor–integrated valve stems. Rubber press-fit stems are normally replaced when new tires are mounted on the rims. Screw-in valve stems normally are not replaced, but may have rubber washers or O-rings that may require replacement when new tires are installed. TPMS-integrated valve stems may be made as part of the valve stem or may screw onto, or snap into, the valve stem, which is replaceable. It is important to know whether the TPMS sensor is connected to the valve stem, so you do not damage it during service and repair. Many TPMS valve stems have valve cores with coated threads and should not be replaced with noncoated threads. They should always be torqued with a valve core torque wrench.

Valve stems can also be damaged by trauma from loose objects on the road, from scraping into curbs or other low structures, or through improper tire changes. Luckily, valve stems are generally inexpensive and relatively easy to replace. To change the valve stem, the wheel must be removed from the vehicle. The tire is then deflated and the top bead broken. On some tires, the sidewall can be pushed down by hand to gain access to the valve stem. In other instances, the tire must be removed from the rim.

After removing the old rubber valve stem, a new one is put into place, with some lubricant around the base of the stem. The stem is then pulled into place using the proper installing tool. The tire is reassembled and inflated to the recommended pressure. Spray some soapy water around the valve stem and core to detect air leaks.

For the valve stems that have a threaded shank with a locking nut and sealing washers, the process is similar. The wheel is removed from the vehicle and deflated, and the top bead is broken. The nut on the valve stem is then unscrewed, and the stem is removed from the inside of the rim. The old sealing washers can be replaced with new ones. Then the valve stem can be reinstalled in the rim. In some cases, only the valve stem core requires replacement. In this case, the old core is screwed out, the new core is installed, and the tire pressure is re-adjusted to vehicle specifications.

To replace a rubber press-fit valve stem, follow the steps in **SKILL DRILL 37-6**.

To replace a screw-in valve stem, follow the steps in **SKILL DRILL 37-7**.

SKILL DRILL 37-6 Replacing a Rubber Press-Fit Valve Stem

1. Remove the wheel from the vehicle, deflate the tire, and break the top bead using the tire machine.

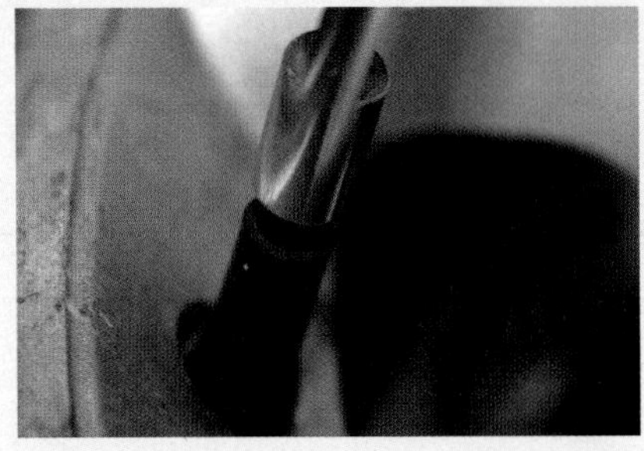

2. Screw the valve stem tool onto the old rubber press-fit stem.

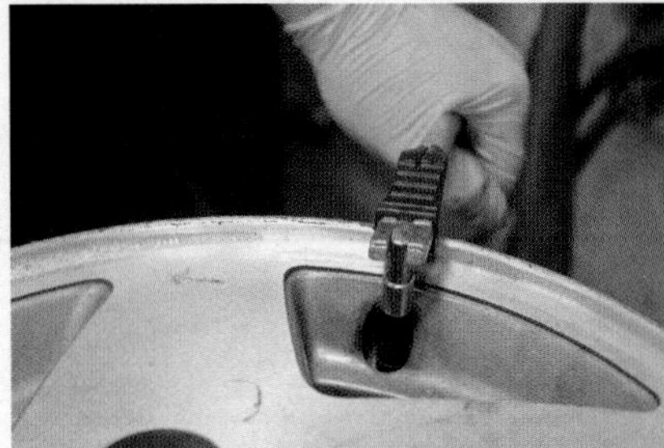

3. Pry the rubber press-fit valve stem from the rim. Use one hand to hold onto the portion of the valve stem in case it breaks off.

4. Clean and inspect the hole in the rim. Clean any rust or corrosion with some sandpaper or other appropriate tool. Lubricate the new valve stem with tire lube.

5. Insert the valve stem into the hole in the rim from the inside, remove the cap, and screw the valve stem tool onto the threaded end of the stem.

6. Use the handle of the valve stem tool as a lever to pull the retaining ridge through the hole in the rim, and verify that the valve stem is properly installed.

SKILL DRILL 37-7 Replacing a Screw-In Valve Stem

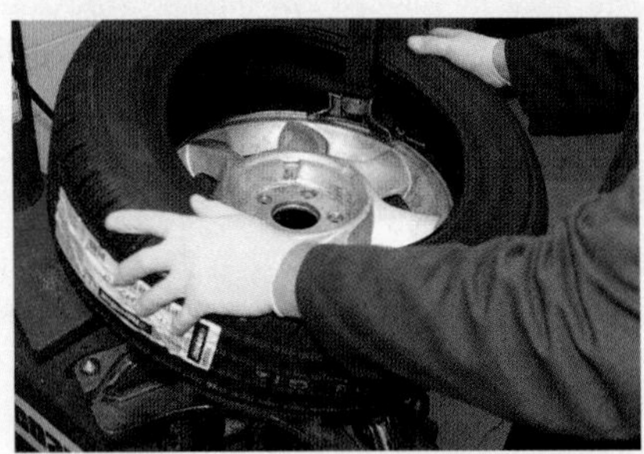

1. Remove the wheel from the vehicle, deflate the tire, break the top bead using the tire machine, and mount the wheel on the machine.

2. Unscrew the nut that holds the valve stem to the rim. Remove the valve stem from the inside of the rim.

3. Discard the old sealing washers, and replace them with new ones. Place the screw-in valve stem with one new sealing washer through the hole on the inside of the rim.

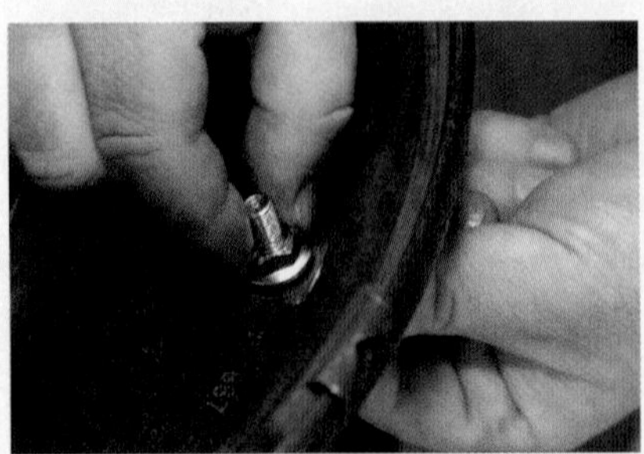

4. Place a new sealing washer over the valve stem, and thread the nut on by hand. Tighten the nut to the specified torque.

Mounting a Tire

In today's vehicles, mounting a tire means more than just installing a properly sized tire onto the rim. Because tires do not come perfectly round and balanced, most tire manufacturers indicate the tire's highest point with a red dot and the tire's lightest point with a yellow dot. The red dot should be lined up with the rim's lowest point, which is called **match mounting**. The yellow dot should be matched up with the rim's heaviest point, which is called **weight matching**. These points occur innately as part of the manufacturing process because it is impossible to create an absolutely perfect, even, round tire. Slight variations in density and shape of the tire are unavoidable. Matching up the tire and rim using one of these methods will help avoid "tolerance stacking," which is when tolerances in mating parts are aligned in such a way that the tolerances are added together rather than canceling each other out. In the case of wheels and tires, it would occur when the tire's heavy or high points are aligned with the rim's heavy or high points.

In addition, the type of tread—directional, symmetric, and asymmetric—affects the direction in which tires are mounted on the rims and which side of the vehicle they are installed on. Directional tires are designed to operate better in one direction, requiring that they be mounted accordingly. Symmetric tires can be operated in either direction, so they can be mounted in either direction. Asymmetric tires are generally side specific, but they may also be directional, in which case specific tires must be mounted on certain rims. Do not fit a tire that is too wide or too narrow for the rim. Check the tire manufacturer's recommendation for the correct range of rim sizes for a particular tire.

The turntable jaws on the tire machine can hold the rim by grasping it from the outside or the inside. When mounting an alloy rim, it is normal that it be clamped from the outside, whereas steel rims are usually clamped from the inside. If the tire is clamped from the outside on the tire machine, it is necessary to release the clamps before fully inflating the tire. Always check the instruction manual for the tire machine you are using for the correct method of clamping a rim.

Inflating a tire for the first time is always dangerous. Some types of tires, such as split rims, must be inflated inside of a tire cage. Doing so contains any pieces if the tire blows up while being inflated. Always follow the manufacturer's guidelines and shop policies when inflating a newly installed tire.

To mount a tire, follow the steps in **SKILL DRILL 37-8**.

SKILL DRILL 37-8 Mounting a Tire

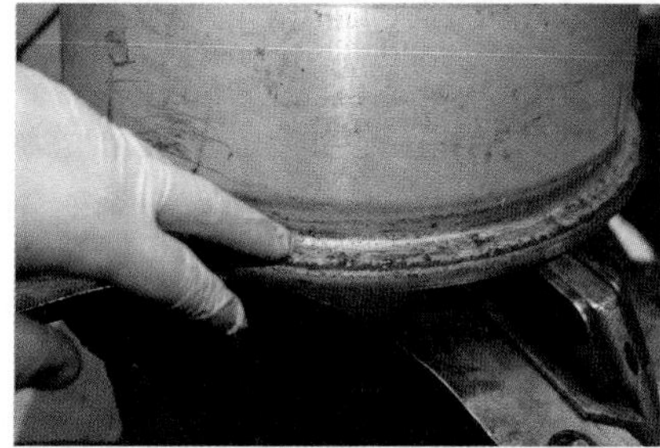

1. Mount the wheel to the tire machine. Examine the wheel, and remove any rust or dirt from the rim bead seat.

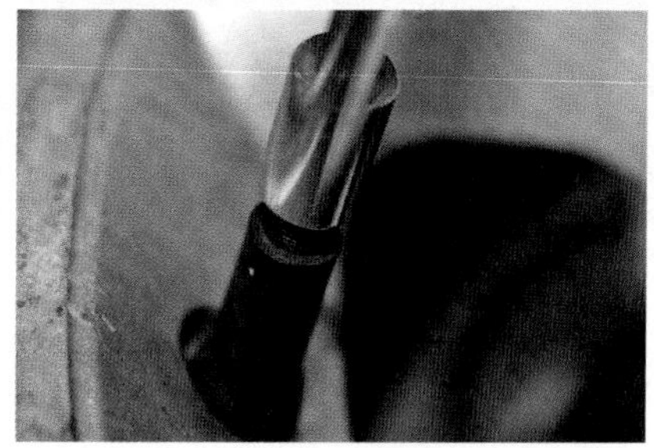

2. Select the correct type of tubeless valve stem, lube it with tire lube, and insert it through the hole in the rim from the inside. Using the valve stem tool, pull the stem through until its groove locates in the hole. Use the valve core tool to unscrew the valve core from the valve stem.

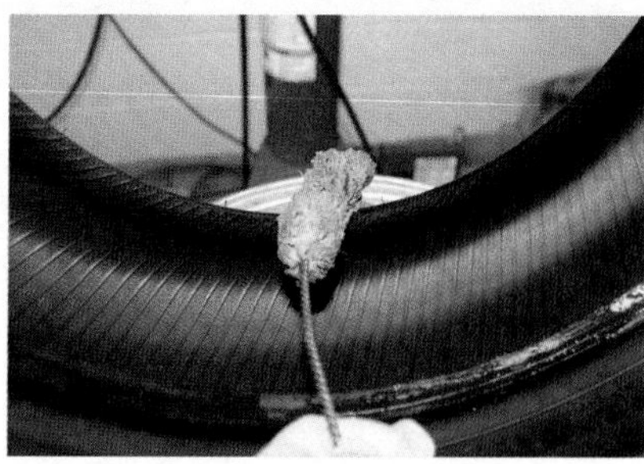

3. Apply some lubricant to the tire bead and rim ridges.

4. Position the tire on top of the rim so that a portion of the lower bead is positioned in the drop center while keeping the lower bead in the tire machine guide.

5. Activate the turntable and guide the lower tire bead onto the rim.

6. Once the lower bead is fitted, position the upper bead into the guide while holding the other side of the upper bead in the drop center.

SKILL DRILL 37-8 Mounting a Tire (Continued)

7. Activate the turntable and guide the tire onto the rim. As the turntable rotates, push the sidewall down, keeping your fingers clear of the rim, so that the tire bead is guided below the safety ridge into the drop center.

8. Attach the tire inflator chuck to the valve stem. Stand clear of the tire and inflate it, being careful to not exceed 30 psi (207 kPa) if both beads have not seated against the rim. If they have not seated by 30 psi, deflate the tire and inspect the rim and tire for damage. If they are okay, relube the tire and rim and reattempt to inflate the tire. If it still will not seat the beads by 30 psi, inform your supervisor.

9. Check the location of the bead indicator ridge to make sure the bead is fully seated. If the rim is clamped from the outside, it will be necessary to release the clamps so the tire can inflate fully.

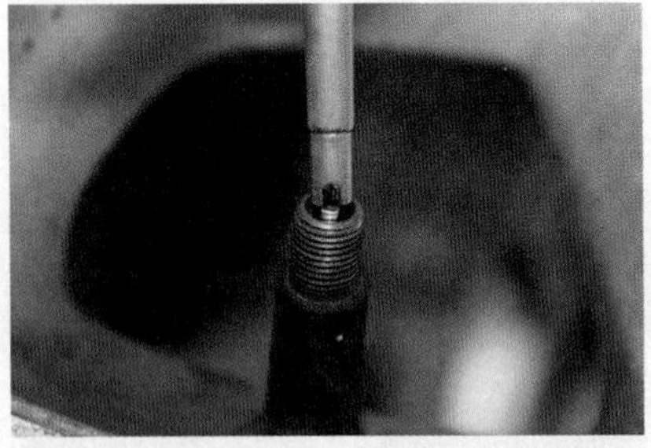

10. When the beads are properly seated, remove the tire inflator chuck, keeping your hands and face clear of the valve stem opening. Once the tire has completely deflated, screw the valve core into the valve stem, using the valve core tool.

11. Reattach the inflator, stand clear, and inflate the tire to the correct pressure as listed on the vehicle's tire placard or in the owner's manual. Be careful never to exceed the maximum tire pressure listed on the tire sidewall.

12. Use a soft brush to apply a small amount of soapy water to the bead. If there are any air leaks, they will be indicated by bubbles.

▶ TECHNICIAN TIP

Always remove any balance weights from both sides of the rim before mounting it on the tire changer. If they are not removed, the bead remover could drag the weights around the rim, causing damage to the rim face. This is particularly important with alloy rims. The damage done by the balance weights is not repairable and usually requires the rim to be replaced, costing the shop a lot of money.

SAFETY TIP

Over-inflated tires can explode. Do not inflate the tire to a pressure greater than what is listed on the sidewall. The possibility of an explosion is why tire inflators have a spring-loaded (dead man's) trigger that does not lock into position. When the tire is being inflated, use a tire cage, if required, or an inflator that allows you to stand clear of the tire. Keep hands and body well away from the tire. When a tire explodes, the tire, rim, or components from the tire changer may cause serious injury or death to any person nearby.

Dismounting, Inspecting, and Remounting a Tire on a Wheel Equipped with a TPMS Sensor

N37004, N37005

Dismounting and mounting tires equipped with a TPMS requires special care to avoid damaging the expensive sensors mounted inside the wheels. There are two general ways that TPMS sensors are mounted in the wheel. The first way involves either attaching it to the valve stem or making it integral to the valve stem, so care must be exercised around the valve stem area. If the TPMS sensor is integrated into the valve stem with a threaded locknut on the valve stem, some manufacturers recommend unscrewing the locknut and pushing the valve stem and TPMS sensor into the tire for safekeeping during disassembly. The second way is by using a band that fits all the way around the drop center of the rim to hold the TPMS sensor to the inside of the wheel. In many cases, the band-style TPMS sensor is positioned 180 degrees away from the valve stem, so care must be exercised on the side opposite of the valve stem. Always check the service information to verify the manufacturer's specified tire dismounting and mounting procedure.

To dismount, inspect, and remount a tire on a wheel equipped with a TPMS sensor, follow the steps in **SKILL DRILL 37-9**.

Dynamic Balancing a Tire

A tire that is dynamically in balance is in balance when it is spinning as opposed to when it is stationary. Dynamic imbalance occurs when a spot on either the inside or the outside of the tire's centerline is heavy. This induces a side-to-side imbalance in the tire as it rotates, which can cause a vibration as well as the steering wheel to shimmy. Dynamic imbalance is usually

SKILL DRILL 37-9 Dismounting, Inspecting, and Remounting a Tire with a TPMS Sensor

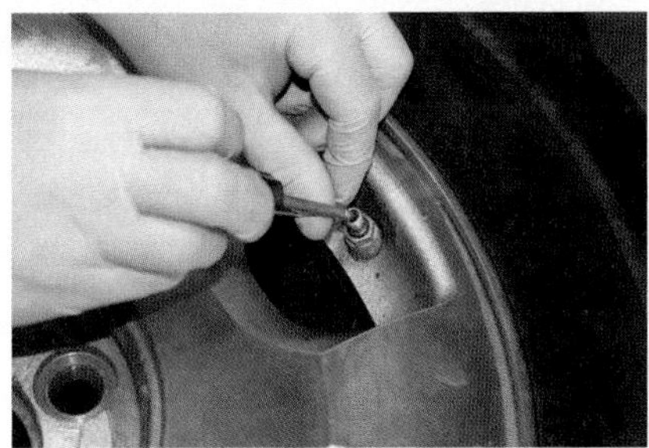

1. Following the specified procedure, remove the wheel from the vehicle, and deflate the tire by removing the valve core. Break the tire beads in the positions specified.

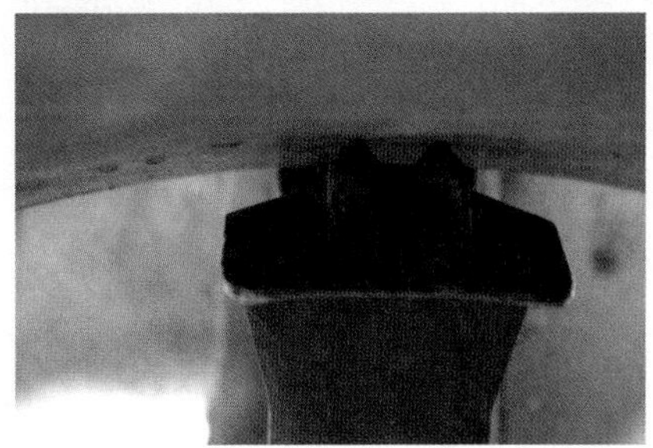

2. Position the tire assembly on the turntable with the shallow side of the wheel up, and engage the jaws to lock the wheel in place.

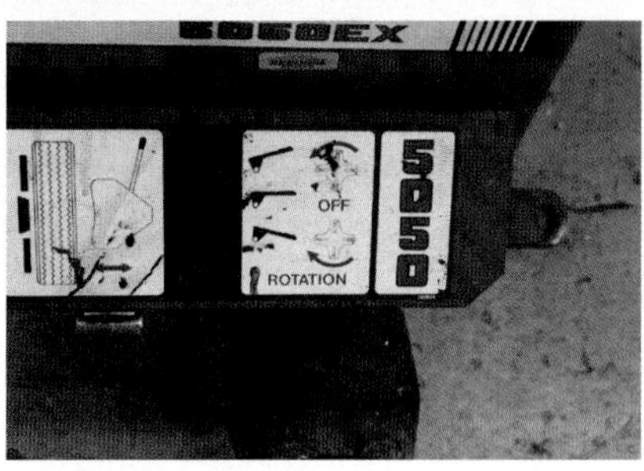

3. Activate the turntable to verify that the wheel is centered and securely held. Stop it in the specified position for removing the top bead.

4. Lubricate both beads with tire lubricant. Make sure you get the lube on the flat bead seat, not just the side of the bead.

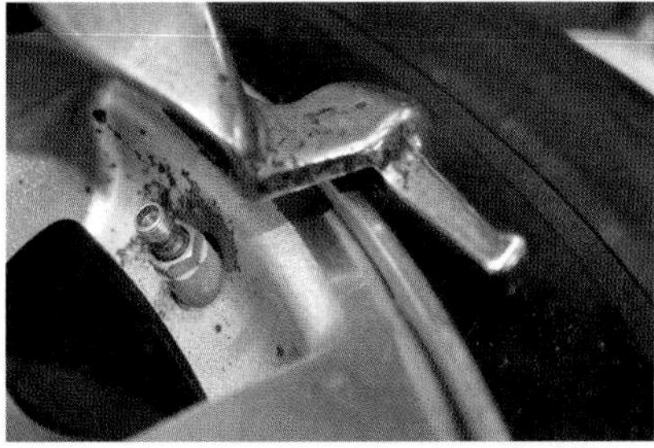

5. Position the bead remover against the edge of the rim; if necessary, adjust it so it has the proper clearance between the rim and the roller. Use the tire lever to pry the tire bead over the bead remover knuckle while pushing down on the sidewall on the opposite side of the tire.

6. Activate the turntable while lifting up on the tire directly behind the remover knuckle to help work the top bead over the flange of the rim.

7. Rotate the turntable to the position specified for removing the lower bead.

8. Carefully pry the lower bead over the knuckle while holding the other side of the tire up in the drop center.

9. Activate the turntable while lifting up on the tire directly behind the remover knuckle to help work the bottom bead over the flange of the rim. Inspect the tire, rim, and TPMS sensor according to the manufacturer's procedure.

SKILL DRILL 37-9 Dismounting, Inspecting, and Remounting a Tire with a TPMS Sensor (Continued)

10. To install the tire, adequately lube both bead seats of the tire with tire lube. Activate the turntable to position the valve stem in the specified position for installing the lower bead.

11. Position the lower bead into the drop center while holding the lower bead in the tire machine shoe.

12. Activate the turntable while helping keep the lower bead in the drop center.

13. Activate the turntable to position the valve stem in the specified position for installing the upper bead.

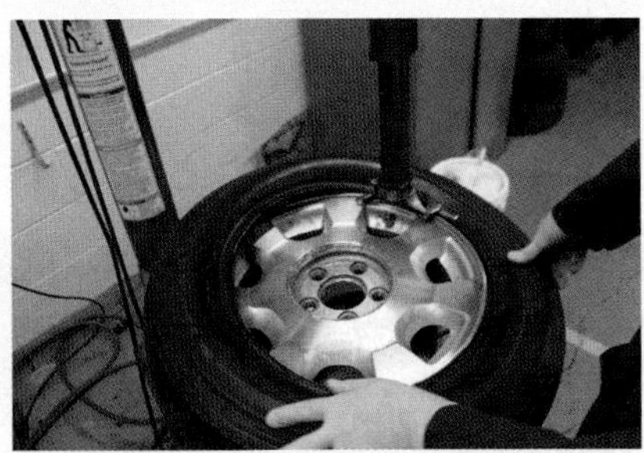

14. Place the upper bead into the drop center while holding the lower bead in the tire machine shoe.

15. Activate the turntable while pushing the top bead into the drop center.

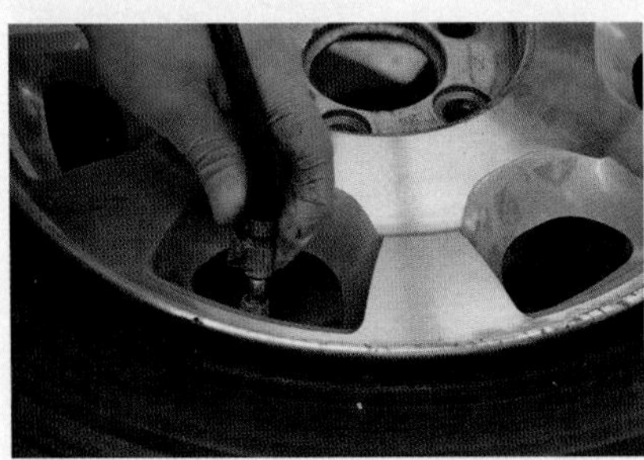

16. Inflate the tire as a non-TPMS wheel, and check for leaks.

a result of manufacturing variations, but it can be caused by a damaged tire or wheel. Obviously, dynamic imbalance reduces ride quality and tends to increase wear on the tires and on steering and suspension system components. Dynamic balancing of the tires should be performed when new tires are installed on the vehicle as well as any time that tire imbalance is suspected.

Dynamic balancing is performed on a tire balancer that is capable of spinning the tire and then measuring the location of any dynamic imbalance. Some balancers are spun at low speed, but others are driven by the balancer at higher speeds. If the tire is spun by the balancer, embedded objects may fly off the tire, so it is important to wear safety glasses. If the wheel balancer is fitted with a safety hood, ensure that it is in place when the wheel is being rotated to further protect against flying objects.

All wheels require one of several specific designs of wheel weights. If the weights fitted to the wheel are not the correct type, they can fly off when the vehicle is driven down the road, causing possible injury or damage. It is good practice to use new wheel weights when balancing a wheel for the same reason. If the vehicle has directional tires, ensure that the wheels are reinstalled in their correct position when balancing is complete.

To balance a tire, follow the steps in **SKILL DRILL 37-10**.

SKILL DRILL 37-10 Balancing a Tire

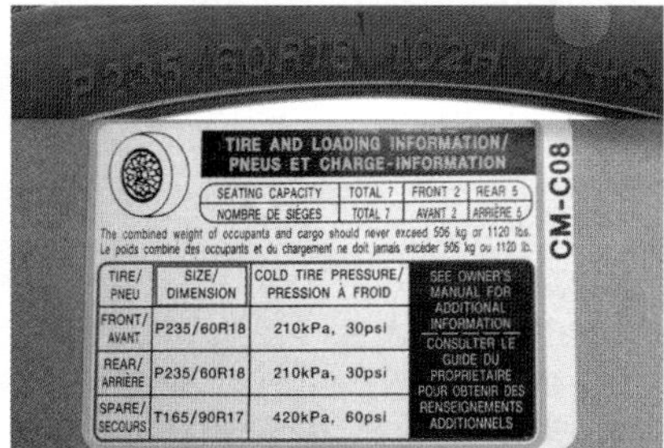

1. If using hand tools, prepare the vehicle by loosening the lug nuts, and then raise the vehicle into a comfortable working position. Check that the tires fitted to the wheels on the vehicle are the appropriate size and rating for the vehicle.

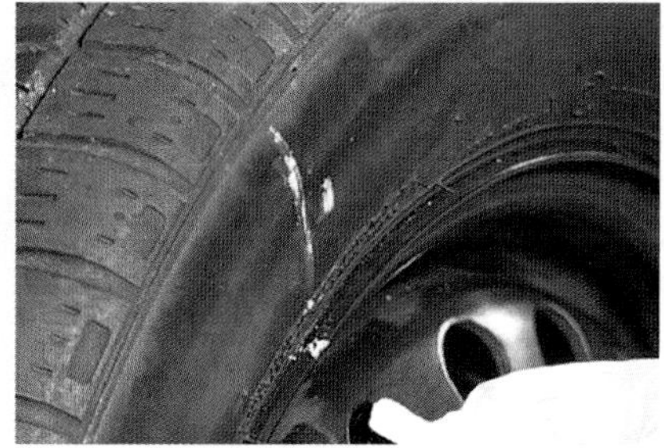

2. Mark the inside of the wheel or tire in relation to its location on the vehicle, and then remove it.

3. Check and adjust the tire pressure before balancing the tire. Mount the wheel and tire on the balancer, putting the inside part of the wheel toward the balancer in most cases. Secure the wheel by screwing the hub nut assembly on the balancer shaft.

4. Calibrate the balancer to the wheel by measuring the width of the rim with a rim caliper, using the gauge on the balancer to determine the offset location of the flange on the wheel, and the diameter of the wheel as listed on the tire. Input these data into the balancer's computer, if fitted. If no computer is fitted, set the balancer adjustments manually according to the instruction manual.

5. If equipped, lower the safety hood over the wheel. Spin the wheel.

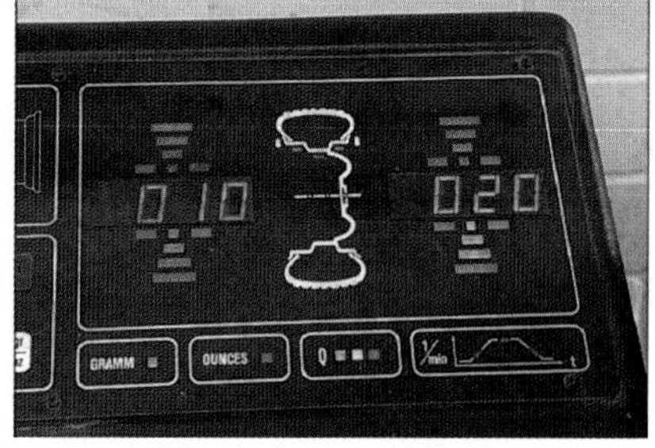

6. Read the balancer's analysis. If the wheel is out of balance, you should remove the old weights and recheck the balance of the wheel before adding new weights.

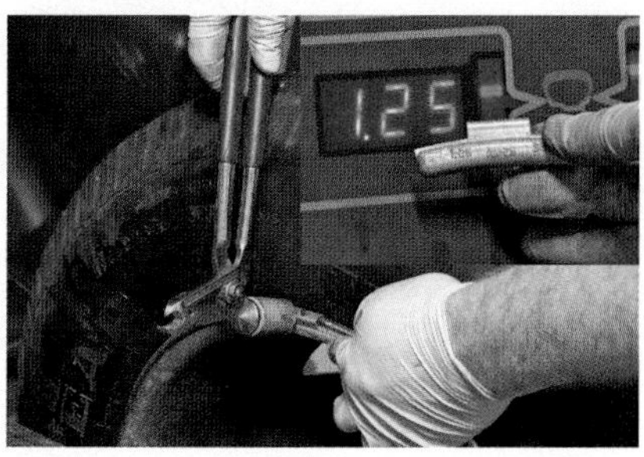

7. Install new weights as recommended by the machine's display.

8. Respin the wheel to check for accuracy of the balancing job and to confirm that balance has been achieved. Repeat the process for the rest of the wheels and tires. Reinstall the wheels and tires to the vehicle.

▶ Diagnosis

S37002, N37006, N37007

Diagnosis of wheel and tire faults is a fairly common task for automotive technicians, as tires are subjected to much abuse compared to some of the other components on a vehicle. Following the strategy-based diagnostic process is key to identifying the cause of the fault. The first step is to verify the customer concern. It is best to make sure you have a good understanding of the concern. You may be able to get enough information from the repair order or the shop's customer concern questionnaire. If not, then discuss the concern further with either the service advisor or the customer.

Next, take the vehicle for a test drive if the vehicle is safe to do so. That way you can verify the concern and get clues to what is causing it, such as location and type of a noise or vibration. You will likely have to test-drive the vehicle in variety of ways to glean as many clues as possible. Just don't violate any driving laws or company policies.

The next step is to research the possible faults. This includes researching TSBs and the service information. Use the clues gathered from the previous step to guide your research. Once the research is complete, you will need to use that plus the knowledge of the tests and repairs listed right after this topic to build an attack plan for diagnosing the fault. Your plan will have to take into account the following considerations:

- The most likely cause based on your test drive and research
- The test most likely to give you the most results in narrowing down the cause of the fault
- The order in which you will perform the tests based on ease and cost of performing
- How you will verify the cause once it's source is found or suspected

Once you have a plan in place, it is time to perform focused testing to locate the root cause of the concern. This could be achieved by performing one or more of the tests that follow in this section. Each time you perform a test, you should know what you are expecting to find, whether the part you are testing is the cause of the fault or not. This expectation comes from understanding the service information, as well as by the experience you are gaining from performing the maintenance tasks.

Once you have identified the root cause of the fault, get authorization to perform the repair. This could be replacement of damaged or worn parts or repairs such as balancing a tire, etc. Just make sure you perform the repair properly. There is very little room for errors when working on tires and wheels. Shortcuts or errors put passengers, others, and property at risk of severe injury.

Once the repair has been performed, verify that it corrected the fault. Initially, you can do this by performing the same test that identified the fault. If that checks out okay, then test-drive the vehicle to verify that it is operating properly. In some cases, you will have fixed the fault only to discover a secondary fault. For example, a vehicle may have a severe vibration caused by a tire with a broken belt. But when you test-drive it after the repair, the vehicle may pull to one side because of an alignment issue. The faulty tire may have been masking the alignment issue. This is another reason why vehicles should be test-driven after the repair. Next, we start to cover some of the tests and repairs for which you will need to gain experience.

Inspecting the Wheel Assembly for Air Loss

N37008

One of the most frustrating complaints from drivers is that they constantly have to pump their tires up as the result of an air loss somewhere. To efficiently check the suspect tire, it must be removed from the vehicle and aired up to its recommended pressure. A preliminary check can be carried out with a spray bottle of soapy water. You need to spray the soapy water around the valve stem and core. In addition, spray around the bead area. Also spray the entire tread area. If there is any air leakage, soapy air bubbles will indicate where the problem is. Mark the place on the tire where the air bubbles are coming from so that the repairs can be carried out.

Another good method for checking where air loss is coming from is to immerse the wheel assembly in a container of water. The container must be large enough to immerse the tire either on its side, with the tire completely underwater, or upright, with approximately half the wheel assembly underwater. If there is a leak, it will be obvious by the discharge of air bubbles in the water (**FIGURE 37-16**). Mark the source of the air bubbles. Remove the wheel from the water tank, and carry out the appropriate repairs.

To inspect the wheel assembly for air loss, using the spray bottle method, perform the steps in **SKILL DRILL 37-11**.

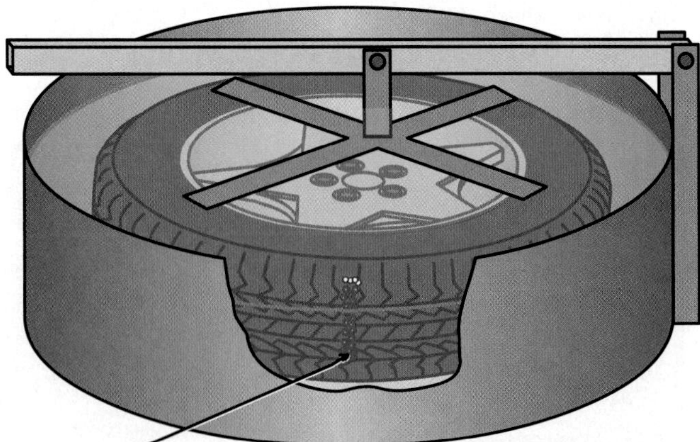
Air Bubbles show leak location, mark this spot for repair when the tire is removed from the tank.

FIGURE 37-16 A good method for identifying the location of an air leak is to immerse the tire assembly in water and watch for air bubbles.

SKILL DRILL 37-11 Inspecting the Wheel Assembly for Air Loss, Using the Spray Bottle Method

1. Remove the tire from the vehicle, and inflate it to the specified pressure. Spray soapy water on the valve core and stem, and look for bubbles.

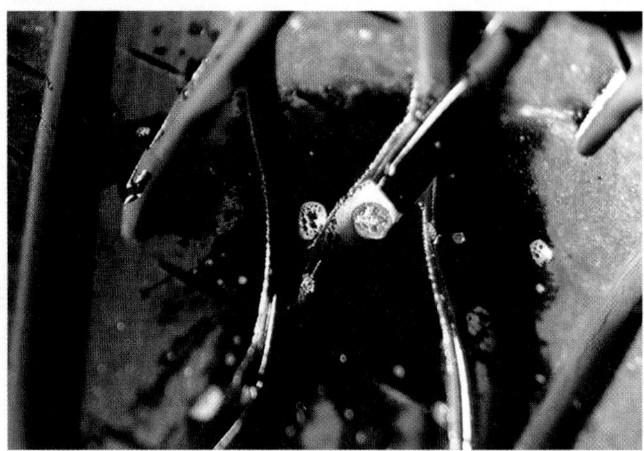

2. Spray soapy water on the tread, and look for bubbles.

3. Spray soapy water around the bead of the tire, and look for bubbles.

Tire Repair

N37009

Driving a short distance on a tire while it is severely under-inflated will cause the tire to overheat as well as weaken the sidewall belts, creating a dangerous, nonrepairable condition. The damage is not visible from the outside, so every tire requiring repairs must be removed from the wheel for inspection and to assess its reparability. The tire needs to be inspected externally first, then internally for any signs of serious damage, such as chafing of the sidewalls or cords sticking through the inner liner. If the inspection shows no signs of nonrepairable failure, then the tire can be repaired in accordance with the procedures recommended by the tire associations, including the Rubber Manufacturers Association.

Repairs from any nail or similar object should be limited to the actual tread area. Among the criteria to perform a proper repair are:

- Repairs are limited to the tread area only.
- Puncture injury cannot be greater than 1/4" (6 mm) in diameter.
- Repairs must be performed by removing the tire from the rim/wheel assembly to perform a complete inspection to assess all damage that may be present.
- Repairs cannot overlap. This means that if there are two or more repairs required to fix the tire, then each repair patch must not overlap.

- A rubber stem, or plug, must be applied to fill the puncture injury, and a patch must be applied to seal the inner liner. A common repair unit is a one-piece plug patch with a stem and patch portion. A plug by itself is an unacceptable repair.

To patch a tire, you need a tire machine to remove the tire from the wheel, a buffer to clean the inside of the tire, glue for attaching the patch, and a tire plug patch. The tire plug patch is a superior product to the old-fashioned tire plug, and in many states it is the only legal way to repair a hole in a tire. It is more expensive but is a much safer alternative for repairing a tire because it patches the inside of the tire in addition to filling the hole in the tire.

To patch a tire, follow the steps in **SKILL DRILL 37-12**.

SKILL DRILL 37-12 Patching a Tire

1. After marking the location of the air leak on the tread of the tire and the position of the tire and weights on the rim, remove the tire from the rim assembly.

2. Mount the tire in a tire spreader so that the hole can be accessed from both the inside and the outside of the tire.

3. Use an air die grinder with the properly sized pointy bit that matches the plug patch, and drill into the tire where the leak was located.

4. Use a tire buffer to smooth the area inside the tire around the hole. Smooth the area approximately 1/2" (13 mm) beyond the expected patch area.

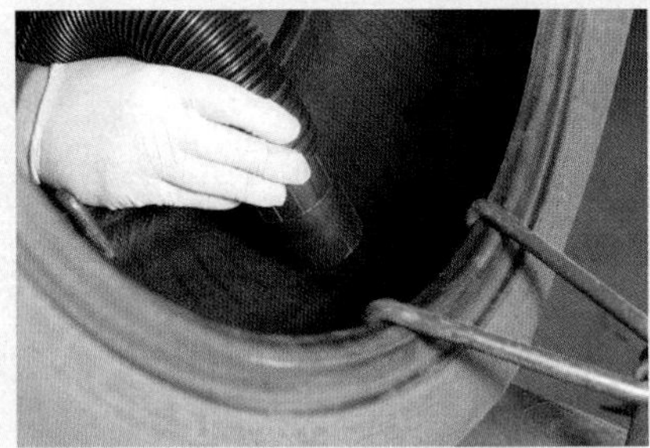

5. After completing the buffing process, clean out all the accumulated debris with a vacuum.

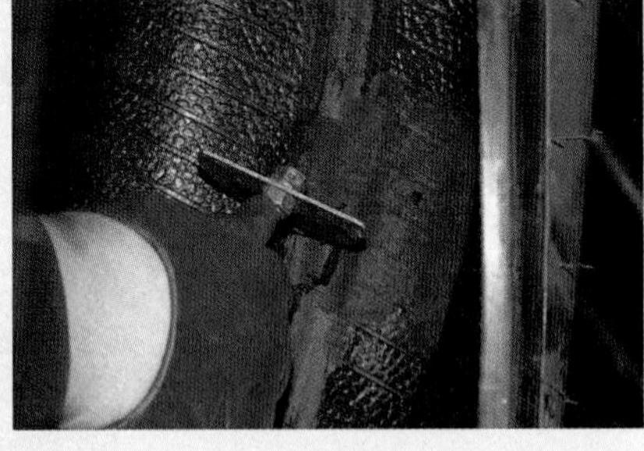

6. Liberally apply the liquid buffing solution to a clean rag, and scrub the area just buffed. Or use cleaner and a scraper to clean the area. Repeat this step once or twice, as needed.

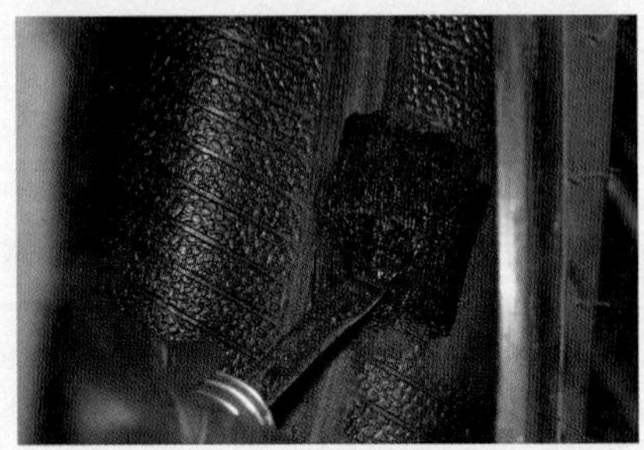

7. Apply vulcanizing cement evenly to the inner buffed surface of the tire. The cement needs to stand until it is relatively dry and is only tacky to touch.

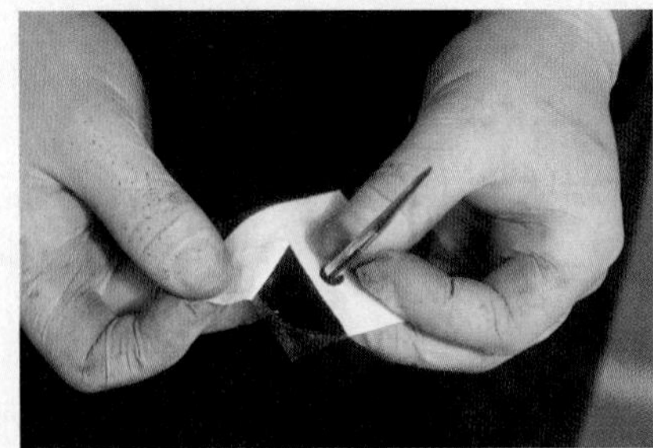

8. After selecting the appropriate tire patch, remove the plastic protective cover on the sticky side of the tire patch without getting your fingerprints on the sticky side.

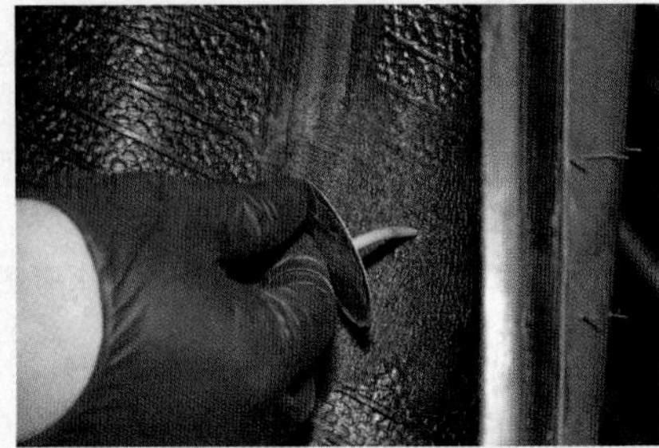

9. Take the pointed part of the patch and push it through the inner side of the tire's hole that was roughed previously, pushing it through to the outside of the tire.

SKILL DRILL 37-12 Patching a Tire (Continued)

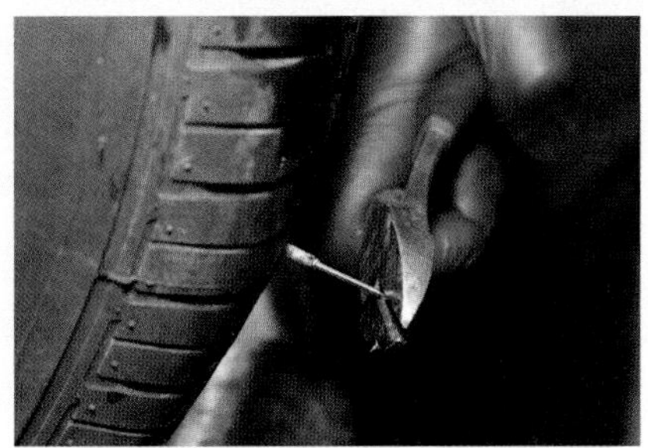

10. Using a pair of pliers, grip the stem of the patch and pull it out so that the disc portion of the patch comes into contact with the cemented area. Pull this pointy part of the patch away from the tire's tread. The sticky side of the patch has now been tightly pressed onto the buffed surface.

11. Use a stitching tool and roll the inner side of the tire patch tight onto the inner surface of the tire. Start at the center of the patch and work outward. Cover the buffed area and newly applied patch with a rubber patch sealant. Trim the plug material even with or just above the tread, using a utility knife or other appropriate cutting tool.

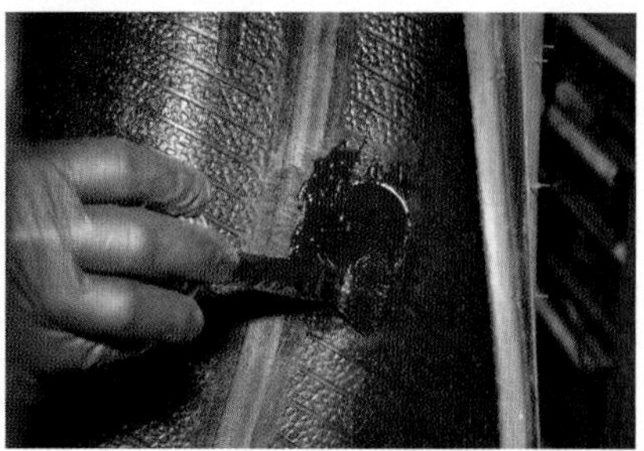

12. After the rubber patch sealant has dried, the tire can be reassembled onto the rim in its original position and inflated to the recommended pressure. Check the wheel assembly for balance before it is reinstalled on the vehicle.

Measuring Wheel, Tire, Axle Flange, and Hub Runout

N37010

Runout is the side-to-side or up-and-down variation in a part in the wheel assembly. In many cases, runout issues cannot be observed when the vehicle is stationary on the ground. Instead, runout problems can be felt as a vibration while the vehicle is being driven, usually getting more noticeable as speed is increased. If the vibration is primarily observed in the steering column or hood, the runout is most likely in one or both of the front wheel assemblies. Vibration felt throughout the entire vehicle suggests that the problem is with one or both of the back wheels. Runout can be defined as radial or lateral. Radial runout occurs when the component is out of round or off center and is felt more as a vertical vibration. Lateral runout occurs when the component is bent or improperly manufactured and causes the wheels to jiggle side to side, creating a horizontal vibration that feels like a shimmy. Lateral runout on front wheels is felt in the steering wheel even at slower speeds. Runout of the wheel, tire, axle, and hub are all measured in the same way, using either a dial indicator or a special runout gauge inserted onto the surface of the component being measured.

To measure tire runout, follow the steps in **SKILL DRILL 37-13**.

To measure wheel runout, follow the steps in **SKILL DRILL 37-14**.

To measure axle flange or hub runout, follow the steps in **SKILL DRILL 37-15**.

SKILL DRILL 37-13 Measuring Tire Runout

1. Research the procedure and specifications. Raise the vehicle on a hoist or place a jack under the vehicle at a suitable lifting point and raise the vehicle.

2. Select the runout gauge or dial indicator, attachment, and bracket that fit the tire. Mount the dial indicator on a firm surface to keep it still.

3. Adjust the dial indicator so the plunger is 90 degrees to the tread of the tire. Press the dial indicator gently against the tire and rotate the tire one full turn. Keep pressing until the plunger settles about halfway into the indicator.

SKILL DRILL 37-13 Measuring Tire Runout (Continued)

4. Verify that the plunger is still 90 degrees to the tire, and lock the indicator assembly into position. Carefully rotate the tire a couple of times while observing the dial readings. If the pointer hovers around a single graduation on the dial, the part has minimal runout or surface distortion, and the test is complete. If the pointer moves significantly left and right, note the variations.

5. Find the point of maximum movement to the left, and move the dial so that zero is over this point.

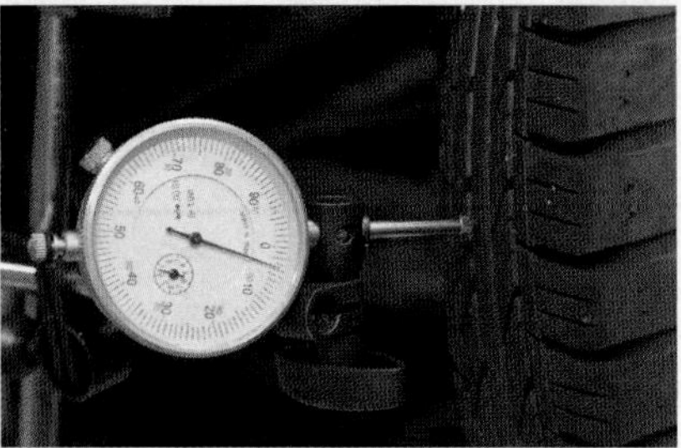

6. Continue to rotate the tire. Find the point of maximum movement to the right, and note the reading. Confirm this value by rotating the tire several more times to verify the zero point and high point. Compare these values to the manufacturer's specifications. If the deviation is greater than the specifications, the wheel and/or hub runout must be measured.

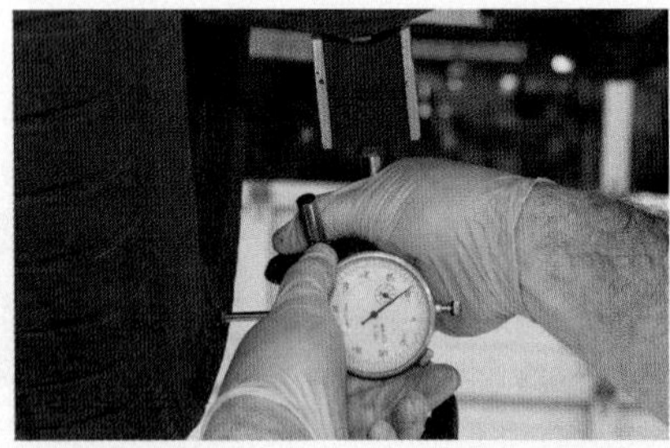

7. Reposition the dial indicator or runout gauge so it is 90 degrees to the sidewall of the tire. Perform steps 3 through 6 to obtain the lateral runout measurement for the tire, and compare to the vehicle's specifications.

SKILL DRILL 37-14 Measuring Wheel Runout

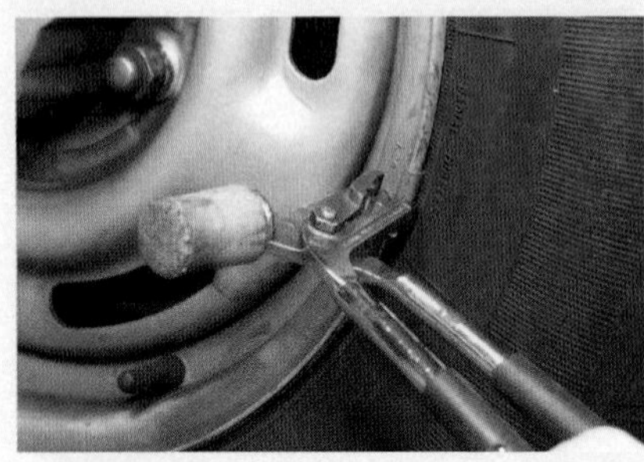

1. If equipped, remove the hubcap and use a wheel weight tool to remove any wheel weights.

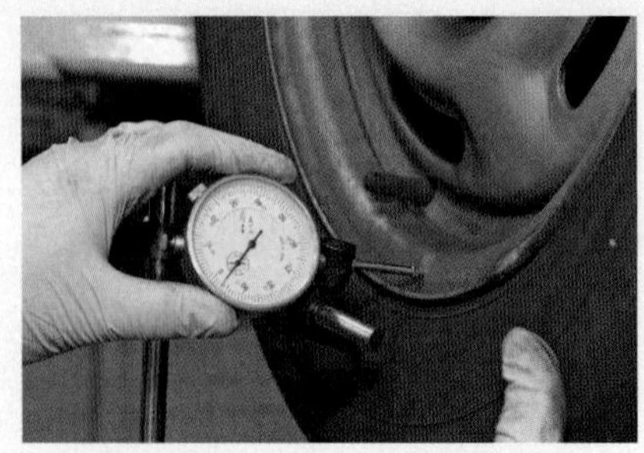

2. Select the runout gauge or dial indicator, attachment, and bracket that fit the wheel. Mount the dial indicator on a firm surface to keep it still. Adjust the dial indicator so the plunger is 90 degrees to the flat portion of the wheel where the wheel weights are installed.

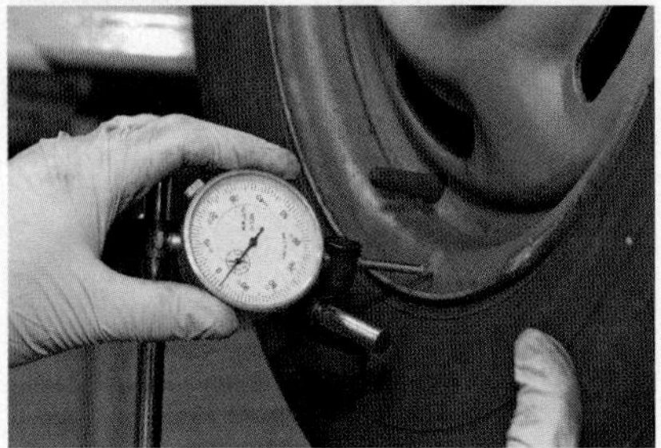

3. Press the dial indicator gently against the wheel, and rotate the wheel one full turn. Keep pressing until the plunger settles about halfway into the indicator. Verify that the plunger is still 90 degrees to the wheel, and lock the indicator assembly into position.

SKILL DRILL 37-14 Measuring Wheel Runout (Continued)

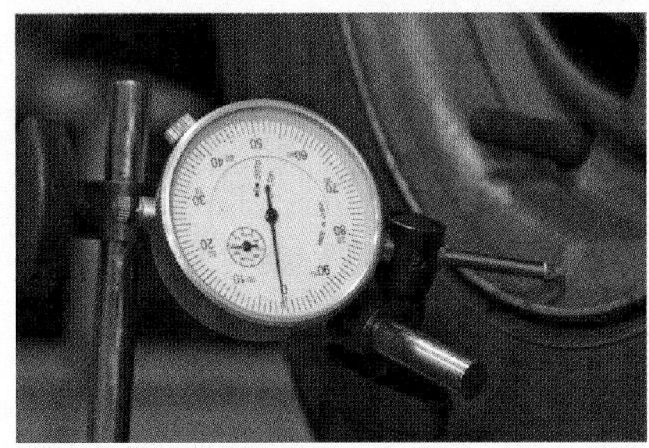

4. Rotate the wheel a couple of times while observing the dial readings. If the pointer hovers around a single graduation on the dial, the wheel has minimal runout or surface distortion and the test is complete. If the pointer moves significantly left and right, note the variations.

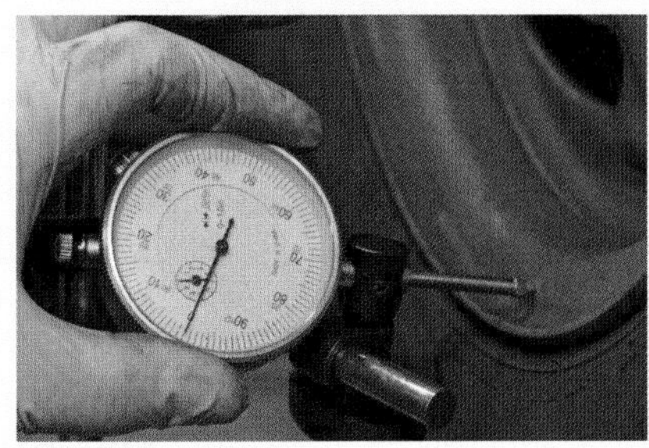

5. Find the point of maximum movement to the left and move the dial so that zero is over this point.

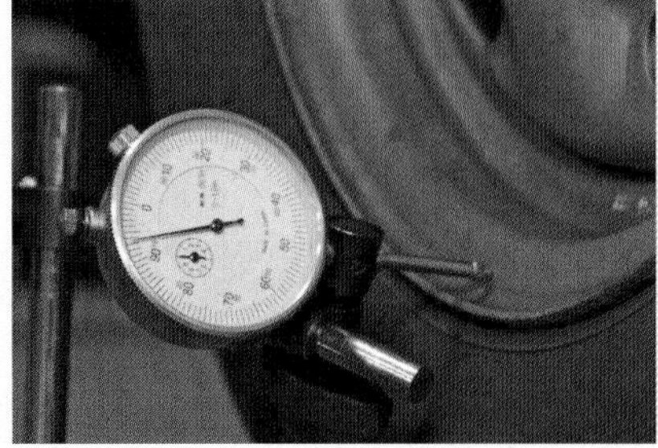

6. Continue to rotate the wheel. Find the point of maximum movement to the right and note the reading. Confirm this value by rotating the wheel to verify the zero point and high point. Compare these values to the manufacturer's specifications. If the deviation is greater than the specifications, the axle or hub runout must be measured.

SKILL DRILL 37-15 Measuring Axle Flange or Hub Runout

1. Prepare the vehicle by removing any hubcaps or lug nut covers. With the vehicle on the ground, loosen the lug or wheel nuts. Raise the vehicle, and remove the lug nuts and tire.

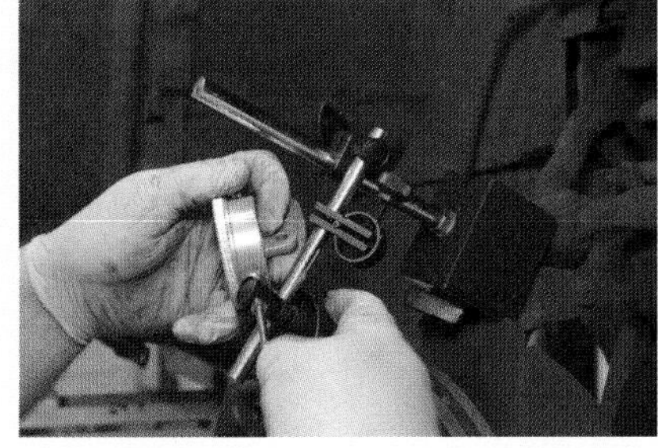

2. Select the runout gauge or dial indicator, attachment, and bracket that fit the axle flange or hub. Mount the dial indicator on a firm surface to keep it still.

3. Adjust the plunger so it is 90 degrees to the axle flange or hub. Press the dial indicator gently against the flat surface of the axle flange or hub, and rotate it one full turn. Keep pressing until the plunger settles about halfway into the indicator. Verify that the plunger is still 90 degrees to the tire, and lock the indicator assembly into position. Rotate the axle flange or hub while observing the dial readings.

4. If the pointer hovers around a single graduation on the dial, the axle flange or hub has minimal runout or surface distortion, and the test is complete. If the pointer moves significantly left and right, note the variations. Find the point of maximum movement to the left, and move the dial so that zero is over this point.

5. Continue to rotate the axle flange or hub. Find the point of maximum movement to the right and note the reading. Confirm this value by rotating the axle several more times to verify the zero point and high point. Compare these values to the manufacturer's specifications. If the deviation is greater than the specifications, the axle or hub must be discarded. Reinstall the wheels and torque the lug nuts following the specified procedure and torque specifications.

Inspecting, Diagnosing, and Calibrating the TPMS

N37011

Inspection, diagnosis, and calibration of the TPMS is required in several general situations. Inspection is needed whenever the tires have been dismounted from the wheels. In some cases, the TPMS sensor batteries need to be replaced and the sensor mounts need to be inspected, or the sensors need to be replaced if they aren't replaceable batteries.

Diagnosis of system faults is required when the system detects a fault and turns on the warning light. A quick check of the system is made by turning the key to "run" and observing the tire pressure warning light or indicator. If the warning light is on, this could be caused by one or more tires that are not at the proper pressure, requiring an inspection for leaks or incorrect pressures, or it could be a system fault requiring a scan tool capable of communicating with the TPMS to interrogate the system to determine what is causing the fault. Calibration of the system is needed on some systems whenever a tire rotation is performed or a sensor is replaced so that the TPMS knows on which wheel each sensor is located.

To inspect, diagnose, and calibrate the TPMS, follow the steps in **SKILL DRILL 37-16**.

SKILL DRILL 37-16 Inspecting, Diagnosing, and Calibrating the TPMS

1. Turn the ignition key to the run position, and observe the TPMS warning light. If the light indicates low tire pressure at any wheel, check the pressure of that wheel with an accurate tire pressure gauge. If incorrect, adjust the pressure, using compressed air.

2. If the light indicates a system fault, connect an appropriate scan tool to the system and read any DTCs. Research DTCs in the appropriate service information, and follow the diagnostic steps listed to identify the cause of the fault.

3. Once diagnosis and repairs are complete, follow the specified relearn procedure. Here, a magnet is used to close the centrifugal switch in the sensor, identifying its location to the PCM.

▶ Wrap-Up

Ready for Review

- The tire placard lists the recommended tire size and pressure (cold) for the vehicle and is typically located on the driver's door pillar.
- Tire manufacturers list the maximum safe inflation pressure and load on the tire sidewall.
- Nitrogen is used to fill some tires because it is slower to leak out, it contains no moisture, and reduces oxidation of the rubber.
- To meet the standards for proper nitrogen fill, the nitrogen level in the tire must be at 95% or higher.
- Tires can be rotated in either a four-tire or five-tire rotation.
- Tires are generally rotated approximately every 5000 miles (8000 km) or with every oil change.
- There are two ways wheels can be balanced: off-car and on-car.
- Static imbalance is when the imbalance is centered across the width of the tire and results in an up and down movement of the tire.
- Dynamic imbalance is when the imbalance is off-center of the tire width and results in a side-to-side wiggling of the tire, along with an up-and-down movement.
- Wheel weights made of lead have been outlawed in some states and replaced with steel weights that are coated with zinc or other protective material.
- Because TPMS sensors are located inside the wheel/tire assembly, they can easily be damaged during tire removal or installation. Follow the manufacturer's procedure.
- Many TPMS sensors use nickel-coated valve cores inside aluminum valve stems to minimize corrosion issues.
- TPMS sensors are a sealed component powered by a small battery (some are replaceable), which is designed to last for 5 to 10 years.

- ▶ Some common issues with tires and wheels are loss of air pressure, imbalance, excessive radial or lateral runout, and pulling while driving.
- ▶ The three most popular types of pocket tire pressure gauges are the pencil type, the dial type, and the digital type.
- ▶ Vehicles that are 2008 model year and newer are required to be equipped with a TPMS.
- ▶ Common irregular tire wear patterns encountered include feathering, one-sided wear, cupping, center wear, and edge wear.
- ▶ When reinstalling the lug nuts thread them on by hand at least two full turns to avoid cross-threading.
- ▶ Tighten the lug nuts to the correct torque in the proper tightening sequence, as specified in the service information.
- ▶ Dismounting a tire is usually performed on a tire machine, which is very powerful, so use extreme care and closely follow the manufacturer's procedure.
- ▶ Dismounting and mounting tires equipped with a TPMS sensor requires special care to avoid damaging the expensive sensors mounted inside the wheels.
- ▶ Some manufacturers recommend unscrewing the locknut and pushing the valve stem and TPMS sensor into the tire for safekeeping during disassembly.
- ▶ Following the strategy-based diagnostic process is key to identifying the cause of tire and wheel faults.
- ▶ The source of tire pressure loss can be located by spraying soapy water on the tire and looking for bubbles. Another method involves dunking the tire and wheel in a dunk tank and looking for bubbles.
- ▶ Air loss in tires generally occur due to foreign objects penetrating the tire, a leaky tire bead, or leaky valve stem or core.
- ▶ Among the criteria to perform a tire proper repair are these: repairs are limited to the tread area only; puncture injury cannot be greater than 1/4" (6 mm) in diameter; repairs must be performed by removing the tire from the rim/wheel assembly to perform a complete inspection, to assess all damage that may be present; repairs cannot overlap, and a rubber stem, or plug, must be applied to fill the puncture injury, and a patch must be applied to seal the inner liner.
- ▶ A common repair unit is a one-piece plug patch unit with a stem and patch portion. A plug by itself is an unacceptable repair.
- ▶ Runout of the wheel, tire, axle, and hub are all measured in the same way using either a dial indicator or a special runout gauge.
- ▶ If the TPMS warning light is on, this could be caused by one or more tires that are not at the proper pressure, or it could be a system fault.

Key Terms

bar A metric unit of measure for pressure.

dynamic imbalance A tire imbalance that causes the wheel assembly to turn inward and outward with each half revolution.

match mounting The process of matching up the tire's highest point with the rim's lowest point for the purpose of reducing the tire's radial runout.

road force imbalance Occurs when the wheel or tire is not concentric or when the tire's sidewall has uneven stiffness.

static imbalance A tire imbalance resulting from a heavy spot on a tire; it will vibrate vertically with the heavy area slapping the road surface with each turn of the wheel.

tire inflation pressure The amount of air pressure in the tire that provides it with load-carrying capacity and affects the overall performance of the vehicle.

tire pressure gauges A gauge used to measure the air pressure within a tire.

weight matching The process of matching the tire's lightest point with the rim's heaviest point (generally at the valve stem) for the purpose of reducing the tire's radial imbalance.

Review Questions

1. Which of the following is a clear indicator that the tire is at the end of its legal life?
   a. Out-of-balance conditions
   b. Tread depth above the level of built-in wear indicators
   c. Low tire pressure
   d. Tread depth below the level of built-in wear indicators
2. Wheel alignment is normally checked and adjusted under all of the following conditions *except*:
   a. when there is abnormal tire wear.
   b. when the steering parts are replaced.
   c. when the oil is changed.
   d. when new tires are put on the vehicle.
3. The cold inflation pressure of a tire is determined based on the:
   a. type of transmission in the vehicle.
   b. type of gas in the tire.
   c. vehicle's design load limit.
   d. engine capacity of the vehicle.
4. The typical procedure to remove any existing oxygen from an air-filled tire before filling nitrogen into it is to:
   a. completely deflate it and then inflate it to high pressure.
   b. inflate and deflate it more than once to remove oxygen.
   c. inflate the tire with air.
   d. inflate the tire once is all that is needed.
5. Jack needs to perform tire rotation on a car that is an all-wheel drive with directional, differently sized tires. Which of these methods should he follow?
   a. Forward-cross pattern
   b. X pattern
   c. Front-to-rear pattern
   d. Dismount and remount on the wheels from the other side
6. Which among the following tire and wheel balancing methods gives the best ride quality?
   a. Road force balancing
   b. Dynamic balancing
   c. Static balancing
   d. Bubble balancing

7. Which of the following valve cores are recommended for use in TPMS sensors?
   a. Unplated brass caps
   b. Brass valve cores
   c. Nickel-plated valve cores
   d. Aluminum valve cores
8. Which of the following tools can act as a safety measure in the event of a tire explosion during inflation?
   a. Tire inflation cage
   b. TPMS torque wrench
   c. Air tire buffer
   d. Tire dunk tank
9. All of the following statements are true *except*:
   a. excessive loaded radial runout on the tire can be corrected by replacing the tire.
   b. the only method for fixing excessive lateral runout on the tire is to replace the tire.
   c. a wheel balancer is used to identify and correct wheel imbalances.
   d. the most common issue with tires is a broken belt.
10. Heavy pulling of a vehicle to either the left or the right could be due to any of the following *except*:
   a. mismatched tire sizes or pressures.
   b. out-of-alignment wheels.
   c. Nitrogen filled tires.
   d. a dragging front brake assembly.

ASE Technician A/Technician B Style Questions

1. Tech A says that when checking tire pressure, the tires should be "cold." Tech B says that the tires should be driven more than 3 miles before checking tire pressure. Who is correct?
   a. Tech A
   b. Tech B
   c. Both Tech A and B
   d. Neither Tech A nor B
2. Tech A says that tires should be rotated every 20,000 miles. Tech B says that when performing a tire rotation, each tire should be moved to the next clockwise position on the vehicle. Who is correct?
   a. Tech A
   b. Tech B
   c. Both Tech A and B
   d. Neither Tech A nor B
3. Tech A says that a static imbalance condition is centered within the width of the tire. Tech B says that a dynamic imbalance condition can be anywhere within the width of the tire. Who is correct?
   a. Tech A
   b. Tech B
   c. Both Tech A and B
   d. Neither Tech A nor B
4. Tech A says that a tire with more wear on the center of the tread is caused by under-inflation of the tire. Tech B says that feathering of the tire tread is most commonly a result of excessive toe-in or toe-out. Who is correct?
   a. Tech A
   b. Tech B
   c. Both Tech A and B
   d. Neither Tech A nor B
5. Tech A says that you should apply tire lubricant to the bead of the tire when removing the tire from the rim. Tech B says that the turntable jaws on many tire changers can hold the rim by grasping it from the outside or the inside. Who is correct?
   a. Tech A
   b. Tech B
   c. Both Tech A and B
   d. Neither Tech A nor B
6. Tech A says that lead wheel weights have been outlawed in some States. Tech B says that when balancing a tire, it is good practice to reuse the wheel weights to cut down on waste. Who is correct?
   a. Tech A
   b. Tech B
   c. Both Tech A and B
   d. Neither Tech A nor B
7. Tech A says that some TPMS sensors are part of the tire's valve stem. Tech B says that some TPMS sensors are held on by a large band that fits around the center of the rim. Who is correct?
   a. Tech A
   b. Tech B
   c. Both Tech A and B
   d. Neither Tech A nor B
8. Tech A says that when repairing a leaky tire with a plug patch, the patch goes on the outside of the tire. Tech B says that the plug patch must be applied while the glue is still wet. Who is correct?
   a. Tech A
   b. Tech B
   c. Both Tech A and B
   d. Neither Tech A nor B
9. Tech A says that it is a good practice to cover the buffed area and newly applied plug patch with a rubber patch sealant. Tech B says that it is good practice to use the mechanical buffer tool to smooth the plug patch after it has been applied. Who is correct?
   a. Tech A
   b. Tech B
   c. Both Tech A and B
   d. Neither Tech A nor B
10. Tech A says that an outside micrometer is used to measure axle flange runout. Tech B says that lateral runout of the tire, wheel, or axle flange causes the wheels to jiggle side to side. Who is correct?
   a. Tech A
   b. Tech B
   c. Both Tech A and B
   d. Neither Tech A nor B

CHAPTER 38

Steering System Theory

NATEF Tasks

- **N38001** Describe the function of suspension and steering control systems and components, (i.e. active suspension and stability control). (MLR/AST/MAST)
- **N38002** Identify hybrid vehicle power steering system electrical circuits and safety precautions. (MLR/AST/MAST)

Knowledge Objectives

After reading this chapter, you will be able to:

- **K38001** Describe the purpose, function, and main types of steering systems.
- **K38002** Describe the purpose of steering geometry and explain how the Ackermann principle applies.
- **K38003** Describe the purpose and function of rack-and-pinion steering components.
- **K38004** Describe the purpose and function of parallelogram steering components.
- **K38005** Describe the purpose and function of steering columns.
- **K38006** Describe the purpose, function, and variety of four-wheel steering systems.
- **K38007** Describe the purpose, function, and variations of steering boxes.
- **K38008** Describe the purpose, function, and components of the rack-and-pinion gearbox.
- **K38009** Describe the purpose, function, and variety of worm gearboxes.
- **K38010** Describe the purpose and function of a worm-and-sector gearbox.
- **K38011** Describe the purpose and function of the worm-and-roller gearbox.
- **K38012** Describe the purpose and function of a recirculating ball gearbox.
- **K38013** Describe the purpose, function, and variety of power steering systems.
- **K38014** Describe the purpose and function of the hydraulically assisted power steering system.
- **K38015** Describe the purpose and function of power steering fluid and hoses.
- **K38016** Describe the hydraulically controlled steering process.
- **K38017** Describe the purpose and function of the flow control valve.
- **K38018** Describe the purpose, function, and variety of electric power steering systems.
- **K38019** Describe the basic operation of basic electric power steering operation.

Skills Objectives

There are no Skills Objectives in this chapter.

▶ Introduction

Every driver knows that turning the steering wheel steers the wheels, which steer the vehicle. Not readily apparent are the many components that are installed between the steering wheel and the wheels that relay the signal and make the wheels steer. This chapter discusses the different steering systems, how they operate, and the purpose and function of each part within them. This understanding will give you a solid foundation for the next chapter, which discusses the steps and tools needed to inspect and service steering systems. Maintaining a well-functioning steering system is necessary for the safety of the driver, passengers, and others sharing the road. But our journey starts with understanding what makes the steering system work.

▶ Steering System Overview

K38001

The **steering system** provides control over the vehicle's direction of travel, good maneuverability for parking the vehicle, smooth recovery from turns as the driver releases the steering wheel, and minimal transmission of road shocks from the road surface through the steering wheel. As vehicle technology has progressed, steering systems have gone through a number of refinements to enhance vehicle safety, performance, and even fuel economy.

A basic steering system has four main parts: a steering column, a steering box, a steering linkage, and a steering knuckle (**FIGURE 38-1**). Added to the basic steering system is a power assist system that makes it easier for the driver to steer the vehicle. The power steering system can be either a hydraulic type or electric type. In the case of electric power steering, in addition to providing power assist, the system can be integrated with the electronic stability control system, such that the powertrain control module (PCM) can take command of the steering, if needed, to prevent loss of control of the vehicle. Electric power steering also makes four-wheel steering more of a feasible option.

The **steering column** transmits the driver's steering effort from the steering wheel down to the steering box. The **steering box** converts the rotary motion of the steering wheel to the **linear motion** needed to pivot the wheels. The steering box also uses principles of gear reduction to give the driver mechanical advantage over the wheels, making it easier to steer them. The **steering linkage** transfers the linear steering effort to the wheels by connecting the steering box to the **steering arm** on each of the steering knuckles, which pivot on the ball joints, allowing the wheels to steer the vehicle.

There are two main types of steering systems used on vehicles today: rack and pinion and parallelogram. Each one has its strengths and weaknesses, but rack-and-pinion systems are the most commonly used. The rack-and-pinion system gets its name because a rotating pinion gear is used to move a flat, toothed rack (**FIGURE 38-2**). The rack is connected through pivoting socket ends and a tie rod directly to the steering arms. It is a simple, compact system with only a few moving and pivoting parts. It fits well into most engine compartments and takes up less space than other systems, which means it is less likely to get in the way of other components. And because of the fewer parts and the orientation of the pinion gear to the rack, it gives a more precise steering feel, making it more responsive.

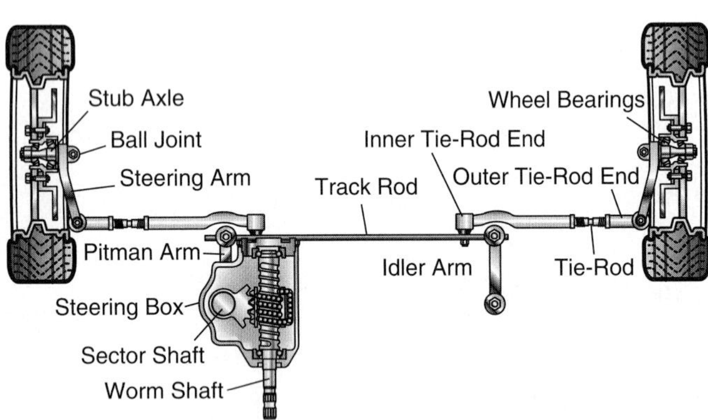

FIGURE 38-1 The components of a basic steering system.

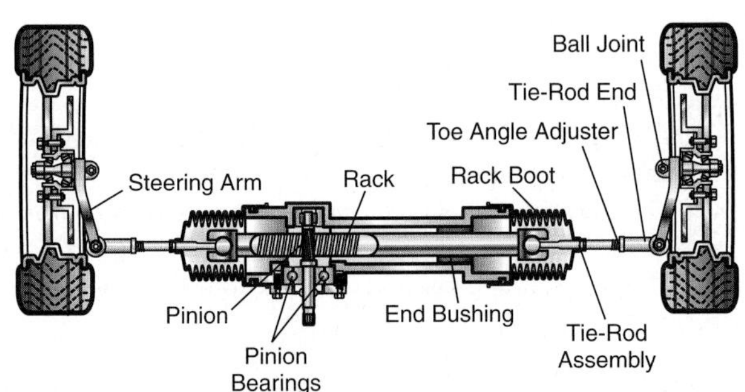

FIGURE 38-2 Rack-and-pinion steering system.

You Are the Automotive Technician

A regular customer who performs some of the maintenance on her car brings it in to you for a wheel alignment after she replaced some steering components. She has been reading up on steering system theory and has some questions for you.

1. What is the Ackermann principle, and how does it affect the steering system?
2. What are the differences between rack-and-pinion and parallelogram steering systems?
3. What do the balls do in a recirculating ball steering gear?
4. What does the power steering switch do?

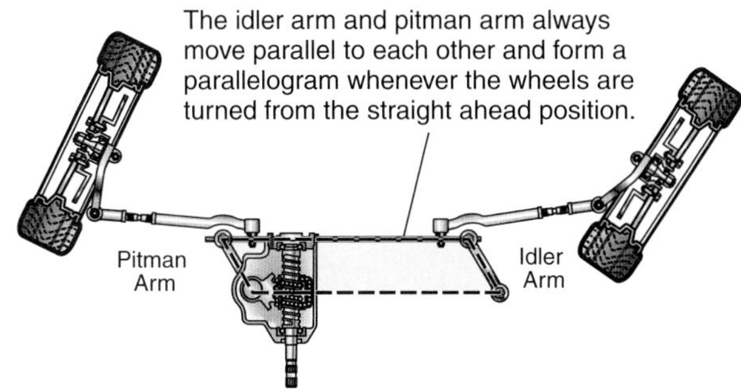
The idler arm and pitman arm always move parallel to each other and form a parallelogram whenever the wheels are turned from the straight ahead position.
Pitman Arm
Idler Arm

FIGURE 38-3 Parallelogram steering system.

The **parallelogram steering system** gets its name because the center link and axle, if equipped, along with the pitman arm and idler arm, always move parallel to each other (**FIGURE 38-3**). Pivoting tie rods connect the parallelogram to the steering arms near the wheels. Parallelogram steering uses a **worm** gearbox, which changes the direction of steering wheel rotation 90 degrees, and a pitman arm and center link to turn the rotary motion into lateral motion. The worm gear design reduces the road shock that is transmitted to the steering wheel, so a parallelogram design provides an advantage, especially in off-road four-wheel drive vehicles or non-sporty vehicles.

The **steering knuckles** are stout components that firmly connect the wheels to the suspension and steering systems (**FIGURE 38-4**). They provide a stub axle upon which the wheel bearings ride, or a hub-style wheel bearing assembly that is pressed or bolted onto the steering knuckle. The steering knuckle pivots on one or two ball joints, depending on the type of suspension, which is covered in the Servicing Suspension Systems chapter. The steering arm transmits the steering force from the steering linkage to the wheel and tire assembly.

Steering Geometry

K38002

The relationships between the steering system, the wheel positions, and the suspension system form what is called the steering geometry. Steering geometry is a geometric arrangement of linkages in the steering of a vehicle designed to solve the problem of keeping the wheels properly oriented through various positions of the steering and suspension systems. As the wheels move up and down relative to the body, the steering linkage swings vertically through an arc. The wheel would thus turn in and out as the vehicle goes over bumps, if it were not for steering geometry. In this case, the pivots for the suspension components cause the wheel to go through a similar arc as the steering components, allowing the wheel to track straight ahead, or in a consistent direction if it is in a turn.

Also, when rounding a corner, the inner and outer wheels must trace circles of different radii; otherwise, the tires would be drug across the road surface, which is called scrub. This is a challenge because no matter which way the vehicle turns, the inside wheel must always turn more sharply than the outside wheel, which is called toe-out on turns. The Ackermann principle, named after the man who patented it in 1818, provides the

FIGURE 38-4 A typical steering knuckle that supports the wheel and brake assembly. It also is the connection point between the wheel and the steering and suspension system.

needed geometry. The Ackermann principle angles the steering arms toward the center of the vehicle such that imaginary lines drawn from the center of the steering knuckle pivot points, through the center of the outer tie-rod ends, intersect at the center of the rear axle (**FIGURE 38-5**). This angle is what allows the wheels to navigate a corner around a common center point, which minimizes tire scrub (see Figure 38-3 for an example of the different wheel angles when turned to the right). Steering geometry and suspension geometry must be within the manufacturer's specifications for the vehicle to operate correctly. If it is not correct, it is usually because of worn or bent components that the technician will have to diagnose and ultimately repair to restore proper operation.

Rack-and-Pinion Steering Linkage

K38003

The **rack-and-pinion steering system** is used on the majority of vehicles today because of the restriction of space under the hood along with their compact and lightweight design (**FIGURE 38-6**). Steering response is very sharp because the rack operates directly on the steering knuckle, and there is very little sliding and rotational resistance, which gives lighter operation. The primary components of the rack-and-pinion steering system are listed here:

- **Pinion**: A toothed helical cut gear that meshes with the rack (**FIGURE 38-7**). The pinion is connected to the steering column. As the driver turns the steering wheel, the forces are transferred to the pinion, causing the rack to move in either direction. This is achieved by having the pinion in **constant mesh** (constant contact) with the rack.
- **Rack**: A toothed, straight piece of metal that meshes with the pinion in the middle of the rack and has tie rods on each end that fasten to the steering knuckle. The rack slides in the housing and is moved by the action of the **meshed pinion**. It normally has a spring-loaded or **adjustable bushing** positioned opposite the pinion to control the components' meshing, and it has a

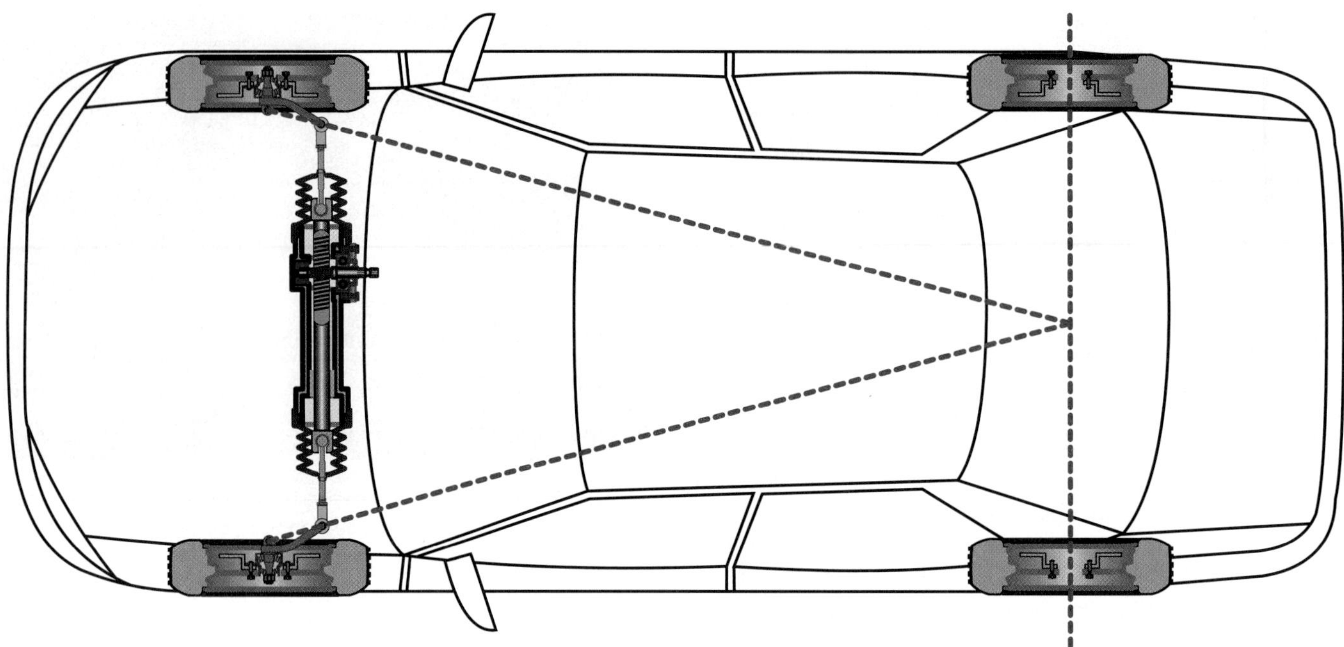

FIGURE 38-5 Angling the steering arms inward allows the inner wheel to turn more sharply than the outer wheel.

FIGURE 38-6 A typical rack-and-pinion steering gear.

FIGURE 38-7 Pinion and rack.

nylon bushing at the other end. This adjustable bushing allows the technician to adjust play out of the rack and pinion. The function of the rack is to transfer motion to the tie rod.

- **Inner tie rod or socket**: The inner tie rod has an inline ball-and-socket joint on one end of a shaft and is threaded on the other end of the shaft. The ball-and-socket joint threads onto one end of the rack (**FIGURE 38-8**). The other end of the inner tie rod threads into the outer tie-rod end. These two joints allow for suspension and steering angle movement. The ball-and-socket joint is arranged in line with the rack, so the tie rod is free to spin in the ball and socket when making adjustments in wheel alignment.
- **Outer tie-rod end**: An outer tie-rod end is attached between the inner tie-rod shaft and the steering arm (**FIGURE 38-9**). It transfers the movement of the rack to the steering arm. The outer tie-rod end is threaded onto the end of the inner tie-rod shaft, and a jam nut locks the shaft and outer tie rod in place. The jam nut allows the length of the tie rod to be adjusted longer and shorter when making adjustments in wheel alignment.
- **Rubber bellows**: The rubber bellows protects the inner joints from dirt and contaminants (**FIGURE 38-10**). In addition, it retains the grease lubricant inside the rack-and-pinion housing. There are two ends of the rack. Each side contains an identical bellows.

Parallelogram Steering Linkage

K38004

The parallelogram steering system may be used on larger vehicles where ride comfort is more important than sporty

FIGURE 38-8 Inner tie rod end or socket.

FIGURE 38-9 Outer tie rod end.

FIGURE 38-10 The rubber bellows seals the end of the rack-and-pinion assembly. Underneath is the inner tie-rod end socket.

FIGURE 38-11 The pitman arm is bolted to the gear box. It supports one end of the center link. It also transmits steering force to the center link.

handling. Parallelogram steering linkage is more complicated than the rack-and-pinion linkage, so there are more wear points to know about and inspect. This is why most vehicles today use rack-and-pinion steering. The primary components of the parallelogram steering system, from the gearbox on, are as follows:

- **Pitman arm**: The pitman arm transfers movement from the steering box to the center link (**FIGURE 38-11**). It is attached to the steering box by a **spline** and nut. Splines are ridges or teeth on a driveshaft that mesh with grooves in a mating piece and transfer torque to it and maintain the angular correspondence between the components. As the driver turns the steering wheel, the steering box mechanism moves the steering linkages via the pitman arm either left or right, depending on the direction in which the steering wheel is turned. The steering box provides the change of angle at 90 degrees to the steering linkage and also provides a steering ratio (the number of turns of the steering wheel compared to the number of turns of the sector shaft) to assist in the ease of turning the steering wheel and linkages.
- **Idler arm**: The idler arm is attached to the **chassis** (the frame of the vehicle) and is positioned parallel to the pitman arm (**FIGURE 38-12**). The idler arm assembly is the pivoting support for the steering linkage. It consists of a rod that pivots on a bracket bolted to the frame of the vehicle on one end and supports a ball socket on the other end. Generally, an idler arm is attached between the opposite side of the center link from the pitman arm and the vehicle's frame, to hold the center link at the proper height so it can accurately relay the pitman arm's movement. Idler arms are generally more vulnerable to wear than pitman arms because of the pivot function built into them.
- **Center link**: The center link connects the pitman arm to the idler arm (**FIGURE 38-13**). In this way, any movement in the pitman arm is directly applied to the idler arm. It also may be called a track rod or drag link.
- **Tie rod**: The tie rods connect the center link to the steering arms that are located on the steering knuckles (**FIGURE 38-14**). Tie rods use a pivoting tie-rod end at each end, which allows the steering linkage to move only as needed.

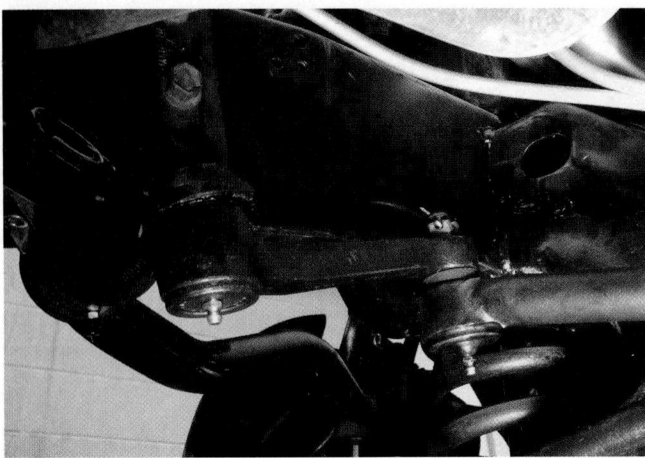

FIGURE 38-12 The idler arm supports the other end of the center link and travels in parallel with the pitman arm.

FIGURE 38-13 The center link connects the pitman arm to the idler arm, so they both move together. Also, each end of the center link is connected to one of the tire rods.

FIGURE 38-14 Tie rods connect the center link to the steering arms. They have pivoting tie-rod ends at each end.

- **Tie-rod end:** Tie-rod ends are attached to each end of each tie-rod shaft. The inner tie-rod ends attach each tie rod to the center link and serve as pivot points. The outer tie-rod ends attach each tie rod to the steering arm. Tie-rod ends pivot as the steering linkage is extended or retracted when the vehicle is negotiating turns. Tie rods and tie-rod ends are left- or right-hand threaded so that when used with an adjustment sleeve, the length of the tie rod can be adjusted during a wheel alignment.
- **Adjustment sleeve:** The adjustment sleeve connects the tie rod to the tie-rod end (**FIGURE 38-15**). The adjustment sleeve is split lengthwise (either partially or completely) with an internal thread that matches the external threads on the tie rods and tie-rod ends. It provides the adjustment point for adjusting the toe setting and is similar to a turnbuckle. When a technician turns the sleeve, the tie-rod threads shorter or longer, which is used to adjust the wheel alignment.

▶ TECHNICIAN TIP

Each type of steering system makes provision for adjustment of the linkage to achieve the manufacturer's recommended **toe setting**. The toe setting is the symmetric angle that each wheel makes with the longitudinal axis of the vehicle. The track rods or tie-rod ends are threaded to provide for their lengthening or shortening.

Each connection point in the steering linkage system has a pivoting joint, which is a ball-and-socket construction (**FIGURE 38-16**). Flexible joints on the center link and on the ends of the tie rods allow for steering and suspension movement. Ball sockets on the tie rods allow steering movement and movement of the suspension. The tie-rod end swivel joints are at equal heights and pivot so that movement in the suspension does not affect steering operation. Movement of the suspension that does affect steering is called bump steer. **Bump steer** is the undesired condition produced when, upon hitting a bump, the vehicle darts to one side as the steering linkage is pushed or pulled as a result of the travel of the suspension. Typically, bump steer is due to a bent tie-rod end or another bent component that would cause unequal heights of the

FIGURE 38-15 The adjustment sleeve is threaded to connect each tie rod to the tie-rod end. One end has right-hand threads, and the other end has left-hand threads for adjustment purposes.

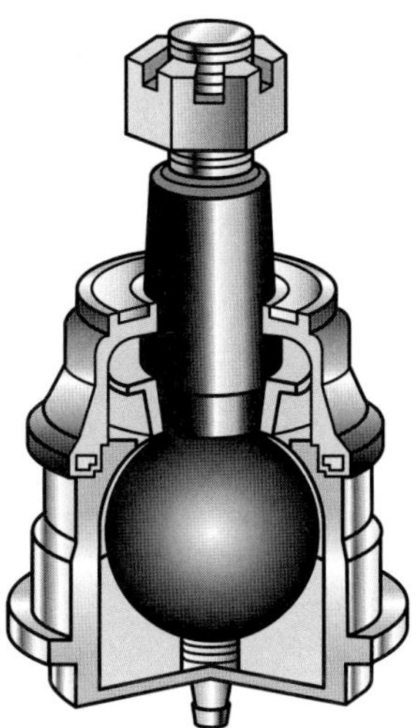

FIGURE 38-16 The flexible joints of the steering linkage are a ball-and-socket construction, which allows movement of the linkage with no play in the joint.

tie-rod ends. The tie-rod end swivel joints allow the suspension to move up and down without affecting vehicle steering.

In four-wheel drive vehicles with a **beam axle**, a single tie rod connects the steering arms on each wheel assembly. In this design, the drag link is connected to an arm on the front of the left-hand wheel assembly. Movement of the pitman arm is transferred through the drag link to the left-hand wheel and through the single tie rod to the right-hand wheel (**FIGURE 38-17**). The steering box is offset from the steering column, so two universal joints and an intermediate steering shaft are used. Four-wheel drive vehicles of this type often use a **steering damper** on the single tie rod. It resembles a shock absorber and operates on a similar principle. Specifically, it keeps the steering wheel from shaking when the driver hits a pothole or other road irregularity. The steering damper is mounted between the tie rod and either the rigid axle or the vehicle frame. When the vehicle is driven over rough terrain, its purpose is to prevent shock forces being transmitted through the steering linkage and back to the steering wheel.

Steering Columns

K38005

Effort applied to the steering wheel is transferred down the steering column, or shaft, to a steering box. In early vehicles, the steering column was a straight shaft, running inside a hollow tube with bushings. The steering wheel was attached to one end, and the steering box to the other. In many frontal collisions, however, this design caused serious injury to the driver, partly because the steering wheel was forced back toward the driver's head and chest, and partly because the sudden stop forced the driver onto the wheel. To reduce this hazard, all steering columns are now fitted with collapsible sections that help protect the driver. During a collision, two forces are applied to the steering column. The first is the force of the steering box being forced toward the steering column and toward the driver. Plastic shear pins allow the lower shaft to move over the upper shaft. The second force is the mass of the driver striking the steering wheel. This force breaks the brackets on the upper part of the column, driving the upper column into the lower column (**FIGURE 38-18**).

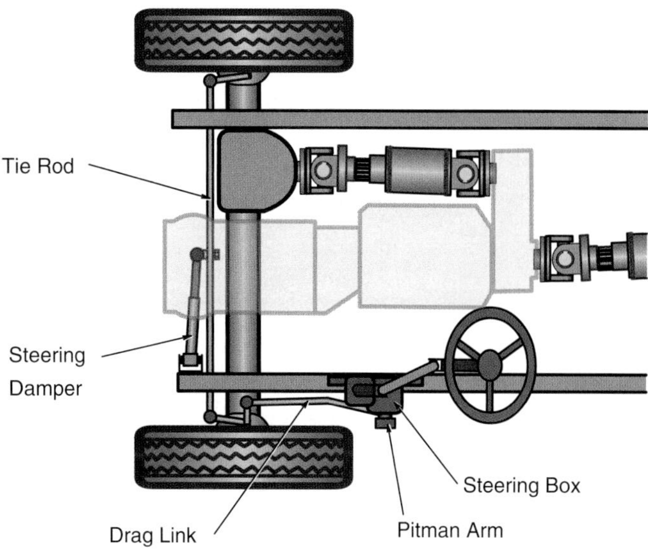

FIGURE 38-17 Steering linkage layout on a solid-axle four-wheel drive vehicle.

FIGURE 38-18 Collapsible steering column.

Some steering columns use an **intermediate shaft**, which runs at an angle from the column to the steering gear. In a collision, the universal joints on the intermediate shaft allow the steering gear to fold under, preventing the impact force from being transferred directly to the column (**FIGURE 38-19**). As added protection in a collision, most vehicles also integrate a driver's side airbag into the steering wheel. This system is discussed in greater detail in a later section.

The steering column is connected to the input shaft of the steering gear by a flexible joint. This flexible joint allows for smooth turning of the steering wheel because the steering column sits at a different angle than the steering box. It also reduces the transmission of road shocks to the driver.

Some manufacturers fit sensors and an electric motor to the steering column. The sensors in the steering column provide information to a steering control module for electric power steering assist or for the electronic stability control system. The sensors are able to sense rotational direction and speed exerted

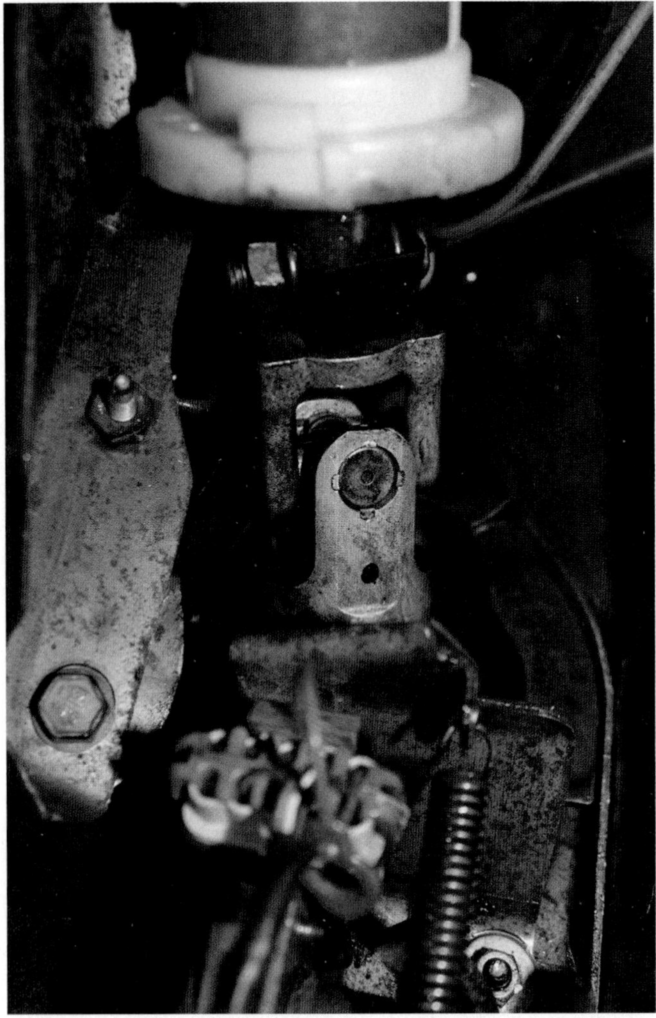

FIGURE 38-19 Steering column with an intermediate shaft for enhanced protection in an accident.

FIGURE 38-20 **A.** Multifunction switch. **B.** Steering wheel entertainment controls.

by the driver and control both the speed and the direction of the electric motor, which is used to provide the electric power steering assist. This system will be discussed further in a later section.

The steering column generally accommodates for a horn button and an ignition switch and lock assembly, although more new vehicles are coming with a push-button start switch located in the dash. The steering column also incorporates a multifunction switch (or separate switches) that may include switches for lights, turn signal indicators, wipers, washer, and cruise control. The steering wheel itself may have integrated control switches for the entertainment system, a cell phone, and instrument panel display options (**FIGURE 38-20**).

Tilt/Telescoping Mechanism

To compensate for variations in driving positions, many manufacturers have included a steering column tilting and/or telescoping mechanism to their vehicles. These features allow drivers to control the steering wheel position so that it best suits their preferences. The tilting mechanism allows drivers to raise or lower the steering column position, and the telescoping mechanism allows the driver to move the steering wheel closer or farther away. However, the amount of movement in any direction does have its limitations.

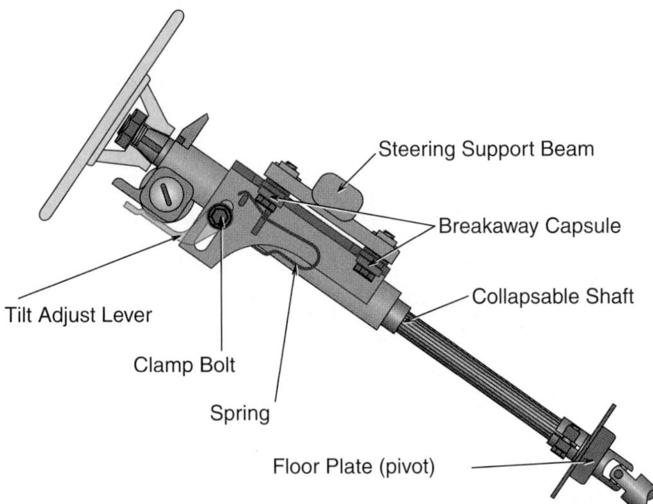

FIGURE 38-21 Typical tilt steering wheel mechanism.

The tilt mechanism uses a heavy spring, a pivoting joint, and a ratchet mechanism to operate (**FIGURE 38-21**). These added components make the steering column more complicated to repair. The telescoping mechanism includes a slip joint and locking mechanism. Both of these features can also put the wires in the steering column under additional stress due to more frequent movement, so remember that when diagnosing wiring faults.

Driver's Side Airbag

The driver's side airbag is designed to provide a collapsing cushion that decelerates a driver's head and chest during an accident. One of the leading causes of fatalities in accidents is the rapid deceleration forces that occur within a body during the accident. The body's internal organs cannot handle the force of sudden deceleration, so engineers have devised a variety of ways to slow the deceleration during an accident. The use of crumple zones in the vehicle, which fold and crush accordion-like, provides some deceleration. Restraint systems such as tear-away seat belts also are designed to slow the deceleration. Airbags also decelerate the driver and prevent the head from contacting the steering wheel or dash. And knee airbags provide protection for knees and lower legs. Side curtain airbags ensure that the driver does not hit the side rail of the roof when involved in a side collision. Next to the seat belt, airbags are the biggest lifesaver in accidents. In most countries, government regulation requires that a vehicle equipped with airbags be capable of having the airbags replaced as part of the repair process of a crashed vehicle. If the airbag has been deployed in an accident, a new airbag and cover must be installed to meet legislative requirements in most states. At the same time, manufacturers may require that other specified components be inspected or replaced in the event of an accident or air bag deployment.

When a technician needs to carry out a servicing procedure on the steering column, the driver's side airbag, which is located in the top cover of the steering wheel, must be disabled. If not properly disabled, the airbag could trigger accidentally, injuring or killing the technician because of the large deployment force created by the ignited propellant. Unintended deployment also creates the need to replace the airbag assembly and other required components, which is expensive. Generally, to disable a driver's side airbag, the correct supplemental restraint system (SRS) fuse must be located and removed, and then verified by turning the ignition switch to the run position and observing the SRS light, which should remain on. Once the fuse is removed, the negative battery terminal is disconnected, and the vehicle is allowed to sit for up to 15 minutes to discharge the capacitors, which store power to deploy the SRS devices in the event power is severed in an accident. Note that a memory minder *should not* be used when working on, or around, the SRS system. Also, do not use a battery, voltmeter, ohmmeter, test light, or any other type of test equipment not specified in the service manual on any airbag, airbag squib, or airbag circuit, in order to avoid accidental deployment of the airbag, which could result in injury or death.

Some manufacturers have you disconnect the airbag connector located under the dash near the steering column. Disconnecting this connector generally activates a shorting bar in the airbag side of the connector, which makes it much harder for the airbag to accidentally deploy (**FIGURE 38-22**). With the airbag disabled, you can carry out any needed diagnosis or repair on the steering column. Just remember to enable the airbag once service is completed. Follow the vehicle's specific service information to properly disable and enable the airbag.

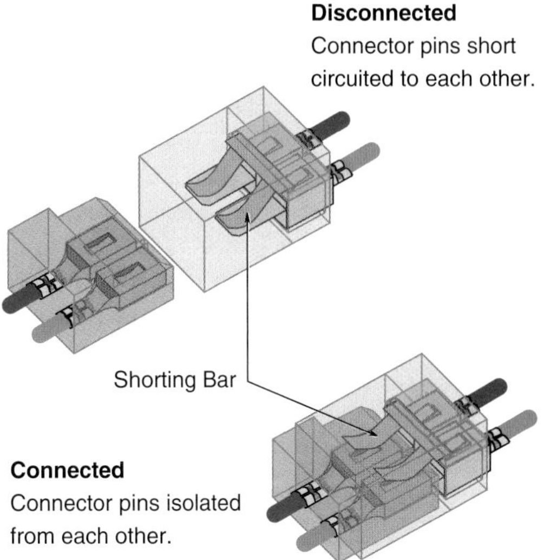
Disconnected
Connector pins short circuited to each other.
Shorting Bar
Connected
Connector pins isolated from each other.

FIGURE 38-22 A shorting bar inside most air bag connectors shorts out the terminals when the connector is disconnected. This helps avoid accidental deployment when the air bag is removed from the vehicle.

SAFETY TIP

Never lean over or in front of an armed airbag when working on the dash, instrument panel, or airbag system. Doing so can cause serious injury or death if it accidentally deploys. Always follow the manufacturer's procedure when disarming and arming the airbag.

Because of the critical nature of the airbag in an accident, it is imperative that it is always connected electrically to its control module, so it can be deployed when needed. Because the airbag is mounted in the steering wheel, a method of maintaining the electrical connection to the airbag is not as easy. This is overcome by incorporating a clock spring into the steering column. The **clock spring**, which is also called a spiral cable, is a special rotary electrical connector located between the steering wheel and the steering column that maintains a constant electrical connection with the wiring while the vehicle's steering wheel is being turned. The wires from the airbag and its electrical system connect through the clock spring, which coils and uncoils as the wheel is turned, and maintains constant electrical contact. When removing the airbag assembly from the steering wheel, inspect the wiring and connector of the airbag to the clock spring to ensure there is no damage. When the steering wheel is removed and the clock spring is easily visible, inspect it to make sure that it and the clock spring wire are not damaged. Typically, if the electrical wiring inside the clock spring is damaged, an SRS system code will set, and the SRS light will turn on. If the wiring has been damaged in any way, it must be replaced and cannot be repaired.

TECHNICIAN TIP

When removing the rack-and-pinion or steering gear, make sure the wheels are pointing straight ahead before disassembly. Also, either lock the steering column with the ignition key or use a steering wheel lock tool to prevent the clock spring from being moved off center. Make sure to also reassemble the parts in the straight-ahead position.

Four-Wheel Steering Systems

K38006

Some vehicles have **four-wheel steering**, meaning the rear wheels can be steered independently of or in conjunction with the front wheels (**FIGURE 38-23**). The ability to steer the rear wheels improves high-speed handling and increases maneuverability during driving tasks such as U-turns and parallel parking. There are two types of four-wheel steering systems—active and passive. Passive systems use compliant rubber bushings that allow a limited amount of rear-wheel steering, typically toward the inside of the corner, under cornering maneuvers (**FIGURE 38-24**). The passive system operates independently of the steering wheel and driver input, which is why it is called a passive system.

In active four-wheel steering systems, the front wheels are controlled normally, and the rear wheels are typically steered through use of a computer and electric motors. In past designs, rear-wheel steering systems could be controlled mechanically through a direct connection between the front and the rear steering boxes.

In modern computer-controlled four-wheel steering, an actuator similar to a front rack-and-pinion assembly attaches to the rear steering knuckles with tie-rod ends. The actuator turns the wheels when commanded by the steering control module (**FIGURE 38-25**). Originally, some actuators were powered hydraulically by the power steering pump and electronic control valves. Most systems today use a rack-and-pinion assembly driven by an electric motor.

Some manufacturers have adopted the standard that at low speeds the rear wheels will turn in the opposite direction to the front wheels, providing a substantially reduced turning radius; at higher speeds the rear wheels will turn in the same direction as the front wheels, producing smaller turning forces (yaw) during turning maneuvers. This standard provides better control of a vehicle at high speeds and better handling at low speeds, which is useful for functions like parallel parking. Other manufacturers use systems that disable rear steering at higher speeds and only allow rear steering at slow speeds, to allow tighter cornering, such as when backing up a trailer.

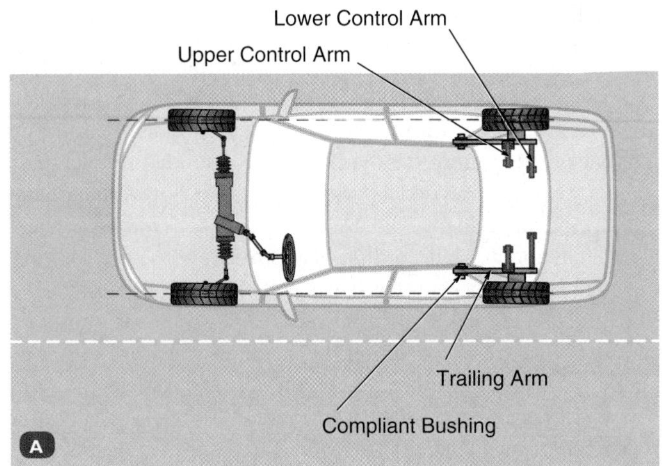

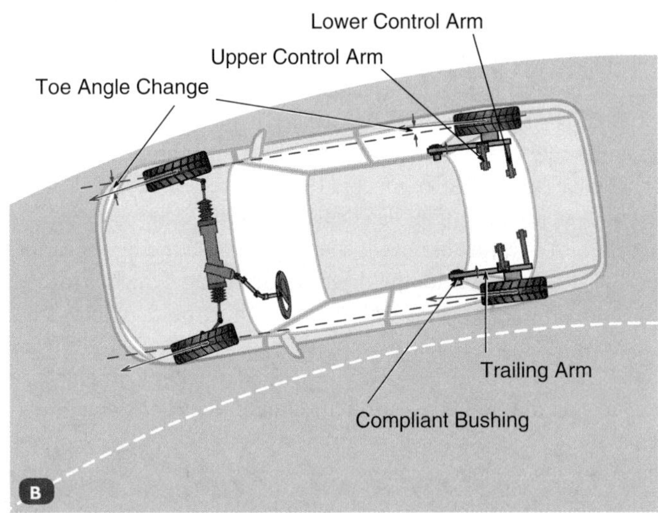

FIGURE 38-24 A passive four-wheel steering system uses compliant rubber bushings that allow a limited amount of rear-wheel steering during cornering.

FIGURE 38-23 Four-wheel steering showing the rear wheels being steered in conjunction with the front wheels.

FIGURE 38-25 The rear actuator for an active four-wheel steering system turning the rear wheels.

▶ Steering Boxes

K38007

There are a number of variations for steering boxes (steering gears). They are either manual steering or power-assisted steering. Manual steering has no power source to help make the wheel easier for the driver of the vehicle to turn, whereas power-assisted steering uses either a hydraulic pump, driven by the engine or an electric motor, or an electric motor that directly aids the driver in turning the wheel. Older vehicles were available with manual steering, and power steering was an expensive option. This is not the case now, as any vehicle purchased comes standard with a form of power steering. The function of all steering boxes is the same, whether manual or power—to transfer the rotary motion of the steering wheel into the side-to-side motion needed to make the wheels pivot left and right.

There are two basic types of steering boxes: those with rack-and-pinion gearing (**FIGURE 38-26**) and those with worm gearing and a sector shaft (**FIGURE 38-27**). In both cases, the gearing in the steering box makes it easier for the driver to turn the steering wheel, and hence the wheels. Some variations of steering boxes are listed here:

- The rack-and-pinion gearbox
- The worm gearbox, consisting of
  - worm and sector;
  - worm and roller; or
  - worm and nut (commonly referred to as the recirculating ball).

Rack-and-Pinion Gearbox

K38008

The rack-and-pinion gearbox has a pinion connected to the bottom of the steering column. This pinion runs in mesh with a rack that is connected to the tie rods. This connection gives more direct operation, which enables the driver to feel the road better. Turning the steering wheel rotates the pinion and moves the rack from side to side. On end take-off racks, ball sockets at the end of the rack locate the tie rods and allow movement in the steering and suspension (**FIGURE 38-28**). On center take-off racks, the tie rods connect to the center of the rack and pinion (**FIGURE 38-29**). The center of the rack is what moves on this style of rack and pinion.

The teeth on both the pinion and the rack are helical gears. The particular shape and positioning of the teeth in a helical gearing enable smooth and quiet operation for the driver.

Mechanical advantage is gained by the **reduction ratio**, the ratio between the turns of the steering wheel and the angle turn of the wheel (both measured in degrees). The value of this ratio depends on the size of the pinion. A small pinion gives easy steering, but it requires many turns of the steering wheel to

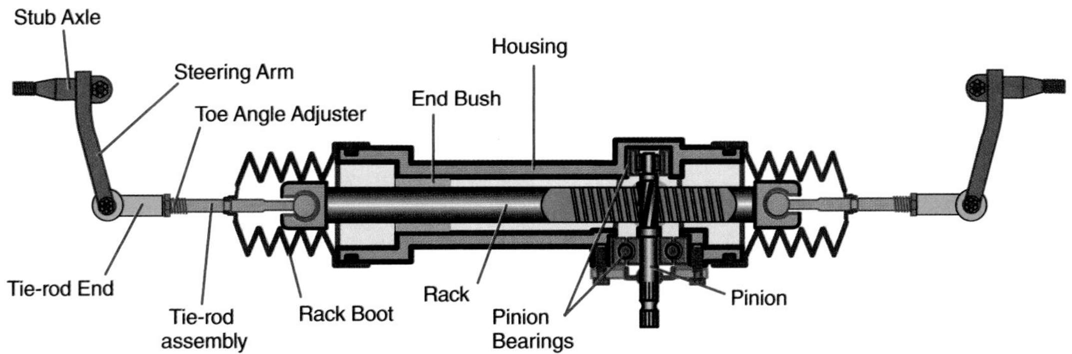

FIGURE 38-26 A rack-and-pinion steering gear.

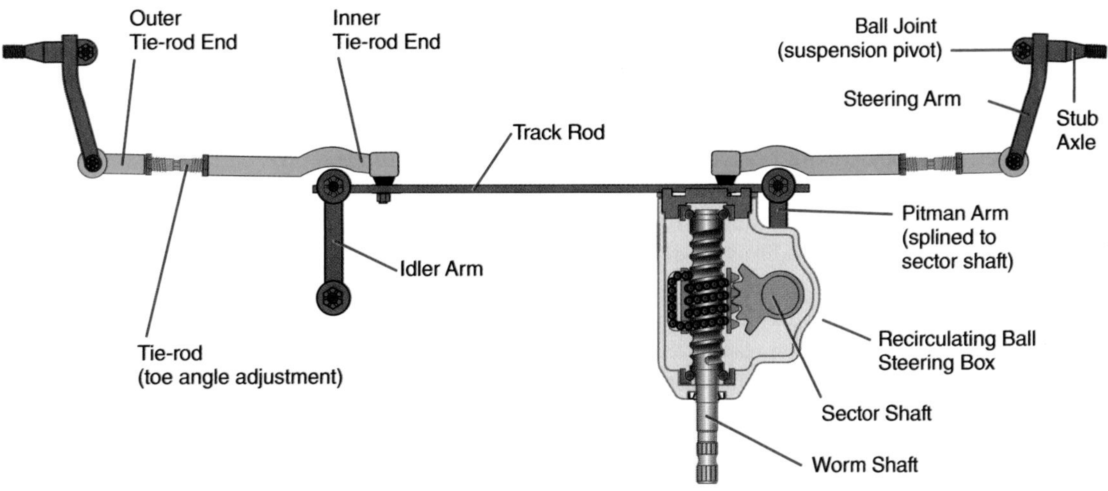

FIGURE 38-27 A worm steering gear.

FIGURE 38-28 The end take-off rack uses ball sockets at each end of the rack.

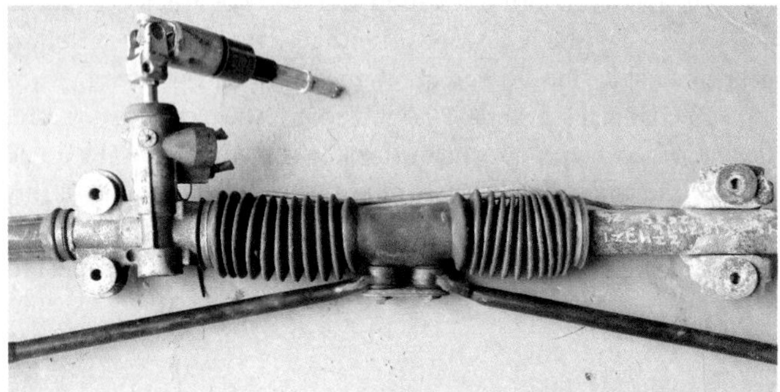

FIGURE 38-29 The center take-off rack uses ball sockets at the center of the rack.

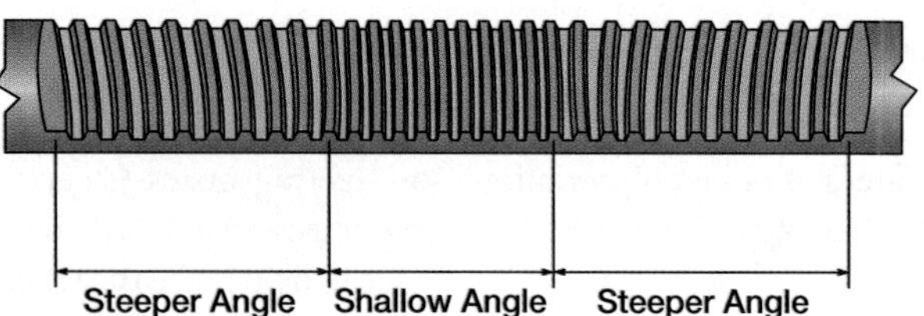

FIGURE 38-30 A variable-ratio rack showing the different angles on the teeth.

travel from lock to lock, which is as far as the steering wheel can be turned from one side to the other. So, instead of one single hard rotation of the steering wheel, the driver rotates the wheel several times with less force. A large pinion means the number of turns of the steering wheel is reduced, but the steering is harder to rotate.

Reduction ratios vary depending on the type of vehicle, such as a passenger vehicle versus a tractor trailer. But in each case, in fixed-ratio rack-and-pinion systems, the ratio is the same for all positions of the wheels. It is a fixed ratio, meaning that no matter where the steering wheel is positioned, the rack-and-pinion system produces a single ratio from lock to lock. The ratio is called steering ratio and is typically calculated by turning the steering wheel one time and checking the number of degrees that the wheel assembly moves. For example, if one turn of the steering wheel produces 20 degrees of movement at the wheel, then dividing 360 degrees of steering wheel rotation by 20 degrees of wheel movement results in an 18:1 ratio. Steering ratio is not the same on all vehicles and is designed by engineers to provide the best steering response for the vehicle.

In some applications, a variable-ratio gear is used, which allows a slow turn rate when the steering wheel is centered and a quicker rate when turned toward the steering lock. This feature provides easier and quicker turning, but keeps the vehicle very stable when traveling straight at high speeds. The variable-ratio rack-and-pinion system uses a specially designed pinion gear tooth, which is more rounded. The rack has teeth that are narrowly spaced (at a shallow angle) in the center and further apart (at a steeper angle) toward the end (**FIGURE 38-30**). Closer (shallow angle) teeth provide slower movement of the rack in relation to the turns of the pinion.

On other vehicles, it is preferable that the steering be more responsive near the center point, which is ideal for maneuvering at slower speeds. This is accomplished by having widely spaced (at a steeper angle) teeth in the center of the rack and closer spaced (at a shallow angle) teeth near the ends. When designed this way, the steering wheel moves the rack further for each rotation of the pinion when in the center position than when at the ends. This also helps prevent stiffer steering toward the full lock positions, because the steering arm leverage is reduced when the wheels are turned sharply.

The rack-and-pinion steering system has the advantages of a large degree of feedback and direct steering feel, meaning the driver can feel the road well. A disadvantage of the rack-and-pinion steering system is that it typically is not manually adjustable. Over time it does wear and develop gear backlash between the pinion teeth and the rack teeth, which is felt as play, or movement, in the steering wheel that does not move the wheels. In many cases, replacement of the rack and pinion is the only cure for excessive gear backlash.

In the rack-and-pinion steering system, the steering rack is supported at the pinion end by being sandwiched between the pinion and a **spring-loaded rack guide yoke**, sometimes called the rack bearing (**FIGURE 38-31**). Spring-loaded rack guide yokes are made of metal, nylon, or other durable material and have a spring that pushes on the back side of the rack to help reduce the play between the rack and the pinion while still allowing for relative movement. There may be an adjuster plug that screws into the body of the rack to put pressure on the rack guide yoke. This adjustment produces the correct mesh of the rack teeth to the pinion teeth. It also affects the amount of torque required to turn the pinion. If the setting is incorrect, such as if you are trying to eliminate play in the rack and pinion, turning force of the pinion will become greater and may result in a steering wheel that will not return properly, a condition sometimes referred to as memory steer. Do not adjust the mesh of the rack-and-pinion gears unless directed by service information.

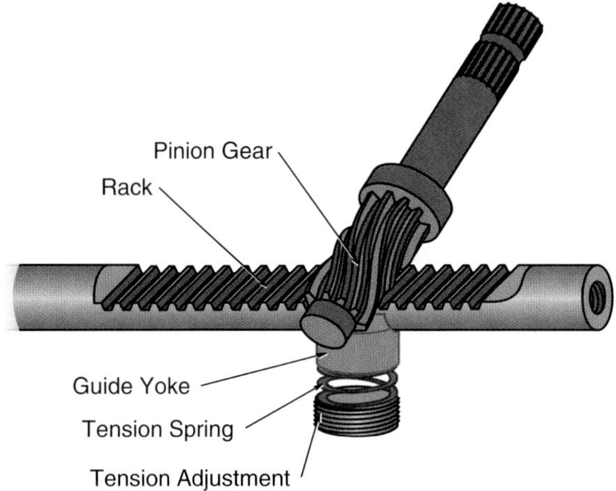

FIGURE 38-31 A spring-loaded rack guide yoke holds the rack against the pinion gear.

The rack is typically supported at both ends of the **rack housing** or tube by a bushing, a nylon piece that keeps the rack from wearing on the metal housing (see Figure 38-28). Nylon is used because it has a low coefficient of friction and low wear rates. The bushings are an integral part of any steering and suspension system. The bushings enable the flexibility needed to accommodate slight radial or lateral movement to assist in **antibinding**, or release when something gets stuck. However, when the bushings wear beyond the limits, they become a liability because they no longer are effective at eliminating play in the steering rack. Regular maintenance and inspection of the steering system is necessary to ensure the safety of the vehicle and its occupants.

The pinion is supported by two bearings in the rack housing (see Figure 38-29). The bearings must be **preloaded** (lightly compressed) to ensure the pinion is in the correct position, relative to the rack, and to eliminate free play.

A rack-and-pinion steering box is normally lubricated by grease. Each end of the rack is protected from dirt and water by a flexible synthetic rubber bellows attached to the rack housing and to the tie rod. The rubber bellows extends and collapses as the tie rods move away from and toward the rack housing as the rack moves. On some vehicles, the rubber bellows are interconnected by a tube so that as the steering wheel is moved from side to side, air is transferred from the collapsing bellows side to the expanding bellows side. This process keeps the rubber bellows from collapsing.

With rack-and-pinion steering, the rack is directly connected to the **tie-rod assembly**, which is attached to the steering knuckle. The tie-rod assembly consists of an inner and an outer tie-rod end that are threaded together.

Worm Gearbox

K38009

Worm gear steering boxes made the process of turning the front wheels an easier task for drivers of the early automobile. The worm gearbox uses two gears: a worm and a worm gear (also called a worm wheel) (**FIGURE 38-33**). The worm has teeth cut in a helical

Applied Math

AM-52: Probability: The technician can relate problem symptoms to the probability of the malfunction of a specific part or system.

A common problem with steering systems, particularly in older vehicles, is excessive free play. Typically, the customer will describe the problem as the steering feeling "loose" or "sloppy." A road test will quickly verify the concern; there will be excessive movement of the steering wheel in the straight-ahead position, before the wheels even begin to turn.

With a pickup truck that has a basic steering system, the excessive free play could be caused by wear in a number of points within the system. There is a higher probability of the fault being caused within the steering linkage than within the steering gearbox, so components such as tie-rod ends, idler arms and bushings, and pitman arms and bushings should be checked and eliminated first, before inspecting the steering gearbox.

The inner tie rod is threaded onto the rack and has a ball-and-socket swivel joint to allow movement in any direction. The other end of the inner tie rod is threaded to allow attachment of the outer tie-rod end, which provides the method used to adjust the toe angle. The steering arm, the **stub axle** knuckle, and the **stub-axle carrier** (the body of the stub axle knuckle) can be forged as one piece and referred to as a steering **knuckle** (**FIGURE 38-32**). They can also be made as separate units and assembled to form one piece.

Figure 38-32 Two-piece steering knuckle and steering arm.

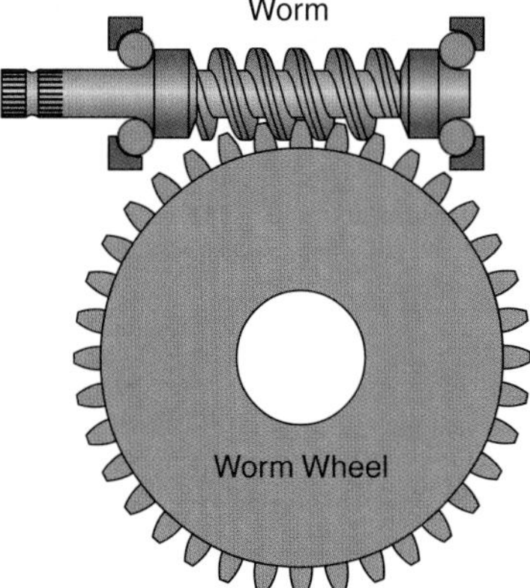

FIGURE 38-33 The parts of a worm gear assembly: **A.** Worm. **B.** Worm gear (or worm wheel).

(spiral) shape and operates the same as a screw. In this case, the helix on the worm moves the worm wheel one tooth for each revolution of the worm. It converts the rotary motion of the steering wheel to the linear motion needed to control the wheels. This system also produces a large gear reduction. The gear ratio of the gearbox is a comparison of the angles of movement between the steering wheel and the wheel assembly and is expressed as a ratio, for example, 18:1. The gear ratio of a worm gearbox increases output torque and reduces the effort the driver has to apply. Wheels require a large force to be steered because of the weight of the vehicle pushing the tire into contact with the road.

Worm gears have another benefit: They do not transmit nearly as much road shock back to the driver as a rack-and-pinion gear assembly (**FIGURE 38-34**). When the worm is driven (by the steering wheel), the output gear moves easily because each turn of the worm advances the worm wheel one tooth. But if the worm wheel is driven (such as by road shock), the worm wheel teeth butt up against the teeth on the worm, preventing most of the force from being transmitted as rotary motion to the worm. This occurs because most of the force from the output gear is nearly perpendicular to the rotation of the worm gear.

Several forms of the worm gearbox have been used throughout the history of the automobile, although they all use the same component—the worm gear arrangement. Three popular styles have included the worm-and-sector, worm-and-roller, and recirculating ball gearboxes.

Worm-and-Sector Gearbox

K38010

The first worm gearbox was of the worm-and-sector style. The worm is meshed with a sector, or portion of a gear, mounted on its own shaft (called a sector shaft) at right angles to the worm (**FIGURE 38-35**). The outer end of the sector shaft has a tapered or straight spline that mates with an internal spline on the pitman arm. As the steering wheel rotates, the worm causes the sector to move through an arc and transfer the motion through the pitman arm to the steering linkage. As the driver turned the steering wheel, this style of gearbox produced more friction between the gears than later worm gearboxes, which made steering difficult. Variations include the worm-and-roller and the recirculating ball worm gear.

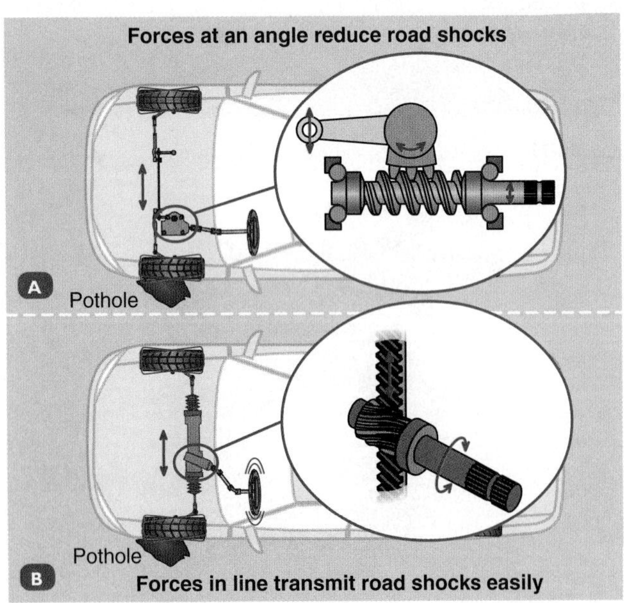

FIGURE 38-34 A. Road shock is minimized in a worm gear arrangement. **B.** Road shock is much larger in a rack-and-pinion gear.

Worm-and-Roller Gearbox

K38011

The worm gear in a worm-and-roller steering box has an hourglass shape, and it meshes with a double-track roller, mounted on bearings, on a pin attached to the pitman shaft (**FIGURE 38-36**). This gearbox was an improvement to the worm-and-sector style and reduced the friction due to the roller, which reduced the friction between the two gears. This design makes turning the steering wheel easier. As the driver turns the steering wheel, the worm rotates, and the roller rolls against the worm, moving in an arc, following the hourglass shape, and transferring the motion to the pitman shaft. The hourglass shape changes the steering ratio slightly as the steering wheel is turned from the central position toward each lock.

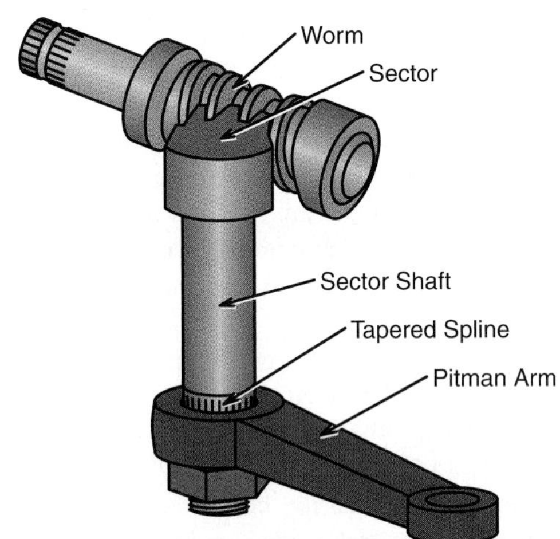

FIGURE 38-35 A worm-and-sector type of worm gear steering box.

FIGURE 38-36 The worm-and-roller steering gearbox.

Recirculating Ball Gearbox

K38012

The worm-and-nut steering gear is also known as a recirculating ball steering gear. The **recirculating ball steering box** contains a worm gear inside a block (nut) with a threaded hole in it and gear teeth cut into its outside that engage the sector shaft to move the pitman arm. It was developed in 1940 and is an improved version of the worm-and-sector and worm-and-roller types of worm gear. In the recirculating ball gearbox, both ends of the **worm shaft** are supported in the housing by angular bearings, which are preloaded to reduce lateral movements known as end play and side thrust movements of the worm gear when it is under load. A ball nut rides on the worm gear, supported on the spiral grooves of the worm and the inside of the nut by many balls (**FIGURE 38-37**). The balls form a low friction internal thread, which causes the nut to thread up or down on the worm as it rotates. With rotation, the balls are rolled along the grooves, partly in the worm and partly in the nut. When they get to the end of the groove, they circulate by passing through **ball-return guides** to the other end of the nut. Ball-return guides are simply special passages through which the balls move. External teeth on one side of the nut mesh with the teeth of the sector gear formed on the sector shaft, and these transfer the motion through the pitman arm to the steering linkage, to produce the side-to-side motion at the front wheels.

> **TECHNICIAN TIP**
>
> On passenger vehicles with independent suspension and a recirculating ball-type steering box, the steering box may be mounted so that the steering linkage is in front of (front steer) or behind (rear steer) the suspension cross member. When the steering linkage is behind the suspension cross member, it is protected by the cross member from possible damage, and the position of the steering box reduces the length of the steering column. This protects the steering box from damage by road debris.

The sector gear and nut teeth are designed so that when the teeth are in the straight-ahead position, they have minimum clearance. This is called the high point. This design reduces free play when the steering wheel is straight. The pitman shaft is supported by two caged needle roller bearings in the steering box housing. The sector teeth are angled, and an adjustment screw on the steering housing cover allows for adjustment of the sector gear height, which provides proper engagement of the sector gear and nut teeth (**FIGURE 38-38**).

> **TECHNICIAN TIP**
>
> In forward-control vehicles, the steering system is mounted in front of the engine and wheels. The steering box is mounted on the frame, with the pitman arm vertical. A **drag link** is a steel rod that transfers movement of the pitman arm to a **relay lever**, which is another steel rod that is used to transfer the movement from the drag link to an idler arm. The relay lever has two arms—one connected to the drag link and the other to the idler arm by the track rod. The longitudinal movement of the drag link pivots the relay lever. As a result of these connections, motion is transferred through the track rod to the idler arm and through the tie rods to the wheels.

▶ Power Steering

K38013

As vehicles became heavier in the front end due to the increased use of front-wheel drive and wider low-profile tires, more steering effort became necessary. As a result, **power steering** was introduced. One type has an **engine-driven hydraulic pump** that delivers hydraulic fluid to the power unit at the steering box or rack-and-pinion assembly through connecting hoses and pipes (**FIGURE 38-39**). When the driver moves the wheel, the hydraulic pressure is sent to the rack or steering box. The power steering system is designed so that the vehicle can still be controlled even if the engine or the power steering system were to fail—though greater effort would be required.

There are three types of power steering:

1. Hydraulically assisted power steering
2. Electrically powered hydraulic steering
3. Fully electric power steering

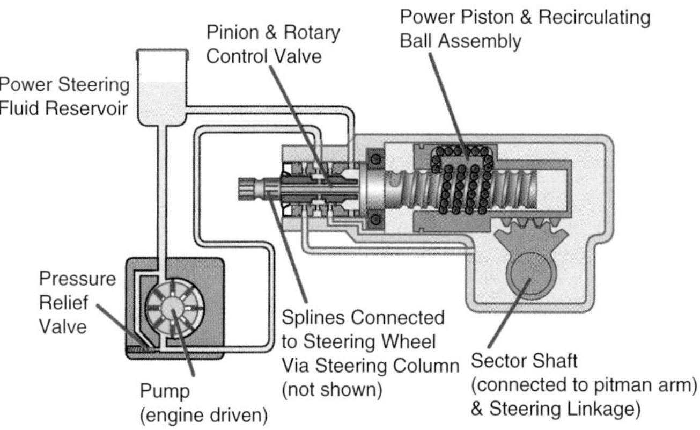

FIGURE 38-37 The recirculating ball steering system uses a worm gear that moves a ball nut by turning against ball bearings, reducing the force needed to turn the wheels.

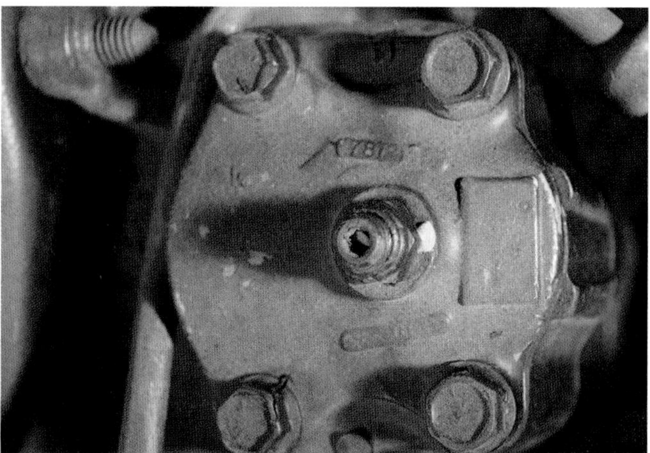

FIGURE 38-38 A sector gear adjustment screw and lock nut maintains the proper engagement between the gears.

FIGURE 38-39 A power steering pump mounted on a vehicle with hydraulically assisted power steering.

FIGURE 38-40 A typical power steering fluid cooler mounted in the high-pressure line.

Hydraulically Assisted Power Steering

K38014

Hydraulically assisted power steering uses hydraulic fluid under pressure to assist the driver in steering the wheels. This design is especially helpful at slower vehicle speeds, when steering wheel turning effort is much higher. An engine-driven hydraulic pump delivers hydraulic fluid to the power unit at the steering gearbox, or rack-and-pinion assembly, through the control valve and connecting hoses and pipes. The fluid reservoir can be mounted on the pump, or it can be remotely mounted. With the engine running, fluid flows continuously from the **power steering pump** to the steering gear control valve and back to the power steering pump. With the steering wheel in the neutral position, minimal pressure is needed to maintain fluid flow, and minimal engine power is needed to operate the system.

Some manufacturers task the power steering system with double duty. The pressure from power steering pumps has been used to power hydraulic brake boosters on hydroboost-equipped vehicles. It has also been used to turn the radiator fan on a number of vehicles. When working on those systems, be aware that the power steering system is integrated into those systems on some vehicles.

Power Steering Fluid and Hoses

K38015

The hydraulic fluid in a power steering system transmits the pressure from the power steering pump to the working chambers in the power steering gear or rack. The fluid must withstand high temperatures and pressures while lubricating the pump and steering gear and also preserving the system seals and pressure hoses. The fluid must also flow freely at very cold temperatures. Many manufacturers have specified certain types of automatic transmission fluid for their power steering systems over the years; others call for specific formulations of a separate power steering fluid. Because there is typically only a quart or two of power steering fluid in the system, it gets worked pretty hard. The harder it is worked, the hotter it gets. The hotter it gets, the shorter its life. For this reason, some vehicles are equipped with a power steering fluid

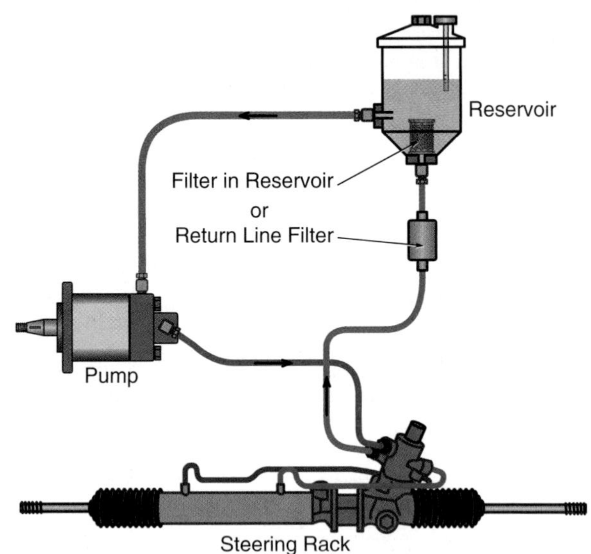

FIGURE 38-41 Some vehicles use a power steering system filter mounted either in the reservoir or in the return line.

cooler (**FIGURE 38-40**). This is usually built into the high-pressure line and helps cool the fluid before it gets to the steering gear.

Power steering fluid also gets contaminated with rubber and metallic particles from internal wear in the system. Because of this, some manufacturers install a replaceable filter in the power steering system (**FIGURE 38-41**). Although many manufacturers do not specify a power steering fluid change interval, the fluid does degrade over time and becomes contaminated. Many technicians recommend flushing power steering fluid every 50,000 to 100,000 miles, depending on the vehicle and how it is driven.

Power steering hoses are used to carry power steering fluid from the pump to the steering gear. This is the high-pressure hose and is usually made of flexible, high-pressure hose material. The return hose runs from the steering gear back to the pump reservoir and carries fluid under much lower pressure. These hoses must also allow movement between the engine and chassis, so they can't be too stiff. Over time, power steering hoses can become weak or damaged. If they leak, they can quickly cause the system to run dry. And because the pressure can get very high, if the high-pressure

FIGURE 38-42 A high-pressure power steering hose using an O-ring to seal the hose to the pump.

hose leaks, it can spray hot power steering fluid all over the engine and exhaust. Power steering fluid is flammable, so leaks can cause extreme fire hazards. Because of this, they should be inspected for seepage or wear during each oil change. Also, most high-pressure power steering hoses use an O-ring to seal the end of each hose to the pump and steering gear (**FIGURE 38-42**). These seals can wear or leak, requiring replacement of the O-ring.

Steering Process

K38016

The hydraulic pressure is controlled by a rotary valve located on the input shaft of the steering gear. When the steering wheel is turned, the rotary valve directs fluid to one side or the other of a piston attached to the steering gear. Pressure then increases as required to provide steering assistance.

In a rack-and-pinion steering gear, the piston is formed centrally on the steering rack, and the rack housing provides the working cylinder (**FIGURE 38-43**). Pressure seals at each end of the cylinder isolate the **power section** from the rest of the rack and pinion. Seals in the rotary valve section at the pinion input prevent internal and external fluid leakage.

In a recirculating ball steering box, the power piston and seals slide in a cylinder in the housing (**FIGURE 38-44**). The power piston has an extension formed on one side, with teeth that engage teeth on the sector shaft. Pressure applied to either side of the power piston produces a force, which is transferred through the teeth, to help turn the sector shaft.

Connecting pipes transfer fluid from the rotary valve housing to one side of the piston or the other to provide assistance, which acts directly on the steering gear. The rotary valve is located between the steering gear input shaft and the pinion gear, or worm. It consists of an inner member that forms part of the input shaft, and a surrounding sleeve member fixed to the pinion gear or worm. It appears as a shaft within a shaft.

Turning the steering wheel makes both members rotate in the steering gear housing, but it is the slight rotary displacement of the inner member and the sleeve member that controls and directs the power steering fluid flow. This slight rotary displacement is allowed by a **torsion bar**, which is a spring-loaded piece of steel connected to the pinion gear or worm at its bottom end and the input shaft at its top end (**FIGURE 38-45**).

When the steering wheel is turned, there is resistance from the front wheels at the road surface. This resistance is transmitted through the steering gear so that the input shaft twists slightly on the torsion bar. Because the inner member is also attached to the input shaft, this twisting provides a relative rotary displacement of the inner and outer members. It is this displacement that lets fluid flow through the valve to act on the

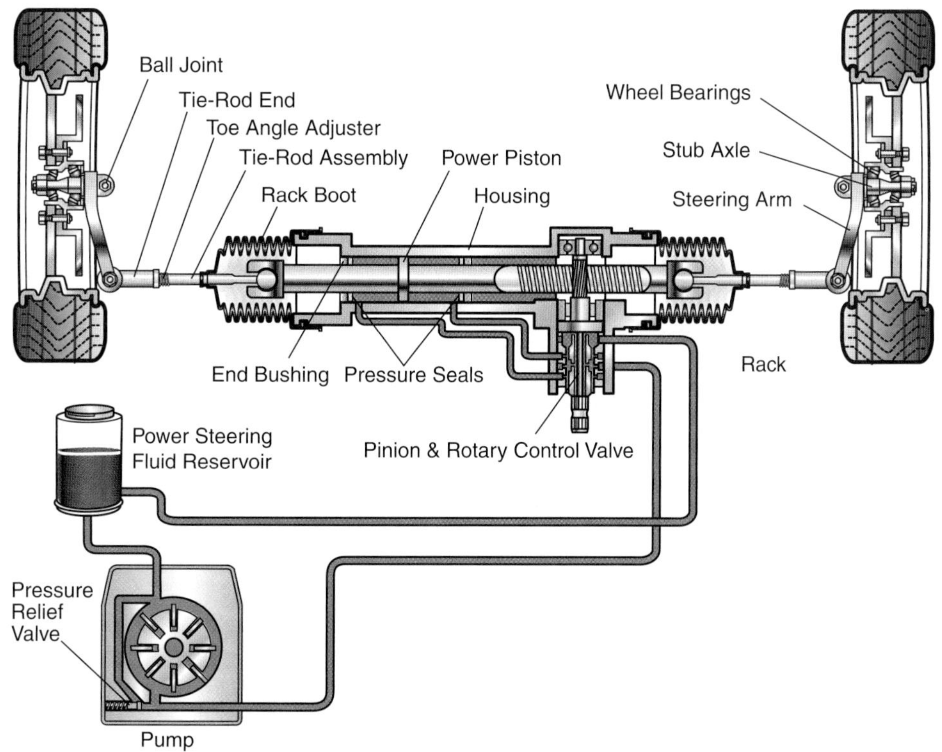

FIGURE 38-43 Power-assisted rack-and-pinion system.

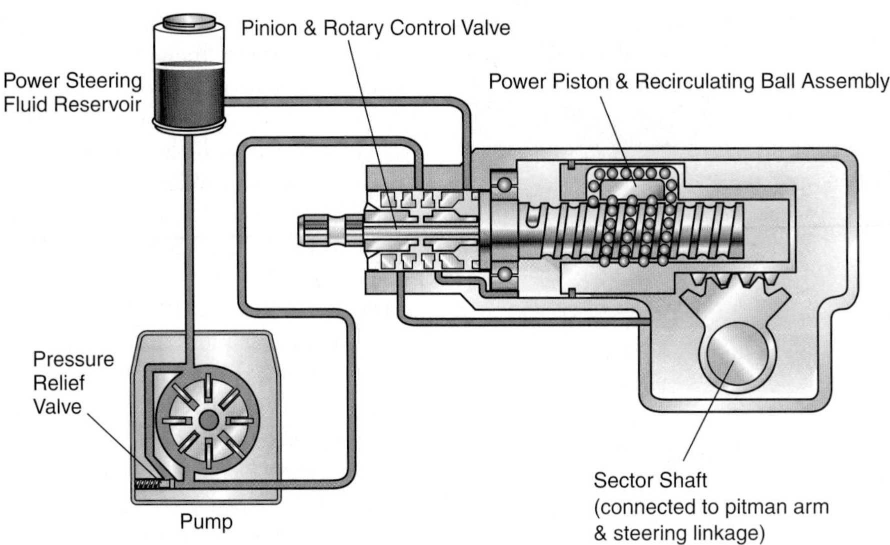

FIGURE 38-44 Power-assisted recirculating ball gearbox.

FIGURE 38-45 Power steering rotary valve and torsion bar.

piston at the steering gear. The input shaft can twist through only a small angle before it contacts a stop on the pinion gear or worm. This is needed as a fail-safe to provide manual steering when power assistance is not available.

With the engine running and the steering in the neutral position, fluid flow is directed into the valve assembly through drilled holes in the outer sleeve (**FIGURE 38-46**). As soon as the steering is turned to the left or right, the slight relative movement occurs between the inner and the outer members. In the neutral position, the inner member lets fluid pass equally to both sides of the rack piston and return to the fluid reservoir. Equal, but low, pressure is applied to both sides of the power piston. No power assistance is needed.

When the steering wheel is turned, fluid is restricted from making a free return to the reservoir. It is now directed to the side that matches the turning action. At the same time, fluid on

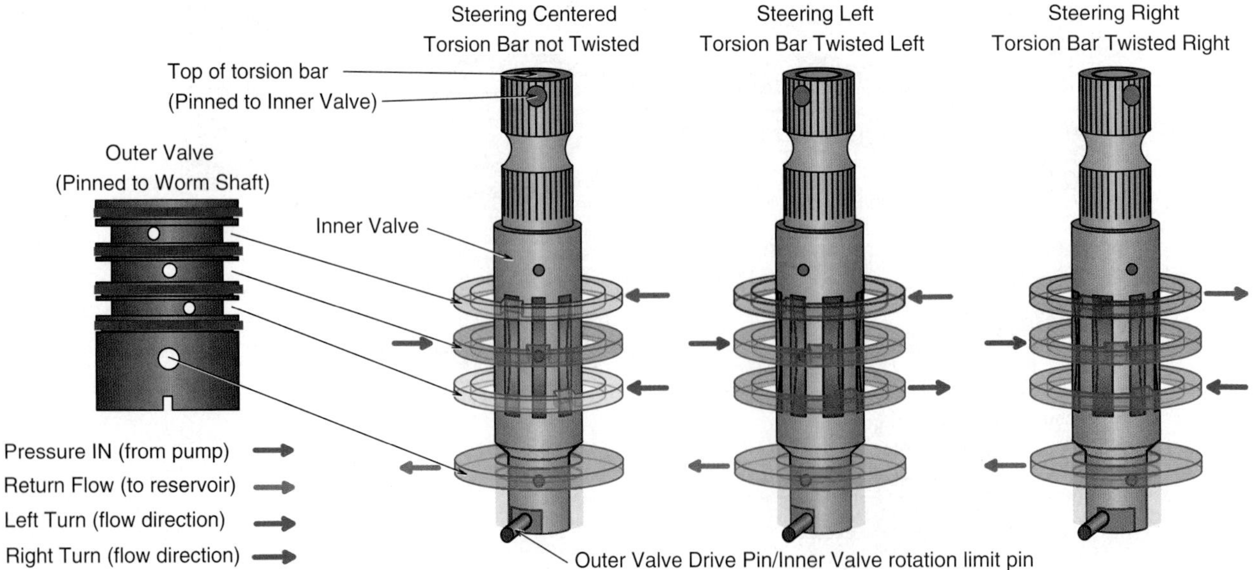

FIGURE 38-46 Slight flexing of the torsion bar allows fluid to be increased or decreased to either side of the power piston in the steering gear.

the opposite side is directed to the return circuit, back to the reservoir. Slight rotation of the valve gives a small amount of assistance, which becomes progressively greater as the torsion bar flexes and more assistance is needed. The grooves of the inner member are precisely shaped to meter the flow of fluid between the apply and release passageways.

Flow Control Valve

K38017

All power steering pumps have a **flow-control valve** to vary fluid flow and power steering system pressures. A pressure relief valve prevents excessive pressures from developing when the steering is on full lock (turned all the way to one side) and is held against its stops in that position. The flow-control valve is located at the outlet fitting of the power steering pump and regulates pressures to as high as 1200 to 1500 pounds per square inch (psi), or 8274 to 10,342 kilopascals (kPa). When the flow-control valve is forced open at a steering stop, there is usually an audible whine from this very high pressure.

During slow cornering or when parking, power steering pump speeds are normally low. There is low fluid flow at this time because the engine speed is low, but high pressure is still needed to provide the required assistance. Discharge ports direct fluid to the pressure side outlet and then to the steering gear. The outlet fluid pressure is slightly lower than the internal high pressure coming from the power steering pump. This drop in pressure occurs as the fluid flow passes the needle and orifice in the outlet fitting. This lower pressure is transmitted through a bypass fluid passage to the spring end of the control valve. The pressure difference on the valve causes it to move away from the outlet fitting, but the force of the spring prevents it moving far enough to uncover a return port back to the pump inlet. Movement of the control valve controls the position of the needle valve in the outlet fitting. This controls the fluid flow to the steering gear (**FIGURE 38-47**).

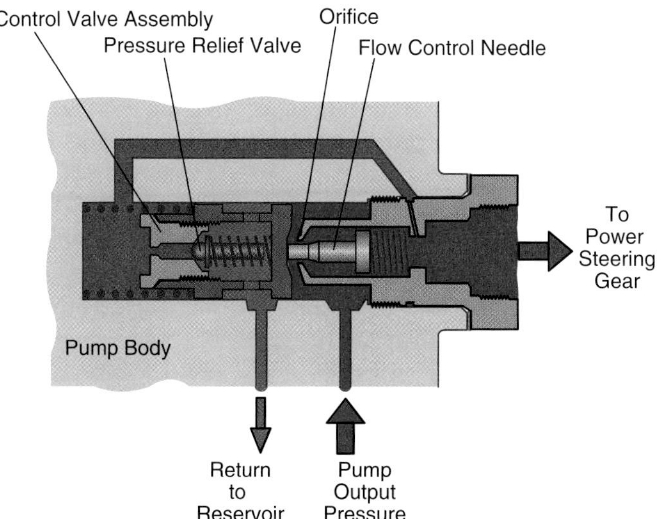

FIGURE 38-47 Movement of the control valve controls the position of the needle valve in the outlet fitting; this controls the fluid flow to the steering gear.

At higher speeds with no steering maneuvers, fluid flow is increased. Because the steering is in the neutral position, there is reduced pressure at the outlet. The lower pressure is transmitted to the spring end of the control valve. The valve moves and opens the return port back to the pump inlet. Movement of the control valve also controls the movement of the flow control needle in the outlet fitting. The needle closes in the orifice, and fluid flow to the steering gear reduces. With the steering wheel held at full lock, the steering rack power piston chamber becomes fully pressurized, and fluid flow through the steering gear stops.

This high pressure is transmitted back to the spring end of the control valve, opening the pressure relief valve. A small amount of fluid passes through the pressure relief orifice, providing a pressure drop. The valve moves and uncovers the return port to the pump inlet (**FIGURE 38-48**). A predetermined relief pressure is thus maintained.

SAFETY TIP

On vehicles with high-voltage systems (50 volts or higher), you may be required to wear insulated gloves appropriate for the system voltage you are working on. Always make sure the gloves have been tested for electrical leaks within the required time frame. And always test the gloves before using them by rolling up the sleeve, trapping air inside, and verifying that there are no air leaks. If there are any air leaks, *do not* use them, as electricity could travel through the hole.

Idle Speed Strategy

N38001

When maneuvering a vehicle at slow speeds, the power steering demand will be high, and the engine speed will be low. If the idle speed is not maintained during this condition, the engine could possibly stall. To prevent stalling, engineers program a strategy into the PCM that raises or maintains the idle speed under these conditions. A power steering switch or sensor inputs power steering pressure information to the engine control module, and based on this input and the actual engine idle revolutions per

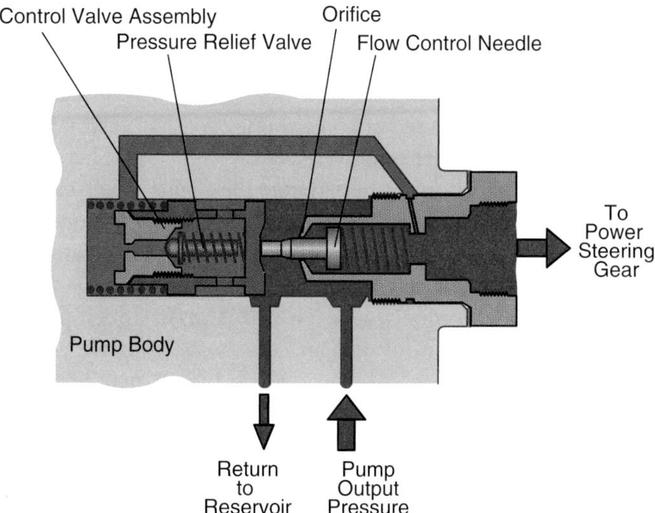

FIGURE 38-48 Flow-control valve—steering wheel at full lock.

minute (rpm), the PCM is able to determine when additional airflow is needed to raise or maintain the desired engine rpm. When the power steering pressure is reduced, the PCM reduces the amount of idle air flow.

Electric Power Steering

K38018

The use of electronics in automotive steering systems enables much more sophisticated control to be achieved. Electric steering is more economical to run and easier to package and install than conventional hydraulic power steering systems and reacts faster to quick steering changes from the driver. Typically, electric and electrohydraulic power steering systems are also lighter and more compact than conventional hydraulic power steering systems. Both the electric power steering system and the electrohydraulic power steering systems are now considered as viable alternatives to conventional hydraulic power steering systems because of their energy efficiency and size.

In **electrically powered hydraulic steering (EPHS)**, a brushless motor replaces the customary drive belts and pulleys that drive a power steering pump in a conventional rack-and-pinion steering system. This system still uses a pump, but it is driven by an electric motor to reduce power drawn from the engine. Unlike an engine driven pump, which runs continuously at engine speed, EPHS only operates on demand. Pump speed is regulated by an electronic controller to vary pump pressure and flow. This provides steering efforts tailored for different driving situations. The pump can be run at low speed or shut off to provide energy savings during straight-ahead driving. An EPHS system is said to use only 20% as much of the engine power used by a standard belt-driven pump, which improves fuel mileage substantially. The engine still contributes power to the steering system through electrical demand on the alternator of the vehicle, but it is greatly reduced from that of hydraulic power steering systems.

Electrically assisted steering (EAS) is a completely electrically powered power-assist system that eliminates all hydraulic components and fluid. An electric motor replaces the hydraulic pump. EAS, or direct electric power steering, completely eliminates hydraulic fluid and the accompanying hardware from the power steering system, creating a fully **electric power steering (EPS) system**. The reduction of components saves weight and reduces drag on the engine, vastly improving fuel mileage. The EPS system is said to require only about 2% of the engine power that a standard belt-driven power steering pump uses. There is still a small amount of power required from the engine, which supplies the electrical demand, but as you can see, it is greatly reduced.

An EPS steering system uses an electric motor attached either to the steering rack or to the steering column via a gear mechanism and torque sensor (**FIGURE 38-49**). A microprocessor or electronic control unit and diagnostic software control the steering dynamics and driver effort. Inputs include vehicle speed, steering wheel torque, angular position, and turning rate.

There are four primary types of electric power assist steering systems:

- **Column-assist type**: In this system, the **power assist unit**, controller, and **torque sensor** are attached to the steering column. The power assist unit is the electric motor; the controller is the electronic control unit; and the torque sensor measures the load on the steering wheel.
- **Pinion-assist type**: In this system, the power assist unit is attached to the steering gear pinion shaft. The power assist unit sits outside the vehicle passenger compartment, allowing assist torque to be increased greatly without raising interior compartment noise.
- **Rack-assist type**: In this system, the power assist unit is attached to the steering gear rack. It is located on the rack to allow for greater flexibility in the layout design (**FIGURE 38-50**).
- **Direct-drive type**: In this system, the steering gear rack and power assist unit form a single unit. The steering system is compact and fits easily into the engine compartment layout. Direct assistance to the rack enables low friction and **inertia** (resistance to a change in motion), which in turn gives an ideal steering feel.

In these systems, **active control** provides constant feedback from sensors in the vehicle to the **control unit**, which calculates sophisticated computer algorithms. These features allow the steering system to react to the road, the weather, and even the type of driver, and provide assistance to the front or rear road wheels independent of direct driver input. For example, if the electronic stability control system detects the start of an oversteer condition, the control unit can actually steer the wheels in the opposite direction, trying to prevent loss of control of the vehicle. Thus, an EPS system might steer the vehicle opposite of the driver's input if that input would lessen the control of the vehicle.

Basic EPS Operation

K38019

A steering sensor is located on the input shaft, where it is bolted to the gearbox housing.

The **steering sensor** performs two functions: First, as a torque sensor, it converts steering torque input and direction into voltage signals for the engine control unit (ECU) to monitor. Second, as a rotation sensor, it converts the rotation speed and direction into voltage signals for the ECU to monitor. An interfaced ECU circuit converts the voltage signals from the torque and rotation sensor into signals that the PCM can process, and ultimately provides the proper output signal to the

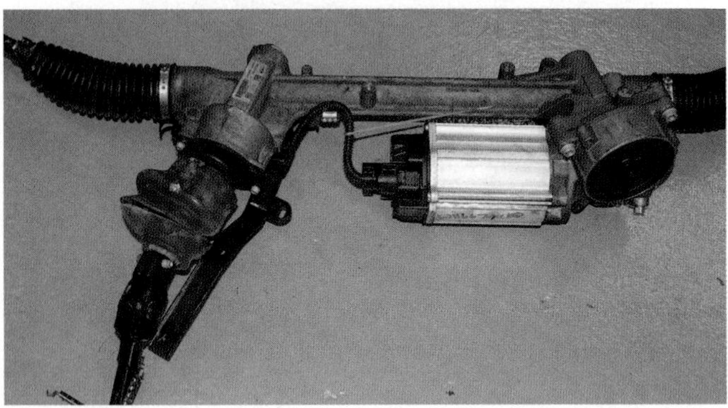

FIGURE 38-49 A typical electronic power steering rack-and-pinion assembly.

FIGURE 38-50 Rack assist steering type.

EPS assembly. The microprocessor control unit also analyzes inputs from the vehicle's speed and wheel speed sensors. The sensor inputs are then compared to determine how much power assistance is required according to the **forces capability map data** stored in the ECU's memory. These map data are preprogrammed by the manufacturer. The ECU sends the appropriate command to the **power unit**, which supplies the electric motor with the necessary current to operate as commanded. The electric motor then pushes the rack to either the right or the left. The direction of rack movement depends on which way the current flows; reversing the current flow reverses directional rotation of the electric motor. Increasing current to the electric motor increases the amount of power assist.

The EPS system has three operating modes:

- Normal control mode—provides left or right power assist in response to input from the torque and rotation sensor's inputs.
- Return control mode—assists steering return after completing a turn.
- Damper control mode—adjusts the amount of assist according to the vehicle speed to improve road feel and dampen kickback.

If the steering wheel is turned and held in the full-lock position—which is as far as the wheel can turn in one direction—and steering assist reaches maximum, the power unit reduces current to the electric motor to prevent an overload situation that might damage the electric motor. The power unit is also designed to protect the electric motor against voltage surges from a faulty alternator or charging system problem.

The electronic steering control unit is capable of self-diagnosing faults by monitoring the system's inputs and outputs and the driving current of the electric motor. If a problem occurs, the electronic steering control unit turns the system off by **actuating** (turning on) a fail-safe relay in the power unit. This eliminates all power assist, causing the system to revert to manual steering. An in-dash EPS warning light is also illuminated to alert the driver.

Higher Voltage Electrically Assisted Power Steering

N38002

Higher voltage electrically assisted power steering systems were developed for modern hybrid vehicles. In hybrid vehicles, the engine is typically off during idling, deceleration, and braking; thus, an EPS system was needed. The higher voltage battery system of the hybrid vehicle provides all the power, with no reliance on engine or hydraulic power. The parts are similar to an electrically assisted power steering system. The electric steering system does not use the full battery voltage from the high-voltage battery pack; instead, it is reduced to a lower voltage through a DC-to-DC converter, which basically just steps the voltage down. Some early hybrids simply used the 12-volt system that was used on gasoline vehicles, but a few models used voltage as high as 46 volts. On more recent vehicles and with the advent of more powerful electric motors, the manufacturers seem to be going back to 12-volt power steering motors, which are simpler systems to design and install on vehicles.

At the time, there were several advantages of using higher voltage to operate electric power steering motors. One was to create more power in the same size motor that was thought to be needed. Smaller gauge wires could be used in the electric steering system, reducing weight in the vehicle, and these lighter weight components would increase fuel economy slightly.

One thing to remember if you work on one of these higher voltage power steering systems is that although voltages under 50 volts are considered relatively safe against electric shock hazards, if the 46 volt system is live, so is the high voltage system that you will be working around. And that requires high-voltage gloves, a high-voltage meter, special insulated tools, and special training. So higher voltage electric power steering systems should only be serviced by highly trained individuals.

Wrap-Up

Ready for Review

- The power steering system can be either a hydraulic or an electric type.
- A basic steering system has four main parts: a steering column, a steering box, a steering linkage, and a steering knuckle.
- Steering geometry is a geometric arrangement of linkages in the steering of a vehicle, designed to keep the wheels properly oriented through various positions of the steering and suspension systems.
- The rack-and-pinion steering system is used on the majority of front-wheel drive vehicles because of the space restriction under the hood.

- The parallelogram steering system is used on larger vehicles where ride comfort is more important than sporty handling.
- When servicing the steering column, it is good practice to disarm the triggering system for the driver's side airbag. If not properly disarmed, it could trigger accidentally.
- Because of the critical nature of the airbag in an accident, it is imperative that it is always connected electrically to its control module, so it can be deployed when needed.
- The function of all steering boxes, whether manual or power, is the same: to transfer the rotary motion of the steering wheel into the side-to-side motion needed to make the wheels pivot left and right.
- There are two basic types of steering boxes: those with rack-and-pinion gearing and those with worm gearing and a sector shaft. In both types, the gearing in the steering box makes it easier for the driver to turn the steering wheel and hence the wheels.
- Four-wheel steering means the rear wheels can be steered independently of or in conjunction with the front wheels. There are two types: active and passive.
- There are three types of power steering: hydraulically assisted power steering, electrically powered hydraulic steering, and fully electric power steering.
- Electrically powered hydraulic steering replaces the customary drive belts and pulleys that drive a power steering pump in a conventional rack-and pinion steering system with a brushless motor.
- The steering sensor performs two functions: First, as a torque sensor, it converts steering torque input and direction into voltage signals for the ECU to monitor. Second, as a rotation sensor, it converts the rotation speed and direction into voltage signals for the ECU to monitor.
- Higher voltage electrically assisted power-steering systems were developed for modern hybrid vehicles. The higher voltage battery system of the hybrid vehicle provides all the power, with no reliance on engine or hydraulic power.

Key Terms

active control A system of providing constant feedback from sensors in the vehicle to the control unit.

actuating The act of making something move or work.

adjustable bushing A brace or nylon part that pushes against the rack to adjust the mesh of the rack teeth to the pinion teeth.

adjustment sleeve A component that connects the tie rods together and to the center link on some applications, providing the adjustment point for toe-in or toe-out, depending on the manufacturer's specifications.

antibinding The release of something when it gets stuck.

ball-return guide A special passage or metal tube through which the balls move in recirculating ball steering boxes.

beam axle A suspension system in which one set of wheels is connected laterally by a single beam or shaft.

bump steer The undesired condition produced when hitting a bump, where the vehicle darts to one side as the steering linkage is pushed or pulled as a result of the travel of the suspension.

chassis The frame of a vehicle, to which the suspension pieces attach.

clock spring A special rotary electrical connector located between the steering wheel and the steering column that maintains a constant electrical connection with the wiring system while the vehicle's steering wheel is being turned.

constant mesh A term used to describe two or more parts, such as gears, that are in constant contact with each other.

control unit Any device that controls another object such as a computer.

drag link A steel or iron rod that transfers movement of the pitman arm to a relay lever.

electric power steering system (EPS) A steering system that uses an electric motor and sensors to provide feedback to the vehicle's computer systems to decrease steering effort.

electrically assisted steering (EAS) A power-assist system that uses an electric motor to replace the hydraulic pump to decrease steering effort.

electrically powered hydraulic steering (EPHS) A steering system that uses an electric motor to produce hydraulic assist for steering.

engine-driven hydraulic pump A power steering pump driven by a belt or gear off the crankshaft.

flow-control valve A valve used in power steering pumps to control the amount of flow out of the power steering pump.

forces capability map data Data preprogrammed into the electronic control unit's memory by the manufacturer and used to determine how much power assistance is needed based on input from the vehicle's speed sensor and steering sensor.

four-wheel steering system A steering system in which the front wheels are controlled normally, and the rear wheels use a computer and electric motors to turn the rear linkage.

inertia The resistance to a change in motion.

inner tie rod or socket Attached to the end of the rack, the inner tie rod allows for suspension movement and slight changes in steering angles.

intermediate shaft A steel rod positioned at an angle from the steering column to the steering gear that functions in transferring movement from one to the other.

knuckle The part that contains the wheel hub or spindle and attaches to the suspension components.

linear motion Movement in a straight line.

meshed pinion A pinion when it is mated with the rack.

outer tie rod The tie rod attached between the tie rod and the steering arm. It transfers the movement of the rack, pivoting as the rack is extended or retracted when the vehicle is negotiating turns.

parallelogram steering system A non–rack-and-pinion system that uses a series of parts, consisting of the pitman arm, idler arm, center link, and tie-rod assemblies, to relay movement from the steering gearbox to the wheel assembly.

pinion A gear located at the end of the steering shaft connected to the rack. It moves the rack from side to side as the pinion rotates, controlling the direction of the wheels.

power assist unit The electric motor in electric power assist steering systems.

power section A chamber in the rack, where pressurized fluid acts upon pistons that assist in steering.

power steering An option on a vehicle that allows movement of the steering wheel with decreased driver effort.

power steering pump A small hydraulic pump that provides assistance to the driver when turning the steering wheel.

power unit A belt- or gear-driven pump that produces hydraulic pressure for use in the steering box or rack.

preloaded A part that is already compressed from pressure.

rack A steel rod driven by the pinion, with tie rods on each end or tie rods connected to the center of the rack.

rack housing The outer shell of the rack-and-pinion steering system that is mounted to the chassis.

rack-and-pinion steering system A steering system composed of a steering wheel, a main shaft, universal joints, and an intermediate shaft. When the steering wheel is turned, movement is transferred by the main shaft and intermediate shaft to the pinion.

recirculating ball steering box A steering box that has worm gear inside a block with a threaded hole in it and gear teeth cut into its outside that engage the sector shaft to move the pitman arm; generally used on trucks and heavy vehicles.

reduction ratio The ratio between the turn of the steering wheel and the turn of the wheel, both measured in degrees.

relay lever A steel rod that transfers movement from the drag link to an idler arm.

rubber bellows Rubber pieces positioned on each end of the rack to protect the inner joints from dirt and contaminants and retain the grease lubricant inside the rack-and-pinion housing.

spline Ridges or teeth on a shaft that mesh with grooves in a mating piece and transfer torque to it, maintaining the angular correspondence between them.

spring-loaded rack guide yoke A spring-containing part that pushes on the back side of the rack to help reduce the play between the rack and the pinion while still allowing for relative movement.

steering arm An arm that extends from the steering knuckle. The tie rods connect to these arms in order to steer the wheels.

steering box (steering gear) A device that converts the rotary motion of the steering wheel to the linear motion needed to steer the vehicle.

steering column A column affixed between the steering wheel and the steering box, usually made to collapse during a crash.

steering damper A device used to prevent shocks from irregular roads from being transmitted through the steering linkage and back to the steering wheel.

steering knuckle The knuckle located between the lower control arm and MacPherson strut or upper control arm, which has a spindle made or bolted onto it, to which the wheel hub is attached.

steering linkage Steel rods that connect the steering box to the steering arms on the steering knuckle.

steering sensor A sensor that can read both torque and rotation from the steering wheel.

steering system A term used to describe all of the components and parts involved in steering a vehicle.

stub axle An axle used for one wheel.

stub-axle carrier The body of the stub-axle knuckle.

tie rod A steering component that transfers linear motion from the steering box to the steering arms at the front wheels.

tie-rod assembly The part that fits between the rack and the steering arms and transfers the movement of the rack.

tie-rod end The close-fitting ball and socket on the end of a tie rod that allows for movement of the steering components.

toe setting Setting of the toe-in or toe-out of the tires to the centerline of the vehicle.

torque sensor A device used to measure the load on the steering wheel.

torsion bar A spring-loaded piece of steel connected to the pinion gear at its bottom end and the input shaft at its top. Also, a torsion bar is a type of spring that some vehicles use to hold up a corner of the vehicle.

wheel assembly A term used to encompass all components of the wheel and tire.

worm A gear with a helical, threaded shaft that is attached to the steering column and meshes with a worm wheel that transfers motion from the steering wheel to the steering linkage.

worm gear steering A robust steering system frequently used on heavier vehicles that uses a worm to turn a meshed worm wheel to provide gear reduction, making steering easier for the driver.

worm shaft The protrusion of the worm gear that serves as the point of attachment to the steering column.

Review Questions

1. All of the following statements are true *except*:
   - a. the steering box converts the rotary motion of the steering wheel to angular motion.
   - b. steering linkage connects the steering box to the steering arm.
   - c. the steering system provides control over the vehicle's direction of travel.
   - d. the steering knuckle pivots on bushings.
2. In an electronic power steering system, PCM stands for:
   - a. powertrain control module.
   - b. portable control mechanism.
   - c. power cutting mechanism.
   - d. power control module.
3. The Ackermann principle is applied in steering geometry to:
   - a. reduce vibration of the vehicle.
   - b. reduce wobbling of the vehicle.
   - c. navigate a corner without scrubbing.
   - d. make steering response sharper.

4. Which of the following components is the inner tie rod socket connected to in a rack-and-pinion steering system?
   a. Rack
   b. Pinion
   c. Nylon brushes
   d. Intermediate shaft
5. The part of the parallelogram steering mechanism that provides the adjustment point for adjusting the toe setting is the:
   a. pitman arm.
   b. tie rod.
   c. idler arm.
   d. adjustment sleeve.
6. In modern computer-controlled four-wheel steering, an actuator similar to a front rack-and-pinion assembly attaches to the rear steering knuckles with:
   a. center arm ends.
   b. tie-rod ends.
   c. bushes.
   d. clock spring.
7. The steering box converts the rotary motion of the steering wheel into:
   a. reciprocating motion.
   b. angular motion.
   c. up and down motion.
   d. tangential motion.
8. Choose the correct statement.
   a. Turning the steering wheel rotates the pinion and moves the rack up and down.
   b. The teeth on both the pinion and the rack are helical gears.
   c. A large pinion means the number of turns of the steering wheel is increased.
   d. The steering ratio is the same on all vehicles.
9. Which type of gear box uses a recirculating ball steering box in it?
   a. Worm-and-sector
   b. Worm-and-roller
   c. Worm-and-nut
   d. Worm-and-steering
10. Which of the following electric power assist steering systems has the power assist unit, controller, and torque sensor connected to the steering column?
    a. Pinion-assist type
    b. Rack-assist type
    c. Column-assist type
    d. Direct drive type

ASE Technician A/Technician B Style Questions

1. Tech A says that the steering column uses one or more flexible joints to connect to the steering gearbox. Tech B says that the pitman arm is bolted to the frame and relays the steering linkage movement to the opposite wheel from the steering gearbox. Who is correct?
   a. Tech A
   b. Tech B
   c. Both A and B
   d. Neither A nor B
2. Tech A says that the clock spring (spiral cable) assists the turning of the steering wheel on vehicles with EPS. Tech B says that the clock spring is used to transmit an electrical signal to the driver's side airbag. Who is correct?
   a. Tech A
   b. Tech B
   c. Both A and B
   d. Neither A nor B
3. Tech A says that some manufacturers specify automatic transmission fluid in their power steering system. Tech B says that in a power steering system, the force needed to steer the wheels is created by the hydraulic pump or electric motor. Who is correct?
   a. Tech A
   b. Tech B
   c. Both A and B
   d. Neither A nor B
4. Tech A says that rack-and-pinion steering systems generally do not use power steering due to their lighter duty construction. Tech B says that rack and-pinion steering systems use a worm gear arrangement to move the rack. Who is correct?
   a. Tech A
   b. Tech B
   c. Both A and B
   d. Neither A nor B
5. Tech A says that a torsion bar is used to operate the control valve in a hydraulically controlled power steering system. Tech B says that in the neutral position, the pressure in the power steering system is very high. Who is correct?
   a. Tech A
   b. Tech B
   c. Both A and B
   d. Neither A nor B
6. Tech A says that most high-pressure power steering hoses use an O-ring to seal the end of each hose to the pump and steering gear. Tech B says that the return hose goes from the steering gear to the power steering pump reservoir. Who is correct?
   a. Tech A
   b. Tech B
   c. Both A and B
   d. Neither A nor B
7. Tech A says that when the vehicle is being driven straight ahead, a belt-driven power steering pump will pump fluid continuously, placing a minimal load on the engine. Tech B says that a belt-driven power steering pump is activated by an electromagnetic clutch, so it only pumps fluid when the wheels are being steered. Who is correct?
   a. Tech A
   b. Tech B
   c. Both A and B
   d. Neither A nor B

8. Tech A says that the pressure relief valve maintains a preset minimum pressure in the system. Tech B says that the pressure relief valve prevents excessive pressure. Who is correct?
   a. Tech A
   b. Tech B
   c. Both A and B
   d. Neither A nor B
9. Tech A says that on a hydraulic power steering system the power steering switch turns off the power steering pump if the pressure gets too high. Tech B says that on the same system, the power steering switch is used to turn on the pump if the pressure gets too low. Who is correct?
   a. Tech A
   b. Tech B
   c. Both A and B
   d. Neither A nor B
10. Tech A says that no matter which way the vehicle turns, the inside wheel must always turn less sharply than the outside wheel. Tech B says that the Ackermann principle angles the steering arms toward the center of the rear axle. Who is correct?
   a. Tech A
   b. Tech B
   c. Both A and B
   d. Neither A nor B

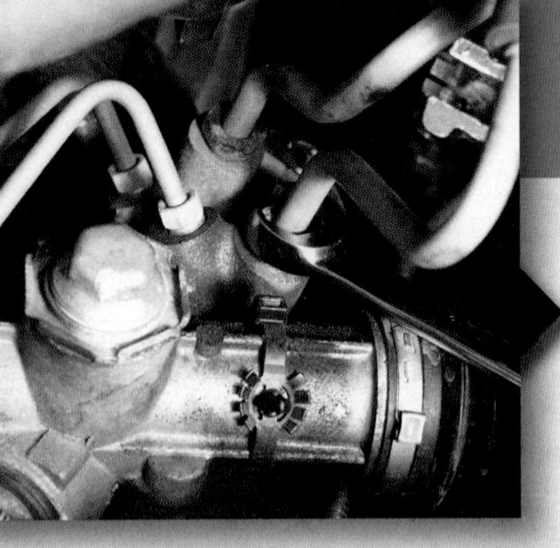

CHAPTER 39
Servicing Steering Systems

NATEF Tasks

- **N39001** Inspect for power steering fluid leakage; determine needed action. (MLR/AST/MAST)
- **N39002** Diagnose power steering gear (non–rack-and-pinion) binding, uneven turning effort, looseness, hard steering, and noise concerns; determine needed action. (AST/MAST)
- **N39003** Diagnose power steering gear (rack-and-pinion) binding, uneven turning effort, looseness, hard steering, and noise concerns; determine needed action. (AST/MAST)
- **N39004** Diagnose steering column noises, looseness, and binding concerns (including tilt/telescoping mechanisms); determine needed action. (AST/MAST)
- **N39005** Flush, fill, and bleed power steering system; use proper fluid type per manufacturer specification. (MLR/AST/MAST)
- **N39006** Remove and reinstall power steering pump. (AST/MAST)
- **N39007** Remove and reinstall press-fit power steering pump pulley; check pulley and belt alignment. (AST/MAST)
- **N39008** Inspect, remove, and/or replace power steering hoses and fittings. (MLR/AST/MAST)
- **N39009** Remove and replace rack-and-pinion steering gear; inspect mounting bushings and brackets. (AST/MAST)
- **N39010** Inspect rack-and-pinion steering gear inner tie-rod ends (sockets) and bellows boots; replace as needed. (MLR/AST/MAST)
- **N39011** Inspect, remove and/or replace pitman arm, relay (center link/intermediate) rod, idler arm, mountings, and steering linkage damper. (MLR/AST/MAST)
- **N39012** Inspect, replace, and/or adjust tie-rod ends (sockets), tie-rod sleeves, and clamps. (MLR/AST/MAST)
- **N39013** Inspect electric power steering assist system. (MLR/AST)
- **N39014** Inspect, test and diagnose electrically- assisted power steering systems (including using a scan tool); determine needed action. (MAST)
- **N39015** Identify hybrid vehicle power steering system electrical circuits and service and safety precautions. (MLR/AST/MAST)
- **N39016** Disable and enable supplemental restraint system (SRS); verify indicator lamp operation. (MLR/AST/MAST)
- **N39017** Remove and replace steering wheel; center/time supplemental restraint system SRS coil (clock spring). (AST/MAST)
- **N39018** Inspect steering shaft universal joint(s), flexible coupling(s), collapsible column, lock cylinder mechanism, and steering wheel; determine needed action. (AST/MAST)

Knowledge Objectives

After reading this chapter, you will be able to:

- **K39001** Describe the purpose and function of tools used in steering system service.

Skills Objectives

After reading this chapter, you will be able to:

- **S39001** Perform basic steering system diagnosis.
- **S39002** Perform maintenance and repair tasks on the power steering system.
- **S39003** Pressure test a power steering system.
- **S39004** Perform maintenance and repair on rack-and-pinion steering systems.
- **S39005** Perform maintenance and repair on parallelogram steering systems.
- **S39006** Inspect and service electric power steering.
- **S39007** Perform maintenance and repair on steering columns equipped with SRS.

▶ Introduction

Because of the critical nature the steering system plays in controlling the vehicle, it is important that all scheduled maintenance and inspection tasks be performed as required. Also, any problems that appear must be diagnosed and repaired as soon as they crop up. There is not much room for procrastination when dealing with steering system faults, as they can cause a loss of control of the vehicle, leading to an accident and injuries, if the repair was performed improperly. So please pay close attention to the procedures listed in this chapter, and always follow the manufacturer's recommended procedure when performing any repairs relating to steering concerns. This is especially true if the manufacturer's procedure is different from the ones presented here, as manufacturers configure their vehicles in a variety of ways.

Steering systems are mostly mechanical, so visual inspections and physical measurements make up a good portion of maintenance and diagnosis tasks. Mechanical joints and bushings, as well as motors, pumps, and switches, wear over time, decreasing in functionality and eventually failing. But as more electronic power steering systems are produced, electrical faults are becoming more common. Sensors also are subject to wear, resulting in inaccurate readings and faulty reactions. The wires are subject to damage and breaks as well as loose or corroded connections.

When confronted with a steering system problem and performing a diagnosis, it is important to consider both the mechanical and electrical components. Also, because the suspension system is directly related to the steering system and helps to hold the wheels in the proper geometry, any problems there can affect the steering system. So ultimately, you need to have an understanding of the suspension system too when diagnosing steering-related issues. Suspension systems are discussed fully in the Suspension Systems Theory and the Servicing Suspension Systems chapters. For now, we focus primarily on steering system maintenance, diagnosis, and repair.

▶ Tools and Equipment

K39001

Some of the tools used in diagnosing and repairing steering system problems can be broken into two categories: tools for mechanical diagnosis and tools for electrical diagnosis. Tools for electrical diagnosis include the factory scanner and the digital multimeter (DMM). The factory scan tool is recommended by most manufacturers because it can perform tests and identify faults with the electrical portion of steering systems. The scan tool can read almost every system on a vehicle and provides valuable data to assist the technician in identifying problems. It is particularly useful in assessing electronic steering systems for wiring and sensor malfunctions. Typically, electronic steering system faults set a diagnostic trouble code that can be retrieved by the scan tool. As well as a DMM use of the service information to diagnose the fault is necessary. The DMM is used to measure voltage and to check the continuity of the electronic circuit. Breaks in the wiring of an EPS system are not uncommon. When this occurs, the system opens or shorts out. The continuity test can pick up on any opens or shorts and help determine where the problem is located.

Tools and equipment used on the mechanical portion of the steering system include a power steering system pressure tester and various measuring devices, such as dial indicators and belt tension gauges. The mechanical components have to be removed and some may need to be replaced, requiring tools such as a tie-rod puller, pickle forks, a pitman arm puller, tie-rod sleeve adjusting tools, inner tie-rod socket tools, inch-pound and foot-pound torque wrenches, hammers, air hammers, and pry bars (**FIGURE 39-1**).

The following tools and parts are used in servicing the power steering system:

- **Floor jacks and safety stands:** Jacks are used to lift a vehicle, and the safety stands provide stability while working on a raised vehicle.
- **Pry bar:** The pry bar is a lever used to apply pressure for testing purposes or to move various components.
- **Dial indicators:** Dial indicators are used to measure the runout or movement on different parts of the steering system, such as play in tie-rod ends.
- **Pitman arm puller:** This heavy-duty puller is made specially for removing the pressed-on pitman arm from the sector shaft.
- **Tie-rod end puller:** This tool is used to pull the tapered shaft on a tie-rod end from its mating steering component.

You Are the Automotive Technician

A customer is driving on the highway and notices a burning oil smell and smoke coming from the front of his six-year-old vehicle with power rack-and-pinion steering. He is able to pull into a visitor's center and check under the vehicle's hood. He identifies a large amount of power steering fluid sprayed all over the engine compartment, and the power steering reservoir is nearly empty. He calls you for advice and wants to know if he can drive it to your shop 35 miles away.

1. What would you advise him to do, and why?
2. What are the most likely leak points?
3. While you are working on the vehicle, what other steering-related inspections would you perform?
4. If a vehicle with power steering comes into your shop with hard steering, what things would you check, and why?

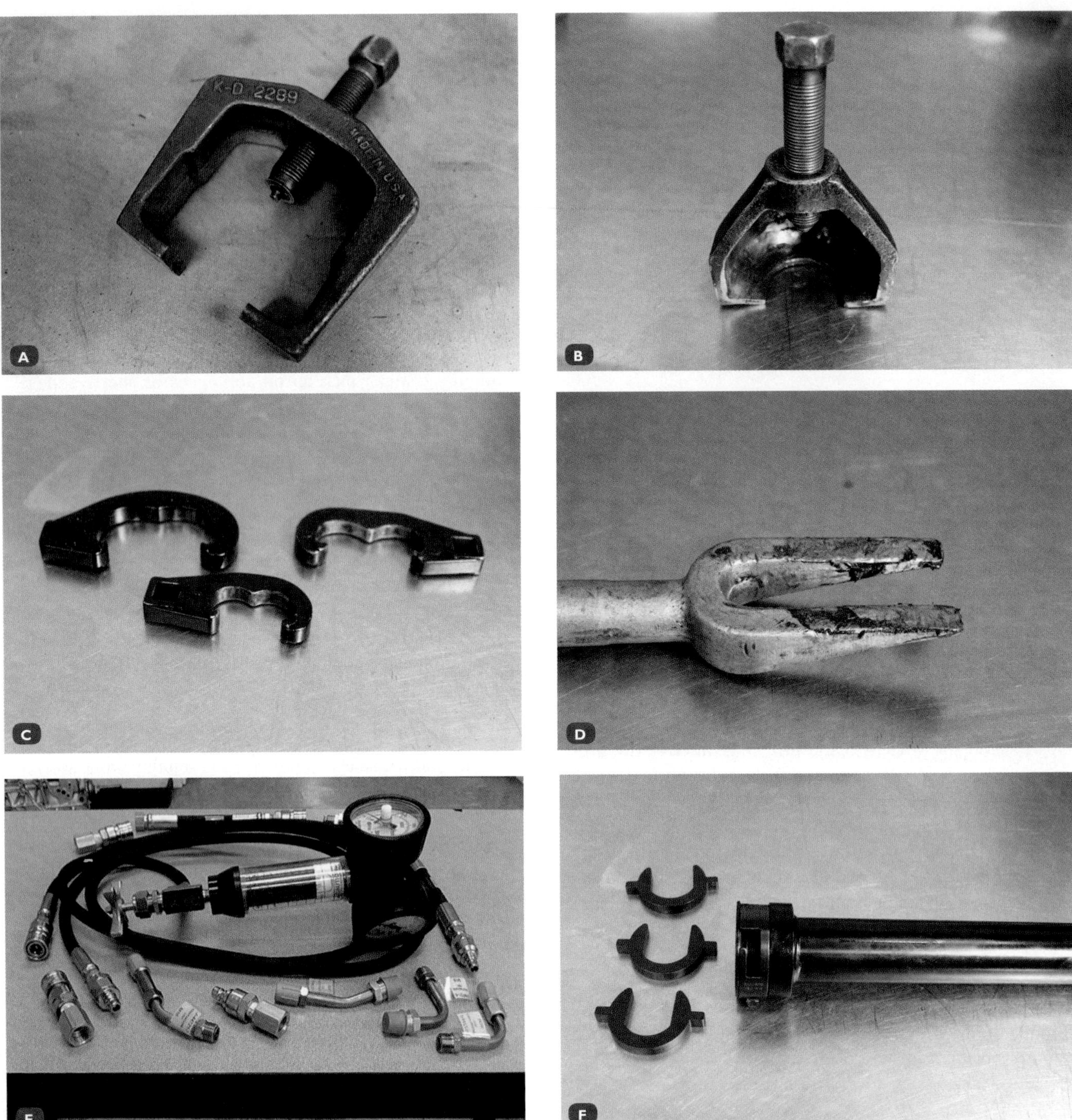

FIGURE 39-1 Steering system specialty tools. **A.** Pitman arm puller. **B.** Tie-rod end puller. **C.** Tie-rod sleeve tool. **D.** Pickle fork. **E.** Power steering system analyzer. **F.** Inner tie-rod tool.

- **Tie-rod sleeve adjusting tool:** This tool has a tab designed to grab the slot in the sleeve and is used to turn the sleeve when adjusting the toe setting.
- **Pickle fork:** This U-shaped wedge is used for separating tie-rod ends and is operated by hammer or air hammer. A pickle fork will usually destroy the dust boot during the process, so it is not the best tool to use on tie rods that will be reused.
- **Inner tie-rod end tool:** This tool is used to loosen and tighten inner tie-rod ends.
- **Manufacturer-specific checking tools:** The following tools are made specifically for auto dealerships. A wide variety of these tools exist, designed for many different purposes:
  - *Power steering system analyzer:* This tool includes a pressure and flow gauge set with various fittings to attach in line with the power steering pump. It is used to check volume of fluid flow, maximum pressure, and leaks internal to the steering gear.

- *Scan tool:* This tool is used to read codes and data from the vehicle's PCM when diagnosing the vehicle's computer-controlled systems.
- *DMM:* The DMM which stands for digital multimeter, is used to measure voltage and ohms in a vehicle's electrical systems.
- *Various jumper leads:* These are small wires with clips on the ends, used to diagnose electrical problems.
- *Circuit tester:* This is a test light used to diagnose electrical problems.
- *Belt tension gauge:* This tool is designed to measure the amount of flex in a drive belt. Some manufacturers require the use of this tool to adjust proper belt tension.
- *Black light and dye kit:* This tool is used to pinpoint fluid leaks. Dye is added to the fluid of the leaking system, and the ultraviolet light makes the dye glow fluorescent.

▶ Diagnosis

Diagnosing Steering Systems

S39001

Steering systems can generate several different customer concerns, but the main problems that arise are play, unusual noises, vibrations/shimmies, and hard or inconsistent steering. The common culprit for these problems is wear, poor lubrication, or damaged parts. Play is the increased movement of parts that results from wear in the connections between parts. In the steering system, play generally results from worn ball sockets and bushings or too much clearance in the steering gearbox.

Hard or inconsistent steering is when it is difficult to turn the steering wheel, either all the time or intermittently. There are many faults that can cause hard steering, from low tire pressure to overly tight adjustments in the steering gearbox, to power steering that is not working properly. It is important to thoroughly check the entire steering system to identify all of the faults causing the hard steering issue. The best way to check the steering system is by physically inspecting each of the components, checking or measuring play, looking for bent or damaged linkage, and testing the operation of the power steering system (**FIGURE 39-2**). We cover each of these tasks in this chapter.

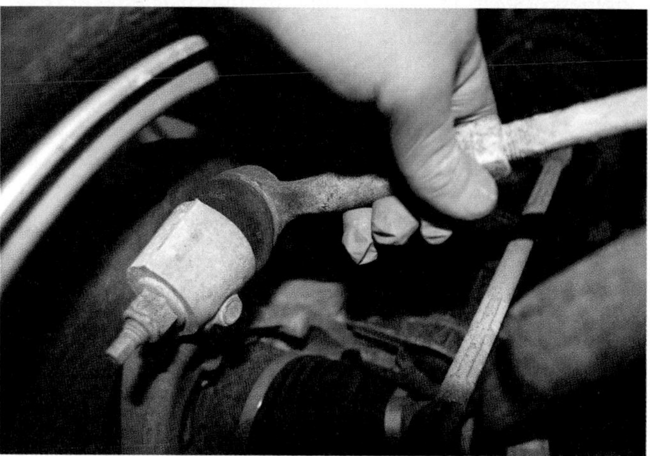

FIGURE 39-2 A technician inspecting a tie-rod end for excessive play.

Diagnosing Power Steering Fluid Leakage

N39001

Power steering fluid is critical to the proper functioning of the entire steering system. Any power steering fluid leak, no matter how small, is due for repair because of the fire hazard it presents. Because a leak can occur anywhere throughout the system, it is necessary to thoroughly inspect it all. One method for checking for leaks is to clean all power steering fluid off the vehicle and then run it for a little while. Turn off the vehicle and reexamine for new leaks. Any drops of fluid are indicative of a leak. Another method uses standard fluorescent dyes. Following the dye manufacturer's instructions, the appropriate amount of dye is added to the power steering system. After running the vehicle for the recommended amount of time, inspect the steering components with an ultraviolet light, which will illuminate the fluorescent dye and make it easier to spot leaks.

Diagnosing Power Steering Gear Issues (Non–Rack-and-Pinion)

N39002

Steering gear issues in a non–rack-and-pinion system can be one of the following: uneven effort needed for turning, looseness of steering, hard steering, unusual noises when steering, or leaks. If the issue is related to the power steering system, such as leaks, noise, or hard steering, you will need to perform a visual inspection of the drive belt and the fluid level and condition, as well as pressure test the power steering system if necessary.

If the issue is related to the steering linkage, you will need to perform a visual inspection of the linkage, joints, and steering gear. With the vehicle jacked up and placed on safety stands, have someone turn the steering wheel as you visually inspect the steering linkage. Check for binding, wear or loose connections, bent or binding components, and any other abnormal conditions. To check for binding issues, it may be necessary to separate the steering linkage to verify where the binding is located. It is possible for binding to happen in the steering gearbox, linkage, ball joints, or steering column. For looseness concerns, do not forget to check the universal joint or coupler between the intermediate shaft and the steering gearbox. These components are prone to wear, which creates play. Refer to the manufacturer's service information for diagnostic information and tests for the concern you are working on.

Diagnosing Power Steering Gear Issues (Rack-and-Pinion)

N39003

Diagnosing steering gear issues in a rack-and-pinion system is similar to a non–rack-and-pinion system and includes the same common complaints, such as uneven effort needed for turning, looseness of steering, hard steering, and unusual noises or vibrations when steering. If the issue is related to the power steering system, perform the inspection and tests associated with the power steering. If the issue is related to the steering linkage, perform a visual and hands-on inspection of the linkage, joints, and steering gear. With the vehicle raised on a hoist or jacked up and placed on safety stands, have someone turn the steering wheel as you visually inspect the steering linkage. Check for binding, wear or loose connections, bent components, debris, or any other anomaly.

Diagnosing Steering Column Issues

N39004

Common steering column issues include unusual noises, looseness, and binding. These problems generally arise from wear or debris lodged in the steering column. The process of diagnosing each of these problems is the same: The vehicle is raised on a hoist or jacked up and placed on safety stands, and one person turns the steering wheel as the technician watches for looseness or binding and listens for noises. Once the location and cause are identified, the best repair plan can be created based on the information in the service information.

Because of the number of electrical components, such as radio control switches and cruise control switches mounted to the steering wheel of modern vehicles, a device called a clock spring must be used in the steering column. The clock spring is a flexible ribbon that is coiled inside of a plastic holder. It carries all the electrical signals that flow to the components on the steering wheel and allows the steering wheel to rotate. The clock spring must be centered when steering column work has been done so that the steering wheel can be turned from lock to lock without damaging the clock spring. Refer to service information to be sure that the clock spring is centered correctly, as the position may vary by manufacturer.

The typical clock spring will have four or so turns of movement before it runs out of travel (turn very gently or you will break the wires). The clock spring is normally marked to show the center position, or where the clock spring should be positioned when the wheels are straight ahead. If you do not know where the center point of the clock spring is, you can carefully turn the clock spring in the correct direction, counting the number of total turns until it reaches the end, then move back half that number of turns. The clock spring should now be centered. Also, be careful not to wind the clock spring in the opposite direction it was designed to be wound as that will greatly shorten its life.

The steering column of a vehicle with EPS or stability control is equipped with a torque sensor and steering angle sensor. If the steering column is being serviced, a steering angle sensor calibration is typically required. Refer to the service information for the correct method of recalibrating the steering angle sensor. Typically, a scan tool is required to perform this function.

TECHNICIAN TIP

A technician who has removed the steering wheel and is removing the steering shaft must be extremely careful not to damage or impair the operation of any multifunction switch or the clock spring.

▶ Maintenance and Repair

S39002

Although the steering system is very critical to the safe operation of the vehicle, it does not normally require a lot of maintenance. In older vehicles, each of the steering and suspension joints must be lubricated with grease during each oil change. Most, but not all, of these joints have been replaced with greased-for-life joints, so most vehicles do not have grease zerks to lubricate the joints (**FIGURE 39-3**). However, always be on the lookout, as many replacement joints use zerk fittings to keep them lubricated. Another area of scheduled maintenance on the steering system is replacement of the power steering belt and also possibly the power steering fluid.

To maintain a safe-operating vehicle, regular inspection of the steering components is essential. This can be performed both by visually inspecting the components and joints for damage, looseness, or leakage and by operating the steering system and looking or feeling for play in the components, and if found, then measuring the play and comparing it to specifications. Also, listening and feeling the steering components can help

FIGURE 39-3 A. A greased-for-life tie-rod end. **B.** Tie-rod end with a zerk fitting used to lubricate the joint.

identify any needed maintenance or repair. Manufacturers list the most appropriate method of inspecting and testing their systems. Always refer to the manufacturer's service information for these methods and wear specifications.

Checking and Adjusting Power Steering Fluid

The power steering fluid transmits pressure throughout a vehicle's power steering system. It also provides lubrication to the moving parts. It must be able to perform these functions in any weather conditions, maintaining its chemical integrity in the face of subfreezing temperatures or scorching heat. Given its importance, it is critical that the power steering fluid be inspected regularly for both purity (no contamination with debris) and level. The power steering fluid usually has both a hot and a cold indicator on the dipstick to ensure an accurate reading of the fluid level in both conditions. Always refer to the service information for the recommended type of power steering fluid specified for the vehicle.

To check and adjust power steering fluid, follow the steps in **SKILL DRILL 39-1**.

Replacing Power Steering Pump Filter(s)

Some vehicles use a replaceable power steering pump filter, which is located in the fluid reservoir or in the return line from the steering gear (**FIGURE 39-4**). If it is in the reservoir, it may be a separate filter that sits in the bottom of the reservoir, and it can be carefully fished out of the reservoir and replaced. Other manufacturers build the filter into the reservoir such that the reservoir and filter are one unit and changed together. If it is an inline filter, it is usually installed in the return line because of the lower pressure and in order to protect the pump. It may be spliced into the hose between the steering gear and the pump, or it could fit inside the return hose where the return hose connects onto the pump. When changing the filter, you may want to consider flushing the power steering system first to remove as much of the old fluid as possible before installing the new filter.

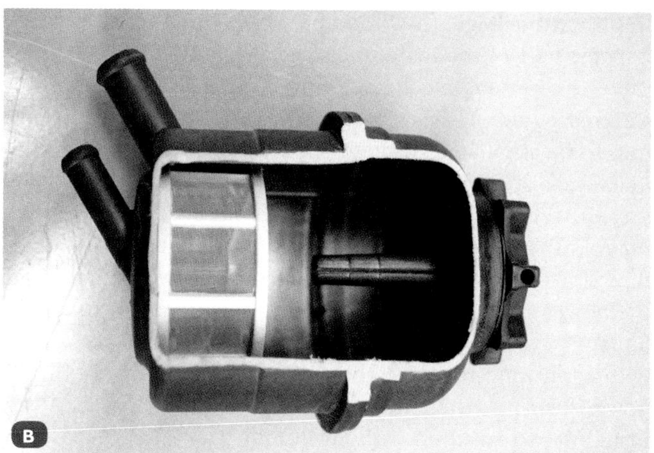

FIGURE 39-4 A. Inline power steering fluid filter. **B.** Reservoir style filter.

Always check the service information to determine the scheduled service interval and the location of the filter, if equipped.

To replace the power steering pump filter(s), follow the steps in **SKILL DRILL 39-2**.

SKILL DRILL 39-1 Checking and Adjusting Power Steering Fluid

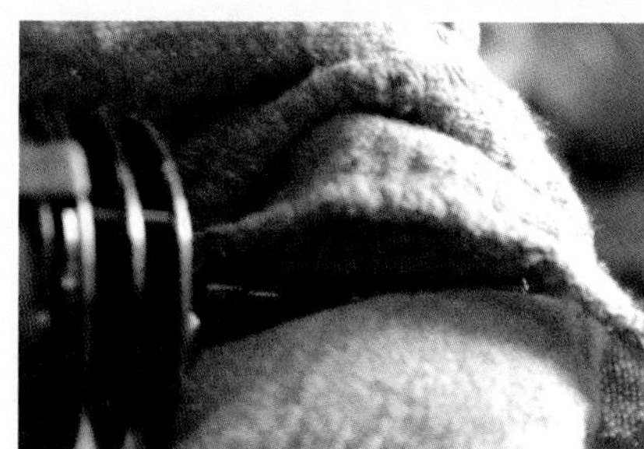

1. Locate the power steering reservoir. Clean around the cap if dirt is present. Unscrew the cap on top of the reservoir, and wipe the dipstick clean.

2. Reinstall the cap for a few seconds, then remove it again and check the fluid level on the dipstick. Verify that the level matches the temperature of the system—hot or cold.

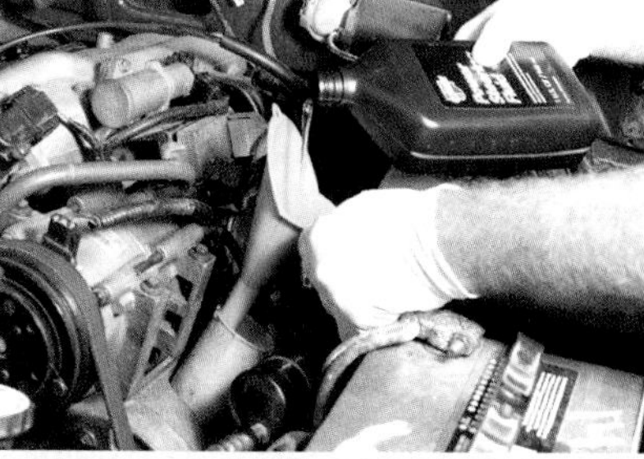

3. If the level is low, top if off with the specified fluid. If the vehicle has a plastic reservoir, check the marks to see if any fluid needs to be added.

SKILL DRILL 39-2 Replacing the Power Steering Pump Filter(s)

1. Research the vehicle you are working on to determine whether it has a power steering pump filter, what its change interval should be, and its location.
2. Following the specified procedure, remove the filter.
3. Inspect any hoses or fittings for cracks and damage.
4. Install the new filter, top off the reservoir with the correct fluid, and check for proper operation and leaks.

Flushing a Power Steering System

N39005

Flushing the power steering system is the process of removing all the old power steering fluid and replacing it with new. This is performed whenever the manufacturer specifies a fluid change, the fluid appears contaminated or dirty, a major part of the hydraulic steering system is replaced, or there is a serious mechanical problem involving the power steering pump or steering gear that could cause metal shavings to be circulated through the system. Only use the manufacturer's specified power steering fluid when flushing the system. When the service is complete, inspect for leaks after turning the steering wheel to the full-lock position in both directions and back to the center. Also ensure that none of the power steering hoses make contact with any of the components. Typically a power steering system flush is performed with a flushing machine that is connected in line with the power steering system pump to remove all old dirty fluid and install new clean fluid without introduction of air. Note that air introduced into the power steering pump can be harmful to the pump, as the pump is lubricated by the power steering fluid, and scoring of the pump can result if the pump is run dry. There are several types of flushing machines; if one is not available, flushing will have to be performed as shown in Skill Drill 39-3. Be sure to follow the manufacturer's procedure for flushing power steering fluid.

To flush a power steering system, follow the steps in **SKILL DRILL 39-3.**

Pressure Testing a Power Steering System

S39003

Pressure testing of the power steering system is performed when the driver complains of hard or inconsistent steering or when there are repeated hose failures. The test checks the operation of the power steering pump, pressure relief valve, control valve, power piston, and hoses. You will be testing the operating pressure, maximum pressure, and fluid flow. All of these tests and readings, along with the service information, will allow you to determine whether there is a fault in the power steering system.

When the service is complete, check for leaks; check operation by turning the steering wheel to full lock in both directions and back to the center; and check the fluid level. Ensure that none of the power steering hoses makes contact with any linkage or other obstructions, which can cause leaks later on. Dispose of discarded fluid properly in accordance with the relevant EPA and government legislation. Use only the manufacturer's specified power steering fluid for the vehicle.

To pressure test a power steering system, follow the steps in **SKILL DRILL 39-4.**

Applied Science

AS-49: Pressure: The technician can demonstrate an understanding of pressure in reaction to the concept of force.

A vehicle is in the shop with a customer concern of extremely hard steering. The technician confirms the concern as lack of power steering assist. After doing a visual inspection for the fluid level, leaks, and belt tension, the technician connects a pressure gauge. Using the manufacturer's service information, the technician conducts the pressure test to discover a reading of only 200 psi (1380 kPa). The specifications for this test are 1500 psi (10,343 kPa).

The technician understands that the lack of steering pressure is causing the lack of steering assist, which results in the extremely hard steering of the vehicle. Pascal's law states that force is equal to pressure multiplied by area. The area in the formula is the size of the piston in square inches or square millimeters. After a new power steering pump is installed, the result will be a greater amount of force needed to steer the vehicle properly.

This is a basic hydraulic application of Pascal's law, which applies to many systems on the automobile. In our example, we consider the power steering, but the same principles apply to the hydraulic lift that raises the vehicle for service.

Removing and Reinstalling the Power Steering Pump

N39006

The power steering pump drives the flow of power steering fluid throughout the steering system. As with any mechanical device, it can fail. Failure is usually described by the driver as difficult or noisy steering, which will have to be fully diagnosed before condemning the power steering pump. For example, excessive noise can occur when the power steering fluid level is low, which can allow air into the system. In this case, replacing the power steering pump is not the correct repair.

Assuming that proper pressure testing and noise diagnosis have confirmed a faulty pump, the pump will have to be replaced. Because there are so many ways that power steering pumps can be installed in different vehicles, always refer to the manufacturer's service information before beginning this operation.

SKILL DRILL 39-3 Flushing a Power Steering System

1. Raise the vehicle, making sure the front wheels are off the ground. Using a suction gun, remove as much power steering fluid from the reservoir as possible.

2. Disconnect the power steering return hose from the reservoir, and stick the end in a suitable container.

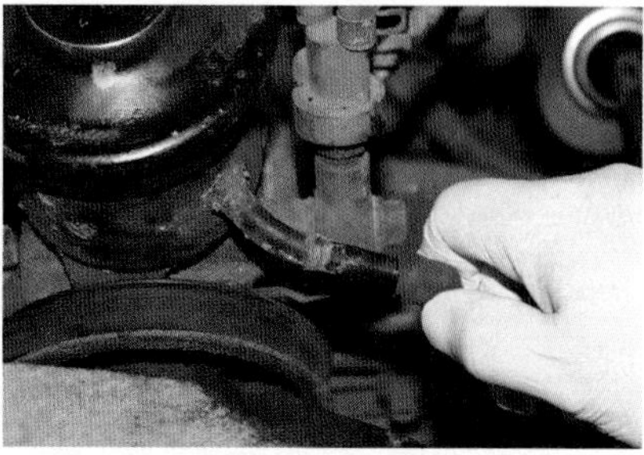

3. Temporarily plug the return fitting on the reservoir with a snug-fitting rubber cap.

4. Add the recommended fluid into the reservoir until it is at the correct level.

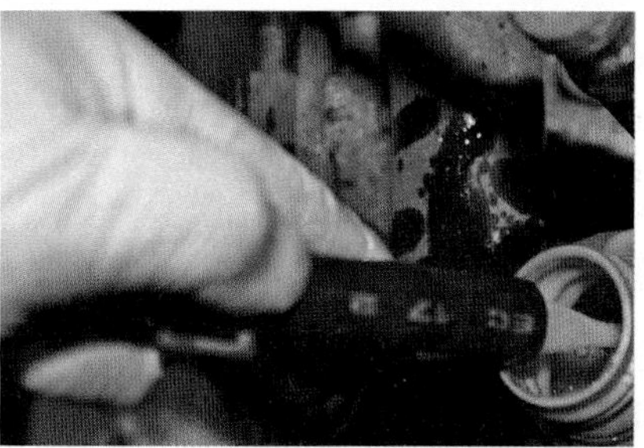

5. Start the engine, and have an assistant turn the steering wheel from lock to lock while keeping the reservoir at or near the full mark. When the fluid coming out of the return hose is clear, stop adding fluid, and keep running the engine until the level in the reservoir is just below the return line inlet. Turn off the engine.

6. Reinstall the return hose on the reservoir, top off the fluid level, and start the engine. Turn the steering wheel lock to lock a few more times, turn off the engine, and check the fluid level. Top off if necessary.

7. Lower the vehicle, start the engine, and turn the steering wheel from lock to lock to check that the system is working correctly with the vehicle weight on the tires. Dispose of the waste power steering fluid in an environmentally approved manner.

SAFETY TIP

- Be aware of moving drive belts and accessories when the engine is running.
- Power steering fluid can become very hot in operation. Take precautions to avoid burns.
- Power steering systems can reach a working pressure of over 1400 psi, or 9650 kPa, during operation. Always follow the vehicle or equipment manufacturer's procedure when working on the system.
- Power steering fluid is flammable. Always keep it away from hot exhaust manifolds, pipes, or catalytic converters.
- Clean up any spills immediately.
- Wear appropriate heatproof chemical gloves when working on hot systems.

To remove and reinstall a power steering pump, follow the steps in **SKILL DRILL 39-5**.

Removing and Reinstalling the Press-Fit Power Steering Pump Pulley

N39007

Most power steering pumps pressurize and pump the power steering fluid throughout the power steering system by a belt-driven pulley. Whenever the power steering pump is removed or replaced, the pulley typically must also be removed from the old pump and installed on the new pump. Power steering

SKILL DRILL 39-4 Pressure Testing a Power Steering System

1. Raise the vehicle, and position a drain pan under the power steering system. Following the manufacturer's instructions for the pressure gauge or analyzer you are using, disconnect the high-pressure hose from the power steering pump.

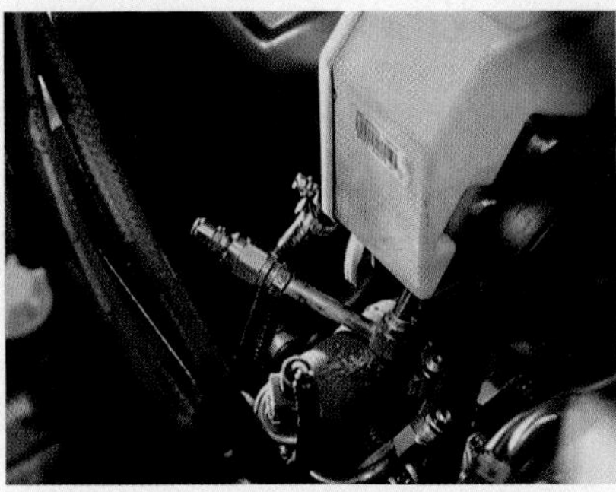

2. Connect the inlet hose from the gauge to the high-pressure connection on the power steering pump, using the appropriate adapter.

3. Connect the high-pressure hose to the outlet valve on the pressure gauge, using the appropriate adapter.

4. With the pressure gauge control valve open, start the engine, and run it at idle until fluid circulates through the tester and hoses, and the air is purged from the fluid. Turn the engine off. Check the power steering fluid level, and add the specified fluid as necessary.

5. Start the engine again. Record the initial pressure and flow readings on the pressure gauge. Close and open the pressure gauge control valve two or three times, making sure the valve is not left closed for more than 5 seconds. Record the highest pressure indicated.

6. With the pressure tester control valve open, turn the steering wheel to the extreme left and to the extreme right. Record the highest indicated pressure at lock. Turn off the engine. Compare these readings to the specifications, and take the results to your supervisor.

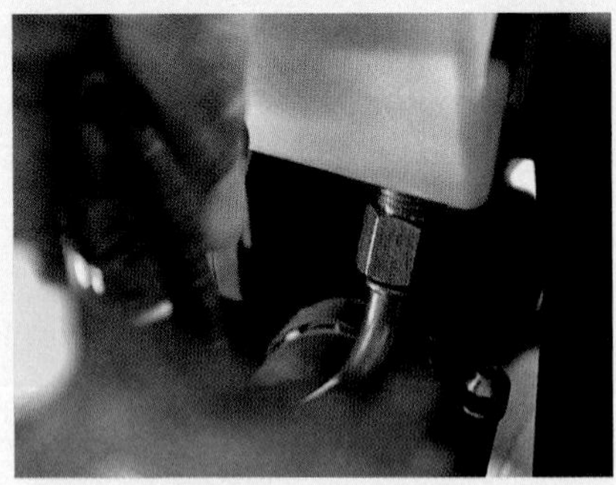

7. Be careful when disconnecting the analyzer hose from the high-pressure connection point on the power steering pump, as the fluid will be extremely hot. Replace any O-rings, and reconnect the vehicle's power steering system high-pressure hose to the power steering pump outlet.

8. Top off the power steering reservoir, and follow the manufacturer's procedure to bleed any air from the system. Some systems require the use of an adapter on the reservoir and a vacuum pump to remove all of the air.

SKILL DRILL 39-5 Removing and Reinstalling a Power Steering Pump

1. After researching the specified procedure for removing the power steering pump, raise the vehicle on a lift or support it on safety stands. Drain the power steering system's fluid by disconnecting the power steering return line, and dispose of the fluid in accordance with environmental legislation.

2. Disconnect the power steering hoses from the pump, and check for damage or leaks. Replace hoses if needed. New O-rings should be used on hose ends on reassembly, if the vehicle is so equipped.

3. Disconnect the drive belt(s) and inspect for cracking or groove depth with a belt groove depth gauge. Replace any belt that is cracked or worn beyond specifications.

4. Remove the pump retaining bolts, and remove the power steering pump. If replacing the pump with a new one, follow Skill Drill 39-6 to remove and replace the pulley.

5. To reinstall the new or overhauled power steering pump, reinstall in the reverse order. Fill and bleed the system with the specified power steering fluid.

pump pulleys are installed differently from most other pulleys; they are typically held in place by a press fit. As such, the pulley will have to be pulled off the power steering pump shaft with a special puller and pressed onto the new pump with the same tool. Some pulleys are now being made of plastic and pose an additional challenge, especially if they have become brittle. On some vehicles, the pulley has to be removed before the pump can be removed. On others, the pump can be removed before the pulley by accessing the pump bolts through holes in the pulley. Because there are many pump and pulley arrangements available today, always refer to the manufacturer's service information before beginning this operation.

To remove and reinstall a press-fit power steering pump pulley, follow the steps in **SKILL DRILL 39-6**.

Inspecting and Replacing Power Steering Hoses and Fittings

N39008

Any leakage of power steering fluid out of a power steering pressure hose can be catastrophic. Because it is being actively pumped, the fluid could spray out of the hose at high speed and spray all over the hot engine and exhaust system. The heat of the engine parts will cause the power steering fluid to smoke and billow out from under the hood, and even catch on fire. Small leaks or leaks in the return line result in low fluid level in the power steering reservoir and leave spots on the floor or parking space. If the fluid level gets low enough, it can cause a buzzing noise when driving. This noise is louder when steering the vehicle. The level of power assist may also fluctuate. If any of these symptoms are present, or to catch issues before they

SKILL DRILL 39-6 Removing and Reinstalling a Press-Fit Power Steering Pump Pulley

1. Assuming the pump is removed from the vehicle, install the correct power steering pulley puller on the pulley, making sure the puller is installed in the correct direction.

2. Tighten the center bolt (while holding the outer nut) to press off the old drive pulley. If replacing the pump, remove the old one and install the new one.

3. Place the pulley onto the power steering pump shaft, with the removal flange facing forward. Using the install portion of the pulley puller tool kit, thread the bolt portion of the tool all the way into the shaft of the new power steering pump.

4. Thread the outer nut down so that the bearing or washer is against the pulley.

5. Tighten the outer nut (while holding the center bolt) until the pulley bottoms on the shaft. Remove the puller.

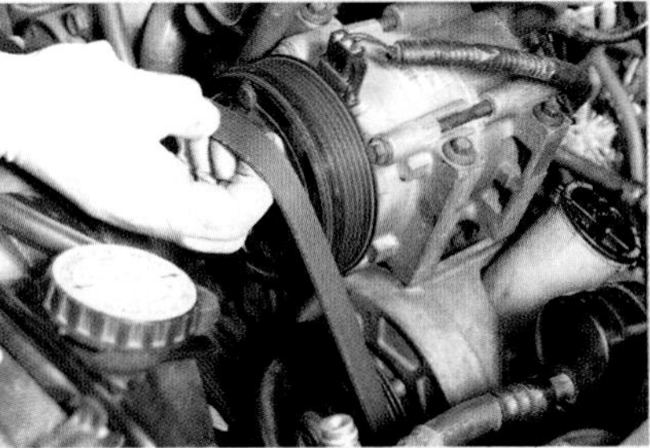

6. Reinstall the drive belt, or replace it with a new one if needed. Refill and bleed the power steering system with the specified fluid.

become this severe, it is important that the power steering hoses be inspected regularly and replaced at any sign of wear. Always refer to the manufacturer's service information before beginning this operation.

To inspect and replace power steering hoses and fittings, follow the steps in **SKILL DRILL 39-7**.

Inspecting Mounting Bushings and Brackets

S39004, N39009

During the normal wear and tear of vehicle use, the rack-and-pinion mounting bushings become compressed, brittle, or torn,

SKILL DRILL 39-7 Inspecting and Replacing Power Steering Hoses and Fittings

1. Raise the vehicle on a lift or support it on safety stands. Inspect each of the power steering hoses and fittings for leaks and damage.

2. Place a drain pan under the fitting to be disconnected. Use a flare nut wrench to loosen the fitting while you are holding the nut on the pump or steering gear with a wrench.

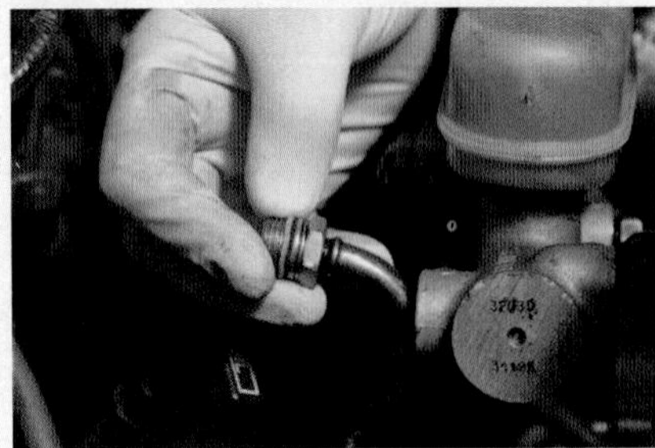

3. If replacing the O-ring, inspect the mating surfaces of the fitting for damage. If no damage is found, replace the O-ring and reinstall the fitting, being careful not to cross-thread or over-tighten.

SKILL DRILL 39-7 Inspecting and Replacing Power Steering Hoses and Fittings (Continued)

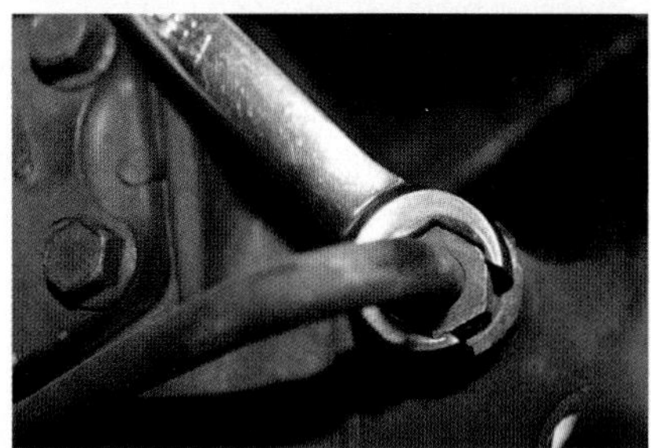

4. If replacing the hose, disconnect the fitting on the other end of the hose. Compare the old and new hoses to verify the correct replacement part.

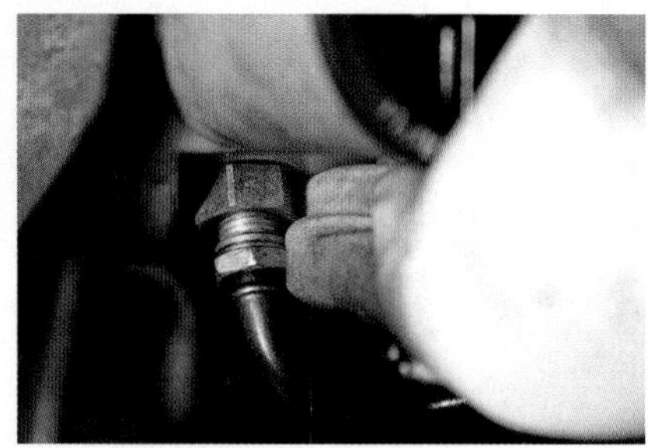

5. Carefully start the fitting on each end of the hose by hand, making sure they are not cross-threaded.

6. Check the routing of the new hoses, ensuring that they do not make contact with any components that could cause a failure; then tighten the fittings to the proper torque.

7. Top off the reservoir with the specified fluid, start the engine, and turn the steering wheel from lock to lock a few times. Check for fluid leaks. Refill the fluid as necessary, and repeat the bleeding process as needed.

and brackets may be damaged, worn, or lost. If the vehicle has an oil or power steering fluid leak, the fluid may leak onto the mounting bushings. The oil tends to soften and degrade the bushings. When this happens, the driver typically complains of either loose steering or bumping noises when driving or turning. In either scenario, a quick inspection of these bushings is warranted.

To inspect mounting bushings and brackets, follow the steps in **SKILL DRILL 39-8**.

Removing and Replacing the Rack-and-Pinion Steering Gear

Removing and replacing the rack-and-pinion steering gear is not to be undertaken lightly. It should be considered only after carefully examining all parts of the steering system and performing a full diagnosis to ensure the problem is not a result of some other fault. Many symptoms can point to a worn rack-and-pinion assembly, such as tire wear, wandering, and hard or uneven steering.

SKILL DRILL 39-8 Inspecting Mounting Bushings and Brackets

1. Raise the vehicle on a lift, keeping the weight of the vehicle on the wheels, if possible. Inspect the bushings and brackets for any faults.

2. Try moving the rack-and-pinion assembly up and down by hand.

3. Have an assistant rock the steering wheel back and forth while you look for movement in the rack-and-pinion bushings. If the rack moves significantly, place a torque wrench on the mounting bolts, and tighten them to specification. If the rack still moves, then bushing replacement will be necessary.

SKILL DRILL 39-8 Inspecting Mounting Bushings and Brackets (Continued)

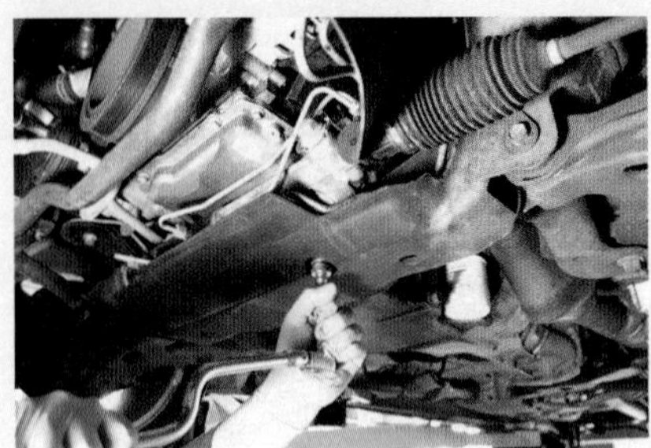

4. Remove the bracket bolts.

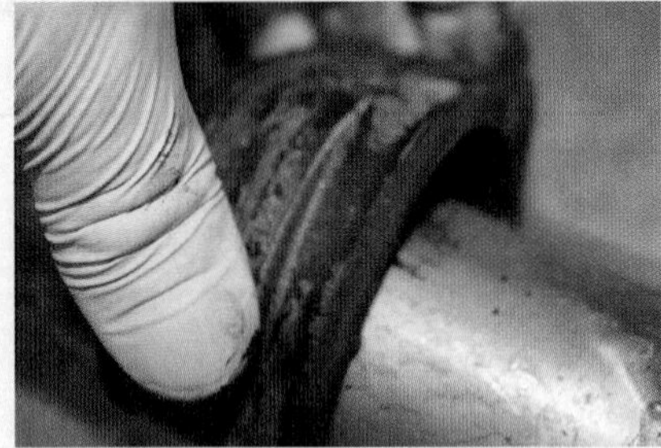

5. Remove the bushings from the rack.

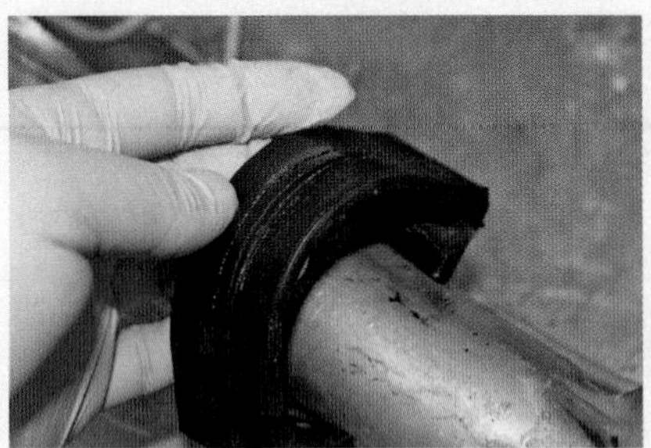

6. Install new bushings in reverse order of removal.

Regardless of the symptoms, all other steering problems should be considered and eliminated before undertaking the removal or replacement of the rack-and-pinion steering gear. Because these systems vary across vehicle types and manufacturer designs, there is no single set procedure for removal and replacement; therefore, the steps that follow are nonspecific. Consult the service information for the specific steps for the vehicle you are working on.

To remove and replace a rack-and-pinion steering gear, follow the steps in **SKILL DRILL 39-9**.

Inspecting and Replacing Rack-and-Pinion Steering Gear Inner Tie-Rod Ends and Bellows Boots

N39010

Over time, the inner tie-rod ends wear and can cause excessive play in the steering linkage. Also, bellows boots may become torn or dislodged from their seat. When bellows boots are torn, dirt or abrasives may enter the unit, accelerating wear

SKILL DRILL 39-9 Removing and Replacing a Rack-and-Pinion Steering Gear

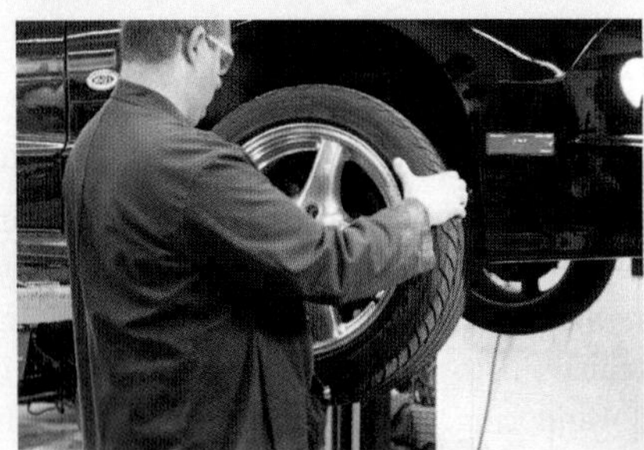

1. After researching the service information, start the engine and straighten the wheels. Turn the engine off and lock the steering column. This will keep the clock spring centered. Remove the front tires.

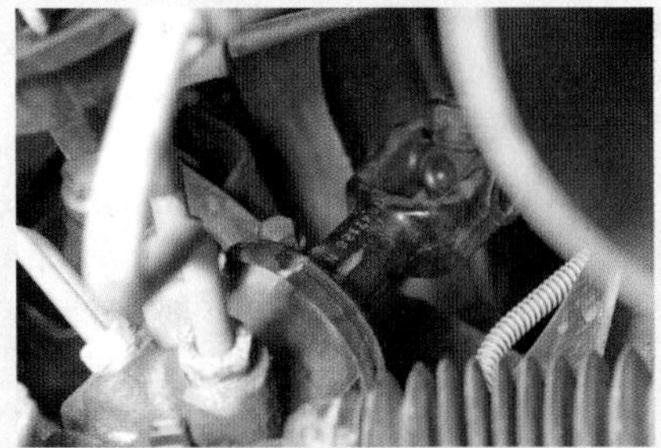

2. Locate the universal joint coupler from the intermediate shaft to the pinion shaft of the rack-and-pinion assembly. Remove the bolt from this coupler. Place a paint line on the coupler and the pinion shaft if reinstalling the same rack.

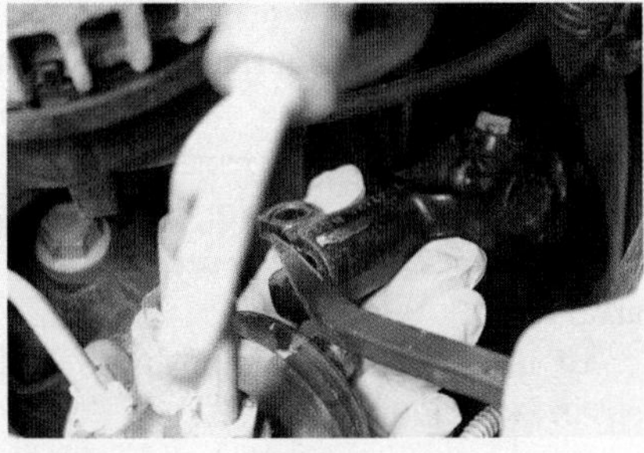

3. Pry the universal joint coupler off the pinion shaft.

4. Place an oil catch container under the power steering lines. Use flare nut wrenches to disconnect the power steering lines from the rack, and plug them to keep fluid from leaking.

5. Remove the cotter pins and nuts from the outer tie-rod ends, and separate the tie-rod ends with an approved tie-rod end puller, hammer method, or pickle fork.

6. Remove bolts from the rack-and-pinion bushing brackets, and remove brackets, if the vehicle is so equipped.

SKILL DRILL 39-9 Removing and Replacing a Rack-and-Pinion Steering Gear (Continued)

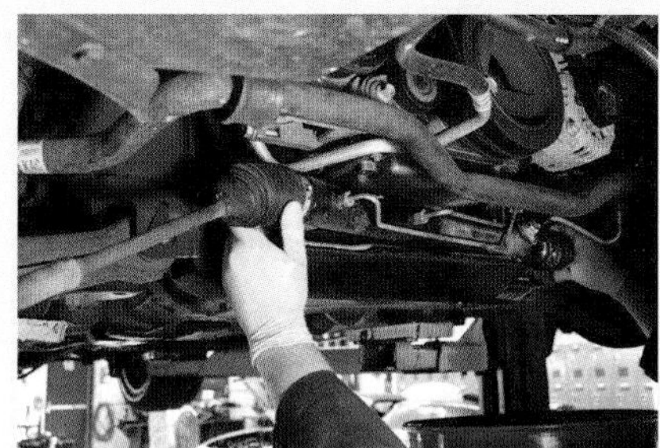

7. Slide the rack-and-pinion assembly out through the vehicle's wheel well.

8. To install the rack-and-pinion steering gear, reverse the removal steps. Be careful not to damage any lines, tubes, or fittings. Also, make sure the rack is centered in its travel so that the clock spring will be indexed to the rack.

9. Torque all fasteners, and install new cotter pins.

10. Check the fluid level in the power steering reservoir. Top off with the proper fluid, and check for leaks from the rack and pinion. Follow the manufacturer's instructions to bleed any air from the power steering system. Top off fluid as necessary.

11. With the engine running, turn the steering wheel from lock to lock. Check for binding, excessive steering effort, and uneven steering effort. After a major repair like this, it is recommended to perform an alignment to ensure all alignment angles are correct.

of rack-and-pinion seals and bushings. Routine inspection can identify damage early, before additional problems arise. Refer to the appropriate service information when replacing the inner tie-rod ends and bellows boots.

To inspect and replace rack-and-pinion steering gear inner tie-rod ends and bellows boots, follow the steps in **SKILL DRILL 39-10**.

Inspecting and Replacing the Pitman Arm, Relay (Center Link/Intermediate), Rod, Idler Arm and Mountings, and Steering Linkage Damper

S39005, N39011

With any steering complaints, it is necessary to check the components of the steering linkage for wear and damage. Slight problems in any of these components may result in significant steering problems and tire wear. Looseness in the idler arm may cause excessive toe change on rough road surfaces, leading to a wandering condition. Looseness in and of the tie-rod ends may be felt as loose steering and is frequently mistaken as a steering gearbox problem. Because of these issues, each of the steering system joints needs to be inspected for excessive wear or damage. If the vehicle is equipped with a steering damper, a faulty one can cause a shimmy in the steering wheel after hitting a bump in the road. Refer to the manufacturer's service information for the procedure and specifications for inspecting and replacing these components.

To inspect and replace the idler arm, follow the steps in **SKILL DRILL 39-11**.

To inspect and replace the pitman arm, follow the steps in **SKILL DRILL 39-12**.

To inspect and replace the center link (relay rod/intermediate rod), follow the steps in **SKILL DRILL 39-13**.

To inspect and replace the steering linkage damper, follow the steps in **SKILL DRILL 39-14**.

SKILL DRILL 39-10 Inspecting and Replacing Rack-and-Pinion Steering Gear Inner Tie-Rod Ends and Bellows Boots

1. Raise the vehicle on a lift. Inspect the rubber bellows for any signs of leaks, tears, or damage.

2. With the vehicle raised, have an assistant turn the steering wheel to one side or the other and rock the steering wheel from side to side. On the side farthest out, squeeze the bellows until you make contact with the inner ball joint, and feel for any play in the inner tie-rod joint. Repeat this procedure for the other side. If play is found, replacement of the inner tie-rod ends will be necessary.

3. Remove the front wheel for the side being replaced, and loosen the locknut on the tie-rod end.

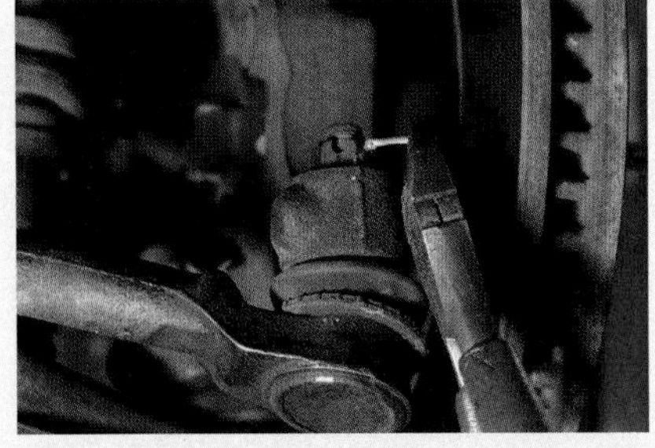

4. Remove the cotter pin and nut holding the outer tie-rod end to the steering arm, and separate the tie-rod end from the knuckle, using a tie-rod removal tool, the double-hammer method, or a pickle fork.

5. Count the number of turns to remove the outer tie-rod end from the threaded sleeve.

6. Remove the spring clamp from the bellows boot end to the inner tie-rod shaft, and remove the crimp clamp from the bellows boot to the rack-and-pinion housing. A new crimp clamp will have to be used on replacement of the boot.

7. Remove the bellows boot.

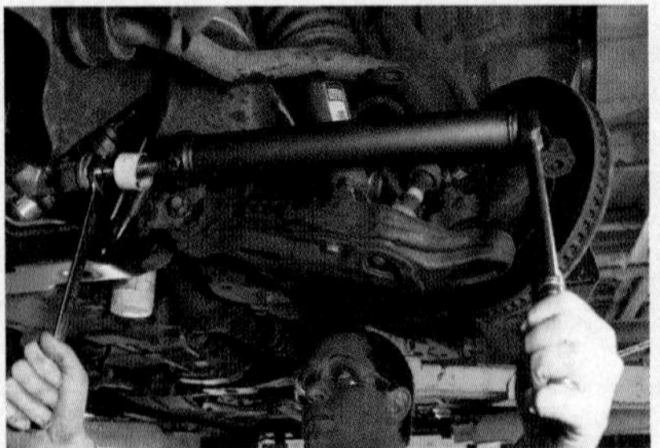

8. Using an inner tie-rod tool and the specified wrench to hold the rack, loosen the inner tie rod from the rack, and remove the inner tie-rod end.

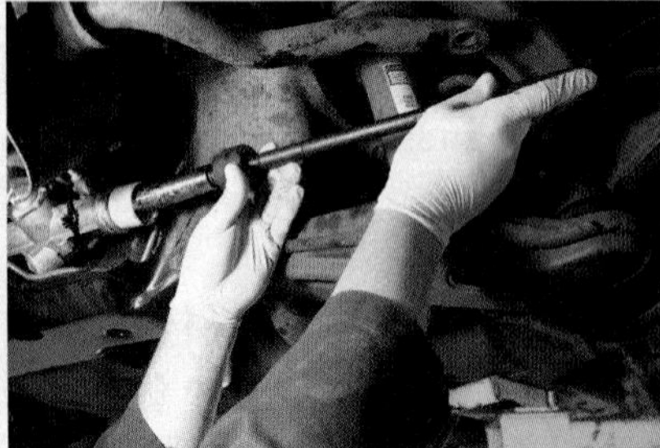

9. Install the new inner tie rod in reverse order of removal, and verify that the play is gone. Perform alignment to reset toe after replacement is performed.

SKILL DRILL 39-11 Inspecting and Replacing the Idler Arm

1. Raise the vehicle with the lift set at the manufacturer's suggested lifting points. Push the center link at the idler arm up and down, and watch the idler arm for excessive movement.

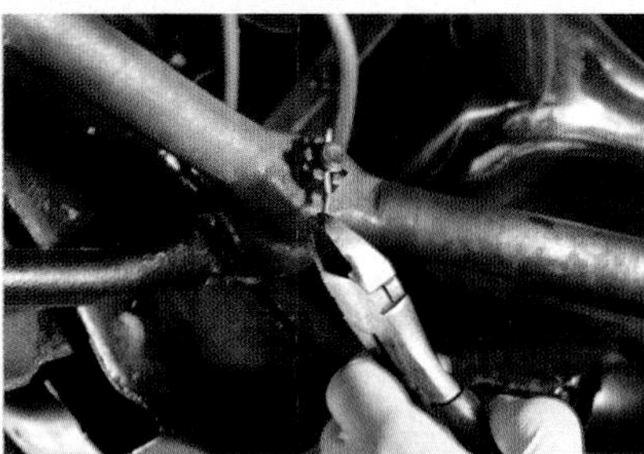

2. If the movement is out of specifications, the idler arm will have to be replaced. Remove the cotter pin, and loosen the nut connecting the idler arm taper stud to the center link.

3. Separate the idler arm taper stud, using the double-hammer method, a pickle fork, or the appropriate puller.

4. Remove the bolts holding the idler arm to the frame.

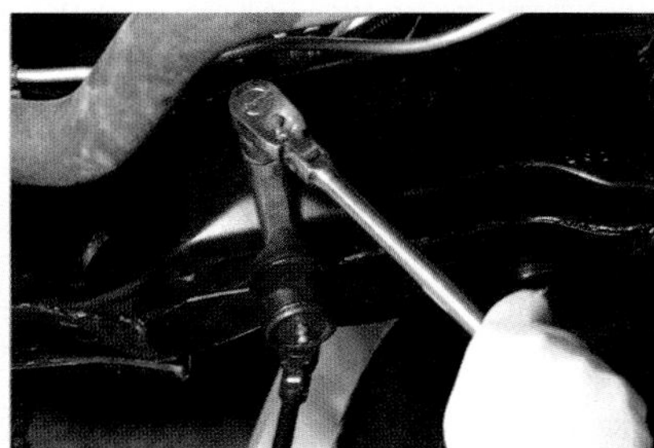

5. Install the new idler arm in reverse order of removal. Note that some idler arms must be installed with the steering centered to avoid inducing a twisting force on the steering linkage, causing the vehicle to pull to one side.

SKILL DRILL 39-12 Inspecting and Replacing the Pitman Arm

1. Push and pull side to side on the front driver's side tire while watching for looseness in the pitman arm joint. If the movement is out of specifications, the joint will have to be replaced. This joint can be located on the pitman arm or the center link.

2. If the center link needs to be replaced, remove the cotter pin and nut holding the pitman arm taper stud to the center link.

3. If the pitman arm needs to be replaced, place an alignment mark on the pitman arm to the sector shaft, with white paint or a punch, to ensure correct positioning on reassembly.

SKILL DRILL 39-12 Inspecting and Replacing the Pitman Arm (Continued)

4. Remove the pitman arm nut holding the arm to the sector shaft of the steering gear.

5. Using a pitman arm puller, pull the pitman arm free from the sector shaft.

6. Install the new pitman arm in reverse order of removal.

SKILL DRILL 39-13 Inspecting and Replacing the Center Link

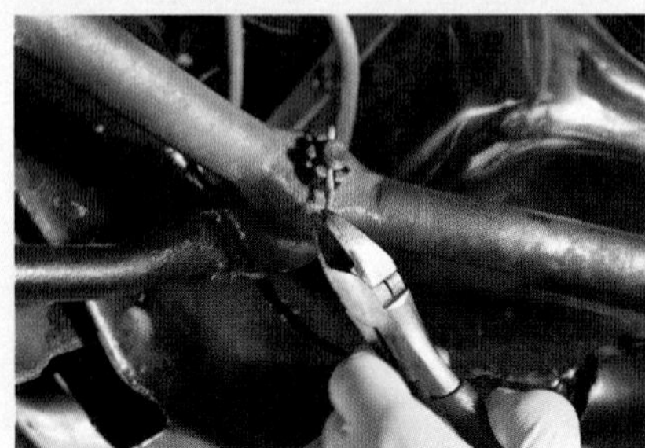

1. Push and pull the tire/wheel assembly from side to side, checking each of the center link joints for excess movement. If the movement is out of specifications, the joint(s) will have to be replaced. Remove the cotter pin and nuts from the taper studs of the pitman arm, idler arm, and inner tie-rod ends.

2. Separate the taper studs, using the double-hammer method, a pickle fork, or the approved puller. Remove the center link.

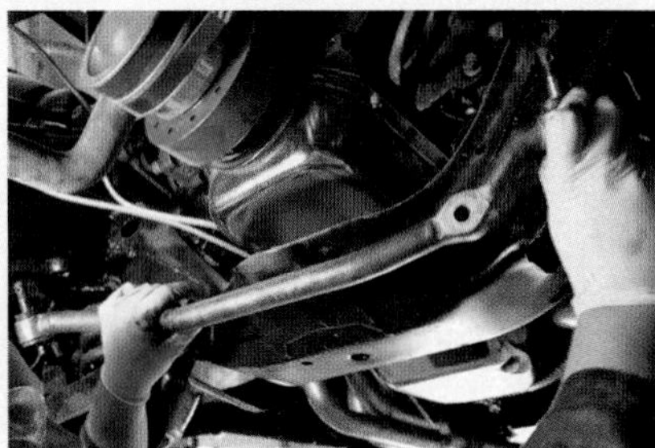

3. Install the new center link in reverse order of removal.

SKILL DRILL 39-14 Inspecting and Replacing the Steering Linkage Damper

1. Safely raise the vehicle with the lift set at the manufacturer's suggested lifting points. Look for fluid leaking out of the damper or a bent rod in the damper, and check that the bushings are tight in the damper.

2. Grab the wheels and turn them right and left. If the damper is working properly, there should be a fair amount of resistance when trying to turn the wheels quickly. If any of these conditions are found, replacement will be necessary. Remove the cotter pin and nut from the center link and frame attaching point, and remove the steering damper.

3. Install the new damper in reverse order of disassembly.

Inspecting and Replacing Tie-Rod Ends, Tie-Rod Sleeves, and Clamps

N39012

Tie rods make the final connection between the steering linkage and the steering arms. The point of connection with the steering linkage is considered the inner tie-rod end, and the end that connects to the steering arm is considered the outer tie-rod end. Checking tie rods is important in identifying steering problems, because the ends are frequently damaged or worn. There are two basic types of tie-rod ends: spring-loaded and preloaded. Each type has its own procedure for inspection and replacement of the ends. Replacement of tie-rod ends requires an alignment to be performed or rapid tire wear will occur. *Note:* If removing a tie rod from an aluminum steering knuckle, do not use a pickle fork, as it will damage the soft aluminum. Use the approved tie-rod end puller to separate the end from an aluminum knuckle or steering arm. Be sure to follow the manufacturer's service information for testing and replacement of tie-rod ends.

To inspect and replace preloaded tie-rod ends, follow the steps in **SKILL DRILL 39-15**.

Adjusting Non–Rack-and-Pinion Worm Bearing Preload and Sector Lash

The gearbox in non–rack-and-pinion steering systems may require adjustment over time, particularly if the driver complains of loose steering. The two adjustments that should be checked are the worm bearing preload and the sector lash. The goal for the worm bearing preload adjustment is to remove play at the worm bearings while maintaining a low resistance to worm shaft movement. The goal for the sector lash adjustment is to remove play from between the worm-and-sector gear teeth in the centered position so that they do not bind up at any point in the travel. For the adjustments to be accurate, the steering shaft and the pitman arm should be disconnected so that turning effort can be measured with an inch-pound torque wrench. Thus it is easier to perform this task when the steering gear has been removed from the vehicle, although it is possible to do it with the steering gear on the vehicle on some vehicles. The adjustment process should be performed according to the manufacturer's service information and specifications.

To adjust non–rack-and-pinion worm bearing preload and sector lash, follow the steps in **SKILL DRILL 39-16**.

Inspecting and Testing Electric Power Assist Steering

S39006, N39013

On vehicles equipped with EPS, the electric power assist system should be tested any time a driver complains of steering difficulties, either hard or loose. The first step is to verify the concern by starting the vehicle and operating the steering wheel while observing the EPS warning lamp. If the lamp goes out as it should and the power steering feels normal, you may need to perform a test drive and check the operation of the steering system under driving conditions. The test drive should confirm the customer's concern and give you valuable information regarding the fault. If the EPS lamp is off, you will want to perform a visual inspection of the mechanical components such as tire pressure, tires, tie-rod ends, rack bushings, and steering column for excessive play or damage. If the EPS lamp indicates a fault,

SKILL DRILL 39-15 Inspecting and Replacing Preloaded Tie-Rod Ends

1. With the vehicle's weight on the tires, have an assistant gently rock the steering wheel between the 10 o'clock and the 1 o'clock positions. Note any side-to-side or up-and-down movement in the tie-rod ends. If the ball and socket is worn, replacement will be necessary. Loosen the tie-rod locknut found where the inner tie-rod end threads to the outer tie-rod end.

2. Remove the cotter pin and nut from the tapered stud of the tie-rod end, and separate it from the steering knuckle with the double-hammer method, a pickle fork, or the approved puller. Count the number of turns the tie-rod end rotates until it is backed out of the sleeve.

3. Install the new tie-rod end, using the number of turns counted in the previous step; this will get the toe angle close. Reinstall the new tie-rod end in reverse order of removal. Perform a test drive to verify repair.

SKILL DRILL 39-16 Adjusting Non–Rack-and-Pinion Worm Bearing Preload and Sector Lash

1. Research the specified procedure and specifications for the adjustments. With the steering gear removed from the vehicle and mounted in a vise, loosen the locknut and back off the sector shaft adjustment screw.

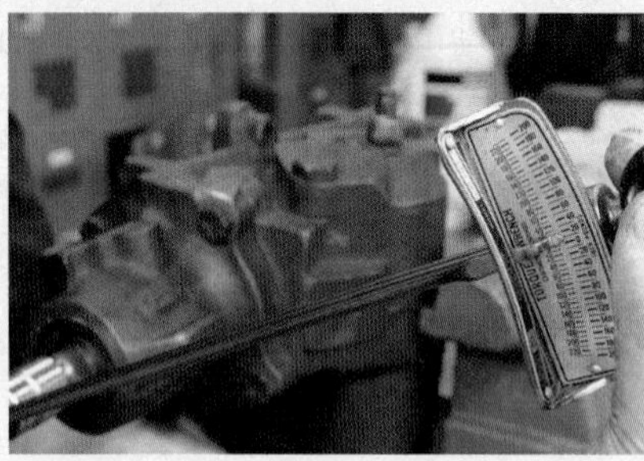

2. Fit a socket onto the input shaft, and use an inch-pound torque wrench to measure the rotating torque from lock to lock. If the rotating torque is too low, loosen the locknut on the worm bearing adjusting plug.

3. Tighten the adjusting plug slightly tighter, and retest the rotating torque. Make adjustments until the rotating torque is within specifications. Tighten the locknut without allowing the adjusting plug to move. Verify that the rotating torque has not changed.

a scan tool will have to be hooked up to the vehicle's data link connector. The scanner accesses and retrieves data from the vehicle's onboard computer and identifies the diagnostic trouble codes (DTCs). The DTCs are used with the manufacturer's flowchart for that code (found in the service information) to identify the exact problem. Consult the service information for the exact steps for diagnosis, repair, or replacement.

To inspect and test the electric power assist steering, follow the steps in **SKILL DRILL 39-17**.

Testing and Diagnosing Electrically-Assisted Power Steering Systems with a Scan Tool

N39014

Once a problem in an electrically assisted steering system has been confirmed and codes have been pulled, it is time to research the codes, technical service bulletins (TSBs), and diagnostic procedure in the appropriate service information. With the proper diagnostic procedure decided, a scan tool updated to the latest calibrations, and a DMM you are ready to start tracking down the fault. Depending on the system and the fault, the service information or TSB will lead you through the diagnosis one step at a time. Just be careful that you understand how the particular system is designed to operate, and read through the procedure carefully so that you don't become confused or start down the wrong diagnostic path.

Adjusting, Repairing, or Replacing Components of Electrically Assisted Power Steering Systems

Electrically assisted steering systems are compact units, with little allowance for adjustment or repair in the face of problems. Generally, the only course of action available is a complete

SKILL DRILL 39-17 Inspecting and Testing the Electric Power Assist Steering

1. Verify the concern by operating the steering wheel and observing the EPS warning lamp.

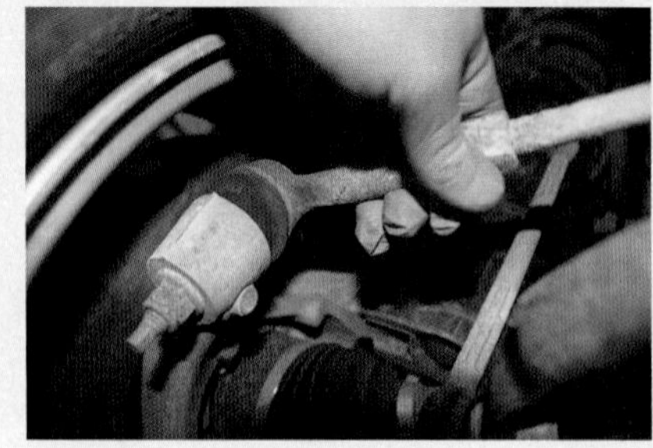

2. If the EPS lamp is off, inspect the steering system, including the tires.

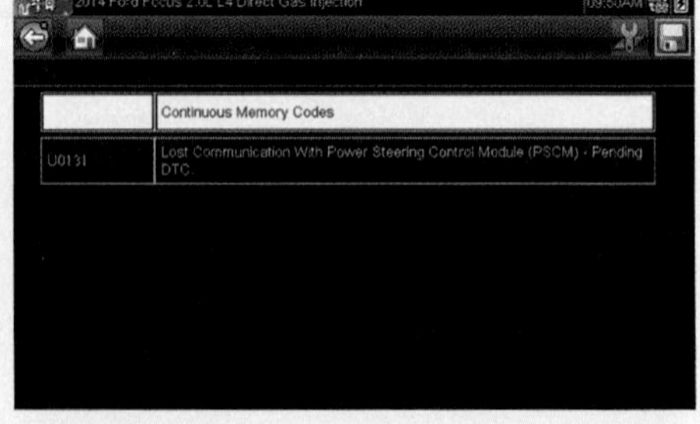

3. If the EPS lamp is on, connect a scan tool and read any EPS-related codes. Diagnose the fault by following the service procedure and using the strategy-based diagnostic process.

replacement with a new unit, replacement of the steering wheel position sensor, or a wiring repair on the system's harness. Each system, including how it fits into the various makes and models of vehicles, is unique. The technician must research the specific instructions supplied by the manufacturer and follow them step-by-step. Once repairs or replacement is complete, the PCM generally must be accessed with a scan tool to perform the process for recentering the steering wheel position sensor.

Identifying Hybrid Vehicle Power Steering System Electrical Circuits

N39015

A technician may be called upon to identify the electrical circuits within the power steering system of a hybrid vehicle, particularly when an electrical problem is suspected. Because the technology in hybrid vehicles varies and is continuously evolving, and because some of these vehicles run on very high voltage, the best procedure is to refer to the manufacturer's service information. This information includes a section devoted to diagnosis, along with the wiring diagrams for the various systems. The technician should examine the diagrams and compare them to the systems in the vehicle. Most manufacturers identify their high-voltage wires with specific colors, but there is not a uniform color designation across all manufacturers (**FIGURE 39-5**). Currently, orange, yellow, and blue are used by manufacturers for circuit voltages higher than 12 volts. Always research the color of high-voltage wiring in the service information for any hybrid or electric vehicles prior to working on them.

Disabling and Enabling the SRS

S39007, N39016

The SRS system must be disabled and enabled while working on or around the steering column, any of the airbags or other pyrotechnic devices, and any of the sensors. Failure to disable the SRS system could cause one or more of the SRS devices to deploy, which can cause serious injury or death, along with an expensive repair. It is important to know that most airbags are inflated by igniting a solid fuel similar to rocket fuel.

Every manufacturer has its own procedures for disabling and enabling the SRS system on its vehicles, and those procedures can be different for each of the vehicle models they sell. Always follow the manufacturer's specified procedure for the model of vehicle you are working on. Also, this is one time that you *do not* use a memory minder or auxiliary power supply on the vehicle. Doing so could supply power to the SRS system after it has been disabled, allowing it to be accidentally deployed.

To disable and enable the SRS, follow the steps in **SKILL DRILL 39-18**.

FIGURE 39-5 Manufacturers identify their high-voltage circuits with special colored wires. Here is one example using orange wire. Note that there isn't a consistent color code used by all manufacturers.

SAFETY TIP

Many airbags are now of the two-stage variety, so they can deploy with the proper amount of force depending on the severity of the accident, approximate weight of the occupant, etc. This means that even a deployed airbag is still potentially dangerous and needs to be treated with caution during removal. It should be deployed soon after removal, following the manufacturer's specified procedure, which will render it safe for disposal.

Removing and Replacing the Steering Wheel and Center/Time the SRS Coil (Clock Spring)

N39017

The clock spring assists in supplying a constant electrical connection to a vehicle's SRS and horn, and in some vehicles, the cruise control, message center controls, and entertainment controls. The clock spring coils and uncoils as needed during turning of the steering wheel, to ensure that the airbag is always ready to be deployed in the event of a collision. Because of this winding and unwinding, the wires in the clock spring can break over time, illuminating the SRS light and disabling the SRS system. Because it is considered an important safety feature, it must be replaced if it is damaged in any way. Be sure to follow the manufacturer's service information to replace a clock spring. Failure to properly disable the airbag could result in serious injury or death. This Skill Drill uses an example vehicle only. *Always* follow the service information for the vehicle you are working on.

To remove and replace the steering wheel and center/time the SRS coil, follow the steps in **SKILL DRILL 39-19**.

SKILL DRILL 39-18 Disabling and Enabling the SRS

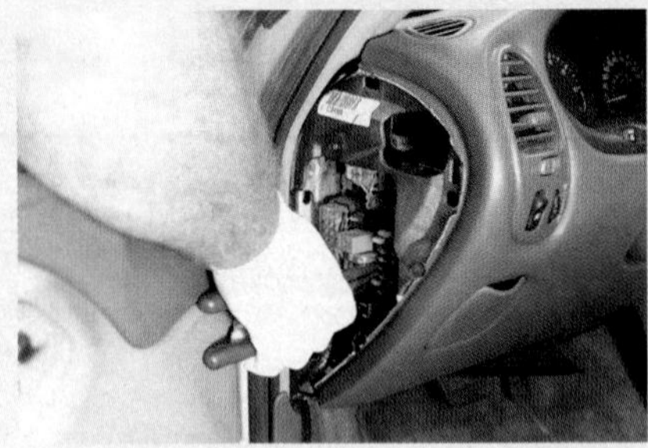

1. Find and remove the SRS fuse. Verify by turning the key on and observing that the SRS light remains lit for at least 30 seconds. If it goes out, you did not remove the correct fuse or all of the required fuses. Make sure the wheels are steered straight ahead. Turn the key off.

2. Remove the negative battery cable, and allow a minimum of 15 minutes to pass to let the SRS system capacitor's discharge. Note any radio presets or other memory features of the vehicle that will be erased when the battery is disconnected. *Do not use a memory minder or auxiliary power source!*

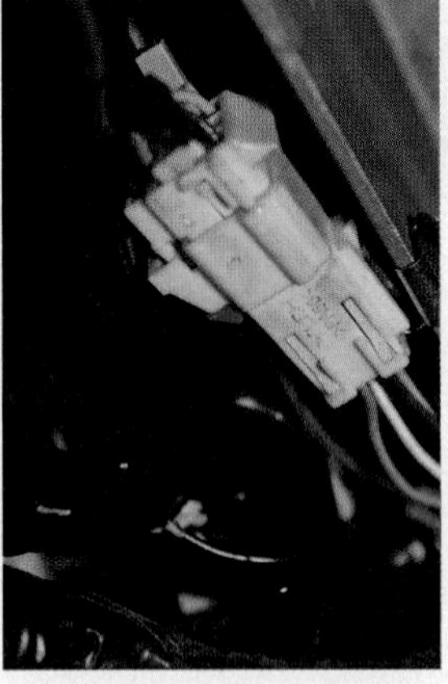

3. To enable the SRS system, verify that all SRS modules, components, and connectors are installed and connected properly.

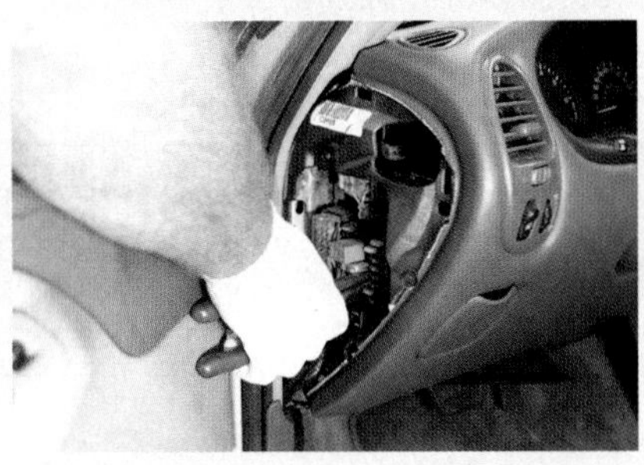

4. Reinstall the SRS fuse.

5. Reconnect the negative battery terminal, and tighten properly.

6. Without being in front of or reaching across the driver's side airbag, turn on the ignition switch and observe the SRS light. It should illuminate briefly and then go out and stay out. If so, the SRS system should be ready to be placed back into service.

SKILL DRILL 39-19 Removing and Replacing the Steering Wheel and Centering/Timing the SRS Coil

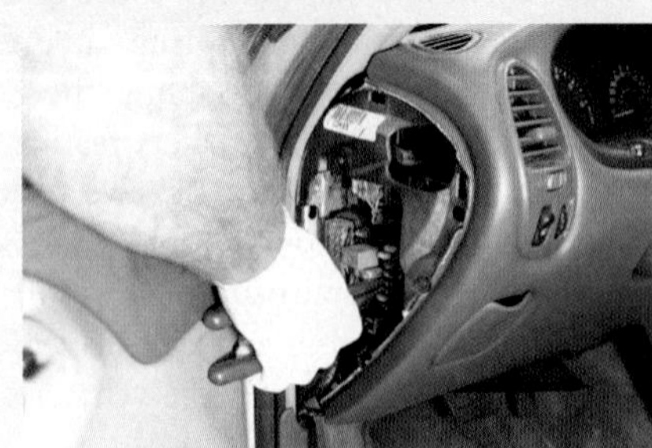

1. Find and remove the SRS fuse. Verify by turning the key on and observing that the SRS light remains lit for at least 30 seconds and does not go out. Make sure the wheels are straight ahead. Turn the key off.

2. Remove the negative battery cable, and allow a minimum of 5 minutes to pass to let the SRS system's capacitors discharge. *Do not use a memory minder or auxiliary power source!*

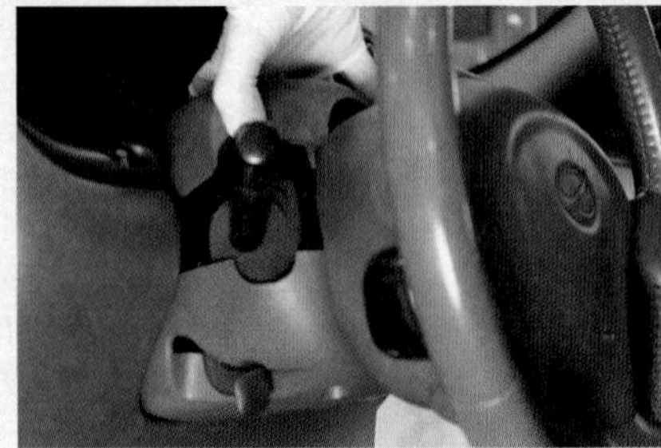

3. Locate the SRS connector at the bottom of the steering column and disconnect. Remove upper and lower trim panels.

SKILL DRILL 39-19 Removing and Replacing the Steering Wheel and Centering/Timing the SRS Coil (Continued)

4. Locate the bolts or spring clips on the back of the steering wheel that hold the airbag to the steering wheel, and remove or release them.

5. Lift the airbag from the steering wheel, and disconnect the airbag connector to the steering wheel harness.

6. Sit the airbag on a bench, face up, in a safe place.

7. Remove the fasteners that hold the steering wheel on.

8. Remove the steering wheel with the manufacturer's recommended puller (this information can be found in the service information).

9. Remove the clock spring screws or snap ring, and lift the clock spring from the steering column shaft. Attach a thin wire to the clock spring connector (airbag connector was disconnected earlier) at the base of the steering column, and pull the clock spring and connector up through the steering column.

10. Gently pull the clock spring harness down through the steering column.

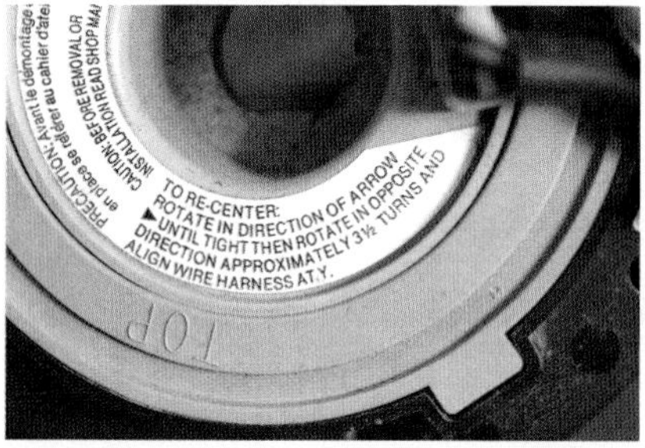

11. If the antirotation key is still installed, skip to step 13. If the antirotation key is not installed, center the clock spring, following the manufacturer's procedure. In this case, we gently rotate the inner rotor counterclockwise until it stops.

12. Rotate the inner rotor clockwise the required number of turns—in this case, four full turns. Verify that all of the marks are aligned in their specified positions.

13. Reinstall the clock spring assembly in the proper orientation, and secure it with the screws or snap ring. Reinstall components in the reverse order of removal, ensuring that torque specifications are followed. Also remove the antirotation key at the appropriate step.

Inspecting the Steering Shaft Universal Joint(s), Flexible Coupling(s), Collapsible Column, Lock Cylinder Mechanism, and Steering Wheel

N39018

The steering wheel and column are responsible for transferring the driver's steering input to the steering gear and on to the wheels, dictating the direction in which the vehicle travels. Any damage or failures of the steering column and steering wheel can cause the vehicle to move in unexpected directions, putting the lives of the driver and those around him or her at risk. Regular inspection of the steering column and its components can identify problems before they become dangerous.

To inspect the steering shaft universal joint (U-joint) or flexible coupler for play, visually inspect the steering shaft U-joint or coupler as another person turns the steering wheel from side to side. The coupler or U-joint should have zero play. If play is noted, replacement will be necessary. Follow service information for replacement of the coupler or U-joint.

Steering columns are designed to collapse in the event of an accident so that the steering main shaft in the column does not injure or kill the driver. The column has a sliding shaft that slides into itself in the event of an accident, and the column housing mounting points are designed to allow the column to slip downward as the driver makes contact with the steering wheel. An inspection of the collapsible steering column normally takes place if the vehicle has been in an accident and the column is suspected to have moved. To inspect the collapsible column for play, position yourself inside the vehicle on the driver's side, looking under the dashboard, while an assistant pulls and pushes on the column. Movement of the column at the mounting locations will be visible if the mounting points have collapsed.

To inspect the lock cylinder, insert the key and ensure that the lock cylinder rotates freely without binding and locks the steering wheel in the lock position. Binding of the lock cylinder may require the cylinder to be removed and replaced or require further teardown of the steering column to find the binding concern. Follow the manufacturer's service information for this concern. To inspect the steering wheel, pull and push on the steering wheel, checking for looseness or play. If play is found in the steering wheel, it may be due to wear in the tilt mechanism or play in the steering column upper bearing. Always check for any TSBs on diagnostic concerns, as there may be a known condition and revised parts. Teardown of the steering column or replacement may be necessary to correct play in the steering column assembly.

To inspect the steering shaft U-joint(s), flexible coupling(s), collapsible column, lock cylinder mechanism, and steering wheel, follow the steps in **SKILL DRILL 39-20**.

SKILL DRILL 39-20 Inspecting the Steering Shaft U-joint(s), Flexible Coupling(s), Collapsible Column, Lock Cylinder Mechanism, and Steering Wheel

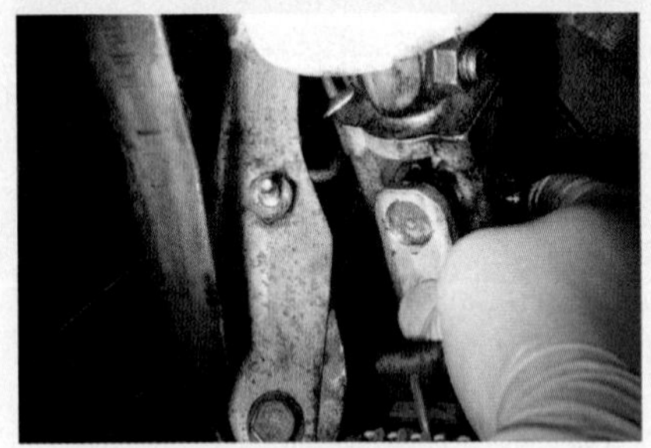

1. Inspect the steering shaft flexible couplings for obvious damage or wear. Have an assistant wiggle the steering wheel back and forth as you watch for play or improper movement. If any is noted, the joint must be replaced.

2. To inspect the collapsible column, push and pull on the steering wheel to check for any looseness of the steering shaft or column assembly.

3. Look under the dashboard at the steering column, and inspect the mounting bolts. Any broken bolts indicate a collapsed steering column.

SKILL DRILL 39-20 Inspecting the Steering Shaft U-joint(s), Flexible Coupling(s), Collapsible Column, Lock Cylinder Mechanism, and Steering Wheel (Continued)

4. Look on the column for the plastic inserts that are typically used to connect the upper and lower halves of the steering shaft. Any damage indicates a collapsed steering column.

5. To inspect the lock cylinder mechanism, insert and remove the key from the lock cylinder, checking for any resistance or difficulties.

6. Place the key inside the lock cylinder, and turn the key to each position, checking for smooth and proper operation.

7. Place the key in the lock position, and check that the steering wheel is locked. If any faults are found, follow the service information for the repair procedures.

▶ Wrap-Up

Ready for Review

- The tools used in diagnosing and repairing steering system problems can be broken into two categories: tools for mechanical diagnosis and tools for electrical diagnosis.
- Many of the problems associated with steering system are mechanical, but as more EPS systems are produced, electrical faults are becoming more common.
- The main problems that arise in the steering system are play, unusual noises, vibrations/shimmies, and hard or inconsistent steering. Common causes are wear, poor lubrication, or damaged parts.
- Power steering fluid is critical to the proper functioning of the entire steering system. Any leak, no matter how small, is cause for repair.
- To maintain a safe-operating vehicle, regular inspection of the steering components is essential.
- Pressure testing the power steering system is performed when the driver complains of hard steering or when there are repeated hose failures.
- Power steering pump pulleys are typically held in place by a press fit.
- The clock spring must be centered when steering column work has been done.
- If the steering column is being serviced, a steering angle sensor calibration will typically be required.
- Removing and replacing the rack-and-pinion steering gear should be considered only after examining all parts of the steering system and performing a full diagnosis to ensure the problem is not the result of some other fault.

- Checking tie-rod ends is important in identifying steering problems because the ends are frequently damaged or worn.
- If removing a tie rod from an aluminum steering knuckle, do not use a pickle fork, as it will damage the soft aluminum. Use the approved tie-rod end puller.
- Looseness in and of the tie-rod ends may be felt as loose steering and is frequently mistaken as a steering gearbox problem.
- A pitman arm puller is used to pull the pitman arm free from the sector shaft.
- On vehicles equipped with EPS, the electric power assist system should be tested any time the driver complains of steering difficulties, either hard or loose, or the warning lamp is on.
- The SRS system must be disabled and enabled while working on or around the steering column, any of the airbags or other pyrotechnic devices, and any of the sensors.
- *Do not use a memory minder or auxiliary power supply* on the vehicle when disabling an airbag.
- The steering column coupler or U-joint should have zero play.

Review Questions

1. When confronted with a steering system problem, all of the following should be taken into consideration *except*:
   a. mechanical components.
   b. the suspension system.
   c. electrical components.
   d. the cooling system.
2. Choose the correct statement.
   a. Steering systems are mostly automatic.
   b. Sensors do not cause steering faults.
   c. As more electronic power steering systems are produced, electrical faults are becoming more common.
   d. The suspension system does not interfere with the steering system.
3. Which tool is particularly useful in performing a continuity test on any opens or shorts in an EPS system and help determine where the problem is located?
   a. TPMS tool.
   b. Factory scanner
   c. DMM
   d. Digital SPM tool
4. The pickle fork tool can be used for:
   a. removing rod ends.
   b. checking the continuity of the electronic circuit.
   c. removing pitman arms.
   d. centering the clock spring.
5. Which of the following devices is used to measure the runout or movement on different parts of the steering system, such as play in tie-rod ends?
   a. Dial indicator
   b. Pickle fork
   c. Pitman arm puller
   d. Tie-rod sleeve adjusting tool
6. All of the following statements with respect to power steering fluid are true *except*:
   a. It transmits pressure throughout a vehicle's power steering system.
   b. It provides lubrication to the moving parts.
   c. It does not get contaminated so only its level needs checking.
   d. It must be able to perform in any weather conditions.
7. Pressure testing a power steering system checks all of the following *except*:
   a. operating pressure.
   b. maximum pressure.
   c. fluid flow.
   d. steering linkage faults.
8. All of the following can create play in the steering wheel *except*:
   a. worn steering gears.
   b. worn tie rod ends.
   c. bent steering arm.
   d. looseness in the idler arm.
9. If the EPS lamp indicates a fault:
   a. check the electrical components of the system.
   b. dismantle the steering system to check for issues.
   c. hook up a scan tool to the vehicle's data link connector.
   d. check the mechanical components of the system.
10. Which of the following systems should be disabled while working on or around the steering column, any of the airbags, or other pyrotechnic devices?
    a. EPS
    b. ABS
    c. SRS
    d. TRW

ASE Technician A/Technician B Style Questions

1. Tech A says that power steering fluid leaks can cause a vehicle to catch fire. Tech B says that fluorescent dye can be added to power steering fluid to locate the source of a leak. Who is correct?
   a. Tech A
   b. Tech B
   c. Both A and B
   d. Neither A nor B
2. Tech A says that in the steering system, play generally results from worn ball sockets and bushings. Tech B says that hard steering is typically caused by tire pressure that is too high. Who is correct?
   a. Tech A
   b. Tech B
   c. Both A and B
   d. Neither A nor B
3. Tech A says that when flushing power steering fluid, you should run the engine for a few minutes after all of the old fluid has drained out of the system. Tech B says that power steering fluid should be flushed if a major part of the hydraulic steering system is being replaced. Who is correct?

a. Tech A
b. Tech B
c. Both A and B
d. Neither A nor B

4. Tech A says that many power steering pump pulleys are typically held in place by a press fit. Tech B says that some power steering pump pulleys are made out of plastic. Who is correct?
   a. Tech A
   b. Tech B
   c. Both A and B
   d. Neither A nor B
5. Tech A says that to properly check a tie-rod end, the technician should twist the tie-rod end, and any rotational movement means the joint is bad. Tech B says that when checking a tie rod, any side-to-side movement in the joint means the tie rod is bad. Who is correct?
   a. Tech A
   b. Tech B
   c. Both A and B
   d. Neither A nor B
6. Tech A says that worn rack-and-pinion mount bushings can cause excessive play in the steering system. Tech B says that a worn idler arm can cause excessive play in the steering system. Who is correct?
   a. Tech A
   b. Tech B
   c. Both A and B
   d. Neither A nor B
7. Tech A says that a pickle fork is used to hold a tie rod while it is being tightened. Tech B says that an outer tie-rod end should have a noticeable side-to-side play if it is okay. Who is correct?
   a. Tech A
   b. Tech B
   c. Both A and B
   d. Neither A nor B
8. Tech A says that worn tie-rod ends can cause a steering wandering complaint. Tech B says that a hard steering complaint could be caused by a worn power steering pump. Who is correct?
   a. Tech A
   b. Tech B
   c. Both A and B
   d. Neither A nor B
9. Tech A says that when disabling the SRS system, verify that the correct fuse was removed by turning the key on and observing that the SRS light remains lit for at least 30 seconds and does not go out. Tech B says that after enabling the SRS system, you should test the system by turning the key to the run position while sitting in the driver's seat. Who is correct?
   a. Tech A
   b. Tech B
   c. Both A and B
   d. Neither A nor B
10. Tech A says that when replacing the clock spring, you should turn it all the way to either end and then install it in the steering column. Tech B says that clock springs are used to return the steering wheel to its centered position. Who is correct?
    a. Tech A
    b. Tech B
    c. Both A and B
    d. Neither A nor B

CHAPTER 40
Suspension System Theory

NATEF Tasks

There are no NATEF tasks in this chapter.

Knowledge Objectives

After reading this chapter, you will be able to:

- **K40001** Describe the purpose, function, and forces acting on suspension systems.
- **K40002** Explain the principles of sprung and unsprung weight and how dampening is affected by them.
- **K40003** Describe the function of the suspension system.
- **K40004** Describe the forces in suspension systems and how they are controlled.
- **K40005** Describe yaw, pitch, and roll as they relate to the vehicle.
- **K40006** Describe the purpose and function of suspension system components.
- **K40007** Describe the purpose, types, and function of springs.
- **K40008** Describe the purpose and function of shock absorbers and struts.
- **K40009** Describe the purpose and function of control arms and rods.
- **K40010** Describe the purpose and function of the steering knuckle.
- **K40011** Describe the purpose and function of ball joints.
- **K40012** Describe the purpose and function of bushings.
- **K40013** Describe the various types of suspension systems.
- **K40014** Describe the purpose and function of dead and live axles.
- **K40015** Describe the purpose and function of a solid axle.
- **K40016** Describe the purpose and types of an independent suspension system.
- **K40017** Describe the main types of front suspension systems.
- **K40018** Describe the components and layout of a strut suspension.
- **K40019** Describe the components and layout of a modified strut suspension.
- **K40020** Describe the components and layout of the SLA suspension.
- **K40021** Describe the components and layout of a twin I-beam suspension.
- **K40022** Describe the main types of rear suspension systems.
- **K40023** Describe the components and layout of rigid-axle leaf-spring suspensions.
- **K40024** Describe the components and layout of rigid-axle coil-spring suspensions.
- **K40025** Describe the components and layout of rigid-axle suspensions.
- **K40026** Describe the components and layout of rear independent suspension (dead axle) suspensions.
- **K40027** Describe the components and layout of rear-wheel drive independent suspensions.
- **K40028** Describe the purpose and function of active and adaptive suspension systems.
- **K40029** Describe the operation of the adaptive air suspension system.

Skills Objectives

There are no Skills Objectives in this chapter.

▶ Introduction

The suspension system of a vehicle is designed to follow the road surface for the purpose of maintaining traction and control of the vehicle while also providing a smooth ride for the passengers. It is composed of a network of springs, arms, struts, and shocks that work together to achieve these purposes (**FIGURE 40-1**).

Wear from extended use is the main cause of suspension problems, though impact from potholes and accidents may cause bent or broken components in the suspension and steering systems. Driver complaints of a bouncy ride, a reduction in steering control, or unusual noises when going over bumps indicate problems in the suspension system. Regular inspection of each component in the suspension system is helpful in identifying wear before it causes severe problems for the owner. Also, know that the suspension system interacts very closely with the steering system. Thus, part of the challenge in accurately diagnosing the vehicle is knowing how both systems operate.

Even if all of the suspension and steering system components are in good working condition, they also have to be aligned correctly with one another and with the vehicle's centerline for the vehicle to operate properly and safely. For this reason, the wheels must be in proper alignment from the factory and throughout the life of the vehicle. Technicians should verify proper alignment, typically when the tires are replaced, when parts are replaced, or if a suspension-related concern is being diagnosed. This chapter helps you become familiar with the various types of suspension systems, their components, and how they operate.

▶ Suspension System Principles

K40001

Sprung and Unsprung Weight

K40002

The **suspension system** reduces the road shocks and vibrations that would otherwise be transferred through the chassis to the driver and passengers. It also must keep the tires in contact

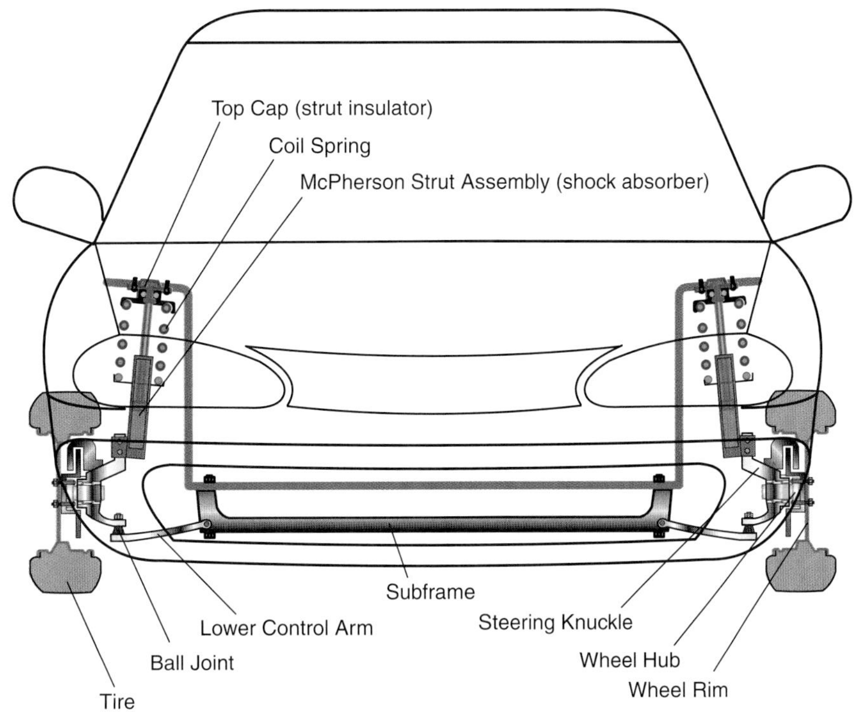

FIGURE 40-1 The suspension system.

You Are the Automotive Technician

A customer with a sporty import vehicle comes into your shop, asking what she can do to make her vehicle handle better on the local autocross track. She says others have suggested that she reduce the sprung and unsprung weight of the vehicle, add a rear sway bar, and add some adjustable shock absorbers. How would you answer these questions she has?

1. What components are part of the unsprung weight of a vehicle? And what would be the easiest way to reduce it?
2. What components are part of the sprung weight of a vehicle? And what are some things she could do to reduce it on race day?
3. What are adjustable shocks, and how might they help her?
4. How would adding a rear sway bar affect the vehicle's handling?

with the road. And because a suspension system must be strong enough to withstand loads imposed by the vehicle's mass during cornering, accelerating, braking, and uneven road surfaces, it can be quite heavy.

When a tire hits a speed bump, it creates a **reaction force**, meaning the tire will move in response to hitting the obstruction. The size of the reaction force generated depends on the **unsprung mass**, or **unsprung weight**. Unsprung weight includes all of the parts of a vehicle that are not supported by the springs, including the wheels, tires, brakes, axles, and any steering and suspension parts not supported by the springs (**FIGURE 40-2**). Sprung weight is the body, drivetrain, and associated parts that are supported by the springs. The heavier the weight of the unsprung components, the greater the reaction force they will generate. When large and heavy wheel assemblies encounter a speed bump, the reaction force may be large enough to make the tires keep moving upward after hitting the bump, which means that the tires lose contact with the road surface. If the tires are not in contact with the road, they have no traction and cannot control the vehicle's direction of travel, a potentially dangerous situation. So the heavier the unsprung components, the harder the vehicle is to control. If

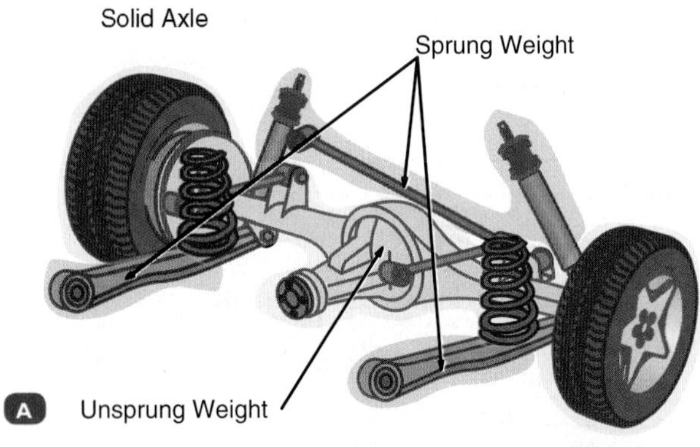

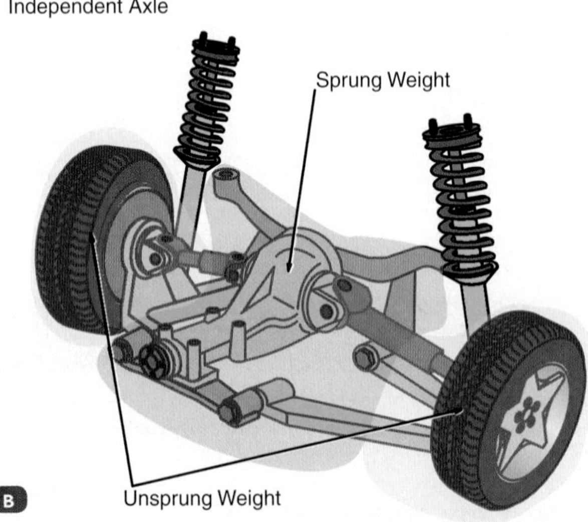

FIGURE 40-2 **A.** Unsprung weight includes all of the components between the spring and the ground. **B.** Sprung weight includes all of the components that are supported by the springs.

the unsprung weight is very light, it will not generate as much upward reaction force, so it is easier for the spring to push it back down, maintaining contact with the road and therefore control of the vehicle. To get an idea of this concept, imagine how much more difficult it would be to stop a fast-moving bowling ball than a fast-moving beach ball. It's the same way with heavy unsprung components: Once in motion, they want to keep moving in that direction.

Reaction forces also transfer a lot of momentum to the vehicle body, causing it to move upward as it absorbs the energy from the wheel assembly. This movement gets transferred to the occupants in the vehicle. So lighter unsprung weight means that less force is transmitted to the body and makes for a more comfortable ride. As a general rule, unsprung mass should be kept as light as possible while maintaining enough strength so that the components will handle the stresses under various driving conditions.

Suspension Function

K40003

Suspension systems must absorb the large road forces generated while driving on nonperfect roads, so those forces won't be passed on to the occupants. And at the same time, the wheels and tires need to be held in the proper orientation to provide traction for acceleration, braking, and steering so the vehicle can be driven safely down the road. On top of all of that, the suspension system has to support the weight of the chassis, drivetrain, passenger compartment, occupants, and any additional load. And it has to do all of this while absorbing the abnormal road forces from potholes, speed bumps, and other hazards (**FIGURE 40-3**). To accomplish this, the suspension system must be the flexible assembly between the wheels and the body. It has to be able to both flex and return to its original shape. This characteristic is called **elasticity**. Automotive suspension systems generally use the elastic properties of special metals formed into springs to provide the elastic medium.

There are three primary types of springs used in suspension systems: leaf springs, coil springs, and torsion bars (**FIGURE 40-4**). **Leaf springs** are located between the frame and the axle assemblies and are typically semi-elliptical in shape. They absorb the **applied force** (pressure of the load) by flattening out under load. They are often used at the rear end of a car or truck to help the vehicle carry large loads. **Coil springs** are formed in a spiral from a single steel rod. They absorb the force of impact by twisting and compressing. They are generally used on smaller vehicles to smooth the ride and improve handling. Torsion bars are held rigid at one end and twist around their center as the other end is deflected. Torsion bars return to their original shape when the **deflecting force** (a force that moves an object in a different direction or into a different shape) is removed. They are typically used in the front of some pickup trucks because they handle better than leaf springs.

Nonmetallic materials, such as rubber, provide cushioning action and are more commonly used as **stops** to limit extreme suspension movement. Stops, called jounce stops, bump stops, bumper stops, or rebound stops, are rubber parts used to control

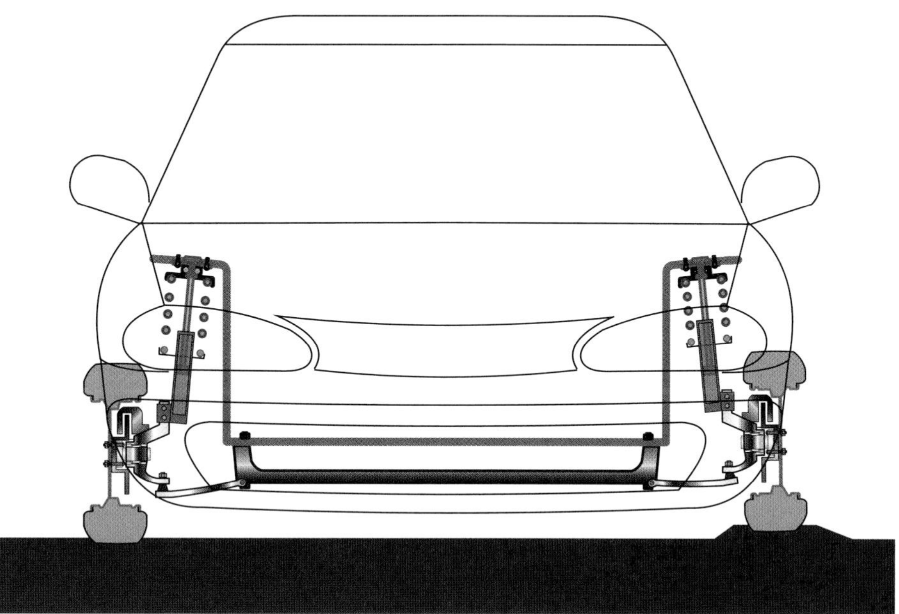

FIGURE 40-3 A suspension system subjected to a speed bump.

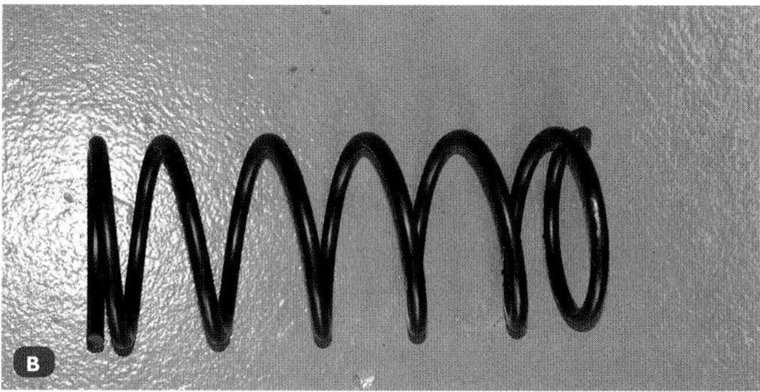

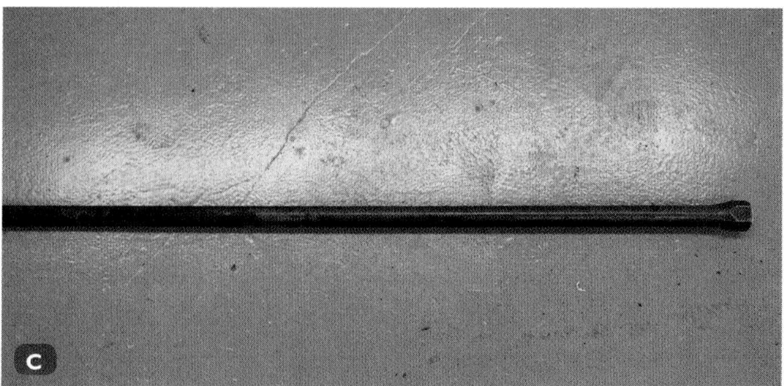

FIGURE 40-4 Three types of springs. **A.** Leaf spring. **B.** Coil spring. **C.** Torsion bar.

FIGURE 40-5 Rubber stops.

the movement of suspension components so they do not bang against each other or the frame of the vehicle, damaging the component over time (**FIGURE 40-5**). These stops are typically a triangle or cone shape to provide a collapsible cushion.

In light vehicle applications, air is used primarily for **ride height** control. Rubber bags filled with air are placed at either the rear of some vehicles to help support additional weight, or front and rear to give additional ground clearance. This is predominately a function of SUVs for when they are driven off-road, where ground clearance can be a problem.

When springs are supporting the weight of the vehicle, they are in a partially compressed condition. When a wheel strikes a bump, the springs are compressed further. But they also impart an upward force on the mass of the vehicle, which causes it to move upward. Once past the bump the spring starts to lengthen, pushing the wheel down but still pushing the body up. The spring tends to **overshoot**, or spring back, past its original

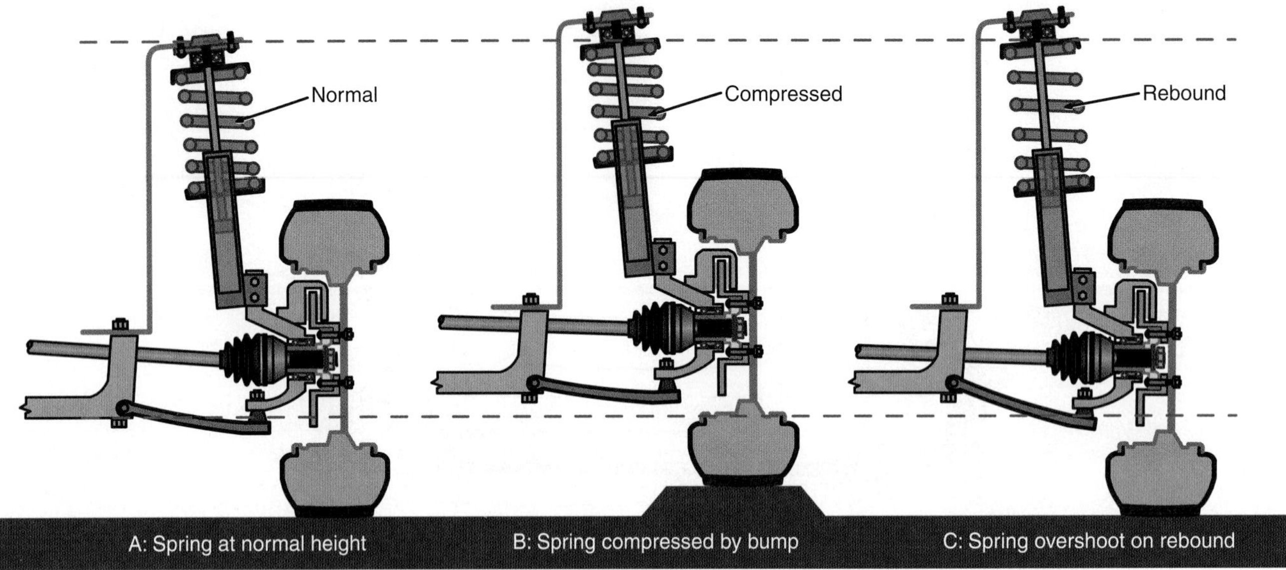

FIGURE 40-6 **A.** Spring at normal height. **B.** Spring compressed by hitting a bump and pushing the body upward. **C.** Spring over shoot on rebound.

length. And because the body is moving upward, there isn't as much weight on the spring, so this overshooting of the spring is magnified. Once the body gets as high as it will go, the spring cannot hold it there, as it is higher than its normal compressed height, so the body falls, compressing the spring once again, producing oscillations in the spring and body (**FIGURE 40-6**). **Oscillations** refer to the fluctuating of an object between two states, basically meaning the spring compresses and rebounds over and over again. As a result, the vehicle bounces up and down, making the ride uncomfortable and unstable. It can even produce forces that make the tires bounce off the ground and lose their grip on the road.

Different materials have different levels of elasticity. Up to a certain point, they can be deformed and released, and they will try to return to their original condition. Beyond that point, they stay deformed. With some materials, returning to their original state too quickly can produce a bouncing effect (oscillation). Preventing or reducing oscillations is called damping, more commonly referred to as dampening. It can occur in many different ways. The dampening material absorbs the energy from the oscillation. In a vehicle suspension system, a hydraulic shock absorber dampens oscillations in the spring. And rubber bushings dampen road shock. These concepts are explored in more depth later in this chapter.

Controlling Forces in Suspension Systems

K40004

When a vehicle is in motion, certain forces exert pressure against the wheel units: driving thrust, braking torque, and cornering force (**FIGURE 40-7**). **Driving thrust** is the force transferred from the tire contact patch through the axle housing. It places a twisting pressure on the suspension members that push the vehicle along the road. It also tends to try to move the wheel forward relative to the body. **Braking torque** places a twisting pressure on the axle housing in the opposite direction around its center during braking. It also tends to try to move the wheel backward relative to the body. Cornering force refers to the lateral movement of the axle housing during turning. It tries to push the wheel to one side or the other. It also tries to fold the wheel over.

These forces are transferred to the frame of the vehicle, but while they act, the wheel units must stay aligned with each other and with the frame. The wheels must be securely located longitudinally, laterally, and vertically while still having the freedom to move vertically to allow for suspension travel, and to pivot to allow for steering. When a vehicle drives over a bump, the tires are forcibly moved. The suspension system must absorb these forces while maintaining precise control of the wheels. Each of the suspension system components contributes to controlling these forces and keeping the wheel units in proper alignment. But the forces do put stress on the joints and pivots, which tend to wear them out over time.

Yaw, Pitch, and Roll

K40005

The terms "yaw," "pitch," and "roll" describe the movement of a vehicle around three axes (**FIGURE 40-8**). The x-axis is the imaginary line drawn down the center of the vehicle from front to back. The y-axis is the imaginary line across the vehicle from left to right. The z-axis is the vertical line that runs through the center of the vehicle from top to bottom. **Roll** is vehicular movement along its x-axis. It is the rolling motion you feel when making a sharp corner and is generally what causes rollovers. **Pitch** is movement around the vehicle's y-axis, commonly felt during hard braking or fast acceleration, when the front of the vehicle noses down or rises up slightly. **Yaw** is movement around the z-axis, felt when the vehicle deviates from its straight path, as when the rear wheels slide out during drifting. Movement

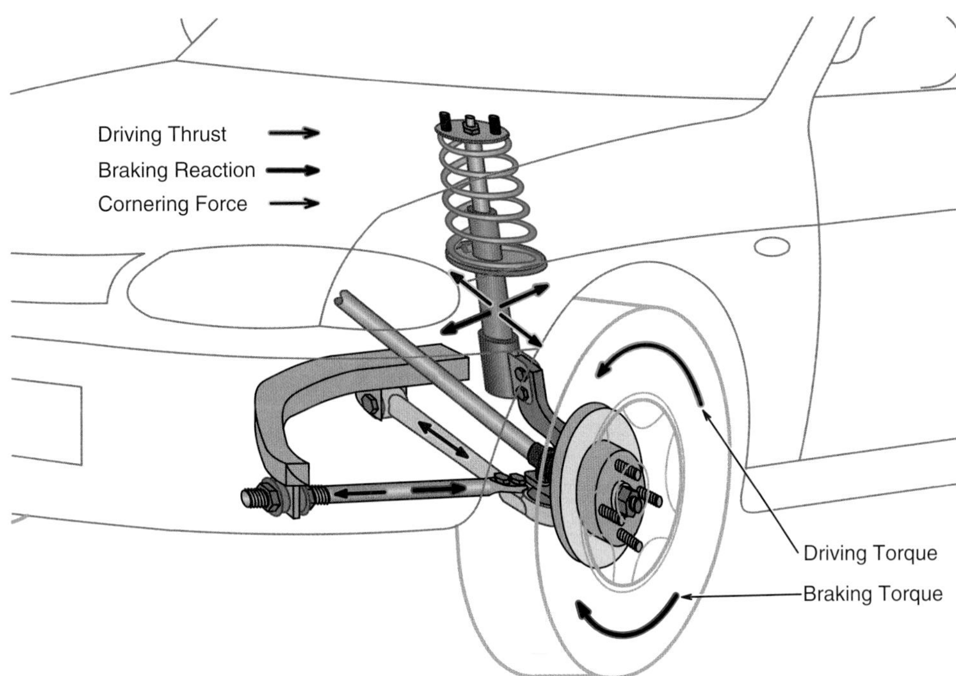

FIGURE 40-7 There are three common forces acting on suspension systems: driving thrust, braking torque, and cornering force.

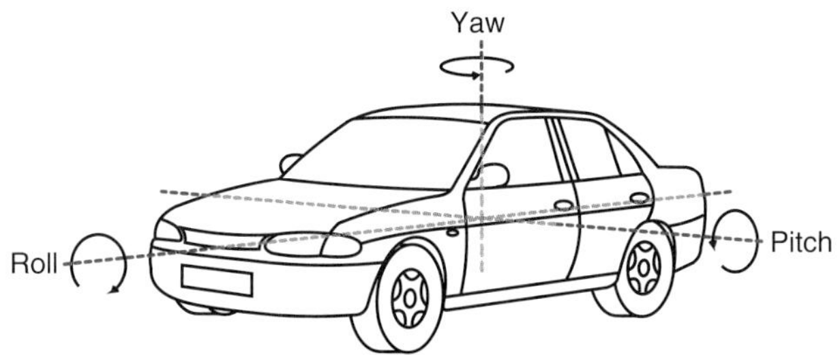

FIGURE 40-8 Yaw, pitch, and roll.

around each of these axes must be controlled during all of the maneuvers of the vehicle for it to be safe. Many safety systems, such as electronic stability control, are designed to keep vehicles within the safe limits of these axes.

▶ Suspension System Components

K40006

In order for the suspension system to perform its many functions, it is made up of many components. A basic suspension system consists of the following parts:

- **Springs**—the flexible components of the suspension. Basic types are leaf springs, coil springs, and torsion bars. Modern passenger vehicles usually use light coil springs.
- **Axles**— used to drive and/or support the wheels.
- **Shock absorbers**—dampen spring oscillations by forcing oil through small holes in a piston. The oil heats up as it absorbs the energy of the motion. This heat is then transferred through the body of the shock absorber to the atmosphere.
- **Control arms**—the primary load-bearing elements of a vehicle's suspension system. They are isolated from the chassis with rubber bushings and pivots that allow the up-and-down movement of the tire and wheel assembly.
- **Rods**—straight (or precisely formed) pieces of steel used to control motion within the vehicle's suspension system. They typically have pivoting or flexible mounts on each end.
- **Ball joints**—swivel connections mounted in the outer ends of the control arms.

FIGURE 40-9 shows the components of the suspension system.

Springs

K40007

Springs suspend the weight of the vehicle on the axles in such a manner that allows the wheels to follow unevenness in the road surface; therefore, springs must be elastic. Most springs are made out of spring steel and sag over time, requiring replacement. Some manufacturers have used composite materials for their leaf springs, which are said to resist sagging better than spring steel, as well as substantially reduce weight. The most common spring configurations are coil springs, leaf springs, and torsion bars.

Coil Springs

Coil springs, also known as helical springs, are used on the front suspension of most modern light vehicles, and in many cases they have replaced leaf springs in the rear suspension (**FIGURE 40-10**).

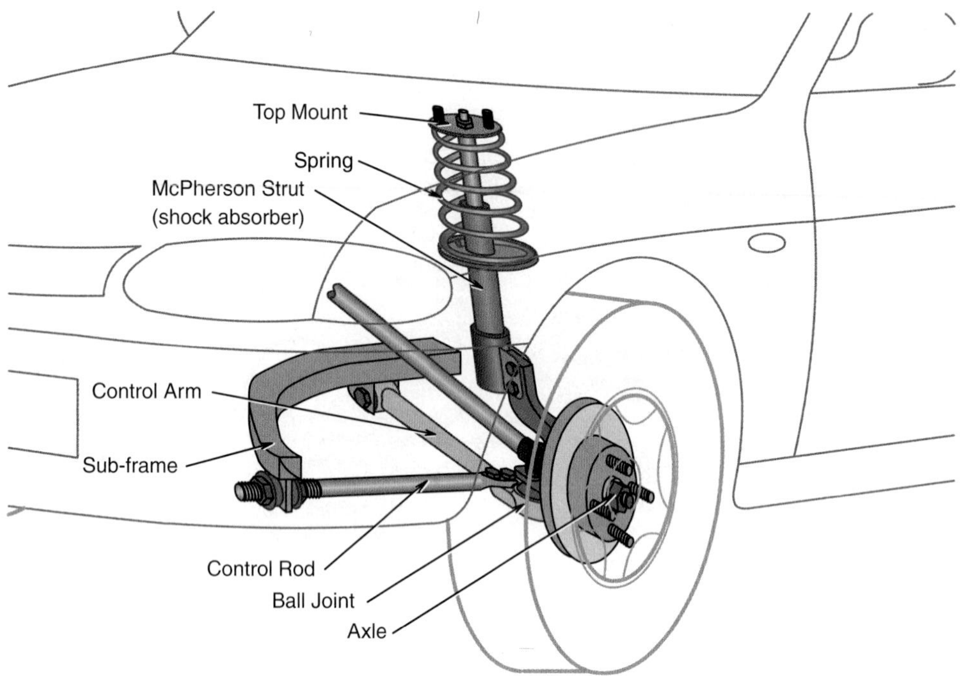

FIGURE 40-9 The components of a suspension system.

FIGURE 40-10 Coil springs, also known as helical springs, are used on the front suspension of most modern light vehicles.

A coil spring is made from a single length of special wire, which is heated and wound on a former in the shape of a coil to produce the required shape. The load-carrying ability of the spring depends on the diameter of the wire, the overall diameter of the spring, its shape, and the spacing of the coils. On a small passenger car, they are lighter and more flexible than springs on a light commercial vehicle, which are more robust and stiff. The stiffer springs are designed to withstand heavier loads, but also make the ride much rougher.

The pitch of a spring is the distance from the center of one coil to the center of the adjacent coil. The coils may be evenly spaced, called **uniform pitch**, or unevenly spaced, called variable pitch. The wire can be the same thickness throughout, or it may taper toward the end of the spring. The spring itself may be cylindrical, barrel shaped, or conical. Generally, a cylindrical spring with uniform wire diameter and uniform pitch has a constant rate of deflection, meaning it takes the same amount of force to compress each coil. Its length reduces in direct proportion to the load applied. When the pitch is varied, the deflection rate varies too (**FIGURE 40-11**). The spring is then said to have a **progressive rate of deflection**, meaning it deflects easily under a light load, but its resistance increases as the load increases. This provides a softer ride when the vehicle is lightly loaded than if the vehicle has heavier constant-rate springs. Coil springs can look alike but have very different load ratings, which are often color coded for identification. They also normally use rubber pads at their seats to prevent the transmission of noise and vibration.

As a cylindrical coil compresses, it can become coil bound, which limits its travel and creates a harsh ride when heavily loaded. As conical and barrel-shaped springs compress, they have the ability to collapse into themselves (**FIGURE 40-12**). This creates a longer suspension travel for a given length of spring than for a cylindrical spring. It also means the spring is less likely to bottom out when heavily loaded, which makes these springs useful in the rear of some pickup trucks or vans.

Leaf Springs

The leaf spring is one of the oldest forms of spring. It is usually used on rear-wheel drive vehicles and is mounted longitudinally. Leaf springs consist of one or more flat springs, commonly made of tempered steel (**FIGURE 40-13**). A number of leaves of different length are used to form a multi-leaf spring. Multi-leaf springs are used in vehicles that haul heavy loads, such as dump trucks and large pickup trucks.

FIGURE 40-11 A constant rate versus a progressive rate spring.

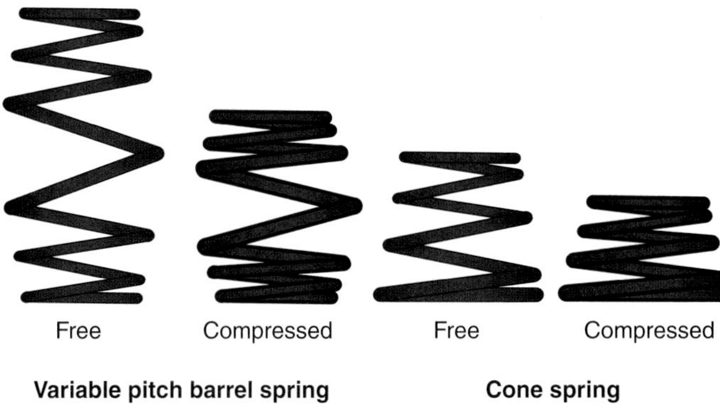

FIGURE 40-12 As conical and barrel-shaped springs compress, they have the ability to collapse into themselves.

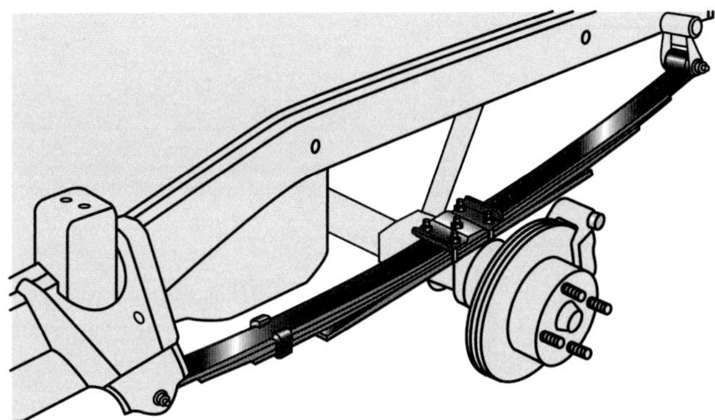

FIGURE 40-13 Leaf springs and components.

The multi-leaves are held together by a center bolt that passes through a hole in the center of each leaf. The center bolt is also used to locate the spring on the axle. The axle is then clamped to the multi-leaf spring by U-bolts that wrap around the axle housing and through a spring plate underneath the spring. **Rebound clips** are metal straps wrapped at intervals around the leaf spring to prevent excessive flexing of the main leaf during rebound, and also keep the end of the leaves in alignment. The longest leaf, called the main leaf, is rolled at both ends to form **spring eyes**. These eyes are used to mount the spring to the frame of the vehicle. Some multi-leaf springs have the ends of the second leaf rolled around the eyes of the main leaf as reinforcement. This leaf is called the **wrap leaf**.

The front of the multi-leaf spring is attached to a **rigid spring hanger** on the vehicle frame. This rigid hanger holds the spring and ultimately the axle in position with the frame. The rear of the multi-leaf is connected to the frame by a **swinging shackle**, which provides a link between the spring eye and a bracket on the frame. This swinging link is needed because the front of the spring is held rigidly to the frame, and as the spring flexes and flattens out under load, the distance between the spring eyes increases. The swinging shackle allows for this lengthening and shortening.

Some multi-leaf springs have inserts between the leaves of plastic, nylon, or rubber. They act as insulators to reduce noise transfer and friction as the leaves move across each other under load. Some older vehicles completely enclose the leaf springs in grease for this same reason. The spring eyes are fitted with replaceable bushings, usually with a flexible rubber section, but nylon and urethane bushings are also used, and sometimes bronze for heavy-duty applications. Rubber insulating pads between the spring mounting pad and the spring, and between the spring plate and the spring, also act as insulators. Most springs are arched upward at the ends, but reverse-arch springs are used occasionally, most often when lowering a vehicle. No matter what, leaf springs support the weight of the vehicle while holding the rear axle in line with the frame.

Torsion Bars

A torsion bar is a long alloy-steel bar that is fixed rigidly to the chassis or subframe at one end and to a control arm at the other end (**FIGURE 40-14**). The torsion bar is connected

FIGURE 40-14 A torsion bar is a long alloy-steel bar that is attached rigidly to the chassis or subframe at one end and to a control arm at the other.

to the control arm in the unloaded condition (no pressure on the bar), and as the suspension control arm is raised, the torsion bar twists around its center, which places it under a **torsional load**. Torsional load is a twisting force that is applied by anchoring one end of an object and then applying a twisting force to the other end.

When the vehicle is placed on the road, with the control arm connected to the suspension assembly, the torsion bar supports the vehicle load and twists around its center to provide the springing action. Spring rate depends on the length of the torsion bar and its diameter. The shorter and thicker the torsion bar, the stiffer its spring rate. This is useful in heavier vehicles or vehicles that regularly haul loads.

Torsion bars can be used across the chassis frame in a **trailing arm suspension**, or as part of the connecting link between two axle assemblies on a semi-rigid axle beam. After a lot of use, all springs can sag, including the torsion bar, meaning the ride height on one or both sides of the vehicle will be lower than specified. On many vehicles with torsion bars, the vehicle can be brought back to the proper ride height by tightening the bolt on the torsion bar adjuster.

Sway Bars

A bar similar to the torsion bar is the **sway bar**, or antiroll bar. The sway bar is used in light vehicles as a stabilizer, or antiroll, bar. The center of the bar is connected to the chassis, and each end is connected to the lower control arm on each side of the suspension system (**FIGURE 40-15**). Sway bars are most common on the front suspension because it is most prone to sway, but are also installed on the rear suspension in sporty vehicles.

When the vehicle is turning, centrifugal force acts on the body and tends to make it lean outward (roll). This tends to push up on that lower control arm, which lifts up on that end of the sway bar. The sway bar then transfers that twisting force to the other control arm and tries to lift it against that spring's downward pressure. Because the car is leaning, this helps keep the inside spring compressed while transferring uses its connections to each side of the suspension to resist this roll tendency. In this way, the spring on the inside of the corner can assist the spring on the outside of the corner to bear the additional load on the outside wheels. This tends to make the vehicle corner more flatly. But because the sway bar can pivot in bushings on the frame, the sway bar just pivots when both wheels go over a speed bump. Also, stiffer roll bars can be installed in a vehicle to enhance its cornering ability.

Rubber Stops

Rubber is used in most suspension systems as stops. If the suspension reaches its limit of travel, stops prevent direct metal-to-metal contact, thereby reducing jarring of the suspension components (**FIGURE 40-16**). The rubber stops protect the suspension parts when they bottom out. The stops also reduce the shock to the driver as the suspension bottoms out. Stops can be shaped to provide an auxiliary springing function, increasing their resistance progressively with suspension contact. Rubber spring stops can be found on most vehicles. For vehicles with MacPherson **struts**, there is a cone-shaped rubber stop located on the strut below the strut bearing surface. In other applications, these stops are found on the frame of the vehicle where the suspension components will make contact.

> **TECHNICIAN TIP**
>
> Rubber has a number of advantages: It doesn't need to be lubricated, it can be made into any shape, as required, and it is silent during use.

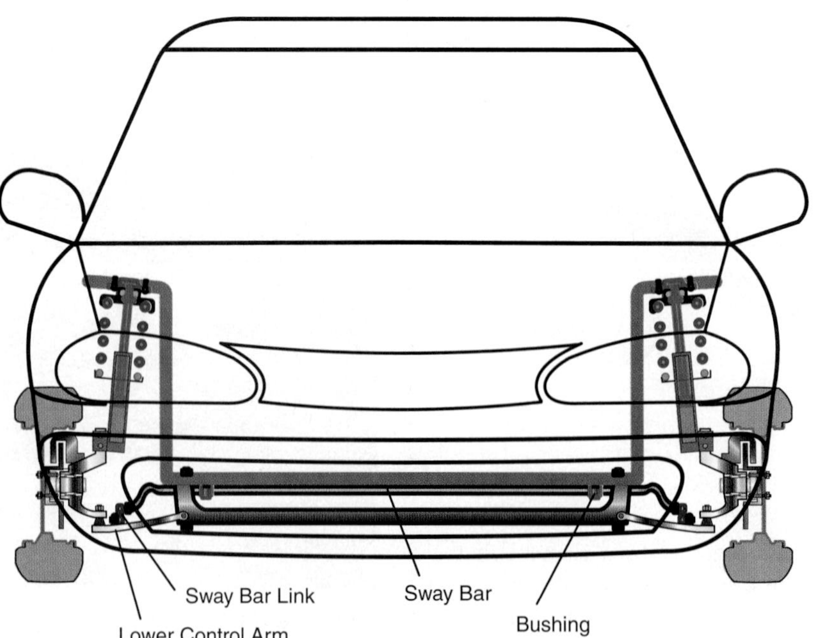

FIGURE 40-15 A sway bar reduces body roll when cornering.

FIGURE 40-16 Variety of rubber stops.

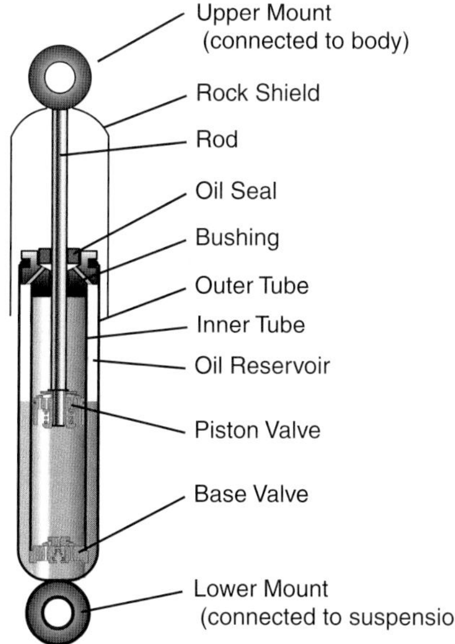

FIGURE 40-17 A shock absorber is a device on a vehicle designed to absorb shock loads caused from driving on irregular surfaces.

Shock Absorbers and Struts

K40008

A shock absorber is a device designed to dampen spring oscillations and absorb shock loads while driving on irregular surfaces. The most widely used hydraulic shock absorber is the direct-acting telescopic type (**FIGURE 40-17**). It can be fitted to the suspension as a self-contained unit that is designed to control only spring oscillations, and referred to as a shock absorber. Or it can be manufactured with a more heavy-duty housing and used to support the top of the steering knuckle as well as dampen spring oscillations. In this configuration, it is referred to as a strut. We look at shock absorbers first.

Most **direct-acting telescopic shock absorbers** are the twin-tube type. The outer tube is normally attached to the suspension member at its base, and the inner tube provides a working cylinder for a piston that is attached to a piston rod. The piston rod is connected to the vehicle's frame at its outer end, and a bushing in the top of the outer tube keeps the rod in alignment as it moves in and out of the shock absorber, with **suspension action** (movement of the chassis up and down). A seal above the bushing prevents external oil leakage and keeps out dirt and moisture. A **shroud** made of steel, plastic, or rubber is typically placed over the shock rod to protect it from damage.

During compression (jounce), which happens when the tire travels over a bump in the road, the rod and its piston move inward inside the shock absorber. In extension (rebound), which is when the tire travels over a dip in the road, the rod and piston move outward inside of the shock absorber. For dampening to be effective, resistance is needed in both directions, although not in the same proportions.

▶ TECHNICIAN TIP

A way to think about the basic function of a shock absorber is to think of a washer fastened to a rod through its center hole. A small hole is drilled through the flat of the washer so that it will let oil move from one side of the washer to the other. Some oil is added to a capped tube, and the rod and washer is inserted into the tube. Adding oil to the top, and capping the tube with a sealed cap allows the rod to move in and out of the tube, but the oil does not leak. Moving the rod up and down in the oil produces resistance, as oil will be forced through the small hole in the washer. Resistance is felt in both directions. If the size of the hole in the washer is increased, there will be less resistance. This is the basic operation of the shock absorber, except in a shock absorber, one-way valves can be used to allow different rates of flow on compression and rebound.

This resistance is provided by forcing oil past disc valves in the piston inside of the shock absorber. Valves control the flow through the piston when it moves in one direction, and different valves control for the other direction, so resistance on compression will vary as compared to rebound (**FIGURE 40-18**). Because of this, manufacturers can design the desired ride characteristics into the shock absorber.

The valves in the shock absorber provide control over the amount of force required to pass fluid through them at any given piston velocity. They can be made to open in stages, according to fluid pressure. This allows light resistance to motion when the piston moves slowly, and heavy resistance when piston velocity is high.

The rapid movement of the piston, continually forcing the oil backward and forward through the valves, causes the oil to heat up as it absorbs the energy of motion of suspension movement. The heat is transferred through the outer tube to the outside air. However, the hotter the oil becomes, the greater its tendency to aerate. Aeration occurs because of the high velocity of the oil as it passes through the small passages in the valves. If the velocity is high enough, air dissolved in the oil comes out of the solution as small bubbles and forms foam. As we know from Pascal's law, fluid cannot be compressed, but air can. Air in the shock absorber results

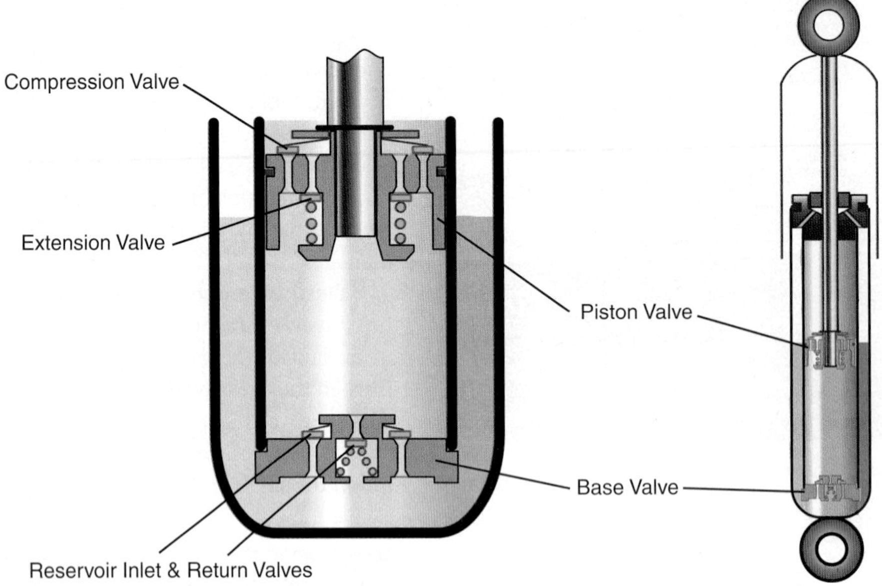

FIGURE 40-18 Shock absorber valves.

in a lack of dampening, which creates a soft, bouncy ride; if this allows the suspension members to contact the stops, a harsh ride results.

Strut-Type Shock Absorbers

A **suspension strut** functions exactly like a shock, except it is much stronger because it functions as a structural part of the suspension when it is integrated into a MacPherson strut suspension. On a MacPherson strut suspension system, the top of the hydraulic strut supports the top of the steering knuckle, so it must be much stronger than an ordinary shock absorber (**FIGURE 40-19**). The hydraulic shock absorber and the hydraulic strut provide their dampening action by transferring oil under pressure through valves in the piston, which restrict the oil flow.

> ▶ TECHNICIAN TIP
>
> Aerated oil has a certain amount of compressibility, so it is unable to provide the dampening force previously achieved in the non-aerated condition, just as a brake pedal will feel spongy when air is present in the brake's hydraulic system. The performance of the shock absorber is thus considerably reduced. This effect is called shock absorber dissolve.

Gas-Pressurized Shock Absorbers

Fluid fills the chambers above and below the piston of the hydraulic shock absorber. As the piston moves in the cylinder, valves control the movement of oil from one chamber to the other. In a gas shock, pressure on the oil is provided by nitrogen gas at the base of the cylinder, acting on a free-floating separation piston

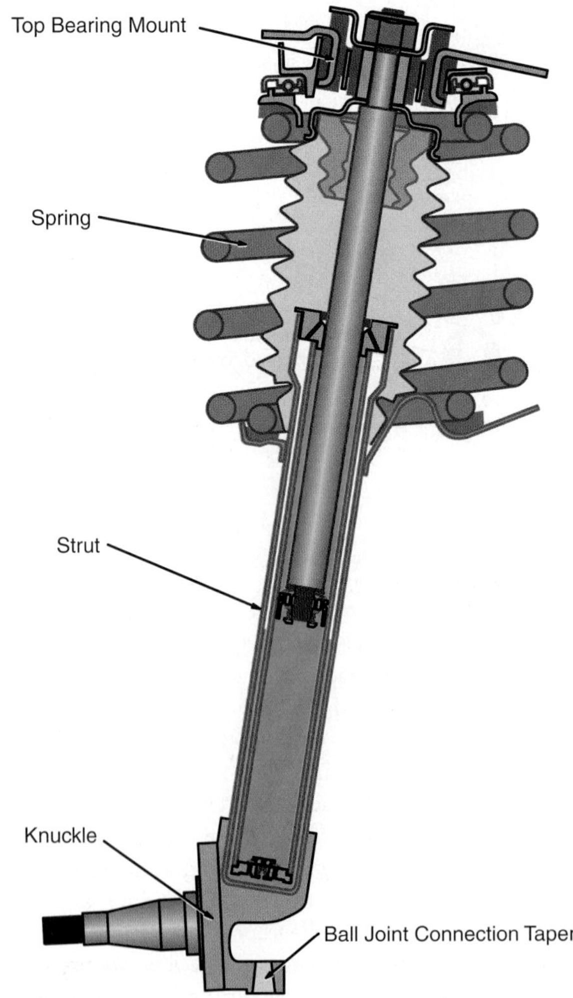

FIGURE 40-19 Cutaway view of a strut.

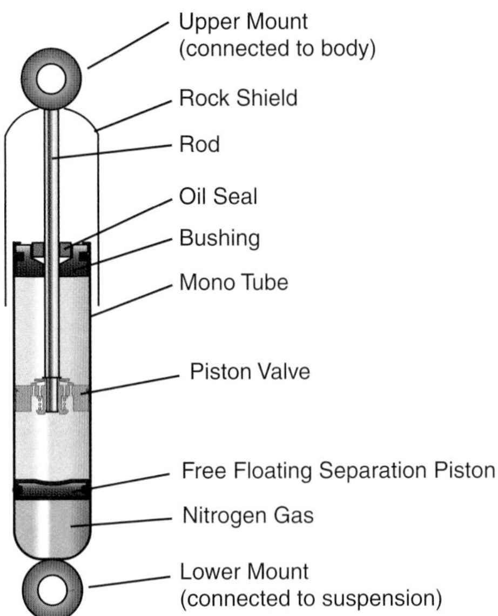

FIGURE 40-20 Gas-pressurized shock absorber.

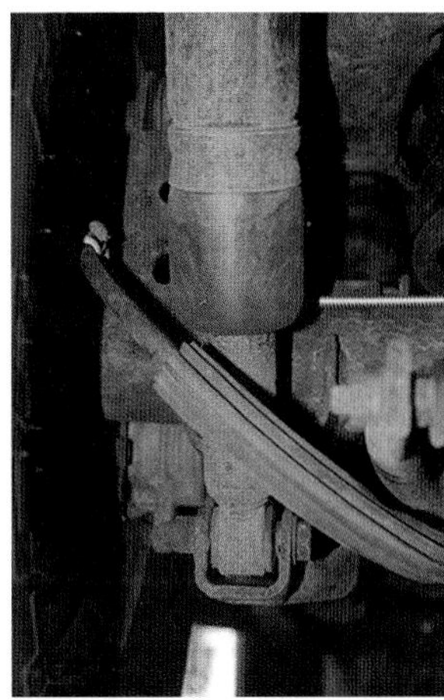

FIGURE 40-21 The air pressure can be adjusted in an air shock to help support heavier loads.

that separates the gas from the oil (**FIGURE 40-20**). On jounce, the piston moves downward, and the penetration of the piston rod displaces a quantity of oil equal to its volume. The separation piston is displaced accordingly, and gas pressure increases slightly. On rebound, the piston and rod move upward, and gas pressure reduces slightly as the separation piston makes up for the rod moving out of the shock. Pressure on the oil is maintained, even when the piston and rod are at the top of their stroke. The pressure applied to the oil keeps air bubbles (or aeration) from forming as easily and thus provides a more consistent job of dampening.

Adjustable Shock Absorbers

In the search for a comfortable ride while maintaining some level of sporty handling or the ability to handle occasional heavy loads, manufacturers have developed a variety of shock absorbers with enhanced capabilities. As you recall, having a few smaller passageways in a shock absorber causes it to provide a high amount of resistance to jounce and rebound, creating a sporty feel. And larger passageways provide less resistance, creating a smoother ride. Manufacturers have machined more passageways into their shock absorber pistons but then provide either manual, electric, or automatic selection and blockage of certain passageways with valves, depending on the desired ride characteristics. As far as extra loads, some shocks incorporate an expandable air bladder that can be used to support some of the load.

Load-Adjustable Shock Absorbers When vehicles carry heavy loads, their suspension is compressed, causing the rear of the vehicle to be lower than normal. As a result, steering can become lighter (less responsive); the alignment of the headlights becomes too high; and the compression length of travel of the suspension over bumps is reduced, causing the suspension to bottom out more easily, which is uncomfortable for passengers. To reduce these effects, a **manually adjustable air spring** can be incorporated into two or all four shock absorbers. This type of shock absorber is commonly referred to as an air shock (**FIGURE 40-21**). The **air spring** consists of a flexible rubber bladder, which seals the outside of the upper and lower halves of the shock absorber. When inflated, the bladder pushes the halves apart.

The shock absorber is a standard hydraulic type, providing normal dampening action, but when a heavy load is placed on the rear of the vehicle, the rubber air bladder can be pressurized with compressed air to assist the vehicle's springs. By changing the air pressure in the bladder, the ride height can be adjusted, as well as the stiffness of the suspension. Compressed air in the bladder can absorb smaller road shocks and provide better ride characteristics than stiff springs alone.

The rubber air cylinder is connected to a filling valve, which looks similar to a tire valve stem, by a flexible plastic hose. Air from a tire pump or air compressor forces more air into the rubber cylinder, allowing the suspension to support more weight. The maximum air pressure setting must not be exceeded, as excessive air pressure can damage the shock absorber's air spring. When the load is removed, the extra air can be released through the filling valve, which allows the suspension to return to its original settings, just as you would release air from a tire. However, a minimum air pressure must be maintained in the cylinder to prevent the rubber from chafing on itself as it collapses internally with shock absorber action. Refer to the manufacturer's specifications for the air shock's minimum and maximum air pressure.

Another type of load-adjustable shock absorber uses a coil spring around the outside of the shock, called a coil-over shock. The spring is installed from the factory, under tension, and connects to both the upper half and the lower half. This tends to push the halves apart, assisting the regular springs in supporting the load (**FIGURE 40-22**).

Manual Adjustable-Rate Shock Absorbers A **manual adjustable-rate shock absorber** has a manual, external damper rate adjustment. The number of valves and size of the

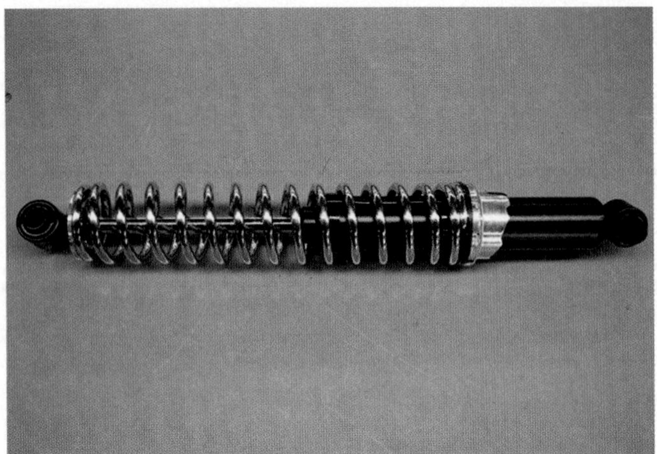

FIGURE 40-22 A coil-over shock acts as a helper spring in supporting a load.

FIGURE 40-23 On some models, the position of the valve is adjusted by turning a spindle located on the shock plunger rod. This style uses a selector knob near the bottom of the shock that can be adjusted at any time.

passageways in the piston can then be selected to vary the amount of restriction of the flow of oil through the piston and to vary the force needed to open the valves.

When all of the orifices are open, a small dampening effect is applied to the oil. The spring force applied to the valve can also be reduced to allow the valve to open more easily. This means the oil can flow through the valves more easily, which gives a softer ride but can also allow more rolling and pitching of the body of the vehicle. Closing some of the orifices and increasing the spring force applied to the valves makes it harder for fluid to flow through the piston. This increases the dampening effect of the shock absorber.

The method of changing the position of the valve varies. On some models, it is adjusted by turning a spindle located on the shock plunger rod. Turning the spindle moves the valves and changes the size of the orifices in the valve. Another style uses a selector knob near the bottom of the shock that can be adjusted at any time (**FIGURE 40-23**). On other models, when the shock absorber is extended to its maximum length, a pin is depressed, locking in an adjusting slide on the piston assembly. Twisting the two halves of the shock absorber changes the number of orifices and the spring force on the valves. The first two types of adjustable shock absorbers can be adjusted with the shock on the vehicle, whereas the last type must be removed to change the dampening.

Electronic Adjustable-Rate Shock Absorbers Some vehicles equipped with electronic ride control systems provide driver-selected control of the ride quality. Typically, a switch is located inside the passenger compartment that allows the driver to select between options such as sport, touring, or automatic. For example, if the driver selects sport, the ride will become firmer by adjusting the shock dampening. Electronic ride control can also be automatic, with no provision for driver control. In automatic ride control, the vehicle's electronic module chooses a softer or firmer ride, sometimes based on the speed of the vehicle or the driving that is being done.

In order to change dampening of the shocks, the shock must be adjustable. Some vehicles use a rotary stepper motor, called an actuator, to change the dampening. The shock works very similar to the manual adjustment shock that has the spindle that is turned. In the manual adjustable shock, you turn this spindle with an Allen wrench or other tool to make dampening changes. In the electronic-controlled shock, the actuator is mounted to the top of the shock and turns the spindle as needed. The actuator changes the number of restrictions or orifices that the oil must pass through. When all orifices are open, oil can flow more easily through the passageways in the piston. Only a small dampening effect is applied to the oil. This provides a dampening force that emphasizes ride comfort when traveling at low speeds. Closing some orifices makes it harder for fluid to flow through the piston. This increases the dampening effect of the shock absorber, providing a firmer ride that is more suitable for higher speeds and faster cornering. The need to increase dampening at high speeds becomes evident when we think about hitting a bump at high speeds, which could cause the wheels to come off the ground, disrupting our ability to steer the vehicle.

Another type of electronic adjustable shock is the solenoid-controlled shock. The solenoid is located in the side of the shock absorber or strut and opens a passageway internal to the shock, to allow more fluid to bypass the piston. If the solenoid is commanded closed, the fluid will not be able to bypass, which will result in more dampening. This type of shock is still controlled by the engine control unit (ECU).

The newest style of electronic adjustable shock uses a special type of fluid called **magneto-rheological fluid**. This fluid has the unique characteristic of changing viscosity when exposed to a magnetic field. General Motors has been using this fluid in vehicles equipped with MagneRide® suspension systems. The fluid is mixed with very small ferrous particles that react to magnetic fields. As the fluid is subjected to a magnetic field, the ferrous particles bind together, effectively increasing the viscosity of the fluid. The stronger the magnetic field, the thicker the fluid becomes. The ability to increase and decrease the viscosity of the fluid eliminates the

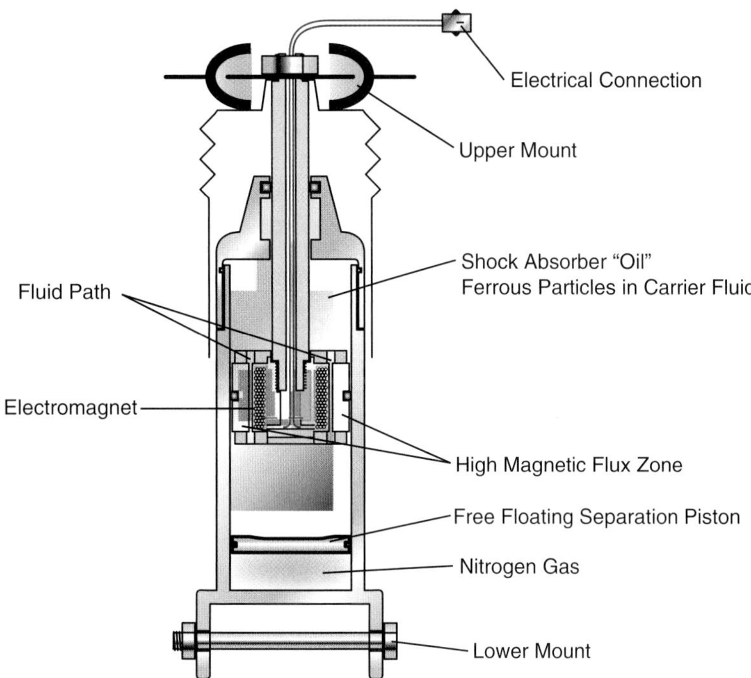

FIGURE 40-24 Electromagnetic shock absorber.

need to change orifice sizes as it is harder to pass a thicker fluid through a hole than a thinner one. Simply varying the magnetic field at the shock absorber causes varied dampening of the suspension (**FIGURE 40-24**). In fact, the viscosity can be changed in a millisecond or less, allowing active dampening of each individual shock absorber as the vehicle is being driven. This means that the powertrain control module (PCM) can stiffen the appropriate shocks in a continuously active manner as the steering wheel is being turned or the brakes are being applied. On the other hand, a particular shock can be softened as a wheel goes over a bump to prevent the shock from being transmitted as fully to the vehicle. Other manufacturers are beginning to use this technology because it is a faster-acting type of electronic dampening suspension. This technology is also used in some motor mounts and other applications.

Automatic Adjustable-Rate Shock Absorbers **Automatic load-adjustable shock absorbers** are also called **self-leveling**, meaning they have a sensor that measures the ride height and uses that information to adjust the self-leveling shocks automatically. An automatic load-adjustable suspension system controls the vehicle ride height automatically, according to the load placed over the rear axle. It consists of air-adjustable shock absorbers fitted to the rear suspension, an electrically driven compressor and air-dryer assembly, a ride height sensor, a control unit, and associated wiring and tubing (**FIGURE 40-25**).

The ride height sensor is mounted to the cross member over the rear axle, and a moveable link connects it to a rear suspension member. As the vehicle is loaded, the normal suspension springs are compressed, which lowers the height of the vehicle. When the ignition is switched on, the control unit senses the lowered ride height and switches on the air compressor. Air is directed to the shock absorbers, causing the airbag around them to expand and raise the suspension back to the normal ride height. If the load is removed, the suspension springs expand, raising the height of the vehicle. The control unit senses the raised ride height, and air is exhausted from the shock absorbers, causing the airbag to deflate and thereby lowering the suspension to the normal ride height.

During normal suspension operation, continual adjustment of vehicle ride height is prevented by a time delay in the control unit. This delay allows the trim height to be adjusted only when the ECU reads an out-of-trim signal for a short period of time—for example, 5 to 15 seconds. Thus, the system does not try to compensate for bumps in the road or weight transfer during braking. The compressor run time or exhaust time is limited to a few minutes. Limiting the operational time prevents it from continuing to operate if the system develops an air leak or if an exhaust vent remains open. If a fault like this develops, most self-leveling systems will set a diagnostic code.

Control Arms and Rods

K40009

Control arms are components that serve as a primary load-bearing element of a vehicle's suspension system. Control arms can be formed in different ways. In the A-arm style, sometimes referred to as a **wishbone control arm**, the arm is a relatively flat triangular part that mounts to the frame or subframe at each leg of the A (**FIGURE 40-26**). These widely spaced mounting points prevent forward or backward movement of the steering knuckle. The mounting points typically use rubber bushings that allow the arm to pivot up and down while dampening or

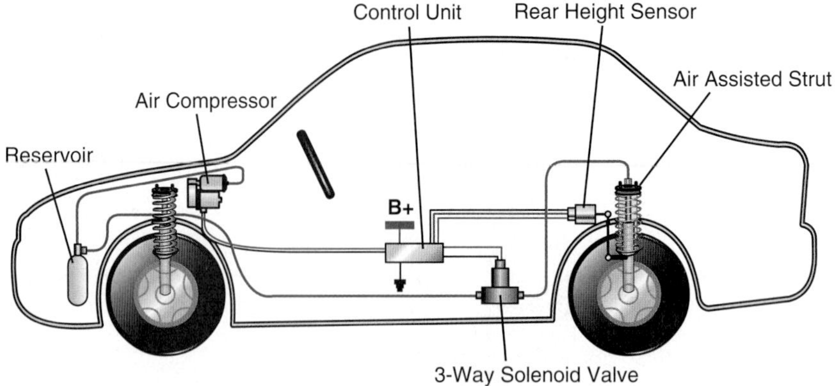

FIGURE 40-25 An automatic load-adjustable suspension system controls the vehicle ride height automatically, according to the load placed over the rear axle.

FIGURE 40-26 Wishbone control arm.

FIGURE 40-27 Single-point control arm.

isolating the road shock and vibrations from the rest of the vehicle. The other end of the control arm has a ball joint that connects the control arm to the steering knuckle assembly. This arrangement provides a smooth yet stable ride for the vehicle.

Another type of control arm uses only one contact point at the frame or body (**FIGURE 40-27**). With this style of control arm, another supporting piece, called a strut rod or radius rod, must be used to keep the control arm from pivoting forward and backward with changes in braking or acceleration. Other types of control arms are made from round tube material and can be used in independent suspensions. These control arms may be called transverse or trailing arms.

Rods

Rods are typically straight (or precisely formed) pieces of steel used to either transfer motion or prevent motion within a vehicle's suspension system. More specific names are given to rods based on their location or attachments. For example, a suspension system may use tie rods, lateral rods, tension rods, control rods, Panhard rods (track bars), steering track rods, or strut rods (**FIGURE 40-28**). Many of the rods use either a bushing or a joint on one or both ends of the rod. Each of these components is discussed elsewhere in this chapter, but it is important to remember that the exact number and types of rods within a vehicle vary greatly, depending on the make and model of vehicle as well as its intended use.

Steering Knuckle

K40010

The steering knuckle, also known as a stub axle or spindle assembly, can be found in many variations. One type uses a forged piece containing the wheel hub or spindle and attaches to the suspension components. Other knuckles may provide a hole for the axle to pass through and the wheel hub to mount to (**FIGURE 40-29**). Some knuckles are cast iron, but newer vehicles may have a cast aluminum steering knuckle to reduce unsprung weight as well as total vehicle weight.

Typically, the steering knuckle also has a steering arm, which either is cast as part of the knuckle or is a separate piece bolted to the knuckle. The steering arm connects to the steering system and in either form serves to transmit the steering force to the steering knuckle when the driver turns the steering wheel. The steering knuckle pivots on a ball joint on the bottom and either a ball joint or a strut bearing on the top.

FIGURE 40-28 A strut rod is one type of rod used in steering and suspension systems.

FIGURE 40-29 A. Steering knuckle—front-wheel drive vehicle. **B.** Steering knuckle—dead axle.

Ball Joints

K40011

Ball joints are swivel connections mounted in the outer ends of the control arms. Ball joints are typically constructed of a ball and socket (**FIGURE 40-30**). And although ball-and-socket joints are used in most tie-rod ends, the term "ball joints" is typically reserved for the primary joints that the steering knuckle pivots on, and "tie-rod ends" is the term used for the ball-and-socket joints on the steering linkage.

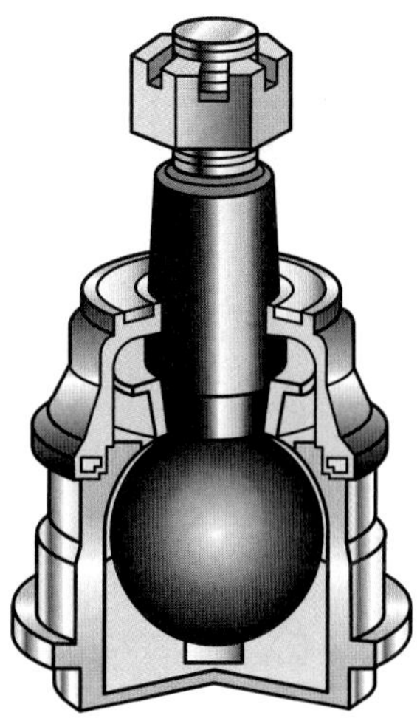

FIGURE 40-30 Ball joint.

Because control arms move up and down with suspension deflection, suspension ball joints allow the knuckle assembly to pivot up and down as well as rotate for steering. The ball joint in most modern vehicles is a sealed, self-contained unit that is replaced as a unit when it is worn out. The ball joint can be fastened to the control arm in several ways. In the past, a ball joint housing was threaded and then screwed into the control arm. This type of ball joint is no longer used, so is only found on classic cars. The ball joint on modern vehicles is either press-fit into the control arm or held by rivets or bolts. If the ball joint is pressed into the control arm, a special tool, called a ball joint press tool, is used to remove and install the joint.

A ball joint is made up of a pressed-steel housing fitted with **sintered** (bonded using pressure and heat) iron seats and a hardened ball stud. Some ball joints use a Belleville spring to hold tension on the joint, which has to be compressed when tested. Typically, a taper on the stud locates in a mating taper on the steering knuckle, although some use a straight stud with a crescent-shaped relief that allows a clamping bolt to orient the stud and clamp it securely (**FIGURE 40-31**).

A rubber seal retains grease and keeps out dirt and water. Some ball joints have grease fittings (grease zerks) installed, which allow for periodic lubrication of the moving ball and stationary socket inside the ball joint (**FIGURE 40-32**). Grease fittings can sometimes be found on other suspension components and need to be lubricated as part of a preventive maintenance program provided by the manufacturer. Most light-duty vehicles manufactured today have maintenance-free suspension components that do not provide access for lubrication. When inspecting the suspension, check to see whether grease fittings are present on the suspension, steering, and drive line components.

Ball joints can be referred to as loaded or unloaded. A loaded ball joint supports the weight of the vehicle. An unloaded ball joint does not support any weight; it just holds the steering

FIGURE 40-31 **A.** Tapered stud ball joint. **B.** Straight stud ball joint with crescent relief.

FIGURE 40-32 Some ball joints have grease fittings (grease zerks) installed, which allow for periodic lubrication of the moving ball and stationary socket inside the ball joint.

knuckle in position and is referred to as a follower ball joint. For example, in a short/long-arm (SLA) suspension system that has an upper and lower control arm, and with the coil spring located between the frame and the lower control arm, the lower ball joint is the loaded ball joint (**FIGURE 40-33**). This is because the force of the spring is placed against the lower control arm and ball joint. In this situation, the upper ball joint is the follower joint.

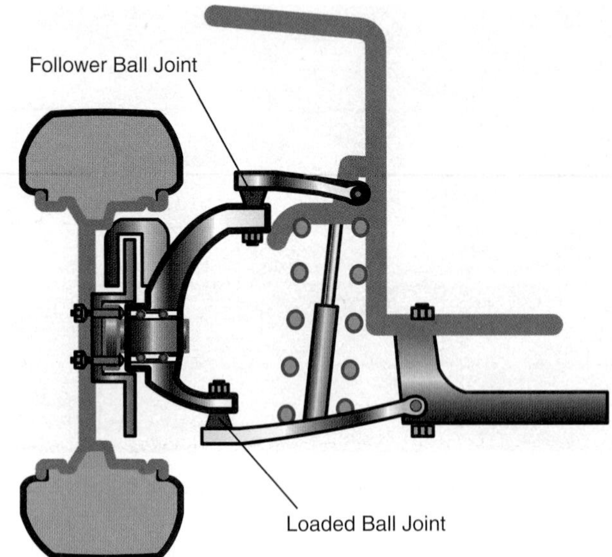

FIGURE 40-33 Loaded versus follower ball joints.

If the spring is located between the frame and the upper control arm, then the upper ball joint is the loaded joint, and the lower ball joint is the follower joint. This distinction becomes important when testing the ball joints for play, as the technician has to unload the joint to accurately test for play. Also, the ball joints can be designed to primarily carry either compression loads or tension loads (**FIGURE 40-34**). Compression forces tend to push the ball into the socket, so those types of joints have most of their bearing surface and socket strength near the base of the ball. Tension ball joints are being pulled apart, so they have most of their bearing surface and socket strength near the stud end of the ball. Never mix up the positioning of ball joints; doing so will cause them to wear out quickly and fail.

Bushings

K40012

Bushings act as pivot points and cushions at suspension fulcrum points such as control arm bushings and strut rod bushings, to allow for limited movement of the component while maintaining its alignment (**FIGURE 40-35**). They can be metallic or made of rubber, nylon, or urethane. Many rubber bushings have a metal inner and outer housing to allow for wear surfaces or for being pressed into place. Rubber bushings isolate noise and harshness and dampen unwanted vibrations. The rubber absorbs small impacts from the suspension action without transmitting them directly to the driver. Rubber requires no lubrication. Rubber bushings can also be used on strut rods (**FIGURE 40-36**). Rubber-bonded bushings can be used to mount the steering rack to the vehicle frame.

Spring shackle bushings can be molded to form two halves to fit into each side of the spring eye on the swinging shackle,

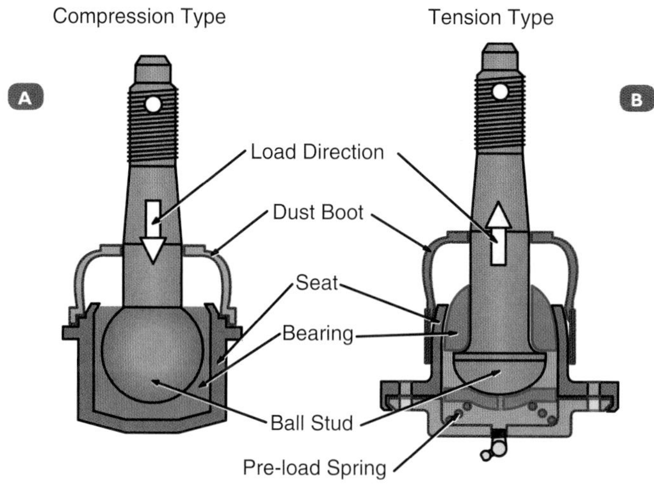

FIGURE 40-34 A. Ball joint made for primarily compression applications. **B.** Ball joint made for primarily tension applications.

FIGURE 40-35 Control arm bushing.

FIGURE 40-36 Strut rod bushings.

which is located on the vehicle frame. With the spring loaded and the shackle plates tightened, the rubber is compressed in the eye and at the face of the plates. As the spring deflects, the rubber deflects without tearing.

FIGURE 40-37 Rubber-bonded bushing used in control arm bushings.

Rubber-bonded bushings are normally used for the front eye of the spring at the fixed shackle point, and also in control arm applications. The rubber-bonded bushing has a steel outer housing and inner sleeve (**FIGURE 40-37**). The rubber medium is bonded between both to provide flexibility between them. The outer casing is normally pressed into place in the component. Relative movement between the casing and the inner sleeve causes the rubber to flex without tearing.

In control arm applications, particularly at the rear of a vehicle, the rubber bushing may be molded with a voided section (**FIGURE 40-38**). This is known as a **compliance bushing**, because it allows the unit or component to comply with a controlled amount of movement in the direction of the void. This movement relative to the vehicle frame may be designed to allow compliance or deflection steer of the road wheels when cornering. Because this influences the steering behavior of the vehicle, it is very important that the voided section be in its correct relative position. It is easy to forget how a bushing was positioned after it is removed. Marking the position with a paint mark is a good way to ensure that the bushing is replaced correctly.

▶ Types of Suspension Systems

K40013

Manufacturers use various types of suspension systems, depending on the intended use of the vehicle, cost of manufacturing, and layout of the drive train. Each system comes in various configurations, which we examine in this chapter. Each design has pros and cons that you need to be aware of when diagnosing and servicing suspension systems. But the function of each suspension system is to maintain the proper positioning of the wheels through all driving conditions while isolating the occupants from the surface of the road as much as possible. Suspension systems also need to accommodate systems that power, brake, and steer some or all of the wheels. Understanding the different suspension types and configurations gives you a good foundation for diagnosing faults in them when they appear. We cover diagnosis in the next chapter.

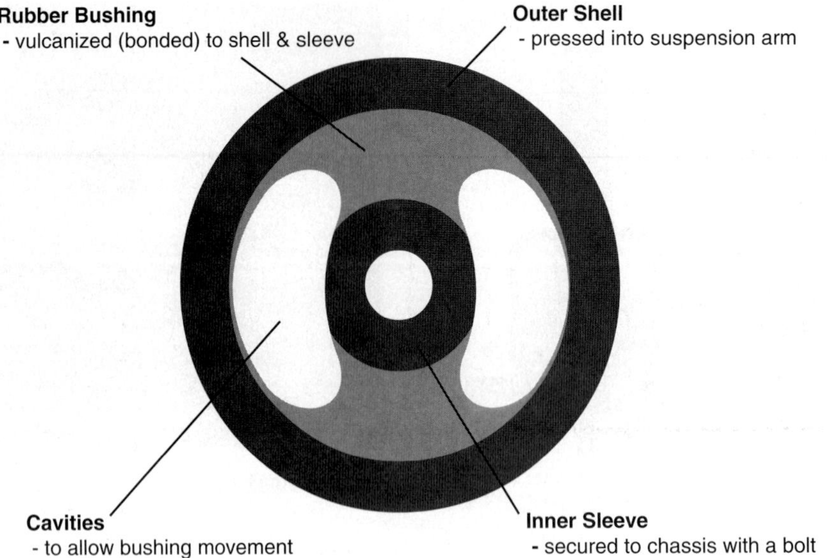

FIGURE 40-38 Compliance bushing.

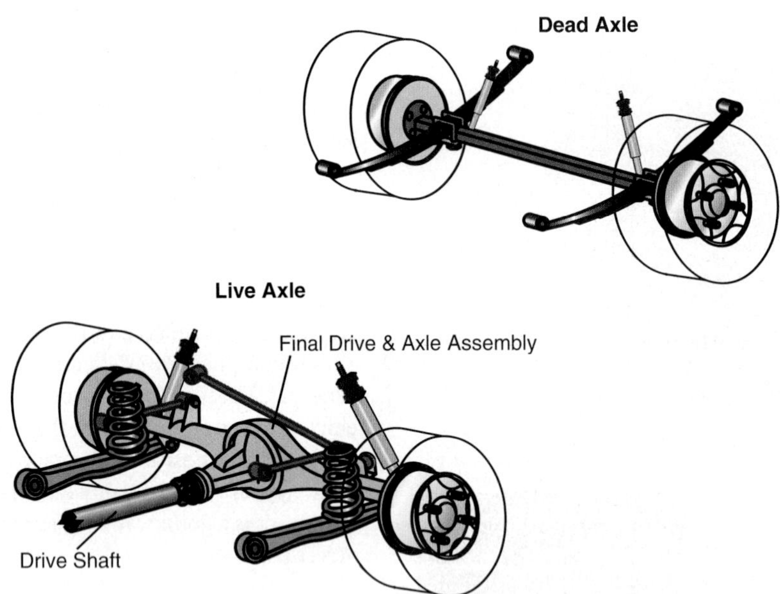

FIGURE 40-39 The terms "dead axle" and "live axle" refer to whether the axle is a driving axle or not.

Dead Axle/Live Axle

K40014

The terms **dead axle** and **live axle** refer to whether the axle is a driving axle or not (**FIGURE 40-39**). A dead axle simply holds the wheels in their proper orientation. A good example of a dead axle is a trailer (**FIGURE 40-40**). It does not power the wheels at all, just holds them in place. A live axle not only holds the wheels in position but also drives them. Live axles are connected to the drive train in such a way that power can be transferred through the components in the live axle (**FIGURE 40-41**). Live axles can be part of either a solid axle suspension system or an independent suspension system. On front-wheel drive vehicles, a simple

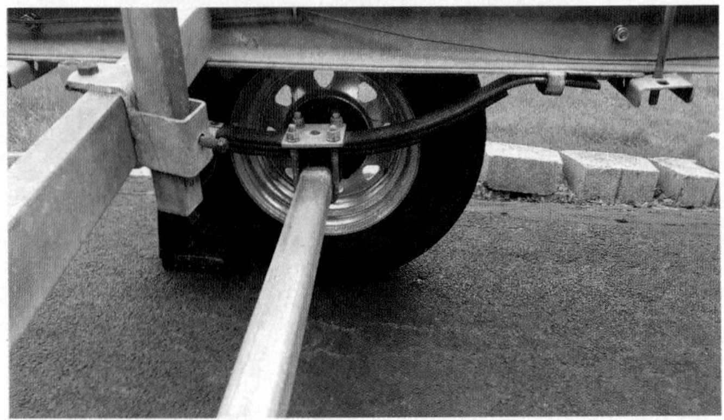

FIGURE 40-40 Trailers are a good example of dead axles, where the axle simply holds the wheels in place.

FIGURE 40-41 A live rear axle not only holds the wheels in position, but it drives them as well.

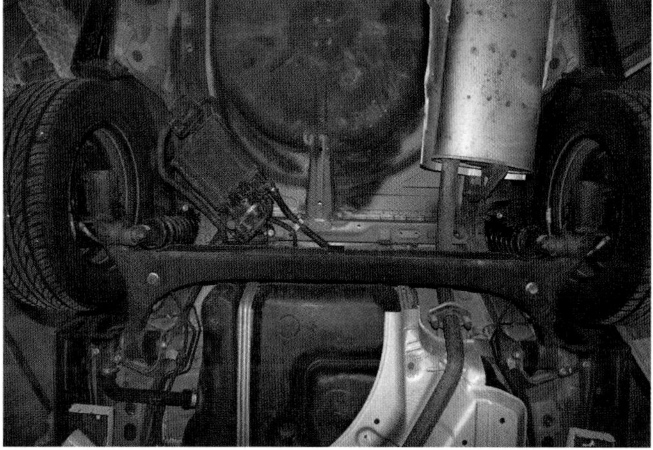

FIGURE 40-42 A solid axle is a non-independent suspension because the wheels on both sides of the axle are connected together.

dead axle is used on the rear wheels. On a rear-wheel drive vehicle, the front axle is a dead axle. And on four-wheel drive or all-wheel drive vehicles, both the front and rear axles are live. Again, they can be of the independent or solid axle styles.

Solid Axle

K40015

The **solid axle** (beam axle) provides a simple means of mounting the hub and wheel assembly. Together with leaf or coil springs, it forms an effective, dependent suspension system (**FIGURE 40-42**). It is used in the rear suspension of many front-engine, rear-wheel drive vehicles and light commercial vehicles and as the front suspension on many heavy commercial vehicles.

A solid axle form a dependent suspension because the wheels on both sides of the axle are connected together. This means when one wheel goes over a bump, the other wheel tilts (**FIGURE 40-43**). This tilting reduces the tire-to-road contact patch, reducing friction. Thus, vehicles with solid axles are said not to handle as well as vehicles with independent suspensions. But they are inexpensive and good for hauling heavy loads.

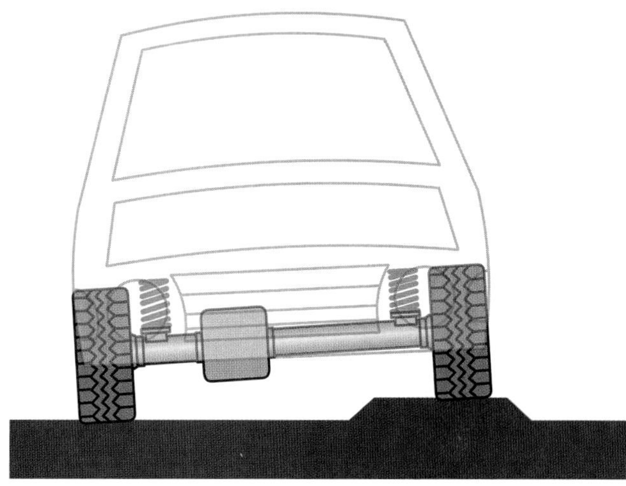

FIGURE 40-43 When one wheel on a solid axle goes over a bump, it affects the other wheel.

Independent Suspension

K40016

An independent suspension allows the wheel on each side of the axle to move up and down independently of the other. In this arrangement, if a wheel on one side hits a road irregularity, it will not upset the wheel on the other side on the same axle (**FIGURE 40-44**). One of the main benefits claimed for **independent suspension** is that unsprung mass can be kept low, as a heavy center axle housing is eliminated. One of the simplest and most common independent suspension systems is the **MacPherson strut** (**FIGURE 40-45**). It can be used on either the front or the rear of the vehicle and consists of a spring and heavy-duty shock absorber unit called a strut. On the front of the vehicle, the lower end of the strut is connected to the knuckle and located by a ball joint fitted to the end of the control arm. Its upper end is located in a strut tower formed in the unibody. It sits in a molded rubber mounting. If the strut is on the front, the upper mounting includes a bearing to allow the complete strut assembly to rotate with the steering.

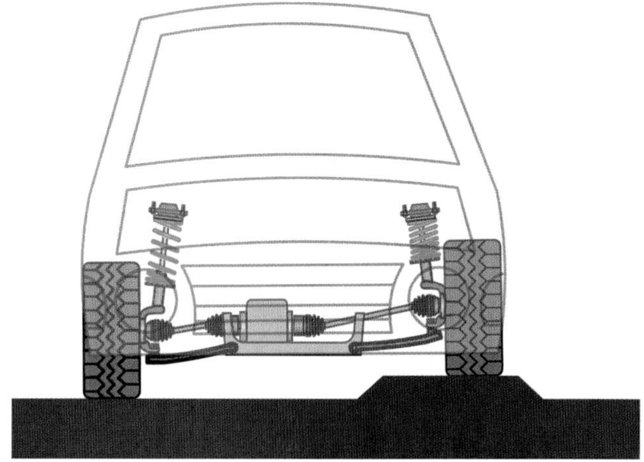

FIGURE 40-44 When one wheel on an independent suspension system goes over a bump, the other wheel is not affected.

FIGURE 40-45 The MacPherson strut suspension system. **A.** Front. **B.** Rear.

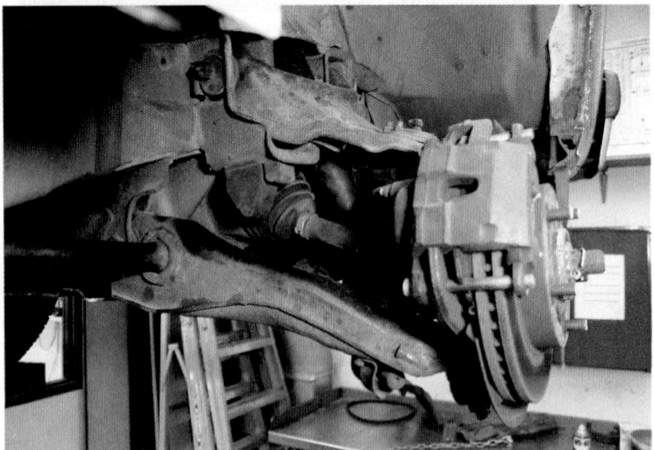

FIGURE 40-46 The SLA suspension system.

A slightly more complicated, yet common type of independent suspension system is the short-long arm (SLA) suspension system (**FIGURE 40-46**). It uses upper and lower control arms to control the movement of the knuckle. It typically uses either a coil spring or torsion bar connected to one of the control arms to support the weight of the vehicle. This system is discussed further in a following section.

Front Suspension

K40017

The front wheels of a vehicle can have the same type or a different type of suspension system than the rear wheels. The exact type of system used in the front can vary depending on the vehicle type, such as whether it is front-wheel or rear-wheel drive. The major difference between front and rear suspension systems is that the front suspension must accommodate steering, whereas most rear suspension systems don't. Another difference is that most front suspension systems incorporate an anti-sway system, typically a sway bar. The two main types of front suspension systems are the independent and the solid axle systems. Regardless of design, the primary function of the front suspension system is to keep the tires in constant contact with the road.

Strut Suspension

K40018

In strut suspension, the shock absorber is contained inside the strut (**FIGURE 40-47**). It is a direct-acting telescopic-type shock absorber and is a type of hydraulic shock absorber. A coil spring is mounted over the strut, inside the suspension tower of the front wheel housing. The strut has an upper mounting point in the suspension tower.

When used on the front suspension, the strut's upper mounting is bearing mounted to allow for the steering movement. The lower control arm is mounted (or held in place) to the frame or subframe by control arm bushings. The outer end of the control arm contains a ball joint, which connects the steering knuckle to the control arm. The steering knuckle can then pivot on the ball joint on the bottom of the knuckle, and the strut bearing on the top. Note that the ball joint on a MacPherson strut is a follower joint, not a loaded joint. This is a compact, responsive, and efficient type of suspension system.

Modified Strut Suspension

K40019

The modified strut system uses one control arm with the strut mounted in the same manner as a MacPherson strut. Unlike a MacPherson strut suspension, the modified strut suspension does not include a spring as part of the assembly and is used in front on some vehicles and on the rear of others. The modified strut system (**FIGURE 40-48**) uses one control arm with the strut mounted in the same manner as a MacPherson strut. The main difference is that the coil spring is mounted on the lower control arm, which is usually an "A" arm. This allows for placement of the coil spring and controls fore and aft movements.

Double-Wishbone Modified Strut

The double-wishbone modified strut suspension uses a high-mount upper arm with a lot of forged aluminum components. **FIGURE 40-49** shows a coil over shock absorber (green arrow), where the upper end of the stabilizer link (yellow arrow) attaches directly to the aluminum hub carrier (steering knuckle

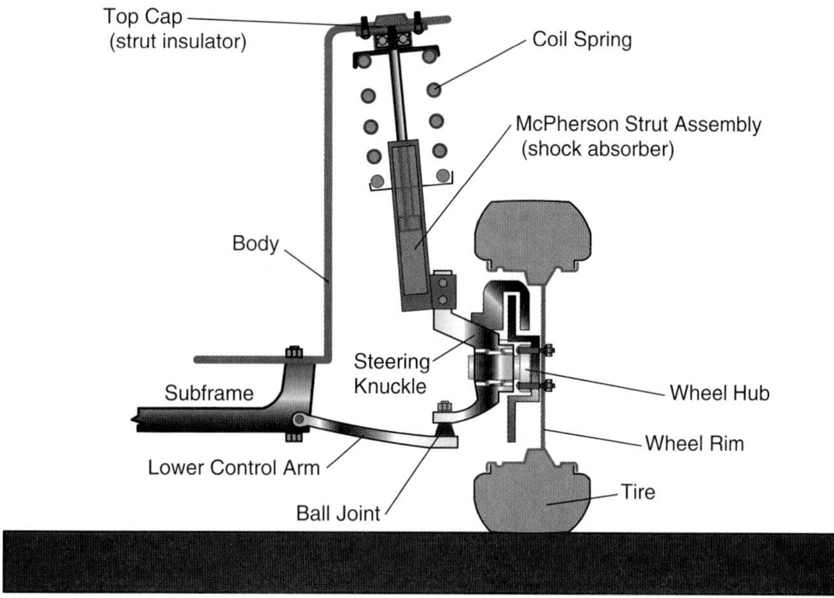

FIGURE 40-47 Strut suspension.

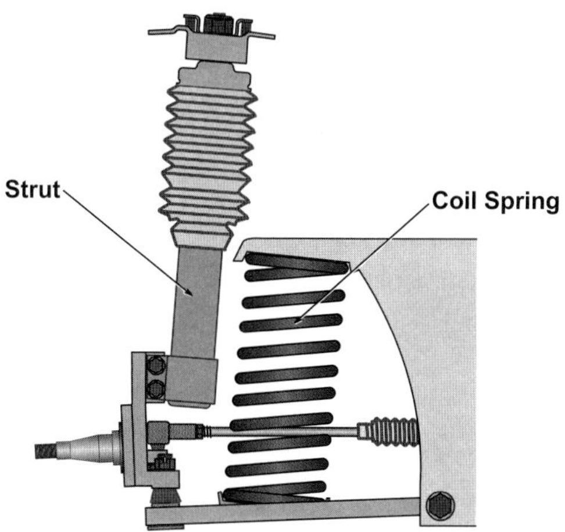

FIGURE 40-48 Modified strut suspension.

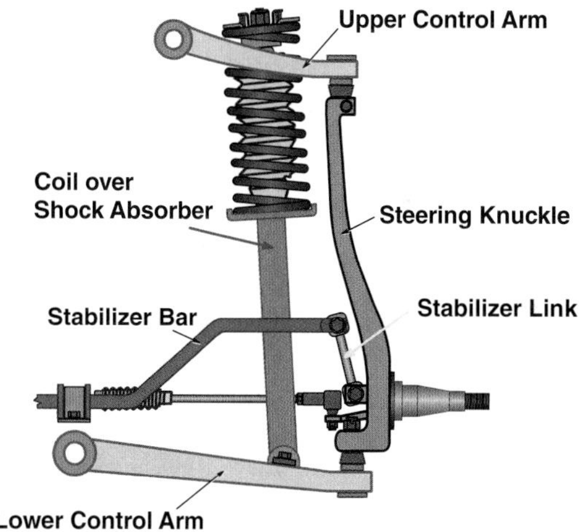

FIGURE 40-49 Double-wishbone modified strut.

or upright) for a 1:1 motion ratio. A 1:1 ratio means they can call this a direct-acting stabilizer bar. Because the bar can give everything it's got, it can be smaller and lighter and be a more responsive suspension. The Cadillac CTS and ATS along with BMW use this type of suspension.

SLA Suspension

K40020

The **short-/long-arm (SLA) suspension** gets its name from using two different-length control arms: one short upper control arm and one long lower control arm. The primary reason that the SLA suspension system was designed was to ensure correct wheel alignment as the vehicle corners. If the arms were the same length, then the wheel would stay perfectly straight as the suspension moved up and down over bumps. This is okay for straight ahead driving, but not when cornering. Having arms of the same length would result in incorrect positioning of the tire in relation to the road surface, as the body tends to roll. The body roll would tilt the top of the wheel outward (toward positive camber), reducing the tire-to-road contact patch, which reduces traction. The SLA system, with the short arm on top, tends to pull the top of the tire inward toward negative camber, keeping the tire-to-road contact patch as large as possible. The shock absorber is located between the frame and the lower control arm and is a direct-acting telescopic-type shock absorber. The coil spring or torsion bar can be mounted either between the frame and the lower control arm (type 1) or between the shock tower (or frame with a torsion bar) and the upper control arm (type 2) (**FIGURE 40-50**).

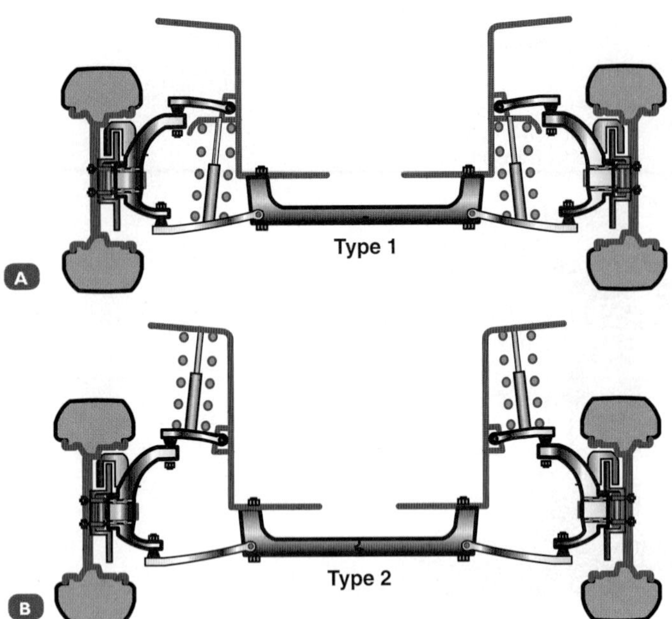

FIGURE 40-50 The coil spring can be mounted two different ways. **A.** Between the frame and the lower control arm (type 1). **B.** Between the shock tower and the upper control arm (type 2).

FIGURE 40-51 Steering knuckles. **A.** Long knuckle SLA. **B.** Short knuckle SLA.

FIGURE 40-52 Twin I-beam suspension system.

Both control arms pivot on control arm bushings. These bushings twist on the control arm pins, which are bolted to the cross member or subframe of the vehicle. Rubber jounce and rebound stops are used to prevent direct metal-to-metal contact between the control arms and the frame if the suspension should reach its maximum limit of travel.

The steering knuckle is mounted at the ends of the control arms by ball joints and allows both up-and-down and steering movement of the tire. The ball joints can be designed to be mounted so they are under compression or tension forces. Compression forces tend to push the ball into the socket, whereas tension forces pull the ball away from the socket.

There are two arrangements of steering knuckles used on SLA suspensions: short knuckle and long knuckle (**FIGURE 40-51**). The short knuckle design locates the upper ball joint inside of the wheel. The long knuckle design moves the upper ball joint above the tire, meaning the knuckle may even partially wrap around the tire. The long knuckle design affords the manufacturer a more ideal king pin geometry as well as better leverage against braking and cornering forces. King pin geometry is covered in the Wheel Alignment chapter.

Twin I-Beam Suspension

K40021

The twin I-beam suspension is a type of independent suspension. For each front wheel, it uses separate I-beams that pivot from the opposite side of the vehicle's frame or cross member (**FIGURE 40-52**). This system gives a wide radius that the wheel assembly swings through as the suspension compresses and rebounds. Most twin I-beam systems use a coil spring to support the weight of the vehicle. This means each I-beam must be supported longitudinally, which is accomplished by use of a radius rod.

The radius rod connects to the I-beam from the rear and is attached to the vehicle's frame with bushings, which allow the radius rod to pivot slightly as the suspension operates.

FIGURE 40-53 In front-wheel drive vehicles, the rear suspension system serves to keep the rear tires in contact with the road and aligned with the front tires.

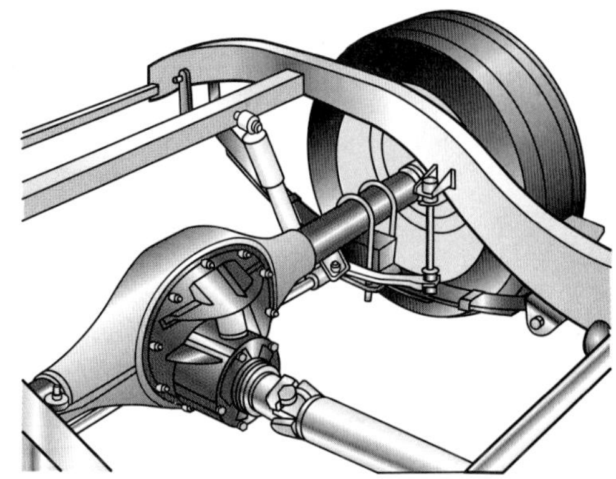

FIGURE 40-54 A rigid-axle leaf-spring suspension.

Rear Suspension

K40022

The main function of the rear suspension system is to keep the rear tires in contact with the road and aligned with the front tires (**FIGURE 40-53**). However, rear-wheel drive or all-wheel drive rear suspension systems are a bit more complicated. In these vehicles, the rear suspension system must not only keep the rear tires in contact with the road and aligned with the front tires, but it must also be engineered to transfer engine torque to the rear wheels. And in vehicles with four-wheel steering, the rear suspension must allow the rear wheels to be steered in a similar manner as the front wheels.

Rear-wheel suspension systems can be of either the independent or solid axle design. It is purely up to the preference of the designer. Many, but not all, rear-wheel drive vehicles use a solid axle, dependent suspension system. On front-wheel drive vehicles, the type of rear suspension system leans toward a higher percentage of independent systems, but solid axle systems are also common.

Rigid-Axle Leaf-Spring Suspension

K40023

A rigid-axle leaf-spring suspension can be used in both dead axles and live axles. The front of the leaf spring is attached to the chassis at the rigid spring hanger (**FIGURE 40-54**). The spring eyes typically use rubber bushings to connect with the vehicle's frame. The axle housing is rigid between each road wheel. Thus, any deflection to one side is transmitted to the other side. The rear of the leaf spring is attached to the swinging shackle, which allows for suspension movement by allowing the spring to extend or reduce in length, as the vehicle moves over uneven ground.

The top of the direct-acting shock absorber is attached to the chassis and to the spring pad at the bottom. The U-bolts attach the axle housing to the leaf spring. They have a clamping force that helps to keep the leaf spring together. Leaf springs hold the axle in position, both laterally and longitudinally, eliminating the need for track bars. The leaf spring is usually made up of a number of leaves of different lengths. The top, or longest, leaf is normally referred to as the main leaf. Most leaf-spring suspensions rely only on the sideways stiffness of the leaf spring to keep the axle in position when turning a corner to help keep the axle from shifting sideways through turns.

Rigid-Axle Coil-Spring Suspension

K40024

In **rigid-axle coil-spring suspensions**, the coil spring is mounted between the axle housing and the vehicle body (**FIGURE 40-55**). One drawback of a coil spring is that it cannot provide any side-to-side or front-to-back stability to the axle. All it can do is suspend the body above the axle. Therefore, unlike the with leaf-spring suspension, control rods must be used to control this potential axle movement during braking, acceleration, and cornering. Manufacturers use a variety of configurations to address this concern.

The first style uses lower control arms near each coil spring that are parallel with the **centerline** of the vehicle and connect between the frame and the axle. They maintain the longitudinal position of the axle. The upper control arms are angled toward the center of the vehicle to counteract any lateral forces as well as twisting forces (**FIGURE 40-56**). Another style uses upper and lower control arms that all are parallel with the centerline of the vehicle. They control the twisting force of the axle but cannot control any lateral forces during cornering. One of two methods is used to control the lateral forces—a Panhard rod or a Watt's linkage.

A **Panhard rod**, also referred to as a track bar, sits parallel with the axle. One end connects to the frame of the vehicle, and the other end connects to the axle (**FIGURE 40-57**). The Panhard rod uses bushings on each end, so the joints can pivot as the suspension compresses and rebounds. The position of the axle is maintained laterally by the rod. A **Watt's linkage** is a bit more complex but functions in a similar way (**FIGURE 40-58**). A lever that is able to pivot in the middle is mounted vertically to the rear axle near its center. The top of the lever is connected to a rod that is parallel to the axle and mounted high on one side of the frame. The bottom of the lever is connected to a similar rod, which is also parallel to the axle but mounted low on the other side of the frame. These two rods allow up-and-down movement of the axle but prevent lateral movement.

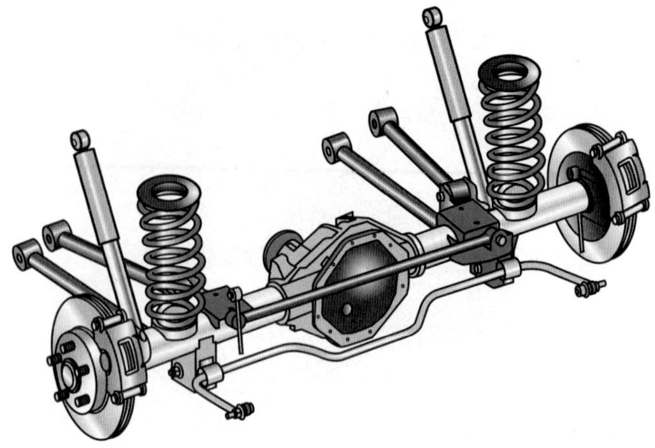

FIGURE 40-55 Rigid-axle coil-spring suspension.

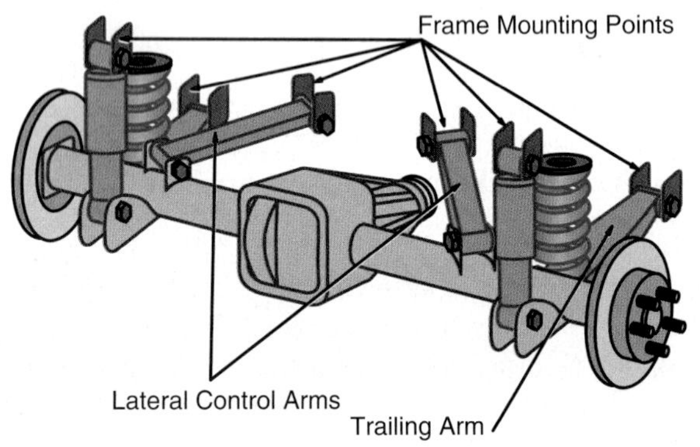

FIGURE 40-56 Lower control arms prevent any front–rear movement of the axle. Upper control arms prevent side-to-side and twisting movement of the axle.

FIGURE 40-57 A Panhard rod sits parallel with the axle.

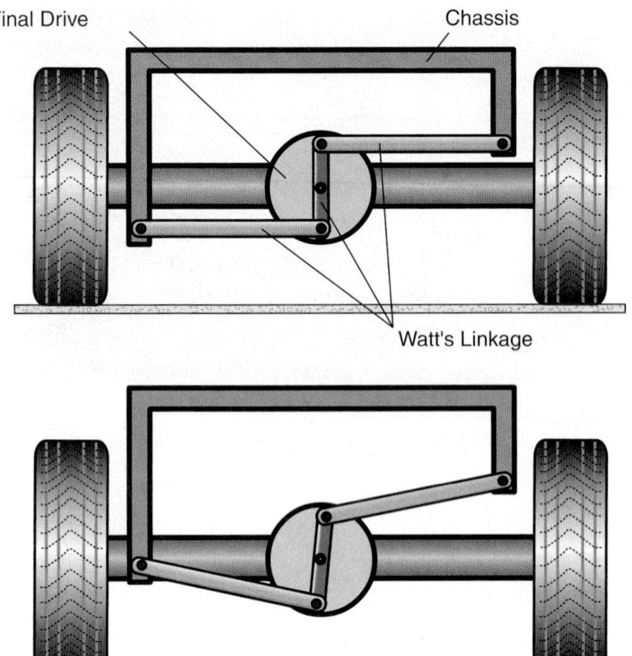

FIGURE 40-58 A Watt's linkage.

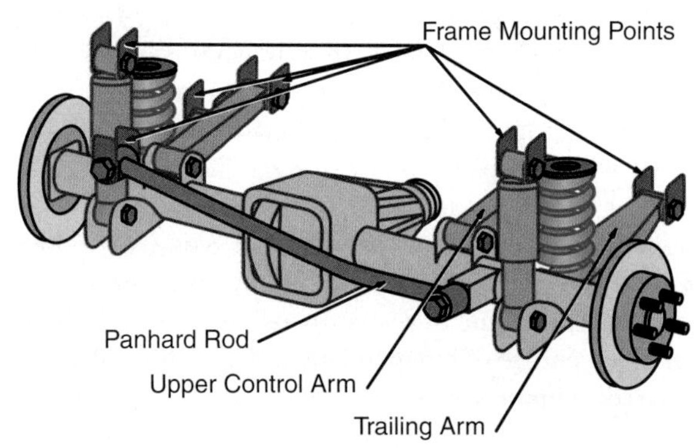

FIGURE 40-59 A rigid dead axle using a strut assembly, trailing arm, and Panhard rod to control axle movement.

Rigid Dead Axle Suspension

K40025

A rigid dead axle is sometimes referred to as a beam axle. It can come in a variety of configurations. With **rigid dead axle suspension**, the longitudinal and lateral position of the axle must be maintained as in all axles, but because it is a dead axle, it typically has to withstand only braking forces. One common rigid dead axle suspension uses lower trailing arms connecting the vehicle frame and axle with a strut assembly on top. These trailing arms maintain the longitudinal position of the axle (**FIGURE 40-59**), and the springs and strut support the weight of the vehicle and assist in controlling any braking forces. The struts have an upper mounting point in the suspension tower. They are nonsteerable and therefore do not require an upper strut bearing. This type of suspension typically uses a Panhard rod to control lateral forces.

Another style, called a torsion beam axle, is a dead axle that uses a U-shaped axle beam with a torsion bar mounted inside it (**FIGURE 40-60**). A trailing arm on each side holds the axle longitudinally; a Panhard rod may be used to hold it laterally; and the

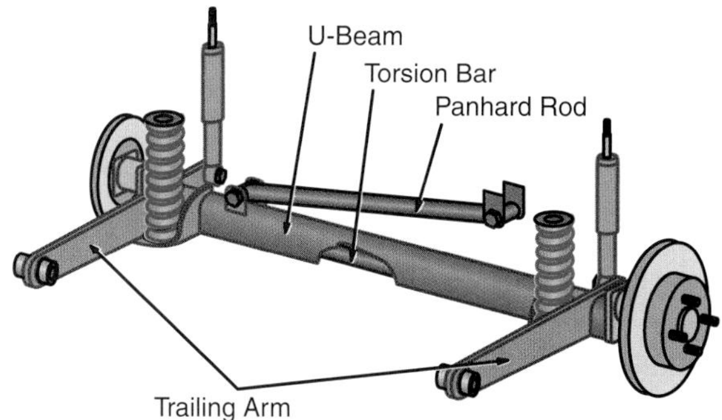

FIGURE 40-60 Torsion beam axle with internal torsion bar.

FIGURE 40-61 A typical MacPherson strut type IRS.

strut and spring support the weight of the vehicle. The torsion bar provides a measure of resistance to twisting forces as one wheel goes over a bump. Manufacturers use a variety of other types of rigid dead axles, but they all tend to use leaf springs, coil springs, or torsion bars, along with control arms, rods, and bushings. As you work out in the shop, be on the lookout for the different arrangements of these systems.

FIGURE 40-62 Ford's Control Blade IRS.

Rear Independent Suspension (Dead Axle)

K40026

The kind of dead independent suspension used on the rear of a vehicle can be fairly simple, as the wheels typically do not steer or drive the vehicle. In this case, the suspension system only has to hold the wheels in the proper orientation while the vehicle is being driven. Because it is an independent type of system, each wheel is not connected to the other, so they can move separately. A MacPherson strut system is commonly used at the rear, which can be similar to the front suspension system, using either a wishbone-shaped lower control arm or two control arms, along with a strut assembly (**FIGURE 40-61**). It can also be similar to Ford's Control Blade® system that uses a trailing arm with a lower control arm (**FIGURE 40-62**). A separate spring sits between the frame and control arm. A long travel shock absorber sits outside of the spring.

FIGURE 40-63 Live independent rear suspension. Note how the final drive is mounted directly to the vehicle frame.

Rear-Wheel Drive Independent Suspension

K40027

On rear-wheel drive vehicles with independent rear suspension, the final drive unit is fixed to the vehicle frame, thereby greatly reducing unsprung weight (**FIGURE 40-63**). Drive is transmitted to each wheel by **external driveshafts**, shafts that transfer power from the final drive to the live axle. Suspension is normally provided by coil springs, and each wheel unit is located by a combination of lateral and longitudinal control arms or by trailing arms to the frame. The final drive assembly is normally bolted to the chassis, and because it must absorb the torque reaction, it must be securely fastened. Driveshafts, with either conventional universal joints or constant-velocity joints, rotate and transmit the drive to the wheels. Universal joints are placed on the end of a driveshaft to compensate for the up-and-down movement of the rear wheels. When conventional universal joints are used, each driveshaft may have a **splined section** that can move in or out to vary the shaft's length due to changes in the suspension action (**FIGURE 40-64**).

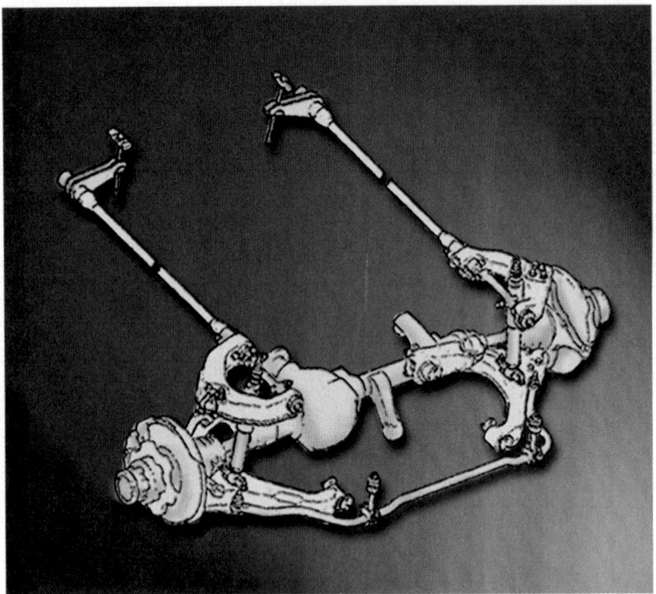

FIGURE 40-64 The independent rear suspension may have upper and lower control arms.

FIGURE 40-65 On some vehicles equipped with an independent rear suspension, the driveshafts or half shafts themselves can be used as the upper link of the suspension, providing the upper pivot point.

On some vehicles equipped with an independent rear suspension, the driveshafts or half shafts themselves can be used as the upper link of the suspension, providing the upper pivot point (**FIGURE 40-65**). In this case, the half shafts are fixed, meaning they do not extend or collapse. This makes the splined section unnecessary, and the shaft can be made as a one-piece unit. The outer wheel bearing hub is held in position laterally and longitudinally at the bottom by a pivot on the end of the lower control arm, and the half shaft holds the top of the hub in position vertically. Because the hub pivots on both the lower control arm and the half shaft, the wheel can move up and down. The lower control arm has widely spaced pivots to provide stability. It is longer than the half shaft, which allows the camber to move slightly positive during cornering, just like an SLA suspension.

> **TECHNICIAN TIP**
>
> External forces, such as curb impact or a collision, can damage control arms or linkages and move the wheel units from their correct position. This can make a vehicle pull to one side, cause abnormal tire wear, and make the vehicle difficult to control.

▶ Active and Adaptive Suspension Systems

K40028

There are two main categories of suspension system control; active and adaptive, also called semi-active. An active suspension system acts independently of the driver, to fully control the actions and responses of each individual wheel while the vehicle is being driven. This means that it can do several things:

- An active suspension system can respond to irregularities in the road surface and raise or lower each wheel so it will follow the irregularities without causing much of a change in the vehicle's body height.
- It can automatically raise or lower the vehicle with changes in speed for aerodynamic or road condition reasons.
- It can prevent the vehicle from leaning in corners by raising the outside wheels when cornering. This maintains a flatter stance and keeps the tires perpendicular to the road surface, which enhances traction.
- It can also prevent the nose and tail of the vehicle from diving or rising during braking and accelerating maneuvers.
- At the same time, it can provide active dampening power to each individual shock absorber, based on the driver's preferred setting, road surface, and driving conditions.

Most active suspension systems are hydraulically actuated, which means that the servos are operated by high-pressure fluid from a hydraulic pump. The pressure is directed to or vented from the servos by a bank of electrically operated solenoid valves. Sensors monitor suspension height, vertical and horizontal acceleration, pitch, and roll of the vehicle body so that adjustments can be made almost instantaneously. Some systems can provide up to 3000 adjustments per second. So, you can see that this type of system can be very responsive. At the same time, these systems are very expensive and have large power requirements. Therefore, few manufacturers include them in their vehicles. Also, because of the amount of work they do and their constant usage, these systems tend to have high-maintenance repair needs, and costs can be prohibitive for most people.

Adaptive air suspension systems are less sophisticated than active systems. They provide electronically controlled dampening at all four shock absorbers, meaning they are able to automatically change the dampening power of the shocks when the road conditions change. At the same time, they can only change the dampening effect but cannot actively raise or lower the wheel in response to road irregularities. An adaptive air suspension system combines the ability to have sporty handling with a high level of ride comfort. Additionally, some systems include speed-dependent lowering of the body; this change in ride height means a low center of gravity, resulting in significantly increased directional stability as well as increased aerodynamic efficiency. The ride height is typically obtained by inflating or deflating air bags at each corner of the suspension. Note that the adjustments in ride height are relatively slow to obtain. They don't happen in the same way that active suspension systems operate.

The information obtained from sensors on the axles for ride height, and acceleration sensors on the body for vehicle pitch, is evaluated in the adaptive suspension's central control unit. This computer can adjust the dampening power of the individual shock absorbers within milliseconds, depending on driving situations. As long as no higher dampening forces are required—for instance, when driving straight ahead on good roads—the damper settings remain comfortably soft.

Specific adjustments to the dampening force at individual wheels eliminate body movement, which enhance handling and could affect occupant comfort and vehicle control. In some cases, when cornering, braking, or accelerating, adaptive dampening can automatically reduce (not eliminate) rolling or pitching movements. Although adaptive suspension systems are not as refined as active suspension systems, the differences between them are shrinking.

Adaptive Air Suspension Operation

K40029

On some vehicles, when the ignition is switched on or when the vehicle's door is opened before ignition, the control system is activated. The height sensor uses the induction principle to constantly monitor the distance between the vehicle's axle and its chassis. If the control system determines that the ride height is too low, it will command that air be added to the appropriate adaptive shock absorbers. When the vehicle is being loaded, unloaded, or lowered because of driver command or vehicle speed, the height sensor monitors the changes and reports them to the control unit. The ECU compares this information to the stored reference values. The ECU activates either the electric motor of the compressor or the exhaust solenoid valve, which is located on the top of each adaptive shock absorber (**FIGURE 40-66**). This also requires the solenoid valve to be actuated or moved to maintain the required level, once reached. The adaptive shock absorber solenoid valves are subject to stringent leakage requirements in order to maintain the vehicle's height even when the system is not being operated.

When the vehicle is being loaded, the compressor delivers air into the four air suspension bellows, through fill solenoids or gate solenoids, until the normal level has once again been reached. For additional air delivery or rapid response, the reservoir solenoid valve is opened, and air flows directly from the reservoir. When the vehicle is being unloaded, the exhaust solenoid valve is activated. Activation of this valve results in airflow from the air suspension bellows being removed via the air dryer solenoid valve in the compressor, then via the relay valve. The air is then exhausted into the atmosphere through a computer-controlled vent valve. Any dynamic air spring movement while the vehicle is in motion is ignored and does not cause the control system to respond. However, the system can adjust the air pressure based on other parameters such as vehicle speed, driver selection of a different mode, or any automatic adjustments that the control system determines are necessary.

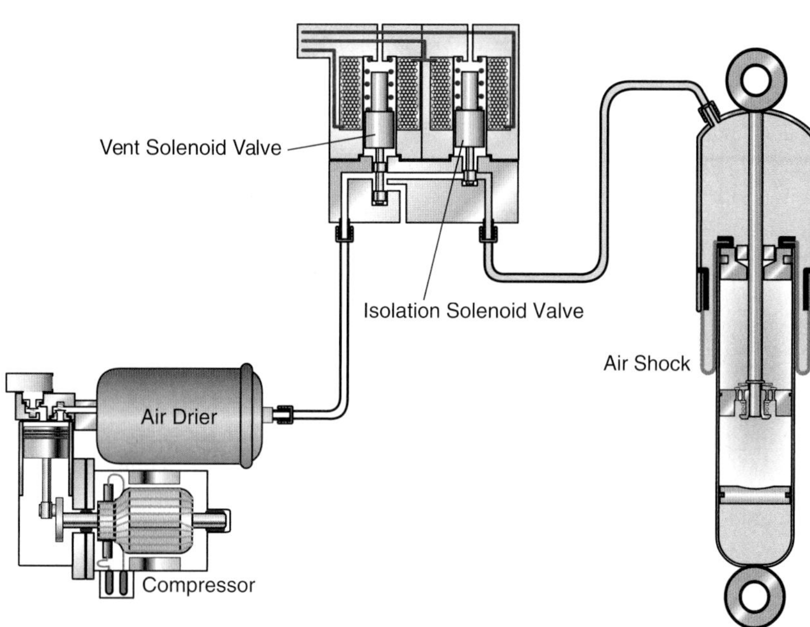

FIGURE 40-66 Adaptive shock absorber.

Wrap-Up

Ready for Review

- The suspension system is designed to absorb road shock and vibrations.
- Unsprung weight refers to the vehicle parts not supported by springs (wheels, tires, brake and steering assemblies); this weight should be kept as low as possible.
- Metal springs, rubber, and air all work to provide suspension system support by absorbing some amount of force.
- Springs react to road shock by not immediately returning to their original state. They may oscillate—that is, fluctuate in length—until the energy from the road shock is dissipated.
- Vehicle movement may be due to yaw, pitch, or roll.
- Suspension system components include the springs, axles, shock absorbers, arms, rods, sway bars, steering knuckle, bushings, and ball joints.
- Types of springs include coil springs, leaf springs, and torsion bars.
- Coil springs can be cylindrical, barrel shaped, or conical and are used on the front suspension of most modern light vehicles.
- Leaf springs are primarily used on rear-wheel drive vehicles and consist of multiple flat springs made of tempered steel.
- Functions of axles include helping support vehicle weight, maintaining wheel position, providing forward propulsion of the vehicle, and transmitting torque to the wheel.
- Axle types include straight, dead, full floating, or semi-floating.
- Types of shock absorbers are hydraulic, gas pressurized, and adjustable (which can be load adjustable, annual adjustable rate, electronic adjustable rate, and automatic load adjustable).
- Shock absorbers use the resistance of a rod and piston, and oil and disc valves, to provide a dampening effect on the force created by the bumps and jolts of driving.
- Manually adjustable air springs can be incorporated into rear shock absorbers for vehicles designed to carry heavy loads.
- The dampening rate of electronic adjustable rate shock absorbers can be manually or automatically altered.
- The primary load-bearing elements of an SLA suspension system are known as control arms, which attach to the wheel assembly on one end and the chassis on the other.
- The primary load-bearing element on a MacPherson strut suspension system is the strut itself.
- Bushings act as bearings at suspension fulcrum points, allowing movement of the component without losing its alignment.
- Types of suspension systems are solid (beam) axles, independent, rear, front, adaptive air, and computer controlled.
- A live axle transfers power from the engine to the wheels; a dead axle does not transmit any drive.
- Independent suspension allows for lower unsprung mass.
- Struts are commonly used in independent suspension systems; the most common type is the MacPherson strut.
- Types of rear suspension systems are rear independent, rear-wheel drive independent, rigid-axle leaf spring, rigid-axle coil spring, and rigid (dead).
- Rear suspension systems in front-wheel drive vehicles are designed to keep the rear tires in contact with the road and aligned with front tires.
- The rear suspension system on a rear-wheel drive vehicle must allow for swiveling of the front wheels during steering; on four-wheel steering vehicles, the system must allow for swiveling of rear wheels as well.
- Front suspensions are generally either independent (front wheels move independently) or solid (dependent).
- Front suspension system types include MacPherson strut, short/long arm (SLA), and torsion bar.
- Adaptive air suspension systems electronically control the height of all four wheels, based on vehicle speed and load.

Key Terms

adaptive air suspension A suspension system that uses rubber bags or bladders filled with air to support the weight of the vehicle.

air spring A part that provides the springing action or auxiliary spring. It is typically used in air suspension systems or heavy truck applications.

applied force Pressure placed on something.

automatic load-adjustable shock absorber Typically, an air shock absorber used in an automatic load-sensing system that adjusts ride height (ground clearance) automatically, such as when additional weight is added to the vehicle; also called self-leveling.

axle The shaft of the suspension system to which the tires and wheels are attached; they are used to drive or support the wheels.

ball joint A swivel connection mounted in the outer end of the front control arm. These swivels are typically constructed with a ball and socket to allow pivoting.

braking torque The torque acting to twist the axle housing around its center during braking.

bushing A rubber, nylon, or urethane part that allows for movement while maintaining alignment.

centerline The imaginary line drawn down the exact center of the vehicle from front to back.

coil spring Spring steel wire, heated and wound into a coil, that is used to support the weight of a vehicle.

compliance bushing A rubber bushing with a voided section molded in it that allows component movement under torque application. It is typically used in control arms on front-wheel drive vehicles to minimize torque steer issues.

control arm The primary load-bearing element of a vehicle's suspension system, commonly referred to as an A-arm or wishbone. These arms may be used as an upper and lower pivot point for the wheel assembly. They attach to the chassis with rubber bushings that allow up-and-down movement of the tire and wheel assembly.

dead axle An axle, used on a rear- or front-wheel drive vehicle that does not drive the vehicle.

deflecting force A force that moves an object in a different direction or into a different shape.

direct-acting telescopic shock absorber A shock absorber designed to reduce spring oscillations.

elasticity The ability to deform and reform into the same shape.

external driveshaft A shaft used to transfer power from the transmission to the live axle.

independent suspension A system for allowing the up-and-down movement of one tire without affecting the other tire on that axle.

leaf spring A spring, made of one or more flat, tempered steel springs bracketed together, that is used in the suspension system to support the weight of the vehicle.

live axle An axle with a final drive unit made into it that transfers the power from the engine to the wheels so the vehicle can move.

MacPherson strut A strut used on an independent suspension where the spring and shock are joined together; used on most front-wheel drive vehicles.

magneto-rheological fluid A fluid that has the unique characteristic of changing viscosity when exposed to a magnetic field.

manual adjustable-rate shock absorber A shock absorber that allows manual adjustment of the dampening rate.

manually adjustable air spring A rubber air bag placed inside coil springs to increase the spring's load carrying ability. It is filled manually through a valve similar to a tire valve stem.

oscillation The fluctuation of an object between two states. With regard to suspension springs, it refers to the uncontrolled compression and decompression of the spring following overshoot.

overshoot The amount a spring extends (springs back) past its original length following compression.

Panhard rod A metal rod used to hold the dead axle and keep it from moving from side to side through corners. It is mounted on the body or frame of the vehicle and the axle; also referred to as a track bar.

pitch Movement of a vehicle around its y-axis (the imaginary line across the center of the vehicle from left to right) that causes the vehicle to lower or rise on the front end during quick braking or acceleration.

progressive rate of deflection The change in deflection rate that occurs as the weight of the vehicle changes. The greater the weight, the lower the rate of deflection due to increased resistance.

reaction force A force that acts in the opposite direction to another force.

rebound clip A metal strap that is warped around the leaf spring to prevent excessive flexing of the main leaf during rebound.

ride height The amount of ground clearance a vehicle has, measured from a point on the body or frame, depending on the manufacturer; also known as ride height.

rigid-axle coil-spring suspension A dead axle that uses a coil spring.

rigid dead axle suspension A type of dead axle suspension system that is non-independent and uses a beam or solid axle.

rigid spring hanger The rigid part typically welded to the body or frame of the vehicle to which the front of the leaf spring is attached.

rod A straight piece of steel used to transfer motion within the vehicle's suspension system. It typically has treads cut on one or both ends.

roll Movement of a vehicle around its x-axis (the imaginary line down the center of the vehicle from front to back). It is commonly referred to as body roll or lean; when cornering, the body tries to move to the outside of the corner against the suspension.

rubber-bonded bushing A bushing that has a steel outer housing and inner sleeve with rubber inside; also known as a metalastic bushing.

shock absorber A device on a vehicle designed to absorb bumps and jolts caused from driving on irregular surfaces and to dampen body movement.

short-/long-arm (SLA) suspension A type of control arm suspension system that uses a short control arm on the top and a long control arm on the bottom. This design ensures correct alignment angles when moving through bumps.

shroud A steel or plastic cover placed over the shock rod.

sintering The process of using pressure and heat to bond metal particles.

solid axle A single piece of steel that provides a simple means of mounting the hub and wheel units; also called beam axle or straight axle.

splined section A flat key made into a shaft to accommodate changes in shaft length due to movement in wheel camber with suspension action.

spring A resilient steel part that stores energy when compressed and releases energy when released to its original state; available as a leaf spring, coil spring, or torsion bar.

spring eyes Rolled ends of some springs used to mount springs to the chassis.

spring shackle bushing A bushing that is positioned in the shackle to which the leaf spring mounts. Bushings allow the spring shackle to move as the leaf spring dimensions change over bumps.

stop A rubber part used to control the movement of control arms (suspension arms).

strut A shock absorber used on a MacPherson strut–type suspension.

suspension action Movement of the chassis up and down.

suspension strut A shock absorber designed to reduce spring oscillations.

suspension system A system within a vehicle designed to isolate the vehicle body from road bumps and vibrations.

sway bar A part used in vehicles as a stabilizer, or antiroll, bar. It is connected to the chassis in the center, and each end is connected to one side of the suspension system. It is typically installed on the front, and sometimes the rear, suspension.

swinging shackle A shackle connected to the rear of the multileaf spring that allows the leaf spring to move downward when a load is placed on the rear of the vehicle.

torsional load A force that is applied by clamping one end of an object to another object that is then twisted.

trailing arm suspension A type of suspension system that uses upper and lower control arms.

uniform pitch A spring whose pitch (the distance from the center of one coil to the center of the adjacent coil) is the same distance throughout.

unsprung mass Any part of the steering and suspension system that is not supported by springs. A large amount of unsprung weight causes the tire to hop off the ground when hitting bumps, as the weight overcomes dampening of the shock absorbers.

unsprung weight *See* unsprung mass.

Watt's linkage Another name for a rigid-axle coil-spring suspension that uses two bars similar to a Panhard rod and a pivot point on the axle, to keep the axle from moving in turns.

wishbone control arm Another term for an A-arm.

wrap leaf A spring containing spring eyes.

yaw Movement of a vehicle around its z-axis (vertical axis), felt when the vehicle deviates from its straight path, as when skidding sideways and the rear comes around.

Review Questions

1. All of these act as common indicators for the problems in the suspension system *except*:
   a. a bouncy ride.
   b. a reduction in steering control.
   c. noises when going over a bump.
   d. improper transmission shifting.
2. For the vehicle to operate properly and safely, suspension and steering components should be aligned with each other and also with the:
   a. engine centerline.
   b. wheels' centerline.
   c. vehicle centerline.
   d. axle centerline.
3. The components that are part of the sprung weight of a vehicle are:
   a. axles.
   b. brakes.
   c. steering parts not supported by springs.
   d. suspension parts supported by springs.
4. The suspension system performs all of the following functions *except*:
   a. absorbing the large road forces generated while driving on non-perfect roads.
   b. holding the wheels and tires in the proper orientation.
   c. monitoring the tread wear of each tire.
   d. supporting the weight of the chassis, drivetrain, passenger compartment, occupants, and any additional load.
5. Which of the following places a twisting pressure on the suspension members that push the vehicle along the road?
   a. Driving thrust
   b. Breaking torque
   c. Cornering force
   d. Centrifugal force
6. The movement around an imaginary line drawn down the center of the vehicle from front to back is called:
   a. yaw.
   b. roll.
   c. pitch.
   d. rock.
7. Which of the following dampen spring oscillations by forcing oil through small holes in a piston?
   a. Springs
   b. Shock absorbers
   c. Control arms
   d. Axles
8. When the pitch is varied, the deflection rate varies in the spring. Then, the spring is said to have a:
   a. uniform pitch.
   b. cylindrical pitch.
   c. barrel pitch.
   d. progressive rate of deflection.
9. Which of the following shock absorbers is also called self-leveling?
   a. Automatic load-adjustable shock absorbers
   b. Electronic adjustable-rate shock absorbers
   c. Manual adjustable-rate shock absorbers
   d. Gas-pressurized shock absorbers
10. All of the following statements are true *except*:
    a. the solid axle provides a simple means of mounting the hub and wheel units.
    b. SLA uses upper and lower control arms to control the movement of the knuckle.
    c. in strut suspension, the shock absorber is contained outside the strut.
    d. solid axle systems are inexpensive and good for hauling heavy loads.

ASE Technician A/Technician B Style Questions

1. Tech A says that the wheel and tire are examples of unsprung weight. Tech B says that the exhaust system is an example of sprung weight. Who is correct?
   a. Tech A
   b. Tech B

c. Both A and B
d. Neither A nor B

2. Tech A says that one purpose of the suspension system is to keep the tires in contact with the road. Tech B says that the heavier the weight of the unsprung components, the smaller the reaction force they will generate. Who is correct?
   a. Tech A
   b. Tech B
   c. Both A and B
   d. Neither A nor B
3. Tech A says that yaw is when a vehicle deviates from its straight path. Tech B says that roll is when the front of the vehicle noses down or rises up. Who is correct?
   a. Tech A
   b. Tech B
   c. Both A and B
   d. Neither A nor B
4. Tech A says that leaf springs are made from a single length of special wire, which is heated and wound on a former in the shape of a coil. Tech B says that a progressive-rate spring offers a soft ride but can also carry a heavier load. Who is correct?
   a. Tech A
   b. Tech B
   c. Both A and B
   d. Neither A nor B
5. Tech A says that shock absorbers dampen in the downward direction. Tech B says that shock absorbers dampen in the upward direction. Who is correct?
   a. Tech A
   b. Tech B
   c. Both A and B
   d. Neither A nor B
6. Tech A says that in a MacPherson strut suspension, the hydraulic strut supports the top of the steering knuckle, so it must be much stronger than an ordinary shock absorber. Tech B says that gas-pressurized shock absorbers don't use hydraulic fluid; they use pressurized nitrogen gas only. Who is correct?
   a. Tech A
   b. Tech B
   c. Both A and B
   d. Neither A nor B
7. Tech A says that a loaded ball joint supports the weight of the vehicle Tech B says that a ball joint in a MacPherson strut suspension is a follower joint (not loaded). Who is correct?
   a. Tech A
   b. Tech B
   c. Both A and B
   d. Neither A nor B
8. Tech A says that a dead axle is designed to carry the weight of the vehicle, with no drive capability. Tech B says that live axles transmit power to the wheels. Who is correct?
   a. Tech A
   b. Tech B
   c. Both A and B
   d. Neither A nor B
9. Tech A says that control arms serve as primary load-bearing elements of a vehicle's suspension system. Tech B says that an A-arm is a relatively flat triangular part that mounts to the frame or subframe at each leg of the A. Who is correct?
   a. Tech A
   b. Tech B
   c. Both A and B
   d. Neither A nor B
10. Tech A says that on short-/long-arm (SLA) suspension systems, the upper control arm is the long one. Tech B says that on SLA suspension systems, the control arms are connected to the frame by ball joints. Who is correct?
   a. Tech A
   b. Tech B
   c. Both A and B
   d. Neither A nor B

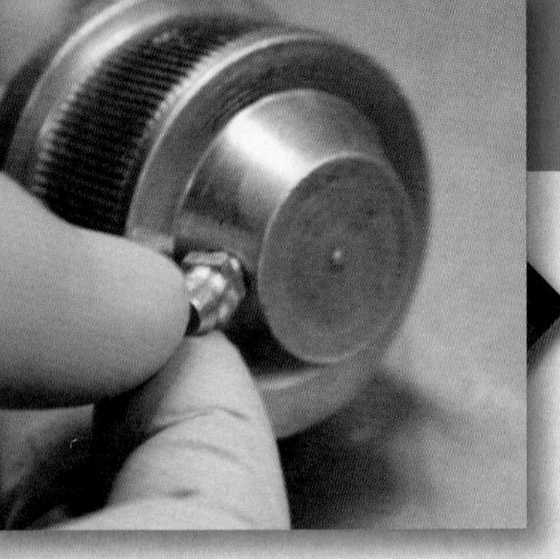

CHAPTER 41
Servicing Suspension Systems

NATEF Tasks

- **N41001** Diagnose vehicle wander, drift, pull, hard steering, bump steer, memory steer, torque steer, and steering return concerns; determine needed action. (AST/MAST)
- **N41002** Diagnose short- and long-arm suspension system noises, body sway, and uneven ride height concerns; determine needed action. (AST/MAST)
- **N41003** Diagnose strut suspension system noises, body sway, and uneven ride height concerns; determine needed action. (AST/MAST)
- **N41004** Inspect, remove, and/or replace front/rear stabilizer bar (sway bar) bushings, brackets, and links. (MLR/ AST/MAST)
- **N41005** Inspect, remove, and/or replace shock absorbers; inspect mounts and bushings. (MLR/AST/MAST)
- **N41006** Inspect, remove, and/or replace short- and long-arm suspension system coil springs and spring insulators. (AST/MAST)
- **N41007** Inspect, remove, and/or replace steering knuckle assemblies. (AST/MAST)
- **N41008** Inspect, remove, and/or replace upper and lower control arms, bushings, shafts, and rebound bumpers. (MLR/AST/MAST)
- **N41009** Inspect, remove, and/or replace upper and/or lower ball joints (with or without wear indicators). (MLR/AST/MAST)
- **N41010** Inspect, remove, and/or replace strut cartridge or assembly, strut coil spring, insulators (silencers), and upper strut bearing mount. (MLR/AST/MAST)
- **N41011** Inspect rear suspension system leaf spring(s), spring insulators (silencers), shackles, brackets, bushings, center pins/ bolts, and mounts. (MLR/AST/MAST)
- **N41012** Inspect, remove, and/or replace strut rods and bushings. (AST/MAST)
- **N41013** Inspect, remove, and/or replace track bar, strut rods/ radius arms, and related mounts and bushings. (MLR/AST/MAST)
- **N41014** Inspect, remove, and/or replace torsion bars and mounts. (MLR/AST/MAST)

Knowledge Objectives

After reading this chapter, you will be able to:

- **K41001** Describe an overview of servicing suspension systems.
- **K41002** Describe the purpose of tools used to service suspension systems.
- **K41003** Describe the diagnostic process for suspension systems.
- **K41004** Describe common issues with suspension systems.

Skills Objectives

After reading this chapter, you will be able to:

- **S41001** Perform maintenance and repair on suspension systems.
- **S41002** Lubricate steering and suspension systems.

▶ Introduction

This chapter covers the diagnosis, maintenance, and repair of modern suspension systems. In the last chapter, you learned that the suspension system is primarily a mechanical system. Electric, electronic, and pneumatic devices may be added to it in some cases, but those devices primarily enhance the function of the mechanical components but do not replace them. You will usually be dealing with mechanical faults such as worn or damaged parts. Faults involving wear are often located during a visual inspection and then confirmed with measurement. Damaged components can typically be identified visually, but not always. In some cases, an alignment machine must be needed to measure the steering and suspension angles and to compare specifications in order to identify damaged parts. As you can see, you need a solid understanding of suspension system components so you can identify faults in them. This knowledge also comes in handy as you replace parts and have to verify that they are operating correctly.

▶ Servicing Suspension Systems

K41001

After any suspension system components are replaced, the vehicle will likely need a wheel alignment to make sure that the suspension system is adjusted properly. Wheel alignment is covered in the next chapter. Some instructors prefer to teach wheel alignment first so students will understand the alignment angles before diving into suspension system parts replacement. So don't be afraid to read ahead and start to learn about the wheel alignment process. But for now, let's get started with an overview of some of the tools you will use during suspension system work.

Tools

K41002

- Dial indicators: Used to measure play in ball joints.
- Pry bars: Used to pry parts to check for play.
- Measuring tapes: Used to check ride height.
- Pitman arm puller: Used to pull the pitman arm from the sector shaft.
- Tie-rod end puller: Should always be used on aluminum steering arms.
- Pickle forks: Used to separate ball joints, but ruins the grease seal.
- Electronic stethoscope: Used to locate unusual noise, vibration, or harshness (NVH); sometime called a chassis ear.
- Coil spring compressor: Used to compress coil springs during removal and installation.
- Scan tool: Used to check for codes in the vehicle's computer.
- Ball joint press tool: Used to remove and replace press-fit ball joints.
- Air chisel: Air-operated hammer that can accept a variety of bits including chisels and punches.
- Strut compressor: Used to compress MacPherson struts to remove the coil spring.
- Strut servicing kit: Used for changing coil springs on struts.
- Universal strut nut wrench kit: Used to remove a strut.
- 24-mm strut rod socket: Used to remove a strut.
- Various lift devices (e.g., jack): Used to raise vehicle off of the ground and safety stands to hold it there; devices include the air jacks found on the alignment lift.

▶ Diagnosis

K41003

Problems in the suspension system generally result in driver complaints of a poor ride, poor handling, or noises. Although these may seem trivial, they could indicate a serious and costly suspension problem that, if left untreated, could lead to additional issues or safety hazards. Diagnosis of suspension system problems follows the same strategy-based diagnostic process as for other systems on the vehicle. It's time we revisit those steps:

Step 1: Verifying the customer's concern
Step 2: Researching possible faults and gathering information
Step 3: Focused testing
Step 4: Performing the repair
Step 5: Verifying the repair

When diagnosing suspension systems, you must always start with a good interview to fully understand the concerns of the driver. Typically the service advisor performs this and then writes up the repair order. In some cases, you may need to perform a road test with the customer to verify the concern.

You Are the Automotive Technician

You work for County Fleet Services in the vehicle service department. One of the drivers wrote up a complaint sheet on a relatively new Dodge full-size delivery van with short-/long-arm suspension on front and leaf-spring suspension on the rear. The truck has a little more than 100,000 miles (160,000 km) on it. The complaint sheet lists several concerns. First, the operator admitted to hitting a curb with the right front wheel. Second, the vehicle pulls hard to the right when driven. Third, it makes an audible clunking noise in the front when going over bumps at around 10 mph (17 kph).

1. What are the potentially damaged components from when the vehicle hit the curb?
2. What will you do to diagnose the pulling condition?
3. What are the possible causes of the clunking noise when hitting a bump?

Once you have gathered information from the customer and performed a road test, you should research the symptoms in the service information to see if there are any related technical service bulletins (TSBs). If so, follow the information listed there.

If there are no related TSBs and you are familiar with the type of suspension system you are working on, take a minute to plan your testing strategy. This likely would involve using logic and your understanding of the system to decide what steps you should take initially—usually a visual inspection of the tires, wheels, steering system, suspension system, as well as a general vehicle inspection. If these initial inspections do not uncover the cause of the concern, you will need to dig deeper, for example, by using an electronic stethoscope to identify any NVH faults, or a wheel alignment machine to check the vehicle's wheel alignment.

Once you identify the root cause of the fault and present that information and the actions needed to correct the fault to the customer for authorization to repair the vehicle, it is time to begin, following the skill drills in this chapter along with the service information. If there is any difference between the Skill Drills in this book and the service information for the vehicle, always follow the service information as it is directly applicable to that vehicle. Finally, once the repair is complete, road test the vehicle to verify that the repair corrected the fault and that no other issues showed up. Doing this helps avoid comebacks and leaves a positive impression with the customer.

SAFETY TIP

Remember that steering and suspension concerns can result in injury or death to the driver and occupants, so extreme attention to safety must be given while diagnosing suspension and steering concerns.

Common Issues

K41004, N41001

The most common problem in the suspension system is play or looseness of the parts. Although many of the parts are designed with special connections that accommodate movement of the wheels, excessive play is actually a bad thing. Excessive play magnifies the feel of road imperfections and makes the steering less responsive to steering wheel input. Excessive play is potentially very damaging, as it causes wear on the connecting parts and tires. Naturally, any amount of play in a fixed part is problematic and is frequently a result of part failure. Remember, unwanted looseness in the suspension system can be extremely dangerous and should be corrected as soon as possible. In some cases, driving the vehicle should not be allowed until the repair is performed. Repair of excessive play typically means replacement of the loose part or parts. At the same time, because the parts are designed to pivot and twist, a small amount of play may be within specifications. Always refer to the manufacturer's service information for proper testing procedures and specifications so that you don't misdiagnose a fault. The upcoming skill drills are designed to work you through the common repair tasks, using common vehicles, but they don't replace the manufacturer's service information. Now let's look at some of the most common customer concerns related to the suspension system.

As discussed earlier, the steering and suspension systems work together to keep the wheels properly positioned. And because each of those systems can have worn parts that cause play, they can have the same or similar symptoms. So you should be familiar with both systems. There are many types of driver complaints relating to the steering and suspension system. Some of the most common are listed here along with their typical cause and the proper diagnostic method:

- **Vehicle wander:** This complaint means the vehicle is not driving in exactly the direction the driver is steering it. It tends to happen when the caster angle is off or there is looseness in the steering/suspension components. A front-end alignment machine is the best tool for diagnosing caster angle problems, and a prealignment inspection would find wear problems on joints, bushings, and components.
- **Drift:** Drift occurs when the driver is holding the steering wheel steady and straight, but the vehicle slowly begins to move either to the right or left. This condition requires correction from the driver to keep the vehicle straight. It generally results from loose or worn steering components, low tire pressure, insufficient caster, and incorrect toe settings. The front-end alignment machine is used to measure toe settings, and a pre-alignment inspection would find wear problems on joints, bushings, and components.
- **Pull:** Pull is felt when the driver feels the steering wheel wanting to go to one side. It can mean the air pressure in one or more of the front tires is low. A simple tire pressure gauge is used to check the tire pressures. Be sure to check the simple things before spending the customer's money on a four-wheel alignment. Pull can occur from a bad tire. Try switching the tires side to side. If the pull is now gone or has moved to the opposite side, then a tire is at fault. Pull can also happen when the camber or caster is out of adjustment from one side to the other by as little as half a degree. A vehicle will pull to the side with the most positive camber. If the caster is off by as little as half a degree from side to side, the vehicle will pull to the side with the most negative caster. A four-wheel alignment is required to verify the alignment angles. Pulls can also result from power steering problems. Power steering problems can be diagnosed by lifting the front tires off the ground and starting the engine while watching for steering wheel movement. If the vehicle pulls only while braking, looseness in steering/suspension components will have to be found and/or problems in the braking system may be diagnosed.
- **Hard steering:** When the act of steering is difficult for the driver, there may be a problem in the power steering system, too much positive caster, or binding in the steering system. A pressure gauge is used to check the power steering system and a front-end alignment machine to check the caster. Hard steering could also be caused by something as simple as under-inflated tires. A tire pressure

gauge would be used to diagnose this. Check for binding in the steering system components by disconnecting components to verify the location of the binding.

- **Bump steer:** Bump steer is when the vehicle turns itself upon hitting a bump. This is caused by incorrect angles in the steering linkage due to a component being bent or an incorrectly positioned steering gear or rack and pinion. The steering linkage must follow the same arc as the suspension. Start with a visual inspection of the steering parts, and if necessary, check the alignment.
- **Torque steer:** Torque steer is a pull that occurs during heavy acceleration. It can be the result of unequal axle lengths as designed, which cannot be repaired by the technician. Compare this vehicle to a known good vehicle to ensure it is a design issue. Abnormal torque steer can result from inner or outer tie-rod ends that are worn. Loose, worn, or broken engine or transmission mounts or control arm bushings can also create this issue. Installing a compliance bushing incorrectly with the void facing the wrong way can also create torque steer. Be sure to look for shifting components under load.
- **Steering return concerns:** This condition is commonly referred to as memory steer. Memory steer can result from binding conditions in any pivot points in the steering or suspension system, including the ball joints, tie-rods, strut bearings, idler arm, binding steering gear, or rack and pinion. Or it can result from unbalanced power steering assist. It can also be caused by insufficient positive caster on the front wheels. Raise the vehicle wheels, and turn the steering while observing components. Disconnecting components one by one may be necessary to find binding in swivel points such as ball joints, strut bearings, or tie-rod ends. If these methods do not uncover the issue, check the caster with an alignment machine.
- **Noises:** Noises can be tracked down through the use of a specialty tool called a chassis ear. This tool involves microphones that can be placed at various locations on the vehicle and a set of earphones worn by the technician. The microphones can be selected one by one by the technician to pinpoint the noise. Some of these tools can connect to a lab scope so that the vibrations/noises can be graphed on the screen, allowing the technician to measure the frequency, which can help identify the suspect component. *Operation of this tool during a road test should be performed by another person, and not the driver, because using the tool can be distracting to the driver.* Some noises can be identified and located while steering the vehicle when it is raised on a lift. Listen and feel for any noises and vibrations. Just be very careful around moving and hot components.

Diagnosing Suspension Noise

N41002, N41003

A noise that is heard coming from the front end whenever a vehicle hits a bump could indicate a problem with a component of the suspension system. Testing the vehicle while turning and hitting bumps may be necessary to pinpoint the source of the noise. Be sure you have complete information on the customer's concern, such as whether the vehicle only makes the noise when cold while going over bumps. In this case, test-driving over bumps while the vehicle is hot will probably not reveal the noise. Suspension system noise can be related to bushings that are worn or broken, ball joints that are worn, spring isolators that have broken or fallen out of place, loose bolts, faulty strut mount or bearing, bushing or joint that has lost its lubrication, or dirt that has gotten into a pivoting point. To diagnose this concern, you made need an assistant to drive the vehicle while you listen from different points inside the vehicle in order to pinpoint whether it is a front or rear suspension issue. The use of chassis ears can provide a quick solution to where the noise is originating (**FIGURE 41-1**). Using an assistant allows one person to focus on finding the problem while the other focuses on safe driving.

To diagnose a suspension noise that can be duplicated when not driving, bounce the vehicle up and down while listening for noise. If there are noises, confirm where the noise is coming from. Have an assistant bounce the vehicle while you check for the noise under the vehicle. You can listen by ear or with a stethoscope, or you can feel by hand for vibrations or clunks. Listen around sway bar bushings, control arm bushings, springs, shock absorbers,

FIGURE 41-1 Chassis ears can be used to help identify the location of suspension noises and vibrations.

component bolts, body panels, steering gear, the MacPherson strut mount, and ball joints. Inspect the component making the noise for damage, and replace if necessary. If the noise is found at a component bolt, check the bolt to see if it is loose, and if so, retighten to the manufacturer's specifications.

Body Sway

To diagnose body sway, first test-drive the vehicle to verify the customer's concern. Because you have to swerve from side to side to perform the test, drive the vehicle in a safe area with little to no traffic, such as in a large empty parking lot. While driving the vehicle, move the steering wheel back and forth to make the vehicle swerve slightly, this will shift the vehicle weight from side to side. Observe how much body sway occurs. It may be necessary to compare the sway of the vehicle to a like vehicle to ensure that the condition is not normal. If there is excessive sway, and the suspension is not electronically controlled, then check the vehicle's front and rear (if equipped) sway bar system. Inspect the sway bar bushings, brackets, and link bushings (**FIGURE 41-2**). Bushings should not be cracked and should have little give when pried. Replacement of faulty components is necessary.

If the vehicle is equipped with an electronically controlled suspension, it is possible that the system is not providing the active control needed to dampen body sway. In this case, the electronically controlled system has to be checked for DTCs and diagnosed according to the manufacturer's specified procedure, which is found in the service information.

Ride Height

The ride height of a vehicle can only be measured if it has matching tires that are properly inflated, and no additional weight in the vehicle. Once these issues are taken care of, the ride height can be measured as specified by the service information. Most ride height specifications require the measurement to be within half an inch side to side. If the measured ride height is greater or less than specified, the vehicle is not in correct alignment and may be causing or contributing to the customer's concern. Check for bent or sagging vehicle springs, bent spring mounting points, a leaking gas-pressurized shock absorber, a faulty electronically controlled suspension system, or a bent frame or axle. Use a measuring tape from a fixed point on the frame or body of the vehicle to the component in question to check for improper ride height. If the ride height issue is related to the coil or leaf spring, then replacement of the spring(s) may be necessary. If the ride height issue is on a vehicle equipped with a torsion bar system, then you may be able to adjust the torsion bar so that ride height is returned to specifications. Refer to the manufacturer's service information for proper adjustment of ride height. Remember that any time ride height is changed, a wheel alignment also must be performed.

To perform ride height diagnosis, follow the steps in **SKILL DRILL 41-1**.

FIGURE 41-2 If the concern is excessive body sway, check the sway bar, bushing, brackets, and link bushings.

Inspecting the Shock Absorbers

Shock absorbers and struts are located near each wheel and dampen body movement from bumps. Common reasons for testing shock absorbers are unusual tire wear, such as tires having a cupped appearance. Also, the driver may complain of a soft or bouncy ride. In some cases, a shock absorber can bind up, creating a very stiff ride. If a vehicle has adjustable shock absorbers, make sure the shock absorber adjustments are the same for the left- and right-hand sides. Some shock absorbers contain pressurized gas, which can leak out, causing uneven ride height and shock absorber performance issues. Many of today's vehicles are equipped with a strut-type suspension instead of conventional shock absorbers, but testing either type of system involves the same procedure, a bounce test. Basically, while the vehicle is stationary, push up and down on a strong point at each corner of the vehicle (not the fenders as they can be dented) several times, and watch how the vehicle responds after you release it. Typically, if you let go at the bottom, a good shock will allow the corner of the vehicle to rise and then settle back into position on most vehicles. On some with softer suspensions, it may allow the corner of the vehicle to rise, fall, and rise back into position. Any more oscillations than that indicate worn shock absorbers.

Pay particular attention to the top strut mounting during the bounce test. Place your hand on top of the mounting during the bounce test. Any noise or movement in the mounting could indicate the need to replace the strut mount. Also have someone turn the steering wheel from lock to lock while feeling and listening to the top strut mount. If you do feel movement or hear noise, report it to your supervisor. Visually inspect the shock absorber mounting points for security and corrosion, and note any wet-looking patches on the sides of the shock absorbers. A wet patch is a common indicator that the strut or shock absorber needs replacing because of a fluid leak. Slight dampness on the shock is typically normal, but drips on the shock are not normal and indicate leaking.

To check shock absorbers, follow the steps in **SKILL DRILL 41-2**.

SKILL DRILL 41-1 Performing Ride Height Diagnosis

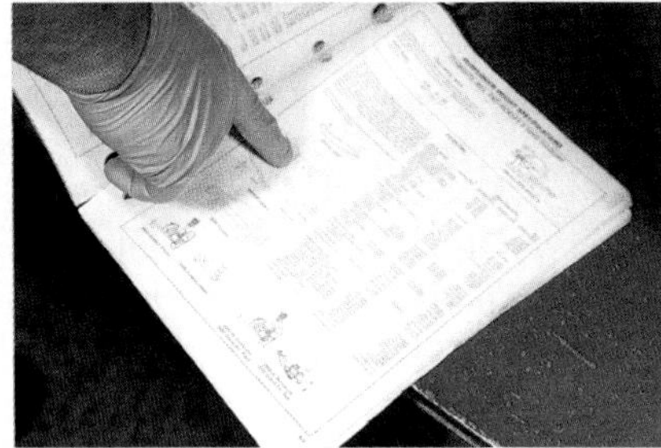

1. Refer to the manufacturer's service information for correct measurement points and specifications.

2. Check for properly sized, matching, and inflated tires. Correct any issues found.

3. Check the vehicle for any nonstandard loads in the trunk or luggage area. Remove them temporarily while measuring ride height.

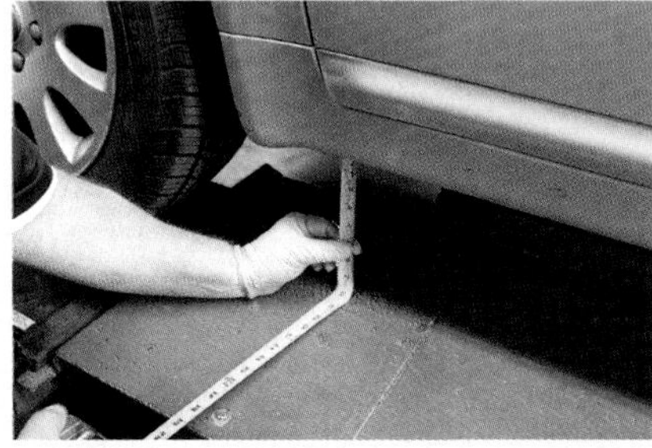

4. Measure from points specified, such as from frame to ground on all four corners of the vehicle, and compare measurements to specifications.

5. Inspect for bent components or a weak or broken spring if any measurements are not correct. If working with a torsion bar suspension, you may have to adjust ride height to correct the condition.

SKILL DRILL 41-2 Checking Shock Absorbers

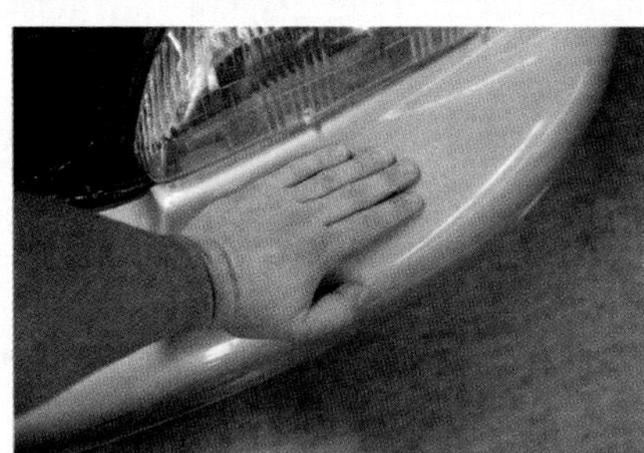

1. Place your weight on a bumper, and begin to bounce the vehicle until it reaches its maximum amount of travel produced by your weight. Stop bouncing at the bottom of the bounce. If the vehicle rebounds and compresses more than twice, replace the shock absorbers. If the shock absorbers are performing well, the vehicle will rebound once or one and a half times, then return to its original position.

2. Pay particular attention to the top strut mounting during the bounce test. Place your hand on top of the mounting during the bounce test. Any noise or looseness in the mounting could indicate the need to replace the mount. While you are driving, the same test can be performed by stopping the vehicle suddenly from a very low speed. If the vehicle bounces up and down when coming to rest, you need to replace the shock absorbers.

3. Visually inspect the shock absorber mounting points for security and corrosion, and note any wet-looking patches on the sides of the shock absorbers. Slight dampness on the shock is typically normal, but a drip on the shock is not normal and indicates leaking.

SAFETY TIP

To prevent personal injury, do not puncture or incinerate gas-charged shock absorbers.

Unloading a Suspension to Measure Ball Joint Play

Play in the suspension system can be damaging to other components or can be a major safety hazard on the road. Testing for play requires the proper technique for the results to be accurate. For play to be measured, the joint must be unloaded. This means the joint cannot be under compression or tension forces. In the case of suspension ball joints, the joint cannot be supporting the weight of the vehicle or the force from the vehicle spring when measuring the play.

The method of unloading the ball joints depends on the layout of the suspension. On a type 1 suspension where the coil spring or torsion bar is pushing against the lower control arm, a floor jack must be placed under the lower control arm and the wheel raised off the ground. The weight of the vehicle is thus supported through the spring to the control arm and then the jack, leaving the ball joint only supporting the wheel, tire, spindle, and upper control arm. In this case, a pry bar can then be used to pry the wheel up and down while measuring the ball joint play. The upper ball joint can be tested by pushing in and out on the top of the tire and measuring any play in the joint.

Some manufacturers specify that their type 1 suspensions be tested with the suspension system left hanging. The vehicle would be supported by the frame in this case. Manufacturers may also specify that only hand pressure be applied when checking for play in the joints. Check the service information before testing a particular vehicle.

On a type 2 suspension where the spring is on the top control arm, it is best to fit a wooden block between the upper control arm and the frame, so the control arm and spring are held in a position as close to normal ride height as possible. This method allows measurement of the play in the joint where maximum wear occurs. The top of the tire can again be pushed in and out, play in the joint can be measured.

In a MacPherson strut suspension with only a lower control arm, testing is performed by raising the vehicle by the frame and allowing the suspension to hang free. This approach tests the point in a position it does not normally operate in, but if there is play, it will likely still be evident. Because it is a follower joint, most manufacturers say if it has any noticeable play, the joint must be replaced.

Be sure to test the ball joint correctly. Refer to the manufacturer's instructions and specifications on how to test these joints, as they are not all tested the same way and may require different tools.

To measure play in the suspension system, follow the steps in **SKILL DRILL 41-3.**

To measure play in the suspension system, follow the steps in **SKILL DRILL 41-4.**

Testing Components of Electronically Controlled Suspension Systems Using a Scan Tool

Electronically controlled suspension systems are tested whenever a driver complains of a suspension problem or the warning light is on. Because it is computerized, diagnosis also requires the use of a scan tool. A diagnostic scan tool is the only way to see inside this type of suspension system and determine what problems exist. The scan tool can retrieve any diagnostic trouble codes (DTCs), as well as provide access to sensor data. In some cases, the scan tool can command the PCM to operate various components in the system so that you can check their operation and any circuit issues. Every manufacturer uses its own system, so it is best to use the service information from that manufacturer to diagnose any related customer concerns.

SKILL DRILL 41-3 Measuring Play in the Suspension System: Loaded Lower Ball Joint

1. Place a dial indicator on the lower control arm and vertically against the steering knuckle.

2. Place a pry bar under the tire and pry it upward, watching the dial indicator reading as you pry and release. Record the total amount of movement in the joint and compare to the manufacturer's specifications.

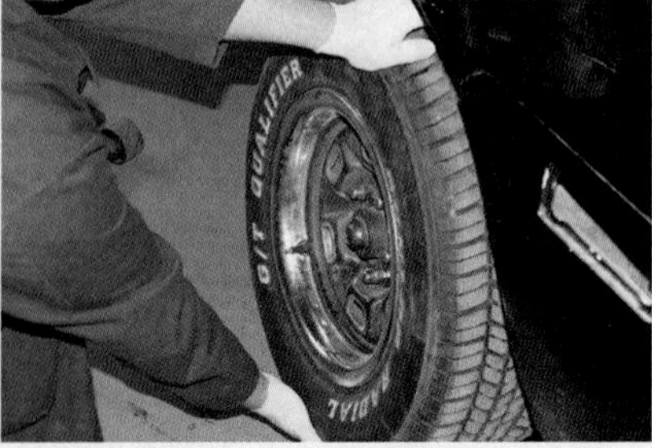

3. Rock the tire in and out at the top, watching for any play in the upper ball joint, and compare to specifications.

SKILL DRILL 41-4 Measuring Play in the Suspension System: Loaded Upper Ball Joint

1. Obtain a properly sized block of wood to fit between the upper control arm and the frame. Place the block of wood between the upper control arm and frame, so it is secure.

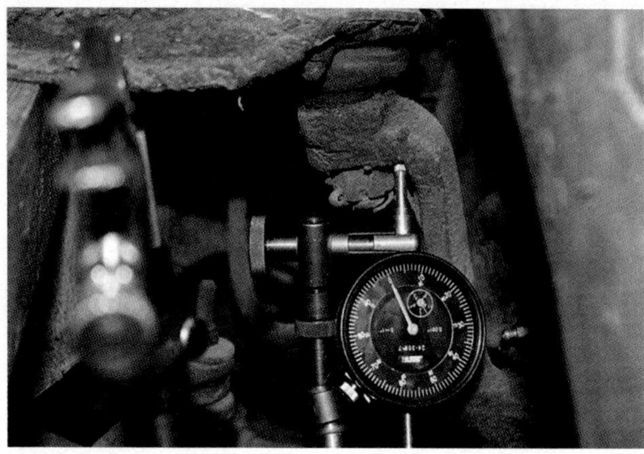

2. Place a dial indicator on the upper control arm and vertically against the steering knuckle.

3. Place a pry bar under the tire and pry it upward, watching the dial indicator reading as you pry and release. Record the total amount of movement in the joint, and compare to specifications.

4. Rock the bottom of the tire in and out, watching for any play in the lower ball joint, and compare to specifications.

▶ Maintenance and Repair

S41001

Removing, Inspecting, and Installing Stabilizer Components

N41004

The stabilizer components (sway bar) help prevent body roll when cornering. The stabilizer bar itself rarely gives any trouble, but the rubber bushings on the bar and links wear out. This usually results in increased body roll as well as a clunking noise in the suspension. The stabilizer components should be checked whenever a vehicle is brought into the shop because of handling concerns or suspension-related noises. Before beginning the procedure, research the manufacturer's procedure and specifications for removing and inspecting the stabilizer components.

Note: We use the same vehicle to perform most of the SLA-related tasks. Also, we address each task as a sequence, as if you were doing the entire series of tasks, one after the other. So the disassembly of each component assumes you have removed the components you have already been told to remove. And reassembly will be held off until all of the related tasks are completed.

To remove and inspect the stabilizer bar bushings and mount brackets, follow the steps in **SKILL DRILL 41-5**.

To remove and inspect the sway bar end links, follow the steps in **SKILL DRILL 41-6**.

Removing and Replacing Shock Absorbers

N41005

Worn shock absorbers cause the vehicle to ride poorly, especially on rough roads. When the tires encounter a bump in the

SKILL DRILL 41-5 Removing and Inspecting the Stabilizer Bar Bushings and Mount Brackets

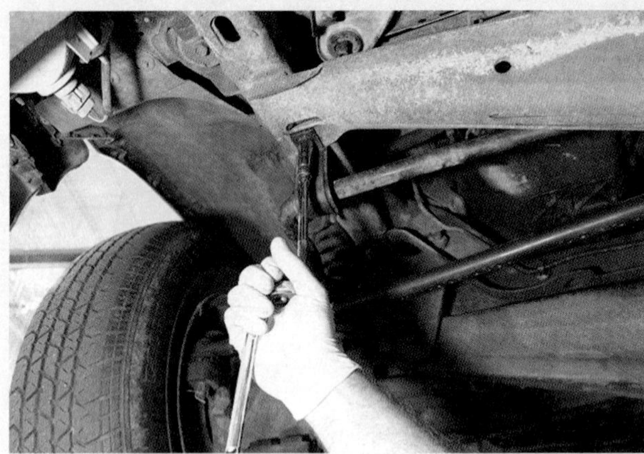

1. Raise the vehicle on a hoist, or use a jack and place safety stands under the frame. Remove the bolts holding the stabilizer bar bushing mount brackets, and inspect the brackets for cracks.

2. Remove the bushings by hand. Inspect the rubber in the bushings for cracks, brittleness, softness, or wear.

SKILL DRILL 41-6 Removing and Inspecting Sway Bar End Links

1. Raise the vehicle on a hoist, or use a jack and place safety stands under the frame. Remove the nut holding the stabilizer bar link.

2. Remove the link by hand.

3. Inspect the rubber link grommets for cracks, softness, brittleness, or wear. Repeat with the other link.

road, a faulty shock absorber cannot dampen the spring oscillations to promote a smoother ride. The suspension continues to rebound and bounce. This action is then transferred to the vehicle frame and ultimately to the driver and passengers. Loose or damaged shock absorbers can be distinguished by abnormal, and in some cases, loud and unusual noises.

▶ TECHNICIAN TIP

Always replace shock absorbers in pairs so the suspension has the same characteristics for the left and right sides. Shock absorbers are rated for "bump" or "jounce," the rate at which they compress, and "rebound," that rate at which they expand.

Visually inspect the shock absorbers for any signs of deterioration, such as oil leaking from the shock absorber shaft seal or damaged body. A slight oil film on the shock is considered normal. The mounting bushings must also be carefully inspected for splits and deteriorated or missing rubber mountings. The mounting supports must be checked for good security and tightness. Many shock absorbers come from the factory pressurized with gas to reduce aeration of the fluid when operating. This pressure tends to expand the shock absorber, making it difficult to install. For this reason, pressurized shocks come compressed with a band holding them together. It is usually easiest to install one end or, in some cases, both ends of the shock before cutting the band.

▶ TECHNICIAN TIP

Shock absorbers use rubber bushings to isolate them from the vehicle body. Always replace these bushings when replacing the shock absorbers.

On some suspension systems (typically non-independent suspensions), the shocks provide the limit for full extension. This means the shocks may be holding the axle up when the shock is fully extended. In this case, removing the shock could cause the axle to slip, pinching fingers or causing the vehicle to shift on the hoist. It is always good practice to place stands under the axle to support it while the shocks are being removed and installed. If the top of the shock is held in by a stud and nut, you may have to use a wrench to hold the top of the stud while you unthread the nut.

If a shock absorber must be replaced, it is industry practice to replace them as pairs, thus ensuring the ride equilibrium of the vehicle. To replace a shock absorber, follow the steps in **SKILL DRILL 41-7**.

Removing, Inspecting, and Installing SLA Suspension Coil Springs and Spring Insulators

N41006

Coil springs absorb the road force by twisting, which compresses them. Whenever a driver complains of the way the vehicle sits or an issue related to the ride quality, the coil springs should be inspected for wear and the vehicle's ride height measured.

Coil springs used in vehicles store very large amounts of energy; thus, extreme caution is required when removing and installing them. Always use an adequately rated spring compressor that is in good working condition when working on a spring. Also, never stand in the direction the spring can fly, should something slip, which is generally in a position near where the tire sits, and beyond. Because the spring can be mounted on either the lower control arm or the upper control arm, you will have to unload it in the same manner as the same type of ball joint. Failure to follow service information can result in serious injury or death. In this Skill Drill, we work with the spring located on the lower control arm.

To remove and inspect SLA suspension system coil springs and spring insulators, follow the steps in **SKILL DRILL 41-8**.

Removing, Inspecting, and Installing the Steering Knuckle Assembly

N41007

The steering knuckle serves as the pivot point for the wheels. Because they rotate on ball joints or tie-rod ends, they are not subject to much wear, but they can be damaged in an accident, requiring replacement. It is more likely that you will have to remove one in order to access other pieces of the steering assembly. This Skill Drill assumes a steering knuckle removal on the vehicle in the previous skill drill. Remember, coil springs are under extreme pressure! Follow the service information to correctly remove any steering knuckle.

To remove and inspect a steering knuckle assembly, follow the steps in **SKILL DRILL 41-9**.

Removing and Inspecting Upper and Lower Control Arms and Components

N41008

Control arms themselves do not generally wear out, but the control arm bushings and ball joints do, so you will likely have to remove them to service those components. However, some manufacturers do not sell bushings and ball joints separately, so you may have to purchase the entire control arm and replace it as a unit. Also, control arms can become bent or damaged as the result of a collision or driving through a deep pothole, so control arms do need to be inspected for damage. Lastly, some customers request that the height of their vehicles be lifted or lowered, which can involve modifying suspension components such as springs and control arms with new upgraded ones. Obviously this changes the ride height, alignment, and handling of the vehicle. Only very experienced and highly trained technicians should diagnose and repair these vehicles. The vehicle used in this Skill Drill has a coil spring

SKILL DRILL 41-7 Replacing a Shock Absorber

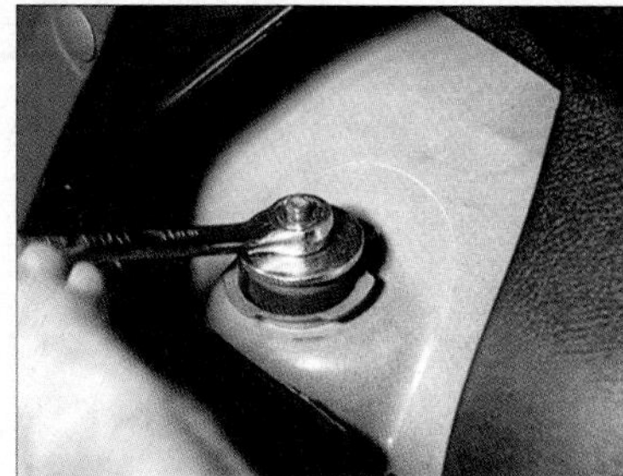

1. Raise the vehicle on a lift, and support the axle/control arm with a jack stand. With a socket and box-end wrench, remove the upper bolts holding the shocks in place.

2. Remove the lower bolts holding the shock in place.

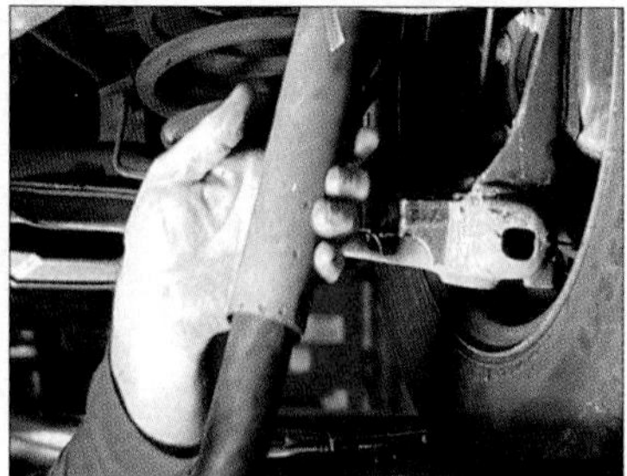

3. Pull the shock out by hand. Replace with new shock. Repeat on the other side.

SKILL DRILL 41-8 Removing and Inspecting SLA Suspension System Coil Springs and Spring Insulators

1. Following the previous Skill Drills, remove the shock absorber and the stabilizer bar end links. Place a spring compressor on the inside or outside of the coil spring to keep the pressure of the spring contained.

2. Place a jacking device under the lower control arm that is being dismantled; maintain pressure on the lower control arm.

3. Remove the cotter pin, and loosen the nut from the upper ball joint.

4. Using a ball joint separator or pickle fork, break loose the ball joint stud taper from the steering knuckle, and remove the nut.

5. Slowly and carefully lower the hydraulic jack supporting the lower control arm, to allow the coil spring to be removed. (Extreme caution must be used here so the spring compressor does not slip and injure you.)

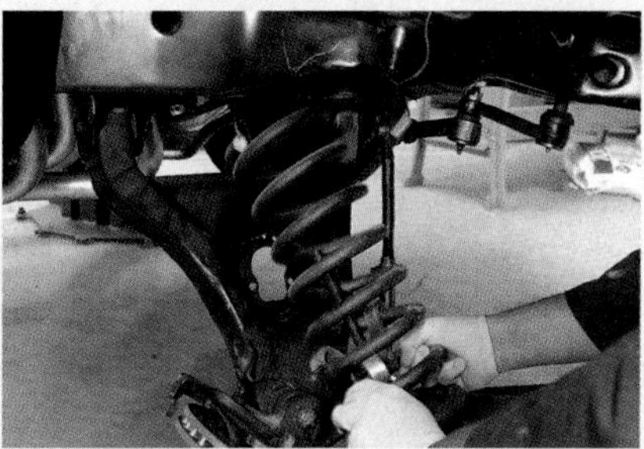

6. Release the spring compressor slowly. Clean and inspect the ball joint stud hole in the steering knuckle boss.

SKILL DRILL 41-9 Removing and Inspecting a Steering Knuckle Assembly

1. Disconnect the lower ball joint, and remove the spring as in the previous Skill Drills. Disconnect the upper ball joint in the same manner as the lower ball joint.

2. Remove the steering knuckle from the ball joint stud. Inspect the tapered portion of the stud for any bright, worn metal, which would indicate a worn steering knuckle boss.

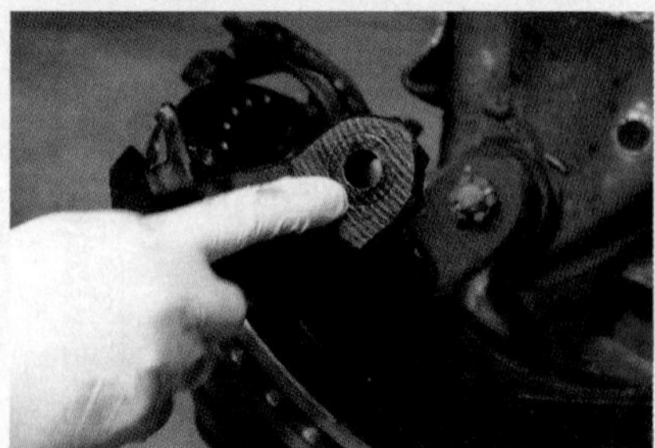

3. Inspect the steering knuckle for any signs of wear or visible signs of bending or twisting.

on the lower control arm. Not all vehicles are the same. Refer to the manufacturer's service information to correctly service the vehicle.

To remove, inspect, and install upper and lower control arms and components, follow the steps in **SKILL DRILL 41-10**.

> **TECHNICIAN TIP**
>
> Remember that modifying a vehicle by raising or lowering the vehicle may be illegal in some states and will affect alignment angles.

Removing, Inspecting, and Installing Upper and Lower Ball Joints

N41009

The ball joint connects the upper and lower control arms to the steering knuckle. Over time, these parts wear and must be replaced. Removal of ball joints varies from vehicle to vehicle. Refer to the manufacturer's service information for the correct procedures. This Skill Drill uses the coil-spring suspension with two control arms and is the more dangerous type that a technician services.

SKILL DRILL 41-10 Removing, Inspecting, and Installing Upper and Lower Control Arms and Components

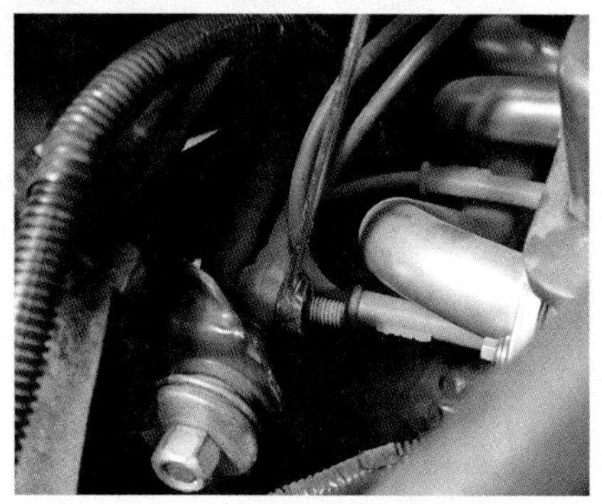

1. Remove the coil spring and steering knuckle by following the steps in previous Skill Drills. Loosen the bolts holding the upper control arm in place.

2. Remove the lower control arm by removing the bolts holding the control arm in place. As you dismantle the system components, inspect the parts being removed to determine whether you need to order replacement parts.

3. Inspect the upper and lower control arm for wear or damage. If the control arm bushings show signs of excessive wear or deterioration of the bushing material, they must be replaced. Replace the worn bushings in the upper and/or lower control arm with a hydraulic press or with an air chisel.

4. *Press method:* Place the control arm in a hydraulic press supported so the bushing is level to the press head. Using the correct-sized adapter, press the bushing out.

5. Using the press, install the new bushing into the control arm. Be careful to start the bushing in straight, or you could bend the control arm. Press it in until the flange bottoms out.

6. *Air chisel method:* Install the control arm in a vise. Use an air chisel to work the bushing out. Be careful not to gouge the mating surface of the control arm.

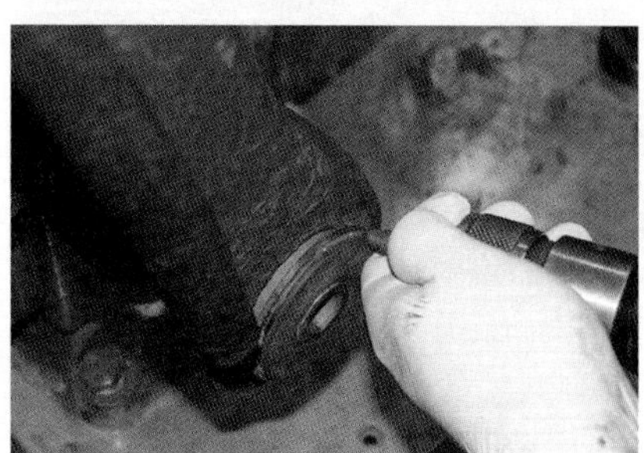

7. Using the air hammer, drive the new bushing into place, making sure it is fully seated.

To remove, inspect, and install upper and lower ball joints, follow the steps in **SKILL DRILL 41-11**.

Lubricating Steering and Suspension Systems

S41002

To function properly, the suspension and steering system must be lubricated. On most modern vehicles, the joints are sealed and lubricated for life, so there are may be no grease fittings present. However, some vehicles do have grease fittings installed, either from the factory or on aftermarket parts. If grease fittings are present, then lubrication must be performed. Lubrication keeps the parts from rubbing and wearing on one another, which extends their life. Add enough grease to see the seal or rubber boot rise slightly. Under no circumstances should you overfill a lubricated joint with grease. Doing so can rupture the seal or rubber boot or bellows.

SKILL DRILL 41-11 Removing, Inspecting, and Installing Upper and Lower Ball Joints

1. Remove the control arms as in the previous Skill Drills. Ball joints are held into the control arm either by a press fit or by rivets or bolts.

Press-fit style: Support the control arm with the correct adapter so there is room for the ball joint to be pressed out of it.

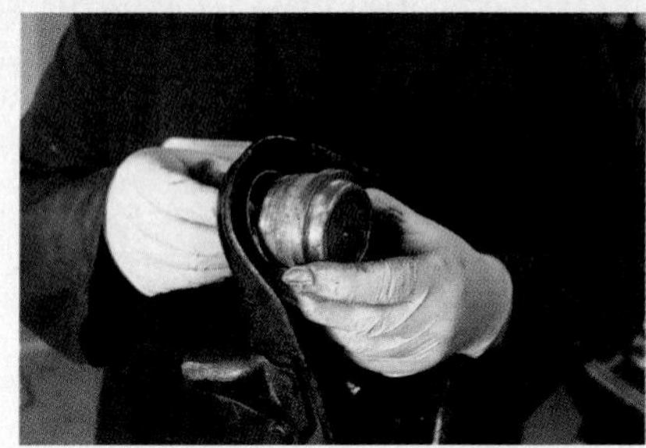

2. Press the joint out of the control arm.

3. Inspect the bore in the control arm to make sure it isn't cracked or damaged.

4. Turn the control arm over and support it, so the ball joint has room to be installed.

5. Press the ball joint into the control arm until it bottoms out.

6. *Riveted style:* Mount the control arm in a vise with the rivet heads facing up or outward.

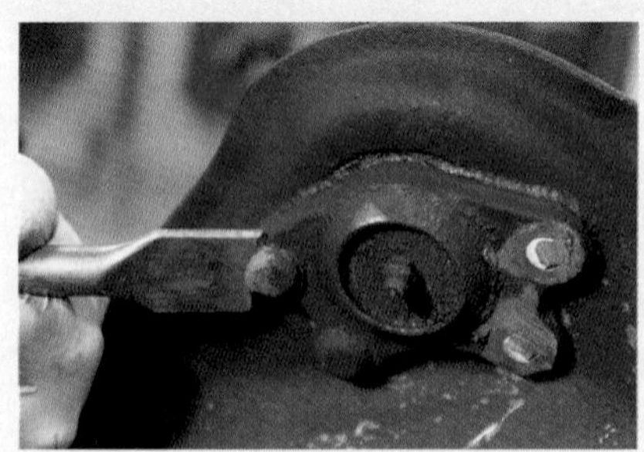

7. Use an air chisel to chisel off the rivet heads, or use a drill to drill approximately three-quarters the diameter of the rivet head to carefully drill the head of the rivet off. Use a hammer and a punch to tap the rest of the rivet out of the ball joint and control arm.

8. Inspect the holes, and remove any burrs with a file. Install the new ball joint with nuts and bolts, and tighten to the proper torque.

9. Install the lower control arm in position, and torque the bolts to specification.

SKILL DRILL 41-11 Removing, Inspecting, and Installing Upper and Lower Ball Joints (Continued)

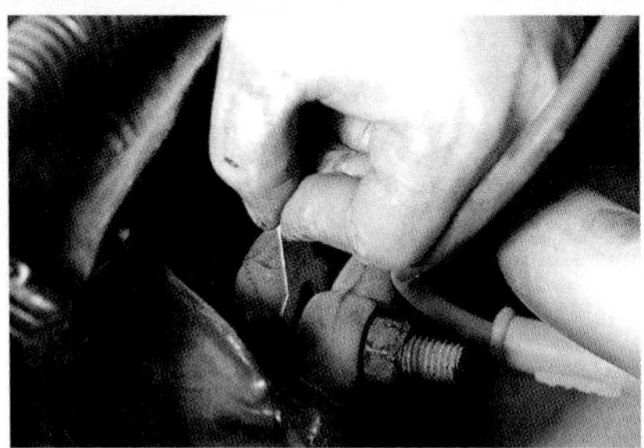

10. Install the upper control arm in position, and screw in the bolts finger tight. Insert shim packs, if equipped. Torque the bolts to specifications.

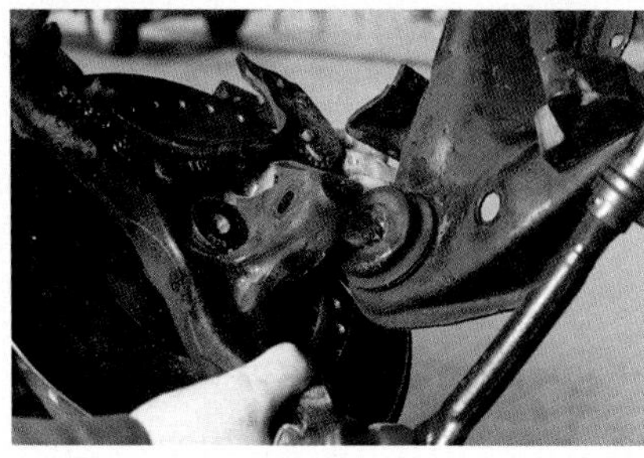

11. Place the steering knuckle in place on the lower ball joint stud. Finger-tighten the ball joint nut.

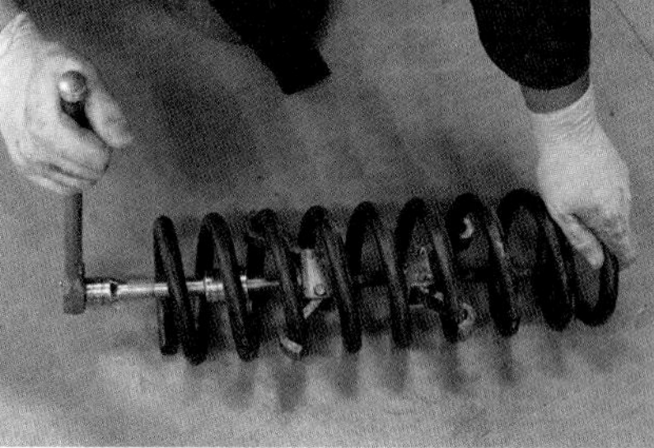

12. Install the spring compressor on the spring so that it can be installed in the vehicle between the lower control arm and the frame.

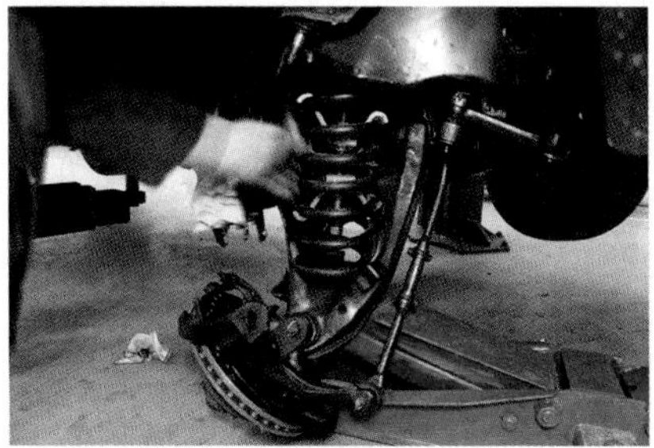

13. Position the spring on the lower control arm, and use a jack to hold it in place.

14. Continue to raise it while guiding the steering knuckle over the upper ball joint stud. Install the nut on the upper ball joint stud finger tight.

15. Use a cotter pin tool to turn the ball joint stud to align the cotter pin hole lengthwise to the vehicle so the cotter pin will be easy to install.

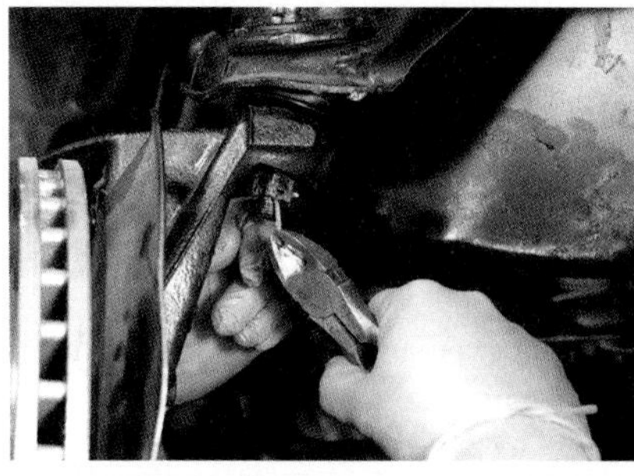

16. Torque both ball joints to specifications, and install the cotter pins.

17. Remove the spring compressor.

18. Install the new washers and bushings onto the new shock, if not already installed. Place the shock in the proper position by hand, and install the other bushings, if needed.

19. Install and properly torque the bolts or nuts to hold the shock in place using a torque wrench. Install the tie-rod end into the steering knuckle, torque the nut, and secure with a cotter pin.

20. Install the new stabilizer bar bushings, placing the bushing around the bar.

21. Install bolts in the mounting bracket, and tighten to specifications.

SKILL DRILL 41-11 Removing, Inspecting, and Installing Upper and Lower Ball Joints (Continued)

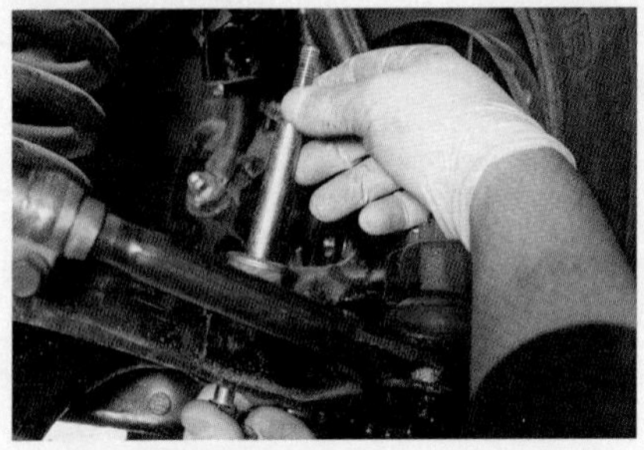

22. Install the stabilizer link to the control arm by assembling it onto the stabilizer bar and control arms with the spacer, washers, and rubber grommets installed in their proper places.

23. Install the nut on the threaded end of the link, then tighten to the manufacturer's specifications.

24. Reinstall the rotor assembly if removed.

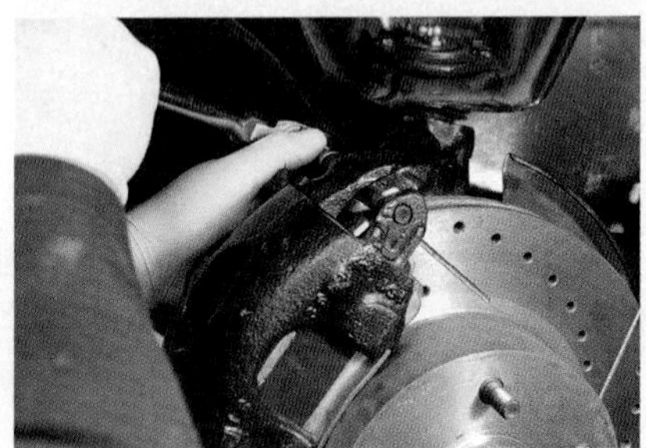

25. Reinstall the brake caliper by placing the caliper over the rotor and tightening to the proper torque.

26. Jack up the lower control arm so the spring is compressed to its normal height, and properly torque the control arm bolts to specifications.

27. Install the wheel and snugly tighten the lug nuts. Lower the vehicle, and torque the wheel nuts to specifications.

To lubricate a suspension system and a steering system, follow the steps in **SKILL DRILL 41-12**.

> **▶ TECHNICIAN TIP**
>
> Clean lubricating equipment very carefully. If you don't thoroughly clean the fitting or nozzle before pumping the grease into the fitting, dirt could be forced into the component. Any dirt entering a component will cause premature failure.

Removing, Inspecting, and Installing the Strut Cartridge or Assembly

N41010

When a vehicle arrives in the shop with chopped tire wear or if the driver complains of ride comfort problems, the parts of the strut assembly should be checked. Most vehicles have a one-piece strut that bolts into the knuckle on the lower end and has the spring seat on the upper end. Some older vehicles have a strut housing that has a replaceable internal strut cartridge. In a strut with an internal strut cartridge, some manufacturers fill the housing with a lightweight oil to allow heat from dampening to be transferred to the outer housing more efficiently. Follow the manufacturer's instructions regarding oil type and the amount to use when replacing inserts. Some parts suppliers are supplying full struts preassembled with the spring and upper strut bearing already installed. This means that the strut has all new parts and is quicker to install, so the customer can get back on the road faster. Be sure to follow manufacturer's service information to replace struts on the vehicle you are working on. This Skill Drill uses a vehicle with a replaceable one-piece strut.

To remove the strut assembly, follow the steps in **SKILL DRILL 41-13**.

Removing, Inspecting, and Installing a Strut Coil Spring, Insulators, and the Upper Strut Bearing Mount

To remove, inspect, and install a strut coil spring, insulators, and the upper strut bearing mount, follow the steps in **SKILL DRILL 41-14**.

> **▶ TECHNICIAN TIP**
>
> If you have to remove the shock absorbers, make sure the vehicle suspension is supported by suitable means. Strut assemblies contain a coil spring that is held under compression. Always use a spring compressor to release the spring pressure before disassembly.

SKILL DRILL 41-12 Lubricating a Suspension System and a Steering System

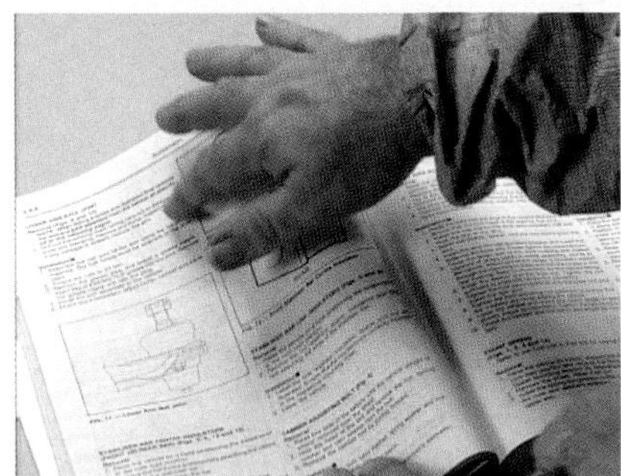

1. Determine the location of lubricating points. Check the shop service manual to determine where the grease points are and the type of grease required. Also look to see if any aftermarket grease fittings are installed. If so, grease them.

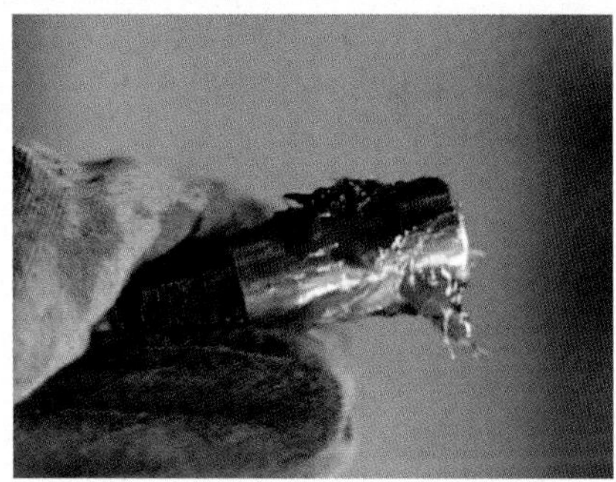

2. Clean each of the lubrication fittings and the grease gun nozzle by wiping all of them with a clean rag. You may need to remove a component's plugs and temporarily install a lubrication fitting. After the component has been lubricated, reinstall the original plug.

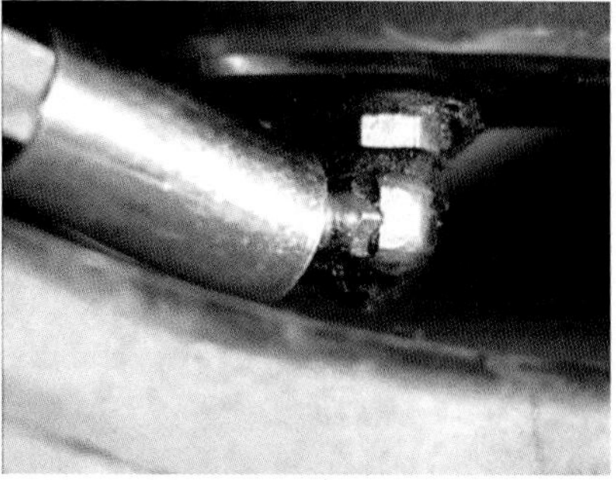

3. Push the grease gun nozzle fully over the fitting. It should snap into place. Add enough grease to see the seal or rubber boot rise slightly. Do not overfill a lubricated joint with grease.

4. If the fitting is clean and will not take grease, remove the grease zerk, and check for blockage. If found, the fitting must be replaced with a new fitting of the same size and angle, and the joint relubricated.

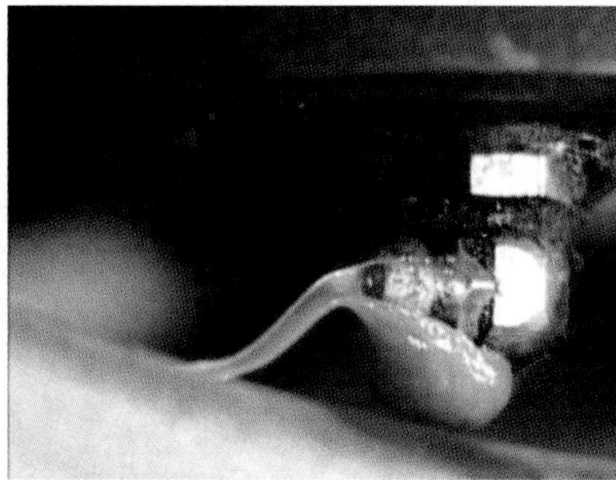

5. Remove the nozzle from the fitting and wipe away any excess grease from it. Repeat the procedure until all the appropriate joints have been lubricated.

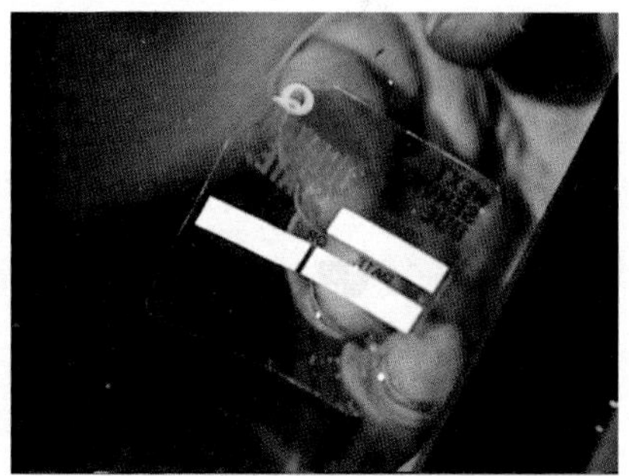

6. After you have completed lubricating all the appropriate joints and cleaned off any excess grease, attach a static cling sticker to the windshield, or reset the maintenance reminder system. Lower the vehicle and remove it from the lifting device.

SKILL DRILL 41-13 Removing the Strut Assembly

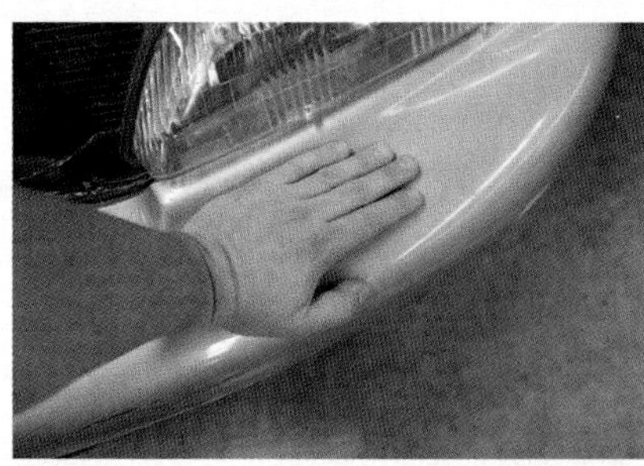

1. Using your hand, push down on one corner of the front bumper continuously until the front of the vehicle is bouncing. When at the bottom, remove your hand and stop pushing. In most vehicles, the corner of the vehicle will rise and the settle into place. A vehicle with softer suspension may rise, fall, and rise to its settling point. If the vehicle bounces more than that, the strut needs to be replaced.

2. Loosen the lug nuts one turn if removing by hand.

3. Safely raise and support the vehicle on a hoist or on jack stands.

SKILL DRILL 41-13 Removing the Strut Assembly (Continued)

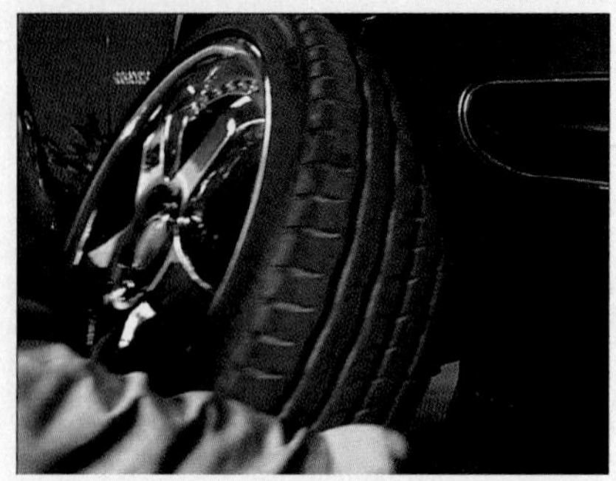

4. Remove the lug nuts and then take off the wheels.

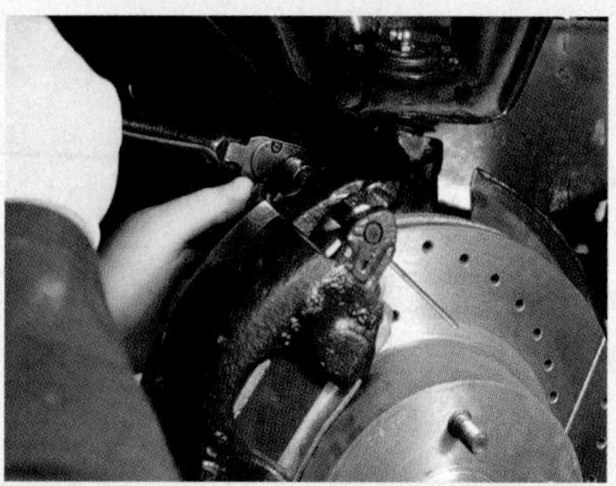

5. Remove the disc brake caliper, and wire it so it does not hang by its brake hose.

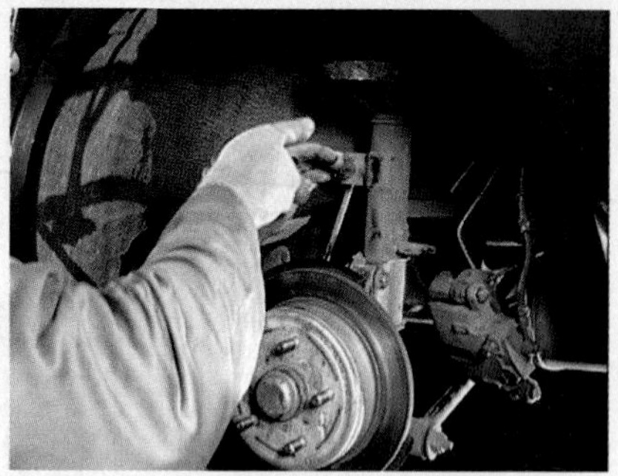

6. Remove the bolts holding the top of the strut to the strut tower (not the center strut bolt).

SKILL DRILL 41-14 Removing, Inspecting, and Installing a Strut Coil Spring, Insulators, and the Upper Strut Bearing Mount

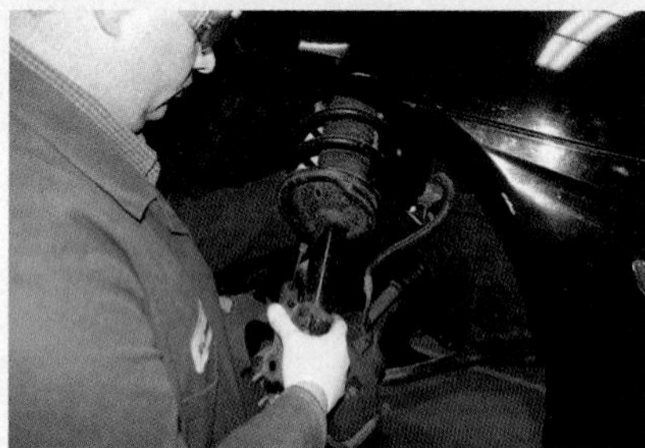

1. Remove the strut from the vehicle, following the previous Skill Drill.

2. Place the strut into the strut compressor. Adjust the lower arms of the strut compressor so the spring is supported properly.

3. Properly place the upper arms of the strut compressor onto the spring.

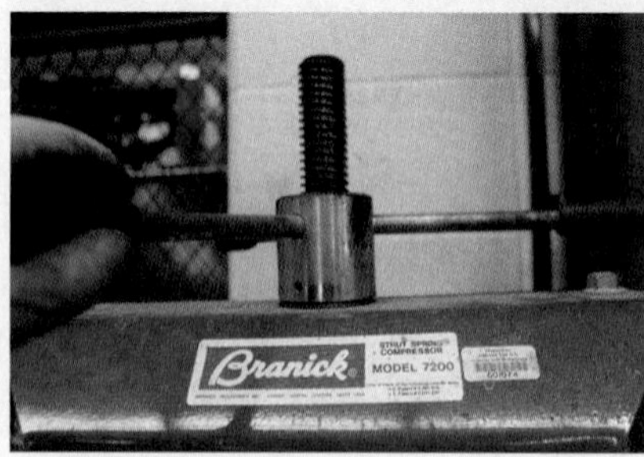

4. Slowly turn the handle of the strut compressor, and compress the spring. Ensure that the strut compressor arms are properly placed and are not sliding.

5. Once the spring is compressed enough that the strut is free from spring pressure, loosen and remove the center strut nut.

6. Remove the old strut from the spring.

SKILL DRILL 41-14 Removing, Inspecting, and Installing a Strut Coil Spring, Insulators, and the Upper Strut Bearing Mount (Continued)

7. Remove from the old strut any strut bellows, bumpers, spring seats that need to be transferred to the new strut. Install them on the new strut.

8. Install the new strut into the spring. Reinstall the upper strut mount in the correct position.

9. Install the upper washer and nut. Torque to the manufacturer's specification. Slowly release the spring compressor, and reinstall the strut into the vehicle.

Removing, Inspecting, and Replacing Leaf Springs

N41011

The leaf springs of a vehicle have to be inspected whenever the ride height does not match the manufacturer's specifications. If they are found to be sagging, bent, or broken, they must be replaced. Removal and inspection is also necessary if noise is found to be coming from the leaf spring or bushings in the leaf spring. Follow the manufacturer's service information to remove and replace a leaf spring. If new leaf spring bushings are needed, the use of proper press tools will ensure proper bushing replacement. Remember that when the leaf springs are removed, the axle will not be supported, and the use of screw jacks to support the axle during removal will be necessary.

To remove, inspect, and install leaf springs, follow the steps in **SKILL DRILL 41-15**.

Removing, Inspecting, and Installing Strut Rods and Bushings

N41012, N41013

The strut rods and bushings hold the control arm and therefore the wheel in position longitudinally. The strut rod bushings wear and degrade over time, requiring replacement whenever they are loose. On some vehicles, the strut rod is the adjustment point for caster. An alignment may have to be performed after performing this task. Refer to the manufacturer's service information for the vehicle you are working on.

To remove, inspect, and install strut rods and bushings, follow the steps in **SKILL DRILL 41-16**.

SKILL DRILL 41-15 Removing, Inspecting, and Installing Leaf Springs

1. Raise the vehicle on a lift and support the rear axle with tall screw jack stands. Check to see if any of the leaves are cracked or broken and that the noise deadening inserts are positioned correctly between the leaves.

2. Test the security of the spring center bolt, and make sure the U-bolts are tight.

3. Check the condition of the bushings or mountings and the spring shackles by placing a lever between the frame and the eye of the spring and levering against the spring.

SKILL DRILL 41-15 Removing, Inspecting, and Installing Leaf Springs (Continued)

4. Remove the bolts holding the leaf spring to the shackle and hanger.

5. Have an assistant help hold the spring while removing the U-bolts, holding the leaf spring to the axle.

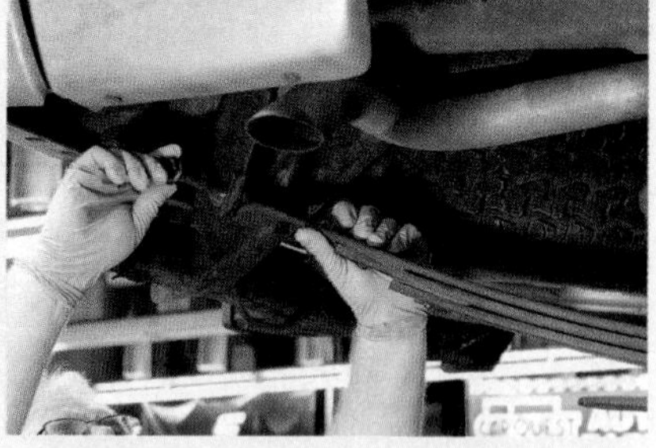

6. To install a new leaf spring, hang the new leaf spring into position by inserting bolts into the hanger and shackle. Use new fasteners whenever specified by service information.

7. Reinstall U-bolts to the axle, and snug them down a little at a time. Tighten all bolts to the manufacturer's specifications.

SKILL DRILL 41-16 Removing, Inspecting, and Installing Strut Rods and Bushings

1. Raise the vehicle on the hoist and inspect the strut rods and bushings. Pry the strut rod side to side and front to back to check for looseness.

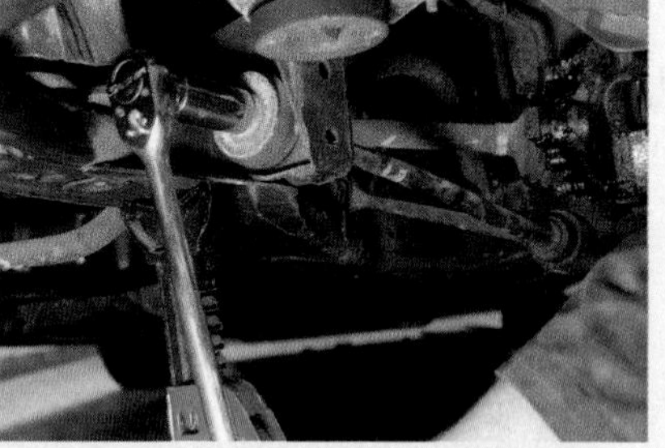

2. Remove the bolts and nuts holding the strut rod to the lower control arm and frame. Remove the rod by hand. The bushing may fall out, or need to be pried out, when the strut rod is removed.

3. To install new strut rod bushings, place half of the bushing onto the strut rod.

4. Slide the rod through the hole in the frame.

5. Place the other half of the bushing onto the rod sticking out of the frame, and start the nut.

Removing, Inspecting, Installing, and Adjusting Torsion Bar Suspension

N41014

The torsion bar provides the spring action in a torsion bar front-end suspension system. In many cases, it is adjustable to allow for corrections in ride height. It has a large bolt in the torque arm that, when adjusted, changes the amount of pressure it places on the control arm to raise or lower the suspension. The torsion bar should be checked whenever the driver complains of suspension problems. Torsion bars are typically stamped or marked with a left or right and must be installed in the proper side. If it has marks for left or right, then it is preset for twisting in that specific direction and will not have the same spring force if twisted the opposite way. If the torsion bars are not marked, then be sure to place a paint mark on them to identify left and right; they have been subjected to torsion in one direction for the life of the vehicle, and installing them backward can result in breakage or incorrect spring rate. The torsion bar may be found on either the lower or upper control arm on certain vehicles. The following Skill Drill will not be the same for all vehicles. Follow service information for proper operation of this task.

To remove, inspect, install, and adjust the torsion bar, follow the steps in **SKILL DRILL 41-17**.

SKILL DRILL 41-17 Removing, Inspecting, Installing, and Adjusting the Torsion Bar

1. Loosen the wheel studs. Safely raise and support the vehicle on a hoist or on jack stands. Remove the wheels.

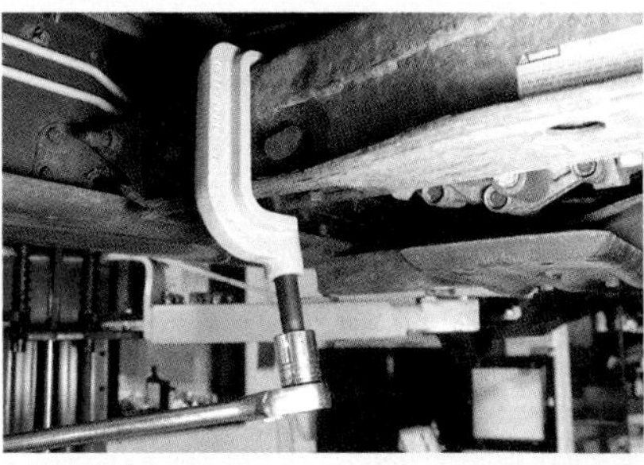

2. Install the correct torsion bar compression tool, and compress the torsion bar.

3. To aid in reinstallation, paint matching marks on the torsion bar spring and its corresponding arm. (Under *no* circumstances should you scribe these marks on the torsion bar and corresponding arm; doing so can seriously damage the bar in the long term, leading to catastrophic failure.)

4. If the torsion bar system has an anchor bolt protrusion, measure the distance from the washer to the end of the stud with a tape measure, and record the value.

5. Remove the adjustment bolt and anchor plate. Before removing these nuts or bolts, ensure that the torsion bar force has been released.

6. Slowly loosen and remove the torsion bar compression tool.

SKILL DRILL 41-17 Removing, Inspecting, Installing, and Adjusting the Torsion Bar (Continued)

7. Remove all the remaining securing bolts using a socket or wrench as required, and remove the torsion bar spring. Release the pressure on the supporting jack, and slowly lower the lower control arm.

8. Ensure that the torsion bar springs were fitted in the correct location by identifying the location marks on the end of the spring ("R" for right side; "L" for left side).

9. Inspect the torsion bar spring for any signs of fretting or cracking, twisting, bends, or wear. Check the serrations of each part for any cracking, twisting, or signs of wear.

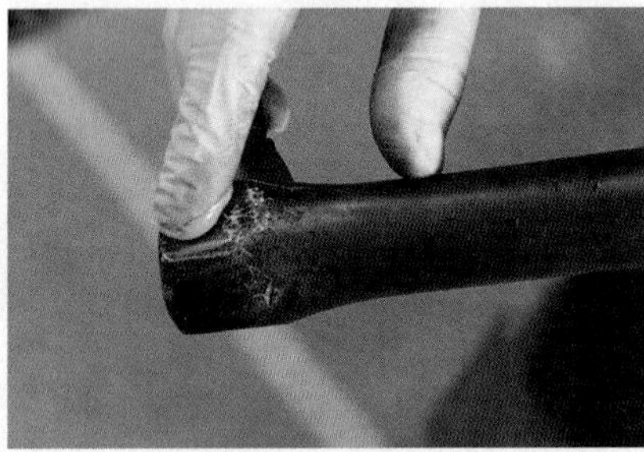

10. Lubricate the ends of the torsion bar with a high-quality, multipurpose grease.

11. Inspect the anchor plate and adjustment bolt for damage or excessive rust.

12. Reinstall the torsion bar spring by sliding it onto the hole with your hands. At this point it is vital that you have ensured that the correct spring (L or R) has been located in its correct side.

13. While aligning the anchor arm with the matching mark, install the anchor arm to the torsion bar. Install the snap ring (circlip) to the anchor arm and dust cover. Snap rings do not always seat in their groove when installed and can pop out as the vehicle is driven, creating a very dangerous situation.

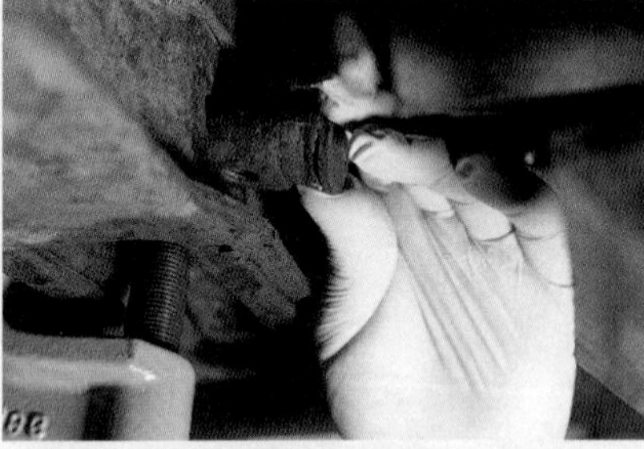

14. Use the correct torsion bar compression tool to compress the bar and install the anchor plate.

15. Install the adjustment bolt and tighten to the correct distance as measured in step 4.

SKILL DRILL 41-17 Removing, Inspecting, Installing, and Adjusting the Torsion Bar (Continued)

16. Carry out recommended inspection procedures to ensure the torsion bar settings are within specifications. Compare the final measurements to the manufacturer's specifications. If needed, adjust the torsion bar height as required by turning the bolt with a socket.

▶ Wrap-Up

Ready for Review

- Common issues related to the suspension system are vehicle wander, drift, pull, hard steering, bump steer, torque steer, and steering return concerns.
- Tools commonly needed to diagnose suspension or wheel alignment issues are the wheel aligner, ramp, turntable, wheel clamps, specification charts, workshop manuals, specialty alignment tools, dial indicators, measuring tapes, scan tools, general hand tools, various lift devices, air ratchets, and wheel braces.
- Common tools for maintenance and repair of suspension systems include those used in diagnosis as well as the following: component compression devices, levers, spring compressor tools, a strut servicing tool kit, a universal strut nut wrench kit, and a 24 mm strut rod socket.
- A common suspension system problem is excessive play in the parts, usually from wear in the joints and bushings.
- Some of the most common customer concerns related to the suspension system include vehicle wander, drift, pull, hard steering, bump steer, torque steer, steering return issues, and noises.

Review Questions

1. Which of these tools is appropriate for changing coil springs on struts?
   - **a.** Strut compressor
   - **b.** Pitman arm puller
   - **c.** Air chisel
   - **d.** Ball joint socket
2. Incorrect angles in the steering linkage due to a component being bent can result in:
   - **a.** noise.
   - **b.** bump steer.
   - **c.** pull.
   - **d.** hard steering.
3. When the driver feels the steering wheel pulling to one side, which of the following should be checked first?
   - **a.** Looseness in the suspension system
   - **b.** Problems in the braking system
   - **c.** Tire pressure
   - **d.** Four-wheel alignment
4. Which of the following can be used to diagnose the source of noise?
   - **a.** Strut servicing kit
   - **b.** Air chisel
   - **c.** Chassis ear
   - **d.** Scan tool
5. All of the following statements with respect to suspension systems are true *except*:
   - **a.** an electronic stethoscope is the best tool for diagnosing caster angle problems.
   - **b.** if one of the front tires is low, it may cause the vehicle to pull to one side.
   - **c.** a four-wheel alignment is needed to verify the alignment angles.
   - **d.** torque steer is a pull that occurs during heavy acceleration.
6. Which of the following statements is *correct*?
   - **a.** Worn shock absorbers will not cause any problems on good roads.
   - **b.** Shock absorbers use plastic bushings to isolate them from the vehicle body.
   - **c.** If there is a slight oil film on the shock absorber, corrective action should be taken.
   - **d.** The stabilizer components help prevent body roll when cornering.

7. When testing a shock absorber, and releasing at the bottom of the stroke, what is the maximum times the vehicle should rebound?
   a. Zero
   b. Once
   c. Twice
   d. Three times
8. The best method for testing the lower ball joint on a MacPherson strut suspension is to:
   a. raise the vehicle by the frame and allow the suspension to hang free.
   b. test it at the floor level.
   c. place a floor jack under the lower control arm and raise the wheel off the ground.
   d. fit a wooden block between the arm and the frame.
9. In the suspension and steering systems of most modern vehicles:
   a. the joints need to be lubricated every month.
   b. there may be no grease fittings present.
   c. lubricated joints should be overfilled with grease.
   d. the joints need to be lubricated every fuel stop.
10. Which of the following should be inspected when the ride height does not match the manufacturer's specifications?
   a. Shock absorbers
   b. Suspension springs
   c. Upper ball joints
   d. Lower ball joints

ASE Technician A/Technician B Style Questions

1. Tech A says that suspension faults involving wear can often times be located during a visual inspection and confirmed as faulty with a measuring tool if necessary. Tech B says that damaged components may require an alignment machine to identify damaged parts. Who is correct?
   a. Tech A
   b. Tech B
   c. Both A and B
   d. Neither A nor B
2. Tech A says that a road test is part of focused testing. Tech B says that a road test is part of verifying the repair. Who is correct?
   a. Tech A
   b. Tech B
   c. Both A and B
   d. Neither A nor B
3. Tech A says that worn sway bar bushings, brackets, and link bushings can cause excessive body sway. Tech B says that unmatched tires can cause ride height to be out of specs. Who is correct?
   a. Tech A
   b. Tech B
   c. Both A and B
   d. Neither A nor B
4. Tech A says that when unloading a suspension where the coil spring or torsion bar is pushing against the lower control arm, a wood block must be placed between the upper control arm and the frame. Tech B says when unloading a MacPherson strut suspension with only a lower control arm, raise the vehicle by the frame, and allow the suspension to hang free. Who is correct?
   a. Tech A
   b. Tech B
   c. Both A and B
   d. Neither A nor B
5. Tech A says that shock absorbers can be tested with a "bounce test." Tech B says that it is industry standard to replace an individual faulty shock absorber. Who is correct?
   a. Tech A
   b. Tech B
   c. Both A and B
   d. Neither A nor B
6. Tech A says that cotter pins should never be installed in ball joints. Tech B says that ball joints are typically welded into the control arm. Who is correct?
   a. Tech A
   b. Tech B
   c. Both A and B
   d. Neither A nor B
7. Tech A says that on most modern vehicles, the joints are sealed and lubricated for life, so there may be no grease fittings present. Tech B says that you should add grease to the ball joint until all the old grease is pushed out past the rubber boot on the joint. Who is correct?
   a. Tech A
   b. Tech B
   c. Both A and B
   d. Neither A nor B
8. Tech A says that both halves of the strut rod bushings go on the back side of the frame. Tech B says that on some vehicles the strut rod is the adjustment point for caster. Who is correct?
   a. Tech A
   b. Tech B
   c. Both A and B
   d. Neither A nor B
9. Tech A says that loose ball joints can cause the vehicle to wander. Tech B says that loose control arm bushings can cause a vehicle to wander. Who is correct?
   a. Tech A
   b. Tech B
   c. Both A and B
   d. Neither A nor B
10. Tech A says that when removing a MacPherson strut assembly from a vehicle, you must use a spring compressor to keep the spring compressed. Tech B says that a spring compressor is used only after the strut assembly has been removed from the vehicle. Who is correct?
   a. Tech A
   b. Tech B
   c. Both A and B
   d. Neither A nor B

CHAPTER 42
Wheel Alignment

NATEF Tasks

- **N42001** Perform pre-alignment inspection; measure vehicle ride height; determine needed action. (MLR/AST/MAST)
- **N42002** Prepare vehicle for wheel alignment on alignment machine; perform four-wheel alignment by checking and adjusting front- and rear-wheel caster, camber and toe as required; center steering wheel. (AST/MAST)
- **N42003** Reset steering angle sensor. (AST/MAST)
- **N42004** Check toe-out on turns (turning radius); determine needed action. (AST/MAST)
- **N42005** Check steering axis inclination (SAI) and included angle; determine needed action. (AST/MAST)
- **N42006** Check rear wheel thrust angle; determine needed action. (AST/MAST)
- **N42007** Check front and/or rear cradle (subframe) alignment; determine needed action. (AST/MAST)

Knowledge Objectives

After reading this chapter, you will be able to:

- **K42001** Explain the purpose and function of wheel alignment and the alignment angles.
- **K42002** Explain camber, its purpose, and effects.
- **K42003** Explain caster, its purpose, and effects.
- **K42004** Explain toe, its purpose, and effects.
- **K42005** Explain toe-out on turns, its purpose, and effects.
- **K42006** Explain steering axis inclination, its purpose, and effects.
- **K42007** Explain included angle and its purpose.
- **K42008** Explain scrub radius, its purpose, and effects.
- **K42009** Explain turning radius, its purpose, and effects.
- **K42010** Explain thrust angle, centerlines, and setbacks, their purpose, and effects.
- **K42011** Explain ride height, its purpose, and effects.
- **K42012** Identify tools used in wheel alignment and their function.

Skills Objectives

After reading this chapter, you will be able to:

- **S42001** Perform a wheel alignment.

▶ Introduction

Even if all of the suspension and steering system components are in good working condition, they also have to be aligned correctly with one another and with the vehicle's **centerline** for the vehicle to operate properly and safely. For this reason, the wheels must be in proper alignment from the factory. Technicians need to verify proper alignment typically when the tires are replaced, when parts are replaced, or if a suspension related concern is being diagnosed. This chapter helps you become familiar with the various types of suspension systems, their components, how they operate, and how to inspect and service them.

▶ Wheel Alignment Fundamentals

K42001

All wheels of a vehicle must be correctly positioned in relation to the vehicle and to one another for the vehicle to drive and steer properly (**FIGURE 42-1**). A driver should not need to keep manipulating the steering wheel to maintain the vehicle in a straight-ahead position on straight, level roads. Similarly, little effort should be needed to turn the vehicle into curves or to let it return to the straight-ahead position once the curve has been negotiated. Wheels are positioned on the suspension at certain angles to provide for easy driving of the vehicle. These angles, taken together, determine the vehicle's **wheel alignment**.

FIGURE 42-1 A technician checking the alignment on a vehicle.

The alignment of the wheels is maintained by the control arms, strut rods, tie rods, knuckles, and the vehicle frame to obtain the proper angles. Alignments should be performed for the following reasons:

- When handling issues related to alignment angles are found
- When tire wear shows any tire wearing angle issue
- When components are replaced that could affect the alignment.
- Whenever new tires are being installed.

But before you can check alignment, you need to know what the alignment angles are and how they affect driving. Alignment angles are split between primary and secondary angles. Primary angles consist of camber, caster, and toe, and are more likely to be adjustable, but not always. Secondary angles consist of toe-out-on-turns, steering axis inclination, included angle, scrub radius, thrust angle, centerlines, setbacks, and ride height. These angles are generally not adjustable, with the exception of thrust line, which is adjustable on most independent rear suspensions. Secondary angles are used primarily for diagnostic purposes. Let's learn more about these alignment angles.

Camber

K42002

Camber is the side-to-side vertical tilt of the wheel. It is viewed from the front of the vehicle and is measured in degrees (**FIGURE 42-2**). A wheel that leans away from the center of the vehicle at the top is said to have **positive camber**. A wheel that leans toward the

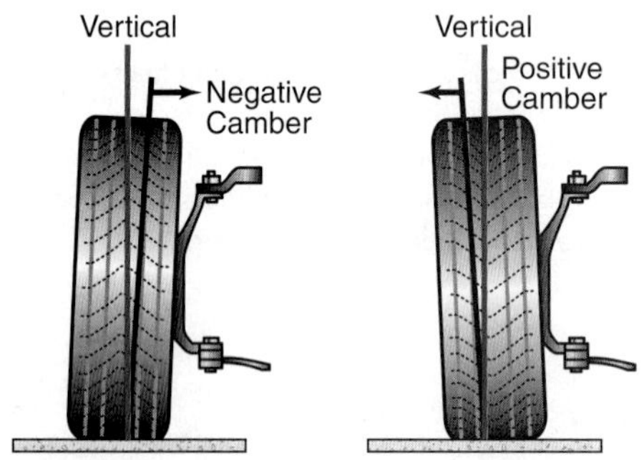

FIGURE 42-2 Camber.

You Are the Automotive Technician

A customer brings his four-year-old vehicle in for some new replacement tires because the ones on the vehicle are worn unevenly. You explain that for the tire warranty to be valid, you will have to perform a wheel alignment on the vehicle. And as part of the alignment, you need to perform a pre-alignment inspection to see if all of the steering and suspension components are within specifications. You warn him that the wear patterns on the tires makes you suspect that there are some worn parts that will have to be replaced. Please answer the customer's following questions.

1. How do worn steering and suspension parts wear out tires?
2. What components will you inspect during the pre-alignment inspection?
3. What are the three types of alignments, and which one do you recommend?
4. What could cause a toe-out on turns measurement to be out of specifications?

center of the vehicle is said to have **negative camber**. Camber tends to pull in the direction of the most positive camber. To picture why it pulls in this direction, think of a paper cup lying on its side. When it is rolled, it turns in the direction of the narrower end of the cup. Tires do the same thing in regard to their camber. Camber is affected by ride height and can be seen by looking at the tilt of the wheel in vehicles that are lowered incorrectly. Camber used to be a heavy tire-wearing angle when tires were of the bias-ply type and had stiff sidewalls. Camber wear is now debatable, given the much more pliable sidewalls in radial tires. It is safe to say that tires can tolerate moderate amounts of improper camber much better than before. There are also studies that say most of what technicians diagnose as camber wear is really toe wear. Pay attention to this debate as it plays out.

On earlier vehicles with narrow, large-diameter tires, large camber angles were used to bring the centerline of road contact closer to the steering axis. Large camber angles also ensured the vehicle weight was carried by the large inner bearing. On modern vehicles, however, tires are much wider and generally smaller in diameter, and large camber angles would cause the tire to ride on the outer edges of the tires. The amount of camber is now reduced so that most vehicles in forward motion have what is called **zero camber**, or no tilt, to provide maximum tire patch contact with the road, as well as long tire life.

▶ TECHNICIAN TIP

Changes in running camber can be caused by driving over road irregularities, load variations, and worn suspension components.

Caster

K42003

Caster is the forward or backward tilt of the steering axis from vertical when viewed from the side of the vehicle (**FIGURE 42-3**). The steering axis is an imaginary line passing through the

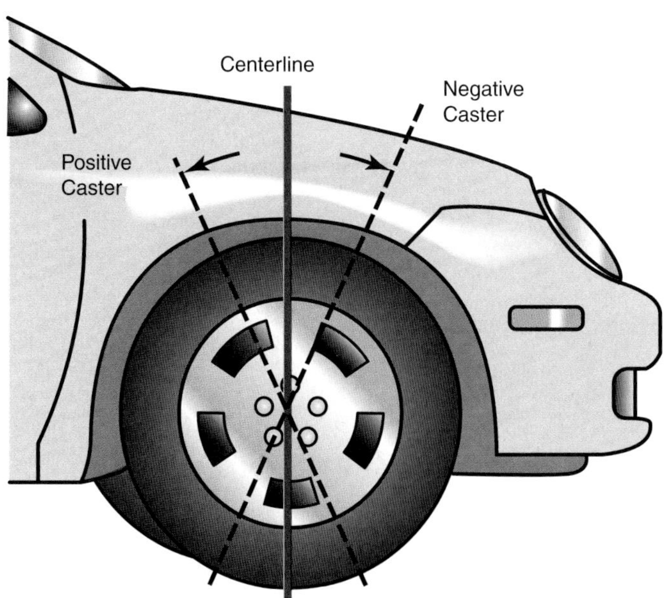

FIGURE 42-3 Caster.

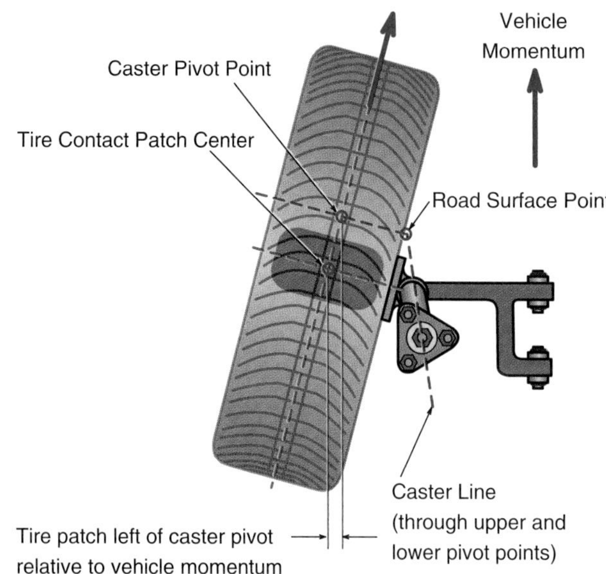

FIGURE 42-4 Positive caster has a self-straightening effect on steering when the wheels are turned.

center of the ball joints on SLA suspensions or passing through the center of the upper strut bearing and lower ball joint on MacPherson strut suspensions. So caster is the tilt of the steering axis from vertical, measured in degrees. Backward tilt from the vertical is **positive caster**. Forward tilt is **negative caster**. When a vehicle has positive caster, a line drawn through the steering axis centerline meets the road surface ahead of the vertical centerline of the wheel. The center of the tire contact point is behind the steering axis centerline.

When the wheel is turned to the right, the tire contact point is moved to the left of the direction of travel (**FIGURE 42-4**). Conversely, when the wheel is turned to the left, the contact point is moved to the right of the direction of travel. In forward motion, this generates a self-centering force that helps return the wheels to the neutral position when the steering wheel is released.

Most vehicles have positive caster, because it causes the tires to travel in a straight line with minimal driver input. However, as positive caster increases, more and more effort is needed to turn the steering wheel to overcome the increased self-straightening force. Also, positive caster causes the spindle to tilt as the wheel is steered, and this places more weight on the inside tire because of the body lifting that occurs with the inside tire. This lift can be seen by turning the steering from lock to lock while the vehicle is stationary.

Differences in caster cause a pull to the side with the most negative caster. So manufacturers generally specify the maximum cross-cater allowed. Many technicians will set a slight amount of negative caster on the driver's side of the vehicle to counteract the natural pull of the road crown, which is toward the outside of the road.

▶ TECHNICIAN TIP

Some vehicles have, by design, an amount of negative caster, which makes the steering light. Negative caster is created by placing the suspension system pivots so that they are tilted forward from the vertical line. Generally, such vehicles operate only at low speeds; vehicles with negative caster can become unstable as speed increases. Steering pulls toward the side of the vehicle with the most negative caster. Caster is affected by ride height and on most modern vehicles is a nonadjustable angle without aftermarket components.

Applied Math

AM-21: Parallel/Perpendicular: The technician can use measurement devices to determine the parallelism or perpendicularity of chassis, suspension, and other vehicle systems requiring the application of geometric alignment principles.

A vehicle was purchased at an auction and has been taken to an automotive repair shop for evaluation. When the vehicle was set up on the alignment machine and alignment readings taken, several of the angles were substantially out of specification, including front-wheel setback, rear thrust angle, and SAI. These angles being out of specification indicate the potential of a bent frame or chassis. Upon further examination, the technician observes cracked undercoat and paint on the subframe components, as well as shifting of the engine cradle bolts. This verifies what the alignment angles are telling him; that the chassis is not parallel and perpendicular. The technician knows that the wheels cannot be brought back into proper alignment with the chassis out of alignment.

Toe-In and Toe-Out

K42004

Toe is the angle of the tires relative to one another when viewed from above (**FIGURE 42-5**). The condition in which the fronts of the wheels are closer together than the rears of the wheels is called **toe-in**. The condition in which the fronts of the wheels are farther apart than the rears is called **toe-out**. Some manufacturers use the terms "positive toe" for toe-in and "negative toe" for toe-out. Toe can be measured in inches, millimeters, or degrees. The **static toe** setting is designed to compensate for slight wear in steering joints and components, which may cause the wheels to splay outward or inward while the vehicle is being driven. This effect is designed into the vehicle based on several factors, such as front-wheel drive versus rear-wheel drive, front/rear brake system split versus diagonal brake system split, and specific desired handling characteristics. Manufacturers specify the proper static toe for the design of the vehicle such that the wheels will be parallel when the vehicle is in forward motion, which avoids scrubbing of the tires. Improper toe settings cause much faster tire wear than camber or caster.

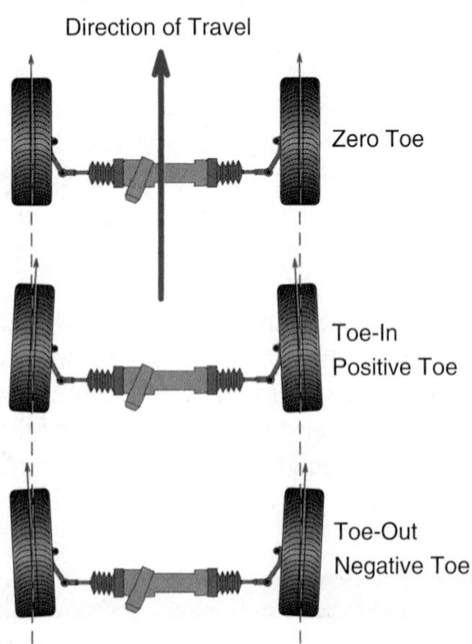

FIGURE 42-5 Toe is viewed from above the tire.

Toe-Out on Turns

K42005

Toe-out on turns is the relative toe setting of the front wheels as the vehicle turns (**FIGURE 42-6**). When a vehicle makes a turn, each wheel should rotate with true rolling motion that is free from tire scrub. True rolling motion is obtained only when each wheel is at 90 degrees to a line drawn between the tire's horizontal centerline and the center point of the turn. Because the rear wheels are fixed, the center of the turn will lie somewhere along the centerline of the rear axle, depending on how far the steering wheel is turned from the straight-ahead position.

To provide true rolling motion, the inner wheel must be turned through a greater angle than the outer wheel. This allows the inner wheel to turn through a smaller turning radius than the outer wheel. This correct positioning of the wheels when steering around a corner is obtained by use of the **Ackermann principle** and layout. The inner wheels must move on a circle with a smaller radius, and the outer wheels move on a circle with a larger radius. The **Ackermann angle** is the angle the steering arms make with the steering axis, projected toward the center of the rear axle (**FIGURE 42-7**). The angling of the steering arms forces the inner wheels to turn through a smaller angle when the steering wheel is turned to that side. When the steering wheel is turned the opposite direction, the wheels that were on the outside are now on the inside of the turn and turn more sharply than the wheels that are now on the outside of the turn. Toe-out

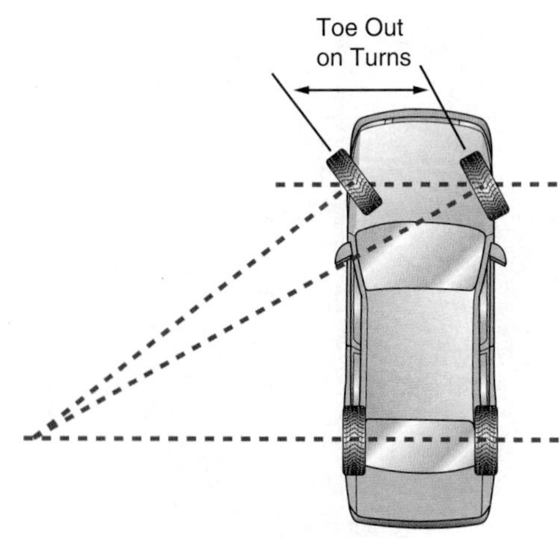

FIGURE 42-6 For toe-out turns to be correct, each wheel must be able to trace its own true arc when turning a corner.

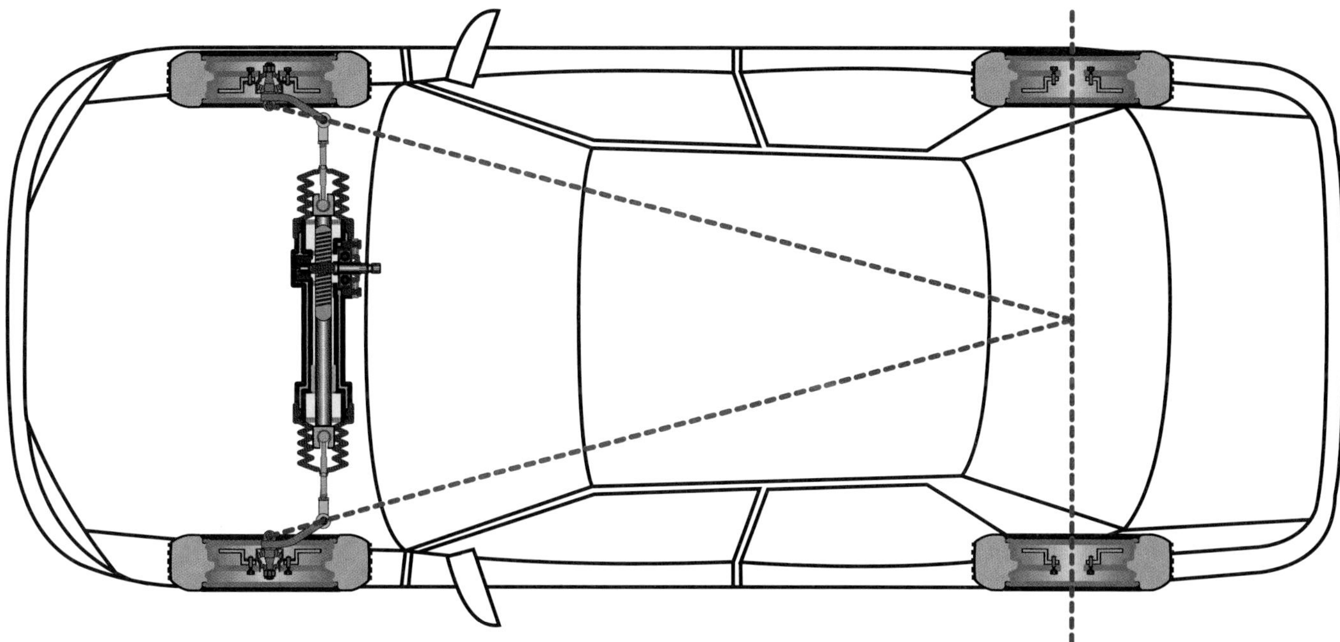

FIGURE 42-7 Ackermann angle.

on turns is a nonadjustable angle. If it is not correct according to the manufacturer's specifications, then you know that a steering linkage is bent, typically the steering arm. This also causes the tires to scrub when turning corners.

Applied Science

AS-15: Force: The technician can use a tension gauge such as a torque wrench to measure the force or tension required to tighten connections to manufacturer's specifications.

A good quality torque wrench is one of the technician's most valuable tools. Proper torque applied to fasteners is critical for the repair job to successful. This is true for chassis work as well as for engines. A torque wrench measures the amount of twisting force, or "torque," applied to the fastener. In physics, torque is sometimes called a rotational force. The formula to calculate the torque applied to a bolt is Torque = Force × Distance. In the study of physics, the distance is known as the moment arm. It takes twice the force at half the distance (or moment arm) to equal a certain torque value. It takes half of the force at twice the distance (or moment arm) to equal the same torque value. When using a torque wrench, the following precautions should be observed:

1. Use manufacturer's service information to obtain the torque specifications.
2. Verify that you have the correct torque wrench for the job that you are doing. The 80/20 rule states that the torque wrench should be used only in the approved range. This would be from 20 to 80 ft.-lb., when using a 0 to 100 ft.-lb. wrench, or 40 to 160 ft.-lb. on a 0 to 200 ft.-lb. wrench.
3. Support the head of the torque wrench with one hand, and pull the wrench handle in a steady motion.
4. Use a general crisscross pattern for tightening fasteners unless a manufacturer's torque pattern is available.
5. Tighten nuts and bolts in at least three stages—one-third recommended torque, two-thirds torque, full torque, and full torque—to double-check once again.

Applied Science

AS-52: Circular: The technician can demonstrate an understanding of circular motion as it relates to toe and camber on turns.

When you steer a vehicle through a turn, the outside front wheel has to travel a wider arc than the inside wheel; therefore, the inside front wheel must steer at a sharper angle than the outside wheel. To provide true rolling motion, the inner wheel must be turned through a greater angle than the outer wheel. This allows the inner wheel to turn through a smaller turning radius than the outer wheel. With the tires on the turntable, toe-out on turns is measured by the turning angle gauges (turn plates) on the wheel alignment machine. The readings are measured electronically and displayed on the screen. Camber is the vertical angle of the wheels relative to the vehicle. This is best viewed from the front of the car. If the vehicle has negative camber, the tops of the wheels will be closer together than the bottom. During cornering, camber compensates for vehicle weight transfer and body roll.

Steering Axis Inclination

K42006

The axis around which the wheel assembly swivels as it turns to the right or left is called the steering axis. It is formed by drawing a line through the center of the upper and lower pivot points of the suspension assembly (**FIGURE 42-8**). Seen from the front of the vehicle, it is tilted inward at the top. The angle formed between this line and the vertical provides the **steering axis inclination (SAI)** angle. Because the SAI is not adjustable, if the camber angle is correct, then the SAI should also be correct; that is, it should match the manufacturer's specifications.

SAI acts with caster to provide a self-centering of the front wheels. When the wheels are in the straight-ahead

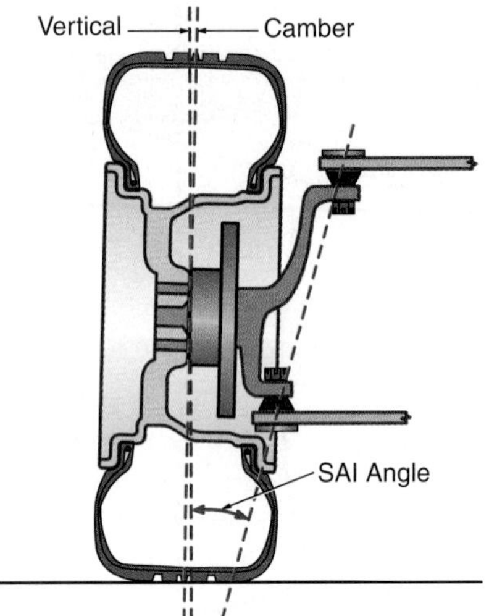

FIGURE 42-8 The axis around which the wheel assembly swivels as it turns to the right or left is called the steering axis. It is formed by drawing a line through the upper and lower pivot points of the suspension assembly.

position, the ends of the stub axles are almost horizontal. When the wheels turn to either side, the effect of SAI is to make the ends of the stub axle tend to move downward, but this tendency is prevented by the wheel and tire on the ground (**FIGURE 42-9**). The stub axle carrier then must move up, which raises the corner of the vehicle. When the steering wheel is released, the weight of the vehicle forces the stub carrier back down, which pushes the wheels back to a central position. When the wheels are turned the other way, the same thing happens. With a perfectly vertical steering axis, no self-centering would occur. The wheel would pivot on a radius, not under the center of the tire patch, but off to the inside of the tire. This would introduce a turning movement on the wheel whenever the tire hit a bump, transmitting road shocks back to the steering wheel, and steering would be difficult to control.

SAI also brings the pivot point close to the center of the tire contact patch at the road surface. For steering purposes, ideally SAI intersects with the camber line (drawn through the center of the tire and the wheel) at the road surface. Any difference in

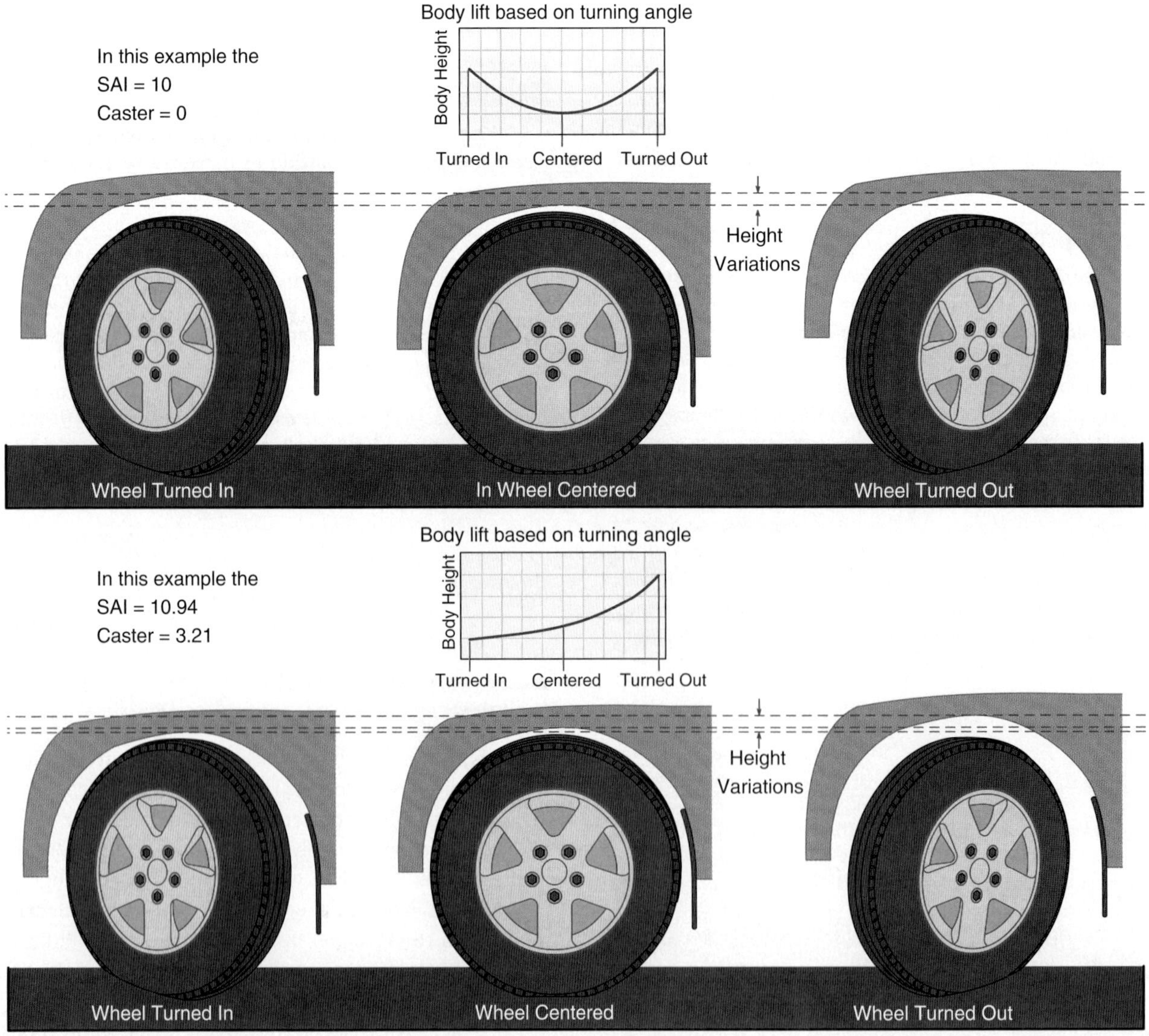

FIGURE 42-9 SAI and caster provide a self-centering effect of the front wheels. SAI tends to push each wheel down when turned either direction. Positive caster tends to push the inside wheel down when turned and lift the outside wheel.

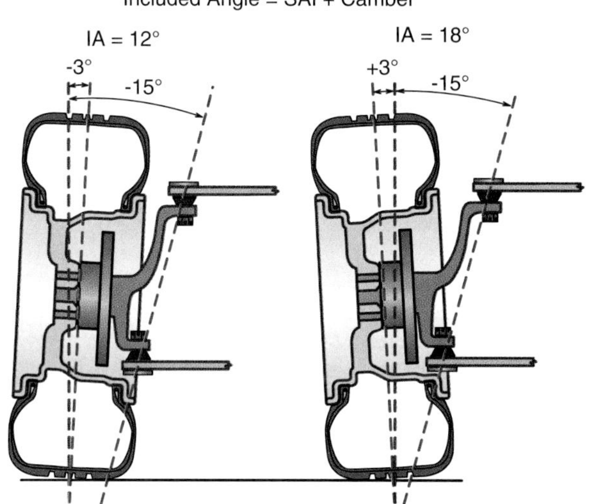

FIGURE 42-10 Included angle is found by adding the positive camber angle to the SAI angle or by subtracting negative camber from the SAI angle.

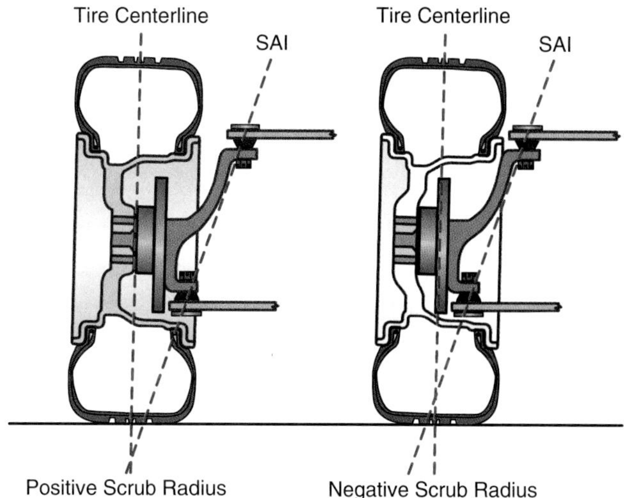

FIGURE 42-11 The scrub radius is the distance between two imaginary points on the road surface: the point of center contact between the road surface and the tire, and the point where the steering axis centerline contacts the road surface.

distance between these two lines produces another suspension angle called the scrub radius, which will be discussed further in a later section.

Included Angle

K42007

The angle formed between the SAI and the camber line is called the **included angle** or diagnostic angle. It is found by adding the SAI angle and the camber angle together (**FIGURE 42-10**). If the camber angle is specified as negative, then it is subtracted from the SAI angle. When an angle is referred to as a diagnostic angle, it means the angle cannot be changed, but is measured to determine if any parts are bent, such as a spindle, control arm, or steering arm.

Scrub Radius

K42008

Scrub radius is also known as steering offset and scrub geometry. It is the distance between two imaginary points on the road surface. One point is the camber line of the tire, and the other point is where the SAI centerline contacts the road surface (**FIGURE 42-11**) If these two lines intersect at the tire patch, then the vehicle is said to have zero offset, or **zero scrub radius**. If the camber line is outside of the SAI line (intersect below the road surface), then it has positive offset, or **positive scrub radius**. If the camber line is inside of the SAI line (intersect above the road surface), then it has negative offset or **negative scrub radius**. Scrub radius can be changed accidentally in a number of ways. Changing the diameter of wheels and tires affects scrub radius, as does the distance that the center of the wheel and tire sits relative to the vehicle body, which is called wheel offset. SAI and camber also affect scrub radius, as we are changing the angles that determine scrub radius.

Incorrect scrub radius can produce undesirable effects. Too much or too little scrub radius causes increased road shock when encountering irregularities in the road surface. Too much positive scrub radius causes the vehicle to dart to the side of greatest braking effort when the brakes are applied. Too much negative scrub radius causes the vehicle to dart when encountering road hazards (not braking) and results in less driver feedback through the steering wheel. Most front-wheel drive vehicles use a negative scrub radius to reduce torque steer and prevent the vehicle from darting when braking in the event that one brake circuit fails, as they use a diagonal split brake system.

On a rear-wheel drive vehicle with positive scrub radius, the vehicle's forward motion and the friction between the tire and the road cause a force that tends to move the front wheels backward. This causes the wheels to toe out. If it has negative scrub radius, the front wheels again tend to move back and the wheels now toe in. On front-wheel drive vehicles, the opposite occurs. Positive scrub radius causes toe-in, and negative causes toe-out.

During braking, if braking effort is greater on one side of the vehicle than the other, positive scrub radius causes the vehicle to veer toward the side with the greater braking effort. Negative scrub radius causes the vehicle to veer away from the side of greatest effort. How much it veers depends on the size of the scrub radius. This is why vehicles with a diagonal-split brake system have negative scrub radius built into the steering geometry. If half of the brake system fails, then the vehicle will tend to pull up in a straighter line.

▶ TECHNICIAN TIP

Because the offset of the wheel rim determines where the centerline of the tire meets the road surface, it is important that the offset is not changed if wheels are being replaced. Changing the rim offset changes the scrub radius and also the predictability of the vehicle handling, especially during brake failure. Also note that if tire size, ride height, or camber adjustment is changed, scrub radius also changes and will affect vehicle handling, potentially creating a safety hazard. Understand that there is a liability risk when modifying the vehicle you are working on.

Turning Radius

K42009

Turning radius is a measure of how small a circle the vehicle can turn in when the steering wheel is turned to the limit (**FIGURE 42-12**). All vehicles have stops to limit how far the front wheels can turn. In some designs, these stops can be adjusted as part of a wheel alignment. If the stops are incorrectly adjusted, they could allow too sharp of a turning angle, and the steering box could bottom out and be damaged.

Thrust Angle, Centerlines, and Setbacks (Tracking)

K42010

On a vehicle with independent rear suspension, undertaking a front-wheel-only alignment is considered an inadequate procedure. The rear tires also need to be aligned or they will experience accelerated tire wear, and the vehicle will not drive properly. The **thrust angle** refers to the relationship between the centerline of the vehicle and the angle of the rear tires. The term **thrust line** refers to the direction in which the rear wheels are pointing (**FIGURE 42-13**). The thrust angle can be adjusted on vehicles with adjustable rear suspensions. On vehicles that do not have adjustable rear suspensions, a small amount of thrust angle can be compensated for by aligning the front wheels to the rear wheels. Referencing the front steering geometry to the rear is very important.

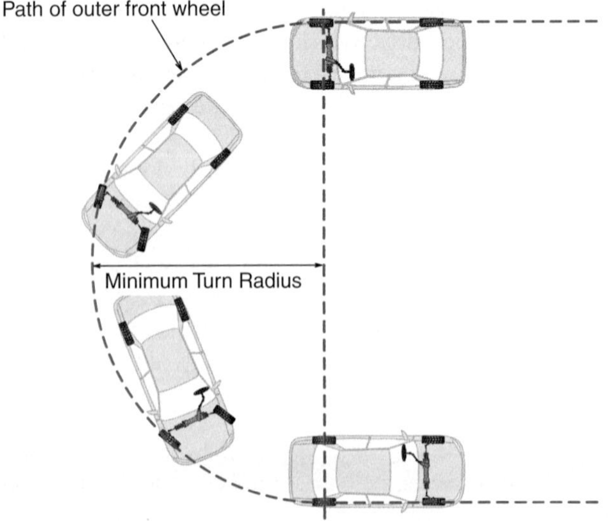

FIGURE 42-12 Turning radius is a measure of how small a circle the vehicle can turn in when the steering wheel is turned to the limit.

Applied Math

AM-26: Visual Perception: The technician can visually perceive the geometric relationships of systems and subsystems requiring alignment.

Bill is an automotive technician trainee who is working with an experienced technician. Bill has an interest in front-end alignment and has performed several alignments with the assistance of his mentor. A vehicle has just come into the shop for an alignment, and Bill has been given permission to do this job on his own. The experienced technician will do a final check to be sure that everything has been done properly. The shop has the latest type of laser-controlled alignment equipment, which makes the job a lot easier for the technician.

Bill drives the vehicle onto the alignment rack, but before attaching the wheel sensors, he looks the car over. He positions himself at the front of the vehicle, sighting from the front tires toward the rear. Bill wants to see the relationship of the front tires to the rear. He observes that there is a small amount of the rear tread showing on each side and that it appears to be an equal amount of tread on each side. He begins the alignment procedure by attaching the sensors on each wheel. The VIN is scanned into the computerized system and the alignment system calculates the thrust angle and geometric centerline. The final result is that a slight toe adjustment is needed, but everything else is within specifications. Bill's visual perception of the relationship of the components helped him to predict this reading before the alignment machine verified it.

Applied Math

AM-51: Trial and Error: The technician is able to solve problems by trying a suggested solution and observing the results.

A vehicle with a MacPherson strut design is in the shop for repairs. The technician who is assigned to this project has done similar jobs in the past. The repair order calls for two new front struts plus an alignment. In the past, the technician has done very little regarding marking of the position of the old strut. In this case, the technician is going to try a new method that was suggested by another technician. The new strategy is to employ careful marking of the old strut's position. This includes the upper and lower positions of the original strut, using a sharp scribe. The technician wants to observe the results of his procedure when the alignment is completed. The strategy is to produce a less involved alignment or perhaps eliminate the need for alignment adjustments to be necessary. The technician understands that this is only a trial-and-error procedure to be tested and could vary from one type of vehicle to another type. He was pleasantly surprised that the alignment was much closer to the specifications than in previous vehicles where he hadn't used this process.

Ideally, the thrust line and the vehicle's geometric centerline should line up closely. The centerline is drawn through points midway between each pair of wheels; however, the thrust line is not always as straightforward to determine. It is normally perpendicular to the rear axle on solid-axle vehicles. In vehicles with an independent rear suspension, it is derived by splitting the toe angle of each of the rear wheels on the vehicle. For instance, if the right rear wheel is toed in 6 degrees, and the left rear wheel is at 0 degrees, the thrust line will veer off 3 degrees

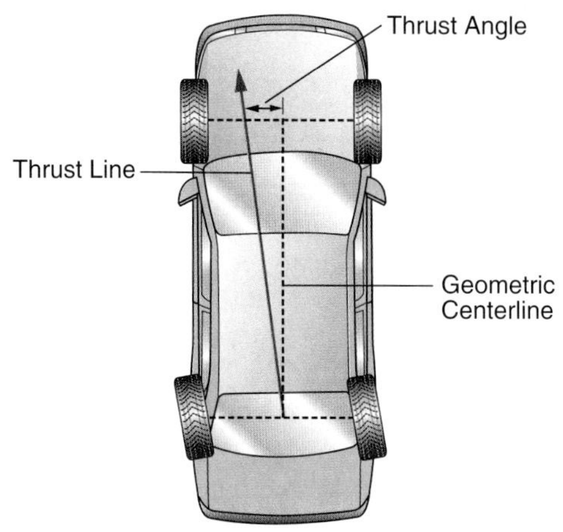

FIGURE 42-13 The thrust angle refers to the average angle of the rear wheels and its relationship to the vehicle's centerline.

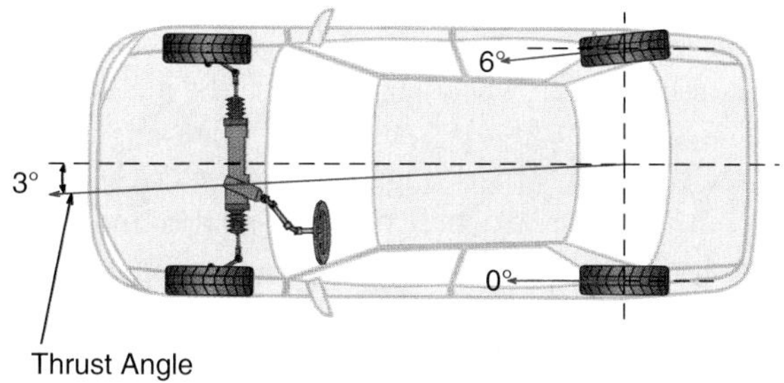

FIGURE 42-14 Rear wheel thrust line can found by splitting the rear toe readings.

to the left of the vehicle's centerline when the vehicle is moving forward (**FIGURE 42-14**).

Ideally, the thrust line and centerline match; however, given the size of a vehicle, the tolerances during manufacture, operational stresses, and component wear, it is rare that they do. If the deviation is very small, then corrective action is normally unnecessary. However, a large deviation must be diagnosed and corrected. Under such conditions, the rear wheels are steering the vehicle away from its centerline, and the driver has to turn the steering wheel to one side to keep the vehicle going in a straight line. Thrust angle issues may be caused by a broken center bolt on a leaf-spring suspension, a bent axle, a bent frame or unibody, worn or damaged suspension parts, or simply misadjusted rear toe.

Setback is a diagnostic measurement because it is generally not adjustable. Setback is the distance one wheel is set back from the wheel on the opposite side of the axle, relative to imaginary lines running through the center of each wheel and perpendicular to the vehicle's centerline (**FIGURE 42-15**) Setback is best measured on an alignment machine (after the wheels are aligned) so it can be measured correctly from the centerline of the vehicle. But a general check of setback can be made by measuring the distance of the wheelbase on one side of the vehicle and comparing it to the wheelbase on the other side.

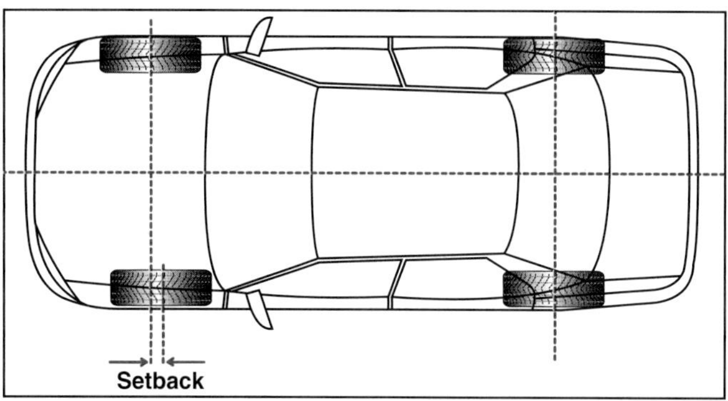

FIGURE 42-15 Setback refers to the distance one wheel is set back from the wheel on the opposite side of the axle, relative to lines running through the center of each wheel and perpendicular to the vehicle's centerline.

If the vehicle has setback, then one wheelbase measurement will typically (but not always) be shorter than the other.

Setback can be incorrect due to damage from a collision or due to a cradle that has been installed incorrectly or has shifted. Setback in the front wheels creates a pull condition for the driver to the side with the most setback or the wheel that is farther toward the rear of the car. A setback issue in the rear will pull toward the side with the least setback. Setback due to a cradle shift may also affect the camber or caster from side to side. On a vehicle that has no adjustment for camber or caster, check for a cradle shift by looking for witness marks around the head of the bolt that holds the cradle to the body. If a partial shiny metal ring surrounds the bolt head, then the cradle has shifted. Loosen the cradle bolts and reposition to achieve proper alignment specifications. Usually, setback should be no more than a quarter of an inch from side to side.

▶ TECHNICIAN TIP

In extreme conditions of setback or thrust angle, the tracks the rear tires make are beside those of the front. This condition is known as dog tracking or crabbing and can cause diagonal tire pattern wear on the rear tires, as well as vehicle instability in some driving conditions. A vehicle that is dog-tracking will appear to be driving slightly sideways down the road.

Ride Height

K42011

Ride height, sometimes referred to as **trim height**, is the amount of distance between the ground and a specified part of the vehicle such as the fender well, upper control arm, or rocker panel (**FIGURE 42-16**). Ride height is measured with the vehicle unloaded. For cars, it is usually given with no cargo or passengers. Changes in ride height alter the position of the control arms, which can have an unfavorable effect on wheel alignment. Ride height that is not within specifications can be caused by improper tires or tire pressure; weak, sagging springs; or bent components such as a control arm, axle, or even a frame

FIGURE 42-16 Ride height measurements.

or unibody. Ride height should typically not vary by more than half an inch from side to side.

Ground clearance is similar to ride height but is the distance from the ground to the lowest part of the chassis, typically a cross member, final drive, or oil pan. Ground clearance is especially important for off-road vehicles. Ground clearance can be increased by replacing stock components with aftermarket performance parts, but increasing ground clearance also increases the vehicle's center of gravity, making it more prone to rollover. It also affects scrub radius and potentially other wheel alignment angles. Thus, increasing ground clearance increases the risk of an accident in several ways; consider that carefully before modifying a vehicle in this way.

Applied Science

AS-88: Pneumatics: The technician can demonstrate an understanding of the forces and motions in pneumatic systems.

A vehicle is in the shop due to an inoperative automatic leveling control, which is part of an air suspension system. This vehicle has an automatic level control (ALC) system that maintains the correct rear suspension trip height. This height will be maintained even if a heavy weight is placed in the trunk. Pneumatics is a system operated by compressed air. In the case of the automatic level system, a battery-powered air compressor is used to power the system. Nylon air lines connect from the compressor to the rear air shocks to lift the rear of the vehicle to the correct height. The forces involved are linear as the weight of the rear of the vehicle is applied to the air shocks. The vehicle height sensor triggers the air compressor to control the amount of run time needed. When weight is removed from the trunk, the rear of the vehicle rises. At this point, the vehicle height sensor gives the appropriate signal to the pressure release solenoid valve to release air from the shocks, and the body of the vehicle returns to the correct height. This type of system is usually found on luxury vehicles. The technician inspects the vehicle and discovers a broken airline. After replacing the line, the vehicle is tested by the addition of some heavy items in the trunk. The air compressor starts, and the rear of the vehicle rises to the correct trim height. When the weight is removed, the vehicle lowers to the proper height.

▶ Performing a Wheel Alignment

S42001

Types of Wheel Alignment

Performing a wheel alignment requires the use of an alignment machine. In the past, simpler devices (such as using a measuring tape or a piece of string to set wheel toe) were used to align the vehicle's wheels. Today's vehicles are more sensitive to the position of the wheels, so using a tape measure to set toe is no longer acceptable. The technician uses a computerized alignment machine to ensure all the angles discussed in the previous sections are correct.

The three basic types of wheel alignment are (1) front-end, two-wheel alignment; (2) thrust-angle alignment; and (3) four-wheel alignment. Two-wheel alignment is outdated and almost never performed on modern vehicles, but we cover it briefly here so you will understand why it would be inappropriate for most vehicles. In front-end, two-wheel alignment, the technician positions the vehicle on the alignment rack and attaches the wheel sensors to the two front wheels. The sensors read the position of both front wheels and provide the measurement to the technician (**FIGURE 42-17**). The two-wheel alignment only compares the front wheel angles to the vehicle's centerline; it does not look at rear wheel position, so it cannot take into account any thrust angle. The technician compares the measurements to the alignment specifications provided by the manufacturer. If they are different, the technician adjusts the angles of the wheels until they match the specifications.

In a thrust-angle alignment, the technician attaches wheel sensors to all four wheels. The front wheels are compared to the angles of the rear wheels and adjusted to them (**FIGURE 42-18**). This technique is typically done on a vehicle with a solid rear axle where no adjustment is possible in the rear. If the thrust is out of specifications, the technician will need to diagnose what is causing the issue, such as a shifted axle or collision damage. If the thrust angle is within specifications, the front wheels are adjusted to compensate for any slight thrust angle of the rear wheels.

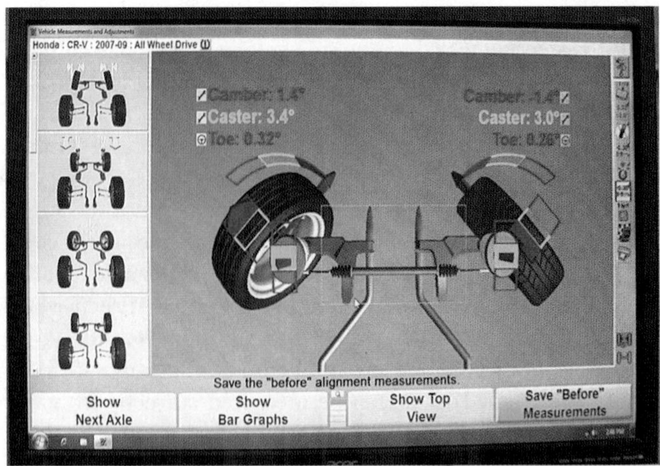

FIGURE 42-17 Readings for a two-wheel alignment.

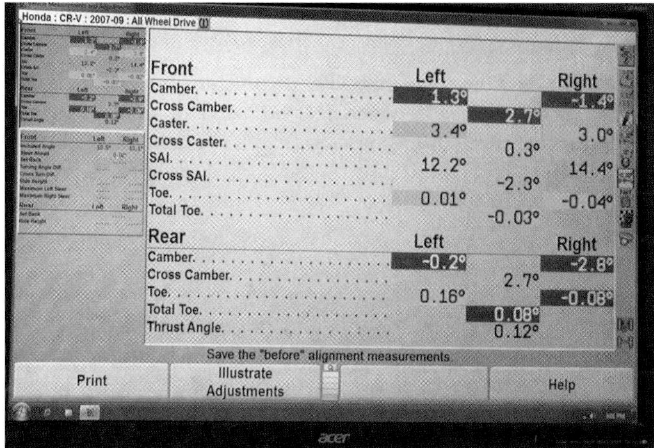

FIGURE 42-18 Readings for a thrust angle alignment.

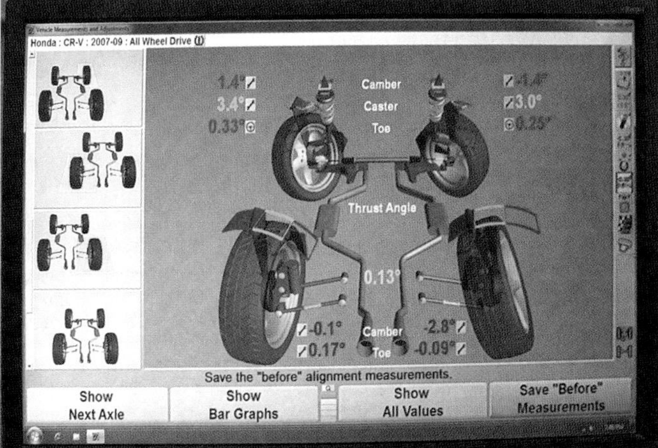

FIGURE 42-19 Readings for a four-wheel alignment.

In a four-wheel alignment, the technician positions the vehicle on the alignment rack and attaches the wheel sensors to all four wheels. The rear wheels are adjusted first so that they conform to the vehicle's centerline; then the front wheels are adjusted to conform to the vehicle's centerline and the position of the rear wheels (**FIGURE 42-19**). The four-wheel alignment provides the most accurate alignment of the wheels if adjustment of the rear suspension is possible.

Remember that any time a worn or damaged steering or suspension component is removed, an alignment should be performed. For example, if the front struts (with or without adjustment slots) are being replaced, an alignment will have to be performed to ensure the correct wheel alignment angles are restored. Remember that any component that holds or allows movement of the wheels is replaced, then it can affect the alignment, so it is good practice to check and align the wheels after these types of repairs.

Wheel Alignment Preliminaries

Performing an alignment can be done for maintenance purposes, or it can be done when tires or other steering and suspension components have been replaced. It can also be a helpful diagnostic step when a driver is complaining of uneven tire wear, pulling, hard steering, or wandering conditions. The alignment machine helps you identify problems that might not be identifiable by a visual inspection, such as bent suspension components, worn parts, or improper repairs that have been performed to the steering or suspension system. Although four-wheel alignments are the norm for current vehicles, many technicians jokingly refer to a proper alignment as a five-wheel alignment. This is because the steering wheel also must be centered at the completion of a wheel alignment; otherwise, the customer will be returning with a crooked steering wheel concern. Assuming you follow the proper procedures, the steering wheel itself will be centered when you are finished performing the alignment adjustments.

But centering the steering wheel physically is not enough on most modern vehicles, as they now use a steering angle sensor that tells the vehicle's stability control module where the steering wheel is pointed. It is possible for the steering wheel to be centered but for the sensor to indicate that it is off center. These vehicles may need to have the steering angle sensor recalibrated if after the alignment the sensor doesn't match the steering wheel position. Typically this process is performed with a scan tool once the alignment angles are within specifications and the steering wheel is physically centered.

Before performing the wheel alignment, you need first to verify the customer concern and then perform a

Applied Math

AM-22: Angles: The technician can use angle measurement equipment and techniques to determine any vehicle angle measurement variance from the manufacturer's specifications.

A front-wheel drive vehicle is in the shop for a routine tire rotation. The technician notices that one of the front tires is slightly worn on one side. He reports his findings to the service advisor, who contacts the vehicle owner and obtains authorization for a four-wheel alignment. The technician starts with a pre-alignment inspection including a check for any loose suspension components. No unusual problems are found, and the technician is now ready to use angle measurement equipment and techniques to determine variance from the manufacturer's specifications.

An electronic sensor is attached to each of the four wheels. The technician scans the VIN, and the alignment system automatically goes to the correct specifications for the vehicle. In addition, the video screen of the alignment system gives a step-by-step approach to the procedure. Because there are many different ways to make alignment corrections, the technician is shown the exact procedure for the vehicle being aligned. A color-coded system—red for out of specifications and green for within specifications—is also helpful.

Computer printouts with before and after adjustments are available to the customer. In this scenario, the alignment equipment indicates that a toe adjustment is needed for the front of the vehicle. As the technician makes the adjustment, the display shows the results on the video screen. When the exact manufacturer's specifications are obtained, the technician locks down the jam nuts to secure the setting.

pre-alignment inspection to ensure that the vehicle can be aligned. If any tire, wheel, suspension, or steering components are found to be worn or damaged, replacement of these components will be necessary before aligning the wheels. Once the pre-alignment inspection is finished and any worn or damaged parts have been replaced, the technician should check the following primary angles: caster (which is not adjustable on all vehicles), camber, and toe.

The secondary angles should also be inspected. They are the SAI, included angle, wheel setback, thrust angle, and toe-out on turns. The secondary angles are typically not adjustable and, if out of specifications, normally mean that a suspension component or the vehicle frame is bent. One exception to this is when the vehicle has an adjustable rear independent suspension. In this case, thrust angle can typically be adjusted by adjusting the toe for each rear wheel, in or out, to correct the thrust angle. Typically, thrust angle is adjusted as close to the vehicle's centerline as possible.

Adjusting wheel alignment angles must be done in this specific order: caster, camber, and then toe. Not all vehicles have adjustment for caster; if they do not, an aftermarket kit must be installed if caster has to be adjusted, or the vehicle may have to go to a body shop for frame straightening. For vehicles equipped from the factory with provisions for adjustment, five common types of adjustment systems are used: shim, eccentric bolts, slots, adjustable strut rod, and ball joint adjusting sleeve. The first type is shim adjustment. Manufacturers have used shims on both the front and rear wheels over time. In this method, shims are added or removed from a shim pack slipped between the control arm pivot shaft and the vehicle frame at each of the two attachment bolts (**FIGURE 42-20**). There are two shim packs used—one at the front attaching bolt and one at the rear attaching bolt. Shims are added or removed to change the position of the pivot shaft and ultimately the control arm and ball joint, affecting the caster and camber angles as desired.

FIGURE 42-20 A shim-type adjustment is probably the most time-consuming alignment adjustment. Moving shims from one attaching bolt to the other affects the caster setting, and changing shim pack size in both packs affects camber.

Applied Math

AM-24: Geometric Figures: The technician can distinguish whether the angles between related parts (e.g., suspension components) are within the manufacturer's specifications.

A three-quarter-ton pickup truck is in the shop with a driveshaft vibration problem. The owner of the vehicle is requesting an inspection of the rear driveshaft, including the measurement of the operating angles. A driveshaft is a component that connects the transmission (or transfer case) to the axles, transmitting torque from the engine to the driving wheels. This component is also known as a propeller shaft. The Cardan-style universal joint is sometimes known as the "cross and caps" type. When the driveline has an operating angle of zero, the universal joint is operating under the best possible conditions for long service life. This "perfect" situation is not realistic for a number of reasons. The suspension system allows the rear axle to move up and down, which changes the operating angle of the driveline. Worn components such as transmission mounts or worn bushings in the spring hangers can change the operating angle.

The technician assigned to the job has a good working knowledge of suspension components and driveline angles. Manufacturer's service information has been accessed, and the technician is aware of the 3-degree maximum operating angle. A special tool called an inclinometer or angle finder is placed on the driveline to determine the angles at the front and rear of the driveshaft. The operating angles on each end of the driveshaft should be equal to or within 1 degree of one another. The technician finds that the driveshaft angles are correct, but the rear universal joint is worn, causing the shaft to be out of alignment and balance and creating the vibration. Replacement of the rear universal joint corrects the problem.

Applied Math

AM-25: Relationships: The technician verifies that the relationship of parallel lines and angles is in conformance with the manufacturer's specifications.

An experienced technician is assigned to assist a technician trainee with an alignment. The shop has recently purchased a state-of-the-art, computerized, laser-operated alignment system. After showing the trainee the basics of a pre-alignment inspection, the vehicle is pulled onto the alignment rack. The technician shows the trainee how to attach a target to each of the four wheels. The target is another name for an electronic device that is attached to each wheel. The VIN number of the vehicle is scanned onto the alignment system. Manufacturer's data will be selected and available for alignment purposes. The experienced technician explained that the alignment system is set up to guide the operator in a step-by-step procedure. On a video screen are instructions that are specific to the exact make and model of the vehicle. There are clear illustrations of all of the adjustments that are needed for each phase of the operation. Concerning the specifications for the vehicle being aligned, caster has a preferred setting of 3.33 degrees with a range of 2.33 to 4.33 degrees. Camber is 0 degrees preferred with a range of −1 to 1 degrees. Toe-in for front is 0.16" preferred, with a range of 0.11" minimum and 0.21" maximum. The technician and trainee then compare the alignment readings with the specifications to determine any needed adjustments.

Because shim adjustment can affect both caster and camber, if the same number of shims is taken out of, or added to, each shim pack, the camber will be changed. If shims are taken from one shim pack and placed in the other, only the caster will be changed. Shims are available in various thicknesses and typically come in 1/64" (0.015 mm) up to 1/4" (0.250 mm). Typically, a shim change of 1/32" moves caster by 0.5 degrees, and camber moves around 0.3 degrees with the same shim change.

Shims are still used on some vehicles to adjust rear camber and toe. This type of shim is slightly wedge shaped and is placed behind the backing plate of the brake assembly or behind the hub of the wheel assembly to move the camber or toe to the correct position (**FIGURE 42-21**). These are aftermarket fixes and typically not used from the factory.

Another type of alignment adjustment for caster and/or camber is the eccentric bolt (**FIGURE 42-22**). The eccentric bolt is slightly egg shaped or has a slightly egg-shaped washer attached to one side of the bolt. This egg shape pushes against ridges on the part being adjusted, which changes caster and/or camber.

If the vehicle has only one attachment point for the control arm to the body or frame, then it will use a strut rod to support the control arm. Some strut rods have an adjustment nut on the front and the rear of the attachment point to the control arm (**FIGURE 42-23**). If both nuts are moved, then the control arm will be pushed forward or pulled backward, which will affect the caster setting.

The last method of adjusting caster and camber involves the ball joint adjusting sleeve (**FIGURE 42-24**). The adjusting sleeve is used on several four-wheel drive vehicles and mounts into a solid axle or a twin I-beam axle. The ball joint tapered stud fits into the sleeve. This sleeve can be turned to change caster and camber slightly. The sleeve has a slightly offset hole so the ball joint tapered stud can fit into it. When turning the adjusting sleeve, the ball joint stud is pushed either forward or backward, or in or out, changing caster or camber. The factory adjusting sleeve typically only gives a half degree of

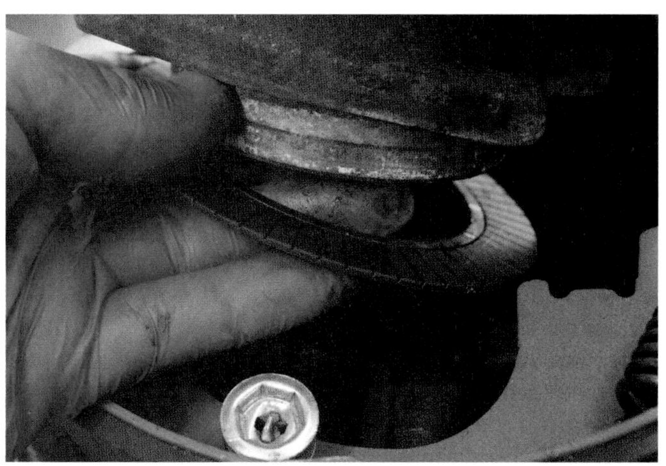

FIGURE 42-21 A typical shim used to adjust the alignment on a rear wheel.

FIGURE 42-23 The strut rod on some vehicles is the adjustment point for caster.

FIGURE 42-22 Eccentric bolt adjustment.

FIGURE 42-24 The adjustable ball joint sleeve gives limited adjustment for caster and camber, and is often replaced with an aftermarket part with more adjustment potential.

alignment change, so an aftermarket adjusting sleeve is often installed if more adjustment is needed.

Adjustment of toe is typically accomplished by lengthening or shortening the tie-rod assembly. Lengthening the tie-rod pushes the steering arm and changes the toe setting; shortening moves the steering arm the opposite direction and changes the toe in the other direction. On a vehicle with an adjusting sleeve, the technician must loosen the two clamp bolts that hold the adjusting sleeve tight and turn the sleeve in the proper direction for toe setting (**FIGURE 42-25**). Don't forget to tighten the clamp bolts when the toe is adjusted properly. On a vehicle with a rack-and-pinion steering system, typically the inner tie-rod threads into the outer tie-rod and no adjusting sleeve is used. To lengthen the tie-rod assembly on this vehicle, loosen the locknut on the outer tie-rod and turn the inner tie-rod in or out to lengthen or shorten the tie-rod assembly (**FIGURE 42-26**). Tighten the locknut when adjustment is finalized.

FIGURE 42-25 The tie-rod assembly comes in two forms and is adjusted by threading the assembly to lengthen or shorten.

FIGURE 42-26 The tie-rod on a rack-and-pinion steering system can be adjusted after loosening the locknut by turning the tie-rod either in or out.

Tools

K42012

The tools commonly used in diagnosing suspension or wheel alignment issues include:

- Wheel alignment machine: Used to measure and view alignment angles on a vehicle.
- Four-post alignment rack: Used to raise and level the vehicle off of the ground so you can align it.
- Turntables: Used on alignment racks so you can turn the front wheels to check caster angle.
- Wheel clamps and extensions: Used to hold alignment targets to the wheels.
- Specification charts: Used to ensure manufacturer-specified angles and measurements for vehicles.
- Workshop manuals or electronic information systems: Used to look up specifications and instructions for replacing parts on the vehicle.
- Specialty alignment tools: Used to align steering angles on a vehicle.
- Steering wheel holder: Used to keep steering wheel from turning when centering steering wheel during an alignment.
- Brake pedal depressor tool: Used to lock the brakes during caster sweep so wheels of vehicle cannot move.
- Steering angle sensor calibration tool or scan tool.

Performing a Pre-Alignment Inspection

N42001

A pre-alignment inspection is performed to ensure that the vehicle is an appropriate candidate for alignment. Worn and loose suspension or steering components, low tire pressure, and heavy objects in the trunk make it impossible to perform an accurate alignment.

As with any diagnosis, verifying the customer's concern is the first step. Normally it is best to test-drive the vehicle prior to the pre-alignment inspection, listening carefully for any unusual noises and noting any improper driving control issues related to the suspension and steering systems. Then use a good pre-alignment inspection form to guide you through the inspection and to document your findings for the customer.

To perform a pre-alignment inspection, follow the steps in **SKILL DRILL 42-1**.

Performing Four-Wheel Alignment

N42002, N42003

The alignment machine measures the angles of the wheels compared to one another and the centerline of the vehicle. These measurements are then compared to specifications so that the proper adjustments can be performed. Because the alignment machine's wheel adapters do not attach to the wheels perfectly straight, the alignment machine has the ability to compensate for wheel adapter runout. Compensation ensures that the alignment machine is reading the angles accurately, no matter where each tire is positioned. Each alignment machine has a different wheel adapter compensation process, so make sure you know

SKILL DRILL 42-1 Preparing a Vehicle for a Wheel Alignment

1. Remove any heavy items from the trunk and passenger compartments.

2. Check the size and condition of all four tires. Adjust the air pressure to specifications.

3. Measure the vehicle's ride height.

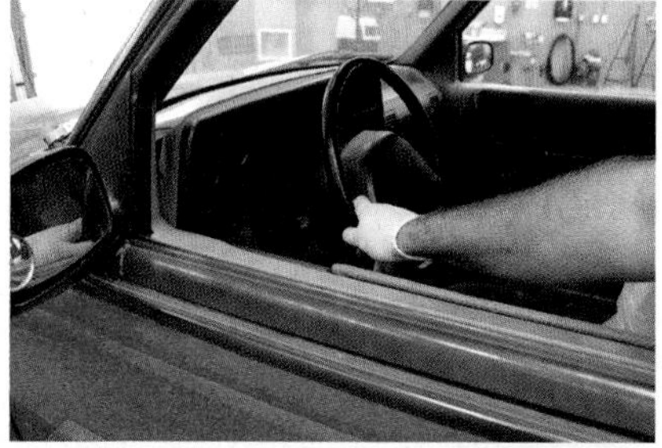

4. Check the play of the steering wheel. Correct any excess play before undertaking the wheel alignment.

5. Bounce each corner of the vehicle to check the correct functioning of the shock absorbers.

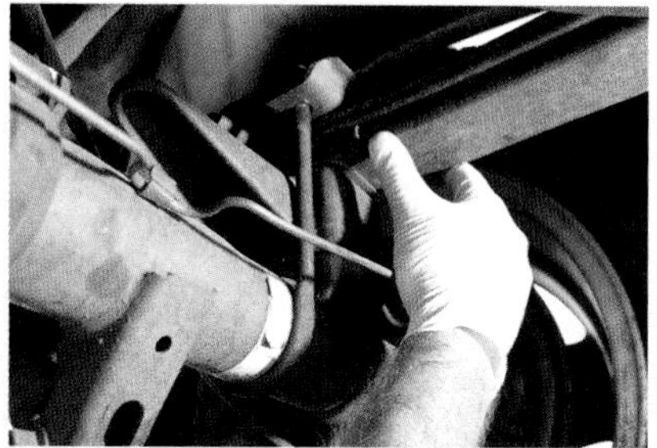

6. With the vehicle raised, inspect all suspension and steering components, including the wheel bearings. Repair or replace all damaged or worn suspension components.

7. Position the vehicle on the wheel alignment ramp making sure the front tires are positioned correctly on the turntables.

8. Position the rear wheels on the slip plates or rear turntables.

9. Attach the wheel units of the wheel alignment machine.

the process for the machine in your shop. Alignments always follow a specific order of adjustments. The rear wheels should be adjusted before the front wheels so that the vehicle can be made to track straight down the road. And you need to confirm that the steering wheel is centered and the steering angle sensor is calibrated. The order is as follows:

(1) rear caster, (2) camber, and (3) toe; then
(4) front caster, (5) camber, and (6) toe; then
(7) verify steering wheel is centered (8) calibrate steering angle sensor.

This sequence is for vehicles with fully adjustable caster, camber, and toe on the front and rear of the vehicle. Performing the alignment adjustment in this way will prevent having to go back and readjust previous settings, which saves you time. Not all vehicles have adjustable caster and camber, so only toe can be set, unless aftermarket components are installed that allow caster and camber adjustment. This Skill Drill is for a vehicle equipped with two eccentric bolts attaching the lower control arm to the frame, allowing caster and camber adjustment. Perform the alignment as the service information instructs or the alignment machine guides you. This Skill Drill is not designed to be an accurate example covering all vehicles.

To perform a four-wheel alignment, follow the steps in **SKILL DRILL 42-2**.

SKILL DRILL 42-2 Performing Four-Wheel Alignment

1. Position the vehicle on the front-end rack. Raise the vehicle to a comfortable working level, and set the rack on its mechanical locks to provide a level surface.

2. Raise the vehicle with the air jacks on the alignment rack.

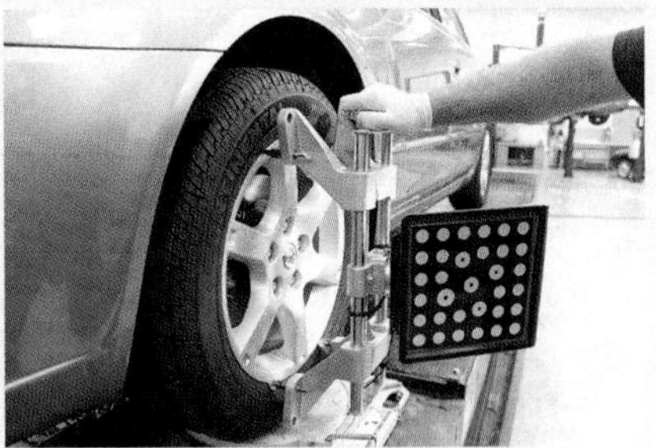

3. Attach sensors and compensate each one.

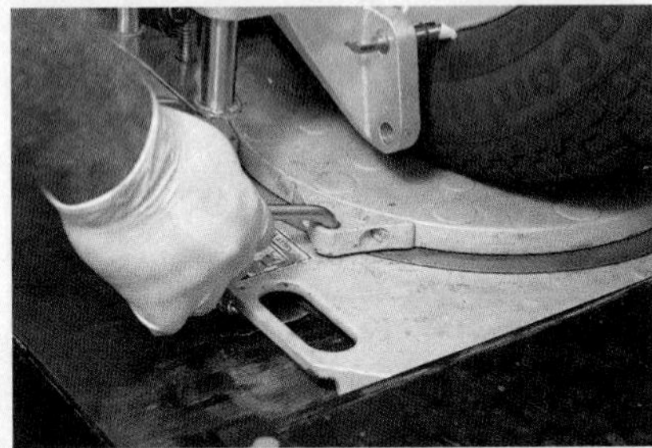

4. Pull the lock pins from the slip plates and turntables.

5. Lower the vehicle as instructed by the machine. Install a brake pedal depressor.

6. Perform a caster sweep by selecting caster sweep on the machine and turning the wheel the number of degrees on the turntable as designated by the machine.

7. Take the alignment readings, and compare them to the vehicle manufacturer's specifications.

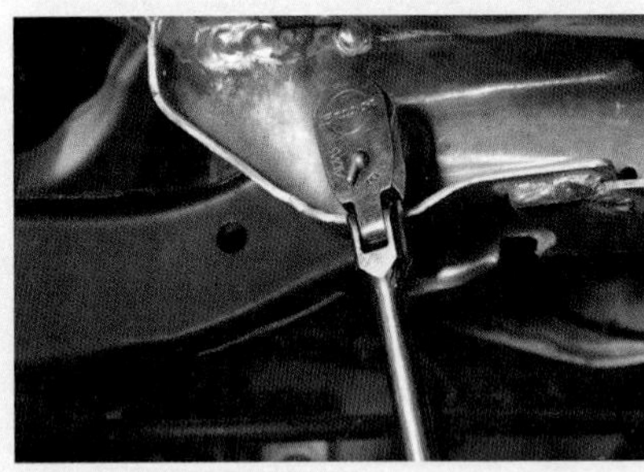

8. Prepare to adjust rear caster, camber, and toe, if possible, by loosening the eccentric bolts.

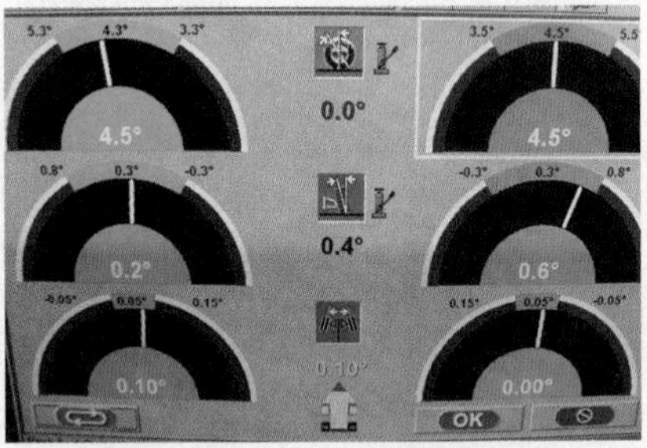

9. Adjust front caster and camber by turning the eccentric bolts attaching the control arm until alignment is within specifications.

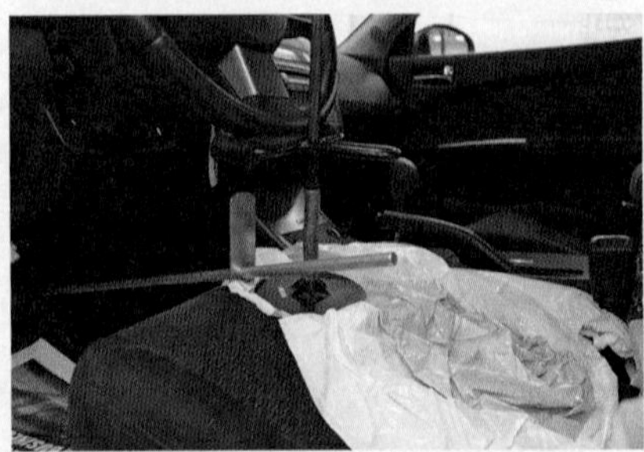

10. Install a steering wheel holder to center the steering wheel.

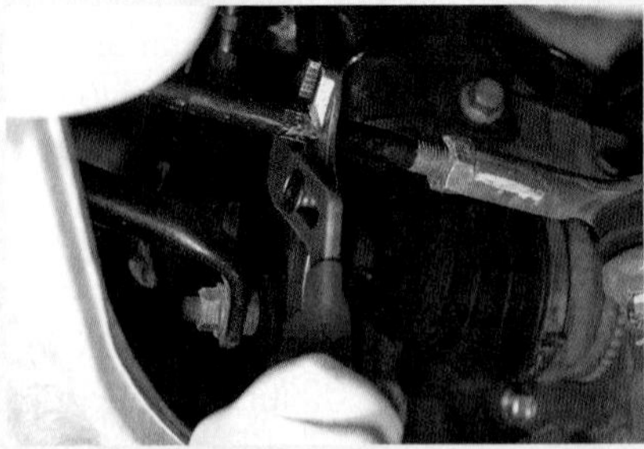

11. Adjust front toe by lengthening or shortening the tie-rod assemblies until toe is within specifications. Tighten the locknuts on the tie-rod assemblies.

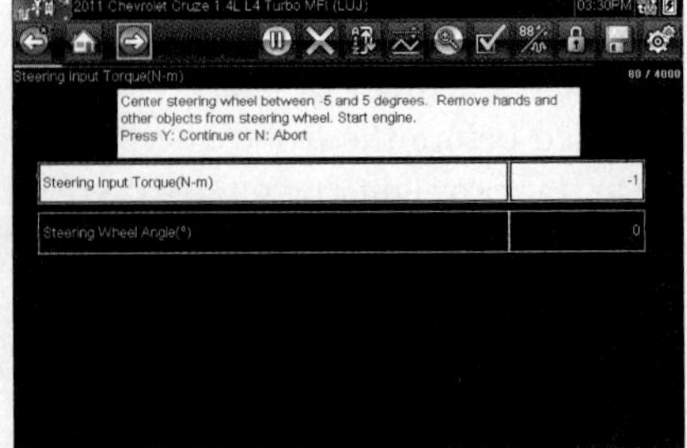

12. Verify that the steering wheel is centered, and calibrate the steering angle sensor. Test-drive the vehicle to make sure the repair was successful.

Checking Toe-Out on Turns

N42004

When you steer a vehicle through a turn, the outside front wheel has to travel a wider arc than the inside wheel. For this reason, the inside front wheel must steer at a sharper angle than the outside wheel to prevent the inside tire from dragging or sliding through a turn. To provide true rolling motion, the inner wheel must be turned through a sharper angle than the outer wheel. Toe-out on turns is sometimes referred to as TOOT, the Ackermann angle, or the track differential angle. Refer to service information to find the specifications for toe-out on turns. With the tires on the turntable, toe-out on turns is measured either electronically and displayed directly on the screen, or by reading the turning angle gauges (turn plates) (**FIGURE 42-27**). Refer to the manufacturer's specifications in the service information for the correct toe-out on turns angles. Typically, a vehicle has an angle of 20 degrees on the inside tire and 18 degrees on the outside tire. This is not the same for every vehicle, so be sure to check specifications. Make sure the readings are at zero on each side when the wheels are straight ahead. Turn the steering wheel so that the inside wheel is at the inside specification, and then check the outside wheel measurement.

If differences from specifications are found, then typically one or both steering arms are bent, but ensure that the tie-rod ends are not bent or the individual toe misadjusted. Be sure that caster, camber, and toe settings are within specifications before checking this angle, as incorrect angles will affect toe-out on turns. Compare the maximum turning angle of each wheel as well. If there are differences from specifications, then check for bent or shifted components such as a rack that is shifted or a steering wheel that was not properly centered during a repair, causing the wheels not to turn as sharply in one direction as the other.

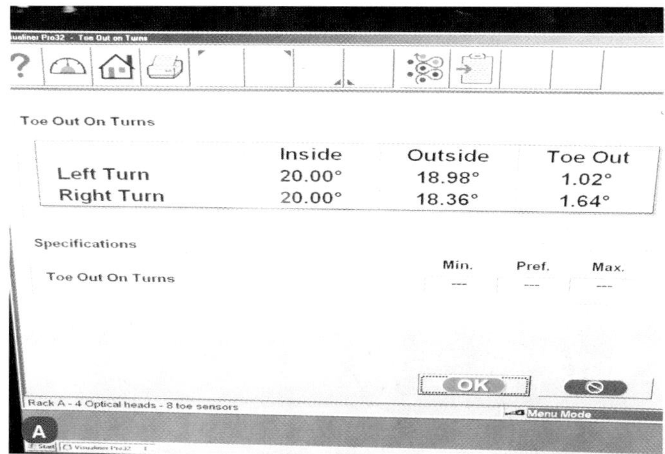

FIGURE 42-27 Displays of toe-out on turns: **A.** Alignment screen. **B.** Turn plates.

Checking SAI and Included Angle

N42005

The SAI is the angle formed by an imaginary line running through the upper and lower steering pivots relative to a vertical line. It cannot be adjusted, though, after an accident or body repair, a technician will on occasion be asked to check the SAI to ensure it is within the manufacturer's specifications. Be sure that caster, camber, and toe settings are within specifications before checking these angles, as incorrect angles will affect SAI and included angle. Generally, SAI and included angles should not vary more than half a degree from side-to-side, plus or minus. If SAI or included angle is incorrect, check for a bent strut, bent control arms, or a bent spindle or steering knuckle. If incorrect, SAI will create an issue with the steering wheel not returning to center after cornering. To check SAI and included angle, follow the steps in **SKILL DRILL 42-3**.

Checking Rear Wheel Thrust Angle

N42006

The rear thrust angle refers to the relationship between the rear wheels and an imaginary centerline down the center of the vehicle. Rear thrust angle problems often result from an accident or other impact that bends the rear axle or axle mounting points. Incorrect rear thrust angle can also be caused by wear on independent rear suspension components or incorrectly set rear toe. Incorrect thrust angle will try to steer the vehicle in the direction the wheels are pointing. Think of a monster truck with rear steering that becomes stuck, as you might have seen on TV. The rear of the vehicle will turn in the direction the rear wheels are pointing and the driver will have to turn the front wheels to make the vehicle crab-walk in a straight line. As this vehicle moves in a straight line, it will look like the body is sitting a bit sideways. Typically, the steering wheel will be off-center while driving because the driver will have to turn the front wheels to compensate for the rear. The ideal thrust angle will be close to zero, but refer to the manufacturer's specifications for correct thrust angle for the vehicle you are working on.

To check the rear wheel thrust angle, follow the steps in **SKILL DRILL 42-4**.

SKILL DRILL 42-3 Checking SAI and Included Angle

1. Position the vehicle on the alignment rack. Attach the wheel sensors on the vehicle to the locations specified by the sensor manufacturer, and compensate.

2. Follow the alignment machine instructions for taking the SAI measurements, and compare them to the vehicle manufacturer's specifications. Typically, the SAI reading will require a caster sweep to be performed. SAI is a nonadjustable angle. No changes can be made; the angle helps the technician to verify that suspension components are bent.

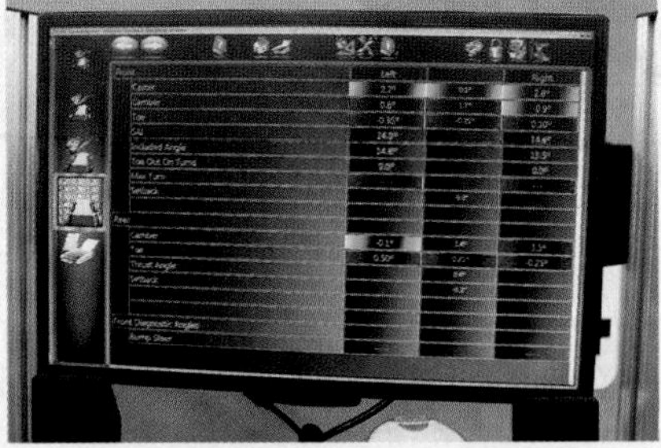

3. Calculate the included angle, if the alignment machine doesn't, by adding the camber reading of each wheel to the SAI of each wheel, and compare to specifications. Remember that if camber is a negative number, you need to subtract the camber from the SAI to get the included angle.

SKILL DRILL 42-4 Checking Rear Wheel Thrust Angle

1. Position the vehicle on the alignment rack. Attach the wheel sensors on the vehicle to the locations specified by the sensor manufacturer, and compensate.

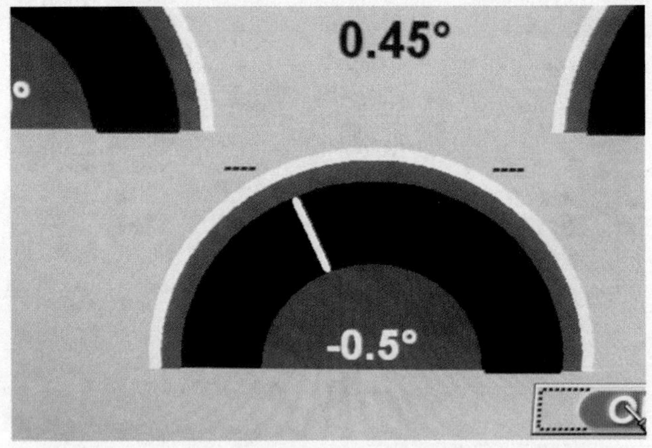

2. Thrust angle can be indicated on most alignment machines, although you may have to go to a special screen. Take the thrust angle reading, and compare it to the vehicle's manufacturer's specifications.

Checking Front and/or Rear Cradle Alignment

N42007

Damage to the cradle of a vehicle can force the wheels out of alignment, possibly changing caster, camber, toe, and set back readings, creating pull and tire wear issues. The cradle should always be perpendicular to the centerline of the chassis. It is possible that after an accident the cradle mounting points have moved (bent), and the vehicle will have to be sent to a frame shop to straighten the body before the alignment can be performed.

It is also important to ensure that the cradle is centered when removing and replacing this major component during service, such as when replacing the engine or transmission. It is recommended on some vehicles that an alignment be performed after cradle removal and reinstallation. Refer to the manufacturer's service information on cradle adjustment.

To check cradle alignment, follow the steps in **SKILL DRILL 42-5**.

SKILL DRILL 42-5 Checking Front and/or Rear Cradle Alignment

1. Position the vehicle on the alignment rack. Attach the wheel sensors on the vehicle, and compensate. Take the alignment readings and compare to the specifications. If camber and caster are incorrect and not adjustable, check for bent parts.

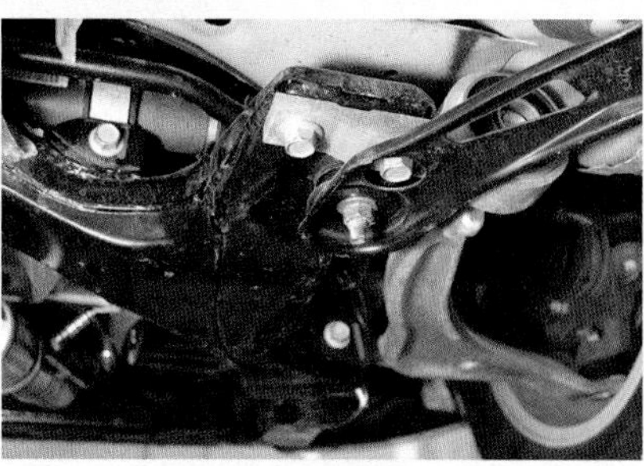

2. If the parts are not bent, check the positioning of the cradle. Loosen cradle bolts and shift in the necessary direction to correct alignment angles.

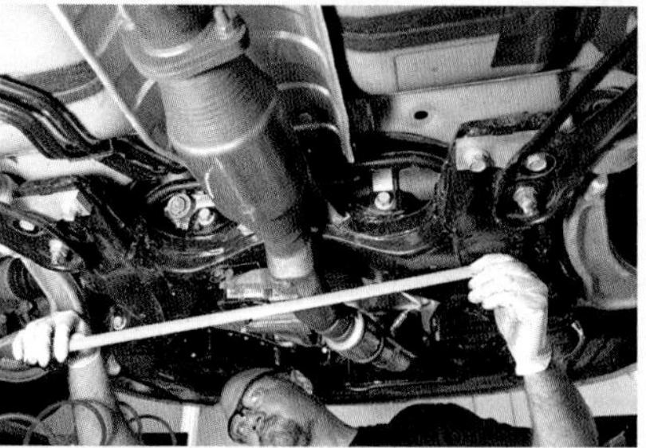

3. If adjustment is still not possible, check for a bent cradle or cradle mounting points by measuring from fixed points on one side compared to the same points on the other side. If they are different, the cradle needs adjusting.

▶ Wrap-Up

Ready for Review

- Wheel alignment is set using the vehicle's control arms, knuckles, and frame.
- Factors affecting wheel alignment include camber, caster, steering axis inclination, toe-in and toe-out, scrub radius, toe-out on turns, turning radius, thrust angle, and ride height.
- Positive camber refers to the top of tires tilting away from the vehicle; negative camber refers to the top of tires tilting toward the vehicle; zero average camber refers to no tire tilt.
- Caster refers to the angle formed between the centerline of steering axis and true vertical, or the forward/backward tilt of the ball joints.
- Steering axis inclination (SAI) provides the front wheels with a self-centering function; it is formed by drawing a line through the upper and lower pivot points of the suspension assembly.
- Scrub radius (steering offset) is the distance between the center point of the tire contact patch at the road surface and the point of steering axis centerline contact with the road.
- The Ackermann principle ensures that the inner wheels turn at a sharper angle than the outer wheels when turning.
- Thrust angle refers to the relationship of the rear wheels to the vehicle's imaginary centerline.
- When performing a wheel alignment, be sure to check front and rear cradle, thrust angle, wheel setback, wheel camber, caster, toe, SAI, and toe-out on turns.
- There are three basic types of wheel alignment: two-wheel alignment, thrust-angle, and four-wheel alignment.
- Tools commonly needed to diagnose suspension or wheel alignment issues are the wheel aligner, ramp, turntable, wheel clamps, specification charts, workshop manuals, specialty alignment tools, dial indicators, measuring tapes, scan tools, general hand tools, various lift devices, air ratchets, and wheel braces.
- Wheel alignment servicing equipment is used to measure the steering and suspension alignment angles.

Key Terms

Ackermann angle The angle the steering arms make with the steering axis, projected toward the center of the rear axle.

Ackermann principle The geometric alignment of linkages in a vehicle's steering such that the wheels on the inside of a turn are able to move in a smaller circle radius than the wheels on the outside.

camber The side-to-side vertical tilt of the wheel. It is viewed from the front of the vehicle and measured in degrees. Negative camber is when the top of the tire is closer to the center of the vehicle than the bottom of the tire.

caster The angle formed through the wheel pivot points when viewed from the side in comparison to a vertical line through the wheel.

centerline The imaginary line drawn down the exact center of the vehicle from front to back.

included angle The angle of camber added or subtracted to the SAI angle. This is the angle of the steering knuckle pivot points in relation to the camber angle of the wheel; also referred to as a diagnostic angle.

negative camber Tilt of the top of the tire toward the centerline of the vehicle.

negative caster Forward tilt of the steering knuckle pivot points from the vertical line.

negative scrub radius A condition in which the camber line is inside the steering axis centerline or when they intersect above the road surface.

positive camber Tilt of the top of the tire is tilted out from the centerline of the vehicle.

positive caster Backward tilt of the steering knuckle pivot points from the vertical line.

positive scrub radius A condition in which the camber line is outside of the SAI line or when they intersect below the road surface.

scrub radius The distance between two imaginary lines—the camber line and the SAI line. Or the point where they intersect above or below the surface of the road.

setback The distance one wheel is set back from the wheel on the opposite side of the axle.

static toe A setting designed to compensate for slight wear in steering components that may cause the wheels to turn slightly outward or inward while the vehicle is in motion.

steering axis inclination (SAI) The angle formed by an imaginary line running through the upper and lower steering pivots relative to vertical as viewed from the front.

thrust angle The angle formed between the perpendicular centerline of the rear axle in comparison to the centerline of the vehicle.

thrust line The imaginary line drawn perpendicular to the rear axle.

toe-in When the front of the wheels, as seen from above, are closer together than the rear of the wheels.

toe-out When the rear of the wheels, as seen from above, are closer together than the front of the wheels.

toe-out on turns The difference in turning angle of the inside tire (smaller turning radius) in comparison to the outside tire. This angle difference allows the tires to roll through the corner rather than the inside tire dragging; also referred to as Ackermann angle.

trim height The height from the ground to a specified part of the vehicle; also known as ride height.

turning radius A measure of how small a circle the outside front wheel on a vehicle can turn around when the steering wheel is turned to the limit.

wheel alignment The practice of aligning the wheels of the vehicle to the centerline of the vehicle and to one another. It ensures that the vehicle will handle correctly and gives best tire wear.

zero camber A tire with no tilt or zero camber angle.

zero scrub radius A condition in which the camber line intersects the SAI line at the road surface.

Review Questions

1. Wheel alignments should be performed for all of the following reasons *except*:
   - **a.** when the vehicle has covered a long distance with low tire pressure.
   - **b.** when tire wear shows any tire wearing angle issue.
   - **c.** when components are replaced that could affect the alignment.
   - **d.** whenever new tires are being installed.
2. All of the following statements with respect to camber are true *except*:
   - **a.** it is the side-to-side vertical tilt of the wheel.
   - **b.** it tends to pull in the opposite direction of the most positive camber.
   - **c.** it is affected by ride height.
   - **d.** in modern vehicles, large positive number angles would cause the tire to ride on the outer edges of the tires.
3. When a vehicle has negative caster:
   - **a.** the center of the tire contact point is in front of the steering axis centerline.
   - **b.** it causes the tires to travel in a straight line with minimal driver input.
   - **c.** it causes the spindle to tilt as the wheel is steered.
   - **d.** it can become unstable as speed increases.
4. All of the following statements are true *except*:
   - **a.** the condition in which the front of the wheels are closer together than the rear of the wheels is called toe-in.
   - **b.** the condition in which the front of the wheels are farther apart than the rear is called toe-out.
   - **c.** toe-out on turns is the relative toe setting of the rear wheels as the vehicle turns.
   - **d.** improper toe settings cause much faster tire wear than camber or caster.
5. The axis around which the wheel assembly swivels as it turns to the right or left is called the:
   - **a.** steering axis.
   - **b.** included angle axis.
   - **c.** diagnostic angle.
   - **d.** inclination angle.
6. On front-wheel drive vehicles, negative scrub radius will cause:
   - **a.** toe-in.
   - **b.** greater torque steer.
   - **c.** toe-out.
   - **d.** the front wheels to move backward.
7. If the camber line is outside of the SAI line (or they intersect below the road surface), then it has:
   - **a.** positive offset.
   - **b.** zero scrub radius.
   - **c.** negative offset.
   - **d.** zero offset.
8. Which of the following statements is correct?
   - **a.** Most front-wheel drive vehicles use a positive scrub radius.
   - **b.** Scrub radius is a measure of how small a circle the vehicle can turn in when the steering wheel is turned to the limit.
   - **c.** Ideally, the thrust line and the vehicle's geometric centerline should line up closely.
   - **d.** A setback issue in the rear will pull toward the side with the maximum setback.

9. A tool that is used to hold alignment targets to the wheels is a:
   a. wheel alignment machine.
   b. turntable.
   c. steering wheel holder.
   d. wheel clamp and extension.
10. Which of the following tools is used to keep the steering wheel from turning when centering the steering wheel during an alignment?
    a. Brake pedal depressor tool
    b. Wheel alignment machine
    c. Steering wheel holder
    d. Scan tool

ASE Technician A/Technician B Style Questions

1. Tech A says that thrust angle refers to the direction the front wheels are pointing. Tech B says that scrub radius refers to the intersection of the tire's camber line in relation to an imaginary line through the steering knuckle pivots. Who is correct?
   a. Tech A
   b. Tech B
   c. Both A and B
   d. Neither A nor B
2. Tech A says that a vehicle will tend to pull toward the side with the most positive camber. Tech B says that camber is the forward and backward tilt of the steering axis from vertical when viewed from the side. Who is correct?
   a. Tech A
   b. Tech B
   c. Both A and B
   d. Neither A nor B
3. Tech A says that caster is when the distance between the fronts of the tires are closer together than the rears of the tires. Tech B says that a vehicle will pull toward the side with the most positive caster. Who is correct?
   a. Tech A
   b. Tech B
   c. Both A and B
   d. Neither A nor B
4. Tech A says that toe-in is when the front of the tires are farther apart than the rear of the tires. Tech B says that excessive toe causes faster tire wear than camber or caster. Who is correct?
   a. Tech A
   b. Tech B
   c. Both A and B
   d. Neither A nor B
5. Tech A says that worn ball joints can affect alignment angles. Tech B says that worn control arm bushings can affect alignment angles. Who is correct?
   a. Tech A
   b. Tech B
   c. Both A and B
   d. Neither A nor B
6. Tech A says that the front wheels should be aligned before the rear wheels. Tech B says that front wheel toe should be adjusted after front wheel caster and camber. Who is correct?
   a. Tech A
   b. Tech B
   c. Both A and B
   d. Neither A nor B
7. Tech A says that when the front wheels are being steered around a corner, the fronts of the tires are further apart than the rears. Tech B says that steering axis inclination is formed by drawing a line through the center of the upper and lower pivot points of the suspension assembly. Who is correct?
   a. Tech A
   b. Tech B
   c. Both A and B
   d. Neither A nor B
8. Tech A says that changing the diameter of the wheels and tires can affect scrub radius. Tech B says that the vehicle's thrust angle is the same as the rear wheels' camber angle. Who is correct?
   a. Tech A
   b. Tech B
   c. Both A and B
   d. Neither A nor B
9. Tech A says that checking tire size and pressure is a part of the pre-alignment inspection. Tech B says that the alignment machine requires a compensation process for the wheel adapters. Who is correct?
   a. Tech A
   b. Tech B
   c. Both A and B
   d. Neither A nor B
10. Tech A says that most all vehicles have adjustable caster and camber. Tech B says that when performing a four-wheel alignment, the rear wheels are adjusted first so they conform to the vehicle's centerline. Who is correct?
    a. Tech A
    b. Tech B
    c. Both A and B
    d. Neither A nor B

brembo

SECTION 6
Brakes

CHAPTER 43
Principles of Braking

NATEF Tasks

- **N3001** Describe the operation of a regenerative braking system. (MLR/AST/MAST)

Knowledge Objectives

After reading this chapter, you will be able to:

- **K43001** Describe the evolution of braking systems over time.
- **K43002** Explain braking system fundamentals.
- **K43003** Describe the factors that affect braking.
- **K43004** Describe kinetic energy.
- **K43005** Describe acceleration and deceleration.
- **K43006** Describe energy transformation.
- **K43007** Describe the purpose and function of friction and friction brakes.
- **K43008** Describe the process of heat transfer.
- **K43009** Explain the types of brake fade.
- **K43010** Describe rotational force and its effect.
- **K43011** Describe levers and mechanical advantage, and provide examples.
- **K43012** Describe the purpose and operation of the various types of brake systems.
- **K43013** Describe the operation of hydraulic brakes.
- **K43014** Describe the operation of air brakes.
- **K43015** Describe the operation of exhaust brakes.
- **K43016** Describe the operation of compression brakes.
- **K43017** Describe the operation of electric brakes.
- **K43018** Describe the operation and components of parking brakes.
- **K43019** Describe the operation of brake-by-wire brake systems.

Skills Objectives

There are no Skills Objectives for this chapter.

▶ Introduction

The brake system is one of the most critical systems on a vehicle. It allows the driver to slow or stop the vehicle as needed. In ideal situations, the driver will have enough time to anticipate the need to slow down well in advance of an event, allowing the vehicle to slow down gradually. However, many situations require the quick use of a very efficient braking system to avoid an accident (**FIGURE 43-1**). In this chapter, we explore the history, theory, and operation of modern braking systems.

▶ The History of Brakes

K43001

Early automobiles evolved from horse-drawn buggies and used a similar scrub braking system. **Scrub brakes** are a simple mechanical system that uses leverage to force a friction block against one or more wheels (**FIGURE 43-2**). In a buggy, for example, the friction between the two surfaces transformed the energy of the moving buggy into heat energy. As heat was created in the friction materials, the buggy slowed down. The scrub braking system was used for more than 2000 years with virtually no change. It worked reasonably well on dry wheel surfaces made of wood or steel but became quickly outdated once rubber tires were developed around 1900, because the scrubbing action on the softer tires significantly decreased tire life.

FIGURE 43-1 Brakes must be fully functional in case of an emergency.

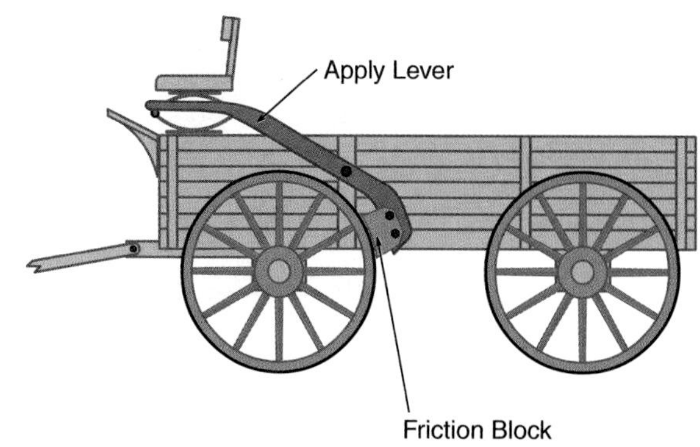

FIGURE 43-2 Scrub brakes operated directly on the wheel.

Faced with the need to replace the scrub braking system with a system that did not apply friction materials directly to the new rubber tires, designers had to consider other options. One option was the **band brake.** It used a metal band lined with friction material that was operated mechanically to clamp around the outside of a small diameter wheel or that was drum mounted to the axle or wheel (**FIGURE 43-3**). This system worked well in the forward direction, but the band would try to unwind in the reverse direction; the system was therefore impractical and was abandoned after just a few short years.

The next major development was the **drum brake**, which is similar to the drum brakes that are used today. The drum brake consisted of two brake shoes that would push against the inside of the brake drum. Early drum brake systems were mechanically operated by rods, links, and levers (**FIGURE 43-4**). This worked fine for applying a single brake unit, but as vehicle operating speeds increased, greater braking demands required brake units to be mounted to each wheel. With this mechanically operated system of rods, links, and levers, it was difficult to maintain equal braking forces at each wheel and caused the vehicle to veer dangerously to one side when braking. This system also required frequent adjustment of the brakes. Over time, the mechanical drum brake system gave way to the modern hydraulic drum brake system because of its ability to automatically equalize braking forces at each wheel.

You Are the Automotive Technician

A customer comes into the shop with a brake concern on her 2009 Ford Taurus. She recently drove down a long mountain pass, using her brakes to stay within the speed limit. As she neared the bottom of the pass, the brakes felt like they weren't working, even though the engine was running and she was pushing very hard on the pedal. You recognize the smell of overheated brakes coming from the car. She asks you why her brakes didn't work correctly.

1. How will you explain what happened to the brakes?
2. What would you advise her to do if she is in that situation again?
3. Which type of brakes are more susceptible to this condition?

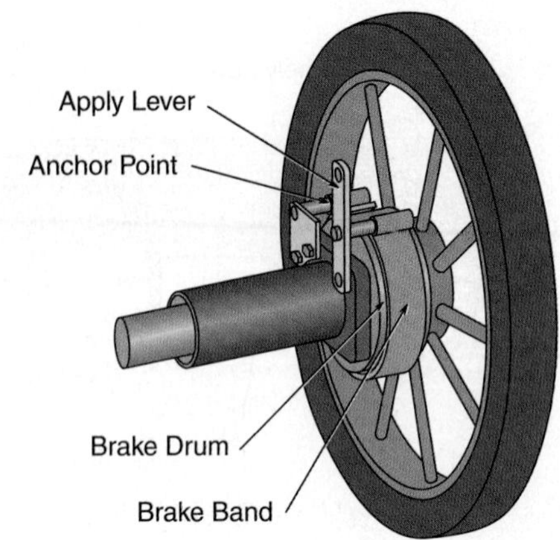

FIGURE 43-3 An old style band brake.

FIGURE 43-4 An early drum brake system.

Disc brakes were originally developed in the early 1900s but didn't find common use until the 1960s. Because the effectiveness of friction brakes depends on the braking components' ability to dissipate heat quickly into the atmosphere, drum brakes with their friction materials on the inside were at a disadvantage. This led to the greater use of **disc brakes** on most vehicles (**FIGURE 43-5**). Disc brakes force brake pads against the outside of the brake rotor and create heat where atmospheric air can quickly remove the heat at its source, making them more efficient under prolonged use.

Advanced Brake Systems

Electronic Brake Control

You might think, "If some brake force is good, then more would be better." But this is not the case. Applying too much brake force can cause the tire to lose traction and skid. If this happens, the driver can lose control of the vehicle. This problem led to the design of electronic brake control (EBC) systems. The first versions of EBC systems were the **antilock brake systems (ABS)**. ABS systems have helped to reduce the number of vehicle accidents each year. They use a computer to monitor each

FIGURE 43-5 Disc brakes force the brake pads against the outside of the brake rotor.

wheel's speed and either hold, release, or apply the hydraulic pressure to each wheel to prevent wheel lockup and maintain the maximum amount of braking power just short of brake lockup. Additional safety demands led to the development of traction control systems (TCS) and electronic stability control (ESC). TCS systems help to prevent tires from slipping during acceleration by reducing engine torque and, if necessary, applying brake pressure to the slipping wheels. ESC adds additional functionality to ABS and TCS to help prevent the tires from losing traction when the vehicle is being steered aggressively or evasive maneuvers are being undertaken. In both TCS and ESC, the control unit can independently apply individual brake units even though the driver is not stepping down on the brake pedal. For more information on ABS brake systems, see the chapter on Electronic Brake Control.

Brake Assist and Brake-by-Wire Systems

Electronic brake control systems have been further enhanced with additional programmed features such as **brake assist (BA)**. BA gives greater control of the braking system to the computer, which can react more quickly and more deliberately than a driver, especially in a panic situation. An example of BA is the **brake-by-wire system**. A full brake-by-wire system does away with the hydraulic portion of the brake system and replaces it with sensors, wires, an electronic control unit, and electrically actuated motors to apply individual brake units at each wheel (**FIGURE 43-6**). The driver applies foot pressure to a **brake pedal emulator**. This tells the computer how firmly the driver intends to brake. The control unit then sends control signals to the appropriate brake actuators, which generate the commanded clamping force and slow the vehicle. All of this is being monitored by sensors reporting data to the control unit, so the desired braking occurs.

Giving greater control of the braking system to the computer increases driving safety. For example, it takes a certain amount of time and stopping distance for the driver to lift his or

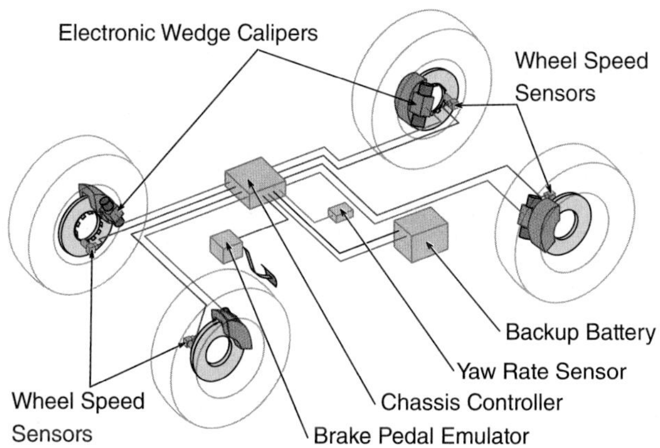

FIGURE 43-6 Schematic of a brake-by-wire system.

her foot from the accelerator pedal, step on the brake pedal, and apply the wheel brake units. In a brake-by-wire system, that time and distance can be reduced. The computer can determine that the driver is quickly releasing the accelerator, which indicates a potential panic stop. As the pedal is being released, the control system immediately applies the brakes lightly, which dries any moisture from the braking surfaces and takes up any clearance in the brake system. This prepares the brakes to be fully applied by the control system if the driver steps on the brake pedal or if the computer detects that a collision is about to happen.

▶ TECHNICIAN TIP

Many auto insurance companies give their customers a discount when the vehicle is equipped with brake safety features such as ABS systems, as they help avoid or minimize accidents.

Regenerative Brake Systems

With the pressure to improve fuel economy, some manufacturers have equipped their hybrid vehicles with regenerative braking. Regenerative braking takes brake-by-wire technology to the next level. Instead of applying friction brakes and losing energy as heat, the brake-by-wire system uses the electric motor as a generator, which slows the vehicle by converting the vehicle's kinetic energy into electrical energy. The amount of stopping power is controlled by how much electricity is being generated (**FIGURE 43-7**). The more stopping power needed, the more electrical output is demanded of the generator (up to its maximum rated output) by the control system. The electricity is stored in the vehicle's high-voltage battery and can then be used later by the electric motor to drive the vehicle. This regeneration process makes the vehicle more fuel efficient, especially in stop-and-go traffic.

▶ Brake Fundamentals

K43002

Service and Parking Brakes

There are two brake systems on all vehicles—a service brake and a parking brake. The **service brake** is used for slowing or stopping the vehicle when it is in motion and is operated by a foot pedal (**FIGURE 43-8**). Service brakes consist of drum and/or disc brakes. Some have disc brakes on the front wheels and drum brakes on the rear wheels; others have disc brakes on all four wheels. The **parking brake** is used for holding the vehicle in place when it is stationary. The parking brake is usually operated by hand, but some vehicles use a foot activated pedal (**FIGURE 43-9**).

FIGURE 43-7 A vehicle display showing regeneration charging the high-voltage battery.

FIGURE 43-8 A vehicle service brake is used for slowing or stopping a moving vehicle.

Modern braking systems are hydraulically operated and have two main sections: the brake assemblies at the wheels and the hydraulic system that applies them. The driver pushes the brake pedal, which applies mechanical force to the pistons in the master cylinder. The pistons apply hydraulic pressure to the fluid in the cylinders. The lines transfer the pressure, which is applied equally in all directions within the confines of the brake lines to the hydraulic cylinders. The hydraulic cylinders at the wheel assemblies apply the brakes (**FIGURE 43-10**).

Force is transmitted hydraulically through the fluid. For cylinders of the same size, the force transmitted from one is the same value as the force applied to the other. By using cylinders of different sizes, forces can be increased or reduced, allowing

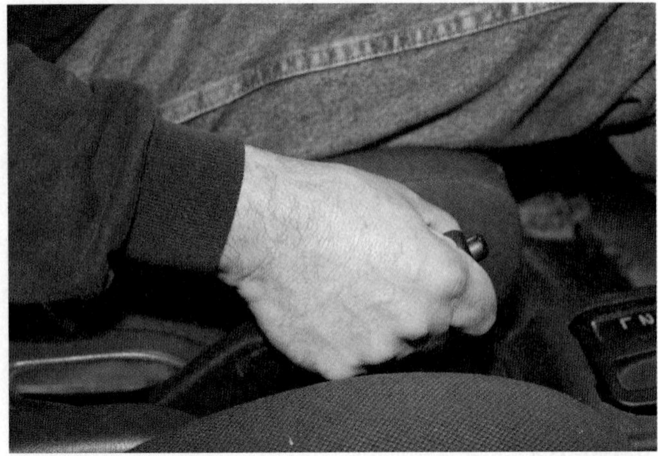

FIGURE 43-9 A vehicle parking brake is used to hold the vehicle in place when it is stationary.

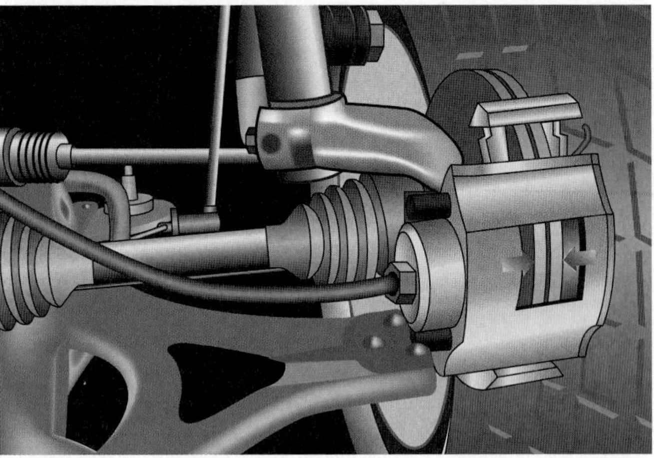

FIGURE 43-12 In disc brakes, pads are forced against the outside of a brake disc.

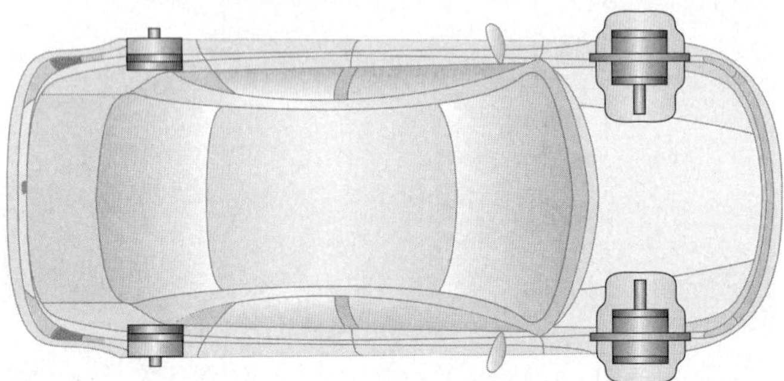

FIGURE 43-10 A braking system in a typical modern vehicle.

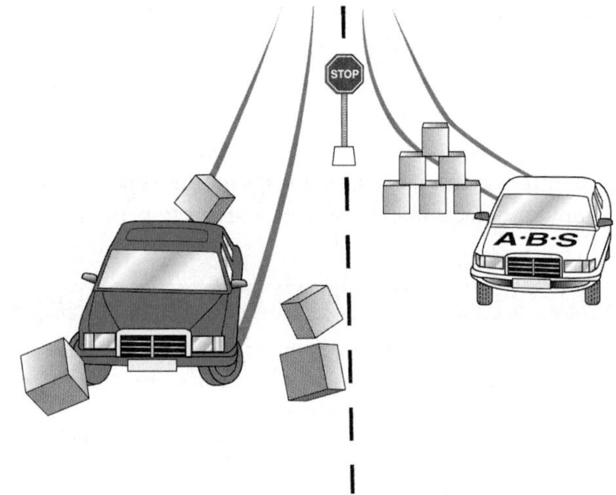

FIGURE 43-13 ABS brakes help to prevent skidding and maintain directional control of the vehicle.

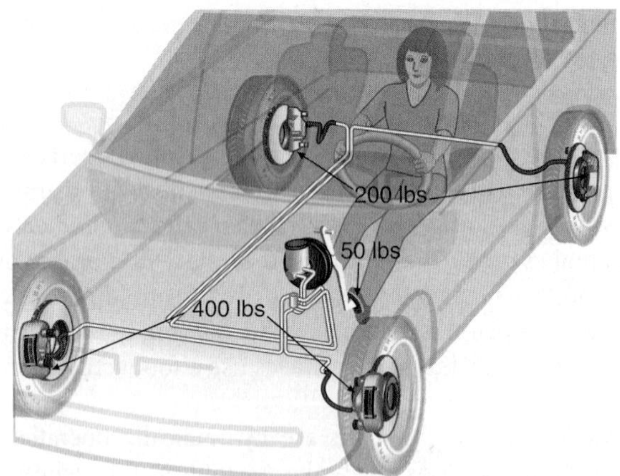

FIGURE 43-11 By using cylinders of different sizes, hydraulic forces can be increased or reduced, allowing designers to obtain the desired braking force for each wheel.

designers to obtain the desired braking force for each wheel (**FIGURE 43-11**). The cylinders force friction linings into contact with the braking surfaces. The resulting friction between the surfaces generates heat energy and slows the vehicle.

In disc brakes, pads are forced against the outside of a brake disc (**FIGURE 43-12**). In both systems, heat spreads into other parts and the atmosphere, so brake linings and drums, pads and discs, and brake fluid must withstand high temperatures and high pressures. On modern vehicles, the basic brake system has some refinements such as a power booster and EBC systems. These help the driver apply the brakes, prevent skidding, and maintain directional control of the vehicle under various driving situations (**FIGURE 43-13**).

Factors That Affect Braking

K43003

A number of factors can influence vehicle braking. An effective braking system takes all of the following factors into account:

- **Road surface:** Generally asphalt and concrete road surfaces allow for good braking while gravel surfaces or dirt roads do not.
- **Road conditions:** Roads that are wet, icy, or covered with loose gravel reduce the tires' traction and result in longer

stopping distances. Extremely hot temperatures on asphalt roads can soften the asphalt, making it slippery.

- **Weight of the vehicle:** Heavier vehicles require more braking force to stop than lighter vehicles and therefore usually have larger wheel brake units. Also, loading down a vehicle increases its stopping distance due to the vehicle's extra mass.
- **Load on the wheel during stopping:** Heavier loads increase the downward force on the wheels, thereby increasing tire traction (**FIGURE 43-14**).
- **Height of the vehicle:** Stopping power is exerted at the point where the tire and the road connect. The centerline of the vehicle's weight is above this tire-to-road contact point. The taller the vehicle, the greater the leverage on the contact point. During braking, this shifts some of the vehicle's weight from the rear wheels to the front wheels. Thus, controlling the vehicle in a panic situation becomes much more difficult.
- **How the vehicle is being driven:** Aggressive driving causes the tires to become hot and possibly overheated, thus reducing the tire's ability to obtain maximum traction. Also, increased speed and aggressive handling force the brakes to work under extreme conditions.

FIGURE 43-14 Tires with differing loads. **A.** The empty pickup truck has low traction on its rear tires. **B.** A heavily loaded truck has higher traction on its rear wheels.

FIGURE 43-15 The condition of the tires affects braking performance. **A.** Tire in excellent condition. **B.** Worn-out tire.

- **The tires on the vehicle:** A tire's composition, tread style, tread condition, and inflation pressure all affect its traction (**FIGURE 43-15**). Manufacturers design tires with different qualities based on vehicle need. Tires are rated for their traction ability. Using the wrong tire will affect the vehicle's stopping power. For example, a tire with tread designed to channel water away from the tire-to-road contact point has greater traction when the road surface is wet than the tires used on drag race cars, which have a slick tread.

Kinetic Energy

K43004

Kinetic energy is the energy of an object in motion. All moving objects have kinetic energy. Heavier objects have more kinetic energy than lighter objects moving at the same speed. If the weight doubles, the kinetic energy doubles. Faster moving objects have more kinetic energy than slower moving objects of the same weight. Kinetic energy increases by the square of the speed. This means that if we double the speed of an object, the kinetic energy will increase by four times. If we triple the speed, the kinetic energy will increase by nine times. Thus, the heavier and faster an object is, the greater its kinetic energy (**FIGURE 43-16**). During braking, all of the kinetic energy in the

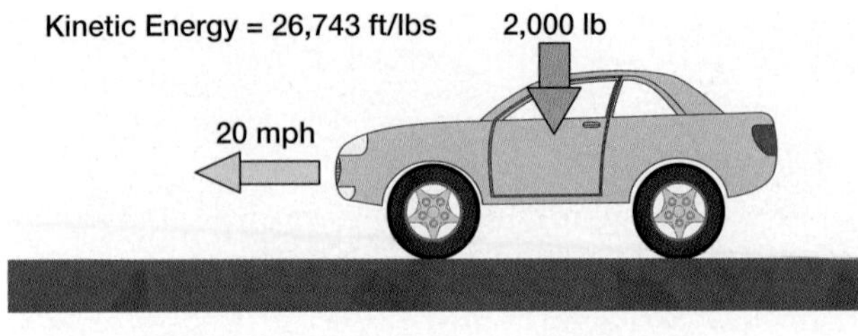

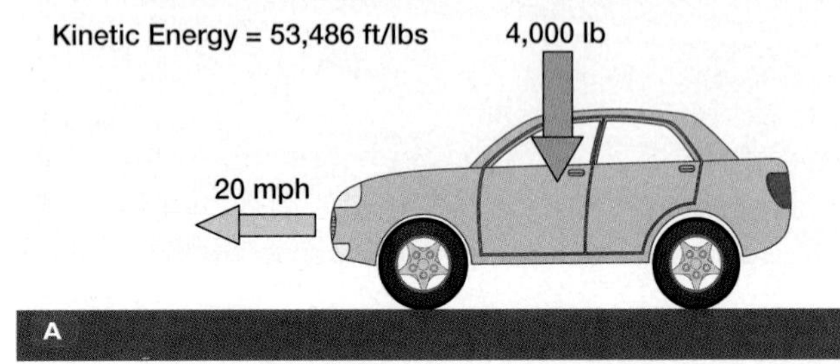

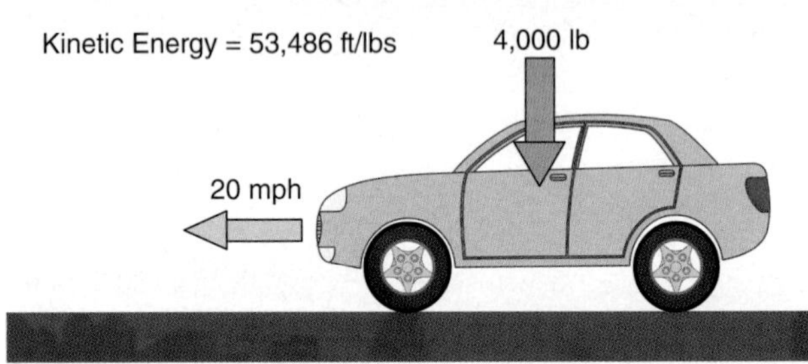

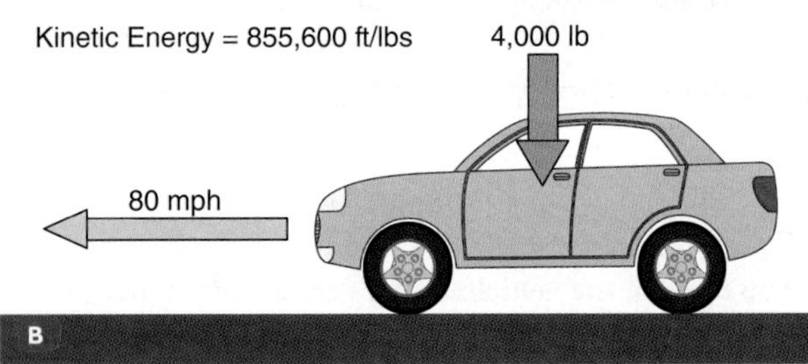

FIGURE 43-16 **A.** If weight doubles, kinetic energy doubles. **B.** If speed increases, kinetic energy increases by the square.

moving vehicle must be converted to another form of energy (in most cases, heat) for the vehicle to stop moving; this is the function of the braking system.

Acceleration and Deceleration

K43005

Newton's first law of motion states that an object will stay at rest or uniform speed unless it is acted upon by an outside force. **Acceleration** refers to an increase in an object's speed. In an automobile, acceleration, or an increase in kinetic energy, is caused by the power from the engine. When the driver steps down on the throttle pedal, the engine's power output is increased, and the vehicle accelerates (**FIGURE 43-17**). This acceleration requires a certain amount of energy. The heavier the vehicle, the more energy required to accelerate it to a given speed. A lighter vehicle requires less energy to accelerate; this is why race cars are stripped of all unnecessary weight.

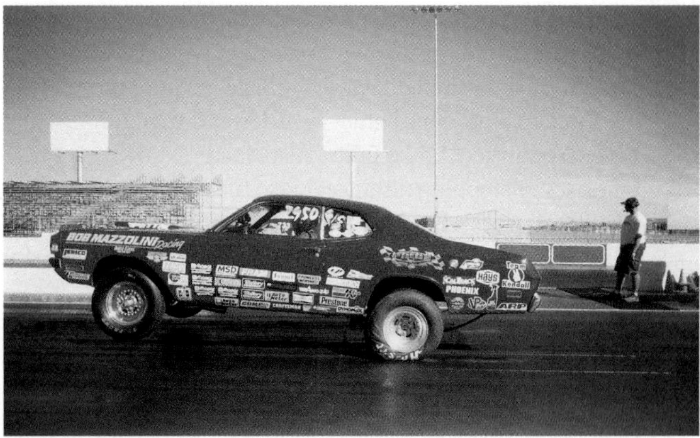

FIGURE 43-17 The engine's horsepower is accelerating this vehicle hard.

Deceleration refers to a decrease in an object's speed. Remember that an outside force is needed for the speed of an object to change. So we need an outside force to act upon the vehicle to cause it to decelerate. That force comes from the mass of the Earth. If you thought it came from the brakes, you would only be partially correct. Imagine traveling at a high speed in a four-wheel drive vehicle and hitting a bit of a jump. Stepping on the brakes in mid-air to slow the vehicle wouldn't do you much good, would it? So the brakes only function when they connect the vehicle to the ground or roadway. In fact, that is what they do; they connect the vehicle to the ground through the rolling wheel and tire assembly. In doing so, they apply a varying amount of force from the ground to the vehicle, thereby causing the vehicle to decelerate. The force of the brakes absorbs the kinetic energy of the vehicle (**FIGURE 43-18**). The heavier the vehicle and the faster it is going, the more kinetic energy must be dissipated and the harder the brakes must work.

Applied Science

AS-51: Acceleration/Deceleration: The technician can demonstrate an understanding of a vehicle's acceleration and deceleration as a function of vehicle weight and power.

Kinetic energy is the energy of an object in motion. All moving objects have kinetic energy. Heavier objects have more kinetic energy than lighter objects moving at the same speed. If the weight doubles, the kinetic energy doubles. Energy is needed to start a vehicle. Heat energy is generated in the engine via chemical energy (fuel); it is then converted via mechanical energy to kinetic energy, putting the vehicle in motion. Kinetic energy is converted to heat energy once again through the operation of the brakes. This heat energy is then dissipated in the surrounding air through the brake system, bringing the vehicle to rest.

Newton's first law of motion states that the greater the weight, the more energy is needed to accelerate and maintain speed. For example, it is easier for three people to push a two-door hatchback than an SUV. Also, the more the vehicle weighs, the more energy it takes to decelerate.

FIGURE 43-18 The vehicle's brakes are decelerating this vehicle hard.

Energy Transformation

K43006

The law of **conservation of energy** states that energy cannot be created or destroyed. This means that the energy used to cause a vehicle to accelerate and decelerate must be transformed from one form of energy to another. Let's follow the cycle of energy transformation in a typical vehicle.

Gasoline or diesel fuels are potential energy in chemical form. A portion of the fuel's chemical energy is transformed within the engine, first into heat energy and then into mechanical energy. The mechanical energy is used to accelerate the vehicle, thus converting the mechanical energy to kinetic energy. Once the vehicle is up to speed, the engine only needs to transform enough chemical energy into kinetic energy to overcome wind resistance, climb hills, and power the vehicle's accessories. This is why most vehicles get better fuel economy while operating at steady speeds than in stop-and-go traffic: it takes a lot more energy to accelerate a vehicle than it does to maintain a particular speed.

Does deceleration require an energy transformation? Yes, it does. The kinetic energy has to be removed from the vehicle for it to decelerate. In other words, the kinetic energy must be transformed into another form of energy for the vehicle to slow down. In a standard vehicle, the braking system transforms the kinetic energy into heat energy (**FIGURE 43-19**). In essence, it takes the same amount of energy to slow a vehicle as it does to accelerate it. For safety's sake, we expect a vehicle to stop from a given speed faster than the time it took to accelerate to that speed. For this reason, the braking system can transform energy faster than the engine. Or in other words, the brake system can absorb energy at a faster rate than the engine can create it.

Friction and Friction Brakes

K43007

Brakes transform kinetic energy to another form of energy. Standard brakes do this through the principle of friction. **Friction** is the resistance created by surfaces in contact. Static friction is resistance between non-moving surfaces and is present in parking brakes. Kinetic friction is resistance between moving surfaces and is present in standard brakes. Just as rubbing sandpaper over a block of wood produces heat, operating the brakes causes the moving friction surfaces to come into contact with each other and generate heat. This transformation of energy converts kinetic energy into heat energy and slows the vehicle.

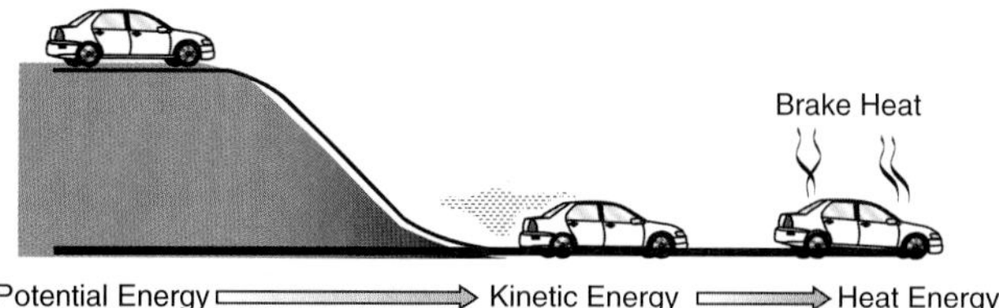

FIGURE 43-19 During deceleration, kinetic energy is transformed into heat energy.

The amount of friction between two moving surfaces in contact with each other is expressed as a ratio called the **coefficient of friction**. It can be found by comparing the amount of force pushing the two surfaces together to the amount of resistive force generated between the two surfaces sliding against each other. For example, a stationary steel surface pushed against a moving steel surface with 100 lb (45.36 kg) of force might generate 20 lb (9.07 kg) of resistive force. This is expressed as 20/100 (9.07/45.36), which equals a coefficient of friction of 0.20 (**FIGURE 43-20**). A stationary block of rubber that is pushed against a moving steel surface with 100 lb of force might generate 125 lb (56.7 kg) of friction. This would be 125/100 (56.7/45.36), which equals a coefficient of friction of 1.25.

▶ TECHNICIAN TIP

A higher coefficient of friction usually results in a faster wearing of the softer material, such as rubber. This is one reason why the brake lining is designed to be made of softer materials than the drum or rotor, which leads to wearing out the brake lining more than the drum or rotor.

Applied Science

AS-90: Friction: The technician can demonstrate an understanding of friction and its effects on linear and rotational motion.

When adjusting drum brakes, the common method is to adjust the brake and turn the wheel at the same time in order to feel how hard it is to turn the wheel while adjusting the brake. As the brake becomes tighter, the friction between the brake drum and the brake shoe increases; therefore, the rotational motion requires more torque or force to produce the same speed of movement.

Linear friction is demonstrated when a heavy object is pulled across a flat surface. The more surface area that is in contact with the object, the harder it is to pull. The larger the surface area that the object is covering, then the greater the force required to move it. For example, try to pull a tool box across a bench. Now lift one end of the toolbox and pull. It is far easier to pull the toolbox with one end lifted because the lifted tool box is in contact with less surface area, reducing the force required to move the box.

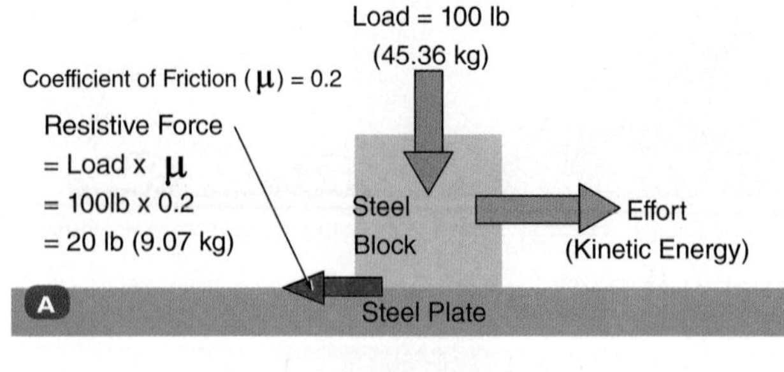

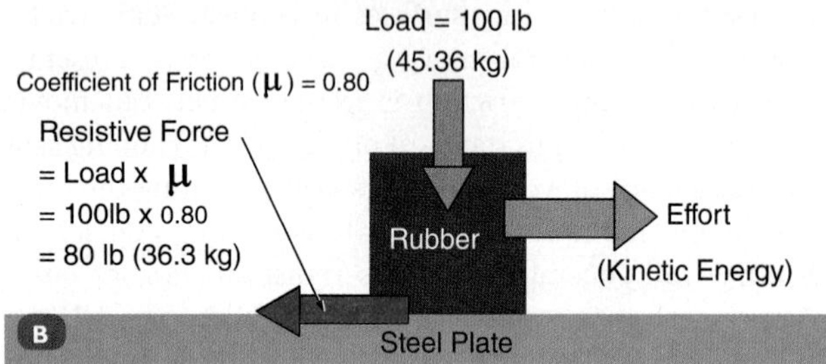

FIGURE 43-20 A. Coefficient of friction = 0.20. **B.** Coefficient of friction = 0.80.

Applied Science

AS-91: Friction: The technician can explain the role that friction plays in acceleration and deceleration.

Friction is very important to acceleration and deceleration. A vehicle's tires must be able to keep friction between the tire and the ground in order to propel the car forward. If the tires are not in contact with the road, they will have nothing to push against. No friction is present and the car will not move forward.

Friction is also very important in deceleration, as the tire again has to maintain friction between the road surface and tire. The braking force on the vehicle also uses friction in order to slow or decelerate the vehicle. As the brake pads press against the brake disc, friction is created between the two surfaces. This friction creates heat and slows the vehicle.

Heat Transfer

K43008

Heat transfers from a hot area to a cool area. Because there is a lot of kinetic energy converted to heat during the braking process, the brakes must be able to dissipate that heat to the atmosphere effectively. Heat transfer is critical to this process. Heat must continually be transferred away from the friction materials so that the brakes can perform their job of transforming the kinetic energy into heat energy.

Ultimately, most of the heat generated by the braking process radiates into the atmosphere. How it radiates into the atmosphere depends on the type of braking system (**FIGURE 43-21**). In drum brakes, the heat is created inside of the drum and transfers through the drum to the outside surface where it radiates into the atmosphere. In disc brakes, the heat is created on the outer surfaces of the rotors, which are in contact with the atmosphere. Disc brakes also may have internal ventilation to help dissipate the heat transferred from the outer surface even faster.

Brake Fade

K43009

In automobiles, **brake fade** is the reduction in stopping power caused by a change in the brake system caused by three factors. The first and most common is **heat fade**. Heat fade is caused by the buildup of heat in the braking surfaces, which get so hot they cannot create any additional heat, leading to a loss of friction (**FIGURE 43-22**). Remember, the brakes must transform kinetic energy into heat energy to decelerate the vehicle. If heat energy cannot be generated, then the kinetic energy cannot be reduced and the vehicle will not decelerate. Heat transfer is used to move heat away from the friction surfaces and allow them to continue generating heat. Once the temperature of the friction materials become so hot that they cannot generate any additional heat, the coefficient of friction drops, and the brakes cannot generate stopping power until some of the heat dissipates.

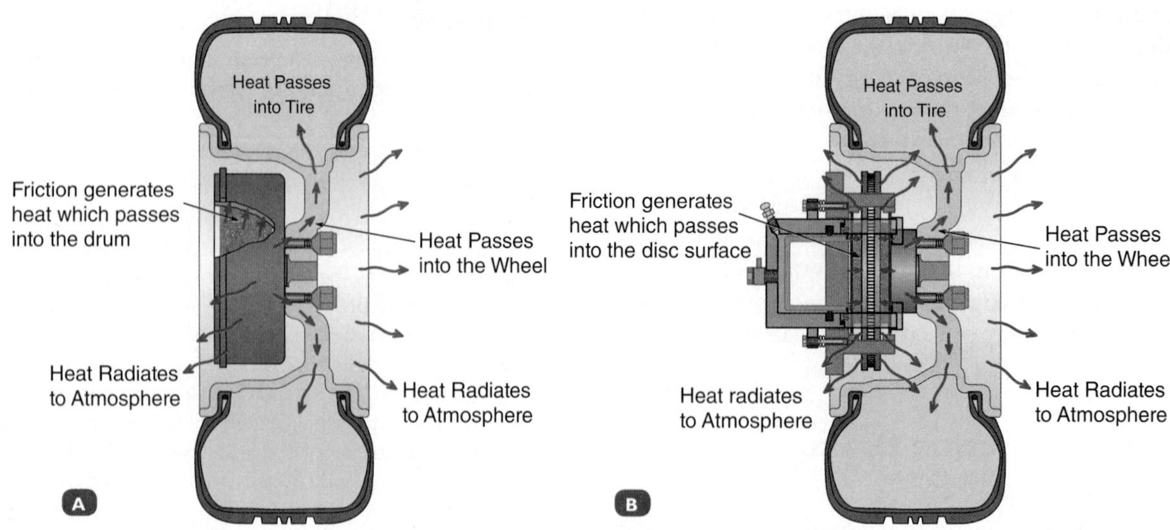

FIGURE 43-21 A. Heat transfer in a drum brake. **B.** Heat transfer in a disc brake.

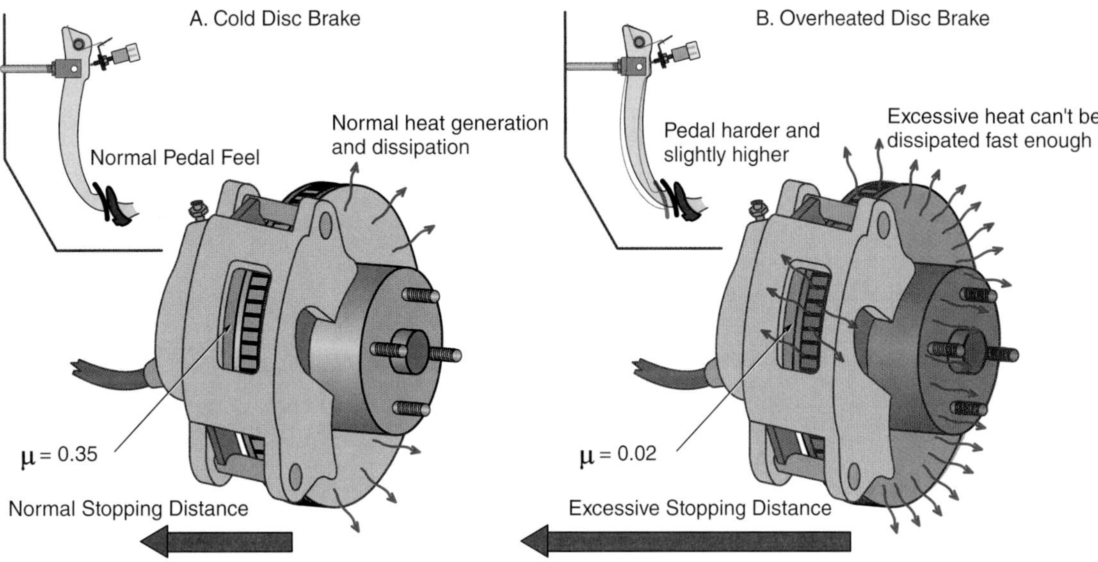

FIGURE 43-22 Heat fade—caused by the brake system reaching the temperature generated by the friction of the brake pads.

▶ TECHNICIAN TIP

Excessive heat can cause warpage of disc brake rotors and brake drums. Brake fade and rotor warping can be reduced through proper braking technique. When traveling on a long downgrade requiring braking, the driver should simply select a lower transmission gear. Periodic rather than continuous application of the brakes allows the brakes to cool between applications. Continuous light application of the brakes, sometimes referred to as riding the brakes, can be particularly destructive in both wear and overheating of the brake components.

A driver experiences heat fade after using the brakes too much, such as during high-performance driving or when going down a long, steep hill, particularly when towing. The brake pedal will be hard, but the braking effect "fades" away, and the vehicle's rate of deceleration decreases. This is a dangerous condition and is why many long hills on freeways have truck escape ramps made of sand or some other soft material to slow a vehicle by absorbing the truck's kinetic energy in the soft material.

The second type of brake fade is called **water fade** and is caused by water-soaked brake linings. The water acts like a lubricant and lessens the coefficient of friction between the braking surfaces (**FIGURE 43-23**). This leads to a hard brake pedal but very little braking power. Once the water is removed from the friction surfaces through evaporation, the normal coefficient of friction will be restored.

The third kind of brake fade is called **hydraulic fade** and is caused by the brake fluid becoming so hot that it boils. Once it boils, it is no longer only a liquid, converting in part to a vapor,

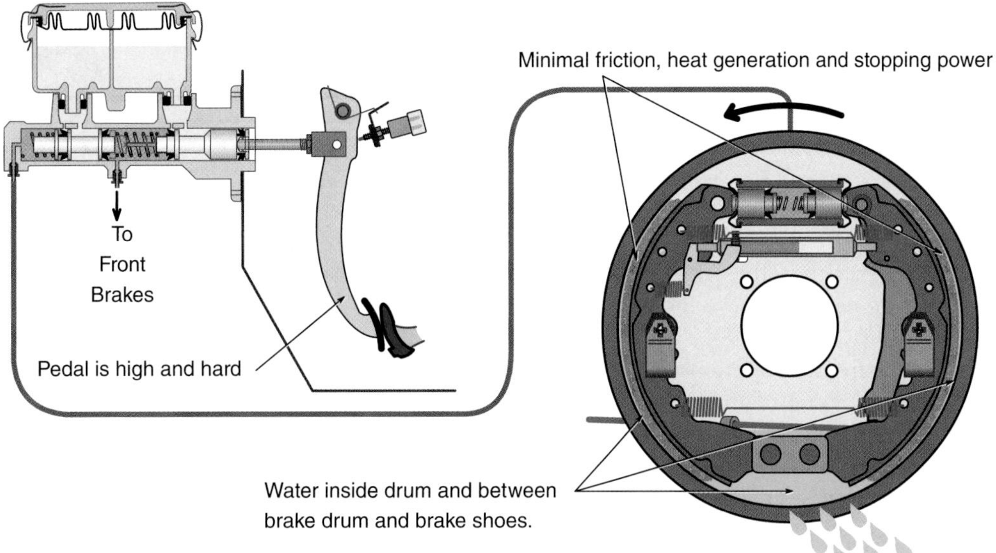

FIGURE 43-23 Water fade—water reduces the friction in brakes, causing a hard pedal with minimal stopping power until the brakes dry out.

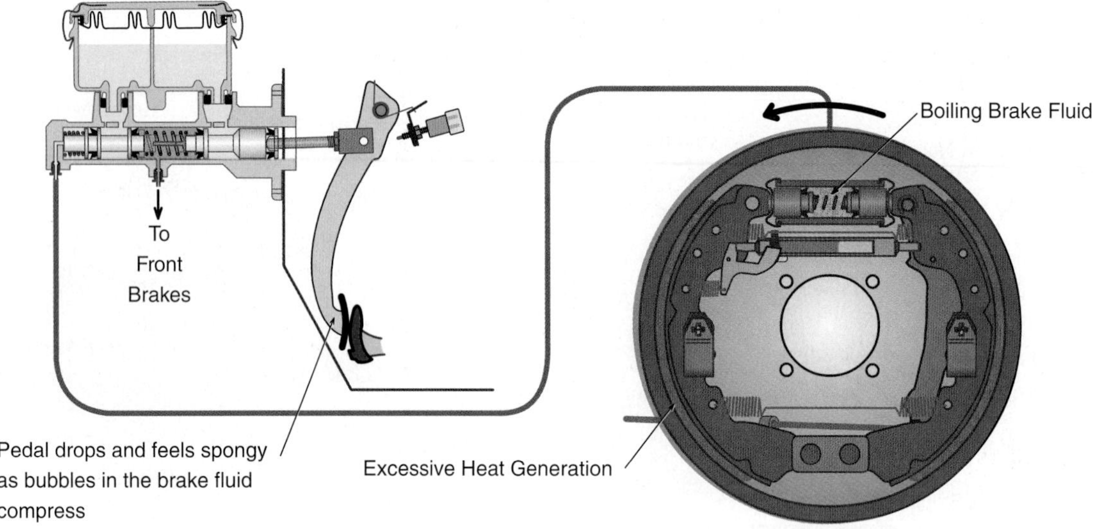

FIGURE 43-24 Hydraulic fade—when the brake fluid boils and can no longer transmit movement in the brake system to apply the brakes. The brake pedal is squishy and goes to the floor.

which can be compressed. The brake fluid can no longer transfer force effectively to the wheel brake units and apply them firmly enough to create friction. Because the boiling fluid can be compressed, hydraulic fade can be recognized by the brake pedal becoming soft and having increased pedal travel during heavy brake usage (**FIGURE 43-24**).

Rotational Force

K43010

When brakes are operated on a moving vehicle, a **rotational force** is generated. As the wheel rotates, the friction between the brake components tends to twist the brake support in the direction of wheel rotation (**FIGURE 43-25**). Because the brake support is ultimately connected to the body of the vehicle, the body too tends to rotate in the same direction. A good example of rotational force is when a motorcycle rider applies the front brake hard enough that the rear wheel is lifted completely off the ground. Rotational forces are usually controlled by the suspension components, but they can become worn and allow movement, which can be felt as a clunk or pop during brake application.

Another result of rotational force is **weight transfer**. The rotational force tends to push the nose of the vehicle down and lift the rear of the vehicle, transferring weight to the front wheels. Rotational force and weight transfer also happen because the centerline of the vehicle is higher than the centerline of the axles; thus, the center of gravity tends to move forward when the brakes are applied firmly (**FIGURE 43-26**).

Weight transfer causes the front wheels to have increased traction, allowing them to bear more of the stopping load, and causes the rear wheels to have less traction, reducing the amount of load they can bear. Engineers take this into account when designing the brakes; otherwise, the front wheels will not get enough stopping power, and the rear wheels will have too much, resulting in rear wheel lockup and loss of control of the vehicle. Lockup is avoided by engineering the system with the proper-sized master cylinder, wheel cylinders, and valving that modifies the hydraulic pressure to the rear wheels under hard braking. This is covered in greater detail in the Hydraulic Components section in the Hydraulics and Power Brakes chapter.

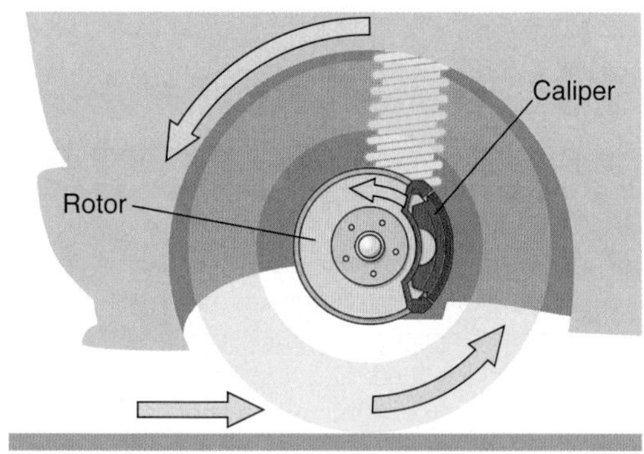

FIGURE 43-25 Braking rotational force occurs when the friction between the brake components twists the brake support in the direction of the wheel rotation.

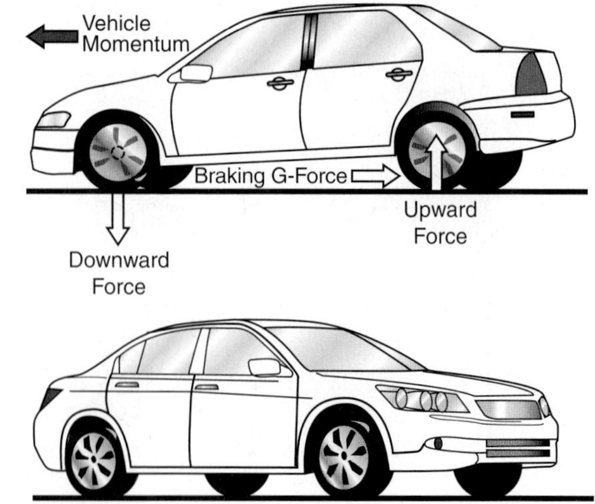

FIGURE 43-26 Weight transfer during braking.

> ▶ TECHNICIAN TIP
>
> Some owners raise their vehicles for better off-road clearance by installing a lift kit and/or large-diameter tires. Doing so raises the vehicle's center of gravity and increases the amount of weight transfer the vehicle experiences, making it more prone to rear wheel lockup and vehicle rollovers. It is important to only use lift kits engineered for the particular vehicle. After installing such a kit, inform the driver of the effect that a higher center of gravity will have on vehicle operation.

Levers and Mechanical Advantage

K43011

Brake systems use levers and mechanical advantage to apply service and parking brakes. A simple example of a **lever** is a bar. The point around which a lever rotates and that supports the lever and the load is called the **fulcrum**. A lever allows the user to lift a large load over a small distance at one end by applying a small force over a greater distance from the other end. The effort distance is from the fulcrum to the point effort is applied. The load distance is from the fulcrum to the point the load is applied.

The effort required to move a load depends on the relative distance of the load and the effort from the fulcrum. The ratio of load and effort is called mechanical advantage. If the effort distance from the fulcrum is greater than the load distance, then the effort required will be less than the load being moved. If the load distance is greater than the effort distance, then the effort required is greater than the load being moved. This is known as a negative mechanical advantage, or **mechanical disadvantage**.

Using the right kind of lever in the right way allows a user to move larger loads with less effort. There are three basic types of levers:

1. **Lever of the first order:** The fulcrum is in the middle, between the load and the effort (**FIGURE 43-27**). Examples are a pry bar or a seesaw. The force applied in this situation is in the opposite direction of the load.
2. **Lever of the second order:** The load is in the middle, between the effort and the fulcrum (**FIGURE 43-28**). An example is a wheelbarrow. The force applied in this situation is in the direction of the load. Brake pedals are usually of the second order. They pivot at the top end (fulcrum). The foot pressure (effort) is applied to the bottom end. And the master cylinder (load) is applied between the two. Mechanical advantage is engineered into the brake pedal to provide the proper brake pedal application and feel (**FIGURE 43-29**).
3. **Lever of the third order:** The effort is in the middle, between the load and the fulcrum (**FIGURE 43-30**). An example is an oar when paddling a canoe, where the hand holding the top of the oar is the fulcrum; the other hand holding the middle of the oar is providing the effort; and the water is the load. The force in this situation is in the direction of the load.

FIGURE 43-27 Lever of the first order.

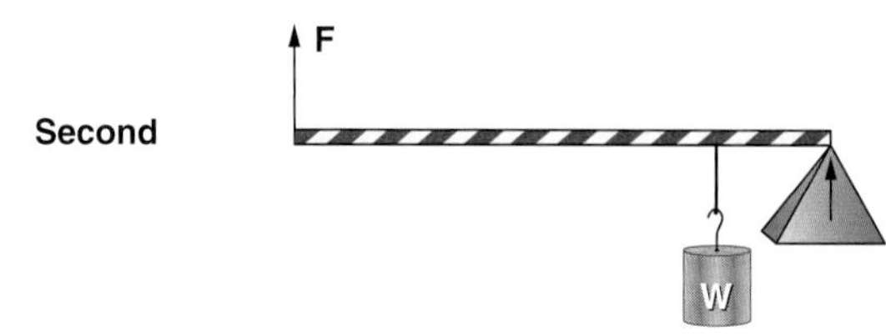

FIGURE 43-28 Lever of the second order.

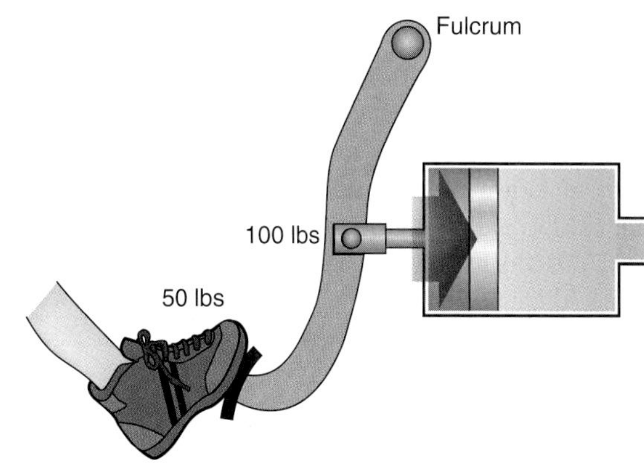

FIGURE 43-29 The brake pedal uses leverage to multiply the force applied to the master cylinder.

FIGURE 43-30 Lever of the third order.

> **Applied Science**
>
> ***AS-12: Levers: The technician can explain how levers can be used to increase an applied force over distance.***
>
> The lever action is applied many times on a daily basis. The simplest example is the use of a wrench. If a bolt or nut is tight, then instead of using a wrench with a short handle, you can use a wrench with a longer handle. The longer handle improves the lever's mechanical advantage by increasing the ratio of the effort distance to the load distance, thus providing more torque to undo the nut or bolt.

▶ Types of Brake Systems

K43012

Now it is time to explore the different types of brakes used on various vehicles. Each vehicle application lends itself to a particular type of brake system. For example, conventional passenger vehicles use hydraulic brakes, which are fairly simple and reliable. Consider trailers. Because they must be able to be hooked up to multiple tow vehicles, connecting a hydraulic brake system becomes difficult because of the leaking of hydraulic fluid as well as air entering the lines. Thus, some heavy trucks use air-operated brakes on both the truck and trailer. That way, when connecting and disconnecting the brake air line, only air will leak, not brake fluid. On passenger-type tow vehicles, trailers are typically equipped with electric brakes. We discuss the various types of brakes in this section.

Hydraulic Brakes

K43013

Hydraulically operated friction brakes use two kinds of wheel brake units. Drum brakes have a drum attached to the wheel hub and rotate with the tire. Braking occurs by means of stationary brake shoes expanding against the inside of the drum, which creates friction and slows the vehicle. Disc brakes have a disc brake rotor attached to the wheel hub and rotate with the tire. The braking occurs by means of stationary pads clamping against the outside of the rotor, creating friction to slow the vehicle. On light vehicles, both of these systems are hydraulically operated, meaning they use hydraulic fluid to transfer the force from the driver. The brake pedal operates a **master cylinder**. Hydraulic lines and hoses connect the master cylinder to the wheel brake units (**FIGURE 43-31**).

Disc brakes require greater force to operate than drum brakes and usually include a power brake booster to assist the driver by increasing the force applied to the master cylinder when the brake is operated (**FIGURE 43-32**). Modern drum and disc brake systems are regularly fitted with an antilock brake system that monitors the speed of each wheel and prevents wheel lockup or skidding, no matter how hard brakes are applied or how slippery the road surface. This allows the driver to better maintain directional control of the vehicle. ABS also generally reduces stopping distances.

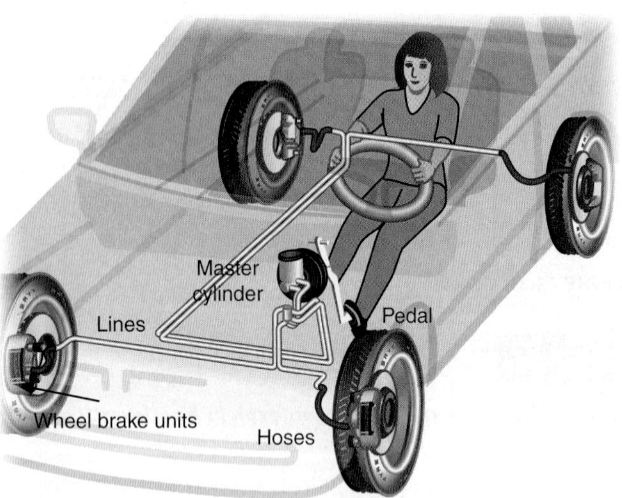

FIGURE 43-31 The hydraulic brake system.

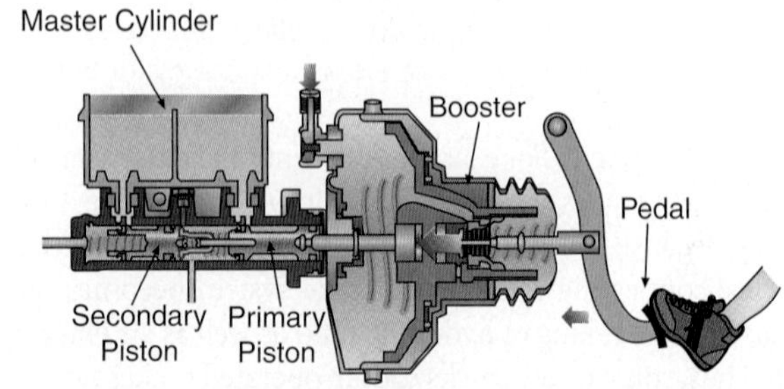

FIGURE 43-32 A power brake booster.

The system consists of a brake pedal, power booster, master cylinder, wheel speed sensors, the electronic control unit (ECU), and the hydraulic control unit, also called a hydraulic modulator.

Air Brakes

K43014

Air-operated braking systems are used on heavy vehicles and are commonly called air brakes (**FIGURE 43-33**). Compressed air, operating on large diameter diaphragms, provides the large forces at the brake assembly that are needed to apply the brakes and slow the vehicle. The wheel brake units can be either drum or disc style.

Some brake canisters incorporate a separate air chamber and large spring assembly (**FIGURE 43-34**) and are referred to as spring brakes. The spring applies the brakes when air pressure is released from the spring brake chamber. Air pressure applied to this chamber compresses the spring and releases the brakes. This system acts as a parking brake when the vehicle or trailer is parked and as a safety measure by applying the brakes in the case of an inadvertent loss of air pressure in the brake system. A vehicle or trailer with this type of air-operated brakes must have a minimum amount of air pressure in the system to compress the spring in order to keep the brakes released.

The air pressure for air brake systems comes from an engine-driven air compressor that pumps air into storage tanks. The tanks provide a reservoir of pressurized air to operate the system. The driver-controlled foot valves then direct the compressed air to different wheel units to operate the wheel brake units. When the driver releases the brakes, the control valve releases the air in the service brake chambers, and the brakes retract.

On articulated vehicles, such as tractor/trailers, any delays in applying the trailer brakes should be minimized. This is achieved by using a relay valve and a separate air tank on the trailer (**FIGURE 43-35**). This arrangement also applies the brakes if the trailer becomes disconnected from the tractor.

Exhaust Brakes

K43015

Medium and heavy-duty vehicles often require increased braking in situations where friction brakes overheat and fail. This can be achieved by using an exhaust brake. An **exhaust brake** works by restricting the flow of exhaust gases through the engine by closing a butterfly valve located in the exhaust manifold. Restricting the exhaust flow results in high pressure in the exhaust manifold and the engine's cylinders. Because the engine acts as an air compressor, restricting the exhaust flow causes the engine speed to slow down, which in turn slows the road wheels through the transmission or powertrain (**FIGURE 43-36**).

Compression Brakes

K43016

In heavy diesel vehicles, engine braking is less effective because more of the compression energy in the cylinders is returned to the crankshaft after the piston reaches top dead center. This results in an engine that wants to freewheel and does not create much force to slow the vehicle. A **jake brake** (or compression brake) uses an extra lobe on the camshaft to control an auxiliary exhaust valve at the top of each cylinder. When engaged, the jake brake releases

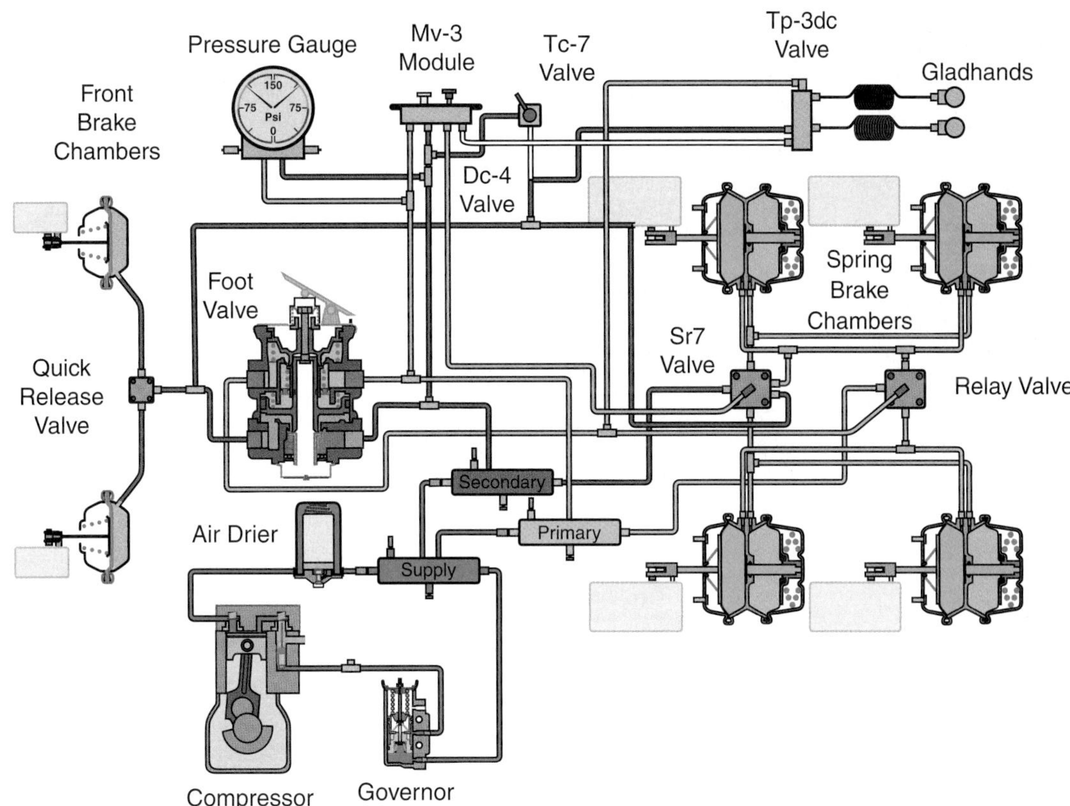

FIGURE 43-33 The air brake system.

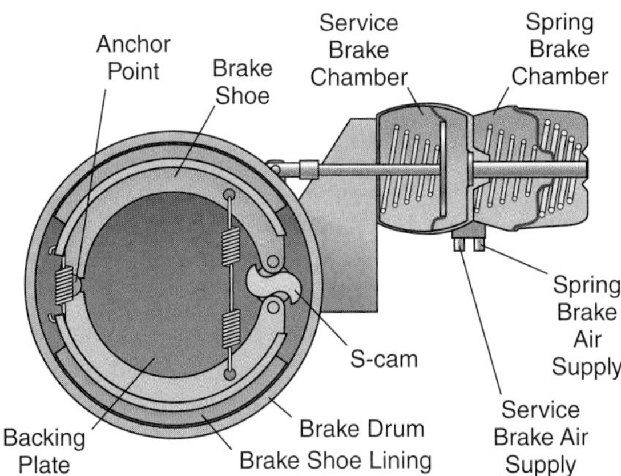

FIGURE 43-34 An air brake canister.

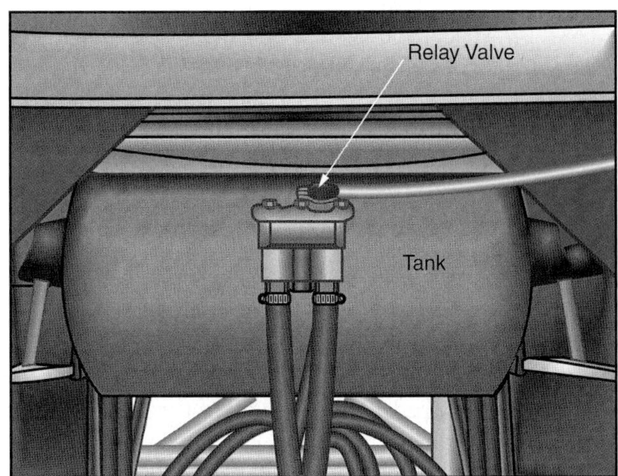

FIGURE 43-35 The relay valve and separate air reservoir.

the compression stroke pressure before it can be transmitted back to the power stroke of the piston. This causes a substantial amount of energy to be used on the compression stroke, which slows the crankshaft and increases the braking effectiveness of the engine. The disadvantage with this form of braking is that it is very noisy, producing a chattering, machine gun sound, usually requiring additional muffling. It is loud enough that some municipalities ban the use of compression brakes in residential areas or other areas where noise would be an issue.

Electric Brakes

K43017

Trailers towed by light vehicles must have a braking system if their gross weight exceeds a certain value. An **electric braking system** is commonly used to activate drum-type friction brakes on the trailer (**FIGURE 43-37**). The driver can increase or reduce the braking effect by adjusting a control unit to suit the load on the trailer. This manual override can be used in certain driving conditions, to dampen a trailer's tendency to sway.

When the brakes in the towing vehicle are applied, the brake light circuit sends the signal to the control unit. The control unit then sends an appropriate current to the trailer brake actuators to operate the trailer brakes at the braking level selected. The electric current is supplied to the brake units at the wheels. The brake assembly uses lever-operated brake shoes. The lever incorporates an electromagnet on the end. The electric current flows through the electromagnet, creating magnetism that draws the magnet

FIGURE 43-36 Exhaust brake butterfly operated by a pneumatic cylinder.

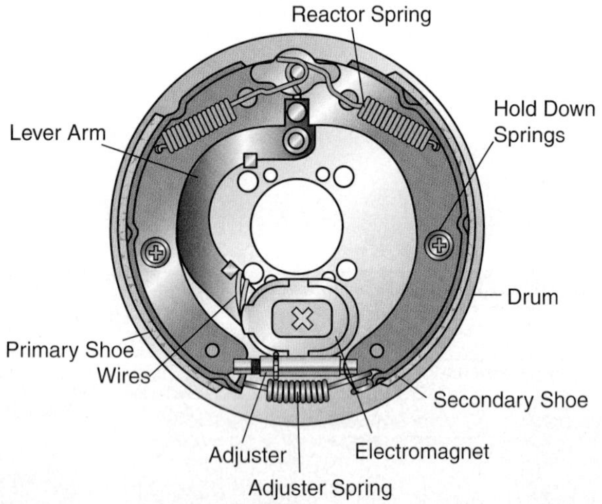

FIGURE 43-37 An electric braking system.

toward the spinning brake drum. The magnet contacts the drum and is pulled around, applying the brake shoes. If the driver steps harder on the brake pedal, the trailer controller sends a stronger electric current to the electromagnet, causing it to be held tighter to the spinning drum, which increases the braking force.

In some places, trailers above a specified weight must also have an auxiliary battery-powered braking system that automatically applies the brakes and keeps them applied if the trailer ever breaks away from its towing vehicle.

▶ Parking Brakes

K43018

As mentioned previously, all vehicles must be fitted with a service brake and a parking brake. The service brake is usually hydraulically operated and is used while the vehicle is being driven. It applies brake units at all four wheels. The parking brake is mechanically operated to hold the vehicle in place when it is parked. It applies the brake units on two wheels only.

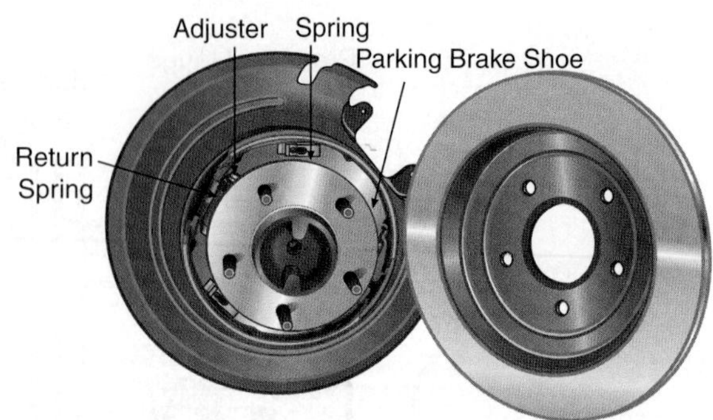

FIGURE 43-38 A top hat design parking brake.

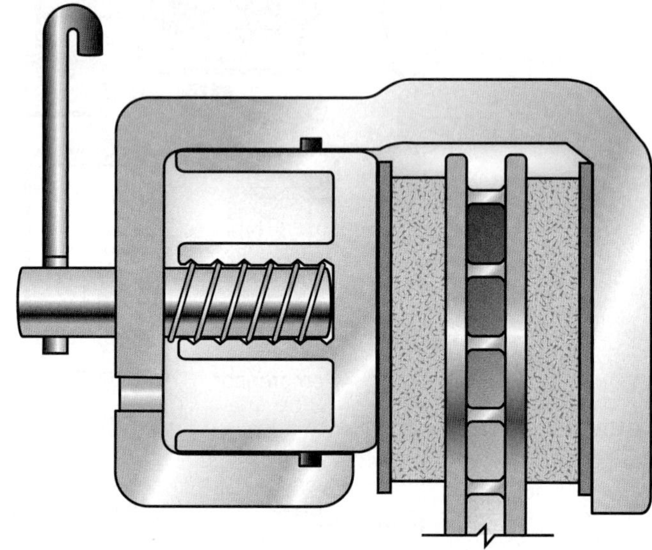

FIGURE 43-39 A disc brake system with an integral parking brake.

There are several types of parking brakes. Some vehicles equipped with disc brakes incorporate a mechanically operated drum-style parking brake, commonly called a **top hat parking brake** design (**FIGURE 43-38**), in the center of the rear disc brake rotors. Other vehicles use mechanical linkage to directly operate the disc-style service brake (**FIGURE 43-39**).

On drum brakes, a **drum-style parking brake** mechanically applies the brake shoes against the drum. The parking brake cable pulls on an actuating lever inside the brake drum assembly (**FIGURE 43-40**). The actuating lever is connected to the secondary brake shoe by a pin or tang, and to the primary shoe by a strut. Movement of the lever forces both shoes against the drum.

Less common, older arrangements of parking brakes include a front wheel–mounted parking brake, which activates each front brake caliper, and a **transmission mounted parking brake**, which uses a small drum brake to prevent the drive shaft from turning (**FIGURE 43-41**). This latter brake is sometimes called a transmission brake.

Parking Brake Cables

Parking brake cables transmit force from the parking brake actuating lever to the brake unit. Part of the cable is inside either a wound steel housing (**FIGURE 43-42**) or a plastic housing,

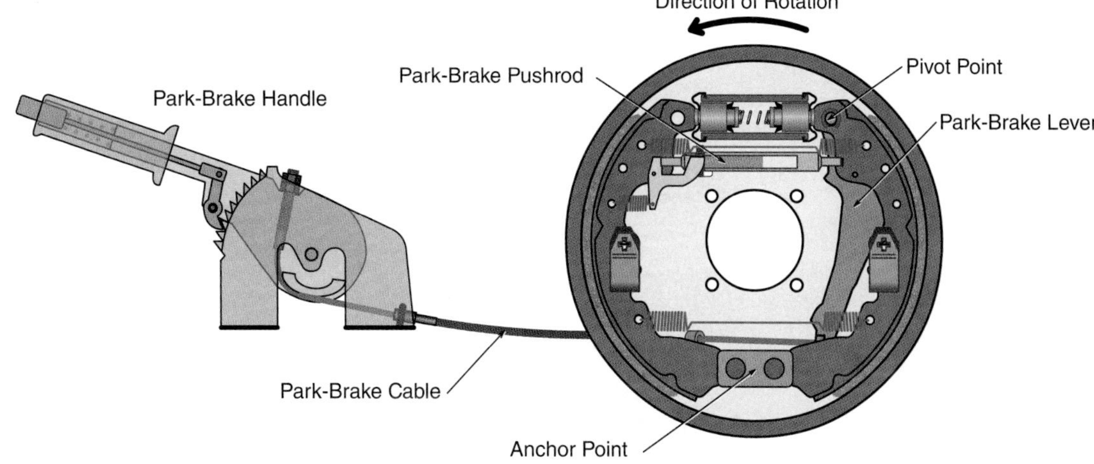

FIGURE 43-40 A drum-style parking brake.

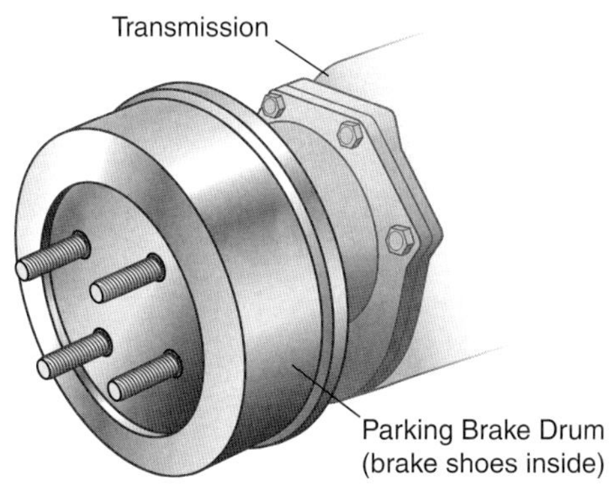

FIGURE 43-41 Parking brake—transmission style.

FIGURE 43-42 A parking brake cable.

which allows it to be somewhat flexible, yet noncompressible, to guide the cable and hold everything in place. Other sections of the cable may be exposed, which can make them susceptible to damage by being caught on road hazards, especially on off-road vehicles. Also, because the cables are steel, they can rust and stick to the inside of the wound steel housing. This is especially true where de-icing chemicals are used on roads. Periodic

FIGURE 43-43 A hand-operated parking brake handle and ratcheting mechanism.

lubrication of the parking brake cables with the lubricant specified by the vehicle manufacturer helps reduce this problem.

Parking Brake Apply Mechanisms

Parking brakes are commonly applied in two ways. The first is a hand-operated lever, usually mounted between the front seats. The lever has its fulcrum on the bottom end, and the cable is attached a few inches from the fulcrum. The driver grasps the handle on the end opposite of the fulcrum, which gives the driver substantial mechanical advantage and ensures that the parking brake can be firmly applied (**FIGURE 43-43**). The handle contains a ratcheting mechanism that engages automatically when the driver pulls up on the parking brake lever. This mechanism holds the handle in the applied position and maintains tension on the parking brake cables and assembly. The driver can release the mechanism by lifting slightly up on the handle and pushing the release button, which retracts the ratcheting tab and allows the driver to lower the parking brake handle.

The second type of operating mechanism works similarly, but is foot operated. It uses mechanical advantage to firmly apply the parking brake, as well as a ratcheting mechanism to hold it once applied (**FIGURE 43-44**). It is usually mounted

FIGURE 43-44 Foot-operated parking brake mechanism.

under the dash near the kick panel toward the outside of the car. Some foot-operated parking brake levers can be released by pulling a release handle, which retracts the ratcheting tang and allows the pedal to return to its rest position. Other pedals can be released by simply pushing the pedal a bit farther down, which automatically retracts the ratchet and allows the pedal to return. The last method uses a vacuum or electric actuator to release the parking brake when the gear selector is moved out of the park position.

Parking Brake Adjustment

Parking brake systems incorporate some method of adjustment, although it is not the same for all vehicles. The method of adjustment could be accomplished by an adjustment nut on the cable under the vehicle, an adjustment on the brake lever assembly, or an adjustment at the rear calipers (**FIGURE 43-45**). When adjusting parking brakes, the order of adjustment is as follows: (1) the parking brake cable, which should have slack in it; (2) the service brakes, which are adjusted according to the manufacturer's procedure; and (3) the parking brake. This sequence ensures that both brake systems are adjusted properly.

Regenerative Braking

N43001

Friction brakes convert the kinetic energy of the moving vehicle into heat, which is dissipated to the atmosphere. This is a waste of good energy. Regenerative braking is technology that allows a vehicle to recapture and store part of that kinetic energy in a reusable form when braking. Regenerative braking is accomplished by causing the hybrid vehicle's electric motor to act as a generator (commonly called a motor/generator) that converts kinetic energy into electrical energy that is stored in the form of chemical energy within the vehicle's high-voltage battery, for future use (**FIGURE 43-46**). The amount of braking force applied by the regeneration system is controlled by a computer. This system still requires traditional friction-based brakes to be used when quick, heavy braking is required. The interaction between the regenerative braking system and the friction braking system must be carefully designed and serviced so that each operates seamlessly with the other.

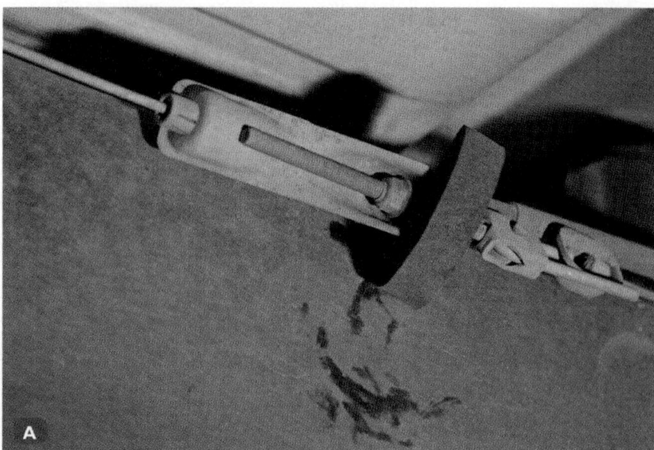

FIGURE 43-45 Methods of adjustment. **A.** Parking brake adjustment—under the car. **B.** Parking brake adjustment—brake lever. **C.** Parking brake adjustment—disc brake caliper.

Brake-by-Wire Braking Systems

K43019

Brake-by-wire means there is no mechanical or hydraulic connection between the brake pedal and the wheel brake units. There are two types of brake-by-wire systems: electric and electrohydraulic (**FIGURE 43-47**). The electric type uses wheel brake units with an integrated motor system that applies clamping force to the disc

FIGURE 43-46 A regenerative braking system.

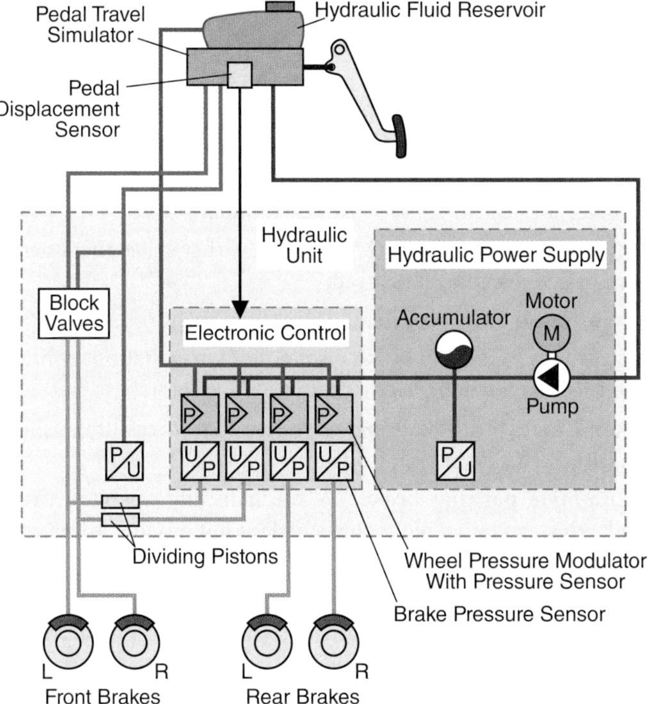

FIGURE 43-47 Electrohydraulic brake-by-wire system.

brake pads. The electrohydraulic type uses a hydraulic controller to apply hydraulic pressure to standard wheel brake units. The brakes, in either case, are computer controlled. They use the brake pedal position and pedal application speed as indicators of the driver's intentions. The computer evaluates that information in light of other data such as vehicle speed and traction conditions. It then compares those data to preloaded information in memory and commands the brake controller to create a specific amount of brake pressure at the appropriate wheel brake units. The driver can modify the programmed braking effort by depressing or releasing the brake pedal an additional amount.

Many vehicles use this same principle to control the engine's throttle (drive-by-wire), meaning there is no mechanical connection between the throttle pedal and the throttle plate. On a throttle, it is easy to replicate throttle pedal feel, as the pedal only works against spring tension. Integrating a spring with similar characteristics results in a throttle pedal that has a similar feel. The feel of a brake pedal is not as easily replicated because, in addition to spring pressure, the driver is also pushing against hydraulic pressure, giving the pedal a much firmer feel. This problem can be overcome by using a brake pedal emulator, which is a sealed cylinder that builds pressure in a manner similar to that of a master cylinder. It also incorporates the sensors that send brake pedal data to the ECU.

There are several benefits to brake-by-wire, including the following:

- Weight is reduced and space is saved.
- Drivers do not feel ABS brake pulsations, making them less likely to remove their foot from the brake pedal during a panic stop.
- The brakes are applied more quickly in an emergency situation.

There are also several drawbacks, such as the risk of a system failure and the initial cost. Some current hybrid vehicles use brake-by-wire technology during regeneration, but with a hydraulic backup brake system. In this case, the brake-by-wire system initiates braking first by increasing generator output, and once that is at its maximum output, it can also apply the hydraulic brakes to get additional stopping power, if needed.

▶ Wrap-Up

Ready for Review

- Braking systems evolved from scrub brakes to band brakes, to the drum and disc brakes that are used today.
- Electronic brake systems use computer technology to assist with braking by monitoring speed and the force needed to stop.
- The development of electronic brake systems has led to improvements in consumer safety and fuel economy.
- Brake-by-wire systems use computer technology in place of the mechanical or hydraulic connection between the brake pedal and wheel brake units.
- Regenerative brake systems (used in hybrid vehicles) convert a vehicle's energy into electrical energy, which is stored in the battery for later use.
- Factors affecting the effectiveness of braking systems include road surface, road conditions, vehicle weight and height, load on wheels, type of tire, and aggressive versus defensive driving.
- Every vehicle has two brake systems: the service brake (for stopping the vehicle in motion) and the parking brake (for holding the vehicle when stationary).

- Hydraulically operated braking systems use a system of cylinders to transfer pressure from the brake pedal to each wheel.
- The braking system converts a vehicle's kinetic energy (the energy of an object in motion) into an alternate form of energy (e.g., heat).
- Acceleration and deceleration determine whether the vehicle's speed is increasing or decreasing. Both require an outside force for action, such as the vehicle's engine (acceleration) or braking system (deceleration).
- The law of conservation of energy requires that energy must be transformed from one form to another; it cannot be created or destroyed.
- Standard brakes use friction to create resistance, thus transforming kinetic energy into heat energy.
- The amount of force pushing two surfaces together compared to the amount of resistive force generated between the two surfaces sliding against each other is called the coefficient of friction.
- Heat energy created by the braking process gets dissipated into the atmosphere.
- Reduction in a vehicle's stopping power (brake fade) can be caused by heat fade, water fade, or hydraulic fade.
- Braking systems create rotational force on the vehicle's suspension, resulting in weight transfer to the front wheels.
- There are three types of levers: lever of the first order (fulcrum in the middle), lever of the second order (load in the middle), and lever of the third order (effort in the middle).
- Disc brakes and drum brakes are two types of friction brakes that are used on lighter vehicles; both can be equipped with antilock braking systems.
- Air-operated braking systems, used on heavier vehicles, use air pressure to apply the brakes (drum or disc style).
- An exhaust brake may be used in addition to friction brakes in medium and heavy vehicles, which creates engine braking by restricting the exhaust flow.
- Jake brakes are used in heavier diesel vehicles and rely on compression to slow the crankshaft and increase braking effectiveness.
- Trailers over a certain weight often use an electric braking system to give the driver braking control.
- Parking brake styles include top hat parking brake, drum-style parking brake, and transmission mounted parking brake.
- Parking brakes use a ratcheting mechanism to maintain tension on the parking brake cables and assembly when applied.

Key Terms

acceleration An increase in a vehicle's speed.

air-operated braking system A braking system that uses compressed air operating on large diameter diaphragms to provide force to the braking assembly; also called air brakes.

antilock brake system (ABS) A safety measure for the braking system that uses a computer to monitor the speed of each wheel and control the hydraulic pressure to each wheel to prevent wheel lockup.

band brake A braking system that uses a metal band lined with friction material to clamp around the outside of a wheel or drum.

brake assist (BA) An enhanced safety system built in to some ABS systems that anticipates a panic stop and applies maximum braking force to slow the vehicle as quickly as possible.

brake-by-wire system A braking system that uses no mechanical connection between the brake pedal and each brake unit. The system uses electrically actuated motors or a separate hydraulic system to apply brake force.

brake fade The reduction in stopping power caused by a change in the brake system such as overheating, water, or overheated brake fluid.

brake pedal emulator A brake pedal assembly used in electronically controlled braking systems to send the driver's braking intention to the computer; it mimics the feel of a standard brake pedal.

brakes A system made up of hydraulic and mechanical components designed to slow or stop a vehicle.

coefficient of friction The amount of friction between two moving surfaces in contact with each other.

conservation of energy A physical law that states that energy cannot be created or destroyed.

deceleration The process of decreasing a vehicle's speed.

disc brakes A type of brake system that forces stationary brake pads against the outside of a rotating brake rotor.

drum brakes A type of brake system that forces brake shoes against the inside of a brake drum.

drum-style parking brake A mechanically operated drum brake that can be set while the vehicle is not moving, to serve as a parking brake.

electric braking system A braking system used to provide braking to trailers; the drum brakes are electrically activated in the trailer when the driver applies the brakes on the tow vehicle.

exhaust brake A brake system that restricts the flow of exhaust gases through the engine by closing a butterfly valve located in the exhaust manifold. Restricting the exhaust flow causes the engine speed to slow down, slowing the vehicle.

friction The resistance created by surfaces in contact. Kinetic friction is resistance to motion when one surface moves over another. Static friction is resistance to motion between two surfaces that are not moving.

fulcrum The point around which a lever rotates and that supports the lever and the load.

heat fade Brake fade caused by the buildup of heat in braking surfaces, which get so hot they cannot create any additional heat, leading to a loss of friction.

hydraulic fade Brake fade caused by boiling brake fluid. Causes a spongy brake pedal.

jake brake A brake system that consists of an extra exhaust valve on a diesel engine, which releases compressed gases from the combustion chamber at the top of the compression stroke; also called a compression brake.

kinetic energy The energy of an object in motion; it doubles with weight, and increases by the square of the speed.

lever A tool that allows the user to move a large load over a small distance at one end by applying a small force over a greater distance from the other end.

master cylinder Converts the brake pedal force into hydraulic pressure, which is then transmitted via brake lines and hoses to one or more pistons at each wheel brake unit.

mechanical disadvantage When the load distance on a lever is greater than the effort distance, which means the effort required to move the load is greater than the load itself.

Newton's first law of motion A physical law that states that an object will stay at rest or uniform speed unless it is acted upon by an outside force.

parking brake A brake system used for holding the vehicle when it is stationary.

parking brake cable A mechanism used to transmit force from the parking brake actuating lever to the brake unit.

rotational force The force created by the rotating wheel when the brakes are applied; it causes the brake components to twist the brake support, and ultimately the vehicle, in the direction of wheel rotation.

scrub brakes A brake system that uses leverage to force a friction block against one or more wheels.

service brake A brake system that is operated while the vehicle is moving, in order to slow or stop the vehicle.

top hat parking brake A drum brake that is located inside a disc brake rotor in order to act as a parking brake.

transmission mounted parking brake A drum brake that is mounted on the drive shaft, just after the transmission, to serve as a parking brake.

water fade Brake fade caused by water-soaked brake linings.

weight transfer Weight moving from one set of wheels to the other set of wheels during braking, acceleration, or cornering.

Review Questions

1. Which of these is a simple mechanical system that uses leverage to force a friction block against one or more wheels?
   a. Drum brake
   b. Band brake
   c. Scrub brake
   d. Disc brake
2. All of the below statements are true *except*:
   a. Service brakes consist of drum and/or disc brakes and are operated by hand.
   b. The parking brake is used for holding the vehicle in place when it is stationary.
   c. In disc brakes, pads are forced against the outside of a brake disc.
   d. Modern braking systems are hydraulically operated.
3. If we triple the speed of an object, the kinetic energy will:
   a. increase by three times.
   b. increase by nine times.
   c. decrease by three times.
   d. increase by six times.
4. The cycle of energy transformation in a typical vehicle is:
   a. Chemical energy→Heat energy→Mechanical energy→Kinetic energy
   b. Kinetic energy→Heat energy→Mechanical energy→Chemical energy
   c. Chemical energy→Mechanical energy→Heat energy→Kinetic energy
   d. Mechanical energy→Heat energy→Chemical energy→Kinetic energy
5. All of the following statements are true *except*:
   a. Heat transfers from a hot area to a cool area.
   b. Friction is the resistance created by surfaces in contact.
   c. Static friction is resistance between non-moving surfaces.
   d. In drum brakes, the heat is created outside of the drum and transferred to the inner surface.
6. The kind of brake fade caused by the brake fluid becoming so hot that it boils is:
   a. heat fade.
   b. water fade.
   c. hydraulic fade.
   d. hard fade.
7. The brake pedal uses leverage:
   a. for weight transfer.
   b. to multiply the force applied to the master cylinder.
   c. to increase traction.
   d. to decrease heat generation.
8. Which of the following is used on heavy vehicles and is commonly called air brake?
   a. Hydraulic control unit
   b. Hydraulic brake
   c. Air operated braking system
   d. Electric brake
9. The force of the brakes absorbs which type of energy from the moving vehicle?
   a. acceleration energy.
   b. kinetic energy.
   c. chemical energy.
   d. hydraulic energy.
10. Vehicles equipped with disc brakes incorporate a mechanically operated drum-style parking brake in the center of the rear disc brake rotors, commonly called a:
    a. regenerative brake.
    b. top hat parking brake.
    c. electric braking system.
    d. transmission mounted parking brake.

ASE Technician A/Technician B Style Questions

1. Tech A says that regenerative braking converts brake heat from friction into electricity. Tech B says that regenerative braking converts electrical energy from the battery into braking energy. Who is correct?
   a. Tech A
   b. Tech B
   c. Both A and B
   d. Neither A nor B
2. Tech A says that an antilock brake system (ABS) allows the front wheels to be steered during a panic stop. Tech B says that ABS sensors apply or release hydraulic pressure to the wheel brake units during ABS brake events. Who is correct?
   a. Tech A
   b. Tech B
   c. Both A and B
   d. Neither A nor B
3. Tech A says that water soaked brake shoes can be a cause of brake fade. Tech B says that disc brakes dissipate heat faster than drum brakes. Who is correct?
   a. Tech A
   b. Tech B
   c. Both A and B
   d. Neither A nor B
4. Tech A says that a disc brake operates by clamping friction materials to the outside of a disc. Tech B says that a drum brake operates by clamping friction materials to the outside of a drum. Who is correct?
   a. Tech A
   b. Tech B
   c. Both A and B
   d. Neither A nor B
5. Tech A says that tire pressure does not affect braking. Tech B says that heavy vehicle loads increase stopping distance. Who is correct?
   a. Tech A
   b. Tech B
   c. Both A and B
   d. Neither A nor B
6. Tech A says that light-duty service brakes are typically applied hydraulically. Tech B says that light-duty parking brakes are typically applied hydraulically. Who is correct?
   a. Tech A
   b. Tech B
   c. Both A and B
   d. Neither A nor B
7. Tech A says that the heavier the vehicle, the more stopping power is needed. Tech B says that the faster a vehicle is moving, the more braking power is needed. Who is correct?
   a. Tech A
   b. Tech B
   c. Both A and B
   d. Neither A nor B
8. Tech A says that kinetic energy is created during braking to stop the vehicle. Tech B says that kinetic energy is converted to heat energy during braking. Who is correct?
   a. Tech A
   b. Tech B
   c. Both A and B
   d. Neither A nor B
9. Tech A says that the brake pedal uses leverage to multiply foot pressure. Tech B says that when braking hard while moving forward, the vehicle's weight transfers to the rear wheels, increasing their traction. Who is correct?
   a. Tech A
   b. Tech B
   c. Both A and B
   d. Neither A nor B
10. Tech A says that friction brakes can fade due to overheating of the brake lining. Tech B says that friction brakes can fade due to overheating of the brake fluid. Who is correct?
    a. Tech A
    b. Tech B
    c. Both A and B
    d. Neither A nor B

CHAPTER 44
Hydraulics and Power Brakes Theory

NATEF Tasks

There are no NATEF tasks in this chapter.

Knowledge Objectives

After reading this chapter, you will be able to:

- **K44001** Describe the principles behind the hydraulic braking system.
- **K44002** Describe the principles of hydraulic pressure and force.
- **K44003** Describe the principles of input force, working pressures, and output force.
- **K44004** Describe the purpose and function of brake system hydraulic components.
- **K44005** Describe the purpose and characteristics of brake fluid.
- **K44006** Describe the types, purpose, and operation of master cylinders.
- **K44007** Describe the operation of single-piston master cylinders.
- **K44008** Describe the operation of tandem master cylinders.
- **K44009** Describe the operation of quick take-up master cylinders.
- **K44010** Describe the operation of ABS master cylinders.
- **K44011** Describe the purpose and operation of master cylinder reservoirs and float switches.
- **K44012** Describe the purpose and function of brake pedals.
- **K44013** Describe the types of divided hydraulic systems and their function.
- **K44014** Describe the purpose, construction, and function of brake lines and hoses.
- **K44015** Describe the purpose and function of sealing washers and fittings.
- **K44016** Describe the purpose and function of hydraulic braking system control components.
- **K44017** Describe the purpose and function of proportioning valves.
- **K44018** Describe the purpose and function of metering valves.
- **K44019** Describe the purpose and function of pressure differential valves.
- **K44020** Describe the purpose and function of combination valves.
- **K44021** Describe the purpose and operation of brake warning lights and stop lights.
- **K44022** Describe the types, purpose, and operation of power brake systems.
- **K44023** Describe the components and operation of vacuum brake boosters.
- **K44024** Describe the components and operation of hydraulic brake boosters.
- **K44025** Describe the components and operation of electrohydraulic brake boosters.

Skills Objectives

There are no Skills Objectives in this chapter.

▶ Introduction

In most modern vehicles, the wheel brake units are applied using the principles of hydraulics. This means the brakes are operated by the force transferred by noncompressible brake fluid flowing through the brake lines and hoses (**FIGURE 44-1**). Pressing down on the brake pedal creates pressure in the brake fluid, which transmits that pressure to the brake units; in turn the brake units apply force to the friction materials, which transform the kinetic energy of the moving vehicle into heat energy and cause the vehicle to decelerate.

To make it easier for the driver to apply the brakes, a power booster is fitted to the brake system to increase the driver's brake pedal force to the master cylinder. This power booster can be operated by engine vacuum or through hydraulic pressure, which is usually generated by the power steering pump or an electric-driven pump. In this chapter, we explore the components of the hydraulic system, their purpose, and how they function. This will give you a solid foundation for understanding the maintenance, repair, and diagnosis process when working on the hydraulic system which is presented in the next chapter.

▶ Principles of Hydraulics

K44001

Pascal's Law(s)

In the 1600s, Blaise Pascal observed the effects of pressure applied to a fluid in a closed system. **Pascal's law** states that pressure applied to a fluid in one part of a closed system will be transmitted without loss to all other areas of the system (**FIGURE 44-2**). This law is the principle behind hydraulic brakes. Pressure created in the master cylinder is transmitted through the hydraulic braking system as long as the system remains closed and has no leaks. In a closed system, hydraulic pressure is transmitted equally in all directions throughout the system. What happens to the pressure levels if there is a leak in the system? According to Pascal's law, a substantial leak prevents the pressure from building up, and therefore the pressure within the system will be equally low. This means that the vehicle may lose some or all of its braking ability if a leak develops.

Pascal's law helps in diagnosing problems with the hydraulic braking system. For example, if the brake pedal is squishy

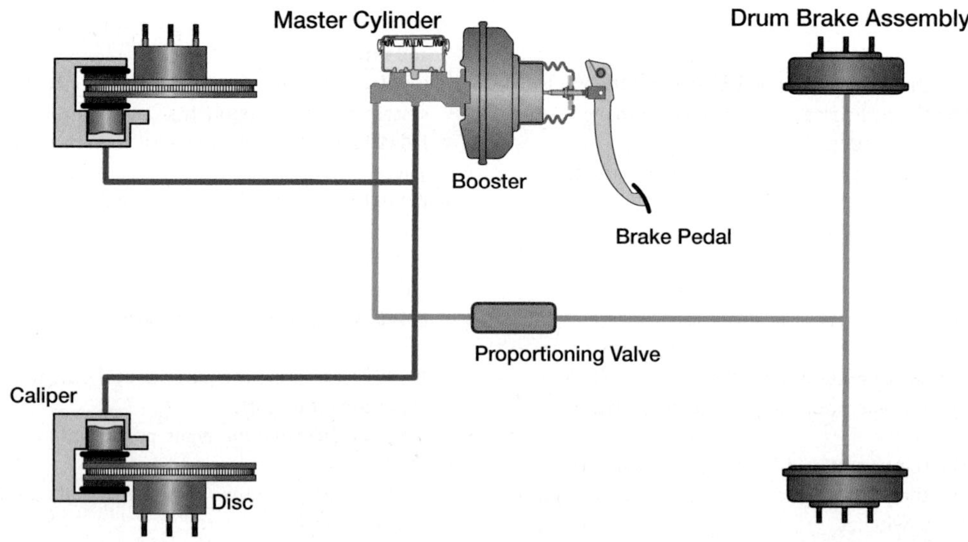

FIGURE 44-1 A schematic view of a hydraulic brake system.

You Are the Automotive Technician

A customer brings his 2011 Chevrolet Tahoe with 96,000 miles to the shop to have a brake concern addressed. The vehicle is regularly driven on the beach and has been pulling to the left, especially after braking. It has gotten worse over the last few weeks. You check the service history and see that the front brake pads were replaced, the rotors refinished, and the brake fluid flushed four years ago, at 54,000 miles. Your mentor technician wants to find out how much you know about braking systems before allowing you to start working on the vehicle. How will you answer his questions?

1. What is Pascal's law, and how can it be used to help you to diagnose this problem?
2. What are the different types of brake fluid, and when should they be used? When should they not be used?
3. What are the purposes of the proportioning, metering, and pressure differential valves?
4. What is the purpose of a tandem master cylinder, and how does it operate?
5. How does a vacuum brake booster operate?

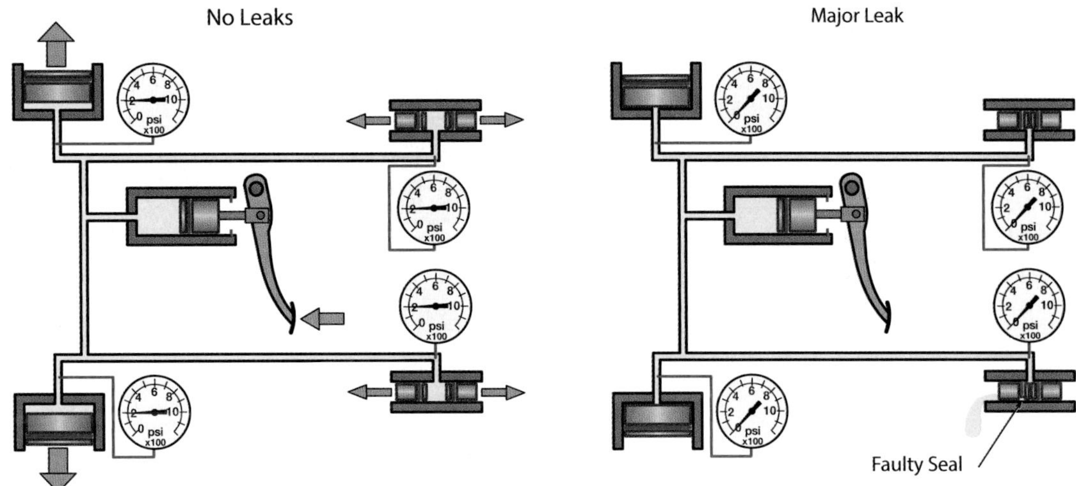

FIGURE 44-2 Pressure is transmitted without loss to all areas of a closed system.

(soft or spongy), there is a good chance that the hydraulic braking system has air in it and has to be bled. If the brake pedal slowly sinks to the floor, there is likely a small leak in the system that must be found. If the vehicle pulls to one side, it could be that a brake hose is plugged up and is not transmitting pressure to one of the brake units. Knowing that pressure should be equal throughout the system helps you identify the cause, based on the reaction of the braking system. Next, we delve more into diagnostics of the hydraulic system later.

Hydraulic Pressure and Force

K44002

Varying amounts of mechanical force can be extracted from a single amount of hydraulic pressure. Because pressure is force per unit area (e.g., 50 lb psi, or 344.7 kPa), the same pressure applied over different-sized surface areas will produce different levels of force (**FIGURE 44-3**). This principle allows engineers to design brakes to have a precise amount of braking force at each wheel. For example, the front wheels on some front-wheel drive vehicles can produce up to 80% of the vehicle's stopping power because of the weight distribution and weight transfer. For these vehicles to brake smoothly, more pressure must be applied to the front brake units than the rear brake units. This is accomplished through the front and rear brake pistons. The larger brake pistons on the front wheels give greater mechanical force and braking power to the front wheels.

Input Force, Working Pressure, and Output Force

K44003

Figure 44-3 illustrates a hydraulic system that has cylinders of different sizes. When the brake pedal is pressed, the force against the piston in the master cylinder applies pressure to the brake fluid. This same pressure is transmitted equally throughout the fluid, but each output piston develops a certain amount of output force depending on its diameter (surface area). The top cylinder is smaller than the master cylinder, so the amount

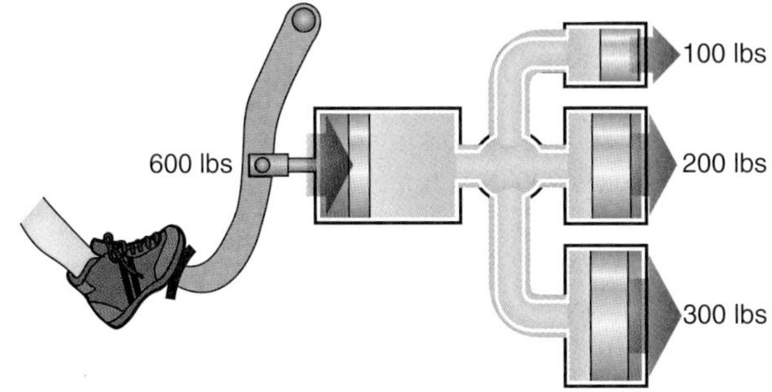

FIGURE 44-3 Engineers apply hydraulic principles to create varying amounts of mechanical force in hydraulic braking systems.

of output force it exerts will be less than the force applied to the master cylinder piston. The middle cylinder is the same size as the master cylinder, so the output force will be the same. The bottom cylinder is larger than the master cylinder, and so its output force will be greater. There are three variables to consider when talking about pressure and force in hydraulic systems (**FIGURE 44-4**):

- **Input force**: The force applied to the input piston is measured in pounds (lb), newtons (N), or kilograms (kg). For example, if 100 lb (45.36 kg) of force were applied to the input piston, this force would be labeled as 100 lb (45.36 kg).
- **Working pressure**: The working pressure of the hydraulic fluid is expressed as the amount of force per specified area. For example, 100 lb of force per square inch is labeled as 100 psi. It could also be expressed as 689.5 kilopascals (kPa). Note, 1 pascal = 1 newton per square meter, or 1 N/m^2. To find the working pressure, divide the input force by the area of the input piston. The example of 100 lb (45.36 kg) of force applied to a 1 sq in. piston creates 100 psi of working pressure. The same 100 lb of force applied to a 0.5 sq in. piston creates 200 psi of working pressure (100/0.5 = 200 psi). Conversely, the 100 lb of force applied to a 2 sq in. piston creates 50 psi of working pressure.

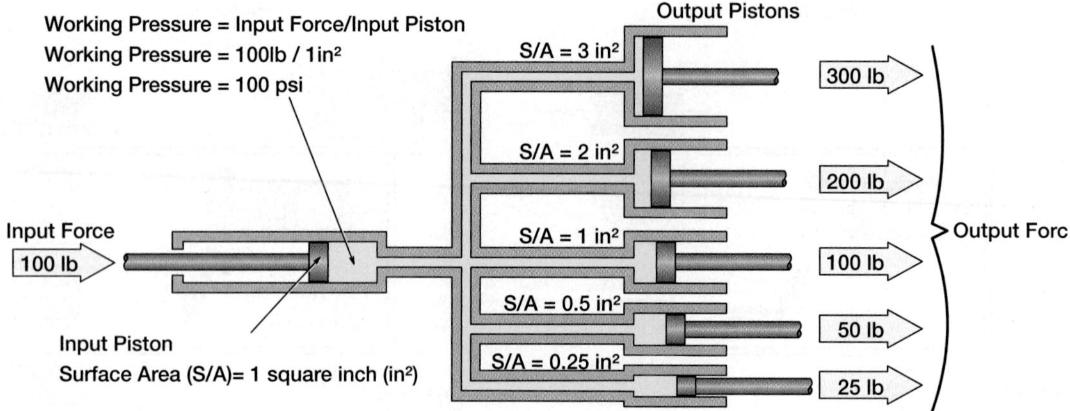

FIGURE 44-4 Working pressure and output force depend on the size of the pistons and the input force.

- **Output force**: Output force is exerted by the output piston and is expressed as pounds, newtons, or kilograms. Finding this measurement is fairly simple: Multiply the working pressure by the surface area of the output piston. For example, 200 psi of working pressure pushing on a 1 sq in. piston exerts 200 lb of force. The same 200 psi of working pressure acting on a 0.5 sq in. output piston creates 100 lb of output force. And if 200 psi of working pressure is applied to a 2 sq in. output piston, 400 lb of force will be created.

SAFETY TIP

It is critical that the hydraulic portion of the brake system not have any leaks or weak spots that could fail and cause leaks. Failure of the braking system could occur, putting the driver, passengers, and others in danger. Vehicles operated in corrosive environments, such as in areas where salt and certain de-icers are used, are susceptible to brake line corrosion. Inspect all vehicles carefully for potential brake fluid leaks.

▶ TECHNICIAN TIP

Input force, output force, and working pressure are optimized during the design of the hydraulic system, based on a specific vehicle application. This is one reason why it is never acceptable to arbitrarily substitute hydraulic components from another vehicle.

▶ Hydraulic Components

K44004

The hydraulic system is made up of a number of components that work together to transmit the driver's effort to the brake pads or shoes. These components must be able to withstand the force, pressure, and temperatures generated in the brake system. They must also be protected from the elements, as they are generally exposed to weather and road hazards. In this section, we explore each of the basic components of the hydraulic braking system.

Brake Fluid Types and Characteristics

K44005

Brake fluid is hydraulic fluid that has specific properties designed for mobile applications. It is used to transfer force, while under pressure, through hydraulic lines to the wheel braking units. Braking applications produce heat, so the brake fluid used must have a high boiling point to remain effective under extreme temperatures. If brake fluid boils, it turns from a liquid to a vapor, which is compressible. This causes a spongy brake pedal and loss of braking ability. Brake fluid must also have a low freezing point so it will not freeze or thicken in cold conditions. If this were to happen, the force from the brake pedal would not be transferred to the wheel brake units.

Standard brake fluid is harmful to painted surfaces because it tends to soften paint; it must therefore be kept off all painted surfaces. Standard brake fluid is also **hygroscopic**, which means it absorbs water. It can absorb water from the atmosphere when it comes into contact with the air in the master cylinder reservoir. Over time, it can even absorb moisture through the flexible brake hoses. Because water boils at a lower temperature than brake fluid, this will gradually reduce the brake fluid's boiling point, making the fluid more likely to boil and cause a hydraulic braking failure. Brake fluid must be flushed periodically to replace old contaminated fluid with new brake fluid to ensure the continued effectiveness of the hydraulic braking system.

▶ TECHNICIAN TIP

Even if they have similar base composition, brake fluids with different Department of Transportation (DOT) ratings should not be mixed.

SAFETY TIP

Because silicone-based fluid tends to **aerate** (tendency to create air bubbles) when forced at high pressure through small passages, it is ***not*** to be used in any vehicle equipped with antilock brakes (ABS). The control valves in ABS brakes would cause DOT 5 brake fluid to aerate under an active ABS stop. This would lead to a spongy pedal and poor brake application.

> ▶ TECHNICIAN TIP
>
> If the brake system needs to be bled, then the brake fluid should be tested to see if it needs to be flushed out and replaced with new brake fluid.

Brake fluids are graded against compliance standards set by the US Department of Transportation (DOT) (**TABLE 44-1**). Brake fluids that meet these standards qualify for the DOT rating and are considered to be quality brake fluids. Brake fluids are tested to ensure they meet the standards for the following:

- pH value
- Viscosity
- Resistance to oxidation
- Stability
- Boiling point

Master Cylinder

K44006

The master cylinder converts the brake pedal force into hydraulic pressure that is transferred to the wheel brake units (**FIGURE 44-5**). Its mounting ears allow it to be mounted to a power booster or the firewall. The cylinder itself has threaded passageways to which the brake lines firmly connect, and a brake fluid reservoir supplies the brake system with an adequate amount of brake fluid. The master cylinder piston is operated by a pushrod from the power booster or the brake pedal. Although all vehicles manufactured for sale in the United States are required to use tandem master cylinders for safety purposes, we start our discussion with the less complicated single-piston master cylinder.

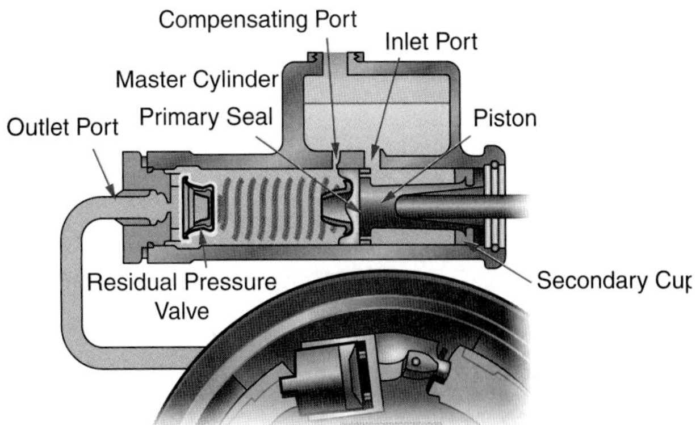

FIGURE 44-5 The master cylinder converts the driver's effort into hydraulic pressure.

Single-Piston Master Cylinder

K44007

Single-piston master cylinders have one piston with two cups: a primary cup and a secondary cup (**FIGURE 44-6**). These cups are also known as seals because they keep the brake fluid from leaking past the piston. When force is applied to the piston by the pushrod, the **primary cup** seals the pressure in the cylinder while the **secondary cup** prevents loss of fluid past the rear end of the piston. An **outlet port** links the cylinder to the brake lines. An **inlet port** connects the reservoir with the space around the piston and between the piston cups. A **compensating port** connects the reservoir to the cylinder, just barely ahead of the primary cup.

With the brake pedal in the released position, the compensating port connects the brake system with the reservoir. The

TABLE 44-1 DOT Ratings for Brake Fluid

Type	Materials	Minimum Dry Boiling Point	Minimum Wet Boiling Point	Additional Specifications
DOT 2	Castor oil based	Not specified	Not specified	• Outdated type of brake fluid that should not be used in any modern vehicles
DOT 3	Various glycol esters and ethers	401°F (205°C)	284°F (140°C)	• Good all-around brake fluid • Harms paint • Absorbs water
DOT 4	Various glycol esters and ethers	446°F (230°C)	311°F (155°C)	• Similar to DOT 3, but higher boiling point. • Compatible with DOT 3 and 5.1
DOT 5	Silicone based	500°F (260°C)	356°F (180°C)	• Not hygroscopic, will not absorb water • Less harmful to painted surfaces than glycol-based brake fluids • Provides better protection against corrosion • More suitable for use in wet driving conditions • *Not* to be used in any vehicle equipped with antilock brakes (ABS)
DOT 5.1	Contains polyalkylene glycol ether	500°F (260°C)	375°F (190.6°C)	• Suitable for ABS-equipped vehicles because of its high boiling point • More expensive than other brake fluids • Compatible with DOT 3, 4, and 5.1

Source: United States Department of Transportation, Federal Motor Carrier Safety Administration: S5.1.2 Wet ERBP.

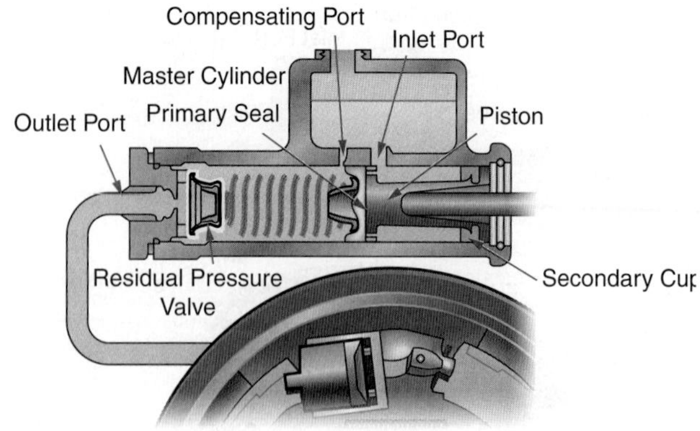

FIGURE 44-6 A single-piston master cylinder with primary and secondary cups.

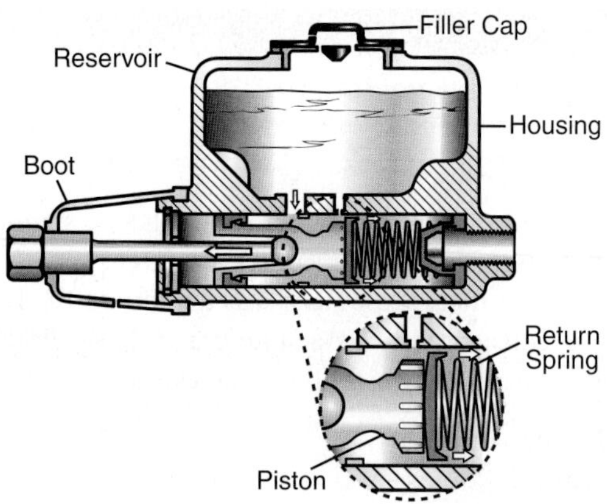

FIGURE 44-7 A single-piston master cylinder with small holes in the piston to allow for recuperation.

compensating port adjusts for changes in the volume of the brake fluid ahead of the piston. This occurs due to the expansion or contraction of the brake fluid as it heats up or cools down. It can also compensate for brake fluid that does not return to the master cylinder due to worn disc brake pads moving the caliper pistons outward, increasing the brake fluid volume in the caliper. In this way, brake fluid can move as needed between the reservoir and the master cylinder bore.

As the pushrod from the brake pedal or power booster moves the piston forward, the compensating port is closed off, trapping brake fluid ahead of the primary cup. Fluid can no longer return to the reservoir. Fluid trapped in the cylinder is then forced from the master cylinder outlet port into the brake lines.

When the brakes are released, the master cylinder piston returns to its original position by action of the spring. When the piston fully returns against its stop, the primary cup uncovers the compensating port. Fluid ahead of the primary cup can now return to the reservoir as needed.

When the brake pedal is released quickly, a spring in the brake pedal pushes the piston back quickly. However, because of the restrictions in the hydraulic system, the brake fluid cannot return as quickly to the cylinder, creating a low-pressure area ahead of the primary cup. As a result, air can be drawn into the system at the wheel cylinders on a drum brake system. To prevent this, small holes are drilled in the piston so that brake fluid from the reservoir can pass through the inlet port and past the edge of the primary cup, thus preventing a vacuum from being created. This is called **recuperation** (**FIGURE 44-7**).

On a drum brake system, when the brake fluid in the lines returns to the master cylinder, brake fluid pressure is held slightly above atmospheric pressure by a valve called the **residual pressure valve**. The residual pressure helps to stop air from entering at the wheel cylinder cups when the brakes are not being applied. The residual pressure valve is located at the outlet end of the master cylinder on single-piston master cylinders, or under the tube seats on tandem master cylinders. Residual pressure valves are not used on disc brake circuits, as the caliper piston seal seals tightly between the piston and bore so that air cannot be easily drawn past it into the hydraulic system. Also, the small residual pressure would keep the brake pads slightly applied against the brake rotors, causing brake drag, premature wear, and reduced fuel economy.

Tandem Master Cylinder

K44008

With a single-piston master cylinder in the braking system, any fluid leak could mean the whole braking system fails. To reduce this risk, modern vehicles must have at least two separate hydraulic braking systems, hence the development of the tandem master cylinder. If one system fails, the other system can still provide a measure of braking ability, although it will not be as effective. For example, the pressure can be 0 psi (0 kPa) in one circuit and normal in the other circuit. **Tandem master cylinders** combine two master cylinders within a common housing that share a common cylinder bore (**FIGURE 44-8**).

Like two single piston cylinders built end to end, a tandem cylinder has a primary piston and a secondary piston. The **primary piston** is in the rear of the cylinder. It is called primary because it is pushed directly by the pushrod. The **secondary piston** is in the front of the cylinder. The secondary piston has a rear-facing seal that seals fluid in the primary chamber and is what ultimately pushes the secondary piston during normal operation. Each half of the cylinder has an inlet port, an outlet port, and a compensating port. There can be two separate reservoirs feeding each half of the cylinder, or just one reservoir divided into separate sections. Dividing the reservoir prevents all of the fluid from draining out due to a leak. This keeps a reserve amount of fluid for the working half of the master cylinder in case of a leak.

When the brakes are applied, the primary piston moves forward and closes its compensating port. Continued movement of the piston causes fluid pressure in front of the primary piston to rise, which acts upon the secondary piston, moving it forward and closing its compensating port (**FIGURE 44-9**). Pressure now builds up equally in both circuits of the master cylinder. Both pistons continue moving forward, displace fluid

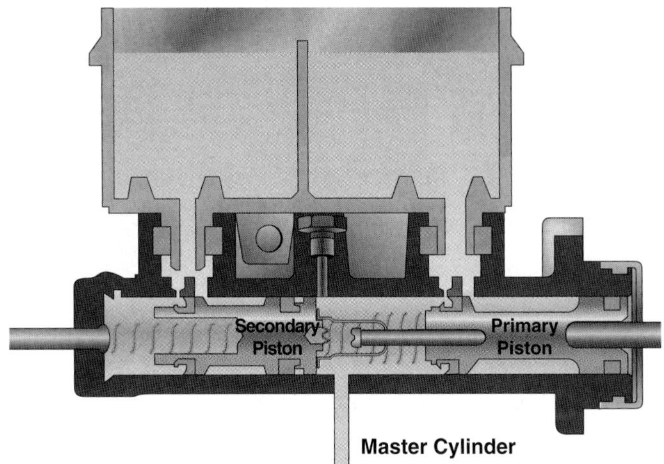

FIGURE 44-8 A tandem master cylinder showing common cylinder.

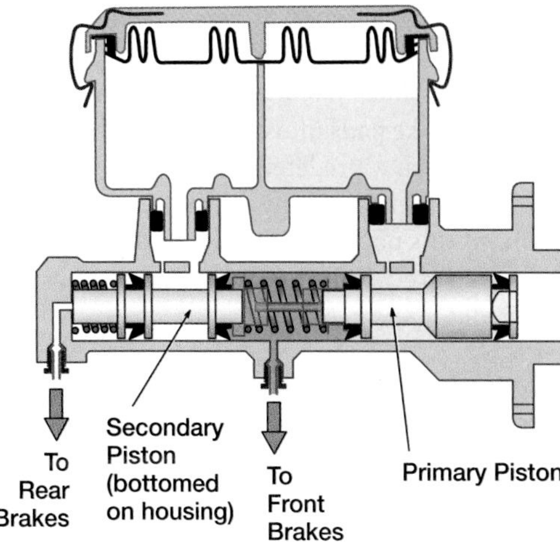

FIGURE 44-10 If there is a leak in the secondary circuit, the secondary piston travels to its stop, and then pressure builds in the primary circuit.

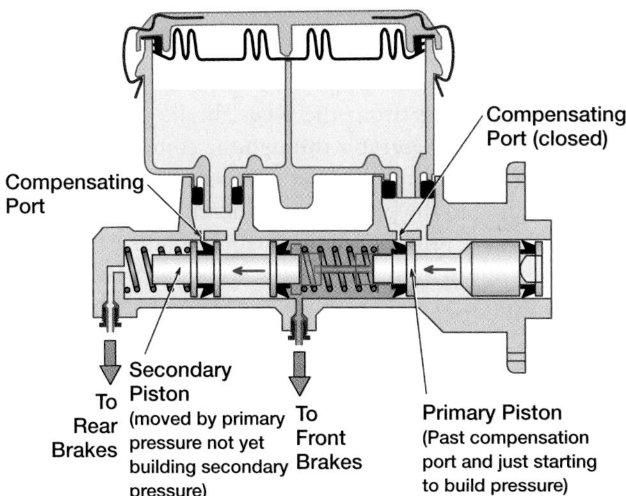

FIGURE 44-9 Moving the primary piston forward closes the compensating port and pushes the secondary piston forward, closing its compensating port.

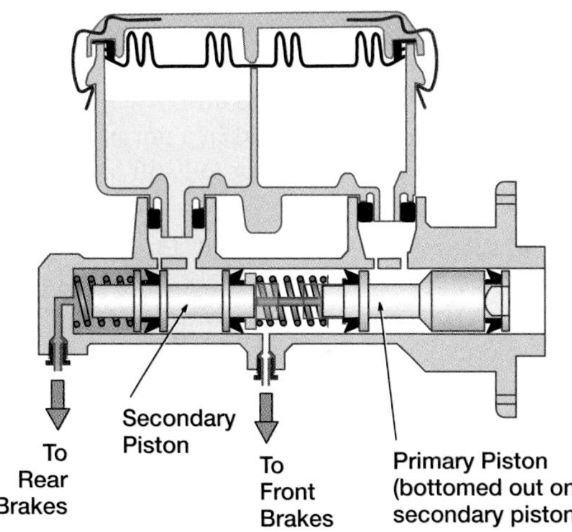

FIGURE 44-11 If there is a leak in the primary circuit, the primary piston travels until it contacts the secondary piston, moving the secondary piston and building pressure in the secondary circuit.

into their separate circuits, and apply the brake units on each wheel.

Just like the single-piston master cylinder, the tandem master cylinder can have problems with a low-pressure area developing when the piston returns quickly but the brake fluid lags. The tandem master cylinder overcomes this by using holes in the piston and grooves in the side of the primary cup. These primary cup grooves allow brake fluid to flow from the inlet port into the low-pressure area, preventing air from entering the system.

If there is a failure in the secondary circuit, the primary system pushes the secondary piston until it contacts the end of the cylinder bore (**FIGURE 44-10**). Once that happens, the primary circuit can start building pressure and operate its brake units. In this situation, it operates, but with increased pedal travel. If the primary circuit fails, no pressure is generated to move the secondary piston; thus a rod attached to the front of the primary piston pushes the secondary piston directly so that it is still able to generate pressure to operate its brake units (**FIGURE 44-11**). This also results in a lower than normal brake pedal. A differential pressure switch in the master cylinder or hydraulic system can illuminate the brake warning light on the instrument panel, alerting the driver to loss of pressure between the two hydraulic circuits, an issue that is covered later.

Quick Take-up Master Cylinders

K44009

Quick take-up master cylinders are used on disc brake systems that are equipped with low-drag brake calipers. These calipers are designed to maintain a larger running clearance between the disc brake pads and rotor. If a standard master cylinder were used, the brake pedal would have to be pushed much farther down before the running clearance would be overcome and the pads would contact the rotor. Hence, a quick take-up master cylinder was designed. It uses a relatively large-diameter piston in the rear of the cylinder to push a large volume of fluid into the hydraulic system at low pressure (**FIGURE 44-12**). This moves the brake pads into contact with the rotor. Once the pressure rises above a predetermined point, a **quick take-up valve** opens and bleeds off any extra pressure created by the large piston. At this point, a smaller diameter piston takes over and builds pressure within the hydraulic system to apply the brakes normally.

ABS Master Cylinders

K44010

The ABS master cylinder is a tandem master cylinder used in divided systems. It has a primary piston and a secondary piston. It may also incorporate the quick take-up principles of operation. In some applications, the compensating port in the secondary chamber is removed, so there is only an inlet port on the secondary piston. The primary chamber still uses a compensating port and an inlet port. The secondary piston incorporates a center valve that controls the opening and closing of a supply port drilled into the piston. At rest, the supply port is open and connects the reservoir with the front brake circuits. This supply port replaces the compensating port in a normal master cylinder (**FIGURE 44-13**). The primary piston still has an inlet port and a compensating port.

When the brake pedal is applied, the primary piston moves and closes its compensating port. Brake fluid pressure in the primary circuit rises and acts with the primary piston spring to move the secondary piston forward, closing the center valve. The pressure builds in the secondary circuit. Pressure keeps building in both circuits and applies force in both circuits. If there is a leak in either circuit, the master cylinder acts like a standard master cylinder and builds pressure in the working circuit.

When the brake pedal is released quickly, recuperation happens, similar to what happens with a tandem master cylinder. When the primary piston is returned fully, any extra brake fluid returning from the wheel brake units displaces brake fluid into the reservoir through the compensating port. In the secondary circuit, the inlet port connects with the supply port drilling in the piston. Any difference in pressure lifts the center valve from its seat and lets the brake fluid enter the chamber ahead of the secondary seal, thus preventing low pressures from developing. When the piston has returned to the "rest" position, the seal is pulled off its seat by the action of the link and spring. This lets the brake fluid still returning from the wheel brake units displace the brake fluid back to the reservoir.

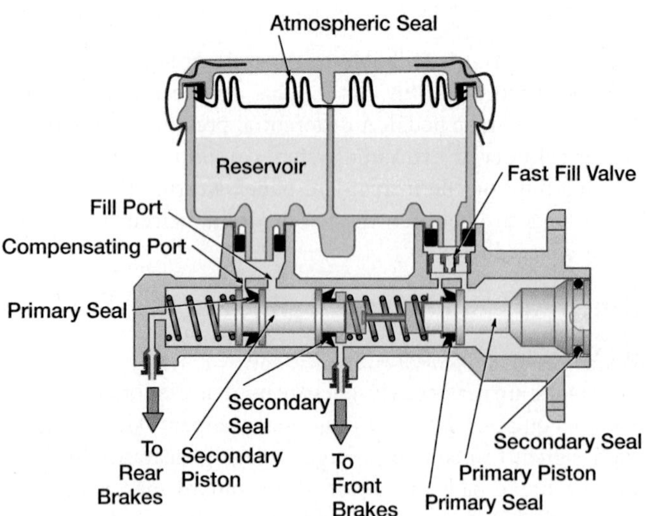

FIGURE 44-12 A quick take-up master cylinder with a larger diameter on the rear of the primary piston.

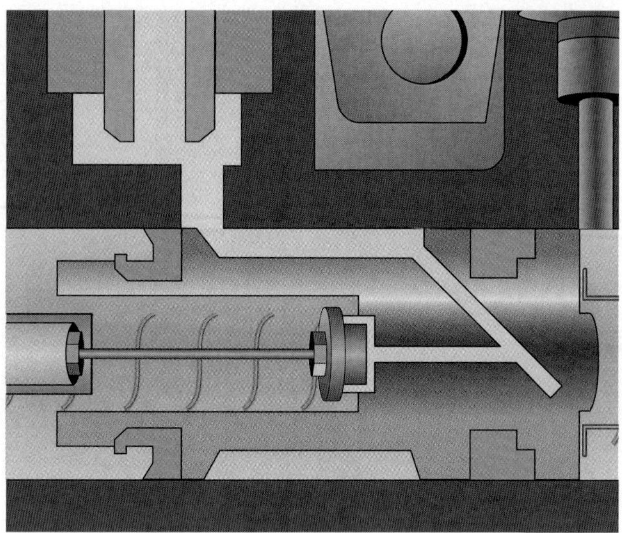

FIGURE 44-13 An ABS master cylinder.

> **TECHNICIAN TIP**
>
> The ABS pedal pulsation has been blamed for actually causing some accidents. Because the ABS system is normally only activated in a panic stop situation, drivers who are unfamiliar with the pedal pulsation have been known to lift their foot off the brake pedal, causing an accident. It is good for drivers to familiarize themselves with the feel of the ABS pedal pulsation by activating the ABS in a safe location, such as an abandoned parking lot. To address this issue, manufacturers have started to move toward electronic braking (brake-by-wire), which prevents hydraulic pulsations from being transmitted to the brake pedal.

If braking conditions are such that the hydraulic modulator must return brake fluid to the master cylinder, then for the front brake circuits, brake fluid is returned to the front section. This forces the secondary piston back against the force of the primary piston spring and the rear brake pressure and pushes both pistons rearward. If enough brake fluid returns, the center valve opens and allows brake fluid to return to the reservoir.

If brake fluid is returned from the rear brake circuit, the secondary and primary pistons tend to be forced apart, causing the primary piston to be driven rearward. If enough brake fluid returns, the compensating port is uncovered and allows brake fluid to return to the reservoir. The amount of brake fluid that returns to the master cylinder is determined by the degree of ABS control. The driver may be aware of a rising brake pedal during this time.

Reservoirs and Float Switches

K44011

Master cylinder reservoirs can be built into the master cylinder housing or can be a separate unit. Built-in- reservoirs are made of the same material as the master cylinder, which is usually aluminum or cast iron, and are formed on top of the master cylinder (**FIGURE 44-14**). The reservoir cover on these master cylinders must be removed to inspect the brake fluid level. The cover uses a rubber diaphragm to isolate the brake fluid from the air. These covers are usually held on by bail clips or tabs molded into the cover.

On two-piece master cylinders, the reservoirs are usually made of a see-through plastic material and use grommets or O-rings to seal them to the master cylinder (**FIGURE 44-15**). Because they are see-through, it is usually unnecessary to remove the cover to check the brake fluid level. The covers can be screw-on caps or clip-on covers. The caps usually incorporate a device to minimize contact of the brake fluid with air. Some systems use a disc that floats on the brake fluid to minimize the surface area of the brake fluid in contact with the air.

To ensure that a leak in one brake circuit does not affect the other circuit, master cylinder reservoirs have two separate chambers. These reservoirs can be two totally separate chambers, or they can use a divider in a common reservoir, which keeps the brake fluid level from falling below a minimum amount.

> **TECHNICIAN TIP**
>
> Master cylinder reservoirs should always have air space at the top of the reservoir to allow for the expansion of brake fluid as it heats up. Never fill master cylinder reservoirs all the way to the top.

FIGURE 44-14 Master cylinder with built-in reservoir.

FIGURE 44-15 Two-piece master reservoir.

Most modern master cylinder reservoirs are equipped with a low brake fluid level float switch that turns on the red brake warning light on the instrument panel and/or sets a notification on the driver information system. The warning system can be activated by the float directly, or a float with an embedded magnet may activate a switch when the float falls to a certain level (**FIGURE 44-16**). The brake fluid level could be low due to worn brake pad linings or a leak in the system; both situations require further investigation. Adding brake fluid without further investigation into what is causing the low fluid condition could lead to a brake failure, putting the driver and occupants in danger.

Brake Pedals

K44012

The brake pedal uses leverage to multiply the effort from the driver's foot to the master cylinder. Different lever designs can be engineered to alter the brake pedal effort required of the driver by using different levels of mechanical advantage. Brake pedals should be mounted securely, free from any excessive sideways movement, and at a height and angle that will allow the driver to quickly move from pressing the accelerator (throttle pedal) to applying the brakes. Brake pedals are covered with a rubber nonslip cover to maintain sure footing. These covers can become worn and lead to slippage, so they need to be inspected periodically.

The brake pedal is usually suspended from a bracket between the dash panel and the firewall (**FIGURE 44-17**). It works as a force-multiplying lever. The pushrod transmits the brake pedal force either directly to the master cylinder or to the power booster. If the power assist fails, the brake pedal's leverage is designed to allow the driver to still generate a reasonable braking force at each wheel brake unit, but with substantially increased foot pressure.

> **TECHNICIAN TIP**
>
> Changes to how far the pedal travels or to its resistance—whether it feels harder or softer than normal—can indicate problems such as a faulty power booster or air in the hydraulic system due to a leak.

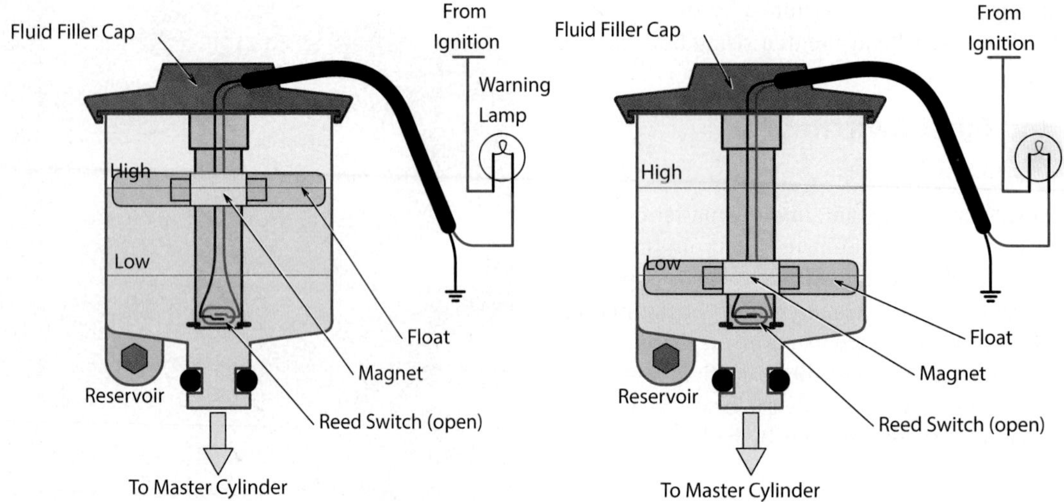

FIGURE 44-16 A magnetic switch is used to activate the brake warning light when the fluid level drops to a specified level.

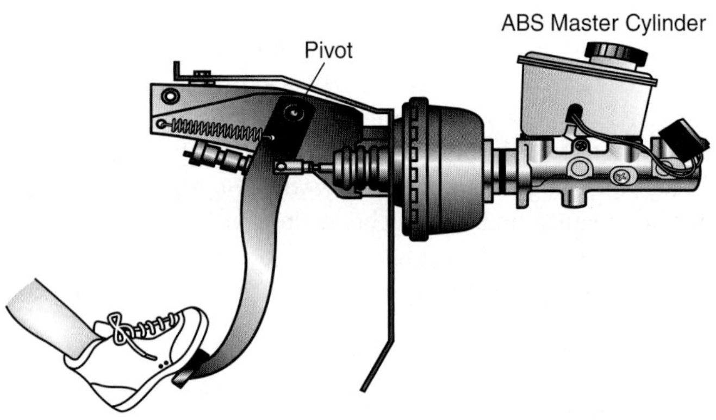

FIGURE 44-17 Brake pedal assembly.

▶ TECHNICIAN TIP

When ABS brakes are activated during heavy braking, the pulsations of the system can be felt by the driver, through the pedal. This is normal. However, a pulsating pedal during normal or light braking can indicate potential braking system problems, such as a rotor that has thickness variation beyond the manufacturer's specifications or that is possibly warped.

Brake pedals must be free to return to their starting position when pressure is removed. This allows the master cylinder piston and pushrod to return to their un-depressed position. The pedal is enabled to return to its starting position by a return spring. The spring action also causes the brake pedal to push the brake light switch open and stop current flow to the brake lights. When the brake pedal is applied, a lighter spring in the brake light switch causes the brake light switch contacts to close and activate the brake lights. Brake light switches are adjustable on some vehicles. Misadjusted brake light switches can cause the brake lights to stay on when they should be off, or not come on when they are supposed to. Both situations can be corrected by adjusting the switch to the proper position. If the switch is nonadjustable, it will likely have to be replaced if it is not operating correctly.

Some vehicles come equipped with adjustable pedal assemblies. Such assemblies allow the driver to raise or lower the brake and throttle pedal assembly for personal comfort. These are usually adjusted by electrically driven motors that are operated by a switch on the steering column or dash (**FIGURE 44-18**).

Types of Divided Hydraulic Systems

K44013

A wheel's braking ability depends on the load it is carrying; therefore, the type of vehicle is a major factor in determining how its system should be divided (**FIGURE 44-19**). A front-engine, rear-wheel drive car has around 40% of its load on its rear wheels and 60% on its front wheels. Its braking system can therefore be divided in a vertical, or front–rear, split. This design puts the front wheels on a different system than the rear wheels. If half of the system fails—either the front or the rear—there is still enough separate braking capability left in the other half to stop the vehicle.

On a front-wheel drive vehicle, a load of about 20% on the rear wheels cannot provide enough braking force to adequately stop the vehicle. Therefore, front-engine, front-wheel drive vehicles use a braking system split in a diagonal, or X, pattern. The left-hand front brake unit is connected to the right-hand rear unit, and the left-hand rear unit is connected to the right-hand front unit. If one system fails, a 50% braking capability is available in the other system.

An alternative arrangement for front-engine, front-wheel drive vehicles is an L split. The front disc brake units have four piston calipers. One inner and one outer piston on each front caliper connect to the right-hand rear brake unit, and the other two pistons of each front caliper connect to the left-hand rear brake unit. As with the diagonal split system, if there is a failure of either half of the system, it still leaves 50% of the braking capability.

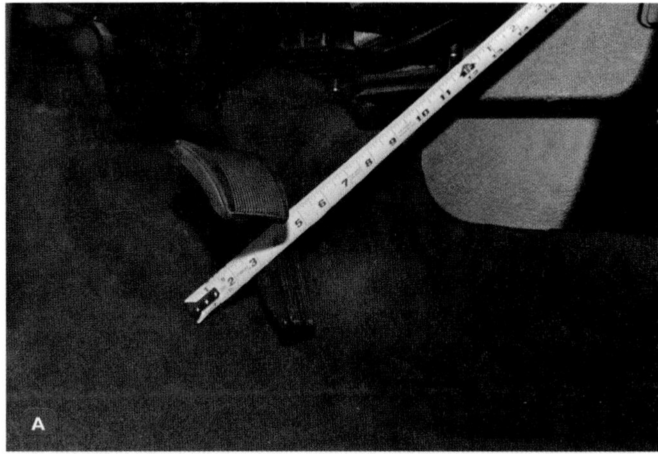

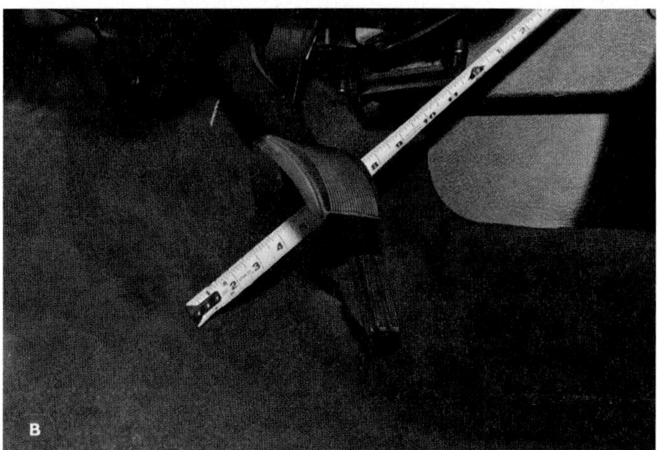

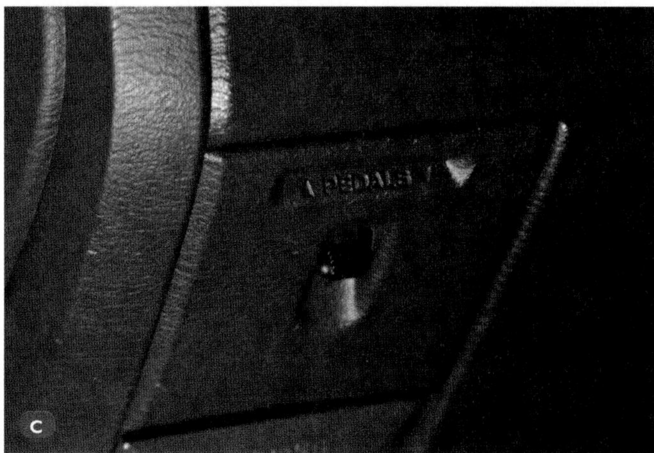

FIGURE 44-18 An adjustable pedal system. **A.** Low position. **B.** High position. **C.** Activating switch.

> ▶ TECHNICIAN TIP
>
> When diagnosing and servicing brakes, it is helpful to know how the hydraulic system is divided. If the vehicle is pulling to one side due to a leak in half of the system, it will generally pull toward the side that is working in the front. Diagnosis and inspection of the other diagonal half usually leads to the leak. In the same way, if air is trapped in half of the hydraulic braking system, then **bleeding** (removing air) the corresponding diagonal half will make it easier to remove the air.

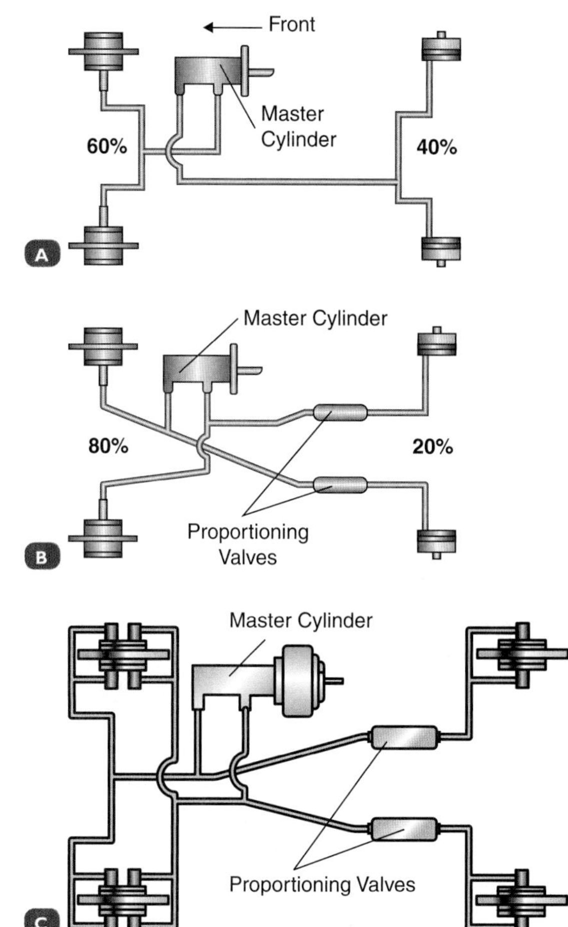

FIGURE 44-19 Divided hydraulic systems. **A.** Vertical, or front–rear, split hydraulic system. **B.** Diagonal, or X, pattern hydraulic system. **C.** L-split hydraulic system.

▶ Brake Lines and Hoses

K44014

Brake Lines

Brake lines and hoses carry brake fluid from the master cylinder to the brake units. They are basically the same on all brake systems and passenger vehicles. For most of their length, they are double-walled steel, coated to resist corrosion, and attached to the body with clips or brackets to minimize damage from vibration (**FIGURE 44-20**). In some vehicles, the brake lines are inside the vehicle to better protect them from corrosion and physical damage. Where the lines must move, flexible brake hoses allow for steering and suspension movement (**FIGURE 44-21**).

Brake Line Materials

The **brake lines** must be able to transmit considerable hydraulic pressure, 1500 psi (10,342 kPa) or more during panic stops. They are made of seamless, double-walled steel rather than a softer, but less corrosive and easier-to-form, material such as copper. They also must conform to applicable standards such as those set by the Society of Automotive Engineers. Only brake lines that meet those standards can be used on vehicles.

FIGURE 44-20 Steel brake line.

FIGURE 44-21 Flexible brake hose.

If a brake line is damaged, it is common practice to replace the entire brake line with a factory replacement rather than repair it. Also, universal brake lines are available in a variety of lengths that are already factory flared with the correct fitting installed. They just need to be formed with the proper tubing bender to the specified shape, following the original brake line routing. Avoid kinks by only using the correct tubing bender. Kinked lines cannot be used or repaired.

Types of Brake Line Flares

Because brake lines connect all of the various hydraulic braking system components, their ends must make leak-proof connections. This is accomplished by using the following types of flared lines and matching fittings:

- **Inverted double flare**: This type of flare is created by first flaring the end of the tube outward in a Y shape. Then about half of the flared end is folded inside of itself (inverted), leaving a double-thick section of brake line on the flared portion of the Y (**FIGURE 44-22**). The flared portion of the tube is clamped between the mating surfaces of the two fittings to provide a secure, leak-proof connection when performed properly.

FIGURE 44-22 **A.** An inverted double-flared line. **B.** A matching fitting.

- **International Standards Organization (ISO) flare**: This type of flare is sometimes called a "bubble flare." The brake line is flared slightly out and then back in, leaving the brake line "bubbled" near the end (**FIGURE 44-23**). The bubble is then clamped between two matching fittings.

SAFETY TIP

Never substitute a copper, aluminum, or non-approved line for the original; doing so could lead to a brake failure, which you could be held liable for.

▶ TECHNICIAN TIP

Many brake and fuel lines screw into an adapter. Always use a double wrench method with flare nut wrenches when loosening and tightening brake and fuel lines. This technique will help prevent twisting of the steel lines (**FIGURE 44-24**).

▶ TECHNICIAN TIP

Flared fittings form a seal by tightly compressing the brake line between the two halves of the fitting. No sealer is needed—nor should it be used—on these types of fittings.

FIGURE 44-23 **A.** An ISO flared line. **B.** A matching fitting. .

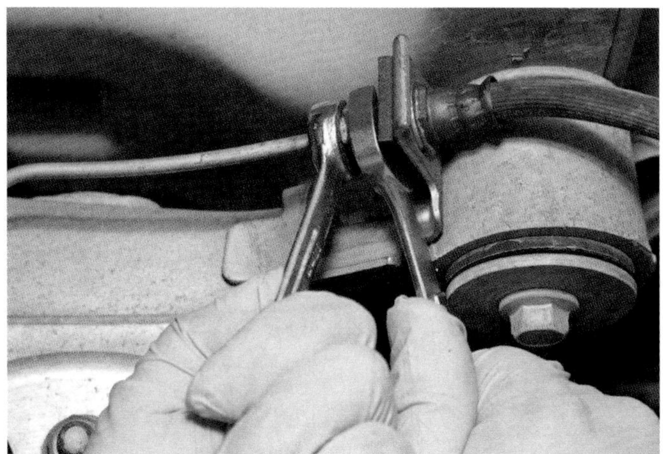

FIGURE 44-24 Double wrench method.

Brake Hoses

A flexible section of the brake lines must be included between the body and suspension to allow for steering and suspension movement. This is accomplished by using flexible hoses made of tough, reinforced tubing. These flexible **brake hoses** transmit the brake system hydraulic pressures to the wheel units (**FIGURE 44-25**). They also must be tough so as not to be damaged easily by objects

FIGURE 44-25 A brake hose transmits hydraulic pressure to the moveable brake unit.

thrown by the tires or other road hazards. When replacing brake hoses, always make sure they are of the proper length. If they are too short, they can be damaged by being stretched. If they are too long, they can contact moving components such as tires or a suspension member, which could weaken or wear a hole in the hose. This is especially common when a vehicle has a lift kit installed.

Brake Hose Materials

Brake hoses are made of several layers of alternating materials. The inside is a liner that helps seal the brake fluid in and any moisture out. The liner is wrapped with two or more layers of flexible webbing, which are usually embedded in a synthetic rubber material and provide reinforcement for the hose. These layers are covered with a tough flexible outer housing jacket designed to resist abrasion and damage (**FIGURE 44-26**).

Although brake hoses are designed to be flexible, they should *never* be pinched, kinked, or bent tighter than a specified radius. Doing so reduces their life and can cause failure of the brake hose. Some technicians mistakenly use vise-grip pliers to crimp a brake hose while disconnected from the caliper or wheel cylinder to prevent brake fluid from leaking out. This practice can damage the brake hose and should be avoided.

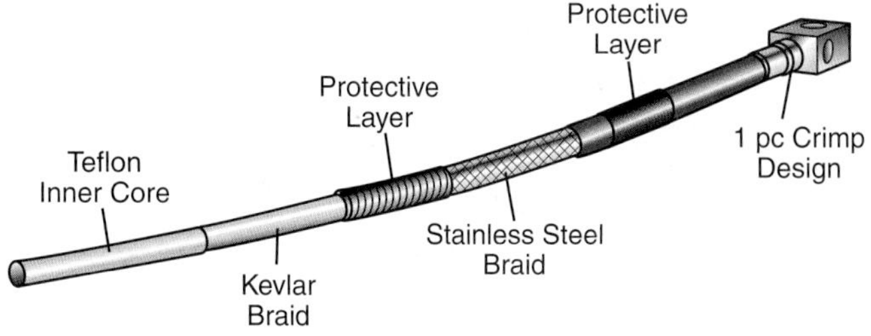

FIGURE 44-26 Flexible brake hose construction.

> **TECHNICIAN TIP**
>
> *Never* hang a disconnected brake caliper by its flexible brake hose. Doing so can damage the brake hose. Always use a piece of wire or other material (e.g., cable or zip tie) to support the weight of the caliper assembly while it is hanging on a suspension or other suitable component.

Brake hoses should be inspected periodically for damage or defects (**FIGURE 44-27**). Some possible issues are:

- **Cracks:** The outer layers become brittle and crack over time. Replacement is required.
- **Bulges:** The reinforcing layers become weak and break over time resulting in bulges in the outer cover. Replacement is required.
- **Abrasion or wear:** This usually happens because the brake hose was routed incorrectly or was too long or short for the application, and it rubbed on a component, resulting, over time, in the abrasion. Replacement and rerouting are required.
- **Kinks:** Kinking usually happens when the brake hose has been pinched with vise grips or twisted on installation. Replacement is required.
- **Internal deterioration causing blockage of the passageway:** Replacement is required.

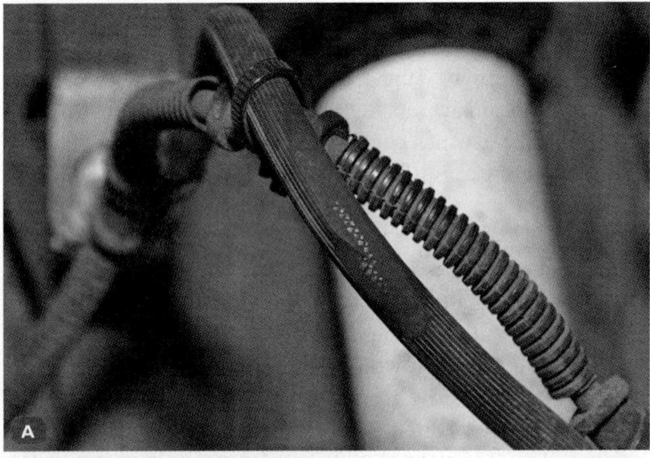

FIGURE 44-27 Possible brake hose issues. **A.** Brake hose with abrasion/wear. **B.** Brake hose with kinks.

FIGURE 44-28 A loosened banjo fitting assembly.

Sealing Washers and Fittings

K44015

Many brake hoses use banjo fittings to connect the hose to the wheel unit. These fittings are comprised of the banjo fitting, banjo bolt, and two copper or aluminum sealing washers (**FIGURE 44-28**). The banjo bolt, banjo fitting, and wheel unit usually have sealing ridges machined in them. These ridges dig into the softer sealing washers to ensure a leak-proof connection. The banjo bolt is hollow and allows the brake fluid through the middle to continue on to the wheel unit. It clamps the banjo fitting to the wheel unit. The sealing washers fit on both sides of the banjo fitting. One sealing washer is between the head of the banjo bolt and the banjo fitting, and the other sealing washer is between the banjo fitting and the wheel unit. Always use the proper torque when tightening banjo bolts so they are not twisted off or left loose to leak.

> **TECHNICIAN TIP**
>
> Because the sealing washers are made of soft metal, they become crushed after use. It is good practice to replace them each time the banjo fittings are removed; otherwise, leaks could occur.

▶ Hydraulic Braking System Control

K44016

The hydraulic braking system must be controlled accurately to maintain adequate control of the vehicle during braking. As we learned earlier, hydraulic working pressure is equally applied throughout a sealed hydraulic braking system. We also learned that the hydraulic braking system can be designed to optimize the output force at each of the wheel brake units. In a perfect scenario, that would work just fine. But because machines are not perfect, you need to be able to modify the hydraulic pressure to accommodate different scenarios. Components such as proportioning valves, metering valves, pressure differential valves, or antilock hydraulic control units are used to modify the pressures within the hydraulic braking system. Let's see how they enhance the braking system.

Proportioning Valves

K44017

Proportioning valves reduce brake pressure to the rear wheels when their load is reduced during moderate to severe braking. Proportioning valves can be pressure sensitive or load sensitive. The pressure-sensitive valve is in the master cylinder or in a separate unit in the rear brake circuit; the load-sensitive type is mounted on the body or on the axle, where it can respond to changes in the vehicle load.

The effectiveness of braking force is determined by tire-to-road friction. The greater the load on the tire, the greater the friction; the greater the friction, the greater the stopping ability. When a vehicle stops abruptly, a portion of the weight on the rear wheels transfers to the front wheels, resulting in greater tire-to-road friction on the front tires and less on the rear. This is called **load transfer**, where weight is being transferred from the rear wheels to the front wheels.

If equal braking force is applied to the front and rear wheels during this load transfer condition, the smaller load in the rear can cause the rear wheels locking up. Skidding tires on the surface of the road (kinetic friction) do not have as much friction as rolling tires, and therefore the stopping distance is increased, which can lead to an accident. A pressure-sensitive proportioning valve reduces the pressure applied to the rear brakes under heavy braking to prevent rear wheel lockup and to help maintain traction.

If the vehicle is equipped with ABS, it may not be equipped with a proportioning valve, as the ABS system can make up for wheel slippage due to the changes in load transfer. This is covered in more detail in the Electronic Brake Control chapter.

Pressure-Sensitive Proportioning Valve Operation

The pressure-sensitive proportioning valve adjusts the braking force to allow for load transfer or variations in loads. During normal braking, the poppet piston of the pressure-sensitive proportioning valve is held in a relaxed position by a large pressure spring. The **poppet valve** is held against its retainer by a light return spring, and brake fluid passes freely through the pressure-sensitive proportioning valve to the rear brakes (**FIGURE 44-29**). In this condition, the rear brakes operate normally without any modification to the pressure.

During heavy braking, master cylinder pressure can reach the poppet valve's crack point. The pressure applied to the two different areas of the poppet piston creates unequal forces, which act on the pressure-sensitive proportioning valve. The higher hydraulic pressure moves the poppet piston against the large pressure spring to close the pressure-sensitive proportioning valve. At a specified inlet pressure, the conical section of the pressure-sensitive proportioning

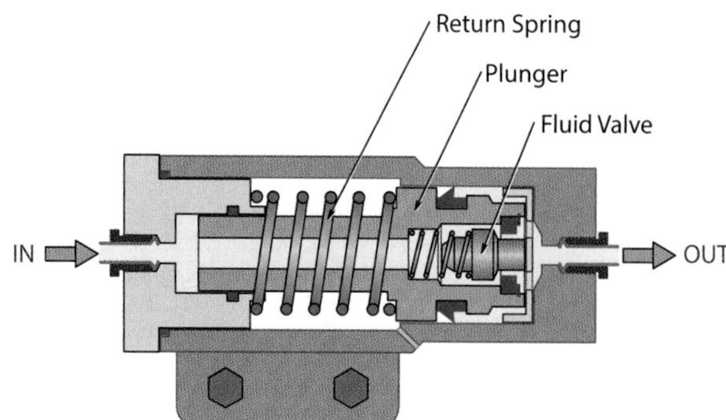

FIGURE 44-29 A pressure-sensitive proportioning valve in the open position.

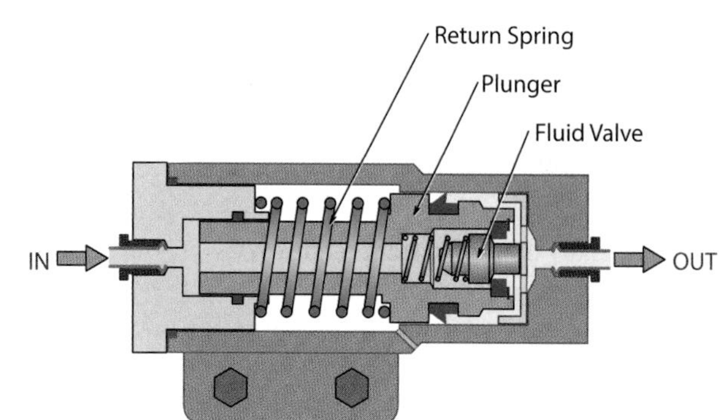

FIGURE 44-30 A pressure-sensitive proportioning valve in the closed position.

valve is held against the seat, which holds the pressure steady to the rear brakes until there is a further change in inlet pressure (**FIGURE 44-30**).

As greater pedal force increases pressure in the master cylinder, brake fluid pressure rises on the smaller end of the poppet piston. This combines with the force of the pressure spring to overcome the lower pressure now on the larger (output) end. As a result, the piston is forced back, opening the poppet valve and allowing pressure to rise to the rear brakes.

The increased pressure now acts on the larger end of the poppet piston and again forces the piston forward to close the poppet valve and hold pressure steady. This repeated action causes a lowering of outlet pressure versus inlet pressure. When the brake pedal is released, the pressure of the rear brake fluid unseats the poppet valve, letting the brake fluid return to the master cylinder. The pressure spring now returns the poppet piston to its relaxed position.

On vehicles that position the proportioning valve in a combination valve, should the front brake system fail, the warning lamp spool moves forward, taking the poppet valve with it (**FIGURE 44-31**). Pressure in the rear brakes rises, and the piston moves forward; however, it cannot close the poppet valve, so full system pressure remains available for the rear brakes in this situation. Should the rear brake system fail, the warning lamp spool will move backward to activate the warning light. The pressure-sensitive proportioning valve cannot operate in this situation.

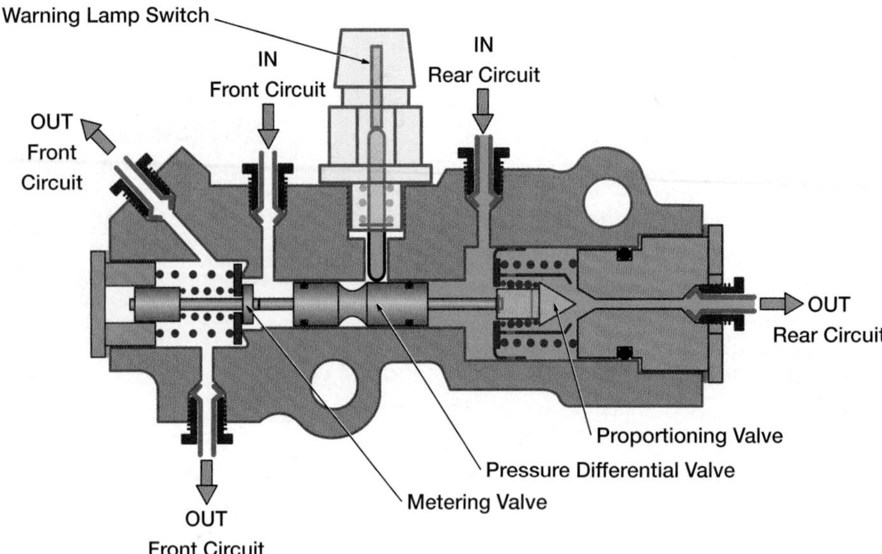

FIGURE 44-31 Proportioning valve action in a combination valve with a front brake circuit failure.

A diagonally divided system requires one pressure-sensitive proportioning valve for each rear wheel, so they are usually located in each rear brake line coming from the master cylinder (**FIGURE 44-32**). Each pressure-sensitive proportioning valve operates in a similar way to the pressure-sensitive proportioning valve in the master cylinder, but without the pressure differential warning light circuit.

Adjustable Proportioning Valves

Aftermarket adjustable proportioning valves are available for performance applications. They have a method of adjusting the crack point of the proportioning valve and can be customized to the specific vehicle. Adjustable proportioning valves are popular with kit car builders, who use components from a variety of vehicles that were not originally designed to work together. Adjustable proportioning valves are not recommended for most applications due to the amount of trial and error necessary to set them properly.

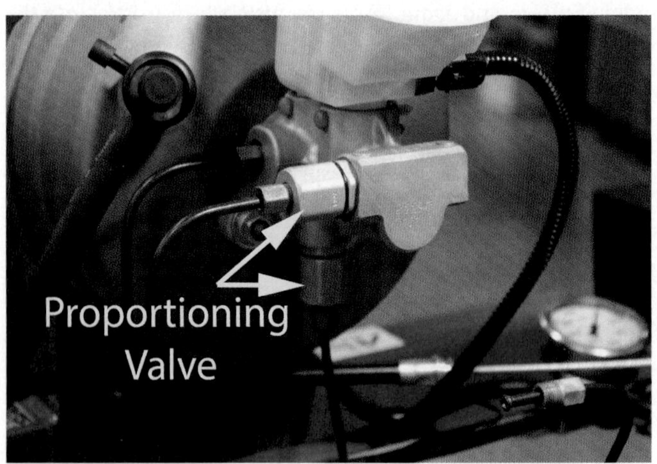

FIGURE 44-32 A pressure-sensitive proportioning valve on a diagonally split system.

Load-Sensitive Proportioning Valve

The load-sensitive proportioning valve reduces rear brake pressure when the vehicle is lightly loaded and allows higher pressure when it is heavily loaded. The load-sensitive proportioning valve is usually located on the chassis and has a lever that is connected to the rear axle (**FIGURE 44-33**). As the vehicle is loaded, the chassis squats on the rear suspension, and the lever is moved, applying more force on the load-sensitive proportioning valve. The lever increases or decreases the point at which pressure is limited by the load-sensitive proportioning valve. Heavier loads allow more pressure through the load-sensitive proportioning valve to the rear brakes, and lighter loads restrict pressure through the load-sensitive proportioning valve. A diagonally split system may have two load-sensitive proportioning valves, one for each rear brake unit. Each load-sensitive proportioning valve is mounted on the chassis, around the rear suspension.

▶ TECHNICIAN TIP

There are very few adjustable components in the hydraulic brake system. One notable exception is the load-sensitive proportioning valve, which is used on some pickup trucks and other load-carrying vehicles. Load-sensing proportioning valves operate so that as load weight increases, more brake pressure is applied to the rear wheels as required.

In the case of a rear-wheel lockup on one of these vehicles, excessive pressure to the rear brakes may be suspected. To check the operation and adjustment of load-sensing proportioning valves, pressure gauges are fitted inline to the front and rear brakes. After the hydraulic brake system is opened to install the gauges, air is bled out, before accurate readings can be obtained.

The weight on the rear axle must be set according to charts or graphs published by the manufacturer, to determine the correct relationship between front and rear pressure at a given load. For example, according to a graph in the service information, if a rear axle is loaded to 1984 lb (900 kg), and front brake pressure is raised to 1138 psi

Brakes Released
Frame Mounted Valve Body
To Rear Brakes
Rear Brake Circuit
Front Brake Circuit
Suspension Connection Link
Brakes Applied
Proportioning Valve
Failsafe Spring
Loaded
Load Spring
Unloaded

FIGURE 44-33 A load-sensitive proportioning valve adjust rear braking force based on changes in rear wheel load.

(7846 kPa), then rear brake pressure should fall within the range of 569 to 711 psi (3923 to 4902 kPa). The rear axle is then loaded to 3700 lb (1678 kg). If the rear pressure is outside the specified range, adjustment of the linkage between the proportioning valve and the rear suspension may be required.

Rear Axle Load lb (kg)	Front Brake Pressure psi (kPa)	Rear Brake Pressure psi (kPa)
1984 (900)	1138 (7846)	569–711 (3923–4902)
3699 (1678)	1707 (11,769)	1323–1493 (9122–10,294)

FIGURE 44-34 A metering valve.

Electronic Brake Proportioning

Many ABS-equipped vehicles integrate an electronic brake proportioning function within the hydraulic control unit on vehicles that require it. This function reduces hydraulic pressure to the rear wheels under heavy braking, similar to the proportioning valve. However, in these vehicles the pressure reduction is handled electronically and can compensate for differences in traction by monitoring wheel slip.

Metering Valves

K44018

Metering valves are used to hold off the application of the front brakes on vehicles with disc brakes on the front wheels and drum brakes on the rear wheels (**FIGURE 44-34**). Drum brakes use springs to return the brake shoes to their rest position. This means that it takes a certain amount of hydraulic pressure to overcome the tension of the return springs and move the shoes to contact the drums. Disc brakes use the much smaller force of the square-cut O-ring to return the caliper piston to its rest position. Thus, very little hydraulic pressure is needed to move the brake pads into contact with the rotor.

Vehicles typically handle better when the rear brakes engage before the front brakes. This helps keep the vehicle tracking straight while the brakes are initially applied. The metering valve keeps the front disc brakes from being applied until the rear drum brakes have had a chance to overcome the tension of the return springs. Vehicles with four-wheel disc brakes don't need metering valves, as the front and rear brakes apply at roughly the same time.

Metering Valve Operation

The metering valve operates like a radiator cap. It has a spring that holds the metering valve closed until a specified pressure is reached (**FIGURE 44-35**). Pressure rises in the hydraulic braking system, overcomes the rear brake return spring tension, and starts to apply the rear brakes. As pressure continues to rise above the metering valve's crack point, brake fluid flows to the front disc brake calipers and starts to apply the front brakes. Because the metering valve only has hydraulic pressure on the inlet side, any pressure above its crack point holds it open.

When the brake pedal is released, the metering valve is pushed closed by the spring, and a fluid return valve opens up,

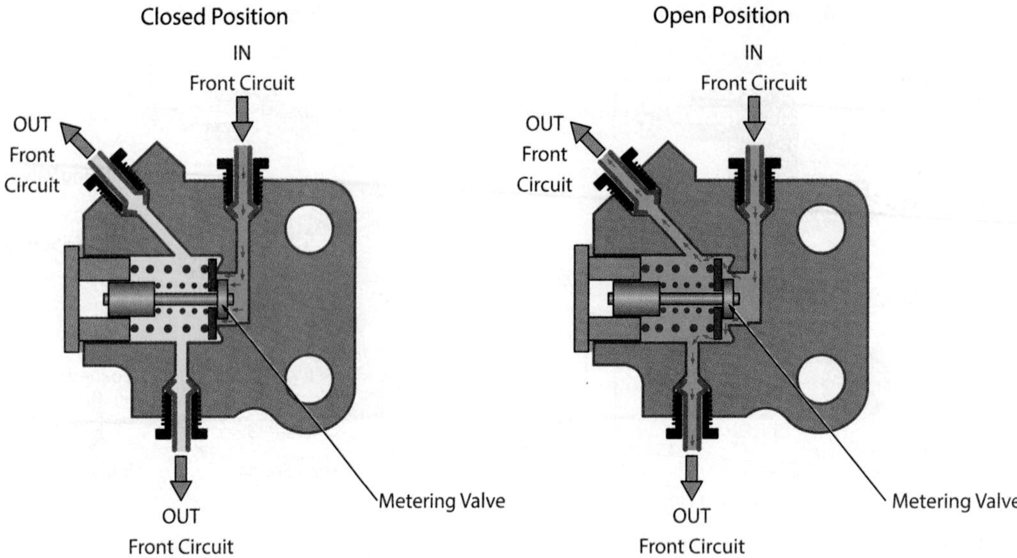

FIGURE 44-35 A metering valve holds pressure from applying the front brakes until a specified pressure is reached.

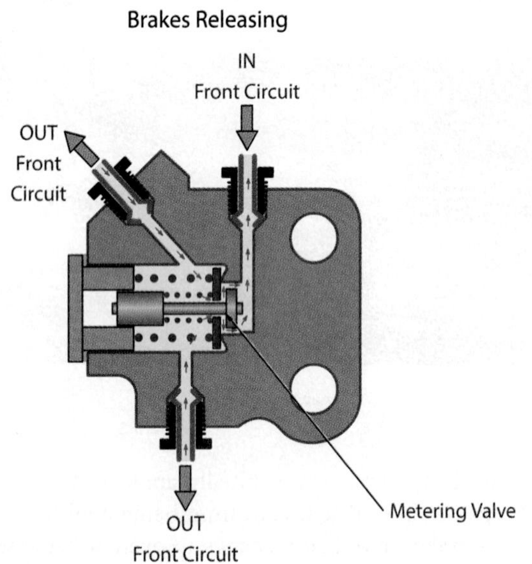

FIGURE 44-36 Brake fluid is free to flow through the fluid return valve back to the master cylinder when the brakes are released.

allowing brake fluid to flow freely back to the master cylinder (**FIGURE 44-36**). During bleeding, if there is air in the lines, it may be difficult to raise the brake fluid pressure enough to open the metering valve. This makes it almost impossible to bleed any air out of the front part of the hydraulic braking system. Most metering valves are designed so that they can be manually held in the open position to allow bleeding to take place. Also, many pressure bleeders operate at an insufficient pressure to open the metering valve, and the valve has to be held open manually.

Vehicles equipped with a diagonally split system are generally front-wheel drive. They usually do not use metering valves for two reasons. First, the torque of the spinning engine during braking compensates for the earlier application of the front brake pads and offsets not having a metering valve. Second, because up to 80% of the braking occurs at the front wheels, they need to be applied as quickly as possible to start braking the vehicle effectively.

SAFETY TIP

After bleeding the hydraulic braking system, always remember to remove the tool that is holding open the metering valve. Failure to do so could cause the vehicle to brake improperly.

Pressure Differential Valve

K44019

A **pressure differential valve** monitors any pressure difference between the two separate hydraulic brake circuits. If there is a moderate leak anywhere in the system, it will illuminate the brake warning light on the instrument panel. Sometimes the light only flickers, or comes on when the brake pedal is pushed, indicating a small leak. The valve can be located in the master cylinder or in the combination valve.

▶ TECHNICIAN TIP

On a front-rear split system, there can also be a pressure difference if the rear drum brake shoes are grossly under-adjusted. When the brake pedal is pushed, the shoes do not make contact. The primary piston moves so far that it contacts the secondary piston, which starts to apply the front brakes. This raises the pressure in the secondary circuit higher than in the primary circuit, moving the pressure differential valve toward the primary circuit and thus illuminating the brake warning light.

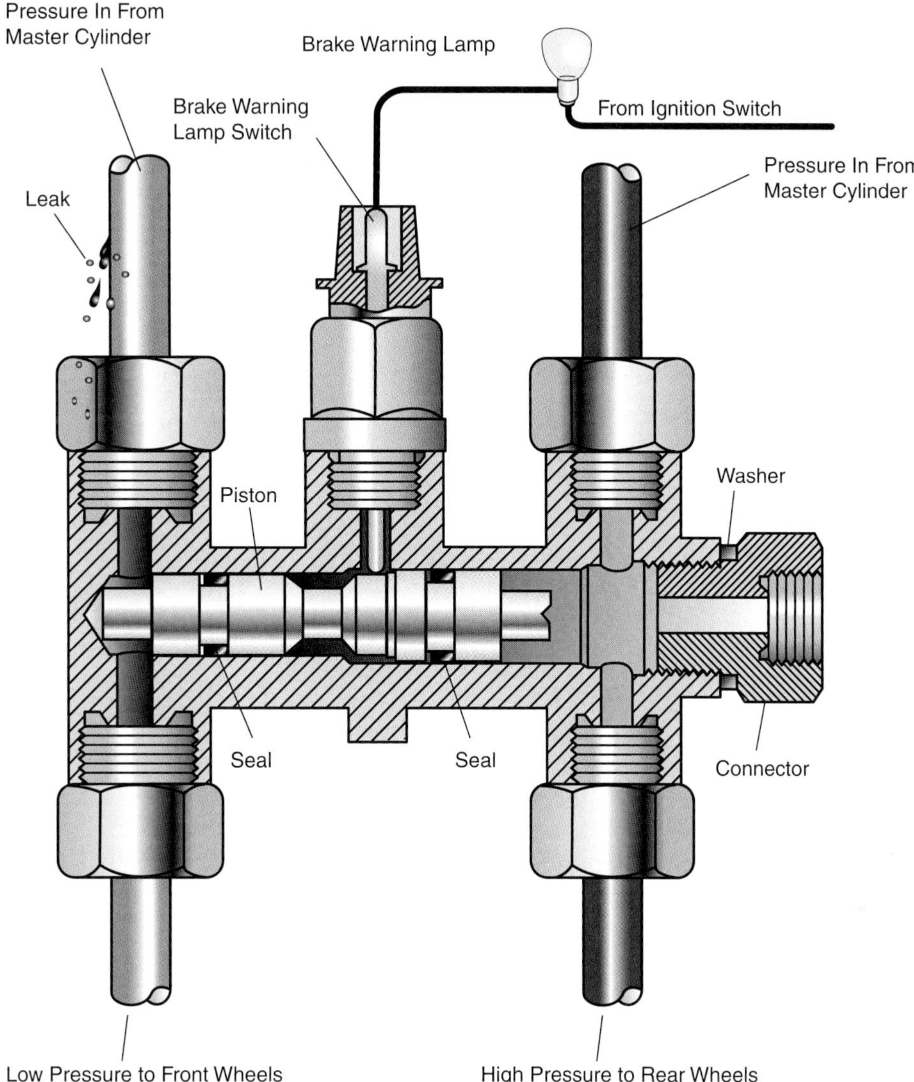

FIGURE 44-37 A pressure differential valve with a leak in the hydraulic braking system.

Pressure Differential Valve Operation

The pressure differential valve is connected between the two halves of the hydraulic braking system so pressure is applied to each end of the pressure differential valve. As long as the pressure stays the same in both circuits, the pressure differential valve remains in the same position and the light stays off. A moderate leak in the hydraulic brake system lowers the pressure on that side of the circuit. This system allows the higher pressure in the non-leaking side to push the pressure differential valve off center toward the side with the leak. The pressure differential switch then closes and illuminates the brake warning light on the instrument panel, telling the driver that there is a serious leak in the hydraulic brake system that has to be diagnosed (**FIGURE 44-37**).

During hydraulic braking system bleeding, the pressure differential valve may need to be centered. Most pressure differential valves have springs that help center them, along with a small amount of clearance between the valve and bore. Applying the brakes firmly causes the pressure to equalize, and the spring returns the pressure differential valve to center. On vehicles without this feature, once the hydraulic braking system has been bled, you must bleed a small amount of brake fluid out of the opposite hydraulic circuit to allow the pressure differential valve to move back to center. Follow the manufacturer's procedure.

Combination Valve

K44020

The combination valve can combine the pressure differential valve, metering valve, and proportioning valve(s) in one unit (**FIGURE 44-38**). Some combination valves combine just the pressure differential valve and proportioning valve(s). On others, just the pressure differential valve and metering valve are combined. Each valve operates individually as it was designed and is contained in one unit. Combination valves are not serviceable. If they become faulty, they must be replaced.

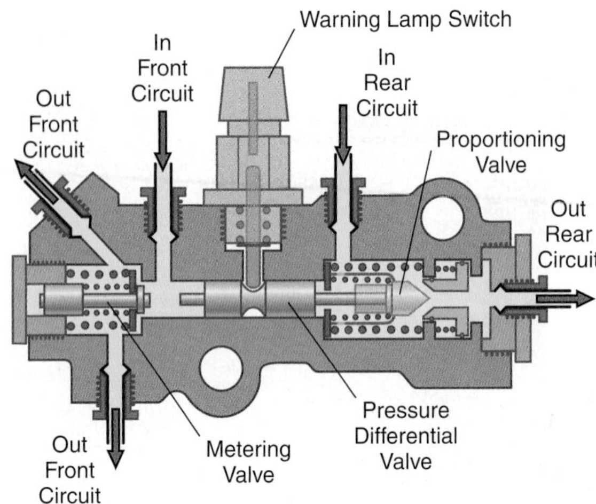

FIGURE 44-38 A combination valve.

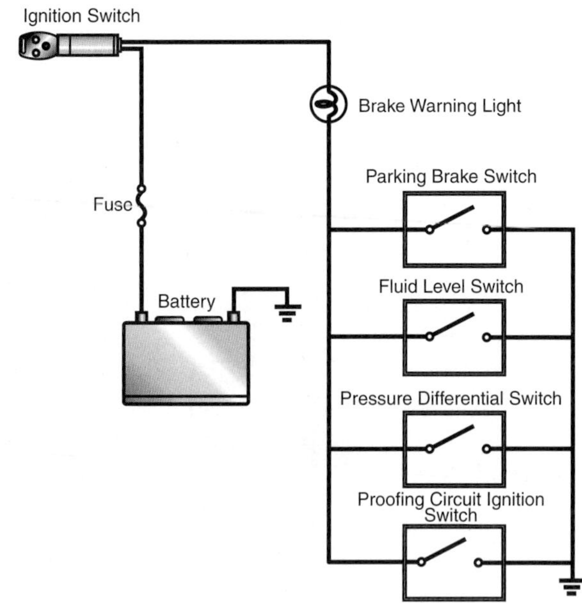

FIGURE 44-40 Brake warning light circuit.

▶ Brake Warning Light and Stop Lights

K44021

Red lights are used on vehicles as warning devices to warn drivers and others of specific conditions. When a red light comes on, the driver should take notice and respond appropriately. There are two general categories; the brake warning light and stop lights. The brake warning light is a single light located in the instrument panel, whereas the stop lights are made up of at least three lights at the rear of the vehicle.

Brake Warning Light

The brake warning light is located on the instrument panel and designed to warn the driver of a condition in the brake system that needs attention (**FIGURE 44-39**). Usually the light can be illuminated by four causes (**FIGURE 44-40**):

- The first is when the parking brake is engaged and the light is turned on by the parking brake lever or pedal. This is to alert the driver that it is on so the driver will release it before driving.
- The second reason the brake warning light comes on is because the brake fluid level is too low in the master cylinder reservoir. This could be caused by a leak in the hydraulic braking system or worn brake pads on the disc brakes.
- On vehicles without ABS, the third cause of the brake warning light coming on is unequal pressure in the hydraulic brake system, which causes the pressure differential valve to activate the brake warning light switch.
- The fourth cause of the brake warning light illuminating is called a "prove out" or "proofing" circuit. On most vehicles designed without controller area network bus (CANbus) system, turning the ignition switch to the Crank position causes the warning light to illuminate, so the driver knows the bulb is good.

FIGURE 44-39 A brake warning light warns the driver of an issue in the braking system.

Stop Lights

Stop lights are designed to warn others that the vehicle is braking. This information is critical to help avoid accidents. The regular stop lights are mounted on the rear of the vehicle and must conform to federal laws for brightness and location. They are activated by a normally closed stop light switch located on the brake pedal assembly. When the driver applies the brake, the brake pedal moves away from the stop light switch. A spring in the switch closes the contacts, allowing current to flow and illuminate the stop lights (**FIGURE 44-41**). Releasing the brakes causes the brake pedal to force the stop light switch open, turning the stop lights off.

In 1986, North America mandated that all new passenger vehicles be equipped with a center high mount stop lamp (CHMSL) (**FIGURE 44-42**). Light trucks and vans were added in 1994. This lamp is located higher than the regular brake lights, near the centerline of the vehicle. It is designed to be more in the driver's line of sight while he or she is looking down the road to anticipate traffic hazards. This higher-mounted lamp helps

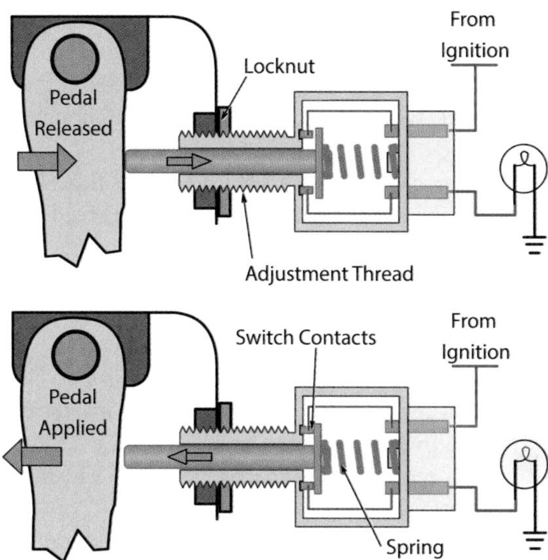

FIGURE 44-41 Brake light switch operation.

FIGURE 44-42 A center high mount stop lamp (CHMSL) mounted on a vehicle.

reduce rear-end collisions by being more visible to drivers, giving them more warning time for stopping.

▶ Power Brakes

K44022

Types and Purpose

A power booster or power brake unit uses an external source of force to multiply the driver's pedal effort and apply that to the master cylinder pistons, thus increasing the hydraulic pressure available from the master cylinder. There are two main types of power brake units: vacuum-assist and hydraulic-assist (**FIGURE 44-43**). The vacuum-assist power booster is the most common.

Units on gasoline engines use the vacuum produced in the intake manifold to power the brake booster. This supplies approximately 20" of vacuum to the booster. Vehicles with diesel engines do not have manifold vacuum, so they are fitted with an engine-driven vacuum pump.

FIGURE 44-43 Power brake boosters. **A.** Vacuum-assist. **B.** Hydraulic-assist.

The vacuum-assisted power booster operates between the brake pedal and the master cylinder. It uses the difference between engine vacuum and atmospheric pressure to increase the force that acts on the master cylinder pistons. The level of assistance this power booster gives depends on the pressure applied to the brake pedal and the pressure difference between the vacuum side of the booster and the atmospheric pressure side.

The hydraulic-assisted power booster operates between the brake pedal and the master cylinder. It usually uses hydraulic pressure from the power steering pump to increase the force that acts on the master cylinder. The level of assistance this power booster gives depends on the pressure applied to the brake pedal and the hydraulic pressure in the system.

Vacuum Booster

K44023

The most common types of vacuum boosters are the single diaphragm and the dual diaphragm. Just as the names imply, they have either one or two diaphragms that are used to extract power from atmospheric pressure. Using two diaphragms allows the diameter of the booster to be smaller, although it is a bit longer than a single-diaphragm unit.

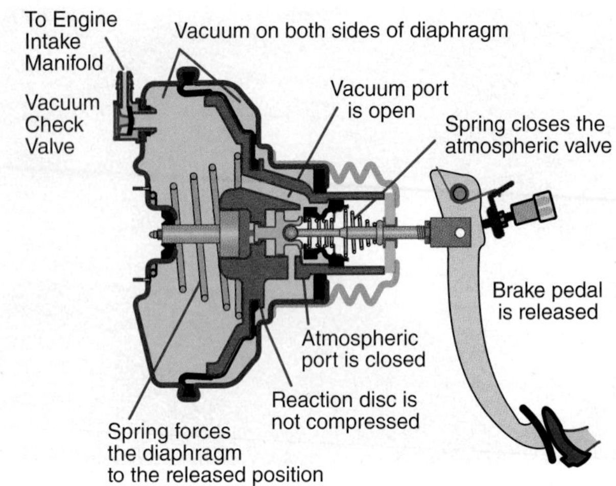

FIGURE 44-44 Vacuum brake booster in the released position.

Vacuum boosters consist of the following (**FIGURE 44-44**):

- A housing, which encases all of the other parts
- One or two sealed diaphragms, which transmit the force to the master cylinder
- A diaphragm return spring, which pushes the diaphragm back when the brake pedal is released
- A control valve, which controls vacuum and atmospheric pressure to each chamber
- A pushrod and reaction disc, which operate in conjunction with the control valve
- A one-way check valve, which holds vacuum in the booster when it is higher than intake manifold vacuum
- Seals, which prevent air leaks

How much force can a vacuum booster create? Let's do the math, using a panic stop situation. The booster receives approximately 20" (51 cm) of vacuum from the intake manifold. That is equal to about 10 psi (69 kPa) of pressure because 2" (5.1 cm) of vacuum equals 1 psi (6.9 kPa). If the diaphragm has a circumference of 12" (31 cm), then it has an area of approximately 113 sq in. (729 cm^2). Therefore, 10 psi (69 kPa) × 113 square inches (729 cm^2) = 1130 lb (513 kg). That means that a 12" (31 cm) booster can add up to 1130 lb (513 kg) of force to whatever the driver's foot effort is after being multiplied through the leverage of the brake pedal. That force is magnified further by the hydraulic system. This is why a vehicle weighing thousands of pounds can be stopped with minimal foot pressure.

Vacuum Booster Operation

When the driver steps on the brake pedal, it moves the brake pedal pushrod forward, which transmits movement through the power unit to the master cylinder piston, to apply the brakes. It also operates a control valve that controls the flow of vacuum and atmospheric pressure to each side of the diaphragm. How it works depends on the position of the control valve.

A hose connects the intake manifold to a vacuum check valve on the power brake unit. With the engine running, the check valve allows air to be evacuated from the booster, but not to return. Vacuum in the intake manifold is used to evacuate the power unit. The valve holds vacuum in the booster in the case of an engine failure, where it allows at least one full boosted brake application.

The booster chambers are separated by a flexible rubber diaphragm attached to the diaphragm plate. It is held in the off position by a large-diaphragm return spring. The master cylinder pushrod and the control valve assembly are centrally located on each side of the plate. The master cylinder pushrod normally incorporates a system to provide an adjustment for pushrod length. The adjustable length provides the proper clearance between the master cylinder pushrod and the master cylinder piston. The length normally does not have to be adjusted, but if adjustment is required, use the proper tools and follow the service information completely.

As the brakes are applied, the pedal pushrod and plunger move forward in the diaphragm plate, which brings the vacuum valve into contact with the vacuum port seat. This closes the vacuum port, sealing off the passage connecting the two chambers and holding the pressure and vacuum steady in each chamber. This is called the hold position (**FIGURE 44-45**).

Further movement of the pushrod and plunger moves the atmospheric valve away from the atmospheric port seat. Air at atmospheric pressure comes in through the air filter in the rear of the unit and enters the chamber behind the diaphragm. The difference in pressure now on both sides of the diaphragm moves the diaphragm plate forward, and it takes the master cylinder pushrod with it, applying greater force to the master cylinder. In this position, called the apply position, the vacuum valve is closed and the atmospheric valve is open.

Once the brake pedal stops moving, the atmospheric pressure continues to build up in the rear chamber. As hydraulic pressure rises, a counterforce acts through the master cylinder pushrod and the reaction disc. This counterforce allows the diaphragm plate and control valve to continue moving forward

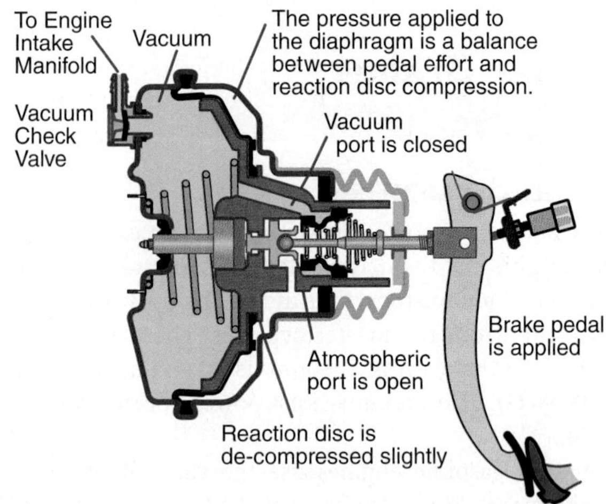

FIGURE 44-45 Vacuum brake booster in the apply position.

a little bit until the movement causes the atmospheric valve to close off the atmospheric port. Closing off the atmospheric port stops the atmospheric pressure from entering the booster, which causes it to hold in that position, ready for increased or decreased pedal pressure from the driver.

During application, the reaction force against the valve plunger works against the driver to close the atmospheric port. With both the atmospheric and the vacuum ports closed, the power unit is in a hold position. It stays this way until increased pedal force reopens the atmospheric port, causing more boost; or a drop in pedal force reopens the vacuum port, reducing boost. When the force on the pedal is held constant, and both the vacuum valve and the atmospheric valve are closed, the valve always returns to the hold position.

When the brake pedal is released, the atmospheric valve is closed, and the vacuum valve opens. As a result, any atmospheric pressure is evacuated from the rear chamber, through the front chamber, out the booster, through the check valve, and into the intake manifold. This reduces the atmospheric pressure pushing on the diaphragm plate. The return spring then pushes the diaphragm plate back to its off position. With the driver's foot off the brake pedal, the vacuum valve remains open, ensuring that there is equal vacuum on both sides of the diaphragm plate ready for the next application.

When the engine is switched off or stops for any reason, no manifold vacuum is available to supply the booster. The vacuum remaining in the booster, held by the one-way check valve, will provide for at least one power-boosted brake application. After this, the brakes will still operate, but without power assistance, which will require much more brake pedal effort from the driver.

Dual-Diaphragm Boosters

Dual-diaphragm power boosters work on the same principle of operation as the single-diaphragm power booster and are smaller in diameter to better fit in the limited space under the hood of some vehicles. However, to keep the surface area of the diaphragm plate large enough to provide the needed boost, two smaller diaphragms in tandem are used, one behind the other. There are separate vacuum and atmospheric chambers for each diaphragm plate. They use the same control valve concepts and therefore apply, hold, and release in the same manner as a single-diaphragm booster.

▶ Hydraulic Brake Booster

K44024

Purpose and Operation

Although not as common as a conventional brake system fitted with a vacuum booster, many vehicles are now equipped with hydraulically assisted brake boosters. The hydraulic booster system uses hydraulic pressure generated by the power steering pump, rather than engine vacuum, to provide the required power (**FIGURE 44-46**). This application is particularly suitable to vehicles with diesel engines, as a separate vacuum source does not have to be provided for the system to operate. Although hydraulic brake boosters use power steering fluid to operate the booster, the master cylinder portion of the system still uses brake fluid. Do not make the mistake of putting power steering fluid in the master cylinder reservoir or brake fluid in the power steering pump reservoir.

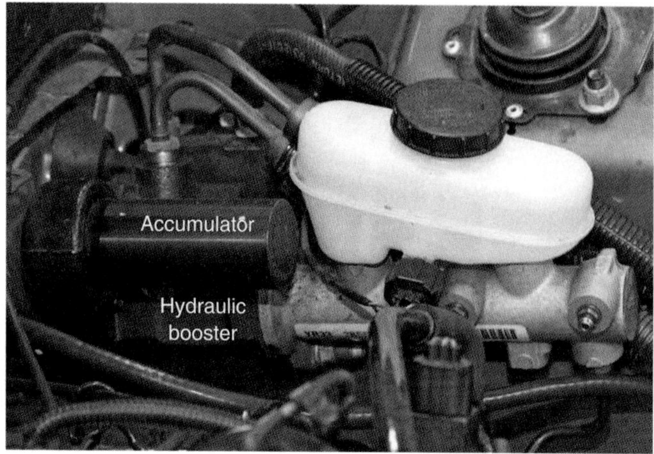

FIGURE 44-46 Hydraulic power brake booster.

Because the hydraulic booster system uses fluid pressure from the existing power steering pump, the booster uses the pressure from the power steering fluid that is always circulating through it as the source of pressure applied against the master cylinder actuating piston. The hydraulic pressure generated by the power steering pump is stored in an accumulator, routed to the hydraulic booster unit, and applied as mechanical force to the master cylinder when the brake pedal is applied (**FIGURE 44-47**).

The booster can generate pressures of 1200 to 2000 psi (8274 to 13790 kPa) to activate the master cylinder. Hydraulic boosters can be completely separate components from the master cylinder, or they can be components integrated with the master cylinder. As a safety measure, part of the hydraulic booster system includes an accumulator, which assists in maintaining a reserve of system pressure. Some are nitrogen pressurized, whereas others are spring loaded, depending on the application. Should pressure in the hydraulic system be lost, such as when the engine stalls or the power steering pump drive belt breaks, the hydraulic booster system's accumulator is designed to store sufficient pressure to provide for a few full-power applications. Once this accumulated pressure is used up, the hydraulic booster system resorts to manual brake function, requiring much higher pedal pressure.

▶ Electrohydraulic Braking

K44025

Purpose and Operation

A less common brake booster system is electrohydraulic braking. **Electrohydraulic braking (EHB)** gets rid of the vacuum

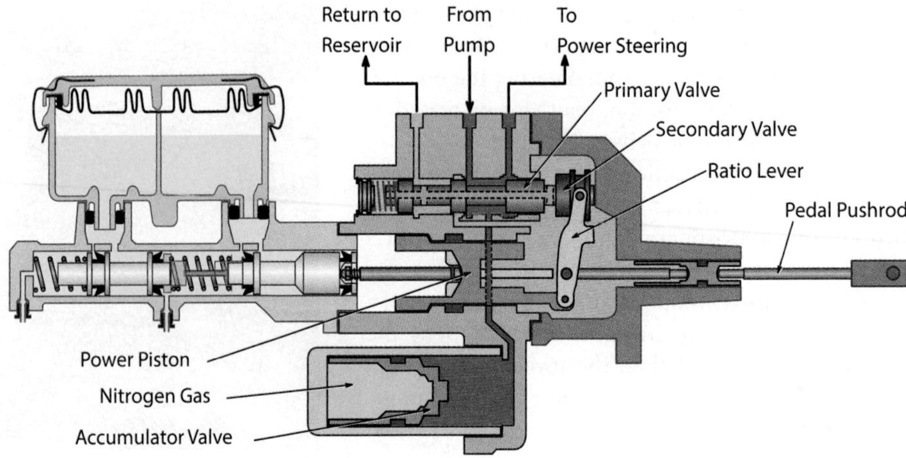

FIGURE 44-47 Cutaway view of a hydraulic power brake booster.

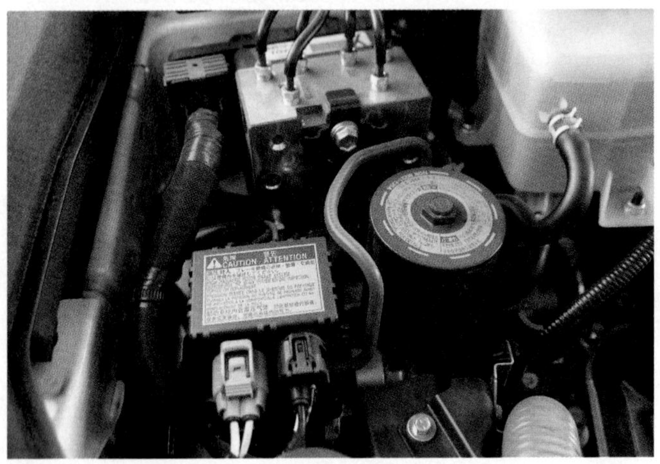

FIGURE 44-48 An electrohydraulic power brake booster.

booster and replaces it with an electrically driven hydraulic pump (**FIGURE 44-48**). Like the hydraulic booster system, it uses a high-pressure accumulator to store the required pressure to activate the master cylinder. However, it uses electric energy to effectively "charge" the accumulator and build sufficient pressure for efficient brake operation. This means that less power is taken away from the engine during operation, as battery power is used only when needed instead of continuously from the power steering pump. Also, there is no chance of problems caused by things such as a worn power steering pump, a slipping or broken pump drive belt, or leaky hose connections.

▶ Wrap-Up

Ready for Review

- The principle behind hydraulic brakes is Pascal's law, which states that pressure applied to a fluid in one part of a closed system will be transmitted without loss to all other areas of the system.
- A substantial leak in the hydraulic braking system will prevent enough pressure from building to exert the necessary braking force.
- Engineers design brakes that have precise (but unequal from front to back) amounts of braking force at each wheel.
- The three variables related to pressure and force in hydraulic systems are input force, working pressure, and output force.
- Main components of the hydraulic braking system are brake pedal, brake fluid, and master cylinder.
- The brake pedal multiplies force from the driver's foot to the master cylinder.
- Brake fluid has a high boiling point and a low freezing point, and is hygroscopic (absorbs water).
- The Department of Transportation (DOT) grades brake fluids on pH value, viscosity, resistance to oxidation, stability, and boiling point.
- Master cylinders convert force exerted from the brake pedal into hydraulic pressure to activate wheel brake units.
- Types of master cylinders are single piston and tandem (required on modern cars).
- Single-piston master cylinders use a primary cup to seal pressure in the cylinder and a secondary cup to prevent fluid loss.
- A single-piston master cylinder traps brake fluid and forces it into the brake lines.
- Residual pressure valves are used on drum brake systems to maintain brake fluid pressure and prevent air entry when the brakes are off.

- Modern vehicles have tandem master cylinders to ensure braking ability in at least one circuit despite a leak.
- Differential pressure switches monitor loss of pressure between the hydraulic circuits.
- Braking units can be split front-to-rear, diagonally, or in an L shape.
- Diagonal and L-shaped braking splits retain 50% braking capability even if half the system fails.
- Quick take-up master cylinders work to compensate for the large running clearance maintained by low-drag brake calipers.
- Hydraulic braking systems use proportioning valves, metering valves, pressure differential valves, or antilock hydraulic control units to modify hydraulic pressure.
- Proportioning valves reduce brake pressure to the rear wheels and are pressure sensitive or load sensitive.
- Load-sensitive proportioning valves adjust rear brake pressure according to the weight of the vehicle's load.
- Pressure-sensitive proportioning valves use a poppet piston to limit the rate of braking pressure increase to the rear brakes.
- Metering valves work to ensure rear brake pressure is applied before front brake pressure.
- The combination valve combines individually operating proportioning valves, metering valve, and pressure differential valve in one unit and cannot be repaired.
- Brake warning lights alert drivers to engagement of the parking brake, low brake fluid intake, and unequal pressure in the hydraulic brake system.
- Stop lights are mounted on the rear of a vehicle and alert other drivers that the vehicle is being braked.
- As of 1986, all vehicles must have a center high mount stop lamp (CHMSL) to reduce incidence of rear-end collisions.
- Power brake units are either vacuum assist (most common) or hydraulic assist.
- Vacuum boosters have single or dual diaphragms to extract power from atmospheric pressure and transmit force to the master cylinder.
- A 12" vacuum booster is capable of generating enough psi to stop a vehicle weighing thousands of pounds.
- A vacuum booster works off the difference between manifold vacuum and atmospheric pressure; a difference in these pressures creates more force on the master cylinder pistons.

Key Terms

aerate The tendency to create air bubbles in a fluid.

bleeding The process of removing air from a hydraulic braking system.

brake fluid Hydraulic fluid that transfers forces under pressure through the hydraulic lines to the wheel braking units.

brake hose A flexible section of the brake lines between the body and suspension that allows for steering and suspension movement.

brake lines Made of seamless, double-walled steel and able to transmit over 1000 psi (6895 kPa) of hydraulic pressure through the hydraulic brake system.

CANbus circuit A two-wire communication network that transmits status and command signals between control modules in a vehicle.

compensating port Connects the brake fluid reservoir to the master cylinder bore when the piston is fully retracted, allowing for expansion and contraction of the brake fluid.

electrohydraulic braking (EHB) A hydraulic braking system that uses an electrically driven hydraulic pump to pressurize fluid for use in the master cylinder.

hygroscopic A substance that attracts and absorbs water (e.g., brake fluid).

inlet port Connects the reservoir with the space around the piston and between the piston cups in a brake master cylinder.

input force The force applied to the input piston, measured in either pounds or kilograms.

International Standards Organization (ISO) flare A method for joining brake lines, also called a bubble flare. Created by flaring the line slightly out and then back in, leaving the line bubbled near the end.

inverted double flare A method for joining brake lines that forms a secure, leakproof connection.

load transfer Weight transfer from one set of wheels to the other set of wheels during braking, acceleration, or cornering.

metering valve A valve used on vehicles equipped with older rear drum/front disc brakes to delay application of the front disc brakes until the rear drum brakes are applied. Located in line with the front disc brakes.

output force Force that equals the working pressure multiplied by the surface area of the output piston, expressed as pounds, newtons, or kilograms.

outlet port Links the cylinder to the brake lines.

Pascal's law The law of physics that states that pressure applied to a fluid in one part of a closed system will be transmitted equally to all other areas of the system.

poppet valve A valve that controls the flow of brake fluid at usually preset pressures.

pressure differential valve A valve that monitors any pressure difference between the two separate hydraulic brake circuits; it usually contains a switch to turn on the brake warning light when there is a pressure difference.

primary cup A seal that holds pressure in the master cylinder when force is applied to the piston.

primary piston A brake piston in the master cylinder moved directly by the pushrod or the power booster; it generates hydraulic pressure to move the secondary piston.

proportioning valves Valves used mostly on older vehicles equipped with rear drum brakes to reduce rear wheel hydraulic brake pressure under hard braking or light loads. Located in line with the rear brakes.

quick take-up master cylinders Cylinders used on disc brake systems that are equipped with low-drag brake calipers to quickly move the brake pads into contact with the brake rotors.

quick take-up valve A valve used to release excess pressure from the larger piston in a quick take-up master cylinder once the brake pads have contacted the brake rotors.

recuperation Process by which brake fluid moves from the reservoir past the edges of the seal into the chamber in front of the piston. This prevents air from being drawn into the hydraulic system caused by low pressure when the brake pedal is released quickly.

residual pressure valve (residual check valve) In drum brake systems, a valve that maintains pressure in the wheel cylinders slightly above atmospheric pressure so that air does not enter the system through the seals in the wheel cylinders.

secondary cup A seal that prevents loss of fluid from the rear of each piston in the master cylinder.

secondary piston A piston that is moved by hydraulic pressure generated by the primary piston in the master cylinder.

single-piston master cylinder A master cylinder with a single piston that creates hydraulic pressure for all wheel units. If there is a leak in the system, there is a loss of pressure for all wheel units.

tandem master cylinder A master cylinder that has two pistons that operate separate braking circuits so that if a leak develops in one circuit, the other circuit can still operate.

working pressure The pressure within a hydraulic system while the system is being operated.

Review Questions

1. All of the statements are true *except*:
   a. Pascal's law is the principle behind hydraulic brakes.
   b. varying amounts of mechanical force can be extracted from a single amount of hydraulic pressure.
   c. the smaller brake caliper pistons on the front wheels give greater mechanical force and braking power to the front wheels.
   d. a power booster is fitted to the brake system to increase the driver's brake pedal force to the master cylinder.
2. If 100 psi of working pressure is applied to a 2 square inch output piston, how much force will be created?
   a. 50 lb
   b. 100 lb
   c. 200 lb
   d. 400 lb
3. Which of the following brake fluids has the highest wet boiling point?
   a. DOT 2
   b. DOT 4
   c. DOT 3
   d. DOT 5
4. The port that connects the reservoir to the cylinder, just barely ahead of the primary cup is the:
   a. compensating port.
   b. inlet port.
   c. outlet port.
   d. primary port.
5. Disc brake systems that are equipped with low-drag brake calipers use:
   a. tandem master cylinders.
   b. quick take-up master cylinders.
   c. single-piston master cylinders.
   d. modern master cylinders.
6. Different lever designs can be engineered to alter the brake pedal effort required of the driver by using different levels of:
   a. input force.
   b. output force.
   c. working pressure.
   d. mechanical advantage.
7. A wheel's braking ability depends most on:
   a. engine speed.
   b. the type of brake fluid used.
   c. the load it is carrying.
   d. the type of transmission used.
8. Which of the following flares is created by flaring the end of the tube outward?
   a. Buggle flare
   b. Bubble flare
   c. Inverted double flare
   d. ISO flare
9. Which of the following is used to hold off the application of the front brakes on vehicles with disc brakes on the front wheels and drum brakes on the rear wheels?
   a. Load-sensitive proportioning valve
   b. Metering valve
   c. Pressure differential valve
   d. Residual pressure valve
10. The valve that reduces rear brake pressure when the vehicle is lightly loaded and allows higher pressure when it is heavily loaded is:
    a. load-sensitive proportioning valve.
    b. metering valve.
    c. pressure differential valve.
    d. residual pressure valve.

ASE Technician A/Technician B Style Questions

1. Tech A says that hydraulic pressure is applied equally in all directions throughout a closed system. Tech B says that air in the hydraulic system will cause the brake pedal to be spongy. Who is correct?
   a. Tech A
   b. Tech B
   c. Both A and B
   d. Neither A nor B
2. Tech A says that a brake pedal that does not return all the way causes the brake warning light on the instrument panel to stay on. Tech B says that a brake pedal that does not return causes the brake lights to stay on. Who is correct?
   a. Tech A
   b. Tech B
   c. Both A and B
   d. Neither A nor B

3. Tech A says that brake fluid should periodically be checked for excessive moisture content. Tech B says that brake fluid should be replaced every 12,000 miles. Who is correct?
   a. Tech A
   b. Tech B
   c. Both A and B
   d. Neither A nor B
4. Tech A says that DOT 5 brake fluid should be used in all vehicles today because it is silicone based and will not absorb water. Tech B says that DOT 4 and DOT 3 are not recommended for use in ABS systems. Who is correct?
   a. Tech A
   b. Tech B
   c. Both A and B
   d. Neither A nor B
5. Tech A says that the secondary piston in a master cylinder is operated by hydraulic force. Tech B says that the brake pedal return spring returns the master cylinder pistons to their original position. Who is correct?
   a. Tech A
   b. Tech B
   c. Both A and B
   d. Neither A nor B
6. Tech A says that the low level brake fluid switch on a master cylinder turns on the brake warning light when the system is low on fluid. Tech B says that the low level switch also monitors the condition of the fluid and activates the warning light when the brake fluid needs to be replaced. Who is correct?
   a. Tech A
   b. Tech B
   c. Both A and B
   d. Neither A nor B
7. Tech A says that the metering valve controls pressure to the rear brakes. Tech B says that the proportioning valve controls pressure to the front brakes. Who is correct?
   a. Tech A
   b. Tech B
   c. Both A and B
   d. Neither A nor B
8. Tech A says that the vacuum booster increases the vacuum in the brake system. Tech B says that the vacuum booster uses vacuum and atmospheric pressure to multiply the driver's foot pressure applied to the master cylinder push rod. Who is correct?
   a. Tech A
   b. Tech B
   c. Both A and B
   d. Neither A nor B
9. Tech A says that input force divided by the area of the input gives working pressure. Tech B says that working pressure multiplied by the surface area of the output piston gives output force. Who is correct?
   a. Tech A
   b. Tech B
   c. Both A and B
   d. Neither A nor B
10. Tech A says that quick take-up master cylinders use two additional pistons to move brake fluid faster. Tech B says that banjo fittings are commonly used to connect brake hoses to wheel brake units. Who is correct?
    a. Tech A
    b. Tech B
    c. Both A and B
    d. Neither A nor B

CHAPTER 45 Servicing Hydraulic Systems and Power Brakes

NATEF Tasks

- **N45001** Select, handle, store, and fill brake fluids to proper level; use proper fluid type per manufacturer specification. (MLR/AST/MAST)
- **N45002** Test brake fluid for contamination. (MLR/AST/MAST)
- **N45003** Bleed and/or flush brake system. (MLR/AST/MAST)
- **N45004** Measure brake pedal height, travel, and free play (as applicable); determine needed action. (AST/MAST)
- **N45005** Check brake pedal travel with and without engine running to verify proper power booster operation. (MLR/AST/MAST)
- **N45006** Diagnose pressure concerns in the brake system using hydraulic principles (Pascal's law). (AST/MAST)
- **N45007** Check master cylinder for internal/external leaks and proper operation; determine needed action. (MLR/AST/MAST)
- **N45008** Remove, bench bleed, and reinstall master cylinder. (AST/MAST)
- **N45009** Measure and adjust master cylinder pushrod length. (AST/MAST)
- **N45010** Inspect vacuum type power booster unit for leaks; inspect the check valve for proper operation; determine needed action. (AST/MAST)
- **N45011** Inspect and test hydraulically assisted power brake system for leaks and proper operation; determine needed action. (AST/MAST)
- **N45012** Inspect brake lines, flexible hoses, and fittings for leaks, dents, kinks, rust, cracks, bulging, wear; and loose fittings/supports; determine needed action. (MLR/AST/MAST)
- **N45013** Replace brake lines, hoses, fittings, and supports. (AST/MAST)
- **N45014** Fabricate brake lines using proper material and flaring procedures (double-flare and ISO types). (AST/MAST)
- **N45015** Inspect, test, and/or replace components of brake warning light system. (AST/MAST)
- **N45016** Check parking brake operation and parking brake indicator light system operation; determine needed action. (MLR/AST/MAST)
- **N45017** Check parking brake system and components for wear, binding, and corrosion; clean, lubricate, adjust, and/or replace as needed. (MLR/AST/MAST)
- **N45018** Check operation of brake stop light system. (MLR/AST/MAST)

Knowledge Objectives

After reading this chapter, you will be able to:

- **K45001** Describe hydraulic system service and liability concerns.

Skills Objectives

After reading this chapter, you will be able to:

- **S45001** Perform hydraulic system maintenance.
- **S45002** Perform hydraulic system component diagnosis and repair.

▶ Introduction

Now that you know about the hydraulic brake components and basic hydraulic principles, we will cover the maintenance, diagnosis, and repair of the hydraulic portion of the brake system (**FIGURE 45-1**). Although the hydraulic system is very dependable overall, it does require periodic maintenance along with occasional diagnosis and repair. Maintenance tasks typically revolve around flushing the brake fluid, regular visual inspection of the hydraulic components, and checking the feel and operation of the brakes. Diagnosis and repair are in order if the maintenance checks on the system indicate faults or if the owner brings the vehicle in with a brake-related concern. But before we cover these processes, we need to discuss the liability issues surrounding brake repair.

▶ Brake Repair Legal Standards and Technician Liability

K45001

Brake repair is right up with steering and suspension repair on the liability scale. Improperly repaired brakes can function reasonably well under normal driving situations but can fail during a panic situation, when they are needed the most. The likelihood of accidents, injury, or death goes up drastically in those situations. Shops and technicians have been successfully sued for improper brake repairs, resulting in large cash settlements. Technicians also risk being found criminally negligent if they are determined to have acted maliciously. Because of this, always follow the manufacturer's procedures when servicing brake systems. Research service information, precautions, and technical service bulletins (TSBs) before performing any brake tasks. Never take shortcuts, which could cause the vehicle to be unsafe.

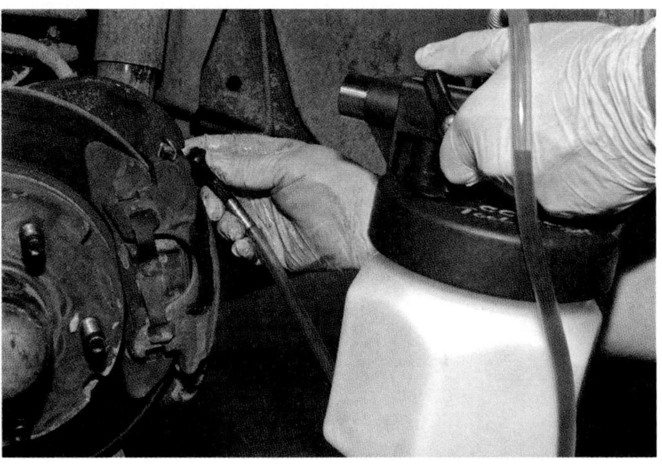

FIGURE 45-1 A technician flushing brake fluid with a vacuum bleeder tool.

Another issue that leads to liability is forgetting to tighten, or failing to properly tighten, components. When you are working on vehicles, it is easy for a distraction to interrupt your work. If that happens, then it is easy to think you already finished a part of the job when you really haven't. A good example is when you are torquing the lug nuts on the wheels. If you get interrupted while torquing the third wheel, it would be easy to come back and think you had torqued all of the wheels and to leave one untorqued and subject to falling off the vehicle down the road. Good technicians reduce liability by creating processes to help ensure that steps of the job are not forgotten. This could be as simple as only installing the lug nuts when you are ready to immediately torque them. Another option if you can't torque them immediately is to only just start them on the lug studs (not running the lug nuts all the way down) so that it is obvious that they haven't been tightened or torqued. The same thing goes with most other parts, if you install them, torque them in place as soon as it is practical. Remember, safety first!

SAFETY TIP

Asbestos is a naturally occurring mineral mined from the earth. Asbestos is a long, very thin fibrous crystal. When asbestos is disturbed, small needlelike fibers can break off and remain airborne, where people inhale them. Because these fibers are so small, they embed themselves deep within lung tissue, causing scarring. Repeated exposure can lead to asbestosis and lung cancer.

Although asbestos has been removed from most brake and clutch materials, it is still present in some replacement and old components. Therefore, you must treat all brake and clutch dust as if it contains asbestos. This is accomplished by using an aqueous brake wash station, or, in some states, an aerosol can of brake cleaning solution to carefully wash down the brake components.

You Are the Automotive Technician

A customer brings in her four-wheel drive vehicle with a brake concern. She describes the problem as the brakes not stopping well unless she pushes very hard on them. The service history shows that the front brakes were replaced and the fluid flushed about five years ago. You test-drive the vehicle and verify the hard brake pedal concern. During the visual inspection, you notice a stick that pushed the flexible brake hose against the right front tire, wearing the hose badly. The brake linings are in serviceable condition.

1. What faults could cause the hard brake pedal concern?
2. What tests would you do to confirm the cause of the hard pedal concern?
3. What fluid service is needed?
4. What do you need to do to repair the brake hose issue?

▶ Hydraulic System Maintenance

S45001

Tools

A number of tools make brake service much easier. Here are some of the most common:

- Brake bleeder wrenches are used to open and close bleeder screws. They come in a variety of configurations and are designed to fit into the tight space where the bleeder screw is located. They are also usually designed with six sides to decrease the likelihood of rounding off the bleeder screw hex head.
- Flare nut wrenches are used to loosen and tighten fittings on brake lines. Their open side slips over the brake line, and the wrench can still grab all six points of the fitting.
- Vacuum brake bleeders can be operated by hand, air, or electricity. The hose from the vacuum bleeder is placed on the end of the bleeder screw, and the bleeder screw is opened. The vacuum is applied and pulls brake fluid from the hydraulic braking system through the bleeder screw; this fluid is captured in a container for later disposal.
- Pressure brake bleeders provide a reservoir of brake fluid under pressure. The pressure bleeder usually mounts to the master cylinder reservoir and supplies a steady stream of clean brake fluid for bleeding or flushing. With the master cylinder pressurized, the bleeder screws on the wheel brake units can be opened one at a time and bled or flushed of old brake fluid and air.
- Proportioning valve/metering valve gauge sets are used to test the operation of the proportioning valve and metering valve. The gauge set is connected to the output line of the appropriate valve, and the brake pedal is activated. The pressures are monitored and compared to specifications.
- Brake fluid testers are used to test the boiling point/moisture content of brake fluid (**FIGURE 45-2**).

Brake Fluid Handling

N45001

You need to understand how to select and handle brake fluid as it is the lifeblood of the hydraulic braking system. Failure to do so could cause damage to the hydraulic braking system

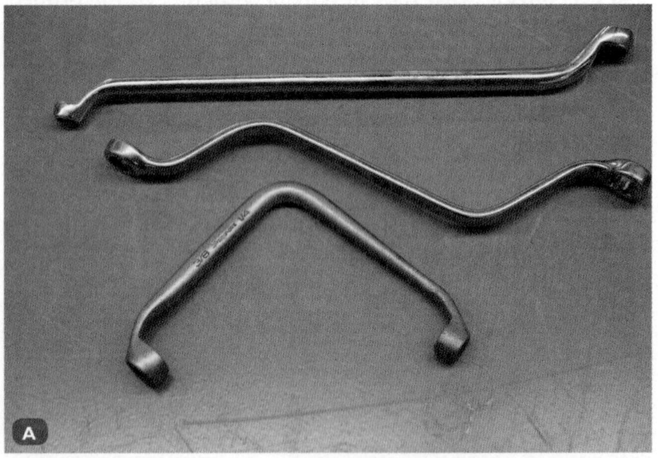
A

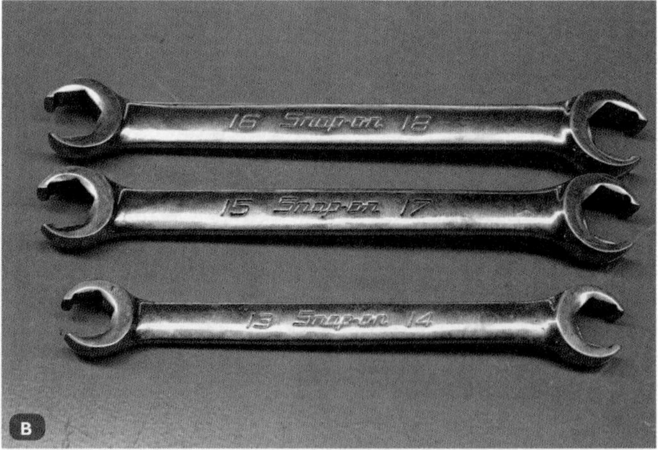

B

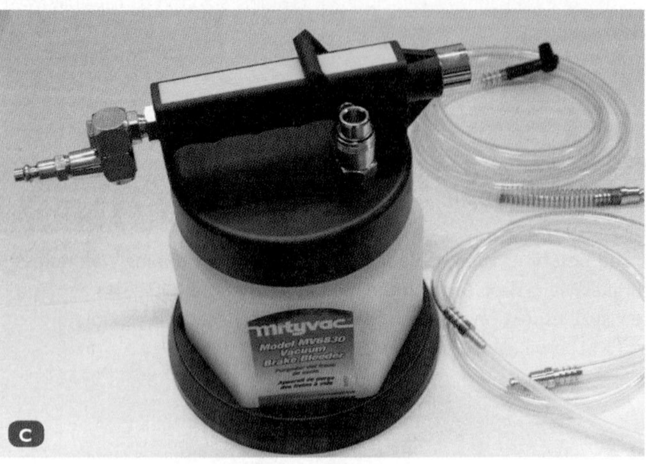

C

D

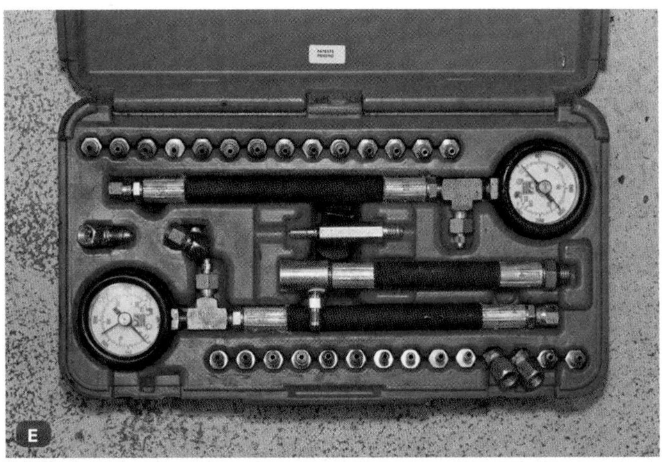

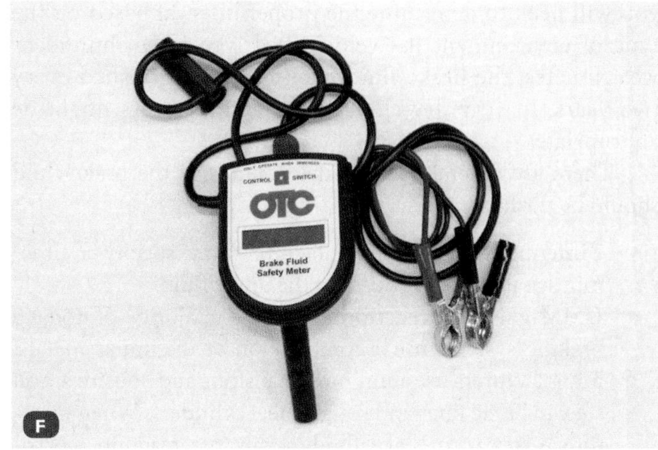

FIGURE 45-2 Common tools used to repair brakes: **A.** Brake bleeder wrenches. **B.** Flare nut wrenches. **C.** Vacuum brake bleeder. **D.** Pressure brake bleeder. **E.** Proportioning valve/metering valve gauge sets. **F.** Brake fluid tester.

and an unsafe situation for the driver and passengers. Brake fluid levels should be inspected during every oil change. The fluid should be near the full mark on the side of the cylinder or within half an inch of the top of each chamber if there are no marks. But because brake fluid is considered to be a non-top-off fluid, if it is near or below the minimum level in the reservoir, the cause of the low brake fluid level must be identified and corrected. A low level could be caused by worn brake linings/pads or a slow leak in the hydraulic braking system. Normally, topping off the brake fluid should only be performed after the brakes have been repaired or serviced. Dilute with fresh clean water any brake fluid that may have been spilled. Do not rub the brake fluid with a cloth, as this could damage any softened paint.

To select, handle, store, and fill brake fluids to proper level, follow the steps in **SKILL DRILL 45-1**.

Brake Fluid Testing

N45002

Brake fluid replacement is a maintenance item for virtually all vehicles. Consult the vehicle's service information to determine the correct intervals for brake fluid flushing. Some manufacturers neglect to specify this time frame, so

SKILL DRILL 45-1 Selecting, Handling, Storing, and Filling Brake Fluid

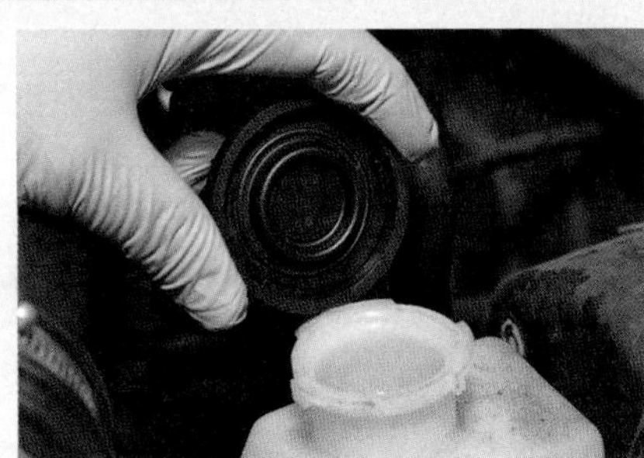

1. Research the specified type of brake fluid in the appropriate service information. Wipe around the master cylinder reservoir cover to prevent any dirt from entering the system. Remove the reservoir cover.

2. Check the fluid level in the reservoir. The fluid should be near the full mark on the side of the cylinder or within half an inch of the top of each chamber if there are no marks.

3. Only add the manufacturer's recommended brake fluid to bring the level to the full mark once all brake faults have been resolved. Replace the cover, and check that it is properly seated. Check for any leaks around the master cylinder. Dilute with fresh clean water any brake fluid that may have been spilled.

you will need to determine the proper interval based on the type of environment the vehicle is driven in. In humid or wet climates, the brake fluid may need to be flushed every two years. In very dry climates, every four years might be appropriate.

There are several ways of determining if the brake fluid should be flushed:

- **Time/mileage:** The manufacturer may specify a time/mileage interval for flushing the brake fluid.
- **DMM-galvanic reaction test:** The majority of today's braking systems use a combination of dissimilar metals. Manufacturers use aluminum in pistons and housings, and steel in brake lines and some wheel cylinders. When moisture mixes with brake fluid, a galvanic reaction (corrosion) can occur. The higher the moisture content in brake fluid, the higher the galvanic reaction and the greater the erosion or corrosion it causes. The DMM-galvanic reaction test uses a DMM to measure the voltage created by the galvanic reaction due to the level of moisture in the fluid (**FIGURE 45-3**).
- **Boiling point test:** Measuring the boiling point of the brake fluid using a special tool can determine the moisture content of the brake fluid (**FIGURE 45-4**).
- **Test strip:** Measuring specific chemicals/metals in the brake fluid can indicate whether there is a chemical breakdown of the brake fluid. Brake fluid test strips contain special color-changing pads that react in the presence of moisture or specific chemicals that have built up in the fluid and that indicate the need for flushing the brake fluid (**FIGURE 45-5**).

To test the brake fluid for contamination, you can use the DMM-galvanic reaction test, the brake fluid tester, or brake fluid test strips. For all three tests, follow these steps first: Brake fluid condition and level should be inspected at every oil change, and the fluid replaced according to the manufacturer's recommended service schedule or when it fails the following tests. Clean around the master cylinder cap to prevent contaminants from entering the reservoir. Remove the master cylinder reservoir cap.

To perform a DMM-galvanic reaction test, follow the steps in **SKILL DRILL 45-2**.

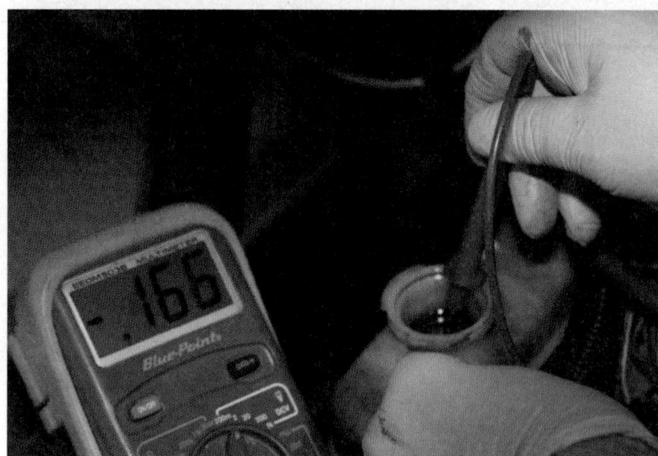

FIGURE 45-3 A DMM measures the voltage created by the galvanic reaction due to the level of moisture in the fluid.

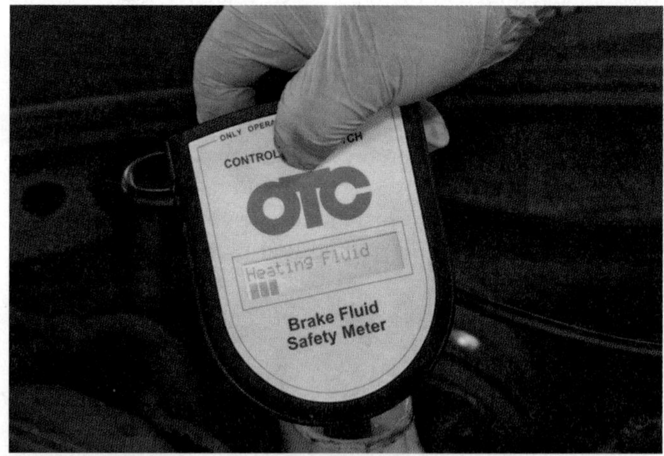

FIGURE 45-4 A brake fluid safety meter boils brake fluid to test for moisture.

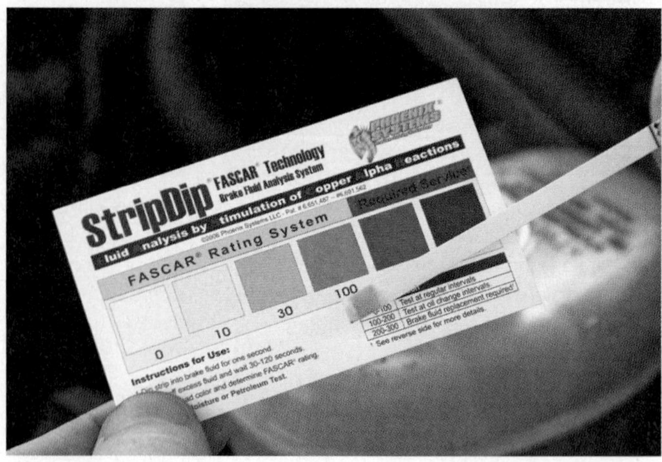

FIGURE 45-5 Brake fluid test strips react to moisture or chemicals in the brake fluid.

Applied Science

***AS-96: Contamination:* The technician can demonstrate an understanding of how a contaminated liquid can cause a chemical reaction that results in the deterioration of performance.**

As moisture levels increase in brake fluid, the boiling point of the brake fluid decreases. This chemical reaction results in the deterioration of the brake fluid's performance. In any liquid when boiling occurs, the liquid turns to vapor. Brakes work on the principle of liquid being incompressible; however, vapor is compressible. The result of brake fluid boiling in brake systems is a spongy pedal and a loss of braking effort at the wheels, including potential for complete brake failure. SAE field tests have documented that the average 1-year-old car has approximately 2% moisture in the brake fluid, making regular brake fluid changes necessary to ensure safety.

SKILL DRILL 45-2 Performing a DMM-Galvanic Reaction Test

1. Set the DMM to DC volts. Insert the red lead in the "v/~" slot and the black lead in the "common" slot.
2. Place the red voltmeter probe in the reservoir brake fluid, and place the black lead on the metal housing of the master cylinder. Make sure to use an unpainted surface of the housing.
3. Compare the voltage reading you obtained to specifications. Less than 0.3 volt is okay. More than 0.3 volt means the fluid needs to be flushed.

SKILL DRILL 45-3 Testing Brake Fluid with a Brake Fluid Tester

1. Follow the directions for the tester you are using.
2. Test the brake fluid. This is usually accomplished by placing an amount of brake fluid from the master cylinder into the tester. Some testers are designed to be placed directly into the brake fluid.
3. Compare the results you obtained to specifications. Most testers will tell you the boiling point of the fluid or the percent of moisture in the fluid.

SKILL DRILL 45-4 Testing Brake Fluid with a Brake Fluid Test Strip

1. Follow the directions for the brake fluid test strips you are using.
2. Test the brake fluid. Most test strips have to be dipped into the brake fluid for a specified amount of time (usually about one or two seconds).
3. Shake off excess fluid. An additional amount of specified time is allowed to pass before comparing the colored pad with the color chart to determine the level of contaminants or pH level.
4. Compare the results you obtained to specifications.

To test brake fluid with a brake fluid tester, follow the steps in **SKILL DRILL 45-3**.

To test brake fluid with a brake fluid test strip, follow the steps in **SKILL DRILL 45-4**.

Bleeding Brake Systems

N45003

Bleeding the brakes means removing air from the hydraulic braking system so that only the brake fluid is left in the system. When pressure is applied to brake fluid in a hydraulic braking system, the brake fluid cannot be compressed into a smaller volume, and therefore the pressure rises within the hydraulic braking system. Pressure is transmitted throughout the hydraulic braking system without loss. If air enters the hydraulic braking system, it can be dangerous. Unlike fluids, gases are compressible; when pressure is applied to air (a gas), the air decreases in volume, and the pressure does not build up as quickly throughout the hydraulic braking system. The brakes feel spongy and are not fully functional. Bleeding the brakes removes any air from the system.

There are a number of different brake bleeding methods. The three most common are described here:

- **Manual bleeding**—Using a helper to manually operate the brake pedal while you open and close the bleeder screws on the wheel units to allow the air and old brake fluid to be pushed out of the hydraulic braking system.
- **Vacuum bleeding**—Using a vacuum bleeder to pull the air and old brake fluid from the hydraulic braking system.
- **Pressure bleeding**—Using clean brake fluid under pressure from an auxiliary tool or piece of equipment to force the air and old brake fluid from the hydraulic braking system.

For all three methods, follow these general steps first: Remove as much of the old brake fluid from the reservoir as possible, using a suction gun, old antifreeze tester, or turkey baster. Refill the master cylinder reservoir, using the specified fluid type. Then determine which method of bleeding you will use. Also research the service information to obtain the bleeding sequence for the vehicle. Following the bleeding sequence will help you perform the task more quickly and with less chance of air entering the system.

The manual bleeding method requires the least amount of equipment and tools. It requires more time if a large percentage of the brake fluid has to be changed. It also is less effective at removing trapped air in systems that have high spots in the brake lines or in vehicles that have a large vertical drop between the master cylinder and the wheel brake units, such as a truck that has been lifted. This method is best when only a small amount of brake fluid has to be bled, such as after replacing front brake calipers or when more expensive equipment is not available.

Using the manual bleeding method requires clear communication between you and the assistant, to prevent air from entering the system. Be sure that you *do not* bleed one wheel so much that the reservoir runs dry and admits air into the hydraulic braking system. If this happens, it will be much harder to bleed the hydraulic braking system, as air is compressible, making it harder to build pressure within the hydraulic braking system.

To perform the manual bleeding method, follow the steps in **SKILL DRILL 45-5**.

Vacuum bleeding uses an air-operated or hand-operated vacuum pump to draw brake fluid out of the bleeder screws at each wheel unit. This method is much faster than using the manual bleeding procedure but easier to set up than the pressure bleeding system, So it is commonly used in shops. Just be sure to watch how much fluid is collecting in the vacuum bottle, so you don't run the master cylinder dry. Also remember that tandem master cylinders use divided reservoirs, so the fluid level could be over half on one side and empty on the other side. Watch the reservoir carefully, and top it off frequently to avoid getting air in the system. Another issue when using a vacuum bleeder is that the tool draws air from around the threads of the bleeder screw while it is operating, which show up as bubbles in the fluid being removed from the brake system. When using a vacuum bleeder, ignore the bubbles and focus more on the color of the fluid. Once the fluid is coming out clear, all air should be removed from the system.

To perform vacuum bleeding, follow the steps in **SKILL DRILL 45-6**.

SKILL DRILL 45-5 Performing Manual Bleeding

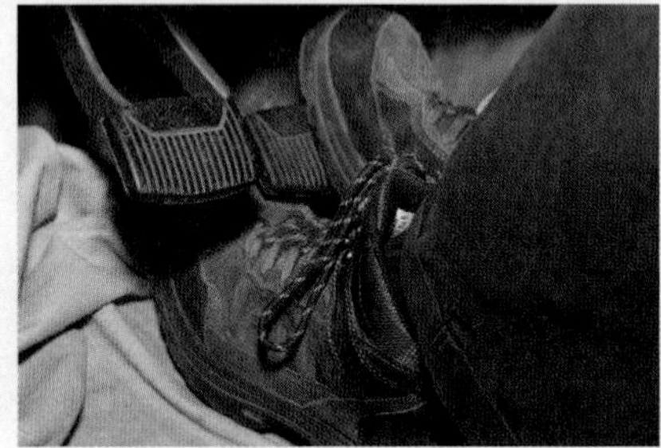

1. Ask an assistant to slowly push the brake pedal down, keeping his or her left foot underneath the pedal to limit full pedal travel.

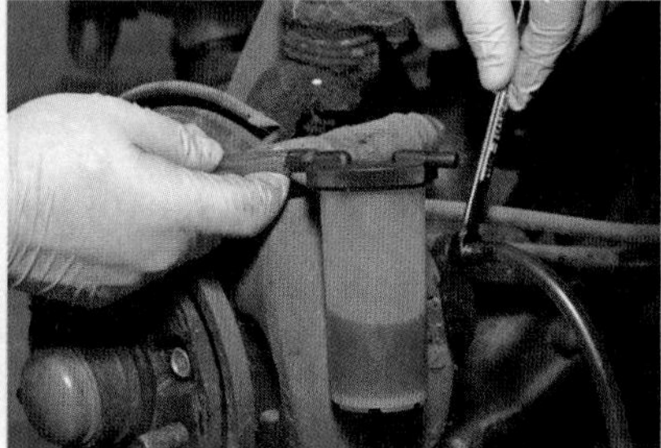

2. Install a clear hose on the farthest bleeder screw. Open the bleeder screw one-quarter to one-half turn. Observe any old brake fluid and air bubbles coming out. When the brake fluid stops, close the bleeder screw lightly, and then have the assistant slowly release the brake pedal. Repeat these steps until there are no more air bubbles or old fluid coming out of the hose.

3. Close off the bleeder screw, and tighten it to the manufacturer's specifications. Check the level in the master cylinder reservoir, top it off, and reinstall the reservoir cap. Repeat this bleeding procedure for each of the brake units, moving closer to the master cylinder, one wheel at a time.

SKILL DRILL 45-6 Performing Vacuum Bleeding

1. Prepare the vacuum bleeder for use. Install the vacuum bleeder on the farthest bleeder screw, and open it one-quarter to one-half turn.

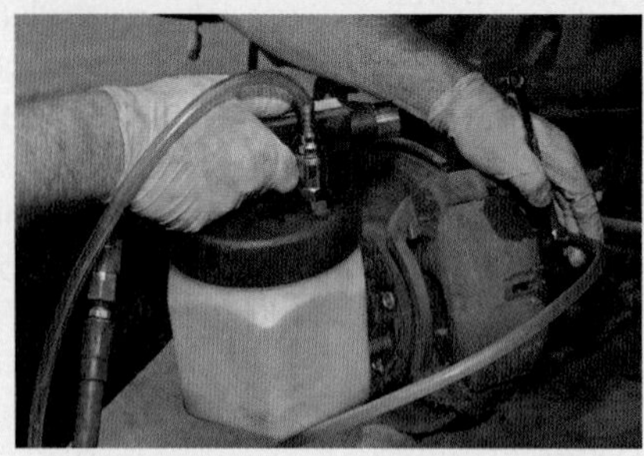

2. Operate the vacuum bleeder to pull brake fluid from the bleeder screw. Observe any old brake fluid and air bubbles coming out. Close off the bleeder screw, and tighten it to the manufacturer's specifications.

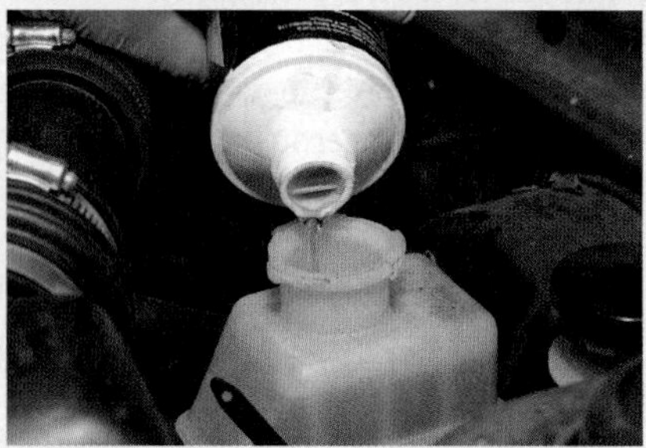

3. Check the level in the master cylinder reservoir, and top it off. Repeat this procedure for each of the brake units, moving closer to the master cylinder, one wheel at a time.

The pressure bleeding method uses equipment that can provide a continuous supply of brake fluid under pressure to the master cylinder reservoir. It may also include small-diameter hoses that attach to the bleeder screws and recover all of the old brake fluid. Because this method takes a bit of work to connect the equipment to the vehicle, it is best used when the hydraulic system needs a full flush or on systems that have a tendency to trap air, making manual bleeding much more difficult. To perform pressure bleeding, follow the steps in **SKILL DRILL 45-7**.

> **TECHNICIAN TIP**
>
> One other method that you should know about is gravity bleeding. This method relies on the master cylinder being mounted higher than the wheel brake units. Because gravity tries to pull the brake fluid down, gravity bleeding can occur just by opening one or more bleeder screws. If the height difference is great enough, the brake fluid will slowly drain out of the system. A technician can use this method to start the bleeding process while working on other wheel brake units. However, gravity bleeding can be a disadvantage. If a technician inadvertently leaves a hose disconnected or a bleeder screw open, *all* of the brake fluid can drain out of the system, leaving it full of air and harder to bleed. Be very careful not to leave the hydraulic braking system open for too long.

> **TECHNICIAN TIP**
>
> Never run the master cylinder dry during bleeding; doing so will make it much more difficult to bleed!

After performing bleeding or flushing brakes, follow these steps:

1. Double-check that all bleeder screws are properly tightened.
2. Replace all bleeder screw dust caps when finished.
3. Refill the master cylinder to the proper level, and reinstall the master cylinder reservoir cap.
4. Start the vehicle, and check for proper brake pedal feel (it should be firm and high).
5. Verify that there are no leaks at each bleeder screw.
6. Dispose of any old brake fluid in an environmentally approved method.

Flushing Brake Systems

Flushing brake fluid is similar to brake bleeding except that it is designed not only to remove any trapped air, but also to replace all of the old brake fluid with new brake fluid. A flush using clean brake fluid can be performed to remove any remaining residue and all air. Usually, clean brake fluid of the proper type for the vehicle is used for flushing a hydraulic braking system. However, if the hydraulic braking system has been contaminated with petroleum products or other improper fluids, some manufacturers specify flushing it with clean alcohol. The alcohol helps dissolve any oil in addition to flushing the system. Once the alcohol flush is complete, dry compressed air can be used to blow out the majority of the alcohol.

> **SAFETY TIP**
>
> Never use any petroleum or mineral-based products, such as gasoline or kerosene, to clean a hydraulic braking system or its components. They are not compatible with the seals used in the braking system and will result in a failure of the hydraulic braking system and its components. This failure may result in injury to the passengers or damage to the vehicle.

Inspecting the Brake Pedal

N45004, N45005

Brake pedal height, free play, and travel are critical for proper brake operation. Having the proper brake pedal height helps to ensure that the brake pedal has enough starting height to fully apply force to the brakes, even if one half of the hydraulic system is rendered useless by a leak. In other words, there has to be a specified distance the brake pedal can travel before it contacts the floor or anything else.

To measure brake pedal height, follow the steps in **SKILL DRILL 45-8**.

SKILL DRILL 45-7 Performing Pressure Bleeding

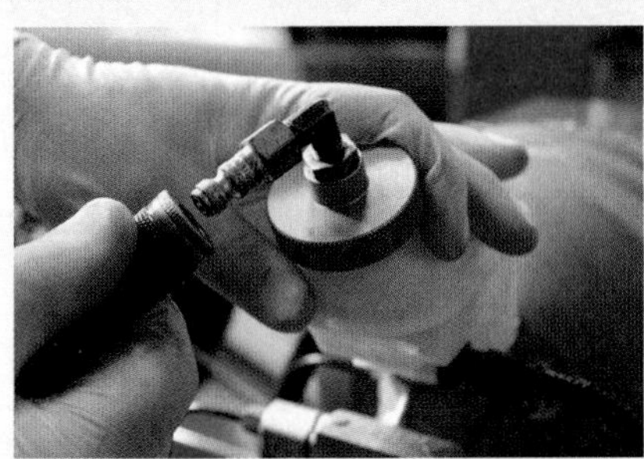

1. Prepare the pressure bleeder, and install it on the vehicle.

2. Install a clear hose on the farthest bleeder screw, and open it one-quarter to one-half turn. Observe any old brake fluid and air bubbles coming out.

3. Close off the bleeder screw when the brake fluid is clear and has no bubbles. Tighten it to the manufacturer's specifications. Repeat this procedure, moving closer to the master cylinder, one wheel at a time.

SKILL DRILL 45-8 Measuring Brake Pedal Height

1. Research the procedure and specifications for measuring brake pedal height, travel, and free play for the vehicle you are working on. Some manufacturers specify how much travel the brake pedal should have, whereas others specify how much reserve pedal should remain when the brake pedal is fully applied. Know which process applies to the vehicle you are working on.
2. Remove any removable floor mats or anything lying on the floor near the brake pedal.
3. With the engine off, measure the brake pedal height between the two specified points, using a measuring stick.
4. Compare this reading to the specifications, and determine any necessary actions.

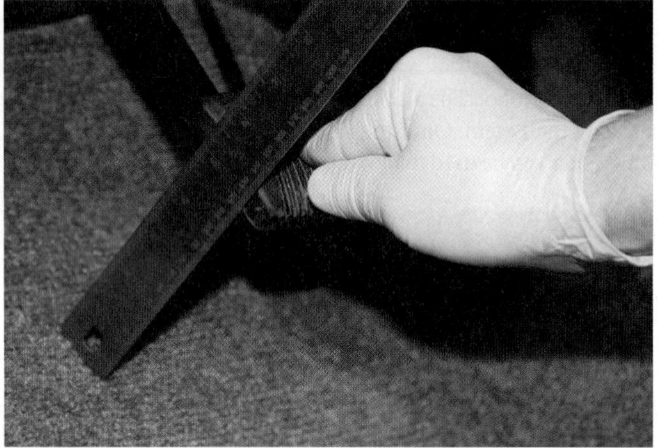

FIGURE 45-6 To measure free play, apply light hand pressure to the brake pedal, and measure how far the pedal travels before you feel resistance.

FIGURE 45-7 Travel is measured by reading brake pedal height and subtracting the height from the floor.

Free play is the amount of clearance between the brake pedal linkage and the master cylinder piston. To measure it, you can apply very light hand pressure to the brake pedal, measuring how far the pedal travels before you start to feel resistance (**FIGURE 45-6**).

To measure brake pedal free play, follow the steps in **SKILL DRILL 45-9**.

Brake pedal travel is sometimes called reserve pedal. Travel is the distance the brake pedal travels from its rest position to its applied height. Travel is measured by reading brake pedal height and subtracting the height from the floor (**FIGURE 45-7**). For example, if the brake pedal height is 8" (203.2 mm) and the travel takes it down to 5" (127 mm) off the floor, then the travel is 3" (76.2 mm). Reserve pedal is the measurement from the floor to the height of the applied brake pedal and represents how much reserve is left for the brake pedal to travel if needed (**FIGURE 45-8**).

To measure brake pedal travel, follow the steps in **SKILL DRILL 45-10**.

FIGURE 45-8 Reserve pedal is the measurement from the floor to the applied brake pedal.

SKILL DRILL 45-9 Measuring Brake Pedal Free Play

1. Research the specified procedure to measure the pedal free play.
2. Use light hand pressure to apply the brake pedal until all clearances are taken up, while measuring the distance the brake pedal moved.
3. Compare this reading to the specifications, and determine any necessary actions.

SKILL DRILL 45-10 Measuring Brake Pedal Travel

1. Research the specified procedure to measure the brake pedal travel or reserve pedal height.
2. Start the engine. This allows the booster to operate normally.
3. Apply the brake pedal with the specified force.
4. Measure the brake pedal travel or reserve height.
5. Compare this reading to the specifications, and determine any necessary actions.

▶ TECHNICIAN TIP

Make sure you understand and observe all legislative and personal safety procedures when carrying out tasks. If you are unsure of what these procedures are, ask your supervisor.

▶ Hydraulic System Component Diagnosis

S45002, N45006

Overview

Diagnosis of hydraulic braking system faults follows the strategy-based diagnostic process. It begins with verifying the concern, which means that you need a solid understanding of the customer concern. It is good practice to gather as much related information as possible about the concern from the customer. The more information you obtain, the better chance you have to diagnose the issue correctly. Ask the customer questions such as these:

- Does the brake pedal sink under hard pressure? Or light pressure?
- Does the vehicle pull to the left when braking? Or the right?
- Does the wheel lock up under hard brake pressure? Or light pressure?

Once you understand how the vehicle is acting, try to determine what conditions are present with the concern by asking questions such as these:

- Does it happen when it is wet? Dry? Or always?
- Does it happen when it is hot? Cold? Or always?
- Does it happen when you are driving at highway speeds? At stop-and-go speeds? Or always?
- Does it happen when you are carrying a heavy load? A light load? Or always?

If the vehicle is safe to drive, it is common practice to test-drive the vehicle to verify the customer concern. If it is an intermittent problem or if it is hard to reproduce, you may need to take the customer along on the test drive to point out what he or she is experiencing. While on the test drive, reproduce the conditions that the customer identified, and observe the concern. Do your best to determine where the fault is located by safely operating the vehicle in a variety of manners.

As you gather information, compare what is happening in the vehicle with your knowledge of the hydraulic braking system and Pascal's law to identify possible causes of the fault. For example, if the vehicle pulls to the left when braking, then suspect that either a brake on the left side is grabbing or a brake on the right side is not operating correctly due to a lack of hydraulic pressure or reduced friction. If the fault is fairly substantial, the cause may be easy to identify. If this situation is not easily identified, then determine whether further test driving will be useful.

Once the test drive has given you as much information as possible, the next step is to research the manufacturer's TSBs to see whether any are related to the concern. If so, follow the information in the TSB to diagnose the vehicle. If there are no related TSBs, then research the manufacturer's diagnostic process in the service information. Once you have researched the service information, it is time to develop a focused testing plan and put it in action. Doing so may involve performing tests in the shop with a scan tool, pressure gauges, or a DMM, or it could require disassembly and visual inspection or measurements of the suspected component or system. If you have been thorough in your diagnostic process, this should lead you to the cause of the fault. If not, take what you learned from the test results and/or inspection, go back to the service information, determine what the next best step is, and perform it. Continue this process until you have identified the cause of the fault.

Once a cause of the fault is identified, consider whether it is the root cause. For example, if the customer concern is that the brake pedal is spongy, you cannot stop your diagnosis when you find low brake fluid level in the master cylinder reservoir. Adding brake fluid and bleeding the brakes might restore proper function to the brake pedal for the time being, but you have to ask, "Why was the brake fluid low?" It is likely that there is a leak in the hydraulic braking system that will cause the customer to return. In this case, the root cause might be a leaky wheel cylinder. Because the brake fluid was low, you should perform a visual inspection of the hydraulic braking system to verify there are no leaks.

Once the repair is complete, you need to verify that the repair corrected the customer concern. This is usually done by retesting the component or system with the test that identified the fault. If it passes the test, then test-drive the vehicle once again to make sure the concern has been corrected and that no other issues are present.

Inspecting the Master Cylinder

N45007

Inspecting the master cylinder for internal and external leaks is usually performed under the following conditions: The brake pedal sinks when the brakes are applied; the brake fluid is low in the reservoir; the brake warning light is on; or the brake pedal reserve height is too low.

Inspect the master cylinder for obvious signs of leakage. Be sure to check all brake line fittings, sensor connections, reservoir seals, and the areas at the rear of the master cylinder near the power booster. Also check the inside of the vacuum hose to the power booster for signs of brake fluid, which would mean brake fluid is leaking past the primary piston in the master cylinder and the front seal on the power booster.

When checking for a sinking brake pedal, start the vehicle, and apply the brake pedal, beginning with a very light pressure. The brake pedal should hold its position without sinking for at least 1 minute. If it continues to sink, the hydraulic brake system may have an external or internal leak, and further inspection of the system is needed. If the brake pedal holds steady, release the brake pedal, and then apply it a bit more firmly. Hold that pressure, and see if the brake pedal holds steady. If it does, repeat this step, increasing the brake pedal pressure until firm pressure is applied. If the pedal sinks and there is no external leak in the brake hydraulic system, including the wheel cylinders and calipers, then you will need to perform one of two tests for confirming whether there is an internal leak in the master cylinder. The first test is the visual method, and the second is the block-off method. Both are explained in the Skill Drill.

To check the master cylinder for internal or external leaks and proper operation, follow the steps in **SKILL DRILL 45-11**.

Master Cylinder Service and Bench Bleeding

N45008

Removing and bench bleeding the master cylinder are usually only performed when the master cylinder is being replaced. Bench bleeding makes bleeding the hydraulic system much easier by removing all of the air from the master cylinder. Failure to bench bleed the master cylinder prior to installing it into the hydraulic brake system will prevent the master cylinder from building much brake fluid pressure. As a result, very little brake fluid will be bled out of each wheel cylinder or caliper, increasing the time it takes to bleed the hydraulic brake system. Also, it will push a lot of air through the brake system, making it more likely that air will get stuck in the system.

To remove, bench bleed, and reinstall the master cylinder, follow the steps in **SKILL DRILL 45-12**.

Measuring and Adjusting Master Cylinder Pushrod Length

N45009

The master cylinder pushrod length is critical for proper brake operation. If it is too short, the driver will have to depress the brake pedal further than specified to operate the brakes. This reduces the amount of reserve pedal available in the case of a hydraulic leak. If the pushrod length is too long, it could prevent

SKILL DRILL 45-11 Inspecting the Master Cylinder for Leakage

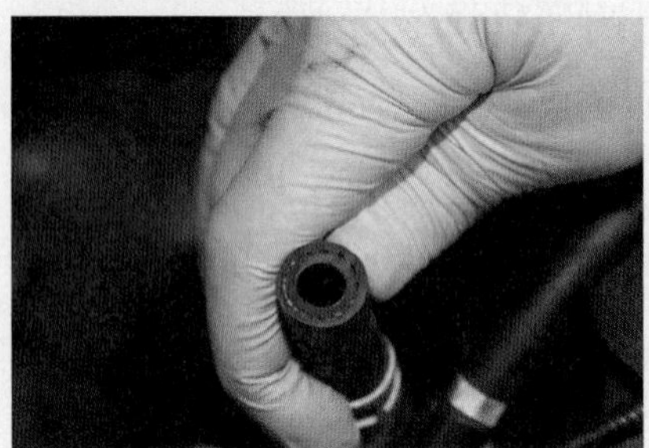

1. Check the brake fluid level to see if it is low. Inspect the master cylinder for obvious signs of leakage. Check the inside of the vacuum hose to the power booster for signs of brake fluid. Start the vehicle, and apply the brake pedal, beginning with a very light pressure. The brake pedal should hold its position without sinking for at least 1 minute. Release the brake pedal, and apply it a bit more firmly. Hold that pressure, and see if the brake pedal holds steady. If it does, repeat this step with increasing pedal pressure until firm pressure is applied.

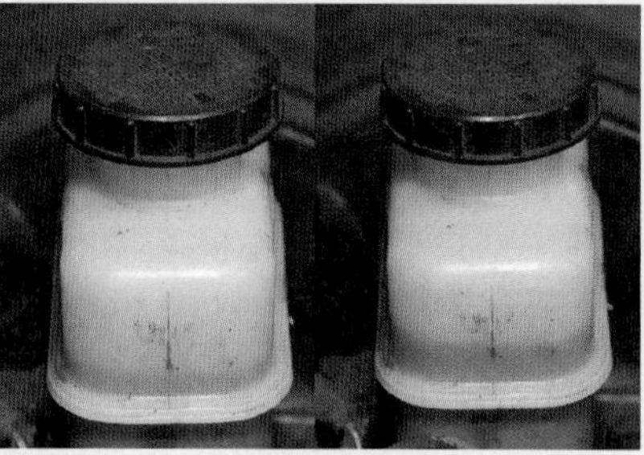

2. Inspect the master cylinder for internal leaks, using the visual method. Only perform this test if the brake pedal sinks.

- Remove the master cylinder reservoir cap.
- Have an assistant apply the brake pedal slowly and hold it firmly.
- Watch the brake fluid in the reservoir; there should be an initial spurt of brake fluid from each of the two compensating ports as the brake pedal is first moved. If it is a continuous stream while the brake pedal is moving and the brake fluid level rises while holding the brake pedal down, then the master cylinder has an internal leak.

3. Inspect the master cylinder for internal leaks, using the blocked-off method. Only perform this test if the brake pedal sinks and the preceding inspection is not conclusive.

- Unscrew the brake line fittings from the master cylinder, and carefully move them out of the way.
- Install block-off fittings into the master cylinder.
- Bleed any air from each master cylinder port, and tighten the block-off fittings properly.
- Have an assistant apply the brake pedal slowly and hold it firmly.
- Verify that the fittings do not leak. If the brake pedal sinks, then the master cylinder has an internal leak.

SKILL DRILL 45-12 Performing Master Cylinder Service and Bench Bleeding

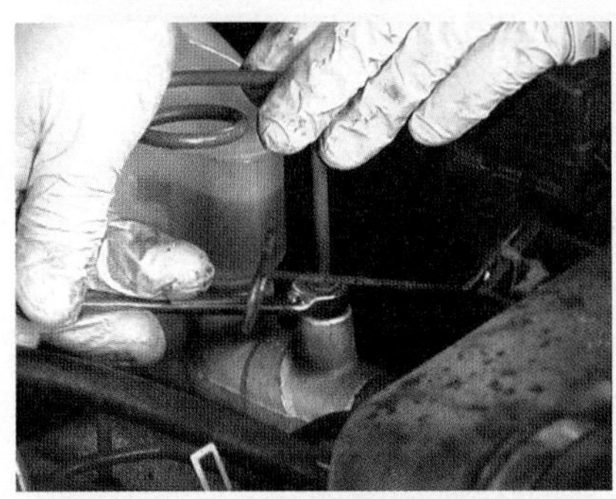

1. Compare the new unit to the old one to verify that it is the correct replacement. Remove all of the old brake fluid from the master cylinder reservoir. Remove the master cylinder brake lines, using a flare wrench and, if necessary, the double wrench method.

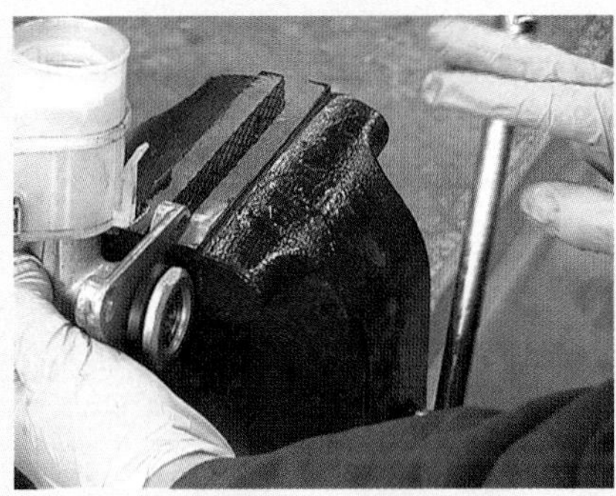

2. Remove the nuts holding the master cylinder to the power brake booster. Remove the master cylinder. Mount the master cylinder in a vise with the reservoir facing up. Now is a good time to perform the next task: measure and adjust pushrod length.

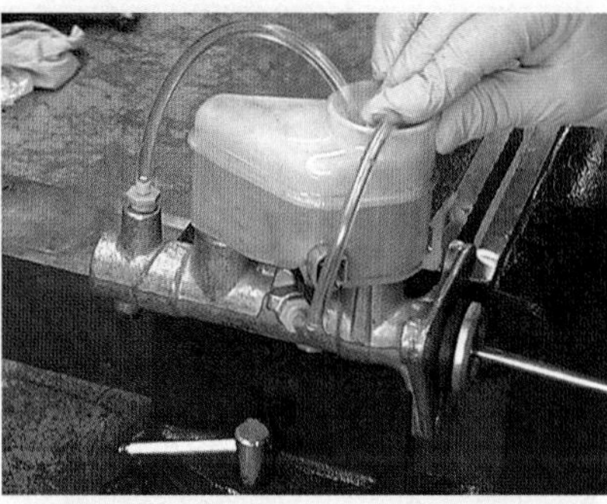

3. Install the bleeder lines into the master cylinder outlet ports with the ends of the lines deep within the master cylinder reservoir. Fill the reservoir about half full with clean brake fluid. With an appropriate tool, slowly push the master cylinder piston into the bore. Allow the piston to return to its rest position. Repeat until all air bubbles have been removed from the master cylinder.

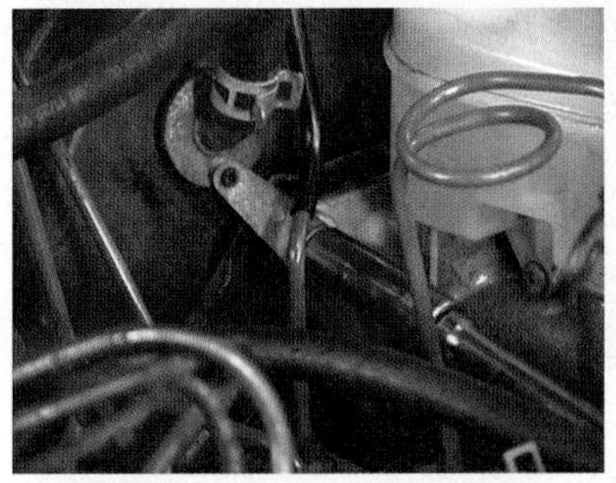

4. Place the master cylinder on the power booster, and install the nuts just far enough to hold it from falling off.

5. Carefully line up the brake lines, and start them by using your fingers to thread them into the master cylinder outlet ports at least four or five threads from when they first catch. *Do not* use a wrench to start the threads.

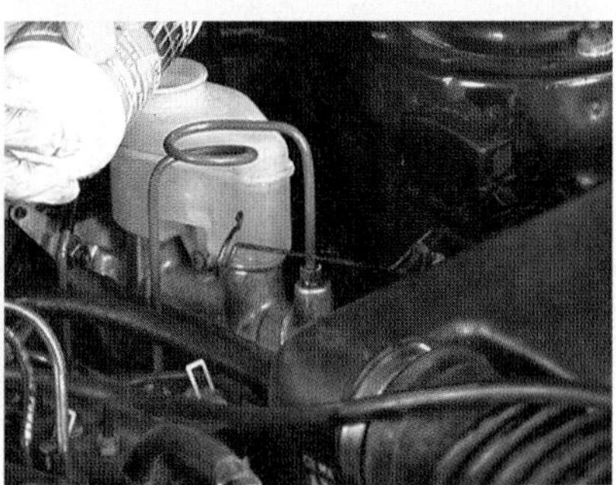

6. Tighten the bolts connecting the master cylinder to the power booster to their proper torque. Tighten the brake line fittings using a flare nut or line wrench. Use the double wrench method if the master cylinder is fitted with an adapter. Top off the master cylinder, and bleed all wheel brake units to remove any remaining air. Start the engine, and check brake pedal feel to verify proper height and firmness. Visually inspect all fittings and bleeder screws for leaks.

the master cylinder piston from returning far enough to uncover the compensating ports, trapping brake fluid pressure in front of the pistons and holding the brakes in the applied position. As important as pushrod length is, normally it does not require adjustment, as it is locked in place. The situations that would call for adjusting it are as follows: Someone changed the adjustment setting; the brake pedal linkage has been repaired or adjusted; or the power booster is being replaced. Just changing the master cylinder does not normally require pushrod length adjustment.

To measure and adjust master cylinder pushrod length, follow the steps in **SKILL DRILL 45-13**.

Diagnosing Power Brake Systems

N45010

All power brake systems should be inspected and tested whenever the customer complains that the brakes are dragging, the brake pedal is harder to push than normal, the pedal height has changed, or if the engine operation changes more than a minimal amount when the brake pedal is applied. Vacuum boosters should also be tested if you determine that the vehicle has an unlocated vacuum leak. Single-diaphragm and dual-diaphragm vacuum brake boosters are diagnosed in the same manner. The

SKILL DRILL 45-13 Measuring and Adjusting Master Cylinder Pushrod Length

1. Remove the master cylinder, following the specified procedure if still installed on the power booster. Make sure the brake pedal is fully released and not bound up. Using the specified tool, measure the master cylinder pushrod length.

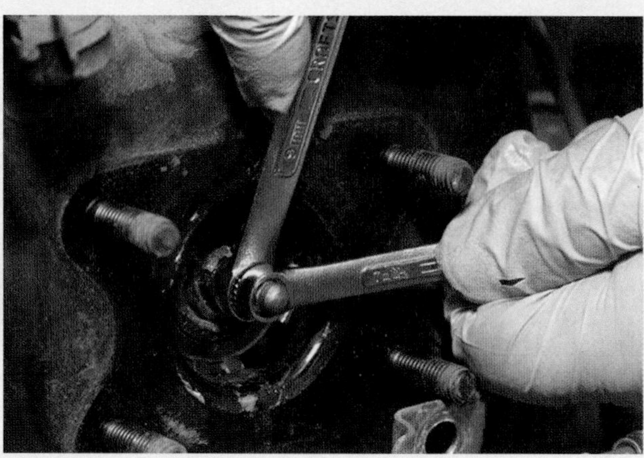

2. If the length is incorrect, loosen the locknut.

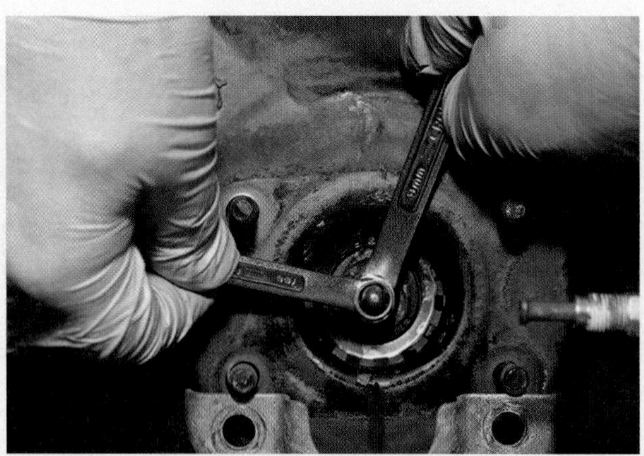

3. Adjust the pushrod to the proper length.

4. Retighten the locknut, and recheck the pushrod length. Reinstall the master cylinder and check pedal free play, pedal height, and reserve pedal.

following tests can be performed on single-diaphragm and dual-diaphragm boosters:

- Brake pedal free travel: Test to see if there is the proper brake pedal linkage clearance.
- Performance/operation test: Test to see if the booster is operational.
- External leak test: Test for leaks to the atmosphere.
- Internal leak test: Test for leaks between the booster chambers.

Testing the Power Booster

Power booster testing starts with a brake pedal free travel test and then follows up with a performance test to see if the booster is operating properly. If not, then an external leak test and internal leak test can determine why the booster isn't working properly. The brake pedal free travel is critical for proper brake operation. The proper amount of free travel ensures that the brake pedal linkage allows the master cylinder pistons to return to their proper rest position and uncover the compensating ports. Insufficient free travel can cause the brakes to drag due to trapped fluid pressure in front of each piston and not being able to return to the reservoir through the blocked compensating ports. Excessive free travel is not good either, as it reduces the amount of reserve pedal for braking in the event of a hydraulic brake system leak.

As important as brake pedal free travel is, it normally does not need to be adjusted because the components are locked in place. The situations that would call for adjusting it include the following: Someone changed the adjustment setting; the brake pedal linkage has been repaired or adjusted; the linkage has worn over time, leading to increased free travel; or the power booster is being replaced. Just changing the master cylinder does not normally require brake pedal free travel adjustment; however, it is good practice to verify that it is within specifications. If adjustment is needed, the manufacturer usually incorporates a locking adjustment rod between the power booster and the brake pedal.

After verifying that the free travel is correct, you need to performance test the power booster. To do so, operate the booster to test its ability to provide boost to the master cylinder. To test pedal free travel and performance test the vacuum booster, follow the steps in **SKILL DRILL 45-14**.

> **TECHNICIAN TIP**
>
> Holding the brake pedal in a steady manner should not affect the operation of the engine. If the engine runs rough when the brake pedal is held down or changes substantially when the brake pedal is released, it could indicate a vacuum leak in the booster. Perform the external and internal leak tests to identify any faults.

Checking Vacuum Supply to Vacuum-Type Power Booster

The brake booster must have an adequate amount of vacuum to operate correctly. Insufficient vacuum requires the driver to increase foot pressure to activate the brakes. Excessive vacuum is not usually a problem, because the booster is designed to work using maximum engine vacuum. Many manufacturers specify a minimum of 16" of mercury (in. Hg; 406 mm Hg) of intake manifold vacuum. If the reading is insufficient, check for vacuum leaks or restrictions in the supply hose or for an improperly tuned engine. To check vacuum supply to a vacuum-type power booster, follow the steps in **SKILL DRILL 45-15**.

> **TECHNICIAN TIP**
>
> Engines with modified camshafts regularly have decreased amounts of manifold vacuum due to the high-duration camshaft. This lowered vacuum results in higher foot pressure required to stop the vehicle. An auxiliary vacuum pump may be needed to provide adequate braking. Also, vehicles operated at high altitude always have less vacuum. The general rule of thumb is that you will lose 1" Hg (25 mm Hg) for every 1000' (305 m) of altitude gained.

Checking Vacuum-Type Power Booster Unit for Leaks and Inspecting the Check Valve

Vacuum leaks in the power booster require increased driver foot pressure to activate the brakes. Because the vacuum normally comes from the intake manifold, a leaky power booster can affect the operation of the engine by changing the air-fuel mixture. Power boosters can leak internally or externally. Perform the external leak test before the internal leak test to avoid confusion in identifying the cause of the leak. To perform an external leak test:

- Start the engine, and allow it to run for at least 10 seconds.
- With your foot *off* the brake pedal, turn the engine off.

SKILL DRILL 45-14 Testing Pedal Free Travel and Performance Testing the Vacuum Booster

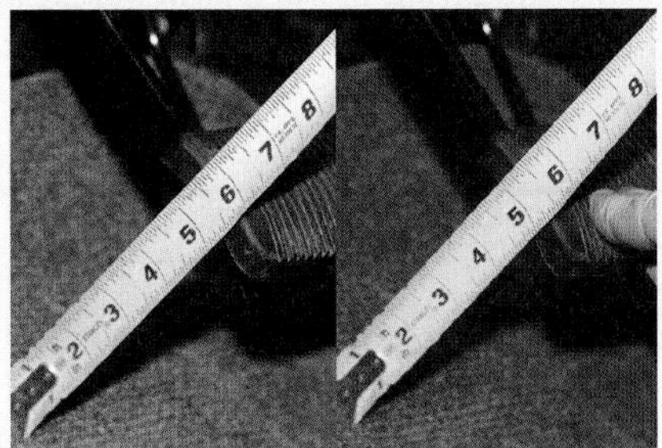

1. To test brake pedal free travel, with the engine off, depress the brake pedal several times to remove any vacuum or hydraulic pressure from the power booster. Measure the distance of the brake pedal free travel by depressing the brake pedal by hand until you just feel all of the slack taken up.

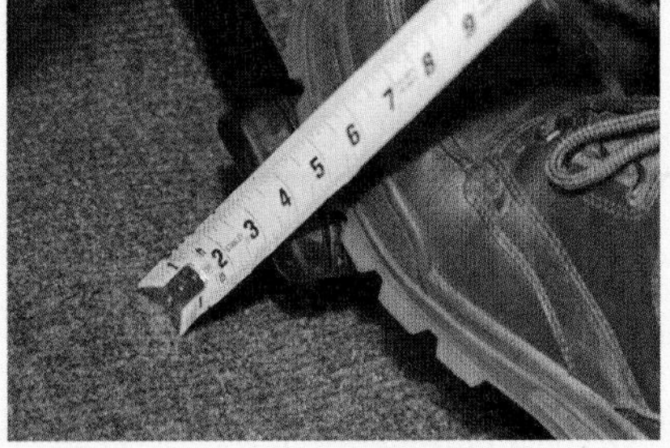

2. Performance test the booster by beginning with the vehicle engine off. Apply and release the brake pedal five or six times to bleed off any vacuum or hydraulic pressure in the power booster. Hold the brake pedal down with moderately firm pressure (20–30 lb [9.1–13.6 kg]).

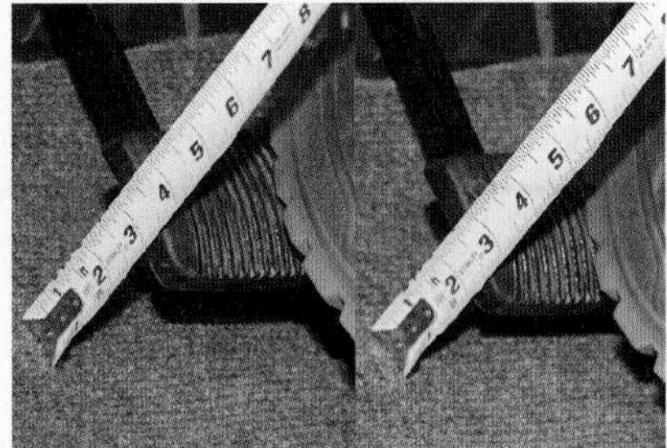

3. Start the engine, and observe the brake pedal. On vacuum-assisted vehicles, if the pedal drops about an inch, the booster is providing boost. If not, the booster is not providing boost, and the following tests must be performed. On hydraulic-assisted vehicles, when starting the engine with your foot on the brake pedal, the pedal should either rise or fall about an inch (depending on the vehicle) if the booster is providing assist.

SKILL DRILL 45-15 Checking Vacuum Supply to Vacuum-Type Power Booster

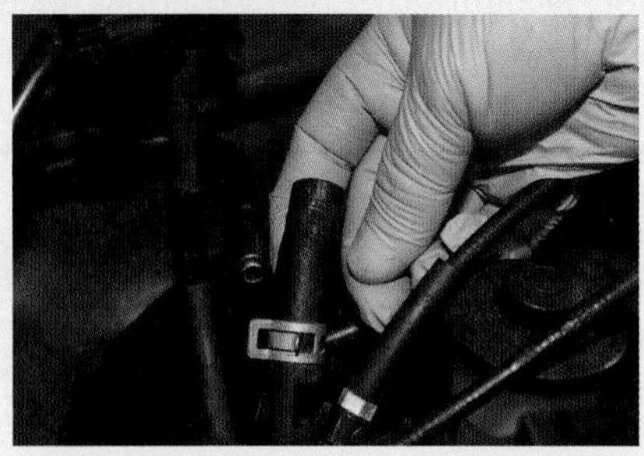

1. With the engine off, remove the inlet hose from the vacuum-type booster.

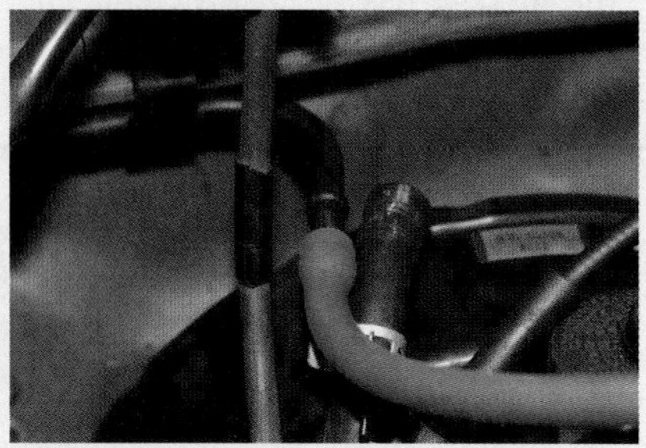

2. Connect a vacuum gauge to the vacuum supply end of the hose.

3. Start the engine, and read the vacuum supply available to the vacuum-type booster. Vacuum should be greater than 16" Hg (406 mm Hg) on most vehicles.

- Wait at least 10 minutes, and then apply the brake pedal with moderately firm pressure (20–30 lb [9.1–13.6 kg]). Note the feel and pedal reserve height.
- Apply the brake pedal a couple more times with the same moderately firm pressure. Each application should result in a higher and firmer brake pedal as the vacuum is released from the booster.

If it does operate properly, then there are no substantial external leaks in the system. If it does not hold vacuum, you need to inspect the booster for external vacuum leaks. The cause could be any of the following:

- The vacuum check valve and/or grommet
- The front or rear seals of the unit
- The atmospheric valve at the rear of the booster
- The case where the halves are crimped together
- A hole worn or rusted through the case

Once you have identified that there are no external leaks, the second step is to perform an internal leak test:

- Begin by starting the engine and letting it idle.
- Apply the brake pedal with firm pressure (30-50 lb [13.6-22.7 kg]).
- Without moving your foot, shut off the engine, and observe the brake pedal for approximately 1 minute.

If the pedal stays steady, there are no internal leaks. If the brake pedal rises, there is an internal leak in the diaphragm, the control valve, or the check valve.

The third step is to perform a check valve operation test:

- Start the engine, and allow it to run for 10 seconds to evacuate the booster. Turn off the engine, wait at least 10 minutes, and then remove the check valve from the booster.

There should be a large rush of air into the booster. If there is, the check valve is holding a vacuum and is okay. If there is not, test the check valve by blowing through it. Air should flow from the booster side of the check valve to the engine side only. If the check valve is okay and the tests indicate a leak, use a stethoscope and listen for leaks around the outside of the booster, including the control valve at the rear of the booster, which is under the dash.

To inspect the vacuum-type power booster unit for leaks and inspect the check valve for proper operation, follow the steps in **SKILL DRILL 45-16**.

Inspecting and Testing a Hydraulically Assisted Power Brake System

N45011

Operation problems can be caused by a number of issues such as leaks inside the booster unit, a worn power steering pump, a slipping or broken pump drive belt, badly contaminated power steering fluid, or leaky hose connections. Most of these issues can be verified with a visual inspection.

To inspect and test a hydraulically assisted power brake system for leaks and proper operation, follow the steps in **SKILL DRILL 45-17**.

Power Booster Testing in the EHB System

Testing the operation of this system can be performed just as the hydraulic booster system was tested. Carry out the same steps to performance test the EHB system. Follow these tests up with the accumulator leak test as described previously. These tests will identify any operational issues with the system. If issues are found, follow the manufacturer's recommended procedure to diagnose the fault.

Inspecting Brake Lines and Hoses

N45012

Inspecting brake lines and brake hoses can show where maintenance is needed to prevent vehicle breakdowns and unsafe vehicle operation. All manufacturers recommend regular inspection

SKILL DRILL 45-16 Checking Vacuum-Type Power Booster Unit for Leaks and Inspecting the Check Valve

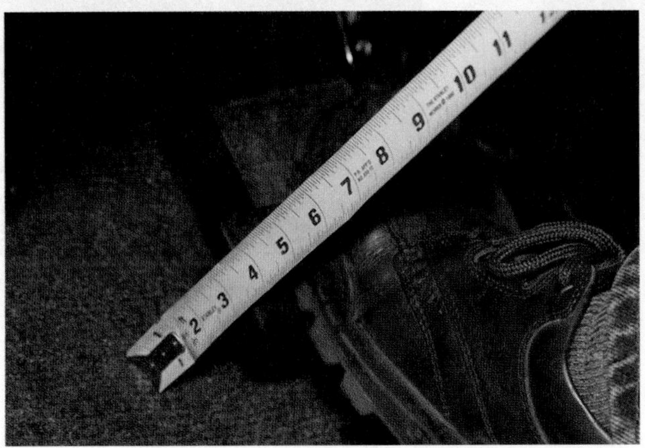

1. The first step is to perform an external leak test. Start the engine, and allow it to run for at least 10 seconds. With your foot *off* the brake pedal, turn the engine off. Wait at least 10 minutes, and then apply the brake pedal with moderately firm pressure (20–30 lb [9.1–13.6 kg]). Note the feel and pedal reserve height. Apply the brake pedal a couple more times with the same moderately firm pressure. Each application should result in a higher brake pedal as the vacuum is released from the booster.

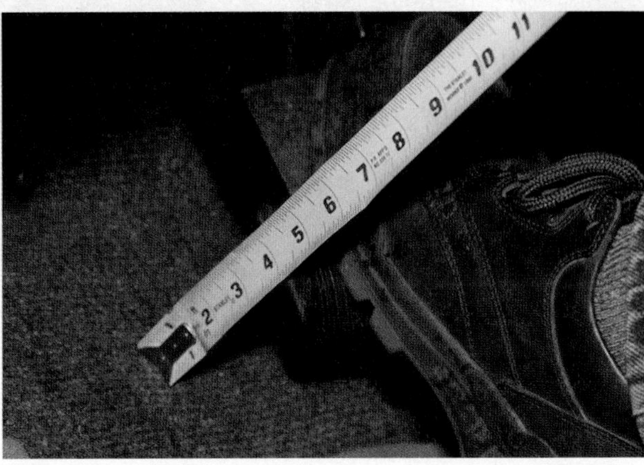

2. The second step is to perform an internal leak test. Begin by starting the engine and letting it idle. Apply the brake pedal with firm pressure (30–50 lb [13.6–22.7 kg]). Without moving your foot, shut off the engine, and observe the brake pedal for approximately 1 minute. If it stays steady, there are no internal leaks. If the brake pedal rises, there is an internal leak.

3. The third step is to perform a check valve operation test. Start the engine, and allow it to run for 10 seconds to evacuate the booster.

4. Turn off the engine, wait at least 10 minutes, and then remove the check valve from the booster. There should be a large rush of air into the booster if the check valve is holding a vacuum properly. If there is not, test the check valve by blowing through it.

of these components. Be sure to take into account regional differences, such as regions that are highly susceptible to corrosion from de-icing chemicals or high humidity. Also, off-road vehicles may be more prone to dents, kinks, and cuts. Inspect all vehicles carefully.

To inspect brake lines, brake hoses, and associated hardware, follow the steps in **SKILL DRILL 45-18.**

Replacing Brake Lines, Brake Hoses, Fittings, and Supports

N45013

When brake lines and hoses have to be replaced, you will need to disconnect them at their fittings. Many times the hose or line fitting is connected very tightly to a matching fitting or

SKILL DRILL 45-17 Inspecting and Testing a Hydraulically Assisted Power Brake System

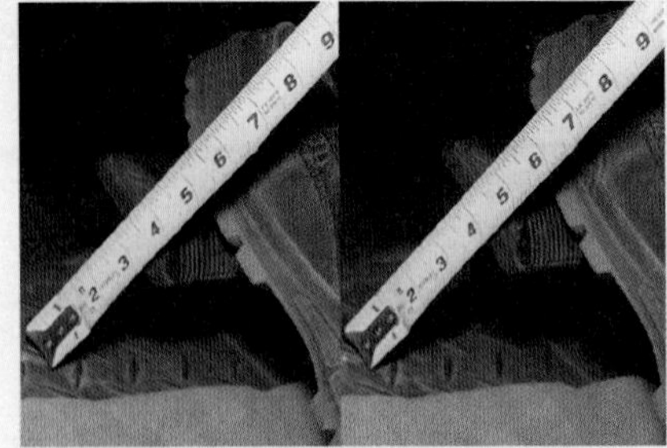

1. First, perform the performance test. Begin with the vehicle engine off. Apply and release the brake pedal five or six times. Hold the brake pedal down with moderately firm pressure (20–30 lb[9.1–13.6 kg]). Start the engine, and observe the brake pedal. It should drop or rise an inch or so if the booster is providing boost.

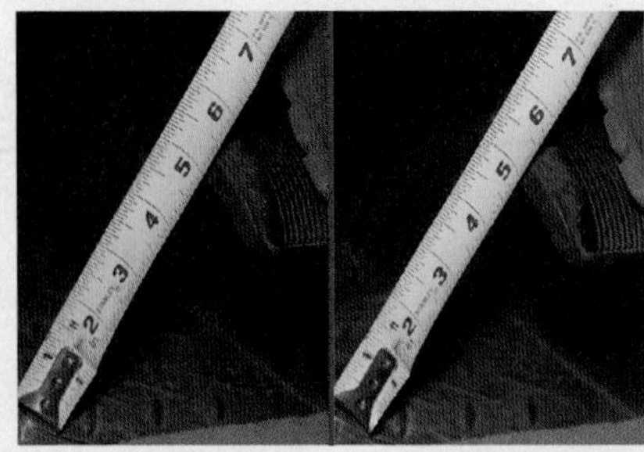

2. Second, perform the accumulator leak test. Start the engine, apply the brake pedal, and note the pedal feel and applied height. Release the brake pedal, turn off the engine, and wait at least 10 minutes.

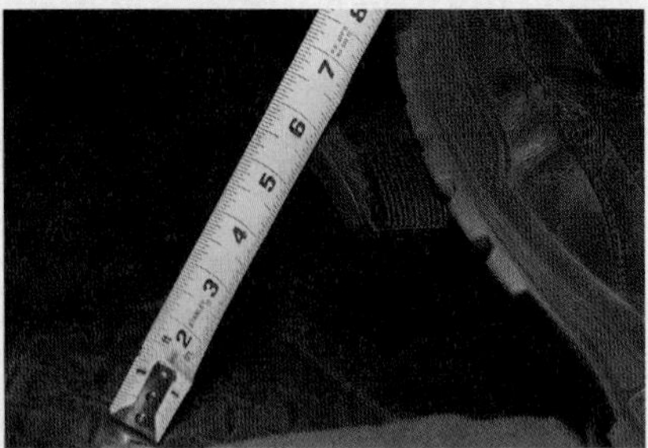

3. Apply the brake pedal with moderately firm pressure (20–30 lb [9.1–13.6 kg]). Note the feel and brake pedal height. Apply the brake pedal a couple more times with the same moderately firm pressure. Each application should result in a higher brake pedal as the accumulator pressure is released from the booster.

SKILL DRILL 45-18 Inspecting Brake Lines, Brake Hoses, and Associated Hardware

1. Safely raise the vehicle on a hoist. Trace all brake lines from the master cylinder to each wheel's brake assembly. Inspect the steel brake lines for leaks, dents, kinks, rust, and cracks.

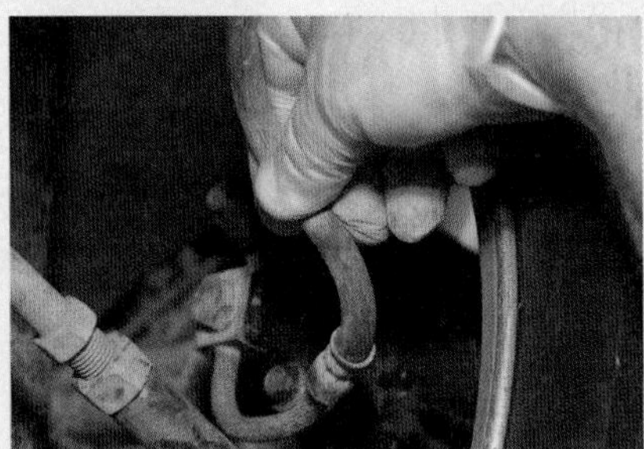

2. Inspect all flexible brake hoses for cracks, bulging, and wear.

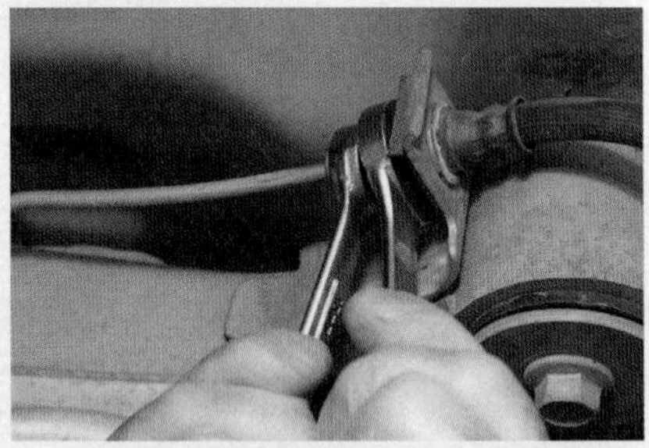

3. Tighten any loose fittings and supports.

adapter on another line, hose, or component. The best way to remove them is with the double wrench method. In this method, you place one flare nut wrench on each of the two fittings, and twist them apart. If you try to use only one flare nut wrench on just one fitting, you will likely kink the brake line or hose, as the line is much weaker than the fittings. So always use the double wrench method when loosening brake line fittings.

To replace brake lines, brake hoses, fittings, and supports, follow the steps in **SKILL DRILL 45-19**.

Brake Line Flaring

N45014

To fabricate brake lines, decide whether you will be using the inverted double flare or ISO method. First select new, double-walled steel brake line of the specified diameter. Make sure the end you are flaring is cut smooth and square, without dings or gouges. Clean up with a file, or recut the end with a tubing cutter. Obtain the flaring tool for the type of flare you are fabricating: double flare or ISO. Also make sure you place

SKILL DRILL 45-19 Replacing Brake Lines, Hoses, Fittings, and Supports

1. Depress the brake pedal with a brake pedal depressor to prevent brake fluid from draining out of the vehicle. Remove the stop light fuse to avoid draining the battery.

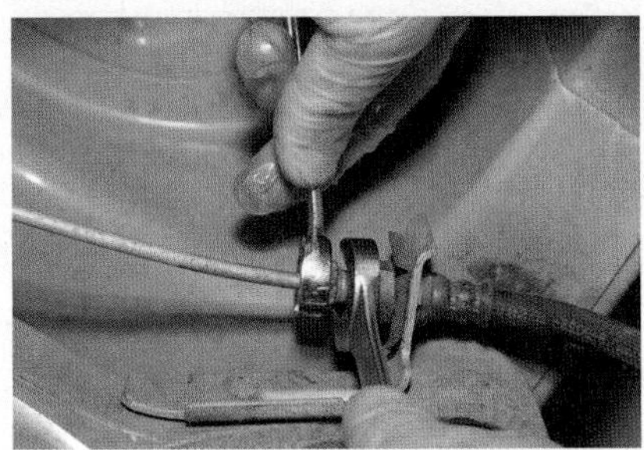

2. Safely raise the vehicle on a hoist. Using flare nut wrenches and the double wrench method, carefully remove any brake lines, hoses, fittings, and supports that are to be replaced. Inspect all components for damage and wear.

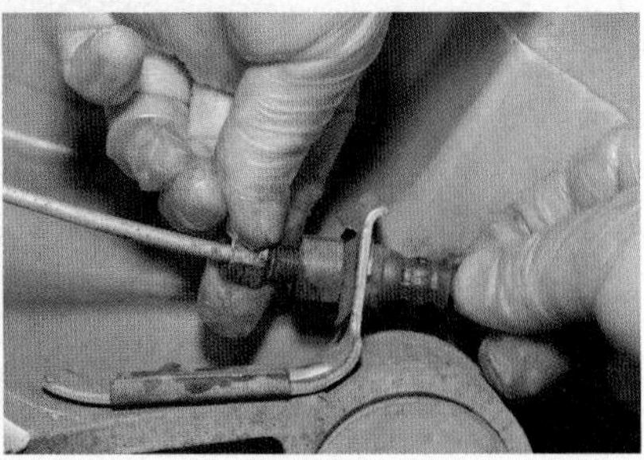

3. Carefully reassemble the removed components by using your fingers only. Once everything is assembled, tighten each fitting and support, using flare nut wrenches and the double wrench method. Bleed any trapped air from the system. Start the vehicle, and verify proper brake pedal height, firmness, and feel. Reinstall the brake light fuse.

Applied Science

AS-87: Hydraulics: The technician can explain how fluid pressure transmits force from one location to another.

Pascal's law states that incompressible liquids transmit pressure in all directions. This is the basic principle underpinning the operation of hydraulic braking systems on motor vehicles. When braking, brake pedal pressure is applied, via the master cylinder, to the brake fluid. The hydraulic fluid under pressure pushes against the caliper or wheel cylinder pistons, forcing friction material against the brake discs or drums to stop the movement of the vehicle.

For the sake of simplicity, let's think of a hydraulic brake system like a tube of toothpaste. Start with a full tube (no air). If you punch a hole in the tube and squeeze, toothpaste (a thick liquid) will be displaced from the hole. If you punch three holes through various parts of the tube and squeeze in only one place, toothpaste will be displaced from all three holes. This illustrates that force applied to a liquid in one area is transferred to all areas within a container.

any fittings required to be in place on the tubing before the tubing is flared.

To perform the inverted double flare method, follow the steps in **SKILL DRILL 45-20**.

To perform the **ISO flare method**, follow the steps in **SKILL DRILL 45-21**.

Diagnosing the Brake Warning Lamp

N45015

A mechanical brake warning lamp system (non-CANbus) circuit is a simple lightbulb in series, with as many as four switches that are connected in parallel with one another (**FIGURE 45-9**). Each switch can illuminate the warning lamp under the right conditions. The warning lamp should be off when the ignition key is in the Run position if the parking brake is released and there are no faults in the hydraulic system. When the ignition is turned to the Crank position, the warning lamp should illuminate.

In the Run position, the warning lamp can be illuminated only when one of the switches is closed. One switch is on the parking brake assembly and illuminates the warning lamp when the parking brake is applied. This alerts the driver that the parking brake should be released before driving the vehicle. The second switch is located on the pressure differential assembly and illuminates the warning lamp if there is a moderate-sized leak in the hydraulic system. The third switch, if the vehicle is so equipped, is located on the master cylinder reservoir and illuminates the warning lamp if the fluid level falls below a certain point. The last switch, if the vehicle is so equipped, is located on the ignition switch and acts as a bulb check feature, which illuminates the warning lamp when the ignition switch is in the Crank position.

When diagnosing the brake warning lamp, it is good to start by verifying that the bulb is operational, using the circuit's bulb check feature if the vehicle is so equipped. To do so, turn the ignition key to the Crank position. The brake warning lamp on the dash should illuminate on most vehicles. If the bulb illuminates, then you know the bulb is good and has power, and the ground circuit through the ignition switch is operating as it should. If it does not illuminate, apply the parking brake, and recheck to see if the warning light is on. If not, check the fuse with a test light (**FIGURE 45-10**). If the fuse is okay, remove the bulb, and verify that it is not burned out by measuring its resistance with an ohmmeter and comparing it to specifications or the resistance of a known good bulb.

SKILL DRILL 45-20 Fabricating Brake Lines, Using the Inverted Double Flare Method

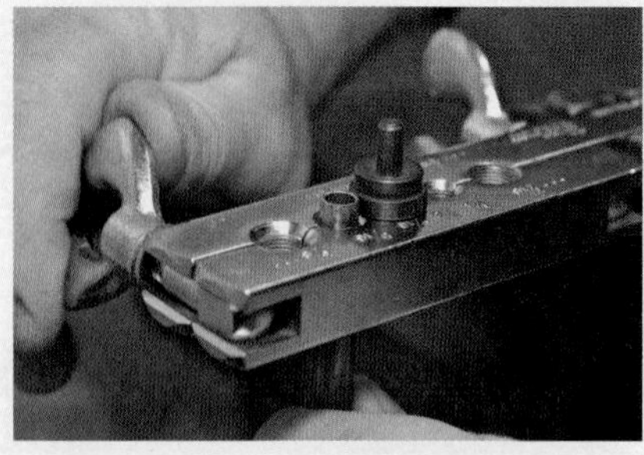

1. Install the proper fitting onto the brake line. Use a bench vise to hold the brake line clamping tool. Select the proper adapter to match the tubing size. Use it to set the height of the tube in the clamp. Tighten down the clamp securely.

2. Insert the adapter into the tubing, and install the flaring tool onto the clamp and over the adapter.

3. Tighten down the flaring tool until the adapter touches the clamp. This forms a widening and narrowing of the tube.

4. Remove the adapter, and reinstall the flaring tool. Tighten down the flaring tool to fold the narrowed portion back into the widened portion of the flare. Stop when the tool starts to get hard to tighten.

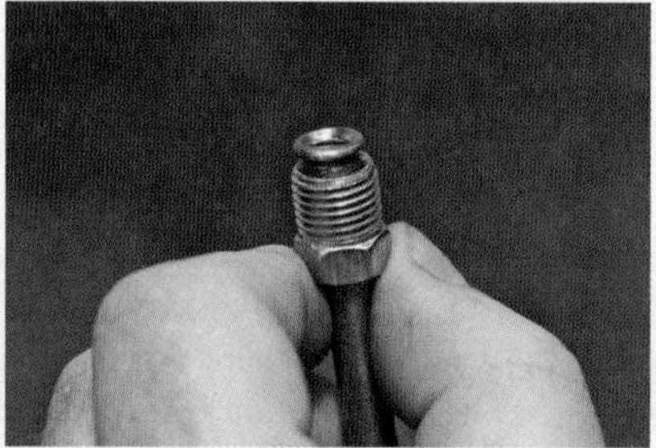

5. Remove the flaring tool, and inspect the flare to see if it has been formed correctly. Remove the flared line from the clamp, and inspect it.

SKILL DRILL 45-21 Performing the ISO Flare Method

1. Install the proper fitting onto the brake line. Use a bench vise to hold the brake line clamping tool. Select the proper adapter to match the tubing size. Use it to set the height of the tube in the clamp. Tighten down the clamp securely.

2. Insert the adapter into the tubing, and install the flaring tool onto the clamp and over the adapter.

SKILL DRILL 45-21 Performing the ISO Flare Method (Continued)

3. Tighten down the flaring tool until the adapter touches the clamp.

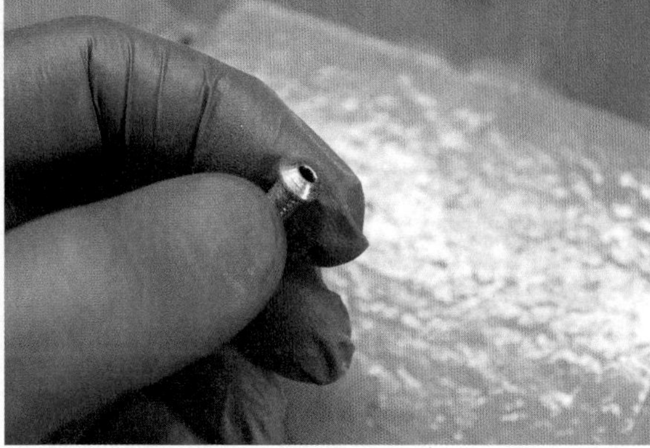

4. Remove the flaring tool, and inspect the flare to see if it has been formed correctly. Remove the flared line from the clamp, and give it a final inspection.

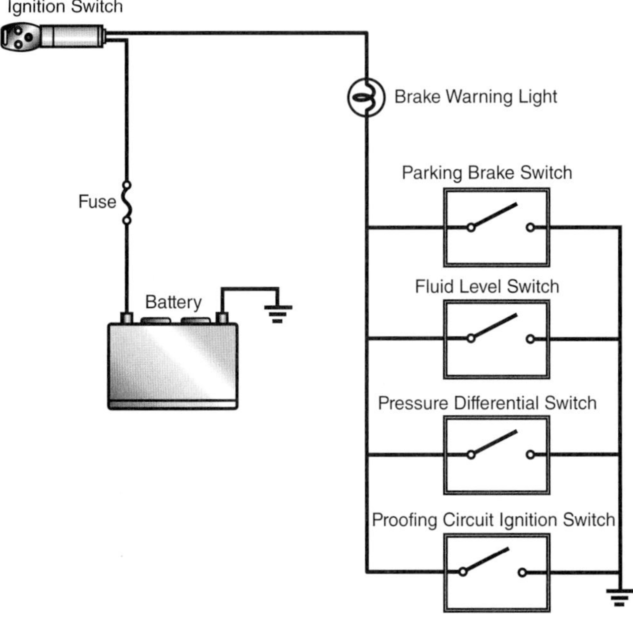

FIGURE 45-9 Brake warning light circuit.

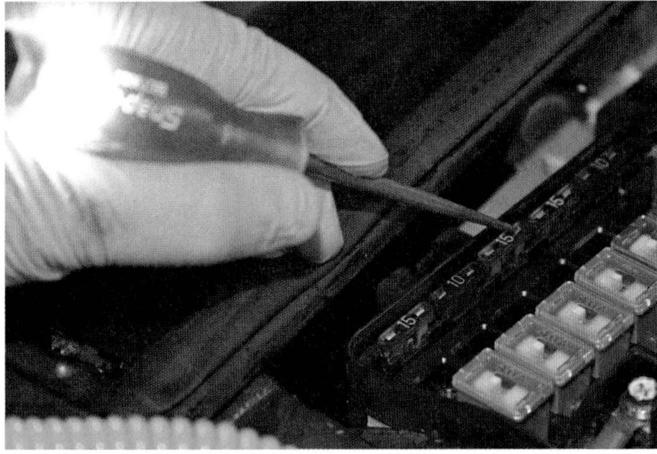

FIGURE 45-10 Checking a brake fuse with a test light.

If the brake warning lamp illuminated in the Crank position, turn the ignition key to the Run position. The brake warning lamp should be off. If it is, apply the parking brake. The warning lamp should illuminate. If it does not illuminate, you need to refer to the vehicle's wiring diagram to determine the best strategy to diagnose the system. If it is similar to the one in Figure 45-9, then locate the parking brake switch, disconnect the wire harness connector from the switch, and jump it to ground with a jumper wire. With the ignition switch on, the brake warning light should now be illuminated. If it is, then the switch needs either adjustment or replacement. If it is not, then there is a problem between the switch and the bulb. Use the same process for the low brake fluid switch, the pressure differential switch, and the ignition switch.

If the brake warning lamp always stays on in the Run position, you need to disconnect each of the four switches, one at a time, checking to see if the warning lamp goes off. If it does, then you need to test each switch and determine whether it is adjusted properly or whether it is shorted. If the switches test okay, then there is likely a short circuit in one of the wires leading to the switches. This is the hardest part of testing the circuit; thankfully, it doesn't happen very often.

In a **CANbus circuit**, the parking brake sensor and the low brake fluid sensors send signals over the CANbus network regarding their status (**FIGURE 45-11**). The appropriate module then signals the instrument panel module to illuminate the brake warning light. Diagnosis involves using a scan tool capable of interrogating the system, a wiring diagram, and a digital multimeter (DMM) to perform specific tests. Refer to the Disc Brake System chapter for more information on parking brakes.

Checking the Brake Warning Light System

To check a brake warning light system on a non-CANbus system, follow the steps in **SKILL DRILL 45-22**.

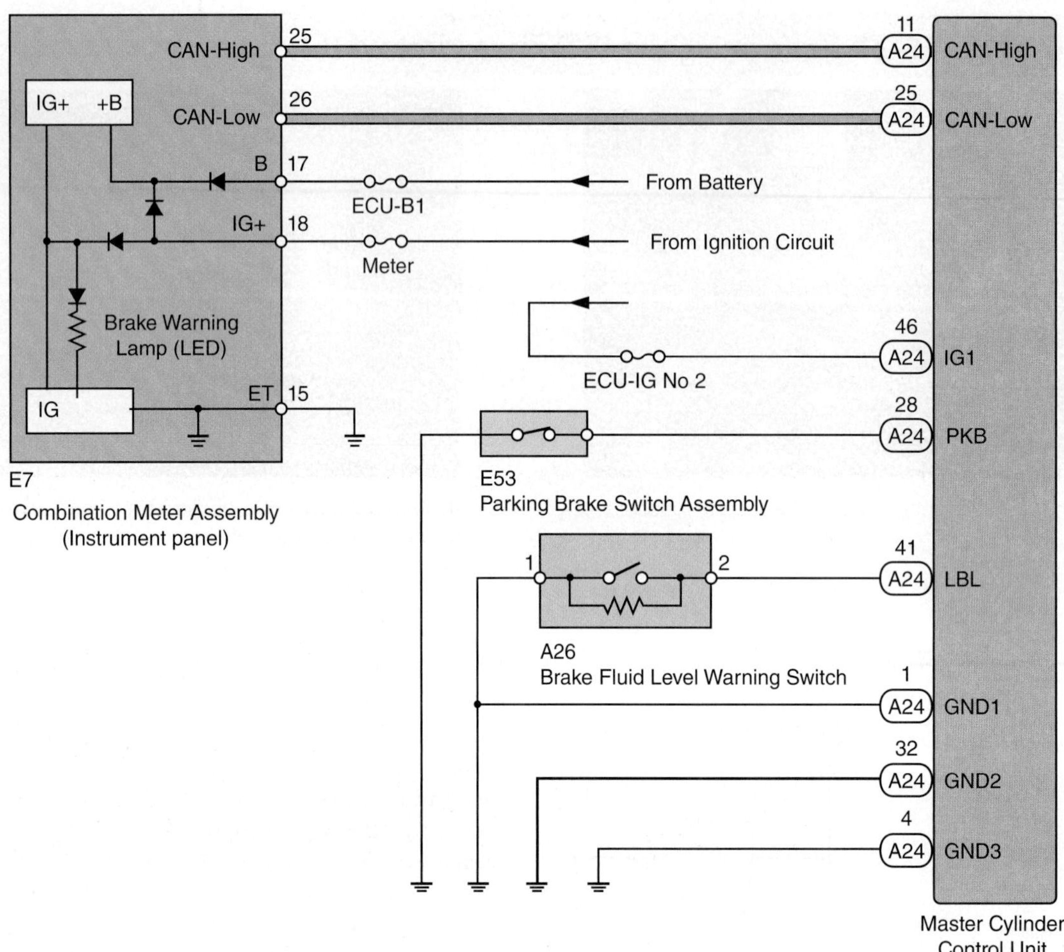

FIGURE 45-11 CANbus circuit for the brake warning light circuit.

Checking the Parking Brake and Indicator Light System

N45016

To check the parking brake and indicator light system operation, follow the steps in **SKILL DRILL 45-23**.

Inspecting and Maintaining Parking Brakes

N45017

To inspect and maintain parking brakes, follow the steps in **SKILL DRILL 45-24**.

Diagnosing Stop Light

N45018

The mechanical system (non-CANbus) circuit is simply a normally closed switch in series with two to six brake lightbulbs connected in parallel with each other (**FIGURE 45-12**). This circuit is also protected by a fuse. Some vehicles use separate fuses for the center high mount stop lamp (CHMSL) and the side stop lights. The switch is turned on and off by the movement of the brake pedal. Diagnosis starts with operating the stop lights, observing their reaction, and comparing that to a wiring diagram to see how the manufacturer has wired the circuit. If they do not illuminate at all, check components common to all of the stop lights such as the fuse, stop light switch, and stop light ground circuit. If individual lights do not illuminate, check those bulbs to see if they are burned out or if they have bad connections, terminals, or wires.

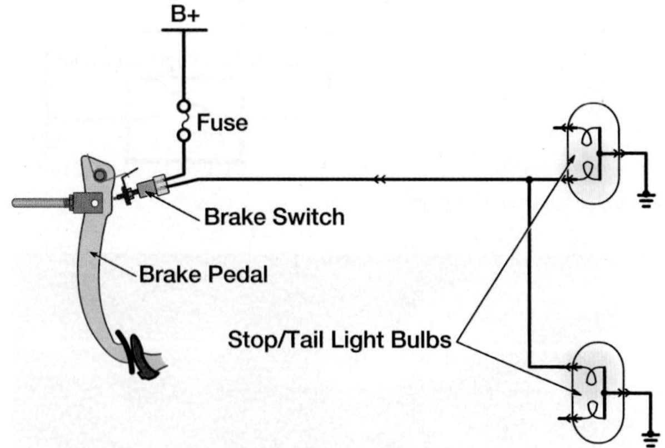

FIGURE 45-12 Typical non-CANbus stop light circuit.

SKILL DRILL 45-22 Checking the Brake Warning Light System in a Non-CANbus System

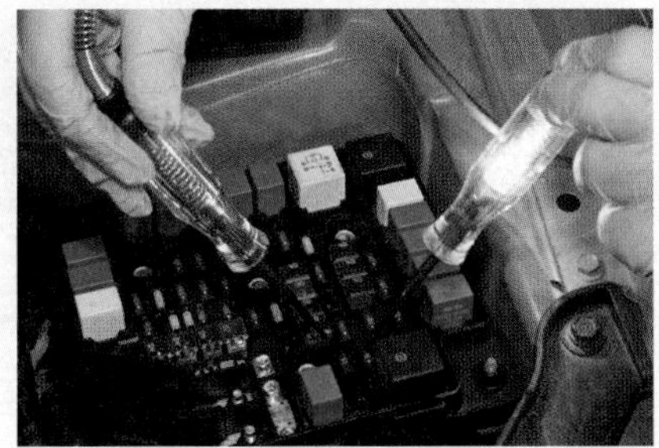

1. Perform a bulb check, and observe the brake warning light. If it is on, go to step 2. If it is off, check the fuse with a test light. If it is okay, check the warning lamp bulb with an ohmmeter.

2. Make sure the parking brake is released, and turn the ignition key to the Run position. The brake warning lamp should be off. If not, go to step 6.

3. If the warning lamp is off, apply the parking brake. The lamp should illuminate. If it does, check the operation of the other switches. If they work fine, there are no faults present in the system.

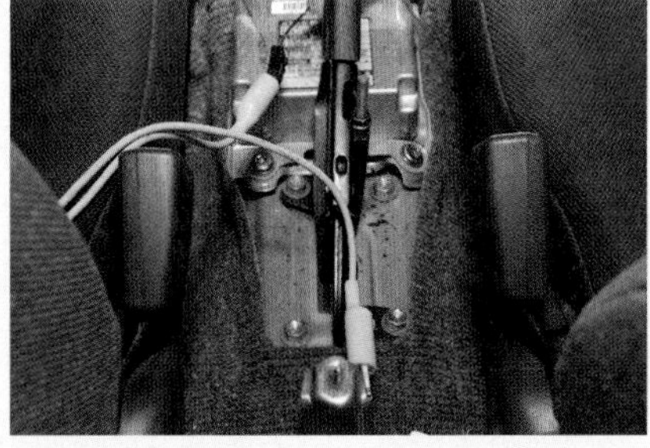

4. If the brake warning light is off when the parking brake is on, disconnect the wire from the parking brake switch, and ground the wire. The brake warning light should illuminate. If it does, test the parking brake switch with a DMM.

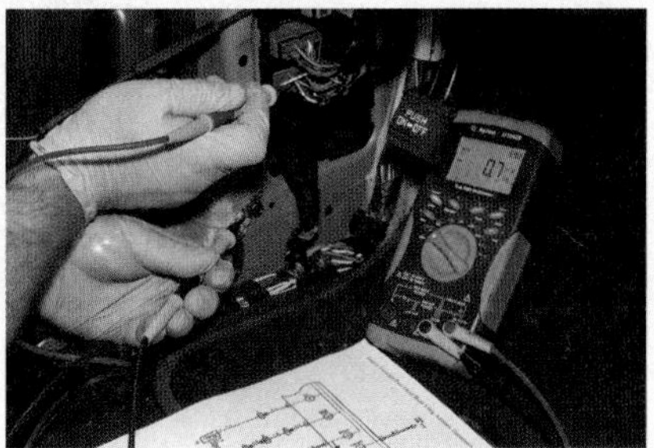

5. If the warning lamp does not illuminate with the parking brake wire grounded, suspect an open circuit between the parking brake wire and the warning lamp. Use a DMM and wiring diagram to diagnose the fault.

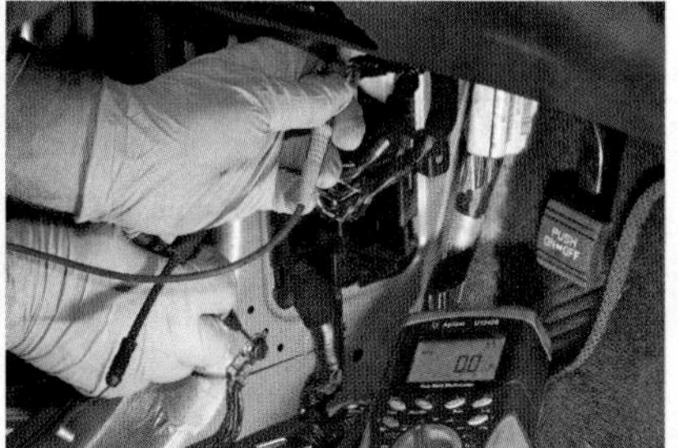

6. If the light stayed on in step 2, apply and release the parking brake several times. If this turns it off, clean or adjust the parking brake mechanism or switch. If it is still on, then disconnect each of the switches that activate it: parking brake, low fluid level, pressure differential, and ignition. If that turns the light off, then test that switch for misadjustment or a shorted condition. If the warning lamp is still on, suspect a short to ground in the wiring harness between the switches and the warning lamp bulb. Use a wiring diagram and DMM to identify the location of the fault.

SKILL DRILL 45-23 Checking Parking Brake and Indicator Light System Operation

1. Count the number of clicks it takes to apply the parking brake, and compare that to the manufacturer's specifications. If incorrect, follow the specified procedure to adjust it.
2. With the parking brake applied, turn the ignition switch to the Run position. The red brake warning lamp should be illuminated.
3. Release the parking brake. The warning lamp should turn off. If it does, the parking brake warning light system is operating properly.
4. If the lamp does not turn off, disconnect the wire to the parking brake switch. If the light turns off, test the parking brake switch.
5. If it does not turn off, check the operation of the other warning lamp switches and circuit.

Checking Operation of a Stop Light System

The stop lights are capable of operating at all times on most vehicles. This means the stop lights should come on whenever the brake pedal is pressed, regardless of whether the key is on or off. Checking the operation of the stop lights involves a visual inspection of each bulb. If the bulbs are on even when the brake pedal is released, suspect a misadjusted brake pedal, a defective brake switch, or a short to power in the circuit. If none of the bulbs light up when the brake pedal is depressed, suspect a fault that is common to all of the bulbs such as the fuse, brake switch, feed wire, or ground wire. If there is a problem with only some of the bulbs, suspect a fault that is common to only the bulbs that do not operate, such as the bulbs

SKILL DRILL 45-24 Inspecting and Maintaining Parking Brakes

1. Safely raise the vehicle slightly off the ground on a lift, and place the transmission in neutral.

2. Count the number of clicks it takes to firmly apply the parking brake, and compare that to specifications.

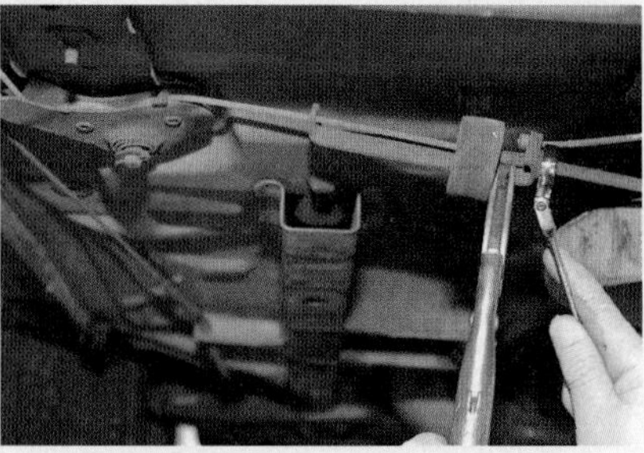

3. If incorrect, follow the specified procedure to adjust it.

4. Release the parking brake, and rotate each wheel that is affected by the parking brake. Feel for abnormal brake drag.

5. If there is no unusual binding, lubricate the cables and pivot points as needed.

or the individual wires. In any case, you will need to use the DMM to locate the open circuit, high resistance, or short circuit.

To check the operation of the brake stop light system and determine any necessary action, follow the steps in **SKILL DRILL 45-25**.

Diagnosing Hydraulic Concerns

Control valves do not normally give technicians trouble. They tend to operate rather dependably. But like any other mechanical component, they can fail. Understanding how they operate will give you a head start on diagnosing concerns associated with them. Generally, the best way to test them is to connect a pressure gauge set designed for the high pressures within a brake system to the input and output of the valve. Operating the brake system and comparing the pressures to specifications will tell you if the control valve is operating correctly.

To perform a metering valve test, follow the steps in **SKILL DRILL 45-26**.

SKILL DRILL 45-25 Checking the Operation of the Brake Stop Light System

1. Verify that the stop lights are off when the brakes are released. If the lights are on, test the operation of the brake pedal and brake switch. Observe the stop lights when the brake pedal is applied. All lights, including the CHMSL, should be illuminated.

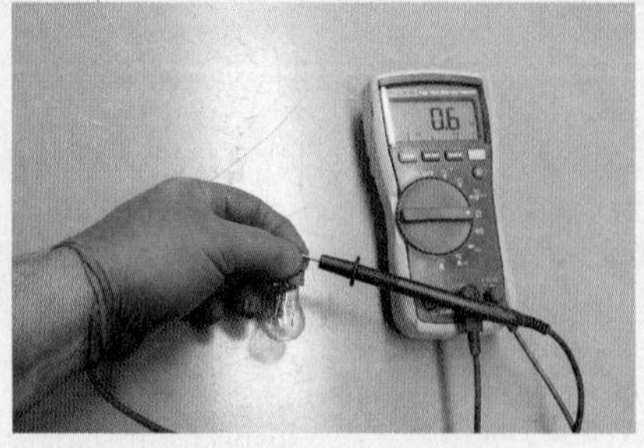

2. If some or all of the stop lamps do not illuminate, test them for an open or high resistance with a DMM.

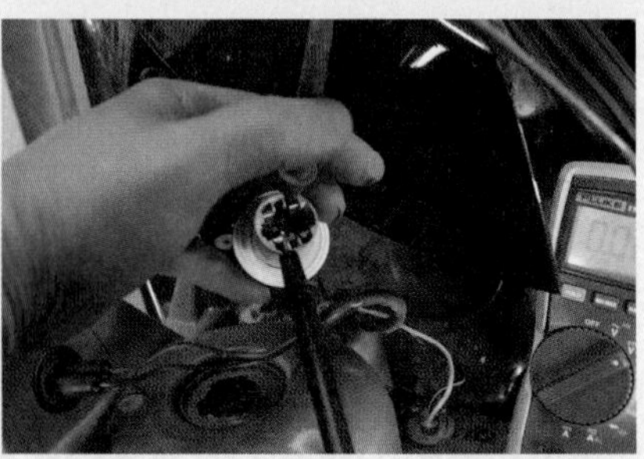

3. If the bulbs are okay, consult the wiring diagram to determine potential causes of the fault in the circuit. Use the wiring diagram and a DMM to identify the cause of the fault.

To perform a proportioning valve test, follow the steps in **SKILL DRILL 45-27**.

To perform a pressure differential valve test, follow the steps in **SKILL DRILL 45-28**.

SKILL DRILL 45-26 Testing Metering Valve

1. Research the testing procedure and specifications in the appropriate service information. Disconnect the inlet line and outlet line from the metering valve.

2. Connect the metering valve pressure tester to the inlet port and outlet port of the metering valve. Reconnect the inlet line to the metering valve, with the pressure gauge teed into it. Operate the brake pedal, and observe both pressure gauges; compare the findings to specifications.

SKILL DRILL 45-27 Performing a Proportioning Valve Test

1. Research the testing procedure and specifications in the appropriate service information.
2. Disconnect the inlet line and outlet line from the proportioning valve. Use a flare nut wrench to accomplish this.
3. Connect the proportioning valve pressure tester to the inlet port and outlet port of the proportioning valve. Reconnect the inlet line to the proportioning valve, with the pressure gauge teed into it.
4. Operate the brake pedal as specified, and observe both pressure gauge readings; compare the findings to specifications.

SKILL DRILL 45-28 Testing Pressure Differential Valve

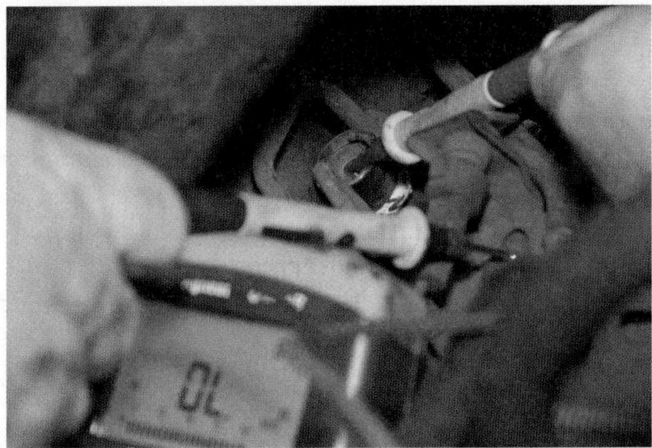

1. Research the testing procedure and specifications in the appropriate service information. Disconnect the electrical connector from the pressure differential valve, and connect an ohmmeter between the terminal on the valve and the body of the valve. Have an assistant lightly apply the brake pedal and hold until the bleeder screw is closed.

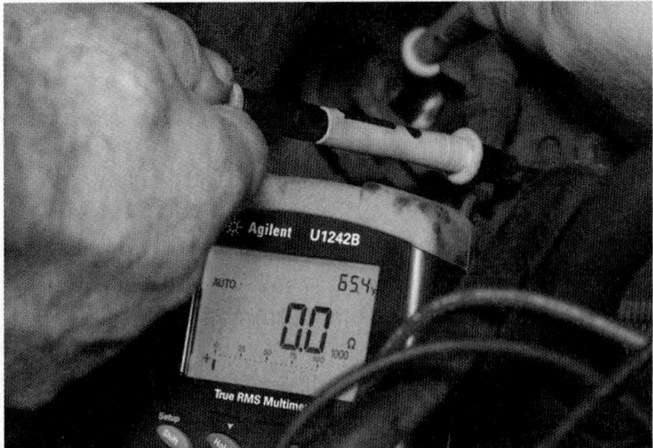

2. Open the front bleeder screw on the driver's side of the vehicle, and watch the ohmmeter reading. It should go to zero. Close the bleeder screw, and move to the rear driver's side wheel. Again, with the assistant applying light pressure to the pedal, open the driver's side rear wheel bleeder screw slightly, and watch the meter reading. It should go to OL and then back to zero as the valve moves to the Off position and then to the On position.

Wrap-Up

Ready for Review

- It can be dangerous to add brake fluid without diagnosing the reason for a low fluid level.
- Always ask customers questions to gather diagnostic information; try a test drive to understand what the customer is experiencing.
- Always try to discern the root cause of a vehicular problem.
- Common tools that are used to repair hydraulic brake systems are brake bleeder wrenches, vacuum brake bleeders, pressure brake bleeders, and valve gauge sets.
- Bleeding the brakes removes air form the hydraulic braking system.
- The three most common brake bleeding methods are manual, pressure, and vacuum (gravity is also used).
- Flushing the brake fluid involves bleeding out the air and replacing the old brake fluid with new.
- Always select the proper grade of brake fluid for the vehicle you are working on.
- Determining if brake fluid should be flushed can be done by time/mileage, DMM-galvanic reaction test, boiling point test, or test strip.
- DMM-galvanic reaction test, brake fluid testers, and brake fluid test strips can all be used to check for brake fluid contamination.
- Manual bleeding requires the fewest number of tools and is best when a small amount of bleeding is needed.
- Vacuum bleeding pulls brake fluid from the bleeder screws. It is faster than manual bleeding and not as complex to set up as pressure bleeding.
- Pressure bleeding requires more equipment and is best when the hydraulic system needs a full flush.
- After bleeding, always check that bleeder screws are properly tightened, with no leaks; refill the master cylinder and reinstall reservoir cup; and properly dispose of brake fluid.
- Brake pedal inspection includes brake pedal height, free play, and travel.
- Free play is the clearance between the brake pedal linkage and master cylinder piston.
- Brake pedal travel is the distance from its rest position to its applied height.
- Check the master cylinder for leaks if: the brake fluid is low in the reservoir, the brake warning light is on or the brake pedal reserve height is too low.
- Inspect the master cylinder for internal leaks only if the brake pedal sinks or is very low.
- Bench bleed the master cylinder prior to installing it, to minimize time needed to bleed the hydraulic brake system.
- Master cylinder pushrod length should only be adjusted if someone changed the adjustment setting; the brake pedal linkage has been repaired or adjusted; or the power booster is being replaced.
- Inspect and test power brake systems whenever the customer complains that the brakes are dragging; the brake pedal is harder to push than normal; the pedal height has changed; or the engine operation changes more than a minimal amount when the brake pedal is applied.
- Power booster testing starts with a brake pedal free travel test and then follows up with a performance test to see if the booster is operating properly.
- Inspect the steel brake lines for leaks, dents, kinks, rust, and cracks.
- Inspect all flexible brake hoses for cracks, bulging, and wear.
- When removing brake hoses and fittings, use two flare nut wrenches and the double wrench method to avoid damaging the brake lines and hoses.
- Double-walled steel brake line must be used when fabricating new brake lines, not copper, or single-wall steel.
- A mechanical brake warning lamp system (non-CANbus) circuit is a simple lightbulb in series with as many as four switches that are connected in parallel with one another.
- When checking parking brakes adjustment, count the number of clicks it takes to firmly apply the parking brake, and compare that to specifications.
- The non-CANbus brake light circuit is simply a normally closed switch in series with two to six brake lightbulbs connected in parallel with each other.

Key Terms

bleeding The process of removing air from a hydraulic braking system.

CANbus circuit A two-wire communication network that transmits status and command signals between control modules in a vehicle.

free play The amount of clearance between the brake pedal linkage and the master cylinder piston

International Standards Organization (ISO) flare method A method for joining brake lines, also called a bubble flare. Created by flaring the line slightly out and then back in, leaving the line bubbled near the end.

manual bleeding A bleeding method where one person manually operates the brake pedal while the other person opens and closes the bleeder screws on the wheel brake units to allow the air and old brake fluid to be pushed out.

pressure bleeding A bleeding method that uses clean brake fluid under pressure from an auxiliary tool or piece of equipment, to force the air and old brake fluid from the hydraulic braking system.

vacuum bleeding Bleeding process that uses a vacuum bleeder to pull the air and old brake fluid from the system.

Review Questions

1. Brake fluid testers are used to test the:
   a. level of the brake fluid.
   b. type of the brake fluid.
   c. current temperature of the brake fluid.
   d. moisture content of the brake fluid.
2. All of the following statements are true with respect to brake fluid handling *except*:
   a. Brake fluid is considered to be a top-off fluid.
   b. Brake fluid levels should be inspected during every oil change.
   c. The fluid should be near the full mark on the side of the cylinder or within half of an inch of the top of each chamber if there are no marks.
   d. The cause of the low brake fluid level needs to be identified and corrected before topping off.
3. All of the following can be used to determine if the brake fluid should be flushed *except*:
   a. manufacturer's time/mileage specification.
   b. boiling point test.
   c. metering valve gauge.
   d. using the test strip.
4. When performing a DMM-galvanic reaction test, the fluid should be flushed if it is:
   a. not equal to 0V.
   b. below 0.3V.
   c. above 0.3V.
   d. above 3V.
5. What happens when bleeding is not done?
   a. The brakes will feel spongy and will not be fully functional.
   b. The brake fluid gets contaminated.
   c. The brake pedal is harder to push than normal.
   d. The brakes will drag.
6. In the case of possible petroleum contamination, which of the following can be used to flush brake fluid?
   a. Compressed air
   b. Alcohol
   c. Gasoline
   d. Mineral-based products
7. The master cylinder should be inspected for internal and external leaks in all of the following scenarios *except*:
   a. the brake fluid is low in the reservoir.
   b. the brake warning light is on.
   c. the brake pedal reserve height is too low.
   d. the brake pedal is harder to push than normal.
8. When a vacuum-type power booster is used, the driver will be required to increase foot pressure to activate the brakes if there is:
   a. insufficient vacuum.
   b. excess air.
   c. excess vacuum.
   d. excess brake fluid.
9. Which of the following tests is used to verify brake booster operation and involves starting the engine while measuring pedal height?
   a. Brake pedal free travel test
   b. Performance test
   c. External leak test
   d. Internal leak test
10. All of the following statements with respect to replacing brake lines, brake hoses, fittings, and supports are true *except*:
    a. depress the brake pedal with a brake pedal depressor to prevent brake fluid from draining out of the vehicle.
    b. using flare nut wrenches and the double wrench method, carefully remove any brake lines, hoses, fittings, and supports that are to be replaced.
    c. reassemble the removed components using tools only.
    d. bleed any trapped air from the system.

ASE Technician A/Technician B Style Questions

1. Tech A says that technicians have been successfully sued for improper brake repairs, resulting in large cash settlements. Tech B says that to reduce liability, good technicians create processes to help ensure steps of the job are not forgotten. Who is correct?
   a. Tech A
   b. Tech B
   c. Both A and B
   d. Neither A nor B
2. Tech A says that a service brake pedal that does not return all the way causes the brake warning light on the instrument panel to stay on. Tech B says that a service brake pedal that does not return causes the brake lights to stay on. Who is correct?
   a. Tech A
   b. Tech B
   c. Both A and B
   d. Neither A nor B
3. Tech A says that brake fluid should periodically be checked for excessive moisture content. Tech B says that brake fluid should be replaced every year. Who is correct?
   a. Tech A
   b. Tech B
   c. Both A and B
   d. Neither A nor B
4. Tech A says that brake pedal free play is the amount of clearance between the brake pedal linkage and the master cylinder piston. Tech B says that brake pedal travel is sometimes called reserve pedal. Who is correct?
   a. Tech A
   b. Tech B

c. Both A and B
d. Neither A nor B

5. Tech A says that a sinking brake pedal is an indication of a faulty vacuum booster. Tech B says that the master cylinder can only have external leaks, not internal leaks. Who is correct?
   a. Tech A
   b. Tech B
   c. Both A and B
   d. Neither A nor B
6. Tech A says that part of the hydraulic system diagnostic process could require disassembly and visual inspection or measurements of the suspected component or system. Tech B says that may involve performing tests in the shop with a scan tool, pressure gauges, or a DMM. Who is correct?
   a. Tech A
   b. Tech B
   c. Both A and B
   d. Neither A nor B
7. Tech A says that to performance test a vacuum booster; start with the engine off, apply the brakes several times, hold the pedal down, start the engine, and observe the brake pedal action. Tech B says that to check a vacuum booster for internal leaks, you should pressurize the booster with compressed air and listen for leaks. Who is correct?
   a. Tech A
   b. Tech B
   c. Both A and B
   d. Neither A nor B
8. Tech A says that when flaring a brake line, you should install the fitting onto the line after the line has been flared. Tech B says that when fabricating a double flare, the adapter widens and narrows the line, and then the flaring tool folds the narrowed portion back into the widened portion of the flare. Who is correct?
   a. Tech A
   b. Tech B
   c. Both A and B
   d. Neither A nor B
9. Tech A says that bench bleeding a master cylinder prevents having to bleed air from the brake lines during replacement. Tech B says that bench bleeding the master cylinder makes bleeding the brakes on the vehicle easier. Who is correct?
   a. Tech A
   b. Tech B
   c. Both A and B
   d. Neither A nor B
10. Tech A says that a faulty vacuum booster can affect engine operation. Tech B says that steel brake line can be replaced with a copper line, as it is easier to bend into shape. Who is correct?
    a. Tech A
    b. Tech B
    c. Both A and B
    d. Neither A nor B

CHAPTER 46
Disc Brake Systems

NATEF Tasks

- **N46001** Identify and interpret brake system concerns; determine needed action. (AST/MAST)
- **N46002** Describe procedure for performing a road test to check brake system operation including an antilock brake system (ABS). (MLR/AST/MAST)
- **N46003** Diagnose poor stopping, noise, vibration, pulling, grabbing, dragging or pedal pulsation concerns; determine needed action. (AST/MAST)
- **N46004** Remove and clean caliper assembly; inspect for leaks, damage, and wear; determine needed action. (MLR/AST/MAST)
- **N46005** Inspect caliper mounting and slides/pins for proper operation, wear, and damage; determine needed action. (MLR/AST/MAST)
- **N46006** Check brake pad wear indicator; determine needed action. (MLR/AST/MAST)
- **N46007** Remove, inspect, and/or replace brake pads and retaining hardware; determine needed action. (MLR/AST/MAST)
- **N46008** Lubricate and reinstall caliper, brake pads, and related hardware; seat brake pads; inspect for leaks. (MLR/AST/MAST)
- **N46009** Retract and readjust caliper piston on an integrated parking brake system. (MLR/AST/MAST)
- **N46010** Clean and inspect rotor and mounting surface; measure rotor thickness, thickness variation, and lateral runout; determine needed action. (MLR/AST/MAST)
- **N46011** Remove and reinstall/replace rotor. (MLR/AST/MAST)
- **N46012** Refinish rotor on vehicle; measure final rotor thickness and compare with specification. (MLR/AST/MAST)
- **N46013** Refinish rotor off vehicle; measure final rotor thickness and compare with specification. (MLR/AST/MAST)
- **N46014** Inspect and replace drive axle wheel studs. (MLR/AST/MAST)
- **N46015** Install wheel and torque lug nuts. (MLR/AST/MAST)
- **N46016** Describe importance of operating vehicle to burnish/break-in replacement brake pads according to manufacturer's recommendations. (MLR/AST/MAST)

Knowledge Objectives

After reading this chapter, you will be able to:

- **K46001** Describe the theory of operation of disc brakes.
- **K46002** Describe the components of the disc brake system.
- **K46003** Describe how the components of the disc brake system work together during braking operations.
- **K46004** Describe the advantages and disadvantages of the disc brake system.
- **K46005** Describe the types, purpose, and function of disc brake calipers.
- **K46006** Describe the purpose, type, and function of disc brake pads.
- **K46007** Describe the principle of the coefficient of friction and how it affects brake lining materials.
- **K46008** Describe the methods used to reduce noise in disc brakes.
- **K46009** Describe the types of wear indicators for disc brakes.
- **K46010** Describe types of disc brake rotors and their function.
- **K46011** Describe the types, purpose, and operation of parking brakes in a disc brake system.

Skills Objectives

After reading this chapter, you will be able to:

- **S46001** Perform disc brake diagnosis, service, and repair.
- **S46002** Perform caliper and pad service.
- **S46003** Perform rotor service.

▶ Introduction

K46001

Disc brakes are so named because they create braking power by forcing flat friction pads against the sides of a rotating disc. This disc is also called a disc brake rotor, or rotor, and is bolted to the wheel (**FIGURE 46-1**). Disc brake calipers straddle the pads. Hydraulic pressure from the master cylinder causes the caliper to create a mechanical clamping action, forcing the brake pads onto the surface of the rotor, creating friction. By forcing friction materials against moving surfaces, the vehicle's kinetic energy is transformed into heat energy. As the vehicle's kinetic energy is transformed into heat energy, the vehicle's speed decreases.

Higher applied forces can be used in disc brakes than in drum brakes because the rotor has to withstand only compressive forces, whereas the drum has to withstand tension forces (**FIGURE 46-2**). Because heat is generated on the outside surfaces of the rotor, it can be easily transferred to the atmosphere. Because of these and other design features, disc brakes are effective at creating substantial braking power using a fairly simple design that is relatively easy for technicians to service and repair.

FIGURE 46-1 Disc brake operation.

> **TECHNICIAN TIP**
>
> The purpose of the disc brake system is to provide an effective means to slow the vehicle under a variety of conditions in an acceptable distance and manner. The better a braking system can do this, the more likely the vehicle will avoid an accident.

▶ Disc Brake System

K46002

Modern passenger vehicles are almost always equipped with disc brakes on at least the front two wheels, and many manufacturers are using them on all four wheels. The primary components of the disc brakes are as follows:

- Rotor
- Caliper
- Brake pads (**FIGURE 46-3**)

The **rotors** are the main rotating part of this brake system. They are durable and resistant to damage by the high

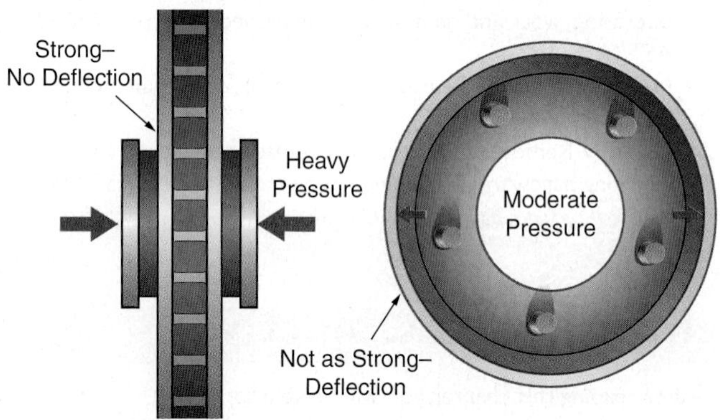

FIGURE 46-2 Disc versus drum brakes.

You Are the Automotive Technician

A customer comes into the service department. She has been experiencing a pulsating brake pedal and high-pitched squealing during braking on her 2013 Town and Country minivan, which has 41,000 miles (69,202 km) on the odometer. After reviewing the vehicle's service history, you notice the vehicle has never had any brake service. You advise the customer that you will need to complete a more thorough inspection to determine what is causing the issues. After completing a visual inspection of the brakes and measuring the rotor's thickness and parallelism, you see that the front brake linings have worn down to the wear indicators, and the rotors are warped beyond specifications. You inform the customer of your findings and recommend that the vehicle needs new front brake pads installed and the rotors refinished. You also recommend that the brake fluid be flushed, as the fluid is beyond the end of its two-year life.

1. What conditions can cause a rotor to become warped?
2. What conditions would require replacement of the rotors rather than just refinishing them?
3. What are the relatively common maximum specifications for rotor runout and thickness variation?
4. What are the benefits of using an on-car brake lathe to refinish the rotors?

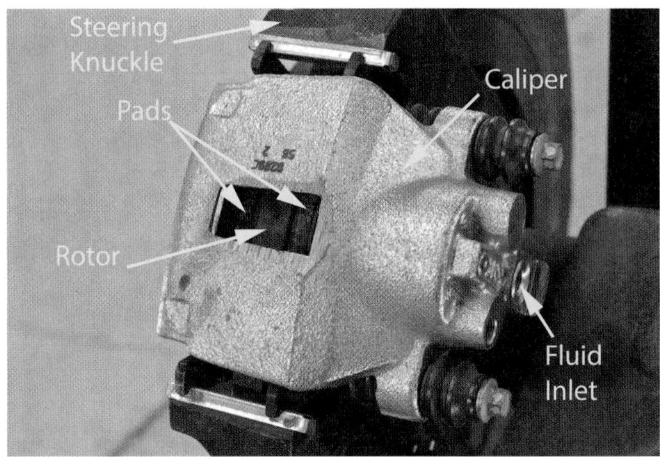

FIGURE 46-3 The disc brake system.

temperatures that occur during braking. In high-performance vehicles, the rotors are made from composite materials, ceramics, or carbon fiber; otherwise, they are usually made of cast iron.

The **caliper** straddles the rotor and houses the disc brake pads and an activating piston(s). The calipers use hydraulic pressure from the master cylinder to apply the brake pads. They are usually bolted to the steering knuckle or, in the case of a non-steering axle, to a suspension component. Calipers should be inspected at the same time as the brake pads.

The **disc brake pads** are located inside the caliper or caliper mounting bracket. The pads are forced against the rotor to slow or stop the vehicle. The disc brake pad consists of a friction material bonded or riveted to a steel backing plate. With this design, the pads will wear out over time and must be replaced periodically.

Disc Brake Operation

K46003

Disc brakes can be used on all four wheels of a vehicle, or on the front wheels, with drum brakes on the rear. When the brake pedal is depressed, a **pushrod** transfers the force through a **brake booster** to a hydraulic master cylinder. The master cylinder converts the pedal force into hydraulic pressure, which is then transmitted via brake lines and hoses to one or more pistons at each brake caliper (**FIGURE 46-4**). The pistons operate on friction pads to provide a clamping force on a rotor that is attached to the wheel hub. This clamping action is designed to stop the rotation of the rotor and the wheel.

The rotors are free to rotate with the wheels, because of wheel bearings and the hubs that contain them. The hub can be part of the brake rotor, called a hub-style rotor. Or it could be separate from the rotor, called a hubless rotor. In this case, the rotor slips over the hub and is bolted to it by the wheel and lug nuts (**FIGURE 46-5**).

The brake caliper assembly is normally bolted to the steering knuckle or vehicle axle housing (**FIGURE 46-6**). In most cases, the brake is positioned as close as possible to the wheel, but there are exceptions. Some high-performance cars with **independent rear suspension (IRS)** use inboard disc brakes on the rear wheels. The calipers are mounted on or next to the differential, which is directly mounted to the vehicle body. This configuration increases vehicle handling because it reduces the vehicle's unsprung weight by taking the differential assembly and brakes from the suspension system and mounting them to the body. Because the wheels and axles are now lighter, the vehicle's springs can do a better job of keeping the wheels on the ground, especially on uneven road surfaces.

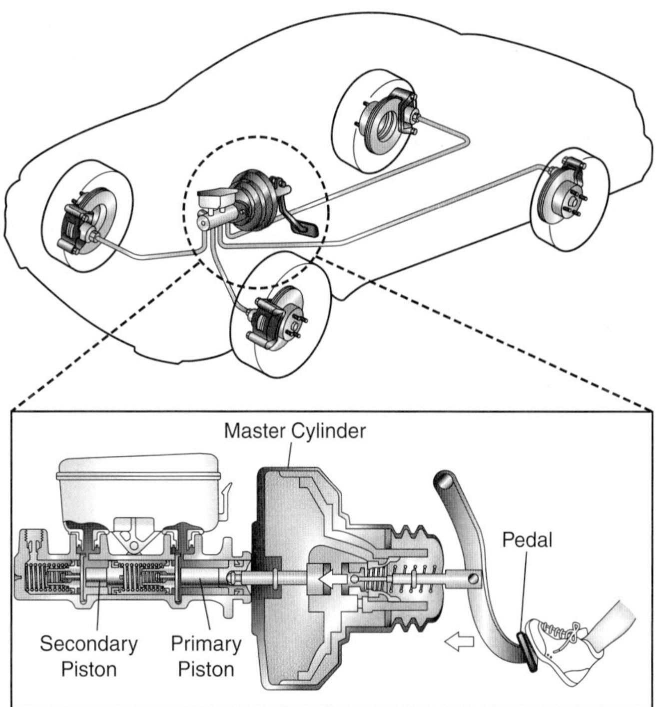

FIGURE 46-4 The master cylinder converts the pedal force into hydraulic pressure, which is then transmitted via brake lines and hoses to one or more pistons at each brake caliper.

FIGURE 46-5 The hub and hubless rotors.

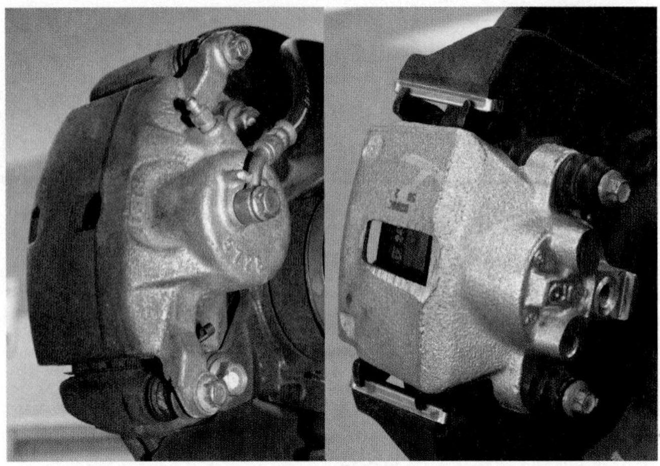

FIGURE 46-6 Floating or sliding caliper mounting methods.

Disc brake pads require much higher application force to operate than drum brake shoes because they are not self-energizing. This additional clamping pressure is created by increasing the diameter of the caliper pistons. Unfortunately, this means the brake pedal travel is lengthened to move the additional fluid being displaced by the larger caliper pistons. Building in more pushrod travel would require more room under the dash for the brake pedal. Manufacturers have overcome this problem by equipping most disc brake systems with a power booster.

Because of the high forces needed to apply a disc brake, using it as a parking brake is more challenging. Some manufacturers have chosen to design more complicated calipers, whereas others have built an auxiliary drum brake assembly into the center of the rear disc brake rotors to provide for parking brake operation. This is referred to as a top hat design.

Advantages and Disadvantages

K46004

Disc brakes have a number of advantages over drum brakes. They also have some disadvantages. In most cases, the advantages outweigh the disadvantages. One of the biggest advantages is that disc brakes can generate and transfer greater amounts of heat to the atmosphere; because most of the friction area of a rotor is exposed to air, cooling is far more rapid than for a drum brake. This faster cooling makes them better suited for high-performance driving or heavy-duty vehicles and reduces the likelihood of brake fade.

Also, because of their shape, rotors tend to scrape off water more effectively. After being driven through water, disc brakes operate at peak performance almost immediately. Further, due to their design, disc brakes are self-adjusting and do not need periodic maintenance or rely on a self-adjusting mechanism that is prone to sticking. Lastly, in most cases, disc brakes are also easier to service than drum brakes.

Although disc brakes have a number of benefits over drum brakes, there are some disadvantages. Probably the most apparent disadvantage is that disc brakes are much more prone to making noise. Their design tends to create squeals and squeaks, which can be very annoying. Many a technician has spent time replacing perfectly functional disc brakes because of excessive noise complaints. Another issue is that the rotors warp more easily than in drum brake systems. Because the brake pads are pressing on each side of the rotor, thickness variations as small as 0.0003" (0.0076 mm) can cause brake pedal pulsations, requiring resurfacing or replacement. The last disadvantage is that because disc brakes are not self-energizing, they need higher clamping forces, which requires a power booster. This also makes it harder to use disc brakes as an effective parking brake.

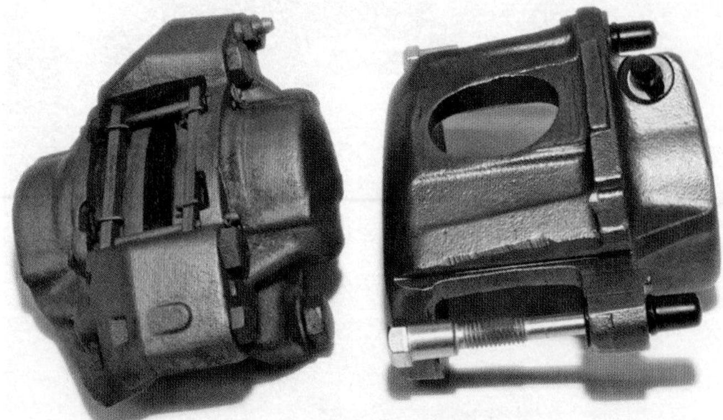

FIGURE 46-7 Fixed and sliding or floating calipers.

▶ TECHNICIAN TIP

Although there may appear to be more disadvantages to disc brakes than drum brakes, the advantages are generally considered much more critical than the disadvantages. Thus, disc brakes are preferred over drum brakes in most applications.

▶ Disc Brake Calipers

K46005

In most applications, the disc brake caliper assembly is bolted to the vehicle axle housing, or steering knuckle, and clamps the brake pads onto the rotors to slow the vehicle. There are two main types of calipers: **fixed calipers** and **sliding or floating calipers** (**FIGURE 46-7**). Sliding or floating calipers are the most common type used in passenger vehicles because they are easier to build and more compact. All calipers are fitted with a **bleeder screw** on the top of the piston bore, to allow for the removal of air within the disc brake system, as well as to help in performing routine brake fluid changes.

Fixed calipers are rigidly bolted in place and cannot move or slide. This makes their application of braking forces more precise than floating calipers. They commonly have one to four pistons on each side of the rotor (**FIGURE 46-8**). When the brakes are applied, hydraulic pressure forces the pistons on both sides of the caliper inward, causing the brake pads to come in contact with the rotor (**FIGURE 46-9**). Once the pad-to-rotor clearance is

FIGURE 46-8 Fixed calipers with multiple pistons.

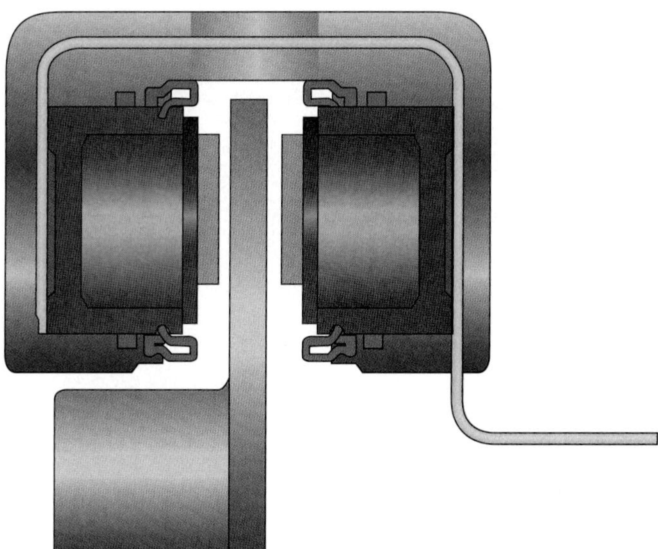

FIGURE 46-9 Fixed caliper application on solid rotor..

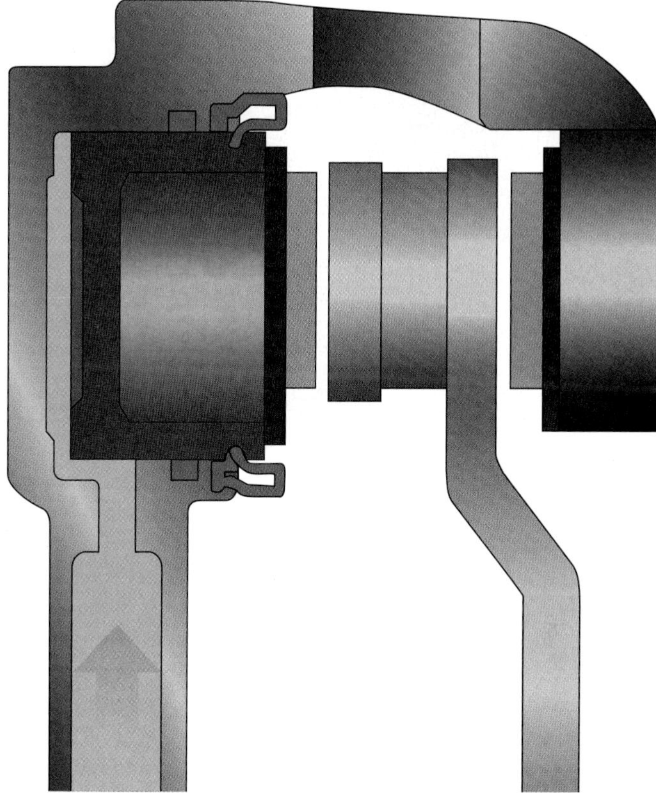

FIGURE 46-10 Sliding or floating caliper application on vented rotor.

FIGURE 46-11 Floating caliper and guide pins.

taken up, the hydraulic pressure rises equally on each side of the caliper, applying both brake pads equally.

The floating or sliding caliper has brake pads located on each side of the rotor, but all of the pistons are only on one side, usually the inboard side of the rotor. Thus, the hydraulic force is generated on one side of the rotor, but the unique design applies equal braking force to both sides of the rotor. This distribution of force is possible because the caliper is mounted on pins, or slides, that allow it to move (float or slide) from side to side, as necessary. This movement allows pistons on one side of the rotor to generate force on both sides of the rotor at the same time.

When the brakes are applied, hydraulic pressure forces the piston toward the rotor. This takes up any clearance between the inboard brake pad and the rotor, and starts to push the pad into the rotor. Because the caliper is free to move on the pins or slides, it gets pushed away from the rotor, pulling the outboard brake pad into contact with the outboard side of the rotor. Once all clearance is taken up on the outboard brake pad, the clamping force increases equally on both brake pads, applying the brakes (**FIGURE 46-10**).

Floating calipers are mounted in place by **guide pins** and bushings (**FIGURE 46-11**). The pins allow the caliper to move in and out as the brakes are operated and as the brake pads wear. Because the calipers move on the pins, the bushings must be lubricated with high-temperature, waterproof disc brake caliper grease when they are serviced. This helps prevent them from binding or sticking. Inspecting, cleaning, and lubricating the pins and pin bores, including any bushings and dust boots, are important steps in disc brake repair.

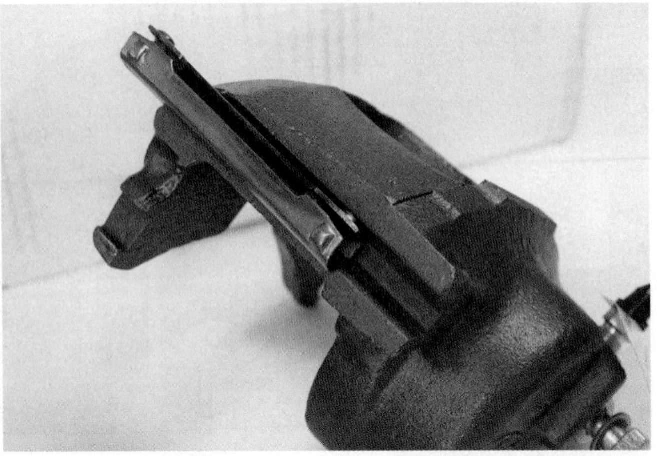

FIGURE 46-12 Sliding caliper.

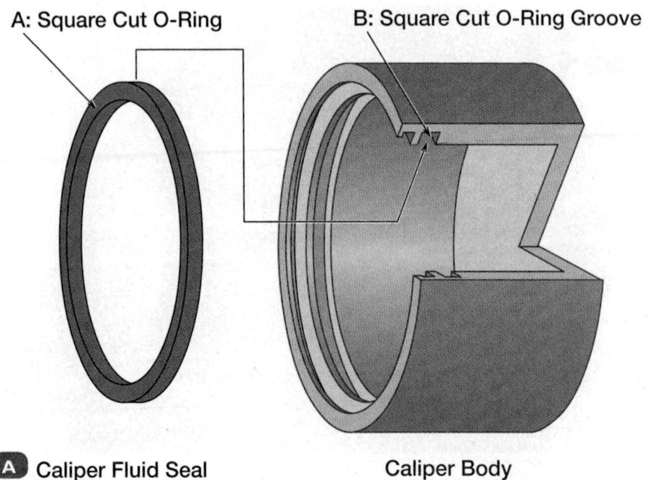

FIGURE 46-13 O-rings. **A.** Square-cut O-ring and O-ring cut to show square section. **B.** Square-cut O-ring groove in caliper.

Sliding calipers have matching machined surfaces on the caliper and caliper mount that allow the caliper to slide in the mount (**FIGURE 46-12**). The sliding mount holds the caliper in position and prevents it from rotating when the brakes are applied. The machined mounts allow the caliper to move side to side, as necessary, to operate the brakes or adjust for brake pad wear. When the calipers are serviced, the surfaces must be cleaned and lubricated with the same high-temperature, waterproof grease as the floating calipers. Sliding calipers are held in place by several methods. Because sliding calipers do not slide as easily as floating calipers, most light-duty vehicles use the floating caliper design.

O-Rings

In disc brake calipers, the piston is sealed by a stationary square section sealing ring, also called a **square-cut O-ring** (**FIGURE 46-13**). This O-ring has a square cross section and is fitted in a machined groove in the caliper. The O-ring is compressed between the piston and caliper housing, creating a positive seal to keep the high-pressure brake fluid from leaking out. It also prevents air from being drawn into the system when the brake pedal is released quickly, creating a low-pressure situation in the hydraulic system.

When the brakes are applied, the piston moves outward, slightly deforming the O-ring seal (**FIGURE 46-14**). When the brakes are released, the elasticity or flexibility of the seal causes it to return to its original shape. This action of the sealing ring retracts the piston to provide a small running clearance between the rotor and pads. As the brake pads wear, the piston must move outward a bit farther than the sealing ring can stretch or flex. The sealing ring is designed to allow the piston to slide through it in this situation, taking up the extra clearance and making the disc brakes self-adjusting.

▶ TECHNICIAN TIP

Because the force generated by the O-ring to retract the piston is fairly small, any corrosion or buildup on the piston or bore will cause the piston to stick and not retract. This holds the brakes in the applied position, causing brake drag, overheated brakes, and poor fuel economy. Technicians can identify this situation by using an infrared temperature gun to measure the temperature of each brake rotor after test-driving the vehicle. The temperatures should be approximately the same on each side. If they are not, suspect a stuck or binding caliper.

Some calipers, sometimes called **low-drag calipers**, are designed to maintain a larger brake pad–to-rotor clearance by retracting the pistons a little bit farther. This is accomplished by modifying the sealing groove in the caliper so the outside of the groove is slightly angled toward the rotor (**FIGURE 46-15**). This position allows the seal to flex a bit farther upon brake application and then retract the piston a greater distance. These systems use a "quick take-up" or "fast-fill" master cylinder to maintain adequate brake pedal reserve height.

The primary sealing surface is the outside diameter of the piston. It is critical that this surface be smooth and free of pitting or rust; therefore, steel pistons are chrome plated. This gives the surface a hard, wear-resistant, and corrosion-resistant finish. Chrome can still rust, but it is much more corrosion resistant than steel. Another way manufacturers have dealt

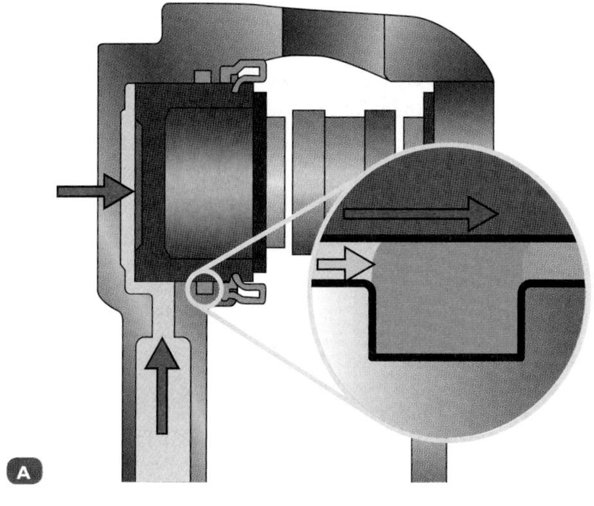

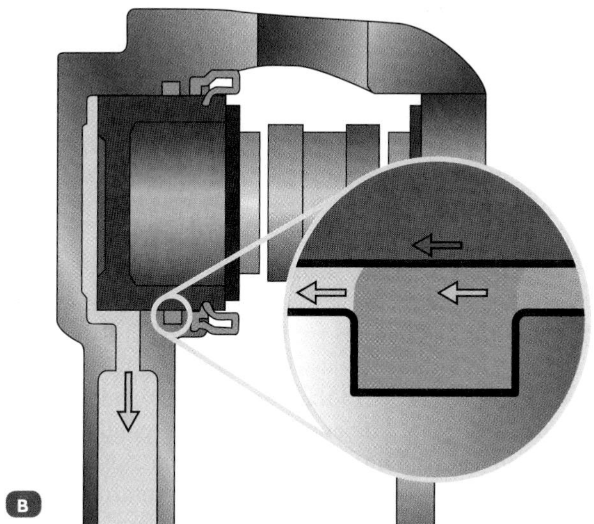

FIGURE 46-14 Square-cut O-ring. **A.** Square-cut O-ring during brake application. **B.** Square-cut O-ring during brake release.

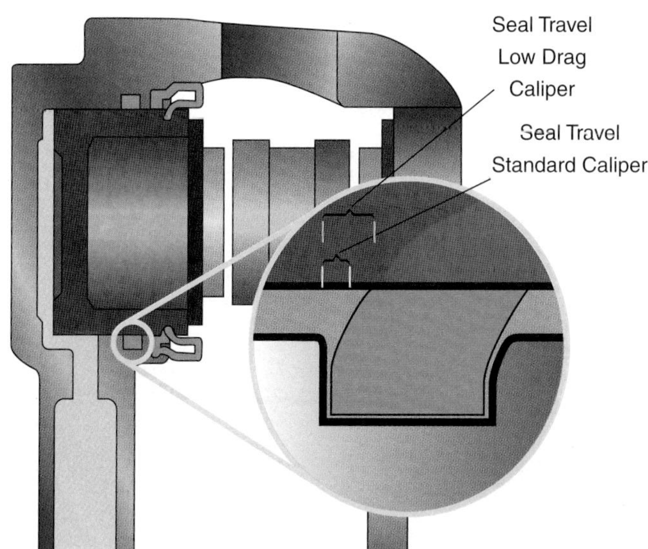

FIGURE 46-15 Low-drag caliper.

FIGURE 46-16 Corroded caliper piston bore.

with the corrosion issue is by making pistons out of a **phenolic resin**. Pool balls also are made from phenolic resin, which is very dense when it hardens and does not corrode or rust, making for a good sealing surface in brake systems. Although the phenolic pistons themselves do not corrode, the cast iron bore of the caliper does corrode and rust and can therefore cause a phenolic piston to seize in the bore (**FIGURE 46-16**).

Phenolic pistons transfer heat more slowly than steel pistons (**FIGURE 46-17**), which is a good thing, because heat transferred through the piston to the brake fluid causes it to boil. Calipers with phenolic pistons are therefore less susceptible to brake failure from boiling brake fluid.

There is also a dust boot that seals the surfaces of the piston and caliper bore from outside dirt and moisture. This seal connects to both the piston and the caliper and must be expandable to allow the piston to move outward as the brake pads wear. It also must be free from cuts and holes; otherwise the piston and bore could corrode and cause the piston to bind in the bore, causing brake drag.

▶ Disc Brake Pads and Friction Materials

K46006

Disc brake pads consist of friction material bonded or riveted onto a steel **backing plate** (**FIGURE 46-18**). **Bonded linings** are more common on light-duty vehicles, as they are less expensive to build and the bonding agent can fail under the very high temperatures of heavy-duty use. **Riveted linings** are used on heavier-duty or high-performance vehicles. Metal rivets provide a mechanical connection to hold the lining to the backing plate that is less susceptible to failure under high temperatures. At the same time, because the rivets actually pinch some of the lining between the rivet head and the backing plate, the linings must be changed sooner than bonded linings. Otherwise, the rivet heads would contact the rotor and wear a groove in the face of the rotor.

The backing plate has **lugs** that correctly position the pad in the caliper assembly and help the backing plate maintain the proper position to the rotor (**FIGURE 46-19**). Disc brakes are

FIGURE 46-17 Heat transfer. **A.** Phenolic piston (slow heat transfer). **B.** Steel piston (fast heat transfer).

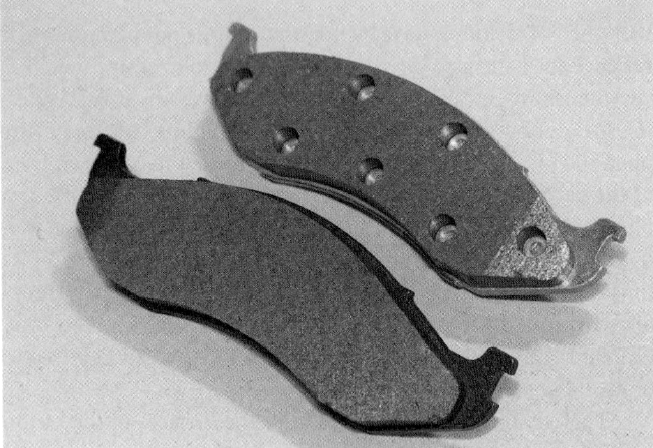

FIGURE 46-18 Bonded and riveted brake pads.

FIGURE 46-19 Brake pad locating lugs.

usually designed so that the thickness of the pads can be checked easily once the wheel has been removed. Most disc brakes also are designed to allow the pads to be replaced with a minimum of disassembly.

The composition of the friction material affects brake operation. Materials that provide good braking with low pedal pressures tend to lose efficiency when they get hot, thus increasing the stopping distance. They also tend to wear out sooner. Materials that maintain a stable friction coefficient over a wide temperature range generally require higher pedal pressures to provide efficient braking. They also have a tendency to put added wear on the disc brake rotor, reducing its useful life (**FIGURE 46-20**).

Brake Friction Materials

K46007

Friction is the force that acts to prevent two surfaces in contact from sliding against each other. The amount of friction between two surfaces is expressed as a ratio and is called the coefficient of friction. When friction occurs, the kinetic energy (motion) of the sliding surfaces is converted into thermal energy (heat). Some combinations of materials, such as a hockey puck on ice, have a very low coefficient of friction. There is very little friction between them and therefore almost no sliding resistance. Rubber tires against a dry, hard road surface have a high coefficient of friction, which means they tend to grip and resist sliding against each other.

Disc brake pads and drum brake linings are made from materials that have a moderate coefficient of friction (**TABLE 46-1**). They also must be able to absorb and disperse large amounts of heat without braking performance being adversely affected. As the heat in brake pads and linings builds up, the coefficient of friction capability of the material—and consequently its stopping power—is reduced. This is called brake fade. Minimizing or overcoming fade is a major factor in the design of brakes and the development of brake friction materials.

Brake friction materials were historically made from asbestos compounds because of the excellent heat resistance of that material. Now that asbestos has been proven to be hazardous, it is generally banned and is not normally used. Today, brakes are

FIGURE 46-20 Brake rotor wear.

TABLE 46-1 Brake Lining Coefficient of Friction (Sliding)

Materials Involved	Coefficient of Friction-Dry Sliding
Rubber and concrete	0.6–0.85
Steel and cast iron	0.23
Copper and cast iron	0.29
Brass and cast iron	0.3
Leather and oak	0.52
Brake lining (FF rating)	0.35–0.45

Applied Math

AM-17: Charts/Tables/Graphs: The technician can interpret charts, tables, and graphs to determine the manufacturer's specifications for a given system.
Algebra (9–12)
C3: Draw reasonable conclusions about a situation being modeled.
Representation
A2: Select, apply, and translate among mathematical representations to solve problems.

Technicians regularly apply math concepts to disc brake diagnosis and repair. For example, a technician measures the amount of rotor thickness variation and runout if there is a brake pedal pulsation problem. Let's say the technician performs these measurements and comes up with a thickness variation of 0.0025" (0.064 mm) and a runout of 0.0015" (0.038 mm) on the left front rotor. Next the technician looks up the manufacturer's specification chart and finds that the maximum allowable thickness variation is 0.0005" (0.013 mm) and the maximum allowable runout is 0.003" (0.076 mm) for the vehicle being worked on. Using the information from the chart, the technician determines that the thickness variation is excessive by 0.002" (0.051 mm) (0.0025" [0.064 mm] – 0.0005" [0.013 mm] = 0.002" [0.051 mm]), and the runout is okay since it is under the maximum allowable specification. But even so, because the thickness variation is out of specifications, the rotor has to be refinished or replaced to bring it back within specification.

manufactured from a variety of different materials, including the following:

- Non-asbestos organic (NAO) materials—organic materials such as Kevlar® and carbon
- Low-metallic NAO materials—small amounts of copper or steel and NOA materials
- Semi-metallic materials—a higher quantity of steel, copper, and/or brass
- Ceramic materials—ceramic fiber materials and possibly a small amount of copper

SAFETY TIP

Don't assume that a brake pad doesn't contain asbestos; it is still found in some applications.

The choice of brake lining compound depends on the application. Lighter passenger vehicles generate less heat in the brakes than heavy or high-performance vehicles. Living in a very hilly region of the country or where there is a lot of stop-and-go traffic puts added demands on the brake pads. The optimum brake composition for any given vehicle or use is a combination of weighted qualities, including the following:

- Stopping power
- Heat absorption and dispersion
- Resistance to fade
- Recovery speed from fade
- Wear rate
- Performance when wet
- Operating noise
- Price

For instance, owners of small economy vehicles tend to value a longer pad life and minimal operating noise rather than resistance to fade in extreme conditions. Owners of high-performance cars, however, may consider fade resistance and stopping power at high speeds more important than noise levels or wear rate.

The Society of Automotive Engineers (SAE) has adopted letter codes to rate the coefficient of friction of brake lining materials. The rating is written on the edge of the friction linings and is called the **edge code** (**FIGURE 46-21**). The lower the letter, the less friction the material has and the harder the brake pedal must be applied to achieve a given amount of stopping power. These code letters represent the following coefficients of friction:

- C: ≤ 0.15
- D: 0.15–0.25
- E: 0.25–0.35
- F: 0.35–0.45
- G: 0.045–0.55
- H: > 0.55
- Z: Unclassified

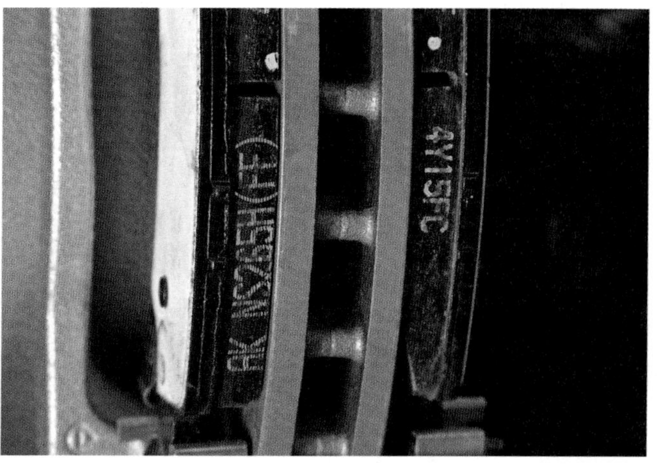

FIGURE 46-21 Brake lining edge code (FF).

The lining is tested both cool and hot. The rating is a two-letter designation such as "FF." The first letter is for the cool performance, and the second letter is for the hot performance. For example, FF has a cool coefficient of friction of 0.35–0.45 and the same coefficient of friction at the hot temperature. It also is very possible that the hot and cold ratings differ from each other. For example, if the rating is FE, the coefficient of friction reduces as the temperature of the lining heats up. Notice that the coefficient of friction range is quite wide for each letter designation. Linings with the same letter ratings may not have the same braking performance as one another. This means that an EE-rated lining from one manufacturer is likely to have different braking characteristics than an EE-rated lining from another manufacturer. Always use high-quality brake lining from reliable companies to help avoid brake issues.

FIGURE 46-23 Example of spring-loaded brake pad retainers.

Antinoise Measures

K46008

Disc brakes are more prone to annoying brake squealing than are drum brakes. Brake squealing is caused by vibrations set up between the brake pad and rotor. Manufacturers have addressed this problem in a number of ways:

1. Using softer linings with a higher coefficient of friction, which are less prone to noise than harder linings with a lower coefficient of friction
2. Adding **brake pad shims and guides** to the brake pads, which help cushion the brake pad and absorb some of the vibration (**FIGURE 46-22**)
3. Using springs to tightly hold the pads in place to minimize vibration (**FIGURE 46-23**)
4. Contouring and grooving the lining material in a way that minimizes vibration. Grooving helps ventilate gases that build up at the surface of the friction material under heavy brake application. It also can help modify the harmonic vibration quality of the friction material to reduce brake squeal. On some pads, the depth of this groove is set so that as the pad wears thinner, the remaining groove gets smaller; when it can no longer be seen, the pad should be replaced (**FIGURE 46-24**).
5. Incorporating **bendable tangs** on the brake pad backing plate that allow technicians to crimp the tangs so they are more firmly mounted in the caliper (**FIGURE 46-25**)

FIGURE 46-24 Brake lining grooves and contouring.

FIGURE 46-22 Brake pad shims and guides.

> **TECHNICIAN TIP**
>
> Some manufacturers claim that refinishing their rotors removes enough material from the rotor that it can cause the brakes to squeal due to less mass, which changes the harmonic vibration qualities of the rotor. They recommend replacing the rotors any time the rotors are worn enough to need resurfacing.

Technicians also can apply noise-reducing compounds to the brake pads. One is a type of high-temperature liquid rubber compound that is applied to the back of the brake pad. When it cures, it stays flexible, absorbs brake pad vibrations, and helps reduce brake noise. Another compound is a specially designed

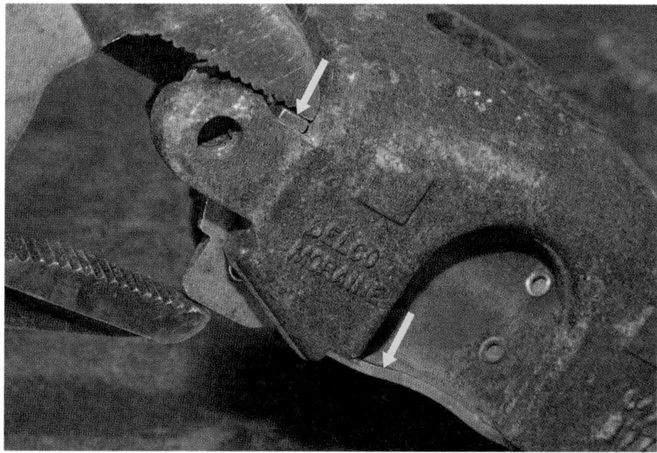

FIGURE 46-25 Brake pad bendable tangs.

liquid that is applied directly to the face of the lining material. This compound helps to modify the lining's coefficient of friction slightly, making it less likely that the lining will squeal. Make sure you apply the correct compound to the correct side of the brake pad.

Applied Science

AS-36: Sound: The technician can demonstrate an understanding of the role sound plays in identifying various problems in the vehicle.

Sound is a series of waves that travel through a gas, liquid, or solid and can often be heard by the human ear. Sound waves are created by vibrating objects, such as a guitar string. Moving objects that are in sliding contact with each other are highly likely to create sound. One example of this is fingernails dragging on a chalkboard. The fingernails vibrate on the surface of the chalkboard and create sound waves that are then heard by the ear.

Technicians commonly use differences in sound to assist in diagnosing disc brake problems. For example, brake pads that are worn down to the metal backing plate make a deep grinding noise when the brakes are applied. Listening to hear which wheel or wheels the noise is coming from helps identify the source of the problem. In the same way, if the disc brakes are equipped with a scratcher style of brake warning system, the brakes will make a high-pitched screeching noise. Many times this noise happens when the vehicle is being driven and the brakes are not being applied. One way to help determine which side the noise is coming from is by driving the vehicle next to a concrete wall or building. If the noise is on that side, it will get much louder than when not near the wall.

▶ Wear Indicators

K46009

Some manufacturers provide a means of notifying the driver that the brake pad linings are worn to their minimum limit. This helps ensure that the brake linings do not wear down to the point that they cannot properly perform their job anymore. Excessively thin brake linings tend to heat up more quickly than thicker linings, which can lead to premature brake fade. Not all manufacturers use a brake lining wear indicator. In those cases, it is especially important to inspect the brakes at regular intervals, usually during tire rotations or oil changes.

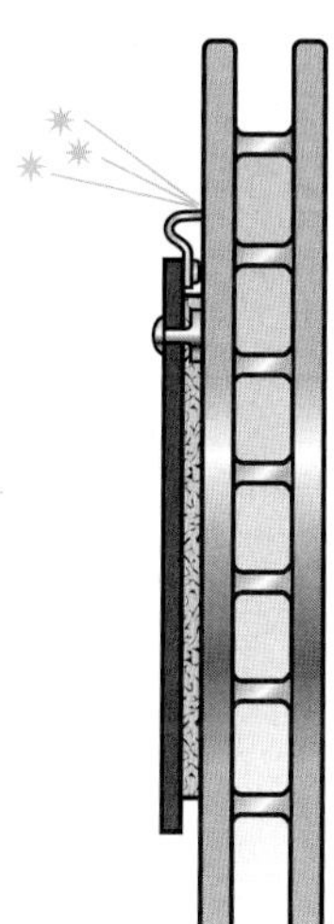

FIGURE 46-26 The scratcher brake wear indicator.

Types

Some manufacturers use a mechanically operated wear indicator to notify the driver that the brake pads are worn to their minimum limit. This is achieved by a spring steel **scratcher** mounted to the brake pad (**FIGURE 46-26**). Part of the scratcher extends below the brake pad backing plate at the lining's minimum wear thickness. When the friction material wears down far enough, the scratcher contacts the surface of the rotor and makes a squealing noise similar to fingernails on a chalkboard. This distinctive noise means the brakes need service right away. When you replace the brake pads, make sure they come equipped with new scratchers set to the correct depth so they can function the next time the pads wear down.

▶ TECHNICIAN TIP

In many cases, the scratchers start to make noise when the brakes are not applied and stop making noise when the brakes are applied, as applying the brakes tends to dampen the vibrations.

Some manufacturers use a warning lamp or warning message on the dash to alert the driver that the lining is worn to its minimum thickness. These systems have an electrical contact installed on the brake pad at the point of the lining's minimum wear thickness. When the pad wears to this minimum thickness, the contact touches the rotor as the brakes are applied, prompting a warning light or warning message that tells the driver the disc brake pads are due for replacement (**FIGURE 46-27**). These contacts can be manufactured into the pad, or they can be clipped onto the pad. The contacts are normally replaced when the pads are replaced. Make sure that either the contacts come with the new pads or are ordered along with the pads.

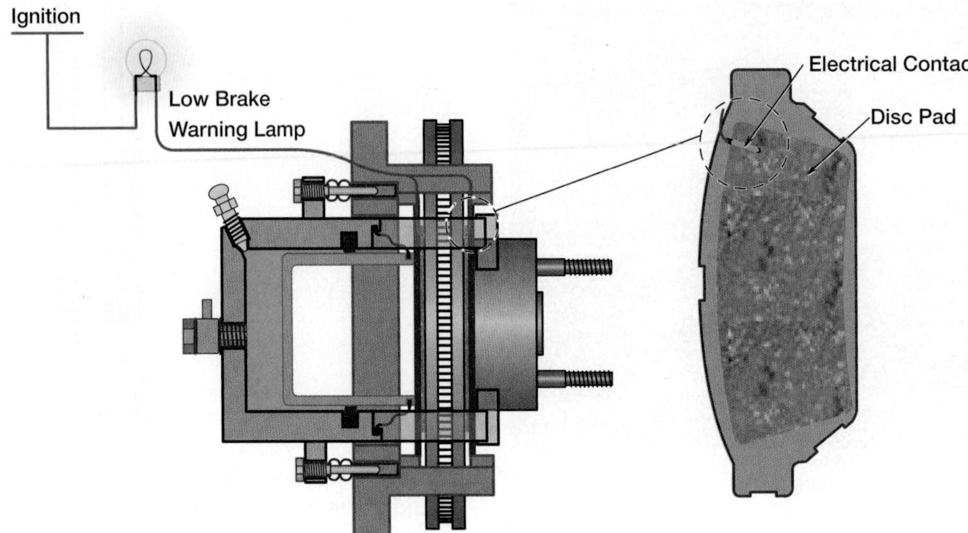

FIGURE 46-27 A brake pad wear indicator system illuminates a warning lamp on the dash when one or more brake pads wear down to a predetermined level.

▶ Disc Brake Rotors

K46010

The brake disc or rotor is the main rotating component of the disc brake unit. The wheel and rotor rotate together, leading some manufacturers to integrate the antilock brake system (ABS) tone wheel into the rotor. Because friction between the rotor and brake pads generates great amounts of heat, rotors must be able to withstand high temperatures. The pads are also forced onto the surface of the rotor with potentially thousands of pounds of force, so the rotor must be strong and have a durable surface. The rotors are usually made of cast iron. To reduce weight, some manufacturers use a two-part rotor with a cast iron disc and a stamped steel center hat (**FIGURE 46-28**). This style of rotor is called a composite rotor. Some heavy-duty and/or high-performance vehicles have rotors made of reinforced carbon, carbon ceramic, or composite ceramic substances to reduce weight and withstand much higher temperatures.

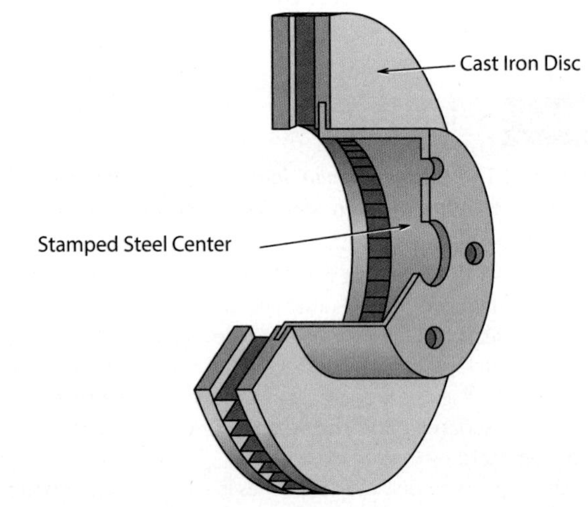

FIGURE 46-28 A composite rotor.

Because the rotor surfaces are squeezed between two brake pads, any unevenness of the rotor surfaces causes pulsation of the brakes as the thicker and thinner portions pass between the brake pads. The rotor surfaces must be parallel to each other to avoid this situation. Rotors can fail in two ways: parallelism, which is also called thickness variation, and lateral runout (**FIGURE 46-29**). **Parallelism** is the most critical condition. If the rotor's thickness varies by as little as 0.0003" (0.0076 mm), the rotor tends to push the brake pads outward at any high spots. This tends to create more pressure on the brake pads and slows the vehicle down faster at that point. It also pushes up on the brake pedal as fluid is being forced back to the master cylinder. The result is a pulsation of the brake pedal and a surging of the vehicle while braking, which is usually more noticeable at lower braking speeds.

Lateral runout, also called warpage, is the side-to-side movement of the rotor surfaces as the rotor turns. A warped rotor can be within specifications for parallelism, but out of specification for lateral runout. Lateral runout tends to move the caliper pistons in the same direction as one another, so brake fluid is not pushed back to the master cylinder. However, the caliper tends to be moved side to side. This movement can cause the steering wheel to shimmy as the warped rotor follows the brake pads, if the lateral runout is greater than about 0.003" (0.076 mm).

Also, runout causes the pads to rub on high spots of the rotor when the brakes are not being applied, causing uneven wear and/or depositing pad material on the rotor, which leads to thickness variation concerns.

▶ TECHNICIAN TIP

Rotors can be warped by improperly torquing the lug nuts. Always use a properly calibrated torque wrench (or the proper torque stick if the shop policy allows) to torque the lug nuts to the manufacturer's specified torque.

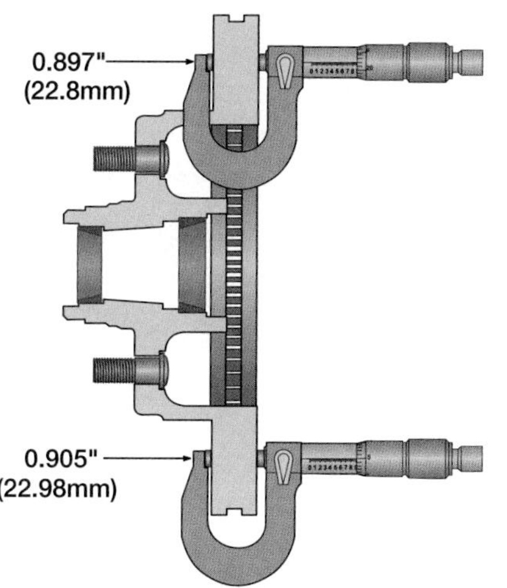

FIGURE 46-29 **A.** Example of thickness variation. **B.** Example of lateral runout.

For proper operation, rotors must maintain their shape and resist warpage under high heat and pressure conditions. Because of these requirements, they are usually made of cast iron. On motorcycles, rotors are often made of stainless steel for cosmetic reasons. Disc brakes also are equipped with a dust shield to help protect the rotor. Dust shields help keep dust, water, and other road debris away from the inside surface of the rotor (**FIGURE 46-30**). They also help direct air flow to the rotor to assist with heat transfer to the atmosphere. Dust shields are commonly made of plastic, but they also can be made of stamped sheet metal. Dust shields can become damaged during brake repair, so always inspect them for proper clearance before installing the wheel assembly.

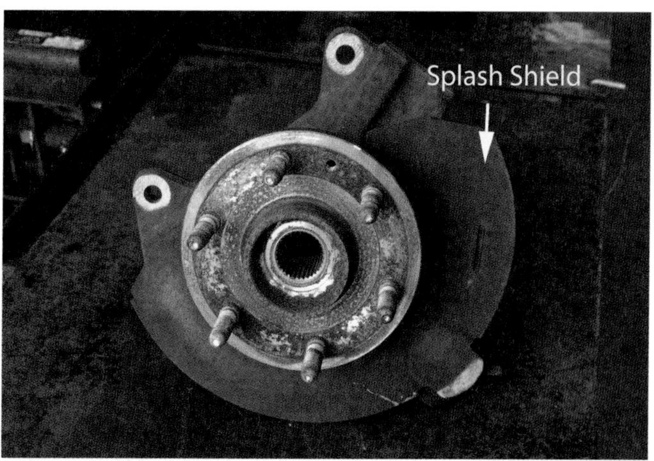

FIGURE 46-30 A typical dust shield.

Solid and Ventilated Rotors

Rotors can be solid or ventilated (**FIGURE 46-31**). **Solid rotors** are less expensive and usually found on smaller vehicles. **Ventilated rotors** are used to improve heat transfer to the atmosphere. These passageways are designed to use centrifugal force to cause air to flow through the center of the rotor when it is rotating. Ventilated rotors are used on heavier vehicles or high-performance vehicles. Some ventilated rotors are directional, meaning they are designed to force air through the rotor in one direction only (**FIGURE 46-32**). If the rotor is rotated in the wrong direction, it will not pump air properly and will overheat easier.

FIGURE 46-31 Rotors. **A.** Standard solid rotors. **B.** Ventilated rotors.

▶ TECHNICIAN TIP

If you hear an unusual scraping or grinding noise when you test-drive a vehicle after servicing the brakes, check to see if the dust shield is contacting one of the rotors. If so, it can make a lot of loud scraping or grinding noises. Because it is thin plastic or sheet metal, chances are good that it is either caught on something or bent. It can usually be put back into place or bent back into shape easily.

Disc brake rotors with holes or slots machined into their surface dissipate heat faster (**FIGURE 46-33**). They also help to remove water quickly from the surface of the pad in wet driving conditions. Because the pads wipe across the holes or slots, the surface of the pad is prevented from becoming hard and

FIGURE 46-32 Directional ventilated rotor.

FIGURE 46-33 Slotted and drilled rotor.

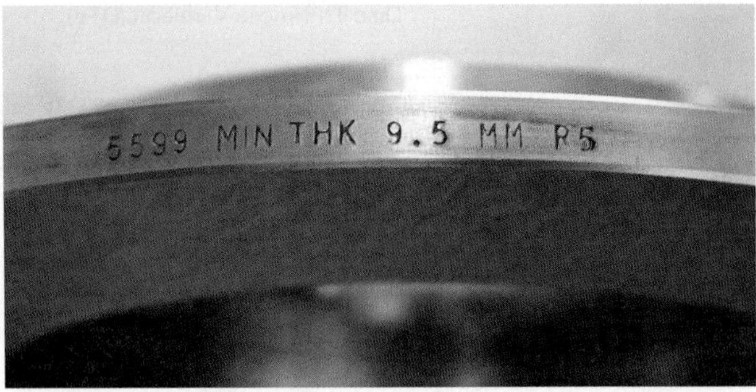

FIGURE 46-34 Rotor with minimum thickness stamped on it.

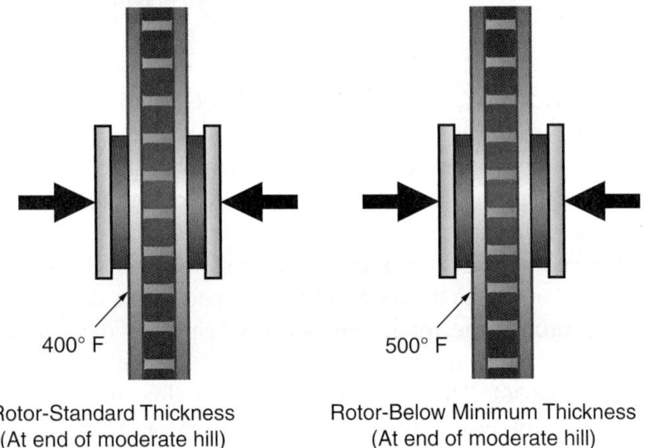

FIGURE 46-35 Rotor thickness and heat capacity.

glassy smooth from the friction and heat of use. However, this scraping action reduces the overall life of the brake pad, so these types of rotors are generally only used in high-performance or heavy-duty vehicles.

Most disc brake rotors are stamped with the manufacturer's minimum thickness specification (**FIGURE 46-34**) This minimum thickness ensures an adequate amount of thermal mass for stopping power. When material is removed, there is not as much material to absorb heat, and the rotor heats up faster (**FIGURE 46-35**). The excess heat can lead to brake fade sooner. Also, when the brake pads wear, if the thickness of the rotor were below this minimum, the piston could be pushed out beyond the edge of the sealing ring, which would cause the brakes to lose hydraulic pressure and fail. Make sure the rotors are always above the manufacturer's minimum thickness before putting them back in service.

▶ Parking Brakes

K46011

Parking brakes are designed to hold the vehicle stationary when parked. Manufacturers are required to design the vehicle so the parking brake will hold the vehicle for a given amount of time on a specified grade in both directions. The parking brake must be separately activated from the service brakes, and the driver must be able to latch it into the applied position. Parking brakes can be foot operated or hand operated. Because disc brakes require higher applied forces to operate, they are a bit more difficult to use as parking brakes. However, manufacturers have overcome this challenge in a couple of ways.

Types of Parking Brakes

Currently, most parking brakes are mechanically applied by use of a cable and ratcheting lever assembly. Parking brakes on disc brake units are primarily of two types: an integrated parking brake caliper and the top hat drum style. Alternatively, electric parking brakes are being used on some vehicles. The electric motor can either pull on a conventional parking brake cable, or the electric motor can be mounted on the caliper and directly drive the caliper piston to apply force to the brake pads.

Integrated Mechanical Parking Brake Calipers

The integrated parking brake mechanically forces the disc brake piston outward, forcing the brake pads to clamp the

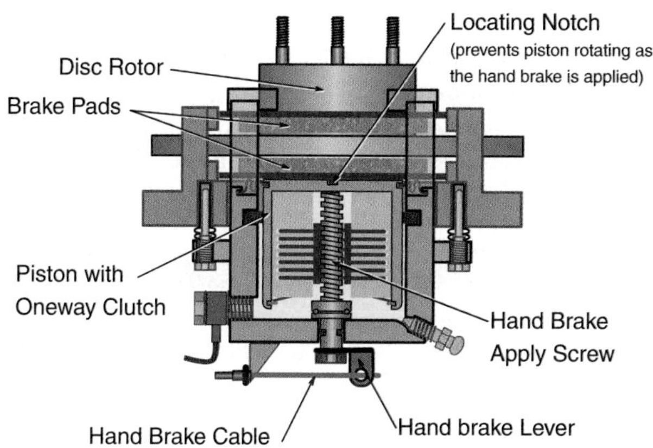

FIGURE 46-36 Integrated parking brake operation.

FIGURE 46-37 A top hat rotor and drum parking brake assembly.

rotor when the parking brake is applied (**FIGURE 46-36**). A lever on the back side of the caliper is pulled by the parking brake cable. The lever converts that motion to rotary motion on a shaft that enters the rear of the caliper cylinder. The shaft uses a seal to prevent fluid leakage from the bore. The shaft has a coarse thread machined into it, which threads into a nut assembly inside the caliper piston. As the shaft is turned by the parking brake lever, the nut causes the piston to be forced outward, applying the brakes. Releasing the parking brake cable allows a spring to unwind the shaft and release the pressure on the brake pads.

Top Hat Design Parking Brake

The top hat design gets its name from the shape of the rotor. The rotor has a deeper offset than normal, giving the appearance of a top hat. The offset portion allows room for a drum surface within the center of the rotor (**FIGURE 46-37**). Drum brake shoes are mechanically forced outward into contact with the inside of the brake drum, which locks the wheel. Releasing the parking brake allows the springs to retract the brake shoes from contact with the drums.

Electric Parking Brake

The electric parking brake uses an electric motor to apply the disc brake assemblies. The cable style uses an electric motor to pull standard parking brake cables, which apply standard integrated mechanical parking brake calipers. The electrically integrated caliper style uses an electric motor mounted on the caliper to directly apply the brakes (**FIGURE 46-38**). Pushing a parking brake button on the dash causes the motor to either tension the cable or directly apply the parking brake. Electric parking brakes also can be integrated with the controller area network bus (CANbus) system to provide additional features beyond just holding the vehicle when it is parked. It can be used to automatically hold the vehicle while it is stopped on a hill, to prevent it from rolling backward or forward. It also can be automatically released by the vehicle's **electronic control module (ECM)** when the throttle is applied for starting to move away from the stop. It also may work with the vehicle's proximity detector when backing up. If the system detects the vehicle getting too close to an object, the ECM can apply the electric parking brake to stop the vehicle and prevent it from striking the object.

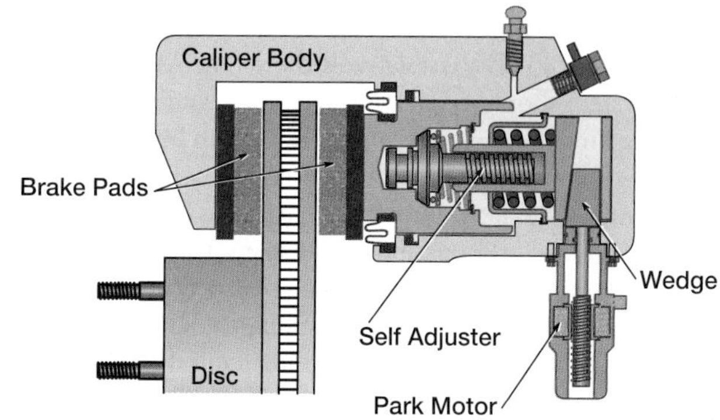

FIGURE 46-38 Electrically integrated parking brake caliper has a built-in motor that mechanically applies the brake pads.

▶ Servicing Disc Brakes

S46001

Diagnosis

N46001, N46002

Disc brake diagnosis starts with understanding the customer's concern. Communicating directly with the customer is the best way to do that, but the customer is not always available. An experienced service advisor will gather the required information, so you should read the service advisor's notes on the repair order carefully or speak with this person directly. Once you understand the customer's concern, a test drive is usually needed to verify the accuracy of the concern. Depending on the concern, it may be as simple as stepping on the brake pedal without moving the vehicle and feeling the pedal sink to the floor, or it could require a more detailed test drive to observe the fault the customer is describing. This is a good opportunity to test the brakes under a variety of conditions. Replicating the customer concern is important in order to address the situation that the customer is experiencing.

During the test drive, find a safe place to operate the brakes at a variety of speeds with a variety of brake pedal pressures, especially trying to mimic the conditions the customer described. It may be necessary to go on a test drive with the customer driving, allowing him or her to operate the vehicle in the way that makes the problem evident and to point out the particular situation he or she is experiencing. Also, it is good to have the customer along in case the problem does not occur, in which case he or she won't think you don't believe there is a problem or are ignoring the issue.

If there is a concern related to the antilock brake system (ABS) that requires a test drive, extreme caution is required so that an accident doesn't happen. Because ABS operates only during extreme braking or poor traction conditions, you run the risk of being rear-ended by a vehicle behind you or losing control of the vehicle if you apply the brakes hard enough to activate the ABS. So make sure the vehicle is being tested away from all other traffic. Remember that if the yellow ABS warning lamp is illuminated, the ABS system is deactivated, and you should not try to activate ABS on a test drive. If the yellow ABS lamp is off, the ABS system should be active and ready to activate. It is helpful when testing ABS to do so on a surface with limited traction, such as wet pavement or a dirt road. When you do apply the brakes firmly, ABS activation can typically be felt as pulsations in the brake pedal and possibly the steering wheel. The vehicle should brake quickly while maintaining steering control. In some cases, a "poor traction" lamp may illuminate, telling you that the ABS system had to activate. This warning lamp will usually turn off after several seconds. If the yellow ABS warning lamp illuminates, it typically means that the vehicle's ECM has observed a fault in the ABS system and should be checked for DTCs.

Once the customer concern has been verified, you will hopefully have enough information to research the concern in the service information and TSBs. Armed with this information, you should be ready to create a plan to begin testing the brake system. This could be as simple as performing a visual inspection of the brake fluid level and condition, removing the wheels to disassemble and inspect the brake units, or measuring the thickness variation of the rotors. It could also involve electrical diagnosis such as locating a short circuit in the brake pad warning indicator system.

Suspension and steering system faults can appear to be brake system faults. An example of this is a pulling condition while braking. If the strut rod bushings on the suspension are worn, then braking the vehicle will cause the wheel to move rearward; at the same time, that will cause the steering angle to change, causing the wheel to point in a direction other than straight down the road, imitating a brake pull. So always inspect the suspension and steering systems when diagnosing concerns related to a vehicle "pull."

The braking system on a vehicle must be restored to its proper operation if one or more braking system faults are present. Diagnosis of any problem must identify all issues that would prevent the brakes from operating normally. Lawyers and technicians have been known to say, "He who touched the brake system last, owns it!" What this means is that if there is a problem in the braking system and you inspected it or worked on it, you are very likely liable for anything that went wrong with it. Brake system failures are more likely to lead to vehicle accidents than failures of most systems. Any diagnosis and subsequent repairs need to be thorough and complete.

Once you have identified the cause of the fault, determine the action that will correct the fault. This information can then be used for an estimate of repairs and given to the customer for authorization to perform the needed work. Once the repair is complete, retest the system to confirm that the concern has been fully addressed and no further issues are present.

TECHNICIAN TIP

One way to identify whether a problem is coming from the front or rear brakes is to test-drive the vehicle in a safe place at a relatively low speed and lightly apply the parking brake. If the condition is still present, the problem is with the rear brakes, because the parking brakes are usually on the rear wheels. If the condition is not present, the problem is likely with the front brakes.

Tools

The tools that are used to diagnose and repair brake systems include those shown in **FIGURE 46-39**:

- **Brake lining thickness gauges**—Used to measure the thickness of the brake lining.
- **Brake wash station**—Used to clean drum and disc brake dust.
- **Caliper piston pliers**—Used to grip caliper pistons when removing them.
- **Disc brake rotor micrometer**—Used to measure the thickness and parallelism of a rotor.
- **Dial indicator**—Used to measure the lateral runout (side to side) of the rotor.
- **Parking brake cable pliers**—Used to install parking brake cables.
- **Caliper piston retracting tool**—Used to retract caliper pistons with integrated parking brakes.
- **C-clamp**—Used to push pistons back into the caliper bore on non-integrated parking brakes.
- **Off-car brake lathe**—Used to machine drums and rotors that are off the vehicle.
- **On-car brake lathe**—Used to machine rotors that are on the vehicle.
- **Caliper dust boot seal driver set**—Consists of a driver and a variety of adapters used to install various sizes of dust boot seals.

Common Disc Brake Concerns

N46003

To diagnose poor stopping, noise, vibration, pulling, grabbing, dragging, or pulsation concerns and determine any necessary actions, follow the steps outlined in the remainder of this chapter. Although braking concerns are generally easier to diagnose than most of the other systems on the vehicle, the large variety of conditions listed makes it imperative that you have a good understanding of disc brake and hydraulic theory. It also helps to use all of your senses to assist you in identifying the location of the fault. As a reminder, **TABLE 46-2** lists some of the common faults to consider for each concern.

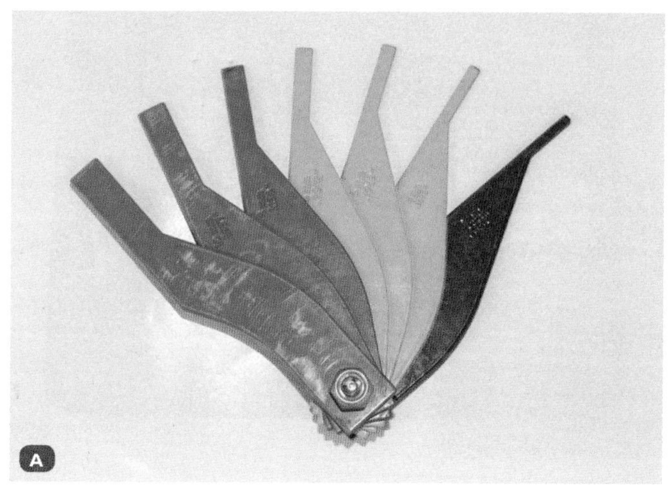

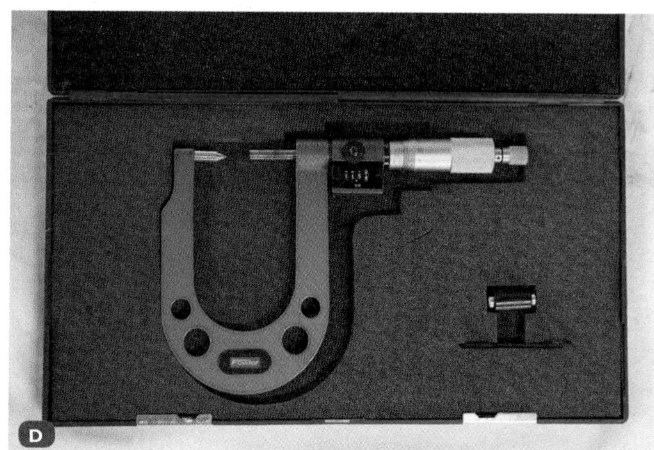

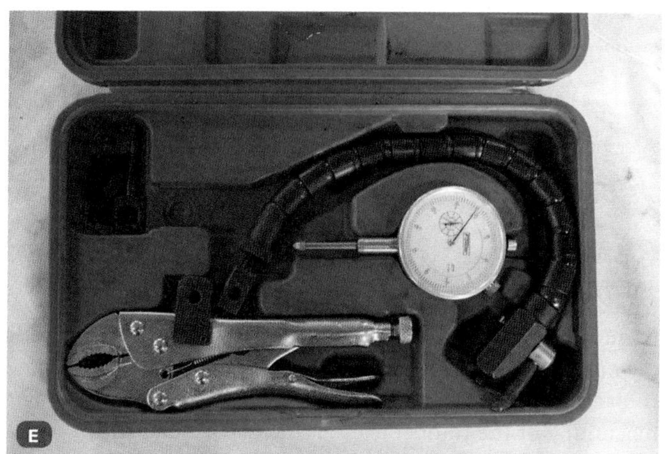

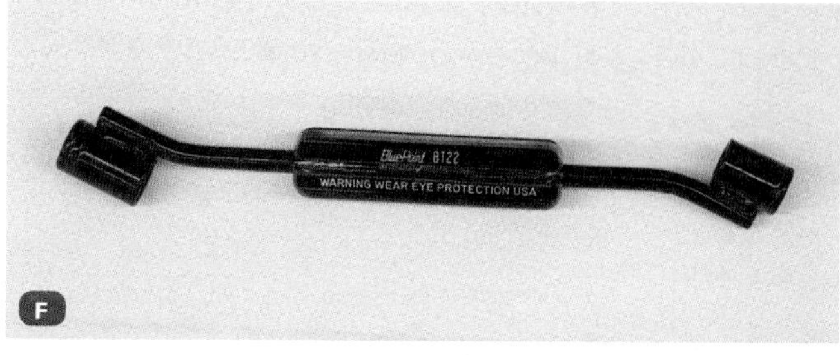

Caption appears on next page.

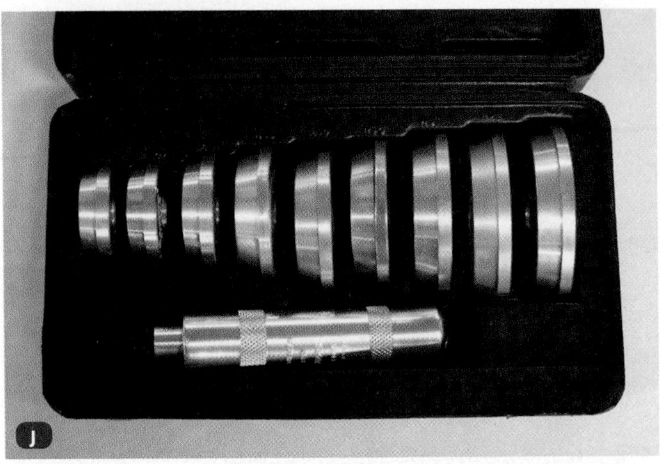

FIGURE 46-39 Disc brake tools. **A.** Brake lining thickness gauges. **B.** Brake wash station. **C.** Caliper piston pliers. **D.** Disc brake rotor micrometer. **E.** Dial indicator. **F.** Parking brake cable tool. **G.** Caliper piston retracting tool. **H.** Off-car brake lathe. **I.** On-car brake lathe. **J.** Dust boot seal/bushing driver set.

TABLE 46-2 Common Brake Issues

Concern	Fault
Poor stopping	• Power booster not operating properly
	• Internal master cylinder leak; air in the hydraulic system
	• Metering valve or proportioning valve blocking fluid flow
	• Improperly adjusted drum brakes
	• Improper friction lining material
	• Contaminated linings
Noise	• Friction lining material too hard
	• Wear indicator touching rotor (worn pads)
	• Lining worn down to metal
	• Worn caliper slides/guide pins
	• Component-specific noises
Vibration	• Improper friction lining material
	• Rotor surface finish not correct
	• Foreign object (mud, rocks, etc.) in rotor
	• Warped rotor/thickness variation excessive
	• ABS operating
Pulling	• Plugged or restricted brake hose
	• Stuck caliper piston
	• Seized caliper guide pins
	• Contaminated lining
	• Lining worn down to metal
	• Air in the hydraulic system
Grabbing	• Contaminated lining
	• Stuck caliper piston
	• Internal master cylinder leak
	• Misadjusted drum brakes
Dragging	• Stuck caliper piston
	• Seized caliper guide pins
	• Misadjusted master cylinder pushrod length
	• Binding brake pedal
	• Plugged restricted brake line or hose
Pulsation	• Warped rotors
	• Rotor parallelism
	• ABS operation

▶ Maintenance and Repair

S46002

Removing and Inspecting Calipers

N46004

Removing the caliper is necessary for replacing the brake pads on most vehicles. It also allows access for machining of the rotors on the vehicle, removal of the rotors for replacement, or machining of the rotors off the vehicle. Further, it allows for a thorough inspection of the caliper, pads, and rotor to determine the cause of a brake concern.

If the caliper is likely to be reinstalled on the vehicle, or rebuilt and reinstalled, it is good practice to loosen the bleeder screws slightly and then retighten them. Doing so ensures that they are not seized in place and that you will be able to loose them later when bleeding the system. Failure to do this now could waste a lot of time later trying to repair a bleeder screw when it breaks off. It is also good practice to flush the old brake fluid from the system at this time so that old brake fluid will not have to be bled through the new or newly rebuilt caliper. Also, leave the master cylinder reservoir level low so that pushing the caliper pistons into their bores will not overflow the reservoir.

To push the caliper pistons back into their bores slightly, using a small pry bar or C-clamp will usually force the piston back just far enough so that the caliper can be removed from the pads and rotor once the caliper is loose from its mount. To remove the caliper assembly, inspect for leaks and damage to the caliper housing, and determine any necessary actions, follow the steps in **SKILL DRILL 46-1**.

▶ TECHNICIAN TIP

Some technicians pinch off the flexible brake hoses with vise-grip pliers. This should be avoided because it crimps the hose, potentially damaging it internally and/or externally. It is a better idea to pull the brake fuse and use the brake pedal holding tool to block off the master cylinder compensating ports, to prevent brake fluid from leaking out of the hose.

Inspecting Caliper Mountings, Slides, and Pins

N46005

The caliper mountings and slides/pins are placed under heavy loads and forces, and they also operate in harsh environments. Therefore, it is common that they experience wear over time. They can also corrode and bind up. Clean caliper mountings and slides/pins thoroughly, and inspect them closely for excessive damage or wear.

To clean and inspect caliper mountings and slides/pins for operation, wear, and damage, and to determine any necessary actions, follow the steps in **SKILL DRILL 46-2**.

SKILL DRILL 46-1 Removing and Inspecting Calipers

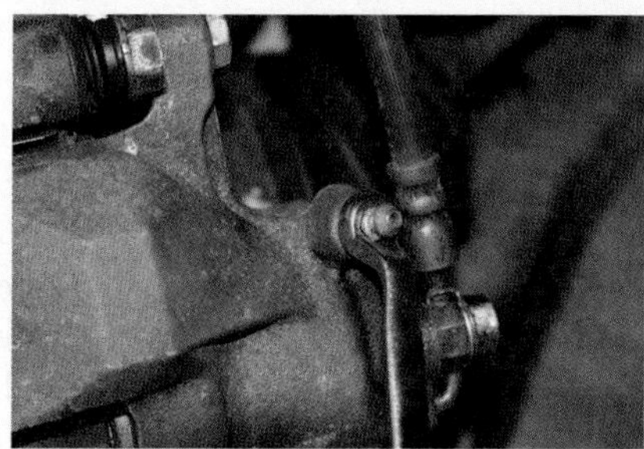

1. Research the procedure for removing the caliper in the appropriate service information. Loosen the bleeder screws slightly, and then retighten them.

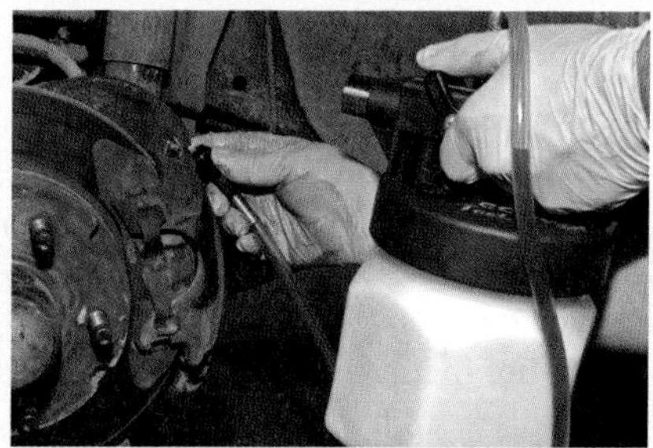

2. If the caliper is being rebuilt or a new caliper will be installed, it is good practice to flush the old brake fluid from the system at this time.

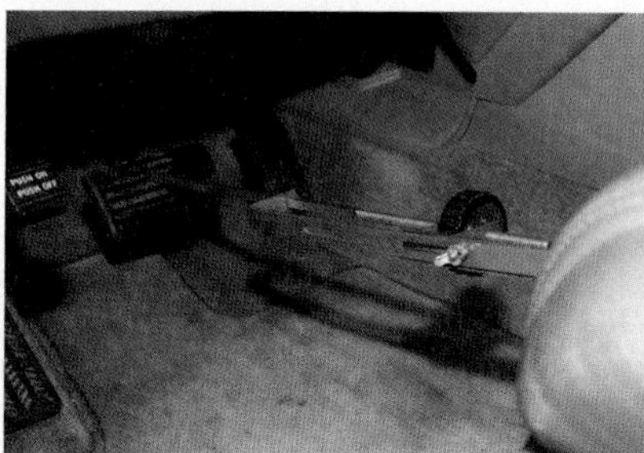

3. Use a brake pedal holding tool to slightly apply the brakes and block off the compensating ports in the master cylinder to avoid excess fluid leakage.

4. Remove the brake line or hose from the caliper. Be careful not to lose any sealing rings.

5. Push the caliper pistons back into their bores slightly. Although many technicians use a screwdriver, as shown, a pry bar or C-clamp is a safer choice.

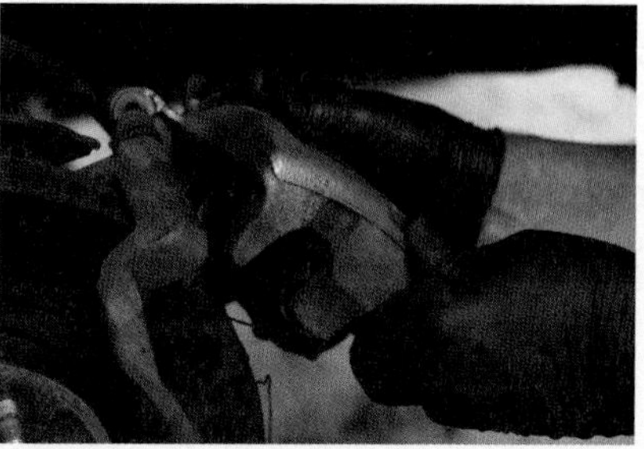

6. Remove the caliper assembly from its mountings.

7. Inspect the caliper, including the piston dust boot for leaks or damage. Determine any necessary actions.

SKILL DRILL 46-2 Inspecting Caliper Mountings, Slides, and Pins

1. Clean the caliper mountings and slides/pins, using equipment/procedures for dealing with asbestos/hazardous dust.

2. Once the dust is taken care of, you may need to use a brake cleaning solvent to clean the components further.

3. Inspect the caliper mountings and slides/pins for wear and damage. Determine any necessary actions.

Inspecting Brake Pads and Wear Indicators

N46006

The wear indicator system could be a scratcher type or a sensor type. It is also very possible that the vehicle does not incorporate any type of wear indicator system. Checking this system usually consists of verifying that the sensor or scratcher is not contacting or nearly contacting the brake rotor. Some manufacturers require that the harness side of the brake pad sensor connector be grounded or jumped with a test lead and verify that the brake pad warning lamp or warning message come on.

To check the operation of the brake pad wear indicator system and determine any necessary actions, follow the steps in **SKILL DRILL 46-3**.

Checking Brake Pads

N46007

During a routine maintenance inspection where there are no customer concerns regarding the brakes, a simple visual inspection along with a measurement of the lining thickness is usually adequate. But if there are customer concerns with the brakes, it may be necessary to remove the pads for a more detailed inspection. Removing the brake pads is the most thorough way to inspect them because it will allow you to inspect not only the thickness of the lining but also the condition of the friction surface. This could show cracks in the lining or other defects that wouldn't be otherwise seen from the outside. It also allows for a more thorough inspection of the retaining hardware. When removing pads, pay attention to the way the brake pads come off, as some pads have slight differences (such as locating nubs) that make it easy to install them incorrectly. To remove, inspect, and replace pads and retaining hardware, and to determine any necessary actions, follow the steps in **SKILL DRILL 46-4**.

Disassembling Calipers

Calipers are disassembled and cleaned for a couple of reasons. The first is to diagnose a brake system concern related to one or more calipers. For example, if the vehicle has a brake pull, it could be caused by a sticking caliper piston.

SKILL DRILL 46-3 Inspecting Brake Pads and Wear Indicators

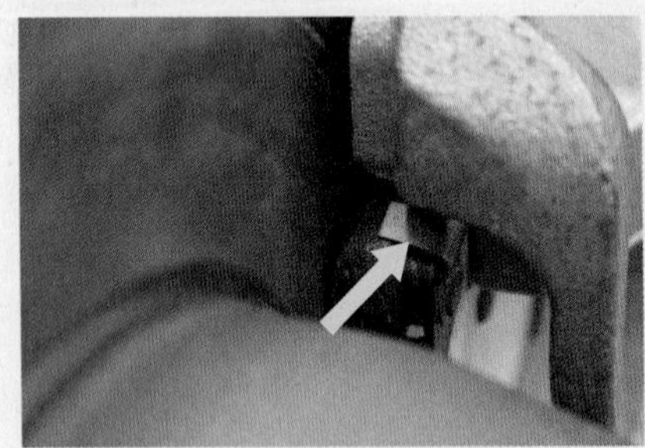

1. Determine the type of wear indicator system utilized. Check that it is not contacting or nearly contacting the rotor.

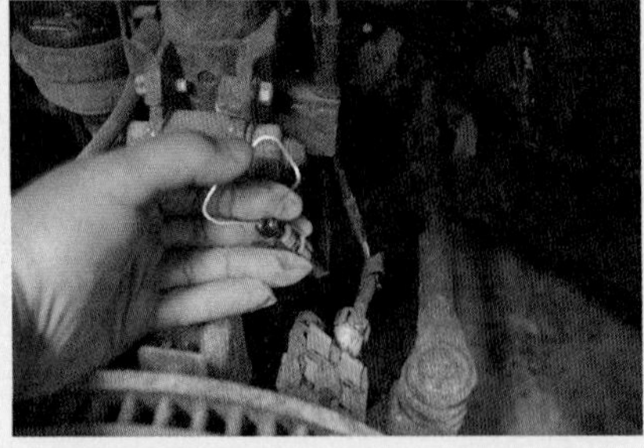

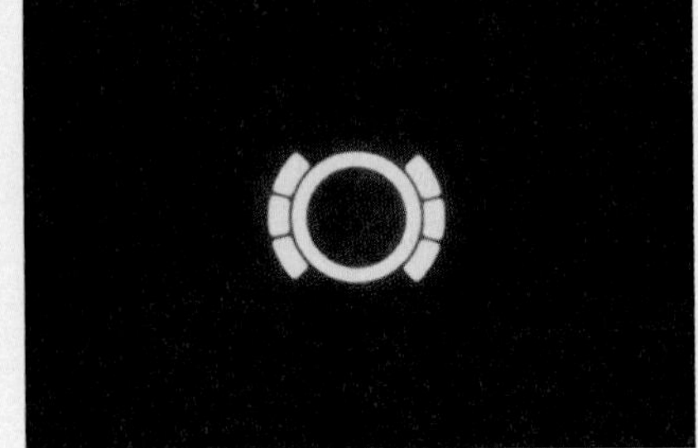

2. If the system is a sensor style, test the system to verify that the system is operational. Determine any necessary actions.

SKILL DRILL 46-4 Checking Brake Pads

1. Remove the pads and retaining hardware.

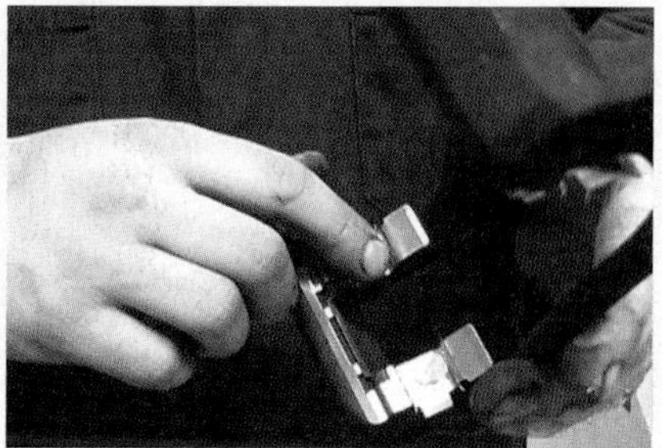
2. Inspect all pads, retaining hardware, and antinoise shims for wear or damage.

3. Measure the remaining brake pad thickness, and compare to specifications. Determine any necessary action(s).

Disassembling the caliper allows you to verify whether that is the case. They also need to be disassembled and cleaned if they are going to be rebuilt. Some shops rebuild the calipers themselves, but most shops just replace them with rebuilt calipers if necessary.

Disc brake calipers are usually side specific, meaning they are designed to be installed on a particular side of the vehicle. Failure to install them on the correct side usually results in the bleeder screws being on the bottom of the caliper cylinder. This prevents air from being bled from the caliper and results in a very spongy brake pedal.

When disassembling a caliper, compressed air is typically used to remove the piston. Be very careful, as the piston can shoot out with great force, enough to break finger bones or pinch fingers off. Always use an approved wood or heavy cardboard cushion between the caliper piston and caliper housing. Keep your fingers away from the area.

Clean all of the caliper parts according to the service manual procedure. Make sure the sealing ring groove is completely clean. This can be performed by scraping the groove with a variety of pick tools, such as dental picks, and then wiping it out with a rag.

▶ TECHNICIAN TIP

Because you have the rotor exposed, skip ahead to the sections on inspecting and servicing rotors, and complete the tasks listed there. This will save you time and effort. Pick up here once you are done with the rotor.

To disassemble and clean the caliper assembly; inspect parts for wear, rust, scoring, and damage; and replace the seal, boot, and damaged or worn parts, follow the steps in **SKILL DRILL 46-5**.

SKILL DRILL 46-5 Disassembling Calipers

1. Disassemble the caliper, following the service manual procedure.

2. Clean all of the caliper parts, including the seal grooves, following the service manual procedure.

SKILL DRILL 46-5 Disassembling Calipers (Continued)

3. Inspect each of the parts for damage, rust, and wear. Also check the caliper pin bores or bushings for wear or damage. Replace if they cannot be cleaned up.

4. Measure the caliper bore-to-piston clearance with a feeler gauge, and compare to specifications. Determine any necessary actions.

Reassembling Caliper and Pad Assembly

N46008

Reassembling the caliper requires patience and attention to detail. Ensure that the sealing ring groove is spotless and that the O-ring gets seated fully in the groove. Use clean brake fluid or approved caliper piston assembly lube on the piston and sealing ring prior to installing it. Be careful not to pinch, twist, or cut the sealing ring and dust boot. Also check the service manual to see when the piston dust boot needs to be installed. Some calipers require the dust boot to be installed in the caliper before installing the piston. In this case, a special technique using air pressure to "balloon" the dust boot is required so the piston can slip inside it. This takes a lot of experience, and it is easy to cut and bruise your fingers or hand. Ask your supervisor to demonstrate this technique. On most calipers, the seal can be installed after the piston is installed. This is usually much easier to do.

When installing shims, springs, and clips, make sure you position them properly. Most of these parts can be installed in a variety of ways, many of which will cause rubbing, wear, noise, or spongy pedal. So to avoid this issue, pay attention to how they came off. Many times you can look at the wear patterns to determine the position the parts were originally in. Once all brakes have been assembled, seat the pads by applying the brake pedal several times. It is a good idea to place your left foot under the brake pedal so that when applying the pedal with your right foot, the pedal does not push the master cylinder pistons farther than normal into the master cylinder bore, which could dislodge sludge or cut the lips of the master cylinder primary seals. Applying the brake forces the brake caliper pistons to adjust to the proper clearance for proper brake application. You may need to start the vehicle to enable the power booster to help you fully apply the brakes, especially if the vehicle is equipped with integrated parking brake calipers.

To reassemble, lubricate, and reinstall the caliper, pads, and related hardware and the seat pads, and to inspect for leaks, follow the steps in **SKILL DRILL 46-6**.

▶ TECHNICIAN TIP

In some cases, the piston cannot be removed with compressed air because it is seized in the bore. If this happens, reinstall the calipers on the vehicle, without the pads; bleed the brakes; and use the brake pedal to force the stuck piston out of the caliper. You may have to block any non-seized pistons so they don't pop out; that way, just the seized piston gets pushed out.

Retracting and Readjusting Pistons on an Integrated Parking Brake

N46009

Retracting the caliper piston on an integrated parking brake system is different than on a standard caliper. Because the integrated parking brake system uses a threaded shaft to force the piston outward from the caliper bore, it cannot just be retracted with a C-clamp. The piston has to be screwed back in on the threaded shaft to retract it into the bore. This is accomplished by using a tool that mates to slots, grooves, or holes in the outer face of the piston. The tool is then turned by hand or wrench to screw the piston back into the bore. Several types of tools work for this purpose. One is a multisided cube with a variety of projections of varying configurations on each side. This cube fits on the end of a ratchet. The ratchet and tool must be held tightly against the piston so it does not slip. There are other more application-specific tools that use the caliper housing to keep the tool from slipping out of the holes in the piston. This style works best if you have access to one, and is shown in the

SKILL DRILL 46-6 Reassembling Calipers

1. Make sure the rotor has been properly installed on the hub/spindle and that the hub surface is free of rust or dirt.

2. Install the piston seal, using brake fluid or brake assembly fluid.

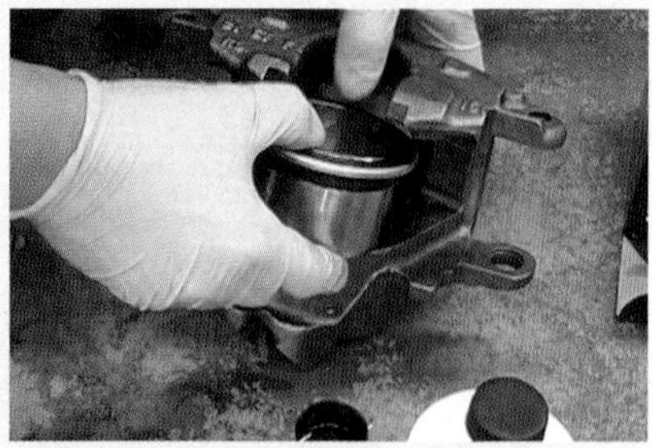

3. Install the piston by hand, being careful not to pinch, twist, or cut the sealing ring and dust boot.

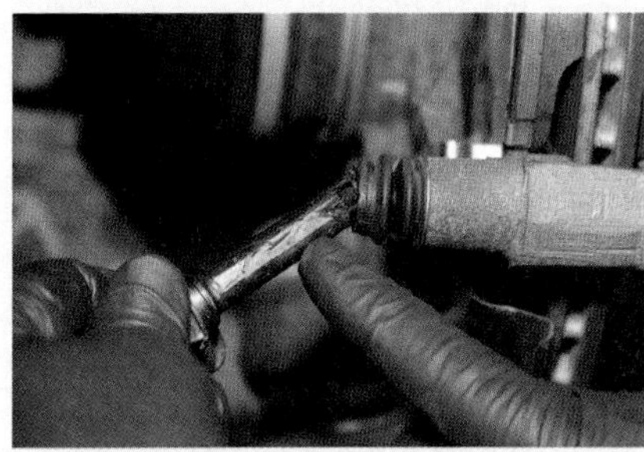

4. Assemble the pads, hardware, and caliper on the caliper mountings, using the specified lubricant. Lubricate all moving parts.

5. Reinstall the brake line fittings, using two new copper washers (if the fittings are so equipped).

6. Tighten the brake line fitting and caliper bolts to the proper torque. Bleed the brakes, following the manufacturer's procedure.

7. Seat the pads by applying the brake pedal several times, not allowing the pedal to go all the way to the floor.

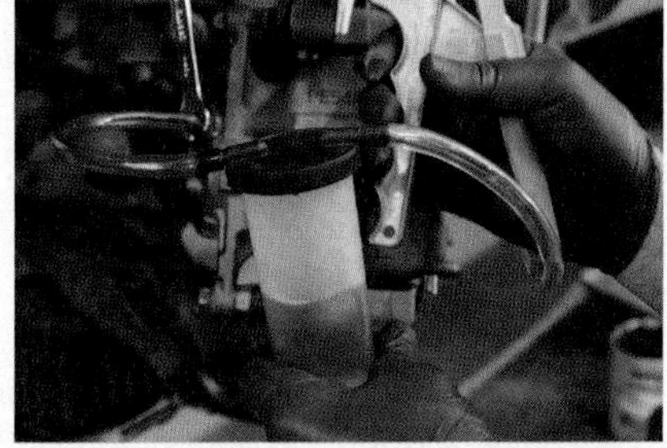

8. If the brake pedal is spongy, you need to bleed the brakes of any remaining trapped air in the system.

9. Inspect the system for any brake fluid leaks, no matter how small.

following Skill Drill. Be careful not to tear the piston dust boot while twisting the piston in.

To retract the caliper piston on an integrated parking brake system, follow the steps in **SKILL DRILL 46-7**.

Inspecting and Measuring Disc Brake Rotors

N46010, S46003

The rotor thickness, lateral runout, and thickness variation must be within specifications for the rotor to function properly. Rotors that are too thin cannot handle as much heat and experience brake fade sooner than thicker rotors do. They also may cause the piston to be pushed out of the caliper bore far enough that the sealing ring no longer seals the piston. This leads to a lack of braking action on at least half of the system.

Excessive thickness variation causes brake pedal pulsation and the vehicle to have a surging feeling while coming to a stop. Excessive lateral runout tends to cause the steering wheel to shimmy. It also can cause an excessive thickness variation problem due to the high spot of the rotor continuously

SKILL DRILL 46-7 Retracting and Readjusting Pistons on an Integrated Parking Brake

1. Research the procedure for retracting the caliper piston. Select the proper adapter or tool to match the caliper piston.

2. Install the tool, and turn it in the direction that causes the piston to retract.

3. Continue turning until the piston is lightly seated at the bottom of its bore.

4. Make sure the dust boot is seated properly in its grooves.

hitting the brake pad while driving down the road. This constant rubbing on the high spot of the rotor wears it slightly, leading to excessive thickness variation across the face of the rotor.

When measuring the rotor for minimum thickness, measure at the deepest groove or thinnest part of the rotor, and compare to specifications. If it is under the minimum thickness, it will have to be replaced. If it needs to be machined and it is above the minimum thickness, check to see how badly it is scored. Remember that removing 0.015" (0.38 mm) on each side of the rotor results in the thickness being reduced by 0.030" (0.76 mm). Many rotors only start with 0.060" (1.52 mm) of machinable material when they are new.

When measuring thickness variation, measure the thickness of the rotor in five to eight places around the face of the rotor. Calculate the maximum thickness variation by subtracting the minimum thickness from the maximum thickness, and compare it to specifications.

To clean, inspect, and measure rotor thickness, lateral runout, and thickness variation and to determine any necessary action(s), follow the steps in **SKILL DRILL 46-8**.

Removing and Reinstalling Rotors

N46011

Removing the rotor is usually required when the rotor needs to be replaced because it is under the specified minimum thickness or would be after machining. It also would need to be removed to be refinished on an off-car brake lathe or to service the wheel bearings or axle shaft. Also, hubless rotors that are being machined with an on-car brake lathe need to be removed to clean off the rust and dirt accumulated between the rotor and the hub.

The hub style has the wheel bearing hub cast into the rotor. This style generally requires disassembly of the wheel bearing hub to remove the rotor from the vehicle. The hubless style uses a wheel bearing hub separate from the rotor, with the rotor held onto the hub by the wheel studs and lug nuts. Some hubs also

SKILL DRILL 46-8 Inspecting and Measuring Disc Brake Rotors

1. Research the procedure and specifications for inspecting the rotor. If you have not already done so, remove the caliper assembly, brake pads, and any hardware. Clean the rotor with approved asbestos removal equipment.

2. Inspect the rotor for hard spots or hot spots, scoring, cracks, and damage.

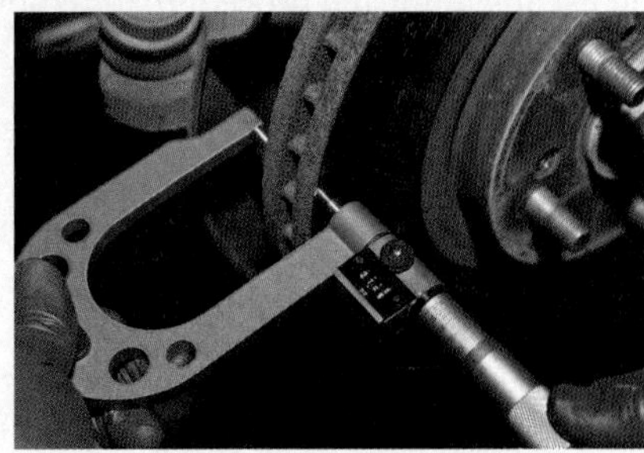

3. Measure the rotor thickness at the deepest groove or thinnest part of the rotor and compare to specifications.

4. Measure the thickness of the rotor in five to eight places around the face of the rotor. Calculate the maximum thickness variation, and compare to specifications.

5. Set up a dial indicator to measure lateral runout. Rotate the rotor and find the lowest spot on the rotor; then zero the dial indicator.

6. Slowly rotate the rotor to find the highest spot on the rotor. Read the dial indicator showing maximum runout.

7. Keep turning the rotor to make sure the dial indicator does not read below zero. If it does, re-zero the dial caliper on the lowest spot. Keep turning the rotor to find the highest spot, and reread the dial indicator. Compare all of your readings to the specifications and determine if the rotor is fit for service, is machinable, or needs to be replaced.

use small screws to hold the rotor on the hub. Removing hubless rotors is generally easier than removing hub-style rotors, although some manufacturers design their rotors to unbolt from the rear side of the bearing hub. In these applications, the wheel bearing hub must be removed before the rotor can be removed from the hub.

When removing hubless rotors, mark the rotor so that it can be reinstalled in the same position. A permanent marker, crayon, or center punch can be used to do this. When removing a hub-style rotor, you need to remove the wheel bearings. First remove the wheel bearing locking mechanism (cotter pin, locknut, or peened washer). Remove the wheel bearing adjusting nut, thrust washer, and outer bearing. If the rear bearing must be removed for service or replacement of the rotor, use the following procedure to remove the inner bearing and grease seal.

- Reinstall the adjusting nut onto the spindle about five turns.
- Grasp the rotor on the top and bottom, and push it toward the center of the vehicle.
- Hold slight downward pressure as you firmly pull the rotor toward yourself. This should cause the adjusting nut to catch the wheel bearing and to pull it and the seal out of the rear of the hub.
- The grease seal will have to be replaced with a new one.
- Now would be a good time to perform any other rotor-related tasks, such as inspecting and measuring a rotor, refinishing a rotor, and servicing wheel bearings. Perform those tasks, and return here to reinstall the rotor.
- If a new rotor is being installed, make sure to clean off any anticorrosion coating it was shipped with. Follow the rotor manufacturer's procedure to remove this coating.

To remove and reinstall the rotor, follow the steps in **SKILL DRILL 46-9**.

SKILL DRILL 46-9 Removing and Reinstalling Rotor

1. Research the procedure for removing and reinstalling the brake rotor. If you have not already done so, remove the caliper assembly, brake pads, and any hardware. If the caliper mount straddles the rotor, remove it.

2. To remove the hubless-style rotor, mark the rotor for proper reinstallation.

3. If there are screws or speed nuts holding the rotor to the hub, remove them.

4. Remove the rotor from the hub.

5. To remove the hub-style rotor, remove the wheel bearing locking mechanism.

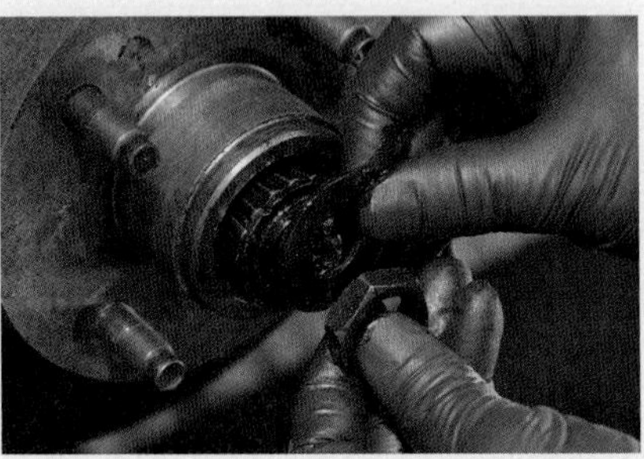

6. Remove the wheel bearing adjusting nut, thrust washer, and outer bearing.

7. Follow the steps of the inner bearing and grease seal removal procedure if the rear bearing must be removed.

8. To reinstall a hubless-style rotor, clean all mounting surfaces on the hub and rotor, and remove any burrs.

9. Slip the rotor over the wheel studs.

SKILL DRILL 46-9 Removing and Reinstalling Rotor (Continued)

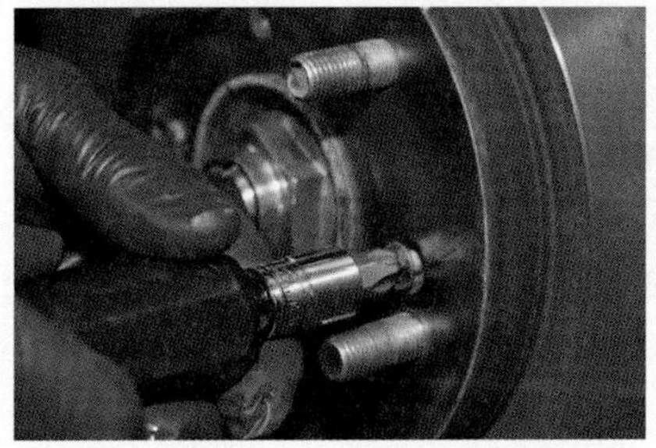

10. If the rotor uses small screws to hold the rotor to the hub, reinstall those and tighten to the proper torque.

11. Spin the rotor to ensure it spins true and does not contact any other components such as the dust shield.

12. To reinstall a hub-style rotor, see the wheel bearing service Skill Drills in the Wheel Bearing chapter. Be sure to install the locking device. Spin the rotor and listen for contact.

Refinishing Rotors on Vehicle

N46012

Rotors need to be refinished when they have excessive runout, thickness variation, or grooving. A brake lathe refinishes the rotor surfaces by removing metal and truing the surfaces. If the grooving or surface defects are too great, the rotor may require the removal of too much metal to satisfactorily refinish the surfaces. The rotor thickness should always be remeasured once the refinishing is complete, to ensure it is above the manufacturer's minimum thickness.

Rotors can be refinished while on the vehicle or off the vehicle. On-vehicle refinishing is preferred by most manufacturers (those who allow refinishing of their rotors) because it minimizes runout issues between the hub and rotor. Because the rotor is being machined as it is mounted on the vehicle, it is being refinished true to the hub and other brake components. This minimizes any lateral runout issues. Because the rotor is still on the vehicle, the brake lathe drives the rotor during the refinishing process. This requires using the proper adapter and mounting it to the lug studs.

Before refinishing can begin, you need to verify that mating surfaces of the hubless rotor and the hub are free from any dirt, rust, and debris. On hub-style rotors, you need to adjust the wheel bearings so that there isn't any end play. Just remember to readjust them after refinishing is complete. When making a cut, set the cutting bits to the proper cutting depth for machining. This is usually between 0.004" and 0.015". Too little removal tends to overheat the edge of the cutting bit. Too much can overload the lathe, as well as make the rotors much thinner.

Many newer machines use an elliptical motion to give a nondirectional finish, so a finish cut or separate nondirectional finish is not needed on these machines. If using an older machine, you may need to give the rotor a nondirectional finish with sandpaper or a special tool.

To refinish a rotor while it is on the vehicle and to measure final rotor thickness, follow the steps in **SKILL DRILL 46-10**.

SKILL DRILL 46-10 Refinishing Hubless Rotors on Vehicle

1. Research the brake lathe manufacturer's procedure for properly refinishing the rotor. Mount the on-car brake lathe to the rotor after cleaning the rust and dirt from between the rotor and hub or adjusting the wheel bearing so there is no end play.

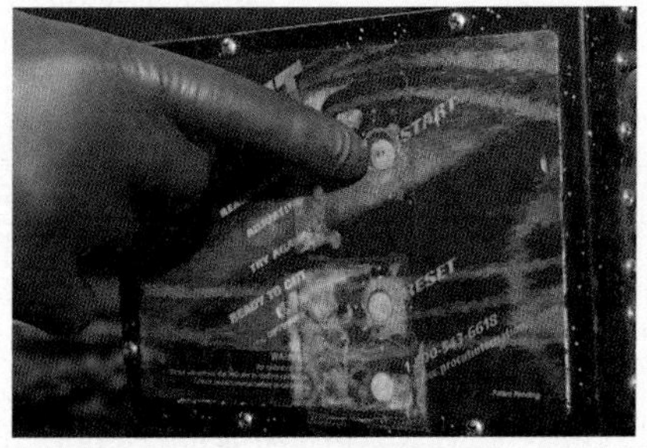

2. Perform the runout calibration on the brake lathe. Some brake lathes require manual compensation, whereas other machines can perform this automatically.

3. Adjust the cutting bits, and cut off any lip at the edge of the rotor.

SKILL DRILL 46-10 Refinishing Hubless Rotors on Vehicle (Continued)

4. Make sure the cutting bits will not contact the rotor face, and move the cutting head toward the inner diameter of the rotor face. Set the cutting bits to the proper cutting depth for machining.

5. Install the antichatter device, if specified.

6. Engage the automatic feed, and watch for proper machining action. If necessary, repeat this step until all damaged surface areas have been removed on both sides of the rotor.

7. If necessary, perform a finish cut on the rotor.

8. Remeasure the rotor thickness to determine if the rotor is above minimum thickness specifications. Readjust the wheel bearings if necessary.

Refinishing Rotors Off Vehicle

N46013

Refinishing a rotor while it is off the vehicle is a bit different than on-vehicle refinishing. The major difference is in the setup. Hub-style and hubless-style rotors each require their own way of being mounted on the brake lathe. Most hub-style rotors use the **bearing races** to drive and center the rotor on the lathe spindle. Thus, bearing adapters of the proper size have to be selected and used. The spindle nut then clamps the rotor onto the spindle through these bearing adapters and races. Hubless rotors are centered using a spring-loaded centering cone to align the rotor's centering hole with the spindle. Clam shell clamps are then used on each side of the rotor to clamp it to the lathe spindle. Composite rotors use a special adapter, which drives the rotor from the center hole while clamping it firmly between solid plates.

To refinish a rotor while it is off the vehicle and measure final rotor thickness, follow the steps in **SKILL DRILL 46-11**.

Inspecting and Replacing Wheel Studs

N46014

Wheel studs have to be replaced when they have been damaged or broken off due to improper installation or normal wear and tear. Wheel studs can be damaged by being over-tightened and stretched. Look for a necked-down or thinned-out section of the stud, which is most likely to happen within the threaded area of the stud. Stretched studs must be replaced. Studs can also have their threads damaged by cross-threading or seizing of the lug nut on the stud. It is best to replace the stud and nut if this occurs. Lastly, wheel studs may even break off if they are over-tightened beyond their stretch point. It is always a good idea to consider replacing all of the studs on a wheel (and maybe the ones on the other wheels also) if one stud is broken off, as it is likely that all of the others have been weakened.

Some vehicles are designed to allow for the removal and replacement of the wheel studs while the hub and wheel flange are still installed on the vehicle. The manufacturer may have provided a recessed spot in the steering knuckle where the studs have enough clearance to be removed; thus, the flange has to be positioned in that particular position. Other vehicles do not have enough clearance on the back side for the stud to be removed; in these cases, the hub and flange must be removed from the vehicle.

There are two primary methods of replacing lug studs: the **drawing-in method** and the **hydraulic press method**. The drawing-in method uses the lug nut to draw the wheel stud into the flange. It is accomplished by inserting the new stud

SKILL DRILL 46-11 Refinishing Rotors Off Vehicle

1. Research the brake lathe manufacturer's procedure for properly refinishing the rotor. Clean nicks, burrs, or debris from the mounting surfaces of the rotor, including the centering hole.

2. Mount the rotor. Check that the rotor is running true on the lathe.

3. Install the antichatter band or antichatter pucks on the rotor.

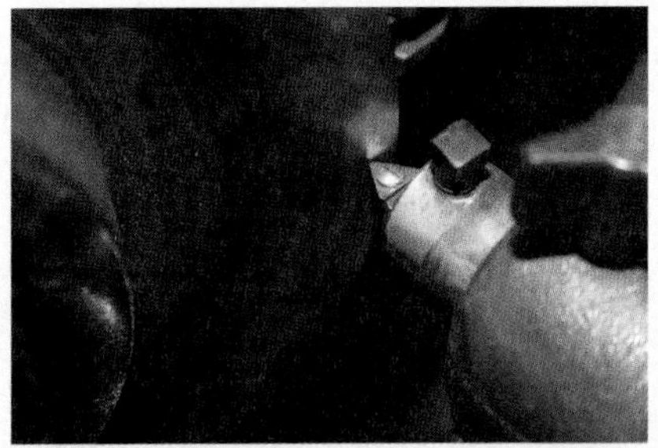

4. Position the cutting head about one-quarter the way in from the outer diameter of the rotor. Turn on the lathe.

5. Set the cutting bits to the proper cutting depth for machining the ridge.

6. By hand, move the cutting head outward toward the ridge. Slowly remove the ridge.

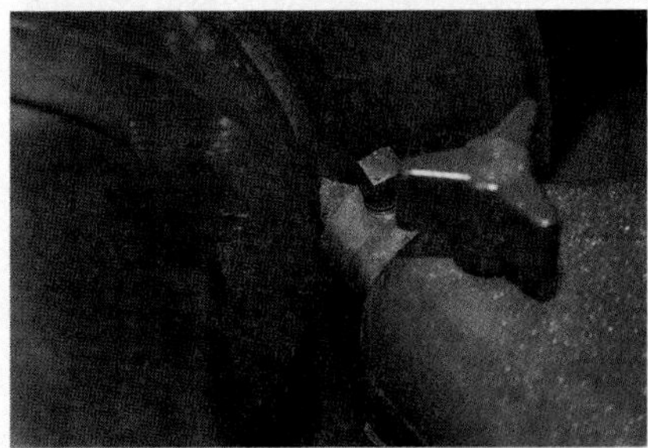

7. Once the ridge is removed, run the cutting head all the way in to the inner face of the rotor.

8. Set the cutting bits to the proper cutting depth for machining the rotor face.

9. Engage the automatic feed on "fast" cut, and watch for proper machining action. Some single-cut machines only have one cutting speed. If necessary, repeat this step until all damaged surface areas have been removed on both sides of the rotor.

10. If necessary, perform a finish cut on the rotor. This is usually done on "slow" speed.

11. Use sandpaper or a drill with a sanding pad to give the rotor faces a nondirectional finish.

12. Remeasure the rotor thickness to determine if the rotor is above minimum thickness specifications. Wash the machined rotor in a hot, soapy water solution or parts washing cabinet to remove any metal particles, and dry.

into the wheel stud hole in the flange and installing enough heavy-duty washers over the stud to allow the lug nut to draw the wheel stud into the flange when tightened. The lug nut is placed on the stud, flat side in, and tightened. Make sure that as the stud is pulled in, the lug nut does not run out of threads on the stud. If it does, remove the nut and add washers. Keep tightening the lug nut until the wheel stud bottoms out in the flange. Verify that the head of the stud is fully seated in the flange.

The hydraulic press method uses a press to force the wheel stud into the flange until it bottoms out. This method requires the hub and flange to be removed from the vehicle. Make sure the flange is positioned and supported properly on the press table. Sometimes it is best to use a short piece of pipe (a bit longer than the wheel stud) to support the flange while the stud is being pressed in. Once the stud is installed, verify that it is fully seated in the flange.

To inspect and replace wheel studs, follow the steps in **SKILL DRILL 46-12**.

Installing Wheels, Torquing Lug Nuts, and Making Final Checks

N46015, N46016

This step, although fairly simple, can result in problems if it is not done properly. Over-tightening the lug nuts can cause the wheel studs to break either immediately or, worse, after the vehicle has been driven for a period of time. Over-tightening also can cause warpage of the rotors, which results in the need to refinish or replace them. Under-tightening can lead to loosening of the lug nuts and result in the wheel working its way off the vehicle. This can cause the driver to lose control of the vehicle and potentially result in an accident. Lug nuts should be tightened to the proper torque, in the specified sequence. All manufacturers specify the sequences for each of their vehicles (**FIGURE 46-40**). The torque pattern is usually in some form of a star or cross.

Be careful which way you install the lug nuts. Many wheels use a tapered hole that matches the tapered end of the lug nut

SKILL DRILL 46-12 Inspecting and Replacing Wheel Studs

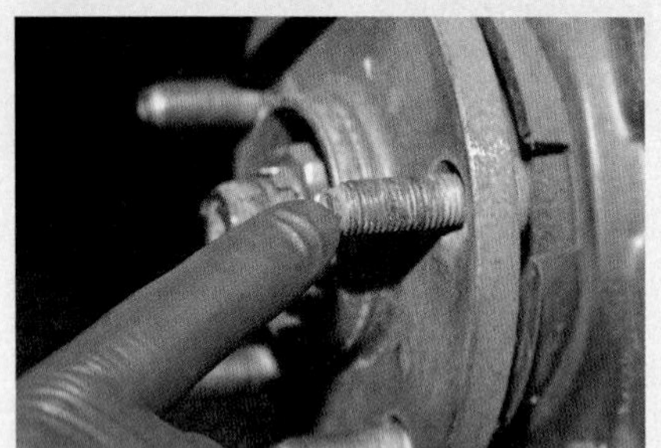

1. Inspect the wheel studs and lug nuts for damage. Look for signs of stretched studs, cross-threaded lug nuts, and broken-off studs. Determine whether the stud can be removed with the flange on or off the vehicle.

2. To perform the drawing-in method, position the flange so the stud has clearance on the back side to be removed.

3. Remove any damaged studs with a hammer. Be careful not to damage any of the surfaces on the flange and hub, including the wheel speed sensor and tone ring.

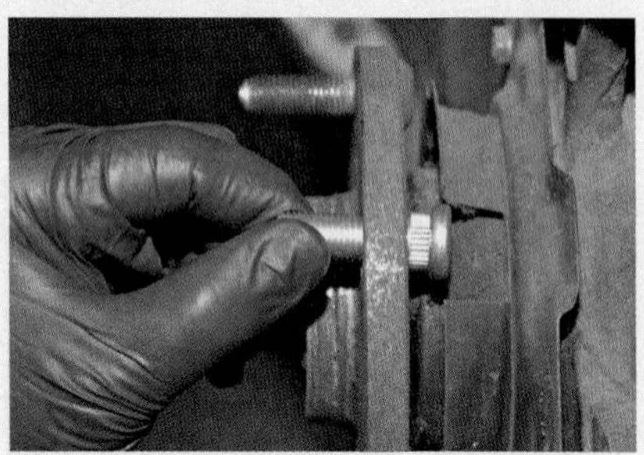

4. Insert the stud in the hole in the flange, and rotate it so that all of the flutes on the stud line up with the notches in the flange.

5. Place enough heavy-duty washers over the stud to prevent the lug nut from bottoming out on the threads.

6. Place the lug nut onto the stud, flat side in. Tighten it until the stud bottoms out in the flange. Inspect the threads on the stud and lug nut to make sure they did not get damaged.

SKILL DRILL 46-12 Inspecting and Replacing Wheel Studs (Continued)

7. To perform the hydraulic press method, remove the hub and wheel flange from the vehicle following the manufacturer's procedure.

8. Use the press to push out any damaged lug studs.

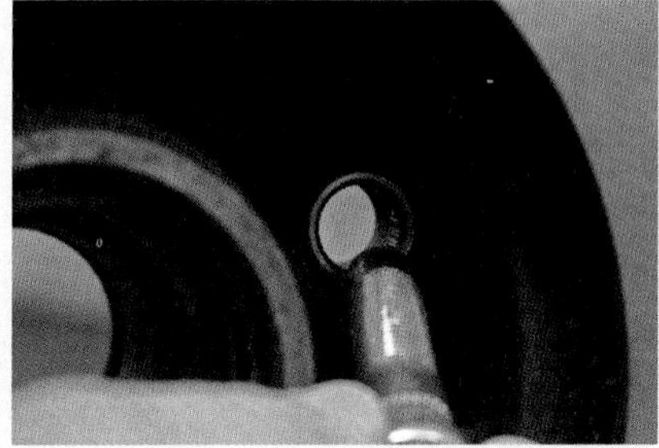

9. Insert the new lug stud into the wheel flange hole. Line up the flutes on the stud with the notches in the flange.

10. Support the hub and flange so the press is pushing the stud straight into the flange.

11. Push the stud in until it bottoms out in the flange.

12. To complete the drawing-in method and the hydraulic press method, verify that each stud is fully seated in the flange.

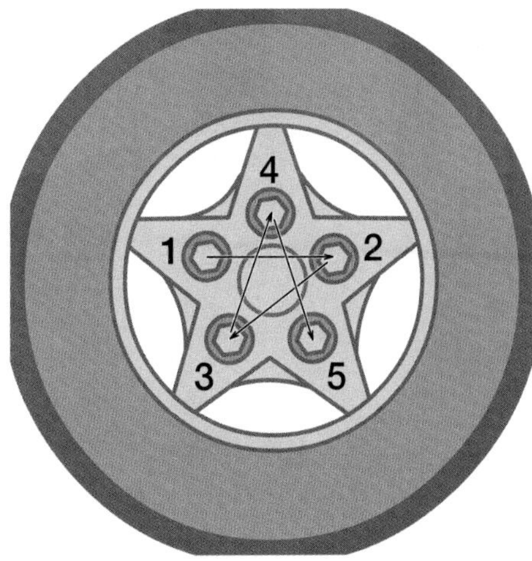

FIGURE 46-40 Common torque sequences for lug nuts.

and centers the wheel on the wheel flange. Other wheels use a flat surface that matches flat surfaces on the lug nuts. But no matter what, the lug nut surface *must* match the mating surface of the wheel. Always check that these surfaces match.

When installing lug nuts, the weight should be off the vehicle so that the wheel is off the ground or is barely touching the ground. The lug nuts should easily center the wheel on the hub. This is especially important on aluminum wheels that use a flat lug nut seating surface. The lug nuts can dig into the sides of the lug nut holes and cause the wheel to not center properly. Once the lug nuts are against the wheel, work the wheel onto the lug nut shafts. Then you can torque them down properly. Once the lug nuts are torqued properly and the brake system checked, it is time for a test drive to verify proper brake operation, and to burnish the new brake pads and rotor surfaces. Burnishing, also called bedding in, is the process of transferring pad material onto the rotor evenly, as well as cooking off the resins that are used to bind the friction material together in the pad. Burnishing results in long, quiet brake life, which results in satisfied customers.

For burnishing to happen properly, the rotor and pad material should be heated slowly and evenly. This is done by making a specified number of stops from a specified speed, with the appropriate wait times between stops. In some cases, brake lining manufacturers want you to perform a series of stops at light to moderate brake application. Other manufacturers specify a series of moderate to heavy stops. Always check the manufacturer's procedure to burnish the brakes once the brake job is complete. This process will help avoid the dreaded disc brake squeal.

> ▶ TECHNICIAN TIP
>
> There is some controversy regarding the use of a lubricant or antiseize on wheel studs and lug nuts. Because the purpose of a lubricant or antiseize is to prevent the components from sticking, CDX errs on the side of not using those products, for fear of the lug nuts coming loose. Also, the torque given for the lug nuts is dry (no lubricant), so torquing them to specifications would lead to over-tightening. At the same time, in areas of the country prone to rust, it is understandable why some want to put a very light amount of antiseize only on the threads of the wheel stud, being careful not to get any on the contact seat of the wheel and lug nut. Doing so prevents rust from building up on the threads. However, the lug nut torque would have to be reduced due to the lubricant. CDX prefers to install lug nuts dry and avoid this issue altogether. Many shops have their customers sign on the repair order that they will return to the shop, so the lug nut torque can be rechecked after 50 to 500 miles of driving to make sure the wheels have not loosened.

To install the wheel, torque lug nuts, and make final checks and adjustments, follow the steps in **SKILL DRILL 46-13**.

SKILL DRILL 46-13 Installing Wheels, Torquing Lug Nuts, and Making Final Checks

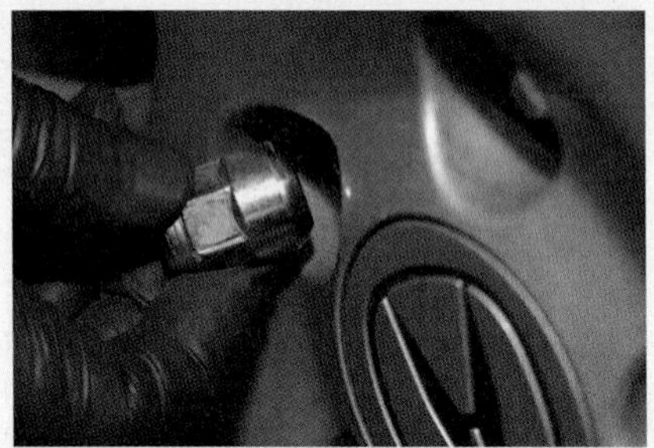

1. Start the lug nuts on the wheel studs, being careful to match up the surfaces.

2. Carefully run all of the lug nuts down so they are seated in the wheel.

3. Lower the vehicle so the tires are partially on the ground to keep them from turning while tightening the lug nuts.

4. Use a torque wrench to tighten each lug nut to the proper torque in the proper sequence.

5. Once all of the lug nuts have been torqued, go around them again, this time in a circular pattern to ensure that you did not miss any in the previous pattern.

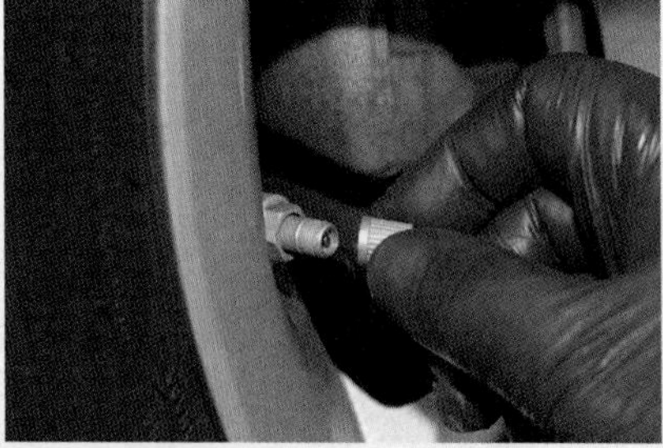

6. If the vehicle was equipped with hubcaps and valve stem caps, reinstall them.

7. Check the brake fluid level in the master cylinder reservoir. Start the vehicle, and check the brake pedal for proper feel and height. Check the parking brake for proper operation. Also inspect the system for any brake fluid leaks and loose or missing fasteners.

Wrap-Up

Ready for Review

- Disc brakes create braking power by forcing flat friction pads against the outer faces of a rotor.
- The vehicle's kinetic energy is transformed into heat energy by the disc brake components, which slow the vehicle when applied.
- Disc brake assemblies consist of a caliper, brake pads, and a rotor.
- Caliper pistons use hydraulic pressure to create a clamping force of the brake pads to the faces of the rotor.
- Disc brake pads require much higher application pressures to operate than drum brake shoes because they are not self-energizing.
- Advantages of disc brakes over drum brakes: They are more effective at transferring heat to the atmosphere, self-adjusting, resistant to water fade, and easier to service.
- Disadvantages of disc brakes compared to drum brakes: They are more prone to noise, more prone to pedal pulsations due to warpage, and more difficult to use as an emergency brake.
- Disc brake calipers come in two main styles: fixed and floating/sliding.
- In disc brake calipers, the piston is sealed by a square-cut O-ring.
- Floating/sliding calipers require clean and lubricated pins, bushings, or guides for proper operation.
- Brake pad lining is either riveted or bonded to the pad backing plate.
- Brake pad lining is available in a variety of materials with varying amounts of coefficient of friction.
- Brake pads may use shims, spacers, guides, and bendable tangs to help minimize squealing.
- Brake pad wear indicators, if used, can be of the mechanical or electronic type.
- Rotors rotate with the wheels and are usually made of durable cast iron with friction surfaces that run true and parallel.
- Brake rotors can be solid or ventilated.
- Disc brake parking brakes can be of the integrated caliper style, top hat drum style, electric pull-cable style, and the integrated electric motor caliper style.
- Diagnosing brake faults requires good information from the customer, an adequate test drive when possible, and a good understanding of brake theory.
- Special precautions are needed when testing ABS during a test drive.
- Removing the caliper is necessary for replacing the brake pads on most vehicles.
- To push the caliper pistons back into their bores slightly, use a small pry bar or C-clamp.
- The caliper mountings and slides/pins experience wear and corrosion over time.
- Removing the brake pads is the most thorough way to inspect them, as it will allow you to not only inspect the thickness of the lining but also the condition of the friction surface.
- Some shops rebuild calipers themselves, but most shops just replace them with rebuilt calipers if necessary.
- Once all brakes have been assembled, seat the pads by applying the brake pedal several times.
- Retracting the caliper piston on an integrated parking brake system requires the piston to be screwed back in on the threaded shaft to retract it into the bore.
- Excessive rotor thickness variation causes brake pedal pulsation and the vehicle to have a surging feeling while coming to a stop.
- Removing 0.015" (0.38 mm) on each side of the rotor results in the thickness being reduced by 0.030" (0.76 mm). Many rotors only start with 0.060" (1.52 mm) of machinable material when they are new.
- On-vehicle refinishing is preferred by most manufacturers because it minimizes runout issues between the hub and rotor.
- Hub-style and hubless-style rotors each require their own way of being mounted on the brake lathe.
- There are two primary methods of replacing lug studs: the drawing-in method and the hydraulic press method.
- Lug nuts should be tightened to the proper torque, in the specified sequence.
- Burnishing, also called bedding in, is the process of transferring pad material onto the rotor evenly, as well as cooking off the resins that are used to bind the friction material together in the pad.

Key Terms

backing plate A metal plate to which the brake lining is fixed.

bearing races Hardened metal surfaces that roller or ball bearings fit into when a bearing is properly assembled.

bendable tangs Small tabs on the brake pad backing plate that are crimped on to the caliper, creating a secure fit and reducing noise.

bleeder screw A screw that allows air and brake fluid to be bled out of a hydraulic brake system when it is loosened and seals the brake fluid in when it is tightened.

bonded linings Brake linings that are essentially glued to the brake pad backing plate; more common on light-duty vehicles.

brake booster A vacuum or hydraulically operated device that increases the driver's braking effort.

brake lining thickness gauge A tool used to measure the thickness of the brake lining.

brake pad shims and guides Small pieces of metal that cushion the brake pad and absorb some of the vibration, helping to cut down on unwanted noise.

brake wash station A piece of equipment designed to safely clean brake dust from drum and disc brake components.

caliper A hydraulic device that uses pressure from the master cylinder to apply the brake pads against the rotor.

caliper dust boot seal driver set A set of drivers used to install metal-backed caliper dust boot seals.

caliper piston pliers A tool used to grip caliper pistons while removing them.

caliper piston retracting tool A tool used to retract caliper pistons on integrated parking brake systems.

C-clamp A tool used to push pistons back into the caliper bore on non-integrated parking brakes.

dial indicator Tool used to measure the lateral runout of the rotor.

disc brake pads Brake pads that consist of a friction material bonded or riveted to a steel backing plate; designed to wear out over time.

disc brake rotor micrometer A specially designed micrometer used to measure the thickness of a rotor.

drawing-in method A method for replacing wheel studs that uses the lug nut to draw the wheel stud into the hub or flange.

edge code A two-digit code printed on the edge of a friction lining that describes its coefficient of friction.

electronic control module (ECM) A computer that receives signals from input sensors, compares that information with preloaded software, and sends an appropriate command signal to output devices; used to manage the antilock brake system (ABS).

fixed caliper A type of brake caliper bolted firmly to the steering knuckle or axle housing, having at least one piston on each side of the rotor.

guide pins Pins that allow the caliper to move in and out as the brakes operate and as the brake pads wear.

hydraulic press method A method for replacing wheel studs that uses a press to force the wheel stud into the flange until it bottoms out.

independent rear suspension (IRS) A type of suspension system where each rear wheel is capable of moving independently of the other.

lateral runout Also called warpage, the side-to-side movement of the rotor surfaces as the rotor turns.

low-drag caliper A caliper designed to maintain a larger brake pad–to-rotor clearance by retracting the pistons farther than normal.

lug A flange that is shaped to assist with aligning objects on other objects.

off-car brake lathe A tool used to machine (refinish) drums and rotors after they have been removed from the vehicle.

on-car brake lathe A tool used to machine (refinish) rotors while they are still attached to the vehicle.

parallelism Also called thickness variation; both surfaces of the rotor should be perfectly parallel to each other so that brake pulsations do not occur.

parking brake cable pliers A tool used to install parking brake cables.

phenolic resin A very dense material used to create some brake pistons that is very resistant to corrosion and heat transfer.

pushrod (braking system) A mechanism used to transmit force from the brake pedal to the master cylinder.

riveted linings Brake linings riveted to the brake pad backing plate with metal rivets and used on heavier-duty or high-performance vehicles.

rotor The main rotating part of a disc brake system.

scratcher A thin, spring steel wear indicator that is fixed to the backing plate of the brake pad; it emits a high-pitched squeal when the brakes are applied if the brake pads have become too thin.

sliding or floating caliper A type of brake caliper that only has piston(s) on the inboard side of the rotor. The caliper is free to slide or float, thus pulling the outboard brake pad into the rotor when braking force is applied.

solid rotor A type of brake rotor made of solid metal, not ventilated.

square-cut O-ring An O-ring with a square cross section that is used to seal the pistons in disc brake calipers.

ventilated rotor A type of brake rotor with passages between the rotor surfaces that are used to improve heat transfer to the atmosphere.

wheel studs Threaded fasteners that are pressed into the wheel hub flange and used to bolt the wheel onto the vehicle.

Review Questions

1. All of the following statements with respect to the disc brake system are true *except*:
   a. the disc brake rotor is bolted to the wheel hub flange.
   b. the hydraulic pressure from the master cylinder causes the caliper to create a mechanical clamping action.
   c. drum brakes require higher applied forces than disc brakes.
   d. the heat generated on the outside surfaces of the rotor is transferred to the atmosphere.
2. Which of the following primary components of the disc brake system uses hydraulic pressure from the master cylinder to apply the brake pads?
   a. Rotor
   b. Caliper
   c. Proportioning valve
   d. Brake pads
3. When the brake pedal is depressed, a pushrod transfers the force to a hydraulic master cylinder through the:
   a. brake booster.
   b. brake liner.
   c. hoses.
   d. pistons.
4. Which of these is an advantage of the disc brake system over drum brakes?
   a. Disc brakes transfer less heat to the atmosphere.
   b. Disc brakes reduce the likelihood of brake fade.
   c. Disc brakes do not need maintenance.
   d. Disc brakes do not create annoying squeals and squeaks.
5. All of the following statements referring to disc brake calipers are true *except*:
   a. disc brake caliper assembly clamps the brake pads onto the rotors to slow the vehicle.
   b. calipers are fitted with a bleeder screw on the top of the piston bore.
   c. the O-ring is compressed between the piston and caliper housing, creating a positive seal to keep the high-pressure brake fluid from leaking out.

d. fixed calipers are the most common type used in passenger vehicles.

6. Which of the following materials have the highest coefficient of friction?
   a. Leather and oak
   b. Brass and cast iron
   c. Rubber and concrete
   d. Copper and cast iron
7. The following edge code letters correspond to four different brake friction materials. Which of these has the lowest coefficient of friction?
   a. EE
   b. FF
   c. GG
   d. HH
8. All of the following statements with respect to wear indicators are true *except*:
   a. a spring steel scratcher is a mechanically operated wear indicator.
   b. when the scratcher makes a squealing noise, it means the brakes must be serviced.
   c. on a sensor type system, contact with the rotor will light a warning light or warning message.
   d. electrical contacts cannot be manufactured into the pad, so they have to clipped to the pad.
9. All of the following statements are true with respect to parallelism *except:*
   a. It results when the rotor's thickness varies by as little as 0.0003".
   b. It causes the brake fluid to boil.
   c. It causes the brake pedal to pulsate.
   d. It should be inspected using a brake micrometer.
10. In which type of parking brake does the rotor have a deeper offset than normal?
    a. Foot-operated parking brake
    b. Integrated parking brake
    c. Top hat design parking brake
    d. Electric parking brake

ASE Technician A/Technician B Style Questions

1. Tech A says that disc brakes operate on the principle of friction. Tech B says that disc brakes operate on the principle of regeneration. Who is correct?
   a. Tech A
   b. Tech B
   c. Both A and B
   d. Neither A nor B
2. Tech A says that some vehicles use fixed calipers. Tech B says that some vehicles use sliding/floating calipers. Who is correct?
   a. Tech A
   b. Tech B
   c. Both A and B
   d. Neither A nor B
3. Tech A says that disc brake pads require higher application force than drum brake shoes. Tech B says that disc brakes are self-energizing. Who is correct?
   a. Tech A
   b. Tech B
   c. Both A and B
   d. Neither A nor B
4. Tech A says that fixed calipers use one or more pistons only on one side of the rotor. Tech B says that sliding/fixed calipers use one or more pistons on both sides of the rotor. Who is correct?
   a. Tech A
   b. Tech B
   c. Both A and B
   d. Neither A nor B
5. Tech A says that calipers use steel piston rings to seal each piston. Tech B says that calipers use a square section O-ring to seal each piston. Who is correct?
   a. Tech A
   b. Tech B
   c. Both A and B
   d. Neither A nor B
6. Tech A says that chrome plated pistons resist rust. Tech B says that pistons made of phenolic resin do not corrode and transmit less heat. Who is correct?
   a. Tech A
   b. Tech B
   c. Both A and B
   d. Neither A nor B
7. Tech A says that some vehicles are equipped with a spring steel brake pad wear indicator that drags on the rotor when the lining thickness is low. Tech B says that some vehicles are equipped with an electric brake pad wear sensor that activates a warning on the dash. Who is correct?
   a. Tech A
   b. Tech B
   c. Both A and B
   d. Neither A nor B
8. Tech A says that rotors should be measured for thickness variation (parallelism). Tech B says that rotors should be measured for lateral runout. Who is correct?
   a. Tech A
   b. Tech B
   c. Both A and B
   d. Neither A nor B
9. Tech A says that rotors that are too thin cannot handle as much heat and will experience brake fade sooner. Tech B says that brake pedal pulsation is the result of air in the hydraulic system. Who is correct?
   a. Tech A
   b. Tech B
   c. Both A and B
   d. Neither A nor B
10. Tech A says that a micrometer is used to measure rotor thickness variation. Tech B says that a micrometer is used to measure rotor lateral runout. Who is correct?
    a. Tech A
    b. Tech B
    c. Both A and B
    d. Neither A nor B

CHAPTER 47
Drum Brake Systems

NATEF Tasks

- **N47001** Diagnose poor stopping, noise, vibration, pulling, grabbing, dragging, or pedal pulsation concerns; determine needed action. (AST/MAST)
- **N47002** Remove, clean, and inspect brake drum; measure brake drum diameter; determine serviceability. (MLR/AST/MAST)
- **N47003** Refinish brake drum and measure final drum diameter; compare with specification. (MLR/AST/MAST)
- **N47004** Remove, clean, inspect, and/or replace brake shoes, springs, pins, clips, levers, adjusters/self-adjusters, other related brake hardware, and backing support plates; lubricate and reassemble. (MLR/AST/MAST)
- **N47005** Inspect wheel cylinders for leaks and proper operation; remove and replace as needed. (MLR/AST/MAST)
- **N47006** Pre-adjust brake shoes and parking brake; install brake drums or drum/hub assemblies and wheel bearings; perform final checks and adjustments. (MLR/AST/MAST)
- **N47007** Install wheel and torque lug nuts. (MLR/AST/MAST)

Knowledge Objectives

After reading this chapter, you will be able to:

- **K47001** Explain the purpose, operation, and components of the drum brake system.
- **K47002** Describe the operation of drum brakes.
- **K47003** Describe the process of self-energization and servo action.
- **K47004** Describe the operation of the different types of drum brakes.
- **K47005** Describe the purpose and function of drum brake components.
- **K47006** Describe the purpose, types, and function of brake drums.
- **K47007** Describe the purpose and function of the backing plate.
- **K47008** Describe the purpose, types, and function of wheel cylinders.
- **K47009** Describe the purpose, types, and function of brake shoes and linings.
- **K47010** Describe the purpose and function of drum brake springs and hardware.
- **K47011** Describe the operation of the drum parking brake system.
- **K47012** Describe the types and purpose of brake tools.

Skills Objectives

After reading this chapter, you will be able to:

- **S47001** Perform maintenance and repair on drum brake systems.

▶ Introduction

Drum brakes get their name from the rotating drum-shaped component called a brake drum (or simply drum). The drums are bolted to the vehicle's axle hubs by the lug nuts. This means the wheels and drums rotate together. Friction lining on brake shoes is forced against the inside of the drums by hydraulic wheel cylinders, causing friction and absorbing the vehicle's kinetic energy. Because heat is created inside the drums, the heat has to transfer through the drum material before it can be transferred to the atmosphere. Situations such as heavy loads or long downhill grades where the brakes are used heavily can lead to overheating of the lining, drums, or brake fluid, resulting in a loss of braking called brake fade.

Most drum brakes require the removal of the drum to inspect the condition of the brake lining and inside drum surface. Drum brake design requires more parts and components, so service and repair can be a bit more complicated than for disc brakes. However, drum brakes are generally less expensive to manufacture and easier to adapt a parking brake to.

▶ Drum Brake System Overview

K47001

Drum brakes get their name from the rotating drum-shaped component called a brake drum (or simply drum). The lug nuts clamp the drum between the wheel and axle flange. This means the wheels and drums rotate together. Friction lining on brake shoes is forced against the inside of the drums by hydraulic wheel cylinders, causing friction and absorbing the vehicle's kinetic energy. Because heat is created inside the drums, the heat has to transfer through the drum material before it can be transferred to the atmosphere. Situations such as heavy loads or long downhill grades, where the brakes are used heavily, can lead to overheating of the lining, drums, or brake fluid, resulting in a loss of braking called brake fade.

Most drum brakes require the removal of the drum to inspect the condition of the brake lining and inside drum surface. Drum brake design requires more parts and components, so service and repair can be a bit more complicated than for disc brakes. However, drum brakes are generally less expensive to manufacture and easier to adapt a parking brake to.

Although many vehicles use disc brakes on all four wheels, drum brakes can be found on the rear wheels of vehicles with a combination of disc and drum brakes, and on all four wheels of older vehicles. Drum brakes can be designed to match the braking requirements of various vehicles. Heavier duty vehicles like pickup trucks use larger-diameter brake drums and shoes as well as wider brake shoes and drum surfaces. These factors allow the drum brakes to create, absorb, and transfer a greater amount of heat energy. The main components of the drum brake system are as follows (**FIGURE 47-1**):

- **Brake drum**: The brake drum fits over the brake linings and forms the braking surface for the brake linings. It is usually made from cast iron and machined so the inside surface rotates true.
- **Backing plate:** The backing plate is made from stamped steel and is bolted to the steering or suspension components. It supports the wheel cylinder(s), brake shoes, and hardware.
- **Wheel cylinder**: The wheel cylinder is attached to the backing plate. The wheel cylinder pistons push the brake shoes into contact with the brake drum to slow or stop the vehicle.
- **Brake shoes**: The brake shoe consists of the steel shoe and the brake lining friction material. The brake shoes are held against the backing plate by hold-down springs and clips.
- **Springs and clips**: Return springs retract the brake shoes when the brakes are released. Other springs work with the self-adjuster and with parking brake linkage operation.

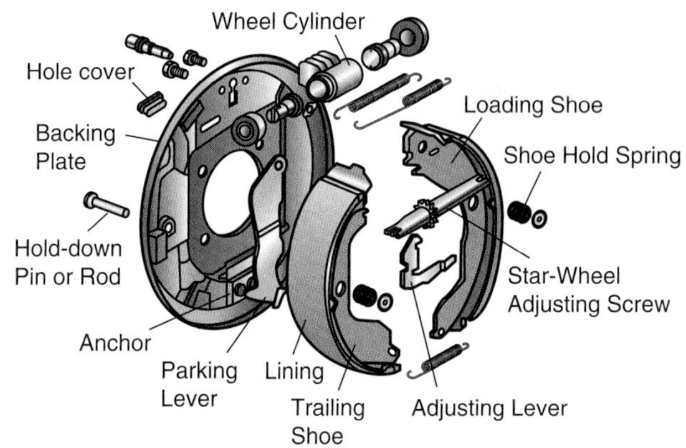

FIGURE 47-1 The main components of a drum brake system.

You Are the Automotive Technician

You are working in San Francisco, California, on your company's fleet vehicles, which are about six years old. Today you are performing a rear brake inspection on a light-duty truck. The driver has indicated that the parking brake isn't able to completely hold the vehicle on some of the steepest hills he has to park on. You operate the parking brake and notice that it can be pushed all the way to the floor before it is completely tight. This means you will have to visually inspect the rear brake assemblies.

1. How is the parking brake similar to and different from the rear drum brakes?
2. What are the possible faults in the service brakes (drum brake style) that could cause the parking brake to be out of adjustment?
3. In what order should the drum brakes and parking brakes be adjusted?

- **Automatic brake self-adjuster**: The automatic self-adjuster automatically adjusts the brakes to maintain a specified amount of running clearance between the shoes and drum. It operates in one of two ways—either when using the brakes while backing up or as part of applying the parking brake. It also makes periodic brake adjustment unnecessary.
- **Parking brake mechanism**: The parking brake linkage mechanically operates the brake shoes (service brakes) to hold the vehicle stationary when the driver applies the parking brake.

Drum Brake Operation

K47002

In drum brake systems, when the brake pedal is depressed, a pushrod transfers the force to a hydraulic master cylinder. The master cylinder converts the brake pedal force into hydraulic pressure, which is then transmitted without loss via the brake lines and hoses to one or two wheel cylinders at each drum brake assembly. The pistons within each wheel cylinder are forced outward by the hydraulic pressure and apply force to both brake shoes, forcing them into each rotating drum (**FIGURE 47-2**). Because the brake shoes are anchored to the backing plate, preventing them from rotating freely with the drum, the friction generated between the moving surfaces slows down the rotation of the drum and the wheel.

Each drum brake has two brake shoes with an attached friction material called a lining. These shoes expand against the inside surface of a brake drum and slow the wheel. The harder the linings are forced against the brake drum, the greater the braking force applied. They can be expanded mechanically or hydraulically. Drum brake systems have to be adjusted to allow for wear of the lining. As the lining wears, the brake shoes must be pushed farther outward to contact the drum. Over time, this wear causes only the top portion of the linings to contact the drum, reducing the surface area of the lining that can dissipate the created heat. Also, if the brakes are not adjusted, the brake pedal reserve height will be too low to be safe. Because of this, most drum brake systems incorporate an automatic adjuster to keep the brakes properly adjusted.

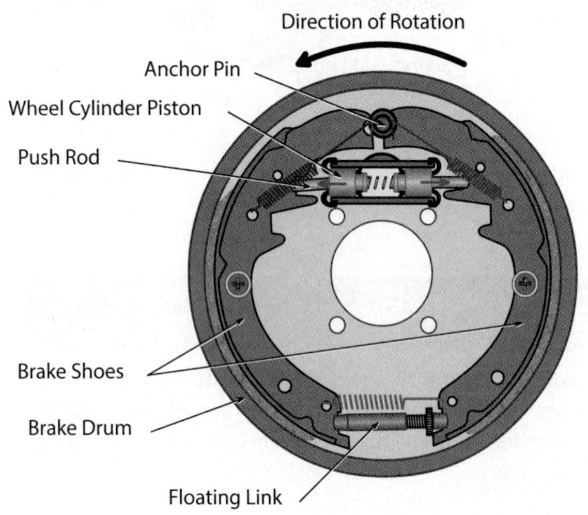

FIGURE 47-2 The pistons within each wheel cylinder are forced outward by the hydraulic pressure and apply force to the brake shoes, forcing them into each rotating drum.

> ▶ TECHNICIAN TIP
>
> Servicing the wheel bearings is a common part of a brake job if the vehicle has serviceable wheel bearings. If the vehicle uses non-serviceable bearings, the bearings should be checked during a brake job and replaced if they are worn out. See the Wheel Bearings chapter for more information on wheel bearings and service.

Self-Energizing and Servo Action

K47003

Drum brakes are **self-energizing**. This means they can increase the force with which they are applied. When brake shoes come into contact with the moving drum, the friction tends to carry them in the direction the drum is rotating. Because the brake shoes are inside the drum and anchored at one end, this has a wedging effect on the brake shoe. This wedging effect assists the driver in applying the brakes, making it so the driver does not have to push so hard on the brake pedal (**FIGURE 47-3**).

The positioning of the brake shoes determines whether a brake shoe is self-energizing. Brake shoes are designed in a leading or trailing manner. **Leading shoes** are installed so the direction they are applied is the same as the forward rotation of the drum. Leading shoes are self-energizing. **Trailing shoes** are installed so the direction they are applied is opposite to the forward rotation of the drum. Trailing shoes are not self-energizing. In fact, they tend to have reduced energization, meaning they are not nearly as efficient at developing braking force as leading shoes are.

Servo action as related to brakes means that one brake shoe, when activated, applies an increased activating force to the other brake shoe, in proportion to the initial activating force. This further enhances the self-energizing feature of some drum brakes. We cover this topic in greater depth later in this chapter.

Types of Drum Brake Systems

K47004

There are three types of drum brake systems: twin leading shoe, leading/trailing shoe (also called single leading shoe), and duo-servo. Each type uses similar drum brake components but functions a bit differently. All three types are self-energizing in at least one direction. The first two types are non-servo brakes. The duo-servo drum brake system uses servo action in both directions. Each of the three types have pros and cons, and designers use the type that best fits the vehicle application.

Twin Leading Shoe Drum Brake Systems

The **twin leading shoe drum brake system** is the least common type in modern automotive use. This system was once popular

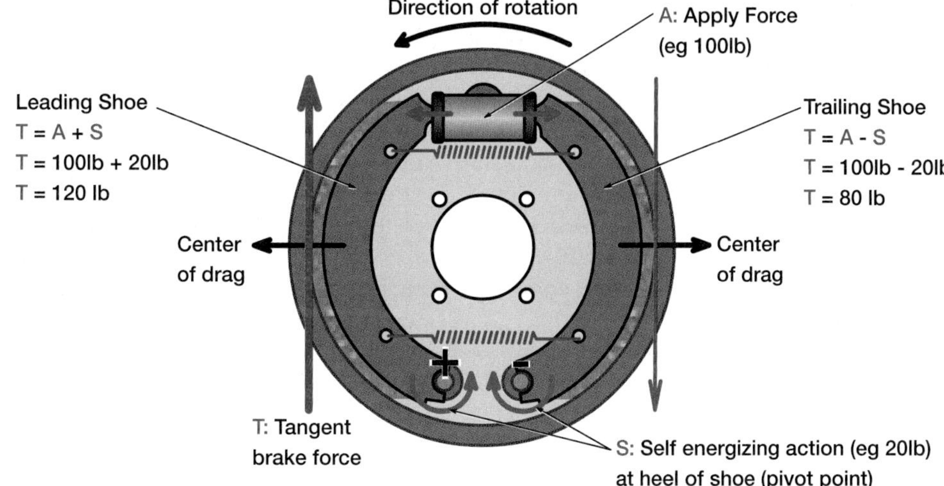

FIGURE 47-3 Drum brake shoes are self-energizing.

on front wheels because it is very efficient at braking in the forward direction. Because the vehicle travels much faster forward than it does in reverse, this matched the braking needs well. The large forward stopping power the twin leading shoe drum brake system generated also allowed the system to operate without a power brake booster.

Twin leading shoe drum brake systems use two single-piston wheel cylinders (also called single-acting wheel cylinders)—one near the top of the backing plate and one near the bottom (**FIGURE 47-4**). Each wheel cylinder activates one of the brake shoes. The brake shoes are anchored at the closed end of the opposite wheel cylinder. It is called a twin leading shoe drum brake system because both shoes are arranged in a leading shoe (self-energizing) configuration in the forward direction. This arrangement gives very good stopping power in the forward direction. When applied in the reverse direction, the braking force is much less, only about 30% as efficient. This type of drum brake system was usually accompanied by one of the other types of brakes on the rear wheels, to be used as a parking brake. The twin leading shoe drum brake system is very well suited for motorcycles because they are driven mostly in the forward direction and rarely in reverse.

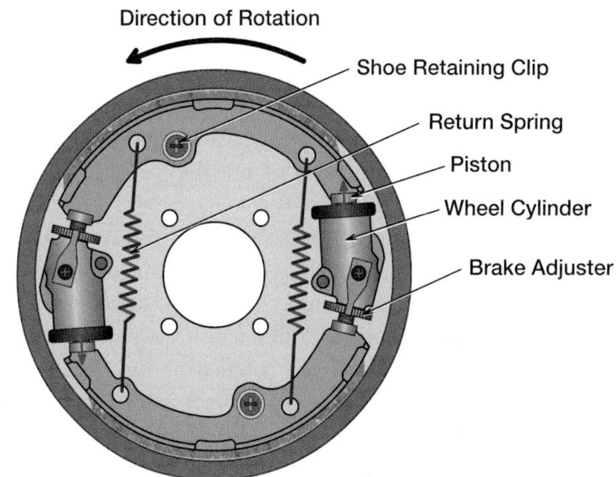

FIGURE 47-4 A twin leading shoe drum brake works very well in the forward direction.

Leading/Trailing Shoe Drum Brake Systems

The **leading/trailing shoe drum brake system** is very common on the rear wheels of front-wheel drive vehicles because of their equal braking forces in both the forward and reverse direction. This system uses a single wheel cylinder with two pistons (also called a double-acting wheel cylinder), usually mounted near the top of the backing plate (**FIGURE 47-5**). Each piston operates one of the brake shoes, and each shoe is anchored at the bottom of the backing plate. This arrangement makes one shoe a leading shoe and the other a trailing shoe. In the forward direction, the front piston forces the front brake shoe into the drum, and it acts like a leading shoe. The rear piston pushes the rear brake

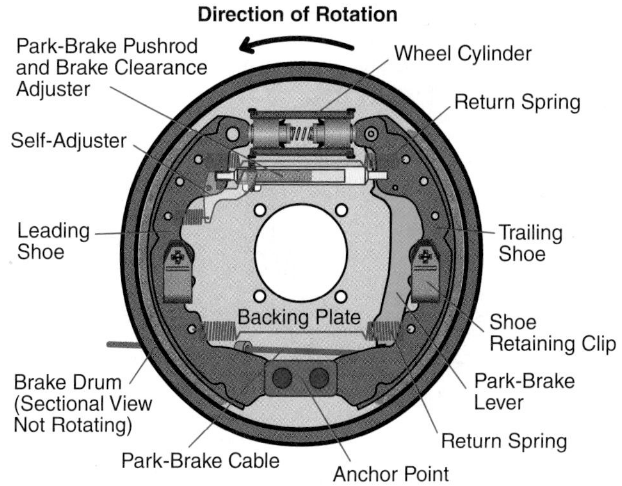

FIGURE 47-5 The leading/trailing shoe drum brake system.

shoe into contact with the drum, but it acts as a trailing shoe, so it does not create as much braking power.

When the car is in reverse or facing uphill and the brakes are applied, the rear shoe becomes a leading shoe and the front shoe becomes a trailing shoe. The leading/trailing shoe drum brake system works equally well in both directions. It is also important to note that it does not provide maximum braking in either direction; it tends to produce a lesser but equal amount of force in both directions. This makes it ideal for the rear wheels of front-wheel drive vehicles because approximately 70–80% of the braking power needed under heavy braking occurs at the front wheels and only 20–30% at the rear wheels.

Duo-Servo Drum Brake Systems

Duo-servo drum brake systems get their name from using the servo action in both the forward and reverse direction. Like the leading/trailing system, the system uses a single wheel cylinder with two pistons (also called a double-acting wheel cylinder), usually mounted near the top of the backing plate (**FIGURE 47-6**). The bottom of each brake shoe is not anchored to the backing plate, but is connected by an adjustable, floating link. This configuration allows the bottom of the brake shoes to move in the direction of the drum. What keeps the shoes from just spinning around with the drum? There is an anchor pin at the top of the backing plate above the wheel cylinder that prevents each shoe from rotating past that point. The shoes can move away from the anchor pin, but they are stopped by it when they rotate toward it.

When the brakes are applied, the front piston overcomes the weaker front return spring tension to move the forward shoe into contact with the drum. When the shoe contacts the drum, the friction causes it to start to rotate with the drum. This puts a small force through the bottom connecting link and applies (servo action) the bottom of the rear shoe, pushing it into contact with the rotating drum. The top of that shoe is thus carried into the anchor pin at the top of the backing plate, causing both shoes to stop rotating. The brakes generate a small amount of force at this point.

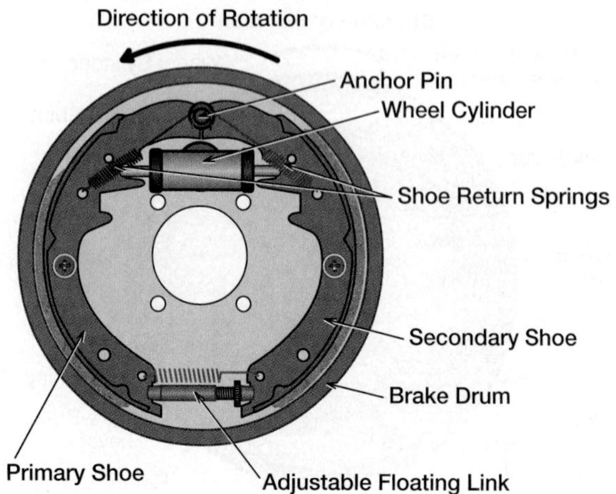

FIGURE 47-6 A duo-servo drum brake system.

As the driver applies more force to the brake pedal, the forward piston in the wheel cylinder pushes the front shoe harder into the rotating drum, which causes the front shoe to apply more force to the rear shoe. This forces the rear shoe harder into the drum as the anchor pin prevents it from rotating. Because hydraulic pressure is the same on both pistons in the wheel cylinder, the rear piston also tends to push the rear shoe outward into the drum, which helps apply it, although not in a completely complimentary direction.

Because the front shoe multiplies the force applying the rear shoe, the rear shoe does more of the braking work. If the linings were the same length front to rear, then the rear one would wear out much faster. This is why manufacturers put more lining on the rear shoe than on the front shoe. They might also use linings with different coefficients of friction for each of the shoes to get the desired braking load between the two shoes. It is important, therefore, to install the correct shoe in the correct position. Failure to do so will cause the brake linings to wear unevenly as well as work improperly.

▶ Drum Brake Components

K47005

Brake Drums

K47006

Brake drums provide the rotating friction surface that the brake lining contacts. They are usually made from cast iron because of its ability to withstand high temperatures, absorb a lot of heat, and maintain its shape. To enhance the cooling of the brake drum, some manufacturers add cooling fins to the outside of the brake drum, and others make the brake drum out of aluminum (**FIGURE 47-7**).

Brake drums are machined to a specific diameter by the manufacturer, which is called its standard diameter. Manufacturers specify the maximum allowable inside diameter a brake

FIGURE 47-7 Brake drums. **A.** Without cooling fins. **B.** With cooling fins.

drum can be worn or machined to and usually stamp or cast that specification on the outside of the brake drum. This specification is commonly 0.060" (1.524 mm) over the standard diameter on many brake drums, but it can be as low as 0.030" (0.762 mm) over standard or as high as 0.090" (2.286 mm) over standard on some passenger vehicles. Always check the manufacturer's specifications. Remember that taking 0.015" (0.381 mm) off one side of the brake drum surface also removes 0.015" (0.381 mm) from the other side, for a total of 0.030" (0.762 mm).

Brake drums that are heavily grooved or warped likely must be replaced. Even if they are not grooved, they have to be measured because they may have been refinished one or more times before, making them oversized. Never put an oversized brake drum back in service because it does not have as much mass to absorb brake heat, and it also is not as strong as a drum that is within specifications.

Types of Brake Drums

Like rotors, brake drums can have an integrated hub, called a hub-style drum, or a separate hub called a hubless-style drum (**FIGURE 47-8**). Hubless drums are the most common in passenger vehicles because of the high percentage of sealed wheel bearings being used. Because the wheel bearings in this case do not have to be packed with grease periodically, they do not have to be removed unless they wear out. Manufacturers design the brake drum to slip over the lug studs on the wheel flange and to be held on by the wheel and lug nuts. Hubless drums are less expensive to replace when they no longer meet the manufacturer's specifications.

Hub-style drums have a one-piece integrated hub/drum assembly. The wheel bearings are housed in the hub and are usually serviceable. If they are serviceable, then it is standard procedure to pack the wheel bearings with the specified grease and replace the grease seals during a brake job. More steps are required to remove hub-style drums from the vehicle than hubless drums, and they are more expensive to replace.

FIGURE 47-8 A. Hubless drum. **B.** Hub-style drum.

▶ Backing Plate

K47007

All of the brake unit components except the brake drum are mounted on a backing plate bolted to the vehicle axle housing or suspension. The backing plate is usually pressed into a very specific configuration from heavy-gauge steel (**FIGURE 47-9**). It uses a labyrinth seal on its outer edge to help keep out dirt and water spray. The drum brake's labyrinth seal is made up of a raised edge on the outer surface of the backing plate that fits into a groove or recess in the brake drum (**FIGURE 47-10**). This makes it so there isn't a direct path for water spray and dirt to

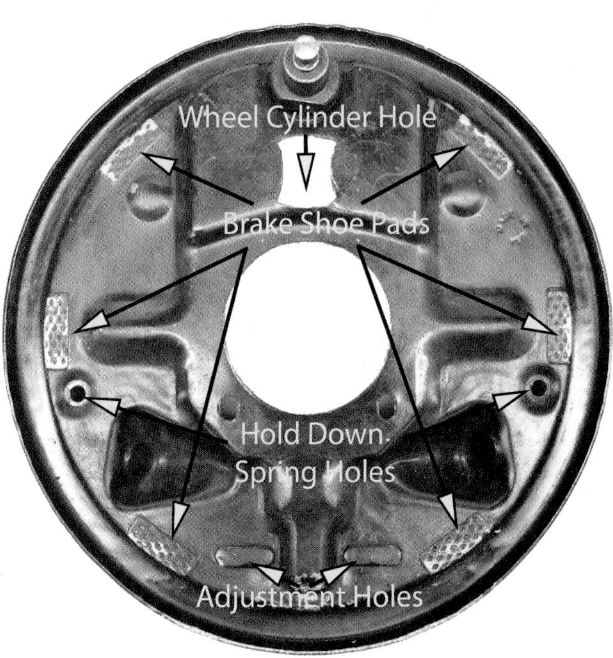

FIGURE 47-9 The backing plate.

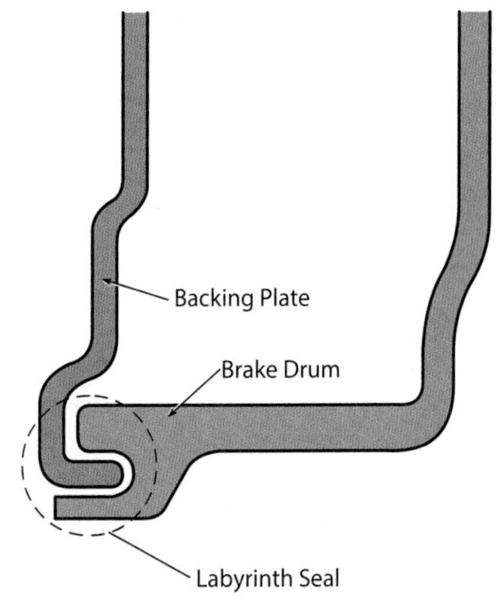

FIGURE 47-10 A labyrinth seal prevents a direct path for water spray and dirt to enter the brake drum.

FIGURE 47-11 Two styles of anchor pins mount on backing plates.

FIGURE 47-12 The wheel cylinder mounted on the backing plate.

enter the drum. The backing plate also has holes stamped in it for the wheel cylinder, hold-down pins, and the parking brake cable to pass through. Some vehicles have one or two openings in the backing plate to allow for manual adjustment of the brake shoes. These holes are plugged with rubber grommets that have to be reinstalled after adjusting the brakes.

One of the most important components of the backing plate is the **anchor pin** (or anchor block). The anchor pin must be able to take all of the braking force when the brakes are applied, so it must be strong and firmly attached to the backing plate (**FIGURE 47-11**). The anchor pin may also hold the shoe guide, return springs, and self-adjuster cable on duo-servo-style brakes.

The inside surface of the backing plate has flat or raised brake shoe contact pads stamped into it. The brake shoes are held against the brake shoe contact pads by spring pressure and are free to move side to side. Over time, the brake shoes can wear a groove in the surface of the contact pad. During brake shoe replacement, these contact pads have to be cleaned and inspected. If there is no wear, then a very light coat of white lithium grease can be applied to the contact pads to lessen any wear during the life of the new brake shoes. If there is light wear, this can be cleaned up with a file or small handheld grinder. If the grooves are too deep, the backing plate will have to be replaced.

▶ Wheel Cylinders

K47008

The wheel cylinder is located inside the brake drum and is either bolted or firmly clipped to the backing plate (**FIGURE 47-12**). It converts hydraulic pressure from the master cylinder into mechanical force that pushes the brake linings against the inside of the brake drum.

Wheel cylinders usually contain one cylinder housing, one or two pistons, a lip seal for each piston, a spring and expander set, a dust boot for each open end of the cylinder, a pushrod for each piston, and a bleeder screw (**FIGURE 47-13**). Wheel cylinder housings are usually made of cast iron or aluminum alloy, and they operate under high pressures and temperatures.

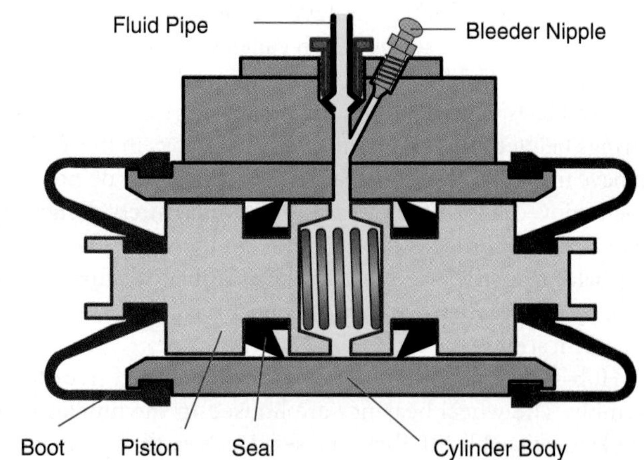

FIGURE 47-13 Wheel cylinder cutaway.

The **cylinder bore**, or inside diameter of the cylinder, is created by drawing a properly sized ball bearing through the bore. This technique gives the surface a hard, smooth finish. It is also why most manufacturers recommend replacing rather than honing the wheel cylinder if it has pits or corrosion. Cylinder bores on aluminum wheel cylinders are usually anodized to help resist corrosion. They also should not be honed, as this would remove the protective finish. Some cylinder bores are sleeved with stainless steel to be longer wearing and more resistant to corrosion.

No matter the type of material used, the cylinder and pistons are manufactured to a precise diameter. This provides the proper amount of clearance between them. It also maintains the proper tension of the piston seal to the cylinder bore, which helps hold pressure in and the air out. These are other reasons why honing the cylinder bore is frowned upon.

Contamination, particularly from water, causes pitting and rusting of the inner surface of the wheel cylinder. Such damage can result in leakage of brake fluid from the cylinder or stuck pistons. The sealing surface in a wheel cylinder is the inside surface of the cylinder bore. This surface must be clean, smooth, and free of pitting. The wheel cylinder cups the seal against the surface of the cylinder bore and prevents the brake fluid from leaking out of the cylinder. Hydraulic pressure forces the lip of the seal into the surface of the cylinder bore even harder when the brakes are applied. The sealing cups are made of materials that are compatible with the type of brake fluid specified for the particular vehicle. Using the wrong seal compound in a brake system or mixing the wrong brake fluid in the brake system can cause seal damage and brake failure. Always verify that the brake fluid and components are compatible with each other.

Wheel cylinder pistons are usually made of anodized aluminum. They have a built-in mating surface for the pushrods or brake shoes to engage. Most pistons use a flat surface that supports the piston seal, but some pistons use a seal that fits within a machined groove at its inner end. Piston-to-cylinder clearance is critical for proper operation, so verify that the cylinder and piston are not worn beyond the specified clearance.

Wheel cylinders may be fitted with a spreader and a light expansion spring to keep the lips of the seal in contact with the cylinder bore during times of low pressure, such as during retraction and while at rest (**FIGURE 47-14**). This helps keep air from being drawn into the cylinder. A flexible dust boot fits over the open ends of the cylinder and allows for piston movement. At the same time, it helps keep brake dust and moisture away from the inside of the cylinder and piston.

Wheel cylinders are fitted with a bleeder screw to allow for bleeding of air and old brake fluid from the hydraulic brake system (**FIGURE 47-15**). The bleeder screw is a hollow screw. It has a taper on the end that mates with a matching tapered seat in the wheel cylinder. These tapered seats seal when the bleeder screw is closed. The bleeder screw has been cross-drilled into the center hole just above the taper. This way, when the bleeder screw is loosened slightly, the tapered seat opens, and brake fluid enters the cross-drilled passage into the center hole and flows out the end of the bleeder screw. Tightening the bleeder screw closes off the tapered seat and holds pressure in the wheel cylinder. Bleeder screws have a small rubber dust cap that fits over the exposed end of the screw to keep water, dirt, and debris from entering the center hole and plugging it up or rusting it in place. Always remember to replace these caps when you finish bleeding the drum brakes.

FIGURE 47-14 Wheel cylinder components.

> **TECHNICIAN TIP**
>
> During a brake inspection, it is good practice to carefully peel back the dust boot from the wheel cylinder to see if there is brake fluid behind the dust boot. If there is, the wheel cylinder is starting to leak and should be replaced.

Types of Wheel Cylinders

Wheel cylinders come in different configurations. They are either single acting or dual acting. Single-acting cylinders use a

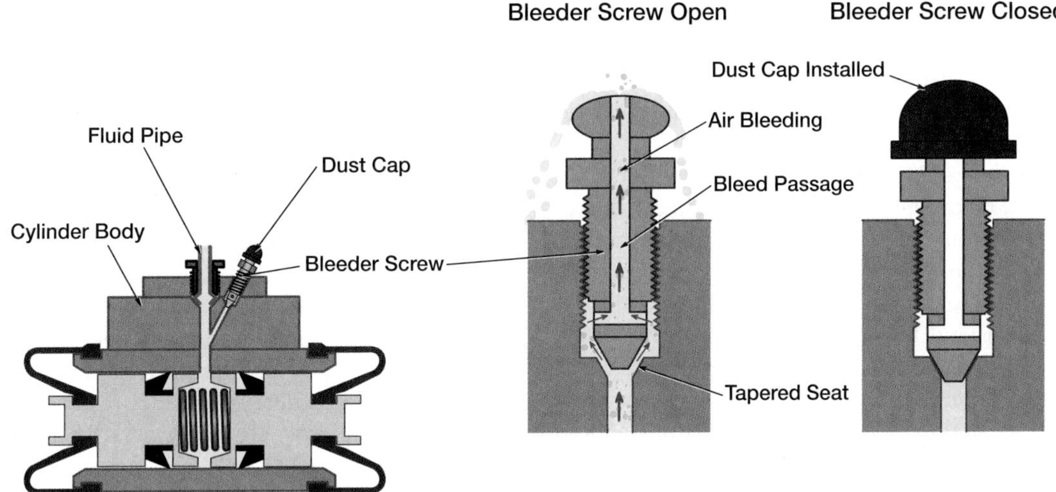

FIGURE 47-15 Bleeder screw allows air and fluid to be bled from the system.

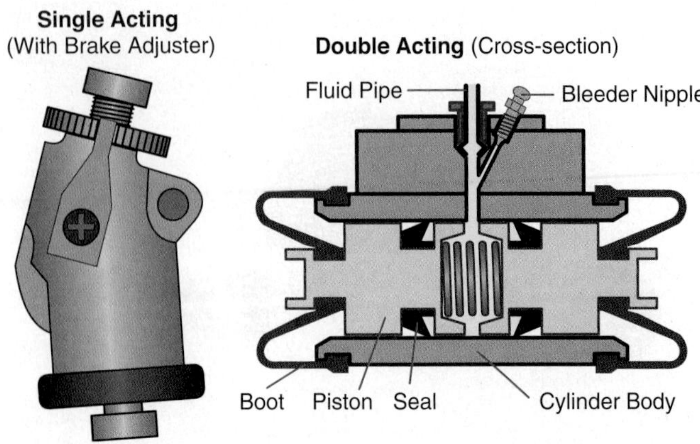

FIGURE 47-16 Types of wheel cylinders.

single piston, meaning that the force is generated in one direction only. Dual-acting cylinders use two pistons opposite of each other, meaning that the force acts in two different directions (**FIGURE 47-16**). Most modern vehicles use double-acting wheel cylinders because they are simpler to design, install, and bleed. Dual-acting wheel cylinders use a common cylinder with a piston and lip seal in each end. There is usually a coil spring with expanders on each end, positioned between the lip seals. The expander helps to hold the seal lips against the cylinder bore when there is little or no hydraulic pressure.

Single-acting wheel cylinders are used on some non-servo drum brakes. Each wheel cylinder only has one piston, so the cylinder bore is closed off at the opposite end. There are two of these cylinders on each wheel assembly, one for each brake shoe. The wheel cylinder is very similar to the double-acting cylinder in that it is made of the same materials and has an aluminum piston, a lip seal, a spring and expander, a dust boot, a pushrod, and a bleeder screw.

▶ Brake Shoes and Linings

K47009

The drum brake system uses metal brake shoes that have holes, slots, and tabs for springs and hardware to attach to. The metal brake shoes have friction material called linings attached to them. Linings can be riveted but are more often bonded to the brake shoes. Even though the brake shoe is technically just the metal portion, technicians generally use the term "brake shoe" to means the metal shoe and lining assembly.

The composition of the lining material affects brake operation. Materials that provide good braking with low pedal pressures tend to lose efficiency when they get hot (**FIGURE 47-17**). This means the stopping distance will be increased. They also tend to wear out more quickly. Materials that maintain a stable friction coefficient over a wide temperature range generally require higher pedal pressures to provide efficient braking. They also tend to put added wear on the brake drum friction surface, reducing its useful life. The Society of Automotive Engineers has adopted codes to rate the coefficient of friction of brake lining materials.

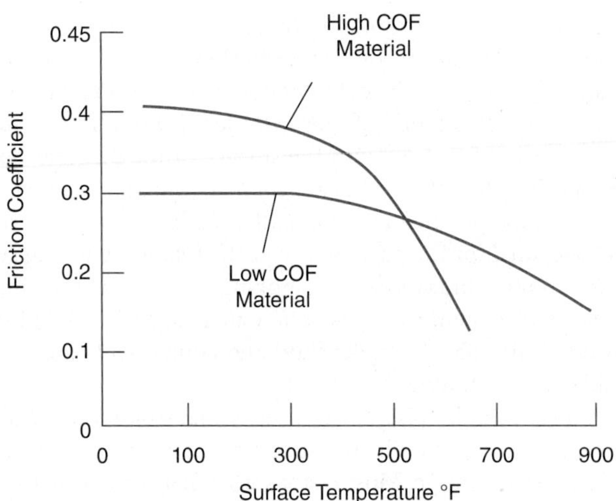

FIGURE 47-17 The brake lining coefficient of friction is affected by temperature.

The rating is written on the edge of the friction linings and is called the edge code. The Disc Brake System chapter discusses these ratings in detail. Drum brakes are usually designed so that the condition of the lining can only be checked once the drum has been removed.

SAFETY TIP

Brake friction materials were historically made from asbestos compounds because of the excellent heat resistance of that material. Asbestos compounds are now known to be hazardous and are generally banned. Do not assume that a brake shoe does not contain asbestos, as it is still found in some applications. Brakes are now manufactured from a variety of different materials that may be non-asbestos organic (NAO), low-metallic NAO, semi-metallic, and ceramic.

▶ TECHNICIAN TIP

As the heat in brake pads and linings builds up, the coefficient of friction capability of the material—and consequently its stopping power—is reduced. This reduction is called brake fade. Minimizing or overcoming brake fade is a major factor in the design of brakes and the development of brake friction materials.

Primary and Secondary Brake Shoes

The terms "primary" and "secondary" refer to the brake shoes in a duo-servo brake system. The primary shoe goes toward the front of the vehicle, and the secondary shoe goes toward the rear of the vehicle (**FIGURE 47-18**). In most cases, the primary and secondary metal shoes are the same, but the linings installed on the brake shoes are of different length. Because the primary shoe applies the secondary shoe, the secondary shoe is responsible for doing most of the braking work. Therefore, the primary shoe lining is shorter in length, and the secondary shoe lining is longer. The primary lining may also have a different coefficient of friction than the secondary lining. New shoes are normally

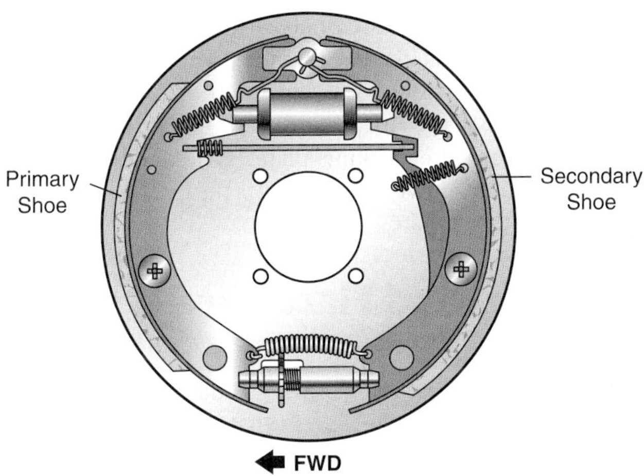

FIGURE 47-18 Primary and secondary brake shoes.

FIGURE 47-19 A common installation error: The primary shoes are on one side of the vehicle, and the secondary shoes are on the other.

labeled with "pri" (for primary) or "sec" (for secondary) on the edge of the lining. Verify that you are installing the lining in the proper position.

> **TECHNICIAN TIP**
>
> New technicians tend to make a couple of rookie mistakes when it comes to duo-servo brake installation. The first relates to the way the new brake shoes are packaged in their box. The manufacturer generally puts the brake shoes in the box in like pairs. What this means is that both primary shoes are in the bottom of the box and both secondary shoes are in the top of the box (or vice versa). When students remove the top two shoes to install them, and compare them to each other, they find that the shoes match (they shouldn't if they are for a duo-servo system). The students assume the matching shoes belong on the same side of the vehicle and install them accordingly (**FIGURE 47-19**). They then do the same with the other two brake shoes (which also match each other) on the other side of the vehicle. So both primary shoes are now installed on one side of the vehicle, and both secondary shoes are installed on the other side.
>
> The second mistake involves installing the shoes in the wrong position on the backing plate. When students install the brake shoes on the first assembly, they may look carefully at the assembly and determine (correctly) that the primary shoe goes on the forward side of the backing plate, which for the driver's side is the left side of the backing plate. They then determine that the secondary shoe goes on the right side of the backing plate. When they move to the passenger side of the vehicle, they believe they have already determined that the primary shoe goes on the left side of the backing plate just like the other side. Unfortunately, that is the *rear* side of the backing plate, and the secondary shoe should be installed in that position. So, the vehicle has the brake shoes installed correctly on one side, but incorrectly on the other side (**FIGURE 47-20**). Watch for these mistakes as you perform brake inspections and service.

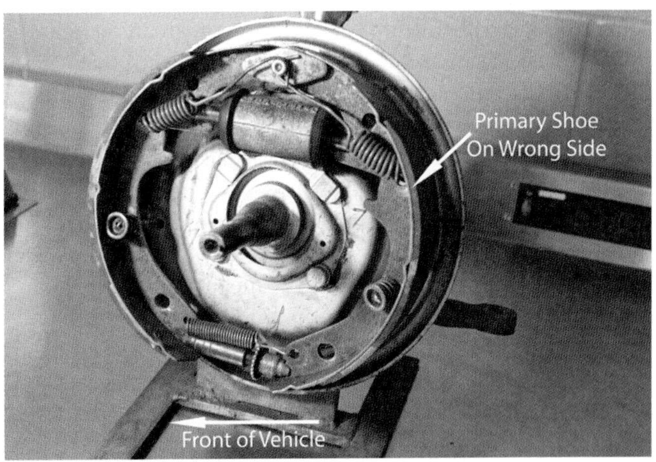

FIGURE 47-20 Another common installation error: The primary and secondary shoes are swapped from side to side.

Riveted and Bonded Friction Materials

Lining on brake shoes is much thinner than on disc brake pads due to the much greater surface area of the brake shoe lining. Drum brake shoes consist of friction material or lining bonded or riveted onto a steel shoe (**FIGURE 47-21**). Bonded brake

FIGURE 47-21 **A.** Bonded brake shoe lining. **B.** Riveted brake shoe lining.

linings are more common on light-duty vehicles because they are less expensive to build. Also, lighter vehicles do not subject the brake linings to as much heat, so the bonding agent is not as likely to fail as it would on heavier vehicles. Bonded linings are glued to the metal brake shoe under high pressure and temperature to ensure that the bonding is as strong as possible.

Riveted linings are used on heavier duty or high-performance vehicles. Metal rivets, usually made of copper or aluminum, provide a mechanical connection to hold the brake lining to the shoe. This means rivets are less susceptible to failure under high temperatures. However, because the rivets actually pinch some of the brake lining between the rivet head and the metal shoe, the brake linings cannot wear down as much before the rivets contact the drum. Riveted linings must be changed sooner than bonded linings to prevent the rivet heads from contacting the drum and wearing a groove in the friction surface.

Drum Brake Noises

Drum brakes are not prone to squealing like disc brakes are, but that does not mean they never make noise. One of the most common sounds that drum brakes make is a groaning noise. It can be caused by excessive brake dust in the drum that causes the brakes shoes to skip and catch, which sounds like a groan. Removing the drum and cleaning the excess brake dust or replacing the brake shoes due to wear usually resolves this issue (**FIGURE 47-22**).

Another noise that drum brakes can make is a grinding noise. If the friction lining wears all the way down to the metal shoe, the shoe and drum can make a metal-on-metal grinding noise. This noise can only be resolved by replacing the brake shoes and refinishing or replacing the drum. Keeping drum brakes from making unwanted noises is pretty simple—keep them clean and inspect the brake lining thickness periodically, replacing it when it gets to the specified minimum thickness.

The last common noise that drum brakes can make is a clicking noise. This can be caused by two situations. First, it could be that the brake shoes have worn grooves in the contact pads on the backing plate and are causing the clicking noise when the shoe moves into and out of the groove. Second, it could be that the brake drum's surface finish cut is too rough and acts like the threads on a screw. When the lining contacts the surface of the drum, the lining follows the threaded finish of the drum, and the linings thread themselves away from the backing plate. When the brakes are released, the hold-down springs snap the brake shoes back against the backing plate.

▶ Springs and Hardware

K47010

Drum brakes use a variety of springs and hardware to control the action of the brake shoes (**FIGURE 47-23**). Each type of brake uses its own arrangement of these components. Pay close attention to how they are installed before removing them. It is good practice to only disassemble and reassemble the brake components on one side of the vehicle at a time. That way you can refer to the other side if you forget how something came apart. Generally speaking, there are several categories of springs used in drum brakes: return, hold-down, and specialty.

Return Springs

Return springs retract the brake shoes when the driver releases the brake pedal. Each return spring either connects to one brake shoe and the backing plate anchor pin or directly to the other brake shoe. In both cases, they pull the brake shoes back into their rest position (**FIGURE 47-24**). Return springs can integrate one or more coil springs in the length of the spring, or it can be a large U-shaped spring. Return springs are generally quite stiff, making them a challenge to install. Brake pliers can make installation easier. Even though return springs may look the same front to rear within a particular brake assembly, be aware that they can have different amounts of strength. On duo-servo brakes, the primary return spring is weaker than the secondary return spring, so the primary shoe can be applied before the secondary shoe.

The coils on return springs should be tightly wound together, meaning that the coils should be touching each

FIGURE 47-22 Brake dust.

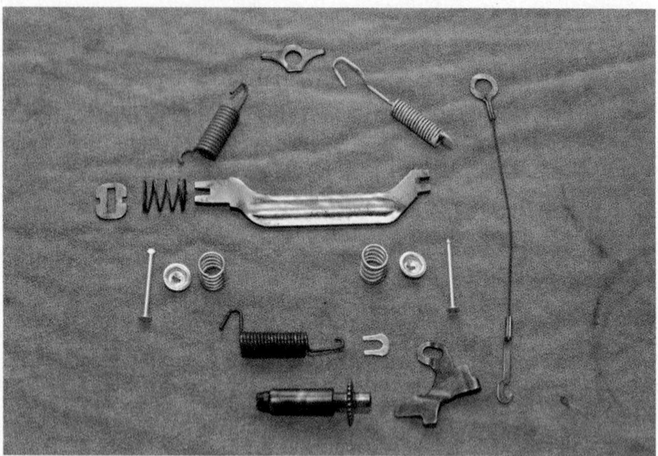

FIGURE 47-23 Examples of brake springs and hardware.

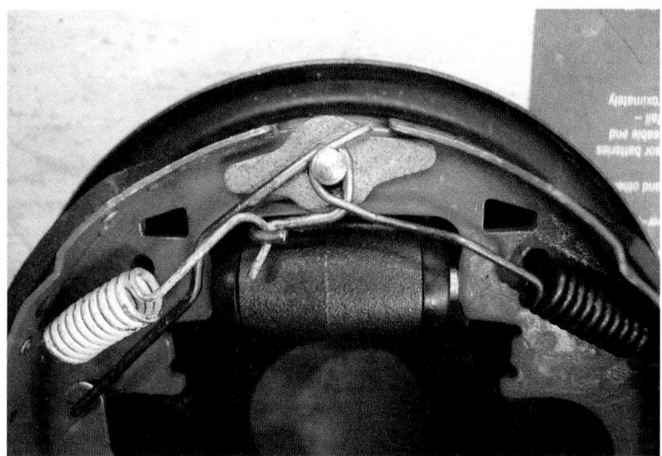

FIGURE 47-24 Return springs pull the shoes back to their rest position.

other, with no space between them. Any space indicates that the return spring has been stretched and must be replaced. Return springs are normally replaced in sets. If one spring has to be replaced, they all should be replaced to avoid mismatched components. Also, because of the high temperatures and the large number of apply and return cycles that drum brake springs must endure, some manufacturers and shops recommend replacing the return springs and hardware during every brake job.

Hold-Down Springs

Hold-down springs do just what their name suggests—they hold the brake shoes against the backing plate (**FIGURE 47-25**). They can be coil springs in combination with a spring retainer and a pin, sometimes referred to as a brake nail, that extends through the backing plate, brake shoe, and center of the spring; or they can be coil springs lying on their side and connected to the pin beside them. They can also use U-shaped spring steel clips in combination with a pin that extends through the backing plate and brake shoe.

Specialty Springs

Specialty springs are used to return links and levers on the parking brake system or the self-adjuster mechanism. Specialty springs can be of all different shapes and sizes and can be used to push or pull components into proper position. Be sure to install each specialty spring in the correct position so that each of the brake systems works as intended. If in doubt, check the other wheel brake unit or the service information to see how it is assembled.

FIGURE 47-25 Hold-down springs. **A.** Coil spring style. **B.** Sideways coil style. **C.** U-shaped clip style.

▶ Self-Adjusters

Brake shoe lining wears over time and increases the clearance between the brake lining and drum. This increased clearance causes the wheel cylinder pistons to travel farther to move the brake shoes out to contact the surface of the drums. It also causes the brake pedal to travel farther before activating the brakes, reducing the reserve pedal height, which leaves the driver vulnerable if a hydraulic failure occurs. To safeguard against these dangers, all manufacturers have been required since 1968 to incorporate a self-adjusting mechanism into their drum brake systems that is capable of maintaining proper shoe-to-drum clearance.

Types of Self-Adjusters

The first type of self-adjuster is used on most duo-servo brakes. It has an adjustable threaded star wheel assembly holding the bottoms of the brake shoes apart. The link has three main parts: the threaded barrel, the non-threaded barrel (and possible thrust washer), and the star wheel, which is threaded on one end and smooth on the other. Both barrels have slots that fit to tabs on the brake shoes, preventing them from turning. The star wheel's threaded end screws into the threaded barrel. The smooth end of the star wheel fits into the non-threaded barrel (**FIGURE 47-26**). As the star wheel is turned, the thread causes the self-adjuster to lengthen. This takes up excess brake shoe clearance, as needed. The question is what turns the star wheel?

A movable self-adjuster link is held against the star wheel (**FIGURE 47-27**). The link is moved up and down by the action of applying the brakes while the vehicle is backing up. If the link moves far enough, it will catch another tooth of the star wheel, and the spring action will turn the star wheel one tooth, adjusting the brakes shoes outward just a bit. This process continues until the brake shoe clearance is taken up enough so that the self-adjuster link cannot move far enough to catch another tooth on the star wheel. As the brake linings wear, the clearance increases, and the movement of the link increases to the point that it catches another tooth. This continues for the life of the brake lining.

▶ TECHNICIAN TIP

Some drivers never use their brakes when backing up. Instead, they back up slowly, put the transmission in first gear, and use their clutch or automatic transmission to stop the vehicle and start going forward at the same time. Backing without braking prevents the self-adjusters from adjusting the drum brakes on vehicles equipped with this type of self-adjuster. Educating customers on how the self-adjuster works will help them keep their brakes adjusted as well as get longer life out of the linings.

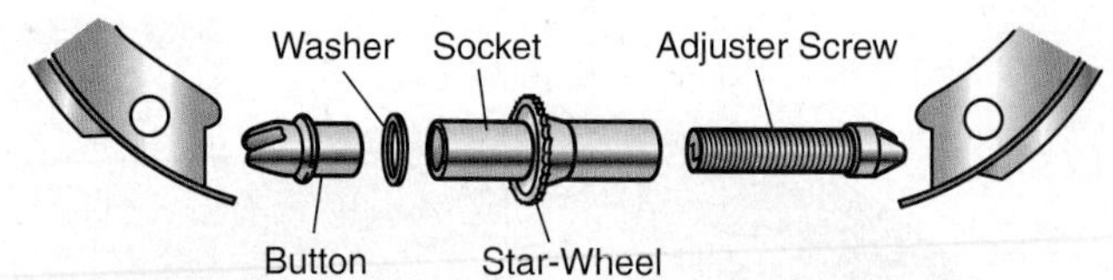

FIGURE 47-26 A star wheel assembly.

▶ TECHNICIAN TIP

Be careful to not switch the self-adjusters from one side of the vehicle to the other. If this happens, the self-adjuster will retract the adjustment, causing the brake shoe clearance to increase as the brakes adjust. The customer will report that the brake pedal keeps getting lower and lower but was fine when he or she picked up the vehicle.

The self-adjuster link is moved by a cable or rod, which is connected to the anchor pin at the top of the backing plate. The movement of the secondary brake shoe, when the brakes are applied while backing up, causes the self-adjuster link to be pulled by the cable or rod. The position of the link and cable or rod is critical to the operation of the self-adjuster assembly. The link should be level with the star wheel and close to its centerline. Also, the star wheel assembly is side specific, meaning that the threads on each star wheel are opposite each other, so they turn in opposite directions.

A similar style of self-adjuster works off movement between one brake shoe and the parking brake strut. It is usually used on non-servo brakes. The star wheel linkage is threaded and is usually mounted between the brake shoes just below the wheel cylinder (**FIGURE 47-28**). It acts as the parking brake strut. There is a pivoting self-adjuster lever attached to one brake shoe, and a spring pulls the lever down when the service brake pedal is applied. The pivoting lever is pushed up by the end of the parking brake strut when the brake pedal is released. If the lever travels far enough

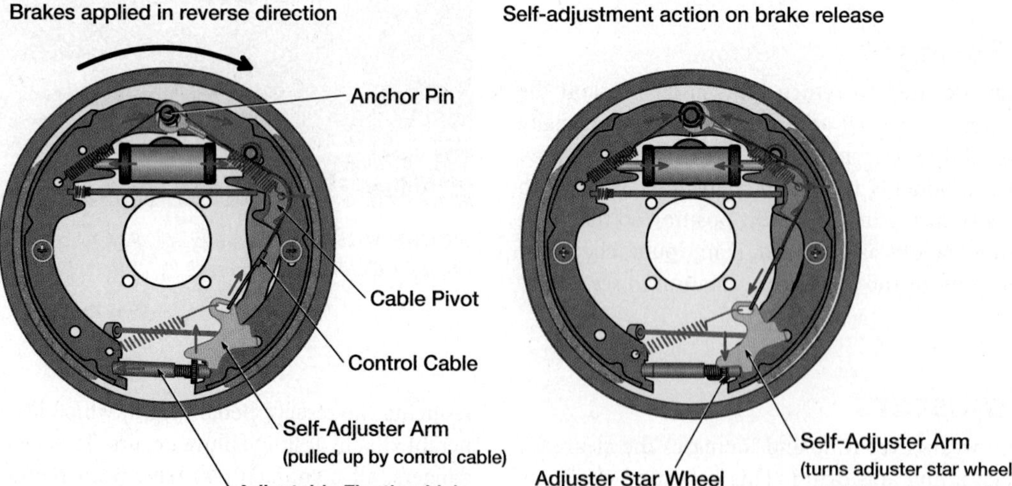

FIGURE 47-27 A self-adjuster on a duo-servo-style brake.

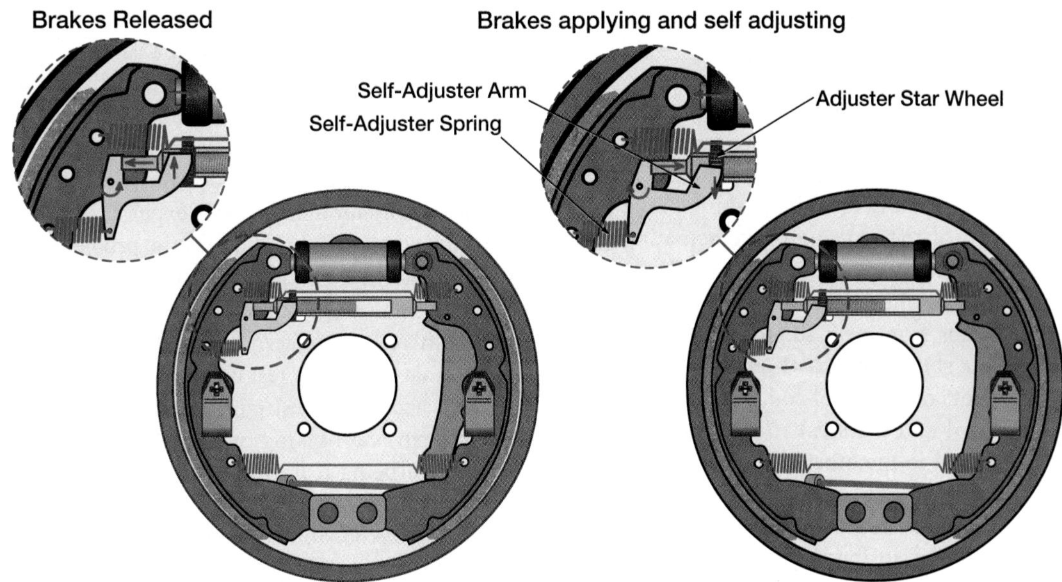

FIGURE 47-28 Alternate style of automatic star wheel brake adjuster.

when the service brake is applied, it will catch the next tooth on the star wheel. As the brake pedal is released, the end of the strut pivots the self-adjuster link upward against spring pressure. This turns the star wheel linkage slightly, lengthens the parking brake strut slightly, and reduces the brake shoe clearance slightly. This process continues as the brake pedal is applied and released until the excess clearance of the brake shoe to the drum is taken up. The self-adjuster lever will then be unable to catch another tooth on the star wheel until the clearance opens a bit more as the brake linings continue to wear.

Some manufacturers use a ratcheting-style self-adjuster, sometimes called a cam-style adjuster (**FIGURE 47-29**). This type uses two toothed pieces held in contact with each other by spring pressure. They can slide over each other in one direction but hold in the other direction. The two pieces allow excessive shoe-to-drum clearance to be taken up when the brake pedal is applied and then prevent the shoes from fully returning to their rest position when the brakes are released. As the service brakes are applied, the shoes move apart. If they travel far enough, the toothed pieces slide over each other until the shoes contact the drum. When the brakes are released, the components allow a small amount of inward brake shoe movement to occur, which creates the proper shoe-to-drum clearance. This style of self-adjuster generally can be adjusted in one brake pedal application, reducing the amount of time spent pre-adjusting the brake shoes when they are being replaced.

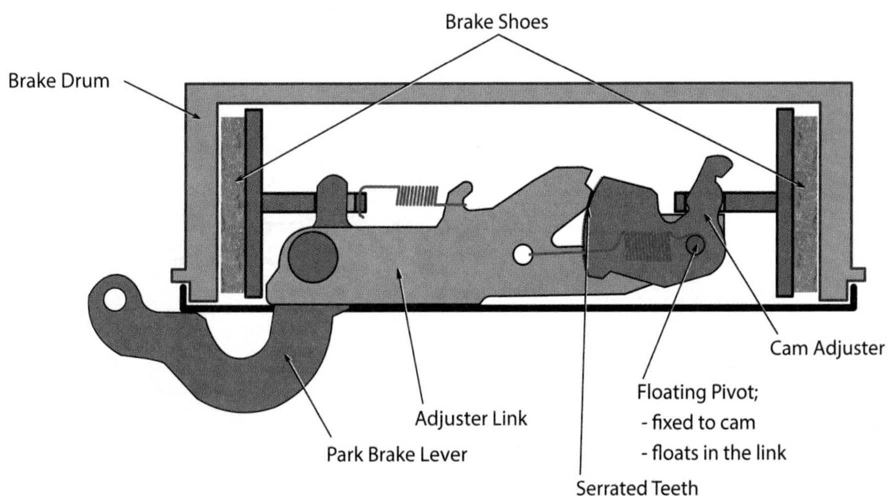

FIGURE 47-29 A ratchet-style adjuster adjusts if excess brake clearance is present when the service brake is applied.

▶ Parking Brake Systems

K47011

Drum parking brake systems mechanically apply the regular service brake shoes. Because drum brakes are self-energizing, it is easier to generate the force needed to apply them mechanically than it is to apply disc brakes. The parking brake cable attaches to the bottom of the parking brake actuating lever in the drum brake assembly. The other end of the lever is attached to the top end of one of the brake shoes with a pin or tang. A strut rod runs from near the top of the actuating lever to the other brake shoe (**FIGURE 47-30**). When the cable is pulled, the lever pivots on the strut rod, pushing the top of the brake shoe rearward and the strut rod forward. Because the strut rod is connected to the other shoe, pulling the lever forces the brake shoes apart and into firm contact with the drum, holding the vehicle stationary. Releasing the cable allows the brake retracting springs to pull the brake shoes away from the drums.

▶ Diagnosis

N47001

Diagnosing Drum Brakes

Drum brake diagnosis starts with understanding the customer's concern. Communicating directly with the customer is the best way to do that, but the customer is not always available. An experienced service advisor will gather the required information, so you should read the service advisor's notes on the repair order carefully or speak with him or her directly. Once you understand the customer's concern, a test drive is usually needed to verify the accuracy of the concern. Depending on the concern, it may be as simple as stepping on the brake pedal without moving the vehicle and feeling the pedal sink to the floor, or it could require a more detailed test-drive to observe the fault the customer is describing. This is a good opportunity to test the brakes under a variety of conditions. Replicating the customer concern is important in order to address the situation that the customer is experiencing.

During the test drive, find a safe place to operate the brakes at a variety of speeds with a variety of brake pedal pressures, especially trying to mimic the conditions the customer described. It may be necessary to go on a test drive with the customer driving, allowing him or her to operate the vehicle in the way that makes the problem evident and to point out the particular situation he or she is experiencing. Also, it is good to have the customer along in case the problem does not occur, in which case the customer won't think you don't believe him or her or are ignoring the issue.

If there is a concern related to the antilock brake system (ABS) that requires a test drive, extreme caution is required so that an accident doesn't happen. Because ABS operates only during extreme braking or poor traction conditions, you run the risk of being rear-ended by a vehicle behind you or losing control of the vehicle if you apply the brakes hard enough to activate the ABS. So make sure the vehicle is being tested away from all other traffic. Remember that if the yellow ABS warning lamp is illuminated, the ABS system is deactivated, and you should not try to activate ABS on a test drive. If the yellow ABS lamp is off, the ABS system should be active and ready to activate. It is helpful when testing ABS to do so on a surface with limited traction, such as wet pavement or a dirt road. When you do apply the brakes firmly, ABS activation can typically be felt as pulsations in the brake pedal and possibly the steering wheel. The vehicle should brake quickly while maintaining steering control. In some cases, a "poor traction" lamp may illuminate, telling you that the ABS system had to activate. This warning lamp usually turns off after several seconds. If the yellow ABS warning lamp illuminates, it typically means that the vehicle's ECM has observed a fault in the ABS system and will have to be checked for DTCs.

Once the customer concern has been verified, you will hopefully have enough information to research the concern in the service information and TSBs. Armed with this information you should be ready to create a plan to begin testing the brake system. This could be as simple as performing a visual inspection of the brake fluid level and condition, removing the wheels to disassemble and inspect the brake units, or measuring the

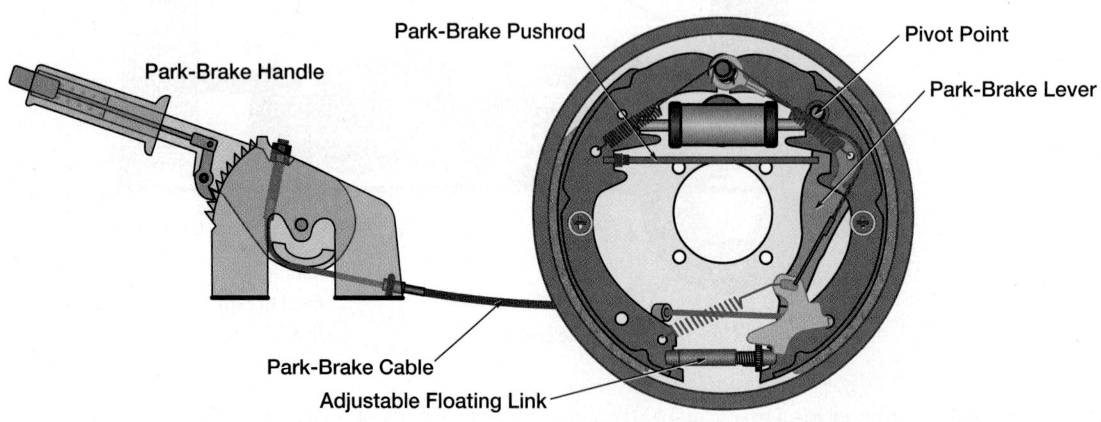

FIGURE 47-30 A parking brake assembly for a drum brake.

runout of the drums. It could also involve electrical diagnosis such as locating a short circuit in the parking brake warning indicator system.

Suspension and steering system faults can appear to be brake system faults. An example of this is a pulling condition while braking. If the control arm bushings on the suspension are worn, then braking the vehicle will cause the wheel to pivot outward, making the wheel point in a direction other than straight down the road, imitating a brake pull. So always inspect the suspension and steering systems when diagnosing concerns related to a vehicle "pull."

The braking system on a vehicle must be restored to its proper operation if one or more braking system faults are present. Diagnosis of any problem must identify all issues that would prevent the brakes from operating normally. Brake system failures are more likely to lead to vehicle accidents than failures of most systems. Any diagnosis and subsequent repairs have to be thorough and complete.

Once you have identified the cause of the fault, determine the action that will correct the fault. This information can then be used to prepare an estimated cost of repairs and given to the customer for authorization to perform the repairs. Once the repair is complete, retest the system to confirm that the concern has been fully addressed and no further issues are present.

▶ TECHNICIAN TIP

It is common to have more than one fault present in a brake system when you diagnose it, such as worn brake shoes, a leaky wheel cylinder, and a leaky seal in the master cylinder. It is good practice to perform a thorough inspection of the brake system whenever one condition is present to determine if any others are present at the same time. This prevents you from having to go back to the customer to get permission to perform the additional work after he or she has agreed to the initial repair, giving the customer cause to doubt your competence or integrity.

▶ TECHNICIAN TIP

One way to help identify whether a problem is coming from the front or rear brakes is to test-drive the vehicle in a safe place at a relatively low speed and lightly apply the parking brake. If the condition is still present, the problem is with the rear brakes, as the parking brakes are usually on the rear wheels. If the condition is not present, the problem is likely with the front brakes.

▶ Maintenance and Repair

S47001

Drum brakes require periodic maintenance and repair as the vehicle ages. As stated previously, visual inspection of the braking system is part of the diagnostic procedure. It is also a critical part of maintenance and repair. A good visual inspection can turn up issues such as lining that isn't wearing evenly, contaminated linings from a leaking wheel cylinder, a warped drum, or even broken springs and hardware. Given the critical role that the braking system plays, a good periodic inspection can address issues before they become life threatening. The following topics and Skill Drills will lead you through the common maintenance and repair procedures of drum brake systems.

Tools

K47012

The tools used to diagnose and repair drum brake systems include (**FIGURE 47-31**):

- **Brake wash station**—Used to clean drum and disc brake dust.
- **Brake spring pliers**—Used to remove and install return springs.
- **Hold-down spring tool**—Used to remove and install hold-down springs.
- **Drum brake micrometer**—Used to measure inside brake drum diameter.
- **Brake shoe adjustment gauge**—Used to pre-adjust brake shoes before installing the drum.
- **Brake spoon**—Used to adjust brake shoes when the drum is installed.
- **Wheel cylinder piston clamp**—Used to hold the pistons in the wheel cylinder while the brake shoes are removed.
- **Off-car brake lathe**—Used to machine drums and rotors that are off the vehicle.
- **Parking brake cable pliers**—Used to install parking brake cables on the parking brake lever.
- **Parking brake cable removal tool**—Used to remove the parking brake cable from the backing plate.

Removing, Cleaning, Inspecting, and Measuring Brake Drums

N47002

Brake drums need to be removed for a variety of reasons, such as inspecting the thickness of the brake linings, checking for wheel cylinder leaks, and packing wheel bearings. While removed, the brake drums should be inspected visually for any damage. They should also be measured with a drum micrometer to verify that they are smaller than the specified maximum diameter. If not, they should be replaced.

Before removing the brake drum, you need to determine if it is a hub style, or hubless (slip-off) style. Most light-duty vehicles now use hubless drums. You can usually tell which style of brake drum you are working on by looking at the holes for the lug studs. If there is clearance between the stud and drum, it is likely a hubless brake drum. You can also look at the clearance between the center hole of the brake drum and the hub. A hubless brake drum has a parting line at the centering hole. A hub-style brake drum appears to be one solid piece.

You can take several steps to ease the drum removal process. First, back off the brake shoe adjustment before removing the

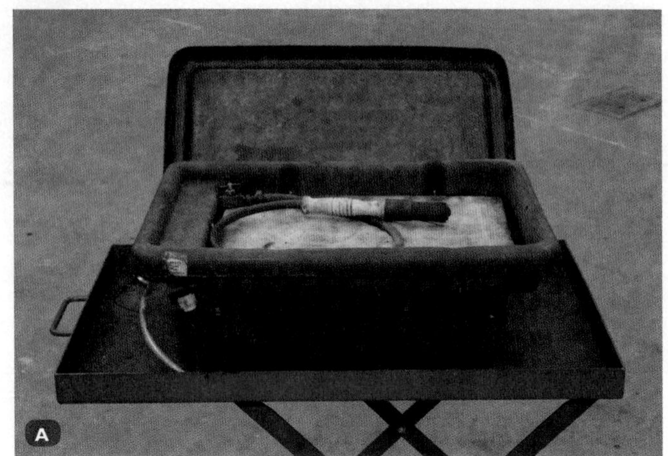

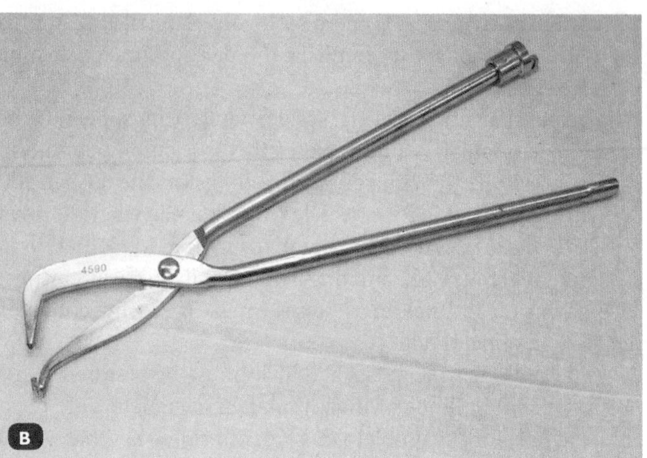

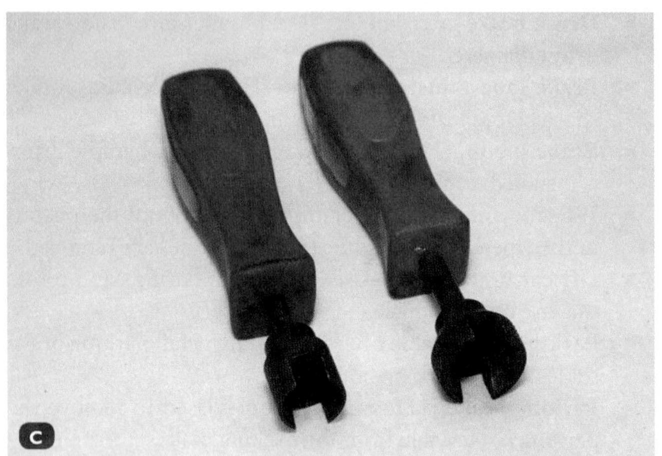

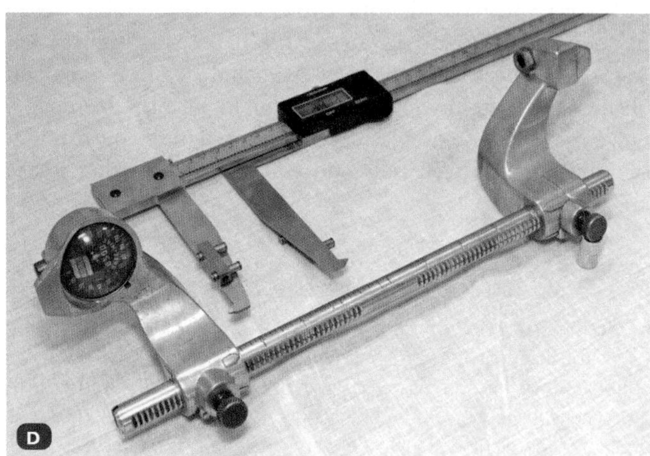

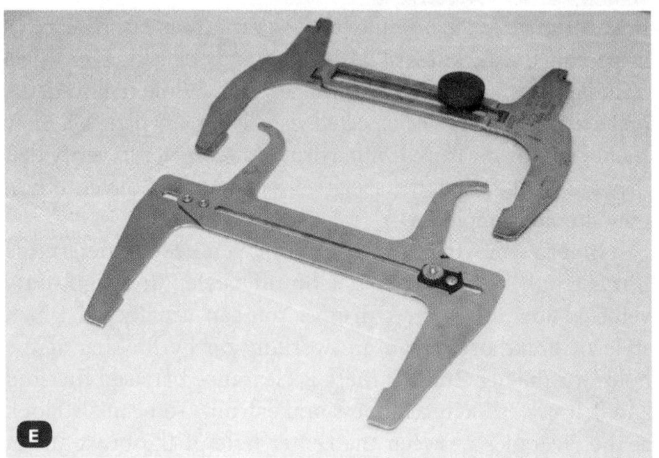

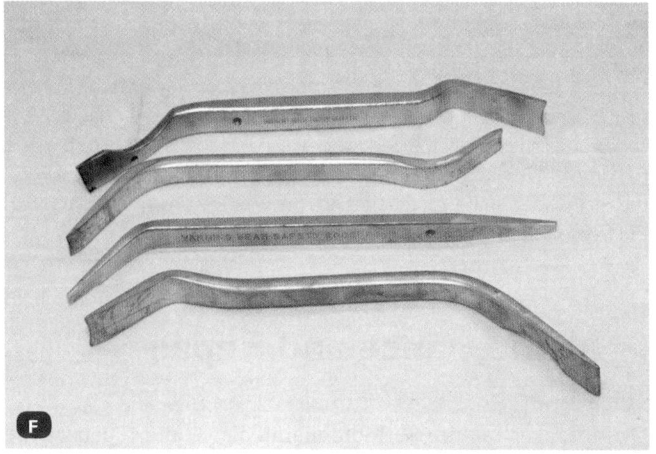

Caption appears on next page.

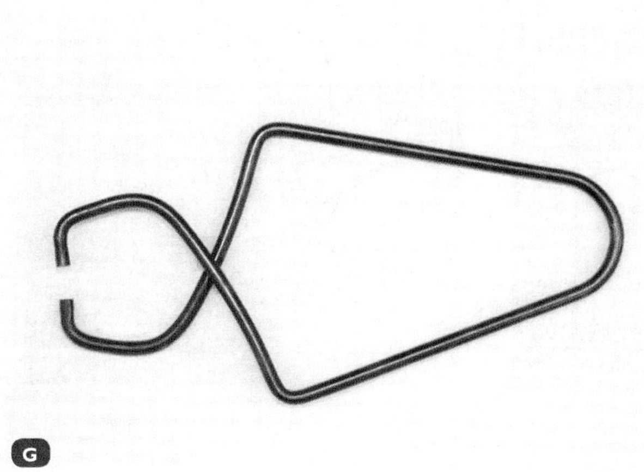

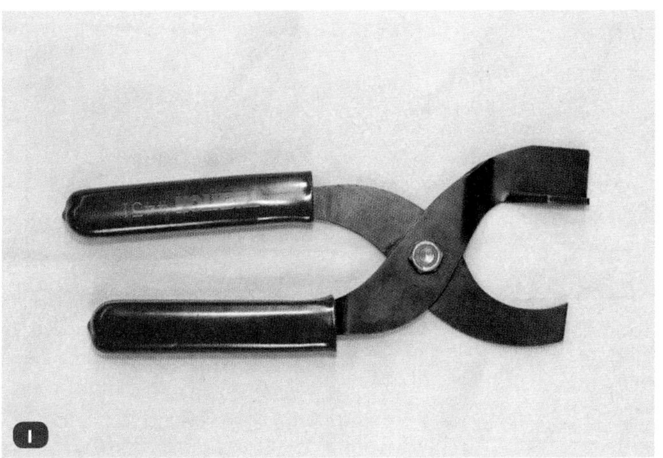

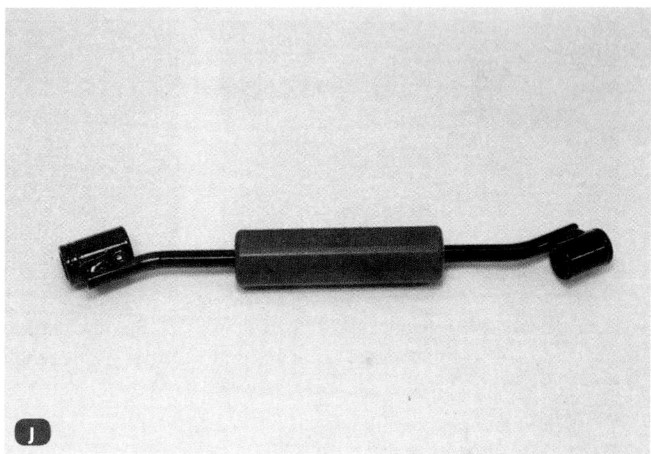

FIGURE 47-31 Drum brake tools. **A.** Brake wash station. **B.** Brake spring pliers. **C.** Hold-down spring tool. **D.** Drum brake micrometer. **E.** Brake shoe adjustment gauge. **F.** Brake spoon. **G.** Wheel cylinder piston clamp. **H.** Off-car brake lathe. **I.** Parking brake cable pliers. **J.** Parking brake cable removal tool.

drum. The shoes and drum are in close proximity to each other, and if you don't back off the adjustment, they can bind during the removal process, especially if there is a ridge on the edge of the drum or the brake shoes are heavily grooved. Second, because of the tight clearances between the hub and brake drum, the surfaces can rust together. You may have to use a medium or large ball-peen hammer to hammer on the drum between the lug studs. Make sure you do not hit the lug studs because you will damage them. Some drums can be rusted on very solidly. Sometimes a torch may be needed to heat up the drum to break the rust bond. Some drum brakes have two threaded holes in the drum that allow bolts to be installed. By tightening the bolts, the drum is forced off the hub. In case the drum still can't be removed, cut the heads off the hold-down pins with a diagonal cutter at the backing plate. This releases the brake shoes from the backing plate and allows the shoes to come off with the drum. If you are having difficulty breaking the drum loose, seek the assistance of your supervisor.

To remove, clean, inspect, and measure brake drums and determine any further actions, follow the steps in **SKILL DRILL 47-1**.

> **TECHNICIAN TIP**
>
> When backing off the self-adjuster, remember that you will have to hold the self-adjuster link away from the star wheel so that the star wheel can be retracted.

Refinishing Brake Drums

N47003

Brake drums must be refinished with a brake lathe when they have excessive grooving or are out-of-round. A **brake lathe** refinishes the drum friction surface by removing metal and making it perfectly round with the proper finish. If the grooving or surface defects are too great, the drum may require the

SKILL DRILL 47-1 Removing, Cleaning, Inspecting, and Measuring Brake Drums

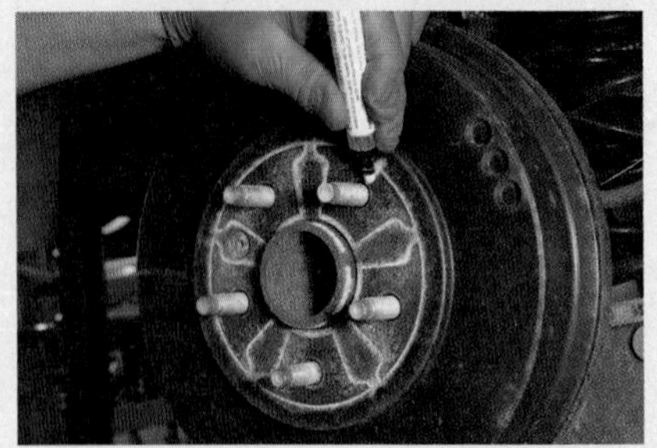

1. To perform this procedure on a hubless-style drum, first make matching marks on the drum and hub for reinstallation in the correct position. A permanent marker, crayon or center punch can be used.

2. If there are screws or speed nuts holding the drum to the hub, remove them following the specified procedure. The speed nuts can be discarded and are not needed upon reassembly. The screws will be reused.

3. Remove the drum from the hub.

4. To perform this procedure on a hub-style drum, remove the wheel bearing locking mechanism (cotter pin, locknut, or peened washer).

5. Remove the wheel bearing adjusting nut, thrust washer, and outer bearing.

6. If the rear bearing needs to be removed for service or replacement of the drum, use the following procedure to remove the inner bearing and grease seal.

- Reinstall the adjusting nut onto the spindle about five turns.
- Grasp the drum/hub assembly on the top and bottom. Push it toward the center of the vehicle. Hold with slight downward pressure as you firmly pull the drum/hub assembly toward yourself. This should cause the adjusting nut to catch the wheel bearing and pull it and the seal out of the rear of the hub.

7. Clean the drum with approved asbestos removal equipment. Inspect the drum for hard spots/hot spots, scoring, cracks, and damage.

8. Measure the drum diameter at the deepest groove or most worn part of the drum, and compare to specifications. If it is over the size limit, it will have to be replaced. If it has to be machined and is below the maximum diameter, check to see how badly it is scored.

9. Find the smallest diameter of the drum, and write down the reading. Find the largest diameter of the drum, and write down that reading. Calculate the difference between the two readings, (out-of-round), and compare it to specifications. Determine any necessary actions.

removal of too much metal to satisfactorily refinish the surface. The drum diameter should always be remeasured once the refinishing is complete to ensure that it is under the manufacturer's maximum diameter. Never put a brake drum that is over the maximum size back in service, because it does not have the ability to absorb as much heat and will experience brake fade much more quickly.

Hub-style and hubless-style drums are mounted on the brake lathe differently. Most hub-style drums use the bearing races to drive and center the drum on the lathe spindle. Bearing adapters of the proper size need to be selected and used. The spindle nut then clamps the drum onto the spindle through these bearing adapters and races.

Hubless drums can be mounted in two ways, composite rotor or clamshell and centering cone. The composite rotor adapter mounts in a similar manner as on a disc brake rotor. The clamshell system uses a spring-loaded centering cone to align the drum's centering hole with the spindle, and clamshell clamps to clamp it to the lathe spindle. In the following Skill Drill, we use the composite rotor adapter method.

TECHNICIAN TIP

Resurfacing or machining the brake drum is an important task that requires precise measurements and the use of a brake lathe. Care must be taken while machining a drum to avoid an inadvertent adjustment that renders it scrap metal with only a quarter turn of the hand wheel.

To refinish a brake drum and measure final drum diameter, follow **SKILL DRILL 47-2**.

Removing, Cleaning, and Inspecting Brake Shoes and Hardware

N47004

Drum brakes need to be disassembled for a variety of reasons such as when the brake lining has worn beyond specifications; the wheel cylinders are faulty; the self-adjuster is stuck or damaged; or the axle seal has failed and leaked gear oil or wheel bearing grease onto the brake lining. Drum brakes are generally more complicated to change than disc brake pads, so care needs

SKILL DRILL 47-2 Refinishing Brake Drums

1. Clean any nicks, burrs, or debris from the mounting surfaces of the drum, including the centering hole, if used.

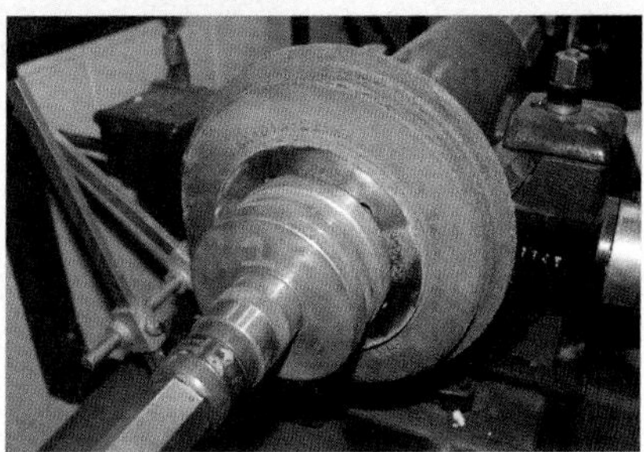

2. Mount the drum on the brake lathe.

3. Visually check to see that the drum is running true on the lathe and is not wobbling. If it wobbles, turn off the lathe and recheck for condition that causes it to wobble.

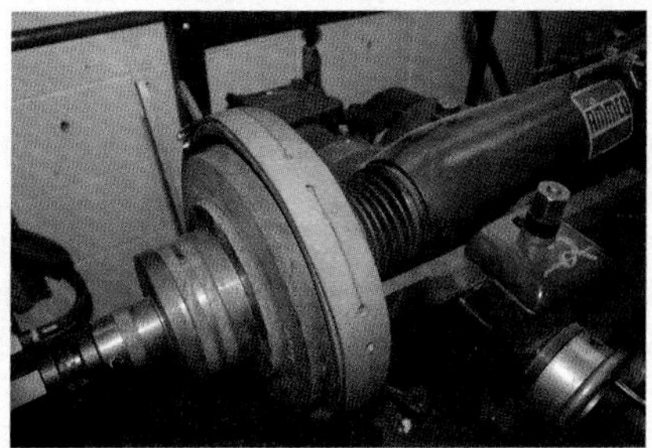

4. Install the anti-chatter band on the drum. It will prevent the drum from vibrating during the machining process, which would create a rough finish and dull the cutting bit.

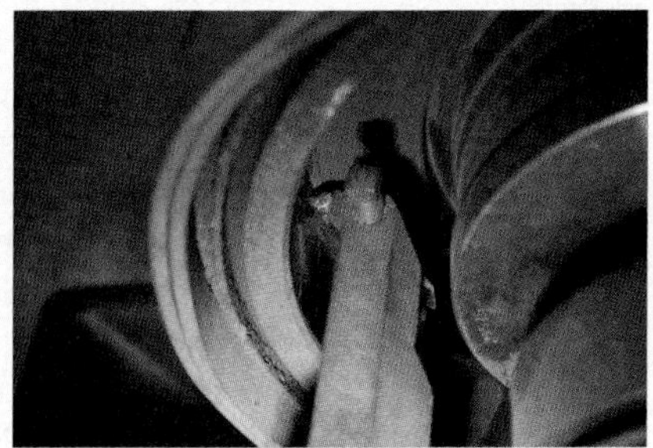

5. Set the position of the cutting tool so that the brake drum is close to the brake lathe when the cutting bit is in the far corner of the drum.

6. Make sure the cutting bits will not contact the face of the drum, and move the cutting head about 0.5" (12.7 mm) in from the outside of the drm.

SKILL DRILL 47-2 Refinishing Brake Drums (Continued)

7. Turn on the brake lathe, and set the depth of the cutting tool so it just touches the surface of the drum. Rotate the brake lathe's handwheel so that the drum moves outward and the cutting bit contacts the ridge. Keep turning slowly to remove the ridge.

8. Once the ridge is removed, run the drum all the way in so the cutting bit is in the inner corner of the drum. Set the cutting bit to the proper depth for machining the surface of the drum, and lock it in place.

9. Engage the automatic spindle feed, and set it to the proper speed; lock it in place, and watch for proper machining action. Some single-cut machines only have one speed.

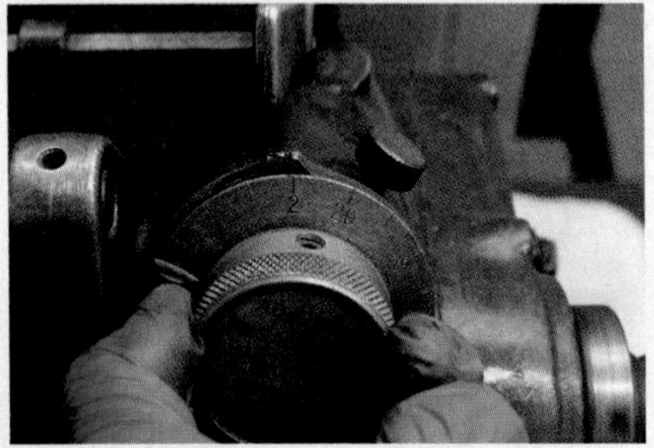

10. If necessary, repeat this step until the worn surface areas have been removed all the way around the surface of the drum. If the brake lathe is not a single-cut machine, perform a finish cut on the drum. This cut is usually done on a slower spindle feed speed.

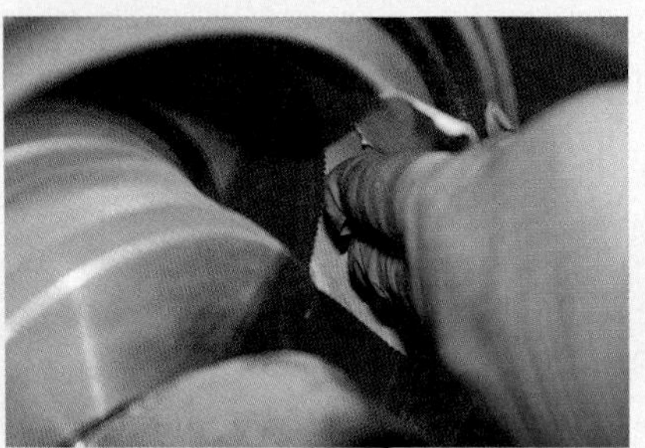

11. Move the drum well away from the cutting bit, and use sandpaper to give the drum surface a nondirectional finish.

12. Remeasure the drum diameter to determine whether the drum is above the maximum-diameter specifications; if so, discard the drum.

Applied Science

AS-37: Vibrations/Waves: The technician can demonstrate an understanding of the types and causes of vibrations caused by out-of-balance or excessively worn systems.

Sound is very important in diagnosing problems in a motor vehicle. Sound is a series of waves that travel through a gas, liquid, or solid, many of which can be heard by the human ear. For example, the solid rumble strips on the sides of the highway make a continuous noise or vibration when driven over by a motor vehicle.

Sound can help you diagnose a worn wheel bearing that needs replacing. As a wheel bearing becomes worn, the case hardening on the bearing race begins to wear away due to constant friction between the bearing race and the bearing rollers. As the case hardening starts to wear, the bearing race develops low spots in the race. As the flat bearing rollers start to roll over the high and low spots in the bearing race, a noise occurs.

To diagnose the noise of a worn bearing in a motor vehicle, you can hoist the vehicle, spin the wheel, and listen for a noise. Or you can test-drive the vehicle and gently turn the steering wheel from side to side. As the vehicle moves from side to side, more weight is placed on each side of the bearing due to load shifting. You will hear a louder noise as you turn away from the side with the faulty bearing.

to be taken when working on them. There is usually a combination of springs, links, levers, guides, and retainers. It is easy to install these items incorrectly. It is good practice to carefully examine each brake assembly before disassembling, to verify that it was assembled correctly previously.

Brake shoes and springs are always replaced in axle sets. If one rear brake assembly has linings contaminated with grease from a failed axle seal, the lining on both wheel brake units will have to be replaced. The same goes with springs. The other parts are left to the discretion of the technician and any shop policies.

Regarding star wheel assemblies, there is some controversy as to whether the threads should be lubricated. The argument on one side is that if you do not lubricate the threads, then they will likely rust and freeze up. The other side argues that any lubricant on the threads will attract brake dust and gum up the threads. There is some truth to both arguments. CDX would suggest using a light coating of lubricant on the threads in areas where the star wheel is likely to come in contact with salt, mud, or water because rust and corrosion are a greater hazard than dust gumming up the threads. In areas where the assembly is unlikely to come in contact with those conditions, use no lubricant on the threads. But always follow the manufacturer's recommended procedure, if specified.

▶ TECHNICIAN TIP

Because it can be hard to remember where all of the parts fit inside of a drum brake assembly, it is good practice to disassemble only one side at a time, leaving the other side as a reference for when you reassemble the first side. Or take a couple of reference pictures before disassembly.

To remove, clean, inspect, and reassemble a duo-servo brake, follow the steps in **SKILL DRILL 47-3**.

To disassemble, clean, inspect, and reassemble a non-servo brake, follow the steps in **SKILL DRILL 47-4**.

SKILL DRILL 47-3 Removing, Cleaning, Inspecting, and Reassembling a Duo-Servo Brake

Note: In the appropriate service information, research and follow the procedure for disassembling the brake assembly. Manufacturers use many drum brake configurations, so the following steps are general in nature and should not be substituted for the manufacturer's procedure.

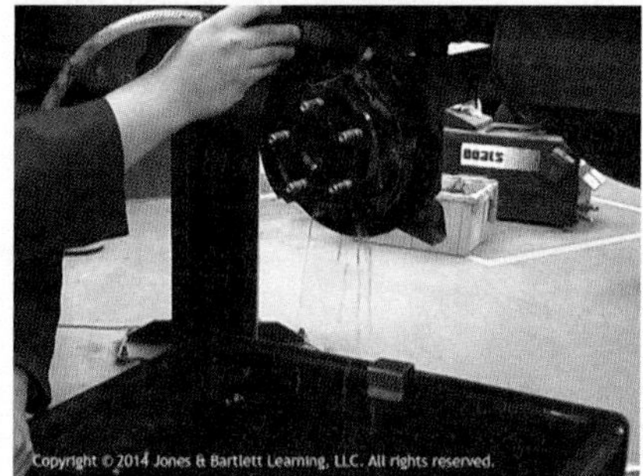

1. Clean the brake shoes, hardware, and backing plates using equipment and procedures for dealing with asbestos and/or dust.

2. To disassemble a duo-servo brake, first remove the return springs, cable guide (if installed), and shoe guide. Set everything aside in the order it will go back on. Remove the parking brake strut and spring.

3. Remove the primary shoe hold-down spring, retainer, and pin.

4. Remove the self-adjuster spring and star wheel assembly and primary shoe.

5. Remove the secondary hold-down spring, retainer, pin, and secondary shoe.

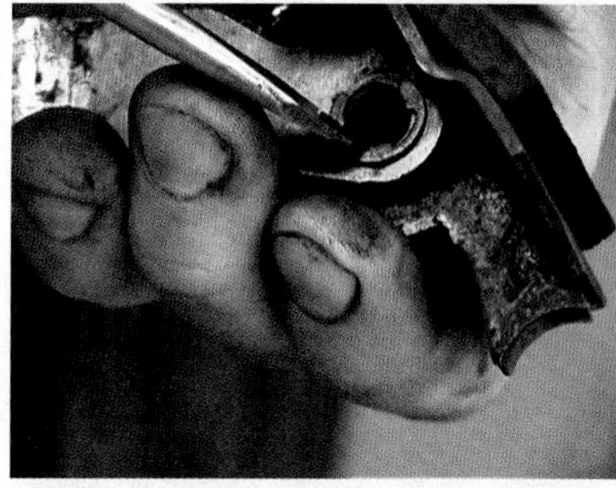

6. Disassemble the parking brake lever from the brake shoe and hardware from the backing plate, being careful not to lose any parts and remembering how they go back together.

7. Finally, clean and inspect all parts according to the manufacturer's procedure.

8. To reassemble a duo-servo brake, first reassemble the parking brake lever on the brake shoe and parking.

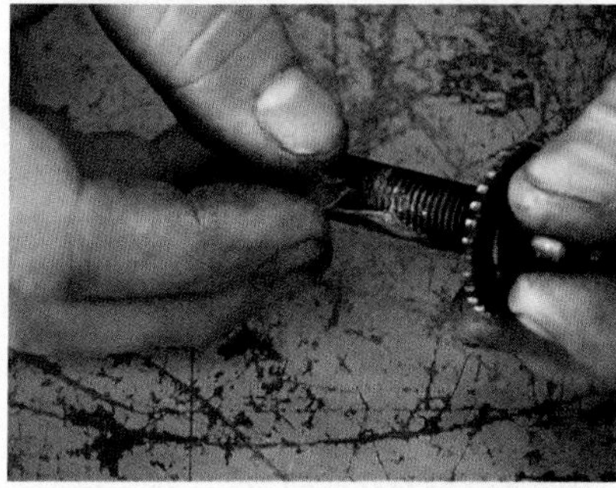

9. Reassemble the star wheel assembly, lubricate the floating end, and set aside. Also lubricate the contact pads on the backing plate.

SKILL DRILL 47-3 Removing, Cleaning, Inspecting, and Reassembling a Duo-Servo Brake (Continued)

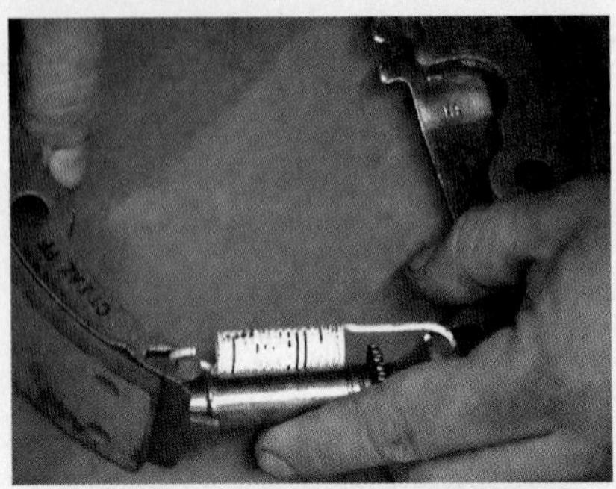

10. Install both shoes to the backing plate with the hold-down spring assemblies.

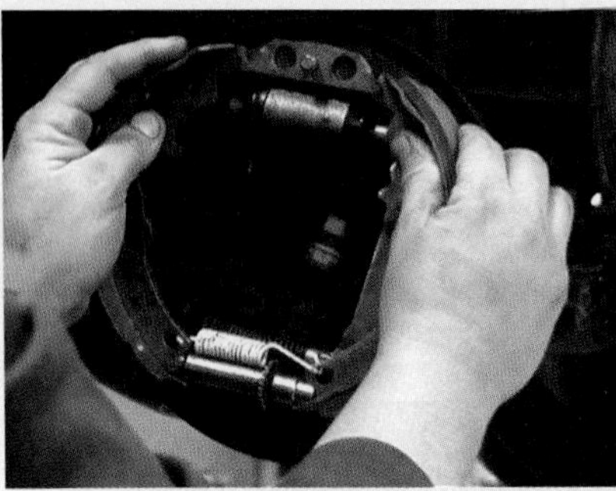

11. Install the shoe guide and self-adjuster cable over the anchor pin.

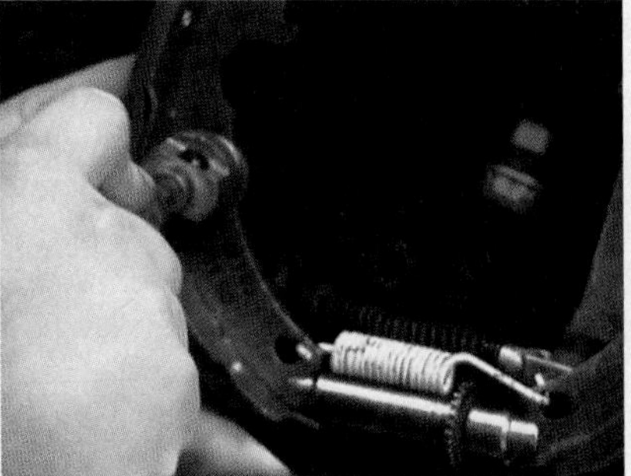

12. Install the cable guide and return spring in the secondary shoe. Also align the wheel cylinder pushrod in the shoe.

13. Position the secondary shoe in place and use brake spring pliers to install the return spring over the anchor pin.

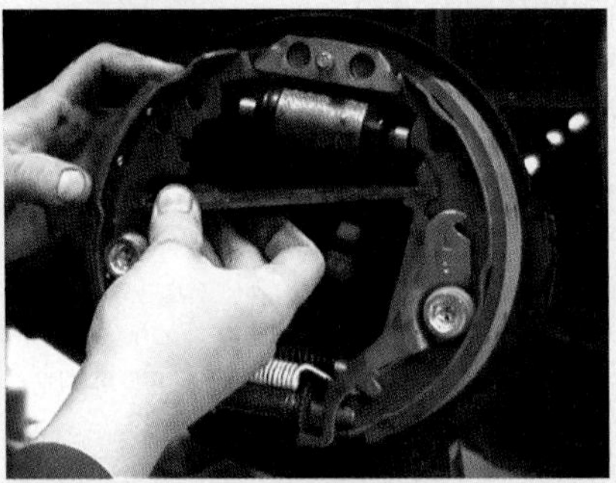

14. Install the parking brake strut rod onto the secondary shoe, and pull the primary shoe engaged with the parking brake strut rod.

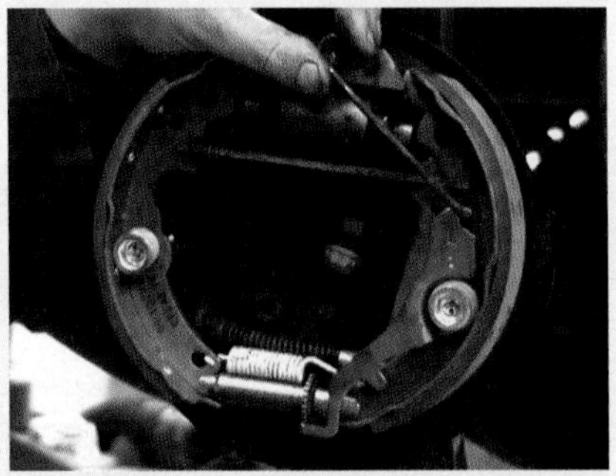

15. Install the return spring in the primary shoe and use brake spring pliers to stretch the return spring over the anchor pin.

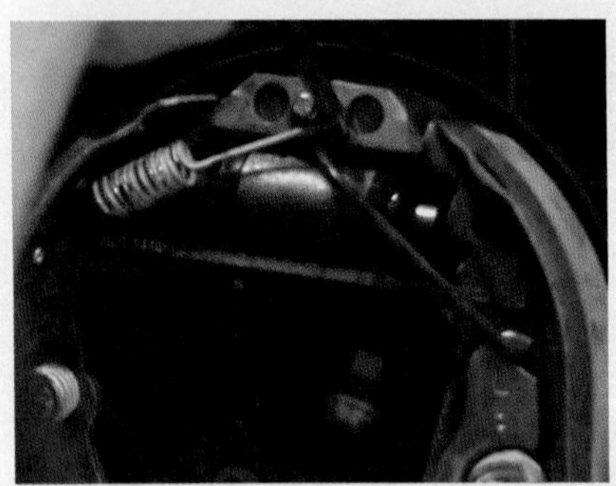

16. Install the self-adjuster link, cable, and spring into position.

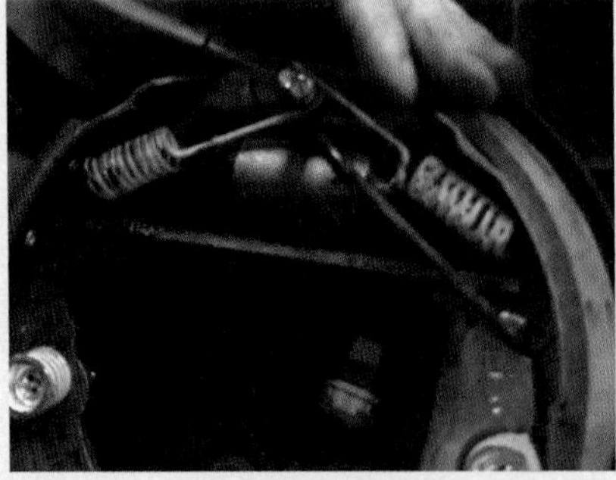

17. Install the star wheel between the bottoms of the two shoes.

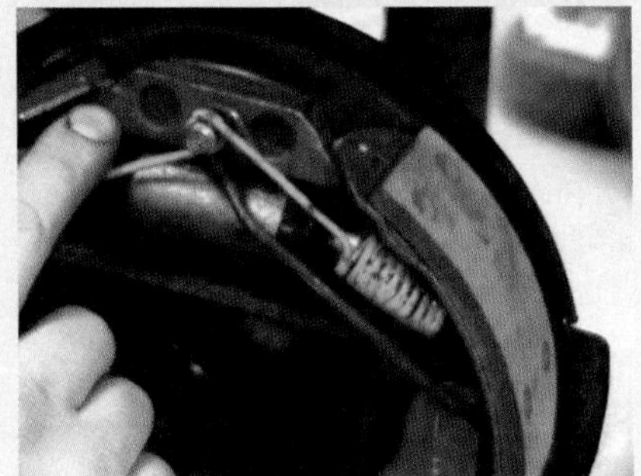

18. Finally, check the fit of all springs, clips, and levers. Operate the self-adjuster cable or lever.

Removing, Inspecting, and Installing Wheel Cylinders

N47005

If a wheel cylinder is leaking or binding, there is a good chance that the cylinder bore is corroded and pitted, requiring removal and replacement of the wheel cylinder. Wheel cylinders also have to be removed if the backing plate is being replaced. Very few shops rebuild wheel cylinders anymore. If you are reusing the wheel cylinder, make sure the bleeder screw can be opened. Because they are hollow, they break off easily, making bleeding virtually impossible.

SKILL DRILL 47-4 Dissembling, Cleaning, Inspecting, and Reassembling a Non-Servo Brake

Note: In the appropriate service information, research and follow the procedure for disassembling the brake assembly. Manufacturers use many drum brake configurations, so the following steps are general in nature and should not be substituted for the manufacturer's procedure.

1. Clean the brake shoes, hardware, and backing plates, using equipment and procedures for dealing with asbestos and/or dust.

2. To disassemble a non-servo brake, first remove the hold-down springs, retainers, and pins.

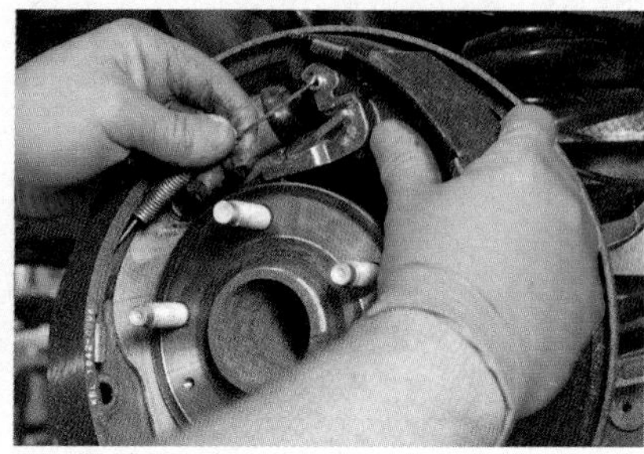

3. Spread the shoes apart and remove the parking brake strut and self-adjuster components.

4. Remove the return springs.

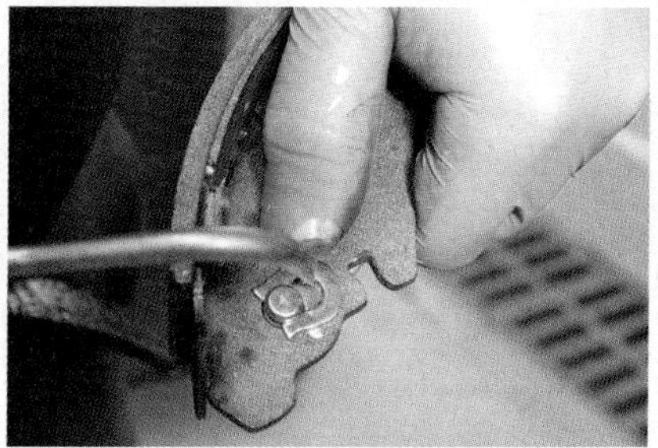

5. Disassemble the parking brake lever from the brake shoe and hardware from the backing plate, being careful not to lose any parts and remembering how they go back together.

6. Clean and inspect all parts according to the manufacturer's procedure.

7. To reassemble a non-servo brake, first assemble and lube the self-adjuster/parking brake strut assembly. Lube the backing plate pads.

8. Place one shoe on the backing plate and install the hold-down spring and pin.

9. Place the self-adjuster/parking brake strut on the installed brake shoe.

SKILL DRILL 47-4 Dissembling, Cleaning, Inspecting, and Reassembling a Non-Servo Brake (Continued)

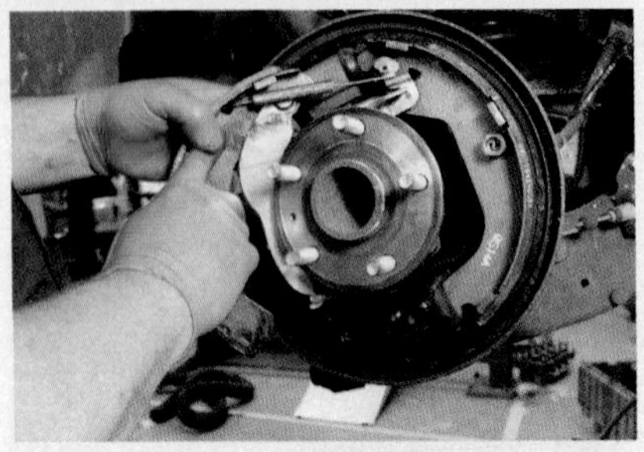

10. Place the retracting springs on both shoes, and fit the loose shoe to the backing plate, being sure to line up the wheel cylinder pushrods, the self-adjuster, and the parking brake mechanism.

11. Install the hold-down spring and pin.

12. Finally, check the fit of all springs, clips, and levers.

> **TECHNICIAN TIP**
>
> If one wheel cylinder is leaking, how long until the other wheel cylinder starts leaking? And if it does leak, will it contaminate the new brake lining? Take into consideration the age and condition of the components when you inspect the system and give the customer your recommendation.

To remove, inspect, and install wheel cylinders, follow the steps in **SKILL DRILL 47-5**.

Pre-Adjusting Brakes and Installing Drums and Wheel Bearings

N47006

Pre-adjusting the brake shoes and parking brake is a routine step in a drum brake job. It saves time because it is faster to adjust the brakes with the drum off than it is when the drum is installed. The pre-adjustment sequence is important. Start by making sure the parking brake is fully released. Then look at the shoes to make sure the parking brake adjustment is not holding the brake shoes in the applied position. You can see this by verifying that the brake shoes are up against their stops on the top and bottom of each shoe. If they are not firmly against them, loosen the parking brake adjustment. Then adjust the service brakes. Once they are adjusted, adjust the parking brake.

To pre-adjust brake shoes and the parking brake and to install brake drums or drum/hub assemblies and wheel bearings, follow the steps in **SKILL DRILL 47-6**.

Installing Wheels, Torquing Lug Nuts, and Making Final Checks

N47007

This procedure, although fairly simple, can result in problems if it is not performed properly. Over-tightening the lug nuts can cause the wheel studs to break either immediately or, worse, after the vehicle has been driven for a period of time. Under-tightening can lead to loosening of the lug nuts and result in the wheel working its way off the vehicle. This can cause the driver to lose control of the vehicle and will potentially result in an accident. Lug nuts should be tightened to the proper torque, in the specified sequence. All manufacturers specify these details for each of their vehicles. The torque pattern is usually in some form of either a star or cross.

Be careful which way you install the lug nuts. Many wheels use a tapered hole that matches the tapered end of the lug nut and centers the wheel on the wheel flange. Other wheels use a flat surface that matches flat surfaces on the lug nuts. No matter what, the lug nut surface *must* match the mating surface of the wheel. Always check that these surfaces match.

When installing lug nuts, the weight should be off the vehicle. The lug nuts should easily center the wheel on the hub. Proper centering is especially important on aluminum wheels that use a flat lug nut sealing surface. The lug nuts can dig into the sides of the lug nut holes and cause the wheel not to center correctly. Once the lug nuts are up against the wheel, work the wheel up onto the lug nut shafts. Then you can torque them down properly.

> **TECHNICIAN TIP**
>
> There is some controversy regarding the use of a lubricant or antiseize on wheel studs and lug nuts. Because the purpose of a lubricant or antiseize is to prevent the components from sticking, CDX errs on the side of not using those products, for fear of the lug nuts coming loose. Also, the torque given for the lug nuts is dry (no lubricant), so torquing them to specifications would lead to over-tightening. At the same time, in areas of the country prone to rust, it is understandable why some want to put a very light amount of antiseize only on the threads of the wheel stud, being careful not to get any on the contact seat of the wheel and lug nut. Doing so would prevent rust from building up on the threads. However, the lug nut torque would need to be reduced due to the lubricant. CDX prefers to install lug nuts dry and avoid this issue altogether. Many shops have their customers sign on the repair order that they will return to the shop so the lug nut torque can be rechecked after 50–500 miles of driving, to make sure the wheels have not loosened.

To install the wheel, torque lug nuts, and make final checks and adjustments, follow the steps in **SKILL DRILL 47-7**.

SKILL DRILL 47-5 Removing, Inspecting, and Installing Wheel Cylinders

1. Peel back the dust boots, and check for brake fluid behind them. Determine any necessary actions.

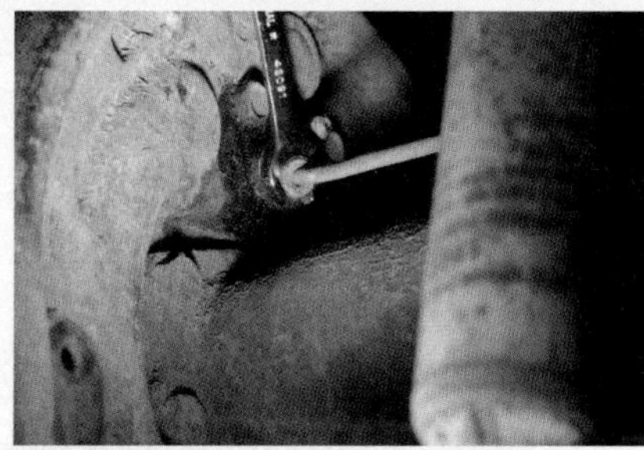
2. Use a flare nut or line wrench to unscrew the brake line from the wheel cylinder.

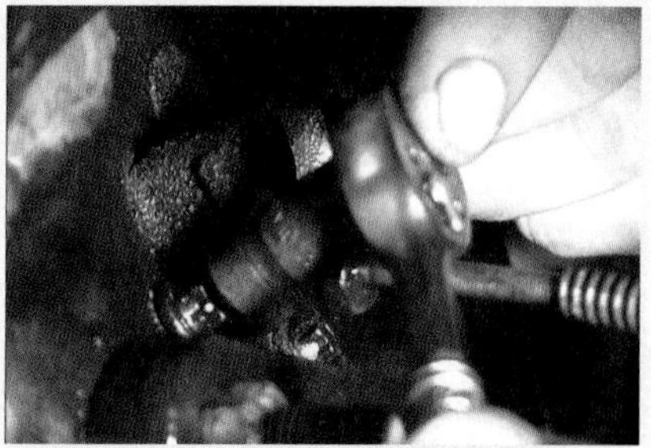
3. Remove wheel cylinder mounting bolts and wheel cylinder from the backing plate.

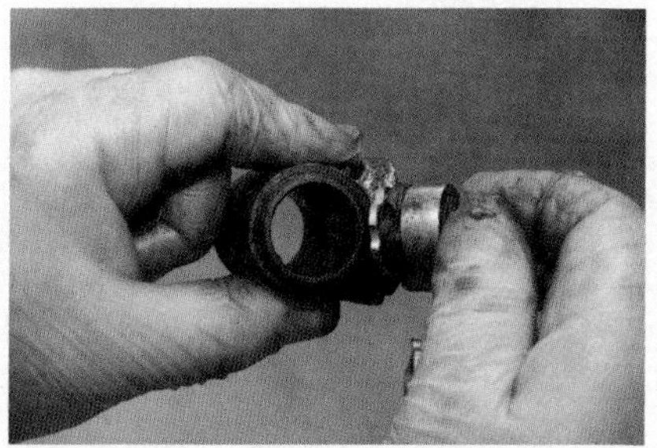
4. Disassemble the wheel cylinder, and inspect each part. Determine any necessary actions.

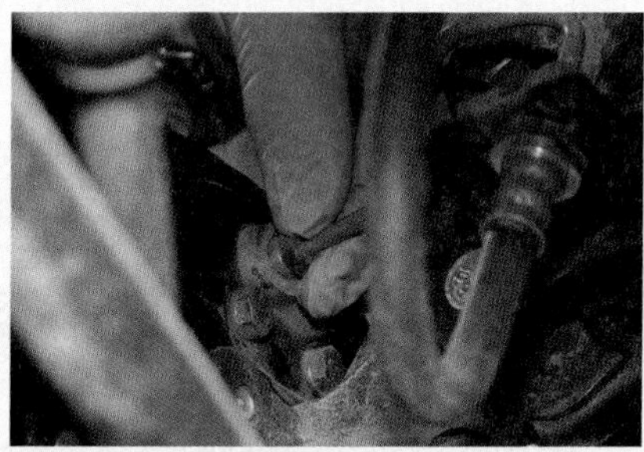
5. If the cylinder can be reused, rebuild it, preferably with new seals and dust boots. If not, replace it with a new wheel cylinder. Reinstall the brake line by hand.

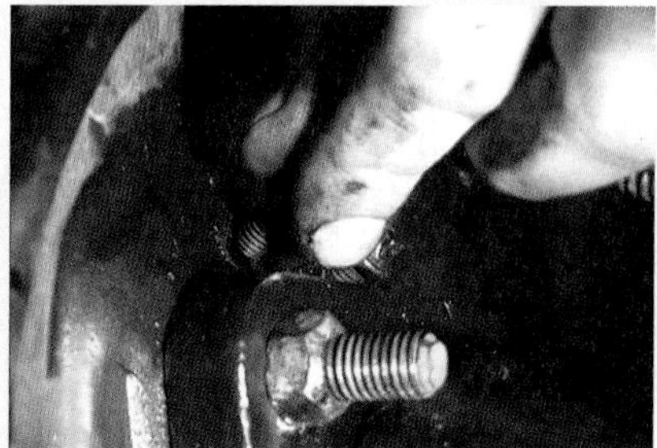
6. Install and tighten any mounting screws. After that, tighten the brake line with a flare nut or line wrench.

SKILL DRILL 47-6 Pre-Adjusting Brakes and Installing Drums and Wheel Bearings

1. Make sure the brake shoes are fully up against their stops and centered on the backing plate.

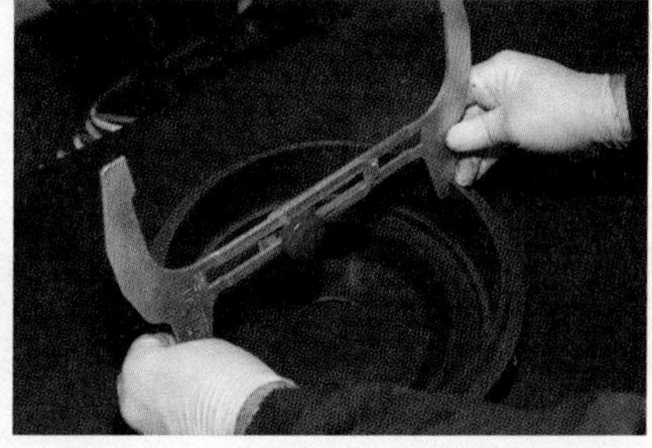
2. Set the pre-adjustment gauge to the drum diameter, and lock it in place.

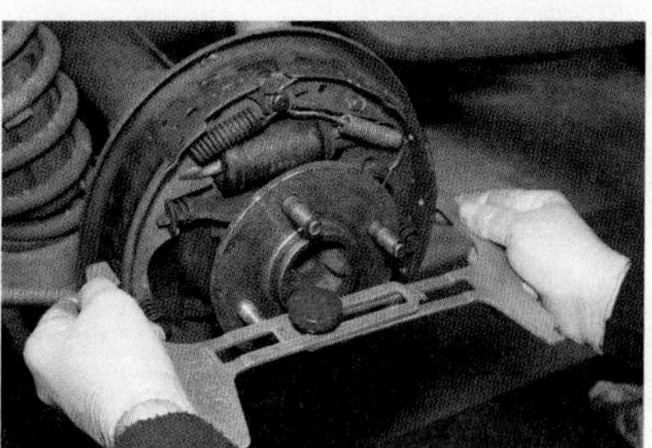
3. Place the pre-adjustment gauge over the center of the brake shoes.

SKILL DRILL 47-6 Pre-Adjusting Brakes and Installing Drums and Wheel Bearings (Continued)

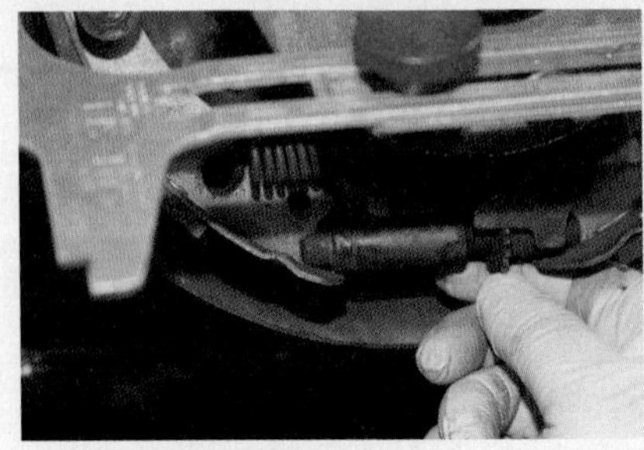

4. Adjust the star wheel until the centers of the brake shoes just contact the pre-adjustment gauge.

5. Test install the brake drum to verify the drum fits. Adjust the shoes as necessary.

6. Adjust the parking brake according to the manufacturer's procedure. If the drum has serviceable wheel bearings, repack, install, adjust, and secure them.

SKILL DRILL 47-7 Installing Wheels, Torquing Lug Nuts, and Making Final Checks

1. Start the lug nuts on the wheel studs by hand. Carefully run all of the lug nuts down so they are seated in the wheel. Lower the vehicle so the tires are partially on the ground. Use a torque wrench to tighten each lug nut to the proper torque in the proper sequence. Once all of the lug nuts have been torqued, go around them again, this time in a circular pattern.

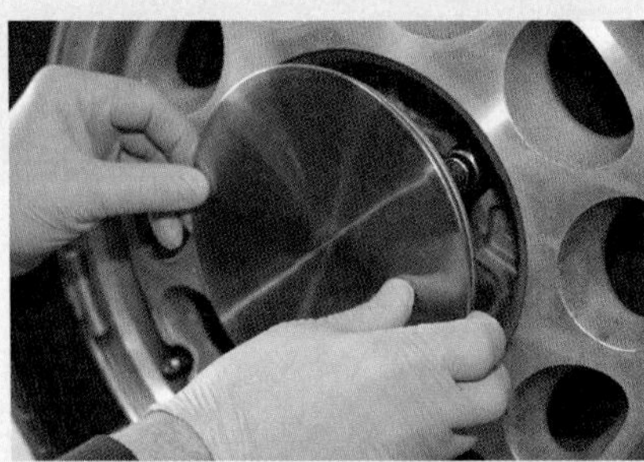

2. Reinstall hubcaps and valve stem caps.

3. Check the brake fluid level in the master cylinder reservoir. Start the vehicle and check the brake pedal for proper feel and height. Check the parking brake for proper operation. Inspect the system for any brake fluid leaks and loose or missing fasteners.

▶ Wrap-Up

Ready for Review

- The main components of the drum brake system are brake drum, backing plate, wheel cylinder, brake shoes, springs and clips, automatic brake self-adjuster, and parking brake mechanism.
- Drum brakes are usually located on the rear wheels of disc-drum applications.
- Hydraulic pressure forces pistons to move the brake shoes into contact with the inside of the brake drums.
- Leading brake shoes are applied in the same direction as the brake drum's forward rotation and are self-energizing; trailing brake shoes are applied in the opposite direction of the brake drum's forward rotation and are not self-energizing.
- Drum brake systems can be twin leading shoe (least common), leading/trailing shoe, and duo-servo.
- Twin leading shoe brake systems are more efficient in forward braking than reverse braking.

- Leading/trailing shoe brake systems provide equal, but not maximum, braking in both directions.
- Duo-servo drum brake systems use servo action for the front brake shoe to multiply the force to the rear shoe, causing rear brake shoe to do more braking work.
- Brake drums are machined to the manufacturer's specified standard diameter and should be replaced once worn or refinished beyond specifications.
- Brake drums styles are hub or hubless (most common in passenger vehicles).
- All brake unit components except the brake drum are mounted to the backing plate.
- Brake shoe contact pads provide a smooth surface for the brake shoes to ride on.
- Wheel cylinders contain a housing, pistons, piston lip seals, spring and expander set, dust boots, push rods, and a bleeder screw.
- Wheel cylinders should be replaced rather than honed if corrosion or pitting occurs.
- Wheel cylinders are single acting (one piston, force generated in one direction) or dual acting (two pistons, force generated in two directions).
- Steel brake shoes have lining material bonded (or riveted) to them to create friction for braking.
- Duo-servo brake systems have primary brake shoes (front of vehicle, shorter shoe lining) and secondary brake shoes (rear of vehicle, longer shoe lining).
- Drum brakes may make the following noises: groaning (excess brake dust in the drum), grinding (worn friction lining), or clicking (grooves worn into backing plate contact pads or rough finish cut on the drum).
- Drum brakes use return, hold-down, and specialty springs.
- Drum brakes must have self-adjusters to maintain proper shoe-to-drum clearance.
- Diagnosis involves understanding the customer concern, test-driving the vehicle, and visually inspecting the brake system.
- Drum brakes require periodic maintenance and repair as the vehicle ages.
- Brake drums should be inspected visually for any damage and measured with a drum micrometer to verify that they are smaller than the specified maximum diameter, and not warped.
- Brake drums need to be refinished when they have excessive grooving or are out-of-round. They are refinished using a brake lathe.
- Brake shoes and springs are always replaced in axle sets.
- It is good practice to disassemble only one side at a time, leaving the other side as a reference for when you reassemble the first side.
- If a wheel cylinder is leaking or binding, the wheel cylinder will have to be replaced.
- Pre-adjusting the brake shoes and parking brake saves time because it is faster to adjust the brakes with the drum off than it is when the drum is installed.
- Lug nuts should be tightened to the proper torque, in the specified sequence.

Key Terms

anchor pin A component of the backing plate that takes all of the braking force from the brake shoes.

automatic brake self-adjuster A system on drum brakes that automatically adjusts the brakes to maintain a specified amount of running clearance between the shoes and drum.

brake drum A short, wide, hollow cylinder that is capped on one end and bolted to a vehicle's wheel; it has an inner friction surface that the brake shoe is forced against.

brake lathe A tool used to refinish the drum surface by removing a small amount of metal and returning it to a concentric, nondirectional finish.

brake shoe A steel shoe and brake lining friction material that apply force to the brake drum during braking.

brake shoe adjustment gauge An adjustable tool used to pre-adjust the brake shoes to the diameter of the brake drum.

brake spoon A tool used to adjust the brake lining-to-drum clearance when the drum is installed on the vehicle.

brake spring pliers A tool used for removing and installing brake return springs.

cylinder bore The inside diameter of a cylinder.

drum brake micrometer A tool used for measuring the inside diameter of the brake drum.

duo-servo drum brake system A system that uses servo action in both the forward and reverse direction.

hold-down springs Springs that hold the brake shoes against the backing plate.

hold-down spring tool A tool used for removing and installing hold-down springs.

leading shoes Brake shoes that are installed so that they are applied in the same direction as the forward rotation of the drum and thus are self-energizing.

leading/trailing shoe drum brake system Type of brake shoe arrangement where one shoe is positioned in a leading manner and the other shoe in a trailing manner.

parking brake cable removal tool A tool used to compress the spring steel fingers of the parking brake cable so that the cable can be removed from the backing plate.

parking brake mechanism A mechanism that operates the brake shoes or pads to hold the vehicle stationary when the parking brake is applied.

return springs Springs that retract the brake shoes to their released position.

self-energizing The property of drum brakes that assists the driver in applying the brakes; when brake shoes come into contact with the moving drum, the friction tends to wedge the shoes against the drum, thus increasing the braking force.

servo action A drum brake design where one brake shoe, when activated, applies an increased activating force to the other brake shoe, in proportion to the initial activating force; further enhances the self-energizing feature of some drum brakes.

specialty springs Springs used to return links and levers on the parking brake system or the self-adjuster mechanism.

springs and clips Various devices that hold the brake shoes in place or return them to their proper place.

trailing shoes Brake shoes installed so that they are applied in the opposite direction to the forward rotation of the brake drum; not self-energizing and less efficient at developing braking force.

twin leading shoe drum brake system Brake shoe arrangement in which both brake shoes are self-energizing in the forward direction.

wheel cylinder A hydraulic cylinder with one or two pistons, seals, dust boots, and a bleeder screw that pushes the brake shoes into contact with the brake drum to slow or stop the vehicle.

wheel cylinder piston clamp A tool that prevents the pistons from being pushed out of the wheel cylinders while the brake shoes are being replaced.

Review Questions

1. The component of the drum brake system attached to the backing plate is the:
   - **a.** automatic self-adjuster.
   - **b.** star wheel.
   - **c.** wheel cylinder.
   - **d.** parking brake lever.
2. All of the following statements with respect to brake systems are true *except*:
   - **a.** drum brakes are self-energizing.
   - **b.** trailing shoes are self-energizing.
   - **c.** leading shoes are self-energizing.
   - **d.** wedging effect assists the driver in applying the brakes.
3. Which of these drum brake systems uses servo action in both directions?
   - **a.** Leading shoe drum brake system
   - **b.** Trailing shoe drum brake system
   - **c.** Twin leading shoe drum brake system
   - **d.** Duo-servo drum brake systems
4. Choose the correct statement.
   - **a.** Brake drums are usually made from fiber.
   - **b.** Hubless drums are more expensive than hub-style drums.
   - **c.** Brake drums can be refinished on the vehicle.
   - **d.** The backing plate has holes stamped in it.
5. The component that prevents a direct path for water spray and dirt to enter the brake drum is the:
   - **a.** labyrinth seal.
   - **b.** wheel bearing.
   - **c.** axle seal.
   - **d.** wheel cylinder dust boot.
6. All of the following statements with respect to a wheel cylinder and piston are true *except*:
   - **a.** It is standard practice to hone out deep pits in the wheel cylinder bore..
   - **b.** The piston is usually made of anodized aluminum.
   - **c.** The wheel cylinder uses a seal to seal the wheel cylinder.
   - **d.** The wheel cylinder is fitted with a bleeder screw to allow bleeding of air.
7. Choose the correct statement.
   - **a.** The composition of the lining material does not affect brake operation.
   - **b.** The primary shoe is responsible for doing most of the braking work.
   - **c.** Riveted linings are used on heavier duty or high-performance vehicles.
   - **d.** Drum brakes create more noise than disc brakes.
8. When the driver releases the brake pedal, the brake shoes are withdrawn by the:
   - **a.** hold-down spring.
   - **b.** return spring.
   - **c.** specialty spring.
   - **d.** coil spring.
9. Among the tools used to diagnose and repair drum brake systems, the wheel cylinder piston clamp is used to:
   - **a.** install parking brake cables on the parking brake lever.
   - **b.** hold the pistons in the wheel cylinder while the brake shoes are removed.
   - **c.** pre-adjust brake shoes before installing the drum.
   - **d.** remove and install hold-down springs.
10. Which of the following brake tools is used to adjust brake shoes when the drum is installed?
    - **a.** Flare nut wrench
    - **b.** Brake spring pliers
    - **c.** Brake spoon
    - **d.** Wheel cylinder piston clamp

ASE Technician A/Technician B Style Questions

1. Tech A says that parking brakes on drum brake vehicles use separate brake shoes for backup in an emergency. Tech B says that parking brakes on drum brake vehicles mechanically operate the standard drum brake shoes. Who is correct?
   - **a.** Tech A
   - **b.** Tech B
   - **c.** Both A and B
   - **d.** Neither A nor B
2. Tech A says that most brake drums are designed to be machined if minor surface issues are present. Tech B says that brake drums can be reused if they are machined larger than specifications, as long as the surface is smooth. Who is correct?
   - **a.** Tech A
   - **b.** Tech B
   - **c.** Both A and B
   - **d.** Neither A nor B

3. Tech A says that duo-servo brake shoes are only anchored on the top. Tech B says that duo-servo brakes typically use single-piston wheel cylinders. Who is correct?
   a. Tech A
   b. Tech B
   c. Both A and B
   d. Neither A nor B
4. Tech A says that both secondary shoes should be installed on the passenger side of the vehicle. Tech B says that the lining on the primary shoe is typically shorter in length than the lining on the secondary shoe. Who is correct?
   a. Tech A
   b. Tech B
   c. Both A and B
   d. Neither A nor B
5. Tech A says that a cleaned-up self-adjuster can be installed on either side of the vehicle. Tech B says that grease seals are meant to be reusable. Who is correct?
   a. Tech A
   b. Tech B
   c. Both A and B
   d. Neither A nor B
6. Tech A says to use an air hose to clean the backing plate of dust and contamination. Tech B says to use a water-based cleaning solution to clean the backing plate of dust and contamination. Who is correct?
   a. Tech A
   b. Tech B
   c. Both A and B
   d. Neither A nor B
7. Tech A says that riveted lining is usually for heavy-duty or high-performance vehicles. Tech B says that bonded lining is glued to the metal brake shoe. Who is correct?
   a. Tech A
   b. Tech B
   c. Both A and B
   d. Neither A nor B
8. Tech A says that a metal-on-metal grinding noise in drum brakes generally requires replacing the brake shoes and resurfacing or replacing the drums. Tech B says that one way to help identify whether a problem is coming from the front or rear brakes is to test-drive the vehicle in a safe place at a relatively low speed and lightly apply the parking brake. Who is correct?
   a. Tech A
   b. Tech B
   c. Both A and B
   d. Neither A nor B
9. Tech A says that brake shoe linings saturated with brake fluid from a leaky wheel cylinder can be cleaned with brake cleaner and reused as long as they aren't worn out. Tech B says that brake shoes should be replaced in axle sets. Who is correct?
   a. Tech A
   b. Tech B
   c. Both A and B
   d. Neither A nor B
10. Tech A says that when performing a brake job on the rear axle of an older vehicle, inspection finds brake fluid under the dust boot; the wheel cylinder should be replaced. Tech B says that when a drum brake return spring has failed, springs on both rear wheels need to be replaced. Who is correct?
    a. Tech A
    b. Tech B
    c. Both A and B
    d. Neither A nor B

CHAPTER 48
Wheel Bearings

NATEF Tasks

- **N48001** Diagnose wheel bearing noises, wheel shimmy, and vibration concerns; determine needed action. (AST/MAST)
- **N48002** Remove, clean, inspect, repack, and install wheel bearings; replace seals; install hub and adjust bearings. (MLR/AST/MAST)
- **N48003** Remove, inspect, service and/or replace front and rear wheel bearings. (AST/MAST)
- **N48004** Replace wheel bearing and race. (MLR/AST/MAST)
- **N48005** Remove, reinstall, and/or replace sealed wheel bearing assembly. (AST/MAST)

Knowledge Objectives

After reading this chapter, you will be able to:

- **K48001** Explain basic wheel bearing theory.
- **K48002** Describe the components of wheel bearings.
- **K48003** Describe the types of wheel bearings.
- **K48004** Describe cylindrical roller bearings.
- **K48005** Describe tapered roller bearings.
- **K48006** Describe ball bearings.
- **K48007** Explain how sealed bearings are different from serviceable bearings.
- **K48008** Explain the purpose and function of grease and axle seals.
- **K48009** Describe the differences between rear drive axle wheel bearing arrangements.
- **K48010** Describe bearing lubricants and their application.
- **K48011** Explain the process of adjusting wheel bearings.

Skills Objectives

After reading this chapter, you will be able to:

- **S48001** Perform wheel bearing maintenance and repair.

▶ Introduction

In this chapter, you learn about each wheel bearing component and the role wheel bearings play in helping to keep the vehicle safe. Because wheel bearings can fail, you learn how to accurately diagnose common issues and recommend the proper service or repair to the customer. You also learn how to properly maintain and, if necessary, replace today's wheel bearings.

▶ Wheel Bearing Theory

K48001

Wheel bearings are a commonly overlooked component. All wheel bearings used to be the serviceable type, requiring periodic maintenance every 24,000–30,000 miles (38,624–48,280 km), consisting of disassembly, cleaning, inspecting, lubricating, reassembling, and adjusting. With the introduction of sealed wheel bearings that do not need periodic service, many manufacturers switched to using them on most of their lighter-duty vehicles. As a result, many technicians do not give the wheel bearings as much consideration as they once did. However, both types of wheel bearings can fail, and vehicles with serviceable wheel bearings still need maintenance, so it is important that you become familiar with wheel bearings.

▶ Wheel Bearings Overview

K48002

Wheel bearings play a critical role in the vehicle by allowing the wheels to roll with a minimum of friction while still maintaining accurate wheel positioning under all driving conditions. Most wheel bearing assemblies include the following components: an **outer race**, an **inner race**, **roller bearings** or **ball bearings**, and a **bearing cage** to hold the rollers or balls in place (**FIGURE 48-1**). The rollers or balls are made of hardened metal and roll between the two races. The races are also made of hardened metal and are carefully formed to match the contour of the rollers or balls, so all of the components roll easily against each other.

Each race either fits firmly within a housing or on a shaft. In many situations, the races are held in place by an **interference fit** with the housing or shaft, which means they must be pressed into place with a high amount of force. In other words, the races are designed so that when they are installed, they should not rotate in or on their respective components. The rollers or balls are specifically designed to provide the rolling function. Because the races, rollers, and balls are made of hardened metals, they are designed to resist wear and damage. However, overloading the vehicle can put the wheel bearings under a greater load than they are rated for, which can cause them to fail. In the same way, using improper or insufficient lubricant can cause them to fail. Wheel bearing failure is discussed in detail later in this chapter.

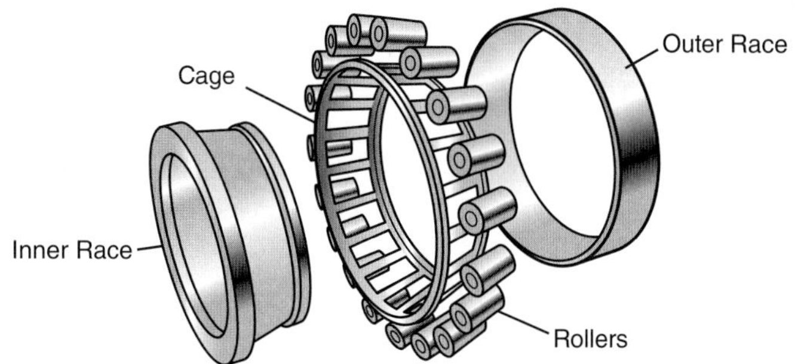

FIGURE 48-1 Components of a typical wheel bearing.

There are two categories of wheel bearings: **serviceable bearings** and **sealed bearings**. Serviceable bearings are designed so they can be disassembled and serviced, whereas sealed bearings are not. Serviceable bearings must be serviced periodically by disassembling, cleaning, inspecting, repacking them with the specified lubricant, reinstalling, and adjusting them. Sealed bearings are designed so they cannot be disassembled or adjusted. Sealed bearings are manufactured with the proper clearance and filled with the specified lubricant from the factory. They are designed to last the life of the vehicle, but bearings do wear out or fail occasionally and have to be replaced. We discuss sealed bearing failure and replacement in detail later in the chapter.

> ▶ TECHNICIAN TIP
>
> Roller and ball bearing assemblies are called **antifriction bearings** because the components are in rolling contact with one another and therefore have minimal friction. A sleeve-type bearing, or bushing, such as a clutch pilot bushing, is called a **friction bearing**, as the components are in sliding contact with one another. Antifriction bearings run much more freely than friction bearings.

You Are the Automotive Technician

A customer pulls her 2008 Ford F250 pickup truck into your shop, complaining that the vehicle is making a howling sound when driving more than about 20 mph. The noise seems to gradually increase as the speed of the vehicle increases. She is concerned that the vehicle might break down and leave her stranded. You ask her where the noise is coming from, and she explains it is coming from the center or rear of the vehicle. You notice that the truck is equipped with a solid live rear axle, and the tires are in good shape, with a fairly smooth highway tread on them.

1. On a test drive, how can you determine whether the noise is coming from a wheel bearing or a transmission bearing?
2. What wheel bearing arrangements can be used on a solid live rear axle, and how do they differ?
3. How are rear wheel bearings lubricated on each type of solid live rear axle?
4. How is a sealed bearing different from a serviceable bearing?

▶ Wheel Bearing Types

K48003

Wheel bearings commonly are of the following types: cylindrical roller bearing, tapered roller bearing, ball bearing, double-row ball bearings, and double-row tapered roller bearings. Each one is designed for a particular application. For example, roller bearings support the load over a larger surface area so they can carry heavier loads than ball bearings. Manufacturers determine the type of wheel bearing they will use based on the particular application and its requirements. You need to be familiar with each type of wheel bearing to successfully service a variety of vehicles.

Cylindrical Roller Bearings

K48004

Cylindrical roller bearing assemblies use rollers that are cylindrical in shape so that the races are parallel to each other with the rollers between them (**FIGURE 48-2**). This type of wheel bearing assembly is used in situations where the wheel bearing is not subject to side loads. It is common in rear axles of rear-wheel drive vehicles, where the rear axles are held from moving side to side by means other than the wheel bearings, such as a differential assembly that uses thrust bearings to prevent the axle from moving side to side. As all side-to-side movement is controlled within the differential assembly, the cylindrical roller bearing assemblies solely support the weight of the vehicle.

In many instances, cylindrical roller bearing assemblies use the surface of the axle as the inner bearing race. In this case, the cylindrical roller bearings ride directly on the axle shaft. If bearing assembly fails, the axle shaft will most likely be damaged and have to be replaced along with the bearing assembly. However, some bearing manufacturers have designed replacement cylindrical roller bearing assemblies that ride farther out on the axle shaft than the original bearing, so the axle may not have to be replaced in this situation.

Cylindrical roller bearing assemblies are manufactured to have the proper running clearance between the rollers and races, so no adjustment of this type of wheel bearing is needed. But like all bearings discussed in this chapter, they do require lubrication to cushion and cool the rollers and races while operating.

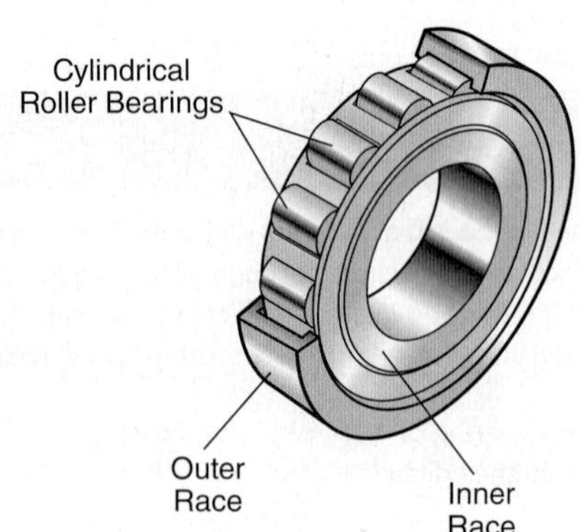

FIGURE 48-2 Cylindrical roller bearing assembly.

Tapered Roller Bearings

K48005

Tapered roller bearings have races and rollers that are tapered in such a manner that all of the tapered angles meet together at a common point (**FIGURE 48-3**). This design allows the tapered rollers to freely roll between the angled inner and outer races. The tapered rollers are contained in a bearing cage that holds the tapered bearings to the inner race as a unit. The inner race is called the cone, and the outer race is called the cup. Together, the cone and cup make up a tapered roller bearing assembly (**FIGURE 48-4**).

Tapered roller bearing assemblies are commonly used where heavy loads and side loads (thrust) have to be supported. Wheel bearing side load conditions occur when the vehicle is cornering. When the wheel is being turned, the vehicle wants to keep going straight. This pushes the bottom of the outside wheel inward and the bottom of the inside wheel outward, which puts them both under a side load condition. In this situation, cylindrical roller bearing assemblies would just allow the wheel and axle to slide sideways, which means they cannot control the side load (thrust) condition. However, tapered roller bearing assemblies can.

Because the components are on an angle to the centerline of (and not parallel to) the axle shaft, they can control side movement (thrust) in one direction. Because of this, tapered roller

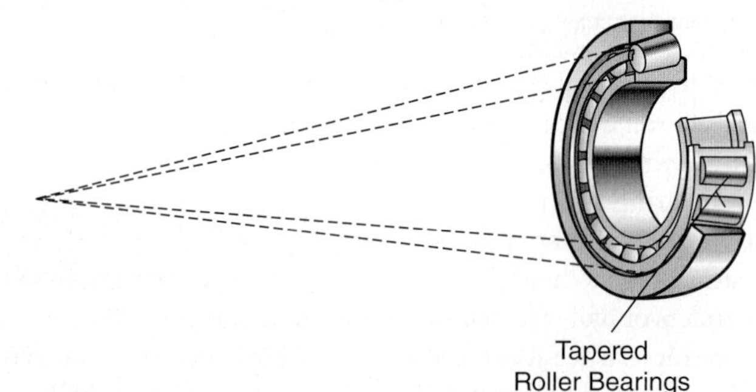

FIGURE 48-3 With tapered roller bearings, the common axis of bearings and races provides minimal rolling resistance.

FIGURE 48-4 **A.** Cone. **B.** Cup.

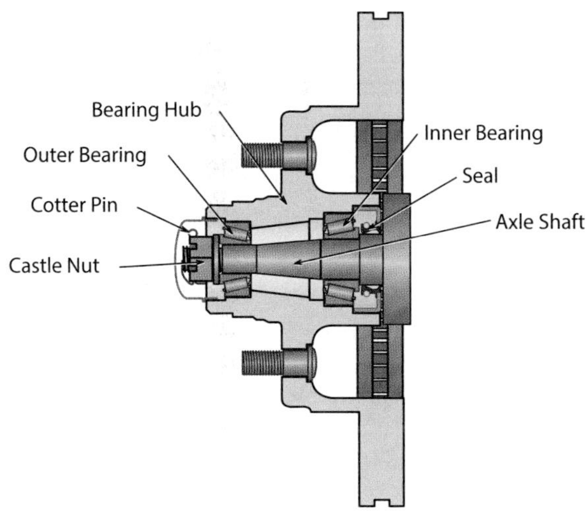

FIGURE 48-5 Tapered roller bearings used to control thrust in both directions.

bearing assemblies are generally used in opposing pairs, so they can control side movement (thrust) in both an inward and outward direction. When used in pairs, the individual tapered roller bearing assemblies are generally referred to as inner (or inboard) and outer (or outboard) bearings (**FIGURE 48-5**). The inner bearing assembly is closer to the centerline of the wheel and supports most of the vehicle weight. Because of this, the inner bearing assembly is typically larger than the outer bearing assembly.

Some manufacturers have designed double-row tapered roller bearing assemblies, which combine two opposing tapered roller bearing assemblies into one sealed unit. This design provides excellent side thrust–carrying capacity and at the same time excellent load-carrying capacity. For more information, see the Sealed Wheel Bearings section.

Because tapered roller bearing assemblies use tapered components, the wheel bearing assembly must be adjusted to have the proper **running clearance** between the tapered rollers and races when the bearing assemblies are installed. The running clearance is the amount of space between components during operation. If the tapered roller bearing clearance is too tight, the components will bind and overheat due to increased pressure and because the rollers squeeze out too much grease, making the lubricating film too thin. If the tapered roller bearing is adjusted with too much clearance, excessive side-to-side and up-and-down movement will occur, which can cause the components to hammer against each other, damaging the surfaces of the tapered roller bearings and races.

Ball Bearings

K48006

Ball bearing assemblies consist of an inner race, an outer race, ball bearings, and a ball bearing cage (**FIGURE 48-6**). The balls roll in deep channels in the races. Deeply grooved ball bearing assemblies are used as wheel bearings on a lot of light-duty vehicles because they have a lower rolling resistance. The much smaller contact area between the balls and races prevents a ball bearing assembly from being used on larger vehicles, which experience higher loads. Side loads are controlled by the balls rolling against the sides of the channels. Because there is much less surface area between the balls and the sides of the channels compared to the tapered roller bearing, ball bearing assemblies are limited in how much side load they can handle. At the same time, the small surface area allows them to roll freer than roller bearings, and they therefore help manufacturers decrease drag on a vehicle, which increases fuel efficiency.

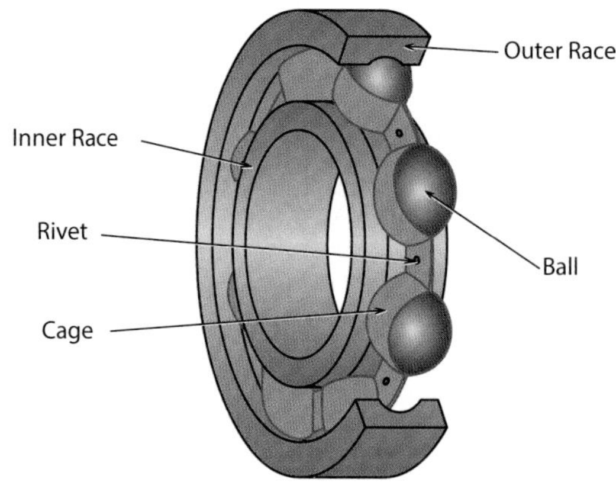

FIGURE 48-6 A typical ball bearing assembly.

Ball bearing assemblies also come in a **double-row ball bearing assembly** configuration (**FIGURE 48-7**). The double-row configuration gives the ball bearing assembly twice the surface contact area, so it can control greater amounts of loads and side loads than a single-row bearing assembly. Virtually all wheel bearings using a ball bearing assembly are of the double-row ball bearing variety, and they are commonly used in automotive light vehicle applications.

The outer race is usually a one-piece unit, whereas the inner race is usually two separate pieces. The inner races are manufactured to butt up against each other to create the correct running clearance when the double-row ball bearing assembly is torqued in place.

Sealed Wheel Bearings

K48007

Some vehicles use sealed wheel bearings. These bearing assemblies are designed and manufactured as completely sealed units. They can be single row or double row, depending on the vehicle application, and can be made up of ball, cylindrical roller, or tapered roller types of bearings. The bearing assemblies are prefilled with

FIGURE 48-7 A double-row ball bearing assembly.

FIGURE 48-8 Sealed bearing.

lubricant and have integrated grease seals built into them to contain the lubricant (**FIGURE 48-8**). They also are manufactured with the proper running clearance, so they do not need to be adjusted. This saves the vehicle manufacturer money by decreasing the time it takes to install them during vehicle assembly. It also makes for a more reliable and consistent installation process, as every unit comes preset at the proper clearance. This style of wheel bearing assembly is also beneficial to vehicle owners because it is designed to last the life of the vehicle during normal use and does not need periodic maintenance, which saves owners money.

Another application for sealed wheel bearings is the **unitized wheel bearing hub** (**FIGURE 48-9**). These usually include either a double-row ball bearing assembly or a double-row tapered roller bearing assembly installed in a housing that is bolted onto the knuckle or axle housing. In many cases, it also has the wheel flange pressed into the inner race of the wheel bearing. In this way, the old bearing assembly can be unbolted and a new one bolted in its place with minimal labor and no adjustments needed. This is virtually a zero maintenance system until it wears out, and then it is usually replaced as a unit. On some vehicles equipped with ABS brakes, the ABS sensor is integrated into the unitized wheel bearing assembly.

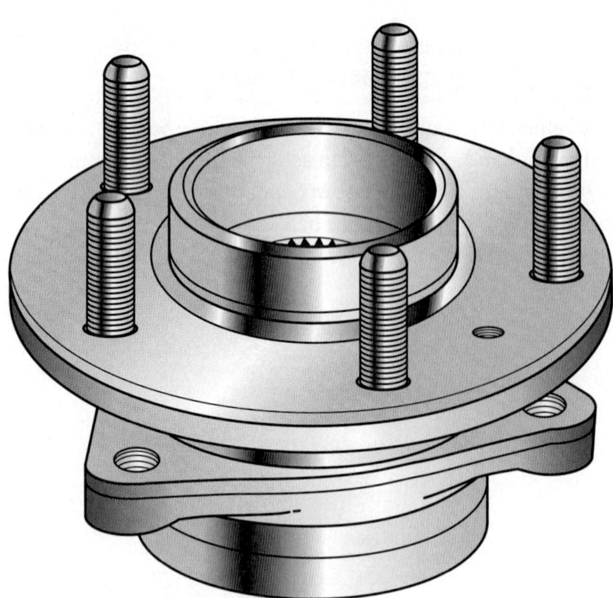

FIGURE 48-9 A unitized wheel bearing hub assembly.

In automotive wheel bearing applications, the double-row ball bearing assembly may come as a unitized wheel bearing hub with the bearing assembly installed in the hub and the wheel flange installed in the center of the bearing assembly.

Grease Seals and Axle Seals

K48008

All wheel bearings rely on some sort of seal to keep the lubricant in and contaminants out. Some wheel bearings have the seal built right into the bearing assembly, such as sealed bearings. Other wheel bearings rely on a completely separate **grease seal**. Wheel bearing grease seals seal against a rotating surface so that lubricants cannot leak out, and dirt and contaminants can't get in. In some applications, the grease seal is a press fit into the axle housing, which is stationary, and seals against the axle shaft, which is rotating. In other applications, the grease seal is press fit into the wheel hub, which rotates and seals against the spindle, which does not rotate.

Most axle seals consist of a stamped sheet metal case and flexible sealing lip with an internal **garter spring** (**FIGURE 48-10**). The metal seal case is constructed to be a press fit when installed in the housing. The outside surface of the metal case usually comes pre-coated with a thin layer of sealer compound to seal minor surface imperfections between the seal case and the housing. The sealing lip is carefully designed and precisely manufactured so that it will seal the specified lubricant. Many seals use a garter spring to help hold the lips of the seal in contact with the shaft it is sealing. The garter spring helps maintain an adequate seal if the parts are slightly out of alignment or if there is a small amount of runout or clearance between the seal and shaft.

Because the garter spring holds the seal against the rotating component, the seal can wear a groove in that component, so always inspect the sealing surface of the rotating component to verify that it doesn't have excessive wear. In some cases, you can purchase a very thin metal repair sleeve that fits snugly over the sealing surface and provides a new surface for the seal to ride

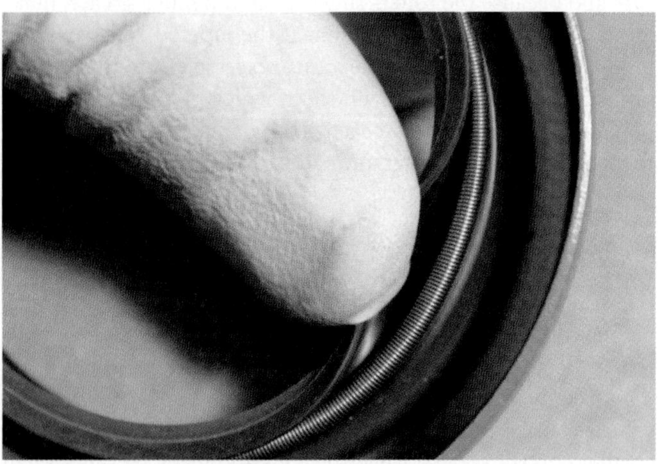

FIGURE 48-10 Components of a typical seal.

against. If a repair sleeve isn't available, the grooved component may have to be replaced if the wear is great enough.

Seals come in a variety of configurations, so it is critical to use the specified seal for the application you are servicing. The sealing lip can also be made from a variety of materials. Always purchase seals from a reputable manufacturer so you can be confident they will do their job and last a long time. When you install a seal, always remember to pre-lube the sealing lip with a small amount of oil or grease. This will prevent it from overheating during the initial use resulting in premature failure of the seal.

> **TECHNICIAN TIP**
>
> Grease seals are designed to be used only one time. If you have to remove a seal for any reason, replace it with a new one. Failure to do so will likely result in a leaky seal. Also, if you damage a seal during installation, you should replace it with a new one. Always use care when installing seals, and use the correct installation tool.

Wheel Bearing Arrangements for Rear Drive Axles

K48009

Rear drive axles come in three different designations: full floating, semi-floating, and ¾ floating. Each designation refers to how the axle and wheel are supported by the wheel bearings. It is also important for you to understand the differences so that you will be able to service each style properly. In a full floating axle arrangement, the axle only carries a twisting force. The weight of the vehicle is fully carried by a pair of tapered roller bearing assemblies, which ride between the hub and axle tube (**FIGURE 48-11**). The axle does not carry any of the vehicle load because the wheel is bolted directly to the bearing hub. The hub also controls side thrust. Full floating axles handle heavy loads better than the other styles, so they are used in heavy-duty applications such as many one-ton pickups, trucks, and vans.

In a semi-floating axle, the wheel flange is part of the axle, which is supported by a single bearing assembly (usually a ball or cylindrical roller bearing style) near the flange end of the axle (**FIGURE 48-12**). The bearing assembly rides between the axle

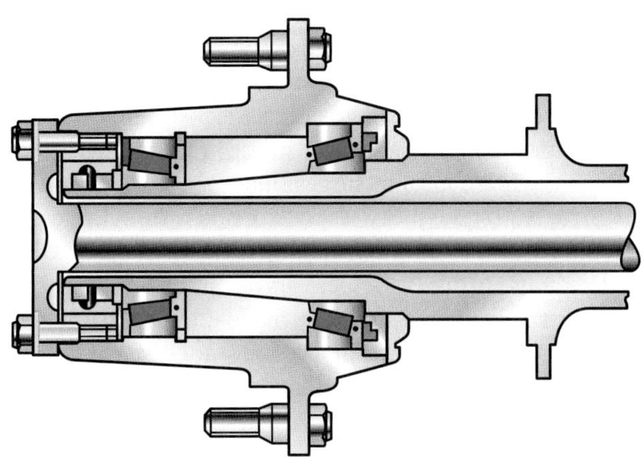

FIGURE 48-11 Full floating axle—two bearings between the axle housing and hub.

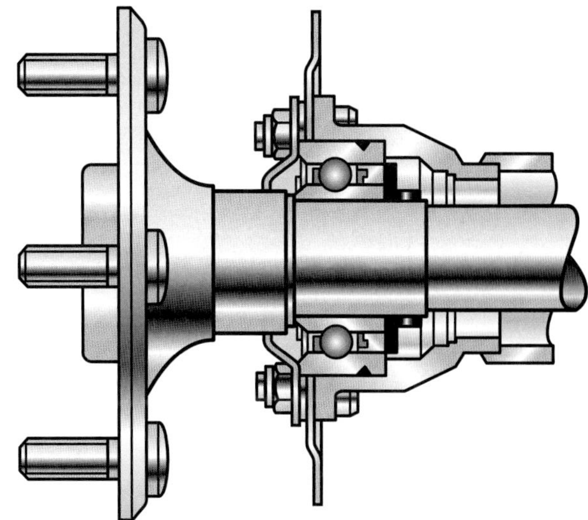

FIGURE 48-12 Semi-floating axle—bearing located between the axle and housing.

and the axle tube, so all of the weight is put on the axle flange, which transfers the weight to the wheel bearing assembly. In this case, the axle and bearing assembly each carry the full weight of the vehicle. The axle also provides the twisting force for the wheel. This arrangement is generally considered the lightest duty of the three types of axle designations.

In a ¾-floating axle design, there is a single bearing assembly (usually a ball or cylindrical roller bearing style) between the outside of the axle tube and the hub (**FIGURE 48-13**). The axle has a wheel flange that bolts to the hub and provides lateral support for the hub and wheel, and the bearing assembly supports the weight of the vehicle. The axle also provides the twisting force for the wheel. This arrangement is generally considered heavier duty than the semi-floating axle, but lighter duty than the full floating axle.

Lubrication

K48010

All wheel bearings require lubrication to extend their useful life. There are two common types of lubricants used—**gear**

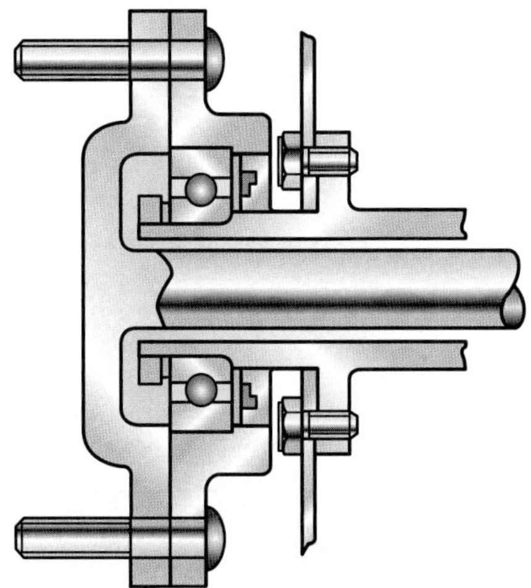

FIGURE 48-13 Three-quarter floating axle—one bearing between the outside of the axle housing and the hub.

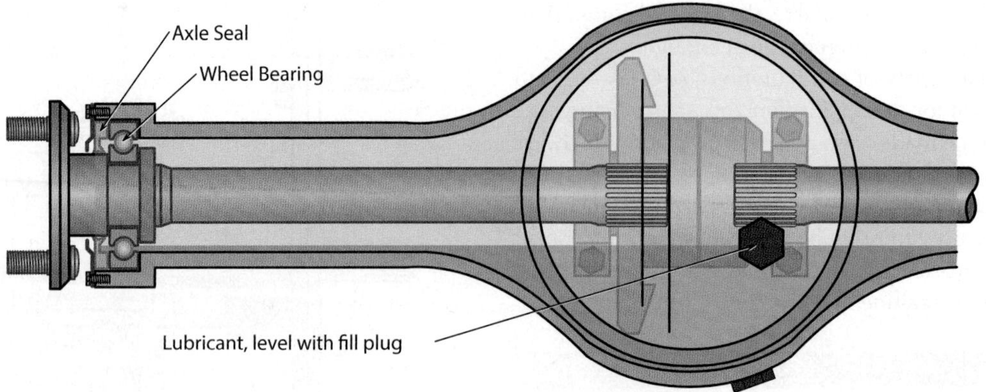

FIGURE 48-14 Gear lube at the proper level lubricates some types of axle bearings.

lube and bearing grease. Each application specifies one or the other. They cannot be substituted, due to the different housing and seal designs each lubricant demands. Gear lube is somewhat thicker than engine oil. Because gear lube flows much more easily than grease, the bearing assembly runs in a housing partially filled with gear lube. The gear lube is thin enough to flow around and between the rollers, lubricating them and helping to prevent overheating of the wheel bearing assemblies.

In many rear-wheel drive vehicles, the wheel bearing assemblies are open to the axle housing, which is partially filled with gear lube. In this case, the gear lube lubricates the wheel bearings and also the final drive assembly (**FIGURE 48-14**). Maintenance of this system involves draining the old gear lube and refilling the system with the specified new gear lube. In most cases, this can be performed without disassembling the wheel bearings by removing a drain plug or bolt and filling the system through the **fill plug** (**FIGURE 48-15**). A fill plug is usually threaded and can be removed to allow the level of a fluid to be checked and filled. Some manufacturers have chosen to use a rubber fill plug that snaps into the fill hole. The level of gear lube should normally be within 0.25" (6.35 mm) of the bottom of the fill plug hole.

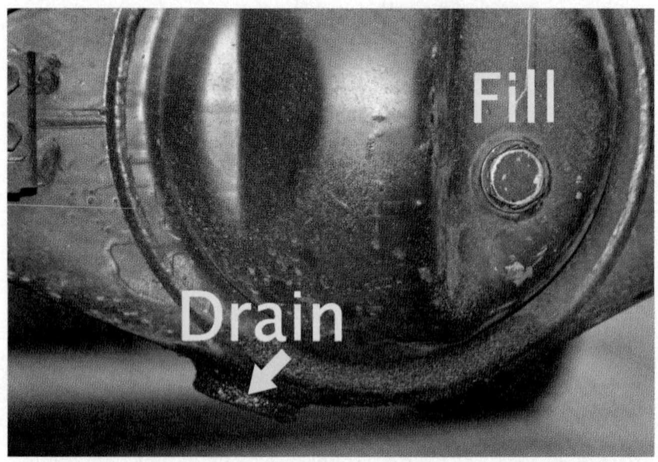

FIGURE 48-15 Drain and fill points for rear axle assembly.

▶ TECHNICIAN TIP

If the level of gear lube is lower than it should be, suspect a leaking axle shaft grease seal. This can usually be verified by looking at the inside of each tire and the back side of the brake backing plate. If gear lube is present, the grease seal is leaking and needs to be replaced. Also check the wheel bearing to make sure it is not faulty; to do so, follow the bearing diagnosis procedure listed later in this chapter.

In order to properly perform maintenance on wheel bearings, you need to understand the different characteristics of gear lube and bearing grease. This helps ensure that you select the proper lubricant for the task you are performing. Gear lube is classified by the Society of Automotive Engineers (SAE) according to its viscosity and by the American Petroleum Institute (API) according to its service grade. **Viscosity** refers to the thickness of the gear lube; the higher the number, the thicker the gear lube. Vehicle manufacturers specify a certain viscosity of gear lube based on the climate the vehicle is operated in or the load it is carrying, so understanding viscosity is very important when servicing vehicles.

Standard viscosities for gear lube are 70W, 75W, 80W, 85W, 90, and 140. The "W" stands for "winter," or the gear lube's cold temperature viscosity. The non-W ratings are the viscosity of the gear lube at a predetermined hot temperature. Similar to engine oil, gear lube is available in multi-viscosity configurations such as 75W-90, 80W-90, or 85-140. Unlike engine oil, the temperature for the "W" rating varies according to the standard being met. For example, 70W has a maximum allowable temperature for its viscosity of –67°F (–55°C), and 85W is –10°F (–23°C). The manufacturers have gone to great lengths to specify the proper gear lube for their vehicles. Make sure you follow their recommendations when choosing the gear lube for a particular vehicle.

Current ratings of the API service grade are GL-4, and GL-5. Generally, the higher the number, the better the lubricant. GL-4 is intended for use with bevel-type gears operating under moderate speeds and loads, such as in many manual transmissions. GL-5 has about twice as much extreme pressure additive as GL-4, so it is intended for most differentials that use hypoid-type gears operating under high-speed/low-speed, high-torque, and shock-load conditions. It can also be used in some manual transmissions. Always check the manufacturer's specifications to determine the proper gear lube for the application you are working on.

Most serviceable wheel bearings require **grease** as their lubricant. Grease is made of a base oil, plus a thickening agent, and specific additives to meet the requirements of the application. **Lithium soap** is a common thickening agent in automotive grease. Other greases use calcium or **molybdenum thickening agents**. Some add small amounts of copper and/or lead to enhance the grease's ability to withstand extreme pressures.

Automotive wheel bearing grease is a thickened lubricant, designated as a plastic solid. This means that although it is thick enough at room temperature to maintain its shape if left undisturbed, it is thin enough to be squeezed into and out of small spaces. Its consistency is similar to a glob of gel toothpaste.

The thickness of grease is graded by the **National Lubricating Grease Institute (NLGI)** (TABLE 48-1). Because it does not flow at room temperature, the grease has to be packed into the spaces around the rollers when the wheel bearing assemblies are installed. This also has to be done when the grease wears out and must be replaced. Packing a wheel bearing assembly can be done by hand or with a **bearing packer**, which is a tool that forces grease into the spaces between the bearing rollers. Packing a bearing assembly should be done only after thoroughly cleaning and inspecting the rollers and races for wear, damage, or corrosion. If any of these are present, the bearing assembly should be replaced.

▶ Diagnosis

N48001

Wheel bearings can be damaged from excessive loads such as overloading the vehicle, shock loads such as hitting a large pot hole, improper adjustment, or just plain wear over time. The wheel bearings must hold up under difficult circumstances, so suspect faulty bearings whenever an unusual rumbling, whirring, howling, or rough sound comes from the wheel areas when the vehicle is being driven. The noise usually can be heard once the vehicle gets up to 15–20 mph (24.1–32.2 kph) and gets louder as the vehicle speeds up.

Loose or worn wheel bearings can also cause the vehicle to wander, shimmy, or vibrate. For these concerns, it is best to lift the wheels off the ground and check the wheel bearings for looseness by grabbing the tire at the 6 o'clock and 12 o'clock positions and lightly wiggling it back and forth. Watch the inside of the wheel to verify that the play is coming from the wheel bearings, and not the ball joints. Then grab the wheel at the 3 o'clock and 9 o'clock positions, and again lightly wiggle the wheel back and forth. Watch to verify that the play is coming from the wheel bearings, and not the tie rod ends. If there is play in the wheel bearings and the vehicle uses sealed bearings, they will have to be replaced. If the vehicle uses serviceable bearings, the maximum allowable play is approximately 0.010" (0.254 mm). If greater than that, they will have to be disassembled, cleaned, inspected, repacked, reinstalled, and readjusted if they are still in good condition.

One way to isolate a wheel bearing noise from a transmission noise is to drive the vehicle at the speed at which it is making the noise and then shift into a higher or lower transmission gear while maintaining the same speed. If the noise speeds up or slows down, then it is a transmission-related issue. If it stays relatively the same, accelerate, coast, and decelerate the vehicle. If the noise changes, suspect the differential or universal joints. If the noise stays relatively the same, it is most likely related to the wheel bearings.

To determine which side the bearing noise is coming from, you can sometimes drive the vehicle at the speed at which it is making the noise, and then lightly rock the car side to side, using the steering wheel. If the noise gets louder when the car is steered right, then it is usually the left side with the bad bearing, and vice versa. Only perform this test in a safe place such as an abandoned parking lot. Another challenge is distinguishing a wheel bearing noise or a tire noise. The best approach to this problem is to drive over different road surfaces, such as asphalt and concrete. If the noise changes, it is likely a tire problem. If the noise does not change, it is likely to be a faulty wheel bearing.

If you cannot determine the source of the noise on a test drive, you might be able to determine the source of the noise by placing the vehicle on a hoist and rotating the wheels. If the vehicle has a MacPherson strut suspension, raise and support the vehicle, and spin the wheel by hand while holding on to the coil spring with the other hand, feeling for a vibration or roughness. The spring tends to magnify the wheel bearing roughness, which can be felt with some practice. If the suspect wheel is a drive wheel, under close guidance of your supervisor have an assistant drive the vehicle in gear while on the hoist, and listen to the wheel bearings with a stethoscope. Because the vehicle is running on the hoist, this can be a hazardous situation and must only be performed under the close guidance of your supervisor.

Failure Analysis

Wheel bearings can be inspected and determinations made about why the wheel bearing failed. Getting to the source of the problem is important to ensure that the same thing does not happen to the new wheel bearings. Failure analysis starts with removing the faulty wheel bearing and cleaning it and the races thoroughly. Once they are clean, visually inspect the wheel bearing components, and compare your findings to the wheel

TABLE 48-1: NLGI Rating System

NLGI Number	Consistency
00	Semifluid
0	Very soft
1	Semisoft
2	Semifirm
3	Soft—common wheel bearing grease
4	Firm
5	Very firm

Courtesy of the National Lubricating Grease Institute

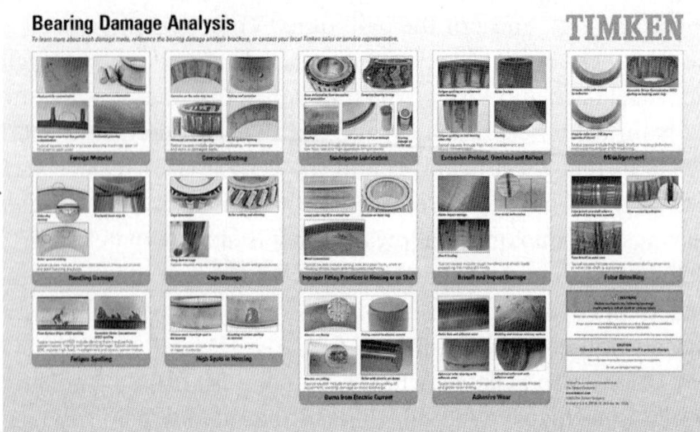

FIGURE 48-16 Wheel bearing failure chart.

bearing manufacturer's failure analysis chart (**FIGURE 48-16**). This information should lead you to what caused the wheel bearing failure.

Once the cause of the wheel bearing failure has been determined, you need to take measures to ensure that the failure does not repeat itself. For example, if the bearing and race surfaces show signs of rust or corrosion, you will want to inspect the sealing surfaces of the hub to see if water is getting past the dust cover or grease seal, along with replacing the wheel bearing assembly. Also, if the wheel bearing is so damaged that there are metal shavings in the wheel hub and grease, a thorough cleaning of the wheel hub will be required so that all of the metal shavings are removed and will not ruin the new bearing assembly.

Some wheel bearing failures involve the races spinning in either the machined bore of the wheel hub or on the surface of the spindle/axle. This can lead to wear of the wheel hub or spindle/axle, which requires replacement of these components. Be sure to inspect the wheel hub and spindle/axle closely for any damage every time the wheel bearings are removed and serviced.

▶ Maintenance and Repair

S48001

Although sealed wheel bearings are used on a higher percentage of vehicles and require no maintenance, serviceable wheel bearings do need periodic maintenance. Maintenance consists of disassembling, cleaning, inspecting, repacking, installing, and adjusting the bearings. This is commonly performed during brake shoe/pad replacement or at intervals specified by the vehicle manufacturer, usually around 24,000–30,000 miles (38,624–48,280 km). A normal part of servicing wheel bearings includes the replacement of the old grease seal with new ones, along with replacement of the **cotter pin** (if used), which retains the wheel bearing adjusting nut. A cotter pin is a soft metal pin that can be bent into shape and is used to retain the bearing adjusting nut. Ensure that the correct replacement parts and grease are available before starting the job.

Tools

Here is a list of common tools used to maintain and repair wheel bearings (**FIGURE 48-17**):

- Bearing packer
- Seal puller
- Wheel bearing race installer/seal installer set
- Wheel bearing **locknut** sockets
- Cotter pin removal tool
- Dust cap pliers

Wheel Bearing Adjustment

K48011

All wheel bearings need the proper end play or preload to operate correctly. End play, in the context of wheel bearings, refers to the amount of inward and outward movement of the hub due to the clearance within the bearing assembly. **Preload** refers to the absence of clearance in the bearing and the specified amount of pressure forcing the bearing components together. Sealed bearings and double-row bearing

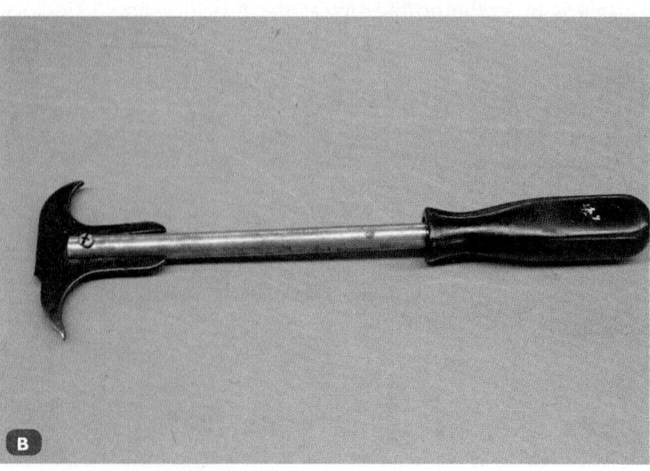

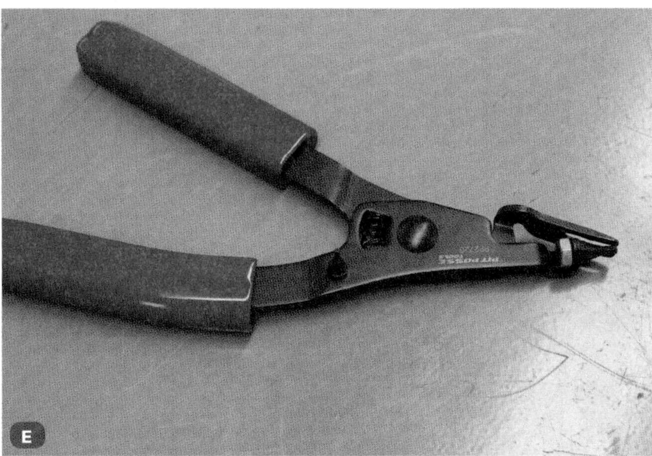

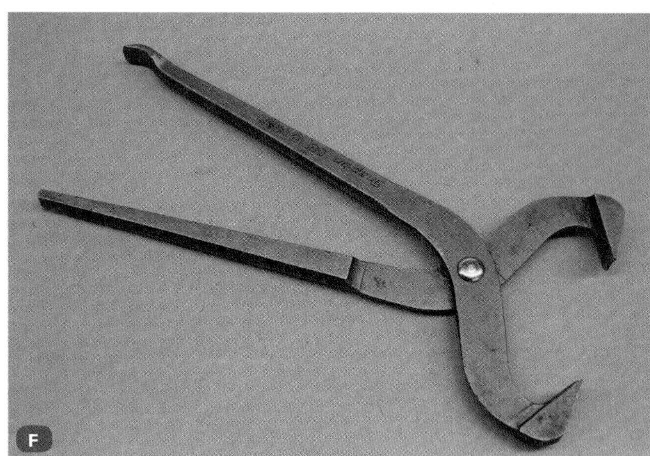

FIGURE 48-17 **A.** Bearing packer. **B.** Seal puller. **C.** Wheel bearing race/seal installer set. **D.** Wheel bearing locknut sockets. **E.** Cotter pin removal tool. **F.** Dust cap pliers.

assemblies come from the manufacturer with the proper clearance machined into them. These wheel bearings are designed so that when the components are tightened together, the races butt up against each other in such a way that the proper clearance is maintained. For these wheel bearings, it is only critical that the retaining bolt or bolts are torqued to the proper specification. This is usually quite high and can typically exceed 200 ft-lb (271.1 N·m). Be sure to check the manufacturer's torque specifications for the application you are working on.

On adjustable wheel bearings, the proper clearance must be set using the **adjusting nut**. In this case, the adjusting nut is initially tightened to about 20 ft-lb (27.1 N·m) to squeeze out any grease between the races and rollers. It is then loosened one-sixth to one-quarter turn, and only tightened lightly (usually about 15–25 in-lb [1.69–2.82 N·m]) so that a small amount of clearance or preload is maintained between the rollers and races, depending on the manufacturer's specifications. The adjusting nut is then locked in place by a locking mechanism so that it cannot loosen. This is critical to prevent the wheel from falling off and causing injury. The most common locking mechanism uses a **keyed washer** (hardened), adjusting nut, **lock cage**, and cotter pin (**FIGURE 48-18**). On four-wheel drive vehicles, the locking mechanism commonly includes a keyed washer (hardened), adjusting nut, **keyed lock washer** (or tang washer), and locknut (**FIGURE 48-19**).

FIGURE 48-18 Typical cotter pin–style wheel bearing locking mechanism.

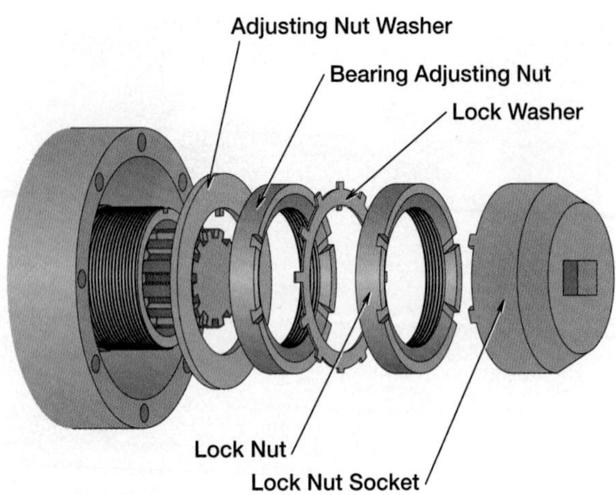

FIGURE 48-19 Typical locknut-style wheel bearing locking mechanism.

FIGURE 48-20 A bearing packer.

Repacking and Adjusting Wheel Bearings

N48002, N48003

Serviceable wheel bearings should be serviced periodically according to the manufacturer's scheduled maintenance chart, whenever brake work is being performed, or if a faulty wheel bearing is suspected. When servicing wheel bearings, it is critical that the proper grease is used. Also avoid mixing different types of grease by thoroughly cleaning all old grease from the wheel bearings. Another option is to use a bearing packer, with the same kind of grease, to force the old grease out of the wheel bearings (**FIGURE 48-20**). Make sure the grease in the bearing packer is not contaminated with dirt or debris and is correct for the vehicle being serviced.

When reinstalling wheel bearings, it is critical to follow the manufacturer's adjustment procedure. Always use new grease seals and cotter pins (if used). Doing so will prevent grease leaks and ensure that the adjusting nut does not back off, causing an unsafe driving situation.

To remove, clean, inspect, repack, and install wheel bearings, and to install the locking mechanism, follow the steps in Skill Drills 48-1 through 48-7. To remove, clean, and inspect the wheel bearings, follow the steps in **SKILL DRILL 48-1**.

To pack grease by hand, follow the steps in **SKILL DRILL 48-2**.

To pack grease with a bearing packer, follow the steps in **SKILL DRILL 48-3**.

To install wheel bearings, follow the steps in **SKILL DRILL 48-4**.

SKILL DRILL 48-1 Removing, Cleaning, and Inspecting Wheel Bearings

1. Remove the wheel bearing dust cap with dust cap pliers or a narrow cold chisel and hammer.

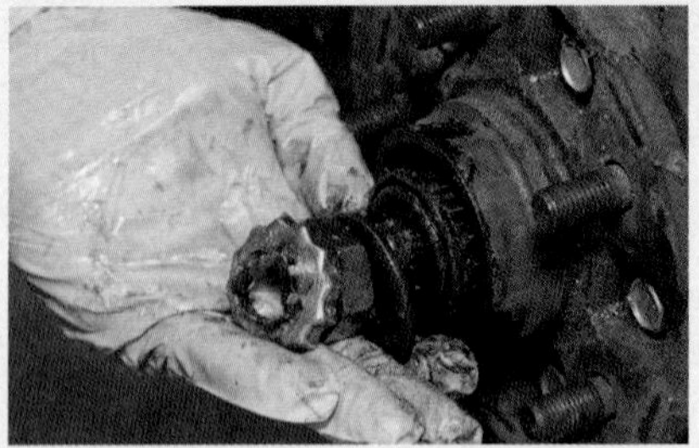

2. Remove the locking mechanism. Remove the adjusting nut, keyed washer, and outer bearing. Reinstall the adjusting nut approximately five turns back onto the spindle.

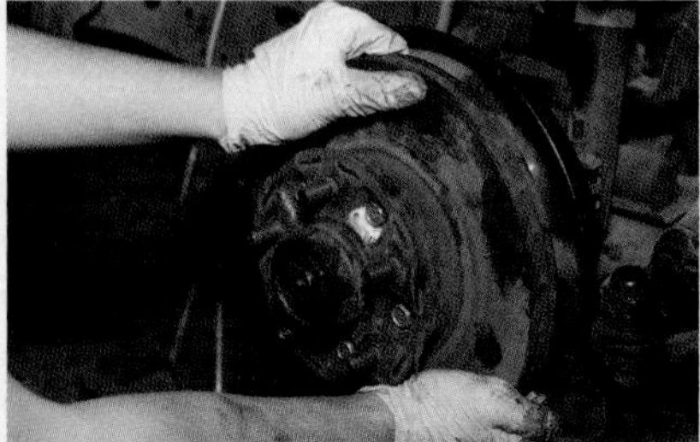

3. Grasp the drum/rotor at the 1 o'clock and 7 o'clock positions or 11 o'clock and 5 o'clock positions. While holding downward pressure, quickly pull the drum/rotor toward you. The adjusting nut should catch the inner bearing race and pop the grease seal and bearing out of the hub, leaving them sitting on the spindle.

SKILL DRILL 48-1 Removing, Cleaning, and Inspecting Wheel Bearings (Continued)

4. Wipe any old grease off of the wheel bearings, races, and spindle with a rag and give them a quick visual inspection. Consult the bearing diagnosis chart to identify any faults.

5. If the wheel bearings are in serviceable condition, completely clean the wheel bearings, races, and hub. If using solvent to clean any of the components, make sure there is no solvent-contaminated grease left on the parts.

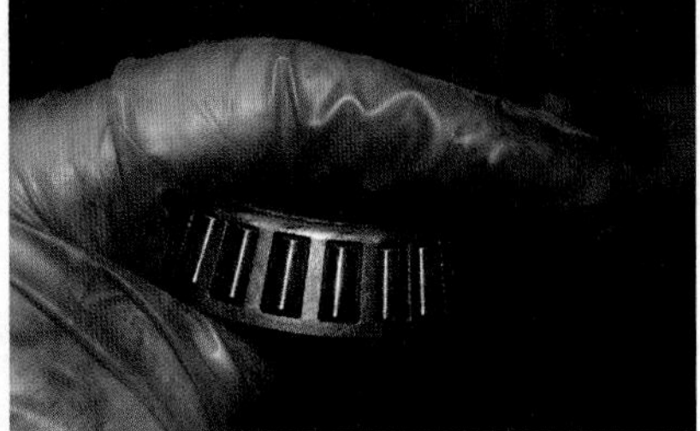

6. Give the parts a final inspection, and consult the bearing diagnosis chart if there are any signs of damage. Using the specified grease, pack both wheel bearings, being careful to keep dirt and debris out of the grease (see Skill Drills 48-2 and 48-3).

SKILL DRILL 48-2 Packing Grease by Hand

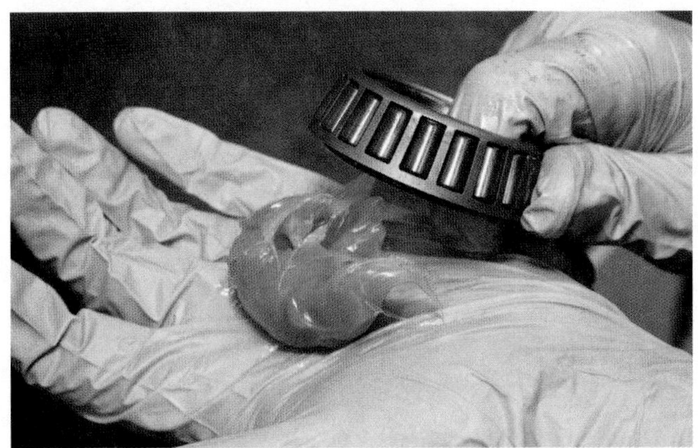

1. Using a pair of latex or nitrile (nitro) gloves, place a small glob of grease in the palm of your nondominant hand. Place the index finger of your other hand through the bearing center hole, with the larger diameter facing down.

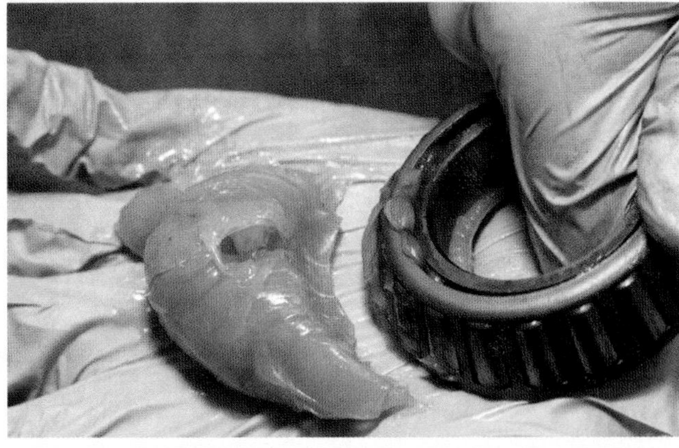

2. Push the large diameter of the bearing down the edge of the grease into your palm. This should force grease into the space between the bearings and races. Continue this process until grease comes out of the top of the bearing.

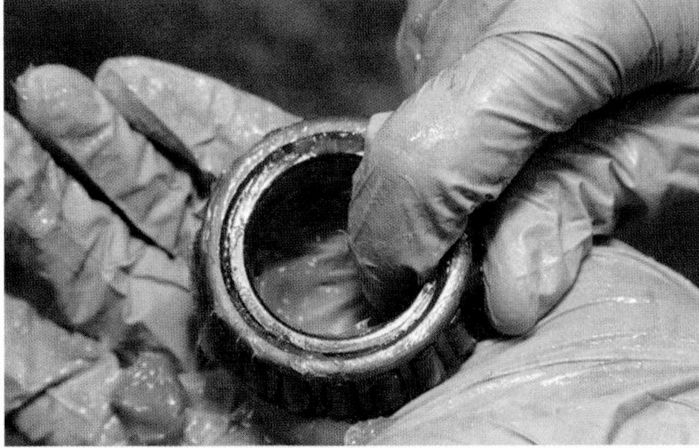

3. Carefully turn the bearing as a unit to a new space, and keep forcing grease between the bearings. Do this until all of the spaces are full.

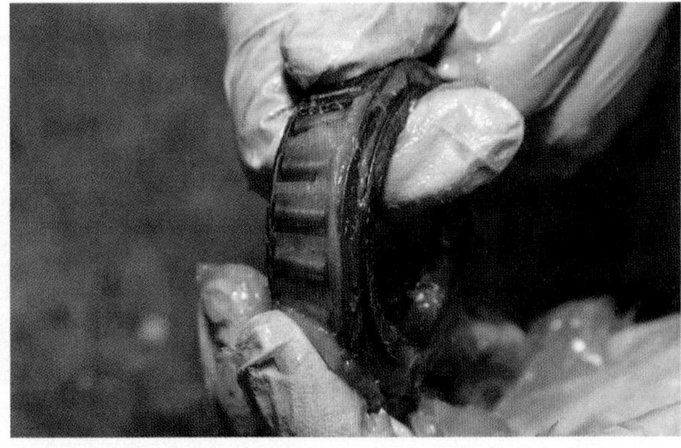

4. Smear some grease around the outside of the bearing. Repeat this process on the other bearing.

SKILL DRILL 48-3 Packing Grease with a Bearing Packer

1. Make sure the bearing packer has the proper grease and that it is uncontaminated. Place the wheel bearing with the narrow side down.

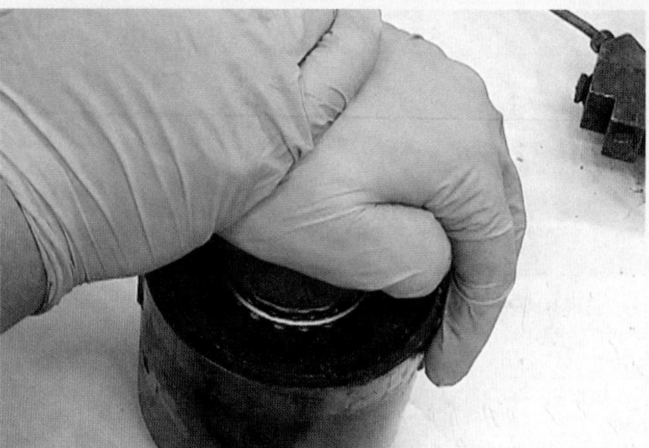

2. Place the packer cone on the top of the wheel bearing. Pack the wheel bearing, following the packer's instructions.

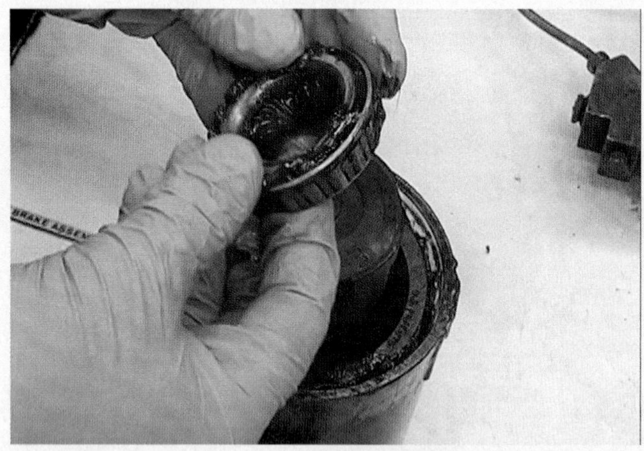

3. Remove the wheel bearing. Wipe off any old grease.

4. Smear some new grease around the outside of the wheel bearing. Repeat this process on the other wheel bearing. Be sure to set a packed wheel bearing down on a clean surface.

SKILL DRILL 48-4 Installing Wheel Bearings

1. Place a small amount of extra grease in the center of the hub. Do not fill it completely. Place the inner wheel bearing in its race, narrow side toward the race.

2. Using a hammer or seal installer, install the new grease seal, being careful not to damage it. Place a small amount of grease on the lip of the seal to provide it with initial lubrication.

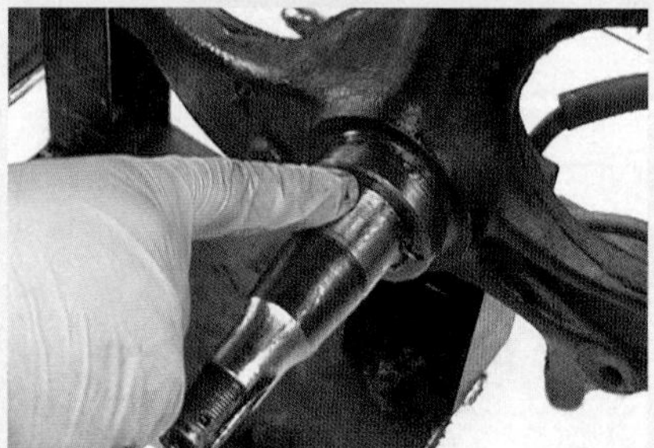

3. Make sure the spindle is clean, including the mating surface for the seal.

SKILL DRILL 48-4 Installing Wheel Bearings (Continued)

4. Without getting grease on the drum/rotor, carefully install it on the spindle, making sure the inner bearing fully seats against the spindle flange.

5. Install the outer bearing on the spindle and into the race.

6. Install the keyed washer and adjusting nut on the spindle, and tighten until finger tight.

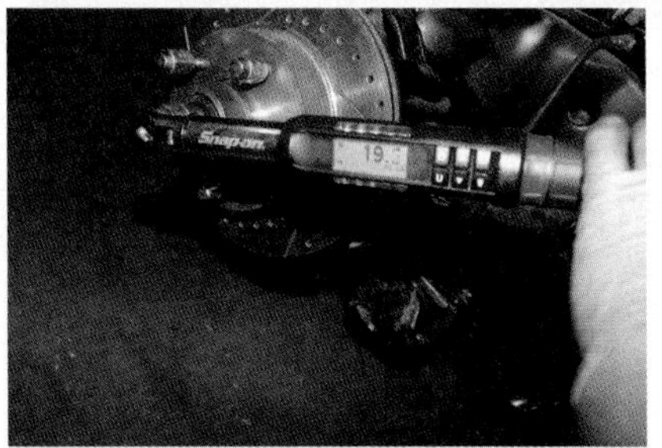

7. Tighten the adjusting nut to the specified seating torque (usually about 20 ft-lb [27.1 N·m]) while turning the drum/rotor. This squeezes the excess grease out from between the wheel bearings and races while seating the bearings. *Do not leave the bearing this tight!*

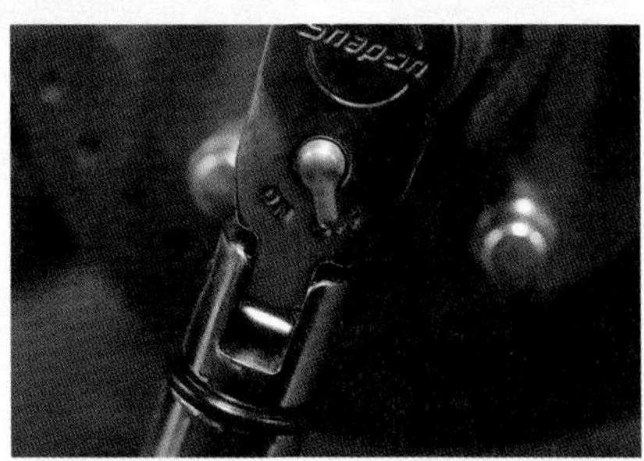

8. Loosen the adjusting nut approximately one-sixth to one-quarter turn without turning the drum/rotor.

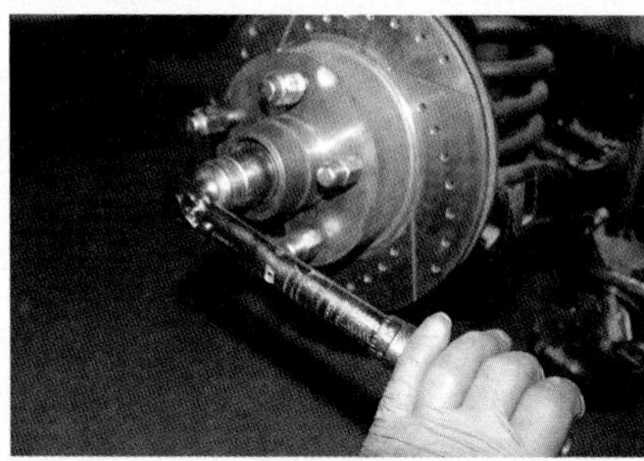

9. Tighten the adjusting nut to the specified preload torque. This is usually about 15–25 in.-lb (1.69–2.82 N·m).

TECHNICIAN TIP

Some technicians simulate the preload torque by placing a 12" crescent wrench on the nut and using the hanging weight of the crescent wrench when it is parallel to the ground. Do not let it drop into position; just lower the handle to where it stops turning the nut. This should be when the handle is approximately level. If not, reposition the crescent wrench on the nut, and allow it to lower until it stops parallel.

To install the locking mechanism, follow the steps in **SKILL DRILL 48-5**.

Replacing Wheel Bearings and Races

N48004

Wheel bearings and races have to be replaced only when they are damaged. If one part of the wheel bearing is damaged, all parts must be replaced. So if the tapered roller bearings are damaged, both the bearing and the race of that bearing have to be replaced at the same time. On serviceable bearings, the inner race, roller bearings, and bearing cage are one unit, which generally slips off of the spindle. The outer race is usually press fit into the hub and has to be driven or pressed out and a new one press fit back in. It is critical that the seat in the hub be spotlessly clean, with no burrs, or the bearing race will not seat properly and the bearing will fail prematurely.

To replace a wheel bearing and race, follow the steps in **SKILL DRILL 48-6**.

Removing and Reinstalling Sealed Wheel Bearings

N48005

Sealed wheel bearings come in two configurations. The first is a replaceable sealed bearing only. On most front wheels, this wheel bearing is pressed between the hub and wheel flange and is the more difficult of the two to replace. The second configuration consists of a unitized wheel bearing hub including a sealed wheel bearing, a removable wheel bearing hub, and possibly the wheel flange. In most cases, this type can be unbolted

SKILL DRILL 48-5 Installing the Locking Mechanism

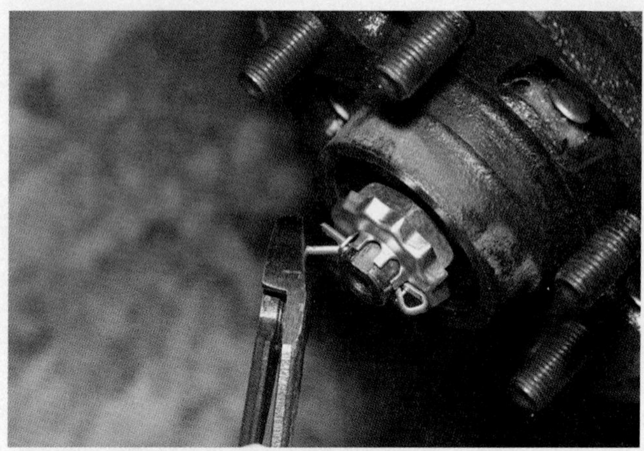

1. If the locking mechanism is a cotter pin, insert the new cotter pin through the **castellated nut** or locking cage, and the spindle. The short leg of the cotter pin should be against the castellated nut, and the long leg should be toward you. With the cotter pin fully engaged in the notch, bend the outer leg toward you and up over the end of the spindle. Cut it off just beyond the spindle. Also cut the short leg off at the nut or cage. Make sure the cotter pin will not hit the inside of the dust cap.

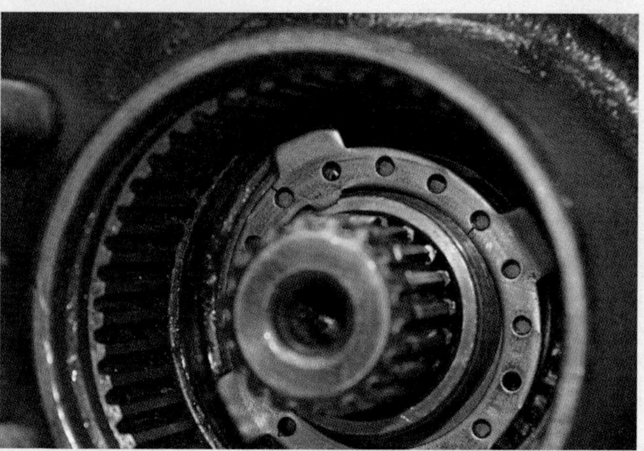

2. If it is a pin and hole style, or a bendable tang locking style, then place the washer against the adjusting nut (with the pin lined up, if that style) and thread the locking nut up against it. Torque the locking nut to the specified torque, and if a bendable tang style, bend the appropriate tang out toward you against the flat side of the locking nut with a small pry bar to lock the adjustment in place.

3. If it is a locking nut style, then tighten the locknut to the specified torque. This is usually a substantial torque of 50 ft-lb (67.79 Nm) or way more.

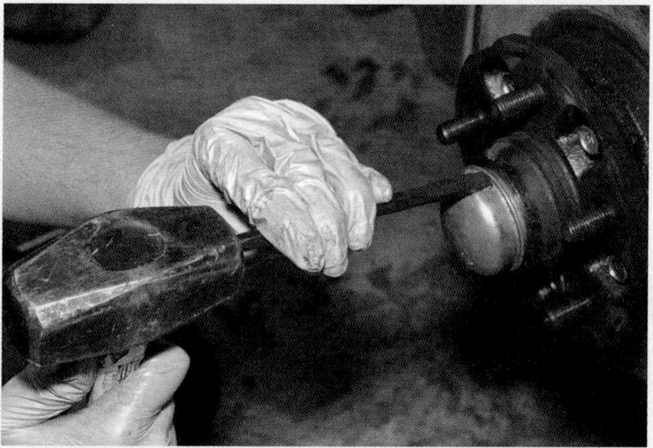

4. Install the dust cap, being sure it is fully seated in the hub. Make sure the drum/rotor turns freely without binding or making any unusual noises.

from the suspension system and a new one bolted in its place, and it is ready to go. This is the most common arrangement on recent vehicles. Clean the inside diameter of the hub in a parts washer.

The replaceable bearing style needs to be pressed apart with a hydraulic press or a special sealed bearing removal/installing tool. If using the hydraulic press method, the steering knuckle will have to be removed from the vehicle so it can be placed on the hydraulic press. If the special sealed bearing tool is used, most bearings can be removed while the steering knuckle is still installed on the vehicle, which can save the technician a fair amount of time. To remove and reinstall a sealed wheel bearing assembly using the unitized wheel bearing hub style, follow the steps in **SKILL DRILL 48-7**.

SKILL DRILL 48-6 Replacing Wheel Bearings and Races

1. With the wheel bearings removed from the wheel hub, clean and inspect the bearing and race for damage. Determine which bearing and race have to be replaced. Using a hydraulic press or a hammer and punch from the opposite side of the hub, carefully force the race from the hub. Keep it as straight as possible while removing it.

2. Remove any burrs with a fine file or Dremel™, and remove any debris from the seat. Lightly lubricate the outside surface of the new race, and set it thick side down in the hub.

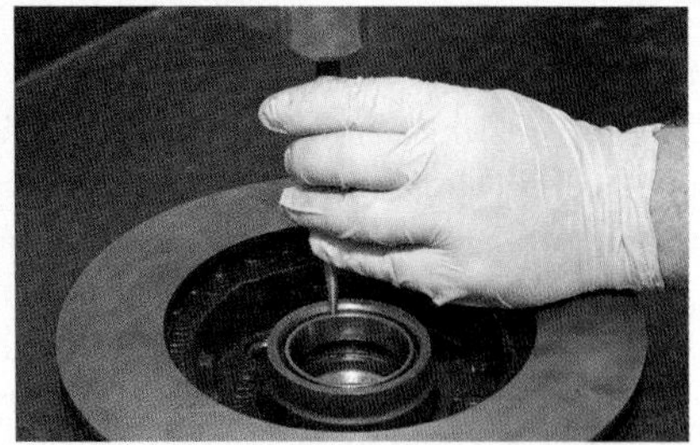

3. Using a hydraulic press or a hammer and bearing race installer, carefully drive the race until it is fully seated in the hub. When you are using a hammer and punch, a distinct sharp metallic sound should be produced when it seats. Inspect the race to verify that it is fully seated. Also check for any damage caused by installation. If everything is good, pack the new bearing and install it according to Skill Drills 48-2 through 48-5.

SKILL DRILL 48-7 Removing and Reinstalling Sealed Wheel Bearings Using the Unitized Wheel Bearing Hub Style

1. Loosen the axle hub nut, if equipped, while the tire is still on the ground. Remove the wheel and brake assembly, following the specified procedure. Also disconnect the ABS connector and/or sensor if mounted to the hub.

2. If the wheel you are working on is a drive wheel, remove the axle hub nut, and tap the drive axle loose with a dead blow hammer.

3. Unbolt and remove the hub assembly from the steering knuckle. Clean the knuckle assembly, and check the hub seat for nicks, burrs, or other damage.

4. Carefully compare the new hub to the old one; then fit the new hub assembly (over the axle shaft, if equipped) to the knuckle, making sure it is fully seated in place, and torque the mounting bolts to the specified torque.

5. Reassemble the brake assembly and ABS sensor, if removed, following the specified procedure; install the wheel, and torque the lug nuts. Be sure the correct ends of the lug nuts are facing the wheel.

6. Install the drive axle nut, if equipped. Use a new hub nut if called for by the manufacturer, and torque to specifications.

▶ Wrap-Up

Ready for Review

- Wheel bearings allow wheels to roll with minimum friction.
- Wheel bearing assemblies include outer race, inner race, roller or ball bearings, and a bearing cage.
- Categories of wheel bearings are serviceable or sealed.
- Types of wheel bearings are cylindrical roller bearing, tapered roller bearing, ball bearing, double-row ball bearing, and double-row tapered roller bearings.
- Cylindrical roller bearing assemblies have parallel races flanking the rollers and are most commonly used in rear axles of rear-wheel drive vehicles.
- Cylindrical roller bearing assemblies rely on lubrication to cushion the rollers and transfer heat to the atmosphere.
- Tapered roller-bearing assemblies have tapered rollers housed in a bearing cage, are used for heavier loads, are used in pairs, and control side movement.
- Tapered roller bearing assemblies must be adjusted for proper running clearance.
- Ball bearings are used in light-duty vehicles, are designed as a manufacturer-sealed bearing assembly, and should be replaced as a unit.
- Wheel bearings are sealed to keep out contaminants and contain the lubricant.
- Serviceable wheel bearings need periodic maintenance, including replacing old grease seals and the cotter pin.
- Wheel bearings are lubricated by gear lube or bearing grease.
- Gear lube is classified by viscosity (thickness) and service grade.
- Bearing grease is graded by thickness; it holds its shape at room temperature.
- For correct operation, wheel bearings must be adjusted for proper end play preload.
- Wheel bearing locking mechanisms include a keyed washer, adjusting nut, lock cage, and cotter pin.
- Damage to wheel bearings may be from vehicle overload, shock loads, improper adjustment, or wear over time.
- Wheel bearing assembly diagnosis may include wiggling the tire (with vehicle lifted) by hand, by test-driving, or by placing on a hoist and rotating the wheels.
- Determine the cause of wheel bearing failure so that the new bearings are not damaged for the same reason.

Key Terms

adjusting nut The nut used to adjust the end play or preload of a wheel bearing.

antifriction bearing Wheel bearing assemblies that use surfaces that are in rolling contact with each other to greatly reduce friction, compared to surfaces in sliding contact.

ball bearings The rolling components of a wheel bearing consisting of hardened balls that roll in matching grooves in the inner and outer races.

bearing cage The component in a wheel bearing that maintains the proper spacing between the roller bearings or ball bearings.

bearing packer A tool that forces grease into the spaces between the bearing rollers.

castellated nut An adjusting nut with slots cut into the top such that it resembles a castle; used with a cotter pin to prevent the nut from loosening.

cotter pin A one-use soft metal pin that can be bent into shape and is used to retain bearing adjusting nuts.

cylindrical roller bearing assembly A type of wheel bearing with races and rollers that are cylindrical in shape and roll between inner and outer races, which are parallel to each other.

double-row ball bearing assembly A single ball bearing assembly using two rows of ball bearings riding in two channels in the races.

fill plug Usually a threaded plug that can be removed to allow the level of a fluid to be checked and filled. This could also be a rubber snap fit plug.

friction bearing A bearing that uses sliding motion between components, such as a clutch pilot bushing.

garter spring A coiled spring that is fitted to the inside of the sealing lip of many seals, used to hold the lip in contact with the shaft.

gear lube A type of lubricant primarily used to lubricate transmission and differential gears but also used to lubricate some wheel bearings.

grease A lubricating liquid thickened to make it suitable for use with many wheel bearings.

grease seal A component that is designed to keep grease from leaking out and contaminants from leaking in.

inner race The inside component of a wheel bearing that has a smooth, hardened surface for rollers or balls to ride on.

interference fit A condition in which two parts are held together by friction because the outside diameter of the inner component is slightly larger than the inside diameter of the outer component.

keyed lock washer The washer that fits between the adjusting nut and the locknut; the face of the washer is drilled with a series of holes that mate to a short pin from the adjusting nut, locking it to the spindle.

keyed washer The washer that fits between the adjusting nut and the wheel bearing and that has the center hole keyed to fit a slot on the spindle or axle tube.

lithium soap A thickening agent for grease to give it the proper consistency.

lock cage The stamped sheet metal cap that fits over the bearing adjustment nut and is secured by a cotter pin going through it and the spindle/axle.

locknut The nut that holds the adjusting nut from turning; usually tightened much tighter than the adjusting nut.

molybdenum thickening agent A compound used in some greases to give it the needed consistency.

National Lubricating Grease Institute (NLGI) An organization that grades the thickness of automotive and industrial grease.

outer race The outside component of a wheel bearing that has a smooth, hardened surface for rollers or balls to ride on.

preload A condition where the wheel bearing components are forced together under pressure and therefore have no end play.

roller bearings The rolling components of a wheel bearing consisting of hardened cylindrical or tapered rollers.

running clearance The amount of space between wheel bearing components while in operation.

sealed bearings Wheel bearings that are assembled by the manufacturer with the proper lubrication and sealed for life; cannot normally be disassembled.

serviceable bearings Wheel bearings that can be disassembled, cleaned, inspected, packed, reinstalled, and adjusted.

tapered roller bearing A type of wheel bearing with races and rollers that are tapered in such a manner that all of the tapered angles meet at a common point, which allows them to roll freely and yet control thrust.

unitized wheel bearing hub An assembly consisting of the hub, wheel bearing(s), and possibly the wheel flange, which is preassembled and ready to be installed on a vehicle.

viscosity The measurement of the thickness of a liquid.

wheel bearing A component that allows the wheels to rotate freely while supporting the weight of the vehicle, made up of an inner race, outer race, rollers and a cage.

Review Questions

1. Which component of wheel bearings holds the rollers or balls in place?
   - **a.** An outer race
   - **b.** An inner race
   - **c.** A bearing cage
   - **d.** Interference fit
2. In most situations, serviceable wheel bearing races are held in place by:
   - **a.** bolts
   - **b.** an interference fit
   - **c.** snap rings
   - **d.** set screws
3. Which type of bearing has an inner race called cone and an outer race called cup?
   - **a.** Cylindrical roller bearings
   - **b.** Double-row ball bearings
   - **c.** Ball bearing
   - **d.** Tapered roller bearing
4. When packing a serviceable wheel bearing:
   - **a.** make sure to fill all of the space in the hub with grease.
   - **b.** make sure to fill the bearing grease cap full of grease.
   - **c.** make sure to fill the spaces between the rollers with grease.
   - **d.** only smear grease on the outside surfaces of the bearings.
5. When installing a cotter pin in a wheel bearing locknut:
   - **a.** adjust the wheel bearing and then install the cotter pin.
   - **b.** install the cotter pin and then adjust the wheel bearing.
   - **c.** install the dust cap and then install the cotter pin.
   - **d.** install the cotter pin and then install the wheel bearing.
6. All of the following statements regarding sealed and serviceable bearings are true *except*:
   - **a.** Sealed bearing assemblies are prefilled with lubricant and have integrated grease seals, unlike serviceable bearings.
   - **b.** In serviceable bearings, proper running clearance must be adjusted, whereas sealed bearings are manufactured with the proper clearance.
   - **c.** Sealed bearing assemblies do not need periodic maintenance, unlike serviceable bearings.
   - **d.** Sealed bearing assemblies never fail whereas serviceable bearings can fail.
7. Choose the correct statement.
   - **a.** All wheel bearings have the seal built right into the bearing assembly.
   - **b.** A garter spring helps maintain an adequate seal if the parts are slightly out of alignment.
   - **c.** Grease seals are designed so they can be reused.
   - **d.** When installing a seal, the sealing lip need not be pre-lubed as it comes with a small amount of oil or grease.
8. In which of the following rear drive axle designations is a pair of tapered roller bearing assemblies used on each side?
   - **a.** Full floating axle
   - **b.** Semi-floating axle
   - **c.** ¾-floating axle
   - **d.** ½-floating axle
9. All of the following statements with respect to lubricants and their application are true *except*:
   - **a.** Gear lube is somewhat thicker than engine oil.
   - **b.** The level of gear lube should normally be within 0.25" (6.35 mm) of the bottom of the fill plug hole.
   - **c.** The higher the viscosity number, the thinner the gear lube.
   - **d.** A bearing packer is a tool that forces grease into the spaces between the bearing rollers.
10. When adjusting serviceable wheel bearings, the final adjustment (2nd adjustment) should be approximately:
    - **a.** 20 ft-lb (27.1 N.m)
    - **b.** 15–25 in-lb (1.69–2.82 N.m)
    - **c.** 0.020" (0.508 mm).
    - **d.** 0.050" (1.27 mm).

ASE Technician A/Technician B Style Questions

1. Tech A says that cylindrical roller bearings can carry more weight than similarly sized ball bearings. Tech B says that tapered roller bearings used in opposing pairs control side thrust in both directions. Who is correct?
   - **a.** Tech A
   - **b.** Tech B
   - **c.** Both A and B
   - **d.** Neither A nor B
2. Tech A says that a tapered roller bearing assembly has less rolling resistance than a similarly sized ball bearing assembly. Tech B says that the bearing assembly in a unitized wheel bearing assembly can normally be disassembled, cleaned, and repacked. Who is correct?
   - **a.** Tech A
   - **b.** Tech B
   - **c.** Both A and B
   - **d.** Neither A nor B
3. Tech A says that over time a grease seal can wear a groove in the sealing surface of the axle or shaft that may require replacement of the axle or shaft. Tech B says that grease seals need to be replaced every time the bearing is removed. Who is correct?
   - **a.** Tech A
   - **b.** Tech B
   - **c.** Both A and B
   - **d.** Neither A nor B
4. Tech A says that in a full floating axle, the axle does not support the weight of the vehicle. Tech B says that when adjusting tapered wheel bearings, the final torque should be about 20 ft-lbs. Who is correct?
   - **a.** Tech A
   - **b.** Tech B
   - **c.** Both A and B
   - **d.** Neither A nor B
5. Tech A says that serviceable wheel bearings can be repacked by removing the dust cap, filling it with grease, and reinstalling it. Tech B says that the cotter pin must be replaced with a new one every time it is removed. Who is correct?
   - **a.** Tech A
   - **b.** Tech B
   - **c.** Both A and B
   - **d.** Neither A nor B
6. Tech A says that the gear lube level in the final drive is okay as long as you can touch the level with your finger. Tech B says that the gear lube level should normally be no more than ¼" below the bottom threads on the fill plug hole. Who is correct?
   - **a.** Tech A
   - **b.** Tech B
   - **c.** Both A and B
   - **d.** Neither A nor B
7. Tech A says that wheel bearings must be replaced as a set, bearing, and race. Tech B says that the wheel bearings and races on both sides of the vehicle must be replaced if one side fails. Who is correct?
   - **a.** Tech A
   - **b.** Tech B
   - **c.** Both A and B
   - **d.** Neither A nor B
8. Tech A says that unitized hubs have a wheel nut with a higher installation torque than serviceable wheel bearings. Tech B says that unitized hubs have the proper bearing end play designed into the assembly once they are torqued properly. Who is correct?
   - **a.** Tech A
   - **b.** Tech B
   - **c.** Both A and B
   - **d.** Neither A nor B
9. Tech A says that when installing a bearing race, it should only take finger pressure to install it. Tech B says that on a tapered roller bearing, the race is also called the cup. Who is correct?
   - **a.** Tech A
   - **b.** Tech B
   - **c.** Both A and B
   - **d.** Neither A nor B
10. Tech A says that when a race is fully seated, a sharp metallic sound is produced when installation is complete. Tech B says that to be sure a race is fully seated, the wheel bearing adjusting nut should be tightened to at least 100 ft-lbs of torque, which will finish seating it. Who is correct?
    - **a.** Tech A
    - **b.** Tech B
    - **c.** Both A and B
    - **d.** Neither A nor B

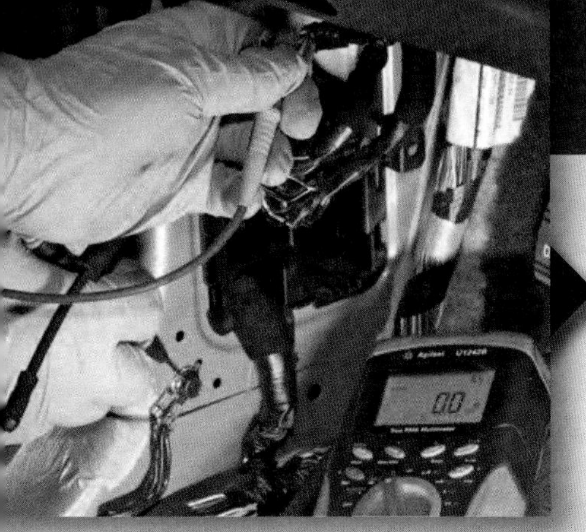

CHAPTER 49

Electronic Brake Control Systems

NATEF Tasks

- **N49001** Bleed the electronic brake control system hydraulic circuits. (MAST)
- **N49002** Depressurize high-pressure components of an electronic brake control system. (MAST)
- **N49003** Diagnose poor stopping, wheel lockup, abnormal pedal feel, unwanted application, and noise concerns associated with the electronic brake control system; determine needed action. (MAST)
- **N49004** Diagnose electronic brake control system electronic control(s) and components by retrieving diagnostic trouble codes, and/or using recommended test equipment; determine needed action. (MAST)
- **N49005** Identify and inspect electronic brake control system components (ABS, TCS, ESC); determine needed action. (AST/MAST)
- **N49006** Test, diagnose, and service electronic brake control system speed sensors (digital and analog), toothed ring (tone wheel), and circuits using a graphing multimeter (GMM)/digital storage oscilloscope (DSO) (includes output signal, resistance, shorts to voltage/ground, and frequency data). (MAST)
- **N49007** Diagnose electronic brake control system braking concerns caused by vehicle modifications (tire size, curb height, final drive ratio, etc.). (MAST)

Knowledge Objectives

After reading this chapter, you will be able to:

- **K49001** Explain the progression and operating principles of electronic brake control systems.
- **K49002** Describe the three most common EBC systems.
- **K49003** Explain the principles of operation of antilock brake systems.
- **K49004** Explain the principles of operation of traction control.
- **K49005** Describe the basic principles of operation of electronic stability control.

Skills Objectives

After reading this chapter, you will be able to:

- **S49001** Perform maintenance and repair tasks on EBC systems.
- **S49002** Remove and install electric and hydraulic EBC components.
- **S49003** Perform diagnostic tasks on electronic brake controlled systems.

Introduction

K49001

Electronic brake control (EBC) systems have greatly increased the safety of vehicles over the years by integrating computer-controlled hydraulics into the braking system (**FIGURE 49-1**). Standard hydraulic brake systems have limitations on how effectively they can stop a vehicle. The driver can only input braking force to the system through the brake pedal, which applies hydraulic pressure at predetermined ratios to the front and rear brakes. In a panic situation, the driver is unable to apply the exact amount of force to maintain the maximum amount of braking. With too little force, the vehicle does not stop as quickly. With too much force, the tires skid, making the vehicle's stopping distance even longer. At the same time, if the front wheels skid, the driver loses the ability to steer the vehicle. If the rear wheels skid, the car could spin out and possibly roll over.

Even if the driver could apply the perfect amount of braking force, there is no way to accommodate different amounts of traction at each wheel under all driving conditions. When tires on one side of the vehicle are on dry pavement and the other tires are on wet pavement, braking the vehicle is likely to lead to a loss of control.

FIGURE 49-1 Brakes must be fully functional in case of an emergency.

In the quest for increased safety, manufacturers developed a series of EBC systems. The first-generation EBC system was the antilock brake system (ABS), which was designed to prevent wheels from locking up under braking conditions, to help the driver maintain steering control of the vehicle. Maintaining steering control allows drivers to be better able to avoid collisions. ABS also shortens most panic stop distances for the same reason.

ABS was a great start, but manufacturers found that by adding a few components to the system and modifying the computer's software, they could provide traction control capability to the vehicle. This helps the driver maintain control of the vehicle while driving, compared to braking the vehicle. Then, with a few more components and advancements in the software, they were able to provide vehicle stability control, which helps prevent loss of driving control and vehicle rollovers. And the enhancements keep coming, with systems that provide engine braking control, trailer sway control, crash avoidance braking, and other features.

Basic Operation of Electronic Brake Control Systems

K49002

ABS systems use a computer that monitors the speed of each wheel as the brakes are applied. If one or more wheels begin to lock up, the computer sends electrical signals to **solenoid valves** that momentarily hold or release hydraulic pressure to that wheel until it speeds up and starts rolling again. Once that happens, the computer allows hydraulic pressure to be applied to that wheel again, slowing it down. This process is repeated very rapidly as the vehicle is brought to a stop. Because the tires remain in rolling contact with the road surface, the vehicle can be steered, allowing the driver to maintain directional control (steerability) (**FIGURE 49-2**). These actions are completely dependent on the driver applying pressure to the brake pedal.

Although the basic ABS system does a good job of managing the braking effort of the driver in a panic stop situation, it is limited to using the hydraulic pressure the driver exerts on the system. This means that the standard ABS system by itself cannot increase the hydraulic pressure in the ABS system; nor can it apply hydraulic pressure separate from the driver. As long as the driver is exerting firm pressure on the brake pedal, ABS can work fully.

The second-generation EBC system was the **traction control system (TCS)**. With the addition of a high-pressure pump

You Are the Automotive Technician

A longtime customer brings his 2012 Ford Explorer into the shop to get his brakes inspected. He claims that the brake pedal pulsates when he applies the brakes. You ask him if he has had any recent work done on the vehicle. He explains that he recently had new tires put on. He also explains that after leaving the shop, a car pulled out in front of him, and he had to lock up the brakes to avoid hitting him. He was startled by very heavy brake pedal pulsations. He says that ever since then, the brakes have a small pulsation that seems to be getting worse. He is wondering if there is a problem with the antilock brakes.

1. Is there a problem with the ABS system? What will you say to the customer?
2. What do you suspect is causing the pedal pulsations, and how did it occur?
3. How would you diagnose a problem when the ABS warning lamp is illuminated?
4. In what two primary ways does traction control reduce wheel slip?

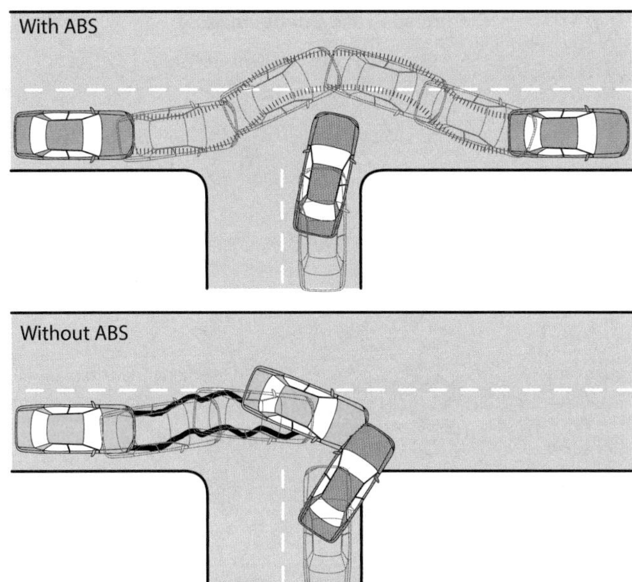

FIGURE 49-2 ABS helps the driver avoid accidents in a panic stop situation by maintaining steering control of the vehicle.

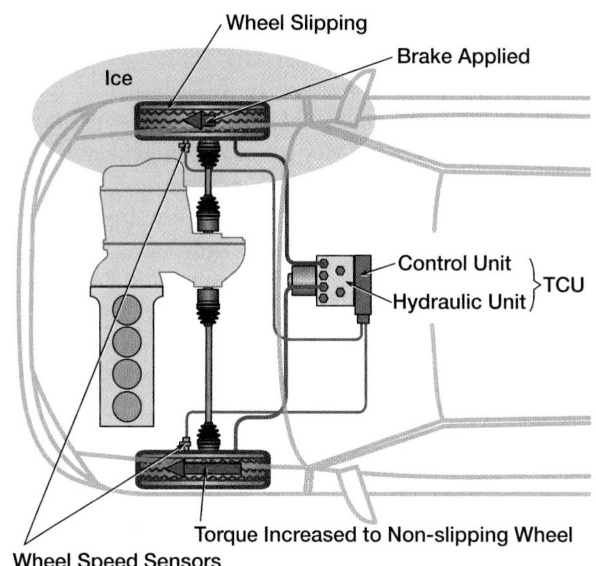

FIGURE 49-3 If a wheel is slipping, the TCS system applies the brake pressure to the slipping wheel. This causes torque to be sent to the wheel with more traction.

and a few **isolation valves** to the basic ABS system, manufacturers found that they could assist the driver in minimizing wheel slip while the vehicle is being accelerated. This is especially effective on slippery road surfaces such as gravel, snow, and ice. In most vehicles, the vehicle's traction is only as good as the traction on the tire with the least traction. So if one tire is on a patch of ice, the vehicle may not have enough traction to move, and will ultimately become stuck. The TCS system applies brake pressure to the slipping tire, which causes more of the engine's torque to be transmitted to the wheel or wheels with the most traction (**FIGURE 49-3**). If necessary, the TCS system can also request that the engine's powertrain control module reduce the power output of the engine to further enhance traction. These actions are controlled by the PCM and do not require any input from the driver.

The next-generation EBC system was the **electronic stability control (ESC) system**. ESC takes the ABS and TCS systems one step further. By adding sensor information regarding the driver's directional intent (from the **steering wheel position sensor**) and sensor information regarding the vehicle's actual direction (from the **yaw sensor**), the EBC module (EBCM) can detect the start of an **understeer, oversteer**, or potential rollover condition. Understeer and oversteer are conditions that happen when a vehicle is traveling too fast for a particular corner. During understeer, the vehicle's front wheels are turned more sharply than the vehicle's path (**FIGURE 49-4**). The front tires are actually sliding somewhat sideways toward the outside of the corner. The greater the understeer, the more the tires slide. Understeer is also referred to as "push," as in "the vehicle is pushing in the corners."

Oversteer is just the opposite. It occurs when the vehicle is turning more sharply than the front wheels are being steered. This happens when the rear tires are sliding sideways toward the outside of the corner. Oversteer is also referred to as "loose," as in "the vehicle is getting loose in the corners." The rear tires lose adhesion while cornering during oversteer. Most passenger

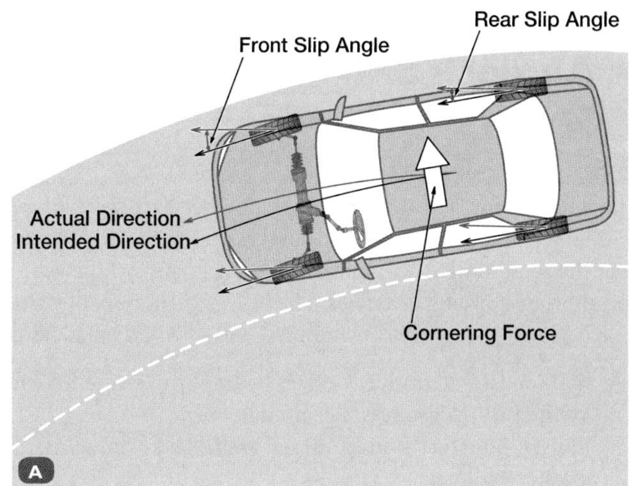

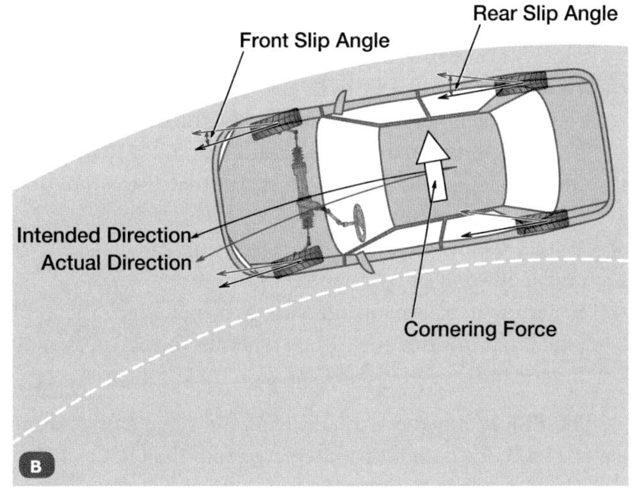

FIGURE 49-4 **A.** Understeer condition. **B.** Oversteer condition.

vehicles are designed to have a bit of understeer because this condition is easier for a driver to recover from than an oversteer.

Using information provided by the sensors of the ESC system, plus the wheel speed sensors, the EBCM monitors the stability of the vehicle and can command individual brakes to be applied and request decreased engine torque, as necessary. For example, if a vehicle is traveling too fast around a right-hand corner and the front wheels are starting to lose traction (understeer), the control system can apply the right rear brake to help pivot the vehicle around the right rear tire. This assists the vehicle in turning and at the same time slowing the vehicle slightly. If additional measures are needed, additional brakes can be applied and the engine torque reduced. The control system performs these functions automatically without any driver input other than steering the vehicle in the desired direction.

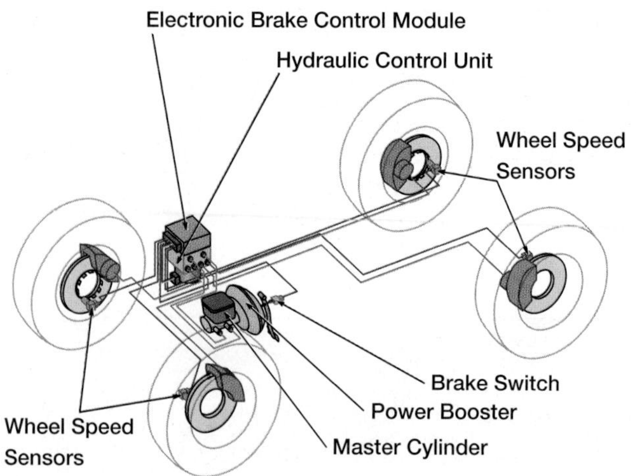

FIGURE 49-5 Typical ABS system.

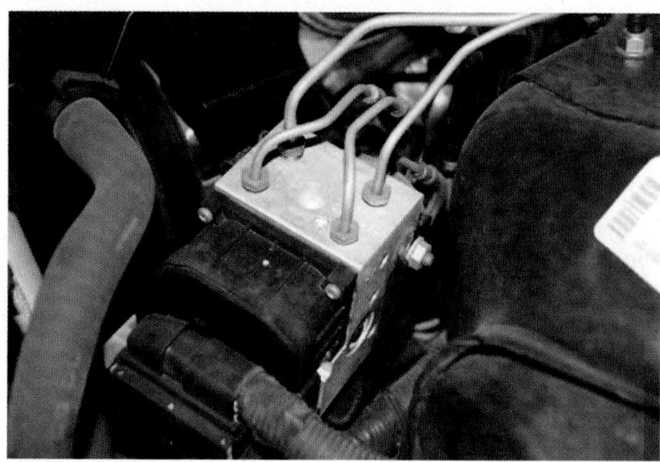

FIGURE 49-6 A hydraulic control unit (HCU).

▶ TECHNICIAN TIP

It is important for customers to know that ABS, TCS, and ESC are not guarantees of avoiding an accident. These systems are designed to help drivers who are driving in a responsible manner to avoid an accident. Drivers can easily exceed the ability of these systems.

▶ Antilock Braking System Overview

K49003

The ABS system is designed to prevent wheels from locking or skidding, no matter how hard the brakes are applied or how slippery the road surface, to maintain steering control of the vehicle and shorten stopping distances. The primary components of the ABS braking system are shown in **FIGURE 49-5** and listed here:

- **ABS master cylinder:** Creates hydraulic pressure for each of the two hydraulic brake circuits.
- **Power booster:** Boosts driver brake pedal force on the master cylinder.
- **EBCM or electronic control unit (ECU):** An onboard computer that is programmed to monitor sensor data and send output control signals to electronic solenoid valves, which modify brake pressure to individual wheel brake units.
- **Hydraulic control unit (HCU)** or modulator: Contains electric solenoid valves controlled by the EBCM to modify hydraulic pressure in each hydraulic circuit (**FIGURE 49-6**). Most systems also contain an **accumulator** to store brake fluid under pressure.
- **Wheel speed sensor**: A device that monitors wheel speed and sends that signal to the EBCM.
- **Brake switch**: An on/off switch mounted at the brake pedal that informs the EBCM whether the driver is applying the brakes.

The EBCM may be located inside the vehicle or mounted near the HCU, or it could be integrated into the HCU. In many cases, it is a separate module from the powertrain control module and may be part of the vehicle's **body control module (BCM)**. The body control module is the computer that controls the electrical system in the body of the vehicle. The EBCM receives input signals from the ABS sensors, compares these data to information stored in its memory, decides what actions are necessary, and sends output commands to the HCU.

The HCU or modulator is connected inline with the brake lines between the master cylinder and the wheel brake units. It houses electric solenoid valves that control the flow of brake fluid to each wheel. The HCU receives operating signals from the EBCM to control the brakes under ABS conditions.

The power booster and master cylinder assembly is mounted on the firewall. In most current applications, these components operate similarly to non-ABS power boosters and tandem master cylinders. Some manufacturers use a portless master cylinder to allow brake fluid to return to the master cylinder reservoir more easily than a master cylinder fitted with a compensating port. When the brakes are operating without ABS action, the brake pressure is controlled by the driver's foot pressure, which is assisted by the power booster. In other words, the ABS system only affects brake pressure when one or more wheels are starting to skid.

The wheel speed sensor consists of a toothed tone wheel (or tone ring) that rotates with the road wheels and a pickup assembly that generates a speed signal. The wheel speed sensor is located near the wheel hub in many applications (**FIGURE 49-7**). The wheel speed sensor sends to the EBCM an electrical signal that varies with the speed of the wheel. Wheel speed sensors can be **variable reluctance sensors** (magnetic induction), generating an analog AC sine wave signal (**FIGURE 49-8**). Wheel speed sensors can also be of the magneto-resistive or Hall effect type, generating a digital square wave signal. These signals can be used by the EBCM to determine the speed of each wheel. We cover the operation of these sensors in much greater depth in the ABS Components section.

Antilock Braking System Operation

When the ignition switch is turned on, the ABS controller illuminates the yellow ABS warning lamp and performs an automatic self-check of the system. If the system check passes, the controller will extinguish the warning lamp, indicating to the driver that the ABS system is functional. Some ABS systems perform an additional self-check once the vehicle is traveling more than approximately 3–5 mph (4.8–8 kph). Failures in the ABS system cause the controller to illuminate the ABS warning light in the instrument panel. If the light is illuminated, the ABS system is shut down and doesn't operate.

FIGURE 49-7 A wheel speed sensor and tone wheel.

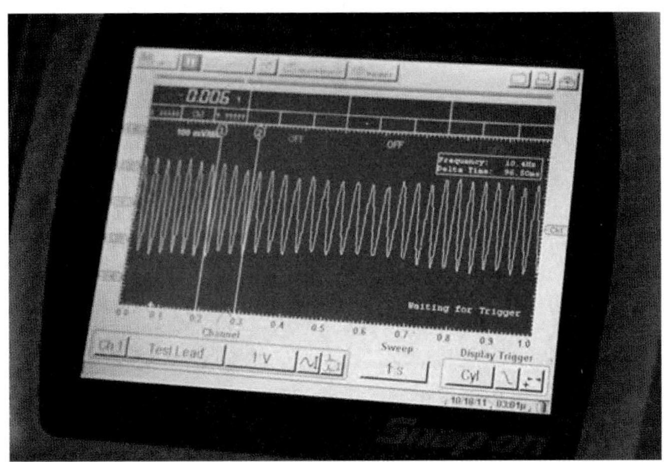

FIGURE 49-8 An oscilloscope pattern from a wheel speed sensor.

As the wheels start to turn, the wheel speed sensors generate small electrical signals and send them to the EBCM. When the brakes are applied, the wheels' rotational speed is reduced. As the speed changes, the signal sent to the EBCM changes in like manner. If the control unit detects that a wheel might be slowing too quickly and starting to lock, it sends output signals to the appropriate solenoid valve in the HCU to modify the hydraulic pressure to the affected wheel brake unit. Let's explore how that works.

Principles of ABS Braking

Braking force and the tendency of the wheels to lock up are affected by a combination of factors such as the friction of the road surface; the type, condition, and loading of each tire; and the difference between the vehicle speed and the speed of the wheels. It should be noted that maximum traction happens with approximately 10–20% tire slip. Thus, maximum braking traction occurs when the wheels are rotating 10–20% slower than the vehicle speed. In the same way, maximum traction during acceleration occurs when the wheels are rotating 10–20% faster than the vehicle speed. At the same time, traction falls off quickly above approximately 20% wheel slip, which is why ABS is so effective. It allows just enough slip to keep the tires at close to their maximum traction. It does so by rapidly modulating the hydraulic pressure in the vehicle's brake system.

During normal braking, as the rotational speed of each wheel falls equally, no ABS intervention is needed. In this condition, the EBCM does not energize the solenoid valves in the hydraulic unit. The master cylinder hydraulic pressure is applied to the wheel brake units, and the ABS is not involved. However, even though the ABS is passive during normal braking, the EBCM is constantly monitoring the speed of each wheel, looking for any wheel that begins to decelerate more rapidly than any of the other wheels.

If one wheel speed sensor signals more severe wheel deceleration, which means the wheel is beginning to slip, the EBCM sends current to the appropriate solenoid valve (**FIGURE 49-9**).

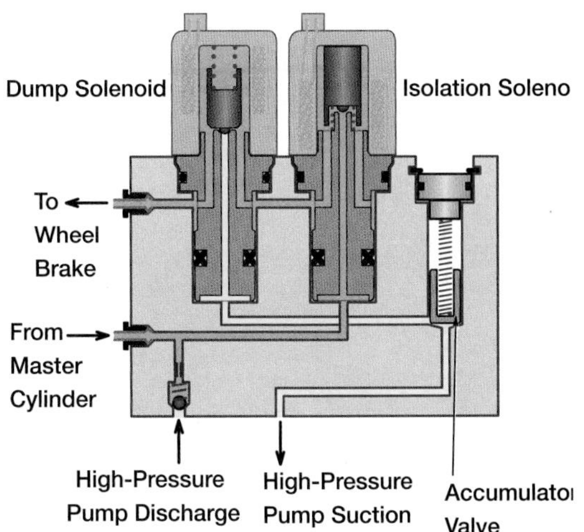

FIGURE 49-9 Hydraulic Control Unit solenoid valve arrangement for modulating ABS hydraulic pressure.

The first level of valve action isolates that brake circuit from the master cylinder. This stops the braking pressure at that wheel from rising and keeps it constant. If the wheel speed sensors indicate that the wheel is still decelerating too rapidly, the EBCM commands the appropriate solenoid valve to release braking pressure. The solenoid valve opens a passage from the brake circuit, releasing the hydraulic pressure to that brake unit. Brake fluid is released from the specific brake circuit back to the master cylinder. Pressure in the brake circuit is reduced so that the wheel is not being braked.

If the wheel speed sensors indicate that reducing the brake pressure is allowing the wheel to roll again, the EBCM de-energizes the solenoid valves. This lets the hydraulic pressure from the master cylinder be applied to the brake again. This cycle repeats itself at up to 16 times per second. It is normal in an ABS system for the valves in the HCU to keep changing position as they modulate the brake pressure that is being applied. These changes in valve position normally cause rapid hydraulic pulsations, which on some vehicles can be felt by the driver through the brake pedal. The solenoid valves also make a fairly loud clicking noise as they cycle on and off.

▶ TECHNICIAN TIP

Drivers should be taught to expect ABS brake pedal pulsation when in a panic stop. Some drivers who have never experienced this actually let up on the brake pedal because of the rapid pulsations and accompanying noise. When in a panic stop, drivers should push hard on the brake pedal and not let up until the vehicle is stopped or out of danger. Brake-by-wire systems are not subject to this brake pedal pulsation situation because the brake pedal is not part of the hydraulic system.

▶ ABS Components

ABS Master Cylinder

ABS master cylinders come in two major configurations: integral and non-integral (**FIGURE 49-10**). **Integral ABS systems** are mostly found on older vehicles. They combine the tandem master cylinder, HCU, and power booster in one unit. The power booster consists of a high-pressure electric pump and accumulator that operates the integrated master cylinder. Brake fluid passes from the master cylinder portion of the assembly to the HCU portion, where pressures are modified by the computer-controlled solenoid valves.

Non-integral ABS systems use a fairly standard tandem master cylinder and a typical vacuum or hydraulic power booster. The booster assists the driver in applying force to the master cylinder. The master cylinder sends fluid under pressure to the HCU, which is a separate assembly that is installed in line with the brake lines between the master cylinder and the wheel brake units. If the pressure has to be modified, the computer-controlled solenoid valves in the HCU will carry out the commands.

FIGURE 49-10 Non-integral and integral master cylinder assemblies.

Purpose and Operation of the ABS Master Cylinder

Non-integral ABS master cylinders are usually identical to non-ABS master cylinders. They both use primary and secondary pistons in a common housing with a **common bore**. Some of these master cylinders utilize a portless ABS master cylinder design, which does not use a compensating port on the secondary circuit. Instead, the secondary piston incorporates a center valve (**FIGURE 49-11**) that controls the opening and closing of a supply port in the piston. At rest, the supply port is open and connects the reservoir with the front brake circuit. The primary piston still uses an inlet port and a compensating port; therefore, the portless design is only used on the secondary circuit.

When the brake is applied, the primary piston moves and closes its compensating port. Fluid pressure in the primary circuit rises. It acts with the primary piston spring to move the secondary piston forward, closing the center valve. Pressure builds in the secondary circuit, keeps building in both circuits, and applies the brakes in both circuits.

If braking conditions are such that the hydraulic modulator must return brake fluid to the master cylinder; then, for the front brake circuits, brake fluid is returned to the front section. This forces the secondary piston back against the force of the primary piston spring and the rear brake pressure. If enough brake fluid returns, the center valve opens and allows the brake fluid to return to the master cylinder reservoir. If brake fluid is returned from the rear brake circuit, the secondary and primary

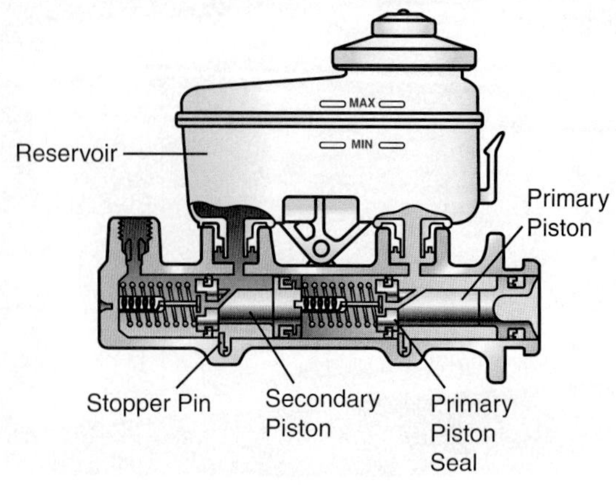

FIGURE 49-11 Portless ABS master cylinder.

pistons tend to be forced apart, which generally moves the primary piston rearward. If it travels far enough, brake fluid will return to the reservoir through the compensating port.

The amount of brake fluid that returns to the master cylinder is determined by the degree of antilock braking control. With as many as 16 ABS control cycles per second, the rapid changes in hydraulic pressure cause brake fluid pulsations to be sent back to the master cylinder; these pulsations can be felt by the driver at the brake pedal.

Control Valve Operation

In a standard ABS system, the HCU houses electrically operated hydraulic control valves (solenoid valves) that control brake pressure to specific wheel brake circuits. Each separate hydraulic circuit within the HCU has one or two solenoid valves that provide three operating conditions: apply, hold, and release (**FIGURE 49-12**). During the apply mode, the solenoid valves allow brake fluid to freely flow through the HCU hydraulic control circuit to the specific brake circuit. In this case, the driver is in full control of the brakes, through the master cylinder.

In the hold mode, a solenoid valve "isolates" the master cylinder from the brake circuit. This prevents brake pressure from building any further. The brake pressure to the wheel is held at that level until the solenoid valve is commanded to change its position. In the release mode, a solenoid valve releases (dumps) the brake circuit pressure to the wheel, allowing it to start rolling again. The solenoid valve opens a passage back to the accumulator, where brake fluid is stored until it can be returned by an electric pump to the master cylinder reservoir. Notice that the brake pressure is held or dumped even though the driver is applying the brake pedal.

> ▶ TECHNICIAN TIP
>
> Many, but not all, HCUs are sealed units and cannot be serviced. If you are working on a vehicle with a sealed HCU and it is faulty, it will have to be replaced. This can be quite costly.

Operation of the Hydraulic Control Unit

The ABS control module (or EBCM) sends commands in the form of electrical signals to the HCU. The HCU executes the commands, using one or two solenoid valves for each hydraulic circuit, depending on the type of HCU. Because the control valves are situated between the master cylinder and the wheel brake units, they can apply, block, or release hydraulic pressure going to the brake units.

In a normal non-ABS braking scenario, brake pedal force is transmitted to the master cylinder and then through the non-energized open isolation valves to the brake units at the wheel. When the signals from the wheel speed sensors show no tendency for the wheels to lock up, the hydraulic pressure from the master cylinder flows freely through the HCU to the brake units at each wheel.

When the control unit detects any lockup tendency, it sends a command current to the isolation solenoid valve for that brake circuit. This current causes the solenoid valve to close, isolating the brake circuit from the master cylinder. That holds the hydraulic pressure between the solenoid valve and the brake circuit constant, regardless of whether the master cylinder hydraulic pressure rises or falls.

If the wheel speed sensor signals that excessive wheel deceleration is continuing, the control module commands the dump valve for that brake circuit to open. This reduces the braking pressure by opening a passage from the brake circuit to the accumulator. A pump in the HCU sends brake fluid back to the master cylinder, pushing one or both pistons rearward in the bore and venting to the reservoir.

If the sensor indicates that the lower pressure has allowed the wheel to speed up, the EBCM signals the dump valve to close and the isolation valve to open. The hydraulic pressure from the master cylinder is again allowed to apply the brakes, and the wheel is again slowed. This process continues until the vehicle comes to a stop or the driver lifts his or her foot from the brake pedal. In most standard ABS systems,

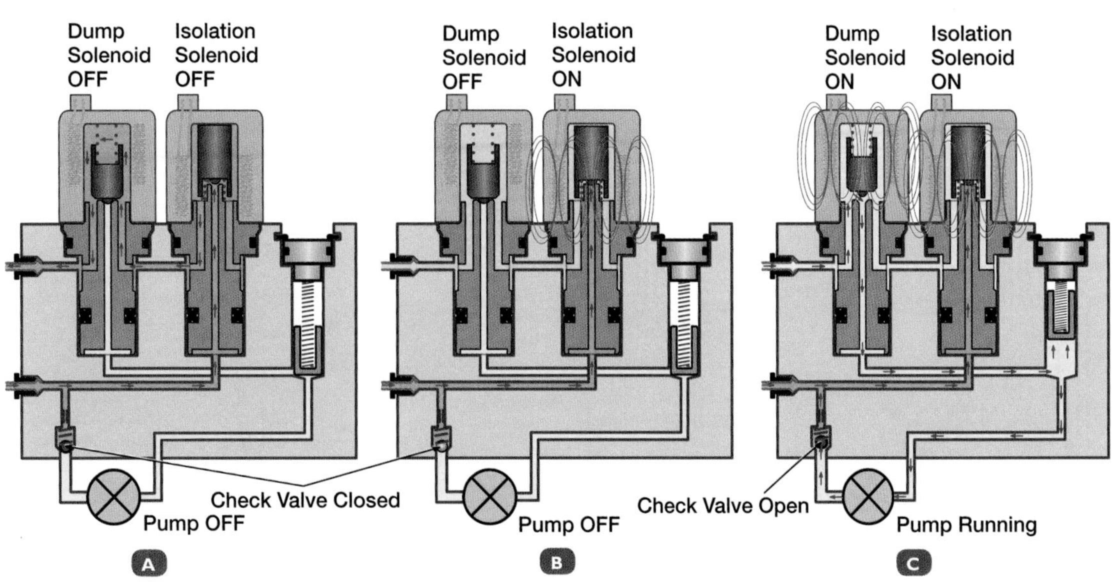

FIGURE 49-12 A. Apply. **B.** Hold. **C.** Release.

the hydraulic pressure in the brake circuits can never rise above the master cylinder pressure.

Types of Hydraulic Control Units

There are a number of HCUs that vehicle manufacturers use, and they generally fall into a few categories. The first category relates to how many channels the system has. A **channel** generally means the number of electrical wheel sensor circuits and hydraulic circuits a system has (**FIGURE 49-13**). A single-channel system uses one sensor circuit, with the speed sensor typically located in the differential and one hydraulic control circuit to control both rear wheels. Since it has only one sensor, the ABS system has nothing to compare the deceleration to. So maximum deceleration is programmed in the control unit. If the deceleration of the differential exceeds the maximum deceleration, then the HCU will be activated.

A two-channel system is similar but uses two separate speed sensors and hydraulic control circuits, one for each rear wheel. The two hydraulic control circuits apply brake pressure separately to the rear wheels. A three-channel system is configured so that each front wheel has its own speed sensor and hydraulic control circuit, whereas the rear brakes use a single speed sensor with a single hydraulic control circuit. A four-channel system uses separate speed sensors and hydraulic control circuits for each of the four wheels. Since these systems have more than one sensor, the control unit not only watches for wheel deceleration that is faster than the programmed maximum, it also watches for individual wheels decelerating faster than others. This gives a more robust ABS function.

Another difference among types of HCUs is the number of solenoid valves per hydraulic control circuit. Some HCU units use a single, three-position solenoid valve per circuit; others use dual, two-position valves per hydraulic circuit (**FIGURE 49-14**). The first position of the single, three-position valve allows brake fluid to flow through the apply port while blocking the release port. The second position blocks the apply port and the release port. The third position blocks the apply port and opens the release port. Thus, the single, three-position valve has all three conditions: apply, hold, and release.

The dual, two-position valve style of HCU uses one solenoid valve to open and close the apply port. This is commonly called the isolation valve. When this valve is not energized, the

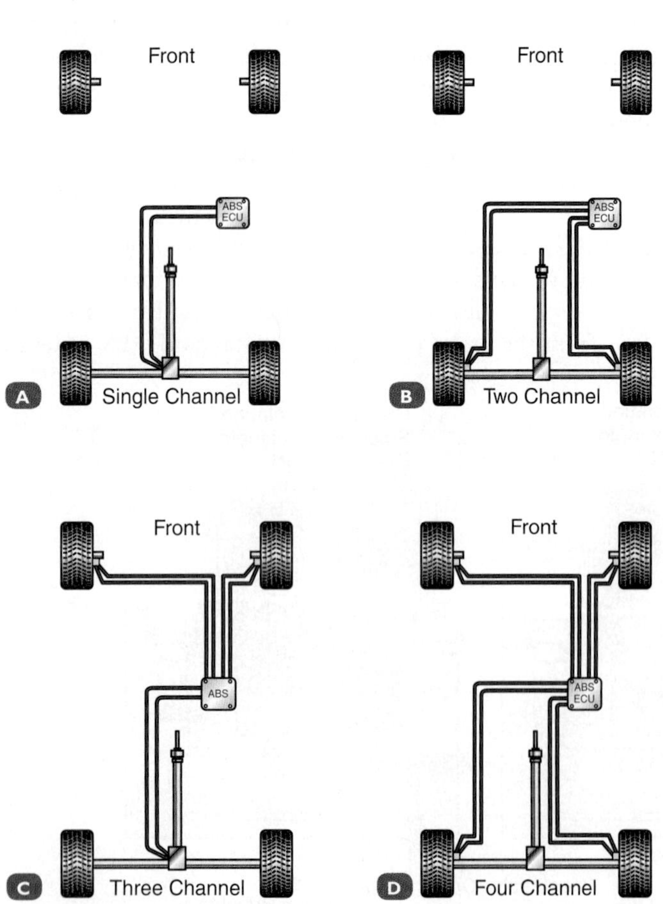

FIGURE 49-13 The four types of ABS channels. **A.** Single-channel system. **B.** Two-channel system. **C.** Three-channel system. **D.** Four-channel system.

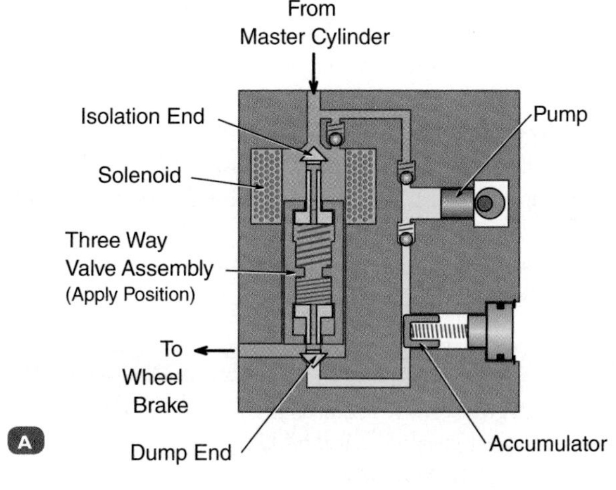

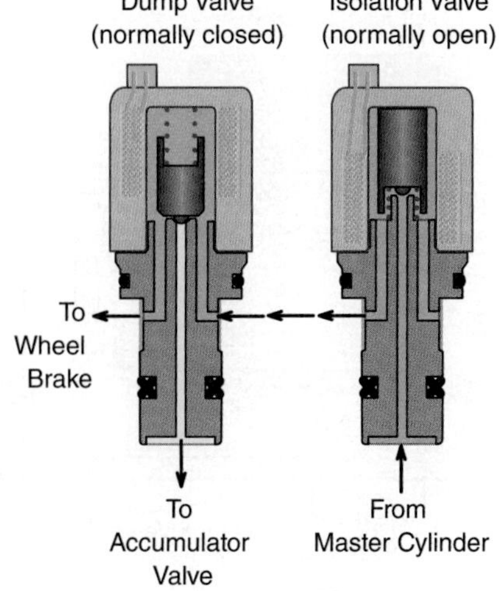

FIGURE 49-14 A. Single three-position valve. **B.** Dual, two-position valves.

apply port is open. The second solenoid valve opens and closes the release port. When this valve is not energized, the release port is blocked. The EBCM operates each of these valves independently to obtain apply, hold, and release functions. Because there are twice as many solenoid valves, and each valve needs its own electrical control circuit, the EBCM is more complicated and costly to build. Therefore, EBCMs are specifically designed to work with only the specified type of HCU, and EBCMs and HCUs cannot be randomly interchanged.

Another difference between HCUs is the type of accumulator used—low pressure or high pressure. **Low-pressure accumulators** hold brake fluid in a spring-loaded chamber when it is released by the dump valves during an EBC event (**FIGURE 49-15**). The accumulator used to supply brake pressure to the HCU also uses a high-pressure pump, pressure switch, and relay to maintain an operating pressure of approximately 1200–2700 psi (8274–18,616 kPa), depending on the system. The hydraulic pressure remains fairly low because an electric pump returns the released brake fluid to the master cylinder when brake fluid in the accumulator reaches a certain point. When the electric pump turns on, the fluid returning to the master cylinder pushes the brake pedal toward the driver's foot, causing the brake pedal to rise. This can be confusing to drivers because it feels like someone is under the dash, pushing the brake pedal back toward them.

High-pressure accumulators are used to store brake fluid under high pressure for one of two purposes: to be used as a power booster for applying the integrated master cylinder or to be used to independently to apply the wheel brake units when the EBCM commands it. When used as a power booster, pressure in the accumulator is maintained by a high-pressure electric pump. The pump is activated by a pressure switch and relay when the hydraulic pressure falls below a certain point. When the pressure reaches the specified upper pressure limit, the pressure switch opens and deactivates the electric pump. The hydraulic pressure is then used to boost the driver's foot pressure on the master cylinder when the driver depresses the brake pedal. If the high-pressure pump fails for any reason, the accumulator holds enough brake fluid at high pressure to apply the brakes 10 to 20 times before the boost is used up. If that occurs, the brakes will still operate but will require much higher foot pressure.

SAFETY TIP

Be careful to follow the manufacturer's procedures when working on EBC systems. High-pressure brake fluid stored in the accumulator is dangerous.

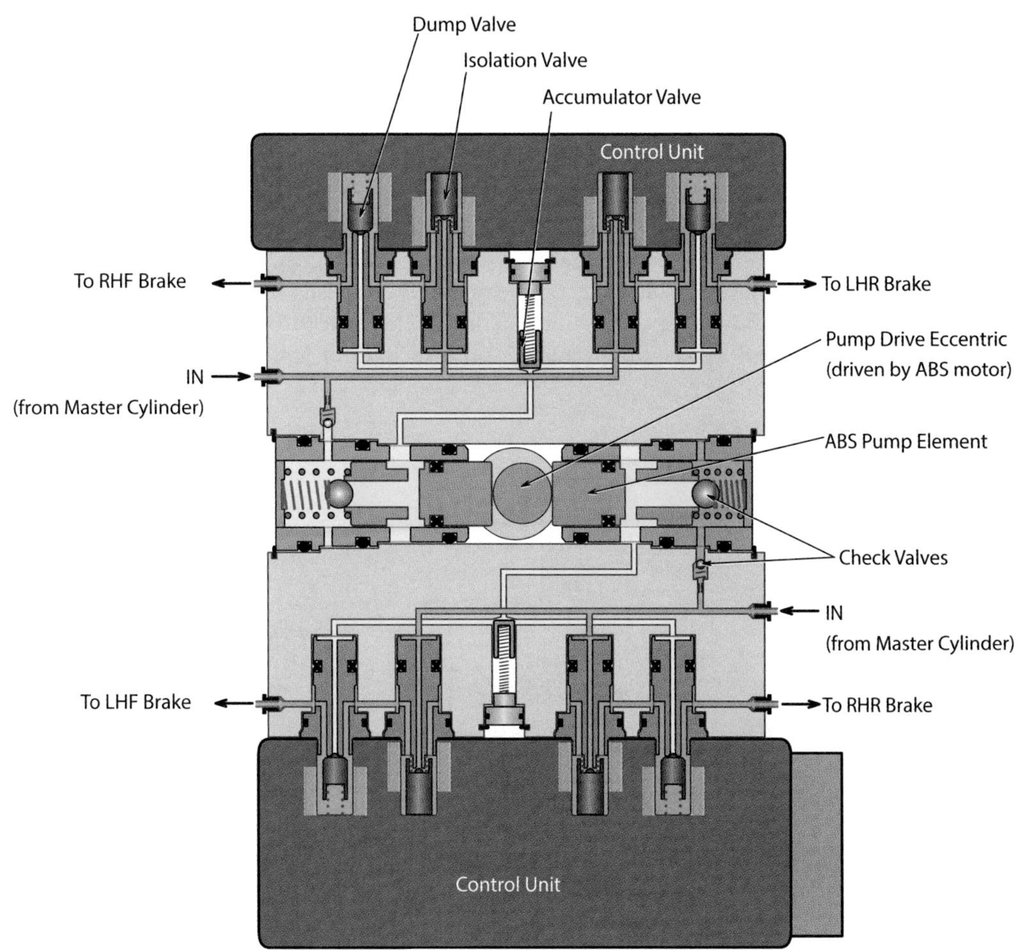

FIGURE 49-15 Low pressure accumulator.

The high-pressure pump pushes the brake fluid against a high-pressure nitrogen chamber, which holds pressure on the brake fluid. The hydraulic pressure is used to independently apply the brakes during a TCS or ESC event. If the high-pressure pump fails while driving and the hydraulic pressure falls below the pump's specified "on" pressure, the EBCM will disable the ABS system and illuminate the yellow warning lamp, alerting the driver to an ABS system fault.

Wheel Speed Sensors

Wheel speed sensors create electrical signals based on the rotational speed of each wheel they monitor. Wheel speed sensors do so by using principles of electromagnetism to generate an analog or digital electrical signal. This signal is read by the EBCM to determine the speed of each wheel as well as the rate of deceleration of each wheel. This information is used to determine whether a wheel is starting to lock up and skid.

A wheel sensor assembly consists of a toothed tone wheel (or tone ring) that rotates with the wheels, and a stationary pickup assembly attached to the hub or axle housing. The **pickup assembly** and **tone wheel** do not touch each other; a small gap, called an **air gap**, must be maintained at the specified clearance. Because there is no mechanical connection, there is virtually no wear unless a foreign object gets between them.

Types of Wheel Speed Sensors

The three most common types of wheel speed sensors are the variable reluctance (magnetic induction style), **magneto-resistive**, and **Hall effect** styles. The variable reluctance type is simpler and usually less expensive for manufacturers to use. This style is sometimes called a passive system because it is self-contained and needs no outside power to function. Magnetic induction occurs when the teeth on the tone wheel pass the sensor, creating an analog AC voltage signal. As each tooth of the tone wheel approaches the pickup, the magnetic field creates a small voltage that pushes current flow in one direction inside the pickup assembly. As each tooth leaves the pickup assembly, voltage is generated that pushes current flow in the opposite direction. This process creates a full-cycle sine wave for each tooth on the tone wheel (**FIGURE 49-16**). The faster the wheel is turned, the faster the sine wave rises and falls. The speed at which the sine wave rises and falls is referred to as frequency. Frequency is measured in hertz, where one hertz equals one full-cycle sine wave per second.

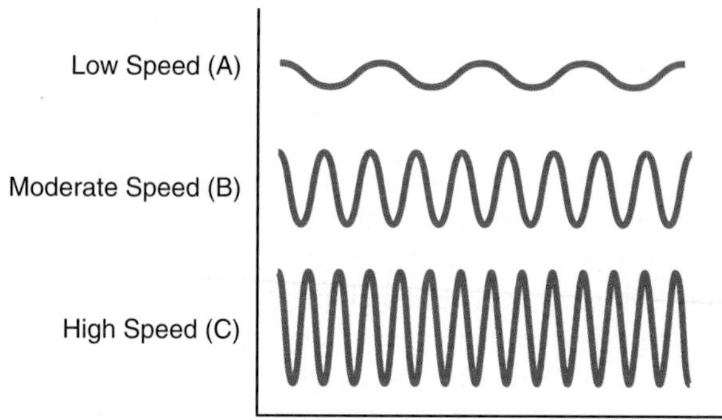

FIGURE 49-16 Wheel speed sensor sine wave. **A.** Signal during low vehicle speed. **B.** Signal during moderate vehicle speed. **C.** Signal during high vehicle speed.

The height of the sine wave, called its amplitude, also tends to change with speed. At very slow vehicle speeds, when the vehicle is just creeping along, the amplitude of the sine wave is very low. As the speed increases, so does the amplitude, along with the frequency. This AC signal is sent to the ECU, where it is processed and then compared to the AC signals from the other wheels to determine wheel lockup.

Most variable reluctance wheel speed sensors are two-wire sensors, which complete the circuit back to the ECU. The variable reluctance sensor assembly consists of a coil of wire around a permanent magnet, with each end of the coil connected to one of the wheel speed sensor terminals, which connect directly into the EBCM (**FIGURE 49-17**). Because this type of sensor operates on principles of magnetism, the air gap between the toothed

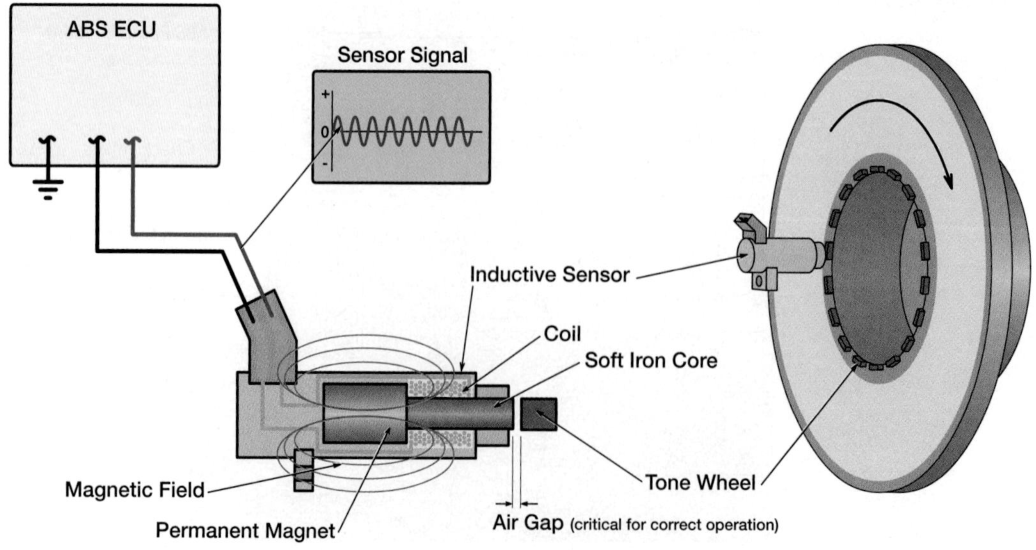

FIGURE 49-17 Variable reluctance wheel speed sensor assembly.

tone wheel and sensor is critical. If the air gap is too small, the parts could contact each other, damaging them. If the air gap is too large, the sensor output signal to the ECU could be too weak and trigger a code or cause the sensor to work intermittently.

One drawback to the variable reluctance sensor is that because it depends on the speed of movement of the tone wheel to create a signal, it does not function effectively below vehicle speeds of around 5 mph (8 kph). In other words, the amplitude of the sine wave it creates at slow speeds is not high enough for the EBCM to read it. This can prevent the ABS from functioning during the last part of a braking event. On a very slippery road surface such as ice, the lack of ABS functionality at that speed could lengthen the stopping distance significantly.

The magneto-resistive and Hall effect sensor systems are called active systems because they require an outside power source to operate. If the sensor loses power or ground, it cannot generate an output signal. The power wire originates from the EBCM and normally supplies the magneto-resistive or Hall effect sensor systems with a reference voltage of between 5 and 12 volts, depending on the manufacturer. This helps ensure that the sensor is not affected by changes in the vehicle's electrical system voltage. A signal wire transmits the output signal from the sensor to the EBCM. The magneto-resistive and Hall effect sensors can be a three-wire arrangement, with the third wire being a dedicated ground, or a two-wire arrangement with ground being provided by the chassis.

The magneto-resistive speed sensor and Hall effect wheel speed sensor types operate similarly to all Hall effect sensors. A reference voltage and ground are supplied to the sensor assembly, where internal circuitry causes a small current to flow across the semiconductor bridge/Hall material (**FIGURE 49-18**). If the bridge/Hall material is exposed to a magnetic force, the magnetism forces the current to flow to one side of the bridge/Hall material. This produces a small difference in voltage across the sides of the bridge/Hall material (**FIGURE 49-19**). Voltage is then amplified and processed into a digital "on" signal (circuit is pulled to ground) and sent to the EBCM. As the magnetic field is removed, the small signal voltage across the bridge/Hall material falls to 0V. The signal sent to the EBCM will be a digital "off" signal (reference voltage). As the magnetic field is alternately applied and removed, the sensor will send a digital square wave on/off signal corresponding to the changes in the magnetic field. Because the magnetic field does not have to be moving for the bridge/Hall effect voltage to be created, the sensor works all the way down to 0 mph (0 kph). This allows the ABS to continue functioning until the vehicle comes to a virtual full stop.

▶ TECHNICIAN TIP

Testing wheel speed sensors is dependent on knowing which kind of sensor the vehicle uses. Do not assume all two-wire sensors are of the variable reluctance style. With the ignition switch set to the Run position and the wheels stationary, use your digital multimeter (DMM) to properly back-probe both sensor wires for voltage. If neither wire has voltage, suspect a variable reluctance sensor. If one of the two wires has a reference voltage, you are likely dealing with a magneto-resistive or Hall effect sensor.

Brake Switch

In addition to activating the rear brake lights, the brake switch sends an electrical input signal to the EBCM, telling it whether the driver is applying the brakes. If the brakes are being applied, the EBCM will activate the appropriate solenoid valves if the wheel speed sensors signal that the wheels are starting to lock up. If the brake switch indicates that the brakes are not being applied and the wheel speed sensors are showing unequal speeds indicating a slippery road surface, the EBCM on some vehicles illuminates a low traction warning lamp, alerting the driver to the low traction condition.

The brake switch is a normally closed switch, meaning that if the switch is not affected by any outside force, electrical current will flow through it. The brake pedal pushes the brake switch

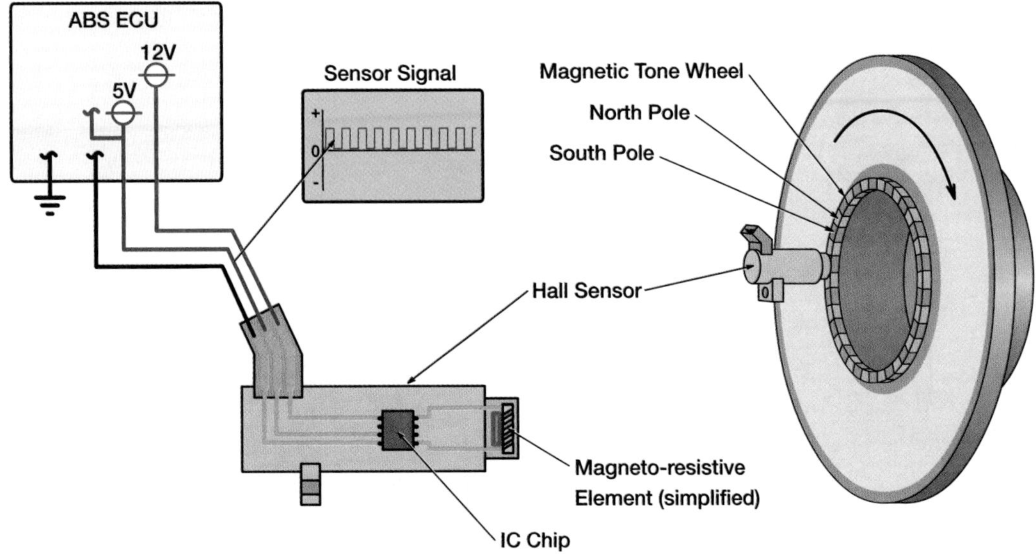

FIGURE 49-18 Hall effect wheel speed sensor assembly.

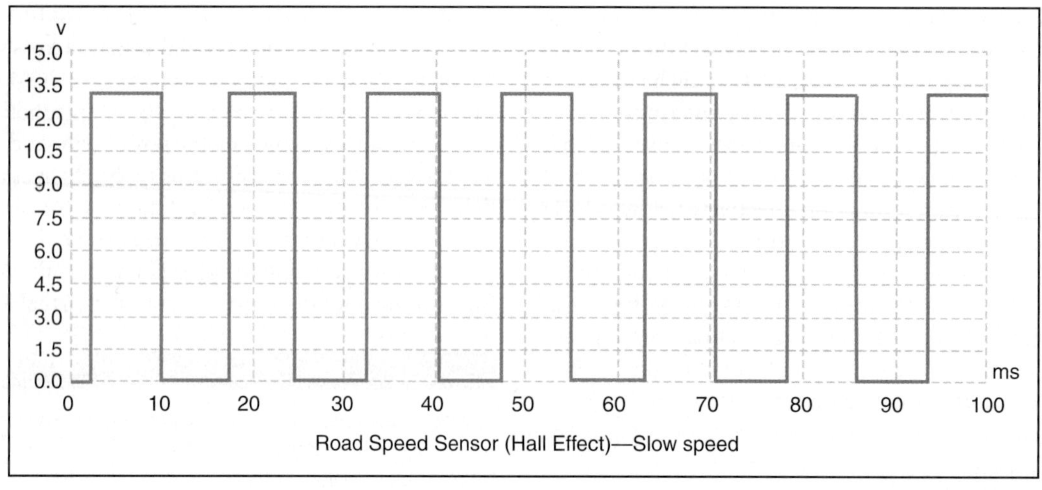

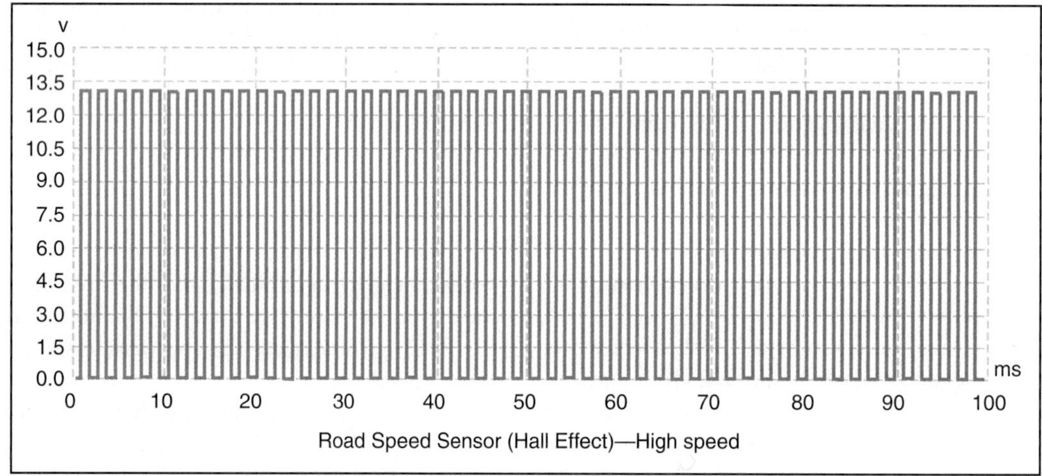

FIGURE 49-19 Hall effect operation.

open when the brakes are released (**FIGURE 49-20**). As soon as the driver steps on the brake pedal, the spring in the brake switch closes the contacts and sends electrical current (signal) to activate the brake lights. This electrical signal is also sent to the EBCM, signaling it that the driver is applying the brakes. In some systems, the brake switch is only used to signal the body control module, which then sends current to illuminate the brake lights.

FIGURE 49-20 The brake light switch is an important input for the ABS. The switch is held open by the released brake pedal.

More advanced EBC systems may use a brake pedal position sensor, which indicates how far and fast the brake pedal is being pushed (**FIGURE 49-21**). It sends a variable signal based

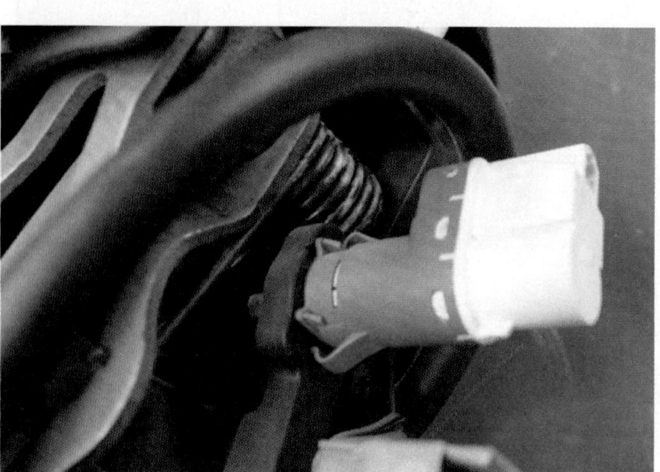

FIGURE 49-21 Typical brake pedal position sensor.

on the application of the brakes, which the EBCM uses to determine brake pedal travel and speed. This gives the EBCM additional information about the type of braking that is being performed, which can be used to modify the ABS intervention.

ABS Electronic Brake Control Module (EBCM)

The EBCM is made up of electronic circuitry to process input signals, an electronic data processor, computer memory, and output drivers to control the output devices such as the electric solenoid valves. The EBCM is programmed from the manufacturer to make brake control decisions and send output commands to the controlled devices based on sensor input data, which are compared to the data maps in its memory. These maps are designed to account for all of the reasonable braking conditions that the vehicle could experience. The EBCM continuously monitors the sensor data for any indication that one or more wheels are about to lock up.

The EBCM receives signals from several sources (**FIGURE 49-22**). A switch at the brake pedal provides a brake on/off condition or, on some vehicles, a brake pedal position signal.

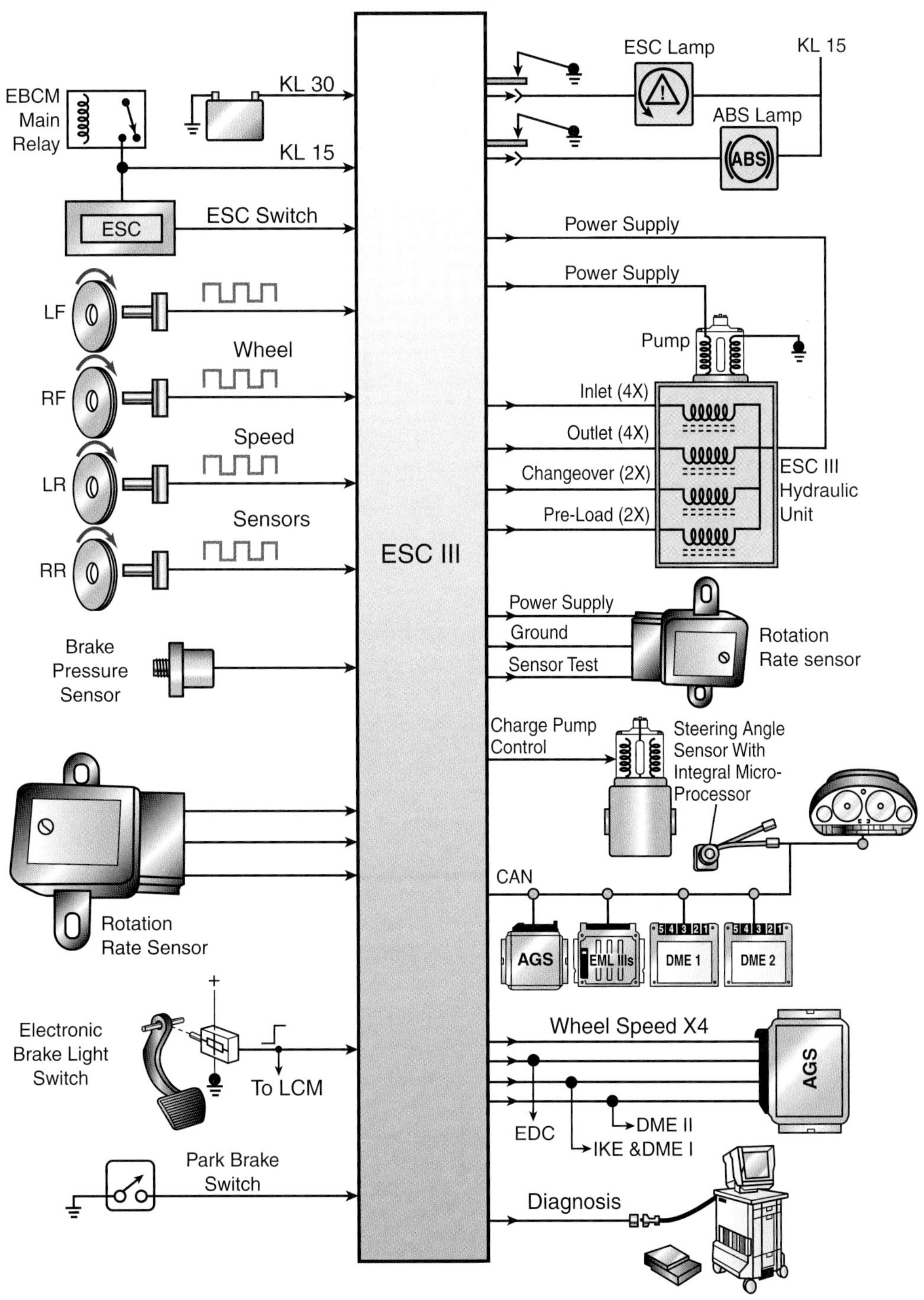

FIGURE 49-22 EBCM circuit.

An input from the ignition switch signals that the driver has turned the ignition on. Some control units monitor the battery voltage and use the rise in charging system voltage to indicate that the engine is actually running. The **vehicle speed sensor** reports the speed of the vehicle. Each of these input signals is used by the EBCM to know whether a wheel starts to lock while the driver is applying the brakes and which ABS actions are necessary to prevent a full skid condition.

Some ABS control modules have additional functionality designed into them. One example is electronic brake proportioning. This feature does away with the mechanical proportioning valve and duplicates that action electronically by using the ABS valves to reduce rear brake hydraulic pressure under moderate brake pedal application. The EBCM restricts pressure to the rear wheels based on how hard the brake pedal is being applied. In this case, the EBCM does not wait until a wheel sensor reports that one or both rear wheels are locking up; instead, it reduces rear brake pressure slightly before lockup occurs. It does this during moderate and heavy braking because of weight being transferred from the rear wheels to the front wheels, reducing the traction at the rear wheels.

The EBCM performs an automatic system self-check and warning lamp bulb check on the ABS system every time the key is turned to the Run position. If the EBCM detects a fault, the ABS warning lamp will remain on and most, but not all, systems store the fault in the EBCM memory. The faults are stored as diagnostic trouble codes (DTCs) for retrieval by technicians when diagnosing ABS system faults.

Some older ABS systems provide **blink codes**, also known as flash codes, through the ABS warning lamp when a specific terminal is grounded or two specific terminals are shorted together. If a code is stored in the EBCM memory, the EBCM will blink the ABS warning lamp in a manner that indicates a particular trouble code. For example, a code 12 would be one blink followed by a short pause, then two rapid blinks followed by a long pause. Each code is usually displayed three times before the next code is displayed. Once all codes have been displayed, the codes start at the beginning again.

Most ABS systems require a scan tool that connects to the EBCM or powertrain control module data link connector to read the **fault codes**. Fault codes indicate which circuit is experiencing a fault, such as an open left front wheel speed sensor circuit. The fault codes also can indicate if there is a condition the EBCM determines is out of acceptable tolerances, such as wheel speeds that do not match within the specified tolerance. The cause could be as simple as having a tire of the wrong size installed on the vehicle or having properly sized tires that are not inflated to the same pressure. Once the codes have been retrieved, technicians use service information to diagnose the problem and locate the cause of the fault.

▶ TECHNICIAN TIP

It is important for the customer to know that the ABS system is disabled when the ABS yellow warning lamp is on. The brakes will work normally, but without ABS function.

▶ Traction Control System (TCS) Overview

K49004

Although a basic ABS system can prevent skidding by holding or releasing individual brake circuit pressure, it has no ability to apply the brakes apart from the driver-created hydraulic pressure. This system works fine as long as the tire slippage is a result of the driver applying the brakes. However, tires also slip because the engine torque accelerating them exceeds their traction with the road surface; in this scenario, they can slip, spin, or break loose, causing a loss of control of the vehicle. The TCS system was developed to prevent the drive wheels from slipping while the vehicle is being accelerated. It is active up to a manufacturer-specified speed. Above that speed, traction control is deactivated by the EBCM because further acceleration is unlikely to cause the wheels to lose traction.

To obtain traction control capabilities, manufacturers have added a few design features to the basic ABS system, one of which is the high-pressure pump and accumulator that was discussed earlier. This pressure is used to activate brake units on the drive wheels independently of the driver. The sensors are the same as in the ABS system, but the ability to apply the individual drive wheel brakes is needed; thus, two to four extra solenoid valves, called **boost valves**, are added to the HCU (**FIGURE 49-23**). These boost valves direct hydraulic pressure from the accumulator to the ABS solenoid valves so that individual wheel brake units can be applied independently. Additional programming is added to the EBCM to control the high-pressure pump and extra HCU valves, and to decide when each of them needs to be activated.

Operation of the TCS

When the TCS is active, the EBCM monitors the speed of the individual drive and non-drive wheels along with the vehicle speed from the vehicle speed sensor. If the driven wheels are accelerating at different speeds from each other or the non-driven wheels, the EBCM can identify which wheel or wheels are slipping. It will then take action to reduce the torque to the appropriate wheels by first applying the brake to any wheels that are slipping. It does this by activating the isolation valve to close off the supply port from the master cylinder. It then activates the boost valve to pressurize the brake circuit on the spinning wheel to slow it down. If that is not enough to prevent the slippage, the EBCM will request reduced power from the engine. This can be accomplished by reducing the throttle plate opening, shutting down one or more fuel injectors, reducing the engine timing, or selecting a higher gear in the transmission. Once the wheel speeds return to proper parameters, the EBCM will return the TCS system to normal and continue to monitor the wheels for slippage.

Some TCS systems can be temporarily deactivated by a TCS function switch located on the dash or center console (**FIGURE 49-24**). If the driver deactivates TCS, the system will not intervene during wheel slip. Drivers will disable the TCS for a variety of reasons. They might be climbing a long hill on a rough

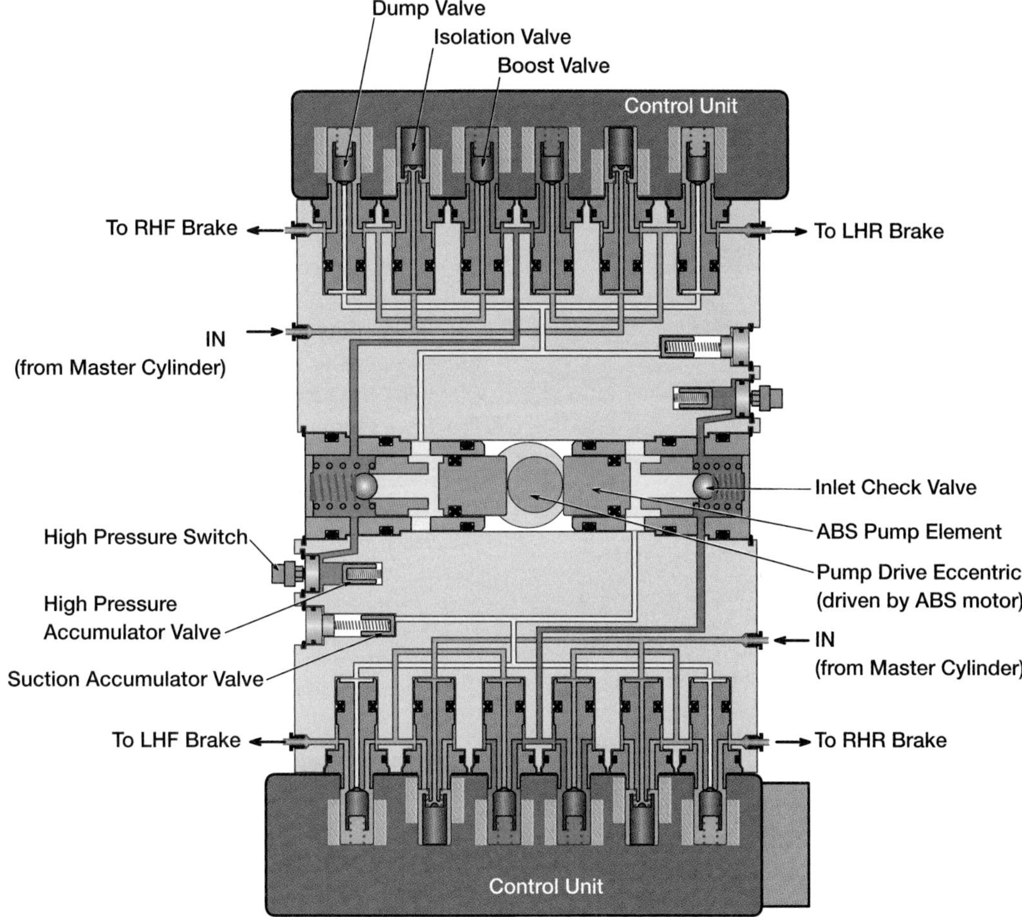

FIGURE 49-23 An HCU with boost valves.

FIGURE 49-24 A switch for deactivating the traction control system.

gravel road, which would continuously activate traction control, overheating the brakes. Or they might want to show off by "roasting" the tires or to experience driving without traction control as they would on a racetrack. The TCS will automatically default back to On during the next ignition switch cycle. In most cases, if the TCS is deactivated, the ABS system will still be active.

▶ TECHNICIAN TIP

Manufacturers use various strategies in their TCS systems, and not all of them apply the brakes as a first step. Some of them reduce engine power first. Even so, EBCMs today operate very fast, so there may only be a few milliseconds between each action.

▶ Electronic Stability Control (ESC) Overview

K49005

ABS does a good job of preventing wheels from locking up under hard braking or poor traction conditions and allows the driver to maintain directional control of the vehicle. TCS also does a good job of maintaining traction when the vehicle is driven in a relatively straight line. However, drivers can lose directional control of the vehicle while driving aggressively, taking emergency steering actions, or if there are sudden changes in the traction of the road surface while in a turn. These situations can cause the vehicle to understeer (push) or oversteer (loose, or fishtail). It can also cause vehicles with a high center of gravity, such as an SUV, to roll over (**FIGURE 49-25**). All of these situations can lead to serious accidents.

FIGURE 49-25 Electronic stability control helps the driver maintain control of the vehicle when driving.

If any of these situations are imminent, the ESC system can independently activate individual wheel brake units as necessary to help keep the driver from losing control of the vehicle. ESC utilizes the ABS and TCS systems, but with a few enhancements to more actively interface with the vehicle's operation in maintaining directional stability while the vehicle is being steered. A US Insurance Institute for Highway Safety 2006 study estimated that if all vehicles were equipped with ESC, approximately 10,000 fatal accidents in the United States could be avoided each year. This finding led the Department of Transportation to require that all vehicles of less than 10,000 lb (4536 kg) gross vehicle weight, and manufactured after September 1, 2011, be equipped with an ESC system that meets their minimum specifications.

The ESC system integrates a yaw sensor, **steering angle sensor**, and sometimes a **roll-rate sensor** into the basic ABS and TCS systems (**FIGURE 49-26**). It also adds new programming parameters into the EBCM to monitor the vehicle's stability, as well as added output command capabilities to apply individual drive wheel and non–drive wheel brake units independent of the driver.

The yaw sensor measures the amount of directional rotation of the vehicle on its vertical axis. In other words, it tells the EBCM the rate at which the vehicle is turning. The steering angle sensor tells the computer what the driver's directional intent is. If equipped, the roll-rate sensor tells the computer the rate of roll and the amount of roll that the vehicle is experiencing. The EBCM continuously monitors these signals and compares them to preprogrammed scenarios and decides which, if any, brake units need to be applied, and if engine torque needs to be reduced to keep the vehicle stable. This process is a good example of the computer feedback loop Input → Control Logic Process → Output.

SAFETY TIP

Most standard passenger vehicles are designed with a bias toward understeer. It is generally agreed that understeer is easier for the average driver to recover from. However, many performance vehicles are designed with a slight bias toward oversteer, which can be managed by an experienced driver while driving aggressively.

▶ TECHNICIAN TIP

The yaw sensor operates similarly to a Wii or other video game controllers. Its internal circuitry senses movement and sends a signal directly related to the movement it senses.

Electronic Stability Control (ESC) Operation

If the ESC system is activated, the ECBM monitors the yaw sensor signal, the steering angle sensor signal, and the roll-rate sensor signal as well as the wheel speed sensor signals. If the vehicle is beginning to understeer, oversteer, or roll, the EBCM detects it in the signal values. It then applies up to three wheel brake units to help bring the vehicle back within proper stability parameters.

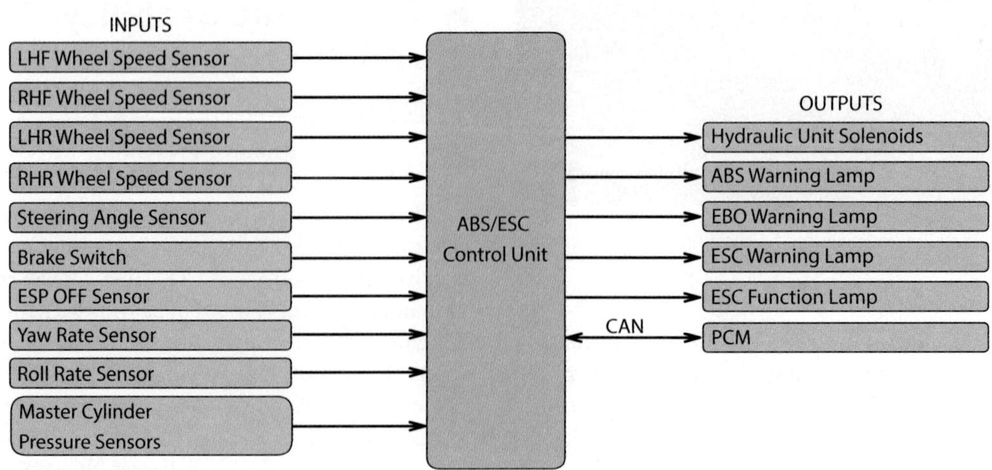

FIGURE 49-26 Typical ESC schematic.

If that does not stop the stability issue, the EBCM will request a reduction in engine power through the powertrain control module to help slow the vehicle further. On most vehicles, the EBCM illuminates a warning lamp on the dash or sound a beeper signifying when the ESC system has detected the start of a skid and reacted to it. This way the driver will be informed that he or she is on the verge of losing control of the vehicle.

On most vehicles, the ESC system defaults to "on" so it is always active. Some vehicles have a switch on the dash or center console to temporarily deactivate the system. This can be useful when driving in mud or sand when traction is virtually nonexistent and the ESC system cannot function effectively. Even though some ESC systems can be turned off, they may still monitor the operation of the vehicle and reactivate the ESC system under certain situations, such as driving above a specified speed or if a spin is detected while the brakes are being applied.

Some ESC systems incorporate a switch that allows the driver to select one or more varying levels of assist from the ESC, such as "touring," "track," or "sport" (**FIGURE 49-27**). This option allows the driver to experience differing levels of wheel slip by being able to push the vehicle closer to the edge of control than when ESC is fully activated, while still having the ESC system available as a backup, but with limited assistance. When driving on a racetrack, for example, the driver may want full control of the vehicle instead of being limited by the ESC system.

In the continuous search for new bells and whistles to impress customers and enhance safety, manufacturers have designed other features into ESC systems, such as these:

- **Hill assist:** Holds the brake pressure until the throttle is depressed and the vehicle starts to move forward.
- **All-wheel drive traction control:** Applies brake pressure as needed to any of the four individual wheels that may be slipping to maintain power to the wheels with the most traction.
- **Engine braking control:** Increases the engine torque if the ESC system detects wheel slippage during deceleration.
- **Panic stop assist:** Detects a driver's rapid throttle release and lightly applies the brakes to dry the rotors and prepare the brakes for a panic stop.

FIGURE 49-27 Switch to select the level of ESC assist desired.

- **Accident avoidance:** Works in conjunction with adaptive cruise control to monitor objects in front of the vehicle. If the ESC system detects an imminent collision, it can apply the brakes or boost the brake pressure above driver pressure.
- **Hill descent control:** Works in conjunction with the ESC system to control the speed of the vehicle when going down loose, rough, or slippery slopes.
- **Trailer sway control:** Detects trailer sway and uses the ESC system to keep it under control.
- **Optimized hydraulic braking:** Monitors brake pressure in each brake circuit and increases it above boosted pressure if deemed necessary.

▶ TECHNICIAN TIP

Many ESC-equipped vehicles monitor signals from other sensors as well to help prevent a loss of control of the vehicle; these sensors include the throttle position sensor, vehicle speed sensor, and brake pedal position sensor. When diagnosing an ESC system fault, research the sensors monitored by the EBCM.

▶ Maintenance and Repair

S49001

Bleeding the EBC System

N49001

Bleeding the EBC system can be easy or difficult. If care is taken to never allow air to enter the hydraulic system, most EBC systems can be bled just like a non-EBC system. However, if air is allowed into the EBC system, it can become trapped in the HCU, requiring a scan tool to operate the solenoid valves to help bleed the air from the EBC system.

Many EBC systems also require that a detailed step-by-step process be followed to successfully purge the air. The best advice is to always make sure that air never enters the EBC system while bleeding the brakes. The easiest way to do so is to check the level of brake fluid in the master cylinder reservoir, often while bleeding the brakes, and to always add fluid well before it gets low enough to allow air to enter the master cylinder. Also, cap any open lines or components so the brake fluid does not drain out of the lines while the system is open. Remember that DOT 5, the silicone brake fluid, should never be used in an EBC system unless the manufacturer clearly specifies it, as it is prone to aeration in the HCU.

To bleed the EBC system's hydraulic circuits, follow the steps in **SKILL DRILL 49-1**.

Depressurizing High-Pressure Brake Components

N49002

Depressurization of an EBC system's high-pressure components is needed on some systems when specific tasks are being performed such as during the removal or disassembly of certain

SKILL DRILL 49-1 Bleeding the EBC System's Hydraulic Circuits

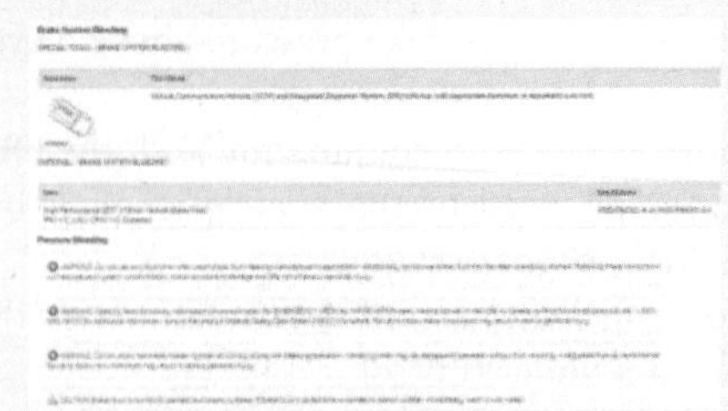

1. Research the bleeding procedure in the service information. Follow the specified procedure precisely. The following steps are given as an example of one manufacturer's bleeding process.

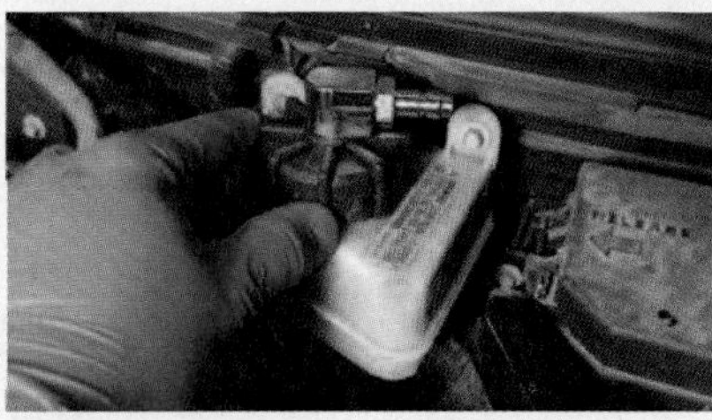

2. Connect a pressure bleeder to the brake fluid reservoir, but do not pressurize it yet.

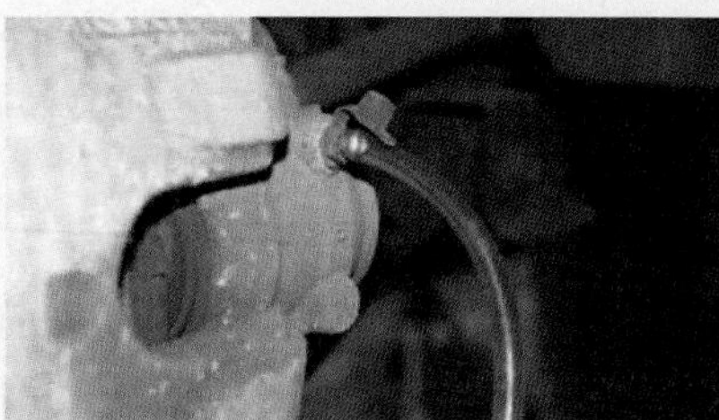

3. Install bleeder hoses over the bleeder screws at each wheel. If applicable, submerse the other end of each hose in a container partially filled with clean, specified brake fluid.

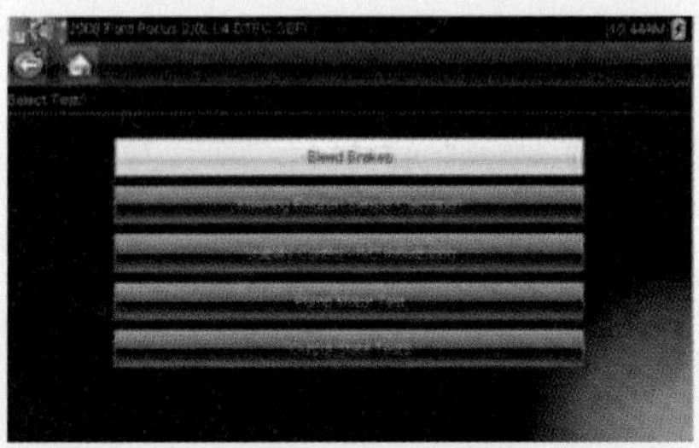

4. Connect the scan tool to the data link connector, and access the ABS bleed function.

5. Apply the specified amount of pressure from the pressure bleeder.

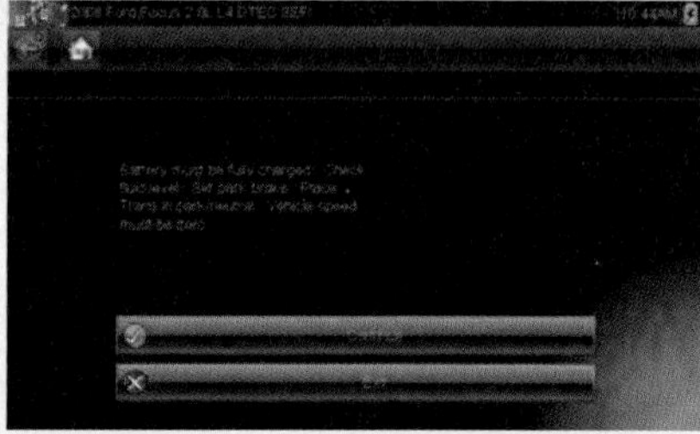

6. Follow the instructions on the screen of the scan tool.

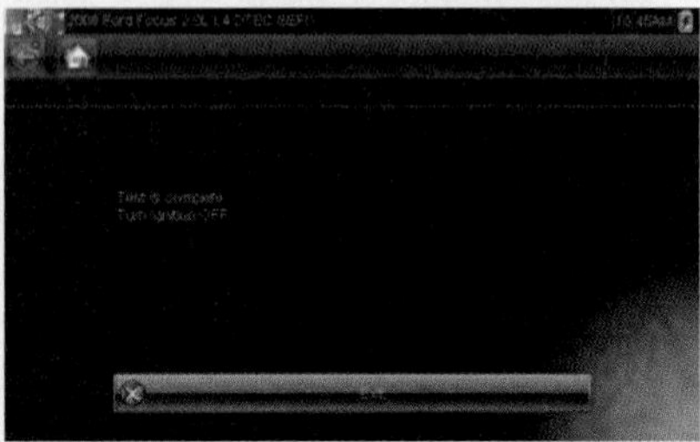

7. Once the scan tool indicates that the HCU is bled, it may direct you to bleed the brakes manually to remove any remaining air from the system.

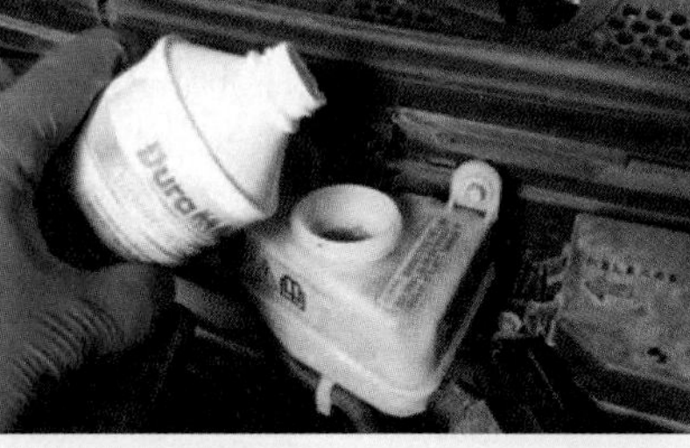

8. Remove the pressure bleeder, and top off the reservoir.

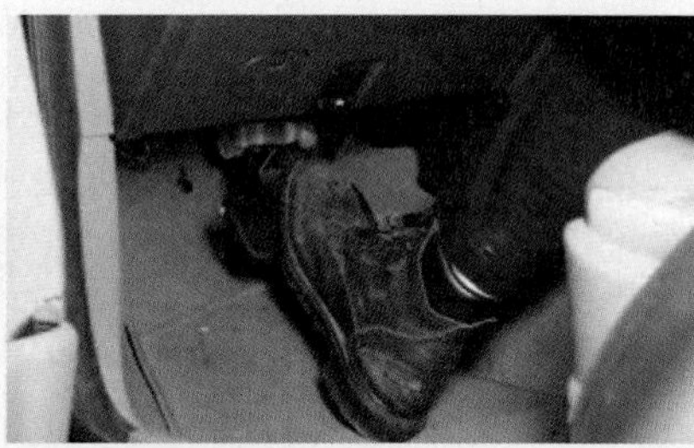

9. Test the brake pedal feel.

EBC system components. Always check the service manual to determine whether depressurization is needed.

In many vehicles, depressurization of the accumulator can be accomplished in one of the following ways:

- Verifying that the ignition switch is in the Off position (to disable the electric pump) and depressing the brake pedal 30–50 times
- Pulling the ABS fuse or relay to prevent the electric pump from charging the system
- Using a scan tool to depressurize the system
- Following the manufacturer's specific timeout process, which depressurizes the system automatically

To depressurize high-pressure components of the EBC system, follow the steps in **SKILL DRILL 49-2**.

SKILL DRILL 49-2 Depressurizing High-Pressure Components of the EBC System

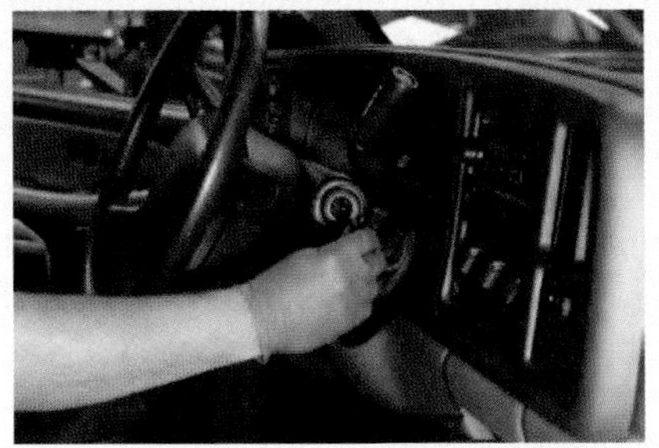

1. Research the procedure for depressurizing the accumulator in the service information. Be sure to carefully follow all instructions. The following steps are given as an example of one manufacturer's process. Verify that the ignition switch is turned to the Off position.

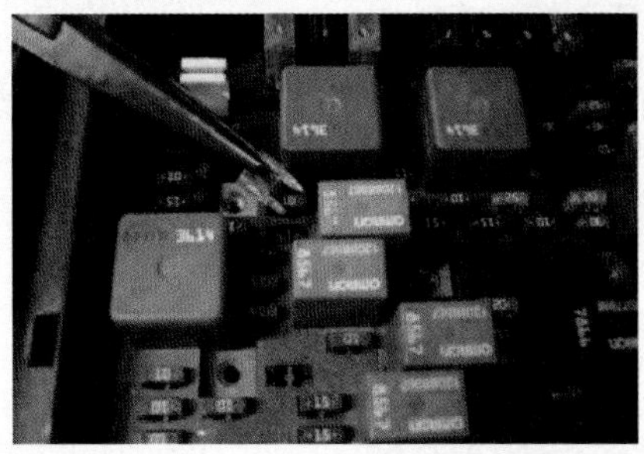

2. Remove the ABS fuse from the fuse box.

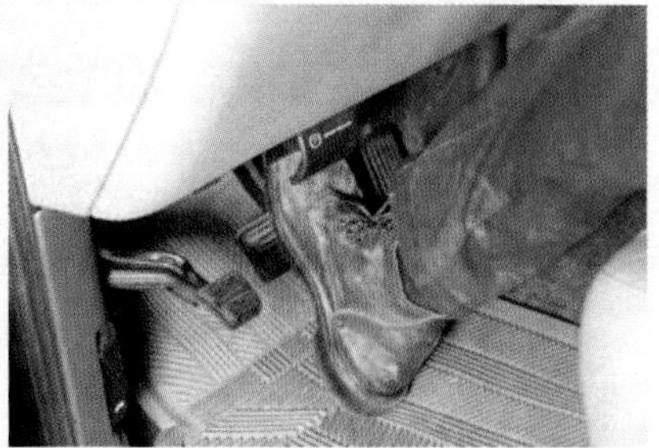

3. Apply the brakes firmly at least 40 times. Verify that there is no power assist. If there is, pump the brakes an additional 10 times, and verify that there is no assist remaining.

Removing and Installing Electric and Hydraulic Components of the EBC System

S49002

Removing and installing electric and hydraulic components of the EBC system is normally only required when there is a fault within the HCU. Because many HCUs are non-serviceable, they will have to be replaced as a unit. This may require depressurizing the high-pressure accumulator on some of these units. If the HCU is serviceable and the faulty components are available, follow the manufacturer's service procedure to remove and install the faulty component. Failure to do so could damage the HCU and render it inoperable.

To remove and install EBC system electrical/electronic and hydraulic components, follow the steps in **SKILL DRILL 49-3**.

▶ Diagnosis Overview

N49003, N49004, N49005, S49003

The diagnosis of ABS, TCS, and ESC systems starts with a thorough understanding of the particular type of system you are working on. Refer to the service information to familiarize yourself with the manufacturer's description and operation of the system. Also check for any DTCs related to the customer concern. In cases where the system warning lamp is illuminated, access the diagnostic trouble codes (DTCs) with a compatible scan tool or code retrieval key. DTCs can be diagnosed following the steps laid out in the service information pertaining to the specific DTCs indicated. In cases where the system warning lamp is not illuminated, indicating there are no faults stored in memory, you can sometimes follow the symptom charts listed in the service information. Just remember that the EBCM makes decisions based on the sensor information and software programming. Those decisions are then sent as output signals to the HCU unit, the vehicle's PCM, and other controlled devices. The controlled devices then need to carry out the commands. So knowing that information, along with the verified customer concern and the service information, you will be able to diagnose the cause of EBC faults.

Tools

The tools used to diagnose and repair ABS, TCS, and ESC systems include these, shown in **FIGURE 49-28**:

- **ABS code retrieval key**: A stamped sheet metal key used to access the ABS blink codes on earlier GM vehicles.

SKILL DRILL 49-3 Removing and Installing EBC System Electrical/Electronic and Hydraulic Components

1. Research the manufacturer's procedure for removing and installing the component on the vehicle you are working on.
2. Verify that the system has been depressurized.

3. Follow all instructions carefully, and be sure to replace all components and tighten all fasteners according to specifications.

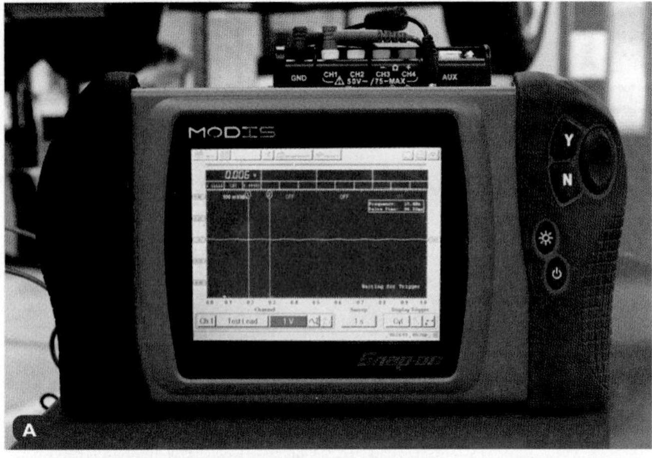

FIGURE 49-28 Tools used to diagnose electronic brake systems. **A.** A digital storage oscilloscope/diagnostic scan tool. **B.** A handheld diagnostic oscilloscope. **C.** A digital multimeter.

- **Scan tool**: Handheld electronic tool used for accessing ABS codes and live data from the EBCM. Some have bidirectional abilities used to command certain outputs, such as activating a solenoid during a diagnostic routine or when bleeding the brakes. They are also used to clear codes after repairs are completed.
- **Oscilloscope**: Handheld electronic tool used to display electrical waveforms relative to voltage signals over divisions of time, such as wheel speed sensor patterns.
- **DMM**: Handheld meter used to measure volts, ohms, and amps in electrical circuits.
- **ABS proportioning valve depressor**: Device that depresses the proportioning valve on some ABS systems while bleeding the brakes.
- **ABS pressure tester**: Device that allows the technician to test accumulator and HCU pressure issues.

Diagnosing All EBC Systems

If the yellow warning lamp is illuminated, indicating there is a fault in the system, retrieve any DTCs following the procedure listed in the service information. If the yellow warning lamp is not illuminated and the brakes are not functioning properly on a vehicle with an EBC system, suspect a problem that is not monitored by the EBCM. This could be from a fault in the base brake system, such as warped rotors, contaminated friction lining, or a seized caliper piston. Diagnose these problems like you would on a non-EBC system. It could also be a problem in the hydraulic system, including the HCU, such as a dump valve that is stuck partially open or an isolation valve that is stuck closed. Even systems that are monitored by the EBCM can cause operating concerns without the yellow warning lamp being illuminated. An example would be a wheel speed sensor that does not have the correct air gap. Using the manufacturer's diagnostic symptom charts and system diagrams and having a good understanding of brake, hydraulic, and system theory will help you diagnose the cause of the concern.

Most EBC-related faults set DTCs and store them in memory. A good starting point for diagnosis is retrieving any stored DTCs with a scan tool or code retrieval key. The service information provides a testing procedure for each of the fault codes. Following the steps listed and using an understanding of brakes, electricity, and the circuit being tested will help lead you to the cause of the fault.

To diagnose EBC system electronic control(s) and components by retrieving DTCs, first connect the scan tool to the data link connector. Navigate to the code retrieval screen and record any DTCs. Research the TSBs and diagnostic procedure for the stored DTCs in the service information. Follow the diagnostic chart to diagnose the cause of the fault. Once the fault has been corrected, clear the diagnostic code, and verify that it does not reset. This may involve a test drive.

Diagnosing Wheel Speed Sensors and Tone Wheels

N49006

Wheel speed sensors are a high-probability cause of illuminated EBC warning lamps and stored DTCs, because they are usually mounted where they are exposed to the elements and in harm's way. Watching their speeds during a test drive allows you to see if they are reading the same speed or not. If not, then check for mismatched or improperly inflated tires. If the tires are not the cause, then testing the wheel speed sensors for electrical faults is a common step in EBC diagnosis. Measuring the output signal with digital storage oscilloscope (DSO) or a graphing multimeter (GMM) and comparing it to known good signals is the most conclusive method of testing the sensors.

Variable reluctance–type sensors (passive) create an analog AC sine wave. Magneto-resistive-type sensors (active) and Hall effect sensors (active) create a digital square wave signal. The sensor electrical circuits can be tested with a DMM for opens, shorts, high resistance, and grounds.

▶ TECHNICIAN TIP

When checking variable reluctance sensors, it is important to use a true-rms meter because the sensor outputs an alternating current (AC) signal. A true-rms meter is designed to accurately represent this signal. Non-true rms meters can only accurately represent clean AC signals, not faulty ones like true-rms meters can. If the meter is a true-rms meter, it typically is labeled as such on its face.

The tone wheel can usually be visually inspected for any faults, such as broken or damaged teeth. Just make sure you inspect the teeth all the way around the tone wheel, not just in one section. Also verify the air gap is within specifications.

To test and diagnose EBC system speed sensors (digital and analog), tone wheel, and circuits, first read any diagnostic codes with a scan tool or other code retrieval method. Using the DTCs, determine which wheel speed sensor or sensor circuit is at fault. Research the service information to determine what type of speed sensors the vehicle is equipped with and the specified testing procedure.

To test variable reluctance sensors, follow the steps in **SKILL DRILL 49-4**.

To test magneto-resistive sensors, follow the steps in **SKILL DRILL 49-5**.

SKILL DRILL 49-4 Testing Variable Reluctance Sensors

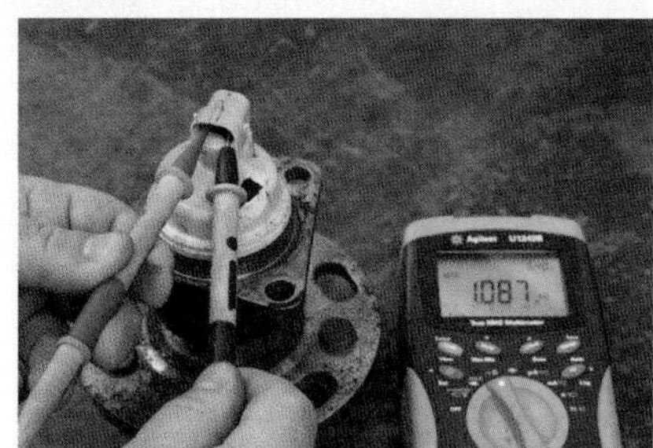

1. Disconnect the suspect sensor, measure its resistance, and compare the reading to the specifications. If the resistance does not meet the manufacturer's specifications, replace the sensor.

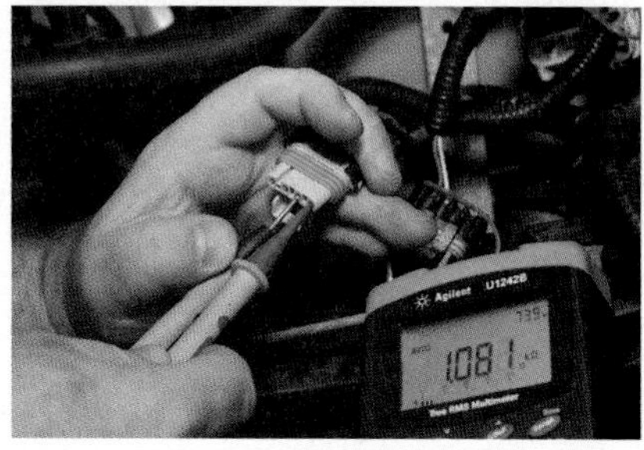

2. If it is within specifications, test the two-wire circuit back to the EBCM for opens, shorts, high resistance, and grounds. Repair as necessary.

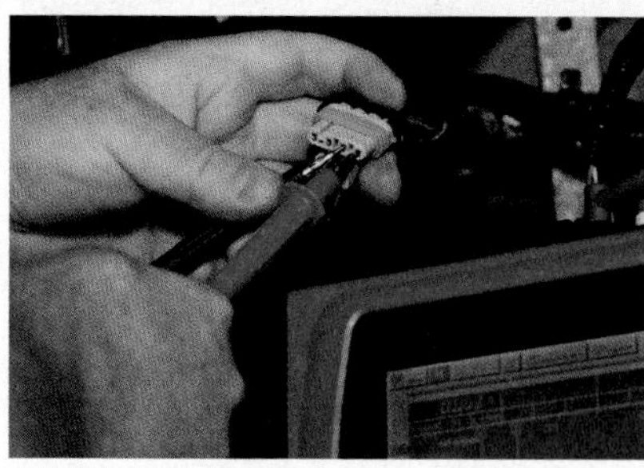

3. If the circuit is good, reconnect the speed sensor connector and attach a GMM or DSO to one of the sensor wires.

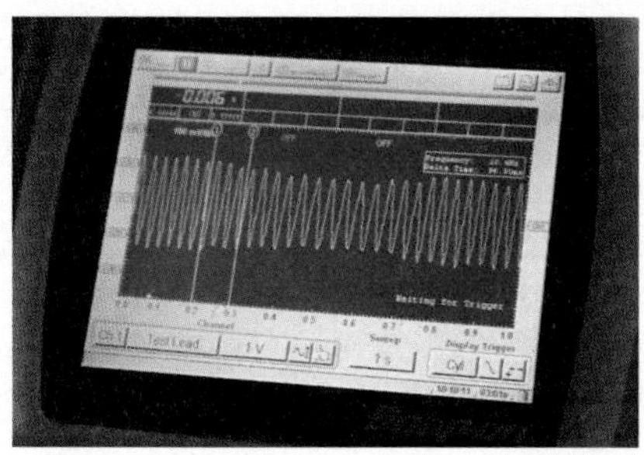

4. Spin the tone wheel and observe the pattern. It should be a clean AC analog pattern of sufficient amplitude (voltage).

5. If the pattern is not correct, inspect the tone wheel for damage, and replace as necessary.

6. If the tone wheel is good, replace the wheel speed sensor. If the pattern looks okay, you may have to compare it to the other wheel speed sensor signals while driving the vehicle. After repair, clear the diagnostic codes, if directed by the service information.

SKILL DRILL 49-5 Testing Magneto-Resistive Sensors

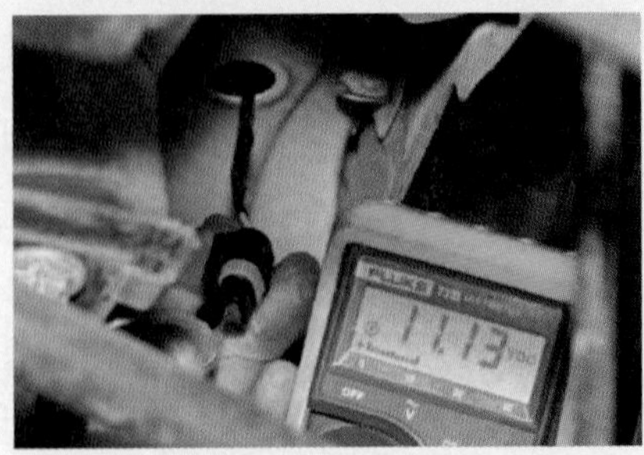

1. Make sure the ignition switch is in the Run position, and measure the available voltage at the suspect sensor. Also check the ground for voltage drop.

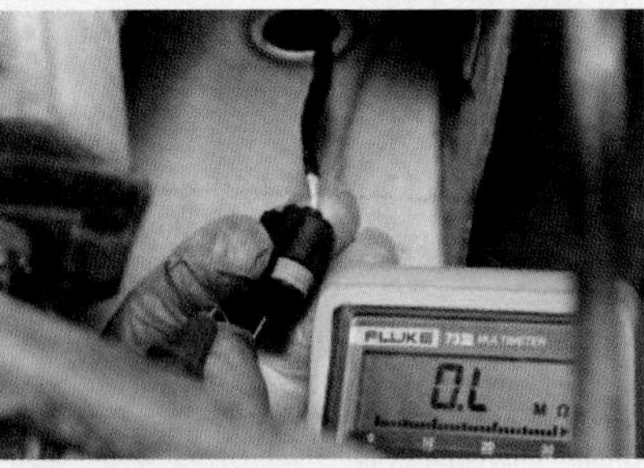

2. If the voltage does not meet the manufacturer's specifications, test the circuit for opens, shorts, high resistance, or grounds. Repair as necessary.

3. If the voltage to the sensor is within the manufacturer's specifications, connect a GMM or DSO to the signal wire.

4. With the key in the Run position, spin the tone wheel, and observe the pattern. It should be a clean digital square wave signal of the appropriate height and shape.

5. If the pattern is not correct, inspect the tone wheel for damage, and replace as necessary. If the tone wheel is good, replace the wheel speed sensor.

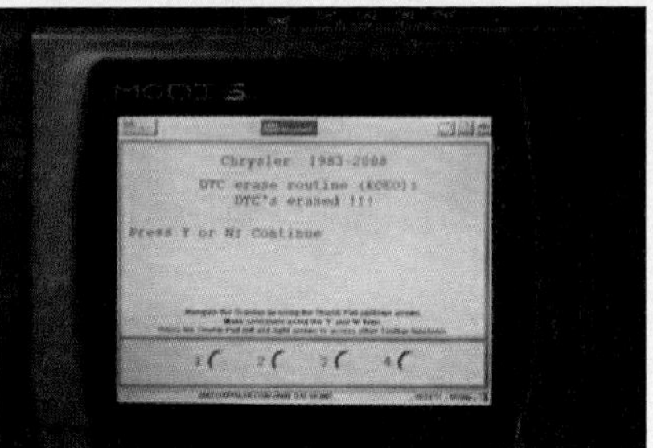

6. After repair, clear the codes if directed by the service information.

Applied Science

AS-85: Problem Solving: The technician can use computers, scan tools, and onboard data to diagnose problems.

Vehicles presenting with ABS warning lights illuminated are now almost an everyday occurrence in most workshops. The first step in the diagnostic process is to read the DTCs stored on the ABS module. The most common cause of ABS malfunctions is faults in wheel speed sensor circuits. Common causes of wheel speed sensor circuit faults include corroded or poorly tensioned connectors, mismatched tire rolling diameters, and physical sensor failures.

Possibly the most effective way to diagnose wheel speed sensor circuit faults is to utilize the live data function of your scan tool. Start in the shop, looking at the speed readings and making sure they are all equal before you even test-drive the car. An inconsistency indicates a fault. During the road test, look at the graphed data outputs from all of the sensors; they should all look consistent. A graph that differs from the others indicates a faulty circuit. Once you have identified the affected circuit, the next step is to test at the sensor with a graphing multimeter or an oscilloscope.

Diagnosing Braking Concerns Caused by Vehicle Modifications

N49007

The EBC systems are engineered so that all of the parts and the programming work together for the purpose of assisting the driver in maintaining control of the vehicle. If any of the original components get replaced with nonstandard parts, the EBC systems will not perform as designed, putting the occupants at risk. Examples of this could be as simple as installing a smaller spare tire on the vehicle due to the original (correctly sized) tire going flat. The smaller tire turns at a faster rotation than the original tire, confusing the EBC system; consequently, the yellow warning lamp illuminates and the EBC system is disabled. In the same way, installing larger tires with a greater circumference will cause the wheel speed sensors to report a slower wheel speed than the vehicle is actually traveling. Because the EBCM is not calibrated to match this parameter, the system cannot respond appropriately, and again a DTC will be reported and the EBC system will shut down.

Changing curb height causes the vehicle's center of gravity to change. Taller vehicles tend to transfer weight more easily. In the case of braking, the weight is transferred to the front wheels, unloading the rear wheels. This makes the rear wheels more prone to lock up. In the case of cornering, the weight transfers to the outside wheels, unloading the inside wheels. This makes the vehicle more susceptible to rollover. In both of these cases, the EBCM is not programmed to take into account the nonstock height of the vehicle. Therefore, the EBCM would not be able to properly anticipate the vehicle's actions in these scenarios.

Changing final drive ratios causes the same situation as changing tire size. The vehicle wheel speed sensors will report a different speed than the vehicle speed sensor, and the EBCM will set a code and disable the system. Also, mismatched final drive ratios on four-wheel drive vehicles can affect the EBC systems, as they can cause the wheels to rotate at slightly different speeds, front to rear.

In some cases, the EBCM can be reprogrammed with the new vehicle information, or an EBCM that does match the new parameters of the vehicle may be available for installation. In any case, technicians need to be aware of how these factors affect the EBC systems. They also need to know how to verify that the components match the information programmed into the EBCM.

▶ Wrap-Up

Ready for Review

- Electronic brake control systems integrate computer controls to prevent wheel lockup, shorten panic stop distances, help drivers maintain steering control, and improve vehicle stability.
- Basic antilock brake systems control hydraulic pressure by holding and releasing via solenoid valves, but cannot function independently of the driver's applied brake pressure.
- Traction control systems minimize wheel slip by automatically applying brake pressure to a slipping wheel's brake unit and reducing engine output.
- Electronic stability control systems use steering wheel position sensors, yaw sensors, roll rate sensors, and wheel speed sensors to independently monitor vehicle stability and apply brakes as necessary.
- Primary components of an antilock braking system are ABS master cylinder, electronic brake control module/electronic control unit, hydraulic control unit/modulator, power booster, wheel speed sensor, and brake switch.
- Braking force and wheel lockup are affected by friction of road surface and type, condition, and loading of each tire.
- Maximum traction occurs with 10–20% tire slip.
- Wheel speed sensors signal the EBCM, which sends current to the solenoid valve, which then holds or releases hydraulic braking pressure.
- ABS master cylinders are integral (mainly in older vehicles) or non-integral with the HCU.
- Solenoid valves provide three operating conditions: apply, hold, and release.
- The hydraulic control unit executes the commands of the ABS control module.
- Hydraulic control units differ by number of channels (one, two, three, or four), number of solenoid valves (single or dual), and type of accumulator (low or high pressure).
- Wheel speed sensors send electric signals to the EBCM to determine speed and rate of deceleration for each wheel.
- Wheel sensor assemblies are comprised of a toothed tone wheel and a pickup assembly, separated by an air gap.
- Types of wheel speed sensors are: variable reluctance, magneto-resistive, and Hall effect.
- The EBCM is comprised of electronic circuitry, electronic data processor, computer memory, and output drivers.
- The EBCM receives input signals from: the brake switch, ignition switch, vehicle speed sensor, wheel speed sensors, and sometimes the battery.
- The ESC system includes a yaw sensor (directional rotation), steering angle sensor (driver's directional intent), and roll-rate sensor (rate and amount of vehicle roll).
- Some TCS and ESC systems can be manually deactivated by the driver.
- EBCM systems can self-diagnose and store faults as diagnostic trouble codes (DTCs) for technicians to retrieve.
- It is best not to allow air into an EBC hydraulic system as they can be difficult to bleed.
- If the HCU needs to be bled, you will likely need a scan tool to perform that that task.
- The EBC system should be depressurized before servicing the hydraulic system.
- Follow manufacturer's procedures when depressurizing and servicing EBC systems.
- When diagnosing EBC systems, don't overlook faults that could be in the base brake system.
- Exercise extreme care when test-driving the vehicle while diagnosing the EBC system.
- Most faults in the EBC system will store a DTC. Use a scan tool to retrieve the DTC, and then research the service information and Technical Service Bulletins.
- Variable reluctance-type sensors (passive) create an analog AC sine wave.

- Magneto-resistive-type sensors (active) and Hall effect sensors (active) create a digital square wave signal.
- Variable reluctance–type sensor circuits can be checked for opens, shorts, and grounds on the connecting wires. And resistance and signal on the sensor.
- Magneto-resistive-type sensor circuits can be checked for, reference voltage, signal quality, and ground voltage.
- EBC systems are affected by vehicle modifications such as tire sizes, curb height, and drive ratios and need to be checked during diagnosis.

Key Terms

ABS code retrieval key A specially shaped metal key that is inserted into the data link connector to retrieve DTCs.

ABS pressure tester A high-pressure gauge designed to connect to the HCU and used to measure high hydraulic pressures in the system.

ABS proportioning valve depressor A tool used to hold the proportioning valve open on some ABS HCUs.

accumulator A storage container that holds pressurized brake fluid.

air gap The space or clearance between two components, such as the space between the tone wheel and the pickup coil in a wheel speed sensor.

blink codes Codes used to communicate DTCs; they are given by the EBCM as a series of blinks illuminated by the ABS warning lamp.

body control module (BCM) The computer that controls the electrical system in the body of a vehicle, including power windows, door locks, heating and A/C systems, and in some cases the EBC system.

boost valve A valve located in the HCU that is controlled by the EBCM; it allows brake fluid under high pressure to flow into the HCU hydraulic circuits to apply the brakes when commanded.

brake switch The electrical switch that is activated by the brake pedal; it turns on the brake lights and signals the EBCM that the brakes are being applied.

channel The number of wheel speed sensor circuits and hydraulic circuits the EBCM monitors and controls.

common bore When a single cylinder is used for two pistons. A tandem master cylinder would be an example of two pistons in one bore.

DMM Handheld meter used to measure volts, ohms, and amps in electrical circuits

electric solenoids An electrically operated valve, which in brake systems is used to control the flow of brake fluid in the hydraulic system.

electronic brake control (EBC) system A hydraulic brake system that has integrated electronic components for the purpose of closely controlling hydraulic pressure in the brake system.

electronic stability control (ESC) system A computer-controlled system added to ABS and TCS to assist the driver in maintaining vehicle stability while steering.

fault codes An alphanumeric code system used to identify potential problems in a vehicle system.

Hall effect An electrical effect where electrons tend to flow on one side of a special material when exposed to a magnetic field, causing a difference in voltage across the special material. When the magnetic fields is removed, the electrons flow normally, and there is no difference of voltage across the special material. This effect can be used to determine the position or speed of an object.

high-pressure accumulator A storage container designed to contain high-pressure liquids such as brake fluid.

hydraulic control unit (HCU) An assembly that houses electrically operated solenoid valves used in electronic braking systems; also called a modulator.

integral ABS system A brake system in which the master cylinder, power booster, and HCU are all combined in a common unit.

isolation valve The valve in the HCU that either allows or blocks brake fluid that comes from the master cylinder from entering the HCU hydraulic circuit.

low-pressure accumulator A storage container for brake fluid coming from the release valves, which is under relatively low pressure.

magneto-resistive sensor A type of wheel speed sensor that uses an effect similar to a Hall effect sensor to create its signal.

non-integral ABS systems A brake system in which the master cylinder, power booster, and HCU are all separate units.

oscilloscope A tool that shows graphically what is happening to voltage over a period of time; it is used to diagnose electrical faults.

oversteer A condition in which the rear wheels are slipping sideways toward the outside of the turn.

pickup assembly A component with a wire coil wrapped around a ferrous metal core; it is used to generate an electrical signal when a magnetic field passes through it.

roll-rate sensor A sensor that measures the amount of roll around the vehicle's horizontal axis that a vehicle is experiencing.

scan tool A tool used to read codes, access live data, and communicate with the vehicle's computers.

solenoid valve An electrically operated valve that when used in brake systems is designed to control the flow of brake fluid in the hydraulic system.

steering angle sensor A sensor that measures the amount of turning a driver desires. This information is used by the ESC system to know the driver's directional intent.

steering wheel position sensor A sensor that signals to the EBCM both the position and speed of the steering wheel.

tone wheel The part of the wheel speed sensor that has ribs and valleys used to create an electrical signal inside of the pickup assembly.

traction control system (TCS) A computer-controlled system added to ABS to help prevent loss of traction while the vehicle is accelerating.

understeer A condition in which the front wheels are turned further than the direction the vehicle is moving and the front tires are slipping sideways toward the outside of the turn.

variable reluctance sensor A type of wheel speed sensor that uses the principle of magnetic induction to create its signal.

vehicle speed sensor The component that creates an electrical signal based on the speed of the vehicle, which is sent to the EBCM.

wheel speed sensor A device that creates an analog or digital signal according to the speed of the wheel.

yaw sensor A sensor that measures the amount a vehicle is turning around its vertical axis. This information is used by the ESC system to know how much a vehicle is turning.

Review Questions

1. Magnetic induction type wheel speed sensors create which type of signal?
   - **a.** 12 volt DC voltage
   - **b.** Analog AC voltage
   - **c.** Digital DC voltage
   - **d.** 12 volt AC voltage
2. Which EBC system uses sensor information regarding the vehicle's actual direction?
   - **a.** ABS system
   - **b.** TCS system
   - **c.** ESC system
   - **d.** Base brake system
3. Choose the correct statement regarding ABS solenoids.
   - **a.** During "hold," the brake pressure is prevented from returning to the master cylinder.
   - **b.** During "release," the brake pressure is directed to the wheel brake unit.
   - **c.** During "hold," the brake pressure is being increased to the wheel brake unit.
   - **d.** During "release," the brake pressure is held steady.
4. Which primary component in an ABS system sends output control signals to electronic solenoid valves?
   - **a.** Brake pedal switch
   - **b.** Hydraulic control unit
   - **c.** Wheel speed sensor
   - **d.** Electronic control unit
5. All of the following statements referring to the principles of ABS braking are true *except*:
   - **a.** During normal braking, the EBCM does not energize the solenoid valves in the hydraulic unit.
   - **b.** If the wheel speed sensors indicate that the wheel is still decelerating too rapidly after the pressure has been held, the EBCM commands the appropriate solenoid valve to release braking pressure.
   - **c.** If the wheels are not skidding, the EBCM does not monitor the speed of each wheel.
   - **d.** Changes in HCU valve position normally cause rapid hydraulic pulsations.
6. Choose the correct statement with respect to the principles of operation of traction control.
   - **a.** The TCS system was developed to prevent the drive wheels from slipping while the vehicle is being accelerated.
   - **b.** Traction control is active at all speeds.
   - **c.** When the TCS is active, the EBCM monitors only the speed of the drive wheels.
   - **d.** If the driver deactivates the TCS, the ABS system will also be deactivated.
7. Which sensor in the ESC system tells the EBCM the rate at which the vehicle is actually turning?
   - **a.** Steering angle sensor
   - **b.** Roll-rate sensor
   - **c.** Yaw sensor
   - **d.** Wheel speed sensor
8. All of the following statements with respect to the operation of the ESC system are true *except*:
   - **a.** The ECBM monitors the yaw sensor signal, the steering angle sensor signal, and the roll-rate sensor signal as well as the wheel speed sensor signals.
   - **b.** The EBCM detects understeer, oversteer, or roll in the signal values and then applies up to three wheel brake units to help bring the vehicle back within proper stability parameters.
   - **c.** The EBCM requests a reduction in engine power through the powertrain control module (PCM) to help slow the vehicle further.
   - **d.** No matter the driving conditions or driver actions, ESC will prevent a loss of control of the vehicle.
9. Which of the following features designed into ESC systems holds the brake pressure until the throttle is depressed and the vehicle starts to move forward?
   - **a.** Hill assist
   - **b.** Hill descent control
   - **c.** Engine braking control
   - **d.** Panic stop assist
10. All of the following are likely to turn the ABS light on except:
    - **a.** replacing a faulty wheel speed sensor
    - **b.** mismatched tires
    - **c.** low tire pressure on one or more tires
    - **d.** installing a different ratio final drive

ASE Technician A/Technician B Style Questions

1. Tech A says that an antilock brake system (ABS) helps shorten the stopping distance during a panic stop. Tech B says that antilock brake systems work by increasing the hydraulic pressure in the brake system so the brakes can be applied harder. Who is correct?
   - **a.** Tech A
   - **b.** Tech B
   - **c.** Both A and B
   - **d.** Neither A nor B

2. Tech A says that traction control can reduce the power output of the engine to increase traction. Tech B says that electronic stability control increases the risk of rollover. Who is correct?
   a. Tech A
   b. Tech B
   c. Both A and B
   d. Neither A nor B
3. Tech A says that an electronic braking system has sensors that monitor wheel speed. Tech B says that understeer is generally easier to recover from than oversteer. Who is correct?
   a. Tech A
   b. Tech B
   c. Both A and B
   d. Neither A nor B
4. Tech A says that ABS controls braking every time the brakes are applied. Tech B says that during an ABS event, it is normal for the brake pedal to pulsate. Who is correct?
   a. Tech A
   b. Tech B
   c. Both A and B
   d. Neither A nor B
5. Tech A says that during antilock braking, brake fluid may be returned to the master cylinder. Tech B says that solenoid valves in the hydraulic control unit will isolate the master cylinder from the brake circuit when it is in the Hold mode. Who is correct?
   a. Tech A
   b. Tech B
   c. Both A and B
   d. Neither A nor B
6. Tech A says that mismatched tires may cause the ABS system to register a fault code. Tech B says that a traction control system may automatically apply brake pressure to a wheel brake unit even if the vehicle is not being braked. Who is correct?
   a. Tech A
   b. Tech B
   c. Both A and B
   d. Neither A nor B

7. Tech A says that an ABS key-on system test checks for faults in the vehicle's base brake system. Tech B says that on most vehicles ABS DTCs are stored in memory for later retrieval. Who is correct?
   a. Tech A
   b. Tech B
   c. Both A and B
   d. Neither A nor B
8. Tech A says that a yaw sensor tells the computer the vehicle's actual direction. Tech B says that raising a vehicle's curb height has no effect on the electronic stability control system. Who is correct?
   a. Tech A
   b. Tech B
   c. Both A and B
   d. Neither A nor B
9. Tech A says that a scan tool may be required to bleed air from the ABS brake system. Tech B says that some vehicles have a high-pressure accumulator that may need to have the pressure bled off before hydraulic brake repairs are made. Who is correct?
   a. Tech A
   b. Tech B
   c. Both A and B
   d. Neither A nor B
10. Tech A says that some ABS wheel speed sensors create a square wave digital pattern. Tech B says that some wheel speed sensors create an AC sine wave pattern. Who is correct?
   a. Tech A
   b. Tech B
   c. Both A and B
   d. Neither A nor B

REC
-13.26
ON
OFF
REL
MIN/MAX
HOLD
ACA/DCA

SECTION 7
Electrical

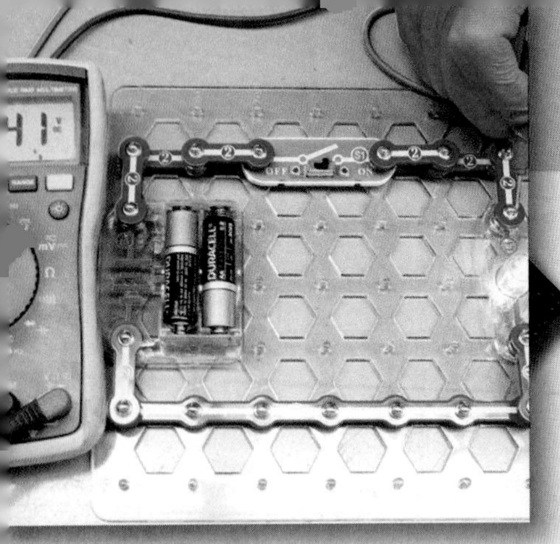

CHAPTER 50

Principles of Electrical Systems

NATEF Tasks

- **N50001** Demonstrate knowledge of the causes and effects from shorts, grounds, opens, and resistance problems in electrical/electronic circuits. (MLR/AST/MAST)
- **N50002** Demonstrate knowledge of electrical/electronic series, parallel, and series-parallel circuits using principles of electricity (Ohm's law). (MLR/AST/MAST)

Knowledge Objectives

After reading this chapter, you will be able to:

- **K50001** Explain the basic fundamentals of electricity.
- **K50002** Define units of electrical measurement.
- **K50003** Define electrical circuit terminology.
- **K50004** Explain the sources and effects of electricity.
- **K50005** Explain the application of basic electrical laws.

Skills Objectives

There are no Skills Objectives for this chapter.

▶ Introduction

The application of electrical principles in the repair of modern vehicles has become increasingly important for technicians as the electrical and electronic complexity of vehicles has increased. As hydrocarbons become scarcer and more expensive, the increased use of sophisticated electrical and electronic systems to improve efficiency and economy will continue into the future. This trend is supported by the increasing popularity of hybrid vehicles and the investment by manufacturers in future technology, such as electric vehicles and fuel cell technology (**FIGURE 50-1**). To service current and future vehicles, it is increasingly important for the technician to have a sound understanding of electrical terminology, the behavior of electricity, and circuit theory. Your success as a technician depends largely on your ability to apply these electrical principles and understand how they relate to the operation of virtually every system in the vehicle.

▶ TECHNICIAN TIP

Many students find that going through electrical theory one time isn't enough to fully understand it. Successful students study a portion of the content and then go back and review it a few times so that they can build a strong understanding of the concepts. Electricity is understandable, but you have to apply yourself. At the same time, becoming a successful technician will be much easier with a sound knowledge of this topic. Please consider this.

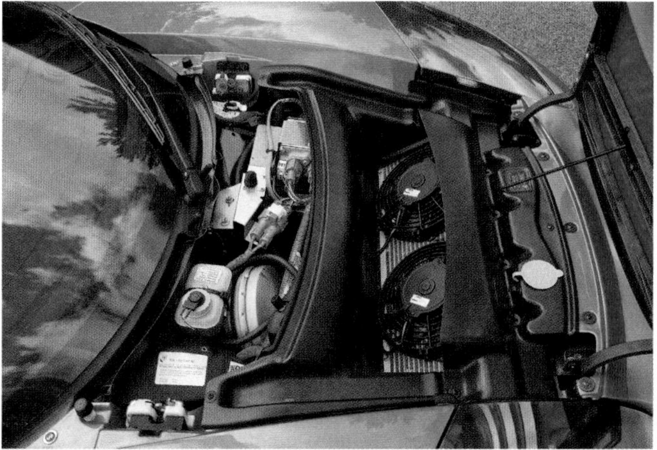

FIGURE 50-1 Manufacturers are relying on electricity and electronics more and more to meet stringent government and customer requirements.

▶ Electrical Fundamentals

K50001

Understanding the behavior of electricity can be more difficult than understanding other concepts such as four-stroke engine theory or the operation of disc brakes. For all intents and purposes, electricity cannot be seen, so you have to use your imagination and visualize what it is doing. At the same time, electricity is governed by the laws of physics, so learning how electricity behaves can be approached in a logical manner, as with any other science.

In this chapter, we explore the various types of circuits and how electricity behaves within each type. We will also explore the various ways that electricity can be created as well as the ways it can be used.

To get started, it is helpful to know that electricity is made up of tangible objects. Even if we cannot normally see them with our eyes because these particles are so small, we can imagine them in our minds. In fact, it may be helpful to think of electricity as nothing more than the movement of specific particles from one point to another. Imagine a line of marbles rolling through a tube, or drops of water through a pipe (**FIGURE 50-2**). The moving marbles or drops of water can perform work if they are directed against another object with force. In the same way, electricity can perform work if it is directed at objects that can extract energy from the moving particles, such as lights and electric motors.

So what makes the marbles move? That is where some of electricity's magic comes in. Remember back to grade school: Unlike charges attract, and like charges repel (**FIGURE 50-3**). These attracting and repelling forces are what cause the electrical particles to move and perform work. As we continue, just remember that electricity is the movement of these particles from one place to another. The next question is, where do these charges come from?

Basic Electricity

All questions about the nature of electricity lead to the composition of matter. All matter is made up of atoms (**FIGURE 50-4**),

You Are the Automotive Technician

You are supervising an intern at the shop where you work. She has been a great help to you, increasing your productivity by doing the tasks she has learned so far. She will be starting electrical training next semester and has been told it can be difficult to learn, so she wants to get a head start on it. She asks you the following questions. How would you respond?

1. What exactly are volts, ohms, and amps?
2. What do "continuity," "open," and "short" mean?
3. What is Ohm's law?
4. What is the difference between a series circuit and a parallel circuit?

FIGURE 50-2 Electricity is like marbles moving through a tube, whose movement can do work.

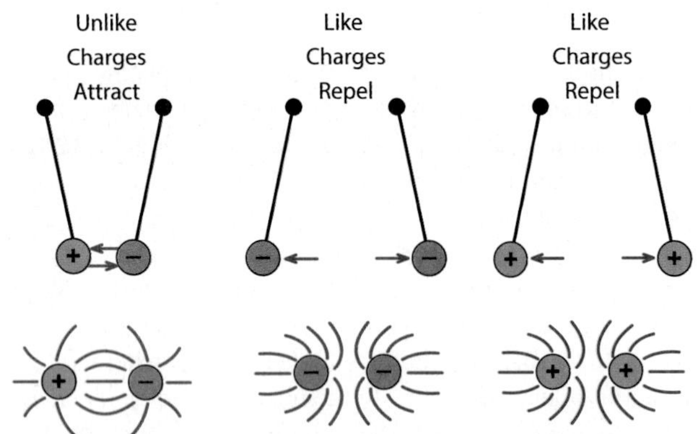

FIGURE 50-3 **A.** Unlike charges attract. **B.** Like charges repel.

which in turn are made up of positively, negatively, and neutrally charged particles. The nucleus has at least one positively charged proton and, in most atoms, at least one neutron that has no charge (neutral). Moving around the nucleus are one or more negatively charged electrons. Electrons travel in different rings, or shells, around the nucleus. Each ring, or shell, can contain a specific maximum number of electrons. Any additional electrons must fit into the next higher ring, or shell. With equal numbers of protons and electrons, the charges within an atom cancel each other out, leaving the atom in a balanced state, with no overall charge. In this state, the electrons and protons are content to stay in the atom just as they are.

Not all atoms are balanced. This means that some atoms are unbalanced. An atom with more electrons than protons has an overall negative charge and is called a negative ion (**FIGURE 50-5**). "Ion" simply means the atom has an imbalance of electrical charges due to the gain or loss of electrons. This negative ion is not balanced. Because the electrons have the same negative charge, they repel one another, and the repelling force wants to push one of the electrons away from this atom.

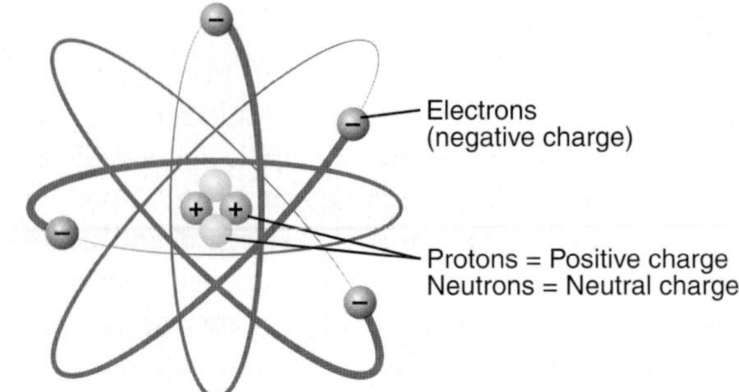

FIGURE 50-4 Parts of an atom.

A deficiency of electrons gives the atom an overall positive charge, and this atom is called a positive ion. It is also not balanced. In this case, it is exerting an attracting force on electrons to try to pull one into its atom from another atom. If a negative ion and positive ion are close enough, the negative charge of the negative ion exerts a repelling force on the extra electron, causing it to be pushed away from its atom; at the same time, the positive ion exerts an attracting force on the extra electron, pulling it from the negative atom, balancing both atoms out. The flow of electrons from atom to atom is called current flow and is the basic concept of electricity (**FIGURE 50-6**).

Not all atoms can give up or accept electrons easily. Materials that can do so easily are called conductors, and those that cannot do so easily are called insulators (**FIGURE 50-7**). The explanation of what makes a good conductor or a good insulator is quite complex and is found in the theories of quantum mechanics, which address the arrangement and behavior of electrons around the nuclei of atoms. To simplify matters, it is safe to say that in some materials there are electrons, called **free electrons**, located on the outer ring, called the valence ring. These electrons are only loosely held by the nucleus and are free to move from one atom to another when an electrical potential (pressure) is applied. In fact, atoms with fewer electrons in the valence ring are the best **conductors**, one electron being the best conductive material. This is because the single electron by itself in the valence ring is held the most loosely

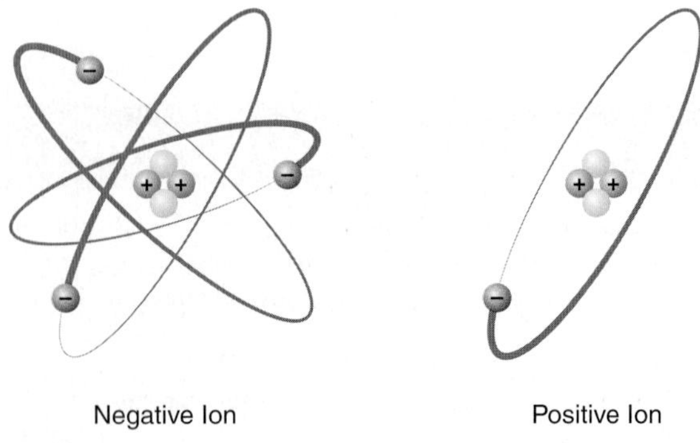

FIGURE 50-5 Ions. **A.** Negative ion. **B.** Positive ion.

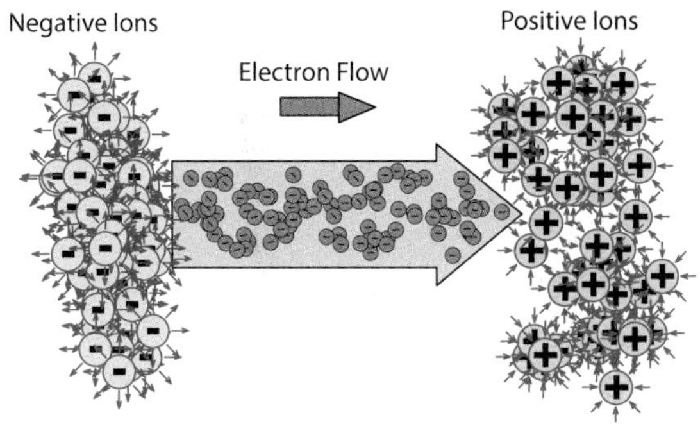

FIGURE 50-6 Electrons being pushed away from the negative side and pulled to the positive side. This movement is current flow.

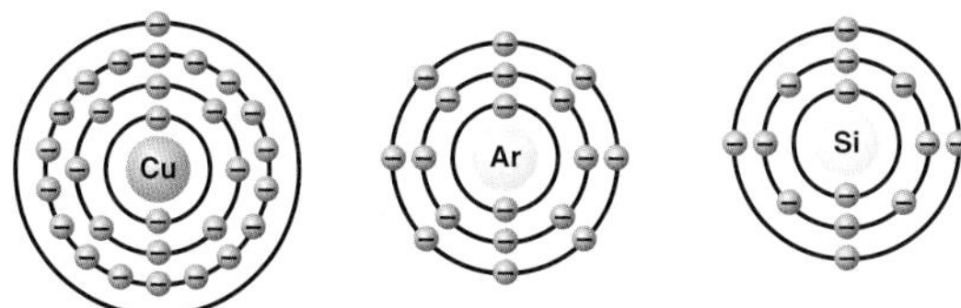

FIGURE 50-7 A. Conductor. **B.** Insulator. **C.** Semiconductor.

by the nucleus. Materials made up of atoms with one to three valance ring electrons are considered conductors. The more atoms with free electrons a particular material has, the better it can conduct electrons. Metals typically have lots of free electrons because of the atoms' structure and are therefore good conductors.

Every substance, even air, will conduct an electrical current if enough electrical pressure (voltage) is applied to it, but the word "conductor" normally is used for materials that allow current flow with little resistance. Most metals are good conductors. The most common conductor used in automobiles is copper. It is used in virtually all of the wiring that connects automotive components together. The more electrons a conductor must carry, the heavier the gauge or thickness the wire needs to be.

Materials that do not conduct electrons easily are called **insulators**. Most plastics are good insulators. The plastic covering on a wire is an example of this. The ceramic portion of a spark plug is also a good insulator. In insulators, electrons in the valence ring are bound much more tightly to the nucleus. A good insulator does not allow current flow because it has no or very few free electrons, and the electrons it does have cannot move freely; therefore, an insulator prevents the movement of electrons when an electrical potential is applied. Insulators are made up of atoms that have five to eight valence ring electrons. The greater the number of valence ring electrons, the better the insulator (**FIGURE 50-7**).

Semiconductors are materials that conduct electricity more easily than insulators, but not as well as conductors. Semiconductors such as silicon are crucial in electronics. They are used to make electronic components, such as transistors and microchips, that can switch the material from a conductor to an insulator and back again very quickly and without mechanical means. Atoms that have four valance ring electrons are considered semiconductors (**FIGURE 50-7**). Note that because the semiconductor material has precisely four electrons, it is only one electron away from becoming either an insulator or a conductor. If an electron is added, it becomes an insulator; if an electron is removed, it becomes a conductor. Thus, the semiconductor material can be used as a switch to control whether electrons flow though the semiconductor material or are stopped by it. All we have to do is add or subtract electrons from the semiconductor material, which we explore further in a later topic.

Movement of Free Electrons

Free electrons are necessary for electrical current, but for the free electrons to move easily, they need two things: a complete pathway, called a circuit, and a force that makes them move. The force from a battery can cause electrons to move. Like charges repel, so the negative electrons repel each other and are forced from the negative terminal of the battery. Unlike charges attract, so the electrons are attracted toward the positive protons in the positive side of the battery. In this case, free electrons flow from the negative terminal to the positive terminal (**FIGURE 50-8**). This flow of electrons is called current flow, or amperage, and is measured in amps. Note that the current is flowing in one direction only, a condition called direct current. Most, but not all, circuits in passenger vehicles operate on direct current. The larger the charge between the negative and the positive terminal, the more strongly the positive terminal attracts and the negative terminal repels the free electrons. This is called **electromotive force** and is typically referred to as voltage. So you could say that **voltage** is the force that motivates electrons to move through a circuit. The unit of measurement for voltage is volts. The greater the voltage, the stronger the amperage.

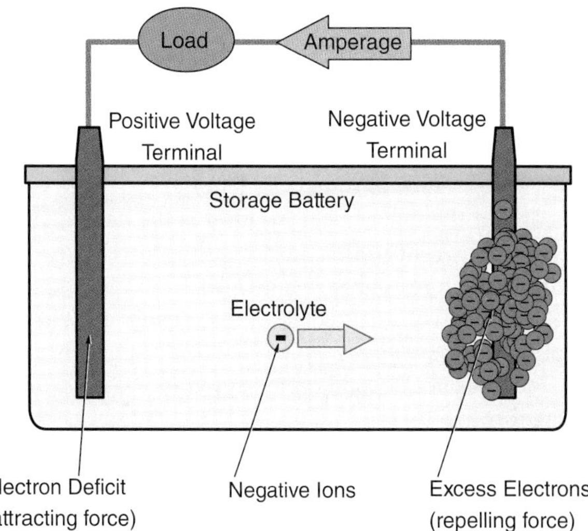

FIGURE 50-8 An excess of electrons on the negative side of the battery pushes electrons toward the positive side of the battery which is pulling on the electrons as well. The force is called voltage. The current flow is called amperage.

> **TECHNICIAN TIP**
>
> Because electrical current is the flow of electrons, it is natural to say that the direction of current is the direction in which the free electrons move—from negative to positive—which is called the **electron theory**. However, before the discovery that electrons are negatively charged, it was thought that the natural way for them to flow was from positive to negative, which is called the **conventional theory**. Most wiring diagrams are written from the conventional theory perspective, whereas electronic circuits are typically designed and operate on the electron theory perspective. Thus, both concepts are still in use. In fact, a third theory exists, closely mirroring the conventional theory, called the **hole theory**. It states that although negative electrons do move from negative to positive, holes move from positive to negative as electrons move from atom to atom; in this case, holes (current) flow from positive to negative, and are sometimes called "positive holes." What's most important to remember is that voltage causes current to flow through conductive paths (resistance). *We use the conventional and hole theories when explaining the electrical concepts throughout this text, unless otherwise noted.*

> **Applied Science**
>
> ***AS-57: Conductors: The technician can explain the difference between an electrical conductor and an insulator.***
>
> The electrical conductivity of a material refers to the freedom of the electrons within the atoms of the material to move around. Materials with high electron mobility are known as conductors, whereas materials with low electron mobility are called insulators. High electron mobility allows electrons to freely flow when a voltage is applied across two points of the material.
>
> Most familiar conductors are metallic, although there are non-metallic conductors such as graphite and salt solutions. Silver is the best conductor, but copper is most commonly used for electrical wiring due to many desirable properties, including high tensile strength and ductility.
>
> An insulator is a material that resists the flow of an electrical charge. Glass and paper are both very good insulators, but polymers and plastics are more commonly used to insulate wiring. The purpose of an insulator in electrical applications is to support or separate conductors without allowing current to flow through them.

Resistance to Current Flow

Also affecting the current flow in a circuit is **electrical resistance**, measured in **ohms (Ω)**. All materials have some resistance—even good conductors. There are four factors that determine the level of electrical resistance:

1. **Type of material:** This refers to how many free electrons are in a material.
2. **Length of the conductor:** As length increases, so does resistance.
3. **Diameter of the conductor:** The larger the conductor, the greater the amount of current it can carry.
4. **Temperature of the conductor:** The higher the temperature, the harder it is for free electrons to pass through and the higher the electrical resistance.

Although all materials have some resistance to current flow, a **resistor** is a component designed to extract energy from the current flow as it flows through the resistor. A typical resistor has a set resistance, usually marked or coded on its surface. Electrical resistance is somewhat like the electrical equivalent of friction in the mechanical world: It is the degree to which a material opposes, or resists, the passage of current flow. Current flowing through resistance creates heat.

Resistance is measured in ohms. Under most conditions except temperature change, the resistance of an object is a constant and does not depend on the amount of voltage or current passing through it.

Electrical Circuits

Electrical circuits are designed to perform electrical work in a controlled manner. They can be compared to a small city; the roads are like wires; the stoplights are like switches; businesses are like electrical devices where work happens; and cars are like electrons that deliver the workers to the workplace. Electrical circuits can be very basic, consisting of a power supply, a fuse, a switch, a component that performs work, and wires connecting them all together (**FIGURE 50-9**). The battery is the power source and creates a potential difference across its terminals, measured in volts. It pushes a flow of electrons (current flow), measured in amps, through the circuit when the switch is in the closed position completing the path. The current flows through the fuse into the circuit wires and to the lamp, where the resistance of the lamp filament causes it to glow, producing light. The current continues to flow through the lamp filament and the return pathway, through the wires and back to the battery, to complete the circuit. When the switch is moved to the open position, the current path is broken, and current flow stops, turning the lamp filament off.

A circuit can be much more complex than the one just described. But even then, most circuits contain a power source, circuit protection device, control mechanism, load, and connecting wires. When you understand the principles of electricity and how it behaves, you will be able to understand more complicated circuits.

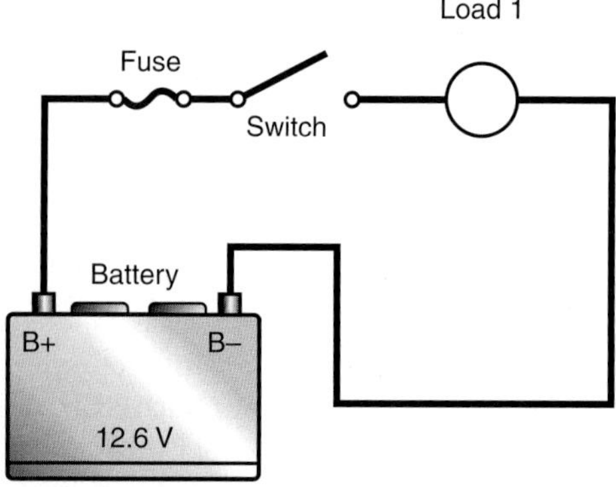

FIGURE 50-9 Simple circuit.

> **TECHNICIAN TIP**
>
> One thing to know about current flow in a circuit: It doesn't flow like dominoes, with a little bit of delay between each domino. Current flow is more like the close fitting marbles inside a tube that we described earlier. As you push on a marble at one end, the marble at the other end falls out. Current flow works the same way, except the effect of current flow travels at the speed of light. Thus, if you add an electron at one end of a wire that is 186,000 miles (300,000 km) long, an electron at the other end of that wire will be pushed out 1 second later. That is why we say that current flow stays the same throughout a series circuit. As one electron is moved, all of the others move with it.

Volts, Amps, and Ohms

K50002

Volts, amps, and ohms are three basic units of electrical measurement. Voltage is the potential or electrical pressure difference between two points in an electrical circuit and is measured in **volts**. For example, the voltage of a typical car battery is 12 volts. This is the potential difference, or electrical pressure, between the positive and the negative battery terminal. It can be measured with a voltmeter or **multimeter** set to read voltage. It might be easier to understand if you think of voltage as the electrical force or pressure in a circuit or battery, just like the water pressure that exists in the bottom of a full tank of water or a home plumbing service (**FIGURE 50-10**). To measure voltage, a voltmeter is used, which measures the difference in voltage (electrical pressure) between the two meter leads.

The ampere, or **amp**, is the unit used to describe how much current flow or how many electrons are flowing at a given point in 1 second when work is being performed—for example, when a lamp is operating. An amp is equal to 6.28 billion billion electrons past a given point in 1 second. Yes, billion billion is correct. To say it another way, think of a pile of 1 billion electrons. One amp would equal the number of elections in 6.28 billion of those piles travelling past a given point in a circuit in 1 second (**FIGURE 50-11**). That is impressive! And just to stretch your thinking, a starter motor may draw about 200 amps. Amperage, or current flow, can be thought of as a faucet being turned on and water flowing. Each drop of water is like an electron (**FIGURE 50-12**). Current flow is measured in amps by placing an **ammeter** into the circuit so that the current flows through the meter.

The ohm is the unit used to describe the amount of electrical resistance in a circuit or component. The higher the resistance, the less current (amps) that will flow in the circuit for any particular voltage. The lower the resistance, the higher the current that will flow in the circuit. Using the water analogy, if you kink a hose with the water running, less water will come out of the hose for a given pressure (**FIGURE 50-13**). If you kink the hose more, more resistance will be added, and even less water will flow through the hose. Lessening the kink lowers the resistance and allows more water to flow. Resistance in a simple electrical circuit works the same way.

An ohmmeter is used to measure the amount of resistance in a component or circuit. The component or wire must be disconnected from the rest of the circuit. The ohmmeter pushes a small amount of current through the part being tested. The amount of resistance in the component changes the amount of current that the ohmmeter can push through the component. The more current the ohmmeter can push through the component, the lower the resistance will read on the ohmmeter.

Power (Source or Feed) and Ground

K50003

"Power" and "ground" are terms to describe the beginning and end of a circuit. Power, source, and feed signify the supply side (beginning) of the circuit, where the electricity originates. Ground signifies the return side of the circuit (**FIGURE 50-14**). In conventional theory, the supply side is the positive side of

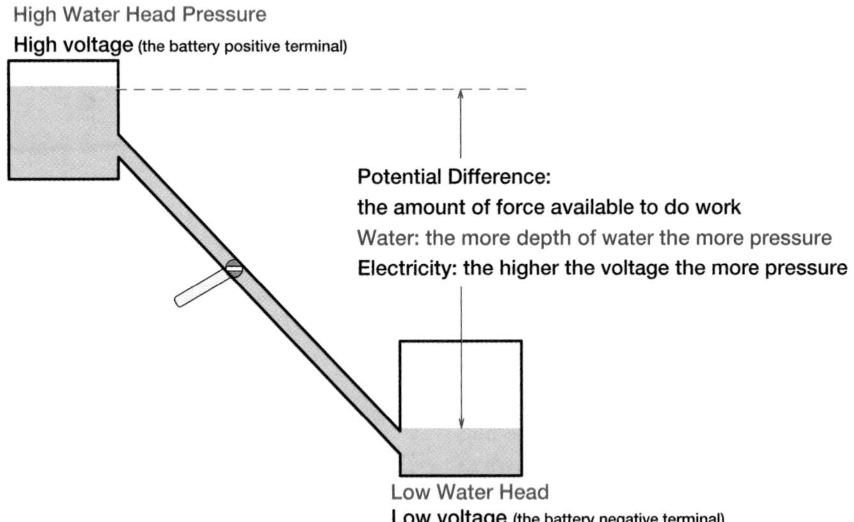
High Water Head Pressure
High voltage (the battery positive terminal)
Potential Difference:
the amount of force available to do work
Water: the more depth of water the more pressure
Electricity: the higher the voltage the more pressure
Low Water Head
Low voltage (the battery negative terminal)

FIGURE 50-10 Pressure (volts) in electrical systems is similar to pressure in a water system. The pressure wants to even out.

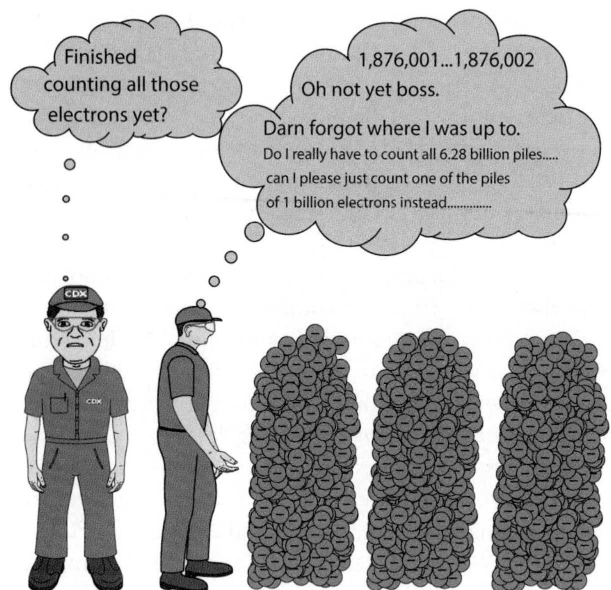

FIGURE 50-11 One amp equals 6.28 billion piles of electrons, each pile containing a billion electrons.

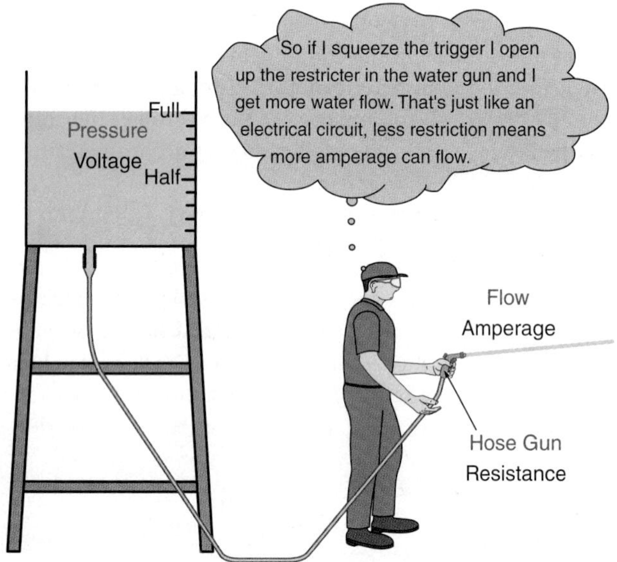

FIGURE 50-12 Current flow (amps) in an electrical circuit is like water flowing through a pipe.

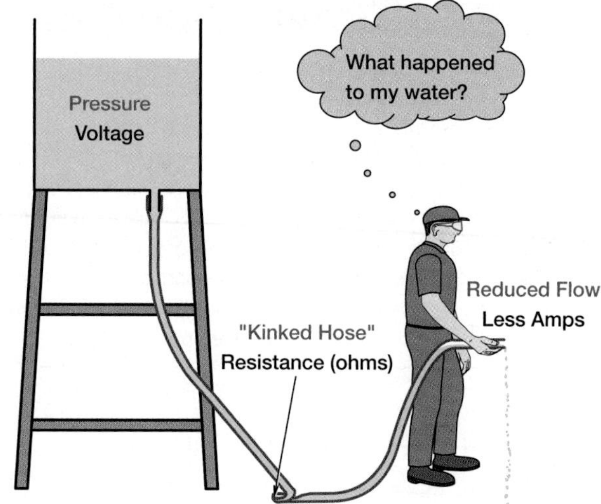

FIGURE 50-13 Resistance (ohms) in an electrical circuit is like a restriction in a water hose.

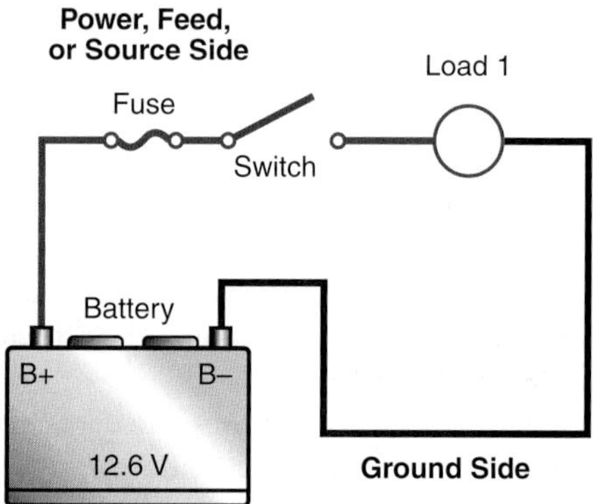

FIGURE 50-14 Power and ground.

the circuit, and the return side is the negative side of the circuit, which we will be using throughout the rest of this text. In a vehicle, the positive battery post is considered the source. The power or feed side of a circuit refers to the wires and components that originate at the positive post of the battery and end at either a switch or a load, whichever occurs first.

Ground is a term used by technicians to indicate the portion of the circuit that returns current flow to the negative side of the battery. The ground side of the circuit starts at the negative post of the battery and ends at either a load or a switch, whichever occurs first. The term "ground" is also used to mean a direct connection with the negative side of the circuit, as in "the switch is grounded." Also, if you are told to ground something, it means that object has to be connected to the negative side of

Applied Science

AS-60: Ground: The technician can demonstrate an understanding of the problems associated with having an electrical circuit inadequately grounded.

Ground problems can present some of the most challenging issues encountered in automotive diagnostics. Let's illustrate why ground points are so important on motor vehicles. In a simple circuit, we might have a battery, a fuse, and a load, all connected by wire. In automotive applications, manufacturers take advantage of the conductivity of metal car and truck bodies by connecting the ground side of electrical components to the vehicle's body or chassis. The negative side of the battery is also connected to the body, so instead of electrical current returning to ground via wiring, it is conducted through the body.

If we have a component with no ground connection, we have an open circuit and no current flow. Loose or dirty ground connections create a point of high resistance in the circuit, resulting in lower current flow and excessive voltage drop. This results in lamps illuminating dimly, electric motors not working or not working at full speed, and running problems in computer-controlled systems. In some circuits, issues may be encountered where electricity finds alternative paths to ground. You may sometimes see vehicles with brake lights flashing along with, or opposite to, turn signal lights. This is commonly caused by a poor ground in the taillight circuit.

the circuit. Many vehicles connect the chassis, body, and engine block to the negative battery terminal, which means most of the metal components on the vehicle are grounded. As a result, many manufacturers use the chassis as the return path to the negative battery terminal because this cuts down the amount of wire needed in the vehicle. At the same time, many computer circuits use dedicated ground wires from their sensors back to the computer so that the electrical signal is accurate and not affected by any stray electrical signals on the ground circuit.

Direct Current and Alternating Current

Electrons must flow in a circuit for work or action to be undertaken. For example, the action of a lamp glowing brightly is caused by the flow of electrons through the filament, heating it up and causing it to glow. There are two fundamental types of **current flow: direct current** (DC) and **alternating current** (AC) (**FIGURE 50-15**). DC is produced by a battery. The battery maintains the same positive and negative **polarity**; therefore, the current flows in one direction only. The characteristics of DC are the fixed polarity of the applied voltage and the flow of charges in only one direction. It is possible to have varying DC; however, the charges always flow in one direction, and the applied voltage polarity remains the same.

AC is the type of current in your home electricity supply. The alternating voltage repeatedly reverses or alternates its polarity. Thus, the current flow moves back and forth within a circuit. AC is produced in what is called a **sine wave**. It operates on a cycle, gradually building to a maximum current flow in one direction (positive value), then gradually reducing to zero, then gradually building to a maximum current flow in the other direction (negative value), and finally gradually reducing back to zero current. In most cases, this cycle can occur many times a second. For example, the AC flow in the house supply has a cycle rate of 60 times per second. **Hertz** is the measurement of frequency and indicates the number of cycles per second. So, Hertz simply means "cycles per second."

AC is used in vehicles to a lesser extent than DC. Alternators use it to create current flow to charge the battery and run the electrical accessories. The AC is first transformed to DC before it leaves the alternator, so it can be effectively put to use in the DC electrical system.

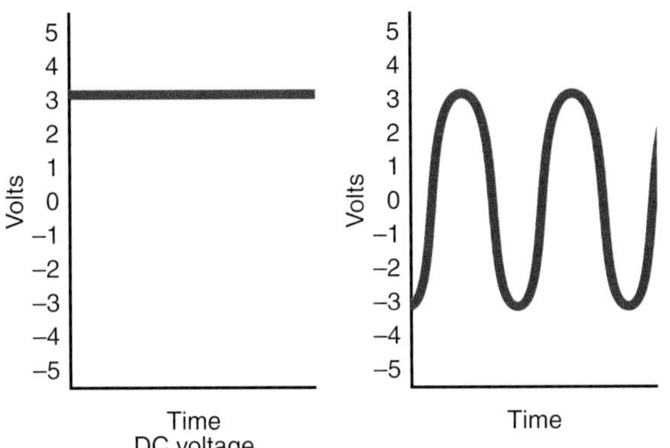

FIGURE 50-15 Waveforms. **A.** Direct current. **B.** Alternating current.

Applied Science

AS-58: AC/DC: The technician can explain the difference between direct and alternating current.

Direct current (DC) and alternating current (AC) are two forms of electricity that are produced differently and have different uses. DC electricity is the simpler form. It starts in one place and then flows in the same direction to its destination. AC electricity flows in one direction for a period of time, then changes direction over and over again continuously.

In modern automotive applications, AC electricity is generated by the alternator using electromagnets. The AC electricity is converted to DC, or rectified, by diodes in the alternator before being supplied to the vehicle's electrical systems or being stored in the battery.

AC is used in the electric motors on most hybrid vehicles. Because those motors generally require high amounts of electrical power, and because AC is more efficient than DC, AC is more advantageous for that application. The AC is created by a sophisticated electronic inverter. In general, electrical components are designed to work on either AC or DC, but not both. For example, a DC motor will not work on AC, and vice versa.

Continuity, Open, Short, and High Resistance (Voltage Drop)

N50001

The terms "continuity," "open," "short," and "high resistance" are often used to describe a circuit's or component's condition. For example, the wiring harness could be described as having a short or the connector as having an open circuit. **Continuity** is achieved when an electrical circuit has a continuous and uninterrupted electrical connection and is thereby capable of conducting current and working as designed. No continuity means there is a break in the circuit, and current cannot flow past the break.

▶ TECHNICIAN TIP

Circuit continuity can be measured with a digital volt-ohmmeter (DVOM) using the ohms setting between two points in the circuit. For example, if a technician suspects that a circuit has a break or bad connections in the wiring, the continuity could be measured between two points. A circuit with no continuity has no path for current to flow. A circuit with continuity has a path for current to flow. But be careful, just because a circuit has continuity doesn't mean it is okay. It is possible that although the circuit has continuity, it may have only one small strand of wire instead of the original 6–50 strands it had. This would affect the operation of the electrical device. Thus, continuity testing is limited in what it can indicate. It is a great test for determining whether a wire or component is open (no continuity) or whether two circuits are connected together (shorted).

The term **open** describes a low-voltage circuit that does not have a complete circuit and therefore cannot conduct current. In other words, the circuit does not have continuity. For example, when a switch is turned off, the circuit is open, and no current can flow. "Open" can also be used to describe a fault in a circuit.

For example, if the fuse is open-circuited, which describes a blown fuse, no current will be able to flow in that circuit (see **FIGURE 50-17**). A multimeter or test lamp can be used to test for an open circuit. An open circuit has infinite resistance—so much resistance that it is not measurable.

The term **short**, in its purest definition, describes a circuit fault in which current takes a shorter path, in terms of resistance, through an accidental or unintended route. The short causes abnormally high current flow in the circuit and may cause the circuit protection devices, such as fuses or circuit breakers, to open the circuit. An example of a pure short would be insulation on the windings within a relay coil that has worn through and is allowing current to bypass many of the windings (**FIGURE 50-16**). In this case, resistance decreases, amperage increases, and the magnetic field created by the winding becomes weaker, potentially causing the relay not to operate.

A pure short circuit is not the only type of short circuit. There are three additional types of short circuits that you need to understand. The first is a short to ground. In this case, the circuit has an unintended path directly from a powered wire to ground. For example, if the wire from the brake switch to the brake lights rubs through the wire insulation on a sharp edge of a body panel, the bare wire may make contact with the metal panel and cause a short to ground when the brake pedal is pressed (**FIGURE 50-17**). A short to ground causes increased current flow and will typically blow the circuit fuse.

Another type of short is a short to power. In this case, the circuit has an unintended path directly to a power source. An example is two wires in a harness that have melted together, one that supplies power to the blower motor and one that feeds power from the brake light switch to the brake lights (**FIGURE 50-18**). In this example, turning on the ignition switch would cause the brake lights to come on because power would be sent to the blower fuse and then, due to the short in the wiring, to the brake lights.

The third type of short is a short from a switched ground wire directly to ground (**FIGURE 50-19**). This can happen on circuits like fuel injectors, where the switch provides a ground for the injector. If the wire between the injector and the switch shorts out to ground, the injector will continuously spray fuel, even if the switch is open.

Other types of electrical faults include unintended **high resistance** in a circuit, which causes a reduction in current flow in the circuit as well as a drop in voltage at the resistance. Both of these result in the intended circuit device not operating effectively or at all. Unintended high resistance can also cause an overheating condition at the area of resistance, which can melt wire insulation or plastic connectors. This condition is caused by a number of faults, including corroded or loose harness connectors, wire that is too thin for the circuit current flow, incorrectly connected terminals, and poorly soldered joints (**FIGURE 50-20**).

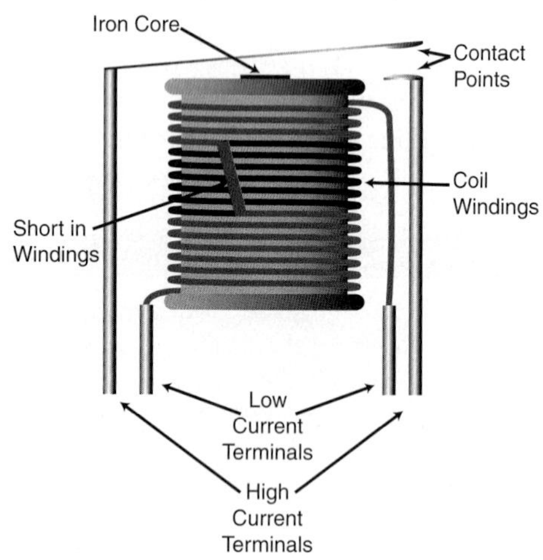

FIGURE 50-16 Short circuit.

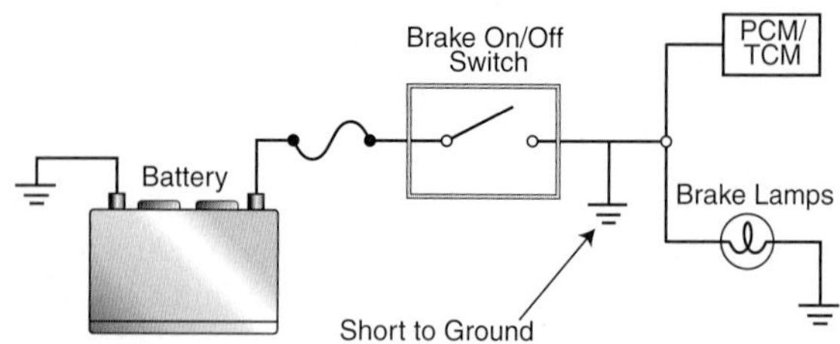

FIGURE 50-17 Short to ground.

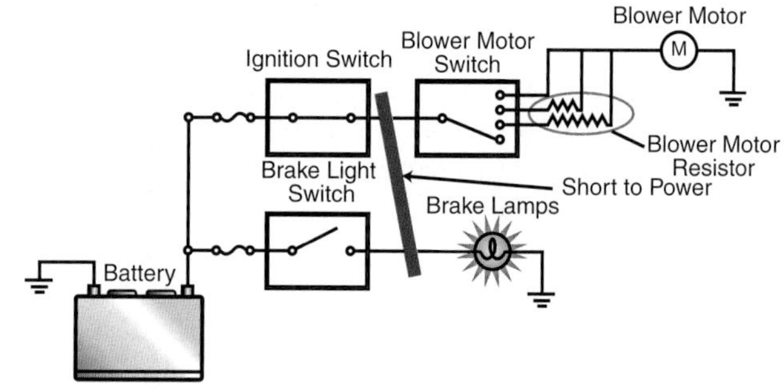

FIGURE 50-18 Short to power.

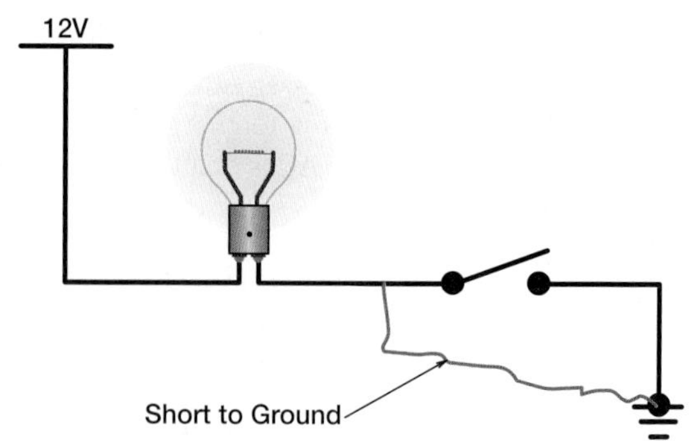

FIGURE 50-19 Short circuit from a switched ground wire to ground.

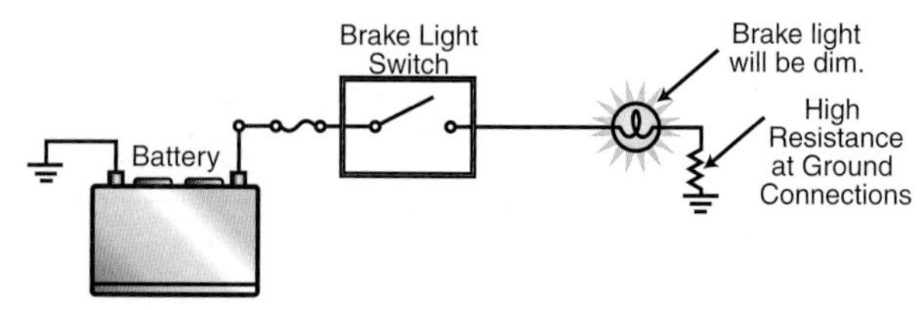

FIGURE 50-20 High-resistance fault.

Applied Science

AS-62: Short Circuit: The technician can demonstrate an understanding of the processes used to locate a short circuit in an electrical/electronic system.

Two types of short circuit can occur in electrical systems. In a short to ground, electrical current finds its way to ground before it was intended, usually due to compromised wiring insulation that allows wiring to touch the metal vehicle body. In a short to power, a circuit is exposed to voltage flowing in another circuit, generally also due to broken wiring insulation. The effect of either type of short will depend on the layout of the circuit and where the short occurs in relation to the load.

The most common indicator of a short circuit is a fuse blowing due to excessive current flow. Traditionally, diagnosing the location of the short has required being able to segment the circuit by disconnecting fuses, components, switches, or harness plugs and checking individual sections of the circuit for integrity. Electronic short-circuit finders are now available that send an electrical signal along the circuit. A receiver is run along the circuit until it stops receiving a signal, at which point the location of the short has been found.

Just like the resistance in a load results in a voltage drop, unintended resistance results in a voltage drop as well. To locate an unintended voltage drop, two conditions must be present: There must be excessive resistance, and the circuit must be activated. Thus, a voltage drop in a circuit indicates that there is resistance present and current is flowing, or trying to flow, through the resistance. If the resistance is excessive, it will reduce both the voltage and the current flow in the circuit, affecting its performance. For example, an unintended voltage drop in the headlight circuit reduces the available voltage for the light, causing the lights to be dim. A high resistance in the main battery cable to the starter motor causes the starter to crank the engine over slowly or not at all. The voltage drop can be measured in a circuit by placing a voltmeter across two different points in a circuit while the circuit is being operated. For example, to measure the voltage drop in the main positive battery cable to the starter motor, the voltmeter's black lead is placed on the positive battery lead, and the red lead is placed on the main battery lead of the starter motor; the voltage drop is then read on the meter while the engine is cranked (**FIGURE 50-21**). If excessive, the cable or connections would have to be replaced or repaired.

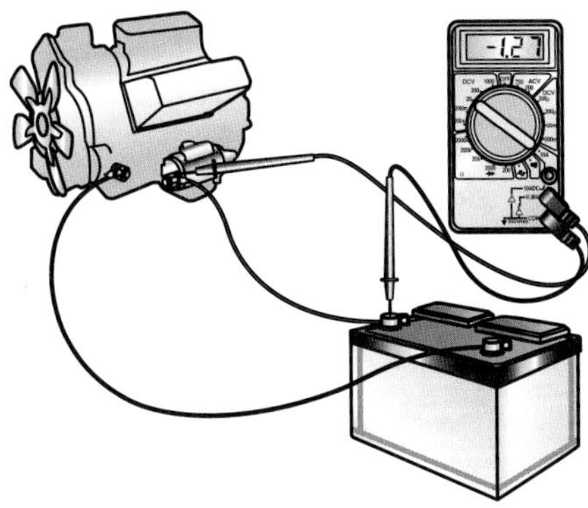
-1.2 1

FIGURE 50-21 Voltage drop testing the positive battery cable.

▶ TECHNICIAN TIP

Because current flowing through a resistance causes heat, you can sometimes locate the high resistance in the circuit just by feeling the wires and connections. For example, if one of the battery posts is corroded between the post and the battery cable, a voltage drop will be present when the starter is engaged. If you operate the starter for several seconds and then feel each battery terminal, the one with excessive resistance will be warmer than the other terminal.

▶ Sources of Electricity

K50004

Electricity is a unique source of energy. Not only can it be transformed into a variety of other kinds of energy, such as thermal energy, light energy, chemical energy, and mechanical energy, it can be created easily in a variety of ways. Think of other types of energy, such as coal and oil. They are much harder to create, generally relying on natural processes over a long period of time. This section explores some of the major ways that electricity can be produced.

A common and easily observable source of electricity is the static electricity produced in thunderstorms. Other ways that electricity can be produced include the movement of a conductor through a magnetic field, the application of pressure to a special type of crystal, the conversion of sunlight by solar cells, and chemical reactions. Regardless of the way it is produced, electricity is always the movement of electrons in a conducting circuit. Let's see how we can get those electrons moving so they can do work for us.

Electrostatic Energy

Static electricity can be induced by rubbing two insulators together. During this process, one material loses electrons to the other. The insulator losing electrons becomes positively charged. The other insulator gains electrons to become negatively charged (**FIGURE 50-22**).

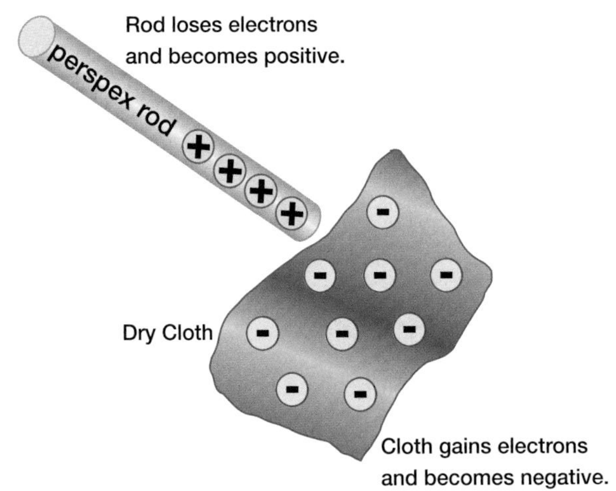
Rod loses electrons
and becomes positive.
perspex rod
Dry Cloth
Cloth gains electrons
and becomes negative.

FIGURE 50-22 Rubbing two insulators together pulls electrons from one surface to the other, creating an electrical charge.

When these two highly charged surfaces are brought close enough together, electrons leap across the gap, causing a spark and canceling out the charge imbalance. This electron transfer can be experienced as an electric shock. It applies to any charged surfaces where the imbalance in charge is great enough to make the electrons leap the gap. The spark can be dangerous. When it is near fuel vapor at a gas station, it can even cause an explosion.

SAFETY TIP

Many fires have been caused by the operator starting the refueling process, getting back into the vehicle to check on a child or put a credit card away, sliding back out across the seat of the vehicle (without touching the metal of the vehicle), and then grabbing the refueling nozzle, at which time a spark ignites the fumes escaping from the filling tank. Always touch the metal of the vehicle or fuel pump before touching the refueling nozzle. Also, fires have reportedly resulted from customers sliding their plastic gas cans out of the back of a plastic-lined pickup bed, creating a spark when the nozzle touched the can or when the pour spout touched the gas tank.

Thermoelectric Energy

If two different metals are joined and heated, a small electrical current can be generated. For a temperature rise of around 392°F (200°C), the potential difference created is about 9 millivolts. The point that is heated is called a **hot junction**, and the whole system is called a **thermocouple** (**FIGURE 50-23**). In developing engine designs, manufacturers use thermocouples to measure the high temperatures of components such as spark plugs and exhaust systems. Thermocouples are also used to measure exhaust gas temperature on race cars to determine the best air-fuel ratio so that the engine parts will not overheat. Pilots in general aviation use thermocouples to measure the exhaust gas temperature to adjust the fuel mixture as the airplane changes altitude.

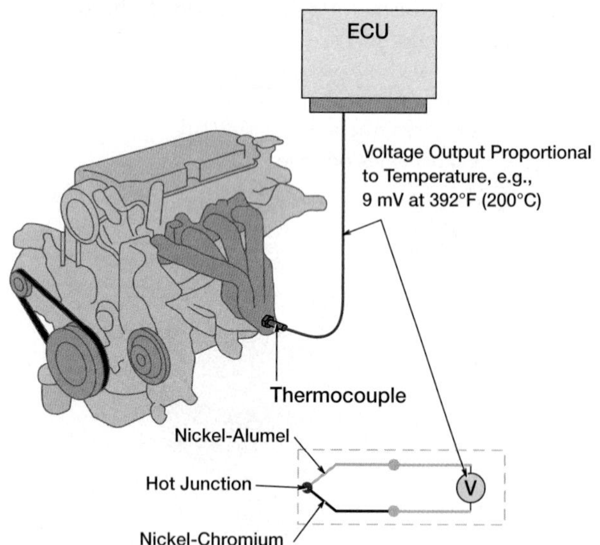

FIGURE 50-23 A thermocouple creates a small electrical current flow that can be used to indicate temperature, for example, that of exhaust gases.

Electrochemical Energy

When two dissimilar metals are immersed in an acidic liquid called an **electrolyte**, the breakdown of chemicals into charged particles, called **ions**, results in a flow of electricity (**FIGURE 50-24**). The process is **electrolysis**. This principle is applied in the standard lead-acid battery used in most vehicles, and is covered more thoroughly in the Batteries, Starting, and Charging Systems chapter.

Photovoltaic Energy

Solar cells convert sunlight directly into electricity and are used in a wide variety of applications to supply electricity. They are made of semiconducting materials similar to those used in computer chips. When sunlight is absorbed by these materials, the photons knock electrons loose from their atoms, causing the electrons to flow through the material to produce electricity. This process of converting light (photons) to electricity (voltage, which moves electrons) is called the **photovoltaic (PV) effect** (**FIGURE 50-25**). The principle is used in heating, ventilation, and air-conditioning (HVAC) systems to determine the sun load on the vehicle. It is also used to sense the amount of

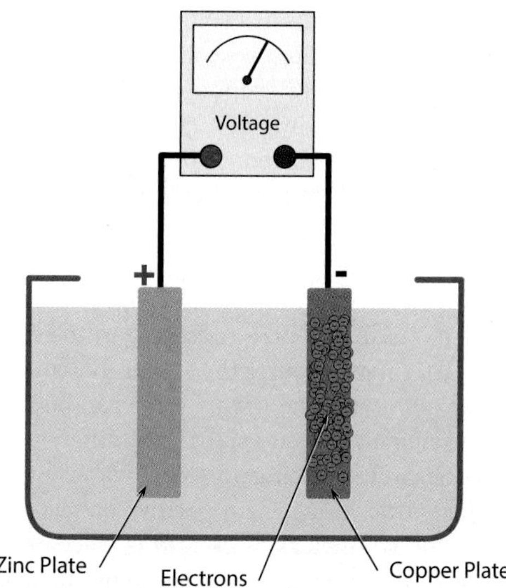

FIGURE 50-24 Electrolysis chemically separates electrons to one plate, creating an electrical charge.

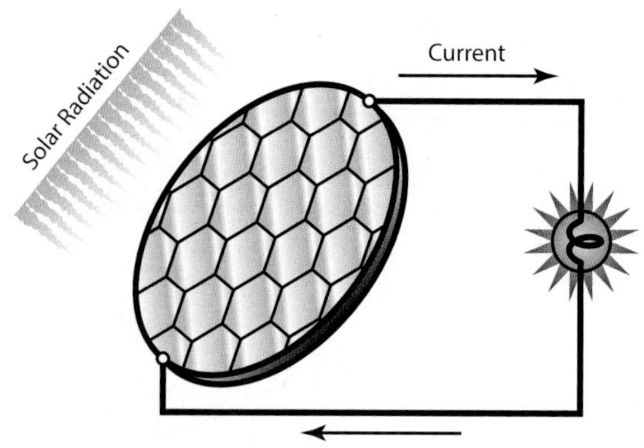

FIGURE 50-25 Photovoltaic effect.

daylight to operate automatic headlights. In addition, it is the principle of the solar cell.

> ▶ TECHNICIAN TIP
>
> The term "photovoltaic" comes from the Greek *photos*, meaning light, and the name of the Italian physicist Volta, after whom the volt (and consequently voltage) is named. Photovoltaic means literally "of light and electricity."

Piezoelectric Energy

When crystals of certain materials, such as quartz, are subjected to mechanical stress, an electrical potential is produced across the crystals (**FIGURE 50-26**). The process is also reversible. A potential electrical difference across the crystal physically distorts the crystal. The best-known simple application of **piezoelectric** energy is the sparking lighters that ignite gas in grills, barbecues, or cooktops, but the piezoelectric principle is also used in automobiles in knock sensors and pressure sensors and in some electronic fuel injectors.

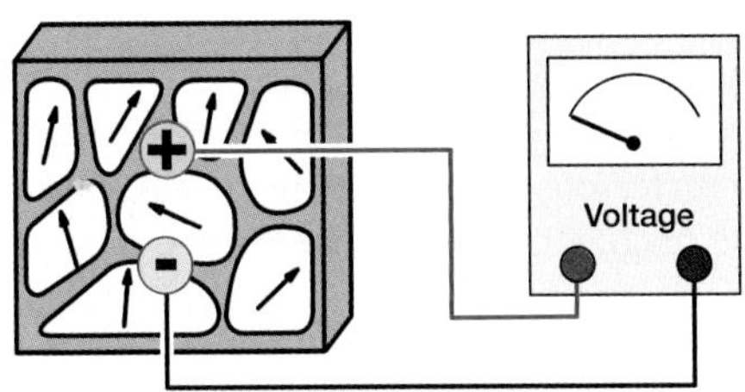
Piezoelectric crystal at rest does not produce any voltage.
Voltage

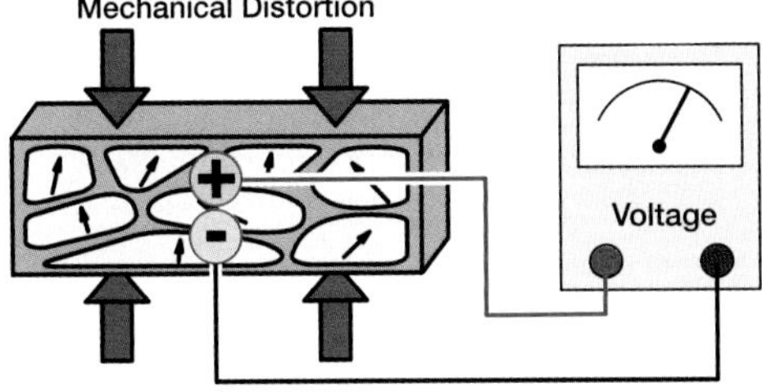
Mechanical distortion applied to a piezoelectric crystal produces voltage.
Mechanical Distortion
Voltage

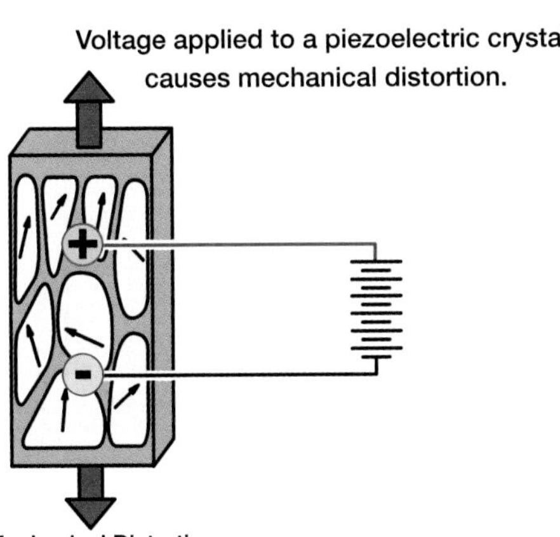
Voltage applied to a piezoelectric crystal causes mechanical distortion.
Mechanical Distortion

FIGURE 50-26 Piezoelectric energy.

Electromagnetic Induction

Electromagnetic induction is the creation of an electrical voltage—or the potential difference—across a conductor within a changing magnetic field. When a conductor cuts across a magnetic field, current flows in the conductor. It flows one way when the conductor cuts the field in one direction, then reverses as it cuts the field in the opposite direction. Thus, it creates alternating current, or AC.

Moving a wire inside a magnetic field produces a current flow. Similarly, moving a magnet inside a stationary coil of wire produces the same effect. For example, to generate electricity, a magnet can be rotated next to a winding, which indicates current flow (**FIGURE 50-27**). As the magnet rotates, the ammeter deflects for current flow. For every half revolution, current flow reverses. Increasing the speed of the magnet increases the amount of electrical energy produced. Electromagnetic induction is applied in alternators, ignition coils, and some sensors on the vehicle.

To have induction, you have to have three things: a winding, a magnet, and relative movement (i.e., movement of one

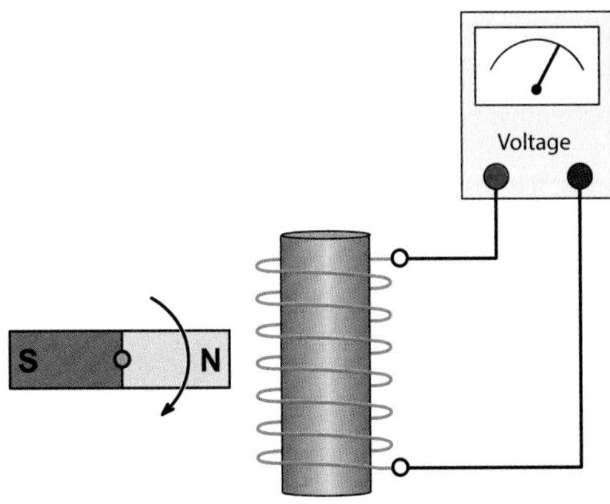
Voltage
S
N

FIGURE 50-27 Electromagnetic induction using a magnet rotating next to a winding.

> **Applied Science**
>
> ***AS-75: Electromagnetism: The technician can explain the relationship between current in a conductor and strength of the magnetic field.***
>
> Faraday's law of induction describes the ways that a voltage can be generated within a conductor moved through a magnetic field. Simply stated, whenever there is a change in the number of magnetic field lines passing through a conductor, whether the conductor is moved or changes occur in the magnetic field, a voltage is generated in the conductor. When the strength of the magnetic field is increased, there is an increase in the number of magnetic field lines. If we have a conductor moving through the magnetic field and the rate of movement remains constant, the increased intensity of the magnetic field will result in more magnetic field lines passing through the conductor, increasing the voltage generated.

past, or through, the other). The amount of induction is dependent upon the strength of the magnetic field, the number of windings, the speed of the movement, and the relative distance between the field and the winding.

▶ Effects of Electricity

Whenever electricity flows, it produces effects. For example, light, heat, chemical reactions, and magnetism are all effects of the flow of electricity. Sometimes those effects are the primary reason for the activity. For example, magnetic effects are used to generate electricity. At other times, the effects are just by-products of the activity. For example, current flowing in the conductors of a generator produces heat as a by-product and is wasted energy. No transformation of energy from one form to another is 100% efficient. There is always waste, which usually shows up as heat. For example, a typical incandescent lightbulb creates light, but only about 10% of the electricity is converted to light; about 90% is wasted as heat.

Heating Effects

Heat is not always bad. Sometimes we need to create heat, such as in circuit breakers and rear window defoggers (**FIGURE 50-28**). As current flows through motor vehicle circuits, most of the electrical energy is transformed into other kinds of energy. Headlights transform the energy into intense heat that then makes the bulb filaments glow white-hot and produce light. The energy can also defog a window or can be used to provide circuit protection from excessive current flow.

Chemical Effects

Chemical effects of electricity depend on ions. Recall that ions are electrically charged atoms or groups of atoms. Atoms that gain electrons become negatively charged; atoms that lose electrons become positively charged.

When two different metal plates are immersed in an electrolyte, one loses electrons and becomes positive. The other gains them and becomes negative. Negative ions in a solution are attracted to the positive **plate** and positive ions to the negative plate, and a chemical reaction can occur.

In a lead-acid battery, the electrical and chemical differences between the sets of plates create a potential difference, or voltage, which makes the current flow in a circuit. It flows in one direction only, so it is called direct current, or DC. As electrons move through the electrical circuits in the vehicle from one set of plates to the other, the same compounds form on the plates. If this happens for too long, there will be no potential difference between the plates, and current will stop flowing, leaving a discharged battery. Of course, a battery can be recharged, which reverses the process so that current is pushed backward through the battery, reversing the chemical reaction and restoring the potential difference between its sets of plates.

Light Effects

The production of light can also be an effect of electricity. An example of this is the LED. The **light-emitting diode (LED)** is a semiconductor diode that creates light by emitting photons from its PN junction when a current flows through it (**FIGURE 50-29**). At one time, LEDs were available only in red, green, and yellow tints. More recently, they have become available in many different colors and have grown in light intensity. They are now replacing traditional bulbs as lighting sources and produce as much or more light with less heat loss, using much less energy. LEDs are also being used as headlights on some top-end cars, and it is likely that as LED technology improves and prices come down, they will be installed on more new vehicles.

Electromagnetic Effects

Electricity can also create magnetic effects, referred to as electromagnetism. An **electromagnet** creates magnetic forces that attract both ferrous metals and unlike magnetic charges and that repel like magnetic charges. These forces can be used to create mechanical movement, such as in a relay or electric motor. In electrical systems, **magnetism** occurs when a current passes through a conductor, and a magnetic field is created around it. If the wire is wound into a coil, the magnetic fields combine to create a stronger and denser magnetic field, with a north and a

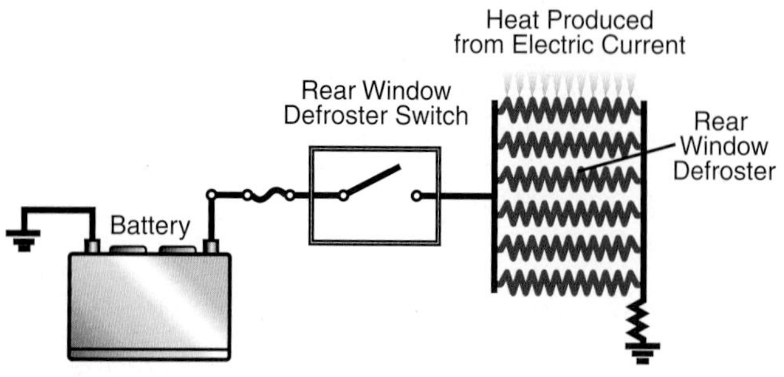

FIGURE 50-28 Heating effect of electricity.

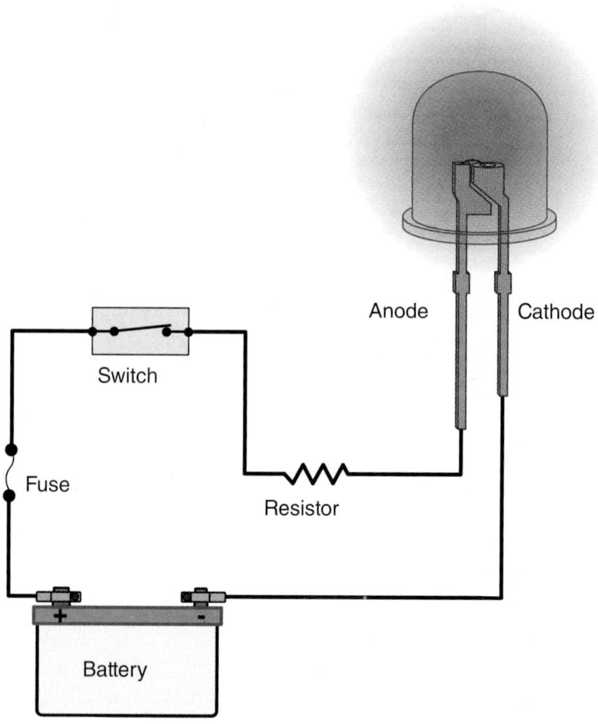

FIGURE 50-29 Light created by an LED.

south pole, just like a permanent magnet. Turn the current flow off, and the magnetic field collapses and disappears. By turning current to the coil on and off, this magnetic effect can be turned into a mechanical movement, pulling a switch open or closed (**FIGURE 50-30**). This is the principle behind a **relay**.

Electromagnets are constructed by winding a conductor wire, many hundreds or thousands of times, around a soft iron or metal core and passing a current through the coil. The strength of the magnetic field produced is determined by the number of turns, or coils, and the value of the current flow through the conductor. The metal core uniformly aligns the magnetic fields, which strengthens the magnetic effect. The size of the electromagnet is determined by its application. A fuel injector contains a small electromagnetic coil using very fine wire, whereas a starter motor uses heavier wire in larger coils. Also, reversing the current flow through the winding reverses the north and south ends of the electromagnet.

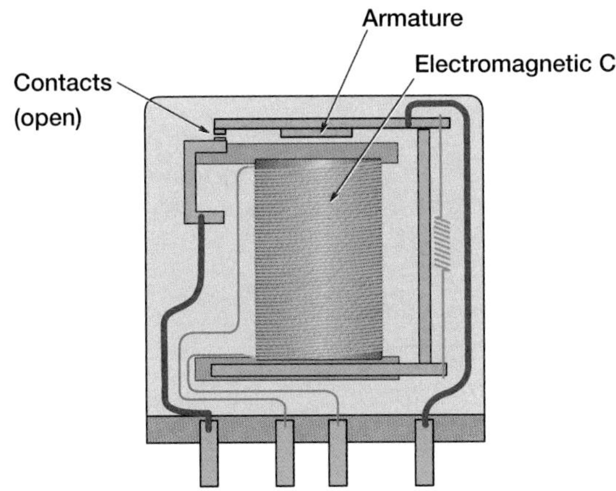

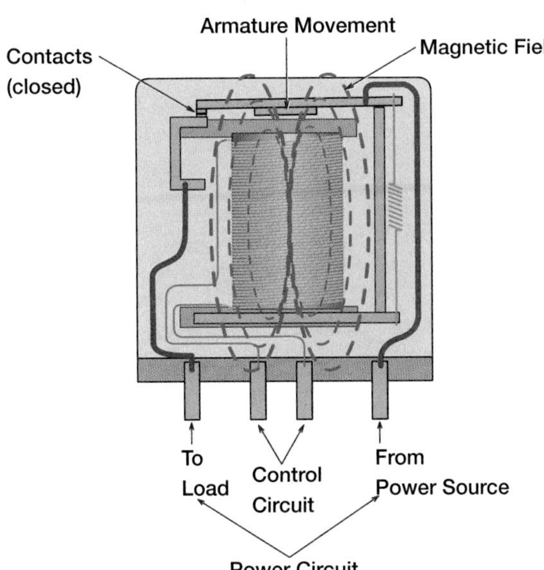

FIGURE 50-30 A relay uses electromagnetic force to open and close a switch.

Applied Science

AS-63: Generators: The technician can explain how the movement of a conductor in a magnetic field generates electricity.

When a conductor is moved through a magnetic field, an electrical current is generated within the conductor. This phenomenon is known as electromagnetic induction. Alternatively, the conductor can remain stationery while the magnetic field is moved in relation to it. The generation of a voltage depends directly on the movement of either the conductor or the magnetic field. Without movement there is no voltage generated.

Electromagnetic induction works by exposing the electrons within the conductor to a magnetic force acting perpendicular to both their motion and the magnetic field, which forces them to move through the conductor, creating a charge.

▶ Ohm's Law and Circuits

K50005

Ohm's law helps us to understand the relationship between volts, amps, and ohms. Understanding how they are related to each other will make electrical diagnosis much easier. Ohm's law shows us that if one of the three units changes, then at least one of the other two must also change, and in specific ways, as they always have to balance out mathematically.

Here's how Ohm's law works. It tells us that it takes 1 volt to push 1 amp through 1 ohm of resistance. That gives us the relationship among the three units. Volts and resistance are physical things. Volts is the electrical pressure in a circuit. Resistance is the physical restriction in the circuit. Amps are the amount of electrons moved. This means that amps are the result of both the voltage and resistance. If voltage or resistance change, then amperage will also change. For example, if voltage stays the same and resistance doubles, half as much amperage can be pushed through the resistance (**FIGURE 50-31**). Conversely, if resistance stays the same but voltage doubles, then the greater force pushes twice as much current through the resistance. Or, if voltage is cut in half and the resistance stays the same, then amperage will be cut in half. Knowing how amperage responds

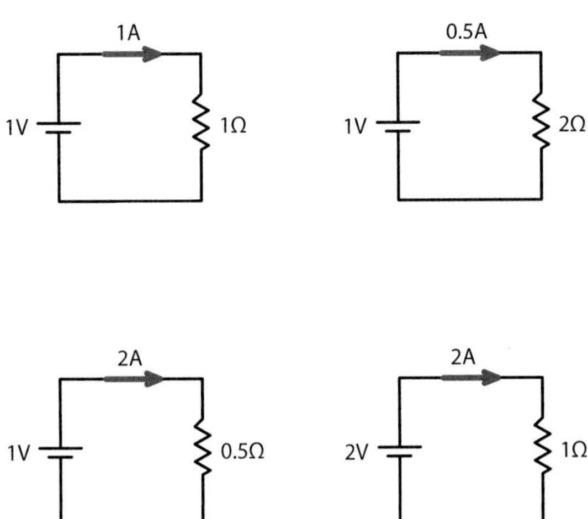

FIGURE 50-31 Current flow is the result of voltage and resistance.

to changes in voltage and resistance will help you know what tests to perform to identify circuit faults. We discuss this further in the chapter on meters.

Ohm's Law and Ohm's Law Calculations

Because Ohm's law is a relationship between volts, amps, and ohms, and because they must always balance out, if we know any two of the values, then we can calculate the third. In calculating Ohm's law, R stands for resistance, V for voltage, and A for amps. (Note that some sources use E for voltage and I for current. Also, we want to highlight the fact that amperage is *always* the result (product) of voltage and resistance. If the amperage is not correct (the device isn't operating correctly), it is because either the voltage or resistance is wrong.) Depending on which value you wish to solve for, you apply one of the following three formulas:

- $A = V/R$
- $V = A \times R$
- $R = V \div A$

Using the Ohm's law circle will help you remember which math operation to use (**FIGURE 50-32**). All you have to do is place your finger over the value you are looking for. If you place your finger on the top value (volts), then you would multiply amps by resistance. If you place your finger on one of the side values, then you would divide volts by the other value. This means two things. First, the values always have to balance. And second, as long as you know any two values, the third can be calculated. For example, battery voltage can be measured. Let's use 12 volts. The value of the resistor, 4 ohms, is on its casing. Current then equals voltage (12 volts) divided by resistance (4 ohms). We can quickly calculate that there should be 3 amps of current flowing through every point in the circuit (**FIGURE 50-33**).

Using the formulas of Ohm's law gives an accurate method of determining expected values in an electrical circuit. We can then compare these expected values to measured values to determine if an electrical fault exists, and if so, where it is located. To assist with these problems, use the Ohm's law circle.

- If the value of V and R are known, then to find A, V is divided by R. Place your thumb over A and the circle shows you this formula.

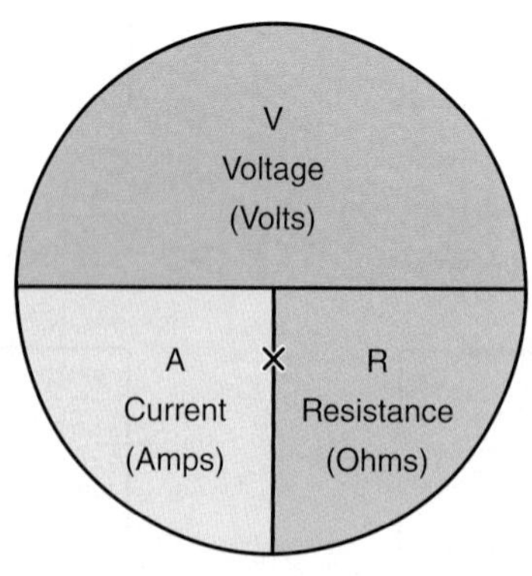

FIGURE 50-32 Ohm's law circle.

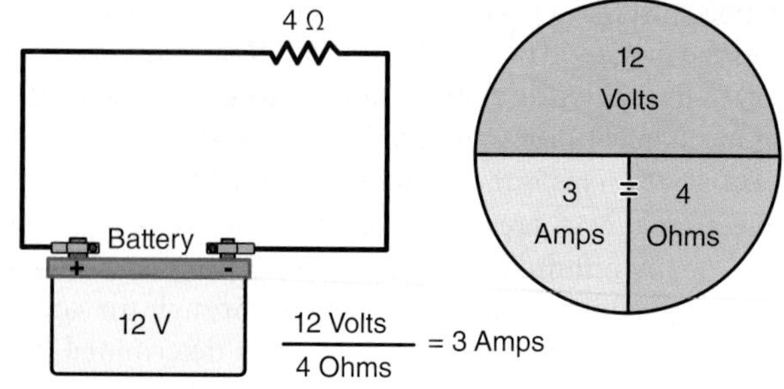

FIGURE 50-33 Calculating current flow in a circuit.

- Similarly, if V and A are known, then R can be found by dividing V by A.
- If A and R, are known, then V is found by multiplying A by R.

In the circuit diagram in **FIGURE 50-34**, the value of the applied voltage is not known, but the amperage and resistance are. To find the value of V, multiply A by R, or 6 amps by 4 ohms. The answer is 24 volts.

In the circuit diagram in **FIGURE 50-35**, the value of the current flow is unknown. To find the value of A, divide V by R, or in this case 12 volts divided by 3 ohms. The value of A is 4 amps.

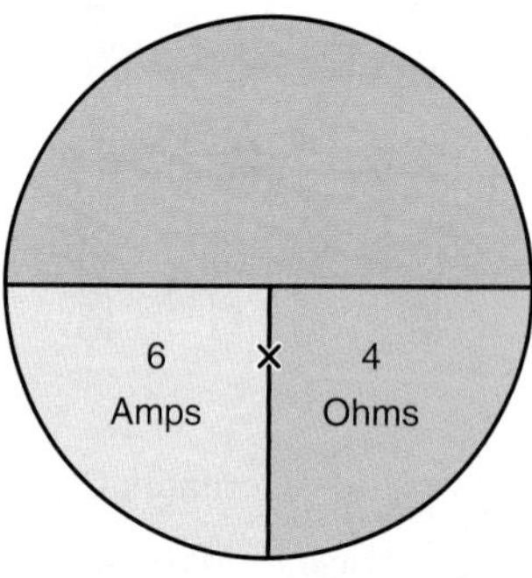

FIGURE 50-34 To find the value of V, multiply A by R, or 6 amps by 4 ohms.

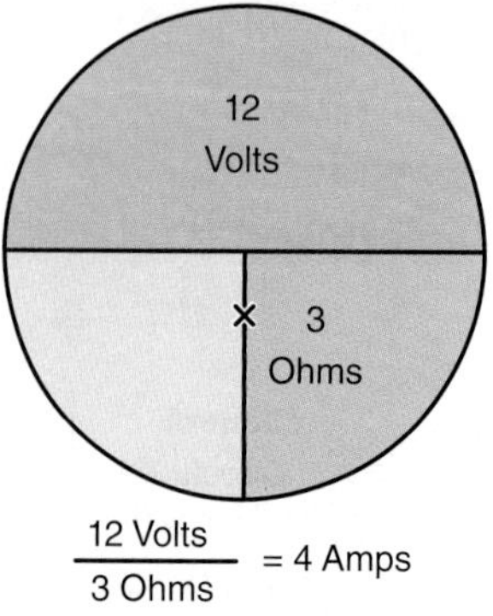

FIGURE 50-35 To find the value of A, divide V by R, or in this case 12 volts divided by 3 ohms.

Applied Science

AS-70: Ohm's Law: The technician can demonstrate an understanding of and explain the use of Ohm's law in verifying circuit parameters (resistance, voltage, amperage).

Ohm's law quantifies the relationship between amperage, voltage, and resistance in any given circuit. It provides a means to calculate any one of these parameters from the values of the other two. Ohm's law is applied most easily through the use of a circle diagram like the one shown here:

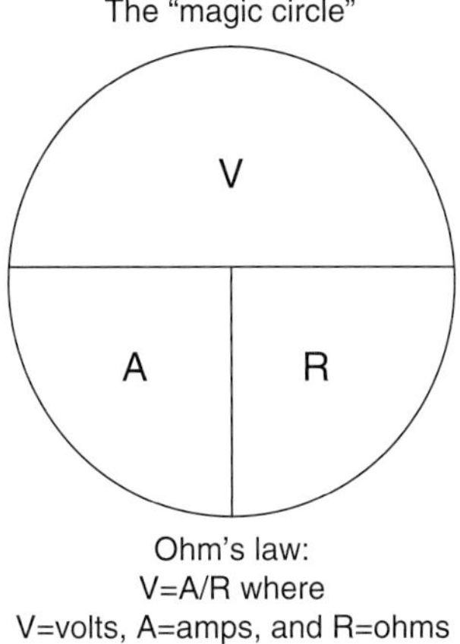
The "magic circle"
V
A
R
Ohm's law:
V=A/R where
V=volts, A=amps, and R=ohms

These diagrams are used by placing your finger over the parameter you are trying to calculate, then reading the remainder of the diagram to identify the calculation you need to perform. For example, if you need to work out the amperage flowing in the circuit, block off the A, and your calculation is V ÷ R, or voltage divided by resistance. For a 12-volt automotive circuit with 20 ohms of resistance, your calculation is 12 ÷ 20, to arrive at 0.6 amp.

Electrical Power and the Power Equation

Energy is the potential to do work. However, work is done only when energy is released. A disconnected battery is not doing work, but it has the potential to do work and is therefore a source of energy. The difference in electron supply at the battery terminals creates electrical force and is sometimes called the potential difference. In the case of a standard charged automotive battery, it has a potential difference of approximately 12 volts. Tapping this potential means turning one form of energy, the battery's chemical energy, into another form of energy, electrical energy.

Turning one form of energy into another is called energy transformation. The amount of energy transformed is the amount of work done. When a person's legs turn the pedals of a bicycle, chemical energy (from oxygen and food) is being turned into mechanical energy. A motorcycle engine turns chemical energy into thermal energy, and then into mechanical energy. In each case, **work** is being done, but there is a difference with the motorcycle—it does the work more quickly than the bicycle, delivering more mechanical energy faster. That difference is called **power**. Power is the rate at which work is performed. It is also known as the rate of transforming energy. In an electrical circuit, power refers to the rate at which electrical energy is transformed into another kind of energy.

The unit of **electrical power** is the **watt**. One watt of electrical power is produced when 1 volt is used while pushing 1 amp of current through a resistance. From this comes the power equation:

- P (the power in watts), equals A (the current in amps), multiplied by V (the voltage in volts). This looks like $P = V \times A$

This calculation is applied similarly to Ohm's law and is typically represented as a triangle (**FIGURE 50-36**).

When current flows in a circuit with a resistor in it, the resistor becomes hotter as it converts electrical energy into heat energy. The greater the voltage used by, or the amperage pushed through, the resistor, the greater the wattage. A lightbulb uses a certain amount of electrical power, but the power used is not an indication of brightness; it is a measure of power consumption only. If a 12-volt circuit with a single light has a current flow of 5 amps, then applying the formula will yield:

- P (watts) = A × V
- P (watts) = 5 A × 12 V
- P (watts) = 60

Applied Science

AS-71: Resistance: The technician can demonstrate an understanding of the relationship of resistance to heat, voltage drop, and circuit parameters.

The electrical resistance of a component is the component's opposition to the flow of electrical current. Any resistance in a circuit creates a voltage drop across the resistive component, be it a resistor, a bulb, an electric motor, or merely a wire. The amount of voltage drop is proportional to the amount of resistance. Resistors work by dissipating energy in the form of heat; therefore, for a given amount of current flowing in a circuit, the higher the resistance of a component, the more heat it will produce.

Let's look at an example using a rear window defogger. The defogger elements form a resistor in the circuit. In this case, heat is produced by the resistor grid to defog the window. The resistance of each grid line creates a steady voltage drop across the length of the grid line. Current flowing through the grid line then creates heat. Broken grid lines cannot conduct current flow, so there is no progressive voltage drop, nor heat generated in that grid line. Battery voltage (12 volts) is applied across the element when you switch on the defogger. If you were to measure the voltage at the center of one of the grid lines in the element, you should expect to see a voltage of approximately 6 volts. If the grid is burned open on the ground side, then the voltage would read about 12 volts, as there cannot be a voltage drop if there is no current flow. If the grid was burned open on the power side, the voltage would be near 0 volts because the current flow cannot get past the open circuit to the middle of the grid line. Moving the voltmeter toward the side with the break can pinpoint the exact location of the broken grid line so that repairs can be made.

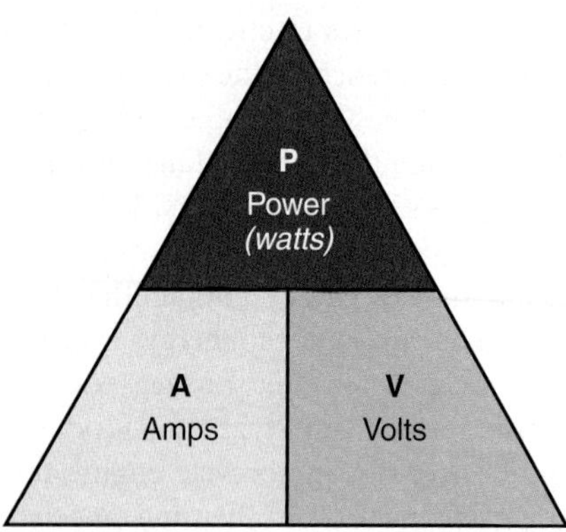

FIGURE 50-36 Watt's law triangle.

The power consumed by the circuit is 60 watts, and the bulb will carry a rating of 60 watts.

If a circuit is powered by a 12-volt battery with a load using 20 amps, using the power equation (P = A × V), we can determine that the load is using 240 watts of power. For reference purposes, 746 watts equal 1 horsepower. So 240W is about 1/3 horsepower.

It is also possible to simplify and transpose the power equation. If power = voltage × amps, then:

- Voltage equals power divided by amps: V = P ÷ A
- Amps equals power divided by voltage: A = P ÷ V

Applied Math

AM-48: Algebraic Expressions: The technician can use Ohm's law and the power law to determine circuit parameters that are out of tolerance.

The power law, or Watt's law, is a mathematical equation describing the relationship of voltage, current, and power in electrical circuits. Power may be defined as the amount of work done in a given period of time. Understanding this equation is useful in calculating characteristics within circuits. The power law is written as: Power (watts) = Voltage (volts) × Current (amps).

An unknown parameter can be calculated from any two known parameters. Like Ohm's law, the power law is commonly represented in a triangle diagram, as shown here:

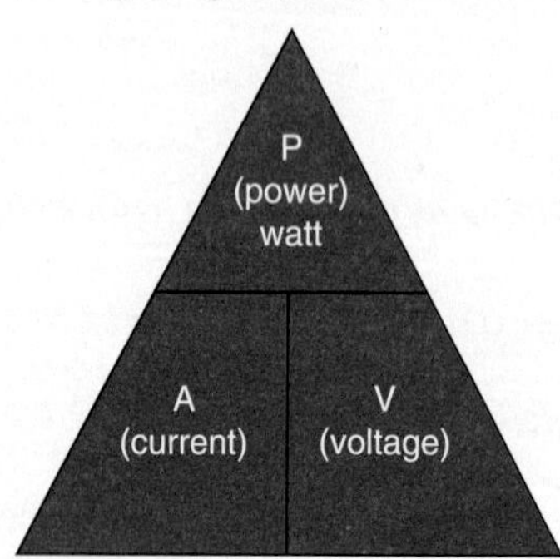

For example, we have an electrical circuit driving a power window motor in a passenger car. The vehicle operates on a 12-volt electrical system. The wattage of the window motor is 22 watts. To calculate the current flow in the circuit, we divide the power (or the motor's wattage) by the voltage, or P/V. In this case the amperage would be 22/12, or 1.8 amps.

Applied Math

AM-44: Formulas: Using formulas, the technician can predict the outcome(s) under different variations.

Ohm's law, a mathematical formula, can be applied to predict the outcome of changes made to electrical systems. Say we have a 12-volt automotive electrical circuit to drive a bulb in an interior light. A customer has complained that the interior light in his car is not bright enough and has requested that a higher wattage bulb be fitted. (*Caution*: We are only using this as an example. Additional wattage creates additional heat, which can cause the fuse to blow or, in an extreme case, a fire. It is recommended that you never modify a vehicle from the manufacturer's original condition.) The resistance of bulbs varies according to their wattage. Generally, the higher the wattage, the lower the resistance.

The original 5-watt bulb fitted to the vehicle had a resistance of 30 ohms. Through Ohm's law, we know that current flow equals voltage divided by resistance, or in this case 0.4 amp. The new 10-watt bulb to be fitted has half the resistance, 15 ohms. Current flow in the circuit is now calculated as 12 volts divided by 15 ohms, which is 0.8 amp, which provides twice the wattage.

You can see that increasing the amps from 5A to 20A increased the watts in the same proportion, from 60W to 240W. The same thing would happen if the voltage were increased.

This formula can be applied to any circuit where the voltage and current flow are known. However, if the values of voltage or current flow are not known, then Ohm's law can be used to determine the missing value.

Series Circuits

N50002

A series circuit is the simplest type of electrical circuit. In a series circuit, there is only one path for current to flow. All of the current flows to each component in turn. It also means that all the electrons flow at the same rate (amps) throughout all parts of the circuit. Amperage is equal everywhere within a series circuit.

In a series circuit, if there is more than one resistance in the circuit, those resistances are connected one after the other; thus the resistances add up in series. The total resistance in a series circuit is the sum of all of the individual resistances. For example, imagine a circuit with three resistors in series, each having 4 ohms of resistance. Total resistance of the three 4-ohm resistors is 4 + 4 + 4 =12 ohms, as resistance adds up in a series circuit.

However, the voltage drops from a potential difference of 12 volts as it leaves the battery to virtually no voltage at all as it returns to the battery. At each point in the circuit where current flows through a resistance, a drop in voltage occurs, which is called a **voltage drop**. Voltage drops are good when they occur inside of an intended load. They are bad when they occur where they are not wanted.

In **FIGURE 50-37**, after the first resistor, voltage has dropped from 12 to 8 volts. After the second, it is down to 4 volts. After the third, it is 0 volts. So all of the source voltage has been used up by the end of the circuit.

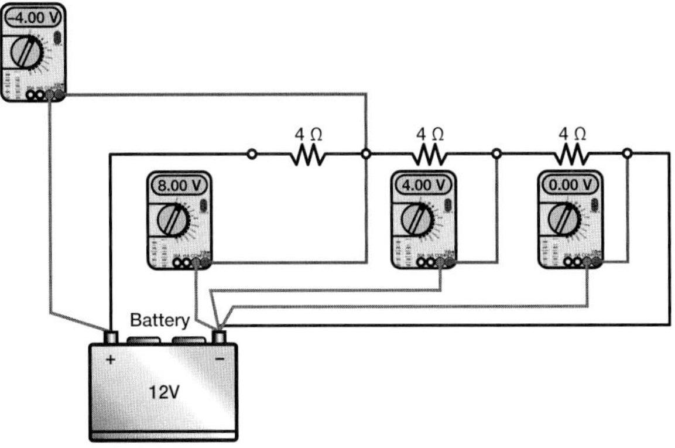

FIGURE 50-37 Voltage drop in a series circuit.

Ohm's law can be used in series circuits to calculate voltage, resistance, and current. Any one of these can be calculated as long as the values of the other two are known.

The series circuit laws listed here provide a summary of how electricity behaves in a series circuit, which is defined as a circuit with one or more loads, but only one path for current to flow:

- Current flow stays the same in a series circuit. Current flow is the same in all parts of the circuit. It doesn't add up or drop. The pressure pushing the amperage gets used up.
- Voltage drops (gets used up) as current goes through resistance(s) in series. The source voltage is equal to the sum of the individual voltage drops in the circuit (Kirchhoff's voltage law).
- Resistance adds up in series. Total circuit resistance is equal to the sum of the individual resistances; for example, $R_T = R_1 + R_2 + R_3$, and so on.

We can use those laws along with Ohm's law not only to help us understand the behavior of electricity but also to diagnose circuit faults. For example, in a circuit with three 2-ohm bulbs in series and a 12-volt battery, we can figure out what is happening electrically in the circuit (**FIGURE 50-38**). First, we have to calculate the total resistance in the circuit, which is 6 ohms. We know the circuit has 12 volts, so 12V (total voltage) ÷ 6Ω (total resistance) = 2A (total amps).

Because current flow stays the same in a series circuit, then each of the three bulbs have 2 amps flowing through them. And because each bulb has 2 ohms of resistance, we can calculate that the voltage used by each one (voltage drop) is 2A × 2 Ω = 4-volt drop across each bulb.

We can do it again with slightly different resistances. So this example has three 8-ohm bulbs in series and a 12-volt battery (**FIGURE 50-39**). Total resistance is 24 ohms. Total volts is 12V. So total amps are 0.5 amp.

Because current flow stays the same in a series circuit, each of the three bulbs have 0.5 amp flowing through it. And because each bulb has 8 ohms of resistance, then we can calculate the voltage used by each one (voltage drop) is 0.5A × 8 Ω = 4-volt drop across each bulb.

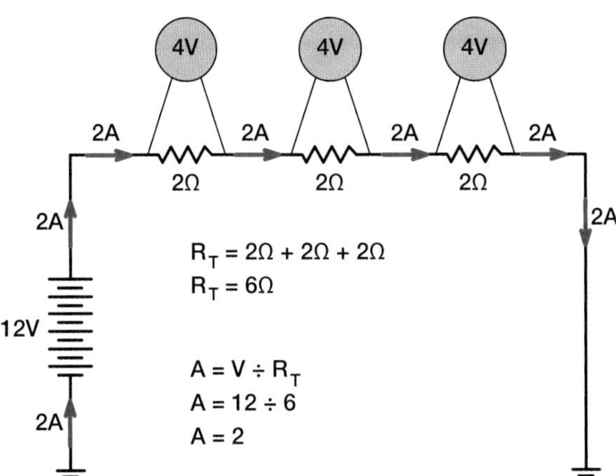

FIGURE 50-38 Using the series circuit laws and Ohm's law to understand how electricity behaves in this circuit.

Parallel Circuits

In a series circuit, components are connected like links in a chain. If any link opens, current to all of the components stops flowing. In a parallel circuit, there is more than one path for current to flow (**FIGURE 50-40**). If any connection or component fails in one branch of a parallel circuit, current continues to flow normally through the remaining branches. This is one reason why parallel circuits are used in automotive headlight and taillight systems. If one lamp fails, current continues to flow

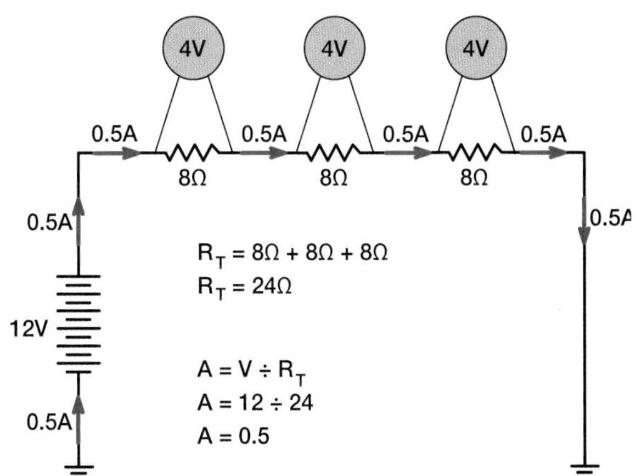

FIGURE 50-39 Notice that increasing the resistance of each of the three bulbs decreased the current flow, but didn't change the voltage drop.

Applied Math

AM-14: Mentally: The technician can determine the proper mathematical operation (addition, subtraction, multiplication, or division) and mentally arrive at the solution.

Determining the operating parameters of electrical circuits can require a selection of different mathematical operations. Some examples may include adding the resistance of individual components in a series circuit to determine the total resistance of the circuit, dividing the voltage by the resistance to determine the current flow in a parallel circuit, or multiplying the amperage of a circuit by its resistance to determine the voltage a device is using.

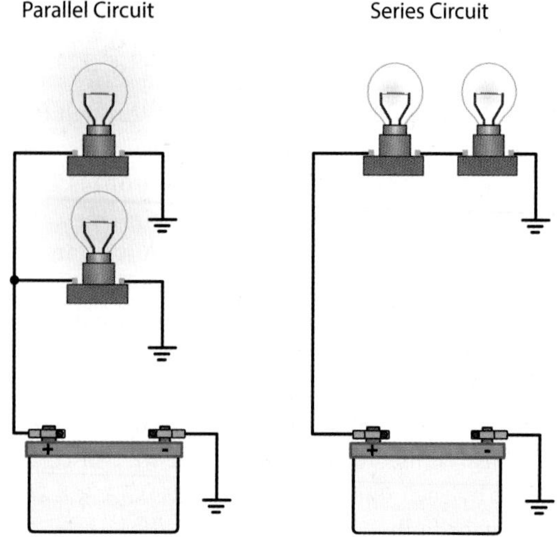

FIGURE 50-40 Typical parallel circuit compared to a series circuit.

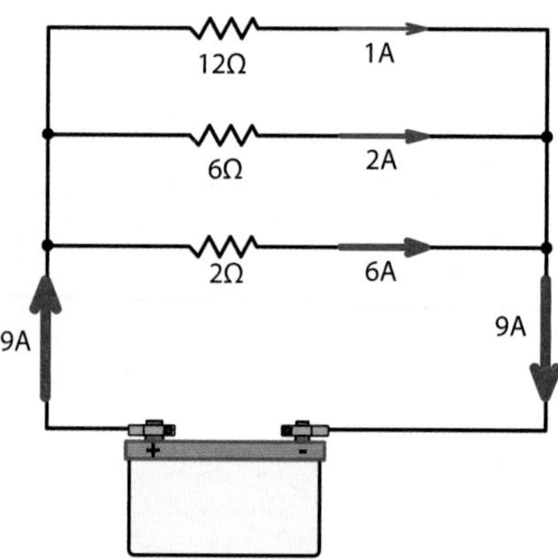

FIGURE 50-41 Parallel circuit with unequal resistances.

through the other lamps in parallel. In a series circuit, all would go out, which could be disastrous.

Although electricity always behaves according to the laws of physics, when it operates in a parallel circuit, there are some additional laws you need to know. The parallel circuit laws listed here should be applied when working with parallel circuits:

- The voltage is the same at the input of all branches of a parallel circuit. The voltage drop is the same across branches in parallel.
- The total current in a parallel circuit equals the sum of the current flowing in each branch of the circuit. Current flow adds up in parallel.
- The total resistance of a parallel circuit decreases as more branches are added. Total parallel circuit resistance will always be less than the branch with the lowest resistance.

Let's look at those laws a bit more closely. A feature of a properly working parallel circuit is that the voltage drop across each branch is the same as the other branches. No matter how many branches are added, or removed, as long as they are in parallel, the voltage across them will be the same as across each of the other parallel branches. Another feature of a parallel circuit is that the current flowing in each branch is determined by the resistance of that branch along with the voltage used by that branch.

In a parallel circuit where the resistors in each branch are the same, the current flowing in each branch is also the same. The sum of their individual current flows is equal to the total current flowing in the entire parallel circuit. When the resistances are not equal, the current divides in accordance with the resistance of each branch, but the total current flow is still the sum of the currents flowing in each branch (**FIGURE 50-41**).

Parallel Circuit Resistance

Resistance in a parallel circuit is not as easily calculated as it is in a series circuit, because as branches are added, another path for current to flow to ground is added. Adding extra paths reduces the circuit's total resistance to current flow. For example, if you have a 12-volt parallel circuit with three branches, each branch having a 12-ohm resistor that allows 1 amp of current flow, the total current flow would be 3 amps. If you then add a fourth parallel branch with a 12-ohm resistor to the circuit, the current increases from 3 amps to 4 amps. This is because in a parallel circuit adding more branches provides more pathways for current to flow, which decreases the overall resistance to current flow; thus, current flow increases. This is like having a freeway with two lanes and bumper-to-bumper traffic. If you add a third lane, the resistance to traffic flow decreases, and cars can move more easily. Another visual example: You have a bucket full of water with three holes in the bottom. If you add another hole, the total flow out of the bucket would speed up, which means resistance to water flow decreases. To calculate resistance, use one of the two formulas in **FIGURE 50-42**, using the ohm value of each resistor in the place of R_1, R_2, etc.

For 2 resistor in parallel

$$R_T = \frac{R1 \times R2}{R1 + R2}$$

For more than 2 resistor

$$R_T = \frac{1}{\frac{1}{R1} + \frac{1}{R2} + \frac{1}{R3}}$$

For example if R1 = 2Ω and R2 = 4Ω and R3 = 6Ω

$$R_T = \frac{2\Omega \times 4\Omega}{2\Omega + 4\Omega}$$

$$R_T = \frac{8\Omega}{6\Omega}$$

$$R_T = 1.33\Omega$$

$$R_T = \frac{1}{\frac{1}{R1} + \frac{1}{R2} + \frac{1}{R3}}$$

$$R_T = \frac{1}{\frac{1}{2\Omega} + \frac{1}{4\Omega} + \frac{1}{6\Omega}}$$

$$R_T = \frac{1}{0.5\Omega + 0.25\Omega + 0.16\Omega}$$

$$R_T = \frac{1}{0.91\Omega}$$

$$R_T = 1.09\Omega$$

FIGURE 50-42 Calculating total resistance in parallel circuits.

Applied Science

AS-61: Parallel/Series Circuits: The technician can explain current flow and voltage in series and parallel circuits.

The terms "series" and "parallel" describe different ways in which components in a circuit can be connected. In a series circuit, all components are connected along a single path. Electricity can flow only one way, so the flow of electrons, or the current flow, within the circuit is the same at all points. The voltage in the system changes at different points due to voltage drops at various resistors. In a parallel circuit, components are connected so that the circuit divides into two or more paths before recombining. All components are connected to the same voltage supply, so the same voltage is applied to all components. Current flow in different branches of a parallel circuit changes depending on the resistance in each branch; therefore, different branches of the circuit may experience differing amperage if their resistance values are different.

Series-Parallel Circuits

When electrical components are wired together one after another so that there is only one path for current flow, they are said to be wired "in series." When they are wired together side by side, so there is more than one path for current to flow, they are said to be wired "in parallel." A **series-parallel circuit** is made of both a series circuit and a parallel circuit. The series circuit portion can be before or after the parallel portion of the circuit. Series-parallel circuits can be analyzed using the same electrical laws that apply to separate series or parallel circuits; you just have to apply the series laws to the series portion of the circuit and the parallel laws to the parallel portion.

A circuit for dash lights is an example of a series-parallel circuit in an automotive application. In this situation, a variable resistor is connected in series with a number of dash lights, which are connected in parallel to each other (**FIGURE 50-43**). When the variable resistor is turned, the resistance changes. As the resistance increases, the voltage drop across the potentiometer increases and the current flow decreases. This lowers the voltage and current flow to each of the dash lights, making the dash lights dimmer. When the variable resistor is turned such that its resistance is lowered, the voltage drop across it is reduced, and the current flow in the circuit increases, making the dash lights brighter. Because the dash lights are in parallel, they each receive the same amount of voltage, and the total circuit current flow is divided equally between them, making them equally bright as one another.

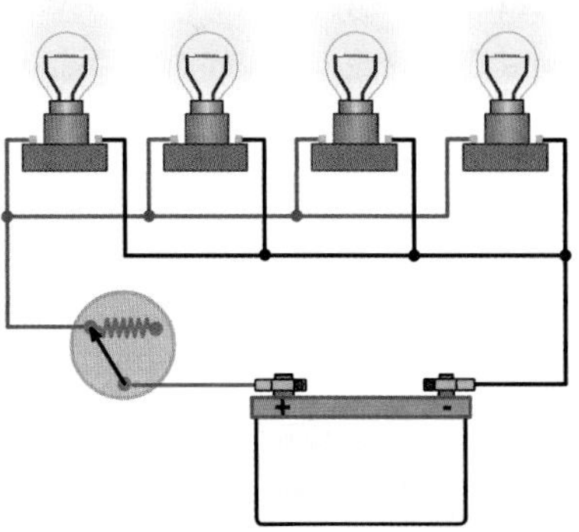

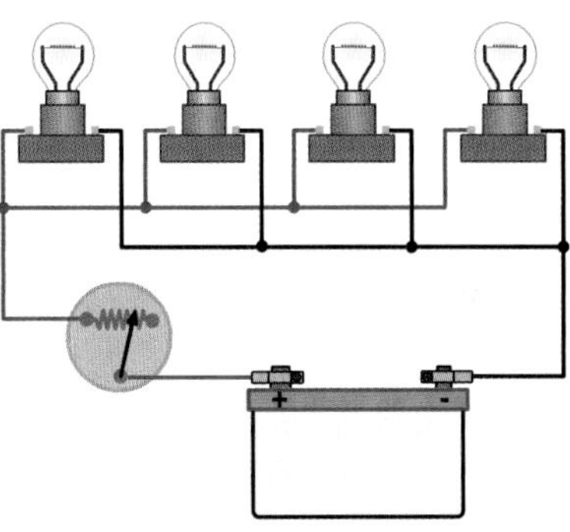

FIGURE 50-43 A dash light circuit is a good example of a series-parallel circuit.

Kirchhoff's Current Law

Kirchhoff's current law describes a fundamental electrical principle and is used by technicians in understanding how all electrical circuits work. Many technicians understand the principle behind the law without necessarily associating it with Kirchhoff. Simply stated, the law is that current entering any junction is equal to the sum of current flowing out of the junction. For example, in a junction of three conductors, current is flowing in at 10 amps from conductor A and out by the other two conductors, B and C (**FIGURE 50-44**). According to Kirchhoff's current law, the sum of the current in conductors B and C equals the current flowing into the junction at conductor A, or 10 amps.

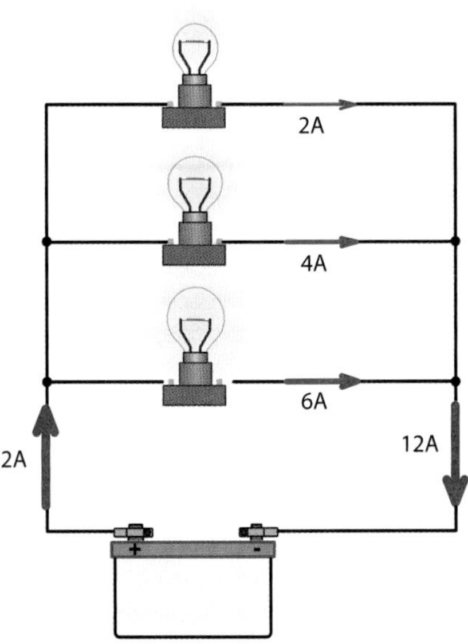

FIGURE 50-44 Kirchhoff's current law: Current entering a junction is equal to the sum of current flowing out of the junction.

Using Electrical Concepts to Solve Problems

If you understand the preceding concepts, then you are well on your way to diagnosing electrical problems. Here is a question: If the amps in a circuit are lower than they should be, what are the two possible causes? If you said that either the voltage is too low or the resistance is too high, you would be correct. If the available voltage to the device is low (charging system not working), then the full amount of current will not flow. In the same way, if the resistance is too high (dirty, loose connection), then current will also be too low.

Second question: If the amps in the circuit are higher than they should be, what are the two possible causes? If you said that the voltage is too high or the resistance is too low, you would be correct. If the voltage is too high (charging system is putting out too much voltage), then the higher voltage pushes more current flow through the resistor. If the resistance is too low, (short circuit) then there will be too much current flow.

As you can see, if you keep your eye on the current flow, you will have a good clue as to what type of fault you are looking for. In many cases, that doesn't require using a meter. You can see it with your eye. For example, a light is dim or even off. In this case, you know that the current flow is too low because work isn't being performed like it should be. And that means there is either not enough available voltage or too much resistance. Quickly testing the available voltage at the device will indicate whether you need to perform a voltage drop test on both sides of the circuit. And a resistance measurement of the electrical device will indicate if a high resistance or open condition is present in the component. Knowing the specifications that go along with these steps will enable you to repair more than half of the electrical problems on a vehicle.

Applied Math

AM-46: Equivalent Form: The technician can write or rewrite an algebraic equation to solve for any unknown variables.

Ohm's law is commonly used to calculate the value of an unknown variable from known values in electrical circuits. The equation is commonly stated as voltage (V) equals current flow (A) multiplied by resistance (R), or V = AR. To calculate current flow or resistance, instead of voltage, the equation must be rewritten. To apply the rules of basic algebra, anything we do to one side of the equation we must also do to the other. To calculate current flow, we need the A by itself on one side of the equation, so we divide both sides by R to arrive at A = V/R. To calculate resistance, the R must stand alone, so we divide both sides of the equation by A to arrive at R = V/A. Ohm's law is commonly represented in a circle format, making the correct algebraic equation very simple to find.

Applied Science

AS-53: Electricity: The technician can demonstrate an understanding of and explain the properties of electricity as they relate to lighting, engine management, and other electrical systems in the vehicle.

Most automotive electrical systems use 12-volt DC electricity supplied from the vehicle's battery to power items such as lightbulbs, electric motors, and heater elements. Some circuits may use switching with varying degrees of resistance to control the output from the circuit's load; examples may include interior blower fan motors that run at varying speeds and the vehicle's fuel gauge, which uses a variable resistor to determine the position of the gauge.

Engine and power train management systems are also powered from the vehicle's 12-volt battery but use various other voltages to monitor operating parameters via sensors. Some sensors operate using a reference voltage, generally 5 volts, which is modified to determine operating parameters. Examples could be a digital crankshaft position/speed sensor or a digital mass airflow meter. The reference voltage is effectively switched "on" or "off" by the digital sensor, producing a square-wave output with a frequency that changes according to the engine speed or airflow rate.

Some sensors generate their own voltage; examples include zirconia oxygen sensors; analog crankshaft or camshaft position sensors, which generate AC voltage; and piezoelectric knock sensors. Voltages generated from these sensors are monitored by the electronic control unit (ECU) to determine engine operating parameters.

Some computer-controlled devices can simply be turned on and off. A cooling fan may be turned on when the coolant reaches a temperature of 219°F (104°C), for example, and turned back off when the coolant temperature falls below 201°F (94°C) for example. Pulse-width modulation provides a means for a computer to control a device with variable operating parameters. Fuel injectors, for example, need to be open for differing amounts of time, depending on the fueling requirements of the engine. An ECU can control injector opening time by sending a pulsed signal with shorter or longer duration.

Wrap-Up

Ready for Review

- Increasingly, vehicle technicians must have an understanding of the electrical principles involved in vehicle system operation.
- Atoms contain negatively charged electrons moving around a nucleus in which there are positively charged protons and neutrons with no charge.
- Atoms with excess electrons have a negative charge and create a negative ion; those deficient in electrons have a positive charge and create a positive ion.
- Free electrons can move from one atom to another if an electrical potential is applied.
- Materials with many free electrons are good electrical conductors.
- Copper is the most common conductor.
- Insulators are materials that do not conduct current easily; an example is plastic.
- Semiconductor refers to a material that conducts electricity more easily than an insulator, but not as well as a conductor.
- Free electrons require a pathway or circuit, and a force to act upon them, such as a battery.
- Like charges repel, and unlike charges attract.
- The attraction of free electrons that creates a force is called voltage.
- The four factors that determine electrical resistance level are the type of material and the length, size, and temperature of the conductor.
- Electrical resistance refers to the degree to which a material opposes the passage of an electrical current.
- Resistance is measured in ohms and is constant in an object unless the temperature changes.
- A semiconductor's ability to conduct electricity depends on negative electrons and holes.
- The number of charge carriers in a semiconductor can be changed by adding small quantities of impurities (doping).
- The PN junction of a semiconductor is located at the depletion layer.
- Semiconductors can prevent or allow current flow, depending on connection to a current source.
- Semiconductor materials include silicon, germanium, gallium-arsenide, and silicon carbide.
- Electrical circuits contain a power supply, a current flow on/off switch, a functional component, a conductive pathway, and a protection device (e.g., a fuse).
- Voltage is the electrical pressure difference between two points in an electrical circuit.
- The ampere (amp) is the unit used to describe how much current is flowing at a given point within a circuit when the functional component is operational.
- The ohm is the unit used to describe electrical resistance in a circuit or component.
- Direct current (DC) flows in one direction only; alternating current (AC) continuously changes its direction of flow.
- Electrical components can work only on AC or DC, but not both.
- Circuits may be described in terms of continuity, open, short, and high resistance.
- Electrostatic energy occurs when two insulators are rubbed together, with one losing electrons to become positively charged and the other gaining electrons to become negatively charged.
- Thermoelectric energy is produced by joining and heating two different metals.
- Electrochemical energy is produced via electrolysis, which is the immersion of two dissimilar metals in a conducting liquid to break down chemicals into ions.
- Photovoltaic energy is produced via solar energy cells.
- Piezoelectric energy is produced when certain crystals are subjected to mechanical stress.
- Electromagnetic induction is created when a conductor cuts across a magnetic field.
- The effects of electricity include light (LED bulbs), heat (headlights), chemical reactions (lead-acid battery), and magnetism (electric motors).
- Electromagnets are used in relays, solenoids, and motors, while electromagnetic induction is used in ignition coils and transformers.
- Relays are used to control circuits that carry high current flow; they can be normally open (NO) or normally closed (NC).
- Solenoids operate similarly to a relay, but create lateral movement rather than closing a circuit.
- Electric motors rely on magnetic fields to create rotary movement.
- Ohm's law states that the total resistance of a circuit always equals the voltage divided by the amperage.
- The term "work" refers to transforming one form of energy into another.
- Power refers to the rate at which work is done, or the rate of transforming energy.
- The watt is the unit of power.
- Kirchhoff's current law states that electrical current entering any junction is equal to the sum of the current flowing out of the junction.
- In a series circuit, current can flow in only one path, and all electrons flow at the same rate.
- Voltage drop refers to the pressure lost by driving the current through a resistor.
- The electrical properties of a series circuit are as follows: current flow is the same in all parts of the circuit; the applied voltage is equal to the sum of the individual voltage drops; and total circuit resistance is equal to the sum of the individual resistances.

- All components in a parallel circuit are directly connected to the voltage supply; hence the voltage across each component is equal to battery voltage.
- Parallel circuit laws are as follows: the voltage across all branches of a parallel circuit are the same; the total current equals the sum of the current flowing in each branch; the amount of current in each branch is inversely proportional to the resistance of the branch; and the total resistance of a parallel circuit can be calculated as:

$$R_T = \frac{(R_1 \times R_2)}{(R_1 + R_2).}$$

Or

$$RT = \frac{1}{\frac{1}{R_1} + \frac{1}{R_2} + \frac{1}{R_3}}$$

- Series-parallel circuits contain both a series circuit and a parallel circuit.

Key Terms

alternating current (AC) A type of current flow that flows back and forth.

ammeter A device used to measure current flow.

amp An abbreviation for amperes, the unit for current measurement.

conductor A material that allows electricity to flow through it easily. It is made up of atoms with one to three valance ring electrons.

continuity A conductive path between two points.

conventional theory The theory that electrons flow from positive to negative.

current flow The flow of electrons, typically within a circuit or component.

direct current (DC) Movement of current that flows in one direction only.

electrical power A measurement of the rate at which electricity is consumed or created.

electrical resistance A material's property that slows down the flow of electrical current.

electrolysis A method of using electrical current to create a chemical reaction.

electrolyte A mixture of water and acid that contains free ions that make it electrically conductive.

electromagnet A conductor wound in a coil that produces a magnetic field when current flows through it.

electromagnetic induction The production of an electrical current in a conductor when it moves through a magnetic field or a magnetic field moves past it.

electromotive force An electrical pressure or voltage.

electron theory The theory that electrons, being negatively charged, repel other electrons and are attracted to positively charged objects; thus electrons flow from negative to positive.

energy The ability to do work.

free electron An electron located on the outer ring, called the valence ring, that is only loosely held by the nucleus and that is free to move from one atom to another when an electrical potential (pressure) is applied.

ground The return path for electrical current in a vehicle chassis, other metal of the vehicle, or dedicated wire.

hertz The unit for electrical frequency measurement.

high resistance The resistance of a component or circuit relative to a low resistance. It can also refer to a faulty circuit where a section or component has excess unwanted resistance.

hole theory The theory that as electrons flow from negative to positive, holes flow from positive to negative.

hot junction The heating point of a thermocouple.

insulator A material that has properties that prevent the easy flow of electricity. These materials are made up of atoms with five to eight electrons in the valance ring.

ions An atom that has fewer electrons than protons (positive) or that has more electrons than protons (negative).

Kirchhoff's current law An electrical law stating that the sum of current flowing into a junction is the same as current flowing out of the junction.

light-emitting diode (LED) A diode that emits light when current flows through it.

magnetism The force that attracts or repels magnetic charges; or the property of a material to respond to a magnetic field.

multimeter A test instrument used to measure volts, ohms, and amps. A digital multimeter may also be called a digital volt-ohmmeter (DVOM).

ohm The unit for measuring electrical resistance.

Ohm's law A law that defines the relationship between current, resistance, and voltage.

open A term used to describe a circuit that does not have a complete path for current to flow.

photovoltaic (PV) effect The conversion of sunlight into electricity.

piezoelectric A type of electricity in which a material such as a quartz crystal produces voltage when mechanical pressure distorts it.

plate A flat, thin square or rectangular sheet of material often used to describe the arrangement of positive and negative electrodes in batteries.

polarity The state of charge, positive or negative.

power The rate at which work is done; electrical power is measured in watts.

relay An electromechanical switching device whereby the magnetism from a coil winding acts on a lever that switches a set of contacts.

resistor A component designed to have a fixed resistance.

semiconductor A material used to make microchips, transistors, and diodes.

series-parallel circuit A circuit that has both a series and a parallel circuit combined into one circuit.

short Also called a short circuit, the flow of current along an unintended route.

sine wave A mathematical function that describes a repetitive waveform such as an alternating current signal.

thermocouple A temperature-sensing component that consists of two dissimilar metals that produce voltage proportional to temperature.

transformer action The transfer of electrical energy from one coil to another through induction in a transformer.

volt The unit used to measure potential difference or electrical pressure.

voltage The electrical pressure that causes current to flow in a circuit.

voltage drop The amount of potential difference between two points in a circuit.

watt The unit for measuring electrical power.

work The process by which one type of energy is transformed into another type of energy.

Review Questions

1. All of the following statements describing the basic principles of electricity are true *except*:
   a. An atom with less/deficiency of electrons than protons has an overall positive charge and is called a positive ion.
   b. An atom with more electrons than protons has an overall negative charge and is called a negative ion.
   c. A positive ion exerts a repelling force on the extra electron.
   d. The flow of electrons from atom to atom is called current flow.
2. Which of the following materials is typically used in the construction of semiconductors?
   a. Silicon
   b. Copper
   c. Plastic
   d. Ceramic
3. Electromotive force can be best described as the:
   a. force of electrons repelling each other.
   b. repelling force of the negative terminal.
   c. attracting force of the positive terminal.
   d. force of attraction/repelling which drives the electrons to move along.
4. The degree to which a material opposes the passage of electrical current through it is called:
   a. voltage.
   b. resistance.
   c. discharge.
   d. amperage.
5. Which of the following is a measure of the number of electrons flowing past a given point in 1 second?
   a. Volt
   b. Watt
   c. Ohm
   d. Amp
6. The device which is used to measure the amount of resistance in a component or a circuit is a(n):
   a. ohmmeter.
   b. ammeter.
   c. voltmeter.
   d. refractometer.
7. The force with which the positive terminal pulls the free electrons toward it in a circuit is measured in:
   a. volts.
   b. amperage.
   c. current.
   d. ohms.
8. All of the following statements describing an electrical circuit are true *except*:
   a. An open circuit does not have continuity.
   b. Current flows through the fuse into the circuit wires.
   c. Electrical circuits consist of a power source, a fuse, a switch, a component that performs work, and wires connecting them all together.
   d. When the switch is moved to closed position, the current path is broken and current flow stops.
9. Which side of a circuit starts at the output of a load and ends at the negative battery terminal?
   a. Positive side
   b. Ground side
   c. Supply side
   d. Power side
10. When two dissimilar metals are immersed in an acidic liquid, the breakdown of chemicals into charged particles results in a flow of electricity. This principle is used in a(n):
    a. HVAC system.
    b. battery.
    c. transistor.
    d. LED.

ASE Technician A/Technician B Style Questions

1. Tech A says that if the specified fuse keeps blowing, it is generally okay to replace it with a larger fuse. Tech B says that a fusible link is one type of circuit protection device. Who is correct?
   a. Tech A
   b. Tech B
   c. Both A and B
   d. Neither A nor B
2. Tech A says that the movement of electrons in a circuit is called current flow. Tech B says that the movement of electrons in a circuit is measured in amps. Who is correct?
   a. Tech A
   b. Tech B
   c. Both A and B
   d. Neither A nor B

3. Tech A says that electromotive force is also known as voltage. Tech B says that when electrons flow in one direction only, this is DC. Who is correct?
   a. Tech A
   b. Tech B
   c. Both A and B
   d. Neither A nor B
4. Tech A says that an unintended resistance results in a voltage drop. Tech B says that one of the factors that determines the level of electrical resistance is the weight of the conductor. Who is correct?
   a. Tech A
   b. Tech B
   c. Both A and B
   d. Neither A nor B
5. Tech A says that if resistance stays the same and voltage goes up, then amperage goes down. Tech B says that if voltage stays the same and resistance goes down, then amperage goes down. Who is correct?
   a. Tech A
   b. Tech B
   c. Both A and B
   d. Neither A nor B
6. Tech A says that a series circuit has only one path for current to flow. Tech B says that a parallel circuit current flows through one resistor before getting to the next resistor. Who is correct?
   a. Tech A
   b. Tech B
   c. Both A and B
   d. Neither A nor B
7. Two technicians are discussing a series circuit with four resistors of various resistances. Tech A says that current flow is different in each resistor. Tech B says that current flow is the same in each resistor. Who is correct?
   a. Tech A
   b. Tech B
   c. Both A and B
   d. Neither A nor B
8. Tech A says that an ohm is a unit of measurement of resistance. Tech B says that high resistance creates heat at the point of resistance in the circuit. Who is correct?
   a. Tech A
   b. Tech B
   c. Both A and B
   d. Neither A nor B
9. Tech A says that in a series circuit with two resistors of 120 ohms each, the total circuit resistance is 240 ohms. Tech B says that in a parallel circuit with two resistors of 120 ohms each, the total circuit resistance is 240 ohms. Who is correct?
   a. Tech A
   b. Tech B
   c. Both A and B
   d. Neither A nor B
10. Tech A says that if a 12-volt light has 10 amps flowing through it, it is using 22 watts of electricity. Tech B says that watts are units of electrical power. Who is correct?
   a. Tech A
   b. Tech B
   c. Both A and B
   d. Neither A nor B

CHAPTER 51

Electrical Components and Wiring

NATEF Tasks

- **N51001** Use wiring diagrams during the diagnosis (troubleshooting) of electrical/electronic circuit problems. (AST/MAST)
- **N51002** Use wiring diagrams to trace electrical/electronic circuits. (MLR)
- **N51003** Inspect, test, repair, and/or replace components, connectors, terminals, harnesses, and wiring in electrical/electronic systems (including solder repairs); determine needed action. (MLR/AST/MAST)
- **N51004** Repair data bus wiring harness. (MAST)

Knowledge Objectives

After reading this chapter, you will be able to:

- **K51001** Describe the function and purpose of electronic components.
- **K51002** Describe wires and wire harnesses.
- **K51003** Repair wiring harness.

Skills Objectives

There are no Skills Objectives for this chapter.

▶ Introduction

Electrical components such as switches, fuses, circuit breakers, resistors, capacitors, and relays are all used in a circuit to modify or manage the flow of current. Each component performs a specific task within the circuit and will be connected into its particular circuit. To ensure the components' correct operation, terminals are often numbered or marked with the connections and referenced in wiring diagrams or schematics.

Some components are polarity sensitive, meaning they must be connected into the circuit with the correct polarity to each lead. For example, some capacitors, and most semiconductor components, are polarity sensitive. In polarity-sensitive components, the polarity is marked on at least one connection to ensure that the user connects them into the circuit correctly.

Electrical components are connected together by wires in wire harnesses. These circuits carry electrical signals and commands that are used in virtually every system on the vehicle. Knowing how these components and circuits are configured and how they operate will help you diagnose and repair them.

▶ Electrical Components

Switches

A **switch** is an electrical device used to turn the current on and off in a circuit. When turned off, switches open the circuit, stopping current flow; when turned on, they close the circuit, allowing current to flow. For example, switch off—light goes out; switch on—light comes on. There are many different types and configurations of switches, including toggle switches, push-button switches, and specialty switches such as for turn signal indicators and windshield wipers (**FIGURE 51-1**). The most basic switches are two-terminal toggle switches that are simply on or off, such as a transmission overdrive switch. Momentary switches are spring loaded to one position (on or off) and can be pushed to the other position. A horn switch would be a good example of a momentary switch.

More complex switches have many terminals and contacts inside them to switch a number of circuits at the same time. The **turn signal switch** is an example of a more complex switch. It has three positions: center for off, moved down for the left turn signals, moved up for the right turn signals. **Circuit or schematic diagrams**, called wiring diagrams, typically show switches, their contacts, and the surrounding circuits so that the technician can identify how they operate in a circuit. The terminals on the switch are often numbered or lettered to indicate the correct way of connecting them into their circuits and to show the mating connectors in the wiring diagrams.

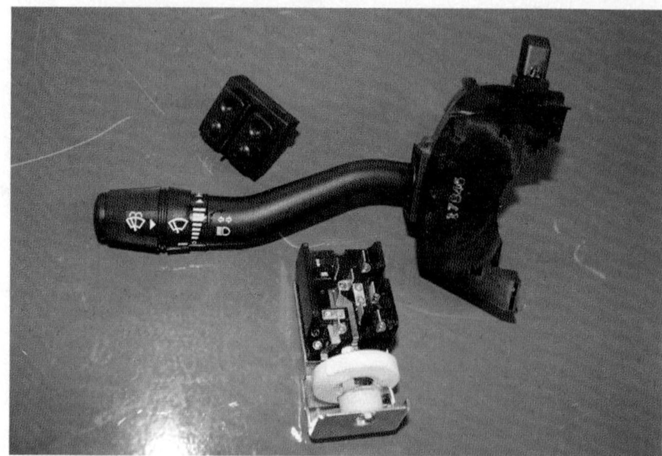

FIGURE 51-1 Typical automotive switches.

Circuit Protection Devices

Fuses and **circuit breakers** are designed to protect electrical circuits by opening the circuit if the current flow is excessive. The most common kinds of circuit protection devices are fuses, fusible links, circuit breakers, and positive temperature coefficient (PTC) thermistor protection devices. Fuses and circuit breakers are rated in amps, and their ratings are usually marked on them. Fusible links are typically rated by their wire size.

Usually, a fuse contains a metal strip that is designed to overheat and melt when subjected to a specified excessive level of current flow, breaking the circuit and stopping the excessive current flow from potentially damaging the wiring harness and more valuable components. Fuses come in a variety of configurations, from cylindrical glass cartridge fuses to plastic blade fuses (**FIGURE 51-2**). Blade fuses are the most commonly used on recent vehicles. They also come in a variety of amp ratings (**TABLE 51-1**). Blade fuses also come in a variety of physical sizes including micro, mini, regular, and maxi. Fuses are typically housed in fuse boxes located around the vehicle, typically under the hood and/or dash (**FIGURE 51-3**). Fuse boxes typically include relays and sometimes diodes as well.

You Are the Automotive Technician

You are preparing to be a guest speaker at the local automotive technician training school. Because you specialize in electrical diagnosis and repair, the instructor would like you to explain some of the things the students need to know to do your job. You need to prepare for the following questions.

1. What does a relay do, and how does it work?
2. What does a diode do, and how does it work?
3. How do you go about tracing wires on a wiring diagram to help you understand how the circuit operates?
4. How do you perform a solder repair on a broken wire in a wire harness?

TABLE 51-1 Blade Fuse Color Codes for Common Fuse Sizes

Color	Amp Rating
Tan	5 A
Brown	7.5 A
Red	10 A
Blue	15 A
Yellow	20 A
Clear	25 A
Green	30 A

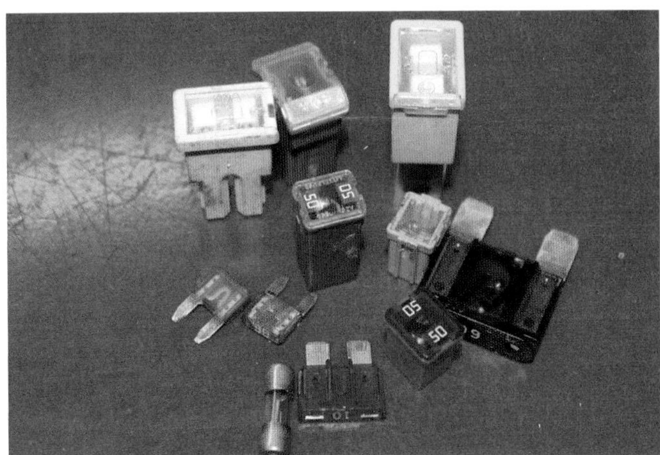

FIGURE 51-2 Various fuses.

FIGURE 51-3 Typical underhood fuse box.

A fusible link is made of a short length (usually 6" [15 cm] or less) of smaller diameter wire that has a lower melting point than standard wire, and insulation that is fire resistant. Fusible links are typically placed near the battery to protect the wiring harness between the battery and any fuse boxes. In most cases, they are used to carry higher current flows than fuses and typically feed power to one or more circuits (**FIGURE 51-4**). Fusible links are fairly durable and do not fail very often, unless there is

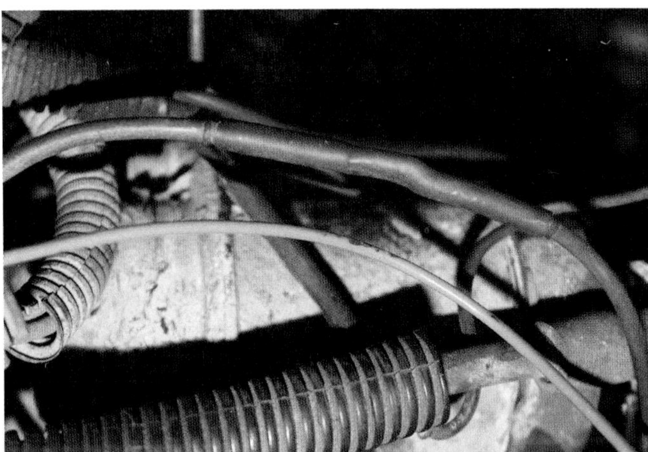

FIGURE 51-4 Fusible links are typically placed near the battery and carry the current needed to power an individual circuit or a range of circuits.

a substantial short circuit in the system or the fusible link wire is abused by excessive flexing or pulling on it. Some newer vehicles use maxi-fuses, which are large blade-type fuses, instead of fusible links.

Circuit breakers are different from fuses and fusible links in two ways. First, they are not destroyed by excess current. And second, they can be reset, either automatically or manually. In a circuit breaker, a bimetallic strip heats up and bends, opening a set of contacts and breaking the circuit when current flow becomes excessive (**FIGURE 51-5**). In most types, as the strip cools, it returns to its original shape. The contacts then close, completing the circuit once more. These are called self-resetting circuit breakers. Manual breakers must be reset by hand, which could involve flipping a lever or inserting a small rod to reset the bimetal spring once it cools down.

PTC thermistors are also used as circuit protection devices. They have very low resistance at room temperature, but increase in resistance as the temperature increases. If too much current starts to flow through a PTC, the small voltage drop creates heat in the PTC. The increased heat produces increased resistance, which further increases the voltage drop. This cycle continues quickly until the PTC reaches its maximum resistance. This heightened resistance effectively shuts off most of the current

Applied Science

AS-69: Fuse: The technician can explain the role of a fuse or fusible link as a protective device in an electrical or electronic circuit.

A fuse, or fusible link, is a form of overcurrent protection device. It consists of a conductive metal strip that melts when subjected to more than a specified level of current. When the fuse "blows," the circuit is broken, and no current flows in the remainder of the circuit. Fuses are installed for two reasons: They can prevent excessive current flow from damaging more-expensive components elsewhere in the circuit, and they can prevent overheating within wiring and components, which could potentially cause a fire.

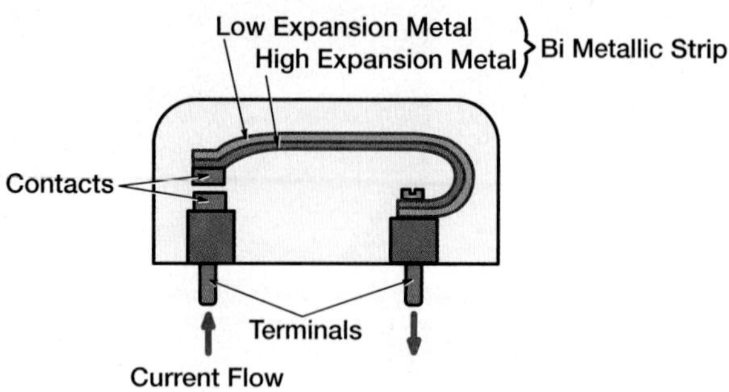

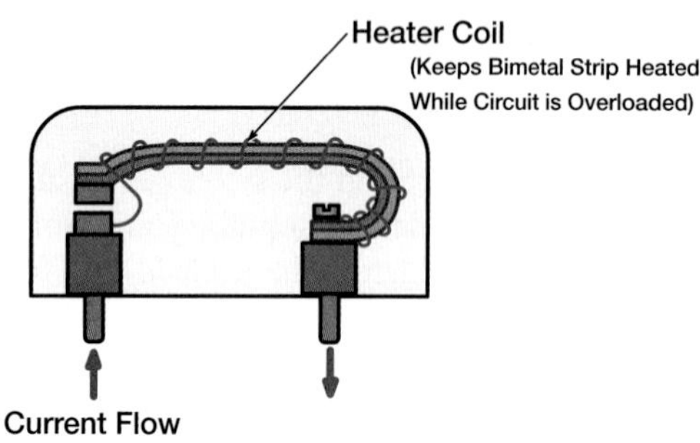

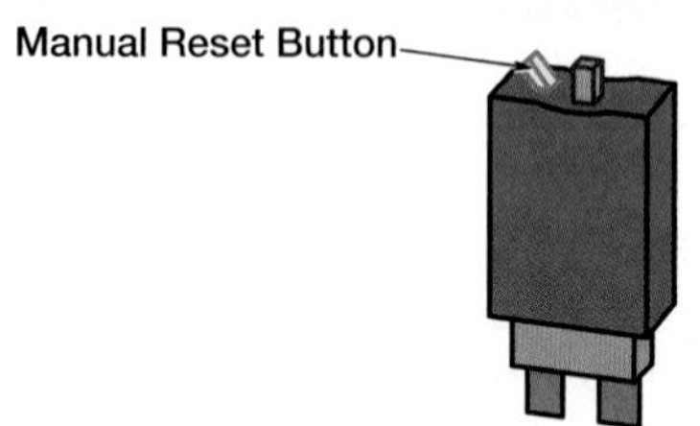

FIGURE 51-5 A circuit breaker opens the circuit if the current goes too high. It can usually be reset either automatically or manually.

flow to the protected device. PTCs generally reset once power is removed and they are allowed to cool. They are typically integrated into components such as power window motors and door locks.

Flasher Can/Control

The **flasher can** is the control mechanism for the turn signal lights on the vehicle. As the name suggests, it flashes or turns the turn signal lights on and off at a regular rate to indicate the driver's intent to change the vehicle's direction. Flasher cans are mechanical devices, whereas flasher controls are generally electronic devices (**FIGURE 51-6**). Both types perform the same job, but the electronic version is more reliable and consistent. Flasher cans operate like an automatically resetting circuit breaker, meaning they use a bimetallic strip to open and close the switch contacts. This opening and closing gives them a distinctive clicking sound that, along with the turn signal indicator lights on the dash, tells the driver when the turn signals are on so that he or she will remember turn them off when needed.

FIGURE 51-6 Flasher control and flasher can.

Electronic flasher controls control the on/off function electronically, which means they can be designed to operate over a wider range of current flow, which makes them ideal for trailer towing. In many cases, the hazard lamps are also operated by a flasher can or flasher control. In newer vehicles, the flasher operation is activated by the body control module, which sends signals over the communication network to the taillight module, which actually turns the turn signal on and off.

Relays

Relays are switches that are turned on and off by a small electrical current. They are ideal for using a small current to control a larger current. An example is the horn circuit in the steering wheel, which can turn a small current on and off, using a light-duty switch and wires. The small current activates the relay, which sends the larger current to the horn. Thus, larger wires are needed only up front, where they can be shorter, and smaller wires can be used to run up the steering column. Most electrical components found in a motor vehicle are controlled by relays. ECUs use relays to control components such as fuel pumps, headlights, and the cooling fan. All of these circuits carry large electrical loads.

The relay is made up of an electromagnet, a set of switch contacts, terminals, and the case (**FIGURE 51-7**). The electromagnet is a winding of fine metal-insulated wire wrapped around an iron core. Each end of the winding is connected to one of the terminals. The contacts usually consist of three contacts—two fixed and one movable. The movable contact is fixed to a spring-loaded armature blade and held against one of the fixed contacts, which is called the **normally closed (NC)** contact. When the relay coil is activated, the electromagnet pulls the movable armature blade contact away from the NC contact and against the **normally open (NO)** contact, sending power out to the controlled device. It also opens the NC contact points on that type of relay.

When the contact points are closed, current flows across the contact points and out to the electrical device to the rest of the circuit. As long as the small control current flows through

FIGURE 51-7 Inside a typical relay.

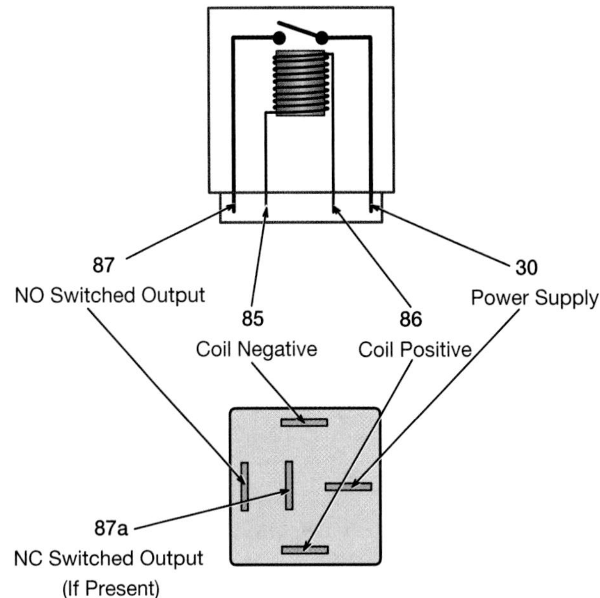

FIGURE 51-9 Relay schematic.

the relay windings, the much larger current for the load will flow through the relay's contact points. A **solid-state relay** acts like a mechanical relay but does not have any moving parts (**FIGURE 51-8**). This means that electronic relays do not use mechanical switches, but instead use transistorized circuitry to turn the circuit on and off. They also do not make any sound, unless they were specifically designed with components to do that.

Automotive relay terminals use one of two standard labeling systems: 85, 86, 30, 87a, and 87 or 1, 2, 3, 4, 5 (**FIGURE 51-9**).

- 85 or 1 = One end of the relay winding
- 86 or 2 = The other end of the relay winding
- 30 or 3 = Common-movable switch contact
- 87a or 4 = NC fixed contact to the common
- 87 or 5 = NO fixed contact to the common

When electromagnetic relays are de-energized, the collapse of the magnetic field induces a large voltage spike in the relay coil windings. If unchecked, this voltage can be transmitted back into the circuit, where it can damage electronic components. Some relays deal with this danger by placing either a suppression diode or a resistor in parallel with the winding

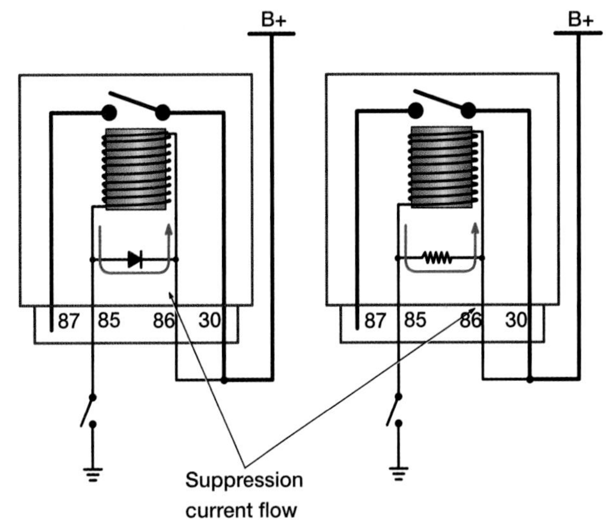

FIGURE 51-10 Spike-protected relays.

(**FIGURE 51-10**). Doing so reduces the voltage spike by shunting it from the output side of the coil back to the input side of the coil, where it is dissipated in the loop. This design prevents the voltage from spiking so high when the relay is de-energized. When replacing a relay that has spike protection, make sure to use a new relay that is specified for the application.

FIGURE 51-8 Inside a typical electronic relay.

Solenoids

A **solenoid** is an electromechanical device that coverts electrical energy into mechanical linear (back-and-forth) movement. Solenoids can be used to pull or push. In a simple solenoid, insulated wire is wound many times around a hollow tube. A sliding mild steel core is made to fit inside the tube. When the winding is energized, the resulting magnetic field attracts the mild steel, drawing it into the tube and producing linear motion. Solenoids can be very strong and produce a lot of mechanical energy. Solenoid design may incorporate return springs, multiple windings, electrical contacts, and mechanical connections.

Fuel injectors and starter motor solenoids are two of the many solenoid-type components used in a motor vehicle (**FIGURE 51-11**). The operation of a solenoid is similar to a relay, but where a relay uses a magnetic field to close an electrical circuit, a solenoid uses a magnetic field to create lateral movement and, in the case of a starter solenoid, to also close heavy electrical contacts. The metal core used by the electromagnet to strengthen the magnetic field is referred to as an **armature**. It is spring loaded so that it is positioned partially outside the electromagnetic coil and is free to move in and out. When the coil is energized, the magnetic field draws the armature into the center of the coil. If the armature is attached to a lever or plunger, it will be forced to move as well. Stopping current flow causes the electromagnet to de-energize and the spring pushes the armature out again.

Another device that uses a solenoid is a vehicle horn. When the armature is drawn in by the electromagnetic coil, it opens a set of electrical contacts so that current flow through the coil is stopped. This stoppage causes the armature to move out again, closing the contacts, drawing the armature back in. This process occurs at very high speeds. The vibration caused by the rapid movement is transferred to a diaphragm, and the familiar horn sound is produced (**FIGURE 51-12**). Refer to the Batteries, Starting, and Charging Systems chapter for a detailed description of the starter solenoid operation.

Motors

Although solenoids use magnetic fields to create lateral movement, electric motors use magnetic fields to create rotary movement. Motors consist of two main components: the armature and the field. The armature contains electromagnetic coils. The field contains either electromagnets or permanent magnets.

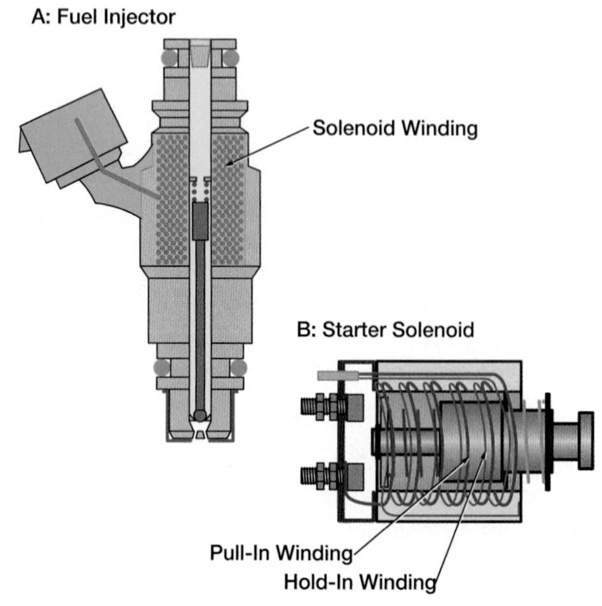

FIGURE 51-11 Examples of solenoids: **A.** Fuel injector. **B.** Starter motor solenoid.

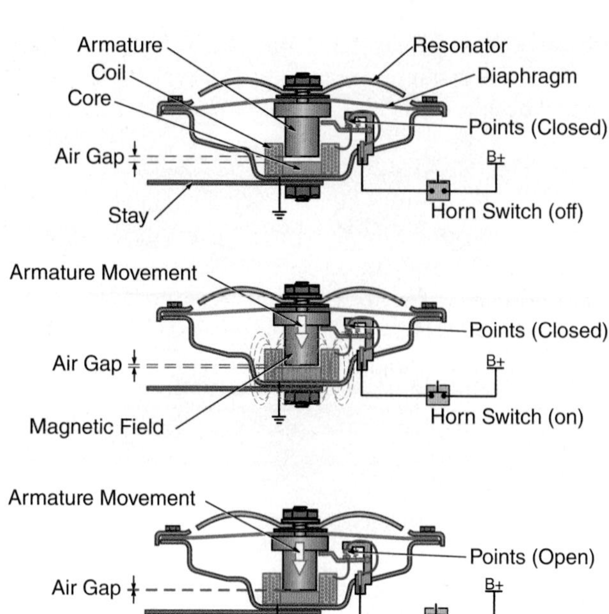

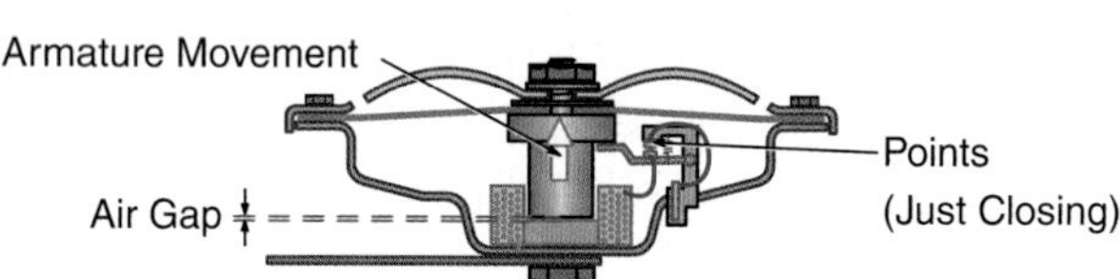

FIGURE 51-12 A horn is a type of solenoid designed to create noise as its output.

The interaction between the magnetic fields of the stationary field coils and the magnetic fields of the moveable armature coils causes the armature to rotate. Because the armature coils can be energized and create electromagnetic fields, those fields can be turned on and off so that they attract and repel the fields of the stationary magnetic fields. The design and orientation of the electromagnetic coils ensure that the armature continues to turn as it rotates through various positions. The armature is connected to the electrical supply by a set of carbon brushes that contact a **commutator**, which is a segmented component of the armature. The commutator and brushes act as switches to control the current flow through the windings of the armature (**FIGURE 51-13**).

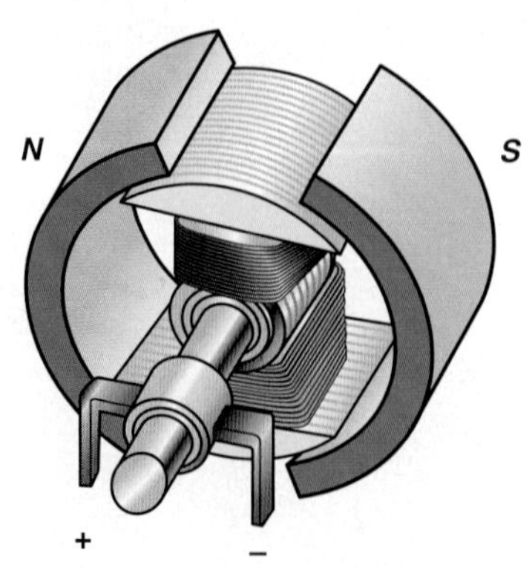

FIGURE 51-13 Simplified electric motor diagram.

The brushes allow the electrical connection to occur even when the armature is spinning. Refer to the Batteries, Starting, and Charging Systems chapter for more information on motors.

Ignition Coils and Transformers

Ignition coils and transformers both operate under the principles of electromagnetic induction. They use electromagnetism to produce electricity rather than mechanical movement. An ignition coil can be described as a **step-up transformer**. This is because the output can be 60,000 volts (60 kv) or more, which is much higher than the input, nominally 12 volts.

Two sets of coil windings are used. One coil winding, referred to as a **primary winding**, is wound around a second winding, the **secondary winding** (**FIGURE 51-14**). The primary coil typically has 200 to 300 turns of light-gauge wire, and the secondary has approximately 30,000 to 60,000 turns of very fine wire. When current is passed through the primary winding, the magnetic field builds surrounding both the windings. When the current is turned off, the magnetic field collapses with enough speed to induce high voltage in the primary winding (self-induction) and very high voltage in the secondary winding (mutual induction) by the rapidly moving (collapsing) magnetic field. Voltage is induced into each of the thousands of windings of the secondary coil. This voltage is strong enough to overcome the infinite resistance of the spark plug gap and push current across the gap, causing a spark with enough heat to ignite the air-fuel mixture in the cylinder.

The **transformer action** causes heat to be produced. In the past, the internal coils were immersed in cooling oil, allowing the heat to be conducted to the case. Modern ignition coils do not use oil. They are usually constructed using a heat-conducting hard resin and are cooled by their location on a heat sink or in a stream of air. The advent of computer controls has allowed the time that current flows through the primary windings to be minimized to reduce heat and electrical loads while still providing enough spark to ignite the air-fuel mixture.

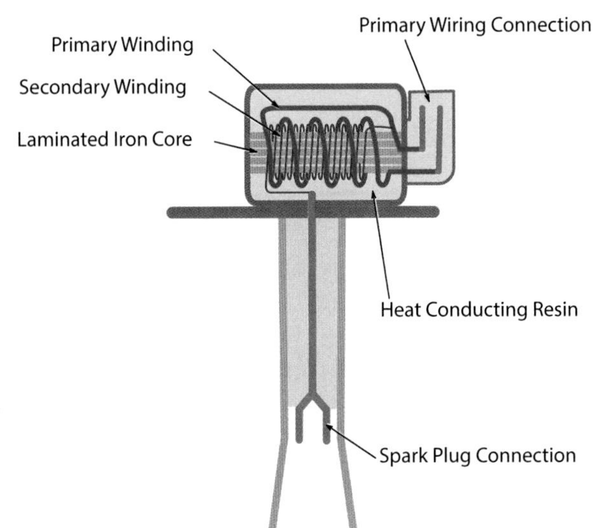

FIGURE 51-14 An ignition coil uses induction to step up the voltage from one coil to much higher voltage in the other coil.

Step-down transformers operate under the same operating principles. The only difference is that the secondary coil has fewer turns than the primary, providing a lower induced output. These transformers are used on power poles to lower the voltage to your house or school, and also used on low-voltage devices in your home plugged into 110-volt outlets, such as cell phone chargers.

> **TECHNICIAN TIP**
>
> You might expect a transformer to be a great way to boost the amount of electrical power that is transmitted, but it isn't. The amount of power after the transformation is relatively the same as before the transformation. For example, if we raise the voltage from 12 volts to 120 volts, the amperage will decrease from, say, 10 amps to 1 amp. Thus, the wattage stays the same:
>
> 12 volts × 10 amps = 120 watts
>
> 120 volts × 1 amp = 120 watts
>
> Also, remember that transformers are not 100% efficient, so some of the power is lost as heat.

Resistors

Resistors are electrical components that resist a current running through them. Putting a resistor in a circuit causes a drop in voltage across the resistor. It also reduces the amperage in the circuit. Resistors are commonly used to control the voltage and amperage that reach various components. It is important to remember that each electrical component has its own resistance. Most high-wattage resistors that can carry large amperage contain a coil of high-resistance wire wound around a ceramic form, to dissipate heat.

Fixed Resistors

Fixed resistors are generally cylinders with connecting metal leads projecting along the axis of the cylinder at each end. Most axial resistors are marked with a series of colored stripes to indicate their resistance and tolerance levels (**FIGURE 51-15**). Fixed resistors can be manufactured as very tiny devices without leads and can be built into integrated circuits with many other miniaturized components.

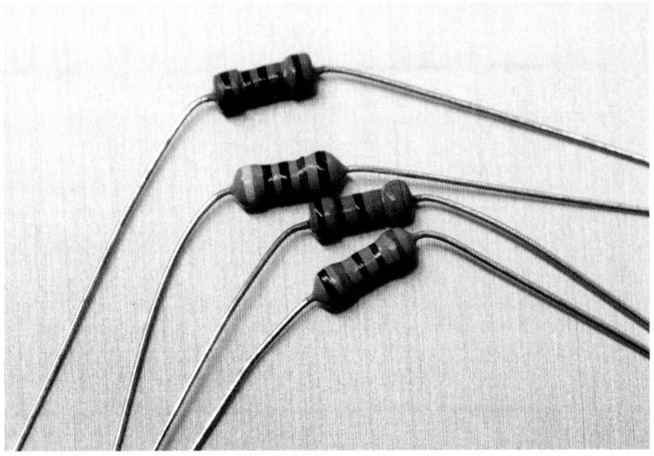

FIGURE 51-15 Typical resistors.

Variable Resistors

Resistors found on circuit boards are normally fixed in value. Some resistors found in the vehicle are variable. The value of **variable resistors** can be altered by movement of a slide or by temperature change. The three types of variable resistors are rheostats, potentiometers, and thermistors. Variable resistors can be linear, meaning their resistance value varies proportionally with movement or temperature change; or nonlinear, meaning the resistance change is not proportional with movement.

Rheostats

A **rheostat** is a mechanical variable resistor with two connections. It consists of a resistance wire wrapped in a loose coil connected to the supply at one end only. A moveable wiper is connected to the other circuit connection and is made to move over the wire manually (**FIGURE 51-16**). When the wiper is close to the beginning of the coil, the total resistance value is very small. As the wiper is positioned closer to the end of the coil, the resistance value increases. Rheostats are commonly used in dash light dimmer circuits and some fuel gauge sender units to alter the current flow and voltage in a circuit.

Potentiometers

Potentiometers are mechanical variable resistors with three connections, two fixed and one moveable. They act as voltage dividers and as such alter the voltage in a circuit. A resistance wire is wrapped between two fixed connections. One fixed connection is attached to the electrical supply, the other to a ground. The third moveable connection is moved across the coil by a wiper, in a similar fashion to a rheostat (**FIGURE 51-17**). The variable voltage output is taken from this point. Throttle position sensors are potentiometers.

Thermistors

As mentioned earlier, **thermistors** are conductors in which resistance value is affected by temperature. There are two types: **negative temperature coefficient** (NTC) and **positive temperature coefficient** (PTC) (**FIGURE 51-18**). The resistance value of NTC thermistors is altered inversely to temperature: As temperature increases, resistance value decreases. In PTC thermistors, resistance value alters proportionally to temperature: As temperature rises, so does resistance value. NTC thermistors are the most common and are used in circuits for ECU inputs as the sensing elements of devices such as coolant and air temperature sensors.

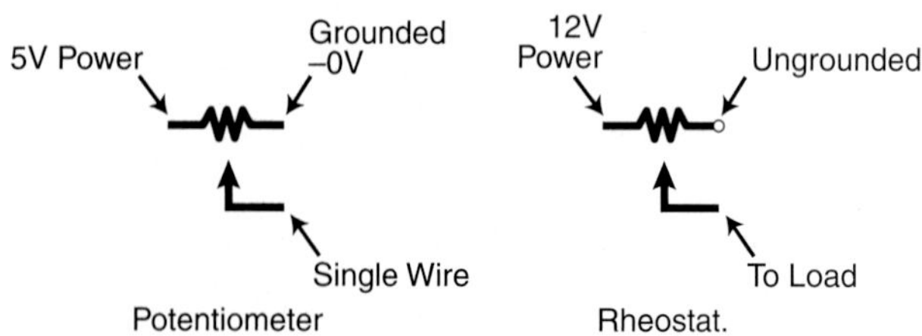

FIGURE 51-16 Rheostat.

FIGURE 51-17 Potentiometer.

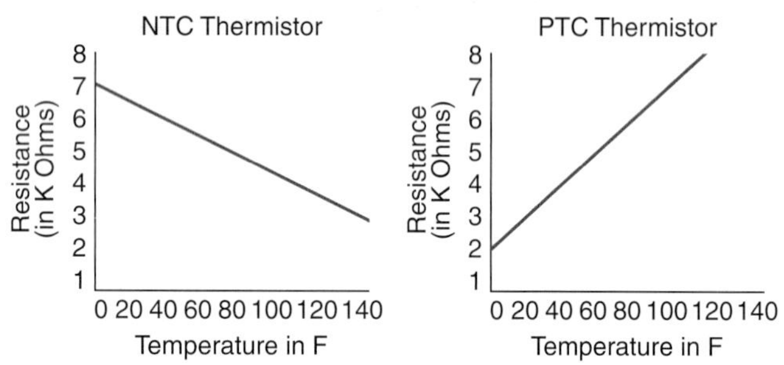

FIGURE 51-18 Thermistors. **A.** NTC thermistor resistance. **B.** PTC thermistor resistance.

Resistor Ratings

Resistance is measured in ohms, represented by the Greek letter omega (Ω). Resistors are rated in ohms as well, to indicate how strongly they will oppose any current flowing through them. Because resistors work by converting some of the electrical energy passing through them into heat, they also have a power rating. Only the resistance value is marked. The resistor's power rating is determined by its size.

Regardless of their power rating, resistors are small, so identification by numbers is impractical. To identify their value, many resistors are marked with four or five colored bands. Each color represents a numeric resistance value. The color bands are set close to each other and read from left to right. The last band, or tolerance band, is spaced farther apart and shows the resistor colors (**FIGURE 51-19**).

Capacitors

A **capacitor** can quickly store a small amount of electrical energy, at which point it is charged. Capacitors are used for several purposes: voltage suppression, to store electricity for safety systems in case of power loss, and in delay circuits for timing purposes. There are two conductive surfaces inside the capacitor, separated by insulating material. When the capacitor is charged, one surface is positively charged, and the other is negatively charged. When a circuit is closed between its terminals, the capacitor releases its charge. At this point, it is discharged.

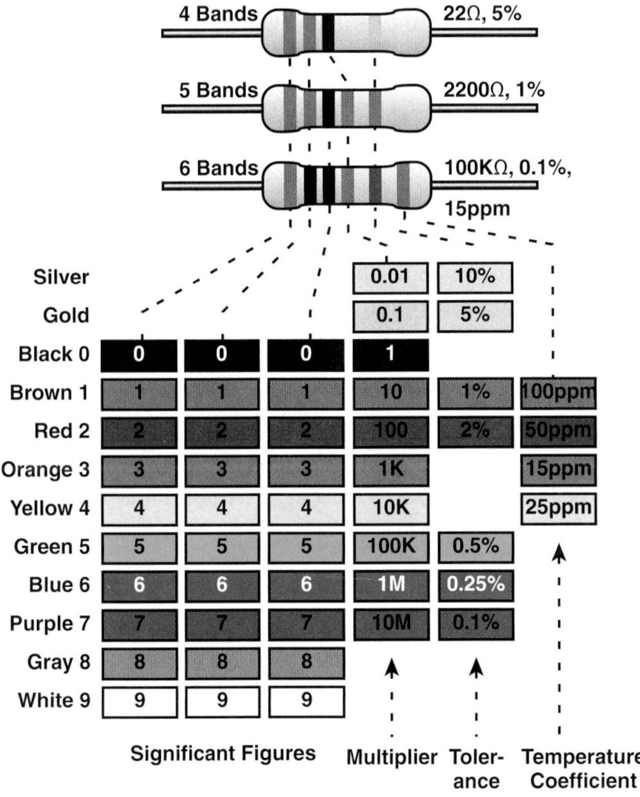

FIGURE 51-19 Resistor colors.

A typical automotive capacitor stores the charge on thin sheets of foil, with sheets of insulation between them (**FIGURE 51-20**). These are rolled together to form a cylindrical shape and placed in a protective canister. A capacitor's **capacitance** (C) is a measure of the amount of **charge** (Q) stored on each plate for a given potential difference or voltage (V) that appears between the plates. In general, as the capacitance and voltage rating of the capacitor increase, the physical size of the capacitor increases. Some DVOMs can measure the capacitance of a capacitor. That way they can be tested if they are suspected of being faulty.

▶ Electronic Components

K51001

The term "electronic components" tends to be used for those components that have no moving parts, such as diodes, transistors, and integrated circuits (**FIGURE 51-21**). Each component has a particular function and, when arranged in circuits, interacts with other components to control electrical and electronic functions. Individual components may be mounted in a housing to operate as a sensor, for example, an engine temperature sensor. Or, many components can be arranged in circuits and mounted on circuit boards to perform various functions. Examples of this are the many electronic components used in control units to manage virtually all of the systems on a vehicle, including ignition, fuel, and emission systems as well as entertainment, lighting, safety, and security systems.

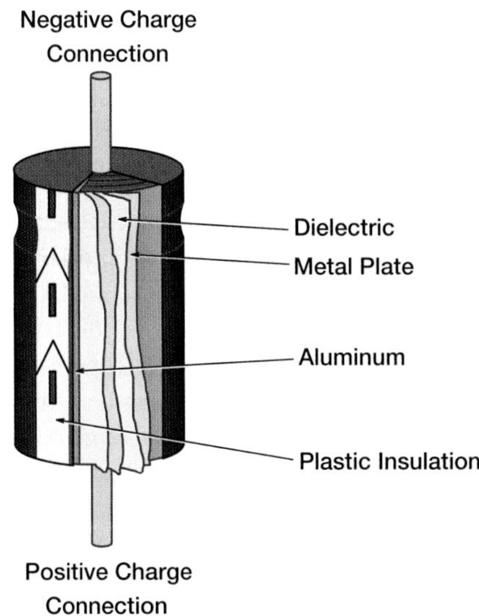

FIGURE 51-20 Cutaway view of the inside of a capacitor.

Semiconductors

The term "electronics" usually refers to devices in which electrons are conducted through a vacuum, gas, or semiconductor. Semiconductors are very versatile substances and are widely used to make various electronic components. Semiconductor devices are replacing many other kinds of switching devices such as mechanical switches and relays. They are small, light, and use low operating voltages. They are also reliable, require no maintenance, and are relatively easy to manufacture, but they are sensitive to heat and voltage spikes.

Automotive applications such as electronic control units (ECUs) mostly use semiconductors such as diodes, transistors, and power transistors. A semiconductor's electrical material has exactly four electrons in its valance ring, which is one higher than that of most conductors, but one lower than that of most insulators. A semiconductor's conducting ability depends on two kinds of **charge carriers**:

- The first type of charge carrier is the negative electron. This kind of semiconductor contains an excess of free electrons that can be made to flow and carry charge.
- The second type of charge carrier is the hole. Holes occur when an electron moves from its existing place to a new

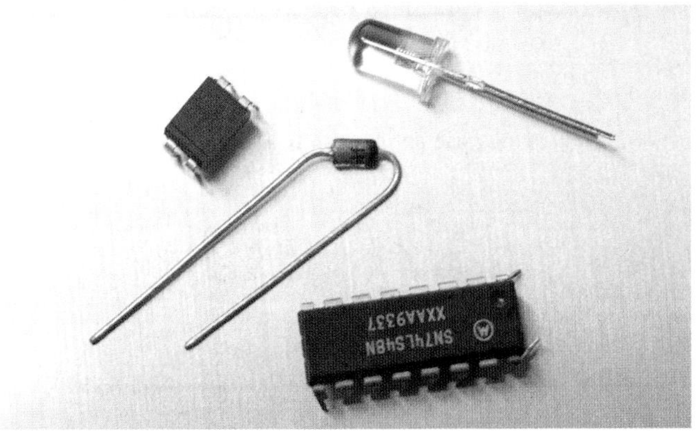

FIGURE 51-21 Typical electronic components.

place, leaving a void where it was. Because the holes are positive, a voltage can make them move toward the negative pole. When connected in a circuit, both the electrons and the holes move, but in opposite directions.

The number of charge carriers, either an excess of electrons or deficiency of electrons (causing holes in a material), can be altered by **doping**, or adding very small quantities of impurities to a semiconductor material. A doped semiconductor always has an excess of one type of charge carrier (**FIGURE 51-22**). For example, excess electrons make it an **N-type** semiconductor. "N" stands for negative. Excess holes make it a **P-type** semiconductor. "P" stands for positive. Each of these materials responds to positive or negative current flow oppositely.

Many different materials have an atomic structure that can be used to make semiconductors. **Silicon** is the most widely used semiconductor material because it has a useful temperature range and is abundant, cheap, and easy to manufacture. **Germanium** was among the first semiconductor materials to be developed and is less widely used than silicon, but it is useful in very high-speed devices when alloyed with silicon. High-speed semiconductors widely use **gallium-arsenide**; however, it is more expensive and more difficult to manufacture. **Silicon carbide** has been used to create blue light-emitting diodes (LEDs), and it can withstand high operating temperatures.

Semiconductor Operation

Most electronic components combine P-type and N-type semiconductors. The point where they join is called the **P–N junction**. In this area, a **depletion layer** occurs: Some electrons and holes cancel each other out, and few charge carriers are present. The depletion layer acts like a weak insulator in this condition (**FIGURE 51-23**).

A **diode** is one P-type semiconductor material joined to one N-type material. Diodes have a single P–N junction (**FIGURE 51-24**). If the diode is connected to a current source, with the P region connected to a negative pole and the N region

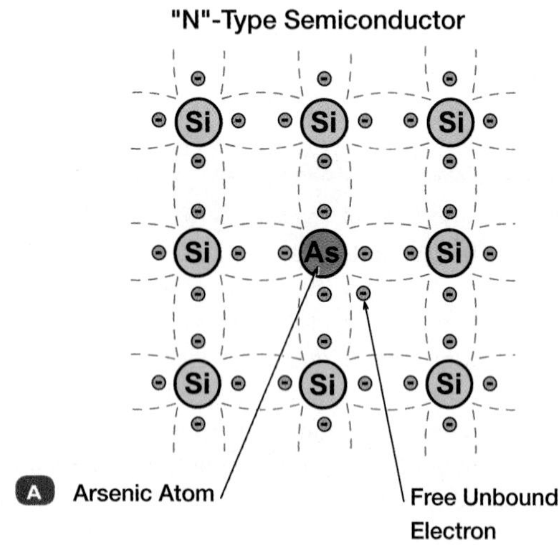

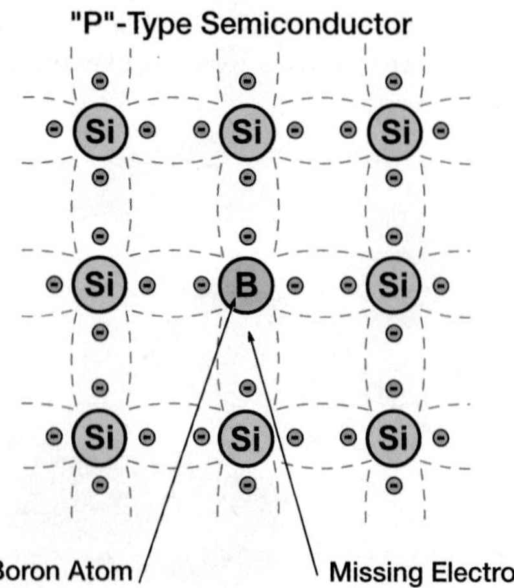

FIGURE 51-22 A. N-type semiconductor material. **B.** P-type semiconductor material.

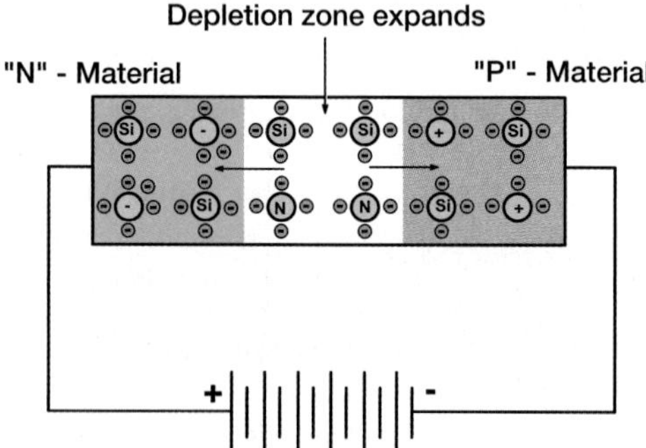

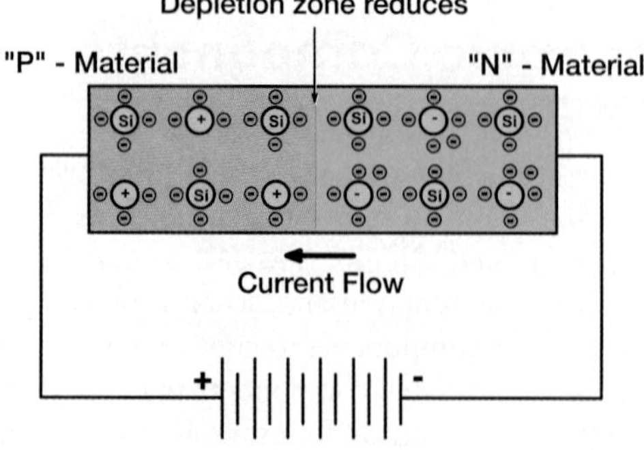

FIGURE 51-23 The P–N junction creates a depletion area that changes its charge carrying ability based on polarity.

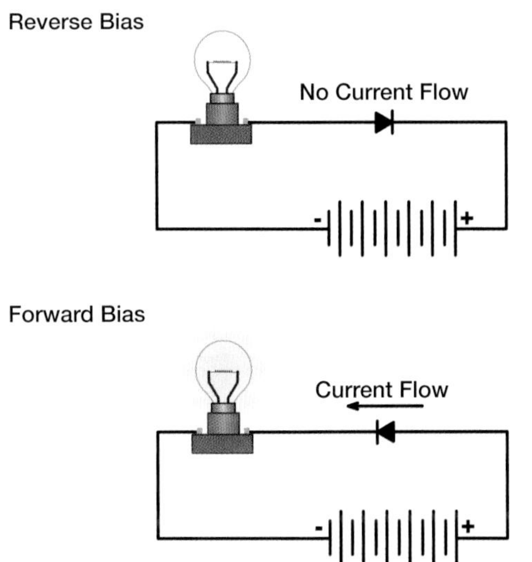

FIGURE 51-24 Diode operation. **A.** Reversed bias. **B.** Forward bias.

to a positive pole, the holes will be attracted toward the negative pole and the electrons to the positive pole. This movement enlarges the depletion layer, which makes the insulated space larger, stopping current flow across the junction. Such diodes are referred to as "reverse biased."

When the voltage source is reversed, lots of holes flow across the junction toward the negative pole, and electrons travel in the opposite direction toward the positive pole. The P–N junction floods with charge carriers; the depletion layer disappears; and with it, the insulator effect disappears. In this direction, the diode allows current to flow. Such diodes are referred to as "forward biased."

Using conventional current flow, a diode lets a low-voltage current flow through it if current flows from its P side to its N side, but stops current flowing through it from its N side to its P side. A diode can be thought of as the electronic version of a one-way check valve.

A **Zener diode** is designed to block current flow through it, but if the voltage is large enough, it can force current to flow through the diode. This is called **breakdown**. As breakdown voltage is reached, the Zener diode's resistance suddenly collapses, letting a large current flow through it without damage. Because Zener diodes respond to certain voltage changes, similar to switches, they are used in voltage regulators.

LEDs, another type of diode, are available in a range of colors and are often used as indicator lights on electrical and electronic appliances. High-powered white LEDs are increasingly being used to replace traditional incandescent and fluorescent lightbulbs; they emit light when they are connected in a forward direction (**FIGURE 51-25**).

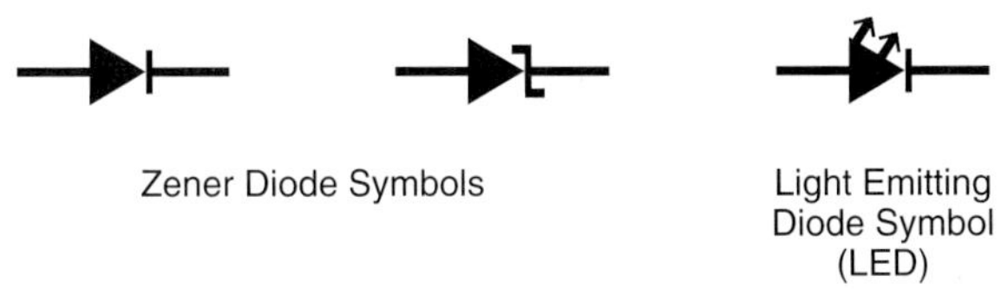

FIGURE 51-25 Diode symbols. **A.** Zener diode symbol. **B.** LED symbol.

Applied Science

AS-81: Photocells: The technician can demonstrate an understanding of the purpose of photocells.

An example of a photocell is a sun load sensor, which is usually mounted on top of the dashboard. The sensor is a part of the heating, ventilation, and air-conditioning (HVAC) system. Its purpose is to send a signal to the programmer based upon the amount of sunlight that is striking it. The signal information causes the programmer to adjust the system output according to sensor output voltage.

Transistors

A **transistor** is a semiconductor device used as a switch and to amplify currents. The transistor, a key component in almost any electronic device, is made up of three sections of semiconductor materials: two of one polarity and one of the opposite polarity. This means that there are two kinds of transistor arrangements: NP–N and P–NP. The **NP–N transistor** has a P-type semiconductor between two N-type semiconductors. A **P– NP transistor** has an N-type semiconductor between two P-types (**FIGURE 51-26**). This arrangement means that transistors have two P–N junctions.

Each of the three regions has a terminal. On a bipolar transistor, the center region is always called the base. The outer regions are the collector and the emitter. In the symbol for a transistor, the emitter is the terminal with the arrow. It always points toward the negative material. On a field-effect transistor, the center region is the gate, and the outer regions are the source and drain.

In a circuit, transistors can act as an electronic switch. If an open control switch is connected to the base, the depletion layer at one P–N junction will block current from flowing through the transistor and powering the load.

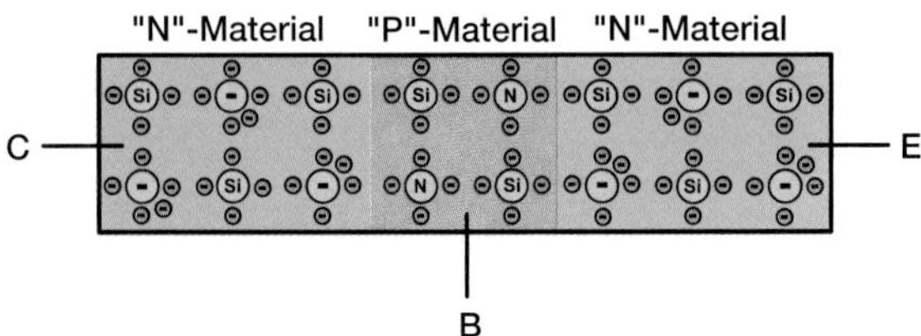

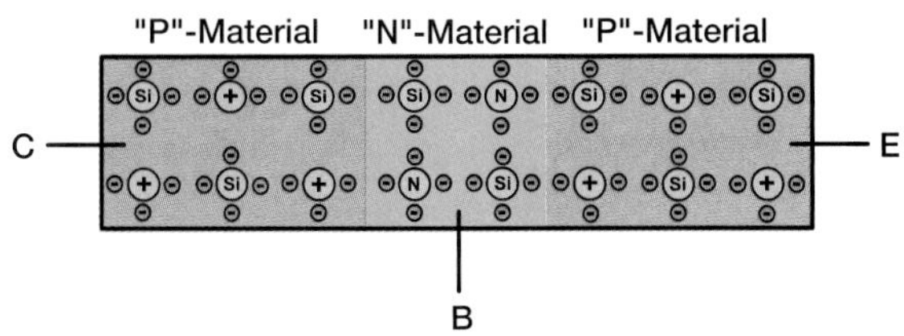

FIGURE 51-26 Transistors. **A.** The NP–N transistor has a P-type semiconductor between two N-type semiconductors. **B.** A P–NP transistor has an N-type semiconductor between two P-type semiconductors.

When the control switch is closed, a small current flows through the emitter-base P–N junction. This minimizes the depletion zone, allowing additional charge carriers to flow across the emitter-collector P–N junction, letting current operate the load. The transistor then operates as a low-resistance conductor between the emitter and collector. Most transistors have about 0.5 volt of voltage drop across the emitter-collector junction while current is flowing. So don't condemn it as an excessive voltage drop. When thinking about the operation of a transistor, think of it as a small current flowing through the base, which lets larger current flow across the emitter-collector junction. The transistor is then said to be "on." Removing the current from the base stops the emitter-collector current flow. The transistor is then said to be "off."

One difference between P–NP and NP–N transistors is whether the base is activated by positive current flow or negative current flow (**FIGURE 51-27**). In the case of an NP–N transistor, positive current, through a resistor (to limit current flow to the base), is applied to the base. In a P–NP transistor, a ground (through a resistor) is applied to the base. Transistors can be placed in the feed side of a circuit or on the ground side of a circuit.

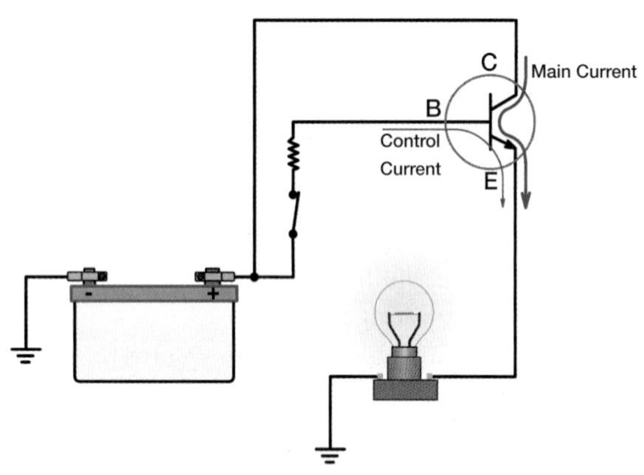

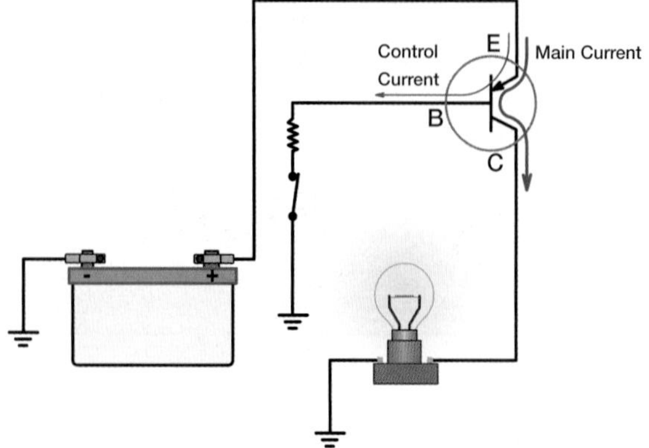

FIGURE 51-27 NP–N versus P–NP transistors. **A.** NP–N transistor connected to power, switch (with resistor), and load. **B.** P–NP transistor connected to power, switch (with resistor), and load.

Applied Science

AS-79: Semiconductor Devices: The technician can demonstrate an understanding of the capability of semiconductor devices to rapidly modify engine operation parameters.

Semiconductor devices are the foundation of modern electronics, having replaced vacuum tubes in nearly all applications. Common semiconductor devices include transistors and many types of diodes. They are used in all types of electronic devices, from radios and televisions to mobile phones and computers, including those used in automotive applications.

Semiconductor devices have enabled the development of automotive engine management systems due to their small size, high resilience to shock and vibration, and high reliability. They can also operate at very high speed, which allows increasingly accurate control of an engine's running parameters. ECUs in modern vehicles may now run at sampling rates exceeding 1.5 million times per second.

Control Modules

Control module and ECU are generic terms that identify an electronic unit that controls one or more electrical systems in the vehicle (**FIGURE 51-28**). Vehicles may have many control units installed, each with a specific purpose. For example, there are engine, transmission, body and suspension, entertainment, antitheft, and many other control units. Control modules monitor a number of inputs from sensors and circuits, process those data according to the information programmed into their memory, and send output commands to actuators and electrical devices, which control the operation of various components and systems on the vehicle. On modern vehicles, the control units share information on a data **bus**, which is a common set of wires that connects the modules together so that information or data can be shared across the network. For example, the transmission control unit needs to know road speed, as does the control unit that operates the instrument cluster and the one for the antilock brake system. The information from the one vehicle speed sensor is shared across the data bus to all three control units. Refer

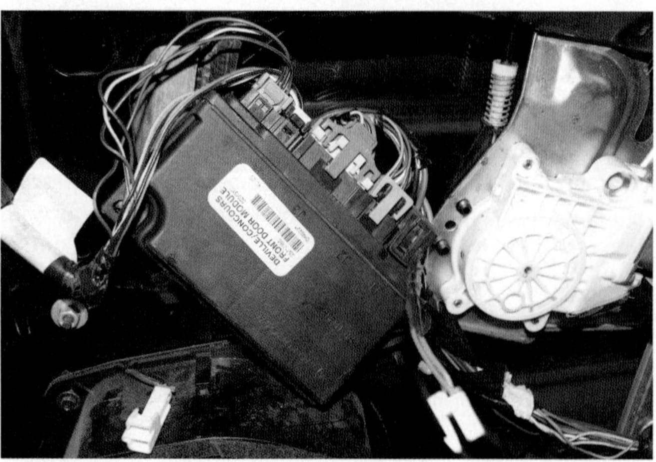

FIGURE 51-28 A door control module.

to the On-Board Diagnostics chapter for more information on control modules.

Microprocessors, Sensors, and Voltage-Generating Devices

Microprocessors, or central processing units (CPUs), are programmable devices that are found in vehicle ECUs. They are computers that monitor information from inputs such as sensors and make decisions based on their program, then output signals to actuators and electrical devices on the vehicle. Microprocessors control and monitor most electrical systems on the modern vehicle. For example, there are microprocessor control units for the engine, transmission, suspension, brakes, steering, and main body control functions. Modern diagnostic equipment connects into the microprocessors on the vehicle via a data link connector to provide information on system performance to the technician. It is not unusual for the microprocessors on a vehicle to require programming updates. Updating is performed by reflashing or reprogramming the memory in the microprocessors with an update supplied by the manufacturer.

Many different types of sensors are installed into the modern vehicle and are used to provide information to the microprocessors on the vehicle. A sensor essentially converts a mechanical action or condition into electrical signals that can be read by the microprocessors in the control units. For example, the sensor installed near the crankshaft converts the rotary movement of the crankshaft into electrical signals that tell the microprocessor each time the crankshaft has rotated through a certain number of degrees and where it is in relation to top dead center of a cylinder, typically cylinder number 1. The temperature sensor converts the temperature of the engine coolant into an electrical signal that can be read by a microprocessor. A number of other sensors are used on vehicles to gather the information needed by the ECM. Refer to the On-Board Diagnostics chapter for more information on sensors.

Delay Circuits

Delay circuits are added to vehicles to build in specific time delays in turning electrical devices on or off. For example, most modern vehicles have a time delay built into the interior lights on vehicles (**FIGURE 51-29**). After closing the door, the lamp will often stay on for a predetermined time before slowly turning off. The delay circuit can be constructed from electronic components, or it can be programmed into an ECU that controls the circuit. Modern vehicles are increasingly using ECUs with programmed timers to perform time-delay functions.

Speed Control Circuits

Speed control circuits are used on the vehicle to control the speed of motors or accessories. One common use is to control the fan speed in the vehicle's HVAC system and the electric radiator fan. Essentially, the speed control circuit manages the speed of the fan motor to control the amount of air movement through the system.

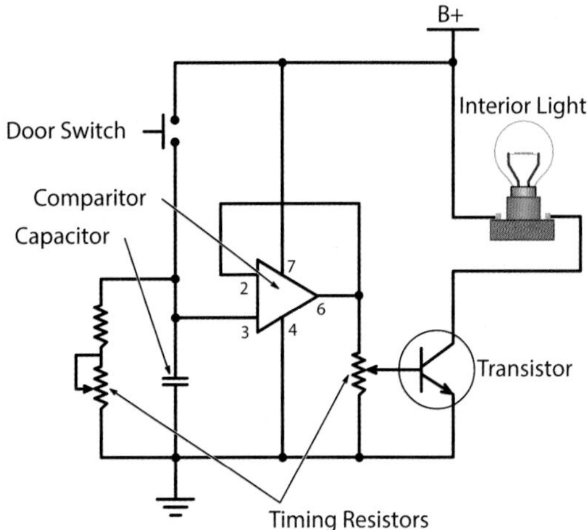

FIGURE 51-29 Typical time-delay circuitry for interior lights.

Most speed control systems use a form of pulse width modulation called duty cycle to control motor speed. Duty cycle refers to the percentage of time a circuit is fully on versus fully off. A pulse width of 50% means that the circuit is on 50% of the time and off 50% of the time (**FIGURE 51-30**). This provides about 50% of the electrical power to the motor, which allows it to operate at about 50% power. Likewise, a 25% pulse width means that the electrical power is on 25% of the time. Because the pulse width is cycled on and off very fast (sometimes exceeding 10,000 cycles per second), power to the motor appears very smooth, and the motor power is controlled very accurately. Like the time-delay circuit, speed control circuits are constructed from electronic components that can be in dedicated modules or integrated into the vehicle's powertrain control module that controls the circuit.

Printed/Integrated Circuits

Printed circuits, or circuit boards, are insulated boards approximately $^1/_{16}$ (25 mm) thick that are designed to hold electronic components such as microprocessors and their associated components (**FIGURE 51-31**). Conductive tracks on the surface of the printed circuits connect the leads of the electronic components to each other to form complex circuits. Printed circuit boards can be very complex, with many high-density integrated circuit components installed. Many boards have more than one layer of conductive tracks. They often have two layers and as many as 16 to accommodate the complex circuitry required to connect many integrated circuits.

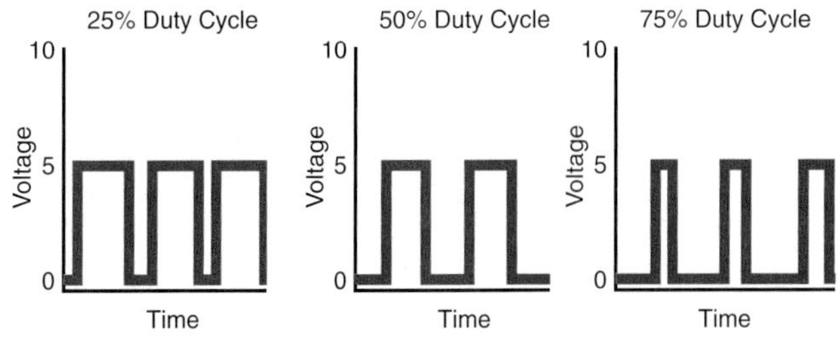

FIGURE 51-30 Pulse width modulation, ground side controlled. **A.** 25% duty cycle. **B.** 50% duty cycle, **C.** 75% duty cycle.

FIGURE 51-31 Typical printed circuit board/integrated circuits.

▶ Wires and Wiring Harnesses

K51002

Wires and wiring harnesses are the arteries of the vehicle's electrical system, and as such they need to be kept in good condition, free of any damage or corrosion. They carry the electrical power and signals through the vehicle to control virtually all of the systems on a vehicle (**FIGURE 51-32**). As technology in vehicles has increased, so too has the number of wires and cables installed on these vehicles. Although wireless communication is being used in some vehicle security, entertainment, and tire pressure monitoring systems, wires are still the dominant signal carriers in a vehicle. To help protect wires and keep them organized, they are bundled together in a wiring harness, a number of which are located throughout the vehicle.

Wires

Electrical wires are used to conduct current around the vehicle. Wire can also be referred to as cable, which typically refers to large-diameter wire. Automotive **wire** is commonly a multi-stranded copper core wrapped with seamless plastic insulation (**FIGURE 51-33**). Copper is typically used as it offers low electrical resistance and remains flexible even after years of use. The insulation is designed to protect the wire and prevent leakage of the current flow so that it can get to its intended destination. **Ribbon cable** is a series of wires that are formed side by side and joined along the wire insulation; they are flat like a ribbon (**FIGURE 51-34**). Ribbon cable works well when several wires run from one component to another. The ribbon design groups them so they can be routed neatly and easily. Ribbon cable is often found inside computers and other electronic components. It is used for connecting between printed circuits or between printed circuits and other components. Some wires, especially signal wires and communication wires, are shielded, which helps to prevent electromagnetic interference, also referred to as "noise."

FIGURE 51-32 Wires carry information and commands throughout the vehicle.

Wire Sizes

Wire size is very important for the correct operation of electrical circuits. Selecting a wire gauge that is too small for an application causes an excessive voltage drop, and in extreme cases

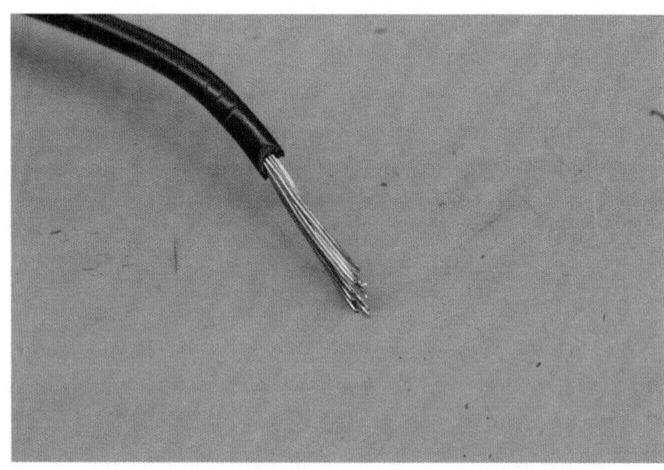

FIGURE 51-33 Multi-strand wire (stripped).

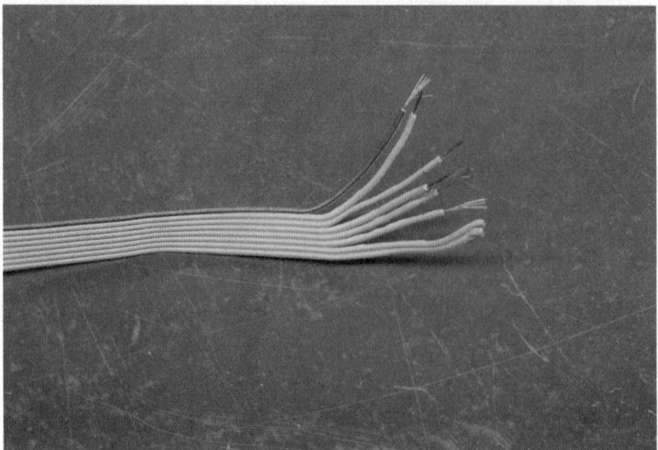

FIGURE 51-34 Ribbon cable (stripped).

the wire will get hot enough to melt the insulation. Selecting a wire gauge that is too large increases the cost, weight, and size of wiring harnesses.

The diameter of a wire affects its resistance and therefore how much current it can carry. Even good conductors have a slight amount of resistance. The resistance of a wire is determined by its length, diameter, construction material, and temperature. The longer the wire and the smaller the diameter, the higher the resistance. The shorter the wire and the larger the diameter, the lower the resistance.

Two scales are used to measure the sizes of wires: the metric wire gauge, which measures the cross-sectional area of the conductor in square millimeters (**TABLE 51-2**), and the **American wire gauge (AWG)**. The AWG system uses a rating number; the larger the rating number, the smaller the wire and the lower its current-carrying capability (**FIGURE 51-35**). Most countries use the metric scale. American manufacturers are split, with some using AWG and others using the metric scale. Manufacturers and standards bodies use wire gauge charts to define how much current each wire gauge can carry safely and efficiently. A vehicle uses a variety of wire sizes depending on the requirements of each particular circuit.

The correct wire size for an application can be looked up on a wire size chart if you know the amperage of the circuit and the length of the wire. For example, if a 12-volt circuit is designed for a maximum current flow of 10 amps and is approximately 15' (6.1 m) long, using the AWG table as a reference, you can determine that the correct wire gauge to use is 12 AWG (**TABLE 51-3**). But be careful of the chart you use, as

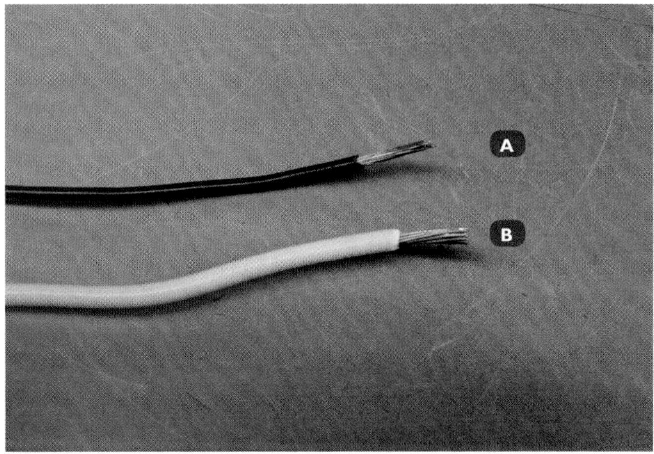

FIGURE 51-35 In the AWG system, the larger the wire number size, the smaller the diameter of the wire. **A.** 16 gauge. **B.** 12 gauge.

TABLE 51-2 Metric Wire Size and American Wire Gauge (AWG) Comparison

Metric Wire Sizes (Square mm)	AWG Wire Sizes (Gauge)
0.22	24
0.35	22
0.5	20
0.8	18
1.0	16
2.0	14
3.0	12
5.0	10
8.0	8
13.0	6
19.0	4
32.0	2

TABLE 51-3 AWG Wire Sizes Based on Amperage and Wire Length*

	Wire Length from Battery to Load						
Circuit Amps	2'	5'	7.5'	10'	15'	20'	25'
2	20	20	20	18	18	18	16
5	18	18	18	18	16	14	14
8	18	16	16	14	14	12	12
10	16	16	16	14	12	12	10
12	16	16	14	14	12	12	10
15	16	16	14	12	10	10	8
18	16	14	12	12	10	8	8
20	14	14	12	10	10	8	8
25	14	12	12	10	8	8	6
30	12	12	10	10	8	6	6

* Chart is based on a maximum 0.4-volt drop per wire size; shorter distances are less than a 0.4-volt drop.

many of them allow up to a 10% voltage drop over the length of the wire, which is significantly more than is allowed in most automotive circuits.

Two different methods describe the conductor size within these standards: A wire may be described in metric size as 5.0, indicating it has a cross-sectional area of 5.0 millimeters squared (mm^2). It can also be expressed as 10/0.5, indicating there are 10 strands of wire, each with a cross-sectional area of 0.5 mm^2. The same system can be applied to the AWG rating using the gauge size.

Terminals and Connectors

Terminals are installed to the ends of wires to provide low-resistance termination. They allow electricity to be conducted from the end of one wire to the end of another wire. In many cases, they allow the wires to be disconnected and reconnected. They come in many different types and sizes to suit various wire sizes and termination requirements (**FIGURE 51-36**). For example, there are **push-on spade terminals, eye ring terminals** to accommodate screws, **butt connectors**, and **male and female terminals** that are designed to be separated and reconnected. Most terminals are the crimp type, which require the use of special tools to crimp the terminal to the end of the wire. They can be insulated or non-insulated. Some **solder-type terminals** are still in use and require the use of electric or gas soldering irons, flux, and solder to make the connection. When soldering wiring, always use a rosin or rosin-core solder; never use an acid-core solder, as acid can cause corrosion and high resistance over time.

Terminals can be installed as a single terminal on a wire or grouped together in a wiring harness with a connector housing, also called **wiring harness connectors**. **Connector** housings have male and female sides and are usually shaped so that they can be connected in only one way (**FIGURE 51-37**). They often incorporate a locking mechanism so the plug cannot accidentally work loose. Many of these connectors are weatherproofed to keep moisture out. Special tools are usually needed to insert and remove the terminals from the connector housing.

FIGURE 51-37 Male and female wire connector.

Wiring Harnesses

Wiring harnesses, also known as wiring looms or cable harnesses, are used throughout the vehicle to group two or more wires together within a sheath of either insulating tape or tubing (**FIGURE 51-38**). Often, harnesses on modern vehicles contain many wires, each terminating at crimped terminals inserted into connector or harness plugs. There are usually a number of harnesses within the vehicle, interconnecting with various connector plugs as required to form the wiring system of the vehicle. Wiring harnesses run around the engine bay, through the dash and interior cabin, and to the rear of the vehicle. They are attached to the vehicle with harness fasteners such as body clips or wire ties, and rubber sealing grommets are used when the harness passes through the metal bodywork.

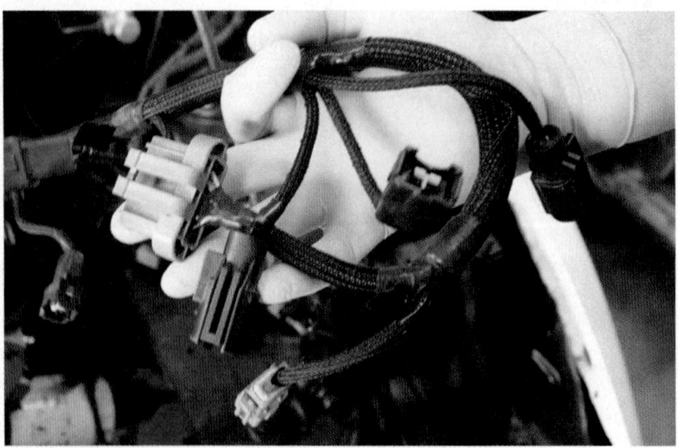

FIGURE 51-36 Terminals and connectors are installed to the ends of wires to provide low-resistance termination of the wires.

FIGURE 51-38 Typical wiring harness.

Shielding

In certain locations within a vehicle and in environments where strong electromagnetic interference (EMI) is present, wiring harnesses are subject to a situation where unwanted electromagnetic induction occurs. This interference is referred to as electrical noise or EMI noise. To prevent noise, many vehicles use **shielded wiring harnesses**. The type of shielding used can be one of three forms: twisted pair, Mylar tape, or drain lines.

Twisted Pair

Twisted pair uses two wires delivering signals between common components. The wires are uniformly twisted through the entire length of the harness and end at a terminating resistor (**FIGURE 51-39**). The twisted wires, along with the terminating resistor, have the effect of canceling any noise that occurs in the wires, reducing the loss of data in the transmitted signals. The controller area network bus, or CANbus, in a modern vehicle may use one or more twisted pairs to connect all the vehicle control units with common data line(s) to share information. Refer to the On-Board Diagnostics chapter for more information on control with the CANbus.

Mylar Tape

Mylar tape is an electrically conductive material that is wrapped around a wiring harness inside the outer harness layer (**FIGURE 51-40**). Any noise that attempts to reach the wires inside the shield is absorbed by the Mylar, where it will be conducted to ground via a ground connection. The shielding is important to prevent electrical noise penetrating into the electrical wiring. If the harness is exposed, the Mylar will have to be rewrapped so that noise cannot penetrate into the harness.

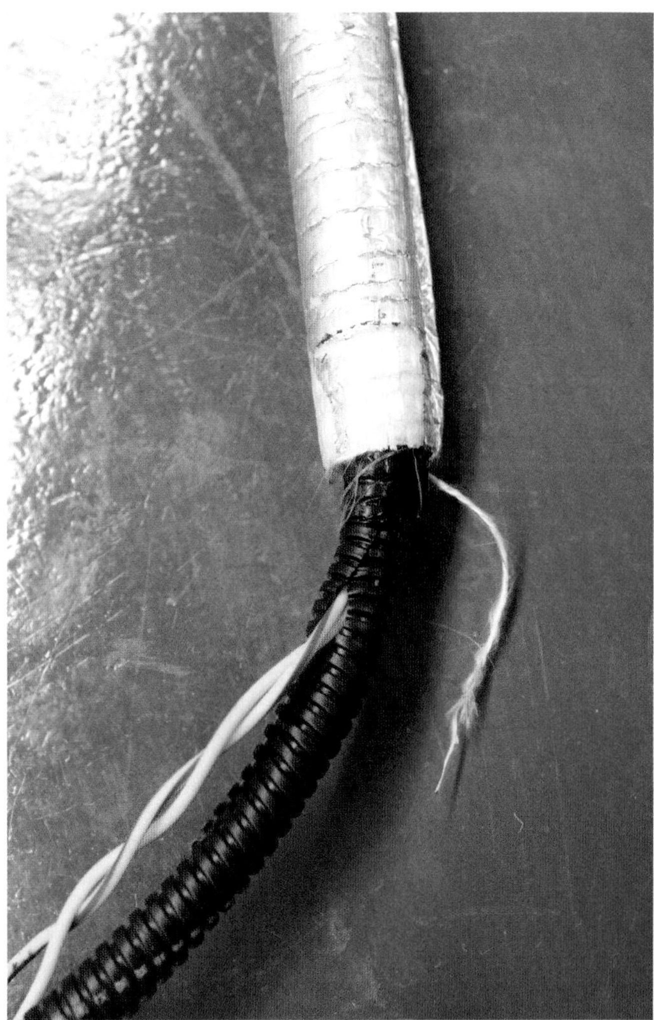

FIGURE 51-40 Mylar tape is used around wiring harnesses to shield them from electrical noise.

Drain Lines

A **drain line** is a non-insulated wire that is wrapped within a wiring harness (**FIGURE 51-41**). The drain wire is connected to ground at the harness source end and conducts any noise to

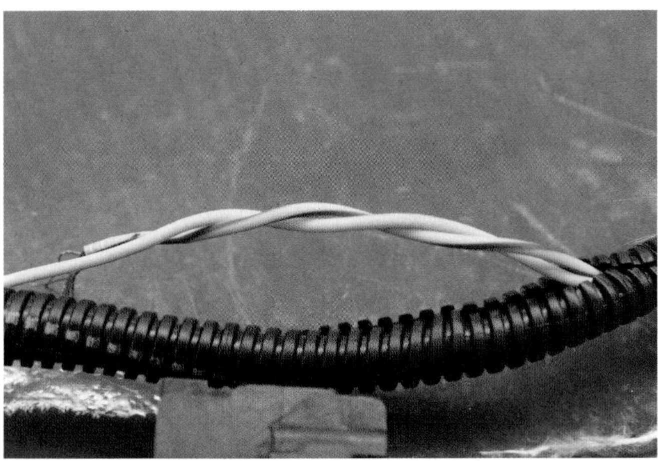

FIGURE 51-39 A twisted pair used to carry signals between components, and also act together to cancel any electrical noise.

FIGURE 51-41 A drain line inside a wire harness, used to drain electrical noise to ground.

Applied Science

AS-78: Magnetic Fields/Forces: The technician can explain the effect of magnetic fields on unshielded circuits in control modules.

Shielding is applied to electrical circuits to prevent unwanted electromagnetic induction, where errant voltages may be generated in unshielded circuits due to electromagnetic interference. Some systems within control modules, particularly sensor inputs, work with high-sensitivity and very low voltages. In these cases, even a small induced voltage, known as "noise," can have serious effects on the running of the vehicle or the functionality of safety systems. The control unit may effectively respond to the noise as if it is actually a signal from a sensor.

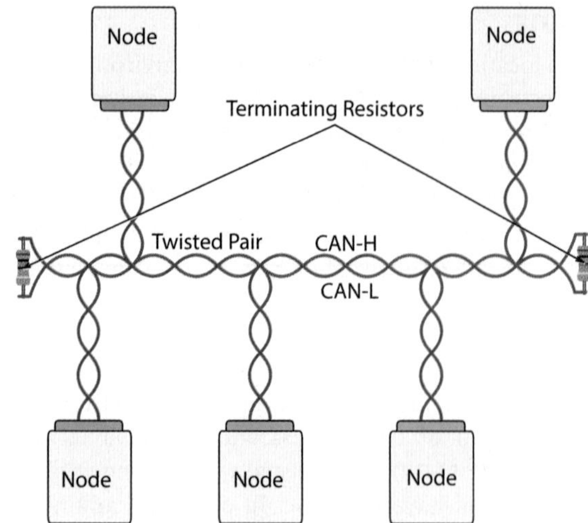

FIGURE 51-42 Typical CANbus network system.

ground, negating the noise effect. If the drain wire is cut, it will be inoperative, so it is important that the wire not be cut or left disconnected.

Networking and Multiplexing

Even the most basic vehicles include many electronically controlled systems. If each electronic system had its own ECU, harness, and sensors, the weight of the added components would negate much of the efficiency the system provided. A vehicle's multiple electronic systems could require more than 1 mile, or 1.6 kilometers, of insulated wiring, consisting of around 1000 individual wires and many terminals.

One solution to the problem is the use of a system that integrates sensors into a common wiring harness by combining many of the individual systems, where possible, into a multiplexed serial communications network so that they can share information. An added advantage of such a system is that if there is less wire and fewer connections, there is less chance of poor connections causing faults.

One system is referred to as a **controller area network bus (CANbus)**, and it uses two thin wires (called a twisted pair) to connect, or multiplex, many of the control units and their sensors to each other (**FIGURE 51-42**). The output devices are referred to as nodes.

The advantage of a multiplex network is that it enables a decreased number of dedicated wires for each function; therefore, there is a reduction in the number of wires in the wiring harness; reduced system cost and weight; and improved reliability, serviceability, and easier installation. In addition, common sensor data, such as vehicle speed and engine temperature, are available on the network, so data can be shared, thereby reducing the number of sensors.

Networking makes it more efficient to add or modify system functionality through the addition of control units and the use of software updates. These updates allow the manufacturer to provide fixes for issues that may crop up after vehicles have been delivered to customers and put into service. Other control units can be added to the system by simply connecting them to the network and the new controlled device. Software can then be updated, if necessary, to control the new device.

A diagnostic scan tool can be connected to the CANbus circuit to both extract operational information and command controlled devices to operate, which assists greatly in diagnosis and fault finding.

Fiber Optics

Fiber optic cables are long, thin strands of very pure glass or plastic about the diameter of a human hair (**FIGURE 51-43**). They can be arranged in bundles and are then called **optical cables.** Light signals can be transmitted along the cable over very long distances. Light is transmitted along the cable, bouncing from wall to wall along the way as it is transmitted through the cable. It can do this because the inner wall of the optical fiber has a mirror finish. This means the light continues to travel along the cable even when it bends around corners.

Optical fiber can be used to transmit digital data for computers or simply to light up dash gauges in a vehicle. The optical fiber is especially suited to reliably transmitting a large amount

FIGURE 51-43 Fiber optic cable made up of individual strands, transmits light and data.

of digital data. In a vehicle, optical fiber cables could be used to replace existing copper data bus systems.

Optical fiber has several advantages over other available technology. It is less expensive and thinner; it can carry much more data than copper wires; it has less signal degradation over length; and it is less affected by interference. Unlike electrical signals, light in one fiber does not interfere with light in another fiber. Optical fiber is nonflammable, lightweight, and flexible. One disadvantage to optical fiber is that it requires more complex splicing procedures.

▶ Wiring Diagram Fundamentals

N51001

Wiring diagrams, also known as electrical schematics or electrical diagrams, use abstract graphical symbols to represent electrical circuits and their connection or relationship to other components in the system. They are essentially a map of all of the electrical components and their connections (**FIGURE 51-44**). The wiring on modern vehicles is very complex, with many wires and components interconnected. A single wiring diagram of the whole vehicle would be very difficult to read. To make it easier, the wiring diagrams are split up into systems and subsystems to reduce the complexity on each page. For example, there may be a wiring diagram for the starter system, a separate one for the charging system, and others for the engine, transmission, antilock brakes, headlights, taillights, and so on.

To assist in understanding the wiring diagrams, manufacturers supply keys for the diagrams, one of which is a list of the component symbols and their names (**FIGURE 51-45**). Other keys refer to the color codes of the wires, wire size, harness connector identification, and pin numbers (**FIGURE 51-46**). Many of the symbols are standardized and used universally by manufacturers, although in some cases, variations may exist. In many cases, wires use two colors; the first one is the solid color, and the second one is the stripe. So "BLK/WHT" would mean the wire uses a solid color of black, with a white tracer (stripe). In many cases, wiring diagrams are set up with power on the top of the diagram and ground on the bottom. This makes it easier to follow the flow of electricity in the diagram. Armed with all of this information, the technician can read the wiring diagram and identify circuits as they relate to the actual components and circuits on the vehicle.

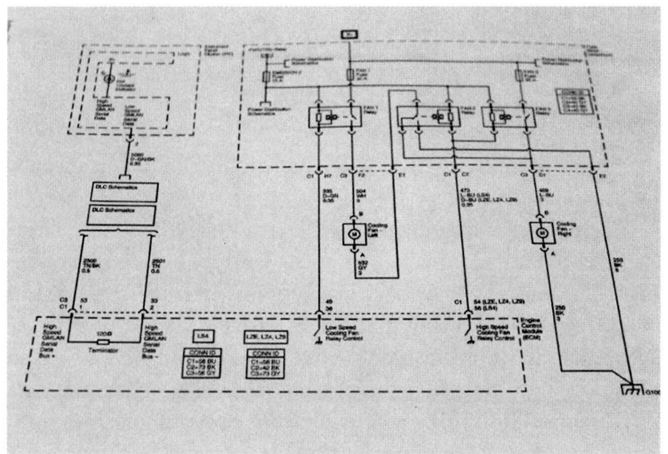

FIGURE 51-44 Typical wiring diagram.

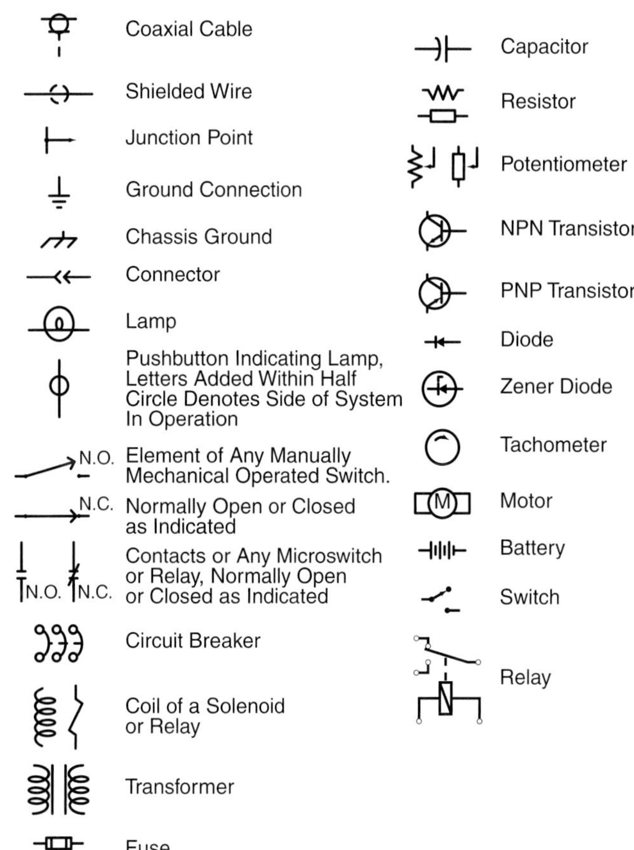

FIGURE 51-45 Every electrical device and component has a corresponding electrical symbol.

Wiring diagrams are a critical tool when diagnosing electrical circuit faults. In the strategy-based diagnostic process, it falls under step 2—Research. Wiring diagrams allow you to understand how the circuit was designed. And by tracing the circuits, as you will learn next, you will be able to determine how electricity flows through the components. Using that information and applying Ohm's law and the circuit laws allows you to predict what voltages should be present at the various connection points. Knowing the expected voltages then allows you to measure the actual voltages present. Any substantial difference between actual voltage and predicted voltage will point you in the direction of the fault. For example, if the customer concern is that the blower motor runs very slowly on all speeds, and the wiring diagram shows no resistance between the battery positive terminal and the blower motor, you would predict that there should be 12 volts at the blower motor (**FIGURE 51-47**). If you check the voltage at that point and find that the actual voltage is only 6 volts, then you know you need to check voltage at upstream points in the circuit (blower switch, fuse, etc.) until you locate the high-resistance fault. These concepts are covered in much more depth in the next chapter.

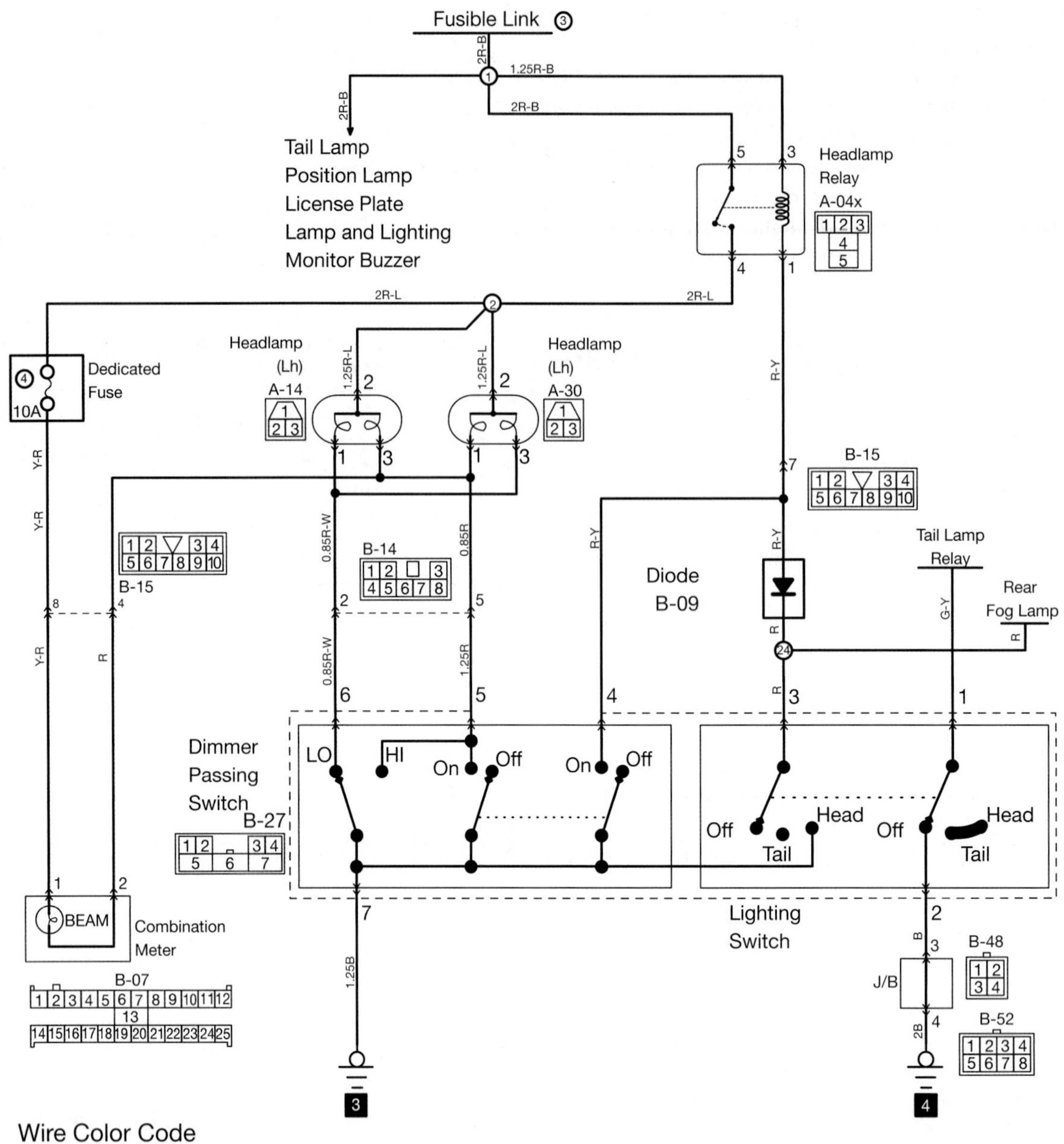

Wire Color Code

B: Black LG: Light Green G: Green L: Blue W: White Y: Yellow SB: Sky Blue
BR: Brown O: Orange GR: Gray R: Red P: Pink V: Violet

FIGURE 51-46 Typical key to wire identification codes.

▶ Using Wiring Diagrams

N51002

Vehicle wiring diagrams, or schematics, may be available as paper-based manuals, computer programs, or online resources. They are produced by manufacturers and some aftermarket publishing companies. Increasingly, repair information is accessed via the Internet, using subscription services that are regularly updated. To use wiring diagrams, an understanding of the symbols, abbreviations, and connector coding used in the diagrams is required. These are usually found on the diagram or in information pages.

Reading a wiring diagram is like reading a road map. There are a lot of interconnected circuits, wires, and components to decipher. Learning to read wiring diagrams takes a bit of time and experience, but knowing that circuits usually consist of a power source, circuit protection device, a switch, a load, and a ground is a good start. Jorge Menchu of AESWave has been promoting a novel approach of using colored crayons or highlighters to trace out wires on the diagram, to help understand how a particular

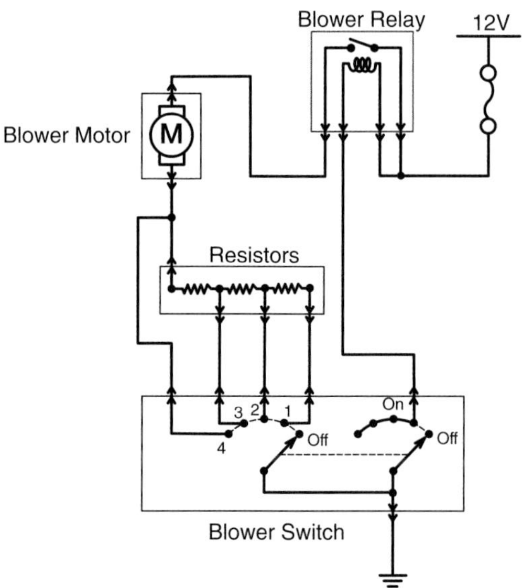

FIGURE 51-47 Heater blower motor circuit wiring diagram being used for understanding the circuit and predicting voltages.

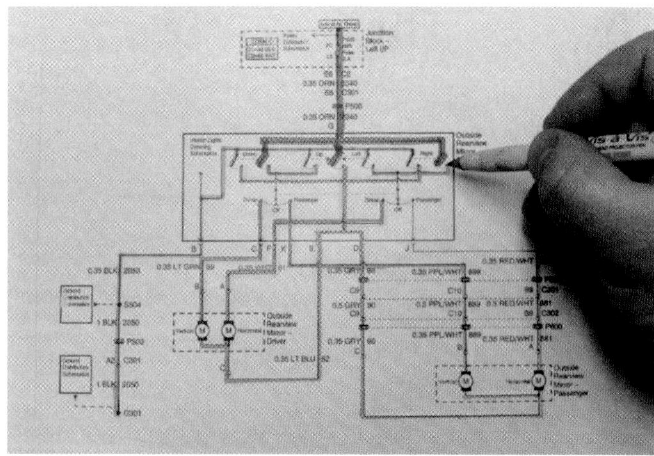

FIGURE 51-48 Color coding of the wiring diagram helps to understand the circuit.

circuit operates (**FIGURE 51-48**). The following is a paraphrased version of that process.

Begin by printing out a copy of the wiring diagram for the circuit being diagnosed. You then identify each wire in the circuit and color it according to the following designations:

- Color all of the wires green that are directly connected to "ground."
- Color all of the wires red that are "hot" at all times.
- Color all of the wires orange that are "switched to power."
- Color all of the wires yellow that are "switched to ground."
- If there are any wires that reverse polarity, such as power window motor wires, mark those with side-by-side orange and yellow lines.
- Finally, color any variable wires, such as signal wires, blue.

Coloring the wires on the wiring diagram in this way does several things. First, it forces you to determine what each wire in the diagram does, which helps you get a total picture of the circuit. Second, it helps to organize your thoughts so that you can understand how electricity flows through the circuit. Third, it helps to keep you from losing your place or forgetting what a particular wire does. And fourth, when you know why the circuit is not working properly and exactly where the problem is located, it can give you confidence that you have properly diagnosed the problem.

To use wiring diagrams to diagnose electrical circuits, follow the steps in **SKILL DRILL 51-1**.

▶ Wire Maintenance and Repair

K51003

Wires are generally trouble free and long lasting. But they can be damaged. Generally speaking, any issues with wiring are more likely to be with the terminals than with the wires themselves (**FIGURE 51-49**). Terminals can corrode, lose their tension, or push back up inside the connector, leading to poor connections and voltage drops. If you suspect a problem with a wire, first inspect the ends. If no problems are found, look for mechanical damage to the wire or wiring harness itself. When a wire is damaged, it is usually due to one of several conditions. One possibility is that the wire has been physically broken, such as a wire that was not disconnected when removing a major component such as the engine or transmission. Another issue is when a wire gets pinched between components, such as between the engine and the transaxle when replacing a clutch. The pinched wire can cause either a short circuit or an open circuit. Wires

SKILL DRILL 51-1 Using Wiring Diagrams to Diagnose Electrical Circuits

1. Identify the correct wiring diagram for the vehicle and system circuit being repaired, and print a copy.
2. Color each wire (using the wiring diagram's color code key) on the wiring diagram for the circuit that requires diagnosis. Note components, wire coding, and harness connectors.
3. Determine circuit test points and their location on the wiring diagram. Find the same test point on the vehicle, and perform the appropriate electrical test.
4. Depending on the results of the test, continue to use the wiring diagram to guide you in performing additional tests on the circuit until the fault has been located.

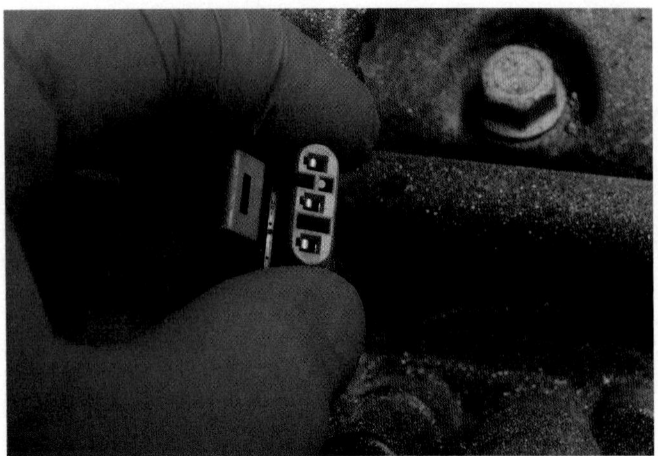

FIGURE 51-49 Inspecting wire terminals for faults.

can also be misrouted so that they lie on a hot surface such as the exhaust manifold, which melts the insulation and causes the wires to short. In all of these cases, the problem can typically be spotted visually. One problem that may be harder to spot is if a circuit shorts out and melts one or more wires together within a harness. In this case, you may have to open up the wiring harness and inspect the wires. Next, we cover how to repair wires, terminals, and connectors.

Stripping Wire Insulation

An insulating layer of plastic covers electrical wire used in automotive wiring harnesses. When electrical wire is joined to other wires or connected to a terminal, the insulation has to be removed. Wire stripping tools come in various configurations, but they all perform the same task. The type of tool you use or purchase will depend on personal preference and the amount of electrical wire repairs you perform. A good pair of wire strippers removes the insulation without damaging the wire strands. Never use a knife or other type of sharp tool to cut away the insulation, as it often cuts away some of the strands of wire as well. This is known as ringing the wire, which effectively reduces the current-carrying capacity of the wire. Only remove as much insulation as is necessary to do the job. Insufficient bare wire may not achieve a good connection, and excessive bare wire may expose the wire to a potential short circuit with other circuits or to ground. Also, removing more than 0.5" (12.7 mm) of insulation at a time can stretch and damage the wire core.

To strip wire insulation, follow the steps in **SKILL DRILL 51-2**.

Installing a Solderless Terminal

N51003

Factories use solderless terminals throughout the vehicle, primarily at connectors. If a wire itself needs to be repaired, it should be soldered back together instead of using solderless terminals to reconnect the wires. Solderless terminals are quick to install and effective at conducting electricity across joints that are designed to be disconnected. Solderless terminals require a clean, tight connection. It is important to make sure the wire and the connection are clean before attaching any terminals. You should use connections that match the size of the wire. Many solderless connectors are color coded for the size of wire they are designed to work with, such as yellow 12–10 AWG, blue 16–14 AWG, and red 22–18 AWG. Use the correct wire stripper to strip only as much insulation off as needed to allow the wire to fully engage the terminal. To keep the wires together after stripping them, give them a slight twist. Do not twist the wire too much; otherwise you risk a poor wire-to-terminal connection. Use the correct crimping tool for the connection, as many manufacturers use special crimping tools. Using the wrong type of tool will cause the connection to have a poor grip on the wire, or it may damage the terminal.

To install a universal solderless terminal, follow the steps in **SKILL DRILL 51-3**.

SKILL DRILL 51-2 Stripping Wire Insulation

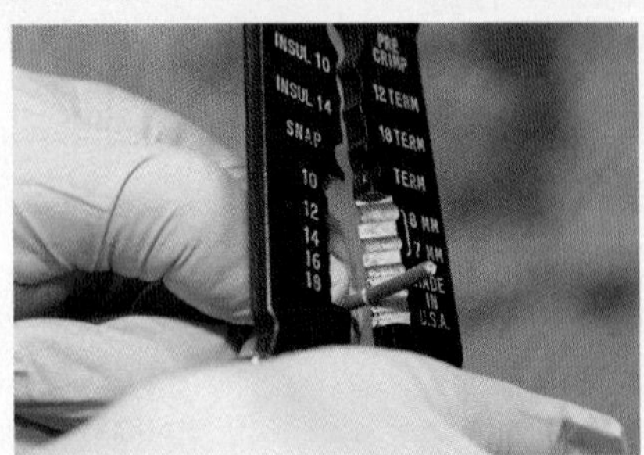

1. Choose the correct stripping tool.

2. Select the hole that matches the diameter of the wire to be stripped. Place the wire in the hole, and close the jaws firmly around it to cut the insulation.

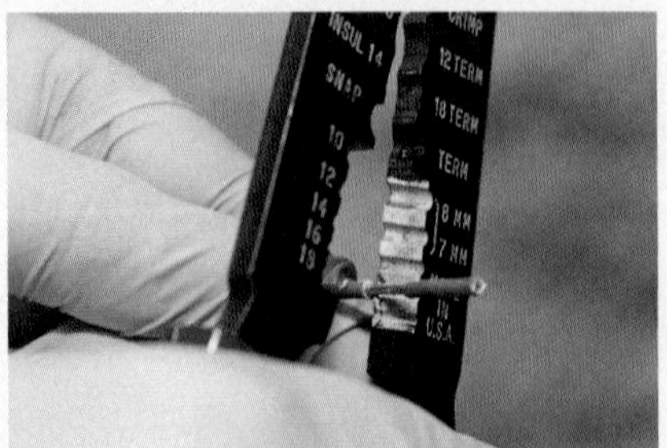

3. Remove the insulation. To keep the strands together, give them a light twist.

SKILL DRILL 51-3 Installing a Solderless Terminal

1. Make sure you have the correct size of terminal for the wire to be terminated and that the terminal has the correct volt/amp rating. Remove an appropriate amount of the protective insulation from the wire.

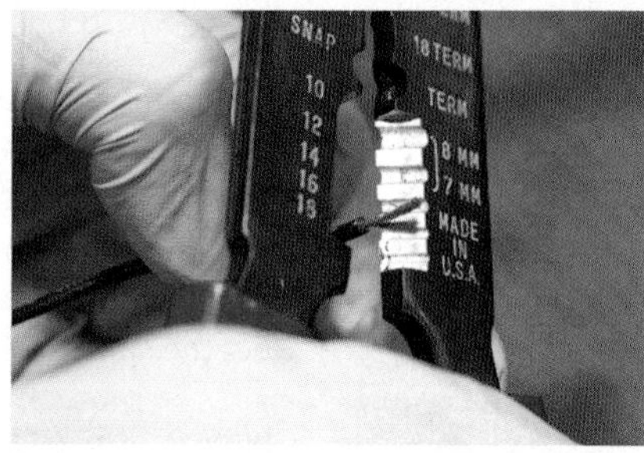

2. Lightly twist the wire strands, and place the terminal onto the wire.

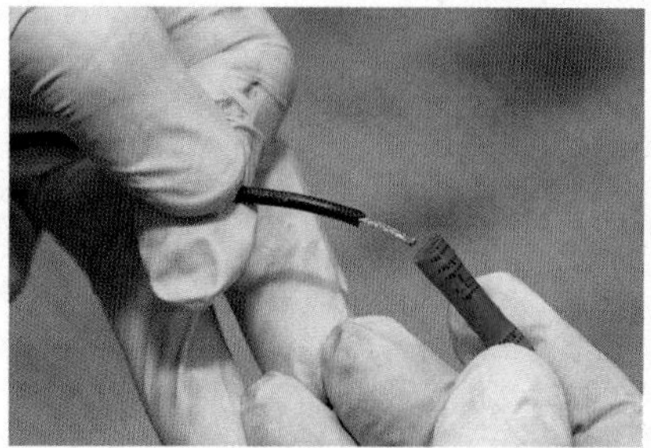

3. Use a proper crimping tool for the terminal you are crimping. Do *not* use pliers, as they have a tendency to cut through the connection. Select the proper anvil.

4. Crimp the core section first. Use firm pressure so that a good electrical contact will be made, but do not use excessive force, as this can bend the pin or terminal.

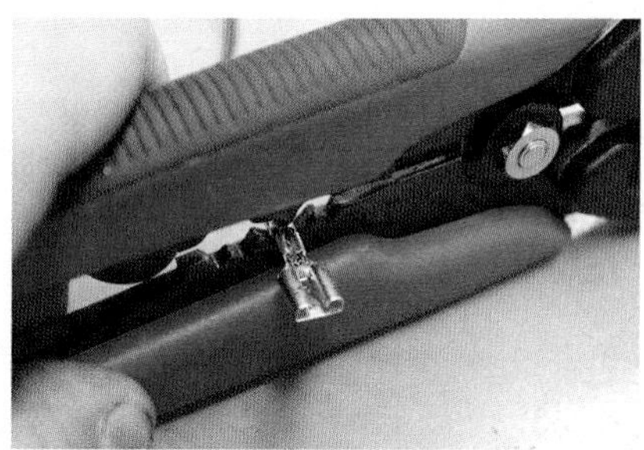

5. If crimping an uninsulated terminal, lightly crimp the insulation tabs so that they hold the insulation firmly.

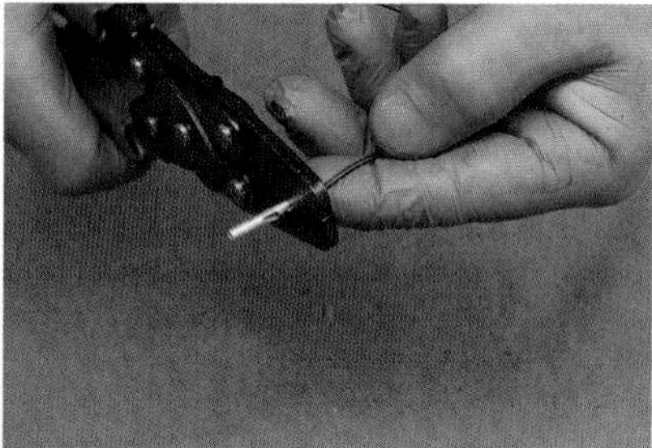

6. If crimping a factory terminal, use the proper tool and follow the instructions.

Soldering Wires and Connectors

Solder used in automotive electrical applications is an alloy typically made up of 60% tin and 40% lead. Solder needs to change from a solid state into liquid easily and return to its solid state quickly as it is heated and cooled. Solder is available as solid or flux cored. Flux is needed to prevent the metals from being joined due to oxidization when they are heated. Solid solder requires an external flux to be applied in the soldering process. Flux-cored solder has a bead of flux within the center of the solder. Flux in flux-cored solder can have either an acid base or a rosin base (**FIGURE 51-50**). Acid flux is designed to be used on non-electrical metal joints such as radiators and must be removed after the soldering process so that the joint does not corrode. Rosin flux solder is used on electrical connections because it is much less likely to corrode the metals than acid flux. Acid flux and rosin flux also come in paste form that can be brushed onto the joint if using solid-core solder.

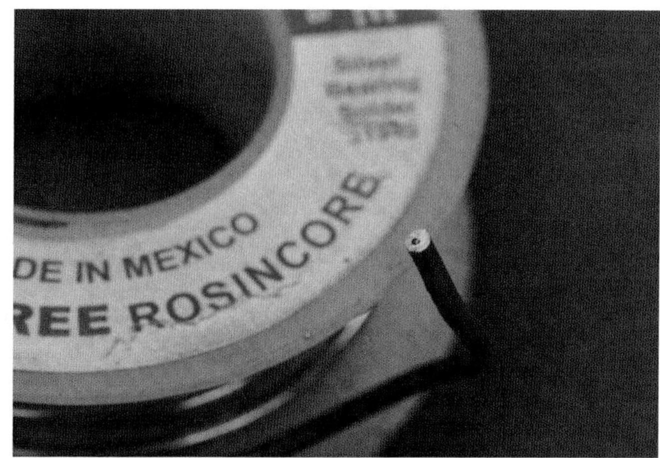

FIGURE 51-50 Rosin core solder is used for soldering electrical components and wires.

Solder is applied with a hot soldering iron. The soldering iron is heated electrically or by an external source such as a butane or oxyacetylene torch (**FIGURE 51-51**). The soldering iron tip absorbs heat that is then applied to the materials to be joined. Once they are hot enough, solder can be melted between the components. It solidifies as it cools, "gluing" the metal pieces together.

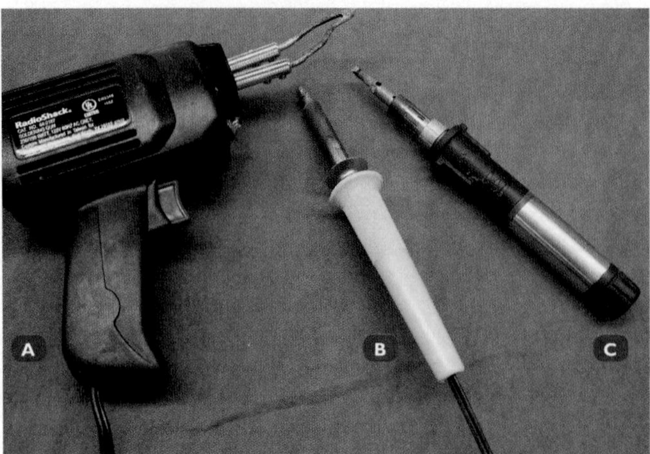

FIGURE 51-51 **A.** Electric soldering gun. **B.** Pencil type soldering iron. **C.** Butane soldering iron.

For a connection to be successful, the soldering iron has to be clean and "tinned." Cleaning may be as simple as heating the tip and wiping it on a damp cloth. Or, with the soldering iron cold, you may need to use a file to remove oxidized metal and reshape it so it can effectively transfer heat to the wires. The tinning process assists in transferring heat to the wire by leaving a small amount of liquid solder on the tip, which increases the surface area where the tip contacts the wires. To tin the soldering iron, the tip is heated, and a small amount of solder is applied to the tip. Excess solder is removed with a cloth rag.

The soldering iron tip is heated and then applied to the wire so heat is transferred to the wire. The solder is then applied to the wire opposite the soldering iron. Once the wire is up to soldering temperature, it will melt the solder and pull the solder into the strands of wire, producing a strong, effective joint. Do not apply too much heat to the wire, or two things will happen. First, the solder will be drawn too far up the strands of wire, making a very long, nonflexible joint that is subject to breaking. The second problem is that the insulation may overheat and melt.

Once the joint is soldered, it has to be protected. This is best done with heat shrink tubing which shrinks when heat is applied to it with a heat gun. Most types of heat shrink tubing are hollow. Another type contains a glue, which, when heated with the heat gun, melts into and seals the joint. If there is no heat shrink tubing available, then it is possible to seal and protect the splice with electrical insulating tape.

SAFETY TIP

Although soldering is generally thought of as a simple process, it can be very dangerous. The solder, soldering iron, and wires are very hot and can cause severe burns. Be careful what you grab or where you set hot items. Molten solder can be flicked by a springy wire up into your eyes, so always wear safety glasses or goggles.

TECHNICIAN TIP

One mistake students make is trying to apply the solder directly to the tip of the soldering iron while the iron is heating up the wires. This does melt the solder, but it is likely that the wire is not hot enough for the solder to stick to it; instead the solder just globs on top of the wires, leading to what is called a cold joint. A cold joint has high resistance, and the wires are likely to break loose from the solder. One sign that the solder joint is good is that you can clearly see the outline of the wires on the surface of the solder all the way around the joint.

When using a soldering iron, you must be careful not to burn yourself or any part of the vehicle you are working on. The tip of the soldering iron has to be hot enough to melt metal solder, so make sure it is in a safe position and not touching anything while it is heating up. To solder wires and connectors, follow the steps in **SKILL DRILL 51-4**.

SKILL DRILL 51-4 Soldering Wires and Connectors

1. Safely position the soldering iron while it is heating up. While the soldering iron is heating, use wire strippers to remove an appropriate amount of the protective insulation from the wires.

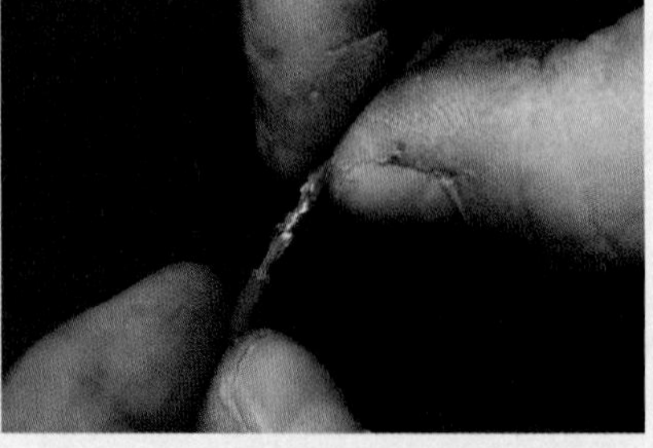

2. Twist the wires together to make a good mechanical connection between them.

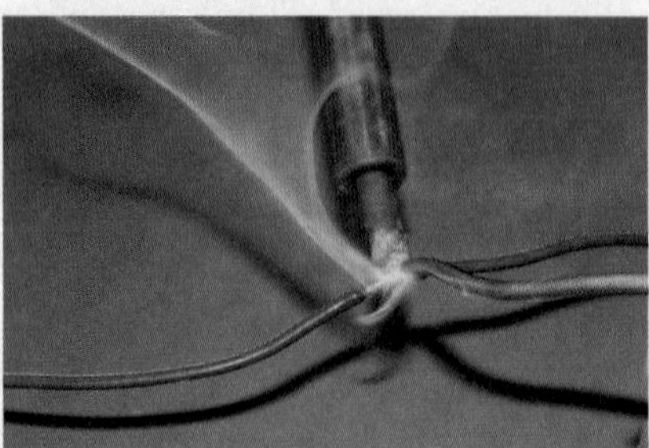

3. Tin the soldering iron tip, and gently heat up the wires while placing the solder opposite of the soldering iron. Allow the solder to be drawn into the joint.

SKILL DRILL 51-4 Soldering Wires and Connectors (Continued)

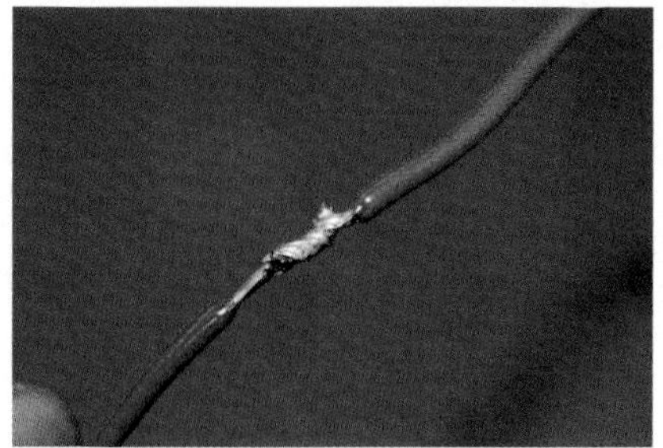

4. Pictured is a good solder joint where the solder has been drawn in.

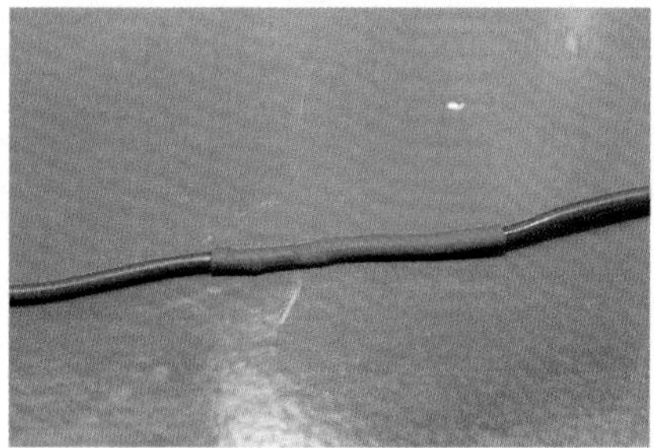

5. Once the electrical connection has been made and it has cooled enough for you to handle it, slide the insulator sleeve cover over the joint, and use a heat gun to shrink the tubing around the joint.

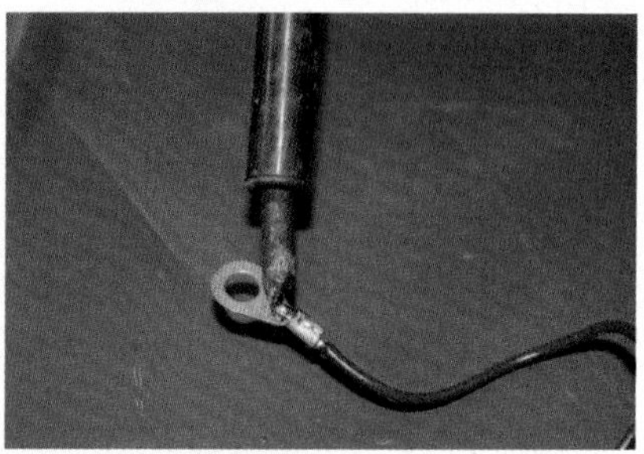

6. To solder a wire to a terminal connector, it is best to crimp it in place as before and use the solder to "glue" the joint together. Place the heated iron onto the terminal to get it hot enough to melt the solder applied to the end of the crimped wire tabs. Some solder will be pulled between the terminal and the wire. Cover the terminal with heat shrink tubing.

Repairing a CANbus Harness

N51004

Always check manufacturer information for the correct procedures when repairing wiring. For a repair on a CANbus system, extreme caution must be taken. The CANbus system carries high-speed data information and is susceptible to communication errors if the repair is not completed according to the manufacturer's recommendations. When measuring resistance or voltage on a CANbus, always use a digital multimeter, and disconnect the negative battery terminal when measuring resistance.

CANbus wires are usually a twisted pair with terminating resistors to reduce noise and interference. Use the correct-sized wires and terminals (if necessary) when making repairs. The challenge in repairing a twisted pair is that you need to perform the solder repair on wires that are twisted together. Simply put, trying to solder one wire back together when the other wire is very close to it is tough. Also, it is harder to use shrink tubing because the wires are twisted. It is possible to make these repairs; it just takes a bit of practice and patience. If both wires in the twisted pair must be repaired, it is best to stagger the joints so that both joints are not side by side, which could lead to a short circuit. To prevent damage, ensure that all insulation is replaced and the harness is secured correctly. If using a soldering iron and solder to make repairs, use a rosin-core flux, and clean the joint before insulating, to finalize the repair. Heat shrink tubing of the appropriate size ensures a protected and insulated joint.

To repair CANbus wires, follow the steps in **SKILL DRILL 51-5**.

SKILL DRILL 51-5 Repairing CANbus Wires

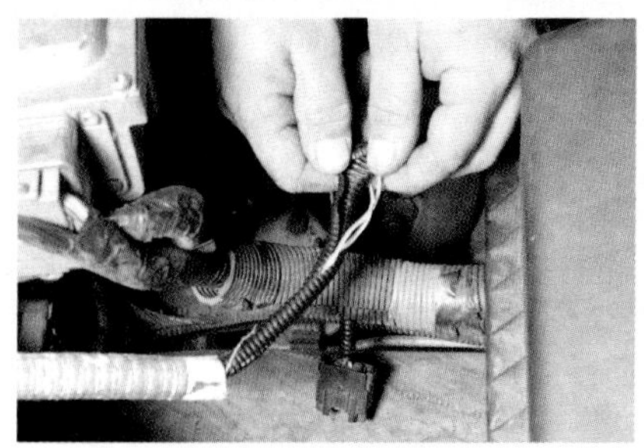

1. Carefully peel back any protective covering and shielding from the area to be repaired.

2. In most cases, you need to splice in a section of one or both wires to maintain equal length after repair. Cut out the bad section, and strip the ends.

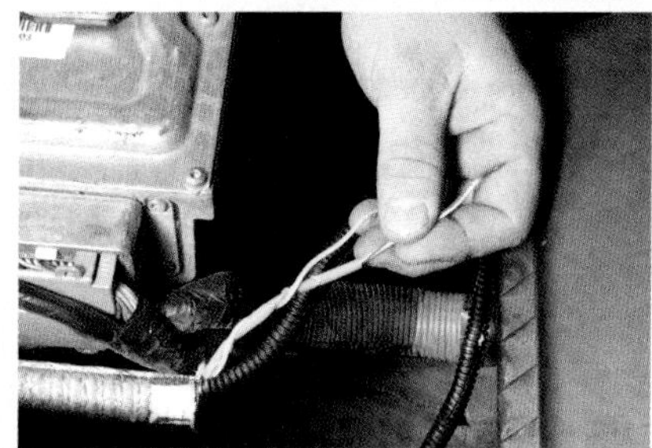

3. Carefully work a piece of shrink tube up the damaged wire.

SKILL DRILL 51-5 Repairing CANbus Wires (Continued)

4. Make sure the wires remain twisted at about one complete twist per inch (25.4 mm); the untwisted section should end up less than 2' (50 mm) after repair.

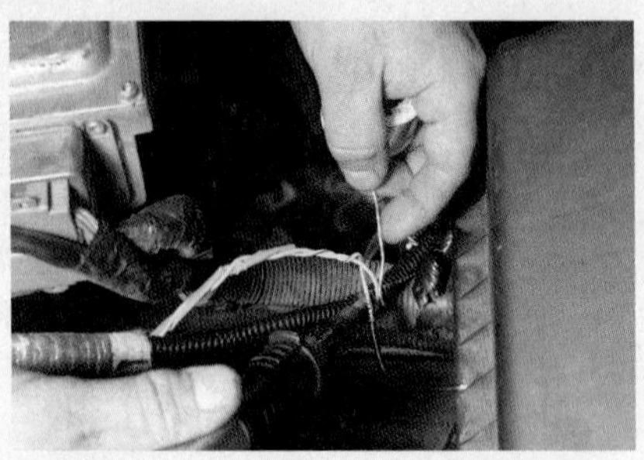

5. Using the correct wire size, twist the strands of the new wire to the old wire on both ends, and solder the wires.

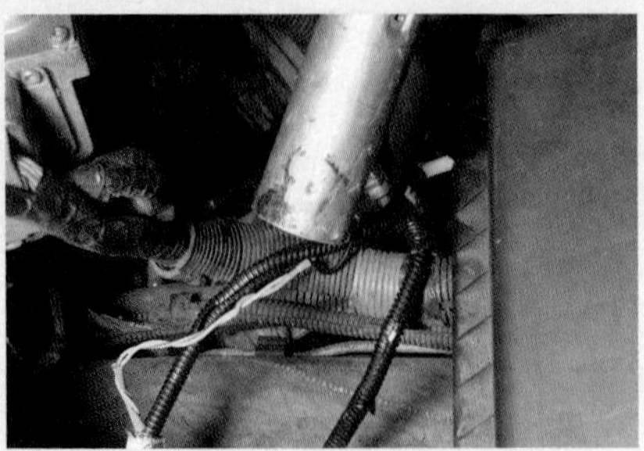

6. Slip the heat shrink tubing in place, and shrink it around the joint. Replace any protective covering, and secure appropriately.

▶ Wrap-Up

Ready for Review

- Electrical components must be correctly connected onto circuits and may have numbered or marked terminals to ensure proper connection.
- Circuit protection devices, which break the circuit during excessive current flow, are fuses, fusible links, and circuit breakers.
- Resistors are used to control voltage that reaches various components because they resist the current running through them.
- Resistor types include fixed, variable, thermistors, metal oxide varistors, and ballast resistors.
- Resistors are rated by both resistance value and power rating.
- Resistance value is indicated by colored bands, and tolerance is indicated by the number of identifying bands.
- Variable resistor types are rheostats, potentiometers, and thermistors.
- Thermistors are a type of conductor in which resistance value is affected by temperature.
- Capacitors are used to store electrical energy and are rated by capacitance, which is the amount of change stored in each plate for a given potential voltage between the plates.
- Diodes are used to restrict current flow to one direction only.
- Transistors (NP–N and P–NP types) are used as switches and to amplify currents.
- Control modules are designed to monitor multiple inputs from sensors and circuits and respond to those inputs.
- Delay circuits can turn an electrical device on or off after a specified time delay.
- Microprocessors are designed to monitor and control most electrical systems on a modern vehicle.
- Most vehicle wires are braided copper with plastic insulation, but other types are shielded wires and ribbon.
- Wire shielding to prevent noise (unwanted electromagnetic induction) can be twisted pair, Mylar tape, or drain lines.
- Proper operation of electrical circuits requires correct wire size.
- Length and diameter of a cable determine its resistance.
- Copper has low resistance value, although as the wire length increases, so does resistance within the wire; therefore, the cross-sectional area needs to increase to overcome resistance.
- Terminals and connectors are fitted to the ends of cables to provide low-resistance cable termination.
- Terminals can be push-on spade terminals, eye ring terminals, and solder-type terminals.
- Connectors can be permanent cable joiners, wiring harness connectors, or male or female connectors.
- Wiring harnesses are used to bind wires together within a sheath of insulating tape or tubing.
- Wiring diagrams are generally split up into systems and subsystems.
- Microprocessors and their assorted components are mounted on circuit boards with conductive tracks designed to connect electronic component leads with one another.
- Electronically controlled systems are integrated to a multiplexed serial communications network, such as a controller area network bus (CANbus).
- Fiber-optic cables can transmit light signals over very long distances.
- Electrical measurement tools include the ammeter, voltmeter, ohmmeter, and digital multimeter.

- Tracing wires on a wiring diagram help you:
  - determine what each wire in the diagram does.
  - organize your thoughts so that you can understand how electricity flows through the circuit.
  - keep from losing your place or forgetting what a particular wire does.
  - have confidence that you have properly diagnosed the problem.
- Use the proper crimping tools when replacing terminals.
- Use only rosin core solder on electrical connections.
- Don't overheat or underheat solder joints.
- On CANbus repairs, maintain the proper twist before and after the joint(s), and make sure the untwisted joint is no more than 2' (50 mm).

Key Terms

American wire gauge (AWG) A standard used to identify different wire sizes.

armature The rotating wire coils in motors and generators. It is also the moving part of a solenoid or relay, and the pole piece in a permanent magnet generator.

breakdown The loss of electrical insulation properties.

bus An electrical circuit for distributing electrical signals bidirectionally.

butt connector A crimp or solder joint that creates a permanent connection.

capacitance The ability of a capacitor to store an electrical charge.

capacitor A device that can quickly store a small amount of electrical energy, at which point it is charged.

charge In measuring capacitance, the amount of electrical energy present.

charge carrier A mobile particle that has a positive or negative electrical charge.

circuit or schematic diagram A pictorial representation or road map of the wiring and electrical components.

circuit breaker A device that trips and opens a circuit, preventing excessive current flow in a circuit. It can be reset to allow for reuse.

commutator A device made on armatures of electric generators and motors to control the direction of current flow in the armature windings.

connector The plastic housing on the end of a wiring harness that holds the wire terminals in place. It can also refer to a type of wire terminal that connects wires together or to a common point such as a bolt.

control module A generic term that identifies an electronic unit that controls one or more electrical systems in the vehicle; also called a control unit.

controller area network bus (CANbus) A wire data network found in vehicles to connect control modules so they can communicate quickly and easily.

delay circuit A combination of electrical and electronic components that provides a time delay for switching an electrical circuit.

depletion layer An area of neutral charge in semiconductors.

diode A two-lead electronic component that allows current flow in one direction only.

doping The introduction of impurities to pure semiconductor materials to provide N- and P-type semiconductors.

drain line A wire included in a harness, with one end grounded to reduce interference or noise being induced into the harness.

eye ring terminal A type of crimp or solder terminal that has an enclosed eyelet to connect the terminal with a bolt or screw.

fiber optics Fine glass fibers through which light is transmitted.

fixed resistor A resistor that has a fixed value.

flasher can A mechanism that turns the vehicle's turn signal and hazard flasher bulbs on and off.

gallium-arsenide A semiconductor used in high-frequency circuits.

germanium A type of semiconductor.

male and female terminal A crimp or solder terminal on which the male and female ends join to create a removable low-resistance connection.

microprocessor An electronic control unit that can process data and control one or more devices.

Mylar tape Polyester film that may be metalized and incorporated into a wiring harness to provide electrical shielding.

negative temperature coefficient (NTC) A characteristic of materials whereby resistance decreases as temperature increases.

normally closed (NC) An electrical contact that is closed in the at-rest position.

normally open (NO) An electrical contact that is open in the at-rest position.

NP–N transistor A transistor in which P-type material is sandwiched between two layers of N-type material.

N-type Semiconductor material with a small amount of extra electrons.

optical cable A cable that contains a number of optical fibers.

P–N junction The junction between N- and P-type semiconductor materials.

P–NP transistor A transistor in which N-type material is sandwiched between two layers of P-type material. This type of semiconductor material has holes, meaning it is missing electrons.

polarity sensitive A term used to describe a component that must be connected into a circuit with the correct polarity to its terminals.

positive temperature coefficient (PTC) A characteristic of materials whereby resistance increases as temperature increases.

potentiometer Also called a pot, a three-terminal resistive device with one terminal connected to the input of the resistor, one terminal connected to the output of the resistor, and the third terminal connected to a movable wiper arm that moves up and down the resistor.

primary winding The coil of wire in the low voltage circuit that creates the magnetic field in a step-up transformer.

printed circuit The fine copper strip or track attached to an insulated board for mounting and connecting electronic components.

P-type Semiconductor material with holes where electrons are missing.

push-on spade terminal A disconnectable type of crimp or solder terminal used to terminate electrical wires.

rheostat An adjustable resistor that varies current flow through a circuit.

ribbon cable A type of flat harness in which cables are insulated from each other but joined together side by side.

secondary winding The coil of wire in which high voltage is induced in a step-up transformer.

shielded wiring harness A wiring harness that has shielding built into it to protect it from induced electrical interference.

silicon A material commonly used to make semiconductors.

silicon carbide A type of material used to make semiconductors.

solder-type terminal A terminal that requires soldering to fasten the terminal to the cable or wire.

solenoid An electromagnet with a moving iron core that is used to cause mechanical motion.

solid-state relay A relay that performs the function of a mechanical relay but uses only electronic components.

speed control circuit A circuit that controls the speed of a motor.

step-down transformer A transformer used to reduce the voltage, such as to allow a battery charger operated on 120 volts to charge a 12-volt battery.

step-up transformer A transformer used to increase the voltage from a lower input voltage to a higher output, such as an ignition coil.

switch An electrical device with contacts that turns current flow on and off.

terminal A means of providing a low-resistance connection or termination at the end of a wire.

thermistor A temperature-sensitive resistor. Its resistance varies substantially with temperature change.

transformer action The transfer of electrical energy from one coil to another through induction in a transformer.

transistor A semiconductor device that allows a small current in the base lead to control a larger current through the emitter collector leads.

turn signal switch A switch that turns the left and right turn signal lights on and off.

twisted pair Two conductors that are twisted together to reduce electrical interference.

variable resistor A component that has a mechanism for varying resistance.

wire A conductor usually made of multi-stranded copper with an external insulated coating.

wiring diagram A schematic drawing and symbol representation of the wiring and components; also called an electrical schematic.

wiring harness A collection of wires or cables insulated from each other but bound together.

wiring harness connector A plug that contains multiple terminals with male and female ends.

Zener diode A diode that forward biases when a certain voltage is reached.

Review Questions

1. Which of the following components can be programmed and are an integral part of other electrical components?
   - **a.** Sensors
   - **b.** Zener diodes
   - **c.** Microprocessors
   - **d.** Transistors
2. Choose the correct statement.
   - **a.** Semiconductor devices use high operating voltages.
   - **b.** Semiconductor devices are sensitive to heat and voltage spikes.
   - **c.** Semiconductor devices are not reliable.
   - **d.** Semiconductor devices need regular maintenance.
3. Which of the following materials is widely used to build high-speed semiconductors?
   - **a.** Silicon
   - **b.** Silver oxide
   - **c.** Gallium-arsenide
   - **d.** Gold-iridium
4. All of the following statements about transistors are correct *except*:
   - **a.** Transistors are used as a switch and to amplify currents.
   - **b.** They are made up of two sections of different polarity.
   - **c.** Transistors have two PN junctions.
   - **d.** Transistor symbols always have an emitter, base, and a collector.
5. What is a common use of speed control units in vehicles?
   - **a.** To control the fan speed in the HVAC system and the radiator fan
   - **b.** To control the speed of the turn signal flashes
   - **c.** To control the speed of the car
   - **d.** To control the speed of the engine
6. Which of the following types of shielding is used in wiring to prevent electromagnetic interference (EMI)?
   - **a.** Special plastic coated wires
   - **b.** Silicon tape
   - **c.** Nylon tape
   - **d.** Mylar tape
7. All of the following statements are true *except*:
   - **a.** Wires are the dominant signal carriers in a vehicle.
   - **b.** A number of wiring harnesses are located throughout the vehicle.
   - **c.** On newer vehicles, most of the wires have been replaced with fiber optic cables.
   - **d.** Wires and wiring harnesses are the arteries of the vehicle's electrical system.

8. Fiber optic cable has all of the following advantages over a conventional copper cable *except*:
   a. It can carry much more data than copper wires.
   b. It is less affected by interference.
   c. Light in one fiber does not interfere with light in another fiber.
   d. It is expensive when compared to conventional cables.
9. If both wires in the twisted pair CANbus need to be repaired, it is best to:
   a. replace both wires.
   b. replace one of the wires.
   c. stagger the joints.
   d. repair without removing the insulation.
10. When installing a terminal on a wire, excessive bare wire may:
   a. expose the wire to a potential short circuit.
   b. reduce the current-carrying capacity of the wire.
   c. not achieve a good connection.
   d. increase the current-carrying capacitiy of the wire.

ASE Technician A/Technician B Style Questions

1. Tech A says that if the specified fuse keeps blowing, it is generally okay to replace it with a larger fuse. Tech B says that a fusible link is one type of circuit protection device. Who is correct?
   a. Tech A
   b. Tech B
   c. Both A and B
   d. Neither A nor B
2. Tech A says that 18 AWG wire is larger than 12 AWG wire. Tech B says that the larger the diameter of the conductor, the more electrical resistance it has. Who is correct?
   a. Tech A
   b. Tech B
   c. Both A and B
   d. Neither A nor B
3. Tech A says that some relays are equipped with a suppression diode in parallel with the winding. Tech B says that some relays are equipped with a resistor in parallel with the winding. Who is correct?
   a. Tech A
   b. Tech B
   c. Both A and B
   d. Neither A nor B
4. Tech A says that NTC thermistors alter their resistance value inversely to temperature (as the temperature increases, resistance value decreases). Tech B says that PTC thermistors are used as throttle position sensors. Who is correct?
   a. Tech A
   b. Tech B
   c. Both A and B
   d. Neither A nor B
5. Tech A says that a transistor has a single P–N junction. Tech B says that a transistor is a semiconductor device used as a switch and to amplify currents. Who is correct?
   a. Tech A
   b. Tech B
   c. Both A and B
   d. Neither A nor B
6. Tech A says that wiring diagrams are essentially a map of all of the electrical components and their connections. Tech B says that in many cases, each wire in wire harnesses use two colors; the first one is the solid color, and the second one is the stripe. Who is correct?
   a. Tech A
   b. Tech B
   c. Both A and B
   d. Neither A nor B
7. Tech A says that a twisted pair is two wires, twisted together, that deliver signals between common components. Tech B says that in many cases, wiring diagrams are set up with power on the top of the diagram and ground on the bottom. Who is correct?
   a. Tech A
   b. Tech B
   c. Both A and B
   d. Neither A nor B
8. Tech A says that it is best to use a knife or other type of sharp tool to cut away the insulation when stripping a wire. Tech B says that any issues with wiring are more likely to be with the terminals than with the wires themselves. Who is correct?
   a. Tech A
   b. Tech B
   c. Both A and B
   d. Neither A nor B
9. Tech A says that while soldering wires, apply the solder to the wire opposite the soldering iron. Tech B says that you should heat the wire up until the plastic insulation melts. Who is correct?
   a. Tech A
   b. Tech B
   c. Both A and B
   d. Neither A nor B
10. Tech A says that a relay is a one-way electrical check valve used in alternators to change AC into DC. Tech B says that a relay uses electromagnetism to open or close a switch. Who is correct?
   a. Tech A
   b. Tech B
   c. Both A and B
   d. Neither A nor B

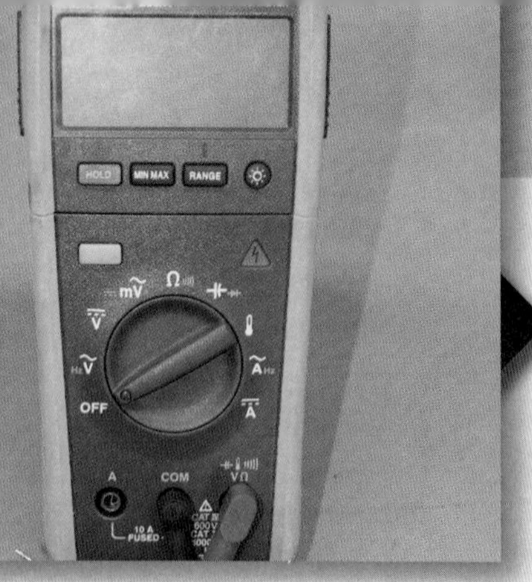

CHAPTER 52

Meter Usage and Circuit Diagnosis

NATEF Tasks

- **N52001** Demonstrate the proper use of a digital multimeter (DMM) when measuring source voltage, voltage drop (including grounds), current flow, and resistance. (MLR/AST/MAST)
- **N52002** Demonstrate knowledge of electrical/electronic series, parallel, and series-parallel circuits using principles of electricity (Ohm's law). (MLR/AST/MAST)
- **N52003** Demonstrate proper use of a test light on an electrical circuit. (MLR/AST/MAST)
- **N52004** Use fused jumper wires to check operation of electrical circuits. (MLR/AST/MAST)
- **N52005** Inspect and test fusible links, circuit breakers, and fuses; determine needed action. (MLR/AST/MAST)
- **N52006** Inspect and test switches, connectors, and wires of starter control circuits; determine needed action. (AST/MAST)
- **N52007** Check electrical/electronic circuit waveforms; interpret readings and determine needed action. (MAST)

Knowledge Objectives

After reading this chapter, you will be able to:

- **K52001** Explain the purpose, function, and layout of a DMM.
- **K52002** Describe the ranges, settings, and setup of a DMM.
- **K52003** Explain the process of measuring volts, ohms, and amps.

Skills Objectives

After reading this chapter, you will be able to:

- **S52001** Perform series circuit measurements.
- **S52002** Perform parallel circuit measurements.
- **S52003** Perform series-parallel circuit measurements.
- **S52004** Perform basic electrical circuit testing and diagnosis.
- **S52005** Locate opens, shorts, grounds, and high resistance.
- **S52006** Inspect a circuit with jumper leads and/or a test light.
- **S52007** Inspect circuit protection and control devices.

▶ Introduction

Digital multimeters (DMMs) and oscilloscopes are electrical measuring tools frequently used to diagnose and repair electrical faults (**FIGURE 52-1**). Like many diagnostic tools, practice is required to understand how the DMM and oscilloscope are used to take electrical measurements and how to connect them into electrical circuits to ensure correct readings are obtained. Once a reading is obtained, it has to be interpreted and applied, in conjunction with knowledge of electrical theory, to diagnose the circuit being tested. This chapter provides an explanation of how to set up and use a DMM for measuring voltage, amperage, and resistance.

This chapter also includes a number of exercises that will expand your knowledge on using a DMM, allowing you to practice taking readings and then apply the results. It also relates the practical circuit examples and DMM readings to Ohm's law calculations. Basic oscilloscope use and typical waveforms are also covered. Knowing how to properly use DMMs and oscilloscopes, as well as how to interpret and apply their readings, will allow you to diagnose electrical faults, making you very valuable to your employer.

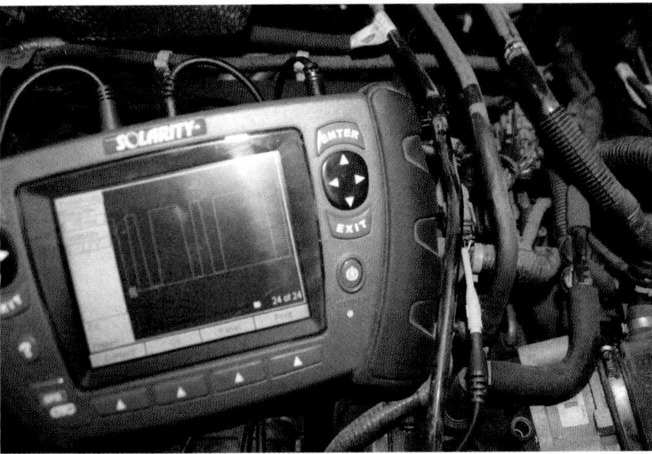

FIGURE 52-1 Electrical measuring tools allow technicians to look inside electrical circuits to see what is happening, a critical step in diagnosing problems.

▶ Digital Multimeters

K52001

A DMM or a digital volt-ohmmeter (DVOM) is a versatile and useful piece of test equipment (**FIGURE 52-2**). They are called digital meters because they give a numerical reading on a digital display. An analog meter, by comparison, uses a needle that hovers over a series of scales, requiring the technician to determine the numerical value of the reading. DMMs are easier to read, which means that a technician is less likely to get the wrong reading. The DMM tends to be the first test tool selected

FIGURE 52-2 A digital multimeter (DMM) is a versatile and useful piece of test equipment.

You Are the Automotive Technician

Your supervisor requested that you train the shop apprentice on how to use a DMM. You will be using the DMM to teach the apprentice how to take simple measurements that will come in handy when doing vehicle inspections such as measuring battery and charging system voltage, testing a fuse to see if it is blown, and performing a voltage drop on a simple circuit. You show him how to set up the meter for each of these tests. Once he is competent in setting up the meter, you show him how to take each type of reading. The following is the list of questions your supervisor has asked you to review with the new employee after the hands-on training:

1. What are the three most common measurements taken by a DMM, and how is the meter connected to the circuit for each one?
2. What are the steps in setting up a meter to take a reading?
3. What is "available voltage," and how is it measured?
4. Describe an unwanted voltage drop and what causes it.
5. Describe the process of measuring voltage drop.
6. What two conditions can cause the current flow in a circuit to be too high?

for electrical diagnosis and repairs. Basic DMMs can measure alternating current (AC) and direct current (DC) voltage, AC and DC amperage, and resistance. Most modern DMMs can also measure frequency and temperature and have a dedicated diode test capability.

DMMs come in a variety of layouts and qualities. You need to get used to the meters in your shop so you will know their capabilities and how to use them. Most

DMMs of average quality or better are "fused," meaning that one or more "fast-blow" fuses are included inside the DMM. If the amperage is too high, the fuse will blow, protecting the meter. If the meter is unfused, it will not be protected and could be damaged if used incorrectly when measuring amperage.

DMMs and test leads also should have a CAT rating listed on the front. CAT is short for "category." Each level, or CAT, is designed to work safely on higher-powered electrical systems (**TABLE 52-1**). CAT ratings were not designed initially for automotive meters because most vehicles use low voltage, but with an increasing number of hybrid and electric vehicles on the road, operating on very high voltages, CAT ratings are becoming important for automotive technicians. Hybrid vehicles typically require meters and test leads rated as CAT III or CAT IV. Another thing to keep in mind if you are working on high-voltage systems is that you need to wear a pair of certified and tested rubber-insulated gloves, most likely with leather protectors over them. Always use the proper CAT-rated meter and leads along with the proper personal protective equipment when working on high-voltage systems.

Digital Multimeter Purpose

DMMs are used to take many different electrical circuit measurements and are one of the first tools used when conducting electrical repair or diagnosis work. As a voltmeter, the DMM can measure electrical voltage within circuits, for example, the voltage available at a fuse, switch, or lamp. The DMM can also measure the resistance of a component, connector, or cable, such as the resistance of an ignition coil to check against specifications. DMMs can also measure current flow in circuits, such as the amount of current flow through an air conditioning compressor clutch, to determine if it meets specifications. Clearly, a DMM is a very versatile tool, which explains why it is the most commonly used electrical diagnostic tool. In the next several sections, we further explore DMMs and how to use them.

Digital Multimeter Layout and Accessories

There are two main components of a DMM: the main instrument body and the test leads that connect the DMM to the circuit being tested. The DMM main instrument body has a function switch to choose the type of electrical measurement to be taken, a digital display to report the readings, and slots to connect test leads (**FIGURE 52-3**). Test leads are used to connect the DMM to the circuit being tested and come in pairs: one red, the other black. Basic leads have a probe on one end for making the connection with the electrical circuit being tested and a connector on the other end for plugging into the slots of the DMM. A wide variety of test leads and adapters are available to make it easier to use the DMM; for example, alligator clips enable hands-free connection of the leads. Accessories such as temperature probes and inductive current clamps connect to the input slots of the DMM and convert temperature or current flow into a voltage that can be measured by the DMM (**FIGURE 52-4**).

Ranges and Scales

K52002

For resistance measurements, DMMs read in a range from very small quantities, 1/10,000 of a unit, up to very large quantities in the range of millions of units. It is not possible for DMMs to effectively and accurately measure such ranges with only a single range or scale; they must have multiple ranges or scales. But before we can talk about those ranges, we need to understand that the DMM screen can only display four or five digits. This means that symbols must be used to substitute for some of

TABLE 52-1: Meter CAT Ratings

Overvoltage Category	Short Description	Examples
CAT I	Electronics	Low-power electronic equipment such as copiers, etc.
CAT II	Single-phase plug-in tools and equipment	Portable tools, appliances, etc.
CAT III	Three-phase fixed equipment and single-phase commercial lighting	Equipment in fixed installations (this is the minimum rating required for hybrid vehicles)
CAT IV	Three-phase utility connection, any outdoor wires	Main power wires from the utility company; outside wire for lighting (this rating is for heavier-duty meters and leads for hybrid vehicles)

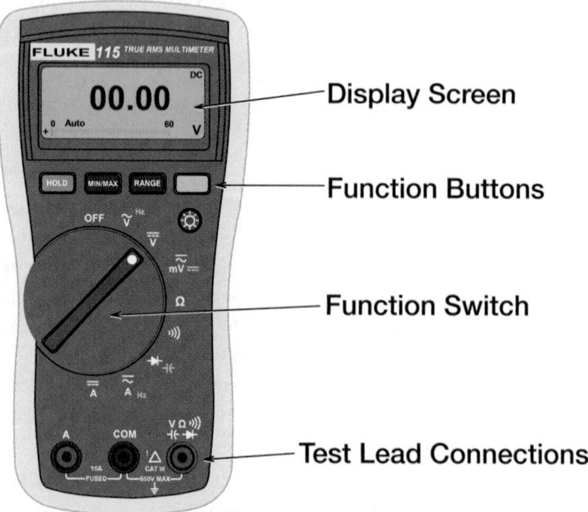

FIGURE 52-3 The body contains the display screen, function switch, function buttons, and slots to connect the leads.

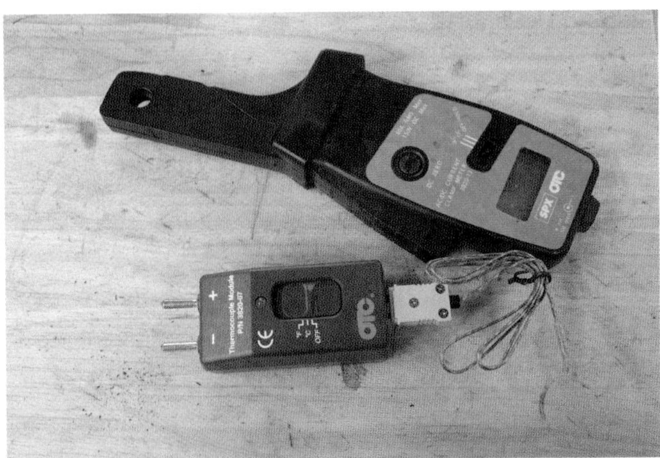

FIGURE 52-4 Accessories such as temperature probes and inductive current clamps are available for DMMs, extending the meter's usefulness.

TABLE 52-3: DMM Ranges

Function	Range	Resolution
mV DC	0–600.0 mV	0.1 mV
V DC	0–6.000 V	0.001 V
	0–60.00 V	0.01 V
	0–600.0 V	0.1 V
	0–1000 V	1 V
Ohms	0–600.0 Ω	0.1 Ω
	0–6.000 kΩ	0.0001 kΩ
	0–60.00 kΩ	0.01 kΩ
	0–600.0 kΩ	0.0001 MΩ
	0–40.00 MΩ	0.01 MΩ

the digits. **TABLE 52-2** shows the common symbols, their prefix, and the factor they represent. You will have to place the appropriate electrical symbol—V, A, or Ω—behind the factor symbol, based on what you are measuring. For example, 2168 mV would be the same as 2168 millivolts. It could also be called 2.168 volts, as there are 1000 millivolts in 1 volt. Either designation is correct. The challenge is taking the meter reading and making sense of it, which takes practice.

Once you understand the symbols and what value each represents, you are ready to decide which range to set the meter to. **TABLE 52-3** lists a typical set of DMM ranges; however, there is no single range or scale value used by DMM manufacturers. The resolution indicates the accuracy of the count within any given range and is different for each range. To achieve the most accurate reading, always select the lowest range possible for the value being measured. For example, if you are measuring 12 volts, you should select the 60-volt range, because 6 volts would be too low and 600 volts would be less accurate.

Most modern DMMs have an automatic ranging capability while maintaining the ability to be used in a manually selected range, which means the user can determine the range. When used in the auto range, the DMM selects the best range for the value being measured so that the technician does not have to be concerned with manually setting the range (**FIGURE 52-5**). But be careful! The meter does not give you flashing light warnings that it has changed ranges, so it is extremely easy to miss that. Many a technician has been led down the wrong diagnostic path by thinking the 12.6 on the meter was volts when in fact the meter had auto-ranged to millivolts; so instead of having full power, the circuit being tested had almost *no* power. To prevent this mistake, many instructors require their students to use only manual ranging when using their meter.

TABLE 52-2: DMM Values

Factor	Prefix	Symbol
1,000,000	Mega	M
1000	Kilo	k
1	No prefix	
0.001	milli	m
0.000001	Micro	μ

Min/Max and Hold Setting

Special settings are incorporated into the design of many DMMs, to assist you in taking measurements of rapidly changing values or to freeze the display so that an individual reading is not lost. In the **min/max setting**, the DMM records in memory the maximum and minimum reading obtained during the time the DMM is connected to a source to take a reading (**FIGURE 52-6**). The min/max setting is often used to measure vehicle battery voltage while the engine is cranking or the battery is charging. During cranking of the starter motor, the battery voltage is at its lowest. In measuring voltage in min/max mode, a DMM captures the minimum and maximum battery voltage.

A limitation in the use of the DMM is the sample rate, which is the speed at which the DMM can sample the voltage. The DMM does not continuously sample the voltage; rather, it checks the voltage at regular intervals, or at a sample rate. Although this occurs quickly—for example, every 100 milliseconds—it does

FIGURE 52-5 Meter set to auto range and reading mV.

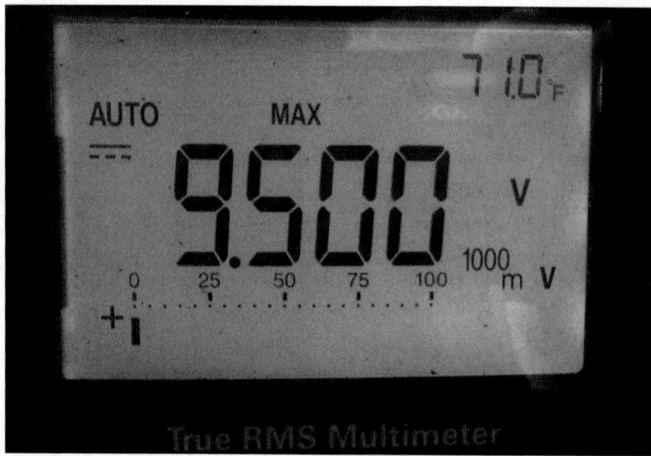

FIGURE 52-6 Meter showing MAX reading.

mean that if a transient voltage occurs between samples, it will not be recorded by the DMM. Where quicker sample rates are required, other tools such as oscilloscopes can be used.

The **hold function** allows the display to be frozen. When the hold function is activated, the display will hold the value on the display until the function or the DMM is turned off (FIGURE 52-7). A variation of the hold function is the auto-hold function found on some DMMs. When activated, the auto-hold function takes a measurement and freezes, or holds, the display until the function or DMM is turned off. This function can be useful when taking measurements in difficult locations, such as underneath a dash, where you may not be able to watch the meter display while making the meter connections.

Setting Up a DMM

To set up a DMM to take accurate measurements, you need to know whether you will be measuring resistance, voltage, or current. You should also know the reading that you are expecting so you can be sure to set up the meter appropriately. If very high voltages are to be measured, it is important to make sure the DMM and leads match the appropriate CAT rating for use at the voltages you will be testing. All of this information determines the way in which you set up the DMM, including the connections you need to make on the DMM and the range you select. Resistance measurements on wires should be undertaken with the circuit unpowered. If measuring the resistance of components, they should be removed from the circuit.

The following steps describe how to set up a DMM:

1. Know what you are testing—volts, amps, or ohms.
2. Know the value you expect to be reading (specification).
3. Select the meter, leads, and probes with the appropriate CAT rating to suit the measuring task.
4. Connect the leads to the DMM.
5. Use the function switch to select the type of measurement to be undertaken (e.g., resistance, volts, or amps; DC or AC).
6. Select the correct meter range if you are using a manual ranging meter.
7. Connect the leads to the circuit being tested.
8. Read the meter display; remember to look at the factor symbol and apply that to the reading.

Test Leads: Common and Probing

N52001

Many people incorrectly label the red lead as positive and the black lead as negative. However, if you look at your meter near the test lead slots, you will not see a "+" or a "–" anywhere. What you will see is "A" (typically 10 A), "mA," "common," and "V/Ω" (FIGURE 52-8). "Common" just means that the slot is common to all of the functions of the meter. In other words, this lead does not need to be moved when different functions of the meter are typically accessed. On the other hand, the red lead does have to move, depending on what function of the meter is being used. That is why it is labeled with the V/Ω symbol, and not "+." If you find this distinction questionable, consider the following: Later on, when we measure various electrical signals at the same time

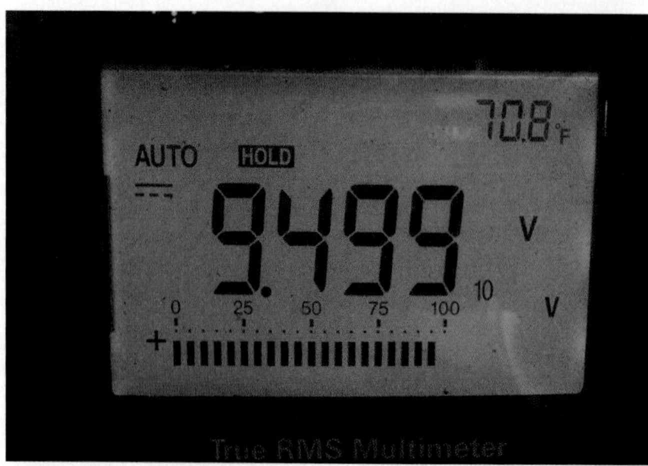

FIGURE 52-7 Meter showing HOLD reading.

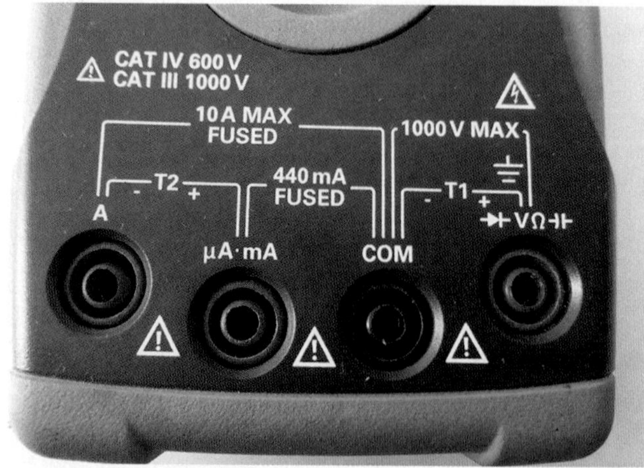

FIGURE 52-8 Slots for meter leads.

on an oscilloscope, we need more than just the red lead. In fact, we also typically use a yellow, a blue, and a green test lead. In all of these situations, the test lead (no matter the color) acts as a probe into the circuit. So rather than referring to the red lead as the positive lead, it is more accurate to refer to it as the probing lead for the DMM. We can then introduce probing leads of other colors when we use an oscilloscope, red being one of them.

The other important note is that the meter screen will always read what the probing lead is touching as compared to what the common lead is touching. For example, if the common lead is touching the battery's negative post and the probing lead is touching the positive post, the meter screen may, or may not, display a "+" before the reading (**FIGURE 52-9**). That means that the probing lead is touching something more positive than the common lead. If we reverse the leads, the meter screen will display a "–" before the number, meaning that the probing lead is touching something more negative than the common lead (**FIGURE 52-10**). When you understand this concept, rather than jumping to the conclusion that "the meter leads are hooked up backward," you will be ready to start diagnosing all kinds of strange electrical problems, especially with ground issues and charging system issues.

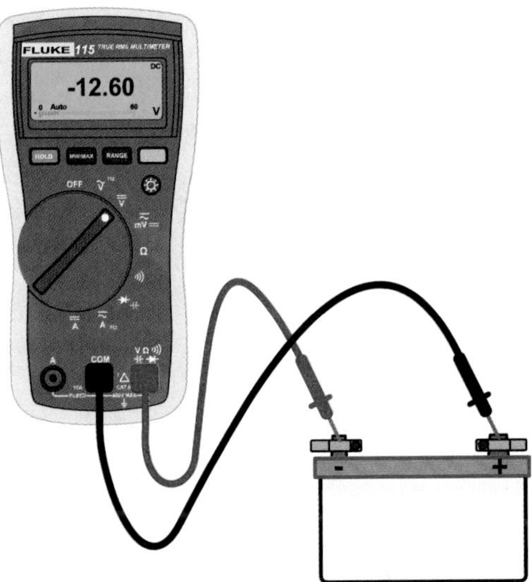

FIGURE 52-10 In this situation, the probing lead is connected to something more negative than the common lead, and is indicated by the "–" in front of the reading.

Probes and Probing Techniques

Gaining access to test points in a circuit requires a variety of probes and **probing techniques**. Some examples of probes are alligator clips, fine-pin probes, and insulation piercing clips (**FIGURE 52-11**). Make sure you know the voltage limits of the probes you use, because high-voltage measurements require special probes that are designed for that purpose. The standard probe leads that are supplied with a DMM are basic straight metal probes useful for making quick measurements in circuits, but they do require the use of both hands to hold them in place. Leads with alligator clips, which come in various sizes, allow the DMM leads to be clipped onto the circuit and held in place by themselves, freeing up your hands for other tasks. These clips are particularly useful for connecting to larger terminals, such as battery terminals.

Back-probing occurs when the probe is pushed in from the backside of a connector to make contact with the terminal. To perform this task, very fine pins are used to reduce the possibility of damage. The pins are designed to slip into the back of connectors and provide contact without causing damage (**FIGURE 52-12**). Insulation-piercing probes are also available but should rarely be used. They have sharp fine pins that pierce the insulation on conductors to create a connection with the wire. If used, remember to always reinsulate the hole that the probe makes to prevent any corrosion. Use liquid insulation or a similar product to reinsulate; do not use room temperature vulcanizing (RTV) silicone, which attracts moisture as it cures, potentially causing corrosion. Because it may result in damage

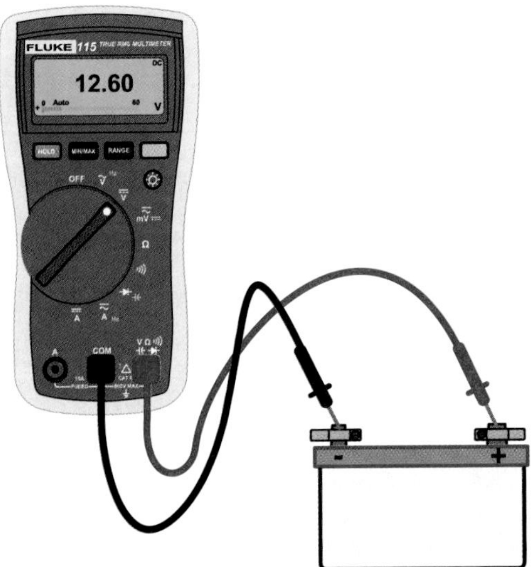

FIGURE 52-9 The screen reads what the red lead is measuring as compared to the black lead. If the black lead is on the negative battery post and the red lead is on the positive battery post, the screen may, or may not, have a "+" in front of the reading.

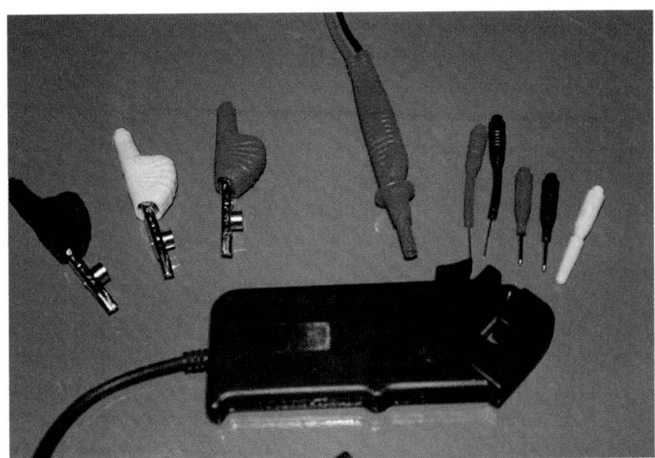

FIGURE 52-11 There are many different leads and probes you can use, depending on the circuit being tested.

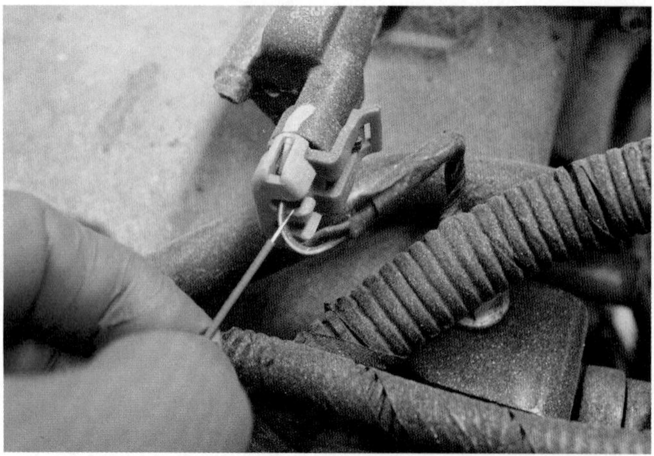

FIGURE 52-12 Back-probing a connector to take a voltage reading.

to the insulation or conductor, many vehicle manufacturers do *not* allow this type of probe.

Measuring Volts, Ohms, and Amps

K52003

The most common measurements taken with DMMs are voltage, resistance, and amperage. To take voltage measurements, the probing lead (red) is connected to the volts/ohms, or V/Ω, slot, and the common lead (black) is connected to the common, or COM, slot of the DMM. An appropriate range or auto range is selected on either AC or DC voltage, depending on the voltage to be measured. The probing lead is typically connected to the positive side of the circuit being tested, and the common lead to the negative post of the battery, or good ground. Connecting the voltmeter in this way is used to check the voltage available at various places in the circuit. This test is called checking available voltage (**FIGURE 52-13**). Watch your screen. If the "+" or "–"

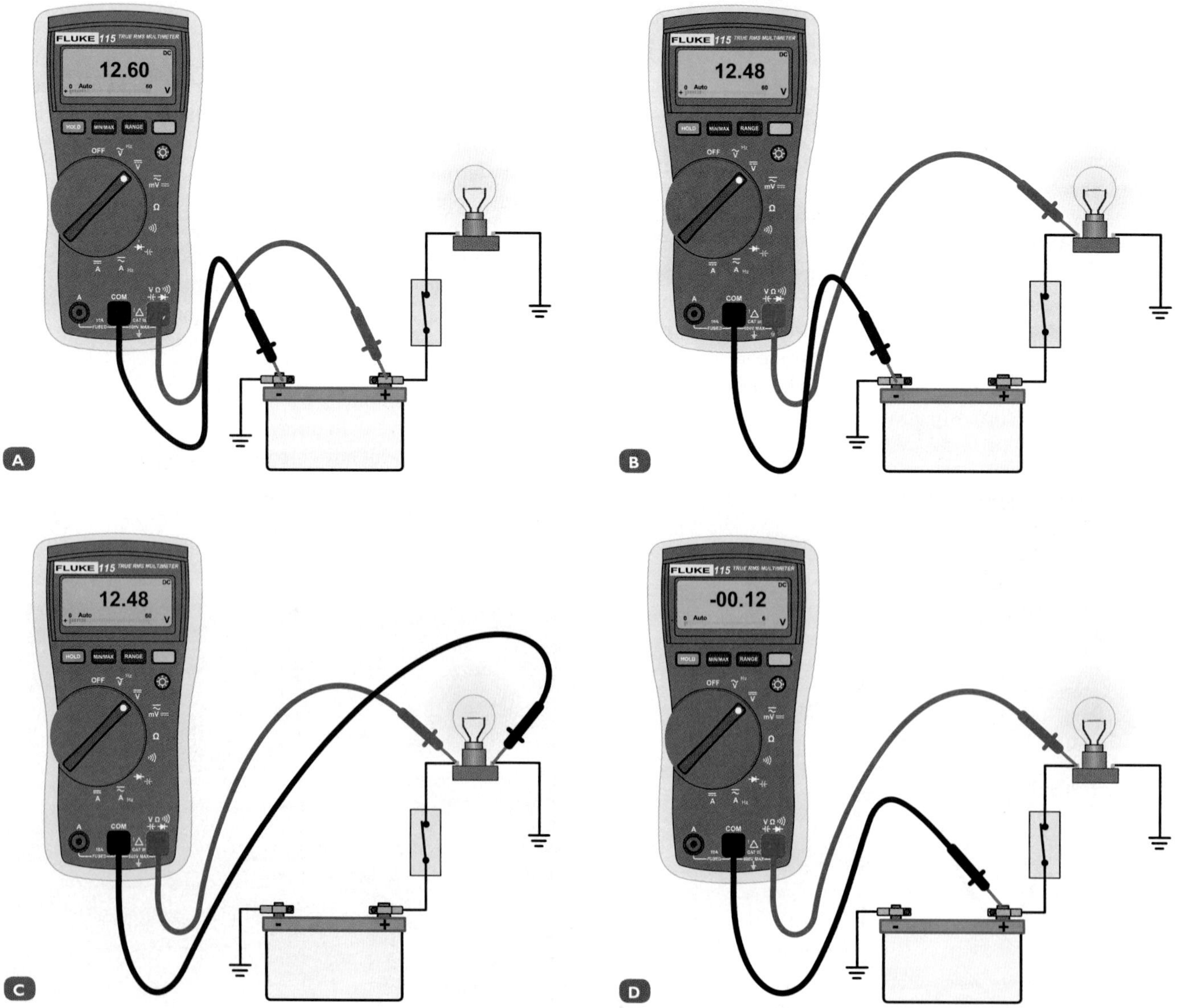

FIGURE 52-13 When connecting a voltmeter in the circuit, the voltmeter leads can be placed anywhere, which will then give you the difference in voltage between those points. Note that **A** and **B** are available voltage tests. **C** and **D** are voltage drop tests.

is not what you were expecting, check the leads to verify they are connected the way you intended. If you still get an unexpected reading, stop and analyze the situation. Ask yourself, what could cause the meter to read that way? Then brainstorm the options.

> **TECHNICIAN TIP**
>
> A technician recently posted an electrical problem on a technical forum. He said that he had hooked up a voltmeter with the black (common) lead on the negative battery terminal and the red (probing) lead on the vehicle engine ground, with the engine running. The meter read a negative number. He asked the forum if he had the meter leads hooked up backward. He received several comments saying that yes, he had hooked them up backward. However, those technicians were not correct. His probing lead was registering a reading that was more negative than the common lead. But what could be more negative than the negative post of the battery? When the engine is running, the alternator can be more negative than the negative battery post. So what his DMM was trying to tell him was that there was a voltage drop between the negative battery post and the alternator frame. If he had understood that the probing lead was not lying to him, that it was reporting exactly what it was touching compared to what the common lead was touching, then he could have started down the path to diagnosing what it indicated. In this case, he should have been looking for a dirty ground connection between the negative battery post and the engine block.

The amps (A) or milliamps (mA) flowing through a circuit can be measured with an ammeter. Most DMMs can measure milliamps or 10 to 20 amps directly through the meter. The correct range needs to be selected, along with AC or DC. The red probe is connected to the A or mA slot (always start high and work your way down if necessary), and the black lead is connected to the COM slot. To measure current, the DMM is connected in series with the circuit, with the probing lead closest to the positive terminal of the power supply or battery. This means that one wire in the circuit has to be disconnected, and the meter leads connected to each disconnected end so that the meter is in series with the load (**FIGURE 52-14**). Quality DMMs typically have an internal fuse that will blow if excessive current flows through it. This fuse is designed to help prevent damage to the meter.

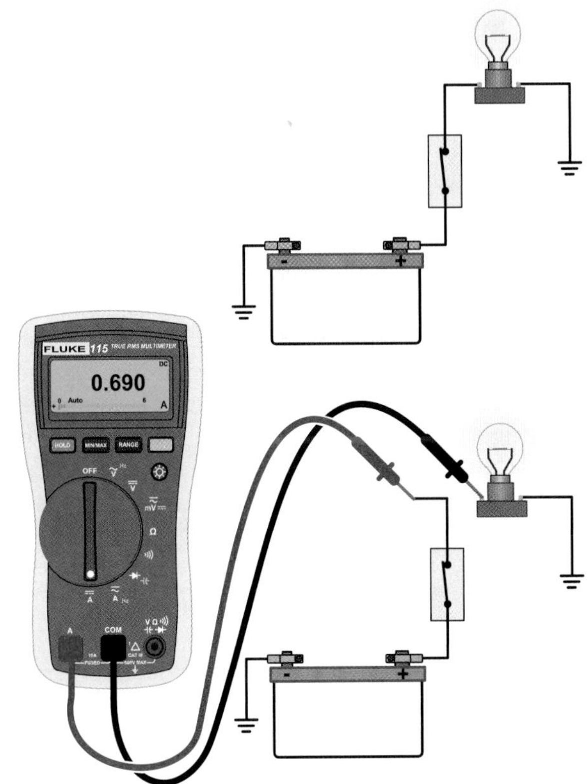

FIGURE 52-14 When measuring amps, one wire in series with the load has to be disconnected, and the ammeter leads connected between the disconnected ends.

> **TECHNICIAN TIP**
>
> There is one big caution to remember when you finish using an ammeter. *Always* move the red test lead from the A slot to the V/Ω slot immediately after finishing. If you leave the red lead in the A slot and then change the selector to measure volts, there is still a straight path through the A slot to the common slot, which means the DMM will create a short circuit when hooked up to measure voltage. This will blow the fuse in the meter as well as potentially damage the circuit.

If larger amperage must be measured, then current clamps can be connected to the DMM, and depending on their range, they can measure high currents, such as starter motor current draw of 400 amps or more. Current clamps are available in a variety of current-measuring ranges. The current clamp fastens around the conductor, measures the strength of the magnetic field produced from current flowing through the conductor, and outputs a voltage that the DMM reads as voltage, which is directly related to the amount of current flowing in amps. When using current clamps, the DMM is set to read volts. Current clamps also have the advantage that they clamp around the conductor, so the circuit does not need to be broken into to insert the DMM in series, as you would have to do with standard probes on an ammeter (**FIGURE 52-15**).

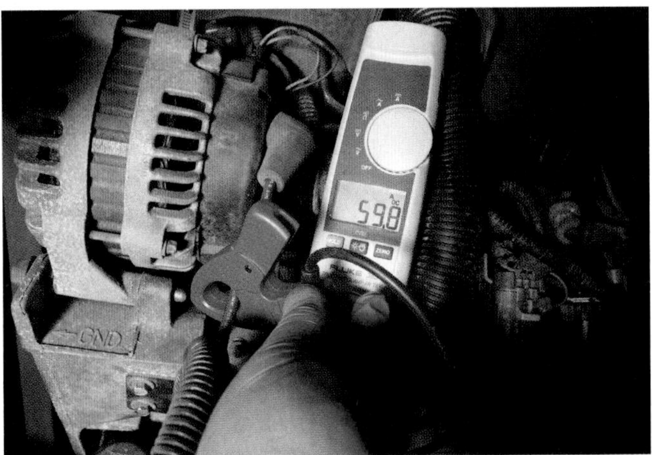

FIGURE 52-15 A current clamp being used with a DMM. Note that the current clamp fits around the wire and is not connected in series.

If you need to measure resistance in a circuit, always make sure the power is disconnected. In order to read resistance, batteries inside the DMM supply the circuit with power to take the measurement. If power is not removed from the circuit being tested, it disrupts the measurement and can provide a false reading or potentially damage the DMM. To take resistance measurements, the red (probing) lead is connected to the V/Ω slot, and the black (common) lead is connected to the COM slot of the DMM. You need to select an appropriate range or auto range to measure resistance. To accurately measure the resistance of a component, you should remove or isolate the component from the circuit. Doing so eliminates the possibility of any parallel circuit resistance affecting the resistance measurement. The red probing lead is connected to the input side of the component being tested, and the black common lead is connected to the output side (**FIGURE 52-16**).

▶ TECHNICIAN TIP

DMMs come in many forms. Always follow the specific manufacturer's instructions in the use of the DMM, or serious damage either to the DMM and/or to the electrical circuit could result.

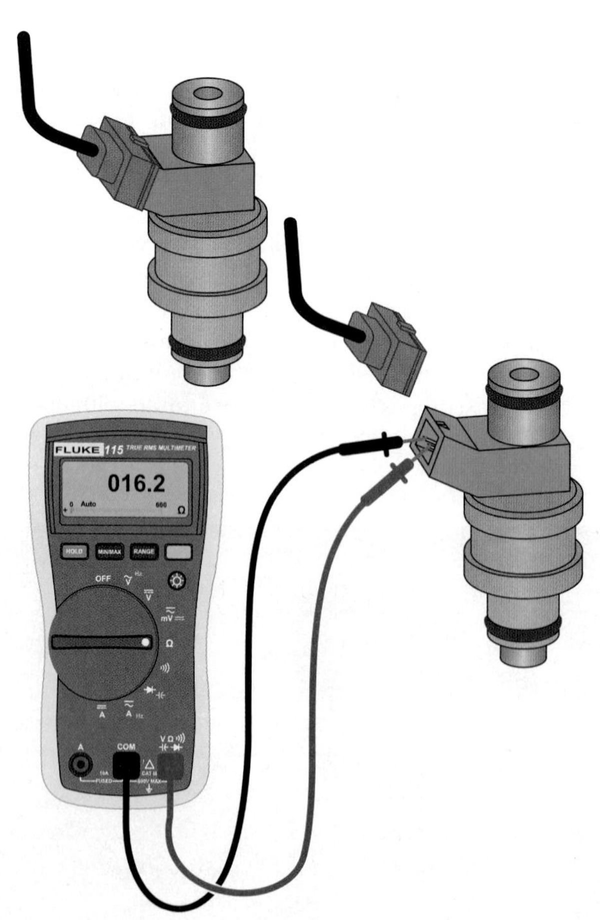

FIGURE 52-16 When measuring resistance, always isolate the component from the circuit, and place the meter leads across the input and output terminals.

Applied Math

AM-47: Specified Symbols: The technician can use conventional symbols (E for voltage, etc.) to solve problems using formulas such as Ohm's law, E = IR.

Voltage, current, and resistance have a specific relationship to one another. This concept forms the basis for electrical diagnosis.

George Ohm discovered that it takes 1 volt to push 1 amp through 1 ohm of resistance. Ohm's law is E = I × R, or Voltage = Amps × Resistance. If you know two of the three values for an electrical circuit, you can find the missing one. To find current flow, the formula is volts divided by resistance. To find voltage, the formula is amps multiplied by resistance. To find resistance, the formula is volts divided by amps.

In this example, we have an electrical circuit that has a resistance of 6 ohms with a 12-volt supply. We want to find the amperage. Our formula will be A = V/R:

A = V/R
A = 12 volts/6 ohms
A = 2 amps, which is our current flow.

▶ Voltage Exercises

The following voltage exercises are designed to explain the use of the DMM in taking DC voltage measurements. Examples are given to show the use of different ranges on the meter display as well as to measure available voltage and voltage drops in the series circuits with equal and unequal loads. As you will see, it is important to understand that the sum of the voltage drops in a series circuit equals the supply voltage, as explained in Kirchhoff's voltage law.

Voltage Ranges

Typically, a DMM has both an auto range and a manual range capability. The way in which you select auto range and manual range varies depending on the DMM. Different DMMs have different range settings. For example, one DMM's setting could be 6 V, 60 V, and 600 V, and another's 4 V, 40 V, and 400 V. **FIGURE 52-17** shows a circuit with two resistors in a series with a 12-volt DC supply. Various DMM ranges can be compared by measuring the voltage at the input and output of each resistor. The sample DMM we used has the following ranges: 600.0 mV, 6.000 V, 60.00 V, 600.0 V, and 1000 V.

Key Learnings: Using the lowest range that will still read gives the most accurate reading. Also note that the voltage drops every time current goes through a resistor.

Measuring Voltage

FIGURE 52-18 shows a series circuit of two resistors with a 12-volt battery and switch. In this example of how to measure voltage, the voltage in various parts of the circuit will be measured with the switch in the open position. Notice what happens to the voltage in the circuit. No voltage got past the open switch, so the loads (resistances) didn't get any current flow because

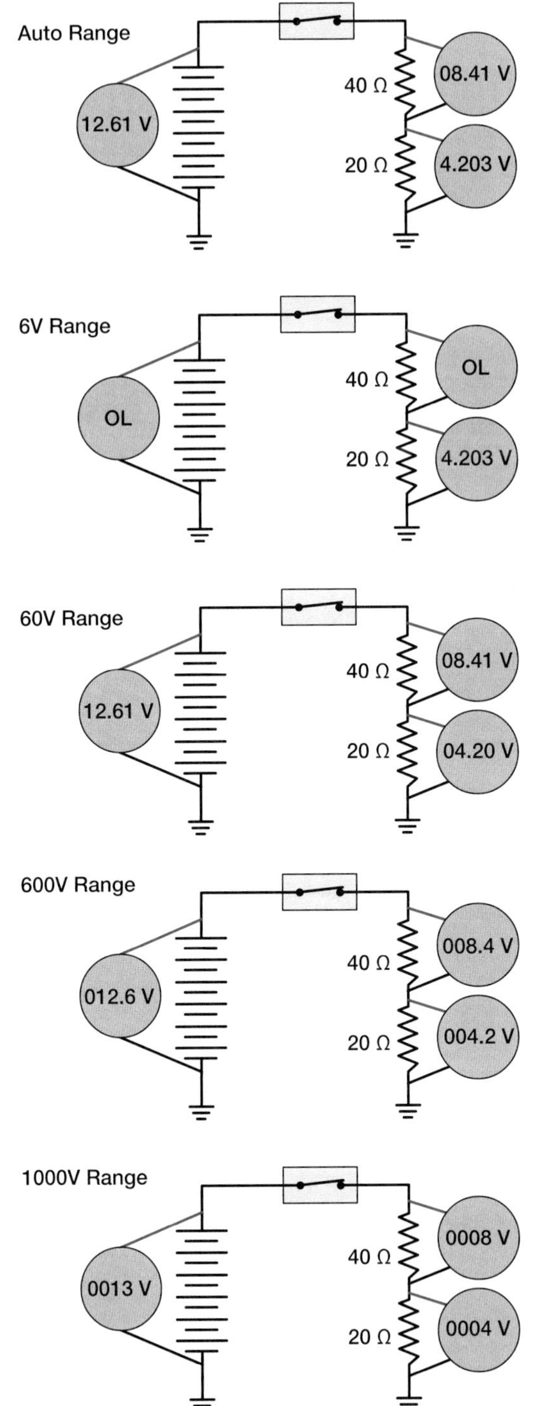

FIGURE 52-17 Measuring voltage using various ranges, in a circuit with two unequal resistors connected in series with a 12-volt DC supply.

there was no voltage after the switch. Open circuits prevent current from flowing.

Key Learnings: With the switch open, no current can flow anywhere in the circuit. With no current flow, there is no voltage used by the resistors. The only place that has a drop in voltage is from the input to the output of the open switch.

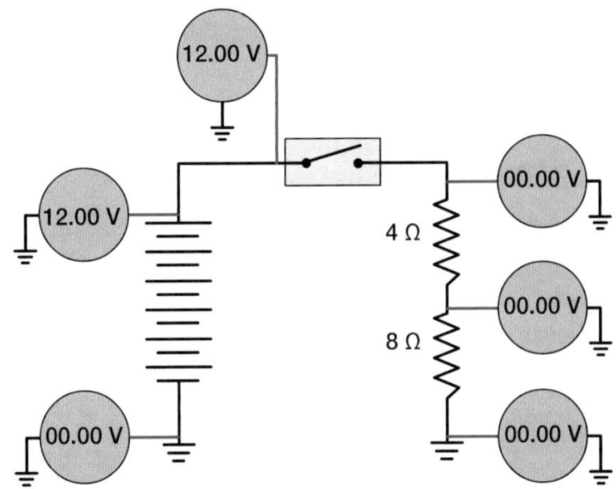

FIGURE 52-18 Measuring voltage at each load in a circuit with two unequal resistors connected in series with a 12-volt battery and with the switch open.

FIGURE 52-19 shows the same series circuit, but with the switch in the closed position. Notice that voltage passed through the closed switch to the first resistor. But not as much voltage made it to the second resistor. Remember that when current flows through a resistor, it uses up voltage. In fact, if we subtract the amount of voltage present after a resistor from the amount before the resistor, we can see how much voltage the resistor used. In this case, the first resistor had 12 V at the input, and 8 V at the output. That means that this first resistor used 4 V. The second resistor had 8 V at the input and 0 V at the output. That means that this second resistor used 8V. Also note that after the second resistor, all of the voltage has been used up.

Because R_1 and R_2 are of different resistances, each will have a voltage drop proportional to its resistance. In a series circuit, the higher resistance has a greater proportion of the voltage drop. Notice that R_2 has twice the amount of resistance that R_1 has; thus it takes twice the amount of voltage to push the current through R_2 as it does through R_1.

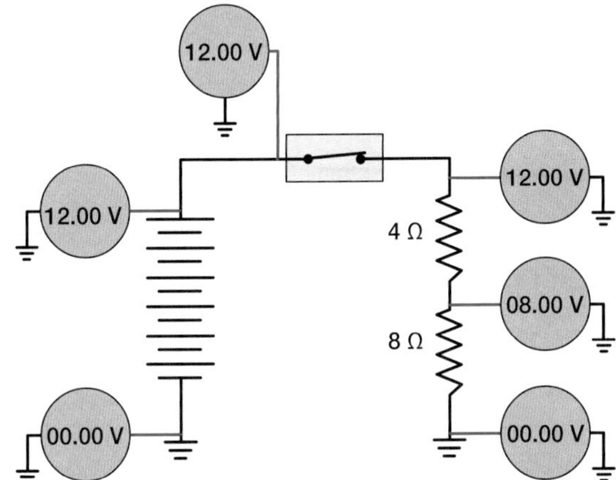

FIGURE 52-19 Measuring voltage in a circuit with two unequal resistors connected in series, with a 12-volt battery and the switch closed.

Key Learnings: Current is now flowing through the entire circuit. Here are important points:

- All of the voltage gets used up by the end of the circuit.
- Each load uses some of the voltage.
- The higher the resistance, the more voltage used.

If we change the resistance of each resistor, the voltage used by the resistors also changes. In this case, both resistances are higher than in the previous example. Notice how the voltage used by each resistance changes, but it is still in proportion to the size of each resistance. Also notice how putting the larger resistance first means the first resistor now uses the most voltage (**FIGURE 52-20**).

Key Learnings: The resistor with the highest resistance always uses the most voltage in a series circuit. But the sum of the voltage drops still equals source voltage.

What happens to voltage in a circuit when we change the number of resistors? In this example, we add one more resistor so there is a total of three resistors in series. Notice how the voltage changes, but once again the voltage used by each resistor changes in proportion to the other's (**FIGURE 52-21**).

Key Learnings: Adding a third resistor splits the voltage up three ways instead of two. The highest-resistance resistor still uses the most voltage. And the sum of all three voltage drops equals source voltage.

Voltage Drop

Voltage drop is the difference in voltage between two points in a circuit. Voltage drop occurs when current flows through a resistance. The voltage coming out of the resistance is less than the voltage going into the resistance. The higher the resistance, the higher the voltage drop, which also lowers the current flow.

In a properly operating circuit, the battery voltage is typically used up in the load (**FIGURE 52-22**). Because most loads are designed to operate on full battery voltage, any voltage drops before or after the load reduce the available voltage at the load, and therefore the voltage used by the load. This reduces the power available to be used by the load. You will remember that electrical power (W) = V × A. Because voltage drops tend to reduce both the voltage and amperage available to the load, power used by the load is also reduced, making it operate less effectively.

Voltage drops become a problem when they become excessive and occur in parts of the circuit other than the load. In reality, a small amount of resistance exists in each cable, connector, and switch within the circuit. A problem arises when the voltage drops become excessive in these components, starving the load.

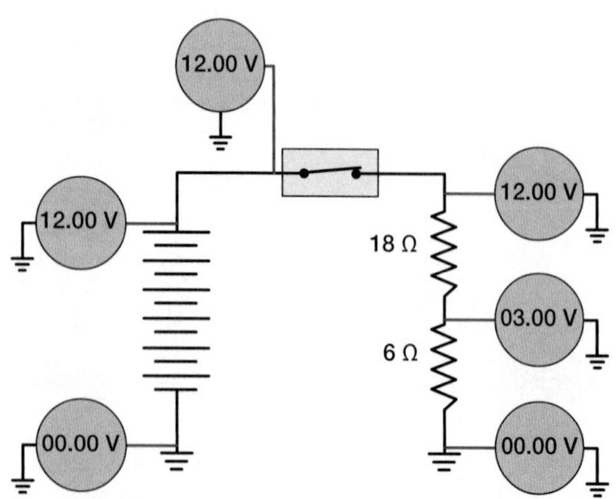

FIGURE 52-20 Another example of measuring voltage in a circuit with two resistors of unequal resistance connected in series.

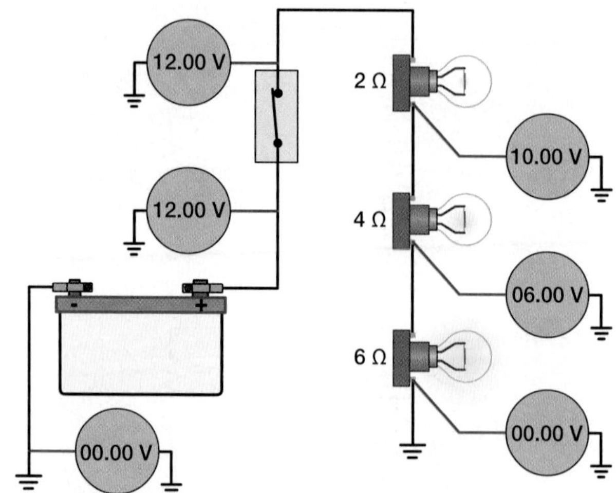

FIGURE 52-21 Notice how the voltage splits up when we have three resistors rather than just two.

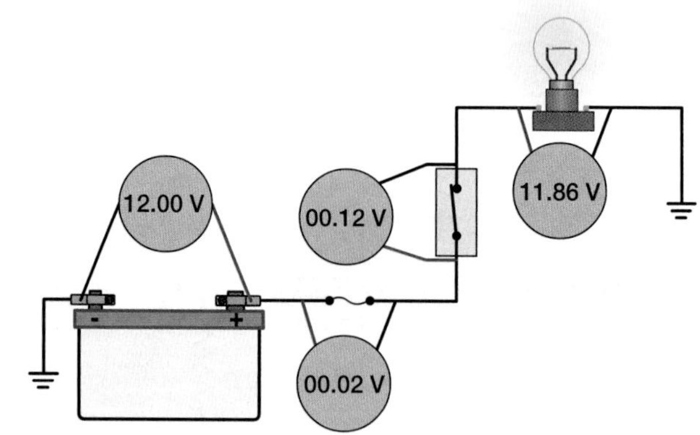

FIGURE 52-22 Typical voltage drops in a properly functioning circuit.

Measuring Voltage Drop

In the earlier voltage exercises, we measured the voltage available at each side of the resistor and used math to calculate how much voltage each resistor used. To take these measurements we placed the common lead on a good ground and the red probing lead in the circuit to find out how much voltage was available at that point. Using the meter this way is called doing an **available voltage test**: we measured the voltage available both before and after each resistor (**FIGURE 52-23**). Because a voltmeter measures the difference in voltage between the black lead and the red lead, we can use the meter to measure voltage drop directly. This can be done by placing one voltmeter lead on the input of a resistance and the other lead on the output. The meter will read

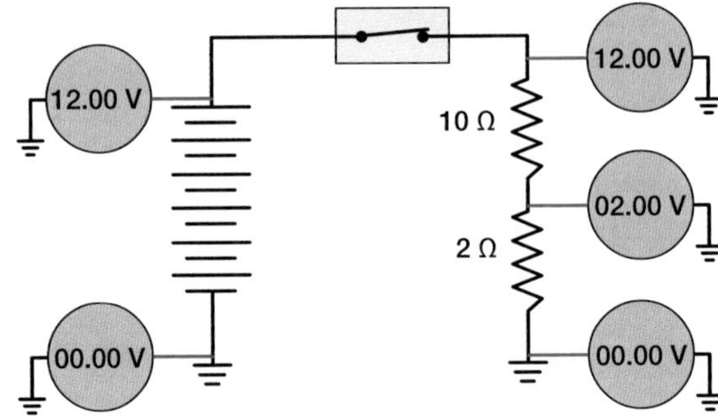

FIGURE 52-23 Available voltage testing.

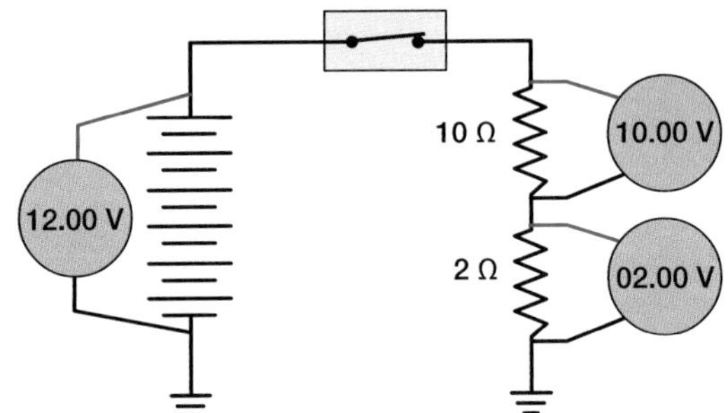

FIGURE 52-24 Voltage drop testing.

the difference in voltage between the two (**FIGURE 52-24**). This is how we perform a **voltage drop test**.

▶ Resistance Exercises

In this section, the exercises are designed to explain the use of the DMM in measuring resistance. Resistance measurements are used to check components or circuits against the manufacturer's specifications; for example, the resistance of sensors should be within the specified range. Examples are given to demonstrate measuring resistance. It is important to understand that Ohm's law tells us that current flow is inversely proportional to resistance. The higher the resistance, the less current will flow. The reverse is also true: The lower the resistance, the higher the current flow.

Measuring Resistance

In this exercise, a total circuit resistance measurement is taken. For resistance measurements, you need to select "auto range Ω" or the correct "manual-range" on the DMM, and connect the red lead to V/Ω slot and the black lead to COM slot. If using "manual range," select an appropriate range by starting at the highest range and working your way down. Resistance measurements should only be taken with power disconnected, and ideally, with the component disconnected from the circuit (**FIGURE 52-25**).

What happens to total circuit resistance when we add another resistor in series with the first one? Our series circuit law for resistance tells us that the resistances add up. So let's see if that happens (**FIGURE 52-26**).

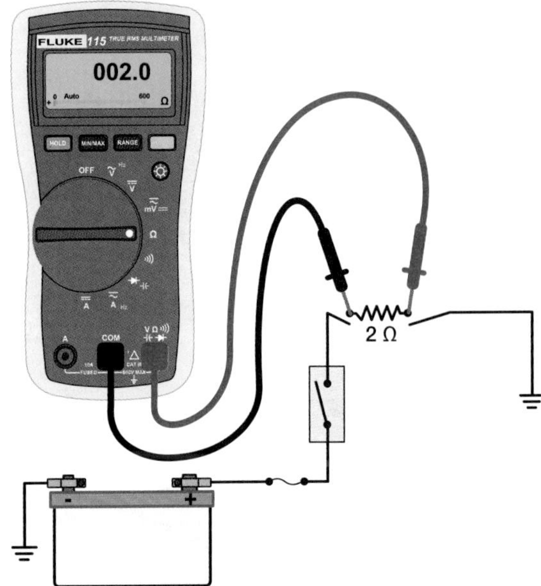

FIGURE 52-25 Measuring total resistance in a circuit with a single resistor.

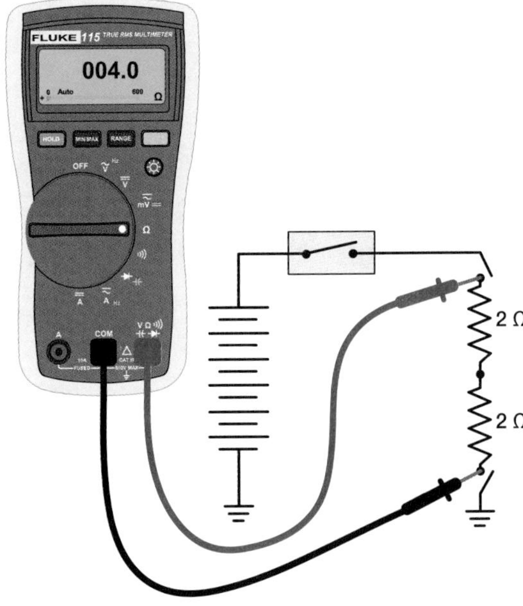

FIGURE 52-26 Measuring total resistance in a circuit with two equal-resistance resistors.

Let's add one more resistor of equal resistance in series with the other two resistors in the circuit. Notice that the total circuit resistance just continues to add up, just like the series circuit law tells us (**FIGURE 52-27**).

Let's try one last experiment. What happens to total circuit resistance if we have three unequal-resistance resistors in series with each other? Note that the total resistance is just the sum of the individual resistances, just like the series law tells us (**FIGURE 52-28**).

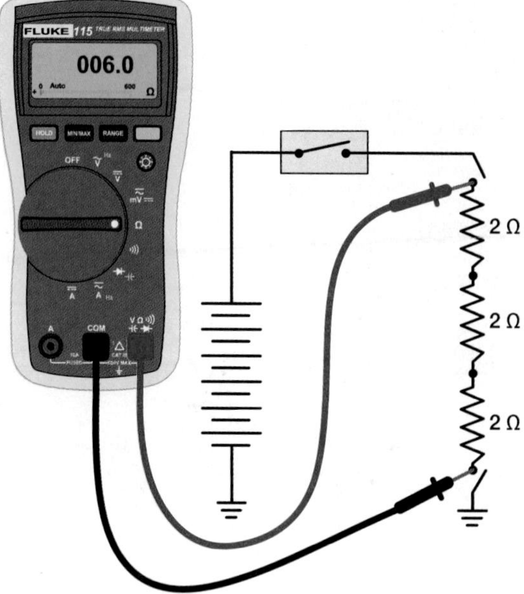

FIGURE 52-27 Measuring total resistance in a circuit with three equal-resistance resistors.

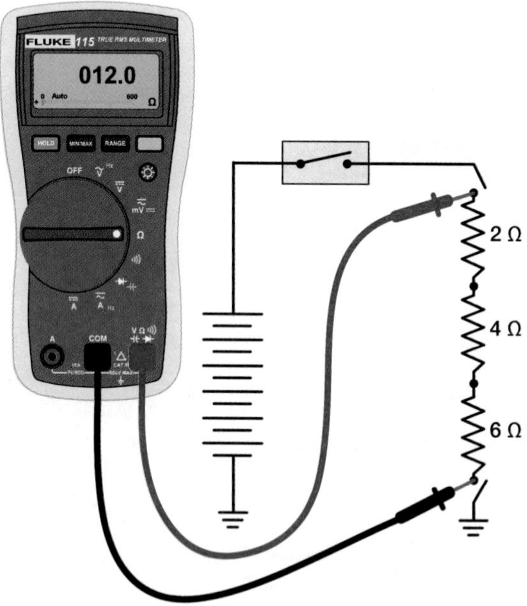

FIGURE 52-28 Measuring total resistance in a circuit with three unequal-resistance resistors.

Applied Math

AM-3: Mentally: The technician can mentally add two or more numbers to determine conformance with the manufacturer's specifications.

AM-6: Mentally: The technician can mentally subtract decimal and whole numbers to arrive at a difference for comparison with the manufacturer's specifications.

In this scenario, a technician is using a DMM to test the resistance of a coil pack for a V-6 engine. The ignition coil pack is suspected to be faulty due to a failed power balance test in which the cylinder was not properly contributing to the performance of the engine. Because coil designs are different, manufacturers' testing procedures vary. With the key off and the battery lead to the coil disconnected, the ohmmeter function of the DMM is used to check resistance. The technician measures the resistance of the primary and secondary windings. In this case, the specifications for the primary windings are 0.3–1.0 ohm. Before taking this reading, the technician checks the resistance in the leads of his DMM and determines there is 0.2 ohm of resistance, which will be subtracted (mentally) from the primary resistance reading. Using service information, the technician places the leads of the DMM in proper slots of the component and obtains a reading of 0.8 ohm. He mentally subtracts 0.2 ohm for the resistance in the meter leads, for a corrected reading of 0.6 ohm. This is within specifications for the primary windings.

The next step is the readings for the secondary windings of the coil pack. The manufacturer's specifications are 8000–9000 ohms. The technician places the meter leads as shown in the service information and obtains a reading of 6200 ohms. Mentally, the technician subtracts this and records the information on the repair order. The coil pack is 1800 ohms below minimum specifications. The technician will install a known good component (coil pack) for a test to see how the engine performs.

▶ Current Exercises

In this section, the exercises are designed to explain the use of the DMM when taking DC current measurements. Undertaking the exercises will improve your understanding of Ohm's law and current measurements. Examples are given to demonstrate measuring current and to show the magnetic fields produced around a conductor when current flows. It is important to understand that current is the same in all parts of a properly working series circuit. Always remember that an ammeter must be connected in series within the circuit. That means that the circuit must be broken in two, and each end of the ammeter should be connected to one of the two broken ends. This method ensures that all of the current flowing through the circuit flows through the ammeter (**FIGURE 52-29**).

Our series circuit law for current flow tells us that current flow stays the same throughout all parts of the circuit. Watch what happens when we place a second resistor of equal resistance in the series circuit. The current flow stays the same at all parts of the circuit. But also note that the current went down from the previous example. If you said that is because the resistance went up, you would be correct (**FIGURE 52-30**).

What happens to current flow if we have a circuit with unequal-resistance resistors? According to our series circuit laws, the current flow should stay the same throughout the circuit. Let's see if that is true (**FIGURE 52-31**).

Current and Magnetic Fields

In this example, a relay controlled by a switch is used to switch the current through a lightbulb. The compass is used to demonstrate that a magnetic field is produced around the relay winding when the current flows through it. To conduct this experiment,

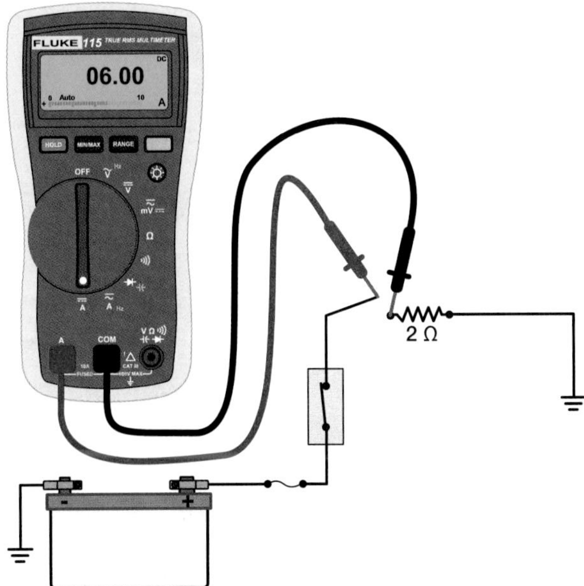

FIGURE 52-29 Measuring current flow in a circuit with a single resistor.

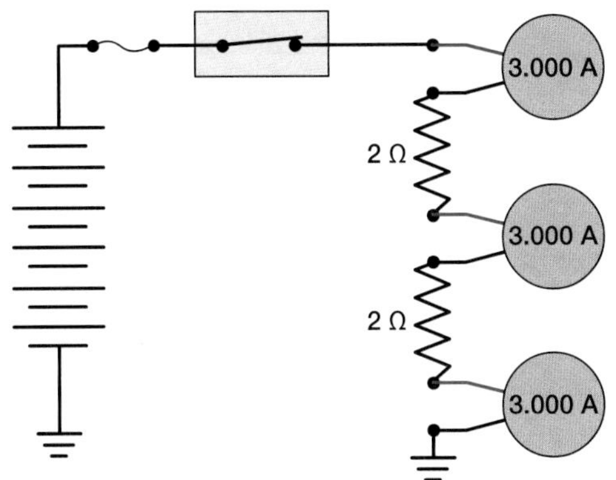

FIGURE 52-30 Measuring current flow in a circuit with two equal-resistance resistors in series.

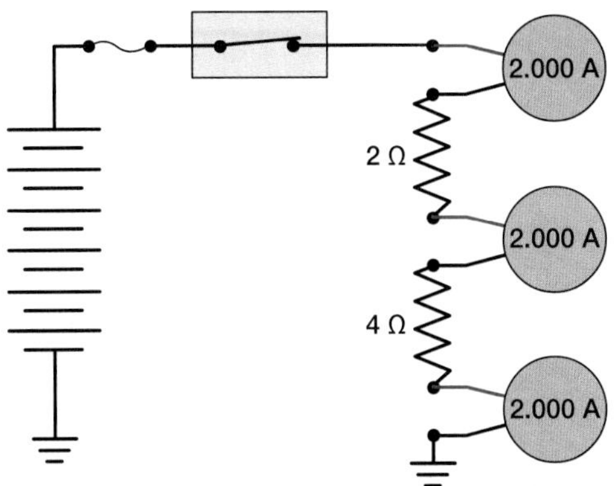

FIGURE 52-31 Measuring current flow in a circuit with two unequal-resistance resistors in series.

set the DMM to measure "DC amps." Connect the red lead to the A slot and the black lead to the COM slot. If using a manual-range DMM, select an appropriate range. **FIGURE 52-32** shows a circuit with a relay controlled by a switch and a single lightbulb with a 12-volt DC supply. Notice that the control current (relay winding) is much smaller than the lightbulb current. Lastly, the compass is used to show that when energized, the relay winding produces a magnetic field. This is one way to check whether the relay winding is operating without removing the relay to test it.

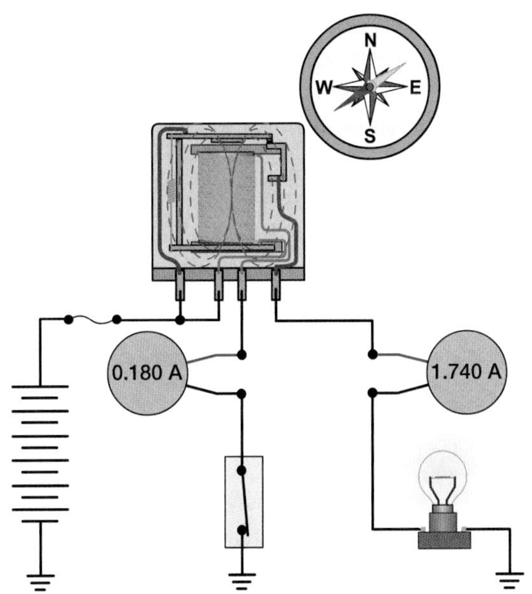

FIGURE 52-32 Measuring amperage in a circuit with a relay controlled by a switch and a single lightbulb with a 12-volt DC supply.

Applied Math

AM-9: Mentally: The technician can mentally divide decimal and whole numbers to determine conformance with the manufacturer's specifications.

In this scenario, the technician is using a DMM to measure cooling fan current flow in amps, and will determine circuit resistance in ohms, using Ohm's law.

The first step is to locate proper service information for the vehicle, covering the following items: fuse block diagram, cooling fan circuit diagram, and specifications or current flow for cooling fan motor. Next, the cooling fan relay is removed, and the DMM is set up to measure current by placing the red lead in the 10A slot on the meter. The black lead should be placed in the COM slot. In this testing procedure, the DMM is connected in series with the red lead in socket 30 and the black lead in socket 87. With the cooling fan motor in operation, the technician observes a reading of 3 amps. The specifications for this vehicle call for a range of 2.3–4.6 amps to operate the cooling fan motor.

To determine the resistance of this circuit, we use the formula, R = V/A. This is a form of Ohm's law (Voltage = Amps × Resistance); R = 12 volts/3 amps = 4 ohms.

▶ Perform Series Circuit Measurements

N52002, S52001

These series circuit exercises are designed to explain the use of the DMM for voltage, resistance, and amperage in a series circuit. If current flow is wrong, it is because either the voltage is wrong or the resistance is wrong, so understanding how volts, amps, and resistance behave in a series circuit allows you to measure them to determine which one is causing the fault. The examples given here demonstrate the use of measuring voltage, resistance, and amperage and describe how current flow is affected by voltage and resistance in a series circuit.

It is important to remember that current flow is the same in all parts of a properly operating series circuit, and the sum of the voltage drops across individual resistors in a series circuit is equal to the supply voltage. The exercise also examines how changing resistance in a series circuit affects the current flow and voltage drops.

One *very* important point should be noted in the following exercises. We are using lightbulbs to represent the loads so that you can see the relative brightness of each bulb to represent the work being done in the circuit. One thing to know about lightbulbs is that the resistance of the filament increases greatly when they light up by becoming very hot. We are *not* taking that increase in resistance into consideration in these exercises. So if you were to measure the current flow in a 12-volt circuit with a 6-ohm bulb (cold resistance), it wouldn't have 2 amps flowing through it; it would probably have about 0.25 amp due to the increase in resistance from the filament being white hot. This has a minimal effect on the exercises listed below. But we want you to be fully aware that if you measure the resistance of a lightbulb cold, then measure the voltage it uses and the current flowing through it, and calculate the resistance, you will get a *much* larger resistance than you measured cold. This just demonstrates the effect that heat has on resistance, and doesn't affect the concepts presented below.

Series Circuit Exercise 1

In this exercise, voltage measurements are taken from the series circuit. To measure voltage drop, set the DMM on the voltage range. Select "auto range volts DC" or the correct "manual range" on the DMM, and connect the black lead to the COM slot and the red lead to the V/Ω slot. You can measure voltage drop across components, connectors, or cables, but the current has to be flowing to get accurate measurements. Remember, when checking voltage, the leads can be placed in either direction. Just remember which way you placed them so you understand what the reading means. Also, when measuring current flow, always break the circuit open and insert the ammeter in series with the load in the circuit. **FIGURE 52-33** shows a typical circuit with a battery supply, fuse, switch, and lightbulb.

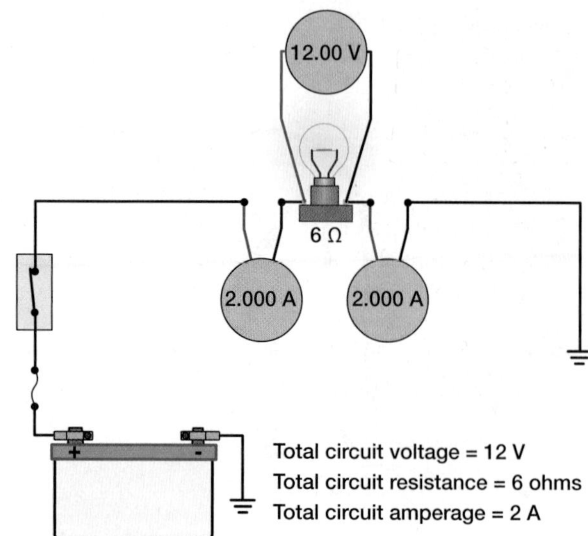

FIGURE 52-33 Electrical behavior in a typical series circuit with a battery supply, fuse, switch, and lightbulb.

Key Learnings: Note that the lightbulb operates at full brightness. Also note the following:

- All of the voltage is used up as current flows through the circuit.
- Virtually all of the voltage is used up in the load on a properly operating circuit.
- The current flow stays the same throughout the circuit.
- Because there is only one resistance in the circuit, that is also the total circuit resistance.

Series Circuit Exercise 2

In this exercise, we add a second equal-resistance bulb in series with the first bulb. Let's look at how this affects resistance, current flow, and voltage drop in a series circuit (**FIGURE 52-34**).

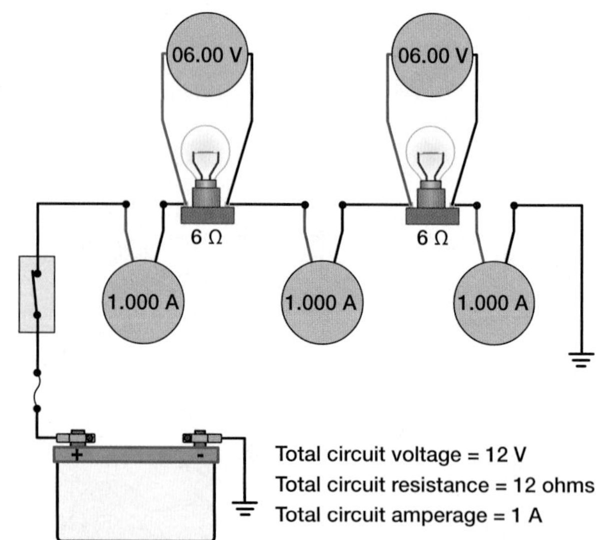

FIGURE 52-34 Electrical behavior in a circuit with two lightbulbs of equal value connected in series.

Key Learnings: Note that having two equal-resistance bulbs means the bulbs are equally bright, but much dimmer than with only one bulb. Also note the following:

- The voltage split up evenly (voltage drop) between the two bulbs due to their equal-resistance value.
- The total resistance doubled, which reduced total current flow in the circuit by half.
- But the current flow stayed at that lower level at every point in the circuit.
- If one bulb burns out, it creates an open circuit, which means that no current flow will flow through the circuit, including the other bulb.

Series Circuit Exercise 3

In this exercise, we have two unequal-resistance bulbs in series. This changes how electricity behaves in the circuit, but that behavior can be predicted by applying the appropriate Ohm's laws and series circuit laws. Let's see how this new arrangement operates (**FIGURE 52-35**).

Key Learnings: Note that having two unequal bulbs in series makes the bulbs operate differently from each other. Please also note the following:

- The voltage split up unevenly (voltage drop) between the two bulbs due to their unequal-resistance value.
- The sum of the voltage drops still equals source voltage.
- The first bulb is the brightest. This is because it has the larger voltage drop due to its higher resistance.
- The total current flow stayed the same throughout the entire circuit, even though the resistances are unequal.
- Total resistance is the sum of the individual resistances.

Series Circuit Exercise 4

In this exercise, we use the same two bulbs as in the last exercise, but this time we reverse their order. In the previous exercise, the first bulb (the small one) was relatively bright, whereas the second bulb (the larger one) was relatively dim. Which bulb do you think will be bright this time? The larger first bulb? Or the smaller second bulb? Let's take a look to see which one uses the most voltage (**FIGURE 52-36**).

Key Learnings: Note that swapping the position of the bulbs in the circuit didn't change the operation of the bulbs; they still operate just like they did in the previous exercise. Please note the following:

- The brightest bulb is still the highest-resistance bulb, even though it is last in the circuit.
- It uses the most voltage, making it brighter.
- Current flow stays the same throughout the circuit, even though the resistances of each bulb are not the same.
- The sum of the voltage drops still equals source voltage.
- The resistance still adds up in a series circuit.

Series Circuit Exercise 5

In this exercise, we place three unequal-resistance lightbulbs in series to see how that affects the behavior of electricity. Again, apply Ohm's law and the series circuit laws to predict the meter readings and calculations shown in **FIGURE 52-37**.

Key Learnings: Note that there are now three bulbs in the series instead of two. Adding the extra bulb in this manner affected the circuit in the following ways:

- The voltage splits up three ways, but the sum of the individual voltage drops still equals source voltage.
- The sum of the individual resistances add up to total circuit resistance, so adding more bulbs increases the resistance.
- The higher total resistance reduced the total current flow from the previous circuits.
- The current flow stays the same at all points in the circuit.

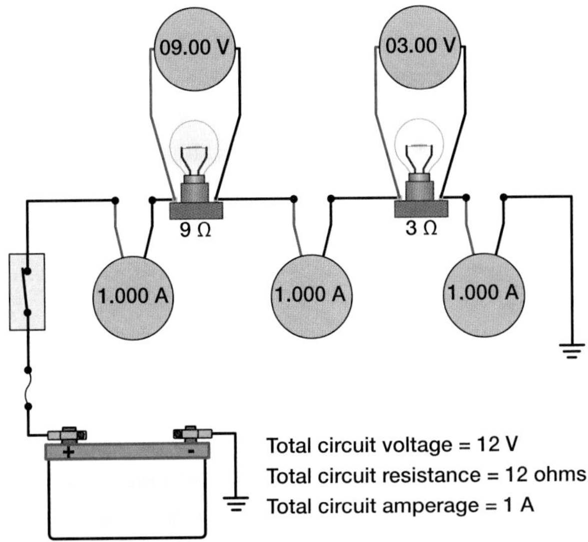

FIGURE 52-35 Electrical behavior in a circuit with two unequal lightbulbs connected in series.

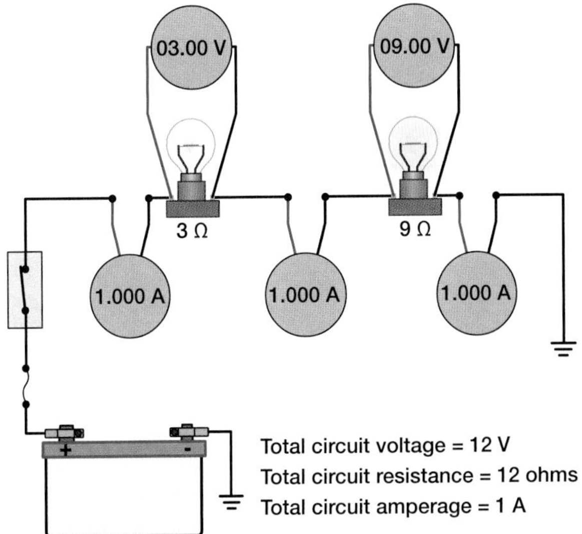

FIGURE 52-36 Electrical behavior in a series circuit with the two unequal-resistance lightbulbs reversed.

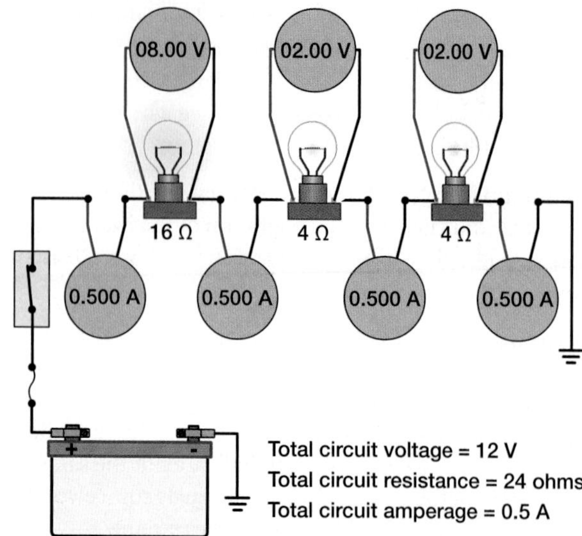

FIGURE 52-37 Electrical behavior in a circuit with three unequal-resistance lightbulbs in series.

▶ Perform Parallel Circuit Measurements

S52002

In this section, the exercises are designed to explain the use of the DMM in measuring volts, amps, and ohms in a parallel circuit. Parallel circuits are commonly used in the vehicle's electrical system, especially for lights. Understanding how they work and the relationships between voltage, amperage, and resistance in parallel circuits helps you to diagnose electrical faults. Examples are given to demonstrate how to measure volts, amps, and ohms and to show how current flows and voltage drops in a parallel circuit. It is important to remember the laws for a parallel circuit: resistance goes down when more parallel paths are added; current flow from individual legs adds up in parallel; and voltage stays the same at all common parallel circuit inputs. The following exercises help to reinforce the understanding of these laws.

One of the main differences in a parallel circuit compared to a series circuit is that resistance goes down as more parallel loads are added, and the total parallel resistance is always lower than the smallest parallel resistance. So we can't just add up the resistance values to find total parallel circuit resistance—we need a new way of doing that. There are actually two ways. The first is the easiest if there are only two resistors in parallel. It can't calculate the resistance for more than two. Here it is, using R_1 = 6 ohms and R_2 = 3 ohms as an example:

$$R_T = \frac{R_1 \times R_2}{R_1 + R_2} = \frac{6 \times 3}{6 + 3} = \frac{18}{9} = 2, \text{ so } R_T = 2 \text{ ohms}$$

So the total resistance for a 6-ohm resistor in parallel with a 3-ohm resistor is 2 ohms.

The second way of calculating parallel resistance is a bit more involved, but it can accommodate any number of parallel resistances; you just have to add more values to the right-hand side of the equation.

$$R_T = \frac{1}{\frac{1}{R_1}+\frac{1}{R_2}+\frac{1}{R_3}}$$

$$R_T = \frac{1}{\frac{1}{2}+\frac{1}{4}+\frac{1}{2}}$$

Parallel Circuits—Exercise 1

In this exercise, we explore how adding an equal-resistance bulb in parallel with an existing bulb affects resistance, current flow, and voltage drops. **FIGURE 52-38** shows two circuits—one series circuit and one parallel circuit—to explore the behavior of electricity in a parallel circuit. The additional resistor in parallel causes an increase in circuit current flow and a decrease in total circuit resistance.

Key Learnings: Adding an equal-resistance bulb in parallel allows both bulbs to be lit to full brightness. Also note the following:

- Adding another bulb in parallel provides another path for current flow. This reduces the total circuit resistance, which increases total circuit amperage.
- Total resistance of the bulbs in parallel is lower than the resistance in the lowest resistance bulb, just like the parallel circuit law for resistance claimed.
- Each bulb uses the full 12 volts, so each bulb is lit to full brightness.
- If one bulb burns out, the other bulb will still operate because each bulb has its own power and ground.

Parallel Circuits—Exercise 2

In this exercise, we see how uneven resistance bulbs affect voltage, resistance, and current flow in a parallel circuit. **FIGURE 52-39** shows two unequal-resistance bulbs in parallel. Notice how the unequal resistance changes the current flow in each leg of the circuit.

Key Learnings: Notice that the current flow in each leg of the parallel circuit is proportional to its resistance. Each leg acts independently in a parallel circuit. Also note the following:

- Total circuit resistance is still lower than the lowest resistance.
- Voltage stayed the same at both parallel inputs.
- Total circuit current flow is found by adding up the current flow for each leg.
- From the above, we can determine that the voltage drop across one parallel branch is the same for all parallel branches, regardless of their resistance.

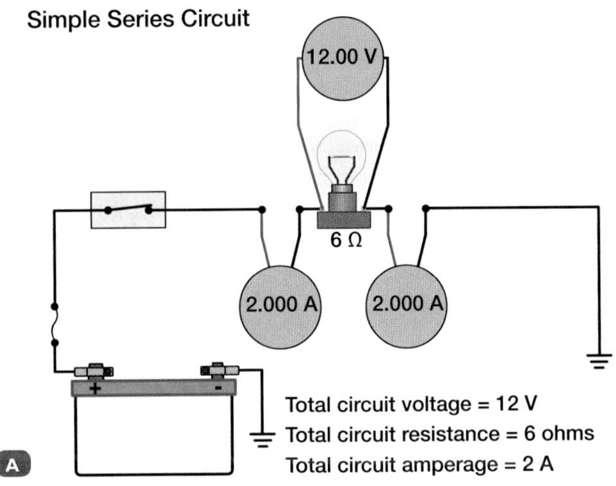

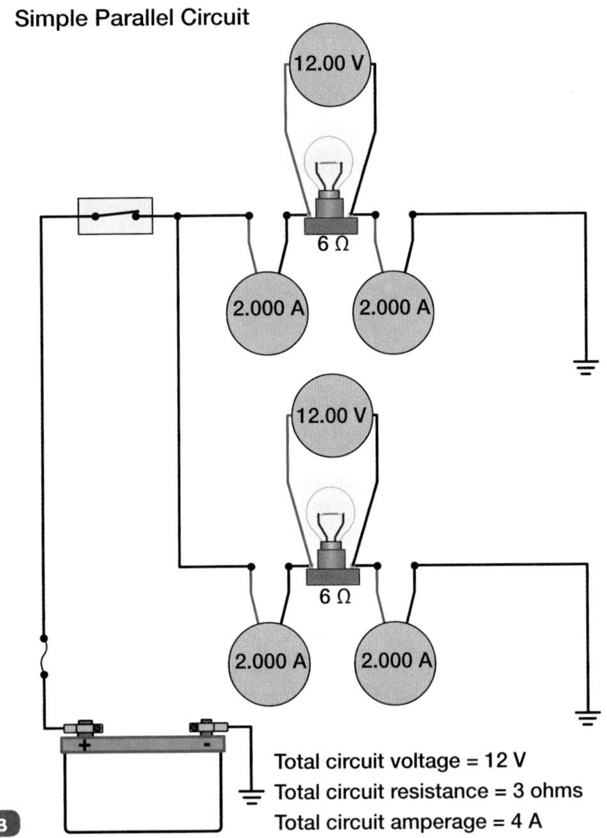

FIGURE 52-38 A. Electrical behavior in a series circuit. **B.** Electrical values in a parallel circuit with equal-resistance bulbs.

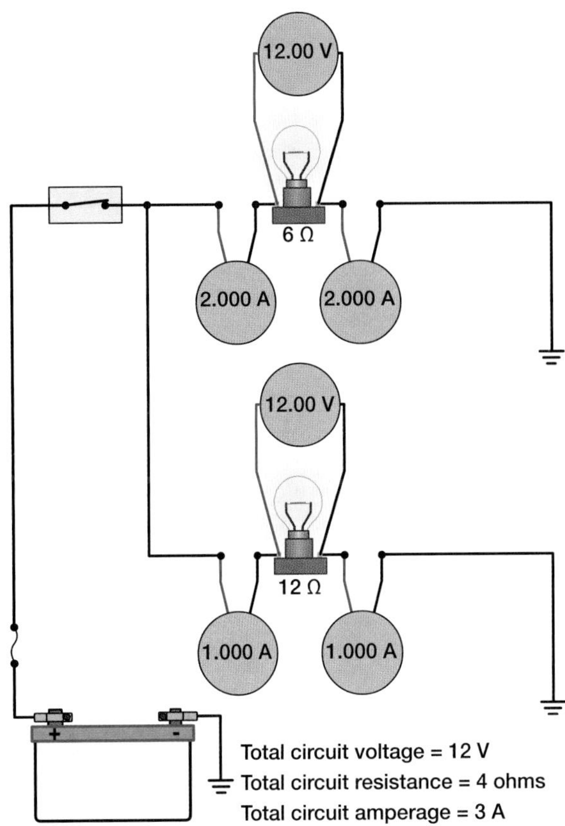

FIGURE 52-39 Electrical behavior in a parallel circuit with unequal-resistance bulbs.

Parallel Circuits—Exercise 3

In this exercise, we add one more bulb in parallel with the existing two bulbs, which are also in parallel. **FIGURE 52-40** has three unequal resistors in parallel. Notice how the electrical values change in this exercise.

Key Learnings: Adding another bulb in parallel made for three bright lights instead of two. Also note the following:

- Total circuit resistance has gone down further and is still lower than the lowest individual resistance.
- Total circuit current flow has increased with the new parallel bulb.
- Each bulb is using full source voltage.

SAFETY TIP

Notice how the current in the circuit increases with each additional load. This is also what happens with a power strip; each time an additional item is plugged in, the current increases. Overloading the circuit may cause the protection device to trip or in extreme cases could cause a fire. Never overload an electrical circuit, whether on a vehicle or in the shop.

Applied Math

AM-45: <, >, =, e.g.: The technician can interpret symbols to determine conformance with the manufacturer's specifications.

In this scenario, we look at three of the most common symbols used in technical manuals.

<, >: These symbols are lesser than (<) and greater than (>). An example of their use is describing the specification for the maximum limit for an AC voltage from an alternator (< 0.5 volt AC).

=: This symbol is a mathematical symbol used to indicate equality. An example is the formula for Ohm's law, which is E = I × R.

e.g.: This means "for example." It comes from the Latin expression *exempli gratia*, or "for the sake of an example." An example of this is "The technician uses a precision measuring instrument (e.g., micrometer) to measure the part."

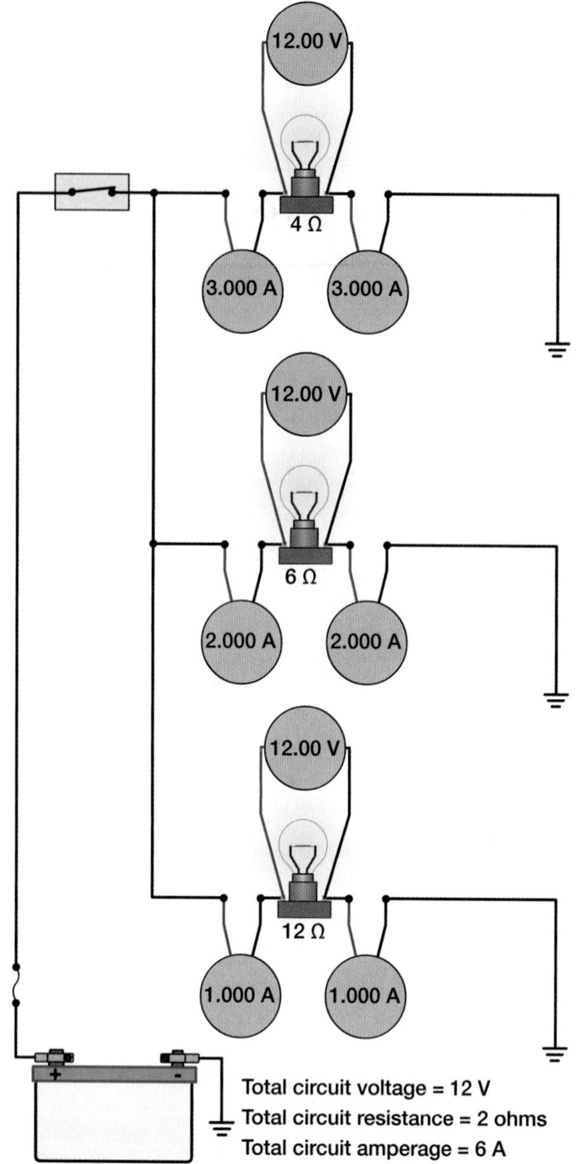

FIGURE 52-40 Electrical behavior in a parallel circuit with three unequal-resistance bulbs.

▶ Perform Series-Parallel Circuit Measurements

S52003

In this section, the exercises are designed to explain the use of the DMM in measuring current and voltage in a series-parallel circuit. Series-parallel circuits are found in vehicles, such as in dash light dimmer circuits, although they are not as common as parallel circuits. Typically, a series-parallel circuit occurs when unwanted resistance shows up in series with a parallel circuit. For example, if the brake light switch contacts become worn out, they can create a voltage drop in series with the brake lights, causing them all to be dimmer than they should be. Understanding how series-parallel circuits work and the relationships between current flow, voltage drops, and resistance helps you diagnose these types of electrical faults. Examples are given to demonstrate measuring voltage and current and to show how current flow and voltage drop are affected by resistance in a series-parallel circuit. It is important to understand that to analyze and calculate current flow and voltage drop, the total resistance of the circuit has to be considered. The voltage drop will be the same across each parallel branch, and the sum of the current flow in each branch is equal to the total parallel circuit current flow. These exercises also examine how the addition of resistors in series to a parallel circuit affects the circuit current flow, voltage drops, and circuit resistance.

Series Parallel—Exercise 1

In this exercise, voltage, resistance, and current measurements are taken from the series-parallel circuit formed by resistors R_1, R_2, and R_3, to observe how the electricity will behave (**FIGURE 52-41**). To measure voltage drop, the DMM must be set on the voltage range. Select "auto range volts DC" on the DMM, and connect the black lead to the COM slot and the red lead to the V/Ω slot. Voltage drop can be measured across components, connectors, or cables as long as current is flowing in the circuit. The red lead of the DMM is normally connected to the positive side of the component. To measure current, the DMM must be set to read DC amps. Connect the red lead to the A slot and the black lead to the COM slot. If using a manual-range DMM, select an appropriate range. Please note that calculations may have been rounded in some instances.

Key Learnings: We used meters to measure the electrical values in the circuit, but you can calculate them as well by using the circuit laws along with Ohm's law. To calculate the values, you need to first effectively convert them to a series circuit. You do this by finding the total resistance of the parallel circuit. Because the parallel portion is in series with the series circuit, the parallel resistance can be added to the series resistance, which gives you the total circuit resistance. Then you can calculate the total current flow by dividing the total circuit voltage by the total circuit resistance. We know the total circuit current flows through the series portion of the circuit. We can then calculate the voltage drop of the series circuit by multiplying the current flowing through R_1 times the resistance of (R_1). We can then subtract the voltage drop of R_1 from source voltage to get the voltage applied to the parallel circuit. We know that each branch of the parallel circuit gets that full voltage, and each branch is grounded, so each branch drops that full applied voltage. Then we can calculate the current flow through each parallel resistor by dividing the parallel circuit voltage drop by each parallel resistance (R_2 and R_3). Also note these findings:

- The total circuit current flows through the series bulb.
- The total source voltage is shared between the series and parallel portions of the circuit.
- Each of the parallel bulbs receives the full parallel circuit voltage.

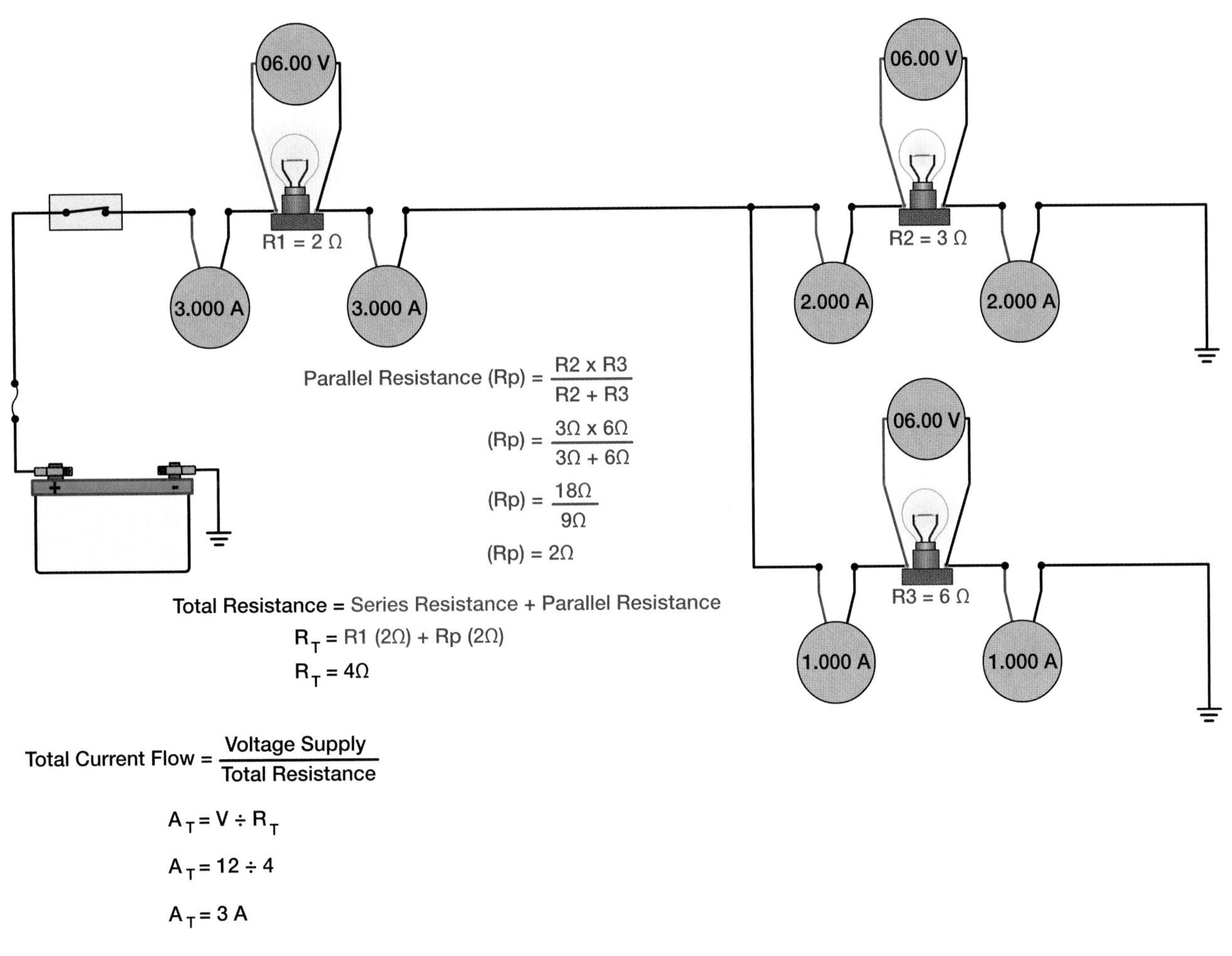
06.00 V
R1 = 2 Ω
3.000 A
3.000 A
06.00 V
R2 = 3 Ω
2.000 A
2.000 A
06.00 V
R3 = 6 Ω
1.000 A
1.000 A
Parallel Resistance (Rp) = R2 x R3 / R2 + R3
(Rp) = 3Ω x 6Ω / 3Ω + 6Ω
(Rp) = 18Ω / 9Ω
(Rp) = 2Ω
Total Resistance = Series Resistance + Parallel Resistance
RT = R1 (2Ω) + Rp (2Ω)
RT = 4Ω
Total Current Flow = Voltage Supply / Total Resistance
AT = V ÷ RT
AT = 12 ÷ 4
AT = 3 A
Voltage Drop = A x R
Series Voltage Drop = 3 A x 2 Ω
Series Voltage Drop = 6 V
Parallel Voltage Drop = 3 A x 2 Ω
Parallel Voltage Drop = 6 V
Parallel Current Flow = Parallel Voltage Drop ÷ R
R2 Current = 6 V ÷ 3 Ω
R2 Current = 2 A
R3 Current = 6 V ÷ 6 Ω
R3 Current = 1 A

FIGURE 52-41 A series-parallel circuit formed by bulbs R_1, R_2, and R_3.

- The total circuit amperage is split up between each branch of the parallel portion of the circuit according to the resistance of each bulb.
- The total resistance of the parallel circuit is lower than any single parallel branch resistance.

Series Parallel Exercise 2

In this exercise, we have added another bulb in the parallel portion of the circuit. Voltage and current measurements will be taken from the series-parallel circuit formed by resistors R_1, R_2, R_3, and R_4 to see how the electricity behaves differently from the last circuit (**FIGURE 52-42**).

Key Learnings: Adding the additional bulb in the parallel portion of the circuit reduced the total resistance of the parallel circuit, which also lowered the resistance of the entire circuit. This increased the total circuit current flow. The increased current flow through R_1 increased the voltage used (voltage drop) of (R_1). This increased voltage drop reduced the voltage to the parallel portion of the circuit slightly. Also, the parallel branches had to share the total parallel current flow with the additional branch, so each branch received slightly less current flow than the previous exercise. Also note these findings:

- Adding the additional bulb in parallel reduced the total circuit resistance slightly.
- The reduced resistance allowed a bit more total circuit current to flow.
- The increased current flow increased the voltage drop across R_1, making it a bit brighter.
- The parallel bulbs got dimmer for two reasons. First, the voltage they received was less. And second, each branch got less current flow. Because Power is V × A, and the bulbs used less of each of these, they operated on lower power.

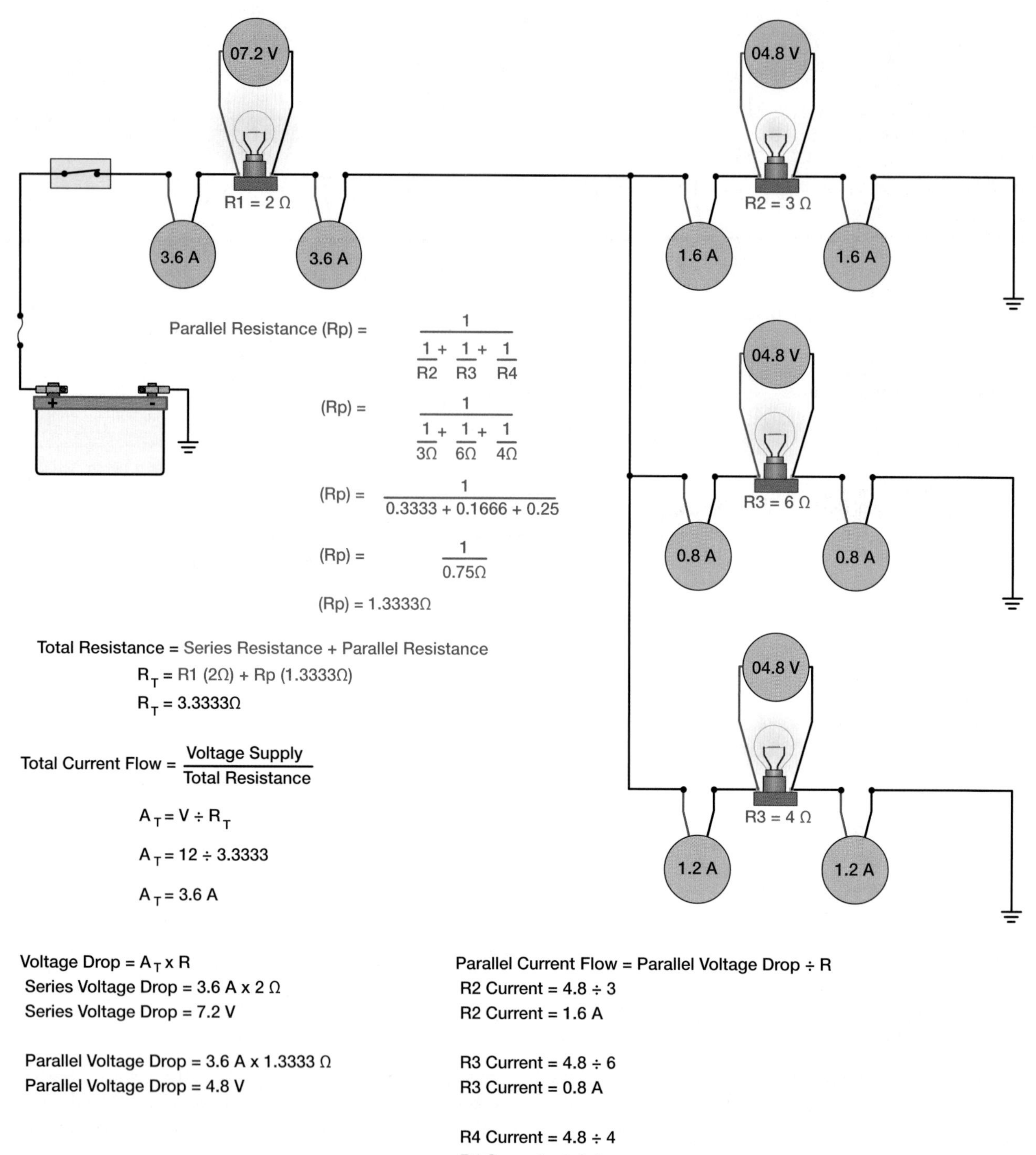
07.2 V
R1 = 2 Ω
3.6 A
3.6 A
04.8 V
R2 = 3 Ω
1.6 A
1.6 A
04.8 V
R3 = 6 Ω
0.8 A
0.8 A
04.8 V
R3 = 4 Ω
1.2 A
1.2 A
Parallel Resistance (Rp) = 1 / (1/R2 + 1/R3 + 1/R4)
(Rp) = 1 / (1/3Ω + 1/6Ω + 1/4Ω)
(Rp) = 1 / (0.3333 + 0.1666 + 0.25)
(Rp) = 1 / 0.75Ω
(Rp) = 1.3333Ω
Total Resistance = Series Resistance + Parallel Resistance
R_T = R1 (2Ω) + Rp (1.3333Ω)
R_T = 3.3333Ω
Total Current Flow = Voltage Supply / Total Resistance
A_T = V ÷ R_T
A_T = 12 ÷ 3.3333
A_T = 3.6 A
Voltage Drop = A_T x R
Series Voltage Drop = 3.6 A x 2 Ω
Series Voltage Drop = 7.2 V
Parallel Voltage Drop = 3.6 A x 1.3333 Ω
Parallel Voltage Drop = 4.8 V
Parallel Current Flow = Parallel Voltage Drop ÷ R
R2 Current = 4.8 ÷ 3
R2 Current = 1.6 A
R3 Current = 4.8 ÷ 6
R3 Current = 0.8 A
R4 Current = 4.8 ÷ 4
R4 Current = 1.2 A

FIGURE 52-42 A series-parallel circuit formed by bulbs R_1, R_2, R_3, and R_4.

▶ Variable Resistors

In this section, the exercises are designed to explain the use of the DMM in measuring voltage and amperage in a circuit with a variable resistor, or potentiometer. Understanding the relationships between voltage, resistance, and current as the potentiometer is adjusted helps you diagnose electrical faults. Examples are given to demonstrate measuring voltage and current and to show how current flows and how voltage drop and current are affected by the position of the wiper of the potentiometer. It is important to understand that as the position of the wiper is changed, so too is the voltage and current flow to a load connected to the potentiometer.

Variable Resistors Exercise

In this exercise, a variable resistor is used as a potentiometer, or voltage divider. For voltage measurements, you need to select "auto range volts DC" on the DMM and connect the red lead to the V/Ω slot and the black lead to the COM slot. For current measurements, select "auto range milliamps DC" on the DMM. Connect the red lead to the A slot and the black lead to the COM slot. If using a manual-range DMM, you will need to select an appropriate range.

FIGURE 52-43 shows circuits with a 250 Ω variable resistor in the circuit as a voltage divider with a 12-volt DC supply. In Figure 52-43A, the wiper of the variable resistor is set so that minimum voltage will occur. In Figure 52-43B, the variable resistor is set so that maximum voltage can occur. The variable resistor is continuously variable between these points, and the voltage varies depending on the position of the wiper (Figure 52-43C). When the wiper in Figure 52-43 is at the upper or lower limits, in practice there may be a small amount of the variable resistor left on the outer edges, so the voltage may not quite reach the full 12 volts or 0 volts.

Key Learnings: A variable resistor can provide an infinitely variable voltage reading from ground to input voltage, depending on the position of the wiper. This makes it ideal for use as a position sensor.

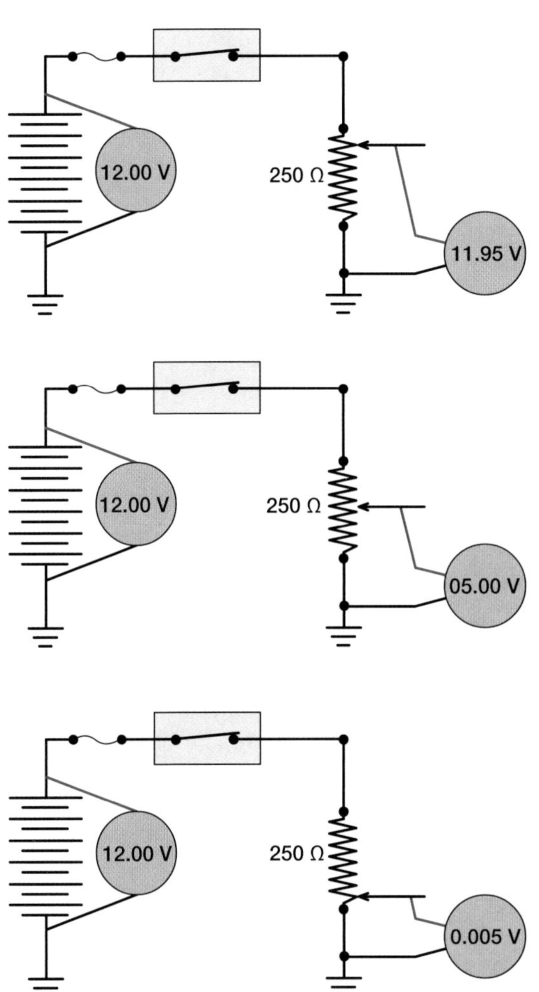

FIGURE 52-43 Measuring voltage in circuits with a 250 Ω variable resistor and a 12-volt DC supply.

▶ Capacitors

In this section, the exercises are designed to explain the use of the DMM in current and voltage in a circuit with a capacitor connected in series with a lamp, which gives you an understanding of how voltage and current change as the capacitor charges. Understanding how capacitors work in a circuit helps you to diagnose electrical faults. Examples are given to demonstrate measuring voltage and current and to show how current flow and voltage drop change as a capacitor charges. It is important to understand that as the capacitor charges, the voltage drop across it increases and the current flow decreases. Capacitors are also used for other purposes in vehicles, such as to filter electrical noise from the charging system. Capacitors can also be tested for the amount of capacitance they have on some DMMs that have that function. If the capacitor is functioning properly, it will have close to its rated capacitance.

Capacitors-Exercise 1

In this exercise, a capacitor is connected in series with a lamp. For voltage measurements, select "auto range volts DC" on the DMM and connect the red lead to V/Ω and the black lead to COM. For current measurements, select "auto range milliamps DC" on the DMM. Connect the red lead to the A socket and the black lead to the COM socket. If using a manual-range DMM, select an appropriate range.

FIGURE 52-44 shows a capacitor connected in series with a lamp with a 12-volt DC supply. The DMM will be used to measure voltage across the capacitor and lamp when the switch is turned on.

Applied Science

AS-77: Capacitance: The technician can demonstrate an understanding of the role of capacitance in timer circuits such as RC timers or a MAP sensor.

Sensors are the components of the system providing input to the computer, making it possible for it to carry out its operations.

The MAP, or manifold absolute pressure, sensor, plays a big role in proper engine performance. There are several different types of MAP sensors. Some of the popular styles are the variable voltage MAP sensor, the variable-inductance MAP sensor, and the variable-capacitance MAP sensor. Here, we focus on the variable-capacitance MAP sensor.

The variable-capacitance MAP sensor consists of two aluminum oxide plates in a chamber that is connected by tubing to the engine's intake manifold. This sensor is capable of generating an output signal in hertz proportional to the change in manifold pressure. This is the key point that the technician should understand. The electrical output signal to the computer is directly proportional to engine load as the manifold pressure changes. At engine idle, the vacuum is at approximately 17–21 inches of mercury (in. Hg) at sea level. Under this condition, the MAP sensor hertz is approximately 95. When the engine is at wide open throttle, the vacuum is almost zero in Hg, which is converted to a hertz reading of approximately 160. The MAP sensor sends the proportional signal to the computer as a result of changes in engine load. The computer responds by providing more or less fuel to the injectors as well as by performing a number of other vital tasks.

1: Switch OFF capacitor discharged

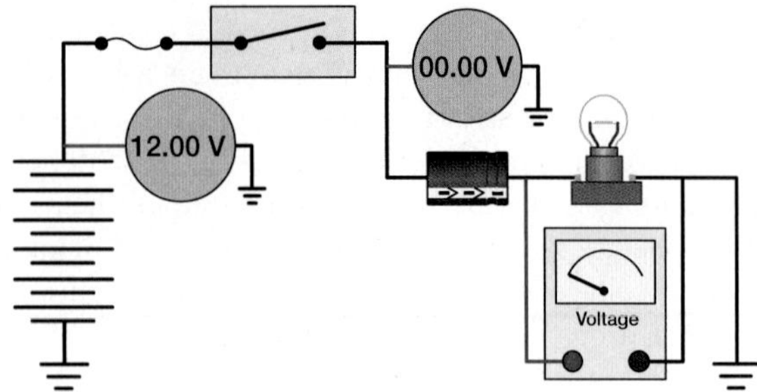

2: Switch ON capacitor starting to charge

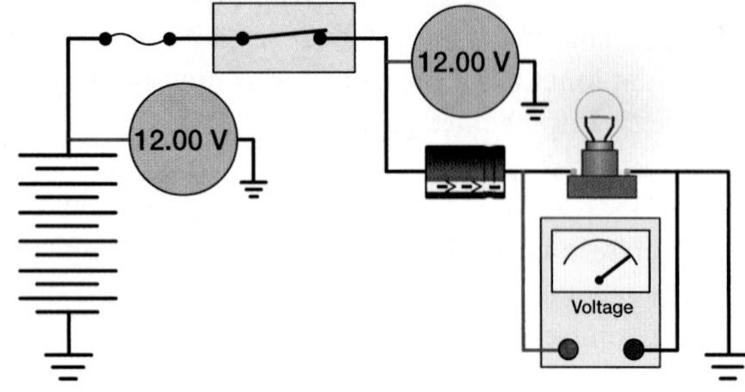

3: Switch ON capacitor charging

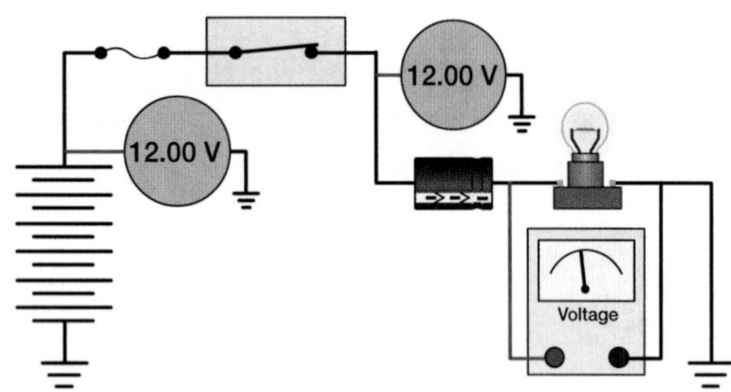

4: Switch ON capacitor charged

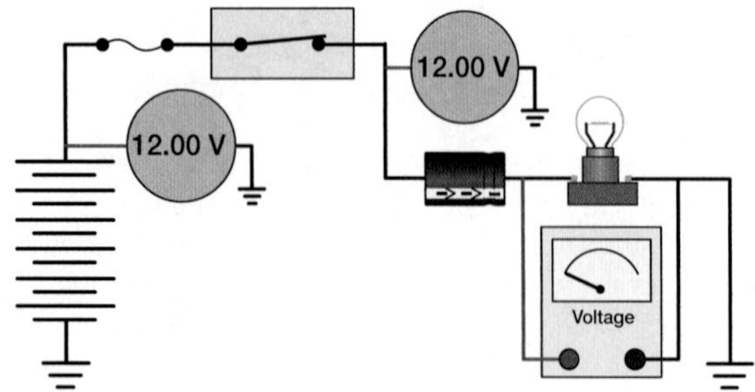

FIGURE 52-44 Measuring voltage in a circuit with a capacitor connected in series with a lamp and a 12-volt DC supply.

Key Learnings: The instant the switch is in the closed position, current can flow in the circuit because the capacitor is fully discharged. Current flow is initially high and gradually reduces as the capacitor charges. Because the lamp is in series, it lights up brightly initially. As the capacitor charges, the current decreases, and the lamp becomes dim before going out as the capacitor fully charges and the current flow stops. Voltage drop across the bulb is initially high because high current through the lamp produces a high voltage drop. As the capacitor charges, the current flow reduces, as does the voltage drop across the lamp.

▶ Electrical Circuit Testing

S52004

Electrical circuit testing begins with understanding circuit types and how electricity behaves within them. Add to that the ability to use meters and oscilloscopes to measure the values of voltage, amperage, and resistance, along with understanding how to read wiring diagrams so you will know how the circuits are constructed, and you will be well on your way to diagnosing electrical faults successfully. Those are the concepts we explore in this section. Feel free to refer back to the previous sections in this chapter to help you remember how electricity behaves as well as how meters are hooked up for specific measurements. Let's kick this off by seeing how Ohm's law can help us predict the behavior of electricity.

Using Ohm's Law to Diagnose Circuits

Ohm's law can be used in two ways to help diagnose electrical circuit faults. The first is by using it to perform the math to predict and verify measurements. The second way is by using the relationships between volts, ohms, and amps that Ohm's law demonstrates to guide you through the diagnosis process.

When using Ohm's law in the first way, you calculate electrical quantities in a circuit, and it is valuable in cross-checking actual measured results within the circuit. For example, if the resistance and voltage of a circuit are known, then the theoretical current can be calculated using Ohm's law. The calculated result can then be compared to the specified or measured amperage to determine if the circuit is functioning correctly. Technicians often do a quick calculation, sometimes just in their head, to obtain an approximate value of an electrical quantity before they take actual measurements. Doing so allows them to anticipate what they will be measuring and to set the measuring tool to the correct range.

Always remember that a calculation may only yield an approximate value; in actual circuits, variations or tolerances exist in components, causing differences between calculated values and actual measurements.

Using Ohm's law the second way helps you understand the relationships between volts, amps, and ohms. For example, if voltage stays the same but resistance decreases, amperage must increase (**FIGURE 52-45**). In the case of a short circuit, the resistance decreases and the amperage increases, potentially blowing the fuse. In the opposite scenario, where resistance increases, current flow decreases (**FIGURE 52-46**). This is the case when a corroded or loose connection introduces excessive resistance into the circuit. It also results in less electrical power (volts and amps) to operate the intended load.

What does Ohm's law tell us to expect when the voltage changes? If voltage decreases and the resistance stays

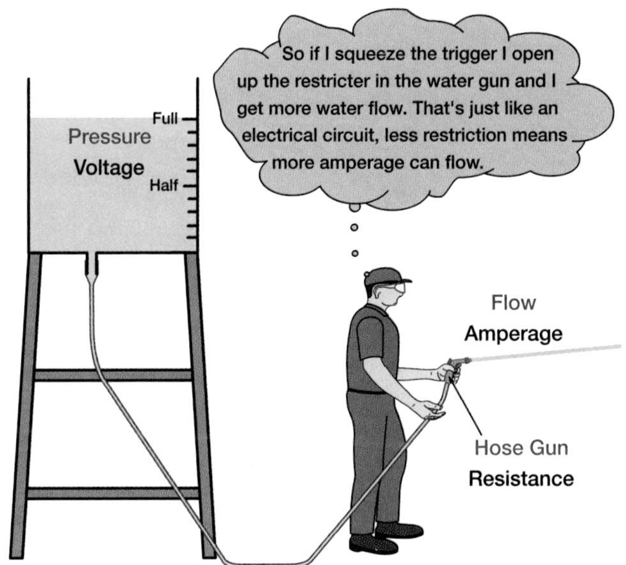

FIGURE 52-45 If voltage stays the same but resistance decreases, amperage must increase.

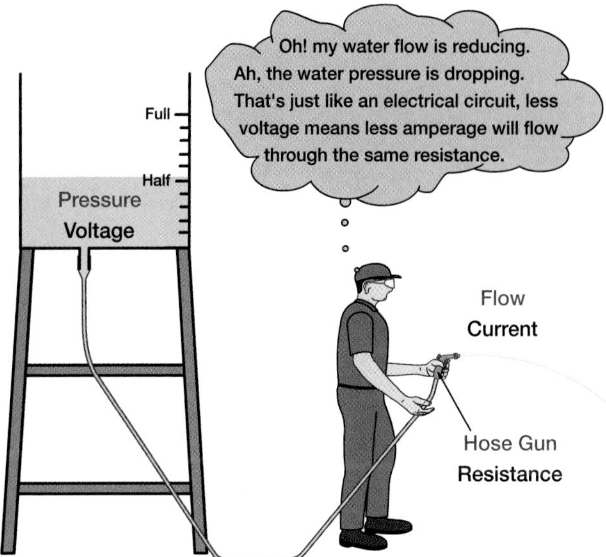

FIGURE 52-47 If voltage decreases but resistance stays the same, amperage must decrease.

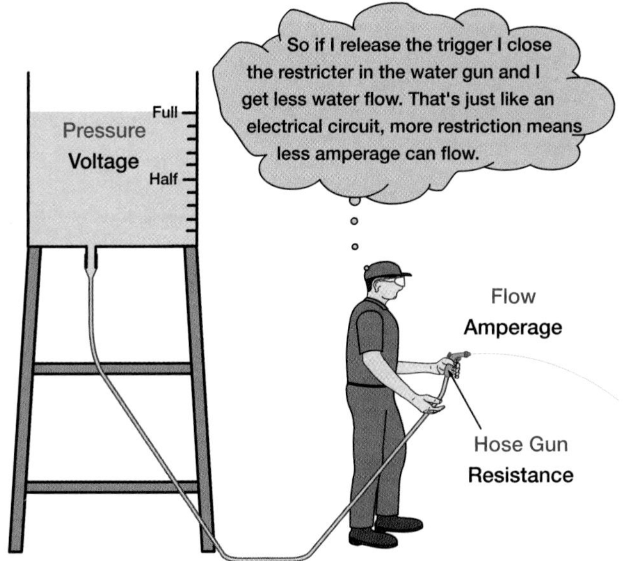

FIGURE 52-46 If voltage stays the same but resistance increases, amperage must decrease.

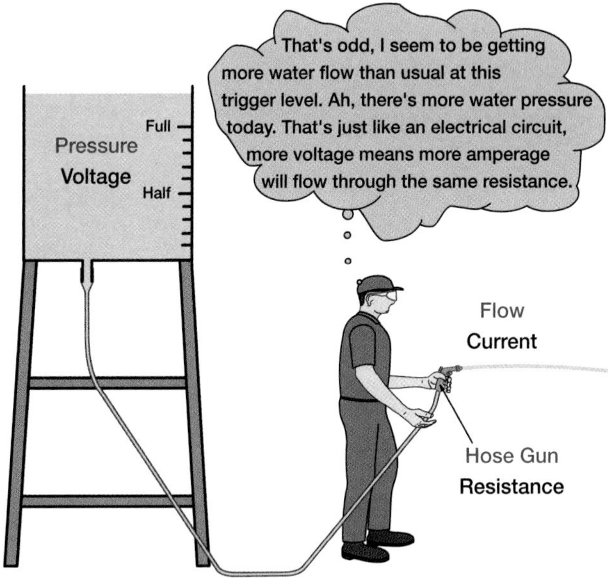

FIGURE 52-48 If voltage increases but resistance stays the same, amperage must increase.

the same, then amperage will decrease (**FIGURE 52-47**). This results in less power being able to operate the load. If the voltage increases and the resistance stays the same, then amperage will increase (**FIGURE 52-48**). If amperage and voltage both increase, then the electrical power operating the load is also increased. This condition can shorten the life of, or even burn out, the load.

So, how do amperage changes affect volts and amps? That is a good question. But if you think about it, amperage is a result of, or product of, the voltage and resistance. Amperage does not exist without both voltage and resistance—a means of pushing the amperage (voltage) and a path for the amperage to flow (resistance). If you ask yourself, "What is the amperage doing in a circuit?" the answer will always be that it is doing whatever the voltage and resistance allow it to do. If the amperage is low, then you know that one of two conditions is present—either the voltage is low or the resistance is high (**FIGURE 52-49**). If the amperage is high, then either the voltage is high or the resistance is low (**FIGURE 52-50**). Understanding the relationships between volts, amps, and ohms helps you know what test you need to perform next during diagnosis.

Because amperage is a product of voltage and resistance, it is a good idea to keep your eye on the amperage. What this means is that if you have a circuit fault, you can generally see

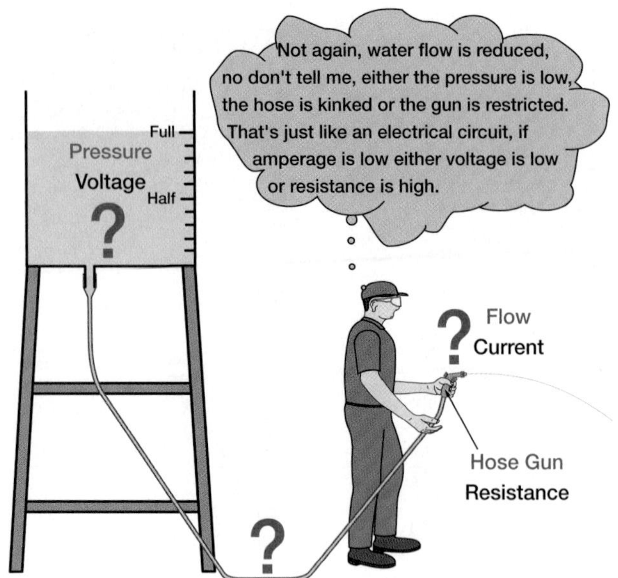

FIGURE 52-49 If amperage is low, then either the voltage is low or the resistance is high.

FIGURE 52-51 You can "see" that the fuse is blown: It most likely blew because of high current flow.

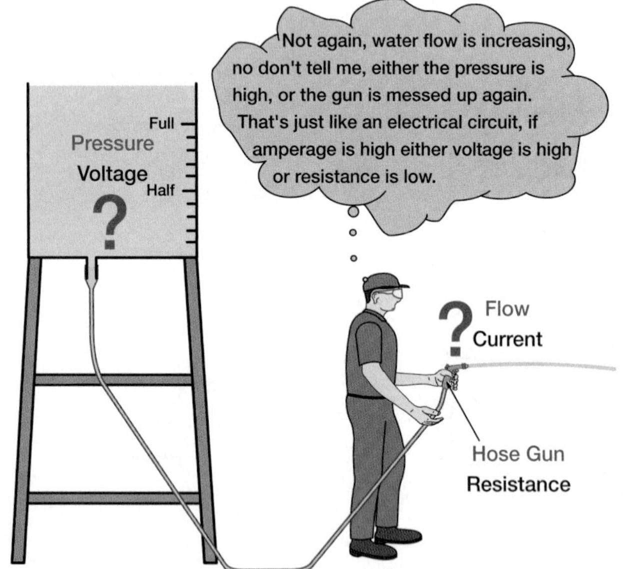

FIGURE 52-50 If amperage is high, then either the voltage is high or the resistance is low.

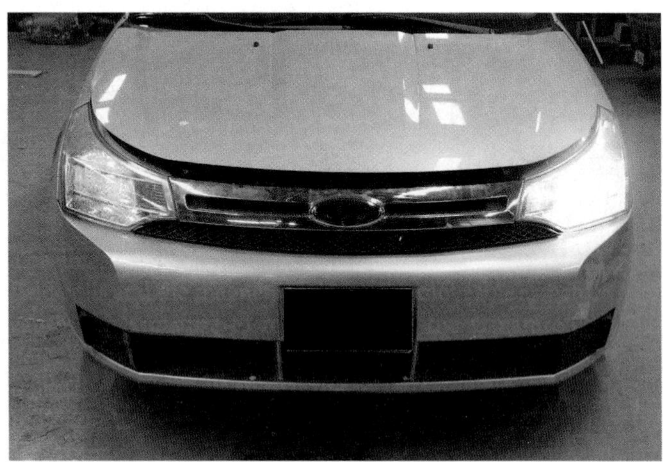

FIGURE 52-52 Our eye can determine that the left front headlight is dim due to low current flow.

what the amperage is doing; it is either high or low. If you truly cannot "see" what the current is doing in the load, like in an internal solenoid, you will have to measure it and compare it to specifications. In most cases, you can see it, and you can do so without measuring it. For example, you can "see" that the fuse blew (**FIGURE 52-51**): It blew because the current was too high. You can "see" that the light is dim: It is dim because the current is too low (most common scenario) because either the voltage is low or the resistance is high. On the flip side, if your eye determines that the current is high, then either the voltage is high or the resistance is low.

For example, let's say that the left front headlight is dim, but the right one is bright (**FIGURE 52-52**). Your eye determines that the current in the left front circuit is low. Thus, either the voltage is low or the resistance is high. Now you just have to test for those two things. Because the right side is bright, the battery voltage is okay and not causing the dim light. But we want to know battery voltage so we can use it for comparison purposes in the next step. Use a voltmeter to measure the voltage at the battery with the lights on, and write it down (**FIGURE 52-53**). If it is low, determine why, such as a faulty charging system. Next, check the voltage drop across the input and output terminals of the headlight with the circuit on (**FIGURE 52-54**). The voltage drop should be well within 1.0 volt of the battery voltage. If not, then switch to looking for the high resistance in the circuit. You do this by first measuring

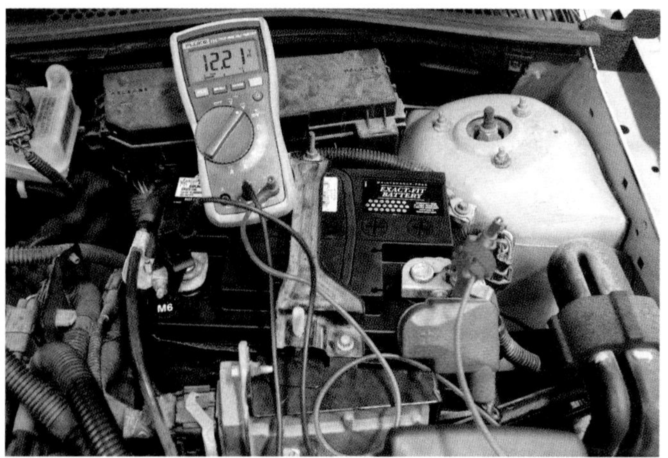

FIGURE 52-53 Measure battery voltage with the headlights on.

FIGURE 52-54 Measure the available voltage at the dim headlight with the light on. It should be well within 1.0 volt of battery voltage.

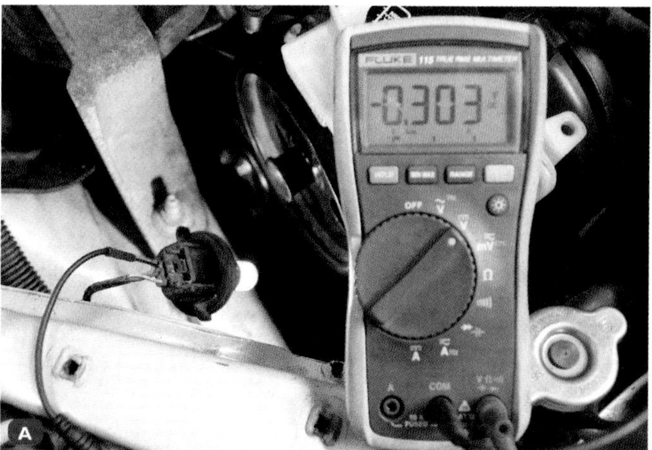

FIGURE 52-55 Measure the voltage drop on each side of the headlight back to the battery. **A.** Power side shows an acceptable voltage drop. **B.** Ground side measurement indicates a fault.

FIGURE 52-56 Here, most of the excessive voltage drop is on the dirty and loose ground connection.

the voltage drop on each side of the headlight circuit back to the battery (**FIGURE 52-55**). If there is an excessive voltage drop on one or both sides, continue to take voltage drop readings by moving the leads closer together in the circuit (**FIGURE 52-56**). If the voltage drop across the headlight is well within 1.0 volt of battery voltage, the problem is most likely the headlight itself. You may be able to check the resistance of the headlight filament and compare it to a known good bulb. Or you may have to measure the current flow through the bulb and compare that to specifications, as its resistance increases greatly due to heat when it is illuminated.

▶ TECHNICIAN TIP

I once ran into a vehicle that had very dim headlights on both sides of the vehicle and on both the low beams and high beams, so I suspected a low-voltage or high-resistance problem in the wiring harness. But when I opened the hood, there was a lot of light from the headlights shining into the engine compartment. A closer look showed that the reflective surface inside the headlights had flaked off and was sitting on the bottom of the headlights. So most of the light from the filament was shining backward into the engine compartment, and not being reflected forward. Replacing the headlights fixed the problem.

If the current flow appears too high in a circuit, then it may have too much voltage or too little resistance. Measuring the battery voltage is an easy way to check for too much voltage. If the voltage is okay, then the resistance must be too low. Resistance of loads can be checked using an ohmmeter and comparing the reading to specifications. An ohms reading that is too low indicates that it is shorted. A high ohms reading indicates a high-resistance fault. The ohmmeter can also be used to check the wire harness for any short circuit conditions as well. We cover this in much more detail later in this chapter.

Using a DMM to Measure Voltage and Voltage Drop

The electrical system is becoming increasingly complex on modern vehicles, and measuring voltages with a DMM is a very common task when diagnosing electrical faults. Ensure that you do not exceed the maximum allowable voltage or current for the CAT rating of the DMM. If you are measuring high voltages, wear appropriate personal protective equipment, such as high-voltage safety gloves, long-sleeved shirts and pants, and protective eyewear, and remove any personal jewelry or items that may cause an accidental short circuit. For most measurements, set the DMM to auto range for ease of use. Select DMM leads and probe ends to match the task at hand; for example, if you need to take a measurement but require both hands to be free, use probe ends with alligator clips.

When using a voltmeter to measure voltage, you have a couple of options. One is the available voltage test. This typically involves placing the common lead on a good ground and the red probing lead anywhere in the circuit to find out how much voltage is available at that point (**FIGURE 52-57**). This gives you a reading of how much more voltage is at the probing lead than is

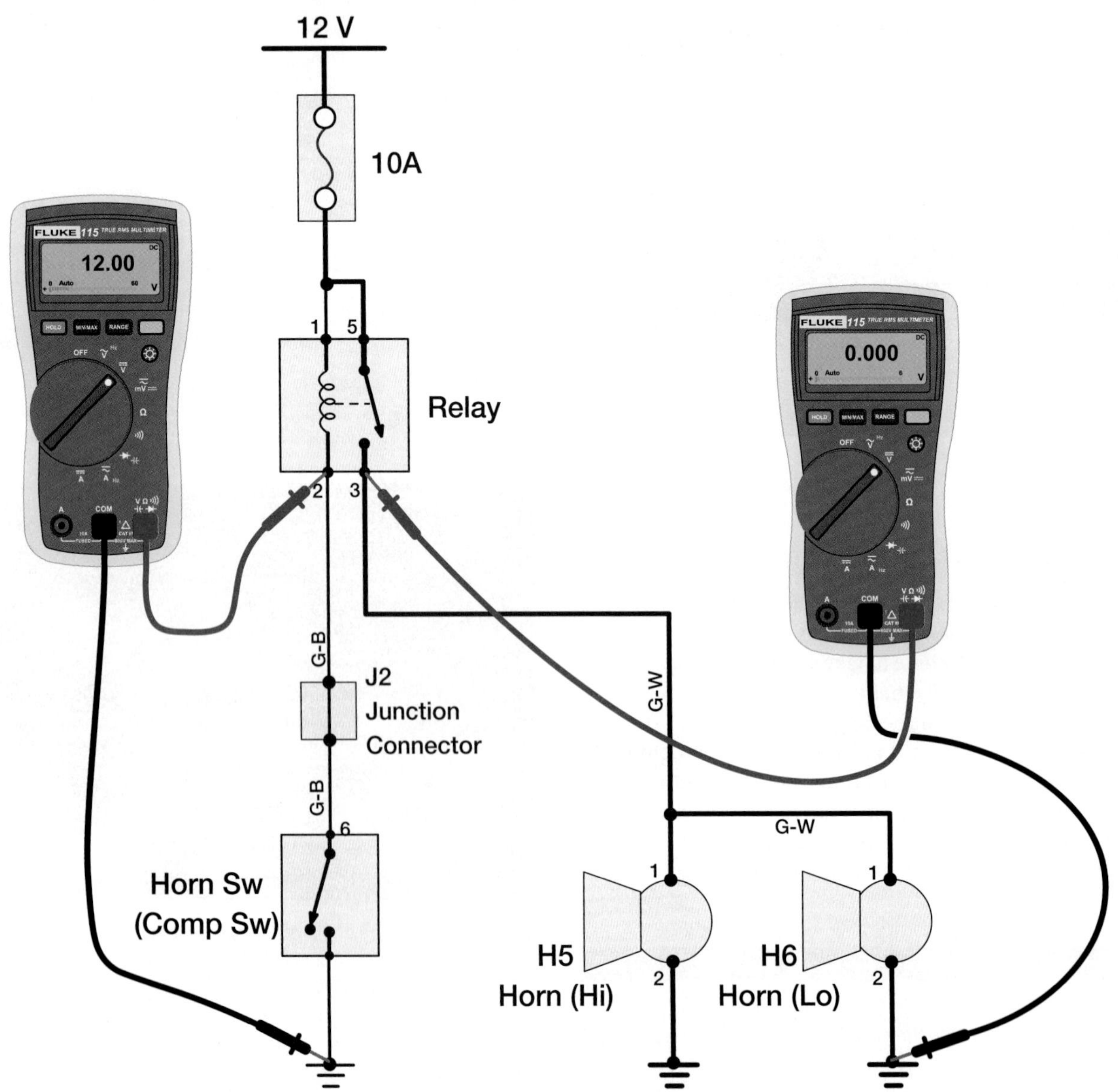

FIGURE 52-57 An available voltage test is performed by placing the black lead on a good ground and the red probe in the circuit where you want to know how much voltage is available.

at the common lead. Although that is helpful, it only gives you an indication of voltage. It does not tell you how much voltage we started with or how much voltage was lost getting to the test point. To do that, you need to either measure the available voltage at the battery and compare that voltage reading to voltage at the component, or perform a voltage drop test.

A voltage drop occurs when current flows through a resistance. The higher the resistance, the higher the voltage drop. We could say that a voltage drop test measures the voltage drop due to excessive resistance in a circuit. When testing for a voltage drop, the circuit must be turned on. That way, the circuit will have current flowing, thereby making voltage drops evident. Just to be clear, in a complete circuit, no current flow means an open circuit, so the only voltage drop will be across the open, and it will be a voltage drop of full battery voltage. Otherwise, if current is flowing, then there will be voltage drops at all resistances in the circuit, wanted and unwanted.

To test for unwanted voltage drop of the conductors, switches, and connectors, measure the voltage across each of these parts of the circuit, and compare them to specs (**FIGURE 52-58**). In general, in a 12-volt system, the individual voltage drops across an individual wire, connection, or common switch should be less than 0.2 volt. Additionally, the total voltage drop across each side of the whole circuit (from battery + to load, or from load to battery –) should not exceed 0.5 volt on 12 V circuits, or 1.0 volt on 24-volt circuits.

Although you could measure the voltage drop on each connection and wire on each side of the circuit, and add them together to see if they are more than 0.5 volt, it is faster if you measure the total voltage drop on each side of the circuit first. This allows you to identify which side of the circuit the excessive voltage drop is on (**FIGURE 52-59**). For the positive side, you can do this easily by placing the black voltmeter lead on the Bat + terminal and the red lead on the input of the load. If the voltage reading is 0.5 volt or lower, that side of the circuit is fine. If it is larger than –0.5 volt, then you need to check each individual connection and wire in that side of the circuit.

The negative side of the circuit can be checked by placing the black voltmeter lead on the Bat – terminal, and the red lead on the output side of the load. Again, the reading should be less than 0.5 volt. If the voltage drop on each side of the circuit is okay, and the load is not working properly, suspect that the load has a fault. You learn how to check loads later in this chapter.

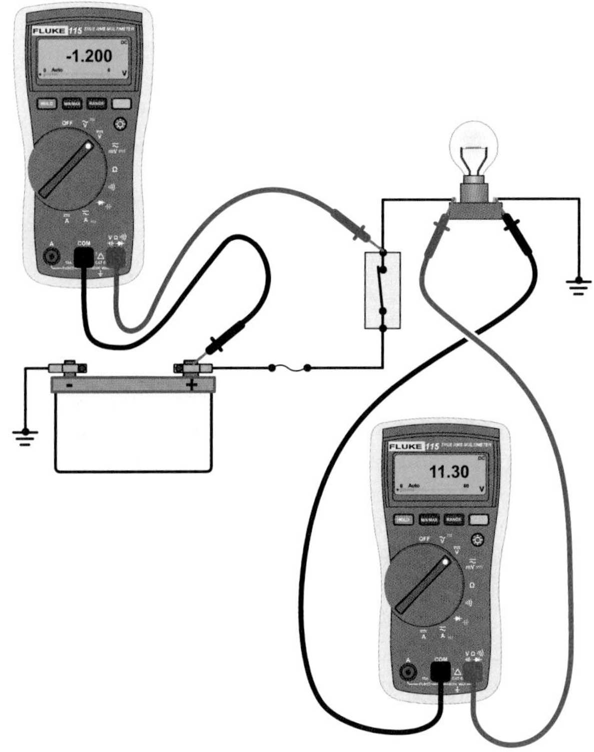

FIGURE 52-58 Excessive voltage drops in an improperly functioning circuit.

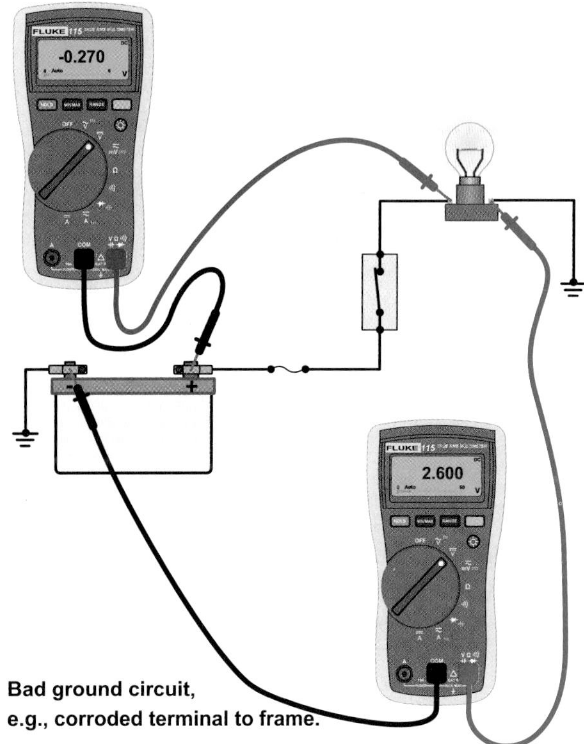

FIGURE 52-59 Measuring voltage drop on each side of the circuit.

Performing a Voltage Drop Test

To measure voltage drop, the DMM needs to be set to the voltage position. You need to set the function switch to either "auto range volts DC" on the DMM or the correct manual range. Then connect the black lead to COM slot and the red (probing) lead to the V/Ω slot.

For this example, the customer concern is that the horn isn't very loud. You verify the concern by activating the horn and hearing that it is not nearly as loud as it should be. You print out the wiring diagram for the horn circuit to familiarize yourself with it. You measure the source voltage at the battery terminals with the horn operating, and it reads 12.48 volts. You then connect the meter across the horn input and output terminals to measure

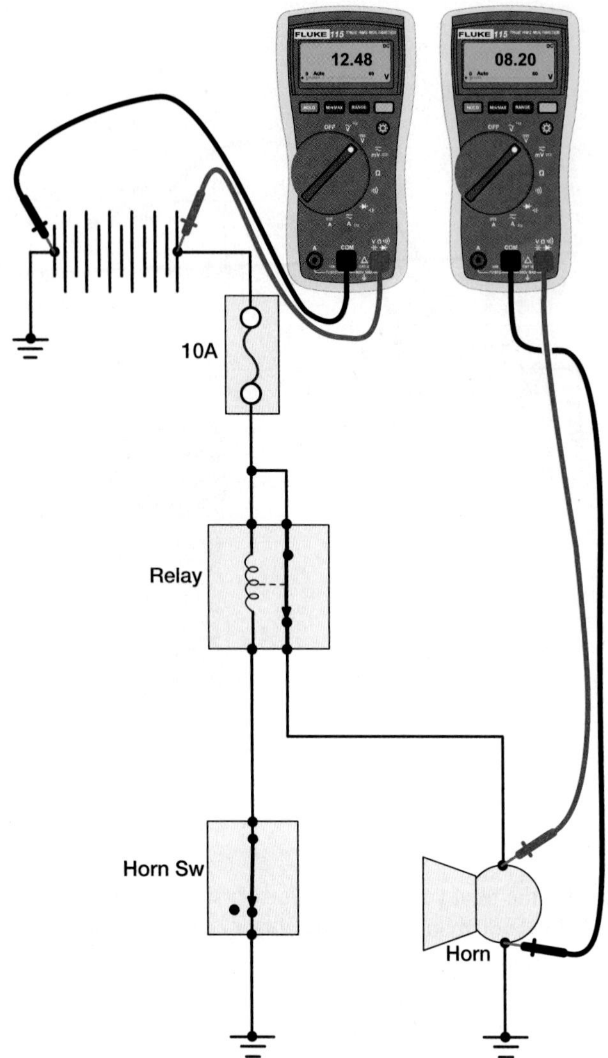

FIGURE 52-60 Comparing available voltage at the battery to the available voltage at the horn shows that the horn isn't operating on full battery voltage.

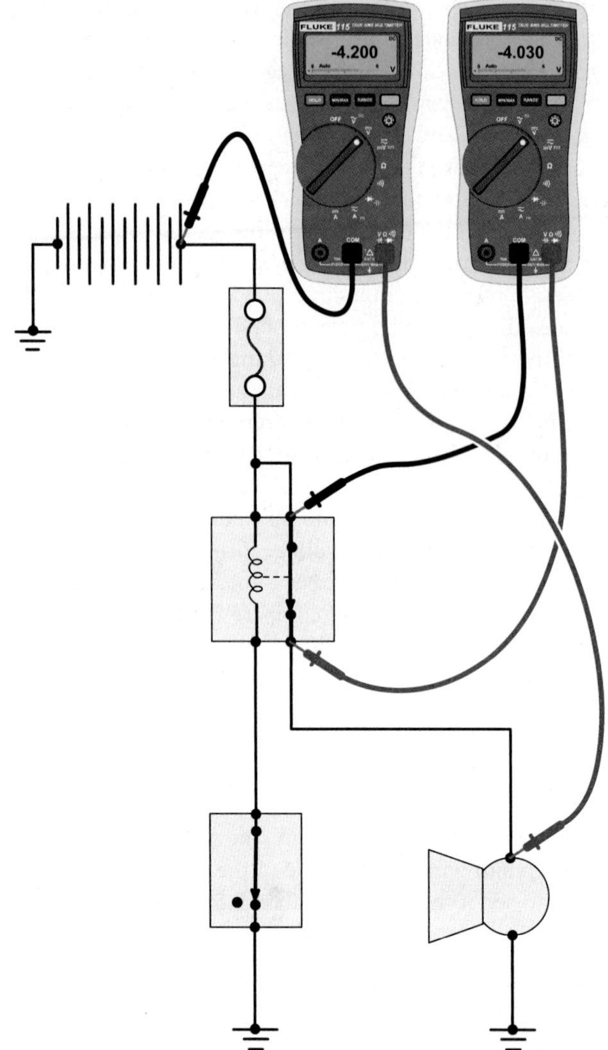

FIGURE 52-61 Excessive voltage drop on the feed side of the circuit due to worn horn relay contacts.

the voltage drop of the horn. Activating the horn shows a voltage drop across the horn of 8.28 volts (**FIGURE 52-60**). Because the circuit started with 12.48 volts, but the horn is only using 8.28 volts, you know that there is another voltage drop of 4.2 volts split between the feed side and ground side of the circuit.

You decide to start on the feed side of the horn circuit, so you connect the black lead to the positive terminal of the battery and the red lead to the input wire of the horn (the wire connected to the horn). When you activate the horn, the voltmeter read –4.2 volts, meaning that there is a voltage drop of –4.2 volts in the feed side of the circuit. In this case, the "–" means "less than." This means that the voltage is 4.2 volts less at the input of the horn (red lead) than it is at the positive battery post (black lead) (**FIGURE 52-61**). Because the voltage drop is more than 0.5 volt, this is an excessive voltage drop in that side of the circuit, and the test leads have to be moved, wire by wire, closer together in the circuit until the point of the voltage drop is located. In this case, there is a –4.03-volt drop across the relay contacts.

You could make the same measurement with the DMM leads reversed. If you place the red lead on the positive battery post and the black lead on the input of the horn, the meter would then read 4.2 volts. In this case, it shows positive. This is because the red lead is on the positive post of the battery, which is 4.2 volts higher than the horn input where the black lead is connected (**FIGURE 52-62**). As you can see, voltmeter leads can be hooked up in a couple of ways. Just remember that the meter always reads what the red lead is touching as compared to what the black lead is touching.

Don't forget that excessive voltage drops can happen on the ground side as well. For example, a corroded or bad chassis ground can cause a voltage drop that reduces the voltage and current available to loads. **FIGURE 52-63** shows a simple circuit with a bulb connected via a switch across a 12-volt circuit. In this circuit, a corroded ground connection has caused a resistance that is dropping 2.6 volts across it. Because the excessive ground circuit voltage drop is in series with the bulb, it reduces the voltage drop across the bulb, which in turn causes poor illumination.

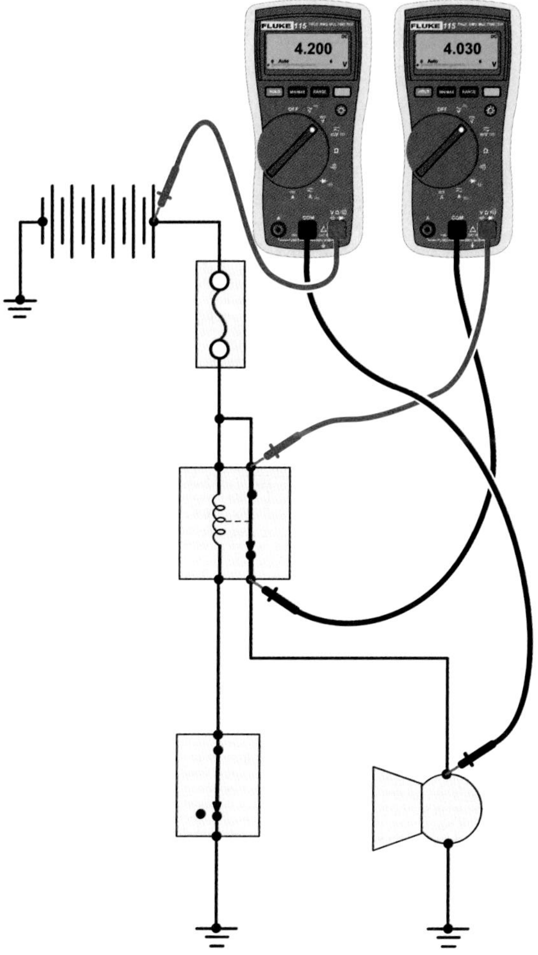

FIGURE 52-62 Reversing the meter leads in the circuit shows 4.2 more volts at the battery + terminal than at the input of the horn.

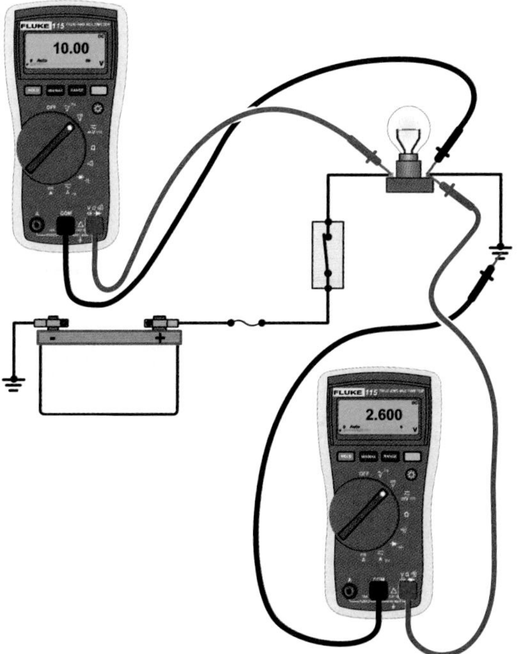

FIGURE 52-63 Voltage drops can occur on the negative side of the circuit, so don't neglect checking it when loads do not operate correctly.

Applied Math

AM-12: Mentally: The technician can mentally multiply numbers that include decimal numbers to determine conformance with the manufacturer's specifications.

In this scenario, a technician is conducting a voltage drop test on the positive side of the starter circuit. As described in the text, a voltage drop occurs when current flows through a resistance. The higher the resistance, the higher the voltage drops. So we could say that a test checks for excessive resistance in a circuit. When testing for a voltage drop, always have the circuit operating. That way, the circuit will have current flowing, so any voltage drops will be evident.

The technician begins by placing the DMM selector on the 20V DC scale and then starts testing at the battery, checking step by step all the way to the starter motor and using the direct method of voltage drop testing, which uses both test leads on the same side of the circuit. The tests include from the battery post to the cable clamp, from the cable clamp to the end of the cable at the solenoid, and from the solenoid to the starter motor terminal. At each of the test points, an assistant cranks the engine over to load the circuit. Each of the three connection points has a reading of 0.30 volt. The technician mentally multiplies those decimal numbers to obtain a total of 0.90 volt (3 × 0.30 volt = 0.90 volt). This reading is in excess of the manufacturer's specifications of 0.5 volt. Next, the technician will clean each of the connection points and retest.

Locating Opens, Shorts, Grounds, and High Resistance

S52005

DMMs, test lamps, and simulated loads tend to be the tools used most often for locating opens, shorts, grounds, and high-resistance faults. Refer to the chapter Principles of Electrical Systems for more information on opens, shorts, grounds, and high-resistance faults. An **open circuit** is a break in the electrical circuit where either the power supply or ground circuit has been interrupted. A systematic check of the circuit is required by first performing a voltage drop check on each side of the affected circuit to determine which side is open (**FIGURE 52-64**). An open circuit causes a voltage drop equal to the source voltage. Once the voltage drop is isolated to one side of the circuit, voltage drop testing can continue on that side, working the leads closer together in steps.

Also, use your understanding of electrical systems to consider the most likely places for the open circuit, such as a blown fuse or a faulty switch. And don't forget that the load could also be open. If the voltage drop test on each side of the circuit is within specifications, use an ohmmeter to check whether the load is open. Some loads such as diodes cannot be tested with a standard ohmmeter. In this case, follow the manufacturer's diagnostic procedure.

High resistance refers to a circuit where there is unintended resistance, which then causes the circuit not to perform properly. It can be caused by a number of faults, including corroded or loose harness connectors, incorrectly sized cable for the circuit current flow, incorrectly fitted terminals, and poorly soldered joints. The high resistance causes an unintended voltage drop in

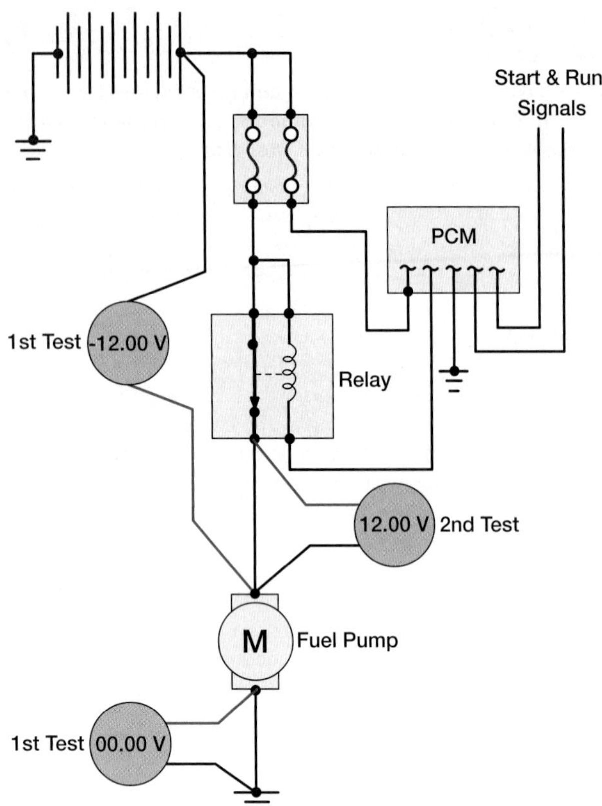

FIGURE 52-64 Locating an open circuit fault starts with a voltage drop test on each side of the circuit, then isolation tests by moving the voltmeter leads together.

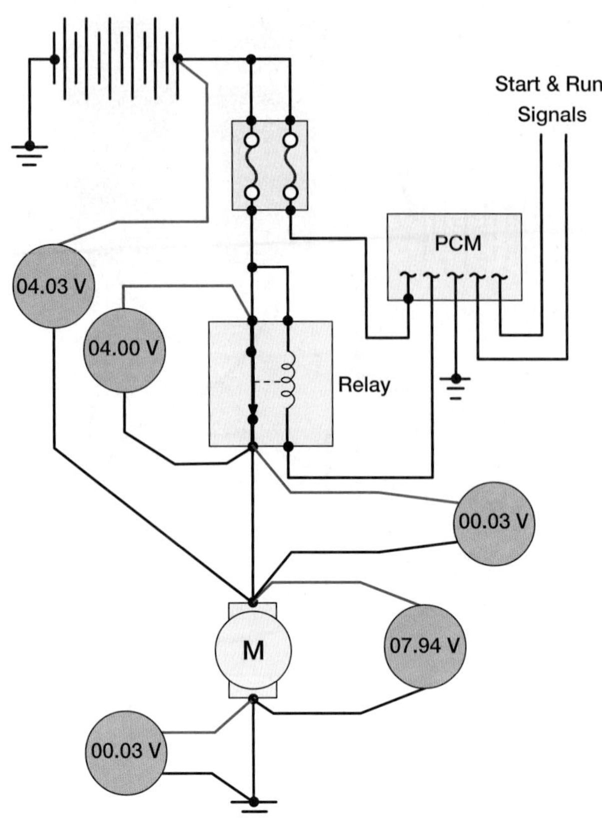

FIGURE 52-65 Example of a high-resistance fault causing an excessive voltage drop.

a circuit when the current flows. This drop reduces the amount of voltage that can be used by the load. The high-resistance fault also reduces the current flow in the circuit. The reduction in voltage and current to the load reduces the amount of electrical power to load (Power = Voltage × Amperage), affecting its performance. Unwanted high resistance can best be located by conducting a voltage drop test in the power and ground circuits just like you did for an open circuit; the only difference is that the voltage drop will be less than battery voltage (**FIGURE 52-65**).

If the high resistance is within the load, such as a relay coil, then the resistance can be checked with an ohmmeter and compared to specifications. Some devices, such as a fuel injector or ignition coil, may need further testing using an oscilloscope. In this way, the waveform can be evaluated, which can indicate issues that an ohmmeter cannot identify as easily.

Shorts, or **short circuits**, can occur anywhere in the circuit and can be difficult to locate, especially if they are intermittent. A short is a circuit fault in which current travels along an accidental or unintended route and can be thought of as a shorter path for current to flow. The short may occur within the load, such as shorted relay windings. It can also be in the wiring, where a wire is shorted to ground or to supply voltage. A short typically causes lower than normal circuit resistance. The low-resistance fault causes an abnormally high current flow in the circuit and may cause the circuit protection devices, such as fuses or circuit breakers, to open the circuit.

A short to power may cause the circuit to remain live even after the switch is turned off. For example, a short between a wire with power on all the time and a wire switched by the ignition switch would cause the circuit controlled by the ignition switch to remain on even after the switch is turned off. Just remember that shorts can be caused by faulty components or damaged wiring.

Shorts that happen within components, such as a relay coil, can usually best be tested by comparing the reading of an ohmmeter to specifications (**FIGURE 52-66**). Shorts that occur in wire harnesses are usually best tested by disconnecting each end of the affected harness and using an ohmmeter to test for unwanted continuity between various wires. A reading on the ohmmeter when connected to two separate wires indicates a short circuit between them. A true short between wires is indicated by a very low ohm reading, typically around 1 ohm or less.

Short to ground refers to a situation where a point in the circuit is unintentionally connected to ground. This can happen on the feed side of the circuit or the ground side. If on the feed side, the current flow will be very high and likely blow a fuse. If on the ground side, then the load will likely run anytime the circuit has power. An initial test can be conducted by disconnecting the fuse and the loads on that fuse's circuit and checking for continuity between the output of the fuse and ground (**FIGURE 52-67**). If there is continuity, you will need to find a connector between the fuse and load, disconnect it to see

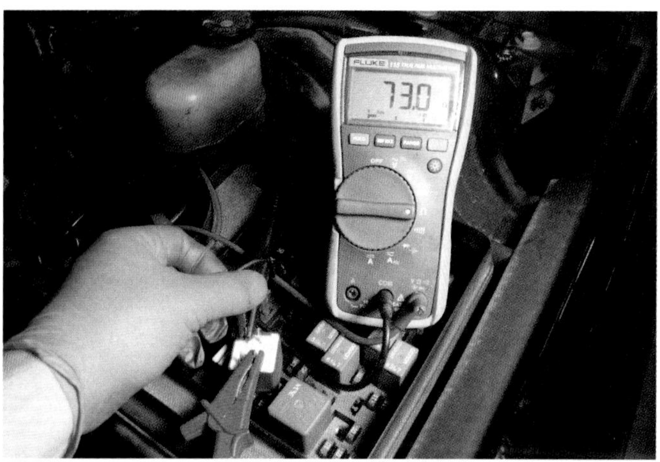

FIGURE 52-66 Shorts within components are usually best tested with an ohmmeter and compared to specifications.

FIGURE 52-68 A test light being used in place of a fuse to indicate current is flowing through the circuit. Disconnect wires until the light goes off.

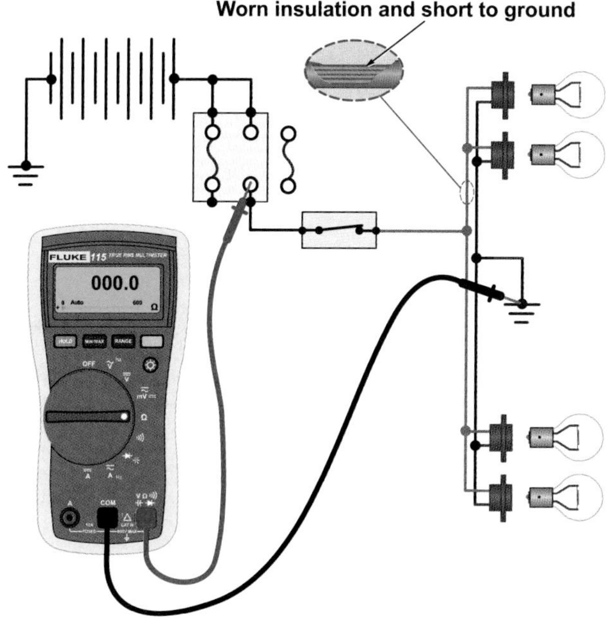

FIGURE 52-67 To find a short to ground, disconnect the loads on the circuit, and check for continuity between the output of the fuse and ground.

which side of the harness has continuity to ground, and keep tracing it until you locate the fault.

If, after disconnecting the loads, there was no continuity between the harness and ground, you need to check the resistance of each load to see if any are shorted internally. For example, if testing the blower motor circuit, first disconnect the blower motor and check for continuity between the output of the fuse and ground. If the short is still in place, then the wiring between the fuse and the load must be at fault. To further narrow down the site of the short to ground, inspect the wiring harness, looking for obvious signs of damage.

Another test can be conducted by connecting a test lamp or buzzer in place of a fuse (**FIGURE 52-68**). Current will flow through the test lamp or buzzer and find a ground through the short. Parts of the circuit can then be disconnected along the wiring harness to narrow down the location of the short. Specialized short-circuit detection tools are also available. They work by sending a signal through the wiring harness. A receiving device is then moved along the wire loom and indicates when a short is located. This type of device can be very useful in situations where it is difficult to access the wiring, such as within large wire looms or under vehicle trim.

Short to power refers to a condition where power from one circuit leaks into another circuit. A short-to-power situation usually causes strange electrical issues. In some cases, one or more circuits operate when they should not. Or in the case of sensor wires, the incorrect signals caused by the short to power can cause the computer to make very wrong decisions based on the faulty data (**FIGURE 52-69**). In this case, the engine, transmission, or other computer-controlled component can react strangely. Shorts to power are diagnosed first with a voltmeter to check for the unwanted voltage. Next, an ohmmeter is used to isolate the problem in the wire harness.

Checking Circuits with a Test Light

N52003, S52006

Non-powered test lamps are useful in determining whether electrical power is present in a part of a circuit. But you should always first test the test light on a known good power and ground before using it to test a circuit. It is possible that the bulb in the test light is burned out. You want to know that before performing any tests. If the test light illuminates, the two ends of the test light are touching both a power and a ground. If the light does not illuminate, the circuit is missing one or both of those elements, or the test light is faulty. Test lights are great for performing simple tests such as testing fuses. The test light lead can be quickly grounded and the probe end touched to each end of the suspect fuse. If both ends light, the fuse itself is good (but the fuse box terminal could be loose). If only one side of the fuse lights the test lamp, then the fuse is blown, and the circuit will have to be diagnosed to find out why it blew.

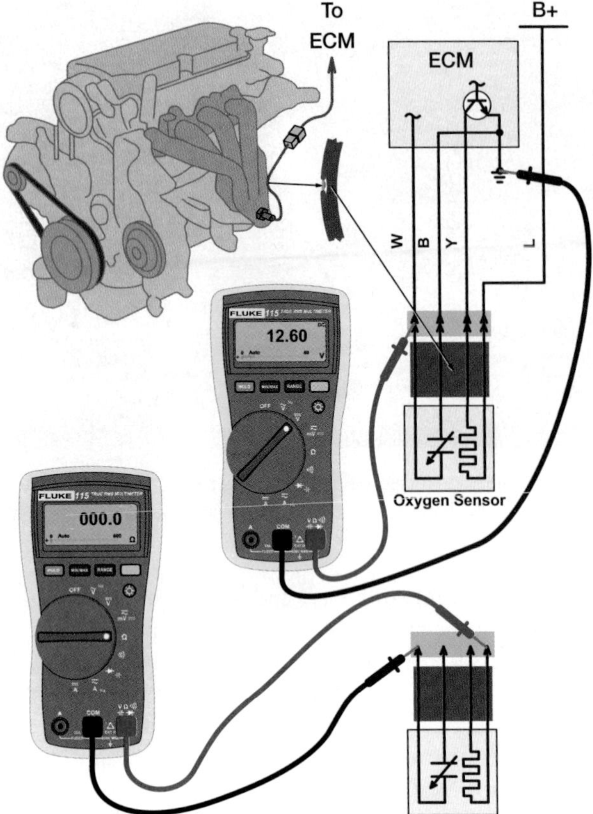

FIGURE 52-69 Short to power: The oxygen sensor heater wire melted to the signal wire, sending battery voltage to the PCM. This can be verified with a voltmeter; then an ohmmeter is used to show continuity between the wires.

To avoid damaging the test light, make sure the circuit voltage you are testing does not exceed the test light's rating. Most test lights are rated for 6- or 12-volt systems, and using the light in a 24-volt system will blow the bulb. You should *not* use a test light to test SRS (supplemental restraint systems), as unintended deployment of the airbags could result, a very dangerous and costly mistake. Also, using a test light on a computer circuit designed for very small amounts of current flow can damage the electrons inside the modules.

To check circuits with a test light, follow the steps in **SKILL DRILL 52-1**.

Checking Circuits with Fused Jumper Leads

N52004

Jumper leads can be used in a number of ways to assist in checking circuits. They can be created by the technician or purchased in a range of sizes, lengths, and fittings, or connectors. They are used to extend connections to allow circuit readings or tests to be undertaken with a DMM. They are also used to jump across terminals on fuses, relays, and other components. In some circumstances, jumper leads may provide an alternate current or ground source for components being tested. Regardless of their application, it is important that the circuit remains protected by a fuse of the correct size. To determine the correct size of fuse for any particular application, refer to the manufacturer's amperage specifications for the components being tested.

To check circuits with fused jumper leads, follow the steps in **SKILL DRILL 52-2**.

SAFETY TIP

Be very careful how you hook up any type of jumper leads, fused or unfused. If you hook them up to the wrong branch of a circuit, especially electronic circuitry, damage can be extensive. There is the old "magic smoke" saying: "Electrical and electronic components work off of the principle of magic smoke. Once the magic smoke is allowed to escape from the component, the component will never function again." Don't use jumper leads in a way that would let the "magic smoke" out of the circuit.

SKILL DRILL 52-1 Checking a Circuit with a Test Light

1. Connect the end of the light with the clip on it to the negative battery terminal. Touch the probe end of the test light to the positive battery terminal. The light should come on.

2. Connect the clip to any known good ground. A typical known good ground is any unpainted metal surface on the vehicle that is directly attached to the battery ground return system.

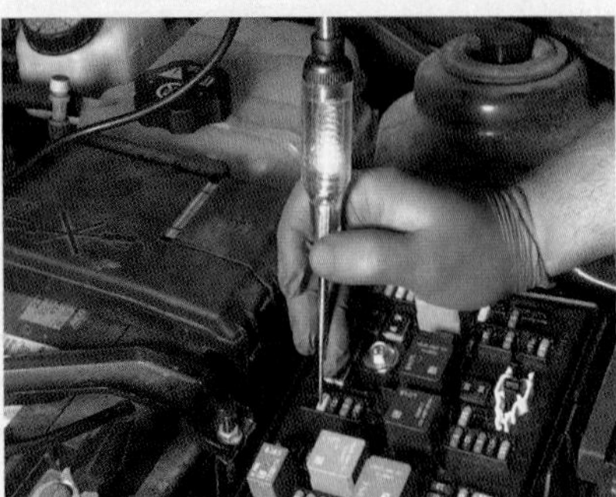

3. Place the probe on the terminal to be tested. If voltage is present, the light will come on.

SKILL DRILL 52-2 Checking Circuits with Fused Jumper Leads

1. Identify the circuit to be checked, and determine the fuse rating for the circuit.

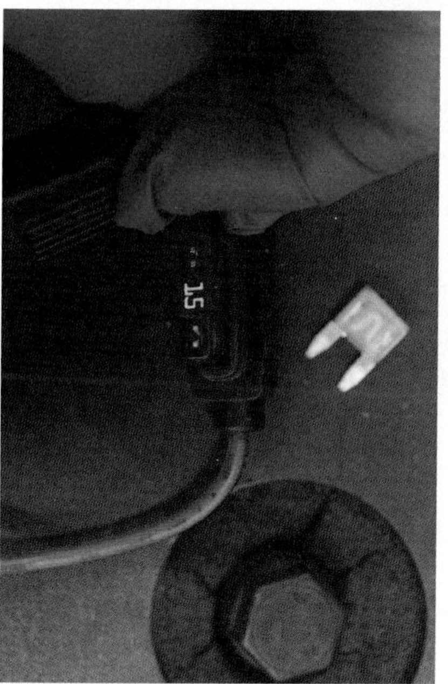

2. Select the appropriate jumper lead, install the correct fuse in it, and connect one end to the battery positive terminal.

3. Touch the jumper lead quickly to the horn input terminal. Never jump across a load; doing so bypasses circuit resistance, causing excessive current in the circuit and damaging it.

Inspecting and Testing Circuit Protection Devices

N52005, S52007

Protection devices are designed to prevent excessive current from flowing in the circuit. Protection devices like fuses and fusible links are sacrificial, meaning that if excessive current flows, they will blow or burn open and have to be replaced. Circuit breakers can be reset. Once they trip, they either reset automatically or require a manual reset by pushing a button or moving a lever. Fuses, fusible links, and circuit breakers are available in various ratings, types, and sizes, and must always be replaced with the same rating and type.

In most vehicles, protection devices are situated in the power or feed side of the circuit. A blown or faulty fuse can be tested using a DMM or test lamp. A good fuse has virtually the same voltage on both sides. A blown fuse typically has battery voltage on one side and 0 volts on the other side (**FIGURE 52-70**). If there are 0 volts on both sides of the fuse, either the ignition may have to be turned on, or a fusible link or maxi-fuse supplying the fuse box may be burned open. Fusible links can be checked for voltage as well. Voltage on both ends means it is good. Voltage only on one end means it is open.

Fuses can typically be visually inspected (**FIGURE 52-71**). This may require the removal of the fuse from the fuse holder. The fusible element should be intact and, if measured by an ohmmeter, should have a very low resistance. The contacts on both the fuse and the fuse holder should be clean and free of corrosion and should fit snugly together.

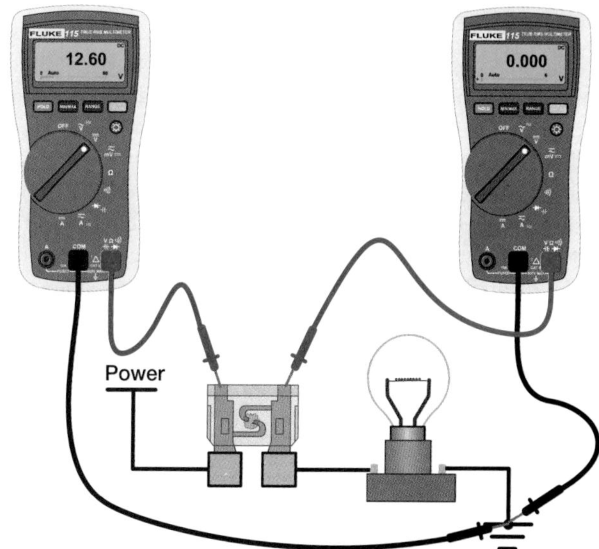

FIGURE 52-70 A blown fuse has battery voltage on one side and 0 volts on the other side.

FIGURE 52-71 Burned open fuse.

▶ TECHNICIAN TIP

It is fine to condemn fuses if they are obviously blown, but if they appear intact, do not rely on your eyes. Over time, fuses heat slightly and cool. This heating and cooling process can cause the fuse to become brittle and crack. The crack can be very fine, almost invisible, yet not conduct electricity, leading your diagnosis astray. So check fuses with a test light or voltmeter.

To inspect and test circuit protection devices, follow the steps in **SKILL DRILL 52-3**.

Inspecting and Testing Switches, Connectors, Relays, Solenoid Solid-State Devices, and Wires

N52006

Manufacturers produce wiring diagrams and diagnostic flowcharts to guide the technician through a diagnostic sequence based on the customer concern and your specific test results. In complex circuits, it is good practice to gather as much information as possible about the operation of the circuit and the customer concern. With that information, the wiring diagram, and the diagnostic flowchart, formulate a testing sequence for diagnosing the fault. Going through this process helps you to understand the problem and to identify possible causes, and a potential sequence of testing.

Inspection of electrical devices and wires usually starts with a visual inspection of the electrical circuit and is followed up with electrical testing. The visual inspection looks for breakage, corrosion, or deformity and includes examination of the insulation for any worn or melted spots. In the case of switches, solenoid contacts, and relay contacts, an electrical inspection is necessary. For example, switches would require voltage drop testing to see if they have excessive resistance (**FIGURE 52-72**). Additionally, solenoid and relay contacts can wear out and have excessive resistance, so performing a voltage drop test on them will determine if they are faulty. Some solenoids can be disassembled and visually inspected. In this case, the solenoid end cap may be removed and the contacts visually inspected (**FIGURE 52-73**). Typically, if there is an excessive voltage drop

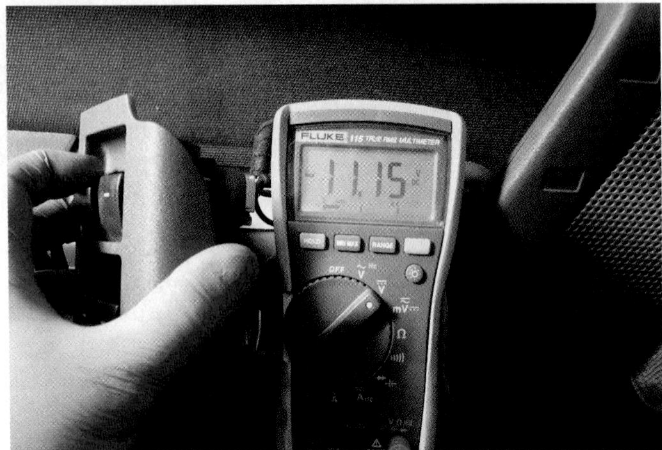

FIGURE 52-72 Switches can be tested for excessive resistance with a voltage drop test.

FIGURE 52-73 Some solenoids can be disassembled and the contacts inspected or replaced.

SKILL DRILL 52-3 Inspecting and Testing Circuit Protection Devices

1. Identify the protection device to be inspected and tested.
2. Conduct a visual inspection.
3. Set up a DMM to read volts, or use a test lamp.
4. Energize the affected circuit, if necessary.
5. Test for voltage on both sides of the circuit protection device. Determine and perform any necessary actions.

across the contacts, the contacts will be pitted and burned. Measuring resistance also comes into play when a shorted relay or solenoid winding is suspected.

DMMs and test lamps are used for most basic testing, with more specialized test equipment, such as oscilloscopes, being used if necessary. It is important to note that test lamps should not be used on electronic circuits due to their higher current draw, which could overpower the electronic components. Some tests, such as resistance tests, can be conducted on components in or out of the circuit. In-circuit tests are often preferred, as they usually provide the opportunity to test the component under load or during operational conditions. After in-circuit testing, components can be removed for individual testing, if required.

▶ Waveforms and Scope Testing

N52007

Oscilloscopes are commonly referred to as lab scopes, or just scopes, and are a very important test instrument for vehicle diagnostics. An oscilloscope provides a graph of voltage values over a time period, typically called a **waveform**, and displays them on a screen (**FIGURE 52-74**). An oscilloscope has a screen, a number of control knobs or buttons, and test sockets to connect the probing leads into the scope (**FIGURE 52-75**).

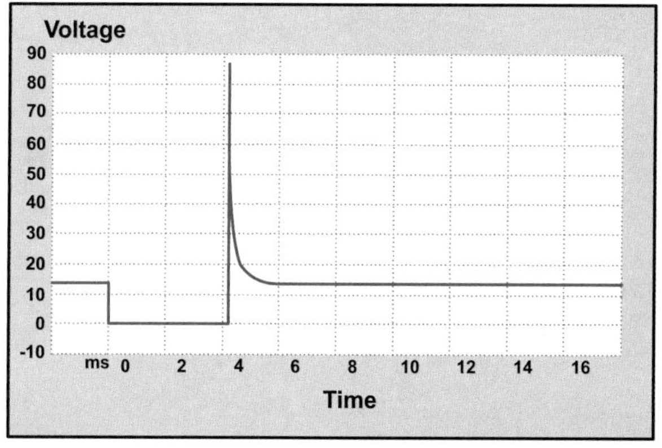

FIGURE 52-74 Oscilloscopes capture what happens to voltage over time.

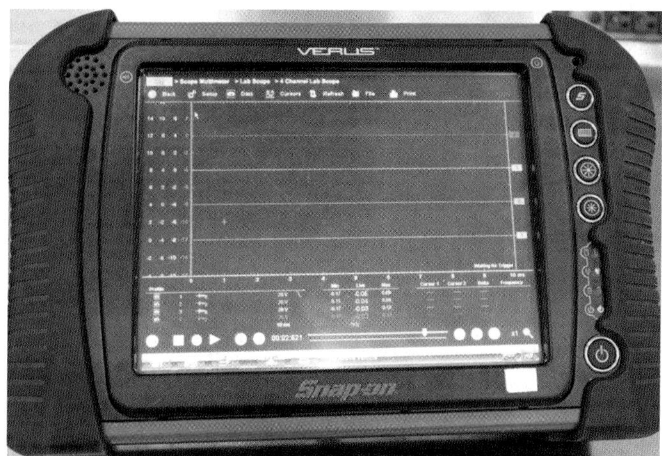

FIGURE 52-75 Typical oscilloscope.

There are many manufacturers of oscilloscopes, and the design and specifications of each vary based on the speed and voltage of waveforms to be measured. Oscilloscopes can be designed to do general electronic measurements or specifically designed for automotive work. Automotive oscilloscopes have special ranges and leads to perform specific automotive measurements; for example, high-voltage probes measure ignition waveforms, and current probes measure fuel pump waveforms.

Oscilloscopes may be analog, displaying waveforms as they occur, or digital. Digital storage oscilloscopes (DSOs) can display waveforms as they occur and store the waveform for later analysis. DSOs are increasingly popular for automotive use because stored waveforms can be compared with known good or bad waveforms. Scopes may incorporate their own screen, usually an LCD display, or connect to a computer by a USB cable and use the computer screen.

Oscilloscopes have a very fast sampling rate, which means they can take many samples per second and display the results on the screen. For example, some DSOs can take a sample every millionth of a second or even faster. Fast sample rates allow the oscilloscope to capture glitches in signals that occur for a very short time (**FIGURE 52-76**). The oscilloscope's sampling speed is a huge advantage over a DMM, which may take a sample every 200,000ths of a second. The oscilloscope can also display the waveform, which provides much more diagnostic information than a numerical reading on a DMM.

Scopes usually have two or more channels. Four channels are common and are used to display up to four different waveforms at the same time (**FIGURE 52-77**). Multiple channels allow the timing of various waveforms to be compared to each other. The timing and relationships of each waveform can be compared to each other and to the manufacturer's specifications.

Voltage is displayed on the vertical axis, and time is displayed on the horizontal axis, or from left to right across the screen. The technician selects the appropriate voltage and time ranges for each axis to capture the waveform on the screen. The pattern can then be read as voltage changes that happen over a period of time. The divisions on the sides and bottom of the

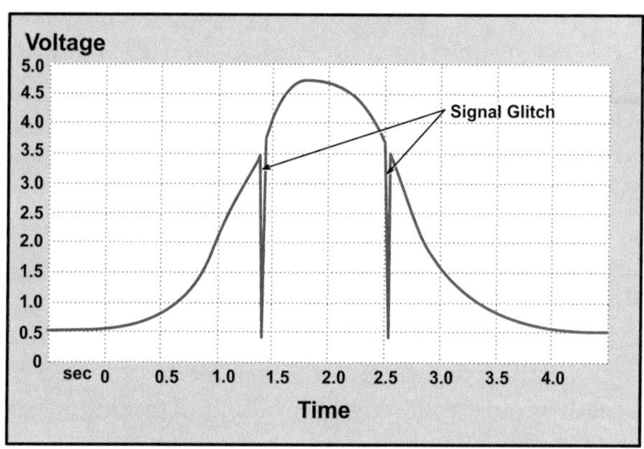

FIGURE 52-76 Oscilloscopes are ideal for finding intermittent glitches in electrical signals. Here, a TPS sensor signal falls out of specification for a very brief amount of time, affecting the drivability of the vehicle.

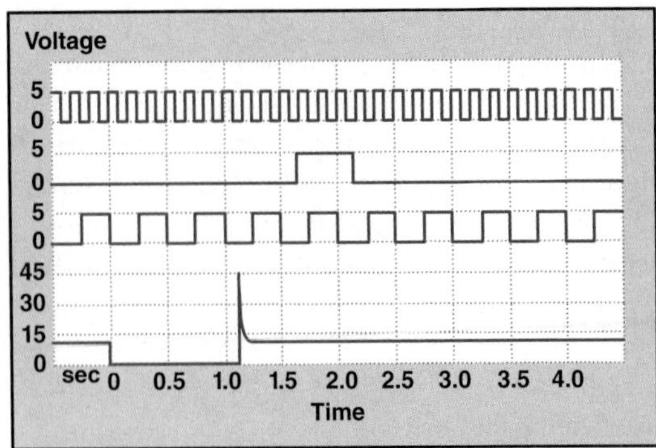

FIGURE 52-77 A four-channel lab scope showing four wave forms.

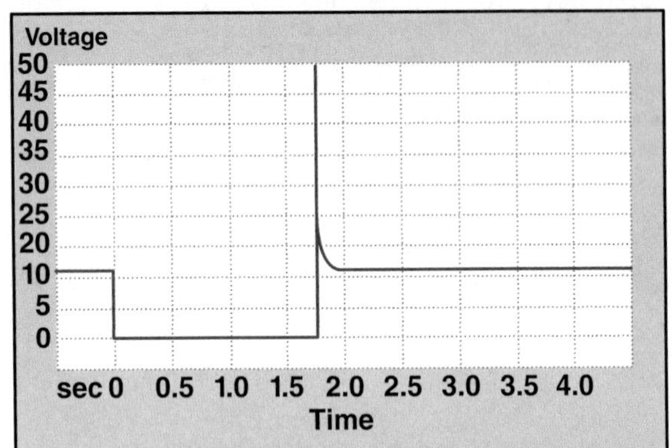

FIGURE 52-78 The divisions on the side and bottom allow voltage and time to be measured.

screen allow for the measurement of each section of the waveform (**FIGURE 52-78**).

Checking Circuit Waveforms

Automotive oscilloscopes tend to come in a kit with specialized leads and probes for connection into automotive-specific circuits. For example, there are ignition probe, fuel injection probes, current probes, temperature sensors, and pressure sensors (**FIGURE 52-79**). Correct setup of an oscilloscope is essential to accurately read circuit waveforms without damage to the circuit being tested, or to the oscilloscope. The two critical settings are the voltage per division and the time base (time per division). Similar to any meter, the voltage scale of the expected reading must be set. It can be set manually, or the oscilloscope may have an automatic range function. Time base is the amount of time being read per division on the screen. The scope screen is typically split horizontally, with 10 divisions for time (**FIGURE 52-80**). Oscilloscopes can measure very fast signals, and it is not unusual to have time base ranges from 100 nanoseconds per division to 200 seconds per division.

The scope screen is also split vertically for voltage, each division representing a user-selected specific amount of voltage, called volts per division. Oscilloscopes are capable of reading voltages as low as 50 millivolts up to maximum voltages of 50 or 100 volts. Sometimes on vehicles, higher voltages have to be measured. For example, the primary voltage of an ignition coil requires a higher voltage reading of about 300 volts. In these circumstances,

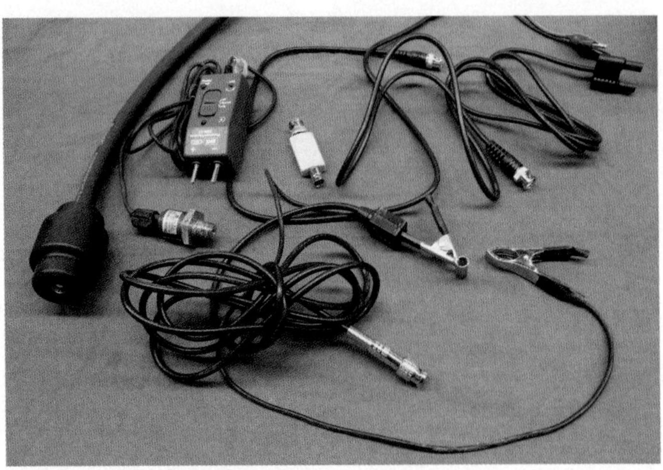

FIGURE 52-79 Various lab scope probes and sensors.

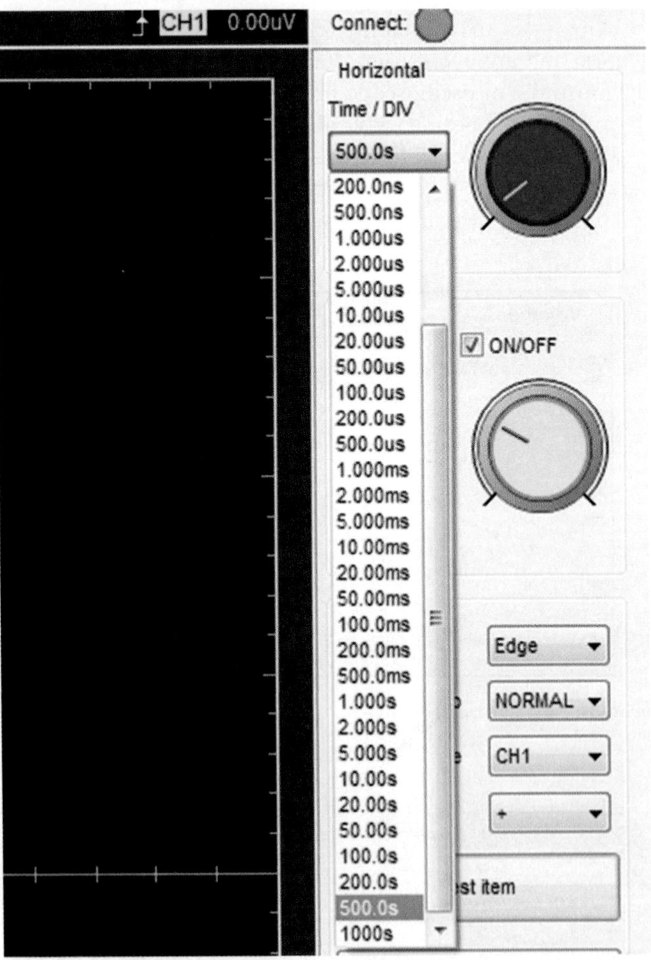

FIGURE 52-80 Typically there are 10 time-based divisions. The lab scope can be set for differing amounts of time per division, allowing you to evaluate a variety of waveforms.

attenuators can be fitted to the probe leads to reduce the maximum voltage to safe levels for the oscilloscope to measure (**FIGURE 52-81**). An attenuator is a device that reduces the amount of voltage passing through it. For example, a 20:1 attenuator divides the amount of voltage passing through it by a factor of 20. Thus, a 20-volt input will appear as an output of 1 volt.

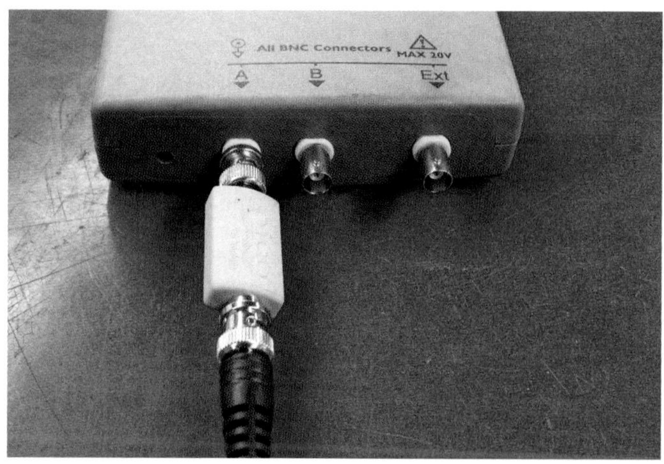

FIGURE 52-81 An attenuator lowers the voltage being sampled by a specified factor, such as 20:1.

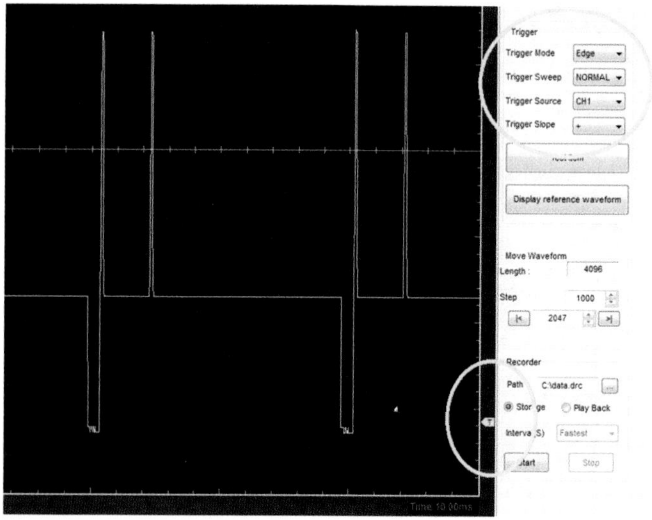

FIGURE 52-82 The trigger sets the specific rising or falling voltage that the waveform begins on the screen.

Most lab scopes have a setting called a trigger, which sets the point that the waveform begins to be displayed. On many scopes, it can be set to start at a specific voltage that is either rising or falling (**FIGURE 52-82**). This allows you to position the pattern on the scope for better viewing.

Always connect the oscilloscope leads according to the manufacturer's specifications. The correct connection of ground leads is important to ensure minimal interference in the waveform. In most cases, the ground lead should be connected directly to the negative battery terminal.

To check circuit waveforms, follow the steps in **SKILL DRILL 52-4**.

Applied Math

AM-41: Proportion/Congruence: The technician can distinguish the congruence of measurements with tolerances specified by the manufacturer.

The word "congruence" comes from the word "congruent," which means to agree or correspond with. A definition is "equal in size and shape."

In this segment, we consider the throttle position sensor (TPS), which sends vital throttle plate position information to the powertrain control module (PCM). A TPS is a potentiometer that is a three-wire variable resistor used to monitor movement. The movement is the throttle shaft rotation in degrees. The PCM sends a 5-volt reference signal to the TPS sensor. There is also a ground wire and a signal return wire to the PCM.

A TPS can be tested to some degree with a DMM, but the best test instrument is the oscilloscope. With this instrument, the technician can watch the signal. It should sweep up and down smoothly with no dropouts or spikes. This is where the congruence comes in. The performance of the TPS signal should be in a congruent style. The slope of the graph should agree or correspond with tolerances specified by the manufacturer. If so, we have congruence of measurements to meet manufacturer's specifications.

SKILL DRILL 52-4 Checking Circuit Waveforms

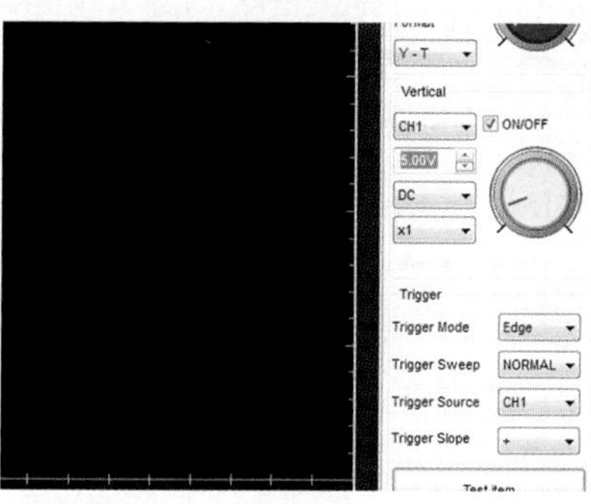

1. Determine the circuit to be tested and the likely voltages and frequency of the waveform to be measured. Set the voltage level per division and time base.

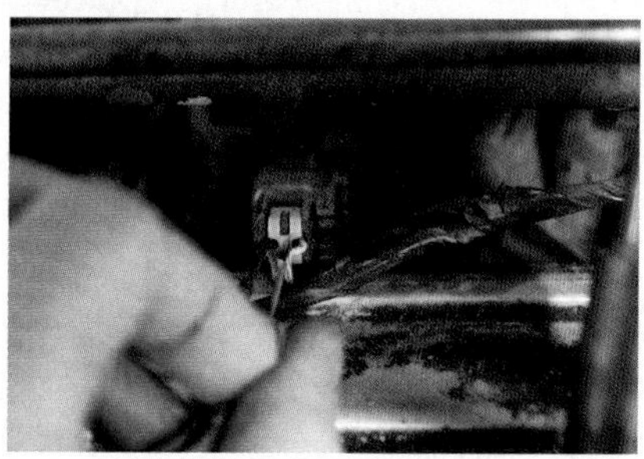

2. Connect the leads to the points in the circuit to be measured.

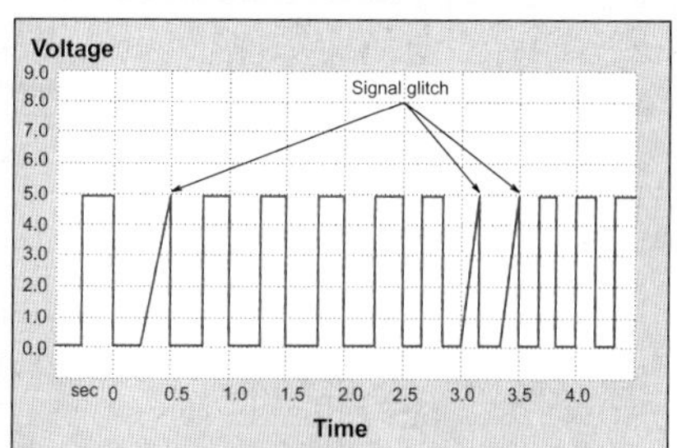

3. Capture waveforms from the circuit being tested. Analyze the waveform, comparing it to the manufacturer's specifications or known good waveforms.

Wrap-Up

Ready for Review

- The digital volt-ohmmeter (DMM) or digital multimeter (DMM) is an electrical measurement tool used to diagnose and repair electrical faults.
- Average or better DMMs are fused to protect the meter. But that doesn't protect the vehicle circuit.
- DMMs have a CAT rating. Each level, or CAT, is designed to work safely on higher-powered electrical systems.
- To properly use a DMM requires time and effort to learn the parts and how they work.
- The DMM can measure volts, ohms, and amps in a circuit.
- An advanced DMM can measure frequency and temperature and has a dedicated diode test capability.
- A DMM is the first tool used to take electrical measurements.
- The DMM allows the technician to see the movement of electrical impulses that cannot be seen without some type of electrical test equipment.
- The DMM can measure electrical volts within circuits.
- The DMM can measure ohms, which is the resistance of a circuit.
- The DMM can measure amps, which is the current flow of a circuit.
- The main parts of the DMM are the main body and the two current leads.
- The main body has a function switch, a connection point for the leads, and a digital display to show values.
- The leads are black for common and red for probing.
- There is a wide selection of leads for the DMM, to enhance the testing capabilities.
- Before you can use a DMM, you need to know the quantity of the measurement (volt, ohm, or amp).
- The DMM can read a wide range of scales, depending what position is selected.
- The DMM can read from low to high values. But always start high and work lower.
- The lower the range selected that will provide a reading, the more accurate the reading.
- The DMM in auto range will select the best value for the range being measured.
- The min/max setting gives the technician the ability to capture the highest and lowest reading.
- The hold function freezes the value measured.
- There are many different ways to probe a circuit, depending on the circuit being tested.
- The probes should never be forced as this could damage the circuit being tested and the probes being used.
- If you use the back probe method, use the proper pins so you won't damage the wire, connector, or terminal.
- The most common measurements taken with the DMM are voltage, current, and resistance.
- Depending on measurements taken, the leads have to be in the correct slots on the body of the DMM.
- If the leads are connected in the wrong place on the DMM, it could cause a fuse to blow, or damage the circuit being tested.
- When voltage is measured, the leads are placed parallel to the circuit being measured.
- When current is measured, the leads are placed in series with the circuit being measured.
- When resistance is measured, the component should be isolated from the circuit so no power is present, and no alternate paths are possible.
- The DMM is very useful in finding opens, shorts, grounds, and high resistance.

Key Terms

attenuator A device that reduces the power of a signal without distorting its waveform.

available voltage test Measurement of voltage at various points in a circuit with the black meter lead on ground and the red lead probing the circuit.

hold function A setting on a DMM to store the present reading.

min/max setting A setting on a DMM to display the maximum and minimum readings.

open circuit A circuit that has a break that prevents current from flowing.

probing technique The way in which test probes are connected to a circuit.

short circuit A condition in which the current flows along an unintended route.

short to ground Fault conditions in a circuit where the circuit is unintentionally contacting a grounded component or wire. This may result in a short, in the case of a power wire; or it could cause a circuit to stay live in the case of a switched ground circuit.

short to power A condition in which current flows from one circuit into another.

voltage drop test Measurement of the difference in voltage between two points in a circuit, the black lead on the end point being tested and the red lead on the beginning point.

waveform A graphical plot of voltage over time that is displayed on an oscilloscope.

Review Questions

1. DMM is used to measure all of the following in a circuit *except*:
   - **a.** voltage.
   - **b.** resistance.
   - **c.** current.
   - **d.** specific gravity.

2. All of the following statements are true *except*:
   a. Measuring source voltage is accomplished with both meter leads on the positive terminal of the battery.
   b. The DMM on the volt setting is measuring the difference in voltage between the positive test lead and the ground.
   c. Measuring available voltage and voltage drop in the circuit are performed as part of the diagnostic process.
   d. Most modern DMMs can display minimum and maximum readings.
3. Which of the following is the recommended range to be set on a DMM to get the most accurate reading?
   a. Highest range possible for the value being measured
   b. Lowest range possible for the value being measured
   c. Always the highest range on the DMM
   d. Always the lowest range on the DMM
4. Which category of DMM is the minimum required when working on hybrid vehicles?
   a. CAT I
   b. CAT II
   c. CAT III
   d. CAT IV
5. Choose the correct statement.
   a. The COM slot is always used for the black lead.
   b. The meter screen will always read what the COM lead is touching.
   c. The standard probe leads that are supplied with a DMM are both black.
   d. To take voltage measurements, the probing lead (red) is connected to the "A" slot.
6. Which of the following is used to designate a reading of 400 megavolts?
   a. 400 mv
   b. 400 MV
   c. 400 kV
   d. 400 μV
7. Which of the following precautions should be taken while measuring the resistance in a series circuit using a DMM?
   a. The switch should be on.
   b. The components should be disconnected from the circuit.
   c. Class "0" gloves shouldn't be worn.
   d. Start by using the lowest meter range.
8. Conducting a voltage drop test on the power and ground circuits is best for locating:
   a. high resistance faults.
   b. capacitance faults.
   c. short circuit.
   d. parasitic draw.
9. What is the total resistance for two 2-ohm resistors connected in parallel?
   a. 1 ohm
   b. 2 ohms
   c. 3 ohms
   d. 4 ohms
10. Which of the following can be fitted to probe leads to reduce the maximum voltage to safe levels for an oscilloscope to measure?
   a. TPS
   b. Attenuator
   c. Capacitor
   d. Resistor

ASE Technician A/Technician B Style Questions

1. Tech A says that total resistance goes up as more parallel paths are added. Tech B says that total amperage goes up as more parallel paths are added. Who is correct?
   a. Tech A
   b. Tech B
   c. Both A and B
   d. Neither A nor B
2. Tech A says that when reading DC voltage on a meter, a "+" before the number means that there is a higher voltage at the red lead than the black lead. Tech B says that a "–" before the number means that there is a lower voltage at the red lead than the black lead. Who is correct?
   a. Tech A
   b. Tech B
   c. Both A and B
   d. Neither A nor B
3. Tech A says that to read amperage, the meter must be hooked up in series in a circuit. Tech B says that to read amperage at a load, you should place one lead of the ammeter on the input side of the load and the other lead on the output. Who is correct?
   a. Tech A
   b. Tech B
   c. Both A and B
   d. Neither A nor B
4. Tech A says that when measuring available voltage, the common (black) lead should be on a good ground. Tech B says that a resistance reading on a lightbulb requires the DMM to be hooked to each side of the bulb and the switch turned on. Who is correct?
   a. Tech A
   b. Tech B
   c. Both A and B
   d. Neither A nor B
5. Tech A says that when checking a voltage drop across an open switch, a measurement of 12 volts is normal. Tech B says that when checking voltage drop across an open switch, a measurement of 12 volts means there are 12 volts on each side of the switch. Who is correct?
   a. Tech A
   b. Tech B
   c. Both A and B
   d. Neither A nor B

6. A customer complains of a weak horn. Tech A says that the horn is faulty and should be replaced. Tech B says that performing a voltage drop test on the horn circuit is a valid test in this situation. Who is correct?
   a. Tech A
   b. Tech B
   c. Both A and B
   d. Neither A nor B
7. Tech A says that when resistance increases, current flow decreases. Tech B says that when voltage decreases, current increases. Who is correct?
   a. Tech A
   b. Tech B
   c. Both A and B
   d. Neither A nor B
8. Tech A says that an open circuit typically causes higher current flow, which will blow the fuse. Tech B says that a short to ground can typically be found by checking voltage at different points in the circuit. Who is correct?
   a. Tech A
   b. Tech B
   c. Both A and B
   d. Neither A nor B
9. Tech A says that when a fuse has blown, it just has to be replaced, because it performed its job. Tech B says that when a fuse has blown, the circuit has to be diagnosed because the fuse was probably not the problem. Who is correct?
   a. Tech A
   b. Tech B
   c. Both A and B
   d. Neither A nor B
10. Two technicians are discussing scope testing. Tech A says that the vertical axis of the screen shows voltage. Tech B says that the horizontal axis shows time. Who is correct?
   a. Tech A
   b. Tech B
   c. Both A and B
   d. Neither A nor B

CHAPTER 53
Battery Systems

NATEF Tasks

- **N53001** Inspect and clean battery; fill battery cells; check battery cables, connectors, clamps, and hold-downs. (MLR/AST/MAST)
- **N53002** Perform slow/fast battery charge according to manufacturer's recommendations. (MLR/AST/MAST)
- **N53003** Jump-start vehicle using jumper cables and a booster battery or an auxiliary power supply. (MLR/AST/MAST)
- **N53004** Perform battery state-of-charge test; determine needed action. (MLR/AST/MAST)
- **N53005** Confirm proper battery capacity for vehicle application; perform battery capacity and load test; determine needed action. (MLR/AST/MAST)
- **N53006** Identify electrical/ electronic modules, security systems, radios, and other accessories that require reinitialization or code entry after reconnecting vehicle battery. (MLR/AST/MAST)
- **N53007** Maintain or restore electronic memory functions. (MLR/AST/MAST)
- **N53008** Diagnose the cause(s) of excessive key-off battery drain (parasitic draw); determine needed action. (AST/MAST)

Knowledge Objectives

After reading this chapter, you will be able to:

- **K53001** Describe battery types and construction.
- **K53002** Explain battery sizing, configurations, and ratings.
- **K53003** Describe lead/acid batteries' charging and discharging process.

Skills Objectives

After reading this chapter, you will be able to:

- **S53001** Perform battery-related service.
- **S53002** Perform battery tests while maintaining memory.

▶ Introduction

The electrical system on modern vehicles is becoming increasingly complex, with an increasing reliance on electrical power to assist in the management and control of not only the engine and emissions system but also almost all aspects of the vehicle, including brakes and suspension, navigation, entertainment, information retrieval, and much more. Almost every system on modern vehicles relies on electrical and electronic components and theory, electronic control modules, and networking systems to connect everything together.

Fundamental to the automobile is the 12-volt storage battery. It provides the initial electrical power for the engine starting system, and once the engine is started, all of the vehicle's electrical systems rely on the charging system to provide sufficient power output and precise voltage control for the vehicle's onboard electrical and electronic circuits. Let's start with the battery, which provides the necessary standby (and reserve) power to get things going.

▶ The Battery

K53001

As you learned in the Principles of Electrical Systems chapter, electricity is a very flexible and useful energy that can be used easily in a variety of ways. But one drawback to electricity is that it cannot be stored easily in its electrical form for later use. This means that electricity must be stored in another form of energy and reconverted to electricity when needed. This is where batteries enter the discussion.

Batteries were developed in the early 1800s, and since that time many varieties and designs have been developed. The battery is part of everyday life and is widely used in modern electrical and electronic devices. Batteries store electricity in chemical form, which is possible because electricity causes a chemical reaction within the battery. In other words, the electrical energy is transformed into chemical energy. The chemical reactions change the composition of the chemicals, which then are stored until the electrical energy is needed. When electricity is needed, the chemicals react with each other, transforming the chemical energy back into electrical energy.

A battery consists of two dissimilar metals, an insulator material separating the metals, and an electrolyte, which is an electrically conductive solution (**FIGURE 53-1**). The strength of the battery depends on the materials used.

The traditional automotive battery type is the lead-acid battery, which is available in many different shapes, sizes, and designs to meet the requirements for various applications. For example, the battery used for starting a vehicle's engine is different from the marine deep cycle battery. Each requires different design characteristics for obviously different applications. Vehicle batteries are designed to provide high current draws for short periods of time, whereas deep cycle batteries supply smaller, continuous loads over longer periods of time. Although the size, case configuration, and design may change, the fundamental components and their operation remain the same.

Lead-Acid Flooded Cell Batteries

The wet cell lead-acid battery is the main storage device in automotive use. It is called a flooded cell battery because the lead plates are immersed in a water-acid electrolyte solution. An automotive battery can supply very high discharge currents while maintaining a high voltage, which is useful when cold starting. It gives a high power output for its compact size, and it is rechargeable.

The standard 12-volt car battery consists of six cells connected in series. Each cell has a nominal 2.1 volts, for a total of 12.6 volts for a fully charged "12-volt" battery. Each cell contains

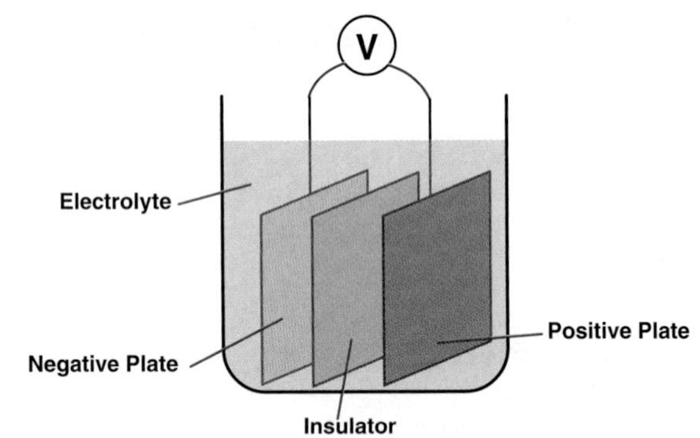

FIGURE 53-1 Components of a simple battery.

You Are the Automotive Technician

A seven-year-old vehicle has just been towed into the shop with a no-crank condition. The driver of the vehicle informs you that he forgot to turn the lights off while he was in a store for about 30 minutes. When he came out, the car cranked very slowly for a second or two and then stopped. He is surprised that it died so quickly. After writing up the repair order and looking up the service history, you inspect the vehicle visually. You notice that the battery is the original one, and the terminals are quite dirty. After using a battery terminal tool to clean the battery terminals, you test the battery's capacity and find it has degraded from its original 650 CCAs to 185 CCAs. With a booster battery connected, the vehicle starts and runs normally.

1. What test is used to measure the battery's CCA capacity?
2. Which parts need to be replaced and why?
3. How would you maintain any adaptive memories while disconnecting the battery?
4. What is parasitic draw, and how is it measured?

two sets of electrodes (called plates)—one set of sponge lead (Pb) and the other set of lead dioxide (PbO_2)—in an electrolyte solution of diluted sulfuric acid (H_2SO_4) and water (about 36% sulfuric acid and 64% water). As the battery discharges, the sulfuric acid is absorbed into the lead plates, and both of the plates slowly turn into lead sulfate. At the same time, the strength of the electrolyte becomes less acidic as the acid is absorbed into the plates. Recharging the battery reverses this process.

In automotive storage batteries, the main storage device is the wet cell. The nominal 2.1 volts of each cell does not depend on the size of the cell; however, its current capacity does. The surface area of the plates in a cell determines the cell's current capacity. In a lead-acid battery, positively and negatively charged plates are assembled so that they are alternated. Inside of a cell, all of the positive plates are connected to each other in parallel, and all of the negative plates are connected to each other in parallel (**FIGURE 53-2**). The more plates, the greater the current capacity of the cell. Because the plates are arranged alternately, they have to be close to each other, but not touching. If they touch, an internal short circuit in the cell will occur, which is typically what has happened in a battery with a "dead cell." The positive and negative plates have shorted together and therefore discharge that cell completely. Normally, the plates are kept from touching each other by separators, usually made of plastic.

The battery's six 2.1-volt cells are connected in series to form the battery, which gives the battery a nominal voltage of 12.6 volts. The cells are sealed from each other and filled with dilute sulfuric acid. The battery case is usually made of plastic or hard rubber. With the cells in series, one end of the battery is connected to the negative post, and the other end is connected to the positive post (**FIGURE 53-3**).

Low-Maintenance and Maintenance-Free Batteries and Cells

Many types of batteries with variations of cell design are available for vehicles. Some batteries are specifically designed for starting, some for deep-cycle or marine usage, some for low-maintenance or maintenance-free applications. "Starting batteries" is another name for flooded cell batteries that were covered earlier. Deep-cycle or marine batteries are made with heavier lead plates that tolerate deep discharging better than starting batteries. But they are heavier and bulkier than starting batteries and have a lower output per pound. Low-maintenance batteries require little, if any, topping off the water in the electrolyte. The plates and venting system are designed so that they do not normally gas and release water vapor to the atmosphere. Low-maintenance batteries still have removable caps so that the electrolyte can be checked and topped off if necessary.

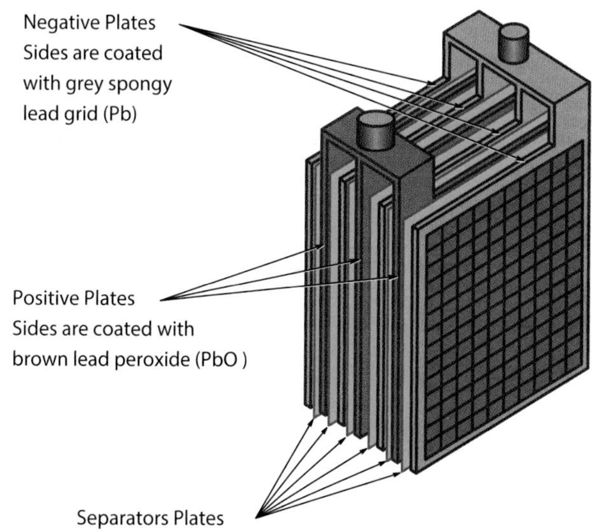

FIGURE 53-2 Typical plate arrangement in a wet cell battery.

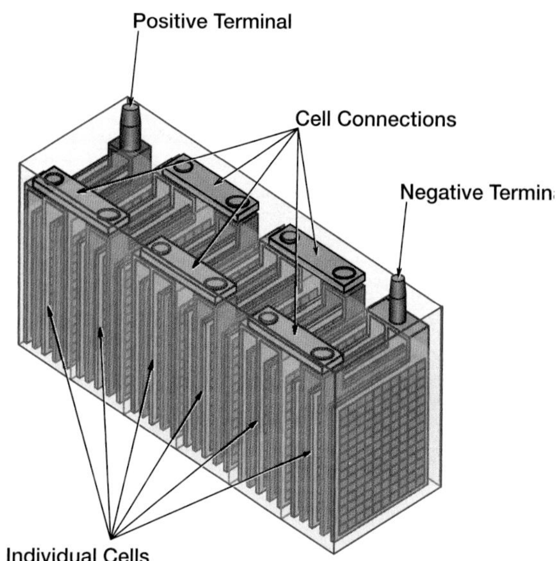

FIGURE 53-3 Interconnections between all six cells in a battery, showing the most negative and positive points of the battery.

> **TECHNICIAN TIP**
>
> Because lead is relatively soft, when it is used in the lead plates of a battery, it is prone to bending and stretching. Many years ago, antimony was added to the plates to strengthen them, but this caused the cells to gas off water, which caused the electrolyte level to fall over time, requiring periodic topping off. Calcium is now added to one or both plates to reduce water usage.

Maintenance-free batteries, also called valve-regulated lead-acid batteries, are fully sealed, but equipped with a pressure relief valve, and do not require the electrolyte to be topped off. In some cases they use a gel-type electrolyte instead of a liquid. **Absorbed glass mat** batteries have the electrolyte absorbed within a mat of fine glass fibers. The plates in this type of fully sealed battery can be made flat or wound in a cylindrical cell (**FIGURE 53-4**). Because the electrolyte in absorbed glass mat batteries is a gel, which does not spill, this type of battery is especially handy for rough handling or tipping. In fact, this type of battery can even be mounted on its side and will still perform well. Thus the absorbed glass mat battery is especially suited for off-road and racing vehicles.

A sealed or low-maintenance battery typically has no removable cell covers, so you cannot adjust or test the fluid levels

FIGURE 53-4 AGM battery.

inside. However, some of these batteries do have a visual indicator (a single-cell hydrometer float) that provides information on the status of the charge and condition of one of the battery cells (**FIGURE 53-5**). Each manufacturer provides details of these visual indicators; refer to these when performing an inspection.

Advanced Batteries

Batteries and cells continue to be developed to create more efficient, higher-density batteries. Consumer electronic devices such as mobile phones and tablet computers have been driving the development of new battery technologies such as nickel-cadmium (Ni-Cd), nickel-metal hydride (Ni-MH), and lithium ion (Li-ion). All of these batteries are types of rechargeable cell batteries, as are lead-acid batteries. Nickel-cadmium batteries contain older technology and have been replaced by nickel-metal hydride batteries, as both have a nominal cell voltage of 1.2 volts. The nickel-metal hydride battery can have two to three times the energy density of a nickel-cadmium battery, which means that for the same size case, it can store two to three times the energy. This made nickel-metal hydride batteries especially useful for drive motor applications such as in early hybrid electric, plug-in hybrid, and battery electric vehicles. They also tended not to have the memory effect that plagued the nickel-cadmium batteries of the past. The memory effect required nickel-cadmium batteries to be fully discharged between charge cycles to ensure that the maximum performance of the battery was maintained.

The lithium-ion battery is a newer type of rechargeable cell and is now used in many consumer electronic devices, such as cell phones and tablet computers. Such applications paved the way for further breakthroughs in the use of lithium-ion batteries in hybrid-electric or battery-electric vehicles; these vehicles use what is known as the rechargeable energy storage system (RESS). A lithium-ion battery has one of the highest energy density ratios of common batteries in production today. This high energy density means the battery can store more energy than comparable batteries of other types. This is a real advantage for vehicle applications. They also have a low self-discharge rate, which means they can sit for long periods without discharging.

The actual cell voltage depends on the final materials used to make the cell. The typical cell voltage for a lithium-ion battery is 3.6 volts. In contrast, cell voltage of a nickel-cadmium or nickel-metal hydride battery is 1.2 volts, and a typical lead-acid battery is 2.1 volts. Like all batteries, the lithium-ion cell has an anode, a cathode, and an electrolyte. When discharging, the lithium ions are removed from the anode and added into the cathode. When the cell is charging, the reverse occurs.

Lithium-ion batteries may suffer from **thermal runaway** and cell rupture if overheated or overcharged. In extreme cases, thermal runaway may result in an explosion. Extreme care should be taken when handling or charging lithium-ion batteries; always follow the manufacturer's recommendations. In hybrid or electric vehicle applications, many small individual cells are connected in series and in parallel arrangements to form a battery pack that delivers the power requirements of the vehicle (**FIGURE 53-6**). Following safety precautions is extremely important because RESS battery packs develop voltages in excess of 200 volts.

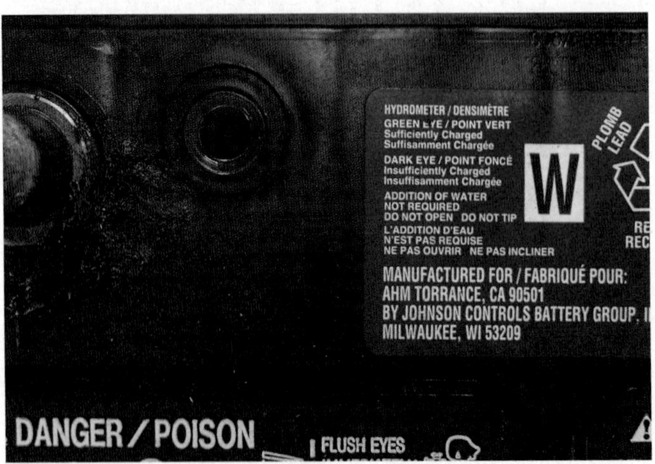

FIGURE 53-5 A status indicator is built into some batteries.

FIGURE 53-6 A typical hybrid vehicle battery stack is made of many small cells.

Advantages of lithium-ion batteries include:

- High energy density: There is more power per pound.
- Low self-discharge: Self-discharge is typically less than half of that of nickel-cadmium.
- Low maintenance: No periodic discharge is required.
- No memory
- Low internal resistance: They are good for high current requirements.

Disadvantages of lithium-ion batteries include:

- Need for circuit protection to ensure current and voltage are within safe limits
- Sensitivity to high temperatures
- Increased cost of manufacturing (however, these costs are being reduced as research improves the technology)
- Potential for damage if completely discharged

It should be noted that battery research is extremely intensive, driven by the search for evermore energy-dense renewable energy storage systems. As new developments find their way to market, expect to see even more variations of lithium-type batteries being used in hybrid electric and battery electric vehicles.

Battery Sizing and Configurations

K53002

Batteries have different sizing and configuration requirements to fit the space allotted for the battery and its location in the vehicle (**FIGURE 53-7**). Physical attributes of the automotive battery include the size of the battery case, the location of the battery terminals (top or side mounted and left and right), and the size or type of battery terminal (round-tapered posts, screw-on terminals, or both) (**FIGURE 53-8**).

Batteries sizes are designated by the Battery Council International (BCI) and given a group number (**FIGURE 53-9**). Individual groups are specified by dimension in length, width, and height. Note that a physically larger battery does not necessarily mean a higher electrical capacity, so always check the battery's ratings in addition to the BCI group size. Other designations relate to the **battery terminal configuration**, which refers to the placing of the positive and negative terminals on the battery.

Battery Cables and Terminals

Battery cables and terminals are designed to carry high discharge currents that are required during cranking of the engine. Battery cable terminals are usually made of lead or zinc-plated

FIGURE 53-7 Vehicle manufacturers locate the battery in a variety of places in the vehicle. **A.** Underhood placement. **B.** Under-seat placement.

FIGURE 53-8 Battery posts. **A.** Side terminal. **B.** Standard post.

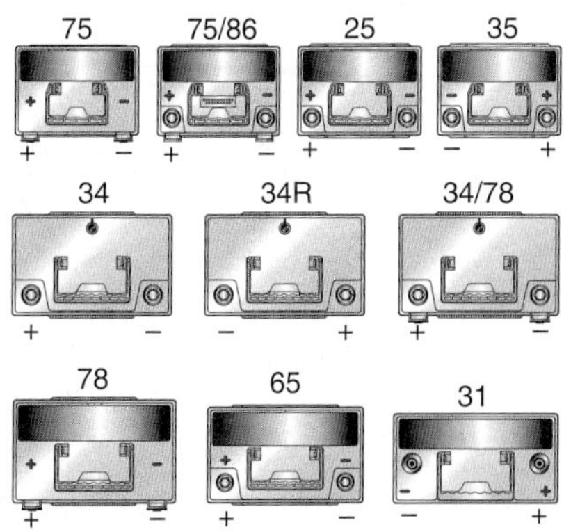

FIGURE 53-9 Typical types of layouts that use a lettering system for identification purposes.

brass. There are a number of designs, the most common being a cone design that provides a large surface contact area, with the ability to tighten the terminal onto the battery post using a nut and bolt. The tapered cone allows easy removal of the clamp when the bolt is loosened. The larger positive and smaller negative terminals are slightly different in size, to ensure correct connection of the cables when installing or servicing the battery (**FIGURE 53-10**).

FIGURE 53-10 A. Tapered post battery terminal. **B.** Side-post cable terminal. **C.** Lug terminal and cable.

Another type of battery terminal is the side terminal. It gets its name because the battery connection is on the side of the battery. The terminal is a flat circle with a center bolt that bolts the terminal tightly to the lead battery connection (see **FIGURE 53-10B**). Side terminal batteries seem to avoid some, but not all, of the oxidation issues of a top-post battery. One shortcoming of a side terminal battery is that it is harder to get a good connection when using jumper cables or a jumper box to jump-start the vehicle.

One less common battery terminal is a flat terminal (lug) with a hole through the center. The post on the battery sticks up vertically, with at least one side flat and a matching hole through the post. The battery cable end butts up against the flat post, and a bolt inserted through the battery post and terminal holds them firmly together (see **FIGURE 53-10C**).

> **TECHNICIAN TIP**
>
> Installing a battery backward into a vehicle instantly destroys expensive onboard electronics, as some technicians and do-it-yourself vehicle owners have unfortunately discovered. Always be careful to connect the cables to the correct polarity, and don't assume that a red cable means positive or a black cable means negative.

Battery cables are usually made of many fine strands of copper wire bound tightly together and insulated to make a cable with high current capacity. A number of cable sizes are available to handle various current capacities. For example, a small-displacement spark-ignition engine requires far less current to crank than a high-compression diesel (compression-ignition) engine. So make sure you use the proper sized cable when replacing them.

Battery terminals are usually crimped or soldered onto the battery cables to ensure strong, low-resistance connections. Often, heat-shrink tubing with sealing adhesive is used over the joint to ensure it is kept clean and protected from corrosion. Battery cables should be protected from chafing or damage, and terminals should be kept clean and free of corrosion.

> **TECHNICIAN TIP**
>
> Corrosion from the battery can creep into and destroy a battery cable beneath the insulation and may not be visible. A cranking voltage drop test on the battery cable is the most reliable way to detect such high resistance caused by corrosion.

> **TECHNICIAN TIP**
>
> When buying battery cables or jumper cables, be careful to verify the wire size. Some wire manufacturers place very thick insulation over a small wire size, making the cable look bigger than it is. Remember, it is the size of the wire that determines current flow, not the size of the insulation.

Battery Ratings

The **electrical capacity**, or the amount of charge a typical lead-acid battery can store, is determined primarily by the total surface area of the plates, but also to some extent by the thickness of the plates. The more plate surface area there is, the higher the electrical capacity of the battery. Automotive battery plates tend to be manufactured in a standard size, so a common method of increasing surface area is to increase the number of plates per cell. For example, a 12-volt, 11-plate cell has a higher capacity than a 12-volt, 9-plate cell.

There are several methods used to rate automotive battery capacity. The three most common are **cold cranking amps (CCAs), cranking amps (CAs)**, and **reserve capacity (RC)**.

- CCA measures the load in amps that a battery can deliver for 30 seconds while maintaining a voltage of 1.2 volts per cell (7.2 volts for a 12-volt battery) or higher at 0°F (−18°C).
- CA measures the same thing, but at a higher temperature-32°F (0°C). This can cause confusion because a 500 CCA battery has about 20% more capacity than a 500 CA battery, so it is important to keep the ratings straight.
- RC is the time in minutes that a new fully charged battery at 80°F (26.7°C) will supply a constant load of 25 amps without its voltage dropping below 10.5 volts for a 12-volt battery. This rating approximates the amount of time that a vehicle can be driven before the battery dies, if the charging system fails completely.

The CCA, CA, and RC are typically marked on automotive batteries, and technicians use these ratings when testing battery performance and when selecting batteries for particular vehicle applications (**FIGURE 53-11**).

Battery Discharging and Charging Cycle

K53003

As the battery is creating current flow to operate electrical devices, it is being discharged (**FIGURE 53-12**). As it discharges, the sulfuric acid in the electrolyte joins with lead dioxide to form lead sulfate, and oxygen from the plate joins the hydrogen from the electrolyte to form water. Lead sulfate also forms at the negative plate, as sponge lead reacts with sulfate from the electrolyte. In a discharged lead-acid cell, the active material of both plates becomes lead sulfate, which means they are no longer dissimilar metals. At the same time, the electrolyte becomes mostly water as the acid leaves the electrolyte and is absorbed into the plates. The result is a very weak sulfuric acid solution in the electrolyte. This, along with the plate material becoming similar, creates a weak electrical output of the battery, and ultimately once the battery is discharged, it is considered "dead."

When being charged, electrical pressure (voltage) higher than that of the battery's total cell voltage pushes electricity back into the battery, reversing the chemical process (**FIGURE 53-13**). The charging device acts like an electron pump, forcing electrons to move from the positive plates back to the negative plates in the battery. At the negative plates, sulfate is forced out of the plate and back into the electrolyte. This changes the negative plates back into sponge lead. It also creates a stronger solution of sulfuric acid in the electrolyte. At the same time, lead peroxide is formed at the positive plates as the sulfuric acid is forced back into the electrolyte. This process restores the cell's electrical potential or voltage.

The charging process increases the amount of acid in the electrolyte, making the electrolyte stronger. When further charging no longer makes the electrolyte stronger, charging is complete. Connecting a lead-acid battery to a load causes chemical changes as the battery discharges.

FIGURE 53-11 Battery ratings.

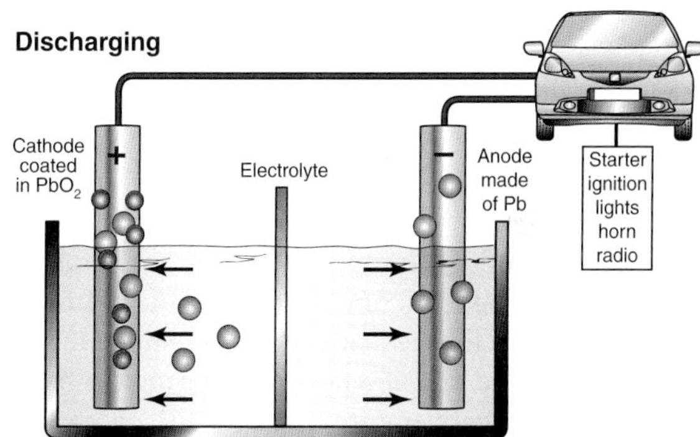

FIGURE 53-12 Discharging cycle.

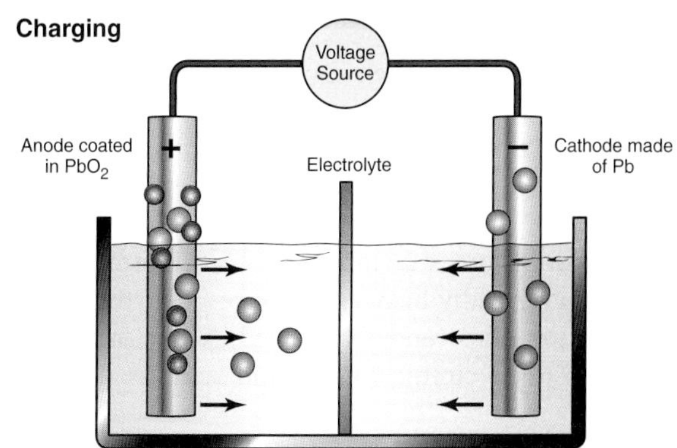

FIGURE 53-13 Charging cycle.

Applied Math

AM-10: Whole Numbers: The technician can multiply whole numbers to determine differences for comparison with the manufacturer's specifications.

An automotive battery is composed of six individual cell compartments within a case. The individual cells are joined by cell connectors in series with one another. Each cell has an open circuit voltage of approximately 2.1 volts at 80°F (26.7°C). When fully charged, a 12-volt automotive battery has an actual open circuit voltage of 12.6 volts at 80°F (26.7°C). For easy calculations with whole numbers, we have 2 volts per cell × 6 cells = 12 volts, which is why we refer to it as a 12-volt battery.

A technician is in the process of checking valve springs for proper tension during the overhaul of a V-8 engine. The valve springs do not meet manufacturer's specifications. and all new valve springs are needed. There are four valve springs per cylinder × eight cylinders. We have whole numbers to multiply, which can be written as 4 × 8 = 32. A quantity of 32 new valve springs is needed to complete the cylinder head work.

Battery Temperature Monitoring

Battery temperature plays an important role in the performance of a battery, and every battery has an ideal operating temperature range. In cold weather, battery performance drops dramatically due to a slowing of the chemical reaction process. At hotter temperatures, battery performance increases, but if the temperature increases too much, battery life may be compromised. Some vehicles monitor the battery temperature by placing a temperature probe in contact with the battery case (**FIGURE 53-14**). The powertrain control module (PCM) then controls the rate at which the battery is charged. Generally speaking, the colder the battery temperature, the higher the rate of charging; and the hotter the battery temperature, the slower the rate of charging. Smart chargers and battery management systems use the feedback from the temperature probe to manage the charge rate of a battery, ensuring the battery is at capacity and the temperature is within the optimal range, thereby promoting longer battery life.

Battery Life

S53001

Batteries do not last forever. Over time and use, the plates start to lose effectiveness as lead paste falls off the plates and collects in the bottom of the battery case. This type of damage is especially likely to develop in a battery subjected to high vibration and shock, as when subjected to off-road travel. As plate material becomes deposited in the bottom of the battery case, it eventually bridges the plates, shorting them out (**FIGURE 53-15**). This renders that cell useless, which lowers the battery's open circuit voltage and reduces the battery's capacity.

A battery may also lose its capacity as it becomes sulfated, meaning the surfaces of the plates harden due to the acid sitting

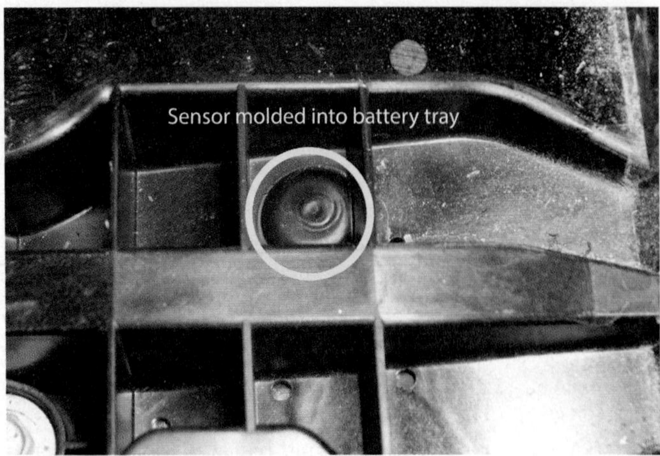

FIGURE 53-14 Some vehicle manufacturers monitor battery temperature to control the rate of charge.

Applied Science

AS-54: Batteries: The technician can demonstrate an understanding of the electrochemical reactions that occur in wet cell and dry cell batteries.

A vehicle that has a no crank condition is towed into a dealership. The technician assigned to the job has completed his diagnosis of the problem and determined that the vehicle has a faulty battery. An apprentice technician working under his supervision has also discovered that the hold-down clamp on the battery is missing.

At break time, the apprentice asks the technician why he thinks the battery failed. The technician explains his thoughts by going back to the basics of battery construction and the electrochemical reactions that occur during various events. The battery case can be thought of as a container. Inside this container are six separate compartments, or cells, which are the energy-producing units. Each cell has a number of plates soaked in an electrolyte solution of about 64% water and 36% sulfuric acid. The plates are separated into the anode, or negative, side and the cathode, or positive, side.

As stated in our text, connecting a lead-acid battery to a load causes chemical changes as the battery discharges. At the positive plate, sulfate from the electrolyte joins with lead to form lead sulfate, and oxygen from the plate joins the hydrogen from the electrolyte to form water. Lead sulfate also forms at the negative plate, as sponge lead forms with sulfate from the electrolyte. Overall, the percentage of acid in the electrolyte falls and the percentage of water rises, which reduces the strength of the electrolyte. As the cell discharges, the plates develop the same composition, which reduces the potential of the cell. Recharging the battery again restores the difference between its sets of plates.

The technician explains his theory of the failure of the battery, attributing it to the missing hold-down clamp. With the battery not being securely fastened, plate material breaks off and drops down into the bottom of the battery case. This eventually causes the plates to short out. He goes on to explain that all batteries will eventually wear out, but excessive vibration shortened the life of this battery.

Applied Science

AS-55: Acids/Bases: The technician can identify the effects of the pH of a solution on various systems.

A technician has replaced a heater core in a vehicle, a task that required many hours of labor due to its location. The technician filled the cooling system with the type of antifreeze recommended by the manufacturer, with a 50/50 mix of distilled water. After running the engine to check for leaks, the technician used a coolant test strip to verify the proper pH level of the coolant. The test strip indicated a pH level of 10, which is in the correct range of 9.8 to 10.2 pH. Any reading below 9.0 indicates that the coolant may cause damage to the water pump, radiator, heater core, or other components. This condition is known as electrolysis.

An acid is a substance that produces hydrogen ions. A base is a substance that produces hydroxide ions. The pH scale is used to tell how acidic or base a substance is. A handheld refractometer is a precision optical device used to measure the concentration, or mix ratio, of liquids. This instrument is more accurate than test strips for pH testing.

In addition to pH testing of the cooling system, brake fluid can be also checked with test strips or a refractometer. The safe range for brake fluid ranges from 7.0 to 11.5 on the pH scale. Battery electrolyte can be tested with a hydrometer or a refractometer. When a battery is discharging, the specific gravity of the electrolyte is dropping. When a battery is being charged, the situation is reversed. Electric current is forced into the battery by the vehicle's alternator or a battery charger. The specific gravity of the battery goes up as a result.

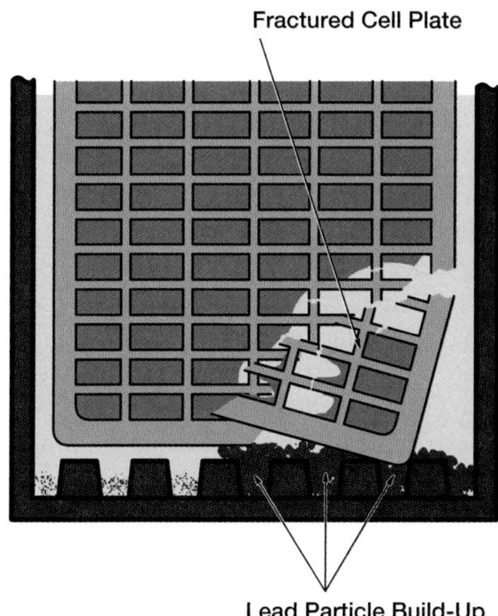

FIGURE 53-15 Over time, lead from the plates can flake off and build up high enough to bridge the gap between the plates, shorting out the cell.

in the plates for extended times. This makes it more difficult for the acid to be absorbed deeper into the plate, creating less current flow. As a result, the battery can no longer readily accept a charge and cannot adequately discharge under load. Leaving a battery on a slow charger overnight and then finding that it fails a load test in the morning is an indication of a sulfated battery. Another indicator is when a conductance test shows that the battery's capacity in CCAs is significantly lower than specified. Sulfated batteries cannot create the necessary current flow and should be replaced and recycled. Leaving a battery less than 80% charged greatly speeds up the process of sulfation. The further discharged, the quicker the battery sulfates. Keeping a battery fully charged reduces the problem of sulfation.

Battery Life Factors

Batteries operate under a range of conditions that affect their performance and life. Batteries should be kept clean, dry, and fully charged; they should have minimal vibration and the correct level of electrolyte; they should be kept at a moderate temperature (e.g., 77°F [25°C]); and they should be well secured. A battery's lifetime is shortened by the following conditions:

- Being fully discharged or having deep discharge cycles
- Remaining over- or undercharged
- Experiencing high discharge rates for extended periods
- Experiencing excessive vibration
- Being exposed to extremes of temperature
- Having dirt or moisture on the case
- Developing corrosion

As you can see, poor maintenance and operating conditions reduce the battery's operating life.

▶ TECHNICIAN TIP

Back in the day when many people worked on their own cars, a very handy do-it-yourselfer bought for his vehicle a top-shelf battery from the local parts store. After a couple of weeks, the customer brought the battery back, saying it had a dead cell. The store employee checked it out, and sure enough, it had a dead cell, so they gave him a new battery. A few weeks later, the customer returned with the same problem. The store employee apologized profusely and gave him another battery. Once again, the customer was back within the month with another bad battery. By this time, the store owner figured something had to be going on, so he went to the customer's home to check it out. When he got there, he saw a beautiful '57 Chevy with a tilt front end. One quick look told him everything he needed to know. When the tilt front end was installed, the battery tray had to be removed. The customer found a nice new spot to mount the battery tray, bolting it right to the top of the upper control arm for the suspension. So every bump in the road bounced the battery and literally shook it to death. Needless to say, the customer ended up paying for a few ruined batteries.

Battery Maintenance

Batteries require regular maintenance consisting of inspection, cleaning, testing, and charging when discharged. Batteries should be checked during scheduled oil changes. They should also be inspected, and any needed maintenance should be performed in the fall, prior to cold weather setting in, which makes the battery work harder.

When inspecting the battery, check that the battery electrolyte level is within the markings on the case, or look in each cell to ensure the plates are well covered (**FIGURE 53-16**). Make sure the exterior case is dry and free of dirt. Dirt on top of the battery can actually cause premature self-discharge of the battery as current "leaks" across the path of dirt or grime. This mixture becomes conductive and drains the battery over time.

To verify whether the surface of the battery has to be cleaned, use a digital multimeter (DMM) set to volts to measure the voltage on the surface of the top of the battery. You can do this by placing the black lead on the negative battery post and rubbing the red lead around the top of the battery, measuring the voltage present there (**FIGURE 53-17**). Any voltage over about 0.2 volt means the battery should be cleaned with a mixture of baking soda and water. Just make sure not to get any of that mixture down inside of the battery, as it will tend to neutralize the electrolyte, damaging the battery.

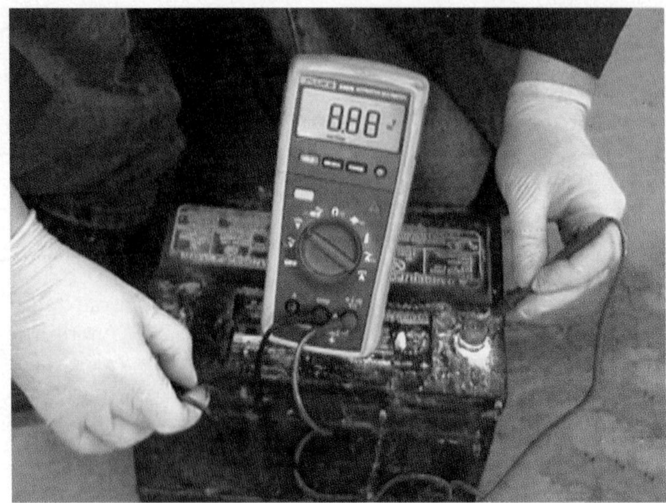

FIGURE 53-17 Use a DMM to measure the voltage on the top of the battery. Any voltage higher than about 0.2 volt means the battery should be cleaned.

Keeping the battery and terminals clean is one of the best maintenance tasks for the money. The lead in the battery posts and terminals oxidizes over time. Unfortunately, lead oxide is an insulator. Therefore, if the lead surfaces of the post and terminal oxidize, the insulator effect can cause the vehicle not to crank over, stranding the driver. Lead oxide can be identified by its dark gray or black color (**FIGURE 53-18**). The only effective way of removing it is with a special battery terminal scraper or wire brush, which removes the oxidized lead and restores the lead to its shiny silver color. Once the battery terminal is reinstalled and tightened, the terminals should be coated with a battery oxidation inhibitor to help slow down the oxidation process.

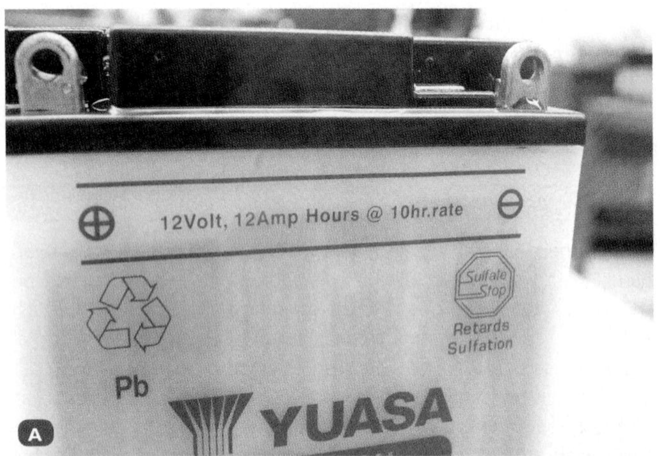

FIGURE 53-16 The electrolyte level should be **A.** between the lines on the battery case or **B.** near or touching the bottom of the fill hole.

Batteries sometimes have to be recharged when lights are left on (a parasitic drain) or when the vehicle has not been driven for several weeks or months. Performing a battery state-of-charge test with a hydrometer or refractometer is a good indicator of whether the battery has to be charged. Another battery maintenance task is the conductance test, which indicates the capacity of the battery. This can be used to determine how much life is left in the battery. These tests are explained in the following topics.

SAFETY TIP

To ensure safety when working on batteries, take the following precautions:

- Make sure the hood is secured with a hood prop rod.
- Wear the appropriate personal protection equipment (PPE) before starting the job. Remember that batteries contain acid; if this acid is splashed in your eyes, it can blind you or seriously damage your eyes.
- Follow the manufacturer's personal safety instructions to prevent personal injury or damage to the vehicle you are working on.

Battery Service Precautions

- When servicing batteries, always ensure that you have the right PPE (e.g., safety eyewear, gloves, and shop clothing).

FIGURE 53-18 A. Heavily oxidized battery post. **B.** Clean battery post.

- Never wear any jewelry, such as neck chains, watches, or rings, when working on or near batteries as jewelry may provide an accidental short-circuit path for high currents.
- If the battery is being, or has been recently, charged, ensure that the space around the battery is well ventilated.
- Avoid making any sparks, which may ignite the **gassing** hydrogen-air mixture. Never create a low-resistance connection or short across the battery terminals.
- Always remove the negative or ground terminal first when disconnecting battery cables. This is to reduce the possibility of a wrench creating an accidental short to ground because you are removing the negative terminal first. If the wrench connects from the negative battery terminal to ground, nothing will happen. And once the negative terminal is off the battery, then if the wrench touches ground while on the positive post, there is not a complete circuit back to the negative battery terminal.

SAFETY TIP

The author has been in the vicinity of two batteries when they blew up. One was from across the shop and sounded like a bomb going off. A student had been welding a van door when a spark from the welding ignited the hydrogen gases around the battery. Fortunately, the battery was on the opposite side of the door from where the student was working. The battery case was blown apart and acid was thrown 5 to 10 feet in all directions. Luckily, no students were within that range. Always keep sparks away from batteries.

Battery Recycling

Batteries contain many environmentally damaging chemicals and metals. If they find their way into a landfill, the acid and metals can contaminate the soil and waterways. Correct disposal of batteries by recycling them is good for the environment, and the precious metals can be reclaimed for reuse. Many municipalities require battery recycling and levy a "core charge" on every new automotive battery sold. The core charge is refunded if an old battery is brought in and exchanged for the new one. This process helps prevent batteries from being discarded in the trash or left lying around. Check local laws and regulations to ensure that batteries are disposed of correctly.

Inspecting, Cleaning, Filling, and Replacing the Battery and Cables

N53001

As noted earlier, batteries last longer if they are properly maintained. In fact, one of the most common causes of vehicle no-starts is dirty or corroded battery cables. Inspecting, cleaning, and filling (if batteries are not maintenance free) are common tasks that should be performed every six months to one year on top-post batteries, and every one to two years on side-post batteries.

Automotive batteries can look lighter than they really are. Always lift them with care, and get help if needed. When reconnecting the battery, be sure you do not connect the battery incorrectly (with reverse polarity); doing so will send current in the reverse direction through the electrical system. The reverse current flow will likely damage some or all of the electronic control units (ECUs) throughout the vehicle so that what started out as a simple maintenance task becomes an expensive diagnosis and repair job that the shop has to perform at no charge to the customer.

SAFETY TIP

When disconnecting the battery terminals, always remove the negative terminal completely first, and install the negative terminal only after the positive terminal is fully tightened. Once the negative terminal is removed, if the wrench touches the chassis when it is on the positive terminal, no spark will be created because there is no path back to the negative terminal. You still need to be careful that the wrench does not bridge straight from the positive battery terminal to the negative battery terminal, as that would cause a huge spark.

To inspect, clean, fill, or replace the battery, battery cables, clamps, connectors, and hold-downs, follow the steps in **SKILL DRILL 53-1**.

SKILL DRILL 53-1 Inspecting, Cleaning, Filling, and Replacing the Battery and Cables

1. Measure the voltage on the top of the battery with a DMM. Place the black lead on the negative post, and move the red lead across the top of the battery to find the highest reading.

2. Remove the cable clamp from the negative terminal first. Then remove the positive terminal. Bend the cables back out of the way so that they cannot fall back and touch the battery terminals accidentally.

3. Remove the battery hold-downs or other hardware securing the battery.

4. Keeping it upright, remove the battery from its tray, and place it on a clean work surface. Inspect the battery for damage.

5. Carefully clean the battery case and the battery tray.

6. Clean the battery posts with a battery terminal tool. Clean the cable terminals with the same battery terminal tool. Examine the battery cables for fraying or corrosion.

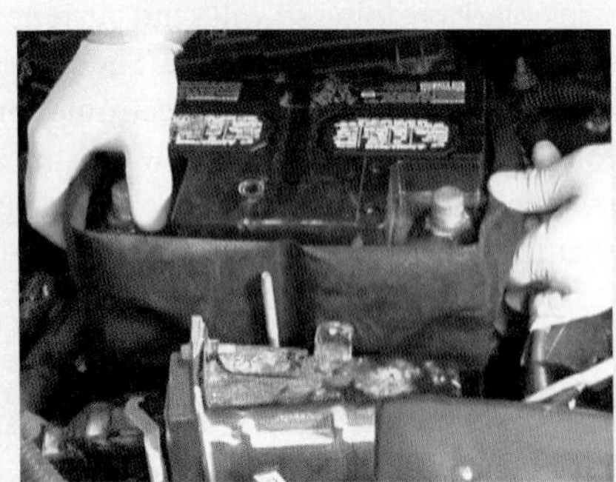

7. Reinstall the cleaned and serviced battery. Replace the hold-downs, and make sure the battery is securely held in position. If installing a new battery, ensure that it meets the original manufacturer's specifications.

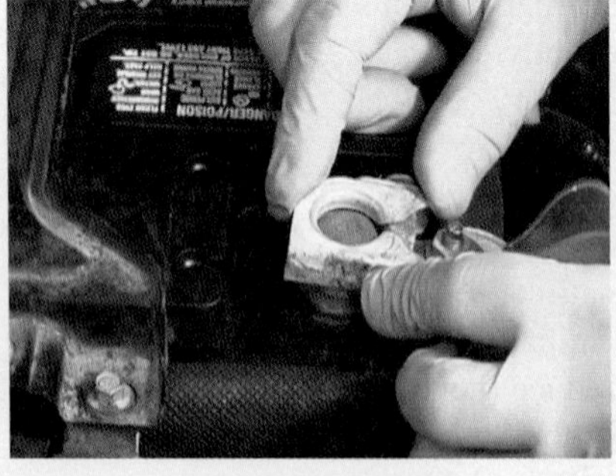

8. Reconnect the positive battery terminal and tighten it in place. Once the positive terminal is finished, reconnect the negative terminal and tighten it.

9. Coat the terminal connections with anticorrosive paste or spray to keep oxygen from the terminal connections. Test that you have a good electrical connection by starting the vehicle.

Applied Science

AS-74: Activity of Metals: The technician can explain the conductivity problems in a circuit when connectors corrode due to electrochemical reactions.

A dealership technician has a do-it-yourself neighbor who enjoys working on vehicles in his driveway. The neighbor has questions from time to time for the professional technician. The do-it-yourselfer was restoring a fairly old vehicle in his spare time. He had just purchased a new battery and a new starter, but the engine would not crank over. The technician demonstrated how to do a voltage drop test using a DMM (digital multimeter). The problem was in a battery cable connection at the starter solenoid mounted on the fender well of the vehicle. When the cable end was removed from the solenoid, corrosion was found. After removing the corrosion from both surfaces and reattaching the cable, the engine cranked over properly. The do-it-yourself neighbor realized that he did not actually need a new battery and starter.

This is an example of a conductivity problem when connectors corrode due to electrochemical reactions. All metals on the vehicle can rust. Battery terminal corrosion is one of the main concerns regarding vehicle maintenance. The terminal corrosion can be removed with a wire brush, but periodic maintenance is still needed. External corrosion on battery terminals is very easily spotted, but internal corrosion between the battery post and cable clamp will not be seen. It is necessary to take the connection apart and clean each surface with a battery terminal cleaning brush. Connections with corrosion problems increases resistance at the battery terminals, which reduces the voltage to the electrical system.

Charging the Battery

N53002

Vehicle batteries may become discharged and require charging, particularly if the vehicle has been sitting idle without starting for more than a couple of weeks or if the charging system is faulty. Also, before testing a battery, it should ideally be fully charged. If charging is needed, you have to determine the charge rate for the battery. But remember that slow charging is less stressful on a battery than fast charging.

- To determine the ideal charging rate: CCAs divided by 70. Example: 500 CCA = 7A
- To determine the max charging rate: CCAs divided by 40. Example: 500 CCA = 12.5A

A slow charger usually charges at a rate of less than 5 amperes. A fast charger charges at a much higher ampere rate, depending on the original battery's state of charge, and should only be used under constant observation. To fully charge a battery can take more than 20 hours, depending on the state of charge.

Some manufacturers recommend removing the negative battery terminal while charging a battery, to reduce risk to the vehicle's electronics. If the battery has to be disconnected during charging, remember to verify whether the vehicle's adaptive memory has to be maintained. If the battery terminals are heavily corroded, you may need to clean them before charging the battery. Also, monitor the battery voltage during charging.

- *Never exceed 15.5 volts* when charging a 12-volt flooded cell battery.
- *Never exceed 14.8 volts* when charging a 12-volt AGM battery.
- *Never exceed 14.3 volts* when charging a 12-volt gel cell battery.

After charging the battery, it is good practice to clean the battery posts, cable terminals, and battery case.

To charge a battery, follow the steps in **SKILL DRILL 53-2**.

SKILL DRILL 53-2 Charging a Battery

1. Ensure that the battery has not been frozen. Install a memory saver into the DLC, and disconnect the negative terminal. Verify that the charger is unplugged and off. Connect the red lead to the positive terminal and the black lead to the negative terminal.

2. Calculate the ideal and maximum charge rates. Plug in the charger, and turn it to the appropriate charge rate. Monitor the battery voltage while charging.

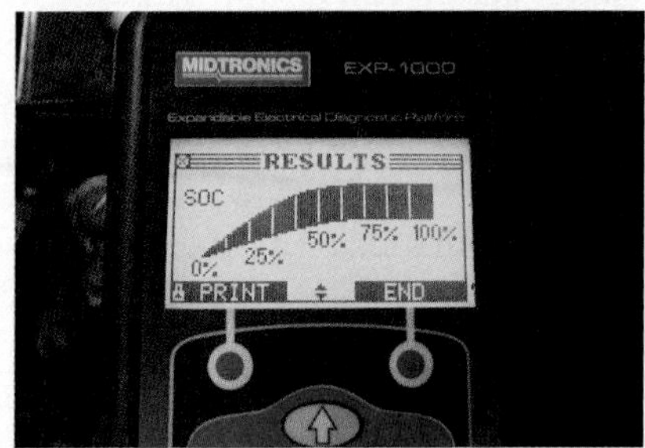

3. Once the battery is charged, turn the charger off. Remove the surface charge by turning on the headlights for three to five minutes. Using a conductance tester, load tester, DMM, refractometer, or hydrometer, test the battery's state of charge and capacity.

Applied Science

AS-73: Electrochemical Reactions: The technician can demonstrate an understanding of the ion transfer process that occurs in an automotive battery.

A vehicle is in the shop for routine maintenance, and a general inspection is being performed. The technician observes that the battery appears to be old and finds a code on the battery, indicating it was manufactured six years ago.

At lunch time, the technician discusses the topic of batteries with another technician. He has the opinion that an automotive battery should be replaced every five years to prevent possible starting problems. Jim does not agree as he believes there are too many variables to make this determination. This leads to a discussion of how a lead-acid battery works.

As stated in our text, in a discharged lead-acid cell, the active material of both plates becomes lead sulfate, and the electrolyte becomes mostly water as the acid is absorbed into the plates. The result is a very weak sulfuric acid solution in the electrolyte. When being charged, the battery is connected to a DC electrical supply with electrical pressure (voltage) higher than that of the battery's total cell voltage. The charging device acts like an electron pump, forcing electrons to move from the positive plates to the negative plates in the battery.

At the negative plates, sulfate is discharged, which changes the chemical composition of the plates into sponge lead and also creates a stronger solution of sulfuric acid in the electrolyte. At the same time, lead peroxide is formed at the positive plates, which helps to restore the cell's electrical potential or voltage.

The charging process increases the amount of acid in the electrolyte, making the electrolyte stronger. When further charging no longer makes the electrolyte stronger, charging is complete.

Jump-Starting the Vehicle

N53003

Jump-starting a vehicle is the process of using one vehicle with a charged battery to provide electrical energy to start another vehicle that has a discharged or dead battery. Because starting a vehicle requires a high amount of electrical energy, jump-starting a vehicle can put stresses on both vehicles. When the discharged vehicle is being cranked, the battery voltage tends to fall very low because the battery is already discharged. This causes the alternator on the running vehicle to put out its maximum current, putting it under heavy load. But as soon as the jumped vehicle stops cranking, the voltage shoots up quickly, potentially high enough to damage electronic components in either vehicle. The same voltage spike can happen when the jumper cables are being disconnected.

Another risk is overheating and damaging the dead vehicle's alternator. With all of the electrical draws on the vehicle, today's alternators work harder than ever. Adding the job of fully recharging a discharged battery on top of the regular electrical loads can push an alternator into the danger zone. Because of the risk of damaging the electronics and alternator, many manufacturers (and tow companies) are now refusing to jump-start newer vehicles or authorize that practice. They now recommend either externally slow charging the battery or replacing the battery with a charged battery.

If you do decide to jump-start a vehicle, always read the owner's manual for both vehicles, and follow their jump-starting guidelines. And never attempt to jump-start a frozen battery. It is usually best to let the running vehicle charge the battery on the other vehicle for 5 to 10 minutes before trying to start the vehicle. Once the dead vehicle's engine has started, let both vehicles come to an idle for a moment. Do not turn off the vehicle with the discharged battery; it needs an extended amount of time to recharge. Before you disconnect the service battery from the discharged battery, it is good practice to place a load on the charged battery by turning on an accessory such as the headlights. This helps absorb any sudden rise in voltage that may occur as the load on the alternator is suddenly decreased. Another method of reducing the risk of damage to sensitive electronic devices is by using jumper leads that have a built-in or auxiliary surge protector.

An option that is less risky when jump-starting a vehicle is the use of a jump box. A jump box does not involve a charging system, which lessens the risk of voltage spikes. This device contains a relatively high-capacity battery and has short cables and spring-loaded clamps that can be connected directly to the dead battery. It provides a substantial additional boost of electrical energy to start the vehicle.

SAFETY TIP

When connecting jumper cables, a spark almost always occurs on the last connection you make. That is why it is critical that you make the last connection on the engine block away from the battery and any other flammables. A spark also occurs when you disconnect the first jumper cable connection, so that should also be the connection at the engine block.

To jump-start a vehicle, follow the steps in **SKILL DRILL 53-3**.

▶ TECHNICIAN TIP

Alternators are not generally designed to charge a dead battery while running several accessories. If a battery must be jump-started, it is always best to recharge the battery using a battery charger. That way, the alternator will not have to work so hard. Some technicians say it takes only 15 minutes to burn up an alternator when charging a dead battery. And if you must drive the vehicle after being jump-started, turn off as many accessories as possible, which will lessen the load on the alternator.

SKILL DRILL 53-3 Jump-Starting a Vehicle

1. Position the charged battery close enough to the discharged battery that it is within comfortable range of your jumper cables. If the charged battery is in another vehicle, make sure the two vehicles are not touching.

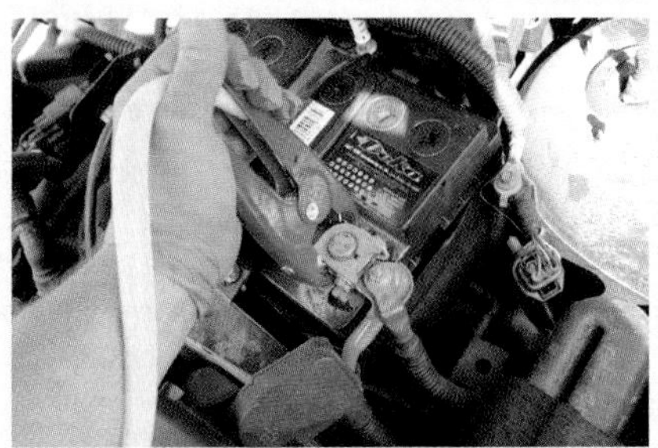

2. First, connect the red jumper lead to the positive terminal of the discharged battery in the vehicle you are trying to start.

3. Next, connect the other end of this lead to the positive terminal of the charged battery or the remote terminal.

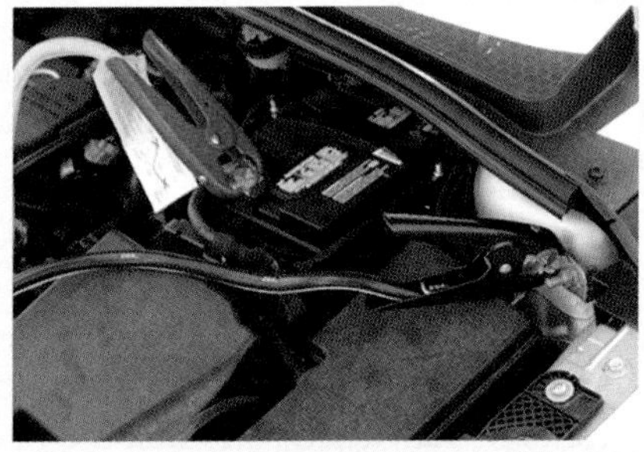

4. Then connect the black jumper lead to the negative terminal of the charged battery or the battery remote terminal.

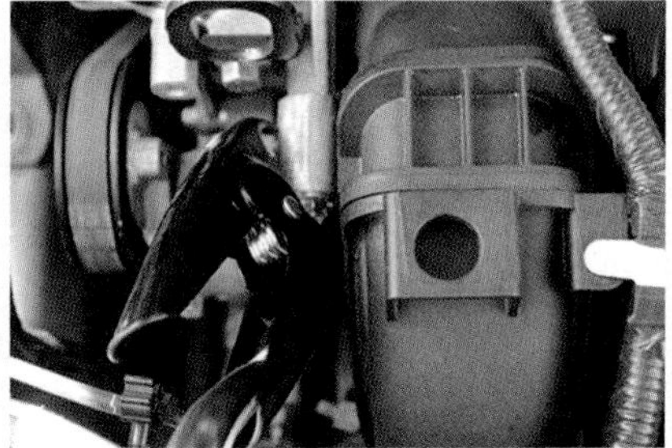

5. Connect the other end of the negative lead to a good ground on the engine block of the vehicle with the discharged battery, and as far away as possible from the battery.

6. Do not connect the lead to the negative terminal of the discharged battery itself; doing so may cause a dangerous spark. Also, do not connect the negative lead to the body or chassis as the ground wire from the body back to the negative battery terminal is usually too small to carry current needed for jumpstarting the vehicle.

7. Try to start the vehicle with the discharged battery. If the booster battery does not have enough charge or the jumper cables are too small in diameter to do this, start the engine in the booster vehicle, and allow it to partially charge the discharged battery for several minutes. Turn on the headlights on the booster vehicle to reduce the possibility of a voltage spike damaging electronic equipment, and try starting the first vehicle again.

8. Disconnect the leads in the reverse order of connecting them.

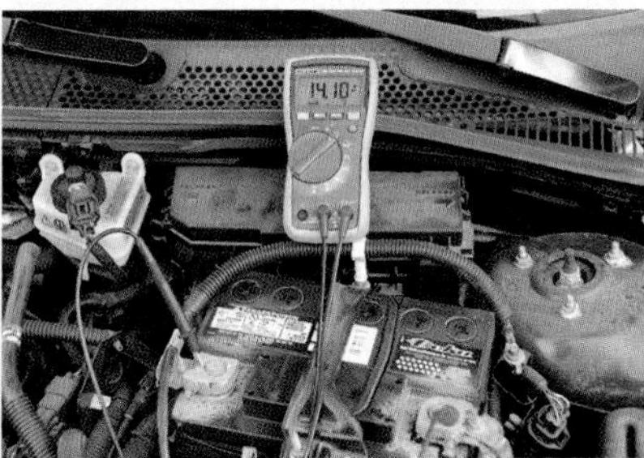

9. If the charging system is working correctly and the battery is in good condition, the battery will be recharged while the engine is running, although it could end up overheating and damaging the alternator.

Testing Battery State of Charge and Specific Gravity

N53004, S53002

Although the conductance test is the preferred test of determining battery condition, other tests may still be used to give us some information. One of those tests is the state-of-charge test. State-of-charge testing tells us how charged or discharged a battery is, not how much capacity it has. The degree of the battery's charge is handy to know when testing starting and charging system issues, as well as just about any other electrical issue.

There are two tests for determining the battery's state of charge: the specific gravity test and the open circuit voltage test. The specific gravity test measures the electrolyte's specific gravity, which indicates the acid content and therefore the state of charge. Although specific gravity testing is the most accurate, most batteries are the low-maintenance or no-maintenance type and may not provide access to the electrolyte in the cells for specific gravity testing. Older batteries with removable caps may be tested for specific gravity. Recall that the acid concentration drops as the battery becomes discharged; this lowers the electrolyte's specific gravity. The higher the specific gravity, the higher the percentage of acid in the electrolyte, which corresponds to a high battery state of charge (**TABLE 53-1**). *Note:* If there is a difference in specific gravity reading of 0.050 or more between the cells, the battery must be replaced.

▶ TECHNICIAN TIP

When performing a state-of-charge test, keep these tips in mind:

- When filling a battery that is not fully charged, never fill it to the top of the full line, as charging the battery raises the electrolyte level.
- Small amounts of electrolyte in the hydrometer may leak out and damage the vehicle's paint or your clothing.
- Do not inadvertently remove electrolyte from one cell or add it to another cell when testing; doing so causes incorrect readings.

The hydrometer's reading, unlike that of the refractometer, must be corrected for electrolyte temperature, as temperature affects the specific gravity. When drawing electrolyte into the hydrometer, draw in enough so that the float is floating, not sitting on the bottom or hitting the top. If the electrolyte level is too low to do this, then you will have to add distilled water, and the battery will have to be fully charged to mix the water and acid.

A very quick and reasonably accurate indicator of battery state of charge is the open circuit voltage test. This test uses a DMM to accurately measure the voltage of a battery that has been sitting with no charge for at least four hours; or, if recently charged, remove the surface charge by turning on the headlights for three to five minutes. Once the surface charge is removed, the headlights can be turned off and the open circuit voltage measured. The voltage reading will then roughly indicate the battery's state of charge. For example, 12.6 volts or above indicates a fully charged battery; 12.4 volts indicates a charge of approximately 75%; and 12.0 indicates a battery with very little charge (**TABLE 53-2**).

To perform a battery state-of-charge test, follow the steps in **SKILL DRILL 53-4**.

Applied Science

AS-56: Density/Specific Gravity: The technician can explain the role of specific gravity in determining the condition of the system.

A technician is assigned to perform a battery state-of-charge (SOC) test on an automotive battery with removable filler caps. This procedure is a specific gravity test using a battery hydrometer.

The technician would normally use an electronic battery tester for a conductance test, but because the battery test was dropped, it is out for repairs. Although an electronic battery tester is often used on today's vehicles, sometimes other methods are used, such as the battery hydrometer.

Observing the safety precautions, the technician uses a hydrometer to draw some electrolyte from each cell, and records the information on a sheet of paper. As stated in our text, a very low overall reading of 1.150 or below indicates a low state of charge. A high overall reading of about 1.280 indicates a high state of charge. The reading from each cell should be the same. If one or two cells are 0.050 different from the rest, it indicates there is something wrong with the battery. See the temperature corrections chart to adjust for temperature variations that are above or below 80°F (26.7°C) for the battery electrolyte temperature.

TABLE 53-1: Specific Gravity and the Corresponding State of Charge

State of Charge	Specific Gravity @ 80°F (26.7°C)
100%	1.265 or higher
75%	1.225
50%	1.190
25%	1.155
0%	1.120

TABLE 53-2: State of Charge as Indicated by Voltage Reading

Voltage	Percentage of Charge
12.6 or greater	100
12.4–12.6	75–100
12.2–12.4	50–75
12.0–12.2	25–50
11.7–12.0	> 0–25
0.0–11.7	0 (no charge)

SKILL DRILL 53-4 Testing the Battery's State of Charge

1. Test the specific gravity of each of the cells by using a hydrometer designed for battery testing. Draw some of the electrolyte into the tester, and read the scale. Compensate for the temperature of the electrolyte if necessary.

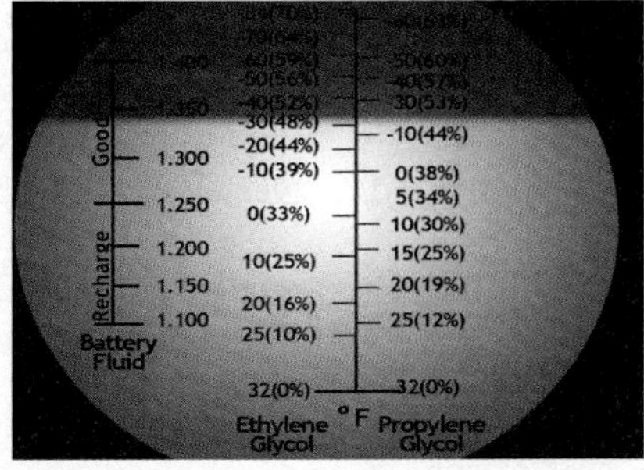

2. To test the specific gravity with a refractometer, place a drop or two of electrolyte on the specimen window, and lower the cover plate. Look into the eyepiece with the refractometer under a bright light. Read the scale for battery acid. The point where the dark area meets the light area is the reading.

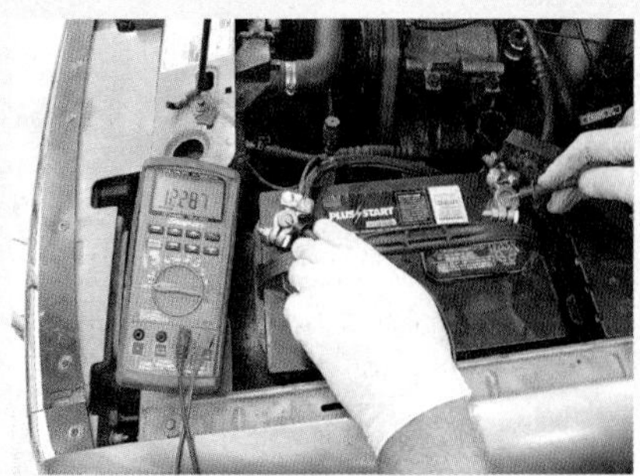

3. To conduct an open circuit voltage test, select the "volts DC" position on your DMM, and attach the probes to the battery terminals (red to positive, black to negative). Compare the reading to those given in **TABLE 53-2**.

Applied Math

AM-5: Decimals: The technician can subtract decimal numbers to determine conformance with the manufacturer's specifications.

A technician is performing a battery state-of-charge test, using a hydrometer. This procedure can be used on automotive batteries that have removable individual filler caps. The tube of the hydrometer is inserted into one of the battery cells. The technician gently draws electrolyte into the hydrometer. The float indicator rises and floats freely. The specific gravity can be obtained by reading the indicator at eye level. When doing this procedure, it is important to consider the hazards of battery electrolyte, which contains sulfuric acid. Eye protection and rubber gloves are needed for personal protection. After reading the hydrometer, the electrolyte is returned to the cell from which it was withdrawn. The same procedure is repeated for the other five cells.

Manufacturer's specifications state that a fully charged battery should show a specific gravity of approximately 1.260 with electrolyte temperature at 80°F (26.7°C). The results of the readings of each cell are be compared to a chart that shows specific gravity as compared to state of charge. If the electrolyte temperature is above or below 80°F (26.7°C), corrections must be made by one of several methods. In this case, the technician consulted a temperature correction chart.

The technician observes that at 40°F (4.4°C), the correction factor is to subtract 0.016 points. The reading for cell 1 is 1.240. To compensate for the huge temperature difference, corrections are needed. In this example, we have 1.240 minus (–) 0.016 = 1.224 for the adjusted specific gravity of one of the battery cells. The next step would be to look at the readings for the other five cells of the battery.

Testing Battery Capacity

N53005

When in doubt of how well a battery can meet the demands placed upon it, its capacity should be tested. But before testing begins, you need to make sure that the battery meets the capacity rating for the vehicle. Look up the service information, and compare it to the rating tag on the battery. Using a battery with too little capacity will overstress the battery itself, plus the starting system. Using a battery with substantially too much capacity puts more stress on the battery tray, as it weighs more, plus the hold-downs are not likely to fit it. It also may put a larger load on the charging system. So always verify the battery is correct for the application.

Conductance Testing for Capacity

Once the application has been verified, you can go about testing the battery. The use of electronic battery testers has by and large replaced the need for hydrometer and other invasive types of battery testing. In fact, many manufacturers are now requiring the use of a conductance test instead of a high-amperage load test in order for warranty coverage to be considered (**FIGURE 53-19**). Many conductance testers have integrated printers so that a printout of the test can accompany the warranty paperwork.

The conductance tester sends low-frequency signals into the battery to determine the battery's ability to conduct current. The greater the ability to conduct current, the higher the CCA capacity of the battery. Because batteries deteriorate over time, their CCA capacity also reduces. The conductance tester is able to measure a battery's CCA capacity by measuring

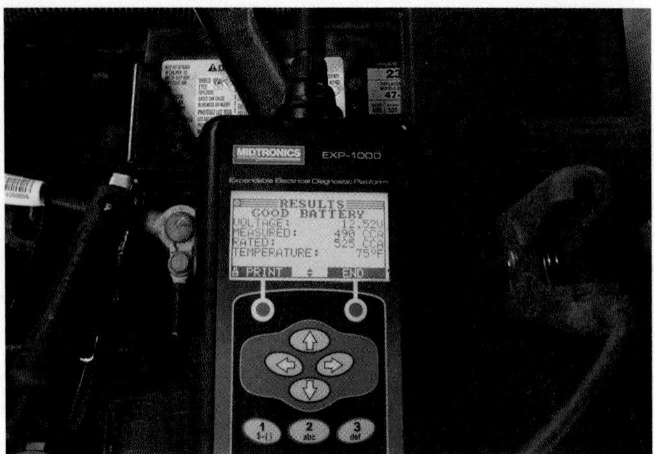

FIGURE 53-19 Conductance tester being used to measure a battery's existing capacity.

its conductance, which provides a near linear comparison between the two. Dirty battery terminals can affect the conductivity, so the terminals may have to be cleaned before taking the reading.

Because conductance testing takes only a minute or two to complete, it is a good way to show customers the status of their battery and how that condition degrades over time. Once the conductance test shows that the battery no longer meets specifications, then customers will have confidence that the battery is at the end of its useful life and must be replaced to avoid it failing and leaving them stranded.

Battery Load Testing

The load test has been used for years to test a battery's capacity and internal condition, but some manufacturers have been saying that their batteries should *not* be load tested and instead should be conductance tested. They claim that load testing can damage the battery. Therefore, always check the service information before performing a load test.

As the name suggests, the load test subjects the battery to a high rate of discharge, and the voltage is then measured at the end of a set time to see how well the battery creates that current flow. In other words, if the battery can maintain a high rate of discharge for a specified time and the voltage is still relatively high, then you know the battery has good capacity and is in relatively good condition. Conversely, if the voltage falls off fairly quickly, the battery's capacity is low. It is kind of like two people running a mile sprint. The one who does it in 5 minutes is likely in pretty good shape; the one who takes 20 minutes is likely in poor shape. It is similar with batteries: The faster the rate at which a battery can create current flow, the higher its voltage, indicating a higher capacity. Another way to think of a load test is that we are testing the battery's ability to produce the high starting current while maintaining enough voltage to operate the ignition and electronic control systems, and to do it a number of times, if needed.

You will remember that CCAs reflect the load in amps that a battery can deliver for 30 seconds while maintaining a voltage of 7.2 volts or higher at 0°F (−18°C). Because vehicle and battery manufacturers specify the CCA rating for every vehicle and battery, that rating is used to calculate the load placed on the battery when load testing. The battery can be either in or out of the vehicle but must be at, or near, a full state of charge for the test to be accurate. The electrolyte temperature should be approximately 70°F (21°C) for the most accurate results, because a cold battery cannot produce current flow as efficiently and will show a false fail result. When load testing a battery, use the following testing parameters:

Test load = Half the CCA of the battery (verify it is sized correctly for the vehicle)
Load test time = 15 seconds
Results: Pass = 9.6 volts or higher; Fail = less than 9.6 volts

If the battery fails the load test, one further test is required before condemning the battery. The battery needs to be slow charged until it is fully charged (can take up to 20 hours), then the load test repeated. If it still fails, the battery is sulfated and has to be replaced. If it passes, test the vehicle's charging system to see if it is charging the battery properly.

▶ TECHNICIAN TIP

Technicians used to perform a three-minute charge test on batteries if they failed the load test. But battery and vehicle manufacturers are discouraging use of this test because of its potential of damaging the battery or vehicle. Thus, we do not discuss this test method further.

To load test a battery, follow the steps in **SKILL DRILL 53-5**.

Applied Math

AM-4: Whole Numbers: The technician can divide whole numbers to determine differences for comparison with the manufacturer's specifications.
AM-7: Whole Numbers: The technician can divide whole numbers to determine differences for comparison with the manufacturer's specifications.

Battery load testing is a procedure that uses a simulated starter current draw on a battery. The results of this test can be compared to manufacturer's specifications for the condition of the battery. There are a number of steps to complete regarding a battery load test using a carbon pile unit. One of the steps is to use one-half CCA rating as the load to be applied to the battery. CCA stands for "cold cranking amps," which is one of the battery's primary ratings. In this example, we have a 770 CCA battery, and the load to be applied is one-half of that amount. To determine this, we can divide by two to give us the 385 CCA load to apply. Another way to describe this would be: 770 CCA/2 = 385 CCA (test load).

In this example, we are dividing whole numbers. Division is the opposite of multiplication.

SKILL DRILL 53-5 Load Testing a Battery

1. With the tester controls off and the load control turned to the Off position, connect the tester leads to the battery. Place the inductive amps clamp around either the black or the red tester cables in the correct orientation.

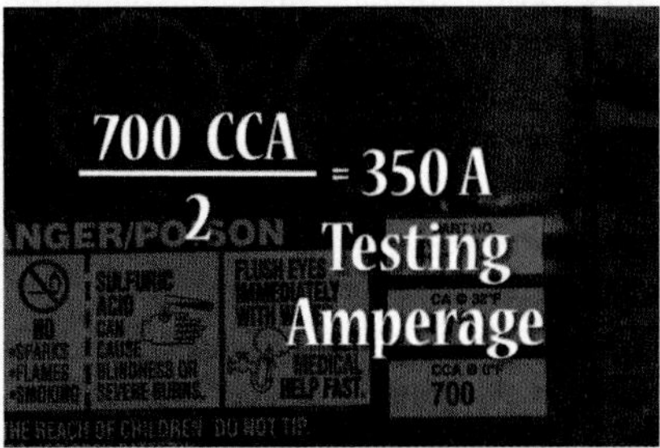

2. Verify that the temperature of the battery is within the testing parameters. Use an infrared temperature gun to determine the temperature by measuring the temperature of the side of the battery. If you are using an automatic load tester, enter the battery's CCA and select "Test" or "Start." If you are using a manual load tester, calculate the test load, which is usually half of the CCA.

3. Maintain this load for 15 seconds while watching the voltmeter. Read the voltmeter, and immediately turn the control knob off. At room temperature, the voltage should be 9.6 volts or higher at the end of the 15-second draw. If the battery is colder than room temperature, look up the compensated minimum voltage. Determine any necessary action.

Identifying Modules that Lose Their Initialization During Battery Removal

N53006, N53007

Many electronic modules in vehicles, such as radio and driver-specific presets, require a small amount of power to maintain their keep alive memory (KAM). When the battery is disconnected from the vehicle, the memory of specific electronic systems is usually lost. For some systems, this can be annoying. The PCM may lose its adaptive learning data, which means the vehicle will have to relearn this information during a period of driving that could take several days. For other systems, such as the security system, loss of memory may prevent the vehicle from being restarted or the radio from being used. At the very least, it may require the dealer or manufacturer to be contacted for vehicle-specific codes to reinitiate the vehicle systems. Check the manufacturers' and owners' information to determine what systems will be affected by the power loss. You need to identify any system that requires security or initialization codes to be reentered, and ensure the procedure and equipment are available to reinitialize systems or modules.

In some cases, it may be possible to use a 9-volt memory minder or memory saver to maintain the vehicle's memory while the vehicle battery is disconnected. This should supply enough power to maintain the memory for small jobs such as changing the battery—provided the vehicle doors or trunk are not opened, causing the interior lights to illuminate and essentially drain the 9-volt battery. Many technicians advocate using an external 12-volt DC power supply connected to the data link control with a suitable cable (**FIGURE 53-20**). If the cigarette lighter socket is always powered on, this too can be used for supplying power to the vehicle while the battery cables and connectors are being serviced. Just remember that providing power back into the circuit makes the system susceptible to short circuits if you ground to any of the powered wires or terminals, including the battery positive terminal.

To identify electronic modules, security systems, radios, and other accessories that require reinitialization or code

FIGURE 53-20 Memory saver and cable.

entry following battery disconnect, follow the steps in **SKILL DRILL 53-6**.

To maintain or restore electronic memory functions, follow the steps in **SKILL DRILL 53-7**.

Measuring Parasitic Draw

N53008

All modern vehicles have a small amount of current draw when the ignition is turned off, called **parasitic draw**. This parasitic current draw is used by the vehicle's **keep alive memory** (KAM) circuits. KAM systems maintain memory functions and monitor systems. For example, the vehicle's theft-deterrent system is part of this parasitic current draw, which should be a relatively small amount of current because excessive draw will discharge the battery over a short amount of time.

Parasitic current draw does not necessarily immediately drop to its lowest level the instant the ignition is turned off. Typically, when the ignition is switched off, the vehicle's engine stops, and the vehicle systems begin to shut down, which is called "going to sleep" or "entering sleep mode." Vehicle systems

SKILL DRILL 53-6 Identifying Accessories that Require Reinitialization

WORK STEPS REQUIRED AFTER RECONNECTING THE BATTERY

Work steps
Switch the ignition on using the ignition key and switch it off again.
Read the DTC memory with the VAS 5051B.
Clock: Check the clock time and set if necessary.
Electrical window regulators: Open and close all the windows. Window regulator comfort closing: The window must close all the way without having to hold the switch.
Perform function test of all electrical consumers.

1. In the appropriate service information or owner's manual, find the section that lists the information on the electronic modules, security systems, radios, and other accessories.

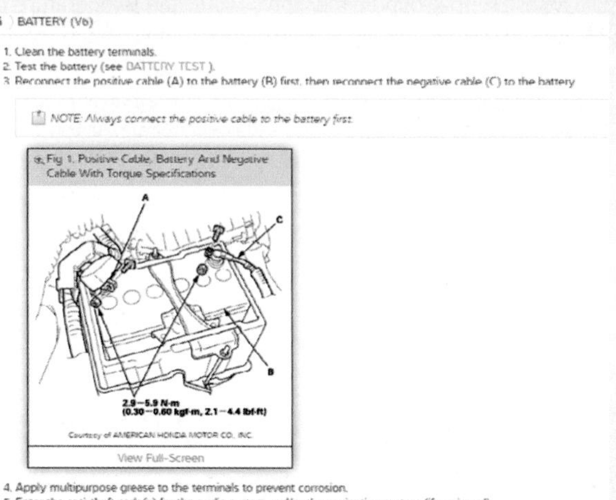
BATTERY (V6)

1. Clean the battery terminals.
2. Test the battery (see BATTERY TEST).
3. Reconnect the positive cable (A) to the battery (B) first, then reconnect the negative cable (C) to the battery

NOTE: Always connect the positive cable to the battery first.

Fig 1. Positive Cable, Battery And Negative Cable With Torque Specifications

4. Apply multipurpose grease to the terminals to prevent corrosion.
5. Enter the anti-theft code(s) for the audio system and/or the navigation system (if equipped).
6. Set the clock (for vehicles without navigation)

2. From the service information, list the systems and modules that may require initialization.

SKILL DRILL 53-7 Maintaining or Restoring Electronic Memory Functions

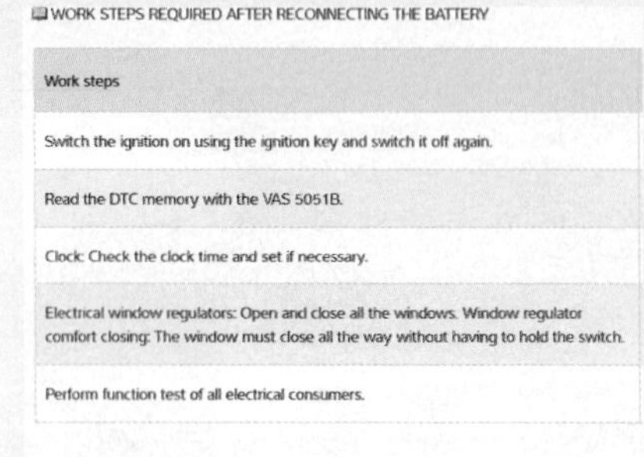
WORK STEPS REQUIRED AFTER RECONNECTING THE BATTERY

Work steps
Switch the ignition on using the ignition key and switch it off again.
Read the DTC memory with the VAS 5051B.
Clock: Check the clock time and set if necessary.
Electrical window regulators: Open and close all the windows. Window regulator comfort closing: The window must close all the way without having to hold the switch.
Perform function test of all electrical consumers.

1. Identify which modules, if any, require reinitialization or code entry when the battery is disconnected, following Skill Drill 53-6.

2. Identify the correct procedure and any needed tools, and verify that initialization codes are available.

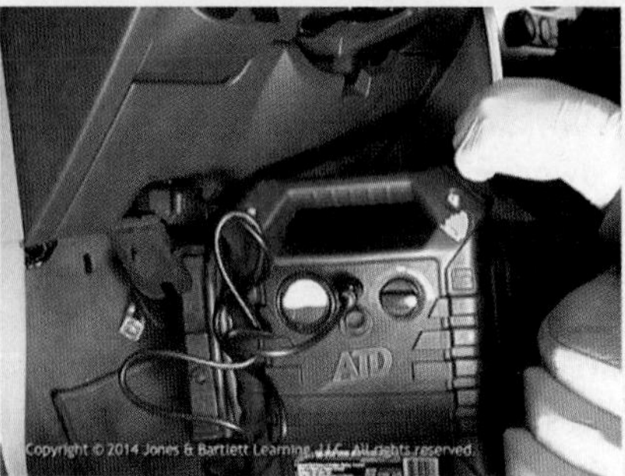

3. If maintaining memory function, install a memory minder prior to the vehicle battery being disconnected. If reinitializing the electronic systems, use the correct codes to reinitialize the modules if required.

will turn off at different rates. For example, some vehicles may take up to a couple of hours before the last system goes to sleep and the parasitic draw reaches its minimum value. Consult the manufacturer's service information to determine the maximum allowable parasitic current draw and the time period after the ignition is turned off, and the time it takes the modules to go to sleep.

Parasitic draw that is higher than specified discharges the battery prematurely, potentially leaving the passengers stranded. Parasitic draw can be measured in several ways, the most common being the process of using an ammeter capable of measuring milliamps and inserting it in series between one of the battery posts and the battery terminal. If the vehicle is equipped with systems or modules that require electronic memory to be maintained, follow the procedure for identifying modules that lose their initialization during battery removal, and maintain or restore electronic memory functions. Note that the modules may reset during the process of disconnecting the battery terminal and connecting the ammeter in series, so you may have to wait for the modules to go back to sleep. If excessive parasitic draw is measured, disconnect fuses or systems one at a time while monitoring parasitic current draw to determine the systems causing excessive draw.

Disconnecting the battery can be avoided if a sensitive low-current (i.e., milliamps) clamp is available. The low-amp **current clamp** measures the magnetic field generated by a very small current flow through a wire or cable. Placing the low-amp current clamp around the negative battery cable allows you to measure the parasitic draw. If excessive parasitic draw is measured, disconnect fuses or systems one at a time while monitoring parasitic current draw to determine the systems causing the excessive draw. If parasitic draw does not decrease after removing all of the fuses, suspect an unfused circuit such as the alternator diodes, the ignition circuit on some vehicles, or control module feed wires.

The last way to measure a parasitic draw is a bit controversial, but it is well worth trying out as it can save a lot of time and requires no special tools other than a DMM. It is named the Chesney parasitic load test after its creator, Sean Chesney. Instead of using an ammeter to measure the draw, an ohmmeter is used, which sounds wrong—but stay with me.

Before doing anything, set the ohmmeter to ohms (the lowest scale), touch the meter leads together, and read the screen. The reading is the resistance of the meter leads and is called the "delta" value, which is the meter's true zero when used with those leads. Typically an ohmmeter will read about 0.1 ohm when the leads are touched together. Remember this number for later. Some meters have a delta feature that recalibrates the ohmmeter to zero when the leads are placed together and the delta button is pushed. If your meter has this

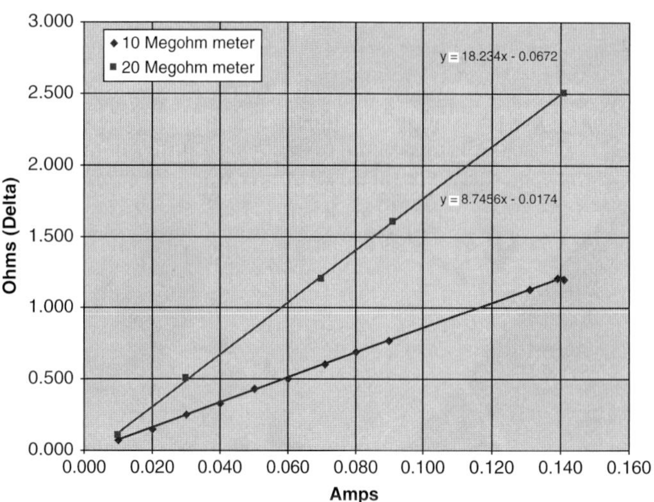

FIGURE 53-21 Chesney parasitic load test-ratio graph.

delta feature, you can use it so that you will not have to remember the delta reading.

Next, with the battery terminals still connected to the battery, place the black lead on the negative post of the battery and the red lead first on an unpainted surface of the alternator housing. Read the ohmmeter, and subtract the delta value from the reading. This reading corresponds to the relative parasitic draw on the system, which we describe next. Then, take the same reading, but with the red lead on a good body ground, and remember the reading.

Through testing, Chesney found that a draw of about 35 milliamps equaled an ohm reading of about 0.3 ohm delta (above the delta value) on a DMM with 10 megohms of impedance, and about 0.6 ohm delta on a DMM with 20 megohms of impedance (**FIGURE 53-21**). Anything above those readings indicates an excessive parasitic draw. You may be skeptical of this method, so try it on any vehicle. Simulate a parasitic draw by opening the driver's door, which illuminates the dome light, and watch the ohmmeter. It went up, right? Close the door. As soon as the light went off, the ohmmeter reading went back down, right? If you use this test and find an excessive draw, you can pull fuses one at a time, watching for the ohmmeter reading to decrease. If it does not decrease after removing all of the fuses, suspect an unfused circuit such as the alternator diodes or the ignition circuit on some vehicles.

To measure parasitic draw with a standard parasitic load test, follow the steps in **SKILL DRILL 53-8**.

To measure parasitic draw with a Chesney parasitic load test, follow the steps in **SKILL DRILL 53-9**.

SKILL DRILL 53-8 Measuring Parasitic Draw—Standard Test

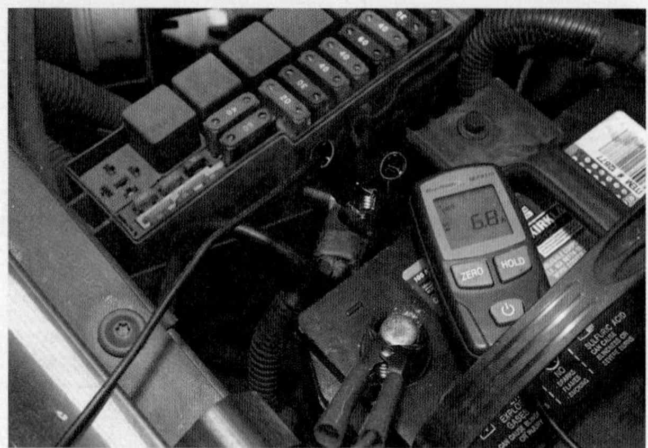

1. Research the parasitic draw specifications in the service information. Connect the low-current clamp around the negative battery cable, and measure the parasitic draw. Compare the parasitic draw to specifications.

2. Disconnect the circuit fuses one at a time to determine which circuit has the excessive parasitic current draw. Determine any necessary actions.

SKILL DRILL 53-9 Measuring Parasitic Draw- Chesney Parasitic Load Test

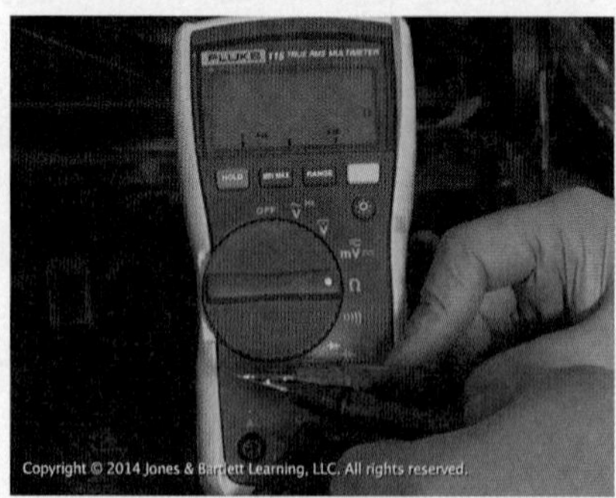

1. Set the DMM to read ohms (lowest scale if available). Connect the leads together and read the meter screen. This is the meter's delta reading.

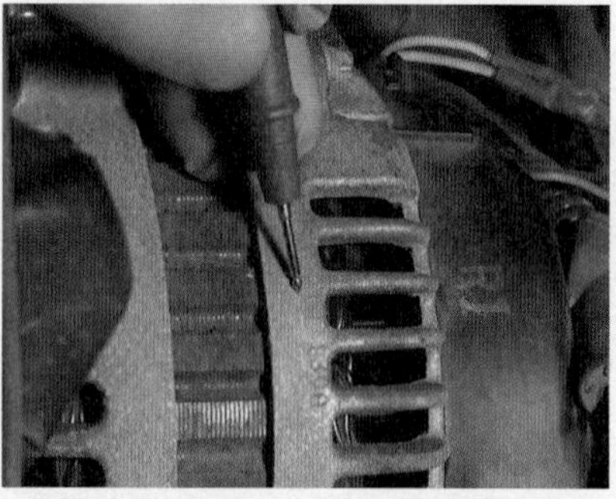

2. Place the black meter lead on the negative battery post and the red lead first on the alternator case and then on a good body ground.

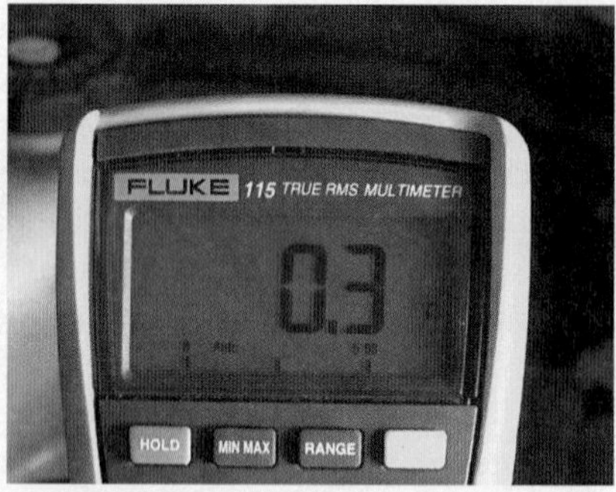

3. Read the meter in both places, and compare the reading to the Chesney parasitic load ratio graph. If there is excessive parasitic load, pull fuses one at a time to identify the faulty circuit.

▶ Wrap-Up

Ready for Review

- The primary components of the battery are the case, cover, vent caps, plates, separators, electrolyte, and terminals.
- Batteries operate by storing electrical energy in chemical form.
- Electrolyte is made up of sulfuric acid and water.
- The standard automotive battery has 2.1 volts per cell, with six cells creating 12.6 volts on a fully charged battery.
- During discharge, sulfuric acid is going into the plates creating current flow and turning the plates into lead sulfate
- During charging, sulfuric acid is pushed out of the plates returning them to sponge lead and lead dioxide.
- The state of charge of a battery is indicated by its specific gravity. The higher the specific gravity, the greater the state of charge.
- Open circuit voltage can also indicate a battery's state of charge. The higher the voltage, the higher the state of charge.
- Batteries are classified into groups, and only certain groups and layouts will fit certain vehicle applications.

- Batteries are rated by their electrical capacity; cold cranking amps, amp hour, and reserve capacity ratings.
- There are two chemicals that make the battery a safety hazard: hydrogen gas and sulfuric acid.
- The absorbed glass mat battery is good for rough handling and can be mounted on its side, unlike the electrolyte battery.
- Battery temperature plays a critical part in battery life. When a battery is cold, it doesn't create current flow as easily, plus it takes more power to crank the engine over; when a battery is too hot, its life is diminished significantly.
- Lithium-ion batteries are the future and are especially being used in hybrids. However, they have not replaced the conventional battery in most applications.
- There are two primary ways to connect the battery to the vehicle: the top-post design and the side-post design.
- Batteries must be kept free of corrosion, as corrosion is the most likely cause of a no-start condition.
- If a vehicle has a dead battery after sitting, the battery must be checked for an excessive parasitic draw.
- Jump-starting a vehicle is one of the simplest automotive tasks to perform, but done incorrectly, it can cause battery and possible vehicle damage. Always ensure the proper polarity when connecting the dead and good batteries together.
- When servicing a battery, always wear the proper personal protective equipment.
- Battery maintenance includes inspecting, cleaning, state of charge testing, and charging. Always follow all applicable technical data when performing battery maintenance.
- Batteries have to be recycled once their life cycle is used up. They contain hazardous materials that cannot be put in a landfill.
- The best way to test a battery's capacity and internal condition is through a conductance test.
- Battery load testing also tests a battery to see if its capacity is okay, but some manufacturers forbid load testing of their batteries.
- The load for a battery load test is half the CCA rating. The voltage should be 9.6 volts or above at the end of 15 seconds.
- A memory minder can be used to maintain electronic memory function when the battery is disconnected.
- A parasitic load test involves measuring the current flowing out of the battery when everything is off, and the modules have gone to sleep. The reading should be below specifications.
- A low amps clamp is used to measure a parasitic draw.
- An ohmmeter can be used to perform a Chesney parasitic load test.

Key Terms

absorbed glass mat A type of lead-acid battery.

battery terminal configuration The placement of positive and negative battery terminals.

cold cranking amps (CCAs) A standard for rating the ability of a vehicle battery to supply high current under cold operating conditions.

cranking amps (CA) A standard similar to CCA, but that measures the battery's function at a higher temperature-32°F (0°C)

current clamp A device that clamps around a conductor to measure current flow. It is often used in conjunction with a DMM.

electrical capacity The ability of a circuit or component to carry electrical loads.

gassing The escape of gas from the battery.

keep alive memory (KAM) A certain minimum amount of parasitic current draw that is used by the vehicle's circuits to maintain memory functions and monitor systems.

parasitic draw Unwanted drain on the vehicle battery when the vehicle is off.

reserve capacity A standard used to specify the time in minutes that a battery will supply a load of 25 amps at 80°F (26.7°C) without its voltage dropping below 10.5 volts.

thermal runaway A cycle during battery charging in which heating lowers resistance, which in turn increases current flow, which in turn further increases heat created. During this cycle, dangerous gases may build up, creating the potential for an explosion or damage through excessive current flow.

Review Questions

1. Which of the following does a battery convert into electrical energy?
   a. Heat energy
   b. Mechanical energy
   c. Chemical energy
   d. Light energy
2. The type of battery that is most common in automobiles is a:
   a. lithium-ion battery.
   b. lead-acid battery.
   c. nickel-cadmium battery.
   d. zinc-bromine battery.
3. All of the following statements are true *except*:
   a. A battery consists of two similar metals.
   b. An automotive battery can supply very high discharge currents while maintaining a high voltage.
   c. The fundamental components and operation of the battery remain the same irrespective of the size and usage.
   d. Batteries store electricity in chemical form.
4. Which of the following batteries are most suited for off-road and all-terrain vehicles?
   a. Low-maintenance batteries
   b. Lead-acid batteries
   c. Lithium-ion batteries
   d. Absorbed glass mat batteries
5. Automotive battery capacity is rated based on all of the following *except*:
   a. cold cranking amps (CCA).
   b. cranking amps (CA).
   c. reserve capacity.
   d. terminal style.

6. The lead plates of the lead-acid batteries are immersed in:
   a. diluted sulfuric acid solution.
   b. distilled water.
   c. mineral water.
   d. brine solution.
7. All of the following are factors in shortening battery life *except*:
   a. having deep discharge cycles.
   b. excessive vibration.
   c. a moderate external temperature.
   d. developing corrosion.
8. Status-of-the charge of a no-maintenance battery can be determined using a(n):
   a. single cell hydrometer float.
   b. TPS.
   c. oscilloscope.
   d. hygrometer.
9. Which of the following is the best and safest method to clean the battery when there is a voltage of more than 0.2 volts on top of the battery?
   a. Clean it in a solvent tank.
   b. Clean it in a hot-water cleaning machine.
   c. Clean it with a steam cleaner.
   d. Clean with baking soda and water mixture
10. Which the following tests is performed to determine the capacity and life left in the battery?
   a. Leak test
   b. Hydrometer test
   c. Conductance test
   d. Vacuum test

ASE Technician A/Technician B Style Questions

1. Tech A says that a battery stores electrical energy in chemical form. Tech B says that a battery creates direct current. Who is correct?
   a. Tech A
   b. Tech B
   c. Both A and B
   d. Neither A nor B
2. Tech A says that a 12-volt battery has six cells. Tech B says that the more plates a cell in a battery has, the more voltage it creates. Who is correct?
   a. Tech A
   b. Tech B
   c. Both A and B
   d. Neither A nor B
3. Tech A says that a parasitic draw is measured in volts. Tech B says that pulling fuses one at a time can help locate a parasitic draw. Who is correct?
   a. Tech A
   b. Tech B
   c. Both A and B
   d. Neither A nor B
4. Tech A says that when disconnecting the battery, the negative terminal should be disconnected first. Tech B says that baking soda and water will remove lead oxide from battery terminals. Who is correct?
   a. Tech A
   b. Tech B
   c. Both A and B
   d. Neither A nor B
5. Tech A says that checking the specific gravity will indicate the battery's cold cranking amps. Tech B says that a battery load test should be performed when the battery is heavily discharged. Who is correct?
   a. Tech A
   b. Tech B
   c. Both A and B
   d. Neither A nor B
6. Tech A says that lead oxide acts as an insulator on the battery posts and has to be scraped away. Tech B says that current can leak across the dirt on the surface of the battery. Who is correct?
   a. Tech A
   b. Tech B
   c. Both A and B
   d. Neither A nor B
7. Tech A says that batteries should be charged as fast as possible. Tech B says that the ideal charging rate is the CCA divided by 70. Who is correct?
   a. Tech A
   b. Tech B
   c. Both A and B
   d. Neither A nor B
8. Tech A says that the surface of the battery can be cleaned with baking soda and water. Tech B says that the electrolyte level should be below the top surface of the plates. Who is correct?
   a. Tech A
   b. Tech B
   c. Both A and B
   d. Neither A nor B
9. Tech A says that the specific gravity of the electrolyte can be checked with a hydrometer. Tech B says that the specific gravity of the electrolyte can be checked with a refractometer. Who is correct?
   a. Tech A
   b. Tech B
   c. Both A and B
   d. Neither A nor B
10. Tech A says that all vehicles have a specified CCA rating for the battery. Tech B says that reserve capacity indicates how many times the engine can be started by the battery. Who is correct?
   a. Tech A
   b. Tech B
   c. Both A and B
   d. Neither A nor B

CHAPTER 54

Starting and Charging Systems

NATEF Tasks

- **N54001** Demonstrate knowledge of an automatic idle-stop/start-stop system. (MLR/AST/MAST)
- **N54002** Perform starter current draw tests; determine needed action. (MLR/AST/MAST)
- **N54003** Perform starter circuit voltage drop tests; determine needed action. (MLR/AST/MAST)
- **N54004** Inspect and test switches, connectors, and wires of starter control circuits; determine needed action. (MLR/AST/MAST)
- **N54005** Inspect and test starter relays and solenoids; determine needed action. (MLR/AST/MAST)
- **N54006** Differentiate between electrical and engine mechanical problems that cause a slow-crank or a no-crank condition. (AST/MAST)
- **N54007** Remove and install starter in a vehicle. (MLR/AST/MAST)
- **N54008** Perform charging system output test; determine needed action. (MLR/AST/MAST)
- **N54009** Perform charging circuit voltage drop tests; determine needed action. (MLR/AST/MAST)
- **N54010** Diagnose (troubleshoot) charging system for causes of undercharge, no-charge, or overcharge conditions. (AST/MAST)
- **N54011** Remove, inspect, and/or replace generator (alternator). (MLR/AST/MAST)

Knowledge Objectives

After reading this chapter, you will be able to:

- **K54001** Describe the purpose, operation, and types of starter systems.
- **K54002** Explain the function of starter motor components.
- **K54003** Explain the operation of the starter solenoid and control circuit.
- **K54004** Describe the purpose and function of the charging system.
- **K54005** Explain the purpose and function of alternator components.
- **K54006** Explain rectification and the operation of the rectifier.
- **K54007** Explain the process of controlling charging system output.

Skills Objectives

After reading this chapter, you will be able to:

- **S54001** Perform starting system inspections.
- **S54002** Perform charging system inspections.

▶ Introduction

The starting system provides a method of rotating (cranking) the vehicle's internal combustion engine (ICE) to begin the combustion cycle. In early vehicles, this was done by the use of a hand-crank handle (**FIGURE 54-1**). Modern vehicles use an electric starter motor that draws its electrical power from the vehicle's battery (**FIGURE 54-2**). The starter is designed to work for short periods of time and must crank the engine at sufficient speed in order for it to start. Modern starting systems are very effective provided that they, and the battery, are well maintained.

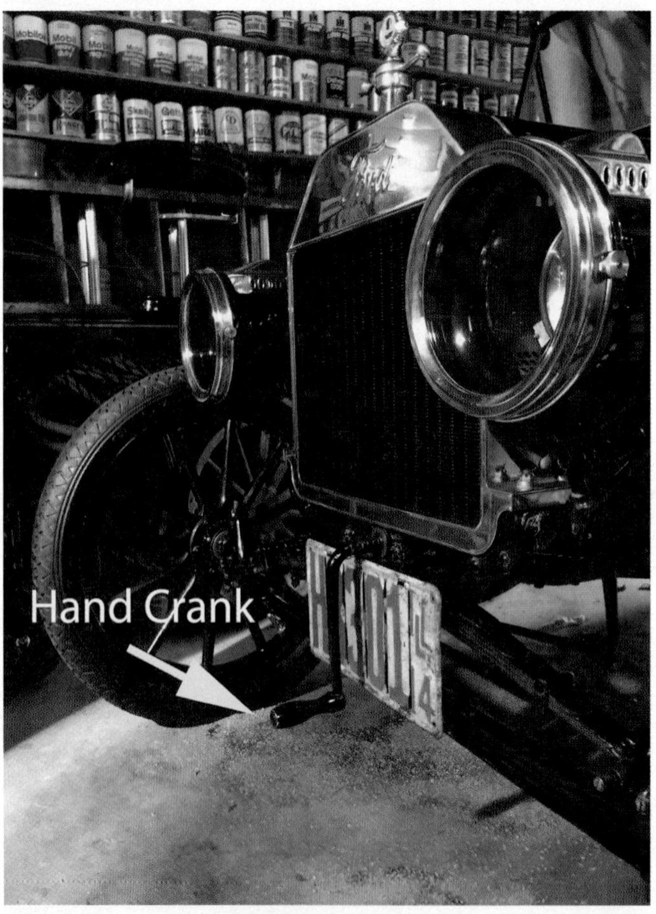

FIGURE 54-1 Early engines were hand cranked to start them.

FIGURE 54-2 Modern electric starter bolted to the transmission.

▶ Engine Starting (Cranking) Systems

K54001

The starting/cranking system consists of two electrical circuits; high amperage and low amperage. The high amperage circuit powers the starter for cranking the engine over. The low amperage circuit is used to control the high amperage circuit. The high amperage circuit consists of the battery, high amperage side of the solenoid, and starter motor assembly. The control circuit consists of the battery, ignition switch, safety switch (clutch switch or neutral safety switch), and the low amperage side of the solenoid. On PCM-activated starting systems, there is also the PCM, a relay, and all of the related sensors that feed information to the PCM. The onboard computer (PCM) and a security/antitheft system may determine when and if the cranking circuit will function.

During the cranking process, two actions occur. The pinion of the starter motor engages with the flywheel ring gear, and the starter motor then rotates to turn over, or crank, the engine. The starter motor is an electric motor mounted on the engine block or transmission. It is typically powered by the 12-volt storage battery, although some hybrid vehicles use the high-voltage battery to operate the starter motor. Starters are designed to have high turning effort (torque) at low speeds. The starter cables are the heaviest in the vehicle because they carry the high current

You Are the Automotive Technician

The owner of an eight-year-old Ford F-150 4X4 has her vehicle towed to your shop because it won't start. She tells you that she went into a store, and when she came out, it wouldn't restart. She agrees to let you diagnose the problem while she waits in the customer lounge. You verify the problem by trying to crank the engine over. The starter motor clicks once, and the starter current is very low.

1. What are the most likely faults that would cause an engine not to crank over?
2. What starting system components are likely *not* causing this problem?
3. What tests will you perform to diagnose the problem, and what results will you be expecting?
4. After repairing the starting concern, how would you verify that the charging system is working properly?

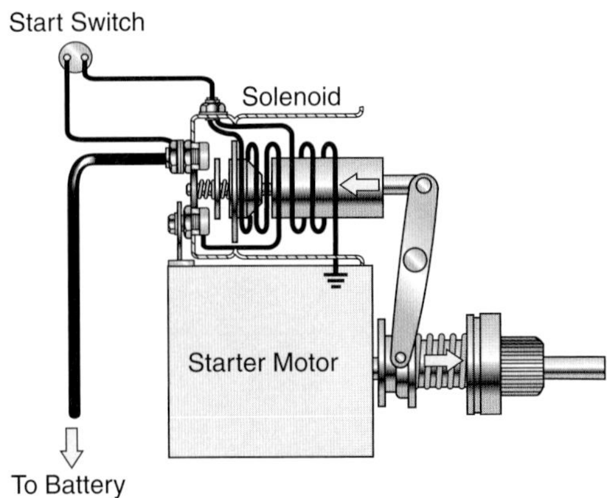

FIGURE 54-3 Typical starter motor, solenoid, and starter drive.

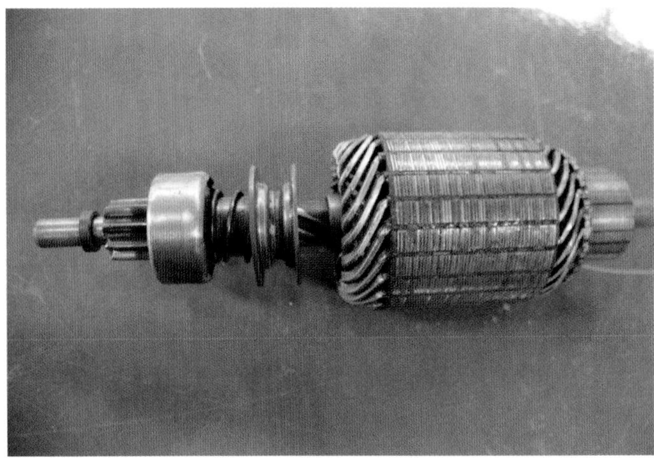

FIGURE 54-4 Armature and starter drive from a direct-drive starter.

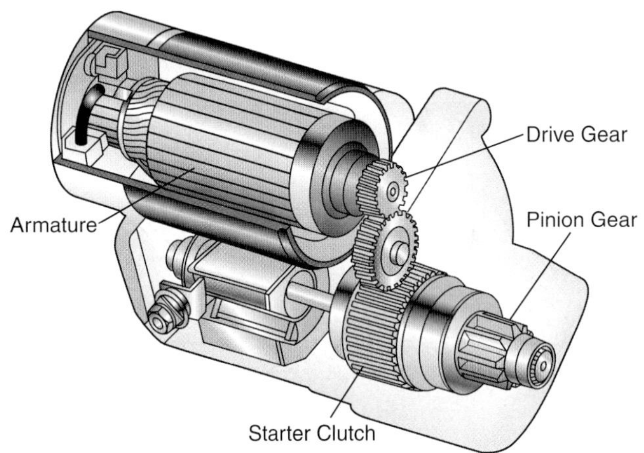

FIGURE 54-5 Gear reduction starter.

needed by the starter motor. The starter motor causes the engine flywheel and crankshaft to rotate from a resting position and keeps them turning until the engine fires and runs on its own.

Starter Motor Principles

The starter motor converts electrical energy to mechanical energy for the purpose of cranking the engine over. There are three sections to the typical starter—the electric motor, the drive mechanism, and the solenoid (**FIGURE 54-3**). The starter motor is mounted on the transmission or cylinder block in a position to engage a ring gear around the outside edge of the engine flywheel, flexplate, or torque converter. Starting is usually accomplished by the operator activating a starter switch built into the ignition lock assembly, or a start button. A relatively small current flows through a neutral safety switch or clutch switch to a starter relay that controls a larger current to operate the starter solenoid, which is typically mounted atop the starter motor. The solenoid plunger moves the drive pinion gear into engagement with the ring gear and also closes a set of heavy-duty contacts. This allows a very large current to flow from the battery to the starter motor, rotating the armature and drive pinion gear, causing the crankshaft to rotate. When the engine starts and is able to run on its own, the operator usually releases the key and the solenoid spring withdraws the pinion gear from the ring gear and brings the armature to a halt. On many modern vehicles, the PCM signals the relay to continue the cranking process once a crank signal is received from the operator until the vehicle starts or the starting operation times out.

Starter Types—Direct-Drive/Gear Reduction

Starter motors can be designed to drive the ring gear in one of two ways: either direct-drive or gear reduction. In the direct-drive system, the starter drive is mounted directly on one end of the armature shaft. The starter drive transfers the rotating force of the armature directly to the engine flywheel (**FIGURE 54-4**). In this arrangement, the only gear reduction is the reduction between the pinion gear and the ring gear.

Gear reduction starters use an extra gear between the armature and the starter drive mechanism. They have a reduction of about 4:1. The gear reduction allows the starter to spin at a higher speed with lower current (**FIGURE 54-5**). It also enables the starter to be downsized yet create a higher torque output. Two types of gearing systems are normally used: spur gears or planetary gears. Spur gears require the armature to be offset via a gear housing that holds the starter drive. A gear reduction system using planetary gears does not require an offset housing; the planetary gears are housed in the drive-end housing in line with the starter drive.

Starter Motor Construction

K54002

A starter motor normally consists of the following components: field coils or large permanent magnets, an armature, a commutator, brushes, a drive pinion with an overrunning clutch, and a drive pinion engagement solenoid and shift fork (**FIGURE 54-6**). The armature is the revolving component of the DC motor. The armature shaft is supported at each end by bushings or bearings pressed into end frames, which locate the armature centrally in the outer casing (i.e., the "barrel") of the motor and between the field coils or permanent magnets. Direct drive starters are equipped with a brake washer that is used to slow the armature when the starter is disengaged after the engine starts.

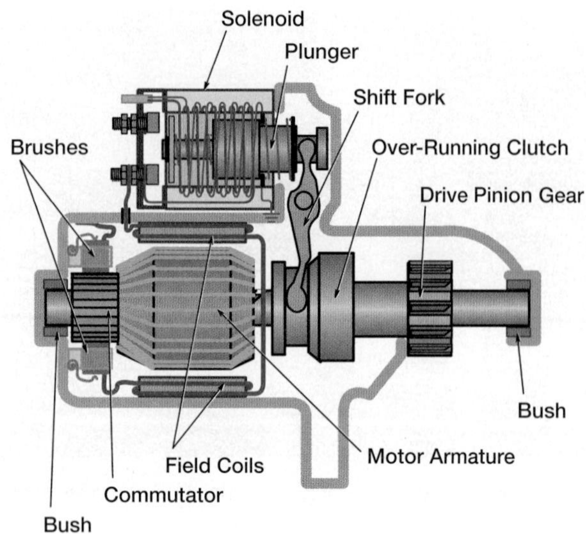

FIGURE 54-6 Cutaway view of a starter motor.

FIGURE 54-7 The brushes transfer electricity to the rotating commutator.

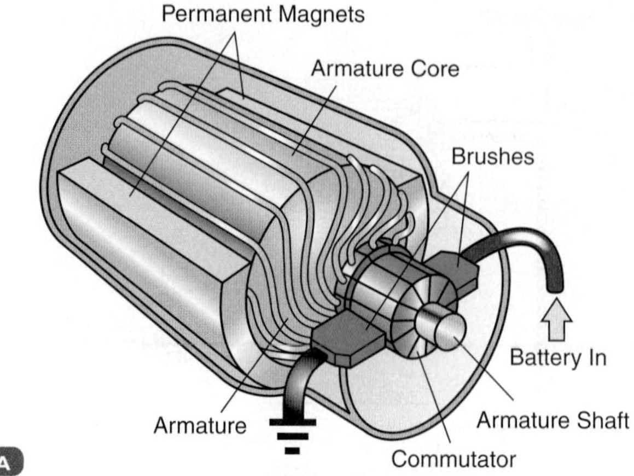

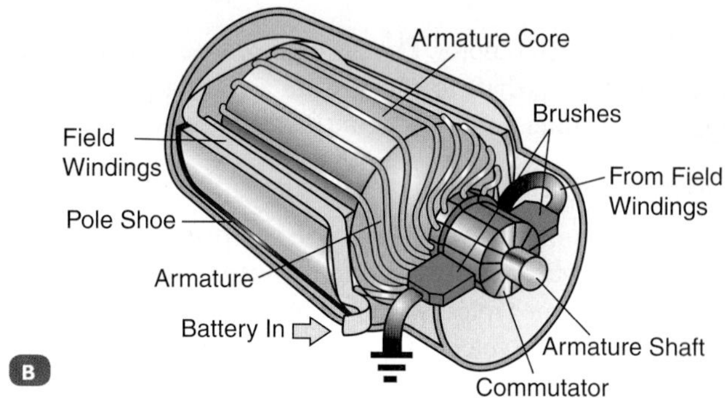

FIGURE 54-8 Electrical schematic of the power flow. **A.** Permanent magnet starter. **B.** Series-wound starter.

The commutator end frame carries copper-impregnated carbon brushes, which conduct current through the armature when it is being rotated in operation. The brushes are mounted in brush holders and are kept in contact with the commutator by tensioned spiral springs (**FIGURE 54-7**).

Half of the brushes are connected directly to the end frame and ground the armature windings via the engine block to the negative battery terminal. The other brushes are insulated from the end frame and connect to the positive battery terminal via the main starter solenoid input terminal. This connection is direct from the solenoid in the case of a permanent magnet starter and is indirect via the electromagnetic field poles in a series-wound motor (**FIGURE 54-8**).

Starter Magnet Types

Starter motors use two magnet types: electromagnetic and permanent magnet (**FIGURE 54-9**). Electromagnetic fields are formed by current flow through heavy strip copper windings, wound around iron pole shoes, which are then fastened to the starter case/barrel. Permanent magnets are located similarly but do not need electricity and therefore occupy less space. The case is made of iron and serves to concentrate the magnetic field produced by the field magnets.

Starter motors with electromagnetic field windings for light vehicle applications are typically series-wound motors. Because the resistance of the field and armature windings is low, the current flow is high when the motor starts under load, and this generates a strong magnetic field that will produce high torque at low speeds. This high initial torque drops sharply as the motor speed increases because of the **counter-electromotive force (CEMF)**, or voltage generated in the armature windings as the motor armature spins in the magnetic field. The CEMF increases with armature speed and opposes current flow. It reduces both current flow and torque output. The faster the motor turns, the less current it draws and the less torque it develops. For example, on a typical V6 engine with standard compression, the initial surge through the starter is usually over 500 amps, but as the armature starts to spin, the CEMF opposes the current flow, reducing it to around 120–150 amps (average) within about a second. This initial surge and subsequent drop of amperage can be observed on an oscilloscope when testing the starter motor.

Some series-wound motors have parallel-wired field windings, but they are still wired in series with the armature. These starters are referred to as series-parallel-wound starter motors.

FIGURE 54-9 **A.** Electromagnetic fields. **B.** Permanent magnetic fields.

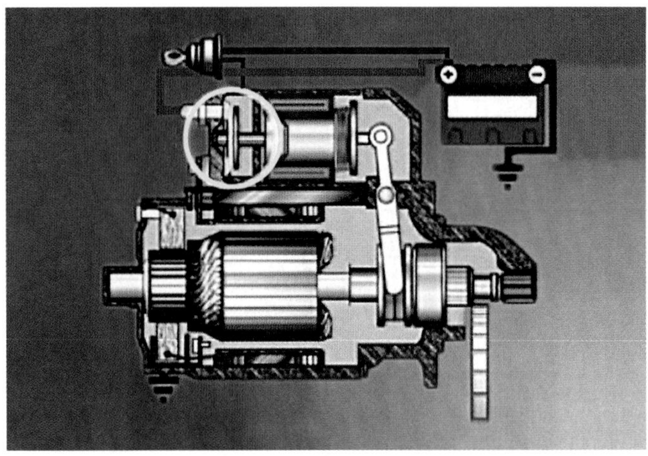

FIGURE 54-10 Cutaway view of starter showing actuating assembly in the engaged position.

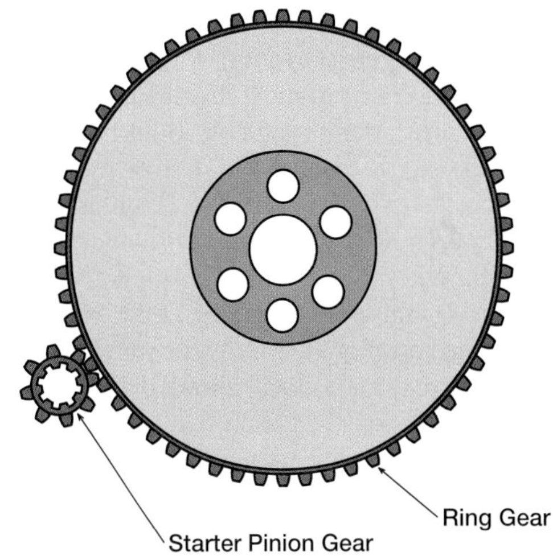

FIGURE 54-11 The starter gear is much smaller than the flywheel ring gear, typically about 17:1.

By connecting the field windings in this way, more current can flow in the circuit, and an overall increase in torque is obtained.

Starter Motor Engagement

Engagement is provided by operation of the ignition switch in the start position or when commanded by the PCM, which activates a starter-mounted solenoid whose plunger is engaged with the end of a pinion shift lever and operating fork. Solenoid operation moves the operating fork, causing the pinion to engage with the ring gear and also causing the plunger contacts to bridge with the main starter terminals (**FIGURE 54-10**). The fork is located in a guide ring on the pinion drive, which is coupled to the pinion gear via a roller-type overrunning clutch. It is designed to transmit drive in one direction only and freewheel in the opposite direction.

The pinion drive is mounted on a slight helix that is machined onto the armature shaft to form a very coarse thread. This arrangement allows the pinion drive to rotate slightly when it is moved toward the ring gear. This feature, together with a chamfer on the leading edge of the ring gear and pinion gear teeth, is designed to assist the teeth in meshing. However, if the pinion gear teeth butt against the ring gear teeth and engagement is prevented, the guide ring continues its movement by sliding over the sleeve of the drive and compressing a meshing spring until the solenoid plunger contacts bridge the main terminals, and the armature begins to turn.

Slight armature rotation and the force from the meshing spring push the pinion teeth into mesh with the ring gear. The meshing spring forces the pinion farther into the ring gear until the pinion contacts a stop ring on the armature shaft. This prevents further axial movement. The starter drive is locked to the shaft via the helix. The one-way clutch drives the pinion gear and transfers the armature rotation to the ring gear.

The pinion has only a small number of teeth compared to the ring gear, usually around 17:1, meaning the armature will rotate 17 times for each revolution of the flywheel (**FIGURE 54-11**). The gear reduction with this ratio also multiplies the torque from the starter motor 17 times, allowing a relatively small electric motor to turn the much larger ICE. If a gear reduction starter is used, then the torque is multiplied further, giving more cranking power from the same size starter motor.

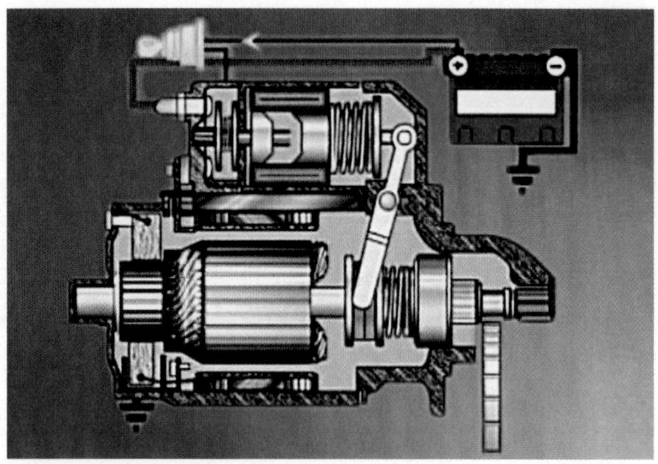

FIGURE 54-12 Cutaway view of starter showing actuating assembly in the released position.

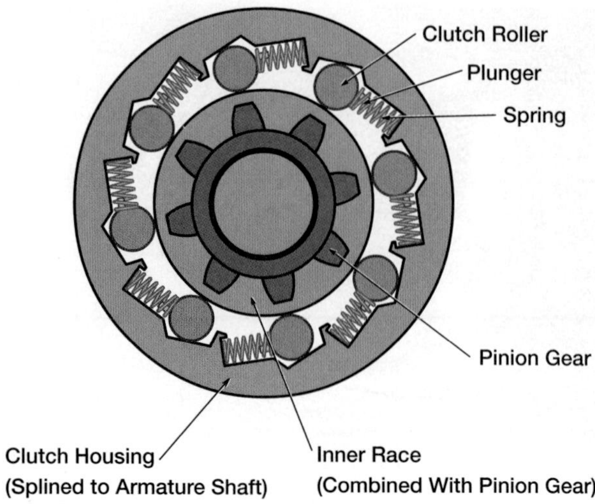

FIGURE 54-13 Starter drive one-way clutch.

As soon as the engine starts, it may easily run at 1000 revolutions per minute (rpm) or more. If still engaged, the engine would then turn the starter pinion gear about 17 times faster, or 17,000 rpm. Turning that fast would destroy the armature. At this instant, the free-wheeling of the overrunning clutch prevents the armature from turning too fast should the driver not release the ignition key from the start position.

The pinion remains meshed as long as the engaging lever is held in the engaged position. Releasing the starter switch (or the PCM terminating the crank signal) allows the solenoid plunger return spring to disengage the pinion gear from the ring gear by returning the engaging lever, starter drive, and pinion gear to their original position (**FIGURE 54-12**).

Starter Drives and the Ring Gear

The starter drive transmits the rotational drive from the starter armature to the engine via the ring gear that is mounted on the engine flywheel, flexplate, or torque converter. The starter drive is composed of a pinion gear, an internal spline that mates with the slightly curved external spline on the armature shaft, an overrunning clutch, and a meshing spring. The pinion gear is small in comparison to the ring gear, which means the starter turns many times faster than the engine ring gear. This also gives a large amount of mechanical advantage to the starter motor, allowing it to crank over the much larger ICE.

The overrunning clutch drives the pinion gear in one direction while allowing it to freewheel in the opposite direction. The overrunning clutch uses roller bearings housed between an inner and an outer shell. The roller bearings ride in tapered ramps built into the outer shell (**FIGURE 54-13**), and springs push the rollers toward the tapered ends of the ramps. In the forward direction, the rollers roll slightly between the tapered ramps and are pinched between the inner and outer shells, thereby locking the assembly and driving the pinion gear. In the opposite direction (once the engine starts), the rollers roll up the inclined ramps against spring pressure, unlocking the rollers and allowing the pinion gear to freewheel. The overrunning clutch prevents the starter motor from being driven by the engine once the engine starts, which would spin the armature faster than it could handle. Without the overrunning clutch, the vehicle engine would overspeed the starter motor, causing significant damage.

Armature Windings and Commutator

When current flows in a conductor, an electromagnetic field is generated around it. If the conductor is placed so that it cuts across a stationary magnetic field, the conductor will be forced out of the stationary field. This occurs when the lines of force of the stationary field are distorted by the electromagnetic field around the conductor and try to return to a straight-line condition. Reversing the direction of current flow in the conductor will cause the conductor to move in the opposite direction. This is known as the motor effect and is greatest when the current-carrying conductor and the stationary magnetic field are at right angles to each other.

A conductor loop that can freely rotate within the magnetic field is the most efficient motor design. In this position, when current flows through the loop, the stationary magnetic field is distorted and the lines of force try to straighten. This forces one side of the loop up and the other side of the loop down, thus turning the loop (**FIGURE 54-14**). The turning motion is called the motor effect and causes the loop to rotate until it is at 90 degrees to the magnetic field. To continue rotation, the direction of current flow in the conductor must be reversed at this static neutral point.

A commutator is used to continually reverse the current flow to maintain rotation of the loop (**FIGURE 54-15**). For example, a commutator consists of two semicircular segments that are connected to the two ends of the loop and are insulated from each other. Carbon-impregnated brushes provide a sliding connection to the commutator to complete the circuit and allow current to flow through the loop.

Rotation begins with both sides of the conductor loop cutting the stationary field. When the loop passes the point where the field is no longer being cut, the momentum of rotation carries the loop and the commutator segments over so that the

Applied Science

AS-75: Electromagnetism: The technician can explain the relationship between current in a conductor and strength of the magnetic field.

Electromagnetism can be defined as the physical relationship between a magnet and electricity. Michael Faraday is best known for his discoveries of electromagnetism back in 1831. His biggest breakthrough was his invention of the electric motor.

Today's automotive electrical systems use a variety of electromagnetic devices to accomplish a wide range of tasks. We have both AC and DC electric motors/generators when hybrid vehicles are taken into account. Automotive starters and generators (alternators) operate on the electromagnetic principle. Electric fuel pumps, cooling fan motors, and windshield wiper motors all produce a turning force based on the principles of magnetic fields.

The basic concept of the electromagnetic principle is to place a rotating armature inside field windings. The field windings are stationary insulated electrical wires formed into a circular pattern. A magnetic field is created when current flows through the field winding. The armature spins as a result of the magnetic field. An example of this is a starter motor.

In the case of a charging system, we can see the relationship of the output of the generator (alternator) voltage by increasing the number of turns, or windings, in the stator. We could also increase the output voltage by increasing the rotor's magnetic field.

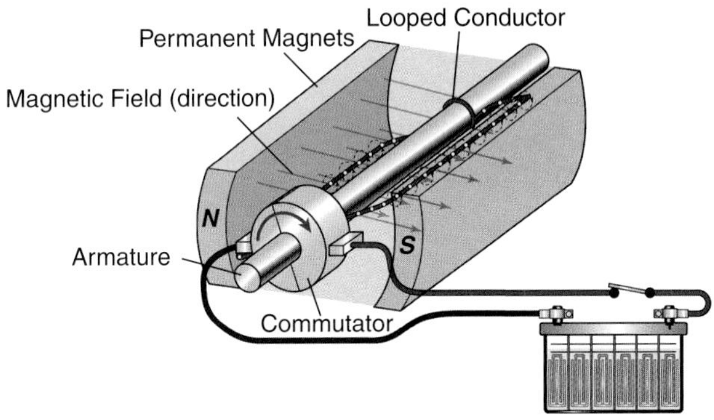

FIGURE 54-14 Simple single-loop motor and electromagnetic fields—with commutator and brushes.

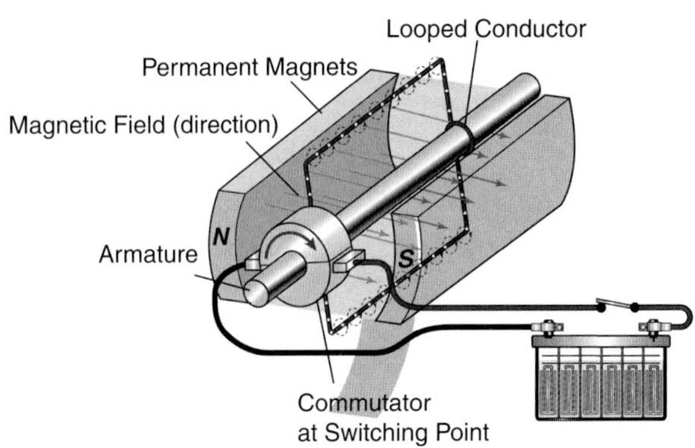

FIGURE 54-15 Simple single-loop motor and electromagnetic fields at the switching point of the commutator.

brushes maintain current flow in the same direction in each side of the loop relative to the stationary field.

This process maintains a consistent direction of rotation of the loop. In order to achieve a uniform motion and torque output, the number of loops must be increased. The additional loops smooth out the rotational forces. A starter motor armature has a large number of conductor loops and therefore many segments on the commutator (**FIGURE 54-16**).

Solenoid Operation

K54003

The solenoid on the starter motor performs two main functions: It switches the high current flow required by the starter motor on and off, and it engages the starter drive with the ring gear. The solenoid is typically a cylindrical device mounted on the starter motor (**FIGURE 54-17**). It is constructed with two electrical windings, a **pull-in winding** and a **hold-in winding**,

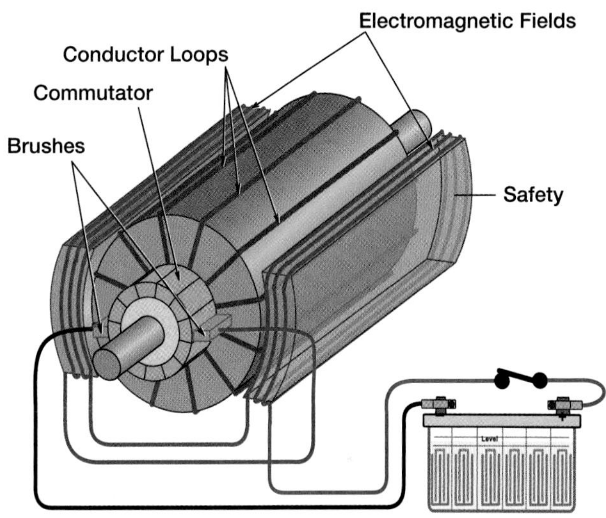

FIGURE 54-16 Simple multiloop motor and electromagnetic fields—with commutator and brushes.

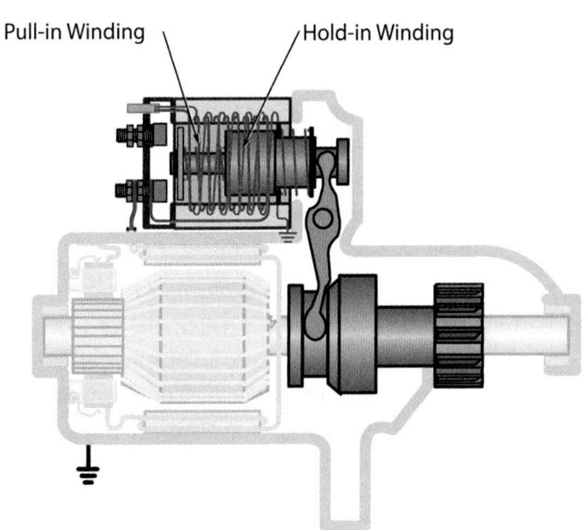

FIGURE 54-17 The solenoid uses two electrical windings: a hold-in winding and a pull-in winding.

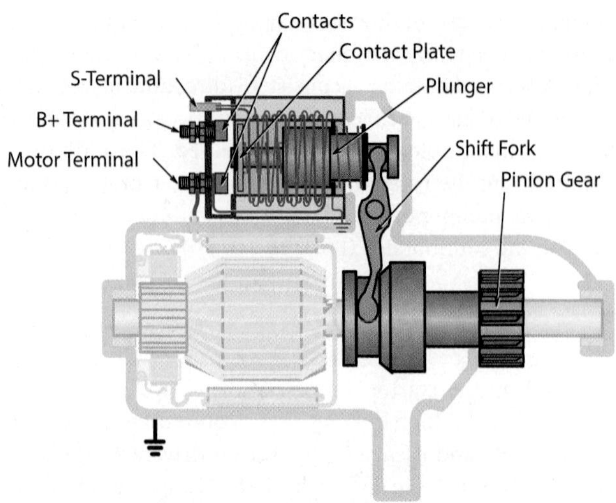

FIGURE 54-18 Solenoid starter contacts and starter drive linkage.

which will be explained in a moment. One end of the solenoid has a moving soft iron plunger, which is connected to a lever that moves the starter drive. The other end has an insulated cap with electrical connections to the solenoid windings along with a set of high-current contacts and a movable copper disc that completes the cranking circuit when the solenoid plunger is drawn forward (**FIGURE 54-18**). Once the cranking circuit is completed, current flows from the battery to the starter fields and armature.

The starter control circuit activates the solenoid winding to draw the plunger forward. The control circuit can be activated directly by the ignition switch on some vehicles or by the PCM on other vehicles. When the control circuit is activated, it supplies battery power to the two windings in the starter solenoid. One of these is a pull-in winding, which draws a higher current and creates a stronger magnetic field than the other winding—the hold-in winding. The inputs of both windings are connected to the S-terminal (control circuit) on the solenoid. Both windings are wound in the same direction in the solenoid housing. The output of the pull-in winding is connected to the main starter terminal leading to the field and armature windings, which provides ground to the pull-in winding until the solenoid contacts close. The output of the hold-in winding is connected to ground on the starter casing, which provides ground to the hold-in winding at all times.

When the starter is activated, current passes through both starter windings. The magnetic fields from both windings work together to attract the solenoid plunger toward the main starter terminals in the solenoid cap (**FIGURE 54-19**). Plunger movement also operates the shift fork lever, thus engaging the drive pinion with the flywheel ring gear. The plunger contacts a switching pin, which transfers the motion through a contact spring, closing the main solenoid terminals. This allows a large current to flow from the battery through the starter motor windings, causing armature and pinion rotation, and rotation of the engine crankshaft. However, closing the contacts does another very important task; it shorts-to-power the output wire of the pull-in winding, resulting in two sequential actions. First, the short-to-power means that the pull-in winding has battery voltage applied to both the input and the output of the winding, stopping the current flow through the pull-in winding and stopping that magnetic field. But the hold-in winding still has power from the control circuit, so it continues to hold the plunger in place while the starter cranks. In fact, during engine cranking, the action of the helix on the rotating armature shaft causes the pinion gear to be held firmly in mesh with the flywheel ring gear. The hold-in winding is used only to ensure that the moving contact continues to bridge the main starter terminals (**FIGURE 54-20**).

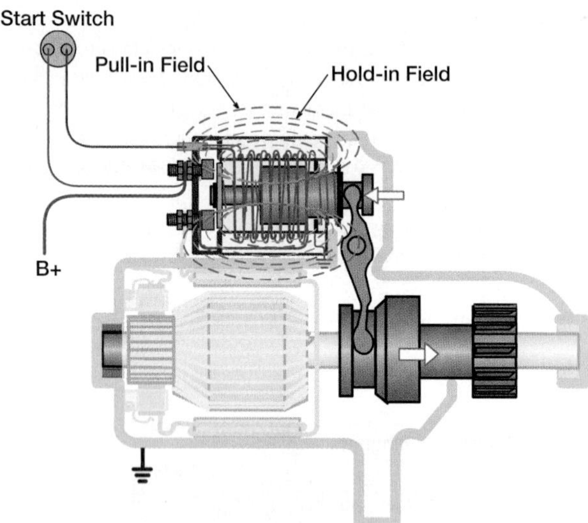

FIGURE 54-19 Both windings energized, creating maximum magnetism, and solenoid plunger starting to move toward the cap.

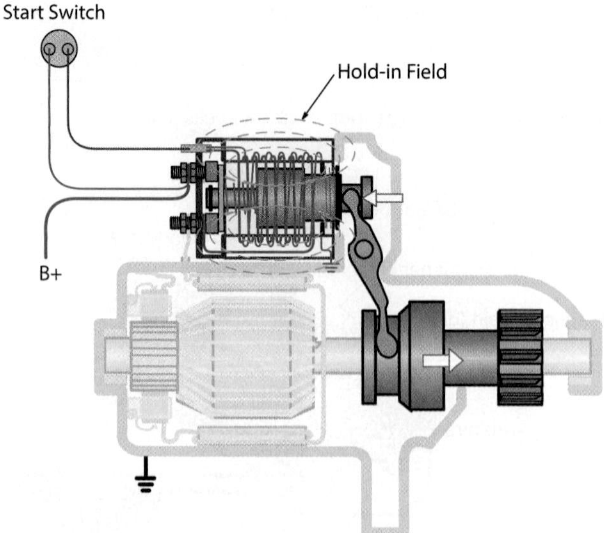

FIGURE 54-20 The solenoid plunger bridging the contacts and causing the motor to turn and shorting-to-power the output of the pull-in winding, stopping current flow through the pull-in winding, and stopping its magnetic field.

Once the engine starts, the control circuit is deactivated, and now is when the second action comes into play. The current stops flowing through the control circuit to supply the input of the hold-in and pull-in windings, but the output of the pull-in winding is still activated through the bridged solenoid contacts, so current flows backward through the pull-in winding and then forward through the hold-in winding (**FIGURE 54-21**). Because the current flows in the two windings are opposite to each other, the two magnetic fields oppose each other and tend to cancel each other out, thereby allowing the plunger return spring to retract the plunger. This disconnects the power from the pull-in and hold-in windings as well as the starter motor, causing it to stop cranking the engine. As the return spring in the solenoid returns the plunger, it also retracts the pinion gear to its rest position.

Ford Starter with Moveable Pole Shoe

Some Ford vehicles use a separate starter relay in the engine compartment, instead of a solenoid, to control the high current for the starter motor. A spring-loaded pole shoe on the starter itself is magnetically pulled into position when current is supplied to the starter (**FIGURE 54-22**). As the pole shoe moves into position, a shift fork attached to it engages the starter drive with the flywheel ring gear, thereby negating the need for a separate starter solenoid. Other functions of this type of starter are similar to those described previously.

Starter Control Circuit

The starter control circuit provides a means of operating the starter motor only within certain parameters, such as when the transmission is in Park, the clutch is depressed, the brake pedal is applied, or the proper ignition key is being used. These requirements help prevent accidentally starting the vehicle in gear (**FIGURE 54-23**), which could cause an accident or injury, and also help prevent the vehicle from being stolen. For many years, manufacturers have placed switches in series with the starter solenoid windings, which prevents the starter from being activated unless each of the switches is closed. If the vehicle was equipped with an automatic transmission, then a neutral safety switch was incorporated into the shifter linkage so that the switch was only closed when the transmission was in Park or

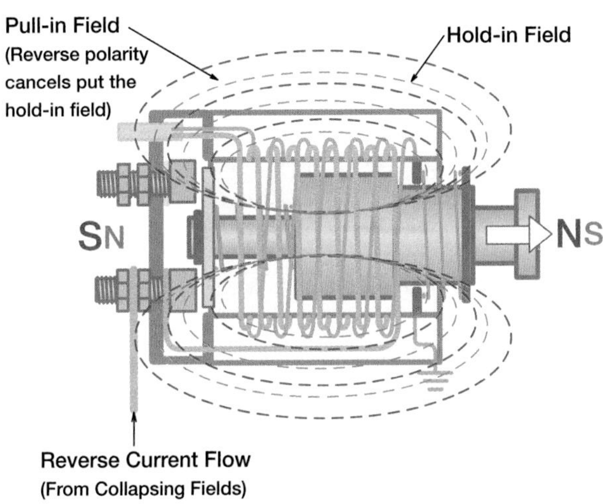

FIGURE 54-21 Current flow goes backward through the pull-in winding when the ignition key is turned from the crank position. This creates an opposite magnetic field, which cancels out the magnetic field from the hold-in winding.

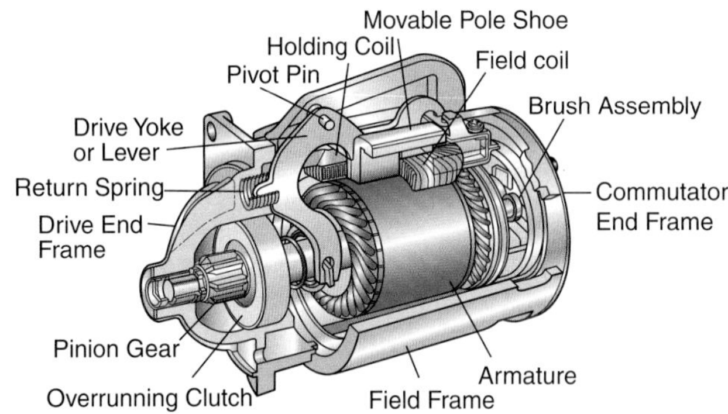

FIGURE 54-22 Ford moveable pole shoe starter.

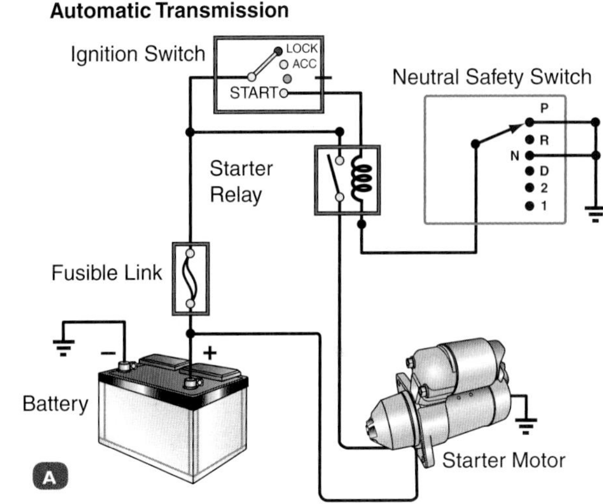

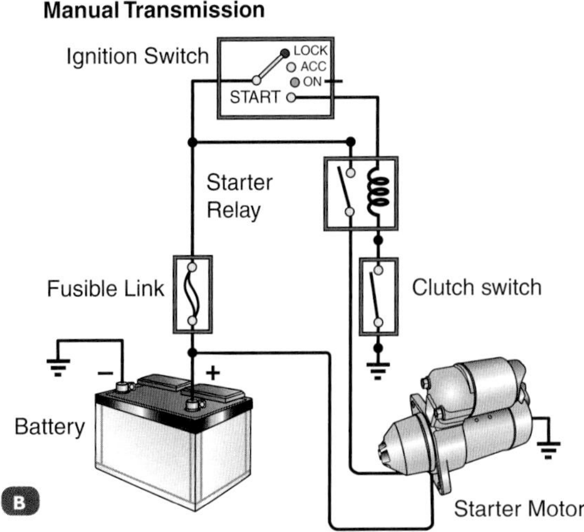

FIGURE 54-23 Basic starter control circuit. **A.** Neutral safety switch circuit. **B.** Clutch switch circuit.

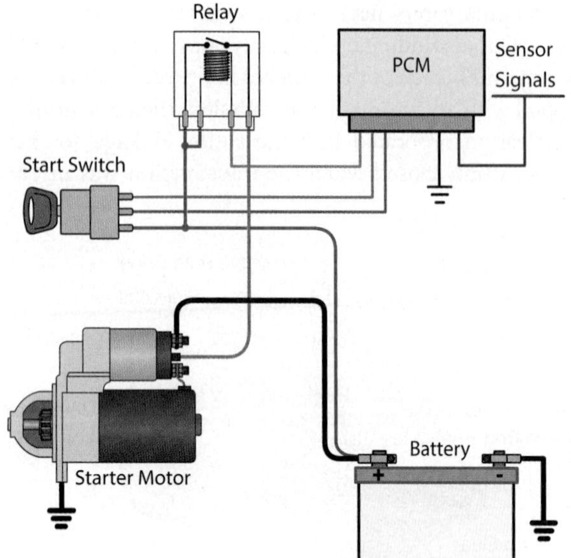

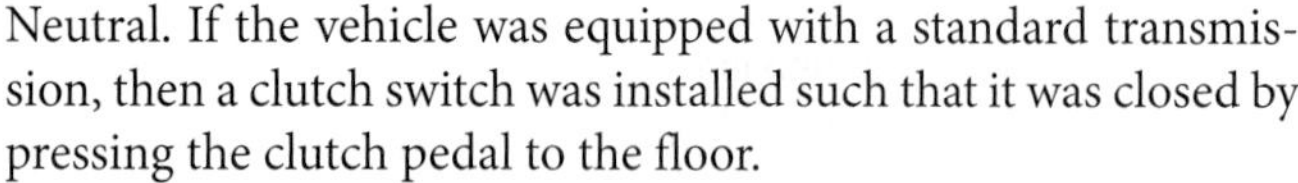

FIGURE 54-24 PCM controlled starter circuit.

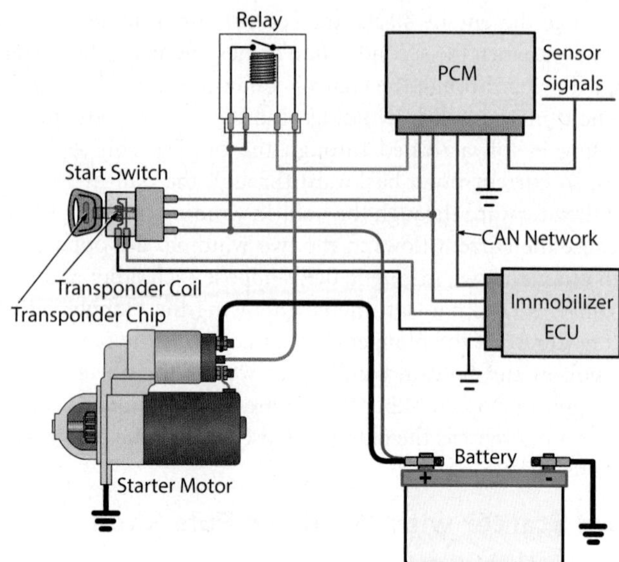

FIGURE 54-25 A typical vehicle immobilization system.

Neutral. If the vehicle was equipped with a standard transmission, then a clutch switch was installed such that it was closed by pressing the clutch pedal to the floor.

Newer vehicles with their computer controls can monitor the same information, and more, so that the starter will not activate until all of the required parameters are met. Once they are met, the PCM either activates a starter relay, which activates the starter solenoid, or the PCM activates the solenoid directly (**FIGURE 54-24**). With PCM-controlled starters, the starter control circuit becomes part of the vehicle theft-deterrent system, since the starter action can be disabled to prevent the vehicle from being started and stolen, as you will see in the next section.

Vehicle Immobilization Systems

There are many different names given to vehicle immobilization systems. Each manufacturer produces its own version, each with subtle differences. Vehicle immobilizers generally comprise a computer-managed security system that disables the vehicle starter and engine systems by using an electronic system to uniquely identify each vehicle key by a security code system. Some keys have a built-in electronic circuit board used to store the code. This system makes it very difficult to start the vehicle with anything other than a correctly coded vehicle key. It also means that the key not only needs to be cut to fit the lock, but it must also be coded electronically to match the vehicle. Key coding is usually done with a scan tool with the correct software and a pass-through device that enables the body control module (BCM) to be programmed from the Internet.

Another type of immobilizer system uses a static code programmed into circuitry built into the key. The circuit is powered by electromagnetic induction from a small coil surrounding the key lock. Once powered up, the circuit in the key sends the static binary code wirelessly to the receiving coil where it is sent to the immobilizer module and compared to the registered code (**FIGURE 54-25**). Newer systems use an immobilizer control unit that generates rolling codes that change each time the system is activated. In this case, the circuitry in the key works with the control unit to determine the rolling codes.

The key identification system has two states of operation: mobilized and immobilized (or secure). When mobilized, the vehicle and engine components are allowed to operate normally. While in the immobilized or secure state, the key identification system is activated to protect the vehicle and prevent the engine from starting and/or cranking. The system is set to immobilize when the engine is switched off and the key removed. Usually the system will incorporate a flashing warning lamp on the dash to identify that it is immobilized. To mobilize the system, the key needs to be inserted into the ignition switch and turned to the on position, or the operator has to be in possession of the electronic smart key, step on the brake pedal, and push the start button.

Some General Motors vehicles have used a key with a built-in resistor. Only when the key with the correct value of resistance is inserted into the ignition switch will the immobilizer module enable the starter control circuit to operate.

Keyless Starting/Remote Starting

Immobilizer systems now use keyless starting. The vehicle has a start button on the dash and does not require the key to be inserted into an ignition switch. In this type of wireless system, the start button will start the vehicle only if the key is in the proximity of the vehicle-for example, in the driver's pocket-and if the driver is stepping on the brake pedal. The vehicle detects the key wirelessly and mobilizes the system so the engine can start if the button is pressed (**FIGURE 54-26**). A further variation of this is where the vehicle can be started remotely (e.g., inside the house) by pressing a start button on the key fob

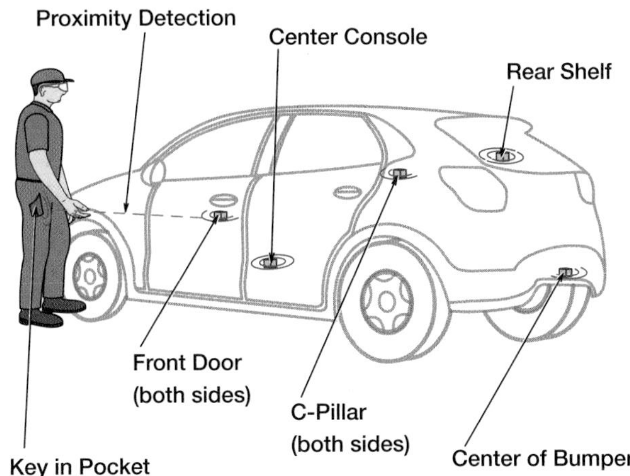

FIGURE 54-26 The vehicle has a number of receivers that detect the key wirelessly, allowing the driver to start the engine with the key in their pocket or purse.

instead of a button on the dash. This is handy in hot or cold climates when warming or cooling the vehicle is desired prior to entering the vehicle.

Starter Location Variation

Engine design usually dictates the location of the vehicle starter motor. Traditionally, most starter motors are mounted on one side of the engine so they can mesh with the ring gear. Starter motors can be difficult to see by just looking into the engine bay; they are often tucked up under exhaust manifolds or engine covers. On trans-verse engines, they are often mounted on the side closest to the firewall, which can make them difficult to locate. Starter motors are sometimes more easily accessed from underneath the vehicle. Often engine covers and components will need to be removed to gain access to difficult starter motor locations. On some engines, such as the General Motors Northstar engine, the starter motor is mounted at the top of the engine under the intake manifold (**FIGURE 54-27**). You may have to check the service information to locate the starter motor on some vehicles.

▶ TECHNICIAN TIP

Hybrid vehicles use both an ICE and electric motors to power the vehicle's drivetrain, although some manufacturers have released (so-called) hybrid vehicles that are driven entirely by an ICE and use only an idle-stop feature. Most hybrid vehicles use one or two high-voltage electric motors for engine start-up, auxiliary power, and regenerative braking functions. Regardless of the hybrid configuration, the ICE still requires a method to crank it over. On most hybrid vehicles, this task is undertaken by one of the hybrid's main electric drive motors, which can spin the ICE at a much faster and smoother rate, causing it to start almost instantly (**FIGURE 54-28**). On some hybrids, a more conventional ICE starter is also provided to start the engine if the main high-voltage battery bank is discharged.

FIGURE 54-27 Starter location on some engines in under the intake manifold.

FIGURE 54-28 Most hybrid vehicles use a high voltage electric motor/generator to start the ICE. This one replaces the engine's flywheel.

Idle-Stop/Start-Stop Systems

N54001

Starting systems have not changed a lot over the years. But one recent change is the addition of idle-stop/start-stop functionality. This system automatically shuts the engine off at times, typically when the vehicle is stopped, and then restarts the engine when needed. This significantly reduces CO_2 emissions and fuel consumption. For the most part, existing components are used, although in some cases they may be upgraded to handle the greater frequency of use. We will look at the three most common methodologies used in idle-stop systems, also called start-stop systems.

The simplest system, using unmodified components already on the vehicle, is on full hybrid vehicles that use a flywheel mounted motor/generator, or a geared motor/generator. This motor/generator is much more powerful than a regular starter motor, so it spins the engine over faster, which is ideal for idle-stop systems. In this case, the hybrid control module activates the motor/generator to spin the engine over when it determines that the engine needs to run. This provides a very smooth, quick start of the engine.

The second type of system uses an upgraded starter motor that is more powerful and robust than a regular starter. It is computer controlled so it can be activated to start the engine on demand. With its higher power, it can spin the engine over faster, making the engine start quickly. Some vehicles equipped this way use an integrated starter/alternator instead of just a starter. In this case, the alternator function may also be used for regenerative braking which recaptures some of the wasted braking energy.

The third type of system is unique since it doesn't rely solely on a starter motor. Instead, it ignites a combustible mixture in an appropriate cylinder to aid in starting the engine spinning. It can only do this if the engine is equipped with a gas direct injection system. Through crankshaft and camshaft position sensors, the PCM knows where each cylinder is at in the firing order as the engine is shutting down. To make sure the pistons stop in the same position every time, some systems manipulate the throttle plate (increase/decrease engine vacuum against pistons) and the alternator output (to increase/decrease the load on the crankshaft), as the engine shuts down and stops rotating. These functions cause the engine to stop at the same position every time. The PCM then knows which cylinder is on the power stroke, injects fuel into that cylinder, and then causes the ignition system to generate a spark in that cylinder. The burning air-fuel mixture pushes the piston down the cylinder, effectively starting to crank the engine over and fire other cylinders. At the same time as the first cylinder is being ignited, the starter motor is also engaged. Both processes work in tandem to crank the engine over faster than could be done separately, and ultimately start it faster. In this arrangement, the starter motor is also used for cranking the engine when it is cold.

No matter how the engine is cranked over, the process for stopping and starting the engine is similar. When the engine is up to operating temperature (including the catalytic converter), and the vehicle is being driven to a stop, the PCM shuts off the fuel injectors. This can happen on some vehicles as it is braking to a stop, but many vehicles wait until the vehicle has actually stopped for a few seconds. While the driver holds the brake pedal down, the engine stays off. On a non-hybrid vehicle, as soon as the brake pedal is released, the PCM commands the engine to be started, and the appropriate modules take over to crank and start the engine. On hybrid vehicles, when the driver releases the brake pedal, the engine may not need to be started, and the electric traction motor may power the vehicle away from the stop. The PCM will determine when the engine needs to be started, which may be down the road a ways.

Some standard transmission vehicles also incorporate stop/start functionality. One manufacturer controls the system in this way. As the vehicle is approaching a stop, the driver depresses the clutch pedal and moves the gear shift to neutral, and then releases the clutch. The engine goes into idle stop if the vehicle is either below a minimum speed, or stopped. The engine stays stopped until the driver depresses the clutch, which then signals the system to restart, allowing the driver to select a gear and drive away from the stop.

Starter Draw Testing

N54002, S54001

Testing starter motor current draw is a good indicator of overall starter motor performance. Manufacturers will specify the current draw for starter motors. When performing the test, it must be performed with a fully charged and correct capacity battery for the vehicle. Starter motors can be tested in two ways: on vehicle or off vehicle. The on-vehicle test is usually called a starter draw test, while the off-vehicle test is called a starter no-load test. Manufacturers will provide specifications for one or both of the tests.

Ideally, the starter motor should be tested under load, which gives you the best indication of any starter issues. This is called the starter draw test and can be accomplished while the starter motor is mounted in the vehicle. Starter current draw is at its highest when the engine is just starting to first rotate. As the starter motor and engine cranking speed increase, the current draw decreases and quickly stabilizes once the engine reaches full cranking speed. It is at this point that the amperage is read and then compared to specifications.

There is a variety of starter test equipment used to test starter draw. Each device will operate slightly differently, but all should have an inductive high-current ammeter to measure the cranking current flow and a voltmeter to measure the cranking voltage. The inductive ammeter is easier to use than a standard ammeter because it does not require any battery cables to be removed; it is quickly clamped around the main starter cable. When conducting the test, the engine must be disabled so it will crank but not start. The current flow and voltage will be measured during cranking and compared to specifications.

To test the starter draw, follow the steps in **SKILL DRILL 54-1**.

Testing Starter Circuit Voltage Drop

N54003

The electrical circuit of the starter motor consists of a high-current circuit and a control circuit. The control circuit activates the solenoid and can either be PCM controlled or non-PCM controlled. Voltage drop can occur across both the high-current and control circuits; however, the high-current circuit is more susceptible to voltage drop due to the much larger amount of current flowing in the circuit. We will look at testing the control circuit voltage drop in the next topic. One thing to remember is that when testing the voltage drop on the high-current side, for the measurement to be meaningful, the starter must be activated by the solenoid. The starter does not necessarily have to spin, but the solenoid must at least click when the ignition switch is engaged. If the solenoid does not click, then you will need to test the control side of the starter circuit.

A voltmeter or digital multimeter (DMM) is used to measure voltage drop across all parts of the circuit. A voltmeter with a minimum/maximum range setting is very useful when measuring voltage drop because it will record and hold the maximum voltage drop that occurs for a particular operation cycle.

SKILL DRILL 54-1 Testing the Starter Draw

1. Research the specifications for the starter draw test. Prepare the starter tester by setting it up to measure starter current. Connect the red lead to the positive terminal of the battery and the black lead to the negative terminal of the battery.

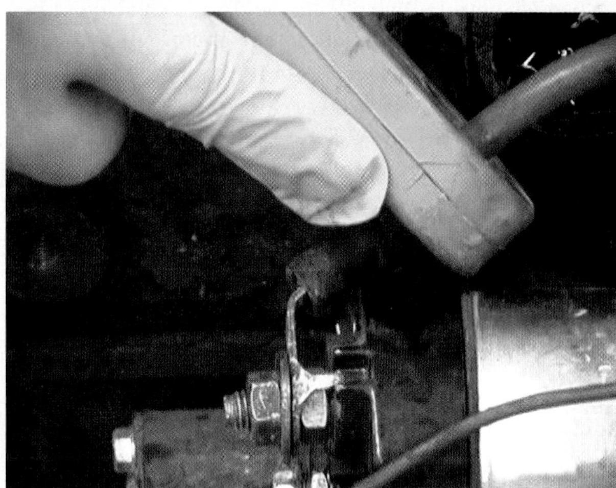

2. Connect the amps clamp around either the positive or the negative battery lead in the correct orientation. Make sure all of the appropriate wires are inside the clamp and the clamp is completely closed.

3. Disable the engine from starting by one of the following methods:

- Clear flood mode: This mode is programmed by manufacturers on many electronic fuel injection vehicles. It can be activated by holding the throttle down to the floor before turning the key. If the engine starts, lift your foot off the throttle and try another method.
- Pull the fuel pump relay and run the engine until it dies.
- Disconnect the fuel injectors or ignition coils.
- Disconnect the spark plug wires.

4. With the engine disabled, crank the engine and read the amps and volts as soon as the amps stabilize. Compare the readings to specifications and determine any necessary actions.

Refer to the Meter Usage and Circuit Diagnosis chapter for more information on DMMs.

Voltage drop is tested while the circuit is under load. The DMM is connected in parallel across the component or part of the circuit that is to be tested for voltage drop. Usually the most efficient method is to test large sections of the circuit first and, if required, narrow down the test to individual components to identify precisely where excessive voltage drop is located. For example, you can connect the black probe to the positive battery terminal and the red probe to the heavy gauge input wire leading into the starter motor, operate the starter with the engine disabled, and record the voltage drop across the positive side of the circuit while the starter motor is energized. Manufacturers will specify the maximum allowable

Applied Math

AM-13: Add/Subtract/Divide/Multiply: The technician can estimate the results of basic arithmetic operations and can accurately round numbers up or down.

The following scenario will give us a review of the application of basic arithmetic operations related to starter current draw. An automobile with a V8 engine has the manufacturer's specifications for starter current draw at 180 amps. The specifications allow for a tolerance of plus or minus 10%. The technician must determine whether the starter circuit will fall within manufacturer's specifications.

Let's determine what those numbers will be in terms of amps. 180 × .90 will give us the minus 10% factor. 180 × .90 = 162 amps. Now, for the plus 10% factor. 180 × 1.10 = 198 will give us the plus 10% factor. Now we know that the specifications for starter draw are 162-198 amps with 180 amps considered the ideal current draw, which could be rounded off to 160 and 200.

When the starter current draw is measured, it reads 310 amps. Our specifications were 198 amps maximum. If we subtract 198 from 310, we have 112 amps over the specifications, indicating a faulty starter motor or partially seized engine.

To accurately round numbers up or down, here are some tips:

- You can round numbers to give a rough idea of an amount or quantity.
- 862 bolts in stock are closer to 900 than to 800, so you could round to 900.
- A population of 76,310 could be rounded to 76,000.

voltage drop, but a rule of thumb is no more than 0.5 volts (500 millivolts) for a 12-volt circuit.

▶ TECHNICIAN TIP

On most vehicles, the starter cable is connected to the input of the solenoid, while the output of the solenoid is connected to the input of the starter motor. Since the heavy contacts are located between the input and the output terminals of the solenoid, where possible, it is best to measure the voltage drop from the positive battery post to the starter motor input (not the solenoid input). That way you are measuring any voltage drop across the solenoid contacts, which are a high-probability failure point.

The same test should also be performed on the negative side of the circuit. To do so, connect the black meter lead to the negative battery terminal, and the red lead to the starter housing, crank the engine, and read the voltmeter. Check the service information for safe methods of disabling the engine so it will not start.

To test starter circuit voltage drop, follow the steps in **SKILL DRILL 54-2.**

▶ TECHNICIAN TIP

A faulty battery will affect voltage drop tests, so always ensure that the battery is fully charged and in good condition before performing starter tests.

Inspecting and Testing the Starter Control Circuit

N54004

The starter control circuit activates the starter solenoid, which activates the starter motor. If there is a problem in the starter control circuit, the vehicle will likely not crank over at all, or maybe intermittently. The control circuit is made up of the battery, fusible link, ignition switch, neutral safety switch (automatic vehicles), clutch switch (manual vehicles) starter relay, and solenoid windings. If the starter is controlled by the PCM, then you must be aware of all of the circuits, such as the immobilizer circuit and the PCM itself.

SKILL DRILL 54-2 Testing Starter Circuit Voltage Drop

1. Set the DMM to volts. Connect the black lead to the positive battery post and the red lead to the input of the starter (not the input of the solenoid unless that is the only accessible terminal).

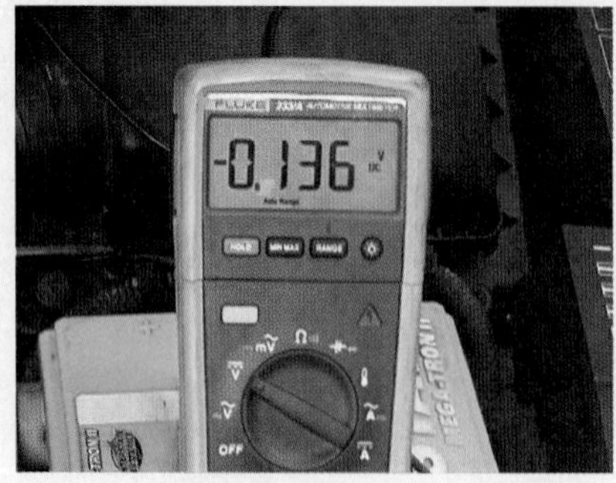

2. Crank the engine and read the maximum voltage drop for the positive side of the circuit.

3. Connect the black lead to the negative battery post and the red lead to the starter housing. Crank the engine and read the voltage drop. If the voltage drop is more than 0.5 volts on either side of the circuit, use the voltmeter and wiring diagram to isolate the voltage drop. Determine any necessary actions.

Before performing any tests, you should know and confirm the customer's concern. The manufacturer's wiring diagrams should be consulted to determine the circuit operation and to identify all components in the starter control circuit.

Once an understanding of the customer concern and circuit operation is obtained, it is time to test. Start by placing the DMM's red lead on the starter input terminal and the black lead on the battery positive terminal. Measure the voltage with the key in the crank position. At that point, assuming a fault in the control circuit is present, the voltage at the control circuit will be more than about 0.5 volts. If it is, you will need to perform voltage drop tests on the power side of the control circuit to determine which side(s) of the circuit the voltage drop is located on. If the voltage drop is less than 0.5 volts, then measure the voltage drop on the starter ground circuit. If the voltage drop is excessive, perform individual voltage drops on the ground leg. If both the control circuit power and the ground circuit voltage are within specifications, the resistance of the solenoid pull-in and hold-in windings will need to be measured. If out of specifications, the solenoid or starter motor and solenoid will need to be replaced.

To inspect and test the starter control circuit, follow the steps in **SKILL DRILL 54-3**.

Inspecting and Testing Relays and Solenoids

N54005

The starting system typically contains solenoids and relays that activate the control circuit. The solenoid is mounted on the starter motor, while the starter circuit relay is usually in or near the main fuse box with other similar devices.

Before performing any tests, ensure that the vehicle battery is charged and in good condition. The manufacturer's wiring diagrams should be checked to determine the circuit operation, identification, and location of all components in the starter circuit.

Relays must be tested in two or three ways depending on the relay. The simplest test is to measure the resistance of the relay winding. If it is out of specifications, the relay will need to be replaced. If it is OK, the contacts will need to be tested for an excessive voltage drop. The best way to do this is by using an adapter that fits between the relay and the relay socket. This will allow the normal circuit current flow to flow through the contacts so that a voltage drop measurement can be taken. Any excessive voltage drop across the relay contacts will require the replacement of the relay. The last test is used only on relays with a suppression diode in parallel with the relay winding. Connect a reasonably fresh 9-volt battery across the relay winding terminals in one direction, and then switch polarity by turning the battery around. If the diode is good, the relay should click in one direction and not in the other (**FIGURE 54-29**). If it clicks in both directions, the diode

FIGURE 54-29 Testing a relay with a 9v battery.

SKILL DRILL 54-3 Testing Starter Circuit Voltage Drop

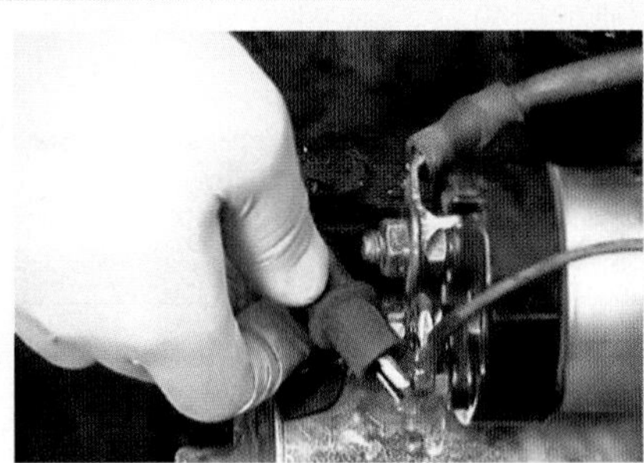

1. Use a DMM to measure the voltage drop on the positive side of the control circuit. Connect the red lead to the control circuit terminal on the solenoid and the positive battery terminal.

2. Crank the engine over and read the meter.

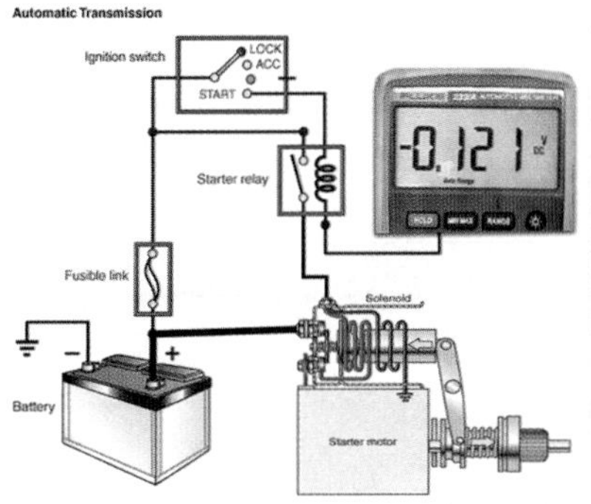

3. Measure the voltage drop on the ground side of the circuit like you did in step 3 of the previous skill drill. If the voltage drop is more than 0.5 volts on either side, use a wiring diagram to determine where to measure the individual voltage drops on that side of the circuit.

is open. If it does not click in either direction, the relay winding is open or the diode is shorted.

Solenoids can be difficult to test on the vehicle due to poor access, and tests will usually be limited to voltage and voltage drop tests on the main contacts. For other tests, such as pull-in and hold-in winding tests, the starter motor will usually need to be removed. Care should be taken when testing relays and solenoids to ensure that cables are not shorted to ground and the engine is not accidentally cranked over.

The first test to perform is a voltage drop test across the solenoid contacts. Place the red lead on the solenoid B positive input and the black lead on the solenoid B positive output. The voltage drop should be less than 0.5 volts. If not, replace the starter assembly.

Testing of the solenoid windings requires partial disassembly of the solenoid. Therefore, it is usually best to disconnect the control circuit connector from the solenoid and use a jumper wire to activate the solenoid. If the solenoid and starter operate, there is probably a fault in the vehicle's control circuit that needs further testing.

Use a jumper wire to apply battery voltage to the control circuit terminal on the solenoid and see if the solenoid clicks. If it does, then there is likely a fault in the control circuit wiring. If the solenoid still does NOT click (and the ground circuit is good), then the solenoid windings or starter brushes are likely worn (sometimes tapping on the starter while the key is turned to the crank position will free up the brushes enough that the pull-in winding can operate). If the solenoid or starter still does not work, then the starter is likely faulty and will need to be replaced.

To inspect and test relays and solenoids, follow the steps in **SKILL DRILL 54-4**.

SKILL DRILL 54-4 Inspecting and Testing Relays and Solenoids

1. To test a relay, measure the resistance of the relay winding and compare to specifications. If out of specifications, replace the relay.

2. Use a relay adapter to mount the relay on top of the relay socket so you can check the control circuit wiring and perform voltage drop tests on the contacts. Activate the relay while measuring the voltage across the relay winding. If it is near battery voltage, the control circuit wiring is okay.

3. Measure the voltage across the contacts with the relay NOT activated. This should read near battery voltage if both sides of the switched circuit are OK. If not, perform voltage drop tests on each side of the switch circuit. Activate the relay while measuring the voltage drop across the contacts. If it is more than 0.2 volts, the relay will need to be replaced.

4. To test a starter solenoid, measure the voltage drop across the solenoid contact terminals with the key in the crank position. If more than 0.5 volts, replace the solenoid or starter assembly.

5. If the solenoid does NOT click with the key in the crank position, remove the electrical connection for the control circuit at the solenoid.

6. Use a jumper wire to apply battery voltage to the control circuit terminal on the solenoid and see if the solenoid clicks. Determine any necessary actions.

Differentiating Between Electrical and Mechanical Problems

N54006

Failure to crank over properly, whether a slow-crank or a no-crank condition, can be caused by electrical or mechanical problems. For example, slow cranking could result from an electrical fault such as high resistance in the solenoid contacts. This problem could be resolved by replacing the starter with a new or remanufactured unit. But the slow-crank condition could also be caused by a mechanical engine fault such as a spun main bearing that is causing a lot of drag on the crankshaft, preventing the starter from cranking it over at normal speed. In this case, the entire engine will need to be rebuilt. As you can imagine, telling customers that they need a new starter motor when in fact they need a new engine (costing 20–40 times as much money) will not make them very happy with you. It is important to be able to differentiate between the two types of faults so that a wrong diagnosis can be avoided and the problem fixed appropriately the first time.

Typical electrical problems that can cause starting system problems include loose, dirty, or corroded terminals and connectors, a discharged or faulty battery, a faulty starter motor, or a faulty control circuit. Mechanical problems that may cause starting system problems include seized pistons or bearings, hydrostatic lock from liquid in the cylinder(s) (e.g., a leaky fuel pressure regulator or water ingestion during off-road operation), incorrect ignition or valve timing, a seized alternator or other belt-driven device, and so on. Gathering as much customer and vehicle information as possible will assist in narrowing down the options of what is causing the fault.

The order in which various tests are undertaken is determined by the likelihood of a particular fault occurring given the facts you gather. For example, if the vehicle is in a parking lot, there is a better chance that the fault is an electrical fault in the starting system than if the vehicle is out in the woods, in the middle of a huge puddle of mud and water, which would tend to move the odds toward a hydrolocked or seized engine. You will also need to consider the ease of conducting a particular test or performing a visual inspection for determining the fault. For example, it is easier to perform a starter draw test than it is to pull the oil pan and inspect the main bearings. Always check for the most common and easiest faults first.

If possible, measure starter draw.

- A slow crank accompanied with high starter draw may indicate low electrical resistance (short circuit) in the starter circuit or the starter itself, or high mechanical resistance from an engine fault such as a seized piston.
- A slow crank accompanied with low starter draw typically points to high electrical resistance in the starter circuit or the starter itself, or low mechanical resistance such as that resulting from a slipped timing belt reducing engine compression.

If an electrical resistance fault is suspected, check for excessive voltage drops or short circuits. Excessive voltage drops reduce current flow in the starter circuit. Locating voltage drops can be done by placing the voltmeter leads systematically in the starter circuit to narrow down the circuit to the actual fault. Use wiring diagrams to identify where you need to take your readings.

Short circuits can cause the starter draw reading to be higher than specifications. Use an ohmmeter to check the resistance of any starter circuit components, and compare them to specs. Also check for grounded wires by isolating them from components and checking to see if there is any resistance between them and ground. If so, electricity has found a path to ground and will have to be located. This can happen by disconnecting the harness connectors and continue checking for grounded wires. If the starter circuit wires are not grounded, suspect that the starter motor has an internal short circuit and will have to be replaced.

If a mechanical fault is suspected, check the oil and coolant for signs of contamination. If the coolant and oil are mixing, suspect a head gasket or cracked head/block issue. If the oil and coolant are not contaminated, turn the engine over by hand to see if it is tight compared to a similar known good engine. If it is harder to turn than it should be, remove the accessory drive belt, spin each of the accessories, and try to turn the engine over again. If still hard to turn over, you will have to go deeper in your visual inspection and start disassembling components based on the information you have gathered along the way. For example, if the crankshaft cannot be turned a complete revolution, remove the spark plugs and check whether liquid is ejected out of one or more spark plug holes. If so, the engine was hydrolocked, and you will have to determine the cause. If no liquids are ejected, then you will have to disassemble the engine further until you determine the cause of the mechanical resistance.

If the engine is easier to turn over than normal, then you will probably need to check the compression on a few cylinders and compare that to specs. If low, perform a wet compression test to see if that is causing the low compression. If the wet test doesn't bring the compression back to normal, check the valve timing to see if it is causing the low compression.

The important thing to remember is that slow-crank and no-crank conditions can be caused by both electrical and mechanical faults, so don't jump to conclusions. You need to identify the root cause of the fault so you can advise the customer on what is needed to repair the vehicle.

Removing and Installing a Starter

N54007

Starter motors are usually located close to the flywheel end of the engine. They can be in difficult-to-reach locations, and some engine components or covers may need to be removed to gain access. In most cases, the starter can be accessed more easily from underneath the vehicle. Always disconnect the negative battery lead before attempting to remove or install a starter motor, and be sure to use a memory minder if specified by the manufacturer.

To remove and install a starter in a vehicle, follow the steps in **SKILL DRILL 54-5**.

Applied Math

AM-2: Decimals: The technician can add decimal numbers to determine conformance with the manufacturer's specifications.

A starter has been rebuilt, and the technician wants to check the pinion clearance. A feeler gauge is used to determine this clearance. Manufacturer's specifications for this clearance are from 0.010" to 0.140", with 0.070" considered the midpoint. The technician's feeler gauge set only goes to 0.045", which fits too loosely in the gap. He places a 0.025" blade next to the 0.045" blade, which, together, fit in the gap with just the right tension. The selected gauges are 0.025" plus 0.045", to equal a total of 0.070".

In this example, we are working with decimal numbers. Decimals are numbers that are expressed using a decimal point.

▶ Charging Systems

K54004

Compared to older vehicles, modern vehicles are increasingly dependent on electronic and electrical systems that require a constant and reliable supply of electrical power. Alternators (sometimes called AC generators, or just generators) supply the electrical energy required for modern vehicles whenever the engine is operating (**FIGURE 54-30**). The terms "generator" and "alternator" are often used interchangeably to describe the electrical generating component. Strictly speaking, the DC generator has not been used on most vehicles since the 1960s. Thanks to solid-state electronics and circuitry, alternators have taken over due to their superior operating characteristics, including the following:

- Greater wattage output at lower rpm
- Higher rpm

SKILL DRILL 54-5 Removing and Installing a Starter

1. Disconnect the negative terminal of the battery after determining whether a memory minder is required.

2. Remove any engine covers or components required to gain access to the starter. Remove starter motor electrical connections, noting how the wires were routed. In some vehicles, the wires cannot be accessed until the starter is removed.

3. Loosen the starter motor mounting bolts, and remove them while holding the starter so it does not fall.

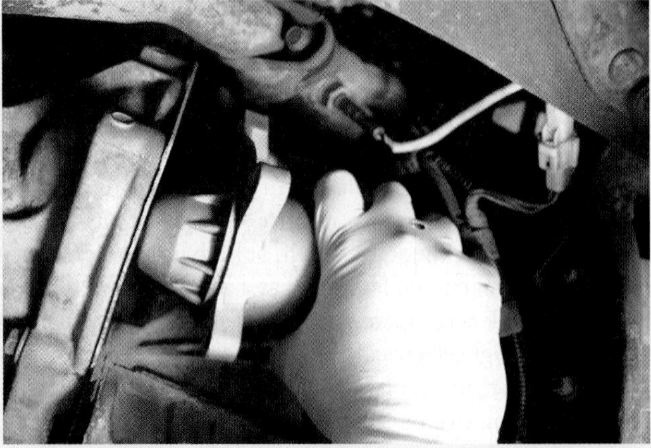

4. Remove the starter motor, being careful to catch any shims that might be between the starter and the block. Reinstall the starter motor by reversing these steps, and verify its proper operation.

FIGURE 54-30 Typical alternator installed on engine.

- Smaller physical size and weight for a given output
- Greater reliability and longer service life (brushes)

Both DC generators and alternators produce electricity by relative movement between conductors and a magnetic field, which induces an electrical potential or voltage within the conductors. The key difference between an alternator and a DC generator is which component rotates or moves to generate electricity. In the DC generator, the conductors that generate power rotate as part of the armature, which rotates within a magnetic field created by the stationary pole shoes (**FIGURE 54-31**). In the alternator, the magnetic field is created by the rotor, which rotates within the stationary stator windings to generate electricity there. In both cases, there is relative movement between the magnetic field and the conductors.

The main parts of the charging system include the battery, the alternator, the voltage regulator (internal or external), a charge warning light, and wiring that completes the circuits. The battery stores an electrical energy in chemical form, acts as an electrical dampening device for variations in voltage or voltage spikes, and provides the electrical energy for cranking the engine. Once the engine is running, the alternator—which is connected to the engine and driven by a drive belt—converts some of the engine's mechanical energy into electrical energy to operate all the electrical components of the vehicle. The alternator also charges the battery to replace the energy used to start the engine. The voltage regulator circuit maintains optimal battery state of charge by sensing battery voltage and then controlling the amount of electrical energy the alternator creates.

Older vehicles have separate (discrete) regulators mounted on the firewall (**FIGURE 54-32**). The next generation of charging systems included regulators that were incorporated inside the alternator. More recently, the PCM is used to control the charging system more efficiently by controlling alternator output based on a number of parameters such as electrical load, engine load and rpm, alternator capability, battery temperature, fuel economy benefits, and more.

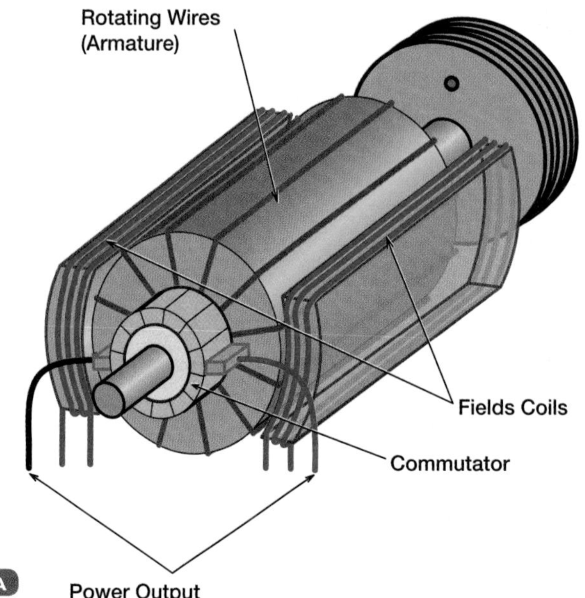

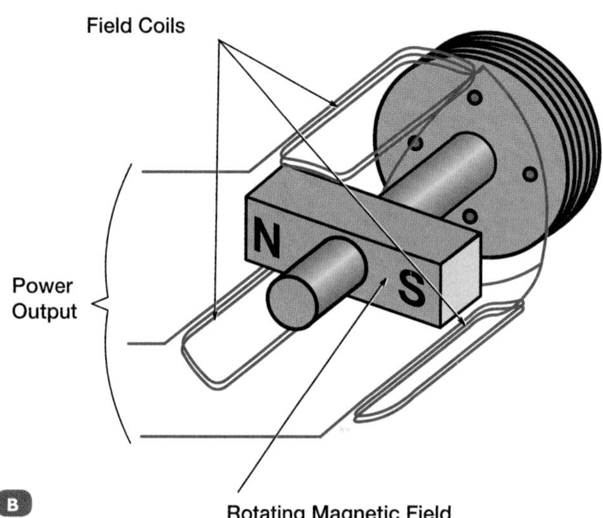

FIGURE 54-31 A. DC generators use rotating wires in a stationary magnetic field. **B.** Alternators use a rotating magnetic field in stationary wires.

Electromagnetic Induction in Alternators

The alternator converts mechanical energy into electrical energy by electromagnetic induction. In a simplified version, a bar magnet rotates in an iron yoke, which concentrates the magnetic field. A coil of wire is wound around the stem of the yoke. As the magnet turns, voltage is induced in the coil, producing a current flow (**FIGURE 54-33**). When the north pole is to the right and south is to the left, voltage is induced in the coil, producing current flow in one direction. As the magnet rotates, and the positions of the poles reverse, the polarity of the voltage reverses as well, and as a result, so does the direction of current flow. Current that changes direction in this way is called alternating current, or AC. In this example, the change in direction occurs twice for every complete revolution of the magnet.

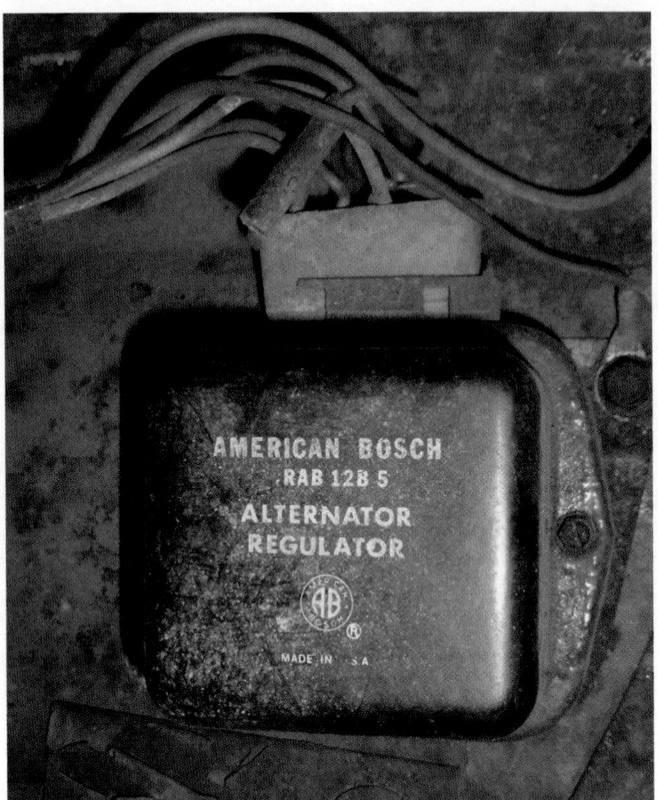

FIGURE 54-32 External voltage regulator.

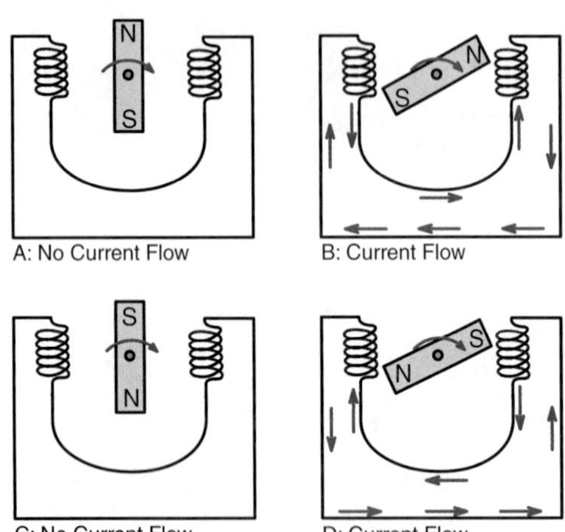

FIGURE 54-33 Electromagnetic induction.

The value of the electromotive force (EMF) or voltage potential induced by an AC generator depends on four factors:

- The strength of the magnetic field—increasing the strength of the magnetic field increases the value (voltage output) of the induced EMF
- The speed at which the magnet rotates
- The relative distance between the magnet and conductors
- The number of turns of wire on the stationary coil

A single-phase AC generator has only one stationary coil, which creates a single sine wave (**FIGURE 54-34**). In a typical automobile alternator, three or possibly four separate coils of wire, or phase windings, are common. The windings are arranged so that when the magnet is rotated, it generates a three-phase (or four-phase) output. The phases are equally spaced in time, and this results in a phase shift of, in the case of a three-phase AC generator, one phase every 120 degrees (**FIGURE 54-35**).

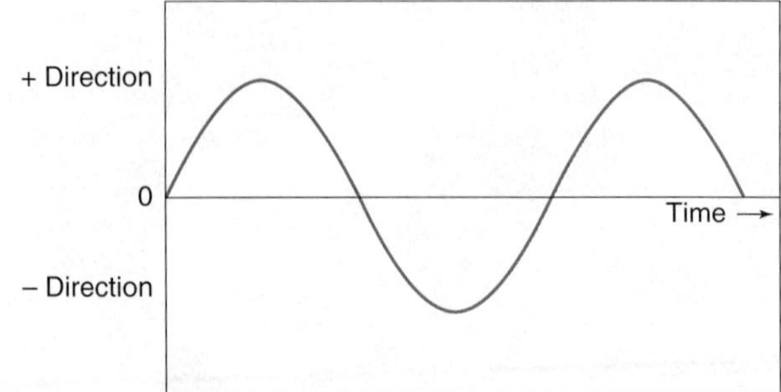

FIGURE 54-34 Single-phase AC signal.

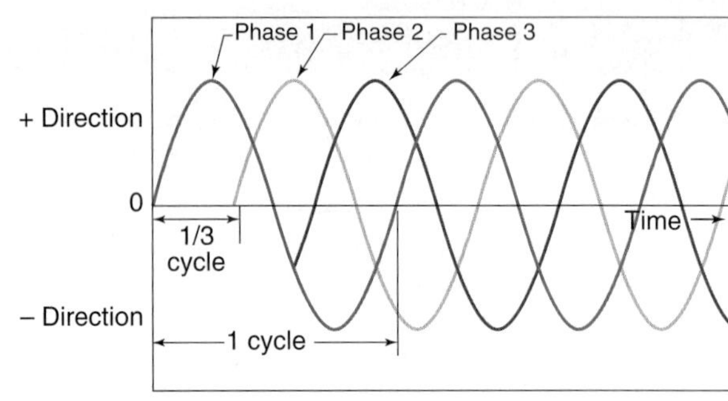

FIGURE 54-35 Three-phase AC signal.

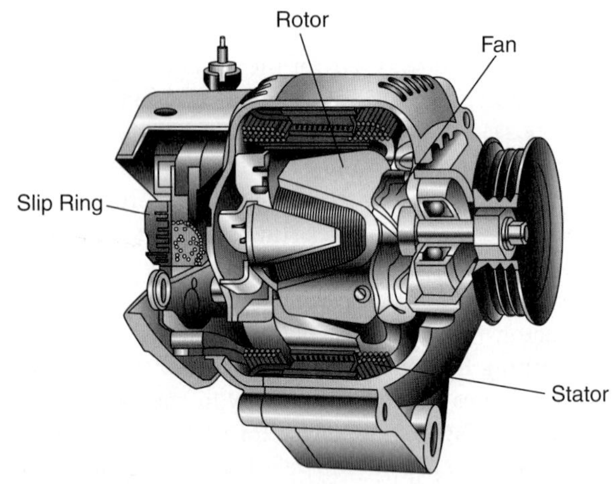

FIGURE 54-36 The alternator.

Alternator Component Overview

K54005

The alternator consists of a stationary winding assembly called the stator, a rotating electromagnet called the rotor with a slip ring, a brush assembly, a rectifier assembly, two end frames, and a cooling fan and drive pulley (**FIGURE 54-36**). A voltage regulator monitors battery voltage and varies current flow through the rotor field circuit, thus controlling the strength of the magnetic field of the rotating magnet. Field current varies as required to perhaps 5 or more amps to keep up with the output demand of the electrical system of perhaps 150 or more amps for a typical automobile. The voltage regulator's job then is to control the output of the alternator so that the system voltage is maintained within specified limits. Regardless of the make or model, the fundamental construction and components of alternators are very similar.

Rotor

The rotor is an electromagnet that rotates freely in the alternator. The rotor is supported on each end by ball bearings. It consists of a coil of insulated wire wound around an iron core and pressed onto a steel shaft. An iron pole piece is fastened on each side of the coil assembly so that the projections or claws (which are bent) interlace. The ends of the rotor coil winding are connected to insulated slip rings mounted on the shaft. The spring-loaded brushes maintain contact with the slip rings at all times so that current can flow into and out of the rotor winding.

When a current is passed through the slip rings and the coil winding, it establishes strong north and south poles at the ends of the iron core and the shaft (**FIGURE 54-37**). The projections then take on the same polarity as the end of the shaft on which they are mounted. This forms pairs of north and south poles that are alternately spaced around the rotor circumference. The rotor usually has 8 to 12 poles, which are tapered to reduce noise and create a smooth AC sine wave output as the rotor rotates. When the rotor is fully energized, the magnetic field is at its strongest and induces maximum current flow in the stator. When the rotor is fully energized, it requires a fair amount of mechanical energy to rotate the rotor inside of the stator. In fact, a high-current alternator can use more than 5 horsepower from the crankshaft to operate. Thus, any sizable electrical load affects fuel economy. Also, a fan is either pressed or slipped onto the rotor shaft to assist in cooling the rotor and stator windings and rectifier assembly.

Slip Rings and Brush Assembly

Slip rings and brushes aid in making an electrical connection to the rotating rotor assembly (**FIGURE 54-38**). The slip rings are normally copper bands, molded onto an insulating material, and then pressed onto the steel shaft of the rotor. Each end of the rotor winding is connected to one of the copper bands, so that as the rotor rotates, the brushes maintain a constant connection with each end of the winding; the brushes carry the current from the stationary end frame to the moving rotor. Brushes are made of a combination of copper and carbon and are carried in brush holders mounted in the end frame of the alternator. They are spring loaded to maintain contact with the slip rings, and they wear out over time.

Brushless Alternators

Because alternator brushes are a wear item that can cause breakdowns when they wear out, some manufacturers have designed brushless alternators. Brushless alternators induce current flow into the rotor through one stationary field winding in the housing and a separate armature on the rotor (**FIGURE 54-39**). The induced exciter current flow that is created in the rotor armature is AC. A rectifier on the rotor turns the AC into DC, which is supplied to the main rotor winding, where it is used to create the rotor's electromagnetic force. The main current in the rotor is controlled by the strength of the exciter field. The greater the current flow in the exciter field, the greater the exciter magnetic field, and the greater the current flow in the exciter armature; and the greater the current flow in the main rotor winding, the greater the strength of the rotor's magnetic field, and the greater the alternator output.

Stator

The stator consists of a cylindrical, laminated iron core, which carries the three (or four) phase windings in slots on the inside (**FIGURE 54-40**). The windings are insulated from each other and also from the iron core. They form a large number of conductor loops, which are each subjected to the rotating magnetic fields. The stator is mounted between the two end housings, and it holds the stator windings stationary so that the rotating magnetic fields can permeate the windings.

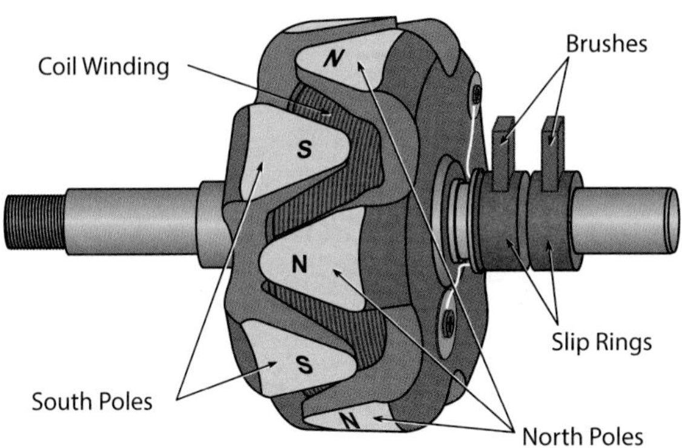

FIGURE 54-37 The rotor has projections that create alternating north and south magnetic fields.

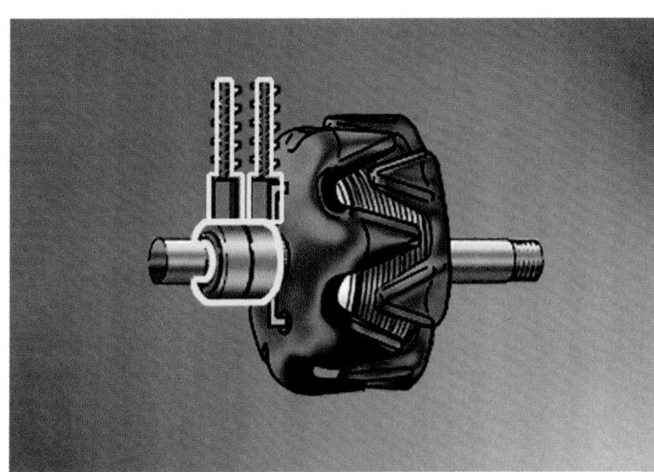

FIGURE 54-38 Slip rings and brushes aid in making an electrical connection to the rotating rotor winding.

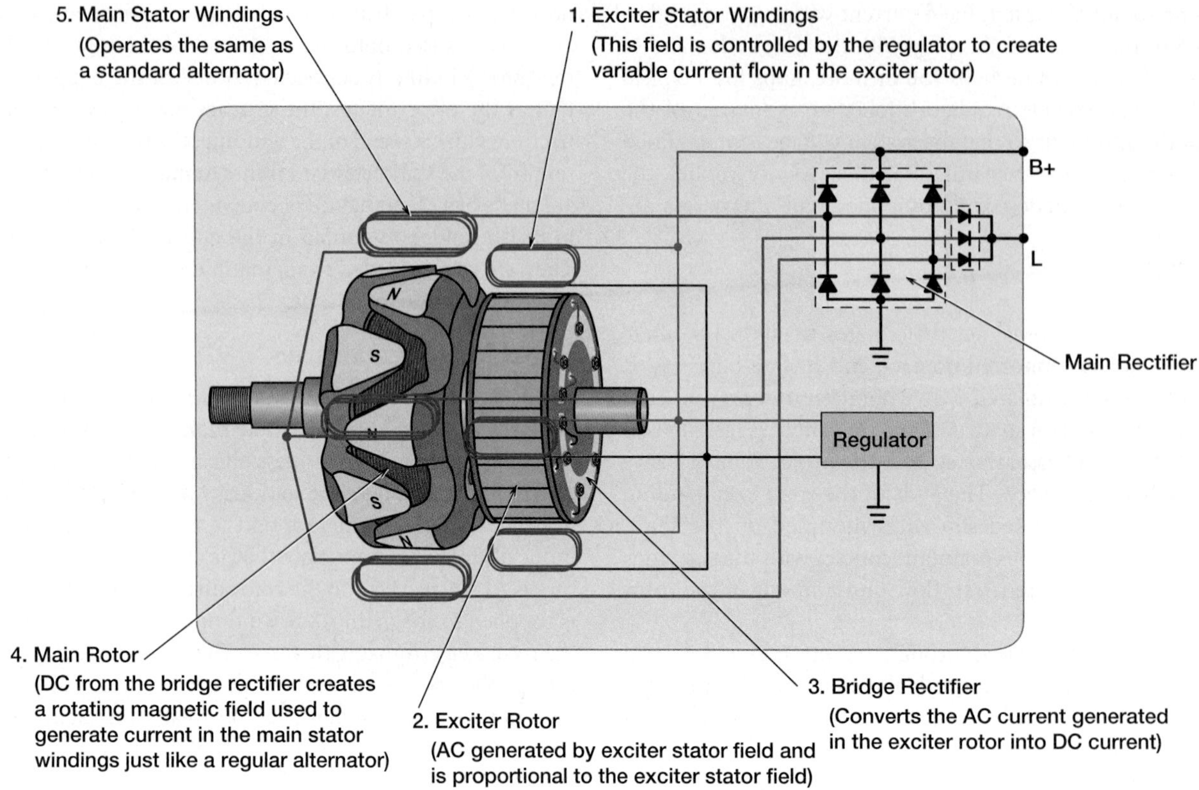

FIGURE 54-39 Brushless alternators use a separate stationary field winding in the alternator housing and a separate armature on the rotor to induce current flow into the rotor. This current flow then controls alternator output just like in a regular alternator.

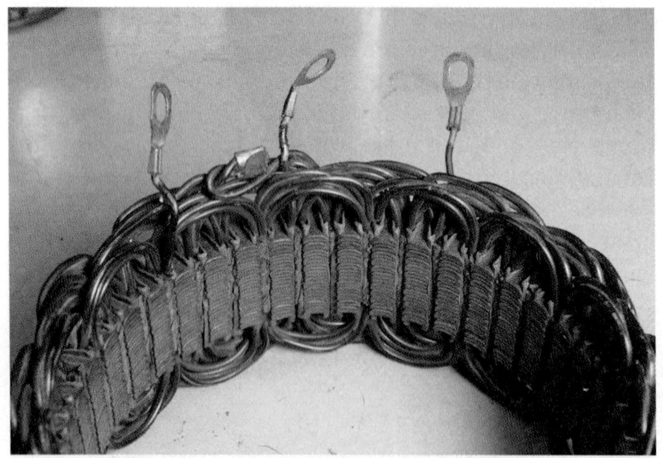

FIGURE 54-40 The stator consists of a cylindrical, laminated iron core, which carries the three-phase windings in slots on the inside.

Two methods of connection are typically used for the stator or phase windings: wye and delta configurations (**FIGURE 54-41**). It all depends on how much alternator output is required. In the wye method of connection, one end of each phase winding is taken to a central point where the ends are connected together. This is known as the star, or neutral, point. The other end of each winding is connected in the bridge rectifier circuit between a positive and a negative diode. Each winding is then always part of a complete circuit,

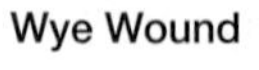

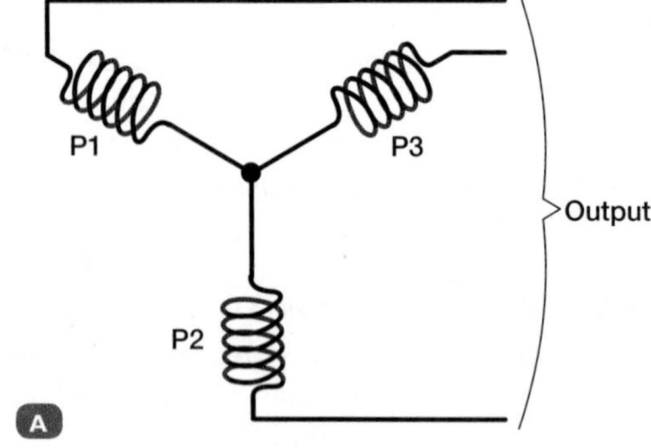

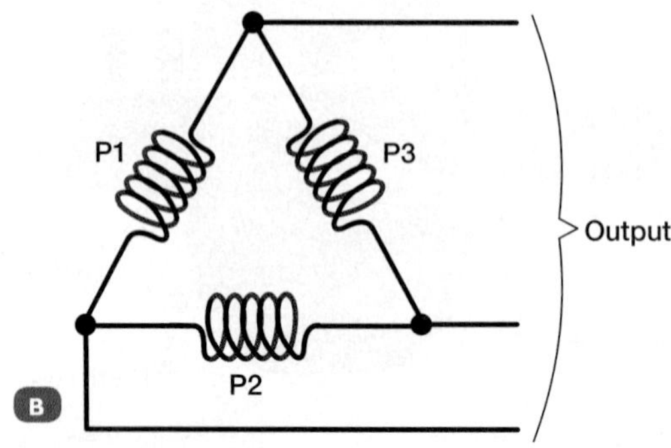

FIGURE 54-41 Phase windings. **A.** Wye. **B.** Delta.

and when current is flowing, it is always flowing through two series windings at the same time.

In the delta method, the windings are connected in the shape of a triangle. Connections are then taken from each point of the triangle, straight to the bridge circuit, between the positive and negative diodes. In this arrangement, only one winding sits between two legs of the rectifier.

Alternator End Frames and Bearings

The alternator housings are typically constructed from aluminum with vents within the frames to provide for a large amount of airflow to assist in dissipating heat (**FIGURE 54-42**). The housings accept the bearing assemblies, which support the rotor at the drive and slip ring ends. A pulley that is driven by a belt is mounted at the drive end of the alternator. Some housings also accept some or all of the diodes, which are pressed into holes in the housing.

Alternator Cooling Fan and Pulley

The alternator's cooling fan is a powerful centrifugal type of fan (**FIGURE 54-43**). It is mounted on the rotor shaft and may be an integral part of the drive pulley or part of the rotor. It is essential to maintain a cooling stream of air over the diodes and stator. Due to the short length of the rotor, the cooling air needs to be spiraled in. This creates a long cooling path for the air and helps maintain component temperatures within the manufacturer's specifications. To achieve this spiraling effect, the cooling fins on the plate have different openings—small and large. To get the maximum cooling effect, the alternator fan must be driven in the correct direction. Refer to the manufacturer's specifications to determine the correct direction if removing and replacing the fan.

A new feature that has been added to alternators over the past number of years is the overrunning alternator pulley (OAP) and overrunning alternator decoupler (OAD) (**FIGURE 54-44**). They are also known as alternator decoupling pulleys (ADPs). These devices are in place of standard pulleys on alternators. They lock in one direction to drive the alternator while the engine is accelerating or at cruise, and they freewheel when the engine is decelerated. They provide the following functions:

- Reduce belt noise and vibration
- Reduce stress placed on the tensioner and belt
- Extend belt and tensioner life
- Improve fuel economy

ADP's are prone to wear just like the tensioner and belt, so they need to be inspected and replaced when faulty. It is also good practice to replace the ADP for maintenance purposes when the tensioner and belt are replaced.

To diagnose an ADP, a couple of tests can be performed. First, you can run the engine at approximately 2000 rpm and turn the engine off while listening for a clicking, popping, or buzzing noise coming from the ADP. The second test involves using special tools to turn the ADP (with the belt still on) in both directions. It should turn smoothly in the freewheeling direction with less than the maximum specified torque. Also, it should *not* turn in the other direction. If any faults are found with the ADP, replace it along with the belt and tensioner. Make sure you replace the ADP with the specified replacement part. They are *not* interchangeable.

FIGURE 54-42 End frames and bearings.

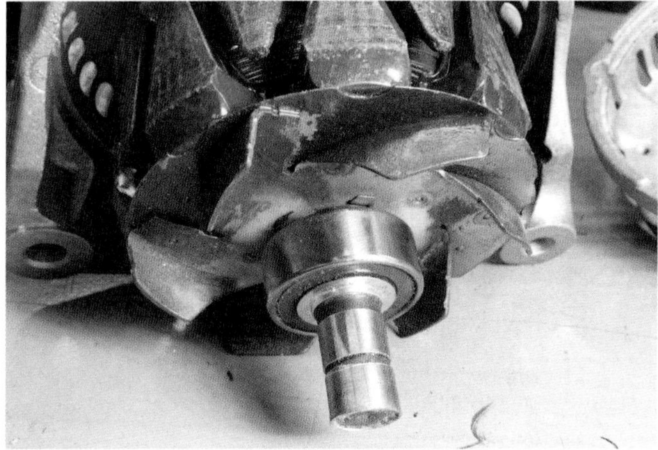

FIGURE 54-43 Alternator fan and pulley.

FIGURE 54-44 Typical overrunning alternator decoupler (OAD).

Rectification

K54006

Rectification is a process of converting AC into the DC that is required by the battery and nearly all of the automobile systems. To change AC to DC, automotive alternators use a rectifier assembly consisting of diodes in a specific configuration. Remember that a diode allows current to flow in one direction but blocks the flow of current in the other direction. A so-called three-phase "bridge" rectifier has a minimum of six diodes (three positive and three negative) to rectify the AC output of the stator windings to DC output from the rectifier. A diode bridge gets its name from two diodes in series, bridged with a wire. If one end of a stator winding is connected between the diodes on the bridge, then the current flows through the ground diode into the stator winding and out of the other end of the winding and through a positive diode. In a three-phase alternator, one side of each stator winding is connected to one of the bridges. Because the other end of each winding is connected to at least one other winding, a complete path from the negative diode through a stator winding(s) and the positive diode can be completed as needed.

As the rotor rotates, each time the magnetic field changes from north to south and south to north, the polarity of each phase winding reverses, and as a result, the current changes direction. No matter what direction the current is flowing in the stator windings, the diodes in the rectifier only allow current to flow into the rectifier and out of the rectifier in one direction (DC). So the diodes take the current flowing no matter what direction through the stator windings and use the current flow (even if it is reversed) to push electrons out of the rectifier's positive terminal. For example, as the rotor turns, it induces a voltage in a stator winding, which generates current flow in one direction. In this position, and with this polarity, the current path is as follows: output of winding A, positive diode A, alternator terminal B-positive, positive battery terminal, battery ground (B-negative), alternator ground, negative diode B, output of winding B, neutral or star point (**FIGURE 54-45**).

When the rotor's magnet rotates further, windings B and C come under the influence of the magnetic field. The current path then is as follows: winding C, at star point, winding B, positive diode B, alternator terminal B-positive, positive battery terminal, battery ground, alternator ground, negative diode C, output of winding C (**FIGURE 54-46**). As the rotor moves through its various positions, individual phase currents change in magnitude and polarity, but the output current to the battery and the electrical circuits remains in one direction only. This is because the individual phase windings are set 120 degrees apart (three-phase system).

Thus, two things are happening. First, the three phases are split so that as the voltage in one phase is falling, the voltage in the next phase is still rising. As the second phase is falling, the third phase is rising. And as the third phase is falling, the first phase is rising again (**FIGURE 54-47**).

Second, by redirecting the current so that it is flowing in the positive direction, the diodes effectively flip all of the negative current flow activity, from below the neutral point of the graph to the positive side, which smoothes out the power flow even further (**FIGURE 54-48**).

The top of the waveform is called the alternator ripple. A ripple that is consistent across each winding indicates that the stator windings and diodes are each creating current flow and voltage consistently. Inconsistent ripple indicates a high-resistance fault in either the diodes or the stator windings. Study the illustrations carefully so that you understand the role the diode bridge plays in providing the relatively smooth DC output required by the automobile's systems.

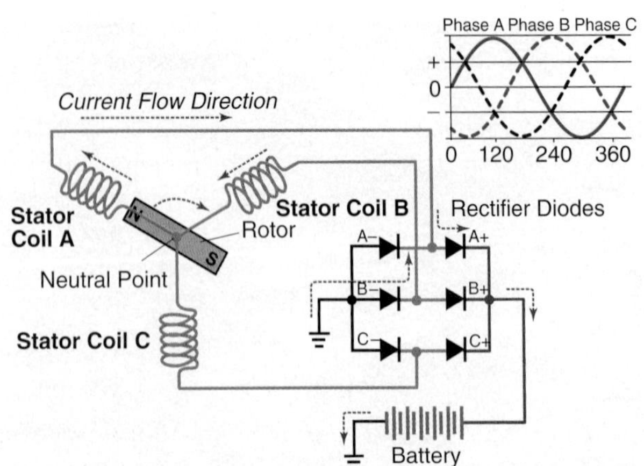

FIGURE 54-45 Current flow through a single phase in the forward direction.

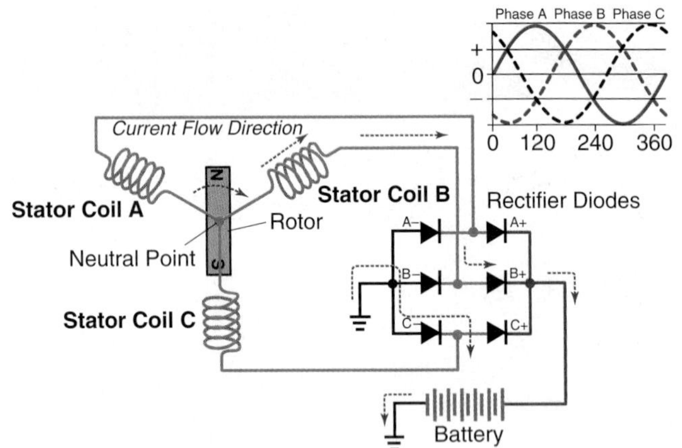

FIGURE 54-46 Current flow through a single phase in the reverse direction.

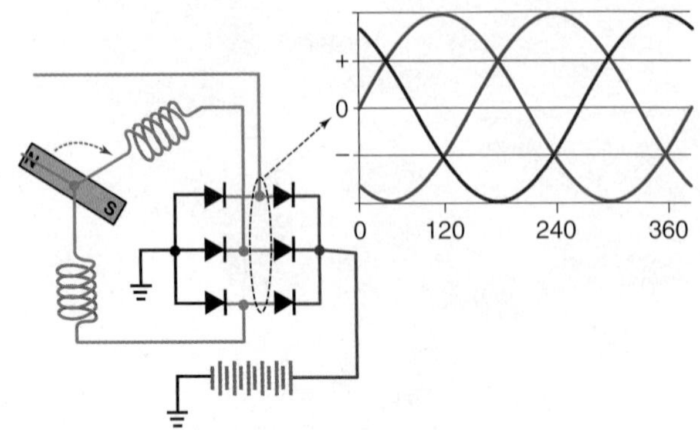

FIGURE 54-47 Three phases—not rectified.

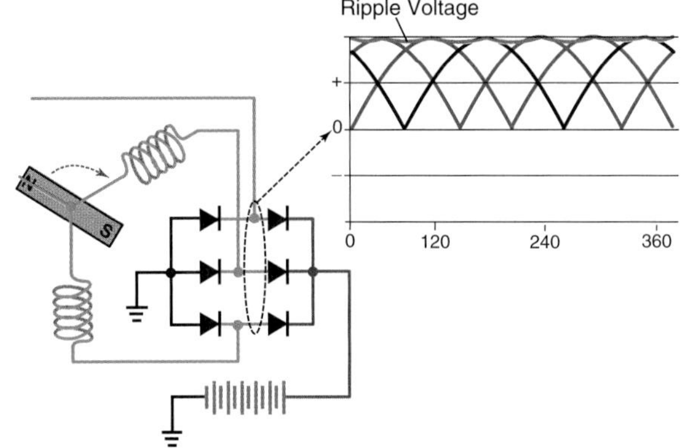

FIGURE 54-48 Three phases—rectified.

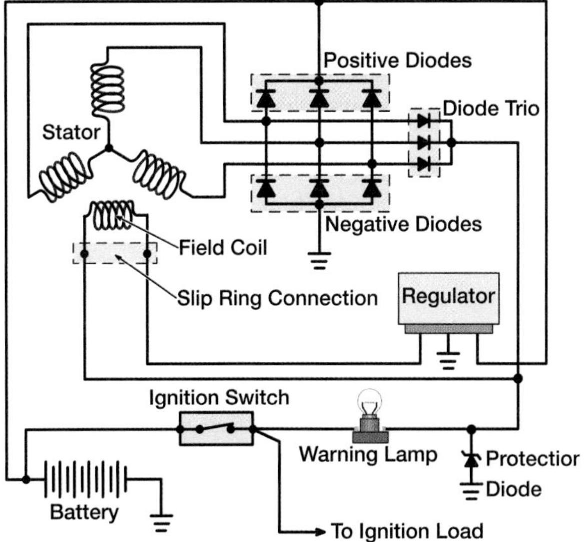

FIGURE 54-49 Positive and negative diode schematic, including diode trio.

Rectifier Assembly

The diodes for rectification (converting AC to DC) are mounted on heat sinks to assist in dissipating the heat generated in the diodes. Each diode drops approximately 0.7 volt and thus tends to run hot. The diodes must be properly cooled to avoid premature failure. Three or more diodes are mounted on each heat sink. One heat sink has the positive diodes, and the other has the negative diodes (**FIGURE 54-49**). The positive diode heat sink is insulated from the frame and is connected through the output terminal to the positive battery terminal. The negative diode heat sink is connected to the frame, which allows the return circuit, via the negative battery terminal, to be completed. Some alternators use a diode trio assembly to supply power to the field windings once the alternator is charging. Each of these diodes is connected to one end of a stator winding.

Voltage Regulation

K54007

The alternator's output is determined by the amount of current flowing through the rotor. The greater the current flow through the rotor, the stronger the magnetic field generated. The stronger the magnetic field, the stronger the alternator output. The weaker the magnetic field, the weaker the alternator output. The voltage regulator monitors battery voltage and adjusts the current flow through the rotor appropriately. This controls the strength of the electromagnet.

When the engine is running and voltage output is low, the regulator allows more current to flow through the rotor field winding. This increased current flow strengthens the magnetic field, which raises the induced voltage in the stator windings, and alternator output then increases. As the output voltage increases to the maximum regulated voltage, the voltage regulator reduces the current flow through the rotor, reducing magnetic strength and alternator output. Regulator switching takes place in milliseconds, so the voltage to the battery is fairly constant. This process occurs quickly enough to maintain consistent system voltage even as loads are switched on or off. Today's field circuits are electronically controlled by a pulse-width-modulated signal, for even smoother regulation and quicker control.

A-Type or B-Type Regulating Circuit

Switching the field current may be done on the positive side of the rotor windings (B type) or on the ground side (A type) (**FIGURE 54-50**).

- In an A-type regulating circuit, alternator B-positive output is fed directly to the rotor; voltage regulation is done on the ground side of the field.
- In a B-type circuit, the voltage regulator is on the positive side of the field; the ground is constant.

Knowing the difference is necessary in order to properly bypass a voltage regulator (full-fielding) when diagnosing some charging systems. For example, many alternators are designed with A-type circuits and a grounding tab on the regulator. The grounding tab provides a method of full-fielding the charging system with the engine running. The tab is grounded by placing a small metal rod or screwdriver on the metal tab and then shorting it out against the alternator frame (**FIGURE 54-51**). This grounds the A-type field, causing full current to flow through the rotor, creating full alternator output. *Caution:* This procedure puts the charging system in an unregulated condition, which can cause the system voltage to increase high enough to damage electronic components. Only perform this test at idle, and then only for a few seconds.

If the charging system is a B type, then battery power has to be fed to the field. This is usually accomplished by disconnecting the regulator connector and then applying power with a jumper lead to the correct terminal. Make sure you do not mix up the types of charging systems or supply power or ground to the wrong place, as you can damage the electronic components.

Rotor Circuit Control

Current is provided to the rotor by means of the copper slip rings on the rotor and brushes. The rotor can be supplied with power in a variety of ways. On older vehicles, rotors on many vehicles were supplied power by means of three extra diodes,

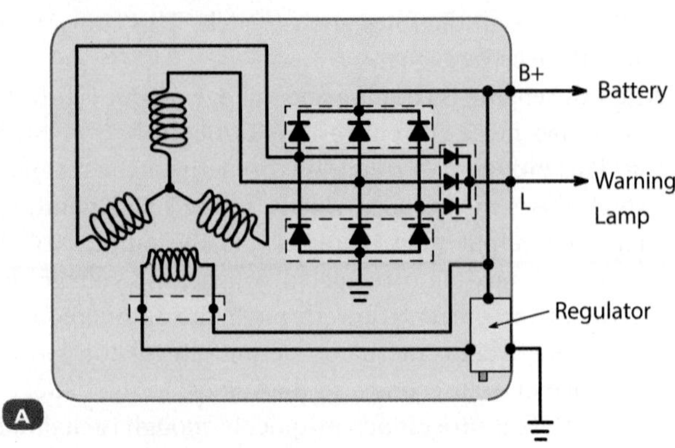

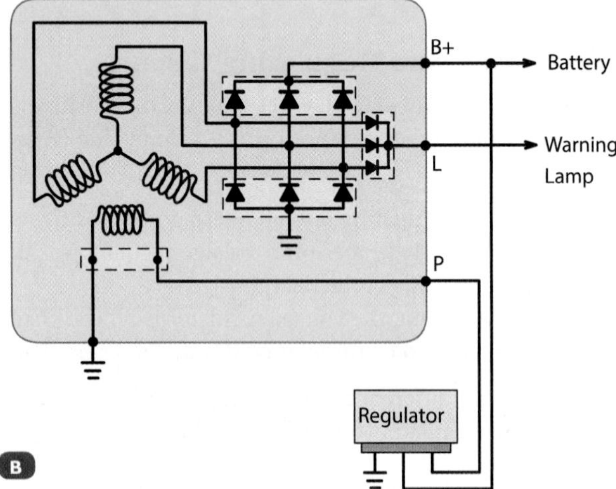

FIGURE 54-50 **A.** Regulator circuit—A Type. **B.** Regulator circuit—B type.

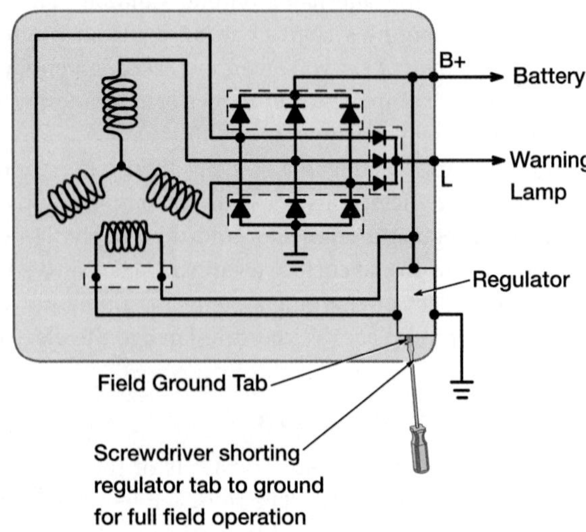

FIGURE 54-51 Full-fielding an A-type regulator circuit.

called a diode trio, and connected to the bridge rectifier circuit (**FIGURE 54-52**). These extra diodes are known as field diodes or exciter diodes, and the alternator is said to be self-exciting. However, this self-excitation can only occur when the alternator is producing an output.

When the ignition is on, current flows from the positive battery terminal through the ignition switch and charge indicator lamp to the L terminal of the alternator (**FIGURE 54-53**). The circuit is completed through the slip rings and rotor field winding, and through the voltage regulator, to ground on the vehicle frame.

The small amount of current flowing in the circuit illuminates the indicator lamp and provides the initial excitation of the field winding. This slightly magnetizes the rotor iron claw pole shoes and produces a weak magnetic field to get the generation process started.

When the rotor is driven (turned by the pulley) from the engine crankshaft, the rotating magnetic field induces a voltage in the stator phase windings, which is then fed to the bridge rectifier to create DC. From there, DC goes to the B-positive alternator output terminal. Stator output is also fed to the exciter diodes so that DC current can flow to the field circuit, restricted only by the resistance of the field winding (and the regulator circuit). This strengthens the magnetic field, and the output voltage rises quickly.

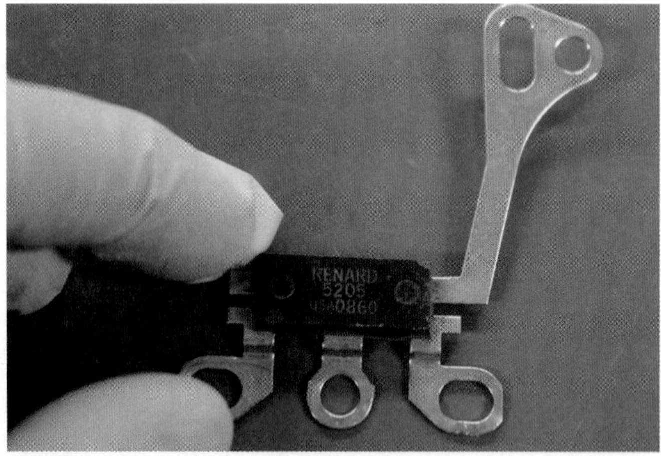

FIGURE 54-52 Diode trio supplies power to the rotor in some alternators.

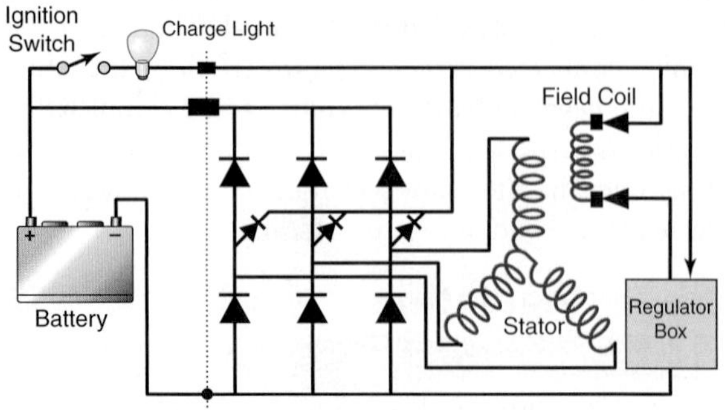

FIGURE 54-53 Current flow from the charge indicator lamp to start the alternator charging.

Applied Science

AS-72: Voltage: The technician can demonstrate an understanding of and explain system voltage generation, uses, and characteristics.

Voltage generation in the majority of modern automotive applications relies on an engine-driven alternator. Using the principle of electromagnetic induction, alternators generate alternating current (AC) electricity through the use of a moving magnet within a coil of wire. AC electricity must be converted to direct current (DC) before being used in automotive systems or stored in a battery. The AC electricity is "rectified," or converted to DC, by being passed through a bridge of diodes mounted in the alternator. Diodes are small solid-state devices that allow current flow in only one direction. The resulting DC electricity can be used to power various vehicle systems, including processing modules, electric motors, and lights, and can also be used to charge the vehicle's battery.

Generators, which produce DC electricity, were once commonly used in automotive applications but were replaced by alternators when small, inexpensive diodes became available. Alternators are more practical, simpler to produce, and require less-intensive maintenance.

The voltage regulator takes control of the field circuit to limit field current to maintain a preset regulated output voltage of approximately 14 volts at the B-positive terminal. As the voltage on each side of the charge indicator lamp is now equal (between the lamp and the diode trio), there is no current flow through the lamp, and the lamp is turned off. The alternator is now charging, and because the output voltage at the B-positive alternator terminal is greater than that of the battery, current flows to the battery to begin the recharging process.

Voltage Regulator

The voltage regulator in modern vehicles is a solid-state electronic device. It may be installed inside the alternator or, for older vehicles, mounted on the fender well as a separate component. The internal regulator as a whole is a replacement part and is not usually serviceable (**FIGURE 54-54**). The regulator's electronic circuit senses the battery voltage and switches the rotor circuit on and off rapidly in order to maintain a constant voltage output up to the alternator's maximum output current.

FIGURE 54-54 Internal regulator mounted inside the alternator.

Modern vehicles control the alternator output voltage from within the PCM or BCM. The control principle is the same: Output voltage is controlled by switching the rotor's field circuit on and off. However, the PCM makes voltage adjustments based on a larger number of inputs, such as engine and ambient temperature, battery temperature, engine cranking, engine load, desired regeneration during deceleration, and electrical load. In most cases, an alternator that is PCM controlled also uses an internal regulator. The PCM works in conjunction with the regulator to control the output of the alternator. In some cases, the PCM communicates the desired charge rate to the regulator, and the regulator adjusts the current flow through the rotor appropriately. The regulator then reports the alternator load and any faults to the PCM. If any faults develop, the PCM will store one or more pertinent codes and will illuminate the malfunction indicator light on the instrument cluster.

▶ TECHNICIAN TIP

As the number of electrical components increases, vehicles require increasing amounts of available electricity. Vehicles have traditionally used 12-volt batteries and a 14-volt charging system, but to meet the increased demand from today's systems, such as electric drives, higher voltage batteries, and battery packs, an appropriately higher voltage electrical system is necessary.

The advantages of using higher voltage systems include increased power for systems and accessories, fewer concerns about voltage drop across the system, and the use of smaller diameter wires for a higher power draw, which reduces the overall weight of the vehicle.

If we apply Ohm's law, we can appreciate the differences between the voltage systems. For example, consider a vehicle with a conventional 14-volt system that has a charging current of 100 amps. Using Ohm's law (Voltage × Current = Power), we can calculate that 14 volts times 100 amps equals 1400 watts. In a higher voltage system, less current is needed to deliver the same amount of power. By comparison, using Ohm's law again, but this time with a 42-volt system, we find that 1400 watts divided by 42 volts equals 33.3 amps. For the same amount of power (i.e., work), the current has been decreased by two-thirds. This means that for any given current flow, the cable diameter on a 42-volt system can be reduced in size and many relays and associated wiring could be eliminated.

Despite these virtues, 42-volt systems have not taken a very strong hold of the market yet. But as electrical demands in vehicles increase, and given the fact that reasonably sized, affordable, and reliable alternators above about 160 amps are more difficult to manufacture, the 42-volt charging system may yet be recognized as an option that can supply the needed power.

Charging System Output Test

N54008, S54002

Vehicle charging systems are voltage regulated, which means that the charging system will try to maintain a set output voltage from the alternator. As electrical load current increases in the vehicle systems, voltage starts to drop. The voltage regulator senses this voltage drop and increases the current output of the alternator, which in turn increases system voltage to try to maintain the correct voltage in the system. The testing of an alternator output initially involves the testing of the system's regulated voltage using a voltmeter. Regulated voltage is the voltage at which the regulator is limiting the alternator output too, once the battery is relatively charged, as evidenced by the greatly reduced current output. Regulated voltage should be between the manufacturer's specified minimum and maximum regulated voltage. If it is incorrect, verify that there are no voltage drops on the alternator and regulator. If no voltage drops are found, the regulator is likely faulty and will have to be replaced.

Once the regulated voltage is confirmed, it's time to check the charging system output. This is done by using an external electrical load, such as a carbon pile, to reduce the battery voltage, thereby tricking the regulator into full-fielding the alternator, making it produce maximum amperage output. This output is read using an inductive ammeter and should then be compared to the manufacturer's rated output specifications. An alternator that puts out within 10% of its rated output is okay. Less than that indicates a faulty regulator, faulty alternator, or excessive voltage drop(s) on the alternator output, ground, or control circuits.

TECHNICIAN TIP

Back in the day of DC generators, it was acceptable to test the charging system by removing one of the battery terminals from the battery post while the engine was running. If the engine kept running, the charging system was working. If the engine died, the charging system was not working. On alternator-equipped vehicles, doing this test is very risky, as any voltage spike caused from disconnecting the battery terminal can destroy any of the electronics in the vehicle. Thus, this test is no longer valid on virtually all vehicles on the road today.

To perform a charging system output test, follow the steps in **SKILL DRILL 54-6**.

Applied Math

AM-1: Whole Numbers: The technician can add whole numbers to determine measurement conformance with the manufacturer's specifications.

An alternator is being tested to determine if it meets manufacturer's specifications. If an alternator is damaged due to a blown diode or similar problem, it is usually out of specifications by a wide margin. For this type of alternator, the output specification is 95 amperes. The technician tests the alternator that puts out 65 amps. The service material states a good alternator will provide an output that is within 15 amps of its rated value. The technician adds 15 amps to the original 65 amps, for a total of 80 amps. This is below the specification of 95 amps for this type of alternator. In this example, we are working with whole numbers. If a number has a negative sign, a decimal point, or a part that's a fraction, it is not considered a whole number.

SKILL DRILL 54-6 Performing a Charging System Output Test

1. Connect a charging system tester to the battery with the red lead to the positive post, the black lead to the negative post, and the amps clamp around the alternator output wire.

2. Start the engine, turn off all accessories, and measure the regulated voltage at around 1500 rpm. The regulated voltage is the highest voltage the system achieves once the battery is relatively charged, as evidenced by the ammeter reading less than about 15–20 amps. Typical regulated voltage specifications are wider than they used to be because of the ability of the PCM to adjust the output voltage for a wide range of conditions. Always check the specifications.

3. Operate the engine at about 1500 rpm, and either manually or automatically load down the battery just enough to obtain the maximum amperage output without pulling battery voltage below 12.0 volts. This reading should be compared against the alternator's rated output. Normally, readings more than 10% out of specifications indicate a problem.

Testing Charging System Circuit Voltage Drop

N54009

An excessive voltage drop in the charging system output and ground circuit tends to cause one of two problems: (1) The battery will not be fully charged because, although the alternator is creating the specified voltage, the voltage drop is reducing the amount of voltage to the battery; or (2) the battery is fully charged, but the alternator is working at a higher voltage to do so, potentially overheating it. Which of the two issues is occurring depends on where the voltage is sensed. If it is sensed at the alternator, then the battery will generally be undercharged. If the voltage is sensed at the battery, then the alternator will work at the higher voltage. Knowing the system will help you diagnose voltage drop issues in the output and ground circuits of the charging system.

The external alternator output circuit components consist of the wires and fuse or fusible link between the positive battery post and the alternator output terminal, and the ground wire back to the battery from the engine and chassis. Voltage drop may occur anywhere in the output current circuit and ground circuit but is especially common at the terminals and connectors because of the high charging system current flowing through them. Even a small amount of resistance can cause significant voltage drop when conducting these higher levels of current flow.

As with testing for voltage drop at the starter circuit, a voltmeter or DMM is used to measure voltage drop across all parts of the circuit. A voltmeter with a minimum/maximum range setting is very useful when measuring voltage drop, as it will record and hold the maximum voltage drop that occurs for a particular operation cycle. Voltage drop tests are only valid when the circuit is under load, because voltage drops can only occur when current is flowing. The greater the flow, the greater the voltage drop, if resistance is present. Therefore, when testing for voltage drop, always perform the test when the circuit is being operated.

To measure for voltage drop, the DMM is connected in parallel across the component, cable, or connection that is to be tested. Usually it is most efficient to first measure the voltage drop on both the entire positive side and the entire negative side of the circuit and, if required, narrow down the test to individual components to identify precisely where excessive voltage drop is located (individual voltage drops in a simple circuit add up). For example, you can connect the black probe to the output side of the alternator and the red probe to the positive post of the battery. Then operate the charging system under a heavy load by turning on as many electrical items as possible or by using a load tester to load down the battery. Record the voltage drop across the output side of the circuit while the alternator is fully charging. Then perform the same test on the ground side of the circuit from the negative battery terminal to the frame of the alternator. Manufacturers typically specify the maximum allowable voltage drop, but a rule of thumb is no more than 0.5 volt (500 millivolts) for each side of a 12-volt circuit.

To perform a charging circuit voltage drop test, follow the steps in **SKILL DRILL 54-7**.

Diagnosing Undercharge and Overcharge

N54010

Undercharging and overcharging are both bad for a vehicle. Undercharging leads to electrical systems that do not function fully, and the battery may not fully charge, leading to an early death due to sulfation. Overcharging can lead to short life of bulbs and other electrical devices, while at the same time overcharging the battery, which can increase gassing and

SKILL DRILL 54-7 Testing Charging Circuit Voltage Drop

1. Set the DMM up to measure voltage, and select min/max if available. Connect the red probe of the DMM to the output terminal of the alternator and the black probe on the positive post of the battery.

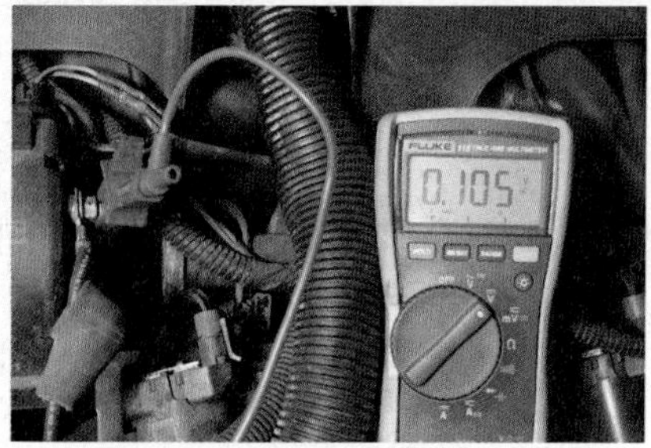

2. Start the engine and turn on as many electrical loads as possible, or use an external load bank to load the battery. Read the maximum voltage drop for the output circuit.

3. Move the leads to measure the voltage drop on the ground circuit by placing the red probe on the alternator case and the black probe on the negative terminal of the battery. Read the maximum voltage drop for the ground circuit. Determine any necessary actions.

Applied Science

AS-64: Motors: The technician can demonstrate an understanding of the role of the generator/alternator in maintaining battery and system voltage.

A vehicle is in the shop for an illuminated battery light on the instrument panel. The repair order states that the customer had to jump-start the vehicle in order to drive it to the shop. An apprentice technician has been given the job of diagnosing this vehicle with the help of an experienced technician, if required. The apprentice starts by verifying the customer's concerns and finds the vehicle once again requires a jump start. When the engine starts, the battery light on the dash remains on. Checking under the hood, the apprentice finds what appears to be a new battery. Reading further on the repair order, the customer states that he replaced the battery two days ago. The apprentice connects his DMM and obtains a reading of 10.2 volts. It is obvious to the apprentice that the alternator is not charging at all. The experienced technician performs a voltage drop test and agrees that a new or remanufactured alternator should be installed.

As stated in our text, the alternator converts mechanical energy into electrical energy by electromagnetic induction. In a simple version, a bar magnet rotates in an iron yoke, which concentrates the magnetic field. A coil of wire is wound around the stem of the yoke. As the magnet turns, voltage is induced in the coil, producing a current flow. A typical charging voltage will be from 13.8 to 14.2 volts DC, which keeps the battery fully charged at 12.6 volts at 70°F (21°C).

loss of water from the electrolyte. Maintaining a proper charge is critical to long life and proper operation of the electrical system.

The voltage regulator maintains a constant voltage output from the alternator up to the maximum output current. Always check the manufacturer's specifications for alternator regulated voltages; more and more manufacturers are using PCM-controlled charging systems, which means the voltages can have a wider range than non-PCM-controlled systems.

As a rule of thumb, for a typical non-PCM-controlled 12-volt charging system, the regulated alternator output voltage should be between 13.8 and 14.5 volts. If the voltage measured is outside of the manufacturer's specifications, further testing will have to be performed to determine the cause.

For this task, we assume that the charging system is charging (not a no-charge condition) and that it is either not charging fully or is overcharging. If the system is overcharging, it can almost always be tracked to either a faulty voltage regulator or a faulty regulator ground, both of which can cause the alternator to overcharge.

An undercharge condition has many more potential causes, such as a loose drive belt, voltage drop in the charging system wiring, faulty regulator, worn brushes in the alternator, high resistance in the rotor, open or shorted diodes, open or shorted stator windings, or even a shorted cell on a battery. Always check for the easiest and most common faults first. For example, check that the drive belts are not slipping, particularly under load. Also verify that the battery does not have a shorted cell and that its capacity is adequate.

If nothing obvious can be found wrong, then you will need to dig deeper. One way to do this is to use an oscilloscope to check the alternator ripple. The ripple indicates faulty stator windings or diodes, meaning the alternator needs to be replaced. If the ripple is okay and the voltage output is low, then testing for voltage drops on the output and ground circuit is a good step. If voltage drops are found, then they will have to be repaired and the system retested. If there are no voltage drops, then the fault is likely to be located inside of the alternator, such as a faulty regulator, worn brushes, or a faulty rotor, which would also require replacement of the alternator.

Replacing an Alternator

N54011

Alternators have to be replaced whenever they are electrically or mechanically faulty. Electrical faults include no-charge, undercharge, or overcharge conditions. Mechanical faults include worn and noisy bearings, or other internal or external mechanical damage. When replacing an alternator, there are some precautions, the first being that battery voltage is always present at the output terminal of the alternator. This means that if the battery negative terminal is not removed from the battery, and a wrench is placed on the output terminal of the alternator, a very large spark could occur if the wrench touches something metal. So always disconnect at least one battery terminal prior to removing the alternator. Also, check the manufacturer's information regarding maintaining the adaptive memory, if necessary.

The alternator must not be operated with the battery disconnected or with the terminals at the back of the alternator disconnected. Even though most vehicle electrical systems are described as being 12 volt, they typically operate at between 13.8 to 14.5 volts. If the system is not operating in this range, the drive belt may have become loose, or excessive voltage drops in the charging or voltage-sensing circuit may be the cause. If all of these components check out, the alternator, external voltage regulator, internal regulator, or even the built-in voltage-regulating circuit in the ECM/BCM may be faulty. Make sure you have identified the cause of the fault before removing the alternator.

To replace an alternator, follow the steps in **SKILL DRILL 54-8.**

SKILL DRILL 54-8 Replacing an Alternator

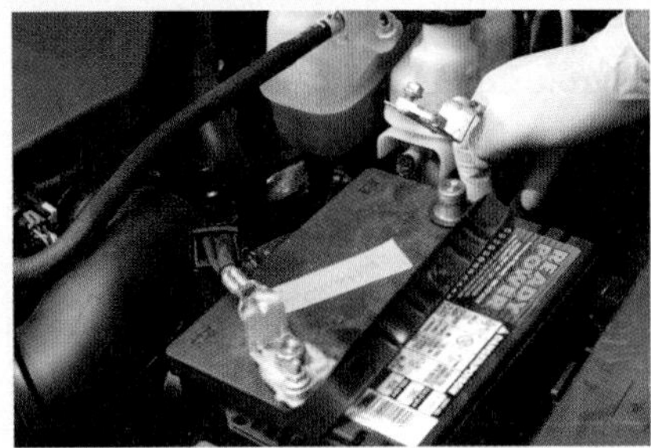

1. Install fender covers. Verify any memory issues, and remove the negative terminal of the battery.

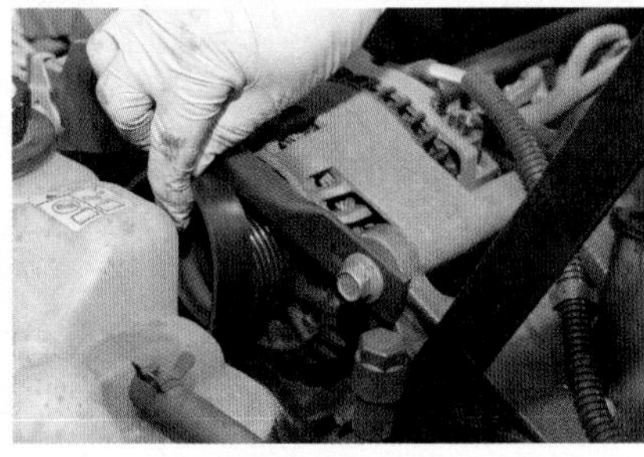

2. Loosen the drive belt and remove it from the alternator pulley. Check the condition of the belt to see if it is still serviceable.

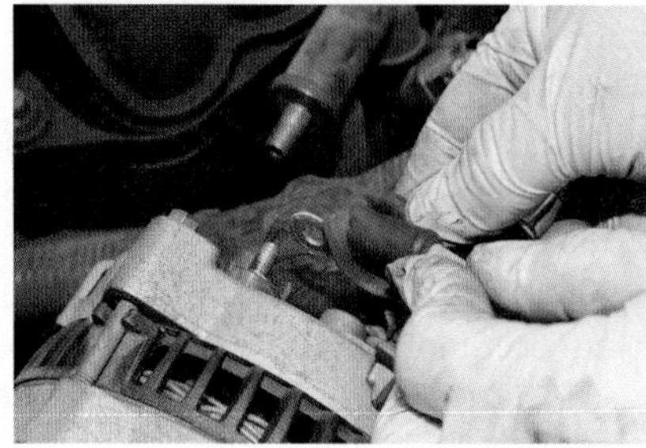

3. Locate the electrical connections at the rear of the alternator, and note their positions. Loosen any securing fasteners or covers, and remove terminals one at a time.

4. Loosen the securing fasteners that hold the alternator to its mounting bracket(s), making sure the alternator is supported. Remove the alternator.

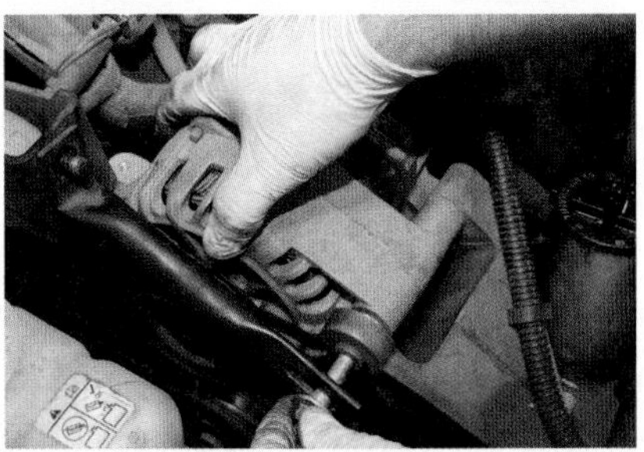

5. Reinstall the alternator. Situate the alternator in the mounting bracket(s) and, while still supporting the alternator, loosely start the securing fasteners that hold the alternator to its mounting bracket(s).

6. Reinstall the electrical wires to their correct terminals, referring to the manufacturer's information. Check the security of any fastening devices.

7. Install the drive belt over the alternator drive pulley and, using the correct tools, adjust the belt to the correct tension.

8. Reattach to the negative post of the battery. Make sure the fastener is tight, and replace any battery terminal covers.

9. Turn the ignition to the On position, and make sure the charge light on the instrument panel illuminates. Start the engine: the charge light should go off. Measure the regulated voltage and maximum alternator amperage output. Remove the fender covers, and return any tools used to their correct place.

▶ Wrap-Up

Ready for Review

- The starting system provides a method of rotating (cranking) the vehicle's internal combustion engine (ICE) to begin the combustion cycle.
- The starting system consists of a battery, cables, a solenoid, a starter motor, a ring gear, and an ignition switch.
- The starter motor converts electrical energy to mechanical energy.
- Some starters operate through gear reduction, giving the same amount of torque at less size and weight.
- A starter motor is basically an electromagnet.
- The starter motor pinion must mesh with the engine ring gear to turn the engine over and allow the vehicle to start.
- Once the engine starts running on its own, the starter must disengage from the ring rear to avoid damaging the starter or the ring gear.
- The solenoid on the starter motor performs two main functions: It switches the high current flow required by the starter motor, and it engages the starter drive with the ring gear.
- Vehicle immobilizers generally comprise a computer-managed security system that disables the start and engine systems by using an electronic system to uniquely identify each vehicle key by a security code system.
- Hybrid vehicles use both an ICE and electric motors to power the vehicle's drivetrain.
- Hybrid vehicles typically use one of the high voltage motors to crank the ICE over, faster and smoother than a 12-volt starter.
- Testing starter motor current draw is a good indicator of overall starter motor performance.
- Compared to older vehicles, newer vehicles put a higher demand on the charging system, as many systems are required not only to operate the vehicle but also to operate an array of entertainment systems.
- The DC generator has not been used on vehicles since the 1960s.
- The AC generator, or alternator, is used on today's vehicles both to recharge the vehicle battery and to run numerous other electrical accessories on the vehicle.
- The alternator converts mechanical energy into electrical energy.
- The rotor creates the rotating north and south magnetic poles. It has slip rings connected to each end of the winding.
- The stronger the magnetic field, the higher the alternator output.
- Brushes ride on the slip rings to supply current flow to the rotor.
- Brushless alternators use magnetic induction to induce current flow into the rotor.
- The pulley spins the rotor, and the fan forces air through the alternator to cool it.
- New alternator decoupling pulleys freewheel during deceleration and drive the rotor during acceleration and cruise.
- The alternator changes AC to DC by using a rectifier with diodes to allow electrical current to flow in one direction only.
- The alternator has a regulator, either built in or external, to keep the alternator from overcharging the battery.
- The regulator controls the amount of current flow to the rotor, which then controls the output of the alternator.
- The more amps the alternator can produce, the more accessories the vehicle electrical system can handle.
- One of the first items that should be checked if an alternator is not charging according to the manufacturer's specifications is the belt tension and condition.
- Overcharging tends to shorten the life of lightbulbs and other electric and electronic devices.
- Undercharging tends to cause the battery plates to sulfate, reducing its life.

Key Terms

counter-electromotive force (CEMF) Voltage created in the field windings as the motor rotates, which opposes battery voltage and limits motor speed.

hold-in winding A low-current winding found in starter solenoids that holds the plunger in the activated position.

pull-in winding A high-current winding found in starter solenoids that pulls the solenoid plunger into the activated position.

rectification A process of converting AC into DC required by the battery and nearly all of the automobile systems.

Review Questions

1. What is the main purpose of the starter system in a vehicle?
   - **a.** To provide a method of rotating the vehicle's internal combustion engine to begin the combustion cycle
   - **b.** To start the vehicle's charging system.
   - **c.** To start all the other systems of the vehicle
   - **d.** To start the conversion of chemical energy into electrical energy
2. A starter draw test measures:
   - **a.** the amount of voltage dropped across the solenoid contacts.
   - **b.** the amount of current a starter draws during cranking.
   - **c.** the amount of current draw in the starter control circuit.
   - **d.** the amount of time it takes for the starter to crank the engine over.

3. To test the charging system output do all of the following EXCEPT:
   a. Connect a voltmeter to the battery, and the amps clamp around the alternator output wire.
   b. Raise the engine RPM to approximately 1.500 RPM.
   c. manually or automatically load down the battery just enough to obtain the maximum amperage output without pulling battery voltage below 12.0 volts.
   d. disconnect the negative battery terminal to see if the vehicle dies.
4. Which component slows down the armature when the starter motor is disengaged?
   a. Shift fork
   b. Copper-impregnated carbon brushes
   c. Tensioned spiral springs
   d. Brake washer
5. What is the purpose of the diodes in an alternator rectifier?
   a. to control the voltage output of the alternator.
   b. to control the amount of current going through the rotor winding.
   c. to transfer the current from the voltage regulator to the slip rings on the rotor.
   d. to transform the alternating current into direct current.
6. What is the main purpose of the fork in starter motor engagement?
   a. It activates a starter-mounted solenoid.
   b. It transmits drive in one direction only and freewheels in the opposite direction.
   c. It helps more current to flow in the circuit.
   d. It engages the starter drive.
7. All of the following statements with respect to the operational function of a starter solenoid are true *except*:
   a. It switches the high current flow required by the starter motor on and off.
   b. It engages the starter drive with the ring gear.
   c. It is typically a cylindrical device mounted on the starter motor.
   d. It conducts electricity from the starter motor to the control circuit.
8. All of the following statements with respect to the operation of a starter solenoid are true *except*:
   a. A pull-in winding draws a higher current and creates a stronger magnetic field than the other winding—the hold-in winding.
   b. The output of the pull-in winding is connected to the main starter terminal.
   c. The movement of the solenoid windings engages the drive pinion with the flywheel ring gear.
   d. The output of the hold-in winding is connected to ground on the starter casing.
9. Alternators have all of the following characteristics *except*:
   a. greater wattage output at lower rpm.
   b. lower maximum RPM.
   c. smaller physical size and weight for a given output.
   d. greater reliability and longer service life.
10. Which component in the alternator controls the strength of the magnetic field of the rotating magnet?
   a. Voltage regulator
   b. Diode assembly
   c. Rectifier assembly
   d. A cooling fan

ASE Technician A/Technician B Style Questions

1. Tech A says that some starters use gear reduction to improve efficiency. Tech B says that a starter converts electrical energy to mechanical energy. Who is correct?
   a. Tech A
   b. Tech B
   c. Both A and B
   d. Neither A nor B
2. Tech A says that the pull-in winding is short-circuited when the solenoid is fully engaged. Tech B says that the starter drive has a built-in one-way clutch. Who is correct?
   a. Tech A
   b. Tech B
   c. Both A and B
   d. Neither A nor B
3. Tech A says that a voltage drop of 0.8 volt on the starter ground circuit is within specifications. Tech B says that high starter draw current could be caused by a spun main bearing in the engine. Who is correct?
   a. Tech A
   b. Tech B
   c. Both A and B
   d. Neither A nor B
4. Tech A says that when testing a relay winding with an ohmmeter, it should show the winding is open. Tech B says that the solenoid closes a set of heavy contacts that send current flow to the starter motor. Who is correct?
   a. Tech A
   b. Tech B
   c. Both A and B
   d. Neither A nor B
5. Tech A says that most hybrid vehicles use the high-voltage electric motor to start the engine. Tech B says that most hybrid engines need to crank over more slowly than regular engines. Who is correct?
   a. Tech A
   b. Tech B
   c. Both A and B
   d. Neither A nor B
6. Tech A says that some alternator pulleys are designed to freewheel in one direction. Tech B says that the alternator output terminal is connected to one of the slip rings on the rotor. Who is correct?
   a. Tech A
   b. Tech B
   c. Both A and B
   d. Neither A nor B

7. Tech A says that the stator rotates inside a magnetic field. Tech B says that the rotor is bolted between the two end housings. Who is correct?
   a. Tech A
   b. Tech B
   c. Both A and B
   d. Neither A nor B
8. Tech A says that the voltage regulator controls the strength of the rotor's magnetic field. Tech B says that the voltage regulator is installed between the output terminal of the alternator and the positive terminal of the battery. Who is correct?
   a. Tech A
   b. Tech B
   c. Both A and B
   d. Neither A nor B
9. Tech A says that overcharging can lead to short life of bulbs and other electrical devices. Tech B says that the charging system regulated voltage is checked with maximum load on the battery. Who is correct?
   a. Tech A
   b. Tech B
   c. Both A and B
   d. Neither A nor B
10. Tech A says that the rectifier assembly controls the output of the alternator. Tech B says that the alternator output circuit voltage drop must be checked with the charging system under a heavy load. Who is correct?
   a. Tech A
   b. Tech B
   c. Both A and B
   d. Neither A nor B

CHAPTER 55

Lighting Systems

NATEF Tasks

- **N55001** Identify system voltage and safety precautions associated with high-intensity discharge headlights. (MLR/AST/MAST)
- **N55002** Aim headlights. (MLR/AST/MAST)
- **N55003** Diagnose (troubleshoot) the causes of incorrect operation of warning devices and other driver information systems; determine needed action. (AST/MAST)

Knowledge Objectives

After reading this chapter, you will be able to:

- **K55001** Describe the function of the lighting systems and types of lightbulbs.
- **K55002** Explain the purpose of each of the convenience and safety light systems.
- **K55003** Explain the purpose of each of the driver's intention indicator light systems.
- **K55004** Explain the function of common headlight systems.
- **K55005** Describe electronically controlled lighting systems.

Skills Objectives

After reading this chapter, you will be able to:

- **S55001** Perform lighting system inspection, testing, and service.
- **S55002** Measure headlight intensity.
- **S55003** Diagnose failures in driver information display systems.

▶ Introduction

Well-designed vehicle lighting systems enhance vehicle safety by increasing the driver's visibility while he or she is operating the vehicle and by clearly signaling the driver's intent to those around the vehicle. Lighting systems are also used inside the vehicle to indicate messages to the driver and provide convenience to any occupants. Manufacturers continue to improve and develop their lighting systems as new technology becomes available. Modern lighting system features include the increased use of light-emitting diode (LED) and xenon lighting, electronic body control units to manage lighting, and high-intensity discharge (HID) lamps, which are much brighter than the traditional sealed-beam or halogen units. Many vehicles also have systems that warn you if lamps are not working properly.

▶ Lighting Systems

K55001

Lighting systems improve visibility at night and make a vehicle visible to other road users. A lighting switch activates taillights, park lights, and headlights to allow the driver to see ahead and be seen from the side and rear. A beam selector switch allows the driver to change the beams from high to low, or vice versa, as required. Brake lights operate when the brake pedal is depressed or when a vehicle control module is automatically applying the brakes. Red or amber turn signals alert other drivers of a change in direction and are mounted so they can be seen from the front, the rear, and sometimes the sides of the automobile. The emergency flasher system operates both front and rear turn signals at the same time as a warning to others. Other circuits operate courtesy, or convenience, lights, backup (reverse) lights, fog lights, and fault indicators.

Types of Lamps

Modern vehicles use many different kinds and sizes of lamps, also known in some places as lightbulbs or light globes. There are several lamp types available, including standard incandescent lamps, halogen lamps, vacuum tube fluorescent (VTF) lighting, HID xenon gas systems, LEDs, and more. Conventional incandescent lamps are being replaced in many applications by these other more efficient types of lights.

Incandescent and Halogen

Incandescent lamps consist of one or more filaments that heat up to approximately 5000°F (2760°C) and glow white-hot (**FIGURE 55-1**). The filament material does not burn because most of the oxygen in the bulb has been replaced by inert gases that stop combustion from occurring. The power in watts consumed is often marked on the lamp; wattage (work performed) is found by multiplying the voltage used by the lamp by the current flowing through it. The higher the wattage, the more light that can be created when compared to a similar type of bulb. Incandescent bulbs are inefficient, converting only about 10% of the electricity to visible light.

Halogen lamps are another type of incandescent lamp, but they are filled with a halogen gas such as bromine or iodine

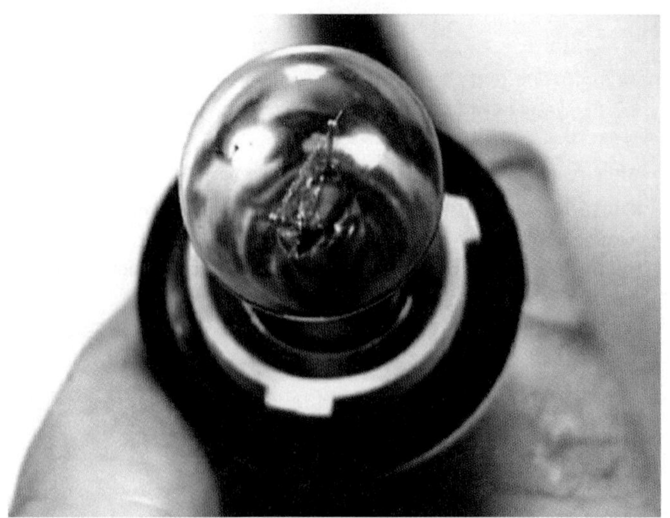

FIGURE 55-1 Incandescent bulb with single filament.

You Are the Automotive Technician

A customer brings his 2014 Dodge Avenger into the auto dealership for a 60,000-mile service checkup. You take the vehicle back to the service bay. Part of the 60,000-mile service is to check the lighting and peripheral systems. Your coworker operates each of the lights while you check the front lights and then the rear lights. You notice that the center high mount stop light (CHMSL) on the trunk lid is not functioning, but the brake lights are working normally. The customer agrees to allow you to diagnose it. After referring to the manufacturer's information and wiring diagrams, you see that the CHMSL is on the same fuse and brake switch as the brake lights, so they do not need to be investigated further. Based on your reading of the wiring diagram, you determine that the problem is either between the brake switch and CHMSL or between the CHMSL and the ground, or the CHMSL bulbs or unit are faulty. You test the power and ground at the CHMSL terminals with the brake pedal applied. You find 12.2 volts available. Because the CHMSL is a sealed LED-style assembly, it will have to be replaced. The customer approves the repair, and you install the new CHMSL and verify that the light is working properly. You review the work and invoice with the customer, thank him for his business, and set him up in the automated email system to send a reminder for his next scheduled maintenance visit.

1. Why didn't you have to check the fuse or brake switch?
2. What did the 12.2-volt reading at the CHMSL terminals tell you?
3. How do wiring diagrams save time?
4. How would you locate the fault if the voltage drop across the CHMSL was 8.3 volts?

FIGURE 55-2 Halogen bulb.

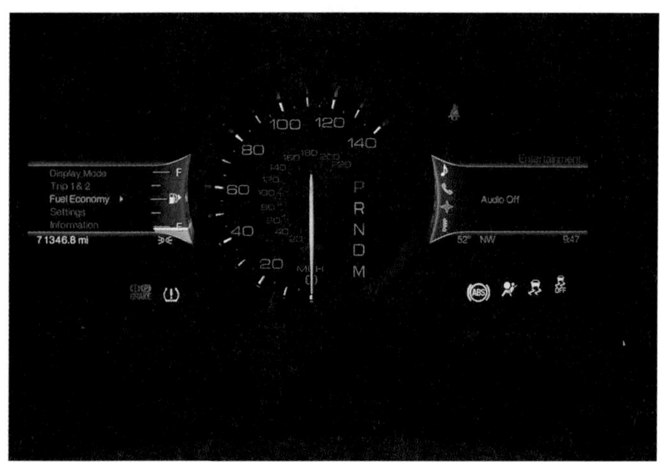

FIGURE 55-3 Typical vacuum fluorescent display.

(**FIGURE 55-2**). These lamps have a much longer life and are generally brighter and produce more light per unit of power consumed. However, they become very hot in use. They are manufactured from highly heat-resistant materials, and the bulbs must be handled carefully because they are sensitive and can be damaged even by finger oil residue left by fingerprints.

Vacuum Tube Fluorescent

Vacuum tube fluorescent (VTF), also called vacuum fluorescent display (VFD), is used for instrumentation displays on vehicle instrument panel clusters. This type of lighting emits a very bright light with high contrast and can display in various colors (**FIGURE 55-3**). Usually VTF displays are of bar graphs, seven-segment numerals, multi-segment alphanumeric characters, or a dot-matrix pattern. VTF displays include different kinds of alphanumeric characters and symbols to alert drivers to various conditions.

High-Intensity Discharge

High-intensity discharge (HID) headlamps produce light with an electric arc rather than a glowing filament (**FIGURE 55-4**). The high intensity of the arc comes from metallic salts that are vaporized within an arc chamber. HIDs produce more light for a given level of power consumption than ordinary tungsten or halogen bulbs. Automotive HID lamps are commonly called xenon headlamps, though they are actually metal halide lamps that contain xenon gas. The light from HID headlamps exhibits a distinct bluish tint as compared with the yellow-white color of tungsten-filament headlamps.

HID headlamp bulbs do not run on low-voltage direct current; they require a **ballast** with an internal or an external igniter that is either integrated into the bulb or included as a separate unit or part of the ballast. The ballast increases the voltage substantially and controls the current to the bulb.

HID headlamps produce between 2800 and 3500 lumens from between 35 and 38 watts, whereas halogen filament headlamp bulbs produce between 700 and 2100 lumens from between 40 and 72 watts at 12.8 volts. The advantage of using HID headlight systems is that they offer substantially greater luminance

FIGURE 55-4 HID headlamp assembly.

than halogen bulbs (about 3000 lumens versus 1400 lumens for comparable halogen bulbs). If the higher-output HID light source is used in a well-engineered headlamp optic, the driver gets more usable light. Studies indicate that drivers react faster and more accurately to roadway obstacles when using good HID headlamps than when using halogen headlamps; therefore, good HID headlamps contribute to driving safety.

The contrary argument is that HID headlamps can impact negatively the vision of oncoming traffic due to their high intensity and the "flashing" effect caused by the rapid transition between low and high illumination in the field of illumination. This potential distraction increases the risk of a head-on collision between a vehicle using HID headlamps and a blinded oncoming driver. Scientific studies of headlamp glare from HID systems has shown that for any given intensity level, the light from HID headlamps is 40% more glaring than the light from tungsten-halogen headlamps.

Some countries mandate that HID headlamps may only be installed on vehicles (except motorcycles) with lens-cleaning systems (to reduce glare) and automatic self-leveling systems (which prevent dazzling oncoming traffic). These systems are usually absent on vehicles not originally equipped with HID lamps, so if a halogen headlamp is retrofitted with an HID bulb, light distribution with illegal levels of glare will be produced.

Another disadvantage of HID headlamps is that they are significantly more costly to produce, install, purchase, and repair. However, some of this cost is offset by the longer lifespan of the HID burner relative to halogen bulbs.

Light-Emitting Diode

Light-emitting diodes (LEDs) have been used for some time in various automotive applications, such as warning indicators and alphanumeric displays. More recent developments in LED technology have seen the production of a wider range of colors and LEDs that are brighter than previous types. It is now possible to get LEDs that emit bright red, green, blue, yellow, and clear or white light. This has made it possible to use LEDs for many new applications, such as more general lighting applications. For example, LEDs are now often used for stop lights, turn signals, and interior lighting on vehicles (**FIGURE 55-5**).

One of the advantages of LEDs is that they turn on instantly. This is particularly useful in brake lights, as they can reduce the braking light response time by two-tenths of a second. This translates to an extra 16' (4.9 meters) of stopping distance for vehicles traveling at highway speeds. LEDs also have better visibility in inclement weather, operate at cooler temperatures, consume less energy, are much smaller, and can last up to 100 times longer, reducing the cost of servicing. LEDs can be specifically designed for LED lighting and also as LED replacement bulbs for more traditional bulb holders.

For automotive applications, a number of LEDs are grouped together to provide the amount of light required for the application. Additionally, LED light lenses are specifically designed with light-focusing prisms and lenses to focus the light generated by the LEDs. A typical LED has a voltage drop of 1.2 to 3.5 volts across it, depending upon the color, when it is forward biased and emitting light.

When used in automotive lighting, many LEDs are required to give off a specified amount of light. To do so, they are usually connected in groups called series strings. A number of series strings are then connected in parallel until enough LEDs are connected to give off the required amount of light.

LEDs work best when the voltage to them and the current flow through them remains constant at a preset level. There are two main ways to achieve this: The first is via a resistor; the second and more preferred way is through the use of a voltage regulation circuit.

Some LED lights are multivoltage, which means they can work on both 12- and 24-volt systems. These lights are normally used in aftermarket products, which can be installed in a wide range of vehicles.

Lamp/Lightbulb Information

All lamps or lightbulbs have letters and numbers stamped on them that typically indicate their part number and often the operating voltage and power consumed. For instance, in a bulb marked 12V/21W, the filament will consume 21 watts of power when 12 volts is applied across the filament. While the wattage is not necessarily an indication of light output, it can be generally assumed that the higher the wattage, the greater the light output.

Lamps and lightbulbs come in a variety of configurations to fit the various applications within a vehicle. One designation is how many filaments the bulb has (**FIGURE 55-6**). Single-filament bulbs are common for use as courtesy lights, dash lights, and warning lights. Dual-filament bulbs have two filaments of different wattage; one filament emits a small amount of light, and the second filament emits more light. These bulbs work well as a combination taillight and brake light. Headlights can also be dual-filament bulbs. In some cases, the low beam filament is lower wattage than the high beam filament, but not always. In a headlight, the filaments are positioned to give a different profile of light. Low beams emit light closer to the vehicle and angled slightly toward the side of the road, whereas high beams tend to focus farther down the road and straight ahead.

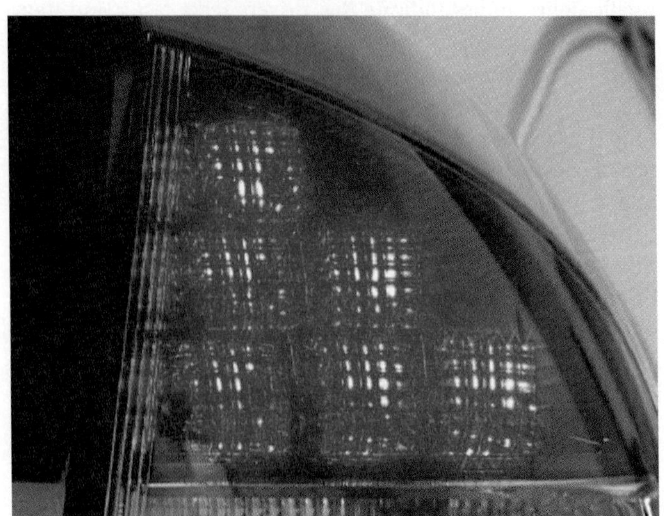

FIGURE 55-5 LED lights.

FIGURE 55-6 Dual-filament bulb.

Another feature that differs among lights is the type of base on the lamp—in other words, what type of socket it is retained in. Bayonet-style bulbs have been around for a long time. They get their name from the two retaining pins on the side of the base (**FIGURE 55-7**). The pins follow slots in the side socket and at the bottom, and the slots turn sideways into a small pocket. The pins are retained in the pocket by the spring-loaded base in the bottom of the socket, which pushes the bulb upward. This design resists vibration very well. Removal of the bulb requires carefully pushing in the bulb, rotating the bulb slightly counterclockwise, and pulling it out. One or two electrical contacts are built into the bottom of the bulb's base. If the bulb is a dual-filament bulb with two contacts on the base, then the pins will be unequal height so that the contacts will be registered properly with the contacts in the socket.

Many newer bulbs use a wedge base either made from the glass bulb itself or with a built-in plastic base (**FIGURE 55-8**). The bulbs are pushed straight into the socket, and tension from the socket retains the bulb. The electrical contact on the glass-based bulbs is made by wires extending from the base of the bulb and bent over opposite sides of the wedge.

FIGURE 55-7 Bayonet-style bulbs.

FIGURE 55-8 Wedge-style bulbs.

FIGURE 55-9 Festoon-style bulbs.

Some dome lights use festoon lights, which have a base on each end of a cylindrical lightbulb (**FIGURE 55-9**). Each end of the filament is connected to one of the bases. Generally, the bases fit in spring steel contacts, which hold the lightbulb securely in place.

▶ Types of Lighting Systems

K55002

There are many different lighting systems or circuits. Each style and type is designed to perform specific roles. For example, warning lamps, turn indicators, stop lights, taillights, courtesy lamps, and headlamps all perform different roles. Lamp locations, color, and brightness are governed by regulations to ensure consistency and safety in the application of lighting on vehicles. Lighting regulations should always be consulted before modifying or adding to any of the vehicle lighting systems. Well-maintained lighting systems improve road safety for all drivers.

Park/Tail/Marker/License Lights

Park, tail, and marker lights are all low-intensity or low-wattage bulbs used to mark the outline or width of the vehicle. Park and tail lamps tend to be installed close to the corners of the vehicle. Park lamps are placed to the front of the vehicle, or in some cases they are incorporated in the headlight assembly and are white or yellow in color. Tail lamps are red and usually installed in a cluster assembly with the stops lamps at the rear of the vehicle.

License plate lamps produce a white light and are designed to illuminate the lettering on the license plate at night without the white light itself being seen from the rear. The bulbs are connected in parallel to each other so that the failure of one filament will not cause a total circuit failure. A license plate illumination lamp or lamps are usually connected in parallel to the taillights and operate whenever the taillights are on.

Taillights are usually incorporated in a cluster assembly at the rear of the vehicle. Government regulations control the height of the lamps and their brightness.

Park lights are located at the front of the vehicle and are used at night when the vehicle is parked on the side of the road and are also on anytime the headlights are on. They use low-wattage bulbs and may have a lens or diffuser that makes the emitted light widespread. In some cases, park lights are incorporated in the headlight assembly. Park lights operate when the light switch is moved to the park light position. For safety reasons, park lights and taillights continue to operate when the light switch is moved to the headlight position. The bulbs are connected in parallel with each other.

▶ TECHNICIAN TIP

In many vehicles produced before the mid-1960s, front park lights turn off when the headlights are on. The thinking was that if the headlights are on, then the park lights were unnecessary. The problem arose when a headlight failed and only one light illuminated on the front of the vehicle, mimicking a motorcycle, which gave the impression of the vehicle being narrower than it is. Designing the park lights to stay on with the headlights helped to prevent this unsafe situation.

On computer-controlled lighting systems, the park lights and taillights are controlled by the body control module (BCM). The lights are supplied with power and ground through a networked light controller. The controller is connected to a network/bus system (discussed later in this chapter) by the twisted pair of communication wires. The park light switch is hardwired to a module. When the park light switch is activated, the module sends a park light request out on the network, where the appropriate controllers pick up the message and supply power or ground to the appropriate lightbulb filaments in order to illuminate the park lights.

Marker lights, used to mark the sides of some vehicles, are often located down the sides of the vehicle or trailer. They can be located on the front and rear fenders. On newer vehicles, they are sometimes placed on side-view mirrors or between the front and the rear doors on large SUVs or pickups (**FIGURE 55-10**). Red marker lamps face toward the rear, and yellow lamps face toward the front of the vehicle. These lights are designed to work when the park lights or headlights are selected, and sometimes operate as turn signals, so a driver can warn others of his or her intent to switch lanes or turn a corner.

Government regulations control the positioning of lights, including the height of the lamps' beam and their brightness. The park, tail, marker, and license plate lamps operate when the headlight switch is in both the park and the headlight-on positions. The bulbs are connected in parallel to each other so that the failure of one filament will not cause a total circuit failure. Tail and park lamps may use separate fuses, so if one circuit fails, the other will continue to operate.

Driving Lights

Driving lights are used to supplement vehicle headlight systems. The driving lights are installed on the front of the vehicle and provide higher intensity illumination over longer distances than standard headlight systems (**FIGURE 55-11**). Vehicle design rules and regulations specify the limitations in relation to the positioning and lens configuration of driving lights. It is essential that the local regulations are adhered to when mounting or adjusting driving lights.

Many types of driving lights are available. They come in different sizes, shapes, lens pattern, and bulb wattage. In some instances, a single driving light can be installed to suit particular applications, but lights are normally installed in pairs.

Most driving lights use quartz halogen bulbs in the 55- to 120-watt range. The quality of the reflector is extremely important in driving lights to get optimum performance. Driving lights are wired so that they operate only when the high beam is operating. This safety feature ensures that driving lights turn off when the headlights are switched from high to low beam, thus ensuring that oncoming traffic is not blinded by excessive

FIGURE 55-10 Marker lights on an SUV.

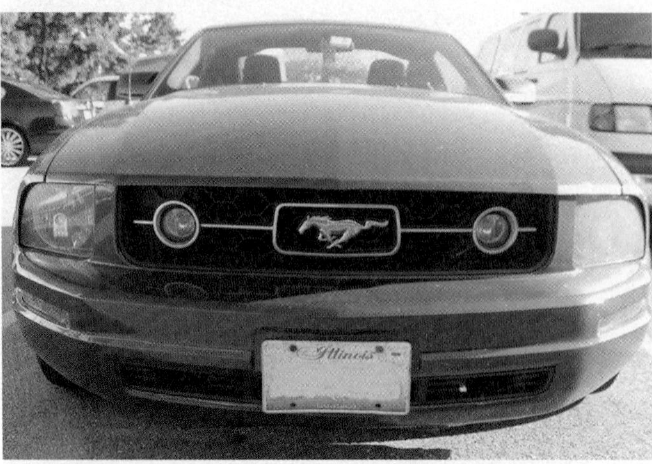

FIGURE 55-11 Vehicle with factory driving lights.

light. Although many performance vehicles come equipped with driving lights, they can be added to almost any vehicle. If they are, a relay and circuit breaker should always be used for circuit protection reasons.

Fog Lights

Fog lights are used with other vehicle lighting in poor weather such as thick fog, driving rain, or blowing snow. Because fog is made up of water droplets suspended in the air, it can reflect headlights back into the driver's eyes at night. In such conditions, fog lights can help drivers see farther ahead and illuminate the road's edges at reasonable speeds. They are used with park lights and low beam headlights, but not with high beams.

Most older fog lights have yellow-colored reflectors, although more recently, white fog lights have become more widely used because yellow lenses reduce fog light brilliance by about 30%. Fog lights typically use quartz halogen bulbs and are available in different shapes and sizes. Fog lights are usually mounted lower than headlights and tend to be aimed straight forward and low (**FIGURE 55-12**). Fog light lenses have a sharp cutoff pattern so that most of the light projected remains below the driver's eye level.

Fog lights are typically wired with a relay and circuit breaker. The method of connection of fog lights depends on local regulations. They may be wired to work only with park lights and to turn off when headlights are used or to work when low beams are used. The BCM normally controls the function of the fog lights if they are installed as original equipment.

Cornering Lights

To improve visibility during night driving, some vehicle manufacturers provide cornering lights (**FIGURE 55-13**). Cornering lights are white lights usually installed into the bumper or fender and are designed to provide side lighting when the vehicle is turning corners. The additional lighting provided by cornering

FIGURE 55-12 Vehicle with factory fog lights.

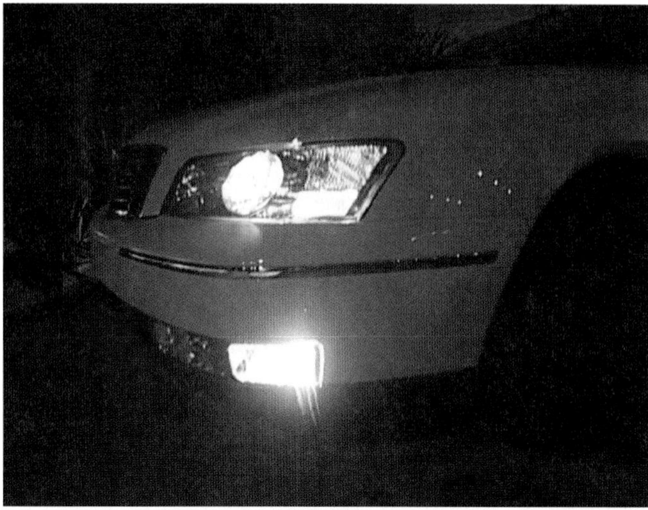

FIGURE 55-13 Cornering lights.

FIGURE 55-14 Courtesy lights provide ambient light where needed.

lights helps the driver to see the curb and any obstacles that may not be illuminated by the headlights. Cornering lights turn on only when the headlights and turn signal switches are both on, and turn off automatically when the turn signal cancels. Even though they come on with the turn signals, their light is steady; that is, they don't blink.

Courtesy Lights

Courtesy lights or lamps are used to provide ambient lighting in the cabin (**FIGURE 55-14**). They are usually low-intensity or low-wattage bulbs and can be overhead and in the doors, glove compartment, and the trunk. Door and latch switches control the various lamps for the trunk and glove compartment. Courtesy lights are usually controlled by the vehicle body computer, with inputs from the ignition and door switches either in the handle, latch, or door pillar. Courtesy lights may be designed with timing circuits that allow for the cabin lamps to stay on for a short period after all the doors are shut or the vehicle is locked.

FIGURE 55-15 Center high mount stop light (CHMSL).

FIGURE 55-16 Typical backup lights.

Brake Lights and CHMSL

K55003

Brake lights, which may also be called stop lights, are red lights mounted to the rear of the vehicle. They are usually incorporated in the taillight cluster. Many vehicles by law now have a higher additional third brake light mounted on top of the trunk lid or on the rear window. This light is called a center high mount stop light (CHMSL), or "chimsul" (**FIGURE 55-15**). The brake lights are activated whenever the driver operates the foot brake to slow or to stop the vehicle or when a control module automatically applies the brakes. The lighting circuit consists of the battery, fusible links and fuses, a brake light switch, brake lightbulbs, wiring to connect the components, and the ground circuit to return current from the light to the battery. It may also include a BCM to command the lights on when the proper inputs are present.

On older vehicle models, when the operator of the vehicle depresses the brake pedal, a switch mounted on the pedal support closes. This allows the electrical current to flow from the battery through the fuse, through the switch, to the brake lamp and to return to the battery by the ground circuit. When the driver releases the pedal, it returns to the rest position and opens the brake switch. The flow of electrical current stops and the brake lamps are extinguished. Today, computer-controlled brake lights are activated by the BCM when the computer sees an input from the brake pedal switch.

Backup Lights

The backup lights, also called reverse lights, are white lights mounted at the rear of a vehicle (**FIGURE 55-16**). They provide the driver with vision behind the vehicle at night and also alert other drivers to the fact that the vehicle is in reverse. The lighting circuit consists of the battery, fuses and fusible links, the ignition switch, the backup light (reverse) switch or a combined transmission backup light/neutral safety switch on the transmission, backup lamps, wiring to connect the components,

Applied Science

AS-33: Refraction: The technician can demonstrate an understanding of refraction as it occurs in systems that employ fiber optics.

A vehicle is in the shop for repairs to the lighting system. The repair order cites as the customer concern that the illumination for the console-mounted shifter indicator is not working. The driver is unable to determine which gear is selected for the automatic transmission at night.

Ethan is the apprentice technician at the dealership who was given the assignment of solving this concern. The first step is to verify the customer concern by covering the windows of the vehicle to simulate night conditions. Ethan looks up the manufacturer's information on a shop computer. Expecting to find an illustration showing the location of a bulb for illumination, he finds the procedure for removal of the illumination control. He then begins the disassembly procedure to get to the components involved. Rather than a conventional light socket and bulb, Ethan finds a fiber-optic lighting assembly that provides illumination for the shifter indicator.

Fiber optics components have been used by Mercedes-Benz, BMW, Daimler, Audi, Porsche, Volvo, and Cadillac. Some of the applications include dashboard lights, interior lights, taillights, and entertainment and information systems.

Refraction is defined as the change in direction of light due to a change in speed as it passes from one medium to another. A reference for the path of light is described in respect to the normal path of light. If light enters a new medium and is slowed down, it bends to the normal path. If it speeds up, it bends away from the normal path.

Aftermarket lighting accessories that are available today include fiber-optic taillights with reverse lights for certain vehicles. These lights are advertised to be considerably brighter than stock units and to light up faster and run cooler. Due to the use of fiber rather than lightbulbs, fiber-optic lights are said to be vibration-proof.

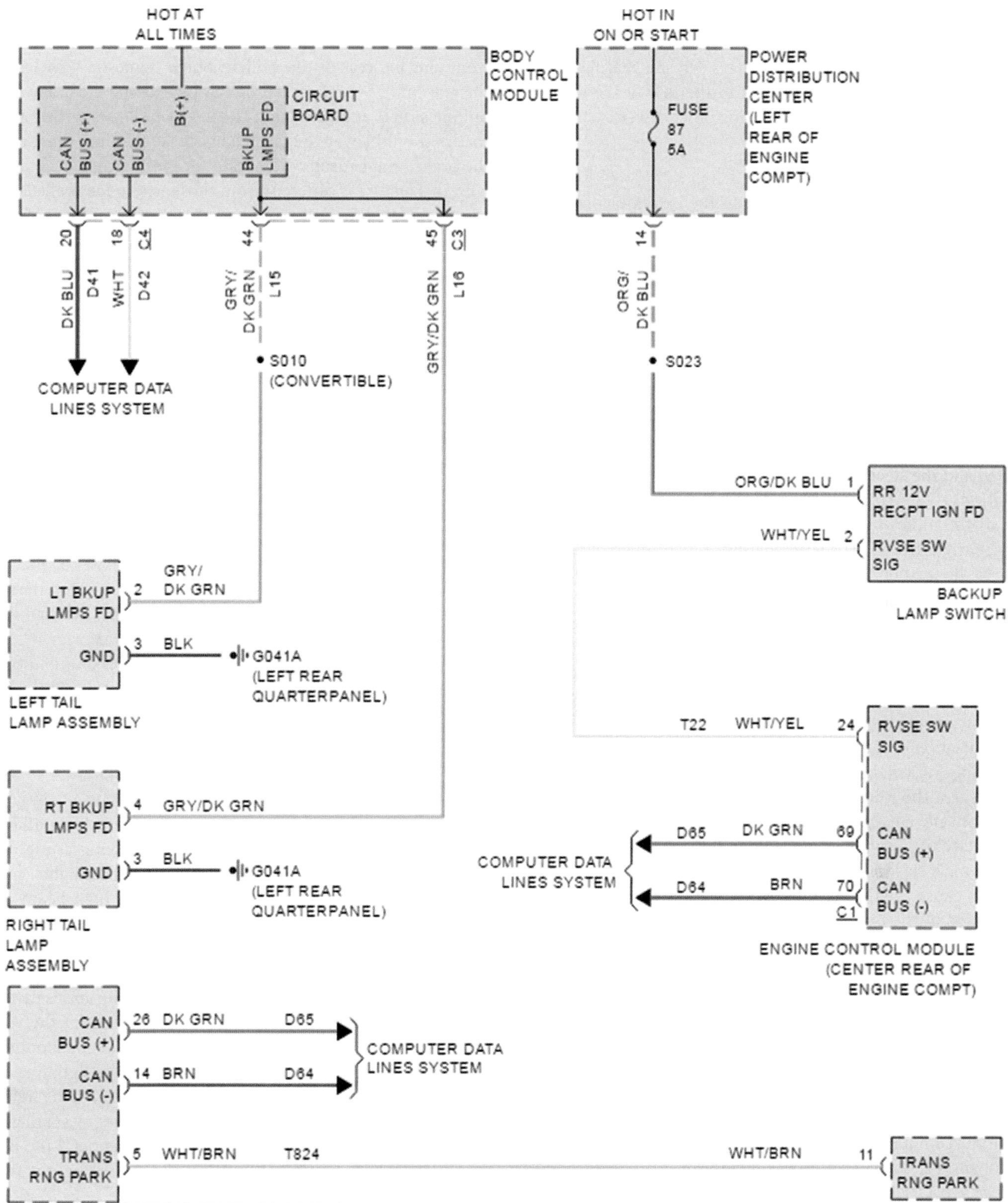

FIGURE 55-17 Typical wiring diagram for a backup light circuit.

and the ground circuit to allow current to return to the battery through the vehicle chassis.

When the ignition is on, and the vehicle is placed in reverse gear, the current flows from the battery, through the ignition switch, and through the closed reverse lamp or transmission position switch on the transmission. Electrical current then flows out of the closed switch to the backup lamps and returns to the battery by the vehicle chassis ground circuit. Modern vehicles use network/bus systems and the BCM to command the backup lights to come (**FIGURE 55-17**).

Backup lights are the only white lights on the rear of the vehicle. Because they operate only in reverse, other drivers can tell that the driver is backing up. This is why it is illegal to have broken tail/brake/turn signal lenses in the rear of the vehicle: White light would be visible, confusing other drivers.

Turn Signal Lights

Turn signal indicators are located on the extreme corners of the vehicle. They are usually amber in the front and can be either red or amber in the rear. A column-mounted switch, operated by the driver, commands a pulsing current to the indicator lights on one side of the vehicle or the other. These pulsing lights warn other road users of the driver's intended change of direction.

Once activated, they continue until the switch is cancelled either by the operator or by a cancelling mechanism in the switch. The cancelling mechanism operates to return the switch to its central or Off position after a turn has been completed and the steering wheel is returned to the straight-ahead position.

If the indicator switch is turned to indicate a right-hand turn, current from the battery typically flows through the fusible link to the ignition switch, where it is directed through a fuse to the flasher unit. The flasher unit uses a timing circuit to pulse the current flowing out of the flasher unit 60 to 120 times per minute. This pulsing current is directed through the indicator switch to the right-hand indicator lights at the front and rear of the vehicle, causing the lamps to flash on and off. An indicator light on the instrument cluster also blinks in sync with the turn signals. The operation of the flasher unit also produces a clicking sound to audibly inform the driver that the indicators are in operation.

When the turn signal switch is returned to the Off position, no current flows through the flasher unit, so the timer circuit is switched off. Older vehicles used a thermo-mechanical flasher unit that relied on heat from the current flow to cause the flasher unit to work. It is very important to use bulbs of the proper wattage on all types of flasher units, as the speed of the flash may be incorrect if incorrect bulbs are used.

These indicators can also be computer controlled. The BCM commands the appropriate turn indicators to come on as it sees an input from the turn signal switch, and flashes it at the proper rate. The computer turns off the turn signals when a steering angle sensor signals the steering wheel is being centered. It can also cancel the turn indicators if the vehicle is driven for a programmed amount of time or distance without the steering wheel being turned. Often a chime will alert the driver if the turn indicator has been left on for too long. On many computer-controlled turn signals, the turn signals will not work at all when the computer senses the wrong amperage flow in the circuit.

Turn Signal Lights-Domestic and Import Systems

All turn signal lamps flash, but there are variations in layout and design. This is most noticeable when comparing some domestic and imported vehicles. Imported vehicles tend to have separate amber-colored turn signal lamps on both the front and the rear of the vehicle. Some domestic vehicles use the rear brake lamps as turn signals by flashing the brake lamp on one side to indicate the turn (**FIGURE 55-18**). Wiring schematics should always be checked because the dual function of the brake lamps means the wiring for brake lamps on vehicles with this feature is different from those where the brake lamps are only related to braking. In such cases, current for the brake lights must flow first from the brake switch through the turn signal switch and on to the individual brake lights. This design tends to make the study of the current path more complicated, because there are two inputs—one for the turn signals and one for the brake lights. So the turn signal switch is much more complicated on a domestic-style turn signal switch than an import-style switch.

Hazard Warning Lights

All modern vehicles are equipped with hazard warning lights. This circuit connects into the turn indicator lights such that it pulses all exterior turn indicator lights and both indicator lights on the instrument panel (**FIGURE 55-19**). Hazard lights can warn other road users that a hazardous condition exists or that the vehicle is standing or parked in a dangerous position on the side of the road. The hazards also use a flasher unit that can be a separate unit or the same as that used for the turn signals. The BCM may control the hazard lights when it sees an input from the hazard switch or an input from the restraints control module indicating a vehicle crash.

Headlights

K55004

Headlights are built into the front of a vehicle to illuminate the road ahead of the vehicle when driving at night or in other conditions of reduced visibility. In headlights, most vehicles require two beams to provide for a high-beam and low-beam operation. The beams are created by separate filaments included either in one lightbulb or in separate bulbs. These filaments must be positioned correctly in relation to the highly polished reflector. This is called focusing and is carried out when designing the light assembly and lens for the vehicle. The high beam filament is positioned at the focal point of the reflector to project the maximum amount of light forward and parallel to the reflector axis (**FIGURE 55-20**). This light is then shaped by the lens, which is made up of many small glass or plastic prisms fused together. These prisms bend the light horizontally and vertically to achieve the desired light pattern for road illumination.

The low-beam filament is often placed above and slightly to one side of the high-beam filament. Mounting the low-beam filament in this position produces a beam of light that is projected slightly downward and toward the curb side (**FIGURE 55-21**). With this arrangement, the high-beam filament produces the most concentrated light output, whereas the low-beam filament gives a downward and dispersed beam that is less likely to blind oncoming drivers.

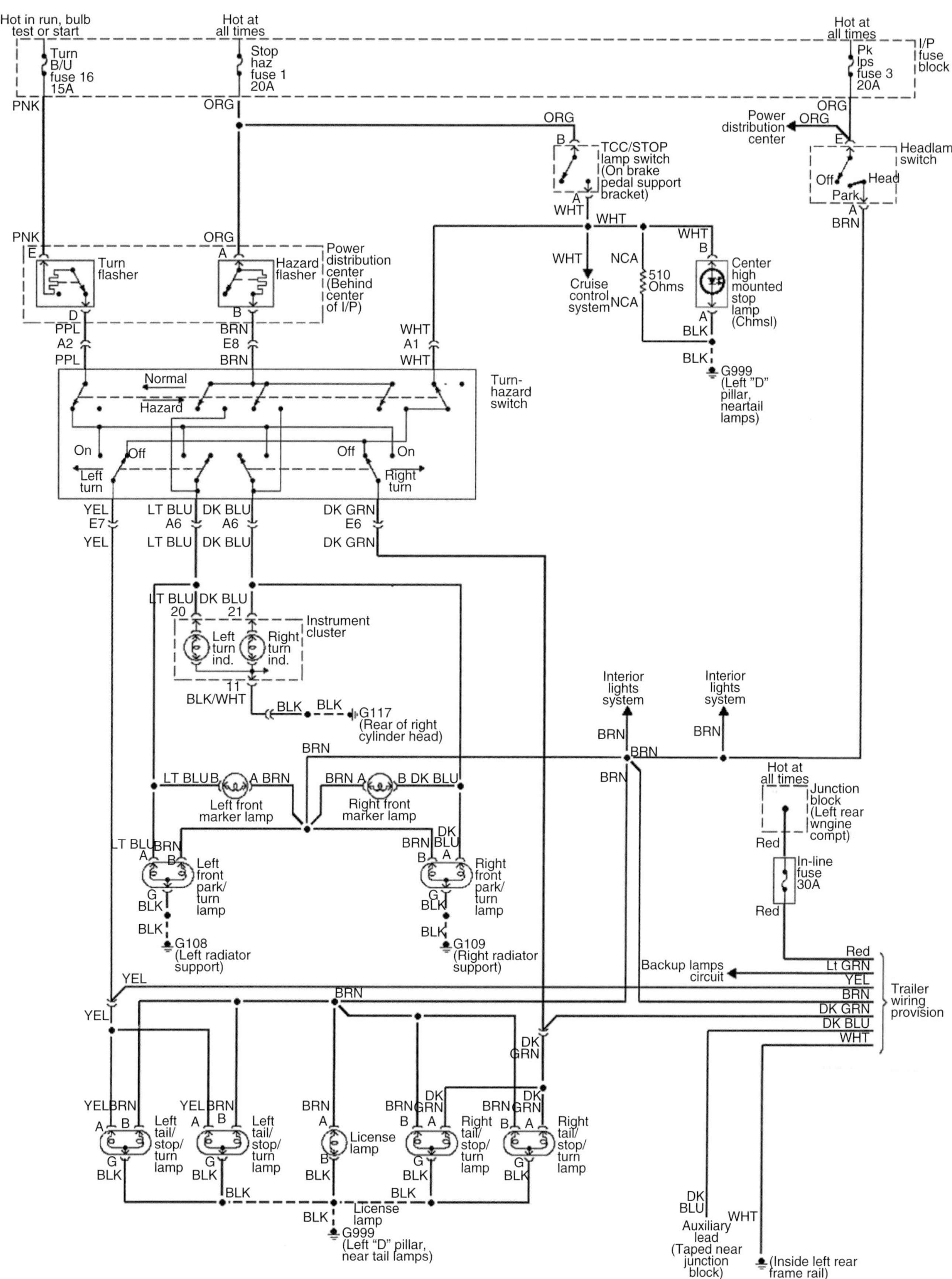

FIGURE 55-18 Typical wiring diagram showing domestic-type turn signals.

FIGURE 55-19 Import-style turn signal circuit with integrated hazard warning lights.

FIGURE 55-20 High- and low-beam filament position in relation to the reflector.

FIGURE 55-21 Low-beam filament position in relation to the reflector.

Headlight Design

Generally speaking, the headlight circuit consists of the battery, the fusible link or maxi-fuse, headlight fuses, the headlight switch, the headlight relay(s), the beam selector switch, headlights, the high-beam indicator light, wiring of a suitable size to carry the electrical current through the circuit, and the ground circuit. Modern vehicles use the BCM to manage headlight functions.

On older, non-computer-controlled systems, when the headlights are switched on, current is supplied from the battery and proceeds through the fusible link or maxi-fuse, headlight fuse(s), headlight switch, and dimmer switch, reaching either the low or the high beams. In more recent, non-computer-controlled systems, some vehicles use relays to control the current through the low- or high-beam circuits. In this configuration, the beam selector switch either powers or grounds the relay windings in each of the relays, depending on the position of the beam selector switch. Activating the beam selector switch in the low-beam position creates a magnetic field inside the low-beam relay that closes the relay contacts, which allows electrical current to flow to the low beams. Activating the beam selector switch in the high-beam position creates a magnetic field inside the high-beam relay that closes the relay contacts, turning on the high beams.

The beam selector switch is a single-pole, double-throw switch, meaning it has one movable pole but makes contact in two positions. In the high-beam position, the current is switched to the high-beam circuit. In the low-beam position, the current is switched to the low-beam circuit. The beam selector switch can be set to switch a common power input to two outputs, or it can switch a common ground input to two outputs.

As with other advanced-technology onboard systems, modern vehicles use a network/bus system to manage these and other headlight and dimmer functions. On many vehicles, the headlight switch and dimmer switch are just inputs to a control module that sends a headlight "on" or "off" request over the network. The appropriate light controller processes the request and actually commands the lights on or off, sometimes with the help of relays-especially for high beams.

A sealed-beam headlight has a highly polished aluminized glass reflector that is fused to the optically designed lens. It is a completely sealed unit that has the filaments accurately positioned in relation to the reflector. Most older vehicles used two dual-filament 7" (178 mm) round lamps; for a time, four single-filament 5.25" (133 mm) round lamps were used. Other older vehicles used two large dual-filament 7.875" (200 mm) rectangular lamps, or four smaller rectangular single-filament lamps. Regardless of size, when a filament fails in a sealed-beam light, the whole sealed unit must be replaced.

A semi-sealed beam headlight uses a replaceable bulb with a prefocus collar. The collar locates the bulb in the headlight and also controls the correct positioning of the filaments to the reflector and lens (**FIGURE 55-22**). Some replaceable headlight bulbs have a partial shield below the low beam filament. This shield prevents light from the filament from striking the lower part of the reflector, which would be reflected higher than the midpoint of the lamp. The shield provides the primary shape of the low beam. The final shaping of the beam is carried out by small cylindrical prisms in the headlight lens. This provides a low beam that is asymmetrical. The asymmetrical lens pattern causes light to be thrown upward at a 15-degree angle on the curb side to better illuminate objects, persons, or animals close to the road.

Types of Headlights

Headlight systems have traditionally used reflector-type lighting systems. In a reflector headlight, the light from the bulb is reflected forward by a specially shaped reflector. An alternative is a projection-type headlight system. This type of headlight often has a smaller front lens; however, it produces a high-intensity forward beam. It uses a lens system, rather than the traditional reflector system, to project the light forward (**FIGURE 55-23**). A projector-style light can use a standard incandescent bulb or, more commonly, an HID light.

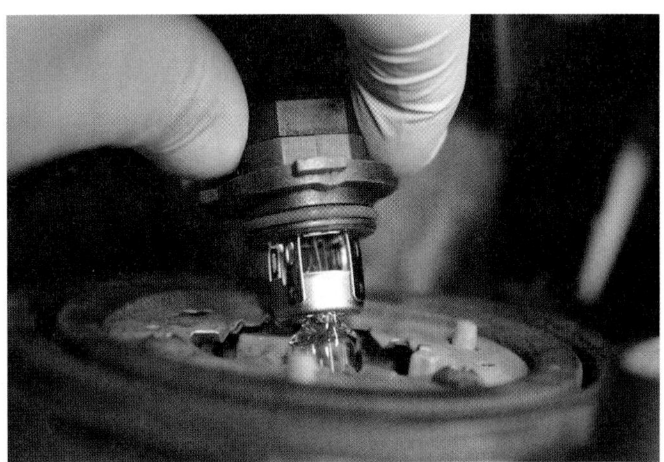

FIGURE 55-22 Replaceable halogen bulb.

FIGURE 55-23 Projector bulb assembly.

FIGURE 55-24 HID headlight assembly.

HID lights use light from an electric arc rather than heating up a filament until it glows (**FIGURE 55-24**). High voltage is applied to tungsten electrodes. Xenon gas inside the bulb is then ionized and creates an electrical path between the electrodes, which lowers the resistance of the gap. As the temperature rises, metallic salts are vaporized and provide a stable arc, emitting much light. To initially jump the gap, a ballast and igniter raise the vehicle's low-voltage direct current (DC) to as much as 25,000 volts alternating current (AC). Once the light reaches full operation, voltage is maintained between approximately 40 and 85 volts AC, depending on the system. HID lights give off a brighter and bluer light than halogen bulbs with less electrical energy, so they help fuel efficiency slightly. They last two to four times as long as a halogen bulb but are quite a bit more expensive. Also, they take up more room in the engine compartment.

Some vehicles use LED lights as headlights. They are not currently as bright as HID lights but are at least as bright as halogen lights. LED lights are known for low power consumption, but for LEDs to create enough light to function as headlights, they must consume almost as much energy as HID lights. Still, LED technology continues to improve. One drawback of current LED headlights is that high temperatures degrade or damage them. Thus, heat sinks and cooling measures are needed, which add complexity, cost, and space.

Night vision is another relatively new technology that enhances a driver's visual perception in dark or poor weather conditions. There are two types of night vision systems: active and passive. Active systems use an infrared light generator that projects infrared light in front of and to the side of the roadway ahead. A special camera picks up the reflected infrared radiation. The image is then displayed either on the windshield, using a heads-up display, or on an LCD screen on the dash or navigation system (**FIGURE 55-25**).

Passive night vision systems use a heat-sensing camera (thermal imaging) to pick up thermal radiation emitted by objects. This system does not have an infrared light source on

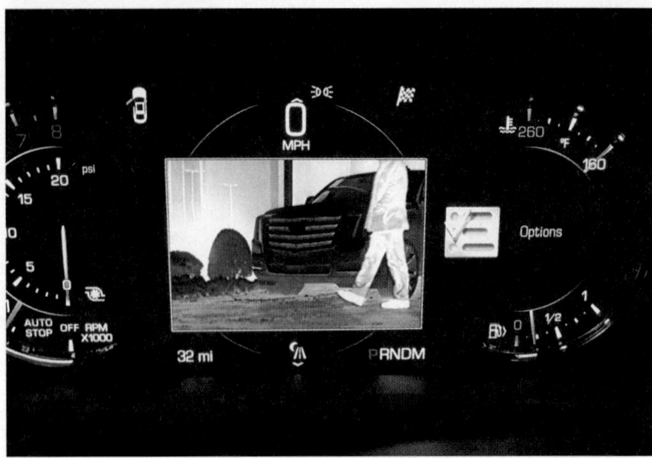

FIGURE 55-25 Night vision enhances visibility under dark or poor weather conditions.

the vehicle. The captured thermal image is then displayed either on a heads-up display or on an LCD screen. One benefit of the thermal system is that it can be programmed to recognize pedestrians and animals and can then either place an outline around them on the display or flash a warning symbol.

Daytime Running Lights

Daytime running lights (DRLs) are an additional safety feature designed on vehicles to improve the vehicle's visibility to other drivers in all weather conditions. They are existing lights that turn on when the vehicle is running and turn off when the engine stops. The system operates the headlights in the front, which are typically operated at about a 60% power level, providing light without excessively decreasing bulb life or using full electrical power. The system operates the taillights on the rear.

DRLs are mandatory on modern vehicles licensed in Canada and some other countries. In the United States, DRLs are permitted, but not required. Their use is somewhat controversial. First, if they are too bright, they can cause daytime glare. Second, they tend to mask the visibility of turn signals, making it harder for other drivers to determine a vehicle's intent. Overall, they have not been proven to reduce accidents or increase safety. They also require energy to operate, which reduces fuel economy and increases carbon dioxide emissions.

Smart Lighting

Different lighting technologies can be mounted individually, or together, to form comprehensive adaptive lighting systems, otherwise known as smart lighting. Automatic headlight leveling systems ensure that the headlights are always correctly aligned, regardless of the load the vehicle is carrying. When a vehicle is loaded, the suspension settles, changing the angle of the headlights to the road. The resulting angle can cause glare into oncoming traffic. Auto-leveling headlights use a sensor to monitor the angle of the vehicle to the road (**FIGURE 55-26**). This information is then used to automatically adjust the headlights. They adjust up or down, so the headlights always stay correctly adjusted. In this system, the load sensors provide information to a control module that operates servomotors installed on the headlights, causing the headlights to tilt up and down.

Another function of adaptive headlights is the ability to adjust the length of the headlight beam based on the distance of oncoming traffic. In this system, a dash-mounted camera can determine the distance of oncoming traffic and relay that information to the BCM, which then continuously adjusts the length of the headlight beam. Some manufacturers add the ability to shade or block light within the beam, based on the position of oncoming traffic. Either shields are moved to block light from specific areas or, in the case of LED lights, specific LEDs are switched off to provide the shaded area. In either case, maximum illumination is maintained without blinding oncoming drivers.

Some types of headlights can improve forward lighting while cornering. This type of headlight swivels slightly through the horizontal plane, so that the headlight turns into the corner as the driver turns the steering wheel (**FIGURE 55-27**). This provides better illumination of the road as the vehicle turns through the corner. In this system, sensors from the steering wheel provide information to a BCM that causes servomotors installed on the headlights to operate, causing the lights to swivel in the desired direction.

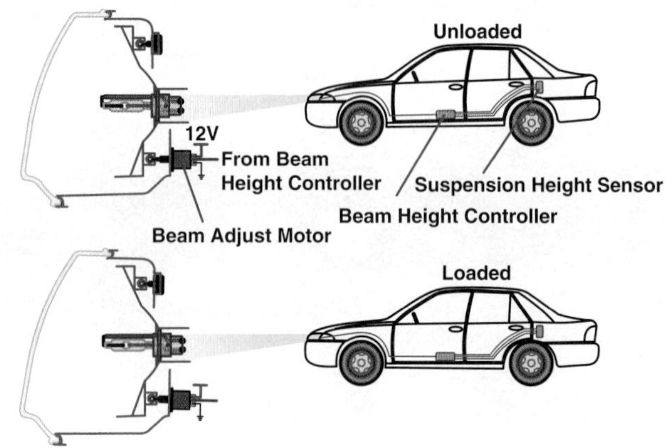

FIGURE 55-26 Auto-leveling headlights.

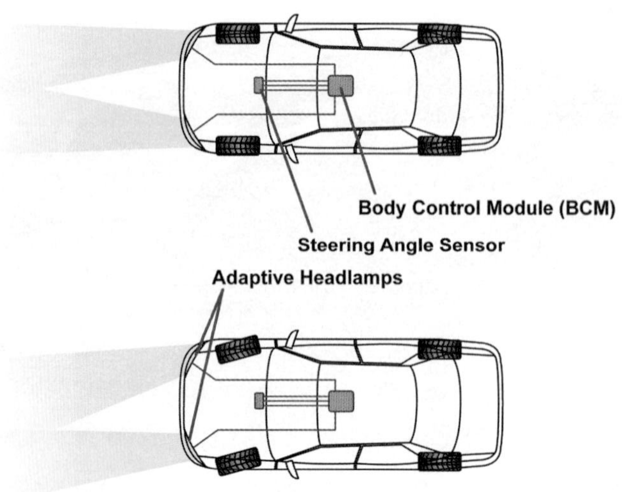

FIGURE 55-27 Headlights that shine around a corner.

Computer Controlled Lighting Systems

K55005

When driving at dusk, the ambient light can fade slowly. In some vehicles, an ambient light sensor determines when light levels are low. It is read by the BCM, which turns the headlights on, when set on "Automatic" (**FIGURE 55-28**). Automatic "dimming" headlights use a light sensor located at the front of the vehicle or on the dash to detect light from oncoming vehicles. If the headlights are on, low beam is selected automatically. This prevents oncoming traffic from being dazzled by the vehicle's high beam. If the light sensor is not signaling that a vehicle is approaching within the defined range, the BCM will select high beams. As a vehicle approaches, the light sensor signals the BCM, which selects the low beams again. Most vehicles have an adjustable knob that sets the sensitivity of the light sensor. Making it less sensitive causes the BCM to switch to low beam when the approaching vehicle is closer in range. More sensitivity causes the BCM to select low beams when the vehicle is farther away.

A delayed "off" feature allows the headlights to be left on after the engine has stopped and the doors are closed or locked. This feature allows the driver to safely see his or her way from the vehicle. The headlights automatically turn off after a period of time. On some vehicles, the BCM turns the headlights off if the headlights are left on after the ignition is off and the driver's door is left opened. On other vehicles, headlight warning alarms sound if the headlights are left on after the engine has been turned off and a door is opened. This feature reduces the risk of the headlights draining the battery. Many vehicles have an adjuster or can be programmed to tailor the length of this delay time up to a few minutes.

On some vehicles, the BCM, in concert with a lamp-out module, illuminates a warning light in the instrument panel cluster when an exterior lightbulb such as a stop, tail, or turn signal light has failed (**FIGURE 55-29**). To detect this malfunction, the lamp-out module compares current flowing through these circuits to values stored in the computer's memory. If the current flowing is outside the set parameters, the warning light will illuminate. Besides providing the correct amount of light, this is another reason to always use bulbs of the proper wattage.

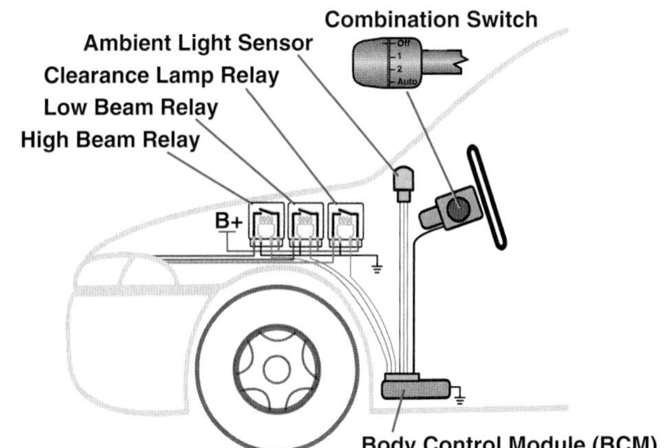

FIGURE 55-28 Automatic headlight system.

FIGURE 55-29 A typical light-out warning system.

The BCM can be programmed to command the interior lighting to remain on after the doors have been closed for a set period of time, to allow the occupants to locate a seat belt or insert the ignition key. It can also turn on the interior lighting when the ignition is turned off to allow the occupants to locate door handles and other items.

Networking and Multiplexing

Even the most basic modern vehicles include many electronically controlled systems. If each electronic system had its own ECU, harness, and sensors, the weight of the added components would negate some of the efficiency it provides and would add cost to the vehicle. A vehicle's multiple electronic systems could require miles of insulated wiring, splices, and terminals.

One solution to reduce the amount of wiring required is the use of a system that integrates sensors into a common wiring harness, called a bus. The bus combines all the individual systems wherever possible into a **multiplexed** serial communications network. An added advantage of such a system is that with less wire and fewer connections, there is less chance of dirty connections causing faults. Of course, if a network fails, it can shut down the entire network and with it the controlled devices. Also, troubleshooting a networked system requires more training and more tools than a simple digital volt-ohmmeter (DMM).

With networking, a single (or multiple) network consisting of one or more twisted pairs of wires, with terminating resistors at the end of each pair, carries digital information throughout the entire vehicle. In many cases, there is a medium- or low-speed bus for non-safety systems and a high-speed bus for systems such as antilock brakes, supplemental restraint systems, and many of the engine controls. All components "listen" to the signals carried on the network and respond when requested to do so. All networked data/information is shared throughout the vehicle. Such a system is often referred to as a controlled area network bus (CANbus) (**FIGURE 55-30**). The thin twisted pair of wires connects all the onboard control modules to each other. Thus, the powertrain control module (PCM), BCM, electronic brake control module, transmission control module, and

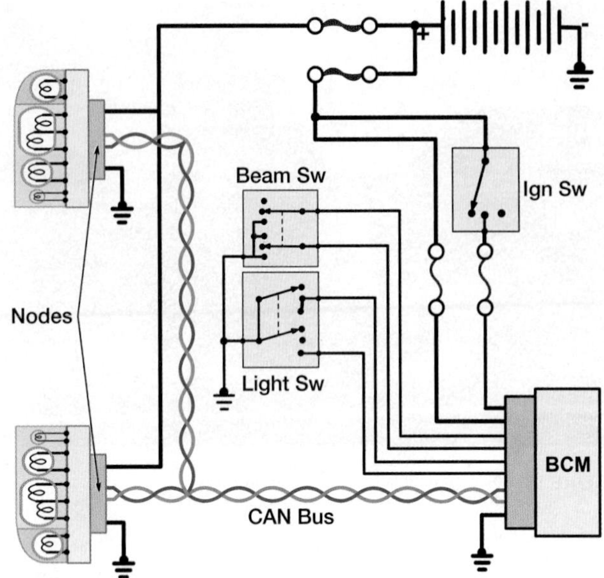

FIGURE 55-30 Typical CANbus diagram.

many others are networked together and communicate, much like "party lines" on old-fashioned telephone systems. Output devices that respond to commands on the network are referred to as nodes. All nodes hear all data on the network but only respond when called upon to perform some action or diagnostic function.

The advantage of a multiplex network is that it enables a decreased number of dedicated wires for each function, and therefore a reduction in the number of wires in the wiring harness. This network also reduces system weight and improves reliability, serviceability, and installation.

Sensor data multiplexed on the network are prioritized. For example, such information as vehicle speed and engine temperature data is shared on the network for other systems such as the instrument panel cluster to access and use for their purposes. This design helps to reduce the number of sensors needed. Some information being shared takes priority over others. For example, a message related to a loss of engine oil pressure would take precedence over a lamp-out message. In such a case, the vehicle operator must be immediately alerted to the urgent fault.

Networking of data also allows greater vehicle content flexibility because functions can be added or modified through software changes. Other control units, such as those related to trailer towing, can be added as required to the system by simply connecting them into the network and updating the software to integrate it into the network. You can think of this process as being similar to adding a printer to your home computer system; plug it in, configure it, and start using it.

A diagnostic scan tool is connected to the CANbus network to both monitor and extract operational information to assist in diagnosis and fault finding. Factory scan tools and other so-called bidirectional scanners are used to command engine and vehicle components to operate in order to confirm their correct function.

▶ Lighting Circuit Testing and Service

S55001

Each lighting circuit has particular operating characteristics based on its purpose and the system design. Knowing how a circuit is designed to operate and how it interacts with other related circuits helps you to better know how to go about diagnosing the system. A typical lighting circuit may be made up of the battery, fuses or circuit breakers, switches, relays, lamps, and wiring to connect the components. Manufacturers design the circuits to operate in specific ways and then create schematics or wiring diagrams, which are a diagrammatic layout of the entire circuit. Learning how to read wiring diagrams takes time and practice. Understanding circuit design, operation, and wiring diagram information can then be used to diagnose circuit faults.

Lighting System Wiring Diagrams

The layouts of electrical circuits and their components are shown as diagrams made up of symbols and connecting lines. Being able to read a wiring diagram is probably the most important skill when trying to trace and correct a fault in an electrical system. For detailed explanation of wiring diagrams, see the Electrical Components and Wiring Diagrams chapter.

Not all wiring diagrams use the same symbols or the same numbering system. It is helpful to refer to the manufacturer's service information for specific details on how to read a particular wiring diagram.

Some of the common symbols and what they represent are shown in **TABLE 55-1**.

HID Safety Precautions

N55001

HID lamps produce a very bright white light. Manufacturers use various designs of HID lamps. Regardless of the design, they generally require a high-voltage spark of up to 25,000 volts to start and a high operating voltage (e.g., 40–85 volts AC) to maintain light. To generate the voltages required to operate, a transformer and electronic circuitry called a ballast are used to supply electrical components.

Several safety precautions should be taken when working on HID systems. There is a risk of electrocution, burns, or shock from the high voltages generated by the HID system. If diagnosing the HID system, be very careful when working on the system when it is live. You should wear safety glasses, high-voltage safety gloves, and safety boots, and you should ensure that the vehicle, engine compartment, and ground under the vehicle are dry. Do not touch the ballast while it is operating; it will often be generating a lot of heat. Persons with active electronic implants, such as heart pacemakers, should not work on HID headlamps. If changing out the bulb, make sure the headlights are turned off. The manufacturer may specify that the battery be disconnected when replacing the bulb. If so, follow the manufacturer's instructions.

TABLE 55-1: Common Symbols

Battery and fuse symbols	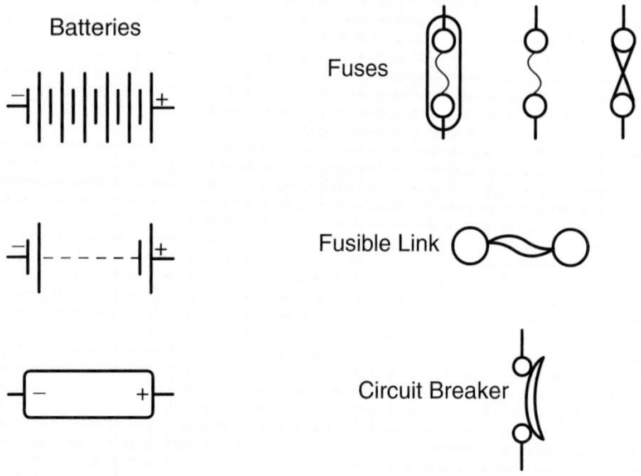 Batteries; Fuses; Fusible Link; Circuit Breaker
Ground and connector symbols	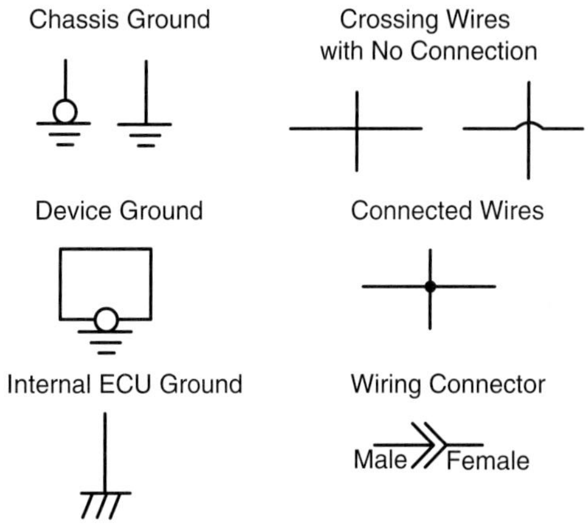 Chassis Ground; Crossing Wires with No Connection; Device Ground; Connected Wires; Internal ECU Ground; Wiring Connector (Male, Female)
Switch symbols	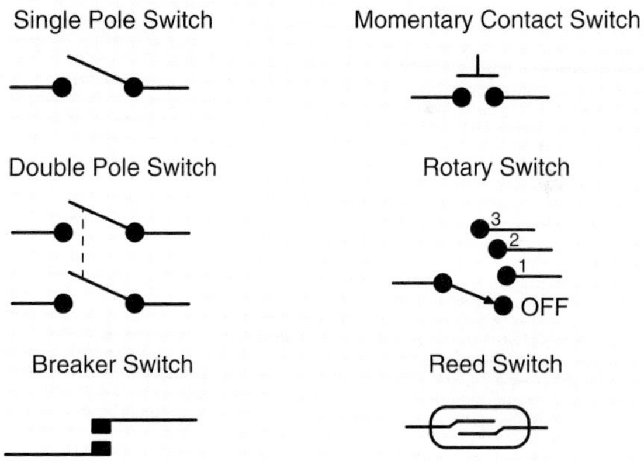 Single Pole Switch; Momentary Contact Switch; Double Pole Switch; Rotary Switch (3, 2, 1, OFF); Breaker Switch; Reed Switch
Resistors, coils, and relay symbols	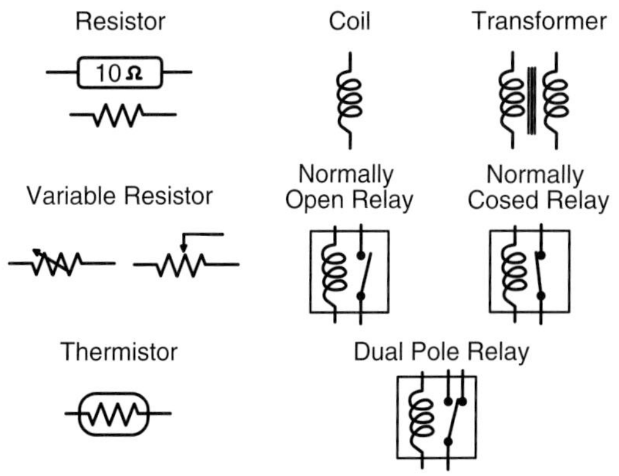 Resistor (10 Ω); Coil; Transformer; Variable Resistor; Normally Open Relay; Normally Cosed Relay; Thermistor; Dual Pole Relay

(Continues)

TABLE 55-1: Common Symbols (*Continued*)

Semiconductor symbols	Diode; Zener Diode; NPN Transistor; Light Emitting Diode (LED); PNP Transistor; Photo Diode; Photo Transistor
Capacitor and device symbols	Piezoelectric Device; Single Filament Globe; Capacitor; Dual Filament Globe; Electrolytic Capacitor; Spark Gap
Motors, generators, and solenoid symbols	Motor; Alternator (ALT); Motor with Permanent Magnet (M); Pulse Generator; Motor with Wound Magnet (M); Solenoid
Gauges and warning device symbols	Voltmeter (V); Chime; Ammeter (A); Speaker; Buzzer; Horn

There is also a risk of injury caused by exposure to ultraviolet light produced by the HID lamp if the lamp is operated outside of its housing. Once ignited, the pressure inside an HID bulb can build up to a very high pressure (around 220 pounds per square inch, 1517 kilopascals) due to the high operating temperature (about 1500°F [816°C]). This pressure creates a potential explosion hazard, so do not attempt to power an HID bulb outside of the headlamp assembly to test it, or operate it near flammable gases or liquids. Also, the bulb must be maintained in a horizontal position when it is on; otherwise, it may overheat and fail. HID headlamps use various heavy metals in their construction; therefore, it is important to always dispose of the bulbs in an environmentally friendly way. Avoid breaking bulbs, as there is also a risk of poisoning caused by inhalation or skin contact of heavy metal vapors and toxic salts.

Checking Lighting and Peripheral Systems

When checking lighting and peripheral systems, be sure to work in a systematic manner to avoid missing a faulty bulb or other component. A vehicle may have warning lights that activate only if that circuit is in use. You may need to turn that circuit on to see the warning light. If you are unsure of where these are, ask your supervisor.

To check lighting and peripheral systems, follow the steps in **SKILL DRILL 55-1**.

SKILL DRILL 55-1 Checking Lighting and Peripheral Systems

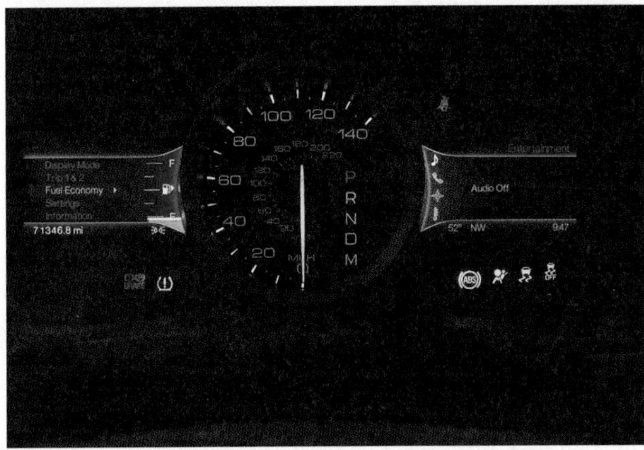

1. In a darkened area, turn on or activate the ignition. The dash warning lights should be displayed. Start the engine. If any warning light stays on after the engine is started, it could indicate a problem in one of the vehicle's safety or mechanical systems. If you are unsure about what any of the warning lights mean, consult the owner's manual.

2. Have someone stand behind the vehicle to report any problems. Turn the ignition to run. Switch on the park lights, taillights, marker lights, and license plate lights. Do the same for left and right turn indicator lights. Depress the brake pedal to make sure the brake lights work and have the proper intensity. Also, place the transmission in reverse and check the backup lights.

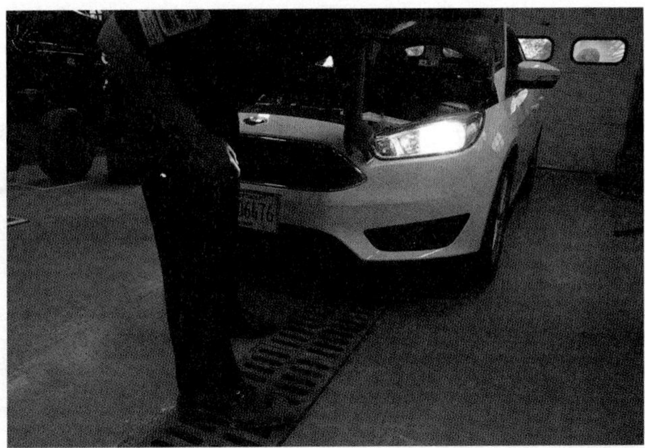

3. With someone in front of the vehicle, make sure the high and low headlight beams, park lights, turn signals, driving lights, fog lights, and cornering lights are all working properly.

4. With the interior light switch in the correct position, open the driver's side door to make sure the interior lights work. If any of these lights do not operate, you may need to replace a bulb or diagnose the fault.

Inspecting and Changing an Exterior Lightbulb

Lightbulbs have a limited life span and burn out occasionally. Inspecting the lighting system's operation periodically, such as during oil changes, will help identify any lightbulb issues. If none of the bulbs in a particular circuit are working, there may be a bigger electrical problem to resolve, requiring diagnosis. Many tail, brake, and signal lightbulbs have more than one filament inside them. These bulbs normally have offset pins to ensure proper orientation of the electrical contacts in the socket. Be sure to look carefully at the bulb you are replacing to make sure you do not try to force the bulb in the wrong way. Also, make sure you don't replace a dual-filament bulb with a single-filament bulb because that can short both lighting circuits together, causing the lights to work very strangely. Some bulbs have a colored glass envelope that enables them to be used with a clear lens. If you replace a bulb of this type, make sure you replace it with one of the same color.

SAFETY TIP

The glass in a lightbulb is thin, and if it breaks, it is extremely sharp. Because bayonet-style lights need to be pushed in to be released, and they can become corroded in place, it may take a lot of force to push them in. Placed under such high pressure, the glass can break and severely cut the skin and even the tendons in fingers. If a bulb does not come out easily, stop, use a folded rag, and try to remove it again. If it shatters, the rag will help to protect your fingers.

To check and change an exterior lightbulb, follow the steps in **SKILL DRILL 55-2**.

Checking and Changing a Headlight Bulb

There are many types of headlight bulbs available. Always make sure you replace a bulb with one of exactly the same type. Sealed-beam units require that the whole light be replaced when one filament has failed. If the reflector in the sealed-beam light shows signs of degradation, it also indicates that you must change the unit. If both lights operate but are not bright when switched on, start the engine to see if this solves the problem; the battery may be in a poor state of charge. When replacing a halogen bulb, avoid touching it with your fingers, which can leave a residue from your fingers on the outer surface. This residue can cause the bulb to crack or shatter and burn out after a short time of operation. If you inadvertently touch the bulb, clean it with alcohol and a lint-free cloth. Do not use gasoline or paraffin to clean the bulb.

To check and change a headlight bulb, follow the steps in **SKILL DRILL 55-3**.

▶ TECHNICIAN TIP

A customer with sealed-beam headlights once complained of poor headlight performance. When the technician turned on the lights, they both worked but didn't seem to create much light. As the technician opened the hood to investigate further, he noticed something strange. The headlights were shining light out the back of the lights, illuminating the engine compartment. He inspected the lights and found that the metallic reflector in the sealed beams had pretty much all flaked away, allowing most of the light to exit the rear of the bulb, instead of being reflected forward. Replacing the sealed-beam headlights with new ones fixed the customer's concern.

Aiming Headlights

N55002

Although the principle of aiming headlights is the same in the majority of cases, the legal rules can differ from region to region. Be sure to check the requirements for your location. If you are unsure of what these are, ask your supervisor. Some manufacturers may suggest that the headlights be aimed on high beam; others, on low beam; and in cases of separate high and low beams, both beams may need to be aimed independently. The manufacturer may also suggest that a load be

SKILL DRILL 55-2 Checking and Changing an Exterior Lightbulb

1. Remove the cover to expose the bulb. If the bulb is pin mounted, gently grip the bulb and push it inward. Turn the bulb slightly counterclockwise, and remove it from the bulb holder. Some bulbs pull straight out.

2. Inspect the bulb holder to make sure there is no corrosion. If there is, clean it with a bulb socket wire brush or emery cloth.

3. Insert the new bulb into the bulb holder, depress it fully, turn it slightly clockwise, and release it. Test it by switching it on and off. Then replace the cover, and test it again.

SKILL DRILL 55-3 Checking and Changing a Headlight Bulb

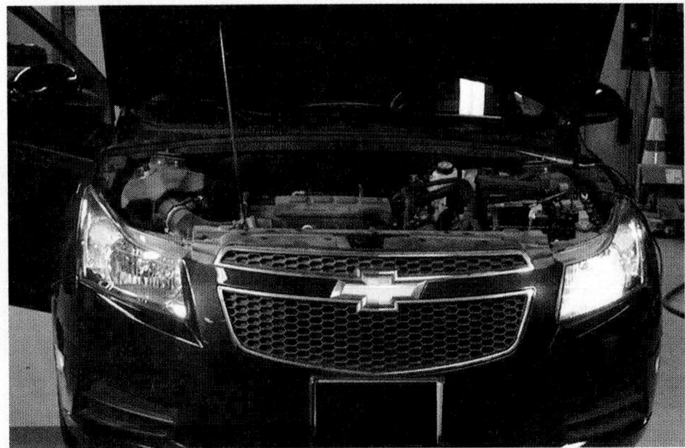

1. Switch the headlights on at low beam, then to high beam. Check that the high-beam indicator is operating. If one of the lights does not operate or is dim, that headlight will have to be diagnosed and potentially replaced.

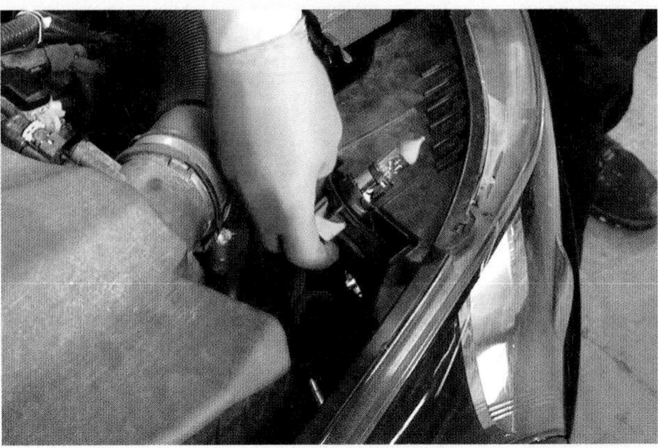

2. Test the vehicle headlights. Obtain the replacement lamp for the vehicle. Unplug the electrical connector at the back of the lamp unit.

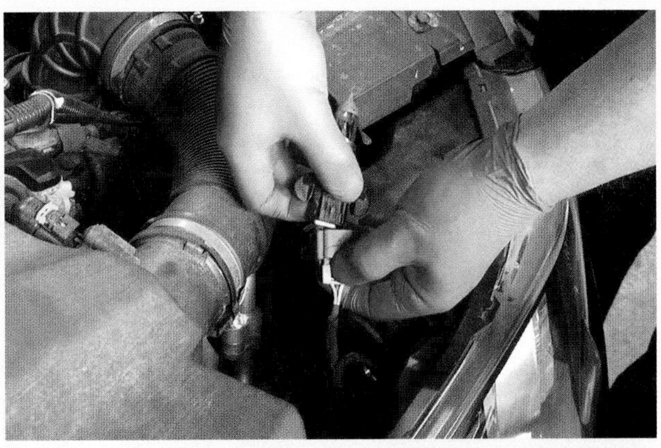

3. Remove the old bulb, and replace it with the new one. Handle the new bulb only by its base or, if supplied, by the card cover.

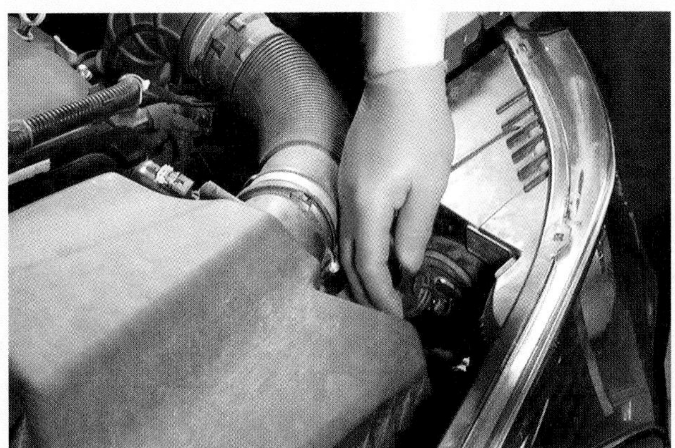

4. Replace the unit and the retaining ring or bulb assembly, and then plug in the connector. Switch on the lights again to confirm that they are both operating correctly.

placed in the vehicle to simulate the ride height of the vehicle when it is travelling down the road. Headlights are typically aligned both vertically and horizontally so that as much of the road is illuminated as possible without blinding oncoming traffic. Some vehicles have headlights that adjust automatically; do not try to adjust those unless you are diagnosing a fault in the system.

When aiming or aligning headlights, there are a couple of possible methods to utilize. The first, while no longer commonly used, involves an on-vehicle headlight aligner. This method was used extensively on sealed-beam headlights. The aligners are calibrated to the floor at the point where the front and rear wheels contact the concrete. Then the aligners are installed on the headlights with a built-in suction cup, which holds them in place. The aligners use a set of mirrors that can be calibrated to local regulations so that the headlights can be aligned properly.

With the introduction of aerodynamic headlights using removable halogen bulbs, off-vehicle headlight aligners became more popular. These aligners sit in front of the vehicle after being calibrated for the floor slope and orientation with the vehicle. With the headlights on, adjustments can be made by following the instructions on the aligner. Refer to the manufacturer's information for specific information regarding headlight aiming.

To aim headlights, follow the steps in **SKILL DRILL 55-4**.

SKILL DRILL 55-4 Aiming Headlights

1. Make sure the tires are inflated properly, the wheels point straight ahead, and there is no extra weight in the vehicle. Position the vehicle correctly in relation to the headlamp aligner unit, following the equipment manufacturer's instructions. Calibrate the aligner for any floor slope and for the vehicle being tested.

2. On the types of aligners that require the headlights to be on during alignment, turn the headlights on to a low-beam setting. The center of the illuminating beams should being the lower right quadrants of the chart or wall markings or as specified by the manufacturer.

3. The high beam should be centered, falling on the intersections of the horizontal and vertical marks or as specified by the manufacturer. If necessary, turn the adjustment screws on the headlight so the lights point to the correct places or bubbles on the levels are centered, depending on the type of aligner equipment you are using.

Headlight Brightness

S55002

Headlight brightness is critical to safe driving. Too dim, and the road ahead will not be adequately illuminated. Too bright, and the life of the headlight will be shortened, making it prone to early burnout. In most cases, the headlights are more likely to be too dim than too bright. Bulbs that are too bright indicate that either the charging system voltage is too high and needs to be measured with a DMM, or someone may have replaced the headlights with an aftermarket set, which may be illegal. Checking bulb numbers will verify this condition.

If the headlights are too dim, there can be several causes. The most common one is high resistance in the light circuit. This can be checked by measuring the voltage drop on both the power side of the bulb and the ground side. The voltage drop should be less than 0.5 volt on each side. If an excessive voltage drop is indicated, use a wiring diagram to research the circuit. Once you know how it is wired, use the DMM to isolate the voltage drop by using the meter to measure each segment of the circuit. When the high resistance is located, perform the needed repair, whether that is a wire repair or a connector, switch, or relay replacement. If the voltage drop is less than 0.5 volt on each side, suspect the lightbulb is wearing out. One way to verify this is with a light intensity meter. This meter can measure the amount of light energy the lamp is producing, so you can compare that to specifications (**FIGURE 55-31**). If the light is being supplied with the specified voltage, but it is not creating enough light, the bulb will have to be replaced.

FIGURE 55-31 A light meter being used to measure light intensity.

Warning Lights

S55003

Warning lamps are usually part of the instrument cluster and are controlled by electronic control units in the vehicle on a modern vehicle. Warning lamps provide the vehicle operator with information regarding the operation of the main vehicle systems, such as the vehicle battery charge level, oil pressure level, and airbag system status (**FIGURE 55-32**). When the ignition initially comes on, the warning lamps usually light up as part of a self-test to show that they are in working order. Once the engine starts, the warning lamps should extinguish and not come on again while the engine is running, unless there is a system fault. Some vehicles have warning lights that self-test while the vehicle is being cranked. This function is performed by the

FIGURE 55-32 Warning lamp.

"proofing circuit" in the ignition switch, which grounds certain warning lights when turned to the "crank" position.

Checking Digit\al Instrument Circuits

N55003

Instrument systems on modern vehicles are controlled by electronic circuits and sensors. Electronic control modules receive information from various sensors and send digital signals to various instruments to display the information. The data from sensors and instruments are sent and received on the CANbus vehicle network.

A diagnostic scan tool connected to the vehicle greatly assists in diagnosing digital instruments. Use the scan tool to check fault codes and vehicle PIDs to check if sensors and instruments are operating correctly. A DMM and oscilloscope can also be used to check sensor outputs and confirm correct operation. Research the diagnostic procedures and sensor outputs in the manufacturer's information. Always disconnect sensors before conducting resistance checks. Automotive test lamps should not be used on digital circuits because they draw too much current and can overload the circuit. Be careful not to accidentally short wires or terminals to ground or power, as doing so may damage the circuits.

You should always inspect sensors, connectors, and wires for corrosion or damage. Connector plugs and pins should be tight, with straight pins that have good connectivity and the proper tension. If back-probing wires or connectors is required, ensure that the pins are not damaged.

To check digital instrument circuits, follow the steps in **SKILL DRILL 55-5**.

Checking Warning Device Operation

Warning devices are installed on the vehicle to provide the driver with information about critical vehicle operating and safety systems, including the battery, supplemental restraint system (SRS), antilock brake system (ABS), park brake, oil pressure, and engine temperature. In some cases, the systems use sensors to relay information to the warning device, such as low oil pressure. In this case, the oil pressure may need to be measured with an external oil pressure gauge. If the oil pressure is okay, then the oil pressure switch may be faulty and will have to be replaced. In other circuits, such as the ABS system, the PCM may send a signal through each wheel speed sensor to measure its continuity. In this case, you need to measure the resistance of the sensor or the continuity of the wiring harness. The warning devices may be controlled by an electronic control unit (ECU).

SRS systems are very dangerous; for example, they control the deployment of airbags by using pyrotechnical charges. Check the manufacturer's procedures and properly disable the SRS system before working on it. Research vehicle fault codes and PIDs in the service information. A DMM or oscilloscope may be used to check signal voltages from sensors used on the vehicle warning devices.

To check warning device operation, follow the steps in **SKILL DRILL 55-6**.

SKILL DRILL 55-5 Checking Digital Instrument Circuits

1. Identify the digital circuit to be tested. Research the correct test procedure from the manufacturer's information.
2. Check fault codes and circuit PIDs, using a scan tool.
3. Inspect the sensor, wires, and connectors.
4. Test the sensor with a DMM or oscilloscope to determine correct output.
5. If the sensor tests okay, check the wiring harness for continuity back to the instrument panel, and determine necessary action.

SKILL DRILL 55-6 Checking Warning Device Operation

1. Identify the circuit to be tested. Research the correct test procedure from the manufacturer's information.
2. Check fault codes and circuit PIDs, using a scan tool. Inspect the sensors, wires, and connectors.
3. Test the affected circuit and any sensors with a DMM or oscilloscope to determine the correct output.
4. If sensors or devices test okay, check the individual circuits for continuity.
5. Identify the circuit to be tested. Research the proper test procedure from the service information.

Wrap-Up

Ready for Review

- Incandescent lamps consist of one or more filaments, which heat up until they glow.
- Halogen lamps have a much longer life and are generally brighter and produce more light per unit of power consumed, but they become very hot during use.
- High-intensity discharge (HID) headlamps produce light with an electric arc rather than a glowing filament.
- Light-emitting diodes (LEDs) have been used for some time in various automotive applications, such as warning indicators and alphanumeric displays.
- One advantage of LEDs is that they turn on instantly.
- All lamps or lightbulbs have letters and numbers stamped on them that indicate the power consumed by bulb operation at the nominal operating voltage.
- Lighting regulations should always be consulted before modifying or adding to any of the vehicle lighting systems.
- Park, tail, and marker lamps are all low-intensity or low-wattage lamps used to mark the outline or width of the vehicle.
- Brake lights, which may also be called stop lights, are red lights mounted to the rear of the vehicle.
- Reverse lights, also called backup lights, are the bright white lights on the back of the vehicle that allow the operator to see behind the vehicle when backing up.
- Hazard lights can warn other road users that a hazardous condition exists or that the vehicle is standing or parked in a dangerous position on the side of the road; they normally use the same bulbs as the turn signal indicators.
- The dim or low beam is usually set to the outside of the front, and the high beams are closer to the front center of the vehicle.
- The driving lights are installed on the front of the vehicle and provide higher intensity illumination over longer distances than standard headlight systems.
- Different lighting technologies can be mounted individually or together to form comprehensive adaptive lighting systems, otherwise known as smart lighting.
- Auto-leveling headlights use a sensor to monitor the angle of the vehicle to the road.
- In some vehicles, an ambient light sensor determines when light levels are low; this then allows the body control computer to turn the exterior lights on.
- Networking, CANbus, or multiplexing light systems eliminate the amount of wire required for the lighting system and are controlled through a main computer.
- Warning lamps provide the vehicle operator with information regarding the operation of the main vehicle systems, such as the vehicle battery charge, oil pressure level, and airbag system status.
- A typical lighting circuit is made up of the battery, fuses or circuit breakers, switches, lamps, and wiring to connect the electrical circuit.
- For motor vehicles and trailers, two or more red tail lamps operate when the headlight switch is in the park position and the headlight position.
- Daytime running lights (DRLs) automatically operate when the vehicle is running and turn off when the engine stops.
- The layouts of electrical circuits and their components are shown as diagrams made up of symbols and connecting lines.
- Being able to read and interpret electrical circuit layouts can cut a technician's troubleshooting time; this is a skill all technicians need to learn.
- When checking lighting and peripheral systems, be sure to work in a systematic manner, or you could miss a faulty bulb or other component.
- If all of the bulbs on one fuse do not work, check the fuse.
- When replacing a bulb, always ensure there is no corrosion in the bulb socket, as this will shorten the life of the new bulb and add resistance in the circuit.
- Always make sure you replace a bulb with one of exactly the same type.
- It is critical that headlights be aimed correctly so as not to impair an oncoming vehicle and to allow correct visibility for the driver.

Key Terms

ballast A device that increases lighting voltage substantially and controls the current to the bulb.

halogen lamp A type of bulb that produces a bright white light.

high-intensity discharge (HID) A type of lighting that produces light with an electric arc rather than a glowing filament.

incandescent lamp The traditional bulb that uses a heated filament to produce light.

multiplex The carrying of multiple signals on one wiring circuit. Digital signals are multiplexed on a dedicated CANbus or similar network; also referred to by the abbreviation MUX.

vacuum tube fluorescent (VTF) A type of lighting used for instrumentation displays on vehicle instrument panel clusters. This type of lighting emits a very bright light with high contrast and can display in various colors; also called vacuum fluorescent display (VFD).

Review Questions

1. Which of the following alert(s) the other drivers of a change in direction?
   a. Red or amber turn signals
   b. Beam selector
   c. Brake lights
   d. Emergency flasher light

2. Which lightbulbs are very efficient and sensitive, and can be damaged even by finger oil residue?
   a. Incandescent lamps
   b. Halogen lamps
   c. Xenon lamps
   d. VTF lamps
3. Which of the following lamps produce more lumens with a bluish tinge for the given wattage when compared with all other lamps?
   a. Incandescent lamps
   b. Halogen lamps
   c. High intensity discharge lamps
   d. VTF lamps
4. Which lights are wired in parallel with the taillights and operate whenever the taillights are switched on?
   a. Headlights
   b. Turn signal lights
   c. Backup lights
   d. License plate lights
5. All of the following statements referring to driving lights are true *except*:
   a. Driving lights typically use quartz halogen bulbs in the 55- to 120-watt range.
   b. The quality of reflector is extremely important in driving lights to get the optimum performance.
   c. Driving lights are wired so that they operate only when the high beam is operating.
   d. Driving lights turn off when the headlights are switched from low to high beam.
6. Which of the following statements describes the purpose of backup lights?
   a. It indicates to other drivers that the vehicle is stopping.
   b. It provides the driver with vision behind the vehicle.
   c. It alerts other drivers that the vehicle is about to turn.
   d. It provides vision to other drivers who are approaching.
7. Choose the correct statement with respect to turn signal lights.
   a. They serve a decorative purpose.
   b. They warn the other drivers that the vehicle is about to stop.
   c. They warn the drivers in front to slow down.
   d. They warn other road users of the driver's intended change of direction.
8. Which of the following lights use flashing control and also warn other road users if the vehicle is parked in a dangerous position on the side of the road?
   a. High intensity flashing beam lights
   b. Turn signal lights
   c. Hazard warning lights
   d. Parking lights
9. All of the following statements with respect to the function of headlights are true *except*:
   a. They illuminate the road ahead.
   b. They help drivers at the time of reduced visibility.
   c. They provide two beams, high and low, to serve different purposes.
   d. They are connected in series with each other.
10. All of the following statements with respect to electronically controlled lighting systems are true *except*:
   a. The bus combines all the individual systems wherever possible into a multiplexed serial communications network.
   b. The common wiring harness increases the weight of the electronic system used.
   c. Troubleshooting a networked system requires more training and more tools than a simple digital multimeter.
   d. An advantage of such a system is less wire and fewer connections.

ASE Technician A/Technician B Style Questions

1. Tech A says that a voltage drop on the ground side of a bulb won't affect its brightness because the electricity has already been used up. Tech B says that a voltage drop is a good way to determine if there is unwanted resistance in a circuit. Who is correct?
   a. Tech A
   b. Tech B
   c. Both A and B
   d. Neither A nor B
2. Tech A says that incandescent bulbs resist vibration well. Tech B says that HID headlamps require up to approximately 25,000 volts to start. Who is correct?
   a. Tech A
   b. Tech B
   c. Both A and B
   d. Neither A nor B
3. Tech A says that some automotive lightbulbs have more than one filament inside. Tech B says that before aligning headlights, make sure the tires are inflated properly. Who is correct?
   a. Tech A
   b. Tech B
   c. Both A and B
   d. Neither A nor B
4. Tech A says that LED brake lights illuminate faster than incandescent bulbs. Tech B says that LED brake lights have more visibility and last longer. Who is correct?
   a. Tech A
   b. Tech B
   c. Both A and B
   d. Neither A nor B
5. Tech A says that if the brake light switch is open, neither brake light will illuminate. Tech B says that the backup lights operate when taillights operate. Who is correct?
   a. Tech A
   b. Tech B
   c. Both A and B
   d. Neither A nor B

6. Tech A says that some brake lights get power from the brake switch through the turn signal switch. Tech B says many turn signals use amber lights. Who is correct?
   a. Tech A
   b. Tech B
   c. Both A and B
   d. Neither A nor B
7. Tech A says that some vehicles automatically "dim" the headlights when light is detected from oncoming vehicles. Tech B says that the CHMSL is illuminated when the taillights are activated. Who is correct?
   a. Tech A
   b. Tech B
   c. Both A and B
   d. Neither A nor B
8. Tech A says that some turn signals are flashed by a flasher can. Tech B says that some turn signals are flashed by the BCM. Who is correct?
   a. Tech A
   b. Tech B
   c. Both A and B
   d. Neither A nor B
9. Tech A says that a light intensity meter is used to measure the brightness of headlights. Tech B says that some headlights can automatically adjust side to side when the vehicle is cornering. Who is correct?
   a. Tech A
   b. Tech B
   c. Both A and B
   d. Neither A nor B
10. Tech A says that if the oil pressure warning light is on, the first thing you should do is replace the oil pressure sending unit. Tech B says that if a bulb is dim, you should perform a voltage drop test on the power and ground side of the bulb. Who is correct?
    a. Tech A
    b. Tech B
    c. Both A and B
    d. Neither A nor B

CHAPTER 56

Network Communications and Body Accessories

NATEF Tasks

- **N56001** Diagnose body electronic systems circuits using a scan tool; check for module communication errors (data communication bus systems); determine needed action. (AST/MAST)
- **N56002** Describe the process for software transfer, software updates, or reprogramming of electronic modules. (AST/MAST)
- **N56003** Diagnose operation of comfort and convenience accessories and related circuits (such as: power window, power seats, pedal height, power locks, truck locks, remote start, moon roof, sun roof, sun shade, remote keyless entry, voice activation, steering wheel controls, back-up camera, park assist, cruise control, and auto dimming headlamps); determine needed repairs. (AST/MAST)
- **N56004** Diagnose operation of safety systems and related circuits (such as: horn, airbags, seat belt pretensioners, occupancy classification, wipers, washers, speed control/collision avoidance, heads-up display, park assist, and back-up camera); determine needed repairs. (AST/MAST)
- **N56005** Diagnose operation of security/anti-theft systems and related circuits (such as: theft deterrent, door locks, remote keyless entry, remote start, and starter/fuel disable); determine needed repairs. (AST/MAST)
- **N56006** Remove and reinstall door panel. (MLR)

Knowledge Objectives

After reading this chapter, you will be able to:

- **K56001** Explain the purpose, function, and types of vehicle networks.
- **K56002** Explain the layout and operation of a LIN-bus network.
- **K56003** Explain the layout and operation of a CANbus network.
- **K56004** Explain the purpose and types of DC accessory motors.

Skills Objectives

After reading this chapter, you will be able to:

- **S56001** Diagnose CANbus network faults.
- **S56002** Diagnose a DC accessory motor fault.
- **S56003** Diagnose horn systems failure.
- **S56004** Diagnose faults in the power door lock systems.
- **S56005** Diagnose faults in the wiper/washer systems.
- **S56006** Diagnose faults in heated accessory circuits.

Introduction

Electrical systems on modern vehicles are becoming increasingly complex with the addition of a range of electronic and accessory systems, such as global positioning systems, entertainment systems, security systems, heated seats, and heated glass. Many of these systems are controlled by body computer systems. Increasingly, the electrical and electronic assemblies are interconnected through vehicle data networks that require diagnosis. The technician requires an in-depth knowledge of these systems, including how they operate and are interconnected through the vehicle network. Diagnosing a vehicle fault, requires the use of dedicated diagnostic tools that communicate with the vehicle's electronic control units. Connecting a diagnostic scan tool to a vehicle allows the technician to monitor data from sensors or control certain components actions.

Vehicle Communications Networks

K56001

The additional features and accessories have increased the need for interaction between the vehicle's systems and components. For example, many systems must know the speed of the vehicle, instrument panel for the speedometer, transmission for gear selection, radio for speed-sensitive volume control, and so on (**FIGURE 56-1**). It is the same way for most of the other sensors and systems as well. So you can see that if we needed separate wires to connect all of the sensors and actuators together, it would require a lot of wires.

Manufacturers solved this issue by using a multiplexed system where information is transmitted across one or two wires that run to all of the control modules. The control modules interpret the information and act on it, if needed, to control the operation of actuators or other electrical/electronic devices. The modules are also supplied with power and ground to both operate the module as well as supply power and ground to the controlled devices. This makes for a much simpler wiring harness with fewer wires.

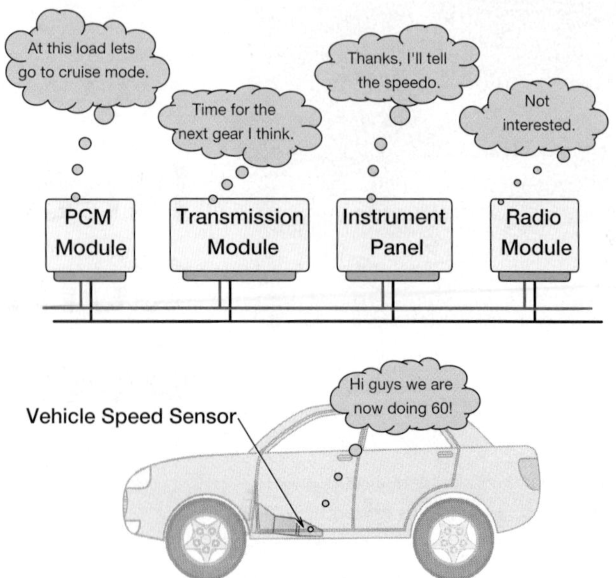

FIGURE 56-1 A networked system allows data to be sent over the system to all modules. The ones that need to know that information act on it, and the rest ignore it.

Networking makes it more efficient to add or modify system functionality through the addition of control units and the use of software updates. These updates allow the manufacturer to provide fixes for issues that may crop up after vehicles have been delivered to customers and put into service. On some networks, control units (modules) can be added to the system by simply connecting them to the network and the new controlled device. Software can then be updated if necessary to control the new device.

There are several types of networks; LIN, LAN, CAN, CAN FD, FlexRay, MOST, and Ethernet-based systems, each with specific benefits and limitations. Many vehicles incorporate more than one type of network due to the varying requirements of the different systems. For example, systems like power mirrors and power seats do not require high-speed networks

You Are the Automotive Technician

A 2010 minivan comes into the dealership because the power window is not working properly. The customer informs you that the "one-touch" function works sometimes, but not at others. After performing a VIN check for service history, you look up technical service bulletins (TSBs) and notice one related to the "one-touch" feature on this vehicle. The issue may be related to a binding window track, plus there is a software update to help address the issue down the road.

The first step is to verify whether there is an overcurrent condition on the window motor. You connect the scan tool and see a code indicating an overcurrent shutdown on the driver's window. Next, you clean out the exposed window track with a clean shop rag. You then spray silicone lubricant into the window track and operate the window several times. Because the window is moving smoothly, the software can be updated. You follow the flash reprogramming steps detailed in the manufacturer's guide, which allow a slightly higher current flow in the circuit that will prevent the computer from disabling the feature. Once the update is complete, you verify that the "one-touch" feature is functioning properly.

1. What are the various types of automotive networks?
2. What is a CANbus system?
3. How can software updates correct problems in the vehicle?
4. What are the precautions before updating software on a vehicle?

to operate effectively, whereas fuel injector control and airbag deployment do. Also, typically, the faster the network, the more expensive it is. So using lower speed/lower cost networks to handle the less-data-intensive systems while using a higher speed/higher cost network for fuel injector control and airbag deployment systems makes sense. This does require that the different networks be able to share information where necessary.

Local Interconnect Network (LIN)

K56002

We look at each one briefly, starting with LIN, which stands for local interconnect network and, like most other networks, is used to send serial data between modules, called nodes. In the LIN system, there is one master node and up to 15 slave nodes that communicate over a single wire at speeds up to 20 kbits/s (**FIGURE 56-2**). This network works well for body-controlled systems such as wipers, mirrors, and lights.

Controller Area Network (CAN)

This network can accommodate up to 100 nodes, with high-end vehicles typically having up to about 70 nodes. Each node is a master node capable of receiving, transmitting, and processing the signals independently. The nodes communicate over a two-wire twisted pair, which reduces electrical interference (**FIGURE 56-3**). One wire designated as CAN–H (CAN–High, or CAN+) carries the high voltage signal, and CAN–L (CAN–Low, or CAN–) carries the mirrored low-voltage signal. This pairing of wires with opposite voltage signals is used to verify the accuracy of the signal being sent by comparing the individual signals to one another. The high-speed portion of the system sends 8 bits per frame and operates at speeds of up to 1 Mbit/s. And the low-speed portion (called fault-tolerant CAN) operates between 40 kbits/s and 125 kbits/s. This dual-speed system works well for high-speed dependent network systems such as engine management and supplemental restraint systems, along

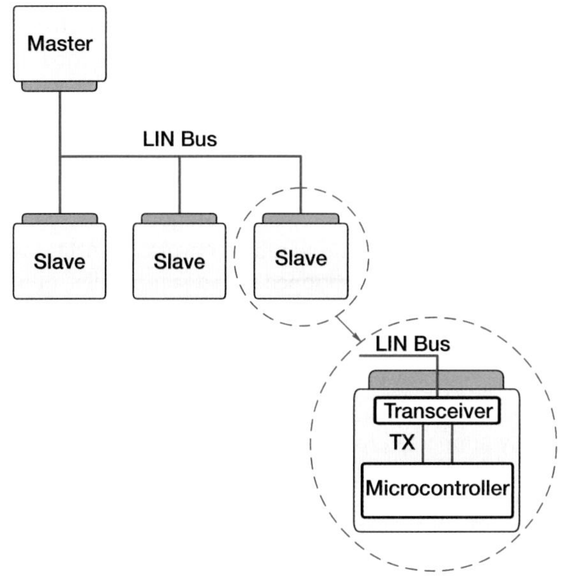

FIGURE 56-2 A typical LIN network using a single data communication wire.

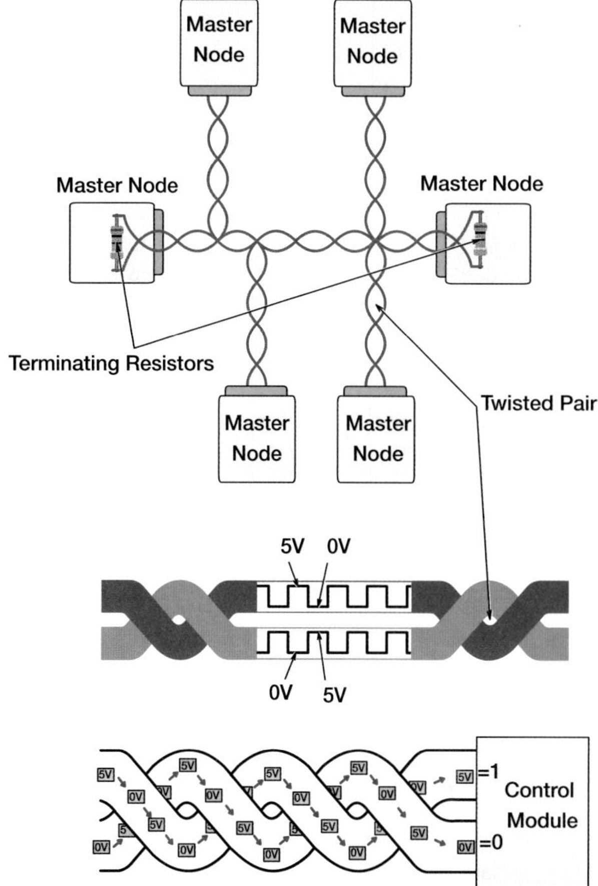

FIGURE 56-3 A typical CAN system with several nodes, each of which is a master node.

with lower speed systems such as the body-controlled devices. This cuts down on the number of sensors required if two separate systems were used.

Controller Area Network with Flexible Data-Rate (CAN FD)

This is an enhanced version of CAN that operates on up to 64 data bytes per frame, compared to 8 bytes per frame, and at a speed of up to 5 Mbits/s. Because of the increased speed and data bit size, the network can handle data traffic more effectively.

FlexRay Networks

FlexRay networks operate at higher speeds of up to 10 Mbits/s. They use either one twisted pair of wires (single channel), or two twisted pairs of wires (dual channel), which increases the fault tolerance of the network (**FIGURE 56-4**). Because this network has a high transfer rate, it works well in data-intensive systems such as engine management.

Media Oriented Systems Transport Networks

Media Oriented Systems Transport (MOST) networks are designed to transmit high-speed audio, voice, video, and data signals. There can be up to 64 MOST devices, and devices are

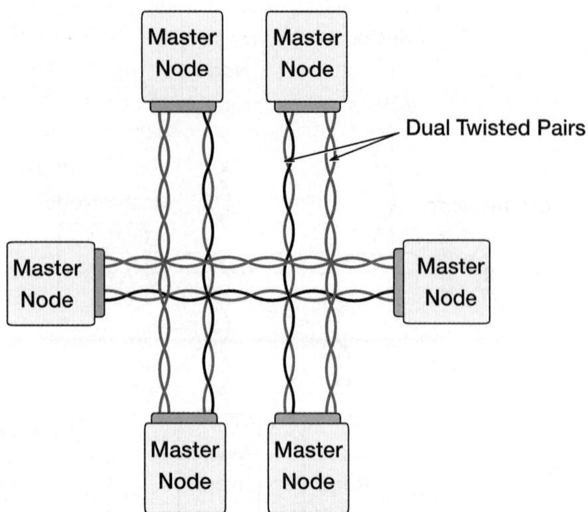

FIGURE 56-4 A typical FlexRay network with a dual channel system.

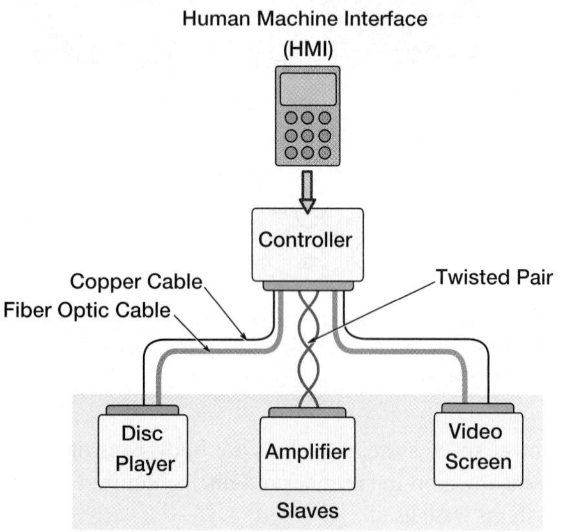

FIGURE 56-5 A typical MOST network with both a fiber optic and electric communication systems.

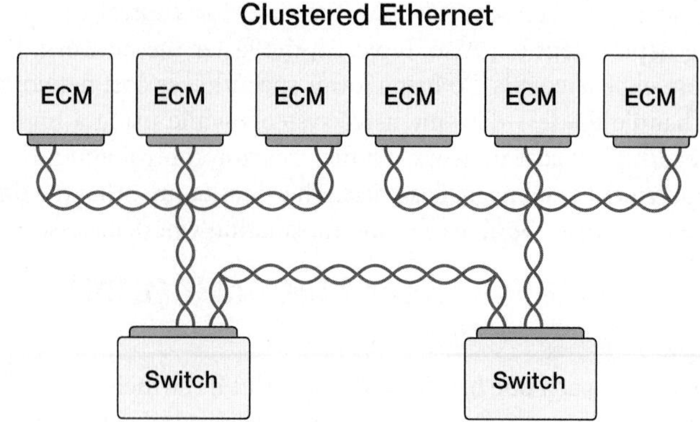

FIGURE 56-6 Ethernet networks are being used as in-vehicle networks.

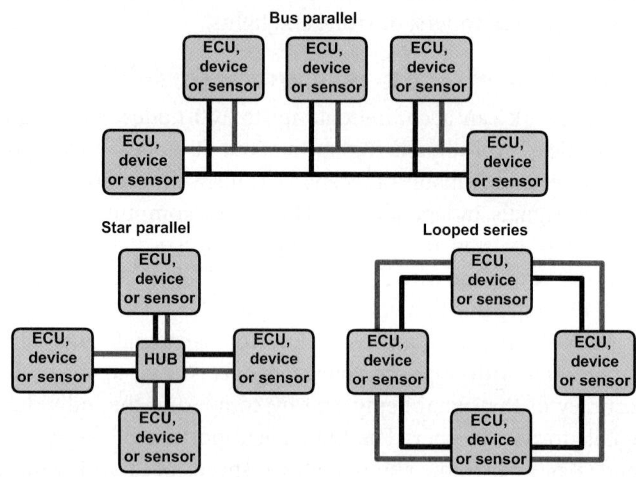

FIGURE 56-7 Network configurations.

plug-and-play, so devices can be added to this system easily, even after the vehicle is in service. There are several versions of MOST systems, each having increased bandwidth over the previous version (25 Mbit/s, 50 Mbit/s, and 150 Mbit/s). MOST systems send data signals either optically, using plastic optic fiber, or electrically, using electrical conductors (**FIGURE 56-5**).

Ethernet Network Systems

Ethernet network systems have been used previously to connect into the vehicle's onboard diagnostics system and to perform software upgrades when accessing vehicle network systems, but are now just being implemented as in-vehicle network systems (**FIGURE 56-6**). Ethernet currently operates at 10 Mbit/s and 100 Mbit/s, with work underway to enhance it to 1 Gbit/s. This makes it ideal for video and entertainment systems and for with connecting to handheld electronic devices such as cell phones and tablet computers. Also, it is being investigated for use in 360-degree camera systems used in drive-by-wire systems. Currently, Ethernet systems use four data wires, but there is an industry push to adopt a single twisted-pair wire for Ethernet networks in automotive applications. This would cut down in weight, wire harness size, and cost.

The network configurations, or how the network connects to each module, vary with each manufacturer and type of network. Some use a simple series configuration; others use variations of series networks and parallel network connections with descriptive names, such as looped series, star parallel, or bussed parallel (**FIGURE 56-7**).

Most vehicles are equipped with more than one type of network because slower networks are less costly and faster networks needed for critical systems are more expensive. Networks are usually connected together so that sensor information can be shared. This means that duplicate sensors for each network are not needed. But because the networks operate differently, there has to be an interface between them so the data can be transferred properly. This function is provided by gateway modules (**FIGURE 56-8**). A vehicle will have one gateway module between each different type of network.

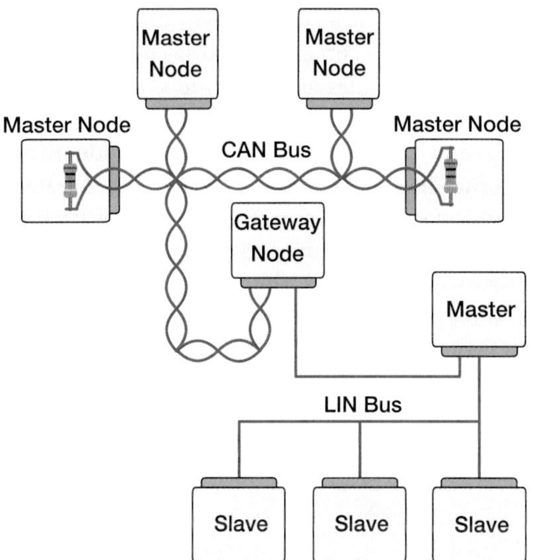

FIGURE 56-8 A gateway node connects two types of networks together so that information can be shared on both.

▶ CANbus System

K56003

The most common network system right now is the CANbus system. We are going to look at it in more depth. The other network systems operate on similar concepts. The terms controller area network (CAN) and bus are used together (often as "CANbus") to describe the data-sharing network in vehicles. The many sensors and electronic control modules can share information across a common network using the CANbus protocol to communicate efficiently and effectively. In electrical and electronic systems, a bus is a means of connecting many electrical or electronic components together for either data or power sharing. Several signal protocols have been developed for use with second-generation, onboard diagnostics (OBDII), including SAE J1850 PWM, SAE J1850 VPM, ISO 9141-2, ISO 14230 KW, and ISO 15765 CANbus (C&B).

Within the CANbus system design, there are several different network layouts or designs, and there may be a gateway or hub controller, but no single dominant electronic control module; each receives the same data and simply chooses the data it needs to operate effectively. It is a robust system, and many modules can send and receive data without corruption or interference. The CANbus system continues to develop and, although standards are in place, manufacturers may use different terminology to describe their particular systems.

As CANbus has developed, so too have the network speeds. CANbus A is a low-speed network (called fault-tolerant CAN), CANbus B is a medium-speed network, and CANbus C (CAN FD) is a high-speed network. Networks with higher speeds are used for critical data such as supplemental restraint systems, antilock brake systems, and engine controls. They are also used for transmitting data-heavy video and voice in the entertainment system. Lower-speed networks are used for less-prioritized data such as data related to accessory systems. CANbus systems continue to evolve, and new standards will be developed; however, the fundamental operating principles are likely to remain the same. And because CANbus-compliant diagnostic systems have been required on all vehicles sold in the United States since 2008, it is even more important that you be familiar with them.

Principles of Operation

The CANbus system uses digital data, called binary data, which have only two states: 0 or 1. All the data can be coded in binary form using a series of 0s and 1s. Each 0 or 1 is called a bit of information, and a group of eight bits is called a byte. The 0s and 1s are represented on the data bus as low and high voltages (**FIGURE 56-9**). Each module on the data bus sees all data, so along with the actual data, other information like message headers is required to identify sender and recipient modules as well as message priority, wake-up commands, and state of health messages. These details are all coded in binary form and sent along the data bus. The data are grouped into blocks, each with a beginning and an end, and each module is capable of reading the binary data, interpreting the required data, and sending its own messages along the data bus.

LIN networks transmit and receive signals over a single wire, but most other types of networks have dual data transfer wires in the form of a twisted pair. In the CANbus network, they are commonly called "CANbus H" (high, or CAN +) and "CANbus L" (low, or CAN –), with the same message sent on both lines (**FIGURE 56-10**). This system provides redundancy in the case of a fault or interference on a single line. The signal on each CANbus line is a differential signal, meaning the twisted pair are mirror images, or opposite polarity, of each other. Typically, both CAN L and CAN H lines are biased at 2.5 volts. When a node sends a "1", CAN L pulls the voltage down 1 volt, and CAN H raises the voltage 1 volt. A "0" shows up as bias voltage. The actual voltages used on other network systems vary with each manufacturer.

Configurations and Terminal Resistors

Regardless of how the network is physically connected, the CANbus-H and CANbus-L system have terminating resistors connected to the data lines to form a loop through the resistors. The value of resistors can vary by manufacturer, but high-speed CANbus systems tend to use dual 120-ohm resistors, giving a

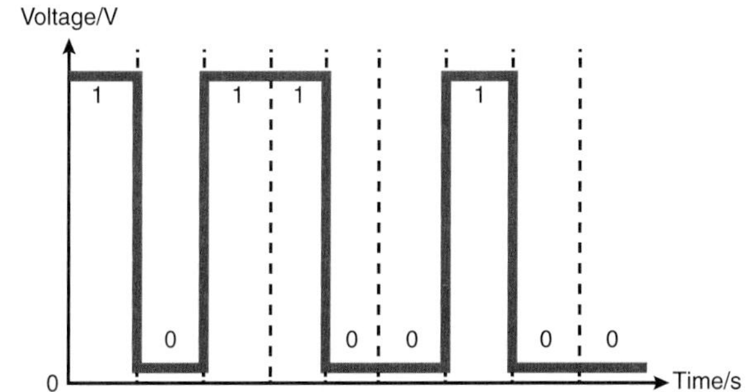

FIGURE 56-9 The CANbus system uses digital data, called binary data, which have only two states: 0 or 1. All the data can be coded in binary form using a series of 0s and 1s.

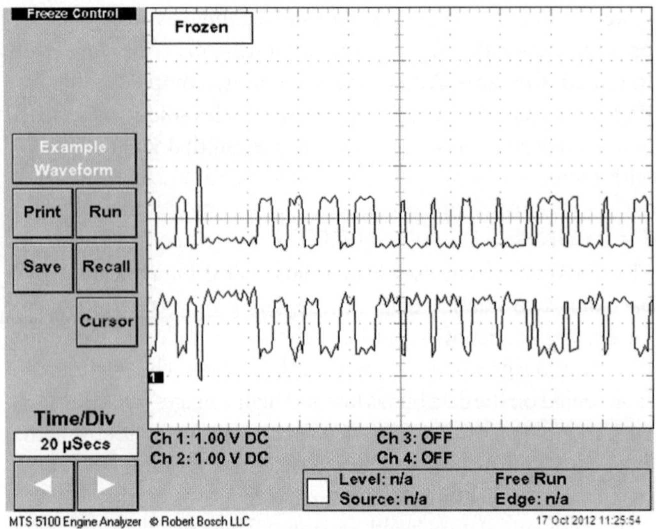

FIGURE 56-10 Typical CANbus-H and CANbus-L signal.

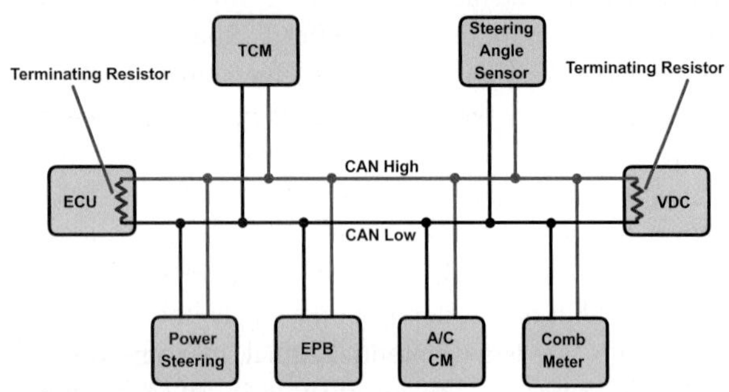

FIGURE 56-11 A network diagram with terminating resistors.

total of 60 ohms per pair because they are connected in parallel, but typically at opposite ends of the bus (**FIGURE 56-11**).

The resistors can be connected externally in the wiring harness or connectors, but they can also be integrated internally into modules. The data lines are also twisted together with a specific twist rotation per length. This twisting, in combination with the terminating resistors, reduces the amount of interference on the data lines. The CANbus system has a **data link connector (DLC)** that provides a connection point into the data network. The DLC is also used to connect scan tools into the network. It can also be used to conduct voltage and waveform tests on the CANbus system.

▶ Testing the CANbus System

S56001

Testing the CANbus system starts just like all diagnostic procedures: verifying the customer concern and then researching the fault in the service information. Once you do that, you can build a plan to focus your testing in the most logical way possible. When there is an issue with the CANbus system, a diagnostic trouble code (DTC) with a "U" designation is typically set, indicating the nature of the fault. Sorting out what is causing the fault requires using a combination of diagnostic tools, including scan tools, digital multimeter (DMM), digital storage oscilloscope (DSO), and DLC breakout boxes. The scan tool connects to the DLC and provides access to the DTCs, and depending on the functionality of the scan tool, it may provide access to the data stream. The scan tool can be used to identify any fault codes that may have been logged by the vehicle's systems. Fault codes may provide good information on what particular part of the system or type of fault needs to be diagnosed. The DMM can be used to measure CANbus data line resistances, system voltages, and grounds to ensure basic circuit integrity.

A DLC breakout box is a tool that connects to the DLC and provides easy access to the terminals in the DLC socket to connect DMMs and DSOs. Some breakout boxes also provide a visual indication of pin voltages with colored LEDs on the sockets. A capable dual-channel DSO can display on the screen the high- and low-voltage data signals present on the CANbus. The CANbus-H and CANbus-L data signals can be compared to ensure that they are mirror images of each other. The signal can also be examined for any interference on the data line that may cause faults or loss of data. Some DSOs are capable of decoding the CANbus data to provide a more accurate comparison of the high and low signals.

Depending on the nature of the fault, it may be necessary to disconnect various modules within the CANbus. Disconnecting modules that are equipped with the terminating resistors during diagnosis or testing may affect the communications on the network due to increased interference. Always refer to the manufacturer's information for test and repair procedures.

▶ TECHNICIAN TIP

Knowing the configuration of the network can help you diagnose it. For example, you will use a scan tool to look at the active modules on the network; if any are missing, knowing the sequence of the control modules can lead you to the area of concern. Checking for continuity of CAN-H or CAN-L between the last active module and the first missing module will show you if the CAN wire is open. Also, measuring the resistance between CAN-H and CAN-L will indicate the condition of the terminating resistors. If the resistance is 60 ohms, both terminating resistors and the CAN lines between them are not open. Resistance of 120 ohms indicates an open line or resistor in one side of the circuit. Using the diagrams and schematics will help you pinpoint faults in the system.

Using a Scan Tool

Many different types and brands of scan tools are available, and the vehicle and system coverage varies with each scan tool. Some scan tools require hardware keys to be inserted for each different type of vehicle; others may have provisions for vehicle manufacturer and model to be selected using software. Always ensure that you understand how to connect and set up the specific scan tool for the vehicle you are working on.

Using the scan tool may require the vehicle's ignition system to be on for an extended period of time without the engine running. The electrical load may be such that the vehicle's battery will discharge; a battery support unit may need to be connected to ensure the vehicle's battery does not discharge.

Battery support units are high-current output power supplies with a steady voltage supply used to support the battery when the engine is not running. Do not use a standard battery charger to perform this role, as many do not have a steady direct current (DC) voltage supply, and they may cause interference on the CANbus communication circuits or damage components.

Checking for Module Communication Errors Using a Scan Tool

N56001

Modern vehicles have multiple modules to control the various vehicle systems. The number of modules and the purpose of each vary among manufacturers and even vehicle models. Modules communicate using the network on the vehicle. If there are communication errors on the network, some or all of the modules will not be able to operate. Also, one or more U codes will typically be set, indicating a fault in the communication network. If the network is active or partially active, a scan tool may be able to help you diagnose the fault.

One of the best things you can do is connect a scan tool to the network DLC and check to see if there are any modules that are missing or not communicating. Typically, if a device is not showing as present, you need to physically locate that module and first check it for power and ground. If those are correct, then use the scan tool to command the module to operate the device, and check the communication wire to see if it is receiving the signal. For example, use a scan tool to *actively test* the output component to see if it operates. If it operates correctly then that means there is nothing wrong on the output side of the circuit.

Also, use a scan tool to check the data list for a "state of change" in signal such as a switch while turn the circuit on and off. If the data corresponds to the state of change signal then the input circuit is working properly.

If the power and ground are sufficient, and if the signal wire(s) is communicating the proper signal and the module is not responding, chances are good that the module is faulty and has to be replaced.

To check for module communication (including CANbus systems) errors using a scan tool, follow the steps in **SKILL DRILL 56-1**.

Performing Software Transfers, Updates, or Flash Reprogramming on Electronic Modules

N56002

Diagnosing and repairing vehicles may require software transfers, software updates, or flash reprogramming of the vehicle's electronic modules with the manufacturer's latest update. Manufacturers periodically update the software for vehicle modules to enhance vehicle operation. Flash reprogramming tools are available and connect to the vehicle's DLC to update vehicle software, and manufacturers make updates available over the Internet, usually through a subscription service.

Using the flash program tool may require the vehicle's ignition system to be on for an extended period of time without the engine running. The electrical load may be such that the vehicle battery will discharge, which may cause a fault in the reprogramming. A battery support unit needs to be connected to ensure that the vehicle's battery does not discharge. Battery

Applied Science

AS-66: Electricity/Measurement: The technician can demonstrate an understanding of the correct procedure to measure the electrical parameters of voltage, current, and resistance.

A vehicle is in the shop for multiple interior electrical problems. The systems affected are the power seats, power windows, power door locks, and interior lighting. This problem was initially intermittent, but at this point the affected systems are not functioning.

In diagnosing the vehicle, the technician performs a voltage drop test to determine if there is a bad ground causing the multiple circuit problems. He is using a long jumper wire along with his DMM to accomplish the test. The technician attaches a long jumper lead to the negative battery post, which connects to the COM or black test lead of the DMM. The DMM is set to read DC volts, and the red lead is used to check the groundside of the components in question. This test will be performed with each separate component under load, such as when attempting to operate a window. A good ground should have a reading of 0.50 volt or less. The technician finds a very high reading for each of the affected circuits and determines that the problem is in a corroded multiple ground connection point. Using the service information, the technician locates the four ground connections secured by a single bolt behind the driver's side kick panel. He uses emery cloth to clean each ground connector on both sides. then cleans the frame location, and tightens the bolt securely. The DMM reading is now 0.07 volt, and all circuits are working properly.

AS-67: Ammeter/Voltmeter: The technician can demonstrate an understanding of how to correctly measure electrical current and voltage in a circuit.

To measure electrical current with the DMM, it is helpful to know that current is the volume of electrons flowing in a circuit. Amps and milliamps are a measure of the volume of electrons. To measure current below 10 amps, place the meter in series with the load. If the current is expected to be above 10 amps, an inductive pickup must be used.

Turn off the power to the circuit, and determine where the test leads should be attached. Put the black test lead into the COM slot. Plug the red test lead into the 10 amps or 300 milliamps slot, and select DC amps. Open the circuit at a connector and connect the test probe's tips to each connector so current will flow through the meter in series. Turn on the circuit and read the meter.

To measure voltage with a DMM, it is helpful to know that voltage is the electrical pressure that causes current to flow. Voltage is the difference in electrical potential between two points in a circuit. To make a voltage measurement, the meter leads must be connected across the load or component being measured. To measure battery voltage, place the red lead on the positive post of the battery. Place the black lead on the negative post of the battery. A fully charged automotive battery should be approximately 12.66 volts at 70°F (21°C).

SKILL DRILL 56-1 Checking for Module Communication Errors Using a Scan Tool

1. After researching the correct procedure in the manufacturer's information, locate the DLC, and connect the scan tool. Power on the scan tool and turn the ignition on. Establish scan tool communications with the vehicle.

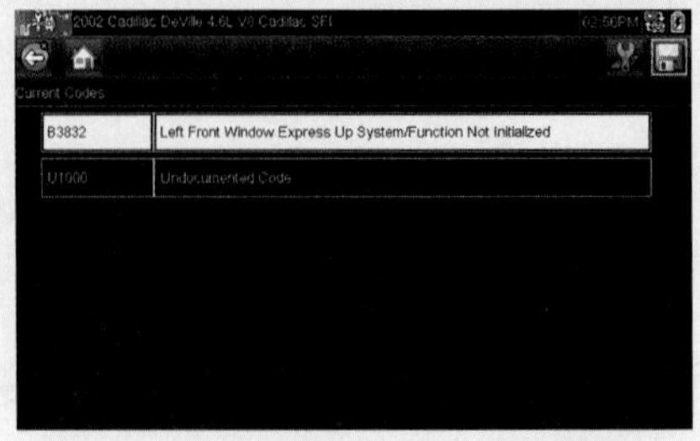

2. Check fault codes using the scan tool, and follow the diagnostic procedure in the service information. Check to see if any modules are inactive, or not listed as active, that should be.

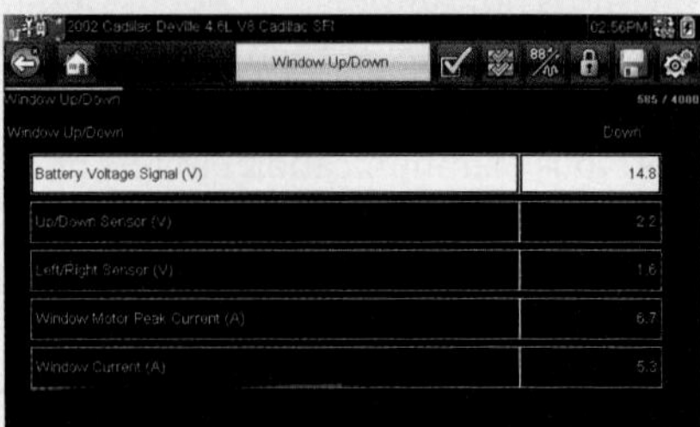

3. Command the module to take the appropriate action, and observe the response. If you do not receive the proper response, refer to the service information for the diagnostic procedure.

support units are high-current output power supplies with a steady voltage supply used to support the battery when the engine is not running. Do not use a standard battery charger to perform this role, as many do not have a steady DC voltage supply, and they may cause interference or damage components resulting in damage to the vehicle. Never shut the key off or lose power when flashing a module, as damage can occur.

To perform software transfers, software updates, or flash reprogramming on electronic modules, follow the steps in **SKILL DRILL 56-2**.

SKILL DRILL 56-2 Performing Software Transfers, Software Updates, or Flash Reprogramming on Modules

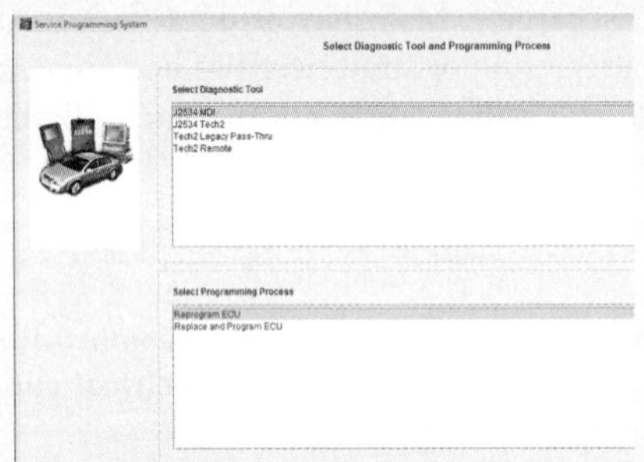

1. Obtain the latest vehicle software program updates from the manufacturer, and load the updates into the flash programmer. Connect a battery support unit to the vehicle.

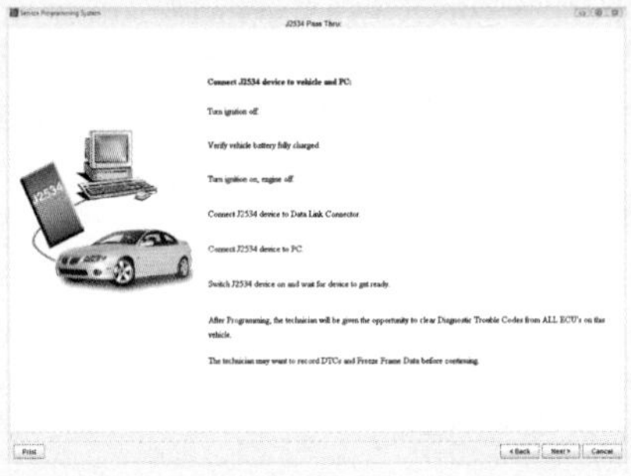

2. Locate the DLC. Connect and power on the flash program tool. Turn the ignition on.

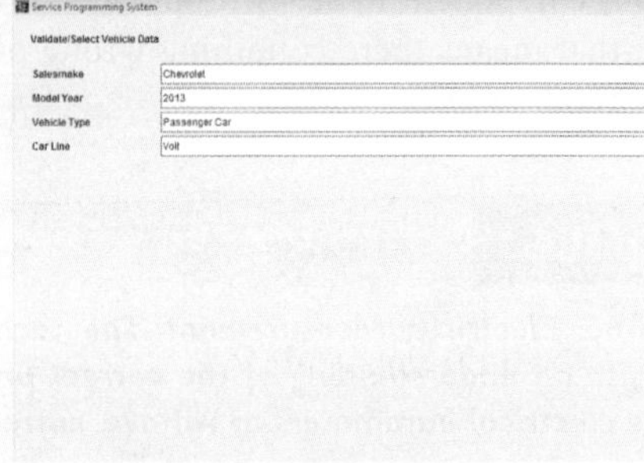

3. Establish flash program tool communication with the vehicle.

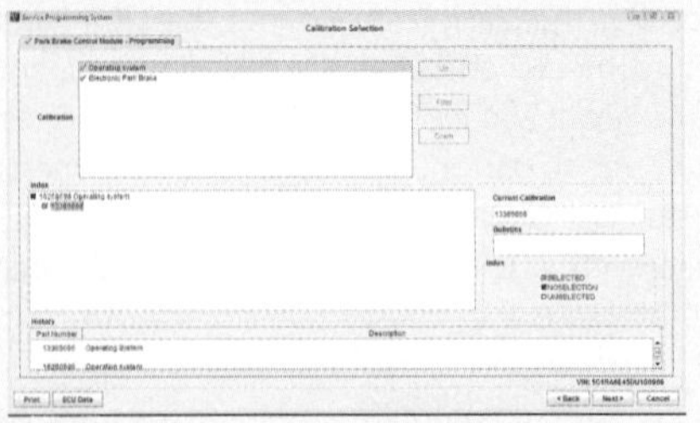

4. Identify the vehicle modules to be updated, and use the flash program tool to perform software transfers, software updates, or flash reprogramming.

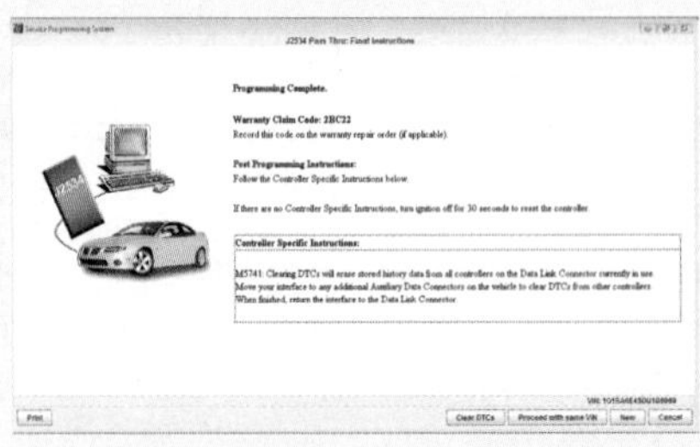

5. Disconnect the flash programmer, and check the vehicle's functionality.

Electric Accessory Motors

K56004

Electric motors rotate as a result of the interaction of magnetic fields; that is, magnetic poles repel each other, and opposite poles attract. Two magnetic fields are required for motor action: one in the casing and the other in the rotating armature. The magnetic field in the casing can be generated by permanent magnets or electromagnets made up of many windings of fine copper wire wound in loops called field coils. The armature's magnetic field is generated in loops of wire that form the armature windings. Motor action occurs through the interaction of the magnetic fields of the field coils and the armature, which causes a rotational force to act on the armature, creating the turning motion (**FIGURE 56-12**).

The rotary motion can be used to drive a variety of devices. Many times, the speed of the motor must be geared down so that more torque can be obtained, such as with windshield wipers and many actuators. Most of the time, the gears are located inside the motor housing, so if one or more gears go bad (e.g., stripped teeth), then the motor may still spin, but the output gear or lever will not move as designed. Thus, listening for motor rotation can help you diagnose the problem. For example, if the windshield wipers do not move when activated, yet you can hear the motor running, you know that the electrical circuitry is working, but the movement isn't getting from the motor to the wipers. This example could be an internal gear problem in the motor, or it could be a problem with the wiper linkage. But at least you will have a good head start on diagnosing the fault.

Motor Types-Brush, Brushless, Permanent Magnet, and Stepper Type

Electric motors are designed to perform specific functions. They all work on the same fundamental principle of the interaction of magnetic fields; however, their construction varies. One of the most common types of DC electric motors is the **permanent magnet type**. In this type of motor, the magnetic field in the casing is produced by permanent magnets, and the armature has an electromagnetic field generated by passing electrical current through loops or windings to produce the motor action (**FIGURE 56-13**). Power is supplied to the armature loops by brushes riding on the commutator, which continually changes the current flow in the armature loops as it rotates. This changes the polarity of the magnetic field in the armature, producing the motor action effect as the armature magnetic field interacts with the permanent magnets in the casing.

A **brushless DC motor**, as the name suggests, does not have any brushes and is sometimes called an "electronically commutated motor." In this type of motor, an electronic control module replaces the brushes and commutator. A brushless motor has permanent magnets rotating around a fixed armature or stator (**FIGURE 56-14**). Having a fixed stator eliminates the need to have brushes. The current flow still needs to be continually alternated in the stationary stator to produce motor action, and this is

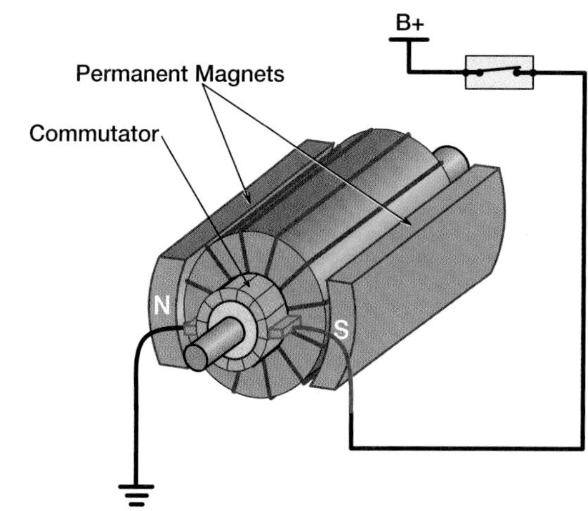

FIGURE 56-13 A typical permanent magnet electric motor.

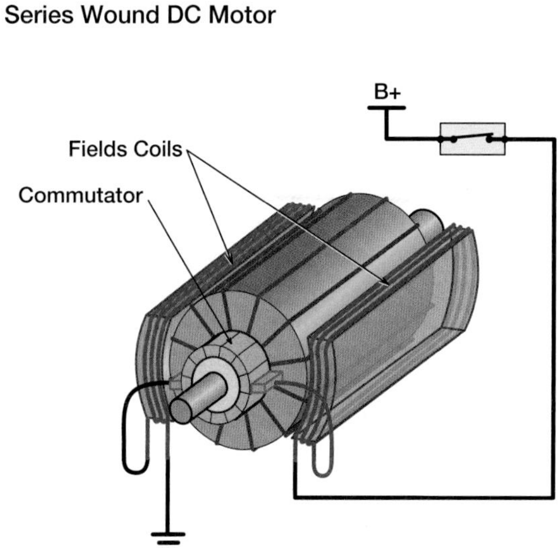

FIGURE 56-12 Typical brush-type motor.

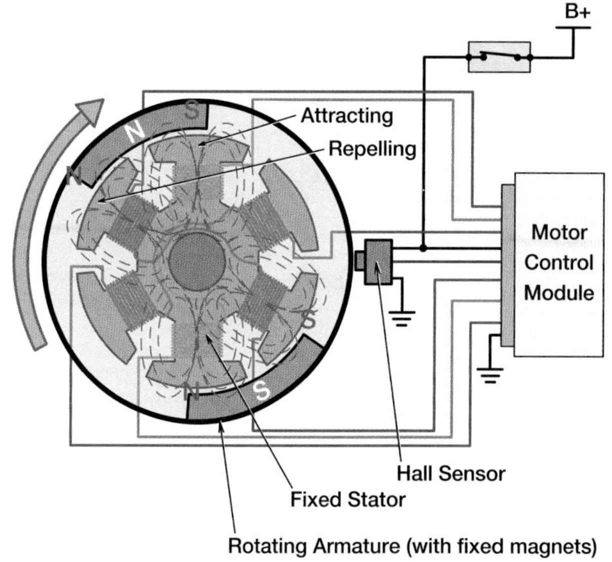

FIGURE 56-14 A typical brushless electric motor.

achieved by an electronic control module. A brushless DC motor has several advantages over brush motors, including improved efficiency and reliability. In addition, brushless motors create less electrical interference because there are no sparks created by brushes making and breaking contact on the commutator.

A **stepper motor** is a type of brushless motor with a key difference: It is designed to rotate in fixed steps, each step being a set number of degrees. The stepper motor uses a rotor with teeth of alternating polarity, and a number of wound coils (typically four in automotive use). Each pair of coils is controlled by an electronic microcontroller. Each time the controller energizes a pair of coils for a short period of time, the motor armature turns a set number of degrees so that two teeth line up with that set of coils (and other teeth are halfway between alignment with the other set of coils) and stops (**FIGURE 56-15**). For example, if a motor were to turn 10 degrees for each step, it would require 18 teeth on the rotor, as each step moves one-half tooth. Energizing each coil in rapid succession creates rotational movement.

Stepper motors can also be used to rotate a short distance and then stop or go in the reverse direction for a set number of degrees by controlling each coil individually through the microcontroller. Stepper motors are increasingly used in vehicles; for example, the electronic throttle body uses a stepper motor to open and close the throttle plate one step at a time, as directed by the power train control module (PCM).

Blower Motor and Circuits

S56002

The **blower motor** in a vehicle is usually the permanent magnet type and moves air over the air-conditioning evaporator and heater core. The blower motor circuit incorporates a speed control for the motor. Speed can be controlled by using a number of resistors connected in series and a switch to select between combinations of resistors or a more complex electronic speed control module. The blower motor circuit consists of the power supply, fuse, on/off switch, speed control, relay, and the motor itself (**FIGURE 56-16**). In vehicles equipped with air conditioning, the blower motor control circuit is incorporated into the control circuits for the air-conditioning system.

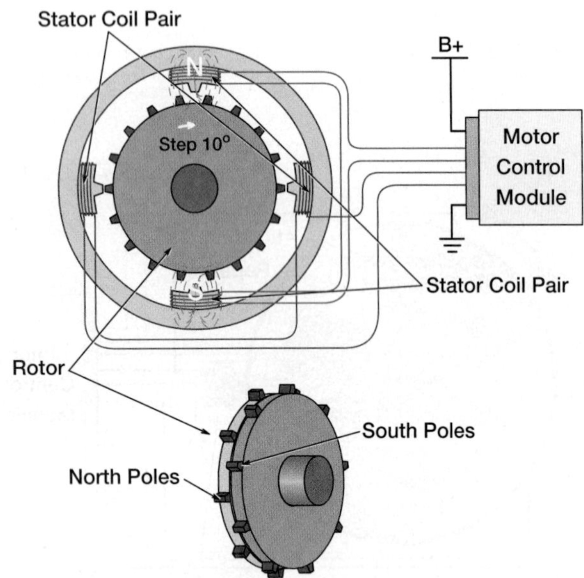

FIGURE 56-15 Typical stepper motor.

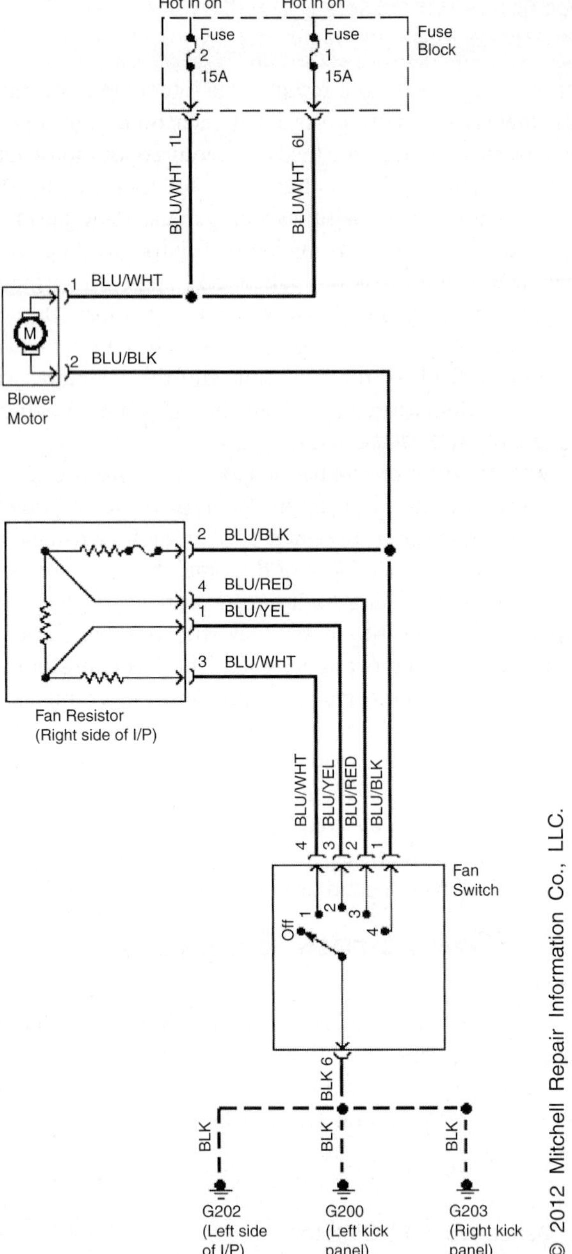

FIGURE 56-16 A blower motor circuit (non-computer control).

Cooling Fans and Circuits

Electric cooling fans for the radiator are increasingly common, particularly for transversely mounted engines and as supplementary cooling fans for vehicles equipped with air conditioning. Cooling fans in modern vehicles are usually switched by relays, which are controlled by an electronic control unit based on information from the coolant temperature sensor. Control units can be quite complex and can incorporate variable speed and on/off control. The control circuits are designed to ensure that the operating temperature of the engine is kept as close as possible to the temperature at which the engine operates most efficiently.

The basic electric cooling fan circuit contains the battery or power supply, fuse, relay, fan control circuits, and fans (**FIGURE 56-17**). The actual circuit and its complexity will vary by manufacturer and vehicle model. The circuit diagram or schematic for the fan system should be examined before attempting any diagnosis.

Power Mirrors

Power mirrors use permanent magnet electric motors to control the up-and-down, in-and-out movement of the passenger and driver side mirrors. Most mechanisms use two motors for each mirror, one each for up-and-down and in-and-out movements (**FIGURE 56-18**). The motors are reversible so that they can provide the movements in either direction. The control circuit turns the motors on and off, controls direction of rotation, and provides dynamic braking. Reversal of the motors is achieved by reversing current flow through the armature. The mechanical arrangements for connection of the motors to the mirrors will vary by manufacturer and vehicle model.

Dynamic braking is critical in this application because it ensures that the motor stops quickly without overrun when the power is removed by the driver operating the switch. Dynamic braking is achieved by providing a resistive path between the two brushes at the time the switch is turned off. The circuit in **FIGURE 56-19** shows a basic control circuit that provides dynamic braking. The typical power mirror circuit includes the battery supply, fuse or circuit breaker, control switches, relays, and motors. The control switches usually incorporate a switch to select between passenger and driver side mirrors and a joystick-type switch with east, west, north, and south positions to control the up-and-down, in-and-out movements.

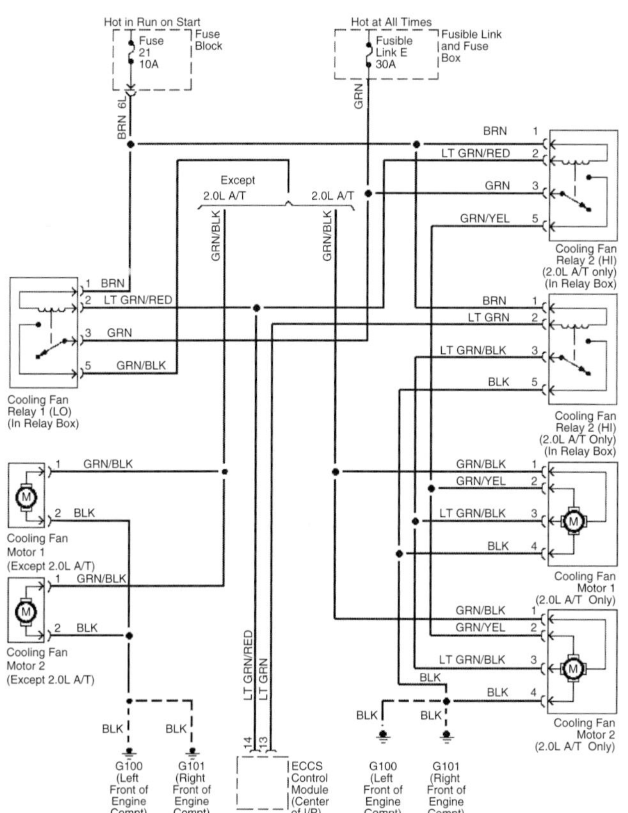

FIGURE 56-17 An electric cooling fan circuit.

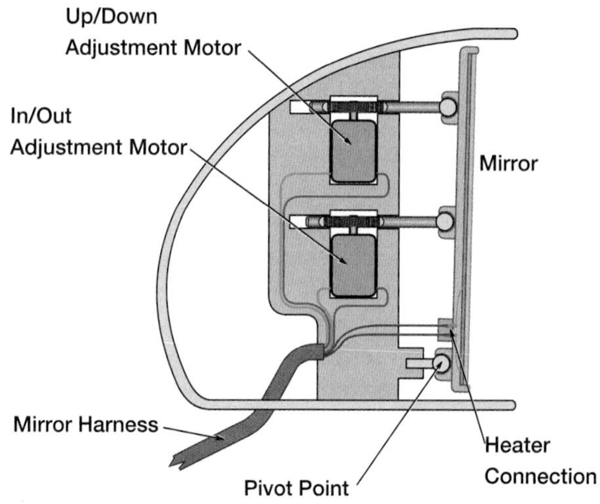

FIGURE 56-18 A basic power mirror showing the position of both motors.

▶ TECHNICIAN TIP

On many modern vehicles, the mirrors are operated by a body computer: when the vehicle is put into reverse, one or both of the mirrors rotate down so that the driver can see the ground or curb. Also, many luxury vehicles have driver presets and allow different drivers to set preferences such as seat position, tilt/telescoping of the steering wheel, radio station, pedal position, and mirror position. These presets are tied to the individual key each driver uses. Also, many mirrors incorporate directional indicators that flash when the turn signals or hazards are on. All of these features are more easily integrated into the vehicle through a networked computer system.

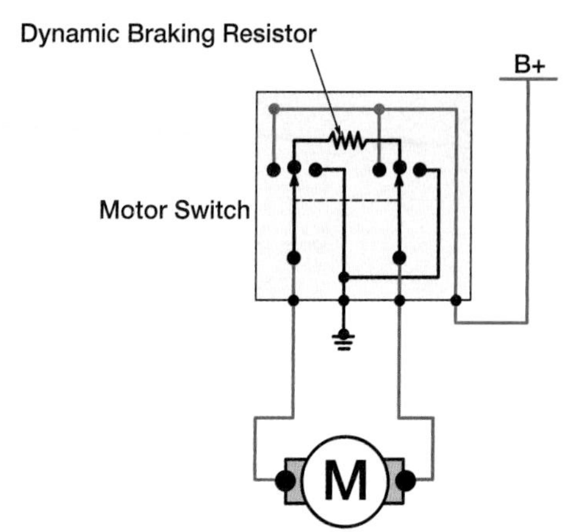

FIGURE 56-19 Dynamic braking of a DC motor.

Power Windows

Electric windows use permanent magnet motors located in each of the passenger doors and sometimes in a rear door to control the up-and-down movement of vehicle windows. The motors are reversible to provide the movement in either direction. The control circuit turns the motors on and off, controls direction of rotation, and provides dynamic braking. Reversal of the motors is achieved by reversing current flow through the armature. A switch on each door controls the direction of travel, and a master switch panel to control all windows is usually located on the driver's side front door (**FIGURE 56-20**). If equipped, a lockout switch in the master switch panel disables the switches for the other window switches.

The mechanical arrangements for connecting the motors to the windows vary by manufacturer and vehicle model. They usually comprise a window regulator made up of gears and scissor-acting levers or a cable and lever system (**FIGURE 56-21**).

Modern vehicles often have more sophisticated electric window control systems to improve safety and usability. They include one-touch up and down, where a single full push on the control switch in either direction enables the window to go all the way up or down without the need to hold the switch. The control system monitors current flow and can reverse the window winding if it senses excessive resistance such as would occur if an arm were caught or trapped in the window, called Jam Protection" on some vehicles. The electric window system includes the battery supply, fuses and circuit breakers, switches at each door and the master panel in the driver's side door, relays, and the DC motor (**FIGURE 56-22**).

FIGURE 56-20 Power window master switch on the driver's door.

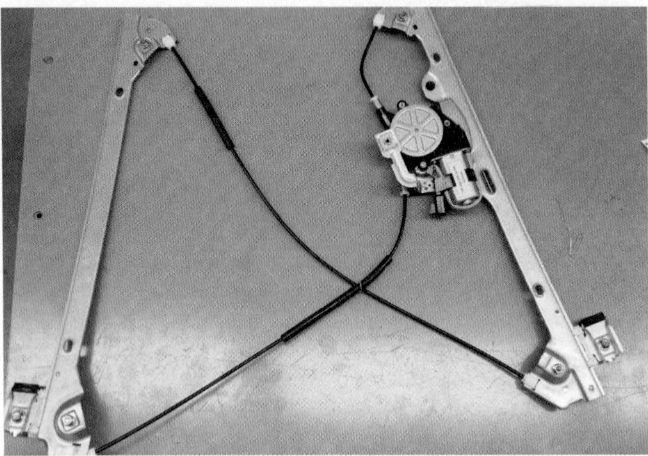

FIGURE 56-21 The mechanical components for a window regulator system.

> **TECHNICIAN TIP**
>
> On some one-touch systems, if power is lost on a control module, most will have to be re-programmed or set back to the HOME position for it to learn where full up is before the one-touch function will work.

Power Seats

Electric or powered seats are becoming standard equipment on modern vehicles. They have a number of electric motors to provide adjustment to the relative position of the seat front to rear, seat height, tilt of the seat base, and often lumbar support. There can be four or more electric motors, which are DC-permanent magnet types and are reversible to provide the movements in either direction. The control circuit turns the motors on and off, controls direction of rotation, and provides dynamic braking. Reversal of the motors is achieved by reversing current flow through the armature. The mechanical arrangements for connection of the motors to the seat mechanisms vary by manufacturer and vehicle model. More sophisticated electric seat systems are controlled by an electronic control unit and can "remember" seat and mirror positions for each driver. The electrical system includes the battery supply, fuses or circuit breakers, switches, relays, electronic control unit (if equipped), and DC motors (**FIGURE 56-23**). In some systems, the switches control relays, which in turn control the motors.

Electric Height Adjustment of Control Pedals

Electric pedal height adjustment allows the repositioning of the brake and accelerator pedals in and out from the firewall. This provides the driver with additional ergonomic comfort by allowing him or her to more effectively position the driving controls in relation to seat and steering position. The movement of the pedals is achieved by a reversible motor-driven screw (**FIGURE 56-24**). Operating the switch activates the drive motor, which in turn rotates an adjustment screw, moving the pedals in toward the firewall or out toward the driver. The accelerator pedal and brake pedal move together and the same distance, so the pedal relationship to each other is maintained. The circuit consists of the fuse or circuit breaker, an adjustment switch, and the reversible motor. Additionally, the system is likely to be controlled by an electronic control unit and may incorporate memory functions and additional safety features, such as inhibitor circuits, that prevent it from operating when the vehicle is in gear or cruise control is enabled.

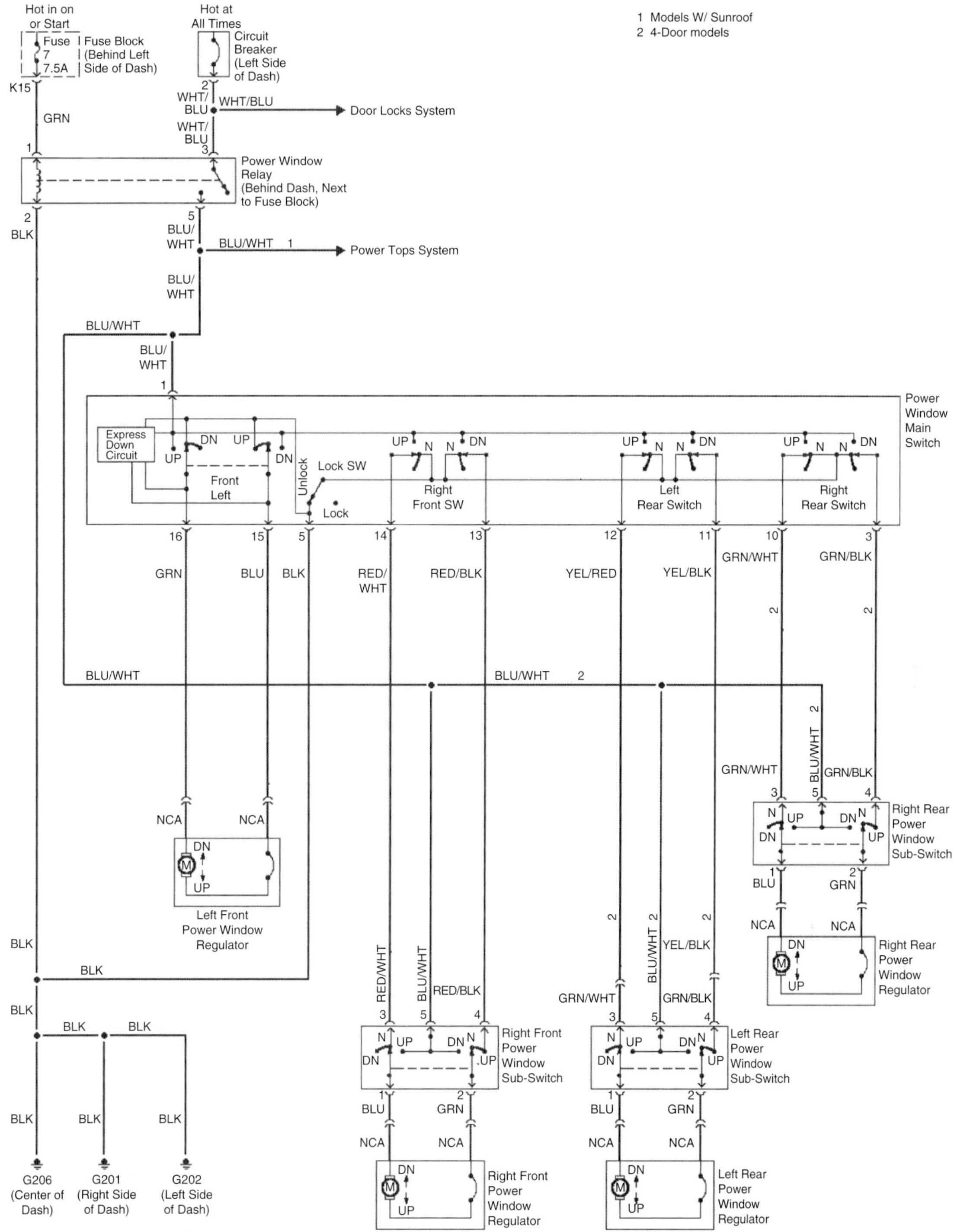

FIGURE 56-22 A circuit diagram for electric windows.

Testing Electric Motors

N56003

Electric motors wear out and fail, but they can also be affected by faulty bearings, bushings, gears and levers, or electrical circuits and control devices. DMMs and automotive test lights will be used to measure voltages and test fuses, relays, switches, and power and ground connection. You will also need to visually inspect the external portion of the motor and any drive mechanism for damage, as they can be the cause of no motor operation.

The DC motors may be located in doors or in locations that require the removal of vehicle trim components, which are prone to breakage, so disassemble them carefully. Permanent magnets within DC motors are brittle and may be damaged if the outer casing is struck with a hammer or other hard object. This activity requires you to work with motors, gears, and mechanical levers, which may create a crush injury hazard; keep fingers away from

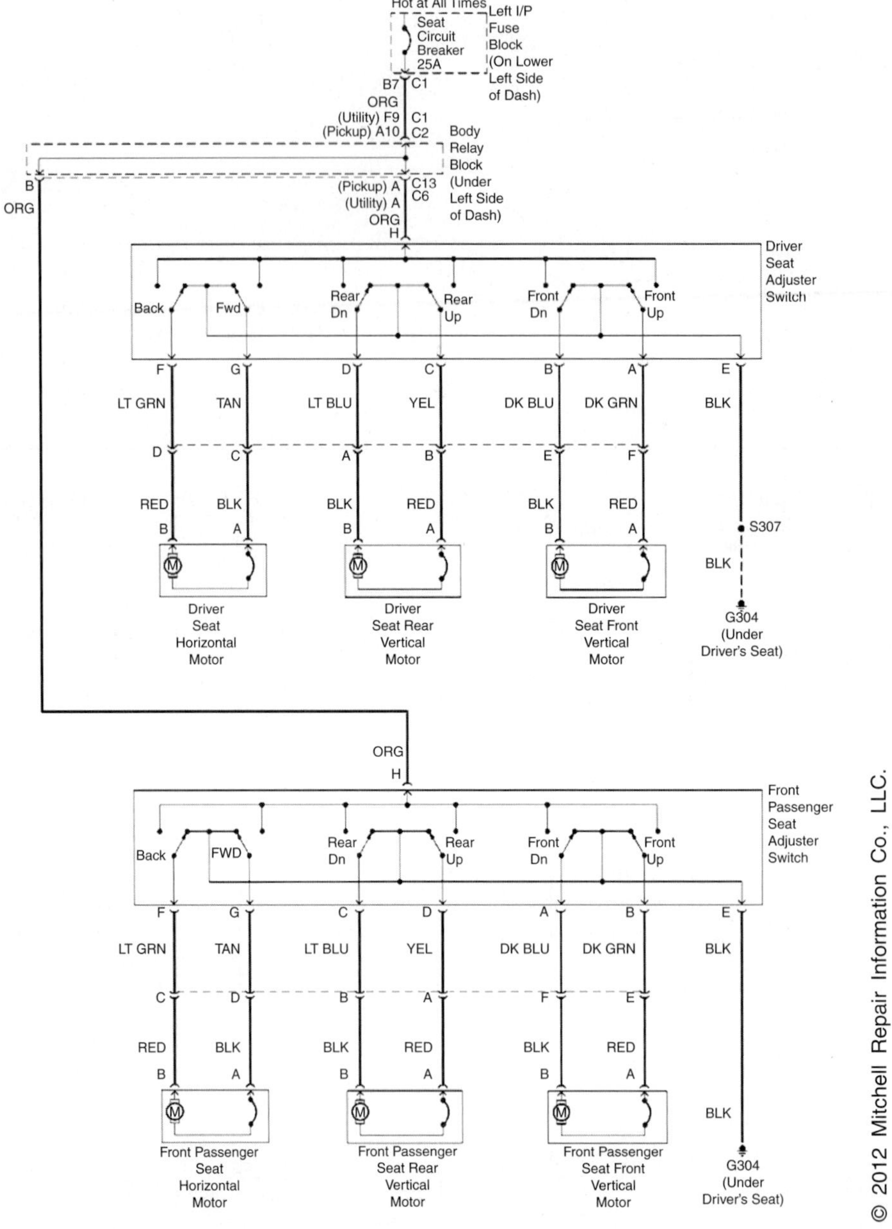

FIGURE 56-23 A circuit diagram for electric seats.

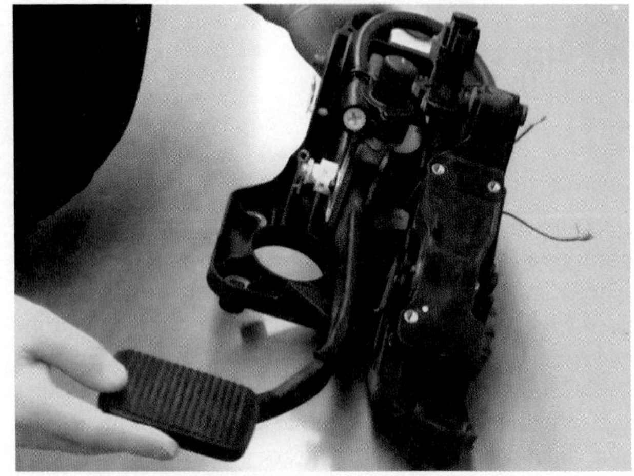

FIGURE 56-24 Adjustable pedal assembly.

mechanisms. Always wear the correct protective eyewear and clothing; use the appropriate safety equipment; and apply fender covers, seat protectors, and floor mat protectors.

To test the electric motor, follow the steps in **SKILL DRILL 56-3**.

▶ Horn Systems

S56003

The vehicle horn is a sound-generating safety device to warn others of a vehicle's presence. Legislation requires vehicles to have a working horn and in most cases also prescribes the sound level required to be made by the horn. Single horns may be installed, but often horns are installed in pairs, one having a high tone and the other with a low tone, to produce a harmonious sound and to give redundancy if one horn fails.

SKILL DRILL 56-3 Testing the Electric Motor

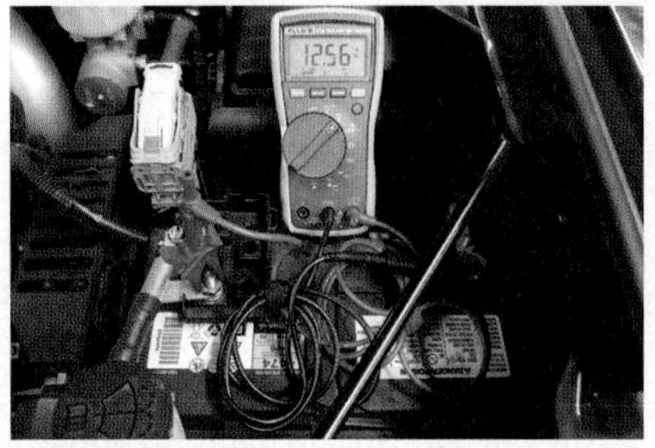

1. Research circuit operation and circuit diagrams for the motor-driven accessory requiring repair, from the manufacturer's information. Measure and record the voltage at the battery.

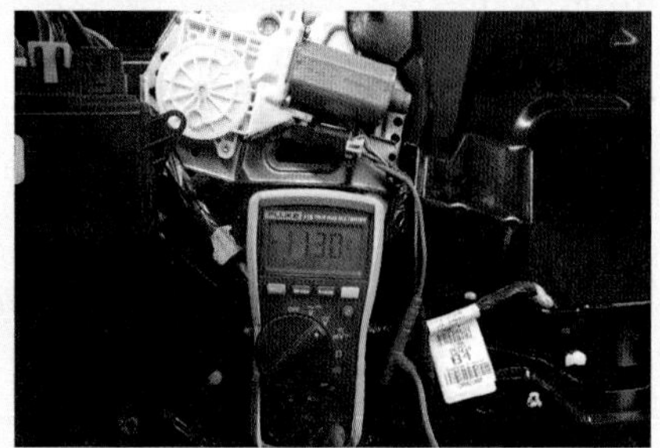

2. Use a DMM to measure the power and ground at the motor while it is being operated. If the voltage reading is more than 1.0 volt lower than battery voltage, perform voltage drop tests on each side of the circuit to determine which side has the fault.

3. Once the faulty side has been identified, measure the voltage drop on each section of the circuit to locate the high resistance. If the total power and ground voltage to the motor was within 1.0 volt of battery voltage, there is a fault in the motor. Determine any necessary actions.

Horn, Relay, Switch, and Clock Spring

The sound of a horn is produced by the vibration of a metal diaphragm, which is operated by an electromagnet switched by a set of contacts. The diaphragm is connected to an armature that moves inside an electromagnet. When the electromagnet is on, the diaphragm is pulled by the armature toward the electromagnet; as this occurs, a trip ring opens the contacts, allowing the diaphragm to spring back to its original position. This cycle happens many times a second, producing sound through the rapid vibration of the diaphragm.

The horn switch is usually mounted in the steering wheel and requires a method to maintain an electrical connection to the circuit as the steering wheel rotates. On vehicles with a driver's side airbag, the clock spring, also called a spiral cable, performs this function. In vehicles without a driver's side airbag, a slip ring and brush assembly is used. The clock spring is also used to carry the electrical signals for the driver airbag circuit and may be used for sound system or cruise control circuits if they are built into the steering wheel. The clock spring is a rotating device with flexible ribbon cable and rotates as the steering wheel rotates (**FIGURE 56-25**). It is designed to turn two or three times in each direction from the center point. It cannot rotate endlessly in a single direction, so it must be centered (timed) to match the position of the steering system when it is installed, to avoid it reaching the end of its travel when the steering wheel is turned in either direction.

The horn circuit is made up of the battery power supply, a fuse or circuit breaker, a relay, the horn switch, and the clock spring. The horn switch is usually a grounding circuit. The switch grounds the relay winding, which in turn supplies power through the relay contacts to operate the horn (**FIGURE 56-26**). The horns on many newer vehicles can also be operated by pressing the panic button on a fob. They may chirp when locking or unlocking, and honk when a breach of security has happened. In most cases, these horns are operated by a body computer when the correct inputs arrive.

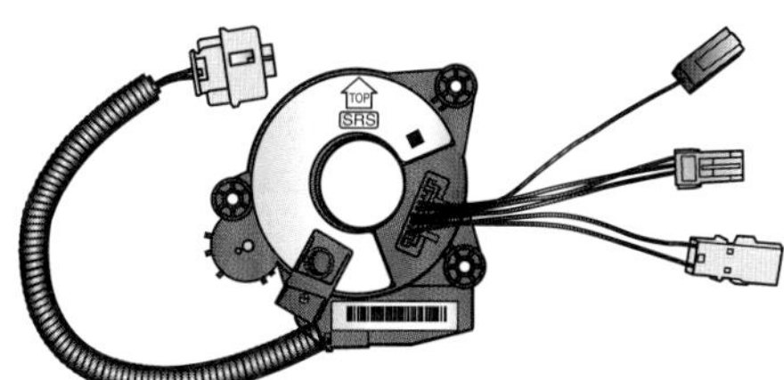

FIGURE 56-25 A clock spring.

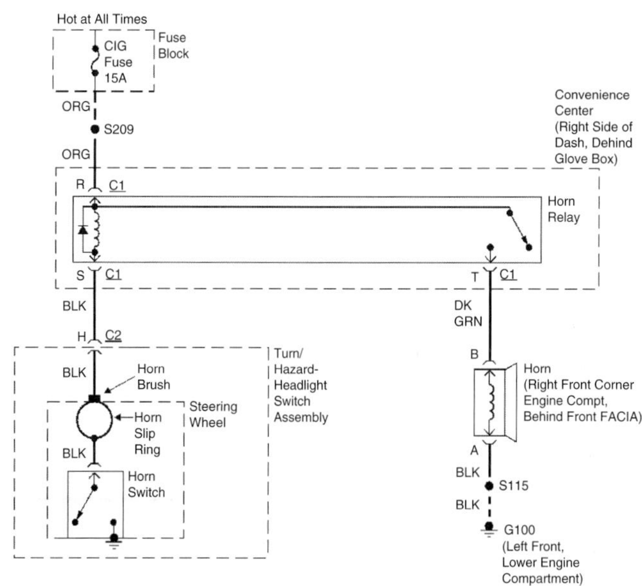

FIGURE 56-26 A typical horn circuit.

Testing the Horn System

N56004

The horn circuit can fail in many ways. Any of the components in the circuit can go bad, as can the wiring. Using a DMM to measure voltages and test voltage drops on fuses, relays, switches, and power and ground connections will tell you how to proceed when testing the circuit. The horn switch is usually mounted within the steering wheel and is usually connected through the steering column by a clock spring. The horn switch usually grounds the horn relay winding, which in turn sends power to the horn.

When testing the horn, it may be easiest to first test to see if power is getting to the horn when the horn is activated, as it is not uncommon for horns to go bad, especially if they are rarely used or are overused. If power is getting to the horn, check the voltage drop on the ground side. If the voltage drop is fine, the horn is bad. In this case, tap the horn while it is energized. It is possible the contacts inside the horn are dirty or rusty, and tapping on the horn may free them up. If so, you will know that the rest of the circuit is operational, and you will probably want to replace the horn as it is likely to go bad again soon.

If the horn does not have power when activated, it is good to go to the horn relay to perform the next tests. At the relay, you have access to each of the legs in the circuit; power from the horn fuse, the output terminal to the horn, power from the fuse feeding the winding, and the horn switch terminal. Using the DMM and a jumper lead, you can check each of these legs. Remove the relay and using the DMM, check for power from each fuse. If present, use the jumper lead from the horn fuse input to the horn output (at the relay base) and see if the horn honks. If it does not honk, check the wiring between the relay and the horn. If it does honk, use a DMM to check the continuity of the horn switch leg with the horn switch depressed. There should be continuity between the horn switch terminal on the relay block and ground. If there is, then the relay is faulty.

To diagnose the horn system, follow the steps in **SKILL DRILL 56-4**.

▶ Power Door Locks

S56004

Power or electric door locks are a common option in modern vehicles and are installed in each door. They are often integrated with the vehicle's security system and remote locking systems and are called central locking systems. Many vehicles also have trunk locks incorporated into the system. The driver's side door is usually a master door lock, which means that when it is locked or unlocked with the key (or the remote fob), the other locks follow suit.

Power door locks use an electric actuator to move the mechanical door lock mechanism. There are two main types of actuators: a DC motor actuator and a solenoid type. The DC reversible permanent magnet motor type is enclosed in a single casing with a number of gears for speed reduction; they operate a rack-and-pinion gear set which converts the rotational motion into linear motion to operate the lock (**FIGURE 56-27**). A centrifugal clutch built into the motor main gear allows the door locks to be operated manually.

In a solenoid-type actuator, there is a permanent magnet plunger with a stationary coil wound around the outside. The north and south poles on the plunger react with the magnetic field created in the coil, producing linear movement of the plunger (**FIGURE 56-28**). The plunger's direction of movement can be reversed by controlling the direction of current flow through the coil. Changing directions of the current flow changes the north and south magnetic fields that are created, which either attract or repel the plunger. The connecting rod of the plunger is connected to the door locking mechanism. The electric door lock circuit consists of the power supply, fuses or circuit breakers, the actuator (motor or solenoid), relays, switches, and cabling. Relays may be used to control the motors or solenoids (**FIGURE 56-29**).

SKILL DRILL 56-4 Testing the Horn System

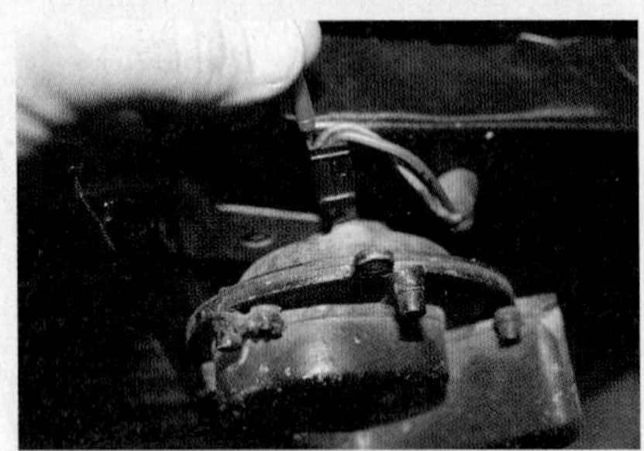

1. Confirm that the horn does not operate. Research circuit operation and circuit diagrams for the horn from the service information. Check for power and ground at horn when the horn switch is operated. If power and ground are present, the horn is faulty and needs to be replaced.

2. If there is no power to the horn, remove the relay, and check for battery voltage at terminal 30 (3). If battery voltage is not present, check the horn fuse and the rest of the feed circuit.

3. If battery voltage is present, jump terminal 30 (3) to terminal 87 (5). The horn should honk. If it doesn't, check for an open wire between terminal 87 (5) and the horn with a DMM.

SKILL DRILL 56-4 Testing the Horn System (Continued)

4. If the horn now honks, check that either terminal 85 (1) or 86 (2) has battery voltage. If 85 has battery voltage then 86 is controlled by the horn switch, or vice-versa.

5. Connect a DMM between the switch leg terminal and ground. Operate the horn switch; resistance should decrease to less than an ohm. If it doesn't, search out the high resistance or open circuit back to the horn switch. Suspect either the clock spring or the horn switch itself.

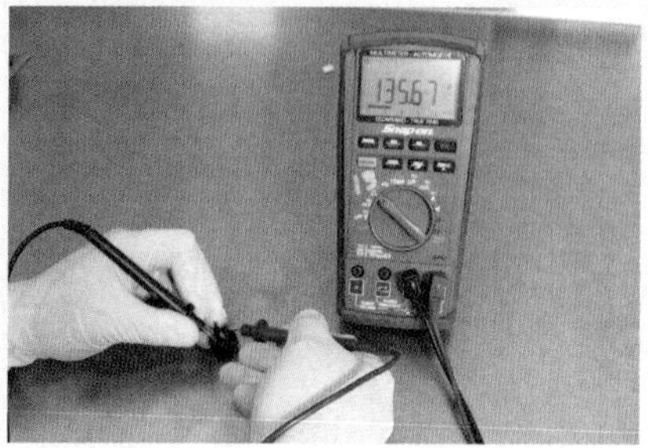

6. If the switch leg is good, measure the resistance between terminals 85 (1) and 86 (2) on the horn relay itself. The resistance should be between about 40 and 100 ohms. If not, replace the relay.

7. If the resistance is good, either measure the voltage drop across the switch contacts of the relay 30 (3) and 87 (5) or substitute a known good relay.

Electric Lock and Keyless Entry Systems

Keyless entry systems and the engine immobilizer or security systems are separate systems, although they are usually linked so that the security system is enabled or disabled as the vehicle is locked and unlocked. Keyless entry is also known as remote key locking, and additional features such as alarms and warning devices may be incorporated into the system. The system is usually controlled by a wireless fob and is used to lock and unlock the vehicle, provided it is near enough (**FIGURE 56-30**).

On networked vehicles, many systems are programmed to perform actions automatically, such as locking the vehicle when

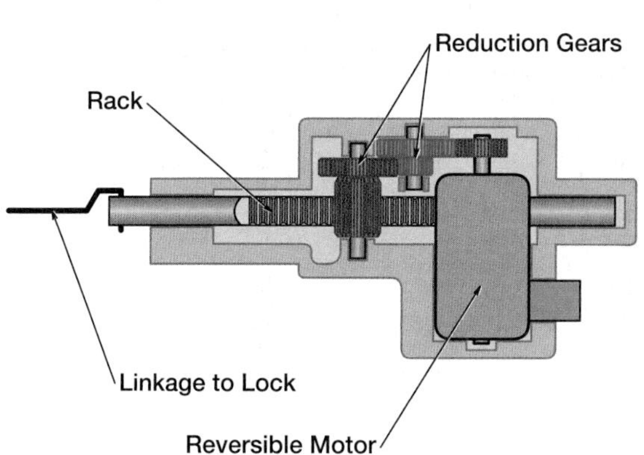

FIGURE 56-27 Power door lock actuator—DC motor type.

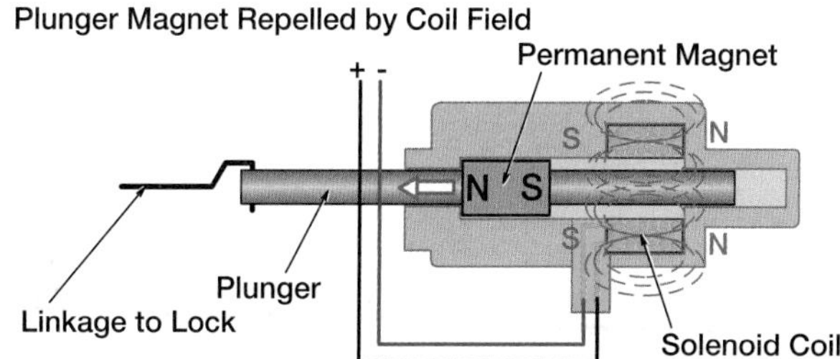

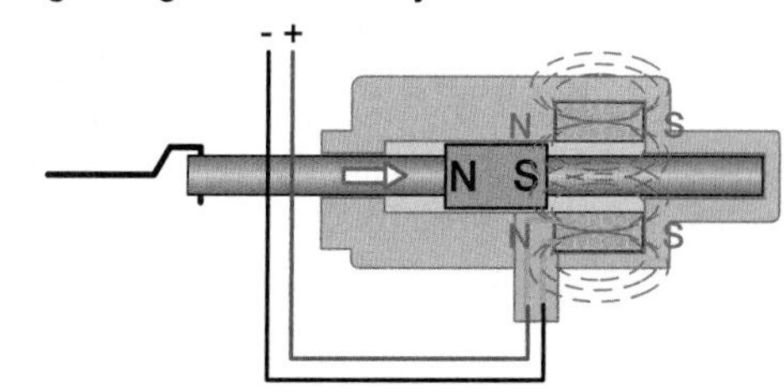

FIGURE 56-28 Power door lock actuator—solenoid type.

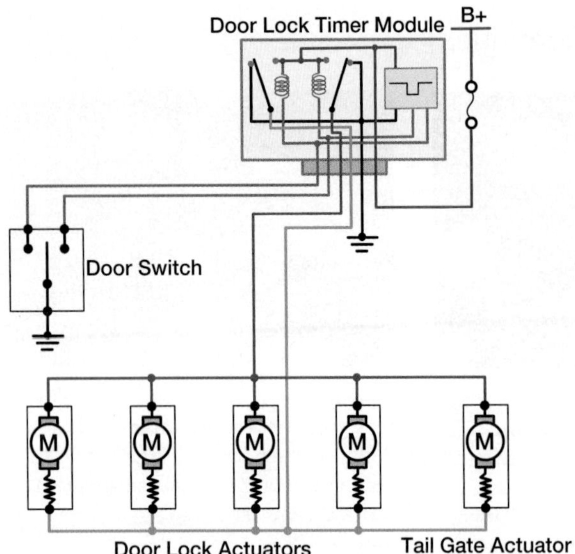

FIGURE 56-29 Typical power door lock circuit.

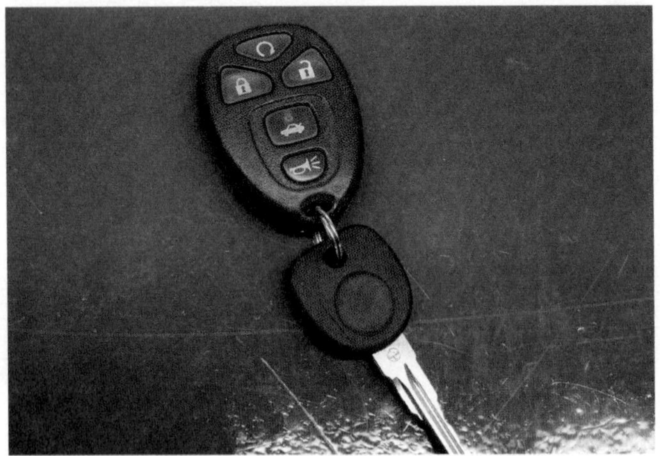

FIGURE 56-30 Wireless key fob with buttons.

it is put into gear (automatic transmission) or when it reaches a certain miles per hour (mph), and unlocking it when put back into park. The driver can program most of these systems to enable or disable these features. Some vehicles have stepped unlocking, which allows one push of the remote fob to unlock the driver's door only and two pushes to unlock all doors. This feature can be programmed on some vehicles as well.

Types of Systems

Manufacturers continue to develop various systems for controlling vehicle access and starting operations; each has its own subtle differences. Despite the differences, the systems tend to fall into two common types: **remote keyless entry (RKE)** and **passive keyless entry (PKE)**. The RKE system uses a fob transmitter that is usually attached to the key or key ring. The fob has a number of buttons or keys for remotely locking and unlocking the doors or trunk. It may also have a panic button that can activate the horn when pushed. The PKE system uses a fob transmitter but does not require any action by the user. The vehicle senses if the fob is within range of the vehicle and automatically unlocks the door when the person touches the handle, without the need to press any buttons (**FIGURE 56-31**). It can also lock the doors automatically once the driver walks away from the vehicle.

Principles of Operation

The RKE system consists of a fob transmitter that the driver has and a receiver inside the vehicle. When a key or button is pressed, the fob transmits a coded signal that is received by the vehicle's receiver, provided it is within range. The **body control module (BCM)** recognizes the key's code and activates the vehicle's electric door locks.

The PKE system consists of an electronic key that is carried on the person's body. A series of receivers determine where the electronic key is located once it is in close proximity to the vehicle (**FIGURE 56-32**). Communication is then established between the vehicle's system and the key. If the key is authenticated, the vehicle unlocks the door when the closest door handle is touched, without the need for any buttons or keys to be pressed. The same authentication system can be used to control the vehicle's start system, security system, and engine immobilizer.

FIGURE 56-31 Passive keyless entry fob automatically communicates with the vehicle. Notice that the remote key is extended.

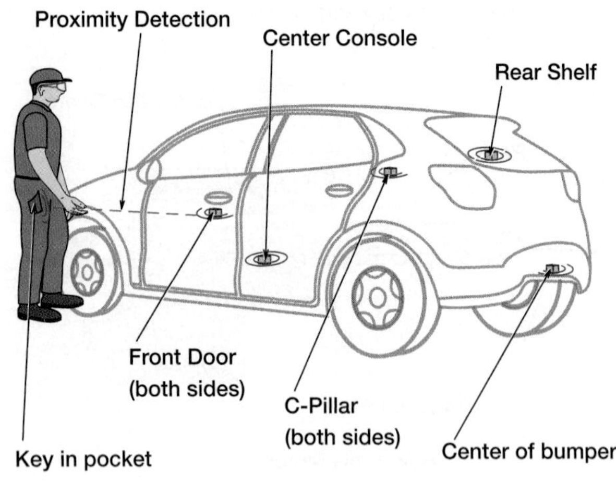

FIGURE 56-32 Typical layout for a passive keyless entry system.

Testing Electric Locks

N56005

Door lock systems can fail for a variety of reasons, from a dead battery in the remote fob to worn-out door lock switches and actuators. If the system is a CANbus system, the network or control module could be faulty. When starting a diagnosis on the electric lock system, it is good to verify what works and what doesn't. For example, if the system works from the lock button on the driver's door, but not on the passenger door, you probably want to focus on the passenger door switch.

Once you have an idea of how the system is functioning, it is a good idea to consult the system's wiring diagram so you can understand how the circuit is supposed to work. Then you can build your diagnostic strategy. At that point, it is probably time to break out the DMM, automotive test light, scan tool, or lab scope to help trace how electricity is flowing (or not flowing) in the circuit. Having an inductive ammeter connected around a jumper lead in the power door lock fuse holder allows you to measure the current flowing through the circuit when the switches are activated (**FIGURE 56-33**). You will be looking for opens, shorts, high resistance, and improper signals in the electrical circuit and for mechanical faults in the motors, solenoids, and linkage.

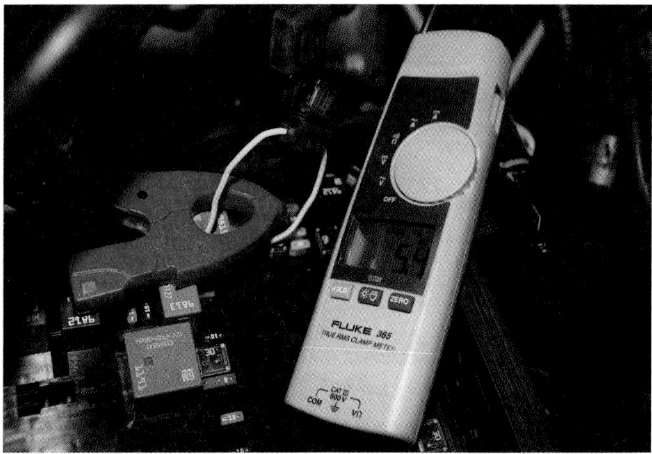

FIGURE 56-33 Using an inductive ammeter around a jumper lead in the fuse holder allows you to measuer the current flowing when the door lock switches are being activated.

Removing the Door Panel

N56006

Removing door panels may require the removal of the arm rest, door lever, window crank (if equipped), and any switch panels mounted on the door. In most cases, the door panel itself will be attached to the door using clips, which need to either be pried out of their holes, or slid off of their stops. Some manufacturers use a few screws to provide extra security. To pry the door panel from the door, it is best to use a door panel tool to prevent damage to the door or paint finishes. Special tools may be required to release the clip that holds the manual window handles in place, if equipped. Preserve the inner lining for reuse by carefully peeling it back from the door frame. Carefully examine the door panel for screws and fastenings that may need to be removed before attempting to pry off the door panel.

To remove the door panel, follow the steps in **SKILL DRILL 56-5**.

▶ Wiper/Washer System

S56005

The wiper and washer system is an important vehicle safety system. The wipers ensure that the driver has a clear line of sight through the windshield by removing any excess moisture from the glass. The wipers usually have a number of speeds and intermittent operation for varying rainfall conditions. The washers provide a cleaning spray that, when used in conjunction with the wipers, helps to clean road grime from the windshield. Some vehicles use a wiper and washer for the rear glass, and even some headlights as well (**FIGURE 56-34**).

Wiper and Delay Circuits

The wiper motor usually has at least a high and low speed and a time delay or intermittent operation. Intermittent operation is controlled by a timer circuit, which may be a discrete electronic

SKILL DRILL 56-5 Removing the Door Panel

1. Research removal and reinstallation of the door panel from the manufacturer's service information. Remove fixtures such as the arm rest.

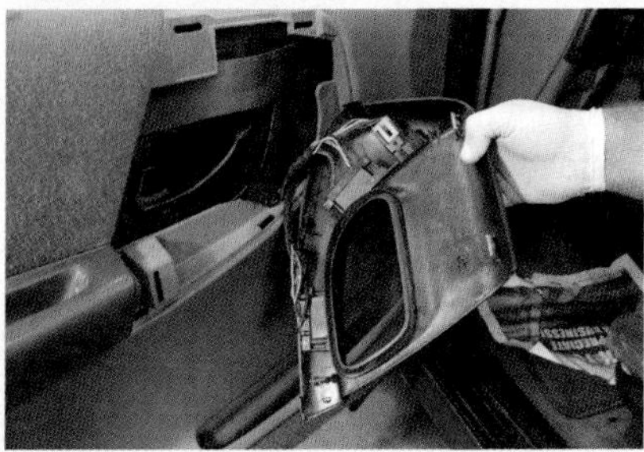

2. Remove the switch panel.

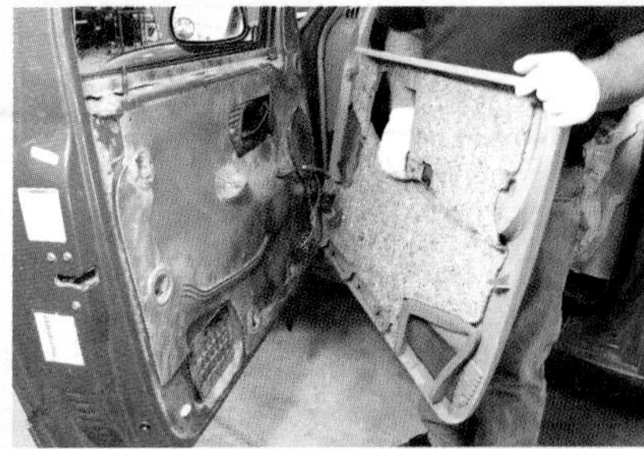

3. Remove the cables, and carefully pry the panel from the frame.

SKILL DRILL 56-5 Removing the Door Panel (Continued)

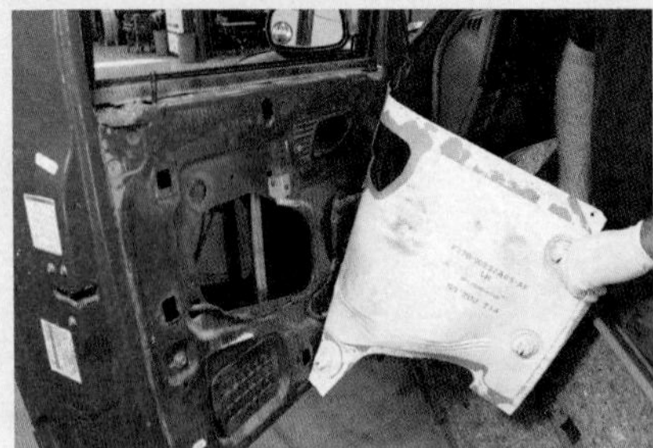

4. Peel back the inner liner from the frame, preserving it for reuse.

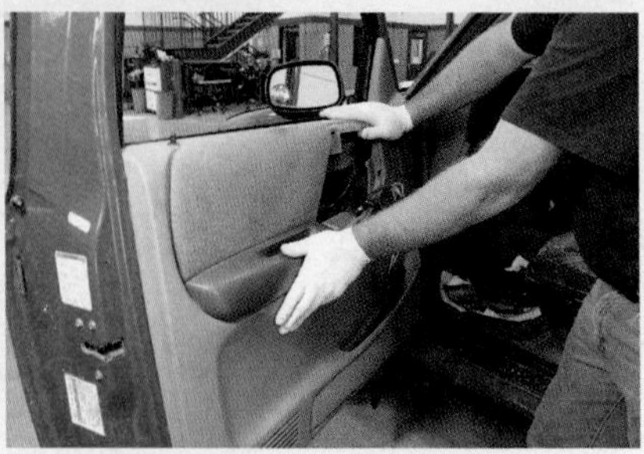

5. Reinstall the door panel using the reverse procedure while ensuring all electrical connections are reinstalled.

FIGURE 56-34 Some headlights use a wiper and washer to keep them clean.

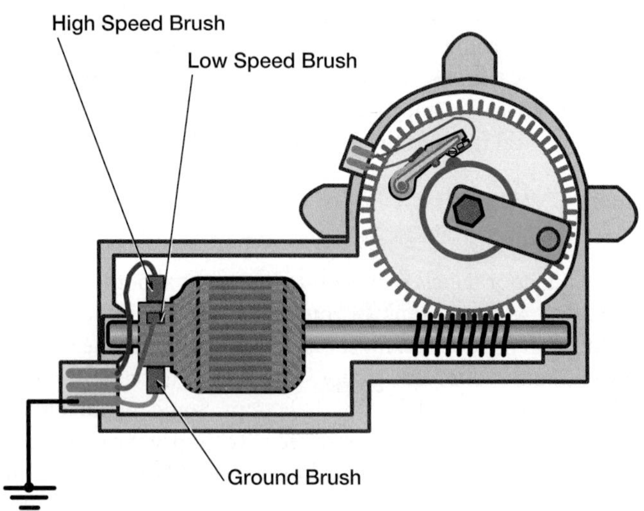

FIGURE 56-35 Many wiper motors use three brushes to get both a low and high speed.

timer or controlled by the vehicle body computer. Incorporated into the circuit and activated by the motor gear is a parking switch to ensure that the motor stops when the wiper blades are in the Park position. The DC motor is dynamically braked to make sure the motor stops instantly, without overrun, in the Park position. To provide high and low speed, the wiper motor is equipped with three brushes. Two brushes are placed 180 degrees apart, as they are in a standard motor, and the additional or third brush is located off-center (**FIGURE 56-35**). When current is switched from the low-speed brush to the high-speed brush, it provides higher rotational speed of the armature due to the different interaction of the magnetic fields.

The wiper circuit consists of the battery, a fuse or circuit breaker, the ignition switch, the wiper switch, an intermittent timer and/or body computer, and the motor assembly with park switch (**FIGURE 56-36**). The park switch is built inside the motor and gearbox housing and switches a power feed to the low-speed brush when the wiper switch is in the Off position and the wiper blades move to the Park position (**FIGURE 56-37**). Once in the Park position, dynamic braking of the motor occurs, bringing the motor to a quick stop. Intermittent operation is achieved by

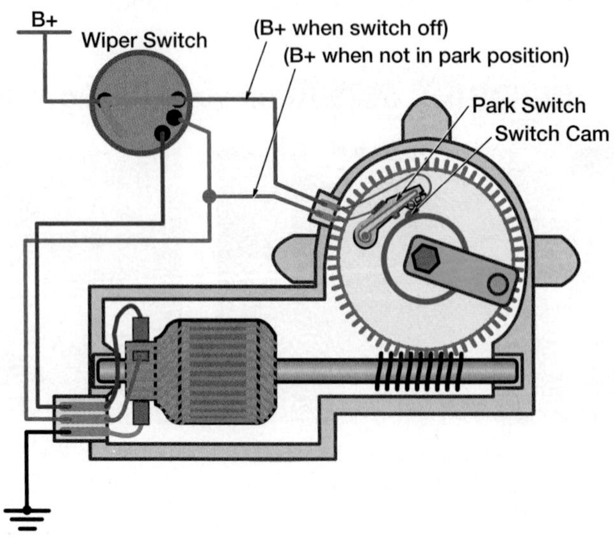

FIGURE 56-36 A wiper circuit diagram.

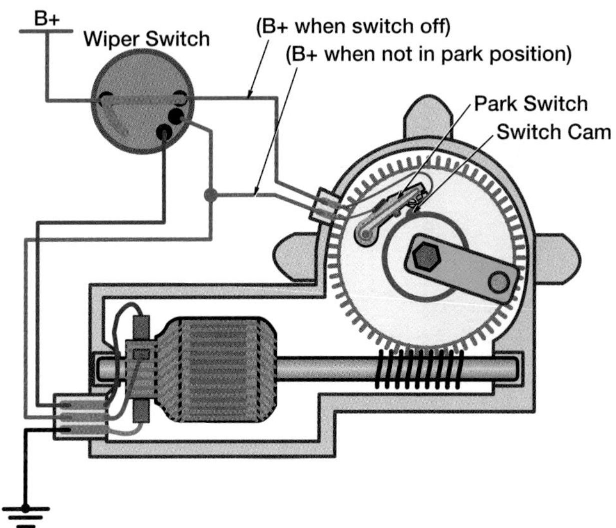

FIGURE 56-37 The park switch allows the wipers to be powered until they reach the Park position, and then they stop.

pulsing the low-speed circuit with a temporary power, causing the motor to move off the Park position. With the power pulse removed, the motor continues on a single wipe of the windshield through the park switch until it again reaches the Park position.

Wiper circuits can also be ground-switched circuits. For example, the ignition switch provides a power feed to the motor through a fuse or circuit breaker, and the wiper switch, depending on the switch position, switches the high- or low-speed brush to ground. Some vehicles use a rain sensor mounted in or near the windshield, which signals the body computer to operate the wipers accordingly.

Testing the Washer System

The windshield and back window washer systems are made up of a washer fluid bottle to hold the washer fluid, a DC electric pump, nozzles to spray the glass, tubing to carry the washer fluid from the pump to the nozzles, and the electric circuitry. The DC motor on the pump is a permanent magnet motor and has power supplied to one brush and ground by the other brush. The switch for the washer is usually located on the same switch assembly as the wipers. Operating the washer switch usually also activates the wipers for at least one cycle; however, some systems allow for multiple wipe cycles before parking the wiper blades.

Both the electrical circuit and the pump discharge circuit need to be tested for correct operation. The electrical circuit consists of the power supply, fuse, switch, and DC pump motor, and can be tested with a test lamp or DMM. The pump or washer fluid discharge circuit is made up of the pump, the washer fluid supply or tank, the tubes that carry the washer fluid to the washer nozzles, and the nozzles. You may find that blowing air through the discharge circuit and nozzles cleans out any debris that may be plugging the system. Just be careful that you don't over pressurize the system and cause an inaccessible hose to be blown off of a fitting. Also, ensure adequate washer fluid in the tank before conducting any tests.

To test the washer system, follow the steps in **SKILL DRILL 56-6**.

Testing the Wiper System

The windshield wipers can fail in a number of ways. First, they may not work at all, such as when a fuse or motor is bad. Second, they may work on only some speeds, such as when the switch or one of the brushes in the motor is faulty. Third, they may not work on the intermittent speed, such as when the delay module is faulty. Fourth, they may not park, as when the park contacts in the motor are faulty or the wiper switch is bad. Each of these faults can be diagnosed by using a DMM and referring to the wiring diagram to locate the fault. Another fault with wiper systems is that the linkage wears out, which can prevent one or both of the wiper arms from operating, or operating properly. To diagnose this, you have to inspect the linkage for worn or disconnected joints.

SKILL DRILL 56-6 Testing the Washer System

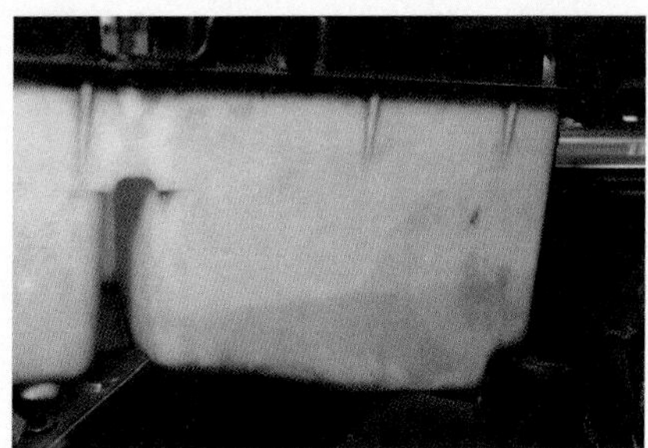

1. Check for the correct solution level in tanks, and listen for washer pump operation.

2. If an electrical fault is indicated by lack of a washer motor sound, check for battery voltage and ground at the pump when it is activated.

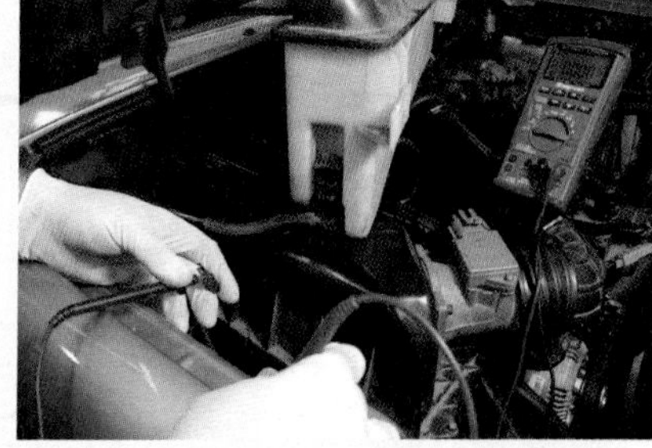

3. If the power or ground circuit is faulty, perform voltage drop tests on the faulty side until the fault is located.

SKILL DRILL 56-6 Testing the Washer System (Continued)

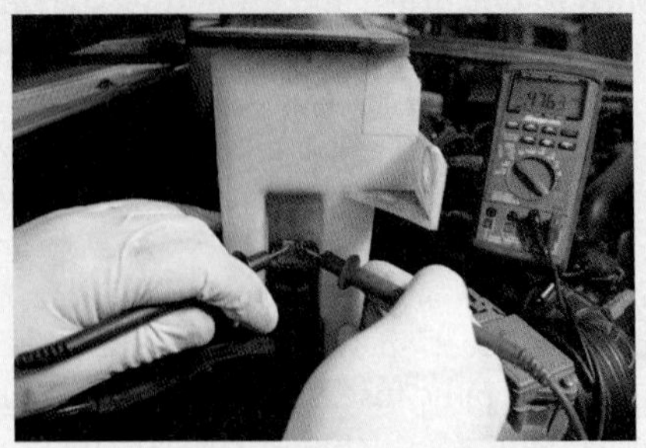

4. If the power and ground circuits are good, measure the resistance of the pump motor, and compare to specifications.

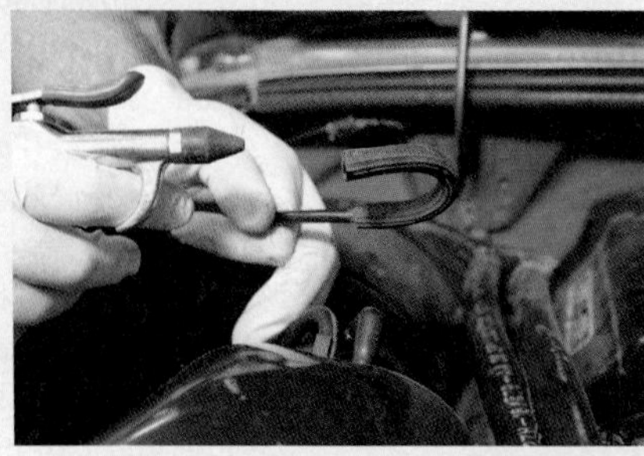

5. If the pump spins when activated but does not pump washer fluid, check the hoses and nozzles for obstructions. Blow out with compressed air if necessary.

This activity requires you to work with motors, gears, and mechanical levers, which may create a crush injury hazard; keep fingers away from mechanisms. Also realize that if you remove the wiper blades and allow the wiper arms to slam back in place, it can break the windshield. Always place a folded up fender cover or heavy cardboard on the windshield when removing wiper blades.

▶ Heated Glass, Mirrors, and Seats

S56006

Any glass surfaces in the vehicle that are colder than the air in the vehicle can cause moisture to condense on the glass and fog it up. Heating the glass is an effective way to reduce the moisture buildup and improve visibility. Both the rear window and vehicle mirrors can be electrically heated to reduce moisture condensation. Most vehicle manufacturers use an electric heater element attached to the glass to provide a heat source (**FIGURE 56-38**). Mirrors have heater elements attached to the rear of the glass in a zigzag pattern to provide coverage across the mirror surface. The rear cabin window has a fine element heating foil attached to the inside of the glass. The element is made fine enough that it does not impair driver vision but creates sufficient heat to ensure that the window remains free of condensed moisture. The heated glass circuit has a power supply, a fuse, a switch, and usually a relay to switch the high current draws of the heating elements.

Heated seats provide additional passenger comfort, particularly in cold climates. Seats are heated with an electric element made inside the seat base and sometimes the seat back (**FIGURE 56-39**).

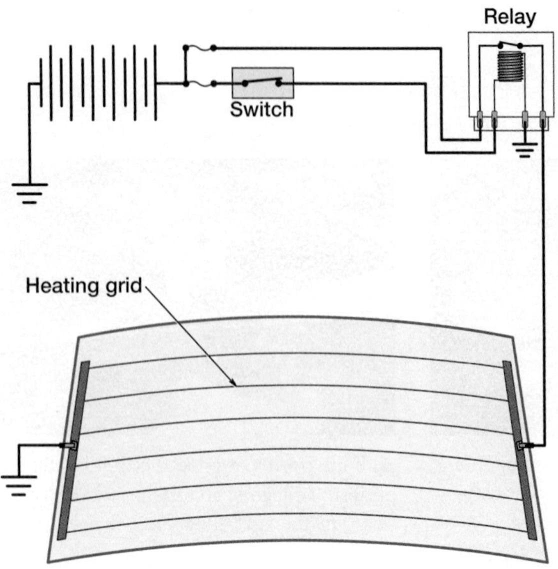

FIGURE 56-38 A heated window grid.

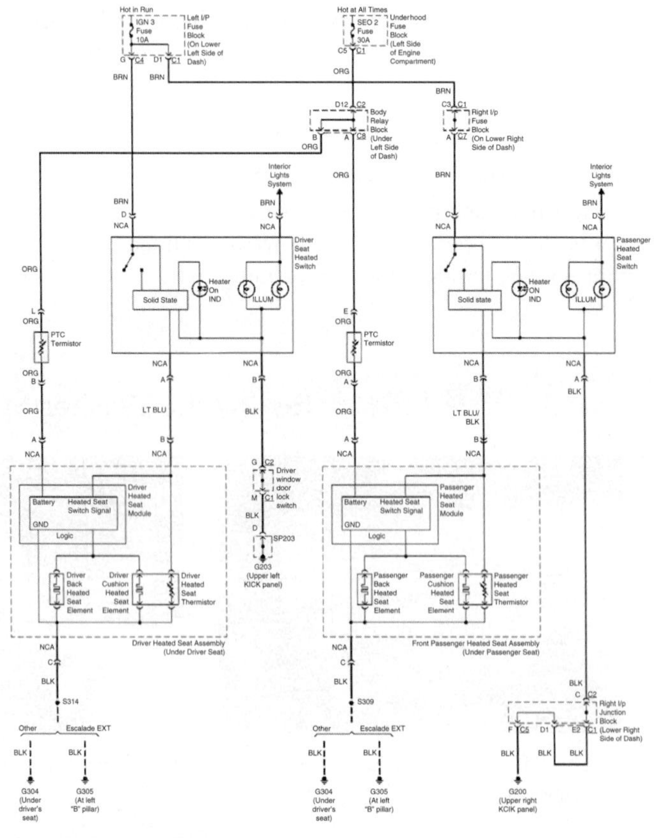

FIGURE 56-39 Heated seat circuit diagram.

The element is controlled by a dash switch and relay. The circuit usually has a relay connected to a thermostat to ensure that the seat does not become overheated. Some systems may also have temperature controls with high and low settings to further provide passenger comfort. Some seats are also cooled. This is covered in the air conditioning section of this book.

Testing Heated Accessories

DMMs and automotive test lights are used to measure voltages and test fuses, relays, switches, and power and ground connections. They are also used to measure resistance of heating elements. Refer to the wiring diagrams and service information to understand the operation of the system as well as its layout. Methodically work through the circuit, testing voltage, resistance, and current against the manufacturer's specifications. An infrared thermometer may be useful in determining the effectiveness of heating elements (**FIGURE 56-40**). Always ensure that power is disconnected before connecting ammeters, and ensure that the ammeter is capable of measuring the expected current draw. The heat generated by the heating elements may cause burns, so always test the surface temperature before touching.

When testing window defogger grids, use a voltmeter or test light to test the individual grid lines to see if they are broken (**FIGURE 56-41**). If they are, some grid lines can be patched with special conducting compound that is brushed across the break and allowed to dry.

FIGURE 56-40 An infrared temp gun being used to measure the temperature of a heated mirror.

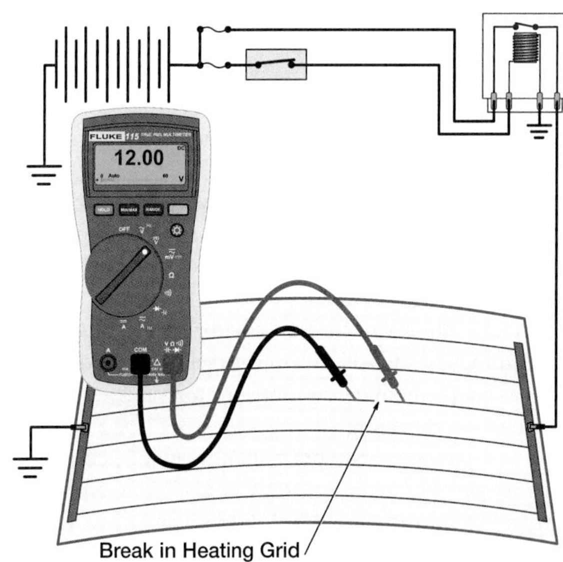

FIGURE 56-41 Testing a heated window grid.

▶ Wrap-Up

Ready for Review

- Electrical systems on modern vehicles are becoming increasingly complex with the addition of a range of electronic and accessory systems, such as global positioning systems, entertainment systems, security systems, electric seats, and heated glass. Many of these systems are controlled by body computer systems.
- The controller area network (CAN) and the bus system are used together in vehicle terminology (often called "CANbus") to describe the data-sharing network in vehicles. The many sensors and engine control units or modules can share information across a common network, using the CANbus protocol to communicate efficiently and effectively with low data loss.
- Diagnosing and repairing vehicles may require software transfers, software updates, or flash reprogramming of the vehicle's electronic modules with the manufacturer's latest update.
- Electric motors are used to drive a variety of accessories on vehicles.
- There are several types of electric motors: brush type, brushless type, and stepper type.
- Electric cooling fans for the radiator are increasingly common, particularly for transversely mounted engines and as supplementary cooling fans for vehicles equipped with air conditioning.
- Power mirrors use two permanent magnet electric motors: one to control the up-and-down movement, and the second to control the in-and-out movement.
- Electric windows use permanent magnet motors located in each of the passenger doors and sometimes in a rear door to control the up-and-down movement of vehicle windows.
- The sound of a horn is produced by the vibration of a metal diaphragm, which is operated by an electromagnet switched by a set of contacts.
- Technicians today need to know how to service all the accessories behind the door panels. This critical skill must be mastered without damaging the accessory, wiring, or door panel.
- The wiper and washer system is an important vehicle safety system. The wipers ensure that the driver has a clear line of sight through the windshield by removing any excess moisture from the glass.

Key Terms

blower motor An electric motor, usually the permanent magnet type, that moves air over the air-conditioning evaporator and heater core.

body control module (BCM) An electronic control module for body electrical systems.

brushless DC motor An electric motor that does not have any brushes and is sometimes called an "electronically commutated motor." In this type of motor, an electronic control module replaces the brushes and commutator.

data link connector (DLC) The port to which a scan tool can be connected.

passive keyless entry (PKE) An active system that senses the proximity of a fob and locks or unlocks the vehicle.

permanent magnet electric motor An electric motor in which the magnetic field in the casing is produced by permanent magnets, and the armature has an electromagnetic field generated by passing electrical current through loops or windings, thereby producing the motor action.

remote keyless entry (RKE) A system that remotely unlocks and locks the vehicle without the use of a traditional key.

stepper motor A type of brushless motor with a key difference: It is designed to rotate in fixed steps through a set number of degrees.

Review Questions

1. What is the function of a multiplexed system in a vehicle?
   a. It transmits information across one or two wires.
   b. It interprets information.
   c. It sends the information only to the module that needs it.
   d. It processes the information travelling through it.
2. Choose the correct statement.
   a. Many vehicles use only one type of networking system.
   b. Usually more than one type of networking system is incorporated in vehicles.
   c. The faster the network, the less expensive it is.
   d. Different networks do not share information between them.
3. The communication network system most likely used to transmit high-speed audio and video is:
   a. CAN network.
   b. LIN network.
   c. MOST network.
   d. CAN FD network.
4. All of the following statements with respect to MOST networks are true *except*:
   a. They can be plug and play devices that can be added even after servicing.
   b. They can use electrical conductors for sending data signals.
   c. They can use plastic optic fiber for sending data signals.
   d. The highest transfer rate in this network is 25 kbps.
5. The horn switch is typically connected to:
   a. the ground side of the relay winding.
   b. the power side of the relay winding.
   c. the power side of the horn.
   d. the ground side of the horn.
6. One of the best things you can do when checking for module communication on a network is to:
   a. trace out the circuit starting at the battery.
   b. connect a scan tool to the network DLC and check to see if there are any modules that are missing or not communicating.

c. measure the voltage on CAN-H.
d. measure the voltage on CAN-L

7. When updating the software in the vehicle's PCM:
   a. the vehicle should be running at full operating temperature.
   b. a battery support unit needs to be connected to ensure that the vehicle's battery does not discharge.
   c. the PCM should be disconnected from the vehilce.
   d. the vehicle's negative battery terminal must be disconnected.
8. The communication line system used for high-speed controller area network bus (CAN-bus) systems is:
   a. two-wire.
   b. twisted-pair two-wire.
   c. two-wire shielded.
   d. single-wire.
9. The network system that is ideal for the entertainment systems on a vehicle is the:
   a. CAN.
   b. LIN.
   c. MOST.
   d. Ethernet.
10. Which of these motors is used to open and close the throttle plate one step at a time, as directed by the power train control module (PCM)?
   a. Brushless motor
   b. Permanent magnet motor
   c. Stepper motor
   d. Blower motor

ASE Technician A/Technician B Style Questions

1. Tech A says that CAN A is a low-speed data network. Tech B says that CAN C is a high-speed network. Who is correct?
   a. Tech A
   b. Tech B
   c. Both A and B
   d. Neither A nor B
2. Tech A says that CAN C typically uses two 120-ohm terminating resistors with a total circuit resistance of 60 ohms. Tech B says that CAN C uses two 120-ohm terminating resistors with a total circuit resistance of 240 ohms. Who is correct?
   a. Tech A
   b. Tech B
   c. Both A and B
   d. Neither A nor B
3. Tech A says that a twisted pair of wires in a wire loom makes it easier to route wires to their destination. Tech B says that the twisted pair of wires transmit differential signals of opposite polarity on the two lines. Who is correct?
   a. Tech A
   b. Tech B
   c. Both A and B
   d. Neither A nor B
4. Tech A says that flashing an electronic control module with updated software can solve customer concerns. Tech B says that when flashing an electronic control unit or module, make sure no power loss or disconnect occurs, as the electronic control unit or module may be damaged. Who is correct?
   a. Tech A
   b. Tech B
   c. Both A and B
   d. Neither A nor B
5. Tech A says that stepper motors are used as blower motors to achieve various speeds controlled by the driver. Tech B says that stepper motors are used where precise movement must occur, such as in electronic throttle controls. Who is correct?
   a. Tech A
   b. Tech B
   c. Both A and B
   d. Neither A nor B
6. Tech A says that resistors are used to control the speed of the blower motor on some vehicles. Tech B says that high speed is obtained when current flows through all of the resistors, to the motor. Who is correct?
   a. Tech A
   b. Tech B
   c. Both A and B
   d. Neither A nor B
7. Tech A says that the clock spring is a device used in the steering wheel to transmit signals across the rotating electrical connection. Tech B says that the clock spring must be centered during installation, as it only allows turning both ways a certain number of rotations. Who is correct?
   a. Tech A
   b. Tech B
   c. Both A and B
   d. Neither A nor B
8. Tech A says that the horn typically gets its power directly from the horn switch. Tech B says that the horn switch activates a relay. Who is correct?
   a. Tech A
   b. Tech B
   c. Both A and B
   d. Neither A nor B
9. Tech A says that if the washer pump operates, but no washer fluid sprays out the nozzles, the windshield wiper motor is faulty. Tech B says that many wiper motors use a low speed brush and a high-speed brush to control wiper speed. Who is correct?
   a. Tech A
   b. Tech B
   c. Both A and B
   d. Neither A nor B
10. Tech A says that a heated window grid can be tested for breaks with a voltmeter or test light. Tech B says that heated seats have special tubes in which heated coolant is circulated to heat them up, just like the vehicle's regular heater. Who is correct?
   a. Tech A
   b. Tech B
   c. Both A and B
   d. Neither A nor B

CHAPTER 57

Safety, Entertainment, and Antitheft Systems

NATEF Tasks

- **N57001** Diagnose operation of comfort and convenience accessories and related circuits (such as: power window, power seats, pedal height, power locks, truck locks, remote start, moon roof, sun roof, sun shade, remote keyless entry, voice activation, steering wheel controls, back-up camera, park assist, cruise control, and auto dimming headlamps); determine needed repairs. (AST/MAST)
- **N57002** Diagnose operation of safety systems and related circuits (such as: horn, airbags, seat belt pretensioners, occupancy classification, wipers, washers, speed control/collision avoidance, heads-up display, park assist, and back-up camera); determine needed repairs. (AST/MAST)
- **N57003** Diagnose operation of entertainment and related circuits (such as: radio, DVD, remote CD changer, navigation, amplifiers, speakers, antennas, and voice-activated accessories); determine needed repairs. (AST/MAST)
- **N57004** Diagnose operation of security/anti-theft systems and related circuits (such as: theft deterrent, door locks, remote keyless entry, remote start, and starter/fuel disable); determine needed repairs. (AST/MAST)

Knowledge Objectives

After reading this chapter, you will be able to:

- **K57001** Describe the purpose, function, and types of cruise control systems.
- **K57002** Describe the purpose and function of various collision avoidance systems.
- **K57003** Explain the purpose, function, and operation of supplemental restraint systems and components.
- **K57004** Describe the types of components and the features that make up the entertainment system.
- **K57005** Describe the process of diagnosing the entertainment system.
- **K57006** Describe the purpose and function of mobile Global Positioning Systems.
- **K57007** Describe the types and function of vehicle antitheft systems.

Skills Objectives

There are no Skills Objectives for this chapter.

▶ Introduction

Over time, occupant safety has become increasingly important, so manufacturers have come up with more greatly enhanced safety systems. We explored antilock braking, traction control, and electronic stability control in the Electronic Brake System chapter. But those are only a few of the safety systems used to protect occupants. Vehicles are equipped with several other systems: Some, such as lane departure warning systems or night vision systems (**FIGURE 57-1**), help prevent an accident; others, such as the supplemental restraint system (SRS), help protect the occupants in the event that an accident does occur.

With drivers and occupants spending more time than ever in their vehicles, entertainment systems have become much more widespread and sophisticated over time (**FIGURE 57-2**). These add to the complexity of the vehicle and the need for technicians trained to diagnose and repair them. Additionally, the complexity and many features of vehicles mean they have become much more valuable, thus more subject to theft. Because of this, manufacturers have introduced a wide variety of antitheft devices and features. We explore each of these systems more fully in this chapter.

▶ Cruise Control Systems

N57001, K57001

Cruise control is not automatically thought of as a safety system. But even in its most basic form, it lowers driver fatigue on trips by allowing the driver not to have to keep a close watch on vehicle speed, so he or she can then focus more effectively on other driving responsibilities. Cruise control is increasingly installed as a standard accessory on vehicles. Cruise control allows the driver to set a cruise speed for the vehicle, which it then memorizes and maintains without throttle input from the driver (**FIGURE 57-3**). For safety reasons, the cruise control has a minimum speed that must be met before it can be set and is deactivated when the driver touches the brake pedal or, in the case of manual transmissions, the clutch pedal. Driver controls built into the steering wheel or column turn the system on and off, allow the speed to be set, and provide deceleration and acceleration controls.

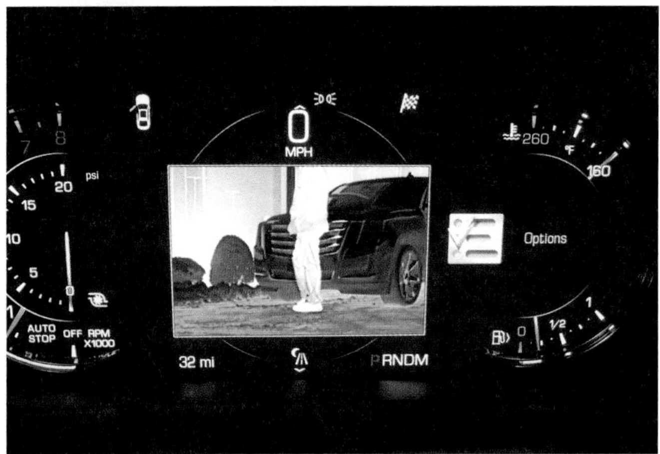

FIGURE 57-1 Night vision systems help drivers see hidden or obscure objects in low visibility situations.

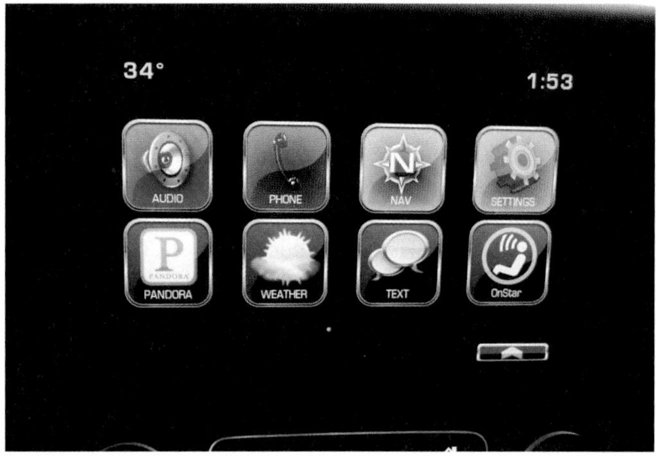

FIGURE 57-2 Manufacturers are adding more entertainment options on vehicles making the ride more enjoyable, but more complicated for technicians to diagnose.

FIGURE 57-3 Cruise control allows the driver to set the vehicle speed, and the system maintains that speed.

You Are the Automotive Technician

Because you are an automotive technician, people ask you regularly for your opinion on various vehicle issues. One of your friends needs to purchase a vehicle for his wife to drive. He is very concerned about her safety and security, so he asks you the following questions.

1. What are the various types of collision avoidance systems, and what do they each do?
2. Besides a driver's side airbag, what other supplemental restraint system devices are available, and what do they do?
3. How does a global positioning system work? And how could it help save her time while driving?
4. What antitheft options are available, and how do they work?

Types of Systems

Cruise control systems are mostly installed as an original equipment accessory by the manufacturer, but they can also be installed as an aftermarket device. Manufacturer-equipped systems enable better integration of the cruise control into the overall vehicle systems and may be linked to stability control and crash avoidance systems. One of three basic types of main actuators controls the throttle plate: electric, vacuum, or electronic operated (**FIGURE 57-4**). The electric actuator is a reversible motor that uses a reduction gear to pull a cable, which moves the throttle plate. The vacuum-operated system uses vacuum to operate a vacuum servo, which is connected to the throttle linkage by a cable. In the electronic style, commonly called a drive-by-wire system, the electronic throttle body is not mechanically connected to the throttle pedal. Instead, this system uses a DC stepper motor that is directly connected to the throttle shaft.

Principles of Operation

The basic cruise control system works by comparing actual vehicle speed to a preset target speed. The driver programs a set target speed, and the cruise control system's control module attempts to maintain the target speed by commanding a throttle actuator to open and close the throttle valve as needed. Cruise control maintains the speed of the vehicle until one of several actions happen: the driver turns the system off; the driver steps on the brake pedal or clutch pedal; or the system detects a loss of traction vehicles with traction control only. Any of those actions will disengage the cruise control system.

Cruise control components can be spilt into two main groups: driver controls and system components. Driver controls include the On/Off switch, dash indicator lamp, and control switches, such as Set, Coast, Resume, and Accelerate. System components include the electronic control module, speed sensor, clutch switch (for manual transmissions) or transmission range sensor (for automatic transmissions), brake pedal switch, and throttle actuator (**FIGURE 57-5**). Note that these components can be integrated into a vehicle network system, and the information shared across the network.

The cruise control system is activated by turning on the cruise control switch. This should illuminate a cruise control indicator in the dash. The driver then accelerates the vehicle to the desired speed and the "Set" button is pushed. The vehicle will maintain that set speed. If the driver wishes to accelerate, the "Accel" button is pushed in short bursts until the desired speed is reached. If the driver wishes to decelerate, the "Decel" button is pushed in short bursts until the desired speed is reached.

▶ TECHNICIAN TIP

Cruise control should *not* be engaged when there is a chance of water puddling on the roadway. Hitting a patch of water can cause the tire to hydroplane, causing its speed to slow down quickly. The cruise control system sees this and applies more engine power, making the tires spin. This can make it harder for the tires to regain traction, and can cause a loss of control of the vehicle.

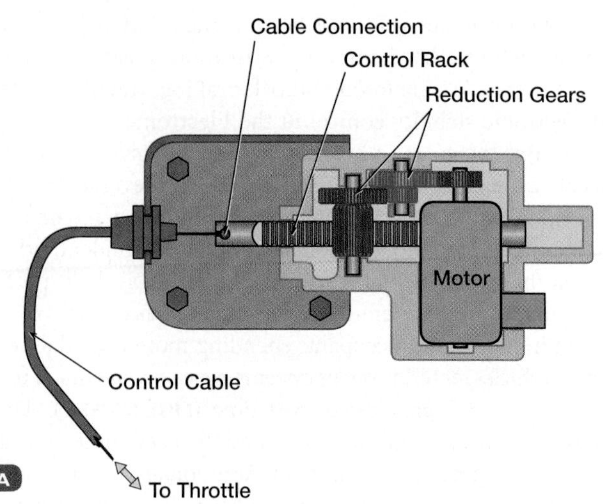

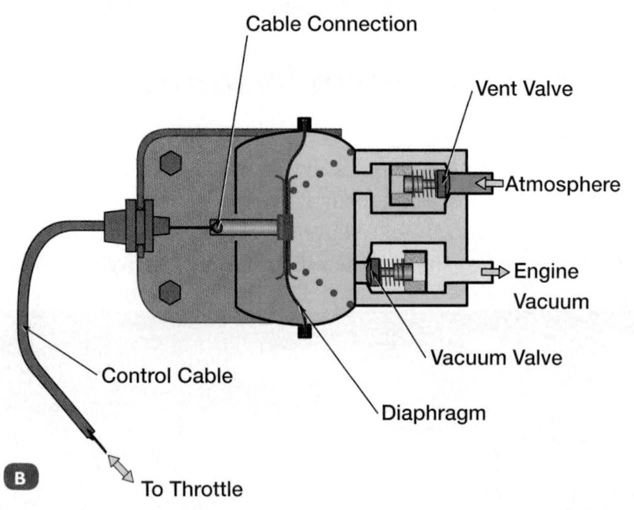

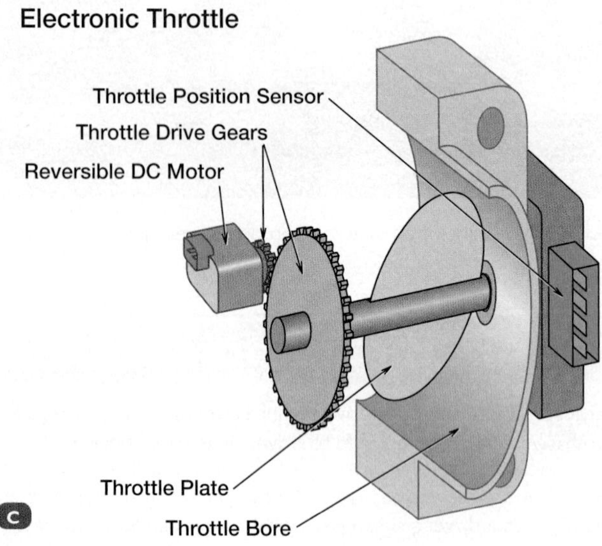

FIGURE 57-4 Cruise control actuators. **A.** Electric. **B.** Vacuum **C.** Electronic throttle.

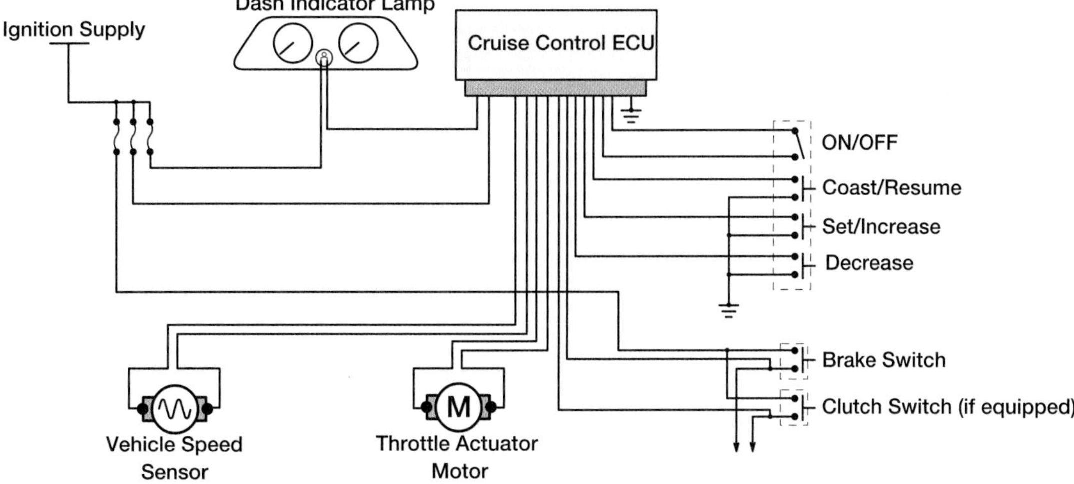

FIGURE 57-5 Wiring schematic for cruise control.

Electronic Throttle Control

Where a vehicle is equipped with electronic throttle control (drive by wire systems),, the cruise control is integrated into the engine control system and does not require an additional throttle actuator. Electronic throttle control systems have no physical connection between the accelerator pedal and the throttle. Instead, an electrical sensor (accelerator pedal position sensor) is built into the accelerator pedal and monitored by the PCM. The PCM controls the position of the throttle based on the position of the accelerator pedal by sending signals to a stepper motor connected to the throttle body (**FIGURE 57-6**). A stepper motor is a special type of electric motor designed to move, by a fixed number of degrees for each step, in either direction. The PCM adjusts the stepper motor in fine increments to control the throttle valve opening based on the position of the accelerator pedal and vehicle speed.

FIGURE 57-6 Electronic throttle control.

Adaptive Cruise Control

Adaptive cruise control systems adapt to the driving environment by sensing and adjusting vehicle speed based on changing conditions. For example, they can maintain the desired speed where possible while keeping a safe distance from the vehicle in front. The systems also incorporate driver warnings for potential collision or the requirement for driver intervention. The advantage of these systems is that the driver does not constantly have to adjust the cruise control in heavier traffic conditions where traffic speed may be variable. Adaptive cruise control requires additional sensors, such as radar or laser systems, to sense the distance and speed of the vehicle in front (**FIGURE 57-7**). In many of these systems, the driver can adjust the desired following distance to best fit the driving conditions. Steering angle sensors are used to determine the vehicle direction. Some active cruise control systems also incorporate automatic brake control to avoid collisions, if necessary. These systems, in addition to controlling the throttle, can apply the brakes if necessary.

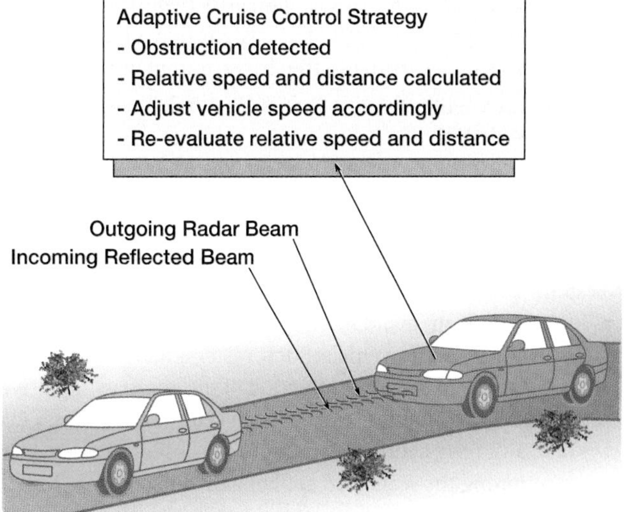

FIGURE 57-7 Adaptive cruise control adjusts the speed of the vehicle so that a safe distance is maintained between your vehicle and the vehicle ahead.

Testing the Cruise Control System

When testing a cruise control system you need to start by verifying the customer concern. Once, you do that, researching the service information will allow you to understand the type of system, how it operates, and the process you need to follow to diagnose the customer concern. For a vacuum-operated system, part of the testing involves performing a visual check of the vacuum hoses and controls, in addition to the electric circuitry.

For diagnosing a fully electronic throttle system, you need to use a scan tool to access any DTCs in the computer system. Once you have retrieved the codes, research them in the service information. This, along with TSBs, will help you develop a focused testing strategy that will guide you through the diagnosis. Typically, a DMM is used to test the wiring and components of the system. If there are not any DTCs stored, you will have to use the symptom charts in the service information to guide your testing. Also, when diagnosing the fault, make sure you identify the root cause of the concern. Once the system is repaired, perform a test-drive, and operate the cruise control system to verify that the concern has been resolved.

▶ Collision Avoidance Systems

N57002, K57002

Collision avoidance systems are designed into vehicles to help avoid collisions, or minimize the severity of a collision, should one happen. There are a number of collision avoidance features that manufacturers use on some of their vehicles. And as time moves on, more of these features are developed and installed on vehicles. Here is a listing of some of the most common collision avoidance features and systems:

- Adaptive cruise control is used in many collision avoidance systems to avoid rear-end collisions. It primarily focuses on other vehicles in front of the car.
- Pedestrian detection systems identify pedestrians and animals that pose a hazard. The system may give an audible and/or visual warning to the driver, and may also have the ability to apply the brakes or steer the vehicle away from the pedestrian.
- Side collision avoidance systems use radar, laser, or cameras to monitor the space beside the vehicle (**FIGURE 57-8**). If the steering wheel is turned in the direction of an obstacle, or the obstacle is moving toward the vehicle, the system will emit a warning to the driver.
- Rear collision avoidance systems monitor vehicles and objects behind the vehicle. Some systems activate the hazard lights to try to catch the attention of the approaching driver. If the system determines a collision is imminent, it may take precautionary actions such as pre-tensioning the seat belts, positioning the seats, and applying the brakes more firmly.
- Lane departure warning systems monitor the lines on the road and either warn the driver of imminent departure or, on some vehicles, make steering adjustments so the vehicle stays within the lane (**FIGURE 57-9**).

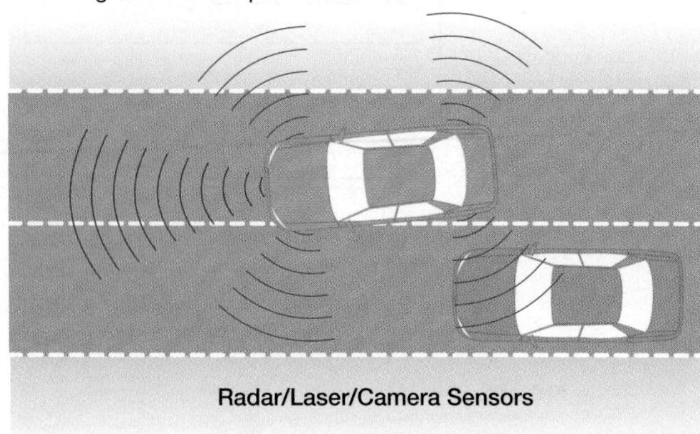
Audible Alert......Obstruction Left...Steer right
Visual Alert.........Flashing stop symbol on dash or HUD
Direct Intervention
1. Steering commanded to veer slightly right
2. Pre-prime braking system for emergency stop
3. Tighten seatbelt pre tensioners for imminent collision
Radar/Laser/Camera Sensors

FIGURE 57-8 Typical side collision avoidance system.

x1000rpm
ECO
C
H
Door
Lights
Sound

FIGURE 57-9 Typical lane departure system warning lamp indicating potential lane departure.

- Driver fatigue monitoring systems use a camera and other sensor feedback to assess driver alertness. If fatigue is detected, audio and visual warnings will be given. Some vehicles tug on the seat belt to alert the driver.
- Night vision systems are usually combined with a heads-up display. Special infrared or low-light cameras are mounted on the front of the vehicle and capture a view of the road in front of the vehicle. The system enhances the image and displays it on the windshield or dash so that the driver can be more aware of hazards, especially in low light or limited visibility conditions such as fog or rain (**FIGURE 57-10**).
- Active headlight systems pivot in the direction the front wheels are turned so that the road around the corner is illuminated and objects are more likely to be seen and avoided. In most cases, an electric motor is used to pivot each headlight based on how tightly the vehicle is being steered.

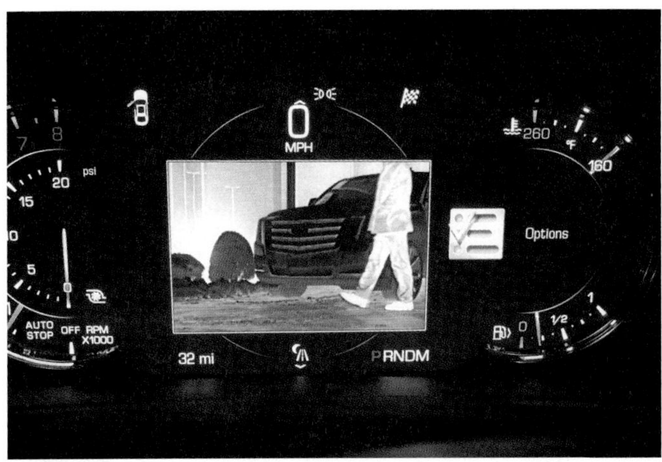

FIGURE 57-10 A night vision system displays hazards in front of and off to the side of the vehicle.

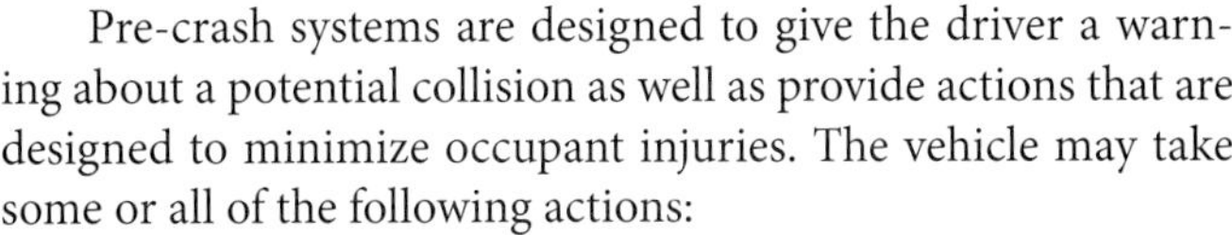

FIGURE 57-11 Proximity sensors.

Pre-crash systems are designed to give the driver a warning about a potential collision as well as provide actions that are designed to minimize occupant injuries. The vehicle may take some or all of the following actions:

- Sense where the driver's attention is focused or whether the driver is drowsy, and if so, take more aggressive actions.
- Audibly and visually warn the driver of the potential collision.
- Pre-tension the seat belts.
- Pre-position the seats and seat backs.
- Roll up the windows and close the sunroof.
- Apply the brakes to precondition them for panic braking.
- Apply the brakes automatically if the driver doesn't take evasive action.
- Steer the vehicle away from the impact zone if there is safe space

Proximity Sensors and Back-Up Displays

To allow for safer driving and parking, proximity sensors can be mounted in the front or rear bumpers and the side if the vehicle (FIGURE 57-11). These color-matched piezoelectric sensors emit ultrasonic sound waves that the human ear cannot detect, and are used to transmit and receive coded sound waves. The control module determines the distance between the sensor and an obstacle by measuring the time taken for the sound wave to leave and return to the sensor. Normally, four sensors are used to allow for full coverage across the width of the vehicle. Proximity to obstacles can be indicated by separate audible and visual alarms, or by integrating warning sounds with the vehicle's audio system. If a trailer is attached to a coupling, the rear sensors are disabled automatically when the harness plug is inserted. Front sensors can be disabled manually in stop-start or stop-and-go traffic. Side sensors are used in collision avoidance systems to warn the driver of a vehicle in the next lane when turning the steering wheel or activating the turn signal.

FIGURE 57-12 **A.** Back-up camera (in tailgate handle). **B.** Display.

A system that provides a similar but more detailed warning than proximity sensors involves back-up cameras and displays (FIGURE 57-12). This system uses a camera mounted at the rear of the vehicle that is activated when the vehicle is put in reverse. The camera's image is then displayed either on the rearview mirror or on the entertainment system's LCD panel, showing the driver what is behind the vehicle. This feature greatly reduces the risk of backing over a child or something such as a bicycle that was left behind the vehicle.

▶ Supplemental Restraint Systems

K57003

The **supplemental restraint system (SRS)** is used to describe passenger safety devices such as airbags (including side curtains, and knee bags) and seat belt pretensioners (**FIGURE 57-13**) installed on vehicles. Manufacturers continue to develop and improve SRS technology to make vehicles safer. When first introduced, a single airbag was usually installed on the driver's side. Now, it is not unusual for modern vehicles to have more than six airbags, along with side-impact protection known as side curtains (**FIGURE 57-14**). One of the newest additions is the pedestrian airbag. These devices are explained in the following sections.

Purpose and Operation of the SRS

Vehicle safety systems are designed to protect occupants during accidents and can be classified as primary (passive) safety systems and secondary (active) safety systems. Primary safety systems are ready to operate in any accident. They include bumper bars, body panels, seat belts, crumple zones, and collapsible steering columns. A secondary safety system has to be activated to work and is only necessary in severe accidents (**FIGURE 57-15**). The two most common types of secondary systems are airbags and seat belt pretensioners.

Seat belts secure and restrain the occupant within the seat and vehicle cabin, and in minor collisions perform their task well. In a more severe impact, inertia causes the occupant to move forward with greater force. This increases the possibility of injury caused by the restraining force exerted by the seat belt or from the occupant striking interior equipment. In these situations, airbags provide additional protection for the occupants.

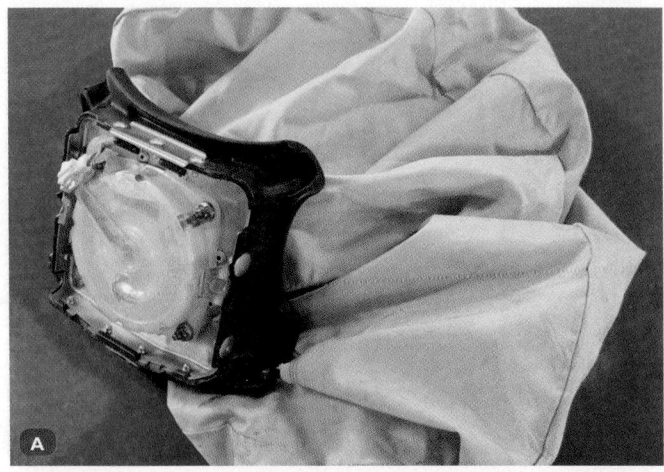

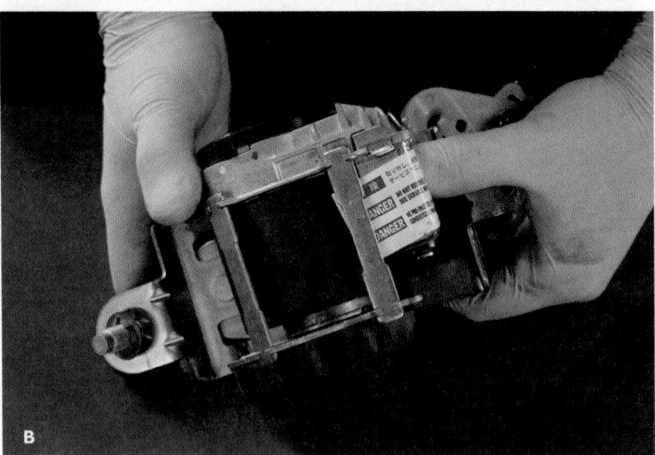

FIGURE 57-13 A. Airbag. **B.** Seat belt pretensioner.

In a vehicle equipped with airbags, they deploy during a severe collision, offering a greater degree of protection from injury. Airbags provide cushioning against the effects of inertia. The bag deploys toward the occupant's approaching body,

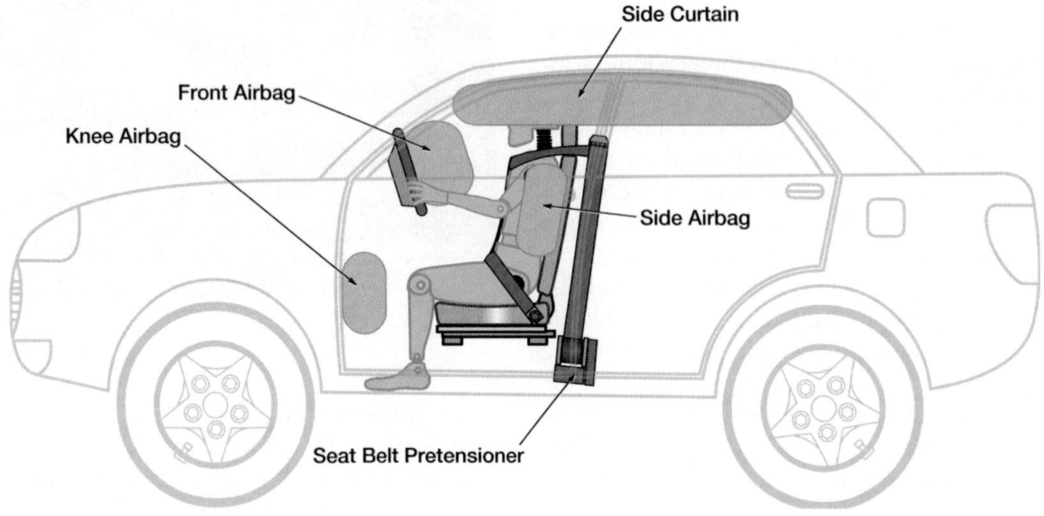

FIGURE 57-14 Typical SRS injury reduction devices.

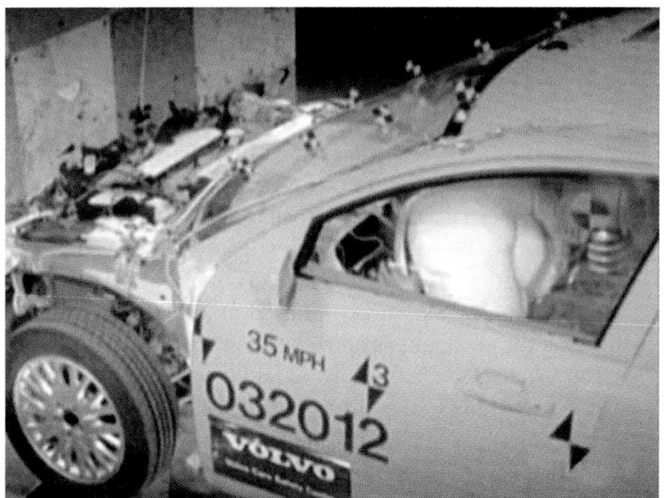

FIGURE 57-15 SRS in action during a collision.

inflated rapidly by pressurized nitrogen gas. Typically, inflation takes no longer than about three-hundredths (0.03) of a second. The airbag is not a nice soft pillow, but a strong counterforce to react against the inertia of the occupants. It is not designed to be comfortable; it is designed to minimize injury. Immediately after absorbing the momentum, the airbag deflates, having done its job.

Airbags and Pre-tensioners

Airbags are usually described as being "supplemental," but in some countries, wearing seat belts is not mandatory. In this case, airbags become the primary restraint mechanism and must be able to trigger at lower speeds and be larger in volume. There are a number of different types of airbags whose size and location are determined by the type of protection they offer (**FIGURE 57-16**). However, just because a vehicle has airbags doesn't mean its passengers don't need to wear seat belts. Seat belts are a primary safety system, meaning they are very important and should be worn. Airbags are designed to supplement seat belts, not replace them.

The most common location for an airbag is in the center of the steering wheel. Here, it protects the driver from frontal impacts. Airbags are also commonly installed on the passenger side of dashes for the same reason. Side-impact airbags are located in the sides of front seats to protect the occupants from side impacts. Curtain side airbags are located in the side edge of roof linings to protect the occupant's head from side impacts. Knee airbags are located under the dash and protect the lower legs. The airbag assembly consists of a nylon bag, **squib**, igniter, gas generator, and airbag triggering mechanism (**FIGURE 57-17**).

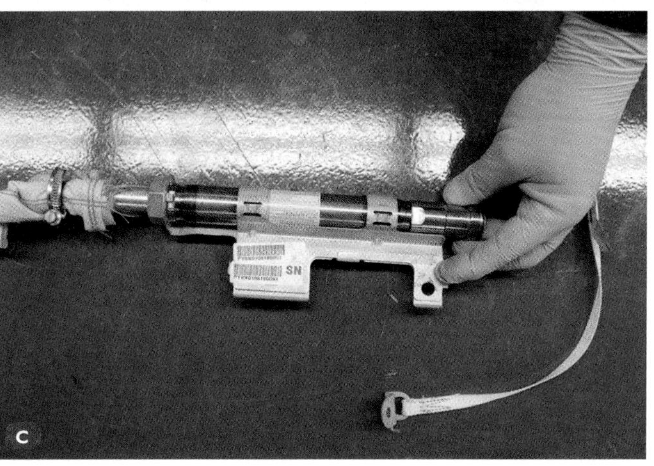

FIGURE 57-16 A. Driver's side airbag. **B.** Passenger's side airbag. **C.** Side airbag. **D.** Knee airbag.

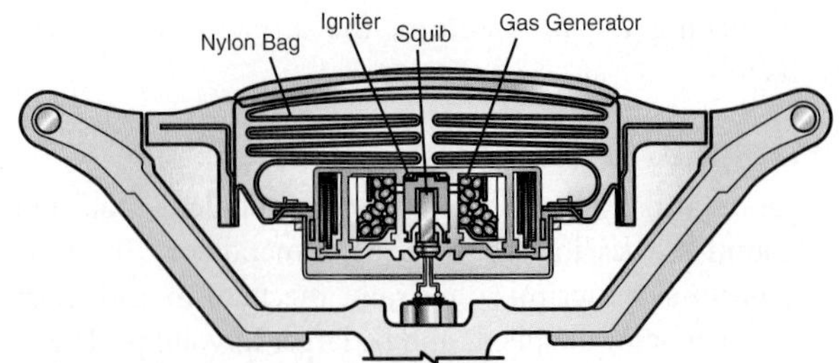

FIGURE 57-17 Parts of an airbag assembly.

There are two types of airbag triggering mechanisms: electrical and mechanical. Most airbags are triggered electrically, with a small electrical current delivered from a remote SRS control module. Mechanically activated systems use inertia to move a triggering pin. Regardless of the type of triggering mechanism, the airbag deploys due to simultaneous actions occurring within the squib, the igniter, and the gas generator. These three parts are located in a metal housing attached to the back of the airbag assembly.

When the control module determines that the airbag should be deployed, the electrical current triggers the squib. The heat generated causes the igniter to burn, which in turn ignites the gas generator. High-pressure nitrogen gas is produced, and the airbag rapidly inflates. When the airbag assembly is mounted, it sits behind a pad, which may have a fracture line cast into the inner face. The force of the gas generated when the airbag deploys causes the cover pad to rupture, allowing the bag to fully inflate (**FIGURE 57-18**).

SAFETY TIP

Airbags, when they deploy, inflate toward the occupant at more than 100 mph (160 kph). Because of this force and speed, they have been known to break wrists and even cause death when not used properly. Always keep the seat positioned so you are at approximately arm's length from the steering wheel; that is, the steering wheel should be at least 10" to 12" (25 to 50 cm) from your chest. Also make sure your hands are in the 9 o'clock and 3 o'clock positions. The old "10 and 2" position can lead to your hands being thrown back toward you or outward (remember that 100+ mph), both of which can injure you.

FIGURE 57-18 You can see the fracture line on this airbag, where it will rupture if deployed.

Mechanically deployed airbags do not have any electrical circuitry. The squib is ignited with a firing pin. Under severe deceleration, inertia causes a steel ball to release a firing pin into the squib. Once the squib has been triggered, the deployment process is identical to electrically triggered airbags. The airbag is fully inflated within three hundredths (0.03) of a second, cushioning the occupant as his or her body moves forward.

The airbag is made from nylon and is folded into the front face of the assembly. Older airbags were coated in cornflower or talcum powder, which acted as a lubricant for the fabric during deployment. Newer ones use either more slippery fabric or silicone to allow the airbag to inflate without binding. Relatively large holes are usually located in the rear face of the airbag to allow the nitrogen gas to escape once the airbag has deployed (**FIGURE 57-19**).

A new type of airbag recently released is called a pedestrian airbag. It is located under the rear edge of the hood, typically in the cowl area. When a collision with a pedestrian is imminent, the hood retracts slightly, and the airbag inflates to cover a portion of the windshield (**FIGURE 57-20**), thus protecting pedestrians from direct impact with the windshield.

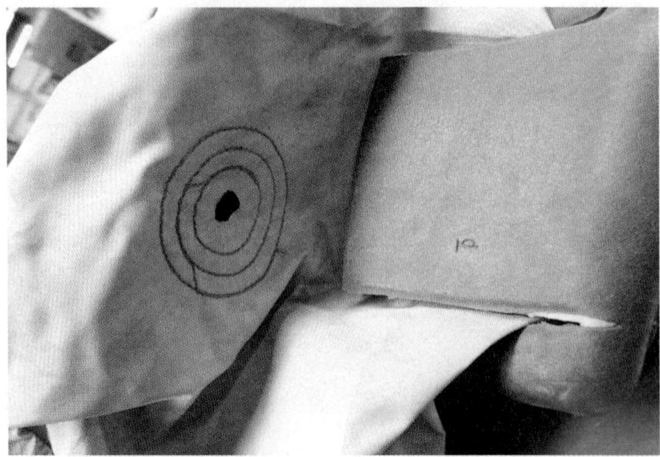

FIGURE 57-19 Large vent holes for deflating the airbag quickly after deployment.

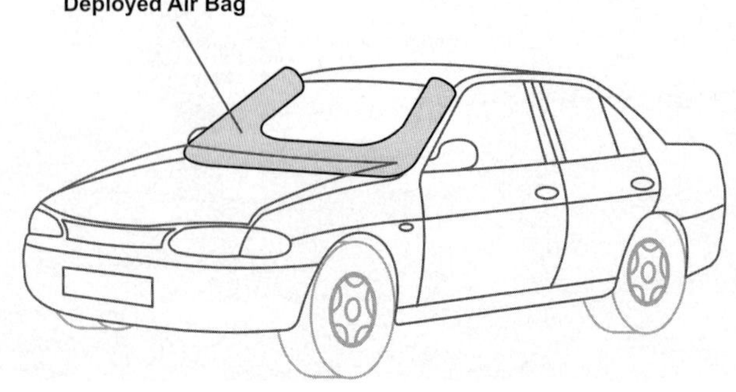

FIGURE 57-20 Pedestrian airbag.

Seat Belt Pretensioners

Seat belt pretensioners are used to tighten the seat belt in a severe frontal accident. There are 3 types of pretensioners, electric, mechanical and pyrotechnic, which are described in the text below. The most common type, sometimes called a ballistic pretensioner, deploys in an actual accident. It relies on an explosive charge that is detonated electronically by an actuator within the seat belt tensioning mechanism (**FIGURE 57-21**). This explosion moves a piston that pulls on a steel cable, causing the belt to tighten by approximately 4" (100 mm). The design allows for the belt to tension before the occupant has moved forward in the seat.

Mechanical systems rely on inertia to move a sensing mass. During the beginning phase of an accident, the movement of the sensing mass releases a spring to pull on a cable, thus tightening the belt. Once the pretensioner has triggered, a ratchet prevents the seat belt from loosening. When the seat belt is removed from the buckle, it cannot be reinserted, and the assembly will have to be replaced.

A newer style of seat belt pretensioner uses an electric motor to pull in the slack on the seat belts (**FIGURE 57-22**). This type is reusable, meaning it can be activated repeatedly. Because of this, it can be deployed much earlier in a potential accident. On some vehicles, the system will tug on the seat belt to get the driver's attention, along with preparing for a collision.

Rip stitching is used on seat belts, in conjunction with an airbag and seat belt pretensioners. During a collision, the pretensioners initially pull the seat belt tight; however, the stitching gradually tears, allowing the occupant to move forward into the airbag at a controlled rate.

For safety reasons, these belts must be replaced once they have had their stitching ripped. Manufacturers generally fit warning labels within the fold to indicate the belt is to be replaced when the label is revealed (**FIGURE 57-23**). Ripping the stitching exposes the warning label.

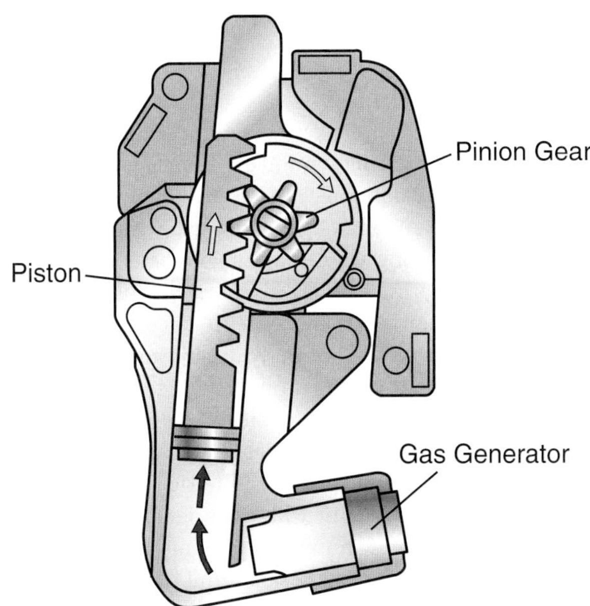

FIGURE 57-21 Seat belt pretensioner.

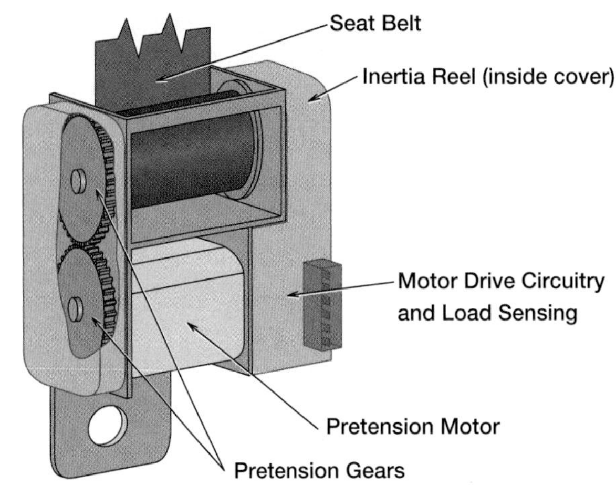

FIGURE 57-22 An electric motor–style seat belt pretensioner assembly.

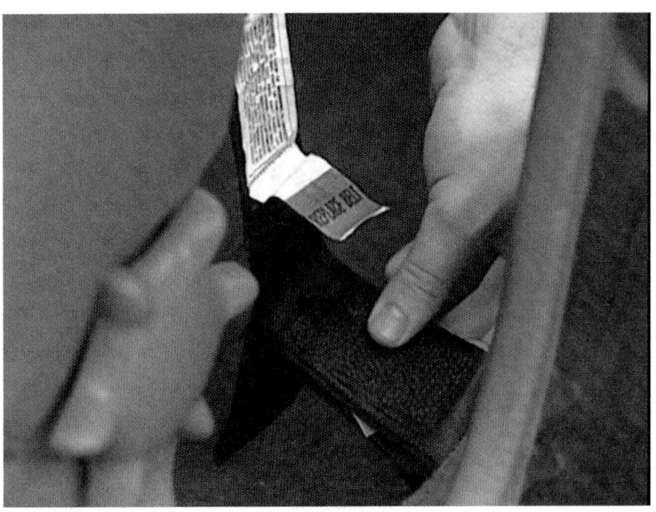

FIGURE 57-23 Rip stitching rips away at a controlled rate so the deceleration rate of the occupant is not as severe.

Sensors, Control Module, and Circuitry

Crash sensors can be installed in various positions throughout the vehicle. Their location depends upon the direction of deceleration they are designed to detect. Some manufacturers place the sensors within the PCM. Others are located behind the front bumper, headlights, and dash (**FIGURE 57-24**).

Side-impact sensors are located in the doorsills or B-pillar. They inform the SRS control module of a side impact and whether to deploy the left- or right-side airbags. When the sensors indicate that a predetermined deceleration rate has been exceeded and it is from the appropriate direction, the SRS control module deploys the relevant airbags.

If the collision is from the front, the driver and passenger airbags will deploy. If the collision is from the side, the sensors determine whether the seat-mounted airbag or the curtain airbags for one side of the vehicle will deploy. With more refined designs, the passenger airbag deploys only if there is an occupant in the seat. Deployment can also depend on the weight

FIGURE 57-24 Typical crash sensor bolted to the front radiator support on the vehicle.

FIGURE 57-25 The safing sensor is typically located inside the SRS control module.

of the occupant and whether the passenger airbag switch, if equipped, is turned on.

To prevent incorrect and unnecessary deployment, some systems include a safing sensor mounted within the SRS control module (**FIGURE 57-25**). The SRS control module will only pass current through the squib if both the safing sensor and a crash sensor indicate simultaneously that a predetermined deceleration rate has been exceeded.

Capacitors within the SRS control module are used to store electricity and act as a backup power supply if the vehicle has its battery destroyed or disconnected in an accident. The capacitors supply the electricity required to keep the SRS system operational for a short time. If a fault is detected in the system, the SRS warning light is illuminated and stays on.

Some seat-mounted side-impact airbags also operate without electricity. When the side of a vehicle is crushed inward, a detonator mounted on the lower outside edge of the seat is activated. Pyrotechnic tubes connect the detonator to the airbags, which in turn ignite the squib. Many vehicles use two-stage side-impact bags. This design provides protection to an occupant's upper torso over a more extended time.

If any of the SRS-related devices are deployed for any reason, most manufacturers require all of the deployed devices, the crash sensors, and the control module to be replaced. This can be quite expensive, especially if you accidentally deployed one or more airbags while working on the vehicle.

Smart Airbag Technology

Manufacturers continue to improve the safety devices installed on vehicles, including airbags. Smart airbag technology uses intelligence to vary the control of the airbag deployment in the event of an accident rather than deploying it with maximum force. The improvements include the use of information regarding the severity of the accident, seat position, and weight of the occupant as inputs for determining how much force is needed during airbag deployment. This information comes from sensors that monitor each of the variables (**FIGURE 57-26**).

Airbags in systems with smart technology are capable of varying the deployment force of the airbag as needed. These adjustments can be performed in one of several ways: diverting some of the gas from the airbag inflation system into the dash, using a dual-chamber airbag, or using a dual-stage airbag that has two different-sized charges that can be deployed individually or together (**FIGURE 57-27**). This makes it more possible for the control systems to deploy the airbags with a more appropriate amount of force, customizing the protection to the occupant.

SAFETY TIP

Just because an airbag has been deployed doesn't mean it is safe. Dual-stage airbags may be designed with two ballistic charges, and only one of them may have been deployed. Airbags must have all charges deployed before disposal. But only do so according to the manufacturer's service information.

Diagnosing the SRS

Working on the SRS can be very dangerous. Consult the manufacturer's service information for the vehicle you are working on, and ensure all safety precautions are followed. Accidental deployment of the airbag system could happen if you inadvertently probe the wrong wire. Many manufacturers use yellow wiring to denote wiring for the airbag system, but don't assume all do. Always be aware of the system/ circuit you are working on, and make sure you properly disable any SRS devices as directed by the service information.

Whenever the SRS light is steadily illuminated on the instrument panel, one or more codes related to the fault is likely stored in the SRS module's memory. Retrieving the codes will give you a very good idea of where the fault is located in the system. Using the manufacturer's service information to further diagnose the fault is required to safely and accurately locate the cause of the fault. Refer to the Servicing the Steering System chapter for detailed steps on disabling the SRS.

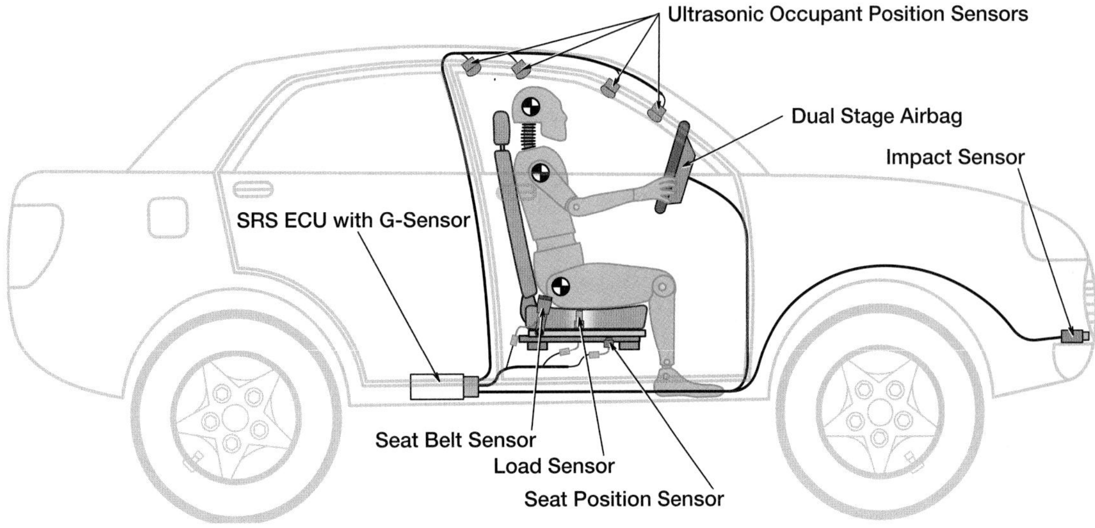

FIGURE 57-26 Schematic of a Smart Airbag system.

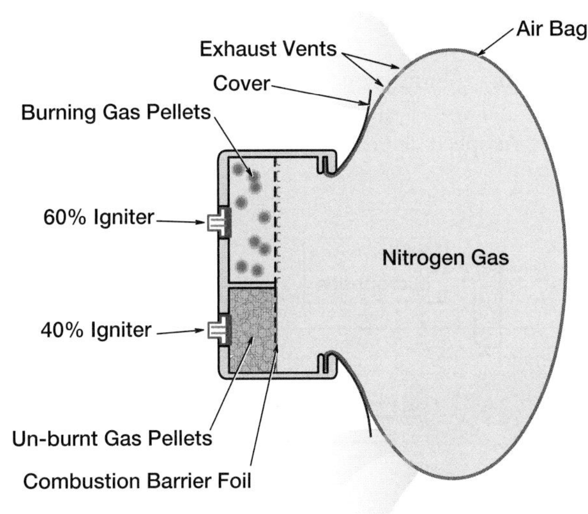

FIGURE 57-27 A typical dual-stage airbag used to tailor the airbag deployment based on the amount of counterforce needed.

FIGURE 57-28 Modern entertainment system.

SAFETY TIP

Although an airbag is powerful enough to counteract the weight of a moving body during an accident, it doesn't take much electricity to ignite one. In fact, a fresh 9-volt battery will provide more than enough electrical energy to set it off. In the same way, an ordinary test lamp connected to 12 volts can ignite the airbag if the wrong wire is probed. Always be careful when working around SRS systems.

▶ Entertainment System

N57003, K57004

Entertainment systems in vehicles are becoming more refined and complex as the technology becomes more accessible and affordable. The basic system incorporates a radio, CD player, MP3 player or iPod control, and multiple speakers with faders, bass control, and treble control. Increasingly, vehicles are being equipped with touchscreen LCD displays, DVD players for passengers, and satellite entertainment systems (**FIGURE 57-28**). Vehicles also often have integrated Bluetooth for cell phone connection and hands-free communication.

Types of Systems

Entertainment systems can range from quite basic to expensive audiovisual systems that are highly integrated. The basic system contains a head unit for control of the radio and CD player, along with inputs for MP3 players. More complex systems incorporate satellite entertainment, DVD players for passengers, game consoles, multiple LCD screens, amplifiers, and complex speaker systems with bass drivers (**FIGURE 57-29**). The head unit in the dash is the main control module for controlling the subsystems installed on the vehicle. It is also quite common on modern vehicles for the navigation system to be built into the entertainment system and share the head unit LCD screen and controls.

FIGURE 57-29 Modern entertainment systems include screens for showing movies or allowing video games to be played.

Principles of Operation

Modern vehicles integrate audio, video, and communication systems into an in-vehicle network. This allows for a high-quality, compact, and ergonomic system that combines entertainment features with simple operation. Controls are centralized with hardware such as CD stackers and DVD players located remotely.

Communication between components requires a combination of hard wiring and data buses (**FIGURE 57-30**). With data buses, warning messages can be broadcast over the vehicle's audio system that relate to other vehicle systems. For example, a voice message such as "The park brake is on," or "Left rear tire is under-inflated," can be broadcast through the audio system. This system allows for features such as the ability to interrupt or replace audio entertainment when there is an incoming phone call, or simply to mute the audio, to allow a hands-free phone conversation.

Audio control functions are usually located on the central control pod or on the component's head unit. Vehicles can also integrate the common audio controls on the steering wheel, and some respond to verbal commands from the driver to allow for safer driving.

The music played on a system usually comes from one of several sources: magnetically on a cassette tape, optically on a CD or DVD, by radio frequency from radio stations or satellites, or by Bluetooth from other portable devices. The information is decoded or processed by the control module and then transmitted to speakers located throughout the vehicle.

Another function provided by the body control module is that of speed-dependent volume. An input from the vehicle speed sensor that allows the control module to gradually increase the audio system's volume proportionally with road speed. As speed increases, and with it engine and road noise, the audio volume increases. As speed decreases, the audio volume decreases.

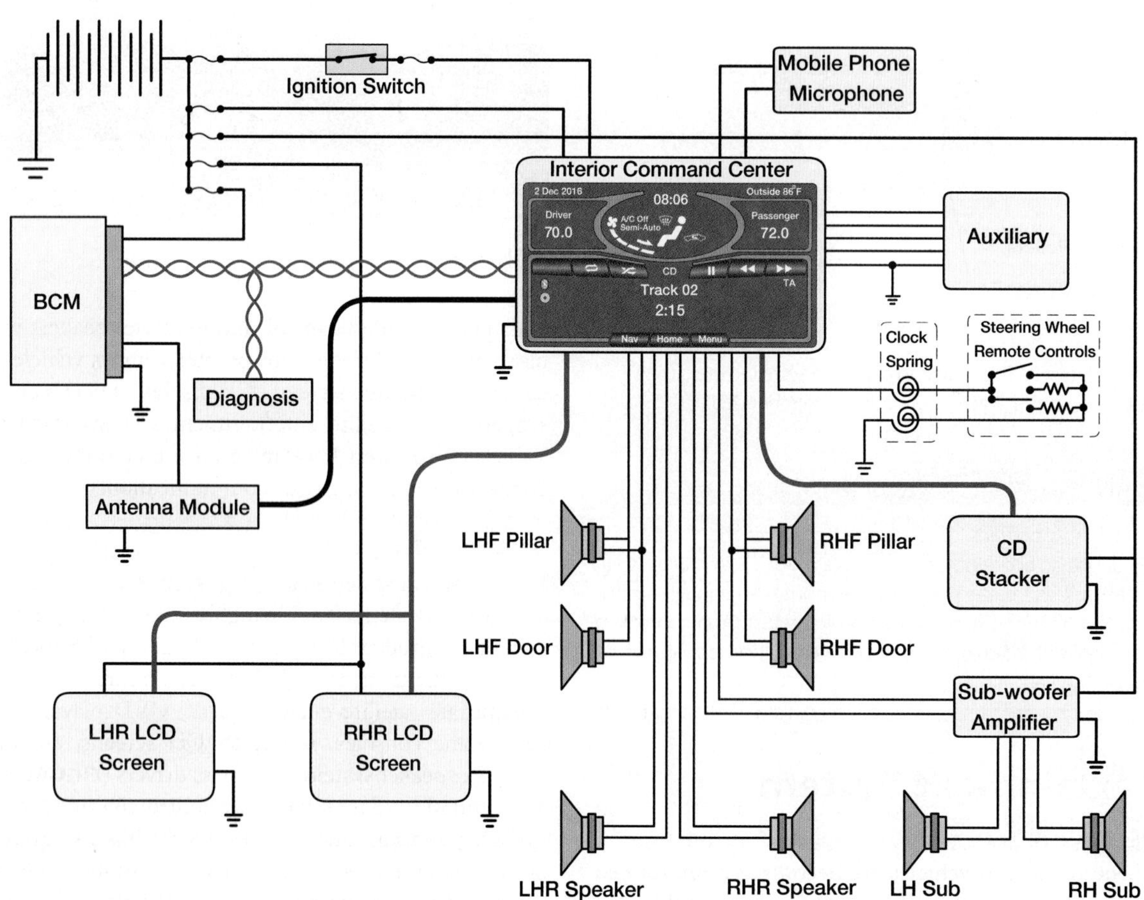

FIGURE 57-30 A typical wiring diagram for a networked entertainment system.

Viewing screens for onboard TV, DVD, and games can be located in the dash; however, if the screen is viewable by the driver, it must be disabled when the vehicle is in motion. This may be done automatically in most vehicles. Other mounting points are from the roof or integrated into the rear of seats or headrests. Control modules can be hardwired or wireless, using an infrared remote control.

Diagnosing the Entertainment System

N57002, K57005

The entertainment system is comprised of many different functions and components built into one system. So testing will have to be guided by the manufacturer's service information for the fault that is present. Problems can range from poor radio reception to failure of the global positioning system (GPS). Always verify the concern by checking what is and is not working properly. Then research the service information to develop your focused testing plan. Here are some things to consider: If there is a radio reception issue, check the radio by selecting a station that you know comes in well in your location. Check for good power and ground connections to the entertainment system components by checking the circuits under load. A lab scope may be useful to detect any unwanted ripple or noise in the power circuit. Ensure that power connections are clean and tight. If interference is suspected, attempt to isolate the cause by turning off or disconnecting electrical items until interference is reduced. Check the antenna for corrosion-free connections and faults; try a substitute antenna to confirm performance. Also check all vehicle grounds.

If the problem is with a video device, you need to verify the concern just like any other diagnosis. Then research the concern in the service information to determine a focused testing process for identifying the cause of the fault. Once the fault has been located and corrected, make sure you verify that the system is operating correctly.

Satellite Communication and Global Positioning System

K57006

Thousands of satellites, serving a variety of purposes, are continually in orbit high above the Earth. The use of satellite technology in vehicle systems provides an ever-increasing array of options for vehicle manufacturers. It is an area of automotive technology that increases the flexibility and power of many onboard systems. Satellite technology is used in navigation, vehicle tracking, vehicle theft recovery, communication, and Internet access. A Global Positioning System (GPS) is used to determine the exact location of a vehicle on the Earth's surface (**FIGURE 57-31**).

Most vehicle tracking systems can work very accurately for short periods of time without satellite communication once their initial position has been determined. Onboard sensors such as vehicle speed and steering wheel position sensors can be used to keep track of nearly the exact vehicle location until satellite communication is reestablished.

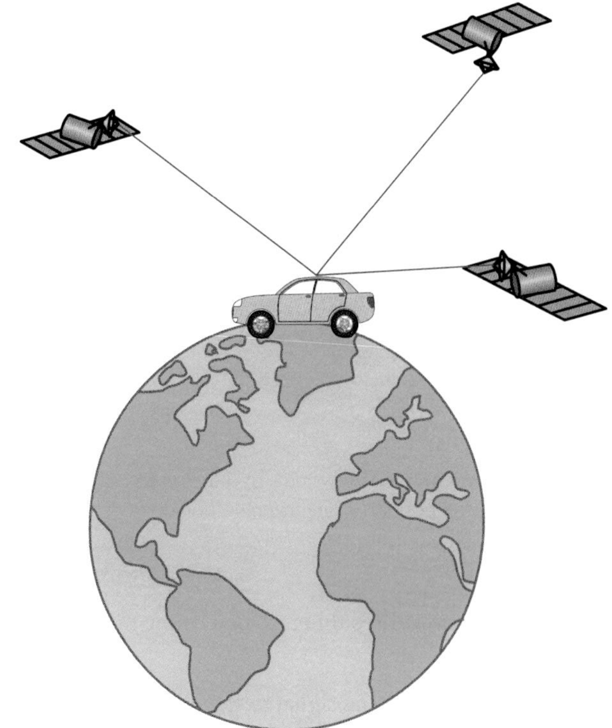

FIGURE 57-31 GPS mapping system.

Triangulation/Trilateration

The GPS relies on a group of at least 24 satellites (and as many as 32) orbiting approximately 12,650 miles (20,350 km) above the Earth. Each satellite circles the Earth two times each day in one of six orbits. Each satellite is equipped with an atomic clock onboard and regularly transmits a unique radio frequency (RF) signal simultaneously with all other global positioning satellites. The RF signals travel out across space in all directions. The vehicle is equipped with a receiving antenna and a computer system that also knows the precise time. The GPS receiver on the vehicle has to receive signals from four or more of these satellites. Once it receives the signals, the time it took for each signal to arrive indicates the relative distance from each satellite.

The GPS unit uses this information to establish its own location. This operation is based on a mathematical principle called **trilateration** in three-dimensional space and is quite complex. For ease of understanding, the term generally used in the automotive industry to describe how the GPS positioning system operates is **triangulation**, which is the process of finding the position of an unknown point based on forming triangles with known points. The more points that are known, then more triangles can be formed, increasing the accuracy of the calculated position. Correspondingly, any fixed point on the surface of the Earth will triangulate with the satellites (**FIGURE 57-32**).

The speed at which an RF signal travels in space is approximately 186,000 miles per second (300,000 km/sec), the speed of light in a vacuum. Each of these transmitted signals will reach the GPS antenna of the vehicle in a certain amount of time depending on the distance the signal travels. The greater the distance, the greater the time taken. The vehicle's onboard GPS

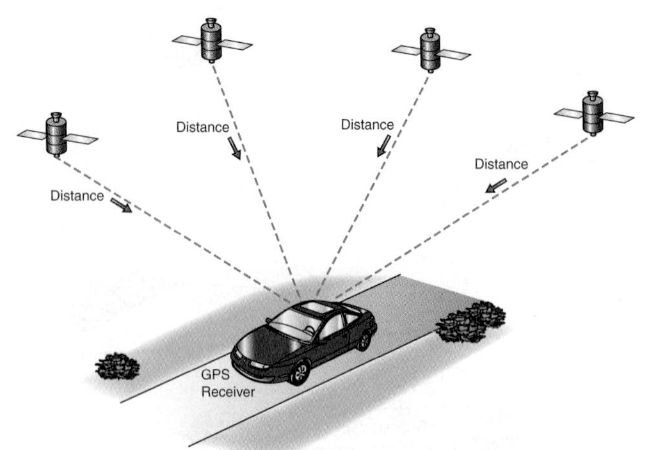

FIGURE 57-32 GPS positioning systems operate on triangulation, the process of finding the position of an unknown point based on forming triangles with known points.

FIGURE 57-33 GPS mapping screen.

system needs to know three things to determine the location of the vehicle:

1. The time it took for the signal to travel from each satellite to the vehicle
2. The position of each satellite
3. The accurate time

Given these facts, enough information is available to form a three-dimensional figure of a pyramid. The base of the pyramid is formed by the location of the satellites, and the apex of the pyramid is the location of the vehicle on the Earth, a point derived from triangulation of the known points of the base.

The GPS equipment knows that all of the apexes (the position of the car) of each triangle must be in the same literal position. Adding more satellites makes this location more accurate because it gives the GPS more triangles to work with. Satellites are positioned so that virtually every location on the Earth's surface has access to at least four satellites 24 hours a day. The system is monitored by Earth-based facilities that adjust the orbit of each satellite and perform any updates that are needed to keep the system accurate.

Inaccuracies caused by weather conditions do occur as the RF signal moves into the Earth's atmosphere and travels through air. The density of air varies, and traveling through denser air slows the RF signal. Inaccuracies also occur when RF signals are reflected off objects such as large buildings. Solar radiation can also cause inaccuracies. But all in all, the GPS system is typically accurate to within 25 feet (7.8 meters).

Satellite-Based Navigation

Once trilateration has been determined, mapping software stored in memory can be used to generate an overlay map and data for the driver on a display screen (**FIGURE 57-33**). As the vehicle moves, the GPS continues to provide the necessary information to allow for plotting of position on the map. Various maps are available to cover the different continents.

Accurate and reliable navigation is possible even when satellite signals become unavailable. This is achieved by using electronic sensors to monitor vehicle variables such as pitch, roll, yaw, road speed, steering angle, acceleration, and deceleration. By using the information from these sensors, the navigation system is not continually and totally reliant on satellites.

During normal operation, the computer program compares vehicle position data derived from the satellites and onboard sensor information to ensure high levels of accuracy. In addition, ground-based stations may be used in suitable locations as an absolute reference point.

Satellite navigation features may include multiple languages and a choice of voice gender. With destination and journey plotting, the most suitable route is provided, and deviating from the recommended route causes the system to provide an alternate route. It may also include directional information provided with a combination of screen icons, maps, and audible instructions; a self-learning route memory function; congestion avoidance, which can warn of the latest traffic bottlenecks and suggest alternative routes; various monitor color display settings; infrared remote control; trip computer; speed-dependent setting; and telephone mute for sound systems.

Telematics

Automotive **telematics** is a satellite-based system that combines two-way communication and information technology within the vehicle. The vehicle is equipped with a satellite transceiver enabling data to be sent to and from the vehicle (**FIGURE 57-34**).

This system allows for vehicle tracking and security features, monitoring of onboard systems, messaging, delivery of travel information, entertainment functions, and use of safety and fleet management systems that monitor information such as location, distance traveled, speed, stops, and fuel usage.

A vehicle manufacturer may offer telematics as a service to its customers. The benefits of this can include the location and immobilization of a stolen or lost vehicle, notification to emergency services after SRS deployment, engine shutdown and door unlocking in the event of a severe accident, roadside assistance, and even remote diagnosis of some system faults.

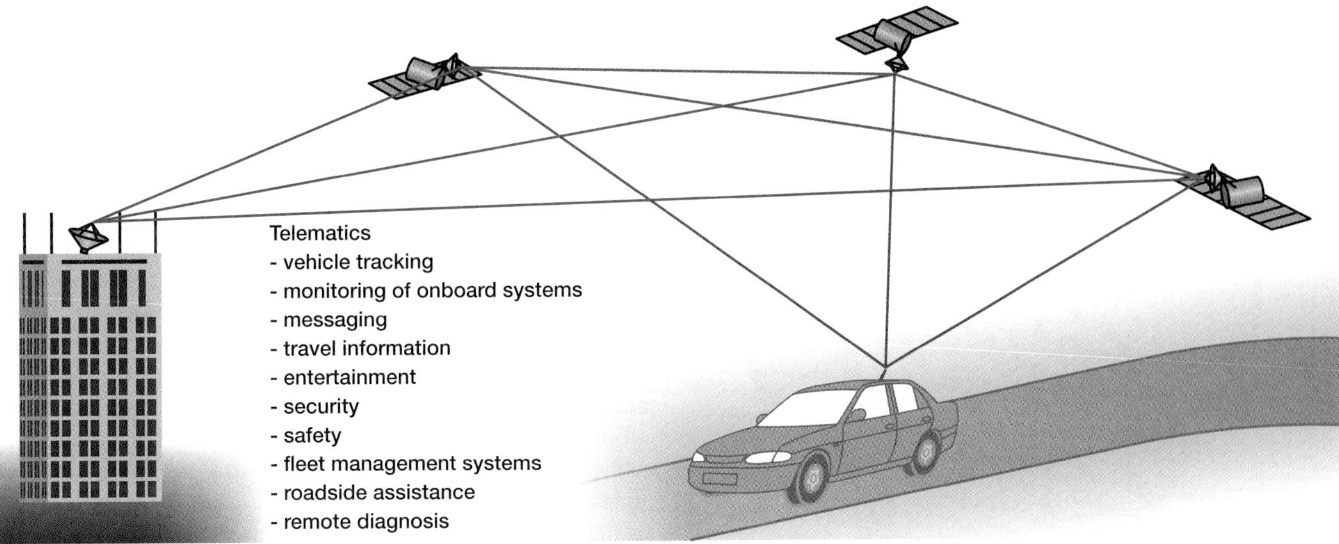

FIGURE 57-34 Automotive telematics is a satellite-based system that combines two-way communication and information technology within the vehicle.

▶ Antitheft Systems

N57004, K57007

Modern vehicle antitheft systems are devices or methods used to prevent the unauthorized use of the vehicle or its component parts. Manufacturers incorporate a number of technologies to discourage the theft of vehicles (**FIGURE 57-35**), and they continue to develop cost-effective means to reduce the risk of vehicle theft through a combination of physical and electronic protections. They also try to discourage theft, removing incentive by making component parts worthless through tagging or marking.

Types of Antitheft Systems

Theft-deterrent systems aim to prevent the vehicle being entered, started, or driven. The actuators used to achieve this are the electric door locks and windows, the starter motor relay, engine management systems, transmission shift solenoids, and an audible alarm.

In an effort to reduce theft, some vehicle manufacturers integrate sophisticated theft-deterrent systems to their vehicles. Theft-deterrent systems can be divided into four categories:

- Component identification
- Vehicle locking (including keyless entry and remote starting)
- Engine and transmission immobilization
- Audible and visible alarms

FIGURE 57-35 Antitheft devices and features help deter vehicle theft.

Component Identification

Component identification, or visible identification of major components, can be a deterrent for theft. Some manufacturers etch the vehicle identification number (VIN) onto labels that are attached to various components on the vehicle, such as fenders, doors, the hood and trunk, and major mechanical components such as the engine and transmission. Etching the VIN onto all windows with a small sand blaster is unobtrusive and makes the vehicle less attractive to steal. If the vehicle is resold or used for parts, the monetary value is greatly reduced.

Microdots are small plastic particles that can have either the VIN or a unique number printed on them (**FIGURE 57-36**). They are practically invisible to the naked eye but can be seen by using an ultraviolet light and a magnifying glass. They are mixed with an adhesive and sprayed onto the underside of the vehicle into the wheel wells and engine bay. A small transfer affixed to the vehicle displays that such a system has been used, making the vehicle less attractive to a thief.

Vehicle Locking

Door lock systems have come a long way. Initially, all locks were operated manually. A key was inserted in the door lock, and when turned, it would unlock only that door. The remaining doors needed to be unlocked manually, via the lock knob or lever. Releasing the key would return it to the same position it was in when inserted so it could be removed. When central locking was introduced, the lock mechanism on the driver's door had a microswitch

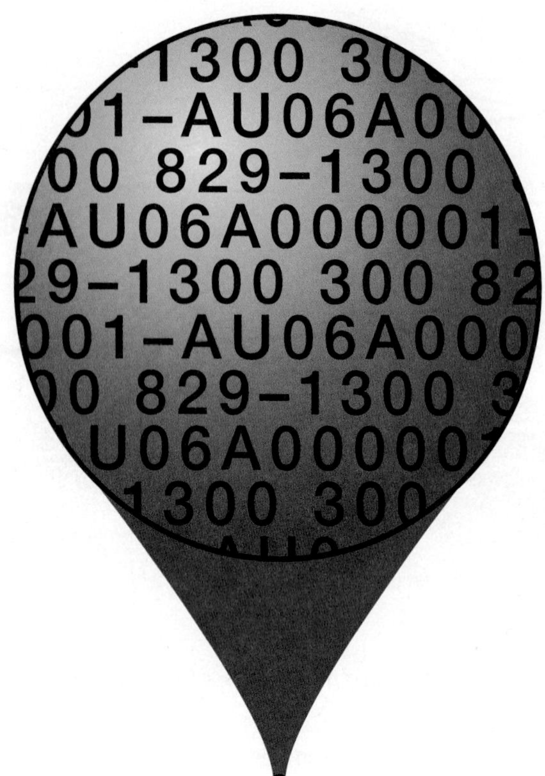

FIGURE 57-36 Microdot component identification.

that, when turned, activated electric solenoids on the remaining doors and trunk and unlocked or locked them. Further enhancement allowed manufacturers to introduce a security system that automatically locked the doors once the vehicle had reached a preset speed. This feature is mandatory in some countries.

Deadlocking adds a further level of security. If the key is turned a quarter turn and removed, a second microswitch activates small electric motors in each door lock mechanism, mechanically locking them, adding additional security.

In keyless systems, remote control keys and key fobs transmit a coded signal that is received by the vehicle's theft-deterrent module. If the code meets preset criteria, the module closes a switch that enables either the driver's door or all doors to be locked and unlocked as required. Pressing the button a second time when locking some vehicles activates the dead lock actuators, deadlocking the vehicle, and turning on the alarm system.

For the remote key and computer to exchange information, wireless communication is needed. High-frequency electromagnetic fields (i.e., radio frequency [RF] signals) are used. This system relies on the same basic technology as cell phones, TV, and radio, but at much lower power.

The remote fob can have one button for locking and unlocking; two buttons, where one locks and the other unlocks; three buttons, where the third button activates the trunk or tailgate; four buttons, where the fourth activates a panic alarm; or five buttons, where the fifth remotely starts the vehicle. Some vehicles actively transmit a radio code looking for a specific key. As the driver approaches the vehicle, the vehicle locates the key, and the theft-deterrent module prepares to unlock the vehicle. On some vehicles, when the driver leaves the vehicle and the key is beyond a certain range, the theft-deterrent module locks the vehicle.

When keyless entry is used, personalization of systems is possible. They can be programmed to recognize different keys, with each driver having his or her own specific key. This allows for different settings to be made that are unique to each driver. When a specific key is identified, the theft-deterrent module communicates with the body control module. Specific settings are remembered from that key, so the seat, steering wheel, and mirror position automatically adjust to the driver as well as climate control settings for cabin temperature and system modes. Entertainment system settings such as radio station presets, volume and audio settings, and transmission shift point preferences can all change to the driver's individual preferences. This system also allows for a valet key, or even a "teen driver" key to be used, limiting engine power, vehicle speed, and maximum radio volume, as well as preventing other vehicle settings from being changed.

Engine and Transmission Immobilization

Immobilization occurs when the theft-deterrent system prevents the vehicle's engine from starting or the transmission from operating without the properly authorized key. Audible and visible alarms may be activated when the locking or immobilization systems are tampered with. The immobilization system is enabled when the lock button is pressed on the remote key, when the doors are locked manually with the door key, or when a period of time has elapsed, typically 15 seconds, after the engine has stopped. The immobilization system is disabled by pressing the unlock button on the remote fob.

The key fob and theft-deterrent modules have to be capable of transmitting and receiving coded information. Operational characteristics vary greatly among manufacturers, and because it is a theft-deterrent system, details are a closely kept secret.

Most systems use a coded system known as "rolling codes." The coded data transmitted and received between the key fob and the theft-deterrent computer randomly change. Thus, the code used to lock and immobilize a vehicle is different from the one used to unlock and mobilize that vehicle. Manufacturers who use this system have seen a dramatic reduction in theft, noticeably from joy riders. The code is transmitted as a digital number, typically containing 16 digits, giving the possibility of billions of different code numbers. When the lock button is pressed, the control module and key preset the agreed-on code, which will be used to unlock the vehicle and deactivate the theft deterrent.

Pressing any button on the remote fob transmits an RF signal, even if the vehicle cannot read it, and changes the rolling code unit from that exchanged between the control module and the fob. The system allows an error factor of 25 steps either side of the agreed-on code for deactivation before the fob transmits a "false code," which causes the control module to disregard the key.

Alarms

Vehicles may also be equipped with alarm systems designed to sound high-decibel sirens or horns if an intruder attempts to break into or move a vehicle. Such systems can be independent of other vehicle systems, although on modern vehicles they are often integrated into the vehicle security system. The alarm system is typically wirelessly armed when the vehicle is locked and disarmed when it is unlocked. Often the system provides an audible chirp, and/or a flash of the headlights, when armed and disarmed so that the vehicle driver knows the alarm status. A basic alarm system is triggered by the door switches, which also turn on the interior lights; more complex systems may incorporate more complex sensors, such as ultrasonic movement sensors. If triggered, the alarm usually sounds for 30 seconds or so before resetting and sounding again if the trigger is still present.

A higher level of antitheft involves monitoring the vehicle for unauthorized opening of the doors, hood, trunk, and fuel filler door, and for activity that is occurring in the vehicle that should not be. The computer that controls the system monitors input signals from various devices. Switches are located at each door, the hood and trunk openings, and the fuel filler door. Vibration sensors detect any unusual vehicle movement such as lifting, jacking, or towing. Ultrasonic sensors detect any movement inside the cabin. Voltage monitoring sensors check for operation of the starter motor, ignition system, or fuel pump. Any of these conditions will cause the alarm to be activated.

Diagnosing the Antitheft System

The antitheft system is likely to be controlled by the vehicle's body control module (BCM) and may include immobilizer systems. Modern vehicle keys are coded to the BCM, and replacement keys have to be programmed for the vehicle. Systems may require security codes to be entered for initialization if the vehicle battery is disconnected, so always verify that you have the codes before disconnecting the vehicle's battery. Vehicle antitheft systems may be connected to sirens, horns, and lights.

As with all diagnosis, start by verifying the concern. Then research the concern in the service information and TSBs. Once you are familiar with the specified testing procedure, use that and your understanding of the system to develop a logical plan to diagnose the system. Make sure you find the root cause of the fault. Once the fault has been corrected, verify that the system operates as specified.

Because the antitheft system is computer controlled and has self-diagnostics built into it, using a scan tool to retrieve any codes from the system is a good place to start your diagnosis. Once you have the codes, you can follow the service information to diagnose the specific components or circuits related to the fault. Also, with the scan tool you will most likely be able to command actions to happen, such as locking the doors. If the system is a networked system, you will also likely be able to observe the confirmation signals on the data bus. The service information will guide you through the process of checking the individual circuits or devices.

Applied Science

AS-68: Electrical: The technician can use precision electrical test equipment to measure current, voltage, and resistance.

A vehicle that was traded in at the dealership is being checked to determine if it can qualify as a certified pre-owned vehicle. The checklist for this certification includes over 120 items to be inspected. Some of the electrical tests can be performed by using a DMM. This testing includes measuring current in amps, voltage in DC volts, and resistance in ohms. The DMM is capable of performing these tests; however, some shops may have electrical test equipment that would work more efficiently than the DMM for specific tests.

The measure of current in amps is the alternator output in amps. For this test, an auxiliary item called an amp clamp will be used because of the very high amperage output by many alternators. The amp clamp is placed around the heavy output wire from the alternator. It is best to follow the specific instructions given by the vehicle manufacturer and the test equipment or DMM manufacturer. Alternator outputs of over 100 amps are not unusual.

The charging system voltage can be checked with a DMM set to the DC 20 volts position. Place the red lead on the positive battery post and the black lead on the negative battery post. The charging system voltage is approximately 13.8 to 14.8 volts DC.

An example of a resistance measurement is testing an electronic ignition coil pack. Set the DMM to the ohms position. With the engine off, place the meter leads on the primary coil terminals. If you do not have an auto-ranging meter, try the lowest ohms setting. All coil packs should have close to the same resistance values. The manufacturer's service information will provide exact specifications and further testing instructions.

▶ Wrap-Up

Ready for Review

- Electrical systems on modern vehicles are becoming increasingly complex with the addition of a range of electronic and accessory systems, such as global positioning systems, entertainment systems, security systems, electric seats, and heated glass. Many of these systems are controlled by body computer systems.
- The controller area network (CANbus) and the bus system are used together in vehicle terminology (often called "CANbus") to describe one type of data-sharing network in a vehicle. The many sensors and engine control modules or modules can share information across a common network using the CANbus protocol to communicate efficiently and effectively with low data loss.
- In the basic cruise control system, the driver programs a set target speed, and the cruise control system's electronic control module attempts to maintain the target speed by commanding a throttle actuator to open or close the throttle plate.

- Adaptive or dynamic cruise control systems monitor the distance of the vehicle in front and adjust the vehicle speed to maintain a safe distance.
- Vehicle safety systems are designed to protect occupants during accidents; they can be classified as primary, or passive, safety systems and secondary, or active, safety systems.
- Collision avoidance systems are designed into vehicles to help avoid collisions or minimize the severity of a collision should one happen.
- Night vision enhances the image of the road and projects it on the windshield or instrument panel so that the driver can be more aware of hazards, especially in low light or limited visibility conditions such as fog or rain.
- Pre-crash systems are designed to give the driver a warning about a potential collision as well as provide actions that are designed to minimize occupant injuries.
- Seat belt pretensioners are used to tighten the seat belt in a severe frontal accident; both mechanical and electronic control systems are available.
- Crash sensors can be installed in various positions throughout the vehicle. Their location depends upon the direction of deceleration they are designed to detect.
- The global positioning system (GPS) relies on a group of at least 24 satellites orbiting approximately 12,650 miles (20,350 km) above the Earth.
- Automotive telematics is a satellite-based system that combines two-way communication and information technology within the vehicle.
- Theft-deterrent systems aim to prevent the vehicle being entered, started, or driven.
- Theft-deterrent systems can be divided into four categories:
  - Component identification
  - Vehicle locking (including keyless entry and remote starting)
  - Engine and transmission immobilization
  - Audible and visible alarms
- Microdots are small plastic particles that can have either the VIN or a unique number printed on them and attached to vehicle parts.

Key Terms

squib The component inside the airbag inflator that triggers the airbag deployment.

supplemental restraint system (SRS) A passenger safety system, such as airbags and seat belt pretensioners.

telematics Satellite-based, two-way communication with the vehicle's electronic systems.

triangulation A method of locating an object using mathematics based on forming a triangle with two known points.

trilateration A method of locating points by the measurement of distances using geometry.

Review Questions

1. Choose the correct statement with respect to cruise control.
  a. It is not an integral part of the safety system.
  b. It controls the vehicle's steering operation.
  c. It sets a cruise speed for the vehicle, which it then memorizes and maintains without throttle input from the driver.
  d. Once the speed of the vehicle is set, it cannot be changed manually.

2. All of the following are types of cruise control acutators *except*:
  a. hydraulic actuator
  b. electric actuator
  c. elcectronic throttle actuator
  d. vacuum actuator

3. All of the following actions will be performed by a vehicle pre-crash system *except*:
  a. pretensioning the seat belts.
  b. prepositioning the seats.
  c. warning audio for the incoming vehicle.
  d. rolling up the windows and sunroof.

4. Which of the following is the purpose of active headlight systems in a vehicle?
  a. They warn the driver of imminent departure.
  b. They illuminate the road around the corner.
  c. They monitor vehicles and objects behind the vehicle.
  d. They enhance the image of the road and project it on the windshield.

5. Which of the following components are a part of secondary (active) safety systems designed to protect occupants?
  a. Seat belt pretensioners
  b. Body panels
  c. Crumple zones
  d. Collapsible steering columns

6. Choose the correct statement with respect to the deployment of air bags in the vehicle.
  a. All airbags are triggered electrically.
  b. A steel ball releases a firing pin into the squib in the case of an electrically deployed airbag.
  c. The air bags are filled with air.
  d. the airbag deploys due to simultaneous actions occurring within the squib, the igniter, and the gas generator.

7. Each of the following are categories of theft-deterrent systems *except*:
  a. Component identification
  b. Vehicle locking
  c. Seat belt pretensioners.
  d. Engine and transmission immobilization

8. The radio reception issue is to be checked by verifying all of the following *except* by:
  a. selecting a station that you know comes in well in your location.
  b. checking for good power and ground connections.
  c. detecting any unwanted ripple or noise in the power circuit.
  d. checking Internet connectivity of the vehicle.

9. The GPS unit establishes its own location:
  a. by transmitting signals to satellites.
  b. by the time it takes for signals to arrive from satellites.
  c. by contacting the nearest radio tower.
  d. with the help of the CPU.

10. Dead locking adds a further level of security by:
    a. requiring manual locking in addition to automatic locking.
    b. locking each door with a different code.
    c. activating small electric motors in each door lock mechanism, mechanically locking them.
    d. requiring a password be entered before the doors unlock.

ASE Technician A/Technician B Style Questions

1. Tech A says that the cruise control disengages when the driver steps on the brake pedal. Tech B says that adaptive cruise control uses radar or laser systems to sense the distance and speed of the vehicle in front. Who is correct?
    a. Tech A
    b. Tech B
    c. Both A and B
    d. Neither A nor B
2. Tech A says that that some night vision systems display an enhanced image of the road, in front of the vehicle, on the windshield. Tech B says that pre-crash systems are designed to make it faster for a person to exit a vehicle after a crash. Who is correct?
    a. Tech A
    b. Tech B
    c. Both A and B
    d. Neither A nor B
3. Tech A says that SRS air bags stay inflated until rescue people arrive on the scene. Tech B says that in severe collisions, wearing seat belts is not important if the vehicle is equipped with SRS airbags. Who is correct?
    a. Tech A
    b. Tech B
    c. Both A and B
    d. Neither A nor B
4. Tech A says that in a collision, SRS air bags are quickly inflated by an electric air compressor. Tech B says that SRS air bags can be repacked and reused after a collision. Who is correct?
    a. Tech A
    b. Tech B
    c. Both A and B
    d. Neither A nor B
5. Tech A says that some entertainment systems have speed-dependent volume, which changes the volume of the radio based on vehicle speed. Tech B says that some warning messages can be broadcast over the audio system. Who is correct?
    a. Tech A
    b. Tech B
    c. Both A and B
    d. Neither A nor B
6. Tech A says that microdots are used as a theft deterrent in some vehicles. Tech B says that microdot is the name given to a single datum transmitted over a vehicle network. Who is correct?
    a. Tech A
    b. Tech B
    c. Both A and B
    d. Neither A nor B
7. Tech A says that immobilization occurs when the theft-deterrent system prevents the vehicle's engine from starting or the transmission from operating without the properly authorized key. Tech B says that on some vehicles, when the driver leaves the vehicle and the key is beyond a certain range, the theft-deterrent module automatically locks the vehicle. Who is correct?
    a. Tech A
    b. Tech B
    c. Both A and B
    d. Neither A nor B
8. Tech A says that most SRS systems use a safing sensor to reduce the possibility of accidental deployment of the airbags. Tech B says that airbags are designed to act like a nice, soft pillow in an accident. Who is correct?
    a. Tech A
    b. Tech B
    c. Both A and B
    d. Neither A nor B
9. Tech A says that If any of the SRS-related devices are deployed for any reason, most manufacturers require all of the deployed devices, the crash sensors, and the control module to be replaced. Tech B says that there are relatively large holes in the rear of an airbag that allow it to quickly deflate after deployment. Who is correct?
    a. Tech A
    b. Tech B
    c. Both A and B
    d. Neither A nor B
10. Tech A says that in some vehicles, airbag deployment can be based on the speed of the impact and the weight of the seat occupant. Tech B says that the SRS seat belt pretensioners tighten the seat belt when you connect the seat belt. Who is correct?
    a. Tech A
    b. Tech B
    c. Both A and B
    d. Neither A nor B

SECTION 8
Heating and Air Conditioning

CHAPTER 58

Principles of Heating and Air-Conditioning Systems

NATEF Tasks

There are no NATEF Tasks for this chapter.

Knowledge Objectives

After reading this chapter, you will be able to:

- **K58001** Describe the history and regulations of automotive HVAC systems.
- **K58002** Explain the basic principles of HVAC systems.
- **K58003** Explain the purpose and function of AC components and controls.
- **K58004** Explain the purpose, function, and types of HVAC restrictions.
- **K58005** Explain the purpose, function, and types of switches and control circuits.
- **K58006** Explain the purpose and function of hoses, pipes, lines, and O-rings.
- **K58007** Explain the purpose and function of driers.
- **K58008** Explain the purpose, function, and types of HVAC compressors and clutches.
- **K58009** Explain the purpose function and types of HVAC heat exchangers (condenser and evaporator).
- **K58010** Explain the purpose, function, and types of refrigerants and refrigerant oils.
- **K58011** Explain the operation of each type of AC system.
- **K58012** Explain the operation of the fixed orifice tube system.
- **K58013** Explain the operation of the thermal expansion valve system.
- **K58014** Explain the purpose and function of heating system.
- **K58015** Describe the purpose and function of the heating and ventilation system components.

Skills Objectives

There are no Skills Objectives in this chapter.

▶ Introduction

This chapter explains the processes involved in regulating the temperature inside the passenger compartment of a vehicle. The quality and temperature of the surrounding air strongly influence a driver's performance and concentration. Within a passenger vehicle, the heating, ventilation, and air-conditioning (HVAC) system is designed to maintain a comfortable temperature in the vehicle cabin and also provide fresh, filtered, conditioned air on a constant basis (**FIGURE 58-1**). Depending on the outside temperature and the desired cabin temperature, the supplied air is either heated, cooled, or both.

The HVAC system of most passenger vehicles uses the heat created by normal engine operation to raise the temperature in the passenger compartment. Refrigerant is used to cool the air in the passenger compartment. To ensure that the air brought into the vehicle is free from dust and debris, many vehicles pass the air through a cabin air filter that removes foreign particles. Without proper maintenance, the HVAC system can degrade, resulting in a less comfortable ride for the passengers, particularly during the summer and winter months. Accordingly, it is critical that automotive technicians be familiar with the theory of operation of the system and its components so that the system can be maintained, diagnosed, and repaired when necessary.

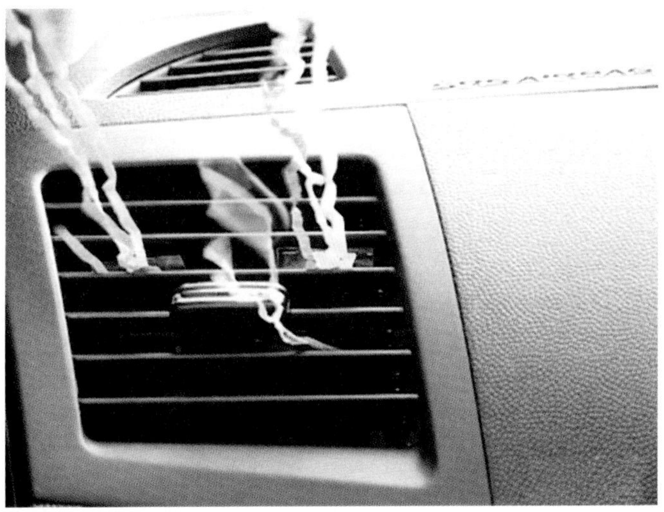

FIGURE 58-1 The HVAC system provides fresh, filtered, and conditioned air to the passenger compartment.

▶ History of Automotive Heating and Cooling

K58001

There is some debate about the origins of systems for heating the passenger compartment of vehicles. What has been confirmed is the use of gas lamps in the early 1900s for heat. Passengers would bring a small gas lamp into the compartment for heat, but it was not connected to the vehicle in any way. Around 1917–1920, some vehicles transferred exhaust gas through pipes into the compartment to provide heat. The first use of air heated from normal engine operation to warm the passenger compartment dates from the late 1920s. These vehicles had rudimentary blowers that simply funneled air from around the engine into the cabin. By the 1930s, the first true heating systems using engine coolant, similar to what is still used today, were developed.

Automotive air-conditioning has a long history, beginning with a discovery by Michael Faraday. In 1820, Faraday discovered how to create cool air by passing air over an evaporating coil of compressed ammonia. Although this was a great advance in comfort, ammonia is a caustic agent and hazardous to humans, causing skin burns, lung damage, and possibly asphyxiation. In 1902, Willis Carrier invented and patented the first electric air conditioner. A great advance was made in 1928 when a research division of General Motors produced **chlorofluorocarbons (CFCs)**, a chemical compound containing chlorine, fluorine, and carbon. CFCs made it possible for air-conditioning systems to be used by consumers with no apparent harm, unlike the ammonia systems.

In 1939, Packard introduced the first automotive air-conditioning system as an add-on luxury unit (**FIGURE 58-2**). The add-on was offered for $274, which in today's money would be more than $4500. Automotive air-conditioning is now standard on most production vehicles because of customer demand, not only for the comfort it provides but also for the increased fuel economy that comes from increased aerodynamic efficiency.

You Are the Automotive Technician

A customer brings his vehicle to the auto shop for a routine maintenance check. He will be driving his vehicle from New York to Florida this fall and won't be returning until the following spring. He is worried about how his vehicle's HVAC system will perform during his long trip and during his stay in sunny Florida. He isn't familiar with auto shops in Florida, so he wants you to inspect the system and suggest any repairs it may need.

1. What are the two types of automotive air-conditioning systems?
2. Refrigerants are another important part of the HVAC system. Why is identifying refrigerants an important part of repairing air-conditioning systems?
3. What license is required for technicians to work with refrigerants?
4. What are the four main components in an air-conditioning system?

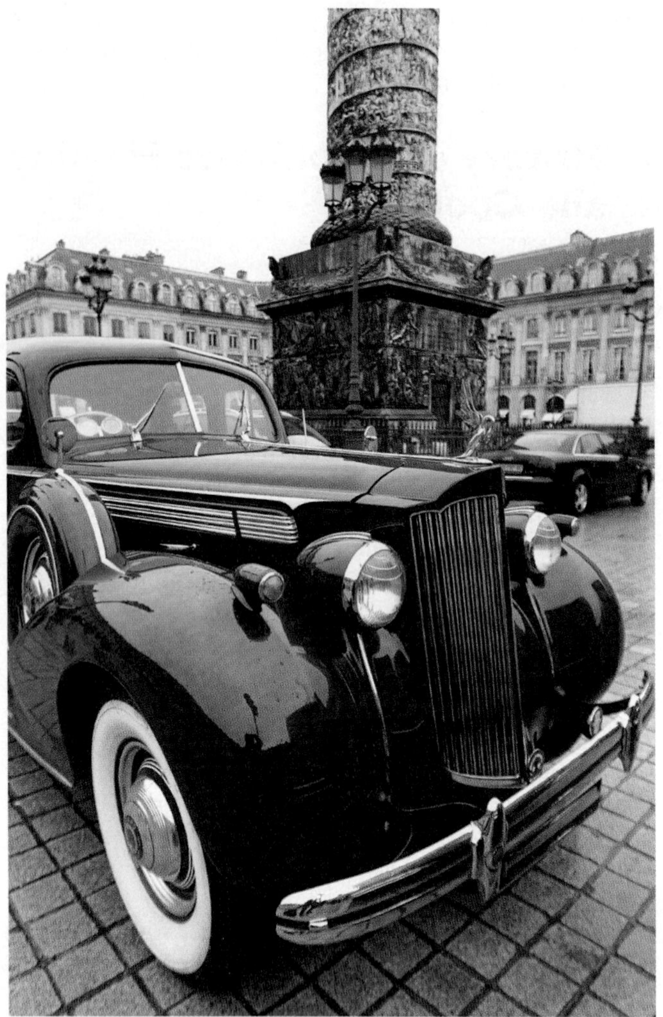

FIGURE 58-2 In 1939, Packard introduced the first automotive air-conditioning system as an add-on luxury unit.

Having the windows rolled up while driving at moderate and high speeds reduces wind drag and increases aerodynamic efficiency in most vehicles.

> ▶ TECHNICIAN TIP
>
> Almost all manufacturers rely on the air conditioner's ability to remove moisture from the air to meet the defrost time standard set by the U. S. Department of Transportation.

▶ HVAC Regulation

Automotive air-conditioning is federally regulated by the **Environmental Protection Agency (EPA)**. Section 609 of the **Clean Air Act (CAA)**—the federal law that regulates air emissions from stationary and mobile sources—contains all of the motor vehicle air-conditioning (MVAC) requirements and laws. Among other things, this law authorizes the EPA to establish National Ambient Air Quality Standards (NAAQS) to protect public health and public welfare and to regulate emissions of hazardous air pollutants.

This act sets the regulations for shops that do air-conditioning work. It mandates that each shop maintain a current government-issued license and stipulates what certifications are required for technicians doing MVAC service. It also outlines the specific refrigerants that can be safely used for each application and how refrigerants must be stored and reclaimed, even citing the specific approved equipment technicians must use.

Licensure

Automotive air-conditioning service and repair technicians need to have a special license (**FIGURE 58-3**) that is granted after passing the 609 test, which is a self-study course that can be found online through companies such as Automotive Service Excellence (ASE), Mobile Air-conditioning Society (MACS), and Mainstream Engineering. Passing the test results in a lifelong license. Because the license never expires, the technician is responsible for keeping up to date with all new regulations and technology. The 609 license also allows the technician to purchase a few select refrigerants, such as R-12, R-134a, and HFC 1234 (refrigerants are identified with the letter "R," followed by a series of numbers that indicate their chemical composition). The 609 license does not mean that the technician knows how to repair the automotive air-conditioning systems, but that the technician knows the laws and will comply with all regulations.

> ▶ TECHNICIAN TIP
>
> The equipment used to recover refrigerants from MVAC systems is also covered in the 609 MVAC laws. The technician is responsible for using properly certified equipment when servicing MVAC systems. The current fine for not complying with the 609 laws is $37,500 per day per incident. This fine is applicable to the person or entity not in compliance with the 609 regulations.

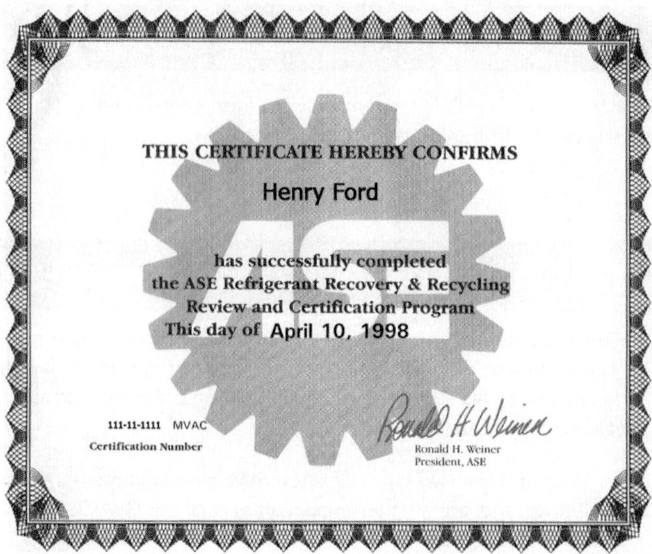

THIS CERTIFICATE HEREBY CONFIRMS

Henry Ford

has successfully completed
the ASE Refrigerant Recovery & Recycling
Review and Certification Program
This day of April 10, 1998

111-11-1111 MVAC
Certification Number

Ronald H. Weiner
President, ASE

FIGURE 58-3 EPA 609 license.

▶ HVAC Principles

K58002

The function of air-conditioning in the HVAC system is to reduce the temperature and humidity inside the passenger compartment to a level that ensures passenger comfort (**FIGURE 58-4**). The HVAC system achieves this by removing excess heat from the passenger compartment through a series of thermal and chemical transformations. In other words, heat is removed from the passenger compartment and transferred by the air-conditioning system to the outside air.

The HVAC system is governed by the principles of physics just like all of the other systems on a vehicle. Understanding the physics will help you to know when a system is operating properly and when it needs to be serviced. One of the simplest concepts concerns heat transfer: Heat always transfers from hot objects to cold objects (**FIGURE 58-5**). Another concept concerns density: Hot liquids and gases are less dense and tend to rise (**FIGURE 58-6**). Thus cool liquids and gases tend to be pulled to the bottom by gravity, and hot liquids and gases tend to float to the top. Another simple concept is the relationship between temperature and

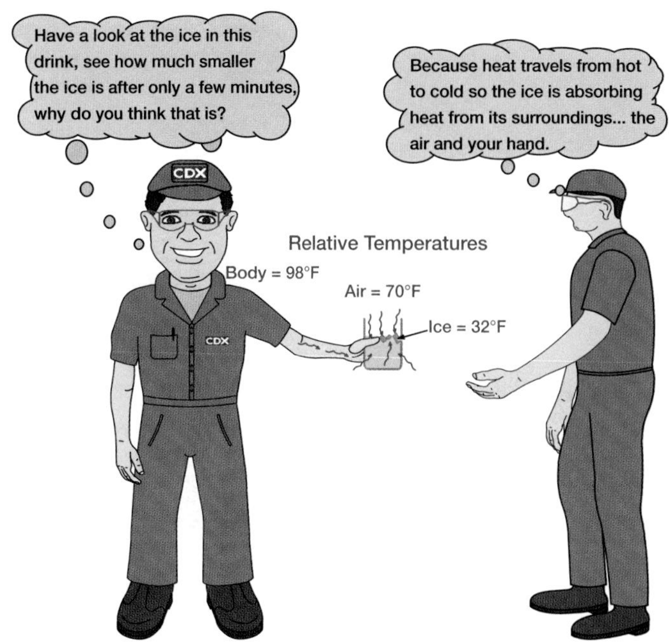

FIGURE 58-5 Heat always travels from hot places to cold places.

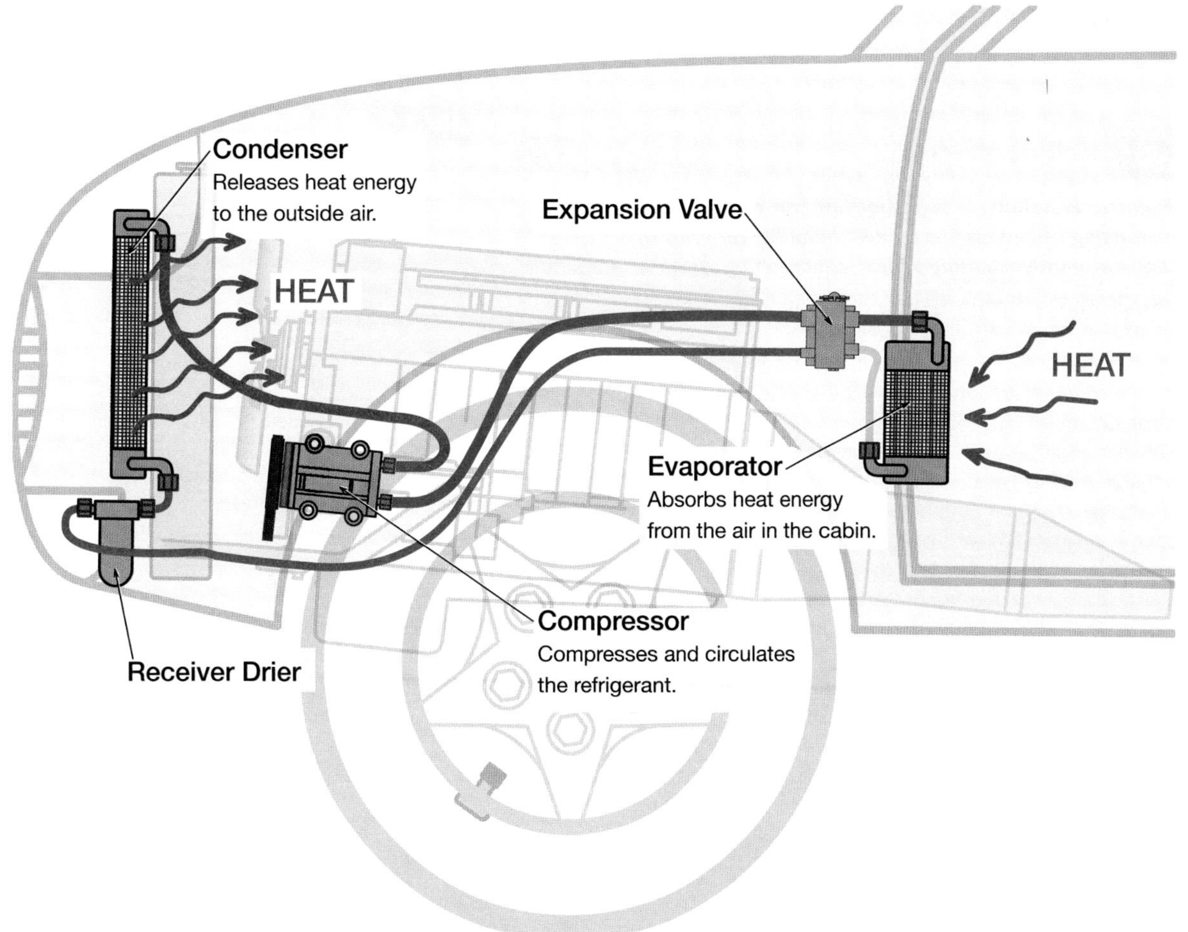

FIGURE 58-4 The HVAC system reduces the temperature inside the passenger compartment by removing excess heat through a series of thermal and chemical transformations.

FIGURE 58-6 Hot things tend to rise, and cold things tend to fall.

Applied Science

AS-84: Relative Humidity: The technician can demonstrate an understanding of and discuss relative humidity in terms of its effect on automotive heating and air-conditioning systems.

Relative humidity is essentially a measure of how moist the air is, or in more detail, the amount of water vapor in the air expressed as a percentage of the amount that the air can "hold" at a given temperature. The warmer air is, the more moisture it is able to hold.

In a car air-conditioning system, warm air is forced through a cold (below dew point) evaporator core. This condenses the water vapor in the air into its liquid form and drains it away, much like the way water condenses on the outside of a cold soda can. Thus, water vapor in the incoming air is removed, and the relative humidity in the vehicle's cabin is lowered. On humid days, more water vapor is in the air, so you may notice larger than usual amounts of water draining from the evaporator onto the ground underneath the vehicle.

Although you may expect heating the incoming air to have the opposite effect, in reality this is not the case. Cool air is passed through a hot heater core (generally heated by coolant from the engine's cooling system), heating the air up, which increases the water carrying capacity of the air. But because no water is being added, heating up the air lowers its relative humidity in the cabin.

pressure: When pressure is raised on a gas, its temperature also rises; when pressure is lowered, so does its temperature. These and other principles will be explored in greater depth in the following sections.

Cold and Hot

The conditions "cold" and "hot" are more of a feeling than a science. Cold is simply a relative term for the absence of heat, or it can be regarded as a condition that exists after heat has been removed. Heat energy is in all matter, from gases (air), to liquids (water), to solids (metal). One might think that 32°F (0°C) is cold, but more heat energy is present there than at –19°F (–28°C). Heat energy cannot be created or destroyed; it can only be transferred from one object to another. Heat always travels from a warmer object to a cooler object.

Consider this age-old question: Does ice make your hand cold, or does your hand warm up the ice? The answer is that your hand is warming up the ice, and the extreme transfer of heat energy makes your hand feel cold. The warming of the ice is known as **heat transfer**. Heat transfer can take place by one of three methods: conduction, convection, or radiation (**FIGURE 58-7**). The HVAC system relies on all three types of

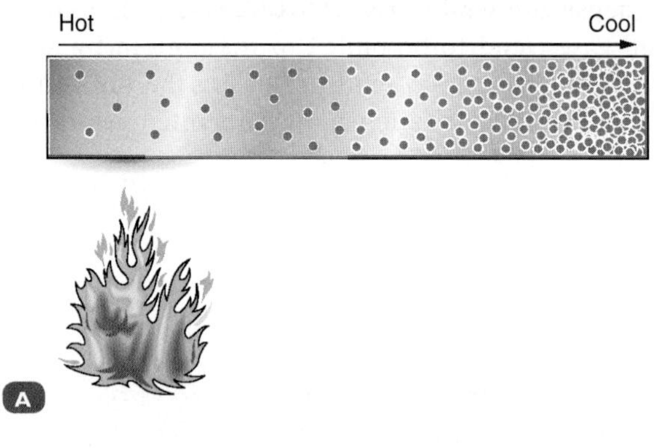

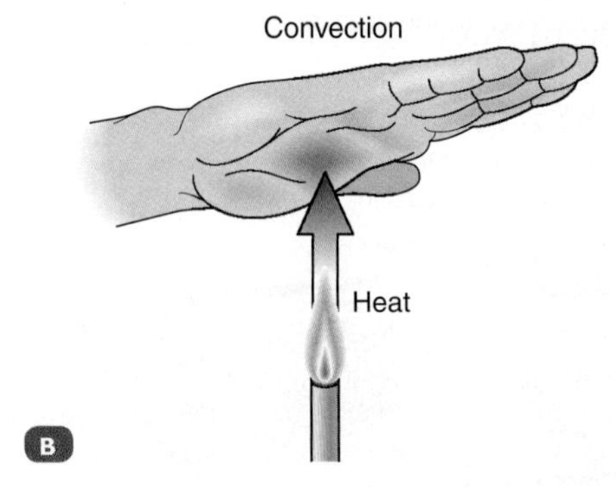

FIGURE 58-7 Heat transfer. **A.** Conduction. **B.** Convection. **C.** Radiation.

heat transfer. **Conduction** is the process of transferring heat through matter by the movement of heat energy through solids from one particle to another. For example, if you hold a lighter up to a steel pipe, the heat from the flame will travel through the steel, eventually heating the entire pipe.

Heat transfer by **convection** is the circulatory movement that occurs in a gas or fluid with areas of differing temperatures due to the variation of the density and the action of gravity. Consider a container of water being heated on a stove. If you put a drop of food coloring in one part of the water and start to heat the water, you can readily see the circulation of the water within the container. As the water heats up, it expands and becomes lighter than the surrounding water. This causes a column of water to rise to the top; the cooler water then falls to the bottom. Thus, a cycle occurs of warmer water rising to the top, pushing cooler water to the bottom. The convection currents have the same effect of a small pump pushing the water around in a set pathway.

Radiation is the transfer of heat through the emission of energy in the form of invisible waves. The sun radiates energy to the Earth over the vast miles of outer space; the electromagnetic radiation readily travels through the vacuum of space. When this energy is absorbed, it becomes converted to heat—that is, the heat you feel as the sun touches your body on a warm day. Of course, the direction in which the radiation travels can be changed or redirected, as when a sheet of shiny aluminum foil reflects the sun's rays and bounces the energy in another direction.

Let's see how heat transfer is used in cooling the passenger compartment. Radiation of heat from the rays of the sun penetrate the windows and heat up the interior components and air. The driver turns on the air-conditioning function of the HVAC system. Convection currents from the heated air in the passenger compartment are pushed through an evaporator core filled with refrigerant, causing heat to be conducted through the metal coils of the evaporator. The metal coils of the evaporator then transfer the heat by conduction to the refrigerant, causing convection currents to transfer heat through the refrigerant. The refrigerant is then circulated by the compressor and carried to the condenser, where conduction and convection dissipate the heat to the surrounding air, and ultimately to the atmosphere.

Heat Energy

Heat energy is measured in **British thermal units (Btus)**. The Btu is a standard unit of measurement of heat quantity. It describes the amount of heat it takes to move temperature up or down in degrees. One pound of water at **atmospheric pressure** and at a temperature of 32°F (0°C) would require 1 Btu of energy added to increase the temperature 1 degree Fahrenheit. A change in the temperature of 1 pound of water from 32°F (the freezing point of water) to 212°F (100°C) (the boiling point of water) would require 180 Btus of heat energy.

Water cannot get any hotter than 212°F at atmospheric pressure. The temperature will not rise because the molecules have reached their maximum speed of movement at atmospheric pressure. Temperature is increased only by moving the molecules faster. If the water is exposed to any more energy, it will start to change its state of matter from a liquid to a gas: steam. Turning a pound of water at 212°F at atmospheric pressure into steam requires 970 Btu. The heat required to change the water at its maximum temperature from a liquid to a gas is called **latent heat of evaporation** (**FIGURE 58-8**). This latent heat is called "hidden heat" because it cannot be felt or measured with a thermometer. Heat that can be felt and measured with a thermometer is called sensible heat. The amount of heat given up by steam to turn back into a liquid is the same 970 Btu and is called **latent heat of condensation**.

When you take the same pound of water from a liquid to a solid, you freeze it. To do this, you must remove heat from the liquid to make ice. At 32°F, the molecules in the water cannot move any slower without changing state. At 32°F, the **latent heat of freezing** begins. It takes the removal of 140 Btu from 1 pound of 32°F water to change its state into ice.

The changing of state in liquids and gases, referred to as vaporizing and condensing, is the basis of air-conditioning. Because it takes more heat to evaporate a liquid than to freeze a liquid, air-conditioning uses the process of evaporating a liquid to absorb a large amount of heat energy. By evaporating liquids, the air conditioner can remove the heat that causes vaporization from the cabin air, thereby making passengers feel cooler. It is the same principle by which sweat cools human beings. Heat is absorbed by the evaporating sweat droplets on the skin, thereby removing heat from the body and cooling it. Similarly, by evaporating liquids, the air conditioner can remove large amounts of heat from the air and make the occupant feel cooler.

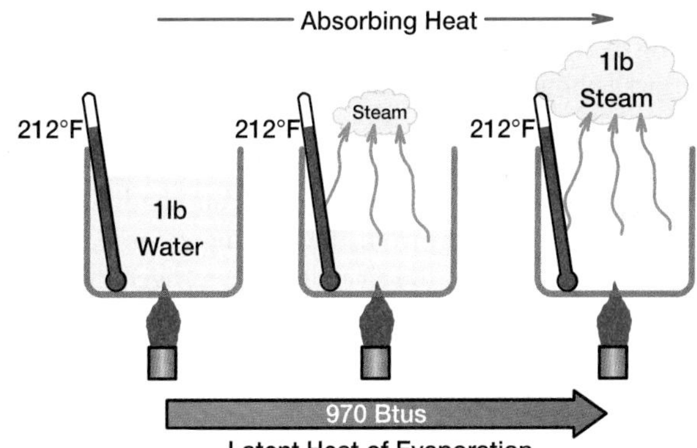

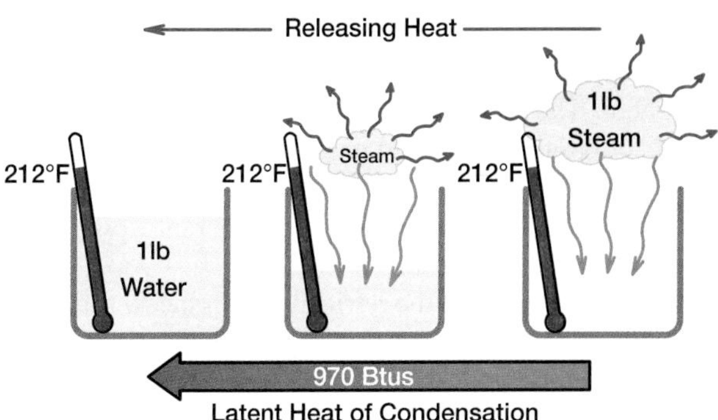

FIGURE 58-8 It takes 970 Btus of heat to turn 1 pound of water into steam, and vice versa.

Refrigerant Principles

Manipulating nature's law of **vaporization** and **condensation** of liquids and gases is how the air conditioner forces ambient passenger compartment air to become cool and less humid for the driver and passengers of the vehicle. For the cycle of vaporization and condensation to occur, the air conditioner must have a fluid that vaporizes and condenses at the correct temperature and pressure, and is also able to change its state in the continuous cycle of vaporization and condensation without breaking down. The fluid used for these purposes is **refrigerant**. The refrigerant in the air-conditioning system must have a boiling point well below freezing so that its boiling and condensing points can be manipulated by changing pressure. The super low boiling point of most refrigerants allows the system to control the vaporization and condensation point of the refrigerant at normal ambient temperatures. A commonly used refrigerant is R-134a (also known as tetrafluoroethane), a hydrofluorocarbon that contains none of the ozone-depleting chlorine found in previous refrigerants.

With the use of refrigerants, maintaining vaporization and condensing points at normal ambient temperatures is simply done by raising or lowering the pressure of the refrigerant. At atmospheric pressure, R-134a boils at about –15°F (–26°C), so in most normal conditions it is in its gaseous state. Adding a small amount of pressure to the refrigerant causes the boiling point of the refrigerant to rise. Indeed, a low pressure (approximately 15 psi [103 kPa]) in a refrigerant can force its boiling point to be just above freezing (32°F [0°C]). Likewise, to allow the R-134a to condense back into a liquid, its boiling point can be increased well above the ambient temperature by further increasing its pressure. Forcing the refrigerant's boiling point so that it is well above ambient temperatures allows it to be cooled by ambient air so that it will condense back into a liquid.

With the refrigerant boiling below the temperature of the air in the passenger compartment, the air conditioner will forcefully remove the heat from the passenger compartment's air by using the heat from the air to boil the refrigerant. R-134a under low pressure (15–25 psi [103–172 kPa]) boils at about 32°F (0°C). Warm air from the passenger compartment is circulated from the passenger cabin across the evaporator containing R-134a, causing it to boil. As R-134a boils, it absorbs that heat and pulls it along with the moving R-134a away from the evaporator. The process continually repeats as the fan circulates the air, and the operating air-conditioning system continually circulates refrigerant, resulting in a continual supply of cool air to the cabin. The air returning to the passenger compartment will feel cool to the driver and passengers because the heat was removed from the air. The heat from the air is absorbed by the refrigerant, changing the state of the refrigerant from liquid to gas while also cooling the air.

If the temperature of the refrigerant falls below the freezing point of water during this process, the air conditioner must shut off. The air passing over the evaporating refrigerant is full of moisture, so the extreme heat removal causes any moisture in the air to condense on the cold surface of the evaporator. The water vapor in the air could freeze on the outside of the evaporator and block the airflow that is needed to cause the refrigerant to vaporize. The air must keep flowing over the evaporator, renewing the heat source to keep the refrigerant evaporating (boiling) efficiently without freezing. The system is designed to prevent freezing if it is operating correctly.

Applied Science

AS-23: Phases/States: The technician can explain in detail the three states of matter.

Solids, liquids, and gases are the three most common states of matter on Earth. The distinction between them is based on characteristics relating to the shape and volume of the piece of matter in question.

Solid matter retains a fixed volume and shape. There are many examples of solid matter used in motor vehicles, from the cast iron, steel, and aluminum used to build engines, to the fabric and plastic materials used in interiors. Liquid matter retains a fixed volume, but the shape adapts to the shape of the container it is held within. Again, there are many liquids used in motor vehicles. In the case of gases, the matter does not retain a fixed volume, but expands to fill whatever volume is available. Air-conditioning refrigerant is an interesting example of a compound that takes the form of both a liquid and a gas in different sections of the air-conditioning system.

Applied Science

AS-28: Conduction/Convection: The technician is able to explain the concept of heat transfer in terms of conduction, radiation, and convection in automotive systems.

Heat can be transferred via three methods: conduction, convection, and radiation. Put simply, conduction is the transfer of heat through solid matter; convection is the transfer of heat through liquid or gas relying on the movement of currents; and radiation is the transfer of heat energy through space by means of electromagnetic waves.

All three of these methods of transfer come into play when trying to keep the inside of a car cool. The heat you feel in the cabin originates from the sun. The sun's heat travels by radiation, and you feel it coming through the glass windows into the vehicle cabin. Heat is transferred through the cabin via convection, with the hot glass panels heating the air inside and causing it to move in currents. When turned on, the air-conditioning system moves the hot air through the cold evaporator core. The heat is moved to the evaporator core via convection and dissipated through the metal evaporator core by conduction.

Differential Gas Pressure (Refrigerant Cycle)

Every vehicle's air conditioner has a low pressure side, also called the low side or suction side (approximately 20–40 psi [138–276 kPa]), and a high pressure side, also called the high side or discharge side (approximately 170–230 psi [1172–1586 kPa]). The low side is designed to allow easy transformation of the refrigerant from a liquid to a gas (evaporation). The high side enables the transfer of heat out of the refrigerant to the atmosphere so it changes back from a gas to liquid (condensation) (**FIGURE 58-9**).

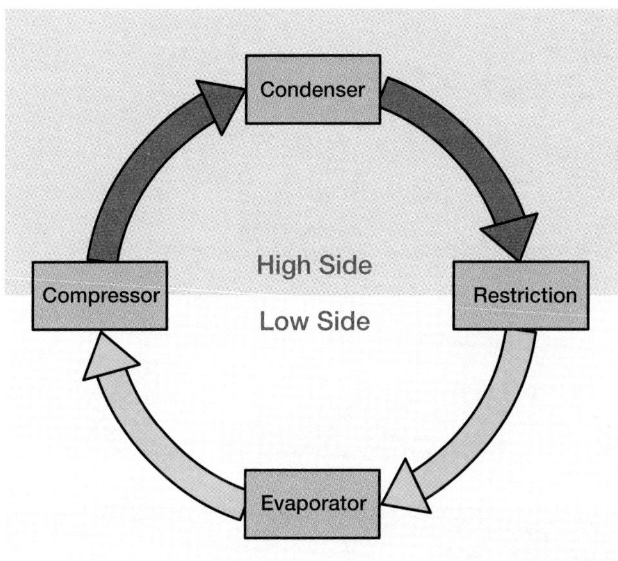

FIGURE 58-9 Low side components and high side components.

Starting at the compressor, a low-pressure gas is compressed into a high-pressure gas and is pumped into the condenser located in front of the vehicle radiator. The gaseous refrigerant is cooled by the air passing through the condenser and leaves the condenser as a high-pressure liquid. The high pressure is required to allow the change from gas to liquid. From there, the liquid travels through a small orifice (restriction) that reduces the pressure of the refrigerant (a fixed orifice tube or thermal expansion valve). After the restriction, the low side of the system begins. The low-pressure liquid leaves the metering device and travels into the evaporator in the passenger compartment. As the low-pressure liquid travels through the evaporator, a fan blows the warm to hot cabin air across the fins, adding heat to the refrigerant until it boils. As it boils, it absorbs heat from the air, and the liquid refrigerant is transformed back into a gas. The gaseous refrigerant then travels back to the compressor to start the cycle over.

Applied Science

AS-29: Expansion/Contraction: The technician is able to demonstrate an understanding of the expansion and contraction of system parts as a result of heat generated during the use of the system.

When matter is heated, it expands, or becomes larger in size. Conversely, when cooled, it contracts, or becomes smaller (assuming the change in temperature does not cause the matter to change state). Thermal expansion (or TX) valves, found in many air-conditioning systems, operate via these principles. The function of a TX valve is to regulate the flow of liquid refrigerant from the evaporator back to the compressor, thereby also regulating the temperature of refrigerant.

The valve is actuated against spring pressure by a temperature-sensing bulb, filled with a similar gas to the refrigerant used in the air-conditioning system. As the refrigerant temperature at the evaporator outlet increases, the pressure in the bulb increases due to expansion of the gas and causes the TX valve to open. When the evaporator outlet temperature drops, the pressure in the bulb decreases as the gas contracts and allows spring pressure to close the valve.

▶ Air-Conditioning Components and Operation

K58003

Each HVAC system uses refrigerant and four major components. Each component has a specific job and purpose. The components are the compressor, condenser, restriction, and evaporator (**FIGURE 58-10**). The air-conditioning cycle starts with the **compressor**, which changes a low-pressure refrigerant gas to a high-pressure gas. The compressor also provides the needed refrigerant movement in the system. Without the movement of refrigerant, the system could not exchange the heat from the condenser and the evaporator. Moving refrigerant also allows for the pressure drop to occur in the restriction. Without movement, the restriction could not drop the pressure. Refrigerant enters the compressor as a low-pressure gas from the evaporator. The compressor's function is to increase the pressure of the gaseous refrigerant and then push it into the condenser.

Forcing a gas to become a liquid is achieved through condensing. So the next component after the compressor in the air-conditioning system is the **condenser**. Hot, high-pressure gas enters the condenser from the compressor. The gas flows through a series of coils in the condenser, and ambient air passes over the outside of the finned coils. The engine's radiator fan and/or dedicated condenser fans move ambient air across the condenser coils and fins. The cooler outside air removes the heat from the hot, high-pressure gas, and the gas begins to condense into a liquid. Emerging from the outlet of the condenser is a relatively warm, high-pressure liquid. The condenser transforms gas to liquid on the high side of the system.

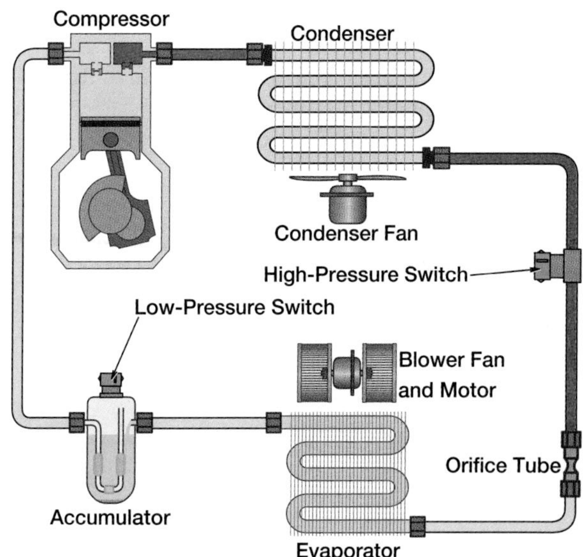

FIGURE 58-10 The components of an air-conditioning unit.

The air conditioner is a **closed loop system**, meaning that nothing else enters or exits the system; the refrigerant just continues to cycle through the system, changing pressure and state to complete the task of cooling the cabin. Before the liquid can be evaporated back into a gas for the compressor, the pressure has to drop so the boiling point can be lowered below the passenger compartment air temperature, thus allowing vaporization to remove the heat from the passenger compartment air. Dropping pressure occurs by flowing a liquid through a **restriction**.

The restriction lowers the pressure by using the Bernoulli principle. Daniel Bernoulli stated that if the flow of a fluid remains constant, the energies of the fluid remain constant; changing one of the two energies (up or down) affects the other energy oppositely. The two energies in fluid are velocity and pressure. These energies react opposite to each other. When velocity increases, the pressure must drop if the flow remains constant. The restriction causes a significant increase in refrigerant velocity because the compressor is constantly moving fluid in the closed loop system.

The fluid flowing through the restriction is a liquid. Liquids are considered incompressible (Pascal's law). Forcing a noncompressible fluid through a small orifice causes the velocity in the restriction to rise. If the velocity of the liquid rises, the pressure of the liquid must drop, according to the Bernoulli principle. Raising the velocity and dropping the pressure is how the restriction transforms the high-pressure liquid refrigerant to a low-pressure liquid refrigerant. The reduced pressure reduces the boiling point of the refrigerant and turns the refrigerant into a low-pressure liquid ready for the evaporator to vaporize into a gas.

The next component in the air-conditioning system is the **evaporator**. The evaporator transforms low-pressure liquid to a low-pressure gas. The transformation from liquid to gas begins when the air from the passenger compartment passes over the evaporator coils and fins. The pressure on the refrigerant is typically held between approximately 20 and 40 psi (138 and 276 kPa); the pressure on the low side of the system reduces the refrigerant's boiling point to a bit over 33°F (1°C). The liquid refrigerant absorbs heat from any source available as it boils. This heat comes from the air flowing through the evaporator core.

The vaporization of the refrigerant is forced to happen just above the freezing point of water so that the evaporator core will not freeze. By moving air across the evaporator coils, the convection currents pass over the evaporator coils, causing conduction of heat to the refrigerant through the tubes of the evaporator. The conduction of heat through the coils then provides the needed energy to boil (evaporate) the liquid refrigerant into a gas. It also strips the air of much of its heat energy and moisture, making the air flowing from the duct cool and crisp.

Restriction

K58004

There are two common types of restrictions: the fixed orifice tube and the thermal expansion valve (**FIGURE 58-11**). A fixed orifice tube restriction is a device with a metered hole in it and a metal or plastic screen to keep out any debris. It allows only a specific amount of liquid refrigerant through at a time. This restriction causes the pressure to drop at that point. No matter what the cabin and evaporator temperatures are, the fixed orifice tube restriction will allow only the specified amount of liquid refrigerant through.

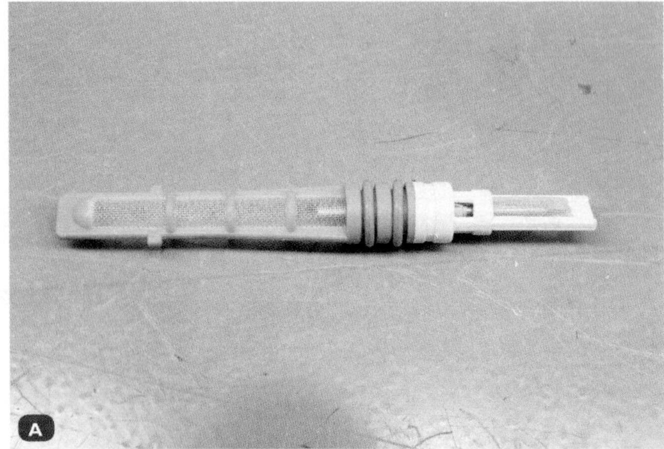

FIGURE 58-11 Restrictions. **A.** Fixed orifice tube. **B.** Thermal expansion valve.

The thermal expansion valve has a sensing device on the outlet side of the evaporator that senses changes in temperature. It also has a valve that can vary the size of the opening to allow more or less liquid refrigerant through. By sensing the temperature coming out of the evaporator, the system adjusts for heat in the cabin. A hot cabin causes the entire refrigerant to boil and creates a high pressure, and the valve opens more to adapt. A cooler cabin generates an incomplete change of state (not all the refrigerant boils), keeping the pressure lower, and the valve closes to reduce the amount of refrigerant in the evaporator. Additionally, as the cabin temperature is cooled by the air conditioner, the valve can vary its output by changing the amount of refrigerant passing through, and improve the boiling process.

Restriction Operation

The restriction lowers the pressure and increases the velocity of the refrigerant. The refrigerant needs pressure in order to be moved through the tight fit of the restriction. The refrigerant

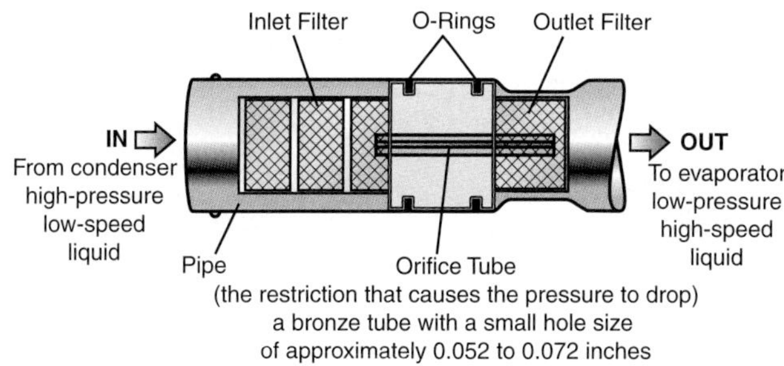

FIGURE 58-12 Pressure drop and speed increase through a restriction.

going through the restriction essentially "uses up" its own pressure to do this, so when it emerges on the other side of the restriction, it is at a lower pressure (**FIGURE 58-12**). Lowering the pressure of the refrigerant also lowers the boiling point of the refrigerant. The boiling point of the refrigerant has to be below the passenger compartment air temperature but slightly above freezing for the heat to be absorbed by the boiling refrigerant in the evaporator. The lower-pressure liquid exiting the restriction is in exactly the correct state and pressure for the evaporator.

Pressure Switches

K58005

High pressures and low pressures can cause severe damage to air-conditioning system components. So pressure switches are used to engage and disengage the compressor clutch. **High-pressure switches** are mounted in the high-pressure side of the air-conditioning system and have their switch contacts in series with the air-conditioning compressor clutch circuit. If the pressure exceeds the switch setting, the switch will open the circuit and the compressor will stop circulating refrigerant. On some electronic HVAC systems, the high-pressure switch or transducer sends a signal to the HVAC module, and the module reduces the output of a variable displacement compressor and/or turns off the compressor clutch by turning off the clutch relay.

Low-pressure switches are mounted in the low-pressure side of the air-conditioning system and are also connected in series with the air-conditioning compressor clutch. Low pressures can occur in air-conditioning systems when the refrigerant has escaped. With little or no refrigerant flowing, lubricating oil flow is also reduced, which can damage the compressor. Low pressure causes the contacts in the low-pressure switch to open and break the electrical supply to the air-conditioning compressor clutch, stopping its operation. Like high-pressure switches or transducers on some electronic HVAC systems, the low pressures can be input to the control module to turn the system off. When this happens, most systems are programmed to output a diagnostic trouble code (DTC), which can be accessed with the proper scan tool.

System Wiring

Early air-conditioning systems did not have much in the way of electrical systems (**FIGURE 58-13**). A power and ground wire to engage the clutch was necessary, and many of the wires had to

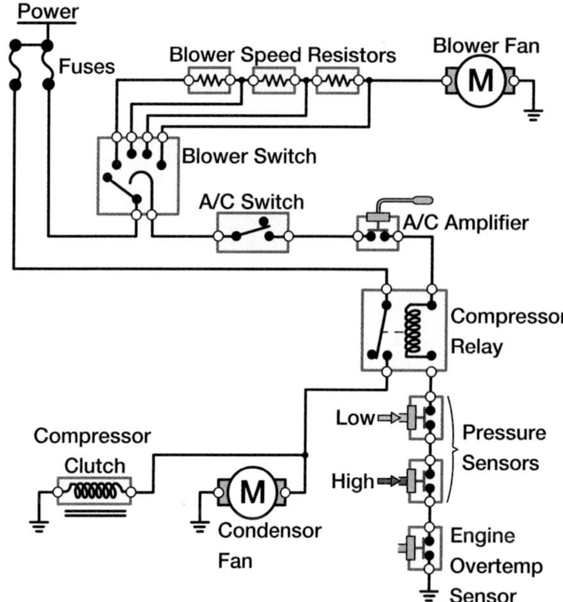

FIGURE 58-13 Complete wiring diagram for a simple mechanical AC system.

pass through a low- and/or high-pressure switch placed in the appropriate line on the system. If the pressure got too low or too high, the switch would open, and no electricity could flow to the clutch. These were safety devices to keep the system from destroying itself. On many systems, the low-pressure switch would open and close often, causing the clutch to cycle on and off. This was used as a control device to keep the pressures where they needed to be. The high-pressure switch was also used to control the condenser fan. The other wiring necessary was for the blower motor in the vehicle. The blower controlled the speed of the air blowing from the vents and was controlled by a switch on the dash and a resistor block designed to control fan speed. Many of the new systems have a lot more wiring to accomplish these and other tasks with relays that are controlled by computers. Refer to the Electronic Climate Control chapter for further information about those systems.

Hoses, Pipes, and Lines

K58006

The terms "pipe" and "line" are used to describe metal tubes. The air-conditioning system and the cooling system do not share any hoses or pipes. The two systems are separated from each other, although many of the hoses and pipes used for air-conditioning and heating run to the same locations and may even look similar. Hoses and pipes are used to carry the refrigerant, both liquid and gas, from component to component. Metal pipes are used for their strength, particularly where no give or movement is necessary. Hoses are used around components that move, such as the compressor, which moves as the engine rocks on its motor mounts (**FIGURE 58-14**). If pipes were used in such applications, they would eventually break from the stress of the movement. However, hoses are not as strong as pipes and tend to be porous.

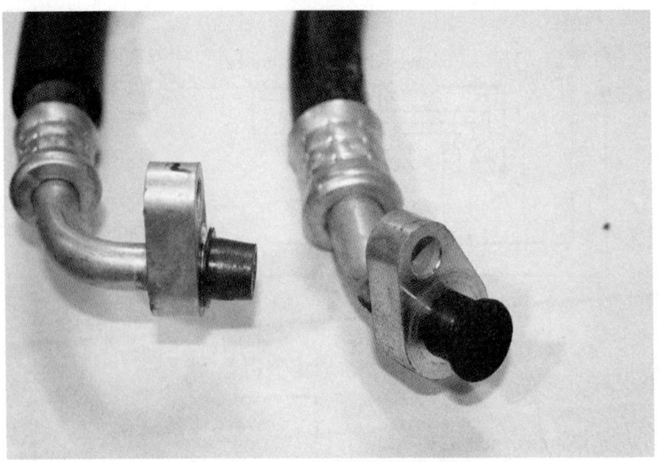

FIGURE 58-14 Flexible hoses allow for small amounts of engine movement and vibration during operation.

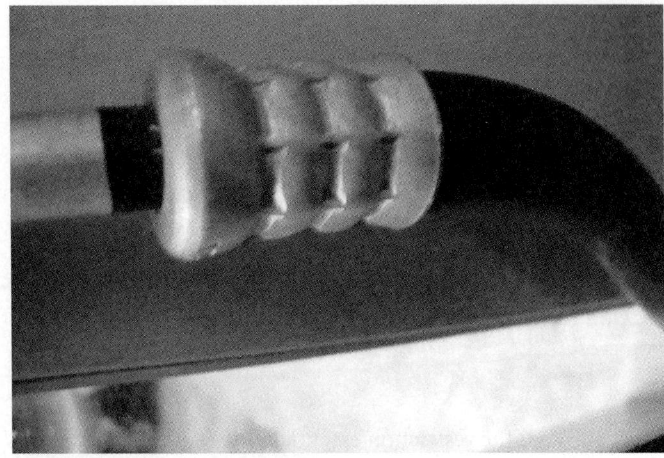

FIGURE 58-15 Hose connections.

The hoses and pipes in an air-conditioning system must be capable of withstanding the high pressures and temperatures found inside the system and must be able to flex and withstand the vibrations and movement of the engine. Air-conditioning hoses must meet SAE (Society of Automotive Engineers) standard J2064. SAE J2064 states that all air-conditioning hoses must be a **barrier-type hose**, which is constructed with a nylon inner core to prevent moisture **ingression** (movement into the system) and permeation of refrigerant from the air-conditioning system. In particular, R-134a has a smaller molecular structure that can pass through smaller holes than R-12 could, so barrier-type hoses are required to keep the R-134a within the closed air-conditioning system. In other words, it must prevent moisture from entering the air-conditioning system through the hose material, and it must prevent refrigerant from leaking through the hose material. Nylon is effective at preventing both of these occurrences.

The hose connections are secured to the hose using special unions known as crimps that must be carefully assembled to avoid leaks (**FIGURE 58-15**). Special crimping tools or machines are needed to correctly crimp a metal pipe to a hose. Most R-134a crimps are a bubble type, named for how they appear when done properly. The metal pipe is slid into the hose, and the crimp is applied over the hose to squeeze it tightly against the pipe. Typically, shops purchase these assemblies already made, as the equipment to do it is expensive and the process must be precisely done by skilled technicians.

TECHNICIAN TIP

The **double crimping** of the hose to fitting ends will cause leaks. Double crimping occurs when the technician over-crimps the fitting in the hose press in an attempt to get a better seal. The best practice for hose crimping is to use the proper dies and hose/fittings for the job.

Pipes are rigid tubes of aluminum or steel often found connecting components that are fixed to the body of the vehicle (**FIGURE 58-16**). Due to the dense makeup of metal pipes, they are less prone to develop leaks. All pipes must have properly

FIGURE 58-16 Typical AC pipes on a vehicle.

formed bends and must be secured to the vehicle body to prevent fatigue cracks or holes made by abrasion.

Hose Size

Air-conditioning hoses are sized by the size of the tube they are designed to fit over. This means that the inside diameter of the hose is slightly smaller than the fitting, allowing a press fit between the hose and fitting. Standard air-conditioning hose is designed to fit over 5/16", 3/8", 1/2", and 5/8" fittings and tubes. The dash system is used by the hydraulic industry, but the automotive industry has applied a number system. This system refers to a hose by how many 16ths of an inch the fitting is that the hose fits over. So a 5/16" hose is referred to as a #5; a 3/8" (6/16ths) hose as a #6; a 1/2" (8/16ths) hose as a #8; and a 5/8" (10/16ths) hose as a #10.

TECHNICIAN TIP

Even though the air-conditioning hoses are built to withstand high pressures and temperatures, they can develop leaks, especially around the fittings or anywhere the hose is repeatedly flexed or chafed. Inspecting hoses and fittings is a critical task during air-conditioning service.

Schrader Valves

Schrader valves are used on many systems both to hold pressure and to allow a gauge or piece of equipment to be attached to the system. For example, the air in the tires is held in, measured, and inflated at the tire's Schrader valve. Schrader valves are placed on both the low and the high side of an air-conditioning system to allow an entry point for checking pressures. Schrader valves are placed in the ports and designed to accept either the low or high side hose from the air-conditioning servicing equipment. When the gauge fittings are inserted, they push down the Schrader valve to allow pressure to be released into the gauge hoses. When the gauge fittings are removed, spring pressure pushes the Schrader valve closed to stop the release of refrigerant (**FIGURE 58-17**). Some Schrader valves are removable with the use of a Schrader valve tool. Once the system pressure has been released, the tool is used to unscrew the Schrader valve and check the O-ring and replace the component, if necessary. Sometimes the valve is not replaceable, and the whole line containing the Schrader valve must be replaced.

> **TECHNICIAN TIP**
>
> Service port caps are used to cap off the service ports for two reasons: (1) to keep dirt and contaminants from collecting in the service port and (2) to provide a secondary seal for the Schrader valve to prevent the loss of refrigerant from the valve. They should always be tightened properly. Also, some repair facilities use tamper-proof caps that cannot be removed without destroying the strap that holds the cap in place. These caps are used to prevent someone from tampering with the air-conditioning system; typically any warranty becomes void if they are broken. So be aware of this if you encounter tamper-proof service caps.

O-Rings

O-rings are an essential part of the air-conditioning system (**FIGURE 58-18**). O-rings are rubber circles that come in many sizes and types of rubber depending on what they are being used to seal. They are placed on the male end of a fitting. The male end fits into the female end, and the O-ring seals between the inside of the female fitting and the outside of the male fitting to keep refrigerant from leaking around the connection. Most R-134a systems use three O-rings in the male pipe to create a good seal. All O-rings must be lubricated with the proper refrigerant oil when installing.

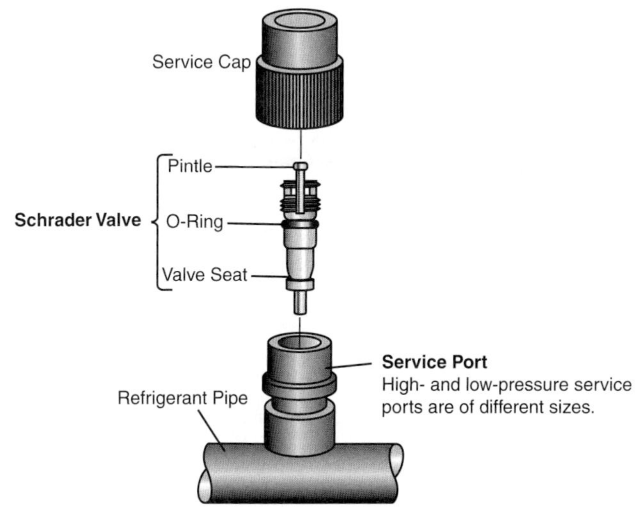

FIGURE 58-17 Air-conditioning Schrader valve.

FIGURE 58-18 O-rings. **A.** R-12-compatible O-rings (nitrile butadiene rubber [NBR]). **B.** R-134a-compatible O-rings (hydrogenated nitrile butadiene rubber [H-NBR]).

> **TECHNICIAN TIP**
>
> Installing the wrong size and type of O-ring can cause problems with achieving the proper vacuum levels while evacuating the air-conditioning system and could cause refrigerant leaks from the air-conditioning system after the vehicle leaves the shop.

Automotive air-conditioning O-rings are made from special materials. Older air-conditioning systems used a **nitrile butadiene rubber (NBR)** compound for the O-rings. This material is not as flexible as pure rubber, but is designed to be compatible with the chemicals in R-12 and mineral oils.

O-rings are also made out of **hydrogenated nitrile butadiene rubber (H-NBR)**. H-NBR was designed to be compatible with the chemical structure of R-134a and PAG oils. The change was required by the federal government in 1992. Like NBR, it is not as flexible as pure rubber. Unlike NBR O-rings, H-NBR O-rings are compatible with R-134a.

> **TECHNICIAN TIP**
>
> When installing O-rings, it is good practice to never use an old O-ring. Using an old O-ring could cause a leak in the system, and the customer to be unsatisfied with your work.

Driers

K58007

Driers, called a **desiccant**, are placed in the receiver dryer or the accumulator (**FIGURE 58-19**). The desiccant is held in a bag and serves as a drying agent to absorb any moisture in the system. The desiccant can become oversaturated if a system is left open and large amounts of moist air enter. If the system has been open for a period of time, then the receiver dryer or accumulator will have to be replaced.

> **TECHNICIAN TIP**
>
> The most common refrigerant desiccants are **XH-7** and **XH-9**, which are crystalline alkali metal alumina-silicate in beaded form. Both are used in R-134a units, but XH-7 is designed for pure R-134a, whereas XH-9 is designed for systems that may not be pure. American-made vehicles use XH-7 because the R-134a is regulated, but many imports do not have the same regulations and may use XH-9.

Sight Glass

The sight glass was an important inspection device on R-12 systems. The sight glass was a glass portal that allowed you to look into the air-conditioning system and see the liquid refrigerant flow (**FIGURE 58-20**). If there were bubbles in the liquid, then the system was low. If no bubbles were seen, then the system was either full or empty (because an empty system would not have any liquid for bubbles to be suspended in). The sight glass is still on some systems but should not be used with R-134a. Because of the chemical makeup of R-134a, some small bubbles are always present. In fact, the appearance of no bubbles in R-134a indicates that the system is overfilled. Because R-12 is no longer used on new vehicles, the sight glass is no longer considered a valid test.

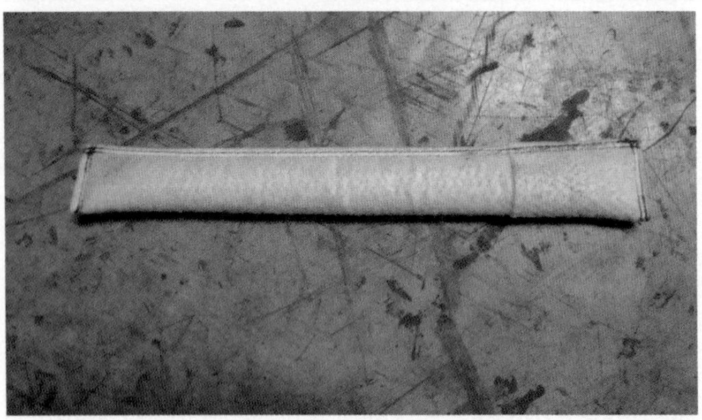

FIGURE 58-19 Air-conditioning system desiccant bag.

FIGURE 58-20 A typical sight glass allowed a visual inspection of refrigerant flow.

Compressor

K58008

The compressor is the heart of the air-conditioning system. It circulates the refrigerant through the system, and it also creates the pressure used to change the refrigerant from a low-pressure gas to a high-pressure gas so the boiling point of the refrigerant is brought above ambient temperature. The pressure from the compressor also keeps the refrigerant moving through the tight restriction of the metering device. There are three general types of compressors: piston, scroll, and rotary vane. Although there may be differences in components, the purpose is the same: to pressurize and circulate refrigerant.

The **axial piston compressor** is one of the most common types of compressor (**FIGURE 58-21**). Cylinders are placed around the drive shaft and parallel to its axis. Each cylinder has a pair of single-ended pistons facing apart, allowing a separate

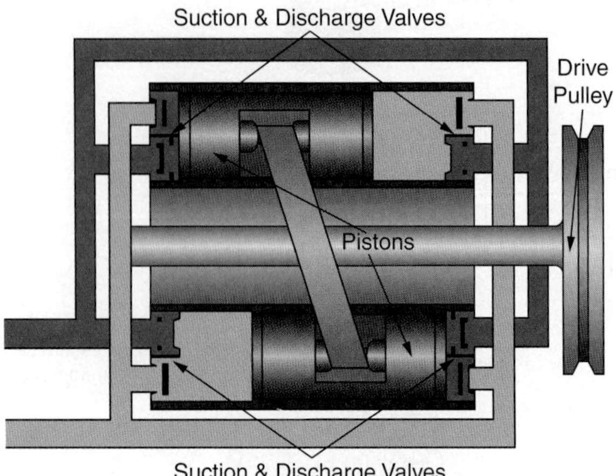

FIGURE 58-21 An axial piston compressor.

pumping chamber at each end of the cylinder. and each pumping chamber has a set of suction and discharge reed valves. The suction reed valves are connected to the inlet port of the compressor by machined passageways. The discharge valves are connected to the discharge port, also by machined passageways. A swash plate is attached to the drive plate at an angle and is positioned between the two rows of opposing pistons. Rotation of the drive shaft causes the outer edge of the swash plate to constantly change its linear position, which moves the pistons back and forth in their respective cylinders.

The pistons are connected to the outer edge of the **swash plate** by swiveling ball joints. The action of the swash plate and ball joints continually moves the pistons back and forth in the cylinder. As the volume above a piston increases, refrigerant is drawn into the cylinder through the **suction reed valves**. When the volume above a piston reduces, the refrigerant is forced out of the cylinder through the **discharge reed valves**.

Compressor Operation

The compressor circulates refrigerant through the air-conditioning system by taking low-pressure gas from the evaporator and moving the gas to the high-pressure side of the system. When the piston is drawn away from the compressor head, the pumping chamber volume increases, and the **inlet reed valve** allows low-pressure refrigerant to enter the pumping chamber. As the piston moves down to the bottom of the chamber, the cavity fills with as much vaporized refrigerant as the chamber can hold at low side pressures. After the piston reaches the bottom of its stroke, it begins to move up toward the head of the compressor. The low side reed valve is forced shut due to the pressure increase as the piston begins to move up in the bore of the cylinder. As the chamber size becomes smaller, the pressure of the refrigerant begins to rise until the high side reed valve opens. The high side reed valve is pushed off its seat because the pressure of the refrigerant in the chamber becomes slightly higher than the pressure in the high side of the system. Compression of the refrigerant also raises its temperature, and hot high-pressure gas is created in the compressor chamber (**FIGURE 58-22**).

▶ TECHNICIAN TIP

The high pressure and volume of gas leaving the compressor is what causes the movement of refrigerant in the air-conditioning system. Increasing or decreasing the revolutions per minute of the pumping chamber can change how fast or slow the refrigerant moves in the air-conditioning system. When testing air-conditioning systems, maintaining a constant rpm can help you isolate the compressor from other components that also affect pressure and temperature.

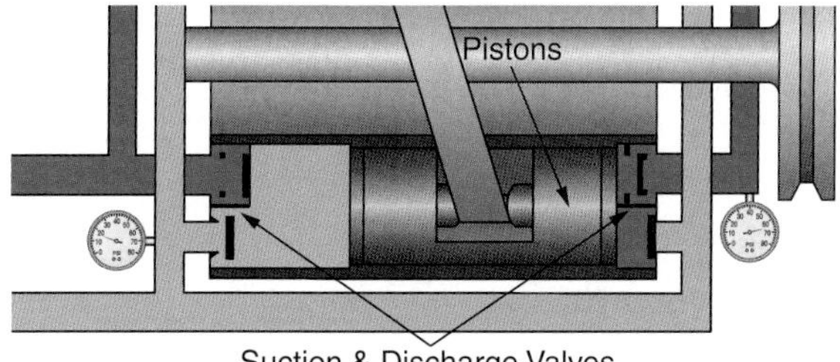

FIGURE 58-22 Compressor operation.

Temperature on the high side of the system will rise 10 to 15 times that of the low side due to the increase in pressure on the refrigerant. The hot high-pressure gas is then forced out of the pumping chamber into a hose connected to the compressor.

Sanden Wobble Plate Compressor

The **Sanden wobble plate compressor** can be found on many different vehicle types. Sanden is one of the major compressor manufacturers, but it is not the only company to manufacture a wobble plate style. The choice of a wobble plate compressor, Sanden's or any other manufacturer's version, is up to the vehicle's designer.

The wobble plate compressor works in much the same way as the axial piston compressor except that it has only a single row of pistons instead of two opposing rows of pistons. Each cylinder has one suction valve (reed valve) and one discharge valve (reed valve). Pistons moving away from the suction valve create a low-pressure area, which draws refrigerant through the suction valve into the cylinder from the low-pressure side of the system. As each piston moves toward the discharge valve, refrigerant is pumped out of the cylinder, through the discharge valve, into the high side of the system.

The Sanden wobble plate compressor is comprised of many moving parts. The wobble plate is connected to the drive shaft of the compressor and is fixed on an angle. As the wobble plate rotates, the cam rotor is forced to move the connecting rods, which are connected between the cam rotor and the pistons in the compressor. The linear movement of the connecting rod moves the piston in a reciprocating (back-and-forth) motion. Each piston has a suction reed valve and a discharge reed valve that allow refrigerant to enter and exit from each chamber.

Rotary Vane Compressor

Rotary vane compressors are used as air-conditioning compressors as well as transmission oil pumps and power steering pumps (**FIGURE 58-23**). Each one works on the same principles. A cylindrical vane cartridge is connected to the drive shaft and rotates within a cylinder. The vane cartridge typically has three or four machined slots that carry spring-loaded sliding vanes. The vanes are free to move in their slots so that each vane rides up against the sides of the cylinder. The vane cartridge is offset in the cylinder so that the vanes create an expanding volume in one half of the cylinder and a decreasing volume in the other half. Rotation of the vane cartridge causes the vanes to move in and out of their slots as the vanes move from the suction side to the discharge side of the cylinder. As the volume increases between the vanes, refrigerant is drawn into the cylinder; and as the volume decreases between the vanes, refrigerant is discharged from the cylinder. This style of pump may not need valves to prevent backflow because the flow is continuous as long as the compressor is turning. There may be a reed valve installed so that the compressor doesn't spin backwards when the compressor clutch is disengaged. As with any compressor, it depends on the vehicle's designers to determine which type of compressor to use.

Sanden Scroll-Type Compressor

The **Sanden scroll-type compressor** uses two inter-woven scrolls (**FIGURE 58-24**). One scroll is fixed and the other oscillates. The oscillating scroll is connected to the drive shaft, which is rotated by the compressor clutch. As the oscillating scroll is rotated around the stationary scroll, the low-pressure gas is drawn into the low side of the compressor pump. The oscillating scroll forces the gas to compress and flow to the center of the scroll. The outlet port is in the center of the scrolls. The gas is then forced from the outlet port to the high side of the system.

Variable Displacement Compressor

Many manufacturers are now using variable displacement air-conditioning compressors on their vehicles. Variable displacement means that the compressor can change the amount of refrigerant it moves from near 0% up to 100%. This variability allows some compressors to do away with the compressor clutch, making it driven all of the time. Most variable compressors are the axial piston or wobble plate style. Variable displacement is accomplished by changing the angle on the swash plate or wobble plate, which lengthens or shortens the stroke of the pistons. The swash plate or wobble plate angle can be adjusted by internal refrigerant pressures pushing against a spring to maintain proper discharge pressure. Commonly, in newer vehicles the angle of the plate is controlled electronically, making for very accurate control of compressor discharge (**FIGURE 58-25**).

Air-Conditioning Compressor Clutch

The **air-conditioning compressor clutch** is an electromagnetic device that uses a metal plate with no friction material to engage and disengage the compressor from the **drive pulley** (**FIGURE 58-26**). When the clutch is disengaged (not connected), the pulley attached via a belt to the crankshaft can spin freely. The clutch assembly uses metal-to-metal friction that allows the components to grab each other when engaged. An electric clutch coil controls the clamping action by engaging and disengaging the clutch assembly. When electricity flows through the clutch coil windings, a magnetic field is created.

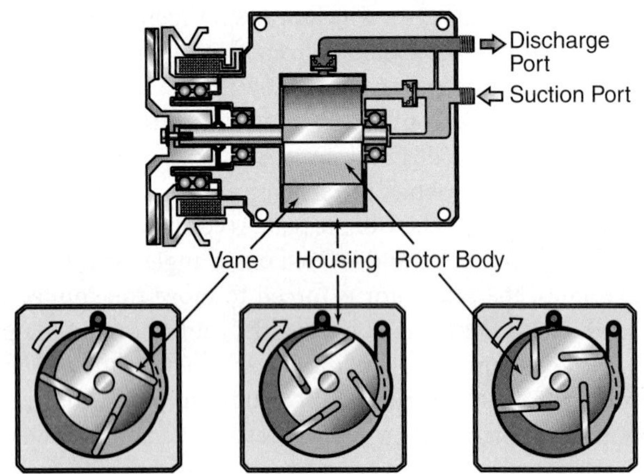

FIGURE 58-23 Rotary vane compressors.

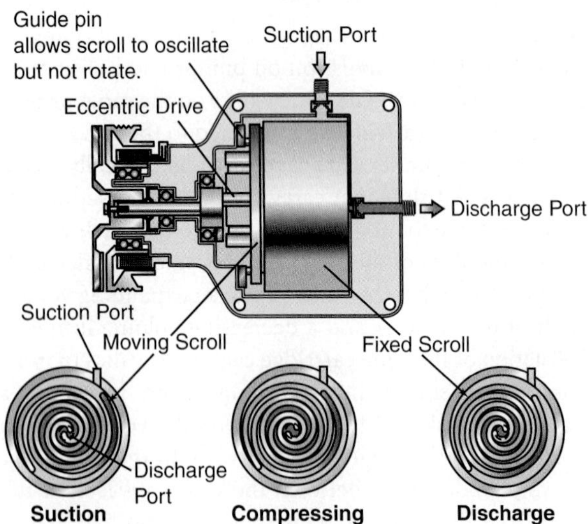

FIGURE 58-24 Sanden scroll-type air-conditioning compressor and clutch.

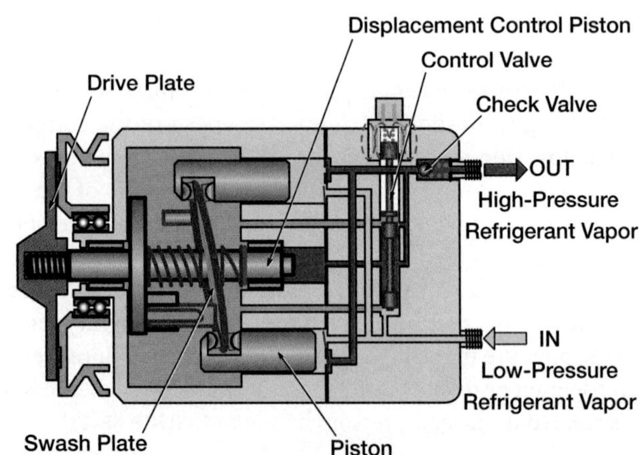

FIGURE 58-25 Variable displacement compressors can change the amount of refrigerant moved from near 0% up to 100%.

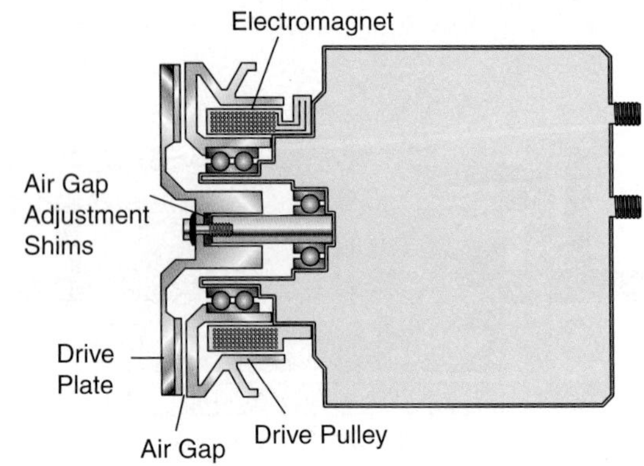

FIGURE 58-26 A compressor clutch.

The magnetic field pulls the metal clutch plate into the spinning pulley and drives the compressor.

The resistance value of the coils determines the amount of electrical current flow through the windings. The lower the resistance value, the more current that can flow. For electromagnets to work there must be relatively low resistance so that enough current can flow through the windings to create a strong magnetic field. Resistance is measured in ohms. The **clutch electrical winding** typically has a resistance value of 3–5 ohms on a 12-volt DC air-conditioning compressor. The electromagnetic clutch is energized by relays, switches, or modules (**FIGURE 58-27**). When the magnetic field is energized, the metal spring-loaded clutch plate is forced to engage the drive pulley being driven by the engine. This engagement drives the compressor and causes refrigerant to move through the air-conditioning system.

Though the air-conditioning compressor clutch is not serviced separately in most air-conditioning repair shops, checking its clearance is important when installing a new compressor. If the gap is too large, the magnet will not be strong enough to pull it in. If the gap is too small, the clutch plate may rub against the pulley when it is not engaged and overheat the clutch plate surfaces, resulting in component failure. The air-conditioning compressor clutch clearance can vary from manufacturer to manufacturer. As a rule of thumb, air-conditioning compressor clutch clearance is typically between 0.015" and 0.025" (between 0.38 and 0.64 mm), but always check the manufacturer's service information.

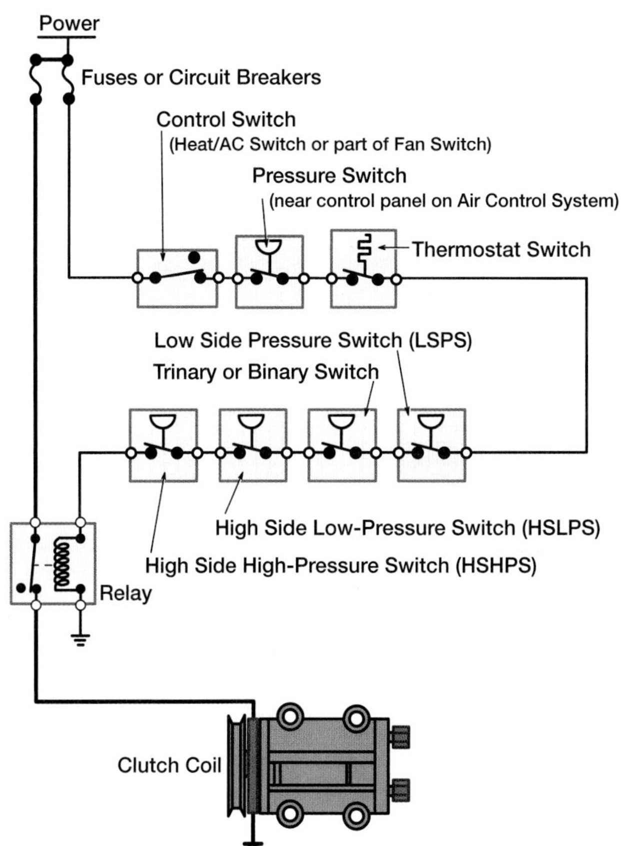

FIGURE 58-27 Typical compressor clutch wiring diagram.

Compressor Clutch Components

The compressor clutch assembly is made up of several components (**FIGURE 58-28**). The pulley rides on a bearing located on the compressor body. The pulley bearing is replaceable and is usually retained by a snap ring. The windings of the clutch coil are mounted behind the pulley and sometimes called the stator. The clutch coil is a replaceable part in many compressors. The clutch plate consists of a metal disc connected to the drive hub by flexible struts. The clutch hub is typically splined and bolted to the compressor shaft. On most compressors, a shim located under the clutch hub allows the technician to adjust the air gap between the two parts of the clutch (the clutch plate and the pulley). If this adjustment is not correct, one of two things may result: The clutch will slip and burn up the friction surface (resulting eventually in no compressor operation), or the compressor will drag when turned off. Be sure to check the service information when replacing a clutch on an air-conditioning compressor.

Muffler

The air-conditioning system may use mufflers to dampen pulsations and noise from the compressor. The muffler is constructed of a hollow cylinder with baffling (**FIGURE 58-29**). As the gaseous refrigerant enters or exits the compressor, the compressor

FIGURE 58-28 Typical compressor clutch components.

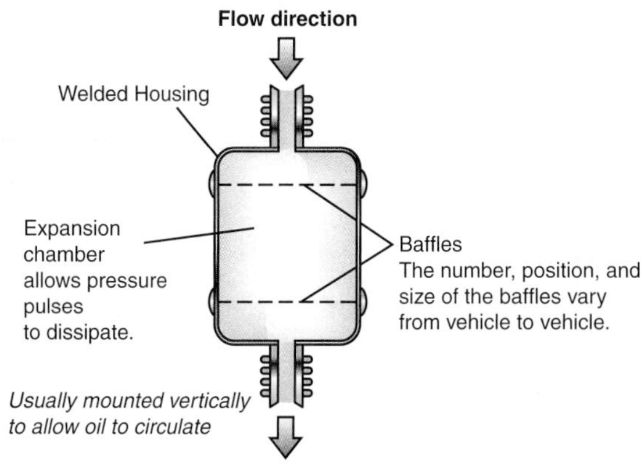

FIGURE 58-29 The muffler is constructed of a hollow cylinder with baffling.

causes pulsations in the gas, which cause unwanted harmonics. The muffler dissipates the pressure pulsations by allowing the gaseous refrigerant in the system to expand into the larger, baffle-lined chamber. This prevents the pulsations from affecting the rest of the air-conditioning system.

Condenser

K58009

The condenser is normally positioned in front of the radiator, where it is exposed to maximum **ram airflow** when the vehicle is in motion (**FIGURE 58-30**). This airflow is assisted by the engine cooling fan or by one or more auxiliary electric fans that create airflow when the vehicle is stationary or at lower speeds. The condenser is usually made with aluminum or copper tubes meshed with a compact series of fins surrounding it. The fins increase the surface area available to maximize the amount of heat transferred from the refrigerant to the outside air. The condenser is a one-piece unit with no serviceable parts.

Applied Science

AS-17: Pressure: The technician can measure pressures in hydraulic or pneumatic systems and compare them to the manufacturer's specifications.

A crucial step in diagnosing air-conditioning system failures is the measurement of the refrigerant pressures. Pressures are measured in two separate places in the system: High side pressure is measured on the discharge side of the compressor, and low side pressure is measured on the suction side of the compressor. Manufacturer service manuals will outline system pressure specifications and the operating requirements to measure them. Some potential scenarios illustrating problems are listed here:

If pressure readings are low on both sides of the system, the most common cause is a low refrigerant charge due to a leak, or leaks, in the system, allowing the loss of refrigerant.

If the pressure readings are the same on both sides of the system, the likelihood is that the compressor is not working.

If the high side pressure is extremely high or the low side pressure is showing vacuum, suspect an internal blockage in the system.

FIGURE 58-30 A typical condenser.

Condenser Operation

The condenser transforms the hot high-pressure gaseous refrigerant on the high side of the system back into a cooler, high-pressure liquid. Gaseous refrigerant enters the top of the condenser. As it passes through the condenser, the airflow across the fins cools the refrigerant, transforming it back into a liquid. Liquid refrigerant leaves the bottom of the condenser.

The gaseous refrigerant in the condenser is at a higher temperature than the outside air flowing across the condenser coils. Putting the refrigerant under pressure raises the boiling point because it now takes more vapor pressure (created by heat) to overcome the pressure in the system. The greater the pressure, the higher the boiling point. The air crossing the condenser is now cooler than the boiling point of the refrigerant. The cool air absorbs the heat of the refrigerant, changing the refrigerant back into a liquid. It also removes a large amount of heat from the refrigerant in doing so.

The refrigerant is able to condense because the pressure of the compressor increased the boiling point of the refrigerant; thus, the ambient air flowing across the condenser causes the refrigerant to condense into a liquid. Just so you understand the difference, relatively cool air from the passenger cab causes the refrigerant to boil, and relatively warm ambient air causes the refrigerant to condense. The only means by which this can happen is the manipulation of the boiling point (pressures) in the system. In other words, because of the pressure drop on the refrigerant, cool cabin air causes the refrigerant to boil. And once the pressure is raised, relatively warm ambient air will cause the refrigerant to condense. Manipulating the pressure on the refrigerant lowers or raises the temperature at which the refrigerant boils and condenses.

Condensers are made in one of two designs: serpentine and parallel flow (**FIGURE 58-31**). **Serpentine condensers** are constructed from a single tube. The tube snakes back and forth

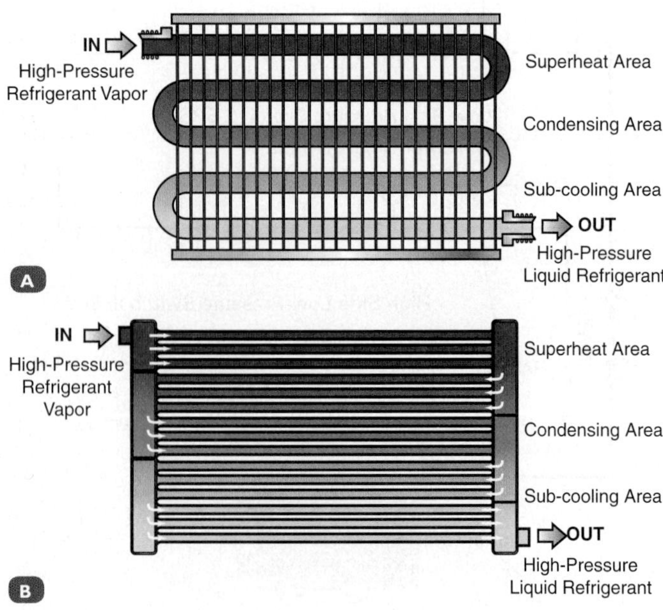

FIGURE 58-31 Condensers. **A.** Serpentine condenser. **B.** Parallel flow condenser.

across the condenser. All of the refrigerant flows along the same path throughout the single tube. This design causes the refrigerant to take a long path from the inlet to the outlet, giving a lot of time for it to transfer its heat to the atmosphere. However, less refrigerant can flow through the single tube, so its efficiency is not high.

Parallel flow condensers are constructed with multiple tubes in parallel with each other and flow refrigerant more like a radiator for engine coolant. Parallel flow condensers allow more refrigerant to flow, but it flows at a slower speed because there are more tubes. By using a parallel flow condenser, the efficiency of the air-conditioning system can be increased by up to 25%, thus allowing the system to be more compact, lightweight, and fuel efficient.

Condenser Fan

The condenser fan is designed to pull or push air from in front of the vehicle through the condenser to remove heat and reduce the high side pressure (**FIGURE 58-32**). In older models, the condenser fan ran off of the belt drive on the front of the engine and was mechanically spun by the speed of the pulley. These systems have been almost completely replaced by newer, higher efficiency, more controllable electric fans. The electric fans are controlled by relays, which turn them on and off as conditions dictate (**FIGURE 58-33**). Turning the air-conditioning system on in many vehicles commands the fan to run. High side pressure switches or transducers input their signals to the HVAC module and can also cause the fan(s) to operate. Many systems vary the speed and number of fans that operate.

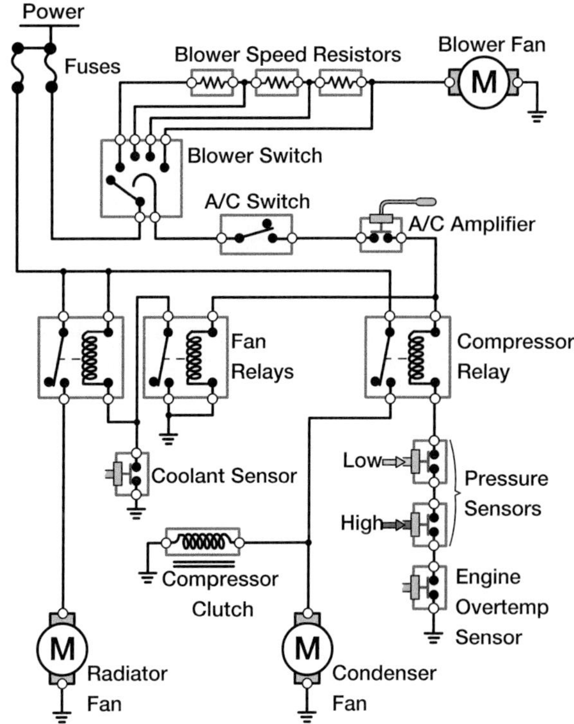

FIGURE 58-33 Typical condenser fan wiring diagram.

Evaporator

The evaporator does not really have components. It is a one-piece unit with no serviceable parts. Like the condenser, it has fins to increase the surface area of the unit and small tubes through which refrigerant flows. It also has an inlet and an outlet pipe through which it connects to the rest of the system. Refrigerant enters the evaporator at the bottom, and primarily vapor exits out the top. It looks similar to a heater core.

The evaporator is usually mounted in the passenger compartment and is positioned so that air from the passenger compartment can be passed across its external surface (**FIGURE 58-34**). It is normally constructed of copper or aluminum tubes with fins attached to increase the surface area. The greater the surface area, the more heat transfer from convection (air) to conduction (evaporator tubes) to convection (refrigerant). More fins increase the surface area of the evaporator and allow a greater area for the air to transfer the energy to the evaporator tubes. This maximizes the rate of heat removal from the air.

Evaporator Operation

The evaporator transforms liquid refrigerant to a gas. The transformation from liquid to gas begins when the air from the passenger compartment passes over the evaporator tubes and fins. The reduced low side pressure (typically held between 20 and 40 psi [138–276 kPa]) from the restriction reduces the

FIGURE 58-32 Condenser fans.

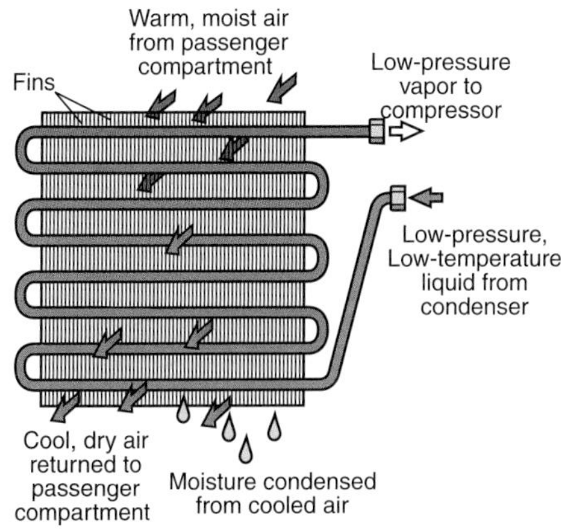

FIGURE 58-34 An evaporator.

refrigerant's boiling point to just above 32°F (0°C). Because of the low boiling point, the warm air inside the cabin is absorbed by the evaporating refrigerant, transforming it from liquid to gas. This heat comes from the cabin air flowing over the evaporator.

The vaporization of the refrigerant is forced to happen at a specific temperature by manipulating the pressure on a liquid that normally boils at negative temperatures. Any heat source will amplify the boiling process. By moving air across the evaporator tubes, the convection currents pass over the evaporator tubes, causing conduction of heat transfer to the refrigerant through the tubes of the evaporator. The conduction of heat through the tubes then provides the needed energy to evaporate the liquid refrigerant to a gas.

The process of transferring heat from the air to the coils and then to the refrigerant is part of the air-conditioning process. After the entire exchange of heat from the air to the refrigerant, the byproduct is what makes the driver and passengers feel cool. The air passing over the evaporator coils is stripped of its heat energy. Some of the moisture and heat are removed. The air flowing from the air ducts is crisp, cool, and low in moisture.

Dual Evaporator System

Some vehicles, such as conversion vans, minivans, and SUVs, use a dual evaporator system, composed of two different units, to cool the vehicle. One unit is placed at the front of the vehicle and the other at the rear (**FIGURE 58-35**). This provides passengers in the rear their own airflow, so they don't have to depend on airflow from the front vents. The front unit is typically placed under the dash, as it usually is, and a rear unit is placed on either side of the back passenger compartment, to hide the unit.

The refrigerant lines for the rear unit run from the front, as it shares the compressor and condenser with the front unit. In many cases, a solenoid valve is used to control refrigerant flow to the rear evaporator. If the rear air-conditioning unit is selected by depressing the control switch, current will flow through the solenoid valve, opening it up so that refrigerant can flow through the rear evaporator. When the rear unit is turned off, the solenoid valve closes. These rear units also have their own blower motor and, in many cases, their own controls.

Types of Refrigerant

K58010

Proper air-conditioning operation requires that all of the components, including the refrigerant, work properly and efficiently. Without proper refrigerant type and amount, or if any one component is not functioning properly, the air conditioner will not cool the passenger compartment effectively. Each component requires that the refrigerant level be correct and that the previous component deliver the refrigerant to it in the correct state. The evaporator requires a low-pressure liquid; the compressor requires a low-pressure gas; the condenser requires a high-pressure gas; and the restriction requires a high-pressure liquid.

There are many types of refrigerants available, but automotive applications generally use one of the three most common: R-12, R-134a, or HFO-1234yf (**FIGURE 58-36**). These refrigerants are not compatible and must not be mixed together. We explore each of these refrigerants next.

Dichlorodifluoromethane (R-12), a member of the chlorofluorocarbon, or CFC, family of gases, was the first common refrigerant to be used in an automotive air conditioner. The first part of the word refers to chlorine, a primary component of CFCs that is banned by the U. S. government in refrigerants because of its detrimental effects on the atmosphere. It is nonflammable, nontoxic, and stable at all temperatures; does not react with aluminum, steel, or copper; and is soluble in mineral oils. It has a boiling point of –21.8°F (–29.9°C). In the early 1990s, R-12 was found to be one of the leading causes of the depletion of the ozone layer and was outlawed for new vehicle installation. Although you will not find R-12 in today's production vehicles, you may still have customers with older vehicles using R-12.

Tetrafluoroethane (R-134a) was the replacement for R-12. It is a hydrofluorocarbon, or HFC. Although R-134a has a boiling point of –15.3°F (–26.3°C), which is nearly the same as R-12's boiling point, R-134a can only be substituted for R-12 (retrofitted) by following a specific procedure. The common refrigerant R-134a uses a polyalkalene glycol (PAG) oil to lubricate the air-conditioning components, and R-12, the less common refrigerant, uses mineral oil. One reason it is so important to always use the correct refrigerant is that PAG oil mixed with

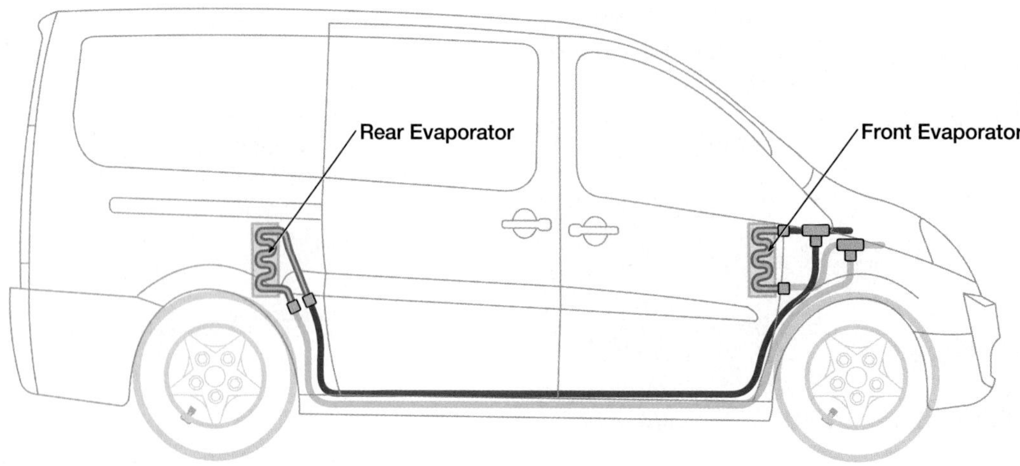

FIGURE 58-35 A typical dual evaporator system.

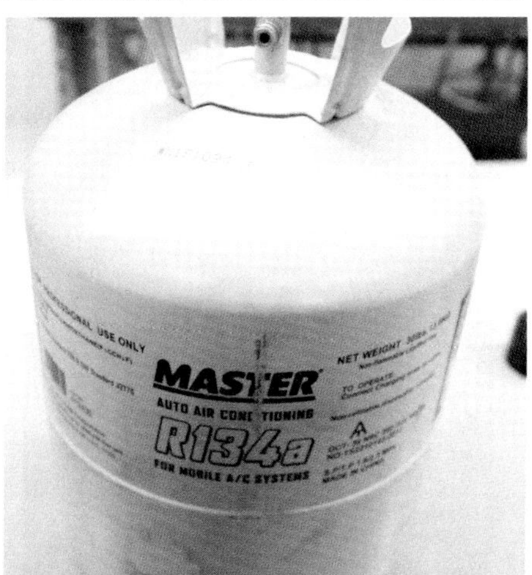

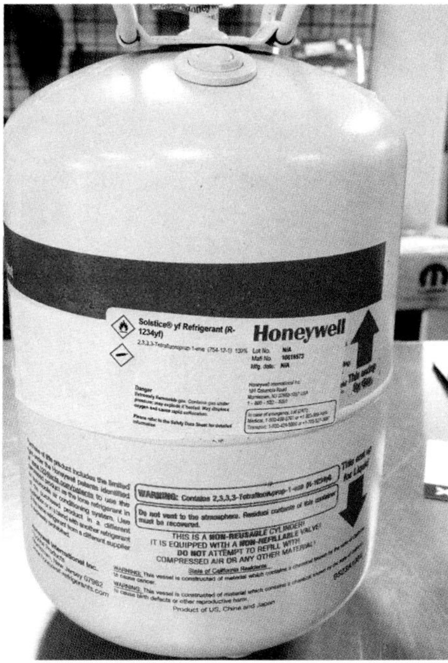

FIGURE 58-36 Types of refrigerant. **A.** R-12. **B.** R-134a. **C.** HFO-1234yf.

mineral oil creates a hazardous gas that can corrode the air-conditioning system. The mixture is also considered a contaminant and must be handled as hazardous waste. To help prevent the wrong lubricant or refrigerant being installed during servicing, the service ports on air-conditioning systems have been changed so that the service equipment for an R-12 system cannot be connected to an R-134a system. Common PAG oils are PAG 46, PAG 100, PAG 133, and PAG 150.

R-134a is nonflammable and is accepted by the automotive world to be safe for automotive air conditioners. R-134a, which is a liquid, boils well below normal room temperatures and in vaporizing will absorb tremendous amounts of heat without increasing its own temperature. Unfortunately, recent discoveries show that R-134a is a major contributor to greenhouse gases, so it is being phased out. Tetrafluoropropene (HFO-1234yf) is now being used in new vehicles.

▶ TECHNICIAN TIP

Identifying refrigerants has become an important part of repairing air-conditioning systems. The accidental mixing of refrigerants or the mixing of different types of refrigerant oil could cause damage to the system and the machines that service air conditioners. Before reclaiming a system, you need to identify the type of refrigerant in the vehicle to prevent the accidental contamination of the refrigerant in your recovering machine.

HFO-1234yf is the latest refrigerant for mobile air-conditioning systems. It has a greatly reduced global warming potential (GWP) of 4, compared to a GWP of 1430 for R-134a refrigerant. Its atmospheric life-time is only 11 days, compared to 13 years for R-134a. It has a boiling point of – 20.2°F (–29°C). It is slightly flammable, but the safety risk has been determined to be significantly less than the risk associated with gasoline. Also, HFO-1234yf is more costly to produce than R-134a, so that will impact customers as their systems need servicing. HFO-1234yf uses a PAG-type oil with different additives on belt-driven compressors and polyolester (POE) oil for electrically driven compressors. HFO-1234yf uses service ports that are different than those used with R-12 or R-134a, so contamination issues can be minimized (**FIGURE 58-37**).

Refrigerant Oils

The oil used in refrigeration systems must be compatible with the refrigerant used. The old refrigerant R-12 used mineral oil. R-134a uses a PAG oil and POE oil is recommended for use with HFO-1234yf (**FIGURE 58-38**). The oil is necessary to keep moving parts in the compressor lubricated and is used on the gaskets and seals to protect and help seal. It must be able to move throughout the system without foaming and be compatible with the refrigerant and the pressure changes taking place. The oil is picked up and carried throughout the system in the refrigerant, so some oil will be found in all the major air-conditioning system components (compressor, evaporator, receiver dryer or accumulator, and condenser). The oil in the compressor is just as important as oil in an engine. Without it, the compressor will overheat and destroy itself.

Mineral oil is clear to light yellow, and PAG oil is usually a light blue color. As the oil picks up dirt or becomes

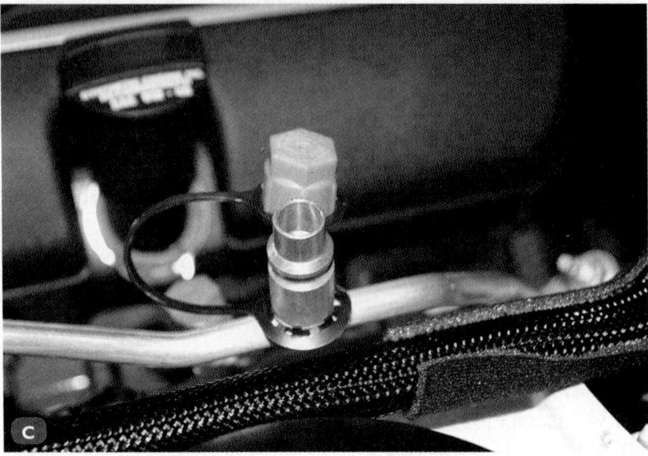

FIGURE 58-37 Service ports. **A.** R-12. **B.** R-134a. **C.** HFO-1234yf.

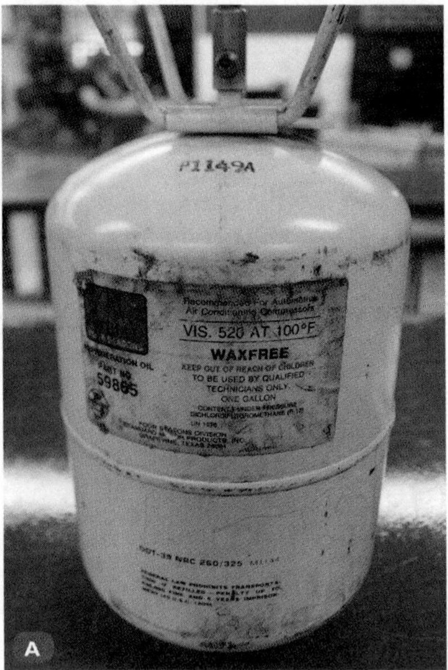

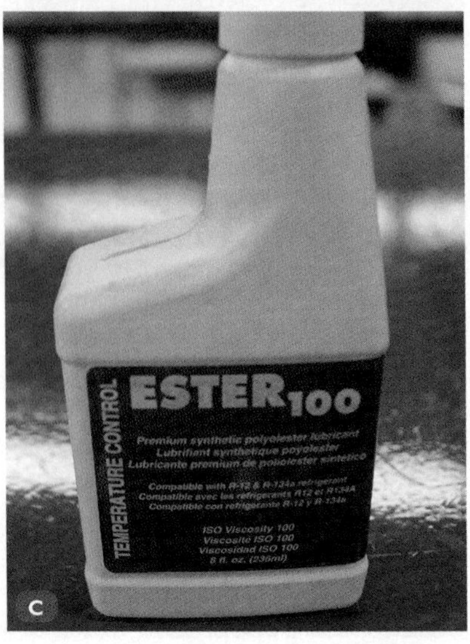

FIGURE 58-38 Refrigerant oils. **A.** Mineral oil. **B.** PAG oil. **C.** POE oil.

contaminated, it will turn brown or black, depending on the level of contamination.

POE oil can sometimes be used with R-134a. It can also be recommended when retrofitting R-134a in R-12 systems because POE oils are not as reactive to small traces of mineral oil residue. Both POE and PAG oils absorb moisture, so be sure to cap the oil bottle whenever you are not pouring it.

▶ Types of Automotive Air-Conditioning Systems

K58011

There are two definitive system types: the **fixed orifice tube system** and the **thermal expansion valve (TXV) system** (**FIGURE 58-39**). The difference is that the TXV is adjustable based on the temperature of the outlet pipe from the evaporator, whereas the fixed orifice tube provides a nonadjustable passage for the refrigerant to pass through. Each type of air-conditioning system contains one subcomponent that is vital to the operation and efficiency of that type of system. The type of restriction being used will determine which subcomponent is required. Fixed orifice tube systems use an accumulator to ensure that a pure gas is delivered to the compressor, and TXV systems use a receiver filter drier to ensure that a pure liquid is delivered to the restriction.

The fixed orifice tube system has the accumulator on the low side line between the evaporator and the compressor. Because this system cannot adapt to temperature changes in the cabin, it is likely that during cooler weather or as the cabin temperature decreases the refrigerant in the evaporator will not fully boil. This means that some liquid refrigerant will still be present as it heads for the compressor. Because liquid cannot be compressed, its presence in the compressor may cause damage from trying to compress a noncompressible substance. The accumulator's job is to prevent liquid from getting into the compressor and causing damage. Liquid and gaseous refrigerant are both able to enter the accumulator. In the accumulator, the liquid falls to the bottom because it is heavier, and the gas rises to the top. The outlet of the accumulator only draws from the top; therefore, it only allows gaseous refrigerant to reach the compressor. The liquid in the bottom is heated by the gas passing through it and heat from the engine compartment, and it boils and rises to be used as a gas.

The TXV system has a receiver dryer in the high side line between the condenser and the TXV. This system can adapt to the temperature changes in the evaporator, so unlike the fixed orifice system, it does not need the safety device for the compressor. The receiver dryer also lets in liquid and gaseous refrigerant, but it actually draws the liquid from the bottom on the outlet side. The receiver stores the accumulated liquid refrigerant in case of a sudden change of temperature in the cabin. If the cabin is cool because the air-conditioning has been on for a while, and then the customer opens a window or door, hot air rushes in, raising the outlet temperature of the evaporator and causing the TXV to rapidly open. This rapid opening stimulates a rush of the extra liquid refrigerant from the receiver dryer to the evaporator to keep it from drawing a vapor and becoming inefficient.

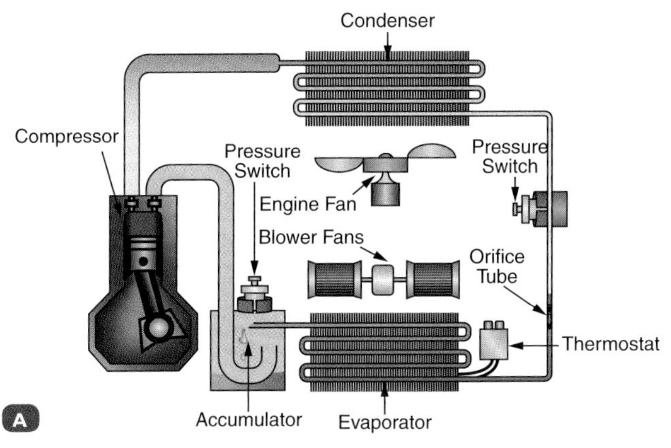

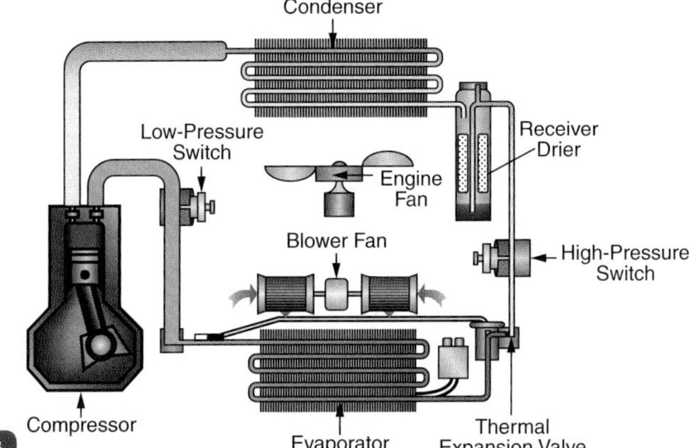

FIGURE 58-39 Types of automotive air-conditioning systems. **A.** Fixed orifice tube system. **B.** Thermal expansion valve system.

Fixed Orifice Tube Air-Conditioning System

K58012

The fixed orifice tube air-conditioning system is a common type of air-conditioning system in the automotive world. The components that make this system unique are the fixed orifice tube, pressure cycling switch, and accumulator. The fixed orifice tube is a metering device that transforms high-pressure liquid into low-pressure liquid. The orifice, or hole, is designed to allow a specific amount of refrigerant through at a designated speed.

Applied Science

AS-19: Fahrenheit/Centigrade: The technician measures system temperatures and converts them to degrees Fahrenheit or degrees Centigrade as required.

A common HVAC-related concern is air-conditioning systems failing to cool effectively. Prior to carrying out any diagnosis, technicians must determine whether the system is operating properly. This is determined by measuring the output temperature at the center vents in the vehicle's dash.

Once the vent temperature has been measured, it should be compared to manufacturer's specifications to determine if the system is operating effectively. Measuring instruments are available reading in either degrees Centigrade or degrees Fahrenheit. The unit of measurement in which specifications are published may vary according to the manufacturer. It may be necessary to convert the measured temperatures from degrees Centigrade to degrees Fahrenheit, or vice versa, to determine if the reading meets specifications.

The pressure cycling switch is normally used in conjunction with the fixed orifice tube. It is installed on the low-pressure line or the accumulator to turn the air-conditioning clutch on and off at certain pressures. The accumulator is placed between the evaporator and the compressor to collect any liquid refrigerant and prevent it from entering the compressor. The accumulator also houses the desiccant.

The fixed orifice tube air-conditioning system is less efficient than the TXV system, as the amount of refrigerant that flows through the restriction cannot be controlled directly. Because it is fixed, it cannot vary the amount of refrigerant flow. It also cannot compensate for engine rpm changes like the TXV system, so refrigerant flows through it continuously while the compressor is running. The only thing that changes the flow is the pressure difference between the high side and the low side. Because the hole of the fixed orifice tube is very small, refrigerant can only pass through it slowly. As the pressure on each side of the tube starts to equalize, meaning the high-pressure side is not too much higher than the low-pressure side, it becomes harder to move the refrigerant through the hole. Once the two pressures are equal, the flow of refrigerant stops. So pressure differential is what controls the amount of cooling that is happening.

Fixed Orifice Tube

The fixed orifice tube is constructed from a plastic outer shell and a small, bronze, fixed-diameter tube with a mesh screen before and after the fixed orifice (**FIGURE 58-40**). The color of the plastic tube is used to identify the diameter of the orifice. For example, one manufacturer uses blue tubes that have an orifice of 0.067" (1.7 mm), red tubes with an orifice of 0.062" (1.6 mm), orange tubes with an orifice of 0.057" (1.4 mm), green tubes with an orifice of 0.052" (1.3 mm), and brown tubes with an orifice of 0.047" (1.2 mm). The different sizes of the orifices are for different flow rates for larger and smaller systems.

The fixed orifice tube is placed in the air-conditioning system between the condenser and the evaporator, near the inlet of the evaporator. A directional arrow on the fixed orifice tubes lets the technician know the proper way the fixed orifice tube should be installed. When removing the fixed orifice tube, note its color, and compare it to specifications. Replace the fixed orifice tube with the proper colored tube, which is usually the color you removed. If a different color is accidently installed, the system will not work properly.

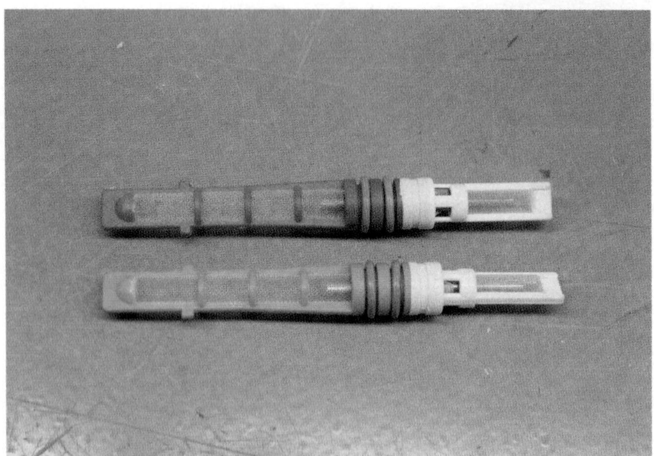

FIGURE 58-40 The fixed orifice tube is constructed from a plastic outer shell and a small, bronze, fixed-diameter tube with a mesh screen before and after the fixed orifice.

Because the fixed orifice tube has no moving parts, flooding the evaporator with liquid is a common occurrence. This happens when the rpm of the engine changes rapidly. Liquid from the evaporator could potentially enter the compressor and cause a seized compressor, which would permanently damage it. This is why fixed orifice tube systems need either a pressure cycling switch and an accumulator or a variable displacement compressor and an accumulator.

> **TECHNICIAN TIP**
>
> When installing the fixed orifice tube, make sure to lube the O-rings on the fixed orifice tube with refrigerant oil, and use the proper orifice tube installation tool. Failure to use caution will lead to breakage of the fixed orifice tube or damage to the O-rings.

Low-Pressure Cycling Switch

The **low-pressure cycling switch** is typically mounted on the accumulator in a fixed orifice tube system. The low-pressure cycling switch also acts as a low-pressure safety switch for the air-conditioning system. If the air-conditioning system loses all refrigerant, the switch will turn off the compressor clutch and possibly save the compressor from failure due to no refrigerant and oil.

The low-pressure cycling switch is a normally open switch that uses system pressure to cycle the compressor on and off during normal operation. The air-conditioning pressure switch is normally open and does not allow electricity to go to the compressor clutch or clutch relay unless low side pressure is between approximately 20 and 46 psi (between 138 and 317 kPa). Below approximately 20 psi (138 kPa), the switch opens. As refrigerant pressure drops on the low side due to the compressor pulling refrigerant from the low side, the low-pressure cycling switch remains closed until the pressure on the low side of the air-conditioning system falls below about 20 psi. Once the pressure falls below 20 psi, the low-pressure cycling switch opens, and the compressor stops pulling refrigerant from the low side. The low-pressure cycling switch will remain open until the pressure rises on the low side of the air-conditioning system to around 46 psi (317 kPa) again. These systems are often referred to as cycling clutch orifice tube (CCOT) systems.

Accumulator

The accumulator is located between the outlet of the evaporator and the inlet of the compressor (**FIGURE 58-41**). It has no serviceable parts, although desiccant is located inside the accumulator. The outlet tube of the accumulator is U-shaped. The gaseous refrigerant rises to the top and is able to exit via the outlet tube. Any liquid refrigerant falls to the bottom of the

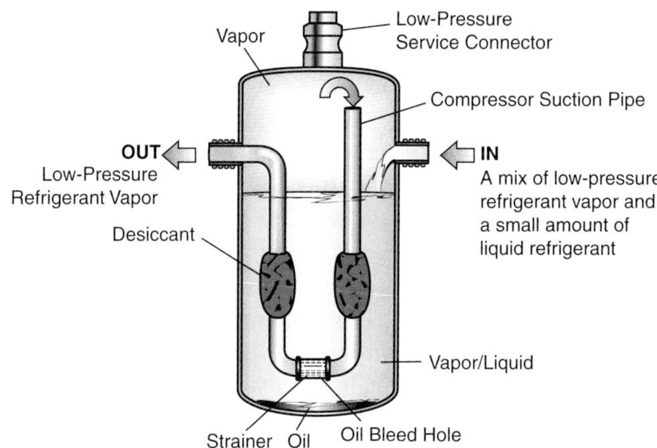

FIGURE 58-41 Accumulator.

accumulator and is heated and converted into a gas as hotter gases pass through the accumulator. The job of the accumulator is to make sure that liquid refrigerant does not enter the compressor. However, some accumulators have a small hole in the bottom of the U-shaped outlet tube so that a small, measured amount of oil can be drawn into the compressor for lubrication purposes.

> **▶ TECHNICIAN TIP**
>
> The outside of the accumulator can frost over when the vehicle comes to a complete stop, because the refrigerant that flooded the evaporator is caught by the accumulator and is boiling off and cooling down, condensing any moisture on the outside of the accumulator. This moisture freezes on the outside of the accumulator, then melts and forms a puddle under the vehicle when it stops, similar to the puddle from the evaporator drain. This puddle can look like the vehicle has a coolant leak, but it is just water that condensed on the accumulator and is nothing to worry about.

Thermal Expansion Valve Air-Conditioning System

K58013

In a TXV air-conditioning system, the components necessary for system operation are the condenser, the evaporator, the compressor, the receiver drier, the TXV, connecting pipes and hoses, a thermostat switch, blower fans, and pressure switches. The TXV is located at the entry to the evaporator and provides a throttling or restricting function to control the quantity of refrigerant entering the evaporator. The TXV controls flow through the use of a passage that can open or close using pressure on a diaphragm. A temperature-sensing bulb, normally containing refrigerant, is attached to the diaphragm via a metal tube. The bulb is attached to the outlet side of the evaporator. The temperature changes in the outlet side cause the refrigerant to expand and contract. As it does so, it adds or removes pressure on the diaphragm to allow it to open and close the valve. As the refrigerant boils completely in the evaporator, it raises the temperature of the outlet pipe. The higher temperature in the pipe raises the pressure in the bulb, which pushes on the diaphragm to open the valve. As the air cools in the passenger compartment, not all the refrigerant passing through the evaporator turns to gas, which means the temperature coming out is cooler, thus lowering the bulb pressure and in turn closing the valve.

The TXV air-conditioning system is charged with refrigerant and a quantity of lubricating oil that circulates with the refrigerant at all times. When the air-conditioning system is switched on, an electromagnetic clutch is energized on the compressor drive plate, and the compressor is driven by the engine crankshaft. Vaporized refrigerant is drawn from the low-pressure side of the system by the intake side of the compressor. It is discharged from the compressor as a high-pressure, high-temperature vapor (compressing the refrigerant raises its temperature) and flows to the condenser, where it is cooled by the air flowing across the condenser coils and fins. The high pressure and the reduction in heat cause the refrigerant vapor to condense and turn into a liquid, which flows to the receiver drier. The liquid refrigerant must pass through the filter and desiccant to reach the pickup tube, where the liquid refrigerant can flow on to the TXV.

Thermal Expansion Valve

At the TXV, the amount of liquid refrigerant flow is determined by the action of a spring-loaded valve, which is controlled by different pressures on each side of a connecting diaphragm (**FIGURE 58-42**). The pressure in chamber A is determined by a temperature-sensing bulb that is taped to the evaporator outlet and insulated from air temperatures. The sensing bulb is filled with refrigerant and is connected to chamber A by a capillary tube. The temperature at the evaporator outlet determines the pressure in the bulb. Higher temperatures give higher pressures, which tends to open the TXV wider and allow more refrigerant to flow through. Lower temperatures give lower pressures, which tends to close the TXV and reduce the flow of refrigerant.

The pressure in chamber B is determined by the refrigerant pressure in the evaporator and by the force supplied from the

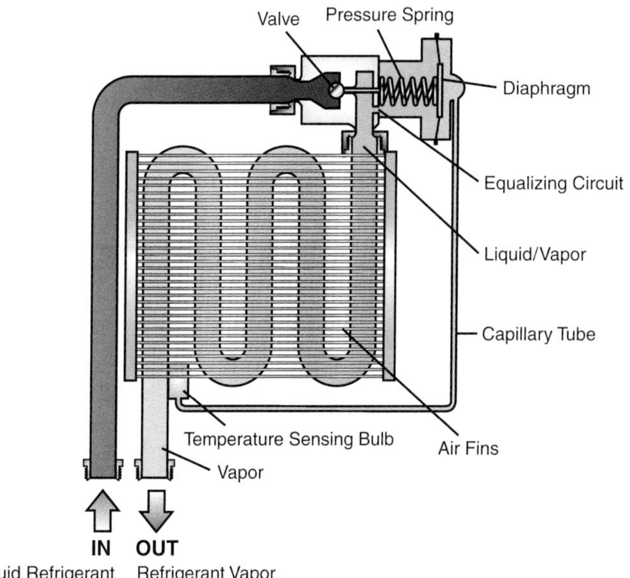

FIGURE 58-42 Thermal expansion valve.

valve spring—often called a **superheat spring**. The superheat spring is designed to ensure that the temperature of the refrigerant leaving the evaporator is between 5°F and 20°F higher than the boiling point of the refrigerant at the current operating pressure. This difference in temperature is called superheat, and it ensures that all liquid refrigerant that enters the evaporator is vaporized before it leaves.

When the compressor switches on, the suction from the compressor removes refrigerant vapor from the evaporator and lowers its pressure. This reduces the pressure in chamber B and allows the pressure in chamber A to move the TXV away from its seat by the action of the diaphragm. High-pressure liquid flows through the orifice and enters the evaporator as a low-pressure liquid. The temperature of the refrigerant leaving the evaporator is interpreted by the internal or external sensing bulb attached to the outlet of the evaporator. If the temperature is low, the refrigerant in the sensing bulb will contract, and pressure in chamber A of the TXV will reduce. The diaphragm moves and the TXV moves toward the valve seat to reduce the quantity of refrigerant entering the evaporator. If the temperature is high, the refrigerant in the sensing bulb expands and exerts more pressure in chamber A, causing the valve to move away from its seat and allowing more refrigerant to enter the evaporator.

The amount of refrigerant flowing depends on the quantity of heat to be removed from the air passing over the evaporator fins. More heat means that more refrigerant is required to remove it; less heat means that less refrigerant is required. A number of control switches are fitted into the system to ensure maximum efficiency, and these operate as described in the section on the fixed orifice system.

The TXV is a self-adjusting restriction. It can be internally or externally compensated. Older TXVs had a long tube with the sensing ball at the end of the tube. The sensing ball was an external unit to the actual TXV body, so it was externally compensated. Most of the new units are known as TXV blocks, or H blocks because of how they look (**FIGURE 58-43**). These units are all encompassed within the block. Both the inlet and the outlet pipe run through the block; thus it is internally compensated because the sensing ball is inside the block. The TXV uses refrigerant trapped in a chamber to control the restriction's opening. If the evaporator starts to flood, the temperature leaving the evaporator drops, and the refrigerant trapped in the TXV chamber contracts, closing the restriction and not allowing any more system refrigerant to enter the evaporator. By shutting off the TXV, the evaporator cannot be flooded, and there is no need for an accumulator in this system. The refrigerant will begin to back up in the system; the receiver drier will act as extra storage for the air-conditioning system while the TXV closes.

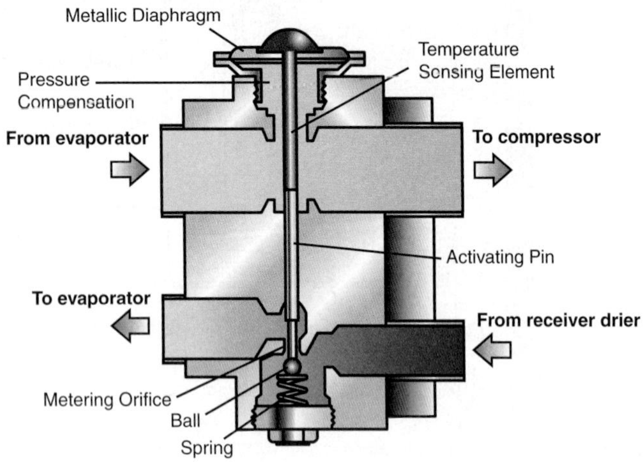

FIGURE 58-43 TXV block.

The TXV is constructed of cadmium-plated steel or aluminum. Three types are commonly used: the internally equalized valve, the externally equalized valve, and the H block valve. Each of these valves has the same job—to open and close the port. The **internally equalized valve** uses pressure from the inlet pipe to overcome the pressure created from the outlet pipe's heat. Lower heat on the outlet pipe creates lower pressure, and the pressure from the inlet pipe is able to overcome it and move the valve. The **externally equalized valve** uses the sensing bulb on the outlet side to overcome spring pressure to open the valve. The **H block valve** uses the outlet side like the external valve, but both the inlet and the outlet pipe go through the block, so the sensing bulb is built into the block.

Thermal Switch Cycling

The **thermal cycle switch** is a normally closed, temperature-controlled switch that controls the compressor clutch. It is located and installed in the fins of the evaporator inside the evaporator housing. The thermal cycle switch is opened and closed by the pressure of the gaseous refrigerant contained in the capillary tube as it expands and contracts with temperature. As the evaporator cools off, the gas contracts in the capillary tube. When the gas in the tube contracts enough, the pressure diaphragm retracts, and the thermal cycle switch opens, shutting the compressor clutch off. As the evaporator warms due to no refrigerant movement, the gas in the capillary tube expands, and the switch closes again, causing the compressor clutch to reengage.

▶ TECHNICIAN TIP

When removing the thermal cycle switch, mark the depth of the old one before removal. This will allow you to install the new one to the same depth.

Receiver Filter Drier

The **receiver filter drier** (**RFD**), often just called a receiver drier, is located between the condenser outlet and the TXV inlet (**FIGURE 58-44**). The RFD has several functions, all of which are important for the operation of the TXV system. The name of the component—receiver, filter, drier—explains what the component does. The main job of the RFD is to ensure that pure liquid refrigerant is being delivered to the expansion valve located close to the evaporator on the inlet side. The delivery of the pure liquid is accomplished by means of the siphon tube on the outlet port of the RFD. The siphon tube extends to the bottom of the RFD, allowing it to siphon pure liquid only. The liquid that flows

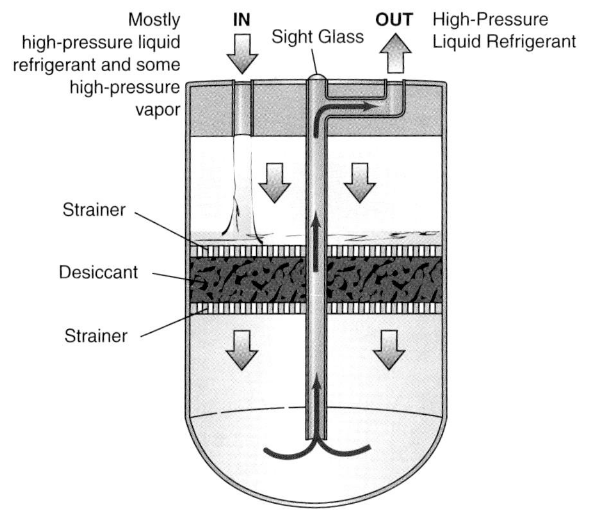

FIGURE 58-44 Receiver filter drier.

from the outlet of the condenser is under high pressure. This high-pressure liquid can include some vapor bubbles, which stay in the top of the RFD, where they condense into liquid.

The next job of the RFD is to act as a surge tank for refrigerant when the TXV closes due to low temperatures and pressures in the evaporator. When the restriction closes, refrigerant backs up in the air-conditioning system because the compressor is still running, and the temperature in the evaporator is not cool enough for the thermal cycle switch to turn off the compressor. The RFD acts as a storage container during these times.

The RFD is constructed with a mesh and cloth filter media. The **filter media** passes the refrigerant through the filter so no debris can enter the TXV and damage the valve. The RFD also contains a desiccant bag to help remove any moisture that might be present in the refrigerant.

TECHNICIAN TIP

It is important not to overcharge the air-conditioning system because the storage capacity of the RFD will be lost, and pressure will tend to rise very quickly in the high side. Overcharging refers to adding more refrigerant to the system than is specified by the manufacturer, such as adding 32 ounces of refrigerant into a system designed to hold a maximum of 28 ounces. With refrigerant capacity being reduced by manufacturers (some systems hold less than 16 ounces when properly filled), the systems cannot tolerate improper charge amounts as well as older vehicles can.

▶ Heating and Ventilation System Overview

K58014

Many of the components used to warm the passenger cabin of a vehicle also function in removing heat from the engine. The key shared components are the radiator, thermostat, water pump, and upper and lower radiator hoses. As the engine heats up from the process of burning fuel to create power, the coolant in the engine is also heated, reaching temperatures of 180–220°F (82–104°C) under normal conditions and 235°F (113°C) or more in hot climates or under heavy vehicle loads. The heat from the hot coolant opens the thermostat, and the water pump pushes the coolant into the radiator, which is located in the front of the vehicle. Airflow from the vehicle moving and/or from the cooling fans passes through the radiator, and the heat is transferred to the air, warming the air and reducing the temperature of the coolant. The coolant is then cycled back through the engine to restart the process. Heated coolant is also pumped to a heater core, where the blower causes air to flow through the heater core and is warmed. The heated air is directed into the passenger compartment.

Just as the pressure inside of the air-conditioning system is manipulated to control the boiling point of the refrigerant, the cooling system is kept under pressure to prevent the coolant from boiling at the normal boiling point of water, which is 212°F (100°C). The cap holds the pressure at approximately 8–20 psi (55–138 kPa) depending on what the manufacturer designed the system to be safely pressurized to. Raising the pressure 1 psi (7 kPa) raises the boiling point of water approximately 3°F, so water in a cooling system with a 15 psi (103 kPa) radiator cap will not boil until 45°F above 212°F, or 257°F (125°C). And if the water is mixed with antifreeze, the boiling point will be even higher.

SAFETY TIP

In the same way that liquid refrigerant boils when its pressure is lowered, so too does coolant. If an engine is overheating, never remove the radiator pressure cap. If you do, and the coolant is above 212°F (100°C), then the coolant will instantly turn into steam and shoot out of the radiator like a geyser, potentially burning you badly. Let all engines cool before opening the radiator pressure cap.

Applied Math

AM-40: Proportion: The technician can solve problems that determine the proportion of variables of a solution and determine if that proportion is within the manufacturer's specifications.

Engine coolant is composed of a mixture of antifreeze, commonly ethylene glycol or propylene glycol, and water. Different operating temperature ranges may require different ratios of antifreeze to water to ensure adequate cooling system performance. A common mixture ratio is 50% antifreeze to 50% water, but other ratios may be specified by antifreeze manufacturers for extreme temperature ranges.

Coolant concentration can be checked with an antifreeze hydrometer, which indicates the proportion of antifreeze in the system. Coolant concentration should be checked during vehicle servicing. Topping up cooling systems with water is a common practice, but the proportion of antifreeze in the system is reduced. If measurement indicates that the proportion of antifreeze in the system is low, some coolant must be drained from the system, which can then be refilled with antifreeze to the correct proportion. Too much antifreeze in the system can cause poor cooling; if this occurs, some coolant must be drained from the system and the system refilled with water to the correct proportion.

In the heating system, two hoses run from the engine to a heater core (small radiator). The heater core is located inside the cabin in the plastic box with the air-conditioning evaporator. As air is moved through the fins of the heater core, hot air is removed and sent out of the vents to heat the passenger cabin. The heater core is made from tubes that have thin metal fins connected to them, which increase the surface area tremendously (**FIGURE 58-45**). Because the tubes are also made of relatively thin metal, they can corrode from the inside and develop pinhole leaks from old acidic antifreeze. Changing a heater core can be a time-consuming and costly job in many of today's vehicles, so regular cooling system maintenance can pay off for a customer.

To adjust the amount of heat delivered to the cabin, some heating systems use a control valve on the inlet of the heater core to control the flow of hot coolant through the heater core. The valve can be mechanically operated by a cable, vacuum operated by a vacuum diaphragm, or electrically operated by an electric motor. Other vehicles use a blend door to control how much airflow goes through the heater core, based on the driver's request, using the temperature settings on the dash. Blend doors can be operated mechanically, by vacuum, or by electric motors. Most vehicles either use a heater control valve or a blend door to control the amount of heat delivered to the passenger compartment (**FIGURE 58-46**).

To improve driver and passenger safety, a supply of fresh air is necessary to maintain comfort and reduce fatigue. Fresh air is drawn from outside, usually through air ducts at the base of the windshield, and directed to the passenger compartment. The air is then delivered through a system of ducts and air doors, which are positioned to direct air to particular points inside the passenger compartment.

Some of the heat generated by the burning fuel in the engine is removed by the cooling system. The hot coolant is pumped through a heater core usually located in the vehicle cabin. As cool air blows across the heater core, the air is heated by the warm coolant. The coolant is then pumped back through the engine to pick up heat from the engine and the cycle continues. In the dash is a control system for the blower fan speed. The blower fan blows the warm or cool air out of the ducts. The control system also has a temperature control that either opens a blend door that determines how much air blows across the heater core or around it, or controls a heater control valve that varies the amount of hot water being pumped through the heater core (**FIGURE 58-47**). In most vehicles, all air flows through the evaporator on its way to the heater core.

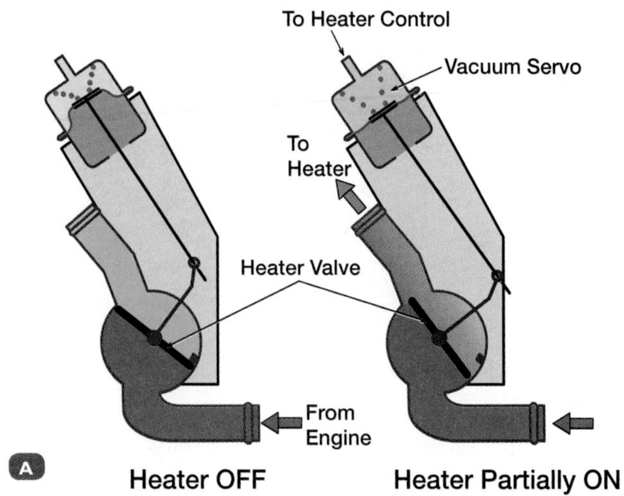

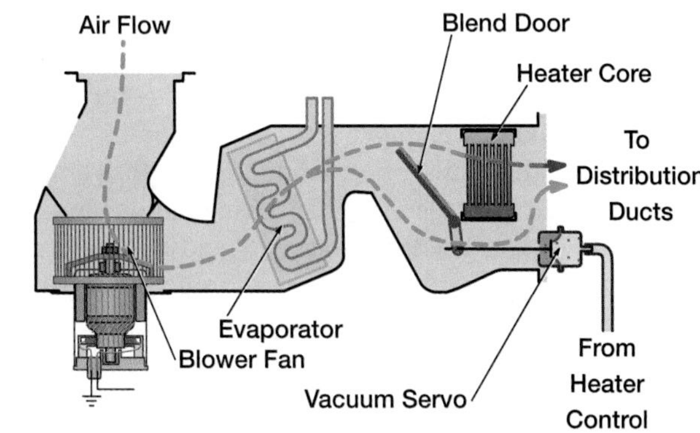

FIGURE 58-46 Most vehicles either use a heater control valve or blend door to control the amount of heat delivered to the passenger compartment. **A.** Typical heater control valve. **B.** Typical blend door.

FIGURE 58-45 Typical heater core.

FIGURE 58-47 A typical temperature control on an HVAC system.

If cold air is desired, then the air-conditioning system is activated to cool the evaporator. If the cool or cold setting is selected, then either the blend door will divert the air around the heater core, or the heater control valve will prevent coolant from flowing through it. If warm or hot air is desired, then the air conditioner will be off, and either the blend door will divert the air through the heater core, or the heater control valve will allow hot coolant to flow through the heater core. If defrost is desired, then both the air conditioner and the heater will operate. The air first goes through the evaporator, and most of the moisture is pulled from the air. Then the air is directed through the heater core where it is heated. This causes hot, dry air to be blown onto the windshield, which defrosts it quickly.

Another control, called a mode switch or selector, regulates location of airflow. Several doors are designed to direct the air to the feet (floor ducts), head (vent ducts), windshield (defrosting ducts), or a combination of locations, such as the feet and windshield (**FIGURE 58-48**). This control also has a vent setting that allows fresh air in, then directs it to the vents. Vehicles with air-conditioning can have a recirculation switch or lever, or they can have a "max" or "maximum" setting. This setting closes the door to the outside air to allow the air-conditioning system to work with only the air already present in the cabin. Using only the air within the cabin produces better cooling because the air-conditioning system will be cooling air that has already been through the air-conditioning system rather than hot, muggy air from outside the vehicle. Recirculation can also be used on some vehicles to help keep the moisture out when using the defroster.

Vehicle Heating and Ventilation Controls

K58015

Air doors can be controlled by cables, **vacuum servos**, or **electric servos** (**FIGURE 58-49**). Servos are devices designed to do work. They are hooked to the ventilation doors and open and close them according to the request from the control assembly. Vacuum servos use a vacuum control at the control head. Vacuum is supplied by the engine and run into the control head

FIGURE 58-48 A typical mode switch on an HVAC system.

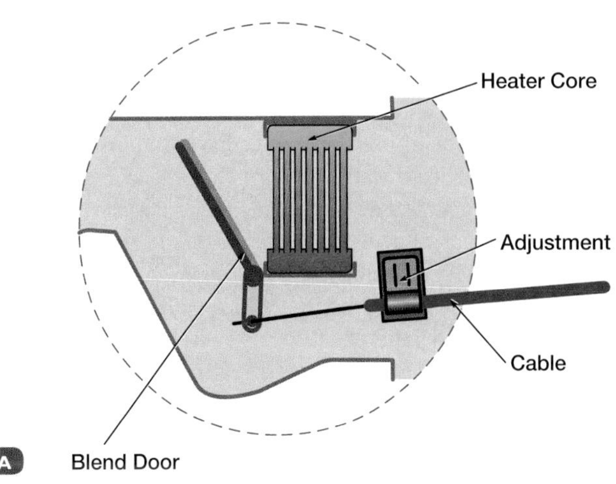

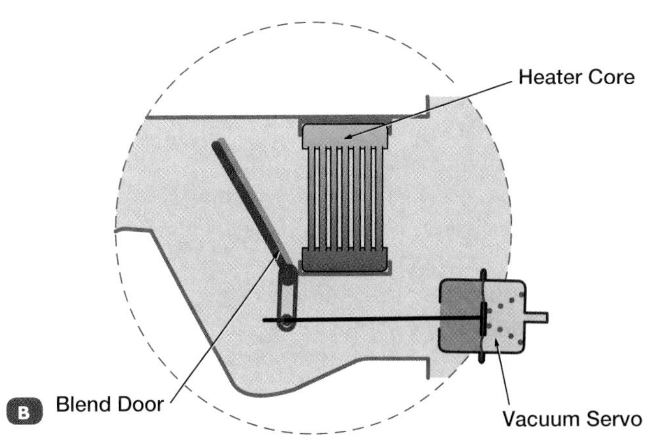

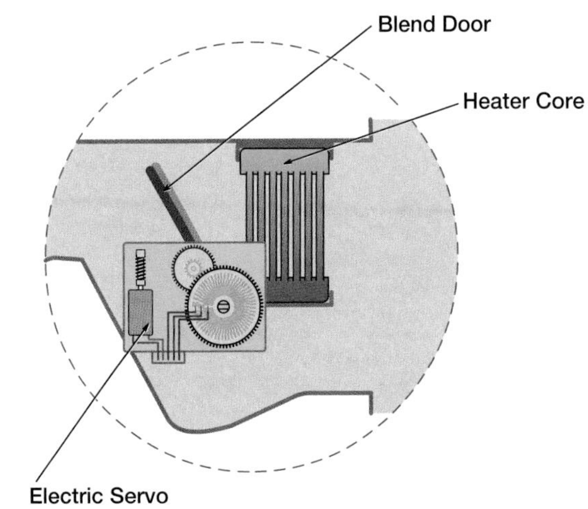

FIGURE 58-49 Air door controls. **A.** Cable. **B.** Vacuum servos. **C.** Electric servos.

with a small hose called a vacuum hose (it is usually black). The control head can route this vacuum to the correct servo by turning the dial on the control head. Each setting has a different-colored hose coming out and going to a separate servo or, in some cases, to a dual servo that can move in two directions. Traditionally, these systems have doors that can be only opened or closed, because they either have vacuum pulling on a diaphragm or they do not. Newer vehicles have electric servos. They operate the same as the old, mechanical servos except, instead of vacuum, they use electric motors to open and close the doors. These can be adapted to only partially open a door. Most of these systems are computer controlled, so the system knows when a door opens and when it does not, and can set trouble codes if a door malfunctions. In this way, the servos can direct air to the vent ducts, floor ducts, windshield ducts, and rear passenger compartment ducts.

A fan, usually driven by an electric motor, is used to assist the air to flow through the radiator and condenser when demands are high or when vehicle speeds are low. When vehicle speeds are low, there may not be enough airflow to cool the coolant and the refrigerant. In these conditions, the fan comes on to pull the air through the radiator and the condenser to keep the vehicle from overheating and also to allow the refrigerant to condense. Some vehicles have two or more fans, typically one for the radiator and one or more for the condenser. They frequently are positioned side by side and attached to the radiator, just as in applications with one fan. Each fan has different controls regulating when it comes on or turns off. Most vehicles with electric fans will turn on at least one fan when the air-conditioning is on, to aid in the cooling of the refrigerant.

The heating system of most passenger vehicles uses hot engine coolant flowing through a **heater core**, which is usually mounted under the dash (**FIGURE 58-50**). The heater core is connected to the engine's cooling system by flexible rubber hoses, which allow for movement and a reduction in the level of engine noise and vibration transmitted into the passenger compartment. The heater core is a small radiator consisting of **tubes**, **fins**, and **tanks**. The tubes are the metal pipes, running side to side or up and down, that the coolant travels through. Fins are small, flat metal pieces placed between the tubes to help transfer heat from the coolant to the air. The fins are in direct contact with the hot pipes and absorb the heat. The heat is then absorbed by the air flowing through the fins, warming the air, which is then directed back into the passenger cabin. Metal or plastic tanks line the sides of the tubes, connecting them together and assisting with coolant flow. The coolant flow is created by the engine water pump directing the coolant through the heater core and then back to the engine. In some applications, a heater control valve is located in one of the heater coolant hoses, to adjust the quantity of coolant flowing through the heater core.

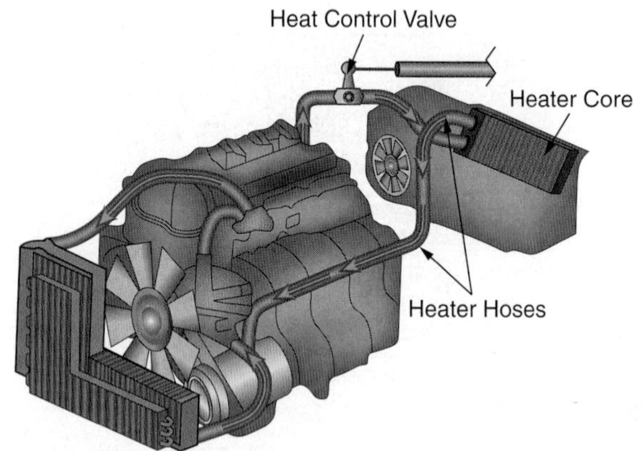

FIGURE 58-50 The heating system.

The coolant enters the heater core from the warm engine, and the air that is passed over the heater core from the passenger compartment cools the coolant in the heater core. The transfer of heat from the hot coolant to the air passing over the heater core is what causes the temperature to rise in the air duct. The air that has been heated by the heater core is warm but still moist because the heat from the heater core cannot condense the moisture out of the air. This condition of warm moist air does not have an adverse effect on the passengers in most cases.

Defroster

When the air conditioner is allowed to operate during the defrost mode, the air is dried from the moisture removal occurring as air passes over the evaporator before passing over the heater core. The result of air passing over both cores is dry, heated air. The hot, dry air from the defrost duct hits the windshield, and the moisture from the glass is quickly drawn into the air. The defrost time of passenger vehicles has decreased considerably because of the air conditioner being switched on during the defrost cycle. Having the air-conditioning operated during defrost also keeps refrigerant and oil circulating through the AC components and

Applied Science

AS-9: Scientific Methods: The technician develops a theory relative to the cause of the problem based on the information provided, then tests the hypothesis to determine the solution.

A vehicle is brought to the shop with an air-conditioning concern. The customer is concerned that all air from the ventilation system is being directed to the windshield and that turning the mode control to either the face or feet positions has no effect.

The vehicle in question uses a vacuum-operated system to control blend-door positions in the HVAC system. The technician forms a theory that the system may have a vacuum leak, based on the fact that, in the event of no vacuum, the system defaults to defroster operation to ensure the driver retains visibility and is able to safely operate the vehicle.

The technician first confirms that the HVAC vacuum line is connected to the engine's intake manifold and then forms a hypothesis that the vacuum leak is occurring within the HVAC system located inside the vehicle's cabin. He is able to confirm this hypothesis by connecting a vacuum pump to the HVAC line in the engine bay and confirming that the system does not hold vacuum. Individual components and sections of vacuum line in the cabin are then individually tested to ensure they hold vacuum. A common failure in this scenario is an internal vacuum leak in the mode control assembly.

hoses during the winter months, when the AC is not likely to be used. This keeps the system active and lengthens its life.

Blower Motors

Blower motors are electric motors that determine the rate of airflow out of the vents. Attached to the spinning shaft of the motor is what is called the "squirrel cage," a plastic cage with fins to pull or push air (**FIGURE 58-51**). The speed is controlled by the driver (except in electronically controlled systems) through the fan switch and a series of resistors to limit the fan speed. Blower motors can be positioned either before the evaporator and heater core or after in the HVAC system. When placed after, they pull the air through the evaporator and heater core in the box. Placing the blower before pushes air through the evaporator and heater core in the box.

Blower Motor Resistor Packs/Speed Controls

The resistor pack or resistor block is a common blower motor control (**FIGURE 58-52**). It uses resistors in series to regulate the speed of the fan (**FIGURE 58-53**). The more resistance there is in the circuit, the less current available to spin the fan. The lowest speed (speed 1) passes the current through several resistors. The highest speed uses no resistors, so the motor has full current flow. The amount of resistors in a pack depends on the number of different speed settings.

FIGURE 58-51 Blower motor and squirrel cage fan.

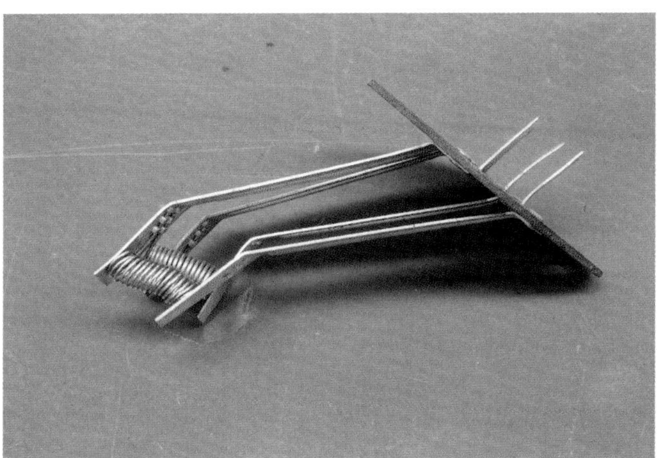

FIGURE 58-52 A typical blower motor resistor pack.

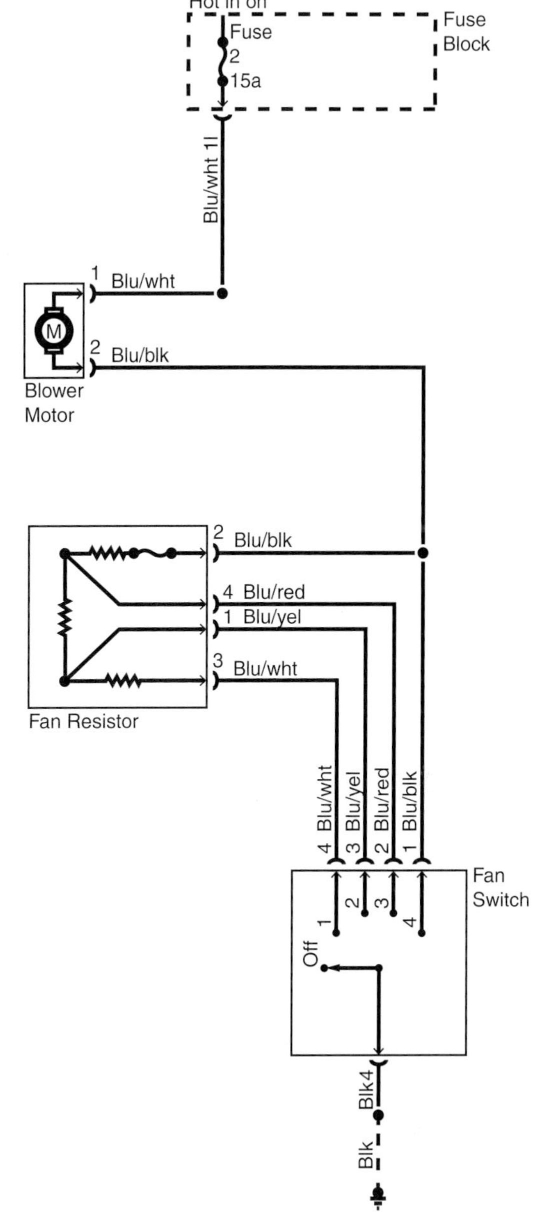

FIGURE 58-53 Blower motor resistor and switch schematic.

On vehicles with electronic air-conditioning systems or automatic systems, where the customer just sets a desired temperature, the fan is controlled by a variable resistor called a rheostat that allows the computer to change fan speed infinitely by varying resistance, like a light dimmer in a house. Other systems use **pulse width modulation (PWM)** to control fan speed. Normally this is done by controlling "on" time compared to "off" time on the ground side of the circuit. The longer the on time, the faster the fan will spin. We cover this in more detail in the Electronic Climate Control chapter.

Cabin Air Filters

Many manufacturers have added cabin air filters to the inlet side of the heating/air-conditioning box where fresh air enters. This filter is designed to catch dust and outside contaminants so they do not get blown into the cabin. Many are paper, just like the air filter for the engine, and as such should be checked regularly and replaced when they begin to clog, or they will cause the system to be inefficient due to poor airflow (**FIGURE 58-54**). Some manufacturers are using activated charcoal cabin filters, which help trap odors and airborne pollutants such as carbon monoxide and oxides of nitrogen. Cabin air filters typically should be replaced once a year or between 12,000 and 15,000 miles.

FIGURE 58-54 Cabin filters. **A.** Clean. **B.** Dirty.

▶ HVAC Developments

With the advent of electric and hybrid passenger cars, the HVAC system continues to evolve. The heater core in your car may soon be a thing of the past; reversing the refrigerant flow through the air-conditioning system would turn your evaporator into a condenser and provide ample heat for the passenger compartment. This is the same principle that home heat pumps operate on. They can reverse the flow of refrigerant, which allows them to be switched easily from an air conditioner to a heater and back.

Another development is electric compressors, which are already taking the place of less efficient belt-driven compressors (**FIGURE 58-55**). The amount of energy loss from the drive belt and electric clutch is too high for hybrid and electric vehicles that need to conserve all the power possible. As passenger cars evolve, so must the air-conditioning system; greater efficiency and less power consumption will be the drivers of new air-conditioning systems.

FIGURE 58-55 A typical electrically driven AC compressor.

▶ Wrap-Up

Ready for Review

- The heating, ventilation, and air-conditioning (HVAC) system maintains a comfortable temperature inside a vehicle and uses an air filter to provide fresh, filtered air.
- The vehicle heating system and automotive air-conditioning system were developed in the 1930s.
- The Environmental Protection Agency regulates automotive air-conditioning via Section 609 of the Clean Air Act.
- Automotive HVAC technicians must obtain a license to verify that they are familiar with federal rules and regulations.
- HVAC systems use heat transfer to ensure temperature comfort inside a vehicle.
- Heat transfer can take place via conduction, convection, or radiation.
- A vehicle's air conditioner transforms the refrigerant from a liquid to a gas and back to liquid.
- Heat energy (the amount of heat needed to move a temperature up or down by degrees) is measured in British thermal units (BTUs).
- The temperature at which a liquid changes to gas is called latent heat of evaporation; likewise, the temperature at which a liquid forms a solid (ice) is called latent heat of freezing.
- Vehicle air conditioners use a refrigerant for the vaporization and condensation process.

- HVAC systems have four major components: compressor, condenser, restriction, and evaporator,.
- The compressor is designed to increase the pressure of the refrigerant (in gas form) and push it into the condenser.
- A vehicle's air conditioner is a closed loop system.
- The condenser is designed to move gas through a series of coils and cool it into a liquid.
- The restriction works to transform high-pressure liquids into low-pressure liquids.
- The evaporator vaporizes liquid into gas as it removes heat energy from the passenger compartment.
- Each HVAC system component requires the refrigerant to be delivered in the proper state and at the correct level.
- Components of a compressor include pulley, clutch, axial plate, piston, reed valves, and body.
- One of the most common types of compressor is an axial piston compressor.
- Refrigerant moves through the air-conditioning system because of the high pressure and volume of gas leaving the compressor.
- The condenser uses ram airflow and the engine cooling fan to cool the refrigerant.
- A restriction can be a fixed orifice tube or a thermostatic expansion valve.
- The velocity and pressure of the refrigerant changes when it moves through the restriction.
- Bernoulli's principle states that the speed/flow of a liquid changes in opposition to the liquid's pressure (when one increases, the other decreases).
- Pascal's law states that liquid cannot be compressed.
- The evaporator is a one-piece unit of copper or aluminum tubes, with no serviceable parts.
- Automotive air conditioners can be fixed orifice tube systems or thermal expansion valve (TXV) systems.
- Fixed orifice tube systems use an accumulator to prevent liquid from entering the compressor.
- TXV systems use a receiver filter dryer to send liquid to the restriction.
- Air doors are controlled by cables, vacuum servos, or electric servos.
- Vehicle heating systems rely on a hot engine coolant flowing through a heater core-a small radiator comprising tubes, fins, and tanks.

Key Terms

air-conditioning compressor clutch The device which connects and disconnects the drive pulley to the compressor shaft and operated by electro-magnetism.

atmospheric pressure The pressure of the air surrounding everything caused by gravity and the weight of air. The higher from sea level, the lower the atmospheric pressure.

axial piston compressor A design of compressor that uses an angled plate (swash plate) to create the piston movement.

barrier-type hose A rubber hose made with a nylon bladder inside to contain substances with small molecular structures, such as R-134a.

British thermal unit (Btu) A measure of heat energy. It takes 1 Btu to raise the temperature of 1 pound of water 1°F.

chlorofluorocarbon (CFC) A manufactured compound designed to be used as a refrigerant. It is now illegal due to the high chlorine content.

Clean Air Act (CAA) A policy signed into law in 1990 that sets standards for air pollution to eliminate ozone-depleting elements.

closed loop system A totally self-contained system with no materials entering or exiting.

clutch electrical winding An electromagnetic control device that creates a magnetic field that pulls in the clutch to make contact with the pulley.

compressor A belt- or electrically driven device designed to increase refrigerant pressure and cause refrigerant to travel through the air-conditioning system.

condensation The changing of a gas into a liquid through cooling.

condenser The air-conditioning component located in the front of the vehicle designed to allow high-pressure refrigerant to change states from gas to liquid.

conduction The process of transferring heat through matter by the movement of heat energy through solids from one particle to another.

convection The process of transferring heat by the circulatory movement that occurs in a gas or fluid as areas of differing temperatures exchange places due to variations in density and the action of gravity.

desiccant A drying agent used in air-conditioning systems to absorb moisture.

dichlorodifluoromethane (R-12) An inert, colorless gas that can be used as a refrigerant. It is stored in white containers.

discharge reed valve A flat, spring-loaded valve used as a check valve on the discharge side of a compressor.

double crimping Overcrimping or recrimping a fitting to keep it from leaking. It normally results in a bigger leak and is not recommended.

drive pulley The disc-shaped device which is driven by the accessory drive belt and used to power the compressor.

electric servo An air-conditioning door actuator controlled by electricity.

Environmental Protection Agency (EPA) A government agency concerned with air quality and pollution issues related to the environment.

evaporator The air-conditioning component normally located in the passenger compartment designed to allow low-pressure refrigerant liquid to change states to a gas.

externally equalized valve A style of TXV where the external valve uses the sensing bulb on the outlet side to overcome spring pressure to open the valve.

filter media A screen designed to keep debris from the TXV that is normally located in the receiver dryer.

fin A small, flat piece of metal placed between the tubes to help with the transfer of heat from the coolant to the air, refrigerant to air, or air to refrigerant. The metal heats up due to contact with the hot pipes in the case of a radiator or heater core and from hot air in the case of an evaporator.

fixed orifice tube system A system with a fixed orifice tube that uses an accumulator between the evaporator and the compressor.

H block valve A style of TXV that has an outlet side like the external valve; however, both the inlet and the outlet pipe go through the block, so the sensing bulb is built into the block.

heat transfer The flow of heat from a hotter part to a cooler part; it can occur in solids, liquids, or gases.

heater core A heat-exchanging device that transfers heat converted by the fuel burning in the engine to the passenger compartment.

high-pressure switch A switch designed to open at a predetermined high pressure to protect the compressor; it is normally located on the high side line near the compressor.

hydrogenated nitrile butadiene rubber (H-NBR) A muffler constructed of a hollow cylinder with baffling.

ingression The passing of foreign bodies, such as moist warm air, through the air-conditioning lines or fittings.

inlet reed valve A flat spring-loaded valve that allows gaseous refrigerant to enter the compressor.

internally equalized valve A style of TXV where the internally equalized valve uses pressure from the inlet pipe to overcome the pressure created from the outlet pipe heat. Lower heat on the outlet pipe creates lower pressure, and the pressure from the inlet pipe is able to overcome it and move the valve.

latent heat of condensation The amount of heat removal necessary to change the state from a gas to a liquid without changing the actual gauge temperature.

latent heat of evaporation The amount of heat required to change the state from a liquid to a gas without raising the actual gauge temperature.

latent heat of freezing The amount of heat removal required to change the state from a liquid to a solid without changing the actual gauge temperature.

low-pressure cycling switch A device installed on the low-pressure line or the accumulator to turn the air-conditioning clutch on and off at specified pressures.

low-pressure switch A device used in nonvariable air-conditioning compressors. It cycles the air-conditioning clutch in response to changes in low side pressure.

nitrile butadiene rubber (NBR) An artificially created rubber that is more resistant to oils and acids than natural rubber; it is compatible with R-12.

O-ring A small donut shaped rubber ring used to seal connections between air-conditioning components.

parallel flow condenser A condenser with multiple parallel tubes flowing from one side tank to the other.

pulse width modulation (PWM) A digital on/off electrical signal used as a variable control for devices such as solenoids.

radiation The transfer of heat through the emission of energy in the form of invisible waves.

ram airflow The flow of air created by a vehicle moving down the road.

receiver filter drier (RFD) The air-conditioning component used on TXV systems to filter and store liquid refrigerant to supply liquid refrigerant to the TXV. It is located between the condenser and the TXV.

refrigerant The name given to a chemical compound designed to meet the needs of the refrigeration system.

restriction A blockage that partially stops or slows the flow of a material such as refrigerant.

rotary vane compressor A compressor design that spins with vanes attached that extend to create the air-conditioning pressure required to run the system.

Sanden scroll-type compressor A compressor design that uses two spiral forms, one spinning and one stationary, to create the pressure needed to drive the air-conditioning system.

Sanden wobble plate compressor A compressor design that utilizes an offset shaft connected to the crankshaft to create the reciprocating motion of the pistons.

serpentine condenser A condenser with one tube snaking back and forth.

suction reed valve A flat, spring-loaded valve used as a check valve on the suction side of a compressor.

superheat spring A spring used as part of the control in the TXV to ensure refrigerant leaving the evaporator is boiled completely or superheated.

swash plate Another name for an axial plate.

tanks Metal or plastic pieces that line the pipes used to connect the tubes together to allow the coolant to continue to flow.

tetrafluoroethane (R-134a) An inert colorless gas that can be used as a refrigerant. It is stored in light blue containers.

thermal cycle switch A normally closed switch located in the fins of the evaporator that controls the air-conditioning compressor clutch. It is opened and closed by the temperature changes in the evaporator.

thermal expansion valve (TXV) system A system with a valve designed to sense evaporator outlet temperature and vary the inlet orifice size accordingly.

tubes Metal pipes running side to side or up and down that the coolant or refrigerant travels through.

vacuum servo An air-conditioning door actuator controlled by a vacuum.

vaporization The changing of a liquid to a gas through boiling.

XH-7 A form of desiccant used with pure R-134a.

XH-9 A form of desiccant used when the refrigerant may not be pure, such as with some imports.

Review Questions

1. Automotive air-conditioning service and repair technicians need to have which special license?
   a. Hazardous materials license
   b. Master Technician license
   c. 609 license
   d. No license is required to work on automotive air-conditioning systems.
2. How does air-conditioning absorb a large amount of heat energy?
   a. By evaporating a liquid
   b. By freezing a liquid
   c. By cooling a liquid
   d. By cooling a gas
3. All of the following are properties of a refrigerant *except*:
   a. It has a super low boiling point.
   b. It changes its state without breaking down.
   c. It vaporizes and condenses at the correct temperature and pressure.
   d. Its pressure remains unchanged during the cooling cycle.
4. Which of the following components provides the needed refrigerant movement in the system?
   a. Compressor
   b. Condenser
   c. Restriction
   d. Evaporator
5. Which of the following components in the AC transforms low-pressure fluid to low-pressure gas?
   a. Compressor
   b. Condenser
   c. Restriction
   d. Evaporator
6. The fixed orifice tube is used for:
   a. sensing changes in temperature.
   b. blocking the flow of liquid refrigerant through the system.
   c. adjusting heat in the cabin.
   d. allowing only a specific amount of liquid refrigerant through at a time.
7. Choose the correct statement with respect to pressure switches.
   a. They are used to display the pressure drop and rise in the system.
   b. They are in parallel connection with the AC compressor.
   c. They send inputs to the control module to vary the compressor speed.
   d. They are used to engage and disengage the compressor clutch.
8. All of the following statements with respect to hoses, pipes, and tubes are true *except*:
   a. Heater hose can substitute for air-conditioning hose if needed.
   b. Pipes are used particularly where no give or movement is necessary.
   c. Hoses are slid into the pipes and a crimp is used to fasten them.
   d. All air-conditioning hoses must be a barrier-type hose.
9. Choose the correct statement with respect to driers.
   a. Driers are circulated in the system.
   b. They absorb any moisture in the system.
   c. They are left open to absorb the moisture.
   d. They are placed in the restriction.
10. What type of compressor uses two interwoven scrolls?
    a. Variable displacement compressor
    b. Sanden scroll-type compressor
    c. Rotary vane compressor
    d. Sanden wobble plate compressor

ASE Technician A/Technician B Style Questions

1. Tech A says that one advantage of an air conditioner is that the system removes water from the air. Tech B says that an AC compressor clutch is activated by a magnetic field. Who is correct?
   a. Tech A
   b. Tech B
   c. Both A and B
   d. Neither A nor B
2. Tech A says that automotive air-conditioning is regulated by the EPA. Tech B says that EPA tests air-conditioning systems every five years. Who is correct?
   a. Tech A
   b. Tech B
   c. Both A and B
   d. Neither A nor B
3. Tech A says that an accumulator creates the pressure needed for the air-conditioning system to operate. Tech B says that the high side refers to refrigerant entering the compressor. Who is correct?
   a. Tech A
   b. Tech B
   c. Both A and B
   d. Neither A nor B
4. Tech A says that heat is removed from the passenger compartment by the heater core. Tech B says that heat in the cab is absorbed by the refrigerant and is transferred to the outside air at the condenser. Who is correct?
   a. Tech A
   b. Tech B
   c. Both A and B
   d. Neither A nor B
5. Tech A says that conduction is when heat transfers through solids. Tech B says that radiation is when heat travels through space. Who is correct?
   a. Tech A
   b. Tech B
   c. Both A and B
   d. Neither A nor B

6. Tech A says that the air-conditioning compressor creates high-pressure liquid. Tech B says that the restriction creates a physical change in refrigerant from a liquid to a gas. Who is correct?
   a. Tech A
   b. Tech B
   c. Both A and B
   d. Neither A nor B
7. Tech A says that when an air-conditioning system freezes up, the refrigerant freezes and stops the cooling process. Tech B says that when an air-conditioning system freezes up, the evaporator coils are covered with frozen water molecules that stop airflow across the coils, thus preventing cooling. Who is correct?
   a. Tech A
   b. Tech B
   c. Both A and B
   d. Neither A nor B
8. Tech A says that the evaporator is on the high side of the system. Tech B says that the condenser transfers heat to the atmosphere. Who is correct?
   a. Tech A
   b. Tech B
   c. Both A and B
   d. Neither A nor B
9. Tech A says that R-134a replaced R-12 as an approved refrigerant. Tech B says that HFO-1234yf refrigerant is flammable. Who is correct?
   a. Tech A
   b. Tech B
   c. Both A and B
   d. Neither A nor B
10. Tech A says that the engine thermostat controls the flow of coolant through the heater system. Tech B says that the heater core can have engine coolant or refrigerant flowing through it, depending on whether heat or cool is commanded. Who is correct?
    a. Tech A
    b. Tech B
    c. Both A and B
    d. Neither A nor B

CHAPTER 59

Heating and Air-Conditioning Systems and Service

NATEF Tasks

- **N59001** Identify and interpret heating and air conditioning problems; determine needed action. (AST/MAST)
- **N59002** Performance test A/C system; identify problems. (AST/MAST)
- **N59003** Identify abnormal operating noises in the A/C system; determine needed action. (AST/MAST)
- **N59004** Inspect A/C condenser for airflow restrictions; perform necessary action. (MLR/AST/MAST)
- **N59005** Inspect evaporator housing water drain; perform needed action. (AST/MAST)
- **N59006** Identify the source of HVAC system odors. (MLR/AST/MAST)
- **N59007** Diagnose A/C system conditions that cause the protection devices (pressure, thermal, and/or control module) to interrupt system operation; determine needed action. (MAST)
- **N59008** Leak test A/C system; determine needed action. (AST/MAST)
- **N59009** Identify refrigerant type; select and connect proper gauge set/test equipment; record temperature and pressure readings. (AST/MAST)
- **N59010** Identify A/C system refrigerant; test for sealants; recover, evacuate, and charge A/C system; add refrigerant oil as required.
- **N59011** Recycle, label, and store refrigerant. (AST/MAST)
- **N59012** Inspect, remove, and/or replace A/C compressor drive belts, pulleys, tensioners, and visually inspect A/C components for signs of leaks; determine needed action. (MLR/AST/MAST)
- **N59013** Remove, inspect, reinstall and/or replace A/C compressor and mountings; determine recommended oil type and quantity. (AST/MAST)
- **N59014** Determine need for an additional A/C system filter; perform needed action. (AST/MAST)
- **N59015** Remove and inspect A/C system mufflers, hoses, lines, fittings, O-rings, seals, and service valves; perform needed action. (AST/MAST)
- **N59016** Remove, inspect, reinstall, and/or replace condenser; determine required oil type and quantity. (MAST)
- **N59017** Remove, inspect, and replace receiver/drier or accumulator/drier; determine recommended oil type and quantity. (AST/MAST)
- **N59018** Remove, inspect, and install expansion valve or orifice (expansion) tube. (AST/MAST)
- **N59019** Determine procedure to remove and reinstall evaporator; determine required oil type and quantity. (AST/MAST)
- **N59020** Inspect condition of refrigerant oil removed from A/C system; determine needed action. (AST/MAST)
- **N59021** Determine recommended oil and oil capacity for system application. (AST/MAST)
- **N59022** Perform correct use and maintenance of refrigerant handling equipment according to equipment manufacturer's standards. (AST/MAST)

Knowledge Objectives

After reading this chapter, you will be able to:

- **K59001** Describe the process of servicing the air-conditioning system.
- **K59002** Describe AC system refrigerant capacity and how to identify overcharge and undercharge conditions.
- **K59003** Identify common AC services and tools.

Skills Objectives

After reading this chapter, you will be able to:

- **S59001** Perform A/C system component inspections.
- **S59002** Inspect and Handle system refrigerant.
- **S59003** Repair compressor failure.
- **S59004** Repair heat exchanger, expansion device, or refrigerant line failure.
- **S59005** Evaluate condition of refrigerant system oil.

▶ Introduction

Having an HVAC system that operates correctly is important to most people, especially in climates where there are temperature extremes, both hot and cold. For this reason, HVAC system diagnosis and repair can be very rewarding, both professionally and financially. Servicing HVAC systems starts with having a good understanding of the theory of operation and components of an HVAC system, which is laid out in the Principles of Heating and Air-conditioning Systems chapter. Once you understand those concepts, it is time to learn about the processes and tools used in the diagnosis and repair of the system, which is the purpose of this chapter. Pay close attention to the details; they will help you learn how to service the system and also how to avoid problems that could damage the components.

The HVAC system is generally a very reliable system. When it does present a problem, it usually results in poor performance, and the customer concern is poor cooling or no cooling. When that is the case, the technician will usually start with a performance test, which tests the function of the system. By performing this test, the technician verifies that there is a problem and will likely understand what is wrong, and what the next steps need to be. As you learn about each of the testing and service procedures, keep in mind the basics you learned in the Principles of Heating and Air-conditioning Systems chapter, and ask yourself how they apply to the task at hand. Doing so will help ensure that you properly perform the basic service and repair procedures necessary to maintain a properly functioning HVAC system.

▶ The Air-Conditioning Service Process

K59001

The HVAC system is a moderately complex system with many principles involved in making the unit function properly. Successful diagnosis and repair of the HVAC system requires a general step-by-step, strategy-based diagnosis process. Following is an overview of that process on a typical vehicle brought in for an air-conditioning repair. Remember, technicians must be EPA 609 certified to handle refrigerants.

1. The way we verify the customer concern on an HVAC system is by carrying out a *performance test* of the system. **Performance testing** is the term used to describe the process of making the system perform to its maximum ability and comparing that to specifications. Typically, it consists of turning on the air conditioner to its maximum cold setting, with the windows closed, and raising the engine revolutions per minute (rpm) to 1200–2000. The vent temperature in the cabin is checked and compared to the temperature and humidity specifications on the manufacturer's temperature chart.
2. After the performance test is complete, the next step is *diagnosing any faults* that are causing the system to operate incorrectly. This starts with researching the results you obtained in the service information, which helps you determine the focused testing steps needed to locate the cause of the fault. It

You Are the Automotive Technician

A new customer stops by the shop with his eight-year-old vehicle and requests that the air-conditioning system be serviced. His concern is that it isn't performing the way it should and isn't cool enough. He says that the AC has worked well up until now and hasn't needed any work. He is wondering if it just needs some extra refrigerant.

You explain that if it is low on refrigerant, then topping it off will make it work better, but that will only last for a short amount of time, and then he will be back in this same situation. Also, it could end up damaging the system if the oil leaks out or moisture gets in. You suggest that he allow you to diagnose the problem, which will tell him why it isn't working properly, and then he will have an opportunity to approve the repair costs. The standard diagnostic fee includes performance testing the vehicle, leak testing the system, up to two pounds of refrigerant, and diagnosis of any faults. He agrees to allow you to diagnose the vehicle.

1. What is involved in performance testing the HVAC system?
2. What are some possible causes of the AC system not providing enough cooling?
3. How would you determine whether the system is low on refrigerant?
4. How would you perform a leak check on a system that is empty due to a relatively large leak?

could be as simple as identifying a slipping belt or performing specified tests on components such as the compressor clutch assembly. It could also be performing a leak test of the system to identify the location of the leak. Just make sure that you find the root cause of the problem, not just the symptom. We cover the most common testing procedures next.

3. If the diagnosis determines that a component of the air-conditioning system is not working properly, such as a faulty cycling switch, then leak testing is not needed, and you will go straight to reclaiming refrigerant. If the diagnosis is that the refrigerant is low, you will need to perform a leak test next. The electronic leak detector (sometimes called a sniffer) and dye leak tester require that the air-conditioning system be relatively full of refrigerant, whereas nitrogen testing requires an empty air-conditioning system. If nitrogen testing is the leak test method to be performed, proceed to the reclaim step.
4. *Leak testing* is the next step. If the system is charged, then electronic leak detectors or fluorescent dye testing are good methods. If the refrigerant has leaked out of the system, then nitrogen testing is a good method of leak testing. Use the three-step method listed later in this chapter to determine whether there is a leak, and take the appropriate action. Do not forget to vent the nitrogen to the atmosphere.
5. *Reclaiming* the refrigerant before any repair or nitrogen testing is required by the Environmental Protection Agency (EPA), Section 609 of the Clean Air Act. Reclaiming involves first checking the system to see if there has been any sealant added to it. The second step is identifying the refrigerant and remove all of it, using an approved machine, then measure and record the amount of refrigerant and oil that was removed from the system. If the oil removed is greater than 1 ounce (30 milliliters [mL]), then the air-conditioning system should be flushed, and the base oil amount should be installed. Compare the amount of reclaimed refrigerant to the under-the-hood sticker to determine whether the air-conditioning system may have been under- or overcharged. If undercharged, suspect a leak, and leak check with nitrogen.
6. *Repair* the air-conditioning problem. Now that the air-conditioning system has been diagnosed and tested, the system fault can be repaired. Take the appropriate action and precautions that the repair requires. At this point, the air-conditioning system can be flushed and the base oil installed if needed. If less than 1 ounce (30 mL) of base oil needs to be added to the system, this step will wait until after the evacuation is complete. Finish the repair and continue to the evacuation.
7. *Evacuation* of the air-conditioning system is the next step after all repairs are made. Evacuation is the process of creating a low pressure or vacuum in the air-conditioning system, using a vacuum pump and gauges. Water boils and turns into a gas at normal ambient temperatures when placed in a strong vacuum, so this process can remove any moisture that entered the system. This process usually requires a strong vacuum for a minimum of 15 minutes for removal of all the moisture. Follow the proper procedure for the evacuation with a mechanical gauge, or use a more accurate micron gauge. After the evacuation is complete, the oil for the air-conditioning system can be added, and then the vehicle is ready for the charging process.
8. *Charging* the air-conditioning system to the proper refrigerant level is an important process of the air-conditioning repair. When adding refrigerant to the system, be sure that you add the correct amount. After the refrigerant has been charged into the air-conditioning system, rotate the compressor shaft by hand to push out any liquid that may have entered the compressor during the charge process.
9. Verifying the repair is performed by *post-testing* of the air-conditioning system. This is recommended to catch any masked problems that did not appear when the original problem was diagnosed. It is a repeat of the performance test to ensure that the system is operating as specified, all moisture is removed, and the pressure gauges show the high- and low side pressures are as specified. All readings should be checked against a diagnostic chart. If you used an electronic leak detector or dye leak testing method, it is a good idea to confirm that there are no leaks. If the vehicle passes the post-test and the air-conditioning system is in good functioning order, then you are ready to return the vehicle to the customer.

▶ Air Conditioner Refrigerant Capacity

K59002

The proper **charge** (amount of refrigerant) of the air-conditioning system is important because if the amount of refrigerant is too low or too high, the resulting change in system operating pressure will change the boiling point of the refrigerant, making it unable to effectively remove heat in the evaporator. Safety features are built into the air-conditioning system. For example, the low-pressure safety switch turns the compressor off if the pressure drops too low due to a low charge of refrigerant, and the high-pressure safety switch turns the compressor off if the pressure gets too high due to an overcharge. Because the system pressures vary greatly during operation, the switches are designed to allow operation within a wide window. Thus, they should not be used as indicators of a proper charge.

The proper charge for the air conditioner can be found under the hood on the air-conditioning identification sticker (**FIGURE 59-1**). The sticker should detail the oil capacity, charge amount, and type of refrigerant. The type of refrigerant must remain the same because cross-contamination will create a hazardous substance.

There are different ways to measure the refrigerant to be added. Some systems give you the charge amount in kilograms, pounds, pounds with ounces, and straight ounces. The machine that you use to charge the system will determine the type of measurement you will need or need to convert to. If the sticker is not under the hood of the vehicle, you will need to turn to

FIGURE 59-1 An air-conditioning identification sticker.

service information or industry literature. Charging a vehicle without knowing the correct amount of refrigerant is playing a guessing game and is likely to lead to a poor-performing system and customer dissatisfaction.

State of Charge

State of charge in the air-conditioning system refers to the amount of refrigerant in the system compared to how much should be in the system. It is best tested by removing all the refrigerant from the air-conditioning system, comparing that amount to the manufacturer's specifications, then recharging the system with the proper amount of refrigerant as specified by the manufacturer. If the air conditioner cannot be discharged for any reason and the state of charge needs to be determined, then you can get a rough indication by performance testing the air-conditioning system on the vehicle (see the Diagnosis section). But the best way to make sure the air-conditioning system is properly charged is to remove all refrigerant from the system, measure it, and reinstall the proper amount back into the system.

Undercharge

An undercharged air-conditioning system contains less refrigerant than the system calls for. For example, if the system is designed to hold 16 ounces of refrigerant, but only 8 ounces are removed during the system evacuation, then the state of charge is half of what it should be. The condition of this air-conditioning system would be referred to as undercharged. Systems become undercharged for two primary reasons: (1) There is a leak in the system, or (2) someone working on the vehicle previously did not properly charge the system. Either way, you should plan on leak testing the system to verify that there are no leaks. The following two sections will discuss how each type of system responds to an undercharged condition.

Orifice Tube System

In a typical orifice tube system, an undercharge of refrigerant causes a rapid cycle time (rapid turning on and off) of the air-conditioning clutch, which itself is caused by the low-pressure cycle switch. This switch, located on the low-pressure line or in the accumulator housing, is designed to turn off at a predetermined pressure, such as 20 pounds psi, or 138 kPa. This prevents the air conditioner from running when there is not enough refrigerant with oil in it passing through the compressor to keep it lubricated, which can cause it to fail. In an undercharged system, it is most likely running fairly close to the predetermined turn-off pressure all the time. When the compressor turns on in this scenario, it draws refrigerant from the low side, creating lower pressure. As a result, the clutch will cycle frequently. The air from the ducts will feel warmer to the hand than it should, and the air-conditioning gauge readings will be lower than normal.

Thermal Expansion Valve System

The thermal expansion valve (TXV) system is more sensitive to a low charge than the orifice tube system. This is because the TXV low-pressure safety switch is located on the high side of the system. In the TXV system, there is not a pressure switch located on the low side of the system. If the charge is below normal, but not completely drained, the air-conditioning system can run below atmospheric pressure (vacuum) on the low side of the air-conditioning system because there is nothing to shut the compressor off. If the air-conditioning system is running below atmospheric pressure, any leak on the low side will allow air and moisture into the system. This may occur due to a blockage in the system or when the system is undercharged. If air and moisture contaminate the air-conditioning system, acid will be created in the oil, corroding the components from the inside out. The air duct will feel slightly cool, and the air-conditioning gauges on both the high and the low sides will read below normal.

Overcharge

Overcharging the air-conditioning system will affect the orifice tube and TXV systems in the same manner. Both the low- and the high side gauges will be running higher than normal. The air in the ducts will feel slightly cool, and the **air-conditioning compressor clutch** will run for a few seconds, then turn off for a long period of time. This behavior is caused by the system pressure running higher than it should and taking more time than it should for the pressure to return to normal levels for the clutch to engage again.

▶ Diagnosis

N59001

Following the strategy-based diagnosis process, start by verifying the concern. Use the results you obtained to research the possible causes and tests in the service information. Use that information and your understanding of the system to develop a focused training plan. Use that to work your way through a thorough testing of the system and components. Here, we cover some generic situations that you are likely to run across, and how to proceed.

The most common air-conditioning complaint that brings vehicles into the shop is simply that the air conditioner is not performing. Usually the root cause is leaks in the system. Other issues that could cause the same complaint are compressor failure, debris causing system blockages on the high or low side, or system blockages caused by moisture freezing in the system. These problems occur far less often than leaks, but they do happen. Normally, a visual check of the system and the gauge pressures will give you clues to what is going on.

- When both gauges read low pressure, the system may be low on refrigerant.
- If the gauges read low pressure on the high side, high pressure on the low side, or about equal pressures, it usually means that the compressor is not working.
- If the low side reads as a vacuum, there is a blockage in the low side of the system that is not allowing refrigerant to pass.
- Extremely high-pressure readings on the high side indicate that there is a blockage on the high side or an airflow restriction across the condenser.

On occasion, the problem of not blowing cold air is not the air-conditioning system at all; rather, the blend door or heater control valve is not closing adequately to keep air or hot coolant from passing through the heater core. This can be identified by checking pressures and determining that the air-conditioning system is performing correctly and moving on to the next logical cause, a heater that is not turning off.

▶ TECHNICIAN TIP

If the duct temperature is warmer than the ambient temperature after running for a minute or two, then the heater control valve is stuck on, or the blend door is not positioned properly. Use hose pinch pliers to shut off the flow to the heater core, and then retest the air-conditioning system.

Applied Science

AS-26: Temperature: The technician can define heat and determine its affect in vehicle components by taking temperature readings.

Heat may be defined as a form of energy that flows between two pieces of matter due to a difference in their temperature. Essentially, it is the condition of something being hot. A complaint that sometimes occurs is that a car's air-conditioning system blows hot air from the vents instead of cooling.

A check of the vent temperature with a thermometer will quickly confirm the complaint. There are two potential scenarios here:

If the vent temperature is not as cold as it should be, but is below ambient temperature, then you need to proceed with diagnosis of the air-conditioning system.

If the vent temperature is above ambient temperature, then you should first suspect the heater system.

Automotive heater systems on water-cooled vehicles function by allowing hot engine coolant to flow through a heater core, placed in the ventilation air stream inside the cabin. A heater valve is fitted to control the flow of coolant through the heater core. The faults described above are most often caused by a faulty heater valve allowing coolant to flow through the heater core while the heater is turned off.

Another driver complaint is unusual smells when running the air conditioner. The air-conditioning box has a drain tube that allows removal of water collected from the moist air as it condenses on the cold evaporator and drips into the box. If the drain is not working correctly, has been knocked off, or is plugged, then the water will provide a breeding ground for bacteria and the smell will be distributed throughout the cabin. This problem is fixed by clearing the drain, spraying the system with a disinfectant and allowing it to sit, then running the system on high to blow out the ducts and vents with all the doors open.

When diagnosing air-conditioning systems, you need to look at all of the factors involved. Always refer to service information, precautions, history, and service bulletins. Some of the variable factors in air-conditioning are the weather, air moisture content, temperature, overall condition of the vehicle, and type of vehicle. Some of these are listed here:

- Obviously the temperature is important, as the hotter or colder the outside air, the harder it is to heat or cool the inside of the vehicle.
- Moisture in the air, through either humidity or precipitation, can stick to the fins of the evaporator and freeze. Because the evaporator is used in air-conditioning and defrost to dry the air, a frozen evaporator can affect both the heating and the cooling systems.
- If a vehicle is running badly, the air-conditioning may be turned off by the powertrain control module (PCM) to try to correct the issue.
- Slipping belts will cause the compressor to spin slowly or not at all.
- Any leaks in passenger doors or windows will cause outside air to enter, which will have to be conditioned.
- The size and color of the vehicle make a difference. Big vehicles have more space, and the materials in the vehicle hold heat or cold; therefore, it takes longer to condition the air. A black vehicle will absorb heat and can be very difficult to cool on a hot sunny day.
- A cloudy day will decrease the heat load on the vehicle, increasing the air-conditioning system's efficiency and possibly giving the impression that the system is cooling fine.

All of these factors, and more, need to be taken into consideration when diagnosing faults in the system.

Performance Testing

N59002

The performance test and visual inspection of the air-conditioning system are the first step in the diagnostic process. *Performance testing* is the term used to describe the standard air-conditioning testing process. You need to look at all of the components and compare how the air-conditioning unit is functioning versus how it is designed to function. When testing the air-conditioning system, make sure to have all of the controls at the maximum settings. The heater fan should be set on the highest speed. The heater control knob should be on the coldest setting. The airflow should be set to the dash vents. Also, it is good practice to use an external fan to supply air to the condenser to simulate driving conditions. Failure to do this typically causes the system to fail the performance test.

The engine's rpm should be at 1200, which is the ideal and average rpm for the majority of air-conditioning systems on vehicles. Before testing the air-conditioning system, let the system stabilize for a few minutes to allow the refrigerant to equalize and the temperatures to reach the proper level.

To performance test an air-conditioning system, follow the steps in **SKILL DRILL 59-1**.

TECHNICIAN TIP

The one rule for diagnosis of an air-conditioning system is "when in doubt, suck it out." This means if you cannot find the problem while diagnosing, remove all refrigerant and start over from a base charge. Starting with a known base charge removes any question as to whether the refrigerant is too low or too high or whether there is any moisture causing a problem in the system.

TECHNICIAN TIP

Performance testing the air-conditioning system must be done before and after the repair. Doing so helps ensure that you identify any faults before starting the repairs, as well as verify that the faults are completely repaired, and nothing new showed up.

Abnormal Noises

N59003, S59001

The compressor is the most common source of abnormal noises arising from the air conditioner. If the compressor fails internally, it may make a knocking noise. Restrictions are another cause of odd noises. Restrictions rattle the air-conditioning pipes, especially if the restriction comes and goes. While performance testing the system, abnormal noises can occur while the air-conditioning system is operating. Listen for noises that may occur when the compressor clutch is engaged. If the noise disappears when the compressor clutch disengages, then the noise is from the air-conditioning system. Be sure to listen for noises under the hood as well as in the vehicle's passenger compartment. Turn the compressor clutch on and off, and move the controls on the dash, listening to determine whether the noises appear and disappear.

Applied Math

AM–33: Temperature: The technician can use standard and metric temperature measurement instruments to measure system temperature and determine conformance to metric specifications.

The most common air-conditioning concern encountered in the workshop environment is the air-conditioning system failing to effectively cool the cabin of the vehicle. Technicians need to determine whether the system is operating effectively before proceeding with diagnosis. This is determined by measuring the vent temperature as the air enters the vehicle cabin.

Once the vent temperature is measured, it must be compared to the manufacturer's specifications to determine if the system is operating effectively. If specifications are published in metric units, either a metric thermometer should be used, or readings must be converted from Fahrenheit to Centigrade, to determine if the measured temperature conforms to specifications

TECHNICIAN TIP

Here are some general rules of thumb when diagnosing an AC system:

- The duct temperature should be no greater than 17°F (–8.3°C) above the evaporator temperature. If so, check the heater control valve or blend door.
- The optimum vent temperature is from 38°F to 48°F (3.3–8.8°C), depending on outside temperature and humidity.
- The air-conditioning clutch cycle should not allow the duct temperature to have greater than a 5° swing from high to low while staying within the range of 38°F to 48°F. If the swing is out of the 5° range, the thermal switch or the pressure cycle switch may have to be replaced, if the vehicle is so equipped.

SKILL DRILL 59-1 Performance Testing an Air-Conditioning System

1. Turn on the vehicle. Place a fan in front of the vehicle to simulate the airflow that occurs when driving.

2. Close all windows. Turn the air conditioner to its maximum cold setting.

3. Raise the engine rpm to 1200–2000. Check the vent temperature using a thermometer. Compare the temperature recorded to the diagnostic chart in the service information.

Inspecting the Condenser for Airflow Restrictions

N59004

The condenser is normally inspected when high side pressures are too high and the system is not cooling well. Road debris, leaves, and animal fur and feathers are common culprits, but anything that sits in front of the condenser can cause this issue. Also, air dams play an important role in directing airflow through the condenser. Make sure they are installed properly and not damaged. This inspection should also be performed with any air-conditioning tune-up or inspection, such as those offered by many shops before summer. Use a flashlight on the back of the condenser. Look in from the outside of the vehicle for the light coming through the condenser (**FIGURE 59-2**). Move the light across the entire condenser. If the light is not showing, the condenser should be cleaned.

When cleaning the condenser for debris, shop air or a water hose may be used to remove the debris, but be careful not to fold over the fins. If the shop air and a water hose do not remove the restriction, the condenser must be removed and professionally cleaned.

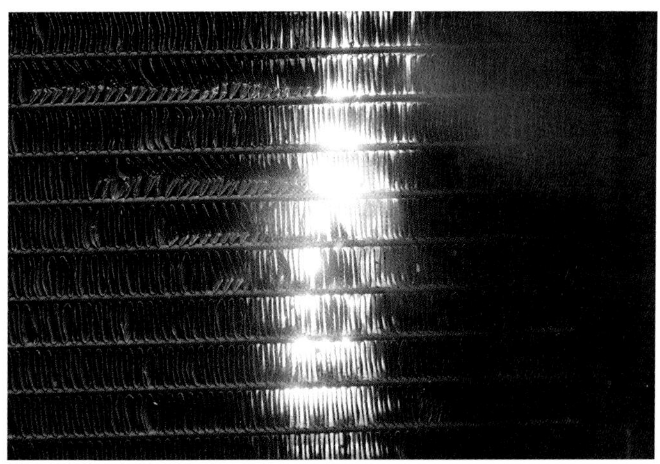

FIGURE 59-2 Using a flashlight to see if the condenser fins are restricted.

> **TECHNICIAN TIP**
>
> There are factory adjustments on both the thermal cycle switch and the pressure cycle switch. The adjustments are for achieving optimum performance at the factory. Trying to adjust either of the switches because of an incorrect temperature swing may mask the faulty switch for a short time. If these switches are incorrectly functioning, replacement is the proper and recommended procedure for the repair.

Inspecting the Evaporator Housing Water Drain

N59005

As the air-conditioning system is running, the evaporator sweats water throughout the day. The water collected has to be drained from the evaporator housing. A drain hose is connected to the evaporator housing and exits through the vehicle's firewall. Checking that the drain is not plugged is a common diagnostic procedure. If it is clogged, try to clear it with a rod or a blast of low-pressure compressed air. Once the water has completely drained, the task has been performed. Advise the customer to periodically check for a water puddle under the vehicle; if one is not present, the clog may have reoccurred, and removal of the air box may be necessary to open the air box and clean out any remaining debris.

To inspect the evaporator housing water drain, follow the steps in **SKILL DRILL 59-2**.

Eliminating Air-Conditioning System Odors

N59006

Air-conditioning systems are prone to unpleasant odors. Because warm, moist air is passed through the cold evaporator, most of the moisture condenses on the surface of the evaporator

SKILL DRILL 59-2 Inspecting the Evaporator Housing Water Drain

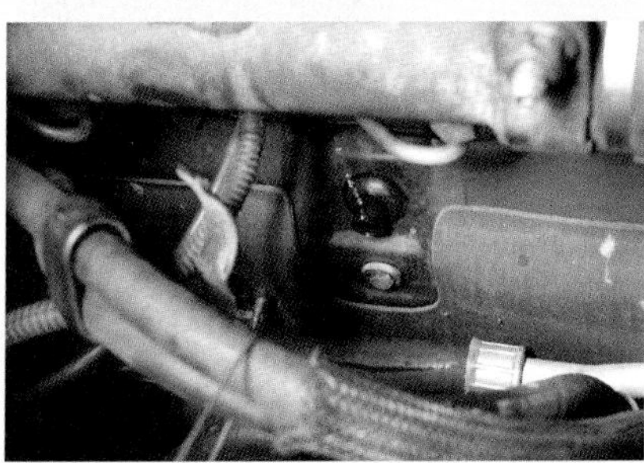

1. Determine that the drain tube is clogged by allowing the air-conditioning system to run while observing the drain tube for water drops. In a plugged drain, no water or very few drops are found.

2. Carefully use a rod to try to clear the inside of the hose.

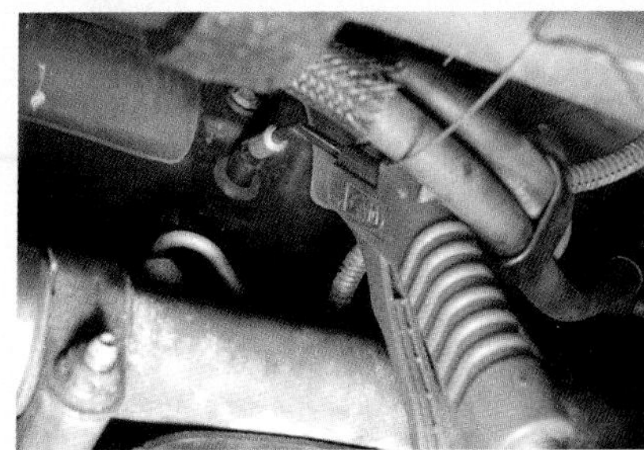

3. If that doesn't work try a quick blast of low-pressure shop air and an air nozzle up into the drain tube. Foul smelling water will come out of the drain tube when the clog is removed; stand back.

and drips to the bottom of the air box. This moisture then sits in the air box and ducts where it is dark, which is an ideal place for mold and bacteria to grow. The mold and bacteria produce odors that are directed at the passengers when the fan is turned on. The ducts also provide an appealing space that mice and other small animals use for building their nests. The nests themselves can give off odors, and sometimes one or more rodents die, giving off a very foul odor.

Identifying and eliminating the cause of the odor usually makes customers very appreciative. If you find that an odor is stronger in one vent position than in the others, the smell is likely coming from that duct. If the odor is in all positions, the problem is likely in the heater/evaporator housing. Keep testing until you can pinpoint the source of the odor. Check for odors as you try each of the following: selecting "fresh" air and turning the fan on high, selecting the dash vent, selecting each vent setting, turning the heater on high, and turning the air conditioner on.

If the drain tube becomes clogged with leaves or debris over time, the water will not be able to drain; it will stagnate and begin growing bacteria, creating an unpleasant odor. The solution to this concern is to unclog the drain tube. If odors continue to be a problem and the drain tube is open, the use of an anti-odor kit may be necessary. These kits require gaining access to the evaporator, either through a specified vent, complete removal of the evaporator, removal of the fan, removal of the resistor pack, or by drilling a small hole into the air box in a precise location. When access is gained to the evaporator, a spray chemical cleaner is used on the evaporator fins and allowed to dry. A follow-up chemical coating is then applied, which keeps bacteria from growing on the fins of the evaporator. If a hole was drilled into the air box according to the directions of the kit, a repair kit is usually included to repair the hole. Sometimes this will be a very strong piece of adhesive tape or a plug that will attach to the hole. Drilling a hole into the air box is not the preferred method and can result in air leaks and noise. Follow the directions of the kit to ensure a quality repair.

To eliminate air-conditioning system odors in the case of a clogged drain, follow the steps in **SKILL DRILL 59-3**.

Cabin Air Filter

Many vehicles now include a cabin air filter in the HVAC system to filter the air before it enters the system. The filter is housed in the air box and can be accessed from one of a variety of positions, depending on the vehicle. The access may be from under the hood near the firewall, under the windshield, or behind the glove box (**FIGURE 59-3**). It is usually fairly easy to remove and replace once you find the access cover. This filter should be inspected during every service and replaced according to the manufacturer's specified interval, typically once a year or every 12,000 to 15,000 miles (19,000 to 24,000 km). When inspecting the cabin air filter, use the same guidelines as for an engine air filter. Hold it up to a light, and look at it to see if the light shines through or if it is blocked from being dirty. Also check it for any cracks, tears, or deformities that would cause it to be ineffective (**FIGURE 59-4**).

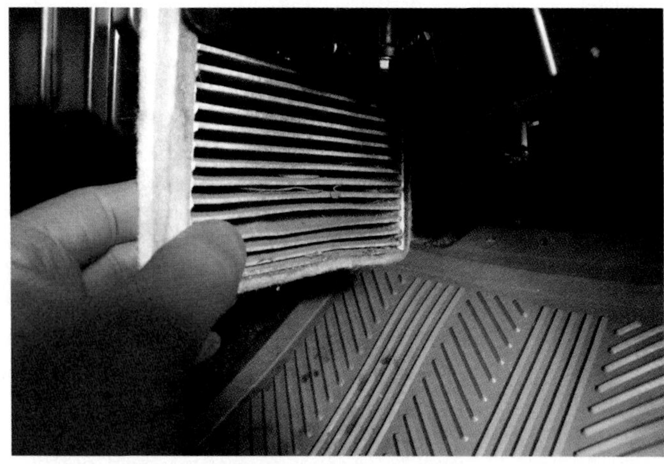

FIGURE 59-3 A typical location of a cabin filter.

SKILL DRILL 59-3 Eliminating Air-Conditioning System Odors

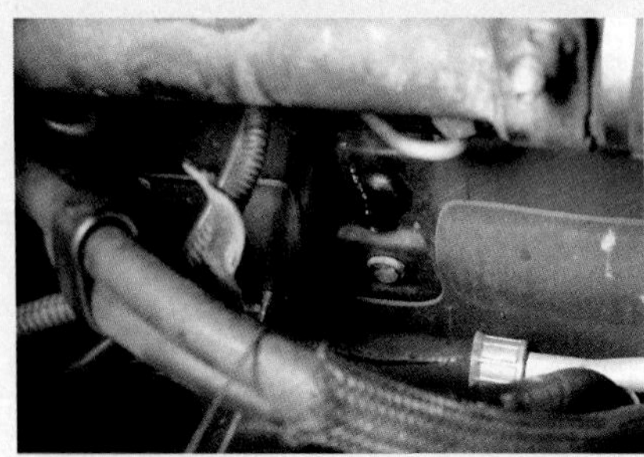

1. Verify that the drain is not plugged. Open all of the doors on the vehicle to allow it to air out a bit.

2. Turn the blower fan on medium speed. Operate the system in all zones to see if the smell is stronger in one position than the others. If the smell is equally strong in all positions, inspect the heater/evaporator housing.

3. If the odor is caused by mold or bacteria buildup, use an anti-odor kit to clean and kill any buildup.

FIGURE 59-4 Cabin filters. **A.** Clean. **B.** Dirty.

Air-Conditioning System Protection Device Issues

N59007

Within the air-conditioning system are protective devices designed to shut down air-conditioning operations to prevent further damage. These protection devices are generally thermal units, fuses, or circuit breakers. Generally, if one of these devices is triggered, the cause is overcurrent, meaning the current flow has become too high. Too much current flow can melt wires and parts. Overcurrents result from shorts in the wires or electrical components, such as the air-conditioning clutch coil. A short to ground on any of these gives a direct path for electricity or current to flow past the device it is supposed to operate and instead pass to the ground side of the battery. Once an overcurrent condition is detected, the source of the short circuit has to be identified. Repairs should be performed based on the specific cause of the overcurrent. This could be a wire that wore through the insulation and shorted out to grounded part; or it could be a component with an internal short. Shorted wires have to be replaced with the proper wire and repair method, and shorted components require replacement.

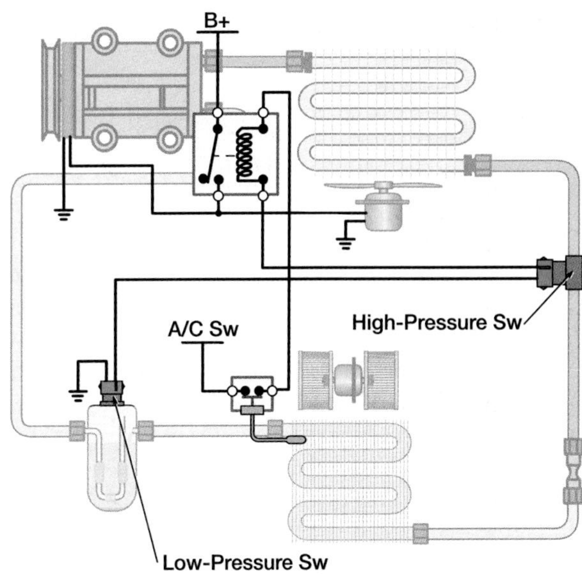

FIGURE 59-5 Typical pressure switches used to protect the AC system if pressures get out of specification.

Pressure switches, whether high or low, have a cutoff pressure to protect the air-conditioning system (**FIGURE 59-5**). A high-pressure switch located on the high side line may have a cutoff pressure of 400 psi (2758 kPa). At that pressure, the switch opens because the pressure is high enough to cause the **hoses** and **lines** to rupture. A low-pressure switch with a cutoff of 15 psi (103 kPa) will open because not enough refrigerant, which carries the lubricating oil, will be able to get to the compressor, and the internal moving parts may fail or overheat.

Computer-controlled air-conditioning systems have the same fail-safe devices as regular systems, but when the pressure switches sense a problem, they send that information to the PCM, which determines whether the system should be shut down until repaired. The PCM may also shut down the air-conditioning if there is a charging system problem or an overheating issue, and it determines that the extra load on the engine created by the air-conditioning system has to be shut down until the other problem is resolved.

If an electrical issue occurs, the problem must be tracked using a digital multimeter (DMM), which is the electrical diagnostic tool used to check voltage, current flow, and resistance of circuits. Once located, the component must be replaced or the wire repaired and the fuse replaced.

Low pressure normally requires refilling the air-conditioning system and finding and repairing the leak. The other issues require diagnosing and repairing the issue that caused the computer to shut off the air-conditioning system, then clearing the codes in the PCM so it will reset and turn the air-conditioning system back on. These problems are found using a scan tool to pull the codes, then diagnosing the code, using the service information.

▶ AC Maintenance and Repair Tools

K59003

To maintain and repair air-conditioning systems, you need the following tools:

- *Sealant detector:* Sealant detectors detect the presence of sealant in the air-conditioning system. Allowing sealant to be drawn into a refrigerant identifier or **air-conditioning machine** can ruin them (**FIGURE 59-6**).
- **Refrigerant identifiers**: Refrigerant identifiers hook to the low side fitting of the air-conditioning system to determine whether the refrigerant is pure or whether it is contaminated with another refrigerant or air. Identifying refrigerant can save the costly procedure of flushing out your reclaim/recycle machine and save you from having to dispose of contaminated refrigerant (**FIGURE 59-7**).
- **Reclaim/recycle machine**: If the air-conditioning machine is used for any air-conditioning work on a customer's vehicle, it is required to meet the EPA requirements. This machine hooks up to the air-conditioning service ports on the vehicle and can check pressures, and it has a storage tank to hold refrigerant (**FIGURE 59-8**). It also has a vacuum pump to evacuate air-conditioning systems. Refrigerant must be recovered or sucked into the machine before repairs are done to the air-conditioning system. Once the repairs are done, the air-conditioning system must be evacuated or vacuumed out to create a low pressure to remove moisture. After this, the machine normally can refill the vehicle's air-conditioning system either from the recycled tank, which contains refrigerant removed from vehicles, sent through a filter, then stored in the tank; or from the virgin tank, which contains new refrigerant. Most shops will draw from the recycle tank until it is too low to charge a vehicle, and then use the virgin tank.
- **Pressure gauge set**: A reliable pressure gauge set is one of the most important air-conditioning tools to obtain. Pressure gauge sets come in both mechanical and digital styles and are used to diagnose and service air-conditioning systems. Both types of gauge sets have a low side and a high side pressure gauge connected to a common manifold that can be used for manual servicing of the air-conditioning system. The hoses connect to each gauge and are connected to the manifold by valves and passageways in the manifold (**FIGURE 59-9**). The two knobs on the front of the manifold

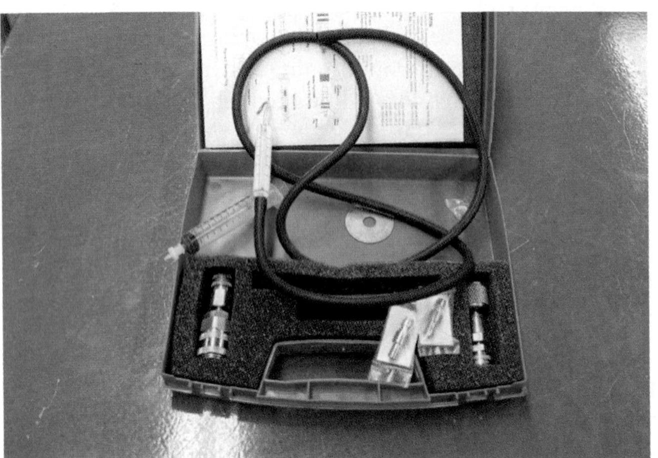

FIGURE 59-6 Sealant detector.

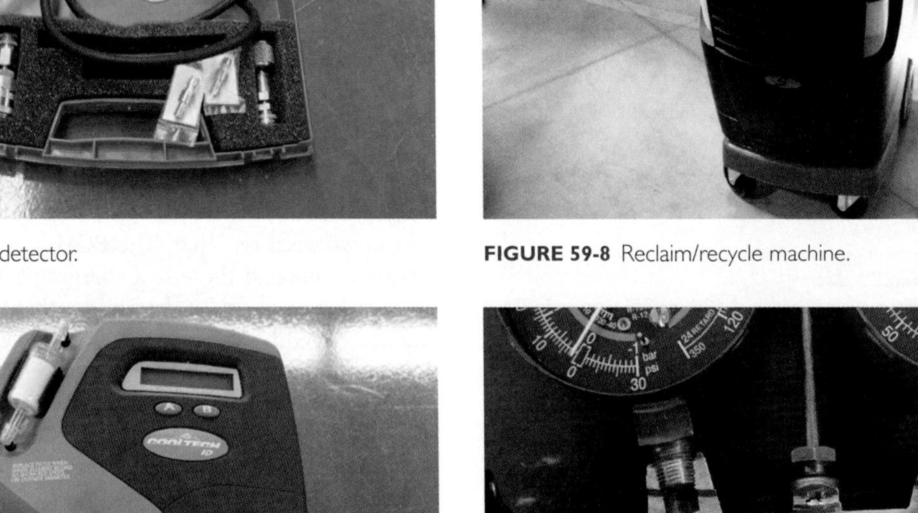

FIGURE 59-8 Reclaim/recycle machine.

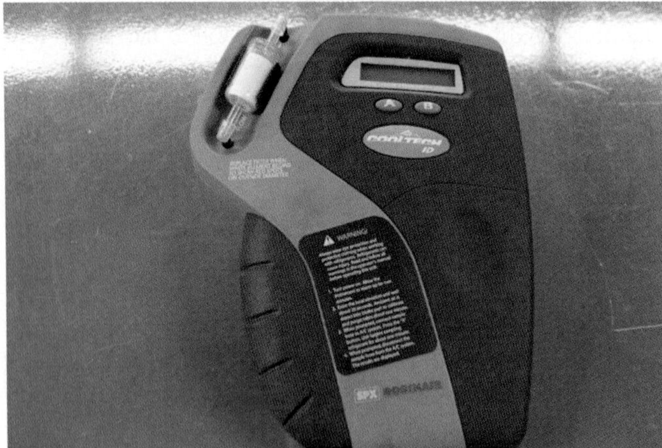

FIGURE 59-7 Refrigerant identifiers.

FIGURE 59-9 Internal passageways of the manifold and gauge set.

connect the high and low hoses to the service port on the manifold. When using the gauge set, make sure both service knobs are in the closed position before connecting them to a system. The hoses on the gauge set are connected to the gauge directly; opening the service valves only connects the passageways from the hoses to the service port. On a digital gauge set, you have to calibrate each gauge to zero so the pressures will read accurately. Follow the instructions in the operator's manual for the procedure on calibrating the gauge to zero.

Both the high-pressure and the low-pressure gauges have to be non-liquid filled if they are mechanical gauges. Oil-filled gauges do not allow the needle to move fast enough for you to see **pressure transients**, which are fluctuations observed as the gauges bounce between pressures, rather than holding at one position or moving smoothly up and down the gauge. Being able to see the needle oscillations provides valuable information when performing certain diagnostic scenarios. For example, a compressor with one bad piston causes the high side pressure to bounce when that piston is trying to push refrigerant. Another example occurs when water is in the system and freezes a line. When it starts to thaw, the gauge on the affected side bounces as the pressure changes from the temporary blockage. Gauge sets are also built into most R/R/R machines (Recover, Recycle, evacuate, leak test, and Recharge).

- **Anemometer**: The air coming out of the vent has to have enough flow to keep fresh air passing over the evaporator. Without air, the refrigerant will not boil. For the air-conditioning system to function properly, the velocity has to be at a minimum of 800 feet per minute (fpm). The most accurate way to test the velocity is to use an anemometer (**FIGURE 59-10**). Anemometers measure the velocity of the air in fpm.
- **Retrofit kit**: This aftermarket kit has the fittings and oil to change an R-12 unit over to R-134a (**FIGURE 59-11**).
- **PT chart**: A PT chart, which stands for pressure-temperature chart, is used to determine what the high- and low side pressures should be at a given outside temperature and humidity in a properly functioning system (**FIGURE 59-12**).

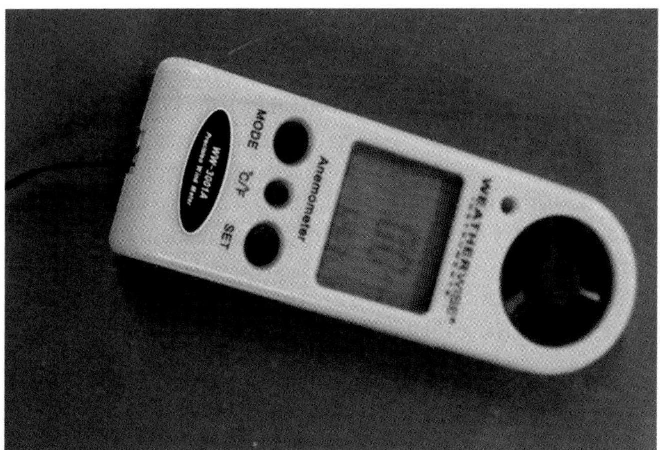

FIGURE 59-10 Anemometer.

FIGURE 59-11 Retrofit kit.

Page 1 of 1

R-134a PressureTemperature Relationship

Pressures shown in the charts are provided as a general guide and may not represent your exact readings. Ambient temperature is measured 2 inches in front of the condenser.

Low Side Readings (PSI)	Evaporator Temperature (Fahrenheit)	High Side Readings (PSI)	Ambient Temperature (Fahrenheit)
16-29	33-50	115-200	70-80
19-39	33-60	140-235	80-90
25-43	40-65	165-270	90-100
37-51	48-65	210-310	100-110

R-12 PressureTemperature Relationship

Pressures shown in the charts are provided as a general guide and may not represent your exact readings. Ambient temperature is measured 2 inches in front of the condenser.

Low Side Readings (PSI)	Evaporator Temperature (Fahrenheit)	High Side Readings (PSI)	Ambient Temperature (Fahrenheit)
10	2	130-160	60
12	6	140-170	65
14	10	150-180	70
16	14	160-190	75
18	18	170-210	80
20	20	180-220	85
22	22	190-230	90
24	24	205-250	95
26	27	220-270	100
28	29	240-290	105
30	32	260-310	110

FIGURE 59-12 PT chart.

- **Vacuum pump**: Vacuum pumps are rated in the total volume of air that can be removed every minute. They are used to create low pressure or vacuum in the system. The volume is measured in cubic feet per minute (CFM). The most common pumps are the 5-CFM and 6-CFM models, although some new air-conditioning machines have a 7-CFM pump.
- *Electronic leak detector:* Sometimes called a **sniffer**, this device is used to locate small refrigerant leaks by electronically reacting to refrigerant gas (**FIGURE 59-13**).
- *Dye leak detector:* This solution is used to detect refrigeration leaks (**FIGURE 59-14**).
- **Oiler**: This device is used to add oil to the air-conditioning system.
- **Vacuum gauge**: This gauge is designed to read negative pressure or vacuum (**FIGURE 59-15**).
- **Micron gauge**: This electronic device is designed to measure vacuum precisely (**FIGURE 59-16**).
- *Line wrenches:* Line wrenches are box-end wrenches or wrenches designed to go all the way around the nut or bolt, with a slot cut out of the box part to allow the technician to slide the wrench over the tube or hose.

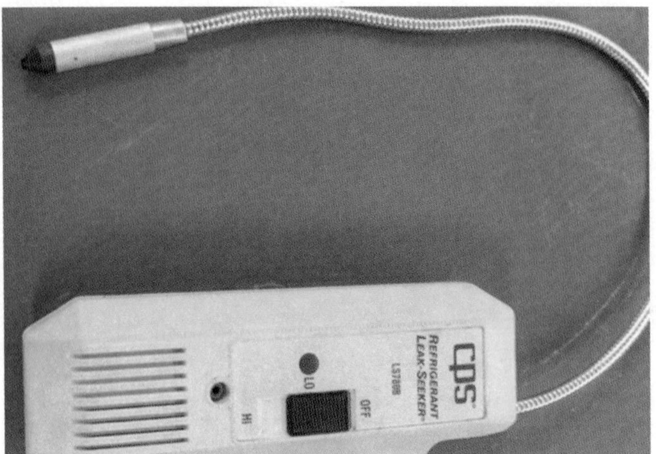

FIGURE 59-13 Electronic leak detector.

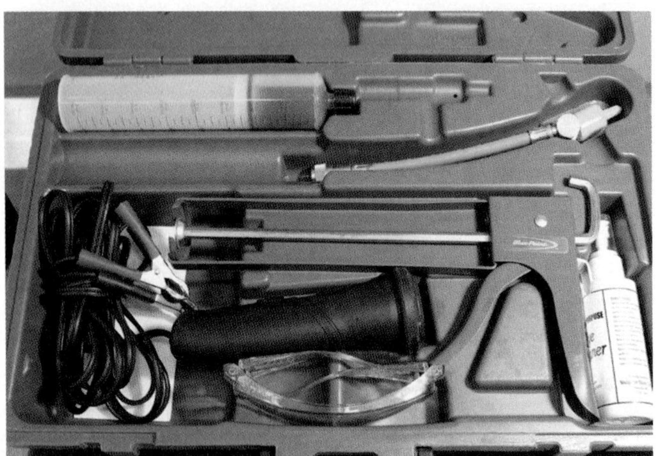

FIGURE 59-14 Dye leak detector.

FIGURE 59-15 Vacuum gauge.

> **TECHNICIAN TIP**
>
> Any air-conditioning machine can be used as a gauge set if the service valves are shut off.

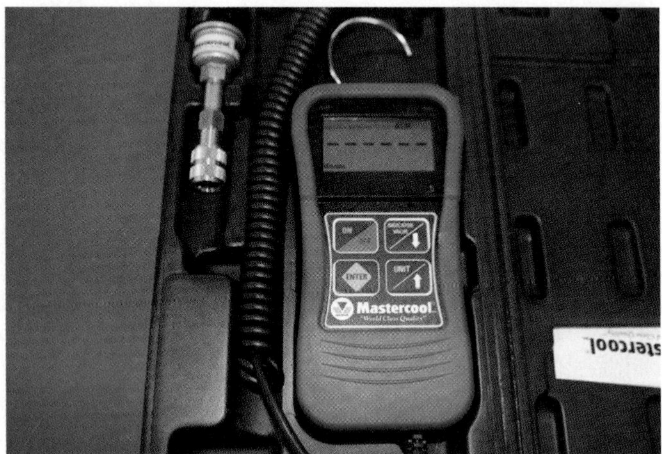

FIGURE 59-16 Micron gauge.

> **TECHNICIAN TIP**
>
> The low-pressure gauge is unique in that it displays both pressure and vacuum. It has a maximum pressure of about 120 psi (827 kPa) and a low pressure of 30" Hg (vacuum measured in inches of mercury, or 102 kPa). The low-pressure gauge displays vacuum readings for diagnostic purposes and to ensure that the proper vacuum levels are achieved when evacuating the system. The high side pressure gauge usually has a maximum pressure of about 500 psi (3447 kPa). Most systems will run with the high side pressure around 120 psi (827 kPa); but if the condenser fan is not functioning, the pressure could rise as high as 400 psi (2758 kPa). This is why you need such a large pressure range on the high side gauge. Otherwise, if the system pressure exceeds the maximum gauge reading, the gauge could be damaged, or it could even explode.

> **TECHNICIAN TIP**
>
> Modifying the air-conditioning machine or repairing the air-conditioning machine should be performed by an authorized dealer, not by the air-conditioning technician, in order to maintain its certification.

Leak Testing

N59008

The air-conditioning system is a sealed system that must not leak; however, leaks do occur and are a common reason that the air-conditioning system is not working properly. Leak testing should be performed on every air-conditioning system that is low on refrigerant or has just been repaired. Leak testing can be done by several methods: sniffer testing, dye testing, and nitrogen testing.

The sniffer and dye leak tests require that the air-conditioning system be relatively full of refrigerant, whereas nitrogen testing requires an empty air-conditioning system. If a nitrogen test is the method being used, first reclaim the air-conditioning system. Let's look closer at each of these types of tests.

Electronic Sniffer Testing

Types of electronic sniffers include **ultrasonic, heated diode**, and **corona suppression**. Ultrasonic sniffers are supersensitive microphones that can pick up the hiss of a leak that is too small to be heard by human ears. Heated diode sniffers use electricity to determine if there is a leak. R-134a passing by the diode is broken down, thereby creating an electrical current charge; the internal mechanics of the diode use this charge to determine how much refrigerant has passed. The diode beeps or chirps at increasing rates as the rate of refrigerant increases. The heated diodes are considered the best sniffers for R-134a. Corona-suppression sniffers function in much the same way as the heated diode except that any particle passing through the device, including dust, will set it off. With corona-suppression, refrigerant gas is used as an insulator to slow current between electrodes. Each of these electronic sniffers requires that the air-conditioning system have a charge of refrigerant for the leak detector to detect a leak.

The sniffer needs to be moved slowly and evenly around the components and lines. You determine where the leak is by watching the sniffer lights and listening to the sniffer sounds. Starting with the sensitivity on a high setting and turning down the sensitivity as the alarms and lights get louder and brighter will help you pinpoint the problem. Moving with a steady hand and taking your time is critical to using this device.

When using the sniffer, be sure there is no wind or draft on the vehicle. A draft or air movement over the vehicle could cause the refrigerant not to enter the sniffer or could mask the noise of the leak, preventing the sniffer from hearing the leak.

To test for leaks using a sniffer device, follow the steps in **SKILL DRILL 59-4**.

Dye Testing

Using ultraviolet (UV) light and refrigerant dye is a common way to test for leaks in an air-conditioning system. Some vehicle manufacturers add dye during the manufacturing process. When using dye for the first time on an air-conditioning system, the dye has to be injected as a concentrate into the system. Most machines have a special port for injecting the dye. Injecting concentrated dye into the low side of the air-conditioning system should be done with the vehicle off. When installing dye for the first time, the air-conditioning system must be operated so that the dye is distributed throughout all of its parts.

If dye has already been installed in the vehicle, or after the vehicle has been test-driven following a new dye injection, locating the leak requires using a UV light. When using a UV light, make sure all safety precautions are followed, including wearing UV safety glasses. Finding leaks with a UV light is easier in the dark; installing a cover over the hood of a vehicle will help you to find the leak.

To perform a dye test to find a leak, follow the steps in **SKILL DRILL 59-5**.

Nitrogen Testing

Nitrogen testing has been used in the home and transport refrigeration repair industry for a long time, but it is not as common in the automotive industry. It works great for testing discharged systems or for prechecking for a leak after repairs have been made and before recharging the system. Nitrogen testing requires a nitrogen tank, pressure regulator, adapters, and a **microleak detector** solution like Refrigeration Technology's BIG BLU. Soapy water is not recommended because of its corrosive properties and inability to show microleaks. This is why a microleak detector solution is recommended. The nitrogen is used to simulate working conditions of the air-conditioning system and provides enough force to make high-pressure leaks leak.

To perform nitrogen testing to find a leak, follow the steps in **SKILL DRILL 59-6**.

SAFETY TIP

Dry nitrogen is the only gas to be used in the air-conditioning system for leak detecting and removing traces of flushing chemicals from the system. Using any other propellant may cause harm and damage. Dry nitrogen is an inert gas and will not explode or ignite like other gases such as compressed air and oxygen.

SKILL DRILL 59-4 Testing for Leaks Using a Sniffer

1. Make sure the air-conditioning system has refrigerant. Slowly move the wand of the leak detector around and under all the air-conditioning lines and components. If a leak is detected, turn the sensitivity down to pinpoint the exact location of the leak.

2. Reclaim the refrigerant and repair the leak.

3. Always recheck for a leak after the system has been properly recharged.

SKILL DRILL 59-5 Testing for Leaks Using Dye

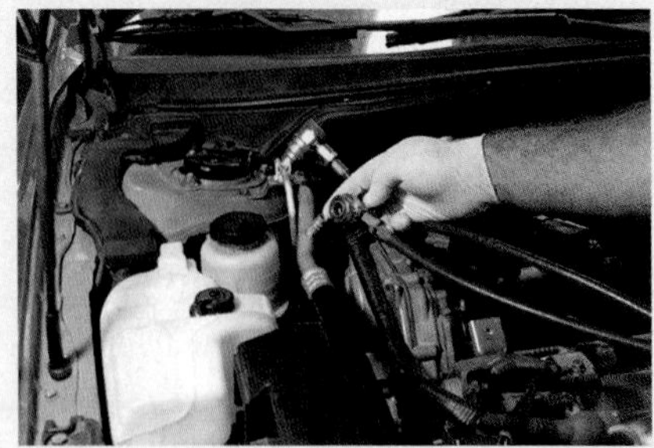

1. Turn on the compressor. If the compressor does not come on, add refrigerant in small increments until the compressor runs.

2. Add the dye through the low-pressure port using a dye injection system. Run the air-conditioning system to circulate the dye. The system may have to run for several minutes or even several days, depending on the size of the leak.

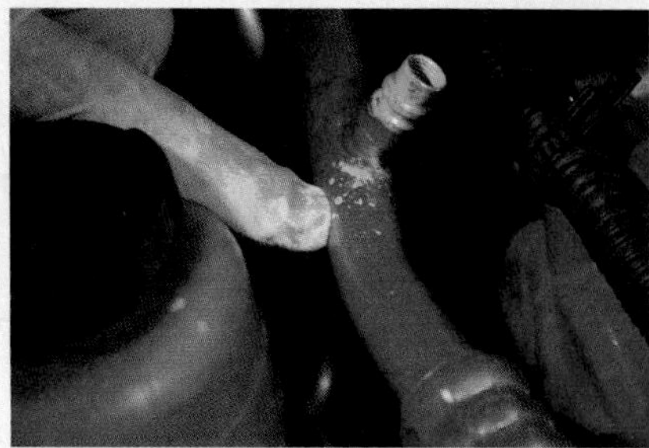

3. Using a black light, follow the lines, hoses, and components, looking for the orange or green glow of the dye, which indicates a leak. If a leak is not found in the lines or hoses, the blower motor resistor may need to be removed to see into the air-conditioning box. Repair the leak and retest.

SKILL DRILL 59-6 Performing Nitrogen Testing to Find a Leak

1. The air-conditioning system has to be reclaimed of all refrigerant. See Skill Drill 59-8 to perform this skill.
2. Attach the nitrogen hose of the nitrogen bottle to the low- or high side service port on the vehicle. Once the quick coupler on the nitrogen hose has been connected and the Schrader valve depressed, start the nitrogen leak test procedure.
3. Turn the regulator on the nitrogen tank to 100 psi (689 kPa). As the pressure in the system rises, look and listen for any leaks. If a leak is located, stop the process and repair the problem.
4. Turn the regulator on the nitrogen tank to 200 psi (1379 kPa). When 200 psi is reached, look for bulging hoses and listen for noises.
5. Turn the regulator on the nitrogen tank to 300 psi (2068 kPa). Once 300 psi is reached, let the system maintain that pressure for 1 minute. Do not exceed 350 psi (2413 kPa); doing so could cause the high-pressure relief valve to vent.
6. After 1 minute, start to spray the leak solution on all of the fittings, hoses, and air-conditioning system components. After spraying the leak solution on the air-conditioning system, it may take upwards of 20 minutes for a leak to present itself.
7. After the nitrogen test is finished, release the nitrogen to the atmosphere. Repair the leak.

SAFETY TIP

When nitrogen testing, safety glasses, gloves, and a face shield are required. Weak components or hoses may burst under nitrogen testing conditions.

Identifying the Refrigerant Type

N59009, S59002

The only way to determine the type of refrigerant in a system is with the aid of a refrigerant identifier. This easy-to-use device connects to the low side air-conditioning fitting and takes a reading of the refrigerant. It will give a readout such as 100% R-134a, or if the system is contaminated, you may see a reading such as 60% R-134a 40% R-12. As long as the refrigerant is pure, it can be recycled using a reclaiming machine and reused. If it is contaminated by the wrong refrigerant or the wrong oil, then it is considered a hazardous waste and must be put in a separate unit and disposed of according to state law. The air-conditioning sticker under the hood lists the type of refrigerant and how much should be in the system.

Refrigerant should be tested to determine its type any time the refrigerant will be recovered. If the refrigerant in the vehicle is contaminated and put in the recovery bottle of the air-conditioning machine, the whole bottle in the machine is contaminated and should be treated as hazardous waste. If it is not caught while in the machine, it will contaminate every vehicle that has air-conditioning refrigerant installed from that machine's recovery bottle.

Another condition that has to be identified is whether sealer has been added to the vehicle's air-conditioning system. Many small cans of refrigerant come with sealant inside. Sealer can ruin refrigerant identifiers and air-conditioning machines. Although not a lot of vehicles have sealer in them, all it takes is one vehicle to contaminate and damage your air-conditioning equipment. Most experts agree that sealant should *never* be used. If there is a leak, fix the leak! Identifying whether the system has sealer in it should be the first step in the refrigerant-identifying process.

To identify the refrigerant type, follow the steps in **SKILL DRILL 59-7**.

SKILL DRILL 59-7 Identifying the Refrigerant Type

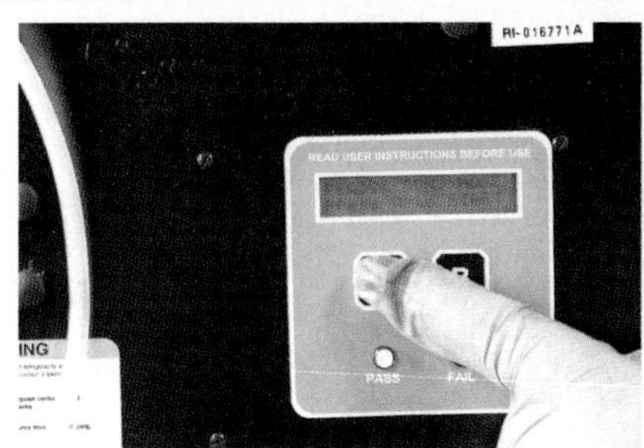

1. Use a sealant identifier to check whether there is any sealant in the system. If there is, notify your supervisor. Turn on the refrigerant identifier, and allow the machine to warm up.

2. Connect the refrigerant identifier to the low side of the air-conditioning system, and open the service valve. Follow the prompts on the refrigerant identifier, and record the refrigerant type and amount of air, if any.

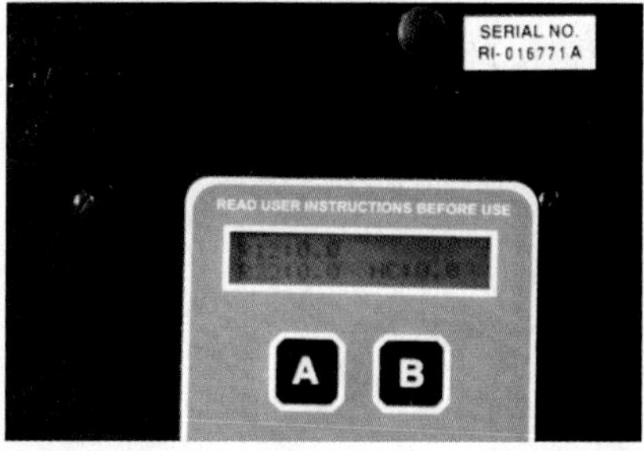

3. Turn off the valve, and disconnect the refrigerant identifier. If the refrigerant does not match the under-the-hood sticker, reclaim the contaminated refrigerant into a contaminated tank that is properly labeled for disposal. If the refrigerant is 100% pure and matches the under-the-hood sticker, continue to the next step.

4. After selecting the pressure gauge set, make sure the service valves on the gauges are in the off or shut position.

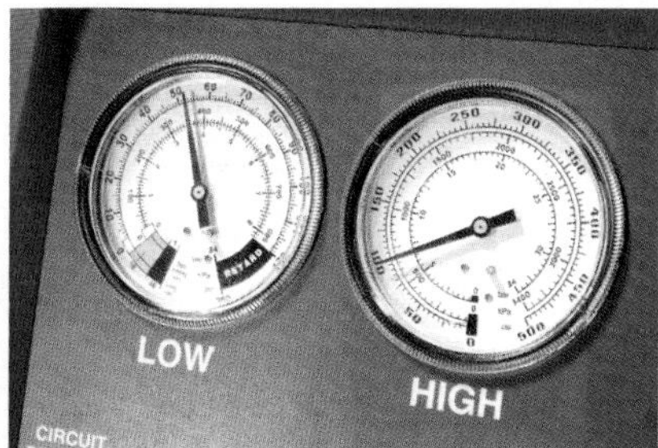

5. Connect the service chuck of the pressure gauge to the air-conditioning system. Open the valves, watch the high- and low-pressure gauges, and record the pressure. If the high- and low side gauges are the same pressure and you could not check for noncondensable gasses, use the PT chart to determine that as follows.

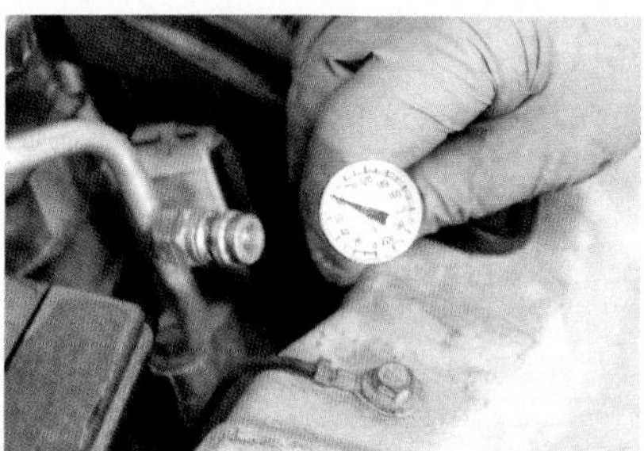

6. Measure the ambient temperature (6–8" [15–20 cm] in front of the condenser) and the pressure on the pressure gauge set. Compare the pressure to the PT chart for that type of refrigerant.

Reclaiming and Recovering the Air-Conditioning System

N59010

Any time the air-conditioning system needs to be opened up, the refrigerant must be removed in such a way that refrigerant will not be released into the atmosphere. Removing refrigerant is called **reclaiming** or **recovering**. The reclaiming process uses the air-conditioning machine to remove refrigerant from the system. It is also a good method of measuring the existing refrigerant charge in a vehicle, which will verify whether the system was undercharged, overcharged, or properly charged. Comparing the amount removed to the specified capacity allows you to determine the charge status. Another reason to recover refrigerant from a vehicle is to remove any noncondensable gases that are mixed in with the refrigerant.

To perform the reclaim process, follow the steps in **SKILL DRILL 59-8**.

Applied Science

AS-20: Liquids/Solids: The technician can measure the volume of a liquid in a system and compare it to the manufacturer's specifications.

Repairing refrigerant leaks in air-conditioning systems is a common workshop activity. Apart from repairing the cause of the leak, it is also important to know that air-conditioning compressors require internal lubrication, thus it is important to ensure that systems are refilled with the correct volume of oil when completing leak repairs.

When working on leak repairs, refrigerant recovery is a necessary step before disconnecting any part of the system. During recovery, any oil removed from the system in held in a graduated container by the recovery machine, to enable the volume of oil removed from the system to be measured. Oil removed from the system must be replaced with the same amount of new oil (if one ounce or less) to ensure adequate compressor lubrication. If the oil removed from the system is found to be contaminated, the system will have to be fully flushed and refilled with the amount of oil specified by the manufacturer.

SKILL DRILL 59-8 Reclaiming and Recovering the Air-Conditioning System

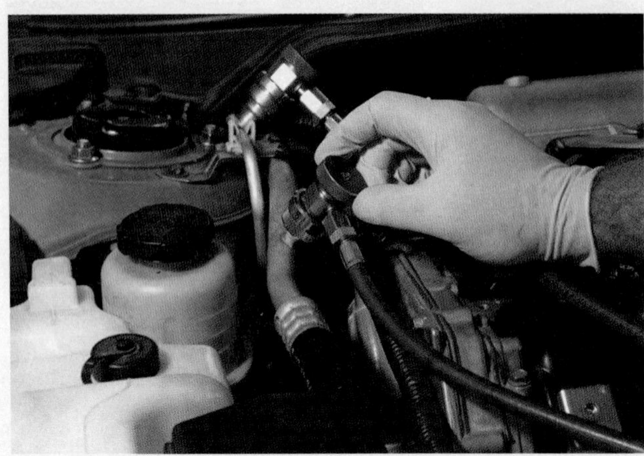

1. Identify the refrigerant using the refrigerant identifier, and verify that there is no sealer in the system. Start the reclaim process by hooking up the air-conditioning machine to the low and high side service ports.

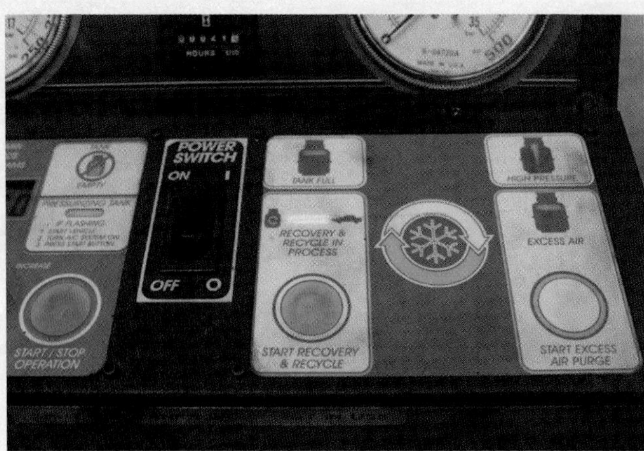

2. Open the valves and turn on the air-conditioning machine. Select the reclaim mode and follow the prompts on the screen.

3. When the refrigerant has been removed, record the amount and compare it to the sticker on the vehicle.

4. Record the amount of oil that is discharged. Install the same amount if the total oil removed is less than 1 ounce (30 mL). If more than 1 ounce, flush the air-conditioning system using a flush machine. Reinstall the proper amount of new oil to the system.

▶ TECHNICIAN TIP

The amount of refrigerant removed will give insight into the condition of the vehicle's air-conditioning system. Compare the weight removed to the under-the-hood sticker amount. If the amount recovered is less than the sticker indicates, there is likely a leak in the system. If there is more than the sticker indicates, then someone else has likely worked on the system, so you will want to give it a good inspection. If it had the correct amount of refrigerant, then the system most likely has no leaks.

Recycling, Labeling, and Storing Refrigerant

N59011

In certain situations, the air-conditioning machine may become full of refrigerant. This will prevent you from recovering any more air-conditioning systems. Removing the refrigerant and storing the refrigerant in an approved container will allow the air-conditioning machine to continue to be used. The refrigerant tank cannot be the virgin disposable tank that the new refrigerant came in. The new virgin refrigerant tanks are a one-use-only tank and have a one-way check valve in the valve stem to prevent refilling of the tank.

To recycle refrigerant, follow the steps in **SKILL DRILL 59-9**.

▶ TECHNICIAN TIP

Recycling refrigerant is one of the processes the air-conditioning machine can perform. After reclaiming the refrigerant, most air-conditioning machines automatically recycle the refrigerant. The filter system in the machine will return the refrigerant to a level of purity accepted by the EPA.

SKILL DRILL 59-9 Recycling Refrigerant

1. Obtain a DOT-approved refrigerant container. There are several sizes, from 15 lb to 500 lb (7 to 227 kg) tanks. The recommended size for most repair shops is a 60 lb (27 kg) container.
2. Connect the air-conditioning machine's hoses to the tank with the appropriate adapters in an air-conditioning retrofit kit. Use the high or low side adapters from the retrofit kit to connect the quick chuck fittings to the 1/4" SAE fittings on the tank.
3. Open the tank valves. Open the quick chuck Schrader depressors on the quick chucks. At this point, you are ready to charge the tank just as you would charge a vehicle.
4. Use the keypad on the air-conditioning machine to select the "charge" mode. Determine the amount of refrigerant you want to transfer, and enter that weight into the display.
5. Begin to charge the refrigerant tank. Do not fill the tank to more than 60% of the total gross capacity of the tank. The total gross capacity is written on the tank, and the technician must mathematically determine what 60% of it is.
6. After the tank is filled with refrigerant, label the tank with the type and the weight of refrigerant. By labeling the tank with the type and weight, the next technician can determine the type and whether there is enough free space to charge any more refrigerant into the tank.
7. Close the tank valves, and disconnect the hoses from the tank.
8. Store the refrigerant in a cool, dry place where the sunlight cannot directly hit the tank.

Evacuating the Air-Conditioning System

After all repairs are made, the air-conditioning system has to be evacuated to remove all of the moisture and air introduced into the lines from opening up the air-conditioning system. Evacuating the air-conditioning system requires either air-conditioning gauges and a vacuum pump or an air-conditioning machine. Either way, the process is the same. To remove moisture from the air-conditioning system, the pressure in the lines has to be lowered with a vacuum pump. This pressure drop lowers the boiling point of the water in the air-conditioning system. When the system at sea level reaches a vacuum of 29.5" Hg (100 kPa) or 10,600 microns, the boiling point will be lowered to 54°F (12°C). Because the boiling point is now lower than ambient air temperature, the moisture in the air-conditioning system's lines and components will boil. The vapor will be evacuated out of the air-conditioning system through the vacuum pump. There are two ways to monitor the evacuation process. We look at both of them next.

Evacuating the Air-Conditioning System, Using a Vacuum Gauge

If an air-conditioning machine is unavailable, the air-conditioning system can be evacuated using air-conditioning gauges and a vacuum pump. The processes are essentially the same. When evacuating the air-conditioning system, the vacuum gauge readings have to be timed and watched. Whether a vacuum gauge or a micron gauge is being used, the time that it takes to achieve water's boiling point pressure from a boiling point chart has to be timed (**TABLE 59-1**). The time it takes for the vacuum pump to pull down the air-conditioning system's pressure is important because there could still be small leaks that were not found during the leak test, and these leaks will cause the time it takes to reach the boiling point to be longer than it should be.

When timing the vacuum gauge readings, you can use the timer on the air-conditioning machine. The time it should take to achieve the boiling point pressure at the ambient temperature with a 5-CFM pump or greater is five minutes. If the vacuum

TABLE 59-1: Typical Boiling Point for Water at Various Vacuum Levels

Temp. F	Temp. C	Microns	Inches of HG Vacuum	Pressure PSIA
212	100	759,968	0.00	14.696
205	96.11	535,000	4.92	12.279
194	90	525,526	9.23	10.162
176	80	355,092	15.94	6.866
158	70	233,680	20.72	4.519
140	60	149,352	24.04	2.888
122	50	92,456	26.28	1.788
104	40	55,118	27.75	1.066
86	30	31,750	28.67	0.614
80	26.67	25,400	28.92	0.491
76	24.44	22,860	29.02	0.442
72	22.22	20,320	29.12	0.393
69	20.56	17,780	29.22	0.344
64	17.78	15,240	29.32	0.295
59	15	12,700	29.42	0.246
53	11.67	10,160	29.52	0.196
45	7.22	7,620	29.62	0.147
32	0	4,572	29.74	0.088
21	-6.11	2,540	29.82	0.049
6	14.44	1,270	29.87	0.0245
-24	31.11	254	29.91	0.0049
-35	37.22	127	29.9150	0.00245
-60	51.11	25.40	29.9190	0.00049
-70	-56.67	12.70	29.9195	0.00024
-90	-67.78	2.54	29.9199	0.00005
		0.00	29.9200	0.000000

Source: http://www.jbind.com/pdf/cross-reference-of-boiling-temps.pdf.

pump cannot pull the pressure in the air-conditioning system to the boiling point pressure in five minutes or less, the air-conditioning system has leaks that must be found.

If using a mechanical vacuum gauge, once the air-conditioning system reaches the boiling point pressure, the vacuum pump needs to be left on the air-conditioning system for an additional 10–15 minutes to fully evacuate the air-conditioning system. You will notice that the default time for most air-conditioning machines is 15 minutes, which allows for the air-conditioning system to reach the boiling point and for the additional 10 minutes for the moisture to be removed. Note that some evacuation equipment and some vehicles require up to 30 minutes of evacuation time. So check the service information for the equipment and vehicle you are using.

To use a vacuum gauge to evacuate an air-conditioning system, follow the steps in **SKILL DRILL 59-10**.

Evacuating the Air-Conditioning System Using a Micron Gauge

The micron gauge is used in conjunction with the mechanical pressure gauge on your air-conditioning machine; they are able to measure vacuum very precisely. Where a vacuum gauge has markings at 29", then 30" of vacuum, the micron gauge can measure accurately between 29" and 30" to determine exactly how well the system is being evacuated. Many vacuum pumps are measured in microns; the fewer microns it can go down to, the better. If you have two pumps, and one can draw down to 500 microns and the other down to 25 microns, the second pump will do a much better job of evacuating the system. Some air-conditioning machines have a micron gauge built in. With the micron gauge, the pressure is measured in microns, or one-thousandths of 1 millimeter. To put it in perspective, in 1 inch, there are 25,400 microns. So, 0 microns equal a perfect vacuum (roughly 30" Hg), and 25,400 microns equal approximately 29" Hg. This super-fine measurement of vacuum allows you to read accurately enough to ensure that all of the moisture has been removed from the air-conditioning system.

When the micron gauge reaches approximately 29" Hg, you still have 25,400 microns of mercury to obtain a perfect vacuum. A mechanical gauge is not accurate enough to see such small pressure movements. If using a micron gauge, watch until it hits 1000 microns or less. This reading means the boiling point of water has dropped to about 0°F (–17.8°C). A reading of 1000 microns or less tells you that all the moisture is gone, as long as it hasn't frozen.

SKILL DRILL 59-10 Using a Vacuum Gauge to Evacuate an Air-Conditioning System

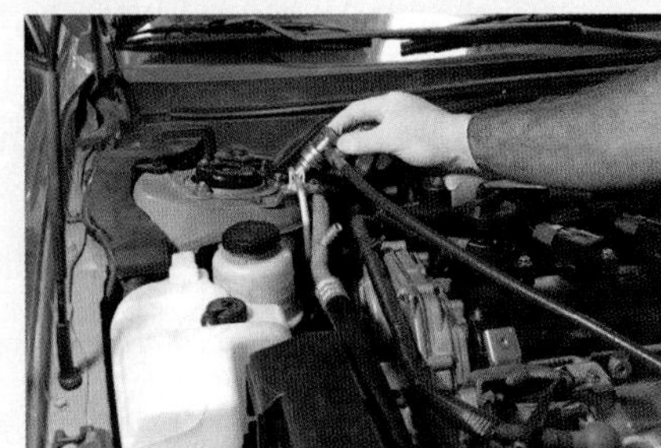

1. Hook up the high and the low side to the air-conditioning gauges.

2. Open the high and low side valves on the gauges.

3. Hook the pump to the center hose (normally a yellow hose), and turn on the pump. Time how long it takes to reach the boiling point.

4. After the required time (typically 10 -15 minutes, but some recommend 30 minutes), turn off the pump and close the valves.

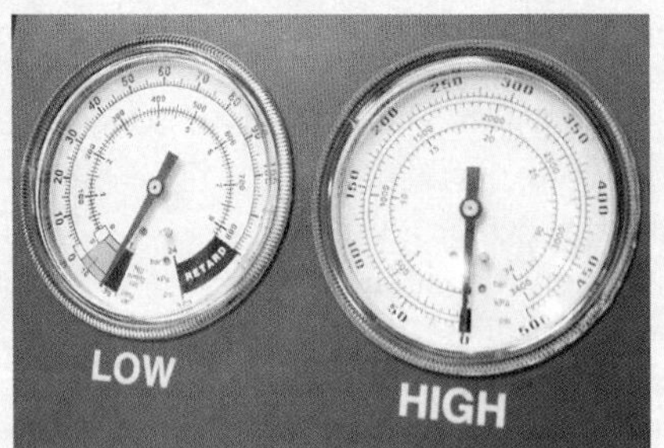

5. Note the vacuum readings. The vacuum should be around 28" Hg or 29" Hg (95 or 98 kPa). Watch the gauges for five minutes to be sure the vacuum holds. If the vacuum drops off, there is a leak in the air-conditioning system.

The micron gauge should reach the pressure of the boiling point of water for the ambient temperature within 5 minutes of turning on the pump. If it does not, then there is still a leak in the system.

To use a micron gauge to evacuate an air-conditioning system, follow the steps in **SKILL DRILL 59-11**.

> **TECHNICIAN TIP**
>
> When using a micron gauge, keep an eye on the numbers as they drop. The micron gauge may hover around 1300. When the numbers hover at 1300 microns, the air-conditioning system is boiling off moisture. Pay close attention at this point. If the numbers drop from 1300–1200 to 1100–1000, the air-conditioning system is okay. If the numbers fall from 1300 directly to 1000 and below, the moisture in the air-conditioning system just froze and will not be evacuated from the system. The freezing of moisture is due to the lack of heat around the air-conditioning system. If the pressure gauge falls from 1300 to 1000 instantly, the vehicle needs to be moved to a warmer environment. The freezing of water in a vacuum cycle becomes a problem around an ambient temperature of 60°F (16°C) and below.

> **TECHNICIAN TIP**
>
> If the micron gauge does not hit 1000 microns within the total vacuum time, then spin the clutch of the compressor and watch the pressure gauge. If the pressure rises, the clutch seal is faulty.

Inspecting and Replacing a Drive Belt

N59012, S59003

Inspecting the drive belt of the compressor is an important part of air-conditioning maintenance and repair. If the belt is worn, the engine cannot drive the compressor properly. On hot days,

SKILL DRILL 59-11 Using a Micron Gauge to Evacuate an Air-Conditioning System

1. Refer to the boiling point chart for the boiling point pressure in microns. Connect the gauge set and vacuum pump.

2. Turn the vacuum pump on to begin the evacuation process. Evacuate for a minimum of 15–20 minutes. Time the pressure gauge reading.

3. The pressure gauge should reach the boiling point pressure from the boiling point chart within 5 minutes.

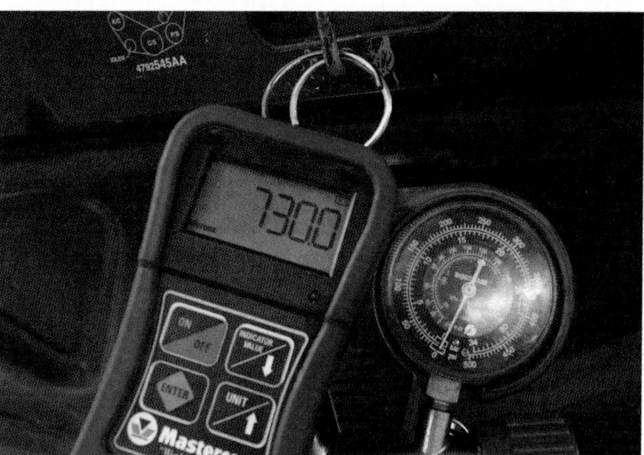

4. If the pressure gauge hits the pressure within five minutes, continue to allow the machine to evacuate the system until the micron gauge drops below 1000 microns. This indicates that the air-conditioning system is moisture free.

the refrigerant pressures are high and the compressor needs an effective drive belt to prevent slipping and causing excessive heat buildup around the clutch. The air-conditioning system requires that the compressor move the refrigerant at the correct velocity for the cooling of the passenger compartment to be at its peak efficiency. The drive belt is the first link in the system and needs to be in good condition and proper tension to ensure efficient air-conditioning. Refer to the Cooling Systems chapter to see the skill drill for inspection and replacement of a drive belt.

Inspecting, Testing, and Replacing the Compressor Clutch

N59013

Testing of the clutch assembly should occur when the air-conditioning pressure is high enough to turn on the clutch, but the clutch is not coming on or is not staying on. Typically, this component is checked using a DMM. The technician needs to determine whether the clutch coil is receiving full power and ground. If it is but still will not turn on, then the clutch coil should be checked for high resistance. If the resistance is too high according to the service information, it has to be replaced. This specification is usually 3–5 ohms. If full power and ground are not present, then wiring and switches need to be inspected and tested according to the wiring diagram (**FIGURE 59-17**).

The air gap is the space between the clutch assembly and the air-conditioning pulley. This gap has to be checked to be sure it is correct according to specifications. If the gap is too small, the clutch will drag on the pulley, creating heat and ruining the clutch. If the gap is too large, the coil will not be able to create a strong enough magnetic force to pull the clutch plate in and hold it against the pulley, or it will pull it in but

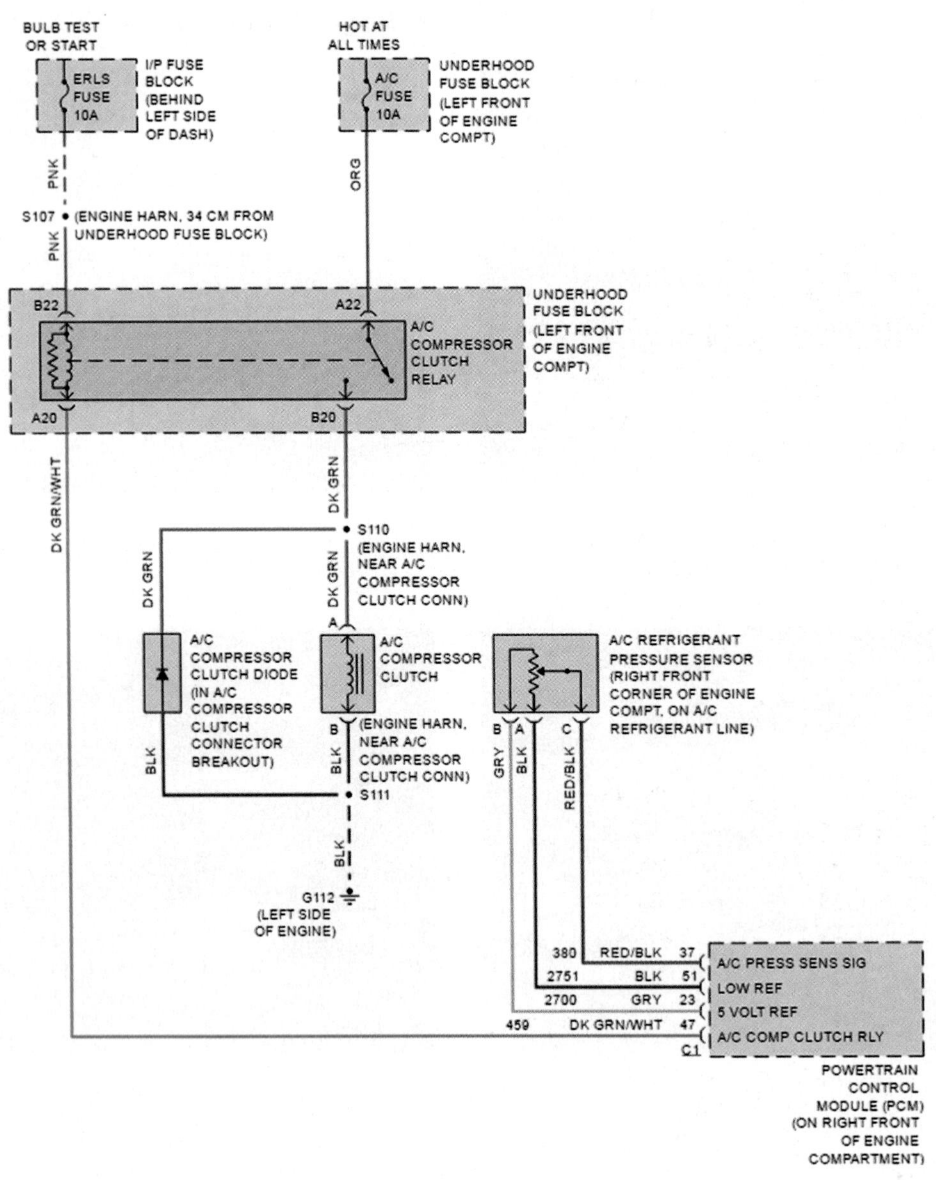

FIGURE 59-17 Typical non-electronic AC compressor clutch wiring diagram.

will not be able to hold it in place when the engine's rpm are raised. A voltage drop in the electrical circuit that energizes the clutch coil may also prevent the compressor clutch from engaging. If it does engage, it will not be able to grip tightly enough and will begin to slip, creating heat and scoring the clutch surface.

An air-conditioning compressor pulley or bearing may need service when noise is detected from the bearing as the engine is running. The pulley spins at all times when the engine is running, and the bearing may fail and make noise. If the bearing has failed, then the pulley will have to be removed. When replacing the compressor clutch, the pulley must be removed as well.

TECHNICIAN TIP

A brown dust on the air-conditioning compressor clutch indicates that there is slippage that could be caused by low voltage and current to the compressor clutch circuit. Slippage could also be caused by the clutch air gap being set too large.

TECHNICIAN TIP

Removing the air-conditioning compressor clutch requires special tools. Each type of compressor clutch has its own mechanism and tools necessary for removal. Before attempting to remove the compressor clutch, make sure you follow the manufacturer's safety procedures, and use the proper tools for the job.

To inspect, test, and replace the compressor clutch, follow the steps in **SKILL DRILL 59-12**.

Removing, Inspecting, and Reinstalling the Compressor

N59013

The compressor needs to be inspected any time there is a compressor failure or leaks are located. Leaks are fairly common around the shaft seal, which can be replaced on most compressors once the clutch is removed. When installing a new compressor, check the tag for instructions on oil; the tag will state

SKILL DRILL 59-12 Inspecting, Testing, and Replacing the Compressor Clutch

1. Inspect the air gap of the clutch. With the clutch disengaged, use feeler gauges to measure the air gap between the drive pulley and the clutch drive plate. If the air gap is too small, follow the manufacturer's procedure to adjust the air gap.

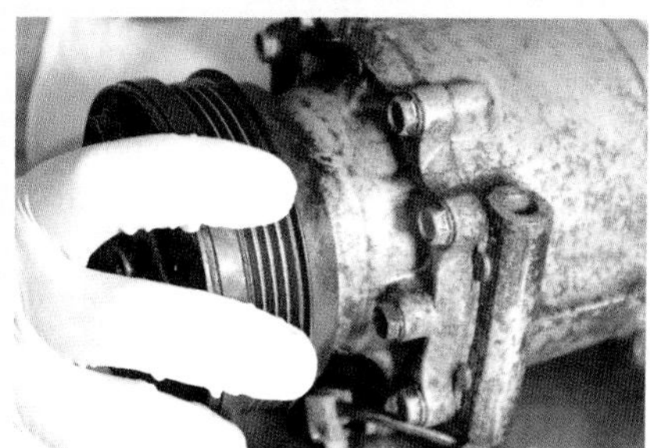

2. If the air gap is too large, check to see whether the pulley bearing is worn, causing the pulley to wobble and make contact with the clutch. Check for excessive plate wear. Replace if necessary.

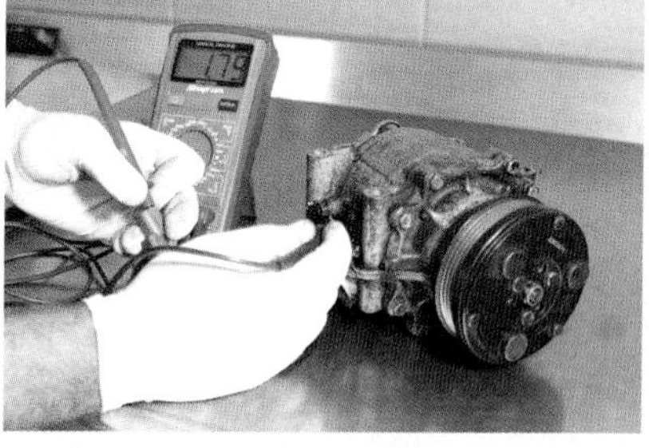

3. Measure the vehicle's voltage and clutch resistance (typically should be 3–5 ohms), and calculate the amperage. Measure the amperage, and compare to the calculation. If the readings are within 5% of each other, adjust the clutch air gap. If out of specification, diagnose the electrical system.

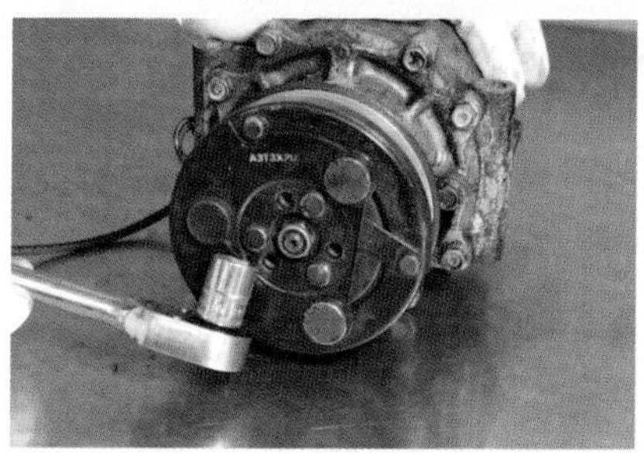

4. Remove the clutch plate center bolt or nut. If the compressor uses shims, be careful to remove these for later use.

5. Using the appropriate puller, remove the clutch plate.

6. Remove the pulley snap ring, and slide the pulley and bearing off the compressor.

SKILL DRILL 59-12 Inspecting, Testing, and Replacing the Compressor Clutch (Continued)

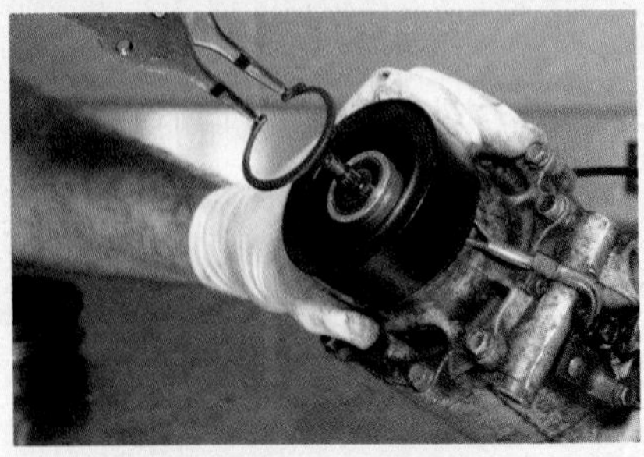

7. Remove the snap ring from the compressor clutch coil, and pull it from the compressor.

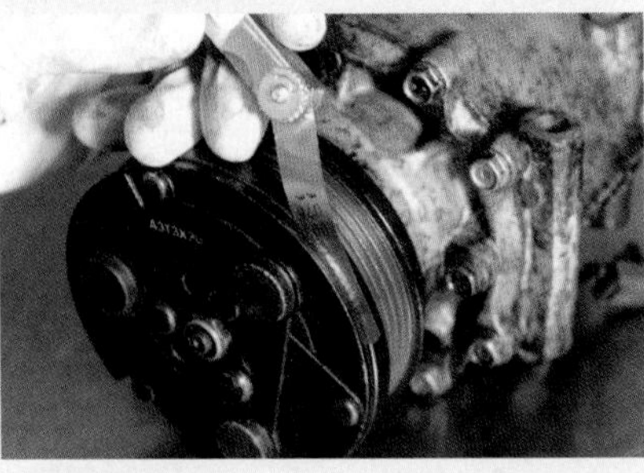

8. Reinstall the components in reverse order. When reinstalling the clutch plate, be sure to adjust the air gap (typical air gap will be 0.012"–0.025" [0.3–0.6 mm]), but check specifications.

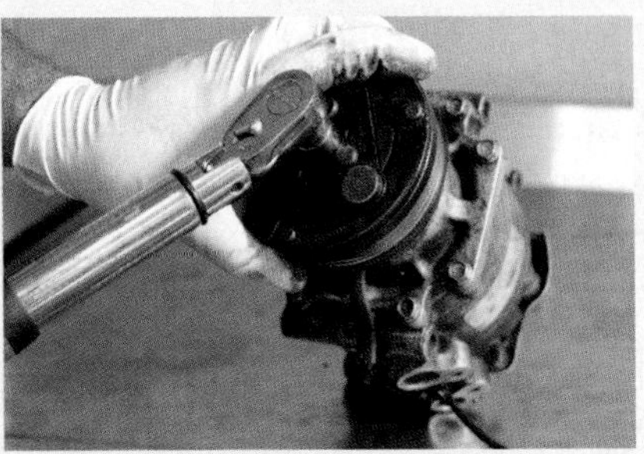

9. Torque the center bolt or nut to specifications.

whether the new compressor is wet (contains oil) or dry (no oil). If it is dry, read the instruction information for the correct amount and type of oil to add.

To remove, inspect, and reinstall the compressor, follow the steps in **SKILL DRILL 59-13**.

Additional Air-Conditioning Filters

N59014

Although thoroughly flushing a system usually removes all debris and contaminants, it is possible that small particles can be wedged or trapped in place. Installing an additional filter provides an extra level of protection for the compressor and expansion valve or orifice tube. Additional filters are designed to catch a small amount of debris that may still be in the system, not to substitute for an adequate flush. Additional filters are generally placed in the liquid line after the condenser, but before the restriction. Any straight section of tubing between these two points can be used. The proper amount of tubing is cut out of the system so that the filter can be installed in the line. Always follow the filter manufacturer's directions when cutting the air-conditioning lines and installing the filter.

To determine the need for an additional air-conditioning system filter and to perform the necessary actions, follow the steps in **SKILL DRILL 59-14**.

SKILL DRILL 59-13 Removing, Inspecting, and Reinstalling the Compressor

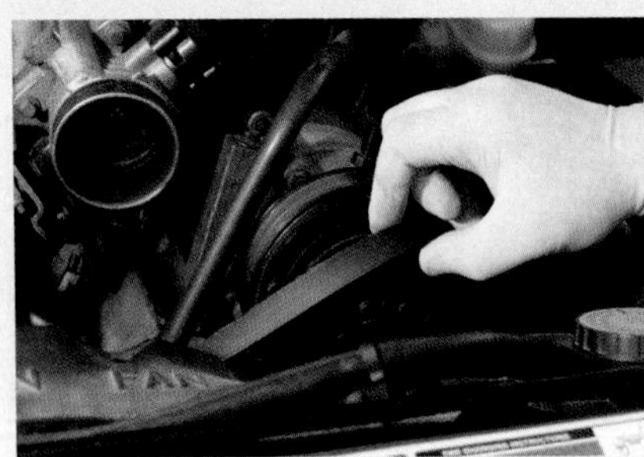

1. Reclaim the refrigerant, making note of the amount removed. Compare this amount to the factory specifications. Remove the belt, using a serpentine belt tool or the proper wrench. Remove tension from the tensioner and slide the belt off, noting routing for reinstallation.

2. Remove the hoses from the air-conditioning system. Cap the lines.

3. Unplug the compressor clutch.

SKILL DRILL 59-13 Removing, Inspecting, and Reinstalling the Compressor (Continued)

4. Remove the mounting bolts, following the manufacturer's specifications. Pour oil from the compressor into a graduated cylinder, and check the oil for acid, using acid test strips.

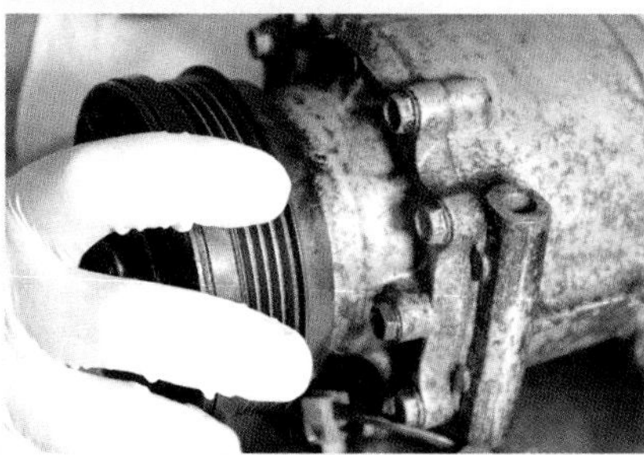

5. Check the resistance of the clutch coil, and check the pulley bearing. Check the terminals for corrosion. Check the tag on the new compressor to see if it came with oil installed. If not, install the specified amount of new oil.

6. Install the new compressor; tighten the mounting hardware to specifications.

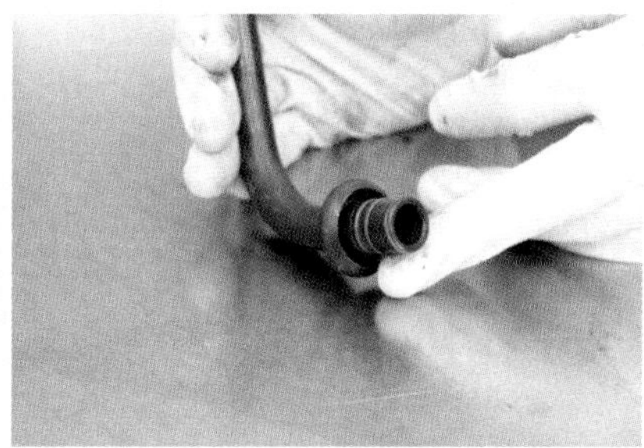

7. Install the air-conditioning lines with new sealing rings.

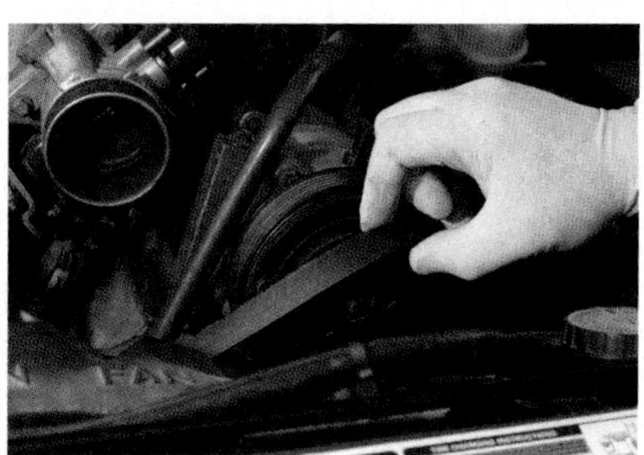

8. Plug in the compressor clutch, and install the drive belt. Proceed to the evacuation procedure.

SKILL DRILL 59-14 Determining the Need for an Additional Air-Conditioning System Filter

1. If the air-conditioning compressor failed, or if there is other debris in the system, an additional air-conditioning filter is recommended. Obtain the correct filter for the application you are repairing, and locate the best position to install the additional filter.
2. Cut the lines using a tubing cutter, carefully deburr the ends of the tubing, and clean out any shavings.
3. Install the filter, seals, and nuts in the proper order and orientation.
4. Tighten the nuts to the specified torque.
5. Evacuate and charge the system, then check for leaks.

Removing and Inspecting Air-Conditioning Mufflers, Hoses, Lines, and Fittings

N59015, S59004

The removal of hoses and lines is done for replacement purposes or to access another component. Some lines have male and female threaded ends that bolt together; others use retainer springs to snap them together. If the hoses are bolted together, line wrenches are used. If they use a spring lock, there is a spring lock release tool that comes in different sizes for different-sized lines. The spring lock tool goes between the fittings and spreads the spring so that the lines can be pulled apart. Any time lines are loosened, the O-rings in the fittings must be replaced or there is potential for leaks. O-rings are usually deformed during hose removal, so new ones of the correct size and type must be used when reinstalling the hoses. Make sure you use the specified O-rings that are compatible with the system you are working on.

When reinstalling hoses, tubes, and fittings, the air-conditioning fittings have to be clocked so there is no rubbing

SKILL DRILL 59-15 Removing and Inspecting Air-Conditioning Mufflers, Hoses, Lines, and Fittings

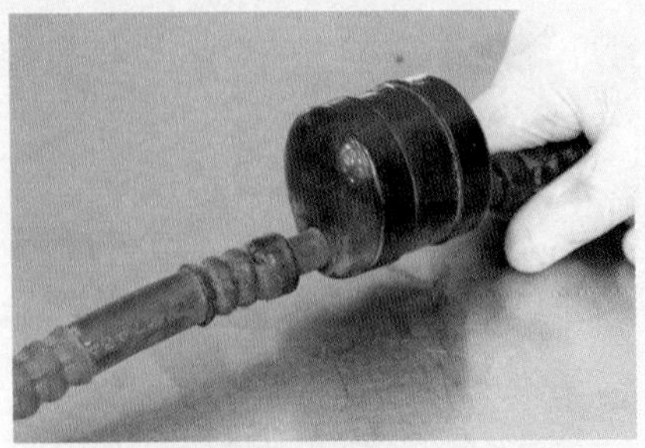

1. After performing a reclaim, remove and inspect the muffler, hoses, lines, and fittings. Shake the muffler and listen for noise. If the muffler sounds like it contains loose debris, it is faulty and has to be replaced.

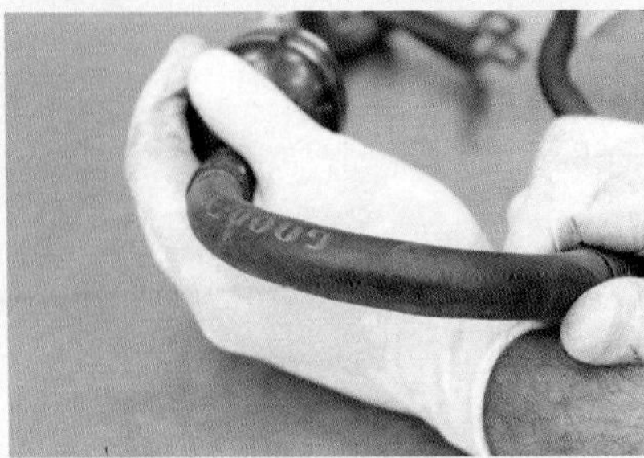

2. Inspect each hose to ensure that it is limber and pliable, not stiff and brittle.

3. Clock all hoses, fittings, and components so they are not rubbing against other components. Replace, lubricate, and install O-rings to prevent refrigerant leaks at fitting points.

with other components. *Clocking* means noting the routing of the hose so it can be reinstalled correctly. The hoses and lines must be kept away from heat sources in the engine that could melt them, and moving parts that could wear a hole in them.

To remove and inspect air-conditioning **mufflers**, hoses, lines, and fittings, follow the steps in **SKILL DRILL 59-15**.

Removing, Inspecting, and Reinstalling the Condenser

N59016

The condenser may be removed to access the radiator or if replacing the condenser because it is plugged, damaged, or leaks. Because the condenser sits in the front of the vehicle, it is damaged quite often in wrecks or by debris flying off the road. It may also have to be removed on some vehicles in order to be able to remove the engine. Same basic hand tools are necessary, as well as wrenches or spring clip removal tools. The process for removing the condenser varies by vehicle type, make, and model. Refer to the service information for specific steps. Also refer to the service information to determine how much oil is needed in the new condenser when it is installed.

To remove, inspect, and reinstall a condenser, follow the steps in **SKILL DRILL 59-16**.

Removing, Inspecting, and Installing the Receiver/Drier or Accumulator

N59017

The receiver/drier or **accumulator** must be removed and inspected whenever there are leaks in the system, or the desiccant is failing to dry the refrigerant. Desiccant failures arise when the system is left open for periods of time, and the desiccant absorbs too much moisture. It can also occur if the

SKILL DRILL 59-16 Removing, Inspecting, and Reinstalling the Condenser

1. After reclaiming the air-conditioning system, remove all necessary components to access the condenser. Refer to the manufacturer's specific procedures on component removal.

2. Remove the inlet and outlet lines on the condenser.

3. Remove the hold-down bolts for the condenser.

SKILL DRILL 59-16 Removing, Inspecting, and Reinstalling the Condenser (Continued)

4. Install the new condenser. Fasten the condenser with the mounting hardware.

5. Install new O-rings on the air-conditioning lines.

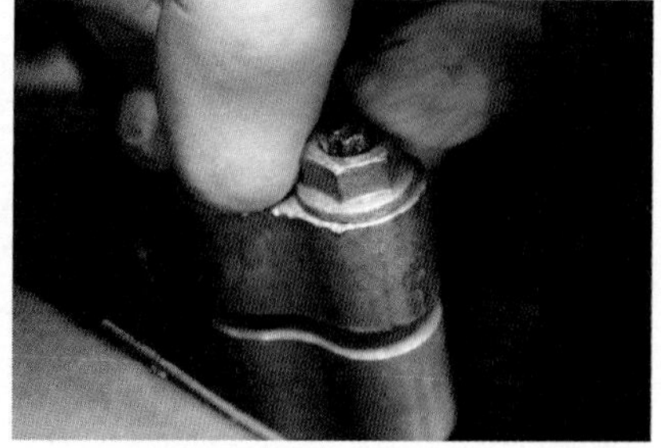

6. Install the inlet and outlet lines. Replace all of the components in reverse order from removal. Evacuate the air-conditioning system if all other air-conditioning repairs are complete.

desiccant bag ruptures, and desiccant has entered the rest of the system or if the compressor has come apart internally. The process for removing the receiver/drier or accumulator differs greatly from vehicle to vehicle. Refer to the service information of the vehicle for the specific procedure. Also refer to the service information to determine the amount of oil needed.

To remove, inspect, and install the receiver/drier or accumulator/drier, follow the steps in **SKILL DRILL 59-17**.

Removing, Inspecting, and Installing a TXV

N59018

A TXV would be removed for inspection if the diagnostic process led to the TXV being stuck open or closed, or restricted by debris, which would be determined by gauge pressures and air-conditioning performance. If stuck open, the high side

SKILL DRILL 59-17 Removing, Inspecting, and Installing the Receiver/Drier or Accumulator

1. After reclaiming the air-conditioning system, remove the inlet and outlet lines from the receiver/drier or the accumulator.

2. Remove the mounting clamps from the receiver/drier or accumulator; remove the receiver/drier or accumulator.

3. Lubricate and install new O-rings on both the inlet and the outlet lines.

4. Install the new receiver/drier or accumulator, and tighten the clamps. Install the inlet and outlet lines.

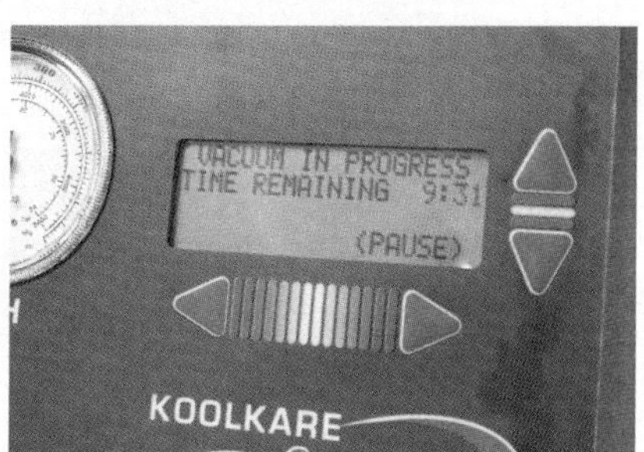

5. Proceed to the evacuation if all other repairs were made.

pressures would be low, and the low side pressures would be high. If stuck closed, the high side pressures would be high, and the low side pressures would be low and could even go into a vacuum.

When working with the expansion valve, caution must be used in all aspects of the removal and installation of the valve. Typically, the TXV is mounted on the front of the evaporator. If the inlet and outlet lines on the evaporator use threaded fittings, a line wrench must be used so the fittings are not damaged or rounded off. A second wrench is used to hold the other end of the line so you do not twist it and break it off. One nut will be on the line; the other nut will be on the TXV. Twist the nut on the line, and hold the nut on the TXV so as not to deform or break the TXV. A backup wrench on the expansion valve must be used also. The lines from the evaporator are aluminum, and the fittings are typically brass, so the use of an open-ended wrench will distort the brass and cause the fitting to leak. The exact procedure for removing, inspecting, and installing a TXV varies widely from vehicle to vehicle and by TXV manufacturer. Refer to the service information for the precise process.

To remove, inspect, and install a TXV, follow the steps in **SKILL DRILL 59-18**.

Removing, Inspecting, and Installing an Orifice Tube

The orifice tube has to be replaced when the same gauge readings as a blocked TXV are observed, as it can block up with debris in the system. It should also be replaced if the compressor comes apart because it can catch debris from the compressor and become clogged. If the orifice is not serviceable, the entire line must be changed with a new one from the factory (there are aftermarket kits for making the orifice tube serviceable on nonserviceable vehicles). The process for removing, inspecting, and installing an orifice tube varies from model to model and vehicle to vehicle. Refer to the service information for the precise procedure.

To remove, inspect, and install an orifice tube, follow the steps in **SKILL DRILL 59-19**.

Removing, Inspecting, and Reinstalling the Evaporator

N59019

Removal of the evaporator is typically done only if it is being replaced (Figure 59-18). Often, the entire dash assembly must be removed before the evaporator box can be accessed;

SKILL DRILL 59-18 Removing, Inspecting, and Installing a TXV

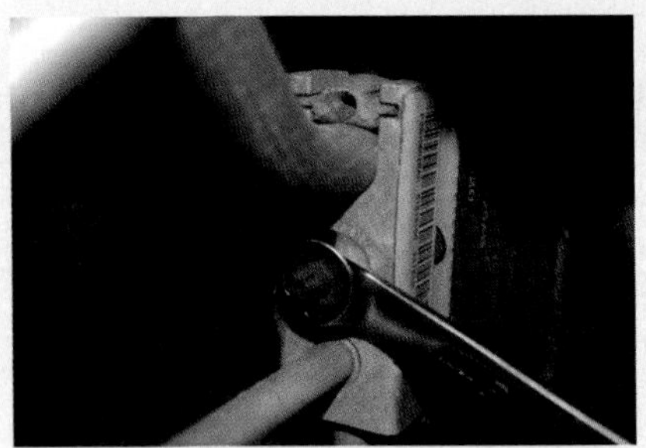

1. After reclaiming the air-conditioning system, use a wrench to loosen the bolt holding the liquid and suction line block to the expansion valve.

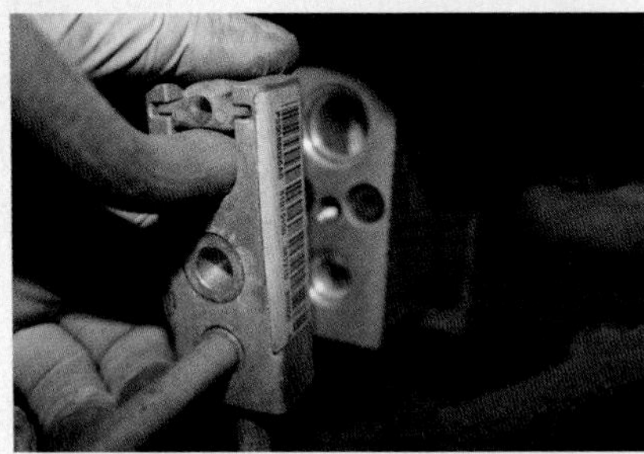

2. Remove the center stud going through the expansion valve. Gently remove the lines from the expansion valve.

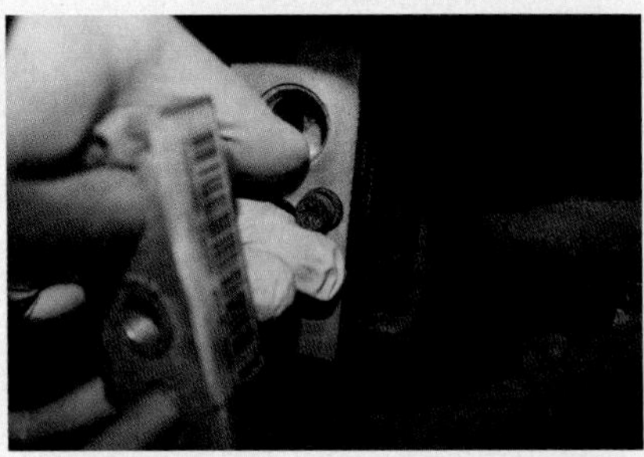

3. Remove the bolts holding the expansion valve to the evaporator inlet.

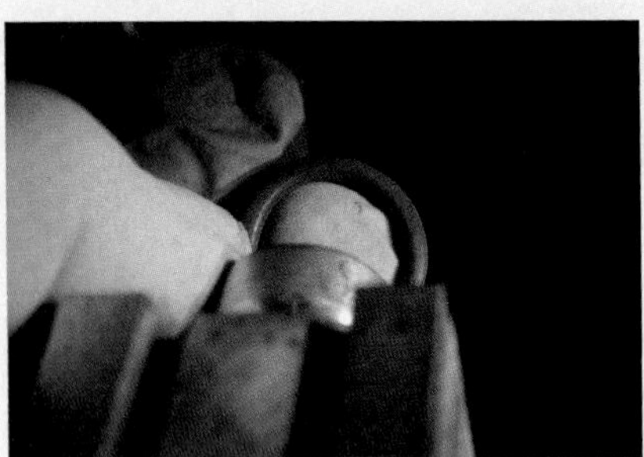

4. Install new gaskets or O-rings onto the new expansion valve and the suction and liquid lines.

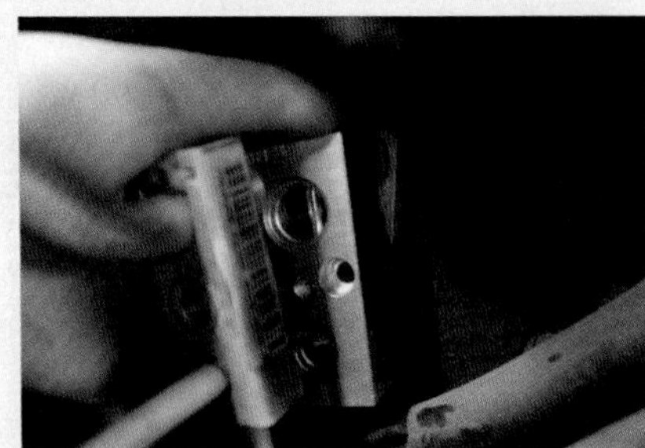

5. Install the new expansion valve, and torque the bolts to specification.

6. Install a new mounting stud. Reinstall the lines to the expansion valve, and tighten the retaining nut. Evacuate the system if all other repairs were made.

SKILL DRILL 59-19 Removing, Inspecting, and Installing an Orifice Tube

1. Locate the orifice tube, and determine whether it is serviceable. After locating the serviceable orifice tube, remove the line where the orifice tube is housed.

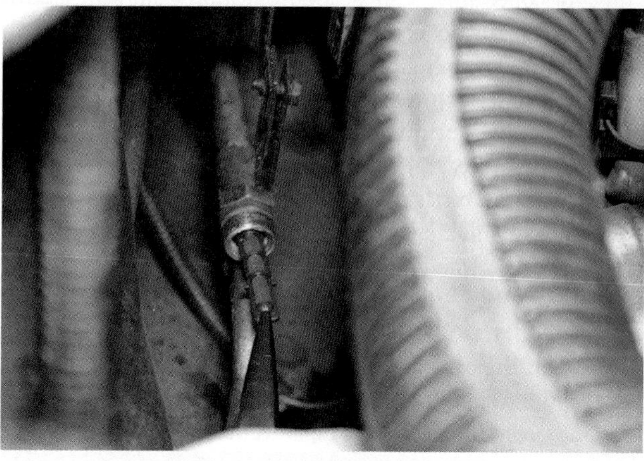

2. Using small needle-nose pliers or the orifice tube removal tool, pull out the old orifice tube from the housing. Note the color of the old orifice tube.

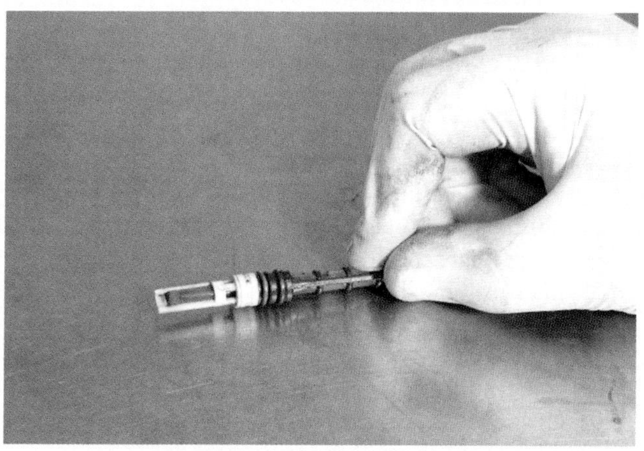

3. Select the new orifice tube, making sure to replace with the properly colored tube. Lubricate the orifice tube O-rings, and install the new one with the direction arrow pointing in the direction of refrigerant flow.

4. Lubricate and install new O-rings on the line, and reinstall the line. Proceed to the evacuation if all other repairs were made.

therefore, many basic hand tools are required. Once the dash is removed, the air-conditioning evaporator box is unbolted from the vehicle, and then the case can be split open by removing more screws to access the evaporator. The lines in and out of the evaporator must be removed before the box can be removed, and this is just like the other lines. The procedure for removing the evaporator varies greatly, as it is situated differently in each type of vehicle make and model. In fact, some late-model Chevy Camaros require that the windshield be removed to provide access to remove the air box. This procedure may even require a new windshield to be installed. Refer to the service information for the particular vehicle for specific instructions. To remove, inspect, and reinstall the evaporator, you should follow the manufacturer's service information.

Inspecting the Condition of Removed Refrigerant Oil

N59020, S59005

After the reclaim process, the air-conditioning machine will deposit any oil removed during the reclaim into a graduated container on the machine. By measuring the amount of oil removed, the technician will know exactly how much oil has to be put back into the system. Only new oil should be installed.

The acid test is used to check refrigerant oil for acidic contaminants and water intrusion. This is a critical part of vehicle HVAC maintenance because acids are capable of destroying the metal lines and compressor; however, it is often overlooked in shops. There are many different makes of acid test kits available from many different companies, though all are composed of either a test strip dipped into the refrigerant oil (**FIGURE 59-18**) or a tube for collecting a sample and then adding a testing chemical. The refrigerant oil changes the strip color or turns color in the sample tube, which is then compared to a color chart supplied by the manufacturer to determine if any acid is present. Be sure to read the test manufacturer's instructions; some tests must be performed with the compressor off the vehicle, whereas others require the refrigerant oil to be drained before testing. There are tests available that can be installed on the service port, where it collects a small amount of the refrigerant oil.

To inspect the condition of removed refrigerant oil, follow the steps in **SKILL DRILL 59-20**.

Now that the air-conditioning system has been diagnosed and tested, the system fault can be repaired. Take the appropriate actions and precautions that the repair requires. At this point, the air-conditioning system can be flushed and the base oil installed if needed. If less than 1 ounce (30 mL) of base oil has to be added to the system, this step will wait until after the evacuation is complete. Finish the repair and continue to the evacuation.

FIGURE 59-18 Typical acid test strip being used.

▶ TECHNICIAN TIP

As a refrigeration technician, it is your responsibility to read the O-ring package and make sure the O-rings you are using are compatible with the type of refrigerant and refrigerant oil you are using.

Air-Conditioning System Flush

The air-conditioning system should be flushed if the compressor comes apart or if the desiccant bag breaks open. In both of these situations, the loose parts or desiccant pellets will start to move through the system, causing blockages. Another reason to flush the air-conditioning system is contaminated oil. A special flush machine is used to flush the air-conditioning system. However, not all of the components are designed to be flushed, such as the compressor, orifice tube or TXV, accumulator, and receiver filter drier. Be sure to check the manufacturer's service information to verify which components can safely be flushed.

After the system has been reclaimed, remove the hoses from the air-conditioning machine, and hook up the flush machine. Several different types of machines are available, but normally a cleaning agent is poured into the machine before an air hose is plugged into the machine. A valve on the machine releases the cleaning agent through the system, pushed by the shop's air pressure. Another valve is used to push air through the system to remove the cleaning agent. Some shops prefer to use nitrogen to blow any residual cleaning agent from the system, as it contains no moisture, unlike air. The system is now ready to have the base oil added to each of the components, reinstalled, and the system recharged. Follow the instructions provided with the flushing machine carefully.

SKILL DRILL 59-20 Inspecting the Condition of Removed Refrigerant Oil

1. Look at the old oil, and check its condition. If the oil is black or acid test strips indicate the presence of contaminants, the air-conditioning system has to be flushed of all the contaminated oil.

2. Check the under-the-hood sticker for the proper system oil and the amount. Flush the air-conditioning system if necessary.

Determining Recommended Oil and Capacity

N59021

As previously discussed, the refrigerant R-12 used a mineral oil, but R-134a uses polyalkylene glycol (PAG) oil. HFO-1234yf uses a different PAG oil, which has a special acid reducer not found in R-134a PAG oil. PAG oil is available in many different viscosities. Each individual air-conditioning system has a capacity of a specific oil. Often, if you replace just one component, the capacity chart will list how much oil should be replaced for that particular component. Unless the entire system is replaced, some oil will remain in each of the system components. Read the manufacturer's service information for specific instructions.

Adding the Proper Oil Amount

Oil should be put back into the air-conditioning system after the evacuation cycle and while pressure is lower than atmospheric pressure in the air-conditioning system. The oiler uses atmospheric pressure to push oil into the air-conditioning system. Because pressure in a vacuum is less than pressure outside, the oil is easily pulled in by the low pressure or vacuum. What actually happens is that the pressures try to equalize. If you have oil in the hose that is allowing outside pressure into the vacuum or lower pressure, the higher pressure pushes the oil in as it tries to equalize pressures. Most air-conditioning machines have a port that a special graduated cylinder screws into. When the port is opened with the cylinder attached, the pressure equalization forces the oil in. You should replace the amount of oil that was removed during the reclaim process if it was less than 1 ounce (30 mL). If the oil removed during the reclaim process was more than 1 ounce, the air-conditioning system needs to be flushed, and a base charge of oil needs to be established.

To add oil to the air-conditioning system, follow the steps in **SKILL DRILL 59-21**.

Charging the Air-Conditioning System

Charging a system is performed after the air-conditioning system has been evacuated and any oil is replaced. The only tool needed is the air-conditioning machine, with the correct type and amount of refrigerant in it. Charge the air-conditioning system to the proper factory level by reading the under-the-hood sticker or obtaining the amount from the manufacturer.

If charging on the high side only, when the charging begins, watch the low side pressure gauge. If the low side gauge does not reach zero before 10 seconds, the TXV or orifice tube is bad. Filling from the high side causes the refrigerant to pass through the TXV or the fixed orifice tube. If it is plugged, it takes a longer time for the refrigerant to pass through it. The system will have to be recovered again and the component replaced.

If the pressure reaches a positive pressure number too rapidly, the reed valves in the compressor are likely bad. The pressure should rise on the low side gauge slowly and smoothly. Refrigerant is passing quickly through the compressor because of a broken valve. The compressor in this situation would need to be replaced. This test is valid only on machines charging from the high side only. Allow the machine to finish charging the refrigerant, and follow any prompts on the screen. Finally, performance test the system and check for leaks if any fittings have been opened.

To charge an air-conditioning system, follow the steps in **SKILL DRILL 59-22**.

SKILL DRILL 59-21 Adding Oil to the Air-Conditioning System

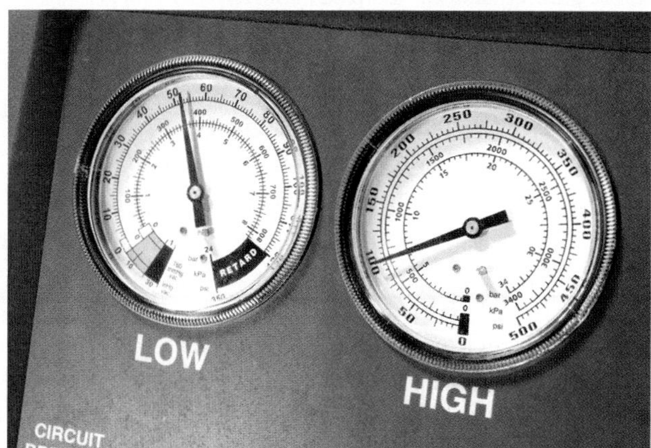

1. Evacuate the air-conditioning system, and watch the vacuum gauge to ensure that it does not bleed up. If no leaks appear after five minutes, install the oil. Open the oiling valve as recommended by the air-conditioning machine information.

2. Shut off the valve when the proper amount of oil has been added. Recharge the air-conditioning system with the amount of refrigerant noted on the under-the-hood air-conditioning sticker or in the service information (see Skill Drill 59-22).

SKILL DRILL 59-22 Charging the Air-Conditioning System

1. Hook the air-conditioning machine to the high- and low side service ports. Open the valves.

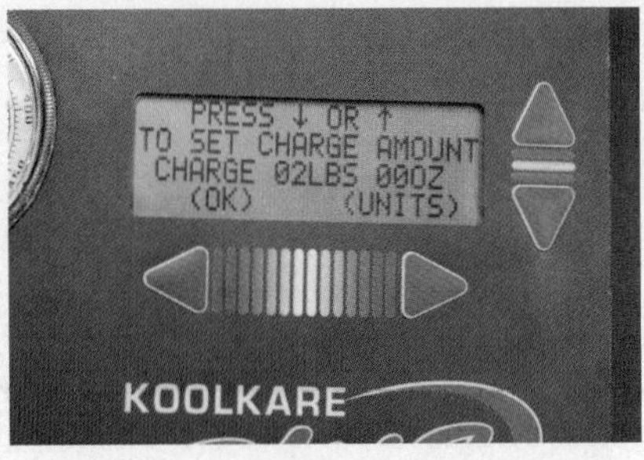

2. Follow the instructions display on the air-conditioning machine for charging. On the display, enter the amount of refrigerant to be charged and select "start" to begin charging the air-conditioning system.

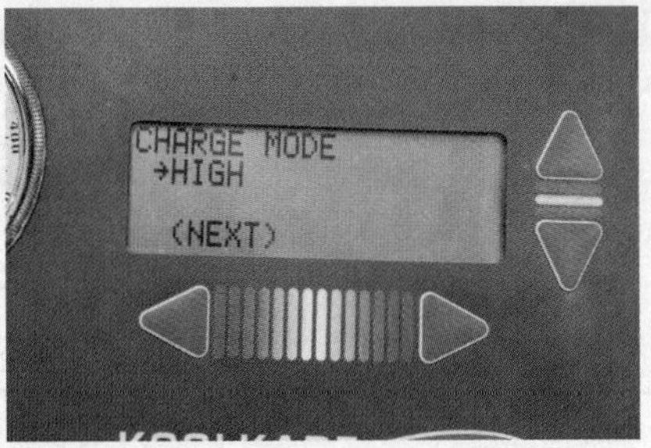

3. If charging on the high side only, watch the low side pressure gauge. The pressure should rise on the low side gauge slowly and smoothly, within 10 seconds. Allow the machine to finish charging the refrigerant, and follow any prompts on the screen. Performance test the system, and check for leaks if any fittings have been opened.

▶ TECHNICIAN TIP

Trying to charge gas to the high side with the air-conditioning system operating will result in the compressor pushing the refrigerant back into the pressure gauge set after the compressor engages. Once the compressor engages, you will be prevented from charging the air-conditioning system due to the high pressure generated from the compressor.

▶ TECHNICIAN TIP

Before charging the air-conditioning system, be sure to have a minimum of 5 lb (2.3 kg) plus your charge amount in the air-conditioning machine. Any less refrigerant than 5 lb will cause an insufficient refrigerant prompt on the air-conditioning machine. If this occurs, you will need to reclaim the air-conditioning system and add refrigerant to the air-conditioning machine.

Refrigerant Equipment Maintenance

N59022

Regular maintenance prompts may appear as you use the air-conditioning machine. If they do, then follow the instructions on the screen. When the air-conditioning machine alerts you that it needs service, you should contact the local service department for the machine.

Vacuum pump oil is one of the regular maintenance items that you will be prompted to change after approximately each 10 hours of vacuum use. Vacuum pump oil can be changed by the shop technician. Failure to change the oil will result in increased pump wear, as well as causing the pump to not be able to draw as deep of a vacuum. Changing the pump oil regularly will extend the life of the pump and make it reach maximum vacuum quicker. You will need a drain pan for the old oil, and new vacuum pump oil for the pump.

When it is time to change the vacuum pump oil, the machine will warm up the oil so it is easier to drain. In most machines, the vacuum pump is behind a cover inside of the air-conditioning machine, which must be removed before use. Remove the drain plug, and drain the oil into the drain pan. The drain plug is near the bottom of the housing. Refer to the air-conditioning machine user manual to find out what type and how much oil to install. Remove the fill plug, install the drain plug, and tighten. The fill plug will be near the top, but not normally all the way on the top. Use a funnel to add the new oil. Turn on the pump, and check that oil is halfway up the sight glass (if the pump has a sight glass). If the level is below half, then add a small amount of oil to bring it to the halfway mark. If it is over half, then drain some oil. Reinstall the fill plug, and check for leaks.

To change the vacuum pump oil in an air-conditioning machine, follow the steps in **SKILL DRILL 59-23**.

Inspecting and Testing the Heater Control Valves

The heater control valves should be tested if the customer complains of no heat or poor air-conditioning. The heater control valve, if equipped, regulates coolant flow to the heater core for heat and limits flow for air-conditioning. Some are vacuum controlled and some are cable controlled. Cable-controlled models must be checked to ensure that the cable is free and working. Vacuum-controlled models should be checked to ensure that the vacuum is supplied to the valve at the right time and that the valve will hold vacuum (**FIGURE 59-19**). If there is a problem with cable or vacuum operation, then the control head in the dash will probably have to be removed, inspected, and repaired or replaced. If the cable or vacuum is working, then the operation of the valve should be checked. Have someone move the temperature lever from hot to cold while you watch the valve (**FIGURE 59-20**). If the

SKILL DRILL 59-23 Performing Refrigerant Equipment Maintenance

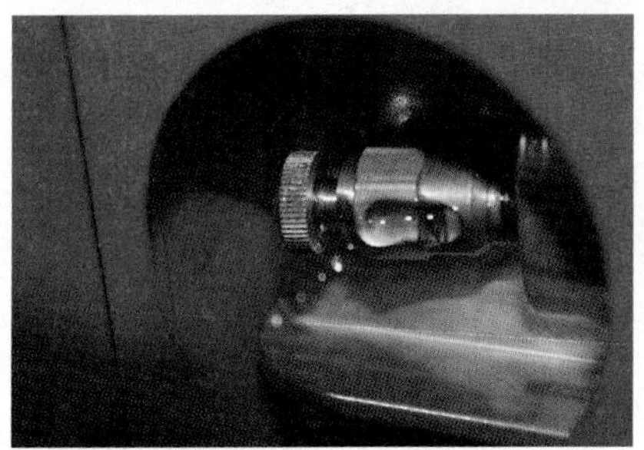

1. Remove the drain plug and drain the oil.

2. Remove the fill plug, install the drain plug, and tighten. Use a funnel to add the new oil.

3. Turn on the pump, and check for leaks. Check that the oil level is halfway up the sight glass. If under the halfway mark, add oil. If over, drain some oil and recheck. Reinstall the fill plug, and check for leaks.

FIGURE 59-19 Checking a vacuum-operated heater control valve diaphragm with a vacuum pump.

FIGURE 59-20 Checking control valve action while an assistant is moving the heat lever.

cable or vacuum servo is moving the control arm on the valve, then the valve is probably okay, and the problem is elsewhere in the system, although the valve could be plugged or slipping on the shaft. Therefore, you may have to remove it from the heater hose and visually inspect that the valve opens and closes.

Removing, Inspecting, and Reinstalling the Heater Core

Heater cores are usually removed only because leaks are present. Leaks from the heater core can result in coolant misting out of the vents or dripping onto the passenger-side carpet, or both. Because this process is complex and different on nearly every model of vehicle, you need to follow the steps in the vehicle manufacturer's service information. One thing to remember is that the heater core inlet and outlet tubes may be made of thin metal and soldered into thin metal tanks. Twisting the heater hoses to get them off almost always results in ripping the heater core tubes out of the tanks. Carefully use a hose slitter or knife to slit the hose lengthwise over the heater core tubes, and then peel the hoses off the tubes. The procedure for removing the heater core varies greatly, as it is situated differently in each type of vehicle make and model. Refer to the service information for the particular vehicle for specific instructions. To remove, inspect, and reinstall the heater core, follow the manufacturer's service information.

Wrap-Up

Ready for Review

- Technicians must be EPA 609 certified to handle refrigerants.
- Performance testing of the air-conditioning system is one of the first steps in diagnosis.
- A performance test involves external cooling for the condenser, running the engine between 1200 and 2000 rpm, placing the air-conditioning on max cold, the fan on high, and measuring the duct temperature, and then comparing the duct temperature to a system performance chart from the manufacturer.
- Servicing the air-conditioning system involves performance testing the system, diagnosing any issues, leak testing, identifying the refrigerant, reclaiming the refrigerant, performing any repairs, evacuating the system, recharging, retesting for leaks, and performance testing.
- Identifying refrigerant helps avoid mixing refrigerants, which cannot be reused and must be stored separately.
- Acid is formed inside an air-conditioning system when water is present.
- Acid test strips can be used to test for the presence of acid.
- An undercharged air-conditioning system contains less refrigerant than the system calls for.
- The one rule for diagnosis of an air-conditioning system is "when in doubt, suck it out."
- The compressor is the most common source of abnormal noises arising from the air conditioner.
- If the high side pressures are too high and the system is not cooling well, check the air flow through the condenser.
- Clearing the drain hose can clear up odors. If this does not work, an anti-odor kit may be required.
- Confirm the refrigerant in the system by using a refrigerant identifier.
- The air distribution system is designed to circulate air through the heating and cooling system, then into the passenger cabin.
- Refrigerant must be tested for acidic contaminants and water intrusion.
- Charge refers to the amount of refrigerant in the air-conditioning system.
- The TXV system is more sensitive to low charge (low refrigerant level) than the fixed orifice tube system.
- Components of a heating system include radiator, thermostat, water pump, upper and lower radiator hoses, heater core, and heater control valve.
- Steps to the diagnosis and repair of air-conditioning systems are performance test and inspection, leak test, reclaiming refrigerant, problem repair, evacuation, charging, and post-testing (performance test).
- The two most common air-conditioning customer complaints are (1) the system is not working or (2) the system has an unusual smell.
- Problems with air-conditioning system performance may be due to system leaks, compressor failure, or system blockages due to debris or frozen moisture.
- Variable factors, such as weather, air moisture content, temperature, condition and type of vehicle, and color and size of vehicle, all affect air-conditioning system functioning.
- Air-conditioning performance testing should be done before and after a repair.
- Air-conditioning system inspection should include possible condenser airflow restrictions, evaporator housing water drain, air filter, hoses and belts, coolant pressure, radiator cap vacuum and pressure, cooling fan, fan clutch, fan shroud, air dams, and heater control valves.
- Air-conditioning systems are equipped with protective devices designed to shut down the system if the pressure becomes elevated.
- Tools needed to maintain and repair air-conditioning systems include: sealant detector, reclaim/recycle machine, refrigerant identifiers, pressure gauge sets, air-conditioning machine, anemometer, retro fit kit, pressure-temperature chart, vacuum pump, electronic leak detector, dye leak detector, oiler, vacuum gauge, micron gauge, and line wrench.
- Methods of leak testing are sniffer testing, dye testing, and nitrogen testing.
- Refrigerant leaks can be detected by sniffers via ultrasonic, heated diode, or corona suppression.
- Any time the air-conditioning system is opened, the refrigerant must be recovered (reclaimed).
- The air-conditioning system should be flushed if the compressor comes apart, the desiccant bag breaks open, or the oil is contaminated.
- Use caution when removing and installing a TXV, to avoid creating leaks.
- Following a repair, the air-conditioning system should be evacuated via an air-conditioning machine or air-conditioning gauges and a vacuum pump.
- Any oil removed from the system has to be replaced with new oil of the correct type (unless more than 1 ounce was removed, in which case the system should be flushed).
- 25,400 microns equal 1 inch of vacuum.

Key Terms

accumulator A device placed between the evaporator and the compressor to collect liquid refrigerant and prevent it from entering the compressor.

air-conditioning compressor clutch An engagement device connected to the compressor crankshaft to engage the crankshaft with a belt-driven pulley.

air-conditioning machine A machine designed to recover, recycle, evacuate, leak test, and recharge (R/R/R) the air-conditioning system.

anemometer A device that measures airflow in feet per minute (fpm).

charge The amount of refrigerant present in the system or the process of installing refrigerant in the system.

corona suppression An electronic sniffer used for detecting refrigerant leaks.

electronic sniffers An electronic leak detector used to locate small refrigerant leaks by electronically reacting to refrigerant gas.

heated diode A sniffer that uses electricity to determine if there is a refrigerant leak; considered the best sniffer for R-134a.

hoses Flexible lines used to direct liquids or gases.

lines Term used interchangeably with pipes or tubes.

microleak detector A solution used to detect refrigeration leaks.

micron gauge A device designed to measure vacuum very precisely.

muffler A device to quiet the pipes of the air-conditioning system with baffles placed inside to deaden the sound of refrigerant moving.

oiler A device used to add oil to the air-conditioning system.

overcharging Overfilling of the air-conditioning system; may result in poor cooling or mechanical failure of the system.

performance testing The process of recreating a driving situation to check air-conditioning performance and vent temperature.

pressure gauge set A set of calibrated gauges for high and low pressure that can show pressure and vacuum readings for use with air-conditioning systems.

pressure transients Minor fluctuations on the gauges that may indicate a problem.

PT chart A pressure-temperature chart that shows the relationship between air-conditioning pressures and evaporator temperature.

reclaim/recycle machine An air-conditioning machine designed to remove and recycle refrigerant for reuse.

reclaiming The process of removing refrigerant from the air-conditioning system by using an air-conditioning machine; also called recovering.

recovering See *reclaiming*.

refrigerant identifiers Devices used to check for impurities in the air-conditioning system.

retrofit kit An aftermarket kit that has the fittings and oil to change an R-12 unit over to an R-134a unit.

sniffer An electronic device used to determine the source of leaks.

state of charge The amount of refrigerant in a system compared to how much should be in it.

ultrasonic A method of leak detection that uses a sensitive microphone to hear small refrigerant leaks by amplifying the hissing noise.

vacuum gauge A gauge designed to read negative pressure or vacuum.

vacuum pump A pump used to evacuate the air-conditioning system and put it into a deep vacuum or low pressure to remove moisture.

Review Questions

1. All of the following statements are true with respect to testing the air-conditioning system *except*:
   - **a.** The heater fan needs to be on the lowest speed.
   - **b.** The heater control knob should be on the coldest setting.
   - **c.** The airflow should be set to the dash vents.
   - **d.** An external fan should be used to supply air to the condenser to simulate driving conditions.
2. Which is the most common source of abnormal noises in an air conditioner?
   - **a.** Evaporator
   - **b.** Condenser
   - **c.** Compressor
   - **d.** Expansion valve
3. Which of the following types of testing requires an empty air-conditioning system?
   - **a.** Corona suppression
   - **b.** Heated diode testing
   - **c.** Dye testing
   - **d.** Nitrogen testing
4. Choose the correct statement.
   - **a.** Experts recommend the usage of a sealant to fix a leak.
   - **b.** A refrigerant identifier is used to determine the type of refrigerant.
   - **c.** A contaminated refrigerant can be recycled using a reclaiming machine and reused.
   - **d.** The reclaimed refrigerant can be stored in the virgin disposable tank that the new refrigerant came in.
5. All of the following can cause the compressor clutch to not engage *except*:
   - **a.** The clutch coil receives full power and ground.
   - **b.** Resistance of clutch coil is more than specifications.
   - **c.** The gap between the clutch assembly and air-conditioning pulley is too large.
   - **d.** A voltage drop is present in the electrical circuit that energizes the clutch coil.
6. Which of the following refrigerants uses mineral oil?
   - **a.** R-22
   - **b.** R-134a
   - **c.** R-12
   - **d.** HFO-1234yf
7. When charging an air-conditioning system, if the pressure reaches a positive pressure number too rapidly, what does it indicate?
   - **a.** The reed valves in the compressor are likely bad.
   - **b.** The TXV or orifice tube is restricted.
   - **c.** The refrigerant is contaminated.
   - **d.** The desiccant bag is broken.
8. When removing the heater core, how should the heater hoses be removed?
   - **a.** By using a wrench
   - **b.** By slitting the hoses lengthwise
   - **c.** By heating to loosen them
   - **d.** By twisting the heater hoses

9. In a typical orifice tube system, which of the following indicates an undercharge of refrigerant?
   a. Both the low and the high side gauges are running higher than normal.
   b. The air-conditioning compressor clutch runs for a long time before it shuts off.
   c. The air in the ducts feels cold.
   d. The air-conditioning clutch has a rapid cycle time.
10. If the pressure gauges read low pressure on the high side, high pressure on the low side, or about equal pressures, what might it indicate?
   a. The system may be low on refrigerant.
   b. The compressor is not working fully.
   c. There is a blockage in the low side of the system that is not allowing refrigerant to pass.
   d. There is a blockage on the high side or an airflow restriction across the condenser.

ASE Technician A/Technician B Style Questions

1. Tech A states that the wider the gap on an air-conditioning clutch, the greater the ohm reading when checking the windings. Tech B states that the air-conditioning clutch is electromagnetically operated. Who is correct?
   a. Tech A
   b. Tech B
   c. Both A and B
   d. Neither A nor B
2. Tech A states that an air-conditioning performance test usually requires that an auxiliary condenser fan be used during the test. Tech B states that a performance test will show if the air-conditioning system is contaminated with sealer. Who is correct?
   a. Tech A
   b. Tech B
   c. Both A and B
   d. Neither A nor B
3. Tech A states that refrigerant in a vehicle should be identified before recovering the refrigerant. Tech B states that refrigerant doesn't need to be identified if you are only topping up a system with refrigerant. Who is correct?
   a. Tech A
   b. Tech B
   c. Both A and B
   d. Neither A nor B
4. Tech A states that when evacuating an air-conditioning system, the vacuum should be maintained for approximately 10–15 minutes (but as long as 30 minutes) after the system reaches the boiling point pressure of water. Tech B states that one main purpose of evacuating an air-conditioning system is to remove any moisture from the system. Who is correct?
   a. Tech A
   b. Tech B
   c. Both A and B
   d. Neither A nor B
5. Tech A states that the system should be flushed if the compressor came apart. Tech B states that the system should be flushed if the oil is contaminated. Who is correct?
   a. Tech A
   b. Tech B
   c. Both A and B
   d. Neither A nor B
6. Tech A states that when using pressurized nitrogen to locate a leak, an electronic sniffer should be used. Tech B states that electronic sniffers are used when the system has at least a minimal refrigerant charge. Who is correct?
   a. Tech A
   b. Tech B
   c. Both A and B
   d. Neither A nor B
7. Tech A states that microns are a much more accurate unit of measuring vacuum than inches of mercury (Hg). Tech B states that microns are a much more accurate measure of time than seconds. Who is correct?
   a. Tech A
   b. Tech B
   c. Both A and B
   d. Neither A nor B
8. Tech A states that to determine how much refrigerant is needed in a system, you must refer to identifying labels on the vehicle or the service information. Tech B states that to determine the amount of refrigerant needed, you just charge the system until the pressures look correct. Who is correct?
   a. Tech A
   b. Tech B
   c. Both A and B
   d. Neither A nor B
9. Tech A states that when removing any component of an air-conditioning system, the oil should be drained from it and measured so that the same amount of new oil can be reinstalled. Tech B states that oil should only be in the compressor, and if any oil is found in any other components, it means that the receiver drier is faulty. Who is correct?
   a. Tech A
   b. Tech B
   c. Both A and B
   d. Neither A nor B
10. Tech A states that if moisture enters the air-conditioning system, acid will be created. Tech B states that evacuating an air-conditioning system will boil moisture, which will be removed from the system as a gas. Who is correct?
   a. Tech A
   b. Tech B
   c. Both A and B
   d. Neither A nor B

CHAPTER 60

Electronic Climate Control

NATEF Tasks

- **N60001** Identify hybrid vehicle A/C system electrical circuits and service/safety precautions. (MLR/AST/MAST)
- **N60002** Check operation of automatic or semiautomatic heating, ventilation, and air-conditioning (HVAC) control systems; determine needed action. (AST/MAST)
- **N60003** Using a scan tool, observe and record related heating, ventilation, and air-conditioning (HVAC) data and trouble codes. (AST/MAST)
- **N60004** Inspect and test HVAC system control panel assembly; determine needed action. (AST/MAST)
- **N60005** Inspect and test HVAC system control cables, motors, and linkages; perform needed action. (AST/MAST)
- **N60006** Diagnose temperature control problems in the HVAC system; determine needed action. (MAST)
- **N60007** Inspect HVAC system ducts, doors, hoses, cabin filters, and outlets; perform needed action. (MLR/AST/MAST)
- **N60008** Inspect and test HVAC system blower motors, resistors, switches, relays, wiring, and protection devices; determine needed action. (AST/MAST)
- **N60009** Diagnose A/C compressor clutch control systems; determine needed action. (AST/MAST)
- **N60010** Diagnose malfunctions in the vacuum, mechanical, and electrical components and controls of the heating, ventilation, and A/C HVAC system; determine needed action. (AST/MAST)

Knowledge Objectives

After reading this chapter, you will be able to:

- **K60001** Describe the purpose and operation of climate control sensors.
- **K60002** Describe the purpose and operation of HVAC control panels.
- **K60003** Describe the operation of climate control system devices.
- **K60004** Describe the unique components of hybrid vehicle HVAC systems.
- **K60005** Describe the testing and diagnosis process on HVAC systems.

Skills Objectives

After reading this chapter, you will be able to:

- **S60001** Diagnose failures in the HVAC system controls.
- **S60002** Diagnose HVAC system control devices and actuators.

▶ Introduction

Electronic climate control systems provide the driver and passengers with a means of automatically moderating the environment inside the vehicle (**FIGURE 60-1**). This chapter offers insight into the operation of these systems and their components. The chapter also includes diagnostic information on determining faults in the climate control system and information on testing the various components. An electronic heating ventilation and air-conditioning system is more technically challenging to diagnose than a mechanical air-conditioning system, and when not functional, many customers will turn to a highly qualified technician to repair it quickly and accurately.

▶ Climate Control Overview

Climate control systems in modern vehicles come in three different forms: **manual, semiautomatic**, and **automatic**. Manual control systems allow the operator to control the direction of airflow, the blower motor speed (to control airflow speed), and the temperature level of the air coming out of the outlets. The manual system gives the operator full control of the climate control system without the use a computer or **electronic control unit** (ECU). But as conditions change, the operator will need to manually adjust the controls to accommodate the changes.

Semiautomatic climate control systems also allow the operator to select the direction of airflow, the blower motor speed, and the temperature. However, the ECU then controls the amount of heat or cold that is discharged to maintain that temperature. The ECU controls heat by one of two methods—a blend door in the air box or a heater control valve in one of the heater hose lines. In a blend door system, the airflow goes through the evaporator and then a blend door moves to direct all of the airflow either across the heater core for maximum heat or around the heater core for cold. The blend door can be positioned anywhere between these two points to "blend" the cold and hot air (**FIGURE 60-2**). The heater control valve controls how much coolant flows through the heater core (it regulates the coolant flow to the heater core) (**FIGURE 60-3**). The further it is open, the hotter the heater core becomes and the more heat that can be transferred to the airstream flowing through it.

Automatic climate control systems automatically maintain the operator-selected climate (temperature) preference within the passenger cabin. The climate control system manages the three subsystems-heating, ventilation, and air conditioning and uses

FIGURE 60-1 Typical control panel for electronic climate control system.

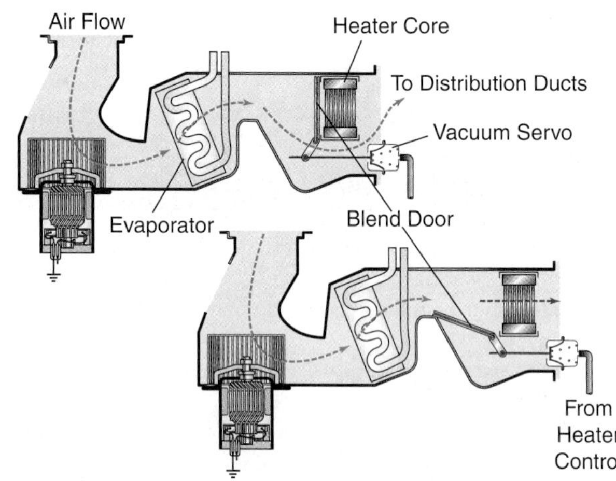

FIGURE 60-2 Blend door.

You Are the Automotive Technician

A woman comes into the shop complaining that her air conditioning isn't cool enough to maintain the temperature she desires in her 2014 Honda Pilot. When she turns her A/C on, it is cool at first, but turns to warm air blowing from the vents after 5 minutes. You perform a system performance test and find that when the vehicle is started up cold, the A/C ducts deliver cool air according to the manufacturer's temperature chart. But over time, the ducted air temperature starts to warm up even though the A/C temperature is at its minimum setting. After about 5 minutes, the ducted air temperature has warmed up to the point that it is warmer than the outside air temperature. You use a pair of hose clamping pliers and clamp off one of the flexible heater hoses. With the engine and A/C still operating, the ducted air temperature starts to fall and after just a couple of minutes, returns to the specified temperature. You now know that this is a heater control problem and not an A/C control problem.

1. During a performance test, when is it normal for the ducted air temperature to be warmer than outside air?
2. Why did you clamp off the flexible heater hose?
3. What caused the ducted air temperature to fall back to its normal temperature?

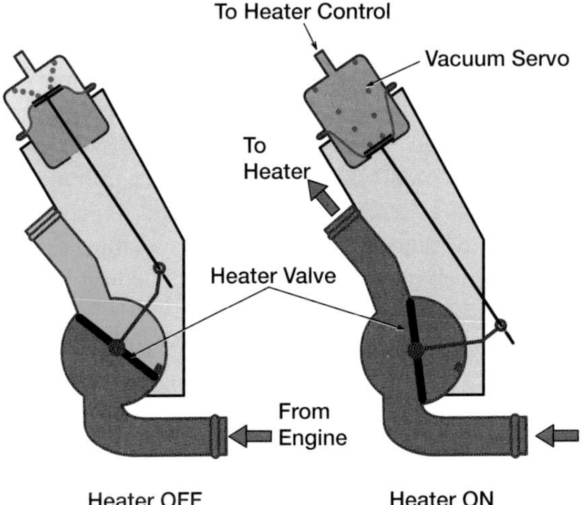

FIGURE 60-3 Heater control valve.

each subsystem to match the climate to the settings selected. If the operator selects auto control and 78°F (26°C), the ECU will check all inputs to determine the temperature outside the vehicle and inside the vehicle. When the temperatures are determined, the ECU will move the doors in the air box to attain and maintain the passenger compartment temperature at the requested value. The ECU may have to turn on or off the air-conditioning compressor to reach the temperature requested.

The automatic climate control system also controls the speed of the blower fan, speeding it up to effect a quick change in climate when the passenger cabin temperature is substantially different from the climate control system setting. Once the climate approaches the desired setting, the fan speed will gradually reduce to maintain the selected temperature.

▶ Electronic Climate Control Components

The electronic climate control system involves many components. Each component fits into one of three categories (**FIGURE 60-4**).

- The first category is the ECU, which is the brains of the system. It controls the system.
- The second category is the HVAC sensors, also called inputs, which provide information to the ECU.
- The last category is comprised of the HVAC-controlled devices, which cause components such as the actuators and the blower motor to function in certain ways.

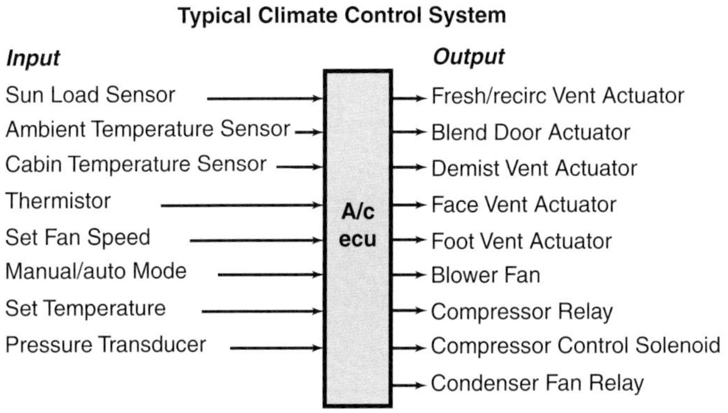

Typical Climate Control System

Input
Sun Load Sensor
Ambient Temperature Sensor
Cabin Temperature Sensor
Thermistor
Set Fan Speed
Manual/auto Mode
Set Temperature
Pressure Transducer

A/c ecu

Output
Fresh/recirc Vent Actuator
Blend Door Actuator
Demist Vent Actuator
Face Vent Actuator
Foot Vent Actuator
Blower Fan
Compressor Relay
Compressor Control Solenoid
Condenser Fan Relay

FIGURE 60-4 Climate control system.

Automatic climate control requires the use of electronic sensors to determine the heat (sun) load on the vehicle, ambient air temperature, cabin air temperature, evaporator temperature, engine coolant temperature, position of various doors, desired temperature setting, and in some vehicles, the temperature of the occupants (**FIGURE 60-5**). The sensors provide the ECU with a sense or feel for temperature inside and outside the vehicle, as well as knowing how much heat from the sun is coming through the windows. By being able to "feel" these conditions, the ECU can decisions based on that information and sends output commands to the **actuators** to control fan speed, heater core coolant flow, air-conditioning clutch activation, compressor displacement, blend doors, and fresh air/recirculation doors. Actuators come in two types: electrically operated and vacuum operated (**FIGURE 60-6**). We look at each of these components in more detail and see how they work together to control the climate within the passenger compartment.

The Electronic Control Unit of an HVAC System

An ECU works with sensors and actuators to control the climate inside the vehicle. It controls the HVAC system. The ECU gathers information from sensors that detect changes in ambient temperature, cabin temperature, coolant temperature, and solar radiation. The ECU evaluates signals from the sensors to determine the necessary operating conditions to meet the desired climate. It then controls actuators to move dampers and valves that control the volume, temperature, and direction of the cool air delivered by the air-conditioning system.

The ECU contains a micro-computer (**FIGURE 60-7**). The starting point of any control module is the printed circuitry. Resistors, capacitors, transistors, and integrated circuits are

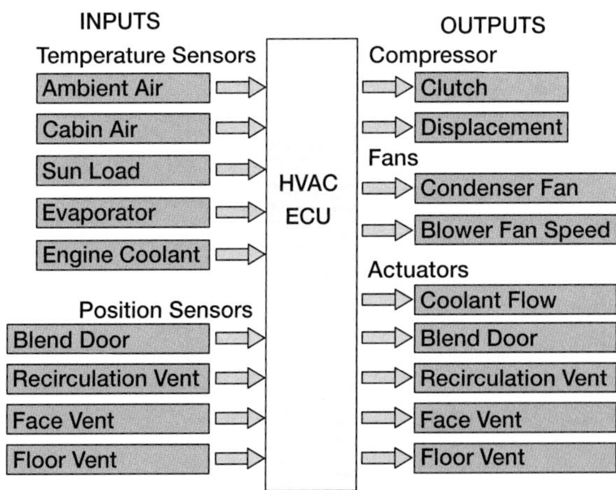

FIGURE 60-5 Automatic climate control systems use a variety of sensors to gather information the ECU needs to determine what actions to take to maintain the desired temperature.

FIGURE 60-6 HVAC actuators. **A.** Electric actuator. **B.** Vacuum actuator.

soldered onto the circuit board to create the control module or computer. The circuitry is mounted inside a metal or plastic housing to protect it from damage. The control module or ECU must be programmed after it is constructed. The programming loaded into the ECU contains all the data to make the system operate properly. One such piece of data programmed is the data map used to determine the voltage versus temperature that is received from interior and exterior temperature sensors. In some automatic climate control systems, the control panel and ECU are one piece and are not located separately.

The ECU receives signals in the form of **voltages** from each of the sensors. The ECU compares the **signal voltage values** from the temperature sensors with values stored in its memory, programmed in when the module is installed into the vehicle. If a difference in voltage values occurs, the ECU determines the action needed to ensure that the temperature requested by the operator is achieved. The ECU may need to change the position of the actuators or increase or decrease blower motor fan speed.

Some actuators provide a **feedback signal** to the ECU in the form of a voltage signal (**FIGURE 60-8**). The voltage signal is used by the ECU to determine what position the actuators are in and whether they are moving. The feedback sensor is a potentiometer, which is a variable resistor that changes resistance values as the position of the wiper contact changes. The actuator's potentiometer works in the same way as the throttle position sensor. As the position of the door changes, the actuator's potentiometer signals its position to the ECU. Refer to the description and operation section of the service information to find out how the vehicle's particular actuator operates.

The ECU also controls the speed of the interior ventilation fan (blower motor fan). On some vehicles, the ECU provides a pulse-width-modulated signal to control the motor's speed, and works by repeatedly turning on and off the current flow to the component. The longer the percent of time the current is turned on, the faster the fan runs (**FIGURE 60-9**). The speed of the interior ventilation fan is infinitely variable between a stopped condition and the fan's maximum speed. This design provides a continuously variable speed capability and therefore

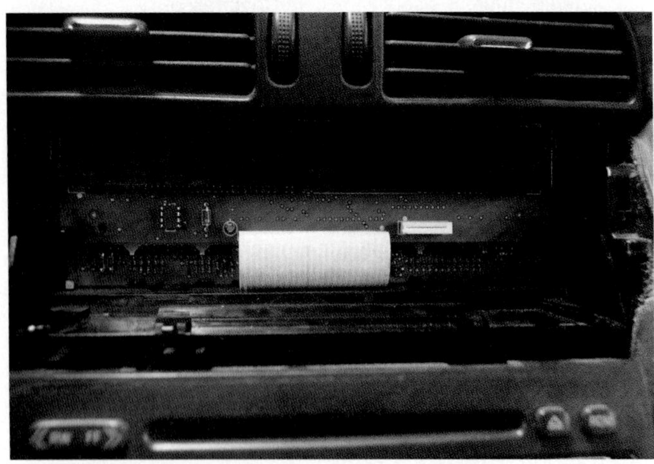

FIGURE 60-7 A typical HVAC electronic control unit.

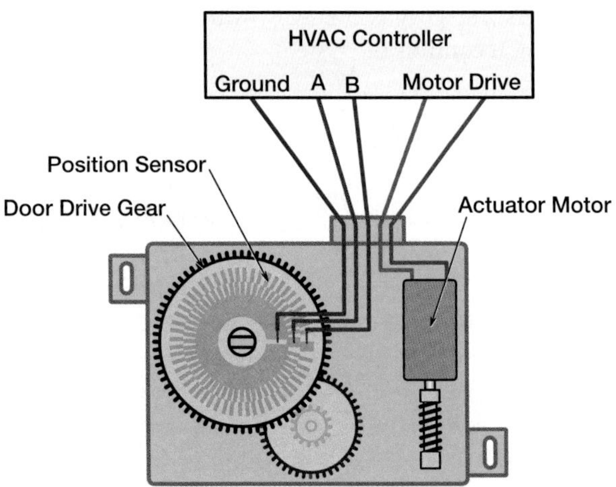

FIGURE 60-8 Some actuators provide position feedback to the ECU in the form of a sensor signal.

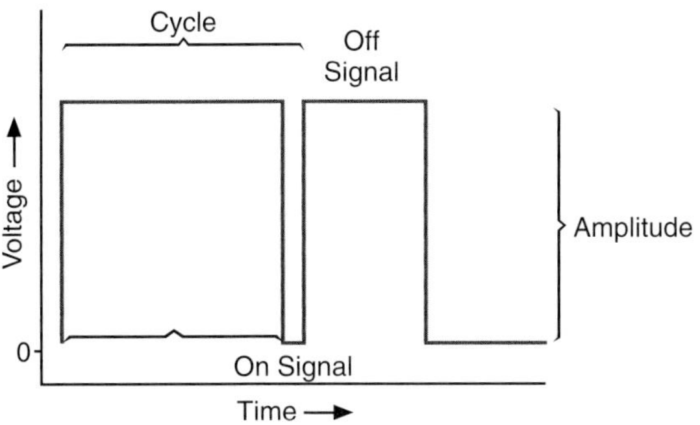

FIGURE 60-9 Pulse-width-modulated (Duty cycle) on the ground side of circuit.

an infinite number of speeds, which allows the fan to operate at a more appropriate speed. A typical resistor pack blower motor, in contrast, has only four speeds. With only four speeds, it is more likely that any single speed is a bit too fast or a bit too slow.

Automatic Climate Control Sensors

K60001

As just discussed, the ECU of an HVAC system needs to know the temperature of the ambient air, the cabin air temperature, the temperature of the air-conditioning evaporator, the temperature of the coolant circulating through the heater core, and whether the vehicle is in bright sunlight. All this information allows the ECU to "feel" its surroundings (interior and exterior temperatures, for example) and to determine from its programming where to position air box doors, how fast to run the blower fan, and whether to turn the air-conditioning compressor on or off.

ECU's depend heavily on temperature sensors to make the right decisions. There are two kinds of temperature sensors, also called **thermistors** because their resistance value changes significantly as temperature changes. One is called a **negative temperature coefficient (NTC) thermistor** because, in contrast to most materials, its resistance goes down as it is heated. The other is the positive temperature coefficient (PTC) thermistor because its resistance goes up as it is heated (**FIGURE 60-10**). Be sure to consult the service information to verify the type of thermistors used in the vehicle you are working with. Most manufacturers use the NTC type, which increases resistance as the temperature goes down. For example, as the temperature sensor is exposed to colder temperatures, the resistance will increase, causing the signal voltage to decrease. Negative temperature coefficient (NTC) thermistors are usually used in automotive applications (**FIGURE 60-11**).

When connected into a circuit, the temperature sensor changes the voltage signal at the ECU. The ECU reads the sensor voltage, interprets it into a temperature, and compares this reading with information stored in its memory (**FIGURE 60-12**). The ECU continuously monitors the signals from all of the sensors so it can make any adjustments as necessary. As an example of the need for the ECU to adjust as necessary, suppose the operator were to leave the air-conditioning controls set to auto with the desired temperature set to 70°F (21°C) while the vehicle is sitting in a garage that is air conditioned to 75°F (24°C). The driver then starts the vehicle and backs it out into the full

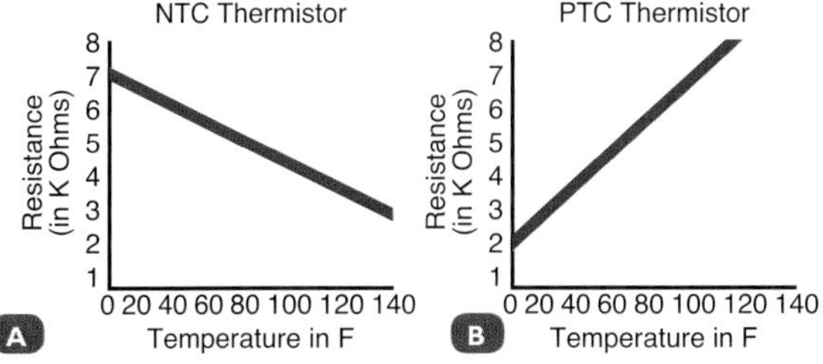

FIGURE 60-10 Thermistors. **A.** Negative temperature coefficient (NTC) thermistor resistance. **B.** Positive temperature coefficient (PTC) thermistor resistance.

FIGURE 60-11 A negative temperature coefficient (NTC) thermistor near the evaporator.

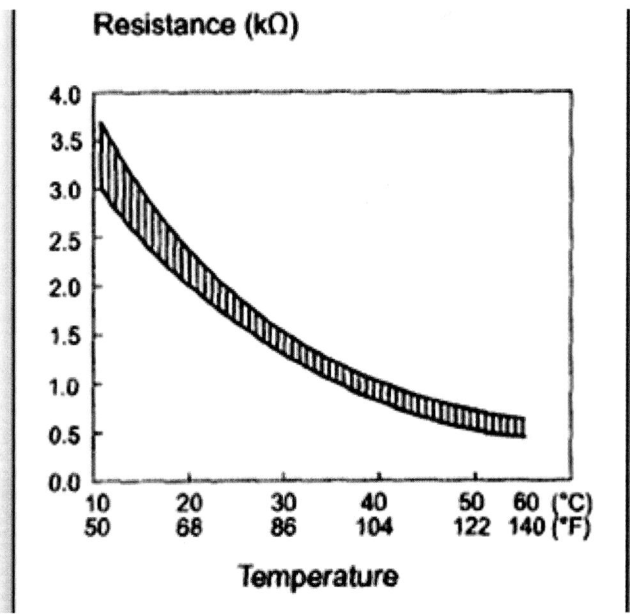

FIGURE 60-12 A typical sensor voltage chart showing the voltage for a corresponding temperature.

sun, where it is 98°F (37°C). The ECU will register the sun load sensor, which detects large amounts of sun, and the exterior sensor, which detects 98°F (37°C). With this information, the air-conditioning compressor will be commanded on for longer periods to cycle cold refrigerant into the evaporator; the blend door will move to full cold; and fan speed will ramp up. All of this is done automatically.

Ambient Air Temperature Sensor

The **ambient air temperature sensor** measures the temperature of the outside air (**FIGURE 60-13**). It is usually located in front of the air-conditioning condenser, where the forward motion of the vehicle and the action of the condenser fans force air over the sensor. The ambient air temperature sensor is a thermistor. The ECU sends a 5-volt reference signal to the thermistor. The resistance of the thermistor affects the current flow through the circuit and the voltage drop across the resistances. The signal is returned to the ECU as a voltage from somewhere near 0.5 up to 4.5 volts. The ECU has an internal voltmeter that looks at the voltage across an internal resistor. If the outside temperature is cold, then the voltage back to the ECU stays high (is not pulled down across the resistor). If the temperature is hot, then the voltage across the internal resistor will be low, and the voltage going back to the ECI is pulled down. If the outside air temperature is low, the ECU will not have to run the compressor as much or move the blend door to as cold a position. The ECU will not allow the air-conditioning compressor to turn on if the ambient air temperature is below a set temperature, typically around 40°F (4°C). The ambient air temperature sensor is also used as an input for the instrument panel display of the outside air temperature.

Cabin Air Temperature Sensor

The **cabin air temperature sensor** may be mounted in a tube-type device called an **aspirator** (**FIGURE 60-14**). The aspirator is a tube that directs airflow to the in-vehicle temperature sensor. One end of the aspirator tube is open to cabin air, and the other is connected to the interior fan. The interior fan creates a flow of air through the aspirator that is directed to the cabin air temperature sensor. The cabin air temperature sensor is a thermistor. A 5-volt reference is sent to the thermistor (cabin air temperature sensor) and is varied depending on the temperature. The cabin air temperature sensor is a valuable input to the ECU to ensure that it knows what the temperature is so it can cause the actuators to respond appropriately.

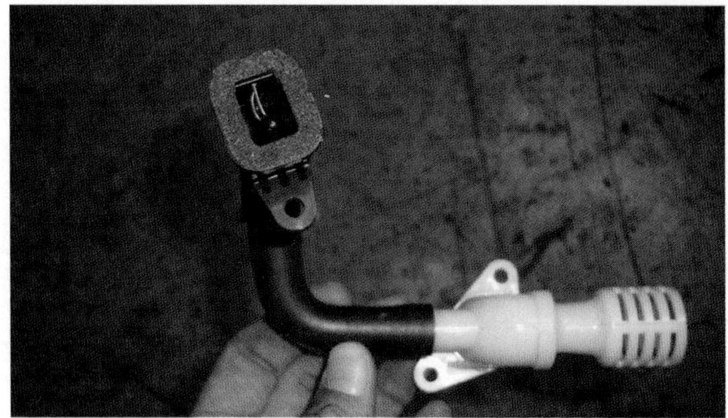

FIGURE 60-14 A cabin air temperature sensor in an aspirator tube.

Engine Coolant Temperature Sensor

The **engine coolant temperature sensor** signals to the ECU the temperature of the coolant that flows through the heater core. Some climate control systems have two of these sensors: the most common one is usually located near the engine thermostat, and the other one may be near the heater core (**FIGURE 60-15**). The engine coolant temperature sensor is a thermistor and works the same as an exterior or interior temperature sensor.

The signal from the engine coolant temperature sensor is used in conjunction with the signals from the ambient air sensor, the sun load sensor, and the cabin air temperature sensor to control heater and fan operation. For example, suppose the ECU registers the following readings: coolant temperature, 60.8°F (16°C); ambient air temperature, 32°F (0°C); sun load, –0; and cabin air temperature, 32°F (0°C). The heater valve will open fully, and the blend door will move to maximum airflow across the heater core; the fan speed will be slow and

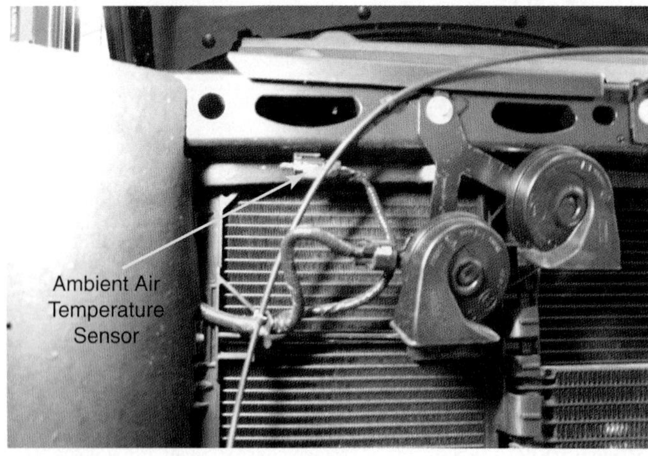

FIGURE 60-13 An ambient air temperature sensor.

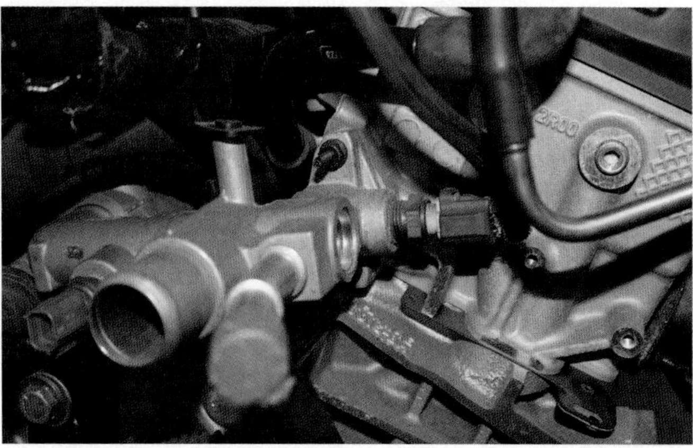

FIGURE 60-15 A typical engine coolant temperature sensor located near the thermostat.

Applied Math

AM-32: Fahrenheit/Centigrade: The technician can identify whether the temperature measurement must be made using a measuring device that measures in degrees Centigrade or whether it should be in degrees Fahrenheit.

Reading in degrees Centigrade or degrees Fahrenheit is largely dependent on the system of measurement used in the publication of specifications for the system.

Ideally, if system specifications are published in metric measurements, an instrument measuring in degrees Centigrade should be used. If standard measurements are used in specifications, an instrument measuring in degrees Fahrenheit would be a better choice. If the appropriate measuring instrument is not available, it is possible to convert Centigrade readings to Fahrenheit, and vice versa for the purpose of comparing measured readings to specifications. Water freezes at 0° and boils at 100° in the Celsius or Centigrade scale. Water freezes at 32° and boils at 212° in the Fahrenheit scale.

steady. As the engine heats up, the fan speed may be increased until the passenger cabin warms up. The ECU monitors the temperatures on the various sensors and takes the appropriate action on the doors, valves, and fan.

Sun Load Sensor

The **sun load**, or **solar, sensor** is located on the dash panel, where it is exposed to light entering through the windshield of the vehicle (**FIGURE 60-16**). As sunlight increases, the passengers get the sensation that it is hotter than it really is inside the passenger compartment. The ECU uses the sun load sensor to increase the fan speed and adjust the blend door. The sun load sensor is a **photodiode**. The photodiode works by varying resistance as sunlight brightness changes. These sensors use a 5-volt reference signal and a signal wire. As sunlight becomes brighter, resistance increases and lowers the voltage signal back to the ECU, similar to how a thermistor works. A table programmed into the memory of the ECU is able to compare the voltage signal to a sun load amount. Photodiodes are used in many other household electronics, including your TV. The photodiode picks up the remote signal when you control the volume or the channel.

When connected to the ECU, a change in light intensity is registered as a change in voltage. The ECU uses this information to gauge the amount of solar heat load the vehicle is experiencing and adjusts the amount of cooling or heating needed.

Air-Conditioning Pressure Sensor

The **air-conditioning pressure sensor** (transducer) is a three-wire sensor that monitors the **air-conditioning system pressure** (**FIGURE 60-17**). Typically, air-conditioning systems will use two pressure sensors to monitor refrigerant pressure - a high side pressure sensor and a low side pressure sensor. Both of these sensors "measure" pressure in pounds per square inch, kPa, or pascals.

The air-conditioning system must have enough refrigerant pressure to ensure that refrigerant oil can be circulated through the air-conditioning compressor. If refrigerant pressure is too low, the air-conditioning compressor will not turn on, which is a precaution designed into the system to keep the compressor from being damaged. The pressure will get too low in the event that there is a leak in the system, or improper refrigerant flow. The low-pressure sensor is located on the low-pressure side of the air-conditioning system to monitor any faults in this side of the system.

The high-pressure sensor is located on the high-pressure side of the air-conditioning system. The high-pressure sensor is designed to keep pressures from going too high and causing the pressure relief valve in the compressor from releasing refrigerant to the air. The relief valve is a safety device built into the compressor by manufacturers to ensure that pressures cannot get too high in the system and rupture a line. The ECU of the HVAC system can turn on the condenser cooling fan (or increase its speed, or turn on an additional fan), to reduce the temperature of the refrigerant, which in turn reduces pressure.

Either the low-pressure sensor or the high-pressure sensor can send a **variable voltage signal** to the ECU, based on the pressure in the system. The air-conditioning pressure sensor's

FIGURE 60-16 A sun load sensor.

FIGURE 60-17 A three-wire air-conditioning pressure sensor.

Applied Science

AS-80: Mechanical Transducers: The technician can demonstrate an understanding of the role mechanical transducers play in sending an electrical control signal to modify the system's operation.

Transducers are electronic devices that convert one form of energy to another. Position sensors are a form of transducer employed in many automotive systems. They indicate the position of components by converting kinetic (motion) energy into electrical energy.

Climate control systems use various doors to control the direction and temperature of airflow into the cabin. These are commonly known as "blend doors." In order for the climate control ECU to effectively manage the operation of the system and achieve the cabin conditions desired by the driver, it must be able to monitor the position of the blend doors. This is accomplished by blend door actuators being fitted with transducers that feed information regarding door positions back to the ECU.

voltage will vary with pressure from about 0.5 to 4.5 volts; this information is again compared to programmed information in memory to determine the pressure in the system and command the correct action.

Evaporator Temperature Sensor

The **evaporator temperature sensor** is located in the airstream where the air leaves the evaporator; it signals the temperature of the air to the ECU (**FIGURE 60-18**). It is usually a two-wire, NTC-type thermistor. The evaporator temperature sensor works in the same manner as the exterior and interior temperature sensors, lowering resistance as the temperature goes up. Information programmed within the ECU recognizes the changes in resistance and translates that into a temperature reading. The evaporator temperature sensor is used to ensure that the evaporator does not get too cold. If the evaporator temperature drops too low, the moisture will freeze on the evaporator and block airflow through the fins. If this happens, cold air will stop flowing to the passenger compartment. When the air conditioner is turned off for a while to let the evaporator warm, the ice melts and airflow is cold and normal when the air conditioner is turned back on. The evaporator temperature sensor ensures that the evaporator does not get too cold by either reducing the compressor on time or shortening the stroke of a variable displacement compressor to reduce the amount of refrigerant flowing in the system. Of course, if it uses a thermal expansion valve, then the ability to modify refrigerant flow will also help keep the evaporator from freezing.

Relative Humidity Sensor

Newer vehicles are being equipped with a **relative humidity sensor** that monitors the relative humidity level in the vehicle. In many cases, the humidity sensor is built into the temperature sensor (**FIGURE 60-19**). Humidity affects the comfort level in the passenger compartment. Both excessively high humidity and low humidity cause discomfort. So relative humidity levels have to be maintained within the comfortable range, which is typically between 30% and 70%. The humidity sensor monitors this. If the humidity is high, then the AC can be commanded on to remove water from the air by condensing it on the evaporator. If the humidity is low, then the system can bring in outside air to raise the humidity level. Because lower levels of humidity tend to make a temperature "feel" cooler to the occupants, keeping the humidity on the low side allows the AC to run less, which increases fuel economy.

HVAC Control Panel

K60002

One other input to the ECU of the HVAC system comes from the HVAC control panel. The operator can tell the ECU where to deliver air and at what temperature, using switches that produce an **on/off-type signal** (**FIGURE 60-20**). The ECU receives either a 12-volt signal or 0 volts. For example, pushing the windshield defrost button sends an "on" 12-volt signal to the ECU. Pushing the floor button cancels the defrost button. The defrost button now sends an "off" 0-volt signal to the control unit. At the same time, the floor button sends an "on" 12-volt signal to the ECU for the floor vent.

FIGURE 60-18 An evaporator temperature sensor.

FIGURE 60-19 A typical relative humidity sensor.

The ECU also receives instructions from the operator regarding the desired cabin temperature via the control panel. The desired temperature selection is used to tell the ECU what temperature the operator wishes the inside of the vehicle to be. The ECU then adjusts the position of valves and doors to achieve the desired temperature and direction of airflow.

Control Panel Types

The control panel buttons, knobs, and dials vary from vehicle to vehicle and from system to system, depending on whether the system is manual, semiautomatic, or fully automatic. There are also dual-zone and rear climate control panels.

If the system is a manual control panel, there will be a rotary or linear switch for fan speed, a rotary or linear switch for modes (floor, defrost, vent, or vent and floor), and a rotary or linear switch for temperature. On some models, an air-conditioning button, as well as a recirculation button, is on the panel (**FIGURE 60-21**).

On a semiautomatic system, the control panel typically has a rotary temperature switch, a rotary fan switch, a rotary mode switch, a recirculation button, and possibly an economy button that limits the air-conditioning compressor operation in an attempt to improve fuel mileage (**FIGURE 60-22**). This system uses an ECU to determine when to run the compressor and the amount of heat required to match and maintain the desired temperature selected by the operator.

The fully automatic system control panel typically has an Auto mode that takes care of the entire system, based on the temperature the driver chooses. The operator can select manual control of this system, so all buttons previously discussed with the manual control panel are present on the automatic panel as well (**FIGURE 60-23**). The control panel usually has a digital display to show the temperature selected.

The dual-zone temperature control panel is designed to allow the driver and passenger to control temperature delivered to their side of the vehicle for a more pleasant climate for each individual (**FIGURE 60-24**). The control panel for this design contains the same buttons used in other systems but has two temperature control switches. With dual controls, the air box must have two blend doors and two blend door actuators to allow for separate temperature control.

FIGURE 60-20 Typical control panel for automatic temperature control system.

FIGURE 60-22 A typical semiautomatic control panel.

FIGURE 60-21 A typical manual control panel with separate AC and recirculation controls.

FIGURE 60-23 A typical automatic control panel with Auto mode along with manual controls.

FIGURE 60-24 A typical dual-control system allows separate control for the driver and passenger side of the vehicle.

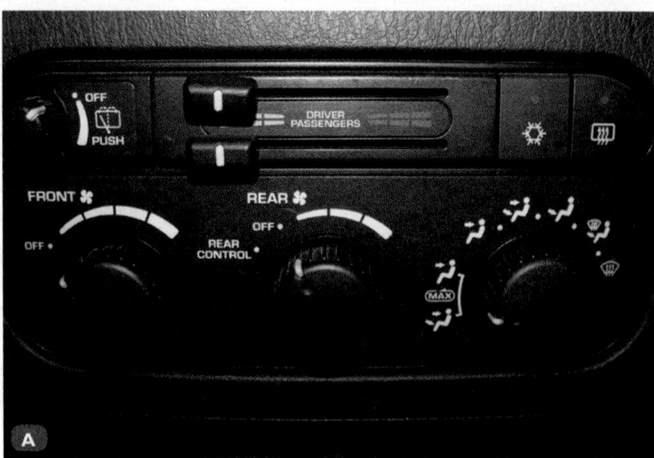

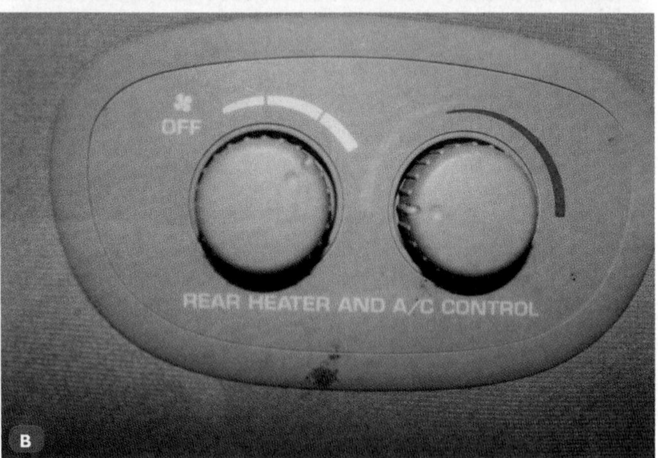

FIGURE 60-25 Rear climate control. **A.** Front control panel with switch to activate the rear air-conditioning. **B.** Rear control panel.

The last system available is the rear climate control panel. The control panel at the front of the vehicle has a rear button in addition to all the other buttons and switches covered here. When the rear button is selected, the rear panel controls located at the back of the vehicle are activated. The driver can turn off control of the panel at the rear by pressing this button on the front panel. With the rear air control panel selected on the front panel, the rear passengers can typically select temperature, fan speed, and mode (**FIGURE 60-25**).

Control Panel Operating Mechanisms

The control panel assembly could be a mechanical system that operates via cables, such as the **heater control cables**, vacuum valves, or electrical switches. It could also be an automatic system that is operated electronically. Cable systems tend to be found only on older vehicles (**FIGURE 60-26**). The cables mechanically move doors and valves. They tend to bind up over time and do not work properly, so most manufacturers have moved away from control cable systems.

Vacuum control systems contain vacuum switches or valves that direct source vacuum from a vacuum canister connected to engine vacuum to different vacuum servos (**FIGURE 60-27**). The vacuum canister is installed on the vacuum system to ensure that vacuum is present for the control panel to use even when the engine is at wide-open throttle (vacuum goes away under wide-open throttle). Thus, the canister provides a reserve of vacuum so the controls can continue to operate. The vacuum switches or valves connect vacuum hoses together at the panel. For example, when the operator wants air to come out at the floor, he or she selects the floor mode, and the switch or valve will connect the source vacuum line to the floor vacuum servo, which moves the mode door to the floor position.

Electronic control panels contain electronic controls that send electrical or electronic inputs to the ECU. In many cases, the ECU is part of the control panel, so it is an all-in-one unit (**FIGURE 60-28**). The ECU sends electrical signals to actuators, which cause the doors or valves to operate.

FIGURE 60-26 An older cable style of HVAC control panel.

HVAC-Controlled Devices

K60003

The controlled devices carry out the commands of the operator. If more airflow is commanded, the operator can move a switch, dial, or lever to select the desired outcome. That information

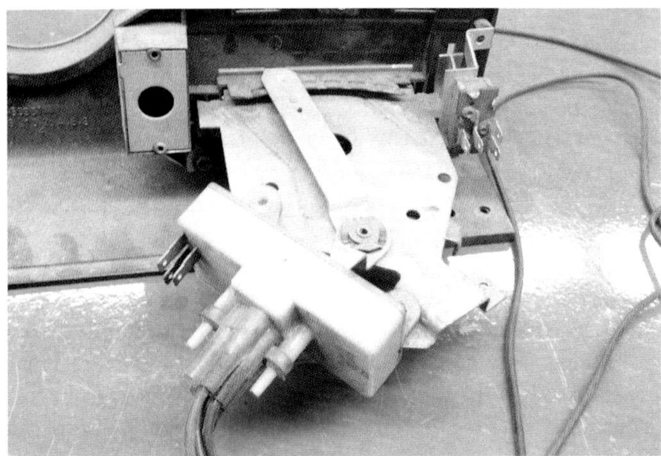

FIGURE 60-27 Vacuum control panel.

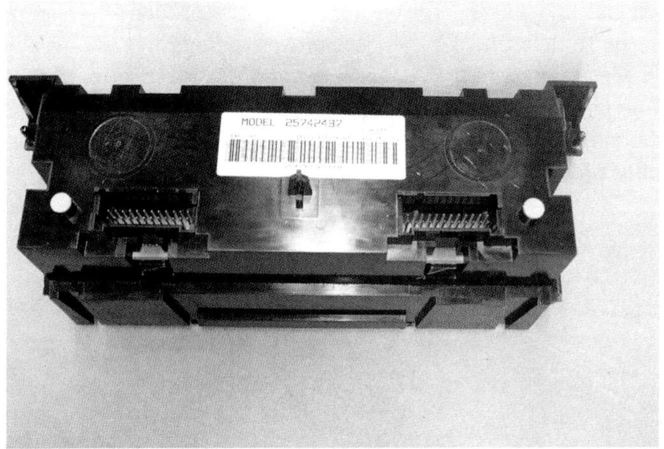

FIGURE 60-28 A typical electronic panel with integrated ECU.

is transmitted as a signal that causes the fan to speed up. The same goes for the preferred location of ducted air: If the operator selects Vent, then the signal will be transmitted to the appropriate actuator and cause an air door to open or close. Each of the controlled devices causes an action to be carried out in the HVAC system. We look at those devices next, one at a time.

Relays

Several different relays can be used in an HVAC system to control electric components such as the **air-conditioning compressor clutch** and condenser fan (**FIGURE 60-29**). Relays control a relatively large current flow in one part of a circuit by a much smaller current flow. Some of these relays are controlled by the ECU of the HVAC system, which sends the smaller current flow signal to the relay when it wants to activate the particular device that the relay operates. The small current in the relay creates a strong magnetic field in the relay windings, which pulls a set of higher current contacts together. The contacts send battery power to the fan or compressor clutch, which is then energized and operates. When the ECU determines that the device no longer has to operate, it stops the signal to the relay, which stops the current flow in the windings and turns off the magnetic field. The spring in the relay now pulls the contacts apart, stopping the current flow to the controlled device and turning it off.

The ECU can control the relay in one of two ways: power or ground. First, I can send power to the relay winding, which would then usually have the ground side connected directly to ground. Or second, it can ground the relay winding, which means that the power side of the winding is most likely connected to power through a fused circuit from the ignition switch. Relay windings are normally equipped with a mechanism to reduce voltage spikes, which are induced in the windings when the relay is deactivated. The voltage suppression can take the form of either a diode or a resistor wired in parallel with the winding (**FIGURE 60-30**). In this way, a path for the voltage spike is provided through the diode or resistor back to the other side of the winding. Voltage spikes can travel back through the circuit and damage electronic devices, so always be sure to use relays with the proper voltage suppression for the vehicle you are repairing.

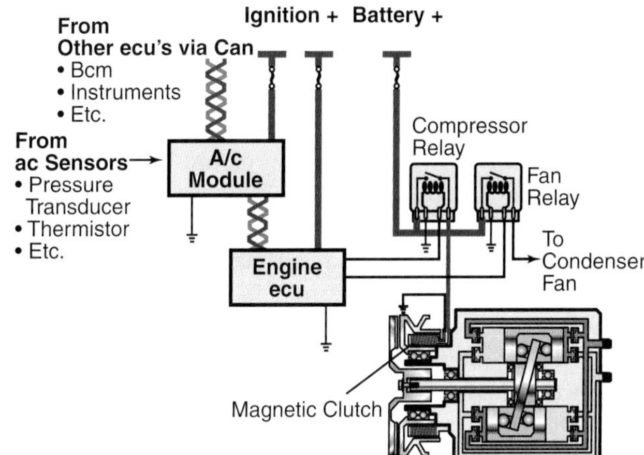

FIGURE 60-29 Relay-activated compressor clutch electrical diagram.

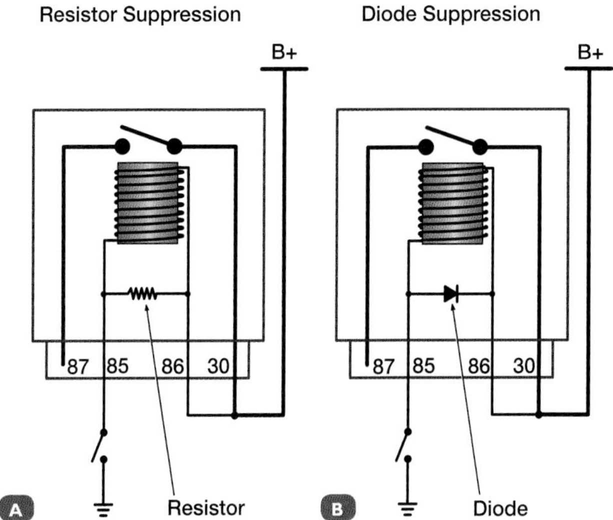

FIGURE 60-30 Some relays use voltage suppression devices across the windings. **A.** Resistor **B.** Diode.

> TECHNICIAN TIP

Diode-style relays are not easily tested for open or shorted diodes. A trick is to use a good 9-volt battery (the little rectangular ones) to activate the relay. Place the battery terminals of the 9-volt battery against the winding terminals (85 and 86).The relay should click when the 9-volt battery is connected in one direction and should not click when it is connected in the other direction. If the relay acts in this way, the diode is operational. If the relay clicks in both directions, the diode is open.

Blower Motor

One of the controlled devices used in the climate control system is the blower motor. The blower motor is the means of moving air into and through the passenger cabin. The blower motor is an electric motor that is attached to a circular fan sometimes referred to by technicians as a squirrel cage fan (**FIGURE 60-31**). The blower motor can be operated at different speeds to provide the desired amount of airflow in the cabin. Blower motors are usually located in the ventilation system before the air-conditioning evaporator and draw air from either the passenger cabin or outside the vehicle.

Blower Motor Speed Control

One way of changing the speed is through the use of a resistor pack in series with the blower motor. A blower motor resistor pack is a set of different resistors oriented in series to one another, but with circuit connection points between them (**FIGURE 60-32**). As additional resistance in the resistor pack is selected, two things happen: Current in the entire circuit drops (Ohm's law), and voltage drop across the resistor pack increases. Both of these conditions reduce the current and voltage to the motor, which makes it spin at a slower speed. Think of a flashlight as the batteries die. The bulb gets dimmer because voltage is lower, but if we insert new batteries, the bulb gets brighter because the higher voltage can push more amperage or current through the bulb. A motor will run slower if the voltage is lower and faster if it is higher.

FIGURE 60-31 Blower motor and squirrel cage fan.

Changing the resistance to the blower motor is accomplished by using the fan control switch to select which resistor(s) the current will pass through. Each time we change the switch position, we change the resistance of the circuit and change the voltage (**FIGURE 60-33**). When we select full speed, the resistors are bypassed, and full voltage is applied to the motor. Most manual air-conditioning systems use three resistors to give four operator-selectable speeds of the blower motor. As we want the motor to spin more slowly, we increase the resistance to the motor (decreasing the voltage and amperage). In the three-resistor pack, we have three speeds that result from resistance added to the circuit. We also have a no resistance speed, which is the fourth speed; this is a circuit with all resistance removed from the circuit except the motor. Resistors can be used on either the positive (B+) side or the negative (B–) side of the motor, depending upon the manufacturer's preference.

When current flows through a resistance, heat is generated. Because the blower motor draws quite a bit of current flow when it is operating, the blower motor resistors generate a lot of heat. To help prevent them from overheating, they are usually made of fairly heavy metal wire, wound in a loose series of coils and placed in the blower airstream. So while the blower motor is operating, airflow from the fan is blown across the resistor coils, thus preventing them from overheating.

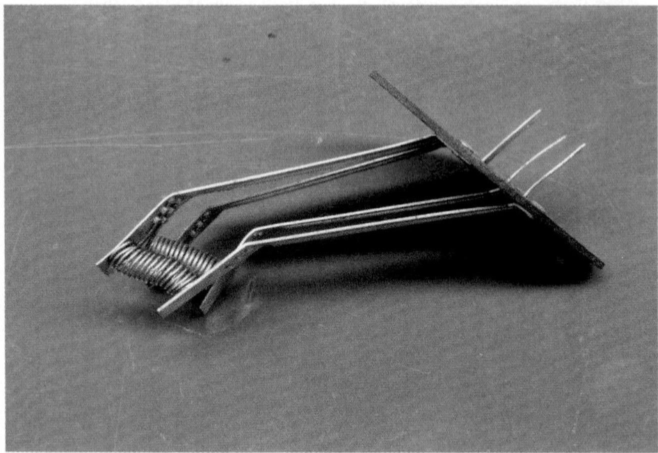

FIGURE 60-32 A typical blower motor resistor pack.

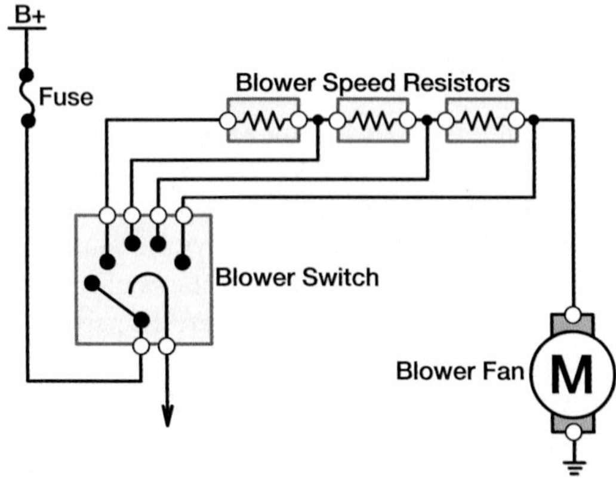

FIGURE 60-33 A typical arrangement of a blower motor resistor pack that gives four speeds.

TECHNICIAN TIP

Some resistor blocks incorporate one or more thermal fuses to protect the circuit if the blower resistors overheat. Because the thermal fuses and resistors are in the blower motor airstream, airflow helps to keep them from overheating. If the cabin air filter becomes clogged or the blower motor becomes bound up, airflow will be reduced and the thermal fuse can overheat and blow. If a thermal fuse, or even a regular blower motor fuse blows, check for excessive blower motor current flow, binding of the fan, and a restricted cabin air filter.

On some vehicles, the highest fan (blower motor) speed may be powered directly from a dedicated fan relay. This design enables full-speed blower motor operation. The fan control switch is connected to a relay. The fan control switch gives 12 volts to the control coil of the relay, which is typically grounded. When the coil gets 12 volts, a magnetic field is produced, and the contacts in the relay close. The relay applies full battery voltage to the blower motor to give high speed. The reason a relay is used is so the fan switch does not have to handle the full blower motor amperage flowing through it. It takes very little amperage to make the magnetic field, but it takes a lot of amperage to run the blower motor at full speed, and this high amperage would damage the fan control switch (**FIGURE 60-34**). In some vehicles, the relay can be used to supply a full ground (B–) to the ground side of the motor instead of the power side (B+).

In automatic climate control systems, the speed of the blower motor can be controlled electronically by the ECU. The ECU controls the speed by pulsing the current flow to the motor. This system is called pulse width modulation and results in a varying duty cycle (**FIGURE 60-35**). The longer the current flow is on, the faster the fan runs. A shorter pulse width signal produces a lower fan speed. The ECU can produce a progressively variable fan speed between zero and maximum speed by varying the pulse width.

Recall that the blower motor resistor design gives us four total fan speeds. By pulsing the current flow longer or shorter, we are able to create an almost infinite numbers of speeds within its design limits. The ECU controls the speed of the blower motor automatically depending on the difference between the interior cabin temperature and the operator-selected temperature. When the difference in temperature is large, the blower motor speed is high to increase air circulation through the cabin. As the difference in temperature decreases, the blower motor speed decreases.

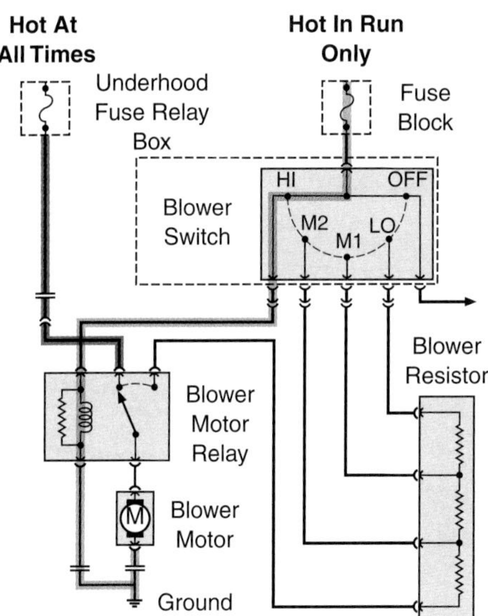

FIGURE 60-34 Blower motor high speed may be controlled by a relay.

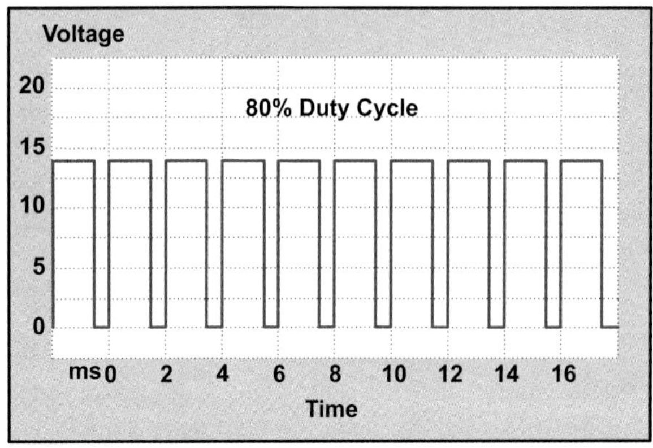

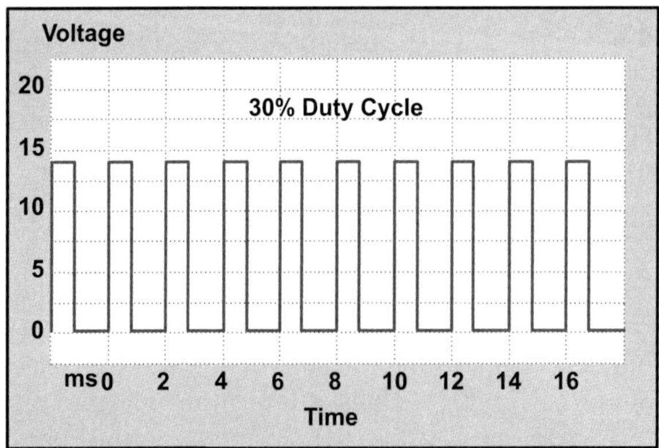

FIGURE 60-35 Using pulse width modulation to vary the duty cycle on the power side.

Air Box and Air Doors

The blower motor is attached to the air box. The air box is a plastic housing that holds other components of the climate control system, such as the heater core and the evaporator. The air box also includes the doors that direct airflow to different discharge areas within the passenger compartment such as the vents, floor, or windshield defroster, and the blend door, which controls air temperature. The blend door directs air through or around the heater core to deliver hot or cold air into the passenger cabin. The mode door changes the location of air delivery to floor, defrost, or vent. The recirculation door switches between outside air and cabin air. It is designed to work with the Max air-conditioning button or switch and will shut airflow off from outside the vehicle and instead recirculate air from inside the passenger compartment back through the evaporator. By moving this air through the evaporator again and again, the temperature inside the vehicle can be reduced further and sometimes quicker (**FIGURE 60-36**). The blower motor and resistor pack are bolted into the air box as well.

Vacuum-Style Servo Motors

Actuators are mechanical devices that create rotary or linear movement. They are sometimes called servos, or servo motors. Actuators move the air box doors to control temperature and position of the airflow from the air distribution system. Actuators can be electric or vacuum powered. Electric actuators come in two styles: geared motor and solenoid. Geared motor actuators are used in HVAC systems, and solenoid actuators are used in systems such as door locks and fuel door latches.

The vacuum servo differs from the electric actuator in that it is a sealed container with a spring and diaphragm inside. The vacuum servo uses vacuum on a diaphragm to pull a link or lever that is connected to the air box door (**FIGURE 60-37**). The diaphragm is connected to one of the air box doors by means of a rod or linkage. When vacuum is released, the spring pushes the diaphragm back to its starting position. Vacuum is applied to the actuator through the control head and is only present or removed when movement is needed.

Vacuum servos can be controlled electrically by using a solenoid to control the vacuum source (**FIGURE 60-38**). The use of a solenoid allows electrical control of a vacuum source. Solenoids are operated by providing power and ground. The ECU of the HVAC system usually controls the ground side. When ground is supplied to the solenoid by the ECU, the solenoid activates, and vacuum is applied through the solenoid valve to the vacuum servo. To release the vacuum servo, the ECU turns the ground off to the solenoid and allows the vacuum to bleed off, which makes the door return to the original position.

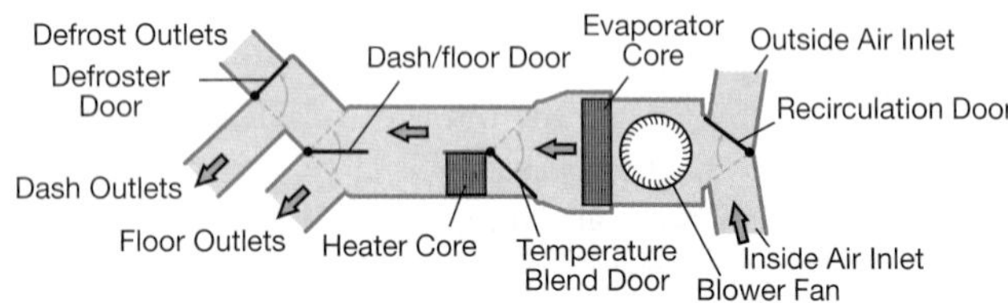

FIGURE 60-36 The recirculation door is designed to shut airflow off from outside the vehicle and instead recirculate air from inside the passenger compartment back through the evaporator.

FIGURE 60-37 Vacuum servo construction.

Electric Servo Motors

Movement of air box doors can also be achieved by using **electric servo motors**, also referred to as actuators. Electric actuators respond to output signals from the ECU of the HVAC system. The electric actuator performs the same job as the vacuum servo, moving the air box doors. Electric actuators use an electric motor and gears to create either rotary or linear motion, which is used to operate the air doors. Older electric actuators may incorporate **limit switches** to stop current to the electric actuator when the door has traveled the maximum distance. This stoppage prevents the electric actuator from burning out. Newer actuators use stepper motors that can be controlled by the ECU precisely one step at a time. This design gives precise movement to the door, and the ECU can monitor the position of the door by knowing how many steps it opened or closed, as commanded by the actuator (**FIGURE 60-39**).

Some electric servo motors also use a potentiometer to signal to the ECU the position of the blend door and, if it is moving, the direction in which it is traveling. The potentiometer is a variable resistor that varies resistance with movement. The potentiometer functions as a position sensor and is used as a throttle position sensor on the engine. The potentiometer is a three-wire sensor that has a 5-volt reference voltage, a

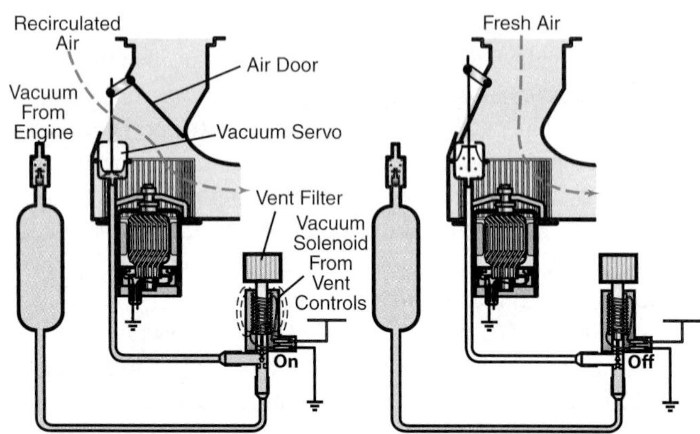

FIGURE 60-38 Vacuum solenoid diagram.

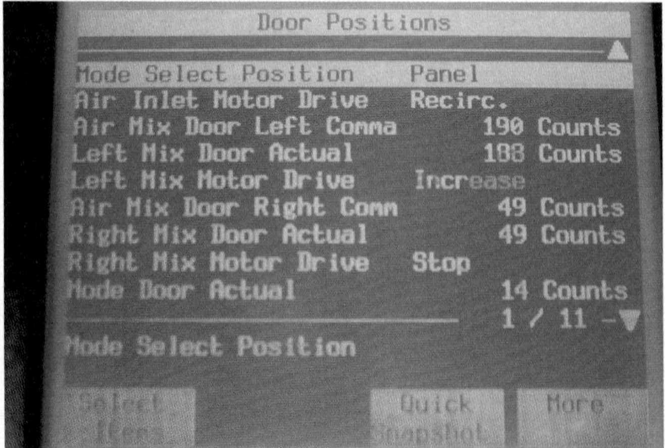

FIGURE 60-39 Scan tool data from an ECU showing the number of steps the actuators are commanded.

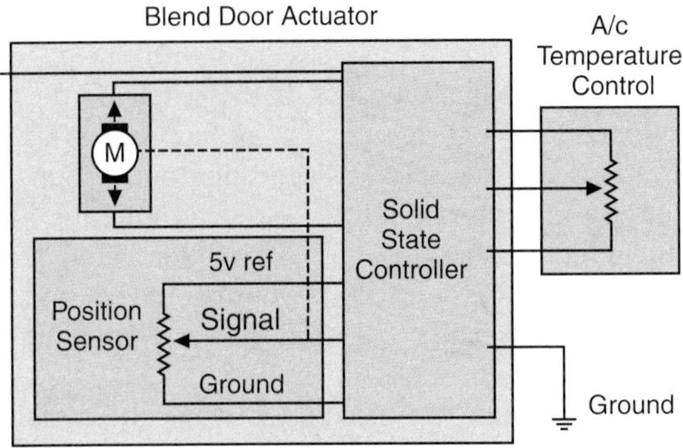

FIGURE 60-40 Electric actuator with potentiometer. The potentiometer is a three-wire sensor that has a 5-volt reference voltage, a feedback wire that signals back to the ECU (in this case), and a ground.

signal wire that provides feedback to the ECU (in this case), and a ground (**FIGURE 60-40**). When the voltage from the actuator sensor matches the programmed information in the memory of the ECU, the door is in the correct position desired, and the actuator's electric motor stops moving. For maximum cooling effect, the blend door is moved to stop airflow through the heater core; in this position, the voltage signal back to the ECU is at its lowest value. As the blend door is moved to allow airflow through the heater core, the voltage at the signal wire progressively increases until at the maximum heat position, where the voltage signal is at its highest. The ECU is thus aware of the blend door position, movement, and direction of travel.

▶ Hybrid Vehicle Air-Conditioning System Principles

N60001, K60004

The air-conditioning system in hybrid vehicles is almost always a climate-controlled HVAC system. This is because there are additional demands put on the HVAC system by the hybrid vehicle. One demand is the need to provide cooling, either directly or indirectly, to the high-voltage battery to prevent it from being overheated. As such, the ECU must be able to take control of the air-conditioning system and operate as needed to keep the battery at the proper temperature. To accomplish this, the ECU monitors the battery temperature with a temperature sensor.

Another difference in some hybrid vehicles is the ability for the driver to remotely activate the HVAC system to cool or warm up the vehicle before entering it. The driver can operate a button on the key fob from inside his or her home or office to activate the HVAC system and begin cooling the passenger compartment (**FIGURE 60-41**). This system requires a higher level of system control. Some hybrid HVAC systems are integrated with a ventilation feature when the vehicle is parked outside in the sun. It may consist of solar cells on the roof of the vehicle, which power a fan to expel hot air from inside the vehicle. This system can be combined with remote HVAC functionality, which allows the driver to turn on the AC system to a few minutes before entering the vehicle.

For the most part, the components in the HVAC system of a hybrid vehicle are the same as in the systems discussed previously, but there are differences. The compressor in the hybrid air-conditioning system may be driven by the internal combustion engine (ICE) through a standard belt and **electromagnetic clutch arrangement**, just like a standard air-conditioning compressor clutch arrangement. Most hybrids now use a high-voltage electric motor that directly drives the compressor (**FIGURE 60-42**). This style does not have a belt driving it. If the compressor has orange, high-voltage wires connected to it, then the compressor uses a high-voltage electric motor to drive the air-conditioning compressor. In this case, the electric air-conditioning compressor motor can be turned on and off as needed by the ECU, without the ICE operating. It can also vary the compressor speed based on the amount of cooling needed. This system greatly improves efficiency and reduces emissions.

SAFETY TIP

The voltage in high-voltage compressor systems is very dangerous. Only technicians certified to work on high-voltage systems should attempt diagnosis and repair on these systems. Voltages can run up to 500 volts. This voltage can be lethal and requires the use of leak-tested, insulated

FIGURE 60-41 A key fob with remote AC functionality.

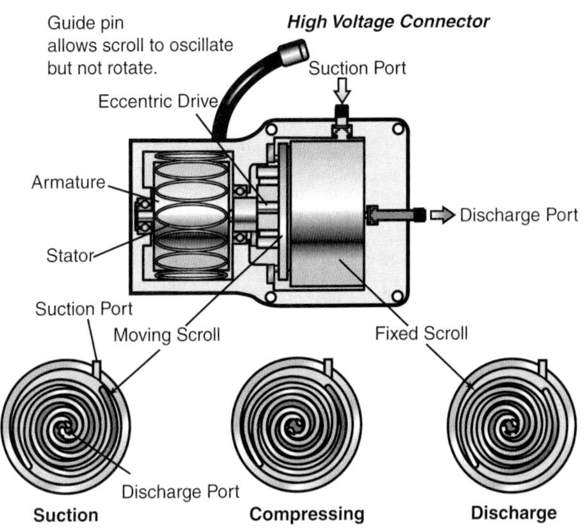

FIGURE 60-42 Electrically driven air-conditioning compressor.

electrical Class 0 1000-volt gloves (and protective covers). It also requires a Cat III-certified digital multimeter (DMM), which is rated to withstand 1000 volts. Also, never attempt to pierce the insulation on high-voltage wires; doing so can cause severe injury or death as well as major damage to the electrical circuit.

Some hybrids use **dual-drive air-conditioning compressors** (**FIGURE 60-43**). These compressors can be belt driven either by the ICE or by a high-voltage motor. The belt may drive a larger-capacity scroll, whereas the electric motor drives a smaller-capacity scroll. These systems can usually be identified by orange high-voltage cables and a belt-driven clutch. This type of system allows the larger compressor scroll to be driven by the ICE whenever it is running. The electric motor drives the smaller scroll when the engine is off and the vehicle is being powered by the storage batteries, or when extra cooling is needed while the ICE is operating. The benefit of this type of system is that the air conditioner continues to operate as needed when the ICE shuts off (such as at a stoplight) to keep the occupants comfortable. When the ICE starts, it powers the compressor instead of pulling power from the high-voltage battery pack. The lower amount of power pulled from the battery pack allows it to recharge more quickly, resulting in better economy.

TECHNICIAN TIP

The compressors in most hybrid vehicles require special oil in their air-conditioning system to protect the insulation on the windings in the high-voltage electric motor. Always check the manufacturer's recommendation for HVAC oil, and do not mix it with the wrong oil. This is critical when reclaiming refrigerant, to prevent cross-contamination with other oils. Even a small amount of contamination can cause the electric compressor motor to short out. Also, manufacturers typically discourage the use of leak detection dye in hybrid vehicles.

▶ Diagnosis

K60005

Diagnosis of HVAC systems follows the same strategy-based diagnosis process as any other vehicle diagnosis. Once you have a good understanding of the customer concern, it is always good practice to perform a system performance test and run the system through all of its operating parameters, noting any differences from the manufacturer's specified operation. Many new vehicles are designed with HVAC systems that have built-in self-diagnostics. In these systems, the ECU is programmed to recognize signals or situations that are out of specifications and to display a diagnostic trouble code (DTC). You can use this information to help determine the cause of the system fault. Using the DTC information, along with any system operational issues, allows you to research and follow the manufacturer's TSB's and service information and locate the cause of the fault.

At the same time, you need to use your understanding of the system's theory of operation and apply that to the way the system is operating. This will help you develop a solid, focused testing plan to guide you through the diagnosis. The more you practice this process, the better you will become at it. We look at the tools and procedures below.

Tools

Several tools are necessary for the service and repair of the HVAC system. Proper training on the use of these tools is required to ensure that you do not hurt yourself, damage the vehicle, or damage the tool itself.

- **Infrared thermometer:** This tool allows you to measure temperatures without opening either the cooling system or AC system. Use extreme caution when working with the cooling and AC systems.
- **Digital multimeter (DMM):** A DMM is absolutely necessary when working with electronics. It has a very high internal resistance, which keeps the meter from pulling too much current from the electronic parts being tested. An older-style analog meter could cause too much current flow during testing and damage components. The DMM is used to test for correct voltages at electrical components in the HVAC system and to test wiring for high resistance. Back probes (small pins to test for voltage at the back side of the connector) are necessary to test electrical components while they are operating (**FIGURE 60-44**).
- **Scan tool:** This tool is necessary for checking DTCs in the ECU of the HVAC system and for performing bidirectional

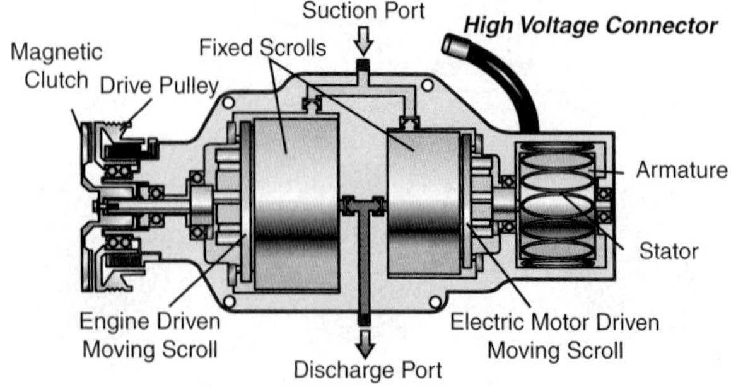

FIGURE 60-43 Dual-drive air-conditioning compressor.

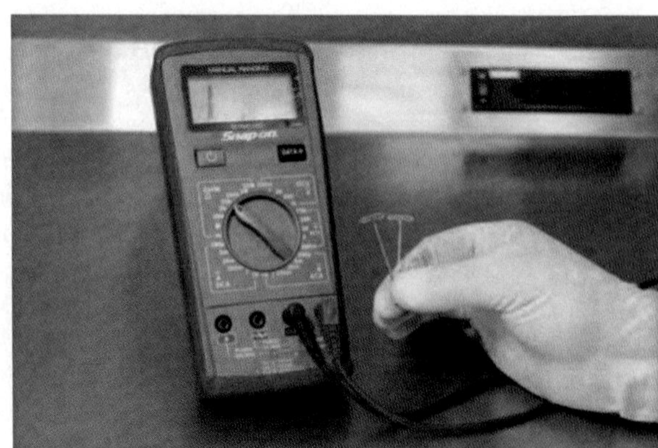

FIGURE 60-44 A DMM with a set of back probes is necessary to test electrical components while they are operating.

controls of electrical loads. Some vehicles have diagnostics built in to the control panel, and in those cases the use of a scan tool is only necessary for bidirectional control of the electrical loads, such as for the cooling fan, blower motor, or electric actuators.

- **Fused jumper wire:** The wire should be fused so that if hooked up incorrectly it is more likely to blow the fuse than to damage the vehicle's wiring. This tool is used to jump suspected bad switches or relays for testing purposes. However, just because it is less likely to damage wiring, does not mean it will not damage electronics. It will. Always verify that you are hooking it up correctly.

SAFETY TIP

Never jump across the load (the electrical part of the circuit doing the work), such as the cooling fan or electric actuator. Doing so will remove the load from the circuit and create a direct short to ground.

Common Issues

Before working on the HVAC system, be sure to check technical service bulletins (TSBs) matching your vehicle for any issues similar to the customer concern. A common issue a technician might encounter with the automatic HVAC system is the failure of electric actuators. Electric actuators have small plastic gears inside that can strip out if the mode doors get jammed or develop drag when turning. Before installing the new actuator, be sure to test the door operation by turning the door by hand and noting the drag on the door. If you are unsure, compare the drag to another door that is easily accessible.

If the vehicle uses vacuum servos to control the mode doors and blend door, failure in the vacuum hose(s) can result in failed operation. Refer to the vacuum chart for the HVAC system you are working on to properly diagnose the concern. Vacuum leaks can be found by using a handheld vacuum pump on one end of the vacuum hose and plugging the other end or by using a smoke machine to blow smoke into the vacuum lines. A visual inspection looking for wisps of smoke leaking out of hoses or vacuum servos will help to determine the exact location of the leak. Vacuum servos fail at times, and they are best tested by using a handheld vacuum pump, pulling a vacuum on the servo, and watching for movement. If no movement is noted, ensure that the door to which the servo is connected is not binding. If the door is not binding, does the servo hold vacuum? If not, replace the leaky servo. If movement is noted, does the servo hold vacuum? If not, replace the leaky servo.

Sensor failure is possible and typically results in a DTC being stored in the ECU. Refer to the manufacturer's service information on details for accessing DTCs on the system you are servicing. Refer to the description of any system you are unfamiliar with to ensure that you understand how the system works. You cannot fix something if you are unsure whether it is broken.

Applied Science

AS-100: Flow Rate: The technician can demonstrate an understanding of how variances in flow rate in airflow sensors and cooling systems can affect engine performance.

Airflow has a fundamental effect on cooling system performance as a steady flow of cool air though the radiator core is essential to remove heat from the system. Radiator fans are fitted to provide additional airflow through the radiator at times when airflow is insufficient. In late model vehicles, radiator fans are controlled by the ECU in response to coolant temperature sensor readings.

When a vehicle is stationery or moving slowly, airflow is generally insufficient to keep the engine cool. In response to high coolant temperatures, the ECU will operate the fan. This occurs in cycles with the fan turned on when a high specified coolant temperature is reached, then turned off when the system temperature returns to a lower specified value. When the vehicle is moving quickly, airflow increases, so little or no fan operation may be required.

Checking Operation of Automatic or Semiautomatic HVAC Control Systems

N60002, S60001

Checking the operation of the automatic or semiautomatic HVAC system is really just a more thorough performance check for the system. It is required when a customer complains of an HVAC issue. This check allows you to verify that there is an issue and to determine the extent of the issue. One of the first steps is to research the system's operation in the service information, which will give you a good idea of how to perform a thorough testing of the system. Some vehicles require auxiliary airflow provided by an external condenser fan, whereas others specify the parameters that have to be met for results to be accurate. The service information also usually provides you with temperature charts to which you can compare your results. Always follow the manufacturer's testing procedure.

To check the operation of automatic or semiautomatic HVAC control systems, follow the steps in **SKILL DRILL 60-1**.

Using a Scan Tool to Observe and Record HVAC-Related Data and Codes

N60003

Scan tools are used to pull codes from a particular computer. Not all HVAC systems use computer controls, so it is important to check the engine computer for issues relating to how the vehicle runs that could affect HVAC performance. If the HVAC system is computer controlled, this check should be done before a performance test, as some codes cause the computer to shut off the HVAC system to protect it. Codes are letters and numbers that can be looked up in a diagnostic chart to see what kind of problem the computer diagnosed. Many different types of scan tools are available, each with its own specific procedure for use. The technician must read and follow the instructions provided with the particular scan tool being used.

SKILL DRILL 60-1 Checking the Operation of Automatic or Semiautomatic HVAC Control Systems

1. Start the engine, and allow the system to normalize. Operate all the controls, and verify the operation of each setting, including fan speeds, air temperature, airflow modes (defrost, vents, floor), and so on.

2. Measure the ambient temperature along with the duct temperature at the center vent.

Ambient Temperature	Relative Humidity	Low Side Service Port Pressure	High Side Service Port Pressure	Maximum Left Center Discharge Air Temperature	Maximum Left Rear Discharge Air Temperature
13-18°C (55-65°F)	0-100%	151-277 kPa (22-33 psi)	1123-1446 kPa (163-210 psi)	8°C (45°F)	18°C (64°F)
19-24°C (66-75°F)	Below 40%	165-261 kPa (24-38 psi)	1150-1481 kPa (167-215 psi)	8°C (46°F)	19°C (66°F)
	Above 40%	179-282 kPa (26-41 psi)	1240-1557 kPa (180-226 psi)	10°C (50°F)	22°C (70°F)
25-29°C (76-85°F)	Below 35%	206-296 kPa (30-43 psi)	1336-1598 kPa (194-232 psi)	10°C (50°F)	23°C (72°F)
	35-50%	220-303 kPa (32-44 psi)	1391-1612 kPa (202-234 psi)	12°C (52°F)	23°C (73°F)
	Above 50%	227-323 kPa (33-47 psi)	1419-1646 kPa (206-239 psi)	13°C (55°F)	25°C (77°F)

3. Compare this temperature to specifications, and confirm that it matches the temperature settings.

To use a scan tool to diagnose HVAC systems, follow the steps in **SKILL DRILL 60-2**.

Inspecting and Testing the Electric Cooling Fan

Inspecting and testing the electric cooling fan is necessary for customer concerns ranging from overheating of the vehicle's engine, to an air-conditioning system that does not get cold enough. The tools required for testing the cooling fan include a scan tool, a DMM, a fused jumper wire, back probes, and service information. It is critical that you have a thorough understanding of how the system works before attempting a repair. This includes whether the circuit is a simple relay-based circuit or a networked circuit.

To inspect and test the electric cooling fan, follow the steps in **SKILL DRILL 60-3**.

Inspecting and Testing the HVAC Control Panel Assembly

N60004

The control panel assembly usually has to be inspected and tested only when the HVAC system does not deliver the selected temperature or placement of the air. This task depends on the type of system you are working on and the manufacturer's diagnostic procedure. There are several types of control panels, including manual, semiautomatic, automatic, dual-temperature control, and rear air-conditioning systems. Be sure to refer to the appropriate service information. The next Skill Drill is conducted with a manual system equipped with a blend door and electric actuators.

To inspect and test the HVAC control panel assembly, follow the steps in **SKILL DRILL 60-4**.

SKILL DRILL 60-2 Using a Scan Tool to Observe and Record HVAC-Related Data and Codes

1. Connect the scanner to the diagnostic connector.

2. Write down all of the air-conditioning codes. Check any freeze-frame data, and write that down.

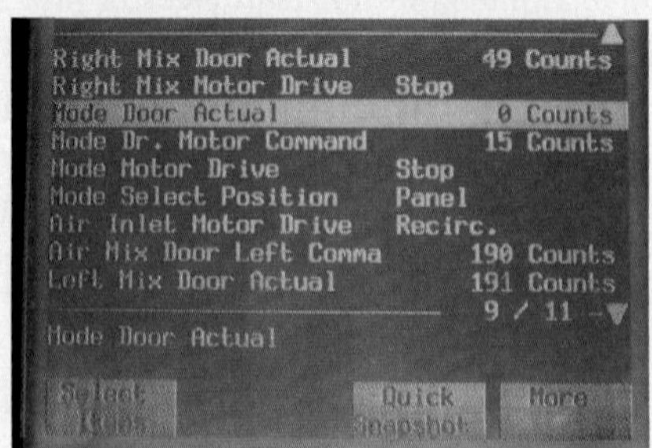

3. Look at the scan data, such as the system temperatures and pressures, and identify any abnormal readings. When repairs are completed, clear the trouble codes.

SKILL DRILL 60-3 Inspecting and Testing the Electric Cooling Fan

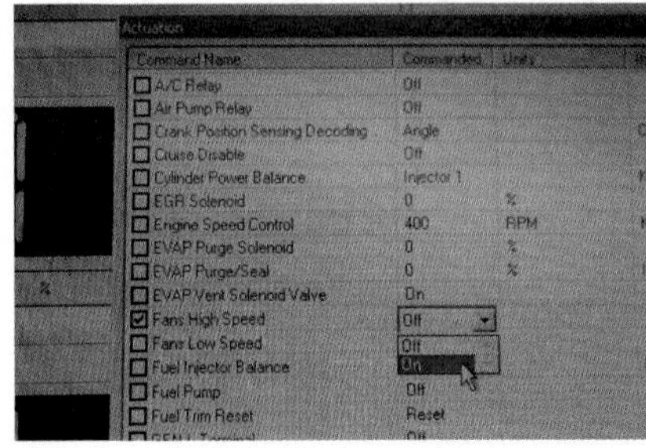

1. If the cooling fan does not come on when the system is turned on, install the scan tool, and find the bidirectional controls for the cooling fan; then turn the cooling fan on.

2. If the cooling fan still does not run, find the coolant fan relay, remove the relay, and find terminals 30 and 87.

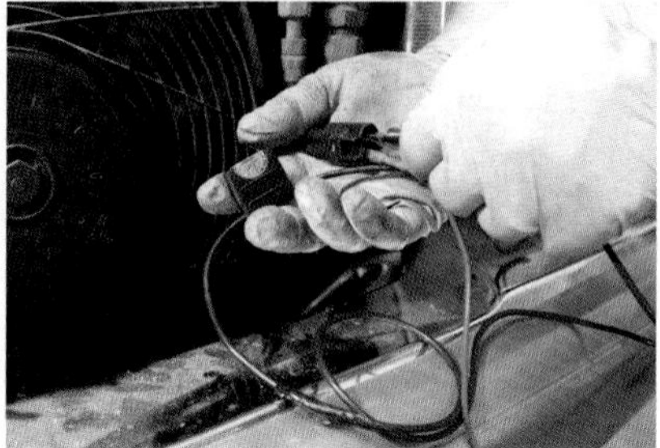

3. Install the fused jumper wire to the connections to provide power to the cooling fan. CAUTION: Make sure you jump the switch circuit (30 and 87) and not the control circuit. Jumping the control circuit can burn up the ECU, even with fused jumper wires!

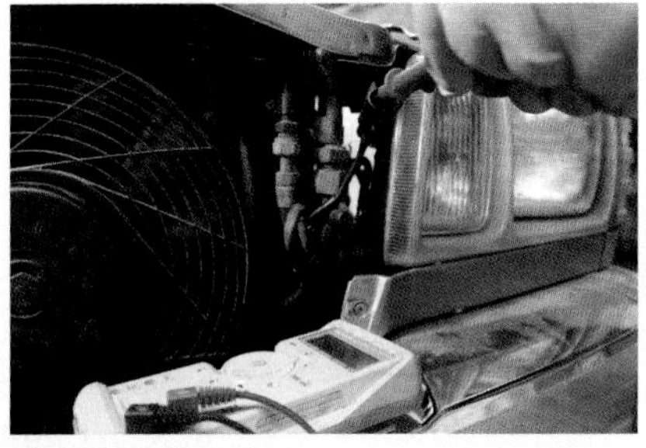

4. Inspect the cooling fan for operation. If the fan is still not operating, back-probe the connector at the cooling fan with the jumper wire still installed from the previous step. Connect the voltmeter leads to both back-probed wires (power and ground); the meter should read battery voltage.

5. If battery voltage is present, replace the faulty cooling fan. If battery voltage is not present, use the voltmeter to locate the open or high resistance in the circuit.

SKILL DRILL 60-4 Inspecting and Testing the HVAC Control Panel Assembly

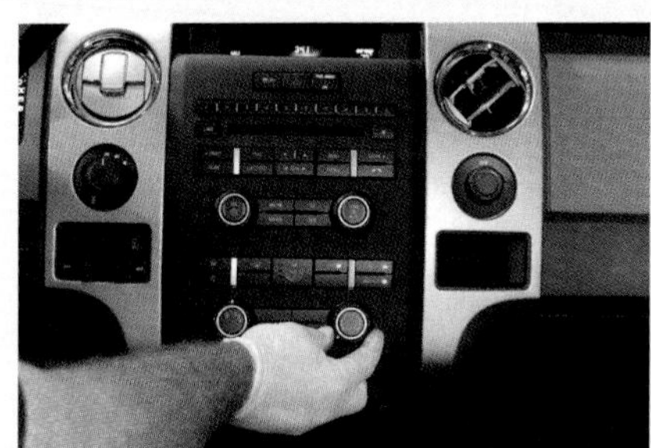

1. After researching the operation and diagnostic procedure for the control panel assembly, start by operating all of the heater and air-conditioning controls. Note any faults. Remove clips and bolts holding hush panels under the dash.

2. Move the temperature control switch from maximum to minimum.

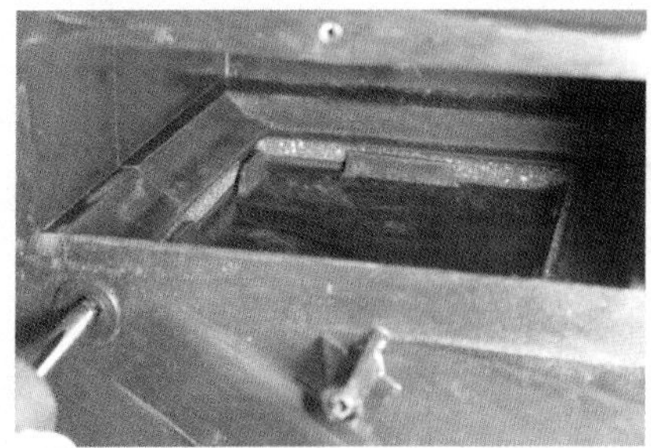

3. Observe the blend door linkage for movement. If no movement is detected, test the actuator system according to service information. If movement is present, remove the actuator by undoing the retainers securing it to the air box.

SKILL DRILL 60-4 Inspecting and Testing the HVAC Control Panel Assembly (Continued)

4. Move the blend door by hand, and check whether it is broken at the shaft or is binding. If the blend door is broken, the entire air box must be removed. Refer to service information for removal and repair.

5. Move each control and watch the appropriate doors for proper movement.

Inspecting and Testing HVAC Control Cables

N60005

Inspecting and testing system control cables and linkages usually will be required only when the air conditioner or heater is not delivering the proper temperature of air or is delivering airflow to a different mode than selected. This will happen when a cable is binding or broken or if it has popped off its linkage at the door or control head. The HVAC control panel will have to be visually inspected and tested, requiring you to gain access to the cables and linkages. Access is different for each vehicle, so research the procedure in the appropriate service information.

To inspect and test the HVAC control cables, follow the steps in **SKILL DRILL 60-5**.

Diagnosing Temperature Control Problems in the HVAC System

N60006

To diagnose temperature control problems, the technician must fully understand how temperature change is controlled in the HVAC system being serviced. Refer to the description and operation in the service information. The need to diagnose temperature control problems arises when the customer is not able to get the vehicle cold enough or hot enough inside the passenger cabin. The cause of such a customer concern can be as simple as a thermostat issue or as complex as an electrical or electronics issue. To diagnose this type of complaint, refer to the manufacturer's service information for troubleshooting charts. The tools required for this job vary greatly depending on the cause of the concern and the type of HVAC control

SKILL DRILL 60-5 Inspecting and Testing the HVAC Control Cables

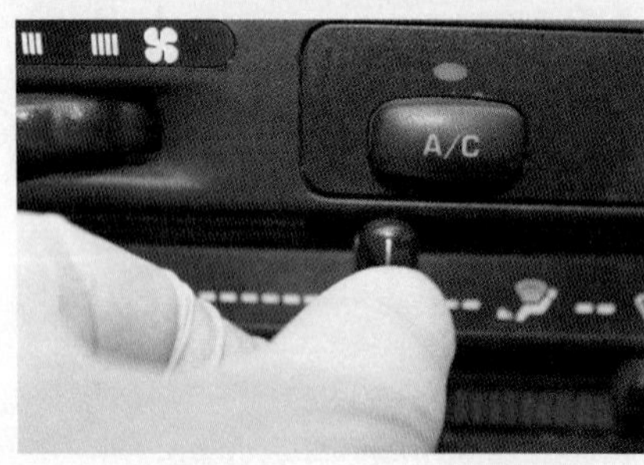

1. Begin by researching the operation and diagnostic procedure for the control panel assembly. Move the control panel switch or levers, and observe the doors on the air box for movement.

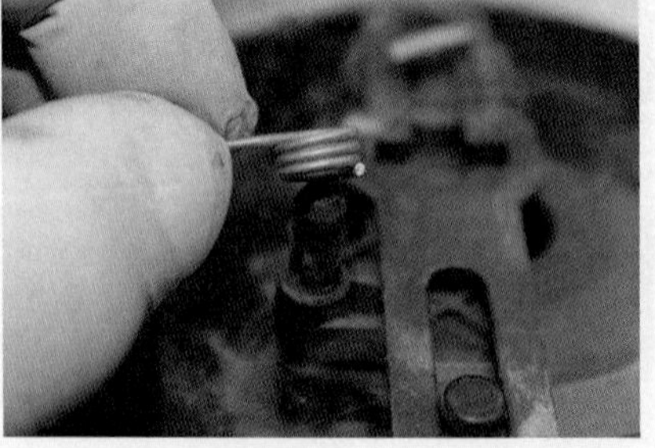

2. If movement is not noted, inspect for a cable that is broken or has popped off the control lever or door.

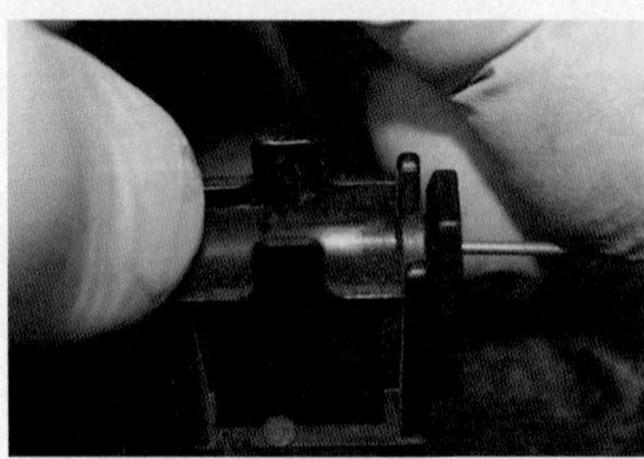

3. If a cable has popped off, reinstall it with the proper retainer, and retest the cable system.

system you are working with. The next Skill Drill assumes use of an automatic HVAC control system with an ECU. The tools required are a scan tool appropriate for the vehicle, a DMM, back probes, an infrared thermometer, and service information.

To diagnose temperature control problems in the HVAC system, follow the steps in **SKILL DRILL 60-6**.

Inspecting the HVAC System: Heater Ducts, Doors, and Cabin Filters

N60007, S60002

Because the HVAC system ducts and doors direct airflow to specific areas within the passenger cabin, it is important that they function as shown on the control panel; for instance, if you select floor mode, air should come out at the floor. Inspecting the HVAC system ducts is accomplished by operating the system in all positions and verifying that it is being controlled appropriately. If not, each position on the HVAC control panel will indicate the air door or duct that is affected.

This inspection of the ducts and doors offers a good opportunity to inspect any cabin filters to see if they have to be replaced. In systems where the cabin air filter is in the fresh air intake, you can use an anemometer to check the difference in the speed of the vent airflow with the system in recirculation mode compared to fresh air mode. Any difference indicates a filter restriction. Also, the average airflow measurement from all vents should be a minimum of 800 fpm. In systems where the cabin air filter filters recirculated air as well, you will need to remove the cabin air filter and hold it up to a light to determine if it needs to be replaced.

To inspect the heater ducts, doors, and cabin filters, follow the steps in **SKILL DRILL 60-7**.

SKILL DRILL 60-6 Diagnosing Temperature Control Problems in the HVAC System

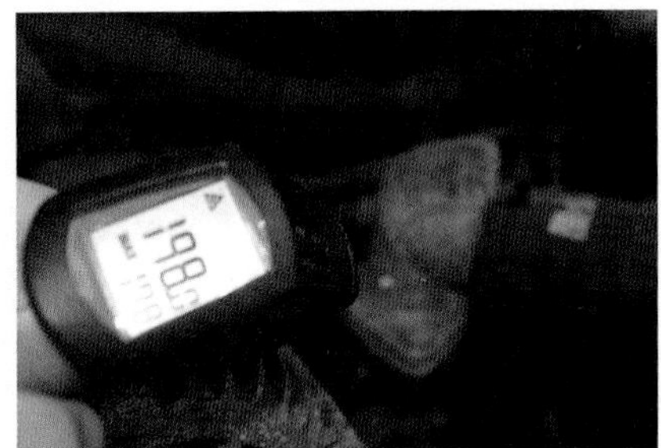

1. Research the operation and diagnostic procedure for the control panel assembly, and verify the customer concern. Check for full operating temperature of the engine by pointing an infrared thermometer at the thermostat housing. If the temperature is not hot enough, the thermostat is likely in need of replacement.

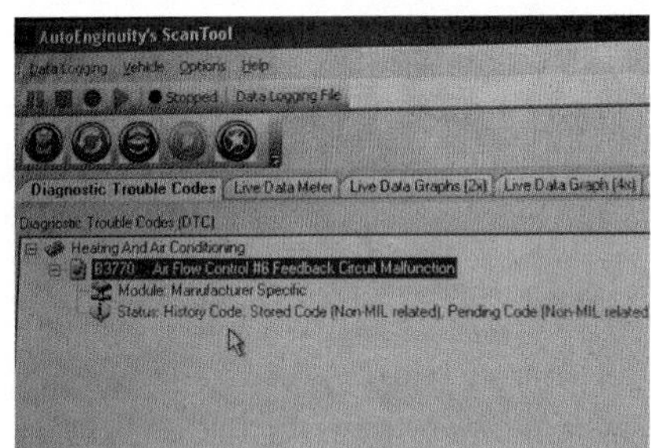

2. Test for codes with the scan tool. If a code is set, follow the troubleshooting chart for the code pulled. If no code is set, follow the symptom-based troubleshooting chart.

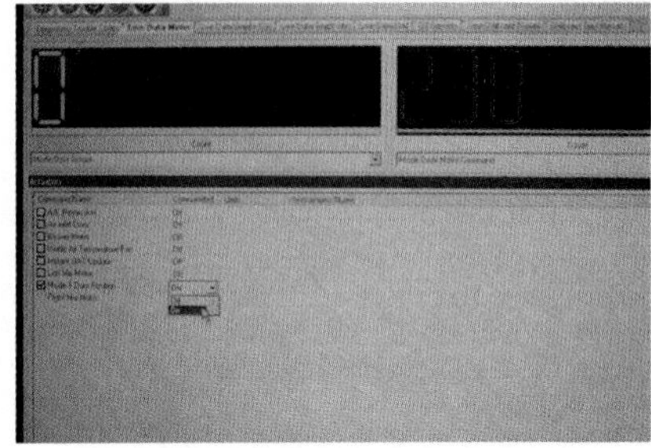

3. If a code is found for the electric actuator for the blend door, pull up bidirectional controls in the scan tool, and command the blend door to operate while visually inspecting the movement of the actuator.

4. If movement is found at the actuator, inspect for a broken or binding door by turning the door by hand. If no movement is found at the actuator, perform the test for voltage (power and ground) at the actuator while commanding it.

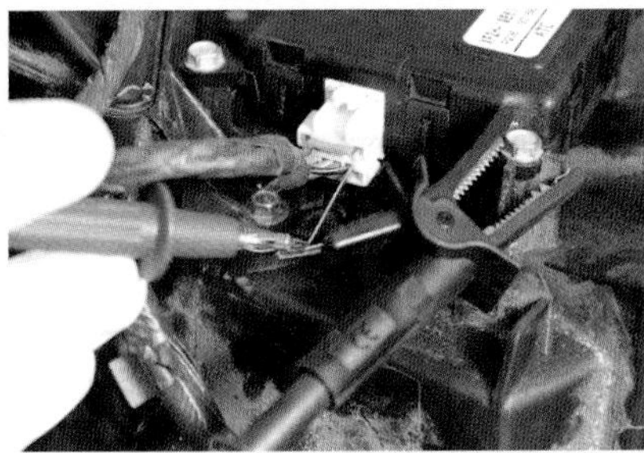

5. If no power and ground are found, backprobe the wires for the actuator at the connector of the ECU voltage (power and ground) while commanding the actuator. If no power and ground are coming from the ECU, the ECU will have to be replaced.

SKILL DRILL 60-7 Inspecting the Heater Ducts, Doors, and Cabin Filters

1. Start the vehicle, and turn the fan switch to the maximum setting.

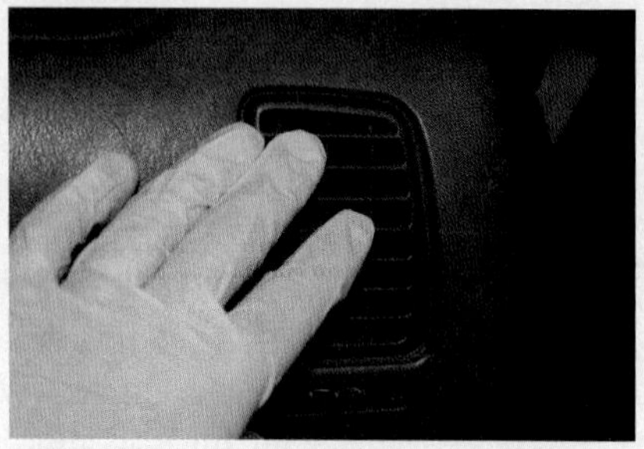

2. Turn the mode selector through all the positions, and make sure the air is distributed through the appropriate vents. Turn the mode selector to dash vents, and recirculate.

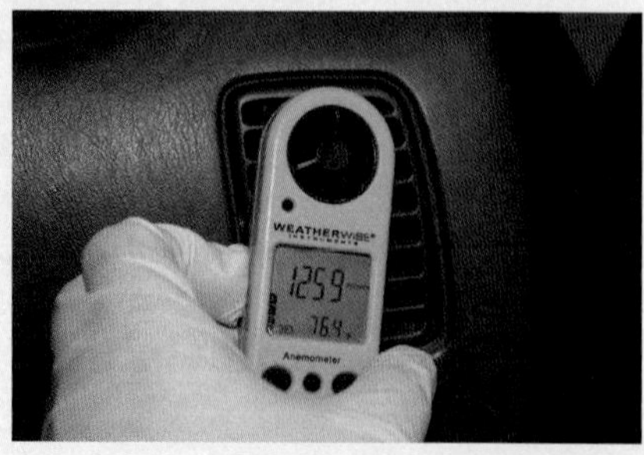

3. Take an average airflow measurement from all vents (800 fpm minimum average). If no flowmeter is available, test the amount of airflow with your hand or a piece of paper, and compare to a similar vehicle.

4. Move the fresh air/recirculate control to the fresh air position, and retest the airflow to see if the cabin filter is plugged. Make sure at least one window or door is open to avoid restricting airflow. The airflow should not drop.

5. If the cabin air filter filters all air in the system, remove the cabin air filter, and hold it up to a light to see if it is plugged. Replace it if it is restricted or damaged.

Inspecting and Testing Blower Motors and Control Circuits

N60008

Blower motors and control circuits (typically, a rotary switch and a resistor pack) are usually not complicated. Basic electrical testing with a DMM and a good wiring diagram will usually isolate the problem fairly quickly. The failure of the blower motor to operate can also be caused by an object falling through one of the defrost vents and finding its way to the blower motor fan, referred to as a squirrel cage fan, and keeping it from turning. A jammed fan could result in a blown blower motor fuse. A good way to check for this problem is to measure the blower motor amperage with the fan switched on. If the ammeter reads zero, the control circuit or the blower motor is open.

Checking the voltage at the blower motor helps you determine what the problem is. Using a DMM, test for voltage at the connector of the blower motor. Be sure to test for power (voltage) and ground; you must have both to make a complete circuit. If there is power (voltage) and ground present, then the blower motor is open. If there is no power (voltage) present, then the control circuit has an open circuit that can be located by voltage drop testing. Voltage drop testing is checking voltage loss across a wire or different components of the circuit.

To perform a voltage drop, place one lead of your voltmeter on one end of the wire or component and the other lead on the other end. With the circuit operating, read the voltage drop. More than 0.2 volt means you have high resistance in that wire, switch, or connector.

If you are using an ammeter, note if it reads near or higher than specifications. If so, the fan will have to be removed and checked for foreign objects (jamming the fan), binding of the fan motor shaft, or a shorted motor. Low resistance creates excessive current flow and slower motor speed.

To inspect and test a blower motor, follow the steps in **SKILL DRILL 60-8.**

Testing Air-Conditioning Compressor Clutch Control Systems

N60009

The air-conditioning compressor clutch control system needs to be checked whenever the clutch does not engage or slips; you will notice that the center of the compressor pulley (the front

SKILL DRILL 60-8 Inspecting and Testing a Blower Motor

1. Turn on the blower motor, and check whether the motor is functioning on each speed. Research the testing procedures for the condition found.

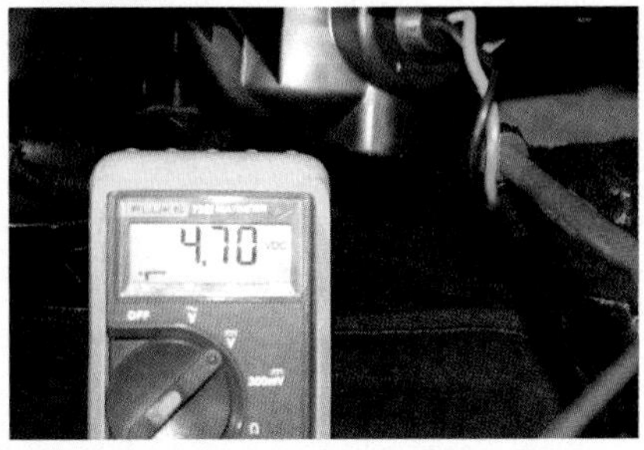

2. Use a DMM to test for power and ground by back-probing the wires at the connector of the blower motor. Turn the ignition switch to Run and the fan switch to the speed that does not operate, and read the meter. If the meter reads more than about 3 volts, the fan should be operating. If it isn't, the blower motor has to be removed. (See photo 4.)

3. If the reading is less than 3 volts, then perform a voltage drop test on the power and ground sides of the circuit. If the voltage drop is excessive, use a wiring diagram and voltmeter to track down the high resistance or opening in the circuit, and repair as necessary.

4. If the control circuit tests okay, but the fan is making unusual noises or is not operating correctly, remove the blower motor by disconnecting the electrical connector and removing any bolts, nuts, or clips retaining the blower motor to the air box.

5. Inspect the fan blades and fins for foreign objects or binding.

6. If the fan is in good operating condition, measure the current flow with the blower motor removed from the vehicle, either (1) with a low-amp clamp (inductive clamp) around the wire to or from the motor or (2) with an ammeter set to 20 amps and wired in series with the motor. Run the motor while reading the amperage. If necessary, replace the blower motor and retest.

portion of the pulley) is not turning or is clicking on and off. This can be caused by a bad clutch coil, an electrical fault in the control circuit, a clutch air gap that is too big, or a low refrigerant level, which causes the pressure to be too low to activate the pressure switch. You will need a DMM and a wiring diagram to locate any control circuit faults.

To test the air-conditioning compressor clutch control system, follow the steps in **SKILL DRILL 60-9**.

Diagnosing Malfunctions in the HVAC Controls

Diagnosis of ECU systems can be complicated because the system components are interrelated. For example, the blend door may be in a position that is causing the duct air temperature to be too hot. This could be caused by a temperature sensor that is reporting the wrong temperature to the ECU, which is commanding the blend door to be in the position it is in. Or maybe the blend door is being held in position by an object that fell into the defrost vents. Or maybe the electric actuator is worn out and cannot respond to the signal it receives from the ECU. Another cause could be that the ECU is faulty and sending an incorrect signal to the blend door actuator. You will need to research the manufacturer's diagnostic procedures and follow the procedure relating to the symptoms present. Be sure to check TSBs as a beginning step for the concern you have.

Many new vehicles are designed with HVAC systems that have built-in self-diagnostics. This means that the ECU is programmed to recognize signals or situations that are out of

SKILL DRILL 60-9 Testing Air-Conditioning Compressor Clutch Control Systems

1. Research the testing procedure and specifications in the appropriate service information. Unplug the compressor clutch electrical two-wire connector located at the front of the air-conditioning compressor. Check the resistance of the clutch windings. If the resistance does not match the manufacturer's specifications, replace the clutch coil.

2. Check supply voltage and ground to the compressor clutch connector. Place the red lead on the positive side of the clutch connector, and place the black lead on the negative side of the clutch connector. Ensure that the air-conditioning compressor is turned on. Your meter should display near battery voltage.

3. Perform a voltage drop test of power and ground. Connect the red lead of your voltmeter to the positive terminal of the clutch coil connector. Connect the black lead to battery positive. Ensure that the air-conditioning clutch is turned on with the control panel. Look for a drop or loss of voltage on the positive wire of no more than 0.4 volt. If more than 0.4 volt is found, the power wire has high resistance in it. Repeat the step for the ground side by connecting the red lead to the negative terminal and the black lead to the battery negative.

specifications and to display a DTC. You can use this information to help determine the cause of the system fault. On some systems, you need to hold down the auto button while pressing the temperature increase button or performing other sequence of button presses. Refer to the specific service information, as this is not standard among vehicles. This process enters the control panel into diagnostic mode. You will be able to display any trouble codes on the screen of the control panel. If the vehicle is set up for scan tool diagnosis, you will be able to pull the code(s) with the scan tool.

For example, suppose you are assigned to work on a particular model of Honda that comes into the shop. The customer complaint is that the rear climate control panel is inoperative and blank (no display). When service information is checked, you notice that body DTCs may be set by faults in the HVAC system. When you pull up DTCs with the scan tool, you find a code for a loss of communication with the rear panel. Following the troubleshooting chart, you perform circuit tests and find a poor connection in the communication wire connector. Repairing the connector and retesting reveals that you have corrected the concern.

Diagnosing Malfunctions in the Electrical Controls of HVAC Systems

Malfunctions in the electrical controls can come in many forms. Be sure to follow the manufacturer's service information on every issue. The use of electrical wiring diagrams is necessary for successful testing. Common issues include failure of electrical components such as actuators, switches, connectors, wiring, or grounds. Pin fit issues in connectors have become one of the more common faults in electrical systems. If the terminal pins and sockets fit loosely with one another, the component may be inoperative or may operate intermittently. Wiggling the testing connectors of the affected circuit is a good way to find pin fit issues.

Diagnosing Malfunctions in the Vacuum and Mechanical Components and Controls of the HVAC System

N60010

Common failures in the vacuum system include vacuum leaks. Refer to the manufacturer's service information to diagnose this system properly. You will need a vacuum chart to show operation of the vacuum controls for the HVAC system. To test for a leaking vacuum line, you have to gain access to the vacuum connections of the vacuum line suspected. Connect one end of a handheld vacuum pump while the other end is plugged. If vacuum is able to be held on the hose, it is not leaking, and the problem lies beyond the line. Vacuum servos can fail by not pulling completely or not at all. Apply vacuum to the servo to see if it is functioning correctly.

Vacuum controls can also be tested with the vacuum pump. Hook up to the port shown in the vacuum chart, and make sure vacuum comes out of the corresponding vacuum port that the chart shows. A blockage, or leak, in the valve or switch will keep the vacuum from passing through, and replacement of the vacuum control will be necessary.

Wrap-Up

Ready for Review

- Climate control systems can be manual, semiautomatic, or automatic.
- Manual climate control systems give the driver full control over blower motor speed, temperature of outlets, and direction of airflow.
- Semiautomatic control systems use an A/C ECU to turn the A/C compressor on or off while the driver controls temperature request, blower motor speed, and airflow position.
- In automatic control systems, the driver selects a temperature, and the A/C ECU manages the subsystems necessary to achieve that temperature (heat, A/C, and air distribution).
- The A/C ECU requires input from electronic sensors to determine vehicle heat load; desired temperature setting; and the temperature of ambient air, cabin air, evaporator, passenger temperature, and engine coolant.
- An electric actuator or a vacuum servo may be used to move air doors.
- Components of a climate control system include blower motor, air box, heater core, evaporator, actuators, and control head.
- The blower motor is attached to a squirrel cage fan and provides the desired amount of airflow to the vehicle cabin.
- The doors that control airflow include blend door, mode door, and recirculation door.
- A potentiometer is a three-wire sensor that signals the position of doors to the A/C ECU.
- A negative temperature coefficient thermistor (temperature sensor) gains resistance as the temperature drops, and a positive temperature coefficient thermistor loses resistance with lowered temperatures.
- Actuator feedback signals may be in the form of voltage signals.
- A pulse width modulation refers to a pulsing current flow (duty cycle) that controls a blower motor's speed and provides an infinite number of speeds within its design limitations.
- The sun load (or solar) sensor is a photodiode that reads light entering the windows of the vehicle.
- The air conditioning pressure sensor monitors refrigerant pressure.
- The evaporator temperatures sensor ensures that the evaporator maintains the correct temperature by controlling compressor cycling.
- Control panels vary according to car type and A/C ECU system. Panels may have rotary switches, recirculation buttons, auto mode, or digital temperature display, and systems may offer dual temperature control or rear-climate control.
- Control panel assemblies may be mechanical or automatic. Mechanical systems may operate cables, vacuum valves, or electrical switches.
- The blower motor speed is determined by resistors in a resistor pack connected to the fan control switch that feeds voltage to the motor.
- In some vehicles, the highest blower motor speed receives a full 12 volts from a dedicated fan relay.
- The A/C compressor is turned on/off by the A/C compressor clutch.
- Components of a compressor clutch include the pulley, clutch plate, clutch coil, compressor shaft, and a shim.
- Hybrid cars may use a standard A/C compressor clutch arrangement or a dual-drive A/C compressor.
- Common HVAC system issues are electric actuator failure, vacuum hose failure, and sensor failure.
- Vehicles with HVAC systems equipped with self-diagnostics will provide a diagnostic trouble code (DTC) that can help determine the cause of system failure.
- Use manufacturers' troubleshooting charts when diagnosing HVAC controls malfunctions.
- Electrical malfunctions in HVAC systems may be due to failure of actuators, switches, connectors, wiring, grounds, or may be due to pin fit issues.
- Consider using an infrared thermometer to check coolant temperature without removing the radiator cap.
- Inspect and test blower motors for foreign objects, shorted motor, or binding of the fan motor shaft.
- Possible causes of the compressor clutch control system not working are bad clutch coil, electrical fault in control circuits, large clutch air gap, or low refrigerant level.
- Inspect and test the heater control panel assembly and control cables if the system is not delivering proper air temperature.
- Air conditioning systems can grow mold and bacteria from trapped condensed moisture, which can then give off an unpleasant odor, necessitating use of an anti-odor kit.
- Tools needed for inspection, maintenance, and repair of an HVAC system include infrared thermometer, digital multimeter (DMM), back probes, fused jumper wire, and a scan tool.

Key Terms

actuator A device that moves the air doors of the air box. It is controlled by input from the climate control panel.

air-conditioning compressor clutch The mechanical coupler that is electromagnetically engaged and provides a way of uncoupling the compressor from the accessory drive belt.

air-conditioning pressure sensor A sensor that gives an input signal of refrigerant pressure in the air-conditioning system to the ECU.

air-conditioning system pressure The refrigerant pressure contained within the air-conditioning system.

ambient air temperature sensor A thermistor that is used to measure the air temperature outside the vehicle.

aspirator A tube that is used to direct airflow across the cabin air temperature sensor.

automatic climate control system A system that automatically adjusts the heating or cooling to meet a specified temperature demanded by the passengers.

cabin air temperature sensor A thermistor that measures air temperature inside the vehicle.

climate control system A system that provides the heating and cooling of air inside the passenger compartment for passenger comfort.

dual-drive air-conditioning compressor An air compressor drive used on some hybrid vehicles. The compressor can be driven by the accessory belt or by an electric motor.

electric servo motor Also referred to as an electric actuator, a motor that provides movement to operate the air doors in an air box to control air temperature and air movement.

electromagnetic clutch arrangement An arrangement used in hybrid air-conditioning systems to drive the compressor. The air-conditioning clutch is an electromagnetic clutch that works by creating a strong magnetic field that pulls the clutch into mesh with the pulley on the air compressor.

electronic control unit (ECU) When referring to the HVAC system, the electronic module that makes the "decisions" of the climate control system settings, based on input sensors and the module programming.

engine coolant temperature sensor A thermistor that measures the temperature of the engine coolant.

evaporator temperature sensor A thermistor that reads the temperature of the evaporator, used to ensure that the evaporator does not freeze.

feedback signal A voltage signal sent back to an electronic control unit. The feedback signal is how the control module is able to interpret temperature or other information from its sensors.

heater control cables Cables that control the air doors in an air box as part of the air distribution system.

limit switches Switches that turn off power flow to an electric motor when a particular limit is reached.

manual climate control system A climate control system fully controlled by the operator.

negative temperature coefficient (NTC) thermistor A thermistor that gains resistance as temperature goes down and loses resistance as temperature goes up.

on/off-type signal A signal that is either a 12-volt signal or zero volts. Typically, switches provide on/off signals. Some systems use a 5-volt reference signal instead of a 12-volt signal.

photodiode An electronic component that creates a varying voltage or current output based on the amount of light striking it.

relative humidity sensor A sensor designed to measure the relative humidity in the passenger compartment.

semiautomatic climate control system A system that provides automatic function of the heater or cooling only, leaving fan speed and mode selection to the operator.

signal voltage value Measured voltage in a signal return circuit that is compared to a specified voltage value published by the manufacturer.

sun load (solar) sensor A photodiode that varies voltage based on light. It is used to determine the radiant heat coming from the sun into the passenger cabin and gives an input signal of sunlight load to the ECU.

thermistor A temperature-controlled variable resistor. As temperature changes, so does resistance.

variable voltage signal A signal that changes based on what the sensor is reading; for example, as temperature varies, so does the voltage signal to the ECU.

voltage The pressure aspect of electricity, measured to show the potential of a circuit to do work.

Review Questions

1. The ECU sends output commands to which type of component to control fan speed, heater core coolant flow, air-conditioning clutch activation, and so on?
   a. Transistor
   b. Actuator
   c. Integrated circuit
   d. Capacitor
2. A potentiometer is a three-wire sensor that signals the ________________ to the A/C ECU.
   a. temperature of the evaporator
   b. amount of sun load
   c. position of doors
   d. ambient temperature
3. Which of the following sensors signals to the ECU the temperature of the coolant which flows through the heater core?
   a. Ambient air temperature sensor
   b. Cabin air temperature sensor
   c. Engine coolant temperature sensor
   d. Sun load, or solar, sensor
4. The ECU receives instructions regarding the desired cabin temperature from the:
   a. relative humidity sensor.
   b. evaporator temperature sensor.
   c. air-conditioning pressure sensor.
   d. HVAC control panel.
5. Which of the following control a relatively large current flow in one part of a circuit by a much smaller current flow?
   a. Relays
   b. Blower motor speed controls
   c. Air boxes
   d. Limit switches
6. Which of these is a function of the blower motor?
   a. Controlling electric components such as the air-conditioning compressor clutch and condenser fan
   b. Providing the desired amount of airflow in the cabin
   c. Controlling temperature and position of the airflow from the air distribution system
   d. Movement of air box doors

7. Which of these is the most common issue a technician might encounter with the automatic HVAC system?
   a. Failure of electric actuators
   b. Vacuum leaks
   c. Sensor failure
   d. Jammed fans
8. Negative temperature coefficient (NTC) thermistors:
   a. increase resistance as they warm up.
   b. are a form of transistor.
   c. decrease resistance as they warm up.
   d. are a form of diode.
9. When there is a difference in the speed of the vent airflow with the system in recirculation mode compared to fresh air mode, may indicate:
   a. restricted cabin air filter.
   b. failure of the blower motor.
   c. a jammed fan.
   d. a bad clutch coil.
10. What should be the first step when diagnosing HVAC ECU systems?
   a. Researching the manufacturer's diagnostic procedures and following the procedure relating to the symptoms present
   b. Recover and recharge the refrigerant in the system.
   c. Changing the settings to manual mode
   d. Checking by replacing with another ECU system

ASE Technician A/Technician B Style Questions

1. Tech A says that a temperature blend air door controls how much air bypasses the heater core. Tech B says that the recirculation blend air door controls fresh air. Who is correct?
   a. Tech A
   b. Tech B
   c. Both A and B
   d. Neither A nor B
2. Tech A says that in a temperature blend air system, the air conditioner runs continuously. Tech B says that in a temperature blend air system, heated coolant always flows in the heater core. Who is correct?
   a. Tech A
   b. Tech B
   c. Both A and B
   d. Neither A nor B
3. Tech A says that air flow in the vents is controlled by adding or removing a second or third motor to increase or decrease air flow. Tech B says that blower motor speed is controlled by controlling voltage and current to make the blower motor run faster or slower. Who is correct?
   a. Tech A
   b. Tech B
   c. Both A and B
   d. Neither A nor B
4. Tech A says that the fresh air setting pulls air from the cabin. Tech B says that when the outside temperature is very hot, the recirculate setting will cool the cabin further. Who is correct?
   a. Tech A
   b. Tech B
   c. Both A and B
   d. Neither A nor B
5. Tech A says that when an air door operated by an electric actuator doesn't move, replace the actuator. Tech B says that when a door operated by a vacuum actuator doesn't move, always verify the door moves freely, and there are no vacuum leaks. Who is correct?
   a. Tech A
   b. Tech B
   c. Both A and B
   d. Neither A nor B
6. Tech A says that some electric actuators are positioned by an A/C ECU which checks the air flow with sensors. Tech B says that electric actuators are positioned by an A/C ECU and some actuators have a potentiometer to tell the A/C ECU the position of the door. Who is correct?
   a. Tech A
   b. Tech B
   c. Both A and B
   d. Neither A nor B
7. Tech A says some newer vehicles use pulse width modulation (duty cycle) to control blower motor speeds, which leads to smooth changes in speeds. Tech B says that the evaporator temperature sensor is used to measure the temperature in the passenger compartment. Who is correct?
   a. Tech A
   b. Tech B
   c. Both A and B
   d. Neither A nor B
8. Tech A says that most hybrids now use a high-voltage electric motor that directly drives the A/C compressor. Tech B says that some hybrids use dual-drive air-conditioning compressors. Who is correct?
   a. Tech A
   b. Tech B
   c. Both A and B
   d. Neither A nor B
9. Tech A says that if the air gap between the compressor clutch and the pulley is too large, the clutch may not engage. Tech B says that if system voltage is too low, the compressor clutch will not engage. Who is correct?
   a. Tech A
   b. Tech B
   c. Both A and B
   d. Neither A nor B
10. Tech A says that when troubleshooting an automatic temperature-controlled A/C or heating problem, one of the first steps should be to check codes for any related faults. Tech B says that codes should only be consulted if you get stuck during diagnosis. Who is correct?
   a. Tech A
   b. Tech B
   c. Both A and B
   d. Neither A nor B

SECTION 9
Engine Performance

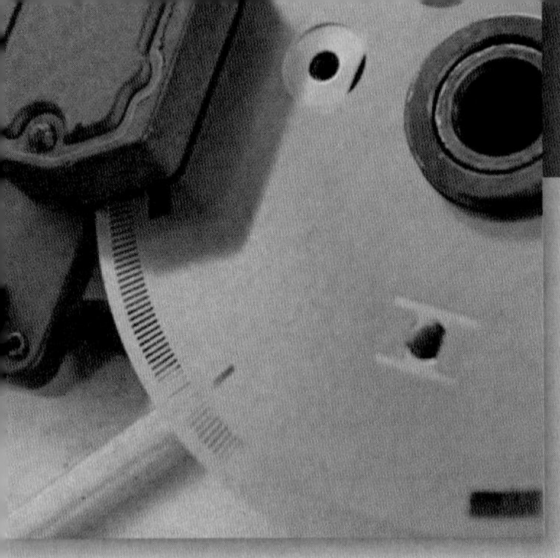

CHAPTER 61
Ignition Systems

NATEF Tasks

- **N61001** Access and use service information to perform step-by-step (troubleshooting) diagnosis. (AST/MAST)
- **N61002** Remove and replace spark plugs; inspect secondary ignition components for wear and damage. (AST/MAST)
- **N61003** Diagnose (troubleshoot) ignition system related problems such as no-starting, hard starting, engine misfire, poor drivability, spark knock, power loss, poor mileage, and emissions concerns; determine needed action. (AST/MAST)
- **N61004** Inspect and test crankshaft and camshaft position sensor(s); determine needed action. (AST/MAST)
- **N61005** Inspect, test, and/or replace ignition control module and powertrain/engine control module; reprogram/initialize as needed. (AST/MAST)

Knowledge Objectives

After reading this chapter, you will be able to:

- **K61001** Explain the basic principles of a modern ignition system.
- **K61002** Explain the need to vary ignition timing during engine operation.
- **K61003** Describe the purpose and functions of the components in an ignition system.
- **K61004** Describe the breaker points ignition system.
- **K61005** Describe the electronic ignition system.
- **K61006** Describe the distributorless ignition system.

Skills Objectives

After reading this chapter, you will be able to:

- **S61001** Perform ignition system maintenance.
- **S61002** Inspect and test ignition system components.
- **S61003** Inspect and test primary and secondary circuits.
- **S61004** Inspect and test ignition coils.
- **S61005** Inspect and test ignition pickup assemblies.
- **S61006** Inspect and test spark plug wires.
- **S61007** Inspect distributor caps and rotors.

▶ Introduction

The air-fuel mixture inside each cylinder must be ignited for it to release its energy and create pressure to force the piston down the cylinder on the power stroke (**FIGURE 61-1**). Gasoline engines use the heat of a high-voltage spark to ignite the mixture. The purpose of the ignition system is to create the high-voltage spark and deliver it at the right time to each cylinder.

The ignition system consists of a primary (low-voltage) circuit and a secondary (high-voltage) circuit. The primary circuit activates the ignition coil, which changes the low voltage of the vehicle battery into the high voltage needed to create the spark that is sent across the spark plug electrodes. The secondary circuit transmits the high voltage from the coil, or coils, to the spark plug at each cylinder. Although there are several types of ignition systems, the components common to all of them include the spark plugs, the ignition coil, and a device for triggering the ignition coil.

For an engine to run smoothly and efficiently, the high-voltage spark must jump across the spark plug electrode as the piston approaches top dead center (TDC) of the **compression stroke**. The ignition system must also be able to advance or retard the timing of the spark, based on engine conditions such as load, speed, and driver input. This chapter explains the principles and operation of modern ignition systems and covers the basics of diagnosis, maintenance, and repair.

▶ Ignition Principles

K61001

When the driver turns the key to the Start position (or presses the electronic start button), an electrical connection is made between the vehicle battery and the primary winding of the ignition coil. As the engine is cranked, a switching circuit turns the primary ignition coil circuit on and off. Each coil amplifies the battery's low voltage and high current signal into a very high voltage and very low current spark. This high voltage is delivered from each ignition coil to each cylinder, where the spark plugs are installed (**FIGURE 61-2**). As the high voltage pushes current across the spark plug air gap crated by the electrodes, the air-fuel mixture is ignited, causing very high cylinder pressure, which in turn pushes the piston down the cylinder on the power stroke.

The ignition system has undergone changes in technology over the decades to meet requirements for dependability, reduced maintenance expectations, and increasingly strict emission standards.

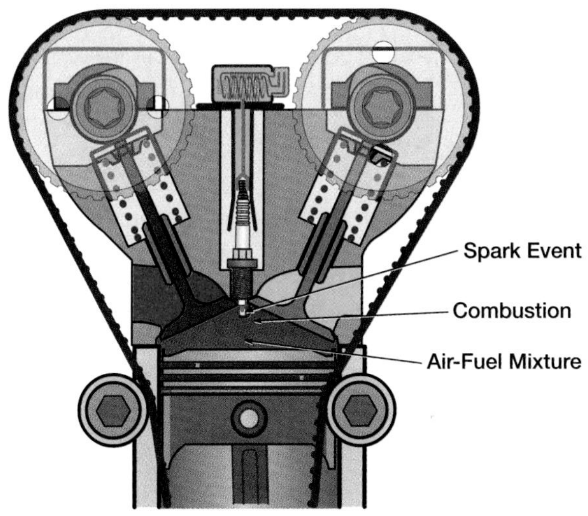

FIGURE 61-1 The purpose of the ignition system is to create a spark to ignite the air-fuel mixture in the combustion chamber.

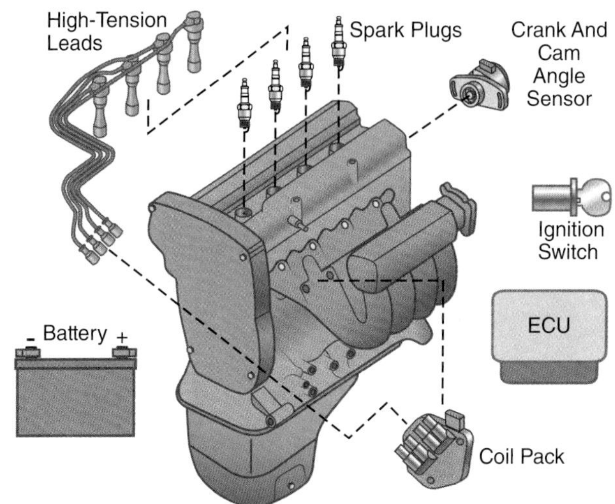

FIGURE 61-2 An ignition system circuit.

You Are the Automotive Technician

Today a customer visits your shop for a scheduled 60,000-mile service on her 2013 vehicle. The service information says to replace the spark plugs and inspect the secondary ignition system, consisting of the ignition coils and spark plug boots. When you twist the coils to pull them off the plugs, several of the boots tear, requiring replacement. After the boots are off, you use compressed air (while wearing safety glasses) to blow out any debris from around the spark plugs. You remove the plugs one at a time and inspect them. The deposits are light and show that they are burning correctly—neither too hot nor too cold. The gaps are worn, as expected. You inspect the ignition coils. The terminals and insulation look good, and you will be replacing all of the boots.

1. How would you prevent the boots from sticking to the spark plugs the next time?
2. How would you test to see if the ignition system was working on a vehicle that cranked but didn't start?
3. If this customer drove only short trips of a mile or two, would you recommend a spark plug with a hotter heat range or colder heat range? Why or why not?
4. How can a spark plug boot cause a cylinder to misfire?

The original system was the **contact breaker point ignition system**. It was a mechanical system with a switch, called **contact breaker points**, that opened and closed as the engine was running (**FIGURE 61-3**). The points turned the primary ignition circuit on and off, which created high voltage in the secondary circuit that was distributed to each spark plug in the firing order.

The first advancement was the replacement of the mechanical switch with an electronic switching device (**FIGURE 61-4**). This device had no moving components, meaning minimal wear on the switching device and very little, if any, maintenance. This ignition system was called an **electronic ignition system-distributor type**, as it still used a single coil and distributor to dispense the spark to the various cylinders.

The next advancement was eliminating the distributor by using dedicated ignition coils—one coil for each pair of cylinders. This system was called a **waste spark ignition system** (**FIGURE 61-5**). A four-cylinder engine would have two ignition coils; a six-cylinder engine would have three coils; and so on. Eliminating the distributor meant doing away with the last mechanical component of the ignition system, which again resulted in increased reliability and reduced maintenance issues.

The latest development has been to give each cylinder its own ignition coil. This system is called a **direct ignition system** or **coil-on-plug** ignition system (**FIGURE 61-6**). This system puts the ignition coil directly on top of the spark plug, thus eliminating the spark plug wires, which are subject to leakage, damage, and wear.

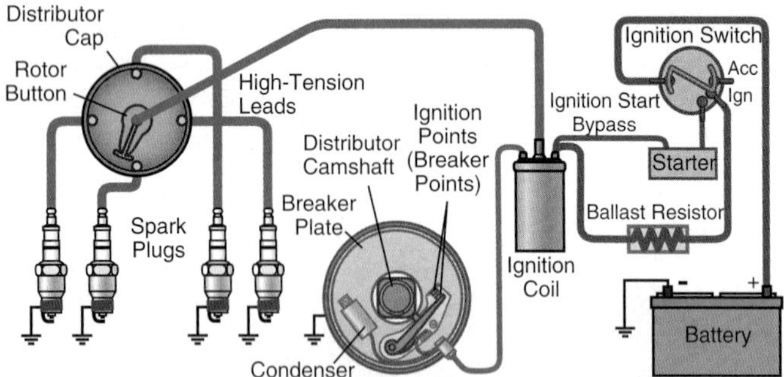

FIGURE 61-3 Breaker point ignition system.

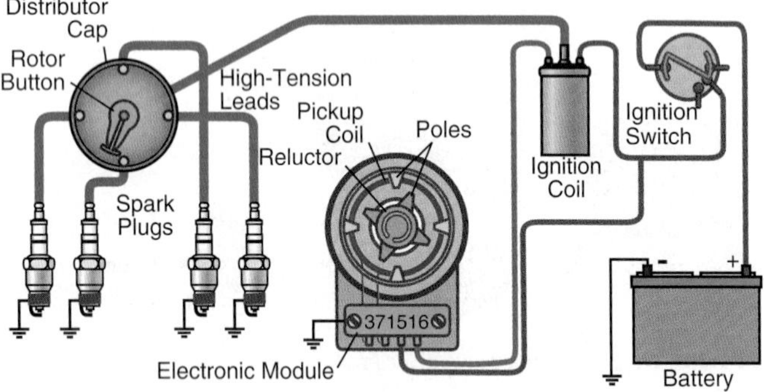

FIGURE 61-4 Electronic ignition system-distributor type.

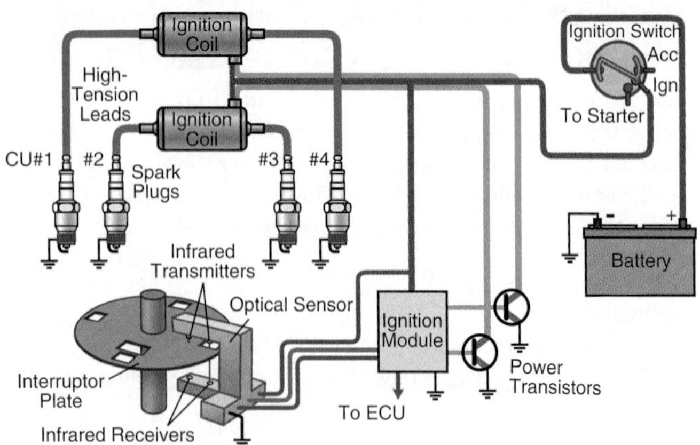

FIGURE 61-5 A waste spark ignition system.

Primary and Secondary Circuits

The ignition system uses an **induction coil** to convert relatively low-voltage and high-current flow into very high-voltage and very low current flow. The low-voltage side is called the **primary circuit**, and the high-voltage side is called the **secondary circuit**. The battery supplies the low-voltage and high-current flow to power the primary circuit. The induction coil steps up the low voltage in the primary circuit to the high voltage in the secondary circuit. A problem in the primary circuit will affect the output of the secondary circuit, so technicians need a complete understanding of both circuits when diagnosing faults relating to the ignition system (**TABLE 61-1**). Notice how the types of primary circuit components stay relatively the same over each type of system, whereas the types of secondary circuit components reduce in number as newer systems are introduced.

Faraday's Law

Most automotive ignition systems use ignition coils that operate on the principles of an induction coil. Automotive induction coils step up the nominal battery voltage of 12 volts to the voltage needed to bridge the gap across the spark plug electrodes, which can be up to 100,000 volts. These induction coils operate according to Faraday's law.

Faraday's law states that relative movement between a conductor and a magnetic field allows four ways by which voltage can be induced in a conductor:

1. Moving a magnet so that the magnetic lines of force cut across a conductor, as in an alternator.
2. Moving a conductor so that it cuts across the stationary magnetic field, as in a generator.

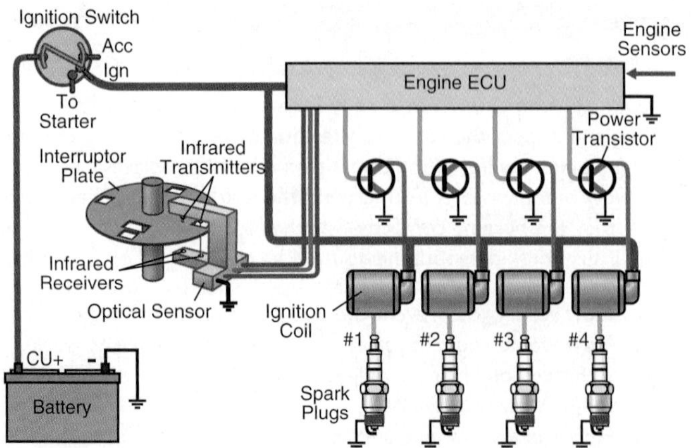

FIGURE 61-6 A coil-on-plug ignition system.

TABLE 61-1: Primary and Secondary Circuit Components

Contact Breaker System	
Primary Circuit Components	**Secondary Circuit Components**
Contact breaker system	Ignition coil—secondary winding
Ignition switch	Coil wire
Ballast resistor	Distributor cap
Ignition coil—primary winding	Rotor
Capacitor	Spark plug wires
Contact breaker points	Spark plugs
Electronic Ignition—Distributor-Style System	
Primary Circuit Components	**Secondary Circuit Components**
Battery	Ignition coil—secondary winding
Ignition switch	Coil wire
Ignition coil—primary winding	Distributor cap
Ignition module	Rotor
Triggering device	Spark plug wires
	Spark plugs
Electronic Ignition—Distributorless-Style System	
Primary Circuit Components	**Secondary Circuit Components**
Battery	Ignition coils—secondary windings
Ignition switch	Spark plug wires
Ignition coils—primary windings	Spark plugs
Ignition module	
Triggering device	
Direct Ignition System	
Primary Circuit Components	**Secondary Circuit Components**
Battery	Ignition coils—secondary windings
Ignition switch	Spark plugs
Ignition coils—primary windings	
Ignition module	
Triggering device	

3. Starting, stopping, or changing the rate of current flow in a conductor. This causes the conductor to induce an electromagnetic field into itself and occurs in the primary windings of an ignition coil. This process is called self-induction.
4. Starting, stopping, or changing the rate of current flow in a conductor that is positioned close to a second conductor. This is called mutual induction. It is used to induce high voltage in the secondary winding of the ignition coil.

When any of these methods are used to induce voltage in a conductor, the value of that voltage depends on the following conditions:

- The density, or strength, of the magnetic field—the stronger the field, the greater the induced voltage
- The number of turns of the windings in the coil—the more turns, the greater the induced voltage
- The speed at which the lines of force are cut—the greater the speed, the greater the induced voltage.

In the induction coil, the secondary winding has many thousands of turns of fine enameled copper wire. The primary winding, with a few hundred turns of relatively heavy wire, is positioned close to the secondary winding. A soft iron core is positioned centrally to concentrate the magnetic field. Current flow through the primary winding establishes a magnetic field around the windings. The higher the current flow, the stronger the field.

Sudden interruption of the primary current effectively disconnects the battery from the coil, and current flow ceases. The magnetic field collapses into the iron core, returning its stored energy to the coil by cutting across the coil's primary and secondary windings. This produces a self-induced voltage in the primary winding and a mutually induced voltage in the secondary winding.

The maximum value of the secondary voltage is partly determined by the ratio of the number of turns in the secondary winding to the number of turns in the primary winding (in this case, approximately 100 to 1) and by the value of the self-induced voltage in the primary winding (in this case, 300 volts). If this coil is 100% efficient, the maximum voltage available from the secondary winding would be 300 volts multiplied by 100, or 30,000 volts.

Because the value of the self-induced voltage in the primary winding is also influenced by the rate of collapse of the magnetic

Applied Science

AS-76: Coil: The technician can explain how a coil can increase the battery voltage needed to fire a spark plug.

An ignition coil is a step-up transformer, increasing the voltage from the battery to a high voltage necessary to fire the spark plugs and ignite the air-fuel mixture in the cylinders. Coils operate via the concept of electromagnetic induction, which dictates that a moving magnetic field (or a change in a stationery magnetic field in the case of an ignition coil) can induce a current in a wire exposed to the field.

The internal makeup of a coil is basically an iron core with two windings of wire, both wound around the core. The secondary winding is wound with considerably more turns than the primary, typically at a ratio of about 100:1.

In operation, battery voltage is applied to the primary winding, creating a magnetic field. This process is called saturation. The secondary winding is exposed to this field. To create a spark, the primary circuit is interrupted, or turned off, by a powertrain control module, ignition module, or breaker points. This interruption causes the magnetic field to collapse very rapidly and a high voltage to be induced into the primary and secondary windings. The increase in voltage is provided by the difference in the number of turns between the windings. Current in the secondary winding is directed to the spark plugs, where it creates a spark as it jumps across the air gap.

field, which is determined by the rate of change of the current flow through the coil, it is essential that the primary current be switched off as quickly as possible. All ignition systems make provisions to ensure that this occurs. This subject is covered in greater detail later in this chapter.

Required Voltage Versus Available Voltage

Understanding the terms "required voltage" and "available voltage" will help you diagnose ignition systems. **Required voltage** is the amount of voltage required to initially get current to jump the spark plug gap. Once the voltage reaches the point where current is flowing in the secondary circuit, the required voltage drops to a much lower level, just sufficient enough to sustain current flow and the spark. This gives more opportunity for the air-fuel mixture to ignite during the duration of the spark.

The other factor is **available voltage**, the maximum amount of voltage available to try to push current to jump the spark plug gap if the gap were infinite. In other words, available voltage is the maximum amount of voltage that the ignition coil can put out. It is very important that the available voltage is always higher than the required voltage (**FIGURE 61-7**). If the available voltage falls below required voltage, or if required voltage is ever higher than the available voltage, there will not be enough voltage to push current across the gap and create a spark. Because the spark is required to ignite the air-fuel mixture, higher required voltage or lower available voltage will prevent the spark from occurring, which causes the cylinder to not fire. This is called a misfire.

As a vehicle is driven over time, typically the required voltage will increase while the available voltage decreases. The required voltage usually increases primarily because of the growing gap of the spark plug as it wears over time. However, it can also increase because of any open circuits in the secondary side of the circuit, such as a broken rotor tip or an open spark plug wire. Manufacturers design their ignition systems to have an amount of reserve voltage above the required voltage so that misfire is prevented. But if the vehicle's ignition system is not maintained, this reserve voltage is reduced, and the engine is prone to misfire, especially when accelerating or climbing a hill. Inspecting all of the components and replacing any that are worn will restore the available voltage to what it should be and reduce the required voltage to within specifications. This will restore the ignition system to its proper operation.

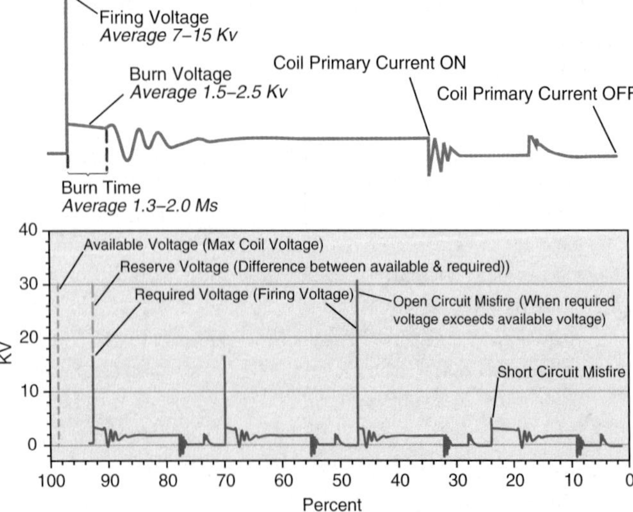

FIGURE 61-7 Available voltage, required voltage, reserve voltage, and misfire.

> **TECHNICIAN TIP**
>
> Required voltage rises by spark plug gaps that are too wide or worn rounded, compression pressures that are higher than normal, or lean mixtures. Available voltage is affected by faults in the primary or secondary circuits such as shorted coil windings or high resistance.

Spark Timing

K61002

The timing of the spark is critical to the smooth and efficient operation of the engine. For any given engine speed and load, the correct **spark timing** varies according to a number of factors, including these:

- Air-fuel ratio
- Detected knock
- Engine speed
- Engine load
- Engine temperature
- Air temperature
- Transmission gear selected
- Throttle position

At higher engine speeds, the time available to burn the air-fuel mixture decreases while the time required to completely burn the mixture remains essentially the same; therefore, the spark must occur sooner in the cycle (advanced) in order to give the air-fuel mixture enough time to burn and create maximum cylinder pressure soon after TDC.

At low engine revolutions per minute (rpm), when the throttle is opened, the engine load increases, which speeds up the burn rate of the air-fuel mixture. The spark does not have to be advanced as much because the engine is not spinning very fast; thus, there is plenty of time for the air-fuel mixture to burn before the piston moves very far in the cycle.

As engine speed increases, there is increasingly less time for the air-fuel mixture to be ignited and for this maximum pressure to develop; therefore, the ignition point must be advanced earlier in the cycle. That is, it must start earlier in relation to the piston's position during the compression stroke. This adjustment must occur automatically in relation to engine speed and engine load.

If the timing is advanced too far, then the engine begins to knock, which can destroy an engine. If it is equipped with a knock sensor located on the engine block, the sensor sends a signal to the **powertrain control module (PCM)**. The PCM may also be called an **engine control module (ECM)** or an electronic control unit (ECU). The PCM then retards the timing to decrease or eliminate the knock. At lower engine temperatures, the fuel does not atomize as quickly and requires the spark to be advanced more than when the engine is fully warmed up.

In early vehicles that use a distributor, base spark timing is set at idle speeds by positioning the distributor body in relation to its rotating cam. The timing is almost always indexed to the **number-one cylinder**, and the contact breaker points are operated in turn by each cam lobe to provide the same timing point for succeeding cylinders in the firing order. This initial setting, which in most, but not all, cases occurs before TDC, allows time for maximum pressure in the cylinder to develop, just as the piston is starting to descend on the power stroke.

Today's vehicles use electronically triggered ignition systems. The timing is engineered into the design of the engine and is not adjustable. The timing components are the crankshaft position sensor and the camshaft position sensor. These sensors are fixed in position on the engine and are triggered by the crankshaft or flywheel (**FIGURE 61-8**) on most engines, or within the distributor (if used). The initial timing specification is programmed into the PCM software.

▶ Components Common to All Ignition Systems

K61003

Most ignition systems include a battery, ignition switch, ignition coil(s), high-tension leads (also called spark plug wires), and spark plugs. The battery is the same one used to start the vehicle and provide electrical power to all electrical loads when the engine is not running. The ignition switch is used to connect the ignition system to the battery during operation and disconnect it from the battery to turn the engine off. The ignition coil is needed to amplify the battery voltage into very high voltage.

The high-tension leads transmit the high voltage from the coil(s) to the spark plugs. The spark plugs are threaded into the top of each cylinder and provide a place for the spark to jump within the combustion chamber.

FIGURE 61-8 A crankshaft position sensor.

Battery

An automotive battery is a chemical-electrical device that provides voltage to all electrical loads while the engine is not running, and when the loads exceed the output of the alternator. The **battery** supplies the electrical energy to the ignition circuit during start-up. Most modern vehicles use a 12-volt lead-acid rechargeable battery located in the engine compartment, in the trunk, or under the rear seats. Once the engine is started, the charging system provides virtually all power for all electrical loads along with charging the battery.

Ignition Switch

The **ignition switch** has more functions than simply starting and stopping the engine. It provides a way to disconnect most electrical accessories from the battery with an ignition key. The common positions of an ignition switch include the following:

- **Lock:** The key can only be removed from the lock position. When this occurs, all nonessential electrical circuits are disabled, and the steering column lock is enabled. If equipped, the engine immobilizer and theft-deterrent system are normally activated at this time.
- **Off:** Turning the key from the lock to the Off position unlocks the steering column, but it does not enable any electrical systems or disable the engine immobilizer or theft-deterrent system.
- **Accessory:** This position allows power to be supplied to the vehicle entertainment system, the blower fan, and in some vehicles the wipers, electric windows, and sunroof. The features enabled by the accessory position are mainly for passenger convenience. Prolonged use of any of them without the engine running will drain the battery.
- **On/Run:** When the switch is turned to the On position, most warning lamps on the instrument panel should illuminate. This is to test the operation of the lamps. On vehicles that are not equipped with engine immobilizers, this position also activates the accessories and ignition system. Vehicles that are equipped with engine immobilizers do not normally activate these systems until the key is turned to the Start position.
- **Start/Crank:** The start position activates the starter motor relay and/or solenoid, which enables the engine to crank and start the engine. Vehicles without engine immobilizers can start immediately, as the required electrical systems will have already been activated when the key was turned to the On position. In vehicles equipped with engine immobilizers, a number of the essential electrical systems are not normally enabled until the key is turned to the start position. When this occurs, communication between various control units determines whether to allow the engine to start or not. Starting is achieved by deactivating the immobilizer and activating the ignition, fuel, and charging systems. There may be a slight delay while this communication is occurring.

The key itself can consist of two parts. The first is a mechanical type that works in the key barrel and turns to unlock the steering column and switch through the various stages of Off, Accessories, On, and Start. The second portion consists of a transponder that is part of an immobilizer system that disables the ignition, fuel, and starter systems unless a transponder device with the correct code is within close range of a receiver in the vehicle. This system was covered in depth in Chapter 57, Safety, Entertainment, and Antitheft Systems.

In many automatic transmission vehicles, the ignition switch also has a transmission shift interlock device connected to it. In these vehicles, the gear selector must be moved into the Park position before the key can be removed from the lock. In the same way, the transmission shift lever cannot be moved out of Park until the key has been turned to the On/Run position and the brake pedal has been depressed.

Ignition Coil

Ignition coils are basically step-up transformers. Ignition coils are electrical devices that amplify the low voltage available from the vehicle's battery to the very high voltage needed for the current to jump the electrodes of the spark plugs, thereby causing spark (**FIGURE 61-9**). There are two sets of windings in an ignition coil: the primary winding and the secondary winding.

When battery voltage pushes current through the primary windings of an ignition coil, a magnetic field is created. When the primary circuit is opened, the collapsing magnetic field induces a voltage of more than 20,000 volts in the secondary winding. This voltage pushes current through the secondary circuit to the spark plug electrodes, creating the needed spark for the power strokes.

High-Tension Leads

A standard ignition coil has a rod-shaped laminated iron core, which is located centrally by an insulator at its base. The secondary winding, with 15,000 to 30,000 turns of very thin **enameled copper wire**, is wound around the core and is insulated from the core by layers of treated insulated paper. The primary

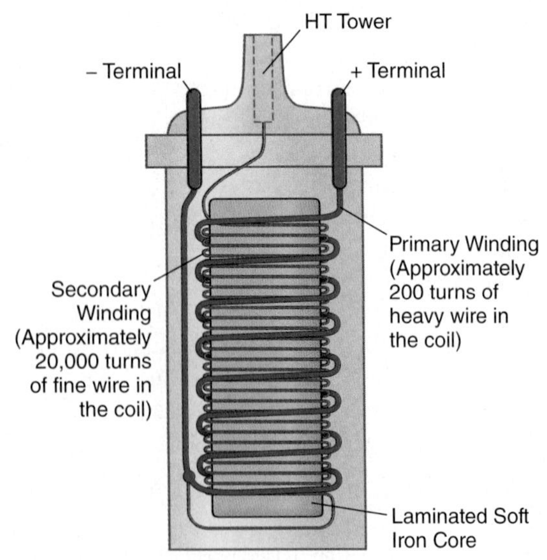

FIGURE 61-9 Cutaway of a basic ignition coil.

Applied Science

AS-65: Transformers: The technician can explain the ignition coil transformer's role in generating the high voltage required to fire a spark plug.

The 12-volt battery voltage used in light vehicles is insufficient to jump spark plug gaps. Much higher voltage is needed to fire the plugs and initiate combustion, especially considering the high pressures and turbulent conditions inside the cylinders. Automotive ignition systems are designed to allow a modest battery voltage to generate the very high voltages required to fire the spark plugs.

The ignition coil is a step-up transformer. The function of a transformer is to transfer voltage from one circuit to another, either as a higher voltage (as in this case) or as a lower voltage (referred to as a step-down transformer).

Depending on the ignition system, the 12 volts applied to the coil may be amplified to between 20,000 and 100,000 volts via a process called electromagnetic induction.

winding, with a few hundred turns of much heavier copper wire, is wound on the outside of the secondary winding. A shield of soft iron surrounds the outer windings, and the complete assembly is inserted into a one-piece steel or aluminum container or encased in an epoxy housing. On older coils, the container is then filled with special oil, which provides good electrical insulation and permits rapid heat dissipation. Epoxy-encased coils are cooled by the surrounding air.

The coil has two terminals, positive and negative, for external connection to the primary circuit. The ends of the primary winding are connected internally to each of these terminals. Provisions are also made for connecting the high-tension coil lead to the coil at a heavy insulated center terminal.

One end of the secondary winding is connected to this center terminal, and the other end is connected either to one end of the primary winding or in some applications to a separate ground connection. On waste spark systems, each end of the secondary winding is attached to a spark plug wire (two plugs per coil). Coil-on-plug and coil-near-plug systems usually have one end of the secondary winding grounded and the other end connected to the high-tension terminal of the spark plug.

High-tension leads, also known as HT leads or spark plug wires, connect the secondary ignition components together, such as the coil to the distributor cap and the distributor cap to the spark plugs. High voltage from the ignition coil pushes current to the distributor cap and on to the spark plugs. The high-tension leads conduct the high output voltage generated in the secondary ignition circuit when each ignition pulse occurs. They link the **high-tension terminal(s)** of the ignition coil, the distributor cap, and spark plugs.

With the higher voltages that today's electronic ignition systems have to endure because of the increased spark plug gaps and leaner air-fuel mixtures, the insulation material on the high-tension leads is now much thicker than in earlier models. The core of the high-tension lead is typically made of carbon-impregnated linen or fiberglass (**FIGURE 61-10**). It has a specific resistance, or opposition, to current flow that helps reduce the radio frequency interference emitted from the

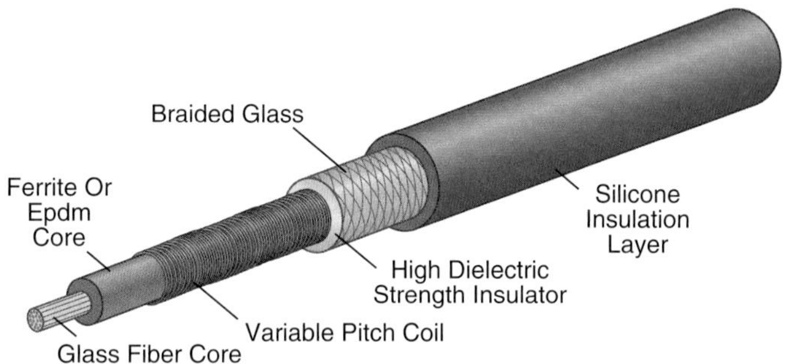

FIGURE 61-10 The parts of a high-tension lead.

high-tension leads. A crimped terminal at each end provides for connection of the components.

If a distributor is used, typically a coil wire transmits the high voltage from the ignition coil to the center terminal of the distributor cap. On systems where the coil is mounted inside the distributor, there is no coil wire. The high voltage is transmitted through a terminal in the distributor cap. On systems that do not use a distributor, there is no coil wire. Instead, the high-tension leads connect the terminals of coil packs directly to the terminals of the spark plugs.

▶ TECHNICIAN TIP

With the high voltage capacity of today's ignition systems, the quality of the spark plug wires, in particular the insulation, is of utmost importance. As the current is conducted down the leads, a magnetic field around the lead is induced. If the insulation around the cable is insufficient, and the leads are not spaced far enough apart and run parallel to each other, then an **induced voltage** can be generated in the other wires. In some instances, this could lead to premature firing of the adjacent spark plug when it is not in sequence. Remember that compression pressures increase required voltage, so the cylinder with lower cylinder pressure (the one not near TDC) is easier to fire than the one that is near TDC. Severe engine damage can result when the firing order of the engine has two side-by-side cylinders fire well before one of them is up on the compression stroke. Make sure the plug wires are routed and secured according to the manufacturer's specifications.

Spark Plug

The **spark plug** consists of a plated metal shell with a ceramic insulator and an electrode extending through the center of the insulator. Threads on the metal shell allow it to be screwed into the cylinder head. A short side electrode is attached to one side and bent toward the center electrode, providing a ground path for the spark. The electrodes are made from a special alloy wire with a manufacturer-recommended gap between them. The spark bridges this gap to ignite the air-fuel mixture in the combustion chamber.

Most modern spark plugs are made of a ceramic insulator, partially covered by a steel shell that threads into the cylinder head at the top of the combustion chamber (**FIGURE 61-11**). All spark plugs include at least one side electrode and may have as many as four side electrodes. Modern spark plugs also include an internal resistor to suppress voltage spikes when the spark plug is energized. The reduced voltage spikes prevent radio frequency interference.

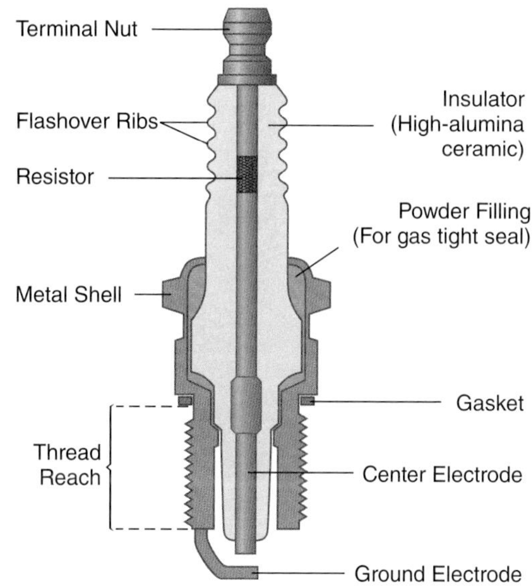

FIGURE 61-11 Spark plug.

Spark plugs are identified by three different features:

- Thread size or diameter
- Reach or length of the thread
- Heat range or operating temperature

Spark Plug Design

The metal case of a spark plug removes heat from the insulator and passes it on to the cylinder head. It also provides structural strength to withstand the torquing force applied when tightening the plug into place. The case also acts as the ground for the current passing through the electrodes.

The insulator covering the center electrode is usually made from an aluminum oxide ceramic with a high tolerance to heat and electrical voltage. The ribbing design, the exact composition of the insulator material, and the length of the insulator can all partly determine the heat range of the plug.

The spark plug seals the combustion chamber when installed, and the seals make sure it is kept airtight. The seal can be a hollow metal washer, which at the correct torque setting is partially crushed between the flat surface of the head and the plug just above the threads. Or the seal can be provided by a carefully machined taper on the spark plug body that mates with a matching machined taper in the spark plug hole in the cylinder head (**FIGURE 61-12**).

A terminal at the outer end of the spark plug connects it electrically to the ignition system. Most ignition leads clip on the spark plug. In a few applications, they may be held in place by a threaded nut. The side electrode runs very hot and is usually made of nickel steel or another high-temperature metal and welded to the side of the metal spark plug case. Some spark plug designs have multiple side electrodes that create the gap between the side of the center electrode and the side electrode (**FIGURE 61-13**).

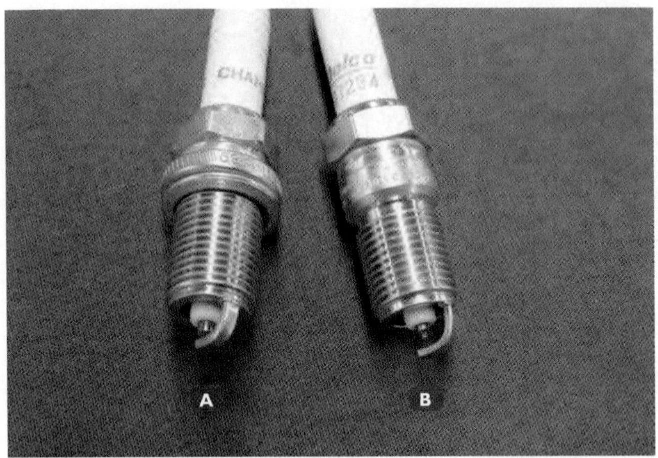

FIGURE 61-12 A. Metal gasket seal. **B.** Tapered seat design.

FIGURE 61-13 Some spark plugs have multiple side electrodes.

Spark plug condition affects required voltage. Some of the factors that increase the required voltage are presented here:

- An excessive gap raises the amount of voltage that is required to jump the gap.
- The electrons tend to stream easily from the sharp edges of the electrodes, and as these edges erode and become less sharp, it becomes more difficult for the electrons to jump the air gap, causing an increase in the required voltage.
- Generally it takes less voltage for a spark to jump from a hot surface to a cooler surface. The **center electrode** is the hottest part of the spark plug, so many older ignition systems were designed to jump from the center electrode to the side electrode. However, with today's high-energy ignition systems, there is enough available voltage to overcome this issue, so some systems are designed so that the spark jumps from the ground electrode to the center electrode.

It used to be necessary to maintain spark plugs by sandblasting built-up deposits from the internal surfaces and filing the electrodes sharp again. Now, low-erosion materials such as platinum and iridium have become available, which last a lot longer (up to 100,000 miles [160,000 km]) and are replaced rather than refurbished.

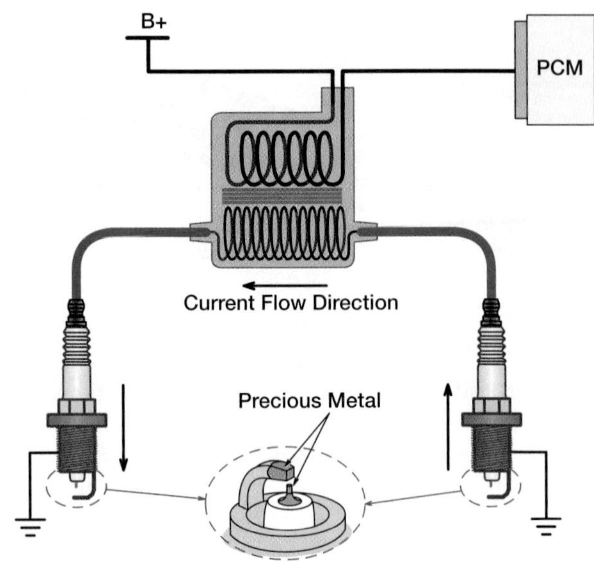

FIGURE 61-14 Polarity-sensitive spark plugs must be installed in the proper orientation so that the spark jumps the intended direction across the air gap.

Be careful when installing plugs on some waste spark systems. Some of these systems use polarity-sensitive spark plugs that have platinum only on one electrode and are designed for the spark to jump from either the center electrode or from the ground electrode. These types of spark plugs have to be installed in the correct spark plug hole so that the spark will jump the correct direction (**FIGURE 61-14**). Manufacturer's service information provides this information.

TECHNICIAN TIP

The same spark plug can sometimes be used in different engines with different gap settings, or sometimes gaps can be damaged during shipment; therefore, the gap should be checked with an accurate spark plug gauge and, if necessary, adjusted according to the specifications when the spark plug is first installed. The gap is critical to engine performance. A narrow gap may give a spark too weak to ignite the air-fuel mixture effectively. A gap too wide may fail to produce any spark at all. Either way, the wrong gap is likely to produce a drivability concern. One exception to the gapping rule is iridium spark plugs. The gap should not be adjusted, as the iridium is brittle and can snap off. Verify that their gap meets specifications. If the gap is wrong, they will have to be replaced.

Spark Plug Size

Spark plug size refers to the diameter of the threads of the spark plug. Nearly all of today's light-duty vehicles use either a 14 mm or an 18 mm spark plug, although some manufacturers use 16 mm "high-thread" spark plugs or even 12 mm long-reach spark plugs (**FIGURE 61-15**). Most manufacturers have been switching to smaller plugs to allow for larger or more valves in the cylinder head. Both the Society of Automotive Engineers (SAE) and the International Organization for Standardization (ISO) have a set of standards that cover length, hex size, thread diameter, and thread pitch.

Spark Plug Reach

Spark plug reach is the distance from the seat of the spark plug to the end of the spark plug threads. The purpose of spark plug reach is to ensure that the spark plug electrodes are in the most efficient position for combustion within the cylinder. If the reach is too long, the piston may strike the spark plug while moving up. If the reach is too short, the spark may occur inside of the threaded spark plug hole, resulting in a misfire. Some spark plugs are only threaded on the bottom half or, in the case of "high thread" plugs, the top half of their reach. Always install the specified spark plug for the engine you are working on (**FIGURE 61-16**).

Heat Range

Spark plugs should operate between average temperatures of 746°F and 1460°F, or 400°C and 800°C. This is referred to as the **heat range**. The operating temperature of a spark plug refers to the temperature at the sparking tip of the spark plug inside a running engine. The temperature that a spark plug will reach depends on the distance the heat must travel from the insulator on the firing end to reach the outer shell of the plug and enter the cylinder head and the water jacket (**FIGURE 61-17**). If the heat path is long, the spark plug will retain more heat and therefore will run at a higher temperature than one with a short heat path.

If a spark plug is too cold, deposits may form on the insulator, providing a path for the spark to travel, leading to misfire. If it is too hot, the heat can ignite the air-fuel mixture itself, causing pre-ignition. A colder plug conducts more of the heat out of the spark plug tip. Using a spark plug with the proper heat range is critical for the operation of the engine. The heat range of a spark plug is affected by a wide range of characteristics, including its design profile and the types of material used in its manufacture, as well as the operating conditions of the engine.

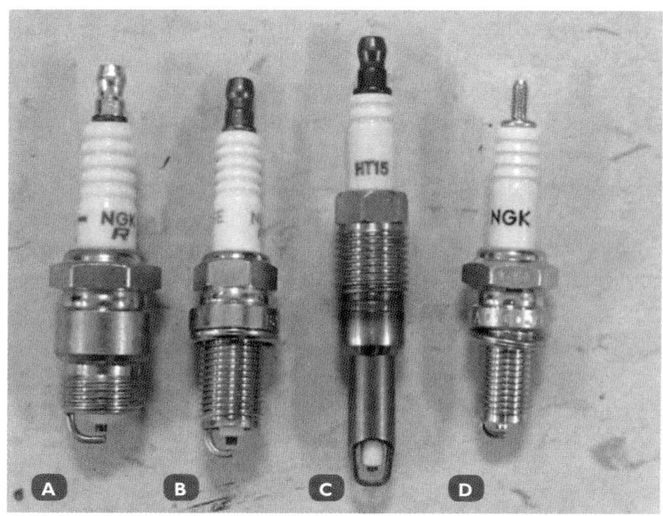

FIGURE 61-15 Common spark plug sizes. **A.** 18 mm. **B.** 14 mm. **C.** 16 mm "high thread." **D.** 12 mm long reach.

▶ TECHNICIAN TIP

Spark plug cooling is affected by spark plug torque. Because almost all of the heat from the spark plug must be transferred through the threads of the spark plug to the cylinder head, properly torquing of the spark plugs keeps the threads in contact with the head, allowing heat to transfer more quickly. This allows the plug to operate at its designed temperature, instead of overheating and causing pre-ignition.

▶ Types of Ignition Systems

K61004

Several types of ignition systems have been used over the years. Contact breaker point ignition systems were used on earlier vehicles and relied on mechanical devices to create the spark. On later vehicles, the mechanical contact breaker was eliminated, and an electronic means of controlling the primary circuit was used. Modern vehicles no longer use a distributor;

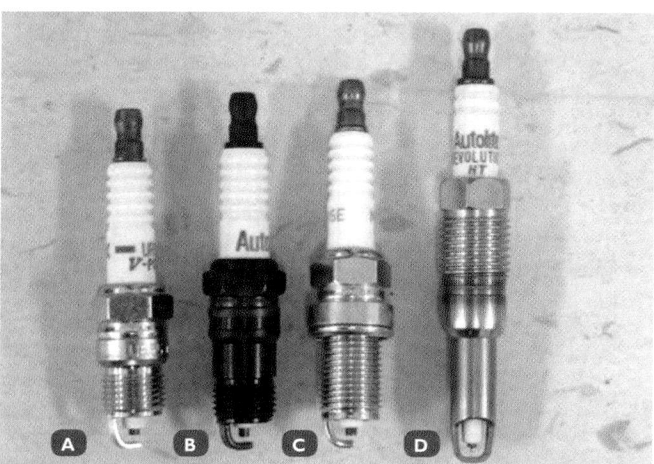

FIGURE 61-16 Spark plug reach. **A.** Short. **B.** Long (half thread). **C.** Long (full thread). **D.** High thread.

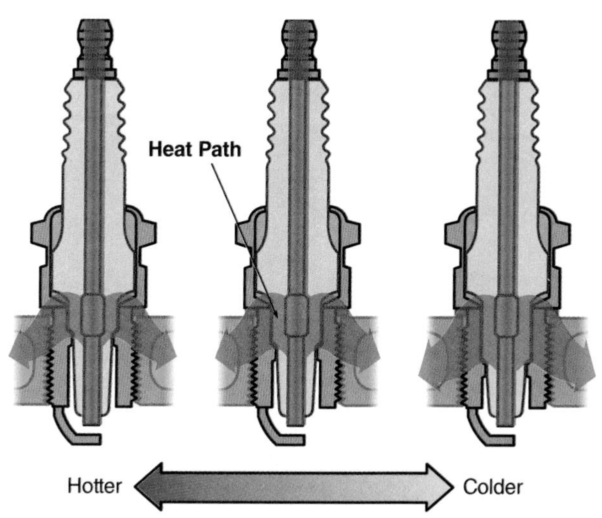

FIGURE 61-17 Heat range of spark plugs.

instead, they send the high voltage from the coil packs or individual coils directly to the spark plugs.

The main difference among these systems is in the way the primary circuits are controlled to produce the secondary spark. The primary circuit is switched on and off, thus creating and then collapsing the primary magnetic field. This switching can be done with breaker points in older vehicles or by electronic switches in modern vehicles.

Contact Breaker Point Ignition Systems

Contact breaker point ignition systems were used on early-model vehicles, providing a mechanical means of quickly connecting and disconnecting the primary side of the ignition coil to ground. The contact breaker is a mechanically operated electrical switch that is fixed to the distributor base plate and opened and closed by the distributor cam with the rotation of the engine (**FIGURE 61-18**). This process builds up and collapses the magnetic field in the ignition coil in order to create a high-voltage output of the ignition coil.

Located in the distributor, the contact points are opened by **cam lobes** on the otherwise round distributor shaft, and closed by a spring. The number of lobes is equal to the number of cylinders, so the contact breaker points open the circuit and activate the ignition coil at the end of each compression stroke for each cylinder.

In order for the spark plugs to fire at the correct moment, the distributor must be installed in the right position. Most manufacturers specify that the ignition **rotor** be aligned with the distributor cap terminal of the number-one cylinder when the number-one piston is at TDC of the compression stroke.

Contact breaker point ignition systems provide a simple means of establishing and interrupting the current flowing in the primary ignition circuit. A basic system consists of the following:

- The battery—provides a source of energy.
- The ignition switch—provides the driver control over system operation.
- An ignition coil—provides step-up transformer action.
- Contact breaker points—opened and closed by lobes on the distributor cam as the engine rotates to make and break the primary circuit at the correct time in the ignition cycle (**FIGURE 61-19**).
- A capacitor—also called a condenser, assists in the rapid collapse of the ignition coil's magnetic field. Any voltage surge across the contacts will charge the capacitor, rather than cause damaging arcing.
- A distributor—rotates at half the speed of the crankshaft to house the contact breaker points and distribute the high voltage from the ignition coil and spark plugs, thus igniting the air-fuel mixture in each cylinder in the correct firing order.
- Connecting wires and leads—suitable for conducting the current flowing in the ignition system, at the appropriate voltage level.

Opening and closing the contact breaker switches the primary current off and on. When the contacts begin to separate, the primary current wants to continue to flow, producing an arc across the contacts. The capacitor absorbs this surge of **inductive current** by providing a very short-term alternative path, in parallel with the opening contacts. The capacitor charges to the peak value of the primary winding voltage almost instantaneously. By the time this has occurred, the gap between the contacts is too wide for a spark to jump between them. This abrupt interruption of the primary circuit assists in the rapid collapse of the magnetic field and hence an increase in the value of the voltage induced in the primary and secondary windings.

As soon as the secondary voltage reaches a value great enough to bridge the gap across the spark plug, a spark occurs, and the voltage in the primary circuit falls to a value below that of the charged capacitor (**FIGURE 61-20**). The capacitor then discharges

FIGURE 61-18 Typical contact breaker point distributor.

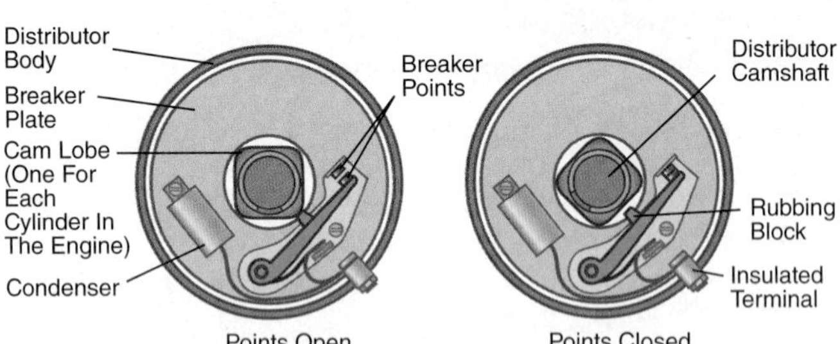

FIGURE 61-19 Contact breaker point operation.

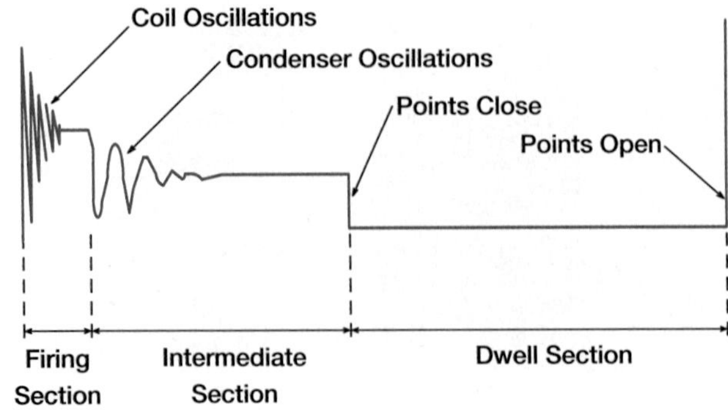

FIGURE 61-20 Primary circuit voltage on an oscilloscope.

back and forth between itself and the primary winding of the coil. When the points close, the capacitor discharges completely.

The spark at the spark plug needs to last for about a minimum of 1 millisecond (0.001 second) to provide enough heat and time to ignite the air-fuel mixture. The time the current is flowing across the electrodes is called the spark duration. Extra-long spark duration causes excessive wear on the spark plug electrodes. Too short, and it may not ignite the air-fuel mixture.

The condenser is made up of two plates constructed from narrow strips of aluminum foil that are insulated from each other by a special waxed paper called a dielectric. The plates and insulating paper are rolled up tightly together and sealed in a metal can by crimping the end over onto a gasket (**FIGURE 61-21**). A spring in the base forces the plates and insulation against the gasket to keep out moisture. One plate is connected to the capacitor case and, through its retaining screw, to ground. The other plate is connected to the external connecting lead. When the capacitor is installed, any voltage surge across the contacts will charge the capacitor rather than erode the contacts on the breaker points. The condenser is typically mounted to either the inside or the outside of the distributor.

▶ TECHNICIAN TIP

The condenser is a critical component of the primary circuit in point-type ignition systems. Without it, the points arc as they are opened. This arcing causes a slow stopping of the primary current flow, which causes a slow collapsing of the magnetic field in the ignition coil, which in turn causes low secondary voltage output of the coil. If you have weak spark output on a point-type ignition system, do not be like a former student and overlook the very simple and inexpensive condenser. This student ended up replacing the spark plugs, spark plug wires, cap, rotor, points, ignition coil, and carburetor before he asked for help. The wise old technician, after listening to his story, pulled out a used condenser and connected it to the negative side of the coil and ground with a couple of test leads and asked the student to try starting the engine. The student was amazed when it started right up and ran like new.

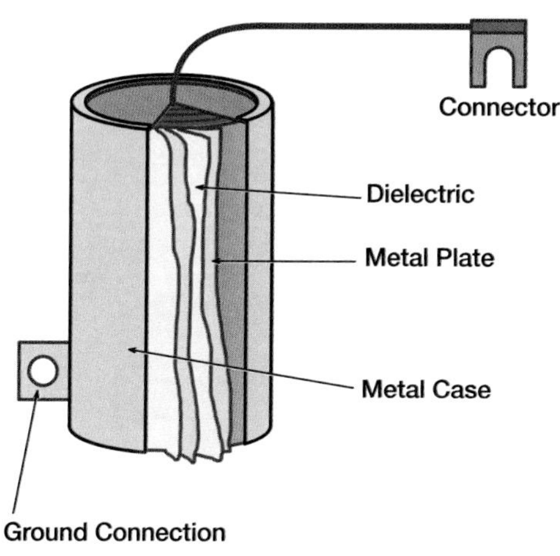

FIGURE 61-21 Typical condenser construction.

▶ TECHNICIAN TIP

The secondary spark occurs when the points just start to open, not when they close. It is the rapid collapse of the magnetic field that induces the high voltage in the secondary windings. Knowing that the spark occurs just as the points open helps a technician static-time an engine after an engine rebuild or when the distributor has been removed.

Dwell Angle

With the primary circuit being switched on and off repeatedly, each ignition coil has to be designed for a particular application so that it operates efficiently. For a four-stroke, four-cylinder engine running at 2000 rpm, 4000 sparks must be supplied every minute. The time available to make and break the primary circuit each time is very short. As engine speed rises, the time available is even shorter; therefore, it is important to ensure that the length of time the current flows through the primary winding is sufficient to create the necessary magnetic field and, therefore, spark.

The condition in which the magnetic field builds to its maximum strength is called coil saturation. In contact breaker systems, it is during the brief time that the contacts are closed and the primary current is flowing that the magnetic field builds. Current flows through the coil primary windings until the next eccentric cam pushes one contact point away from the other. The amount of time, in degrees of distributor rotation, that the contacts are closed is called the **dwell angle** (**FIGURE 61-22**).

The setting of the contact breaker gap influences the dwell angle. On most systems, once it has been set, the dwell angle remains fixed regardless of engine speed. A large gap gives a small dwell angle, as the contact points are closed for a shorter time. A small gap gives a large dwell angle because the contact points are closed for a longer time. The manufacturer's recommended gap provides the specified dwell angle for each application, assuming there is no wear on the distributor cam. The dwell angle should be adjusted to specifications, and then the point gap should be

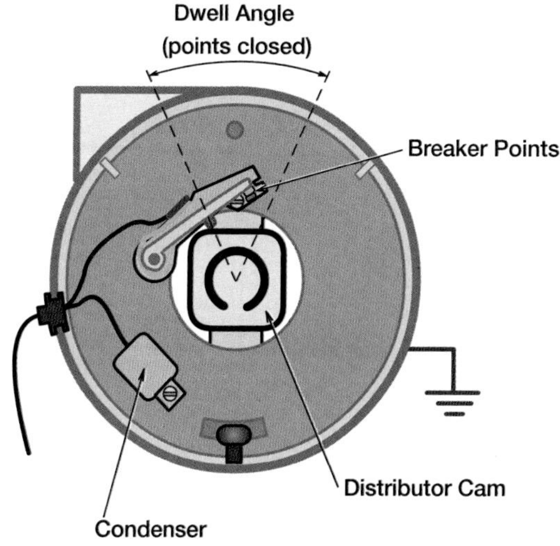

FIGURE 61-22 Dwell is the time that the points are closed and the magnetic field is building in the ignition coil.

checked to identify any cam lobe wear. The proper gap also ensures there is sufficient spacing between the contacts to prevent arcing.

> ▶ TECHNICIAN TIP
>
> As the rubbing block wears, the point gap decreases and the dwell increases. Setting the dwell to the lower end of the specifications allows the dwell to stay within specifications longer. Also, for every degree the dwell increases as the rubbing block wears, the ignition timing retards one degree, reducing fuel mileage and power. As a result, engines need to be "tuned up" every 10,000 to 15,000 miles (16,000 to 24,000 km) to install a new set of points and condenser, set the dwell angle, and adjust the ignition timing.

Ballast Resistor and Bypass Circuit

A **ballast resistor** limits the amount of current flowing in the ignition primary circuit (**FIGURE 61-23**). It is inserted in series between the ignition switch and the positive terminal of the ignition coil. It is usually located near the ignition coil, where it can dissipate its heat into the air.

When the ignition key is turned to the Start position and the starter begins to crank the engine, a heavy load is placed on the battery. This load causes the battery voltage to drop to about 10 volts. Ignition systems must be able to fire the air-fuel mixture on this reduced voltage. Yet, when the engine is running, ignition systems must be able to run on approximately 14 volts, because the charging system is charging the battery.

The ballast resistor and bypass circuit is key to providing the correct voltage in a point-type ignition system. When the voltage is low because the engine is cranking over, a resistor bypass circuit bypasses the resistor, giving full cranking voltage to the coil. When the ignition switch is turned back to the Run position, the resistor bypass circuit is disabled, so current has to flow through the resistor, dropping its voltage from 14 volts to approximately 9–10 volts. More modern solid-state electronic ignition systems do not need a ballast resistor because they have much more reserve voltage, which stays above the required voltage even when primary voltage drops during cranking.

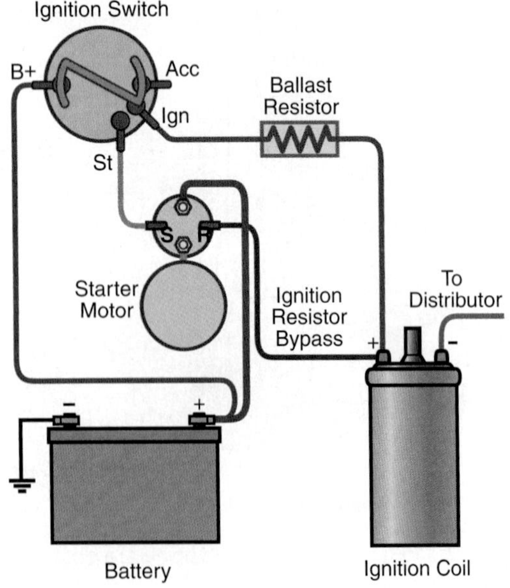

FIGURE 61-23 A ballast resistor in an ignition circuit.

Distributors

The main function of the **distributor** is to distribute the spark to the spark plugs in the correct sequence and at the correct time in the engine cycle (**FIGURE 61-24**). The distributor includes a distributor cap, rotor, switching device, and a shaft with cam lobes for operating the switching device. The distributor also houses the vacuum and mechanical (centrifugal) timing **advance mechanisms** on older engines. These are used to advance the timing of the spark under certain driving conditions.

The **distributor cap** covers the end of the distributor to protect the components inside. It is held onto the distributor by clips or screws. It also provides a connection point between the rotor and the spark plug leads. The rotor is a high-voltage rotating switch that transfers the secondary voltage from the center terminal of the distributor to the side terminals. The distributor distributes the high-voltage pulses to the individual spark plug wires in the correct sequence and at the correct instant in the engine cycle.

In some applications, the ignition coil is mounted inside of the distributor; the cap covers the coil and provides an electrical path from the coil's output terminal to the center of the rotor. In other applications, the ignition coil is bolted to the top of the distributor cap on the outside and covered with a plastic shield (**FIGURE 61-25**). In this instance, the center of the coil sits right above the rotor, and a hole in the cap allows a spring-loaded button to transfer the high voltage directly to the center of the rotor.

An insulated **rotor arm** with a brass or steel electrode is keyed to the shaft directly above the cam and rotates within a molded insulated distributor cap. An internal spring-loaded carbon brush conducts each high-voltage spark from the central terminal of the cap to the center of the rotor as it turns.

The high-voltage spark is timed to occur when the rotor electrode is aligned with a fixed electrode inside the distributor cap. The high voltage bridges the small gap between the electrodes, driving current through the ignition terminal for that cylinder and bridging the gap at the spark plug. As the rotor turns inside the distributor cap, the high-voltage electrical sparks take place successively for each cylinder according to the firing order. For example, an in-line four-cylinder engine

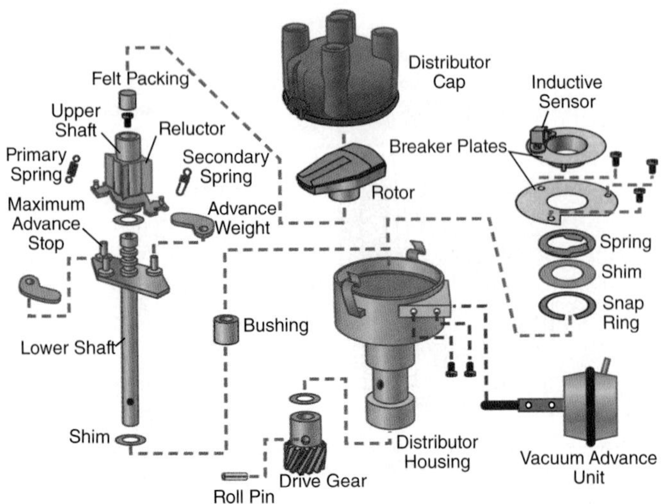

FIGURE 61-24 Distributor.

FIGURE 61-25 Typical distributor with the coil mounted on top of the distributor cap.

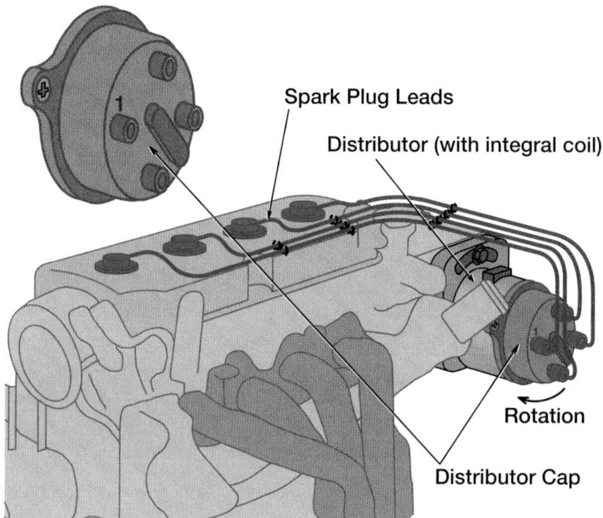

FIGURE 61-26 The spark plug wires are arranged in the firing order on the distributor cap.

typically has a firing order of 1–3–4–2. If the distributor shaft were designed to spin clockwise, the high-tension leads would be arranged clockwise every 90 degrees in the firing order (**FIGURE 61-26**).

Engine Spark Timing

The ignition timing refers to the point at which the spark plug "fires" near the end of the compression stroke (typically 5-10 degrees BTDC). On older vehicles with distributors, the base timing can be adjusted by rotating the position of the distributor housing in relation to the engine. On newer engines, the timing is not adjustable because the computer controls it based on all of the sensor information.

Engine manufacturers include timing marks on many of their engines so that the technician can check and adjust the engine timing if some conditions have changed. There are two types of timing on gasoline engines: valve timing and ignition timing. Valve timing refers to the position of the valves in relation to the piston stroke. During the four-stroke cycle, the valves must open and close at the proper times. Engine manufacturers place marks on the gears, sprockets, or pulleys and the corresponding surfaces of the block and head to allow the valve timing to be properly set whenever the valve timing components are being replaced.

> **TECHNICIAN TIP**
>
> Some newer engines still use a distributor for distributing the high-voltage spark, but they use a crankshaft position sensor to control the timing of the spark. Turning the distributor on this application interferes with the rotor to cap alignment. It does not actually change the timing. So just because it has a distributor does not mean that adjusting it will change the spark timing.

Centrifugal Advance Units

The **centrifugal advance mechanism** (also called mechanical advance) controls ignition timing in relation to engine speed. It is located within the distributor and can be above or below the contact points' base plate. The distributor shaft has two sections that rotate on the same axis. The inner shaft is driven by the camshaft, and the outer shaft has the cam lobes that control the opening time of contact points. Both parts are joined by the centrifugal advance mechanism (**FIGURE 61-27**). Its function is to allow the outer shaft to rotate forward, within limits, in relation to the inner shaft, proportional to engine speed. This forward rotation advances the timing.

The inner shaft has two flyweights attached, each pivoted at opposite ends and controlled by a spring. As the inner shaft is turned, they are thrown outward by the effect of centrifugal force. The faster the shaft turns, the more they move outward.

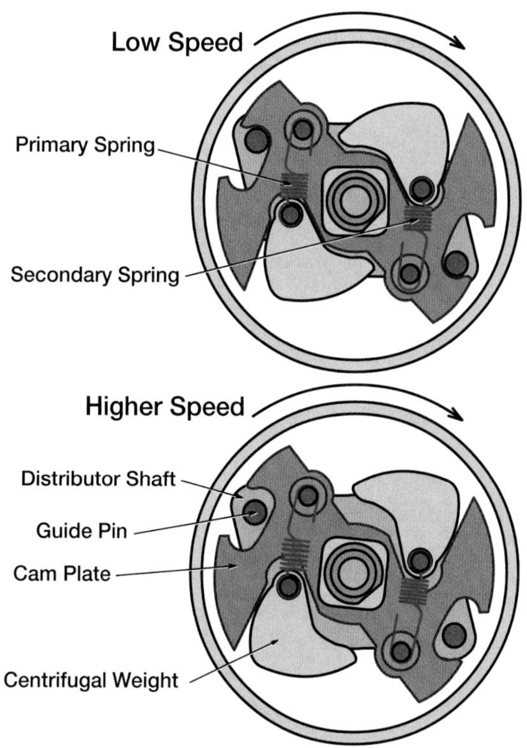

FIGURE 61-27 A typical centrifugal advance unit.

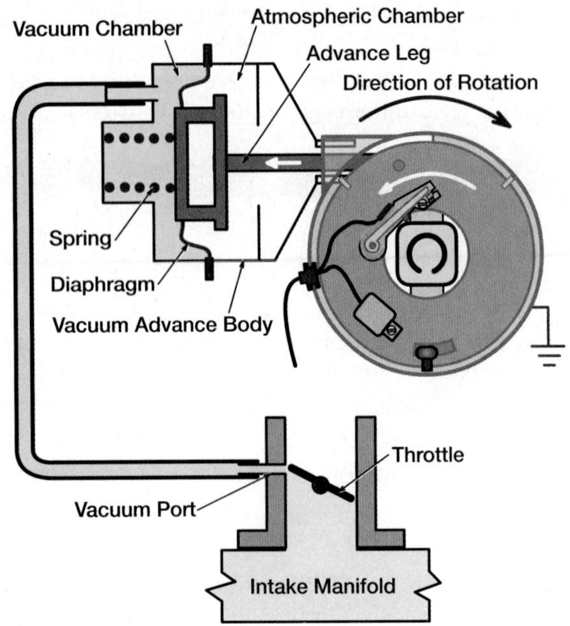

FIGURE 61-28 Typical vacuum advance unit, which advances the ignition timing under light engine loads.

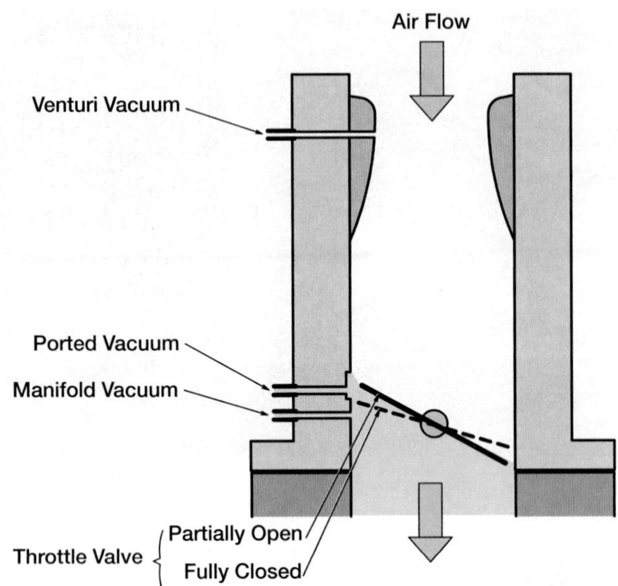

FIGURE 61-29 Vacuum sources: ported, manifold, and venturi.

Slowing the shaft speed reduces the amount of centrifugal force, and spring force pulls them back in.

A slot in each weight is keyed to the outer shaft, and as the weights move outward, they turn the outer shaft forward in relation to the inner shaft. This movement has the effect of advancing the opening of the contact points in relation to engine speed. The relationship between speed and advance is determined by how heavy the weights are and by the tension of the springs. The maximum advance is determined by limiting the movement of the weights with slots and pins or bushings.

Vacuum Advance Units

The vacuum advance mechanism controls **ignition advance** in relation to engine load. At light load, an engine is more efficient when the timing is advanced slightly. Under heavy load, the timing must be retarded slightly to avoid "pinging," or spark knock. The function of ignition advance is to improve fuel economy and, in doing so, reduce exhaust emissions. Vacuum is routed from the carburetor to a spring-loaded diaphragm housing attached to the distributor. The diaphragm is attached to a pull rod that, when operating, acts on the **distributor base plate**, turning it opposite to the direction of distributor shaft rotation (**FIGURE 61-28**). The movement results in the opening time of the contact points being advanced during light engine loads and retarded during heavy engine loads.

Manufacturers designed their vacuum advance to operate on one of three types of vacuum: ported, manifold, or venturi (**FIGURE 61-29**). On older vehicles, ported vacuum is the most common. It is obtained through a small slit in the throttle bore located just above the throttle plate. This slot has no vacuum at idle, creating no vacuum advance at idle. As the throttle is opened, the vacuum to the slit increases, causing some vacuum advance, which tapers off at wide-open throttle.

Manifold vacuum is obtained through a small hole in the throttle bore just below the throttle plate. It provides strong vacuum at idle and less vacuum as the throttle is opened. Venturi vacuum is obtained from a port that is connected to the smallest diameter of the venturi. This system has no vacuum at idle, but as airflow increases through the venturi, the vacuum increases. The vacuum in each of these systems acts on the vacuum diaphragm, causing the timing to advance as vacuum increases, or retard as vacuum decreases. Because the vacuum responds differently in each type of vacuum, the manufacturer would specify the base timing and the advance curve profile for the engine.

Electronic Ignition Systems

K61005

In **electronic ignition systems**, the contact breaker points are eliminated and replaced by both an electronic triggering system and an electronic switching device (**FIGURE 61-30**). The triggering system, called a pickup assembly, generates an electronic signal that turns the electronic switching device on and off. The switching device, called an ignition module, uses a power transistor to turn the primary circuit on and off. This causes the magnetic field to build up and collapse in the primary coil winding in the same way that the point ignition system does.

In general, a pickup assembly has one stationary component and one component that spins with the distributor shaft. When the two components interact as designed, a signal is sent to the ignition module. This is typically accomplished through the use of either an inductive pickup assembly, a **Hall-effect sensor**, or an **optical sensor**, which will be covered later. Many of the components, such as the battery, ignition switch, spark plugs, and wiring, in an electronic ignition system are essentially the same as those in a contact breaker point ignition system. Early electronic ignition systems included an ignition module and a triggering device coupled with a standard distributor.

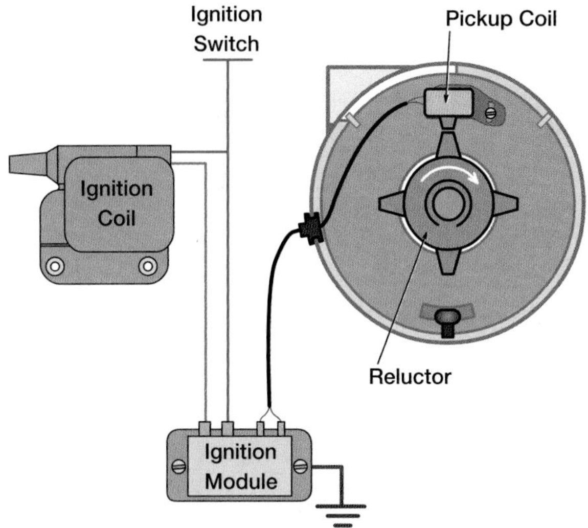

FIGURE 61-30 Electronic ignition systems eliminated the contact breaker points and condenser and replaced them with a pickup assembly and ignition module.

The following sections describe only those components of an electronic ignition system that are not part of a contact breaker point ignition system.

Components of Distributor-Type Electronic Ignition Systems

Ignition Modules

Ignition modules are used in virtually every type of ignition system except the contact breaker point system. In its simplest form, it uses the information from the triggering device to turn the primary circuit on and off. As technology progressed, processing power was added to the module so that the dwell time could be modified. Thus, the dwell time could be shortened during low engine rpm and lengthened as rpm increased. This flexibility helps to prevent overheating of the ignition coil and transistor during low speed while still maintaining adequate time for coil saturation during all engine speeds. Further refinement saw current-limiting circuitry added to the ignition module, which allowed for the use of low-impedance ignition coils.

In the 1980s, manufacturers started to use their PCMs to work along with the ignition module. The module typically provided for base timing while the engine was being started and then switched over and allowed the PCM to control ignition timing through the ignition module once the engine was running.

In early electronic-ignition vehicles, the ignition module was typically located in one of three places: in the distributor, on the distributor, or away from the distributor on the firewall or inner fender (**FIGURE 61-31**). On many waste spark systems, the ignition module is located either under the ignition coils or within the coil assemblies. On newer vehicles, the ignition module functions are either completely controlled by the internal circuitry of the PCM or housed in the ignition coils themselves. In many cases, this eliminates the need for an external ignition module.

Induction-Type Pickup Assemblies

Induction-type systems are electronic systems that use a magnetic pulse generator, also called a variable reluctor sensor (VR sensor), to generate an AC signal. They come in a variety of configurations, but all of them have a **stator** (also called a pickup coil) mounted on the distributor body and a rotor unit (also called a **reluctor** or trigger wheel) attached to the distributor shaft (**FIGURE 61-32**). The reluctor has one tooth for each cylinder. As it spins, the teeth interact with the stator to trigger the ignition module. The stator may have only one projection with a stationary coil of fine enameled wire wound around it, or it may have a circular permanent magnet with a number of projections or teeth corresponding to the number of engine cylinders, and a stationary coil of fine enameled copper wire wound on a plastic reel and positioned inside or around the magnet.

As the **stationary winding** is influenced by the magnetic field, a voltage is induced in the winding each time the magnetic field changes. As the teeth approach, the strength of the magnetic field increases, inducing a voltage and current flow in the winding. The polarity of the voltage is said to be positive as it produces a current flow in a certain direction. When the teeth are in alignment, the magnetic field is at its strongest, but at that point, it is not changing. Voltage and current now fall to zero (**FIGURE 61-33**).

As the teeth move away, the strength of the magnetic field changes again, once more inducing voltage and causing current to flow in the winding. This time, current flow is in the opposite direction, and the polarity is now said to be negative. Because polarity changes every time the teeth approach and leave the stator teeth, the voltage produced is an alternating current voltage, and current flow is also alternating (**FIGURE 61-34**).

After the positive pulse from the pickup coil is received by the ignition module, a **dwell control section** determines how soon in the cycle the primary circuit will be switched on (**FIGURE 61-35**). This also determines how long current will flow in the primary winding before the module turns it off and the spark occurs. The dwell can be varied according to engine speed, improving coil efficiency. At low engine speeds, the dwell is short when measured in degrees, but when measured in time, it is enough to fully saturate the coil. As engine speed increases, the dwell increases, which helps to maintain an adequate amount of time for coil saturation.

Ignition coils employed with electronic systems are referred to as **low-inductance coils** because their primary winding resistance and number of turns are low. Current flow is much higher than with contact breaker point systems and reaches its optimum level sooner. A current-limiting device within the control module limits maximum current flow to a safe value. Ignition advance, according to engine speed and load, may be provided by centrifugal and vacuum advance mechanisms in a distributor, or through the ignition module or PCM on newer vehicles.

The pickup coil winding is connected to the ignition module and forms part of an **internal module control circuit**. This section of the ignition module is responsible for receiving the trigger signal. With the ignition switched on and the engine stationary, battery voltage is applied to the coil's positive terminal.

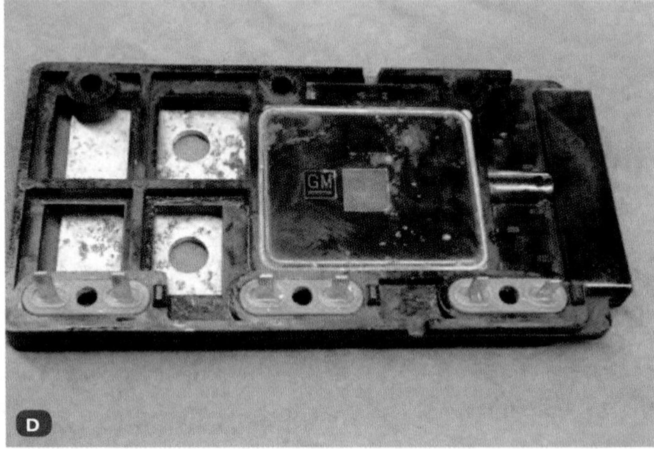

FIGURE 61-31 Externally mounted ignition modules. **A.** In the distributor. **B.** On the side of the distributor. **C.** On the firewall or fender. **D.** Under the coils.

FIGURE 61-32 Types of inductive ignition pickup coils.

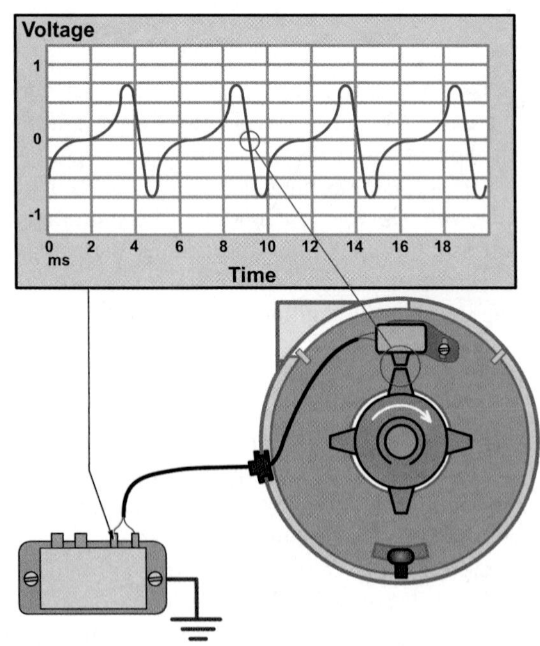

FIGURE 61-33 Inductive pickup coil and the signal it generates.

However, there is no current flow through the primary winding as the switching transistor is in the Off state.

With engine rotation, the reluctor teeth on the distributor shaft approach the stator teeth, and a voltage of positive polarity is induced in the winding. This voltage signals the control circuit to turn the primary current on, and a voltage is applied to the base/emitter of the **switching transistor**. The switching transistor acts as a relay. When activated, it allows current to flow through the ignition coil primary winding and through the collector/emitter of the switching transistor to ground. The magnetic field is built up in the ignition coil. Further rotation of the reluctor causes the teeth to align with the stator teeth or pickup coil, which signals the switching transistor to turn off, stopping current flow through the ignition coil primary winding.

Simply stated, the primary winding in the ignition coil is continuously turned on and off as the reluctor teeth spin past the stator. This causes the primary winding in the ignition coil to build and collapse the magnetic field, which induces high voltage spikes in the secondary winding that is used to create a spark at the end of the compression stroke of each cylinder.

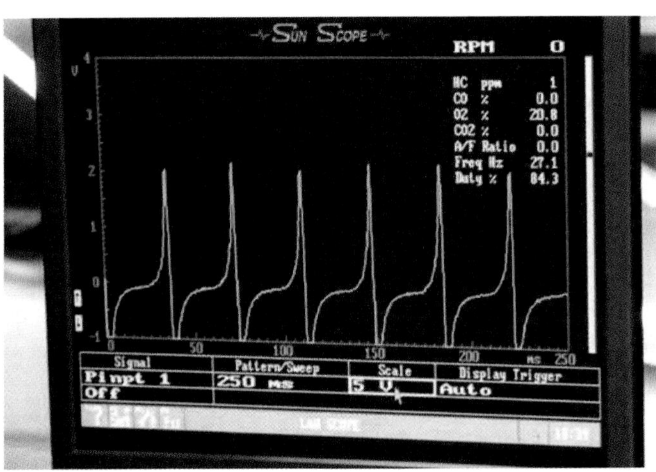

FIGURE 61-34 Typical ignition pickup coil scope pattern.

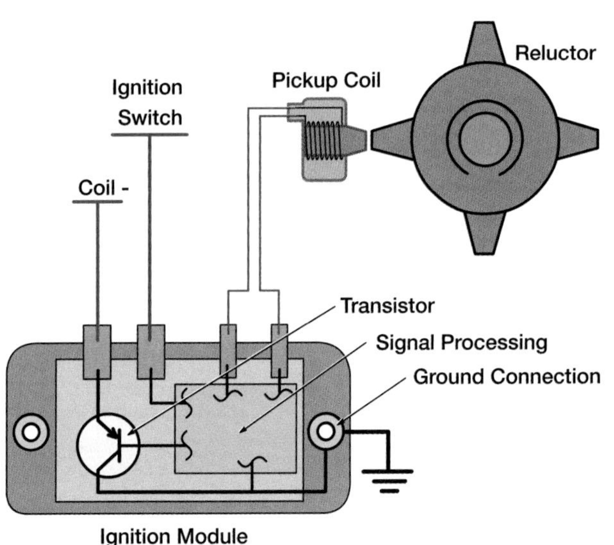

FIGURE 61-35 Schematic of an electronic ignition system that uses an induction-type triggering component.

Hall-Effect Sensors and Operation

Another type of electronic ignition system is the Hall-effect type. In Hall-effect systems, a Hall-effect generator, also called a Hall-effect sensor or switch, can be located inside the distributor to signal an ignition module to turn the primary circuit on and off. Hall-effect generators operate by using a potential difference, or voltage, created when a current-carrying conductor is exposed to a magnetic field. If a magnetic field is applied at right angles to the direction of current flow in a conductor, the lines of magnetic force permeate the conductor, and the electrons flowing in the conductor are deflected to one side of the conductor. This deflection creates a potential difference or voltage across the conductor (**FIGURE 61-36**). The stronger the magnetic field, the higher the voltage. This dynamic is called the Hall-effect voltage, and if the magnetic field is alternately shielded and exposed, it can be used to trigger a switching device. Hall devices are made of semiconductor material, which can quickly respond when shielded from or exposed to a magnetic field.

In a distributor, the Hall-effect generator and its **integrated circuit** are located on one leg of a U-shaped assembly mounted on the distributor base plate (**FIGURE 61-37**). An integrated circuit is a semiconductor chip that contains miniature versions of various electrical components within one housing. A permanent magnet is located on the other leg, and an air gap is formed between them. An **interrupter ring**, which has the same number of blades and windows as engine cylinders, is rotated by the distributor shaft, moving the blades through the air gap. The purpose of an interrupter ring is to systematically block and expose the magnetic field. The ring is shaped like a very shallow cup with slits, or windows, cut into it at evenly spaced intervals. It is made of a ferrous metal.

When a window is aligned with the assembly, the magnetic field is at its strongest, and its lines of magnetic force permeate the Hall generator material. The Hall generator material creates

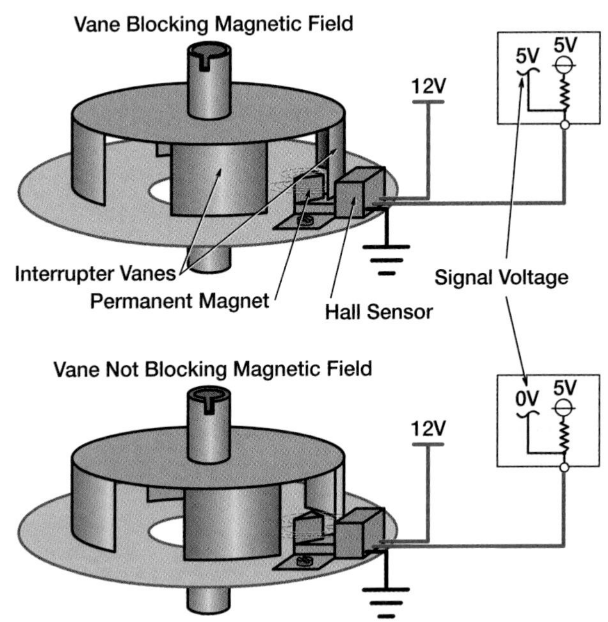

FIGURE 61-36 Hall-effect sensor operation.

FIGURE 61-37 Hall-effect setup in a distributor.

FIGURE 61-38 Optical sensor ignition system.

a voltage that is used to switch the primary circuit to ground. When the interrupter ring rotates so that the blade aligns with the assembly, the magnetic field is shielded from the generator, and the Hall material does not create a signal voltage; thus, the primary circuit is not switched to ground. With continuous rotation, the blades repeatedly move in and out of the air gap, and the signal voltage turns on and off repeatedly. This can be used to control the operation of the ignition coil primary circuit.

Optical-Type Sensors

Another type of electronic ignition system uses a light-emitting diode (LED) and a phototransistor to create an optical sensor. The optical sensors located inside the distributor can be used to sense the position of the crankshaft and send an appropriate voltage signal to the ignition module or PCM. A signal rotor plate is attached to the distributor shaft (**FIGURE 61-38**). Although various designs are used, a typical rotor plate has 360 slits at 1-degree intervals on its outer edge. Inboard of these slits are four slits on a four-cylinder engine, six slits on a six-cylinder engine, and so on. One slit is larger than the others to identify the position of the number-one cylinder to the PCM.

As the rotor plate turns, it passes between an LED above the rotor plate and a phototransistor below the plate (**FIGURE 61-39**). When provided with a suitable voltage, LEDs transmit a fine beam of light. Phototransistors receive this light and use it to make a voltage output signal. When a slit is in alignment, the light beam passes through it, and a signal is transmitted to the control unit. When the slit is out of alignment, the light beam is interrupted, and the signal falls to zero.

The control unit uses the signals it receives at every 1 degree to gauge engine rpm and **crank angle position** in 1-degree increments. It uses the signals from the inner slits to gauge the piston position, with the signature slit identifying the number-one piston. The signals from both sets of diodes are converted to on/off pulses in the sensor's internal circuitry.

After computing data from the various input sensors against the optimum settings for each operating condition recorded in its memory, the control unit determines ignition timing. It then switches the primary circuit on and off by controlling the operation of a power transistor in the ignition module.

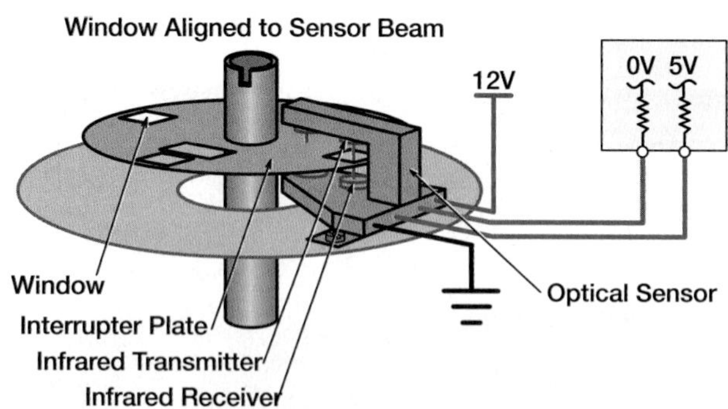

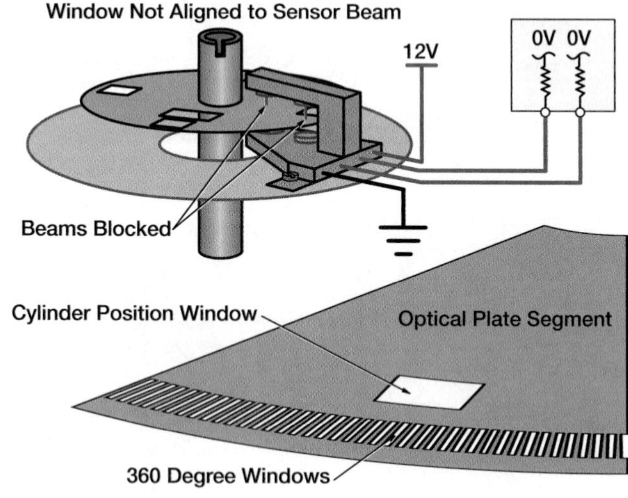

FIGURE 61-39 A typical optical sensor assembly.

> **TECHNICIAN TIP**
>
> The optical sensor is very precise due to the small slits that the light passes through. However, because the slits are so small, they can easily become blocked by dirt, debris, or even oil. Make sure the slits in the rotor plate stay clean when you are working around it.

Components of Distributorless-Type Systems

K61006

In **distributorless ignition systems**, the distributor is eliminated and replaced by multiple ignition coils. The main advantage of the distributorless ignition system is the elimination of all the mechanical parts included in a distributor assembly such as bushings, bearings, the **breaker plate** assembly, and advance mechanisms. Also, the distributorless ignition system has eliminated maintenance items such as a distributor cap, rotor, contact points, and condenser.

There are two main types of distributorless ignition systems: waste spark, and coil-on-plug (COP) ignition systems (**FIGURE 61-40**). In waste spark systems, there is one ignition coil for each pair of companion cylinders. For coil-on-plug systems, there is one ignition coil for each cylinder. Signals from a **crankshaft position sensor** and, if equipped, a **camshaft position sensor** are used by the PCM to control the on/off signal to the primary windings of the coils. Both of these sensors are either Hall-effect or induction-type sensors, to let the PCM know the positions of the crankshaft and the camshaft. Not all engines use a camshaft position sensor; on waste spark ignition engines, the PCM only needs to know when the number-one cylinder is at TDC (compression or exhaust doesn't matter) to give the proper spark timing for all cylinders.

Crank and Cam Sensors

Crankshaft and camshaft position sensors are used in distributorless ignition systems to calculate engine speed and determine engine position. The crankshaft position sensor can be mounted at the front, middle, or back of the engine, and functions like a pulse generator in a distributor. The reluctor or trigger wheel is mounted to the flywheel, the crankshaft, or the crankshaft pulley. The camshaft position sensor monitors the position of the camshaft to let the PCM know which cylinder is approaching its power stroke. The two sensors work together to control the firing of the spark plugs, including advance and retard functions (**FIGURE 61-41**). They are also used for fuel injector control. These sensors provide more accurate readings than contact breaker points, as there are fewer mechanical parts to wear, which affects accuracy.

As an example, in one system the identification of each pair of cylinders and a signal for their triggering is provided by a dual crank sensor fixed to the engine timing cover. This contains two Hall-effect switches sharing a central magnet that forms two air gaps between them. Two **concentric interrupter rings**, which look like one shallow cup placed inside a larger shallow cup, with a number of blades and windows on their outer edges, are mounted on the rear of the crankshaft balancer and rotate through the air gaps (**FIGURE 61-42**). The voltage signals provided allow the ignition module to identify which pair of companion cylinders must be provided with ignition.

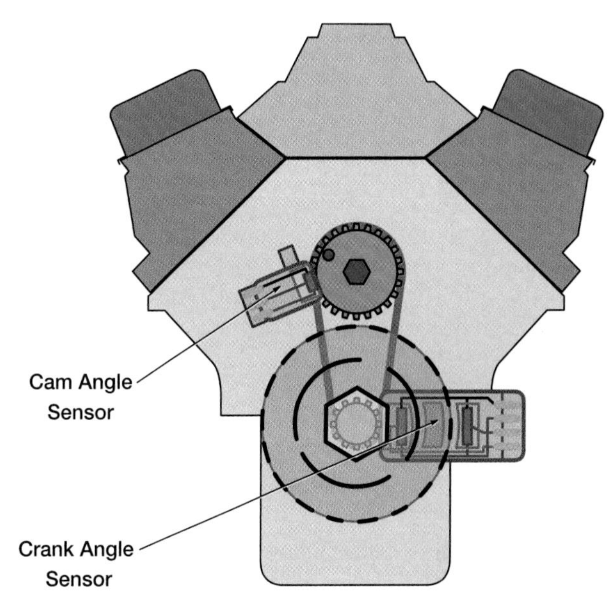

FIGURE 61-41 The crankshaft position sensor and the camshaft position sensor work together to provide timing signals that the PCM uses to fire the spark plugs and control variable valve timing.

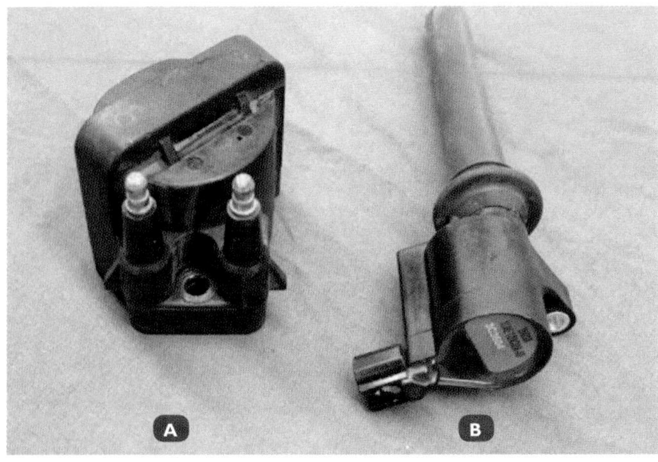

FIGURE 61-40 **A.** Waste spark ignition coil. **B.** Coil-on-plug ignition coil.

FIGURE 61-42 The interrupter rings on an engine's crankshaft pulley.

SAFETY TIP

Crankshaft and camshaft sensors are typically located around the front of the engine near the belts and other moving parts. Always remove the key from the ignition switch before working around the moving parts of the engine.

Waste Spark Ignition System

In a waste spark system, each ignition coil serves two cylinders, with each end of the secondary winding attached by a high-tension lead to a spark plug. These two plugs are on companion cylinders—that is, cylinders where the pistons reach TDC at the same time, but on opposite strokes (**FIGURE 61-43**). The cylinder on the compression stroke is said to be the **event cylinder**, because it is the one getting ready to produce power, and the cylinder on the exhaust stroke is the **waste cylinder**, as the spark has no effect on engine operation.

When the high voltage is induced in the secondary winding, the secondary circuit is completed by current flowing through the high-tension lead to the center electrode on one spark plug, bridging the gap and creating a spark in that cylinder, traveling through the cylinder head, bridging the spark plug gap in the companion cylinder, and flowing back through that high-tension lead to its starting point at the ignition coil. The cylinder on the compression stroke with its charge of fuel and air is "fired" by its spark, driving the piston down on the power stroke, while the spark at the plug of the cylinder on exhaust simply serves to complete the circuit, and is "wasted."

When the crankshaft rotates through one revolution, the roles of the cylinders are reversed, and the waste cylinder becomes the event cylinder. Firing of both spark plugs again takes place, and this cylinder now drives its piston down on the power stroke. This same process occurs on each pair of companion cylinders as they approach the TDC position. The primary circuit in each coil must therefore trigger at the correct time in each crankshaft revolution. Timing is controlled by the PCM based on all of the engine sensor data.

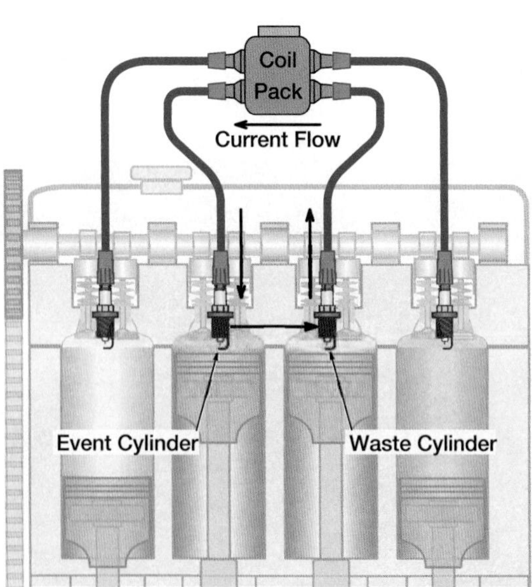

FIGURE 61-43 A waste spark system fires two spark plugs at a time: the event cylinder and the waste cylinder.

Because each coil feeds two cylinders, there are three ignition coils for a six-cylinder engine, each with its own primary and secondary windings. The coils can be combined to form one coil pack, or they can be three separate coils (**FIGURE 61-44**). An eight-cylinder engine has four ignition coils.

Coil-on-Plug Ignition System

Most current production gasoline engines use a single ignition coil for each cylinder (**FIGURE 61-45**). There are two common layouts: coil-on-plug (COP) and coil-near-plug (CNP). The coil-on-plug system positions each coil directly on top of each spark plug. In fact, the coil has a replaceable, integrated spark plug boot on the end of the coil. The coil-near-plug system positions the coils a short distance away from the spark plugs (usually due to the spark plugs being in the side of the cylinder head instead of the top), and uses short spark plug wires to deliver the spark to the spark plugs (**FIGURE 61-46**).

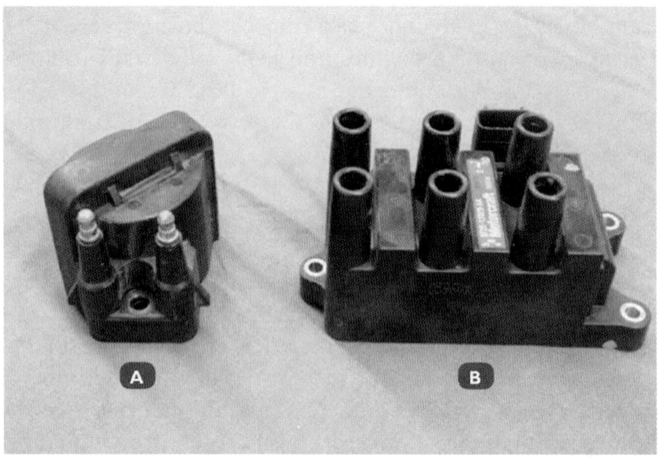

FIGURE 61-44 Typical waste spark ignition coils. **A.** Individual ignition coil. **B.** Coil pack.

FIGURE 61-45 Typical coil-on-plug ignition system with the coils partially removed.

FIGURE 61-46 Typical coil-near-plug ignition system with the spark plugs in the side of the head and the coils nearby.

In many ways, these systems are the simplest ignition system in terms of components. They do away with most of the secondary components and use existing engine management components to replace some of the dedicated primary circuit components in older ignition systems (**FIGURE 61-47**). In fact, in most of these systems, the PCM acts as an ignition module for the primary circuit of each coil. This controls the saturation time of each coil and the time of each spark.

COP and CNP systems have several benefits over other ignition systems:

- Better high rpm performance. Each coil only sparks once per cylinder, per cycle, meaning they have to supply a lot fewer sparks. This allows a greater amount of time for coil saturation between sparks, especially at high rpm.
- Ability to create higher-voltage spark output without overstressing the coils.
- Fewer parts required, which reduces cost, weight, failures, and maintenance.
- Better control of emissions due to a hotter spark.

Coil-on-plug systems come in two physical configurations; the coils can be separate coils that are each mounted on top of one spark plug as mentioned, or they can be molded into an insulated cassette that is mounted directly over the top of the spark plugs (**FIGURE 61-48**). The individual COP coil systems have become more common on recent vehicles.

Coil-on-plug coils typically come in one of three different electrical configurations: two-wire, three-wire, and four-wire. Each one of them operates slightly different from the others. On a two-wire system, one terminal is connected to the power side of the circuit, and the other terminal is connected to the switching circuit (**FIGURE 61-49**). Power is fed to the power terminal (B+) that supplies power to the primary winding. The other end of the primary winding is connected to one end of the secondary winding and the switching terminal. The switching occurs in either the PCM or an external electronic ignition module. This coil acts like the standard ignition coil described previously.

Some manufacturers use a Zener diode in parallel with dwell control circuit in the PCM to verify COP coil operation. The large induced voltage in the primary winding from the collapse of the magnetic field overpowers the Zener diode and provides a signal indicating that the coil fired (**FIGURE 61-50**). The PCM looks for this signal; if it is not present, the PCM will set a DTC for that particular ignition coil.

The three-wire COP coil has the following terminals; power, ground, and triggering. In this case, a power transistor

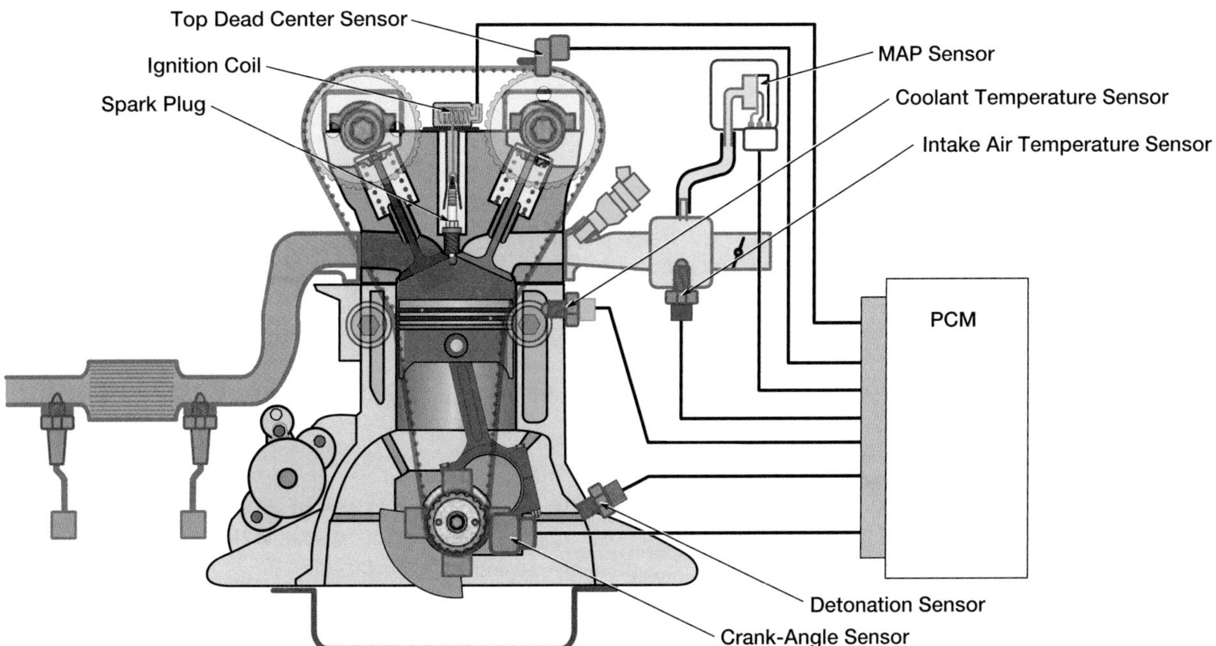

FIGURE 61-47 Schematic of a typical COP ignition system on a four-cylinder engine. Notice that there are many fewer secondary components than a distributor style ignition system.

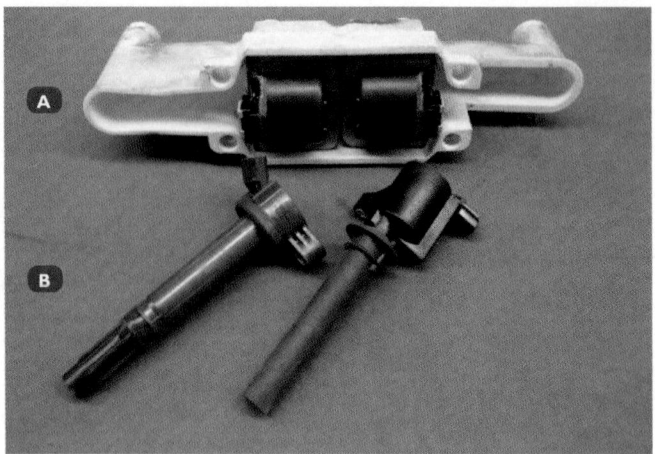

FIGURE 61-48 **A.** Cartridge style coil-on-plug assembly. **B.** Individual coil-on-plug assemblies.

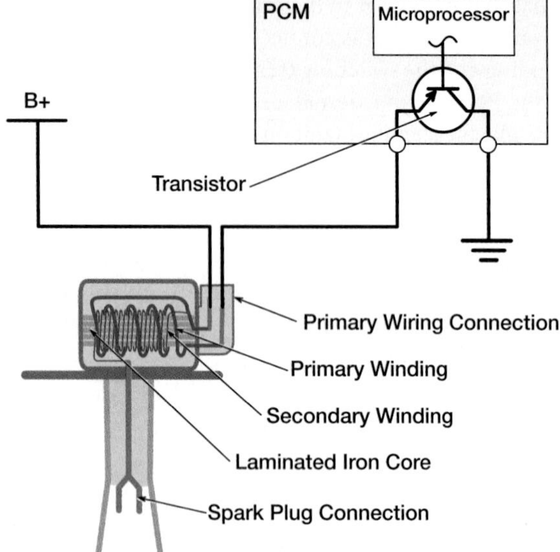

FIGURE 61-49 Typical wiring diagram of a two-wire COP coil system.

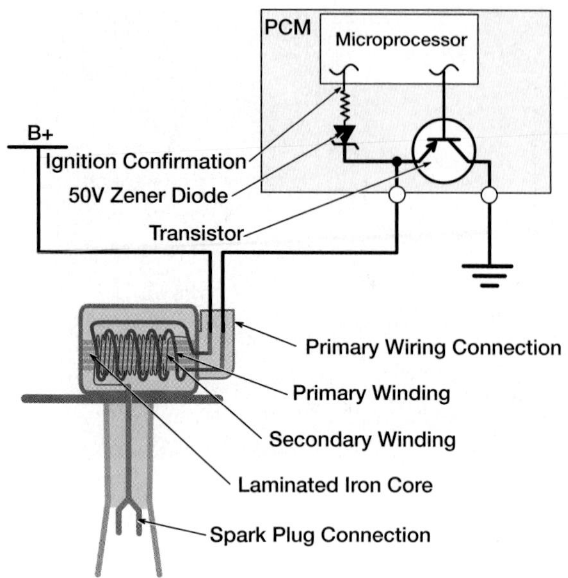

FIGURE 61-50 Some ignition systems use a Zener diode in the PCM to indicate that the magnetic field collapsed in the coil (created a spark).

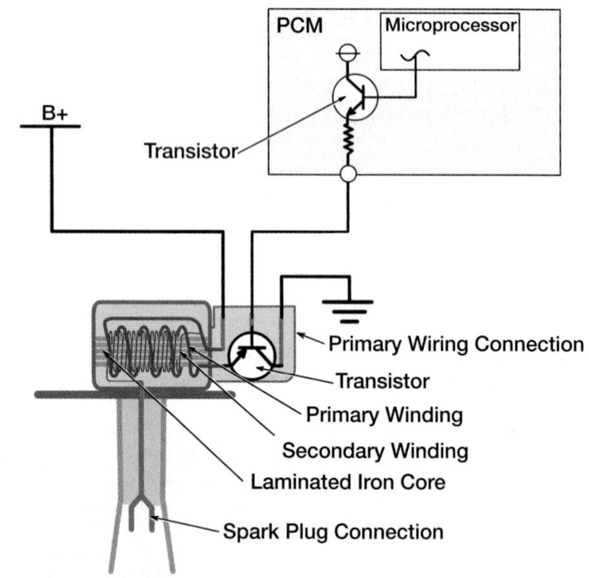

FIGURE 61-51 Typical three-wire COP coil system.

is contained inside of the COP coil. So the power wire is connected to the primary winding and the transistor. The triggering wire activates the power transistor to switch the primary current flow on and off (**FIGURE 61-51**). The ground wire provides ground for the power transistor so it can perform its job.

The four-wire COP coil has the following terminals: power, ground (chassis), triggering, and either a dedicated computer ground or coil feedback circuit (**FIGURE 61-52**). The power, chassis ground, and triggering terminals work similarly to the three-wire system. The fourth wire is used differently by the various manufacturers, but it is generally some sort of feedback wire to the PCM that either indicates whether the coil is operating or the cylinder is actually firing.

Some electronic ignition systems provide more than one spark per cylinder. These types of systems are referred to as multi-strike or multiple-spark discharge. They provide up to three sparks in rapid succession to a single spark plug at low engine rpm (**FIGURE 61-53**). This helps ensure that the air-fuel mixture is ignited during every combustion cycle.

Distributorless Ignition System Timing Control

The spark timing for distributorless ignition systems is controlled by the PCM, which decides the best time fire the spark plugs. It makes this decision based on information received from sensors, for example:

- The **mass airflow sensor**, which measures the airflow into the engine
- The **manifold absolute pressure (MAP) sensor**, which determines the actual pressure in the intake manifold

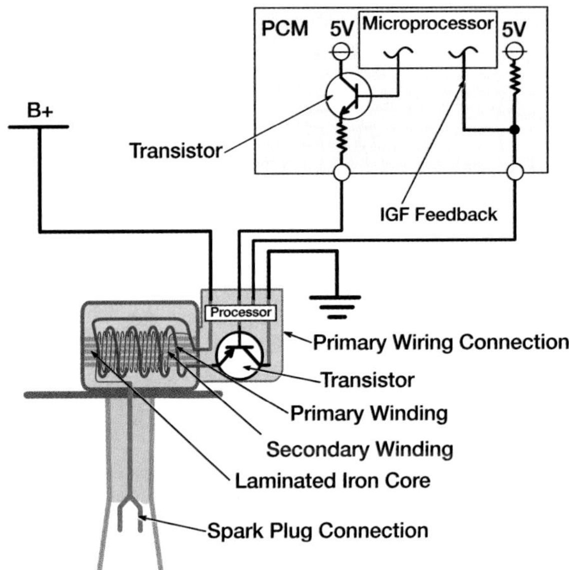

FIGURE 61-52 Typical four-wire COP coil system.

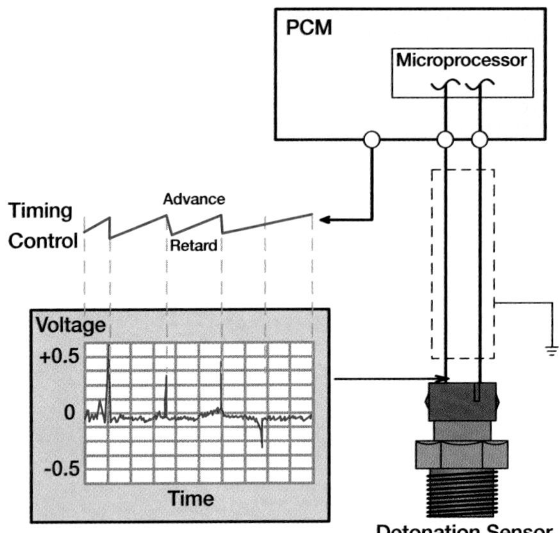

FIGURE 61-54 Knock sensor feedback circuit.

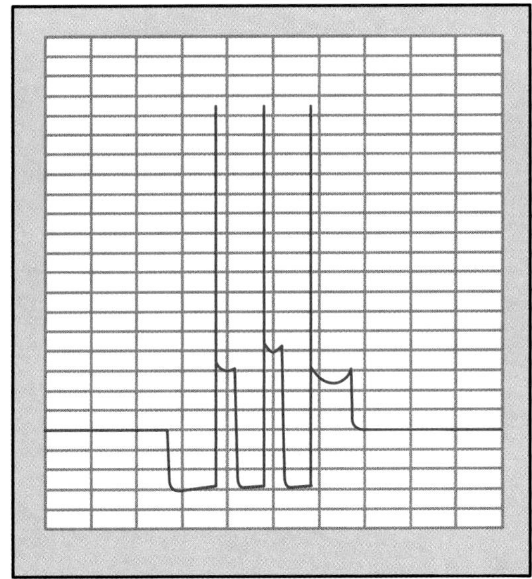

FIGURE 61-53 Some ignition systems can provide up to three sparks in rapid succession to the spark plug. This helps to ensure the air-fuel mixture is ignited at low engine rpms.

- The **throttle position sensor (TPS)**, which indicates how much the throttle plate is open
- The **cam** and **crank angle sensors**
- The coolant temperature sensor (CTS)
- The intake air temperature sensor (IAT)

Once the PCM has evaluated all the sensor information, it decides when to fire the spark plugs based on its stored programming. To do this, the PCM switches the power transistors that control the primary current for each coil. These power transistors can be located inside the PCM or inside the COP coil.

Some systems use a knock sensor, which generates a small voltage signal when the engine is "pinging." When the PCM receives this knock sensor signal, many systems are programmed to retard the ignition timing a few degrees (**FIGURE 61-54**). If the pinging continues, the timing is further retarded in steps until the pinging stops or the timing reaches its limit. If the pinging stops, the PCM advances the timing 1 or 2 degrees and monitors the knock sensor again. If no pinging is detected, the PCM continues to advance the timing a small amount in steps until it detects pinging again or the timing reaches its limit. This process is continuously applied so that the ignition timing is maintained near the most efficient point for fuel economy and emissions purposes.

Another technique some manufacturers use in detecting detonation or pinging, as well as misfires, is called an ion sense or compression sense system. This type of system uses additional circuitry in the ignition module or PCM that monitors the amount of voltage it takes to ionize the space between each spark plug's electrodes (**FIGURE 61-55**). As pressure increases, so does the amount of voltage it takes to ionize the gap. Thus, the monitoring circuitry can tell when the air-fuel mixture has been ignited prior to the spark jumping the gap. It can also tell when the cylinder pressure does not rise in the cylinder, and registers that as a misfire. In the case of detonation or pinging, the PCM will likely retard the timing to try to reduce the pinging. In the case of a misfire, the PCM will set the proper diagnostic trouble code (DTC) when the enabling criteria have been met for that code.

▶ TECHNICIAN TIP

A knock sensor is not very good at distinguishing between pinging and mechanical noise such as the knocking from a loose air-conditioning bracket. A loose vibrating bracket may mimic the frequency of engine knock, and the knock sensor will generate the signal voltage equal to engine knock. This error causes the vehicle to lack power and reduces fuel economy. Ion sense only looks at cylinder pressure, so it is less likely to be fooled.

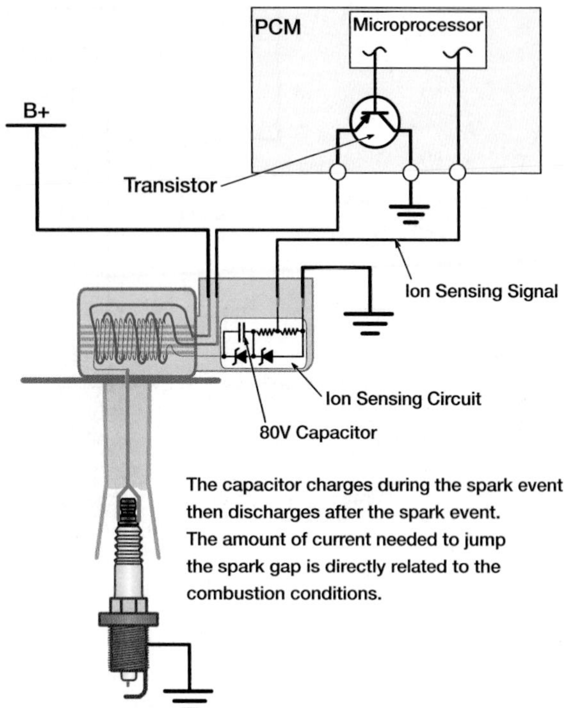

FIGURE 61-55 Ion sense systems detect detonation and misfire situations by monitoring the amount of voltage it takes to ionize the air-fuel mixture across the spark plug gap.

▶ Ignition System Maintenance

Maintenance Tools

S61001

The ignition system is capable of putting out up to 100,000 volts, so the proper tools and safety practices are necessary when working around the ignition system. Here are several tools you will need:

- A special spark plug socket is needed to remove and reinstall the spark plugs. These sockets are deeper than regular sockets, come primarily in three sizes ($^{9}/_{16}$", $^{5}/_{8}$", and $^{13}/_{16}$"), and have a rubber insert. The rubber insert makes removal and installation less likely to damage the spark plug. This insert allows the spark plug socket to grip the spark plug when the spark plug is being installed in a deep well. It also holds the spark plug centrally within the socket, preventing the spark plug from being cocked sideways inside the socket, which can break the ceramic insulator.
- A spark plug gapping tool is used to measure and adjust the spark plug gap before installation (**FIGURE 61-56**).
- A high-tension insulated wire puller makes it easier to remove stuck high-tension wires from spark plugs (**FIGURE 61-57**).
- A tool for removing broken-off high-thread spark plugs (**FIGURE 61-58**).

FIGURE 61-56 Spark plug gapping tool.

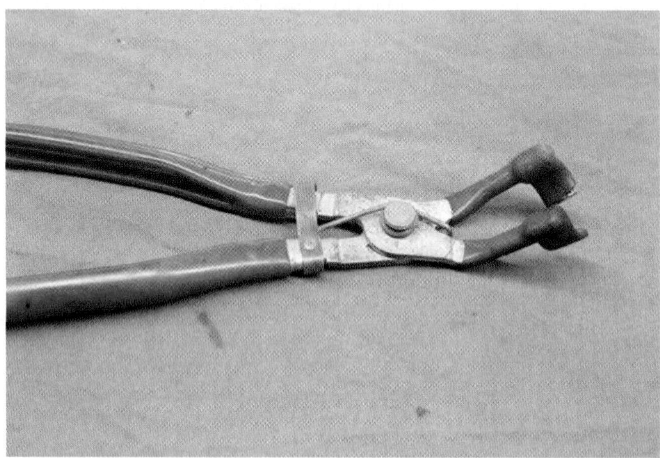

FIGURE 61-57 High-tension insulated wire puller.

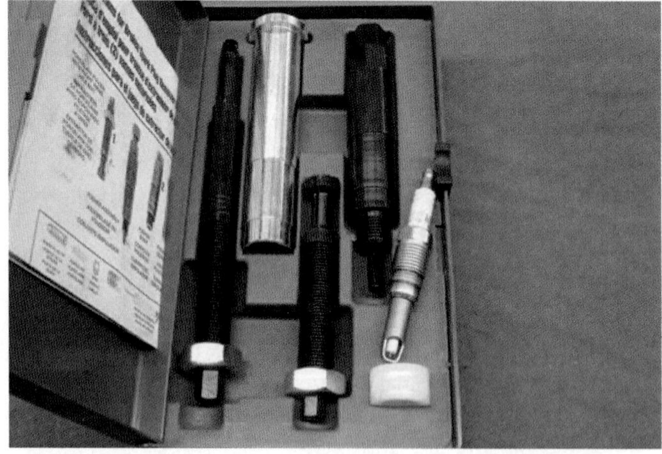

FIGURE 61-58 Tool for removing broken-off high-thread spark plugs.

Using Service Information

N61001

Service information is critical for diagnosing and servicing every system on the car, including the ignition system. This information is normally accessed in an online format from either a subscription service or the manufacturer-supplied service information system. There are three main areas of service information you need to access: specifications, technical service bulletins (TSBs), and diagnostic/repair information.

These topics were covered in the Vehicle Service Information and Diagnostic Process chapter. But for now, when working on ignition systems, you need to research the TSBs whenever you start a diagnosis. You will need to research the specifications when servicing or diagnosing the system. And you will need to research the diagnostic, testing, and repair procedures when you are performing any ignition system work.

To use service information, follow the steps in **SKILL DRILL 61-1**.

Removing and Replacing Secondary Ignition Components

N61002

As the vehicle ages, the components of the secondary ignition system will wear. Periodic replacement of the spark plugs, high-tension leads (spark plug wires) or coil boots, and possibly the distributor cap and rotor, is required to maintain vehicle performance and to meet emission standards. The main components to be replaced on today's vehicles are the spark plugs and the high-tension leads (if equipped) and coil-on-plug coil boots on some applications. Hard starting, misfiring under load, or high tailpipe emissions may indicate a need for service of these ignition components. We explore the process of removing, inspecting, and replacing these secondary ignition components next.

Servicing the Spark Plug

According to most maintenance schedules, the spark plugs should be replaced every 30,000 to 50,000 miles (50,000 to 80,000 km). Some newer vehicles use high-performance spark plugs that are made with platinum, iridium, or gold palladium, and do not require replacement until the engine reaches 100,000 miles (160,000 km) (**FIGURE 61-59**).

FIGURE 61-59 A. Platinum spark plug. **B.** Iridium spark plug.

When service is needed, avoid attaching the spark plug wires to the wrong cylinder, by replacing the spark plugs one at a time. Note that some manufacturers recommend that the spark plugs only be replaced on a cold engine to avoid damaging the threads in the cylinder head. Check the service procedure before attempting to replace the spark plugs on a hot engine. Some manufacturers have very specific procedures for removing their spark plugs, and the procedure should be followed exactly, or broken-off plugs can occur. Even when following exact procedures, some applications still result in broken-off spark plugs, and you will have to use the proper tools and follow the correct procedures for removal of the broken plug.

It is quicker to check the spark plug gap on all of the new spark plugs at one time. Although most spark plugs come pre-gapped, they should be checked to make sure they were not damaged during shipping. If the gap of the spark plug is correct, there will be a slight drag between the gapping tool gauge and the spark plug electrodes as the gauge is installed and removed. If the gap is too big, the outer electrode should be bent down slightly until it is correct. If the gap is set too small, bend the outer electrode away from the center electrode until the gapping tool fits snuggly between the two electrodes. Some spark plugs, such as those made with iridium, are not to be adjusted; if the gap is wrong, they are to be replaced, as they break very easily.

To replace the spark plugs, follow the steps in **SKILL DRILL 61-2**.

Replacing the Spark Plug Wires or COP Boots

The spark plug wires, high-tension leads, or COP boots should be replaced periodically. There is not usually a recommended

SKILL DRILL 61-1 Using Service Information

1. Obtain the website address for the service information system used by your repair shop. There may be a user name and password for the business as a whole or each technician may have individual login information.
2. If not using an online service, obtain the service manual for the vehicle you are servicing.
3. Using the year, make, and model of the vehicle, locate the section that covers the ignition system. This section is usually located within the "Engine Performance" or "Engine Electrical" section; it may also be within the "Engine Controls" section.
4. Locate and record the specifications for the pickup sensors, spark plug wires, ignition coil (primary and secondary windings), or any other related component. The specifications will be in ohms or volts, or sometimes in frequency (Hz).
5. Print out or write down the specifications for use at the vehicle while you are testing.
6. Research the TSBs to see if there are any updated procedures, parts, or related symptoms that can guide you through your task.

SKILL DRILL 61-2 Replacing the Spark Plugs

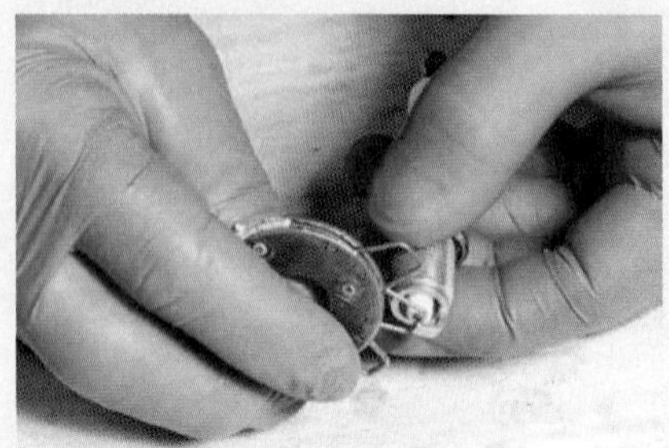

1. Check the electrode gap of each spark plug by finding the correct-sized gauge on the gapping tool, and attempt to place it between the center and the side electrode.

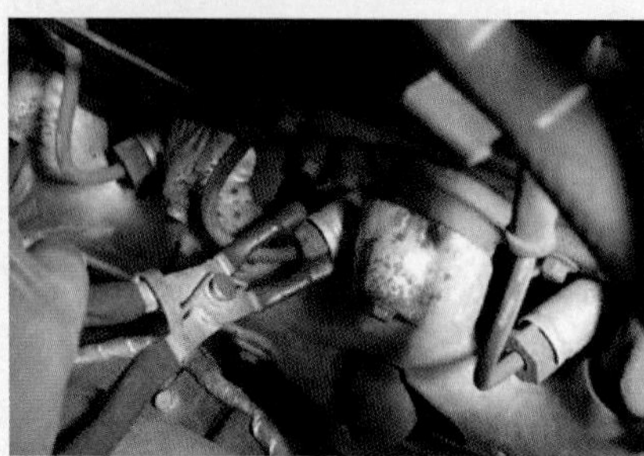

2. To gain access to a spark plug, first remove the spark plug wire by gripping its boot by hand, or with a high-tension wire puller, and gently pulling it off the spark plug.

3. Clean out the spark plug pocket with compressed air.

4. Remove the spark plug, using the correct-sized spark plug socket.

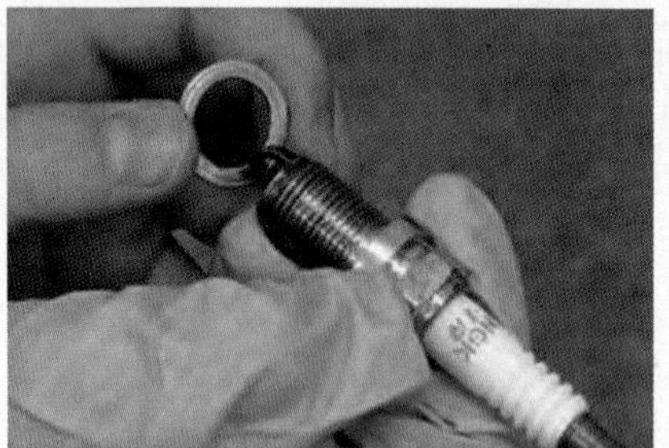

5. If the new plugs are equipped with a gasket, ensure they are installed properly. Install the spark plugs by hand to avoid cross-threading. Use a torque wrench to properly tighten the spark plug.

6. Once the spark plug is installed, reinstall the spark plug wire by pushing it on to the terminal of the spark plug until it snaps into place.

replacement interval stated in the owner's maintenance schedules for spark plug wires. Many shops recommend replacement at about the 100,000-mile (160,000 km) range. When removing spark plug wires, twist the boots to break them loose from the spark plug. Also make sure you remove them by pulling on the boot and not the wire; otherwise, the terminal can pull off the end of the wire. Spark plug boot pliers can help you reach down to grab the boot in arrangements where the boots are obstructed.

When replacing a set of plug wires, replacing one spark plug wire at a time will help prevent you from getting the wires and cylinders mixed up. It is also good to lay out the wires shortest to longest so that you can match up the wires if they are not numbered. When reinstalling the plug wires, some spark plug boots should have a small amount of high-temperature dielectric grease smeared on the inside to prevent the boots from sticking to the spark plug porcelain. In some cases, you need to burp any trapped air from the spark plug wire boots with a small, dull screwdriver or thin wire so that the trapped air does not push the boots back off after they are installed.

To replace the spark plug wires, follow the steps in **SKILL DRILL 61-3**.

▶ Diagnosis

N61003

A healthy ignition system is critical to the proper operation of an engine. Faulty or worn ignition parts may contribute to hard starting, no starting, lack of power, rough running, excessive emissions, and other drivability problems. If a vehicle runs poorly or will not start, a thorough understanding of the customer concern will be necessary. Many shops use a drivability worksheet that the customer fills out, listing the concern and as many of the details about the condition as possible. The worksheet steps the customer through a series of questions designed to help the him or her be thorough in describing the concern. This worksheet can be very helpful to the technician diagnosing the vehicle. Once you understand the concern, you have to verify it, typically by performing a test drive if the vehicle is operational and it is safe to do so.

Most vehicles on the road today are equipped with an OBDII-compliant computer system that can greatly help the technician diagnose the vehicle. If so, then the PCM will monitor the engine for misfires on each cylinder as well as individual ignition coil operation. Therefore, connecting a scan tool to the vehicle's data link connector (DLC) and retrieving any DTCs

SKILL DRILL 61-3 Replacing the Spark Plug Wires

1. Remove one end of the wire (or boot) from the distributor cap, coil pack, or COP coil. Remove the other end from the spark plug with a high-tension wire puller by twisting the boot on the spark plug while gently pulling it off.

2. If dielectric grease is specified for the spark plug boots, use a cotton swab to spread a small amount around the inside of the boot.

3. Install the new spark plug wire by pushing one end onto the spark plug and the other onto the distributor cap, coil pack, or COP coil. The spark plug wires should be routed in their factory positions to prevent damage.

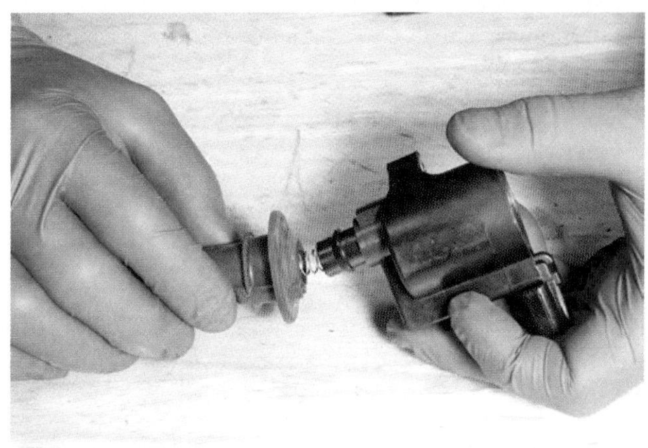

4. If replacing COP boots, carefully work the boot over the secondary terminal until it is fully seated.

will generally give the technician some good clues on where to start. DTCs can be researched in the service information, along with TSBs, to formulate a focused testing plan.

If the vehicle is not equipped with an OBDII system, and an ignition system fault is suspected, it is good to perform a spark output test to see if there is enough voltage to jump the spark plug gap inside the combustion chamber. Realize that it takes a much higher voltage to create a spark that will jump the spark plug gap when it is under the compression forces inside the cylinder than it does outside the cylinder. Just because a spark can be made to jump the gap of a spark plug when it is removed from an engine does not mean it will do so inside the combustion chamber. Use a spark tester to verify that the spark is able to jump a $^{3}/_{8}$" to $^{1}/_{2}$" (9.5 to 12.7 mm) gap. If it can jump that gap outside the engine, it is strong enough to jump across the spark plug gap inside the engine. Just be sure to inspect the spark plug to make sure it is not contaminated (fouled) and that the ceramic center is not broken and falling down, covering up the gap between the electrodes.

If the spark test shows a weak spark, inspect the rotor and the inside of the distributor cap for cracks and carbon tracks. If no cracks are present, you will need to refer to the service information for further testing procedures. Replace any parts that are worn, cracked, or broken.

The condition of the spark plugs, when viewed on the electrodes side, can give a good indication of major engine problems. For example, if the electrodes are covered in heavy, crusty, brown deposits, there may be worn valve seals or oil-control piston rings, allowing oil to be burned in the combustion chamber. If the electrodes are covered with heavy, crusty, black carbon deposits, the engine may be running excessively rich.

These tests and more are covered in the information below. And don't forget to test-drive the vehicle once the repair is completed in order to verify that you have resolved the concern.

Testing Tools and Equipment

- Spark testers are used to test the condition of spark plugs and to see whether there is adequate spark getting to the spark plug. On a coil-on-plug system, a special coil-on-plug tester is used to check the coils for output (**FIGURE 61-60**).
- A standard test light is used to test for power and grounds within the primary circuit. It is a light spliced into a wire and used as an indicator of the presence of voltage. It has a spring-loaded clip on one end and a probe on the other (**FIGURE 61-61**).
- An LED test light is used where circuits are not designed to withstand the current flow that a standard test light allows to flow. This would typically be low current triggering circuits (**FIGURE 61-62**).
- A digital multimeter (DMM) is frequently required to test for power, ground, and continuity within the ignition primary circuit and continuity in secondary circuits (**FIGURE 61-63**).
- A scan tool used to retrieve DTC, freeze-frame data, and data stream information (**FIGURE 61-64**).
- An **oscilloscope** or lab scope is used to observe the electrical patterns created by the triggering devices and ignition coil (**FIGURE 61-65**).

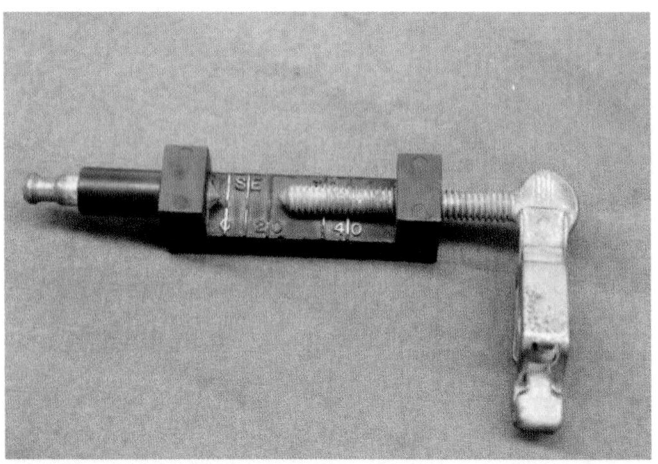

FIGURE 61-60 Spark testers.

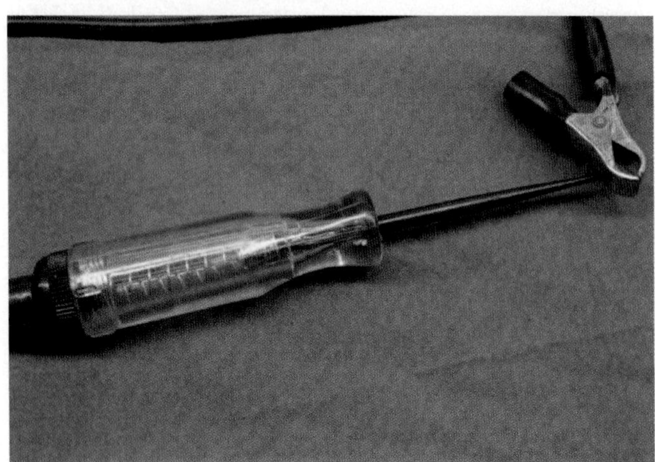

FIGURE 61-61 Standard test light.

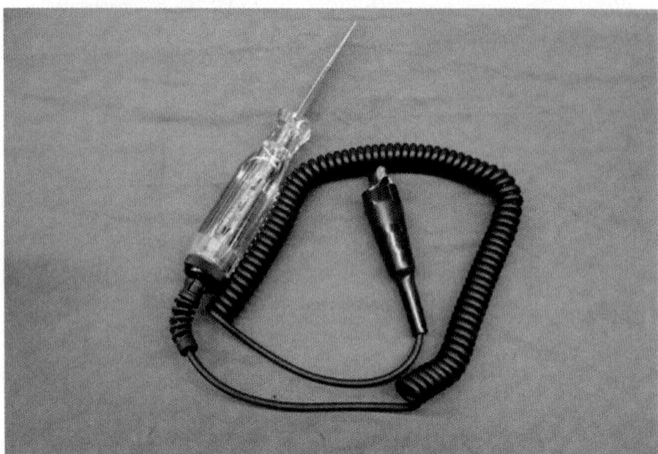

FIGURE 61-62 LED test light.

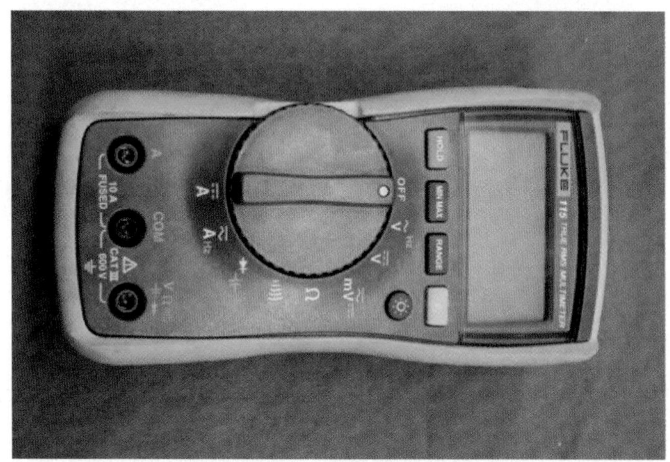

FIGURE 61-63 Digital multimeter (DMM).

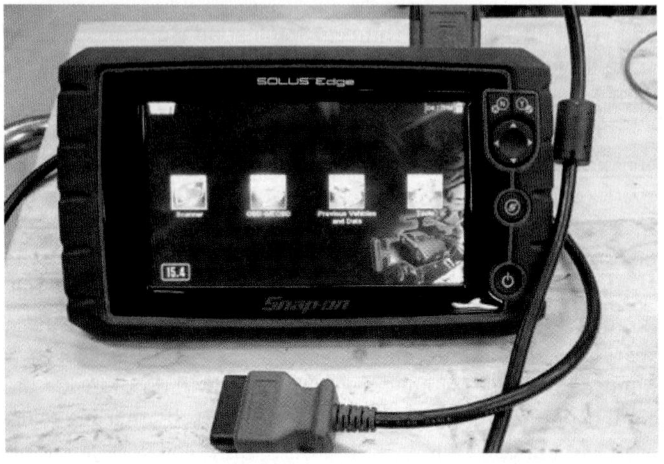

FIGURE 61-64 Scan tool used to retrieve DTC, freeze-frame data, and data stream information.

FIGURE 61-65 Oscilloscope.

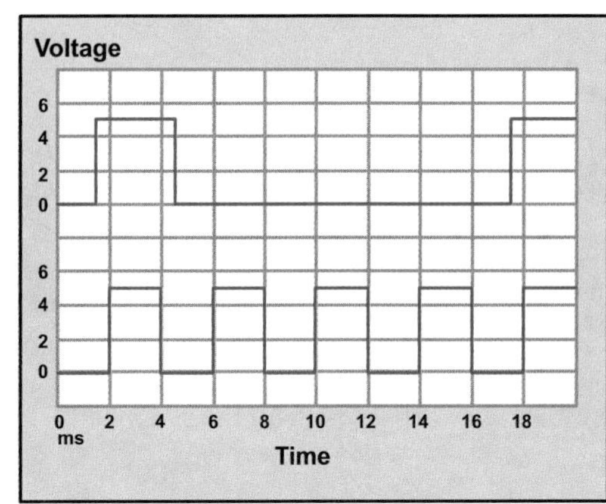

FIGURE 61-66 Lab scopes display. Voltage is displayed vertically. Time is displayed horizontally.

SAFETY TIP

While the secondary voltage is very high (up to 100,000V), the current flow is very low. Because the current flow is so low, an ignition system shock is unlikely to kill you directly (unless you have a pacemaker), but it could hurt you. If you get shocked by the high voltage, it will hurt, and you will likely jerk away from it. This can cause you to get tangled up in the fan belts, or ram your head into the hood safety latch, which could seriously injure you. Always respect secondary ignition systems, and work around them carefully.

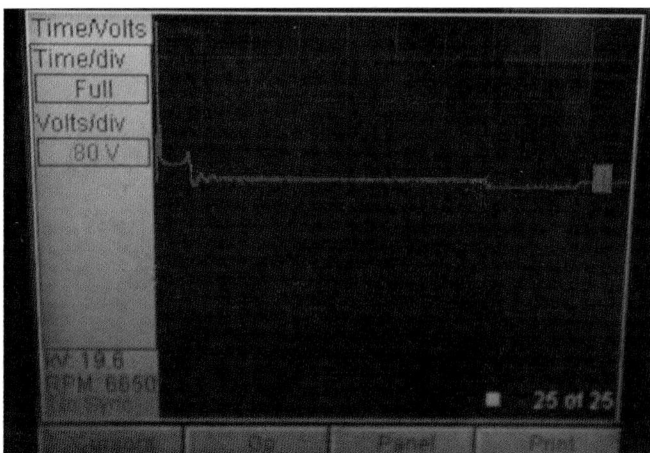

FIGURE 61-67 Typical lab scope primary coil-on-plug pattern.

FIGURE 61-68 Adapters being used to adapt an oscilloscope to a COP ignition coil.

Lab Scope/Oscilloscope Usage

One of the primary tools for evaluating the condition of ignition systems is the oscilloscope or lab scope. These tools display voltage in relation to time on a screen. Voltage is displayed on the vertical axis, and time is displayed on the horizontal (**FIGURE 61-66**). For example, if the voltage needed to maintain the spark across the gap of a number-one spark plug is 2000 volts, and that voltage is maintained for 3 milliseconds (0.003 second), the vertical axis will display the voltage, and the horizontal axis will display the time that the voltage is that high.

The oscilloscope can be used to test both the primary and the secondary circuits, as they tend to mirror each other. Although the oscilloscope historically has been most frequently used to test the condition of the secondary side of the circuit, with the advent of coil-on-plug systems, where the secondary portion of the circuit is not as accessible, the lab scope is now being used more on the primary side of the circuit (**FIGURE 61-67**).

At the same time, new adapters have been developed to adapt oscilloscopes to display secondary COP ignition patterns. They consist of paddles, foils, or probes that are held against the side or top of the COP coil to sense the voltage created in the secondary ignition circuit (**FIGURE 61-68**). This gives a good look at the secondary ignition pattern.

On distributor type vehicles, when testing the secondary side of the circuit, the oscilloscope can display several scope patterns. A **parade pattern** shows all the cylinders firing in sequence (**FIGURE 61-69**). It shows how much voltage is required to fire each spark plug and displays those voltages for every spark plug on the screen. If the voltage to one of the cylinders is too high, too low, or nonexistent, it can quickly be identified by viewing the "parade."

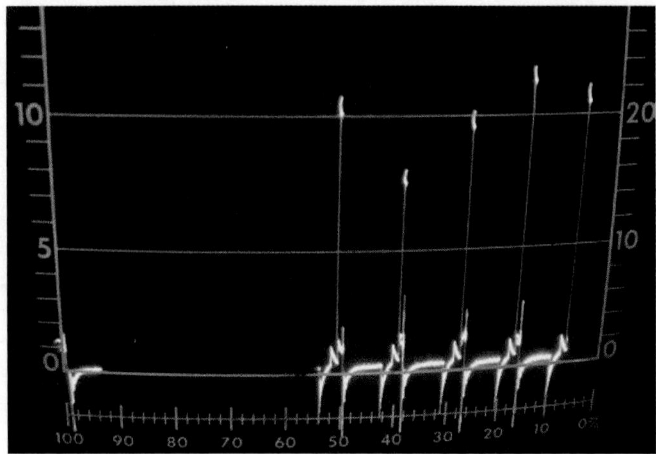

FIGURE 61-69 Typical parade scope pattern.

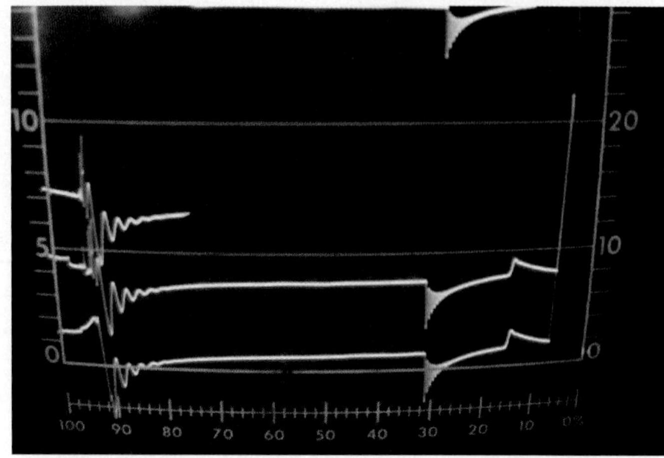

FIGURE 61-71 A raster pattern focusing on the spark line faults.

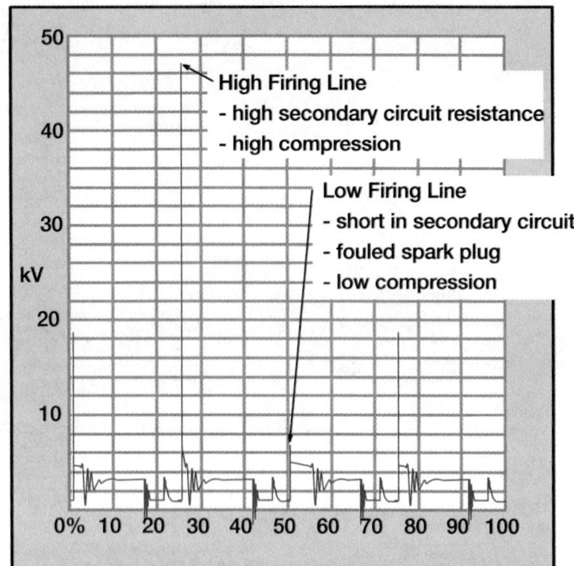

FIGURE 61-70 Differences in the firing lines on Parade pattern indicate high resistance, open circuits, short circuits, and high and low compression.

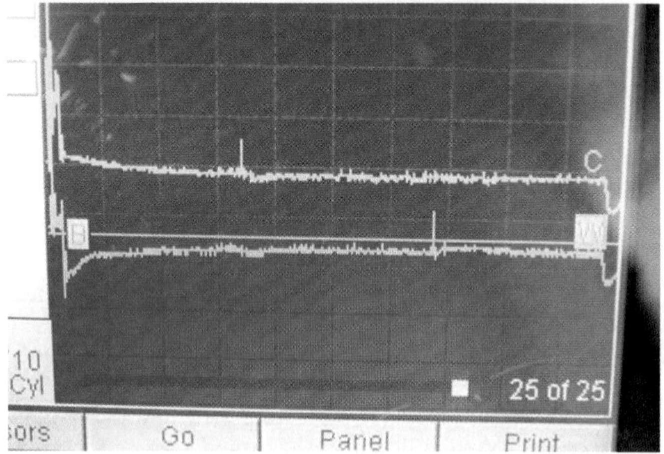

FIGURE 61-72 Oscilloscope display for a secondary pattern on a waste spark ignition.

When viewing the parade pattern, the tall lines indicate the voltage required to ionize the spark plug gap. This is called the **firing line**. A high firing line on just one cylinder indicates high resistance in that spark plug circuit, typically from a wide gap in the secondary side of the circuit (**FIGURE 61-70**). This may also be due to an open or disconnected spark plug wire. A low firing line indicates low resistance, a short in the circuit, or low compression. Low resistance can be caused by a shorted spark plug wire or a fouled spark plug. On ignition systems that use a distributor, if the coil wire or rotor has a large gap, then all of the firing lines will be high.

Many oscilloscopes have the ability to stack each spark plug's ignition pattern vertically, which is called **raster** (**FIGURE 61-71**). Raster allows a technician to compare **spark lines** with one another, which tells how much voltage it takes to keep the spark burning and for how long. Raster is also helpful for viewing the timing of each event to events in the other cylinders, such as the firing line. If everything is good, all of the firing lines should be straight above one another. If there is wear on the timing chain or the distributor shaft is bent, then the firing lines will have an S shape to them. The greater the S, the bigger the problem, as the firing line occurs when ignition occurs in the combustion chamber. Therefore, any difference in firing line alignment means that the cylinders are not firing at the same time in the cycle. In this case, cylinder 1 may be properly timed, but all of the other cylinders may be in an advanced or retarded condition.

Waste spark ignitions can also be displayed on an oscilloscope if you are using the correct adapters. In fact, both the firing lines and spark lines for the event cylinders and waste cylinders can be compared to each other. Faults will be displayed similar to the distributor type firing lines and spark lines (**FIGURE 61-72**).

Spark Testing

S61002

A spark test should be performed when an engine will not start, has one or more cylinders misfiring, or has other drivability problems, such as a lack of power, or stalling. A spark tester can look like a regular spark plug, but it has a large alligator clip attached to its side, or it can be an adjustable tester that allows you to adjust the gap (preferred style). If the spark jumps across the specified gap of the spark tester, then the spark is big enough to jump across the spark plug gap in the combustion chamber. You may need to check more than one spark plug wire to verify that the spark is good in all wires.

To perform a spark test, follow the steps in **SKILL DRILL 61-4**.

Inspecting the Ignition Primary and Secondary Circuits

S61003

Before digging into the primary and secondary winding, verify the condition of the spark directly out of the coil, if this has not already been performed. If the spark is good out of the coil, but not out of the spark plug wire, then check the distributor cap, rotor, and spark plug wire. If there is no spark or the spark is weak, then the primary wiring should be checked for cuts, abrasions, and signs of arcing. The primary wires are the two to four smaller wires attached to the ignition coil(s), and the secondary wires are the larger wires leading from the coil pack to the spark plugs or from the distributor cap to the spark plugs, depending on the type of ignition system. Loose or corroded wires can also cause an ignition system to perform poorly. If the visual inspection checks out okay, then the primary circuit will have to be tested for power, continuity, and switched ground (also called triggering).

To test for power, switch the ignition on, and connect the red lead of the multimeter to the positive terminal of the coil, and the black lead to an engine ground. A typical reading is generally above 12 volts. If using a test light, the light should illuminate brightly. If you are dealing with a no-start condition, you should check that the power remains at that terminal while the engine is cranked over. If no power is present, you will have to consult a wiring diagram and follow the circuit back toward the battery, especially any relays or fuses. If power is available, continue on.

To test whether the coil primary circuit has continuity, place the red meter lead or test light on the negative side of the coil and the black lead on a good ground. With the ignition switch in the Run position, there should be at least 12 volts or a bright test light. If not, either the coil primary is open, or the primary output wire is grounded. Measure the resistance of the coil primary winding to see if it is open. Also compare your reading to specifications. If the primary winding is within specifications, check the primary output wire for a short to ground.

If power is present at the coil negative terminal, then keep the meter or light connected as above, and crank the engine and look for the meter to **oscillate** or the test light to flash rapidly, indicating the coil is being switched on and off. If so, check the secondary circuit of the coil. If the light or meter stays on and doesn't oscillate, then the ignition module, PCM, pickup assembly or crankshaft position sensor, and wiring will have to be tested further.

To inspect the primary and secondary circuits, follow the steps in **SKILL DRILL 61-5**.

Testing the Ignition Coil

S61004

Whether an engine uses a single ignition coil, an **ignition coil pack**, or one coil for each cylinder, it is critical that the ignition coil(s) are operating properly. A faulty ignition coil may cause hesitation under acceleration, prevent an engine from starting, or cause a misfire. Ignition coils can fail in a few ways. One way is that the primary and secondary windings may be open, shorted, or have high resistance. A DMM is used to measure the resistance of the primary and secondary windings and compared to factory specifications. If out of specifications, the coil will need to be replaced.

Another way ignition coils can fail is breakdown of the insulating material. This can happen either internally or externally. A common external location for this is the secondary terminal of the coil. It can become carbon tracked and allow voltage to leak across the carbon track to ground. On COP coils, the spark plug boot can get a small hole burned through it.

SKILL DRILL 61-4 Performing a Spark Test

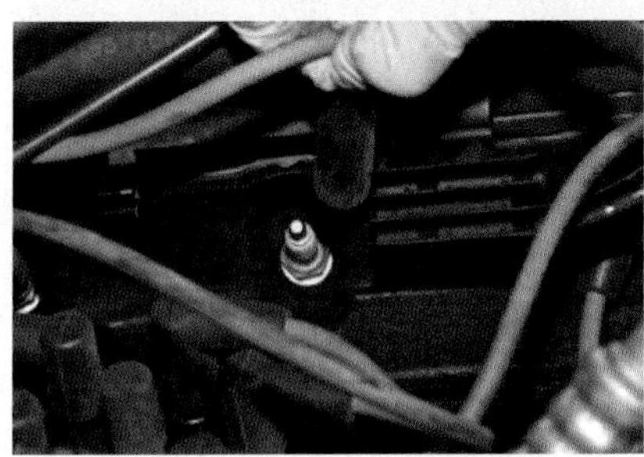

1. Remove the high-tension lead or coil from the end of the spark plug.

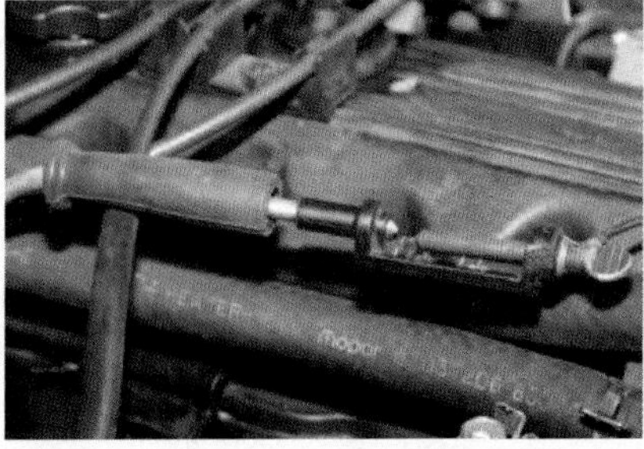

2. Connect a spark tester to the boot of the coil or high-tension lead, and attach the clamp to a good ground.

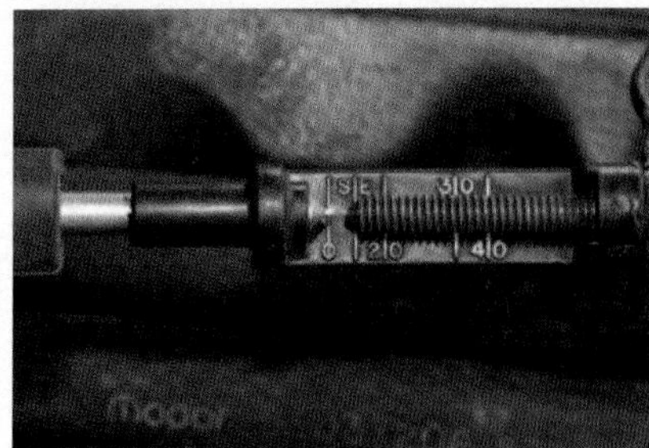

3. Crank the engine, and watch for a blue spark across the tester electrodes. If there is no spark present, continue on to the next skill drill.

SKILL DRILL 61-5 Inspecting the Primary and Secondary Circuits

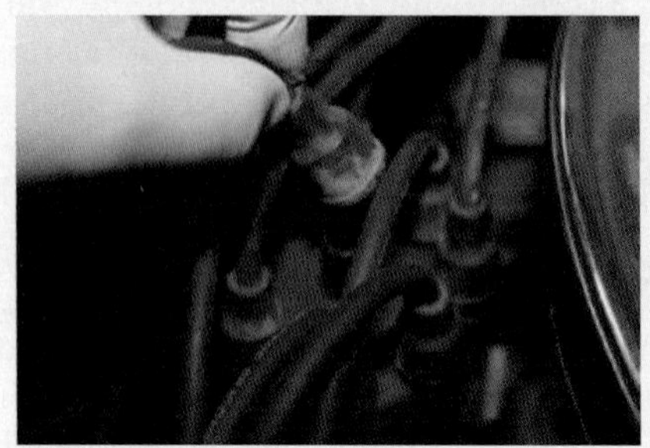

1. Remove the secondary wire from the center terminal of the distributor cap, install a spark tester, and crank the engine to see if there is any spark coming out of the ignition coil.

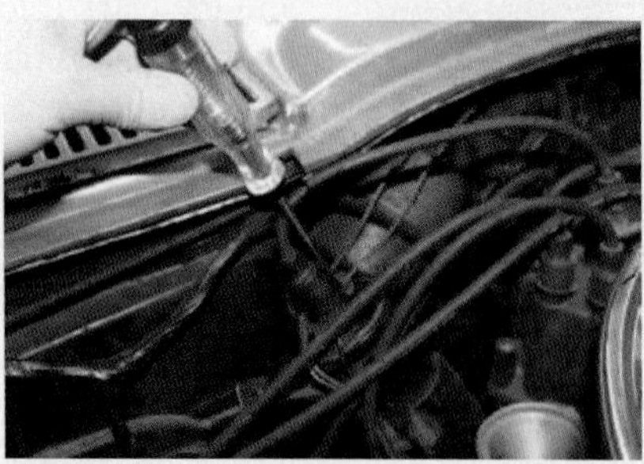

2. Use a test light, or connect the red lead of a DMM to the positive terminal at the coil and the black lead to a good ground with the key switched on. A typical reading is 10 to 12 volts. If using a test light, the light should illuminate brightly.

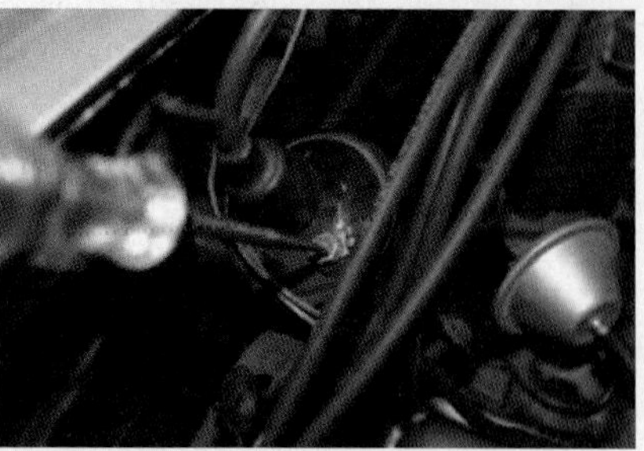

3. Place the red meter lead or the test light probe on the negative side of the coil and the black lead on a good ground. With the key on Run, the light should be on steady. While cranking the engine, the test light should blink on and off, or the meter reading bounce around, indicating the coil is being triggered.

An internal breakdown of the insulation can also occur allowing the voltage to leak to the primary circuit or ground. Break down of the insulating materials results in low available voltage output from the coil. Finding this fault requires the coil to be activated, and the available voltage to be measured on an oscilloscope and compared to specifications. If there is a weak spark or no spark during the spark test, and the spark plug wires and primary circuit have been tested and are found to be satisfactory, the ignition coil may be faulty, and needs to be tested.

To test a two-wire ignition coil primary and secondary winding, you will need to disconnect the wires from the coil and connect an ohmmeter to the coil's positive and negative terminals. A typical reading for a standard ignition coil is less than 5 ohms. A typical reading for a low-impedance coil is 0.7–5.0 ohms. Always research the specifications for the vehicle you are using.

To test the secondary windings, place one ohmmeter lead to the secondary tower terminal and the other to one of the primary terminals or the coil secondary ground, if equipped. This could be the negative terminal of the primary winding, or the case of the ignition coil. In general, the readings should be between about 5000 and 15,000 ohms. But again, research the specifications. If the readings are not within the manufacturer's specifications, replace the ignition coil.

To test a two-wire ignition coil, follow the steps in **SKILL DRILL 61-6**.

To test a waste spark ignition coil, follow the steps in **SKILL DRILL 61-7**.

SKILL DRILL 61-6 Testing the Ignition Coil

1. Visually inspect the ignition coil(s) and wires. All the coil terminals should be clean, secure, and free of corrosion.

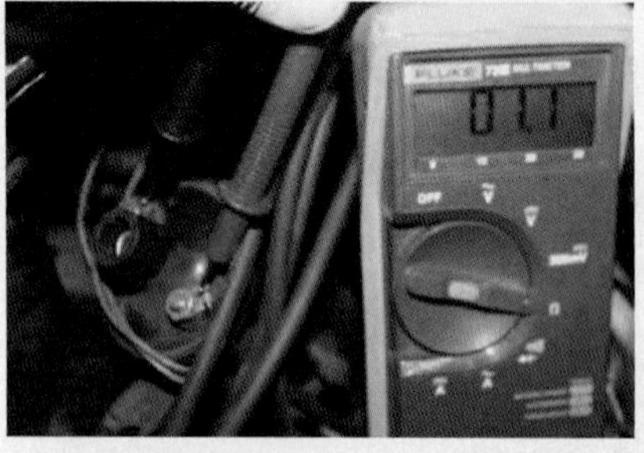

2. To test the resistance of the primary windings, place the ohmmeter leads on each of the two primary winding terminals. If the reading is not within specifications, the coil is faulty and must be replaced.

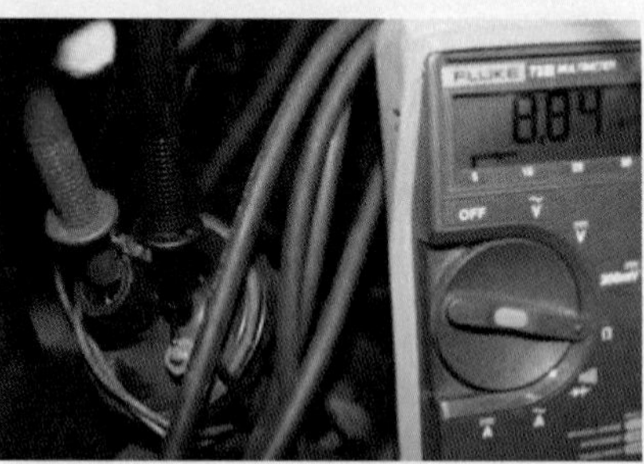

3. To test the secondary windings, place one ohmmeter lead on the secondary tower terminal and the other to the negative primary terminal, or the coil secondary ground, if equipped. If the reading is not within specifications, the coil is faulty and must be replaced.

SKILL DRILL 61-7 Testing a Waste Spark Ignition Coil

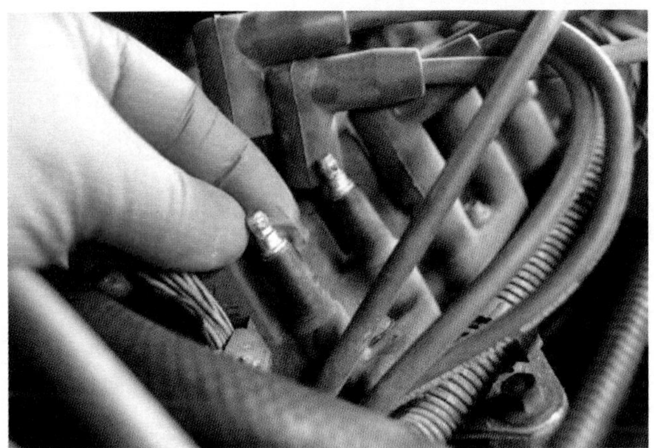

1. Visually inspect the ignition coil(s) and wires. All the coil terminals should be clean, secure, and free of corrosion.

2. Measure the resistance of the secondary winding at the two secondary terminals. If out of specifications, replace the ignition coil.

3. Remove the coil from the ignition module.

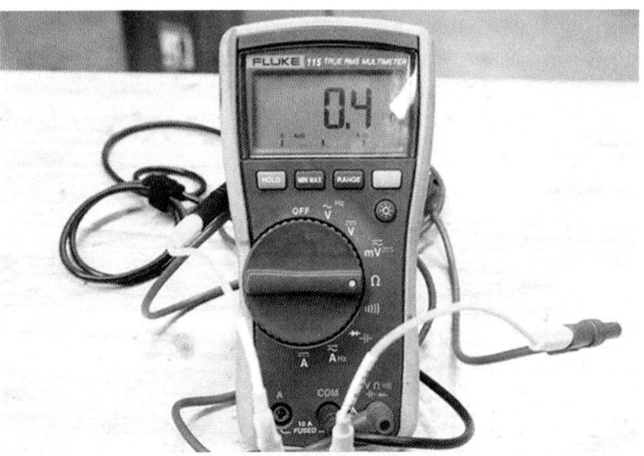

4. Measure the resistance of the primary circuit across the two primary terminals. If out of specifications, replace the ignition coil.

To test a three-wire or four-wire COP ignition coil, follow the steps in **SKILL DRILL 61-8**.

Testing the Pickup Assembly

S61005

A pickup assembly is typically tested when a vehicle will not start because of a lack of spark. You need to know which style of pickup assembly the vehicle is equipped with: magnetic, Hall effect, or optical. There should be an on/off signal being generated at the ignition coil's negative terminal. If there is not, the pickup assembly should be tested.

To check the resistance of a magnetic pickup assembly, disconnect any wires connected to the assembly. Measure the resistance between the two terminals of the pickup coil by placing the leads of an ohmmeter to each terminal of the pickup coil. If the reading is within specifications, wiggle the pickup coil wires inside and outside of the distributor while watching the ohmmeter to see if there is any change in the resistance caused by a broken or partially broken wire.

To check the pickup coil pattern on a lab scope, connect the scope to the pickup coil windings. Crank the engine over, and observe the alternating current pattern. If the reading or pattern is not within specifications, the pickup coil is faulty and should be replaced.

To perform a test of the magnetic pickup assembly, follow the steps in **SKILL DRILL 61-9**.

Testing the Spark Plug Wire

S61006

A faulty spark plug wire may cause a misfire, high exhaust emissions, poor fuel economy, and/or a lack of power. Over time, spark plug wires become brittle and frayed and tend to have higher resistance; internally, they can burn in two, causing an open. This can be checked by placing one lead of an ohmmeter on each end of the spark plug wire. Flex the wire while measuring to detect a broken conductor within the wire. A good wire will typically have a maximum of 3000 to 7000 ohms of resistance per foot of wire, but check the specifications. An open

SKILL DRILL 61-8 Testing a Three-Wire or Four-Wire COP Ignition Coil

1. Visually inspect the ignition coil(s) and wires. All the coil terminals should be clean, secure, and free of corrosion.

2. Measure the voltage at the power and ground terminals of the ignition coil in the run and crank positions. If not within 0.5 volt of battery voltage, research the wiring diagram and perform voltage drop tests on the power and ground circuits.

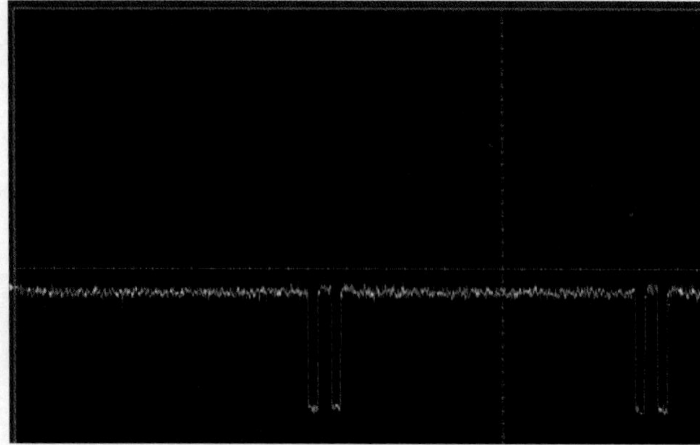

3. Connect a lab scope to the trigger wire, crank or start the engine and inspect the pattern. Compare to a known good pattern. If faulty, check the PCM, crankshaft position sensor, and wiring.

4. If a four-wire COP coil, research the specified testing process for testing the fourth wire, and follow that process carefully.

SKILL DRILL 61-9 Testing a Pickup Coil

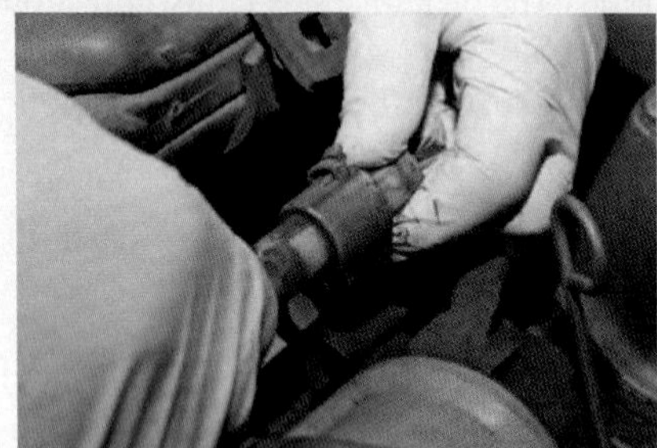

1. Disconnect the electrical connector from the pickup coil.

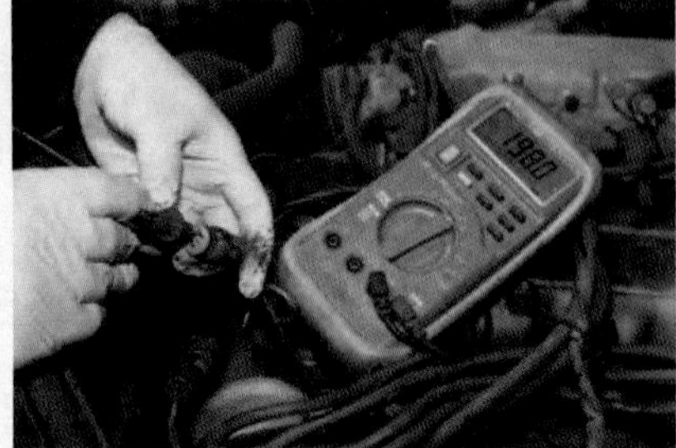

2. Measure the resistance between the two terminals of the pickup coil with an ohmmeter. If the reading is not within specifications, replace the pickup. If the reading is within specifications, wiggle the pickup coil wires while watching the ohmmeter reading. It should not change.

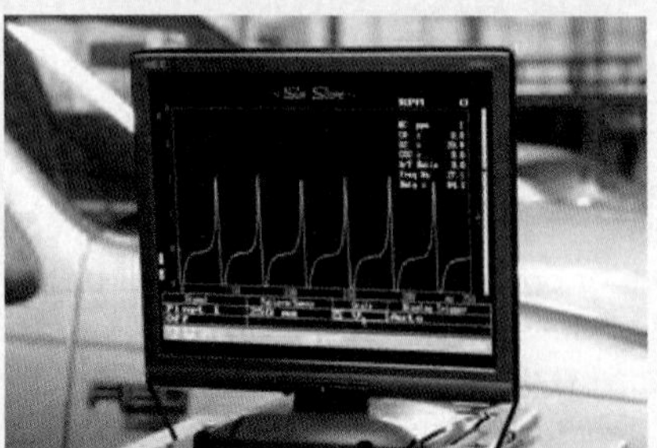

3. Connect a lab scope to the pickup coil windings. Crank the engine over, observe the pattern, and compare it to a known good pattern.

wire will read OL or OR or 1. An open spark plug wire can cause the engine to misfire. Misfires can cause catastrophic damage to the catalytic converters on the vehicle, so misfires are serious issues if left undiagnosed and not repaired.

If the vehicle has an intermittent misfire and is equipped with spark plug wires, you can use a spray bottle with plain water to mist the wires while the engine is running. Or you can use a grounded test lead and run it along the wires with the engine running. If the spark plug wires have a weak spot in the insulation, you should be able to see and hear the spark jumping to another wire, the test lead, or conductive engine parts. If the vehicle is equipped with a coil-on-plug system, or if the spark plug wires fit down inside of a recessed well, then you can use insulated pliers to carefully lift the coil or plug wire boot off the spark plug while the engine is running in order to see whether the spark is jumping through a hole or crack in the boot to the side of the spark plug well or the grounded test lead.

To test a spark plug wire, follow the steps in **SKILL DRILL 61-10**.

Inspecting the Distributor Cap and Rotor

S61007

A defective distributor cap or rotor can also prevent a spark plug from firing or can cause a weak spark. If an engine has a misfire or exhibits a lack of power, poor fuel economy, or high emissions, the distributor cap and rotor should be inspected. Distributor caps and rotors become carbon tracked over time. This means that the high voltage ionizes the surface of the plastic parts and eventually can create a path to ground or to another spark plug terminal. This carbon track allows electricity to flow along that path instead of firing the proper spark plug. Carbon tracks look like small hairline cracks but can become much more noticeable over time. A faulty distributor cap or rotor may also cause an engine to run rough, backfire, or not start at all.

To inspect a distributor cap and rotor, follow the steps in **SKILL DRILL 61-11**.

SKILL DRILL 61-10 Testing a Spark Plug Wire

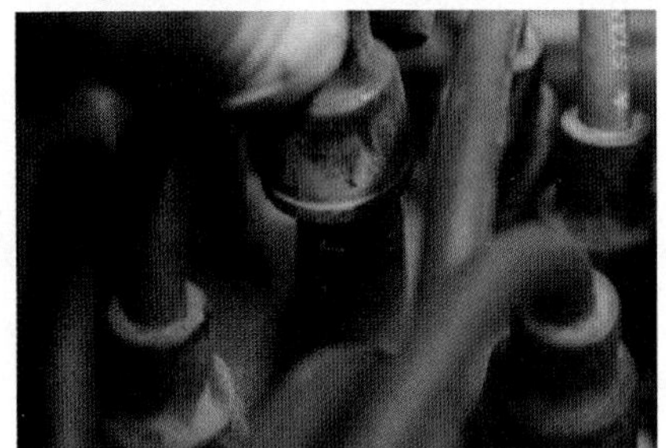

1. Disconnect the spark plug wire at both ends by grasping the boots on the ends and twisting while pulling the wire off. Inspect each wire for cracks, brittleness, or burnt spots.

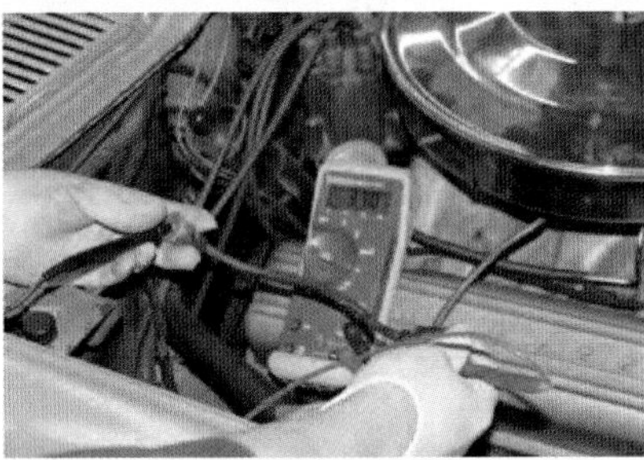

2. Place one lead of the ohmmeter on each end of the spark plug wire. Flex the wire while reading the ohmmeter.

3. Replace the spark plug wires, start the engine, and lightly mist water on the spark plug wires, or run a grounded test lead along each wire. If the wires arc, replace the spark plug wires, making sure to route them in their factory positions to prevent damage.

SKILL DRILL 61-11 Inspecting a Distributor Cap and Rotor

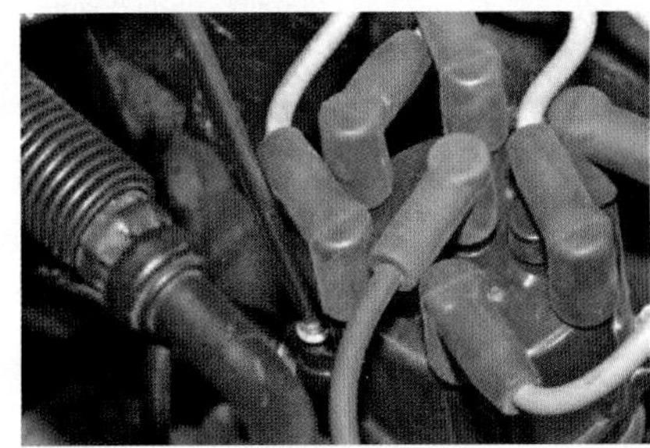

1. Remove the distributor cap from the top of the distributor by unscrewing it or removing the clips.

2. Check for cracks, carbon tracks, and burned terminals, and also check the center terminal for wear.

3. Check the rotor for burned contacts and cracks in the plastic housing. If the rotor is held in place by a retaining screw, make sure it is tight.

Testing Crankshaft and Camshaft Position Sensors

N61004

A defective crankshaft or camshaft position sensor typically prevents an engine from starting. In some cases, it causes the engine to die intermittently. If an engine cranks but will not start, the crankshaft and camshaft position sensors should be tested after performing a spark test. Note that some crankshaft position sensors are adjustable and will not function properly if they are out of adjustment. If a DTC indicates a problem with the crankshaft or camshaft position sensor, do not assume that the sensor is faulty. Test the entire circuit, including the wiring, according to the manufacturer's procedures. The crankshaft and camshaft position sensor can be tested with a lab scope or digital multimeter.

To test the crankshaft and camshaft position sensors, follow the steps in **SKILL DRILL 61-12**.

Testing the Ignition Module

N61005

On modern engines, the spark at the spark plugs is controlled by the ignition module and/or the PCM. If either of these parts is defective, the engine will not start or will run erratically. Some modules are grounded through their mounting hardware, so it is important that the module is clean and fastened securely. The ignition module may be mounted to the side or inside of the distributor, or it may be mounted under the ignition coils with the coils bolted to it. Many manufacturers recommend that heat-conductive silicone grease be applied to the bottom of the ignition module prior to installation to help dissipate heat and prevent overheating the module. Both the ignition module and the PCM are sealed units and cannot be serviced. If they are defective, they must be replaced.

To test an ignition module, follow the steps in **SKILL DRILL 61-13**.

SKILL DRILL 61-12 Testing the Crankshaft and Camshaft Position Sensors

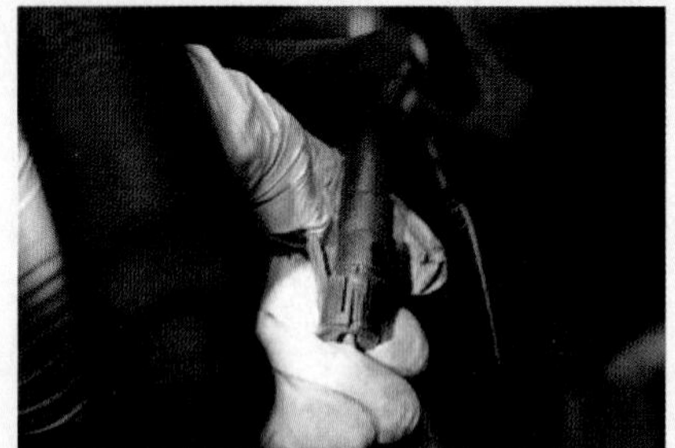

1. Remove the connector from the sensor by removing any lock clips that may be present, and pull the connector off.

2. If it is an inductive-type sensor, use an ohmmeter to read the resistance, and compare the readings to the specifications. Replace the sensor if its resistance is too high or too low, indicating that it is defective.

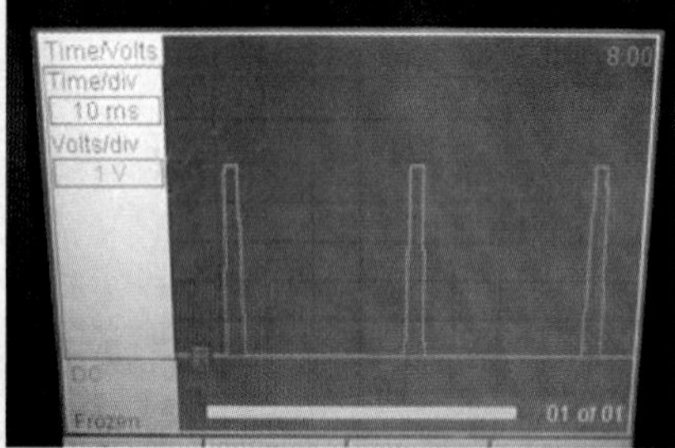

3. If the sensor is a Hall-effect or optical sensor, or the previous reading is normal, use a lab scope to observe the sensor signal, and compare to a known good pattern. If faulty, replace the sensor.

SKILL DRILL 61-13 Testing an Ignition Module

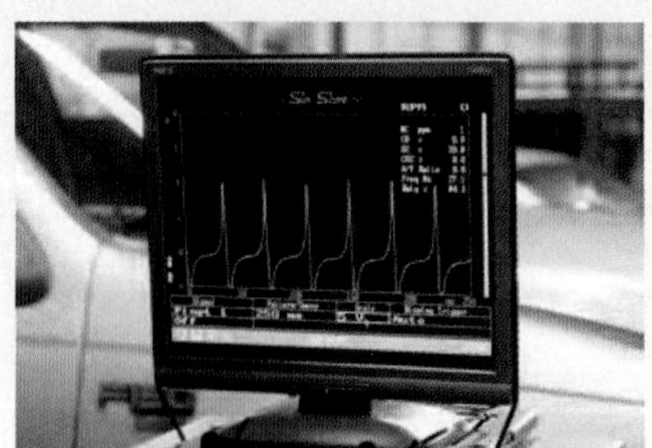

1. Test all the input wires leading to the ignition module for proper voltage and ground signals.

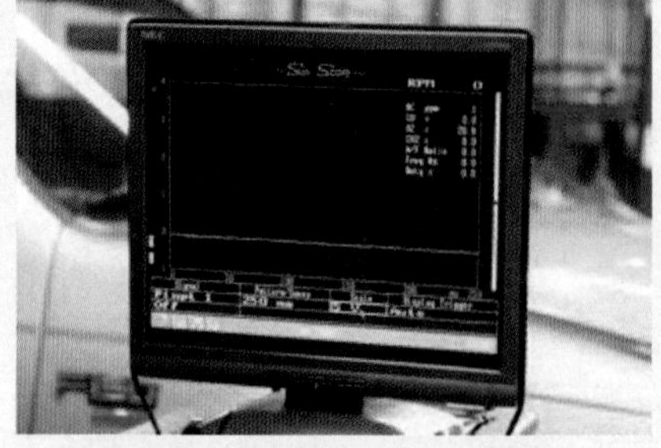

2. Check for proper output signals from the ignition module with the key set to run and while cranking. If the specified inputs are present but the ignition module will not produce the proper output signals, the ignition module should be replaced.

3. Check the voltage drop between the ignition module and the engine ground with a DMM. If the voltage drop is more than 0.1 volt, check for corrosion or rust between the ignition module and the mounting surface. Clean and reinstall the ignition module.

> **TECHNICIAN TIP**
>
> The electronic circuits inside of an ignition module are more prone to failure when hot. Some cars and light trucks die but then start right up once towed to the repair shop. This makes it difficult for the technician to diagnose. If a vehicle dies but will restart after it cools down, the ignition module may be at fault. This intermittent condition can quickly be tested by placing a few ice cubes on the ignition module. If the vehicle starts while the vehicle is still hot with the ice cubes on the ignition module, then use a heat gun to carefully warm the module back up. If the engine dies, the ignition module is faulty and should be replaced.

Wrap-Up

Ready for Review

- Ignition systems have a primary (low-voltage) circuit to connect and disconnect the ignition coil and a secondary (high-voltage) circuit to send voltage from the battery to the spark plug.
- Common components of an ignition system are the spark plugs, high tension leads ignition coil, and ignition-coil triggering device.
- Principles of an ignition system involve a circuit activating the ignition coil, which converts the battery's low voltage to high voltage. The voltage then travels to the spark plug in each cylinder, igniting the air-fuel mixture and creating enough pressure to push down the pistons.
- The original contact breaker ignition system was replaced with an electronic ignition system of the distributor type.
- Modern systems are direct ignition systems (or coil-on-plug), which have a dedicated coil for each cylinder.
- The amount of available voltage should always be higher than the amount of required voltage.
- As ignition components age and become worn, required voltage increases and available voltage decreases.
- Spark timing is affected by any of these factors: air-fuel ratio, detected knock, engine speed, engine load, engine temperature, and throttle position.
- Spark timing must be correct in order to give the air-fuel mixture enough time to burn.
- Modern vehicles have electronically programmed spark timing via the PCM; older vehicles mechanically advance or retard the spark timing.
- Common points on an ignition switch are the lock and off functions and the accessory, on/run, and start/crank positions.
- Automatic transmission vehicles may employ a transmission shift interlock device to ensure that the gear selector is in park before key removal can occur.
- Standard ignition coils contain primary and secondary windings around a rod-shaped laminated iron core.
- The main difference among types of ignition systems is the method used to control the primary circuit to produce the secondary spark.
- In the primary circuit, a magnetic field develops that is interrupted by the ignition triggering device, thus collapsing the field and returning stored energy to ignition coil terminals.
- Because the secondary winding has approximately 100 times as many turns as the primary winding, it can produce 100 times greater voltage.
- Basic contact breaker point ignition systems are comprised of the battery, ignition switch, ignition coil, contact breaker points, capacitor, distributor, and appropriate voltage-connecting wires and leads.
- The purpose of the distributor is to transfer the spark to the spark plugs with the correct sequence and timing.
- The distributor cap provides rotor and spark plug lead connection.
- The distributor controls ignition timing in relation to speed via a centrifugal advance mechanism and ignition timing in relation to vehicle load via a vacuum advance mechanism.
- Vehicles regulate voltage to the ignition system via a ballast resistor, which is inserted in the primary circuit and lowers the voltage.
- Spark plugs all have at least one, and up to four, side electrodes, as well as an internal resistor to suppress voltage spikes and prevent radio frequency interference.
- Spark plugs are identified by thread size or diameter, reach or length of thread, and heat range or operating temperature.
- Spark plug components include the metal case, insulator, terminal, side electrode, and center electrode.
- To work effectively, spark plugs must have proper reach (distance from sealing area to end of spark plug threads) and proper heat range (operating temperature, generally 746°F to 1460°F [400°C to 800°C]).
- High-tension leads carry a high-voltage spark from the ignition coil to the distributor cap and on to the spark plugs.
- Engine timing can refer to valve timing or ignition timing.
- How long the contacts are closed and current is flowing in a contact breaker system is determined by the dwell angle.
- Electronic ignition systems electronically trigger the primary circuit (rather than using a contact breaker).
- The triggering device of an electronic ignition system can be a magnetic pulse generator (pickup coil) or a Hall-effect switch.

- Ignition modules process information from various sensors and then interrupt the signal to the primary winding of the ignition coil.
- Induction-type systems use a magnetic pulse generator with a stator and a reluctor attached to the distributor body and shaft, respectively.
- Induction-type systems produce an alternating current voltage.
- The internal module control circuit receives the trigger signal.
- The reluctor teeth spin past the stator, turning on and off the primary winding and causing the secondary windings to generate a high-voltage spark at the end of each cylinder's compression stroke.
- Hall-effect systems use a potential difference, or voltage, which is used as a switch device.
- In a Hall-effect system, the magnetic field is alternately blocked and exposed by an interrupter ring.
- Some ignition systems use a phototransistor to receive light from an LED and transform it into a voltage output signal.
- In distributorless ignition systems, the distributor is eliminated so crankshaft and camshaft position sensors help determine when to send a signal to the coil's primary windings by monitoring which cylinder is approaching its power stroke.
- In a six-cylinder, waste spark engine, each set of two cylinders is paired with an ignition coil. The cylinders alternate between an event cylinder and a waste cylinder.
- The PCM controls spark timing for distributorless ignition systems via information from the following sensors: mass airflow sensor, manifold absolute pressure sensor, throttle position sensor, cam and crank angle sensors, engine speed sensor, knock sensor, and coolant temperature sensor.
- Special tools and equipment for ignition system diagnostics include the spark plug socket, spark plug gapping tool, high-tension wire puller, spark tester, test light, a digital multimeter, scan tool, and an oscilloscope.
- Oscilloscopes test the condition of the secondary circuit and can show if voltage to any of the cylinders is too high, too low, or nonexistent.
- The first step in diagnosing ignition system issues is to conduct a visual inspection to look for loose connections, broken or corroded terminals, or any components that need replacing.
- Inspect the primary and secondary circuit wiring for cuts, abrasions, and signs of arcing.
- Test the ignition coil for primary and secondary winding resistance.
- If the vehicle will not start, performing a spark test is a good step.
- Test to ensure that the spark plug wire, distributor cap and rotor, crankshaft and camshaft position sensors, and ignition modules are in proper working order.
- Always follow manufacturer-recommended procedure for diagnosing each ignition system component.
- Spark plugs and spark plug wires will wear over time and require replacement; check the manufacturer's recommendations for replacement schedules.

Key Terms

advance mechanism A device used to trigger an earlier spark based on engine conditions.

available voltage The maximum amount of voltage that the induction coil secondary is capable of putting out.

ballast resistor Used to limit the amount of current flowing in the ignition primary circuit.

Battery An electrochemical device used to supply voltage to a vehicle's electrical systems.

breaker plate The movable plate that the breaker points are mounted on and that pivots as the vacuum advance pulls on it.

Cam The egg-shaped lobe, machined to a shaft, used to cause opening and closing of the valves of a four-stroke cycle engine.

cam lobes Raised areas or protrusions on an otherwise round shaft.

camshaft position sensor A sensor mounted near the camshaft and used to send camshaft and valve position information to the PCM.

center electrode The electrode located in the center of a spark plug. It is the hottest part of the spark plug.

centrifugal advance mechanism An ignition timing device, located above or beneath the distributor base plate, that rotates with the distributor cam and is used to advance the spark. As engine speed rises, the flyweights on the advance mechanism are thrown outward by centrifugal force. Since the distributor cam is able to rotate on the distributor shaft, the weights act against their springs and move the distributor cam forward.

coil-on-plug A type of ignition system used on late-model vehicles that uses one coil placed above each spark plug.

compression stroke The stroke in the four-stroke cycle in which the air-fuel mixture is compressed.

concentric interrupter rings Two interrupter rings that have the same center.

contact breaker point ignition system A type of ignition system that uses a mechanical means of turning the primary circuit on and off.

contact breaker points A mechanically operated electrical switch that is fixed to the distributor base plate and opened and closed by the distributor cam with the rotation of the engine. The contacts normally form a self-contained unit, fixed to the base plate by a retaining screw engaged in a slot in the fixed contact.

crank angle position The position of the crankshaft, measured in degrees.

crank angle sensor A sensor that measures crank angle position.

cranking Rotating the engine by turning the ignition key to the start position.

crankshaft position sensor A sensor mounted near the crankshaft and used to send engine speed and position signals to the PCM.

direct ignition system May refer to a waste spark ignition or a coil-on-plug ignition system, in which the coils are directly attached to the spark plugs.

distributor The part of an ignition system that distributes the spark to the spark plugs in the correct sequence and at the correct time. It includes a distributor cap, rotor, shaft, and usually a switching device.

distributor base plate A round metal plate near the top of the distributor that is attached to a distributor housing; also called a breaker plate.

distributor cap The top portion of a distributor, used to make a connection between the spinning rotor and the high-tension leads.

distributorless ignition system An ignition system that does not include a distributor. It uses signals from the crankshaft position sensor and the camshaft position sensor sent to the PCM to determine when to send a signal to the ignition module.

dwell angle The amount of time that the primary circuit is energized, measured in degrees of distributor rotation.

dwell control section Part of the ignition control circuit that determines when the primary circuit will be switched on and for how long current will flow in the primary winding. The dwell period can be varied according to engine speed, improving coil efficiency.

electronic ignition system-distributor type An ignition system that uses a distributor but replaces the contact points with an electronic triggering device and control module.

electronic ignition system An ignition system that uses a nonmechanical (electronic) method of triggering the ignition coil's primary circuit.

enameled copper wire Wire that uses a thin layer of enamel as insulating material. The thinness of the insulation allows the wire to be closely wound in a coil, creating a dense magnetic field when current flows through it.

engine control module (ECM) A computer that controls the ignition and fuel control and emissions control systems on an engine; also called the electronic control unit (ECU) or powertrain control module (PCM).

event cylinder The cylinder that is on compression and ready for the spark to ignite the air-fuel mixture on a waste spark ignition system.

firing line The tall lines on a parade pattern that indicate the voltage required to initially jump the spark plug gap.

Hall-effect sensor The portion of an electronic ignition system used to trigger the ignition system. Hall-effect switches operate by using a potential difference, or voltage, created when a current-carrying conductor is exposed to a magnetic field. If a magnetic field is applied at right angles to the direction of current flow in a conductor, the lines of magnetic force permeate the conductor, and the electrons flowing in the conductor are deflected to one side. This deflection creates a potential difference across the conductor. The stronger the magnetic field, the higher the voltage.

heat range The rating of a spark plug's operating temperature.

high-tension leads The heavy insulated wires used to connect the distributor cap terminals to the spark plugs, and the ignition coil to the distributor cap; or on waste spark systems, the coils to the spark plugs.

high-tension terminals The terminals on the coils and distributor cap that the high-tension leads are connected to.

high-voltage spark The electrical arc that takes place between the center and the side electrode of a spark plug.

ignition advance The means of causing the spark to occur earlier within the compression stroke for better performance and fuel economy during changing engine conditions.

ignition coil A device used to amplify an input voltage into the much higher voltage needed to jump the electrodes of a spark plug.

ignition coil pack A group of two or more ignition coils housed in one assembly.

ignition module An electronic component that electronically controls the ignition coil or coils.

ignition switch A switch operated by a key or start/stop button and used to turn a vehicle's electrical and ignition system on or off.

induced voltage The creation of voltage in a conductor by movement of a magnetic field that is near that conductor.

induction coil An electrical transformer that uses magnetic fields to produce high-voltage pulses from low-voltage direct current.

induction-type system A type of ignition system that uses a magnetic pulse generator to trigger the spark

inductive current The current that has been created across a conductor by moving it through a magnetic field.

integrated circuit A semiconductor chip that contains miniature versions of various electrical components within one housing.

internal module control circuit A section of the ignition module responsible for receiving the trigger signal.

interrupter ring A ferrous metal ring, shaped like a very shallow cup with slits or windows cut into it at evenly spaced intervals. The ring has the same number of blades and windows as engine cylinders and is rotated by the engine moving the blades through an air gap. The purpose of an interrupter ring is to systematically block, and expose, the magnetic field in a Hall-effect sensor in order to turn the primary ignition circuit on and off.

low-inductance coils Ignition coils designed with low primary winding resistance and a low number of turns. Current flow on these coils is much higher than with contact breaker point systems and reaches its optimum level sooner.

manifold absolute pressure sensor A sensor that is used to measure vacuum or pressure in the intake manifold.

mass airflow sensor PCM input that measures the mass of the air entering the engine.

number-one cylinder Typically the cylinder located farthest forward on the engine. It is the first cylinder in the firing order.

optical sensor A sensor that generates a voltage when excited by a beam of light.

Oscillate To cycle above and below a given value.

Oscilloscope A tool that shows graphically what is happening to voltage over a period of time; it is used to diagnose electrical faults.

parade pattern The display of all the cylinders firing in sequence on an oscilloscope.

powertrain control module (PCM) A computer that controls the ignition and fuel control and emissions control systems on an engine; also called the electronic control unit (ECU) or engine control module (ECM).

primary circuit The low-voltage circuit that turns the coil on and off.

raster The scope pattern where all of the ignition firing sequences are stacked vertically on top of each other.

Reluctor A rotating, toothed wheel that changes the reluctance of a material to conduct magnetic lines of force.

required voltage The amount of voltage needed to push current across the electrodes of a spark plug located in the combustion chamber.

rotor A high-voltage rotating switch that transfers voltage from the distributor cap's center terminal to the outer terminals.

rotor arm The portion of the rotor that extends toward, but not touching, the outer distributor cap terminals.

secondary circuit The part of an ignition system that operates on higher voltage and delivers the necessary high voltage to the spark plugs.

spark line The horizontal line immediately after the firing line on an ignition pattern. It shows how much voltage and the length of time the spark is burning.

spark plug A device that provides a gap for the high-voltage spark to occur in each cylinder.

spark plug reach The length of the spark plug from the seat to the end of the threads.

spark timing The point at which a spark occurs at the spark plug relative to the position of the piston.

stationary winding An extended length of wire wrapped into a circle. These windings are fixed as opposed to some types, which are meant to spin.

Stator Portion of an electronic ignition system that is mounted to the base of the distributor. It has a circular permanent magnet with a number of projections or teeth corresponding to the number of engine cylinders, and a stationary coil of fine enameled copper wire wound on a plastic reel and positioned inside the magnet.

switching transistor An electronic device used to control the flow of current through the ignition coil primary winding.

throttle body The housing on an intake manifold that is used to control the amount of filtered air that enters the cylinders.

throttle position sensor A sensor that measures how far the throttle valve is opened.

vacuum advance unit A mechanism that controls ignition timing advance in relation to engine load and causes the spark at the spark plug to occur sooner based on engine conditions. Its function is to improve fuel economy and, in doing so, reduce exhaust emissions.

waste cylinder The cylinder in a waste spark ignition system that receives a spark near the top of its exhaust stroke.

waste spark system An ignition system in which each ignition coil serves two cylinders, with each end of the secondary winding attached by a high-tension lead to a spark plug. The spark is used to ignite the air-fuel mixture in one cylinder and has no effect on the other cylinder.

Review Questions

1. All of the following components are common to all ignition systems *except*:
   a. spark plugs.
   b. the ignition coil.
   c. a device for triggering the ignition coil.
   d. a distributor.
2. Which of the following does the ignition system use to convert relatively low-voltage and high-current flow into very high-voltage and very low-current flow?
   a. Induction coil
   b. Ignition switch
   c. Ignition module
   d. Spark plugs
3. Which of these usually increases primarily because of the growing gap of the spark plug as it wears over time?
   a. Required voltage
   b. Available voltage
   c. Reserve voltage
   d. Primary voltage
4. In electronically triggered ignition systems, the spark timing is engineered into the design of the engine and is not adjustable. Which component measures engine speed?
   a. Throttle position sensor
   b. Manifold absolute pressure sensor
   c. Crankshaft position sensor
   d. Mass airflow sensor
5. Which of the following ignition switch positions unlocks the steering column but does not enable any electrical systems or disable the engine immobilizer or theft-deterrent system?

a. Lock
b. Off
c. Accessory
d. On

6. Which of the following systems has the fewest parts, which reduces failures and maintenance the most?
   a. Waste spark ignition system
   b. Coil-on-plug ignition system
   c. Contact breaker point ignition system
   d. Distributor-type electronic ignition system
7. Which of the following equipment is used to test for voltage drops and high resistance within the primary circuit?
   a. Spark testers
   b. LED test light
   c. Digital multimeter
   d. Standard test light
8. On distributor type vehicles, when testing the secondary side of the circuit, the oscilloscope can display several scope patterns. A high firing line on just one cylinder may indicate:
   a. low resistance.
   b. a short in the circuit.
   c. an open spark plug wire.
   d. low compression.
9. If there is a slight drag between the gapping tool gauge and the spark plug electrodes as the gauge is installed and removed, it means:
   a. the gap of the spark plug is correct.
   b. the gap of the spark plug is too high.
   c. the gap of the spark plug is too low.
   d. the spark plug was not pre-gapped.
10. If the coil primary circuit has continuity, the reading should be 12 volts or more when:
   a. the ignition switch is on, and the red lead of the multimeter is connected to the secondary output terminal of the coil and the black lead to engine ground.
   b. the ignition switch is off, and the red lead of the multimeter is connected to the positive terminal of the coil and the black lead to engine ground.
   c. the ignition switch is in the run position, and the red lead or test light is connected to the negative side of the coil and the black lead to a good ground.
   d. the ignition switch is off, and the red lead or test light is connected to the negative side of the coil and the black lead to a good ground.

ASE Technician A/Technician B Style Questions

1. Tech A says that contact breaker points are a mechanical switch that opens and closes once for every ignition spark that is created. Tech B says that contact breaker points send high voltage directly from the points to the spark plugs. Who is correct?
   a. Tech A
   b. Tech B
   c. Both A and B
   d. Neither A nor B
2. Tech A says that as engines gain miles, the spark plug gap increases, which raises the ignition system's available voltage. Tech B says that misfire occurs when required voltage is higher than available voltage. Who is correct?
   a. Tech A
   b. Tech B
   c. Both A and B
   d. Neither A nor B
3. Tech A says that as engine RPM increases, spark timing generally increases. Tech B says that as engine load increases, spark timing generally decreases. Who is correct?
   a. Tech A
   b. Tech B
   c. Both A and B
   d. Neither A nor B
4. Tech A says that a Hall-effect switch uses light to turn a circuit on and off. Tech B says that waste spark systems don't need distributors. Who is correct?
   a. Tech A
   b. Tech B
   c. Both A and B
   d. Neither A nor B
5. Tech A says that coil-on-plug ignition systems use one coil to fire two cylinders. Tech B says that you should twist spark plug boots before removing them. Who is correct?
   a. Tech A
   b. Tech B
   c. Both A and B
   d. Neither A nor B
6. Tech A says that the ignition system will maintain spark at the spark plug for approximately 23 degrees of crankshaft rotation. Tech B says that the duration of spark in the spark plug only lasts 2 to 3 degrees of crankshaft rotation. Who is correct?
   a. Tech A
   b. Tech B
   c. Both A and B
   d. Neither A nor B
7. Tech A says that the positive side of the coil primary circuit is typically switched by the ignition module. Tech B says that on two-wire COP coils, the coil is switched by an external ignition module or the PCM. Who is correct?
   a. Tech A
   b. Tech B
   c. Both A and B
   d. Neither A nor B
8. Tech A says that one advantage of distributorless ignition systems is no moving parts to maintain. Tech B says that the resistance of the coil primary winding can best be tested with a voltmeter. Who is correct?
   a. Tech A
   b. Tech B
   c. Both A and B
   d. Neither A nor B

9. Tech A says that an inductive pickup coil can be tested for shorts and grounds with an ohmmeter. Tech B says that an inductive pickup coil pattern can be tested with a lab scope. Who is correct?
   a. Tech A
   b. Tech B
   c. Both A and B
   d. Neither A nor B

10. Tech A says that on a secondary ignition pattern, the firing line shows how much voltage it takes to jump the spark plug gap. Tech B says that the spark line on a secondary pattern tells the length of the spark plug reach. Who is correct?
    a. Tech A
    b. Tech B
    c. Both A and B
    d. Neither A nor B

CHAPTER 62
Gasoline Fuel Systems

NATEF Tasks

- **N62001** Replace fuel filter(s) where applicable. (MLR/AST/MAST)
- **N62002** Inspect and test fuel pump(s) and pump control system for pressure, regulation, and volume; determine needed action. (AST/MAST)
- **N62003** Check fuel for contaminants; determine needed action. (AST/MAST)
- **N62004** Inspect, test, and/or replace fuel injectors. (AST/MAST)

Knowledge Objectives

After reading this chapter, you will be able to:

- **K62001** Explain the principles related to gasoline fuel systems.
- **K62002** Explain the purpose and function of the fuel supply system components.
- **K62003** Explain the principles EFI fuel systems.
- **K62004** Explain the types and operation of carbureted fuel systems.

Skills Objectives

After reading this chapter, you will be able to:

- **S62001** Perform maintenance and preliminary testing of fuel systems.

▶ Introduction

Today's gasoline **fuel systems** must meter a precise amount of fuel into the engine under a wide range of operating conditions. The most important job of the fuel system is optimizing engine performance while keeping fuel consumption and emissions to a minimum. Although carburetors were commonly used on vehicles up until about 30 years ago, they didn't perform these jobs adequately (**FIGURE 62-1**). Their limitations prompted the development of fuel injection, starting with mechanically operated systems that typically sprayed a continuous flow of fuel through fuel injectors at each individual intake port in the intake manifold. These systems were sometimes referred to as "bug sprayers" because they put out a continuous fine mist. The amount of fuel sprayed was controlled mechanically by an air-flow-operated fuel distributor. The more air that entered the engine, the more fuel that was sprayed.

FIGURE 62-1 Types of fuel systems. **A.** Carburetor. **B.** Throttle body injection. **C.** Multipoint fuel injection. **D.** Gasoline direct injection.

You Are the Automotive Technician

A new customer brings her late model vehicle into your shop, complaining that it hasn't been "running right" for the last few weeks. She took it to another shop, where the technician said it needed a tune-up and changed the spark plugs, oxygen sensors, fuel filter, and air filter. It improved at first, but then it started getting worse. At first the MIL came on intermittently but now is on most of the time. The vehicle seems to be low on power as well, and yesterday it didn't start. When it started today, she thought she should bring it in. You explain that you will need to diagnose the problem. She authorizes the diagnostic charge. You scan the PCM and find a P0171 (system too lean bank 1), and a P0174 (system too lean bank 2). Because the oxygen sensors are new and look like they are OEM replacements, you decide to perform a fuel pump pressure and volume test. You will also perform an alcohol content test on a small sample of the fuel after the volume test.

1. Why do you suspect a fuel pump problem in this situation?
2. If the fuel pump passes the pressure test, should you perform the volume test? Why or why not?
3. How could too much alcohol in the gasoline cause the engine to run too lean?
4. How can a scan tool and test drive be used to find a low fuel pressure and volume condition?

Mechanical fuel injection still could not meet the increasingly stringent emissions standards, so electronic fuel injection (EFI) was introduced. In this way, fuel delivery could be controlled electronically, allowing continuous adjustments to the air-fuel ratio. Most early EFI systems used throttle body injection (TBI), wherein one or two injectors were mounted in a throttle body. These systems looked like a simplified carburetor. As technology progressed, multipoint fuel injection (MPFI or just PFI) became more prevalent, involving one injector installed in the intake manifold near the intake port of each cylinder. This design allowed more equal distribution of fuel among the cylinders, which improved fuel economy and reduced emissions further. But even that wasn't the end of the line for new innovations. Gasoline direct injection (GDI) places the fuel injectors in the cylinder head near the spark plugs, where they can spray fuel directly into the combustion chamber. This technology allows manufacturers to operate the engine on much leaner mixtures than has otherwise been possible during certain engine operating conditions.

To make EFI function correctly, an electronic control unit (ECU), also called a powertrain control module (PCM), is needed to determine the proper quantity of fuel to be injected. For the PCM to make the proper determination, information from a variety of electronic sensors is needed. It is the wide variety of information that allows the fuel injection system to operate so precisely and efficiently. As a technician, it is critical to have a deep understanding of how these systems and components operate. This understanding will give you a head start on properly inspecting, diagnosing, and repairing fuel system faults for your customers, and thus keeping their vehicles operating at peak efficiency and performance.

▶ Gasoline Fuel System Principles

K62001

The purpose of a fuel system is to provide the ideal air-fuel mixture for the operating conditions of the engine. Liquid fuel will not burn. Fuel has to be vaporized, turning from a liquid to a gas. It then has to be mixed with the proper amount of air. It takes time and temperature for fuel to vaporize fully. The smaller the liquid droplets, the faster they can be vaporized. The process of making the droplets small is called atomization. A fuel system needs to atomize the fuel as small as possible so that it can be vaporized before it is ignited.

As explained in the introduction, older cars and trucks used a carburetor to atomize the fuel and mix it with air. Modern gasoline-powered vehicles use EFI systems, which atomize the fuel into much smaller particles, giving superior efficiency and performance over carburetors.

One of the reasons that carburetors cannot atomize fuel as efficiently as fuel injection systems is because the principles of operation are different. Pressure differential is central to all fuel systems. The greater the differential, the easier it is to atomize the fuel. A carburetor works off pressures below atmospheric pressure (vacuum), which involves a relatively low pressure differential. Fuel injection, in contrast, works off pressures above atmospheric pressure, which generally involves a much higher pressure differential (**FIGURE 62-2**). With a very small pressure difference, as in a carburetor, fuel passages must be much larger. This means the fuel entering the airstream in a carburetor is in much larger droplets than fuel that is sprayed out of a very small hole at high pressure in a fuel injection system.

Modern fuel injection systems have three subsystems:

- **Fuel supply system:** This system provides pressurized, filtered gasoline to the fuel injectors or carburetor (in older vehicles). The fuel supply system draws in gasoline from the gas tank (fuel cell) and delivers it under pressure to a fuel metering device. Today's vehicles typically use an in-tank electric fuel pump. Older vehicles used a mechanical fuel pump that was typically mounted on the engine and driven by the camshaft.
- **Air supply system:** The air supply system, also called the induction system, provides clean, filtered air for combustion in the engine. An air filter is a paper filter or **element** that filters the incoming dirty/dusty air. The element is usually housed in a plastic or metal air cleaner housing located in the engine compartment. Engineers have made numerous configurations to prevent water and dirt intrusion. In addition, air intake sounds have been carefully analyzed, and intake systems have been engineered to give the optimum vehicle performance while providing smooth, quiet operation.
- **Fuel metering system:** This system constantly meters and adjusts the amount of fuel that the engine is burning. The most common type of EFI fuel metering is multipoint fuel injection. Although specific systems vary, many of the systems have similar parts. For example, most vehicles with multipoint fuel injection systems have sensors, fuel injectors, and a PCM.

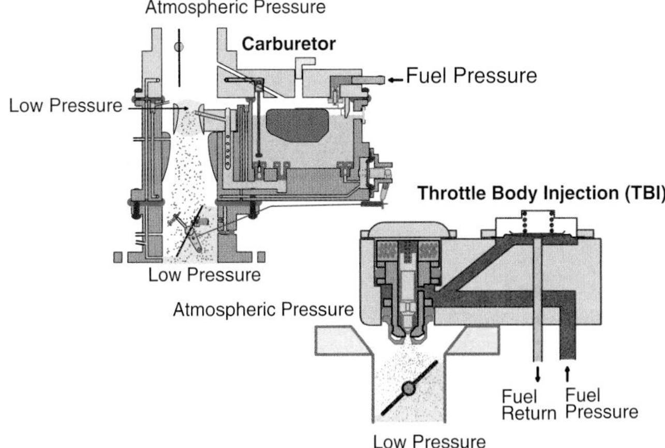

FIGURE 62-2 The main difference between carburetion and fuel injection is that carburetion works off pressures below atmospheric pressure (vacuum), and fuel injection works off pressures substantially above atmospheric pressure.

▶ Gasoline Fuel

Gasoline fuel is derived from crude oil. Crude oil is taken out of the ground as a liquid mixture of highly flammable compounds of hydrogen and carbon, called hydrocarbons, together with impurities. It is then processed into many fuel and lubricant products at an oil refinery through the fractional distillation process in which the crude oil is heated in the base of a tower and allowed to condense at different temperatures (levels) of the tower. More volatile compounds rise to the top of the tower while less volatile compounds stay low in the tower (**FIGURE 62-3**).

Gasoline is very volatile, mixing easily with air to form gas or vapor. The more effectively liquid gasoline is changed into vapor, the more efficiently it burns in the engine. Thus, high volatility is desirable. However, gasoline vapor allowed to mix with air in the open is highly explosive; therefore, it can be very dangerous, and gasoline must be handled with care. High volatility also can create excessive hydrocarbon emissions, which, if vented to the atmosphere, contribute to air pollution and smog. If liquid gasoline is heated, it is even more volatile. If it vaporizes in the fuel pump, vapor is pumped through the system instead of liquid. This causes the engine to run very poorly, and it may even die. This situation is called **vapor lock** and was fairly common on carbureted engines. It rarely happens on fuel-injected engines.

Gasoline is mainly a mixture of paraffins, naphthenes, aromatics, and olefins, together with some other organic compounds and contaminants such as sulfur. Some of these contaminants can cause corrosion, so they must be removed. Tight regulations in some countries limit the allowed proportion of aromatics, olefins, and sulfur in gasoline. Gasoline in its raw processed form is not suitable for use in vehicle engines; it must be enhanced with different additives such as detergents, octane boosters, and oxygenates such as ethanol. Detergents help to keep the fuel system clean, especially the fuel injector nozzles and intake valves. Octane boosters make it harder to ignite the gasoline, which makes it less prone to be ignited before the spark occurs. And oxygenates add oxygen to the fuel, which helps reduce carbon monoxide pollutants by slightly leaning out the air-fuel mixture.

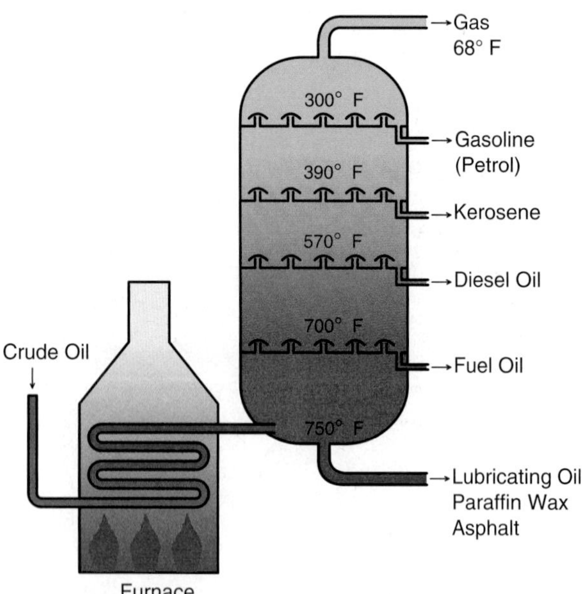

FIGURE 62-3 In processing gasoline from crude oil, more volatile compounds rise to the top of the tower while less volatile compounds stay low in the tower.

When a mixture of gasoline (petrol) and air is compressed inside an engine cylinder, it heats up. If the compression of the engine is high enough, and if the fuel is able to ignite easily enough, the air-fuel mixture may spontaneously ignite before the spark occurs. This is called pre-ignition, also known as **knocking**.

Gasoline fuel can be modified during processing by including additives so that it is less prone to spontaneously ignite. This ignition characteristic is known as its **octane rating**. The less easily the fuel ignites, the higher the octane rating. Higher compression engines are more susceptible to engine knock, so they require fuels with a higher octane rating—that is, fuels that ignite less readily.

A gasoline's octane rating is measured by the producer. Two different methods are used to measure the octane rating of a fuel—the Research Octane Number (RON) and the Motor Octane Number (MON). Depending on the composition of the fuel, the MON of a modern gasoline will be about 8 to 10 points lower than the RON; however, there is no direct link between RON and MON. Both are measurements of a fuel's resistance to knock, but the MON is a better measure of how the fuel behaves when under load.

In most countries, including the whole of Europe and much of the rest of the world, the RON rating is the one that is usually displayed on the pump at filling stations. In the United States and Canada, and some other countries, the displayed fuel rating is an average of the RON and the MON rating: (RON + MON)/2. Consequently, whatever the rating may be called at the pump, the rating number for identical fuels will on average be about 4 to 5 points higher in Europe than it will appear to be in the United States.

There is a popular belief that fuels with higher octane ratings will improve performance in vehicles that are designed to run on fuels with lower octane ratings. This is largely a myth, although some premium fuels do have higher energy ratings. Higher-powered engines usually have a higher compression ratio, so they generally require more expensive higher-octane fuels to prevent pre-ignition and detonation. Higher octane does not in itself mean higher energy output, so a fuel designed for a high-compression engine will not necessarily deliver any more power in a lower compression engine; in fact, it will likely be lower. Engines perform best when used with the octane rating recommended by the engine manufacturer.

Octane can be boosted with additives. Prior to the introduction of catalytic converters, tetraethyl lead was added to boost octane. Lead was also used to lubricate valve faces and seats, which slowed down wear. But lead also coats the catalytic converter and oxygen sensor, rendering them inoperable. So leaded fuel was phased out for general vehicle use, and unleaded fuel took its place. So-called unleaded gasoline may contain small amounts of lead, but maximum levels are tightly controlled. Octane is now boosted with additives of ethanol, aromatic hydrocarbons, and ether chemicals.

▶ Controlling Fuel Burn

For gasoline to burn properly, it must be mixed with the right amount of air. The air-fuel ratio where all of the fuel and all of the oxygen in the air is burned is about 14.7 to 1 by mass. By volume, it is about 11,000 to 1 (**FIGURE 62-4**). This means that it takes a lot of air to burn a small amount of gasoline. A lean air-fuel mixture has more air in proportion to the amount of fuel. A slightly lean mixture gives good fuel economy and low exhaust emissions-suitable for cruising conditions. A mixture that is too lean can make an engine run rough and overheat.

A rich air-fuel mixture has less air in proportion to the amount of fuel (**FIGURE 62-5**). A slightly rich mixture can produce more power, but the extra fuel it uses steps up fuel consumption and greatly increases emissions. A mixture that is slightly rich causes incomplete burning, which reduces efficiency and increases carbon monoxide emissions significantly. And if the mixture is extremely rich, it can foul spark plugs and increase carbon deposits in the combustion chamber and exhaust, as well as reduce fuel economy and increase emissions even further. Controlling the air-fuel ratio so it is at the proper level for each driving condition is impelling changes in fuel system technology.

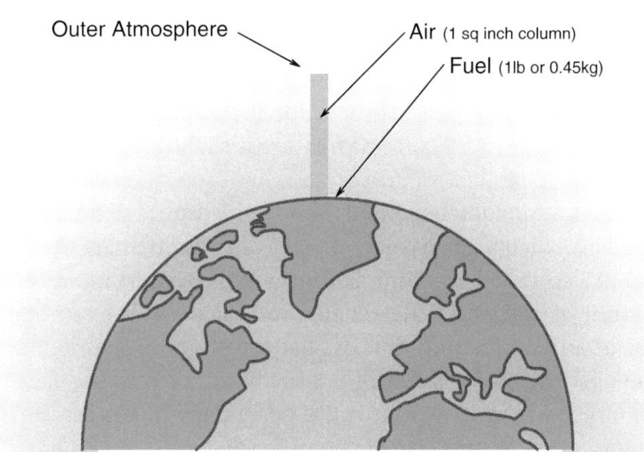

FIGURE 62-4 Air-fuel ratio where all of the fuel and all of the oxygen is burned is about 14.7:1 by mass and 11,000:1 by volume.

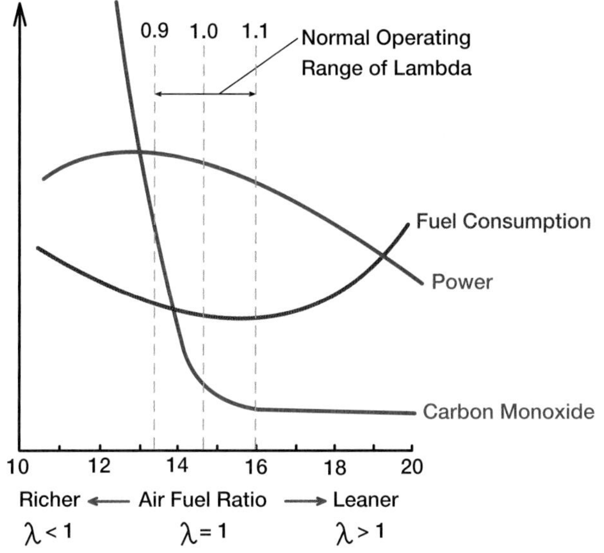

FIGURE 62-5 Rich and lean air-fuel ratios.

Combustion

In normal combustion, the spark plug ignites the mixture, and a small ball of flame forms around the tip of the plug (**FIGURE 62-6**). The piston finishes compressing the mixture. The flame spreads faster and moves evenly to halfway through the mixture, and the piston reaches top dead center. The flame picks up more speed, then shoots out to consume the rest of the mixture. Combustion ends with the piston a short way down the cylinder. Ideally, this would completely burn all of the fuel that entered the cylinder, extracting the maximum thermal energy from the gasoline. As you will see, this process can go wrong.

Detonation/Pre-ignition

Detonation is a violent collision of flame fronts in the cylinder, caused by uncontrolled combustion. It occurs after the spark plug has fired. The sudden rise in pressure can cause the remainder of the mixture to ignite spontaneously, creating a second flame front that collides with the first front, making a knocking sound (**FIGURE 62-7**). Sustained detonations can raise temperatures enough to melt holes in the top of pistons.

Pre-ignition occurs before normal combustion, when something in the combustion chamber heats up enough to ignite the mixture before the spark plug fires. A mixture may also ignite simply because it is unstable under the higher heat and pressure. Detonation and pre-ignition can cause severe damage and must be avoided. Many engine management systems use knock sensors to detect detonation, and the PCM retards ignition timing to certain limits to try to stop it. Knock sensors are discussed in greater detail in the next chapter.

An engine that keeps running after it is switched off is said to be running-on or **dieseling**. That's because, as in a diesel

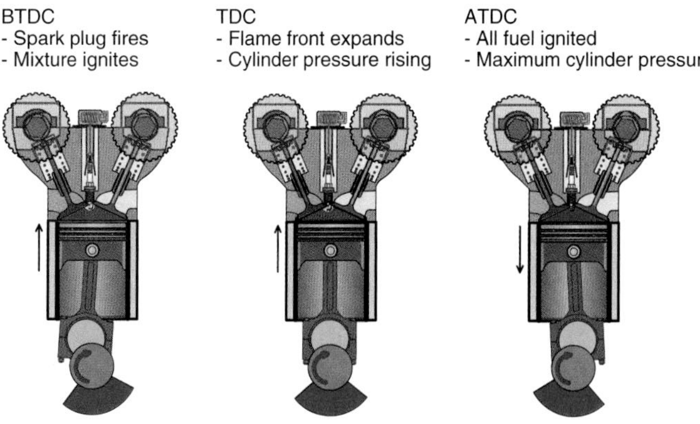

FIGURE 62-6 In normal combustion, the spark plug ignites the mixture, and a small ball of flame forms around the tip of the plug.

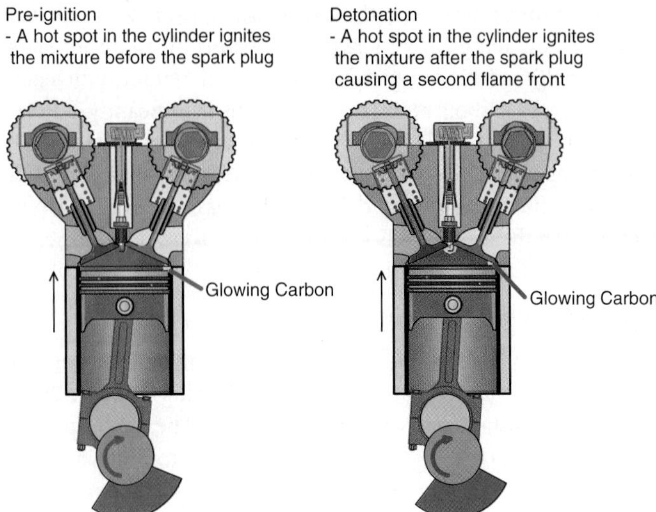
Pre-ignition
- A hot spot in the cylinder ignites the mixture before the spark plug
Glowing Carbon
Detonation
- A hot spot in the cylinder ignites the mixture after the spark plug causing a second flame front
Glowing Carbon

FIGURE 62-7 Pre-ignition occurs before normal ignition. Detonation occurs after normal ignition.

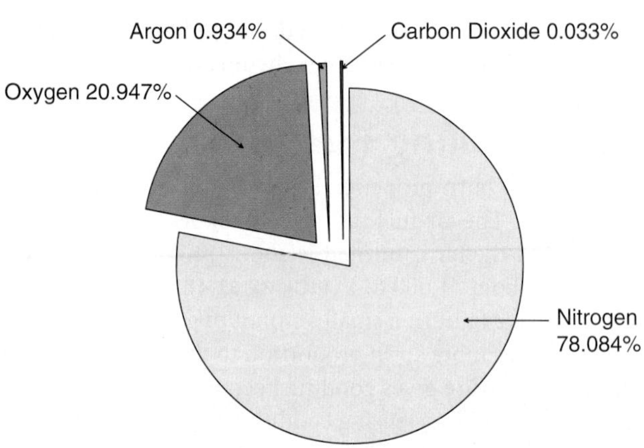
Argon 0.934%
Carbon Dioxide 0.033%
Oxygen 20.947%
Nitrogen 78.084%

FIGURE 62-8 The four primary components of the Earth's atmosphere are nitrogen, oxygen, argon, and carbon dioxide.

engine, the fuel is igniting just from heat in the cylinder, not from a spark from the plug. This situation may even cause an engine to run backwards for a brief time when it comes to a stop. Dieseling can be caused by a high idling speed, an overheated engine, too many carbon deposits in the chamber, or use of a gasoline with an octane rating that is too low. On carbureted engines, it can be prevented by a special valve in the carburetor idle circuit that cuts off fuel when the ignition is turned off, or a solenoid can be used to close the **throttle** to a below-idle position. Fuel-injected vehicles do not experience dieseling unless there is a leaking injector when the ignition is turned off, because the PCM shuts off the injector(s) during shutdown.

Air

Air is one of the essential components of the internal combustion engine. The components of air are a mixture of gases and small particles, which vary in composition. There are four primary components of the Earth's atmosphere: nitrogen, oxygen, argon, and carbon dioxide (**FIGURE 62-8**). Nitrogen, the largest component of air, is considered inert, meaning that it doesn't burn under normal temperatures, so it typically just passes through the engine. Oxygen is a highly reactive gas that promotes combustion of all types of fuels. It is the burnable part of air that mixes with fuel to power the engine. Argon and carbon dioxide are trace elements in air; neither burn so they also pass through the engine. These components are discussed further in the Emission Systems chapter.

Stoichiometric Ratio

The term **stoichiometric ratio**, represented by the Greek letter lambda (λ), describes the chemically correct air-fuel ratio necessary to achieve complete combustion of the fuel and air. In other words, it is when all of the oxygen completely combines with the fuel, leaving no molecules of either remaining. All of the oxygen and fuel have been chemically combined through complete combustion. For gasoline fuel, the stoichiometric air-fuel ratio is 14.7 parts air to 1 part fuel by mass, not volume. By volume, that equals 11,000 gallons (or liters) of air to every 1 gallon (or liter) of fuel—a ratio of 11,000:1. So if the air-fuel mixture is at the stoichiometric air-fuel ratio of 14.7 to 1, then the lambda value is 1 ($\lambda = 1$).

If an air-fuel mixture has a higher figure, say a lambda value of 1.05, there is more air in proportion to the fuel than 14.7 to 1; in fact, it is about 15.5:1, and the mixture is slightly lean. A mixture with a lower lambda value, say 0.95, has proportionately less air than fuel, approximately 14.0:1, and the mixture is slightly rich.

The exhaust gas oxygen sensor (often written as "O_2" sensor) is also called the lambda sensor. It is used to indicate the amount of oxygen in the exhaust so that the PCM can maintain a lambda reading that oscillates just above and below 1. The upstream oxygen sensor is usually installed in the exhaust manifold, where it measures the percentage of oxygen in the exhaust gases. A high percentage of oxygen may mean there isn't enough fuel to burn up all the oxygen, the mixture is too lean, and lambda is greater than 1. The oxygen sensor delivers this information to the PCM, which adjusts the mixture to richen it up slightly. Similarly, a low percentage of oxygen in the exhaust may mean there isn't enough oxygen to burn all of the fuel, and lambda is less than 1. The PCM would then adjust the mixture to lean it out slightly. Different fuels have a different stoichiometric ratio. For instance, methanol's air-fuel ratio is 6.4:1, and ethanol's is 9:1, again measured by mass, not volume.

The density of air is its mass per unit volume. Thus, a volume of air at high density has a higher mass than the same volume at low density. And if there is a larger mass of air, it will contain proportionally more oxygen. The density of air in the atmosphere changes at different temperatures and altitudes. That means the air that enters an engine at different locations could have very different amounts of oxygen. The amount of oxygen in air directly affects how well it supports combustion, so it is very important in determining an air-fuel ratio for an engine.

Pressure

Pressure and negative pressure, or **vacuum**, are terms used daily in the automotive industry. Manufacturers recommend a pressure to which tires should be inflated. And manifold vacuum is used to operate a power brake booster. Gases exert pressure on all bodies they make contact with. This applies also to the air in the Earth's atmosphere. Air has mass, and as a result it exerts pressure, called atmospheric pressure, not only on the Earth's surface, but also on all objects on the Earth's surface. Atmospheric pressure reduces with altitude. At sea level, it is calculated as 14.7 pounds per square inch (psi), or 101 kilopascals (kPa). But if this is so, why does a pressure gauge read zero when it is not in use? This is because the gauge indicates only pressure above atmospheric pressure. This reading is called gauge pressure (**FIGURE 62-9**). If you needed to know absolute pressure, an absolute pressure gauge would be required, and it would read 14.7 psi at sea level. Absolute pressure equals gauge pressure plus atmospheric pressure. Readings on a tire gauge are based on gauge pressure, not absolute pressure, so they would need 14.7 psi added to them (at sea level) for atmospheric pressure. In automotive use, it is most common for pressures to be measured in gauge pressure.

If pressure being measured is below atmospheric pressure, a pressure gauge with its zero reading (gauge pressure) cannot measure those lower pressures. Therefore, pressure below atmospheric pressure is called "vacuum," and a vacuum gauge is normally used to measure it. An example of vacuum in automotive is intake manifold vacuum.

In most gasoline engines, the position of the throttle plate controls the volume of air, or air-fuel mixture, entering the manifold. At **idle** speed, the pistons draw air away from the manifold at a faster rate than it can pass the throttle plate into the manifold, creating a high vacuum, or low absolute pressure. At wide-open throttle, depending on load, the vacuum is much less, and pressure in the manifold rises closer to atmospheric pressure (**FIGURE 62-10**).

A vacuum gauge can be calibrated in inches of mercury in a scale reading from 0" to 30", or in millimeters of mercury, on a scale from 0 to 760 mm. The scale is derived from the fact that atmospheric pressure at sea level supports a column of mercury approximately 30" (760 mm) high. This glass tube, closed at one end, is filled with mercury and then inverted in a bowl of mercury. The space above the mercury is a vacuum, and atmospheric pressure on the exposed surface of the mercury supports the column (**FIGURE 62-11**). At higher altitude, its height falls, which indicates that atmospheric pressure at that altitude is less. Thus, this scale can be used to indicate the amount of vacuum that exists below atmospheric pressure. At idle speeds, a vacuum gauge connected to the intake manifold will indicate a reading of approximately 15" to 21" (308 to 530 mm) of mercury, depending on altitude.

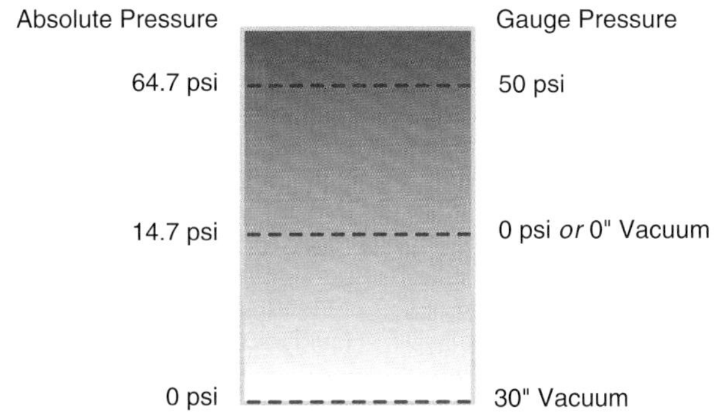

FIGURE 62-9 Absolute pressure versus gauge pressure.

FIGURE 62-10 Engine vacuum. **A.** At idle. **B.** Partial throttle.

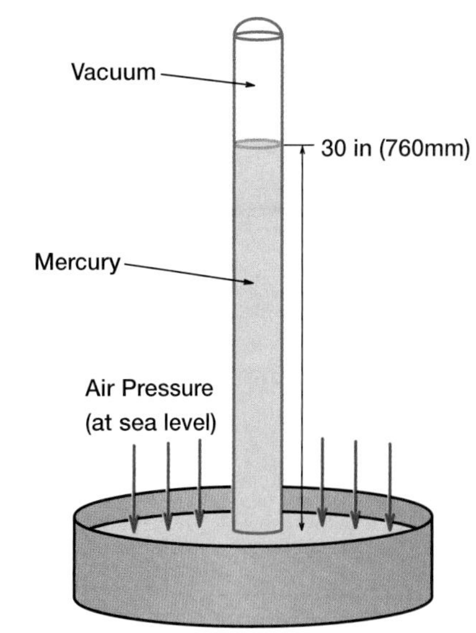

FIGURE 62-11 Atmospheric pressure at sea level supports a column of mercury approximately 30 inches (760 mm) high.

Some engine management systems measure changes in atmospheric pressure by using a barometric pressure sensor in the PCM. This is because, above sea level, air pressure is reduced, which means that it contains less oxygen. To maintain the correct air-fuel ratio, the amount of fuel delivered to the engine must be reduced to match the amount of oxygen present.

▶ Fuel Delivery System Components

K62002

The fuel supply system consists of several components that all play key roles in delivering gasoline to the fuel injection system. We will look much more in depth at each of the following:

- Fuel tank—stores fuel for delivery to the engine as needed.
- Fuel filler neck—allows fuel to be added to the fuel tank.
- Gas cap—prevents leaks into and out of the filler neck.
- Evaporative emission control system (EVAP)—Temporarily stores fuel vapors until they can be burned in the engine.
- Fuel pump relay—sends power to operate the fuel pump on some vehicles.
- Fuel pump—pumps fuel at the proper pressure and volume to the injectors.
- Fuel tank sending unit—provides a method of signaling the level of fuel in the fuel tank to the fuel gauge on the instrument panel.
- Fuel filter—filters the fuel going to the fuel injectors to strain out particles of dirt and debris.
- Fuel lines—provide a route for fuel to flow from the tank to the injectors. Also typically provides a location for a fuel filter.
- Fuel rail—a hollow tube that delivers fuel to the injectors. May have a pressure tap and pressure regulator mounted on it.
- Fuel pressure regulator—the device that maintains the proper pressure in the delivery system. Can be mechanical or electronic.
- Fuel injector(s)—electromechanical solenoid valves that control the flow of fuel into the intake manifold or combustion chamber.

Fuel Tank

The fuel tank (or gas tank, as it is sometimes called) is the primary reservoir of the onboard fuel supply. The typical fuel tank of modern vehicles consists of a gas cap, filler neck, fuel, fuel pump, and **gauge sending unit** (**FIGURE 62-12**). Its primary function is to safely hold an adequate supply of gasoline for prolonged engine operation.

Where the tank is mounted depends on where the engine is and on space and styling. Safety demands that it be positioned well away from heated components and outside the passenger compartment. Tanks are made either of tinned sheet steel that has been pressed into shape or of nonmetallic materials. Aluminum or steel is used on commercial vehicles. The metal tank is usually in two parts, joined by a continuous weld around the flanges where the parts fit together. Baffles make the tanks more rigid. They also prevent the sloshing of fuel and ensure that fuel is available at the pickup tube.

Because fuel expands and contracts as temperature rises and falls, fuel tanks are vented to let them breathe. Modern emission controls prevent tanks from being vented directly to the atmosphere. They must use evaporative control systems. Vapor from the fuel tank is vented through a charcoal canister where fuel vapors are stored until they are burned in the engine (**FIGURE 62-13**). A vapor or vent line with a check valve connects the space above the liquid fuel with the canister. This valve opens above a specified pressure and lets through vapor, but not liquid.

Liquid fuel closes the check valve and blocks the line, which stops it from reaching the charcoal. Some systems have a small container, called a liquid-vapor separator, above the fuel tank. It also prevents liquid fuel from reaching the charcoal.

Fuel Filler Neck

The fuel filler is where fuel enters the tank. The **fuel filler neck** is a pipe that extends above the fuel tank (**FIGURE 62-14**). On

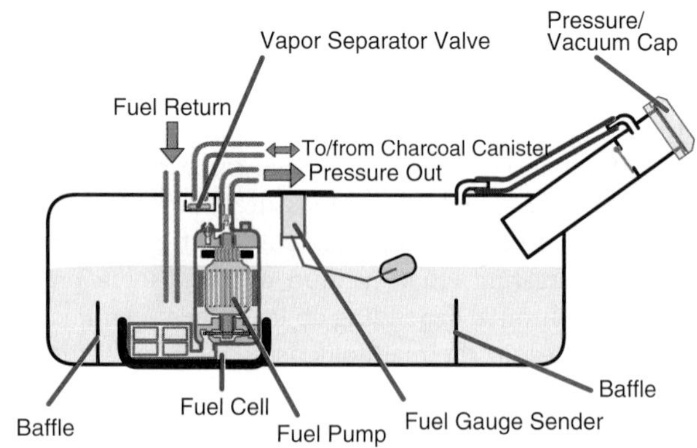

FIGURE 62-12 The typical fuel tank of modern vehicles consists of a gas cap, filler neck, fuel, fuel pump, and gauge sending unit.

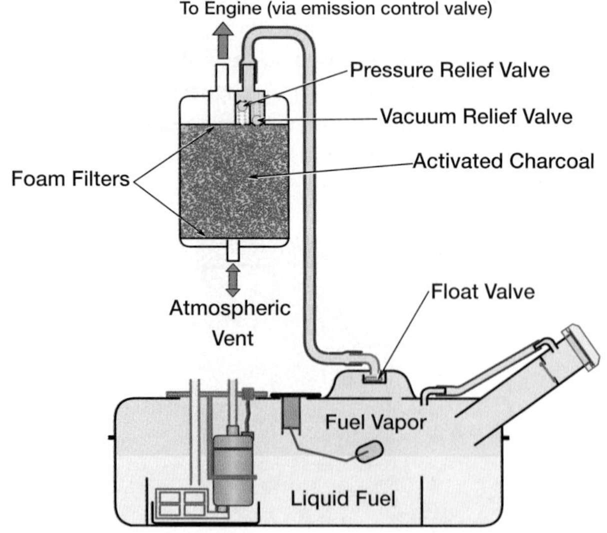

FIGURE 62-13 The fuel tank is vented through the evaporative emissions system.

FIGURE 62-14 The fuel filler neck.

older unleaded gasoline vehicles with catalytic converters, the filler neck was designed to prevent leaded fuel being added. Its diameter is smaller than those on leaded vehicles, and a trapdoor inside the filler can be opened only by the nozzle of the unleaded gasoline spout. The location of the filler neck depends on the design of the vehicle and location of the tank. The filler neck can incorporate the use of a blowback ball valve to prevent fuel from leaking from the vehicle during fill-ups and to deter gas theft.

SAFETY TIP

Never use a hose and your mouth to siphon gasoline from a vehicle. It is easy to swallow or inhale the gasoline, which is poisonous. If gas must be removed from a vehicle, always use a gas caddy with a suction pump.

Gas Cap

Modern vehicles are required by the Environmental Protection Agency (EPA) to have a non-vented gas cap. This non-vented cap prevents fuel vapors from being directly vented to the atmosphere, which wastes gasoline and contributes to air pollution. To help ensure that the cap is tightened properly, many of these caps use a ratchet system. Drivers tighten the cap until it ratchets. Newer Ford vehicles are using an Easy Fuel capless system (**FIGURE 62-15**). There is no gas cap to remove when refueling the vehicle. When the properly sized nozzle is placed on the top of the cap, latches release that open the valve and allow the nozzle to be fully inserted. This cap also reduces the amount of gasoline vapors that escape to the atmosphere when compared to a standard cap.

▶ TECHNICIAN TIP

A loose or leaky gas cap can cause the malfunction indicator lamp (MIL) to illuminate on the dash when the evaporative emission system monitor runs and the leak is detected. Always check the tightness of the cap if an evaporative emission system leak code is present. If the cap is tight, you may want to test the cap for leaks on a cap tester before digging into the evaporative emission system to locate the leak.

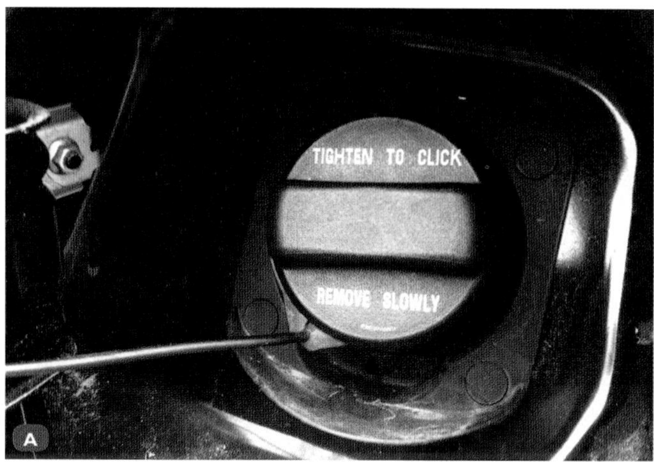

FIGURE 62-15 Gas caps. **A.** Standard gas cap. **B.** New capless system.

Fuel Pump Relay

Fuel pumps can draw a considerable amount of current while they run. To carry that high current load, modern vehicles use a **fuel pump relay**. The relay is an electromagnetically operated switch that basically activates the fuel pump when the ignition is turned on for priming the fuel system; when the engine is cranked; and when the engine is running. As low-amperage current is passed through the winding in the relay, a magnetic field is created that pulls the contacts together, controlling the larger current flow going to the fuel pump (**FIGURE 62-16**). The current operating the fuel pump relay winding is typically controlled by the PCM. In most cases, the PCM grounds the winding when it wants the fuel pump to operate.

Fuel Pump

Most fuel-injected vehicles use one or two electrical pumps to supply the fuel system with pressurized fuel. The pump is typically located either inside the fuel tank, but it can be mounted on the frame. If located in the tank, the pump is the submersible type. Electric fuel pumps can be the low-pressure type or the high-pressure type, depending on the fuel system used on the vehicle.

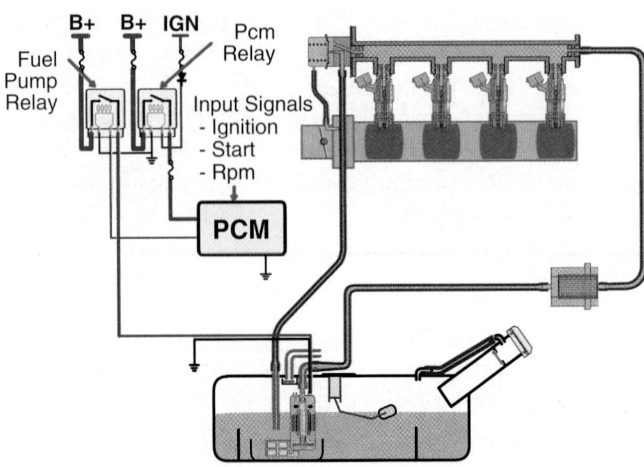

FIGURE 62-16 Fuel pump circuit.

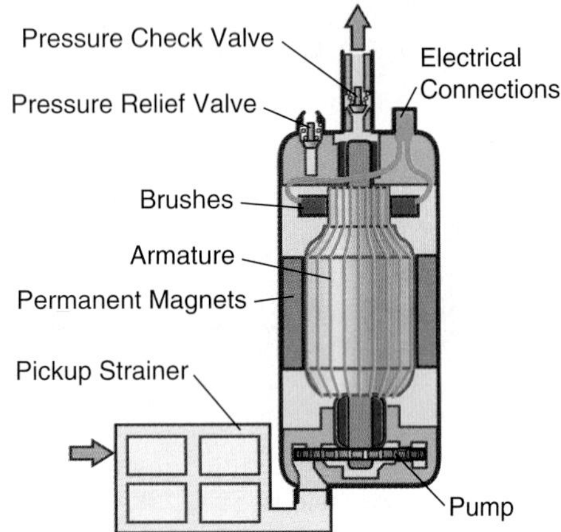

FIGURE 62-17 The fuel pump is driven by a permanent-magnet electric motor, a sealed unit integral with the pump.

The fuel pump is electrically operated and electronically controlled. It is driven by a permanent-magnet electric motor, a sealed unit integral with the pump (**FIGURE 62-17**). Fuel flows through the pump and around the electric motor when it is running. There is never an ignitable mixture (of air and fuel); only fuel is inside the pump housing, so there is virtually no danger of explosion. The pump is designed to deliver more fuel than the maximum requirement of the engine, so pressure in the fuel system is maintained at all times. Either a fuel pressure regulator maintains fuel pressure between the pump and the injectors, or the speed of the pump is varied to maintain the correct pressure.

For a short time after an engine is switched off, the engine temperature keeps rising, which can produce vapor in the fuel lines. A pressure check valve in the pump maintains the fuel pressure in the system after engine shutdown, preventing the fuel from boiling (vaporizing) as it absorbs heat from the engine (heat soak). This pressure diminishes after about 20 to 30 minutes, but it ensures effective hot-starting capability of the engine.

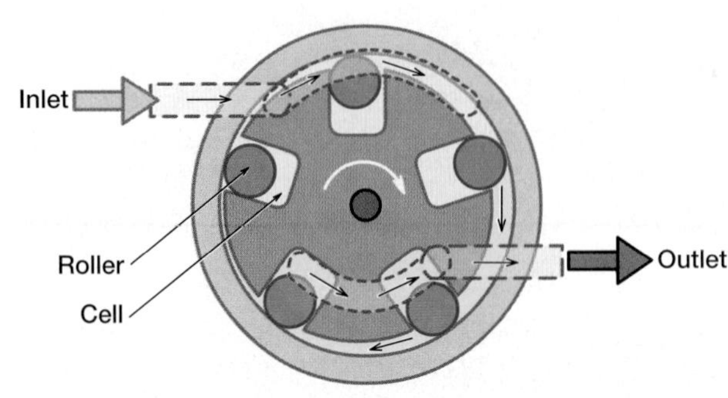

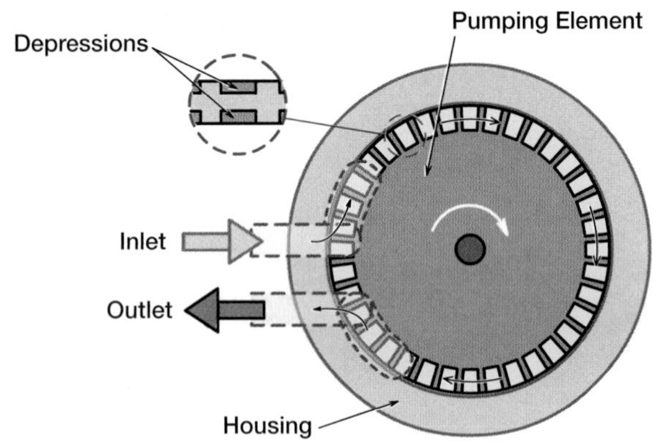

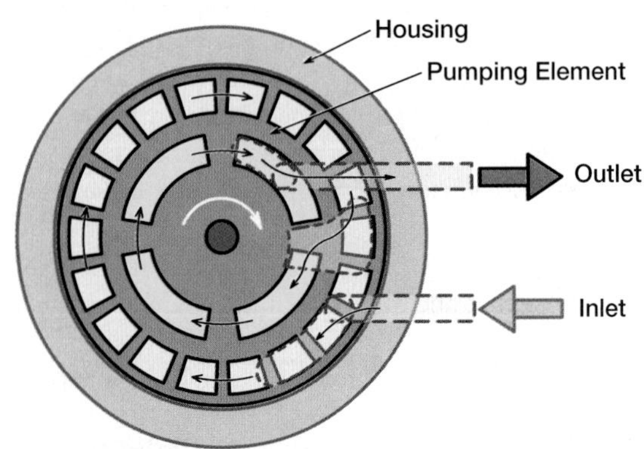

FIGURE 62-18 Types of pumps used in fuel pump assemblies.

Electric fuel pumps use various types of pump chambers, one of which is the roller cell (**FIGURE 62-18**). In this style, rollers float in channels in an offset rotor. As the rotor turns, the rollers are thrown outward, drawing fuel in as the space between the rollers expands and forcing fuel out as the space decreases. Other types are the peripheral type and the side channel pump.

Fuel Tank Sending Unit

For obvious reasons, it is important for the driver to know how much fuel the vehicle has in the tank. The primary job of the **sending unit** is to send constant electrical signals to the gas gauge located in the driver information center or to the BCM, which then controls the gas gauge. The sending unit is a variable resistor that is attached to a float mechanism (**FIGURE 62-19**). As the fuel level changes, so does the resistance and, in turn, the amount of current sent to the fuel gauge. This current directly indicates an accurate amount of fuel in the tank. On today's EFI vehicles, the sending unit is incorporated with the fuel pump and, in some cases, a replaceable or nonreplaceable fuel filter. The entire assembly has three jobs: pick up fuel from the bottom of the tank by way of the fuel pump, strain (filter/clean) the fuel and pressurize it, and of report the level of fuel in the tank at the driver information center. These systems are meant to last at least the life of the warranty.

Filter Sock

Fuel contamination can be extremely detrimental to the fuel supply system as well as the engine itself. To this end, a **filter sock** is the first line of defense and is incorporated into the end of the fuel pickup tube. The sock typically consists of a fine mesh, which prevents most small particles from being drawn into the fuel pump and sent through the rest of the fuel system (**FIGURE 62-20**). It typically has to be replaced when the fuel pump is replaced; otherwise, the manufacturer won't warranty the fuel pump. They can become plugged up due to varnish, water, or dirty fuel.

Fuel Lines

Fuel lines are usually made of metal tubing or synthetic materials. A fuel supply line carries fuel from the tank to the engine. A return line may also be provided to allow excess fuel to return to the tank. Returning the excess fuel helps prevent the formation of vapor that can occur in the fuel supply line during hot conditions. Other lines that run along with the fuel lines include evaporative emission system lines, which connect between the fuel tank, charcoal canister, and intake manifold (**FIGURE 62-21**). These lines can be made of steel or plastic. Steel lines can rust out in parts of the country that use salt on the roads. Both types of lines can be damaged by improperly lifting or supporting a vehicle with a jack, jack stand, or hoist.

Fuel Filter

The **fuel filter** typically consists of a pleated paper filter housed in a sealed container. Its primary function is to prevent contaminants from reaching the injectors. The paper filter can also prevent small amounts of water from getting past it, because the paper absorbs the water, causing the fibers in the filter to swell and restricting liquid from getting past it. Most fuel filters are directional, meaning they must be installed in the proper direction. Usually an arrow on the filter indicates the direction of flow. Fuel filters can typically be found in two places: on the frame rail or near the tank or in the tank as part of the fuel pump assembly (**FIGURE 62-22**).

FIGURE 62-20 A typical fuel sock installed on the bottom of a fuel pump assembly.

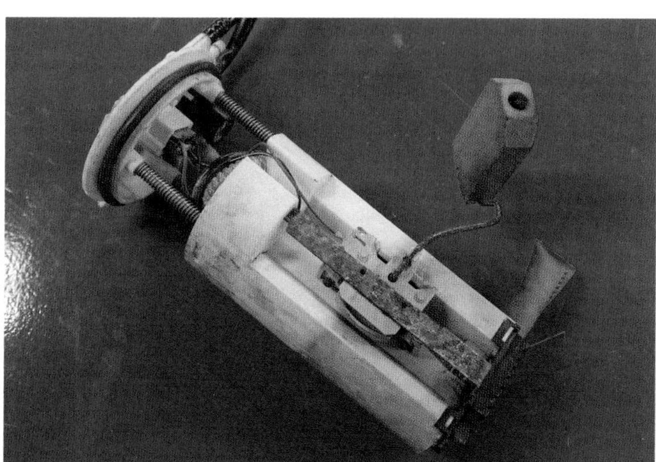

FIGURE 62-19 The sending unit is a variable resistor that is attached to a float mechanism.

FIGURE 62-21 Fuel and evaporative emission lines.

Fuel Rail

A **fuel rail** is a special manifold designed to provide a reservoir of pressurized fuel for the fuel injectors. The fuel rail typically sits on top of the fuel injectors (**FIGURE 62-23**). The gasoline is fed from the fuel tank by way of the fuel lines. The fuel, under pressure of course, is supplied by the fuel pump. The fuel rail can also serve as a mounting place for the fuel pressure regulator. The couplings where the fuel lines meet the fuel rail are usually sealed by flared ends; quick-connect, high-pressure O-rings; or banjo fittings (**FIGURE 62-24**). These methods ensure a sealed, leak-proof joint between the parts. Many, but not all, fuel rails provide a means for connecting a fuel pressure tester to the rail. This design, called a Schrader valve, uses a one-way valve to hold pressure in the system while the pressure tester is connected and disconnected to the fuel rail. Schrader valves usually are covered by a screw-on cap with a seal inside to prevent leaks in the event the Schrader valve goes bad.

Fuel Pressure Regulation

Fuel pressure must be regulated in some manner. On port fuel-injected engines, the fuel pressure has to be maintained a certain amount above manifold pressure. Because manifold pressure changes with engine load, fuel pressure also has to change with it. Some vehicles use a pressure regulator on the fuel rail; some are in the fuel tank; and others control fuel pressure by controlling the speed of the fuel pump.

The EFI pressure regulation system can be either a circulation type or a returnless type. In a circulation system, fuel is drawn from the tank by a fuel pump and delivered to solenoid-operated

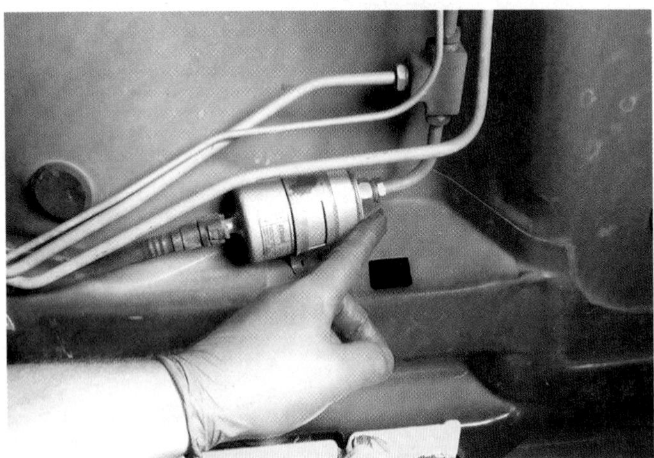

FIGURE 62-22 Typical inline fuel filter.

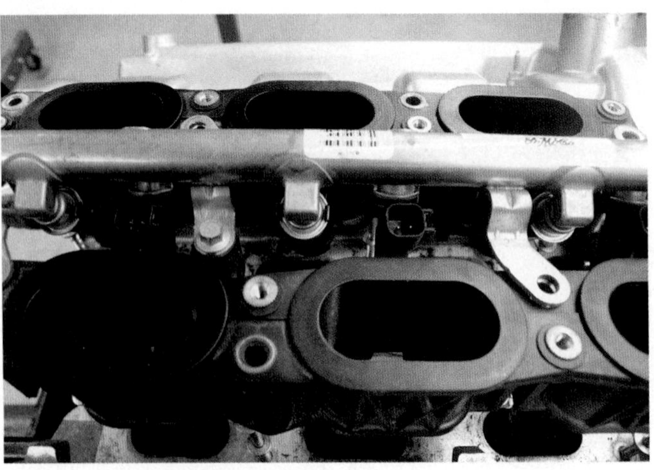

FIGURE 62-23 Typical fuel rail that sits on top of the fuel injectors.

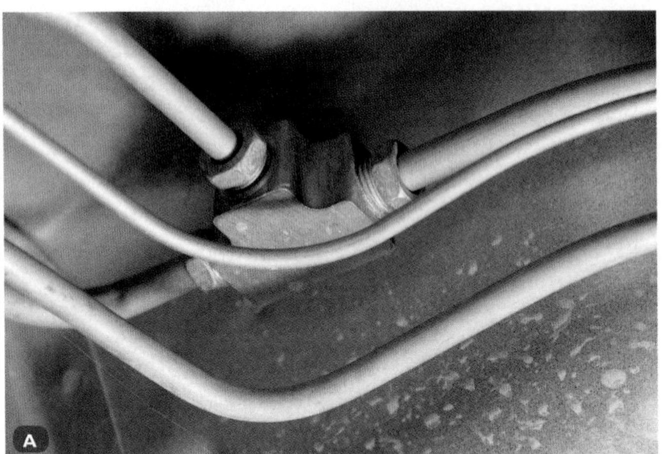

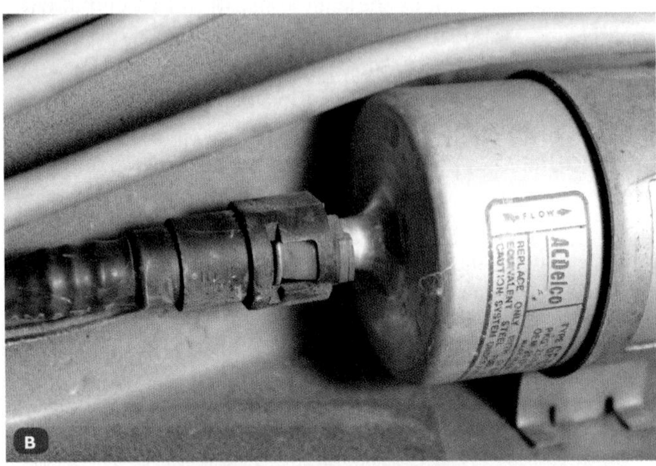

FIGURE 62-24 Types of fuel line connections. **A.** Flare. **B.** Quick couplings. **C.** Banjo fitting.

injection valves, called **injectors**. Fuel pressure at the injector(s) is maintained by a fuel pressure regulator, and excess fuel flows back to the tank through a return line. On most returnless systems, the pressure is controlled by the speed of the pump, and the pump only delivers the amount of fuel being injected.

A **fuel pressure regulator** mounted on the fuel rail incorporates a diaphragm-operated valve (**FIGURE 62-25**). The fuel pressure regulator inlet is connected to the fuel rail, and an outlet lets fuel return to the tank. Between them is a control diaphragm and pressure spring that determines the exposed opening of the outlet and the amount of fuel that can return. Movement of the diaphragm opens and closes the valve, which vents pressure back to the fuel tank through a return line. Fuel pressure builds against the diaphragm, and once it hits the pre-set pressure, the valve opens slightly and bleeds off pressure. This action reduces the pressure on the diaphragm closing the valve. With the valve closed, the pressure rises again until the valve once again opens. Thus, the strength of the pressure spring determines fuel pressure in the fuel rail and keeps it at a fixed value.

However, the pressure in the intake manifold varies considerably with changes in engine speed and load. So for any injection duration, if fuel is held at constant pressure, then as manifold pressure varies, so does the amount of fuel delivered. That means fuel pressure must be continuously adjusted to maintain a constant pressure drop across the injector.

Constant pressure drop across the injector is maintained exposing the spring side of the diaphragm to intake manifold vacuum. The spring-loaded vacuum chamber is connected by a manifold vacuum line to the intake manifold. The vacuum pulls against spring pressure to modify the pressure at which the valve opens to allow fuel to return to the fuel tank. As the throttle is opened, the manifold vacuum reduces. That, in turn, allows more spring pressure to push on the diaphragm, which then requires more fuel pressure to open the valve, providing higher fuel pressure in the fuel rail (**FIGURE 62-26**). This maintains a consistent pressure drop across the fuel injector (fuel rail pressure compared to manifold vacuum) as the throttle is manipulated by the driver.

When manifold vacuum is high (low absolute pressure), as it is at idling, fuel pressure is low. As manifold vacuum is low (high absolute pressure), toward open throttle, fuel pressure rises. Because the injectors are all subjected to the same pressure drop, they all inject an equal amount of fuel for a given "on" time (pulse width). The quantity of fuel delivered is thus controlled accurately by the pulse width of the injector.

Note that manifold pressure is not needed to assist the spring against the diaphragm in TBI systems, as the injection occurs above the throttle plate, at atmospheric pressure. Thus, fuel pressure is determined solely by the force of the regulator spring acting on the diaphragm.

When the engine is running, the circulation of fuel ensures that cool fuel is delivered at all times and that vapor formation in the lines is prevented. The pump control circuit normally allows the pump to operate for only a few seconds when the ignition is switched on, but the engine is not running or cranking; this is for priming the system. The pump control circuit allows the pump to operate during cranking and when the engine is running above a specified minimum revolutions per minute (rpm).

Newer vehicles have been using **returnless fuel injection systems** for the past 15 years or so. Manufacturers turned to

FIGURE 62-25 A fuel pressure regulator mounted on the fuel rail incorporates a diaphragm-operated valve.

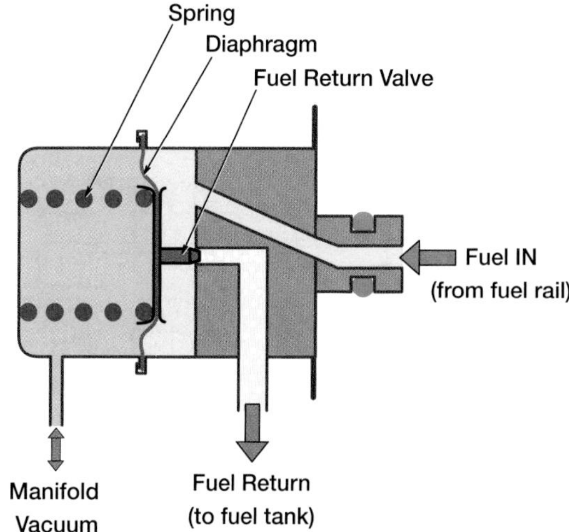

FIGURE 62-26 Typical fuel pressure regulator that uses intake manifold vacuum to maintain a constant pressure drop across the fuel injector.

using returnless fuel systems to reduce evaporative emissions. On a return-type system, the fuel returning to the tank is hot from engine heat. Hot fuel will readily vaporize and make the evaporative system work harder, requiring additional fuel vapor storage capacity. Because no hot fuel is returned to the tank in a returnless system, the fuel in the tank stays relatively cool, thus minimizing vaporization.

There are two types of returnless systems. One uses a pressure regulator in the fuel tank (mechanical), and the other controls the speed of the fuel pump to modify pressure (electronic) (**FIGURE 62-27**). The in-tank regulator uses a spring-loaded pressure regulator similar to the type mounted to the fuel rail, except it does not change with manifold pressure and therefore does not use a vacuum hose. Excess fuel pressure is simply vented into the tank.

On the electronically controlled system, the PCM sends a square wave (digital) signal, which controls the speed of the fuel pump (**FIGURE 62-28**). The faster it turns, the higher the pressure and flow. The lower it turns, the lower the pressure and flow. The PCM monitors the fuel pressure through a fuel pressure sensor mounted on the fuel rail and controls the fuel pump based on engine speed, load, and other factors.

Fuel Injector

The modern fuel injector is simply a spring-loaded, electric-solenoid spray nozzle (**FIGURE 62-29**). It incorporates a filter screen in the inlet and is usually sealed with O-rings to the fuel rail and intake manifold. The nozzle can be one of several types, such as rotating disc style, pintle style, and ball-valve style. Each type has its own pros and cons. Its job is to spray the proper amount of gasoline in the proper pattern either into the intake ports, directly into the combustion chamber, or into a pre-chamber in response to signals from the PCM (**FIGURE 62-30**). The fuel injector is positioned between the fuel rail and the intake

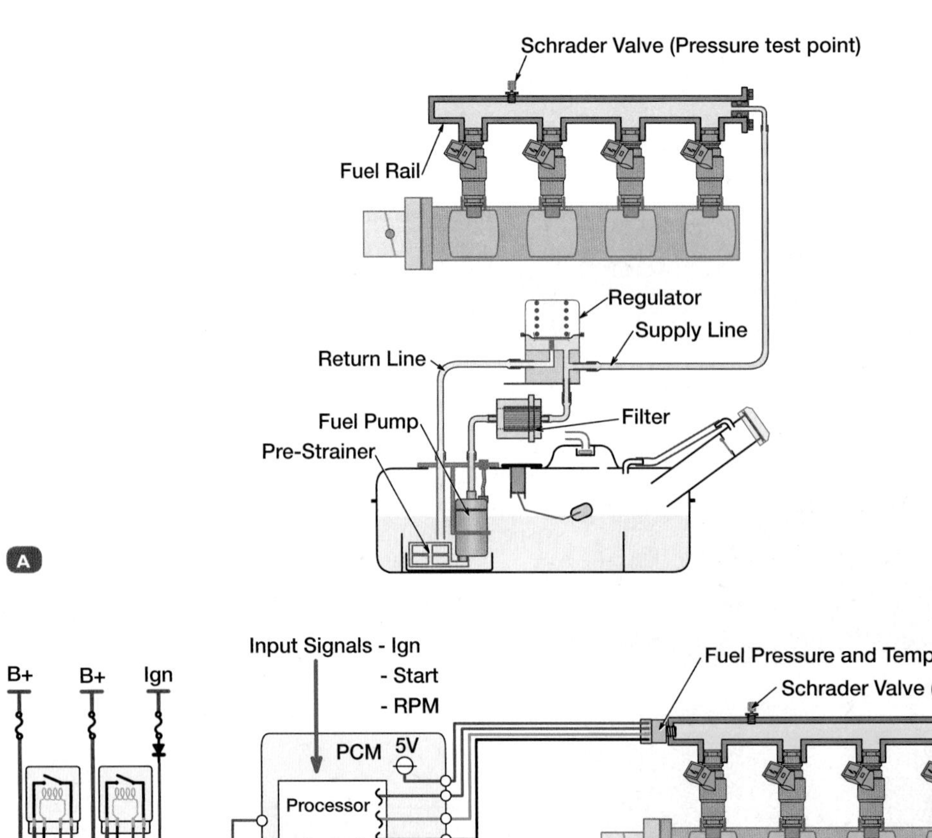

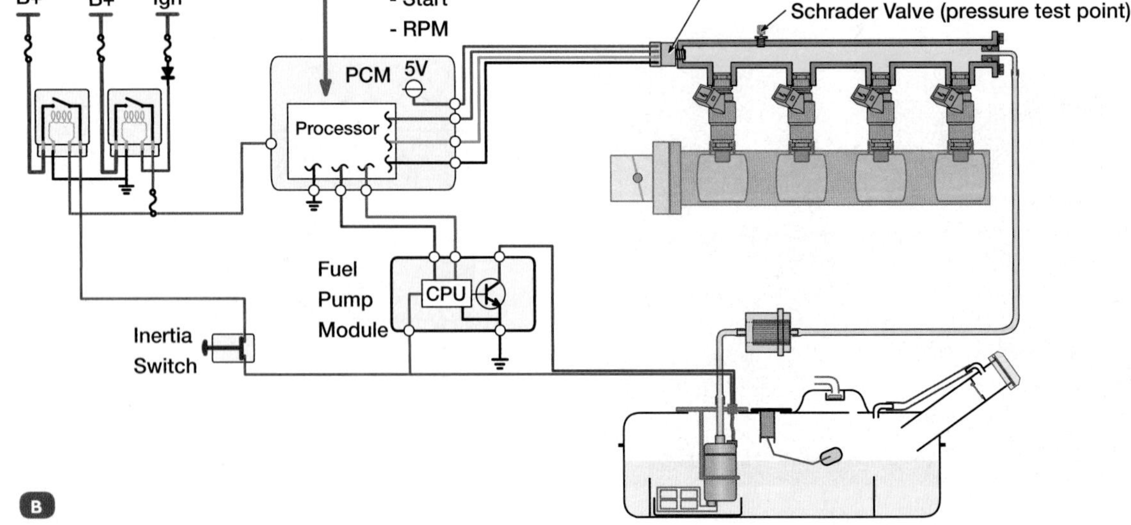

FIGURE 62-27 A. Mechanical returnless fuel system (regulator can be in-tank). **B.** Electronic returnless fuel system.

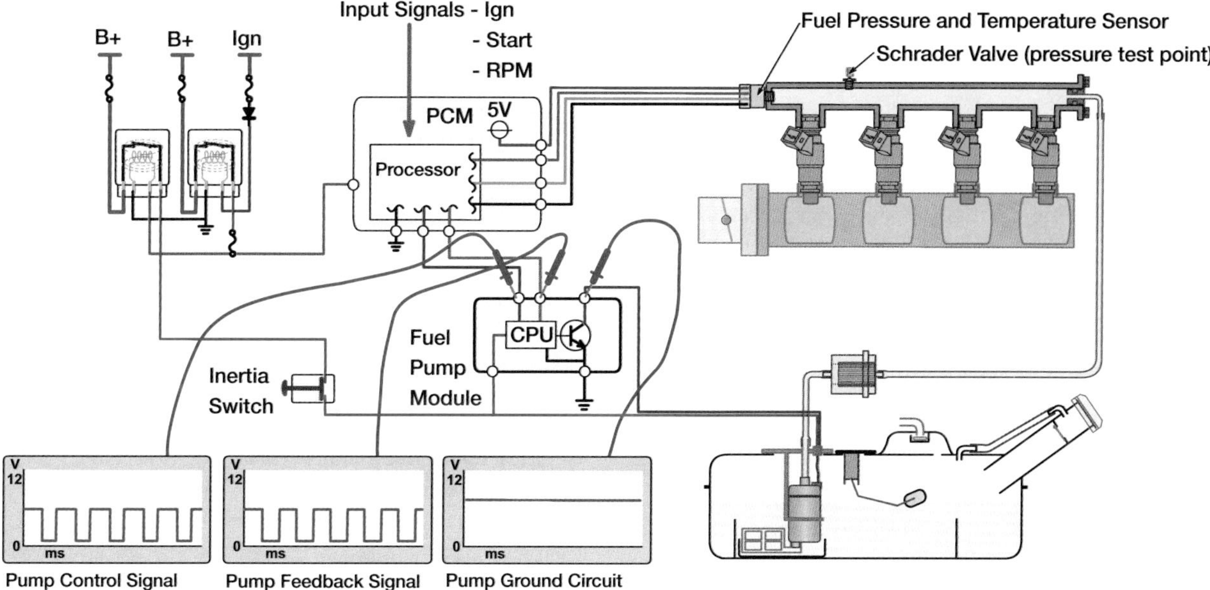

FIGURE 62-28 Electronically controlled fuel pressure.

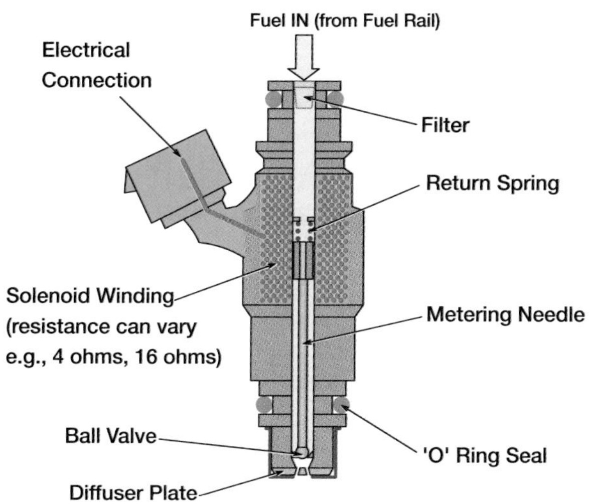

FIGURE 62-29 The inside components of a port fuel injector.

FIGURE 62-30 The modern fuel injector sprays a fine mist of fuel when activated.

manifold, or in the case of a GDI fuel system, between the fuel rail and the combustion chamber.

As you know, fuel is pressurized and waiting in the fuel rail. When the PCM grounds the electrical circuit from the fuel injector, current flows through the windings of the injector and opens the pintle valve. When it is time to close the pintle valve, the PCM opens the circuit to the fuel injector and stops the current flowing through the fuel injector. An internal spring closes the pintle valve, and it waits for the next "on" command. This happens repeatedly hundreds of times per minute at each fuel injector. The fuel injectors are designed and built to very exacting tolerances, giving them a high degree of fuel delivery precision and a long service life.

The response time to lift the injector needle to the fully open position is about 1 millisecond. If battery voltage is low, this response time takes longer, and the cylinders receive less fuel. As the voltage falls, the PCM can compensate for this slower opening time by extending the pulse width of the injector(s).

The injectors are sealed into the manifold by O-rings that prevent air entering at that point. O-rings are also commonly used to seal the injector to the fuel rail. The O-rings, together with plastic caps on the injector nozzles, also act as a barrier to heat being transferred to the injector body (**FIGURE 62-31**).

FIGURE 62-31 Typical O-rings and plastic cap act as heat barriers on a fuel injector.

▶ Electronic Fuel Injection Principles

K62003

EFI systems employ electronically controlled injectors to spray the fuel into either the intake manifold or the directly into the cylinder. The following discussion covers the basic EFI systems you will encounter as an automotive technician, and then describes specifically how these systems operate.

Types of EFI Systems

There are three basic EFI systems: throttle body injection, also called single-point injection; multipoint fuel injection; and gasoline direct injection, or simply direct injection (**FIGURE 62-32**). The specific characteristics of each of these designs will be explained further, but they all operate on similar principles: Fuel is supplied to the injectors at the specified pressure; the PCM sends an electric signal to each injector to cause it to open for a certain amount of time (pulse width); and fuel is injected into the intake manifold or combustion chamber.

Throttle Body Injection Systems

Throttle body injection (TBI), also known as single-point injection or central-point injection, is a system with one or two fuel injectors located centrally on the intake manifold, right above the throttle plates (**FIGURE 62-33**). Fuel is sprayed into the top center of the throttle body and then atomized with the incoming filtered air. This air-fuel mixture is then delivered to each of the engine's cylinders relatively evenly. TBI is the simplest type of EFI system, requiring only one or two injectors. This allows PCM to be a simpler, less powerful design. And because the fuel is sprayed above the throttle plates, it is at atmospheric pressure, so the pressure drop across the injector is always the same. Thus, fuel pressure does not have to change with throttle opening or engine load.

In TBI systems, the central injector is normally triggered on every ignition pulse. However, if there are two injectors, alternate triggering may be used. At idling speeds, the frequency may be less to provide finer control. TBI is the predecessor to modern-day multipoint (or multiport) fuel injection.

In TBI systems, fuel pressure and flow is provided by an electric fuel pump. It is typically located on the frame rail but can also be located in the fuel tank. Most TBI systems operate on fairly low fuel pressure, typically between about 15–20 psi (103–138 kPa).

Multipoint Fuel Injection Systems

In **multipoint fuel injection** (MPFI) systems, a fuel injector is used for each cylinder. Each injector is located in the intake manifold near each intake valve and sprays fuel toward the valve. Each injector is connected to the fuel rail, which supplies the injector with fuel under pressure (**FIGURE 62-34**). Each injector also has an electrical connector that provides it with power and ground. Most electronic fuel injectors are supplied with constant battery voltage when the ignition key is in the Run or Crank position. The PCM then switches the negative side of the injector circuit to ground to turn it on. When the PCM switches the circuit off, the spring inside the injector closes the injector, and fuel stops spraying.

In MPFI systems, fuel pressure and flow is provided by an electric fuel pump, most likely located in the fuel tank. The

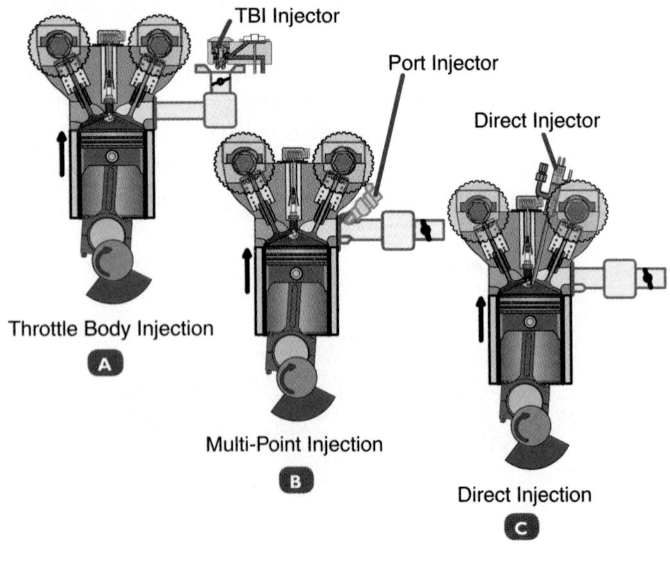

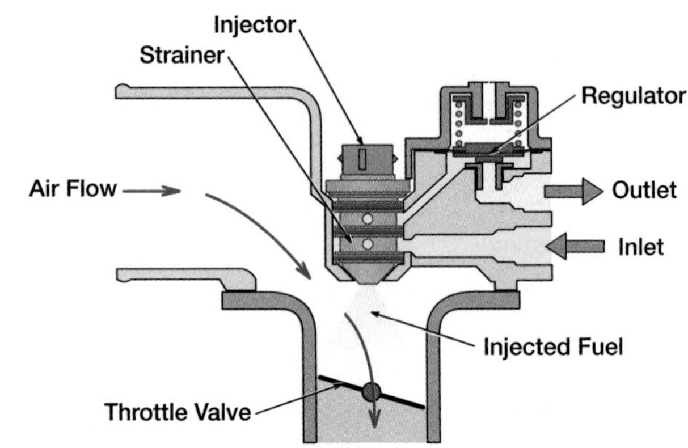

FIGURE 62-33 Typical TBI system.

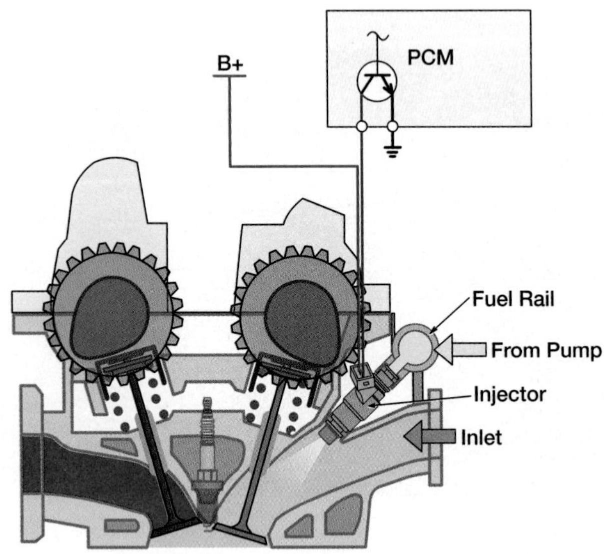

FIGURE 62-34 Typical MPFI system.

FIGURE 62-32 The three basic systems. **A.** Throttle body injection. **B.** Multipoint fuel injection. **C.** Direct injection.

system typically operates on higher fuel pressure than TBI, of approximately 35–70 psi (241–482 kPa).

Simultaneous Fuel Injection

In MPFI, the injectors can be triggered in various ways. The simplest way is to trigger them at the same time or in groups. This is called simultaneous injection, grouped injection, or banked injection. This arrangement operates the injectors twice per cycle—once each crankshaft revolution, each time delivering half the fuel for the cycle. In a six-cylinder engine, groups of three injectors are triggered every third ignition pulse (**FIGURE 62-35**). Grouping the injectors allows for a less powerful processor in the PCM and fewer drivers inside. It also means fewer terminals coming out from the PCM and fewer wires in the harness as there is only one control wire for each group of injectors.

Sequential Fuel Injection

Sequential injection means injection occurs in the sequence of the firing order. The injectors spray fuel following the firing order of the engine. Each injector typically opens only once in each cycle to deliver the fuel needed. Because each injector is controlled separately, the PCM has to have individual drivers to turn each injector on and off. This system requires more computing power to manage each injector circuit. It also requires more wires back to the PCM, as each injector requires its own circuit (**FIGURE 62-36**). Because each injector fires only once per cycle, the timing of the injection pulse is important, and therefore the position of both the number-one cylinder and the camshaft must be known. This also adds to the complexity of the system over a TBI or simultaneous-injection system.

Gasoline Direct Injection Systems

With the pressures to reduce emissions and increase fuel economy, as well as to gain an advantage over the competition and offset the escalating price of crude oil, manufacturers have looked to new technologies in fuel injection. Just as TBI replaced carburetion, **gasoline direct injection (GDI)** is the natural successor to **indirect fuel injection**. In fact, approximately half of all new vehicles sold in the United States are equipped with GDI engines. Although indirect fuel injection atomizes fuel at or near the intake valve, GDI systems take their design cues from diesel technology by spraying the fuel directly into the cylinder (**FIGURE 62-37**). Because of this, direct injection engines can run effectively on extremely lean fuel mixtures—much leaner than stoichiometric (14.7 parts fuel to 1 part air). Their fuel mixtures can be as lean as about 65 to 1! Also, because the fuel can be injected near the top of the compression stroke, detonation can be minimized, meaning that compression ratios can be higher, which also increases efficiency.

The fuel injector for each cylinder is located in the cylinder head. Fuel is directly sprayed into the combustion chamber as a very highly atomized mist at the precise time it is needed, depending on the operating conditions of the engine. There are four typical modes of operation, plus variations of these: stratified charge, stoichiometric, full power, and catalyst heating. The PCM selects the mode best suited for the operating conditions.

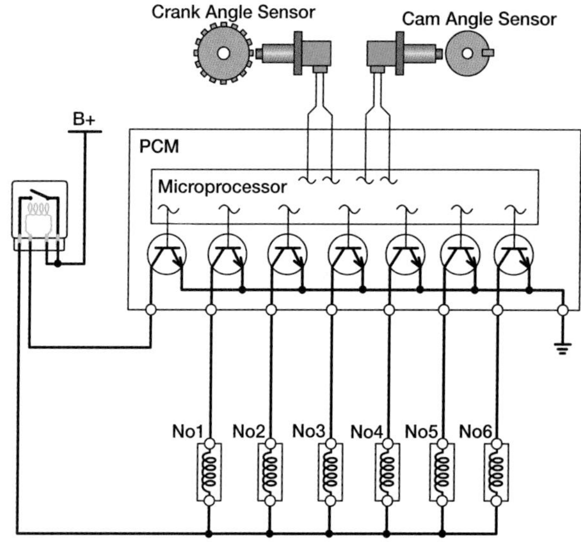

FIGURE 62-36 A typical sequential fuel injection schematic showing how each injector is controlled by the PCM.

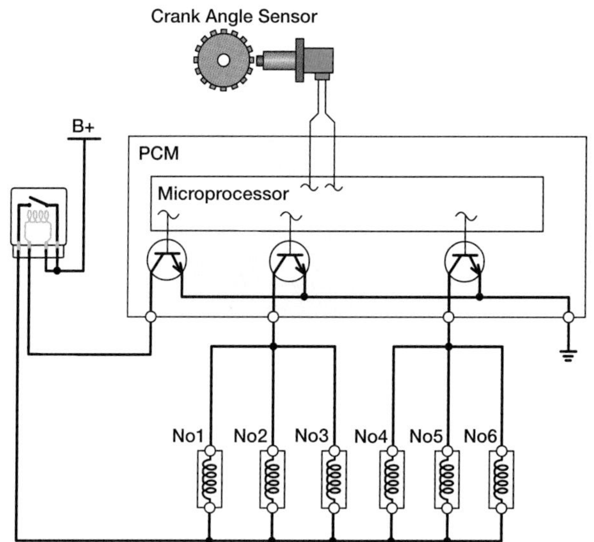

FIGURE 62-35 Typical simultaneous fuel injection circuit showing each group of injectors on a six-cylinder engine.

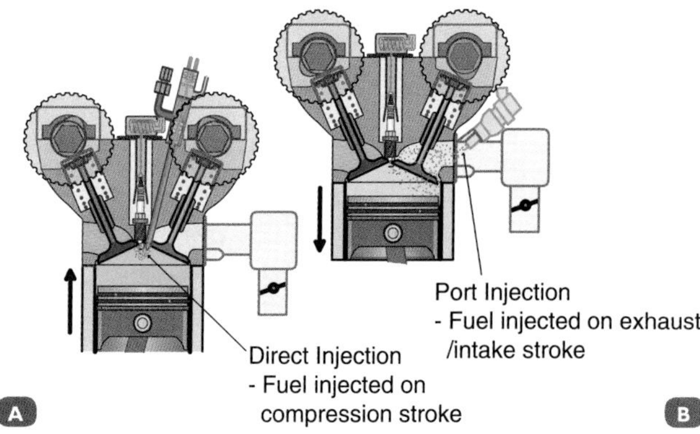

FIGURE 62-37 Injection systems. **A.** Direct injection system. **B.** Indirect injection system.

The **stratified charge** mode is designed to be used during light throttle, cruise conditions. It provides the best fuel economy of all of the modes because it runs so much leaner than the stoichiometric ratio. The reason this type of engine can run so lean is because the injector can inject fuel near the end of the compression stroke, when the piston is near the spark plug. The pistons have a specially designed pocket that helps concentrate the fuel in a small space near the spark plug. Various pocket designs are in use. This makes it so the fuel particles can be close enough together that they all burn (**FIGURE 62-38**). The surrounding area outside this pocket is primarily air and recirculated exhaust gases, which provide a barrier so the flame and heat stay further away from the cylinder walls. Heat loss is reduced in the combustion chamber, which also reduces the load on the cooling system. This is called a stratified charge because a combustible volume of fuel is mixed with air in just a small area, not thoroughly dispersed among the entire amount of air in the combustion chamber. This design greatly increases fuel economy.

The stoichiometric mode is designed for moderate engine load conditions, when a bit more power is needed than for light cruising. During this mode, just enough fuel is injected during the intake stroke to create a stoichiometric air-fuel ratio. Injecting on the intake stroke gives more time for the larger amount of fuel to mix more fully with the air. Because the fuel is injected on the intake stroke, it is able to disperse more evenly throughout the combustion chamber and is not contained by the pocket in the piston (**FIGURE 62-39**). This is called a **homogeneous mixture** and results in good power and a very clean burn.

The **full power mode** is designed for heavy engine load conditions, when full power is needed. During this mode, a slightly rich mixture is injected during the intake stroke in a homogenous mixture, creating maximum power and reduced spark knock (**FIGURE 62-40**). Injecting the fuel on the intake stroke in a very atomized condition causes the fuel to act as a

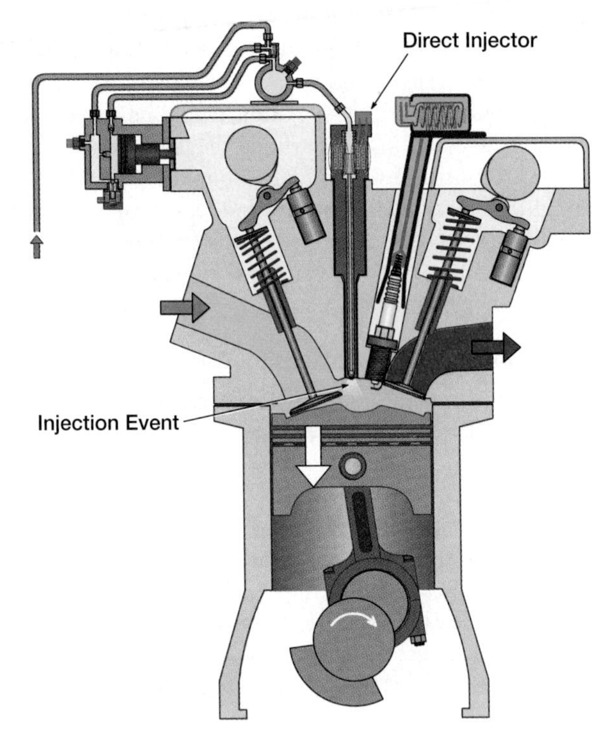

FIGURE 62-39 Stoichiometric mode. Fuel is injected during the intake stroke, fully mixing the air and fuel.

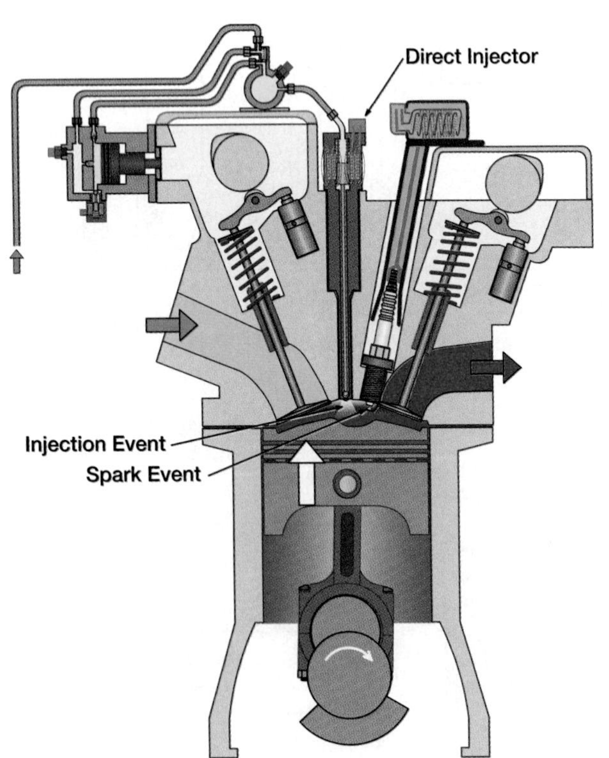

FIGURE 62-38 Stratified charge mode. Fuel is injected near the top of the compression stroke, near the spark plug.

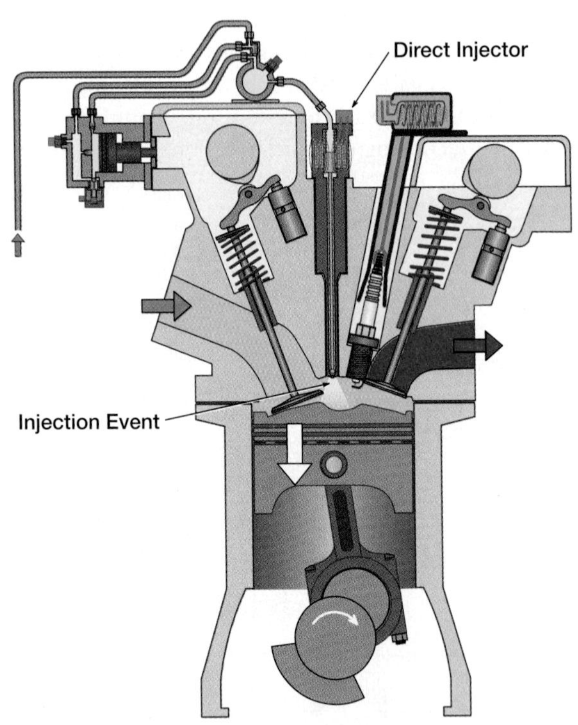

FIGURE 62-40 Full power mode. A slightly rich mixture is injected on the intake stroke. The additional fuel helps cool the charge to reduce detonation.

coolant, absorbing some of the heat in the combustion chamber, reducing detonation, and vaporizing it more fully.

The **catalyst heating mode** is designed to quickly heat the catalytic converter. It injects fuel in the stratified mode near the top of the compression stroke. It also injects a small amount of fuel during the power stroke. This delayed fuel continues to burn for a longer period of time and is used to heat up the catalytic converter.

So far GDI engines still use spark plugs. Using spark plugs allows the GDI system to accommodate the different operating modes listed above. For example, if ignition of the air-fuel mixture depended on the heat of compression, then fuel couldn't be added during the intake stroke as it would ignite prematurely on the compression stroke, making the engine detonate and destroy itself. Using a spark plug allows the ignition timing to be carefully controlled and adjusted as needed.

Speaking of timing, many GDI fuel systems can provide more than one injection event per cycle. For one example, a small amount of fuel may be injected in stratified charge mode at the top of the compression stroke, and a small second injection event following ignition provides a small amount of additional power while still maintaining a lean mixture (**FIGURE 62-41**). If the second injection event is delayed, then it provides the ability to quickly heat the catalytic converter.

Another example of multi-pulse injection events is when a moderately lean fuel mixture is injected during the intake stroke, and a small amount of fuel is injected near the top of the compression stroke to act as a primer charge. This results in a more complete burn of a moderately lean mixture. Some manufacturers accomplish this by using both direct injection fuel injectors in the cylinder head and a second set of common indirect fuel injectors in the intake manifold (**FIGURE 62-42**). The indirect fuel injectors operate when fuel is needed for moderate and heavy power demands. They inject fuel just like a sequential EFI system. The direct fuel injectors operate under light loads or in tandem when full power is needed.

GDI fuel systems can be either a low-pressure variety (only inject during the intake stroke or early in the compression stroke) or a high-pressure variety (can inject during the intake stroke, the compression stroke, and potentially the power stroke). Low pressure systems operate on a fuel pump system, similar to an MPFI system, with about 35–70 psi (241–482 kPa). High-pressure systems generally have a low pressure pump, like the one just described, which feeds a high-pressure pump that raises the fuel pressure up to as much as 3000 psi (20,684 kPa).

Most high-pressure GDI systems include the following components (**FIGURE 62-43**):

- Low-pressure pump—An in-tank electric pump used to supply fuel to the high-pressure pump at the correct pressure and volume. The pressure is typically controlled by the PCM.
- Low-pressure sensor—Used to monitor the fuel pressure in the low-pressure system. The PCM uses this reading to control the output of the low-pressure fuel pump to maintain the proper pressure.
- High-pressure pump—A mechanical pump capable of creating fuel pressure of up to approximately 3000 psi (20,684 kPa) in some vehicles. It is typically operated off an eccentric with two or three cam lobes on the camshaft.

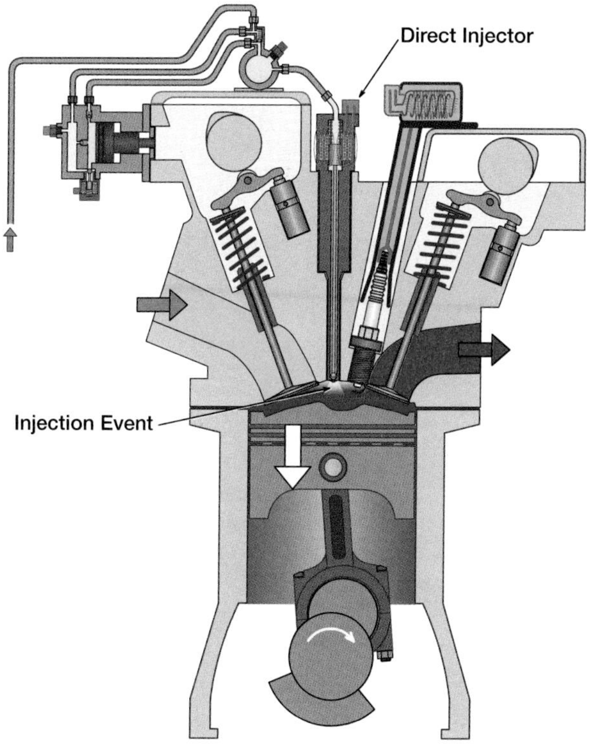

Direct Injector
Injection Event

FIGURE 62-41 A multi-pulse injection event.

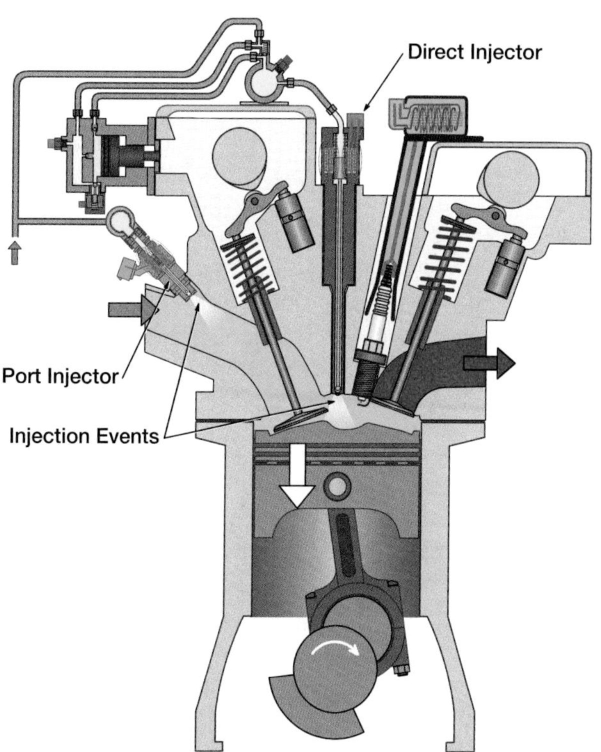

FIGURE 62-42 Some manufacturers use both a direct injection system and an indirect injection system.

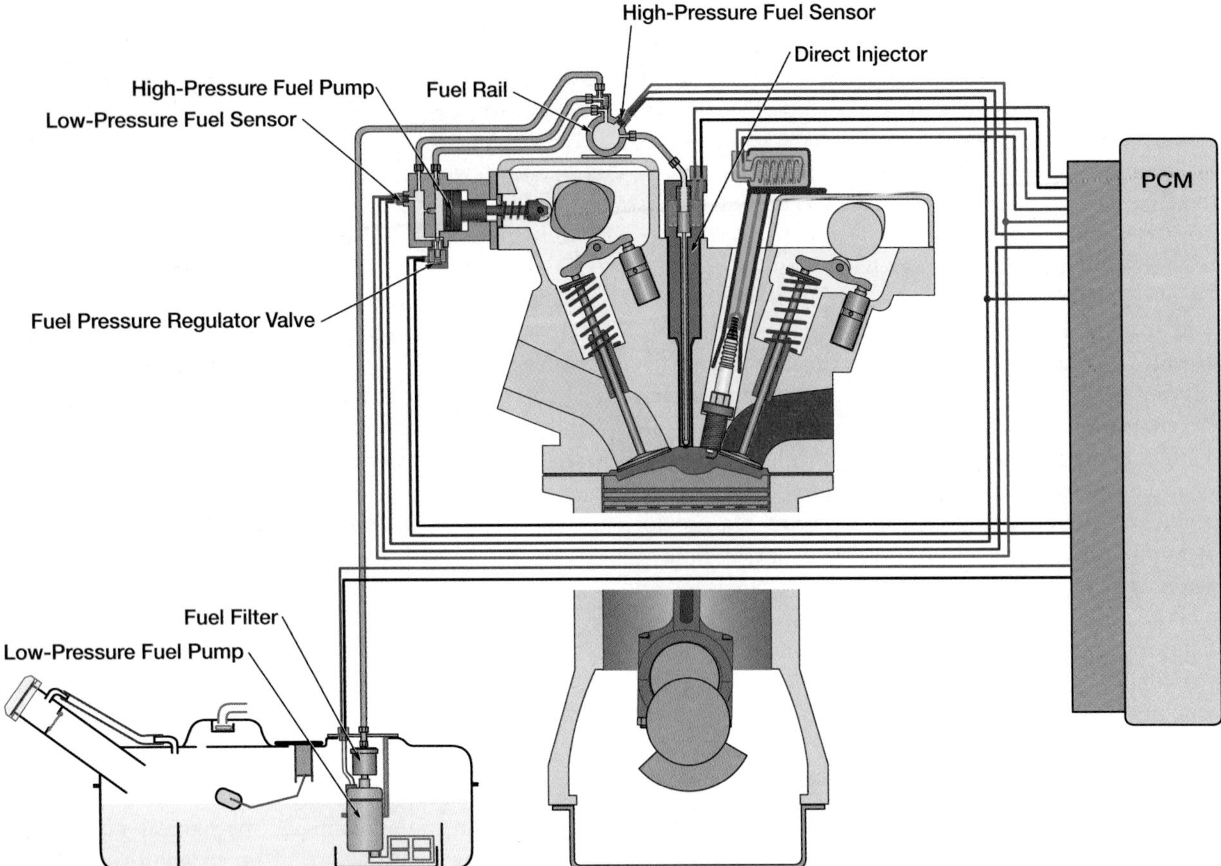

FIGURE 62-43 Typical layout of a GDI system.

A roller rides on the lobes and moves a plunger in a bore, which creates the high pressure needed in GDI systems. The pressure is maintained in the fuel rail, so it is available to all injectors equally.

- High-pressure sensor—Used to monitor the fuel pressure in the high-pressure system. The PCM uses this reading to control the output of the fuel pressure regulator valve, based on the operating condition of the engine.
- Fuel pressure regulator valve—Used to control the fuel pressure in the high-pressure system. This valve is typically electromagnetically controlled by signals from the PCM. The fuel pressure regulator valve uses a check valve in the outlet to hold the high pressure in the fuel rail when the engine is shut down. It is not unusual for the high pressure side to hold substantial pressure for up to two hours.
- High-pressure fuel injectors—These injectors have to accurately control very high fuel pressures, but also the nozzles are subject to the high pressures and temperatures of combustion. These injectors are electromagnetically operated solenoid valves, some of which operate on voltage higher than 12 volts.

GDI Drawbacks

Although GDI systems have a lot of positive qualities such as increased fuel economy, reduced emissions, and more power, there are some negative points as well. Besides being more expensive, complicated, and noisy, one of the most common drawbacks that technicians run into is their tendency for carbon to build up on the intake valves and runners (**FIGURE 62-44**). This reduces airflow into the engine, which reduces volumetric efficiency and engine power. Because of this tendency, it is critical to use gasoline with adequate detergents. One way to do this is to use Top Tier Detergent Gasoline™ fuel, which has much more deposit-reducing detergent in it than the minimum

FIGURE 62-44 GDI engines are prone to carbon buildup on the intake valves and runners.

specified by the EPA. It has also been tested to demonstrate its ability to reduce the buildup of carbon deposits. Another method of reducing carbon buildup is to use a high-quality motor oil specified for the vehicle, and change the oil at the recommended intervals. This will prevent oil vapor from creating additional carbon deposits.

If customers don't take the preventive actions above, and the valves carbon up, there are several ways to remove the carbon buildup, but none of them are quick, easy, and without risk. Here are the options:

- Remove the cylinder head, disassemble it, and clean it. This is absolutely the least risky way to remove the carbon, especially if the buildup is very heavy. But it is also likely to be the most expensive method unless you factor in that the other methods have various levels of catastrophic risk to the engine.
- Remove the intake manifold and use a crushed walnut shell blaster to blast away the carbon. The walnut shells are soft enough not to damage the base metal while blasting away the carbon. There is a risk of walnut shells and debris getting where it shouldn't be and causing problems down the road. But that is minimized if this procedure is carried out carefully.
- Remove the intake manifold, and use a chemical cleaner to soften and dissolve the carbon so it can be removed. This is a fairly safe method as long as all of the cleaner and loosened carbon is removed. There are two primary dangers. First, if there are hard carbon deposits that don't get removed, they can pass into the combustion chamber and get wedged between the top of the piston and cylinder head, creating an engine knock and possibly damaging engine bearings, pistons, or connecting rods. The second way is if chemical cleaner gets caught in an inaccessible cavity and is drawn into the cylinder during start-up. This can cause hydrolocking of the cylinder, which can cause the same damage as just listed.
- Induction cleaning is the least invasive because minimal engine disassembly is required. It is a good option on engines that only have soft carbon buildup, not hard carbon deposits. This is likely to be the case on engines with less than about 30,000 miles (48,280 km). This method has the same dangers as with the previous option—the possibility that there are unknown hard carbon deposits that will be loosened up and damage the engine; or chemical cleaner can pool in cavities and cause hydrolocking. Another problem with this method is on vehicles equipped with turbochargers. The induction cleaning chemicals can overheat the turbo and even the catalytic converter when the vehicle is driven while the chemical is working.

Another drawback to GDI systems is the condition called low-speed pre-ignition, which causes severe pressure waves in the combustion chamber that can catastrophically damage components in just a few revolutions of the engine. It is thought to be caused by very small, poor-quality oil droplets that work their way past the piston rings into the combustion chamber. Once there, they ignite during the compression stroke (pre-ignition) and ignite the air-fuel mixture, causing severe pressure waves as the piston is still moving up. This condition can be reduced by using quality oil that is specified by the manufacturer. In fact, Toyota mandates a specific oil to reduce this condition in some of their GDI vehicles.

▶ Carbureted Fuel Systems

K62004

We are covering carburetion lightly because some states still require the teaching of carburetors. Also, some instructors like to use the mechanical systems of the carburetor to introduce the components of EFI systems that do the same jobs. This section gives you some background on how carburetors operate in case you ever have to work on one. **If you have no interest in carburetors, skip this section.**

Initially, manufacturers used carburetors and mechanical fuel pumps on their vehicles to deliver fuel to the engine. The carburetor system is a supply-and-demand fuel delivery system. The carburetor was usually fed gasoline by an engine-driven mechanical fuel pump. Traditionally, fuel pumps for a carburetor-type fuel system were a diaphragm-type pump mounted to the engine block and driven directly by the camshaft. Whenever the engine was running, the pump would draw fuel from the tank and deliver it to the carburetor with a low amount of fuel pressure. This delivery of fuel was then held up at the carburetor's float bowl, which was kept at bay by a needle and seat (**FIGURE 62-45**). The pressure in the fuel line was low; thus delivery into the carburetor occurred when the float lowered enough to allow fuel to pass by the needle through the seat. Once the bowl was full again, the needle was closed off against the seat. In actual operation at a steady speed, the float dropped just enough to allow the same amount of fuel in that was being drawn out.

Carburetors, when tuned properly, delivered fuel effectively. However, if not properly maintained over time, they would cause drivability and emissions problems. As electronics came into play in the automotive industry, carburetors were eventually phased out, with manufacturers opting for more effective, efficient, and reliable **electronic fuel injection (EFI)** systems.

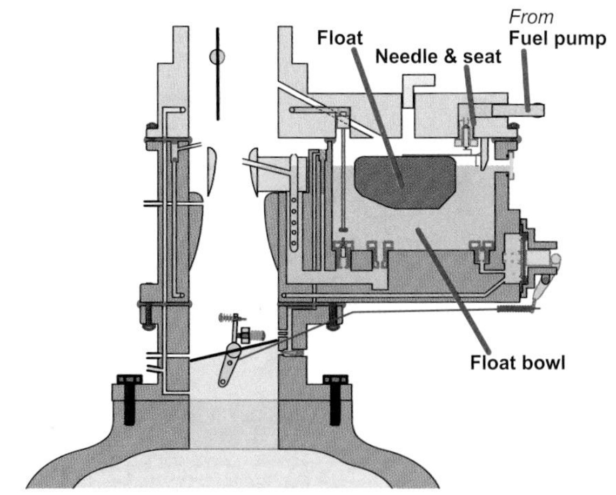

FIGURE 62-45 Float bowl and float system.

The carburetor supplies the engine with the approximately correct air-fuel mixture for all conditions of operation.

- It atomizes the fuel and mixes it with air.
- It controls the delivery of this approximately correct mixture to the intake system.

Carburetors come in different designs (**FIGURE 62-46**). Most carburetors are the downdraft design, but there are also side-draft carburetors and updraft carburetors. These terms refer to the direction air flows through the carburetor. Each type has a float bowl where a float and a needle valve control the fuel level. The air horn and **venturi** are located in the top of the barrel of the carburetor. A throttle valve controls airflow through the venturi and is linked to the throttle pedal.

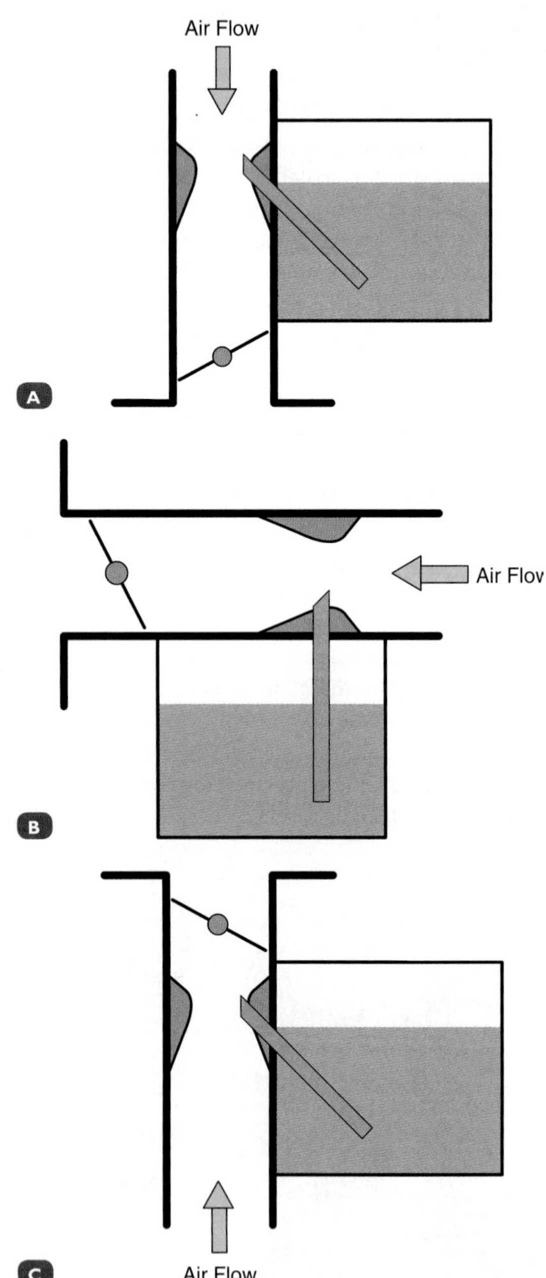

FIGURE 62-46 Types of carburetors. **A.** Downdraft. **B.** Side draft. **C.** Updraft.

As the piston moves through its intake stroke, it creates a low-pressure area, and as a result, air from the atmosphere flows through the venturi (**FIGURE 62-47**). The venturi is narrower than the rest of the barrel, and it is shaped to make the air speed up as it passes through. A similar effect occurs around the wings of aircraft. The shape of the wing section speeds up the airflow over the top of the wing and creates a low-pressure area there, lower than the atmospheric pressure below. The result is an upward force that provides lift for the aircraft.

The shape of the venturi is designed to apply the same principle, known as the Bernoulli effect. It creates a low-pressure area where the end of the nozzle protrudes into the airflow. Atmospheric pressure in the float bowl is greater than the pressure on the end of the nozzle, forcing fuel from the nozzle. The fuel mixes with the passing air, breaking up into droplets, or atomizing.

Depressing the accelerator increases air speed through the carburetor, thus lowering air pressure at the nozzle. Pressure on the fuel in the float bowl stays constant, so more fuel is forced into the venturi to mix with the increased air. This keeps the air-fuel ratio roughly constant for a range of throttle openings. The throttle valve also controls flow of mixture into the engine. Opening it allows more mixture to be delivered, which increases engine power and speed. Closing it has the opposite effect.

Carburetor Operation

The carburetor is a mechanical device that delivers the approximately correct mixture of air and fuel to each of the cylinders for combustion during the power stroke. It is bolted to the intake manifold. While the engine is running, the intake stroke of each piston creates a low-pressure area (vacuum) in the intake manifold. The vacuum created in a mechanically sound gasoline engine at sea level is between 18" and 21" (457 and 533 mm) of mercury. With atmospheric pressure at sea level being 14.7 psi (101 kPa), a moderate pressure differential results. That differential in pressure allows the clean filtered air to enter the carburetor. This pressure is used to meter out the corresponding

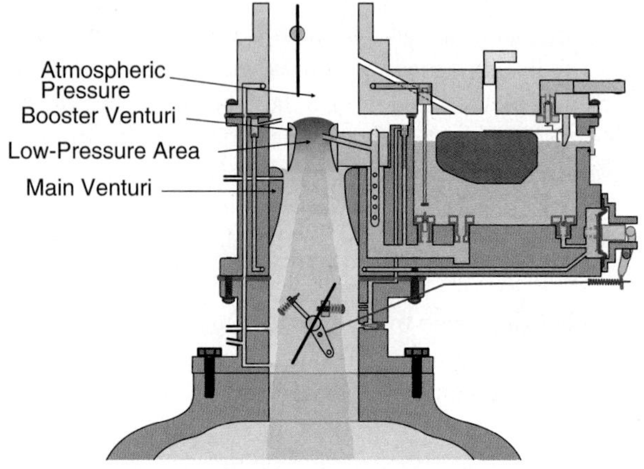

FIGURE 62-47 As air is drawn through the venturi, its speed increases and its pressure drops, creating a low-pressure area.

amount of fuel over varying and constantly changing condition requirements. Ideally, the carburetor governs the air-fuel mixture as precisely as possible.

Carburetor Circuits

The basic carburetor components are housed inside a metal casting called the carburetor body. This carburetor body serves as the mounting point for the various essential carburetor system components, which are grouped together to create carburetor circuits. Each of the carburetor circuits is designed to provide a certain function to the carburetor. The six circuits are the float, idle, main metering, power, accelerator pump, and choke.

Float Circuit

The **float chamber** holds a quantity of ready-to-use fuel at atmospheric pressure. Its supply is refilled by a float-driven valve; as the level drops, the float drops too and opens an inlet, which allows the fuel pump to deliver more fuel to the float chamber. The float rises with the replenished fuel level, closing off the inlet. To allow atmospheric pressure to act on the fuel, the float bowl is open to either the atmosphere (unbalanced carburetor), the air horn above the venturi (balanced carburetor), or the **charcoal canister** (evaporative emission carburetor) (**FIGURE 62-48**). If the float level is too low, more airflow through the venturi will be required to pull out the fuel, leaning out the air-fuel ratio. Consequently, too high a float level will cause the mixture to be too rich. Float adjustment is important when rebuilding a carburetor. Flooding a carburetor also produces rich mixtures. Flooding can be caused by a worn needle and seat, or by dirt trapped between the needle and seat that causes the level in the float bowl to rise and fuel to dribble from the nozzle, resulting in little or no venturi action.

Idle and Off-Idle Circuits

When the throttle valve is closed or nearly closed, the manifold vacuum created behind the throttle is sufficient to pull a small amount of fuel and air through small openings located after the butterfly valve (**FIGURE 62-49**). This design is called the idle circuit, and it enables the engine to keep running when there is not enough air speed through the venturi to create a vacuum. As the throttle valve opens slightly, the manifold vacuum is reduced, so additional small openings are revealed to compensate for this. This design is the "off-idle" circuit.

Main Metering Circuit

The main metering circuit comes into action above fast idle, as airflow through the venturi increases. A main **metering jet** in the float bowl meters fuel passing into the discharge nozzle (**FIGURE 62-50**). How much fuel leaves the nozzle depends on the pressure difference created by the airflow through the venturi. As the throttle opens, and airflow increases and speeds up, more and more fuel is drawn from the discharge nozzle. However, the mass of air does not increase in proportion with the speed, and as a result, high speeds can produce a mixture that is too rich. To correct this, more air can be added. This is called compensation by air correction.

As the throttle opens and engine speed increases, the level in the jet well falls, exposing air bleed holes in the discharge tube. Air can now mix with the fuel and prevent the mixture from becoming too rich. As the throttle opens farther, the fuel level falls too, exposing more air-holes. More air bleeds in to maintain the correct mixture. Main metering fuel flow can typically be adjusted by replacing the removable jets with jets having larger or smaller orifices.

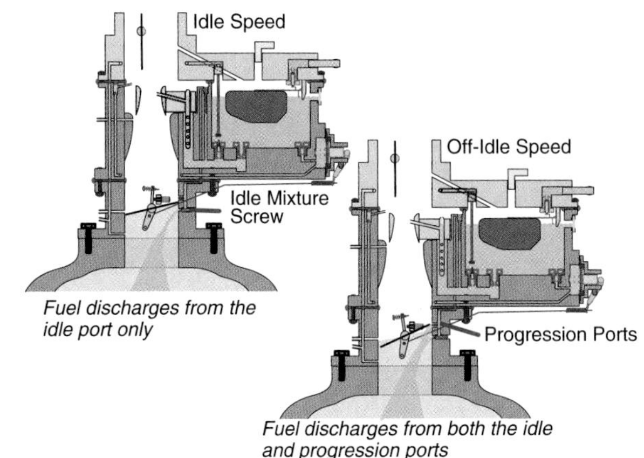

FIGURE 62-49 Idle and off-idle circuit.

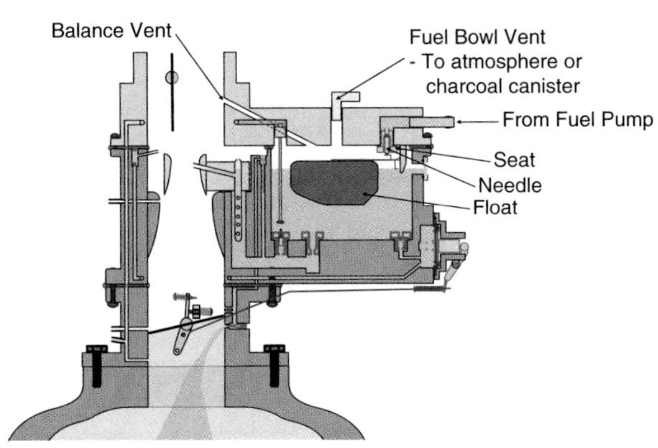

FIGURE 62-48 Float bowl, float, and needle and seat.

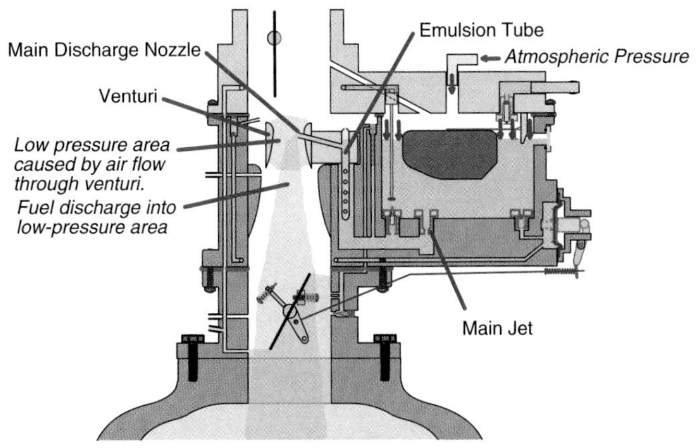

FIGURE 62-50 A main metering jet in the float bowl meters fuel passing into the discharge nozzle.

Power Circuit

The size of the main jet is selected to provide the best mixture for economy under cruising conditions. When the throttle is open wide for maximum power, a richer mixture is required. The extra fuel is provided by a power valve, with a vacuum piston and rod opening it as it is needed. At low speeds, intake manifold vacuum is transferred through a passage to the vacuum piston. This holds the piston up and keeps the power valve closed. With the throttle valve fully open for full engine power, the vacuum in the intake manifold falls. A spring pushes down the vacuum piston and rod to open the power valve. Additional fuel flows through the power valve to enter the fuel well and add to the fuel from the main jet. This provides the extra fuel needed to enrich the mixture for full power. Some carburetors use metering rods instead of a vacuum piston. The metering rods are pulled down into the main jets at idle and cruise to restrict the fuel flow. When manifold vacuum drops under heavy load, springs push the metering rod(s) up, increasing the opening size of the main jet(s). Other carburetors use a diaphragm-type power valve that opens an additional passage when vacuum drops under load.

Accelerator Pump Circuit

Extra fuel is also needed for accelerating. Suddenly opening the throttle increases the airflow, but fuel cannot flow from the discharge nozzle quickly enough to match it. An extra squirt of fuel is needed, which is where the **accelerator pump circuit** comes into play. Depressing the pedal compresses a duration spring that exerts a force on the plunger of a small plunger pump (**FIGURE 62-51**). This pressurizes fuel below the plunger and closes off the inlet valve. Fuel flows past a check valve and enters the airstream from a discharge nozzle above the venturi. The duration spring extends the time for delivering the fuel. Releasing the pedal lets the linkage move the plunger upward. The check valve closes, and the inlet valve opens to let fuel refill the pump chamber from the float bowl, priming it for the next shot of fuel. Thus, whenever the throttle is opened, the **accelerator pump** discharges a small amount of fuel into the throat of the carburetor. So constantly working the throttle pedal when driving down the road wastes fuel.

The Choke

Fuel ignites less readily when cold, and if the engine is also cold, then some fuel vapor can condense out of the air-fuel mixture onto the intake manifold and cylinder walls. This loss of air makes the combustible mixture leaner. To compensate for this, a valve known as the **choke** restricts the flow of air at the entrance to the air horn, lowering the pressure at the venturi and off-idle circuits even though the throttle valve has been opened. In this way, fuel is sucked into the incoming air through all the fuel circuits—idle, off-idle, and main—at the same time.

The choke can be controlled manually by a cable that operates the valve. However, most are controlled automatically so that the valve is closed when the engine is cold and opens progressively as the engine warms up (**FIGURE 62-52**). When the engine is warm, the fuel drawn into the manifold vaporizes readily, and the engine can be started without the aid of a choke. The choke should operate as briefly as possible. Overusing it produces rich mixtures that cause extreme amounts of exhaust pollution and increase fuel consumption. Some later-model carburetors that used a cable-operated choke also used a spring-loaded choke release that turned the choke off after a set amount of time.

Carburetor Barrels

Carburetors can have one, two, three, or four barrels. Extra barrels improve performance, particularly at high speeds, letting more air and fuel enter the cylinders than with a single barrel. However, a carburetor that is too large reduces engine power because not enough airflow can be maintained through the venturi to draw the fuel out properly. Two-barrel carburetors have two outlets to the intake manifold and come in two basic designs. One has a common float chamber, with each barrel having a complete set of all other circuits and the throttles having the ability to open simultaneously. The float chamber may be straddled by two connected floats that almost surround the air passage. This design leaves the float chamber unaffected by cornering, climbing, accelerating, or braking.

In the other basic design, called a progressive carburetor, throttles open in two stages (**FIGURE 62-53**). This design combines the two barrels to act as a single carburetor. The two stages combine good low-speed operation of a single-barrel design with the extra airflow of two barrels.

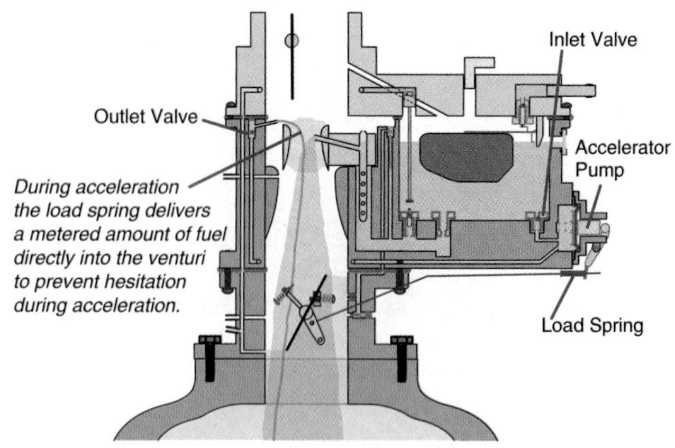

FIGURE 62-51 Accelerator pump circuit.

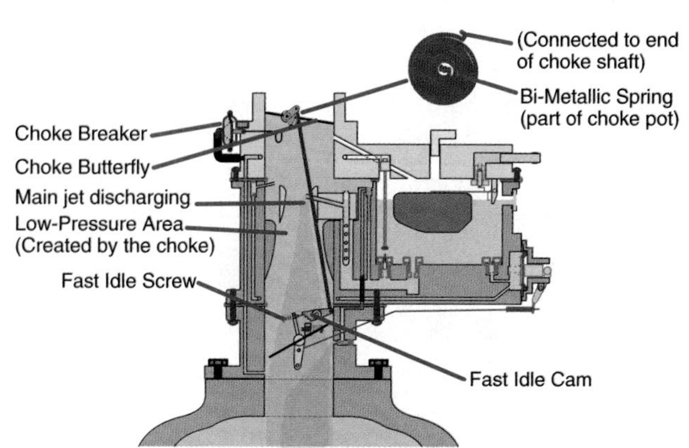

FIGURE 62-52 The choke.

FIGURE 62-53 A progressive carburetor.

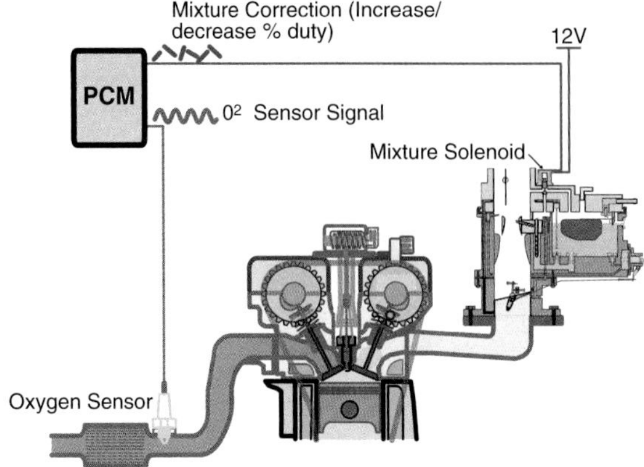

FIGURE 62-54 A computer-controlled carburetor normally uses an electronically controlled solenoid valve.

One of the two barrels of the carburetor has all the circuits needed to supply mixtures for the whole range of operation. It is called the primary side. The other barrel, the secondary side, supplies extra mixture, but only at high speed or full throttle. It normally has a main metering system and a nonadjustable idling system. The primary side has a choke for cold starting. When the engine is being started, the throttle on the secondary side is already closed, so a choke is not needed. From idle to medium speeds, only the primary throttle is open. When engine speed rises to where additional breathing capacity is needed, the secondary throttle opens to admit more air-fuel mixture. By the time the primary throttle is wide open, so is the secondary throttle. Opening the throttles can be controlled mechanically or by a vacuum unit that is connected by a pull rod to a lever on the secondary shaft. When air flows past ports in the venturis, it produces low-pressure areas. A hose transmits this low pressure to a diaphragm chamber. This low pressure acts on the diaphragm and opens the secondary throttle.

Large-capacity V8 engines may use a four-barrel carburetor of two-stage design—effectively two, two-stage carburetors combined, with each side supplying four of the eight cylinders. Some carburettors have a central molded plastic fuel bowl and suspended metering system that is incorporated into the design. The result is lowered fuel temperatures with more precise fuel metering and closer control over air-fuel ratios.

Computer-Controlled Carburetors

A **computer-controlled carburetor** normally uses an electronically controlled solenoid valve, called a mixture control solenoid, to respond to the PCM commands (**FIGURE 62-54**). This type of carburetor system uses various sensors in the exhaust and engine to monitor operating conditions. In turn, these sensors send that information to the PCM. According to the sensor data, the PCM then calculates rich or lean fuel conditions and adjusts the air-fuel mixture accordingly. The computer constantly sends commands to the mixture-control solenoid to open and close air and fuel passages in the carburetor.

Mechanical Fuel Pump

Fuel pumps on carbureted systems can be electric or mechanical. The mechanical fuel pump for the carburetor system is usually mounted on the cylinder head or the engine block. It has a flexible diaphragm, which is a flexible piece of neoprene rubber separating two chambers (**FIGURE 62-55**). This diaphragm is operated by an eccentric on the camshaft. The eccentric rotates, making the rocker arm move. This movement is transferred to the diaphragm, pulling it down. As a result, fuel is drawn into the pumping chamber on the other side of the diaphragm. The diaphragm spring moves the diaphragm up, forcing fuel from the pumping chamber, out of the pump and into the carburetor.

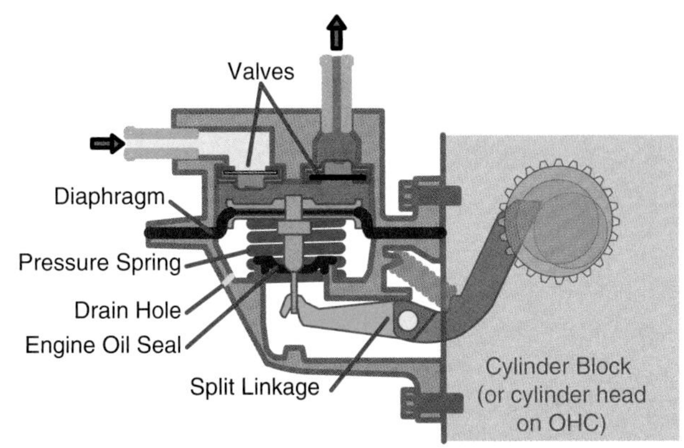

FIGURE 62-55 The mechanical fuel pump has a flexible diaphragm—a flexible piece of neoprene rubber separating two chambers.

When the engine needs more fuel, the diaphragm moves through a long stroke to pump more fuel. When less fuel is needed, pressure builds up in the fuel line to the carburetor and in the pumping chamber. The diaphragm spring cannot push the diaphragm so far, and the pumping stroke is reduced. Some pumps also have a return line to send excess fuel back to the tank. As the fuel circulates, it cools the fuel pump and lines, reducing the chance of vapor lock.

Low-Pressure Electric Fuel Pump

Carburetor systems can also use electric fuel pumps. Most of these pumps are located outside the tank, though some are inside the tank. One widely used pump is the diaphragm type (**FIGURE 62-56**). It has an electrical section and a mechanical section. When the ignition is switched on, current magnetizes the solenoid. The magnetic field energizes an armature, which

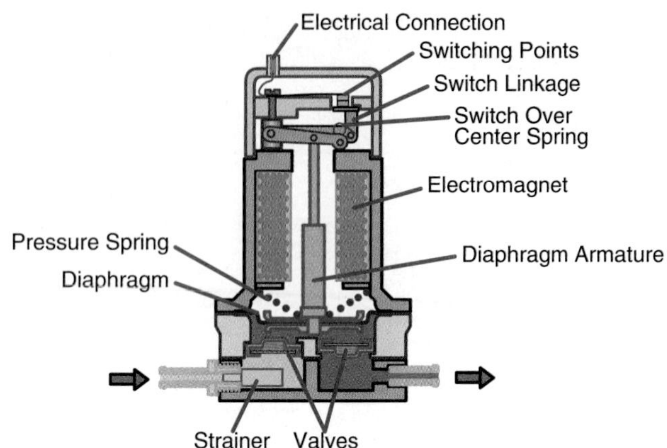

FIGURE 62-56 Diaphragm-type electric fuel pump.

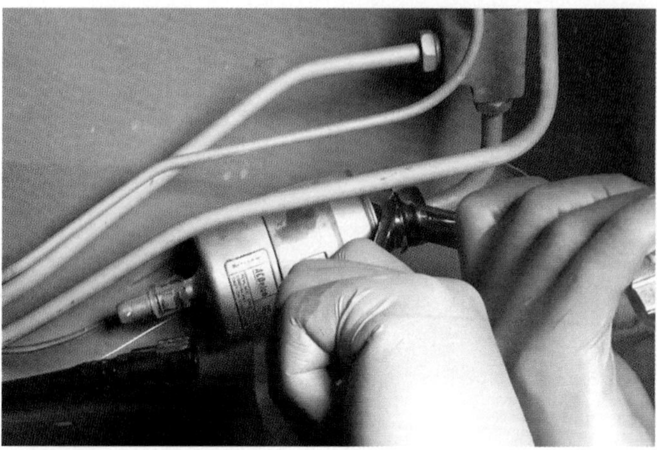

FIGURE 62-57 When removing filters with flared fittings or banjo fittings, use the double wrench method to loosen and tighten them.

pulls down the diaphragm. This creates a low-pressure area that draws fuel into the pump.

Pulling down the diaphragm breaks the circuit and stops the current, causing the solenoid to no longer be an electromagnet. The armature is released, and the diaphragm spring forces up the diaphragm, which forces fuel out of the pump and to the carburetor. This action continues, delivering fuel each time.

When the engine needs less fuel, pressure builds up in the fuel lines to the carburetor and reduces how much fuel is delivered. Some carburetor-equipped vehicles with electric pumps have a safety switch that prevents the pump from continuing to run if the engine stops.

▶ Maintenance and Repair

Replacing a Fuel Filter

N62001, S62001

Fuel filters are a maintenance item and need to be replaced according to the manufacturer's specified replacement schedule, or anytime restricted flow is encountered. Fuel filters can be located in a variety of places, depending on the vehicle. Common locations are under the vehicle in the fuel line, under the hood in the engine compartment, and inside the fuel tank as part of the fuel pump assembly. Because fuel pressure is maintained in all electric fuel pump systems, even when the engine is not running, it is important to release the pressure before opening the system so that fuel does not spray everywhere. There are several ways to relieve the static pressure in the fuel system before removing the fuel lines. One way is to remove the fuel pump relay or the fuel pump fuse, and run the engine until it dies. Another method is to connect a fuel pump pressure gauge (that has a bleed valve) to the fuel rail test point, and release the excess pressure from the system into a gas can. On GDI-equipped engines, you may need to use a scan tool to release the fuel pressure. Just remember that fuel pressure on the high side of a GDI system can be up to 3000 psi (20,684 kPa), so following the manufacturer's specified procedure is critical for safety purposes.

No matter how you release the static pressure in the fuel system, vent the pressure in the gas tank by removing the gas cap. Doing so reduces the amount of gas that dribbles out of the lines when removing the filter. Always refer to the service information for the specific procedure for removing and replacing the fuel filter for the vehicle you are working on.

Some vehicles use metal or plastic lines that bolt onto the filter, using either a flared fitting or a banjo bolt. For either type, you need to use the double wrench method to unbolt the line; otherwise, you may twist or kink the metal or plastic line. The double wrench method uses two wrenches, one on the filter nut and the other on the flare nut or banjo bolt (**FIGURE 62-57**). Use the wrench on the filter nut to prevent the filter from twisting while applying force to the other wrench. Often you can place the wrench handles so they are at a slight angle to each other and squeeze them together to break the fittings loose.

Many newer vehicles use quick couplers to retain the fuel lines on the fuel filter. These couplers make it very quick and easy for the factory to install the filter because the fuel filter lines only have to be pushed into or onto the filter inlet and outlet. The quick coupler may use a plastic clip or a coiled spring to retain the line on the filter. Only use the proper release tool when removing the line; otherwise, you could damage the fitting or line. Also, most quick coupler systems use O-rings to seal the fuel line to the filter. Make sure the O-rings are in good shape, or replace them with new ones.

If the fuel lines connecting the fuel tank to the filter are flexible hoses rather than metal lines, check their condition to determine whether it is necessary to replace the hoses and clamps when you replace the filter. Some replacement filters come with these items; when they are supplied, you should always use them. If they are not supplied, but you need to replace them anyway, obtain a sufficient length of the proper type of new fuel line and suitable clamps. There are different types of clamps for flexible fuel lines, including spring type, worm type, and rolled edge. You will need to use the appropriate tool when installing new clamps on the hoses.

Some manufacturers have been installing the fuel filter in the fuel tank along with the fuel pump assembly (**FIGURE 62-58**). The only way to get to the pump and filter is through the top of the fuel tank. Sometimes manufacturers provide a removable access cover under the backseat that allows you to get to

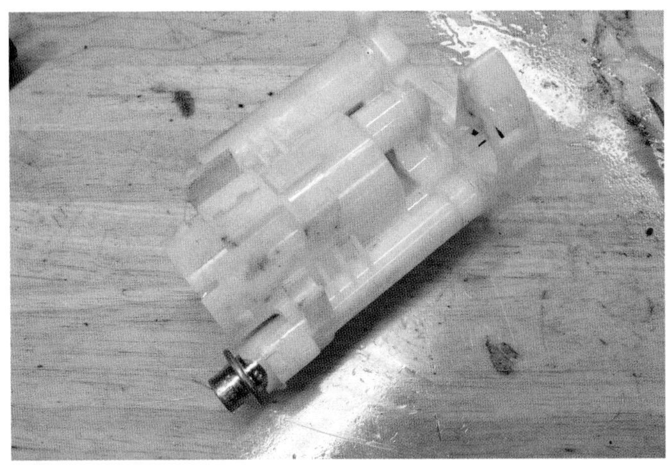

FIGURE 62-58 Typical fuel filter assembly located in the fuel tank.

the top of the tank. Other times the tank has to be removed to gain access. Once the covers are removed, the fuel pump assembly can be carefully removed. On some vehicles, the filter can be replaced separately. On others, the fuel filter is changed as part of the fuel pump assembly. In this case, it is not normally a maintenance item.

SAFETY TIP

Gasoline fuel vapor is explosive and highly flammable. Be careful not to spill any fuel onto a hot engine component, where it could evaporate, ignite, and start a fire. Also take care not to cause any sparks while you are changing a fuel filter. Collect the gasoline waste in a metal container, and dispose of it in an environmentally prescribed way. Always wear the appropriate personal protection equipment before starting the job.

To replace a fuel filter, follow the steps in **SKILL DRILL 62-1**.

SKILL DRILL 62-1 Replacing a Fuel Filter

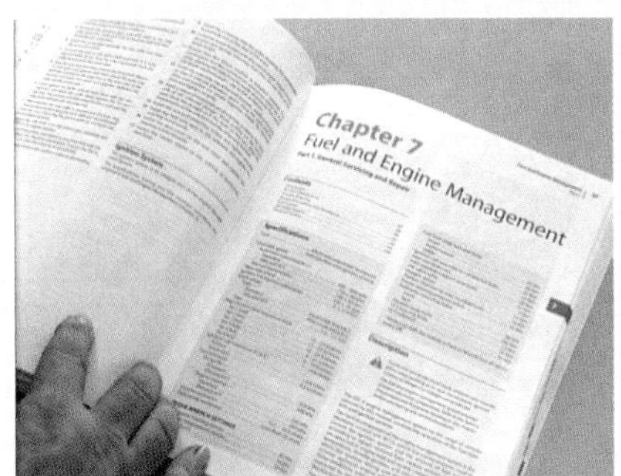

1. Refer to the service information to identify the location, type of fuel filter, and the procedure for removing and replacing it. If it is equipped with an electric fuel pump, release the pressure according to the service information.

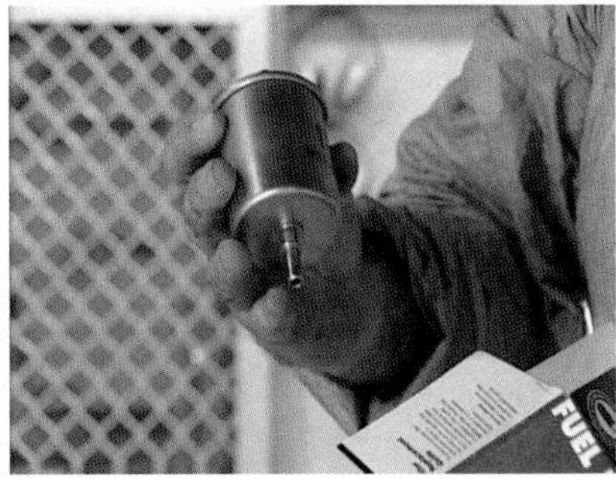

2. Obtain the correct replacement filter and components. Loosen the bracket holding the filter in place, if equipped. Follow the steps below according to the type of filter you are replacing.

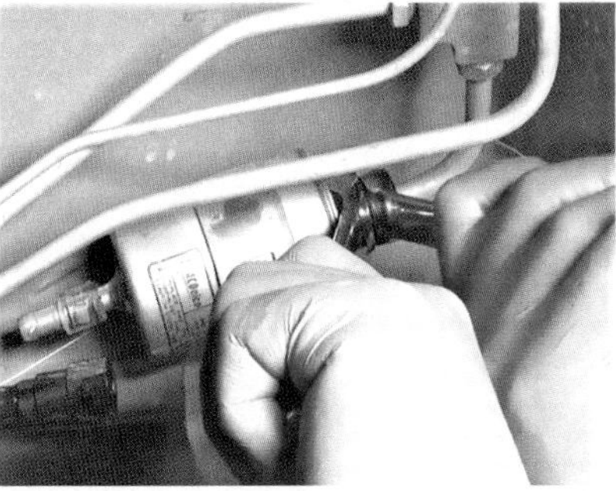

3. For a flared fitting type of filter: A number of connection types are used on filters. On flared fitting types, disconnect the fuel line on the engine side of the filter, using the double wrench method. If necessary, drain any excess fuel into a fuel-proof container.

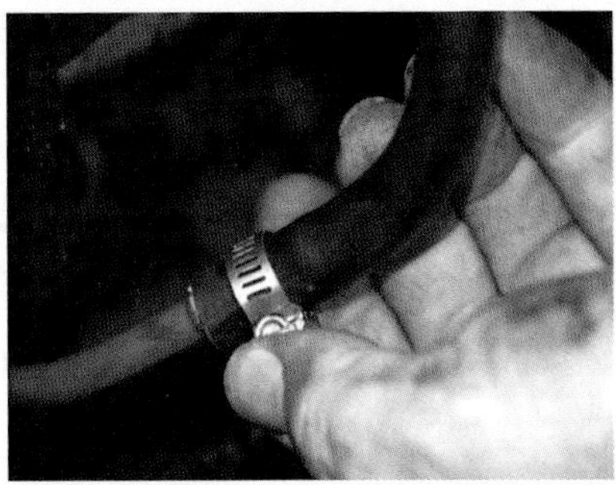

4. Some low pressure types use clamps to seal the connections.

5. Reinstall the filter, making sure you have the filter facing in the right direction, with the flow indicator arrow pointing toward the engine. Tighten the fittings on both ends, using the double wrench method.

6. For a quick disconnect filter: Using the correct tool, release the quick disconnect connectors from the outlet end of the filter, catching any leaking fuel in a fuel-proof container.

SKILL DRILL 62-1 Replacing a Fuel Filter (Continued)

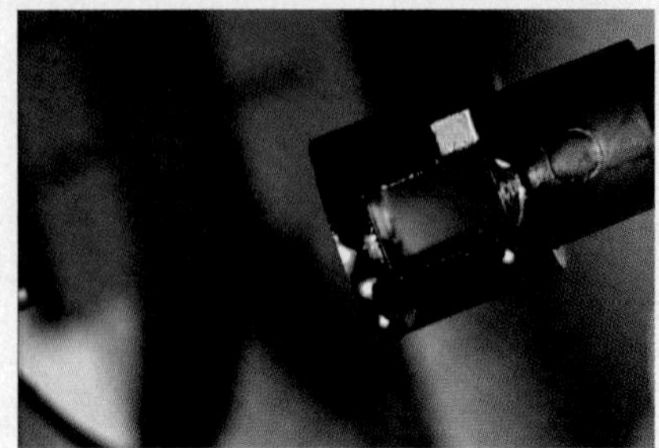

7. Release the quick disconnect connectors from the inlet end of the filter, and remove the filter from the lines.

8. Reinstall the filter, with the flow indicator arrow pointing toward the engine, and fully engage the lines, making sure they are secure.

9. On vehicles with the filter in the tank, remove the fuel pump assembly.

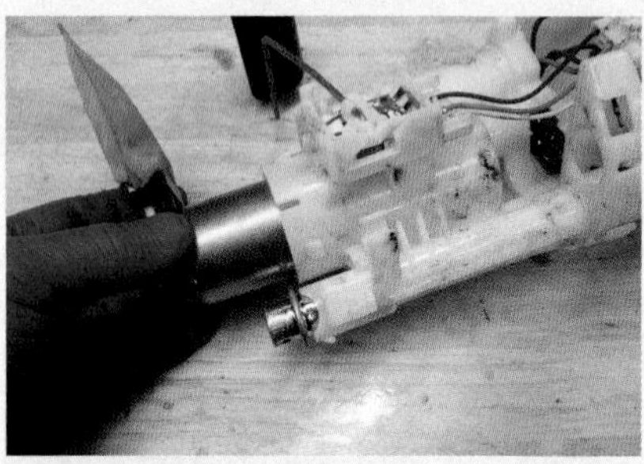

10. Carefully remove the in-tank fuel filter from the pump assembly, and replace it with a new one, if its a replaceable type.

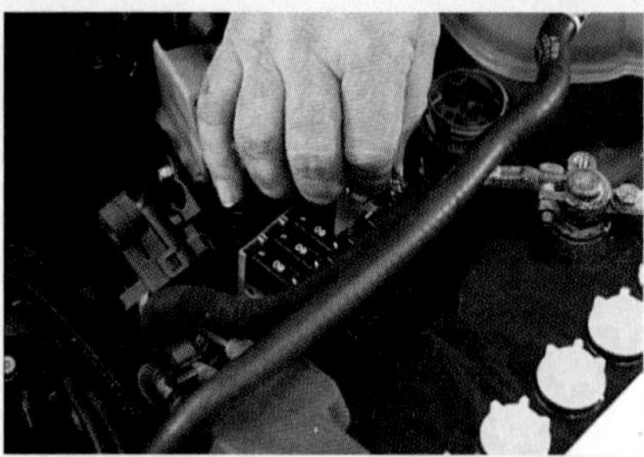

11. Wipe any residual fuel off with a clean shop rag, and write the date and mileage on the filter. Remember to replace the fuel pump fuse, if removed.

12. Turn the key to the On position for a few seconds, but *do not start the engine*, and then turn it back to off. Repeat the process two more times, checking the filter connections for leaks. If no leaks are found, start the vehicle, letting it run for two to three minutes before shutting it off. Recheck the filter connections for leaks.

Inspecting and Testing Fuel Pumps

N62002

Fuel pumps have to be tested if a vehicle experiences low engine power or if the vehicle will not start due to a fuel-related issue. Some shops test a fuel pump's performance, looking to see if it is nearing a failure. There are several ways to test a fuel pump: pressure/volume test, lab scope inductive current flow test, and scan tool data stream test. Each will be described in this section.

The pressure/volume test measures the fuel pressure being delivered to the fuel rail along with the volume of the fuel pump. Many, but not all, fuel-injected engines provide a test port that a fuel pressure tester can connect to. Once connected, the pressure can be measured with either just the key on or with the engine running. Manufacturers give pressure specifications to compare the pressure readings with. If the pressure is low, and the system uses a return line from the pressure regulator on the fuel rail, you may need to pinch off the return line to see if the pressure goes up due to a faulty pressure regulator. If the pressure is within specifications, turn the engine off, but watch the pressure gauge. It should stay steady. If it drops, there is a leak in the system, either at the pressure regulator, the check valve in the fuel pump, or even a fuel injector.

The pressure may be within specifications and hold pressure when the engine is off, but the fuel pump volume may be low. Specifications are given in volume per amount of time, such as 1 pint in 30 seconds or 28 lb per hour, which usually has to be converted to volume per amount of time. The volume is measured with the engine idling. Most EFI fuel pressure gauges have a valve that allows fuel to be taken from the rail without killing the engine and then caught in a calibrated container. If the fuel flow is low, check to see if the fuel filter is plugged and restricting the flow. If the filter is okay, you may have to remove the fuel pump from the fuel tank and inspect the fuel sock at the bottom of the fuel pump to see if it is plugged. If all of the filters are not restricted but the pump volume is low, the pump will have to be replaced.

Some technicians like to use their lab scope to display the trace for the fuel pump's current flow. As each pair of segments of the fuel pump's commutator aligns with the brushes, current flows through the connected windings in the armature. If there is high resistance or a bad connection in one of the commutator segments or armature windings, the current flow will decrease. Because each pair of segments is connected to a different set of windings, the current flow can change for each set if the resistances differ from each other. This will be evident on the lab scope by current pulses of different height in a repeating pattern. The speed of the pump can also be determined on the lab scope by measuring how many revolutions there are in a fixed amount of time. That can then be converted to revolutions per minute by doing a bit of math. The rpms can then be compared to either specifications or a known good pump.

One of the best ways to verify that there is a fuel delivery fault is to use a scan tool and test drive. If possible, set the scan tool to record rpm, VSS, front oxygen sensors, short-term fuel trim, and long-term fuel trim, and if equipped, a fuel pressure sensor, then go on a test drive. In a safe place and manner, briefly drive the vehicle under heavy load and at higher rpm. After returning to the shop, look at the recording; if you see a low fuel pressure, or a continuous lean oxygen signal with an associated large increase in fuel trim, a fuel delivery problem is indicated. This problem could be as simple as a restricted fuel filter or something more severe such as a faulty pump or plugged fuel injectors.

To inspect and test fuel pumps, follow the steps in **SKILL DRILL 62-2**.

Checking Fuel for Contaminants and Quality

N62003

It is easy to forget that fuel by itself can cause drivability issues if it becomes contaminated or old, or even if it is the wrong fuel for the vehicle. In fact, with more people owning diesel-powered vehicles, you may encounter vehicles filled with the wrong fuel, such as a diesel car filled with gasoline, or vice versa. Also, higher blends of gasoline and alcohol make it easy to fill up a standard gasoline vehicle with E-85, which is 85% alcohol and 15% gasoline. A flex fuel vehicle will operate correctly on that blend, but a non-flex fuel vehicle will run very lean because alcohol requires a much richer mixture to run properly. Anytime a vehicle is not running properly and other common causes have not identified the issue, you need to check the fuel for contaminants and quality.

Short of sending fuel out to be tested by a lab, there are some simpler tests that can be done. All require getting a sample from the fuel supply, which is best taken from the fuel rail or where the gas enters the carburetor. That way you will be testing the fuel that the engine is running on. You will want to collect a cup or two of fuel in a clear container, preferably not glass, which breaks easily. Allow the fuel to settle, and then look at it. Is it the right color? Is it cloudy? Does it have a separation line due to being contaminated with water? Does it smell right? If any of those conditions are present, you will need to drain the fuel system, flush it out, and replace the fuel filters. If the fuel passes the visual inspection, it is time to perform an alcohol content test. For this test, you will use a tall 100 mL graduated cylinder. A mixture of 10 mL of clean water and 90 mL of gasoline are carefully agitated for 30 seconds. The mixture is allowed to settle for a minute or two. Any alcohol in the fuel will absorb the water, raising the level above the original 10 mL. Each mL above 10 equals the percentage of alcohol in the gasoline. Most vehicles can tolerate 10% alcohol without causing any drivability issues. And flex fuel vehicles can typically operate normally on 85% alcohol.

To check fuel for contaminants and quality, follow the steps in **SKILL DRILL 62-3**.

SKILL DRILL 62-2 Inspecting and Testing Fuel Pumps

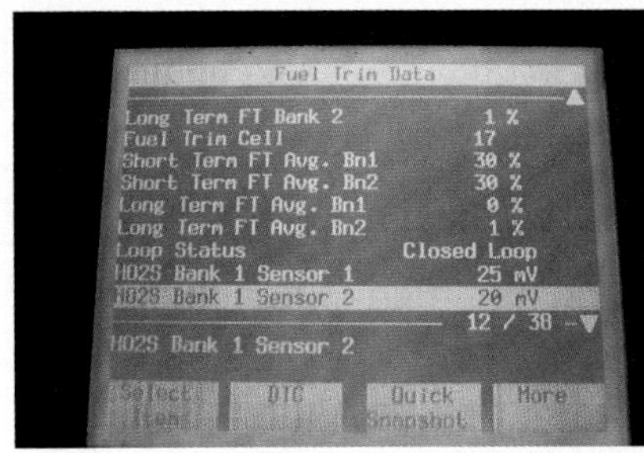

1. Connect a scan tool to the DLC and record rpm, VSS, front oxygen sensors, both fuel trims, and fuel pressure, if equipped, under heavy load and at higher rpm. After returning to the shop, look at your recording; if you see a low fuel pressure or a lean oxygen signal with a large increase in fuel trim, the vehicle has a fuel delivery problem.

2. After researching the procedure for testing the fuel pump in the service information, install a fuel pump gauge on the fuel rail test port.

3. Turn the key to the Run position and measure the pressure. If none, test the fuel pump's electrical circuit.

SKILL DRILL 62-2 Inspecting and Testing Fuel Pumps (Continued)

4. Start the engine, measure the pressure, and compare to specifications. With the engine running and the end of the fuel line from the fuel pressure gauge in a 1-quart plastic bottle, open the valve on the gauge and stop the fuel flow after 15 seconds.

5. Measure the amount of fuel delivered in that time, and compare to specifications.

6. If the pressure test or volume test indicated less fuel than specified, you can use a lab scope and low-amps clamp to graph the current flow going to the fuel pump. Remove the fuel pump fuse, and connect a jumper lead between the terminals in the fuse box. Place the amps clamp around the jumper lead.

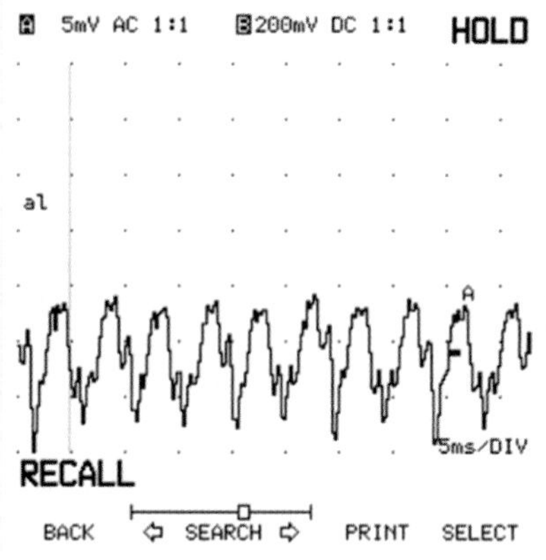

7. Start the engine, look at the lab scope pattern, and compare the pattern to the service information.

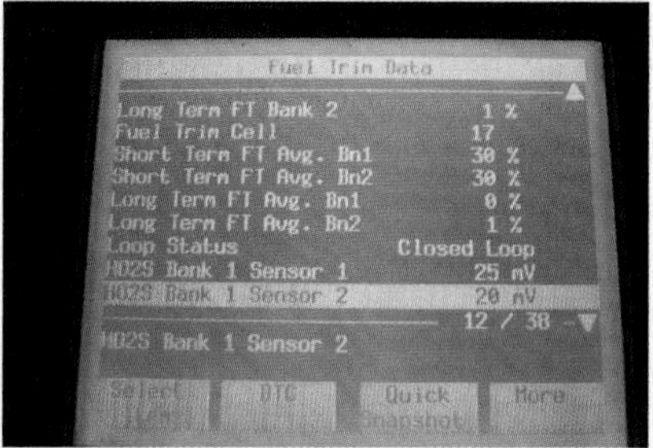

8. After repair, reconnect the scan tool and test drive the vehicle to verify the repair by monitoring the oxygen sensors and fuel trim.

SKILL DRILL 62-3 Checking Fuel for Contaminants and Quality

1. Collect a quantity of fuel from the vehicle's fuel rail in a clear plastic fuel container. Let it settle, and check for contaminants, cloudiness, or improper odor.

2. Pour 10 mL of water into the 100 mL graduated test tube.

3. Add 90 mL of gasoline, bringing the total volume to 100 mL.

SKILL DRILL 62-3 Checking Fuel for Contaminants and Quality (Continued)

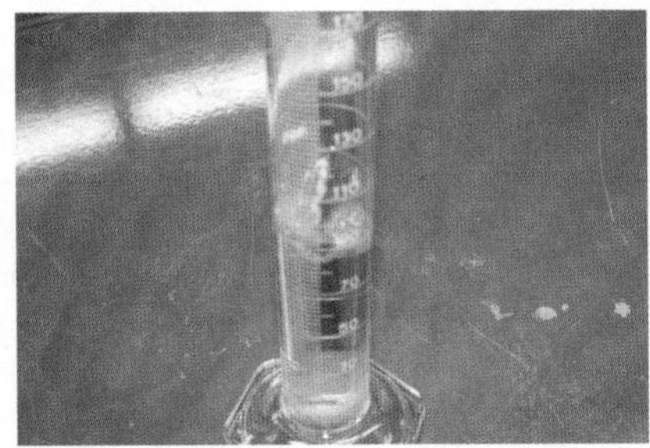

4. Cap the test tube tightly. Slowly agitate the fuel-water mixture for 30 seconds. If there is any alcohol in the fuel, this motion will allow the water to be absorbed by the alcohol.

5. Allow the mixture to settle. Observe the level of the water in the bottom of the test tube. Anything higher than the initial 10 mL is the amount of alcohol in the fuel. List your observations, and determine any necessary actions.

6. Carefully pour off the fuel in the test tube back into the fuel container. Make sure no water leaves the test tube. Properly dispose of the remaining water-fuel mixture.

> **TECHNICIAN TIP**
>
> This author was recently at a Ford dealer catching up on new technologies. The service manager related a story about a four-month-old diesel truck that the customer inadvertently filled with gasoline and drove for a short distance. In this situation, all of the high pressure components had to be replaced at a cost of about $15,000. This was a very expensive mistake that the customer was responsible for. Be careful to use the correct fuel.

Inspecting and Testing Fuel Injectors (Non-GDI)

N62004

Fuel injectors can fail electrically or mechanically. They can also become plugged, or deposits can build up on the nozzle, affecting the spray pattern of the injector. If the injector fails completely, it will cause the cylinder to misfire. If the injectors have a poor spray pattern due to deposits, the engine may experience general roughness or random misfires that are not isolated to one or two cylinders. Before specifically testing an injector, it should be identified as causing a drivability issue. In other words, the ignition system, compression, vacuum leaks, and other engine mechanical faults can cause one or more cylinders not to work correctly, just like a fuel injector problem. Use all of your diagnostic skills and tools to narrow the fault down to the injectors. For example, you may need to retrieve the codes, read the data stream, or access mode 6 data to help you narrow down the area of the fault. If you suspect a fuel injector is causing the fault, you can quickly measure the resistance of the injector winding and compare it to specifications. Any improper resistance will require replacement of the injector.

If the injector resistance and on/off signal from the PCM are good, you can perform an injector pressure drop test to see if the injector is restricted with deposits. This test uses a fuel pressure gauge and injector pulsing tool. The fuel pressure gauge is connected to the test port in the fuel rail, and the pulsing tool is connected to the battery and the suspect injector. The pulse tool pulses the injector a certain number of times for a certain amount of time. The idea is that you will pressurize the fuel rail by turning the key to the On position and charging the fuel rail. Then start the pulse tool, which pulses the injector the set amount of times. Read the pressure on the fuel pressure gauge, and compare this reading to the pressure reading after doing the same test on the other injectors. If the flow is the same between the injectors, the pressure at the end of each test should be within a psi or two of each other. If not, the one with the higher pressure at the end of the test is not allowing as much fuel to flow as the other, which could indicate deposit buildup on the nozzle or a restricted filter screen on the injector. Also note that all of the injectors could be restricted and giving relatively even readings. And lastly, just because the pressure drops are within specifications, the spray pattern can be wrong, which prevents good atomization of the fuel.

To inspect and test fuel injectors, first research the fuel injector testing procedure and specifications for the vehicle in the appropriate service information. Inspect the fuel injectors for leaks and damage; then follow the service information to test the fuel injectors. If the information does not provide a method for testing the injectors, follow the steps in **SKILL DRILL 62-4**.

SKILL DRILL 62-4 Inspecting and Testing Fuel Injectors

1. Measure the resistance of the suspect fuel injector, and compare to specifications. If the resistance is not in specifications, replace the injector.

2. Install the fuel pressure gauge on the fuel rail. Connect the injector pulsing tool to one fuel injector, according to the tool maker's instructions, and set it for the appropriate number and time of pulses.

3. Pressurize the fuel rail by turning on the ignition switch for a few seconds. Then turn the ignition switch off. Record the fuel pressure.

4. Activate the injector pulsing tool for the appropriate amount of time. Watch the pressure gauge, and record the pressure after the injector has been pulsed.

5. Repeat this test on each fuel injector, pressurizing the fuel rail each time before activating the injector pulsing tool. Record the pressure readings before and after each test. Compare all readings to specifications and to one another. Determine any necessary actions.

▶ Wrap-Up

Ready for Review

- The most important job of the fuel system is optimizing engine performance while keeping fuel consumption and emissions to a minimum.
- As a technician, you must be able to properly inspect, diagnose, and repair fuel system components for your customers, thus keeping their vehicles running at optimum efficiency and performance.
- Modern fuel injection has three subsystems: fuel supply, air supply, and fuel metering.
- Vapor lock occurs when vapor forms in the fuel line; the vapor blocks the flow of fuel at the pump and stops the engine.
- Engines perform best when used with the fuel that meets the engine manufacturer's recommended octane rating.
- Combustion is the burning of the air-fuel mixture.
- Detonation is when the entire mixture explodes instead of burning smoothly across the combustion chamber.
- Gasoline must be burnt in a controlled manner, or it will burn either too fast or too slow.
- The term "stoichiometric ratio" describes the chemically correct air-fuel ratio necessary to achieve complete combustion of the fuel and oxygen in the air.
- Air is one of the essential components of the internal combustion engine.
- Air is primarily made up of nitrogen and oxygen. It includes small amounts of argon, CO_2, and other elements.
- Fuel pumps can be electrical or mechanical. The most common today are electrical and are installed in the fuel tanks.

- Most of today's vehicles use electronic fuel injection (EFI), which is better for fuel economy and emission controls.
- The typical fuel tank (or gas tank, as it is sometimes called) of modern vehicles consists of a gas cap, filler neck, fuel, fuel pump, and gauge sending unit. The tank's primary function is to hold an adequate supply of fuel.
- Modern vehicles are required by the Environmental Protection Agency (EPA) to have a non-vented gas cap. This non-vented cap prevents the dangerous fuel vapors from being directly vented to the atmosphere.
- There are three types of electronic fuel injection: throttle body injection (TBI), port fuel injection (PFI), and gas direct injection (GDI).
- Airflow, volume, and density are critical in ensuring the engine has the correct ratio of air to fuel.
- All electronic fuel systems use sensors and other devices to tell the PCM how much fuel is needed for a given engine operating condition.
- If the fuel system is maintained, it will give many years of dependable service to the vehicle's owner.
- Always refer to the manufacturer's technical data for the latest service information on the specific vehicle that is being serviced or repaired.

Key Terms

accelerator pump A small pump usually located inside the carburetor that sprays an extra amount of fuel into the carburetor air horn during acceleration.

accelerator pump circuit The carburetor circuit involved with heavy acceleration, directly connected to the accelerator pump. Also used to prime the engine during cold starting.

air supply system Equipment in a motor vehicle that delivers air to the engine.

catalyst heating mode An operating mode in GDI systems where fuel is injected near the top of the compression stroke, the fuel is ignited, and then an additional amount of fuel is injected during the power stroke to create heat later in the cycle so that the catalytic convert can be warmed up faster.

charcoal canister A device used to trap the fuel vapors. The fuel vapors adhere to the charcoal until the engine is started, and engine vacuum can be used to draw (purge) the vapors into the engine so that they can be burned along with the air-fuel mixture.

choke A device that provides a rich air-fuel mixture until the engine warms up, by restricting the flow of air at the entrance to the carburetor, before the venturi.

computer-controlled carburetor A carburetor that changes and sets air-fuel ratio based on commands from a PCM that uses memory and input from sensors.

detonation The condition in which the remaining fuel charge fires or burns too rapidly after the initial combustion of the air-fuel mixture. It is audible through the combustion chamber walls as a knocking noise.

dieseling A condition in which the engine continues to run after the ignition key is turned off. Also referred to as run-on.

electronic fuel injection (EFI) An injection system in which fuel delivery is controlled electronically, allowing continuous adjustments to the air-fuel ratio.

element The replaceable portion of a filter, such as an air filter element or oil filter element.

filter sock The first line of defense in the fuel supply system. The sock typically consists of a fine mesh, which prevents most small particles from being drawn into the fuel pump and sent through the rest of the fuel system.

float chamber A chamber that holds a quantity of fuel at atmospheric pressure, ready for use.

fuel filler neck The upper end of the fuel filler tube leading down to the fuel tank, which accepts the fuel hose nozzle at the gas station pump.

fuel filter A device that removes impurities (dirt and water) from the fuel before they reach the carburetor or injection system. Filters may be made of metal or plastic screen, paper, or gauze.

fuel metering system Equipment in a motor vehicle that delivers the proper amount of fuel to each cylinder.

fuel pressure regulator A system that controls the pressure of fuel entering the injectors.

fuel pump A mechanically or electrically driven vacuum device used to draw fuel from the tank and force it into the fuel system.

fuel pump relay A relay to turn on or off the high-amperage circuit of the fuel pump.

fuel rail Tubing that connects several injectors to the main fuel line.

fuel supply system Equipment in a motor vehicle that delivers fuel to the engine.

fuel system Equipment in a motor vehicle that delivers fuel to the engine.

full power mode An operating condition in a GDI fuel system where the fuel is injected on the intake stroke in a slightly rich condition and allowed to fully mix with the air in the cylinder. This creates maximum power and reduces detonation.

gasoline A volatile, flammable liquid mixture of hydrocarbons, obtained from crude oil and used as fuel for internal combustion engines.

gasoline direct injection (GDI) A fuel injection system in which fuel is sprayed directly into the combustion chamber.

gauge sending unit A device used for transmitting a signal to control a fuel gauge.

homogenous mixture An air-fuel mixture evenly dispersed throughout the combustion chamber.

idle The speed at which an engine runs without any throttle applied.

indirect fuel injection Any fuel injection that is not sprayed directly into the combustion chamber.

injector A valve that is controlled by a solenoid or spring pressure to inject fuel into the engine.

knocking A noise heard when the air-fuel mixture spontaneously ignites before the spark plug is fired at the optimum ignition moment.

metering jet A calibrated orifice in a carburetor for fuel to flow through. Often it is replaceable for performance or economy desires.

misfire Failure of one or more cylinders to fire or complete combustion.

multipoint fuel injection (MPFI) An injection system in which fuel is injected into the intake ports just upstream of each cylinder's intake valve, rather than at a central point within an intake manifold. Also called multiport injection.

octane rating A standard measure of the performance of a motor or aviation fuel. The higher the octane number, the more heat the fuel can withstand before self-igniting.

pressure The force per unit area applied to the surface of an object.

relay A magnetically operated switch used to make or break current flow in a circuit.

returnless fuel injection system A type of injection system in which no hot fuel is returned to the tank, thus keeping the fuel in the tank relatively cool and minimizing vaporization.

sending unit The component in the fuel supply system responsible for sending constant electrical signals to the gas gauge located in the driver information center.

stoichiometric ratio The optimum ratio of air to fuel for combustion—14.7 parts air to 1 part fuel by weight.

stratified charge A layering of the air-fuel mixture in the combustion chamber. Used when the fuel injector places fuel near the spark plug, but not the surrounding space.

throttle A device used to produce acceleration by controlling the air-fuel mixture.

throttle body injection (TBI) A fuel injection system that uses one or more fuel injectors mounted above or in the throttle body itself. Also called single-point injection.

vacuum A pressure in an enclosed area that is lower than atmospheric pressure.

vapor lock A situation in which vapor forms in the fuel line, and the bubbles of vapor block the flow of fuel and stop the engine.

venturi A restriction (narrowed area) in the air horn.

Review Questions

1. All of the following statements with respect to the fuel system are true *except*:
   - **a.** It provides the ideal air-fuel mixture for the operating conditions of the engine.
   - **b.** Fuel has to be vaporized, turning from a liquid to a gas.
   - **c.** The larger the liquid droplets, the faster they can be vaporized.
   - **d.** It takes time and temperature for fuel to vaporize fully.
2. Which of these air-fuel mixtures gives good fuel economy and low exhaust emissions suitable for cruising conditions?
   - **a.** A slightly lean mixture
   - **b.** A rich mixture
   - **c.** A slightly rich mixture
   - **d.** An extremely rich mixture
3. Which of the following temporarily stores fuel vapors until they can be burned in the engine?
   - **a.** Fuel tank
   - **b.** Fuel pump
   - **c.** Fuel rail
   - **d.** Evaporative emission control system (EVAP)
4. Which of these fuel injection systems operates with fuel injectors located only in the intake manifold near each intake valve and sprays fuel toward the valve?
   - **a.** Central-point injection
   - **b.** Throttle body injection
   - **c.** Multipoint fuel injection
   - **d.** Gasoline direct injection
5. Choose the correct statement with respect to the stoichiometric mode of operation.
   - **a.** It is designed to be used during light throttle, cruise conditions.
   - **b.** It is designed for moderate engine load conditions when a bit more power is needed than for light cruising.
   - **c.** It is designed for heavy engine load conditions when full power is needed.
   - **d.** It is designed to quickly heat the catalytic converter.
6. In GDI engines, which of the following is the least risky way to remove the carbon on intake valves and runners, especially if the buildup is very heavy?
   - **a.** Removing the cylinder head, disassembling it, and cleaning it
   - **b.** Removing the intake manifold and using a crushed walnut shell blaster to blast away the carbon
   - **c.** Removing the intake manifold and using a chemical cleaner to soften and dissolve the carbon so it can be removed
   - **d.** Induction cleaning
7. Which tool is used when measuring fuel pump volume?
   - **a.** Digital multimeter
   - **b.** Fuel pressure gauge
   - **c.** Lab scope
   - **d.** Pulse tester
8. All of the following are valid ways to relieve the static pressure in the fuel system before removing the fuel lines *except*:
   - **a.** removing the fuel pump relay or the fuel pump fuse and running the engine until it dies.
   - **b.** connecting a fuel pump pressure gauge (that has a bleed valve) to the fuel rail test point, and releasing the excess pressure from the system into a gas can.
   - **c.** on GDI equipped engines, using a scan tool to release the fuel pressure.
   - **d.** using fuel filter lines.
9. The alcohol content that can be tolerated by flex vehicle and non-flex vehicles, respectively, without causing any drivability issues are in the range of:
   - **a.** 25%; 45%.
   - **b.** 85%; 10%.
   - **c.** 35%; 95%.
   - **d.** 50%; 50%.

10. An injector pressure drop test to see if the injector is restricted with deposits can be done using a(n):
    a. oscilloscope.
    b. pulse tool.
    c. digital multimeter.
    d. fuel line release tool.

ASE Technician A/Technician B Style Questions

1. Tech A says that most electric fuel pumps are now mounted inside of the fuel tank. Tech B says that fuel flows through the center of the electric pump and is used to cool the pump. Who is correct?
    a. Tech A
    b. Tech B
    c. Both A and B
    d. Neither A nor B
2. Tech A says that the stoichiometric ratio is 14.7 gallons of air to 1 gallon of fuel. Tech B says it is 14.7 lbs of air to 1 lb of fuel. Who is correct?
    a. Tech A
    b. Tech B
    c. Both A and B
    d. Neither A nor B
3. Tech A says that some fuel pressure regulators are connected to the throttle plate. Tech B says that too much alcohol in gasoline can cause the engine to run improperly. Who is correct?
    a. Tech A
    b. Tech B
    c. Both A and B
    d. Neither A nor B
4. Tech A says that high octane fuel is harder to ignite than lower octane. Tech B says that using high octane fuel in an engine designed for lower octane will produce better fuel economy and power. Who is correct?
    a. Tech A
    b. Tech B
    c. Both A and B
    d. Neither A nor B
5. Tech A says that dieseling in a car today indicates a possible leaking injector. Tech B says that fuel filters with flared fittings need to be loosened and tightened with the double wrench method. Who is correct?
    a. Tech A
    b. Tech B
    c. Both A and B
    d. Neither A nor B
6. Tech A says that a lambda greater than 1 means the engine is running rich. Tech B says that the amount of oxygen in the exhaust indicates how rich or lean the mixture is. Who is correct?
    a. Tech A
    b. Tech B
    c. Both A and B
    d. Neither A nor B
7. Tech A says that gauge pressure indicates pressure above atmospheric pressure. Tech B says that atmospheric pressure increases as elevation increases. Who is correct?
    a. Tech A
    b. Tech B
    c. Both A and B
    d. Neither A nor B
8. Tech A says that fuel pressure is regulated by a fuel pressure regulator on some vehicles. Tech B says that fuel pressure is regulated by controlling the speed of the fuel pump on some vehicles. Who is correct?
    a. Tech A
    b. Tech B
    c. Both A and B
    d. Neither A nor B
9. Tech A says that measuring the alcohol content in gasoline involves using water. Tech B says that as long as the fuel pressure is correct, you don't have to worry about fuel pump volume. Who is correct?
    a. Tech A
    b. Tech B
    c. Both A and B
    d. Neither A nor B
10. Tech A says that fuel injectors can be checked for the specified amount of resistance. Tech B says that an injector pressure drop test can determine if an injector is restricted with deposits. Who is correct?
    a. Tech A
    b. Tech B
    c. Both A and B
    d. Neither A nor B

CHAPTER 63
Engine Management System

NATEF Tasks

There are no NATEF tasks in this chapter.

Knowledge Objectives

After reading this chapter, you will be able to:

- **K63001** Explain the purpose and function of engine management sensors.
- **K63002** Explain the purpose and operation of potentiometer-based sensors.
- **K63003** Explain the purpose and operation of thermistor-based sensors.
- **K63004** Explain the purpose and function of inductive and Hall-effect sensors.
- **K63005** Explain the purpose and operation of oxygen sensors.
- **K63006** Explain the purpose and operation of air measurement sensors.
- **K63007** Explain the purpose and operation of pressure sensors.
- **K63008** Explain the purpose and operation of switches as sensors.
- **K63009** Explain the purpose and function of the PCM.
- **K63010** Explain the purpose and function of the controlled devices.
- **K63011** Explain the purpose function and application of relays as controlled devices.
- **K63012** Explain the purpose function and application of solenoids as controlled devices.
- **K63013** Explain the purpose function and application of control modules as controlled devices.
- **K63014** Explain the purpose function and application of electric motors as controlled devices.
- **K63015** Explain the engine management system control strategies.
- **K63016** Describe the operation of speed density and mass airflow systems.
- **K63017** Explain feedback and looping.
- **K63018** Explain short- and long-term fuel trim.
- **K63019** Explain fuel shutoff mode and clear flood mode.

Skills Objectives

There are no Skills Objectives in this chapter.

▶ Introduction

K63001

Previously, in the Motive Power Types chapter, we discussed the inefficiency of gasoline engines from the 1970s and before. These engines were about 25% efficient, meaning that only 25% of the available energy in gasoline was used to power the vehicle down the road, and the other 75% was wasted. Wasting large amounts of gasoline became unacceptable as gasoline became more expensive and environmental concerns heightened. A movement began to improve the efficiency of the gasoline engine, and led to the addition of a whole range of technology and systems.

Although engine management systems were born out of a need for greater efficiency and emission control, they have grown in scope and complexity over time. Here, we won't cover the systems as they evolved, but we look at current engine management systems. For the most part, older systems used many of the same components. The engine management system controls the operation of the engine and all of its related systems, but we focus primarily on the fuel and ignition system portions of the engine management system. We cover the diagnostics portion of the engine management system in the On-Board Diagnostics chapter, and the emission controls portion of engine management in the Emissions Control chapter. For each of these topics, the engine management system contains components categorized in one of three separate groups:

- Sensors monitoring any condition that the powertrain control module (PCM) requires.
- A control module that receives sensor information, processes that information, and compares it with stored data. The module then sends output commands to the controlled devices. This also called input-process-output.
- Controlled devices, also called output devices or actuators, implement the commands sent by the control module.

Today's engine management systems are much more powerful and use many additional sensors and controlled devices to maintain a high degree of control over the engine operation. But before we discuss each of the main components in the system, we need to explore the various types of electrical signals utilized in the system.

▶ Digital and Analog Signals

There are two main types of electrical signals, analog and digital. Analog signals are continuously variable, typically changing in strength and time, whereas a digital signal is a direct on/off with no in-between transition (**FIGURE 63-1**). Both types of signals are useful and can be used as either sensor signals or output signals. Analog signals show how something changes over time, or they can cause something to operate in a variable manner (**FIGURE 63-2**). For example, a throttle position sensor typically

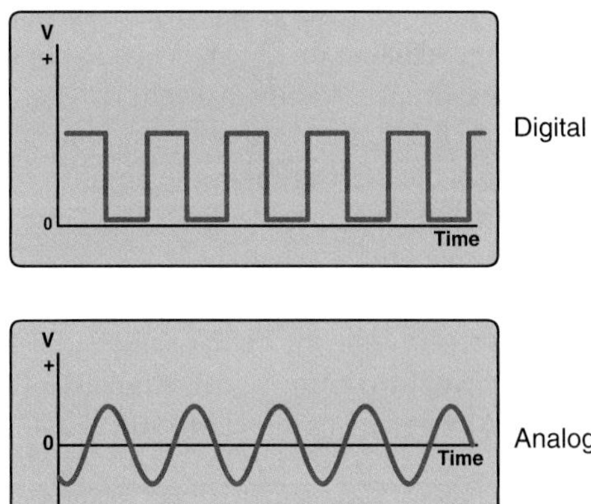

FIGURE 63-1 Typical analog and digital waveforms.

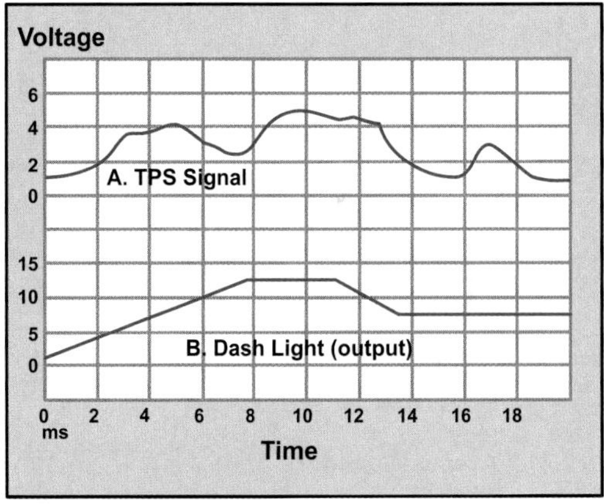

FIGURE 63-2 Analog signals. **A.** TPS sensor signal (analog). **B.** Dash light control output (analog).

You Are the Automotive Technician

The intern at your shop is preparing for an upcoming Skills USA technical competition that will include test questions about engine management systems and related terminology. Each competitor received the following questions to study prior to the competition. She has asked you to give her an overview of each concept to help her get started.

1. What is the difference between digital and analog signals?
2. How do potentiometer-based sensors operate?
3. How do PTC and NTC thermistors operate?
4. How do inductive and Hall-effect sensors operate?
5. How does a wide-band oxygen sensor operate differently than a stoichiometric sensor?
6. What are the four distinct sections of a PCM, and what do they do?
7. How are relays, solenoids, modules, and motors used to control engine operation?

sends an analog signal to the PCM. The signal voltage generated at closed throttle is typically around 1 volt. The signal voltage rises as the throttle is opened, reaching a maximum reading of almost 5 volts. Dashlight control is a good example of an analog output signal. As the dash light dimmer switch is rotated, it turns a rheostat, which sends a proportional output signal to the lights, causing them to dim or brighten.

Digital signals can be used to show whether something is on or off, or to command something on or off (**FIGURE 63-3**). A good example of a digital signal is the brake on/off signal. When the driver steps on the brake pedal, the brake on/off switch sends a full voltage signal to the PCM. When the brake pedal is released, the signal drops instantly to 0 volts. Digital signals can also be sent by the PCM as an output signal to turn devices on or off. For example, the PCM grounds the negative side of a fuel injector to turn it on and allows it to inject fuel into the engine for a certain amount of time. When the PCM opens the ground, the injector turns off.

When cycled very quickly, digital signals can be used to create a varying output control signal, similar to the variable control provided by an analog signal (**FIGURE 63-4**). The PCM creates an alternating on/off signal many times per second. The percentage of time that the signal is off, versus on, can be varied by the PCM. This allows a varying amount of output signal to be sent to the controlled device. This type of pulse width modulation is called duty cycle, as the percent of on-time can be changed by the PCM (**FIGURE 63-5**).

In a varying analog signal, variable resistance is used to control the strength of the output signal. The electrical energy used by the resistance is wasted as heat. When a digital duty cycle signal is used, the circuit alternates between full on and full off, giving a lower average current flow and voltage to the load. The percent of on-time determines the strength of the output control signal (average current flow and voltage); therefore, there is no energy that is wasted as resistance. The increased energy efficiency of digital technology is being used more and more to control and monitor all of the systems a vehicle, including the engine management system.

Another type of signal used to transmit data is frequency. Frequency is measured in cycles per second, where a cycle is the distance between waves or wavelengths. A higher number of cycles per second indicates a higher frequency (**FIGURE 63-6**).

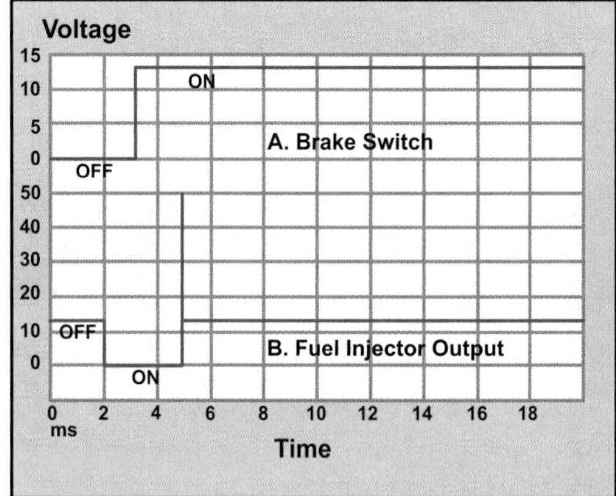

FIGURE 63-3 Digital signals. **A.** Brake on/off switch (digital). **B.** Fuel injector output (digital).

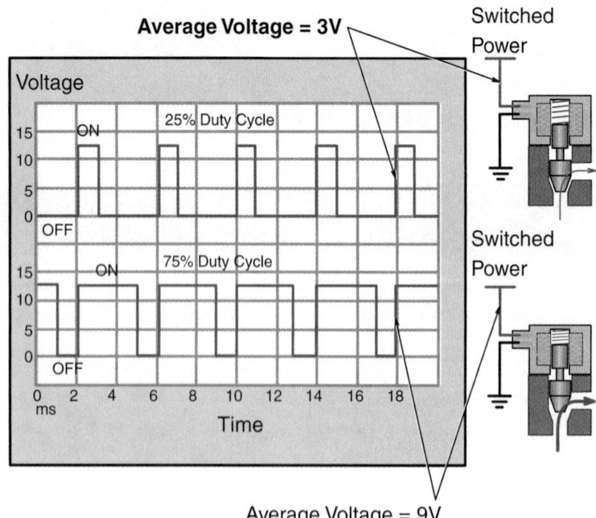

FIGURE 63-4 A digital signal can be used to create a varying output signal, as in this pulse-width-modulated EGR valve.

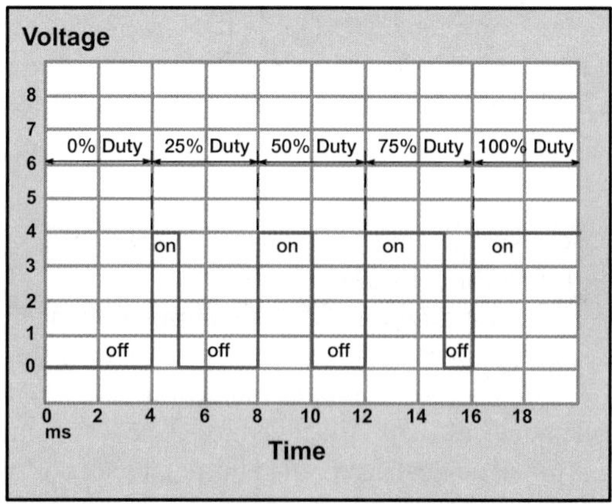

FIGURE 63-5 Duty cycle is a type of digital signal that uses PWM to control the strength of the output signal generated. Duty cycle can generally be controlled from 0% to 100% on-time.

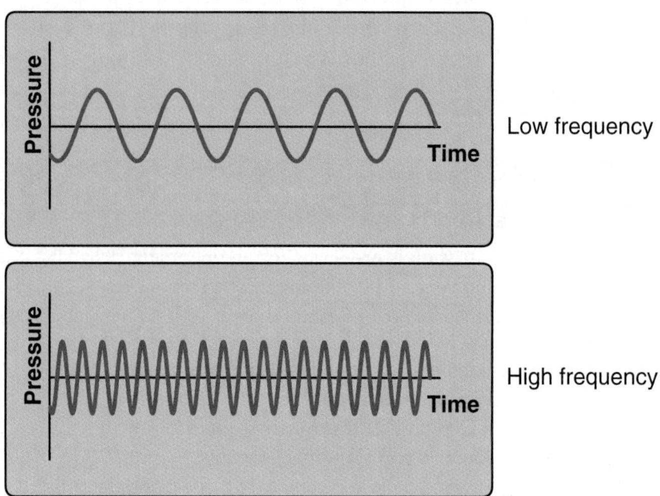

FIGURE 63-6 The higher the frequency, the greater the number of cycles per second.

Both analog and digital signals have frequency because both signals cycle between a high and low state.

Sound waves are a good way to illustrate frequency. Sound travels through the air by producing pressure waves—areas of high pressure and areas of low pressure. The rate at which these waves are created is called **frequency**. The higher the frequency, the higher the pitch of the sound. The range of human hearing is approximately 20 to 20,000 cycles per second. An engine produces sounds across a wide range of frequencies. The mufflers and resonators in the engine exhaust system reduce these sounds to an acceptable level. In the combustion chamber, burning the air-fuel mixture creates sound. Proper burning creates a moderate level of sound. Detonation creates a much louder and sharper level of sound. A sensor can be used to monitor the frequency and volume of sound created in the combustion chamber, which indicates whether fuel is being burned properly. The PCM can then make adjustments when necessary. Frequency is also used in wheel speed sensors to indicate wheel speed and deceleration.

Engine Management Sensors

As mentioned earlier, the engine management system is made up of three main parts, the first of which is input sensors. Each sensor monitors specific conditions and creates a signal that corresponds to the conditions. The signal is then sent to the PCM, where it is evaluated and ultimately used to make decisions. The most common sensors used in recent engine management systems include the following:

- Throttle position sensor (TPS)
- Accelerator pedal position (APP) sensor
- Engine coolant temperature (ECT or CTS) sensor
- Intake air temperature (IAT) sensor
- Crankshaft position (CKP) sensor
- Camshaft position (CMP) sensor
- Oxygen sensor (before and after catalytic converts) (O_2 or HO_2S) sensor
- Manifold absolute pressure (MAP) sensor
- Barometric pressure (BARO) sensor
- Mass airflow (MAF) sensor
- Air-conditioning compressor clutch signal
- Brake on/off (BOO) switch
- Knock (KS) sensor
- Vehicle speed sensor (VSS)
- Fuel pressure sensors
- Ignition pickup assembly

▶ Potentiometers

K63002

Some sensors use a **potentiometer** to create a variable voltage signal that corresponds to a specific condition. Because a potentiometer is a resistor with a mechanically variable wiper that contacts the resistor along its length, it can access an increasing or decreasing level of voltage as the wiper moves (**FIGURE 63-7**). The resistor in engine management sensors is generally a thin film, and it can be linear or circular in construction. It has three

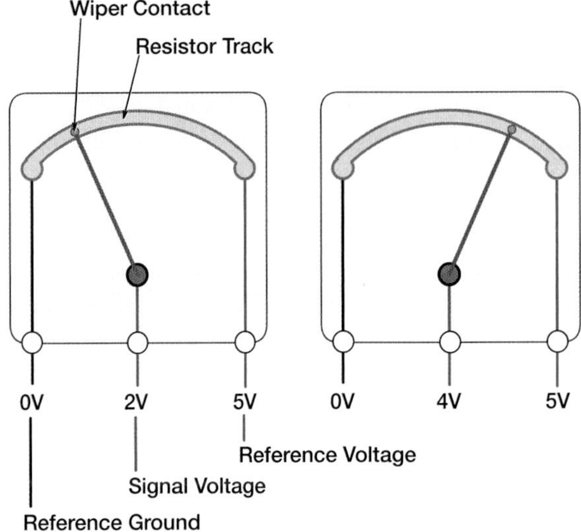

FIGURE 63-7 A potentiometer has a fixed resistor that drops the voltage evenly from input to output. The wiper contacts the resistor and sends the voltage out at that point on the resistor as a voltage signal.

electrical connecting points: one at each end of the resistor, and a third attached to a center sliding contact, called a wiper. The wiper is designed to move across the resistor. A reference voltage (typically less than battery voltage; 5 V is the most common) is applied to the resistor allowing a steady current flows through the resistor. As the wiper contact moves across the resistor, it picks up the voltage at that point and sends it as signal voltage referencing that position.

In automotive applications, the circular form is commonly used as a throttle position sensor, with the center contact attached to the throttle plate. The throttle plate position can then be monitored by the control unit.

Throttle Position Sensor

The throttle position sensor (TPS) gathers information on throttle position to allow the control unit to make adjustments according to operating conditions. It is located on the throttle body and is operated by rotation of the throttle shaft. A potentiometer-type sensor monitors throttle position over its full range. One end has a 5-volt reference voltage from the control unit (**FIGURE 63-8**). The other end is connected to the control unit ground. A third wire runs from a sliding contact in the TPS sensor to the input circuits of the control unit. The sensor is a variable resistor. As the angle of the throttle valve changes, so does the voltage signal along the third wire.

At closed throttle, the reading is usually below 1 volt. As the throttle valve opens, the voltage signal rises. At wide-open throttle, the signal is about 4.5 volts. Ongoing monitoring of throttle position provides accurate data for the control unit, allowing control over a wider range of operating conditions.

Although TPS voltage can be monitored using a scan tool, use of a digital storage oscilloscope (DSO) is recommended because the scan tool's data rate is relatively slow and cannot display the TPS voltage quickly enough to find flaws in the sensor. To test a TPS sensor, measure the voltage on the signal return wire with the vehicle in key on, engine off (KOEO) mode. Using the throttle pedal, slowly open the throttle all the way, and release it. The voltage should rise steadily and return to the closed-throttle level.

Accelerator Pedal Position Sensor

Accelerator pedal position (APP) sensors are used on vehicles with an electronic throttle control. The APP sends signals about the position, direction, and speed of the throttle pedal. The PCM uses this information to determine driver intent, related to engine power needed. The APP is made up of two throttle position sensors that work opposite of each other. One sensor moves from low voltage to high voltage while the other one is moving from high voltage to low voltage (**FIGURE 63-9**). This creates a redundant circuit of opposite readings, allowing the PCM to quickly diagnose a problem with one or both of the APP sensor signals.

Thermistors

K63003

A resistor that changes its resistance with the changes in temperature is called a **thermistor**. Thermistors can be the positive temperature coefficient (PTC) type or negative temperature coefficient (NTC) type (**FIGURE 63-10**). In PTC thermistors, the resistance goes up with the temperature. In NTC thermistors, the resistance goes down as the temperature goes up. Thermistors are used in various temperature-related sensors on modern vehicles. Coolant temperature sensors, fuel temperature sensors, and ambient temperature sensors are just a few types (**FIGURE 63-11**). For example, the coolant temperature sensor is a NTC thermistor, so as the engine temperature warms up, the internal resistance decreases. This resistance change causes changes in a voltage signal used by the PCM to decrease the injector on time as necessary to keep the engine from running too rich as it warms up. As a result, the engine operates with the proper air-fuel ratio for good drivability, fuel efficiency, and emissions purposes.

FIGURE 63-8 A potentiometer-type sensor monitors throttle position over its full range. One end has a 5-volt reference voltage from the control unit.

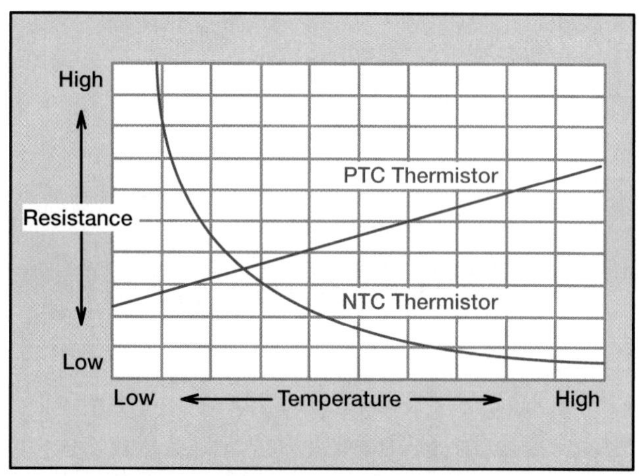

FIGURE 63-10 Typical resistance changes of PTC and NTC thermistors at different temperatures.

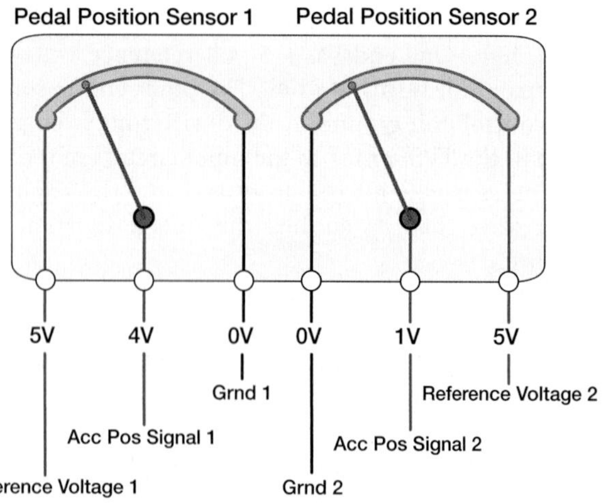

FIGURE 63-9 Typical APP sensor operation.

FIGURE 63-11 A typical thermistor used as a coolant temperature sensor.

Engine Temperature Sensors

To maintain the air-fuel ratio within an optimum range, the control unit must account for coolant temperature and air temperature. Extra fuel is needed when the engine is cold, because the air is colder and therefore denser.

The engine coolant temperature sensor (ECT) is immersed in coolant in the cylinder head, block, or intake manifold. It consists of a hollow threaded pin that has a resistor sealed inside it (**FIGURE 63-12**). This resistor is made of a thermistor semiconductor material whose electrical resistance falls as temperature rises. The signal from the ECT is used by the PCM to control the air-fuel mixture of the engine throughout its operating temperature.

Enrichment of the air-fuel mixture occurs during engine cranking, then slowly reduces as the engine warms up. This design ensures that the engine runs smoothly as soon as it is started. The control unit continually monitors coolant temperature during engine operation. Some manufacturers use a cylinder head temperature sensor instead of a ECT to signal engine temperature. This sensor screws into the metal casting of the cylinder head to sense head temperature.

Air temperature is one factor that changes the density of the air, so air temperature has to be monitored by the PCM to better judge how much fuel needs to be injected. An air temperature sensor does just that. If the air temperature sensor is installed in the airflow sensor, it is positioned in the airstream and is called an **intake air temperature (IAT) sensor** (**FIGURE 63-13**). When it is installed in the intake manifold, it is located in one of the intake runners and is called a manifold air temperature (MAT) sensor. In both cases, it relays information about air temperature and therefore density. The control unit can then vary the injector pulse width accordingly.

Crankshaft Position Sensor

K63004

Crank angle sensing sends information on the speed and position of the crankshaft to the PCM, for the control of ignition timing and injection sequencing. The control unit can then trigger the ignition and injection to suit virtually all operating conditions. The **crankshaft position (CKP) sensor** may be mounted externally on the crankcase housing, or it may be inside the housing of the ignition distributor. There are different kinds of CKP sensors.

Inductive-type sensors sense the movement of a toothed disc on the crank pulley (reluctor), or the ring gear teeth on the flywheel (**FIGURE 63-14**). These sensors do not make physical contact. The sensor is mounted on the crankcase housing. It consists of a stator with a central permanent magnet, and a soft iron core surrounded by an induction winding. The housing around all of these components is insulated from the iron core and windings. The iron core interacts magnetically with the teeth on the reluctor.

The stator is positioned so that it has a very small clearance, or air gap, between the end of the soft iron core and the reluctor teeth. As the reluctor rotates, the teeth approach and leave the stator, and the air gap changes. As this occurs, the strength of the magnetic field changes. The winding is part of a complete circuit, so changing the magnetic field produces an alternating voltage and current in the circuit.

FIGURE 63-13 Intake air temperature (IAT) sensor.

FIGURE 63-12 The engine coolant temperature sensor (ECT) is immersed in coolant in the cylinder head, block, or intake manifold.

FIGURE 63-14 An inductive-type sensor and toothed wheel.

As the tooth approaches the stator, the strength of the magnetic field increases. This induces a voltage, and current flow, in the winding. The polarity of the voltage is said to be positive, as it produces a current flow in a certain direction. When the tooth aligns with the stator, the magnetic field is at its strongest, but at that point it is not changing. Voltage and current flow fall to zero.

As the tooth moves away from the stator, the strength of the magnetic field changes again, and once again voltage and current flow are induced in the winding. This time, current flow is in the opposite direction, and the polarity of the voltage is now said to be negative.

Because polarity changes every time a tooth approaches and leaves the stator, the voltage produced is an alternating current (AC) voltage, and the current flow in the winding is AC. It is the frequency of this alternating voltage that is used by the control unit to calculate engine rpm.

Crankshaft position is detected by a separate inductive sensor. It sends a signal to the control unit when a tooth or bore passes, and generates one pulse per revolution. It signals the control unit that the number-one piston is, for example, 80 degrees before top dead center. Ignition timing is then decided according to the operating conditions and is triggered to occur a certain number of degrees from that point.

When only one sensor is used, the reluctor is shaped to provide information on both crankshaft speed and position. The reluctor has a number of equally spaced teeth around its circumference, but one or two teeth are omitted. The frequency of the pulse from each tooth gives engine speed in rpm, but on each revolution, the pulse alters as the gap from the one or two missing teeth passes the sensor. This, again, gives the position of the number-one piston.

A Hall-effect CKP sensor measures the position of the crankshaft, as well as engine speed, represented in rpm of the crankshaft. The sensor is in very close proximity to a metal disc that is typically attached to the crankshaft or camshaft. Around the circumference of the disc, there are evenly spaced windows and shutters in one or two rows (**FIGURE 63-15**). The sensor is stationary and is typically mounted to the block. It has a magnet on one side and a sensor on the other side. The windows and shutters alternately allow and block the magnetic field from the sensor as the crankshaft rotates. The sensor is made of up Hall-effect material that reacts to the varying magnetic field and creates an on/off digital signal. This is what gives the PCM a measurement of the engine rpm. This sensor may have two rows of windows and shutters. If so, one is used for rpm, and the other is used for crankshaft position, which can also be used for ignition module and injector timing.

Camshaft Position Sensor

The **camshaft position (CMP) sensor** (or cam sensor) sends constant data to the computer to let it know when the number-one piston is on the compression stroke. From there, the position of other pistons on their power strokes can be determined as well. It also identifies the position of the camshaft and valve on variable valve-timing engines. It does this by constantly reading a point on the camshaft or on the distributor housing (**FIGURE 63-16**). The CMP sensor works in conjunction with the CKP and a knock sensor to provide information the computer needs to keep the ignition timing adjusted for all load conditions, thus reducing pinging and knocking.

Ignition Pickup-Style Position Sensor

The ignition pickup is primarily a Ford design. Known as PIP, which stands for profile ignition pickup, it is a Hall-effect vane switch that provides camshaft position data to the ignition module and Ford ECC-IV processor (**FIGURE 63-17**). This data allows the Ford ECC-IV system to make adjustments to the ignition timing and fuel injection according to the driving conditions. The pickup assembly is located inside of the distributor, which rotates at camshaft speed, so it can determine crankshaft position.

Vehicle Speed Sensor (VSS)

The vehicle speed sensor (VSS) is a detection device that sends the vehicle speed information (i.e., how fast the vehicle is traveling) to the PCM. The VSS is sometimes located in the instrument cluster as part of the speedometer; however, most

FIGURE 63-15 A Hall-effect CKP sensor measures engine speed, represented in rpm of the crankshaft.

FIGURE 63-16 The camshaft position (CMP) sensor sends data to the computer to let it know which cylinder is on its power stroke as well as the position of the camshaft and valve.

FIGURE 63-17 A typical Ford EEC-IV ignition pickup assembly.

FIGURE 63-18 A typical VSS magnetic pickup sensor.

manufacturers mount these sensors on the transmission or transfer case.

Most VSSs are a type of sensor called a magnetic pickup, or magnetic reluctance, sensor. The sensor assembly consists of an iron core wrapped with fine copper wiring. There is also a reluctor wheel, which has raised "teeth" located around the circumference of the wheel. **FIGURE 63-18** shows the parts of the magnetic pickup sensor. The reluctor wheel is attached to the output shaft of the transmission and spins with the output shaft. As the teeth of the reluctor wheel get closer to the iron core, a positive voltage is produced. As the tooth moves away, the voltage becomes negative. This process repeats over and over again as the reluctor wheel spins, creating an AC voltage. This signal is sent to the PCM, where the vehicle speed can be determined from the signal frequency. The magnetic pickup sensor produces its own voltage while the reluctor wheel is spun. These sensors do not need a separate wire to supply them with a reference voltage.

Another type of VSS is called a **reed switch**. There is a thin, movable blade switch inside the sensor, and a magnet installed in a rotating part, called the rotor, on the output shaft (**FIGURE 63-19**). As the output shaft spins, the rotor spins. When the magnet comes near the reed switch, the switch closes,

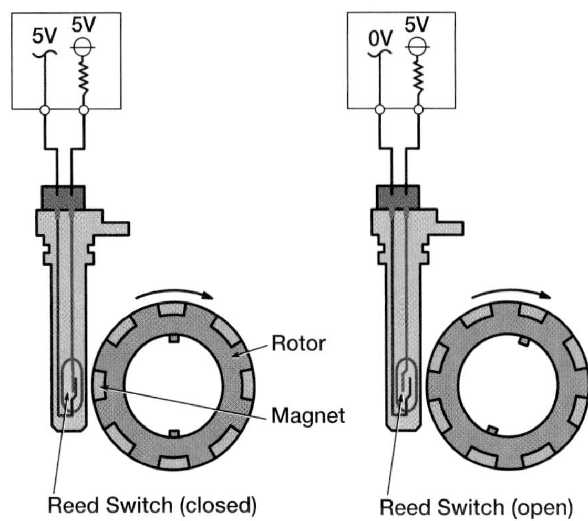

FIGURE 63-19 A reed switch VSS. As the magnet in the rotor passes the reed switch, the switch closes, allowing current to flow.

allowing current to flow through the switch. When the magnet moves away from the switch, the switch opens, breaking the current flow. This opening and closing of the switch creates a DC square wave pattern that the PCM can use to determine vehicle speed. Reed switches have to be supplied with electricity to operate. The amount of voltage supplied to the sensor varies with different vehicle manufacturers. Some of these switches apply a ground to a wire coming from the PCM; others simply allow a voltage to pass from a voltage source, through the reed switch, to the PCM.

Oxygen Sensor

K63005

An **oxygen sensor (before and after catalytic converter)** is positioned in the exhaust pipe and provides the engine management PCM with an electrical signal that relates to the amount of oxygen in the exhaust gas. The PCM uses this and other information to determine the correct amount of fuel to be injected into the engine.

Originally, automotive oxygen sensors were stoichiometric sensors. Although these are still in use today, they indicate only whether the air-fuel ratio is rich (deficient oxygen) or lean (excess oxygen), but they do not indicate how rich or how lean. Their output signal changes almost vertically on either side of an air-fuel ratio of stoichiometric, (14.7:1). Above 0.45 volt indicates rich; below 0.45 volt indicates lean (**FIGURE 63-20**). It is the "Nernst cell" inside the sensor that produces this signal voltage. The Nernst cell operates by comparing the amount of oxygen in the exhaust gas to oxygen levels in the outside air (**FIGURE 63-21**). To operate, this cell has to be hot, approximately 600°F (315°C) or more. Exhaust gas can heat the sensor; however, when an internal heater is added, the sensor becomes operational more quickly and is called a heated oxygen sensor (HO_2S). This is most relevant when the sensor is positioned in cooler parts of the exhaust system, away from the exhaust manifold.

As emission standards become tighter for both gasoline and diesel engines, a more precise signal is required. In these

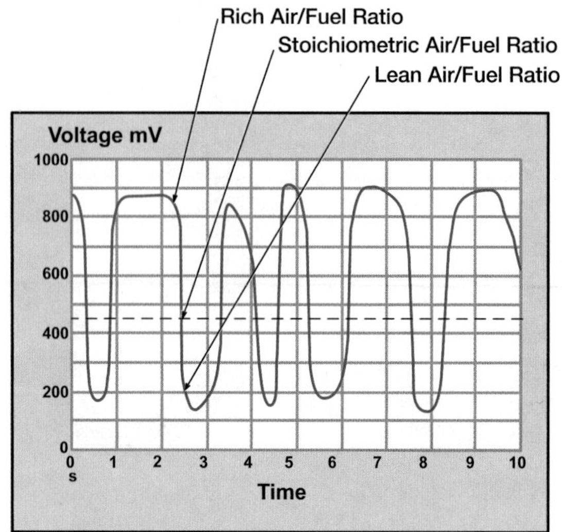

FIGURE 63-20 Many oxygen sensors typically have a switching point of 0.45 volt, indicating a stoichiometric air-fuel ratio.

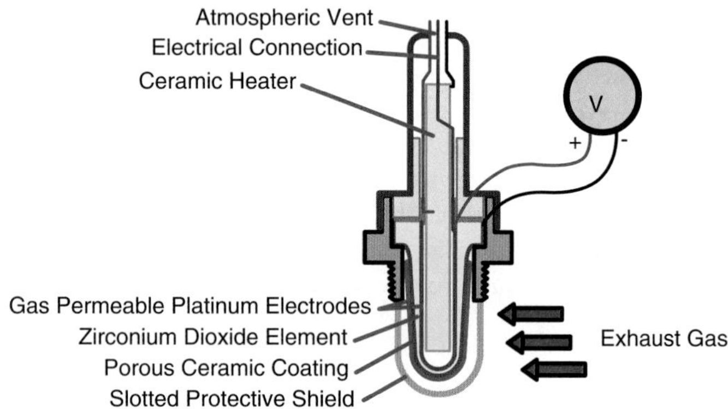

FIGURE 63-21 Oxygen sensors, when at temperature, create an electrical signal based on the difference in oxygen in the exhaust stream compared to ambient air.

systems, the broadband oxygen sensor (or wide-band oxygen sensor) informs the PCM of a range of air-fuel ratios from about 9:1 up to atmospheric air. These systems are ideal for optimum emissions for vehicle gasoline or diesel engines, particularly with the extremely lean-running GDI engines.

The wide-band oxygen sensor is far more sophisticated in operation than the earlier sensors, but it does include some similar parts. The Nernst cell is still used; however, the exhaust gas oxygen levels are referenced to a sealed chamber of air within the sensor and not outside air. These sensors have an electrical heating element that heats the sensor quickly from a cold start, typically in less than 10 seconds. This is a shorter time than that taken by the older sensors. This faster heating is achieved because the sensors have far less material within the sensing element.

Current through the heater is controlled by the PCM, allowing the correct operating temperature to be continuously maintained. A small chamber within the sensor has access to the exhaust gas. It is from this chamber that the Nernst cell samples exhaust gas. This sensor uses a solid-state pump to add or remove oxygen from the exhaust gas chamber (**FIGURE 63-22**).

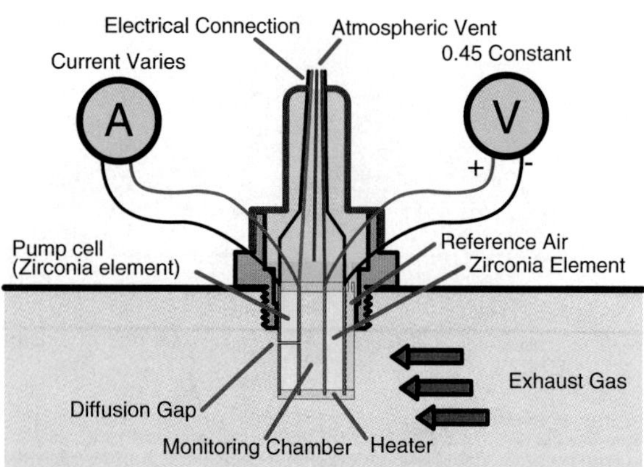

FIGURE 63-22 The PCM sensor uses a solid-state pump to add or remove oxygen from the exhaust gas chamber.

The computer controls the current flowing through the oxygen pump so that the Nernst cell output is maintained at a stoichiometric ratio. Current flowing in one direction through the pump adds oxygen, whereas current in the opposite direction removes oxygen. The value and direction of current required to do this represents the level of oxygen in the exhaust gas. Reading this current allows the PCM to determine the amount of oxygen in the exhaust. This allows the PCM to precisely monitor the air-fuel ratio between approximately 9:1 and atmospheric air, so that the PCM can better control the amount of fuel injected to maintain correct emission levels.

All manufacturers selling road-going production cars must install an oxygen sensor before and after the catalytic converter to test for correct converter operation. For a catalytic converter to change exhaust gases correctly, it has to be capable of storing and releasing oxygen from the catalyst. When this happens, the amount of oxygen entering the converter will differ from that leaving. The PCM uses these two signals, along with response time, to determine if the catalytic converter is functioning properly. If a malfunction is detected for a predetermined period, the MIL will be illuminated. Many manufacturers also use the rear oxygen sensor to fine-tune the fuel trim. If the PCM determines that there is a catalyst-damaging event occurring, most systems will flash the MIL light (aka Check Engine Light) to warn the driver.

▶ Air Supply

K63006

In multiport fuel injection (MPFI) systems, the air required for the combustion of the fuel travels from the air filter, through the throttle valve, and into the common manifold, or plenum chamber. From here, individual intake runners, or pipes, branch off to each cylinder. Ideally, all of these runners are of equal length (**FIGURE 63-23**). The design of the intake system determines how large an air mass can be drawn into a cylinder at any given engine rpm. This is where the most advantage can be taken from fuel injection, as it can achieve uniform distribution of the air delivered to the cylinders.

With unobstructed passages to each cylinder, the cylinder fills with air as efficiently as possible. The breathing of the

FIGURE 63-23 Individual intake runners branch off to each cylinder.

engine, or its volumetric efficiency, is improved. As more air is forced into the cylinder, the air-fuel mixture becomes denser during ignition. This increases the pressure on the piston, as well as engine power output.

The temperature of the air influences density of the air-fuel mixture. Cold air is more dense than hot air, so it has a greater mass in any given volume. Thus, to ensure the entry of cool air, on most intake systems the air entry to the intake is located away from engine heat. Because the manifold carries air only, it does not have to be heated by coolant. Filtered air arrives at the intake port as cool and dense as possible, ready for mixing with the fuel from the injector. Refer to the Induction and Exhaust chapter for more information on intake manifolds.

Air Measurement

The amount of air entering the engine must be measured, or calculated so that the amount of fuel injected forms a mixture to suit the engine operating conditions at that time. The volume of air can be determined in two ways—speed density calculation or direct measurement.

The speed density method calculates the volume of air entering the engine, which varies according to the engine speed and the load. If calculations are used, they are based primarily on the engine rpm as measured by the crankshaft position sensor, and the engine load as measured by the manifold absolute pressure sensor. Then, using the intake air temperature, throttle position, and coolant temperature, the airflow can be calculated using pre-programmed fuel maps. The oxygen sensor is then used to provide feedback about the richness or leanness of the fuel that has been injected.

Mass Airflow Sensor

A mass airflow (MAF) sensor directly measures the mass of filtered air entering the engine in grams per second (gps). The MAF sensor is typically installed in the air intake hose leading to the throttle body assembly. However, some MAF sensors are built into the throttle body assembly, whereas others are mounted on the air cleaner assembly. All air entering the intake flows through the MAF sensor, then through a sealed, flexible hose or duct to the throttle body, after which it enters the engine. Older systems used a vane-type airflow meter. These meters have a spring-loaded vane, or flap, that deflects according to how much air enters (**FIGURE 63-24**). The electrical signal to the PCM varies with this deflection as contacts attached to the vane move across a potentiometer.

A newer style uses a hot wire to measure the mass of air entering the engine. The sensor uses a platinum wire placed in the airstream within the MAF sensor housing (**FIGURE 63-25**).

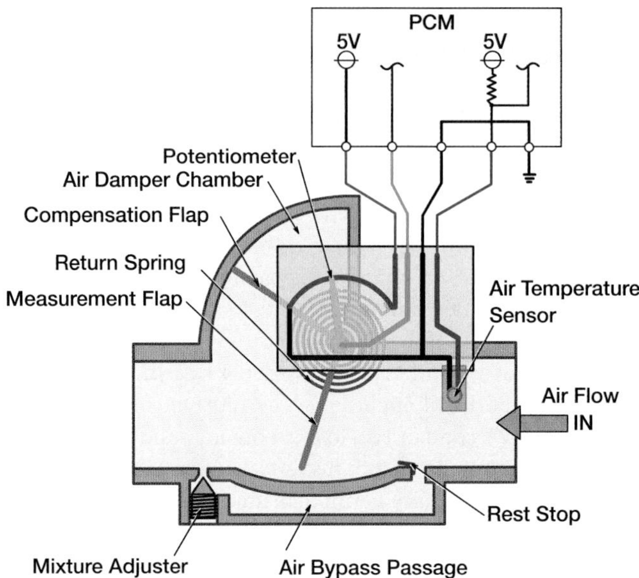

FIGURE 63-24 Typical vane-type airflow meter.

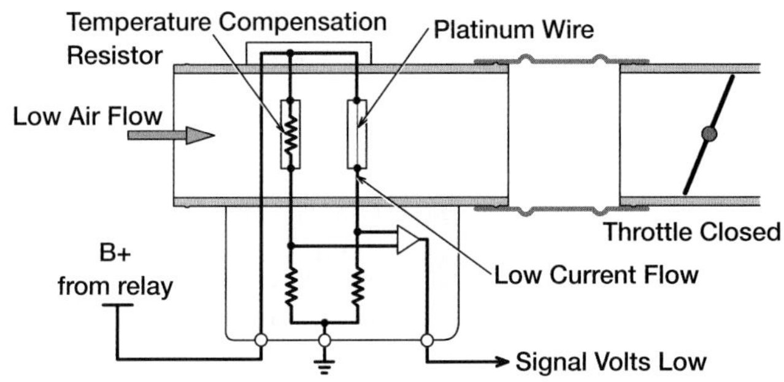

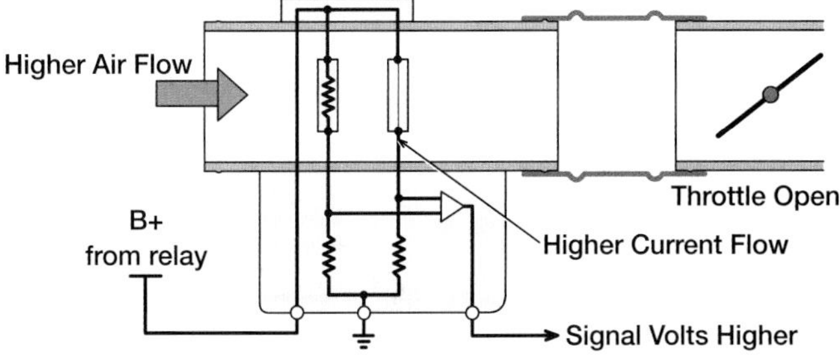

FIGURE 63-25 A hot-wire mass airflow sensor uses the cooling effect of air passing across the heated wire to change the current flow. The change in current flow indicates the mass of the air entering the engine.

A constant voltage drop is maintained across the wire. As the wire heats up, its resistance rises, and current flow drops. As the engine is started and air is drawn across the wire, the air cools the wire, which lowers its resistance and increases the current flow. The more air that flows across it, the higher the current flow. Also, air temperature and density affect the amount of heat removed from the wire. The cooler and denser the air, the more it cools, and the greater the current flow. This cooling effect is directly related to the mass of air passing through the airflow meter, so the airflow meter can send a signal to the PCM that indicates the mass of the airflowing through it.

Some hot-wire MAF sensors have a built-in cleaning feature where the PCM applies a higher current flow across the hot wire for a second or two to burn off any residue or contaminants. This typically occurs just after engine shutdown. Some hot-wire sensors require periodic manual cleaning. Clean these components carefully, using a cleaner recommended for MAF sensor cleaning. Note that some manufacturers specifically say not to clean their MAF sensors and to instead replace them if they are not reading correctly. Be careful to check the manufacturer's recommendation regarding whether cleaning the MAF sensor is possible, and follow the manufacturer's directions precisely if cleaning is allowed.

The Karman vortex airflow sensor measures airflow disruptions. There are three types of vortex sensors, all of which use an air regulator to first smooth out or reduce any turbulence in the airflow entering the sensor; a triangular vortex column then disrupts a portion of the airflow creating vortices or whirlpools in the airflow (**FIGURE 63-26**). These "whirlpools" are then measured by one of three different methods:

1. Optical type—The air vortices cause a mirror to oscillate and interrupt an infrared signal.
2. Ultrasonic type—The whirlpools disrupt an ultrasonic beam that is transmitted through the air stream.
3. Pressure type—The pressure waves caused by the vortex column are measured by what is essentially a pressure sensor, similar to a MAP sensor. This is now the most common type.

Regardless of the method used to measure the whirlpools, all of these sensor types produce a 5-volt digital signal, with a frequency that is proportional to the airflow through the sensor.

The intake system must be airtight, with no leaks downstream of the meter; otherwise inaccurate signals will be sent to the PCM. Any air that leaks into the system without going through the airflow meter is called false air, or unmetered air. If air bypasses the airflow meter, it tends to lean out the mixture because the PCM is not accounting for the extra air. Always be on the lookout for cracked, missing, or unplugged vacuum hoses that may contribute to this problem.

Manifold Absolute Pressure Sensor

K63007

Changes in engine speed and load cause changes in intake manifold pressure. A manifold absolute pressure (MAP) sensor measures these pressure changes and converts them into an electrical signal. The signal may be an output voltage or a frequency. By monitoring output signals, the PCM senses manifold pressure and uses this information to calculate the basic fuel requirement. The MAP sensor can use a piezoelectric crystal. If there is a change in the pressure exerted on this crystal, the resistance changes. This alters its output signal. Some manufacturers also use the MAP sensor to monitor changes in manifold pressure when determining whether the **exhaust gas recirculation (EGR) valve** opened or not.

The sensor is either connected to the intake manifold by a small-diameter, flexible tube or is screwed directly into the manifold. The control unit typically sends a 5-volt reference signal to the sensor. As manifold pressure changes, so does the electrical resistance of the sensor, which in turn produces a change in the output voltage (**FIGURE 63-27**). During idling, manifold pressure is low (high vacuum), which produces a comparatively low MAP output. With a wide-open throttle, manifold pressure is higher, closer to atmospheric pressure (low vacuum), and voltage output is higher. Another type of MAP sensor that has been used by some manufacturers is one

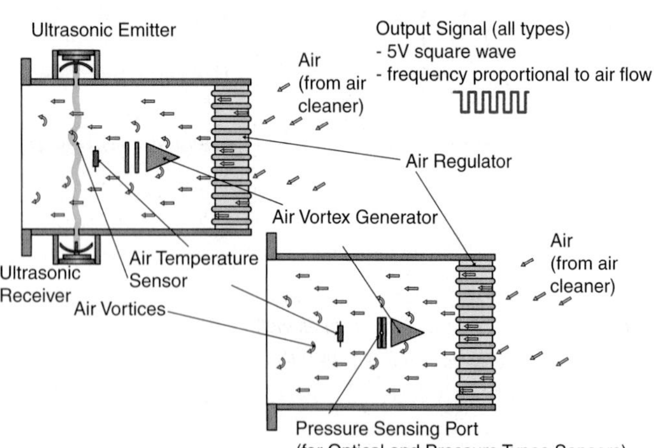

FIGURE 63-26 A triangular, vortex-generating rod disturbs the airflow and causes whirlpools of air to form in the body of the sensor.

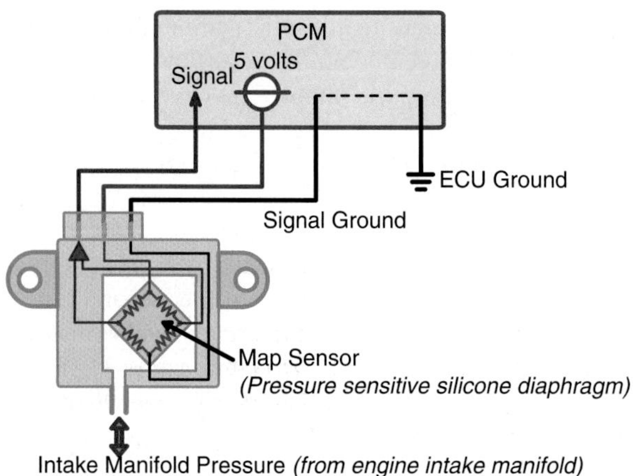

FIGURE 63-27 As manifold pressure changes, so does the electrical resistance of the sensor, which in turn produces a change in the output voltage.

that sends a varying frequency to the PCM. As the sensor measures manifold pressure, it sends a corresponding frequency signal. Frequency can be read on a digital multimeter that has a frequency function.

Barometric Pressure Sensor

A **barometric pressure (BARO) sensor** measures barometric pressure, which helps to calibrate the fuel injection system. The barometric pressures at sea level and, for example, in Denver, Colorado, which is called the mile-high city because of its high altitude, are extremely different. The vehicle must function properly at both locations or anywhere in between. The BARO sensor works in conjunction with the MAP sensor to provide constant pressure difference information to the PCM. This information is critical for proper air-fuel mixture as well as ignition timing.

Older BARO sensors were typically mounted on the firewall or along the fender in the engine bay. Most late-model vehicles use the MAP sensor to take a barometric pressure reading before the vehicle is started, and the sensor is usually mounted on the intake manifold.

Fuel Pressure Sensor

Most vehicles now use a fuel pressure sensor mounted on the fuel rail (**FIGURE 63-28**). This sensor allows the PCM to monitor the fuel pressure in the fuel rail so that it can control the speed of the fuel pump to maintain the correct pressure in the system. GDI systems typically use two fuel pressure sensors, one for the low pressure system and one for the high-pressure system.

FIGURE 63-28 Fuel pressure sensor mounted on the fuel rail.

Knock Sensor

Engine knock occurs in the combustion chamber when there is an unwanted spike in pressure, caused by pre-ignition or detonation. This unwanted and damaging event can be caused in different ways. Some vehicles are equipped with a **knock sensor**, the function of which is to monitor the noise that is created by the pressure spike. This noise, or "knock," can be created by an excessive load on the engine, ignition timing that is too advanced, or overheating the engine. The PCM can adjust the ignition timing to help reduce the knocking, except in the case of engine overheating. In addition, adjustments to the ignition timing and fuel delivery can be made if the vehicle begins to overheat. Following is a brief description of what occurs during excessive engine load and during engine overheating:

- **Excessive load on the engine**
  1. Ignition starts.
  2. The expanding gases create a pressure wave designed to push the piston down.

Applied Science

AS-10: Standard/Metric: The technician can convert measurements taken in the standard or metric system to specifications stated in either system.

An apprentice technician has been assigned to replace an intake manifold and fuel rail on a vehicle. An experienced technician is working with the apprentice on certain key points. Using the manufacturer's service information, the two technicians locate and discuss the proper torque specifications, which are given in both standard and metric units. The experienced technician explains that if only one specification is given, there are conversion factors available to determine the other torque value.

The common English term "foot-pound" (ft-lb) is sometimes interchangeably used with the preferred SAE term "pound-foot" (lb-ft). To convert 1 lb-ft to the metric unit of newton meters (N·m), we multiply the lb-ft value by 1.355 N·m. For example, 10 lb-ft = 13.55 N·m. To convert 1 N·m to lb-ft, we multiply the N·m value by 0.737 lb-ft. For example, 10 N·m = 7.37 lb-ft.

In addition to using these conversion factors, conversion tables are available. A number of interactive calculators are available online to convert various units from standard to metric and metric to standard.

Applied Science

AS-83: Barometric Pressure: The technician can demonstrate an understanding of the relationship of barometric pressure to engine performance.

Atmospheric pressure is a measure of the pressure exerted by the weight of the atmosphere. This is also called barometric pressure because barometers are used to measure it. The term "barometer" comes from the Greek words for "weight" and "measure." A mercury barometer compares atmospheric pressure to a column of mercury and measures the results in inches or millimeters of mercury.

A barometric pressure (BARO) sensor measures barometric pressure and is crucial to most fuel injections systems. The barometric pressure varies by geographic elevation. The vehicle must function properly in any given location. Accurate pressure readings are critical for proper air-fuel mixture as well as ignition timing.

The principle of the operation of barometric sensors is based upon the flexing of a silicon chip. The output voltage signal varies based upon the amount that the chip flexes due to barometric pressure. The PCM uses this information along with other data to determine the proper adjustments that should be made to the fuel system as well as other systems.

3. The forces opposing piston movement are too high.
4. The piston accelerates slowly and maintains a small volume above the piston.
5. The unburnt mixture is compressed by the advancing pressure wave.
6. This fuel self-ignites due to an increase in temperature as the mixture is compressed and creates its own flame front.
7. The two advancing flame fronts create a huge spike in pressure, creating engine knock.
8. This knock has enough energy to badly damage pistons, rings, spark plugs, and rod bearings.

- **Overheated engine—faulty thermostat**

1. Ignition starts.
2. The expanding gases create a pressure wave designed to push the piston down.
3. The temperature of the unburnt fuel is too high due to the overheated engine.
4. The unburnt mixture is compressed by the advancing pressure wave.
5. This fuel self-ignites due to an increase in pressure/temperature and creates its own flame front.
6. The two advancing flame fronts create a huge spike in pressure, causing engine knock.

One way of overcoming engine knock is to allow the fuel to be ignited later in the compression stroke. Doing so requires having less ignition advance. If the fuel is ignited later, then the pressures above the piston will be less during the early stages of the power stroke, reducing engine knock.

▶ TECHNICIAN TIP

Another way of preventing engine knock is to use fuel with a higher octane rating. This makes the fuel harder to ignite and slower to burn, both of which reduce engine knock.

The function of the knock sensor is to produce an electrical signal that the PCM can use to determine whether knock has occurred. If knock is occurring, the PCM will retard the timing a few degrees and listen to detect whether the knock is still present. If so, the PCM will continue to retard the timing in steps, within its limits, until knock is eliminated. If knock is not occurring, the PCM will advance the timing a degree or two and listen for knock. The timing is kept at the optimum point for the driving conditions. V-configured engines often have two knock sensors installed, known as twin knock sensors. This design allows the PCM to have control over knock on separate banks.

The sensor is screwed into the engine block, where it is influenced by all engine vibrations (**FIGURE 63-29**). Using a piezoelectric crystal, the sensor produces a signal voltage proportional to the vibrations applied to it. The PCM interprets the strength, frequency, and timing of the signal to determine if knock has occurred. Although knock sensor technology has improved, it cannot always know the difference between engine knock and a loose air-conditioning compressor bracket that rattles when it is on. In this situation, the rattling air-conditioning compressor bracket causes a noise that the knock sensor picks up as engine knock. But because the "knock" in this case does not go away as the timing is retarded, the PCM gradually reduces ignition advance until it reaches its limits, causing low power, poor fuel economy, and possible engine overheating. Monitoring the knock sensor signal and amount of timing retard on a scan tool will help you identify this issue.

FIGURE 63-29 The sensor is screwed into the engine block, where it is influenced by all engine vibrations.

▶ Switches

K63008

Switches provide a digital on/off signal to the PCM, so they can be used to indicate when devices and circuits are activated that require compensation by the PCM. For example, the brake on/off switch indicates to the PCM that the driver is applying the brakes. This information is useful for reducing or stopping fuel injection pulses, cruise control deactivated, and so on.

While at idle, engine speed can be adversely affected by fairly minor changes in load— from transmissions, power steering pumps, and air-conditioning compressors, for example. For optimum engine control, it is necessary for the PCM to be informed of these loads before the engine speed drops. Switches inform the PCM when the power steering pump pressure is high, automatic transmission drive gear is selected, and the air-conditioning compressor is being engaged. When the PCM receives the signal from the switch, it can immediately start to increase the engine's power to meet the need of the additional load. The result is excellent PCM control of idle speed as loads are placed on the engine.

Applied Math

AM-15: Mean/Median/Mode: The technician can calculate the average (mean) of several measurements to determine any variance from the manufacturer's specifications.

A new vehicle was purchased that was advertised to have a fuel economy rating of 24 mpg city driving and 30 mpg highway driving. The owner decided to calculate the average (mean) of several trips to see if the advertised ratings were correct. The plan was to check both city and highway driving separately, using several tests for each category.

The city mileage test consisted of three separate fill-ups, with miles recorded from the odometer. The first fill-up required 14.1 gallons of fuel for 330 miles of driving, equaling 23.4 mpg. The second fill-up required 13.6 gallons of fuel for 338 miles, equaling 24.9 mpg. The third fill-up required 12.9 gallons for 286 miles, equaling 22.2 mpg. To calculate the average (mean), he added 23.4 mpg + 24.8 mpg + 22.2 mpg and divided by 3. The result is 23.5 mpg city driving, which is very close to the advertised 24 mpg.

The highway driving test consisted of two fill-ups on a road trip. The first fill-up required 9.5 gallons for 260 miles, equaling 27.4 mpg. The second fill-up required 10.1 gallons for 305 miles, equaling 30.2 mpg. To determine the mean, he added 27.4 mpg + 30.2 mpg and divided by 2. The result is 28.8 mpg, which is slightly less than the advertised 30 mpg for highway driving.

Applied Math

AM-18: Standard/Metric: The technician can measure/test with tools designed for standard or metric measurements, and then convert the measurement to the system used by the manufacturer for specifications and tolerances.

A vehicle is in the shop for engine performance concerns. The technician needs to check the fuel pressure. As he looks in his toolbox for the gauge, he remembers that his friend has not returned his fuel pressure gauge. He has another one, but it is calibrated only in the metric unit bar. He connects this pressure gauge to the Schrader valve and starts the engine. With the engine running at idle, the gauge shows a reading of three bars. The technician uses the conversion factor of 1 bar = 14.503 psi multiplied by 3 to obtain the result of 43.5 psi. The service information states that the vehicle should have 44 psi, plus or minus 3 psi. This confirms that the fuel pressure is acceptable, and the technician will now conduct a fuel volume test.

▶ Powertrain Control Module

K63009

The PCM is like the brain of the engine. It gathers information, processes that information based on stored data, makes decisions, and commands components to operate in specified ways. It has the ability to learn from its actions and to use that information to make better decisions in the future. It can also perform self-diagnostics on virtually all of the components in the system, including itself. In basic terms, the PCM is comprised of four distinct parts (**FIGURE 63-30**):

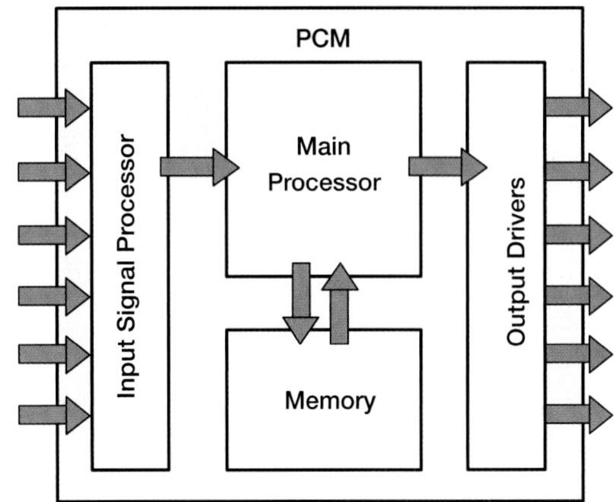

FIGURE 63-30 The sections of a PCM.

- A memory and storage section
- A sensor input/output signal processing section
- A data processing section
- An output drivers section

The memory and storage section contains the software for the computer, as well as a place to retain diagnostic trouble codes (DTCs), freeze-frame data, and learned data. Software is loaded into the memory, which is the basis for processing data. The software contains all of the information needed for the processor to interpret the sensor information and make decisions based on the programmed information. It also contains the instructions for performing self-diagnostic tests on most of the circuits and the PCM itself. In many cases, the software in the PCM's memory can be updated by uploading software updates, called a reflash process. By using software that can be updated, manufacturers can design fixes for many of the issues that may arise once a vehicle goes into operation, and a software update is much easier, quicker, and less expensive for resolving common customer or drivability issues than replacing parts.

The input signal processing section is responsible for sending the proper reference voltage to many of the sensors. It also receives the sensor signals and processes them into information that the main computer processor can use (**FIGURE 63-31**).

The data processor and drivers are hardware related, so they make up the physical portion of the PCM. The processor function receives the data from the input signal processing section and compares it to data maps in the memory. Once the sensor data has been analyzed, the processor looks up the appropriate data map and makes a decision about what actions to take (**FIGURE 63-32**).

The processor then sends the appropriate commands to each of the output drivers, which are electronic switching or control devices that control each of the actuators in the engine management system. The processor also commands the diagnostic tests to be run on the system and evaluates the results compared to the information stored in memory. The output

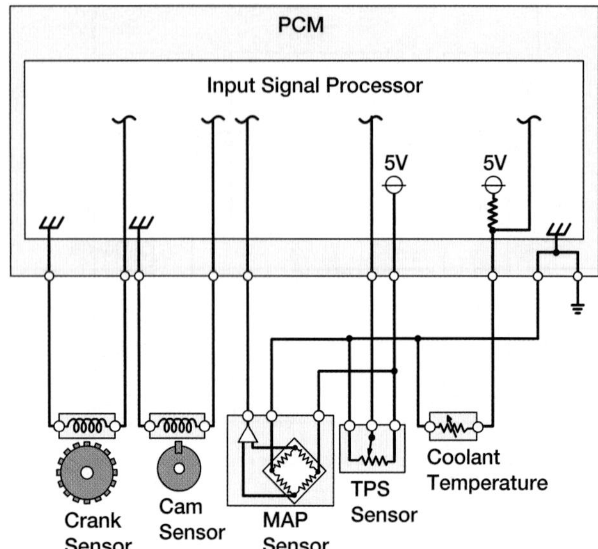

FIGURE 63-31 The input signal processing section sends reference voltage to many of the sensors. It also processes the signals returning from the sensors so the data processor can use the data.

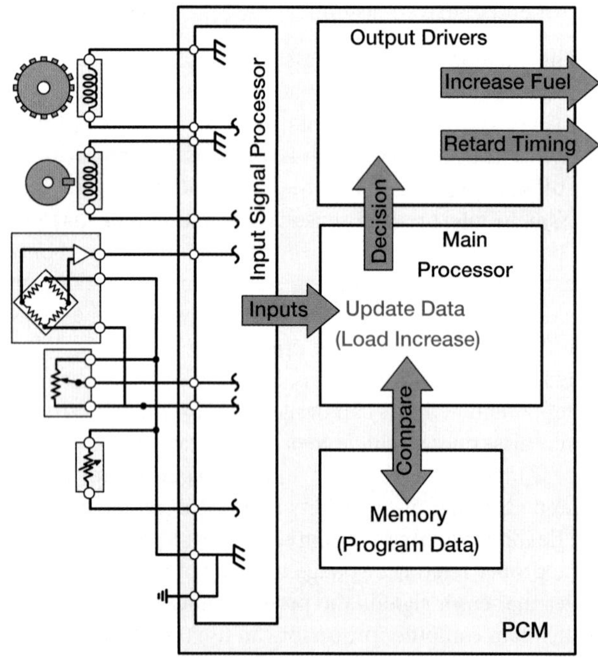

FIGURE 63-32 The data processor evaluates the sensor data and compares it to data maps and decides what actions to take. It then commands the output devices to send output signals to the actuators.

drivers typically send either simple on/off signals or pulse-width-modulated signals to the actuators, depending on the actuator being controlled (**FIGURE 63-33**).

The PCM also communicates with other electronic control modules on the vehicle since the control modules share data from each other's sensors through a communication network. Some of those control modules will be on the same network, and some will be on other networks, because most late-model vehicles use more than one communication network, as outlined

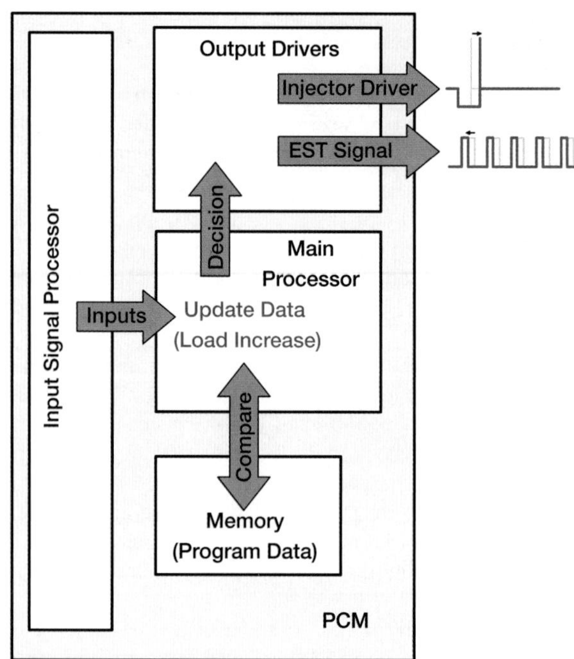

FIGURE 63-33 The output drivers send either on/off signals, pulse-width-modulated signals, or timing signals to the actuators.

in the chapter on Body Electrical System. In the case of multiple communication networks, the PCM sends a command over the network system, and the appropriate module carries out that command when the message is received. The module has its own output drivers that control the actuator.

The PCM is a gateway node that is connected to each of the other networks. A gateway node translates the information from, or to, the other networks so it can be shared between networks. Here is an example of how it would work: Suppose the driver begins to drive the vehicle forward. The PCM monitors the VSS signal for engine management purposes. The body control module (BCM) is programmed to lock the doors at a predetermined speed. The BCM monitors the VSS signal coming across the network from the PCM. At the predetermined speed, the BCM sends the command to lock the doors on its network. Each door control module receives the message and commands its own output driver to power the circuits to lock the door. The doors then lock at the predetermined vehicle speed, and a confirmation may be sent back across the network that the message was received and the action taken (**FIGURE 63-34**).

Controlled Devices

K63010

Remember, sensors monitor conditions and send information to the PCM. Actuators such as relays, solenoids, modules, and motors do the work. When a specific criterion is met, or the vehicle driver gives a certain command, an input signal is sent to the PCM for processing, then an output signal is sent to the corresponding controlled device (relay/solenoid/module/motor) to make an action happen. Next, we look at the most common engine management output devices related to the fuel and ignition systems.

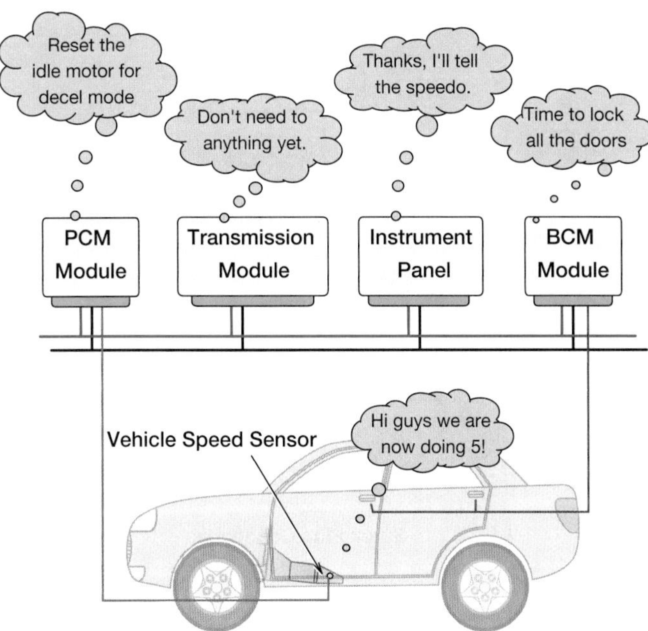

FIGURE 63-34 The PCM shares sensor data on the vehicle networks so that other control modules have access to those data they need.

▶ Relays

K63011

A **relay** is often used as an actuator to control other actuators when the computer must control a high-current load. The computer typically controls the relay ground circuit. When a high-current job is requested or commanded, the computer simply completes the ground circuit, and the relay coil field pulls the mechanical contacts closed, allowing the high current levels to flow to the appropriate load. Therefore relays can be used to control many different devices, such as fuel pumps, starter motor solenoids, and AC compressor clutches.

Most fuel pumps are operated by a fuel pump relay. The relay controls power to the fuel pump and is typically controlled by the PCM (**FIGURE 63-35**). Unlike a mechanical pump that would quit pumping if the vehicle were involved in an accident that resulted in the engine dying, an electric fuel pump will continue running as long as electricity is flowing to it. For safety's sake, the PCM will shut down the fuel pump relay if it does not see indications from the crankshaft position sensor that the engine is running above about 350 rpm. This way, the fuel pump is likely to be shut down in the event of an accident. The PCM can also operate the fuel pump for a few seconds if the ignition switch is turned on to prime the system or while the vehicle is being cranked.

Some vehicles use an inertia switch to turn the fuel pump off in the event of an accident. The inertia switch is held closed by a spring-loaded trigger. A loose weight is free to move within the area of the trigger. If the vehicle experiences enough of a jolt in any direction, the weight will dislodge the trigger, opening the switch. Once the switch opens, it has to be reset manually (**FIGURE 63-36**).

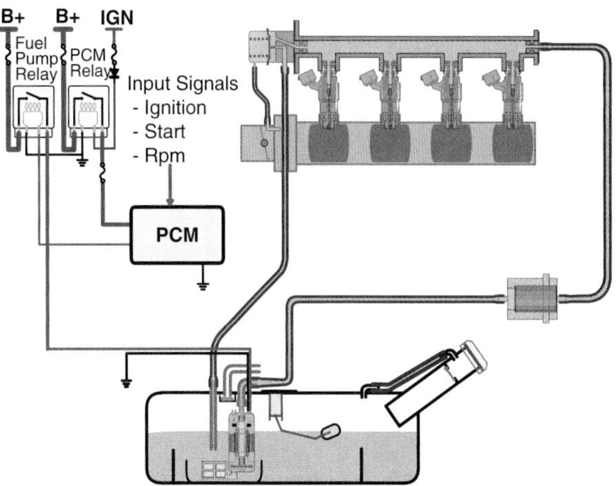

FIGURE 63-35 Relays are one type of controlled device activated by the PCM. Relays then power all kinds of other actuators.

FIGURE 63-36 A typical inertia switch with reset button on top.

▶ Solenoids

K63012

Solenoids are another type of controlled device used to carry out commands from the PCM. A solenoid is an electromechanical device similar to a relay. But instead of moving a switch magnetically, it moves a valve or linkage. When solenoids are used to open and close valves, they can control hydraulic or pneumatic circuits. A few of the common solenoids that are controlled by the PCM are listed here:

- Fuel injectors
- Fuel pressure control valves in pressure regulators
- Idle air control valves
- Vacuum control valves
- Vent valves
- Oil pressure control valves (variable cam drives, etc.)
- EGR valves

Solenoids can be controlled on either the power side or the ground side of the circuit. In most cases, solenoids like vacuum control valves are designed to withstand continuous duty

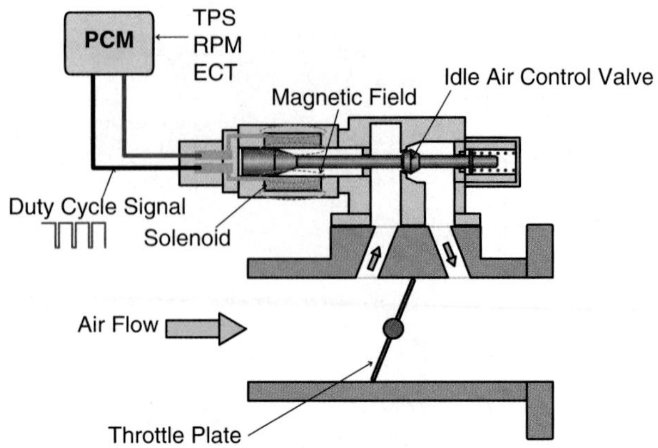

FIGURE 63-37 The solenoid-type air control valve acts on signals from the control unit to bypass a measured airflow around the throttle plate.

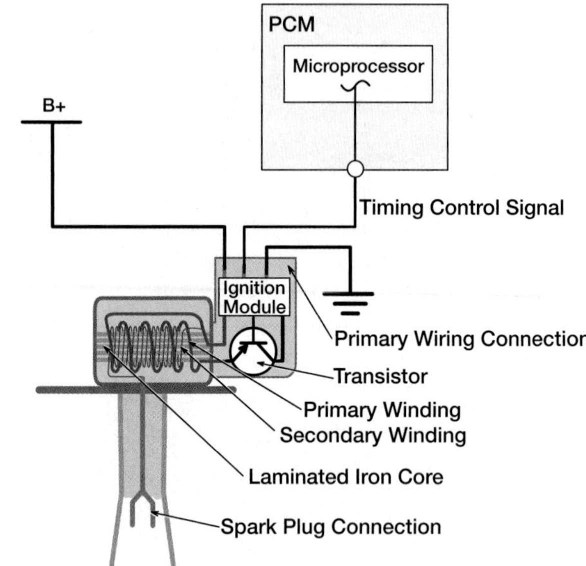

FIGURE 63-38 A coil-on-plug ignition module built into the coil.

cycle, meaning they can be on for long periods of time. In this case, they would receive typical on/off signals from the PCM. Other solenoids, like fuel injectors, are designed to be cycled on and off continuously at a very rapid rate. In this case, they receive a pulse-width-modulated signal from the PCM, as discussed earlier.

The solenoid-type air control valve acts on signals from the PCM to allow a measured amount of airflow to bypass the throttle plate (**FIGURE 63-37**). The position of the valve depends on how much current the control unit applies to the solenoid. Maximum current flow opens the valve fully to give maximum airflow. This is generally the starting position; thereafter, the taper valve is positioned by pulsing the solenoid winding circuit at a predetermined frequency.

The amount of valve opening, and therefore the amount of airflow, depends on the "on" time of the pulse. A long "on" pulse with a short "off" pulse will produce a high average voltage, and therefore a large opening of the valve. A short "on" pulse with a long "off" pulse will produce a low average voltage, and therefore a small valve opening and an increase in airflow. This "on" and "off" time is called the duty cycle, and it is generally expressed as a percentage. A variation in pulse width, at the set frequency, is called pulse width modulation.

▶ Control Modules

K63013

Most electronic modules on vehicles have their own switching circuit built inside the module that controls any associated actuators. Modules just need a timing signal to know when to switch the output circuit on and off. In this case, the **ignition control module** (ICM) controls the ignition coil's primary circuit. When the ignition control module receives a timing signal from the PCM, it then uses that timing signal to know when to switch the ignition coil primary circuit on and off.

There are a wide range of configurations of PCM-controlled ignition control, from the simple to the complex. Examples of simple ignition modules are many coil-on-plug modules built directly into the ignition coils. Most of them are simple switching devices that make no adjustments to spark plug timing. They only receive on/off timing signals from the PCM, and they turn the primary circuit on and off based on those commands (**FIGURE 63-38**). All of the timing calculations are handled by the PCM in this case. And because the PCM has access to all of the engine related sensor data, it is probably the best-suited component to manage ignition system timing.

Some ignition modules have more functionality than coil-on-plug ignition modules and receive signals from the CMP sensor, the CKP sensor, and the PCM (**FIGURE 63-39**). This type of module is capable of processing the CMP and CKP signals to determine crankshaft position and rpm. In this case, the module is able to control timing, within limits, across a broad rpm range without the timing signal from the PCM. These types of modules typically completely control the ignition timing during engine start-up and then switch over to PCM-controlled timing once the engine is running. Because of this, if the PCM timing signal is ever lost, the ignition module can provide enough control over the ignition system to enable the vehicle to be driven to a shop for service. As long as the PCM timing signal is present, its calculations are based on a lot more sensor data, so its timing decisions are the most valid. These types of ignition control modules are usually mounted under or near the ignition coils on a waste spark system or on/in the distributor (if equipped).

Electric Motor Actuators

K63014

Another type of actuator is an electric motor. Electric motors operate many devices on a vehicle from starters to the power windows to the throttle plates. Any type of motor can be controlled by the PCM, it just needs the right type of signal to operate. For example, a power window uses a typical permanent magnet DC motor, so to move the window up and down it only needs an on/off signal that can be reversed. A stepper motor

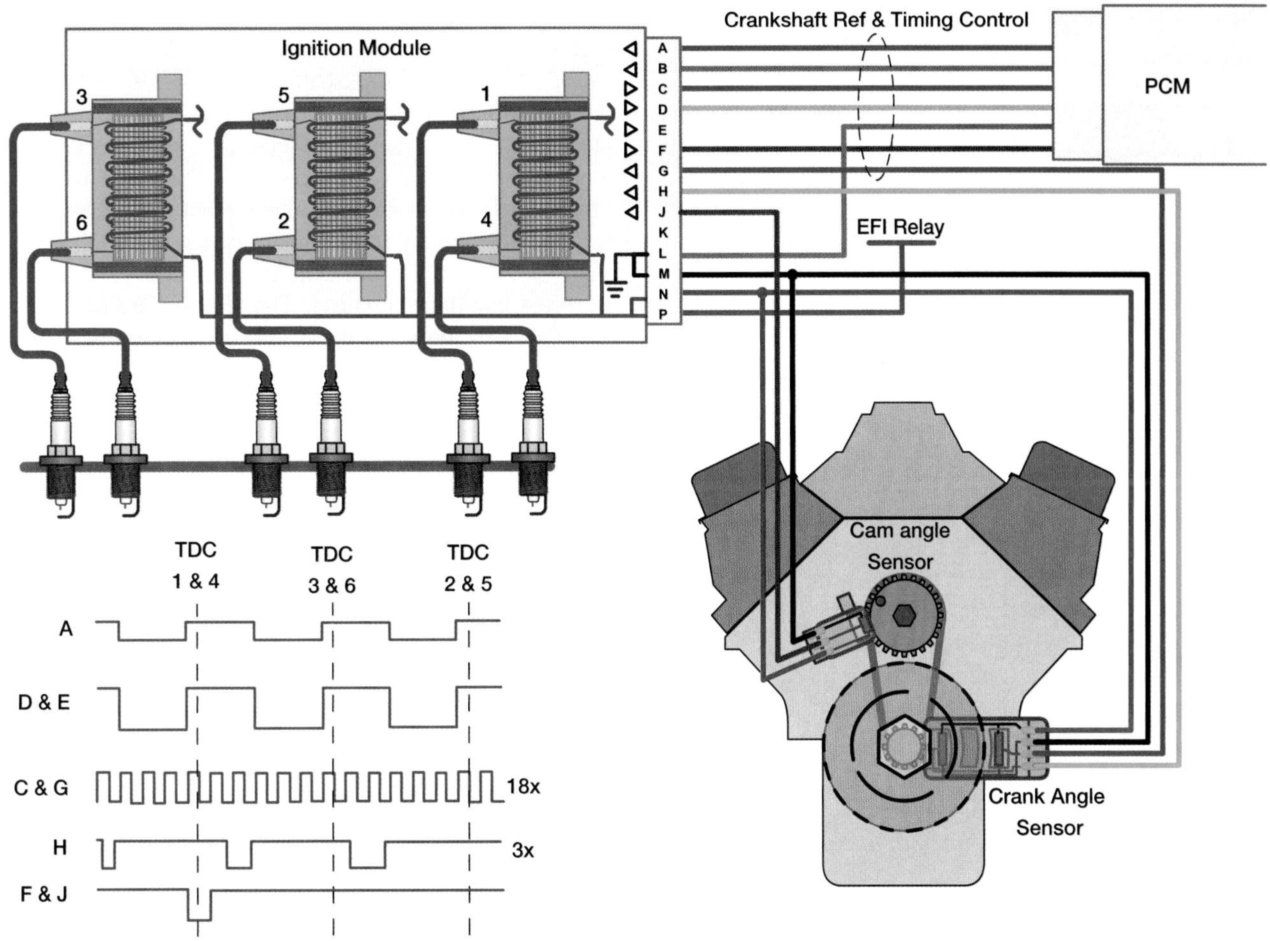

FIGURE 63-39 Typical waste spark ignition module circuit.

FIGURE 63-40 In the stepper motor type of idle control, the tapered pintle is positioned using a screw and nut assembly.

needs two to four alternately reversing signals, which moves the motor one step per signal.

An idle air control valve that uses a stepper motor has a tapered pintle valve that is opened and closed using a screw and nut assembly (**FIGURE 63-40**). The nut is part of a centrally located, permanent magnet armature that engages with the screw on the pintle. Rotating the armature extends or retracts the pintle, which closes or opens the air passage. The further the valve is open, the more air that bypasses the throttle plate, and vice versa. Rotation occurs by switching current flow in two coils. The pintle turns in forward or reverse steps, numbered from zero up to 255, with the bypass air passage fully open for maximum airflow. This large number of steps allows the pintle position to be finely controlled.

When the idle speed is controlled by the position of the throttle plate, a direct current motor may be used as a variable throttle stop. The controlling circuit can let current flow through the motor in either direction, to extend or retract the linear actuator.

Electronically Controlled Throttle

Commonly referred to as drive-by-wire or throttle-by-wire, an **electronically controlled throttle** is the replacement device for the long-standing use of a throttle cable (**FIGURE 63-41**). This system uses a pedal position sensor to calculate the driver's desired throttle position, and the command is sent to the PCM. If the conditions are right, the PCM in turn sends a command to a throttle servomotor to open or close the throttle plates. The computer can regulate cruise consistency, acceleration, and deceleration to improve fuel economy, reduce emissions, and

FIGURE 63-41 An electronically controlled throttle is the replacement device for a throttle cable.

prevent sudden changes in speed that could affect the life of the drivetrain components.

Some late-model, high-end vehicles may have a driver-controlled switch that provides a throttle response choice between "touring" and "sport." Touring has a slower and more conservative throttle response, which increases fuel economy. The sport setting provides a much quicker and more aggressive throttle response, making performance a priority over fuel economy.

▶ Engine Management System Control Strategies

K63015

Speed Density and Mass Airflow Systems

K63016

As we are focusing primarily on fuel management, we need to explore how the PCM determines the quantity of fuel to inject. As you know, the air-fuel ratio for normal vehicle operation should be roughly 14:1, by mass. It is relatively easy for the PCM to measure the amount of fuel being injected because fuel is not compressible. The PCM knows the fuel pressure and injector flow rates, so it knows how much fuel will be injected for a given pulse width. However, because air is compressible and changes density based on elevation, temperature, and humidity, it is much more challenging for the PCM to know how much air is entering each cylinder. If the amount of air entering the engine is not known, the proper amount of fuel to inject to obtain the correct air-fuel mixture is also not known. There are two primary methods that manufacturers use to determine the mass of the air: the speed-density method and the mass airflow sensor method.

The speed-density method calculates the mass of the air by using the engine rpm, manifold absolute pressure, and air temperature. All of these factors affect the mass of the air. As we saw in an earlier chapter, rpm affects volumetric efficiency. Volumetric efficiency varies with engine speed, so rpm is reported to the PCM by the crankshaft position sensor. A table is programmed into the PCM, accounting for volumetric efficiency at all rpms. Manifold absolute pressure affects the density of the air. The higher the absolute pressure in the intake manifold, the greater the mass of air. The intake manifold pressure is measured by the manifold absolute pressure sensor and reported to the PCM. Again, it has a table that indicates air density at all expected manifold pressures. The temperature of the air also affects its density. The warmer the air is, the less dense it is. An intake air temperature sensor measures the temperature of the air in the intake manifold and sends it to the PCM. The PCM uses the data from these sensors and compares it to the data maps in its memory to determine a base amount of fuel needed. It then commands the fuel injector to inject fuel for the required amount of time to obtain that amount of fuel.

In the speed-density system, the other sensors provide data that the PCM uses to refine the injector pulse width slightly from the above calculations. For example, if the TPS indicates that the throttle opened quickly, the PCM will enrich the mixture slightly to provide more power from the engine and avoid detonation. If the vehicle is coasting down a long hill at freeway speeds, the PCM may lean the mixture out, or shut the fuel off completely, as power is not needed.

Speed-density systems are fairly simple, inexpensive, and reliable. They also do not unduly restrict airflow, which means they can create slightly more power than engines equipped with a mass airflow sensor. The drawbacks of this system are that they aren't as accurate, nor can they take into account as many variables, as mass airflow systems. They also don't work as well with engine performance modifications such as bigger camshafts, turbochargers, or superchargers.

Mass airflow systems use a mass airflow (MAF) sensor that relatively accurately measures the mass of the air entering the intake manifold. It is mounted in front of the throttle body, and all air entering the engine should be going through the sensor. The MAF takes into account the temperature, humidity, and density of the air passing through it, so it sends a signal that accurately reflects the mass of the air. We discussed the different types of MAF sensors and their operation earlier in this chapter.

MAF systems are the most accurate type of system for determining the amount of air entering the engine. However, because all air flows through the sensor, MAF systems can restrict airflow. Also, they can be temperamental, causing drivability issues. They are also more expensive than speed-density systems, and they don't replace the other sensors, which are used in the same way to modify the injector pulse width from the programmed amount based on the mass airflow sensor reading. The other sensors cause the PCM to alter the pulse width based on the operating conditions they indicate.

Now, both of these systems are used only to determine the base amount of fuel required for the expected air-fuel ratio. But that doesn't mean that the expected air-fuel ratio is achieved. Each of these systems relies on the feedback of the oxygen sensor to verify that the actual air-fuel mixture burned in the combustion chamber is the correct amount. We explore more of how this is done in the next section.

▶ Feedback and Looping

K63017

For efficient operation of the catalytic converter, the mixture ratio must be maintained close to the stoichiometric ratio—14.7 parts air to 1 part fuel (or leaner during certain operating parameters). Efficiency of combustion can be monitored by measuring the percentage of oxygen in the exhaust gas. A high percentage may mean the mixture entering the cylinder is too lean (not enough fuel).

The oxygen sensor in the exhaust manifold, or in the exhaust pipe, tells the PCM how much oxygen is in the exhaust gas. The PCM then varies injector opening time to achieve the correct mixture ratio. Monitoring the data over time also allows the base fuel settings to be updated in the PCM memory as components age. This is called **adaptive learning**. The PCM memorizes its fuel settings and their results for different operating conditions and stores them for future use. If a fault occurs in a component or part of a system, a fault code will be stored in memory and can be retrieved by connecting a scan tool to the data link connector. Information on the fault can then be analyzed.

Although most injectors operate at the nominal battery voltage of 12 volts, many sensors have a reference voltage of 5 volts. This adjustment is made by a voltage regulator in the control unit. It gives more accurate signaling, as fluctuations due to changes in battery voltage do not occur at that reduced voltage.

During engine operation, three main **pollutants** are created: oxides of nitrogen, carbon monoxide, and hydrocarbons. These three pollutants enter the catalytic converter. How efficiently they are converted depends on the composition of the exhaust gases, and that depends on the air-fuel mixture sent into the cylinder for combustion. The more ideal the mixture, the fewer pollutants produced.

If the air-fuel ratio supplied to the engine is too rich, then the nitrogen oxides are converted efficiently, but the carbon monoxide and the hydrocarbons are not. If the air-fuel mixture supplied to the engine is too lean, the opposite occurs: Carbon monoxide and hydrocarbons in the exhaust gas are converted efficiently, but the nitrogen oxides are not.

Highly efficient conversions of all three pollutants occur only in a narrow range of air-fuel ratios. This range, the stoichiometric point, occurs around the ideal air-fuel ratio of 14.7 to 1 (**FIGURE 63-42**).

If the mixture ratio falls outside this range, the efficiency of conversion of either the nitrogen oxides or the hydrocarbons and carbon monoxide will rapidly decrease. Because of this strict requirement, vehicles with a three-way catalytic converter have a feedback system, called looping. It is made up of the fuel injectors, PCM, and oxygen sensors. When the feedback loop is active, it is called closed loop. When the feedback loop is not active, it is called **open loop**. During closed loop, the PCM adjusts the air-fuel ratio based on the readings from the oxygen sensors, also known as a lambda sensors. This is called a feedback loop because the PCM constantly adjusts the injector "on" time while monitoring the oxygen sensor reading to determine whether the mixture it sent was rich or lean.

For example, if the oxygen sensor indicates a rich condition, the PCM will command a slightly shorter pulse width of the injector. The oxygen sensor should then indicate a lean condition; if so, the PCM will command a slightly longer pulse width of the injector to enrichen it slightly. In stoichiometric mode, the oxygen sensor should toggle back and forth between slightly lean and rich conditions. This is ideal for most catalytic converters (**FIGURE 63-43**).

When an engine is first started, it starts in open loop. This means that the PCM pretty much ignores the oxygen sensor signals and determines pulse width primarily on the speed-density or MAF system, plus information from the other sensors. During normal operation, open loop may also occur

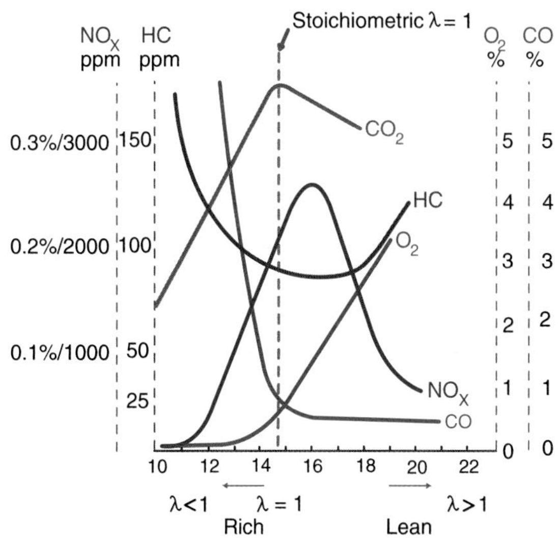

FIGURE 63-42 The stoichiometric point.

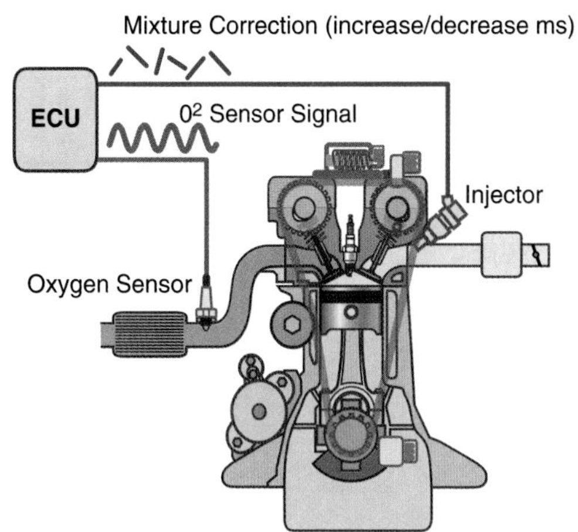

FIGURE 63-43 Closed loop means the control unit is using feedback from the oxygen sensor to alter the injection pulse width.

during extended idle times, or under maximum power. It can also occur if there is a fault that causes the air-fuel ratio to be excessively rich or too lean for long periods.

For the system to be in closed loop, several conditions must be met, but they are different for each manufacturer. Generally, the engine temperature must be above a certain point; the oxygen sensor must be active above a certain voltage; and a minimum amount of time must have passed since the vehicle was started. Until those conditions occur, the vehicle operates in open loop.

When the oxygen sensor reaches its operating temperature, around 600°F (315°C), it sends an output voltage to the control unit to signal whether the mixture is richer or leaner than a lambda value of 1.00—that is, the air-fuel ratio of 14.7 to 1. When the mixture deviates from this point, the output voltage changes sharply above or below its switching point (approximately 0.45 volt) (**FIGURE 63-44**).

The oxygen sensor voltage signal changes sharply when the air-fuel mixture changes from lean to rich. If the control unit sees the voltage as high, then the quantity of fuel injected is reduced. Similarly, if the mixture changes from rich to lean and the voltage is low, more fuel is injected to enrich the mixture.

The control unit then adjusts the pulse width of the injector to ensure the most efficient operation of the catalytic converter. During engine operation, this adjusting is continuous, and almost instantaneous, trying to maintain an air-fuel ratio for lambda very near 1.

Short- and Long-Term Fuel Trim

K63018

The computer is programmed with the expected injector pulse width for nearly all operating conditions, but circumstances such as engine and fuel system wear, or differences in fuel, cause the programmed settings to result in an improper mixture, as indicated by the oxygen sensors. Because the oxygen sensors are the computer's window into the exhaust stream, they are used to validate the accuracy of the computer's pulse width signal. If the mixture is off, then the PCM will add or subtract fuel from the base amount. The difference between the stored pulse width and the actual pulse width required to keep the mixture at the correct ratio is called fuel trim. This means that the PCM is adjusting the amount of fuel injected, plus or minus, from the programmed amount. It is also part of the PCM's adaptive learning.

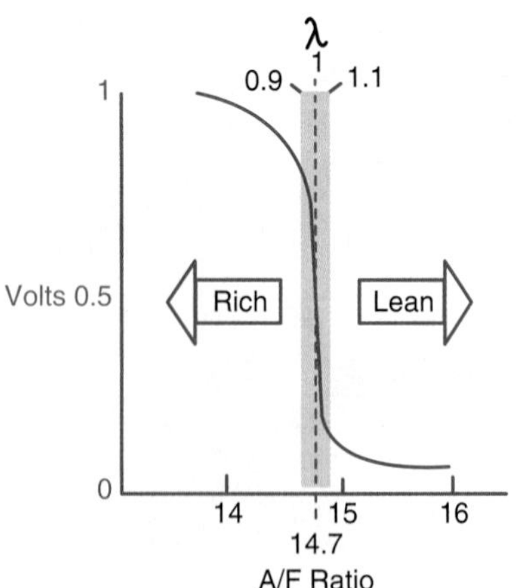

FIGURE 63-44 When the mixture deviates from the stoichiometric point, the output voltage changes sharply above or below its switching point of approximately 0.45 volt.

The PCM watches the average of the enriching and leaning over short periods of time (less than a second), calculates the percentage changed from the calculated value, and displays it as a positive (adding fuel) or negative (subtracting fuel) percentage. This is called short-term fuel trim (STFT). The PCM aims to keep the STFT at its midpoint so that fuel trim adjustments can respond quickly. If it sees the STFT start to creep away from the midpoint, the PCM will increase or decrease its base pulse-width setting (long-term fuel trim [LTFT]) to bring the STFT back to center. It will then record the updated base pulse width and store it in the PCM's memory. It will also store the difference between the original programmed pulse width and current pulse width, as a LTFT percentage, and store it in memory—again, either positive or negative. Note that the STFT is not normally stored in memory when the key is turned off.

Both short- and long-term fuel trims can be displayed on a scan tool connected to the PCM. STFT and LTFT readings are useful when diagnosing drivability concerns and give us a window into how the engine management system is operating. As you can imagine, if readings for both LTFT and STFT are near zero, then it is likely that the system is operating close to its design. If you see the LTFT is more than single digits positive, it means the PCM thinks the vehicle is running too lean and is substantially enriching the mixture. This can indicate a plugged fuel filter, worn fuel pump, restricted fuel injectors, vacuum leak, exhaust manifold leak, or oxygen sensors that are weak and reading too low a voltage. If the LTFT is more than single digits negative, it means the PCM thinks it is running too rich and is leaning it out. This could be caused by a leaky fuel injector, a leaky pressure regulator, a purge valve that is stuck open, or gasoline contaminated oil in the oil pan. It could also be that the oxygen sensors that are malfunctioning by sending too high a voltage (this is less common).

▶ Fuel Shutoff Mode and Clear Flood Mode

K63019

Another operating mode is called the **fuel shutoff mode**. This mode turns the fuel injectors off when the vehicle is coasting above a certain rpm and speed with the throttle closed. Rather than inject fuel into the cylinders during coasting, which would be wasted, the PCM shuts the fuel injectors off, until either the rpm drops below a certain level or the throttle is opened.

Clear flood mode is similar to fuel shutoff mode, but for a different reason and condition. With carbureted vehicles, if the engine is flooded with fuel, the driver would hold the throttle plate open while cranking the engine over to clear the flooded condition. Some manufacturers have built that functionality into their fuel-injected vehicles. Clear flood mode is activated if the PCM sees the throttle position sensor indicating a wide-open throttle when the ignition is first turned on or cranked. If this occurs, the PCM will keep the injectors turned off to clear a flooded engine condition. Technicians use this mode to fool the engine into cranking without starting during a cranking sound diagnosis test. Holding the throttle to the floor, and then turning the key to crank the engine over, allows the engine to crank, but not start. This allows the technician to evaluate the cranking sound without having to disable the ignition system or fuel system manually. Just be aware that not all electronically fuel-injected vehicles are equipped with this functionality.

▶ Wrap-Up

Ready for Review

- The engine management system is made up of components categorized in one of three separate groups: sensors, a control module (PCM), and controlled devices, also called output devices or actuators.
- There are two main types of electrical signals: analog and digital. Analog signals are smooth and gradually changing in strength, whereas a digital signal is a direct on/off with no in-between transition.
- Frequency is measured in cycles per second, where a cycle is the distance between waves or wavelengths. The more cycles per second there are, the higher the frequency.
- Each sensor monitors specific conditions and creates a signal that corresponds to the conditions.
- Several sensors use a potentiometer to create a variable voltage signal that corresponds to a specific condition.
- The throttle position sensor (TPS) gathers information on throttle positions to allow the control unit to make adjustments according to operating conditions.
- Thermistors are used in various temperature-related sensors on modern vehicles and use a resistor that changes its resistance with the changes in temperature. Thermistors can be the positive temperature coefficient (PTC) type or negative temperature coefficient (NTC) type. In PTC thermistors, the resistance goes up as the temperature goes up. In NTC thermistors, the resistance goes down as the temperature goes up.
- Engine coolant temperature maintains the air-fuel ratio within an optimum range. The control unit must take account of coolant temperature and air temperature.
- Several sensors use either inductive pickup sensors or Hall-effect sensors to indicate the speed or position of components.
- The crankshaft position (CKP) sensor uses information on the speed and position of the crankshaft to control ignition timing and injection sequencing.
- The camshaft position (CMP) sensor sends constant data to the computer to let it know which cylinder is on its power stroke, as well as the position of the camshaft and valves.
- The vehicle speed sensor (VSS) is a detection device that sends the vehicle speed information (i.e., how fast the vehicle is traveling) to the electronic control module.
- Manifold absolute pressure (MAP) measures changes in engine speed and load and converts the findings into an electrical signal to control engine operations.
- A barometric pressure (BARO) sensor measures barometric pressure, which helps to calibrate the fuel injection system.
- Some vehicles are equipped with a knock sensor (KS), the function of which is to monitor the noise that is created by the pressure spike.
- Oxygen sensors placed in the exhaust stream indicate the rich/lean condition of the air-fuel mixture.
- Stoichiometric oxygen sensors are good for indicating AF ratios around 14.7:1. Wide-band oxygen sensors can accurately indicate AR ratios from about 9:1 up to atmospheric air.
- Mass airflow (MAF) sensors measure the mass of the air entering the engine. There are several types of MAFs.
- Switches provide a digital on/off signal to the PCM so they can be used to indicate when devices and circuits are activated for which the PCM is required to compensate.
- The PCM is the brain in the system. It gathers information, processes that information based on stored data, makes decisions, and commands components to operate in specified ways.
- When a specific criterion is met or the vehicle driver gives a certain command, an output signal is sent to the corresponding controlled device (relay/solenoid/module/motor) to make an action happen.
- A relay is often used as an actuator to control other actuators when the computer needs to control a high-current load.
- A solenoid is an electromechanical device similar to a relay. But instead of moving a switch magnetically, it moves something else such as a valve or linkage.
- Most electronic modules on vehicles have their own switching circuit built inside the module which control their associated actuators. Modules just need a timing signal to know when to switch the output circuit on and off.
- Electric motors are a type of actuator that operate many devices on a vehicle, from starters, to the power windows, to the throttle plates.
- There are two main strategies for determining the amount of air entering the engine; the speed-density method and the mass airflow sensor method.
- During closed loop, the PCM adjusts the air-fuel ratio based on the readings from the oxygen sensors, also known as a lambda sensors.

- Open loop means that the PCM pretty much ignores the oxygen sensor signals and determines pulse width primarily on the speed-density or MAF system, plus information from the other sensors.
- The difference between the stored pulse width and the actual pulse width required to keep the mixture at the correct ratio is called fuel trim.
- STFT and LTFT readings are useful when diagnosing drivability concerns and give us a window into how the engine management system is operating.
- Clear flood mode is activated if the PCM sees the throttle position sensor indicating a wide-open throttle when the ignition is first turned on or cranked. If this occurs, the PCM will keep the injectors turned off to clear a flooded engine condition.

Key Terms

adaptive learning Monitoring the data over time also allows the base fuel settings to be updated in the PCM memory as components age.

barometric pressure (BARO) sensor A sensor that measures atmospheric pressure.

camshaft position (CMP) sensor A detection device that signals to the PCM the rotational position of the camshaft.

crankshaft position (CKP) sensor A sensor similar to a Hall-effect switch that is used to monitor crankshaft position and speed.

electronically controlled throttle A system that uses electronic, instead of mechanical, signals to control the throttle. Sometimes called drive-by-wire.

exhaust gas recirculation (EGR) valve A valve that allows a controlled amount of exhaust gas into the intake manifold during a certain period of engine operation. Used to lower nitrogen oxide exhaust emissions.

frequency The rate of change in direction, oscillation, or cycles in a given time.

fuel shutoff mode A safety precaution by which fuel is shut off when certain conditions are met during a vehicle crash.

ignition control module (ICM) A general control unit of some electronic ignition systems, usually with current and dwell angle control; driver and output stage; and in some cases, electronic spark timing functions.

inductive-type sensor A sensor mounted on the crankcase housing that is used to sense the movement of the ring gear teeth on the flywheel, or a toothed disc on the crank pulley.

intake air temperature (IAT) sensor A sensor that measures the temperature of the incoming air through the air filtration system.

knock sensor An engine sensor that detects pre-ignition, detonation, and knocking.

open loop When the feedback loop is not active.

oxygen sensor (before and after catalytic convertor) An exhaust sensor used to measure the amount of oxygen in the exhaust gases produced by the engine PCM, used to determine fuel mixture and spark timing.

pollutant A potential threat to human health or the environment resulting from excessive amounts of chemicals and waste.

potentiometer A variable resistor that can be used to adjust voltage in a circuit.

reed switch A type of speed sensor that uses a magnetic field to open and close a movable set of contacts. It is used with a rotating magnet to measure rpm of a shaft and send the signal to the PCM.

relay An electromechanical switching device whereby the magnetism from a coil winding acts on a lever that switches a set of contacts.

thermistor A device that changes its resistance in relation to heat.

Review Questions

1. In the engine management system, which of the following create input signals to the PCM?
   a. Sensors
   b. Relays
   c. Control modules
   d. Solenoids
2. Which of the following is the best tool to measure the voltage output from a throttle position sensor?
   a. Scan tool
   b. DVOM
   c. DMM
   d. Digital storage oscilloscope (DSO)
3. In a vehicle, thermistors are used in various sensors related to:
   a. pressure.
   b. temperature.
   c. voltage.
   d. speed.
4. Which of the following sensors are most involved in the control of ignition timing and injection sequencing?
   a. Fuel temperature sensor
   b. Accelerator pedal position sensor
   c. Crankshaft position sensor
   d. Oxygen sensor
5. Which of the following sensors provides information that is the most critical for proper air-fuel mixture, as well as ignition timing?
   a. Manifold absolute pressure (MAP) sensor
   b. Barometric pressure (BARO) sensor
   c. Fuel pressure sensor
   d. Knock sensor
6. Which part of a powertrain control module retains diagnostic trouble codes (DTCs), freeze-frame data, and learned data?
   a. Memory and storage section
   b. Sensor input/output signal processing section
   c. Data processing section
   d. Output drivers section

7. Which of the following is often used as an actuator to control other actuators when the computer needs to control a high-current load?
   a. Solenoids
   b. Thermistors
   c. Motors
   d. Relays
8. The speed-density method of determining the amount of air entering the engine takes into account all of the following *except*:
   a. engine rpm.
   b. manifold absolute pressure.
   c. air temperature.
   d. spark knock.
9. The feedback looping system in a three-way catalytic converter serves to adjust:
   a. the air-fuel ratio.
   b. engine speed.
   c. the temperature in the passenger cabin.
   d. fuel consumption.
10. Which of the following modes is used by technicians to fool the engine into cranking without starting during a cranking sound diagnosis test?
   a. Fuel shutoff mode
   b. Clear flood mode
   c. Key on, engine off mode
   d. Stoichiometric mode

ASE Technician A/Technician B Style Questions

1. Tech A says that analog signals are continuously variable signals. Tech B says that digital signals are direct on/off signals with no in-between transition. Who is correct?
   a. Tech A
   b. Tech B
   c. Both A and B
   d. Neither A nor B
2. Tech A says that potentiometer type sensors use a variable resistance to create a variable voltage signal that corresponds to a specific condition. Tech B says that oxygen sensors are the potentiometer type. Who is correct?
   a. Tech A
   b. Tech B
   c. Both A and B
   d. Neither A nor B
3. Tech A says that the oxygen sensor measures the amount of oxygen entering the intake manifold. Tech B says that many oxygen sensors have a built-in heater to warm them up more quickly. Who is correct?
   a. Tech A
   b. Tech B
   c. Both A and B
   d. Neither A nor B
4. Tech A says that mass airflow sensors determine mass airflow based on engine speed and manifold pressure. Tech B says that some airflow sensors use a hot wire to determine mass airflow. Who is correct?
   a. Tech A
   b. Tech B
   c. Both A and B
   d. Neither A nor B
5. Tech A says that the output drivers in the PCM typically send either simple on/off signals, or pulse-width-modulated signals to the actuators, depending on the actuator being controlled. Tech B says that in many cases, the software in the PCM's memory can be updated by uploading software updates, called a reflash process. Who is correct?
   a. Tech A
   b. Tech B
   c. Both A and B
   d. Neither A nor B
6. Tech A says that speed density systems use vehicle speed and fuel density to determine injector pulse width. Tech B says that mass airflow systems use a sensor to measure the mass of the air entering the engine. Who is correct?
   a. Tech A
   b. Tech B
   c. Both A and B
   d. Neither A nor B
7. Tech A says that open loop is when the PCM pretty much ignores the oxygen sensor signals. Tech B says that closed loop is when the PCM shuts off the fuel injectors during a collision. Who is correct?
   a. Tech A
   b. Tech B
   c. Both A and B
   d. Neither A nor B
8. Tech A says that the difference between the stored pulse width and the actual pulse width required to keep the mixture at the correct ratio is called fuel trim. Tech B says that long-term fuel trim values can be positive or negative. Who is correct?
   a. Tech A
   b. Tech B
   c. Both A and B
   d. Neither A nor B
9. Tech A says that fuel shutoff mode is when the PCM shuts off the fuel pump when the fuel tank is overfull. Tech B says that clear flood mode is activated by holding the throttle to the floor with the key off and then cranking the engine. Who is correct?
   a. Tech A
   b. Tech B
   c. Both A and B
   d. Neither A nor B
10. Tech A says that the oxygen sensor is part of the feedback system for the engine management system. Tech B says that long-term fuel trims that are positive means that the PCM is leaning out the fuel mixture from the base pulse-width setting. Who is correct?
   a. Tech A
   b. Tech B
   c. Both A and B
   d. Neither A nor B

CHAPTER 64
Onboard Diagnostic Systems

NATEF Tasks

- **N64001** Describe the use of OBD monitors for repair verification. (MLR/AST/MAST)
- **N64002** Retrieve and record diagnostic trouble codes (DTC), OBD monitor status, and freeze-frame data; clear codes when applicable. (MLR/AST/MAST)
- **N64003** Inspect and test computerized engine control system sensors, powertrain/engine control module (PCM/ECM), actuators, and circuits using a graphing multimeter (GMM)/digital storage oscilloscope (DSO); perform needed action. (MAST)
- **N64004** Perform active tests of actuators using a scan tool; determine needed action. (AST/MAST)
- **N64005** Interpret diagnostic trouble codes (DTCs) and scan tool data related to the emissions control systems; determine needed action. (AST/MAST)
- **N64006** Diagnose the causes of emissions or driveability concerns with stored or active diagnostic trouble codes (DTC); obtain, graph, and interpret scan tool data. (MAST)
- **N64007** Diagnose emissions or drivability concerns without stored or active diagnostic trouble codes; determine needed action. (MAST)
- **N64008** Diagnose drive ability and emissions problems resulting from malfunctions of interrelated systems (cruise control, security alarms, suspension controls, traction controls, HVAC, automatic transmissions, non-OEM-installed accessories, or similar systems); determine needed action. (MAST)

Knowledge Objectives

After reading this chapter, you will be able to:

- **K64001** Describe the reasons for onboard diagnostic systems.
- **K64002** Explain the two main generations of onboard diagnostic systems.
- **K64003** Decode and explain diagnostic trouble codes and freeze-frame data.
- **K64004** Explain the purpose and function of the system readiness monitors.
- **K64005** Describe the purpose and function of scan tools.
- **K64006** Explain the role of the scan tool in the diagnostic process.

Skills Objectives

After reading this chapter, you will be able to:

- **S64001** Diagnose emissions or drivability concerns.

▶ Introduction

The automobile we drive today has certainly evolved significantly, even over the last few years. Properly operating vehicles are running cleaner than ever and put out a small amount of pollution compared to vehicles of 30 years ago. Yet despite all the innovations, **emissions** from automobiles can still harm the environment, as well as our health, if something fails. In fact, one badly operating new vehicle can put out as much emissions as hundreds of properly running vehicles combined. Because of this, the engine management system needs to be in proper operating condition, and any faults it reports have to be diagnosed and repaired in a timely manner.

▶ Reasons for Onboard Diagnostic Systems

K64001

In the United States, Congress has passed federal emission regulations, starting with the Clean Air Act in 1963. This program was expanded in 1967 to help curb emissions from automobiles. Major amendments to the law, requiring regulatory controls for air pollution, were enacted in 1970, 1977, and 1990.

In some parts of the country (usually urban areas) where noncompliance with clean air laws exists, auto technicians know that vehicle emission testing and repair remedies are a big part of day-to-day business. Today, when a vehicle's malfunction indicator lamp (MIL)—formerly called a "check engine light"—is illuminated, it means the vehicle has identified a fault that is likely causing it not to comply with clean air regulations (**FIGURE 64-1**). In the interest of public health, keeping cars running clean is a mandate. Low tailpipe emissions also mean a vehicle is running efficiently, which also helps to conserve fuel, in turn reducing carbon dioxide emissions.

FIGURE 64-1 The malfunction indicator lamp (MIL) indicates to the driver that the PCM has identified a fault in the engine management system.

▶ Pollutants

Vehicles emit pollutants to the atmosphere in two major ways: evaporative emissions, and tailpipe emissions. Evaporative emissions are known as **volatile organic compounds (VOCs)**. VOCs may be fuel or oil vapors emitted from the fuel tank, fuel lines, engine crankcase, or elsewhere. In earlier vehicles, engine draft tubes spewed VOCs from the crankcase onto the road, as evidenced by dark center-of-lane deposits on uphill stretches of our highways. Likewise, hot summer days once meant the smell of gasoline fumes around parked cars—but not today!

The tailpipe emits a variety of pollutants, including these:

- **Carbon monoxide (CO)**
- **Hydrocarbons (HCs)**, or VOCs
- **Particulate matter (PM)**
- **Carbon dioxide (CO_2)**
- **Sulfur dioxide (SO_2)**
- **Oxides of nitrogen**—nitric oxide (NO) and nitrogen dioxide (NO_2)

Carbon monoxide is highly toxic, as discussed in the Personal Safety chapter. Oxides of nitrogen and hydrocarbons react together to create ground-level ozone, which is considered a health hazard. Ground-level ozone especially affects children, the elderly, and those with respiratory problems. On "ozone alert days," such people are advised to stay indoors, and the operation of VOC-spewing lawn and garden equipment is discouraged. Carbon dioxide is a greenhouse gas that contributes to the greenhouse effect.

You Are the Automotive Technician

Jim, a new customer, brings his vehicle into your shop for an illuminated malfunction indicator lamp (MIL). He says it has been on for about a week and wants to know if you have one of "those computers" that will tell you what is wrong with his vehicle. You tell him, as you chuckle, that you wish there was a tool you could buy that would do that, but you do have several scan tools and the training to know how to use them. With that and some good technical service information, you can diagnose what is causing the MIL to be on, so it can be fixed right the first time. Jim says that makes sense, as one of his coworkers had a similar problem, and the shop he went to told him that the vehicle needed a particular part, but that part didn't fix the problem. The coworker ended up buying several parts over a couple of visits, and the MIL always came back on. Jim agrees to let you diagnose the problem.

1. What does the MIL being illuminated indicate?
2. What is "freeze-frame" data? And how can it help you diagnose the problem?
3. What do monitors that report "Ready" and "Not Ready" indicate?
4. What is the "data stream," and how is it used during diagnosis?

All of these pollutants are hazardous to one degree or another and are monitored and controlled by efficient onboard electronic systems. The PCM runs the engine management systems through self-tests at engine start-up and continues to monitor the systems once underway. If any of the sensor readings or actuators are out of specifications, the driver is alerted, a snapshot of operating conditions is saved for technicians to examine later, and, if needed, alternate operating strategies are initiated to protect the vehicle and the environment. Knowing how the onboard diagnostic system operates will give you a head start in diagnosing engine management and drivability concerns. We cover those topics in this chapter.

Many strategies have been implemented to clean up vehicle emissions, and in more recent years closed-loop electronic systems have assumed the major responsibility. The following is an abbreviated timeline of the many strategies devised for curtailing light-duty vehicle emissions in the United States since the 1960s:

- 1961—First positive crankcase ventilation (PCV) systems required in California
- 1964—Sealed crankcase systems replace crankcase "draft tubes"
- 1968—Controls for crankcase VOCs, tailpipe CO, and hydrocarbons (PCV, AIR, etc.)
- 1969—Commencement of limiting oxides of nitrogen emissions by retarding ignition and valve timing
- 1971—Introduction of evaporative emission (EVAP) control systems (charcoal canisters)
- 1973—Exhaust gas recirculation (EGR) systems used to control oxides of nitrogen
- 1975—Unleaded gasoline phase-in
- 1975—Oxidizing catalytic converters introduced
- 1981—Closed-loop systems and three-way catalysts introduced; first-generation onboard diagnostics (OBD I) enacted nationwide (OBD I started with California vehicles)
- 1996—Second-generation onboard diagnostics (OBD II) enacted nationwide
- 1998—Commencement of vehicle refueling emission controls
- 2003—Widespread adoption of controller area network (CAN) communication systems (originated in 1998 with Robert Bosch–equipped vehicles like Volkswagens)
- 2006—The FlexRay network was first added to a production vehicle. It is faster and more reliable than CAN.

Let's look further into what these onboard diagnostic (OBD) systems do and what features are available to help technicians diagnose and repair vehicle system **faults**.

▶ Onboard Diagnostic Systems

K64002

OBD systems are a great resource to help the technician work more efficiently and remove some of the guesswork involved in diagnosing problems in today's sophisticated vehicles. These systems can be diagnosed using the powerful onboard computers in harmony with sophisticated diagnostic equipment to pinpoint problems in engine management systems and components. Various onboard computers (sometimes referred to as **modules**) monitor the systems and components and inform the vehicle operator when a fault occurs. The systems are also capable of providing information about the fault, thereby assisting technicians in identifying and repairing malfunctioning circuits or components.

In the United States, two main generations of OBD systems have been used. The first generation of onboard diagnostic (**OBD I**) systems operated under manufacturer standards, starting with California vehicles and becoming nationwide in 1981. The second generation of onboard diagnostic (**OBD II**) systems operate under standards set by the Society of Automotive Engineers (SAE) to conform to U.S. Environmental Protection Agency (EPA) regulations. Some manufacturers started using OBD II with a few of their models in 1994 and 1995, in order to become familiar with the systems, but for the 1996 model year, OBD II was standardized and initiated nationwide.

Where OBD I monitored mainly hard electrical circuit malfunctions, OBD II is an enhanced diagnostic system that identifies faults that may affect the vehicle's emission output and drivability. OBD II constantly monitors and manages the engine management system as it controls the engine and vehicle's emission systems. This in turn enables much tighter emissions standards to be enforced, with cutoff points far lower than those previously possible.

OBD I allowed each vehicle manufacturer to have its own OBD system, with its own diagnostic connector, its own trouble codes, its own component names, and its own scan tool. This made diagnosing and repairing vehicles much more difficult for technicians and repair shops. The OBD II standardized all of this and provided a certain degree of commonality in that all vehicle manufacturers must adhere to certain standards regarding the diagnostic connector, fault code descriptions, and nomenclature (names of parts). This means that any so-called "generic" OBD II faults and background data can be accessed and read by **aftermarket** as well as original equipment manufacturer (OEM) scan tools. This greatly reduced the cost and complexity of repairing these vehicles.

Both OEM and aftermarket OBD II scan tools serve to access OBD information via the data link connector (DLC). The DLC enables the scanner to access data stored in the vehicle's various computers (**FIGURE 64-2**). The DLC itself is a 16-pin connector with a common (SAE J1962) size, shape, and pin layout (**FIGURE 64-3**). Regardless of vehicle make and model, its location is also now fairly standardized as being within 2' of the steering wheel below the driver's side instrument panel.

Data Stream

One of the advantages of OBD II is that the sensor data can be accessed with a scan tool. That means that you can use the scan tool to access the information from virtually any sensor in the system, through the PCM. The scan tool displays the interpreted sensor readings that the computer is receiving. This means that you can examine all of the sensor data while hunting for clues to the cause of the fault. For example, the customer complained

FIGURE 64-2 The data link connector (DLC) enables the scanner to access data stored in the vehicle's various computers.

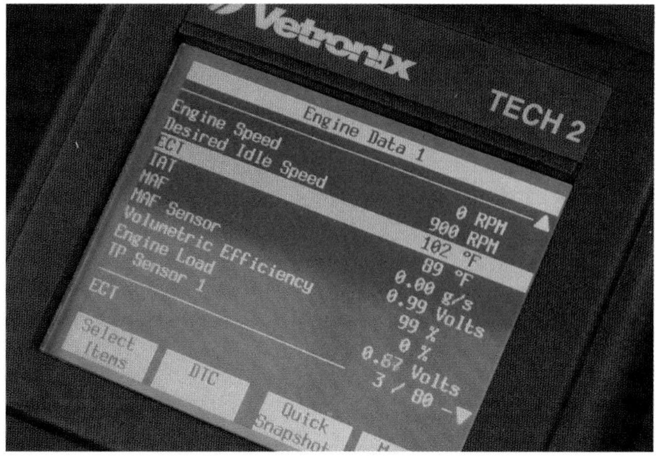

FIGURE 64-4 Data can be accessed by using the appropriate scan tool either to read data or command actuator functions to operate.

1	2	3	4	5	6	7	8
9	10	11	12	13	14	15	16

1 Manufacturer's discretion
2 Bus + Line SAE J1850
3 Manufacturer's discretion
4 Ground
5 Ground
6 Manufacturer's discretion
7 K Line, ISO 9141
8 Manufacturer's discretion
9 Manufacturer's discretion
10 Bus – Line, SAE J1850
11 Manufacturer's discretion
12 Manufacturer's discretion
13 Manufacturer's discretion
14 Manufacturer's discretion
15 L Line, ISO 9141
16 Power supply

FIGURE 64-3 Pin layout for a data link connector (DLC).

about the engine temperature gauge reading way colder than it used to. You can connect the scan tool and access the coolant temperature sensor data. If it shows full engine operating temperature of 195°F (90.5°C), then it is likely that the temperature gauge or sending unit is faulty, so you can focus your further testing there. If the coolant temperature sensor showed the engine operating temperature at 153°F (66.7°C), then you could suspect that the engine thermostat is likely stuck open and pursue that fault.

OBD Terminology

The automotive technician needs to know and understand the terms, or "lingo," used in the diagnosis and repair of OBD systems. Before the SAE published the appropriate "recommended practices," OEMs used nonstandardized nomenclature. The SAE has helped to standardize automotive terms used by engineers and technicians alike. A complete listing of OBD I and OBD II terms is available from the SAE (**SAE J1930**). Some aftermarket and manufacturer service information also list the OBD terms.

Networking

Standard copper wiring bundles on vehicles have become a major weight and cost issue due to the increased interconnectivity required between onboard computers and components. To simplify wiring, a **controller area network-bus (CANbus)** and other types of networks have become commonplace in today's vehicles. A CANbus is a localized (onboard) vehicle network that enables computers and components to send and receive signals across a shielded twisted pair of wires. The wires are shielded by being twisted around one another and carrying balanced (+ and −) signals to reduce the likelihood of radio frequency interference into the network. The shielded pair of wires forms a common loop around the vehicle to connect components and handle encoded messages between components. Thus, the various computers/modules can network, or "talk," to one another. For example, a throttle position sensor or coolant temperature sensor may provide valued information to a number of other onboard modules to serve their respective needs.

The network is constantly monitored by the PCM. Additionally, for diagnostic purposes, data can be accessed by the technician, using the appropriate scan tool either to read data or command certain actuator functions to operate and be tested (**FIGURE 64-4**).

Diagnostic Trouble Codes

K64003

Diagnostic trouble codes (DTCs) indicate the component or circuit in which a fault has been detected. **Codes** are "set" (stored) in one or more of the vehicle's powertrain control modules (PCMs) once a fault is detected. An emission-related fault is generated when values being monitored determine the emissions would be 1.5 times the EPA Federal Test Procedure (FTP). For example, an open in the engine's coolant temperature circuit will cause the fuel system to run rich, increasing the emissions so a DTC will be set. Also, if two sensors send conflicting data, a fault is recorded, and the MIL is illuminated. Some DTCs set due to implausibility and conditions that are out of parameters, but it is not always a case of a failed component or wiring. Understanding how the affected system operates is very important when troubleshooting the OBD II system. If you don't know how it works, how can you fix it? Locating, reading, and understanding the strategy used by the manufacturer to generate the DTC (enabling criteria) is of utmost importance.

Emission-related DTCs are the same (generic DTCs) across all vehicle makes and models, as are the SAE-recommended names used to describe components and systems. Generic DTCs—those that are in any way emission related—are now the same from one vehicle to the next. Refer to **SAE J2012** for more information on interpreting generic DTCs.

Most manufacturers go beyond the generic OBD II requirements and program their own proprietary functionality into the PCM. This can include manufacturer (non-generic) codes, sensor information, and bidirectional tests. Refer to OEM service manuals for accessing and interpreting OEM-specific (non-generic) system codes, and performing manufacturer specific tests. But know that typically, the non-manufacturer side of the OBD II system has more options and information than the generic OBD II system has. Many scan tools can access the manufacturer side of the system, so that is typically a good option. And manufacturer scan tools will generally give you even more options and control over the system.

If a vehicle system reports that a fault exists, the MIL located in the instrument cluster will likely illuminate (**FIGURE 64-5**). Sometimes it takes two consecutive drive cycles, or "trips," to illuminate the MIL. The MIL alerts the driver that there is a problem in the engine management system. Even a loose gas cap can trigger a DTC and illuminate the MIL (or a special gas cap warning light) under certain conditions. If the catalytic converter is at risk, such as from a continuous misfire, the MIL will flash, thus telling the driver to stop the vehicle as soon as safely possible and turn off the engine to avoid serious damage to the converter.

Once the MIL is on, it will remain on until either the technician clears the codes, or the fault has not been detected during three consecutive trips when the **monitor** is run. OBD systems store DTCs in the computer's so-called keep alive memory (KAM). The codes remain in memory until power is disconnected for older vehicles; for newer vehicles, DTCs are saved even if the vehicle's battery is disconnected. DTC information remains stored in the respective control module's long-term memory regardless of whether a "hard" (continuous) or an intermittent fault has set the code.

DTCs are stored according to their status. **Current DTC** means that the test failed during the last one to three drive cycles (**FIGURE 64-6**). There is a good probability that this is a hard fault, meaning that it is not intermittent. **Pending codes** are codes that have not been validated by failing a second test consecutive test, or a current code that was downgraded because it is intermittent. (**FIGURE 64-7**). **History codes** are pending or current codes that have happened in the past but haven't been cleared manually with a scan tool, or automatically after approximately 40 warm-up cycles without the code reoccurring (**FIGURE 64-8**). A **warm-up cycle** is when the engine is started and run until the engine operating temperature reaches 160°F (71.1°C) and has increased in temperature at least 40°F (22°C), and the engine is turned off again.

A DTC may be generated by any of the numerous computer modules on board. It serves not only to inform the driver of a fault but also to enable the automotive technician to determine where the fault has occurred. DTCs are used in conjunction with diagnostic flowcharts found in the manufacturer's service information to assist technicians in determining the likely cause of a failure.

OBD II codes use a series of letters and numbers grouped together to identify which system, component, or circuit is at

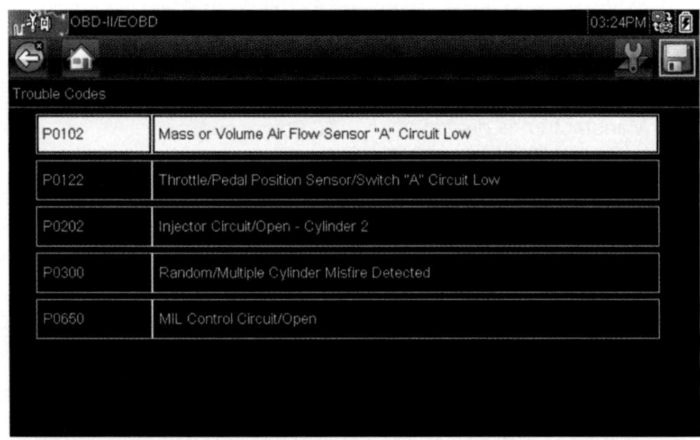

FIGURE 64-6 A scan tool displaying current codes.

FIGURE 64-5 The malfunction indicator lamp (MIL).

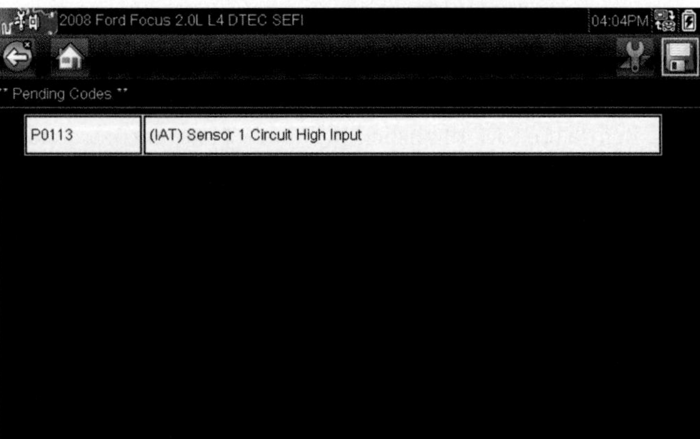

FIGURE 64-7 A scan tool displaying pending codes.

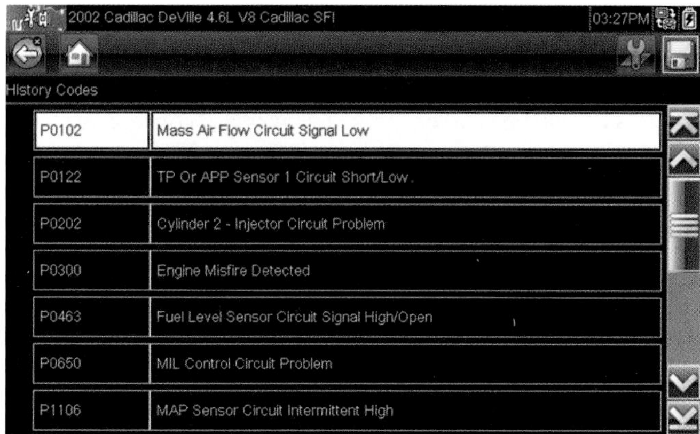

FIGURE 64-8 A scan tool displaying history codes.

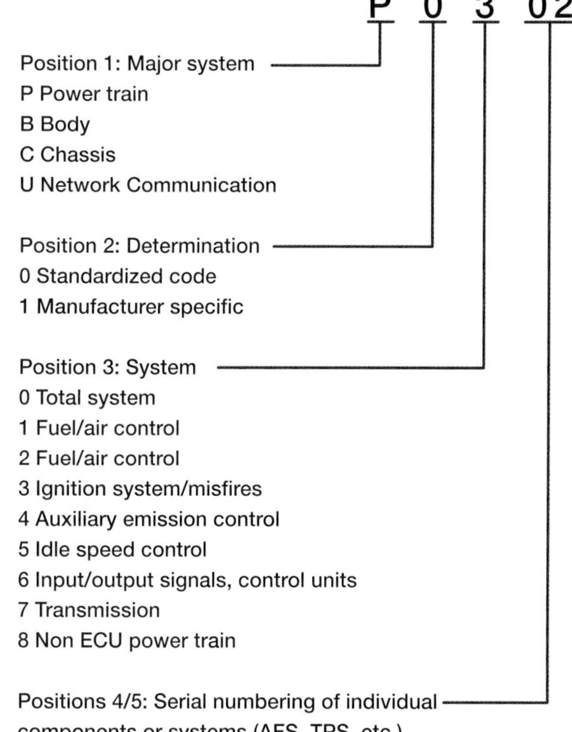

FIGURE 64-9 Decoding a diagnostic trouble code (DTC).

fault (**FIGURE 64-9**). The first character of the code is a letter that identifies whether the fault is located within the powertrain (P), body (B), chassis controller (C), or communication system (U). The 2nd digit indicates SAE controlled (generic) or manufacturer (OEM) specific where a "0" or "2" indicates SAE controlled (generic) code and a "1" indicates manufacturer (OEM) specific code. Additionally, a "3" indicates both SAE controlled (generic) and manufacturer specific (OEM) code. The 3rd digit indicates the powertrain subgroup DTC and usually from "0-9" as listed in the table above. The 4th/5th digit indicates the area or component involved in the fault. Refer to DTC charts (on the Internet or in service literature) for a complete listing of the hundreds of possible codes. The list grows constantly as onboard systems become even more sophisticated and unique, as with hybrid-electric vehicles.

Freeze-Frame Data

Since OBD II came about in 1996, technicians have had the ability to access **freeze-frame data** that is automatically recorded in the vehicle's PCM when a vehicle DTC is stored. Freeze-frame data are a list of sensor data (parameter ID [PID]) that are associated with the specific DTC. Typically, freeze-frame data can only be saved for one DTC, generally the highest priority DTC. If a more serious fault supersedes another previously less serious recorded fault, freeze-frame data are updated. The highest priority DTCs are related to misfires; the next highest are one-trip DTCs; the next are two-trip DTCs, and the lowest priority are non-emissions-related DTCs.

Difficult-to-find intermittent faults can thus be diagnosed by carefully reviewing the freeze-frame data for clues to the location of the fault. For example, if the vehicle set a misfire DTC while crossing a rough set of railroad tracks, but at no other time, the out-of-range data in the freeze frame for a particular circuit would give the technician a good place to start looking for the fault. Modern scan tools can retrieve freeze-frame data and display them for analysis after the fact (**FIGURE 64-10**).

Malfunction Indicator Lamp (MIL) Operation

Under OBD II requirements, the operation of the malfunction indicator lamp and the storing of diagnostic codes is different from OBD I vehicles. The MIL can come on or flash depending on a number of factors. Even when a fault is detected, the MIL and the storing of a DTC will only occur under the following conditions:

- On a one-trip DTC, the MIL will illuminate when the DTC and freeze-frame data are stored.
- On a two-trip DTC, if the same fault is detected in two consecutive drive cycles, the MIL is illuminated, and a DTC and freeze-frame data are stored.
- If a catalyst-damaging fault occurs, the MIL will flash which fault is present.

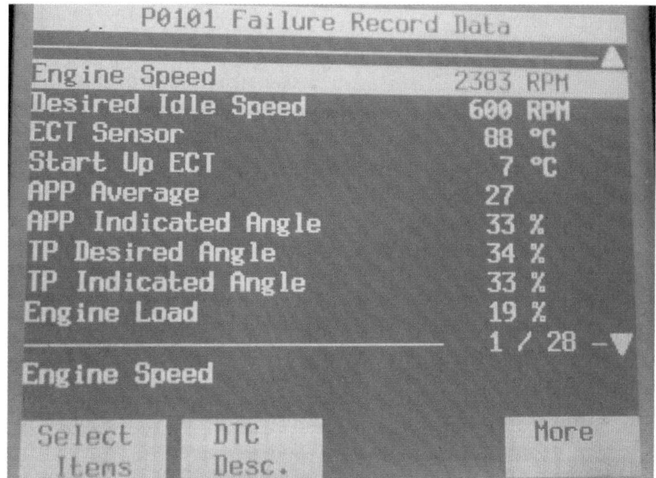

FIGURE 64-10 A scan tool can retrieve and display freeze-frame data.

- If a fault occurs that triggers limp home mode, the warning lamp may be illuminated, and a DTC will be stored even if the fault only occurs in one drive cycle.

The engine warning lamp will go out if no faults are detected in three consecutive drive cycles; the DTC will, however, remain as a history code. To turn off the MIL manually, use a scan tool to erase the codes and freeze-frame data.

Drive Cycles

A **drive cycle** is designed to operate the vehicle in such a manner as to meet the **enabling criteria** to cause the PCM to perform the readiness monitor tests. This typically includes the following events: A vehicle starts, warms up, is accelerated, cruises, slows down, accelerates once more, decelerates, stops, and cools down. The EPA has established a "drive cycle" for emission testing of various kinds of vehicles, and the light-duty vehicle (240-second) drive cycle is the basis for how OBD II readiness monitors (tests) are run. A complete driving cycle should include running the monitors on all systems and should be completed in under 15 minutes; however, sometimes certain monitors cannot be run during a drive cycle if qualifying conditions for running that particular monitor are not met. For example, an evaporative emission (EVAP) monitor will not run if the fuel tank is less than one-quarter full or more than three-quarters full. In some cases, particular monitors cannot run until after other monitors have been run and passed. EVAP monitoring will also not run if the engine coolant temperature monitor has not passed. State-run emission inspection stations in the Unites States will sometimes allow a vehicle to pass an emission inspection, even if one or two monitors have not run.

Warm-up cycle is a term used to determine when automatic clearing of DTCs occurs.

▶ System Readiness Monitors

K64004

A monitor is a computer's test of its system components and circuits for malfunctions that would cause emissions to increase 1½ times more than FTP (Federal Test Procedures) regulations allowed. OBD II standards dictate that a vehicle's PCM must monitor the components and systems with two priorities in mind, high priority (**continuous monitors**) and low priority (**non-continuous monitors**). High priority faults that require continuous monitoring are listed here:

- Misfire monitor—continuously monitors either crankshaft speed or possibly combustion chamber feedback that indicates misfiring cylinders. If the monitor detects misfires between the lower and upper threshold, a "pending DTC" will be set, and freeze-frame data will be stored. If the misfire reoccurs during the next consecutive trip, a "current DTC" will be set and the MIL illuminated. If the monitor detects misfires above the upper threshold, the MIL will blink immediately and a "current DTC" will be set. It typically doesn't require a second trip. This may also cause the engine management system to default to open loop status to better control fuel trim.
- Fuel system monitor—monitors the amount of LTFT correction needed to keep the air-fuel mixture correct. Above or below the specified fuel trim threshold, a "pending DTC" will be set. This typically takes two consecutive trips to turn the MIL on and convert this to a "current DTC."
- Continuous component monitor—continuously monitors the engine management sensors and actuators for faults. Failure on the first trip, during which the failure causes the emissions to exceed the specified limits, results in a DTC set, freeze-frame data stored, and MIL illuminated. Failure on the first trip during which the failure does not cause an emission failure sets a pending DTC and requires a second consecutive failure to turn the MIL on.

Note that continuous monitors do not have a monitor readiness status indicator because they run continuously.

Lower-priority faults are monitored with **noncontinuous monitors** such as these:

- Catalytic converter monitor—monitors the efficiency of the catalytic converter.
- Heated catalyst monitor—monitors the condition of the catalyst heaters
- Oxygen sensor monitor—monitors the activity and response time of the oxygen sensors
- Oxygen sensor heater monitor—monitors the condition of the oxygen sensor heaters
- Evaporative system monitor—monitors the system for leaks as well as the ability to purge the canister of vapors.
- EGR system monitor—monitors the ability of the EGR to allow exhaust gases to recirculate. May also monitor variable valve timing on engines that use that in place of an EGR valve system.
- Secondary air system monitor—monitors the secondary air system, if equipped, to verify that air is directed where specified.

In such cases, noncontinuous monitors are run only once during each engine warm-up cycle, or even less often, depending on certain requirements such as ambient temperatures or fuel level. Because of this, they have readiness monitor status indicators that indicate whether the readiness monitor has run.

Each time the engine is started and the vehicle is driven according to the enabling criteria for a readiness monitor to run, the PCM runs them one at a time to verify whether the emissions system it monitors is functioning correctly. If the system completes the monitor readiness test, then the monitor status will be logged as "Ready" in the PCM's memory. This doesn't mean that the monitored system doesn't have a fault—but only that the test completed. Until the monitor runs and completes the system readiness test, the status will be listed as "Not Ready" or "Not Completed." If a fault is detected, the MIL is illuminated after either the first fail if a one-trip DTC, or the second consecutive fail if a two-trip DTC, indicating that the vehicle needs attention.

Some vehicles can indicate whether the readiness monitors have run and are "ready" without using a scan tool. On those vehicles, this mode can be activated by turning the ignition to On (not Crank) and watching the MIL. The MIL should come on and stay on steady if all of the readiness monitors have passed. If after 15–20 seconds the MIL blinks, it indicates that not all of the readiness monitors have passed.

Misfire Monitor

A misfire is defined as a lack of combustion. It can be caused by poor spark, poor fueling, poor compression, or any other engine related fault that would have a significant impact on normal combustion. When a misfire occurs, raw fuel and excess oxygen enter the exhaust, adversely affecting the oxygen sensor signal, and can cause permanent damage to the catalytic converter in a very short time. In normal operation, catalytic converters get very hot just dealing with a small amount of combustible gases. When misfires occur, the full volume of air and fuel from the misfiring cylinder is burned in the catalyst. This can turn the catalytic converter red hot and even cause it to melt down. Detecting misfire is important to prevent this. In fact, if the PCM detects a cylinder that is misfiring above a certain threshold, it will typically cause the MIL to blink (not just come on steady), warning that there is a catalyst-damaging fault in the system. Manufacturers incorporate misfire monitoring into 1996 and newer vehicles to address this risk.

As an engine rotates through the four-stroke cycle, the speed of the crankshaft varies. This is due to the change in forces placed on the piston and crankshaft primarily during the compression and power strokes. These changes are predictable when the engine is operating. By using a high-definition (many teeth) CKP sensor/reluctor, the PCM can be programmed to monitor the speed of the crankshaft and look for the slowing down and speeding up for each power pulse. Misfire monitoring software identifies unacceptable changes in crankshaft speed, indicating that a power pulse did not occur due to a misfire. The PCM uses the camshaft position sensor (CMP) to tell which cylinder(s) are misfiring or if it is a random misfire. This information allows the PCM to store a cylinder-specific misfire code or a random misfire code. If the monitor detects misfires between the lower and upper threshold, a "pending DTC" will be set, and freeze-frame data may be stored. If the misfire reoccurs during the next consecutive trip, a "current DTC" will be set and the MIL illuminated. If the monitor detects misfires above the upper threshold, the MIL will blink immediately, and a "current DTC" will be set. It typically doesn't require a second trip. This may also cause the engine management system to default to open loop status to better control fuel trim.

▶ TECHNICIAN TIP

If a new PCM is installed into a vehicle that is misfiring, incorrect misfire monitoring may result. Because the PCM has not yet carried out a reference check for "normal operation," it may consider the misfiring engine to be "normal."

Oxygen Sensor Monitor

Oxygen sensor feedback is the most effective method of monitoring and controlling exhaust emissions. The front oxygen sensor is the most important component of this system. It is mounted in the exhaust system between the engine and the catalytic converter, typically as close to the combustion chambers as possible.

When the sensor reaches its operating temperature of approx. 600°F (316°C), the sensor produces a voltage inversely proportional to the amount of oxygen in the exhaust gas. The engine ECU uses this information to calculate the fuel injector pulse width necessary to achieve a 14.7 to 1 air-fuel ratio (stoichiometric).

Normal cycle	100 mV – 900 mV (0.1 V – 0.9 V)
Lean mixture (high exhaust oxygen content)	low voltage output (below 0.45 V) (High HC)
Stoichiometric	450 mV (0.45 V) (Minimal CO and HC output)
Rich mixture (low exhaust oxygen content)	high voltage output (above 0.45 V) (High CO and HC)

When the engine PCM uses the oxygen sensor signal for correction of the A/F ratio, it is called a closed loop operation. When the engine ECU does not use the oxygen sensor signal for correction of the A/F ratio, it is called open loop operation.

A properly operating oxygen sensor must be able to:

- Generate a signal below a specified low voltage and above a specified high voltage.
- Change voltage quickly or, in other words, have a fast response time to changes in the exhaust oxygen level.

An oxygen sensor that has a slow response rate or a reduced voltage output is no longer operating efficiently. This can lead to higher exhaust emissions due to richer mixtures being burnt. Under the OBD I engine management system, fault code 11 typically referred to a sensor that was not functional. It could not identify a low voltage or slow sensor. Under the OBD II engine management system, monitoring ensures that a marginal oxygen sensor is identified early, before exhaust emissions are affected significantly.

Oxygen sensor monitoring is carried out several times during each drive cycle as soon as the oxygen sensor has passed its threshold voltage (0.5 V) and it has been judged that the oxygen sensor has been activated.

Monitoring occurs under specific conditions. Here is an example from a recent vehicle (other vehicles may have different parameters):

Engine coolant temperature	Above 140°F (60°C)
Engine load	25 to 60%
Engine speed	1250–3000 rpm

This is a two-trip DTC. That is to say, the OBD II monitor must judge the sensor faulty in two consecutive drive cycles before a "Current" DTC will be recorded.

The OBD II monitor also measures the rich-lean switching frequency of the oxygen sensor output during a short time period (see example diagram below). The monitoring is carried out a certain number of times during each drive cycle, and the switching frequency is the average of these measurements. The oxygen sensor must switch above and below the switching point a minimum number of times in a given amount of time (**FIGURE 64-11**).

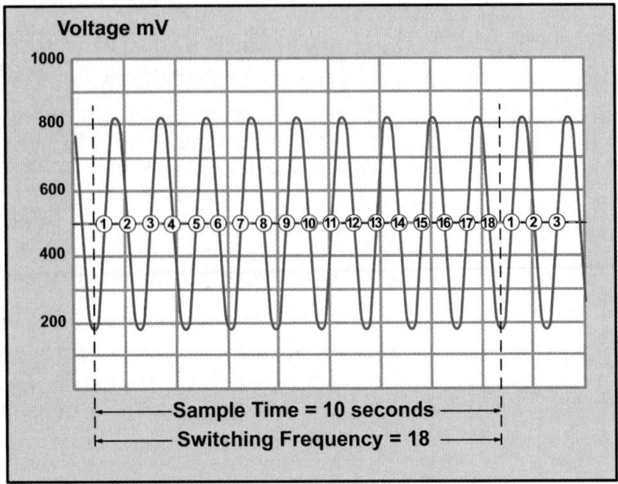

FIGURE 64-11 Example of oxygen sensor switching frequency.

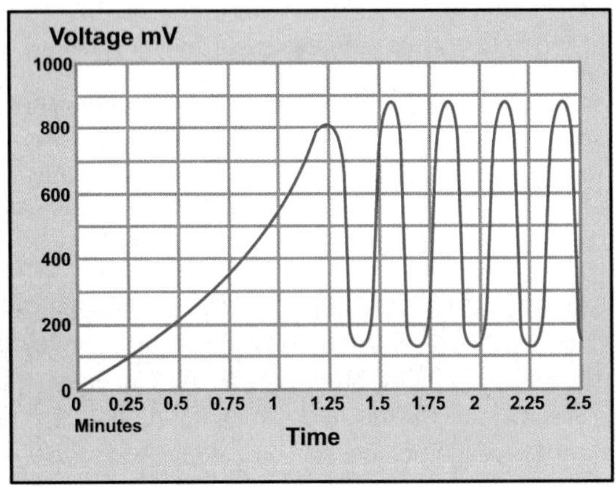

FIGURE 64-12 Oxygen sensor heater monitors the time it takes for the sensor to operate reliably.

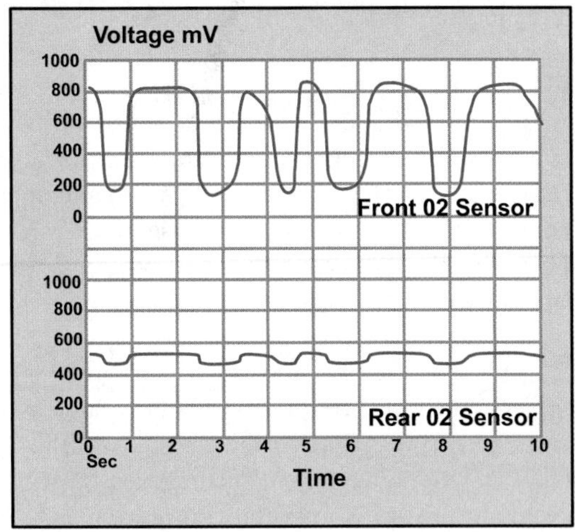

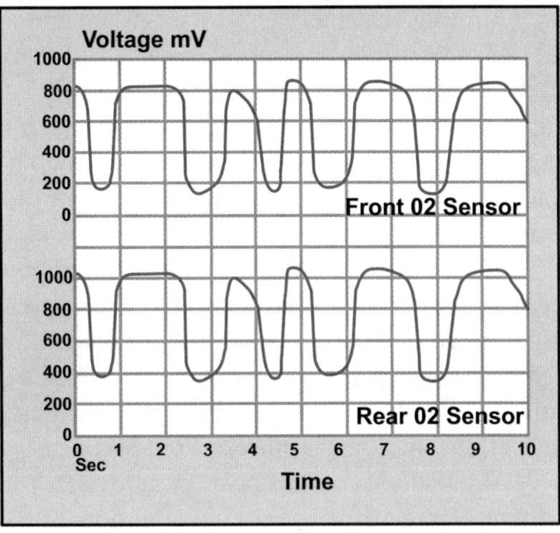

FIGURE 64-13 The catalytic converter monitor compares the post-cat sensor reading to the pre-cat sensor reading to determine catalyst efficiency.

Oxygen Sensor Heater Monitor

Because most emissions from a vehicle happen when the engine and components are cold, heating up the oxygen sensors can get them into operation much sooner. This allows the fuel mixture to be controlled much more accurately. The oxygen sensor heater monitors how long it takes for the oxygen sensor to start operating reliably from cold start-up (**FIGURE 64-12**). If the time is too long on two consecutive drive cycles, a DTC is set.

Catalytic Converter Monitor

The catalytic converter (cat) substantially reduces pollutants in the exhaust stream if it is operating correctly. The catalytic converter monitor uses pre-cat and post-cat oxygen sensors to test the ability of the converter to store and release oxygen as it is operating. Readings that are outside of specified parameters indicate that the catalytic converter has lost some of its efficiency. Typically, a failing cat is indicated by the post-cat oxygen sensor pattern following the pre-cat sensor pattern (**FIGURE 64-13**).

Fuel System Monitor

For emissions output to be low, the fuel system has to be in proper control of the air-fuel mixture. The fuel system monitor monitors the amount of LTFT correction needed to keep the air-fuel mixture correct once the engine is warmed up above a minimum specified temperature. If the LTFT goes above or below the specified threshold, a "pending DTC" will be set. This typically takes two consecutive trips to turn the MIL on and convert this to a "Current DTC."

Evaporative System Monitor

The evaporative emissions system captures and stores fuel vapors from the tank and then directs them to the engine to be burned at the appropriate time. This system can fail in two general ways: leaks and problems with purging the system. So the monitor tests the system for leaks as well as to verify that purge occurs when commanded.

To run the EVAP monitor, the engine must typically be below a specified temperature, and the fuel level must be between

EVAP System Test Sequence
1. PCM turns the normally open Vent Valve ON to close the Vent Valve.
2. PCM switches the Purge Valve to 100% duty cycle to pull the EVAP system into vacuum then switches the valve OFF to isolate the system.
3. PCM monitors the tank pressure via the pressure sensor to test the system for leaks.

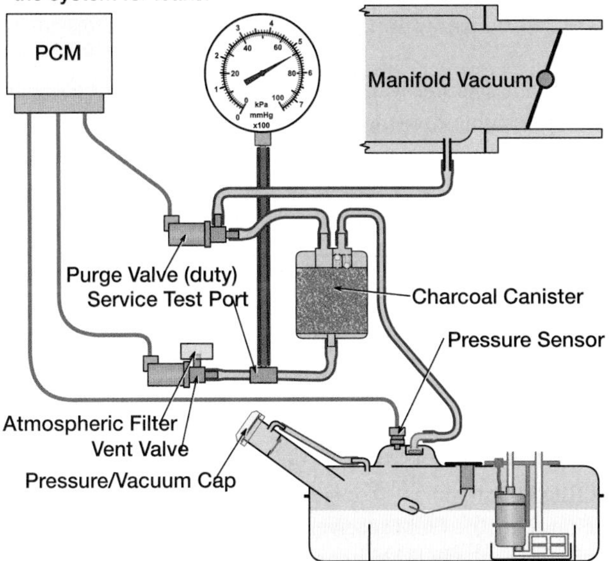

FIGURE 64-14 The EVAP monitor draws a vacuum on the EVAP system, seals it off, and then monitors the pressure to determine whether there are any leaks.

one-quarter and three-quarters full. The monitor commands the EVAP vent valve to close and the purge valve to open so that a specified vacuum can be built up in the EVAP system. Once the specified vacuum level is reached, the purge valve is closed, sealing off the EVAP system. The vacuum in the fuel tank is monitored to see if the pressure drops over a specified time, indicating a leak in the system (**FIGURE 64-14**). Many systems can report two levels of leakage DTCs: a small leak DTC and a large leak DTC.

EGR System Monitor

The EGR system reduces NOx emissions by introducing burned exhaust gases back into the intake manifold to dilute the air-fuel mixture. This slows down the rate of combustion, which lowers its peak exhaust gas temperature and thereby reduces the formation of NOx. The EGR monitor monitors the flow of EGR gases when the EGR valve is commanded open. This monitor is run after the engine is fully warmed up and has met other specified conditions based on the type of monitoring sensor used. Here are the most common EGR monitoring sensors used to determine operation.

- **EGR temperature sensor method:** A temperature sensor is used in the intake manifold side of the EGR passage. When the monitor commands the EGR valve to open, the monitor looks for a resulting temperature increase above a minimum specification.
- **MAP sensor method:** The monitor commands the EGR valve to open and then looks for a resulting pressure drop from the MAP sensor above a minimum specification.
- **Oxygen sensor method:** The monitor commands the EFT valve to open and then looks for a resulting signal change of the oxygen sensor beyond a minimum specified.

▶ Scan Tools

K64005

Essentially, a scan tool ("scanner") is a device able to electronically communicate with and extract data from the vehicle's one or more onboard computers. Onboard computer modules include the PCM, **electronic brake control module (EBCM)**, body control module (BCM), transmission control module (TCM), and perhaps numerous others. Simple scan tools from the 1980s could read and erase fault codes and little more. But as onboard systems became more complex, so too did professional-grade scan tools used in the service bay (**FIGURE 64-15**). Today, adapters are available that work with smartphones. The dongle plugs into the DLC and wirelessly communicates with the smartphone, which acts as a scan tool (**FIGURE 64-16**). These devices are available for $25–$100.

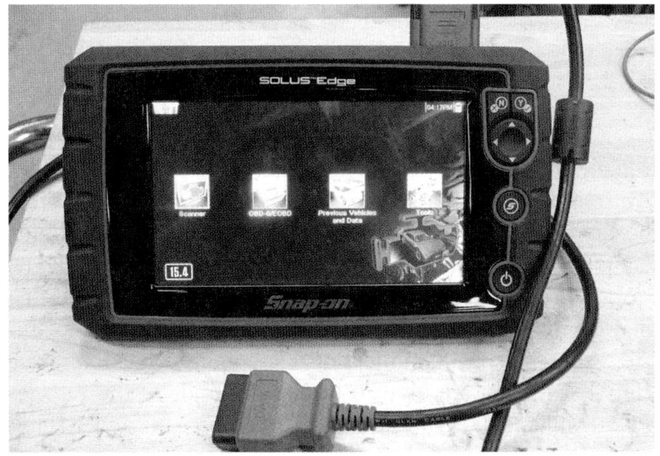

FIGURE 64-15 Typical scan tool used in a shop.

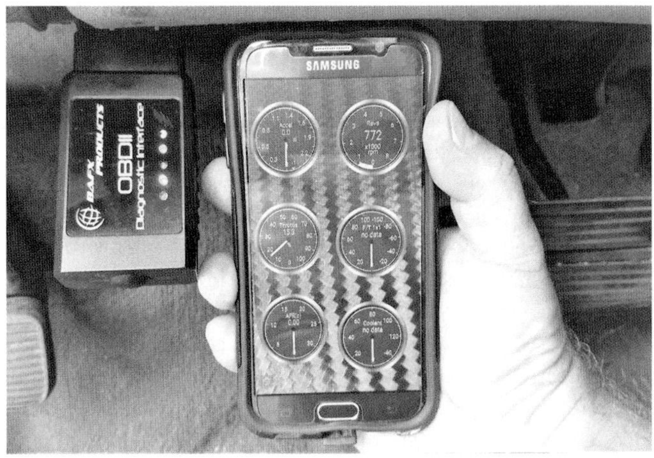

FIGURE 64-16 Typical smartphone dongle which turn the smartphone into a scan tool.

Applied Science

AS-18: Time: The technician uses direct and indirect methods to measure time and compare the results to the manufacturer's specifications.

A technician is troubleshooting a vehicle that has an intermittent problem with hesitation on acceleration. The technician has started his evaluation by conducting a visual inspection. After this, he uses a scan tool to check for codes but does not find any. He checks fuel pressure and volume, and both are within the manufacturer's specifications. The technician checks for service bulletins for this symptom but has not found any leads in this area.

Another technician suggests confirming that the injector pulse width meets the manufacturer's specifications, as he has a similar vehicle that had problems in this area. The technician checks the service manual and finds that the injector pulse width should be 2.5 to 2.7 milliseconds at engine idle. The manual also gives other pulse-width specifications for other engine speeds. Using a scan tool, the technician is able to verify that the pulse width is within the manufacturer's specifications at all published engine speeds. The scan tool data is on a direct-time basis as real-time engine data is being used.

Pulse width is the amount of on-time measured in milliseconds that a fuel injector delivers fuel to the cylinder. The injector pulse width depends on the input signals supplied to the PCM from its various engine sensors. The pulse width expands on acceleration and contracts under lighter loads.

Cost and complexity increase with the level of bells and whistles desired in a scan tool. Automotive technicians today are finding that they must use faster and more accurate diagnostic instruments, such as graphing scan tools, to see hidden faults in component or system waveforms. Scan tools and digital storage oscilloscopes or PC oscilloscopes, are now used to display engine compression and vacuum anomalies, and so forth.

Scan tools with bidirectional capability, called **bidirectional scanners**, are used to send commands to the vehicle's PCM, which will then cause various components and systems to operate for testing purposes. For example, using the scan tool to command the radiator fan to operate gives you lots of information. First, if the fan does run, then you know that the fan is good; power/ground to it is good; and the PCM can cause it to operate. This would then cast suspicion on a faulty coolant temperature sensor or a thermostat stuck open as the cause of the fan not running. If the fan did not operate, then you would know that the PCM is *not* able to activate it, so checking power, ground, wires, and the fan itself would be in order. Scan tools are used along with digital storage oscilloscopes, portable five-gas **emission analyzers**, and other diagnostic equipment for much more effective time-saving diagnostic routines.

The scan tool is not a "magic bullet," however. A fault code found does not necessarily point directly to the problem. For example, a P0171 Fuel System Too Lean code could be caused by a defective oxygen sensor, but it could also be caused by an underreporting mass airflow sensor, a vacuum leak, use of the wrong fuel, and so on, so further testing is often required.

Training and experience play a major role in effectively using these modern information-gathering tools. Understanding the principles of combustion theory, internal combustion engine operation, and emissions systems are essential ingredients for success. Something that is often forgotten in this electronic age is that engines, transmissions, and braking systems are still, by nature, mechanical devices. Electronics can control and monitor them, but they cannot overcome large physical defects. For instance, worn valve guides causing excessive oil consumption may go undiagnosed because a code is not set.

After the scanner is used to read any codes, but before any electronic troubleshooting begins, a visual inspection of the vehicle should be performed. Simple faults like a cracked or loose vacuum hose, or even a loose gas cap, may easily be found and corrected-whether a code is set or not.

Indeed, the field is demanding. It requires extensive training and also a sizeable investment. Opportunities abound for the technician who understands the theory and masters the use of diagnostic tools to efficiently analyze, diagnose, and solve problems on today's advanced vehicle engine systems.

▶ OBD II Modes of Operation

The OBD II system has 10 operating modes that give you access to the raw data the PCM receives. This can give you better information in case the PCM either processes the information wrong when displaying it in the Generic OBD II mode or delays it. Each of the 10 operating modes allows for access to different data. Here is an overview of these data:

- $01 Show current data—also establishes communication between the PCM and scan tool.
- $02 Show freeze-frame data—allows you to access any freeze-frame data that was stored when a DTC was set.
- $03 Show stored DTCs—allows access to all DTCs; this may include non-generic OBD II codes
- $04 Clear DTCs and stored values—clears DTCs, freeze-frame data, and readiness monitor status.
- $05 Test results, oxygen sensor monitoring—allows you to access oxygen sensor data.
- $06 Test results, other component/noncontinuous monitors—allows access to test results from non-continuous monitors and related components. Very useful in diagnosing no-code faults, and for verifying the fault has been repaired.
- $07 Show pending DTCs—displays current and pending DTCs, which is very useful in verifying the repair and ensuring the MIL doesn't reoccur for the same or a related fault.
- $08 Control operation of onboard component/system—allows bidirectional control of some components.
- $09 Request vehicle information—used for verifying calibration and software versions for flash updates.
- $0A (Mode 10) Permanent DTCs (cleared DTCs)—active on 2010 and newer vehicles. Used to store codes, even if they have been erased, for emissions test compliance. These codes can only be cleared by a full readiness monitor check that doesn't reset a code.

▶ Diagnosis and Testing

Diagnosing Engine Concerns

N64001, K64006

Modern vehicle engine systems are complex systems that have developed to provide performance and efficient use of fuel while ensuring that exhaust emissions meet strict environmental regulations. To achieve these demands, engineers have improved the performance of mechanical components and used computer control of the electrical, ignition, and fuel systems to ensure the highest possible efficiency while reducing exhaust emissions. All of these systems must be working together, each interacting with the others correctly to ensure optimal engine operation. A fault in one system affects the performance of other systems. For example, a plugged injector may cause a cylinder misfire, reducing engine power and also affecting the secondary ignition voltages. Diagnosis of the engine requires that you understand each of these systems, how they operate, and how their performance affects other systems and components.

To diagnose correctly, start with gathering information to understand the customer's concern, and then verify the fault. Once the fault is verified, it may be time efficient to retrieve any DTCs and perform a quick visual inspection of the related parts and systems to see if the fault can be identified quickly. If not, research the concern, symptoms, and DTCs in the service information and technical service bulletins (TSBs).

Once your research is complete, formulate a focused testing plan that takes into account the information from your research, as well as tests that will give you the most information, and tests that are quick and easy. Once you have your focused testing plan, perform the tests, gather the resulting information, and evaluate it against the service information specifications.

Determine what faults may cause the symptoms, and conduct mechanical, electrical, ignition, and fuel system checks to correctly identify the location and types of faults. It is also possible that a vehicle may have multiple faults; for example, a plugged injector may have been caused by poor-quality fuel.

You will need to use test equipment to diagnose engine faults. This equipment may include OBD scanners, digital storage oscilloscopes, compression gauges, digital volt-ohmmeters (DVOMs), fuel system pressure gauges, and flowmeters. You will have to interpret the test results, compare the results to the manufacturer's specifications, and then apply the test results to determine any next steps. Continue testing, researching, and retesting until the root cause of the fault is identified.

Once you have identified the cause of the fault, inform the customer so that he or she can authorize the repair. Follow the service information when making the repair. Once the repair is complete, retest the vehicle with the same test that identified the fault, to make sure the repair solved the concern. Also test-drive the vehicle to make sure any other faults don't show up. This requires performing a drive cycle that causes all of the monitors to run. This could mean having to add fuel to the tank and driving in an area that can allow all of the drive cycle operating conditions to be met.

One step that many technicians fail to take when working on the engine management system is to make sure that the system isn't about to turn on the MIL due to pending codes. One way of doing this is to use the scan tool to access Mode 6 and Mode 7, looking at the results of the monitors. Typically on Mode 6, there is a minimum value, maximum value, and the current value for each noncontinuous monitor. The closer the current value is to the maximum, the sooner the monitor will fail. On Mode 7, you can see the status of any "current" or "pending" codes. Remember that many codes are two-trip codes, so if it set a pending code on the test drive, it will likely turn the MIL on when the customer drives away from the shop.

Retrieving and Recording DTCs, OBD Monitor Status, and Freeze-Frame Data

N64002

Modern vehicles are required to meet strict environmental emission regulations over the life of their operation. In order to meet these standards, vehicles are equipped with sophisticated electronic control units and sensors to monitor and adjust the engine fuel and the ignition and emission systems and to provide diagnostic capability to meet the required standards. Retrieving and recording DTCs, monitor status, and freeze-frame data are integral steps in diagnosing faults within vehicle systems. The technician must be able to do so anytime a vehicle illuminates the MIL, sets a DTC, fails emission monitors, or fails an emission test.

Every vehicle sold is required to meet the requirements of the Federal Test Procedure, which is designed to simulate actual driving conditions. The OBD II system constantly tests and analyzes each system's performance over the drive cycle. This is achieved through a set of PCM programs called monitors, which test each system's operation. The monitor sets a ready or not-ready status based on each system test. The PCM will set and store DTCs if a fault is detected. Freeze-frame data are also stored, which provide a snapshot of particular data before, during, and after the time of the fault being detected. Never delete or reset the DTCs until all testing has been completed. And even then, it may be better not to erase the codes after repair and let the PCM turn the MIL off. This way, all of the monitors will not have to rerun to change their status to "ready." If you do need to erase the codes before diagnosing the fault, make sure that any codes and freeze-frame data are recorded and safely stored.

To retrieve and record DTCs, OBD monitor status, and freeze-frame data, follow the steps in **SKILL DRILL 64-1**.

Testing Computerized Engine Control System Sensors, the PCM, Actuators, and Circuits

N64003

Graphing multimeters and digital storage oscilloscopes are used to make precise electrical measurements over time and display the results on a graph as a waveform. Given the complex

SKILL DRILL 64-1 Retrieving and Recording DTCs, OBD Monitor Status, and Freeze-Frame Data

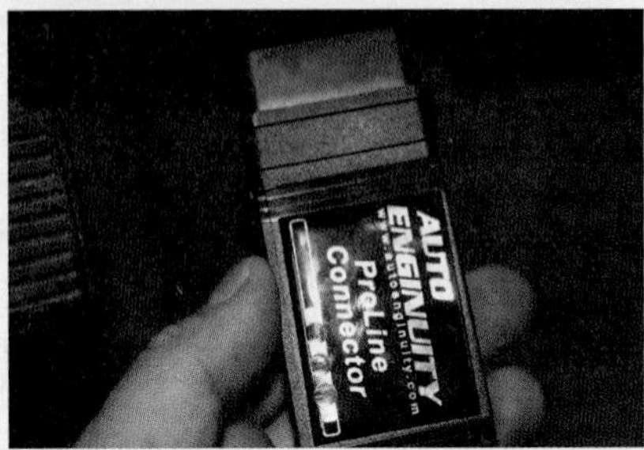

1. Select the scan tool to provide the best coverage for the type and make of vehicle. Locate the DLC and connect the scan tool.

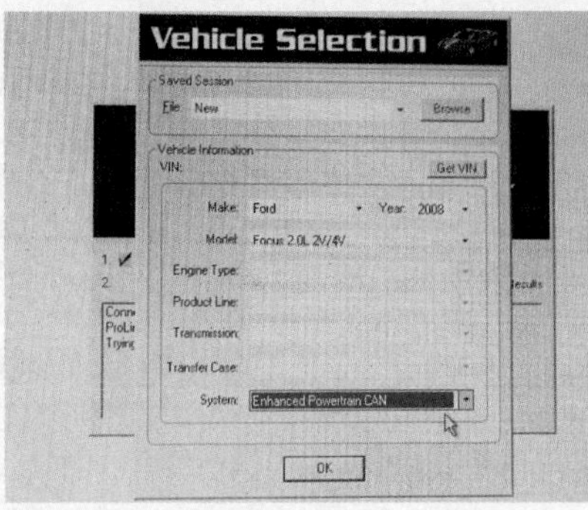

2. Power on the scan tool, and turn the ignition on. Establish scan tool communications with the vehicle.

3. Retrieve and record the DTCs.

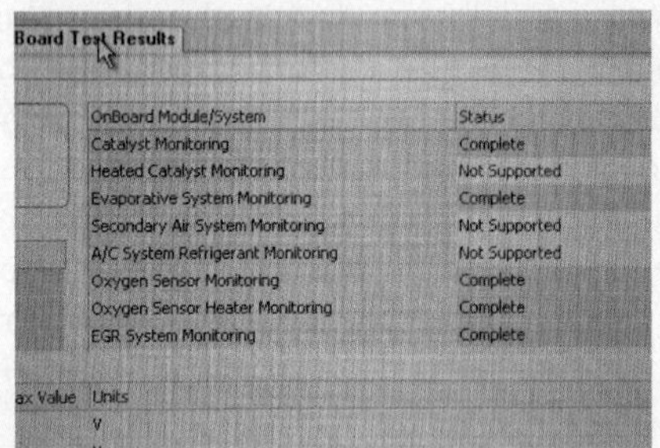

4. Retrieve and record OBD monitor status.

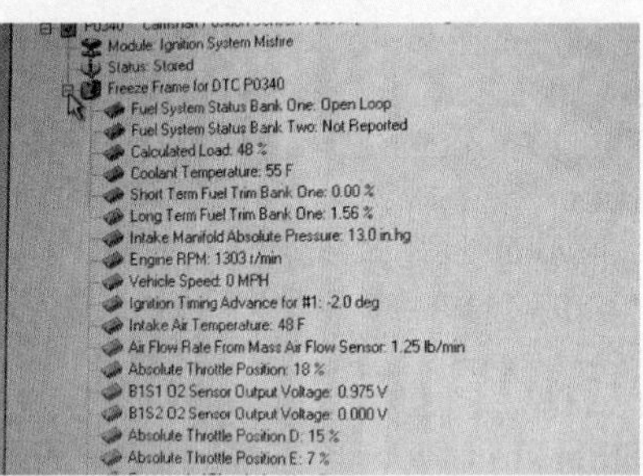

5. Retrieve and record freeze-frame data applicable to DTCs and monitors.

6. Power off the scan tool, turn the ignition off, and disconnect the scan tool.

FIGURE 64-17 An oscilloscope being used to check electrical and electronic signals to and from a sensor.

systems found on vehicles today, these tools are essential in checking electrical and electronic signals to and from sensors and PCMs (**FIGURE 64-17**). Typical graphing multimeters and digital storage oscilloscopes can store the results of captures as data on the instrument and in many cases also on a computer that can connect to the tool.

A graphing multimeter or a digital storage oscilloscope is used to verify signals going to and from electrical and electronic components. To use these tools, you need to know how to hook the test equipment into the electrical circuit, set the test instrument to read accurately, and interpret the results of the capture. There are many different types of leads and adapters available for graphing multimeters and digital storage oscilloscopes, such as current probes, high-voltage probes, and pressure sensors. You need to know when and where to use each lead and adapter to ensure accurate readings are taken.

To inspect and test computerized engine control system sensors, the PCM, actuators, and circuits using a graphing multimeter, follow the steps in **SKILL DRILL 64-2**.

Testing Actuators

N64004

Actuators are devices such as solenoids, fuel injectors, and stepper motors. They are used to convert electrical signals into mechanical movement. A typical vehicle has many actuators that are controlled by the vehicle's computer systems to manage

SKILL DRILL 64-2 Testing Engine Control System Sensors, the PCM, Actuators, and Circuits

1. Determine the circuit to be tested and the likely voltages and frequency of the waveform to be measured. Set up the graphing multimeter or oscilloscope, setting the voltage level and time base to commence measurements.

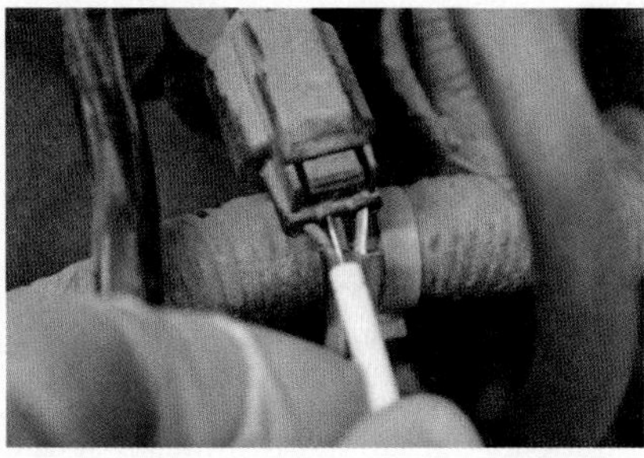

2. Connect the leads to the points to be measured.

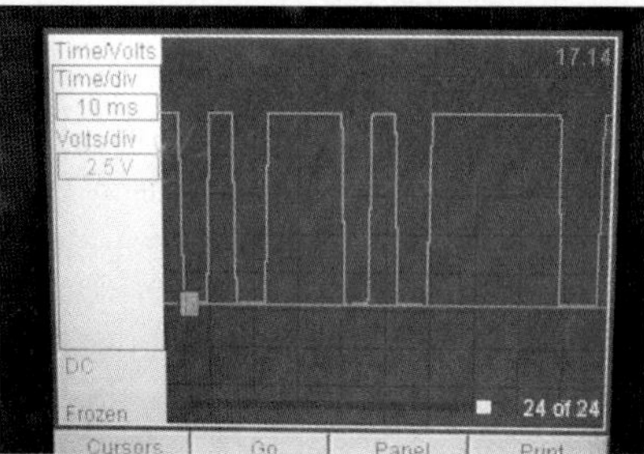

3. Capture the waveforms from the circuit under test. Analyze the waveform, comparing it to the manufacturer's specifications or known good waveforms. Determine the necessary actions.

such functions as fuel injection, throttle position, and emission valves. Actuators have to be tested for correct operation when faults occur in vehicle systems.

The test process usually entails checking for correct signals to the actuator to verify that correct operation is possible and checking the mechanical and electrical operation of the actuator to ensure it is capable of working under operating conditions. Testing actuators may involve the use of test equipment such as DVOMs, digital storage oscilloscopes, and pressure gauges, and may require other tools such as jumper leads and power supplies. Many actuators operate on voltages supplied by the system computer and may operate at voltages other than battery voltage—for example, 5 volts. Always check the manufacturer's information for correct voltages, and always isolate an actuator from its control circuit before applying any external voltage or power supply to test its operation. Never jump battery or power supply voltages to an actuator while it is connected to its control circuit; doing so may result in damage to the system computer.

To test actuators, follow the steps in **SKILL DRILL 64-3.**

Interpreting DTCs and Scan Tool Data Related to the Emission Control Systems

N64005

DTCs and scan tool data must be interpreted by the technician when there is a drivability or emission problem. Scan tools are an important test instrument for modern vehicles, as they allow access to the OBD and data systems. Scan tools fall into two main categories—aftermarket and OEM. OEM scanners are specific to a particular manufacturer's line of vehicles and often provide additional functionality that aftermarket scanners cannot provide. Aftermarket scanners, in contrast, usually work on multiple manufacturers' products, but not necessarily with the same degree of functionality as an OEM scan tool.

The scan tool is connected via the diagnostic plug, which allows access to stored fault codes and data that the vehicle computer systems have stored in regard to vehicle performance. Never erase data or fault codes before they have been accurately recorded. You will need to reference the manufacturer's information or code books to identify fault codes and the systems they relate to.

To interpret DTCs and scan tool data related to the emission control system, follow the steps in **SKILL DRILL 64-4.**

Diagnosing the Cause of Drivability or Emission Concerns with DTCs

N64006, S64001

DTCs often set when there is a fault within vehicle systems, such as with drivability or emission concerns. When one-trip or two-trip DTCs are set, the MIL will illuminate on the dash, indicating a fault. After you verify the concern, the DTCs are a good place to begin the diagnostic process. You should retrieve

SKILL DRILL 64-3 Testing Actuators

1. Reference the manufacturer's information as necessary. Determine the actuators to be tested and the required operating parameters required for correct operation.
2. Select the appropriate test equipment.
3. Conduct actuator tests on and off the vehicle as required.
4. Analyze the test results against the manufacturer's specifications, and make recommendations as required.

SKILL DRILL 64-4 Interpreting DTCs and Scan Tool Data

1. Locate the DLC, and connect the scan tool.
2. Power on the scan tool, and turn the ignition on.
3. Establish scan tool communications with the vehicle.
4. Retrieve and record the DTCs.
5. Look up the DTCs' definitions in code books or in the manufacturer's service information.
6. Interpret the results, referencing the manufacturer's service information.
7. Power off the scan tool, turn the ignition off, and disconnect the scan tool.

all codes and record them electronically or print them out. Researching the service information will provide you with a description of each DTC. It is not unusual for multiple codes to be set. Some codes may relate to the root cause of the problem, and some may be set as an effect of the system not functioning correctly. Follow the procedures listed in the service information to diagnose concerns when DTCs are stored.

To diagnose drivability or emission concerns when there are stored or active DTCs present, and to obtain and interpret scan tool data, follow the steps in **SKILL DRILL 64-5**.

Diagnosing Drivability or Emission Concerns without DTCs

N64007

A vehicle may fail an emission test or a customer may complain about a drivability issue even though DTCs have not been set. Just because a DTC has not been set does not mean the vehicle is running perfectly. The OBD II system is designed to set fault codes if the vehicle exceeds the limits programmed within the system to meet the EPA Federal Test Procedure. Pass and fail points are different for every vehicle and are integrated within the system's operation, which is designed to compensate and keep emissions in check even in the event of a fault.

The PCM can make adjustments to ensure engine emissions continue to meet the standard. The PCM adjustments may have an effect on the vehicle's drivability, but because requirements are still met, no DTC is set. For example, an engine misfire may not set a DTC because the intermittent nature of the misfire is not enough to set a DTC, so the customer complains about a periodic stumble. Additionally, in some cases, codes are not set if a PCM is faulty or a system fault is such that it prevents a code from being set.

To diagnose a vehicle when DTCs have not been set, you need to rely on diagnostic procedures using test equipment such as scan tools to read system parameters, digital oscilloscopes, DVOMs, smoke machines, pressure gauges, and flowmeters. The technician should collect as much information as possible and analyze the result to narrow down the cause of the fault. Some of that information will be stored in the PCM and can be accessed with a scan tool using Mode 6 or Mode 7. This area of the PCM's memory stores the results of its own component tests and readiness monitors. So by looking at this information, you will be able to see data such as the number of misfires individual cylinders have, for example, the number-three cylinder has accumulated a total of 1327 misfires. But if the misfire are random and below the threshold for setting a DTC, there would be no easy way of knowing that. So by looking at this information, you may be able to get clues to identifying the cause of the concern.

To diagnose drivability or emission concerns that do not involve stored or active DTCs, first gather information on the complaint from the customer. Verify the complaint, and identify the symptoms. After referencing the service information, use diagnostic test equipment to conduct tests; then interpret the results and diagnose the drivability or emission concerns. Recommend a course of action for repair.

Diagnosing Drivability and Emission Problems Resulting from Malfunctions of Interrelated Systems

N64008

This task is performed when systems other than the engine management system cause problems with emissions or drivability. Drivability and emission problems can be caused by interrelated systems, such as automatic transmissions, cruise control, security systems, suspension control, air-conditioning, and other non-OEM-installed accessories. Each of the systems, aside from the non-OEM accessories in most vehicles, is controlled by a system computer and has electronic diagnostic capability and

SKILL DRILL 64-5 Diagnosing Drivability or Emission Concerns

1. Locate the DLC, and connect the scan tool.
2. Power on the scan tool, and turn the ignition on.
3. Establish scan tool communications with the vehicle.
4. Retrieve and record DTCs.
5. Look up DTC descriptions in the service information.
6. Formulate a focused testing plan, using the service information and retrieved data.
7. Diagnose drivability or emission concerns, and identify and recommend a course of action for repair.
8. Power off the scan tool, turn the ignition off, and disconnect the scan tool.
9. Test-drive the vehicle to verify the repair. Recheck the status of monitors and DTCs to make sure there are no potential pending faults.

SKILL DRILL 64-6 Diagnosing Drivability and Emission Problems from Malfunctions of Interrelated Systems

1. Verify the customer complaint, and identify the symptoms.
2. Plan diagnostic tests on interrelated systems with reference to the manufacturer's information.
3. Use diagnostic test equipment to conduct tests.
4. Interpret results with reference to the manufacturer's service information.
5. Diagnose drivability and emission problems resulting from malfunctions of interrelated systems, and recommend a course of action for repair.

associated fault codes. The technician needs to understand how each system works and interacts with other systems. This way he or she can understand the cause and effect of faults and their impacts on other systems.

DTCs and customer information are the best place to start gathering information and planning a diagnostic pathway based on suspected systems and components. Non-OEM-installed accessories can make diagnosing difficult due to a lack of information and knowledge about the products, their installation, and how they work. Ensure that as much information as possible is gathered from the customer before commencing work.

To diagnose drivability and emission problems resulting from malfunctions of interrelated systems, follow the steps in **SKILL DRILL 64-6.**

▶ Wrap-Up

Ready for Review

- Second-generation onboard diagnostics (OBD II) started in late 1994 and became standard in 1996. All vehicles use the same generic adapter for a generic scan tool to connect to the vehicle's computer.
- Since California's emission laws of the 1960s, the standards for emissions have gotten more stringent, requiring many changes to subsequent vehicles.
- OBD systems help the technician work more efficiently and help take the guesswork out of diagnosing problems on today's sophisticated vehicles.
- OBD systems are a great resource, using powerful onboard computers in harmony with sophisticated diagnostic equipment to pinpoint problems in vehicle systems and components.
- Both OEM and aftermarket OBD II scan tools serve to access OBD information via the data link connector (DLC).
- The Society of Automotive Engineers (SAE) has standardized the terminology of parts and systems nomenclature (J1930) and across-the-board identification of generic DTCs (J2012).
- Diagnostic trouble codes (DTCs) are either one-trip DTCs or two-trip DTCs.
- Codes on OBD systems are categorized as powertrain (P), body (B), chassis controller (C), and communications system (U).
- Freeze-frame data are the best way for a technician to determine the conditions under which a concern happens, which then helps determine the cause of the DTC.
- System readiness monitors are either continuous or noncontinuous.
- Continuous monitors are misfire monitor, fuel system monitor, and continuous component monitor.
- Noncontinuous monitors are catalytic converter monitor, heated catalyst monitor, oxygen sensor monitor, oxygen sensor heater monitor, evaporative system monitor, EGR system monitor, secondary air system monitor.
- Scan tools communicate with the PCM and computer modules, displays codes and data, and erases codes. Some scan tools have bidirectional capability.
- It is very important to run all monitors after repair so that you can verify that the fault is fixed and the MIL will not illuminate again.
- Scan tools, a lab scope, and a digital volt-ohmmeter (DVOM) are a technician's best tools when it comes to troubleshooting an OBD code.
- Always refer to the manufacturer's service information and follow the diagnostic flowcharts to ensure that the vehicle malfunction is identified and repaired correctly.
- Always use wiring diagrams when diagnosing any electric/electronic circuits in the OBD system, and do not forget about the grounds.

Key Terms

aftermarket That segment of the trade that supplies parts, services, and repair for vehicles outside of the original equipment manufacturer (OEM) or the dealer network.

bidirectional scanners Scanners used to cause various components and systems to operate for test purposes.

carbon dioxide (CO_2) A vehicle emission that is considered a primary greenhouse gas, though not toxic and not yet regulated.

carbon monoxide (CO) A vehicle emission produced by partially burned fuel that is colorless, odorless, and highly toxic and that causes asphyxiation when inhaled.

code Another name for a fault code; see *DTC*.

continuous monitoring A term that describes OBD II monitors that run continuously throughout the drive cycle.

controller area network (CAN) A localized (onboard) vehicle network that enables computers and components to send and receive signals across a shielded twisted pair of wires.

current DTC Condition indicating that the test failed during the last ignition cycle and for two-trip codes, on the previous cycle.

drive cycle A series of prescribed automobile operating conditions during which emissions testing is performed.

electronic brake control module (EBCM) The module that controls and monitors the antilock braking system.

emissions Tailpipe and volatile organic compound pollutants emitted by the automobile.

emission analyzer A service bay or lab device used for detecting/measuring vehicle tailpipe emissions.

enabling criteria The operation conditions required before a monitor is allowed to run.

fault See *diagnostic trouble code (DTC)*.

freeze-frame A feature of OBD II that records sensor data when a fault occurs.

history code A fault code that has occurred but is not current and is saved in the PCM's memory for 40 warm-up cycles.

hydrocarbon (HC) A microscopic unburned fuel particle that contributes to photochemical smog and helps form ground-level ozone.

module An electronic computer or circuit board that controls specific functions.

monitor An OBD II test run to ensure that a specific component or system is working properly.

noncontinuous monitor A monitor that runs only once per drive cycle.

OBD I The first generation of onboard diagnostic systems that originated for California vehicles.

OBD II The second generation of onboard diagnostic systems, which have been in effect for all U.S. vehicles since 1996.

oxides of nitrogen A vehicle emission that contributes to ground-level ozone. Oxides of nitrogen are produced when nitrogen and oxygen react during combustion, given sufficient temperatures and pressures. Forms include nitrogen oxide and nitrogen dioxide.

particulate matter (PM) Unseen microscopic particles of carbon consisting of soot (especially prominent with diesel exhaust). PM clogs the lungs and is carcinogenic.

pending codes Codes that have not been validated by failing a second test.

SAE J1930 An SAE standard for across-the-board standardization of parts and systems nomenclature.

SAE J2012 An SAE standard for across-the-board identification of generic DTCs.

sulfur dioxide (SO_2) A pollutant resulting from sulfur in motor fuel and which contributes to acid rain.

volatile organic compound (VOC) The hydrocarbons in petroleum products that contribute to combustion.

warm-up cycle When the engine is started and run until the engine operating temperature reaches 160°F (71.1°C), has increased in temperature at least 40°F (22°C), and the engine is turned off again.

Review Questions

1. Why is it important to run the monitors after repair?
   a. To ensure that the fault doesn't immediately reoccur.
   b. So that the vehicle will operate in closed loop.
   c. So that the vehicle will operate in open loop.
   d. To ensure that the PCM is updated to the latest software.
2. All of the following statements regarding DLC are true *except*:
   a. Scan tools serve to access OBD information via the DLC.
   b. Its location is also now fairly standardized as being within 2' of the engine.
   c. The DLC enables the scanner to access data stored in the vehicle's various computers.
   d. It is a 16-pin connector with a common (SAE J1962) size, shape, and pin layout.
3. If the MIL flashes, it indicates that the:
   a. engine needs to be serviced within the next 100 miles.
   b. driver should stop the vehicle as soon as safely possible.
   c. vehicle needs refueling.
   d. PCM is malfunctioning.
4. Pending or current codes that have happened in the past but haven't been cleared manually with a scan tool, or automatically after approximately 40 warm-up cycles without the code reoccurring, are known as:
   a. current DTCs.
   b. pending codes.
   c. history codes.
   d. generic codes.
5. If the first character of the DTC is U, it indicates that the fault is located within the:
   a. powertrain.
   b. body.
   c. chassis controller.
   d. network communication system.
6. All of the following statements with respect to freeze-frame data are true *except*:
   a. Freeze-frame data are automatically recorded in the vehicle's PCM when a vehicle DTC is stored.
   b. Freeze-frame data are a list of sensor data that are associated with the specific DTC.
   c. Freeze-frame data is saved for all DTCs stored in memory.
   d. Difficult-to-find intermittent faults can be diagnosed by carefully reviewing the freeze-frame data for clues to the location of the fault.

7. Which of the following is a continuous monitor?
   a. Fuel system monitor
   b. Catalytic converter monitor
   c. Evaporative system monitor
   d. EGR system monitor
8. Bidirectional scanners:
   a. can be used to cause various components to operate on many vehicles.
   b. can only extract data from the vehicle's one or more on-board computers.
   c. are used to place a fault in a computer system.
   d. can only be used to detect engine faults.
9. How can you make sure that the system isn't about to turn on the MIL due to pending codes?
   a. Checking the results of the monitors on Mode 6 and Mode 7
   b. Reassembling the engine
   c. Erasing the codes
   d. Checking freeze-frame data
10. Which of the following are used to make precise electrical measurements over time and display the results on a graph as a waveform?
   a. Digital storage oscilloscopes
   b. Scan tools
   c. Continuous monitors
   d. DMM

ASE Technician A/Technician B Style Questions

1. Tech A says that on OBD II vehicles, it is a good idea to clear the codes before diagnosis and see if they reset. Tech B says that the DTC indicates which part needs to be changed. Who is correct?
   a. Tech A
   b. Tech B
   c. Both A and B
   d. Neither A nor B
2. Tech A says that OBD I and OBD II DLC connectors are different from each other. Tech B says that OBD II standardizes the designations for diagnostic trouble codes (DTCs). Who is correct?
   a. Tech A
   b. Tech B
   c. Both A and B
   d. Neither A nor B
3. Tech A says that monitors are designed to test whether emission systems are working properly. Tech B says that monitors are designed to store sensor data about a fault if a DTC is set. Who is correct?
   a. Tech A
   b. Tech B
   c. Both A and B
   d. Neither A nor B
4. Tech A says that a CANbus network communicates over a bundle of many individual wires, specially designed for fast communication speed. Tech B says that OBD II mandates that a DTC must be set if the emissions are more than 1.5 times the EPA FTP. Who is correct?
   a. Tech A
   b. Tech B
   c. Both A and B
   d. Neither A nor B
5. Tech A says that OBD II allows a technician to hook up a generic scan tool to read DTCs and clear DTCs. Tech B says that when the MIL is on, the vehicle should not be driven. Who is correct?
   a. Tech A
   b. Tech B
   c. Both A and B
   d. Neither A nor B
6. Tech A says that something as simple as a loose gas tank fill cap will turn on the MIL. Tech B says that some DTCs require two trips to set a "current" code. Who is correct?
   a. Tech A
   b. Tech B
   c. Both A and B
   d. Neither A nor B
7. Tech A says that a "P" in the DTC stands for a powertrain code. Tech B says a "U" stands for an undetermined code. Who is correct?
   a. Tech A
   b. Tech B
   c. Both A and B
   d. Neither A nor B
8. Tech A says that if the MIL is blinking, it indicates a catalyst-damaging fault. Tech B says that the EGR readiness monitor can use the MAP sensor to verify the EGR gases are flowing. Who is correct?
   a. Tech A
   b. Tech B
   c. Both A and B
   d. Neither A nor B
9. Tech A says that the MIL can turn off if the fault doesn't reappear in three sequential trips. Tech B says that when an MIL is activated, the PCM stores the DTC until a predetermined number of drive cycles automatically erase the DTC or if cleared manually by a scan tool. Who is correct?
   a. Tech A
   b. Tech B
   c. Both A and B
   d. Neither A nor B
10. Tech A says some mechanical engine problems can cause OBD II DTCs to be set. Tech B says that OBD II codes only monitor non-powertrain components. Who is correct?
   a. Tech A
   b. Tech B
   c. Both A and B
   d. Neither A nor B

CHAPTER 65
Intake and Exhaust Systems

NATEF Tasks

- **N65001** Inspect throttle body, air induction system, intake manifold, and gaskets for vacuum leaks and/or unmetered air. (AST/MAST)
- **N65002** Verify idle control operation. (AST/MAST)
- **N65003** Inspect integrity of the exhaust manifold, exhaust pipes, muffler(s), catalytic converter(s), resonator(s), tailpipe(s), and heat shield(s); perform needed action. (MLR/AST/MAST)
- **N65004** Inspect condition of exhaust system hangers, brackets, clamps, and heat shields; determine needed action (MLR/AST/MAST).
- **N65005** Perform exhaust system back-pressure test; determine needed action. (AST/MAST)
- **N65006** Test the operation of turbocharger/supercharger systems; determine needed action (MAST).

Knowledge Objectives

After reading this chapter, you will be able to:

- **K65001** Explain the purpose and function of the intake system and components.
- **K65002** Explain the components and function of intake manifolds.
- **K65003** Explain forced induction.
- **K65004** Explain the purpose and function of the exhaust system and components.
- **K65005** Explain the purpose and function of the muffler system.

Skills Objectives

After reading this chapter, you will be able to:

- **S65001** Perform intake and exhaust inspection and testing.

▶ Introduction

The intake and exhaust systems are critical parts in the operation of the internal combustion engine. You can think of it as "breathing." The air enters the engine through the intake system and exits out the exhaust system. The intake system ensures that clean air is supplied to the engine, which is then mixed with fuel and burned in the combustion chamber, creating the thermal expansion that pushes the pistons down the cylinder. In fact, the more air an engine can take in, the more power it can make.

Clean air is essential for a proper, long-lasting engine, because dirt would get between the close-moving parts and cause premature wear of the engine. In fact, it only takes a tablespoon or two of dirt entering the engine through the intake system to ruin an engine. The intake system must provide a sealed passageway to the combustion chambers to ensure that no contaminants leak into the system and that no air is allowed to bypass the airflow sensor, which could create a drivability problem. On most vehicles, the intake system also controls the amount of air entering the engine, by use of a throttle plate, which is how engine speed and power are controlled.

The exhaust system provides a path for the burned exhaust gases to exit the engine and travel safely out the rear of the vehicle. In doing so, it provides a method of reducing the noise from the power pulses. It also includes components that help reduce the harmful emissions in the exhaust stream. Getting rid of exhaust gases is just as important as getting air into the engine. If exhaust gases cannot leave the engine easily, then they will back up in the system, and not as much air will be able to enter the engine, reducing power output. An efficient, free-flowing exhaust system assists the engine in creating maximum power with minimal emissions. In this chapter, we explore the operation, diagnosis, and testing of the components in both the intake and the exhaust systems.

▶ The Intake System

K65001

The intake system is primarily designed to control the flow of air delivered to the combustion chamber. In many fuel-injected and carbureted engines, fuel is mixed in the intake system so that air and fuel are being delivered together. In diesel and gasoline direct-injection engines, fuel is injected directly into the combustion chamber, where it is mixed with air; thus, only air (plus any positive crankcase ventilation and evaporative emissions) flows through the intake system. For any fuel to burn efficiently, no matter where it is injected, it must be vaporized and fully mixed with the proper amount of air. Vaporization begins when the fuel system atomizes the fuel by breaking it up into very small particles by using high pressure to spray it into the charge of air. These small particles make it much easier for the fuel to vaporize. Heat and low pressure also may be used to vaporize the atomized fuel.

Intake System Components

The primary components of the automotive intake system are the intake manifold, the throttle body, and the air induction system. The intake manifold is bolted to the cylinder head. Its construction and design depend on the engine for which it is created. The intake manifold directs airflow into each cylinder (**FIGURE 65-1**), and when restricted by the throttle plate, it provides a source of vacuum for systems such as the power brake system.

The intake manifold creates a mounting place for a throttle body assembly. In throttle body injection, one or two fuel injectors, which are mounted in the top of the throttle body, spray fuel down into the intake manifold. In a port fuel–injected engine, a throttle body is bolted to the opening of the intake manifold, and the injectors are mounted in the intake manifold near the intake valves (**FIGURE 65-2**). In gas direct injection, a throttle body is bolted to the opening of the intake manifold, and the injectors are mounted in the cylinder heads.

FIGURE 65-1 The intake manifold.

You Are the Automotive Technician

A customer comes into the dealership complaining that her 12-year-old Ford Mustang doesn't idle as well as it should and that the malfunction indicator lamp (MIL) is illuminated. You scan it for codes and notice that it has a P0131 (Circuit Low Voltage B1S1), a P0171 (Fuel Trim Lean B1), and a P0300 (Engine Misfire) detected. As an apprentice technician, you refer to the Ford Motor Company's service information to look up the process for diagnosing the codes. You notice that there could be several things that cause those codes, including a vacuum leak, weak oxygen sensor, and dirty fuel injectors. You decide to start by looking for a vacuum leak because you hear what sounds like a leak when the vehicle is running.

1. What are the different ways of locating a vacuum leak?
2. Where are the possible places that a vacuum leak can occur?
3. How does a vacuum leak affect a vehicle equipped with a mass airflow sensor?
4. How does a vacuum leak affect the oxygen sensor reading?

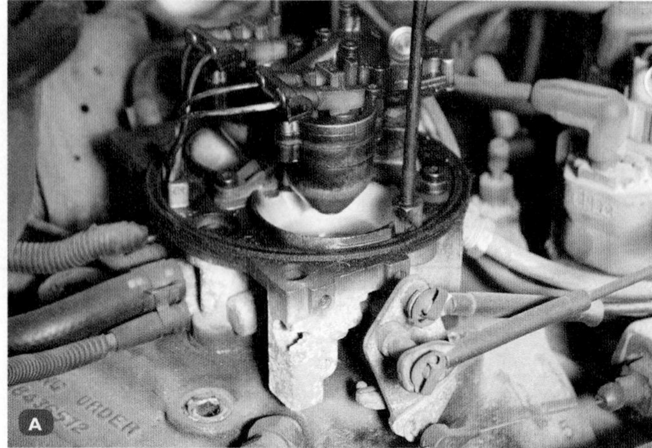

FIGURE 65-2 A. Throttle body injection. **B.** Port fuel injection. **C.** Gas direct injection.

FIGURE 65-3 The throttle plates can be: **A.** Cable operated. **B.** Electric motor operated.

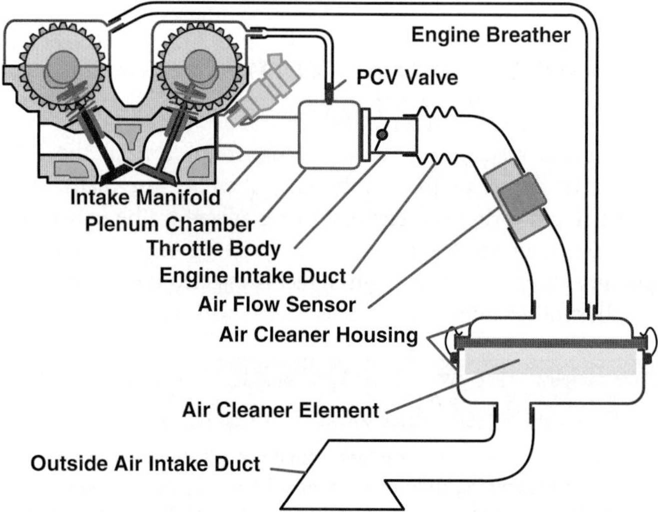

FIGURE 65-4 The air induction system directs air from the inlet through an air filtration device and on to the intake manifold.

The throttle body controls airflow with a butterfly valve or valves, also called throttle valves. The throttle valves are opened and closed either by a throttle cable or, if the engine uses an electronic throttle control, by an electric motor that opens and closes the throttle plates (**FIGURE 65-3**). Attached to the throttle body is a throttle position sensor, which provides throttle position information to the engine computer. The throttle body also includes vacuum ports for the operation of vacuum-controlled devices such as evaporative emission systems and the power brake booster.

In front of the throttle body are the following components: an air cleaner and housing, solid and flexible-duct tubing, connectors, and sometimes a mass airflow sensor (**FIGURE 65-4**). The inlet opening of the induction system may be located in various positions under the hood, depending on the available space for the automotive engineer to work with.

Air Cleaner

The air cleaner assembly filters the incoming air. In past designs, the air cleaner housing was made of stamped metal and housed a round air filter. Most air cleaners are now made of plastic and can vary in shape from square to round. The air filter element may be manufactured from pleated paper or from oil-impregnated cloth or felt; in much older vehicles, it was manufactured in an oil bath configuration (**FIGURE 65-5**). Another function of the air cleaner assembly is to muffle the noise of the intake pulses and the incoming air. The air cleaner can also act as a flame arrester. If a gasoline engine backfires, the air cleaner can contain the flame within the intake manifold or carburetor. The location of the air cleaner depends on the available space and the hood design.

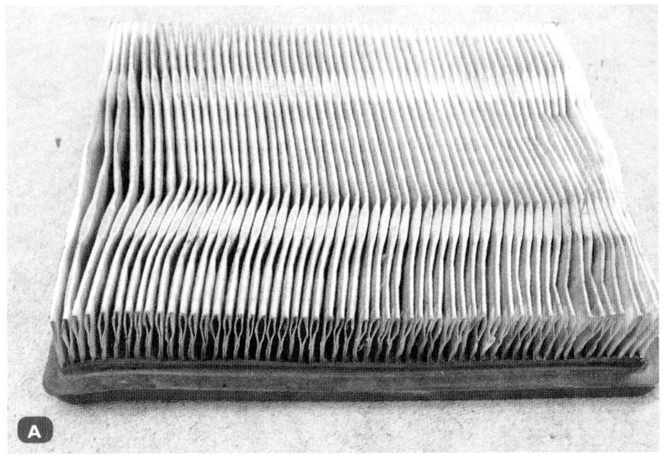

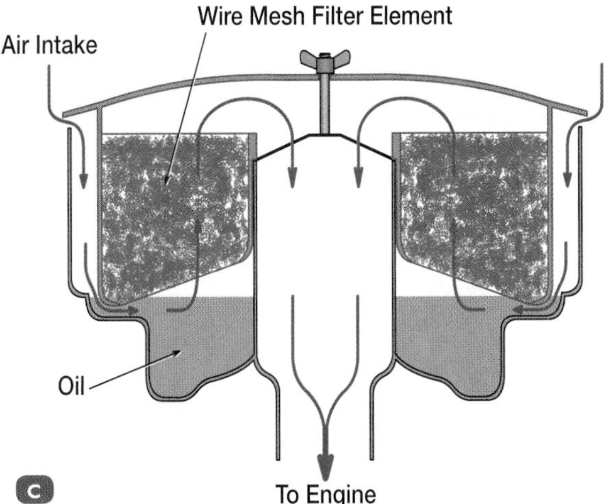

FIGURE 65-5 Types of air filter systems. **A.** Pleated paper. **B.** Oil-impregnated cloth or felt. **C.** Oil bath air cleaner.

A lot of air passes through the intake system into the engine. In a gasoline engine, the air-fuel mixture, by weight, is about 14.7 parts air to 1 part fuel. By volume, that's 10,000 times more air than fuel. When an engine consumes 10 gallons of gas, the air filter will have filtered 100,000 gallons of air. The air-fuel mixture enters the engine, so the air must be clean. Any abrasives that enter the engine can quickly cause wear and damage.

> ▶ TECHNICIAN TIP
>
> One way to tell if a pleated paper air filter has to be changed is by holding it up to a light and seeing how much light passes through it. Just because it is a bit dirty on the outside doesn't mean it has to be replaced. In fact, a lightly plugged filter prevents particles from going through the filter better than a brand new filter. But if it is too plugged, it will restrict airflow to the engine. If you can see a lot of light through it, it is reusable. If it restricts the light very much, it should be replaced.

An air cleaner on a multipoint electronic fuel-injected engine usually has a different shape from that on a carbureted engine, but it serves the same purpose. In many vehicles, the air inlet is mounted where it can obtain cool, clean air. Many electronically fuel-injected systems have an airflow sensor between the air cleaner and the throttle body, which accurately measures all air entering the engine. It is essential that there be no air leaks after the airflow sensor, as leaks will upset the air-fuel mixture. It is interesting to note that a mass airflow (MAF) sensor can measure air entering the engine down to tenths of a gram per second.

On most heavy-duty diesel engines and a few gasoline engines, the air cleaner assembly uses an air filter indicator to identify whether the filter needs to be serviced (**FIGURE 65-6**). The indicator typically has a red band that will show if the filter is restricted. Some indicators lock in place, even when the engine is stopped; others indicate filter condition only when the

FIGURE 65-6 Air cleaner with filter service indicator.

engine is running. The indicator is mounted between the air cleaner and the engine. When the air filter creates enough of a restriction, the vacuum produced in the intake tube causes the indicator to display the warning.

Some heavy-duty air cleaners also incorporate a cyclone-type pre-cleaner, which is mounted directly onto the air cleaner unit (**FIGURE 65-7**). The cyclone system uses angled vanes that give the incoming air a swirling motion. Centrifugal force throws the heavier dirt particles outward, and they collect in a bowl at the bottom of the cleaner, where they can be removed manually. An efficient cyclone pre-cleaner can remove up to 90% of particles before they reach the main filter element, greatly extending its life.

One type of air cleaner system is the long-life filtration system that is currently used on the Ford Focus (**FIGURE 65-8**). This air cleaner assembly is a nonserviceable unit that is designed to last 150,000 miles (240,000 km) or more. It is made to also capture any hydrocarbons being released from the throttle body and then allow them to be pulled back into the engine as air flows across the filter. The filter element is made of specially designed foam. This filter can be serviced only by replacement of the entire assembly, which costs several hundred dollars.

FIGURE 65-7 Cyclone-type pre-cleaner.

Ducting

The ducting is what connects the ridged air cleaner to the throttle body of the intake manifold. Ducting is typically made of hardened plastic with flexible rubber couplings to absorb engine movement. It may have a MAF sensor in line and be retained with worm-style clamps to seal the duct to the sensor. Some ducting also includes air dampening, sometimes referred to as a Helmholtz resonator. A **Helmholtz resonator** is a container that is sealed and specially shaped to cancel noise created by pressure waves. The pressure wave bounces off the walls of the resonator and collides with the incoming waves, cancelling the noise. On the vehicle, this resonator is a tube that is connected to the air ducting and is simply a sealed chamber that is open where it is connected to the air duct (**FIGURE 65-9**). The pressure waves created when the engine pulls air into the air duct produce a loud suction sound. Use of the resonator minimizes the noise produced in the air intake system.

Diesel Induction Systems

The two main components of the four-stroke intake system are the air cleaner and the intake manifold. In a diesel engine, fuel is delivered directly to the combustion chamber. The intake system carries air only and typically does not have a throttle body. Diesels do not need a throttle plate to control engine rpm, since fuel delivery is what controls engine rpm. At the same time, most newer model diesels are equipped with an exhaust gas recirculation (EGR) valve which then may require the use a throttle plate (**FIGURE 65-10**). The throttle plate is then designed to restrict airflow to the engine so that exhaust gases can be drawn into the intake manifold, thus lowering cylinder temperatures and reducing oxides of nitrogen. The throttle plate can also be used in some diesel engines to govern the engine

FIGURE 65-8 Typical long-life air filter that is not serviceable.

FIGURE 65-9 The air duct on some vehicles contains a resonator to cancel air noise as air flows through the intake.

Applied Science

AS-38: Amplification: The technician can explain to a customer how sound can be amplified in a vehicle due to resonant cavities and other physical characteristics of the vehicle.

The noise level of a vehicle plays an important role in the overall satisfaction rating by the owner. A quiet vehicle is expected by the majority of vehicle owners. Some vehicle owners prefer the sound of a high-performance vehicle, and the noise level is not as much of a concern. Air intake and exhaust systems can be manufactured and modified for a variety of different sounds.

Back in 1863, physicist Herman Von Helmholtz did many studies regarding air-dampening devices. He discovered a resonator that uses the principle of sound waves colliding, resulting in the concept of cancelling noise. The resonator can be used in the exhaust system to assist the muffler in further reducing noise. The resonator can also be used in the induction system. The benefit is the ability to muffle airflow noise. The Helmholtz resonator is basically a simple device consisting of a cavity with one or more short narrow tubes. For each application, the device must be precisely tuned. Concerning the air induction system, the resonator is installed between the air filter and the engine inlet. The design of the Helmholtz resonator creates a sound frequency that cancels out some of the intake air noise. On some vehicles, there is a vacuum-activated valve that disables the effects of the resonator at certain rpm.

AS-39: Carriers/Insulators: The technician can demonstrate an understanding of how sound generated in one place can be carried to other parts of the body or engine through metal and materials.

Sound travels through steel approximately 17 times faster than through the air. This is because the molecules in steel are closer together. Sound is produced when an object vibrates. Vibrations can pass from molecule to molecule quickly in materials such as steel.

When a vehicle has a bad wheel bearing that is producing a rumbling sound, it is often difficult to tell which of the four wheels the noise is coming from. In this case, you might try driving the vehicle alongside a building or a concrete block wall. Sound waves travel in all directions from their source. Some of them will strike the wall, and the noise will be reflected back toward the vehicle. A greater proportion of the sound waves will strike the wall when it is near the side of the vehicle with the faulty wheel bearing, meaning more sound waves will be reflected toward the vehicle. Thus, if the noise is louder when the passenger side is near the wall, the technician will know the faulty bearing is on the passenger side.

A mechanic's stethoscope can also be used to locate sounds on vehicles. The tip of the metal probe can be used to trace the sound to the source of the problem. There are now electronic stethoscopes that can be used to locate sounds on vehicles.

FIGURE 65-10 Some late-model diesel engines use a throttle plate to create a low pressure in the intake manifold to draw EGR gases in.

speed by simply closing down the amount of air that can be pulled into the engine.

To get more power on some diesel engines, one or more turbochargers or a supercharger is used to pressurize the air. Diesel engines may have more than one air cleaner, as needed for dealing with severe working conditions. The air cleaner is usually mounted away from the engine to obtain cleaner, cooler air.

Intake Air Heating

Heating of the intake manifold is required in a carbureted or throttle body–injected engine to provide greater vaporization of the fuel and to ensure that fuel does not collect on the cold walls,

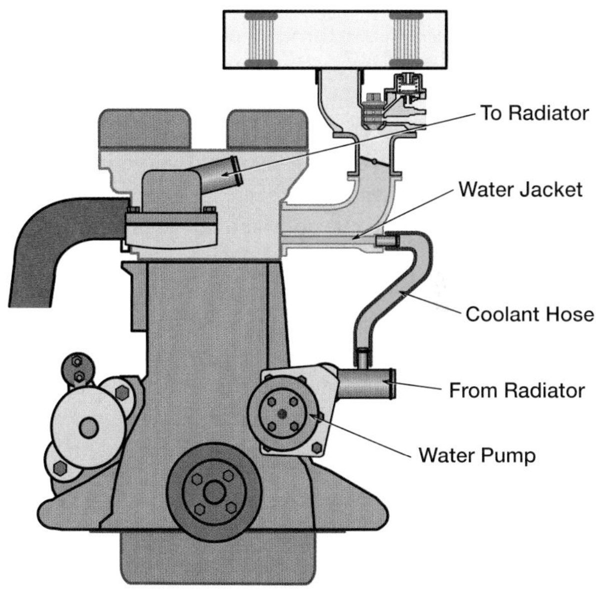

FIGURE 65-11 Using hot engine coolant to heat the manifold.

creating a lean mixture. Port fuel–injected or GDI intake manifolds do not normally have to be heated, because the manifold does not carry fuel.

There are several methods that manufacturers have used to heat intake manifolds, including hot engine coolant, hot exhaust gases, and electric heaters. Flowing hot engine coolant from the cooling system through passageways in the intake manifold heats the manifold (**FIGURE 65-11**). However, it takes awhile for the coolant to become hot, so this is not the fastest way to heat the intake manifold. In this type of system, coolant continuously circulates through the manifold, even when the engine is fully warmed up.

On V-type engines, exhaust gases can also be directed through passageways in the intake manifold during engine warm-up. The hot exhaust gases are hot immediately and therefore heat the manifold up more quickly than coolant does. A heat riser valve at the end of the exhaust manifold blocks most of the exhaust gases from exiting through the exhaust pipe. The exhaust gases are directed through a passageway in the cylinder head, then through a chamber in the bottom of the intake manifold, out the passageway in the other cylinder head, and to the exhaust pipe on the other side of the engine. Once the engine is warm, the heat riser opens the valve in the exhaust, and exhaust gases can flow normally out of the exhaust pipe. If the heat riser valve sticks closed, the engine will run hotter than normal and will likely be low on power.

Electric heaters have also been used to preheat the air. A preheat grid is typically placed between the intake manifold and the throttle body (**FIGURE 65-12**). During cold engine operation, current flows through the heating grid, warming the air passing through it. Once the engine reaches a certain temperature, the current flow to the grid is turned off, and the engine operates normally. If the heater stays on too long, the heater grid can melt down, dropping particles down the intake manifold, which can damage the engine.

Another method of helping the fuel to vaporize in the intake manifold on carbureted and throttle body–injected engines is to warm the incoming air by passing it through a shroud around the exhaust manifold. This method was called an early fuel evaporation (EFE) system (**FIGURE 65-13**). An enormous amount of heat passes through the exhaust manifold and is used to heat the incoming air. The heated air then enters the air cleaner and proceeds through the throttle body and intake manifold while it mixes with the fuel.

Once an engine is hot, the incoming air can become too hot, so the amount of hot air entering the engine has to be controlled. On vehicles with emission controls, air cleaners use a thermostatic valve to control how much hot air enters the air cleaner. These systems are called heated air cleaners (HAC) and heated air systems (HAS). When the engine is started, only heated air

FIGURE 65-12 An electric heater.

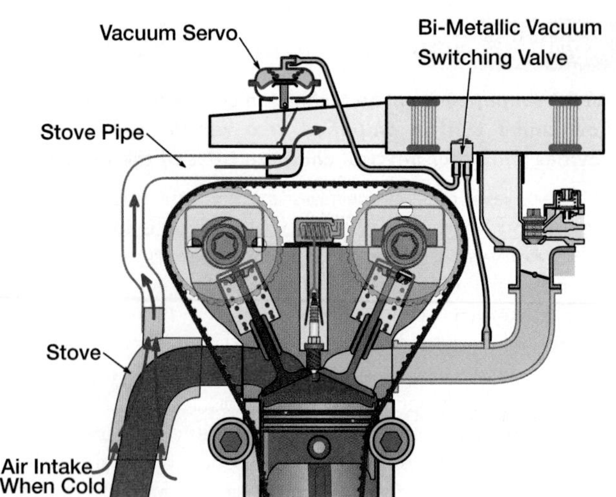

FIGURE 65-13 The heated air intake system, found on carbureted and throttle body–injection systems, is used to ensure that fuel does not collect on the cold intake manifold.

is used. As engine temperature rises, the valve opens and blends cold ambient air with the hot air. This ensures that the temperature of the air supplied to the engine stays fairly constant. This valve can be a simple thermostatic (mechanical) type.

Another way to control the intake air temperature is to use a vacuum control unit and a control valve mounted inside the air cleaner unit. A diaphragm is attached to the control valve of the vacuum control unit. When a cold engine starts, vacuum from the intake manifold moves the diaphragm, opening the valve and allowing hot air to flow into the air cleaner while closing the ambient air intake. A heat-sensitive valve in the air cleaner responds to changes in air temperature. Below a certain level, it opens, letting hot air flow into the air cleaner. As the temperature rises, it slowly closes, reducing the flow of hot air and blending in cooler air. When carburetors and the throttle body–injection units were no longer installed in vehicles, heated intake air was no longer needed.

With a multipoint injection setup, the intake manifolds carry air only, so heating of the intake manifold is not needed. Also, the cross-sectional area of the tubes can be larger. Because more air can flow, the engine will produce more power. Fuel is injected directly into the intake ports of the cylinder head. The other possible fuel delivery method is the gasoline direct injection system, which directs fuel straight into the combustion chamber. This system uses an intake similar to the multipoint injection setup, flowing air only through the intake. The fuel-injected engine manifold has a **plenum chamber** that provides a reservoir of air and helps prevent interference with the flow of air between individual branches. The plenum chamber is a large portion of the intake manifold after the throttle plate and before the intake runner tubes (**FIGURE 65-14**).

Intake Manifolds

K65002

The intake manifold is usually a cast iron, aluminum, or plastic part with several tubular branches. In carbureted or throttle

FIGURE 65-14 The port fuel injection manifold has a large open area, called the plenum chamber, and tubes that deliver air individually to each cylinder called intake runner tubes.

FIGURE 65-15 Plastic intake manifold.

body–injected engines, the intake manifold carries the air-fuel mixture into the cylinder head. In a port injection or gas direct injection engine, the intake manifold carries air only. The cross-sectional area of each tube has to be kept small enough to maintain the high air speeds that improve vaporization, but not so small that it restricts the airflow to the engine at higher speeds.

Intake manifolds were originally made from cast iron, and later from light alloy metal castings. In modern vehicles, the intake manifold is made from special heat-resistant polymers and plastics (**FIGURE 65-15**). Manifolds using this type of construction lighten manifold weight by up to 50% and contribute to higher fuel efficiency. These intake manifolds are normally molded from a glass fiber–reinforced grade of crystalline polymer that consists of a blend of syndiotactic polystyrene and polyamide, a polyamide nylon–based material. This type of material is ideally suited for replacing metal in underhood applications because of its strength, stiffness, and chemical resistance under high-temperature operating conditions. The plastic can also be recycled, similar to cast iron and aluminum. The use of plastic has enabled more efficient airflow into

FIGURE 65-16 Cross-flow head.

the engine due to smoother surfaces of the manifold and has allowed for easier shaping of the manifold, creating better airflow characteristics.

Cylinder heads that have intake and exhaust manifolds on opposite sides of the engine are known as cross-flow heads. That means the intake manifold is on one side and the exhaust manifold is on the other (**FIGURE 65-16**). This design tends to produce more power because the flow of air moves more easily across the head rather than coming in and back out on the same side. Most modern engines are set up with a cross-flow cylinder head. The cross-flow head contains individual branches or ports to carry air and fuel into the combustion chamber. In past designs, two cylinders could share an intake port, but this design did not allow for free flow of air and was replaced with individual ports to allow for more power.

Variable Intake Systems

The air intake manifold for an electronic fuel injection multipoint engine normally has long branches of equal length. The long branches increase the pulsing effect of the airflow at lower engine rpm and help charge the cylinders. The more air drawn into the cylinder, the denser the air-fuel mixture is when the intake valve closes. The increased density of the air-fuel mixture determines how much pressure develops in the combustion chamber and the amount of force on the piston to turn the crankshaft.

The characteristic torque curve of a naturally aspirated engine depends mainly on how the engine's mean pressure changes across the rpm operating range. The design of the intake system largely influences the mass of air that can be drawn into a cylinder at a given engine speed; thus, the intake system largely influences the engine's torque curve. In general, long intake manifold runners produce high torque at lower engine rpm because of the increased amount of time between the air pulses due to the inertia of the long column of air. The increased time between the pulses matches the longer time at lower rpm operation. Shorter intake manifolds produce higher torque at higher engine rpm because of the shorter amount of

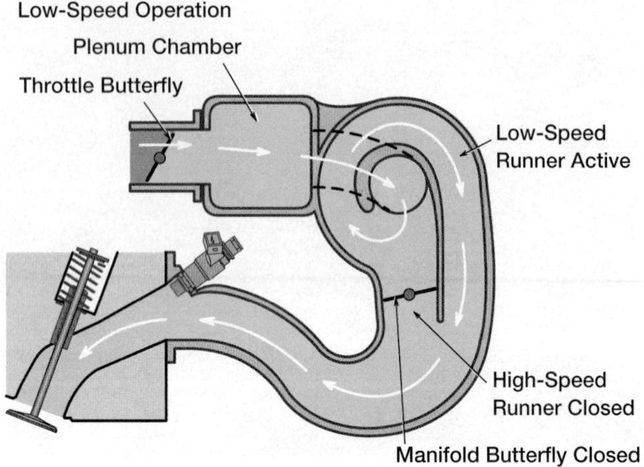

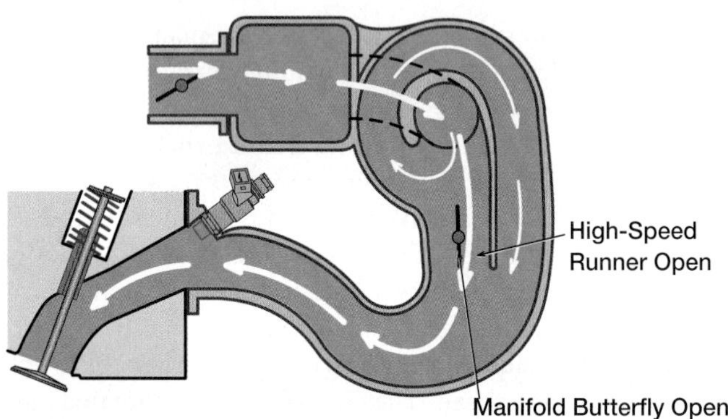

FIGURE 65-17 Typical variable intake manifold with alternate runners.

FIGURE 65-18 Different intake manifold runner lengths can be obtained by opening and closing alternate runners.

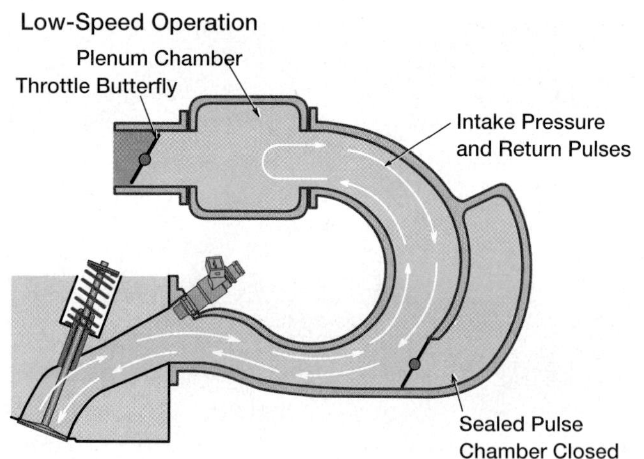

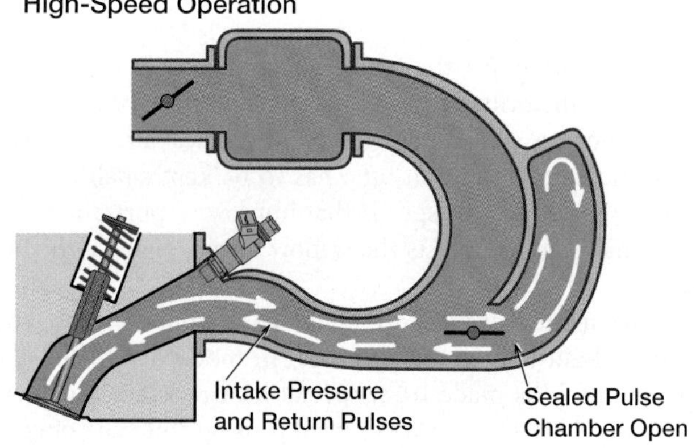

FIGURE 65-19 Typical operation of a pulse chamber intake manifold.

time between the pulses due to the inertia of a shorter column of air. This ram air effect allows for more total airflow at and near the tuned rpm.

Manifolds that respond to changes in engine load and speed by changing their effective length in two or three stages are called variable inertia, or variable intake, manifold systems. The primary section is long and narrow for the low range of rpm. The secondary section is shorter and wider for the high range of rpm (**FIGURE 65-17**). This combination maintains a high-speed airflow in the system. The three-stage manifold extends the torque curve even more, so that the torque curves overlap each other as advantageously as possible.

The different manifold stages are controlled by valves that are computer controlled by the engine management system to open at a specified engine speed and extend the torque output. The system is called intake manifold runner control (IMRC). Specific intake runner stages can be opened or closed off so that air is forced to run through the other set of runners. The runners are closed off by a butterfly-type valve that is moved by a stepper motor or vacuum actuator (**FIGURE 65-18**). The stepper motor or supply of vacuum is controlled by the engine control module.

Some manufacturers are now using sealed pulse chambers for each intake runner that is opened and closed by butterfly valves. The pulse chambers are designed to time the return of intake pressure pulses back to the intake valves above a certain rpm (**FIGURE 65-19**). At this rpm, the butterfly valves open, and the pressure pulses assist in filling the cylinders with more air. Below this range, the butterfly valves are closed, and the pressure pulses are timed to happen at a lower rpm range. This helps to maintain a broader and more powerful torque range across the operating rpm of the engine.

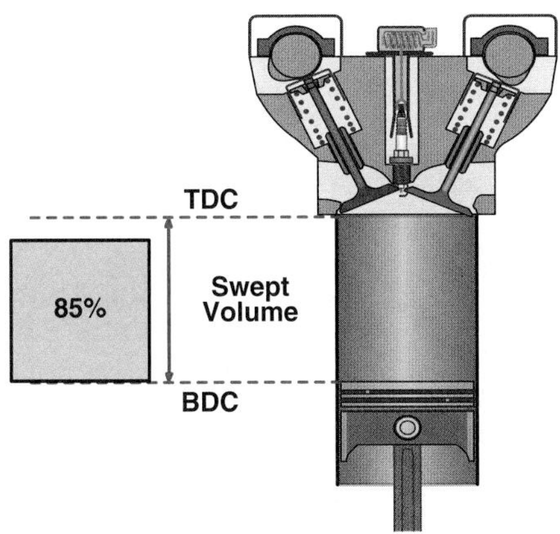

FIGURE 65-20 Volumetric efficiency of a normally aspirated engine.

Volumetric Efficiency

Volumetric efficiency compares the volume of air entering a cylinder during intake to the swept volume of the cylinder when the piston is at bottom dead center. It is usually expressed as a percentage. In a stock naturally aspirated engine, one without forced induction, volumetric efficiency can almost never be 100%. This is because of the resistance to airflow that occurs at the throttle body, manifold, and intake valve (**FIGURE 65-20**). However, some highly modified engines can exceed 100% due to the valve overlap being matched to the intake and exhaust manifold sizes and lengths, which maximizes the ram air effect at the tuned rpm.

To understand volumetric efficiency, you must have a good understanding of the physics of airflow. Air likes to move in a straight manner and is affected by the bends in the intake manifold. Sharp bends force the air to move to the opposite side of the tube and force the air to pile up, slowing the airflow. Air also works like a bearing surface: Air near the surface of the manifold slows as friction of the tube wall resists its motion, while air in the center of the airstream tends to speed up as it moves over top of the slower air. All of these factors slow the airflow and reduce volumetric efficiency.

With forced induction, the incoming air is compressed by some sort of pump or compressor. Because of this greater pressure, a greater volume and mass of air is forced into the cylinder during the intake stroke, which takes the volumetric efficiency to well above 100%, increasing the engine's power output proportionately.

Volumetric efficiency is also affected by back pressure. Back pressure in an exhaust system refers to a buildup of pressure in the system that interferes with the outward flow of exhaust gases. This area of high pressure acts as a kind of wall to stop gas flow. It can be caused by a blockage in a muffler or a similar restriction. One common cause of excessive back pressure is a catalytic converter that has melted down and created a restriction in the exhaust stream. Catalytic converters become hotter as the combustible mixture is increased, such as from a misfire. The catalyst in the converter will melt if it gets too hot, and will essentially block off the small passageways. Back pressure typically should be no more than 3 pounds per square inch (psi), or 20.7 kilopascals (kPa). If it is greater, it can create performance issues, as it will cause exhaust to back up into the combustion chamber, minimizing the amount of fresh air (combustible mixture) flowing into the engine. This reduces volumetric efficiency and engine power output.

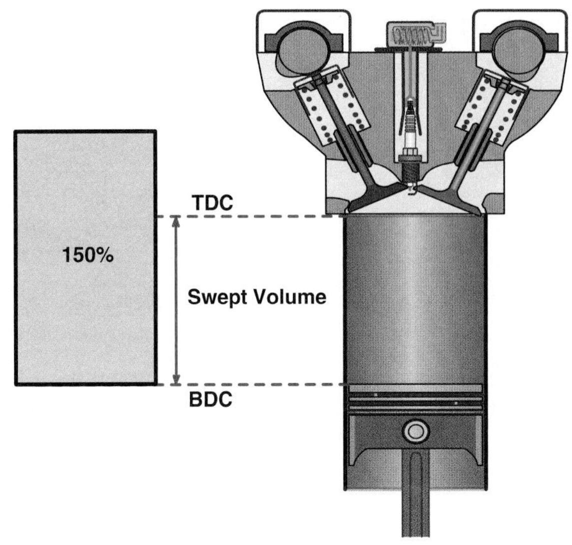

FIGURE 65-21 Volumetric efficiency of a forced induction engine.

Forced Induction

K65003

One way to improve engine output is to increase the amount of the air-fuel mixture that is burned in the cylinder (increasing volumetric efficiency). Volumetric efficiency can be boosted the most by what is called **forced induction**. Forced induction increases air pressure in the intake manifold above atmospheric pressure. Thus, an engine using forced induction can have a volumetric efficiency well above 100% (**FIGURE 65-21**).

One way to achieve forced induction is by using a turbocharger. A turbocharger uses energy that is normally wasted through the exhaust to spin a compressor that forces air into the engine. Another way to achieve forced induction is to use a supercharger. A supercharger is turned by the crankshaft, compresses the air, and forces it into the engine. The faster the engine turns, the more air is moved by the supercharger.

Supercharger Systems

Power is produced when a mixture of air and fuel is burned inside an engine cylinder. If more air is forced into the cylinder, then more fuel can be burned and more power produced with each power stroke. A **supercharger** compresses the air in the intake system to above atmospheric pressure, which increases the density of the air entering the engine. Naturally aspirated engines operate with uncompressed air at atmospheric pressure, 14.7 psi (101 kPa). But a supercharger boosts that pressure another 6 psi (41.4 kPa) or higher.

In a supercharged system, there is a greater air mass flow rate—that is, a higher density and speed of airflow. Air pressure is increased by the compressor on the way into the engine, and

FIGURE 65-22 The supercharger uses power from the crankshaft to force additional air and fuel into the engine, which increases the engine's power output.

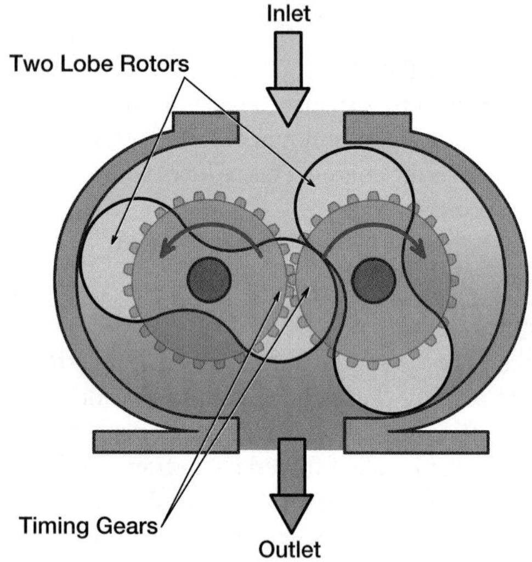

FIGURE 65-23 A Roots-style supercharger moves air around the sides of the lobes.

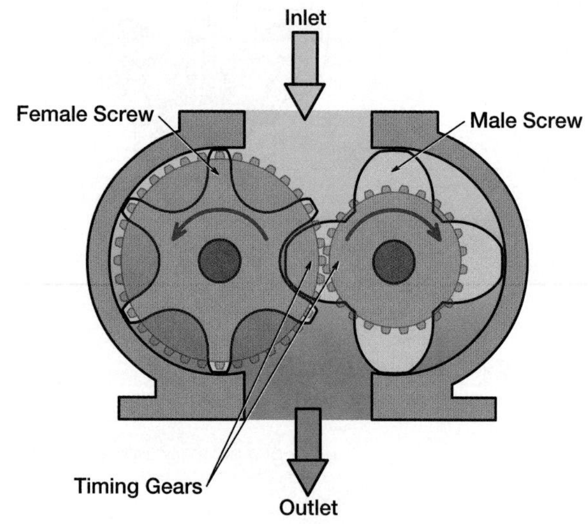

FIGURE 65-24 Typical twin-screw supercharger.

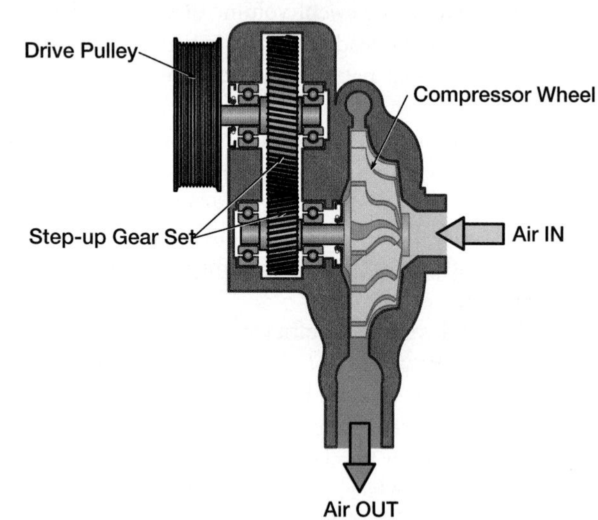

FIGURE 65-25 Typical centrifugal supercharger.

more fuel is added, which creates more cylinder pressure and power. The increased cylinder pressure causes the exhaust gases to exit much more rapidly, making the timing and exhaust sizing less important. Also, any residual gases tend to be pushed out of the cylinder by the airflow entering the combustion chamber, as the supercharger boosts the manifold pressure substantially.

Although some of the extra power produced must be used to drive the supercharger, the net result is more total power from the crankshaft (**FIGURE 65-22**). The supercharger may include a **bypass valve** system that allows the air to be returned back to the supercharger inlet and prevents excessive boost. The bypass valve can be mounted remotely or directly onto the intake port.

Several types of superchargers are manufactured for use on vehicles. The oldest version is the Roots-style supercharger, which uses lobes on two shafts to move air into the engine. This type of supercharger does not draw air between the two shafts, but rather pushes air around the sides of the housing and into the intake (**FIGURE 65-23**). The lobes on both shafts mesh together in the middle, so air cannot exit between them. The Roots-style supercharger uses stacking air against the intake valves to build boost pressure as the lobes come together and prevent air from leaking back between them.

Another version of the supercharger is the twin screw type. The twin-screw supercharger compresses air between the screws and forces it into the intake (**FIGURE 65-24**). This compressor design is very efficient, but very expensive.

Many aftermarket high-performance superchargers are the centrifugal design. This supercharger shaft is turned by a drive belt from the crankshaft and spins a compressor wheel, similar to what is used in the turbocharger (**FIGURE 65-25**). The supercharger shaft is connected by a set of gears to create a higher speed of the compressor. The compressor increased the air pressure, which forces more air into the engine.

Because compressing air heats the air up, many supercharged engines use a heat exchanger, also called an intercooler, located between it and the intake manifold. The heat exchanger

FIGURE 65-26 A typical turbocharger removed from the vehicle.

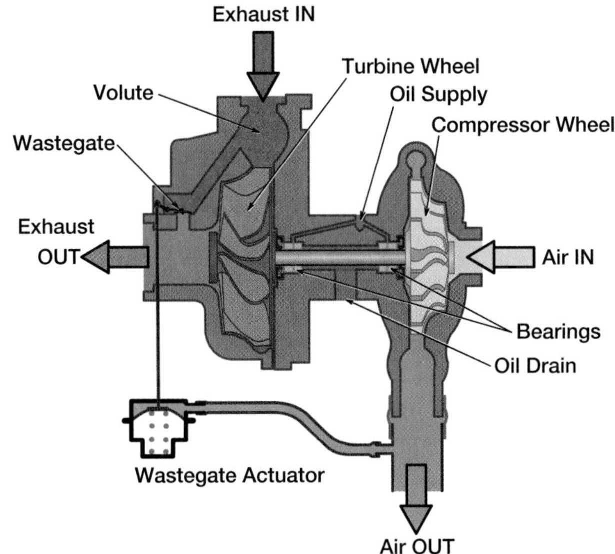

FIGURE 65-27 The turbocharger compresses air to feed to the intake and is powered by exhaust gases.

is like a radiator that the air flows through to cool it before entering the engine. This helps prevent engine-damaging detonation or overheating. Intercoolers are covered in detail later in the chapter.

Turbocharger Systems

A **turbocharger** is a forced induction system that uses wasted kinetic energy from the exhaust gases to increase the volume of air entering the engine (**FIGURE 65-26**). At the same time, it has a negative effect on the flow of exhaust gases out of the engine. This means that for maximum power output, valves, cam timing, and exhaust system design have to be closely matched to the turbocharger and engine.

The turbocharger uses exhaust gases to turn a turbine fan. A shaft connects the turbine rigidly to a centrifugal compressor (**FIGURE 65-27**). The compressor compresses the air and forces it under pressure into the intake manifold. The turbine is turned by exhaust gases, so it runs at very high temperatures. It,

FIGURE 65-28 The turbocharger wastegate actuator controls the amount of boost pressure going to the intake manifold.

along with the compressor, can rotate at well over 100,000 rpm. Because of this, the turbocharger needs a good supply of clean oil to lubricate the bearings and carry excess heat away from the turbocharger. Some engines also supply coolant to the turbocharger body to improve cooling.

Higher engine speeds mean increased cylinder pressure and exhaust gas volume, and that makes the turbocharger spin faster and force more air into the cylinders. Pressures left unchecked can damage the engine if allowed to get too high. When pressure increases in an engine, so does cylinder temperature. As the temperature increases, the possibility of detonation also increases because the gasoline will ignite before the piston reaches the proper position in the compression stroke. To control this risk of detonation, a device called a **wastegate** is installed on the exhaust inlet of the turbocharger. When intake manifold pressure reaches a preset level, the wastegate automatically directs the exhaust gases so they bypass the turbine (**FIGURE 65-28**), thus preventing intake pressure from building further. The wastegate can also be computer controlled so that intake air pressure can be reduced in the case of detonation or knocking.

On a mechanically operated wastegate, the spring pressure in the diaphragm controls the pressure at which the wastegate valve opens. The wastegate actuator is a spring-loaded pressure diaphragm that moves when intake manifold pressure pushes against it. The manifold pressure pushes on the diaphragm, overcoming the spring force at the specified point. This causes the control linkage to open the wastegate, allowing exhaust gases to bypass the turbine wheel, which reduces the boost pressure (**FIGURE 65-29**). As intake manifold pressure drops, the wastegate closes and allows exhaust to flow to the turbine wheel again.

The wastegate can be PCM controlled. The manifold absolute pressure sensor reports the manifold pressure to the PCM. When needed, the PCM signals the wastegate actuator to open the wastegate and reduce turbocharger boost. The actuator can be vacuum operated or electric motor operated.

A pressure relief valve known as a **blow-off valve**, may also be installed to prevent excessive manifold pressure if the

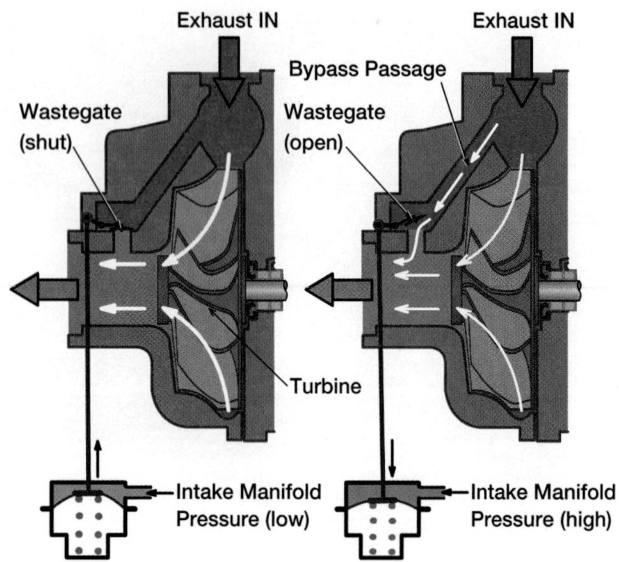

FIGURE 65-29 Typical mechanical wastegate operation.

FIGURE 65-30 The blow-off valve is a spring-loaded pressure relief valve that releases excessive pressure in the air intake tube going to the intake manifold.

wastegate should fail. A blow-off valve works against spring pressure. As boost pressure rises above spring pressure, the pressure opens the valve, allowing excess pressure to "blow off" into the atmosphere.

The blow-off valve may also vent the pressure to the air cleaner box to reduce the turbocharger pressure release noise. The blow-off valve is located in the air intake tubing connected to the intake manifold. It is also used to ensure that when the throttle is abruptly shut, pressure does not rise excessively high between the throttle body and the turbocharger compressor wheel (**FIGURE 65-30**).

If pressure rises excessively, **compressor surge** will occur. Compressor surge is when manifold pressure rises above normal and works against the spinning exhaust turbine wheel, slowing it considerably. The pressure wave can then flow backward out of the compressor, creating a fluttering noise and continued pressure waves. Compressor surge, which is recognizable by a rapid fluttering sound, can be damaging to the compressor wheel, because the pressure wave forces the wheel to slow.

Applied Science

AS-40: Decibels/Intensity: The technician can demonstrate an understanding of how sound intensity can be measured.

The degree of loudness is measured in units called decibels (dB). The average human can hear sounds between 0 and 120 dB. One-time exposure to noises above 120 dB, or sustained exposure above 85 dB, can damage the ears. Decibels can be measured with a sound pressure level meter. These meters are available in analog and digital styles.

As a baseline, the decibel level of some common sounds are as follows:

- Conversational speech at a distance of 3 feet: 60 dB
- A diesel truck engine at a distance of 30 feet: 90 dB
- A chain saw at a distance of 3 feet: 110 dB
- A jet aircraft at takeoff at a distance of 150 feet: 150 dB

Technicians should be concerned about the decibel rating, as well as the exposure time, of sounds in the workplace. The length of time that a person is exposed to the sound is a very important factor. For example, the CDC recommends that a person should be exposed to a sound pressure level of 100 dB for no more than 15 minutes. Ear protection should be worn to protect your hearing, based on the decibel rating and exposure time.

Because the turbocharger uses the energy of the exhaust gases, there is a short delay between when a driver opens the throttle and when maximum power is available. This delay is called turbo lag, and on larger engines, it can be quite noticeable if the turbocharger is not sized correctly.

When the turbocharger housing is large, boost pressure is higher at high speed, but lag is worse for a given exhaust flow. When the housing is small, boost pressure is high at lower speeds, but a restriction occurs in the exhaust system at high rpm. To combat turbo lag and avoid creating a restriction, several designs have been made. The first is the twin scroll turbocharger, which uses two passageways into the turbine housing from the exhaust manifold. The one passageway is controlled by a flap that can be closed to accelerate the exhaust flow, making the turbocharger operate at a higher boost at low engine speeds.

Another way to improve the turbo lag issue is to use twin turbochargers. Many manufacturers use two small turbochargers (one on each bank of the engine) that can spin up more quickly, thereby creating boost sooner. The turbo lag is decreased as boost is created at a lower engine speed, and yet additional boost is created at a higher speed because there are two turbochargers to deal with the total exhaust flow.

The final way to create boost pressure without lag is to use a variable-geometry turbocharger. The variable-geometry turbocharger uses a set of movable vanes or fingers that change the flow of exhaust gases through the turbine housing (**FIGURE 65-31**). Under low exhaust speed, the fingers are positioned so that the exhaust gases spin the turbine faster, creating boost. Under high rpm, the speed of the exhaust gases is very high; therefore, the fingers are moved so they do not over-speed the turbine while also

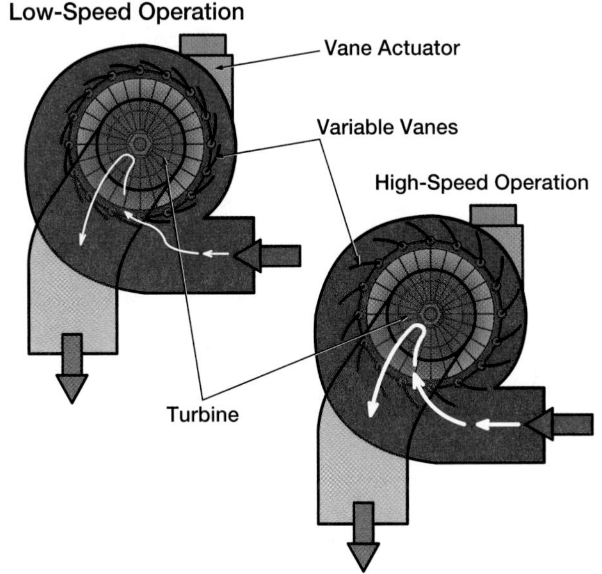

FIGURE 65-31 Typical variable-geometry turbocharger.

not restricting the exhaust flow and creating backup of exhaust, which would lower power. This type of turbocharger offers the best of both worlds. Currently it is used on Ford's Powerstroke series of diesels.

Another new development in turbocharger technology is used on the BMW engine. The engine design uses reverse flow cylinder heads. The center of the V-type engine, which normally was for intake, is now for the exhaust passages (**FIGURE 65-32**). Exhaust leaves the head and moves into the twin turbochargers, one per engine bank. The intake manifold is now on the outside of the head. The different positioning of the turbochargers keeps exhaust temperatures high and produces more turbine speed, which increases the efficiency of the turbocharger. The catalytic converter also gets up to temperature faster, reducing emissions.

Because a turbocharger recycles heat energy that would otherwise be lost, a turbocharged engine can increase an

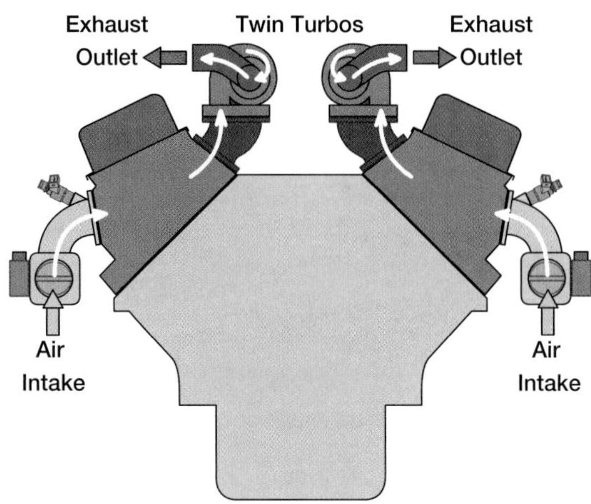

FIGURE 65-32 BMW's reverse flow cylinder heads with twin turbos in the center of the V.

engine's efficiency and fuel economy—as long as the vehicle is being driven conservatively. However, even though a turbocharger may seem to be offering additional energy for nothing, it can introduce problems of its own. The extra heat and power it generates can put an extra load on the engine's cooling and lubrication systems. This is why most turbocharged vehicles have shorter service intervals for oil changes than non-turbocharged vehicles.

As the turbocharger compresses the air, it heats up. Hot air is less dense than cool air, so it tries to expand again, and some of the benefits of compressing it are lost. To stop this expansion and improve efficiency, some engines use an intercooler to cool the compressed air. It fits between the turbocharger and the engine. We look at this topic next.

Intercoolers

A turbocharger or supercharger is used to increase the volume of air entering the engine by compressing the air above atmospheric pressure. However, when air is compressed, it heats up, which causes the volume of the gas to increase, lowering the

Applied Science

AS-41: Frequency/Hertz: The technician can explain how frequency of a sound can be used to identify normal and abnormal operating systems.

A customer complains that his vehicle is making a high-pitched whining sound upon acceleration. The technician test-drives the vehicle to verify the concern.

The technician uses a mechanic's stethoscope to discover that the abnormal sound is coming from the alternator. Further diagnostic testing reveals that the alternator has a bad diode that is causing the whining sound. The recommendation to the customer is to replace the alternator.

Concerning the properties of sound, frequency is the number of waves produced in a given time. The frequency of sound waves determines the audible pitch, either high or low. Hertz is the unit used to measure the frequency of sound waves or pitch.

AS-42: Hearing: The technician can demonstrate an understanding of the role of listening for unusual sounds as part of a troubleshooting process.

A vehicle has been towed to a dealership with a no-start condition. A technician has been assigned to determine the cause of the problem. While an assistant attempts to start the vehicle, the technician listens to the cranking rhythm of the engine. The engine has a very uneven cranking rhythm, which indicates uneven compression between cylinders. Using diagnostic equipment, the technician verifies that the engine has very low compression on several cylinders, caused by a broken timing belt. When the timing belt broke, several of the engine's valves were bent, resulting in the low compression.

Listening for unusual sounds can be an essential part of a troubleshooting process. It is helpful to use other methods as well to verify the diagnosis.

FIGURE 65-33 A typical air-to-air intercooler lowers the temperature of the compressed air before it enters the intake manifold.

air density. Hot air under pressure in a cylinder contains fewer oxygen molecules than cooler air at the same pressure in the same volume.

The purpose of an intercooler is to reduce the intake air temperature up to a few hundred degrees Fahrenheit before it enters the intake manifold. Decreasing air temperature increases the density of the pressurized air and improves engine efficiency. It also lowers the chance of engine-damaging detonation.

The intercooler is most efficient at removing heat when the turbocharger's boost pressure is near its maximum and the compressed air is the hottest. This is typically above 15 psi, or 100 kPa. Most intercoolers operate on an air-to-air principle, by feeding compressed air from the turbocharger through the intercooler and then into the intake manifold.

The intercooler works like a radiator. Inside it, the air passes through small tubes with thin fins attached. The heated compressed air flowing through the intercooler gives up its heat to the fins and the tubes (**FIGURE 65-33**). As the vehicle moves forward, the cool outside air flowing across the fins pulls heat away from the tubes and fins. This heat transfer occurs constantly during engine operation.

Some larger engine applications have a liquid-operated intercooler. In this system, the air is fed through small tubes in a heat exchanger, and the vehicle coolant surrounds the tubes. The coolant absorbs the heat and transfers it to the engine cooling system.

▶ The Exhaust System

K65004

The exhaust system is made up of several components that work together to perform four main functions: remove exhaust gases from the engine, quiet the exhaust noise, ensure that poisonous exhaust gases do not enter the passenger compartment, and reduce harmful emissions in the exhaust stream (**FIGURE 65-34**). Exhaust flow is described as follows: Burned gases exit the cylinder through the exhaust port and pass into the exhaust manifold. The first pipe is usually called the engine pipe or down pipe. The down pipe is connected to the outlet of the manifold, which carries the exhaust gases to the catalytic

FIGURE 65-34 The exhaust system removes exhaust gases from the engine, prevents harmful exhaust gases from entering the passenger compartment, reduces noise, and reduces harmful emissions in the exhaust stream.

converter. Catalytic converters were introduced in vehicles in the mid-1970s and were designed to reduce exhaust emissions. The exhaust exits the converter and continues on through an intermediate pipe to the muffler, which reduces exhaust noise. Exhaust gases are then either passed through a resonator or simply discharged to the atmosphere through a tailpipe, usually at the rear, to the side, or above the vehicle.

During engine operation, each time an exhaust valve opens, a pulse of hot exhaust gases is forced into the exhaust manifold. These hot gases produce a lot of noise, some of it at very high frequency. Even while quieting the exhaust noise, the exhaust system can be designed to enhance engine operation and efficiency. In fact, a well-designed system can improve drivability and performance while still reducing the noise.

Exhaust System Components

Exhaust Manifold

The exhaust manifold is bolted to the engine's cylinder head. It also typically provides a mounting place near its outlet for the oxygen sensor, as it is operates better when positioned as close to the cylinders as possible. This position helps it warm up faster during a cold start. The exhaust manifold is usually made from either cast iron or stainless steel. Because of the extreme temperatures generated at the exhaust manifold, heat shields can be installed to protect other vehicle components from heat damage.

On many current vehicles, the exhaust manifold is often replaced with a header. The **header** is an exhaust manifold made of mandrel-formed tubes that are of equal lengths and join at a common collector. **Mandrel forming** uses a special pipe-bending tool to ensure that piping bends in a smooth arc that is not collapsed, which would create a partial restriction. The equal length of the pipes ensures that the exhaust pulses create a more equal exhaust flow out of each cylinder. Exhaust manifolds can be restricting, forcing the pistons to work harder to

FIGURE 65-35 An exhaust manifold is typically made of cast iron, and a header is made of tubing of equal length.

push the exhaust gases out. A header is freer flowing, allowing gases to leave the engine quickly (**FIGURE 65-35**).

The headers also provide a **scavenging** effect to help remove exhaust gases from the cylinders. The outgoing pulse from one cylinder is timed to arrive at the junction at exactly the right time to help draw out the pulse from another cylinder. This setup is called tuned exhaust, and the lengths of the header tubes determine the rpm range they are tuned to. Tuned exhaust is widely used on high-performance vehicles and race cars. It has also found its way into regular production vehicles as well. Performance gains can be realized by ensuring that exhaust gases flow freely so that the engine can pull air in and push exhaust out efficiently.

Engine Pipe

The engine pipe, or down pipe, is attached to the exhaust manifold and connects to the catalytic converter. The engine pipe is usually made of a nickel chromium material, which resists rust and corrosion to ensure it is long lasting. Some exhaust down pipes may also use stainless steel to ensure long life. The engine pipe may be attached to the exhaust manifold by spring-loaded bolts that allow the exhaust to move slightly as the engine moves in its mounts. Some engine pipes also have flexible connectors that allow movement between the engine pipe and the intermediate pipe (**FIGURE 65-36**). The flexible connector is used close to the gap. Its main functions are to allow engine movement—especially in front-wheel drive vehicles—and to reduce vibration without passing it along the exhaust (**FIGURE 65-37**).

Catalytic Converter

A catalytic converter is used to convert unacceptable exhaust pollutants, such as carbon monoxide, hydrocarbons, and oxides of nitrogen, into less dangerous substances. Three-way converters convert oxides of nitrogen back into nitrogen and oxygen, and the hydrocarbons and carbon monoxide to water and carbon dioxide. Older two-way catalytic converters only converted hydrocarbons and carbon monoxide to water and carbon dioxide but were not able to convert the oxides of nitrogen.

FIGURE 65-36 The engine pipe connects the exhaust manifold to the catalytic converter and may contain a flexible connector.

FIGURE 65-37 The flexible connector is used to allow the pipe to flex as the engine moves.

FIGURE 65-38 The catalytic converter is located after the engine pipe and before the muffler. It changes harmful gases into nonharmful gases to be released to the atmosphere.

A catalytic converter fits in line with the exhaust system. It is located close to the exhaust manifold so that it can reach its operating temperature as soon as possible (**FIGURE 65-38**).

FIGURE 65-39 The exhaust is supported by rubber mounts that ensure that vibration is not felt by the driver.

FIGURE 65-40 A typical intermediate pipe between the catalytic converter and muffler.

Some manufacturers install a catalytic converter in the base of the exhaust manifold and another one downstream before the muffler.

Catalytic converters can become contaminated by lead or silicone. Leaded fuel must not be used in an engine with a catalytic converter, because lead coats the catalyst and prevents it from interacting with the pollutants. Some types of silicone sealer also coat the catalyst. Once coated, the converter will most likely have to be replaced.

The catalytic converter operates by creating and then maintaining a chemical reaction while converting the exhaust pollutants. It usually creates heat as it is converting the harmful gases to less harmful ones, so it can get extremely hot. Because of this, it has a heat shield to prevent heat from radiating to bodywork and other parts. The catalytic converter is covered in greater detail in the Emission Control chapter.

Exhaust Brackets

The exhaust components are supported along the length of the vehicle by brackets suspended from the underbody. The supports are usually rubber mounted and help isolate the vibrations of the exhaust from the main body of the vehicle. Rubber is preferred because of its natural dampening effect (**FIGURE 65-39**).

Intermediate Pipe

The intermediate pipe connects the catalytic converter to the muffler (**FIGURE 65-40**). This pipe can be made to fit inside of the pipe of the catalytic converter and may be sealed with an exhaust clamp. The pipe may also have flanges on the ends that bolt to a flange on the catalytic converter, and has a gasket between the bolted flanges to seal the exhaust in.

The Muffler System

K65005

The purpose of a vehicle's muffler is to manage the sound coming from the exhaust system. Typically this means as much noise reduction as possible. But some customers desire a certain exhaust tone, so the muffler system is designed to provide that sound. Exhaust sounds originate from the combustion process within the engine. Exhaust noise becomes an issue as vehicle systems become generally quieter and as the number of vehicles on our roads increases.

To understand the operation of modern exhaust noise reduction systems, it is helpful to understand what sound is. We sense sound with our eardrum, located within the ear. The eardrum is made to move by variations in air pressure. Variations in air pressure can be created when a force is placed upon an object. An example is clapping your hands. As the two hands collide, they push the air surrounding them away. The moving air creates a wave of air pressure, or a sound wave. This sound wave moves your eardrum, which is interpreted as sound by your brain.

The engine produces noise because each combustion process is a rapid burning of air and fuel-a controlled explosion. These explosions create a great deal of noise if they are not absorbed or canceled. *Noise absorption* refers to using sound-deadening materials to absorb sound waves. This happens in some mufflers by putting a sound-deadening material around a perforated pipe that the exhaust gases flow through. This is similar to placing noise-absorbing materials inside the walls of a house to make the rooms quieter. *Noise cancellation* is a system that prevents the sound waves from leaving the exhaust system by canceling them out inside the muffler. These systems create gas pressures that are equal in force but opposite in direction to the noise source. These generated pressures are known as anti-noise. Any remaining sound is referred to as residual noise.

The muffler is designed to quiet the noises of combustion without restricting exhaust flow to the point of adversely affecting performance. The goal is to produce a vehicle that is smooth and quiet as well as powerful. A muffler may use a dissipative technique, which is sound absorption, or a noise-canceling technique, or both. Exhaust noise can be reduced by various means, including baffles and chambers, variable-flow exhaust, and electronic mufflers.

Applied Science

AS-43: Noise/Acoustics: The technician can demonstrate an understanding of why the acoustics of the vehicle affect specific noises.

The acoustics of a vehicle is related to the behavior of sound waves in an enclosed space. Noise can enter the passenger compartment from a number of sources. Engine noise, tire noise, wind noise, and the noise of other vehicles are some sources. Dampening material around the vehicle cabin is a very good method of noise reduction. Insulation is placed in trunk panels, rear wheel wells, doors, and the roof, as well as under the carpet. All of this insulation reduces unwanted environmental noise.

On some vehicles, to reduce weight and be more price competitive, insulation and dampening materials have been reduced. In comparison, luxury vehicles may have more than 100 pounds of sound-deadening material placed carefully in the most needed areas to ensure a quiet environment.

Technicians should understand that noises may not be the same in all vehicles due to these variations. The acoustics of a particular vehicle may differ widely as compared to another. On one type of vehicle, it may be typical for the driver to hear a squealing water pump as it failed. On another, the engine covers and multiple layers of dampening material between the engine and the passenger compartment may prevent the driver from hearing the same sound.

AS-44: Overtones/Harmonics: The technician can explain that the presence of overtones may indicate changes in vibration in systems.

A technician in a luxury car dealership has replaced a number of water pumps on a certain model of a popular vehicle. The published labor rate for this water pump replacement is 11.2 hours. This greatly exceeds the average water pump replacement labor rate by three or four times.

The technician has observed that the water pump on this particular vehicle tends to have a slightly noisy bearing before it fails. As a preventive measure, the technician uses a stethoscope to listen to the sound of the water pump bearing when vehicles of this model are in for basic service. The pump sounds are checked with the stethoscope at idle, at 1000 rpm, and again at 1500 rpm. Over the years, the technician has developed the listening skills with his stethoscope to evaluate the condition of the water pump bearing on this particular vehicle model. When an abnormal condition exists during the stethoscope evaluation, the technician is aware of the situation and can detect the presence of an overtone, which is a higher tone, above the average tone. This condition may indicate the first sign of a bearing with excessive vibration. The technician will make a record of his findings on the repair order to alert the owner of this concern before the vehicle has a water pump failure.

▶ TECHNICIAN TIP

Pedestrians are faced with the opposite problem when dealing with hybrid and electric vehicles—the vehicles are too quiet. It is easy for pedestrians who are visually or hearing impaired, or who just aren't paying close attention, to walk out in front of one of these vehicles while it is operating on electric power, and get hit. Federal guidelines have just been released that all hybrid and electric vehicles will have to be equipped with a new alert system by September 1, 2019. The regulation also specifies that half of a manufacturer's new hybrid and electric vehicles will have to be compliant one year before that.

Baffles and Chambers

The dissipative-type muffler uses a perforated tube or baffle that is wrapped in fiberglass material, and this absorbs the noise of combustion as gases flow past. The baffle then gets welded or bolted into the exhaust pipe. Typically, aftermarket motorcycle mufflers have removable baffles to allow the owner to rewrap the fiberglass insulation or to remove it if a louder exhaust note is wanted.

The noise-canceling technique uses a large chamber that the exhaust pipe enters. The exhaust flows into the large chamber, and the exhaust gases bounce off the chamber's wall. The exhaust gases bounce back toward the incoming gases, and the pressure waves collide, canceling the noise. The exhaust gases then flow out the opposite pipe to the tailpipe. There are several variations to the design of the noise-canceling muffler, which are intended to produce less back pressure (**FIGURE 65-41**).

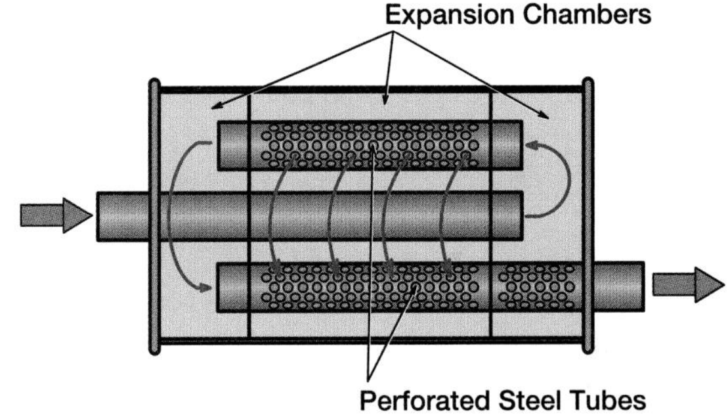

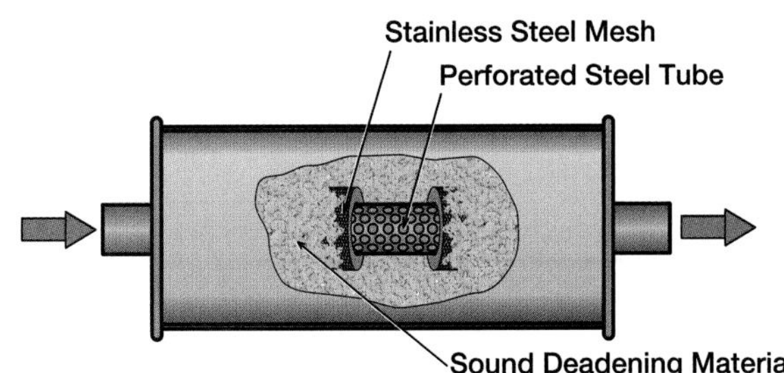

FIGURE 65-41 The muffler is one of two types of designs, either the dissipative type or the noise-canceling type.

Variable-Flow Exhaust

A moveable valve built within the exhaust system is used to change the path for exhaust to flow as well as the amount of exhaust back pressure. This system is used on many high-end performance vehicles to ensure they meet noise restrictions. The system operates in two stages. In the first stage, when the engine rpm and throttle position are low, the exhaust takes a longer route through two or more mufflers/resonators. It also provides a small amount of back pressure, which further lowers

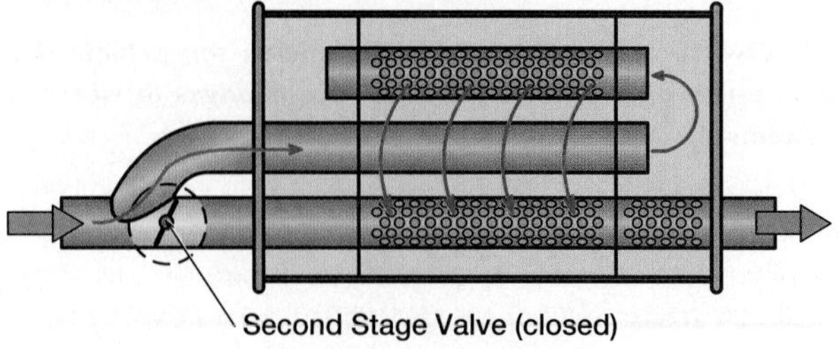

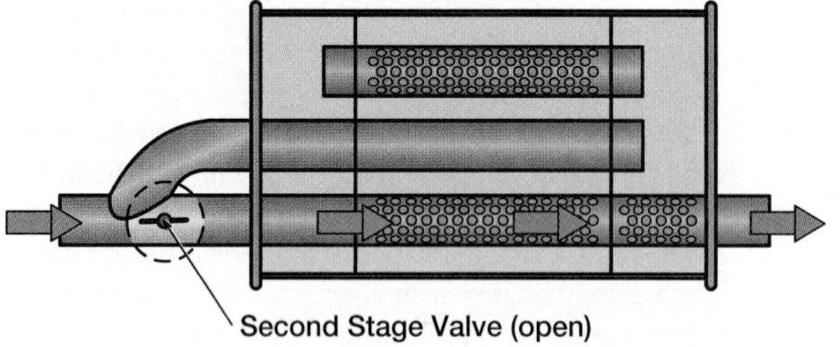

FIGURE 65-42 A typical variable-flow exhaust system.

the exhaust noise as well as the hydrocarbon emissions during valve overlap at low rpm. When the throttle is opened and engine speed increases, the second stage is activated. A valve opens and allows the exhaust to bypass some of the silencers in the system, making the exhaust path more free-flowing (**FIGURE 65-42**). The valve can be operated by the following means:

- Exhaust gas pressure
- Vacuum diaphragm
- Electronic actuator

Adding a variable-flow exhaust to the baffle or chamber system reduces exhaust noise during normal driving but increases power and sound during performance driving. This is because the system can respond to changes in engine speed and load. Variable-flow exhaust was typically limited to high-end exotics such as Ferraris, Aston Martins, or Lamborghinis. This can now be found on domestic vehicles such as Corvettes, Camaros, and Challengers.

Electronic Mufflers

Any restriction to exhaust flow in the exhaust system creates back pressure. Although some back pressure can be beneficial, excessive back pressure reduces an engine's volumetric efficiency. This in turn reduces engine efficiency. Electronic mufflers are designed to produce anti-noise without restricting exhaust flow. This computer-controlled system uses a microphone to detect the sound waves produced within the exhaust system. A computer-driven loudspeaker is operated to generate equal but opposite sound waves to cancel the exhaust noises. This use of opposing sound waves is sometimes called anti-noise. The anti-noise is applied to the exhaust stream in an electronic muffler. The result is a virtually silent exhaust without generating additional and unwanted back pressure across all engine operating conditions. This system increases fuel economy and reduces exhaust emissions. The electronic muffler is a research part at the moment and has never been released on a production vehicle.

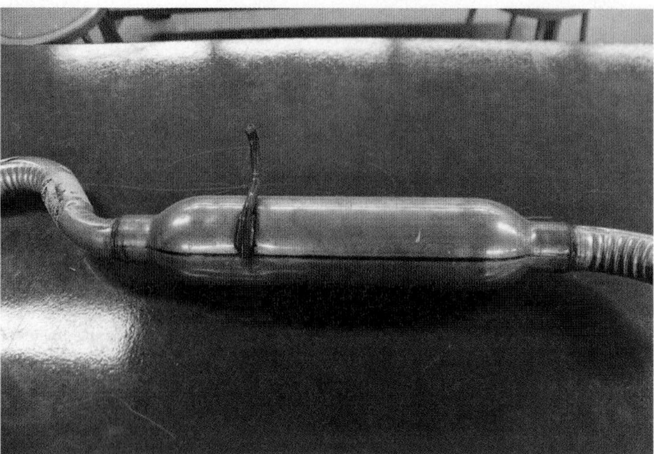

FIGURE 65-43 The resonator is an assistant to the muffler to ensure engine noises are canceled adequately.

▶ TECHNICIAN TIP

Some vehicles use the audio system to cancel noise inside the passenger compartment in order to combat engine noise. This is done in two ways. The first is to use the entertainment system to actively cancel engine and road noise. The second way is to use a speed-sensitive entertainment system that adjusts the volume of the entertainment system according to vehicle speed: low volume at low speed and higher volume at higher speed. This helps overcome objectionable noise.

Resonator

Some manufacturers use a resonator in the exhaust system. It is located between the muffler and the exhaust outlet, and its function is to reduce any resonance levels that the muffler cannot adequately suppress (**FIGURE 65-43**). Some manufacturers may use more than one resonator to quiet the combustion noises even further. Some resonators work on the Helmholtz principle—namely, that air blown across a tube connected to a rounded container creates a vibration, and noise is produced, similar to air blown across a bottle. This principle is then used to cancel noise by bouncing waves off one another, reducing their intensity, which lowers the exhaust noise.

Tailpipe

The tailpipe takes the exhaust gases away from the vehicle. Its exit point must not allow any of the exhaust gases to enter the vehicle. The tailpipe is connected to the muffler or resonator and is held by flexible exhaust mounts. It is made of a noncorrosive and rust-resistant material to ensure long life. It also may include an exhaust tip for appearance purposes. And some exhaust tips include built-in resonators (**FIGURE 65-44**).

Exhaust Gaskets

Exhaust components are sometimes bolted together. If they are, they may need to have a gasket to seal them. Exhaust gaskets can be found between the engine cylinder head and the exhaust

Applied Science

AS-45: Pitch/Frequency: The technician can explain the relationship of pitch to frequency.

Heinrich Hertz, a German physicist, was the first to prove the existence of electromagnetic waves. He did so by engineering a radio wave transmitter. In honor of his work, the unit for measuring frequency, hertz, is named after him. Because it is named after a person, its abbreviation (Hz) begins with a capital letter.

Frequency is the number of sound waves produced in a given time. The frequency of a sound wave determines its pitch, described in terms of how "high" or "low" the sound is. Humans can perceive sound waves ranging from 20 Hz (lowest pitch) to 20,000 Hz (highest pitch).

The concept of sound is based on the principle that a sound wave begins with a vibrating object. The frequency of a sound wave, or Hz, can be expressed as 1 Hz = 1 vibration per second.

AS-46: Resonance: The technician can demonstrate an understanding of what happens when an object resonates.

In physics, resonance refers to the amplitude at which an object vibrates at a given frequency. There are a number of different types of resonance, including mechanical resonance. When a large group of soldiers are marching across a bridge, they are told to break step in order to avoid damaging the bridge's structure. If the soldiers' footsteps were to fall simultaneously, extreme vibrations could result, which the bridge might not be able to withstand.

In automotive applications, resonance issues are often experienced in exhaust systems that produce a droning noise from vibration. Vehicle manufacturers in some cases add mass (weight) to the various exhaust components to change the resonance frequency.

FIGURE 65-44 Exhaust tips are typically used for appearance purposes, but some include resonator tips.

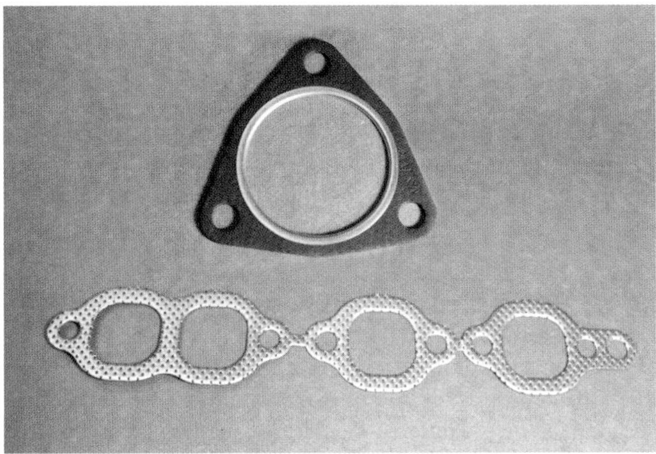

FIGURE 65-45 Exhaust gaskets come in many forms and are used in the extreme temperature of the exhaust system.

manifold, the engine pipe and the catalytic converter, and possibly the catalytic converter and the muffler. The exhaust is extremely hot, so the gaskets must withstand the temperatures without burning (**FIGURE 65-45**). Exhaust gaskets can

FIGURE 65-46 Donut gaskets are used between the exhaust manifold and down pipe on some vehicles.

be multilayered high-temperature alloys that are formed with graphite or mica to provide a good seal as exhaust components move with expansion and contraction. Donut gaskets are another type of exhaust gasket (**FIGURE 65-46**). They use spiral-wound steel and filler material, which is shaped into a round design, that, when installed in ball-shaped pipe, provides seal while allowing the pipes to be joined even if they are slightly out of line. Some of these gaskets also allow for slight movement of the engine.

Ceramic Coatings

High-performance developments are being continuously incorporated into on-road vehicles as efficiency gains are sought. One technology that is now becoming commonplace is the use of metallic ceramic coatings. The most common applications for ceramic coatings are on the exhaust system, intake manifolds, and exhaust headers. When ceramic thermal barrier coatings are applied to exhaust manifolds or headers, they provide two advantages. First, they protect the components from rust and corrosion, and second, they reduce heat loss, which can be translated into higher engine output. If the exhaust headers are internally coated, the hot exhaust gases travel at a higher velocity with less turbulence due to the smoother surface inside the system.

Internally, applying ceramic coating to the cylinder head's combustion chamber and exhaust ports has the effect of creating a thermal barrier from the hot gases to the cylinder head itself. This allows more of the thermal energy to be converted to mechanical energy, increasing an engine's efficiency. It also protects the base metal from excess heat. Listed here are the primary benefits of using metallic ceramic coatings:

- A metallic ceramic coating, when applied to metal surfaces, protects against rust and corrosion and extends component life.
- As a thermal barrier, a ceramic coating enhances engine performance and significantly reduces engine compartment and component temperature.
- A ceramic coating will not chip, crack, or peel, and it can survive bending and thermal shock.
- A ceramic coating is highly resistant to corrosives.
- A ceramic coating can survive base metal temperatures up to 2000°F (1100°C).
- A ceramic coating can come in a range of surface finishes and colors to suit a particular application.
- A ceramic coating is easily cleaned.

▶ Inspection and Testing

Inspecting the Throttle Body and the Induction System

N65001, S65001

The throttle body and the induction system control the airflow into the cylinders. The throttle body controls the amount of air passing entering the induction system, which allows for the control of engine rpm and power. If air leaks are present in the induction system downstream of the MAF sensor, then the sensor will not accurately measure the air coming into the engine. This additional air is sometimes referred to as "false air." As a result, the PCM will not deliver enough fuel, based on the MAF signal, and the engine will tend to run lean.

There are several ways to test for air leaks, including using a substitute fuel to artificially enrich the mixture at the point of the leak, using a smoke machine to fill the intake system and show any external leaks, and using an electronic stethoscope (or piece of heater hose) to listen for and pinpoint the source of the leak.

The use of a substitute fuel such as acetylene, propane, or carburetor cleaner is a quick and accurate method, but you risk catching the engine on fire. The technician carefully directs a small amount of fuel toward any suspected leaking components (**FIGURE 65-47**). If the fuel is passed over an intake leak, the vacuum will pull the gas into the intake manifold and richen the mixture up. This will be noted by smoothing out the engine, or the rpm will increase. If the engine has an idle control system, it may prevent the rpm from increasing, except when you pull the fuel away from the leak. You may need to disable the idle control system or hold the throttle open just enough that it doesn't control the idle. You may also be able to use a scan tool to watch the oxygen sensor, injector pulse width, and short-term fuel trim reading while applying and removing the acetylene or propane.

FIGURE 65-47 To find a leaking intake manifold gasket, a technician can use acetylene, propane, or carburetor cleaner to enrich the mixture.

Leaks can also happen in the ductwork, hoses, and vacuum diaphragms that are connected to the intake manifold, creating a false air condition. Although the acetylene and propane may still be able to find the leak, a smoke machine is invaluable. The smoke machine pushes smoke at low pressure into the intake manifold. The smoke machine can then be used to fill the entire intake system with smoke. If there are any leaks, the smoke will leak from inside the system to the outside. So anywhere the smoke is leaking from is an air leak. A high-intensity light can be used to make the smoke more visible.

SAFETY TIP

Be careful using carburetor cleaner, or any flammable liquid, on a hot engine. Flammable chemicals can start a fire on a hot engine.

To inspect the throttle body, air induction system, and intake manifold for leaks, follow the steps in **SKILL DRILL 65-1**.

Checking Idle Control Operation

N65002

Idle speed of the engine is controlled by the PCM through an idle air control motor, idle air control valve, or an electronically controlled throttle. The PCM is programmed to control idle speed based on the different operating conditions sensed by the input sensors. Common inputs include rpm, the engine coolant temperature, the power steering pressure switch, and the air-conditioning switch. For example, when the PCM sees that the air conditioner is on, the idle control is moved to allow more air to bypass the throttle plate, or the throttle plate is opened more and the rpm will increase slightly or remain at its present rate.

The engine speed can surge if engine sensor inputs are out of range or if a vacuum leak is present. Be sure to follow the

SKILL DRILL 65-1 Inspecting the Throttle Body, Air Induction System, and Intake Manifold

1. Start the engine and let the idle stabilize. Using an acetylene- or propane-testing tool, place the hose near the suspected leak area.

2. Open the fuel valve to slowly release the acetylene or propane.

3. On vehicles without an idle control system, observe the rpm and smoothness of the engine. If the engine speed increases or smoothes out, then a vacuum leak is present. Determine any necessary repairs.

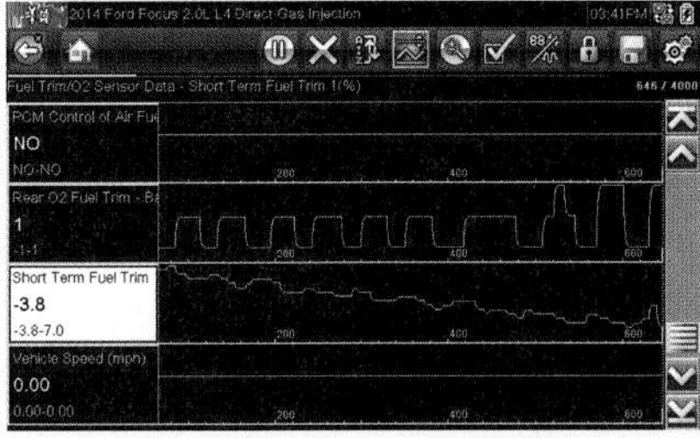

4. On vehicles with an idle control system, connect a scan tool and select data stream. Observe the oxygen, short-term fuel trims, and injector pulse width as you are moving the acetylene or propane around.

5. Shut off the engine and connect a smoke machine to the intake manifold. Start the smoke machine and inject smoke into the intake manifold.

6. Using a bright light, look around the engine compartment for traces of smoke. Any smoke coming from the air cleaner inlet is normal. Determine any necessary action(s).

manufacturer's diagnostic procedures for the particular vehicle you are working on for an idle system concern. The idle control system can also result in an unwanted low or high idle speed if the motor is worn or sticky.

The scan tool allows the technician to see what is happening with the idle control system. The scan tool typically displays the desired idle speed along with the actual idle speed. These two numbers should be very close to each other, as the engine computer is trying to keep the idle to the desired speed.

In vehicles with stepper motor idle control, the scan tool also displays the stepper motor counts from 0 at fully closed, to 140 at fully open. Starting the engine on a vehicle without a vacuum leak should result in the stepper motor count of around 20. If a leak is present, the idle count will be lower, and if the leak is bad enough, it can read 0. If the idle air control valve is gummed up, the count will be higher than 20, depending on how bad it is. When the air conditioner is then turned on, the idle count should step up, and the rpm should increase slightly or remain steady. Most scan tools allow output control of the idle air control, which means the technician can command the idle air control system to increase or decrease the rpm. Doing so allows the technician to know whether the system is capable of controlling idle airflow or not.

Be sure to follow the manufacturer's test information to diagnose the vehicle you are testing.

To test the idle air control operation, follow the steps in **SKILL DRILL 65-2**.

Inspecting the Exhaust System

N65003, N65004

Inspection of the exhaust system is a necessary task as an automotive technician. Exhaust inspection is part of any mandatory state inspection process, as a leak will result in possible injury or death to the passengers due to carbon monoxide poisoning. Special care should be taken to ensure that the exhaust system is not leaking. Exhaust leaks leave a black carbon marking. Also, most municipalities have noise ordinances limiting the amount of noise an exhaust can emit.

A quick test for exhaust leaks is to use your shoe and a rag to block off the exhaust flow coming from the tailpipe; if the exhaust system is not leaking, pressure will build and either leak past the rag or cause the engine to die. If there are leaks in the exhaust, then hissing will be heard from the exhaust system through the leaks. It may be necessary to have an assistant perform this test (block off the tailpipe with a rag) while

SKILL DRILL 65-2 Testing the Idle Air Control Operation

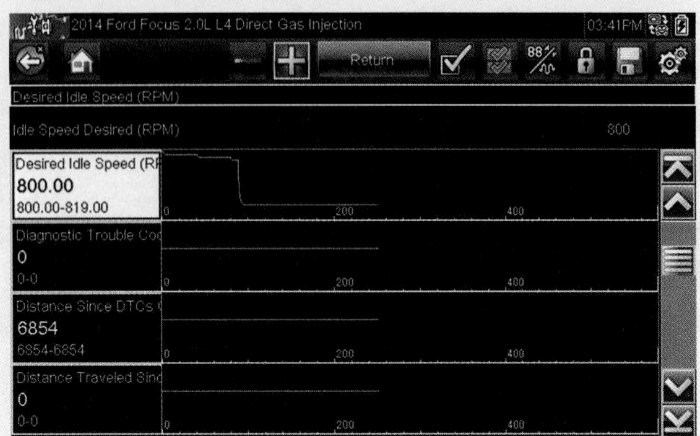

1. Connect a scan tool that is compatible with the vehicle you are working on, and start the engine. Compare the desired idle speed to the actual idle speed. The speeds should be very close to each other.

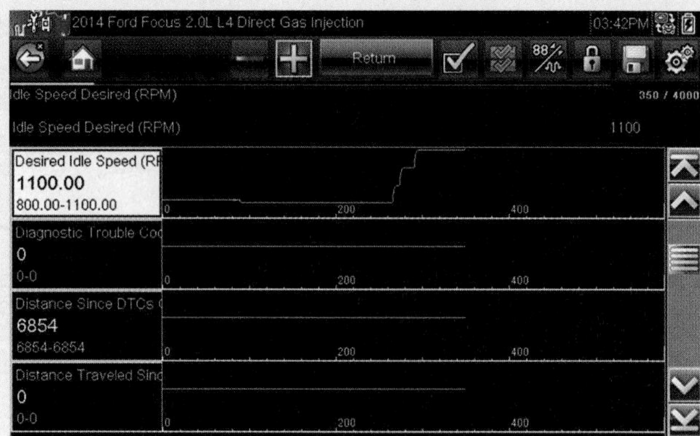

2. Locate the output commands function in the scan tool. Command a step-up of the idle speed and verify that it does so smoothly. Complete the same process for stepping down the idle speed. If the system is not operating correctly, follow the steps in the service information to diagnose the fault. Determine necessary actions.

the technician listens for exhaust leaks. A small hose can help pinpoint small leaks while this test is being performed.

Another testing option is to use a smoke machine with an exhaust pipe adapter to fill the exhaust system with smoke. With the engine off, any signs of smoke will pinpoint leaks in the exhaust system. Remember you may see smoke coming from the air cleaner assembly due to the valves in one or more cylinders being on overlap. This is considered normal and should not be confused as a leak from the exhaust.

The exhaust pipes can rust out over time. The water and corrosive chemicals in the exhaust stream corrode the pipe. A good way to test the integrity of the pipe is to use a large pair of adjustable pliers to squeeze the pipe. If it is springy or crushes easily, the pipe has to be replaced. Make sure you check any low spots in the system along with bends in the pipe, as these are typically the weakest spots.

To inspect the exhaust system for leaks, follow the steps in **SKILL DRILL 65-3**.

Performing an Exhaust Back-Pressure Test

N65005

An exhaust system back-pressure test is used to determine if there is a restriction that will create low power. A technician can use one of two methods to test for back pressure. The most accurate method is to use a pressure gauge installed into the exhaust stream to see how much exhaust pressure is backing up in the system. This can be done by removing an oxygen sensor near the exhaust manifold and screwing the back-pressure gauge into the hole. Another possible test point is the EGR passageway. After installing the back-pressure gauge, the engine is started and accelerated

SKILL DRILL 65-3 Inspecting the Exhaust System for Leaks

1. Safely lift and secure the vehicle on a hoist. Inspect for leaks or rust holes in the exhaust system, including the exhaust manifolds. Inspect brackets, hangers, and clamps.

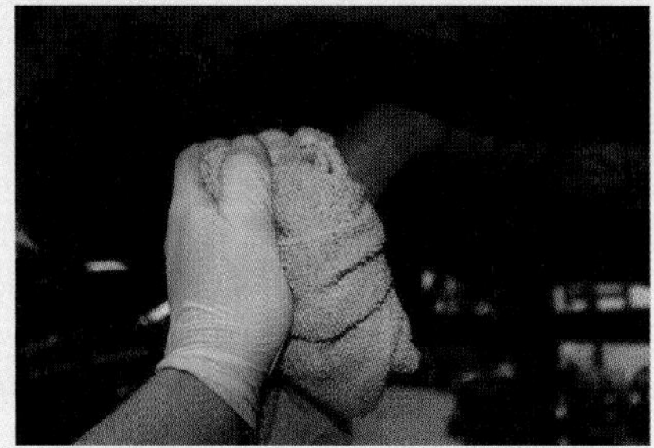

2. Have a helper hold a rag against the exhaust pipe(s) while the engine is idling to increase the pressure in the system. Listen and feel for any leaks.

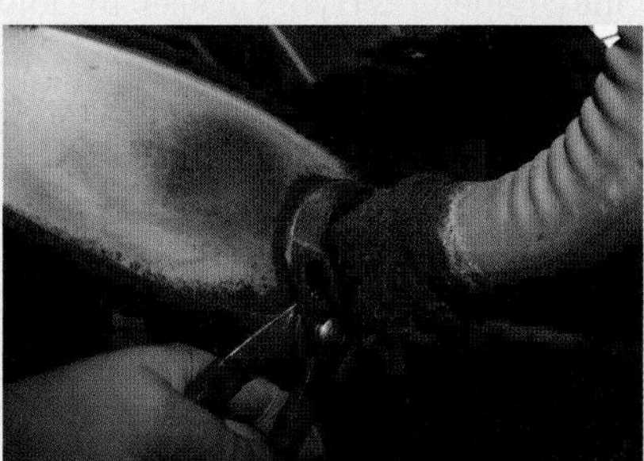

3. Use large adjustable pliers to test the integrity of the pipes by moderately squeezing the pipes. Determine necessary repairs.

to about 2000 rpm, and the pressure is compared to specifications. Refer to the service information for the maximum amount of acceptable pressure in the exhaust system.

The other method of testing the exhaust system for back pressure is to measure intake manifold vacuum (or manifold absolute pressure). You can use an external vacuum gauge or a scan tool to display manifold absolute pressure. Either connect the vacuum gauge to the intake manifold vacuum or watch the scan tool, and raise the engine rpm to 2000 rpm for two minutes. If the vacuum falls as the engine is held at 2000 rpm, then suspect a restricted exhaust system. This is because exhaust will begin to back up at the restriction in the exhaust system, all the way into the combustion chamber. This will make it so the intake stroke will not be able to pull in as much air, since the exhaust cannot fully escape. This excessive back pressure will result in an engine that has low power or may even stall.

The technician may need to remove parts of the exhaust system to locate the restriction. Typically, the technician will disconnect the connection at the front of the catalytic converter and check to see if it is clogged or restricted. If so, the root cause of the converter failure will have to be diagnosed, such as an overly rich-running engine or a cylinder misfire, which can cause it to overheat and melt down. If the converter is okay, then the muffler will have to be inspected.

To perform an exhaust back-pressure test with a pressure gauge, follow the steps in **SKILL DRILL 65-4**.

Performing a Turbocharger/Supercharger System Operation Test

N65006

Performing a turbocharger/supercharger test is a task that should be performed only by certified technicians or under very close oversight of the instructor. Testing the boost pressures that the system can produce requires operating the vehicle at wide-open throttle for a short time, making this an extremely risky test. Refer to the manufacturer's service information for specific turbocharger/supercharger diagnostic testing process for the vehicle you are working on.

The processes are similar when testing either the turbocharger or the supercharger for proper function. You need to be able to measure manifold pressure. This can be accomplished with the scan tool by watching the manifold absolute pressure reading. If this vehicle doesn't have that capability, or if the MAP sensor is suspect, a vacuum gauge that can measure low pressures (use a gauge that will read about 20% more than specifications) is needed.

Both the supercharger and the turbocharger will put out a maximum pressure, but that pressure can be reached safely only while the engine is being loaded down by driving on the road, or better yet a chassis dynamometer. Connect a scan tool to the DLC, and set it up to display MAP sensor pressure. If necessary, connect the vacuum gauge to the intake vacuum port, and route it out the hood and through the passenger door window, or tape it to the outside of the windshield away from the windshield wipers, facing in (**FIGURE 65-48**). Drive the vehicle in a safe place

FIGURE 65-48 Connect a vacuum/pressure gauge to the engine intake manifold and route it under the hood and through the passenger window, or tape it securely to the windshield.

SKILL DRILL 65-4 Testing the Exhaust System for Back Pressure

1. Remove the oxygen sensor. Install an adapter to the oxygen sensor hole, and connect a low-pressure gauge.

2. Start the engine, and observe any exhaust back pressure, which should be no more than 1.5 psi (10.3 kPa) at idle and no more than 3 psi (20.7 kPa) at 2000 rpm. Increase the engine rpm to 2000 rpm, and observe the pressure. Determine any necessary actions.

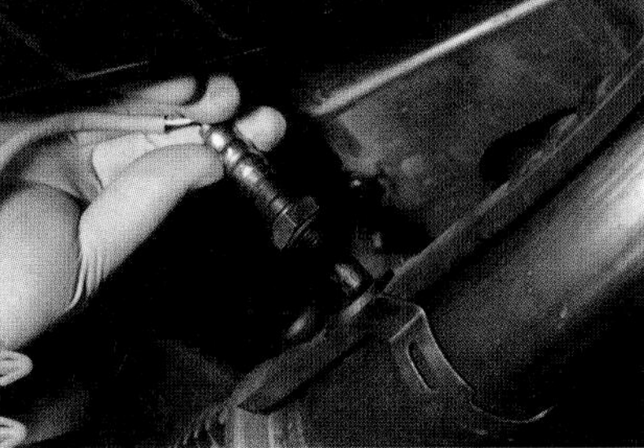

3. Inspect the oxygen sensor gasket, put anti-seize on the sensor threads, and torque to the proper specifications.

and manner. Observe the manifold pressure and compare it to specifications. If the specified pressure isn't reached, check that the wastegate on the turbocharger is not sticking open or that a pressure supply line is not disconnected. If the supercharger is not building pressure, check that the bypass valve is not sticking, a pressure line is not disconnected, or an electrical connection is not disconnected. If no faults are found, follow the service information procedure for further diagnostic steps.

To perform an operation test of a turbocharger/supercharger system, follow the steps in **SKILL DRILL 65-5**.

SKILL DRILL 65-5 Performing an Operation Test of a Turbocharger/Supercharger System

1. Connect a scan tool to the DLC, or a vacuum/pressure gauge to the engine intake manifold, and route it under the hood and through the passenger window, or tape it securely to the windshield.
2. Drive the vehicle in a safe place and in a safe manner.
3. Observe the manifold pressure and compare it to the manufacturer's specifications. Determine any necessary actions.

▶ Wrap-Up

Ready for Review

- The intake system ensures that clean air is delivered to the engine to be mixed with fuel.
- Clean air is essential to a long-lasting engine.
- The intake system ensures that air and fuel are delivered to the engine efficiently.
- The primary components of the intake system are the intake manifold, the throttle body, and the air induction system.
- There are two types of intake manifolds: wet and dry.
- In front of the throttle body are an air cleaner assembly, solid and flexible-duct tubing, and connectors.
- Ducting can be made of plastic and rubber connections for engine movement. Ducting connects the air cleaner to the intake manifold.
- The intake manifold provides a path for air (and fuel) to travel to the engine combustion chambers.
- Volumetric efficiency compares the volume of air entering a cylinder during intake to the internal swept volume of the cylinder when the piston is at bottom dead center.
- Back-pressure typically should never get higher than 3 psi (20.7 kPa). If it is greater, it can create poor performance issues, as it will cause exhaust to back up into the combustion chamber, minimizing the fresh airflow into the engine.
- Forced induction increases air pressure in the intake manifold above atmospheric pressure.
- A supercharger compresses the air intake to above atmospheric pressure, which increases the intake air density to the engine.
- A turbocharger is a forced induction system that uses wasted kinetic energy from the exhaust gases to increase the intake pressure.
- When using a turbocharger, exhaust gases turn the turbine, which in turn pulls in fresh air at a higher rate than normally aspirated engines.
- When air is compressed, it heats up; an intercooler is used to cool the air prior to being forced into the engine.
- The exhaust system involves many components that work together to remove the byproducts of combustion from the engine.
- The primary components of the exhaust system are the exhaust manifold, engine pipe (sometimes called the down pipe), catalytic converter, intermediate pipe, muffler, resonator, tailpipe, and exhaust brackets.
- The exhaust manifold provides a path for exhaust gases to escape the engine. In a high-performance engine, headers are tuned for better engine breathing.
- The flexible connector allows movement between the engine and the rest of the exhaust system, and keeps the pipes aligned under a load.
- A catalytic converter is used to convert unacceptable exhaust pollutants, such as carbon monoxide, certain hydrocarbons, and oxides of nitrogen, into less dangerous substances.
- There may be one or two heated oxygen sensors, one before and one after the catalytic converter.
- The rubber exhaust brackets support the system and allow movement and dampening action in the exhaust system.
- The purpose of a vehicle's muffler is to manage the sounds coming from the exhaust system.
- The baffles in the exhaust system help to muffle the noises that the engine makes during normal operation.
- The tailpipe prevents the harmful exhaust fumes from entering the passenger compartment.
- Gaskets are used throughout the exhaust system to make a tight, leak-free system.
- The throttle body and the induction system ensure that the engine gets airflow into the cylinders and must be inspected for proper operation.
- Refer to applicable manufacturer's information when inspecting the induction system or the exhaust system.
- Exhaust systems have to be inspected for leaks, rust, and restrictions.
- An exhaust back-pressure test will reveal whether there is a blockage in the exhaust system that will affect engine performance or drivability.

Key Terms

blow-off valve A valve that allows the release of excessive boost pressure from the turbocharger when the throttle plate is quickly closed.

bypass valve The pressure control valve of the supercharger. When the bypass valve opens, it lets air move around the supercharger compressor section, reducing the boost pressure.

compressor surge The backup of air against the throttle plate as it is closed. The turbocharger is still spinning, pressurizing air, when the throttle plate is closed. Air will stack up, creating a rapid slowing of the turbocharger compressor wheel. This can damage the compressor wheel.

forced induction The pressurization of airflow going into the cylinder through the use of a turbocharger or supercharger.

header A specially tuned exhaust manifold typically made of exhaust pipes. These pipes are usually made equal length to ensure equal flow between cylinders.

Helmholtz resonator A device that uses the principle of noise cancellation through the collision of sound waves. When necessary, the resonator is used in addition to the muffler to cancel additional sounds. This resonator may also be used on the induction system to muffle noise of airflow through the induction system. It is named after physicist Hermann Von Helmholtz.

mandrel forming Using a pipe bender with mandrels to allow for very tight bends without creating kinks or reducing the size of the pipe.

plenum chamber A large portion of the intake manifold after the throttle plate and before the intake runner tubes. The plenum provides a reservoir of air and helps prevent interference with the flow of air between individual branches.

scavenging The process of exhaust flow pulling fresh air into the cylinder as it creates a low-pressure area in the cylinder as it moves out of the cylinder.

supercharger A device that pressurizes airflow into the engine, working similarly to a turbocharger. The supercharger is driven by the crankshaft through a belt or gears and does require power from the engine.

turbocharger A device that pressurizes airflow into the engine. The turbocharger works similarly to a supercharger but is driven by the exhaust gases.

wastegate A pressure regulator device that allows control of the pressure produced by the turbocharger. The wastegate opens to allow exhaust gases to bypass the turbine wheel of the turbocharger.

Review Questions

1. The intake system does all of the following *except*:
   a. prevent carbon dioxide from entering the engine.
   b. ensure that clean, dry air is supplied to the engine.
   c. provide a sealed passageway to the combustion chambers.
   d. control the amount of air entering the engine.
2. Choose the correct statement with respect to the exhaust system.
   a. It stores the exhaust gases until the catalytic converter heats up.
   b. It blocks the air rushing into the engine.
   c. It helps reduce the harmful emissions.
   d. It is used to warm up the fuel tank.
3. Which of the following is one of the functions of an air cleaner?
   a. Preventing harmful gases from entering the passenger compartment
   b. Muffling the noise of the intake pulses
   c. Restricting the airflow to the engine at higher speeds
   d. Reducing the noise of combustion
4. In cross-flow heads:
   a. the design produces lesser power.
   b. the intake and exhaust manifolds are on the opposite sides of the engine.
   c. air and fuel are carried on the same branch.
   d. the intake and exhaust manifolds are beside each other.
5. Which of the following components in the exhaust system helps in the reduction of exhaust emissions?
   a. Manifold
   b. Catalytic converter
   c. Muffler
   d. Resonator
6. Which of these is attached to the exhaust manifold and connects to the catalytic converter?
   a. Resonator
   b. Heat shield
   c. Muffler
   d. Flex pipe
7. A catalytic converter converts all of the following gases into less harmful gases *except*:
   a. carbon monoxide.
   b. hydrocarbons.
   c. oxides of nitrogen.
   d. sulfur oxides.
8. To inspect an exhaust system, technicians do all of the following except:
   a. block off the exhaust pipe with the engine running and look for leaks.
   b. use large adjustable pliers to test the integrity of the pipes by moderately squeezing the pipes.
   c. spray the exhaust system with dye check and use a florescent light to locate leaks.
   d. inspect for leaks or rust holes in the exhaust system.
9. An exhaust system back-pressure test is used to:
   a. determine whether there is a restriction that will create low power.
   b. test the boost pressures that the system can produce.
   c. test the integrity of the exhaust pipes.
   d. test for exhaust leaks.
10. Air leaks in the induction system can be tested by all of the following methods *except*:
    a. using a substitute fuel to artificially enrich the mixture at the point of the leak.

b. using a smoke machine to fill the intake system and show any external leaks.
c. using an electronic stethoscope to listen for and pinpoint the source of the leak.
d. submerging the system in water and looking for air bubbles.

ASE Technician A/Technician B Style Questions

1. Tech A says that coolant circulates through some intake manifolds to help warm them up. Tech B says that some intake manifolds use an electric heater grid to warm up the intake air. Who is correct?
   a. Tech A
   b. Tech B
   c. Both A and B
   d. Neither A nor B
2. Tech A says that in some gas engines, the intake manifold runner length can be varied. Tech B says that some intake systems use resonators to quiet intake noise. Who is correct?
   a. Tech A
   b. Tech B
   c. Both A and B
   d. Neither A nor B
3. Tech A says that the fuel injectors are mounted in the throttle body on gas direct injection engines. Tech B says that air and fuel are mixed in the intake manifold on gas direct engines. Who is correct?
   a. Tech A
   b. Tech B
   c. Both A and B
   d. Neither A nor B
4. Tech A says that scavenging creates a low pressure in the cylinder by means of the exhaust gases flowing through the exhaust pipe. Tech B says that increased back pressure in the exhaust system increases the scavenging effect. Who is correct?
   a. Tech A
   b. Tech B
   c. Both A and B
   d. Neither A nor B
5. Tech A says that shorter intake manifolds produce higher torque at lower rpm. Tech B says that longer intake manifolds produce higher torque at high engine speeds. Who is correct?
   a. Tech A
   b. Tech B
   c. Both A and B
   d. Neither A nor B
6. Tech A says that air becomes heated as a supercharger compresses it. Tech B says that a turbocharger is designed to pull the exhaust gases out of the engine more effectively. Who is correct?
   a. Tech A
   b. Tech B
   c. Both A and B
   d. Neither A nor B
7. Tech A says that a wastegate is designed to direct waste gases to the intake manifold. Tech B says that a blow-off valve allows excess boost pressure to "blow off" into the atmosphere. Who is correct?
   a. Tech A
   b. Tech B
   c. Both A and B
   d. Neither A nor B
8. Tech A says that most engines with a turbocharger have turbo lag. Tech B says that a variable-geometry turbocharger will overcome most turbo lag. Who is correct?
   a. Tech A
   b. Tech B
   c. Both A and B
   d. Neither A nor B
9. Tech A says that a catalytic converter operates best at cold temperatures. Tech B says that the catalytic converter is typically located before the muffler. Who is correct?
   a. Tech A
   b. Tech B
   c. Both A and B
   d. Neither A nor B
10. Tech A says that small vacuum leaks can be found with a smoke machine. Tech B says that stepper motors are used to control idle speed in some vehicles. Who is correct?
    a. Tech A
    b. Tech B
    c. Both A and B
    d. Neither A nor B

CHAPTER 66
Emission Control Systems

NATEF Tasks

- **N66001** Diagnose emission and driveability concerns caused by catalytic converter system; determine needed action. (AST/MAST)
- **N66002** Diagnose emissions and driveability concerns caused by the evaporative emissions control (EVAP) system; determine needed action. (MAST)
- **N66003** Diagnose oil leaks, emissions, and driveability concerns caused by the positive crankcase ventilation (PCV) system; determine needed action. (AST/MAST)
- **N66004** Inspect, test, service, and/or replace positive crankcase ventilation (PCV) filter/breather, valve, tubes, orifices, and hoses; perform needed action. (MLR/AST/MAST)
- **N66005** Diagnose the cause of excessive oil consumption, coolant consumption, unusual exhaust color, odor, and sound; determine needed action. (AST/MAST)
- **N66006** Diagnose emissions and driveability concerns caused by the exhaust gas recirculation (EGR) system; inspect, test, service and/or replace electrical/electronic sensors, controls, wiring, tubing, exhaust passages, vacuum/pressure controls, filters, and hoses of exhaust gas recirculation (EGR) systems; determine needed action. (AST/MAST)
- **N66007** Diagnose emissions and driveability concerns caused by the secondary air injection system; inspect, test, repair, and/or replace electrical/electronically operated components and circuits of secondary air injection systems; determine needed action. (MAST)

Knowledge Objectives

After reading this chapter, you will be able to:

- **K66001** Explain the composition of air, regulated emissions, and their sources.
- **K66002** Explain each type of vehicle emissions.
- **K66003** Explain the various ways that emissions are controlled in an engine.
- **K66004** Explain the purpose and function of each emission control system.
- **K66005** Diagnose faults related to PCV system or fluid consumption.

Skills Objectives

There are no Skills Objectives in this chapter.

▶ Introduction

Emissions are the release of substances into the atmosphere. They can occur naturally from forest fires, volcanoes, and the decomposition of natural materials, and they can also be human-made, such as from vehicles and industrial sites (**FIGURE 66-1**). In recent automotive applications, most emissions are the by-product of combustion and are emitted from the exhaust system. Not all emissions from combustion are considered hazardous. Those emissions from the automobile that are hazardous can be managed by carefully designing the engine, accurately controlling the air-to-fuel ratio, or converting them to nonharmful or less harmful gases through the use of add-on emission control devices. Today's vehicle technology has reduced the harmful emissions a vehicle produces to almost zero. This chapter details the methods and control devices designed to either reduce those harmful emission gases or convert them into less harmful or nonharmful elements.

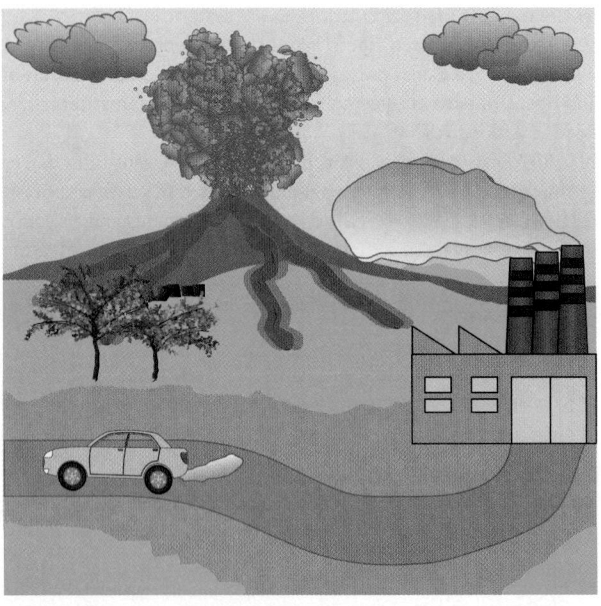

FIGURE 66-1 Emissions occur naturally or are human-made, including stationary and mobile sources.

▶ Composition of Air

K66001

The air we breathe is composed mainly of two gases: nitrogen and oxygen. Nitrogen by far makes up the largest percentage of air (78%). It is an inert gas, which means it does not react easily with other compounds. Oxygen (21%) is the second most abundant element in air. It is a highly reactive gas, which means it combines readily with almost all other elements. In many cases, oxygen reacts through the process of combustion, which is also called oxidation. These two gases account for approximately 99% of the content of air. That leaves 1% for all of the other elements found in air, including carbon dioxide and argon.

The level of oxygen in the atmosphere is critical to humans. The gases in our atmosphere provide a balanced environment for plant, animal, and human life. Humans breathe in oxygen and exhale carbon dioxide; trees and plants take in carbon dioxide and give back oxygen. This process is part of the oxygen cycle (**FIGURE 66-2**). There is also a nitrogen cycle, in which nitrogen from our atmosphere is converted to usable food for plants by bacteria in the ground. There is also a carbon cycle. Carbon fuels are burned and produce carbon dioxide in the air, which is then converted to carbon and oxygen by plants. The carbon is used to grow the plant while oxygen is given back to the air.

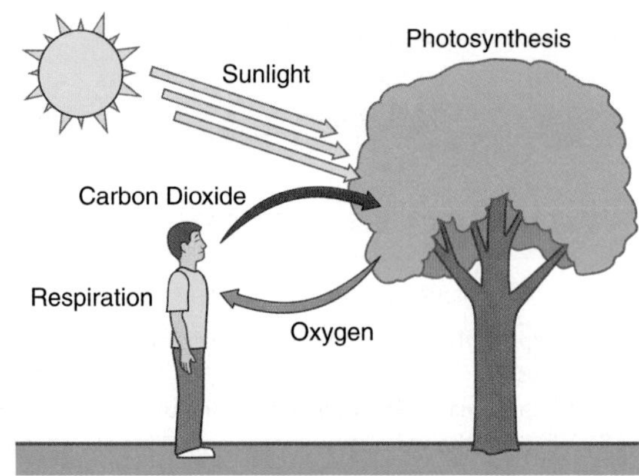

FIGURE 66-2 Humans breathe in oxygen and exhale carbon dioxide; trees and plants take in carbon dioxide and give back oxygen.

You Are the Automotive Technician

A customer brings her car into your shop with the malfunction indicator lamp (MIL) illuminated. You explain to the customer that she did the right thing by bringing it in because some issues can cause further damage if left unchecked. She authorizes you to perform the diagnosis. You first plug in the scan tool and pull the active diagnostic trouble codes (DTCs). There are several: P0133-O2 Sensor Circuit Slow Response B1S1; P0401-EGR Flow Insufficient Detected; P0420-Catalyst Efficiency below Threshold-Bank 1. You know that a slow-responding oxygen sensor can affect the catalytic converter, so you start by diagnosing that DTC and then diagnose the other two DTCs.

1. How can a slow-responding O_2 sensor affect the catalytic converter?
2. Under what conditions should the EGR system operate?
3. Describe two good ways of testing the catalytic converter.
4. How does an "EGR flow insufficient" condition affect emissions?

TABLE 66-1: Percentages of Gases Contained in the Atmosphere

Gas	Percentage
Nitrogen (N_2)	78.08
Oxygen (O_2)	20.95
Water (H_2O)	0 to 4, variable
Argon (Ar)	0.93
Carbon dioxide (CO_2)	0.0360, variable
Neon (Ne)	0.0018
Helium (He)	0.0005
Methane (CH_4)	0.00017, variable
Hydrogen (H_2)	0.00005
Oxides of nitrogen (NO_X)	0.00003, variable
Ozone (O_3)	0.000004

Although nitrogen and oxygen make up 99% of the air, the other 1% contains a variety of gases (**TABLE 66-1**). Some of these gases can be useful in different ways. As a technician, while welding, you may have to use carbon dioxide, argon, or helium to keep oxygen away from the weld. These gases are used because if oxygen is allowed to come in contact with the molten metal of the weld, oxidation will occur and the weld will be weak and could fail. Neon gas is drawn from the atmosphere. It produces light when exposed to electricity and is used as an accent light in some show cars.

▶ Sources of Emissions

The term "emission," when used in automotive circles, normally refers to the pollution produced by a light vehicle during operation or while sitting stationary. All fuels, with the exception of pure hydrogen, produce pollution when stored or burned. **Emission control systems** are designed to limit the pollution caused by the storing and burning of various fuels. Emissions from current gasoline-fueled motor vehicles usually come from three main sources: the fuel system, the crankcase, and the tailpipe (**FIGURE 66-3**).

FIGURE 66-3 Vehicle emissions typically come from three main sources: the fuel system, the crankcase, and the tailpipe.

Liquid fuel stored in the fuel tank tends to evaporate into a gas. If not controlled, those vapors can escape into the atmosphere and are called evaporative emissions. When gasoline or other hydrocarbon fuel is burned in the combustion chamber along with the oxygen and nitrogen from the air, emissions are created. Some of the emissions are pollutants and are released from the exhaust system in the form of gases; these emissions are called tailpipe emissions. Some of the combustion gases leak past the compression rings and cylinder walls and find their way into the crankcase. These blowby gases create pressure in the crankcase and, if not controlled, leak into the atmosphere and are called crankcase emissions.

In a **compression-ignition engine** (diesel engine), emissions are created in the combustion chamber the same way as they are in spark-ignition engines, and escape to the atmosphere through the exhaust and the crankcase breather. Diesel fuel does not evaporate as easily as gasoline, so evaporative emissions are not a major concern on vehicles powered by diesel fuel.

▶ Regulated Emissions

Manufacturers are required by law and the Environmental Protection Agency (EPA) to control the emissions produced by their vehicles. The EPA requires every manufactured powertrain to pass a federal emission test before it can be sold in the United States. This test is called the Federal Test Procedure (FTP) and is designed to determine the emission output of that particular vehicle as classified by year, make, model, and drivetrain. The test uses a dynamometer that duplicates the normal driving pattern of the typical commuter, known as the FTP drive cycle (**FIGURE 66-4**). As the vehicle is operated according to the FTP drive cycle, the emission output is monitored in grams per mile (g/mi). For the vehicle to pass, the emissions must stay under the limits for that type of vehicle. Other countries have other agencies that regulate emissions, but any vehicle entering the United States must pass the U.S. FTP test.

Not only must the vehicles pass the test before they can be sold, they must be capable of detecting an emission failure over their life, as they are being operated. As of 1996, all vehicles must meet the onboard diagnostics II (OBD II) standard. OBD II is an automotive self-diagnostic system mandated by the EPA that requires a warning light to alert the driver of an emission system fault. The EPA requires that when a vehicle's emissions deviate from the FTP limit by 1.5 times, the warning light (malfunction indicator lamp [MIL]) must turn on, and one or more specific diagnostic trouble codes (DTCs) must be set in memory.

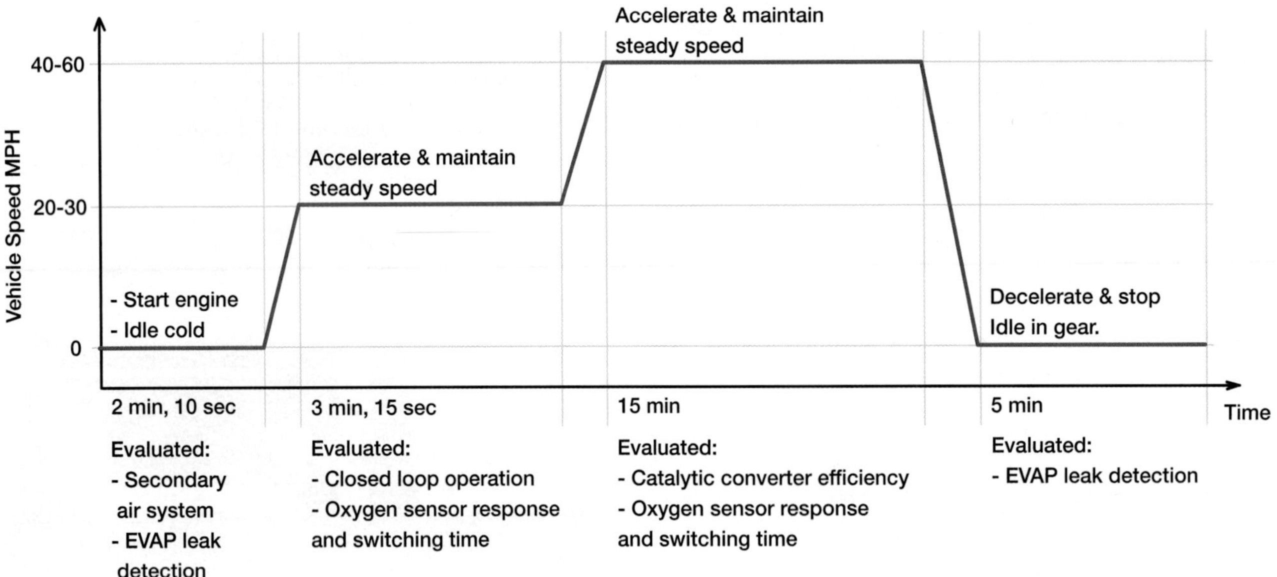

FIGURE 66-4 FTP drive cycle is used to certify the emission output of vehicles and ensure they meet regulatory standards.

If the vehicle begins to run rich—for example, due to a leaking fuel injector—and the emission output exceeds 1.5 times the FTP limit, the engine management system must be able to detect the rich condition. The powertrain control module (PCM) will turn the MIL on to warn the driver that the vehicle has a fault and requires service. The OBD II system was developed purely with the intent of keeping vehicles emission-compliant and warning drivers if they are not. At the same time, it provides a lot of good data and functionality when diagnosing drivability concerns.

Regulated air pollutants can be divided into two groups: gases and particulates. The primary pollutant from vehicles is gases, which include hydrocarbons, carbon monoxide, oxides of nitrogen, and sulfur dioxide. These pollutants can have a damaging effect on people, the atmosphere, and the natural environment. We will explore them in a minute. Particulates, often referred to as particulate matter, are tiny particles of solid or liquid suspended in the air. Particulate matter is graded in a size range from 10 nanometers to 100 micrometers in diameter. Particulates of less than 10 micrometers are dangerous to humans because they can easily become airborne and reach deep into the lungs. Smaller particles also tend to stay airborne longer than larger particles, which settle more quickly.

Many countries have recognized the global effects that the continual release of pollutants into the earth's atmosphere can have. These countries have developed emissions laws or follow an international protocol on emission control standards. Vehicle manufacturers are required to comply with these laws to ensure that the emissions from the vehicles they produce meet the emission control standards.

Manufacturers design emission control systems for their vehicles, which monitor and control emissions so that they do not exceed the emission control limits. A **limit value** is the maximum amount of emissions that a vehicle is permitted to emit. Limit values are assigned to different classifications of vehicles. Passenger and light vehicle limit values are normally given in grams per mile in the United States and grams per kilometer in Europe. These regulations have become progressively more stringent as government regulations become tougher in an effort to further reduce vehicle emissions.

In the quest to meet the tougher emission control standards, many manufacturers are turning to hybrid technologies, which couple two power sources together (electric and ICE). Hybrid vehicles have reduced emission output and meet tougher fuel economy standards. Alternative fuels, such as compressed natural gas (CNG) and propane, are also being used more frequently in the attempt to further reduce pollutions. Ever-stricter emission control is at the forefront of vehicle design and construction.

▶ TECHNICIAN TIP

Historically, the regulated emissions have been hydrocarbons, carbon monoxide, oxides of nitrogen, and particulate matter. In recent years, carbon dioxide has been added to the list of emissions that are being regulated. Not all scientists agree with this regulation because carbon dioxide is necessary for plant life and is given off in human respiration.

▶ Types of Emissions

K66002

Internal combustion engines produce several emission gases as part of the conversion process of turning fuel (chemical energy) into mechanical energy. There are two categories of emission gases: nonharmful and harmful. The nonharmful emissions are water (H_2O), nitrogen (N_2), and oxygen (O_2). The harmful emissions are carbon monoxide (CO), hydrocarbons (HC), carbon dioxide (CO_2), oxides of nitrogen (NOx), sulfur dioxide (SO_2), and particulates. Each of these pollutants will be explained in the following sections.

Water

Water is a natural by-product of complete combustion and is not harmful to the environment or our air quality. Water is formed by the engine when two hydrogen molecules from the fuel are drawn into the combustion chamber and are combined with one oxygen molecule (**FIGURE 66-5**). Water is emitted as a result of complete combustion of hydrogen and oxygen. We could say that water is a by-product of completely burned hydrogen fuel. In fact, burning 1 gallon of gasoline in an engine creates about 1 gallon of water. Thus, water is formed in all internal combustion engines and is discharged along with all normal combustion gases through the exhaust system.

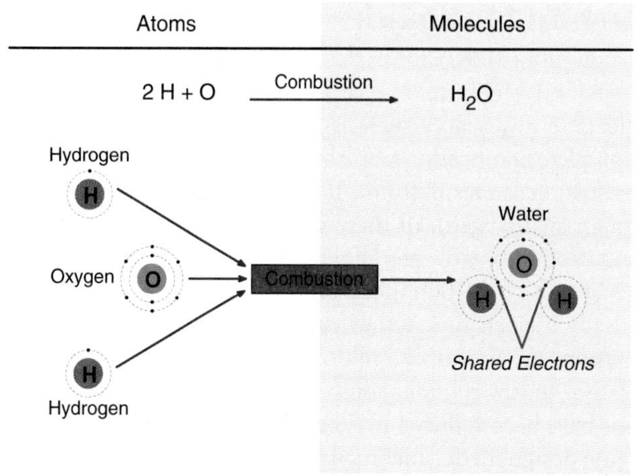

FIGURE 66-5 Water is formed by the engine when two hydrogen molecules from the fuel are drawn into the combustion chamber and are combined with one oxygen molecule.

▶ TECHNICIAN TIP

In a gasoline engine, about 1 gallon of water is created for every 1 gallon of gasoline burned. This water usually leaves the exhaust system as superheated steam because it is produced under very high combustion temperatures. If the exhaust pipe is cool because the engine was not running very long, the steam can condense into water and drip out of the tailpipe. Water also mixes with any sulfur dioxide in the exhaust and creates sulfuric acid. This water and sulfuric acid can sit in low spots in the exhaust system and rust or corrode any steel and iron components. Many manufacturers use exhaust system pipes and mufflers made of stainless steel, which resists rust and corrosion.

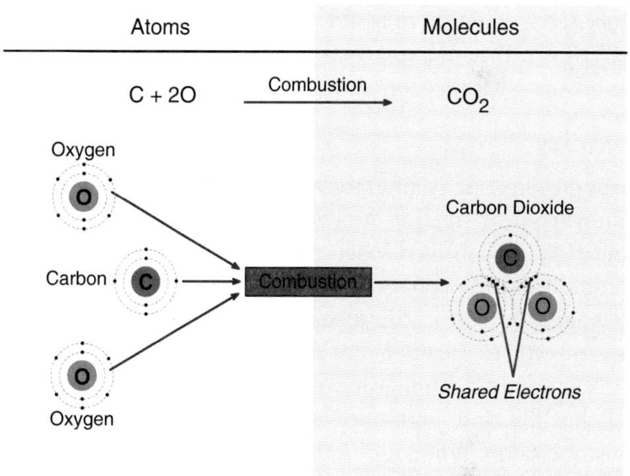

FIGURE 66-6 Complete combustion of air and fuel occurs when one carbon molecule combines with two oxygen molecules to form one carbon dioxide molecule.

Carbon Dioxide

Carbon dioxide is a by-product of complete combustion, along with water. When complete combustion of air and fuel occurs, one carbon molecule combines with two oxygen molecules to form one carbon dioxide molecule (**FIGURE 66-6**). So we could say that carbon dioxide is the result of completely burned carbon fuel. Carbon dioxide is not normally considered a poisonous gas; in fact, humans exhale it as our bodies use the oxygen we breathe in, and plants use the carbon from carbon dioxide for growth while releasing the oxygen back into the atmosphere. Recently, the EPA has labeled carbon dioxide as a pollutant. Because motor vehicles are a major source of carbon dioxide, governments are implementing regulations with the goal of reducing the total amount of carbon dioxide released into the atmosphere.

Carbon dioxide is also formed inside catalytic converters. The catalytic converter completes the combustion process by adding another oxygen molecule to the carbon monoxide (CO) to convert it to carbon dioxide (CO_2). In addition, carbon dioxide is produced by diesel, liquid petroleum gas (LPG), propane, CNG, or any other carbon-based fuel used in a vehicle with an internal combustion engine.

Hydrocarbons

Gasoline, diesel, LPG, and natural gas are all hydrocarbon compounds. Hydrocarbons are created when raw fuel evaporates into the atmosphere or when it does not burn at all in the combustion chamber and is exhausted out the tailpipe. In other words, hydrocarbon emissions are unburned fuel that escapes to the atmosphere.

Gasoline must evaporate easily so that it can mix with air to burn properly in the combustion chamber. But this property also means it evaporates easily into the atmosphere at ordinary temperatures and pressures. For example, when a vehicle is being refueled, hydrocarbon vapors can escape from the filler neck into the atmosphere. When the vehicle is left in the sun, the temperature of the fuel increases, and fuel can evaporate from the tank.

▶ TECHNICIAN TIP

Hydrocarbon emissions react with oxides of nitrogen (NOx) in the presence of sunlight to produce **photochemical smog**, a brown haze that hangs in the sky, typically seen over large cities. Smog is a major health issue to humans because it affects lung tissue. Hydrocarbons are a major component of photochemical smog.

Carbon Monoxide

Carbon monoxide (CO) is a colorless, odorless, tasteless, flammable, and highly toxic gas. It is also heavier than air, so it collects in low spaces such as lube pits. It is a product of the incomplete combustion of carbon and oxygen. The incomplete combustion can result from either too much fuel (carbon) or not enough air (oxygen). In these conditions, the carbon can only bond itself to one oxygen molecule, creating carbon monoxide (**FIGURE 66-7**). Carbon monoxide is formed by rich mixtures or partially burned fuel. Other causes of high carbon monoxide emissions are improper ignition timing, low combustion temperature, or low cylinder compression. Carbon monoxide emissions have been reduced in modern vehicles as a result of better engine design, better fuel management, and the use of catalytic converters in the exhaust system.

Most engine fuels, except pure hydrogen, use carbon-based fuel. Unlike the ideal combustion process, a real-world internal combustion engine operating on a carbon-based fuel tends to produce some carbon monoxide emissions. This is especially true when the combustion temperature is low, such as during a cold start, when there is not enough oxygen present to completely burn the fuel, or when there is insufficient time in the combustion chamber for complete combustion.

Toxicity

Carbon monoxide is an extremely poisonous gas. Because you cannot see it or smell it, it is very dangerous. Inhaling carbon monoxide in a confined space can be lethal. Because it is known to come from the exhaust system, it is important not to allow any engine to run inside a shop without venting the exhaust gases directly to the open air outside.

Victims of **carbon monoxide poisoning** can sometimes look healthy and pink-cheeked. This is because concentrations of carbon monoxide in the bloodstream give the blood a brighter red color than normal. Carbon monoxide binds much more easily than oxygen with hemoglobin, which is the blood component that carries life-giving oxygen around the body. As the hemoglobin becomes saturated with carbon monoxide, it becomes unable to carry oxygen. When the brain is starved of oxygen, the victim becomes unconscious, resulting in brain damage or death. Early signs of carbon monoxide poisoning include headaches, nausea, weakness, and irritability.

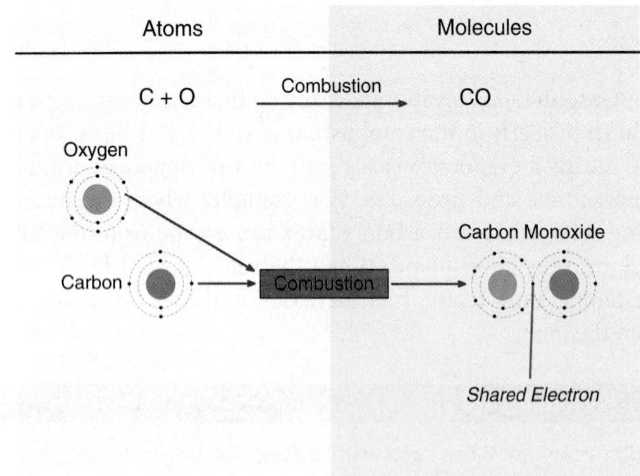

FIGURE 66-7 The incomplete combustion that creates carbon monoxide can result from either too much fuel (carbon) or not enough air (oxygen), causing the carbon to be able to bond to only one oxygen molecule.

Applied Science

AS-24: Chemical Reactions: The technician can demonstrate an understanding of the chemical reactions that occur in the automotive engine that are related to the combustion of fuels and the operation of the catalytic converter.

An automotive teacher asks a student to use a hacksaw to cut apart a failed catalytic converter for evaluation purposes. When the converter is cut apart, it is evident that the converter had melted down. The students find six large pieces of material that resemble rocks. The ceramic catalyst was damaged because of a super-heated condition that resulted from unburned fuel entering the converter. This teaching aid provides an introduction to the catalytic converter and how it works and relates to the combustion of fuels in an automotive engine.

The stoichiometric ratio is the ratio of air to fuel at which all of the oxygen in the air and all of the fuel are completely burned. This ratio is 14.7 parts of air to 1 part of fuel. Maintaining close control of the air-fuel ratio allows an engine to run cleaner than ever. In many cases, precise control of the air-fuel ratio has made the job of hydrocarbon (HC)– and carbon monoxide (CO)–related emission control devices much easier, as they are dealing with much smaller volumes of the pollutants. Catalytic converters use a base structure of ceramic material coated with precious metals to react with and break up or combine gases in the exhaust system. The catalytic converter cleans up the pollutants left over from combustion, and tailpipe emissions are reduced.

In the case of the failed converter that was cut apart for evaluation, the engine had a misfiring cylinder as a result of a faulty spark plug. The vehicle owner ignored the check engine light that was illuminated on the dash for about one month, resulting in the failure of the converter. The unburned fuel from the misfiring cylinder was ignited in the converter, causing overheating of the ceramic monolith.

Because carbon monoxide is absorbed so easily by red blood cells, the following agencies set limits for the permissible exposure limit (PEL) for the workplace:

- OSHA PEL is 50 parts per million (ppm) for an eight-hour shift.
- The National Institute for Occupational Safety and Health (NIOSH) PEL is 35 ppm for an eight-hour shift, with a ceiling of 200 ppm for any length of time.
- The American Conference of Governmental Industrial Hygienists (ACGIH) PEL is 25 ppm for an eight-hour shift and a 40-hour workweek.

Exposure levels of 400 ppm may be fatal in as little 30 minutes. As you can see, it doesn't take much carbon monoxide in the air to be hazardous. To further help you understand the smallness of the PELs, consider that 1% carbon monoxide equals 10,000 ppm of carbon monoxide. Vehicles with an emission-related problem can run perfectly well and still easily have carbon monoxide tailpipe readings between 3% and 5%, which equates to 30,000–50,000 ppm—more than 1000 times the allowable PELs given in the preceding list.

SAFETY TIP

Be sure when working in a shop environment with a vehicle running that you provide adequate ventilation by using an exhaust extraction device (fan system and exhaust hoses designed to draw exhaust gases from the vehicle's exhaust pipe and push them outside of the shop). Small amounts of carbon monoxide can kill if proper ventilation is not provided.

First Aid

First aid for victims of carbon monoxide poisoning starts with recognizing the signs of exposure: shortness of breath, nausea, headache, dizziness, or light-headedness. Awareness will allow you to make the decision to remove the other person, or yourself before severe poisoning occurs. If the victim of carbon monoxide poisoning is unresponsive or demonstrates severe symptoms, call for qualified help, and then assess the environment to determine if you can safely remove the person from further exposure. Remember, CO is heavier than air, so it will collect in low spaces such as lube pits. If you are trained, you may need to perform CPR. If available, give first aid oxygen as well. Continue to treat the victim until medical help arrives.

Oxides of Nitrogen

Air that is drawn from the atmosphere into an engine contains almost 80% nitrogen, which is normally considered to be an inert gas, meaning it does not react easily. But under the high temperature and pressure of combustion, nitrogen may combine with oxygen to produce oxides of nitrogen (NOx) (**FIGURE 66-8**). Because nitrogen is inert, it is typically produced in large amounts only when combustion temperatures of around 2500°F (1400°C) and above occur. A lean mixture results in higher oxides of nitrogen because lean mixtures burn faster than rich mixtures, producing an excessive combustion temperature. The increase of available oxygen can oxidize the nitrogen at those high temperatures.

Oxides of nitrogen are major contributors to photochemical smog, along with hydrocarbons and sunlight. For example, the brown-looking cloud that often hangs over large cities is photochemical smog. Oxides of nitrogen irritate the eyes, nose, and throat. In extreme cases, such as regular exposure to high concentrations, coughing and lung damage can occur.

Sulfur Dioxide

Gasoline and diesel fuels contain sulfur as part of their chemical makeup. When sulfur burns in the combustion chamber, one sulfur molecule combines with two oxygen molecules, creating sulfur dioxide. Sulfur dioxide mixes with water vapor formed during the combustion process and produces sulfuric acid (**FIGURE 66-9**). This corrosive compound is emitted into the atmosphere through the exhaust. Sulfuric acid is a major environmental pollutant, coming back to earth in contaminated rainwater, called acid rain. This acid rain has been responsible for degrading vast areas of arable land. As a result, the reduction or removal of sulfur from motor fuels has become a major part of most countries' vehicle emission control programs.

High sulfur levels in fuel, when combined with water vapor, can also cause corrosive wear on valve guides and cylinder liners, which can lead to premature engine failure. Using proper lubricants and draining oil at the correct intervals help combat this effect and reduce the degree of corrosive damage. Many manufacturers have been using stainless steel in their exhaust pipes and mufflers to resist corrosion caused by sulfur in these components.

Sulfur reduces catalyst efficiency in modern vehicles, and vehicles operating with higher sulfur gasoline have higher emissions than vehicles operating on lower sulfur gasoline. There is evidence that in some instances sulfur in gasoline may also degrade the performance of oxygen sensors, which would then contribute to higher levels of all tailpipe emissions.

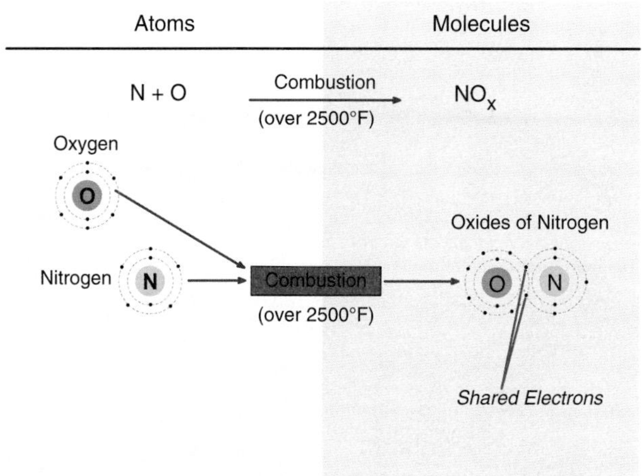

FIGURE 66-8 Under the high temperature and pressure of combustion, nitrogen may combine with oxygen to produce oxides of nitrogen.

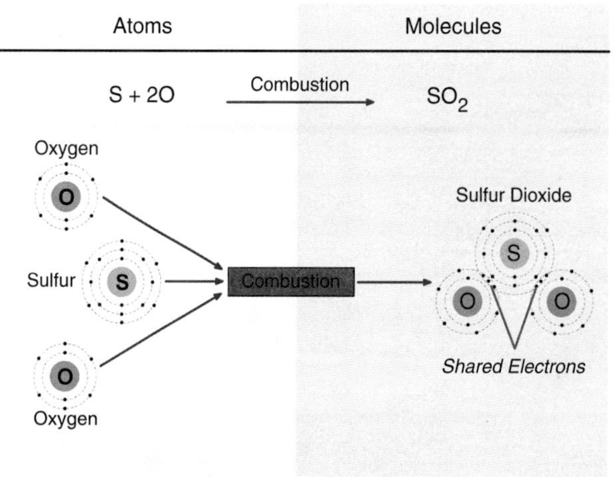

FIGURE 66-9 Sulfur dioxide mixes with water vapor formed during the combustion process and produces sulfuric acid.

Although regulations have reduced the permissible levels of sulfur in fuel, there are some side effects from using low-sulfur diesel fuel. The refining process used to reduce the sulfur level can reduce the natural lubricating properties of the diesel fuel, which is essential for the lubrication and operation of diesel fuel system components such as fuel pumps and injectors. These components either have to be manufactured with higher-quality materials, or special lubricants have to be used to provide the required level of protection.

To reduce the sulfur content, oil companies change the overall chemical composition of the fuel, which can affect fuel pump seals, engine seals, and O-rings, some of which react to changes in fuel composition by swelling or shrinking. This problem can be fixed by performing regular maintenance of the fuel system and by manufacturers replacing the seals with ones made of materials that are less susceptible to swelling and shrinking.

Particulates

Particulates are small particles of solid matter that are suspended in the air. Particulates from modern engines are usually small particles of carbon (**FIGURE 66-10**). In spark-ignition engines, particulates are caused by incomplete combustion of rich air-fuel mixtures. In this case, the hydrogen burns away from the hydrocarbon molecule, leaving the carbon by itself. In compression-ignition engines, particulates are caused by either too rich of a mixture or a poorly mixed air-fuel mixture. This produces very rich areas in the combustion chamber, even though the overall mixture could be very lean. Newer compression-ignition engines use a particulate filter to catch and hold the particulates in the exhaust stream. In some systems, the particulate filter can be regenerated periodically by burning off the carbon particles. The PCM controls the regeneration process automatically, or a technician can command regeneration with a scan tool connected to the PCM (**FIGURE 66-11**).

Smog

Smog is produced in the atmosphere when unburned hydrocarbons and oxides of nitrogen react chemically with sunlight. Smog appears reddish brown on the horizon and can be seen over major cities (**FIGURE 66-12**). Smog is also known as ground-level ozone. Ozone blocks harmful ultraviolet light from the sun in the upper atmosphere; however, in the lower atmosphere, it creates irritation of the lungs and is believed to contribute to asthma.

FIGURE 66-10 Particulates from modern engines are usually small particles of carbon.

▶ Controlling Emissions

K66003

Manufacturers are relatively free to determine how they will meet emission standards, as long as they do meet them. In designing systems and processes that effectively and cost efficiently meet the standards, manufacturers use a couple of main strategies:

- The first is to design the engine in such a way as to reduce emissions as much as possible, such as building turbulence-enhancing strategies and precisely controlling valve and spark timing.
- The second strategy is to closely control the air-fuel mixture under all driving conditions, thereby minimizing emission output.

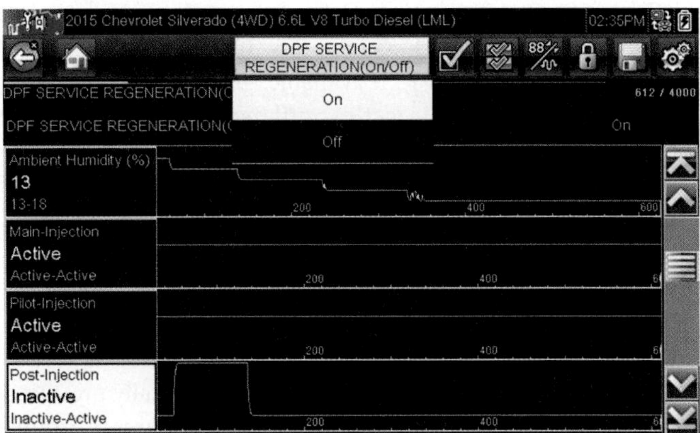

FIGURE 66-11 A technician using a scan tool to command regeneration of the particulate filter.

FIGURE 66-12 Smog over a city.

- The third strategy is to warm up the engine as quickly as possible, as well as operate it at higher temperatures than older engines.
- The final strategy is to integrate precombustion and post-combustion add-on devices that are designed to further reduce emission output.

We look next at each of these strategies in more depth.

▶ TECHNICIAN TIP

Manufacturers continue to produce cleaner and more efficient vehicles. Most vehicles use similar emission controls; however, you should always refer to the service information for a description and operation of the emission system before attempting to diagnose or repair it.

Start Date/Time: 01/05/2007 04:52 PM | Operator 1: AMANDAM AMANDA | Operator 2: LARRYWE LARRY

Gas Cap Test Results — Result: PASS | Cap 1: PASS | Cap 2: N/A | Cap 3: N/A | Cap 4: N/A

This emission test was performed in accordance with federal regulations on emission tests (40CFR 85 Subpart W) by Applus Technologies, Inc. for the Department of Eco

Test Results — Test Type: ASM | Prescribed Type: ASM

Overall ASM Result: PASS | Number of Restarts: 0

	HC (PPM)	CO (%)	CO+CO2 (%)	O2 (%)	RPM
Cruise Limit:	150	1	6	N/A	N/A
Cruise Emissions:	36	0.21	15.51	0.3	N/A
Cruise Result:	PASS	PASS	N/A	N/A	N/A
Idle Limit:	220	1.2	6	N/A	N/A
Idle Emissions:	24	0	15.3	0.5	850
Idle Result:	PASS	PASS	N/A	N/A	N/A

FIGURE 66-13 One indication that an engine is in good working order is a clean emission test.

Importance of Controlled Combustion

Because most emissions are by-products of combustion, accurately controlling combustion within the cylinders can minimize emission output. For fuel to burn efficiently, it has to be fully evaporated and completely mixed with the right proportion of air. This is pretty easy to do in the controlled environment of a laboratory, but doing so repeatedly at a high rate of speed in very short periods of time under a huge variety of conditions is much more difficult. These are the conditions we are dealing with inside automotive combustion chambers. Thus, maintaining low emission output is a challenge and requires all of the engine's systems and components to be working properly for combustion to occur efficiently. In fact, a good general test that will tell you the overall condition of the engine and its control systems is an emission test. If the vehicle passes the emission test with clean results, that is a good indication that all of the engine systems and the engine itself are operating well (**FIGURE 66-13**).

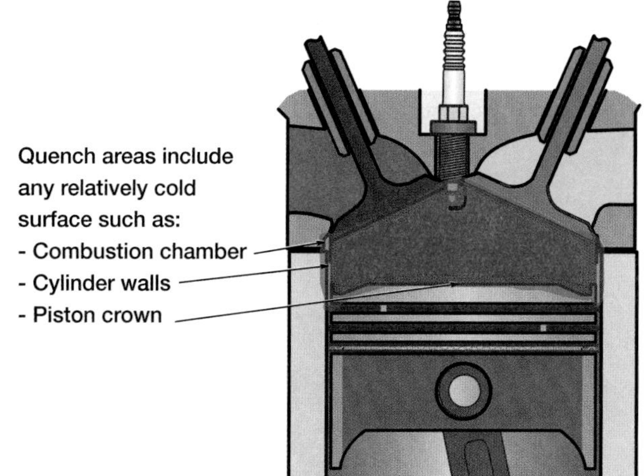
Quench areas include
any relatively cold
surface such as:
- Combustion chamber
- Cylinder walls
- Piston crown

FIGURE 66-14 Quench areas in a combustion chamber.

Combustion Chamber Design

Combustion chamber design can affect the combustion process, which also affects the level of emissions. In the combustion chamber where surface temperatures are low, the combustion flame can be **quenched** in those areas (**FIGURE 66-14**). The flame temperature drops so low that the flame goes out, or is quenched. Fuel left unburned in these zones is then exhausted as hydrocarbon and carbon monoxide emissions.

If the spark plug is positioned near the center of the combustion chamber so that the **flame front** travels evenly through the combustion chamber, combustion is more complete. The flame front is the rapid burning of the air-fuel mixture that moves outward from the spark plug across the cylinder. Combustion chambers where the flame front isn't blocked by protrusions or sharp corners help ensure complete combustion. Proper placement of the spark plug helps to minimize this effect.

The designs of intake ports and cylinder heads can aid in swirling the air-fuel mixture as it travels to the combustion chamber. Swirling of the air and fuel helps to fully mix the two. If fuel does not stay mixed in the air when being drawn into the combustion chamber, the mixture will be distributed unevenly and create drivability problems and excessive emissions. This shows us how important proper movement of air is to combustion.

One factor that affects the ability of air to hold fuel is gas flow rate. Fast-moving air tends to hold fuel better than slow-moving air, which allows fuel to fall out of the airstream (**FIGURE 66-15**). Fast-moving air more easily causes swirling of the air and fuel. Gas flow rate can be increased by the use of smaller intake ports. This design works well at keeping good airflow at low engine revolutions per minute (rpm). But at high rpm, it can restrict the airflow too much and reduce engine power. Some manufacturers overcome this problem by using two or more intake valves in each cylinder. With the extra valves open at high rpm, the overall size of the intake port opening is then increased, and a good gas flow rate is maintained. The more ideal the gas flow rate of air entering the combustion chamber, the more efficient that engine, and with better efficiency comes lower emissions.

Changing valve timing also alters the combustion process. Reducing the valve overlap reduces the **scavenging effect**, which is when the exhaust gases moving out of the combustion chamber create a low-pressure area in the exhaust manifold, helping to pull air and fuel into the cylinder. If the valve overlap is excessive, unburned air and fuel can be pulled

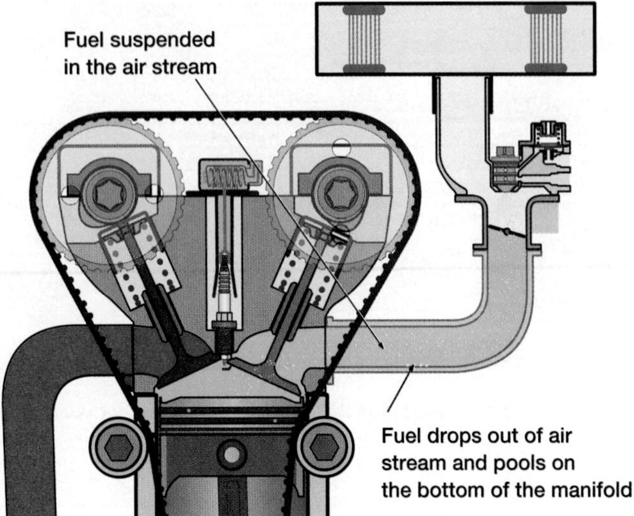
Fuel suspended in the air stream
Fuel drops out of air stream and pools on the bottom of the manifold

FIGURE 66-15 Gas flow rate affects the ability of air to hold fuel. Fast-moving air tends to hold fuel better than slow-moving air, which allows fuel to fall out of the airstream.

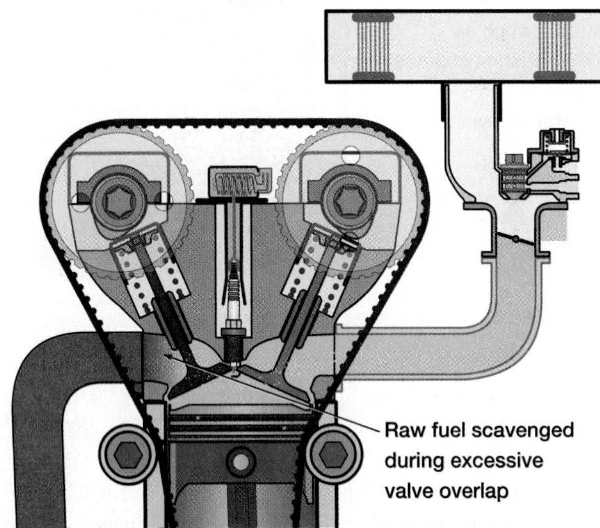
Raw fuel scavenged during excessive valve overlap

FIGURE 66-16 Excessive valve overlap can cause raw fuel to be scavenged out of the cylinder, raising HC emissions.

out of the exhaust valve, which will increase hydrocarbon emissions (**FIGURE 66-16**).

Stoichiometric Ratio

The stoichiometric ratio, also referred to as **lambda**, is the ratio of air to fuel at which all of the oxygen in the air and all of the fuel are completely burned. As you can imagine, complete combustion reduces hydrocarbon and carbon monoxide emissions. The typical stoichiometric ratio for gasoline is 14.7:1 (14.7 parts air to 1 part gasoline by weight or mass). This ratio equates to a lambda reading of 1.0. Each fuel has its own stoichiometric ratio (**TABLE 66-2**). As the air-fuel ratio moves away from the stoichiometric point, the hydrocarbons and carbon monoxide emissions increase. In many cases, precise control of the air-fuel ratio has made the job of hydrocarbon and carbon monoxide emission control devices much easier because they are dealing with much smaller amounts of the pollutants.

TABLE 66-2: Stoichiometric Ratios for Various Fuel Types

Fuel	Ratio
Gasoline	14.7
No. 2 Diesel	14.7
Methanol	6.45
Ethanol	9.00
Propane	15.7
Compressed natural gas	17.2
Hydrogen	34.3

Source: Department of Energy website. Stoichiometric ratio is the ratio of air to fuel, expressed as 14.7:1.

A rich condition means more fuel is introduced than the engine can completely burn. This is anything lower (richer) than 14.7:1 or less than 1.0 lambda. When this condition occurs, hydrocarbons can only partially burn because of the lack of oxygen. This rich condition creates large amounts of carbon monoxide. If excessively rich, some of the hydrocarbons will not burn at all, leaving hydrocarbons to pass through the engine unburned. Elevated amounts of carbon monoxide in the exhaust are a good indicator that the engine is running rich.

A lean condition means more air is introduced than the engine can completely burn. This is anything above (leaner than) 14.7:1 or greater than 1.0 lambda. When this condition occurs, not all of the hydrocarbons burn because the flame cannot reach some of the hydrocarbon molecules due to the excessive amount of oxygen and nitrogen molecules. Thus, the hydrocarbon molecules are pushed out of the exhaust pipe unburned. Elevated hydrocarbon readings (with low carbon monoxide readings) are a good indicator of a lean condition.

Carbon dioxide is a good indicator of whether the fuel is burning completely. For example, high readings of 12% to 15% of carbon dioxide in the exhaust indicate proper combustion efficiency. A smaller percentage of carbon dioxide in the exhaust indicates that combustion is not complete, evidence of a rich or lean mixture, ignition problems, lowered compression, or an exhaust system leak.

▶ TECHNICIAN TIP

Carbon dioxide is a gas that manufacturers are trying to reduce. It is believed to contribute to global warming, and laws are underway that will require manufacturers to reduce its creation. Some of the ways that carbon dioxide can be reduced are listed here:

- Making vehicles smaller, lighter, and more aerodynamic
- Using smaller, more fuel-efficient engines
- Shutting down the engine when the vehicle is decelerating or stopped at a stop light
- Using regenerative braking systems to capture waste energy during braking

Controlling Air-Fuel Ratios

Precise control of the air-fuel ratio is one of the main strategies for minimizing a vehicle's emission output. Keeping the air-fuel ratio close to the stoichiometric point produces minimal hydrocarbon and carbon monoxide emissions. Electronic fuel injection and engine management systems can adjust the air-fuel mixture within certain limits as parts wear and thereby maintain proper control of the air-fuel ratio over a broad range of conditions. These systems closely control the air-fuel ratio entering each cylinder and ensure the ignition timing matches operating conditions.

Sensors around the engine send information about airflow, coolant temperature, throttle position, engine speed, and other inputs to the PCM. The PCM uses this information to set fuel and ignition settings, which change as sensor data change. The PCM has a programmed memory that works like a reference chart. The manufacturer's engineers program the PCM to constantly adjust the fuel and spark, depending on the values received from the sensors. For example, as airflow rises, the fuel increases proportionately.

Fuel delivery is controlled by the fuel injector. The typical fuel injector is simply an electric solenoid. When electrical current flows in the fuel injector winding, a magnetic field is produced (electromagnet), causing the fuel injector to open and allow fuel to flow through it into the intake manifold. The PCM (or powertrain control module) opens the fuel injector for a longer or shorter period of time, creating more or less fuel delivery. The longer the fuel is delivered, the richer the mixture, and vice versa.

▶ TECHNICIAN TIP

The pulse width of the fuel injector is a piece of data that can be read on a scan tool or lab scope to assist the technician in testing fuel delivery. The pulse width is the amount of time the fuel injector is turned on and fuel is sprayed into the intake manifold, and is measured in milliseconds. The longer the pulse width, the more fuel is delivered to the engine.

Changes in operating conditions can change mixture conditions. For instance, if an engine is being driven at moderate speed and the throttle is suddenly closed, any fuel condensed on the intake manifold walls or intake port of the cylinder head will be drawn into the cylinders. A smaller amount of air moves into the combustion chamber with the throttle plates closed, so turbulence tends to be poor, which can lead to incomplete combustion and the release of unburned gases into the exhaust.

To reduce this effect, the PCM detects that the throttle is being closed through the use of the throttle position sensor (TPS) and reduces the fuel spray by reducing the on time of the injectors. If the engine is being used to slow the vehicle, the injectors can be completely shut off to save fuel and reduce pollution. This is called fuel shutoff mode. Newer engines with electronic throttle control can move the throttle plate independent of driver input. This feature allows the computer to control the throttle opening as well as the fuel injection pulse width to create a proper mixture during all driving conditions.

Precombustion/Postcombustion Treatment

The pollution emitted from the internal combustion engine can be reduced by taking actions both prior to combustion and after combustion. Precombustion treatment focuses on assuring that combustion happens in a controlled and complete manner that minimizes postcombustion pollutant gases. There are two primary precombustion emission control systems: the heated air intake system and the exhaust gas recirculation system. The heated air intake system was used on throttle body–injected (and carbureted) engines. It was designed to keep the air entering the engine at a constant temperature of approximately 100–110°F (38–43°C) to help vaporize the fuel in the intake manifold. The better vaporized the fuel, the more likely it could be distributed evenly within the air, and the cleaner the combustion process. This system is discussed further in the Emission Control System Evolution section.

Most port fuel-injection systems and gas direct-injection systems do not need a heated air intake system because the fuel is either sprayed into the intake manifold right next to the intake valve, or into the combustion chamber itself. The injector sprays a fine mist of fuel, which is easier to vaporize. Plus it does not have to sit or travel along a cold intake manifold, where it can condense out of the air. Most new vehicles monitor the intake air temperature and use that to help determine the mass of air entering the engine, so they know exactly how much fuel to inject.

The second precombustion emission control system is the exhaust gas recirculation system. This system recirculates some of the inert exhaust gases back into the combustion chamber when combustion temperatures are high enough to create oxides of nitrogen. The inert gases help lower peak combustion temperatures below the threshold at which oxides of nitrogen are created. Exhaust gas recirculation systems are covered in greater depth in the following Emission Control Systems section.

Postcombustion emission control systems deal with the remaining pollutants that are left over from the combustion process. There are three primary postcombustion systems: the positive crankcase ventilation (PCV) system, the secondary air injection system, and the catalyst system.

The PCV system draws the blowby gases out of the crankcase and routes them to the intake manifold, where they are drawn into the cylinders and burnt in the combustion chamber. The catalyst system reduces oxides of nitrogen in the exhaust stream, converting them back into nitrogen and oxygen, and oxidizes hydrocarbons and carbon monoxide. The secondary air injection system, if equipped, helps to complete the burning of hot hydrocarbons and carbon monoxide in the exhaust system by adding oxygen to the exhaust stream. The oxygen in the air reacts with the hydrocarbons and carbon monoxide and converts them into carbon dioxide and water.

All three of these systems reduce emissions to near zero if the engine and all of the systems are operating as designed. Each of them is covered in greater depth in the following section.

▶ Emission Control System Evolution

K66004

Emission control systems have been created by manufacturers to minimize the amount of pollution a vehicle and its internal combustion engine release into the atmosphere. Before the days of emission control systems, vehicles and engines produced very large amounts of pollutants, as evidenced by the number of smog days in many big cities. Even though there are now many more vehicles and many more miles driven, the number of pollution days in a year has been reduced substantially and continues to decline in most areas.

One of the earliest control devices was the positive crankcase ventilation (PCV) system. The system was first used around 1961 and later became a mandatory emissions device. PCV systems stop the release of hydrocarbons from the crankcase of the engine to the atmosphere. Another control device is the evaporative control system, known as the EVAP system. This system was used beginning around 1972 and later became a mandatory emission system. It is used to stop the release of hydrocarbons from the fuel tank to the atmosphere.

Another early emission control device from the 1960s was the early fuel evaporation system. It was designed to quickly heat up the induction system and intake air on cold engine starts. There were several different ways of doing this, from heat riser valves to electric heater grids.

Exhaust gas recirculation (EGR) systems began to appear on the automobile around 1972 and became a mandatory part of emission controls. These systems control the release of oxides of nitrogen to the atmosphere by controlling the temperature of the combustion process. In the 2000s, variable valve timing (VVT) started to be used to dilute the air-fuel mixture in the combustion chamber and allowed some vehicles to do away with the EGR valve.

The use of a system called secondary air injection began in the late 1960s to limit hydrocarbon release from the engine during cold engine operation. The system injects air into the exhaust stream near the exhaust valve to mix with the hydrocarbons and aid in the conversion process as the exhaust exits the combustion chambers. Later versions of this system sent air down to the catalytic converter once the engine warmed up.

Catalytic converters found their way onto the automobile around 1973. Catalytic converters are used to reduce remaining pollutants in the exhaust stream before they leave the tailpipe. As catalytic converters were introduced, leaded gas was phased out. Lead in the gas would contaminate (coat) the catalyst in the converter and render it inoperative.

Catalytic Converters

Catalytic converters are a primary emission control device. They are installed near the front of the exhaust system (**FIGURE 66-17**). They perform a final cleanup of the tailpipe emissions on a properly running engine equipped with functioning emission control systems. Catalytic converters are not designed to handle the emission output of a poorly running engine or one with faulty emission control systems. Catalytic converters create heat as they convert harmful gases into less harmful gases. Forcing converters to deal with more gases than they are designed to handle can cause them to overheat. In some cases, they may even melt down internally, which ruins them.

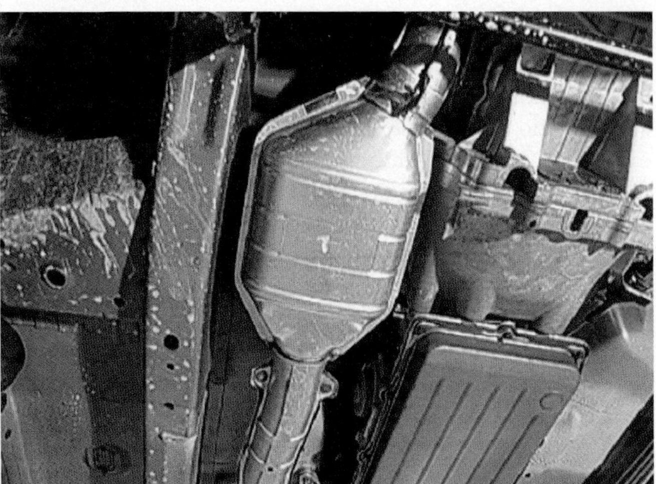

FIGURE 66-17 Catalytic converters use a base structure of ceramic material coated with precious metals to react with and break up or combine gases in the exhaust system.

Catalytic converters use a base structure of ceramic material coated with precious metals to react with and break up or combine gases in the exhaust system. The term "catalyst" refers to a material that causes a chemical reaction without itself being consumed or changed in the reaction. Specific precious metals have the ability to act as a catalyst that can convert harmful exhaust gases to nonharmful or less harmful gases. Catalytic converters come in two types: oxidizing and reduction. Each type is used to react with particular gases.

The reduction catalyst contains a platinum and rhodium coating, which helps to reduce the oxides of nitrogen molecules into their base compounds. When a nitric oxide or nitrogen dioxide molecule comes into contact with the catalyst coating, the coating strips the nitrogen atom out of the molecule and retains it. This frees up the one or two oxygen atoms in the molecule, which combine in pairs to form molecules of oxygen. The nitrogen atoms bond with other nitrogen atoms that are retained in the catalyst and form molecules of nitrogen; thus, two molecules of nitric oxide become one molecule of nitrogen and one molecule of oxygen, or two molecules of nitrogen dioxide become one molecule of nitrogen and two molecules of oxygen (**FIGURE 66-18**).

The second type of catalyst is the oxidizing catalyst, which oxidizes any unburned hydrocarbons and carbon monoxide as they travel over the platinum and palladium coating. This causes the carbon monoxide and hydrocarbons to react with any remaining oxygen in the exhaust gas (**FIGURE 66-19**). Each carbon monoxide (CO) molecule combines with an oxygen (O) molecule to make one less harmful carbon dioxide (CO_2) molecule. And two hydrogen (H) atoms combine with one oxygen (O) molecule to form one nonharmful water (H_2O) molecule.

Catalytic converters are referred to as either two-way or three-way catalytic converters. The **two-way catalytic converter** was the first type of catalytic converter designed. The two-way catalytic converter is an oxidizing catalyst and is named because it converts two gases, hydrocarbons and carbon monoxide, to carbon dioxide and water. The early two-way catalytic converters used a bed of catalytic pellets (pellet-style converter) that the exhaust passed through. More recent versions of the two-way (and three-way) catalytic converters are designed with a honeycomb catalytic element. The honeycomb structure provides long narrow passageways for the exhaust gases to flow through, which ultimately helps the gases to come in contact with the catalyst materials on the sides of the honeycomb walls. This contact between the gases and the catalyst promote oxidation of the hydrocarbons and carbon monoxide once the catalyst reaches its operating temperature.

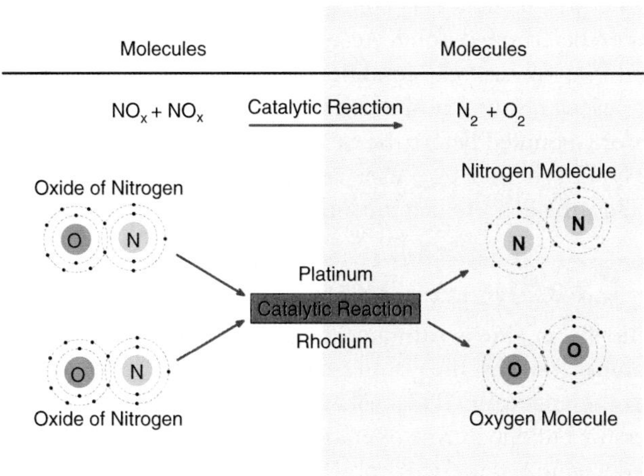

FIGURE 66-18 The reduction catalyst splits nitric oxide and nitrogen dioxide molecules into nitrogen atoms and oxygen atoms. The nitrogen atoms bond to form molecules of nitrogen (N_2), and the oxygen atoms bond to form oxygen (O_2).

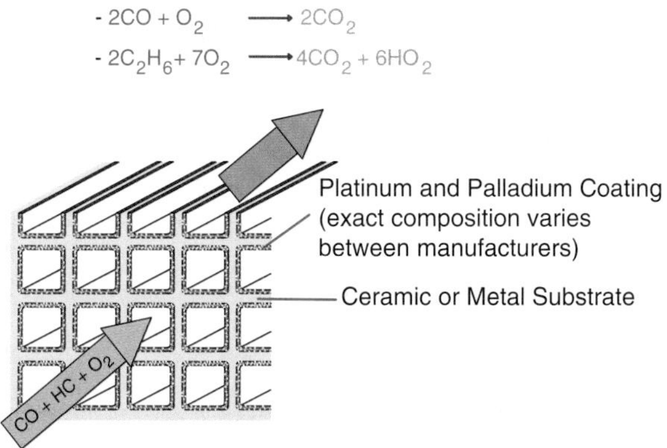

FIGURE 66-19 The oxidizing catalyst oxidizes any unburned hydrocarbons and carbon monoxide as they travel over the platinum and palladium coating.

SAFETY TIP

Catalytic converters must be above approximately 500°F (260°C) to begin conversion. Typically, they operate at 900–1600°F (482–871°C). Catalytic converters can cause serious burns, so use gloves and heat protection sleeves when working around a hot catalytic converter.

TECHNICIAN TIP

Catalytic converters may contain rhodium, palladium, platinum, or cerium. These materials are more valuable than gold and should be recycled. Catalytic converters have become a target for thieves, who cut them out of the exhaust system while the vehicle is parked in a parking lot or driveway.

Modern vehicles operating on petroleum-based fuels are fitted with three-way catalytic converters. **Three-way catalytic converters** contain both a reduction catalyst and an oxidizing catalyst. The term "three-way" is in relation to the three regulated emissions the three-way catalytic converter is designed to convert: carbon monoxide, hydrocarbons, and oxides of nitrogen. The reduction catalyst is in front and converts the oxides of nitrogen, nitric oxide, and nitrogen dioxide back into harmless nitrogen and oxygen molecules (**FIGURE 66-20**). The oxidizing catalyst then uses the oxygen from the reduction catalyst to oxidize the hydrocarbons and carbon monoxide into water and carbon dioxide.

Because of strict emission requirements, vehicles with three-way catalytic converters require a **feedback system** on the fuel system. The PCM monitors the oxygen content in the exhaust stream by using an exhaust gas oxygen (EGO) sensor, also known as a lambda sensor, mounted in the exhaust manifold. This sensor tells the engine computer how much oxygen is in the exhaust. The computer uses this information to control the pulse width of the fuel injectors.

The PCM can increase or decrease the amount of oxygen in the exhaust by adjusting the air-fuel ratio. The ECU ensures that the air-fuel ratio cycles just above and below the stoichiometric

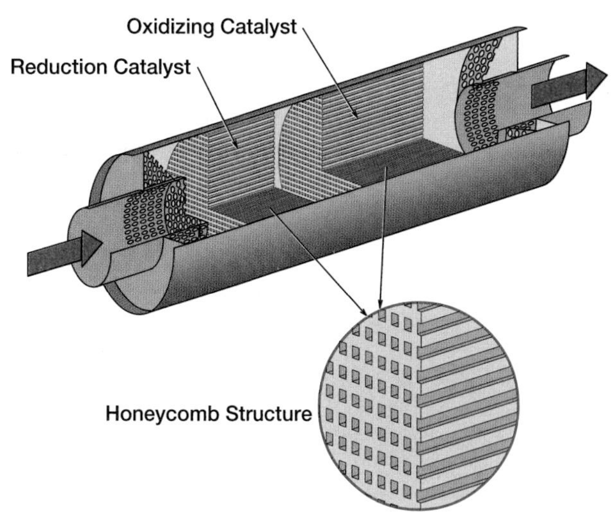

FIGURE 66-20 A typical three-way catalyst.

point in normal driving conditions. This ensures that there is always sufficient oxygen in the exhaust system to allow the oxidization catalyst to deal with unburned hydrocarbons and carbon monoxide.

> ▶ TECHNICIAN TIP
>
> Each style of catalytic converter is designed for a specific vehicle; one size does not fit all. A converter intended for the treatment of a four-cylinder engine will not be efficient on an eight-cylinder engine. Be sure to install the correct catalytic converter in the correct vehicle. Always check the manufacturer's specifications before making a repair. Be careful with aftermarket catalytic converters, which may not use as much catalyst agent (precious metals) as the emission system requires.

Catalyst Monitoring

Any vehicle produced after 1996 monitors the efficiency of the catalyst by comparing the readings of two oxygen sensors (**FIGURE 66-21**). One oxygen sensor is in front of the catalytic converter, as in the past, and indicates the amount of oxygen in the exhaust stream as it leaves the cylinders. The reading from this sensor normally toggles rich and lean as the PCM is controlling the air-fuel ratio. The second oxygen sensor is located after the catalytic converter and measures the oxygen in the exhaust stream after it leaves the catalytic converter.

Because the converter uses oxygen to oxidize the hydrocarbons and carbon monoxide, the oxygen content is normally lower after the converter and results in a higher and steadier oxygen sensor voltage. The catalyst monitor tracks both oxygen sensor readings to ensure that the gases are being effectively converted by the catalytic converter. As the catalytic converter fails, the rear sensor tends to track the oscillations of the front oxygen sensor.

> **Applied Science**
>
> ***AS-89: Catalytic Converter: The technician can explain the principles by which a catalytic converter modifies emission gases at the atomic level to provide a lower level of HC, CO, and NOx in the final exhaust.***
>
> One of the greatest emission devices to be installed on vehicles is the catalytic converter. This component will last for the life of the vehicle if given proper care. Proper care for the converter includes maintaining the engine according to the manufacturer's service information. If the engine is operating normally, the converter won't have to work hard to accomplish its task. Catalytic converters are designed to provide control of hydrocarbons (HC), carbon monoxide (CO), and oxides of nitrogen (NOx). Emissions are passed through the converter's catalytic inlaid materials such as rhodium, palladium, platinum, or cerium. All of these special materials are considered to be precious metals. The reduction catalyst is in front and converts the oxides of nitrogen, nitric oxide, and nitrogen dioxide back into harmless nitrogen and oxygen molecules. The oxidizing catalyst then uses the oxygen from the reduction catalyst to oxidize the hydrocarbons and carbon monoxide into water and carbon dioxide.
>
> A catalytic converter can be checked in several ways we discuss in the text. The manufacturer's service information should always be consulted.

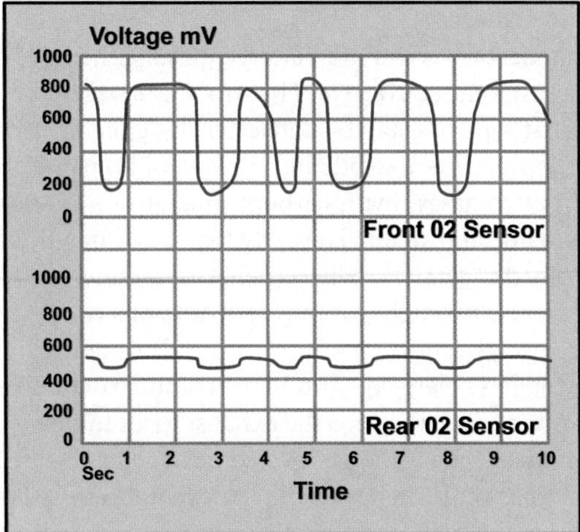

FIGURE 66-21 Any vehicle produced after 1996 monitors catalyst efficiency by using two oxygen sensors.

The PCM monitors these two oxygen sensor signals and turns on the MIL to set a DTC once the catalyst efficiency degrades to a predetermined point. Access to this information allows the technician to verify the condition of the catalytic converter when it is suspect. Note that newer vehicles use an air-fuel ratio (AFR) sensor (mounted before the catalytic converter), which replaces the typical zirconia O_2 sensor, and has a wider range of detection to allow the ECM to better control the air-fuel mixture.

▶ Crankcase Emission Control

While the engine is running, some gases from the combustion chamber leak past the piston rings and the cylinder walls, down into the crankcase. This leakage is called **blowby**. To prevent pressure buildup during operation, the crankcase must be ventilated. Unburned fuel (hydrocarbons) and water from condensation also find their way into the crankcase and sump. When the engine reaches its full operating temperature, the water and fuel in the crankcase evaporate. In older vehicles, crankcase vapors were vented directly to the atmosphere through a **breather tube** or road-draft tube. It was shaped so that air flowing past it while the vehicle was being driven helped draw the vapors from the crankcase. This resulted in blowby gases being vented directly to the atmosphere.

Modern vehicles are required to direct crankcase blowby gases and vapors back into the intake manifold, where they can be returned to the combustion chamber to be burned. A common method of directing gases and vapors back to the intake is through the emission control system called the **positive crankcase ventilation (PCV)** system (**FIGURE 66-22**). The PCV system regulates the flow of blowby gases between the crankcase and the intake manifold.

Types of PCV Systems

There are three main types of PCV systems: the **variable orifice**, **fixed orifice**, and **separator** types. All three types do the same job—ventilate blowby gases back to the intake manifold to be burned in the combustion chamber.

In a variable orifice–type system, a replaceable, spring-loaded **PCV valve** regulates gas flow (**FIGURE 66-23**). The position of the PCV valve is controlled by the pressure in the manifold. With the engine off, the spring holds the PCV valve in the closed position, and air cannot enter the inlet manifold. This allows the engine to start. At idle, relatively high intake manifold vacuum draws the PCV valve to the other end of the PCV valve's housing, where it allows only a small, measured amount of blowby gases and air past the PCV valve. At wider throttle openings, the PCV valve plunger position allows maximum flow through the PCV valve's housing, which gives maximum crankcase ventilation and tends to match the higher amount of blowby gases under that condition.

The PCV system is designed to remove more air than just blowby gases, so there should almost always be more ventilation capacity than the amount of blowby. The additional air that the system removes comes from a fresh air intake hose or tube, usually attached to the air cleaner assembly. The fresh air intake hose directs filtered air to one side or end of the crankcase. This intake point is usually as far as possible from the PCV valve to ensure that as much of the blowby gases are pulled from the crankcase as possible, giving good ventilation.

The PCV system has another purpose: preventing a buildup of pressure in the crankcase. Modern PCV systems are the closed (sealed) type. This means both ends of the PCV system are connected to the intake system. The PCV valve is connected directly to the intake manifold below the throttle plate, whereas the fresh air hose is connected to the air intake above the throttle plate. Having a closed system means any gases that cannot be handled through the vacuum side of the system are directed back through the fresh air connection to the air cleaner assembly, where they are drawn into the intake airstream and burned in the combustion chamber (**FIGURE 66-24**).

The fixed orifice–type PCV system usually involves a screw-in fitting with a small hole drilled in it. The hole creates a predetermined vacuum leak that draws a predetermined amount of crankcase vapors from the crankcase. Engineers have factored the vacuum leak into the engine management system to ensure the correct air-fuel mixture. The fixed orifice is typically mounted on or in the intake manifold (**FIGURE 66-25**).

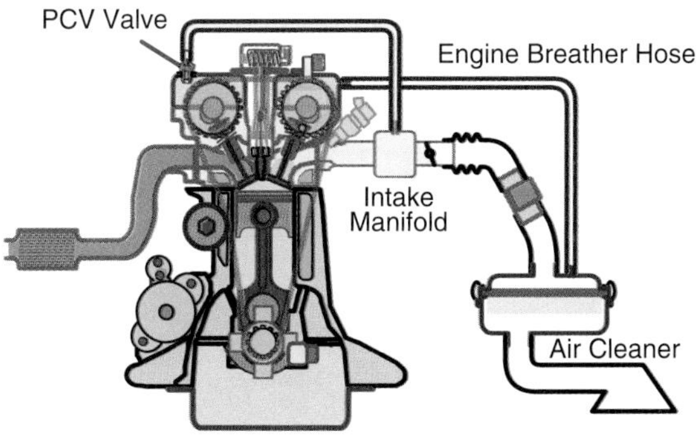

FIGURE 66-22 PCV systems draw blowby gases out of the crankcase to be burned in the combustion chambers.

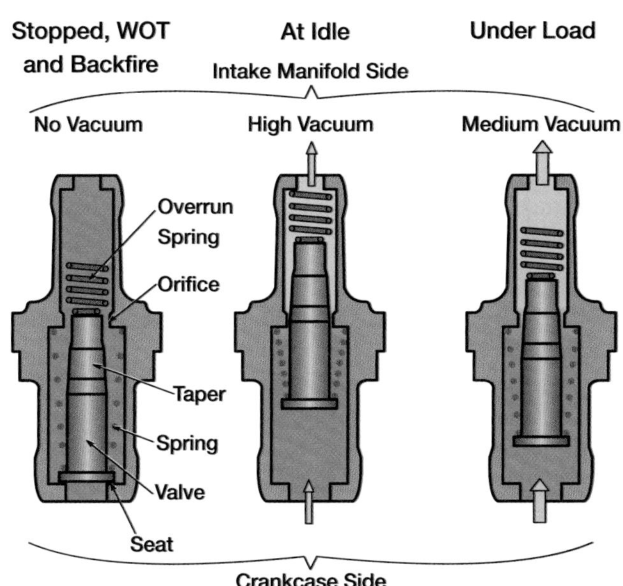

FIGURE 66-23 A variable-orifice PCV valve meters the amount of blowby gases, depending on engine load.

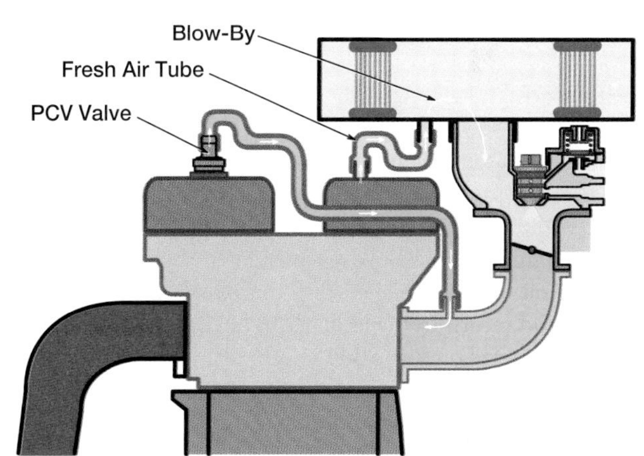

FIGURE 66-24 During heavy acceleration, excess blowby gases vent through the fresh air tube to the air cleaner assembly.

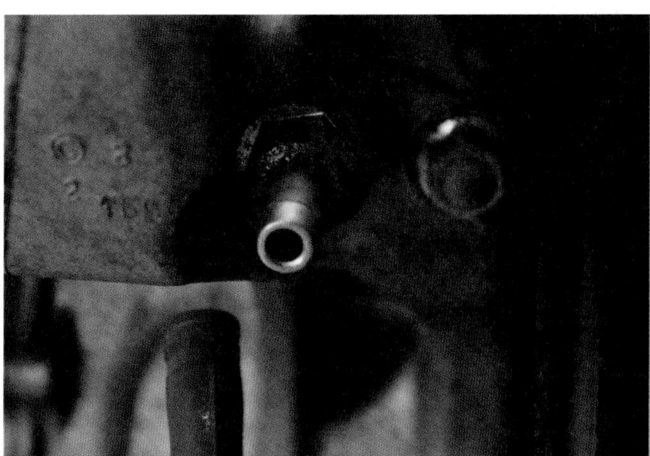

FIGURE 66-25 Typical fixed orifice passageway for that type of PCV system.

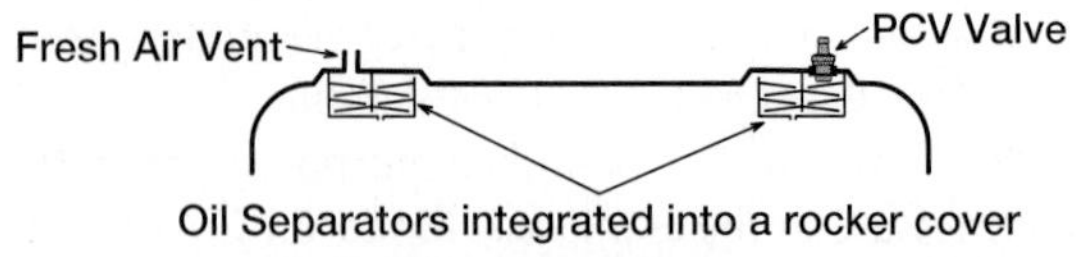

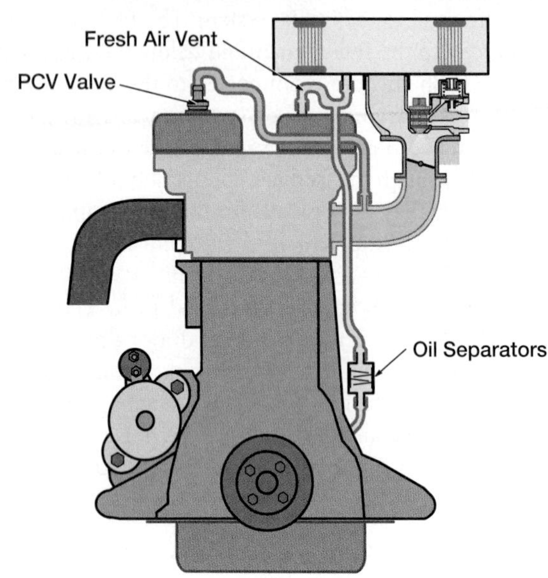

FIGURE 66-26 Separator type of PCV system.

FIGURE 66-27 The exhaust gas recirculation (EGR) system was designed by automotive engineers in the 1970s to control the emission of oxides of nitrogen. Nitrogen is oxidized in large amounts once combustion temperatures reach approximately 2500°F (1371°C).

The separator type of PCV system involves a valve that is hooked to the pressure side of the crankcase, an oil return line at the bottom of the valve, and a suction line on the other side. Oil that is mixed with blowby gases enters the separator, and the heavy oil tends to fall out of the mixture to the bottom of the valve and return to the crankcase (**FIGURE 66-26**). The separator type is used on turbocharged applications because turbocharged systems pressurize the intake system and would close off a standard PCV valve.

▶ Exhaust Gas Recirculation System

The **exhaust gas recirculation (EGR) system** was designed by automotive engineers in the 1970s to control oxides of nitrogen, which are created in large amounts once combustion temperatures reach approximately 2500°F (1371°C). This temperature is reached when the vehicle is under moderate to heavy loads and the air-fuel ratio is at stoichiometric or leaner. EGR systems were designed to operate an EGR valve to allow a measured amount of exhaust gases to enter the intake manifold during those times (**FIGURE 66-27**). As with all vehicle technology, EGR systems have evolved over time. Instead of using an EGR valve, some newer engines use variable valve timing to draw some exhaust gases back into the cylinder through the exhaust valve, which is held open a little longer into the intake stroke than is typical.

Purpose and Operation

An EGR valve connects a passage from the exhaust port, or manifold, to the intake manifold (**FIGURE 66-28**). If engine operating conditions are likely to produce oxides of nitrogen, the EGR valve opens, letting some burned exhaust gases pass from the exhaust into the intake system. During combustion, an amount of fresh air-fuel mixture is displaced by inert exhaust gases that do three things:

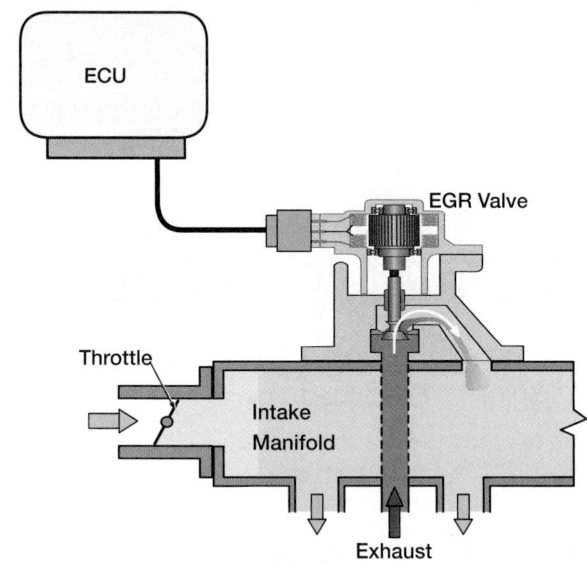

FIGURE 66-28 Typical computer-controlled EGR system.

- First, they make it so there is not quite as much burnable mixture, which helps keep the temperature from rising as high as it otherwise would.
- Second, because they are inert and mix thoroughly with the rest of the air-fuel mixture, they cause the mixture to burn a bit more slowly, and therefore cooler, as the flame front must travel farther to burn around the inert molecules.
- Third, the inert gases absorb some of the heat of the burning gases.

These three situations lower the peak combustion temperature, thus reducing the formation of the oxides of nitrogen.

The EGR valve only opens under the conditions that create oxides of nitrogen. Because oxides of nitrogen are not created in a cold engine, the EGR valve is kept from opening until after the engine achieves its normal operating temperature. The EGR valve also is not opened at idle because there is only a small amount of air-fuel mixture being burned, so the combustion temperature is well below 2500°F (1371°C).

There is one condition when oxides of nitrogen are being created in large amounts and the EGR valve may not operate: when the vehicle is operated at wide-open throttle (WOT). In this condition, the EGR system is typically shut off so that full engine power is available for emergency needs such as avoiding an accident. But most passenger vehicles do not operate very often or for very long at WOT, so the overall amount of oxides of nitrogen created is minimal.

Some manufacturers have been able to accomplish the same reduction in oxides of nitrogen by using variable valve timing on their engines instead of the traditional EGR system. In situations where oxides of nitrogen can be created, the PCM retards the exhaust camshaft so that the exhaust valves are held open during the first part of the intake stroke. As the piston moves down slightly, the lower cylinder pressure pulls some of the exhaust gases back into the combustion chamber (**FIGURE 66-29**). These exhaust gases are used to dilute the air-fuel mixture with inert gases just like an external EGR system does. In this case, the PCM controls how long the exhaust valve remains open, which then affects the amount of exhaust gases that are retained in the cylinder. This is sometimes referred to as an internal EGR system, and the beauty is that it uses only the variable valve timing components, which are used for other fuel economy and drivability reasons. So there are fewer components to install and go bad.

▶ TECHNICIAN TIP

Many people believe that all emission control devices kill engine power and use extra gasoline. Although there may have been a small amount of truth to that when some of the systems were first introduced, that is almost never the case any longer. In the case of the EGR system, exhaust gases are typically recirculated only during part throttle operation. If any power is lost at that time, the throttle can be opened slightly to make up for it. And when the throttle is wide open, the EGR system typically does not operate, so there is no impact at that time. Lastly, it could be argued that the extra inert gases from the EGR system help maintain a higher average cylinder pressure, which makes up for less air-fuel entering the cylinder.

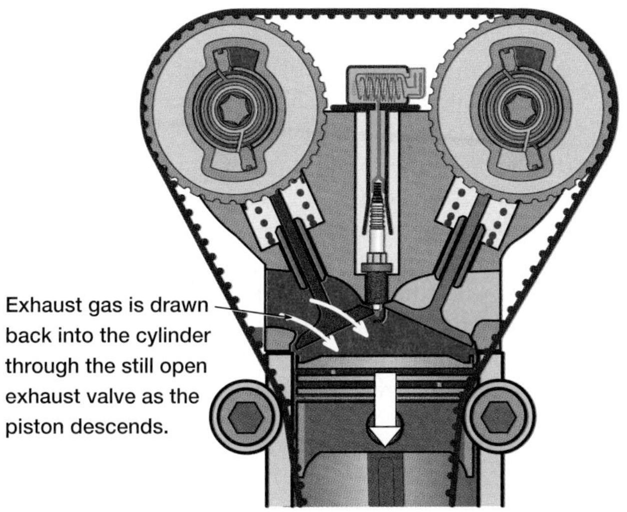

FIGURE 66-29 Variable valve timing can be used instead of an EGR valve to draw exhaust gases back into the combustion chamber.

Types of EGR Valves

There are two main types of EGR valves found on current automobiles: the vacuum diaphragm type and the electrically operated type. Each of these types comes in a variety of styles. For vacuum diaphragm types, the EGR valve is available in three styles: the single-diaphragm valve, the negative back-pressure valve, and the positive back-pressure valve. The single-diaphragm EGR valve has a single, spring-loaded diaphragm that seals off the vacuum chamber (**FIGURE 66-30**). The diaphragm is connected to and operates a **pintle**, which is a tapered valve that seals in a tapered seat and allows exhaust gas to flow into the intake manifold when it is open. When vacuum is applied to the chamber, the diaphragm moves against spring pressure and opens the valve. Spring pressure closes the valve when vacuum is released.

The **negative back-pressure EGR valve** uses two diaphragms to operate (**FIGURE 66-31**). The main diaphragm is connected to the pintle valve and is spring loaded just like the single-diaphragm valve. The second diaphragm is connected through a hole in the pintle shaft to the exhaust side of the EGR valve. The second diaphragm opens and closes a bleed hole in the main diaphragm. A small spring holds this bleed valve in the closed position, allowing the main chamber to be operated by vacuum. Once the EGR valve opens, the pressure in the exhaust system keeps the bleed valve closed as long as there is not a vacuum in the exhaust system. If a vacuum occurs, then the bleed valve opens, which releases the vacuum in the main chamber and allows the main spring to close the EGR valve. This system ensures that the EGR valve is open only when the engine is under a load, as evidenced by the positive back pressure. It is helpful to remember that a negative

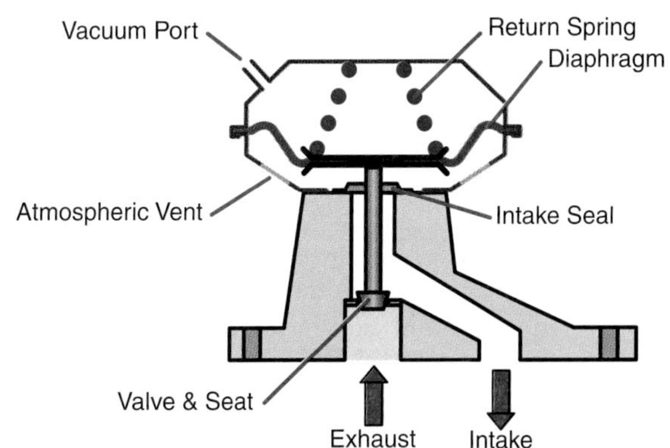

FIGURE 66-30 Single-diaphragm EGR valve.

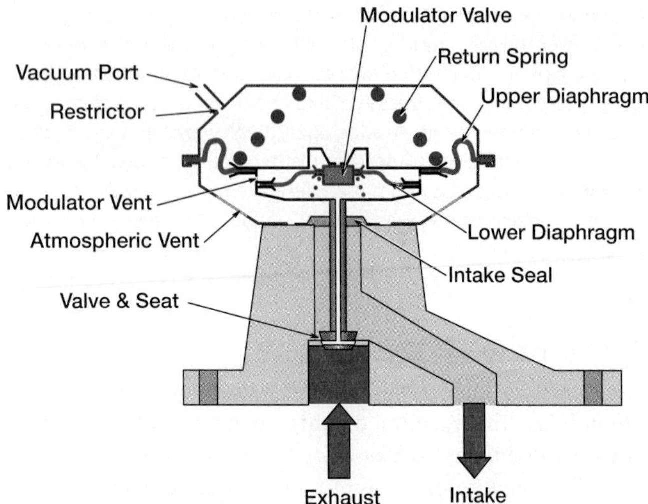

FIGURE 66-31 Negative back-pressure EGR valve.

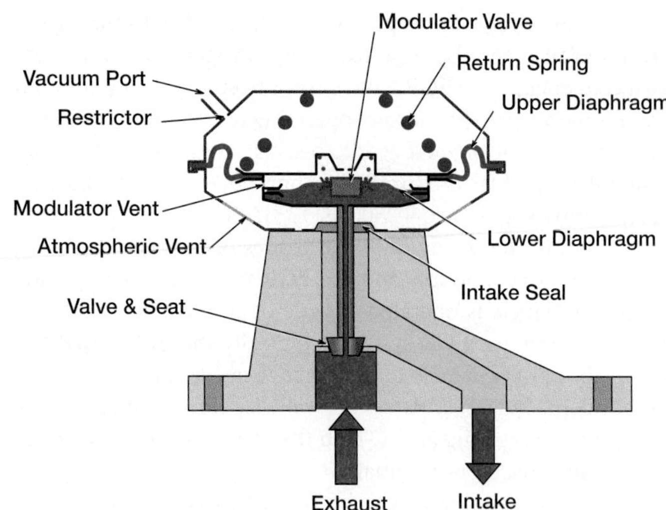

FIGURE 66-32 Positive back-pressure EGR valve.

back-pressure EGR valve is released whenever there is negative exhaust pressure. This valve can be tested with a simple vacuum pump to ensure that the valve and diaphragm lifts, holds, and closes properly. In many cases, a negative back-pressure EGR valve can be identified by the "N" at the end of the part number.

▶ TECHNICIAN TIP

In the combustion chamber, higher combustion temperatures tend to create more power. But high temperatures also create very high amounts of oxides of nitrogen. One way to overcome this is by understanding the difference between high peak combustion temperature and high average combustion temperature. High peak temperatures last for very little time. Although they produce a sharp spike in pressure, it does not last long and tends to give a hammer blow to the piston, which is not very effective in pushing it down the cylinder. High average combustion temperature maintains high average pressure, which is good for a longer, smoother push of the piston. It produces good power while minimizing the creation of oxides of nitrogen.

The **positive back-pressure EGR valve** is very similar to the negative back-pressure EGR valve. However, the small spring holds the bleed valve open until positive exhaust pressure acts on the second diaphragm and closes the bleed valve (**FIGURE 66-32**). Once the bleed valve is closed, vacuum can operate on the main diaphragm to open the EGR pintle. If exhaust pressure falls, the bleed valve will open and bleed off the vacuum in the main diaphragm chamber, and the main spring will close the EGR pintle. Without exhaust gas pressure, the spring-loaded valve will not close and seal off the diaphragm chamber so that intake vacuum cannot pull on the main diaphragm and lift the pintle. Operation of any of these diaphragm types of EGR valves can be controlled through the use of a temperature-controlled vacuum valve or a computer-controlled actuator (EGR solenoid). In many cases, a positive back-pressure EGR valve can be identified by the "P" at the end of the part number.

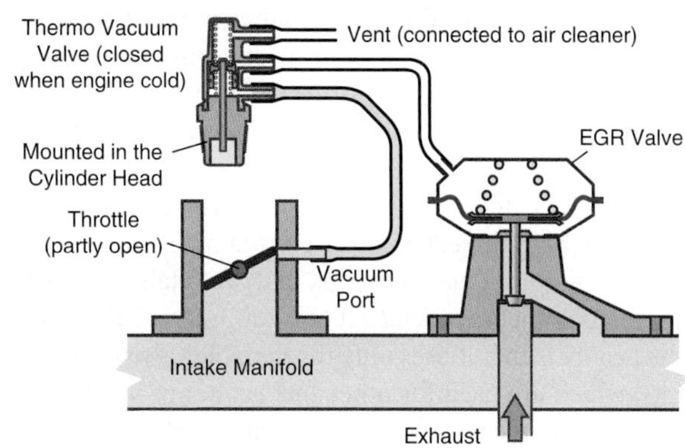

FIGURE 66-33 Typical EGR vacuum control system.

Vacuum Controls in the EGR System

Older vehicles used vacuum controls solely to operate the EGR system (**FIGURE 66-33**). In many cases, the throttle body had a vacuum port for the EGR system that was positioned a little bit above the throttle plate. That way there would not be any vacuum to the port at idle or WOT. But when the throttle was partially open, the port would receive manifold vacuum to send to the EGR valve. Many vacuum-controlled EGR systems also used a temperature-sensitive vacuum valve that prevented vacuum from reaching the EGR valve until the engine reached operating temperature. This design prevented the EGR from operating when the engine was cold. Manifold pressure reaches the EGR valve when the engine is warm and when the EGR port is influenced by the position of the throttle. This is generally at moderate throttle openings, when a lean mixture could increase creation of oxides of nitrogen.

Electronic Controls in the EGR System

Electrically operated EGR valves are commonly used on vehicles today because they can be controlled much more accurately. There are two common types of electric EGR valves: pulse-width-modulated solenoid type and the **stepper motor EGR valve** type (**FIGURE 66-34**). The pulse-width-modulated style uses a spring-loaded valve that is opened by an electric solenoid and closed by the spring. The PCM sends a pulse-width-modulated signal that varies the solenoid on/off times. The longer the pulse width, the higher the duty cycle of the solenoid and the farther the valve is held open. This design allows for reasonably good control of EGR flow.

The stepper motor EGR valve controls EGR flow much more precisely. The valve is moved in and out by means of a threaded rod that is turned in steps by the electric motor. The PCM sends signals to the stepper motor, which can open or close the valve one step at a time. Because the PCM also knows the engine temperature, throttle position, vehicle speed, and engine load, it can operate the valve only when EGR flow is needed and precisely control how much is allowed to flow. Most PCM-controlled EGR systems incorporate a feedback loop, called a monitor, to ensure that the valve is working properly and EGR gases are flowing in the system. One type of EGR monitor is shown (**FIGURE 66-35**). If the PCM detects a fault, it will set a DTC and illuminate the MIL. The EGR monitoring strategies were covered in the Onboard Diagnostics System chapter.

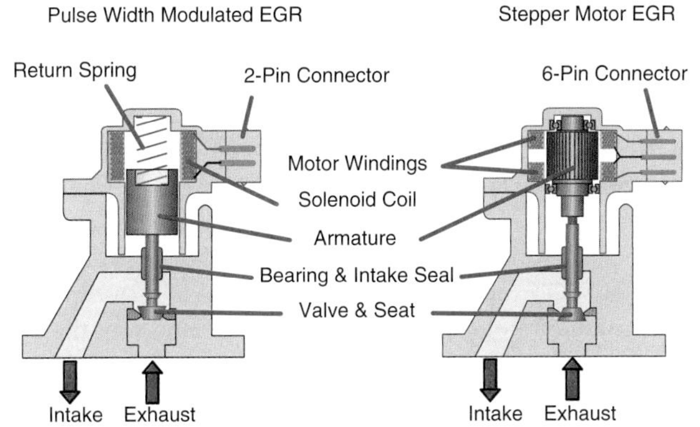

FIGURE 66-34 Pulse-width-modulated EGR valve and stepper motor EGR valve.

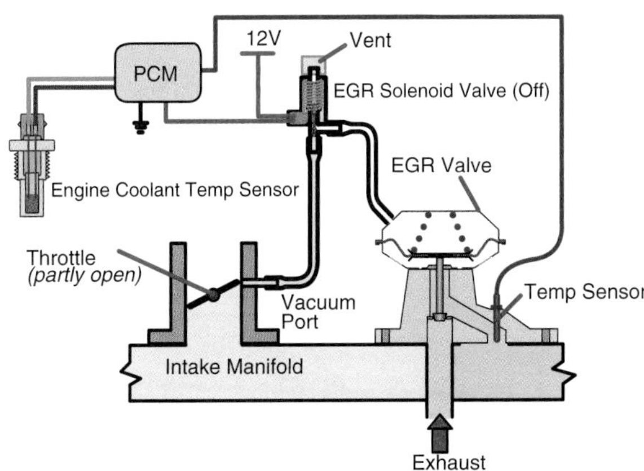

FIGURE 66-35 EGR temperature monitoring system.

> **TECHNICIAN TIP**
>
> It is important to know that the results of monitor tests are erased when DTCs are cleared. Because it can take days, weeks, or in some cases months for the monitors to run, it may be prudent to leave any DTCs in memory and let the PCM turn off the MIL once you have completed the repairs on the vehicle.

▶ Evaporative Emission Control

The **evaporative emission (EVAP) system** is designed to ensure that hydrocarbons are not released into the atmosphere when fuel in the fuel tank begins to vaporize and build pressure. The EVAP system uses a charcoal canister and a series of hoses and valves to capture the hydrocarbon vapors that would normally be lost from the fuel tank and then deliver them to the intake manifold to be burned in the engine's combustion chambers (**FIGURE 66-36**). The EVAP system stores the fuel vapors in the charcoal canister until the PCM determines they can be burned in the engine without affecting the drivability of the vehicle. This is typically when the vehicle is driving down the road between certain speeds and within a certain engine operating temperature range, and Once the PCM determines that the conditions are right for burning off the vapors, it will control the rate of vapors being released to the engine. The EVAP system also incorporates a monitoring system to ensure that there are no leaks in the fuel storage system, as well as to verify that the vapors are being delivered, as designed, to be burned. We will look at those in a minute.

Fuel Storage

Before 1971, vehicles vented the fuel tank through the filler cap into the atmosphere. As fuel in the tank vaporized, vaporous hydrocarbons would escape from the filler cap and the carburetor. This is a waste of perfectly good fuel as well as a pollutant of the air we breathe. Non-vented filler caps were designed to help prevent the release of these vapors. But the system has to accommodate changes in fuel tank level. Today, a **vacuum relief valve** somewhere in the system allows air to enter the fuel tank to relieve the low pressure as the fuel is drawn out of the tank or when the fuel contracts as the temperature drops. The vacuum relief valve also stops the fuel tank from collapsing in these situations.

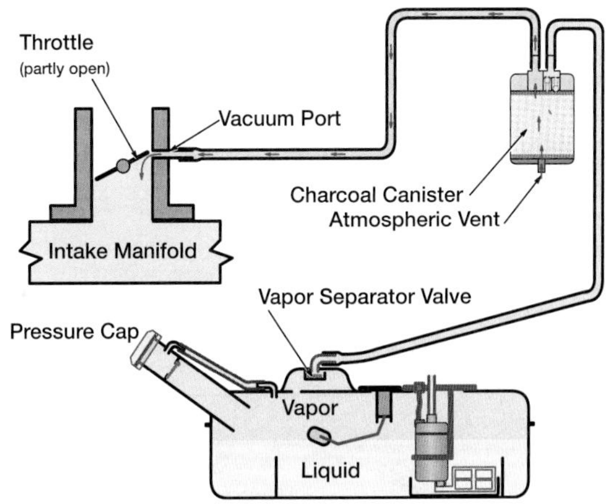

FIGURE 66-36 EVAP system.

The fuel cap may also incorporate a **pressure relief valve.** If the fuel tank's internal pressure exceeds the set value of the relief valve, the pressure relief valve releases a small amount of pressure to prevent the fuel tank from rupturing. Some modern caps have no valves at all and are completely sealed to stop the entry of air and water into the fuel tank and to stop the release of fuel vapor. Modern tanks also contain an **expansion volume,** which is an air chamber that allows for the expansion of liquid fuel on hot days (see Figure 66-36). It can be either directly built into the top of the fuel tank or included as a separate chamber connected to the fuel tank by tubing.

A liquid-vapor separator may be connected to the fuel tank by a number of tubes. This separator allows a space for liquid fuel to separate from the vapors and return to the fuel tank (**FIGURE 66-37**). This separation ensures that the liquid fuel does not get pushed into the charcoal canister and ruin it. A **vapor line** is connected to the vapor space in the fuel tank or the liquid vapor separator and carries fuel vapors from the fuel tank to a storage container called a charcoal canister. This vapor line can incorporate a check valve, which is a valve with a steel or plastic ball in it. The check ball moves into the passageway to block fuel if the vehicle is tilted too far from the horizontal and rolls over. The check valve stops liquid fuel from leaking and creating a fire hazard; thus, it is sometimes called a rollover valve.

Charcoal Storage

A charcoal canister is a canister filled with **activated charcoal** (**FIGURE 66-38**). "Activated" means the charcoal is highly porous with a very large surface area that can absorb and store large quantities of fuel vapor. The charcoal canister is connected to the fuel tank vapor line, and the charcoal absorbs excess fuel vapor from the fuel tank, primarily when the engine is not running. In some earlier carbureted engine designs, the canister also has a vapor line connected to the **carburetor float bowl** to absorb any fuel vapors from it.

The charcoal canister is designed to store the fuel vapors for a limited time before it must be purged. **Purging** is the process of drawing fresh air up through the activated charcoal for the purpose of pulling the fuel vapors from the charcoal and burning them in the engine. Purging is designed to normally occur when the engine is above a minimum temperature and has run for a specified time. This system prevents the engine's air-fuel ratio from being thrown rich or lean by the flow of purged gases while the engine is warming up. Purging is an important process, as it prepares the canister to absorb fuel vapors the next time the vehicle is not running. Because all air that enters the engine must be filtered, most charcoal canisters have an air filter installed on their air intake. This filter is replaceable on some charcoal canisters.

Capturing evaporated fuel vapors is a critical part of reducing one of the major sources of hydrocarbon emissions. The rates of evaporation are not the same for all fuels. Evaporation is higher with gasoline than diesel because gasoline is more volatile. If a fuel is more volatile, it is able to evaporate quickly and easily. Volatility is important for cold weather operation but causes excess vapor in warmer weather. For this reason, gasoline fuels are rated for winter or summer use, as the winter fuel must be more volatile due to the cooler atmospheric air temperature. Thus, a problem arises when a vehicle has a winter blend of gasoline in the tank and an unexpectedly hot spring day happens. The charcoal canister can then become overwhelmed with more vapor than it can handle and allow some of the vapor to escape past the canister to the atmosphere.

Manufacturers size charcoal canisters to a particular amount of fuel tank capacity and fuel volatility. In fact, some vehicles with large fuel tanks require the use of two or more charcoal canisters to provide enough fuel vapor capacity for the vehicle.

Vacuum Controls in the EVAP System

Older vehicles use vacuum to control the opening and closing of the canister **purge valve,** which is spring loaded in the closed position. Vacuum applied to the purge valve opens it so the canister can be purged (**FIGURE 66-39**). The vacuum signal to the purge valve is controlled by a thermal vacuum valve, which allows vacuum to pass through it to the purge valve only when the engine is above the specified operating temperature. The vacuum is provided by a ported vacuum tap just above the throttle plate, so vacuum is provided only when the vehicle is at cruise. At WOT the vacuum in the manifold falls to zero; thus, the canister cannot purge at WOT in the same way the EGR system cannot operate at WOT.

With the advent of early PCMs, manufacturers used an electrically operated **purge solenoid** to turn on and off the control vacuum to the purge valve (**FIGURE 66-40**). In this way, manifold vacuum can be controlled by the purge solenoid when the PCM sees that the engine is above the specified operating temperature and throttle opening. The PCM then sends an

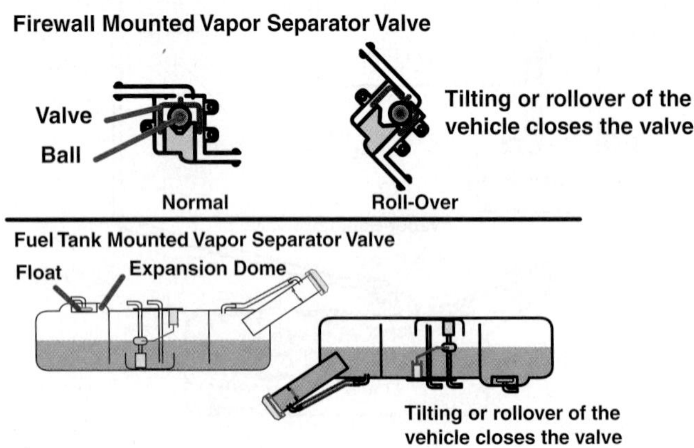

FIGURE 66-37 Liquid vapor separator.

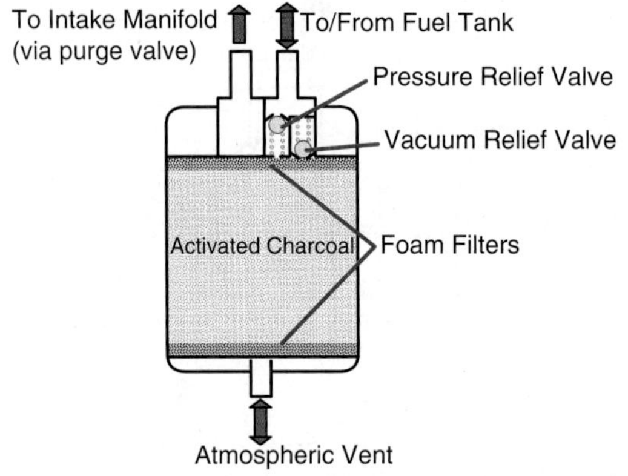

FIGURE 66-38 Charcoal canister.

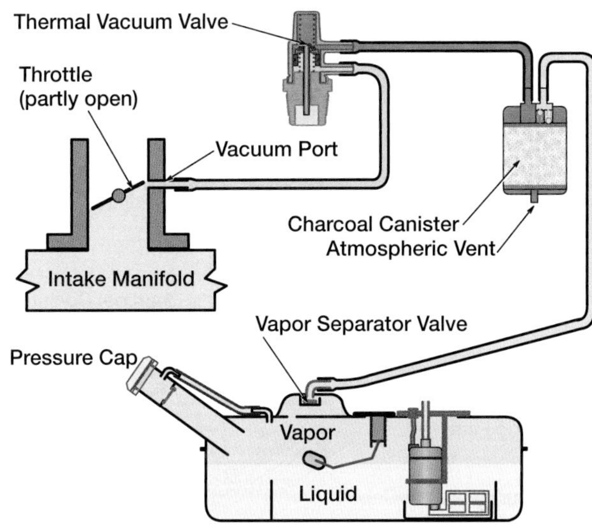

FIGURE 66-39 Vacuum-controlled EVAP system.

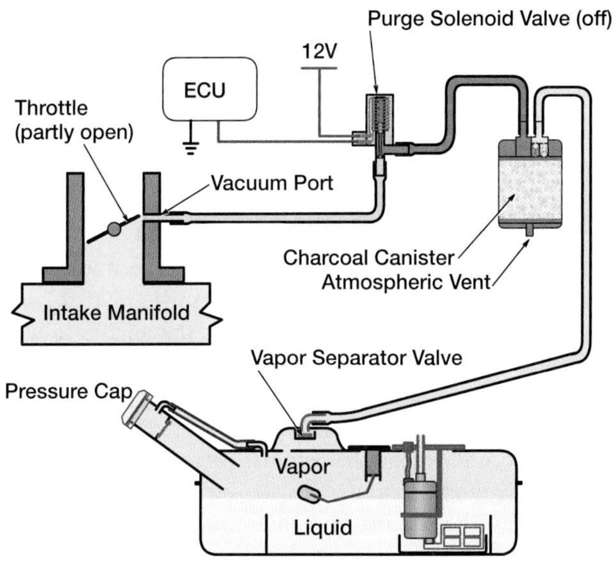

FIGURE 66-40 Electrically operated vacuum purge system.

electrical signal to the purge solenoid, which opens its vacuum valve and allows manifold vacuum to open the purge valve and start purging the canister.

Electronic Controls in the EVAP System

Most recent EVAP systems are electronically controlled to allow for more precise purge control. The PCM takes into account engine temperature, rpm, throttle position, and calculated load to decide when and at what rate the vapors can be purged. The PCM then commands the purge vapor from the charcoal canister (**FIGURE 66-41**).

The purge valve can be opened by an integrated solenoid or stepper motor. If it is a solenoid style, it can be either an on/off solenoid or a pulse-width-modulated solenoid, depending on the vehicle. If it is a stepper motor style, then the PCM can progressively open and close the purge valve one step at a time. Both of these styles allow for a variable amount of purge to occur.

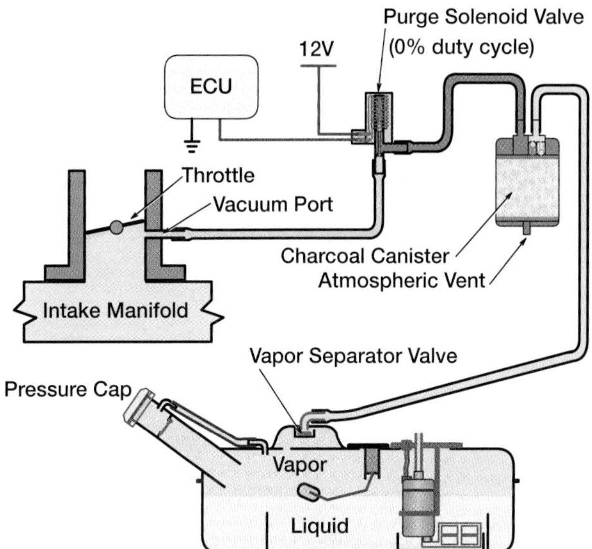

FIGURE 66-41 An electronically controlled EVAP system allows variable purging of stored gases and is controlled by the PCM.

Onboard Refueling Vapor Recovery System

The onboard refueling vapor recovery system is an enhancement to the EVAP system. Its purpose is to capture fuel vapors that would normally escape to the atmosphere during fueling of the vehicle. Instead of venting the vapors out the filler neck, they are directed through the charcoal canister where they are captured and held until they can be burned. The canister vent valve is open when the engine is turned off. This allows the vapors from the tank to vent through the canister where the fuel vapors are trapped and the air passes on, minus most of the hydrocarbons. In fact, the EPA has determined that the ORVR captures approximately 98% of the fuel vapors that would otherwise be vented to the atmosphere.

The system consists of a one-way check valve at the bottom of the fill pipe. The valve opens during refueling to allow fuel to enter the fuel tank. It closes after refueling to keep liquid fuel from splashing up the fuel tank filler pipe. Also, the filler pipe has a section which is reduced in diameter to act as a liquid seal that prevents vapors from traveling back up the filler neck as the tank is being filled. An ORVR vapor control valve is located inside the fuel tank and contains a float that rises with the fuel level and blocks off the vent hose when the tank is full preventing fuel from saturating the canister. A larger diameter than normal vent hose from the fuel tank to the charcoal canister allows the vapors to more easily pass to the canister during refueling. Also, the canister may be of larger than normal capacity to handle the additional vapors during fueling.

Computer Monitoring Strategies in the EVAP System

The EVAP system monitoring is a bit different from the other emission control monitors. EVAP monitoring is required not only to detect faults that prevent proper purging of the canister but also to detect any leaks greater than 0.020" (0.51 mm) in the

FIGURE 66-42 Purge switch.

system. Manufacturers are free to design their own monitoring systems as long as the system can accurately detect the minimum specified fault. Because manufacturers have the ability to design their own systems, a wide variety of strategies and components are used by the manufacturers. We explore some of the more common strategies in this section.

In some cases, manufacturers use dedicated sensors such as a **purge switch**, which is located in the purge line and senses either the flow of gases, or the pressure drop caused by the flow of gases, that are being purged (**FIGURE 66-42**). The purge switch then signals the PCM that purge is actually happening. Other EVAP monitoring strategies use the oxygen sensors to measure the decrease in exhaust oxygen content when the EVAP system is being purged. Another method accurately monitors the fuel trim correction data during purge.

Newer EVAP systems monitor purge system operation by reading a **fuel tank pressure sensor** while pulling a vacuum on the system. The PCM system closes the **vent solenoid**, which seals off the EVAP system from outside air. Either the purge valve is opened to allow manifold vacuum to pull a vacuum on the entire EVAP system. If the purge system is operating correctly, the fuel tank pressure sensor will report an increase in vacuum (decrease in pressure) that is within the specified pressure and time factors for the conditions currently present in that vehicle (**FIGURE 66-43**).

If the system passes the purge flow test, the PCM performs a leak test. The PCM commands the purge valve to close, thus sealing off the EVAP system from manifold vacuum. Because the vent solenoid is still closed, the EVAP system is completely sealed. The PCM monitors the fuel tank pressure sensor for a set time and looks for any loss of vacuum. If noted, the PCM will set a DTC based on the calculated size of the leak. The PCM is required to detect leaks as small as 0.040" (0.102 mm) (vehicles prior to 2000, large leak) and 0.020" (0.51 mm) (2000 and newer vehicles, small leak). A loose gas cap may set a separate DTC.

Some EVAP leak detection monitors are run after the vehicle has been shut down for several hours and outside air temperature is below a specified maximum (example 95°F or 35°C). If the air temperature is still not below the maximum specified at that time, then the PCM will try to run the monitor again after a period of time (example, two hours). In this type

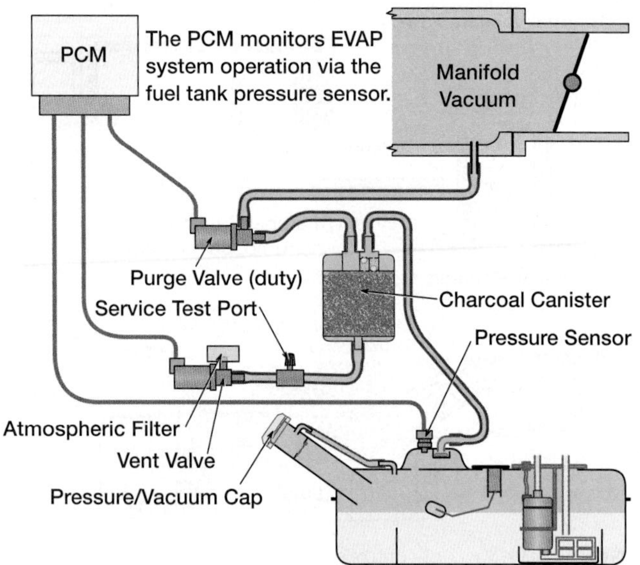

FIGURE 66-43 If the purge system is operating correctly, the fuel tank pressure sensor will report a specified increase in vacuum.

of system, because the engine isn't running, it uses an electric pump to create the vacuum (or pressure) in the system. The fuel tank pressure sensor is used to monitor the system for leaks.

▶ Secondary Air Injection

Secondary air injection is less common now than it was in previous years. There are several reasons for this. The first reason is improved engine design, which causes the engine to warm up much more quickly and therefore emit fewer emissions. Improved engine design also enhances turbulence and mixing of the air and fuel, which has the effect of reducing the emissions. The second reason that engines run much cleaner is due to the much more accurate fuel delivery to the combustion chamber. And the last reason is due to the use of highly efficient three-way catalytic converters. In fact, many new vehicles are able to meet the tightened emission regulations even without the secondary air injection system. Of those vehicles that are equipped with a secondary air injection system, many only operate during engine warm-up to burn off the increased hydrocarbons and carbon monoxide, as well as to help heat up the catalytic converter more quickly.

When older carbureted and throttle body–injected engines were started up cold, they needed a richer mixture to operate due to the cold intake manifold surfaces, which condensed fuel out of the air and created a lean condition. This rich mixture resulted in the excessive release of hydrocarbons and carbon monoxide from the tailpipe until the engine warmed up.

To reduce hydrocarbons and carbon monoxide, the secondary air injection system was added. It injected air (oxygen) into the hot exhaust gases as they left the combustion chamber. The extra oxygen caused the hot hydrocarbons and carbon monoxide to burn, converting them into water and carbon dioxide. The burning of the hydrocarbons and carbon monoxide also added extra heat to the exhaust stream, which helped warm up the catalytic converter more quickly, making it operational sooner.

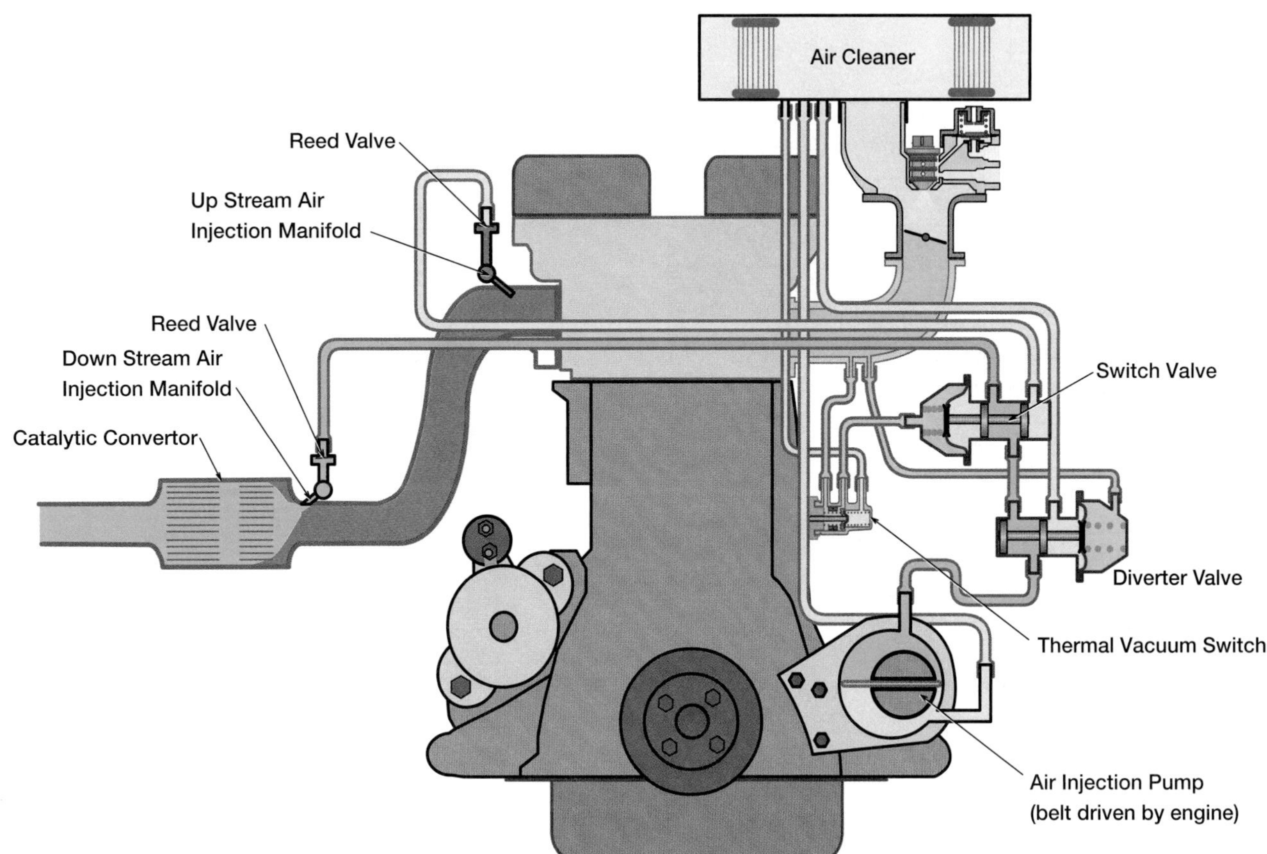

FIGURE 66-44 Secondary air injection system.

Once the catalytic converter heated up, the air (oxygen) was switched from upstream (near the exhaust valves) to downstream, where it entered the oxidizing catalyst in the converter. The extra oxygen was used to help oxidize any remaining hydrocarbons and carbon monoxide in the converter.

Secondary air injection systems use three valves to control airflow in the system. The **air switching valve** is used to switch air from upstream to downstream once the engine reaches the specified temperature. The **air diverter valve** diverts the injection air from the exhaust stream to the air cleaner assembly whenever additional air in the exhaust stream could cause backfiring in the exhaust, such as during deceleration or WOT. The last type of valve used in the system is a check valve (**FIGURE 66-44**). It prevents hot exhaust gases from traveling backward through the hoses and valves to the air pump, which would overheat and ruin the hoses, valves, and pump in short order.

Early secondary air injection systems used a thermal vacuum switch in the vacuum line to control the switching valve. When the engine was cold, the thermal vacuum switch allowed vacuum to the switching valve, which moved the switching valve to the upstream position. Once the specified engine temperature was reached, the thermal vacuum switch would block the vacuum from reaching the switching valve, and a spring moved the switching valve to the downstream position. In later designs, the air switching valve and diverter valve were operated by an electrical actuator controlled by the PCM for more precise control of the air injection system.

Secondary Air Injection Types

There are two main types of air injection systems: pumped and aspirated. Aspirated systems use an aspirator valve instead of an air pump to help facilitate the moving of air in the system. Aspirated systems are also known as pulse air systems. The pumped type of system uses an air pump to move the air and may be belt driven by the engine or electrically driven by an electric motor, only when needed (**FIGURE 66-45**). The air pump is turned on and off by the PCM based on the needs of the emission system. The electrically driven air pump type is the most recent of the secondary air injection systems. Any of these systems do the same job of ensuring that the exhaust stream receives oxygen at the correct locations and time.

The **pulse air system** uses the pulsations of the exhaust gas to open and close a **reed valve** (also called an aspirator valve), which is a flat metal valve that flexes when pressure is applied against it (**FIGURE 66-46**). The reed valve admits air into the exhaust manifold in short bursts but prevents air from moving back through the reed valve. Air drawn from the air filter enters the exhaust manifold through the reed valve. The pulse air injection is suited for engines with four or fewer cylinders, because their exhaust pulsations are further apart and more pronounced, creating more efficient pulling of air into the exhaust stream.

Systems that use an **air pump** are installed on engines with more than four cylinders and supply a steady volume of air to the system. The air pump is commonly a mechanical vane-style pump that moves the air at low pressures, in some cases as low

FIGURE 66-45 There are a two main types of air injection pumps: **A.** Belt driven. **B.** Electrically driven.

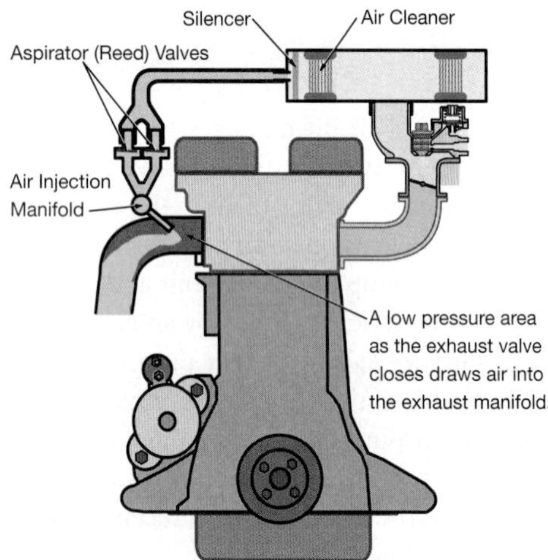

FIGURE 66-46 Pulse air injection system.

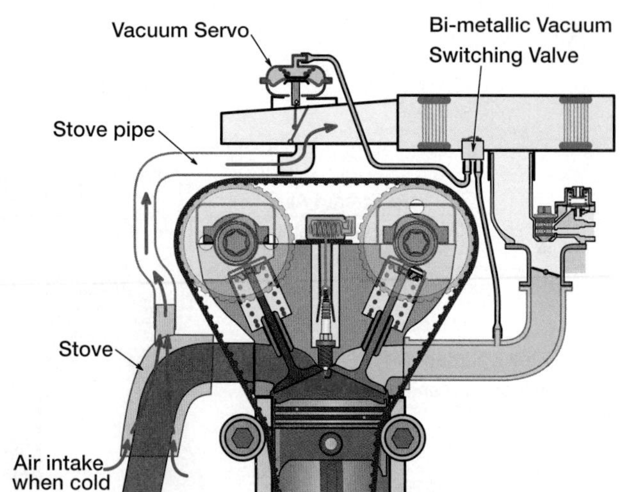

FIGURE 66-47 Heated air intake systems use the air around hot exhaust manifolds to blend with outside air, to supply warm air at a predetermined temperature.

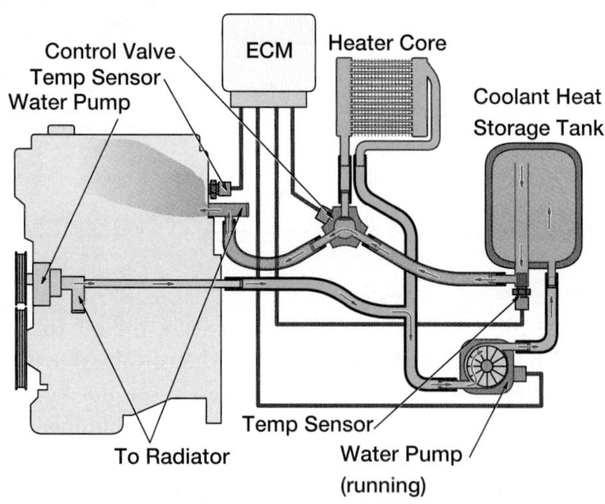

FIGURE 66-48 The Toyota coolant heat recovery system is used to preheat the engine during cold start.

as 2 pounds per square inch (psi), or 13.8 kilopascals (kPa). If the air pump is belt driven, it can be driven by a V-belt or a serpentine belt.

Heated Air Intake Systems

Because liquid fuel will not burn, fuel must be in a vapor form for it to burn. So the challenge is to get liquid fuel to fully vaporize by the time it has to be ignited, but not much sooner, and certainly not later. Intake manifolds carry the air-fuel mixture. Fuel tends to condense on the inside surface of the cold manifold. One method of minimizing this problem is to use a **heated air intake system**. This system collects hot air from around the exhaust manifold and mixes it with outside air entering the air cleaner assembly (**FIGURE 66-47**). The system maintains a specified air temperature of air entering the engine. This simple system uses a temperature-sensitive valve inside the air cleaner. It operates a flap that blends the hot air with cool air so that the intake manifold receives air at about 100–110°F (38–43°C), regardless of the outside air temperature. Maintenance of this consistent temperature assists in the proper **vaporization** of the fuel, particularly when the engine is cold.

Vaporization is also assisted in some engines, such as in the Toyota Prius, by circulating stored hot liquid coolant through passages in the intake manifold and cylinder head, thereby preheating them right before the engine is started (**FIGURE 66-48**). Another past method of preheating fuel and air mixtures before they entered the combustion chamber was to use a heated grid under the throttle body to assist in heating air in order to promote the vaporization of the fuel (**FIGURE 66-49**). Fuel will not mix easily in cold air temperatures, so the use of a heating grid assures a more combustible fuel mixture.

FIGURE 66-49 An electric grid under the throttle body was used on some vehicles to heat the incoming air.

▶ Emission System Diagnosis

Diagnosing emission control system faults can be some of the most demanding work a technician performs. It requires a very good understanding of the following disciplines and more:

- Engine theory and diagnosis
- Fuel system theory and diagnosis
- Ignition system theory and diagnosis
- Electrical/electronic system theory and diagnosis
- Emissions and emission control systems theory and diagnosis
- Strategy-based diagnosis theory and experience

As you can see, that is a lot of content to know and experience to have. At the same time we have covered each of the other systems earlier in this book. So you should have a good head-start on understanding them. We focus mostly on the emission control system diagnosis and testing in the rest of this chapter.

With emission control systems, often the first indication of any trouble is an illuminated MIL on the dash. In many cases, the vehicle may not exhibit any drivability issues. Even so, diagnosis of emission control systems starts with verifying the concern. This could be by doing a test drive, or it could be by observing the MIL and retrieving the DTCs, or both. If there are DTCs, you will most likely want to follow up with those. Once you have any DTCs, research the service information and any related technical service bulletins (TSBs). This information will help you determine which focused tests need to be performed to narrow down the cause of the fault. One additional piece of information is knowing the enabling criteria for the DTCs that you retrieved. Knowing what conditions trigger the setting of a particular DTC will be a big advantage in guiding your diagnosis.

The PCM's freeze-frame data and data stream can also give you good clues as to where to look for the fault. Any data that are out of specification will help inform your diagnostic process. If the vehicle's PCM has bidirectional capability, it may be helpful to operate suspect components and observe the results. For example, if there is a DTC regarding the canister vent solenoid, you might want to command it to close and then check to see if it indeed did close, both electrically and mechanically. You can then determine if it is responding appropriately and again gather information that will point you in the direction of the fault.

FIGURE 66-50 Five-gas analyzer.

If there are no DTCs present, then having a good understanding of the types of emission control systems the vehicle is equipped with is a good place to start. This can best be found by researching the "description and operation" for each of those systems in the appropriate service information. Use that information plus any "symptom diagnosis" procedures to start diagnosing the system. Also, we cover some of the common testing and diagnostic procedures for each of the emission control systems below.

▶ Emission Testing

You must have a thorough understanding of emission control systems in order to test and repair an emission failure. Some states require emission testing to ensure that all emission control devices are working properly and that proper combustion is taking place. This could be as simple as verifying that there are no DTCs set in memory and that certain monitors have run. Or it could require a tailpipe emission test.

Several tools can be used to ensure that emission systems are working properly. One of these tools is the **five-gas analyzer** (**FIGURE 66-50**), which uses sensors to measure the level of gases in the exhaust stream. The technician compares the analyzer readings to the manufacturer's acceptable levels for that particular vehicle. The manufacturer's specifications may be located in the software of the gas analyzer you are using, or your state emission inspection agency may provide the specifications for each vehicle. Some states require the use of the five-gas analyzer with a dynamometer; others require only a five-gas idle test or a combination of an idle and a 2000-rpm test. See **FIGURE 66-51**, which shows a typical state emission test report for a vehicle that passed the exhaust analyzer test.

▶ TECHNICIAN TIP

The emissions standard chart is used to ensure that a vehicle passes the required emissions laws for the state. The chart may be labeled in percentage (%), parts per million (ppm), or grams per mile (g/mi). The EPA provides emissions standards for federal and California emissions, although individual states or municipalities may have their own standards.

Start Date/Time: 01/05/2007 04:52 PM | Operator 1: AMANDAM AMANDA | Operator 2: LARRYWE LARRY

Gas Cap Test Results | Cap 1: PASS | Cap 2: N/A

Result: PASS | Cap 3: N/A | Cap 4: N/A

This emission test was performed in accordance with federal regulations on emission tests (40CFR 85 Subpart W) by Applus Technologies, Inc. for the Department of Ecol

Test Results | Test Type: ASM | Prescribed Type: ASM

Overall ASM Result: PASS | Number of Restarts: 0

	HC (PPM)	CO (%)	CO+CO2 (%)	O2 (%)	RPM
Cruise Limit:	150	1	6	N/A	N/A
Cruise Emissions:	36	0.21	15.51	0.3	N/A
Cruise Result:	PASS	PASS	N/A	N/A	N/A
Idle Limit:	220	1.2	6	N/A	N/A
Idle Emissions:	24	0	15.3	0.5	850
Idle Result:	PASS	PASS	N/A	N/A	N/A

FIGURE 66-51 Typical state emission test report for a vehicle that passed the test.

A rule of thumb for emission testing is found in **TABLE 66-3**. This chart is for engines at idle and is typical of what a five-gas analyzer will display. Be sure to refer to your vehicle's specifications if performing an emission test for your state, as each powertrain option creates differing emissions.

Vehicles produced after 1996 use the OBD II requirements. This system contains additional sensors and testing strategies that have led some states to do away with mandatory five-gas dynamometer testing and instead perform an **onboard diagnostic (OBD)** system test to ensure that all tests (or monitors) have been run and passed by the engine's computer and that there are no DTCs in memory. If an OBD test fails, a related diagnosis code may be set that alerts technicians to the fault. If all tests pass, an **idling five-gas test** may be performed. In an idling five-gas test, a five-gas analyzer takes a sample of gases from the exhaust pipe.

Setting Up a Gas Analyzer

A gas analyzer is a useful tool to determine emissions and drivability problems with a vehicle by analyzing the tailpipe gases. If combustion is not working as designed or an emission control device is malfunctioning, excessive emissions will be released. Some states make this analysis a mandatory test conducted periodically (every one or two years) to certify that the vehicle is not excessively polluting the air.

When setting up an analyzer, there are several details to be aware of. The first is that all emission analyzers require a certain amount of time to warm up. Up to about 10 minutes is required on some machines. Any readings before the analyzer is fully warmed up will not be valid.

Next, the analyzer hose and probe are susceptible to cracking. Cracks would allow ambient air to leak into the sample exhaust stream, diluting it and leading to an incorrect reading. The hose and tube should be leak checked before every test or at least once per day. Leak testing can normally be performed by plugging the holes in the metal sniffer tube and watching for the low flow indicator to illuminate on the analyzer. If it does illuminate, there are no substantial leaks in the sample hose and sniffer. If it does not illuminate, there is a leak that will have to be found and repaired.

The analyzer also has to be zeroed before it can be used. On some analyzers, this step is done automatically; on others, it is manually performed. Just make sure you follow the instructions and zero the analyzer properly. Also, the analyzer should be checked with calibration gases, typically monthly. This process uses calibration gases of specific concentrations to run through the analyzer so that its accuracy can be checked. If the analyzer is operating correctly, it will display the proper readings for the particular calibration gas.

Preparing a five-gas analyzer to obtain exhaust readings requires several steps before the actual exhaust gas reading is obtained. To prepare a five-gas analyzer to obtain exhaust readings, follow the steps in **SKILL DRILL 66-1**.

Catalytic Converter Testing

N66001

A catalytic converter can fail for a number of common reasons, including contamination, overheating, physical damage, or just old age. Contamination happens when the catalyst becomes coated by a substance such as lead from leaded fuel, silicone from improper sealants, or carbon from excessively rich conditions or repeated short driving trips. The contaminant prevents the exhaust gases from interacting with the catalyst, so the gases cannot be converted properly, if at all. Lead and silicone contamination cannot generally be cleaned from the catalyst, so the converter must typically be replaced in this case. Carbon-contaminated catalyst can sometimes be restored first by correcting any conditions that caused the engine to run rich, and second by driving the vehicle for an extended time at freeway speeds under a moderate load. This process will heat up the converter and possibly burn the carbon contaminants from the catalyst, restoring its function.

Overheating the catalyst can cause the catalyst material to burn off the substrate, rendering it inoperative. Overheating can also melt the substrate, which can block the passageways and restrict exhaust gas flow, reducing the engine's power and efficiency. Overheated catalysts cannot be restored and must be replaced.

Physical damage can happen due to an object such as a large rock striking the converter or from thermal shocks, which can happen if the hot converter is immersed in cold water, such as a deep pothole. The substrate, being made of a ceramic material, is quite brittle and can break reasonably easily. Physical damage can usually be found by first visually examining the converter

TABLE 66-3: Typical Emission Test Readings

With Converter	Hydrocarbon	Carbon Monoxide	Oxygen	Carbon Dioxide	Oxides of Nitrogen
	< 100 ppm	< 0.5%	0% to 2%	12% to 15%	< 100 ppm, idle

Note: This chart will not relate to all vehicles; refer to your state's specifications.

Source: Information from Arizona state emissions site.

SKILL DRILL 66-1 Using a Five-Gas Analyzer

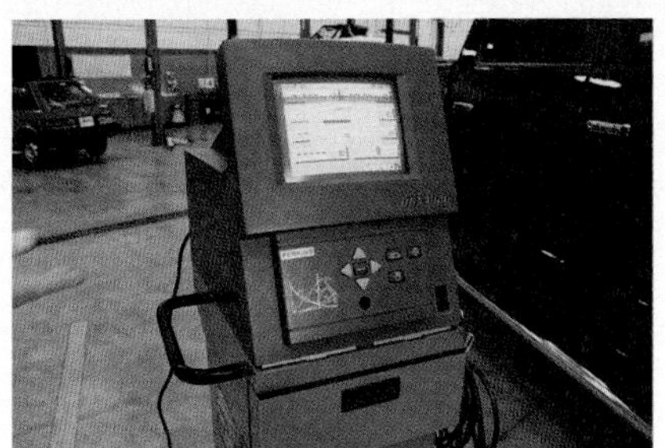

1. Research the local emission limits for this vehicle in the applicable manual or online sources. Warm up both the vehicle and the exhaust gas analyzer. Leak test the analyzer hose and probe.

2. Calibrate the analyzer, if applicable. Insert the exhaust analyzer probe in the vehicle's exhaust pipe at least 18" (46 cm).

3. Allow the vehicle to idle, and record the emission readings. Compare your readings to the specifications, and determine any necessary action(s).

shell for dents indicating abuse. Also, lightly tapping on the converter shell with your hand will cause any broken pieces of substrate to rattle around inside, indicating a broken catalyst. Physically damaged converters must be replaced. Also, if the catalyst is broken, it is possible that the broken pieces may be plugging up the rear of the converter or even the muffler. Check for this problem if the catalyst is broken.

Catalysts that are damaged due to contamination, overheating, or old age can be tested to see if they are still functioning. Testing the catalytic converter should be performed according to the manufacturer's service information. If no procedure is outlined in the service information, then it may be necessary to use one of the following tests.

If the vehicle is OBD II compliant, you can either use a scan tool that can access mode 6 and look at the test results for each of the catalyst monitors, or you can use a lab scope to compare the upstream and downstream oxygen sensor patterns to determine whether the catalyst is working efficiently. When using mode 6, you need to research the specific codes that the manufacturer uses to identify each catalyst monitor, as well as the coding for the actual results. Once you know what you are looking at, you will be able to determine how well each catalyst is performing. For example, if the upper limit for the catalyst monitor in mode 6 is 20, and mode 6 shows it at a 3, you will know that the catalyst is working very well. If the monitor reading is 25, you will know that the converter is not working well. Just make sure the rest of the ignition, fuel, and emission systems are working well before condemning a catalytic converter. Of course, this is true no matter how you test the converter.

Some manufacturers use mode 6 to display how many counts the catalyst monitor has detected oxygen sensor activity outside the specified parameters, indicating a faulty converter. Be sure to research the manufacturer's information when using mode 6 for diagnosis.

When diagnosing cat problems using the O_2 sensors, you need to understand the O_2 sensor numbering scheme. The O_2 sensors should be labeled, for example, $HO_2B_1S_1$ (for heated oxygen sensor, bank one, sensor one) and HO B_1S_2 (for heated oxygen sensor, bank one, sensor two). If the engine is a four-cylinder in-line engine, there is only one bank. If there are two banks, then bank 1 is the bank with the number-one cylinder in it. As for the oxygen sensor number; S1 is the closest to the combustion chamber, S2 is next closest, and so on. So S1 is typically the pre-cat sensor, and S2 is typically the post-cat sensor. But check the service information to verify that, as some vehicles have three sensors in each bank.

When using a lab scope connected to both the pre-cat O_2 sensor (short for "pre-catalytic converter oxygen sensor") and the post-cat O_2 sensor, be sure the catalytic converter is up to operating temperature. If the catalyst is working properly, the downstream oxygen sensor will be relatively stable, whereas the upstream sensor will be constantly toggling between rich and lean. If the catalyst is not working properly, the upstream sensor will be toggling (indicating a properly operating fuel system), and the downstream sensor will also be toggling, similar to the upstream sensor.

For vehicles using an air-fuel ratio sensor mounted before the catalytic converter instead of a zirconia oxygen sensor, you cannot run the same test procedure with this type of air-fuel sensor. It is tested differently based on a 3.3V (0.0 mA) reading at stoichiometry. A reading above 3.3V ($\leq$ 2.2 mA) indicates a lean mixture (high exhaust oxygen content), and a reading below 3.3V ($\geq$ -2.2mA) indicates a rich mixture (low exhaust oxygen content). Check with each specific manufacturer for instructions on how to test these air-fuel ratio sensors. At the same time, most manufacturers use zirconia oxygen sensors after the catalytic converter so you can monitor them for the level of fluctuation, which will be of a lower, but opposite voltage, than the AFR sensors.

A good test to use on a non–OBD II vehicle is called the **intrusive test**. When performing the intrusive test, a hole is drilled in the exhaust pipe in front of the converter so that an emission analyzer tube can be inserted into the exhaust stream entering the converter. That reading is then compared to a reading taken after the converter at the end of the tailpipe. If the catalyst is working properly, a general rule of thumb is that hydrocarbons, carbon monoxide, and oxides of nitrogen will be reduced by around 90%, assuming the engine is operating as designed. The upstream reading indicates whether the engine is operating correctly by giving readings that are consistent with a stoichiometric ratio. The downstream reading indicates whether the converter is operating correctly if the harmful emissions are being reduced by approximately 90%. Once the test is complete, the drilled hole is plugged with a stainless steel rivet.

If the catalytic converter failed, determine the cause of the failure. Replacement of the converter may mask the symptoms, but only for a short time, and then they will show up again as another failed converter. So always verify that the engine management system is functioning properly before installing a new converter.

TECHNICIAN TIP

Because the catalytic converter creates heat as it is operating, it was once thought that you could measure the temperature from the inlet side of the converter and compare that to the temperature at the outlet of the converter. If the converter was working properly, then there would be a 50–100°F increase of temperature at the outlet of the converter. But with fuel systems that run much tighter fuel ratios, that is no longer the case. Thus, this test is no longer considered to be valid on newer vehicles.

To test a catalytic converter for efficiency (lab scope method), follow the steps in **SKILL DRILL 66-2**.

SKILL DRILL 66-2 Testing a Catalytic Converter for Efficiency

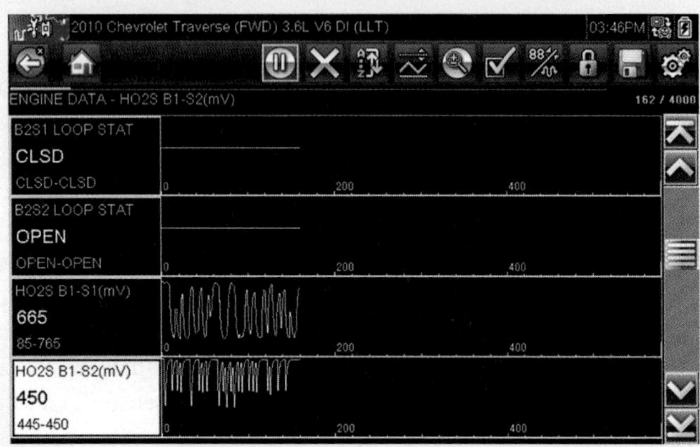

1. First, ensure the vehicle is fully warm by performing a road test. Install a scan tool with graphing capability, and highlight the pre-cat O_2 sensor and the post-cat O_2 sensor.

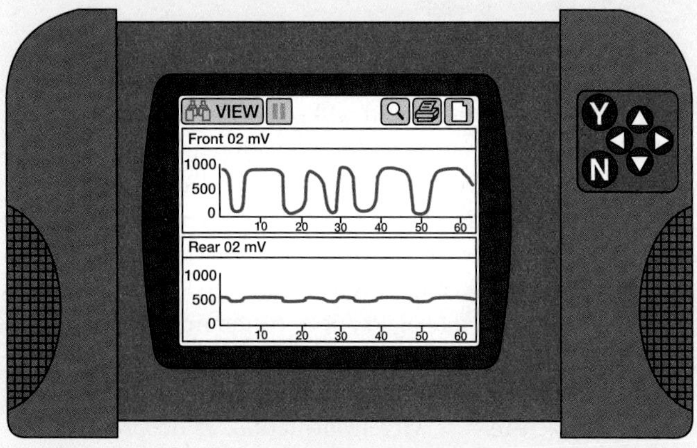

2. Set the scan tool so it will graph both sensor readings. Raise engine rpm to 2000, and watch the oxygen sensor readings. The pre-cat sensor should toggle from rich to lean a certain number of cycles in a specified time, indicating good control of the mixture. The post-cat sensor should tend to run at a steady midpoint reading because the converter is using up much of the leftover oxygen, converting the hydrocarbons and carbon monoxide.

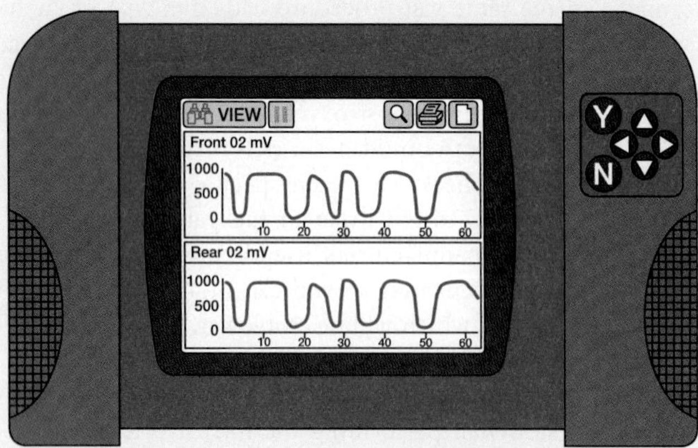

3. If the rear oxygen sensor pattern tracks the front sensor, the catalytic converter is not functioning and will have to be replaced.

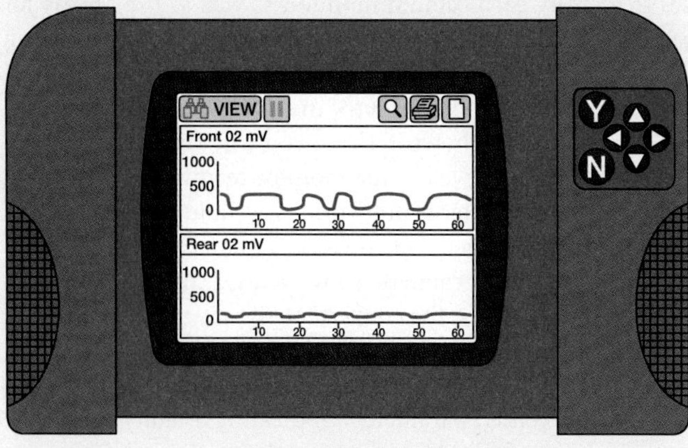

4. If the front sensor is not toggling properly, the engine or fuel system is not operating properly, and you will need to diagnose that condition further. You must keep in mind the symptom-to-system-to-component-to-cause diagnostic process used by most manufacturers.

▶ Diagnosing EVAP System–Related Concerns

N66002

EVAP system faults may or may not create drivability issues. If the EVAP system continues to purge at idle, a rough idle could be created. If the vapor line between the charcoal canister and the fuel tank is ruptured, there are likely to be no drivability issues, but the MIL will probably be illuminated. Because the federal regulations require the OBD II system to monitor each of the emission control systems and turn on the MIL when the emission output is 1.5 times the FTP, emission control system faults normally are accompanied by an illuminated MIL. All vehicles built from 1996 up have enhanced evaporative systems that can detect purge flow and leakage in the evaporative system.

EVAP systems are somewhat prone to failures. One weak area of the EVAP system is the gas cap. Because it is removed and reinstalled often, it is possible that it might not get tightened all the way. It should be tightened until it clicks to ensure there is no leak. Leakage could also result from a damaged gasket or seal. Gas cap leaks are detected by the monitor. A code is then set and the MIL turned on.

Another common EVAP fault relates to the rubber and plastic vacuum lines that run between the components of the EVAP system. A small hole or crack in the line will allow vapors to escape to the atmosphere. This hole will be detected by the EVAP monitor. A DTC will then be set, and the MIL will come on. Many vehicles use a plastic fuel tank, which can crack and leak, resulting in an EVAP system failure. Sometimes, the use of a smoke machine may be necessary to find the leak in the system.

Other failure areas for the EVAP system are the vent and purge solenoids. If the valve fails electrically, it may not open or close when commanded by the PCM. The valves can also fail mechanically and stick open or closed. The vent solenoid may also get plugged with mud or debris. These types of failures typically result in a DTC being set and the MIL turning on.

Inspecting and Servicing the EVAP System

The hoses of the EVAP system carry the hydrocarbon vapors from the fuel tank to the charcoal canister and then to the engine to be burned. Valves and solenoids are used to control the flow of vapors into and out of the charcoal canister, as well as to direct them into the engine depending on operating conditions. If a leak exists in the EVAP system (including the gas cap) or hoses, hydrocarbons can be released into the atmosphere. An EVAP system code will be set by the ECM if a leak is detected. If the leak is large, the PCM will set a P0455 (gross leak) DTC. If the leak is small, the PCM will set a P0456 (small leak) DTC. Other codes could indicate low purge flow or a fault in one of the sensors, valves, or solenoids. The service information will guide you through the proper diagnosis, depending on the DTC. However, a visual inspection of the hoses and components may give you a clue as to the cause of the fault.

To inspect the hoses and components of the EVAP system, follow the steps in **SKILL DRILL 66-3**.

To diagnose EVAP-related concerns, a good understanding of how the system operates is required. Use a scan tool to pull any diagnostic codes. The freeze-frame data stored can give a good indication of the conditions that set the code. Most manufacturers have the technician record the freeze-frame data in order to recreate the operating conditions at the time of the fault. If diagnostic codes are present, follow the manufacturer's diagnostic trouble chart. If there are no diagnostic codes, use the scan tool to view the live sensor and actuator data for the EVAP system sensors and actuators. Compare the data displayed with the service information data. If necessary, command the actuators on one at a time, and watch how the sensor data respond. The data, along with an understanding of the system, will likely lead you to the fault.

SKILL DRILL 66-3 Inspecting and Servicing the EVAP System

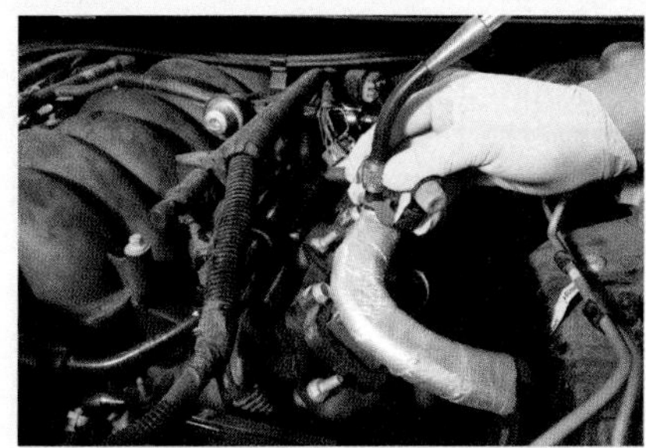

1. Use a scan tool to check for codes related to the EVAP system. If a code is found, follow the diagnostic procedures for that code. If the diagnostic code indicates a leak, connect an EVAP tester to the EVAP test port.

2. Use the EVAP system leak detector to determine whether the system is in fact leaking and, if so, the size of the leak.

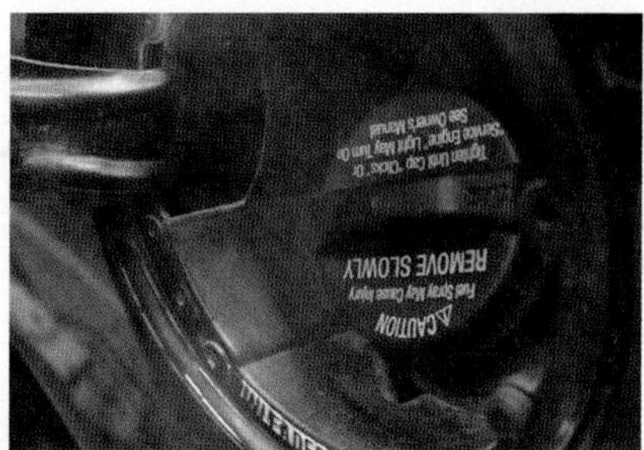

3. Use a smoke machine to help locate the leak. If smoke is detected, look in the area of the smoke to identify the cause of the leak. If no smoke is detected, the system could be leaking internally through a stuck-open purge valve or a leaky vent valve. Repair according to the manufacturer's procedure.

▶ Diagnosing PCV-Related Concerns

N66003, K66005

A PCV system is used to remove blowby gases from the crankcase in an environmentally friendly manner. If blowby gases are not removed from the crankcase, pressure can build up and cause oil seals and gaskets to leak oil. Replacement of seals or gaskets might temporarily fix the problem until excessive pressure pushes the seal or gasket out of place again. Proper testing of the PCV valve system is critical to get to the root cause of the concern and avoid repeat failures.

To diagnose PCV-related concerns, first locate the PCV valve. Refer to the service information for location of the PCV valve and hoses. If the PCV valve is located in the valve cover, remove the PCV valve from the cover, and ensure that a strong vacuum is coming through the valve with the engine running. This ensures the PCV valve and hose are clear. Reinstall the PCV valve into the valve cover, and remove the breather tube from the air cleaner assembly. With the engine still running, check that vacuum builds up in the breather tube (and crankcase), indicating that the breather tube is clear and there are not excessive blowby gases. Inspect all hoses for softening, brittleness, cracks, kinks, and holes. Cracks in hoses can result in vacuum leaks, which can create a poor idle situation. Inspect the PCV valve and ensure it has the correct part number for the vehicle. If the incorrect part is found or the valve is sludged up or sticky, replace it with a new PCV valve.

Inspecting and Servicing the PCV System

N66004

The PCV system draws crankcase gases out of the crankcase and burns them in the engine to reduce crankcase emissions. It is an important system that has to be serviced on a regular maintenance interval determined by the manufacturer and published in the service information.

PCV failures can occur and potentially create drivability problems for the customer. The most common problem is when the PCV system has a vacuum leak. The hose connecting the PCV valve to the intake manifold can crack and create a vacuum leak. Vacuum leaks tend to cause a lean air-fuel mixture and create a rough idle, which is most noticeable when the vehicle is stopped at a stop light. It also causes the LTFT to increase, the worse the leak.

PCV systems can also become clogged. Over time, especially if the vehicle's oil changes are neglected, the PCV valve or hose can plug up with sludge. If this happens, the PCV system cannot ventilate the crankcase, and blowby gases will contaminate the oil further, which creates more sludge throughout the engine. Another common failure from a clogged PCV valve or hose is oil being pushed up into the housing. If oil is found in the air filter housing, the PCV system is either being overwhelmed by the blowby gases (the piston rings are worn out), or the suction side of the PCV system is clogged/restricted. This causes pressure to build up in the crankcase, which then forces oil mist up the fresh air hose to the air cleaner housing. In this case, check that there is strong vacuum present at the PCV valve when the engine is idling and that the valve is not restricted. Also test the crankcase pressure, using a blowby gauge.

Repeat engine oil leaks may be the result of excessive pressure in the crankcase. If this pressure cannot be released, it will continue to build and eventually either push oil past the seals or in some cases even push oil seals and gaskets out of position, causing an oil leak. In the event of multiple gasket or seal failures or repeat failures, be sure to check that the PCV system is operating correctly by testing for the proper PCV system flow.

To test the PCV system, follow the steps in **SKILL DRILL 66-4**.

SKILL DRILL 66-4 Inspecting and Servicing the PCV System

1. Locate the PCV valve. With the engine idling, remove the PCV valve, and check for the presence of a strong vacuum.

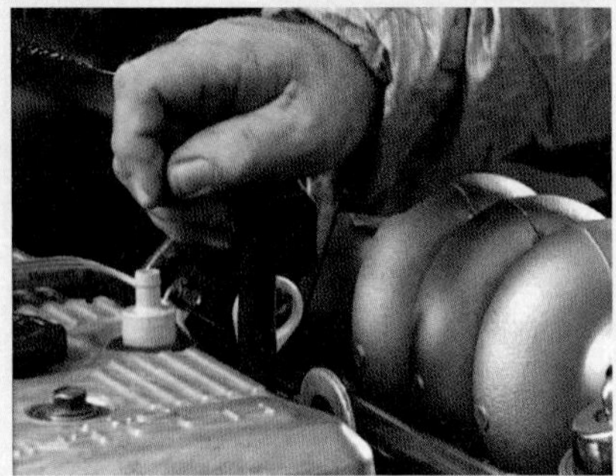

2. Remove the hose, and check that it is still pliable and not clogged with sludge deposits.

3. Remove the PCV valve, and inspect it for deposits. If there are any issues, replace it with a new one of the same type. If fixed orifice style, make sure the passageway is clean.

SKILL DRILL 66-4 Inspecting and Servicing the PCV System (Continued)

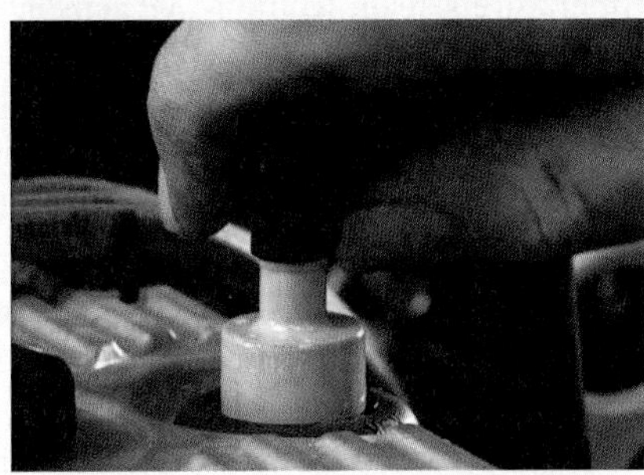

4. Reinstall the PCV valve into the valve cover.

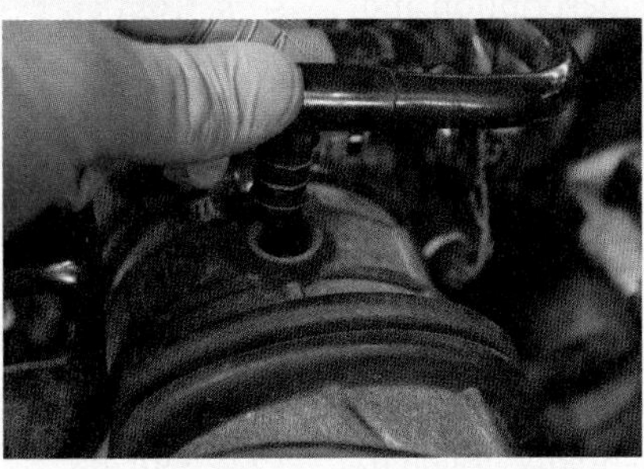

5. Remove the breather hose from the air cleaner assembly.

6. With the engine idling, block off the breather hose and feel for vacuum building up in the breather hose and crankcase, indicating that the PCV system can handle the amount of blowby gases. There should also be a slight amount of measurable vacuum in the crankcase at idle, typically measurable at the dipstick tube as 0.5"–2" H_2O. This would indicate that the PCV system is operating and adequate for the blowby produced by the engine.

▶ Diagnosing Abnormal Exhaust

N66005

To diagnose abnormal exhaust color, you have to understand what each color represents. Exhaust with a bluish tint represents oil that is being burned in the combustion chamber. This condition typically means the piston oil control rings are worn; there is excessive valve stem-to-guide clearance; or the valve seals are worn. Black exhaust represents an overly rich air-fuel mixture. This condition could be caused by a significant number of faults, such as a leaking or sticking fuel injector, a leaky vacuum diaphragm on the fuel pressure regulator, an open coolant temperature sensor circuit, or a MAP sensor reading incorrectly. All of these conditions would supply more fuel than needed, causing the engine to run rich and possibly emit black smoke. White exhaust represents water or coolant burning in the combustion chamber. This condition could be caused by a number of faults, such as a blown head gasket, a leaky intake manifold gasket, or a crack in the head or block that is allowing coolant to leak into the combustion chamber.

To diagnose an abnormal exhaust, gain as much of an understanding of the customer concern relating to exhaust as possible. Understanding the customer's complaint is the first stage of the repair process. Does the customer have additional information to give you, such as whether the issue happens when the engine is cold or hot? Did the engine recently overheat? Verify the customer concern by starting the engine and attempting to duplicate the conditions that the customer described. Research the possible causes of the concern in the appropriate service information or TSBs. Perform any specified tests to verify the cause of the customer's concern.

▶ Diagnosing EGR-Related Concerns

N66006

EGR valves typically cause two types of drivability concerns. One is a rough idle concern, and the other is a preignition or detonation concern known as **pinging**. Pinging can be referred to as the sound of marbles being shaken in a glass jar and is damaging to the engine if it is allowed to continue. Engines ping under moderate and heavy loads, such as when climbing a hill or pulling a heavy load. If the EGR valve is stuck in a closed position or if passageways are clogged, pinging can result, along with excessive oxides of nitrogen emissions. If the EGR valve is stuck in an open position, the engine will idle excessively rough and may stall at stops.

If the vehicle is using a **diaphragm-style EGR valve** (either the positive back-pressure or negative back-pressure type), the diaphragm could develop a leak, a vacuum line could crack or break, or a temperature control valve could stick open or closed. The EGR exhaust passageway could also clog from a buildup of carbon, which would block exhaust gases from entering the intake, and combustion temperatures would continue to climb. If the EGR system uses an actuator or solenoid to control vacuum, the actuator or solenoid could develop an electrical or mechanical fault, resulting in little to no vacuum to the EGR valve or full vacuum at all times. Stepper motor EGR valves can fail electrically or mechanically. In either case, the pintle will not operate correctly, creating idle or drivability problems. The EGR valve can also be the source of a vacuum leak, either externally or internally.

To diagnose EGR-related concerns, first gather the customer concern and as many specific details as possible. Consult the service information, and research the "description and operation" of the EGR system for the vehicle you are working on. With either a scan tool or a vacuum pump (depending on the system; refer to the service information on testing), operate the EGR valve with the vehicle running. Idle should become rough, and the engine may eventually stall as the EGR valve is opened and exhaust gas flows into the intake manifold. If there is no difference, or very little difference, in the idle speed and quality, suspect plugged-up EGR passageways, or a faulty EGR valve. If the idle was substantially affected, perform testing of the control system to ensure that a signal is getting to the EGR valve. Verify that either a vacuum or an electrical signal is getting to the EGR valve when the engine is running at about 2000 rpm. If a signal is not getting to the EGR valve, inspect the remainder of the control circuit to identify the fault. With the information that you have collected, determine any necessary actions.

TECHNICIAN TIP

Many vehicles have individual EGR passageways for individual intake manifold runners. If some of the passageways become clogged, too much EGR gases will go to the remaining open passageways when the EGR valve opens, causing an engine misfire. If the engine experiences rough engine operation when the EGR valve is open, suspect this condition.

Inspecting and Servicing the EGR System

The EGR system is intended to reduce the amount of oxides of nitrogen that are formed during combustion. The introduction of exhaust gases reduces the combustion temperature, which in turn lowers oxides of nitrogen. EGR systems normally are not active on engines when cold, at idle, or at WOT.

To inspect and service the EGR system, start by researching the description and operation in the service information. Then perform an output test on the EGR valve to see if it is operational. Do this with either a scan tool or a vacuum pump (depending on the system; refer to the service information on testing). Operate the EGR valve with the vehicle running. Idle should become rough, and the engine may eventually stall as the EGR valve is opened and exhaust gas flows into the intake manifold. If the idle becomes rough, the EGR valve is capable of operating, and the passageways are relatively clear. Also check the EGR monitoring device (sensor) to see if it is registering EGR flow. This could be monitored by one of a number of sensors, from an EGR temperature sensor, to the MAP sensor, to the O_2 sensor. See more on this in the next Skill Drill.

If there is no change in the engine idle speed when the EGR valve is activated, perform testing of the control system to ensure that a signal is getting to the EGR valve. This could be a vacuum signal or an electrical signal, depending on the type of EGR control system. If a signal is getting to the EGR valve, inspect the valve itself for faults and the EGR passageways to ensure they are not clogged.

With the information you have collected, determine the necessary actions. Remember that the EGR valve will be hot to the touch if the engine is warm. Because there are many types of EGR sensors, one description cannot adequately cover this procedure; therefore, the following steps are generic.

To test the EGR system, follow the steps in **SKILL DRILL 66-5**.

Inspecting and Servicing the EGR Sensors

EGR sensors give feedback to the PCM that the EGR valve is flowing exhaust gas. If gas is not found to be flowing, the MIL will be turned on and a diagnostic code will be set. The method of inspecting EGR sensors depends on the type of sensor system used by the manufacturer. Some systems use sensors to monitor the position of the pintle and calculate EGR flow. Some use a temperature probe on the downstream side of the EGR valve and watch for the temperature rise when the EGR gases are flowing. Some use the MAP sensor to watch for the pressure change in the intake manifold when the EGR valve is commanded on. Many systems used on Ford Motor Company vehicles use a pressure sensor (called a delta pressure feedback EGR [DPFE]) to monitor the pressure difference across a restriction in the EGR tube. When EGR gases flow across the restriction, a pressure difference occurs, which causes the sensor to change its signal, indicating EGR flow.

To diagnose EGR sensor–related concerns, first consult the service information, and research the proper operation of the EGR system on the vehicle. You also need to verify that the EGR valve is operating. Follow the steps in the previous Skill Drill to test the EGR system's operation.

Further diagnosis using the service information will be needed. Because there are many types of EGR sensors, one description cannot adequately cover this procedure; therefore, the following steps are generic.

SKILL DRILL 66-5 Testing the EGR System

1. Determine the type of EGR valve the vehicle is equipped with and the test procedure needed. Assuming you are working with a digital EGR valve, follow the procedures in the service information.
2. Attach a scan tool to the DLC, and retrieve any codes related to the EGR sensor. If diagnostic codes are present, follow the trouble chart in the service information for further diagnosis.
3. Follow the manufacturer's troubleshooting chart. If the troubleshooting chart tells you to inspect passageways, it may be necessary to remove the EGR valve to inspect.
4. Based on the diagnostic chart, determine the path to follow to repair the vehicle.
5. Verify the repair. Perform a drive cycle and use mode 6 to verify a pending EGR code did not reset, before releasing the vehicle to the customer.

SKILL DRILL 66-6 Testing the EGR Sensors

1. Research the procedure for testing the electrical sensors and wiring of the EGR sensor system in the service information.
2. Attach a scan tool to the data link connector (DLC), and retrieve any codes related to the EGR system. If diagnostic codes are present, follow the trouble chart in the service information for further diagnosis.
3. Follow the service information procedure for testing the EGR sensor and wiring. Testing involves the use of a digital volt-ohmmeter (DVOM) and basic electrical testing. It may also require a scan tool to activate the EGR valve for testing purposes. Follow the service information for testing.
4. After following service procedures, determine any necessary actions.

To test the EGR sensors, follow the steps in **SKILL DRILL 66-6.**

▶ Diagnosing Concerns Related to the Secondary Air Injection System

N66007

The secondary air injection system consists of several components: an air pump, either belt driven or, on some newer models, electronic; distribution tubes and hoses; check valves that prevent exhaust flow back into the air pump; and air management valves or solenoids to direct the airflow. The air pump reduces hydrocarbons and carbon monoxide by pumping air into the exhaust stream and assisting the catalytic converter in converting the hydrocarbons and carbon monoxide to water and carbon dioxide. If air is not injected into the exhaust manifold when the engine is cold, the hydrocarbons and carbon monoxide emissions will be excessive. If the air is injected into the exhaust manifold at all times, the oxygen sensor will pick up on the excessive oxygen and tell the PCM a lean condition is present. The PCM will then richen up the fuel mixture, creating more emissions, poor gas mileage, and possibly a drivability issue. Check valves can burn out and separate, or the diaphragm may fail, creating an audible exhaust leak or exhaust flow back into the air management valves and possibly the air pump itself, damaging their plastic components.

To diagnose concerns related to the secondary air injection system, first check for diagnostic codes in the PCM with a scan tool. Pull any trouble codes, and consult the service information. View the oxygen sensor readings on a scan tool. If the oxygen sensor readings indicate a lean condition, verify that the oxygen sensor is functioning properly, and refer to the service information on further testing. Verify that air is not being injected to the wrong location or at the wrong time. You may have to pinch off the rubber supply hose from the pump. If the oxygen sensor now responds correctly, suspect an air injection system fault.

Inspecting and Servicing the Secondary Air Injection System

The secondary air injection system is not as common as it once was due to the more accurate fuel systems, heated oxygen sensors, and faster warm-up. Secondary air injection system failures on older systems were numerous and more common because more mechanical parts were involved. The air pump itself had bearings that could fail and cause a pump and accessory belt failure. The switching and diverter valves could stick and direct air to the wrong place. Vacuum hoses and valves would crack and break, creating a vacuum leak and failure of the entire secondary air injection system. The air distribution hoses to the exhaust manifold required the use of a check valve, which kept exhaust from flowing backward and destroying the switching and diverter valve, hoses and tubing, and air pump.

With the introduction of the PCM, electric controls replaced some of the vacuum-operated parts, which helped reduce the amount of failures. Common failures now include connector terminal pin fit issues that create open circuits or high resistance and other common electrical circuit faults.

Always verify that the vehicle you are working on has a secondary air injection system. Testing the secondary air injection system starts with checking for any related DTCs. If any are present, then refer to the service information, which will provide detailed testing procedures. If no DTCs are present, then you will want to verify that the system is operating according to design; if not, you will have a better idea of where the problem is located.

To verify that the system is operating correctly, check to see that air is being directed upstream during cold start-up, and watch the engine temperature to see when it switches to downstream. Many times you can squeeze the air injection hoses and feel the pulsations of air flowing through it. By feeling each of the hoses, you can determine where and when air is being directed. Because the types of air injection systems are so great, one description cannot adequately cover this procedure; therefore, the following steps are generic.

To inspect the secondary air injection system, follow the steps in **SKILL DRILL 66-7.**

Inspecting and Testing Electrical Components of Secondary Air Injection Systems

The air injection electrical system controls the operation of the air valves and, in some cases, the electric air pump. If the electrical system fails, a check engine light should illuminate and a diagnostic code will set. Start by retrieving any related DTCs. Follow the service information to diagnose the operation of the secondary air electrical system. Because the types of air injection electrical systems are so great, one description cannot adequately cover this procedure; therefore, the following steps are generic.

To test the electrical system of the secondary air injection system, retrieve any related DTCs and research the service information for the diagnostic procedures for a secondary air

SKILL DRILL 66-7 Inspecting the Secondary Air Injection System

1. Check for any stored DTCs related to the secondary air injection system with a scan tool. If any are found, research the service information for proper procedures for diagnosing the codes. Ensure you have a proper understanding of the system's operation before attempting to repair the system.
2. Check the system's operation. Air should be directed toward the exhaust manifold during engine warm-up and to the converter once warm. Also verify that air diverts to the air cleaner or atmosphere during deceleration.
3. Determine necessary actions. If the air pump is locked or any mechanical components are seized, replace with new parts and retest the system.

injection electrical concern. Be sure you understand the operation of the secondary air injection system. Follow the service information to diagnose any secondary air injection system faults, and replace or repair the faulty component.

▶ Wrap-Up

Ready for Review

- In automotive applications, most emissions are the by-product of combustion and are emitted from the exhaust system.
- Not all emissions from combustion are considered hazardous.
- The term "emission," when used in automotive circles, normally refers to the pollution produced by a light vehicle during operation or while sitting stationary.
- The fuel tank allows liquid fuel to evaporate into a gas; if not controlled, those vapors can escape into the atmosphere and are called evaporative emissions.
- In a compression-ignition diesel engine, emissions are created in the combustion chamber the same way as in spark ignition engines and escape to the atmosphere through the exhaust and the crankcase breather.
- Manufacturers are required by law and the Environmental Protection Agency (EPA) to control the emissions produced by their vehicles.
- The federal test procedure (FTP) requires all manufacturers to pass a federal emission test before vehicles can be sold in the United States.
- Not only must the vehicles pass the test before they can be sold, but also they must be capable of detecting an emission failure over the life of the vehicle, as they are being operated.
- As of 1996, all new vehicles most meet the OBD II standard, meaning all vehicles must have the same diagnostic connectors, DTCs, and scan tool communication requirements.
- The EPA requires that when a vehicle's emissions deviate from the FTP limit by 1.5 times, the warning light (malfunction indicator lamp) must turn on and one or more specific diagnostic trouble codes (DTCs) set in memory.
- The OBD II system was developed primarily with the intent of keeping vehicles emission compliant and warning drivers if they aren't.
- Regulated air pollutants can be divided into two groups: gases and particulates.
- Many countries follow an international protocol on emissions requiring vehicle manufacturers to comply with these laws.
- Historically, the regulated emissions have been hydrocarbons, carbon monoxide, oxides of nitrogen, and particulate matter. In recent years, carbon dioxide has been added to the list of emissions that are being regulated.
- There are three categories of emission gases: nonharmful, harmful, and debatable.
- Emissions can be reduced using precombustion systems or postcombustion systems.
- Emissions can be minimized by using a good engine design and closely controlling the air-fuel mixture.
- The catalytic converter reduces NOx, CO, and HC. It is a postcombustion emission device that splits up and combines pollutants so that less harmful gases are created.
- The evaporative emission system reduces HC. It stores fuel vapors and then directs them to the intake manifold to be burned in the engine.
- The positive crankcase ventilation system pulls vapors from the crankcase and directs them to the intake manifold to be burned in the engine.
- The exhaust gas recirculation system reduces oxides of nitrogen by recirculating a small amount of exhaust gases to the intake manifold to dilute the air-fuel mixture with inert gas.
- The secondary air system reduces HC and CO. It directs air to the exhaust manifold during cold start and then switches to the catalytic converter once the engine warms up.

Key Terms

activated charcoal Charcoal that will absorb large amounts of vapor, used in charcoal canisters to store and release fuel vapors.

air diverter valve A valve that changes secondary airflow from the exhaust stream to the air cleaner or vice versa.

air pump A device used to pump air to the exhaust stream to ensure the converter has enough oxygen to perform gas conversions.

air switching valve The valve used to switch secondary air from near the exhaust valve (upstream) to the catalytic converter (downstream) for the purpose of oxidizing hydrocarbons and carbon monoxide.

blowby Pressure that leaks past the compression rings during compression and combustion.

breather tube A tube used in the PCV system to allow fresh air into the engine crankcase.

carbon monoxide poisoning Exposure to higher than tolerable levels of carbon monoxide, resulting in headaches, fatigue, or loss of consciousness, eventually resulting in death.

carburetor float bowl The part of the carburetor that holds fuel to be burned in the engine; the bowl is at a constant level of fuel to ensure adequate fuel is present during driving.

compression-ignition engine An engine that uses the heat of compression to ignite the air-fuel mixture; also known as a diesel engine.

diaphragm-style EGR valve An EGR valve that is operated by a diaphragm that moves when vacuum is applied to it, opening a passageway for exhaust to flow.

emission control system A system of devices that are designed to control or reduce harmful gases released to the atmosphere.

evaporative emission (EVAP) system A system used to capture vapors or gases from an evaporating liquid.

exhaust gas recirculation (EGR) system A system that recirculates a portion of burned gases back into the combustion chamber to displace air and fuel and cool combustion temperatures.

expansion volume An engineered space that allows for growth of the volume of a liquid as it heats and expands.

feedback system A system that uses feedback to adjust what it is doing. Typically, a module gives a command and waits to see if the command was followed by use of a sensor.

five-gas analyzer A tool that uses sensors to measure the level of gases in the exhaust stream.

fixed-orifice PCV system A system in which a hole of a predetermined size is used as a means of pulling crankcase vapors into the intake manifold to be burned.

flame front The rapid burning of the air-fuel mixture that moves outward from the spark plug across the cylinder.

fuel tank pressure sensor A sensitive pressure sensor mounted in the fuel tank or EVAP system used to monitor the system for leaks.

heated air intake system A system that uses hot air from around the exhaust manifold to warm the air going into the intake manifold.

idling five-gas test A five-gas analysis test conducted during vehicle idling.

intrusive test A test in which a gas analyzer hose is placed into the exhaust stream before and after the catalytic converter; it may require drilling an access hole in the exhaust pipe in front of the converter.

lambda The ratio of air to fuel at which all of the oxygen in the air and all of the fuel are completely burned; See also *stoichiometric ratio*.

limit value The maximum amount of emissions that a vehicle is permitted to emit. Values are assigned to different classifications of vehicles.

negative back-pressure EGR valve A dual-diaphragm EGR valve that uses vacuum in the exhaust manifold to operate a control diaphragm that opens and closes a vent in the main diaphragm chamber.

onboard diagnostic (OBD) The first generation of a self-diagnostic system built into the vehicle computer system that provided a means of warning the driver when the computer detected a problem in the system. This system allowed the technician to retrieve codes to assist in diagnosis of the vehicle.

PCV valve A valve that controls the amount of crankcase ventilation flow that is allowed; it varies with changes in manifold pressure.

photochemical smog A brown haze that hangs in the sky, typically seen over large cities. Smog is a major health issue to humans because it affects lung tissue.

pinging A preignition or detonation concern that is damaging to the engine if it is allowed to continue. It is described as the sound of marbles being shaken in a glass jar.

pintle A tapered valve that sits in a tapered seat to seal a passageway for air or fuel.

positive back-pressure EGR valve A dual-diaphragm EGR valve that uses pressure in the exhaust system to act on the control diaphragm, which closes a bleedvalve in the main diaphragm, thus allowing the EGR valve to open only when there is positive exhaust pressure.

positive crankcase ventilation (PCV) system A system that draws blowby gases from the crankcase into the intake to be burned.

pressure relief valve A valve that is designed to release pressure if it gets above a calibrated pressure; used in a gas cap as a safety device to release excessive pressure if pressure gets too high.

pulse air system The use of normal exhaust engine pulses to draw air into the exhaust stream.

purge solenoid A control device that lets fuel vapors move from the charcoal canister to the engine. The solenoid is controlled by the engine control module.

purge switch A device used to show the computer when purge is occurring. It is used as a feedback device to allow the computer to determine whether flow is happening.

purge valve A valve used to control the flow of evaporative emissions from the charcoal canister to the intake manifold.

purging The process of pulling stored fuel vapors from the charcoal canister and moving them into the engine to be burned.

quenched The state in a combustion chamber in which the flame cannot burn due to cold surfaces or poor distribution of the fuel mixture.

reed valve A flexible valve made from spring steel that flexes to open or close, usually due to pressure changes.

scavenging effect A condition caused by moving columns of air, which create a low-pressure area behind them, resulting in a pulling force that is used to pull the remaining burned gases from the combustion chamber. Valve timing affects the amount of scavenging effect an engine has.

separator PCV system A PCV system with a device that uses gravity to allow oil to fall to the bottom of the valve and be returned to the crankcase; the valve prevents liquid from traveling to the intake manifold.

stepper motor EGR valve An EGR valve that uses an electrically controlled motor to open the pintle in steps in order to let exhaust gases flow into the intake manifold.

three-way catalytic converter A converter that changes hydrocarbons, carbon monoxide, and oxides of nitrogen into harmless elements.

two-way catalytic converter A converter that changes only hydrocarbons and carbon monoxide into harmless elements.

vacuum relief valve A mechanical valve used on the gas cap that prevents a low-pressure condition from occurring. It also ensures that the fuel tank will not collapse due to fuel usage or contraction of fuel on cooldown.

vaporization The changing of a liquid to a gas through boiling.

vapor line A rubber or plastic line that carries vapors from the fuel tank to the charcoal canister.

variable-orifice PCV system A system in which a replaceable, spring-loaded PCV valve regulates gas flow. The position of the PCV valve is controlled by the pressure in the manifold.

vent solenoid A solenoid that allows fresh air to enter the evaporative system during a purge event. Also used for an evaporative system monitoring test.

Review Questions

1. All of the following are sources of emissions from current gasoline-fuelled motor vehicles *except*:
   a. the fuel system.
   b. the crankcase.
   c. the HVAC system.
   d. the tailpipe.

2. The EPA requires that the malfunction indicator lamp (MIL) turn on and one or more specific diagnostic trouble codes (DTCs) be set in memory:
   a. whenever emissions are released.
   b. when emissions cross the FTP limit.
   c. when emissions are twice the FTP limit.
   d. when emissions are 1.5 times the FTP limit.

3. Which of the following is a nonharmful emission?
   a. Nitrogen
   b. Carbon monoxide (CO)
   c. Oxides of nitrogen (NOx)
   d. Sulfur dioxide (SO_2)

4. The reduction catalyst in a catalytic converter helps to:
   a. reduce the oxides of nitrogen molecules into their base compounds.
   b. oxidize any unburned hydrocarbons and carbon monoxide.
   c. convert exhaust water to oxygen and hydrogen.
   d. oxidize nitrogen before releasing it into the atmosphere.

5. Which of the following is a function of the positive crankcase ventilation (PCV) system?
   a. It burns fuel vapors from the charcoal canister.
   b. It directs blowby gases and vapors back into the intake manifold.
   c. If prevents exhaust gases from entering into the engine.
   d. It ensures that the ventilation capacity is less than the amount of blowby.

6. Choose the correct statement with respect to the exhaust gas recirculation system.
   a. It recirculates gases from the crankcase to the exhaust system.
   b. The EGR valve opens only during full throttle.
   c. The EGR valve is open during idle.
   d. Most PCM-controlled EGR systems incorporate a feedback loop, called a monitor, to ensure that the valve is working properly and EGR gases are flowing in the system.

7. Which of the following ensures that hydrocarbons are not released into the atmosphere when fuel in the fuel tank begins to vaporize and build pressure?
   a. Catalytic converter
   b. EVAP system
   c. EGR system
   d. PCV system

8. Which of the following uses oxygen sensors to measure the level of gases in the exhaust stream?
   a. Idling five-gas test
   b. Catalytic converter testing
   c. Intrusive test
   d. Purge flow test

9. In which of these cases is the catalytic converter most likely to be saved?
   a. When the catalyst is severely overheated
   b. When contaminated by lead and silicone
   c. When contaminated by carbon
   d. When the converter is physically damaged

10. Black exhaust represents:
   a. a blown head gasket.
   b. oil that is being burned in the combustion chamber.
   c. an overly rich air-fuel mixture.
   d. a leaky intake manifold gasket.

ASE Technician A/Technician B Style Questions

1. Tech A says that the purge valve is part of the evaporative emission system. Tech B says that OBD II systems must illuminate the MIL when the emissions exceed 1.5 times the FTP. Who is correct?
   a. Tech A
   b. Tech B
   c. Both A and B
   d. Neither A nor B
2. Tech A says that carbon monoxide is partially burned fuel. Tech B says that oxides of nitrogen are unburned fuel. Who is correct?
   a. Tech A
   b. Tech B
   c. Both A and B
   d. Neither A nor B
3. Tech A says that hydrocarbons are a result of complete combustion. Tech B says that a catalytic converter creates a chemical reaction, changing carbon monoxide and hydrocarbons to water and carbon dioxide. Who is correct?
   a. Tech A
   b. Tech B
   c. Both A and B
   d. Neither A nor B
4. Tech A says that allowing hot exhaust gases into the engine through the EGR valve helps to warm up the engine during warm-up. Tech B says that rich fuel mixtures create low amounts of carbon monoxide. Who is correct?
   a. Tech A
   b. Tech B
   c. Both A and B
   d. Neither A nor B
5. Tech A says that burning gasoline in an engine creates water as a by-product of combustion. Tech B says that variable valve timing can be used to reduce oxides of nitrogen, eliminating the need for an EGR valve on some engines. Who is correct?
   a. Tech A
   b. Tech B
   c. Both A and B
   d. Neither A nor B
6. Tech A says that the PCV system recirculates exhaust gases into the intake manifold. Tech B says that the PCV system recirculates blowby gases into the intake manifold. Who is correct?
   a. Tech A
   b. Tech B
   c. Both A and B
   d. Neither A nor B
7. Tech A says that the PCM monitors the pre-cat and post-cat oxygen sensors to determine catalytic converter efficiency. Tech B says that a catalytic converter can be tested by graphing the oxygen sensor readings on a scan tool or lab scope and comparing them. Who is correct?
   a. Tech A
   b. Tech B
   c. Both A and B
   d. Neither A nor B
8. Tech A says that the EVAP system stores escaping hydrocarbons until the engine can burn them. Tech B says that catalytic converters were designed for regular leaded gas. Who is correct?
   a. Tech A
   b. Tech B
   c. Both A and B
   d. Neither A nor B
9. Tech A says that the EGR valve is open fully at idle so that the engine will not die. Tech B says that oxides of nitrogen are created in large amounts when the combustion temperature is above 2500°F (1400°C). Who is correct?
   a. Tech A
   b. Tech B
   c. Both A and B
   d. Neither A nor B
10. Tech A says that black exhaust indicates a very rich-running engine. Tech B says that white exhaust can indicate coolant in the exhaust. Who is correct?
    a. Tech A
    b. Tech B
    c. Both A and B
    d. Neither A nor B

CHAPTER 67

Alternative Fuel Systems

NATEF Tasks

- **N67001** Identify safety precautions for high-voltage systems on electric, hybrid, hybrid electric, and diesel vehicles. (MLR/AST/MAST)
- **N67002** Identify service precautions related to service of the internal combustion engine of a hybrid vehicle. (MLR/AST/MAST)
- **N67003** Identify hybrid vehicle A/C system electrical circuits and service/safety precautions. (MLR/AST/MAST)
- **N67004** Identify hybrid vehicle auxiliary (12 V) battery service, repair, and test procedures. (MLR/AST/MAST)

Knowledge Objectives

After reading this chapter, you will be able to:

- **K67001** Describe alternative fuel and the conditions favorable for its use.
- **K67002** Describe common alternative fuels.
- **K67003** Explain the purpose and function of the various types of gaseous fuel storage system.
- **K67004** Explain the applications of biofuels.
- **K67005** Explain fuel cell operation.
- **K67006** Explain battery electric vehicles and their components.
- **K67007** Describe hybrid electric vehicles.
- **K67008** Explain hybrid drive configurations and operation.
- **K67009** Describe the safety and service precautions for servicing hybrid and electric vehicles.

Skills Objectives

There are no Skills Objectives in this chapter.

▶ Introduction

The use of alternative fuels in the internal combustion engine is nothing new. Back between 1890 and 1896, Henry Ford designed one of his first vehicles—the Quadricycle—to run on ethanol derived from local farm crop waste (Ford's Model T also ran on ethanol) (**FIGURE 67-1**). Most of the earliest automobiles and trucks operated either on electricity or by using external combustion steam engines fueled with paraffin or internal combustion engines running on "town gas" produced from coal. But for a number of reasons (both political and practical), once discovered, petroleum replaced almost all other options for motor fuel used in light-duty road vehicles. Petroleum-based fuels have dominated the transportation sector for the last century. Presented with a long-standing record of gasoline use, some people question the logic of switching back to alternative fuels for powering our road vehicles.

▶ Alternative Fuels

K67001

What are "alternative fuels," alternative fuel vehicles (AFVs), and alternative energy sources? An **alternative fuel** is essentially anything other than a petroleum-based motor fuel (gasoline or diesel fuel) that is used to propel a motorized vehicle. Alternative fuels may be liquid or gas (or electric) and contain a variety of latent heat energy for use in the internal combustion engine. An alternative fuel vehicle, then, is a vehicle powered by something other than petroleum. One example is an electric vehicle, which is fueled by electricity generated in a variety of ways (more on this later). Here is a list of some of the recognized alternative fuel vehicles currently produced:

FIGURE 67-1 One of the first alternative fueled vehicles—Henry Ford's Quadricycle.

- Flexible-fuel vehicles (FFVs)
- Dedicated and bi-fuel compressed natural gas vehicles (CNGVs)
- Liquefied natural gas vehicles
- Liquid petroleum gas (LPG, also called propane) vehicles
- Hydrogen-powered internal combustion engine (ICE) vehicles
- Battery-electric vehicles (BEVs)
- Hybrid electric vehicles (HEVs)
- Hydrogen fuel cell (electric) vehicles (FCVs)

Some of the alternative fuels include:

- Ethanol (alcohol)
- Methanol (alcohol)
- Biodiesel
- Methane: compressed natural gas (CNG) and liquefied natural gas (LNG)
- Liquefied petroleum gas (LPG, also called propane) from methane
- Synthetic fuels
- Electricity (not a fuel, per se, but an energy source)

What has prompted the renewed interest in alternative fuels? Reasons for making the switch from time-proven petroleum to "alternatives" are widely debated and include the following:

- Enhancing energy security by using domestically sourced fuels/energy
- Reducing the export of funds to overseas oil producers
- Reducing pollution from vehicle emissions
- Stimulating and bolstering the national and local economies (agri-business)

You Are the Automotive Technician

A Toyota Prius hybrid, which is a high-voltage vehicle, is brought into the shop for a regular 120,000-mile service, requiring replacement of the spark plugs and other maintenance items. You know that you need to follow the latest safety precautions when servicing a high-voltage vehicle. First, you do some research on the Toyota website and locate the service information for this make and model. Next, you create a service plan to prepare the workspace for safety and determine which tools are needed and which technicians are trained to perform the work.

1. Why is it important to read the Toyota service information for the hybrid vehicle prior to servicing?
2. Why is it necessary to inform or work with someone else while you are servicing a high-voltage vehicle?
3. What is the standard PPE and equipment needed to work on high-voltage vehicles?
4. Why must the main high-voltage battery pack be disconnected before servicing a high-voltage vehicle?

Boiled down, alternative fuels advance the following "3E" causes:

- Energy security
- Environmental concerns
- Economy

Let's look at each of these in more detail.

Energy Security

National security is a key issue in the United States, especially after the 9/11 terrorist attacks. For years, the United States has been importing oil from other countries. Oil importation for transportation has been as high as 70%. President George W. Bush said in his 2006 State of the Union Address that "Americans are addicted to oil" (**FIGURE 67-2**). Foreign petroleum stakeholders have large petroleum energy reserves, some of whom are waging war against the United States. In an attempt to shift the playing field after the oil embargo of the 1970s, the United States Congress took action. In 1992, Congress passed the **Energy Policy Act (EPAct92)**, the intent of which was to advance, through mandates, the use of vehicles capable of using alternatives to gasoline and diesel. EPAct92 considered the previously listed vehicles (and fuels) as legitimate AFVs. In 1996, EPAct was amended to include synthetically derived fuels such as coal-to-gas liquids (derived from the Fischer-Tropsch method), bio-based fuels, and more.

Unfortunately, EPAct92 failed to account for the fact that an alternative fuel **infrastructure** did not, for the most part, exist in the United States. As a result, while federal, municipal, and private fleets were mandated to include AFVs as a certain percentage of their new vehicle purchases, alternative fuels, such as ethanol and methanol, were not widely available for use in these flexible-fueled vehicles (meaning they can use either alcohol or gasoline as their prime fuel). The Renewable Fuels Standard and other legislation helped to promote a greater availability of alternative fuels for on-road and off-road vehicles (agricultural, railroads, etc.) and for marine and aviation applications. Today the U.S. military is a driving force in the move toward alternative fuels.

In considering the need to develop alternative fuel sources, another important point is that petroleum is not considered a renewable resource, meaning there is a finite amount of it. According to peak oil advocates, once supplies of quality petroleum become hard enough to extract from proven easily accessible reserves, the cost will become prohibitive. As sweet light crude oil becomes scarce, crude oil will become more expensive. With developing countries like India and China seeing an increased number of automobiles on their roads, the worldwide demand for oil has caused sharp rises in the price of crude oil, which recently have been tempered by the shale oil discoveries in the United States. Oil has become a valuable "black gold" commodity on the world market, making energy a strategic resource.

Environmental Concerns

Consider the pollution created by drilling for oil, refining it, and transporting the end products to the local filling station for purchase. Once burned in the internal combustion engine, the resulting tailpipe emissions must be tightly regulated. "Well-to-wheels" emissions are now being studied and regulated, to some degree or another, in most countries of the world. Greenhouse gas emissions are a worldwide concern. Global climate change is widely discussed, and pollutants are heavily regulated by many countries. Photochemical smog has created health issues for the young as well as for the elderly (**FIGURE 67-3**). Automobiles burning gasoline or petrodiesel fuel create the pollutants responsible for harm to humans and the environment, so transitioning to cleaner alternative energy and fuel choices can help reduce (mitigate) the effects of pollution. Later in the chapter, we explore the various pollutants associated with petroleum burned in automobile engines and why alternative choices make sense.

Economic Concerns

To offset the unfavorable balance of U.S. dollars sent overseas for petroleum, "home grown" bio-based fuels are being used. Congress has passed legislation requiring the use of alternatives to gasoline, and some states require that ethanol be blended at the pump with gasoline to help their farm economies. Nationwide, gasoline is widely blended with 10% ethanol (mostly corn-based ethanol), yet it still does not meet targets of nonpetroleum usage. Congressional targets for alternative

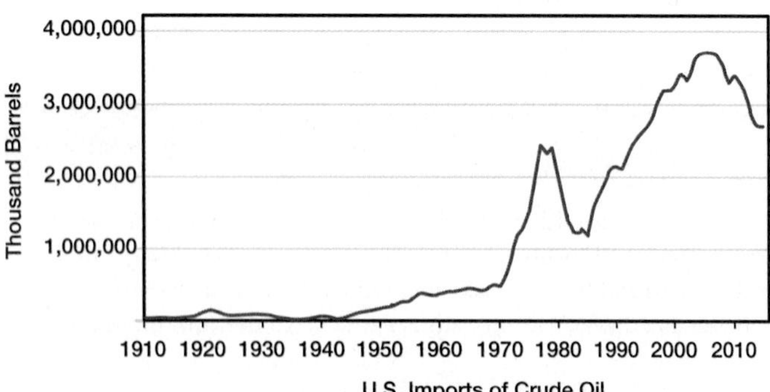

FIGURE 67-2 US imports of crude oil have gone down as domestic oil production and alternative fuel production has gone up.

FIGURE 67-3 Smog is still a problem in parts of the world. Alternative fuels help reduce the problem.

fuel usage are difficult to reach because producers have hit the "blend wall," meaning ethanol supplies now outweigh market demand. The Environmental Protection Agency (EPA) has approved the use of 15% ethanol for 2001 and newer vehicles, and this should help both ethanol suppliers and the environment. Corn prices per bushel have risen due to increased demand for fuel and for export overseas. The co-products of ethanol production, such as dried distiller grains, are sold as feed for livestock.

Other alternatives likewise are helping local economies to prosper. Soybeans and other U.S.-grown feedstocks intended for use as biodiesel are finding their way to Midwestern U.S. markets. Some states mandate the use of 10% or more blends to encourage and assure local economies of a market for their products.

▶ Vehicle Emissions and Standards

As clean as they are by today's standards, left unchecked, internal combustion engines (ICEs) and fuel systems emit pollutants both from without and from within the tailpipe. The pollutants regulated by EPA emission standards are:

- **Carbon monoxide (CO):** This product of partially burned fuel displaces oxygen in the bloodstream and causes asphyxiation.
- **Hydrocarbons (HC):** These volatile organic compounds (VOCs) of unburned fuel contribute to photochemical smog.
- **Carbon dioxide (CO_2):** Although harmless in small amounts, too much carbon dioxide is an atmospheric concern. It is considered a greenhouse gas, although there are disagreements on how damaging it is.
- **Nitrogen oxides (NOx):** The various compounds of nitrogen contribute to the creation of smog in the presence of sunlight.
- **Particulate matter (PM):** Mainly produced by diesel engines, PM is literally fine soot (measured in microns, or millionths of an inch). PM becomes trapped in the lungs and causes health issues. At one time PM-10 (10 microns large) was thought to be harmless; today even PM-2.5 is considered dangerous.

Emission Standards

Emission standards are set by various countries, states, and even localities. Standards affect cutoff points for the amount of pollutants that vehicles may release into the environment. Different countries focus on different pollutants, depending on what environmental issues they are dealing with, but globally, most nations' emission standards have progressively tightened to the point that tailpipe exhaust is sometimes cleaner than what goes into the engine's intake system.

In the United States, the Environmental Protection Agency (EPA) controls federally mandated emission standards, mostly following the lead of California's clean air regulations. Some states, like California, have enacted more stringent emission standards than those of the EPA. The California Air Resources Board (CARB) is a department within California's EPA. California is the only state that is permitted to have such a regulatory agency, because it is the only state that had one before the passage of the federal Clean Air Act. Other states are permitted to follow CARB standards or use the federal ones, but not set their own.

We will not go into the actual cutoff points for each of the pollutants, vehicles, and locations or discuss when they became effective; the regulations are far too varied to list here. In general, however, regardless of where clean air regulations are regulated and enforced, pollution cutoff points have become progressively more stringent over time.

Low-Emission Vehicle Program Emission Standards

California used unique terms to describe vehicle compliance with their clean air regulations, and window stickers seen on vehicles reflected their compliance with California's clean air limits. For example, from 1994 to 1999, California's regulations were stricter than the U.S. national "tier" regulations. The **Low-Emission Vehicle (LEV) standard** included six major emission categories, each with several targets depending on vehicle weight and cargo capacity. Vehicles with a test weight up to 14,000 lb (6350 kg) were covered by these regulations. The major emission categories were as follows:

- TLEV: Transitional Low-Emission Vehicle
- LEV: Low-Emission Vehicle
- ULEV: Ultra-Low-Emission Vehicle
- SULEV: Super-Ultra-Low-Emission Vehicle
- ZEV: Zero-Emission Vehicle

The last category is largely restricted to electric vehicles and hydrogen vehicles, although such vehicles are usually partly reliant on grid-supplied power, which is derived from coal, oil, natural gas, hydro (water), or nuclear sources.

▶ Types of Alternative Fuels

K67002

Alternative fuels are used as a way of either reducing the operating cost of vehicles or reducing pollution from vehicle emissions, or both. Some alternative fuels such as alcohol can be mixed with traditional gasoline and used as a fuel additive to reduce the total amount of gasoline used. For example, some vehicles are designed to operate on two different fuels blended together in various mixture amounts by percentage. These are called **flexible-fuel vehicles (FFVs)**. Ethanol FFVs, for example, are designed to operate on 100% standard gasoline or up to 85% alcohol and 15% gasoline (called **E85**). Other fuels such as **compressed natural gas (CNG)** may replace the use of liquid fuel altogether. Such vehicles require a different fuel system. For example, a dedicated CNG vehicle uses only natural gas; a gasoline vehicle converted for both CNG fuel delivery as well as gasoline is called a bi-fuel vehicle. Let's look at the various alternative fuel and clean fuel vehicles in greater detail.

Liquefied Petroleum Gas

Liquefied petroleum gas (LPG), also called propane, is a liquid under pressure but boils into a gas at atmospheric pressure. LPG is a by-product of the process of refining crude oil or natural gas. It is odorless and colorless and has been used for many years to power specially modified gasoline engine vehicles. It is the third most common fuel used worldwide, and it is nontoxic and nonpoisonous, with a very small flammability range (**FIGURE 67-4**). LPG is heavier than air and is the least carbon-emitting hydrocarbon fuel available. It emits far fewer harmful emissions than gasoline because its gaseous state and less complex chemical makeup (C3H8) make it easier to burn completely. LPG can be used in a wide range of spark-ignition engines as an alternative to gasoline. It is generally cheaper than gasoline; therefore, it is popular for light-duty vehicles, medium-duty trucks, and shuttle vans used in high-mileage fleet applications, for example, taxis and delivery vehicles, where the cost of vehicle modification can be recouped over time more easily through lower fuel costs. Propane is also domestically produced and provides nearly the same power, torque, and towing capacity as gasoline-powered vehicles.

LPG is stored at normal temperatures under moderate pressure (several hundred pounds) as a liquid and is therefore very energy dense for onboard storage. When the pressure is released to atmospheric, propane expands to 270 times its liquid volume and is burnt as a dry gaseous vapor. A fuel pressure regulator is fitted in the system to reduce storage tank pressure. Many engines use a gaseous vapor (fumigation) system, but newer vehicles are using liquid propane injection. A filter is fitted to the system to remove any particles that may be present in the propane. When propane is injected into the engine's intake manifold behind the intake valve, it becomes a clean vapor. For so-called fumigation systems, a special air-fuel mixer is installed at the engine's throttle body to ensure the correct amounts of LPG and air are mixed together to achieve the correct stoichiometric ratio (air-fuel mixture) for clean combustion.

FIGURE 67-4 Propane is easily obtained, being the third most common fuel worldwide.

Applied Science

AS-21: Fractional Distillation: The technician can explain the ignition characteristics of fuels resulting from varying levels of fractional distillation (i.e., fuels with differing chemical makeups).

The type of oil that comes from the ground is known as crude oil. It is a thin to thick, smelly liquid containing a mixture of hydrocarbons. The crude oil is not ready to be used until a separation process is completed. This separation process is known as fractional distillation and is accomplished by using the difference in boiling points.

Crude oil is heated in a furnace before entering a tall steel distillation tower, which acts as a huge heat exchanger. As the crude oil goes up the tower, it becomes progressively cooler. This difference in temperature separates the different fractions from one another. The hydrocarbon gases condense back into liquids as a result of the temperature decrease.

Motor oil is separated into the basic categories of fuels or lubricants at approximately 626°F (330°C); diesel fuel, at approximately 482°F (250°C); and gasoline, at approximately 122°F to 158°F (50°C to 70°C). Additives improve the products in a number of ways. For example, many types of gasoline contain detergents designed to keep fuel systems clean.

Today's light- and medium-duty LPG-powered vehicles are fitted with LPG fuel control processors, which are onboard computers that control the sequential operation of fuel injectors. Many vehicle manufacturers, including Ford, GM, Toyota, and Mitsubishi, offer factory-fitted or tier 1 supplier-approved fumigation or liquid fuel–injected LPG systems. LPG fuel systems may also be installed by aftermarket conversion companies. The EPA mandates that any vehicle converted from gasoline to an alternative fuel must be as clean as, or cleaner than, the original gasoline system.

One drawback to LPG vehicles is the size of the fuel tank required. For this reason, most LPG conversions are payload-type vehicles that can accommodate placement of the onboard propane fuel tank. Still, the LPG fuel tank is far less costly than the cylinders used for CNG.

Natural Gas

Natural gas (sometimes referred to simply as NG) is a fossil fuel often found in offshore gas fields or trapped in coal beds, aquifers, or petroleum domes. Natural gas is primarily methane but may also include ethane, propane, butane, and other gases such as carbon dioxide, nitrogen, and helium. It requires no processing other than filtering physical impurities, water, sulfur, and the other gases mentioned previously. Almost all foreign gases other than methane are removed during processing prior to using natural gas as an automotive fuel. Because methane has no odor, an **odorant** called **mercaptan** is added so that leaks may be detected by humans. Methane is the cheapest and most efficient of all fuels when properly burned, but it requires a large amount of air for combustion. It is one of the most popular forms of alternative fuels for use in vehicles because of its availability, relatively low cost, and low emissions. Methane has a low carbon content, and its simple chemistry (CH_4) offers much cleaner exhaust (lower tailpipe emissions) compared to gasoline.

Methane has an octane number of 125 RON. This gives it a very high resistance to pre-ignition and to knock (abnormal detonation). Thus, gas engines can use a higher compression ratio—as high as 12:1, compared to a traditional gasoline/petroleum engine ratio ranging from 7.5:1 to 10:1. The higher compression ratio results in an increased power output and a greater fuel economy compared to gasoline. It also means natural gas engines have better thermal efficiencies than conventional engines.

Natural gas spark-ignition (SI) engines work on the same basic principles as gasoline-powered SI engines. The fuel is mixed with air and drawn into the cylinder of a four-stroke engine. The air-fuel mixture is ignited by a spark from a spark plug to generate the energy to move the piston and turn the crankshaft. The major changes between conventional SI engines and natural gas engines are in the fuel supply system and the method of mixing and induction of the air-fuel mixture. The difficulties of storing and distributing natural gas have inhibited its more widespread as a fuel for vehicles. The benefits of natural gas can be more easily justified in high-use vehicles, so natural gas in both compressed and liquefied forms is more frequently used by buses, trucks, and taxis. Such vehicles also have a greater storage space for the natural gas fuel tanks.

Compressed Natural Gas

Compressed natural gas vehicles (CNGVs) use methane compressed for storage and use in light-duty passenger vehicles, delivery trucks, and buses (**FIGURE 67-5**). CNGVs carry natural gas typically compressed up to 3600 pounds per square inch (psi), or 25,000 kilopascals (kPa), and stored in heavy-duty cylinders of various types. The cylinders themselves are quite safe, being rated at four times the storage pressure of the fuel system. A fuel pressure regulator reduces natural gas storage pressure to around 100 psi (700 kPa) if fuel injected, and a second stage reduces it to near atmospheric pressure for a fumigation system. A filter is used to trap water or compressor oil "carryover" in the gaseous fuel. Natural gas is used in both bi-fuel (gas or gasoline is used) and dedicated (CNG only) SI engine vehicles. It is also used in dual-fuel compression-ignition (CI) engines.

What looks like white exhaust smoke when a natural gas engine starts up is actually steam caused by the different ratio of hydrogen to carbon in the fuel. This results in a higher ratio of water to carbon dioxide in the exhaust, and water vapor is visible until the exhaust system heats up. In a CNGV, the exhaust is frequently more visible because it is more humid. This steam is not a result of soot, oil blowby, or unburned hydrocarbons.

Under normal operating conditions, oil consumption should be about the same for natural gas engines as it is for traditional gasoline engines, although modern CNG engines require synthetic oils to control higher internal operating temperatures. However, the lubricating oil is generally cleaner because there is less soot formed in combustion, and unburned hydrocarbons cannot contaminate the lubrication oil. The same properties apply to LPG engines.

One drawback of CNGVs is their heavy cylinders and limited range, unless multiple cylinders are used. In that case, a lot of room is needed either under or over the vehicle (bus), or in the valuable payload space inside or outside (e.g., in the pickup truck bed). Another drawback is the incremental cost of the CNG fuel system.

Liquefied Natural Gas

Liquefied natural gas (LNG) is used in heavy-duty long-haul trucks, heavy-duty refuse and delivery vehicles, buses, and so forth. LNG is almost pure methane and has a gasoline per gallon equivalent much closer to gasoline than CNG—a prime consideration for achieving reasonable vehicle range. For the LNG to become a more compact and easily stored liquid, it has to be cooled to an extremely low cryogenic temperature of −263.2°F (−164°C). The liquid must remain very cold at all times, or else it will boil off and greatly expand into a gaseous fuel.

LNG tanks are constructed with two walls of material separated by a layer of insulating material that is placed under a vacuum that enhances the long-term cooling of the fuel—as long as several weeks (**FIGURE 67-6**).

A **float-type fuel gauge** is placed in the tank and raises or lowers in a tube for fuel level calculation. All LNG containers

FIGURE 67-5 CNG is used in buses, delivery trucks, and fleet vehicles.

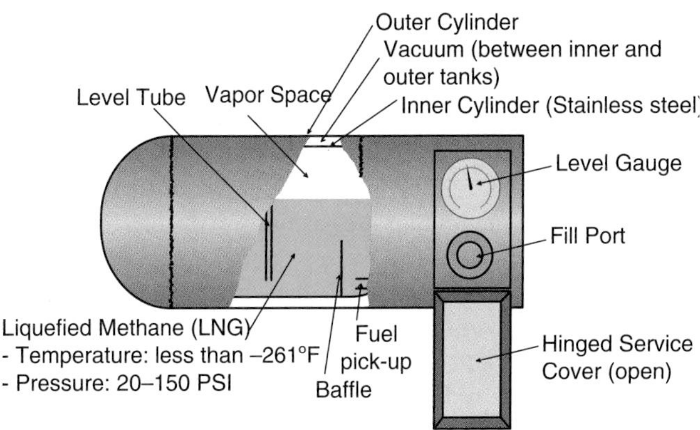

FIGURE 67-6 LNG fuel tank.

must be mounted within the normal extremities of the vehicle and in the correct orientation so that the internal baffles to limit fuel sloshing can be effective when the vehicle is in motion and so the correct evaporation space can be maintained. LNG systems usually circulate engine coolant through the final regulator to maintain the natural gas at a constant pressure and temperature before it is used. This feature is a very important requirement with LNG because of the extremely low temperature at which the gas is stored in its tank.

When refueling LNG tanks, a vapor return pipe prevents accidental overfilling and ensures that there is sufficient vapor space above the liquid in the tank. Excess liquid pumped in will discharge out again through this vapor return connection. As for the LNG infrastructure, the expense of compressor stations, LNG storage tanks, and LNG onboard fuel systems limits widespread acceptance of LNG in vehicles. Exceptions are the vehicles traveling so-called clean corridors. These are specifically designed and equipped U.S. interstate highways with compressor stations and LNG refill locations strategically located to accommodate long-haul trucks. Long-haul global transit of methane is most economically done with LNG ocean-going ships, but establishing a land-based LNG refueling infrastructure comes at a price in terms of the extensive equipment required plus the energy used for compressing methane gas to cryogenic temperatures.

Manufactured Gases

Manufactured gases are derived from other forms of fuel such as wood or coal. These gases generally contain carbon monoxide so are usually poisonous as well as flammable. **Producer gas** is a generic term that refers to a number of manufactured gases such as wood gas, town gas, and syngas. **Wood gas** is a hydrogen and methane gas created by a process of thermal gasification of wood or some other biomass. This gas was used to power ICEs and vehicles fitted with onboard gasifiers or trailer gasifiers, which produced the gas from other feedstocks while the vehicle was in transit. **Town gas**, used for lighting and other purposes, was used for ICE vehicles during the very early days before petroleum became the fuel of choice (thanks to its energy density—Btu per gallon). Town gas, or coal gas, is mostly methane and is made by either the carbonization or the gasification of coal. Originally, this gas was a by-product of turning bituminous coal into the cleaner coke fuel. It is still extensively used for heating and cooking and for some industrial tasks, but it is not usually used to power automobiles.

Syngas, or synthetic gas, consists mainly of varying combinations of hydrogen and carbon monoxide, and it can be produced in a number of different ways, including the Fischer-Tropsch method for the gasification of coal. Syngas can be an intermediate step leading to a catalyzed chemical process to convert the gas to liquid (GTL) hydrocarbons, which can then be used as a synthetic lubricant or as a fuel. **Water gas** is mostly hydrogen and carbon monoxide, and it is created by passing steam over red-hot coke.

Bi-fuel engines and vehicles are capable of running on either gasoline or natural gas, though some engine modifications are required for them to run efficiently. Only one fuel is used at a time, but they may be switched seamlessly while underway. Bi-fuel vehicle engines tend to be limited by the lower compression ratio needed to accommodate gasoline.

Dual-fuel-powered vehicles run literally on two fuels at the same time. For example, in the dual-fuel diesel system, both diesel fuel and natural gas are injected directly into the combustion chamber, almost at the same time. Clean air is drawn into the cylinder from the manifold, and during the compression stroke, a small amount of diesel fuel is "pilot injected" into the combustion chamber. The diesel fuel starts ignition, and the combined fuel mixture burns as it would in an engine fueled with 100% diesel fuel. By using natural gas fuel this way, engine knock is no longer a problem, and the compression ratios for the engine do not have to be altered. In the dual-fuel engine, the liquid fuel is typically injected from a center-located pintle in a dual-fuel injector. Once combustion is initiated, liquid injection is immediately followed by a main spray of gaseous CNG from holes surrounding the liquid fuel pintle. Injectors may be drop-in configured to convert a conventional diesel engine to dual fuel for fuel savings and to lower emissions. In other applications, natural gas is injected through a second injector into the combustion chamber.

▶ Diesel-to-Compressed Natural Gas Conversions

Some fleets use diesel engines converted to run on natural gas alone instead of diesel fuel by converting the CI engine to an SI engine. This is usually done by replacing the diesel injector in the engine's cylinder head with a spark plug and premixing the gaseous fuel with the intake air. The existing diesel fuel injection system is replaced with a high-tension SI system, and the natural gas fuel is mixed with air with port injectors, or with a carburetor-type spray ring or fumigation rail before the throttle body in the intake system. Compression ratios for SI engines have to be significantly lower than for diesel engines, so a spacer is usually fitted between the block and the head on the engine. This design reduces the compression ratio to around the necessary 12:1, but it also reduces the thermal efficiency of the engine, decreasing torque, particularly at low speed, which is a common advantage of CI engines.

In a dedicated natural gas conversion, the diesel injectors are replaced with spark plugs, and the natural gas fuel is premixed with air before intake occurs. This is a relatively inexpensive conversion, and it allows the fleet to use cheaper, cleaner natural gas fuel.

With a dual-fuel CI system, the diesel injectors remain in place, and the engine can still run on 100% diesel fuel if necessary. Natural gas is mixed with air in the inlet manifold, and the amount of diesel fuel injected is correspondingly reduced. Only a relatively small proportion of diesel fuel is necessary for ignition of the whole fuel mixture to occur through engine compression. Combustion in this way is actually more effective than with SI because the diesel fuel injector nozzle spray jets create a number of ignition sources, not just one. This allows for a more complete and rapid combustion.

Natural gas can account for up to 98% of the fuel content, as long as the remaining percentage is injected diesel fuel, although the higher the proportion of natural gas fuel in the mix, the greater the tendency of the engine to knock.

▶ Gaseous Fuel Storage Systems

K67003

LPG Fuel Storage Tanks

Because of the insulation used for LNG fuel storage, these systems are more complicated to design and much more costly to manufacture. Although the weight by mass of LNG fuel is about half that of diesel fuel, an LNG fuel tank is substantially heavier and bulkier.

There are two types of LPG storage container: **portable universal cylinders** and **permanently mounted tanks** (**FIGURE 67-7**). Portable cylinders can be removed from a vehicle and taken to a refueling site to be filled. These are useful for off-road vehicles or remote sites where the vehicle itself cannot easily be taken to a filling station. When not installed on a vehicle, these containers must be stored in a secure outdoor area. They are called universal cylinders because they can be used in a vertical or horizontal position.

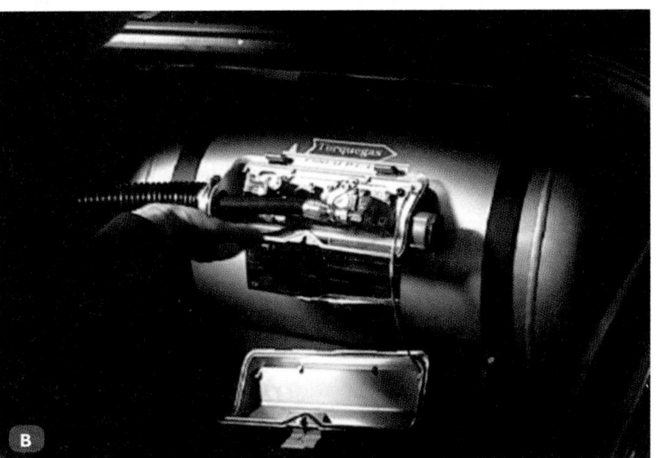

FIGURE 67-7 LPG fuel tanks. **A.** Portable. **B.** Permanent.

When a vehicle is capable of traveling to a refueling site, the tank should be permanently mounted to the vehicle. Permanently installed tanks must be placed in their correct orientation, and they usually carry marks to indicate the top of the tank. LPG containers must be mounted within the extremities of the vehicle. The tanks used in automobiles (such as taxis) are usually found in the trunk of the passenger vehicle. In the case of pickup trucks, the tank sits in or under the bed of the truck. LPG tanks are normally made of a heavy-gauge steel, as are the fuel lines and pressure safety valves used to supply the fuel to the vehicle system.

The normal working pressure of a tank varies depending on ambient temperatures and the quantity of fuel in the tank, but common operating pressures are normally within the range of 102–170 psi (700–1200 kPa). LPG tanks are constructed to withstand much higher pressures of approximately 1000 psi (7000 kPa). This is necessary because LPG expands when its temperature rises.

To make sure the fuel has room to expand, the tank is never filled to capacity. LPG tanks are fitted with a special refueling port and an overfill protection device—an **automatic fuel limiter**—which ensures that the tank cannot be filled past the safe fill limit of 80%. This design allows for fuel expansion on a hot sunny day, and tanks are generally painted white to reflect sunlight. LPG tanks are also equipped with pressure relief valves that can release propane vapors to the atmosphere to prevent the tank from rupturing under abnormally high-pressure conditions. A fuel lock-off lets fuel flow when the engine is running but shuts off the fuel supply as soon as the ignition is turned off or if the engine stops running, such as from a collision.

▶ TECHNICIAN TIP

The basic LPG servicing procedures are limited to visual inspection and leak testing. Because LPG is heavier than air, an electronic leak detector is moved slowly below fittings to detect any leakage. Soap and water or leak detection spray may also be used for leak testing. The tank must be evacuated and purged before disassembling any of the fittings, and no welding is permitted on the container.

CNG Fuel Storage Cylinders and Systems

CNG is compressed and stored at very high pressures (3000–3600 psi [20,700–24,800 kPa]). The volume of space required for CNG storage tanks is almost twice that required for LNG tanks for delivering the equivalent mileage range.

Both fuel storage (depot) and vehicle storage tanks must be extremely robust due to the high pressures required for storing compressed gas. CNG cylinders come in four types (types 1 through 4), ranging from heaviest to lightest (**FIGURE 67-8**). Type 1 cylinders are made of steel; type 2 are made of aluminum; type 3 are made of a combination of metal and an approved composite material; and type 4 (the lightest in weight) are made totally of a carbon composite material wound around a dense rubber core. Cylinders must be permanently marked "CNG" by the manufacturer. Pressurized containers for CNG use have a rated service pressure of not less than 3600 psi at 70°F (24,800 kPa

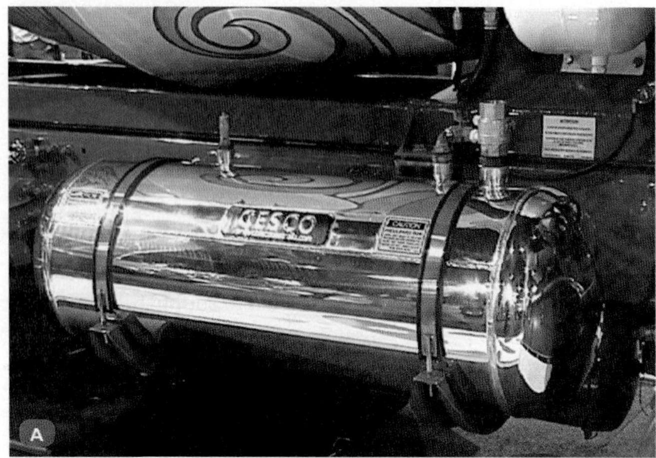

FIGURE 67-8 CNG fuel tanks.

at 21.1°C). They are tested under extreme conditions and must hold pressure to four times their rated service pressure.

Vehicle CNG fuel storage systems consist of one or more cylinders. Each is shock mounted in place using clamps that enable pressurized cylinders to expand when filled while still holding them firmly in place. The straps enable the vehicle chassis to deform as intended in a high-speed collision without deforming the CNG storage tank.

Each cylinder is fitted with a thermal- and/or pressure-activated pressure relief device (PRD), which releases gas in the vent of over-pressurization. Each cylinder is equipped with either a manual safety shutoff valve or a solenoid lock-off device connecting to a power source. If more than one cylinder is used, a seamless steel tubing manifold connects the cylinders. If the cylinders are located under the vehicle, a shield protects the safety valve/lock-off and PRD. If cylinders could be exposed to sunlight, they are protected from ultraviolet damage by a cover.

Because CNG is a gas, not a liquid, it does not matter whether the container is standing up or lying down; however, all CNG containers must be mounted within the extremities of the vehicle, for their protection. CNG cylinders may be found in the trunk of a passenger vehicle, inside a van or SUV, or underneath the vehicle in the rear or between the frame rails. In some applications, such as urban transit passenger buses, as many as eight or more CNG cylinders are mounted on the roof and are encased in a special fiberglass enclosure (**FIGURE 67-9**). There are height restrictions for such applications to avoid serious damage to the cylinders caused by overhead hazards such as low bridges. Similarly, cylinders must be a minimum distance from the road, with the vehicle fully loaded.

FIGURE 67-9 CNG fuel tanks mounted on the roof of a bus.

If the CNG tank is mounted within 8" (200 mm) of a heat source such as the exhaust pipe or muffler, then an effective heat shield must be mounted between the tank and the heat source without touching either. There must also be an insulator such as nylon or neoprene between the tank and its mounting bracket to protect it from movement wear and corrosion.

Note: For the latest CNGV installation and inspection specifications, refer to **National Fire Protection Association (NFPA)** standards published in their **NFPA-52** document.

Service procedures are normally limited to a so-called basic visual inspection for damage and rust, and leak testing. Leaks can be detected with an approved electronic methane leak detector or an approved leak test solution, which is sprayed over the tank, valve, PRD, lines, and so on. Soap and water may also be used, but not ammonia. Bubbles in the solution indicate a leak. Because methane is lighter than air, testing is done above the fittings rather than below (as with propane). If a cylinder is suspected to have a problem, the vehicle should be immediately taken out of service, the cylinder defueled and removed from the vehicle, and a detailed cylinder inspection performed by a trained and certified tank inspector.

▶ TECHNICIAN TIP

LNG tanks may be serviced only by specially trained and certified specialists.

▶ Biofuels

K67004

Biofuels are fuels derived from biological sources, such as corn, sugarcane, soybeans, and algae. Biofuels help to reduce consumption of petroleum fuels, which provide the previously stated benefits. In transportation, biofuels generally come in two major categories: biodiesel and bio-alcohol. Biodiesel is used in CI engines either as a blend with traditional petrodiesel or by itself. Bio-alcohol is used primarily as a blend with gasoline. Biofuel production has continued to increase over the years, although it still contributes only a small percentage of total energy use.

Biodiesel

Biodiesel is renewable fuel made by chemically combining natural oils from soybeans, cottonseeds, canola, animal fats, algae, jatropha seeds, or even recycled cooking oil, with an alcohol such as methanol or ethanol, and a catalyst like lye. Depending on the fuel base stock, biodiesel fuels can be more expensive than diesel made of petroleum products (petrodiesel), but biodiesel burns with less particulate and with no **sulfur** or **aldehydes**, thus producing less harmful and irritating emissions.

The improved **lubricity** and zero sulfur content of biodiesel result in longer maintenance intervals, longer engine and fuel system life, and lower emissions. Biodiesel fuel is compatible with petrodiesel and may be used completely in place of petrodiesel, or in most applications may be blended with petrodiesel. A typical blend would be 20% biodiesel with 80% petrodiesel fuel (**FIGURE 67-10**). Biodiesel tends to clean petrodiesel residues from the fuel system, so fuel filters may require frequent servicing for the first few tank fills.

Biodiesel Use

Biodiesel, depending on the feedstock and process used, is a light-yellow to dark-gold liquid, with a **viscosity** very similar to petrodiesel fuel. Commercial blends of biodiesel and petrodiesel are designated with the letter "B" followed by the volumetric percentage of biodiesel in the blend: **B20 fuel**, the blend most commonly available, contains 20% biodiesel and 80% petrodiesel. **B100 fuel** is pure biodiesel (called "neat" diesel). Commercially produced biofuels must meet ASTM standards. Commercially processed biodiesel-blended fuels can run in any diesel engine designed for petroleum-based diesel fuels. Some diesel engines can run on **straight vegetable oil (SVO)** or on **waste vegetable oil (WVO)**, but only if it has been properly treated to remove the glycerin, which could otherwise gunk up injectors and engine parts like piston rings.

FIGURE 67-10 Biodiesel fueling station.

SVO is a pure form of vegetable oil or canola oil, peanut oil, soybean oil, and corn oil. WVO oil is filtered oil usually from restaurants but may be a combination of many restaurants. The filtered oil is sent to a processing plant, where it is generally mixed with clean pure soybean oil for biofuel processing. The by-product of using this oil is the waste, which becomes greater when large parts of animal fats are present. The cleanest restaurant oil used comes from the frying of non-animal products. The risk of using restaurant oil is a batch that exceeds the heat range of the oil, generally over the 600°F (316°C), where the oils start breaking down structurally. This oil is unusable for the biofuel process.

Biodiesel may be more expensive to produce than petrodiesel, depending in part on governmental taxes and subsidies. Biodiesel blends tend to perform better than today's low-sulfur petrodiesel. Just a 1% blend of biodiesel adds lubricity lost, yet needed to protect injectors from wear. Biodiesel burns cleaner with less smoke and may have a unique odor (like french fries!).

Biodiesel fuel has some drawbacks. It may make the engine harder to start in winter temperatures. High feedstock and production costs tend to limit wholesale commercial implementation, but the U.S. military is committed to the use of alternatives such as biodiesel for defense purposes. Large-scale production demand will inevitably bring down the cost.

As with petrodiesel fuel, contamination with moisture can corrode components such as pumps and injectors, as well as encourage the growth of microbes. Biodiesel fuel can also degrade natural rubber gaskets and hoses. Today's vehicles use synthetics such as Viton to avoid such problems. Even older diesel vehicles can tolerate a blend of B5, and some original equipment manufacturers (OEMs) accept B20 or higher blend usage with no warranty concerns.

Biodiesel Characteristics

Biodiesel is both biodegradable and nontoxic. It produces significantly fewer emissions than petroleum-based diesel fuel. Tests have shown that the use of soybean biodiesel fuel (B100) in urban transit buses reduces net carbon dioxide emissions by over 75%. B100 (neat) biofuel contains about 11% oxygen, which improves combustion, reducing most exhaust pollutants. Drivers also note that the exhaust from biodiesel is less irritating than exhaust from petrodiesel-powered engines.

One of the most important characteristics of any diesel fuel is its ability to auto-ignite. The cetane number (a rating of how easily diesel fuel may be ignited) for biodiesel ranges from 46 to 57 for soybean oil, with an average of around 50. This means that biodiesel fuel ignites more quickly than petrodiesel, which normally has a cetane rating of 40 to 52.

Biodiesel is a better solvent than petrodiesel, so it helps clean the fuel system as it flows through it, and it can break down and flush out residue deposits in vehicles that have been running on petrodiesel. As some ocean-going fishing vessel captains have discovered, it is important to clean the fuel system

FIGURE 67-11 Diesel cloud point.

FIGURE 67-12 E-85 is 85% ethanol and 15% gasoline.

and bring extra filters along when changing over to biodiesel fuel; the solvent action of biodiesel will quickly clog up fuel filters by flushing out accumulated deposits in older and dirty fuel systems.

Although it has some performance advantages, biodiesel does have some disadvantages, with biodiesel fuels made from yellow grease, or recycled cooking oil, performing worse than soy-based biodiesel. At lower temperatures, biodiesel fuel normally clouds and forms wax crystals, which can clog fuel lines and filters in a vehicle's fuel system. To combat this natural formation of restricting crystals, manufacturers add fuel line and filter heaters to the fuel systems. The **cloud point** is the temperature at which diesel fuel starts to appear cloudy, indicating that wax crystals have begun to form (**FIGURE 67-11**). At even lower temperatures, fuel develops a waxy gel that cannot be pumped, thus starving the engine of fuel. The **pour point** is the temperature below which diesel fuel will not flow at all. The cloud point and the pour point temperatures for biodiesel are both higher than those for petrodiesel, so some operators report that it does not perform as well at low temperatures.

The energy efficiency of a vehicle is the percentage of the fuel's energy delivered as engine output. The energy content per unit of volume measure (Btu per gallon) of biodiesel is somewhat lower than that of petrodiesel, so biodiesel may be less economical to use. Fuel efficiency is usually expressed as miles traveled per gallon (mpg) of fuel, or kilometers per liter (km/L) of fuel. Vehicles running on higher blends of biodiesel may be less fuel efficient.

Ethanol

First-generation biofuels like **ethanol** are usually derived from an organic process—for example, the fermentation of sugars that come from plants such as corn, cane sugar, sugar beet, or grains. Second-generation **bio-ethanol** is derived from a variety of cellulose-based plants such as switchgrass. Third-generation biofuels like ethanol are algae based—that is, produced by/from algae. Algae-based ethanol production is the most promising in terms of yield per acre/hectare.

Bio-ethanol is a renewable resource and is also known as grain alcohol. Ethanol is heavier than air and is **hygroscopic**, which means it readily absorbs water vapor from the atmosphere. Ethanol is a high-octane fuel (around 100 RON/MON/2 octane [see the Gasoline Fuel Systems chapter]), depending on various factors. It is a renewable fuel intended for SI engines. It is used in FFVs in the United States in a blend of up to 85% ethanol to 15% gasoline, called E85 (**FIGURE 67-12**). Some countries use 100% ethanol instead of gasoline, which would be E100.

Ethanol is primarily used to reduce the negative emission effects of gasoline and as a gasoline oxygenate to replace harmful methyl tertiary-butyl ether. This mixture may be harmful to gasoline/petrol-fueled vehicles without the proper FFV seals and elastomers. Equipping an OEM vehicle for FFV use is inexpensive, and increasingly more OEMs are doing so in the interest of making their vehicle offerings "green." Vehicle computers calculate the percentage of ethanol blend by a variety of methods and alter injection quantity and spark timing to achieve the desired stoichiometric ratio of fuel and air and the resulting low emissions possible with ethanol.

Ethanol separates and evaporates quickly from water, which makes it less polluting to streams and waterways in the event of a spill. Compared to gasoline, ethanol burns particulate-free and releases fewer toxic emissions than gasoline/petrol. Its luminosity (visibility when it burns) is hard to detect in sunlight, thus making it somewhat dangerous in the event of a vehicle fire. It does release carbon dioxide during the combustion process, but again, less than the equivalent amount of gasoline/petrol. Ethanol holds even greater promise for reducing petroleum usage and providing greater engine power when engines are optimized (through valve timing or stroke variation) for use of higher octane ethanol, as compared to gasoline. When used in plug-in hybrid FFVs, such vehicles are said to achieve a gasoline-per-gallon equivalent as high as 500 mpg.

Methanol

Methanol is produced from wood products, natural gas, acetylene, or other organic materials (plant/animal wastes or garbage). It is highly poisonous and dangerous to the environment. It has a lower energy density than ethanol, and its **burn rate** (measured in feet per second) is not as high as that of gasoline fuel. Methanol is heavier than air and is hygroscopic. However, it has lower evaporative emissions than petrol/gasoline, as well as fewer toxic hydrocarbon emissions.

Methanol is sometimes referred to as wood alcohol because it is possible to produce it by fermenting biomass. **Biomass** is a renewable energy source, plant, or animal material used as a source of fuel. **Fuel methanol** is normally sold as a blend of 85% methanol and 15% unleaded premium gasoline/petrol, known as M85.

Methanol cannot be used in conventional vehicles without significant modifications. **M85 vehicles** are FFVs that can run on any mixture of methanol and gasoline/petrol, though the limit of methanol in the mixture is 85% in the United States. A **fuel composition sensor** tells the engine computer what percentage of methanol is in the fuel, and it adjusts the injection quantity ignition timing accordingly. This means that the FFV can run on conventional gasoline/petrol if methanol is not available; two separate fuel systems are not necessary.

Biobutenol

Biobutenol is produced in a similar fashion to ethanol, but butenol has characteristics similar enough to gasoline that it can be transported through the existing petroleum (pipeline) distribution infrastructure. Butanol has several advantages over ethanol, such as higher energy content, lower water absorption, and better blending ability. It can be used in conventional combustion engines without modification. Like ethanol, it can be produced through fermentation or by petrochemical methods. Butenol is not widely used but holds great promise as an alternative fuel for SI ICE-powered vehicles.

In summary, liquid fuels like biodiesel, ethanol, methanol, and butenol are perhaps the most transparent alternative fuels. With some limitations (petroleum pipelines are off-limits to some of these fuels), they are the least distinguishable from gasoline/petrol in how they are transported, refueled, and used. The fuel system of a car or truck requires only slight modifications in order to run on E85 or M85.

▶ Fuel Cell Operation

K67005

Although most alternative fuel technologies were developed many decades ago, many of them require expensive materials or manufacturing processes to mass produce them affordably. Developments and advancements are constantly being made in alternative fuel technologies. In this section, we explore fuel cell technology, which is still in its infancy but provides a lot of promise if the remaining obstacles can be overcome.

Fuel Cells

Fuel cell technology has been used for many years in the space industry. Recent improvements in the technology and the need to seek alternative fuel technologies for the automotive industry have seen a number of manufacturers develop fuel cell technology for use in automobiles. In a vehicle powered by a fuel cell, the electric motor is powered by electricity generated by the fuel cell.

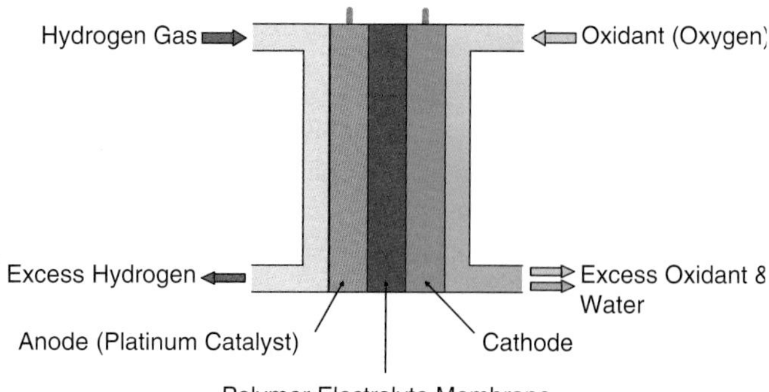

FIGURE 67-13 Components of a fuel cell.

A **fuel cell** is an **electrochemical device** that combines hydrogen and oxygen to produce water, and in the process produces electricity and heat. Fuel cells operate without combustion, so they are virtually pollution free. Another electrochemical device that you are already familiar with is a battery. In a battery, all the chemicals are stored inside, and the battery converts those chemicals into electricity. This means the battery eventually becomes discharged until it is recharged. In a fuel cell, the chemicals oxygen and hydrogen constantly flow through the fuel cell, like fuel through an engine, so it continues to produce electricity as long as oxygen and hydrogen are available.

There are four basic elements to a fuel cell: the **anode**, **cathode**, **electrolyte**, and **catalyst** (**FIGURE 67-13**). Pressurized hydrogen flows into the fuel cell anode. The platinum coating on the anode helps to separate the gas into protons (hydrogen ions) and electrons. The electrolyte in the center allows only the protons to pass through to the cathode side of the fuel cell. The electrons cannot pass through this electrolyte and therefore flow through an external circuit in the form of electrical current. At the same time, oxygen flows into the fuel cell cathode, where another platinum coating helps the oxygen, protons, and electrons combine to produce pure water and heat.

A fuel cell only produces a voltage of about 0.7 volt. To get the required voltage for automotive applications, a fuel cell stack is created. The number of fuel cells in the stack determines the total voltage, and the surface area of the fuel cells determines the total current.

Fuel cells use hydrogen and oxygen to produce electricity. The oxygen can come from the air; however, hydrogen is not readily available in its pure form. Hydrogen is difficult to store and distribute. To address this problem in automotive applications, an additional device called a **reformer** is used to separate hydrogen from other fuels, such as gasoline.

The fuel reforming process can be performed on a small scale on an as-needed basis immediately before its introduction into the fuel cell. One example is for a fuel cell–powered vehicle to have a gasoline tank onboard that uses the existing infrastructure of gasoline delivery. An onboard fuel processor reforms the gasoline into a hydrogen-rich stream that is fed directly to the fuel cell.

At the present time, it is not practical to perform separation of other products of the reforming process from the hydrogen at this small scale. So to make fuel cells practical for automotive applications, vehicle manufacturers are developing better fuel cell systems and technology to improve the efficiency of the system while using readily obtainable fuels.

Hydrogen Fuel Cells

Hydrogen is the simplest and most plentiful element in the universe. It is a colorless, odorless, and tasteless gas. Hydrogen is the lightest element, yet it has the highest energy content per unit weight/mass of all energy-based fuels—three times the energy of gasoline/petrol.

Hydrogen can be extracted from virtually any hydrogen compound and is the ultimate clean energy carrier. Hydrogen is commonly extracted from water by electrolysis, which is the process of separating H_2O (water) into its components of hydrogen and oxygen. It can also be extracted from a variety of other sources, meaming that hydrogen's energy can be harnessed in many pollution-free ways. Hydrogen can be burned directly as a fuel in an ICE instead of gasoline. It varies depending on the source of hydrogen, but total carbon dioxide emissions using hydrogen in the ICE vehicle can be less than half those of a conventional gasoline vehicle. Such vehicles are a slightly modified version of the traditional gasoline-powered engine.

Alternatively, hydrogen can be used to create electrical power in fuel cells to create power for electric motors, which in turn can power a vehicle. A fuel cell consists of two electrodes sandwiched around an electrolyte. Oxygen passes over one electrode and hydrogen over the other to generate electricity, water, and heat.

A fuel cell operates in some ways like a battery, but unlike a battery, it does not run down or require recharging. It produces electrical energy and heat as long as hydrogen fuel and oxygen are supplied. Although a battery is a closed electrical storage system with plates (electrodes) that react and change their chemical configuration as that battery is charged or discharged, fuel cells are very different. The electrodes of a fuel cell are relatively stable, but catalytic, and the fuel cell consumes hydrogen as a reactant that must be continuously replenished. Hydrogen is fed into the anode of the fuel cell. Oxygen (i.e., air) enters the fuel cell through the cathode. The hydrogen atom passes a catalyst, which causes the hydrogen to split into a proton and an electron. The protons pass through the electrolyte to the cathode, but the electrons must take a different path through a circuit that converts them to electrical current. In a **fuel cell (electric) vehicle (FCV)**, the current is used to charge a traction battery or to power electrical devices before they return to the cathode. The hydrogen electrons and protons then recombine with oxygen to become water molecules. Water is the "exhaust" from the fuel cell.

Besides the lack of harmful emissions, one of the main advantages of using hydrogen fuel cells as a generator of energy for traction batteries is their reliability. Fuel cells are compact and lightweight and have no moving parts, and they do not involve any of the explosive combustion processes necessary with conventional fuels.

Fuel Cell Vehicles

Gasoline-powered vehicles waste a lot of thermal energy. Fuel is used to heat and expand air to drive pistons and the crankshaft, with exhaust gases containing pollutants emitted to the atmosphere. Heat energy that has not been fully used by the expansion process is wasted through radiation and from the exhaust. FCVs, however, are cleaner and more efficient. The conversion of fuel to electricity is a direct electrochemical process, with more energy extracted from the fuel source. And because the vehicle has fewer components in the drive system, such as pistons and camshafts, frictional losses and reliability issues associated with these components are eliminated.

Concept cars using hydrogen fuel cells have been in existence for many years, but FCVs have in the past struggled to produce electricity in sufficient quantity (kilowatts) at an acceptable cost. However, the first production line vehicle with a dedicated hydrogen fuel cell platform was made available for public sale in 2008. The Honda FCX Clarity has an effective driving range of approximately 280 miles (450 km) on one tank of hydrogen gas (**FIGURE 67-14**). This is the equivalent to an EPA-certified 74 mpg (30 km/L). Other OEMs have since produced pre-production and commercially available FCVs, like the Toyota Mirai, are now being tested for reliability the world over. The downfall for FCVs occurs if the water in the fuel cell is allowed to freeze. If the water turns to ice, the expanding ice will ruin the fuel cell stack, consisting of many electrodes.

A typical FCV has a fuel cell stack, an air compressor to make sure enough air is being fed into the fuel cell, a radiator with circulating coolant, a hydrogen recirculating system to optimize efficiency, hydrogen storage tanks, a battery, and an electric motor. Either a lithium-ion battery or a supercapacitor may be used to provide supplemental power for the vehicle and to store energy generated during braking.

The hydrogen used comes from other fuel sources, such as from natural gas or by electrolyzing (electrically separating) water (H_2O) into its hydrogen and oxygen elements. Although energy-intensive, even taking this part of the fuel production process into account, FCVs significantly reduce the amount of

FIGURE 67-14 Honda FCX Clarity.

carbon dioxide emissions. Some companies use solar energy to electrolyze sea water, thus producing hydrogen in a truly sustainable fashion.

Systems used for transporting and storing CNG are somewhat similar to those needed for hydrogen. Because of its molecular structure, hydrogen (H_2)—the smallest of all elements— tends to easily leak through ordinary storage media. It also tends to crystallize metal. Thus, one of the limitations in the more widespread use of hydrogen fuel cells in vehicles has been the relatively small number of available hydrogen refueling outlets. One way this may be overcome is by installing a domestic hydrogen generator as a standard element in a home heating system. This generator could be powered by the conventional domestic supply of natural gas, so your vehicle could be economically **slow-fill** refueled overnight in your own garage rather than at a commercial filling station.

Fuel Cell Systems

Fuel cell power-generating systems have separate hydrogen gas fuel and air intake ports, and an outlet for the waste water produced by the fuel cell process. Electrical current is generated by the fuel cell and is recovered through the anode and the cathode plates in the stack. Antifreeze is circulated through the system as a coolant.

The electrochemical reaction that creates electrical power takes place within individual fuel cells. The fuel cell system consists of many fuel cells arranged in a "stack." Together, they output enough electrical energy to drive a vehicle's electric motor. Each cell consists of a **membrane electrode assembly** sandwiched between **plates**. Tiny flow channels in the plates diffuse the hydrogen and oxygen gases onto the surface of the membrane, where a catalytic reaction occurs to split off hydrogen electrons to produce electricity (**FIGURE 67-15**). The two gases then combine to form water, which is the only exhaust.

The reaction also generates significant heat, which must be removed. Coolant is circulated from a radiator between the fuel cells or through the plates without making direct contact with the membrane.

The individual fuel cells are combined in series into a stack so that the whole module will deliver a required output power. For example, suppose a stack produces 20 kilowatts of power (**FIGURE 67-16**). Four such stacks combined will produce 80 kilowatts, enough to fully power an average vehicle. Smaller fuel cell stacks can be used as a range extender—an auxiliary power source used in combination with a battery. Some fuel cells are used to power smaller vehicles or as auxiliary generators for trucks and refrigeration units during overnight layovers or extended idle periods.

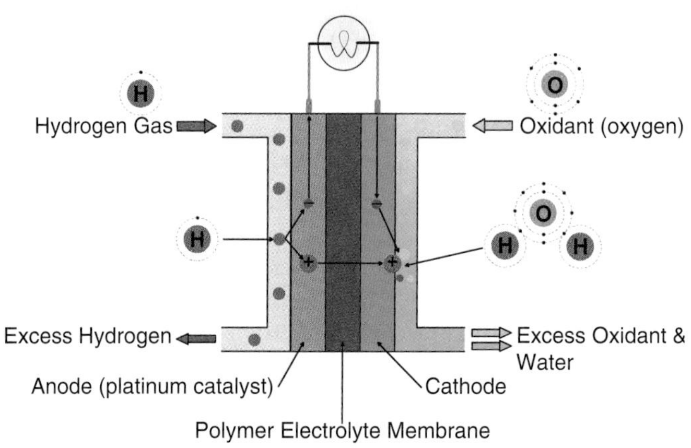

FIGURE 67-15 Fuel cell creating electricity.

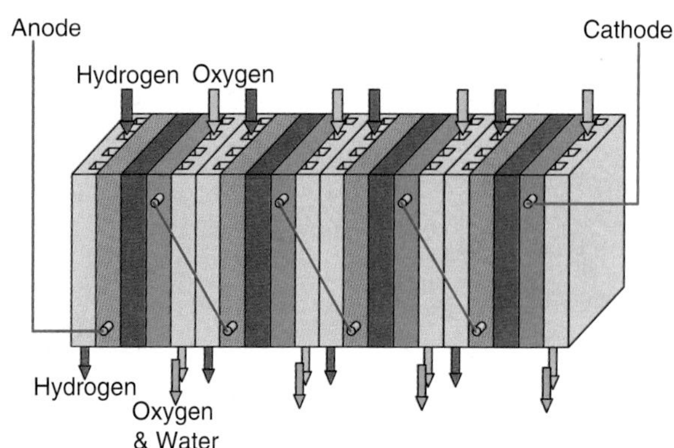

FIGURE 67-16 Fuel cells in a stack.

Battery Electric Vehicles

K67006

Battery electric vehicles (BEVs) are powered entirely by batteries that need recharging when they are low. In the past, lead-acid or nickel–metal hydride batteries were typically used. Now lithium-ion and other types of batteries are used for greater range. Early BEVs include GM's EV1, Ford's EV Ranger, Toyota's RAV4 EV, and Nissan's Altra. Most of these early electric vehicles have been reclaimed and scrapped by the OEMs. Now, a whole new generation of electric vehicles is becoming available, such as the Tesla, Chevy Bolt, and Nissan Leaf (**FIGURE 67-17**). Watch for more BEVs to be released soon.

BEV Development

BEV principles were used more than a century ago, before the advent of gasoline-powered internal combustion vehicles. With modern-day improvements in vehicle systems and battery technology, electric vehicles are returning to the marketplace. Because more than 75% of daily commuters drive no more than 40 miles (about 65 km) a day, electric vehicles are becoming viable, as they can now deliver that range without having to consume any gasoline or diesel fuel at all, and therefore producing no emissions.

BEVs first enjoyed a brief resurgence during the late 1990s in the United States with a number of OEMs offering them for lease. These included GM's EV1, Toyota's RAV4 EV, Ford's Electric Ranger pickup, and Nissan's Altra station wagon/crossover vehicle, to name a few (**FIGURE 67-18**). These vehicles were preproduction or limited production test vehicles mainly for California commuters, police or postal fleets, and other limited-range applications.

FIGURE 67-17 Nissan Leaf.

FIGURE 67-18 Early attempt by manufacturers to develop BEVs.

The BEV limitations included the lack of a recharging infrastructure and lack of a single-charge port configuration; various conductive hookups were used along with the innovative inductive "paddle." The vehicles also suffered from disappointing range issues. Where lead-acid batteries offered a limited range-to-weight ratio, the nickel–metal hydride batteries proved expensive and at the time somewhat unreliable. Thus, most BEVs of that period were recalled by the OEMs and crushed. But the knowledge and technology gains from those electric vehicles propelled a new generation of HEVs and BEVs forward to the marketplace. As a result of improved development in battery technology and propulsion system management algorithms, many manufacturers have now entered the BEV market.

BEVs Today

BEVs have been on a resurgence lately. When coupled with sustainable or renewable energy sources, the BEV can truly serve as an emissions-free vehicle. Sustainable energy may be provided by unlimited sources of supply, including these:

- Solar panels and reflectors (to drive steam turbines)
- Wind generators

FIGURE 67-19 Ford Focus Electric being charged.

- Tidal and wave-power generators
- Geothermal sources

These sources and others offer limitless energy sustainability without the use of fossil fuels like petroleum or coal. The BEV itself normally uses only one power source—the battery. The battery powers traction motors to propel the vehicle. The vehicle is returned to a charging station to maintain the energy (**FIGURE 67-19**). With a BEV, the average distance is limited to the storage capacity of the battery and the efficiency of its motor. Most BEVs, such as the Nissan Leaf, can travel about 60–100 miles (129–161 km) on a single charge, with some exceptions. Because of this, the Leaf has been one of the bestselling BEVs in history, with about 250,000 vehicles sold since it was introduced in 2010.

The Tesla Roadster was reportedly the first production automobile to use lithium-ion battery cells and the first production BEV (all-electric) to travel more than 200 miles (322 km) per charge. Tesla has followed that up with the Model S sedan and the Model X SUV (**FIGURE 67-20**). There is also a Model 3 in development that promises to sell for $35,000, competing with the Nissan Leaf and Chevy Bolt.

With such promising battery and vehicle technology, BEV range will likely extend to more than 300 miles (483 km) in the not-too-distant future. In fact, Elon Musk, owner of Tesla says, "Future models may reach a 500-mile (800 km) range, partially because of a new patented battery system, pairing metal-air and lithium-ion batteries."

Another BEV, the Chevy Bolt, became available in 2016. It uses a "nickel-rich lithium-ion" battery pack that tolerates higher cell temperatures. The battery pack is rated at 60 kW of electrical storage and has a range of approximately 240 miles (380 km).

The Chevy Volt is known as an extended-range electric vehicle. It uses a lithium-ion battery that can be recharged from a domestic electrical power outlet (110 or 220 volts AC). This vehicle, however, is not a pure electric vehicle, but a form of hybrid vehicle, as it also has a small gasoline/petrol-powered engine that drives a generator to provide electric power, should the vehicle travel beyond its battery-only range.

FIGURE 67-20 **A.** Tesla Model S. **B.** Tesla Model X.

Batteries

Battery technology has advanced in recent years, making BEVs and HEVs viable for many more people. From the early days when lead-acid batteries only provided a range of 20–30 miles, lithium-ion batteries have been improved to the point that they now provide over 200 miles of range. Although current battery technology is adequate for many people, it is still expensive, putting this technology out of the reach of the masses. But the progress being made in battery technology indicates that the price will continue to decrease, and the capacity increase, hastening the day when electric vehicles become a majority of vehicles on the road. Here is the progression of battery types used in production vehicles.

- Lead-acid battery—Variation of standard storage batteries used in regular passenger vehicles, but with thicker lead plates, which allow the battery to better withstand the heavy cycling from the charged to the discharged state, and back. This cycling is common in electric vehicles, and the batteries are called deep cycle batteries. Advantages are: availability, cost, and simplicity of charging equipment required. Disadvantages include weight, size, and storage capacity.
- Nickel-cadmium (Ni-Cd)—Rechargeable battery with higher energy density than lead-acid batteries. Advantages are the ability to withstand high discharge rates, ability to withstand deep discharge for long periods of time, ability to withstand more discharge/charge cycles, and its being lighter and smaller than lead-acid batteries. Disadvantages include a higher self-discharge rate, higher cost than lead-acid (equivalent cost to NiMH but less capacity); also, the cadmium is highly toxic, and Ni-Cd requires a special charger because the batteries' internal resistance falls as the cell temperature rises.
- Nickel-metal hydride (NiMH)—Rechargeable battery similar to Ni-Cd but with the cadmium replaced with a metal hydride structure. Advantages are that there are no highly toxic materials; it has two to three times the energy density of Ni-Cd; it is able to withstand high discharge rates; it is less expensive than lithium-ion batteries, and it is in mainstream use in production EVs and HEVs. Disadvantages are that it requires a special charger, has a moderately high self-discharge, and can be damaged if discharged too deeply (polarity reversal).
- Lithium-ion (Li-ion)—Rechargeable battery with the highest energy density currently in use. Li-ion batteries are made from a variety of materials in addition to the lithium salt electrolyte. Advantages include high energy density, minimal self-discharging, ability to withstand large amount of discharge/charge cycles, and its relatively nontoxic materials. Disadvantages are that it has the highest cost of current battery options, is a possible thermal runaway and fire hazard, and requires a special charger.

Many high-voltage battery packs, especially Li-ion batteries, have to operate within a specified temperature range. This requires special cooling for the battery module. Cooling can happen in a couple of ways: The first is with outside air being drawn into the battery pack by a computer-controlled fan (**FIGURE 67-21**); the other is with a separate cooling system connected into the vehicle's HVAC system.

In many cases, the high-voltage controller also has to be cooled. This is typically cooled by a separate cooling system from the ICE's cooling system (**FIGURE 67-22**), although they may share separate sections of a common radiator. The coolant is circulated by an electric water pump as needed, which is controlled by the HV controller. Typically, this coolant is the same type as the ICE coolant but has to be flushed separately.

Electric Motors

Vehicles that include electric drive motors have many advantages over traditional vehicles. Lower noise and gas emissions are among the major benefits. Other benefits include fewer moving parts, fewer maintenance requirements, good low-speed torque output, and increased reliability.

Electric vehicles for road use today use AC motors. The DC motors used in electric concept cars were originally carryovers from the electric forklift industry. Smaller non-road (golf cart style) or low-speed neighborhood electric road vehicles (NEVs) may use DC motors and relays to control the drive system. DC installations tend to be simpler and less expensive than an AC system. But AC motors are more efficient and have a higher

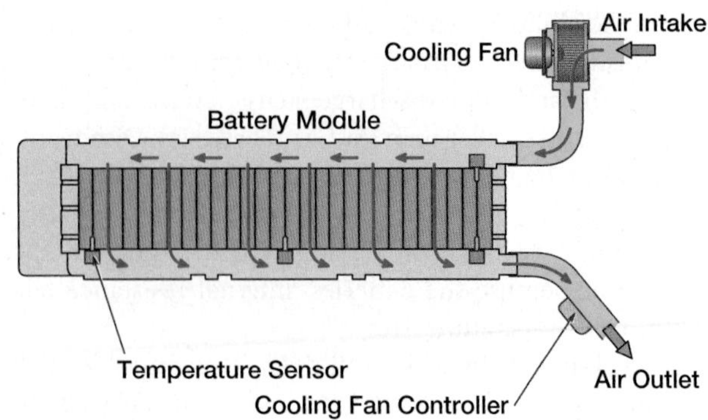

FIGURE 67-21 Typical battery pack cooling system.

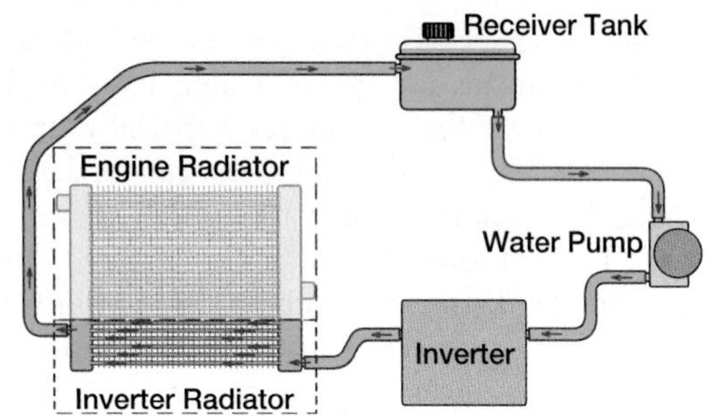

FIGURE 67-22 The high-voltage controller may have its own cooling system to prevent it from overheating.

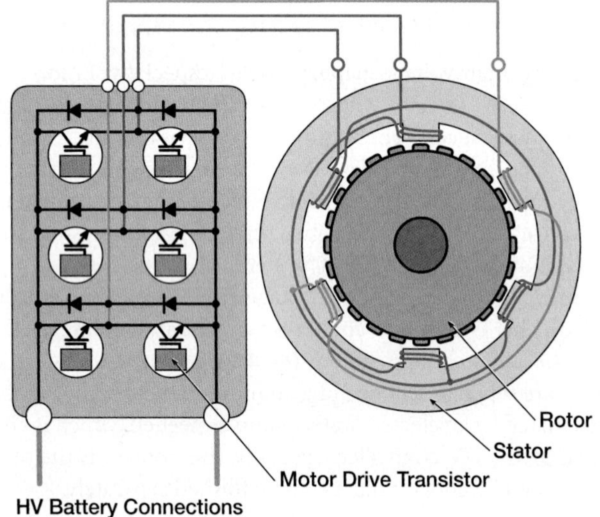

FIGURE 67-23 Typical three-phase motor and controller schematic.

operating speed than DC motors. AC installations allow the use of a three-phase AC motor (**FIGURE 67-23**). AC motors and controllers often incorporate a "regen" feature like that used with many commercial hybrid vehicle applications. This means that during braking, the motor functions as a generator to deliver otherwise lost kinetic energy into power to recharge the batteries.

In vehicles designed for highway use, drive motor systems are computer controlled. The computer controls the direction, frequency, order, and amplitude of current flowing through the motor(s). The magnetic field produced by this current reacts with permanent magnets in the motor, causing the motor's rotor to rotate. However, it is important to note that the AC systems used in on-road vehicles operate on dangerously high voltages, some as high as 500 to 650 volts.

The electric motor is connected to road wheels through a variety of ways, via planetary gears, a CVT, or even direct drive. A few vehicles use so-called wheel motors, which are built into the wheel hub to save valuable cargo-carrying space normally displaced by the drive system.

During deceleration or vehicle braking, the electric motor connected to the drive wheels serves as a generator to create electrical energy (normally lost as heat from the friction brakes) and to provide both regenerative braking, and regenerative charging. This energy, captured as high current during regenerative braking, is routed back into the high-voltage battery for storage. This charging must be strictly controlled by the battery pack controller so the battery pack isn't overcharged, or overheated, which would result in serious damage to the high-voltage battery. Any additional braking force needed beyond regeneration is supplied by friction brakes.

The most popular types of electric motors used are brushless, multiphase, synchronous, and permanent magnet motors.

- **Brushless**—Brushes and commutators are not used.
- **Multiphase**—The stator contains more than one winding, usually three.
- **Synchronous**—The speed of the rotor is synchronized to the frequency of current flowing through the **stator windings**. This design maintains excellent power and efficiency characteristics, even at low speeds.
- **Permanent magnet**—The rotor is made from very powerful rare earth permanent magnet alloys—typically, neodymium-iron-boron or samarium-cobalt.

Because these electric motors are brushless, the only wear components are the bearings that support the rotor shaft. These motors enjoy a long, low-maintenance/zero-maintenance life; no oil changes, no fuel filter changes, no spark plug changes, no coolant changes, and so on.

Types of Hybrids and Electric Vehicles

K67007

There are various kinds of electric and hybrid vehicles on the market today, and they are growing in popularity (**FIGURE 67-24**). They may be classified by body style, but how is the *degree* of hybridization described? There seems to be some confusion. HEVs may be classified by the type of drive configuration used—such as planetary gear sets, multiple electric machines, and continuously variable transmissions. Other descriptions are thrown around with abandon. The following are some widely used descriptions you should be familiar with, but don't be surprised if other designations (often created for marketing appeal) are used for these hybrid vehicles.

Power Sources

HEVs have two power sources, typically a gasoline ICE and high-voltage batteries working in both series and parallel. HEVs

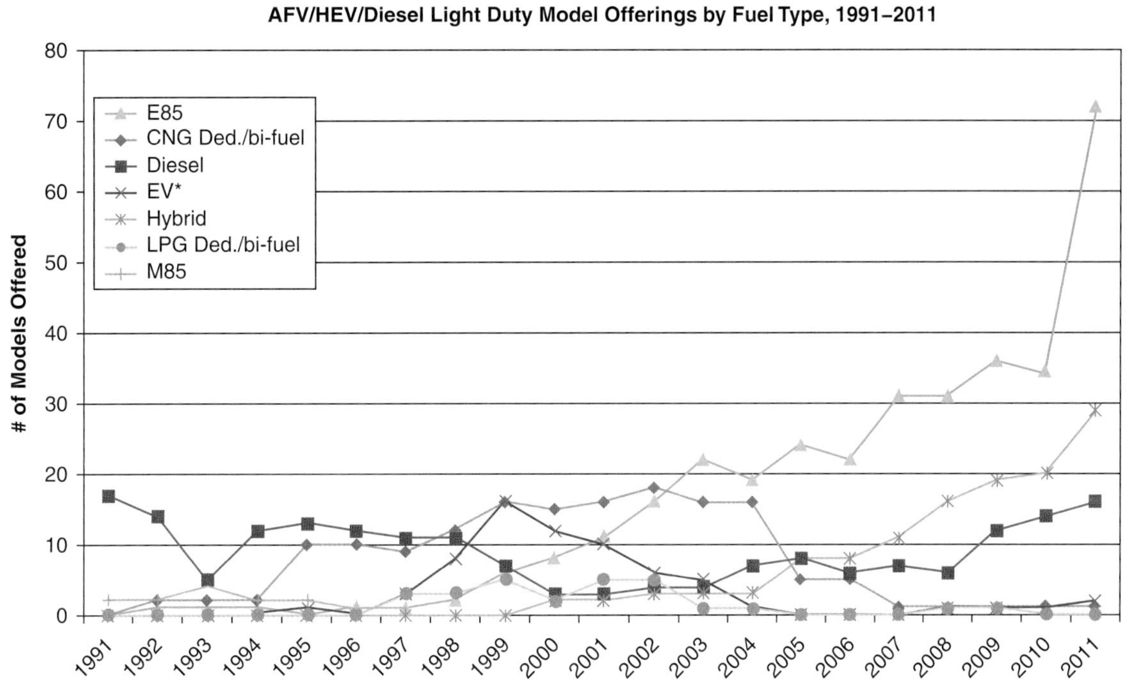

AFV/HEV/Diesel Light Duty Model Offerings by Fuel Type, 1991–2011

FIGURE 67-24 Notice that starting from 1999, close to 3 million hybrids rank second highest among AFVs in the United States. These numbers will continue to climb in the coming years. Note also that as of 2011 there are more than 7 million E85 (85% ethanol) FFVs on U.S. roads. Most owners of these FFVs are unaware that they can handle alcohol fuel.

are not typically electric grid rechargeable, although some have been converted to plug-in HEVs (see the next section) for more than 100 mpg (43 km/L). HEV types include micro, partial, and full hybrid vehicles.

- Micro-hybrids include GM's Silverado and Sierra with start/stop features.
- Partial hybrids include Honda's Civic and Insight, in which the engine (ICE) does not shut off while the vehicle is underway.
- Full hybrids include the Toyota Prius Highlander, and Camry models,, the Ford Escape and Mercury Mariner hybrid, and the GM vehicles with their two-mode hybrid system. These and many others (perhaps 90%) of the market are considered full hybrids, as they can be driven short distances and at low speeds using only the (high-voltage) batteries.

FIGURE 67-25 More plug-in hybrid vehicles are coming to market.

Plug-in Hybrid Electric Vehicles

Many manufacturers are entering the **plug-in hybrid electric vehicle (PHEV)** market. A PHEV is different in that only one power source, the battery, is used to propel the vehicle for a certain distance, limited by the storage capacity of the battery and the efficiency of the motor. A PHEV-20, for example, is designed to go 20 miles (32 km) before the batteries need charging and the ICE is needed; a PHEV-40 can go 40 miles (64 km), and so on (**FIGURE 67-25**). The ICE may also start up for added power before the batteries need a recharge. Ford, Toyota, and other light-duty vehicle OEMs have PHEVs on test. PHEV automobiles achieve great fuel economy, with some getting more than 100 mpg (43 km/L). Medium-duty vehicles from Mercedes-Benz and others offer PHEVs for commercial use. International Truck and Engine sells a PHEV school bus. Combined with FFV capability, PHEVs offer a good miles per gallon of gasoline equivalent (MPG-e).

Extended-Range Electric Vehicles

Extended-range electric vehicles (EREVs) are plug-in electric vehicles that can drive a moderate distance on electric-only power. But they have a relatively small auxiliary ICE to drive a generator for charging the batteries and extending the vehicle's range. The Chevy Volt is a series EREV (**FIGURE 67-26**).

FIGURE 67-26 The Chevy Volt is an extended range electric vehicle.

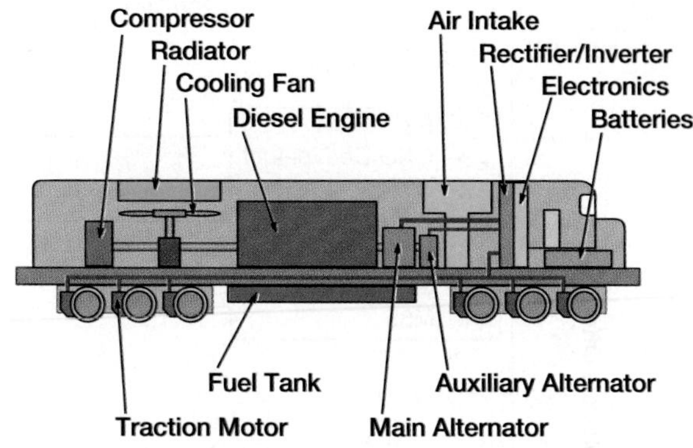

FIGURE 67-27 Hybrid drives have been used for about a hundred years in some applications.

Fuel Cell (Electric) Vehicles

FCVs are considered hybrid vehicles because they have two power sources. The sources typically are a hydrogen fuel cell "stack" and a storage device from which electric power is drawn. Regenerative braking electrical energy may be stored in the vehicle's high-voltage battery pack, or as with Honda's early FCX and others, in ultra- or supercapacitors, which can more quickly be charged or provide acceleration power as needed.

Hybrid Electric Vehicles

If you consider diesel-electric submarines or diesel-electric locomotives, you will realize that the hybrid drive concept is anything but new (FIGURE 67-27). Light- and medium-duty hybrids now come in a variety of configurations. Power from the ICE (running on gasoline, natural gas, diesel, or propane) combines with auxiliary backup or primary power from electric batteries, ultra- or supercapacitors, hydraulic-pneumatic accumulators, or even flywheels operating at super-high revolutions per minute (rpm). Other power options, such as micro-turbines, are used in non-road special-purpose hybrids. U.S. light-duty hybrids are almost exclusively gasoline-electric.

Hybrid electric vehicles (HEVs) use a combination of electric power and an ICE. They combine the different advantages of ICEs and electric motors to provide a vehicle that operates more efficiently. Usually, the engine drives a generator to provide the electricity, which is stored in batteries that then drive an electric motor. The engine may also drive the wheels directly. Hybrid vehicles are designed to drive like conventionally powered vehicles while producing significantly lower emissions and increased fuel economy.

In conventional vehicles that have only an ICE, the engine must be large enough and powerful enough to produce momentary torque sufficient for rapid acceleration at low speeds or from a stationary position. Such an engine is larger than necessary for highway cruising, which requires far less power. The parasitic losses of the larger engine take their toll on fuel economy. A vehicle that has only an electric motor has its own complications. It must carry a large volume of heavy batteries to give it sufficient range to be useful, and these must be recharged or replaced when discharged.

Although electric motors produce their maximum torque at stall speed, a gasoline engine is more efficient at higher rpm. Combining a small but efficient gasoline engine that is designed to operate at its optimum peak torque rpm with an electric motor that can double as a generator to recapture braking energy is both effective and efficient.

Hybrid Drive Configurations

K67008

There are three basic kinds of hybrid drive configurations (FIGURE 67-28):

- **Series hybrid**—The gasoline engine is used only to drive a generator and charge a battery that produces power to drive an electric motor. The electric motor then drives the wheels of the vehicle. The gasoline engine is really only a battery charger inside an electric vehicle.
- **Parallel hybrid**—The engine always drives the wheels and keeps the batteries charged. The electric motor acts like a large starter motor, helping the engine to crank over when more power is needed, particularly when starting from rest or when accelerating rapidly. This allows the engine to be smaller and lighter and more fuel efficient than if it were the sole source of power in the vehicle.
- **Series-parallel hybrid**—This design allows both the engine and the motor to drive the vehicle, or either one by itself, under the control of an onboard computer that determines the optimum combination of power delivered to the wheels at any time. The engine can shut down completely when the batteries are charged. When needed, the gasoline engine fires up again to assist in driving the wheels or to recharge the battery. The engine can be optimized for efficiency within a narrow rpm range, and the overall combination of both power sources takes up little more room than a conventional all-purpose gasoline engine, and yet can deliver lower emissions and much greater fuel economy. This design is used by the Toyota Prius and Ford Escape.

Almost all hybrids (except the Chevy Volt) transfer motive power to the wheels in either the parallel or the series-parallel drive configuration.

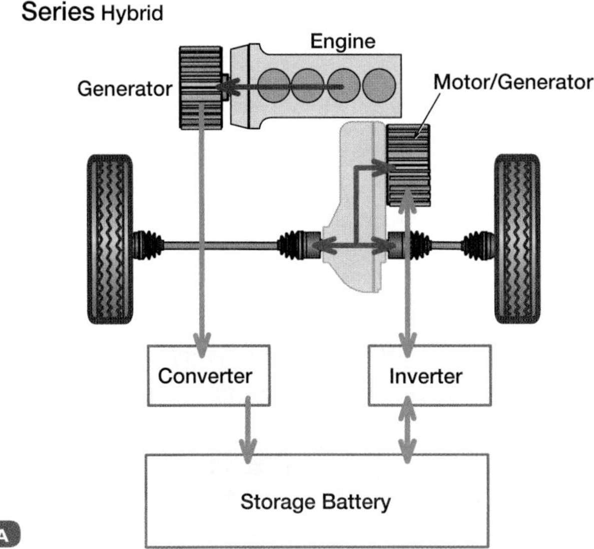

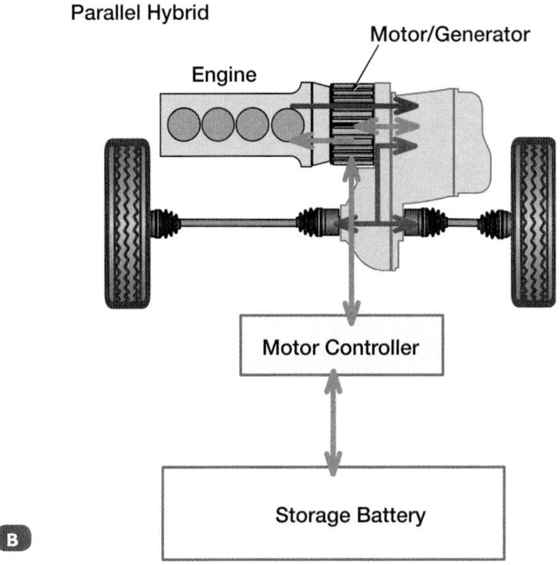

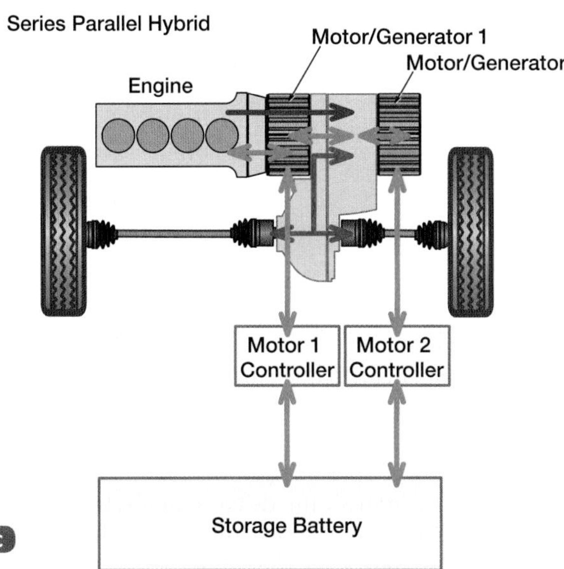

FIGURE 67-28 The three types of hybrid drive configurations.

A variety of transmissions, clutches, and other features are used, depending on the desired amount of performance (or alternatively, fuel economy) and the selling price. Some hybrids come with manual or automatic (up to eight speeds) planetary gear set transmissions; others come with continuously variable transmissions (some with sporty "paddle"-type shifters). High-voltage controls vary also, along with the types of batteries and capacities, type of battery cooling, and so on. There may be one, two, or more high-voltage **electric machines** (drive motors/generators) used. These may be three-phase AC permanent magnet or inductive motors. Usually, the ICE provides power when the vehicle is traveling at higher speeds, and the electric motor is used when slowing to a halt and at lower speeds. The parallel hybrid arrangement is best suited to vehicles that travel greater distances between stops, such as more traditional passenger vehicles and delivery trucks.

Hybrid Vehicle Efficiency Enhancements

HEVs are able to offset ICE inefficiency by using a smaller volumetric engine size and by running the ICE at the optimal torque and speed. Efficiency is also achieved through regenerative braking and by using displacement-on-demand cylinder deactivation, power-on-demand, and idle stop. **Regenerative braking** occurs when the drive motor(s) act as generators to recharge the traction batteries during deceleration or when braking. This feature reportedly saves around 10–15% on fuel consumption. **Displacement-on-demand** deactivates cylinders when operating under cruise or coasting conditions to save fuel. The **power-on-demand** feature temporarily shuts off the ICE when it is not needed, such as when idling or coasting. Drive may also be provided instead only by the electric motor, without burning fuel, thus saving energy and reducing emissions. **Idle stop** turns off the ICE when the vehicle is at a standstill. Expect all of these features to be more widely used as newer vehicles hit the market. Other fuel-saving features include variable valve timing, Atkinson cycle (five-cycle) or Miller cycle (five-cycle with low rpm supercharger boost) engines, off-set pistons, and similarly interesting engine innovations. Refer to the Motive Power Types chapter for more on these engines.

But engine technology isn't the only thing that improves efficiency on hybrid vehicles. Many of them use

- advanced aerodynamic technologies to reduce wind resistance. This can be due to shaping the body and exterior components to slip through the air easier.
- smoothing out the underside of the vehicle with panels that reduce drag.
- low drag tires that are specially made for hybrid vehicles.
- electric power steering, electric air conditioning, and electric water pump to reduce engine load.

As you can see, making vehicles more efficient involves all aspects of the vehicle. And it involves understanding how each of the new technologies operate as well as how to diagnose and service them.

Hybrid vehicle types are classified by body style and by the degree of performance and luxury they offer. But technically speaking, they are also classified by how far they can be driven on batteries alone. For example, a "full hybrid" can cruise a limited distance without the engine running. A "partial hybrid" runs on ICE power at all times except when stopped. A large amount of U.S.-registered full hybrid electric vehicles use drive systems with one or more planetary gear sets licensed from, or similar to, Toyota's Hybrid Synergy Drive design. A number of other OEMs are using the GM-originated two-mode hybrid drive system. See the Hybrid and Continuously Variable Transmission chapter for more information on these drive systems.

Toyota Hybrid System Overview

The ICE in a hybrid vehicle is designed to operate only within its most efficient operating range, typically somewhere between 2000 and 4500 rpm, where engine peak **volumetric efficiency** and peak torque are developed. Its construction has lighter internal components than would otherwise be required, resulting in an efficient engine with low frictional qualities. It may also incorporate other fuel-efficient technologies such as engine cylinders that can be selectively shut down, multi-runner intake manifolds, and throttleless intake manifolds.

A powerful permanent magnet brushless motor/generator is mounted adjacent to the engine (**FIGURE 67-29**). It draws power from the traction battery to drive the vehicle in situations involving low load and speed, and it assists the engine to provide more power during acceleration. During deceleration and braking, it functions as a generator, reclaiming kinetic energy by converting it to electrical energy and storing it in the battery.

Toyota uses a power splitting assembly mounted between the ICE (engine) and the electric motor (**FIGURE 67-30**). This device functions as an **electronically continuously variable transmission** (e-CVT) and transaxle assembly. The e-CVT does not "shift" from one gear to another in the usual sense; rather, it continuously varies the amounts of speed and torque from the engine and both electric motors to the wheels, selecting the best (most efficient) engine and motor rpm balance for any given driving condition.

FIGURE 67-29 Main traction motor.

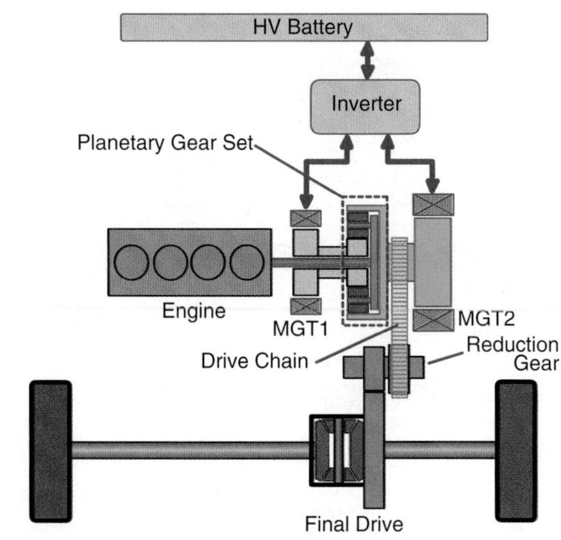

FIGURE 67-30 Power splitting transmission.

FIGURE 67-31 A battery pack installed in a Toyota Prius.

Batteries have transitioned from lead-acid in early electric vehicles to sealed nickel–metal hydride 200- to 300-volt battery packs to sealed lithium-ion battery packs, which provide higher power output and are lighter than other battery types (**FIGURE 67-31**). Their output may be inverted and raised to 650 volts by the power inverter to power the electric drive motors (also called electric machines because they also generate electricity during overrun situations, such as when coasting). Hybrid battery packs are generally made up of a lot of small individual battery cells that are connected in series. They are monitored and managed by a control unit that controls the rate of charge and monitors their temperature.

A small permanent magnet brushless motor/generator replaces the engine flywheel (**FIGURE 67-32**). It functions as the engine's starter motor and battery charger. It also provides additional power for the larger main motor/generator. A motor control unit controls the delivery of electrical current to and from the battery and the electric motors. It has its own cooling system to control its operating temperature. It also contains an AC-to-DC converter used to charge the battery,

FIGURE 67-32 Stator for a flywheel-mounted motor/generator.

and a DC-to-AC power **inverter** to change direct current from the battery into alternating current to power the electric motors.

Hybrid Vehicle Operation

When the vehicle is moving from a stationary position, and when traveling at low to moderate speeds, the main electric motor/generator (MG2) drives the vehicle. At these speeds, the ICE is less efficient and is normally used only to charge the battery.

During normal driving, the ICE starts and drives the generator (MG1) and the power splitter. Power from the generator (MG1) is used to drive the electric motor (MG2). The motor control unit controls the **power splitter** so that the drive remains at its most efficient. When accelerating and when using power from the ICE, the control unit draws power from the battery and directs it to both electric motor/generators, providing more power to the wheels than the ICE could supply on its own.

During deceleration and braking, the ICE is turned off. Inertia from the wheels drives one or both electric motor/generators, using the current produced to charge the high-voltage battery. Normally this energy would be wasted as heat, but in the hybrid, it is recovered and reused. The **retarding effect** (vehicle slowing) caused by this system provides a moderate amount of deceleration. So gentle application of the brake pedal does not apply the brakes. The braking effect is achieved by using the electric motor/generator, which converts the vehicle's inertial momentum into electrical energy to charge the high-voltage battery. This is called "regenerative braking."

When the vehicle is stationary and the brakes are applied, the ICE is stopped, and power is not applied to the electric motor/generator. No fuel is burned and no emissions are produced. Releasing the brakes applies a small amount of power to the electric motor. Moving the accelerator pedal further, more power is sent to the electric motor, and if required, the ICE is started to increase the power further.

▶ Hybrid and Electric Vehicle Service

N67001, N67002, K67009

Maintenance needs for hybrid electric vehicles (HEVs), plug-in hybrid electric vehicles (PHEVs), and all-electric vehicles (EVs) are similar to those of conventional vehicles. However, these vehicles have high-voltage electrical systems that range from 100 to 600 volts. They also may start up at any time if not fully shut down. Consequently, servicing them requires some particular precautions.

Before attempting to perform service operations on any hybrid vehicle, read the manufacturer's service information for the vehicle you are about to service. The service manual will instruct you in how to properly follow all safety procedures. The most important rule to remember is, never work on a high-voltage vehicle without first notifying someone who is trained in dealing with high-voltage electrocution. You need a responsible person in your work environment to check on you as you work around high voltage.

The matter of servicing these vehicles safely is to be taken seriously. Always follow the manufacturer's safety and service procedures to keep out of trouble. There are many safety features built into hybrid and electric vehicles to prevent accidental shock, but once you decide to venture into high-voltage areas, be prepared to spend some money on safety equipment. This is no time to pinch pennies! Some of the standard equipment for servicing HEV systems is listed here:

- OEM or equivalent scan tools
- A high-quality (three-phase CAT III or CAT IV) digital multimeter (DMM) with appropriate high-voltage leads (**FIGURE 67-33**)
- High-voltage insulated gloves rated for 5000 volts and certified to 1000 volts (Class 0) and air-checked on a daily basis; also approved glove covers (**FIGURE 67-34**)
- High-voltage, insulated shoes

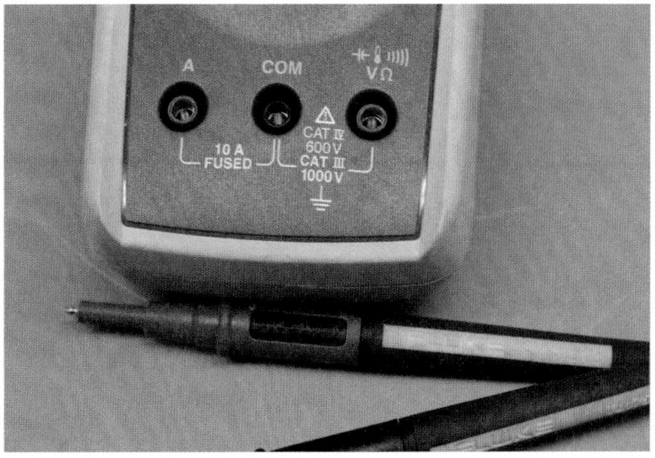

FIGURE 67-33 CAT III, 1,000 V DMM.

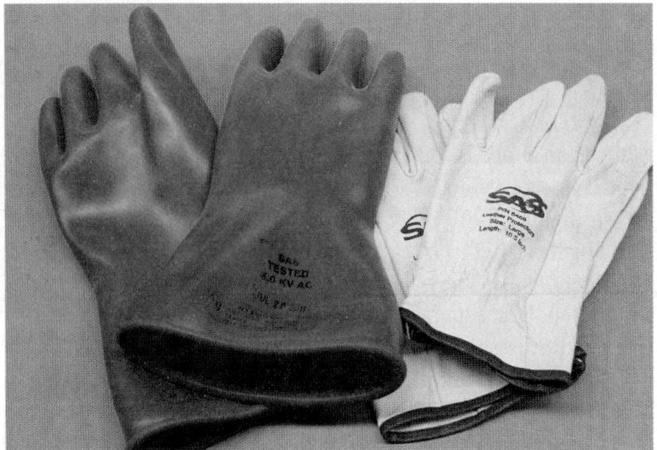

FIGURE 67-34 Class 0 HV insulated gloves and approved covers.

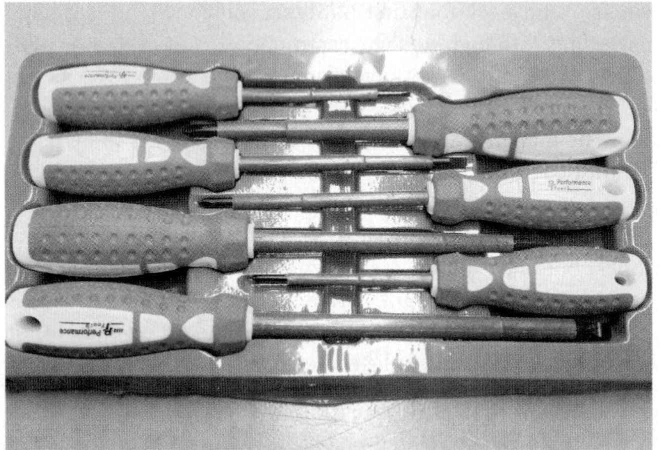

FIGURE 67-35 Insulated tools with high-voltage rating.

- Insulated tools meeting 1000 V/300 A specifications (**FIGURE 67-35**)
- Safety cones and safety tape (to mark off service areas) and other such items
- A retrieval hook in case someone becomes disabled from electrical shock (**FIGURE 67-36**)
- ABC-type fire extinguisher

If you decide to service electric and hybrid vehicles, get the proper training on the high-voltage areas, as well as a full understanding of how they operate. Stay alert and continually practice situational awareness.

Besides the high-voltage dangers of a hybrid or electric vehicle, there are other safety precautions you need to be aware of. One of the first is that the vehicle should be completely powered down before working on it, since the ICE can start at any time the vehicle's PCM deems necessary. Never assume that the vehicle is turned off because it is silent or still. The vehicle may appear to be turned off because the ICE has been shut down, and the vehicle is silent, but the system can still be in "ready" mode (most vehicle models show this on

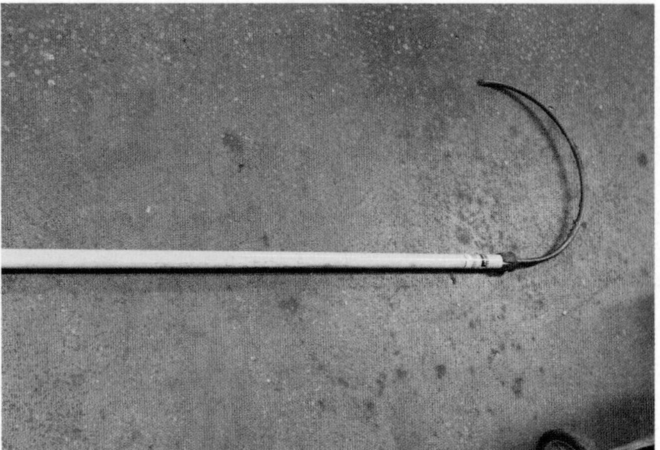

FIGURE 67-36 Retrieval hook for pulling a person away from high voltage.

the vehicle dash display panel), capable of engaging the ICE or the drive motor at any moment. If a technician is changing a belt or working around the ICE and the engine starts up, he or she could be injured or killed. Also, prior to service, the vehicle should be in Park, with the parking brake applied and the power button powered down. The keyless fob, if the vehicle is so equipped, should be more than 15' (4.6 m) from the vehicle. Also remove the hybrid high voltage service plug.

SAFETY TIP

Always make sure there is someone in your work environment who is trained in the safety procedures regarding high-voltage electrocution, to check on you whenever you work around high-voltage systems.

Identifying and Disabling the High-Voltage System

N67003

There are plenty of opportunities to service hybrid and electric without venturing into the high-voltage areas of the vehicle, such as changing oil. But just because you are not working on the main traction motor doesn't mean that you won't come into contact with other high-voltage circuits. In many hybrid and electric vehicles, high voltage is used to operate a variety of accessories, such as power steering and air-conditioning, so you need to be aware that high-voltage wires can be located nearly anywhere in the vehicle. These high-voltage wires are recognized by orange (usually) convolute or wiring (**FIGURE 67-37**). Some vehicles have used other colors, such as blue, to designate high-voltage wiring, so check the service information.

It should be noted that manufacturers of HEVs, PHEVs, and EVs design these vehicles with safety features that deactivate the high-voltage electrical system when they detect a collision, short circuit, or certain HV system faults. Typically the deactivation occurs in the high-voltage battery itself, or the

FIGURE 67-37 High-voltage wiring.

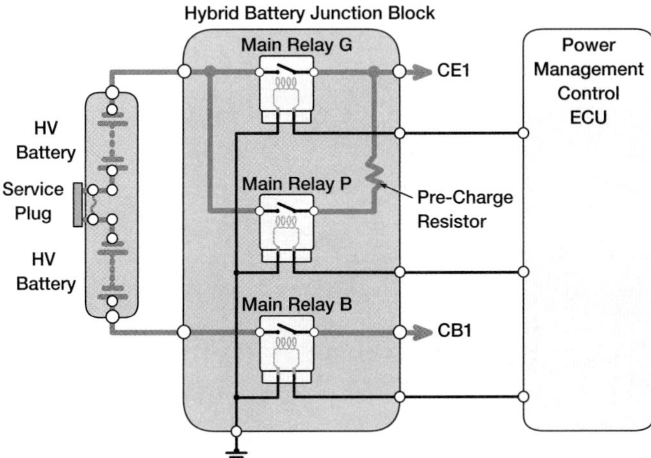

FIGURE 67-38 Typical high-voltage battery safety system on a Toyota.

battery junction box by one to three special relays (Toyota calls them system main relays (SMRs) that open the high-voltage circuit (**FIGURE 67-38**). The relays are controlled by the high-voltage power management control ECU and switches them off if a fault is detected.

One method manufacturers use to detect shorts in the high-voltage wiring is by building electrical leakage detectors within the insulating material. If a leak is detected, the power management controller shuts down the high-voltage system. So never pierce a high-voltage wire when testing for voltage, or you may inadvertently shut down the system and permanently ruin the cable.

Avoid working on these high-voltage areas of a hybrid vehicle without first having the proper training, service information, and experienced help on hand. Also, the main high-voltage battery pack must be disconnected before service is performed on or around the high-voltage components. Every manufacturer has a different way of disconnecting high-voltage batteries from the high-voltage electrical system. Many systems use a service plug that is removed from or near the battery. Removing the service plug opens up the high-voltage circuit inside the battery (**FIGURE 67-39**). Others systems use a battery module switch that is turned to the Off position, which also disconnects the battery from the high-voltage system.

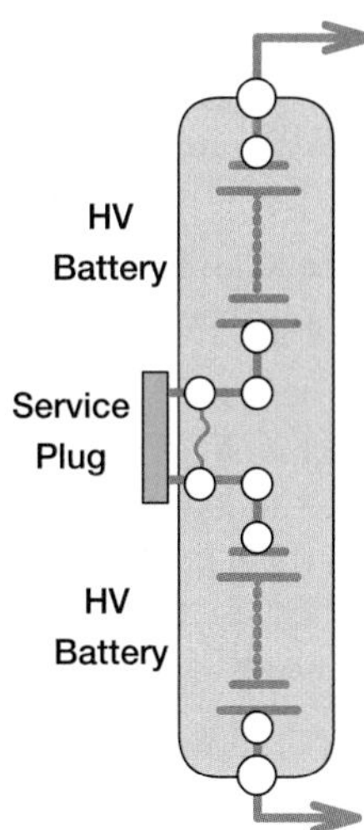

FIGURE 67-39 Removing the service plug opens the high-voltage circuit in the battery.

FIGURE 67-40 Locate the high-voltage battery disconnect (service plug), release the lock, release the lever, and remove the service plug.

Before you disable the high-voltage system, make sure you are wearing the appropriate personal protection equipment (PPE) and are following the manufacturer's procedure. As an example of disconnecting the high-voltage battery using a third-generation Prius, follow these general guidelines:

1. Locate the high-voltage battery disconnect (service plug).
2. Release the lock, release the lever, and remove the service plug (**FIGURE 67-40**).
3. Store it in a safe place, where it cannot be reinserted accidentally. Some manufacturers suggest that you keep it in your pocket so that it can't get reinstalled by anyone.
4. Wait for the specified time for the residual electricity to bleed off. It typically takes 5 to 15 minutes for this to occur.

5. Then, verify that the system is no longer live. To do that, locate the high-voltage access panel, and remove the bolts that secure the cover, using insulated tools (**FIGURE 67-41**).
6. Next, remove the cover and, using the correctly rated meter, measure that there is no voltage at the specified terminals (**FIGURE 67-42**).

To power down the high-voltage system on a Nissan Leaf, there are three possible methods; primary method, alternate method number 1, and alternate method number 2. We cover the primary method here, because the alternative methods are only used in the event the "Ready" indicator won't shut off:

- Verify that the "Ready" indicator is off.
- Keep the key at least 16 feet away from the vehicle.
- Open the hood and disconnect the 12 volt battery's negative cable.
- Insulate the negative battery cable with insulated tape.
- Wait at least 10 minutes for the high-voltage capacitor to discharge.

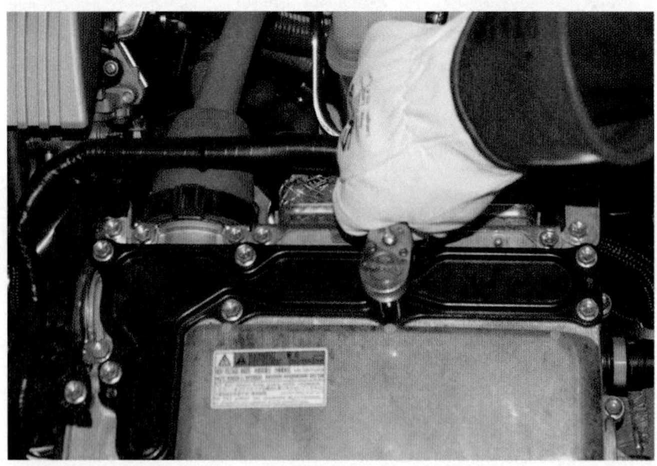

FIGURE 67-41 Locate the high-voltage access panel, and remove the bolts that secure the cover, using high-voltage insulated tools.

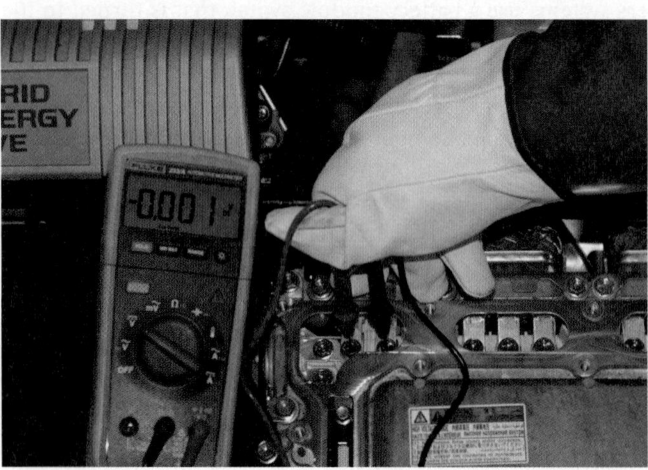

FIGURE 67-42 Remove the cover, and using the correctly rated meter, measure that there is no voltage at the specified terminals.

Hybrid Auxiliary (12 V) Battery Service

N67004

On most hybrid vehicles, the 12-volt auxiliary battery is used to power up the vehicle's computers and run virtually all of the accessories. Because it powers up the computers, it is critical to starting the vehicle, even though it is generally not used to crank the engine over; that is typically done by one of the traction motors and a high-voltage battery. This means that the 12 V battery in a hybrid vehicle is generally of smaller capacity and size than in a standard vehicle, making the battery more susceptible to parasitic current draws, as they don't have as much current capacity. It is also located separately from the traction batteries, typically either under the hood or in the trunk (**FIGURE 67-43**). In most cases, the 12 V battery is charged from an inverter built into the hybrid electronic controller, instead of a separate alternator (**FIGURE 67-44**).

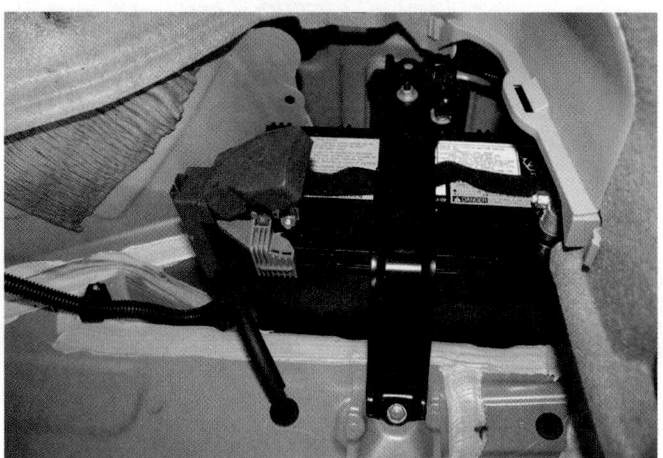

FIGURE 67-43 A 12 V Prius battery installed.

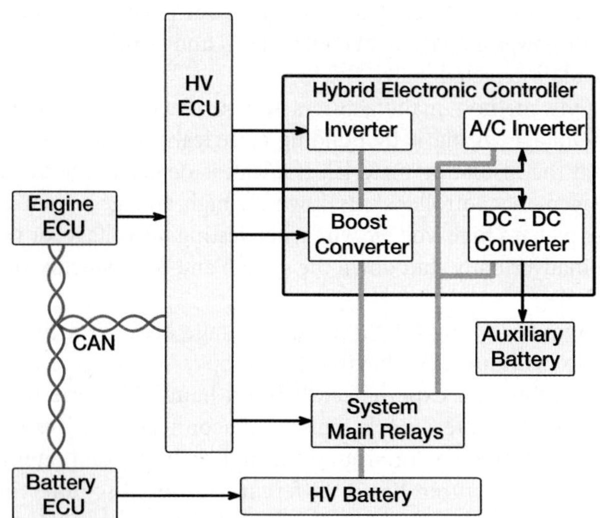

FIGURE 67-44 The 12-volt battery is typically charged from a DC-DC converter built into the hybrid electronic controller.

In most hybrids, the 12 V battery is the absorbed glass mat (AGM) construction. This battery is sensitive to overcharging, which can damage it, so a special smart battery charger is required whenever charging is needed. Some manufacturers recommend removing the 12 V battery from the vehicle while it is charging to prevent any damage to the vehicle's electrical system. But this can erase the computer and entertainment memories, as well as activate the antitheft system, so always follow the manufacturer's procedure when disconnecting any battery. Also, most manufacturers prohibit hybrid vehicles from being used to jump-start another vehicle because of the reduced size of the auxiliary battery and the potential damage to the hybrid electrical system. The 12 V battery is best tested with a conductance tester, as described in the Battery System chapter (**FIGURE 67-45**).

FIGURE 67-45 The 12 V auxiliary battery is best tested with a conductance tester.

▶ Wrap-Up

Ready for Review

- An alternative fuel is any fuel that powers a vehicle that is not petroleum-based motor oil.
- Alternative fuels are not new; they have been around since the 1890s.
- Alternative fuels have to meet the California Air Resources Board (CARB) standards to be certified.
- The push for alternative fuels is driven by the need for enhanced security, environmental concern, and economic factors.
- Alternative fuel is regulated by many different acts, starting in 1992.
- The Environmental Protection Agency (EPA) is the driving force behind alternative energy.
- There are many different standards for low-emission vehicles, beginning in the early 1980s and growing progressively stringent into 2016.
- Alternative vehicles are designed to reduce emissions and be more economical to operate.
- Compressed natural gas (CNG) and liquefied natural gas (LNG) were first introduced in the 1980s, but due to lack of refilling stations, they did not sell well to consumers.
- CNG and LPG must be stored in special cylinders that will withstand pressures and damage.
- CNG and LPG containers are fitted with special pressure relief systems to release pressure as required.
- CNG and LNG must be protected from a heat source. In some designs, special heat shields are installed around the exhaust system to protect the container from excess heat.
- Leaks can be identified by using a special explosion meter that finds methane gas leaks.
- Biofuels are made by chemically combining natural oils from soybeans, cottonseeds, canola, animal fats, algae, jatropha seeds, or even recycled cooking oil, with an alcohol such as methanol or ethanol and a catalyst such as lye.
- There are different qualities of biodiesel fuels, and they should not be confused with the fuel made from old vegetable oil.
- The one main disadvantage to biodiesel is that starting in cold weather may require longer cranking of the engine.
- The military is the driving force in the use of alternative fuels like biodiesel.
- Biodiesel produces less emission than similar diesel fuels.
- Biodiesel has higher water content than traditional diesel fuels.
- Just like with typical diesel fuel, the cetane number of biodiesel must be correct for overall engine performance, typically around 46–57; the higher, the better.
- Ethanol is being used mainly to lower fuel emissions in vehicles and is made from three products: sugars, grains, and algae.
- Methanol or wood alcohol is being used in E85 vehicles, which means 85% of the fuel mixture is methanol.
- The advantage to methanol is the lower output of emissions.
- The main disadvantage with methanol is that it cannot be burned in conventional vehicles without major engine modifications.
- Biobutenol is similar to ethanol, but butenol has characteristics similar enough to gasoline that it can be transported through the existing petroleum (pipeline) distribution infrastructure.
- Fuel cell technology has been used a number of years by NASA; given the current need for alternative fuels, it has pushed for another look at this technology by several manufacturers.
- Fuel cells are complex systems that require oxygen and hydrogen to produce water and electricity.
- Fuel cells have zero pollutants; they exhaust only water.
- There are four elements to a fuel cell: the anode, cathode, electrolyte, and catalyst.

- Fuel cells do not produce enough energy by themselves; to achieve the required power, they must be stacked.
- The electric motor shows great promise because it has high efficiency, fewer moving parts, fewer maintenance requirements, good low-speed torque output, and increased reliability.
- The computer gives the electric motor new life and has some advantages; the biggest disadvantage is the electric motor must have a battery to run it.
- Hydrogen fuel cells are simple and have three times the energy of gasoline. Fuel cell vehicles, however, are more efficient and cleaner.
- Battery electric vehicles (BEVs) are making a comeback with the addition of computer technology and lithium-ion batteries.
- The Chevy Volt was supposed to be the first BEV, but it is really an extended range hybrid, because it has a gasoline engine to recharge the battery.
- The locomotives on the railroad tracks today are diesel electric; the engine charges the batteries and the locomotive runs on the batteries.
- Many city buses are diesel electric, which demonstrates that hybrid technology is not new; it is being redefined, with the new vehicles using computers to control operations.
- The engine in hybrids can be much smaller, as its main purpose is to recharge the hybrid's battery. The hybrid gets most of its power from the battery running the electrical drives.
- When you think of hybrids, think of electrical circuitry. There are series, parallel, and series parallel designs.
- On series hybrids, the engine runs only to supply the electric motor or charge the batteries.
- On parallel hybrids, the engine runs to charge the battery and directly assist propelling the vehicle.
- On series-parallel hybrids, the engine runs to charge the battery, operate the electric motor, and assist with propelling the vehicle all at the same time.
- Hybrids offset the power lost by internal combustion engines (ICEs) in several ways. One of the most efficient is the regenerative braking system.
- There are numerous hybrids, from plug-in to extended-range. They all work basically the same way: They allow the vehicle to be propelled by electrical power. Thus, they are more fuel efficient.
- The hybrid normally starts out on electrical power, and as the need changes, the engine comes on to assist or recharge the batteries, depending on the design.
- The ICE is turned off during deceleration and braking.
- The ICE starts and runs at higher rpm for peak efficiency.
- Hybrids can be dangerous. Always refer to the appropriate repair information.
- In hybrids, it is usually the orange wires that indicate high voltage. Other colors may be used, so check the manufacturer's information.
- Always wear special gloves to remove the service plugs to disable hybrids.
- Always use correctly rated meters and tools when working on hybrid vehicles.
- Anytime service is performed, the technician must ensure that the hybrid is not in the ready mode and that everything is off.

Key Terms

aldehyde A pollutant/irritant in diesel fuel.

alternative fuel A nonpetroleum-based motor fuel.

anode A positively charged electrode or plate.

automatic fuel limiter A special refueling port and an overfill protection device fitted on LPG tanks, which ensure that the tank cannot be filled past the safe fill limit of 80%.

B20 fuel A fuel that is 20% biodiesel blended with 80% petrodiesel.

B100 fuel A fuel that is 100% biodiesel.

battery electric vehicle (BEV) A vehicle powered by battery only.

bi-fuel Two fuels used in a vehicle one at a time, such as compressed natural gas or gasoline.

biodiesel Renewable fuel made from organic feedstocks.

bio-ethanol A renewable fuel derived from a variety of cellulose-based plants.

biomass Organic matter used for fuel.

brushless An electric motor without brushes.

burn rate The rate of flame spread for liquid or gaseous fuel.

catalyst A device that changes chemical composition without self-destruction.

cathode A negatively charged electrode or plate.

cloud point The temperature at which wax (paraffin) in diesel fuel starts to solidify.

compressed natural gas (CNG) Methane compressed for onboard storage in cylinders.

compressed natural gas vehicle (CNGV) A vehicle that uses methane stored in high-pressure cylinders.

displacement-on-demand A feature that allows cylinders to be taken off-line when not needed, such as during vehicle cruise.

dual-fuel-powered system A system that allows an engine to burn two fuels at one time.

E85 A fuel with 85% ethanol and 15% gasoline blend.

electric machine Another name for a traction motor with regenerative capability.

electrochemical device A device such as a battery or fuel cell that uses two forms of energy.

electrolyte The solution in a battery, such as sulfuric acid and water used in a lead-acid battery.

electronically continuously variable transmission (e-CVT) A transmission without individual gears or gear ratios.

emission standards Cutoff points set by governmental agencies to limit tailpipe emissions.

Energy Policy Act (EPAct92) Regulations enacted in 1992 by the U.S. Congress to promote the use of alternative fuel vehicles and fuels.

ethanol Alcohol-based motor fuel made from starches and sugars.

flexible-fuel vehicle (FFV) A vehicle that can operate on two different fuels blended together in various mixture amounts by percentage.

float-type fuel gauge A system for determining the level of liquid fuel in a tank using a float and potentiometer.

fuel cell An electrochemical device that uses hydrogen and oxygen to create electricity.

fuel cell (electric) vehicle (FCV) A vehicle that converts fuel to electricity in a direct electrochemical process, with more energy extracted from the fuel source than in traditional internal combustion engines.

fuel composition sensor A device that measures the percentage of alcohol in gasoline.

fuel methanol A toxic alcohol fuel.

hybrid electric vehicle (HEV) A vehicle that uses two power sources for propulsion, one of which is electricity.

hygroscopic The ability of a substance to attract and hold water molecules.

idle stop A feature that turns off the internal combustion engine when the vehicle is at a standstill.

infrastructure The system for supplying fuel for transportation use; includes ships, pipelines, trucks, etc.

inverter A device that converts direct current to alternating current.

liquefied natural gas (LNG) A cryogenic liquid produced from gaseous methane.

liquefied petroleum gas (LPG) Gaseous fuel at atmospheric pressure; commonly referred to as propane.

Low-Emission Vehicle (LEV) standard A program initiated in California to improve vehicle emission standards.

lubricity The ability to lubricate moving parts.

M85 vehicle A flexible-fuel vehicle, meaning any mixture of methanol and gasoline/petrol in the fuel tank can be used by the engine.

membrane electrode assembly The individual voltage-producing component in a fuel cell stack.

mercaptan An agent that smells like sulfur that is added to liquefied petroleum or natural gas to aid in detecting leaks.

methanol A motor fuel made from wood alcohol.

multiphase An electric motor that operates through more than one phase.

National Fire Protection Association (NFPA) The agency that publishes standards for liquefied petroleum and compressed natural gas vehicles.

natural gas Methane; often referred to simply as NG.

NFPA-52 A document from the NFPA that sets standards for compressed natural gas refueling stations and compressed natural gas vehicle fuel systems.

odorant An agent (mercaptan) that smells like sulfur that is added to liquefied petroleum or natural gas to aid in detecting leaks.

parallel hybrid A hybrid vehicle driven simultaneously by an internal combustion engine and an electric machine.

permanent magnet A material with natural or artificially created constant magnetic properties.

permanently mounted tank A compressed natural gas or propane cylinder not designed to be portable.

plates The parts of a battery or fuel cell that contain materials that enable electrons to flow into or out of them.

plug-in hybrid electric vehicle (PHEV) A hybrid electric vehicle in which only one power source, the battery, is used to propel the vehicle for a certain distance, limited by the storage capacity of the battery and the efficiency of the motor.

portable universal cylinder A liquefied petroleum gas tank that can be moved or transported.

pour point The temperature at which diesel fuel is not gelled.

power splitter A device that receives power from an internal combustion engine and electric machine to power a hybrid electric vehicle.

power-on-demand A feature that shuts down the engine when not needed to save fuel.

producer gas A generic term that refers to a number of manufactured gases such as wood gas, town gas, and syngas.

reformer An onboard device that extracts hydrogen from another source material.

regenerative braking A feature that allows drive motors to act as generators to recharge the traction batteries during deceleration or when braking. Also called regenerative charging.

retarding effect The result of retarding (slowing) the vehicle.

series hybrid A hybrid electric vehicle powered by an internal combustion engine but driven by an electric motor and battery.

series-parallel hybrid A hybrid electric vehicle that uses the internal combustion engine and/or the battery pack for propulsion.

slow-fill The method of filling a compressed natural gas vehicle (such as overnight) so the cylinders are completely filled.

stator winding A winding in an alternator that creates the current output, or a winding in a motor that creates the needed magnetism for the motor to rotate.

straight vegetable oil (SVO) Biodiesel made from plants such as soybeans, jatropha, and palms.

sulfur A regulated by-product of petroleum-based (especially diesel) fuel.

syngas An artificially created motor fuel.

town gas Locally available gas used to power early internal combustion engines.

viscosity The ability of a liquid to be poured.

volumetric efficiency A measure of how well an internal combustion engine can breathe at a given rpm at wide-open throttle.

waste vegetable oil (WVO) Used cooking grease or oil used for producing biodiesel fuel.

water gas A gas made from hydrogen and water.

wood gas A gas made by the carbonization or gasification of coal.

Review Questions

1. Alternative fuels advance all of the following *except*:
   a. energy security.
   b. environmental concerns.
   c. economy.
   d. elegance.
2. Which of the following is derived from other forms of fuel such as wood or coal?
   a. Manufactured gas
   b. Liquefied natural gas
   c. Compressed natural gas
   d. Liquefied petroleum gas
3. Choose the correct statement.
   a. CNG is compressed and stored at low pressures.
   b. The volume of space required for LNG storage tanks is almost twice that required for CNG tanks for delivering the equivalent mileage range.
   c. Because CNG is a gas, not a liquid, it does not matter whether the container is standing up or lying down.
   d. CNG cylinders come in two types (heavy and light).
4. Bio-alcohol is used primarily:
   a. as a blend with traditional petrodiesel.
   b. by itself.
   c. as a blend with liquefied natural gas.
   d. as a blend with gasoline.
5. When compared to petrodiesel, biodiesel:
   a. has a lower cetane rating.
   b. has a lower cloud point and pour point.
   c. produces significantly fewer emissions.
   d. produces more harmful emissions.
6. Choose the correct statement with respect to a fuel cell.
   a. It produces higher emissions than biodiesel-powered vehicles.
   b. It combines hydrogen and oxygen to produce water and in the process it produces electricity and heat.
   c. A fuel cell produces a voltage of about 1000 volts.
   d. Fuel cells are heavy and increase vehicle load.
7. Which of the following is a rechargeable battery with the highest energy density currently in use?
   a. Lead-acid
   b. Nickel-cadmium
   c. Nickel-metal hydride
   d. Lithium-ion
8. Which of the following hybrid drive configurations is where the ICE only drives a generator to charge the batteries, which then power the electric motor to drive the wheels?
   a. Series hybrid
   b. Parallel hybrid
   c. Series-parallel hybrid
   d. Battery-electric hybrid
9. Hybrid vehicles that can be driven short distances and at low speeds using only the (high-voltage) batteries are classified as:
   a. full hybrids.
   b. partial hybrids.
   c. micro hybrids.
   d. macro hybrids.
10. All of the following statements are true with respect to servicing hybrid and electric vehicles *except*:
    a. Pierce a high-voltage wire when testing for voltage.
    b. Avoid working on the high-voltage areas of a hybrid vehicle without first having the proper training.
    c. Disconnect the main high-voltage battery pack before service is performed on or around the high-voltage components.
    d. Never assume that the vehicle is turned off because it is silent or still.

ASE Technician A/Technician B Style Questions

1. Tech A says that fuel cells only emit CO_2 to the atmosphere. Tech B says that fuel cells run on water. Who is correct?
   a. Tech A
   b. Tech B
   c. Both A and B
   d. Neither A nor B
2. Tech A says that ethanol-blended gas at 10% is a nationwide standard. Tech B says that 15% ethanol-blended gas has been approved for 2001 vehicles and newer. Who is correct?
   a. Tech A
   b. Tech B
   c. Both A and B
   d. Neither A nor B
3. Tech A says that high-voltage wires in hybrid and electric vehicles are usually a special color. Tech B says that special high-voltage gloves should be worn when working on high-voltage vehicles. Who is correct?
   a. Tech A
   b. Tech B
   c. Both A and B
   d. Neither A nor B
4. Tech A says that E85 vehicles are designed to use 85% ethanol efficiently. Tech B says that hydrogen cars are not zero-emission vehicles. Who is correct?
   a. Tech A
   b. Tech B
   c. Both A and B
   d. Neither A nor B
5. Tech A says that lithium-ion batteries currently have the highest energy density of batteries used in HEVs. Tech B says that nickel–metal hydride batteries are less expensive than lithium-ion batteries. Who is correct?
   a. Tech A
   b. Tech B

c. Both A and B
d. Neither A nor B

6. Tech A says that CNG fuel tanks hold 3600 psi (24,800 kPa). Tech B says that CNG engines tend to emit white smoke, which is water vapor. Who is correct?
   a. Tech A
   b. Tech B
   c. Both A and B
   d. Neither A nor B
7. Tech A says that some hybrid battery packs are cooled by a cooling system. Tech B says that regenerative braking in a hybrid vehicle uses electricity from the high-voltage battery to slow the vehicle. Who is correct?
   a. Tech A
   b. Tech B
   c. Both A and B
   d. Neither A nor B
8. Tech A says that prior to working on a hybrid vehicle, make sure the "Ready" indicator is on. Tech B says that once the service plug is removed, you can immediately start working on the high-voltage system. Who is correct?
   a. Tech A
   b. Tech B
   c. Both A and B
   d. Neither A nor B
9. Tech A says that a CAT III meter is required when working on the EV or HEV high-voltage system. Tech B says that a HEV can start the ICE on its own if the "Ready" indicator is on. Who is correct?
   a. Tech A
   b. Tech B
   c. Both A and B
   d. Neither A nor B
10. Tech A says that biodiesel mixed with diesel fuel does not reduce emissions. Tech B says that a 1% or more blend of biodiesel extends the life of injectors. Who is correct?
   a. Tech A
   b. Tech B
   c. Both A and B
   d. Neither A nor B

APPENDIX A

2017 NATEF AUTOMOBILE ACCREDITATION TASK LIST CORRELATION GUIDE

Task List	MAST	AST	MLR	Chapter
ENGINE REPAIR				
General: Engine Diagnosis; Removal and Reinstallation (R & R)				
Complete work order to include customer information, vehicle identifying information, customer concern, related service history, cause, and correction.	P-1	P-1		
Research vehicle service information including fluid type, internal engine operation, vehicle service history, service precautions, and technical service bulletins.	P-1	P-1	P-1	
Verify operation of the instrument panel engine warning indicators.	P-1	P-1	P-1	10
Inspect engine assembly for fuel, oil, coolant, and other leaks; determine needed action.	P-1	P-1	P-1	13
Install engine covers using gaskets, seals, and sealers as required.	P-1	P-1	P-1	
Verify engine mechanical timing.	P-1	P-1	P-2	
Perform common fastener and thread repair, to include: remove broken bolt, restore internal and external threads, and repair internal threads with thread insert.	P-1	P-1	P-1	8
Inspect, remove and/or replace engine mounts.	P-2	P-2		
Identify service precautions related to service of the internal combustion engine of a hybrid vehicle.	P-2	P-2	P-2	67
Remove and reinstall engine on a newer vehicle equipped with OBD; reconnect all attaching components and restore the vehicle to running condition.	P-3			18
Cylinder Head and Valve Train Diagnosis and Repair				
Remove cylinder head; inspect gasket condition; install cylinder head and gasket; tighten according to manufacturer's specification and procedure.	P-1	P-1		19
Clean and visually inspect a cylinder head for cracks; check gasket surface areas for warpage and surface finish; check passage condition.	P-1	P-1		19
Inspect pushrods, rocker arms, rocker arm pivots and shafts for wear, bending, cracks, looseness, and blocked oil passages (orifices); determine needed action.	P-2	P-2		19
Adjust valves (mechanical or hydraulic lifters).	P-1	P-1	P-3	
Inspect and replace camshaft and drive belt/chain; includes checking drive gear wear and backlash, end play, sprocket and chain wear, overhead cam drive sprocket(s), drive belt(s), belt tension, tensioners, camshaft reluctor ring/tone-wheel, and valve timing components; verify correct camshaft timing.	P-1	P-1		
Establish camshaft position sensor indexing.	P-1	P-1		
Inspect valve springs for squareness and free height comparison; determine needed action.	P-3			19
Replace valve stem seals on an assembled engine; inspect valve spring retainers, locks/keepers, and valve lock/keeper grooves; determine needed action.	P-3			
Inspect valve guides for wear; check valve stem-to-guide clearance; determine needed action.	P-3			19
Inspect valves and valve seats; determine needed action.	P-3			19
Check valve spring assembled height and valve stem height; determine needed action.	P-3			
Inspect valve lifters; determine needed action.	P-2			
Inspect and/or measure camshaft for runout, journal wear and lobe wear.	P-3			19
Inspect camshaft bearing surface for wear, damage, out-of-round, and alignment; determine needed action.	P-3			19
Identify components of the cylinder head and valve train.			P-1	
Engine Block Assembly Diagnosis and Repair				
Remove, inspect, and/or replace crankshaft vibration damper (harmonic balancer).	P-1	P-2		

Task List	MAST	AST	MLR	Chapter
Disassemble engine block; clean and prepare components for inspection and reassembly.	P-1			20
Inspect engine block for visible cracks, passage condition, core and gallery plug condition, and surface warpage; determine needed action.	P-2			
Inspect and measure cylinder walls/sleeves for damage, wear, and ridges; determine needed action.	P-2			20
Deglaze and clean cylinder walls.	P-2			20
Inspect and measure camshaft bearings for wear, damage, out-of-round, and alignment; determine needed action.	P-3			
Inspect crankshaft for straightness, journal damage, keyway damage, thrust flange and sealing surface condition, and visual surface cracks; check oil passage condition; measure end play and journal wear; check crankshaft position sensor reluctor ring (where applicable); determine needed action.	P-1			20
Inspect main and connecting rod bearings for damage and wear; determine needed action.	P-2			20
Identify piston and bearing wear patterns that indicate connecting rod alignment and main bearing bore problems; determine needed action.	P-3			20
Inspect and measure piston skirts and ring lands; determine needed action.	P-2			20
Determine piston-to-bore clearance.	P-2			
Inspect, measure, and install piston rings.	P-2			
Inspect auxiliary shaft(s) (balance, intermediate, idler, counterbalance and/or silencer); inspect shaft(s) and support bearings for damage and wear; determine needed action; reinstall and time.	P-2			20
Assemble engine block.	P-1			22
Lubrication and Cooling Systems Diagnosis and Repair				
Perform cooling system pressure and dye tests to identify leaks; check coolant condition and level; inspect and test radiator, pressure cap, coolant recovery tank, heater core, and galley plugs; determine needed action.	P-1	P-1	P-1	17
Identify causes of engine overheating.	P-1	P-1		17
Inspect, replace, and/or adjust drive belts, tensioners, and pulleys; check pulley and belt alignment.	P-1	P-1	P-1	17
Inspect and/or test coolant; drain and recover coolant; flush and refill cooling system; use proper fluid type per manufacturer specification; bleed air as required.	P-1	P-1	P-1	17
Inspect, remove, and replace water pump.	P-2	P-2		
Remove and replace radiator.	P-2	P-2		17
Remove, inspect, and replace thermostat and gasket/seal.	P-1	P-1	P-1	17
Inspect and test fan(s), fan clutch (electrical or mechanical), fan shroud, and air dams; determine needed action.	P-1	P-1		17
Perform oil pressure tests; determine needed action.	P-1	P-1		15
Perform engine oil and filter change; use proper fluid type per manufacturer specification.	P-1	P-1	P-1	15
Inspect auxiliary coolers; determine needed action.	P-3	P-3		15
Inspect, test, and replace oil temperature and pressure switches and sensors.	P-2	P-2		15
Inspect oil pump gears or rotors, housing, pressure relief devices, and pump drive; perform needed action.	P-2			
Identify components of the lubrication and cooling systems.			P-1	
AUTOMATIC TRANSMISSION AND TRANSAXLE				
General: Transmission and Transaxle Diagnosis				
Identify and interpret transmission/transaxle concerns, differentiate between engine performance and transmission/transaxle concerns; determine needed action.	P-1	P-1		27
Research vehicle service information including fluid type, vehicle service history, service precautions, and technical service bulletins.	P-1	P-1	P-1	
Diagnose fluid loss and condition concerns; determine needed action.	P-1	P-1		27
Check fluid level in a transmission or a transaxle equipped with a dip-stick.	P-1	P-1	P-1	27
Check fluid level in a transmission or a transaxle not equipped with a dip-stick.	P-1	P-1	P-1	27

(continued)

Task List	MAST	AST	MLR	Chapter
Perform pressure tests (including transmissions/transaxles equipped with electronic pressure control); determine needed action.	P-1			27
Diagnose noise and vibration concerns; determine needed action.	P-2			27
Perform stall test; determine needed action.	P-2	P-2		27
Perform lock-up converter system tests; determine needed action.	P-3	P-3		27
Diagnose transmission/transaxle gear reduction/multiplication concerns using driving, driven, and held member (power flow) principles.	P-1	P-1		24
Diagnose electronic transmission/transaxle control systems using appropriate test equipment and service information.	P-1			27
Diagnose pressure concerns in a transmission using hydraulic principles (Pascal's Law).	P-2	P-2		
Demonstrate knowledge of pressure test including transmissions/transaxles equipped with electronic pressure control.		P-3		
Diagnose electronic transmission/transaxle control systems using appropriate test equipment and service information.		P-2		
Check transmission fluid condition; check for leaks.			P-2	27
Identify drive train components and configuration.			P-1	
In-Vehicle Transmission/Transaxle Maintenance and Repair				
Inspect, adjust, and/or replace external manual valve shift linkage, transmission range sensor/switch, and/or park/neutral position switch.	P-1	P-1	P-2	27
Inspect for leakage; replace external seals, gaskets, and bushings.	P-2	P-2	P-1	27
Inspect, test, adjust, repair, and/or replace electrical/electronic components and circuits including computers, solenoids, sensors, relays, terminals, connectors, switches, and harnesses; demonstrate understanding of the relearn procedure.	P-1	P-1		
Drain and replace fluid and filter(s); use proper fluid type per manufacturer specification.	P-1	P-1	P-1	27
Inspect, replace and align powertrain mounts.	P-2	P-2	P-2	27
Off-Vehicle Transmission and Transaxle Repair				
Remove and reinstall transmission/transaxle and torque converter; inspect engine core plugs, rear crankshaft seal, dowel pins, dowel pin holes, and mounting surfaces.	P-2	P-2		28
Inspect, leak test, flush, and/or replace transmission/transaxle oil cooler, lines, and fittings.	P-1	P-1		28
Inspect converter flex (drive) plate, converter attaching bolts, converter pilot, converter pump drive surfaces, converter end play, and crankshaft pilot bore.	P-2	P-2		28
Describe the operational characteristics of a continuously variable transmission (CVT).	P-3	P-3	P-3	29
Describe the operational characteristics of a hybrid vehicle drive train.	P-3	P-3	P-3	29
Disassemble, clean, and inspect transmission/transaxle.	P-1			28
Inspect, measure, clean, and replace valve body (includes surfaces, bores, springs, valves, switches, solenoids, sleeves, retainers, brackets, check valves/balls, screens, spacers, and gaskets).	P-2			28
Inspect servo and accumulator bores, pistons, seals, pins, springs, and retainers; determine needed action.	P-2			
Assemble transmission/transaxle.	P-1			28
Inspect, measure, and reseal oil pump assembly and components.	P-2			28
Measure transmission/transaxle end play and/or preload; determine needed action.	P-1			28
Inspect, measure, and/or replace thrust washers and bearings.	P-2			28
Inspect oil delivery circuits, including seal rings, ring grooves, and sealing surface areas, feed pipes, orifices, and check valves/balls.	P-2			28
Inspect bushings; determine needed action.	P-2			28
Inspect and measure planetary gear assembly components; determine needed action.	P-2			28
Inspect case bores, passages, bushings, vents, and mating surfaces; determine needed action.	P-2			28

Task List	MAST	AST	MLR	Chapter
Diagnose and inspect transaxle drive, link chains, sprockets, gears, bearings, and bushings; perform needed action.	P-2			
Inspect measure, repair, adjust or replace transaxle final drive components.	P-2			28
Inspect clutch drum, piston, check-balls, springs, retainers, seals, friction plates, pressure plates, and bands; determine needed action.	P-2			28
Measure clutch pack clearance; determine needed action.	P-1			28
Air test operation of clutch and servo assemblies.	P-1			28
Inspect one-way clutches, races, rollers, sprags, springs, cages, retainers; determine needed action.	P-2			28
MANUAL DRIVE TRAIN AND AXLES				
General: Drive Train Diagnosis				
Identify and interpret drive train concerns; determine needed action.	P-1	P-1		
Research vehicle service information including fluid type, vehicle service history, service precautions, and technical service bulletins.	P-1	P-1	P-1	
Check fluid condition; check for leaks; determine needed action.	P-1	P-1	P-2	32, 34
Drain and refill manual transmission/transaxle and final drive unit; use proper fluid type per manufacturer specification.	P-1	P-1	P-1	32
Identify manual drive train and axle components and configuration.			P-1	
Clutch Diagnosis and Repair				
Diagnose clutch noise, binding, slippage, pulsation, and chatter; determine needed action.	P-1	P-1		31
Inspect clutch pedal linkage, cables, automatic adjuster mechanisms, brackets, bushings, pivots, and springs; perform needed action.	P-1	P-1		31
Inspect and/or replace clutch pressure plate assembly, clutch disc, release (throw-out) bearing, linkage, and pilot bearing/bushing (as applicable).	P-1	P-1		31
Bleed clutch hydraulic system.	P-1	P-1		31
Check and adjust clutch master cylinder fluid level; check for leaks; use proper fluid type per manufacturer specification.	P-1	P-1	P-1	
Inspect flywheel and ring gear for wear, cracks, and discoloration; determine needed action.	P-1	P-1		31
Measure flywheel runout and crankshaft end play; determine needed action.	P-2	P-2		31
Describe the operation and service of a system that uses a dual mass flywheel.	P-3	P-3		
Check for hydraulic system leaks.			P-1	
Transmission/Transaxle Diagnosis and Repair				
Inspect, adjust, lubricate, and/or replace shift linkages, brackets, bushings, cables, pivots, and levers.	P-2	P-2		
Describe the operational characteristics of an electronically-controlled manual transmission/transaxle.	P-2	P-2	P-2	32
Diagnose noise concerns through the application of transmission/transaxle powerflow principles.	P-2			32
Diagnose hard shifting and jumping out of gear concerns; determine needed action.	P-2			32
Diagnose transaxle final drive assembly noise and vibration concerns; determine needed action.	P-3			32
Disassemble, inspect, clean, and reassemble internal transmission/transaxle components.	P-2			33
Drive Shaft and Half Shaft, Universal and Constant-Velocity (CV) Joint Diagnosis and Repair (Front, Rear, All-wheel, and Four-wheel drive)				
Diagnose constant-velocity (CV) joint noise and vibration concerns; determine needed action.	P-1	P-1		34
Diagnose universal joint noise and vibration concerns; perform needed action.	P-2	P-2		34
Inspect, remove, and/or replace bearings, hubs, and seals.	P-1	P-1	P-2	
Inspect, service, and/or replace shafts, yokes, boots, and universal/CV joints.	P-1	P-1	P-2	34
Check shaft balance and phasing; measure shaft runout; measure and adjust driveline angles.	P-2	P-2		34
Inspect locking hubs.			P-3	

(continued)

Task List	MAST	AST	MLR	Chapter
Check for leaks at drive assembly and transfer case seals; check vents; check fluid level; use proper fluid type per manufacturer specification.			P-2	
Drive Axle Diagnosis and Repair				
Ring and Pinion Gears and Differential Case Assembly				
Clean and inspect differential case; check for leaks; inspect housing vent.	P-1	P-1	P-1	34
Check and adjust differential case fluid level; use proper fluid type per manufacturer specification.	P-1	P-1	P-1	
Drain and refill differential case; use proper fluid type per manufacturer specification.	P-1	P-1		
Drain and refill differential housing.			P-1	
Diagnose noise and vibration concerns; determine needed action.	P-2			34
Inspect and replace companion flange and/or pinion seal; measure companion flange runout.	P-2	P-2		34
Inspect ring gear and measure runout; determine needed action.	P-3			34
Remove, inspect, and/or reinstall drive pinion and ring gear, spacers, sleeves, and bearings.	P-3			34
Measure and adjust drive pinion depth.	P-3			34
Measure and adjust drive pinion bearing preload.	P-3			34
Measure and adjust side bearing preload and ring and pinion gear total backlash and backlash variation on a differential carrier assembly (threaded cup or shim types).	P-3			
Check ring and pinion tooth contact patterns; perform needed action.	P-3			34
Disassemble, inspect, measure, adjust, and/or replace differential pinion gears (spiders), shaft, side gears, side bearings, thrust washers, and case.	P-3			34
Reassemble and reinstall differential case assembly; measure runout; determine needed action.	P-3			34
Limited Slip Differential				
Diagnose noise, slippage, and chatter concerns; determine needed action.	P-3			34
Measure rotating torque; determine needed action.	P-3			34
Drive Axles				
Inspect and replace drive axle wheel studs.	P-1	P-1	P-1	34, 46
Remove and replace drive axle shafts.	P-1	P-1		34
Inspect and replace drive axle shaft seals, bearings, and retainers.	P-2	P-2		34
Measure drive axle flange runout and shaft end play; determine needed action.	P-2	P-2		34
Diagnose drive axle shafts, bearings, and seals for noise, vibration, and fluid leakage concerns; determine needed action.	P-2			34
Four-wheel Drive/All-wheel Drive Component Diagnosis and Repair				
Inspect, adjust, and repair shifting controls (mechanical, electrical, and vacuum), bushings, mounts, levers, and brackets.	P-3	P-3		35
Inspect locking hubs; determine needed action.	P-3	P-3		35
Check for leaks at drive assembly and transfer case seals; check vents; check fluid level; use proper fluid type per manufacturer specification.	P-3	P-3		
Identify concerns related to variations in tire circumference and/or final drive ratios.	P-2	P-2		35
Diagnose noise, vibration, and unusual steering concerns; determine needed action.	P-3			35
Diagnose, test, adjust, and/or replace electrical/electronic components of four-wheel drive/all-wheel drive systems.	P-2			35
Disassemble, service, and reassemble transfer case and components.	P-2			35
SUSPENSION AND STEERING				
General: Suspension and Steering Systems				
Research vehicle service information including fluid type, vehicle service history, service precautions, and technical service bulletins.	P-1	P-1	P-1	
Identify and interpret suspension and steering system concerns; determine needed action.	P-1	P-2	P-1	

Task List	MAST	AST	MLR	Chapter
Steering Systems Diagnosis and Repair				
Disable and enable supplemental restraint system (SRS); verify indicator lamp operation.	P-1	P-1	P-1	
Remove and replace steering wheel; center/time supplemental restraint system (SRS) coil (clock spring).	P-1	P-1		39
Diagnose steering column noises, looseness, and binding concerns (including tilt/telescoping mechanisms); determine needed action.	P-2	P-2		39
Diagnose power steering gear (non-rack and pinion) binding, uneven turning effort, looseness, hard steering, and noise concerns; determine needed action.	P-2	P-2		39
Diagnose power steering gear (rack and pinion) binding, uneven turning effort, looseness, hard steering, and noise concerns; determine needed action.	P-2	P-2		39
Inspect steering shaft universal-joint(s), flexible coupling(s), collapsible column, lock cylinder mechanism, and steering wheel; determine needed action.	P-2	P-2		39
Remove and replace rack and pinion steering gear; inspect mounting bushings and brackets.	P-2	P-1		39
Inspect rack and pinion steering gear inner tie rod ends (sockets) and bellows boots; replace as needed.	P-1	P-1	P-1	39
Inspect power steering fluid level and condition.	P-1	P-1	P-1	
Flush, fill, and bleed power steering system; use proper fluid type per manufacturer specification.	P-2	P-2	P-2	39
Inspect for power steering fluid leakage; determine needed action.	P-1	P-1	P-1	39
Remove, inspect, replace, and/or adjust power steering pump drive belt.	P-1	P-1	P-1	
Remove and reinstall power steering pump.	P-2	P-2		39
Remove and reinstall press fit power steering pump pulley; check pulley and belt alignment.	P-2	P-2		39
Inspect, remove and/or replace power steering hoses and fittings.	P-2	P-2	P-2	39
Inspect, remove and/or replace pitman arm, relay (centerlink/intermediate) rod, idler arm, mountings, and steering linkage damper.	P-2	P-2	P-1	39
Inspect, replace, and/or adjust tie rod ends (sockets), tie rod sleeves, and clamps.	P-1	P-1	P-1	39
Inspect, test and diagnose electrically- assisted power steering systems (including using a scan tool); determine needed action.	P-2			39
Identify hybrid vehicle power steering system electrical circuits and safety precautions.	P-2	P-2	P-2	38, 39
Test power steering system pressure; determine needed action.	P-2			
Inspect electric power steering assist system.		P-3	P-2	39
Suspension Systems Diagnosis and Repair				
Diagnose short and long arm suspension system noises, body sway, and uneven ride height concerns; determine needed action.	P-1	P-1		41
Diagnose strut suspension system noises, body sway, and uneven ride height concerns; determine needed action.	P-1	P-1		41
Inspect, remove, and/or replace upper and lower control arms, bushings, shafts, and rebound bumpers.	P-3	P-3	P-1	41
Inspect, remove, and/or replace strut rods and bushings.	P-3	P-3		41
Inspect, remove, and/or replace upper and/or lower ball joints (with or without wear indicators).	P-2	P-2	P-1	41
Inspect, remove, and/or replace steering knuckle assemblies.	P-3	P-3		41
Inspect, remove and/or replace short and long arm suspension system coil springs and spring insulators.	P-3	P-3		41
Inspect, remove, and/or replace torsion bars and mounts.	P-3	P-3	P-1	41
Inspect, remove, and/or replace front/rear stabilizer bar (sway bar) bushings, brackets, and links.	P-3	P-3	P-1	41
Inspect, remove, and/or replace strut cartridge or assembly, strut coil spring, insulators (silencers), and upper strut bearing mount.	P-3	P-3	P-2	41
Inspect, remove, and/or replace track bar, strut rods/radius arms, and related mounts and bushings.	P-3	P-3	P-1	41

(continued)

Task List	MAST	AST	MLR	Chapter
Inspect rear suspension system leaf spring(s), spring insulators (silencers), shackles, brackets, bushings, center pins/bolts, and mounts.	P-1	P-1	P-1	41
Inspect front strut bearing and mount.			P-1	
Inspect rear suspension system lateral links/arms (track bars), control (trailing) arms.			P-1	
Related Suspension and Steering Service				
Inspect, remove, and/or replace shock absorbers; inspect mounts and bushings.	P-1	P-1	P-1	41
Remove, inspect, service and/or replace front and rear wheel bearings.	P-1	P-1		48
Describe the function of suspension and steering control systems and components, (i.e. active suspension and stability control).	P-3	P-3	P-3	38
Wheel Alignment Diagnosis, Adjustment, and Repair				
Diagnose vehicle wander, drift, pull, hard steering, bump steer, memory steer, torque steer, and steering return concerns; determine needed action.	P-1	P-1		41
Perform prealignment inspection; measure vehicle ride height; determine needed action.	P-1	P-1	P-1	42
Describe alignment angles (camber, caster, and toe).			P-1	
Prepare vehicle for wheel alignment on alignment machine; perform four-wheel alignment by checking and adjusting front and rear wheel caster, camber and toe as required; center steering wheel.	P-1	P-1		42
Check toe-out-on-turns (turning radius); determine needed action.	P-2	P-2		42
Check steering axis inclination (SAI) and included angle; determine needed action.	P-2	P-2		42
Check rear wheel thrust angle; determine needed action.	P-1	P-1		42
Check for front wheel setback; determine needed action.	P-2	P-2		
Check front and/or rear cradle (subframe) alignment; determine needed action.	P-3	P-3		42
Reset steering angle sensor.	P-2	P-2		42
Wheels and Tires Diagnosis and Repair				
Inspect tire condition; identify tire wear patterns; check for correct tire size, application (load and speed ratings), and air pressure as listed on the tire information placard/label.	P-1	P-1	P-1	37
Diagnose wheel/tire vibration, shimmy, and noise; determine needed action.	P-2	P-2		37
Rotate tires according to manufacturer's recommendation including vehicles equipped with tire pressure monitoring systems (TPMS).	P-1	P-1	P-1	37
Measure wheel, tire, axle flange, and hub runout; determine needed action.	P-2	P-2		37
Diagnose tire pull problems; determine needed action.	P-1	P-1		37
Dismount, inspect, and remount tire on wheel; balance wheel and tire assembly.	P-1	P-1	P-1	37
Dismount, inspect, and remount tire on wheel equipped with tire pressure monitoring system sensor.	P-1	P-1	P-1	37
Inspect tire and wheel assembly for air loss; perform needed action.	P-1	P-1	P-1	37
Repair tire following vehicle manufacturer approved procedure.	P-1	P-1	P-1	37
Identify indirect and direct tire pressure monitoring system (TPMS); calibrate system; verify operation of instrument panel lamps.	P-1	P-1	P-1	37
Demonstrate knowledge of steps required to remove and replace sensors in a tire pressure monitoring system (TPMS) including relearn procedure.	P-1	P-1	P-1	37
BRAKES				
General: Brake Systems Diagnosis				
Identify and interpret brake system concerns; determine needed action.	P-1	P-1		46
Research vehicle service information including fluid type, vehicle service history, service precautions, and technical service bulletins.	P-1	P-1	P-1	
Describe procedure for performing a road test to check brake system operation including an anti-lock brake system (ABS).	P-1	P-1	P-1	46

Task List	MAST	AST	MLR	Chapter
Install wheel and torque lug nuts.	P-1	P-1	P-1	46, 47
Identify brake system components and configuration.			P-1	
Hydraulic System Diagnosis and Repair				
Diagnose pressure concerns in the brake system using hydraulic principles (Pascal's Law).	P-1	P-1		45
Measure brake pedal height, travel, and free play (as applicable); determine needed action.	P-1	P-1		45
Describe proper brake pedal height, travel, and feel.			P-1	
Check master cylinder for internal/external leaks and proper operation; determine needed action.	P-1	P-1	P-1	45
Remove, bench bleed, and reinstall master cylinder.	P-1	P-1		45
Diagnose poor stopping, pulling or dragging concerns caused by malfunctions in the hydraulic system; determine needed action.	P-1	P-3		
Inspect brake lines, flexible hoses, and fittings for leaks, dents, kinks, rust, cracks, bulging, wear; and loose fittings/supports; determine needed action.	P-1	P-1	P-1	45
Replace brake lines, hoses, fittings, and supports.	P-2	P-2		45
Fabricate brake lines using proper material and flaring procedures (double flare and ISO types).	P-2	P-2		45
Select, handle, store, and fill brake fluids to proper level; use proper fluid type per manufacturer specification.	P-1	P-1	P-1	45
Inspect, test, and/or replace components of brake warning light system.	P-3	P-3		45
Identify components of hydraulic brake warning light system.	P-2	P-2	P-3	
Bleed and/or flush brake system.	P-1	P-1	P-1	45
Test brake fluid for contamination.	P-1	P-1	P-1	45
Drum Brake Diagnosis and Repair				
Diagnose poor stopping, noise, vibration, pulling, grabbing, dragging or pedal pulsation concerns; determine needed action.	P-1	P-1		46, 47
Remove, clean, and inspect brake drum; measure brake drum diameter; determine serviceability.	P-1	P-1	P-1	47
Refinish brake drum and measure final drum diameter; compare with specification.	P-1	P-1	P-1	47
Remove, clean, inspect, and/or replace brake shoes, springs, pins, clips, levers, adjusters/self-adjusters, other related brake hardware, and backing support plates; lubricate and reassemble.	P-1	P-1	P-1	47
Inspect wheel cylinders for leaks and proper operation; remove and replace as needed.	P-2	P-2	P-2	47
Pre-adjust brake shoes and parking brake; install brake drums or drum/hub assemblies and wheel bearings; perform final checks and adjustments.	P-1	P-1	P-1	47
Disc Brake Diagnosis and Repair				
Diagnose poor stopping, noise, vibration, pulling, grabbing, dragging, or pulsation concerns; determine needed action.	P-1	P-1		
Remove and clean caliper assembly; inspect for leaks, damage, and wear; determine needed action.	P-1	P-1	P-1	46
Inspect caliper mounting and slides/pins for proper operation, wear, and damage; determine needed action.	P-1	P-1	P-1	46
Remove, inspect, and/or replace brake pads and retaining hardware; determine needed action.	P-1	P-1	P-1	46
Lubricate and reinstall caliper, brake pads, and related hardware; seat brake pads; inspect for leaks.	P-1	P-1	P-1	46
Clean and inspect rotor and mounting surface; measure rotor thickness, thickness variation, and lateral runout; determine needed action.	P-1	P-1	P-1	46
Remove and reinstall/replace rotor.	P-1	P-1	P-1	46
Refinish rotor on vehicle; measure final rotor thickness and compare with specification.	P-1	P-1	P-1	46
Refinish rotor off vehicle; measure final rotor thickness and compare with specification.	P-1	P-1	P-1	46
Retract and re-adjust caliper piston on an integrated parking brake system.	P-2	P-2	P-2	46
Check brake pad wear indicator; determine needed action.	P-1	P-1	P-1	46
Describe importance of operating vehicle to burnish/break-in replacement brake pads according to manufacturer's recommendations.	P-1	P-1	P-1	46

(continued)

Task List	MAST	AST	MLR	Chapter
Power-Assist Units Diagnosis and Repair				
Check brake pedal travel with and without engine running to verify proper power booster operation.	P-2	P-2	P-2	45
Identify components of the brake power assist system (vacuum and hydraulic); check vacuum supply (manifold or auxiliary pump) to vacuum- type power booster.	P-1	P-1	P-1	
Inspect vacuum-type power booster unit for leaks; inspect the check-valve for proper operation; determine needed action.	P-1	P-1		45
Inspect and test hydraulically-assisted power brake system for leaks and proper operation; determine needed action.	P-3	P-3		45
Measure and adjust master cylinder pushrod length.	P-3	P-3		45
Related Systems (i.e. Wheel Bearings, Parking Brakes, Electrical) Diagnosis and Repair				
Diagnose wheel bearing noises, wheel shimmy, and vibration concerns; determine needed action.	P-1	P-2		48
Remove, clean, inspect, repack, and install wheel bearings; replace seals; install hub and adjust bearings.	P-2	P-2	P-1	48
Check parking brake system and components for wear, binding, and corrosion; clean, lubricate, adjust and/or replace as needed.	P-1	P-1	P-2	45
Check parking brake operation and parking brake indicator light system operation; determine needed action.	P-1	P-1	P-1	45
Check operation of brake stop light system.	P-1	P-1	P-1	45
Replace wheel bearing and race.	P-3	P-3	P-2	48
Remove, reinstall, and/or replace sealed wheel bearing assembly.	P-1	P-1		48
Inspect and replace wheel studs.	P-1	P-1	P-1	
Electronic Brake Control Systems: Antilock Brake (ABS), Traction Control (TCS), and Electronic Stability Control (ESC) Systems Diagnosis and Repair				
Identify and inspect electronic brake control system components (ABS, TCS, ESC); determine needed action.	P-1	P-1		49
Identify traction control/vehicle stability control system components.			P-3	
Describe the operation of a regenerative braking system.	P-3	P-3	P-3	43
Diagnose poor stopping, wheel lock-up, abnormal pedal feel, unwanted application, and noise concerns associated with the electronic brake control system; determine needed action.	P-2			49
Diagnose electronic brake control system electronic control(s) and components by retrieving diagnostic trouble codes, and/or using recommended test equipment; determine needed action.	P-2			49
Depressurize high-pressure components of an electronic brake control system.	P-2			49
Bleed the electronic brake control system hydraulic circuits.	P-1			49
Test, diagnose, and service electronic brake control system speed sensors (digital and analog), toothed ring (tone wheel), and circuits using a graphing multimeter (GMM)/digital storage oscilloscope (DSO) (includes output signal, resistance, shorts to voltage/ground, and frequency data).	P-2			49
Diagnose electronic brake control system braking concerns caused by vehicle modifications (tire size, curb height, final drive ratio, etc.).	P-1			49
ELECTRICAL/ELECTRONIC SYSTEMS				
General: Electrical System Diagnosis				
Research vehicle service information including vehicle service history, service precautions, and technical service bulletins.	P-1	P-1	P-1	
Demonstrate knowledge of electrical/electronic series, parallel, and series-parallel circuits using principles of electricity (Ohm's Law).	P-1	P-1	P-1	50, 52
Use wiring diagrams to trace electrical/electronic circuits.			P-1	51
Demonstrate proper use of a digital multimeter (DMM) when measuring source voltage, voltage drop (including grounds), current flow and resistance.	P-1	P-1	P-1	52

Task List	MAST	AST	MLR	Chapter
Demonstrate knowledge of the causes and effects from shorts, grounds, opens, and resistance problems in electrical/electronic circuits.	P-1	P-1	P-1	50
Demonstrate proper use of a test light on an electrical circuit.	P-1	P-1	P-2	52
Use fused jumper wires to check operation of electrical circuits.	P-1	P-1	P-2	52
Use wiring diagrams during the diagnosis (troubleshooting) of electrical/electronic circuit problems.	P-1	P-1		51
Diagnose the cause(s) of excessive key-off battery drain (parasitic draw); determine needed action.	P-1	P-1		53
Measure key-off battery drain (parasitic draw).			P-1	
Inspect and test fusible links, circuit breakers, and fuses; determine needed action.	P-1	P-1	P-1	52
Inspect, test, repair, and/or replace components, connectors, terminals, harnesses, and wiring in electrical/electronic systems (including solder repairs); determine needed action.	P-1	P-1	P-1	51
Check electrical/electronic circuit waveforms; interpret readings and determine needed repairs.	P-2			52
Repair data bus wiring harness.	P-1			51
Identify electrical/electronic system components and configurations.			P-1	
Battery Diagnosis and Service				
Perform battery state-of-charge test; determine needed action.	P-1	P-1	P-1	53
Confirm proper battery capacity for vehicle application; perform battery capacity and load test; determine needed action.	P-1	P-1	P-1	53
Maintain or restore electronic memory functions.	P-1	P-1	P-1	53
Inspect and clean battery; fill battery cells; check battery cables, connectors, clamps, and hold-downs.	P-1	P-1	P-1	53
Perform slow/fast battery charge according to manufacturer's recommendations.	P-1	P-1	P-1	53
Jump-start vehicle using jumper cables and a booster battery or an auxiliary power supply.	P-1	P-1	P-1	53
Identify safety precautions for high voltage systems on electric, hybrid, hybrid-electric, and diesel vehicles.	P-2	P-2	P-2	67
Identify electrical/electronic modules, security systems, radios, and other accessories that require reinitialization or code entry after reconnecting vehicle battery.	P-1	P-1	P-1	53
Identify hybrid vehicle auxiliary (12v) battery service, repair, and test procedures.	P-2	P-2	P-2	67
Starting System Diagnosis and Repair				
Perform starter current draw tests; determine needed action.	P-1	P-1	P-1	54
Perform starter circuit voltage drop tests; determine needed action.	P-1	P-1	P-1	54
Inspect and test starter relays and solenoids; determine needed action.	P-2	P-2	P-2	54
Remove and install starter in a vehicle.	P-1	P-1	P-1	54
Inspect and test switches, connectors, and wires of starter control circuits; determine needed action.	P-2	P-2	P-2	52, 54
Differentiate between electrical and engine mechanical problems that cause a slow-crank or a no-crank condition.	P-2	P-2		54
Demonstrate knowledge of an automatic idle-stop/start-stop system.	P-2	P-2	P-3	54
Charging System Diagnosis and Repair				
Perform charging system output test; determine needed action.	P-1	P-1	P-1	54
Diagnose (troubleshoot) charging system for causes of undercharge, no-charge, or overcharge conditions.	P-1	P-1		54
Inspect, adjust, and/or replace generator (alternator) drive belts; check pulleys and tensioners for wear; check pulley and belt alignment.	P-1	P-1	P-1	
Remove, inspect, and/or replace generator (alternator).	P-1	P-1	P-2	54
Perform charging circuit voltage drop tests; determine needed action.	P-1	P-1	P-2	54
Lighting Systems Diagnosis and Repair				
Diagnose (troubleshoot) the causes of brighter-than-normal, intermittent, dim, or no light operation; determine needed action.	P-1	P-1		

(continued)

Task List	MAST	AST	MLR	Chapter
Inspect interior and exterior lamps and sockets including headlights and auxiliary lights (fog lights/driving lights); replace as needed.	P-1	P-1	P-1	
Aim headlights.	P-2	P-2	P-2	55
Identify system voltage and safety precautions associated with high-intensity discharge headlights.	P-2	P-2	P-2	55
Instrument Cluster and Driver Information Systems Diagnosis and Repair				
Inspect and test gauges and gauge sending units for causes of abnormal readings; determine needed action.	P-2	P-2		
Diagnose (troubleshoot) the causes of incorrect operation of warning devices and other driver information systems; determine needed action.	P-2	P-2		55
Reset maintenance indicators as required.	P-2	P-2		
Body Electrical Systems Diagnosis and Repair				
Diagnose operation of comfort and convenience accessories and related circuits (such as: power window, power seats, pedal height, power locks, truck locks, remote start, moon roof, sun roof, sun shade, remote keyless entry, voice activation, steering wheel controls, back-up camera, park assist, cruise control, and auto dimming headlamps); determine needed repairs.	P-2	P-3		56, 57
Diagnose operation of security/anti-theft systems and related circuits (such as: theft deterrent, door locks, remote keyless entry, remote start, and starter/fuel disable); determine needed repairs.	P-2	P-3		56, 57
Diagnose operation of entertainment and related circuits (such as: radio, DVD, remote CD changer, navigation, amplifiers, speakers, antennas, and voice-activated accessories); determine needed repairs.	P-3	P-3		57
Diagnose operation of safety systems and related circuits (such as: horn, airbags, seat belt pretensioners, occupancy classification, wipers, washers, speed control/collision avoidance, heads-up display, park assist, and back-up camera); determine needed repairs.	P-1	P-3		56, 57
Diagnose body electronic systems circuits using a scan tool; check for module communication errors (data communication bus systems); determine needed action.	P-2	P-3		56
Describe the process for software transfer, software updates, or reprogramming of electronic modules.	P-2	P-3		56
Disable and enable supplemental restraint system (SRS); verify indicator lamp operation.			P-1	39
Remove and reinstall door panel.			P-1	56
Describe the operation of keyless entry/remote-start systems.			P-3	
Verify operation of instrument panel gauges and warning/indicator lights; reset maintenance indicators.			P-1	
Verify windshield wiper and washer operation; replace wiper blades.			P-1	
HEATING, VENTILATION, AND AIR CONDITIONING (HVAC)				
General: A/C System Diagnosis and Repair				
Identify and interpret heating and air conditioning problems; determine needed action.	P-1	P-1		59
Research vehicle service information including refrigerant/oil type, vehicle service history, service precautions, and technical service bulletins.	P-1	P-1	P-1	
Identify heating, ventilation and air conditioning (HVAC) components and configuration.			P-1	
Performance test A/C system; identify problems.	P-1	P-1		59
Identify abnormal operating noises in the A/C system; determine needed action.	P-2	P-2		59
Identify refrigerant type; select and connect proper gauge set/test equipment; record temperature and pressure readings.	P-1	P-1		59
Leak test A/C system; determine needed action.	P-1	P-1		59
Inspect condition of refrigerant oil removed from A/C system; determine needed action.	P-2	P-2		59
Determine recommended oil and oil capacity for system application.	P-1	P-1		59
Using a scan tool, observe and record related HVAC data and trouble codes.	P-3	P-3		60
Refrigeration System Component Diagnosis and Repair				
Inspect, remove, and/or replace A/C compressor drive belts, pulleys, tensioners and visually inspect A/C components for signs of leaks; determine needed action.	P-1	P-1	P-1	59

Task List	MAST	AST	MLR	Chapter
Inspect, test, service and/or replace A/C compressor clutch components and/or assembly; check compressor clutch air gap; adjust as needed.	P-2	P-2		
Remove, inspect, reinstall, and/or replace A/C compressor and mountings; determine recommended oil type and quantity.	P-2	P-2		59
Identify hybrid vehicle A/C system electrical circuits and service/safety precautions.	P-2	P-2	P-2	60, 67
Determine need for an additional A/C system filter; perform needed action.	P-3	P-3		59
Remove and inspect A/C system mufflers, hoses, lines, fittings, O-rings, seals, and service valves; perform needed action.	P-2	P-2		59
Inspect for proper A/C condenser airflow; determine needed action.	P-1	P-1		59
Inspect A/C condenser for airflow restrictions; determine necessary action.			P-1	59
Remove, inspect, and replace receiver/drier or accumulator/drier; determine recommended oil type and quantity.	P-2	P-2		59
Remove, inspect, and install expansion valve or orifice (expansion) tube.	P-1	P-1		59
Inspect evaporator housing water drain; perform needed action.	P-1	P-1		59
Diagnose A/C system conditions that cause the protection devices (pressure, thermal, and/or control module) to interrupt system operation; determine needed action.	P-2			59
Determine procedure to remove and reinstall evaporator; determine required oil type and quantity.	P-2	P-2		59
Remove, inspect, reinstall, and/or replace condenser; determine required oil type and quantity.	P-2			59
Heating, Ventilation, and Engine Cooling Systems Diagnosis and Repair				
Inspect engine cooling and heater systems hoses and pipes; perform needed action.	P-1	P-1	P-1	17
Inspect and test heater control valve(s); perform needed action.	P-2	P-2		17
Diagnose temperature control problems in the HVAC system; determine needed action.	P-2			60
Determine procedure to remove, inspect, reinstall, and/or replace heater core.	P-2	P-2		17
Operating Systems and Related Controls Diagnosis and Repair				
Inspect and test HVAC system blower motors, resistors, switches, relays, wiring, and protection devices; determine needed action.	P-1	P-1		60
Diagnose A/C compressor clutch control systems; determine needed action.	P-2	P-2		60
Diagnose malfunctions in the vacuum, mechanical, and electrical components and controls of the heating, ventilation, and A/C (HVAC) system; determine needed action.	P-2	P-2		60
Inspect and test HVAC system control panel assembly; determine needed action.	P-3	P-3		60
Inspect and test HVAC system control cables, motors, and linkages; perform needed action.	P-3	P-3		60
Inspect HVAC system ducts, doors, hoses, cabin filters, and outlets; perform needed action.	P-1	P-1	P-1	60
Identify the source of HVAC system odors.	P-2	P-2	P-2	59
Check operation of automatic or semi-automatic HVAC control systems; determine needed action.	P-2	P-2		60
Refrigerant Recovery, Recycling, and Handling				
Perform correct use and maintenance of refrigerant handling equipment according to equipment manufacturer's standards.	P-1	P-1		59
Identify A/C system refrigerant; test for sealants; recover, evacuate, and charge A/C system; add refrigerant oil as required.	P-1	P-1		59
Recycle, label, and store refrigerant.	P-1	P-1		59
ENGINE PERFORMANCE				
General: Engine Diagnosis				
Identify and interpret engine performance concerns; determine needed action.	P-1	P-1		
Research vehicle service information including vehicle service history, service precautions, and technical service bulletins.	P-1	P-1	P-1	
Diagnose abnormal engine noises or vibration concerns; determine needed action.	P-3	P-3		13

(continued)

Task List	MAST	AST	MLR	Chapter
Diagnose the cause of excessive oil consumption, coolant consumption, unusual exhaust color, odor, and sound; determine needed action.	P-2	P-2		66
Perform engine absolute manifold pressure tests (vacuum/boost); determine needed action.	P-1	P-1	P-2	13
Perform cylinder power balance test; determine needed action.	P-2	P-2	P-2	13
Perform cylinder cranking and running compression tests; determine needed action.	P-1	P-1	P-2	13
Perform cylinder leakage test; determine needed action.	P-1	P-1	P-2	13
Diagnose engine mechanical, electrical, electronic, fuel, and ignition concerns; determine needed action.	P-2	P-2		
Verify engine operating temperature; determine needed action.	P-1	P-1	P-1	17
Verify correct camshaft timing including engines equipped with variable valve timing systems (VVT).	P-1	P-1		
Remove and replace spark plugs; inspect secondary ignition components for wear and damage.			P-1	
Computerized Controls Diagnosis and Repair				
Retrieve and record diagnostic trouble codes (DTC), OBD monitor status, and freeze frame data; clear codes when applicable.	P-1	P-1	P-1	64
Access and use service information to perform step-by-step (troubleshooting) diagnosis.	P-1	P-1		61
Perform active tests of actuators using a scan tool; determine needed action.	P-1	P-2		64
Describe the use of OBD monitors for repair verification.	P-1	P-1	P-1	64
Diagnose the causes of emissions or driveablility concerns with stored or active diagnostic trouble codes (DTC); obtain, graph, and interpret scan tool data.	P-1			64
Diagnose emissions or driveablility concerns without stored or active diagnostic trouble codes; determine needed action.	P-1			64
Inspect and test computerized engine control system sensors, powertrain/engine control module (PCM/ECM), actuators, and circuits using a graphing multimeter (GMM)/digital storage oscilloscope (DSO); perform needed action.	P-2			64
Diagnose driveablility and emissions problems resulting from malfunctions of interrelated systems (cruise control, security alarms, suspension controls, traction controls, HVAC, automatic transmissions, non-OEM installed accessories, or similar systems); determine needed action.	P-2			64
Ignition System Diagnosis and Repair				
Diagnose (troubleshoot) ignition system related problems such as no-starting, hard starting, engine misfire, poor driveablility, spark knock, power loss, poor mileage, and emissions concerns; determine needed action.	P-2	P-2		61
Inspect and test crankshaft and camshaft position sensor(s); determine needed action.	P-1	P-1		61
Inspect, test, and/or replace ignition control module, powertrain/engine control module; reprogram/initialize as needed.	P-3	P-3		61
Remove and replace spark plugs; inspect secondary ignition components for wear and damage.	P-1	P-1		61
Fuel, Air Induction, and Exhaust Systems Diagnosis and Repair				
Diagnose (troubleshoot) hot or cold no-starting, hard starting, poor driveablility, incorrect idle speed, poor idle, flooding, hesitation, surging, engine misfire, power loss, stalling, poor mileage, dieseling, and emissions problems; determine needed action.	P-2			
Check fuel for contaminants; determine needed action.	P-2	P-2		62
Inspect and test fuel pump(s) and pump control system for pressure, regulation, and volume; perform needed action.	P-1	P-1		62
Replace fuel filter(s) where applicable.	P-2	P-2	P-2	62
Inspect, service, or replace air filters, filter housings, and intake duct work.	P-1	P-1	P-1	10
Inspect throttle body, air induction system, intake manifold and gaskets for vacuum leaks and/or unmetered air.	P-2	P-2		65
Inspect, test, and/or replace fuel injectors.	P-2	P-2		62
Verify idle control operation.	P-1	P-1		65
Inspect integrity of the exhaust manifold, exhaust pipes, muffler(s), catalytic converter(s), resonator(s), tail pipe(s), and heat shields; perform needed action.	P-1	P-1	P-1	65

Task List	MAST	AST	MLR	Chapter
Inspect condition of exhaust system hangers, brackets, clamps, and heat shields; determine needed action.	P-1	P-1	P-1	65
Perform exhaust system back-pressure test; determine needed action.	P-2	P-2		65
Check and refill diesel exhaust fluid (DEF).	P-2	P-2	P-2	10
Test the operation of turbocharger/supercharger systems; determine needed action.	P-2			65
Emissions Control Systems Diagnosis and Repair				
Diagnose oil leaks, emissions, and driveablility concerns caused by the positive crankcase ventilation (PCV) system; determine needed action.	P-3	P-3		66
Inspect, test, service, and/or replace positive crankcase ventilation (PCV) filter/breather, valve, tubes, orifices, and hoses; perform needed action.	P-2	P-2	P-2	66
Diagnose emissions and driveablility concerns caused by the exhaust gas recirculation (EGR) system; inspect, test, service and/or replace electrical/electronic sensors, controls, wiring, tubing, exhaust passages, vacuum/pressure controls, filters, and hoses of exhaust gas recirculation (EGR) systems; determine needed action.	P-2	P-3		66
Inspect and test electrical/electronically-operated components and circuits of secondary air injection systems; determine needed action.		P-3		
Diagnose emissions and driveablility concerns caused by the secondary air injection system; inspect, test, repair, and/or replace electrical/electronically-operated components and circuits of secondary air injection systems; determine needed action.	P-2			66
Diagnose emissions and driveablility concerns caused by the evaporative emissions control (EVAP) system; determine needed action.	P-1			66
Diagnose emission and driveablility concerns caused by catalytic converter system; determine needed action.	P-2	P-3		66
Interpret diagnostic trouble codes (DTCs) and scan tool data related to the emissions control systems; determine needed action.	P-2	P-2		64
REQUIRED SUPPLEMENTAL TASKS				
Shop and Personal Safety				
Identify general shop safety rules and procedures.	R	R	R	3
Utilize safe procedures for handling of tools and equipment.	R	R	R	6
Identify and use proper placement of floor jacks and jack stands.	R	R	R	9
Identify and use proper procedures for safe lift operation.	R	R	R	9
Utilize proper ventilation procedures for working within the lab/shop area.	R	R	R	4
Identify marked safety areas.	R	R	R	3
Identify the location and the types of fire extinguishers and other fire safety equipment; demonstrate knowledge of the procedures for using fire extinguishers and other fire safety equipment.	R	R	R	3
Identify the location and use of eye wash stations.	R	R	R	3
Identify the location of the posted evacuation routes.	R	R	R	3
Comply with the required use of safety glasses, ear protection, gloves, and shoes during lab/shop activities.	R	R	R	4
Identify and wear appropriate clothing for lab/shop activities.	R	R	R	4
Secure hair and jewelry for lab/shop activities.	R	R	R	4
Demonstrate awareness of the safety aspects of supplemental restraint systems (SRS), electronic brake control systems, and hybrid vehicle high voltage circuits.	R	R	R	
Demonstrate awareness of the safety aspects of high voltage circuits (such as high intensity discharge (HID) lamps, ignition systems, injection systems, etc.).	R	R	R	
Locate and demonstrate knowledge of safety data sheets (SDS).	R	R	R	3

(continued)

Task List	MAST	AST	MLR	Chapter
Tools and Equipment				
Identify tools and their usage in automotive applications.	R	R	R	
Identify standard and metric designation.	R	R	R	6
Demonstrate safe handling and use of appropriate tools.	R	R	R	6
Demonstrate proper cleaning, storage, and maintenance of tools and equipment.	R	R	R	6
Demonstrate proper use of precision measuring tools (i.e. micrometer, dial-indicator, dial-caliper).	R	R	R	6
Preparing Vehicle for Service				
Identify information needed and the service requested on a repair order.	R	R	R	5
Identify purpose and demonstrate proper use of fender covers, mats.	R	R	R	9
Demonstrate use of the three Cs (concern, cause, and correction).	R	R	R	5
Review vehicle service history.	R	R	R	5
Complete work order to include customer information, vehicle identifying information, customer concern, related service history, cause, and correction.	R	R	R	5
Preparing Vehicle for Customer				
Ensure vehicle is prepared to return to customer per school/company policy (floor mats, steering wheel cover, etc.).	R	R	R	9
Workplace Employability Skills				
Personal Standards (see Standard 7.9)				
Reports to work daily on time; able to take directions and motivated to accomplish the task at hand.	R	R	R	11
Dresses appropriately and uses language and manners suitable for the workplace.	R	R	R	11
Maintains appropriate personal hygiene.	R	R	R	11
Meets and maintains employment eligibility criteria, such as drug/alcohol-free status, clean driving record, etc.	R	R	R	11
Demonstrates honesty, integrity and reliability.	R	R	R	11
Work Habits/Ethic (see Standard 7.10)				
Complies with workplace policies/laws.	R	R	R	11
Contributes to the success of the team, assists others and requests help when needed.	R	R	R	11
Works well with all customers and coworkers.	R	R	R	11
Negotiates solutions to interpersonal and workplace conflicts.	R	R	R	11
Contributes ideas and initiative.	R	R	R	11
Follows directions.	R	R	R	11
Communicates (written and verbal) effectively with customers and coworkers.	R	R	R	11
Reads and interprets workplace documents; writes clearly and concisely.	R	R	R	11
Analyzes and resolves problems that arise in completing assigned tasks.	R	R	R	11
Organizes and implements a productive plan of work.	R	R	R	11
Uses scientific, technical, engineering and mathematics principles and reasoning to accomplish assigned tasks.	R	R	R	11
Identifies and addresses the needs of all customers, providing helpful, courteous and knowledgeable service and advice as needed.	R	R	R	11

APPENDIX B

NATEF INTEGRATED APPLIED ACADEMIC SKILLS CORRELATION GUIDE

Applied Math	Chapter	Applied Science	Chapter	Applied Communication	Chapter
AM-1	54	AS-1	4	AC-1	11
AM-2	54	AS-2	3	AC-2	11
AM-3	52	AS-3	3	AC-3	11
AM-4	53	AS-4	3	AC-4	11
AM-5	53	AS-5	3	AC-5	11
AM-6	52	AS-6	13	AC-6	11
AM-7	53	AS-7	5	AC-7	11
AM-8		AS-8	5	AC-8	11
AM-9	52	AS-9	58	AC-9	11
AM-10	53	AS-10	63	AC-10	11
AM-11	21	AS-11	5	AC-11	11
AM-12	52	AS-12	43	AC-12	11
AM-13	54	AS-13	19	AC-13	11
AM-14	50	AS-14	19	AC-14	11
AM-15	63	AS-15	42	AC-15	11
AM-16	13	AS-16	21	AC-16	11
AM-17	46	AS-17	58	AC-17	11
AM-18	63	AS-18	64	AC-18	11
AM-19	20	AS-19	58	AC-19	11
AM-20	20	AS-20	59	AC-20	11
AM-21	42	AS-21	67	AC-21	11
AM-22	42	AS-22		AC-22	11
AM-23		AS-23	58	AC-23	5
AM-24	42	AS-24	66	AC-24	11
AM-25	42	AS-25	14	AC-25	11
AM-26	42	AS-26	59	AC-26	11
AM-27	36	AS-27	16	AC-27	11
AM-28	36	AS-28	58	AC-28	11
AM-29	14	AS-29	58	AC-29	11
AM-30	14	AS-30	16	AC-30	11
AM-31	20	AS-31	16	AC-31	11
AM-32	60	AS-32	16	AC-32	11
AM-33	59	AS-33	55	AC-33	11
AM-34		AS-34	13	AC-34	11
AM-35	32	AS-35	13	AC-35	11
AM-36	10	AS-36	46	AC-36	11
AM-37	20	AS-37	47		
AM-38	21	AS-38	65		
AM-39	21	AS-39	65		

(continued)

Applied Math	Chapter	Applied Science	Chapter	Applied Communication	Chapter
AM-40	58	AS-40	65		
AM-41	52	AS-41	65		
AM-42	34	AS-42	65		
AM-43	23	AS-43	65		
AM-44	50	AS-44	65		
AM-45	52	AS-45	65		
AM-46	50	AS-46	65		
AM-47	52	AS-47	37		
AM-48	50	AS-48	17		
AM-49	31	AS-49	24, 39		
AM-50	10	AS-50	12		
AM-51	42	AS-51	43		
AM-52	38	AS-52	42		
AM-53	27	AS-53	50		
		AS-54	53		
		AS-55	53		
		AS-56	53		
		AS-57	50		
		AS-58	50		
		AS-59			
		AS-60	50		
		AS-61	50		
		AS-62	50		
		AS-63	50		
		AS-64	54		
		AS-65	61		
		AS-66	56		
		AS-67	56		
		AS-68	57		
		AS-69	51		
		AS-70	50		
		AS-71	50		
		AS-72	54		
		AS-73	53		
		AS-74	53		
		AS-75	50, 54		
		AS-76	61		
		AS-77	52		
		AS-78	51		
		AS-79	51		
		AS-80	60		
		AS-81	51		
		AS-82			
		AS-83	63		
		AS-84	58		
		AS-85	49		

Applied Math	Chapter	Applied Science	Chapter	Applied Communication	Chapter
		AS-86	24		
		AS-87	45		
		AS-88	42		
		AS-89	66		
		AS-90	43		
		AS-91	43		
		AS-92	14		
		AS-93	19		
		AS-94	25		
		AS-95			
		AS-96	45		
		AS-97	12		
		AS-98	29		
		AS-99	12		
		AS-100	60		
		AS-101	16		
		AS-102	22		
		AS-103	14		

APPENDIX C

SAFETY DATA SHEET

Section 1: Product & Company Identification

Product Name: **Brakleen® Brake Parts Cleaner** (aerosol)

Product Number (s): **05089, 05089-6, 05089T, 75089, 85089, 85089AZ**

Product Use: Brake parts cleaner

Manufactured / Supplier Contact Information:

In United States:
CRC Industries, Inc.
885 Louis Drive
Warminster, PA 18974
www.crcindustries.com
1-215-674-4300(General)
(800) 521-3168 (Technical)
(800) 272-4620 (Customer Service)

In Canada:
CRC Canada Co.
2-1246 Lorimar Drive
Mississauga, Ontario L5S 1R2
www.crc-canada.ca
1-905-670-2291

In Mexico:
CRC Industries Mexico
Av. Benito Juárez 4055 G
Colonia Orquídea
San Luís Potosí, SLP CP 78394
www.crc-mexico.com
52-444-824-1666

24-Hr Emergency – CHEMTREC: (800) 424-9300 or (703) 527-3887

Section 2: Hazards Identification

Emergency Overview

DANGER: Vapor Harmful. Contents Under Pressure.
As defined by OSHA's Hazard Communication Standard, this product is hazardous.
Appearance & Odor: Colorless liquid, irritating odor at high concentrations

Potential Health Effects:

ACUTE EFFECTS:

EYE: May cause slight temporary eye irritation. Vapors may irritate the eyes at concentrations of 100 ppm.

SKIN: Short single exposures may cause skin irritation. Prolonged exposure may cause severe skin irritation, even a burn. A single prolonged exposure is not likely to result in the material being absorbed through skin in harmful amounts.

INHALATION: Dizziness may occur at concentrations of 200 ppm. Progressively higher levels may also cause nasal irritation, nausea, incoordination, and drunkenness. Very high levels or prolonged exposure could lead to unconsciousness and death.

INGESTION: Single dose oral toxicity is considered to be extremely low. Swallowing large amounts may cause injury if aspirated into the lungs. This may be rapidly absorbed through the lungs and result in injury to other body systems.

CHRONIC EFFECTS: Repeated contact with skin may cause drying or flaking of skin. Excessive or long term exposure to vapors may increase sensitivity to epinephrine and increase myocardial irritability.

TARGET ORGANS: Central nervous system. Possibly liver and kidney.

Medical Conditions Aggravated by Exposure: None known.

See Section 11 for toxicology and carcinogenicity information on product ingredients.

Product Name: Brakleen® Brake Parts Cleaner (aerosol)
Product Number (s): 05089,05089-6, 05089T, 75089, 85089, 85089AZ

Section 3: Composition/Information and Ingredients

COMPONENT	CAS NUMBER	% by Wt.
Tetrachloroethylene (PERC)	127-18-4	> 95
Carbon Dioxide	124-38-9	< 5

Section 4: First Aid Measures

Eye Contact: Immediately flush with plenty of water for 15 minutes. Call a physician if irritation persists.

Skin Contact: Remove contaminated clothing and wash affected area with soap and water. Call a physician if irritation persists. Wash contaminated clothing prior to re-use.

Inhalation: Remove person to fresh air. Keep person calm. If not breathing, give artificial respiration. If breathing is difficult give oxygen. Call a physician.

Ingestion: Do NOT induce vomiting. Call a physician immediately.

Note to Physicians: Because rapid absorption may occur through lungs if aspirated and cause systemic effects, the decision of whether to induce vomiting or not should be made by a physician. If lavage is performed, suggest endotracheal and/or esophageal control. If burn is present, treat as any thermal burn, after decontamination. Exposure may increase myocardial irritability. Do not administer sympathomimetic drugs unless absolutely necessary. No specific antidote.

Section 5: Fire-Fighting Measures

Flammable Properties: This product is nonflammable in accordance with aerosol flammability definitions. (See 16 CFR 1500.3(c)(6))

Flash Point: None (TCC) Upper Explosive Limit: None
Autoignition Temperature: None Lower Explosive Limit: None

Fire and Explosion Data:

Suitable Extinguishing Media: This material does not burn. Use extinguishing agent suitable for surrounding fire.

Products of Combustion: Hydrogen chloride, trace amounts of phosgene and chlorine

Explosion Hazards: Aerosol containers, when exposed to heat from fire, may build pressure and explode.

Protection of Fire-Fighters: Firefighters should wear self-contained, NIOSH-approved breathing apparatus for protection against suffocation and possible toxic decomposition products. Proper eye and skin protection should be provided. Use water spray to keep fire-exposed containers cool and to knock down vapors which may result from product decomposition.

Section 6: Accidental Release Measures

Personal Precautions: Use personal protection recommended in Section 8. Do not breathe vapors.

Environmental Precautions: Take precautions to prevent contamination of ground and surface waters. Do not flush into sewers or storm drains.

Methods for Containment & Clean-up: Dike area to contain spill. Ventilate the area with fresh air. If in confined space or limited air circulation area, clean-up workers should wear appropriate

Product Name: Brakleen® Brake Parts Cleaner (aerosol)
Product Number (s): 05089,05089-6, 05089T, 75089, 85089, 85089AZ

respiratory protection. Recover or absorb spilled material using an absorbent designed for chemical spills. Place used absorbents into proper waste containers.

Section 7: Handling and Storage

Handling Procedures: Vapors of this product are heavier than air and will collect in low areas. Make sure ventilation removes vapors from low areas. Do not eat, drink or smoke while using this product. Use caution around energized equipment. The metal container will conduct electricity if it contacts a live source. This may result in injury to the user from electrical shock and/or flash fire. For product use instructions, please see the product label.

Storage Procedures: Store in a cool dry area out of direct sunlight. Aerosol cans must be maintained below 120 F to prevent cans from rupturing.

Aerosol Storage Level: I

Section 8: Exposure Controls/Personal Protection

Exposure Guidelines:

	OSHA		ACGIH		OTHER		
COMPONENT	TWA	STEL	TWA	STEL	TWA	SOURCE	UNIT
Tetrachloroethylene	100	N.E.	25	100	N.E.		ppm
Carbon dioxide	5000	30000 v	5000	30,000	N.E.		ppm
N.E. – Not Established (c) – ceiling (s) – skin (v) – vacated							

Controls and Protection:

Engineering Controls: Area should have ventilation to provide fresh air. Local exhaust ventilation is generally preferred because it can control the emissions of the contaminant at the source, preventing dispersion into the general work area. Use mechanical means if necessary to maintain vapor levels below the exposure guidelines. If working in a confined space, follow applicable OSHA regulations.

Respiratory Protection: None required for normal work where adequate ventilation is provided. If engineering controls are not feasible or if exposure exceeds the applicable exposure limits, use a NIOSH-approved cartridge respirator with organic vapor cartridge. Air monitoring is needed to determine actual employee exposure levels. Use a self-contained breathing apparatus in confined spaces and for emergencies.

Eye/face Protection: For normal conditions, wear safety glasses. Where there is reasonable probability of liquid contact, wear splash-proof goggles.

Skin Protection: Use protective gloves such as PVA, Teflon, or Viton. Also, use full protective clothing if there is prolonged or repeated contact of liquid with skin.

Section 9: Pysical and Chemical Properties

Physical State: liquid
Color: colorless
Odor: irritating odor
Odor Threshold: 50 ppm
Specific Gravity: 1.619

Product Name: Brakleen® Brake Parts Cleaner (aerosol)
Product Number (s): 05089,05089-6, 05089T, 75089, 85089, 85089AZ

Initial Boiling Point: 250 F
Freezing Point: ND
Vapor Pressure: 13 mmHg @ 68 F
Vapor Density: 5.76 (air = 1)
Evaporation Rate: very fast
Solubility: 0.015 g/ 100 g @ 77 F in water
Coefficient of water/oil distribution (log P_{ow}): 2.88
pH: NA
Volatile Organic Compounds: wt %: 0 g/L: 0 lbs./gal: 0

Section 10: Stability and Reactivity

Stability: Stable

Conditions to Avoid: Avoid direct sunlight or ultraviolet sources. Avoid open flames, welding arcs, and other high temperature sources which induce thermal decomposition.

Incompatible Materials: Avoid contact with metals such as: aluminum powders, magnesium powders, potassium, sodium, and zinc powder. Avoid unintended contact with amines. Avoid contact with strong bases and strong oxidizers.

Hazardous Decomposition Products: Hydrogen chloride, trace amounts of chlorine and phosgene

Possibility of Hazardous Reactions: No

Section 11: Toxicological Information

Long-term toxicological studies have not been conducted for this product. The following information is available for components of this product.

Acute Toxicity:

Component	Oral LD50 (rat)	Dermal LD50 (rabbit)	Inhalation LC50 (rat)
Tetrachloroethylene	2629 mg/kg	> 10 g/kg	5200 mg/kg/4H
Carbon dioxide	No data	No data	470,000 ppm/30M

Chronic Toxicity:

Component	OSHA Carcinogen	IARC Carcinogen	NTP Carcinogen	Irritant	Sensitizer
Tetrachloroethylene	No	Group 2A	Reasonably Anticipated to be a Carcinogen	E (mild) / S (severe)	No
Carbon dioxide	No	No	No	None	No

E – Eye S – Skin R - Respiratory

Reproductive Toxicity: No information available
Teratogenicity: No information available
Mutagenicity: Tetrachloroethylene: in vitro studies were negative
animal studies were negative
Synergistic Effects: No information available

Section 12: Ecological Information

Ecological studies have not been conducted for this product. The following information is available for components of this product.

Ecotoxicity: Tetrachloroethylene -- 96 Hr LC50 Rainbow Trout: 5.28 mg/L (static)
96 Hr LC50 Fathead minnow: 13.4 mg/L (flow-through)

Persistence / Degradability: Biodegradation under aerobic conditions is below detectable limits. Biodegradation may occur under anaerobic conditions. Biodegradation rate may increase in soil and/or water with acclimation.

Bioaccumulation / Accumulation: Bioconcentration potential is low (BCF less than 100).

Mobility in Environment: Potential for mobility in soil is medium.

Section 13: Disposal considerations

Waste Classification: The dispensed liquid product is a RCRA hazardous waste for toxicity with the following potential waste codes: U210, F001, F002, D039. Pressurized containers are a D003 reactive waste. (See 40 CFR Part 261.20 – 261.33)
Empty aerosol containers may be recycled. Any liquid product should be managed as a hazardous waste.

All disposal activities must comply with federal, state, provincial and local regulations. Local regulations may be more stringent than state, provincial or national requirements.

Section 14: Transport Information

US DOT (ground): Consumer Commodity, ORM-D

ICAO/IATA (air): Consumer Commodity, ID8000, 9

IMO/IMDG (water): Aerosols, UN1950, 2.2, Limited Quantity

Special Provisions: None

Section 15: Regulatory Information

U.S. Federal Regulations:

Toxic Substances Control Act (TSCA):
All ingredients are either listed on the TSCA inventory or are exempt.

Comprehensive Environmental Response, Compensation and Liability Act (CERCLA):
Reportable Quantities (RQ's) exist for the following ingredients: Tetrachloroethylene (100 lbs)

Spills or releases resulting in the loss of any ingredient at or above its RQ require immediate notification to the National Response Center (800-424-8802) and to your Local Emergency Planning Committee.

Superfund Amendments Reauthorization Act (SARA) Title III:
Section 302 Extremely Hazardous Substances (EHS): None

Section 311/312 Hazard Categories:

Fire Hazard	No
Reactive Hazard	No
Release of Pressure	Yes
Acute Health Hazard	Yes
Chronic Health Hazard	Yes

Section 313 Toxic Chemicals: This product contains the following substances subject to the reporting requirements of Section 313 of Title III of the Superfund Amendments and Reauthorization Act of 1986 and 40 CFR Part 372:

Brakleen® Brake Parts Cleaner (aerosol)

Tetrachloroethylene (97.7%)

Clean Air Act:
Section 112 Hazardous Air Pollutants (HAPs): | Tetrachloroethylene

U.S. State Regulations:

California Safe Drinking Water and Toxic Enforcement Act (Prop 65):
This product may contain the following chemicals known to the state of California to cause cancer, birth defects or other reproductive harm: Tetrachloroethylene

Consumer Products VOC Regulations: This product cannot be sold for use in California and New Jersey. In other states with Consumer Products VOC regulations, this product is compliant as a Brake Cleaner.

State Right to Know:

New Jersey:	127-18-4, 124-38-9
Pennsylvania:	127-18-4, 124-38-9
Massachusetts:	127-18-4, 124-38-9
Rhode Island :	127-18-4, 124-38-9

Canadian Regulations:

Canadian DSL Inventory: All ingredients are either listed on the DSL Inventory or are exempt.

WHMIS Hazard Class: A, D1B, D2A, D2B

European Union Regulations:

RoHS Compliance: This product is compliant with Directive 2002/95/EC of the European Parliament and of the Council of 27 January 2003. This product does not contain any of the restricted substances as listed in Article 4(1) of the RoHS Directive.

Additional Regulatory Information: None

Section 16: Other Information

HMIS® (II)	
Health:	2
Flammability:	0
Reactivity:	0
PPE:	B

Ratings range from 0 (no hazard) to 4 (severe hazard)

NFPA

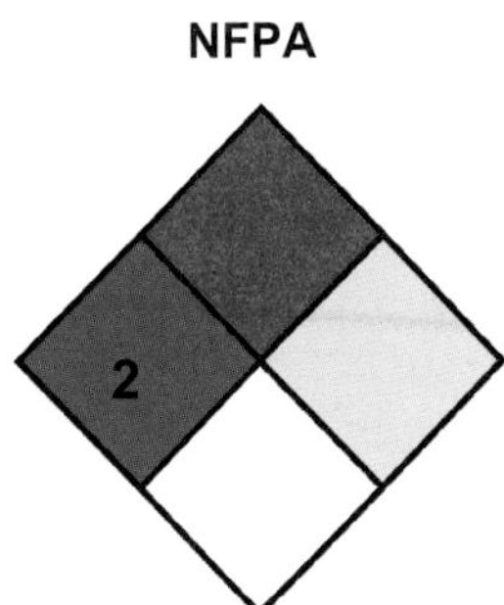

Prepared By: Michelle Rudnick
CRC #: 491G
Revision Date: 01/25/2010

Changes since last revision: MSDS reformatted to meet the requirements of the Canadian Controlled Products Regulations.

Product Name: Brakleen® Brake Parts Cleaner (aerosol)
Product Number (s): 05089,05089-6, 05089T, 75089, 85089, 85089AZ

The information contained in this document applies to this specific material as supplied. It may not be valid for this material if it is used in combination with any other materials. This information is accurate to the best of CRC Industries' knowledge or obtained from sources believed by CRC to be accurate. Before using any product, read all warnings and directions on the label. For further clarification of any information contained on this SDS consult your supervisor, a health & safety professional, or CRC Industries.

ACGIH: American Conference of Governmental Industrial Hygienists
CAS: Chemical Abstract Service
CFR: Code of Federal Regulations
DOT: Department of Transportation
DSL: Domestic Substance List
g/L: grams per Liter
HMIS: Hazardous Materials Identification System
IARC: International Agency for Research on Cancer
IATA: International Air Transport Association
ICAO: International Civil Aviation Organization
IMDG: International Maritime Dangerous Goods
IMO: International Maritime Organization
lbs./gal: pounds per gallon
LC: Lethal Concentration
LD: Lethal Dose
NA: Not Applicable
ND: Not Determined
NIOSH: National Institute of Occupational Safety & Health
NFPA: National Fire Protection Association
NTP: National Toxicology Program
OSHA: Occupational Safety and Health Administration
PMCC: Pensky-Martens Closed Cup
PPE: Personal Protection Equipment
ppm: Parts per Million
RoHS: Restriction of Hazardous Substances
STEL: Short Term Exposure Limit
TCC: Tag Closed Cup
TWA: Time Weighted Average
WHMIS: Workplace Hazardous Materials Information System

Canadian Regulations

Canadian DSL Inventory: All ingredients are either listed on the DSL Inventory or are exempt.

GLOSSARY

abrasive discs (cutting wheels) Abrasive wheels or flat discs fitted to bench, pedestal, and portable grinders.

ABS code retrieval key A specially shaped metal key that is inserted into the data link connector to retrieve DTCs.

ABS pressure tester A high-pressure gauge designed to connect to the HCU and used to measure high hydraulic pressures in the system.

ABS proportioning valve depressor A tool used to hold the proportioning valve open on some ABS HCUs.

absorbed glass mat A type of lead-acid battery.

acceleration An increase in a vehicle's speed.

accelerator pedal The foot-operated pedal used by the driver to increase and decrease the amount of power the engine develops.

accelerator pump A small pump usually located inside the carburetor that sprays an extra amount of fuel into the carburetor air horn during acceleration.

accelerator pump circuit The carburetor circuit involved with heavy acceleration, directly connected to the accelerator pump. Also used to prime the engine during cold starting.

accessory drive pulley A round circular metal disc attached to the front of the engine crankshaft with a rubber/fiber belt that helps operate automotive accessories.

accumulator A device placed between the evaporator and the compressor to collect liquid refrigerant and prevent it from entering the compressor.

Ackermann angle The angle the steering arms make with the steering axis, projected toward the center of the rear axle.

Ackermann principle The geometric alignment of linkages in a vehicle's steering such that the wheels on the inside of a turn are able to move in a smaller circle radius than the wheels on the outside.

activated charcoal Charcoal that will absorb large amounts of vapor, used in charcoal canisters to store and release fuel vapors.

active control A system of providing constant feedback from sensors in the vehicle to the control unit.

actuating The act of making something move or work.

actuator A device that is electrically or vacuum controlled and is used to physically move doors within the heater box to control airflow.

adaptive air suspension A suspension system that uses rubber bags or bladders filled with air to support the weight of the vehicle.

adaptive learning Monitoring the data over time also allows the base fuel settings to be updated in the PCM memory as components age.

adjustable bushing A brace or nylon part that pushes against the rack to adjust the mesh of the rack teeth to the pinion teeth.

adjusting nut The nut used to adjust the end play or preload of a wheel bearing.

adjustment sleeve A component that connects the tie rods together and to the center link on some applications, providing the adjustment point for toe-in or toe-out, depending on the manufacturer's specifications.

advance mechanism A device used to trigger an earlier spark based on engine conditions.

aerate The tendency to create air bubbles in a fluid.

aftermarket That segment of the trade that supplies parts, services, and repair for vehicles outside of the original equipment manufacturer (OEM) or the dealer network.

air checking The use of compressed air to check clutch and servo operation on transmissions during and after assembly.

air diverter valve A valve that changes secondary airflow from the exhaust stream to the air cleaner or vice versa.

air drier A device fitted to compressed air lines to remove moisture.

air drill A compressed air–powered drill.

air gap The space or clearance between two components, such as the space between the tone wheel and the pickup coil in a wheel speed sensor.

air hammer A tool powered by compressed air with various hammer, cutting, punching, or chisel attachments; also called an air chisel.

air impact wrench An impact tool powered by compressed air designed to undo tight fasteners.

air nozzle A compressed air device that emits a fine stream of compressed air for drying or cleaning parts.

air pump A device used to pump air to the exhaust stream to ensure the converter has enough oxygen to perform gas conversions.

air ratchet A ratchet tool for use with sockets, powered by compressed air.

air spring A part that provides the springing action or auxiliary spring. It is typically used in air suspension systems or heavy truck applications.

air supply system Equipment in a motor vehicle that delivers air to the engine.

air switching valve The valve used to switch secondary air from near the exhaust valve (upstream) to the catalytic converter (downstream) for the purpose of oxidizing hydrocarbons and carbon monoxide.

air-conditioning compressor clutch An engagement device connected to the compressor crankshaft to engage the crankshaft with a belt-driven pulley.

air-conditioning machine A machine designed to recover, recycle, evacuate, leak test, and recharge (R/R/R) the air-conditioning system.

air-conditioning pressure sensor A sensor that gives an input signal of refrigerant pressure in the air-conditioning system to the ECU.

air-conditioning system pressure The refrigerant pressure contained within the air-conditioning system.

air-operated braking system A braking system that uses compressed air operating on large diameter diaphragms to provide force to the braking assembly; also called air brakes.

aldehyde A pollutant/irritant in diesel fuel.

all-wheel drive (AWD) A drivetrain arrangement in which all of the wheels drive the vehicle.

Allen wrench A type of hexagonal drive mechanism for fasteners.

alternating current (AC) A type of current flow that flows back and forth.

alternative fuel A nonpetroleum-based motor fuel.

aluminum rod A lightweight, strong, either cast or forged metal shaft that connects the piston assembly to the crankshaft assembly.

ambient the air surrounding the vehicle; atmospheric air.

ambient air temperature sensor A thermistor that is used to measure the air temperature outside the vehicle.

American wire gauge (AWG) A standard used to identify different wire sizes.

ammeter A device used to measure current flow.

amp An abbreviation for amperes, the unit for current measurement.

anchor pin A component of the backing plate that takes all of the braking force from the brake shoes.

anemometer A device that measures airflow in feet per minute (fpm).

angle grinder A portable grinder for grinding or cutting metal.

anisotropic An object that has unequal physical properties, along its various axes. Used in head gaskets to pull heat laterally from the edge surrounding the combustion chamber to the water jacket.

anode A positively charged electrode or plate.

antibinding The release of something when it gets stuck.

antifoaming agents Oil additives that keep oil from foaming as it moves through the engine.

antifriction bearing Wheel bearing assemblies that use surfaces that are in rolling contact with each other to greatly reduce friction, compared to surfaces in sliding contact.

antilock brake system (ABS) A safety measure for the braking system that uses a computer to monitor the speed of each wheel and control the hydraulic pressure to each wheel to prevent wheel lockup.

applied force Pressure placed on something.

arc joint pliers Pliers with parallel slip jaws that can increase in size. Also called Channellocks.

armature The rotating wire coils in motors and generators. It is also the moving part of a solenoid or relay, and the pole piece in a permanent magnet generator.

aspect ratio The ratio of sidewall height to section width of a tire.

aspirator A tube that is used to direct airflow across the cabin air temperature sensor.

asymmetric tread pattern A tread pattern that differs on each side and therefore is usually directional.

Atkinson cycle An engine cycle that uses a longer effective exhaust stroke than intake stroke to reduce exhaust emissions. This type of engine is widely used in hybrid-electric vehicles.

atmospheric pressure The pressure of the air surrounding everything caused by gravity and the weight of air. The higher from sea level, the lower the atmospheric pressure.

attenuator A device that reduces the power of a signal without distorting its waveform.

automatic brake self-adjuster A system on drum brakes that automatically adjusts the brakes to maintain a specified amount of running clearance between the shoes and drum.

automatic climate control system A system that automatically adjusts the heating or cooling to meet a specified temperature demanded by the passengers.

automatic fuel limiter A special refueling port and an overfill protection device fitted on LPG tanks, which ensure that the tank cannot be filled past the safe fill limit of 80%.

automatic load-adjustable shock absorber Typically, an air shock absorber used in an automatic load-sensing system that adjusts ride height (ground clearance) automatically, such as when additional weight is added to the vehicle; also called self-leveling.

automatic oiler A device fitted to compressed air systems to oil air tools.

Automotive Service Excellence (ASE) An independent, nonprofit organization dedicated to the improvement of vehicle repair through the testing and certification of automotive professionals.

Automotive Youth Educational Systems (AYES) An independent, nonprofit organization that is a partnership between automotive manufacturers, their dealerships, and affiliated secondary automotive programs.

available voltage The maximum amount of voltage that the induction coil secondary is capable of putting out.

available voltage test Measurement of voltage at various points in a circuit with the black meter lead on ground and the red lead probing the circuit.

aviation snips A scissor-like tool for cutting sheet metal.

axial load The load applied in line with a shaft. It can be controlled with thrust bearings.

axial piston compressor A design of compressor that uses an angled plate (swash plate) to create the piston movement.

axle A shaft connected to wheels that transmits the driving torque to the wheels.

B100 fuel A fuel that is 100% biodiesel.

B20 fuel A fuel that is 20% biodiesel blended with 80% petrodiesel.

back clearance The area or space behind the piston rings when the rings are in the piston ring grooves.

backing plate A metal plate to which the brake lining is fixed.

backlash The amount of movement between the pinion teeth and the ring gear teeth.

baffles Metal plates that are welded into the oil sump. These plates help to keep the oil at the oil pickup screen when the vehicle is cornering, braking, or accelerating hard. Oil will move during these conditions, and the baffles help to keep it from moving away from the pickup.

balance shaft A metal shaft attached and located inside the engine cylinder block assembly to counteract crankshaft vibrations.

ball bearings The rolling components of a wheel bearing consisting of hardened balls that roll in matching grooves in the inner and outer races.

ball hone An assembly of metal rods, with abrasive stones attached, which is inserted into the cylinder and spun to break the glazing off the cylinder walls and make a new crosshatched pattern.

ball joint A swivel connection mounted in the outer end of the front control arm. These swivels are typically constructed with a ball and socket to allow pivoting.

ball-peen (engineer's) hammer A hammer that has a head that is rounded on one end and flat on the other; designed to work with metal items.

ball-return guide A special passage or metal tube through which the balls move in recirculating ball steering boxes.

ballast resistor Used to limit the amount of current flowing in the ignition primary circuit.

ballast A device that increases lighting voltage substantially and controls the current to the bulb.

band A metal band with friction material bonded to one side. The band is contracted around a drum to stop the drum from spinning.

band brake A braking system that uses a metal band lined with friction material to clamp around the outside of a wheel or drum.

bar A metric unit of measure for pressure.

barometric pressure (BARO) sensor A sensor that measures atmospheric pressure.

barrier cream A cream that looks and feels like a moisturizing cream but has a specific formula to provide extra protection from chemicals and oils.

barrier-type hose A rubber hose made with a nylon bladder inside to contain substances with small molecular structures, such as R-134a.

base circle The rounded bottom part of the camshaft (off the lobe) where the valves remain closed or at rest.

Battery A device that converts and stores electrical energy through chemical reactions.

battery charger A device that charges a battery, reversing the discharge process.

battery electric vehicle (BEV) A vehicle powered by battery only.

battery terminal configuration The placement of positive and negative battery terminals.

baulk-ring synchromesh unit A synchronizer used in all current transmissions. It uses a blocking ring. Also called a blocker ring synchromesh unit.

bead seat The part of the wheel that the tire seals against.

beam axle A suspension system in which one set of wheels is connected laterally by a single beam or shaft.

beam-type axle A rear-wheel drive axle assembly that has a solid tube incorporating the differential gears

bearing cage The component in a wheel bearing that maintains the proper spacing between the roller bearings or ball bearings.

bearing crush The force created to seat the bearing by the extra bearing material when the ends of the bearing inserts touch each other and are forced against each other.

bearing inserts Components made out of soft metal materials that are replaceable and come in pairs; also called half-shell bearings.

bearing packer A tool that forces grease into the spaces between the bearing rollers.

bearing races Hardened metal surfaces that roller or ball bearings fit into when a bearing is properly assembled.

bedding-in The process of a valve wearing into the valve seat and creating a positive seal around the whole diameter.

belt alternator starter (BAS) A type of hybrid drive system that uses a belt-driven alternator/starter that operates on 42 volts.

belt routing label A label that lists a diagram of the routing of the belt(s) for the engine accessories.

bench grinder (pedestal grinder) A grinder that is fixed to a bench or pedestal.

bench vice A device that securely holds material in jaws while it is being worked on.

bendable tangs Small tabs on the brake pad backing plate that are crimped on to the caliper, creating a secure fit and reducing noise.

bi-fuel Two fuels used in a vehicle one at a time, such as compressed natural gas or gasoline.

bias-ply A tire constructed in a latticed, crisscrossing structure, with alternate plies crossing over each other and laid with the cord angles in opposite directions.

bidirectional control The ability to command different solenoids and actuators on and off to check their operation.

bidirectional scanners Scanners used to cause various components and systems to operate for test purposes.

billet The roughly shaped steel piece that has been cast or forged but has not undergone its final machining.

bimetal Aluminum, tin, and silicon alloy metals placed together with steel to form one piece of material and used for bearing materials.

bio-ethanol A renewable fuel derived from a variety of cellulose-based plants.

biodiesel Renewable fuel made from organic feedstocks.

biomass Organic matter used for fuel.

bleeder screw A screw that allows air and brake fluid to be bled out of a hydraulic brake system when it is loosened and seals the brake fluid in when it is tightened.

bleeding The process of removing air from a hydraulic braking system.

blind rivet A rivet that can be installed from its insertion side.

blink codes Codes used to communicate DTCs; they are given by the EBCM as a series of blinks illuminated by the ABS warning lamp.

block deck The "top" of the engine block and cylinder bore where the cylinder head is bolted on.

blocking ring A synchronizer part that increases or decreases a gear's speed to match shaft speed so that the synchronizer sleeve can lock the gear to the shaft. Also called a baulk ring.

blow-off valve A valve that allows the release of excessive boost pressure from the turbocharger when the throttle plate is quickly closed.

blowby Pressure that leaks past the compression rings during compression and combustion.

blowby gas The result of combustion gases leaking past the compression rings and getting into the crankcase.

blower motor An electric motor, usually the permanent magnet type, that moves air over the air-conditioning evaporator and heater core.

bob weights Weights used during the balancing process that are used to mimic the exact weight of the piston and connecting rod weights for that particular rod bearing journal.

body control module (BCM) An onboard computer that controls many vehicle functions, including the vehicle interior and exterior lighting, horn, door locks, power seats, and windows.

body electronic module (BEM) An electronic control module for body electrical systems.

boiling point The temperature at which a substance begins to change from a liquid to a gas.

Bolt A type of threaded fastener with a thread on one end and a hexagonal head on the other.

bolt cutters Strong cutters available in different sizes, designed to cut through non-hardened bolts and other small-stock material.

bonded linings Brake linings that are essentially glued to the brake pad backing plate; more common on light-duty vehicles.

boost valve A valve located in the HCU that is controlled by the EBCM; it allows brake fluid under high pressure to flow into the HCU hydraulic circuits to apply the brakes when commanded.

bore A tool used to make or enlarge a hole in or through an object.

bored out An object that has been bored.

boring bar A long bar used to position and align a single-point tool, such as an engine cylinder block boring machine, for boring operations.

bottom dead center (BDC) The position of the piston at the end of its stroke, when it is closest to the crankshaft.

bottoming tap A thread-cutting tap designed to cut threads to the bottom of a blind hole.

box-end wrench A wrench or spanner with a closed or ring end to grip bolts and nuts.

brake assist (BA) An enhanced safety system built in to some ABS systems that anticipates a panic stop and applies maximum braking force to slow the vehicle as quickly as possible.

brake booster A vacuum or hydraulically operated device that increases the driver's braking effort.

brake drum A short, wide, hollow cylinder that is capped on one end and bolted to a vehicle's wheel; it has an inner friction surface that the brake shoe is forced against.

brake fade The reduction in stopping power caused by a change in the brake system such as overheating, water, or overheated brake fluid.

brake fluid Hydraulic fluid that transfers forces under pressure through the hydraulic lines to the wheel braking units.

brake hose A flexible section of the brake lines between the body and suspension that allows for steering and suspension movement.

brake lathe A tool used to refinish the drum surface by removing a small amount of metal and returning it to a concentric, nondirectional finish.

brake lines Made of seamless, double-walled steel and able to transmit over 1000 psi (6895 kPa) of hydraulic pressure through the hydraulic brake system.

brake lining thickness gauge A tool used to measure the thickness of the brake lining.

brake pad shims and guides Small pieces of metal that cushion the brake pad and absorb some of the vibration, helping to cut down on unwanted noise.

brake pedal emulator A brake pedal assembly used in electronically controlled braking systems to send the driver's braking intention to the computer; it mimics the feel of a standard brake pedal.

brake shoe A steel shoe and brake lining friction material that apply force to the brake drum during braking.

brake shoe adjustment gauge An adjustable tool used to pre-adjust the brake shoes to the diameter of the brake drum.

brake spoon A tool used to adjust the brake lining-to-drum clearance when the drum is installed on the vehicle.

brake spring pliers A tool used for removing and installing brake return springs.

brake switch The electrical switch that is activated by the brake pedal; it turns on the brake lights and signals the EBCM that the brakes are being applied.

brake technician A technician who specializes in working on vehicle brake systems.

brake wash station A piece of equipment designed to safely clean brake dust from drum and disc brake components.

brake-by-wire system A braking system that uses no mechanical connection between the brake pedal and each brake unit. The system uses electrically actuated motors or a separate hydraulic system to apply brake force.

brakes A system made up of hydraulic and mechanical components designed to slow or stop a vehicle.

braking torque The torque acting to twist the axle housing around its center during braking.

breakdown The loss of electrical insulation properties.

breaker plate The movable plate that the breaker points are mounted on and that pivots as the vacuum advance pulls on it.

breather tube A tube used in the PCV system to allow fresh air into the engine crankcase.

British thermal unit (Btu) A measure of heat energy. It takes 1 Btu to raise the temperature of 1 pound of water 1°F.

broach-style surface A tool used in resurfacing the engine block.

brushless An electric motor without brushes.

brushless DC motor An electric motor that does not have any brushes and is sometimes called an "electronically commutated motor." In this type of motor, an electronic control module replaces the brushes and commutator.

bucket-style lifter A bucket-shaped lifter assembly that sits on top of the valves and is operated directly by the camshaft. It can be hydraulic or mechanical.

bump steer The undesired condition produced when hitting a bump, where the vehicle darts to one side as the steering linkage is pushed or pulled as a result of the travel of the suspension.

burn rate The rate of flame spread for liquid or gaseous fuel.

bus An electrical circuit for distributing electrical signals bidirectionally.

bushing A rubber, nylon, or urethane part that allows for movement while maintaining alignment.

butt connector A crimp or solder joint that creates a permanent connection.

bypass filter An oil filter system that only filters some of the oil.

bypass valve The pressure control valve of the supercharger. When the bypass valve opens, it lets air move around the supercharger compressor section, reducing the boost pressure.

C-clamp A clamp shaped like the letter C; it comes in various sizes and can clamp various items.

cabin air temperature sensor A thermistor that measures air temperature inside the vehicle.

caliper A hydraulic device that uses pressure from the master cylinder to apply the brake pads against the rotor.

caliper dust boot seal driver set A set of drivers used to install metal-backed caliper dust boot seals.

caliper piston pliers A tool used to grip caliper pistons while removing them.

caliper piston retracting tool A tool used to retract caliper pistons on integrated parking brake systems.

cam The egg-shaped lobe machined to a shaft, used to cause opening and closing of the valves of a four-stroke cycle engine.

cam lobes Raised areas or protrusions on an otherwise round shaft.

cam lobe centerline The location of the cam lobe in relation to top dead center of the engine in degrees.

cam lobe ramp The rise of the lobe from the base circle to the top of the lobe, which is where the valve starts to lift, on the side opposite of where it starts to close.

cam lobe separation The number of degrees between the centerline of the intake lobe and the centerline of the exhaust lobe; this with cam duration determines the amount of valve overlap.

cam-in-block engine An engine in which the camshaft is located in the engine block rather than on the cylinder head.

camber The side-to-side vertical tilt of the wheel. It is viewed from the front of the vehicle and measured in degrees. Negative camber is when the top of the tire is closer to the center of the vehicle than the bottom of the tire.

camshaft The part of the engine that activates the valve train by using lobes riding against lifters.

camshaft follower A slider or roller placed in direct contact with the lobes of the OHC camshaft that pushes on the tip of the valve to open it.

camshaft lobe The eccentric "egg-shaped" portion of the camshaft that pushes on the valve lifter or camshaft follower.

camshaft position (CMP) sensor A detection device that signals to the PCM the rotational position of the camshaft.

CANbus circuit A two-wire communication network that transmits status and command signals between control modules in a vehicle.

canted valves A valve arrangement in the cylinder head where the valves are at an angle to the cylinder bore. Canting the valves can make for a straighter path for air to flow into and out of the intake and exhaust ports.

capacitance The ability of a capacitor to store an electrical charge.

capacitor A device that can quickly store a small amount of electrical energy, at which point it is charged.

carbon dioxide (CO_2) A vehicle emission that is considered a primary greenhouse gas, though not toxic and not yet regulated.

carbon monoxide (CO) A vehicle emission produced by partially burned fuel that is colorless, odorless, and highly toxic and that causes asphyxiation when inhaled.

carbon monoxide poisoning Exposure to higher than tolerable levels of carbon monoxide, resulting in headaches, fatigue, or loss of consciousness, eventually resulting in death.

carburetor float bowl The part of the carburetor that holds fuel to be burned in the engine; the bowl is at a constant level of fuel to ensure adequate fuel is present during driving.

carrier The part of the throw-out bearing assembly that holds the bearing.

casing plies A network of cords that give the tire shape and strength; also known as casing cords.

cast rod A connecting rod that is created by the metal casting process.

castellated nut An adjusting nut with slots cut into the top such that it resembles a castle; used with a cotter pin to prevent the nut from loosening.

caster The angle formed through the wheel pivot points when viewed from the side in comparison to a vertical line through the wheel.

casting line A by-product of the metal casting process that forms a line found on parts that are cast in metal.

catalyst A device that changes chemical composition without self-destruction.

catalyst heating mode An operating mode in GDI systems where fuel is injected near the top of the compression stroke, the fuel is ignited, and then an additional amount of fuel is injected during the power stroke to create heat later in the cycle so that the catalytic convert can be warmed up faster.

cathode A negatively charged electrode or plate.

center bolt A fastener or bolt found in the middle of the harmonic balancer or cam sprocket that attaches it to the crankshaft or the camshaft.

center electrode The electrode located in the center of a spark plug. It is the hottest part of the spark plug.

center punch Less sharp than a prick punch, the center punch makes a bigger indentation that centers a drill bit at the point where a hole is required to be drilled.

centerline The imaginary line drawn down the exact center of the vehicle from front to back.

centrifugal advance mechanism An ignition timing device, located above or beneath the distributor base plate, that rotates with the distributor cam and is used to advance the spark. As engine speed rises, the flyweights on the advance mechanism are thrown outward by centrifugal force. Since the distributor cam is able to rotate on the distributor shaft, the weights act against their springs and move the distributor cam forward.

centrifugal force A force pulling outward on a rotating body.

centrifugal switch A switch that is only activated when centrifugal forces are placed on a vehicle.

chamfer A rounded or angled edge located on the blocking ring, with ridges machined into it to help grab the cone surface of the gear.

chamfering The process of cutting into the top of the oil passage hole with a specialized chamfering tool or regular drill bit to make the top of the oil hole more like a 45-degree angle, to keep the edge from digging into the bearing the way a 90-degree edge would.

channel The number of wheel speed sensor circuits and hydraulic circuits the EBCM monitors and controls.

charcoal canister A device used to trap the fuel vapors. The fuel vapors adhere to the charcoal until the engine is started, and engine vacuum can be used to draw (purge) the vapors into the engine so that they can be burned along with the air-fuel mixture.

charge The amount of refrigerant present in the system or the process of installing refrigerant in the system.

charge carrier A mobile particle that has a positive or negative electrical charge.

chassis The main support frame in a vehicle. It includes the running gear, such as suspension, the engine, and the drivetrain.

chassis dynamometer A machine with rollers that allows a vehicle to attain road speed and load while sitting still in the shop.

chassis technician A technician who specializes in working on vehicle suspension and steering systems.

check valve Also known as a shuttle valve, a valve that allows the flow of hydraulic oil in one direction only. It is typically used to allow oil to escape quickly from an application device when shifting gears.

chlorofluorocarbon (CFC) A manufactured compound designed to be used as a refrigerant. It is now illegal due to the high chlorine content.

choke A device that provides a rich air-fuel mixture until the engine warms up, by restricting the flow of air at the entrance to the carburetor, before the venturi.

circuit breaker A device that trips and opens a circuit, preventing excessive current flow in a circuit. It can be reset to allow for reuse.

circuit or schematic diagram A pictorial representation or road map of the wiring and electrical components.

Clean Air Act (CAA) A policy signed into law in 1990 that sets standards for air pollution to eliminate ozone-depleting elements.

cleaning gun A device with a nozzle controlled by a trigger fitted to the outlet of pressure cleaners.

climate control system A system that provides the heating and cooling of air inside the passenger compartment for passenger comfort.

clock spring A special rotary electrical connector located between the steering wheel and the steering column that maintains a constant electrical connection with the wiring system while the vehicle's steering wheel is being turned.

closed loop system A totally self-contained system with no materials entering or exiting.

cloud point The temperature at which wax (paraffin) in diesel fuel starts to solidify.

clutch binding A condition in which the clutch disc is dragging, leading to grinding gears during gear shifts and possibly clutch chatter.

clutch chatter A condition in which the clutch shudders when the clutch pedal is released and the vehicle starts to move forward.

clutch disc The center component of the clutch assembly, with friction material riveted on each side. Also called a clutch plate or friction disc.

clutch electrical winding An electromagnetic control device that creates a magnetic field that pulls in the clutch to make contact with the pulley.

clutch fork The part of the clutch linkage that operates the throw-out bearing.

clutch pedal The foot-operated pedal used by the driver to engage and disengage the clutch.

clutch safety switch An electrical switch that is operated by the clutch pedal and keeps the starter motor from cranking the engine over until the clutch is fully depressed.

clutch system A mechanically operated assembly that connects and disconnects the engine from the transmission.

coarse (UNC) Used to describe thread pitch; stands for Unified National Coarse.

code Another name for a fault code; see *DTC*.

coefficient of friction A value assigned to materials to describe the amount of friction when two objects slide against each other.

coil bind A result of excessive valve lift. When the coils of the spring touch each other, excessive wear on the cam lobe and bending the pushrod.

coil spring Spring steel wire, heated and wound into a coil, that is used to support the weight of a vehicle.

coil spring pressure plate A type of pressure plate that uses coil springs to provide the clamping force.

coil-on-plug A type of ignition system used on late-model vehicles that uses one coil placed above each spark plug.

cold chisel The most common type of chisel, used to cut cold metals. The cutting end is tempered and hardened so that it is harder than the metals that need to be cut.

cold cranking amps (CCAs) A standard for rating the ability of a vehicle battery to supply high current under cold operating conditions.

collet A cone-shaped gripping adapter that encloses and grips a rod or shaft when inserted into the sleeve of a lathe or other machine.

column inertia The principle that as a column of air flows, it creates inertia, which keeps air flowing until its inertia energy is spent; sometimes referred to as a "ram effect" when using tuned intake or exhaust systems.

combination pliers A type of pliers for cutting, gripping, and bending.

combination wrench A type of wrench that has a box-end wrench on one side and an open end on the other.

combustion chamber The area of an engine in which the air-fuel mixture is burnt. It consists of the bottom of the cylinder head and the top of the piston.

combustion pressure The force exerted by the expanding gases during the burning process. It is what causes the internal combustion engine to operate.

common bore When a single cylinder is used for two pistons. A tandem master cylinder would be an example of two pistons in one bore.

commutator A device made on armatures of electric generators and motors to control the direction of current flow in the armature windings.

companion flange A splined flange that transmits power from the driveshaft to the pinion gear.

compensating port Connects the brake fluid reservoir to the master cylinder bore when the piston is fully retracted, allowing for expansion and contraction of the brake fluid.

compliance bushing A rubber bushing with a voided section molded in it that allows component movement under torque application. It is typically used in control arms on front-wheel drive vehicles to minimize torque steer issues.

complicated fracture A fracture in which the bone has penetrated a vital organ.

compound planetary gear set Two or more simple planetary gear sets connected together.

compressed natural gas (CNG) Methane compressed for onboard storage in cylinders.

compressed natural gas vehicle (CNGV) A vehicle that uses methane stored in high-pressure cylinders.

compression ratio (CR) The volume of the cylinder with the piston at bottom dead center as compared to the volume of the cylinder at top dead center, given in a ratio such as 9 : 1 CR.

compression ring A metal ring found inside the grooves on the side of the engine piston.

compression ring groove A square groove found on the outside of the piston skirt in which the piston ring fits.

compression stroke The stroke of the piston during which air and fuel is being compressed into a small area prior to ignition.

compression tester A device used to measure the amount of compression pressure a cylinder can generate.

compression-ignition (CI) engine An internal combustion engine that uses the heat of compression to ignite the compressed air-fuel mixture.

compressor A belt- or electrically driven device designed to increase refrigerant pressure and cause refrigerant to travel through the air-conditioning system.

compressor surge The backup of air against the throttle plate as it is closed. The turbocharger is still spinning, pressurizing air, when the throttle plate is closed. Air will stack up, creating a rapid slowing of the turbocharger compressor wheel. This can damage the compressor wheel.

computer-controlled carburetor A carburetor that changes and sets air-fuel ratio based on commands from a PCM that uses memory and input from sensors.

concentric interrupter rings Two interrupter rings that have the same center.

concentricity A term used to describe a valve seat where the valve seat and valve stem share a common center. In this design, the valve can seal against the seat no matter how it is rotated.

concentricity The roundness of a hole. If the hole is not round, it can also be referred to as out of round.

condensation The changing of a gas into a liquid through cooling.

condenser The air-conditioning component located in the front of the vehicle designed to allow high-pressure refrigerant to change states from gas to liquid.

conduction The process of transferring heat through matter by the movement of heat energy through solids from one particle to another.

conductor A material that allows electricity to flow through it easily. It is made up of atoms with one to three valance ring electrons.

connecting rod A cast or forged metal rod that connects the engine pistons to the engine crankshaft.

connector The plastic housing on the end of a wiring harness that holds the wire terminals in place. It can also refer to a type of wire terminal that connects wires together or to a common point such as a bolt.

conservation of energy A physical law that states that energy cannot be created or destroyed.

constant mesh A term used to describe two or more parts, such as gears, that are in constant contact with each other.

constant-velocity (CV) joint A joint used to transmit torque through wider angles and without the change of velocity that occurs in U-joints.

constant-velocity (CV) joints Joints commonly used in FWD vehicles to allow flexibility of the axle while turning.

contact breaker point ignition system A type of ignition system that uses a mechanical means of turning the primary circuit on and off.

contact breaker points A mechanically operated electrical switch that is fixed to the distributor base plate and opened and closed by the distributor cam with the rotation of the engine. The contacts normally form a self-contained unit, fixed to the base plate by a retaining screw engaged in a slot in the fixed contact.

continuity A conductive path between two points.

continuous monitoring A term that describes OBD II monitors that run continuously throughout the drive cycle.

continuously variable transmission (CVT) A transmission that does not have conventional set gear ratios, but is instead infinitely variable between the transmission's lowest gear ratio and highest gear ratio.

control arm The primary load-bearing element of a vehicle's suspension system, commonly referred to as an A-arm or wishbone. These arms may be used as an upper and lower pivot point for the wheel assembly. They attach to the chassis with rubber bushings that allow up-and-down movement of the tire and wheel assembly.

control module A generic term that identifies an electronic unit that controls one or more electrical systems in the vehicle; also called a control unit.

control unit Any device that controls another object such as a computer.

controlled area network bus (CANbus) A wire data network found in vehicles to connect control modules so they can communicate quickly and easily.

controller area network (CAN) A localized (onboard) vehicle network that enables computers and components to send and receive signals across a shielded twisted pair of wires.

convection The process of transferring heat by the circulatory movement that occurs in a gas or fluid as areas of differing temperatures exchange places due to variations in density and the action of gravity.

conventional oil Oil that is processed from crude oil; about 20% of oil is additives.

conventional theory The theory that electrons flow from positive to negative.

convertible A vehicle that converts from having an enclosed top to having an open top by a roof that can be removed, retracted, or folded away.

Coolant An antifreeze concentrate mixed with water, called engine coolant. Most manufacturers recommend a 50/50 mixture.

coolant control valve A valve that blocks off coolant flow to keep hot water from entering the heater core when less heat is requested by the operator.

coolant label A label that lists the type of coolant installed in the cooling system.

cooler flow test The placement of a specialty measuring device into the transmission cooler line to measure fluid flow to the cooler. Low cooler flow can be a sign of other issues with the pump and lubrication system.

core plug A metal cap for the holes that are used to remove core sand used during the casting process.

cored solder Solder that is in the form of a hollow wire. The center is filled with flux, which is used as a cleaning agent while the solder is being applied to the metal surfaces.

cornering force The force between the tread and the road surface as a vehicle turns.

corona suppression An electronic sniffer used for detecting refrigerant leaks.

corrosion inhibitors Oil additives that keep acid from forming in the oil.

cotter pin A one-use soft metal pin that can be bent into shape and is used to retain bearing adjusting nuts.

counter-electromotive force (CEMF) Voltage created in the field windings as the motor rotates, which opposes battery voltage and limits motor speed.

coupe A two-door vehicle that has seating for two people and may have a small rear seat.

crank angle position The position of the crankshaft, measured in degrees.

crank angle sensor A sensor that measures crank angle position.

crank core The rough, unfinished crankshaft assembly that has just left the foundry or forging area.

crankcase The bottom area of the engine cylinder block where the crankshaft is located.

Cranking Rotating the engine by turning the ignition key to the start position.

cranking amps (CA) A standard similar to CCA, but that measures the battery's function at a higher temperature-32°F (0°C).

crankshaft A part mounted in the lower side of the engine block that has offset journals that rotate to change the up-and-down motion of the pistons into rotary motion at the crankshaft.

crankshaft end play The amount of forward and rearward movement of the crankshaft in the main bearings. Crankshaft end play is controlled by the engine main bearing thrust bearing.

crankshaft journal A metal part of the crankshaft that is positioned offset to the main bearing journal on the crankshaft.

crankshaft position (CKP) sensor A sensor used by the PCM to monitor engine speed. It can be one of three types of sensors—Hall effect, magnetic pickup, or optical.

crescent pump An oil pump that uses a crescent-shaped part to separate the oil pump gears from each other, allowing oil to be moved from one side of the pump to the other.

crimp The bent shape of the oil ring expander that allows it to provide outward force on the oil rings.

cross-arm A description for an arm that is set at right angles or 90 degrees to another component.

cross-cut chisel A type of chisel for metal work that cleans out or cuts key ways.

cross-flow radiator A radiator that uses cooling tubes that run horizontal with tanks on each end. This design allows lower hood profile for better vehicle aerodynamics.

crosshatch A pattern of lines placed at angles to each other, appearing as a series of Xs across a surface.

crude oil Material pulled from the earth, originating from organic compounds broken down over time and formed into petroleum. This material is processed in a refinery to break down into various hydrocarbon substances such as diesel, gasoline, and mineral oil, among others.

crush sleeve A collapsible spacer between the bearings that provides a means of maintaining a preset torque on the pinion nut.

cubic boron nitride (CBN) cutter An engine block resurfacing tool used on cast iron parts.

current clamp A device that clamps around a conductor to measure current flow. It is often used in conjunction with a DMM.

current DTC Condition indicating that the test failed during the last ignition cycle and for two-trip codes, on the previous cycle.

current flow The flow of electrons, typically within a circuit or component.

curved file A type of file that has a curved surface for filing holes.

CV joint An abbreviation for constant velocity, a type of universal joint used on the drive axles or half-shafts of a vehicle Usually refers to front-wheel drive vehicles.

cylinder bore The hole in the engine block that the piston fits into.

cylinder head The part of the engine that is bolted to the engine block and caps off the top of the combustion chamber.

cylinder leakage tester A device that pumps air into the cylinder and measures the percentage of air that is leaking out of the cylinder.

cylindrical roller bearing assembly A type of wheel bearing with races and rollers that are cylindrical in shape and roll between inner and outer races, which are parallel to each other.

data link connector (DLC) An under-dash connector for connecting a diagnostic scanner.

data link connector (DLC) The connector through which the scan tool communicates to the vehicle's computers; it displays the readings from the various sensors and can retrieve trouble codes, freeze-frame data, and system monitor data.

dead axle An axle, used on a rear- or front-wheel drive vehicle that does not drive the vehicle.

dead blow hammer A type of hammer that has a cushioned head to reduce the amount of head bounce.

deceleration The process of decreasing a vehicle's speed.

deck The surface area at the top of the engine block against which the cylinder head seals.

deep dish wheel A wheel with negative offset, which gives the outside of the wheel a deep dish appearance. Deep dish refers to the side of the wheel that is farthest from the drop center.

deflecting force A force that moves an object in a different direction or into a different shape.

degree wheel A disc with 360 one-degree markings near its outer edge; it bolts to the front of the crankshaft and is used to check valve and cam timing.

delay circuit A combination of electrical and electronic components that provides a time delay for switching an electrical circuit.

depletion layer An area of neutral charge in semiconductors.

depth filter A hydraulic filter that has thick filter media to trap dirt and other particles of various sizes as they pass through the filter.

depth micrometer A measuring device that accurately measures the depth of a hole.

desiccant A drying agent used in air-conditioning systems to absorb moisture.

detent mechanism The mechanism that holds or helps hold the shift rail into position to ensure that the gear does not pop out when selected and to let the driver feel when a shift is completed.

detent valve See *kickdown valve.*

detergents Oil additives that help to keep carbon from sticking to engine components.

detonation The condition in which the remaining fuel charge fires or burns too rapidly after the initial combustion of the air-fuel mixture. It is audible through the combustion chamber walls as a knocking noise.

diagnostic trouble code (DTC) A code set by the computer indicating system or component malfunction.

diagonal cutting pliers Cutting pliers for small wire or cable.

dial bore gauge An accurate measuring device for inside bores, usually made with a dial indicator attached to it.

dial indicator An accurate measuring device where measurements are read from a dial and needle.

diaphragm pressure plate A slightly conical, spring steel plate used to provide the clamping force for the clutch assembly.

diaphragm-style EGR valve An EGR valve that is operated by a diaphragm that moves when vacuum is applied to it, opening a passageway for exhaust to flow.

dichlorodifluoromethane (R-12) An inert, colorless gas that can be used as a refrigerant. It is stored in white containers.

die Used to cut external threads on a metal shank or bolt.

die stock A handle for securely holding dies to cut threads.

diesel exhaust fluid (DEF) A mixture of urea and water that is injected into the exhaust system of a late-model diesel-powered vehicle to reduce exhaust oxides of nitrogen emissions.

dieseling A condition in which the engine continues to run after the ignition key is turned off. Also referred to as run-on.

differential gears Gears situated in the final drive assembly that are meshed together and with both axles, allowing the wheels to rotate at different speeds when turning a corner.

differential gear set The arrangement of gears between two axles that allows each axle to spin at its own speed when the vehicle is going around a corner.

diode A two-lead electronic component that allows current flow in one direction only.

dipper A type of splash lubricating system used in small engines. It works like a spoon scooping up oil and throwing it upward onto the crankshaft and other wear surfaces.

direct current (DC) Movement of current that flows in one direction only.

direct drive A condition in which the input shaft and output shaft are locked together and turning at the same rate of speed.

direct ignition system May refer to a waste spark ignition or a coil-on-plug ignition system, in which the coils are directly attached to the spark plugs.

direct TPMS A type of automated tire pressure monitoring system that measures tire pressure and possibly temperature via a sensor installed inside each wheel.

direct-acting telescopic shock absorber A shock absorber designed to reduce spring oscillations.

directional and asymmetric tread pattern A tread pattern that is both directional and asymmetric, which means the tire is designed to rotate in only one direction and has one side that must face outward to ensure that the tire performs as designed under operating conditions.

directional tread pattern A tread pattern designed to pump water out from under the tire; each tire must be placed in a particular spot on the vehicle.

disc brake pads Brake pads that consist of a friction material bonded or riveted to a steel backing plate; designed to wear out over time.

disc brake rotor micrometer A specially designed micrometer used to measure the thickness of a rotor.

disc brakes A type of brake system that forces stationary brake pads against the outside of a rotating brake rotor.

discharge reed valve A flat, spring-loaded valve used as a check valve on the discharge side of a compressor.

dislocation The displacement of a joint from its normal position; it is caused by an external force stretching the ligaments beyond their elastic limit.

dispersants Oil additives that keep contaminants held in suspension in the oil, to be removed by the filter or when the oil is changed.

displacement-on-demand A feature that allows cylinders to be taken off-line when not needed, such as during vehicle cruise.

distributor The part of an ignition system that distributes the spark to the spark plugs in the correct sequence and at the correct time. It includes a distributor cap, rotor, shaft, and usually a switching device.

distributor base plate A round metal plate near the top of the distributor that is attached to a distributor housing; also called a breaker plate.

distributor cap The top portion of a distributor, used to make a connection between the spinning rotor and the high-tension leads.

distributorless ignition system An ignition system that does not include a distributor. It uses signals from the crankshaft position sensor and the camshaft position sensor sent to the PCM to determine when to send a signal to the ignition module.

doping The introduction of impurities to pure semiconductor materials to provide N- and P-type semiconductors.

double Cardan joint A type of joint that uses two Cardan joints housed in a short carrier and that reduces the change in velocity of a single Cardan joint by using the second joint to cancel out the changes in velocity of the first joint.

double crimping Overcrimping or recrimping a fitting to keep it from leaking. It normally results in a bigger leak and is not recommended.

double flare A seal that is made at the end of metal tubing or pipe.

double insulated Tools or appliances that are designed in such a way that no single failure can result in a dangerous voltage coming into contact with the outer casing of the device.

double-clutching A technique used to shift gears when a non-synchronized transmission is used. The driver must push the clutch in multiple times to change gears.

double-row ball bearing assembly A single ball bearing assembly using two rows of ball bearings riding in two channels in the races.

down-flow radiator A radiator that uses cooling tubes that run vertical. This design requires a higher hood profile.

drag The slowing down of the crankshaft when it strikes the oil in the crankcase.

drag link A steel or iron rod that transfers movement of the pitman arm to a relay lever.

drain line A wire included in a harness, with one end grounded to reduce interference or noise being induced into the harness.

drawing-in method A method for replacing wheel studs that uses the lug nut to draw the wheel stud into the hub or flange.

drift punch A type of punch used to start pushing roll pins to prevent them from spreading.

drill chuck A device for securely gripping drill bits in a drill.

drill press A device that incorporates a fixed drill with multiple speeds and an adjustable worktable. It can be free standing or fixed to a bench.

drill vice A tool with jaws that can be attached to a drill press table for holding material that is to be drilled.

drivability technician A technician who diagnoses and identifies mechanical and electrical faults that affect vehicle performance and emissions.

drive axle An axle that provides power to a wheel.

drive axle assembly The components that make up the drive axle, including the axles, final drive assembly, bearings, and axle housing.

drive cycle A series of prescribed automobile operating conditions during which emissions testing is performed.

drive pulley The disc-shaped device which is driven by the accessory drive belt and used to power the compressor.

driveline angularity The relationship of the driveshaft to the component that the driveshaft attaches, measured in degrees of angle.

driveline vibrations Rotational fluctuations caused by out-of-balance, misaligned, worn, or bent driveline components.

driven center plate The friction disc that is held firmly against the flywheel by a pressure plate and that transfers power from the flywheel to the transmission input shaft.

driveshaft The shaft or tube fitted with universal couplings that is connected between the transmission and other drivetrain components to transmit torque and rotation.

drivetrain A term used to identify the engine, transmission/transaxle, differential, axles, and wheels.

drop center A wheel design with part of the center section of the wheel a smaller diameter than the rest. It is used for mounting and demounting the tire.

drum brake micrometer A tool used for measuring the inside diameter of the brake drum.

drum brakes A type of brake system that forces brake shoes against the inside of a brake drum.

drum-style parking brake A mechanically operated drum brake that can be set while the vehicle is not moving, to serve as a parking brake.

dry flanged sleeve A metal cylinder with a flange embedded into the block pressed into an engine cylinder to give it a new wear surface. The flanged sleeve is held in by the cylinder head.

dry sleeve A metal cylinder that is pressed into an engine cylinder block that does not come in direct contact with coolant. It is held in place with an interference fit.

dual overhead cam (DOHC) A design that, in a V-engine, includes four cams; also called a twin cam engine.

dual overhead cam (DOHC) engine An engine design that is like the overhead cam engine but with two camshafts used per cylinder head; one operates the intake valves and the other operates the exhaust valves.

dual-clutch transmission A type of automatically shifting manual transmission in which two separate input shafts are connected to their own clutch. Shifting of the gears alternates between the two input shafts.

dual-drive air-conditioning compressor An air compressor drive used on some hybrid vehicles. The compressor can be driven by the accessory belt or by an electric motor.

dual-fuel-powered system A system that allows an engine to burn two fuels at one time.

dual-shaft transmission An automatic transmission that more closely resembles a manual transmission, as they do not use planetary gear sets.

duo-servo drum brake system A system that uses servo action in both the forward and reverse direction.

duration The amount of time the valve stays open, given in degrees of rotation of the crankshaft.

dwell angle The amount of time that the primary circuit is energized, measured in degrees of distributor rotation.

dwell control section Part of the ignition control circuit that determines when the primary circuit will be switched on and for how long current will flow in the primary winding. The dwell period can be varied according to engine speed, improving coil efficiency.

dynamic imbalance A tire imbalance that causes the wheel assembly to turn inward and outward with each half revolution.

E85 A fuel with 85% ethanol and 15% gasoline blend.

ear protection Protective gear worn when the sound levels exceed 85 decibels, when working around operating machinery for any period of time, or when the equipment you are using produces loud noise.

edge code A two-digit code printed on the edge of a friction lining that describes its coefficient of friction.

EH2 rim The specialized rim design that is used with some run-flat tires.

elasticity The ability to deform and reform into the same shape.

elasticity The amount of stretch or give a material has.

electric braking system A braking system used to provide braking to trailers; the drum brakes are electrically activated in the trailer when the driver applies the brakes on the tow vehicle.

electric machine Another name for a traction motor with regenerative capability.

electric power steering system (EPS) A steering system that uses an electric motor and sensors to provide feedback to the vehicle's computer systems to decrease steering effort.

lectric servo motor Also referred to as an electric actuator, a tor that provides movement to operate the air doors in an air control air temperature and air movement.

electric servo An air-conditioning door actuator controlled by electricity.

electric solenoids An electrically operated valve, which in brake systems is used to control the flow of brake fluid in the hydraulic system.

electrical capacity The ability of a circuit or component to carry electrical loads.

electrical power A measurement of the rate at which electricity is consumed or created.

electrical resistance A material's property that slows down the flow of electrical current.

electrical technician A technician who diagnoses, replaces, maintains, identifies fault with, and repairs electrical wiring and computer-based equipment in vehicles.

electrically assisted steering (EAS) A power-assist system that uses an electric motor to replace the hydraulic pump to decrease steering effort.

electrically powered hydraulic steering (EPHS) A steering system that uses an electric motor to produce hydraulic assist for steering.

electrochemical device A device such as a battery or fuel cell that uses two forms of energy.

electrohydraulic braking (EHB) A hydraulic braking system that uses an electrically driven hydraulic pump to pressurize fluid for use in the master cylinder.

electrolysis The process of pulling metals apart by using electricity or by creating electricity through the use of chemicals and dissimilar metals.

electrolyte A mixture of water and acid that contains free ions that make it electrically conductive.

electromagnet A conductor wound in a coil that produces a magnetic field when current flows through it.

electromagnetic clutch arrangement An arrangement used in hybrid air-conditioning systems to drive the compressor. The air-conditioning clutch is an electromagnetic clutch that works by creating a strong magnetic field that pulls the clutch into mesh with the pulley on the air compressor.

electromagnetic induction The production of an electrical current in a conductor when it moves through a magnetic field or a magnetic field moves past it.

electromotive force An electrical pressure or voltage.

electron theory The theory that electrons, being negatively charged, repel other electrons and are attracted to positively charged objects; thus electrons flow from negative to positive.

electronic brake control (EBC) system A hydraulic brake system that has integrated electronic components for the purpose of closely controlling hydraulic pressure in the brake system.

electronic continuously variable transmission (ECVT) A type of hybrid transmission that often uses two electric motors in combination with an ICE. The two electric motors and the ICE transfer power through a planetary gear set, allowing an infinite amount of gear ratios.

electronic control module (ECM) A computer that receives signals from input sensors, compares that information with preloaded software, and sends an appropriate command signal to output devices; used to manage the antilock brake system (ABS).

electronic control unit (ECU) When referring to the HVAC system, the electronic module that makes the "decisions" of the climate control system settings, based on input sensors and the module programming.

electronic fuel injection (EFI) An injection system in which fuel delivery is controlled electronically, allowing continuous adjustments to the air-fuel ratio.

electronic ignition system An ignition system that uses a nonmechanical (electronic) method of triggering the ignition coil's primary circuit.

electronic ignition system-distributor type An ignition system that uses a distributor but replaces the contact points with an electronic triggering device and control module.

electronic pressure control (EPC) solenoid A solenoid used to regulate line pressure on a computerized transmission.

electronic reverse lockout An electronic solenoid used to prevent accidental engagement into reverse gear while the vehicle is moving forward.

electronic sniffers From p. 29.

electronic stability control (ESC) system A computer-controlled system added to ABS and TCS to assist the driver in maintaining vehicle stability while steering.

electronically continuously variable transmission (e-CVT) A transmission without individual gears or gear ratios.

electronically controlled throttle A system that uses electronic, instead of mechanical, signals to control the throttle. Sometimes called drive-by-wire.

element The replaceable portion of a filter, such as an air filter element or oil filter element.

emission analyzer A service bay or lab device used for detecting/measuring vehicle tailpipe emissions.

emission control system A system of devices that are designed to control or reduce harmful gases released to the atmosphere.

emission standards Cutoff points set by governmental agencies to limit tailpipe emissions.

emissions Tailpipe and volatile organic compound pollutants emitted by the automobile.

enabling criteria The operation conditions required before a monitor is allowed to run.

enameled copper wire Wire that uses a thin layer of enamel as insulating material. The thinness of the insulation allows the wire to be closely wound in a coil, creating a dense magnetic field when current flows through it.

end play Referring to the input or output shaft, fore and aft movement in the transmission.

Energy Policy Act (EPAct92) Regulations enacted in 1992 by the U.S. Congress to promote the use of alternative fuel vehicles and fuels.

energy The ability to do work.

engine balancer An expansive bench unit, complete with computer controls, a lathe and drill press for machining metal for the balancing procedure, and all necessary adapters for most applications of crankshaft assemblies.

engine bearing A specially designed metal piece that supports circular moving parts.

engine block deck The portion of the engine cylinder block on which the head gasket lies and to which the cylinder head is bolted.

engine block dowel pin hole The hole where the engine cylinder block dowel pins are fixed.

engine configuration The way engine cylinders are arranged—for example, V, flat, or in-line.

engine control module (ECM) A computer that controls the ignition and fuel control and emissions control systems on an engine; also called the electronic control unit (ECU) or powertrain control module (PCM).

engine coolant temperature (ECT) sensor A sensor that changes resistance based upon coolant temperature; also known as a thermistor.

engine coolant temperature sensor A thermistor that measures the temperature of the engine coolant.

engine displacement The size of the engine given in cubic inches, cubic centimeters, and liters. It is found by multiplying the piston displacement by the number of cylinders the engine has. Sometimes called "swept volume."

engine hoist A small crane used to lift engines.

engine support bar Used to support the engine during engine mount removal in an FWD vehicle or when removing the transaxle without removing the engine.

engine-driven hydraulic pump A power steering pump driven by a belt or gear off the crankshaft.

engineering and work practice controls Systems and procedures required by OSHA and put in place by employers to protect their employees from hazards.

Environmental Protection Agency (EPA) A government agency concerned with air quality and pollution issues related to the environment.

epitrochoid curve The circular movement around the perimeter of another circle. This is the movement that the rotary engine uses to ensure that the rotor stays in contact with the housing.

ethanol Alcohol-based motor fuel made from starches and sugars.

ethylene glycol A chemical used as antifreeze that provides the lower freezing point of coolant and raises the boiling point. It is a toxic antifreeze.

evaporative emission (EVAP) system A system used to capture vapors or gases from an evaporating liquid.

evaporator The air-conditioning component normally located in the passenger compartment designed to allow low-pressure refrigerant liquid to change states to a gas.

evaporator temperature sensor A thermistor that reads the temperature of the evaporator, used to ensure that the evaporator does not freeze.

event cylinder The cylinder that is on compression and ready for the spark to ignite the air-fuel mixture on a waste spark ignition system.

exhaust brake A brake system that restricts the flow of exhaust gases through the engine by closing a butterfly valve located in the exhaust manifold. Restricting the exhaust flow causes the engine speed to slow down, slowing the vehicle.

exhaust gas recirculation (EGR) system A system that recirculates a portion of burned gases back into the combustion chamber to displace air and fuel and cool combustion temperatures.

exhaust gas recirculation (EGR) valve A valve that allows a controlled amount of exhaust gas into the intake manifold during a certain period of engine operation. Used to lower nitrogen oxide exhaust emissions.

exhaust stroke The stroke of piston during which the exhaust valve is open and the piston is moving from bottom dead center to top dead center, to push exhaust gas out of the cylinder.

exhaust valve The valve through which exhaust gases are forced out of the combustion chamber.

expander The part of an oil control ring that holds the ring against the cylinder wall.

expansion volume An engineered space that allows for growth of the volume of a liquid as it heats and expands.

Extended Mobility Technology (EMT) Tires with thick sidewalls that allow the tire to be driven on even when it has no air pressure.

extension housing A component of the automatic transmission housing that covers the output shaft of the transmission. The extension housing also supports the end of the driveshaft and may hold components such as the vehicle speed sensor, speedometer drive assembly, and governor assembly.

external bleeding The loss of blood from an external wound; blood can be seen escaping.

external combustion engine An engine that runs on heat applied externally to the cylinder, for example, the steam engine.

external driveshaft A shaft used to transfer power from the transmission to the live axle.

externally equalized valve A style of TXV where the external valve uses the sensing bulb on the outlet side to overcome spring pressure to open the valve.

extreme loading Large pressure placed on two bearing surfaces. Extreme loading will try to press oil from between bearing surfaces.

extreme-pressure additive An oil additive that ensures that a protective coating is given to moving engine parts and that keeps oil from being forced out under extreme pressure. Helps oil to cushion components.

eye ring terminal A type of crimp or solder terminal that s an enclosed eyelet to connect the terminal with a bolt or

fast chargers A type of battery charger that charges batteries quickly.

fasteners Devices that securely hold items together, such as screws, cotter pins, rivets, and bolts.

fault See *diagnostic trouble code (DTC)*.

fault codes An alphanumeric code system used to identify potential problems in a vehicle system.

feedback signal A voltage signal sent back to an electronic control unit. The feedback signal is how the control module is able to interpret temperature or other information from its sensors.

feedback system A system that uses feedback to adjust what it is doing. Typically, a module gives a command and waits to see if the command was followed by use of a sensor.

feeler gauge A thin blade device for measuring space between two objects.

fiber optics Fine glass fibers through which light is transmitted.

fill plug Usually a threaded plug that can be removed to allow the level of a fluid to be checked and filled. This could also be a rubber snap fit plug.

fillet The radius portion of an inside corner that reduces stress at the corner.

fillet area The area of the crankshaft that meets up with the rod bearing journal or main bearing journal.

filter media A screen designed to keep debris from the TXV that is normally located in the receiver dryer.

filter sock The first line of defense in the fuel supply system. The sock typically consists of a fine mesh, which prevents most small particles from being drawn into the fuel pump and sent through the rest of the fuel system.

fin A small, flat piece of metal placed between the tubes to help with the transfer of heat from the coolant to the air, refrigerant to air, or air to refrigerant. The metal heats up due to contact with the hot pipes in the case of a radiator or heater core and from hot air in the case of an evaporator.

final drive A component that provides a final gear reduction and allows for the difference in speed of each wheel when cornering.

final drive assembly An assembly used to power the drive wheels and allow the wheels to rotate at different speeds as the vehicle turns.

fine (UNF) Used to describe thread pitch; it stands for Unified National Fine.

finished rivet A rivet after the completion of the riveting process.

fire rings Steel rings integrated into the cylinder head gasket nearest the combustion chambers that provide extra sealing to seal in the high combustion pressures.

firing line The tall lines on a parade pattern that indicate the voltage required to initially jump the spark plug gap.

first aid The immediate care given to an injured or suddenly ill person.

first-degree burns Burns that show reddening of the skin and damage to the outer layer of skin only.

five-gas analyzer A tool that uses sensors to measure the level of gases in the exhaust stream.

fixed caliper A type of brake caliper bolted firmly to the steering knuckle or axle housing, having at least one piston on each side of the rotor.

fixed orifice tube system A system with a fixed orifice tube that uses an accumulator between the evaporator and the compressor.

fixed resistor A resistor that has a fixed value.

fixed-orifice PCV system A system in which a hole of a predetermined size is used as a means of pulling crankcase vapors into the intake manifold to be burned.

fixed-type joint A joint that does not slide to allow for shaft lengthening or shortening; it simply allows for angle changes as the suspension moves.

flame front The front edge of the burning air-fuel mixture in the combustion chamber.

flame propagation The movement of the flame through the combustion chamber during the combustion process.

flare nut wrench A type of box-end wrench that has a slot in the box section to allow the wrench to slip through a tube or pipe. Also called a flare tubing wrench.

flashback arrestor A spring-loaded valve installed on oxy-acetylene torches as a safety device to prevent flame from entering the torch hoses.

flasher can A mechanism that turns the vehicle's turn signal and hazard flasher bulbs on and off.

flat blade screwdriver A type of screwdriver that fits a straight slot in screws.

flat seat with washer A type of lug nut that is flat where it bolts to the wheel and has a washer affixed that allows it to turn independent of the hex part of the lug nut.

flat seat without washer A type of lug nut that is flat where it bolts to the wheel.

flat tappet A camshaft specifically designed to push on flat bottom lifters (nonroller types). It typically has a higher rolling resistance than the roller-style camshafts.

flat-head engine An L-head engine with valves in the block.

flat-nose pliers Pliers that are flat and square at the end of the nose.

flexible-fuel vehicle (FFV) A vehicle that can operate on two different fuels blended together in various mixture amounts by percentage.

flexplate A metal part attached to the rear of the crankshaft that connects to the torque converter in a vehicle with an automatic transmission.

float chamber A chamber that holds a quantity of fuel at atmospheric pressure, ready for use.

float-type fuel gauge A system for determining the level of liquid fuel in a tank using a float and potentiometer.

floating pin A round metal pin that has a very small clearance but is free to float in the piston and connecting rod.

flow-control valve A valve used in power steering pumps to control the amount of flow out of the power steering pump.

fluid coupler A type of hydraulic coupling used on vintage vehicles to connect and transfer power from the engine to the transmission.

fluid dissipater A silicone fluid component that allows the mass of a fluid type harmonic balancer to rotate at a more constant speed than the crankshaft, to help even out the torsional vibrations of the crankshaft.

Flux A liquid or paste that protects a soldering or welding joint from oxidization.

flywheel A heavy metal disc bolted to the crankshaft that is used to smooth out the engine's power pulses and keep the engine moving through the non-power strokes. Also provides the mating surface for the clutch disc and pressure plate.

flywheel ring gear Large, round, externally toothed gear that is usually press fit to the outer diameter of the flywheel and used along with the starter to crank the engine over.

force The effort to produce a push or pull action.

forced induction The pressurization of airflow going into the cylinder through the use of a turbocharger or supercharger.

forces capability map data Data preprogrammed into the electronic control unit's memory by the manufacturer and used to determine how much power assistance is needed based on input from the vehicle's speed sensor and steering sensor.

forcing screw The center screw on a gear, bearing, or pulley puller. Also called a jacking screw.

forged connecting rod A strong metal connecting rod hammered into shape by large presses used to connect the piston to the crankshaft.

four-post hoist A type of hoist that the vehicle is driven onto that uses two long, narrow platforms to lift the vehicle.

four-wheel drive (4WD) A drivetrain layout in which the engine drive has either two wheels or four wheels, depending on which mode is selected by the driver.

four-wheel steering system A steering system in which the front wheels are controlled normally, and the rear wheels use a computer and electric motors to turn the rear linkage.

fracture split A method used on powdered metal connecting rods where the big end of the connecting rod is broken in half and can be identified by a ragged finish of its parting surfaces.

free electron An electron located on the outer ring, called the valence ring, that is only loosely held by the nucleus and that is free to move from one atom to another when an electrical potential (pressure) is applied.

free play The amount of clearance between the brake pedal linkage and the master cylinder piston

freewheeling engine An engine that has enough clearance between the piston and the valves so that in the event the timing belt or chain breaks, the valves that are hanging all the way open will not contact the piston, thus preventing engine damage.

freeze-frame A feature of OBD II that records sensor data when a fault occurs.

frequency The rate of change in direction, oscillation, or cycles in a given time.

friction The resistance created by surfaces in contact. Kinetic friction is resistance to motion when one surface moves over another. Static friction is resistance to motion between two surfaces that are not moving.

friction bearing A bearing that uses sliding motion between components, such as a clutch pilot bushing.

friction facing The material riveted to each side of the clutch disc that mates to the flywheel and pressure plate. Used to provide friction and a wear surface for the clutch assembly.

front bearing retainer The housing that bolts the input shaft bearing in place on the front of the transmission.

front pump A hydraulic pump used to supply lubrication oil and hydraulic pressure to the different components inside the automatic transmission.

front-wheel drive (FWD) A drivetrain layout in which the engine drives the front wheels.

fuel cell An electrochemical device that uses hydrogen and oxygen to create electricity.

fuel cell (electric) vehicle (FCV) A vehicle that converts fuel to electricity in a direct electrochemical process, with more energy extracted from the fuel source than in traditional internal combustion engines.

fuel composition sensor A device that measures the percentage of alcohol in gasoline.

fuel filler neck The upper end of the fuel filler tube leading down to the fuel tank, which accepts the fuel hose nozzle at the gas station pump.

fuel metering system Equipment in a motor vehicle that delivers the proper amount of fuel to each cylinder.

fuel methanol A toxic alcohol fuel.

fuel pressure regulator A system that controls the pressure of fuel entering the injectors.

fuel pump A mechanically or electrically driven vacuum device used to draw fuel from the tank and force it into the fuel system.

fuel pump relay A relay to turn on or off the high-amperage circuit of the fuel pump.

fuel rail Tubing that connects several injectors to the main fuel line.

fuel shutoff mode A safety precaution by which fuel is shut off when certain conditions are met during a vehicle crash.

fuel supply system Equipment in a motor vehicle that delivers fuel to the engine.

' system Equipment in a motor vehicle that delivers fuel to gine.

fuel tank pressure sensor A sensitive pressure sensor mounted in the fuel tank or EVAP system used to monitor the system for leaks.

fulcrum A half-round bearing that the rocker moves on as a bearing surface.

fulcrum ring A steel ring that is used as a pivot point for the diaphragm spring in the pressure plate.

full floating axle An axle that does not support any weight; if removed, the vehicle will still roll on its wheels.

full power mode An operating condition in a GDI fuel system where the fuel is injected on the intake stroke in a slightly rich condition and allowed to fully mix with the air in the cylinder. This creates maximum power and reduces detonation.

full-flow filter An oil filter installed on production cars. This oil filter cleans all oil coming from the oil pump on its way to the lubricated components.

FWD engine hoist Useful when removing the engine and transaxle from below the vehicle when dropping the entire cradle.

galleries Passageways drilled or cast into the engine block or head(s), which carry pressurized lubricating oil to various moving parts in the engine, such as the camshaft bearings.

gallium-arsenide A semiconductor used in high-frequency circuits.

garter spring A coiled spring that is fitted to the inside of the sealing lip of many seals, used to hold the lip in contact with the shaft.

garter spring lip seal A seal that has a spring to maintain pressure on the seal's surface to prevent leakage.

gas direct injection (GDI) cylinder head A cylinder head designed to accept a gasoline injector directly in the combustion chamber instead of in the intake manifold.

gas welding goggles Protective gear designed for gas welding; they provide protection against foreign particles entering the eye and are tinted to reduce the glare of the welding flame.

gasket A rubber, cork, or paper spacer that goes between two parts to seal the gap between the parts.

gasket scraper A broad sharp flat blade to assist in removing gaskets and glue.

gasoline A volatile, flammable liquid mixture of hydrocarbons, obtained from crude oil and used as fuel for internal combustion engines.

gasoline direct injection (GDI) A fuel injection system in which fuel is sprayed directly into the combustion chamber.

gassing The escape of gas from the battery.

gauge sending unit A device used for transmitting a signal to control a fuel gauge.

gear A relatively round, rotating part with internal or external teeth that are designed to mesh with another gear for the purpose of transmitting torque.

gear lube A type of lubricant primarily used to lubricate transmission and differential gears but also used to lubricate some wheel bearings.

gear pullers A tool with two or more legs and a cross bar with a center forcing screw to remove gears.

gear pump A type of front pump used on some vehicles that uses two rotating gears to force fluid out of the pump.

gear ratio The ratio of the number of turns that a drive gear must complete to turn the driven gear one turn. Typically, this is a calculation of the driven gear to the drive gear. Ratios are listed, for example, as 2:1 or 4.3:1.

gear reduction The use of a small gear to drive a large gear. The result is in an increase in torque, but a decrease in speed.

gear set Two or more gears that are in mesh with each other.

gear synchronizer An assembly in the transmission that is used to bring two unequally spinning shafts or gears to the same speed when upshifting or downshifting.

geared oil pump An oil pump that has two gears running side by side together to move oil from one side of the pump gears to the other.

gearshift lever The lever that the driver uses to shift the transmission.

gelling A thickening effect of oil in cold weather. This is not a desirable trait for lubricating oil, as it will not flow when it is gelling. Wax content in base stock mineral oil makes gelling worse.

germanium A type of semiconductor.

girdle A metal device connected to the bottom of the engine cylinder block to create strength for the main bearing caps.

governor A mechanical device that creates a pressure using centrifugal force. The pressure is proportional to vehicle speed.

governor pressure The pressure created by the governor, which is used to make the shift valves upshift and is proportional to vehicle speed.

gradient resistance Resistance encountered when a vehicle travels up a hill, requiring torque to be applied to overcome it.

gravity pouring The casting process used for creating metal parts.

grease seal A component that is designed to keep grease from leaking out and contaminants from leaking in.

grease A lubricating liquid thickened to make it suitable for use with many wheel bearings.

grinding wheels and discs Abrasive wheels or flat discs fitted to bench, pedestal, and portable grinders.

ground The return path for electrical current in a vehicle chassis, other metal of the vehicle, or dedicated wire.

guide pins Pins that allow the caliper to move in and out as the brakes operate and as the brake pads wear.

H block valve A style of TXV that has an outlet side like the external valve; however, both the inlet and the outlet pipe go through the block, so the sensing bulb is built into the block.

half-shaft An axle that has CV joints on each end and that fits between the transaxle and wheel. Typically, one is used on each side of a vehicle.

Hall effect An electrical effect where electrons tend to flow on one side of a special material when exposed to a magnetic field, causing a difference in voltage across the special material. When the magnetic fields is removed, the electrons flow normally, and there is no difference of voltage across the special material. This effect can be used to determine the position or speed of an object.

Hall-effect sensor The portion of an electronic ignition system used to trigger the ignition system. Hall-effect switches operate by using a potential difference, or voltage, created when a current-carrying conductor is exposed to a magnetic field. If a magnetic field is applied at right angles to the direction of current flow in a conductor, the lines of magnetic force permeate the conductor, and the electrons flowing in the conductor are deflected to one side. This deflection creates a potential difference across the conductor. The stronger the magnetic field, the higher the voltage.

halogen lamp A type of bulb that produces a bright white light.

hard rubber mallet A special-purpose tool with a head made of hard rubber; often used for moving things into place where it is important not to damage the item being moved.

hard shifting A shifting problem in which the shifter will not move smoothly into the desired gear, requiring excessive force by the driver to set it into gear.

hardened seat A valve seat that has undergone a heat-treating process to become more robust and capable of withstanding the severe demands of today's vehicles and drivers.

harmonic balancer A round metal disc connected to the front of the crankshaft that smoothes out torsional vibrations created by the crankshaft.

hatchback A vehicle that has a shared passenger and cargo area; it typically is available in three- and five-door arrangements.

hazard Anything that could hurt you or someone else.

hazardous environment A place where hazards exist.

hazardous material Any material that poses an unreasonable risk of damage or injury to persons, property, or the environment if it is not properly controlled during handling, storage, manufacture, processing, packaging, use and disposal, or transportation.

head gasket A thin piece of material, often a multilayered, bimetallic sheet used to seal the cylinder head assembly to the engine block.

header A specially tuned exhaust manifold typically made of exhaust pipes. These pipes are usually made equal length to ensure equal flow between cylinders.

headgear Protective gear that includes items like hairnets, caps, or hard hats.

heat buildup A dangerous condition that occurs when t[illegible] glove can no longer absorb or reflect heat, and heat is tra[illegible]ferred to the inside of the glove.

heat dissipation The spreading of heat over a large area to increase heat transfer.

heat fade Brake fade caused by the buildup of heat in braking surfaces, which get so hot they cannot create any additional heat, leading to a loss of friction.

heat range The rating of a spark plug's operating temp-erature.

heat transfer The flow of heat from a hotter part to a cooler part; it can occur in solids, liquids, or gases.

heated air intake system A system that uses hot air from around the exhaust manifold to warm the air going into the intake manifold.

heated diode A sniffer that uses electricity to determine if there is a refrigerant leak; considered the best sniffer for R-134a.

heater control cables Cables that control the air doors in an air box as part of the air distribution system.

heater core A heat-exchanging device that transfers heat converted by the fuel burning in the engine to the passenger compartment.

heavy line technician A technician who undertakes major engine, transmission, and differential overhaul and repair.

helical-cut gear A type of gear in which the teeth are cut in a spiral down the axis of the gear.

helical-geared limited slip differential A type of differential that responds very quickly to changes in traction and that does not bind from friction in turns or lose its effectiveness since there are no clutches.

helix The curve created by a smooth spiral and used in the angle of gear teeth and coil springs.

Helmholtz resonator A device that uses the principle of noise cancellation through the collision of sound waves. When necessary, the resonator is used in addition to the muffler to cancel additional sounds. This resonator may also be used on the induction system to muffle noise of airflow through the induction system. It is named after physicist Hermann Von Helmholtz.

hemispherical cylinder head A combustion chamber that is hemispherical in shape with the valves in a cross-flow arrangement and the spark plug near the center of the cylinder head directly over the top of the piston.

hertz The unit for electrical frequency measurement.

high resistance The resistance of a component or circuit relative to a low resistance. It can also refer to a faulty circuit where a section or component has excess unwanted resistance.

high-intensity discharge (HID) A type of lighting that produces light with an electric arc rather than a glowing ilament.

h-pressure accumulator A storage container designed to ain high-pressure liquids such as brake fluid.

pressure switch A switch designed to open at a predeter- l high pressure to protect the compressor; it is normally n the high side line near the compressor.

high-tension leads The heavy insulated wires used to connect the distributor cap terminals to the spark plugs, and the ignition coil to the distributor cap; or on waste spark systems, the coils to the spark plugs.

high-tension terminals The terminals on the coils and distributor cap that the high-tension leads are connected to.

high-voltage spark The electrical arc that takes place between the center and the side electrode of a spark plug.

history code A fault code that has occurred but is not current and is saved in the PCM's memory for 40 warm-up cycles.

hold function A setting on a DMM to store the present reading.

hold-down spring tool A tool used for removing and installing hold-down springs.

hold-down springs Springs that hold the brake shoes against the backing plate.

hold-in winding A low-current winding found in starter solenoids that holds the plunger in the activated position.

hole theory The theory that as electrons flow from negative to positive, holes flow from positive to negative.

hollow punch A punch with a center hollow for cutting circles in thin materials such as gaskets.

homogenous mixture An air-fuel mixture evenly dispersed throughout the combustion chamber.

honing The smoothing-out process of the cylinder walls performed after the boring process. This is the final step in refinishing the cylinder; it creates the proper crosshatched pattern necessary for the piston rings to seal and seat into to give the combustion chamber the proper seal.

Hooke's joint A joint that consists of a steel cross with four hardened bearing journals, mounted on needle rollers in hardened caps, which locate the cross in the eyes of the yokes. The cross swivels in the yokes as the drive is transferred across the joint.

horizontally opposed engine An engine with two banks of cylinders, 180 degrees apart, on opposite sides of the crankshaft. It is also called a flat engine or a boxer engine.

horsepower An amount of work performed in a given time.

hoses Flexible lines used to direct liquids or gases.

hot junction The heating point of a thermocouple.

hybrid drive system A drive system that uses two or more propulsion systems such as electric motors and an ICE.

hybrid electric vehicle (HEV) A vehicle that uses two power sources for propulsion, one of which is electricity.

hydraulic actuator A hydraulically controlled cylinder that engages or disengages the clutch pedal.

hydraulic control unit (HCU) An assembly that houses electrically operated solenoid valves used in electronic braking systems; also called a modulator.

hydraulic fade Brake fade caused by boiling brake fluid. Causes a spongy brake pedal.

hydraulic jack A type of vehicle jack that uses oil under pressure to lift vehicles.

hydraulic press method A method for replacing wheel studs that uses a press to force the wheel stud into the flange until it bottoms out.

hydraulic pressure test The use of a hydraulic pressure gauge to measure the amount of hydraulic pressure produced in each gear range.

hydraulic valve lifter A small mechanical cylinder with a hydraulically operated internal piston used to automatically take up the slack (valve clearance) in the valve train.

hydrocarbon (HC) A microscopic unburned fuel particle that contributes to photochemical smog and helps form ground-level ozone.

hydrocracking A process in which group 2 and group 3 oils are refined with hydrogen at much higher temperatures and pressures. This process results in a base mineral oil with many of the higher performance characteristics of synthetic oils.

hydrodynamic seal Oil seal flutes or swirls that are part of the oil seal and that create a pumping action to return oil to the transmission as the shaft rotates, enabling lubrication.

hydrogenated nitrile butadiene rubber (H-NBR) A muffler constructed of a hollow cylinder with baffling.

hydrogenating A process used during refining of crude oil. Hydrogen is added to crude oil to create a chemical reaction to take out impurities such as sulfur.

hydrometer A tool that measures the specific gravity of a liquid.

hygroscopic A property of a substance or liquid that causes it to absorb moisture (water), as a sponge absorbs water. Brake fluid absorbs water out of the air; thus it is hygroscopic.

hygroscopic The ability of a substance to attract and hold water molecules.

hypereutectic piston An aluminum piston that is cast with a 16% to 19% silicon content and that has minimal expansion rates.

hypoid bevel gear A special design of spiral bevel gear, with the centerline of the pinion below the centerline of the ring gear.

hypoid gear A type of helical gear used to change the direction of motion 90 degrees. The axis of the input gear does not line up on the centerline of the output gear.

idle The speed at which an engine runs without any throttle applied.

idle stop A feature that turns off the internal combustion engine when the vehicle is at a standstill.

idler gear A gear used in between two gears to change the direction of the rotation of the driveshaft or drive axles in the transmission.

idling A condition in which a gear is spinning, but not moving.

idling five-gas test A five-gas analysis test conducted during vehicle idling.

ignition The lighting of the fuel and air mixture in the combustion chamber.

ignition advance The means of causing the spark to occur earlier within the compression stroke for better performance and fuel economy during changing engine conditions.

ignition coil A device used to amplify an input voltage into the much higher voltage needed to jump the electrodes of a spark plug.

ignition coil pack A group of two or more ignition coils housed in one assembly.

ignition control module (ICM) A general control unit of some electronic ignition systems, usually with current and dwell angle control; driver and output stage; and in some cases, electronic spark timing functions.

ignition module An electronic component that electronically controls the ignition coil or coils.

ignition switch A switch operated by a key or start/stop button and used to turn a vehicle's electrical and ignition system on or off.

impact driver A tool that is struck with a hammer to provide an impact turning force to remove tight fasteners.

in-line engine An engine in which the cylinders are arranged side by side in a single row.

incandescent lamp The traditional bulb that uses a heated filament to produce light.

included angle The angle of camber added or subtracted to the SAI angle. This is the angle of the steering knuckle pivot points in relation to the camber angle of the wheel; also referred to as a diagnostic angle.

independent rear axle A type of rear suspension system that allows each wheel on the axle to move independently of the other.

independent rear suspension (IRS) A type of suspension system where each rear wheel is capable of moving independently of the other.

independent suspension A system for allowing the up-and-down movement of one tire without affecting the other tire on that axle.

independent suspension drive axle A type of suspension that allows each wheel on a drive axle to move independently of the other.

indirect fuel injection Any fuel injection that is not sprayed directly into the combustion chamber.

indirect TPMS A type of automated tire pressure monitoring system that uses the anti-lock braking system of a vehicle to measure the difference in the rotational speed of the four wheels, to determine tire pressure.

induced voltage The creation of voltage in a conductor by movement of a magnetic field that is near that conductor.

induction coil An electrical transformer that uses magne fields to produce high-voltage pulses from low-voltage di current.

induction hardening A metal hardening process that uses an electromagnetic field to quickly heat up the surface of a metal part so it can be quenched to provide an appropriate amount of hardness.

induction-type system A type of ignition system that uses a magnetic pulse generator to trigger the spark.

inductive current The current that has been created across a conductor by moving it through a magnetic field.

inductive-type sensor A sensor mounted on the crankcase housing that is used to sense the movement of the ring gear teeth on the flywheel, or a toothed disc on the crank pulley.

inert gas A gas that will not react chemically.

inertia The resistance to a change in motion.

infrastructure The system for supplying fuel for transportation use; includes ships, pipelines, trucks, etc.

ingression The passing of foreign bodies, such as moist warm air, through the air-conditioning lines or fittings.

injector A valve that is controlled by a solenoid or spring pressure to inject fuel into the engine.

inlet port Connects the reservoir with the space around the piston and between the piston cups in a brake master cylinder.

inlet reed valve A flat spring-loaded valve that allows gaseous refrigerant to enter the compressor.

inner race The inside component of a wheel bearing that has a smooth, hardened surface for rollers or balls to ride on.

inner tie rod or socket Attached to the end of the rack, the inner tie rod allows for suspension movement and slight changes in steering angles.

input force The force applied to the input piston, measured in either pounds or kilograms.

input shaft speed sensor A sensor inside the transmission that measures the rpm of the input shaft; also called a turbine shaft sensor.

inside micrometer A micrometer designed to measure internal diameters.

instrument panel warning lamps Lamps that illuminate to warn a driver of a fault in a system.

insulator A material that has properties that prevent the easy flow of electricity. These materials are made up of atoms with five to eight electrons in the valance ring.

intake air temperature (IAT) sensor A sensor that measures the temperature of the incoming air through the air filtration system.

intake port The port through which the air or air-fuel mixture travels from the throttle body area to the combustion chamber.

ıtake stroke The stroke of the piston from top dead center to ttom dead center, during which the intake valve is open and s pulled into the cylinder.

e valve The valve through which air and fuel enter the stion chamber.

in ABS system A brake system in which the master cylinder, booster, and HCU are all combined in a common unit.

integral valve guide A valve guide machined into the cylinder head during cylinder head construction.

integrated circuit A semiconductor chip that contains miniature versions of various electrical components within one housing.

integrated motor assist (IMA) A Honda hybrid drive system that uses a moderate-sized electric motor installed between the engine and the transmission.

interference angle The built-in differences in the angles between the valve seat and the valve face for the purpose of providing quick wearing-in of the surfaces; there usually must be between 0.5 and 1 degree of difference.

interference engine An engine that has minimum clearance between the valves and the pistons during normal operation; in the event that the timing belt or chain breaks, the open valves will be contacted by the piston and bend the valves, possibly breaking the piston.

interference fit A condition in which two parts are held together by friction because the outside diameter of the inner component is slightly larger than the inside diameter of the outer component.

interlock mechanism A mechanical device that prevents engagement of two different gears at the same time.

intermediate shaft A steel rod positioned at an angle from the steering column to the steering gear that functions in transferring movement from one to the other.

intermediate tap One of a series of taps designed to cut an internal thread. Also called a plug tap.

internal bleeding The loss of blood into the body cavity from a wound; there is no obvious sign of blood.

internal combustion engine An engine that burns a fuel internally and creates movement due to thermal expansion of gases.

internal module control circuit A section of the ignition module responsible for receiving the trigger signal.

internally equalized valve A style of TXV where the internally equalized valve uses pressure from the inlet pipe to overcome the pressure created from the outlet pipe heat. Lower heat on the outlet pipe creates lower pressure, and the pressure from the inlet pipe is able to overcome it and move the valve.

International Standards Organization (ISO) flare method A method for joining brake lines, also called a bubble flare. Created by flaring the line slightly out and then back in, leaving the line bubbled near the end.

interrupter ring A ferrous metal ring, shaped like a very shallow cup with slits or windows cut into it at evenly spaced intervals. The ring has the same number of blades and windows as engine cylinders and is rotated by the engine moving the blades through an air gap. The purpose of an interrupter ring is to systematically block, and expose, the magnetic field in a Hall-effect sensor in order to turn the primary ignition circuit on and off.

intrusive test A test in which a gas analyzer hose is placed into the exhaust stream before and after the catalytic converter; it may require drilling an access hole in the exhaust pipe in front of the converter.

inverted double flare A method for joining brake lines that forms a secure, leakproof connection.

inverter A device that converts direct current to alternating current.

ions An atom that has fewer electrons than protons (positive) or that has more electrons than protons (negative).

isolation valve The valve in the HCU that either allows or blocks brake fluid that comes from the master cylinder from entering the HCU hydraulic circuit.

jack stands Metal stands with adjustable height to hold a vehicle once it has been jacked up.

jake brake A brake system that consists of an extra exhaust valve on a diesel engine, which releases compressed gases from the combustion chamber at the top of the compression stroke; also called a compression brake.

journal saddle A semicircular cut located at the bottom of the engine block that is used to support the engine crankshaft.

keep alive memory (KAM) A certain minimum amount of parasitic current draw that is used by the vehicle's circuits to maintain memory functions and monitor systems.

keeper groove A groove machined into the top of the valve stem near the tip of the valve that is used to "lock in" the valve keepers to help retain the valve spring onto the valve assembly.

keepers Locking devices that keep the valve retained by the valve spring seat.

keyed lock washer The washer that fits between the adjusting nut and the locknut; the face of the washer is drilled with a series of holes that mate to a short pin from the adjusting nut, locking it to the spindle.

keyed washer The washer that fits between the adjusting nut and the wheel bearing and that has the center hole keyed to fit a slot on the spindle or axle tube.

kickdown valve A type of spool valve that is connected to the throttle on the vehicle. The valve is used to force a downshift when the throttle is opened all the way, assuming the governor pressure is below a specified point. Kickdown valves are also called detent valves.

kinetic energy The energy of an object in motion; it doubles with weight, and increases by the square of the speed.

Kirchhoff's current law An electrical law stating that the sum of current flowing into a junction is the same as current flowing out of the junction.

knock sensor An engine sensor that detects pre-ignition, detonation, and knocking.

knocking A noise heard when the air-fuel mixture spontaneously ignites before the spark plug is fired at the optimum ignition moment.

knuckle The part that contains the wheel hub or spindle and attaches to the suspension components.

knurled A process in which a special bit with a spiral groove is threaded through the valve guide to create raised ridges on either side of the groove, effectively shrinking the size of the valve guide.

L-head A type of four-stroke internal combustion engine having both intake and exhaust valves located in one side of the engine block, which are operated by lifters actuated by a single camshaft. It is sometimes called a flat head because there are no valves in the head.

labor guide A guide that provides information to make estimates for repairs.

lambda The ratio of air to fuel at which all of the oxygen in the air and all of the fuel are completely burned; See also *stoichiometric ratio*.

latent heat of condensation The amount of heat removal necessary to change the state from a gas to a liquid without changing the actual gauge temperature.

latent heat of evaporation The amount of heat required to change the state from a liquid to a gas without raising the actual gauge temperature.

latent heat of freezing The amount of heat removal required to change the state from a liquid to a solid without changing the actual gauge temperature.

lateral runout Also called warpage, the side-to-side movement of the rotor surfaces as the rotor turns.

leading shoes Brake shoes that are installed so that they are applied in the same direction as the forward rotation of the drum and thus are self-energizing.

leading/trailing shoe drum brake system Type of brake shoe arrangement where one shoe is positioned in a leading manner and the other shoe in a trailing manner.

leaf spring A spring, made of one or more flat, tempered steel springs bracketed together, that is used in the suspension system to support the weight of the vehicle.

lemon law buyback A consumer protection law used in some states to identify a new vehicle that has undergone several unsuccessful attempts to repair the same fault.

lever A tool that allows the user to move a large load over a small distance at one end by applying a small force over a greater distance from the other end.

lift The amount the valve will open. The more the valve lifts off its seat, the more air can get into and out of the engine.

light line technician A technician who diagnoses and replaces the mechanical and electrical components of motor vehicles.

light-emitting diode (LED) A diode that emits light when current flows through it.

limit switches Switches that turn off power flow to an electr[illegible] motor when a particular limit is reached.

limit value The maximum amount of emissions that a ve[illegible]le is permitted to emit. Values are assigned to different cla[illegible]cations of vehicles.

limited slip differential assembly A differential assembly that uses a clutch assembly or gear assembly to allow a limited amount of slip between the two axles. It is used to increase drive wheel traction in slippery conditions.

limp-in mode A transmission operating mode in which limited computer controls are needed to operate for the purpose of getting the vehicle to a shop.

line bored/line honed A process for correcting a mainline that is out of line.

line pressure A hydraulic pressure that is used to apply bands and clutches, and is regulated by the pressure regulator valve.

line pressure sensor A variable resistor sensor used to monitor line pressure. It sends a signal back to the PCM where it can be translated into a psi reading.

linear motion Movement in a straight line.

lines Term used interchangeably with pipes or tubes.

lip-type dynamic oil seal A seal with a precisely shaped dynamic rubber lip that is held in contact with a moving shaft by a garter spring. An example would be a valve seal or camshaft seal.

liquefied natural gas (LNG) A cryogenic liquid produced from gaseous methane.

liquefied petroleum gas (LPG) Gaseous fuel at atmospheric pressure; commonly referred to as propane.

lithium soap A thickening agent for grease to give it the proper consistency.

live axle An axle with a final drive unit made into it that transfers the power from the engine to the wheels so the vehicle can move.

load transfer Weight transfer from one set of wheels to the other set of wheels during braking, acceleration, or cornering.

lobe The raised portion on a camshaft; used to lift the lifter and open the valve.

lock cage The stamped sheet metal cap that fits over the bearing adjustment nut and is secured by a cotter pin going through it and the spindle/axle.

locking hub Four-wheel drive front axle hubs that are manually locked or unlocked by turning the knob on the hub.

locking pliers A type of pliers where the jaws can be set and locked into position.

locknut The nut that holds the adjusting nut from turning; usually tightened much tighter than the adjusting nut.

lockout solenoid The cam and lock pin operated by an electronic solenoid that keeps the reverse gear in lockout until it is selected.

lockout/tag-out A safety tag system to ensure that faulty equipment or equipment in the middle of repair is not used.

ng block An engine assembly that includes the short block and 's the valve train cylinder heads, timing cover, oil pan, and, in s cases, manifolds.

lo udinal The orientation of the engine in which the front of th ngine is facing the front of the vehicle. It is most commonl und in rear-wheel drive vehicles.

lost foam A metal casting process that uses a foaming process to create the pattern of the part being cast in metal.

low-drag caliper A caliper designed to maintain a larger brake pad–to-rotor clearance by retracting the pistons farther than normal.

Low-Emission Vehicle (LEV) standard A program initiated in California to improve vehicle emission standards.

low-inductance coils Ignition coils designed with low primary winding resistance and a low number of turns. Current flow on these coils is much higher than with contact breaker point systems and reaches its optimum level sooner.

low-pressure accumulator A storage container for brake fluid coming from the release valves, which is under relatively low pressure.

low-pressure cycling switch A device installed on the low-pressure line or the accumulator to turn the air-conditioning clutch on and off at specified pressures.

low-pressure switch A device used in nonvariable air-conditioning compressors. It cycles the air-conditioning clutch in response to changes in low side pressure.

lube technician A technician who carries out scheduled maintenance activities on a range of mechanical and related vehicle components.

lubricating oil Processed crude oil with additives to help it perform well in the engine.

lubrication system A system of parts that work together to deliver lubricating oil to the various moving parts of the engine.

lubricity The ability to lubricate moving parts.

lug A flange that is shaped to assist with aligning objects on other objects.

lug nuts Nuts that secure the wheel onto the wheel studs.

lug wrench A tool designed to remove wheel lugs nuts and commonly shaped like a cross.

M85 vehicle A flexible-fuel vehicle, meaning any mixture of methanol and gasoline/petrol in the fuel tank can be used by the engine.

MacPherson strut A strut used on an independent suspension where the spring and shock are joined together; used on most front-wheel drive vehicles.

magnafluxing An electromagnetic process used to locate cracks in ferrous engine blocks and cylinder heads and other ferrous metal parts.

magnetic pickup tools Useful for grabbing items in tight spaces, it typically is a telescoping stick that has a magnet attached to the end on a swivel joint.

magnetism The force that attracts or repels magnetic charges; or the property of a material to respond to a magnetic field.

magneto-resistive sensor A type of wheel speed sensor that uses an effect similar to a Hall effect sensor to create its signal.

magneto-rheological fluid A fluid that has the unique characteristic of changing viscosity when exposed to a magnetic field.

main bearing Metal bearing insert located in the main saddles that support the main journals.

main bearing caps Sturdy caps placed over the main journals of the crankshaft assembly and fastened with either two or four bolts.

main cap girdle A metal attachment that strengthens and connects all the main bearing caps together. It can be integral from the main bearing caps or separate.

main journal The smooth machined area of the crankshaft assembly that allows the crankshaft to rotate within the main bearings.

mainline The imaginary centerline down the engine cylinder block.

male and female terminal A crimp or solder terminal on which the male and female ends join to create a removable low-resistance connection.

malfunction indicator light (MIL) A dash light usually indicating the presence of a diagnostic trouble code or malfunction.

Mallory metal A tungsten alloy of copper and nickel that is used as a metal substitute added to the crankshaft counterweight for balancing purposes.

mandrel The shaft of a pop rivet.

mandrel forming Using a pipe bender with mandrels to allow for very tight bends without creating kinks or reducing the size of the pipe.

mandrel head The head of the pop rivet that connects to the shaft and causes the rivet body to flare.

manifold absolute pressure (MAP) sensor A vacuum sensor that is attached to the intake manifold by a passageway or vacuum hose. The sensor measures engine intake manifold pressure to determine engine load and sends a corresponding signal to the PCM.

manifold absolute pressure sensor A sensor that is used to measure vacuum or pressure in the intake manifold.

manual adjustable-rate shock absorber A shock absorber that allows manual adjustment of the dampening rate.

manual bleeding A bleeding method where one person manually operates the brake pedal while the other person opens and closes the bleeder screws on the wheel brake units to allow the air and old brake fluid to be pushed out.

manual climate control system A climate control system fully controlled by the operator.

manual lever position (MLP) switch A switch that is used by the PCM to tell which gear range the driver has selected with the shift lever.

manual valve A type of spool valve directly connected to the gearshift mechanism in the vehicle; when the driver selects a drive gear, the manual valve is positioned in the valve body to allow line pressure to push fluid to the proper valves and components inside the transmission.

manually adjustable air spring A rubber air bag placed inside coil springs to increase the spring's load carrying ability. It is filled manually through a valve similar to a tire valve stem.

mass airflow (MAF) sensor A sensor located in the air intake system that is used to measure the mass of the air flowing into the engine.

master cylinder Converts the brake pedal force into hydraulic pressure, which is then transmitted via brake lines and hoses to one or more pistons at each wheel brake unit.

match mounting The process of matching up the tire's highest point with the rim's lowest point for the purpose of reducing the tire's radial runout.

measuring tape A thin measuring blade that rolls up and is contained in a spring-loaded dispenser.

mechanical advantage The process of using a device to get more output force than the amount of input force, with the trade-off being that the input distance is proportionately longer than the output distance.

mechanical disadvantage When the load distance on a lever is greater than the effort distance, which means the effort required to move the load is greater than the load itself.

mechanical fingers Spring-loaded fingers at the end of a flexible shaft that pick up items in tight spaces.

mechanical jack A type of vehicle jack that uses mechanical leverage to lift a vehicle.

mechanical pressure regulator valve A spool valve and spring assembly that is used to control the amount of line pressure in a transmission.

membrane electrode assembly The individual voltage-producing component in a fuel cell stack.

memory saver (memory minder) Battery backup device for vehicle computer systems.

mercaptan An agent that smells like sulfur that is added to liquefied petroleum or natural gas to aid in detecting leaks.

meshed pinion A pinion when it is mated with the rack.

metallurgical bonding A process of sintering metals until they are fused together as one metal.

metering jet A calibrated orifice in a carburetor for fuel to flow through. Often it is replaceable for performance or economy desires.

metering valve A valve used on vehicles equipped with older rear drum/front disc brakes to delay application of the front disc brakes until the rear drum brakes are applied. Located in line with the front disc brakes.

methanol A motor fuel made from wood alcohol.

microleak detector A solution used to detect refrigeration leaks.

micrometer An accurate measuring device for internal and external dimensions. Commonly abbreviated as "mic."

micron gauge A device designed to measure vacuum very precisely.

microprocessor An electronic control unit that can process data and control one or more devices.

Miller cycle An engine cycle that uses a longer exhaust stroke than intake stroke through delayed closing of the intake valve. This engine uses a supercharger to pressurize air into the cylinder when needed.

min/max setting A setting on a DMM to display the maximum and minimum readings.

mineral oil Base stock processed from crude oil in a refinery, used as the base material of all conventional oil.

minivan A lighter-duty van used for carrying six to eight occupants or light cargo.

misfire Failure of one or more cylinders to fire or complete combustion.

modulator pressure A pressure created by the vacuum modulator that is proportional to engine load. Modulator pressure is used to delay transmission upshifting based upon engine load. It may also be used to raise line pressure to more firmly apply bands and clutches under higher engine loads.

module An electronic computer or circuit board that controls specific functions.

moly The abbreviation for the lubricant called molybdenum disulfide.

molybdenum thickening agent A compound used in some greases to give it the needed consistency.

monitor An OBD II test run to ensure that a specific component or system is working properly.

Morse taper A system for quickly changing and securing drill bits to drills.

Morse-style chain A heavy-duty chain constructed of many links and held together with pins. It is used in some transaxles and transfer cases to transmit torque from one component to another.

muffler A device to quiet the pipes of the air-conditioning system with baffles placed inside to deaden the sound of refrigerant moving.

multi-plate clutch A clutch assembly that is comprised of two or more clutch plates and used to increase the torque-carrying capacity of the clutch.

multidisc clutch A type of holding device used by an automatic transmission to stop the movement of one component of a planetary gear set. It uses several thin friction discs and thin steel plates that are squeezed together when hydraulic pressure is applied to a piston in the clutch.

multilayer steel (MLS) head gasket A gasket composed of multiple layers of steel and coated with a rubberlike substance that adheres to metal surfaces. They are typically used between the cylinder head and the cylinder block.

multimeter A test instrument used to measure volts, ohms, and amps. A digital multimeter may also be called a digital volt-ohmmeter (DVOM).

multiphase An electric motor that operates through more [illegible]an one phase.

[illegible]ltiplex The carrying of multiple signals on one wiring ci[illegible]uit. Digital signals are multiplexed on a dedicated CAN-bus[illegible] similar network; also referred to by the abbreviation MU[illegible]

multi[illegible]nt fuel injection (MPFI) An injection system in which [illegible]l is injected into the intake ports just upstream of each cylinder's intake valve, rather than at a central point within an intake manifold. Also called multiport injection.

Mylar tape Polyester film that may be metalized and incorporated into a wiring harness to provide electrical shielding.

N-type Semiconductor material with a small amount of extra electrons.

National Automotive Technicians Education Foundation (NATEF) An accrediting body for secondary and postsecondary automotive training programs; an independent, nonprofit organization under the umbrella of the ASE.

National Fire Protection Association (NFPA) The agency that publishes standards for liquefied petroleum and compressed natural gas vehicles.

National Lubricating Grease Institute (NLGI) An organization that grades the thickness of automotive and industrial grease.

natural gas Methane; often referred to simply as NG.

needle roller bearing Typically a small-diameter pin rolling bearing that can be held in a cage or placed into the inside diameter of a hole.

needle-nose pliers Pliers with long tapered jaws for gripping small items and getting into tight spaces.

negative back-pressure EGR valve A dual-diaphragm EGR valve that uses vacuum in the exhaust manifold to operate a control diaphragm that opens and closes a vent in the main diaphragm chamber.

negative camber Tilt of the top of the tire toward the centerline of the vehicle.

negative caster Forward tilt of the steering knuckle pivot points from the vertical line.

negative offset A condition in which the plane of the hub mounting surface is positioned toward the brake side or back of the wheel centerline.

negative scrub radius A condition in which the camber line is inside the steering axis centerline or when they intersect above the road surface.

negative temperature coefficient (NTC) A characteristic of materials whereby resistance decreases as temperature increases.

negative temperature coefficient (NTC) thermistor A thermistor that gains resistance as temperature goes down and loses resistance as temperature goes up.

neutral steer A condition in which both the front and the rear tires of a vehicle are experiencing the same slip angle.

Newton's first law of motion A physical law that states that an object will stay at rest or uniform speed unless it is acted upon by an outside force.

NFPA-52 A document from the NFPA that sets standards for compressed natural gas refueling stations and compressed natural gas vehicle fuel systems.

nippers (pincer pliers) Pliers designed to cut protruding items level with the surface.

nitriding A metal surfacing process that uses nitrogen ammonia gas to create a very thin, highly hardened surface.

nitridization process The nitriding metal surfacing process in which nitrogen ammonia gas is used inside a furnace and heated up and then cooled.

nitrile butadiene rubber (NBR) An artificially created rubber that is more resistant to oils and acids than natural rubber; it is compatible with R-12.

noise, vibration, and harshness (NVH) test A test to measure for any audible noises, vibrations, and harsh operation. It can be completed by the technician with or without the aid of an NVH tester. The tester is used to pinpoint the exact frequencies of the noise and vibrations.

non-integral ABS systems A brake system in which the master cylinder, power booster, and HCU are all separate units.

non-integral valve guide Also called a replaceable valve guide, a valve guide that is pressed into the cylinder head after the head has been constructed. These guides can be removed and replaced.

noncontinuous monitor A monitor that runs only once per drive cycle.

nondirectional tread pattern A tread pattern that is non-specific, allowing the tire to be placed on any wheel of a vehicle.

normally closed (NC) An electrical contact that is closed in the at-rest position.

normally open (NO) An electrical contact that is open in the at-rest position.

NP-N transistor A transistor in which P-type material is sandwiched between two layers of N-type material.

number-one cylinder Typically the cylinder located farthest forward on the engine. It is the first cylinder in the firing order.

nut A fastener with a hexagonal head and internal threads for screwing on bolts.

O-ring A small donut shaped rubber ring used to seal connections between air-conditioning components.

O-ring seal A seal that is a complete circle. If a cutaway is done, it will also be a complete circle. These seals are usually set into a groove that is machined into the housing or part that does not move and is intended to be sealed.

O-ring valve stem seal A seal used where the valve spring retainer attaches to the valve stem to seal the juncture and prevent excessive oil from leaking into the valve guide.

OBD I The first generation of onboard diagnostic systems that originated for California vehicles.

OBD II The second generation of onboard diagnostic systems, which have been in effect for all U.S. vehicles since 1996.

Occupational Safety and Health Administration (OSHA) Government agency created to provide national leadership in occupational safety and health.

octane rating A standard measure of the performance of a motor or aviation fuel. The higher the octane number, the more heat the fuel can withstand before self-igniting.

odorant An agent (mercaptan) that smells like sulfur that is added to liquefied petroleum or natural gas to aid in detecting leaks.

off-car brake lathe A tool used to machine (refinish) drums and rotors after they have been removed from the vehicle.

offset screwdriver A screwdriver with a 90-degree bend in the shaft for working in tight spaces.

offset vice A vice that allows long objects to be gripped vertically.

ohm The unit for measuring electrical resistance.

Ohm's law A law that defines the relationship between current, resistance, and voltage.

oil control ring The widest of the metal rings on the piston head assembly that controls oil flow to and from the cylinder walls.

oil cooler A device that takes heat away from engine oil by passing it near either engine coolant or outside air. Cooling the engine oil helps to keep it from overheating and breaking down.

oil filter wrench A specialized wrench that allows extra leverage to remove an oil filter when it is tight.

oil galleries Oil passages that are drilled into the engine block and cylinder head(s). These passageways carry oil from the oil pump to critical moving parts.

oil monitoring system A system that alerts the driver when it is time to change engine oil. These systems will need to be reset for the customer after an oil change is performed.

oil pan The metal pan that covers the bottom of the engine, contains oil sump where engine oil is held.

oil pressure relief valve A valve, usually located in the oil pump, that limits the oil pressure. When oil pressure is reached, excessive pressure is bled back to the sump.

oil pump A device that pumps lubricating oil through the engine.

oil pump strainer A screen located on the oil pump pickup that keeps debris from being picked up by the oil pump.

oil ring groove A metal groove cut into the piston head assembly designed to hold the oil control ring, which lubricates the cylinder wall and pistons.

oil seal Any seal used to seal oil in and dirt, moisture, and debris out.

oil slinger A device used on small engines, located on the crankshaft or driven by the camshaft. It works to fling oil up onto moving engine parts.

oil spurt holes Holes drilled into the connecting rod that spray oil up onto the cylinder walls and the piston wrist pins.

oil sump The lower part of the oil pan that holds lubricating oil for the engine. The oil pickup screen sits in this low point.

oil-life monitoring (OLM) system A system that displays the useful life remaining in the engine oil.

oiler A device used to add oil to the air-conditioning system.

on-car brake lathe A tool used to machine (refinish) rotors while they are still attached to the vehicle.

on/off-type signal A signal that is either a 12-volt signal or zero volts. Typically, switches provide on/off signals. Some systems use a 5-volt reference signal instead of a 12-volt signal.

onboard diagnostic (OBD) The first generation of a self-diagnostic system built into the vehicle computer system that provided a means of warning the driver when the computer detected a problem in the system. This system allowed the technician to retrieve codes to assist in diagnosis of the vehicle.

one-way clutch A type of holding device used by an automatic transmission to stop the movement of one component of a planetary gear set. It allows free spinning in one direction but will lock up when the part attempts to spin in the opposite direction.

open A term used to describe a circuit that does not have a complete path for current to flow.

open circuit A circuit that has a break that prevents current from flowing.

open differential assembly A differential assembly that allows both axles to turn at their own speed when turning a corner, but is dependent on the traction of the tires to deliver torque to the ground. If one wheel has no traction, all of the engine's torque will be used at that wheel, causing it to simply spin.

open fracture A fracture in which the bone is protruding through the skin or there is severe bleeding.

open loop When the feedback loop is not active.

open-end wrench A wrench with open jaws to allow side entry to a nut or bolt.

optical cable A cable that contains a number of optical fibers.

optical sensor A sensor that generates a voltage when excited by a beam of light.

orifice A precisely sized hole used to reduce the speed and pressure of hydraulic oil flowing to a clutch or band.

Oscillate To cycle above and below a given value.

oscillation The fluctuation of an object between two states. With regard to suspension springs, it refers to the uncontrolled compression and decompression of the spring following overshoot.

oscilloscope A tool that shows graphically what is happening to voltage over a period of time; it is used to diagnose electrical faults.

outer race The outside component of a wheel bearing that has a smooth, hardened surface for rollers or balls to ride on.

outer tie rod The tie rod attached between the tie rod and the steering arm. It transfers the movement of the rack, pivoting as the rack is extended or retracted when the vehicle is negotiating turns.

outlet port Links the cylinder to the brake lines.

output force Force that equals the working pressure multiplied by the surface area of the output piston, expressed as pounds, newtons, or kilograms.

outside micrometer A precision measuring instrument meant to measure the outside of components. It is usually accurate to 0.0001" (0.0025 mm).

overcharging Overfilling of the air-conditioning system; may result in poor cooling or mechanical failure of the system.

overdrive Any gear ratio that results in a torque reduction with a speed increase. Overdrive is used on vehicles to reduce the engine speed when traveling at highway speed in order to save fuel.

overflow tank A tank used to catch any coolant that is released from the radiator cap (works like a catch can).

overhauling The process of refurbishing the transmission to like-new condition.

overhead cam (OHC) engine Any engine in which the camshaft is housed inside the cylinder head and directly actuates the valves through rocker arms or cam followers or by pushing directly above the valve stem on a bucket.

overhead valve (OHV) engine An engine in which the valves are positioned in the cylinder head assembly, directly over the top of the piston.

overshoot The amount a spring extends (springs back) past its original length following compression.

oversteer A condition in which a vehicle's front slip angles are larger than the rear slip angles. This vehicle is said to be "pushing" in the corners.

owner's manual An informational guide supplied by the manufacturer; it contains basic vehicle operating information.

oxidation inhibitor An oil additive that helps keep hot oil from combining with oxygen to produce sludge or tar.

oxides of nitrogen A vehicle emission that contributes to ground-level ozone. Oxides of nitrogen are produced when nitrogen and oxygen react during combustion, given sufficient temperatures and pressures. Forms include nitrogen oxide and nitrogen dioxide.

oxyacetylene torch A gas welding system that combines oxygen and acetylene.

oxygen sensor (before and after catalytic convertor) An exhaust sensor used to measure the amount of oxygen in the exhaust gases produced by the engine PCM, used to determine fuel mixture and spark timing.

P–N junction The junction between N- and P-type semiconductor materials.

P–NP transistor A transistor in which N-type material is sandwiched between two layers of P-type material. This type of semiconductor material has holes, meaning it is missing electrons.

P-type Semiconductor material with holes where electrons are missing.

paddle A shifting mechanism or electronic control usually attached to the steering wheel.

pan inspection The process of removing the transmission pan to check for clutch material, metal, and other debris or contaminants.

Panhard rod A metal rod used to hold the dead axle and keep it from moving from side to side through corners. It is mounted on the body or frame of the vehicle and the axle; also referred to as a track bar.

parade pattern The display of all the cylinders firing in sequence on an oscilloscope.

parallax error A visual error caused by viewing measurement markers at an incorrect angle.

parallel flow condenser A condenser with multiple parallel tubes flowing from one side tank to the other.

parallel hybrid A hybrid vehicle driven simultaneously by an internal combustion engine and an electric machine.

parallel hybrid drivetrain A type of hybrid transmission in which power can flow from either a gasoline engine or an electric motor or any combination of the two.

parallelism Also called thickness variation; both surfaces of the rotor should be perfectly parallel to each other so that brake pulsations do not occur.

parallelogram steering system A non–rack-and-pinion system that uses a series of parts, consisting of the pitman arm, idler arm, center link, and tie-rod assemblies, to relay movement from the steering gearbox to the wheel assembly.

parasitic draw Unwanted drain on the vehicle battery when the vehicle is off.

parasitic loss A loss of engine efficiency caused by internal friction, inefficient breathing, etc.

parking brake A brake system used for holding the vehicle when it is stationary.

parking brake cable A mechanism used to transmit force from the parking brake actuating lever to the brake unit.

parking brake cable pliers A tool used to install parking brake cables.

parking brake cable removal tool A tool used to compress the spring steel fingers of the parking brake cable so that the cable can be removed from the backing plate.

parking brake mechanism A mechanism that operates the brake shoes or pads to hold the vehicle stationary when the parking brake is applied.

particulate matter (PM) Unseen microscopic particles of carbon consisting of soot (especially prominent with diesel exhaust). PM clogs the lungs and is carcinogenic.

parts program A computer software program for identifying and ordering replacement vehicle parts.

parts specialist The person who serves customers at the parts counter.

Pascal's law One of the fundamental physics principles behind the operation of an automatic transmission, this law states that if fluid is placed in a confined space, pressure will be transferred equally in all directions inside the space.

passive keyless entry (PKE) An active system that senses the proximity of a fob and locks or unlocks the vehicle.

PCV valve A valve that controls the amount of crankcase ventilation flow that is allowed; it varies with changes in manifold pressure.

peening A term used to describe the action of flattening a rivet through a hammering action.

pending codes Codes that have not been validated by failing a second test.

performance testing The process of recreating a driving situation to check air-conditioning performance and vent temperature.

permanent magnet A material with natural or artificially created constant magnetic properties.

permanent magnet electric motor An electric motor in which the magnetic field in the casing is produced by permanent magnets, and the armature has an electromagnetic field generated by passing electrical current through loops or windings, thereby producing the motor action.

permanently mounted tank A compressed natural gas or propane cylinder not designed to be portable.

personal protective equipment (PPE) Safety equipment designed to protect the technician, such as safety boots, gloves, clothing, protective eyewear, and hearing protection.

phenolic resin A very dense material used to create some brake pistons that is very resistant to corrosion and heat transfer.

Phillips head screwdriver A type of screwdriver that fits a head shaped like a cross in screws.

photochemical smog A brown haze that hangs in the sky, typically seen over large cities. Smog is a major health issue to humans because it affects lung tissue.

photodiode An electronic component that creates a varying voltage or current output based on the amount of light striking it.

photovoltaic (PV) effect The conversion of sunlight into electricity.

pickup A vehicle that carries cargo; it has stronger chassis components and suspension than a sedan.

pickup assembly A component with a wire coil wrapped around a ferrous metal core; it is used to generate an electrical signal when a magnetic field passes through it.

pickup tube A tube connected to the oil pump that acts like a straw for the oil pump to pull oil from the sump of the oil pan.

piezoelectric A type of electricity in which a material such as a quartz crystal produces voltage when mechanical pressure distorts it.

pilot bearing The bearing or bushing that supports the front of the transmission input shaft.

pin punch A type of punch in various sizes with a straight or parallel shaft.

pinging A preignition or detonation concern that is damaging to the engine if it is allowed to continue. It is described as the sound of marbles being shaken in a glass jar.

pinion A gear located at the end of the steering shaft connected to the rack. It moves the rack from side to side as the pinion rotates, controlling the direction of the wheels.

pintle A tapered valve that sits in a tapered seat to seal a passageway for air or fuel.

pip mark A small indent or dimple on the piston ring that indicates which side of the ring is installed upward. It is also used on some timing sprockets.

pipe wrench A wrench that grips pipes and can exert a lot of force to turn them. Because the handle pivots slightly, the more pressure put on the handle to turn the wrench, the more the grip tightens.

piston The round metal plug found inside the engine cylinder that moves up and down inside the cylinder.

piston assembly All of the parts of the piston including the piston, piston rings, and piston pin.

piston displacement The volume of air that is moved by the piston from bottom dead center to top dead center.

piston engine An internal combustion engine that uses cylindrical pistons, moving back and forth in a cylinder, to extract mechanical energy from chemical energy.

piston pin (wrist pin) A round circular metal manufactured part that attaches the piston assembly to the connecting rod assembly.

piston ring A metal ring that is placed in a square groove around a piston for sealing purposes.

piston ring groove A square-cut groove located on the piston, designed to hold a metal piston ring.

piston slap An engine noise caused by excessive clearance between the piston skirt area and the cylinder wall.

piston stroke The distance the piston travels through the engine cylinder from top dead center to bottom dead center.

pitch Movement of a vehicle around its y-axis (the imaginary line across the center of the vehicle from left to right) that causes the vehicle to lower or rise on the front end during quick braking or acceleration.

pitch circle diameter (PCD) The diameter of the imaginary circle drawn through the center of the wheel bolt holes.

planet carrier The device that holds the planet gears in place, keeping them equally spaced.

planetary gears The small gears in a planetary gear set that revolve around the sun gear; also known as pinion gears.

plasma cutter A tool that uses electricity and compressed gas to produce a stream of high-temperature gas to cut metal.

Plastigauge® The trademarked name for a plastic gauging material used to check the clearances between two surfaces.

plate A flat, thin square or rectangular sheet of material often used to describe the arrangement of positive and negative electrodes in batteries.

plenum chamber A large portion of the intake manifold after the throttle plate and before the intake runner tubes. The plenum provides a reservoir of air and helps prevent interference with the flow of air between individual branches.

pliers A hand tool with gripping jaws.

plug-in hybrid electric vehicle (PHEV) A hybrid electric vehicle in which only one power source, the battery, is used to propel the vehicle for a certain distance, limited by the storage capacity of the battery and the efficiency of the motor.

plunge-type joint The inner joint on the half shaft that allows for changes in shaft length.

ply rating A rating system that denotes the number of belt layers, or plies, that make up the tire carcass. In radial tires, ply rating denotes the relative strength of the plies, not the actual number of plies.

pneumatic jack A type of vehicle jack that uses compressed gas or air to lift a vehicle.

pocket bearing A roller in the rear of the input shaft that supports the front of the main shaft.

polarity sensitive A term used to describe a component that must be connected into a circuit with the correct polarity to its terminals.

polarity The state of charge, positive or negative.

policy A guiding principle that sets the shop direction.

pollutant A potential threat to human health or the environment resulting from excessive amounts of chemicals and waste.

polyalphaolefin (PAO) An artificially made base stock (synthetic) used in place of mineral oil. Oil molecules are more consistent in size, and no impurities are found in this oil as it is made in a lab.

polycrystalline diamond (PCD) cutter An engine block resurfacing tool used on aluminum parts.

pop rivet gun A hand tool for installing pop rivets.

poppet (mushroom) valve A cam-operated, spring-loaded, mushroom-type valve used to control intake into, and exhaust out of, the combustion chamber.

poppet valve A valve that controls the flow of brake fluid at usually preset pressures.

portable universal cylinder A liquefied petroleum gas tank that can be moved or transported.

positive back-pressure EGR valve A dual-diaphragm EGR valve that uses pressure in the exhaust system to act on the control diaphragm, which closes a bleedvalve in the main diaphragm, thus allowing the EGR valve to open only when there is positive exhaust pressure..

positive camber Tilt of the top of the tire is tilted out from the centerline of the vehicle.

positive caster Backward tilt of the steering knuckle pivot points from the vertical line.

positive crankcase ventilation (PCV) system A system that draws blowby gases from the crankcase into the intake to be burned.

positive offset A condition in which the plane of the hub mounting surface is positioned toward the outside or front of the wheel centerline.

positive scrub radius A condition in which the camber line is outside of the SAI line or when they intersect below the road surface.

positive temperature coefficient (PTC) A characteristic of materials whereby resistance increases as temperature increases.

positive valve stem seal A valve seal located at the top of the valve guide that is a more positive seal than an umbrella or O-ring seal.

potentiometer A type of variable resistor that increases or decreases its resistance as it is turned, which creates a varying voltage signal.

potentiometer Also called a pot, a three-terminal resistive device with one terminal connected to the input of the resistor, one terminal connected to the output of the resistor, and the third terminal connected to a movable wiper arm that moves up and down the resistor.

pour point The temperature at which diesel fuel is not gelled.

pour point depressants Oil additives that keep wax crystals from forming and causing the oil to gel during cold operation.

powdered metal rod A connecting rod manufactured through the process of heat, compression, and forging of powdered metals into a connecting rod.

power The rate at which work is done; electrical power is measured in watts.

power assist unit The electric motor in electric power assist steering systems.

power flow The path that power takes from the beginning of an assembly to the end. In a transmission, power flow changes as different gears are selected by the driver.

power section A chamber in the rack, where pressurized fluid acts upon pistons that assist in steering.

power splitter A device that receives power from an internal combustion engine and electric machine to power a hybrid electric vehicle.

power steering An option on a vehicle that allows movement of the steering wheel with decreased driver effort.

power steering pump A small hydraulic pump that provides assistance to the driver when turning the steering wheel.

power stroke The stroke during which combustion is pushing the piston from top dead center to bottom dead center in the cylinder. This stroke is where power is produced.

power take-off (PTO) A device attached to the transmission that is gear driven and can be used to run accessories such as winches and towing equipment. It can also refer to the gears that send power to the rear axle in a predominantly FWD vehicle.

power tools Tools powered by electricity or compressed air.

power unit A belt- or gear-driven pump that produces hydraulic pressure for use in the steering box or rack.

power-on-demand A feature that shuts down the engine when not needed to save fuel.

power-splitting transmission (PST) A type of hybrid transmission that splits the power flow going to the wheels from one or more electric motors and an internal combustion engine.

powertrain control module (PCM) A computer that controls the ignition and fuel control and emissions control systems on an engine; also called the electronic control unit (ECU) or engine control module (ECM).

powertrain mount A rubber or metal bracket used to secure the engine and transmission into the vehicle. Some vehicles use hydraulic or electrohydraulic powertrain mounts.

preload A condition where the wheel bearing components are forced together under pressure and therefore have no end play.

preloaded A part that is already compressed from pressure.

press-fit pin A description of how a piston pin is placed inside a connecting rod by pressing the pin into the connecting rod.

pressure The force per unit area applied to the surface of an object.

pressure bleeding A bleeding method that uses clean brake fluid under pressure from an auxiliary tool or piece of equipment, to force the air and old brake fluid from the hydraulic braking system.

pressure differential valve A valve that monitors any pressure difference between the two separate hydraulic brake circuits; it usually contains a switch to turn on the brake warning light when there is a pressure difference.

pressure gauge set A set of calibrated gauges for high and low pressure that can show pressure and vacuum readings for use with air-conditioning systems.

pressure plate The assembly that applies and removes the clamping force on the clutch disc.

pressure relief valve A valve that is designed to release pressure if it gets above a calibrated pressure; used in a gas cap as a safety device to release excessive pressure if pressure gets too high.

pressure testing A process for checking the cylinder heads for cracks, using air pressure.

pressure transducer A device used to measure engine vacuum and display it graphically on a lab scope.

pressure transients Minor fluctuations on the gauges that may indicate a problem.

pressure washer/cleaner A cleaning machine that boosts low-pressure tap water to a high-pressure output.

pressure, or force-feed, lubrication system A lubrication system that has a pump to pressurize the lubricating oil and push it through the engine to moving parts.

prick punch A pinch with a sharp point for accurately marking a point on metal.

primary circuit The low-voltage circuit that turns the coil on and off.

primary cup A seal that holds pressure in the master cylinder when force is applied to the piston.

primary piston A brake piston in the master cylinder moved directly by the pushrod or the power booster; it generates hydraulic pressure to move the secondary piston.

primary winding The coil of wire in the low voltage circuit that creates the magnetic field in a step-up transformer.

printed circuit The fine copper strip or track attached to an insulated board for mounting and connecting electronic components.

probing technique The way in which test probes are connected to a circuit.

procedure A list of the steps required to get the same result each time a task or activity is performed.

producer gas A generic term that refers to a number of manufactured gases such as wood gas, town gas, and syngas.

progressive rate of deflection The change in deflection rate that occurs as the weight of the vehicle changes. The greater the weight, the lower the rate of deflection due to increased resistance.

proportioning valves Valves used mostly on older vehicles equipped with rear drum brakes to reduce rear wheel hydraulic brake pressure under hard braking or light loads. Located in line with the rear brakes.

propylene glycol A chemical used as antifreeze. It is labeled as a nontoxic antifreeze.

pry bar A high-strength carbon steel rod with offsets for levering and prying.

PT chart A pressure-temperature chart that shows the relationship between air-conditioning pressures and evaporator temperature.

pull-in winding A high-current winding found in starter solenoids that pulls the solenoid plunger into the activated position.

pullers A generic term to describe hand tools that mechanically assist the removal of bearings, gears, pulleys, and other parts.

pulse air system The use of normal exhaust engine pulses to draw air into the exhaust stream.

pulse-width modulation (PWM) A digital on/off electrical signal used as a variable control for devices such as solenoids.

punches A generic term to describe a high-strength carbon steel shaft with a blunt point for driving. Center and prick punches are exceptions and have a sharp point for marking or making an indentation.

purge solenoid A control device that lets fuel vapors move from the charcoal canister to the engine. The solenoid is controlled by the engine control module.

purge switch A device used to show the computer when purge is occurring. It is used as a feedback device to allow the computer to determine whether flow is happening.

purge valve A valve used to control the flow of evaporative emissions from the charcoal canister to the intake manifold.

purging The process of pulling stored fuel vapors from the charcoal canister and moving them into the engine to be burned.

push-on spade terminal A disconnectable type of crimp or solder terminal used to terminate electrical wires.

push-type clutch A typical clutch system, used in modern vehicles, where the clutch fork pushes the release bearing forward to release the friction facing from the pressure plate.

pushrod (braking system) A mechanism used to transmit force from the brake pedal to the master cylinder.

pushrod A tubular rod that stands between the tappet and the rocker arm in an overhead valve engine; the pushrod transfers cam motion to the rocker arm.

quadrant ratchet The device used in some cable-operated clutches to provide self-adjustment as the clutch disc wears. Some quadrant ratchets adjust if you lift up on the clutch pedal.

quench or squish area The narrow area between the top of the piston at top dead center and the cylinder head. It derives its name from the squishing of the air-fuel mixture into a small "charge."

quenched The state in a combustion chamber in which the flame cannot burn due to cold surfaces or poor distribution of the fuel mixture.

quick take-up master cylinders Cylinders used on disc brake systems that are equipped with low-drag brake calipers to quickly move the brake pads into contact with the brake rotors.

quick take-up valve A valve used to release excess pressure from the larger piston in a quick take-up master cylinder once the brake pads have contacted the brake rotors.

rack A steel rod driven by the pinion, with tie rods on each end or tie rods connected to the center of the rack.

rack housing The outer shell of the rack-and-pinion steering system that is mounted to the chassis.

rack-and-pinion steering system A steering system composed of a steering wheel, a main shaft, universal joints, and an intermediate shaft. When the steering wheel is turned, movement is transferred by the main shaft and intermediate shaft to the pinion.

radial load The load that is perpendicular to a shaft, usually controlled by bearings or bushings.

radial tire A tire with two or more layers of casing plies and cord loops running radially from bead to bead.

radiation The movement of energy through space, such as the movement of energy from the sun to the earth.

radiator A device that takes hot coolant and cools it by passing heat energy to the surrounding air.

radiator hoses Rubber hoses that connect the radiator to the engine. Because they are subject to pressure, they are reinforced with a layer of fabric, typically nylon.

radius A straight line extending from the center of a circle to its edge or from the center of a sphere to its surface.

ram airflow The flow of air created by a vehicle moving down the road.

raster The scope pattern where all of the ignition firing sequences are stacked vertically on top of each other.

ratchet A generic term to describe a handle for sockets that allows the user to select direction of rotation. It can turn sockets in restricted areas without the user having to remove the socket from the fastener.

ratcheting box-end wrench A wrench with an inner piece that is able to rotate within the outer housing, allowing it to be repositioned without being removed.

ratcheting screwdriver A screwdriver with a selectable ratchet mechanism built into the handle that allows the screwdriver tip to ratchet as it is being used.

rattle gun A term used to describe an air impact wrench, based on the noise it makes.

Ravigneaux gear set A type of compound planetary gear set that uses two different sun gears and two different-diameter planets while only using one ring gear.

reaction force A force that acts in the opposite direction to another force.

rear-wheel drive (RWD) A drivetrain layout in which the engine drives the rear wheels.

rebound clip A metal strap that is warped around the leaf spring to prevent excessive flexing of the main leaf during rebound.

receiver filter drier (RFD) The air-conditioning component used on TXV systems to filter and store liquid refrigerant to supply liquid refrigerant to the TXV. It is located between the condenser and the TXV.

reciprocating motion An up-and-down motion within the cylinder.

reciprocating weight The amount of weight that is moving up and down. It is everything from the middle of the connecting rod upward, including the piston, wrist pin, and rings.

reciprocation The back-and-forth movement of the piston assembly inside the cylinder.

recirculating ball steering box A steering box that has worm gear inside a block with a threaded hole in it and gear teeth cut into its outside that engage the sector shaft to move the pitman arm; generally used on trucks and heavy vehicles.

reclaim/recycle machine An air-conditioning machine designed to remove and recycle refrigerant for reuse.

reclaiming The process of removing refrigerant from the air-conditioning system by using an air-conditioning machine; also called recovering.

recovering See *reclaiming*.

rectification A process of converting AC into DC required by the battery and nearly all of the automobile systems.

recuperation Process by which brake fluid moves from the reservoir past the edges of the seal into the chamber in front of the piston. This prevents air from being drawn into the hydraulic system caused by low pressure when the brake pedal is released quickly.

reduction ratio The ratio between the turn of the steering wheel and the turn of the wheel, both measured in degrees.

reed switch A type of speed sensor that uses a magnetic field to open and close a movable set of contacts. It is used with a rotating magnet to measure rpm of a shaft and send the signal to the PCM.

reed valve A small flexible metal plate that covers the inlet port of a two-stroke engine and opens and closes to let air and fuel into the crankcase.

reformer An onboard device that extracts hydrogen from another source material.

refrigerant identifiers Devices used to check for impurities in the air-conditioning system.

refrigerant label A label that lists the type and total capacity of refrigerant that is installed in the A/C system.

refrigerant The name given to a chemical compound designed to meet the needs of the refrigeration system.

regenerative braking A feature that allows drive motors to act as generators to recharge the traction batteries during deceleration or when braking. Also called regenerative charging.

relative humidity sensor A sensor designed to measure the relative humidity in the passenger compartment.

relay An electromechanical switching device whereby the magnetism from a coil winding acts on a lever that switches a set of contacts.

relay lever A steel rod that transfers movement from the drag link to an idler arm.

release mechanisms Components that operate the clutch. Usually included are the throw-out bearing and the clutch fork. Some manufacturers include the operating system.

Reluctor A rotating, toothed wheel that changes the reluctance of a material to conduct magnetic lines of force.

remote keyless entry (RKE) A system that remotely unlocks and locks the vehicle without the use of a traditional key.

repair order A form used by shops to collect information regarding a vehicle coming in for repair, also referred to as a work order.

required voltage The amount of voltage needed to push current across the electrodes of a spark plug located in the combustion chamber.

reserve capacity A standard used to specify the time in minutes that a battery will supply a load of 25 amps at 80°F (26.7°C) without its voltage dropping below 10.5 volts.

residual pressure valve (residual check valve) In drum brake systems, a valve that maintains pressure in the wheel cylinders slightly above atmospheric pressure so that air does not enter the system through the seals in the wheel cylinders.

resistor A component designed to have a fixed resistance.

respirator Protective gear used to protect the wearer from inhaling harmful dusts or gases. Respirators range from single-use disposable masks to types that have replaceable cartridges. The correct types of cartridge must be used for the type of contaminant encountered.

restriction A blockage that partially stops or slows the flow of a material such as refrigerant.

retarding effect The result of retarding (slowing) the vehicle.

retrofit kit An aftermarket kit that has the fittings and oil to change an R-12 unit over to an R-134a unit.

return springs Springs that retract the brake shoes to their released position.

returnless fuel injection system A type of injection system in which no hot fuel is returned to the tank, thus keeping the fuel in the tank relatively cool and minimizing vaporization.

reverse boost valve A component of the pressure regulator valve that increases line pressure when the vehicle is in reverse.

reverse shift fork A shift fork used to engage the reverse gear.

rheostat An adjustable resistor that varies current flow through a circuit.

rib A design feature created in the metal casting/forging process that places extra metal area on a part for strength and durability, yet keeps the weight of the engine block low.

ribbon cable A type of flat harness in which cables are insulated from each other but joined together side by side.

ride height The amount of ground clearance a vehicle has, measured from a point on the body or frame, depending on the manufacturer; also known as ride height.

ridge reamer A tool used to remove the metal lip on top of the cylinder walls caused by engine wear.

rigid dead axle suspension A type of dead axle suspension system that is non-independent and uses a beam or solid axle.

rigid spring hanger The rigid part typically welded to the body or frame of the vehicle to which the front of the leaf spring is attached.

rigid-axle coil-spring suspension A dead axle that uses a coil spring.

rim The outer circular lip of the metal on which the inside edge of the tire is mounted.

rim flanges The outside edge of the wheel that helps keep the tire from popping off the wheel.

rim width The distance across the rim from one rim flange to the other.

ring gap The distance between the ends of the piston rings when the rings are seated against the inside of the engine cylinder.

ring gear A large circular gear that is typically mounted on the flywheel or flexplate and mates to the starter gear, for the purpose of cranking the engine over.

ring land The area between the piston ring grooves found on the piston head assembly.

ring tension The built-in force created inside the piston rings; it is caused by the piston rings being made bigger than the cylinder walls in order to generate a scraping action against the cylinder walls.

riveted linings Brake linings riveted to the brake pad backing plate with metal rivets and used on heavier-duty or high-performance vehicles.

road force imbalance Occurs when the wheel or tire is not concentric or when the tire's sidewall has uneven stiffness.

rocker arm A lever that actuates a valve by pivoting near the center and pushing on the tip of the valve to open it.

rocker arm The fulcrum that transfers pushrod endwise motion to the valve stem.

rod A straight piece of steel used to transfer motion within the vehicle's suspension system. It typically has treads cut on one or both ends.

rod beam The area of a connecting rod between the big and small ends.

rod bearing A bearing insert located in the big end of the connecting rod.

rod cap The very bottom part of the connecting rod that retains the rod bearing inserts.

rod journal Also called the crankshaft rod throw, an area machined to a very smooth finish that contains an oil hole to provide oil to lubricate the surfaces of the crankshaft and rod bearing inserts.

roll Movement of a vehicle around its x-axis (the imaginary line down the center of the vehicle from front to back). It is commonly referred to as body roll or lean; when cornering, the body tries to move to the outside of the corner against the suspension.

roll bar Another type of pry bar, with one end used for prying and the other end for aligning larger holes, such as engine motor mounts.

roll-rate sensor A sensor that measures the amount of roll around the vehicle's horizontal axis that a vehicle is experiencing.

roller bearings The rolling components of a wheel bearing consisting of hardened cylindrical or tapered rollers.

rolling resistance Resistance that is present from tires contacting the road and wind resistance against the vehicle while rolling down the highway.

room temperature vulcanizing (RTV) silicone A silicone adhesive that sets up, or "vulcanizes," at room temperature.

Rosin A type of liquid or paste (flux) that is in solid form contained within solder and is used to prevent oxidization.

rotary engine An engine that uses a triangular rotor turning in a housing instead of conventional pistons.

rotary flow A type of fluid flow in a torque converter in which fluid flows around the centerline of the torque converter in a circle.

rotary vane compressor A compressor design that spins with vanes attached that extend to create the air-conditioning pressure required to run the system.

rotating assembly The assembly of the crankshaft, connecting rod, and piston that is found inside the engine cylinder block.

rotating weight The amount of weight that is moving in a circular motion, including everything from the center of the connecting rod down to the connecting rod cap and all components of the crankshaft.

rotational force The force created by the rotating wheel when the brakes are applied; it causes the brake components to twist the brake support, and ultimately the vehicle, in the direction of wheel rotation.

rotational speed The speed at which an object rotates, measured in revolutions per minute (rpm).

rotor A high-voltage rotating switch that transfers voltage from the distributor cap's center terminal to the outer terminals.

rotor arm The portion of the rotor that extends toward, but not touching, the outer distributor cap terminals.

rotor housing The engine block of the rotary engine. The rotor moves within it.

rotor lobes Lobes or rounded edges on rotors that squeeze oil and create pressure.

rotor pump A type of front pump found in automatic transmissions. The rotor pump has two different-sized rotors that mesh together to force fluid out of the pump.

rotor-type oil pump An oil pump that uses rounded gears to squeeze oil through.

rubber bellows Rubber pieces positioned on each end of the rack to protect the inner joints from dirt and contaminants and retain the grease lubricant inside the rack-and-pinion housing.

rubber-bonded bushing A bushing that has a steel outer housing and inner sleeve with rubber inside; also known as a metalastic bushing.

run-flat technology A tire design that allows the vehicle to keep moving under driver control, following a puncture or rapid loss of pressure.

running clearance The amount of space between wheel bearing components while in operation.

Rzeppa joint A type of fixed constant-velocity joint that has an inner race, six steel ball bearings, a bearing cage, and an outer race.

SAE J1930 An SAE standard for across-the-board standardization of parts and systems nomenclature.

SAE J2012 An SAE standard for across-the-board identification of generic DTCs.

safe working load (SWL) The maximum safe lifting load for lifting equipment.

safety data sheet (SDS) A sheet that provides information about handling, use, and storage of a material that may be hazardous.

safety-type drop-center rim A rim designed with a slight hump at the inside edge of the bead ledges to hold the tire beads in place during a flat tire.

salvage title Also called a branded title; a record that a vehicle has been severely damaged or deemed a total loss by an insurance company.

sand or bead blasters A cleaning system that uses high pressure and fine particles of glass bead or sand.

Sanden scroll-type compressor A compressor design that uses two spiral forms, one spinning and one stationary, to create the pressure needed to drive the air-conditioning system.

Sanden wobble plate compressor A compressor design that utilizes an offset shaft connected to the crankshaft to create the reciprocating motion of the pistons.

sander/polisher A power tool with a rotating disc or head to which polishing or sanding discs can be attached.

scan tool A tool used to read codes, access live data, and communicate with the vehicle's computers.

scavenge pump A pump used with a dry sump oiling system to pull oil from the dry sump pan and move it to an oil tank outside the engine.

scavenging effect A condition caused by moving columns of air, which create a low-pressure area behind them, resulting in a pulling force that is used to pull the remaining burned gases from the combustion chamber. Valve timing affects the amount of scavenging effect an engine has.

scavenging The process of removing burned gases from the cylinder through the use of moving airflow pulling or extracting the gases out.

Schrader valve A one-way valve used in a valve stem.

scratcher A thin, spring steel wear indicator that is fixed to the backing plate of the brake pad; it emits a high-pitched squeal when the brakes are applied if the brake pads have become too thin.

screw extractor A tool for removing broken screws or bolts.

scrub brakes A brake system that uses leverage to force a friction block against one or more wheels.

scrub radius The distance between two imaginary lines—the camber line and the SAI line. Or the point where they intersect above or below the surface of the road.

sealed bearings Wheel bearings that are assembled by the manufacturer with the proper lubrication and sealed for life; cannot normally be disassembled.

second-degree burns Burns that involve blistering and damage to the outer layer of skin.

second-order vibration Vibration that occurs at twice the engine rpm.

secondary circuit The part of an ignition system that operates on higher voltage and delivers the necessary high voltage to the spark plugs.

secondary cup A seal that prevents loss of fluid from the rear of each piston in the master cylinder.

secondary piston A piston that is moved by hydraulic pressure generated by the primary piston in the master cylinder.

secondary winding The coil of wire in which high voltage is induced in a step-up transformer.

sedan A vehicle configuration that has an enclosed body, with a maximum of four doors to allow access to the passenger compartment.

selective thickness shim A shim of a prescribed thickness that is used to control shaft clearances in a transmission.

selector gate The U-shaped cutaway in shift shafts that the shifter lever fits into.

selector shift rail A rod that is attached to the shift fork. These rails move when the shifter lever is moved against them.

self-energizing The property of drum brakes that assists the driver in applying the brakes; when brake shoes come into contact with the moving drum, the friction tends to wedge the shoes against the drum, thus increasing the braking force.

self-sealing tire A tire constructed with a flexible and malleable lining inside the tire around the inner tubeless membrane. The lining can seal small tread-area punctures instantly and permanently.

semi-floating axle An axle that carries the weight of the vehicle; if removed, there is no way to connect the wheel to the vehicle.

semiautomatic climate control system A system that provides automatic function of the heater or cooling only, leaving fan speed and mode selection to the operator.

semiconductor A material used to make microchips, transistors, and diodes.

sending unit The component in the fuel supply system responsible for sending constant electrical signals to the gas gauge located in the driver information center.

separator PCV system A PCV system with a device that uses gravity to allow oil to fall to the bottom of the valve and be returned to the crankcase; the valve prevents liquid from traveling to the intake manifold.

separator plate Sometimes called a spacer plate, a thin sheet metal plate installed between the valve body and the transmission case. Orifices can be installed in the separator plate, and check valves can work with holes in the plate.

series hybrid A hybrid electric vehicle powered by an internal combustion engine but driven by an electric motor and battery.

series hybrid drivetrain A type of hybrid transmission in which power flows from the engine through an electric motor. The electric motor supplements the power from the engine to the wheels.

series-parallel circuit A circuit that has both a series and a parallel circuit combined into one circuit.

series-parallel hybrid A hybrid electric vehicle that uses the internal combustion engine and/or the battery pack for propulsion.

series-parallel hybrid drivetrain A type of hybrid drivetrain that can function as both a series hybrid and parallel hybrid. That means the gasoline engine can turn a generator that can be used to power an electric motor. The gasoline engine can also drive the vehicle directly through the transmission. And the electric motor can work in parallel with the gasoline engine to drive the vehicle.

serpentine condenser A condenser with one tube snaking back and forth.

service brake A brake system that is operated while the vehicle is moving, in order to slow or stop the vehicle.

service campaign and recall A corrective measure conducted by manufacturers when a safety issue is discovered with a particular vehicle.

service consultant/advisor A customer service worker who works with both customers and technicians; the first point of contact for customers seeking vehicle repairs.

service history A complete list of all the servicing and repairs that have been performed on a vehicle.

service manager The shop supervisor who is responsible for the management of the service department.

serviceable bearings Wheel bearings that can be disassembled, cleaned, inspected, packed, reinstalled, and adjusted.

servo A hydraulic device consisting of a piston inside a cylinder that is used to apply a band. The servo typically has a spring that releases the servo when it is no longer **needed.**

servo action A drum brake design where one brake shoe, when activated, applies an increased activating force to the other brake shoe, in proportion to the initial activating force; further enhances the self-energizing feature of some drum brakes.

setback The distance one wheel is set back from the wheel on the opposite side of the axle.

shaft The long, narrow component that carries one or more gears or has gears machined into it.

shielded wiring harness A wiring harness that has shielding built into it to protect it from induced electrical interference.

shift fork A mechanism that moves the synchronizer sleeve to lock the gear to the main shaft.

shift solenoid An electromechanical device used to control oil flow to bands and clutches in an automatic transmission to help shift the transmission.

shift valve A spool valve used to control oil flow to a clutch or band for a particular gear.

shock Inadequate tissue oxygenation resulting from serious injury or illness.

shock absorber A device on a vehicle designed to absorb bumps and jolts caused from driving on irregular surfaces and to dampen body movement.

shop foreman The supervisor in a shop who oversees the work of technicians and staff and communicates with customers and external suppliers.

shop or service manual Manufacturer's or after-market information on the repair and service of vehicles.

short Also called a short circuit, the flow of current along an unintended route.

short block An engine assembly that includes the engine block, camshaft, timing set, pistons, rods, and crankshaft installed.

short circuit A condition in which the current flows along an unintended route.

short to ground Fault conditions in a circuit where the circuit is unintentionally contacting a grounded component or wire. This may result in a short, in the case of a power wire; or it could cause a circuit to stay live in the case of a switched ground circuit.

short to power A condition in which current flows from one circuit into another.

short-/long-arm (SLA) suspension A type of control arm suspension system that uses a short control arm on the top and a long control arm on the bottom. This design ensures correct alignment angles when moving through bumps.

shroud A steel or plastic cover placed over the shock rod.

side force The pressure on the wheel that pushes it toward the outside or inside of the rim as the vehicle makes a turn.

side gear A gear that is splined to the axle shaft and meshes with the spider gears and allows the axles to rotate at their own speeds when cornering and turning.

side rail The thin portion of the oil control ring that is used to scrape oil off the cylinder walls.

signal voltage value Measured voltage in a signal return circuit that is compared to a specified voltage value published by the manufacturer.

silicon A material commonly used to make semiconductors.

silicon carbide A type of material used to make semiconductors.

simple fracture A fracture that involves no open wound or internal or external bleeding.

Simpson gear set A type of gear set with two planetary gear sets that share a common sun gear.

sine wave A mathematical function that describes a repetitive waveform such as an alternating current signal.

single flare A sealing system made on the end of metal tubing.

single-piston master cylinder A master cylinder with a single piston that creates hydraulic pressure for all wheel units. If there is a leak in the system, there is a loss of pressure for all wheel units.

single-post hoist A type of vehicle hoist that uses a single central platform to lift a vehicle.

sintering The process of using pressure and heat to bond metal particles.

sintering process A metal hardening process in which the metal is fused together without melting.

slave cylinder The component in a hydraulically operated clutch that converts hydraulic pressure to mechanical movement at the clutch fork.

sledgehammer A heavy hammer, usually with two flat faces, that provides a strong blow.

sliding or floating caliper A type of brake caliper that only has piston(s) on the inboard side of the rotor. The caliper is free to slide or float, thus pulling the outboard brake pad into the rotor when braking force is applied.

sliding spline driveshaft A two-piece driveshaft that is joined in the middle with splines. The driveshaft can slide on itself to increase or decrease in length.

sliding T-handle A handle fitted at 90 degrees to the main body that can be slid from side to side.

slip angle A tire's sideways distortion that makes the vehicle follow a path at an angle to the direction the road wheel is pointing.

slip yoke Part of a two-piece driveshaft that is splined and allows for a change in length of the shaft as the suspension compresses and rebounds.

slippage A condition in which two surfaces in firm contact with each other slide.

slow charger A battery charger that charges at low current.

slow-fill The method of filling a compressed natural gas vehicle (such as overnight) so the cylinders are completely filled.

smart charger A battery charger with microprocessor-controlled charging rates and times.

snap ring The spring steel, C-shaped ring that is fitted in a groove and holds gears, bearings, and shafts in place.

snap ring pliers A pair of pliers for installing and removing snap rings or circlips.

sniffer An electronic device used to determine the source of leaks.

socket An enclosed metal tube commonly with 6 or 12 points to remove and install bolts and nuts.

soft plug A thin-walled, metal, cup-shaped disc designed to be pressed into a machined passageway in the block for the purpose of plugging it.

Solder A mixture of lead and tin with a low melting point for connecting wires.

solder-type terminal A terminal that requires soldering to fasten the terminal to the cable or wire.

soldering irons A heating tool to heat solder and wires to produce a low-resistance joint.

solenoid An electromagnet with a moving iron core that is used to cause mechanical motion.

solenoid valve An electrically operated valve that when used in brake systems is designed to control the flow of brake fluid in the hydraulic system.

solid axle A single piece of steel that provides a simple means of mounting the hub and wheel units; also called beam axle or straight axle.

solid rear axle A type of axle that has a one-piece axle housing, so that the action of hitting a bump with one wheel affects the other wheel.

solid rotor A type of brake rotor made of solid metal, not ventilated.

solid valve lifter A non-hydraulic valve lifter.

solid-state relay A relay that performs the function of a mechanical relay but uses only electronic components.

solvent tank A tank containing solvents to clean vehicle parts.

spark line The horizontal line immediately after the firing line on an ignition pattern. It shows how much voltage and the length of time the spark is burning.

spark plug A device that provides a gap for the high-voltage spark to occur in each cylinder.

spark plug reach The length of the spark plug from the seat to the end of the threads.

spark timing The point at which a spark occurs at the spark plug relative to the position of the piston.

spark-ignition (SI) engine An engine that relies on an electrical spark to ignite the air and fuel mixture.

specialty springs Springs used to return links and levers on the parking brake system or the self-adjuster mechanism.

speed brace A U-shaped socket wrench that allows high-speed operation. Also called a speeder handle.

speed control circuit A circuit that controls the speed of a motor.

speedy sleeve An aftermarket repair kit that consists of a thin metal sleeve that fits tightly over the seal surface of the axle, providing a new, undamaged surface for the seal to ride against.

splash lubrication A lubrication system that relies on oil being splashed onto moving parts by rotating engine parts striking the oil. These systems are typically used in small engines.

spline Ridges or teeth on a shaft that mesh with grooves in a mating piece and transfer torque to it, maintaining the angular correspondence between them.

splined Typically, a shaft and gear that have parallel grooves machined in them so they mate with each other and lock together rotationally.

splined section A flat key made into a shaft to accommodate changes in shaft length due to movement in wheel camber with suspension action.

split ball gauge A measuring device used to accurately measure small holes.

spool valve A type of valve commonly used in automatic transmissions that resembles the spool that thread or fishing line comes on.

sport utility vehicle (SUV) A passenger vehicle built on a light-truck chassis; it is usually equipped with four-wheel drive and capable of hauling heavier loads than typical passenger vehicles.

sprain An injury in which a joint is forced beyond its natural movement limit.

spray wash cabinet A cleaning cabinet that sprays cleaning solution under pressure to clean vehicle parts.

spring A resilient steel part that stores energy when compressed and releases energy when released to its original state; available as a leaf spring, coil spring, or torsion bar.

spring eyes Rolled ends of some springs used to mount springs to the chassis.

spring pressure Pressure exerted by a metal coil, usually measured in pounds or kilograms.

spring shackle bushing A bushing that is positioned in the shackle to which the leaf spring mounts. Bushings allow the spring shackle to move as the leaf spring dimensions change over bumps.

spring-loaded key A part of the synchronizer that helps hold the synchronizer collar in position.

spring-loaded rack guide yoke A spring-containing part that pushes on the back side of the rack to help reduce the play between the rack and the pinion while still allowing for relative movement.

springs and clips Various devices that hold the brake shoes in place or return them to their proper place.

spur gear A type of gear in which the teeth of the gear are cut in a straight line down the axis of the gear.

spur gears Gears with straight-cut gear teeth.

square file A type of file with a square cross section.

square thread A thread type with square shoulders used to translate rotational to lateral movement.

square-cut O-ring An O-ring with a square cross section that is used to seal the pistons in disc brake calipers.

squib The component inside the airbag inflator that triggers the airbag deployment.

staged governor A governor in which the assembly uses two valves: a primary valve and a secondary valve.

stall Often referred to as stall speed, the maximum rpm difference between the turbine and the impeller in the torque converter.

stall test A test that involves raising the engine rpm to wide-open throttle while the brake is firmly applied and the transmission is in gear. The test is used to check torque convertor and transmission operation on some vehicles.

standard torque converter A hydraulic coupling device consisting of an impeller, turbine, stator, and housing; located between the engine and the transmission.

starter teeth Machined teeth located on the ring gear for meshing with the starter drive teeth located in the starter motor.

state of charge The amount of refrigerant in a system compared to how much should be in it.

static imbalance A tire imbalance resulting from a heavy spot on a tire; it will vibrate vertically with the heavy area slapping the road surface with each turn of the wheel.

static toe A setting designed to compensate for slight wear in steering components that may cause the wheels to turn slightly outward or inward while the vehicle is in motion.

station wagon A vehicle configuration with four doors, with a roof line that continues into the rear cargo area and a rear door for access.

stationary winding An extended length of wire wrapped into a circle. These windings are fixed as opposed to some types, which are meant to spin.

Stator Portion of an electronic ignition system that is mounted to the base of the distributor. It has a circular permanent magnet with a number of projections or teeth corresponding to the number of engine cylinders, and a stationary coil of fine enameled copper wire wound on a plastic reel and positioned inside the magnet.

stator winding A winding in an alternator that creates the current output, or a winding in a motor that creates the needed magnetism for the motor to rotate.

steel hammer A hammer with a head made of hardened steel.

steel rule An accurate measuring ruler made of steel.

steel-disc–type rim A plain steel wheel that is typically covered by a hubcap.

steering angle sensor A sensor that measures the amount of turning a driver desires. This information is used by the ESC system to know the driver's directional intent.

steering arm An arm that extends from the steering knuckle. The tie rods connect to these arms in order to steer the wheels.

steering axis inclination (SAI) The angle formed by an imaginary line running through the upper and lower steering pivots relative to vertical as viewed from the front.

steering box (steering gear) A device that converts the rotary motion of the steering wheel to the linear motion needed to steer the vehicle.

steering column A column affixed between the steering wheel and the steering box, usually made to collapse during a crash.

steering damper A device used to prevent shocks from irregular roads from being transmitted through the steering linkage and back to the steering wheel.

steering knuckle The knuckle located between the lower control arm and MacPherson strut or upper control arm, which has a spindle made or bolted onto it, to which the wheel hub is attached.

steering linkage Steel rods that connect the steering box to the steering arms on the steering knuckle.

steering sensor A sensor that can read both torque and rotation from the steering wheel.

steering system A term used to describe all of the components and parts involved in steering a vehicle.

steering wheel position sensor A sensor that signals to the EBCM both the position and speed of the steering wheel.

step-down transformer A transformer used to reduce the voltage, such as to allow a battery charger operated on 120 volts to charge a 12-volt battery.

step-up transformer A transformer used to increase the voltage from a lower input voltage to a higher output, such as an ignition coil.

stepper motor EGR valve An EGR valve that uses an electrically controlled motor to open the pintle in steps in order to let exhaust gases flow into the intake manifold.

stepper motor A type of brushless motor with a key difference: It is designed to rotate in fixed steps through a set number of degrees.

stoichiometric ratio The optimum ratio of air to fuel for combustion—14.7 parts air to 1 part fuel by weight.

stop A rubber part used to control the movement of control arms (suspension arms).

straight edge A measuring device generally made of steel to check how flat a surface is.

straight grinder A powered grinder with the wheel set at 90 degrees to the shaft.

straight vegetable oil (SVO) Biodiesel made from plants such as soybeans, jatropha, and palms.

strain An injury caused by the overstretching of muscles and tendons.

strategy-based diagnosis The process of logically and systematically (strategically) solving problems.

stratified charge A layering of the air-fuel mixture in the combustion chamber. Used when the fuel injector places fuel near the spark plug, but not the surrounding space.

stroke The movement of an object in a straight line. The piston sees four strokes during one combustion cycle in a four-stroke cycle engine, meaning it moves up and down twice during each cycle.

strut A shock absorber used on a MacPherson strut–type suspension.

stub axle An axle used for one wheel.

stub-axle carrier The body of the stub-axle knuckle.

stud A type of threaded fastener with a thread cut on each end rather than having a bolt head on one end.

subframe A mount attached to the vehicle that is used to support the engine and transaxle assembly.

suction reed valve A flat, spring-loaded valve used as a check valve on the suction side of a compressor.

sulfur A regulated by-product of petroleum-based (especially diesel) fuel.

sulfur dioxide (SO_2) A pollutant resulting from sulfur in motor fuel and which contributes to acid rain.

sulfuric acid A type of acid that when mixed with pure water forms the basis of battery acid or electrolyte.

sun gear The center gear of a planetary gear set around which the other gears rotate.

sun load (solar) sensor A photodiode that varies voltage based on light. It is used to determine the radiant heat coming from the sun into the passenger cabin and gives an input signal of sunlight load to the ECU.

supercharger A device that pressurizes airflow into the engine, working similarly to a turbocharger. The supercharger is driven by the crankshaft through a belt or gears and does require power from the engine.

superheat spring A spring used as part of the control in the TXV to ensure refrigerant leaving the evaporator is boiled completely or superheated.

supplemental restraint system (SRS) A passenger safety system, such as airbags and seat belt pretensioners.

supporting statement A statement that urges the speaker to elaborate on a particular topic.

surface filter A filter that is a simple screen mechanism to catch dirt and other particles in the hydraulic oil as it passes through.

surface roughness average (Ra) A measurement used to classify how rough a particular mating surface is at the microscopic level.

surge protector An electrical protection device for preventing electrical surges.

surge tank A sealed tank that captures coolant coming from the head that has turned to steam and then changes the steam back to coolant to be reused by the cooling system.

suspension action Movement of the chassis up and down.

suspension strut A shock absorber designed to reduce spring oscillations.

suspension system A system within a vehicle designed to isolate the vehicle body from road bumps and vibrations.

swash plate Another name for an axial plate.

sway bar A part used in vehicles as a stabilizer, or antiroll, bar. It is connected to the chassis in the center, and each end is connected to one side of the suspension system. It is typically installed on the front, and sometimes the rear, suspension.

swinging shackle A shackle connected to the rear of the multileaf spring that allows the leaf spring to move downward when a load is placed on the rear of the vehicle.

switch An electrical device with contacts that turns current flow on and off.

switching transistor An electronic device used to control the flow of current through the ignition coil primary winding.

symmetric tread pattern A tread pattern with the same tread pattern on both sides of the tire; typically nondirectional.

synchromesh transmission A modern transmission that uses gear synchronizers to match the speeds of gears and shafts during upshifts and downshifts.

synchronizer An assembly that allows for the selection of gears without grinding by matching the speed of the two assemblies.

synchronizer sleeve The sleeve that slides to lock the selected gear to the main shaft. The synchronizer sleeve is part of the synchronizer assembly.

syngas An artificially created motor fuel.

synthetic blend A blend of conventional engine oil and pure synthetic oil.

synthetic oil Synthetic oil that, in its pure form, uses artificially made base stocks and is not derived from crude oil. This oil lasts longer and performs better than normal oil. The base stock additives are similar to those in conventional oils.

tandem master cylinder A master cylinder that has two pistons that operate separate braking circuits so that if a leak develops in one circuit, the other circuit can still operate.

tang A raised or slightly bent corner on some engine bearing shells that fits into a matching recess in the bearing mounting surface to keep the bearing from rotating in the bore.

tanks Metal or plastic pieces that line the pipes used to connect the tubes together to allow the coolant to continue to flow.

tap A term used to generically describe an internal thread-cutting tool.

tap handle A tool designed to securely hold taps for cutting internal threads.

taper An object that is smaller in diameter at one end.

taper tap A tap with a tapper; it is usually the first of three taps used when cutting internal threads.

tapered roller bearing A type of wheel bearing with races and rollers that are tapered in such a manner that all of the tapered angles meet at a common point, which allows them to roll freely and yet control thrust.

tapered seat A type of lug nut with a tapered end toward the rim that helps center the wheel on the wheel studs.

tappet Another name for a valve lifter. Tappets may be fat or have rollers to ride on the cam lobes.

technical service bulletin (TSB) Information issued by manufacturers to alert technicians of unexpected problems or changes to repair procedures.

telematics Satellite-based, two-way communication with the vehicle's electronic systems.

telescoping gauge A gauge that expands and locks to the internal diameter of bores; a caliper or outside micrometer is used to measure its size.

temperature grade A standardized grading system that indicates the extent to which heat is generated or dissipated by a tire.

tensile strength A measure of how strong a material is as it is being pulled apart.

terminal A means of providing a low-resistance connection or termination at the end of a wire.

test certificate A certificate issued when lifting equipment has been checked and deemed safe.

tetrafluoroethane (R-134a) An inert colorless gas that can be used as a refrigerant. It is stored in light blue containers.

thermal cycle switch A normally closed switch located in the fins of the evaporator that controls the air-conditioning compressor clutch. It is opened and closed by the temperature changes in the evaporator.

thermal expansion valve (TXV) system A system with a valve designed to sense evaporator outlet temperature and vary the inlet orifice size accordingly.

thermal runaway A cycle during battery charging in which heating lowers resistance, which in turn increases current flow, which in turn further increases heat created. During this cycle, dangerous gases may build up, creating the potential for an explosion or damage through excessive current flow.

thermistor A temperature-controlled variable resistor. As temperature changes, so does resistance.

thermo-control switch A temperature-sensitive switch that is mounted into a coolant passage on the engine or into the radiator to control electric fan operation.

thermocouple A temperature-sensing component that consists of two dissimilar metals that produce voltage proportional to temperature.

thermostat Located under the thermostat housing, the thermostat regulates the flow of coolant, allowing coolant to flow from the engine to the radiator when the engine is running at its operating temperature.

third-degree burns Burns that involve white or blackened areas and damage to all skin layers and underlying structures and tissues.

thixotropy The ability of a semisolid grease to flow when agitated or stressed.

thread file A type of file that cleans clogged or distorted threads on bolts and studs.

thread pitch The coarseness or fineness of a thread as measured by either the threads per inch or the distance from the peak of one thread to the next. Metric fasteners are measured in millimeters.

thread repair A generic term to describe a number of processes that can be used to repair threads.

threaded fasteners Bolts, studs, and nuts designed to secure parts that are under various tension and sheer stresses. These include bolts, studs, and nuts, and are designed to secure vehicle parts under stress.

three-angle grind A process of grinding the valve seats on the cylinder head so that air can pass through them more smoothly and quickly. Also used to narrow and position the seat.

three-quarter floating axle An axle on which there is only one wheel bearing that bears the weight of the vehicle, but the axle prevents the wheel from tipping inward or outward.

three-way catalytic converter A converter that changes hydrocarbons, carbon monoxide, and oxides of nitrogen into harmless elements.

threshold limit value (TLV) The maximum allowable concentration of a given material in the surrounding air.

throttle A device used to produce acceleration by controlling the air-fuel mixture.

throttle body The housing on an intake manifold that is used to control the amount of filtered air that enters the cylinders.

throttle body injection (TBI) A fuel injection system that uses one or more fuel injectors mounted above or in the throttle body itself. Also called single-point injection.

throttle position sensor (TPS) A type of potentiometer used by the PCM to measure throttle angle.

throttle position sensor A sensor that measures how far the throttle valve is opened.

throttle valve A type of spool valve that is connected to the throttle on a vehicle. The throttle valve creates a pressure proportional to throttle opening and is used to delay upshifting based on throttle opening.

throttle valve pressure The pressure created by the throttle valve that is proportional to throttle opening.

throw The offset area of the crankshaft where the connecting rod bolts on.

throw The part of the crankshaft that is offset for the connecting rods to mate with; also called the connecting rod journal.

throw-out bearing The part of the clutch release mechanism that imparts clutch pedal force to the rotating pressure plate levers.

thrust angle The angle formed between the perpendicular centerline of the rear axle in comparison to the centerline of the vehicle.

thrust bearing A main bearing insert that either has integrated flanges or separate flanges that provide bearing surfaces that prevent forward or lateral movement of the crankshaft assembly.

thrust line The imaginary line drawn perpendicular to the rear axle.

thrust washer A small steel washer coated with bearing material. The thrust washer is used to control forward and backward movement of a part in an automatic transmission.

thrust-type angular-contact ball bearing A type of bearing that uses a deep groove in the bearing races where the ball bearings ride; this design is for thrust conditions.

tie rod A steering component that transfers linear motion from the steering box to the steering arms at the front wheels.

tie-rod assembly The part that fits between the rack and the steering arms and transfers the movement of the rack.

tie-rod end The close-fitting ball and socket on the end of a tie rod that allows for movement of the steering components.

timing chain A steel chain connecting the crankshaft assembly to the camshaft assembly.

timing gear A sprocket attached to the crankshaft assembly and the camshaft assembly.

tin snips Cutting device for sheet metal, works in a similar fashion to scissors.

tire inflation pressure The amount of air pressure in the tire that provides it with load-carrying capacity and affects the overall performance of the vehicle.

tire inflation pressure The level of air in the tire that provides it with load-carrying capacity and affects overall vehicle performance.

tire placard A metal, vinyl, or paper tag permanently affixed to a vehicle that indicates the appropriate tire size and inflation pressure for the vehicle.

tire pressure gauges A gauge used to measure the air pressure within a tire.

tire pressure monitoring system (TPMS) An automatic sensor system in most modern vehicles that alerts drivers of tire pressure problems.

tire valve The valve through which air is inserted into a tire to inflate it.

title history A detailed account of a vehicle's past.

toe setting Setting of the toe-in or toe-out of the tires to the centerline of the vehicle.

toe-in When the front of the wheels, as seen from above, are closer together than the rear of the wheels.

toe-out When the rear of the wheels, as seen from above, are closer together than the front of the wheels.

toe-out on turns The difference in turning angle of the inside tire (smaller turning radius) in comparison to the outside tire. This angle difference allows the tires to roll through the corner rather than the inside tire dragging; also referred to as Ackermann angle.

tone wheel The part of the wheel speed sensor that has ribs and valleys used to create an electrical signal inside of the pickup assembly.

top dead center (TDC) The position of the piston when it is farthest from the crankshaft.

top hat parking brake A drum brake that is located inside a disc brake rotor in order to act as a parking brake.

toroidal CVT A type of CVT that uses moveable rollers in contact with input and output drive discs. The rollers transfer power from one drive disc to the other. Their position determines the effective gear ratio.

torque angle A method of tightening bolts or nuts based on angles of rotation.

torque Twisting force applied to a shaft that may or may not result in motion.

torque Twisting force.

torque angle A tightening procedure in which a bolt is torqued to a set torque value and then tightened using a measured angle instead of a torque value. Example: A bolt is torqued to 40 ft-lb and then tightened another 90 degrees.

torque assist Use of an electric motor to supplement the engine's torque whenever additional torque is needed, allowing for a smaller ICE to be used.

torque converter A device that is turned by the crankshaft and transmits torque to the input shaft of an automatic transmission.

torque converter clutch (TCC) A hydraulically operated clutch located inside the torque converter that applies at predetermined conditions and stops torque converter slippage.

torque multiplication The increase of torque.

torque plate A 2" (51 mm) thick plate that is bolted where the cylinder head is fastened on the engine block during the boring process, simulating head bolt torque.

torque sensor A device used to measure the load on the steering wheel.

torque smoothing A process that uses an electric motor to smooth out engine power pulses when an ICE is operating at low rpm or when the vehicle is using fuel management techniques such as cylinder deactivation.

torque specifications Supplied by manufacturers and describes the amount of twisting force allowable for a fastener or a specification showing the twisting force from an engine crankshaft.

torque steer A condition in which the vehicle pulls to one side during hard acceleration.

torque wrench A tool used to measure the rotational or twisting force applied to fasteners.

torque-to-yield A tightening procedure in which a bolt is designed to be slightly elastic when tightened; the elastic bolt retains an even pressure on the head gasket.

torque-to-yield A method of tightening bolts close to their yield point or the point at which they will not return to their original length.

torque-to-yield (TTY) bolts Bolts that are tightened using the torque-to-yield method.

torsion bar A spring-loaded piece of steel connected to the pinion gear at its bottom end and the input shaft at its top. Also, a torsion bar is a type of spring that some vehicles use to hold up a corner of the vehicle.

torsional load A force that is applied by clamping one end of an object to another object that is then twisted.

torsional vibrations The speeding up and slowing down of a shaft, which happen at a relatively high frequency. Crankshafts have torsional vibrations due to the power pulses of the pistons.

town gas Locally available gas used to power early internal combustion engines.

toxic dust Any dust that may contain fine particles that could be harmful to humans or the environment.

traction control system (TCS) A computer-controlled system added to ABS to help prevent loss of traction while the vehicle is accelerating.

traction grade A standardized grading system that indicates how well a tire will maintain contact with the road surface when wet.

trailing arm suspension A type of suspension system that uses upper and lower control arms.

trailing shoes Brake shoes installed so that they are applied in the opposite direction to the forward rotation of the brake drum; not self-energizing and less efficient at developing braking force.

transaxle A type of transmission typically used in FWD vehicles in which the transmission also includes the differential and final drive gear assembly.

transfer case An assembly used in FWD vehicles to transmit power to either two wheels only or all four wheels.

transformer action The transfer of electrical energy from one coil to another through induction in a transformer.

transistor A semiconductor device that allows a small current in the base lead to control a larger current through the emitter collector leads.

transmission An assembly that houses a variety of gear sets that allow the vehicle to be driven at a wider range of speeds and terrain conditions than would be possible without a transmission.

transmission control module (TCM) A computer module that controls the transmission operation. It may be integrated into the PCM.

transmission fluid temperature (TFT) sensor A type of variable resistor used inside the transmission to monitor oil temperature.

transmission input shaft The shaft that brings engine torque into the transmission.

transmission mounted parking brake A drum brake that is mounted on the drive shaft, just after the transmission, to serve as a parking brake.

transmission specialist A technician who diagnoses, overhauls, and repairs transmissions.

transverse A term used to describe the side-to-side engine orientation when mounted in the engine compartment.

tread wear grade The number imprinted on the sidewall of a tire by the manufacturer, as required by the National Highway Traffic Safety Administration (NHTSA), that indicates the tread life of a tire's tread.

triangular file A type of file with three sides so it can get into internal corners.

triangulation A method of locating an object using mathematics based on forming a triangle with two known points.

trilateration A method of locating points by the measurement of distances using geometry.

trim height The height from the ground to a specified part of the vehicle; also known as ride height.

trimetal The use of three different types of metals to build up a bearing insert to give it long-lasting wear characteristics.

truck A large, heavy vehicle for carrying cargo.

tubes Metal pipes running side to side or up and down that the coolant or refrigerant travels through.

tube flaring tool A tool that makes a sealing flare on the end of metal tubing.

tubing cutter A hand tool for cutting pipe or tubing squarely.

tulip/tripod joint A constant-velocity joint that has three equally spaced fingers shaped like a star. This configuration allows in-and-out movement of the shaft while allowing flexing.

turbocharger A device that pressurizes airflow into the engine. The turbocharger works similarly to a supercharger but is driven by the exhaust gases.

turn signal switch A switch that turns the left and right turn signal lights on and off.

turning radius A measure of how small a circle the outside front wheel on a vehicle can turn around when the steering wheel is turned to the limit.

twin leading shoe drum brake system Brake shoe arrangement in which both brake shoes are self-energizing in the forward direction.

twist drill A hardened steel drill bit for making holes in metals, plastics, and wood.

twisted pair Two conductors that are twisted together to reduce electrical interference.

two-mode hybrid A type of hybrid drive system in which there are two distinct modes of operation. In one mode, the electric motor can propel the vehicle and be used for regenerative braking; in the second mode, the electric motors can be used to assist the engine while the engine uses fuel management techniques such as cylinder deactivation.

two-post hoist A type of vehicle hoist that uses two parts (one on each side of vehicle) and four arms to lift the vehicle.

two-stroke engine An engine that uses only two strokes to complete its running cycle.

two-way catalytic converter A converter that changes only hydrocarbons and carbon monoxide into harmless elements.

ultrasonic A method of leak detection that uses a sensitive microphone to hear small refrigerant leaks by amplifying the hissing noise.

umbrella-style valve stem seal A seal that surrounds the top of the valve stem to prevent excessive oil from leaking into the valve guide.

understeer A condition in which a vehicle's front wheels turn more sharply than the vehicle's direction when the vehicle is steered around a corner. This vehicle is said to be "loose" in the corners.

unibody design A vehicle design that does not use a rigid frame to support the body. The body panels are designed to provide the strength for the vehicle.

uniform pitch A spring whose pitch (the distance from the center of one coil to the center of the adjacent coil) is the same distance throughout.

Uniform Tire Quality Grading (UTQG) A standardized grading system, established by the National Highway Traffic Safety Administration (NHTSA), designed to provide tire buyers with a comparative measure of a tire's tread life, traction, and temperature characteristics.

unitized wheel bearing hub An assembly consisting of the hub, wheel bearing(s), and possibly the wheel flange, which is preassembled and ready to be installed on a vehicle.

universal joint (U-joint) A cross-shaped flexible joint on which caps fit over the ends of the cross. Needle bearings fit between the ends of the cross and the caps, allowing the caps to rotate smoothly.

universal joint A cross-shaped joint with bearings on each leg, where one set of parallel legs is connected to the end of one shaft and the other set of parallel legs is connected to the end of a second shaft. This arrangement allows the shafts to operate at shallow angles to each other.

unsprung mass Any part of the steering and suspension system that is not supported by springs. A large amount of unsprung weight causes the tire to hop off the ground when hitting bumps, as the weight overcomes dampening of the shock absorbers.

unsprung weight *See* unsprung mass.

U. S. Department of Transportation (DOT) A federal agency that regulates transportation safety in the United States, including vehicles' wheels and tires. The DOT requires a code—a series of letters and numbers—to be stamped into the sidewall of every tire made for public use in the United States. These codes contain information such as the date of manufacture and the plant where the tire was manufactured.

V blocks Metal blocks with a V-shaped cutout for holding shafts while working on them. Also referred to as vee blocks.

V engine A term used to describe an engine configuration that has two banks of cylinders sitting side by side in a V arrangement sharing a common crankshaft.

vacuum A pressure in an enclosed area that is lower than atmospheric pressure.

vacuum advance unit A mechanism that controls ignition timing advance in relation to engine load and causes the spark at the spark plug to occur sooner based on engine conditions. Its function is to improve fuel economy and, in doing so, reduce exhaust emissions.

vacuum bleeding Bleeding process that uses a vacuum bleeder to pull the air and old brake fluid from the system.

vacuum gauge A device used to measure the amount of vacuum an engine can generate during various operating conditions.

vacuum modulator A device on a hydraulically controlled transmission that converts engine manifold vacuum into an engine load signal called modulator pressure.

vacuum pump A pump used to evacuate the air-conditioning system and put it into a deep vacuum or low pressure to remove moisture.

vacuum relief valve A mechanical valve used on the gas cap that prevents a low-pressure condition from occurring. It also ensures that the fuel tank will not collapse due to fuel usage or contraction of fuel on cooldown.

vacuum servo An air-conditioning door actuator controlled by a vacuum.

vacuum tube fluorescent (VTF) A type of lighting used for instrumentation displays on vehicle instrument panel clusters. This type of lighting emits a very bright light with high contrast and can display in various colors; also called vacuum fluorescent display (VFD).

validating statement A statement that shows common interest in the topic being discussed.

valve A device used to control the flow of air and fuel into the combustion chamber and exhaust gases out of the combustion chamber.

valve body An aluminum or cast iron housing inside the transmission that houses the majority of the valves that control transmission operation.

valve core The one-way spring-loaded valve that screws into the valve stem; it allows air to be pumped into a tire and prevents it from flowing out.

valve cover gasket A gasket used to seal the valve cover (also called a rocker arm cover or cylinder head cover) to the cylinder head assembly.

valve face A machined surface on the back of the valve head; this area seals onto the valve seat in the cylinder head.

valve float A condition that occurs when the valves are moving so fast that the spring tension is not great enough to fully seat the valves as designed.

valve guide A hole machined into the cylinder head or a hollow metal tube pressed into the cylinder head in which the valve stem rides, holding the stem in the proper position within the cylinder head.

valve head The portion of the valve that is exposed to the combustion chamber and contains the valve face and margin.

valve keeper A device used to keep the valve spring retainers attached to the valve while in the cylinder head.

valve lifter A device that transfers motion from the cam lobe to a pushrod or directly acts on the valve and spring, depending on whether the cam is in the engine block or the cylinder head; sometimes called a tappet.

valve margin The flat surface on the outer edge of the valve head between the valve head and the valve face.

valve seat An integral part of the head, or circular metal rings that are pressed into the cylinder head, that makes up the mating surface where the valve face sits when it is closed.

valve spring A metal coil spring that returns valves to their fully closed positions after being opened.

valve spring retainer A washer-shaped piece of metal positioned near the top of the valve stem that holds the top of the valve spring to keep pressure on the valve while in the cylinder head.

valve stem The shaft that is attached to the valve head and provides the sliding surface for the valve in its guide as it opens and closes.

valve stem cap A cap that fits tightly onto the valve stem to prevent debris from clogging it and acts as a secondary seal.

valve tip The end of the valve stem against which the rocker arm, cam follower, or hydraulic bucket-style lifter directly presses on to open the valve.

valve train A system encompassing all of the parts used in the opening and closing of the valves.

vane pump A type of front pump in an automatic transmission that uses small vanes placed radially around a center hub. The vanes may be forced out against the housing of the pump by centrifugal force. The volume between the vanes varies as the pump turns, resulting in fluid being forced out of the pump.

vapor line A rubber or plastic line that carries vapors from the fuel tank to the charcoal canister.

vapor lock A situation in which vapor forms in the fuel line, and the bubbles of vapor block the flow of fuel and stop the engine.

vaporization The changing of a liquid to a gas through boiling.

variable displacement vane pump A type of front pump that uses a vane-style pump that can pivot so that its displacement can be varied.

variable reluctance sensor A type of wheel speed sensor that uses the principle of magnetic induction to create its signal.

variable resistor A component that has a mechanism for varying resistance.

variable voltage signal A signal that changes based on what the sensor is reading; for example, as temperature varies, so does the voltage signal to the ECU.

variable-diameter pulley (VDP) A type of CVT that uses two pulleys with moveable sheaves, allowing the effective diameter of the pulleys to change, resulting in variable gear ratios.

variable-orifice PVC system A system in which a replaceable, spring-loaded PCV valve regulates gas flow. The position of the PCV valve is controlled by the pressure in the manifold.

vehicle emission control information (VECI) label A label used by technicians to identify engine and emission control information for the vehicle.

vehicle hoist A type of vehicle lifting tool designed to lift the entire vehicle.

vehicle identification number (VIN) A unique serial number that is assigned to each vehicle produced.

vehicle inspection pit A trench permanently fitted into the floor of the shop to allow easy work access to the vehicle's underside.

vehicle jack A tool for lifting a vehicle.

vehicle safety certification (VSC) label A label certifying that the vehicle meets the Federal Motor Vehicle Safety, Bumper, and Theft Prevention Standards in effect at the time of manufacture.

vehicle speed sensor (VSS) A sensor used by the PCM to measure vehicle speed. It is often located in the transmission extension housing.

vent solenoid A solenoid that allows fresh air to enter the evaporative system during a purge event. Also used for an evaporative system monitoring test.

ventilated rotor A type of brake rotor with passages between the rotor surfaces that are used to improve heat transfer to the atmosphere.

venturi A restriction (narrowed area) in the air horn.

vernier calipers An accurate measuring device for internal, external, and depth measurements that incorporates fixed and adjustable jaws.

viscosity The ability of a liquid to flow.

viscosity index improver An oil additive that resists a change in viscosity over a range of temperatures.

viscous coupler Called a fan clutch, a hub that connects the water pump drive to the cooling fan using a temperature-sensitive viscous fluid to cause the fan to turn faster as the temperature of the air pulled through the radiator increases.

viscous coupling An silicone clutch assembly used in AWD differentials to provide a slight amount of differential action for control of axle rotational speeds.

volt The unit used to measure potential difference or electrical pressure.

volatile organic compound (VOC) The hydrocarbons in petroleum products that contribute to combustion.

voltage The pressure aspect of electricity, measured to show the potential of a circuit to do work.

voltage drop The amount of potential difference between two points in a circuit.

voltage drop test Measurement of the difference in voltage between two points in a circuit, the black lead on the end point being tested and the red lead on the beginning point.

volumetric efficiency A ratio, given as a percentage, of the amount of air actually inducted at a given engine speed at full throttle compared to the internal engine displacement. For a normally aspirated engine (without supercharging or turbocharging), an engine's volumetric efficiency may peak at around 85%. Peak engine torque is developed at peak volumetric efficiency.

vortex flow State in which the fluid in the torque converter is traveling between the turbine, stator, and impeller.

VR engine A term used to describe an engine configuration that uses a single bank of cylinders staggered at a shallow 15-degree V.

W engine A term used to describe an engine configuration consisting of two VR cylinder banks in a deep V arrangement.

wad punch A type of punch that is hollow for cutting circular shapes in soft materials such as gaskets.

warding file A type of thin, flat file with a tapered end.

warm-up cycle When the engine is started and run until the engine operating temperature reaches 160°F (71.1°C), has increased in temperature at least 40°F (22°C), and the engine is turned off again.

waste cylinder The cylinder in a waste spark ignition system that receives a spark near the top of its exhaust stroke.

waste spark system An ignition system in which each ignition coil serves two cylinders, with each end of the secondary winding attached by a high-tension lead to a spark plug. The spark is used to ignite the air-fuel mixture in one cylinder and has no effect on the other cylinder.

waste vegetable oil (WVO) Used cooking grease or oil used for producing biodiesel fuel.

wastegate A pressure regulator device that allows control of the pressure produced by the turbocharger. The wastegate opens to allow exhaust gases to bypass the turbine wheel of the turbocharger.

water fade Brake fade caused by water-soaked brake linings.

water gas A gas made from hydrogen and water.

water jacket A passageway for coolant to flow inside the engine block that is formed when the block is cast.

water jackets The passages in the engine block and cylinder head that surround the cylinders, valves, and ports.

watt The unit for measuring electrical power.

Watt's linkage Another name for a rigid-axle coil-spring suspension that uses two bars similar to a Panhard rod and a pivot point on the axle, to keep the axle from moving in turns.

waveform A graphical plot of voltage over time that is displayed on an oscilloscope.

web A reinforced area of a metal part formed during the casting process to create strength and durability.

wedge combustion chamber A combustion chamber design where the valves are often directly lined up beside each other in a row and positioned at an angle over the pistons, forming a wedge-shaped combustion chamber.

weight matching The process of matching the tire's lightest point with the rim's heaviest point (generally at the valve stem) for the purpose of reducing the tire's radial imbalance.

weight transfer Weight moving from one set of wheels to the other set of wheels during braking, acceleration, or cornering.

Welch plug A soft, round, metal disc pressed into the engine block to plug a water jacket or oil passageway.

welding helmet Protective gear designed for arc welding; it provides protection against foreign particles entering the eye, and the lens is tinted to reduce the glare of the welding arc.

wet sleeve A replaceable steel cylinder installed into the block that provides a wear surface for the piston and rings. It is in direct contact with coolant on its outside surface.

wheel alignment The practice of aligning the wheels of the vehicle to the centerline of the vehicle and to one another. It ensures that the vehicle will handle correctly and gives best tire wear.

wheel assembly A term used to encompass all components of the wheel and tire.

wheel bearing A component that allows the wheels to rotate freely while supporting the weight of the vehicle, made up of an inner race, outer race, rollers and a cage.

wheel center The part of the wheel containing the holes for the lug studs.

wheel cylinder A hydraulic cylinder with one or two pistons, seals, dust boots, and a bleeder screw that pushes the brake shoes into contact with the brake drum to slow or stop the vehicle.

wheel cylinder piston clamp A tool that prevents the pistons from being pushed out of the wheel cylinders while the brake shoes are being replaced.

wheel retaining nuts Lug nuts used to hold the wheel on the hub.

wheel rim The part on which the tire is mounted. Also called a "wheel" or "rim."

wheel speed sensor A device that creates an analog or digital signal according to the speed of the wheel.

wheel studs Threaded fasteners that are pressed into the wheel hub flange and used to bolt the wheel onto the vehicle.

windage tray A plate that bolts onto the bottom of the main bearing saddles inside the oil pan, helping to keep the crankshaft from contacting the engine oil.

wire A conductor usually made of multi-stranded copper with an external insulated coating.

wire feed welder A welding machine that automatically feeds the filler wire by operating a trigger mechanism on a welding gun.

wiring diagram A schematic drawing and symbol representation of the wiring and components; also called an electrical schematic.

wiring harness A collection of wires or cables insulated from each other but bound together.

wiring harness connector A plug that contains multiple terminals with male and female ends.

wishbone control arm Another term for an A-arm.

wood gas A gas made by the carbonization or gasification of coal.

Woodruff key A half-round, rectangular key that is placed into a half-round, rectangular opening in the nose of the crankshaft or other round shafts found on the engine.

work The result of force creating movement.

working pressure The pressure within a hydraulic system while the system is being operated.

worm A gear with a helical, threaded shaft that is attached to the steering column and meshes with a worm wheel that transfers motion from the steering wheel to the steering linkage.

worm gear steering A robust steering system frequently used on heavier vehicles that uses a worm to turn a meshed worm wheel to provide gear reduction, making steering easier for the driver.

worm shaft The protrusion of the worm gear that serves as the point of attachment to the steering column.

wrap leaf A spring containing spring eyes.

wrenches A generic term to describe tools that tighten and loosen fasteners with hexagonal heads.

XH-7 A form of desiccant used with pure R-134a.

XH-9 A form of desiccant used when the refrigerant may not be pure, such as with some imports.

yaw Movement of a vehicle around its z-axis (vertical axis), felt when the vehicle deviates from its straight path, as when skidding sideways and the rear comes around.

yaw sensor A sensor that measures the amount a vehicle is turning around its vertical axis. This information is used by the ESC system to know how much a vehicle is turning.

yield point The point at which a bolt is stretched so hard that it will not return to its original length when loosened; it is measured in pounds per square inch of bolt cross section.

Zener diode A diode that forward biases when a certain voltage is reached.

zero camber A tire with no tilt or zero camber angle.

zero offset A condition in which the plane of the hub mounting surface is even with the centerline of the wheel.

zero scrub radius A condition in which the camber line intersects the SAI line at the road surface.

Zyglo™ A crack detection method used on engine blocks and cylinder heads that uses a liquid fluorescent penetrant and a fluorescent light to find cracks in ferrous metal.

INDEX

Note: The letters 'f' and 't' following locators refer to figures and tables.

A

C

D

E

F

G

S

U

V

W

Y

Z

CREDITS

Chapter	Figure ID	Page Number	Credit Line
2	F02B	19	© Teddy Leung/Shutterstock
2	F04	19	© G Fiume/Getty Images Sport/Getty
2	F05	20	© ZUMA Press, Inc./Alamy Stock Photo
2	F06	20	© Mark Elias/Bloomberg/Getty
2	F09	21	© Car Collection/Alamy Stock Photo
2	F10	21	© Luis Sinco/Los Angeles Times/Getty
3	F03	41	© New York Daily News Archive/Getty
3	F10	44	© ESB Professional/Shutterstock
3	F20	50	© Craig F. Walker/Denver Post/Getty
3	F29	53	© Kurdiansyah Kurdiansyah/EyeEm/Getty
4	F39	77	© pixelaway/Shutterstock
4	F44	79	© Biophoto Associates/Science Source/Getty
6	F38	116	© Joe Giddens/EMPPL PA Wire/AP Photo
16	F01	367	© robcocquyt/iStock/Getty Images Plus/Getty.
20	F03	478	© LostFoam.com
20	F04	478	© LostFoam.com
21	F15	526	© Matthew Richardsons/Alamy stock photo
21	F36	533	Used with permission from Goodson Tools & Supplies for Engine builders
21	F54	539	Used with permission from Gary Thompson Memorial
26	F12	645	© AsianDream/iStock/Getty Images Plus/Getty
36	F01	861	DaveAlan/E+/Getty
43	F17	1046	© Wonderlust Industries/Photonica World/Getty
43	F18	1047	© Sean Murphy/Stone/Getty
48	F16	1188	Used with permission from The Timken Company.
50	F01	1229	© Alex Ramsay/Alamy Stock Photo
67	F01	1737	© Everett Collection Inc/Alamy stock photo
67	F03	138	© Hung Chung Chih/Shutterstock
67	F20A	1751	Used with permission from Tesla Motors
67	F20B	1751	Used with permission from Tesla Motors